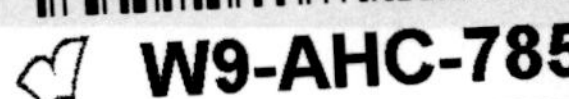

W9-AHC-785

KEY CONCEPTS

Interactive assignments and tools ensure that students succeed in grasping and applying fundamental course concepts.

The optional highlighting tool makes it easy for students to interact with the text on the screen the way they would on a page. ▶

Students are given the opportunity to apply the scientific language they have learned to seminal primary source readings. ▶

Ongoing summaries require critical thinking and ensure that students are internalizing the information. ▶

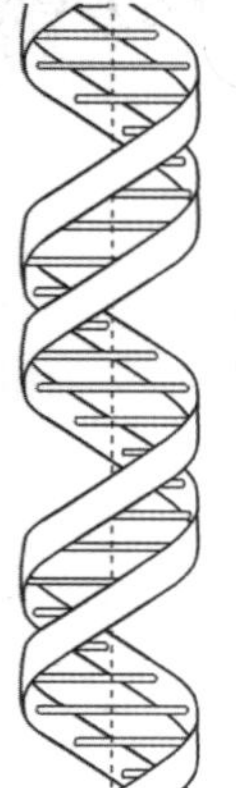

We wish to put forward a radically different structure for the salt of deoxyribose nucleic acid. This structure has two helical chains each coiled round the same axis (see diagram). We have made the usual chemical assumptions, namely, that each chain consists of phosphate diester groups joining deoxyribose residues with __________ linkages.[1] Both chains follow right-handed helices, but the sequences of the atoms in the two chains run in __________ .[2] Each chain loosely resembles Furberg's model No. 1; that is, the __________ are on the inside of the helix and the __________ on the outside. ...

[1][This sentence] describes the sugar-phosphate backbone.

[2][This sentence] describes the [▼]

[Clear this section's highlighting]

The novel feature of the structure is the manner in which the two chains are held together by the purine and pyrimidine bases. The planes of the bases are perpendicular to the fibre axis. They are joined together in pairs, a single base from one chain being __________ to a single base from the other chain, so that the two lie side by side.[3] ... One of the pair must be a purine and the other a pyrimidine for bonding to occur. ...

[3][This sentence] describes the [▼]

It is found that only specific pairs of bases can bond together.

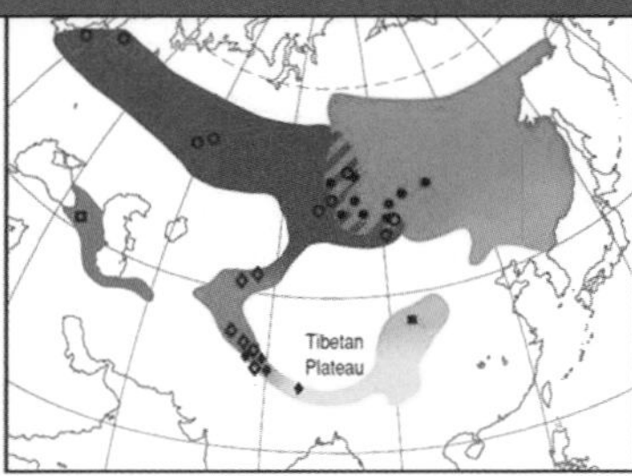

Tibetan
Plateau

Greenish warblers live throughout Asia, as shown on the distribution map on the left. Although they share a broad range in central Siberia, the eastern Siberian greenish warblers (range represented in red) and western Siberian greenish warblers (range represented in purple) do not interbreed.

Source: D. E. Irwin, S. Bensch, J. H. Irwin, and T. D. Price, "Speciation by distance in a ring species," *Science* 307, no. 5708 (January 2005): 414-416. Copyright © 2005 by the American Association for the Advancement of Science. Reprinted by permission.

On the map, a graduated transition in color from one geographic region to another represents gene flow between adjacent populations. Notice that, despite the physical distance between them, the eastern Siberian population does exchange genes with the populations to the south. Based on the information provided, the greenish warblers consist of __________ .

In 2000, Dr. Darren E. Irwin traveled all over Asia and collected data about local green warbler populations. The symbols on the map above refer to locations he visited. These locations are represented on the graphs below by a two-letter designation (for example, MN, which stands for Mongolia). At each site, he quantified characteristics of the local greenish warbler song and properties of the local environment that would have a strong influence on bird song. For example, the density of the forest has direct effects on the acoustics of bird song, or how it is heard.

How the environment varies with latitude

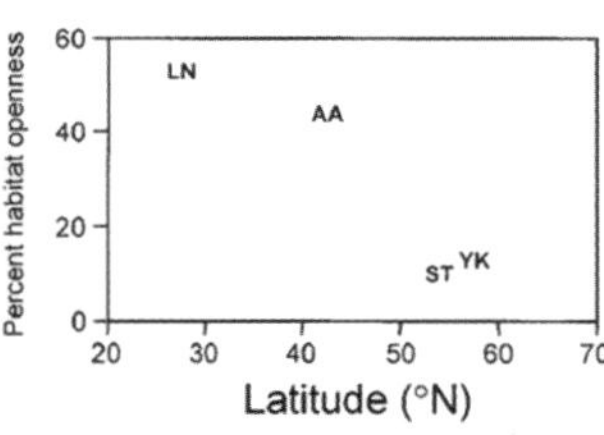

Percent habitat openness
60
40
20
0
LN
AA
ST
YK
20 30 40 50 60 70
Latitude (°N)

How bird song behavior varies with latitude

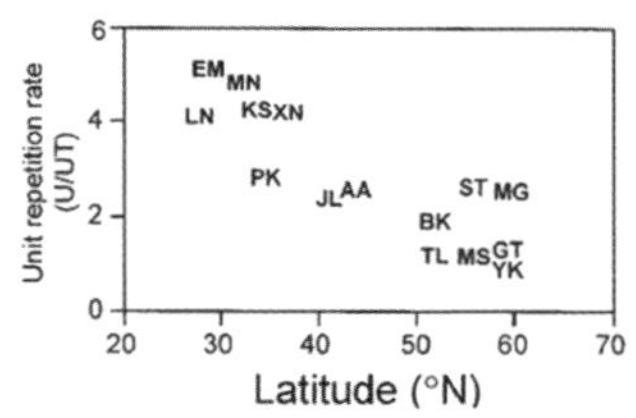

Unit repetition rate (U/UT)
6
4
2
0
EM
MN
LN
KS
XN
PK
JL
AA
ST
MG
BK
TL
MS
GT
YK
20 30 40 50 60 70
Latitude (°N)

Source: Darren E. Irwin, "Song variation in an avian ring species," *Science* 54, no. 3 (2000): 998-1010, Figure 5. Copyright © 2000. Reprinted by permission of John Wiley & Sons, Inc.

As he moved farther north (increase in latitude), what observations did he make about how the environment changed?

- ○ The habitat becomes less open.
- ○ The habitat becomes more open.

As he moved farther north (increase in latitude), what observations did he make about how song behavior changed?

- ○ The warblers repeat themselves more.
- ○ The warblers repeat themselves less.

EMPIRICAL RESEARCH

Aplia's text-specific problems offer a variety of question types and styles and encourage students to think critically.

◀ Students are required to analyze primary research data. Questions are varied to accommodate different student learning styles, such as that of the visual learner.

◀ Graph-based questions test students' quantitative understanding.

INSTRUCTOR TOOLS

Grading and Performance: Aplia keeps you informed about student participation and progress. As students complete assignments, their grades are imported directly into your Aplia gradebook. ▶

Course Management System: Aplia has a full course management system that can be used either independently or in conjunction with other course management systems such as Blackboard and WebCT. ▶

Personalized Support: Aplia has the industry-leading in-house team for the best instructor training, custom course creation, and support. ▶

Instructor Tools

Grading ahd Performance

Aplia keeps you informed about student participation and progress. As students complete assignments, their grades are imported directly into your Aplia gradebook.

- Track class and individual student performance by class distribution and the class average of assignments
- See students' scores and compare them to class averages
- Analyze an individual student's performance on a single assignment
- View and download the gradebook in a simple spreadsheet (Excel compatible)

Course Management System

Aplia has a full course management system that can be used either independently or in conjunction with other course management systems such as Blackboard and WebCT.

- View and download your gradebook
- Switch between instructor and student views of the course
- E-mail students
- Post announcements
- Manage the student discussion board
- Assign, edit, or delete homework and quizzes
- Change assignment due dates and times
- Upload files
- Add in links to other websites
- Share ideas with other instructors in the instructor discussion board

Personalized Support

Aplia has the industry-leading in-house support team. Aplia's support team:

View Grades | Analytics | Import | Export

Name	Aplia					Other	
Find Students	1	2	3	4	5	1	2
Alison, Hoselton	0	0	0	0	0	0	2
Himsielon, Claires	3	5	3	13.1	13	33	1.2
Ho, Charles	0	0	0	0	0	0	0.2
Ho, Mei Han Christine	0	0	0	0	0	11	3.1
Hoehne, Timothy	2	5	3	13.1	13	1	0.3
Hogstead, Allan	2	5	3	13.1	13	33	0.7
Holitski, Robyn	2	0	0	0	0	0	0.9
Hong, Cindy	1	0	0	0	0	11	0.3
Hood, mark	2	5	3	13.1	13	1	0.4
Horbay, timish	3	5	3	13.1	13	33	0.3
Hosking, Bryan	3	5	3	13.1	13	11	0.2
Hou, Xuan	0	0	0	0	0	1	0.3
Houle, Daniel	1	0	0	0	0	33	0.3
Houle, Ryan	2	5	3	13.1	13	0	0.4
Hoy, Alexia	3	5	3	13.1	13	11	0.6
Hrupp, Jennifer	1	0	0	0	0	1	0.6
Hsieh, chiEn	1	0	0	0	0	33	0.6
Hsung, Antoine	2	5	3	13.1	13	0	0.8
Hu, Youyuan	2	5	3	13.1	13	11	0.9
Hua, Vinson	2	0	0	0	0	1	0.9

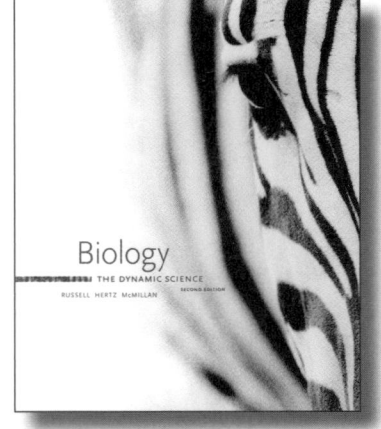

Biology

THE DYNAMIC SCIENCE

SECOND EDITION

RUSSELL HERTZ McMILLAN

BROOKS/COLE
CENGAGE Learning™

Australia • Brazil • Japan • Korea • Mexico • Singapore • Spain • United Kingdom • United States

Biology: The Dynamic Science, Second Edition

Peter J. Russell, Paul E. Hertz, Beverly McMillan

Editor in Chief: Michelle Julet

Publisher: Yolanda Cossio

Senior Developmental Editors: Mary Arbogast, Shelley Parlante

Editorial Project Manager: Jake Warde

Art Developmental Editor: Suzannah Alexander

Assistant Editors: Elizabeth Momb, Shannon Holt, Alexis Glubka

Editorial Assistants: Joshua Taylor, Lauren Crosby

Managing Media Editor: Shelley Ryan

Media Editor: Lauren Oliveira

Marketing Manager: Tom Ziolkowski

Marketing Assistant: Elizabeth Wong

Marketing Communications Manager: Linda Yip

Content Project Manager: Teresa L. Trego

Art Director: John Walker

Print Buyer: Karen Hunt

Rights Acquisitions Account Manager, Text: Bob Kauser

Rights Acquisitions Account Manager, Image: Dean Dauphinais

Production Service: Graphic World Inc.

Text Designer: Jeanne Calabrese

Art Editor: Steve McEntee

Photo Researcher: Chris Althof for Bill Smith Group

Copy Editor: Graphic World Inc.

Illustrators: Dragonfly Media Group, Steve McEntee, Graphic World Inc.

Cover Designer: John Walker

Cover Image: Pat Hermansen/The Image Bank

Compositor: Graphic World Inc.

Aplia for Biology: Peg Knight, Alison Petersen, Qinzi Ji

Selected artwork courtesy of Starr/Taggart/Evers/Starr, *Biology: The Unity and Diversity of Life* (0-495-55792-0).

Library of Congress Control Number: 2010932431

Volume 1: ISBN-13: 978-0-538-49372-7
Volume 1: ISBN-10: 0-538-49372-0

Volume 2: ISBN-13: 978-0-538-49373-4
Volume 2: ISBN-10: 0-538-49373-9

Volume 3: ISBN-13: 978-0-538-49374-1
Volume 3: ISBN-10: 0-538-49374-7

Brooks/Cole
20 Davis Drive
Belmont, CA 94002-3098
USA

Cengage Learning is a leading provider of customized learning solutions with office locations around the globe, including Singapore, the United Kingdom, Australia, Mexico, Brazil, and Japan. Locate your local office at **www.cengage.com/global.**

Cengage Learning products are represented in Canada by Nelson Education, Ltd.

To learn more about Brooks/Cole, visit **www.cengage.com/brookscole**

Purchase any of our products at your local college store or at our preferred online store **www.CengageBrain.com.**

Printed in Canada
1 2 3 4 5 6 7 14 13 12 11 10

Brief Contents

About the Authors

Peter J. Russell received a B.Sc. in Biology from the University of Sussex, England, in 1968 and a Ph.D. in Genetics from Cornell University in 1972. He has been a member of the Biology faculty of Reed College since 1972; he is currently a professor of biology. Peter teaches a section of the introductory biology course, a genetics course, and a research literature course on molecular virology. In 1987 he received the Burlington Northern Faculty Achievement Award from Reed College in recognition of his excellence in teaching. Since 1986, he has been the author of a successful genetics textbook; current editions are *iGenetics: A Molecular Approach, iGenetics: A Mendelian Approach,* and *Essential iGenetics.* Peter's research is in the area of molecular genetics, with a specific interest in characterizing the role of host genes in the replication of the RNA genome of a pathogenic plant virus, and the expression of the genes of the virus; yeast is used as the model host. His research has been funded by agencies including the National Institutes of Health, the National Science Foundation, and the American Cancer Society. He has published his research results in a variety of journals, including *Genetics, Journal of Bacteriology, Molecular and General Genetics, Nucleic Acids Research, Plasmid,* and *Molecular and Cellular Biology.* Peter has a long history of encouraging faculty research involving undergraduates, including cofounding the biology division of the Council on Undergraduate Research in 1985. He was Principal Investigator/Program Director of a National Science Foundation (NSF) Award for the Integration of Research and Education to Reed College, 1998–2002.

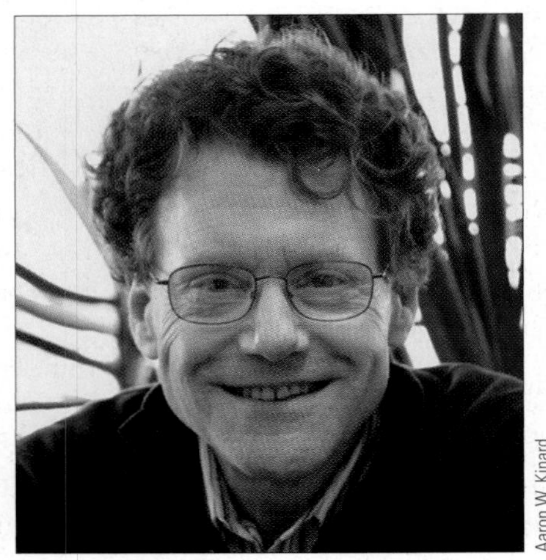

Aaron W. Kinard

Paul E. Hertz was born and raised in New York City. He received a B.S. in Biology from Stanford University in 1972, an A.M. in Biology from Harvard University in 1973, and a Ph.D. in Biology from Harvard University in 1977. While completing field research for the doctorate, he served on the Biology faculty of the University of Puerto Rico at Rio Piedras. After spending two years as an Isaac Walton Killam Postdoctoral Fellow at Dalhousie University, Paul accepted a teaching position at Barnard College, where he has taught since 1979. He was named Ann Whitney Olin Professor of Biology in 2000, and he received The Barnard Award for Excellence in Teaching in 2007. In addition to serving on numerous college committees, Paul chaired Barnard's Biology Department for eight years. He is also the Program Director of the Hughes Science Pipeline Project at Barnard, an undergraduate curriculum and research program that has been funded continuously by the Howard Hughes Medical Institute since 1992. The Pipeline Project includes the Intercollegiate Partnership, a program for local community college students that facilitates their transfer to four-year colleges and universities. He teaches one semester of the introductory sequence for Biology majors and preprofessional students, lecture and laboratory courses in vertebrate zoology and ecology, and a year-long seminar that introduces first-year students to scientific research. Paul is an animal physiological ecologist with a specific research interest in the thermal biology of lizards. He has conducted fieldwork in the West Indies since the mid-1970s, most recently focusing on the lizards of Cuba. His work has been funded by the NSF, and he has published his research in such prestigious journals as *The American Naturalist, Ecology, Nature, Oecologia,* and *Proceedings of the Royal Society.* In 2010, he received funding from NSF for a project designed to detect the effects of global climate warming on the biology of *Anolis* lizards in Puerto Rico.

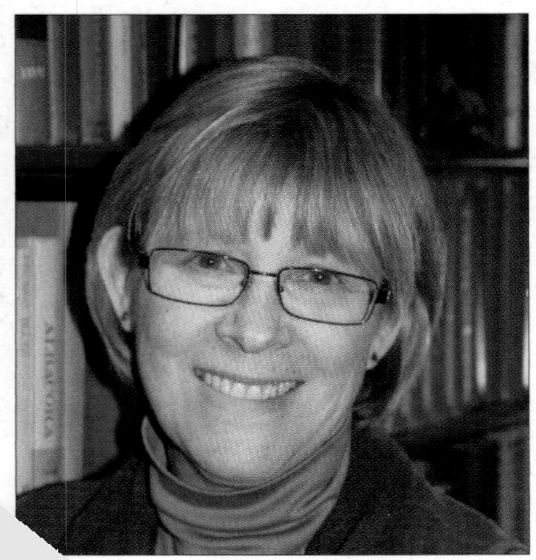

Beverly McMillan has been a science writer for more than 25 years and is coauthor of a college text in human biology, now in its eighth edition. She has worked extensively in educational and commercial publishing, including eight years in editorial management positions in the college divisions of Random House and McGraw-Hill. In a multifaceted freelance career, Bev also has written or coauthored 10 trade books, as well as story panels for exhibitions at the Science Museum of Virginia and the San Francisco Exploratorium. She has worked as a radio producer and speechwriter for the University of California system and as a science writer and media relations advisor for the Virginia Institute of Marine Science of the College of William and Mary. She holds undergraduate and graduate degrees from the University of California, Berkeley.

Preface

Welcome to the second edition of *Biology: The Dynamic Science.* The title of our book reflects the explosive growth in knowledge of the living world over the past few decades—indeed, even over the few years since the first edition appeared. Although the rapid pace of discovery makes biology the most exciting of all the natural sciences, it also makes it the most difficult to teach. How can college instructors—not to mention their students—absorb the ever-growing body of ideas and information? The task is daunting, especially in introductory courses that provide a broad overview of the discipline.

Building on a strong foundation . . .

As scientists and authors, we viewed the first edition of this book as an experiment: we presented a focused view of the essential knowledge that we thought students would need to begin their careers as biologists. Like all good experiments, our first edition generated lots of data: the many instructors and students who used the book offered positive feedback about elements that enhanced students' learning as well as valuable suggestions for possible modifications. We also received input from a small army of expert reviewers as well as media and art advisory boards. As a result of these efforts, every chapter has been revised and updated, and some units have been reorganized. In addition, the second edition includes many new or modified illustrations and photos, and we have taken great pains to make the text and the art even more tightly integrated than they were in the first edition.

Emphasizing the big picture . . .

In this textbook, we have applied our collective experience as college teachers, researchers, and writers to create a readable and understandable introduction to our field of study. We provide straightforward explanations of fundamental concepts, presented from the evolutionary perspective that binds the biological sciences together. Having watched our students struggle to navigate the many arcane details of college-level introductory biology, we constantly remind ourselves and each other to "include fewer facts, provide better explanations, and maintain the narrative flow," thereby enabling students to see the big picture. Clarity of presentation, thoughtful organization, a seamless flow of topics within chapters, and spectacularly useful illustrations are central to our approach.

Focusing on research to help students engage the living world as scientists . . .

A primary goal of this book is to sustain students' curiosity about the living world instead of burying it under a mountain of disconnected facts. We can help students develop the mental habits and fascination of scientists by conveying our passion for biological research. We want to amaze students not only with *what biologists know* about the living world, but also with *how they know it* and *what they still need to learn.* In doing so, we can encourage some students to accept the challenge and become biologists themselves, posing and answering important new questions through their own innovative research. For students who pursue other careers, we hope that they will leave their introductory—and perhaps only—biology course armed with intellectual skills that will enable them to evaluate future discoveries with a critical eye.

In this book, we introduce students to a biologist's "ways of knowing." Scientists constantly integrate new observations, hypotheses, experiments, and insights with existing knowledge and ideas. To help students engage the world as scientists do, we must not simply introduce them to the current state of knowledge. We must also foster an appreciation of the historical context within which those ideas developed, and identify the future directions that biological research is likely to take.

To achieve these goals, our explanations are grounded in the research that established the basic facts and principles of biology. Thus, a substantial proportion of each chapter focuses on studies that define the state of biological knowledge today. When describing research, we first identify the hypothesis or question that inspired the work and then relate it to the broader topic under discussion. Our research-oriented theme teaches students, through example, how to ask scientific questions and pose hypotheses, two key elements of the "scientific process."

Because advances in science occur against a background of past research, we also give students a feeling for how biologists of the past formulated basic knowledge in the field. By fostering an appreciation of such discoveries, given the information and theories available to scientists in their own time, we can help students understand the successes and limitations of what we consider cutting edge today. This historical perspective also encourages students to view biology as a dynamic intellectual enterprise, not just a list of facts and generalities to be memorized.

We have endeavored to make the science of biology come alive by describing how biologists formulate hypotheses and evaluate them using hard-won data; how data sometimes tell only part of a story; and how studies often end up posing more questions than they answer. Although students might prefer simply to learn the "right" answer to a question, they must be encouraged to embrace "the unknown," those gaps in knowledge that create opportunities for further research. An appreciation of what biologists *don't* know will draw more students into the field. And by defining *why* scientists don't understand interesting phenomena, we encourage students to think critically about possible solutions and to follow paths dictated by their own curiosity. We hope that this approach will encourage students to make biology a part of their daily lives—to have informal discussions about new scientific discoveries, just as they do about politics, sports, or entertainment.

Presenting the story line of the research process . . .

In preparing this book, we developed several special features to help students broaden their understanding of the material presented and of the research process itself. A Visual Tour of these features and more begins on page xi.

- The chapter openers, entitled *Why It Matters. . .*, tell the story of how a researcher arrived at a key insight or how biological research solved a major societal problem, explained a fundamental process, or elucidated a phenomenon. These engaging, short vignettes are designed to capture students' imaginations and whet their appetites for the topic that the chapter addresses.
- To complement this historical or practical perspective, each chapter closes with a brief essay, entitled *Unanswered Questions,* prepared by an expert in the field. These essays identify important unresolved issues relating to the chapter topic and describe cutting-edge research that will advance our knowledge in the future.
- Each chapter also includes a short, boxed essay entitled *Insights from the Molecular Revolution,* which describes how molecular tools allow scientists to answer questions that they could not have even posed 20 or 30 years ago. Each *Insight* focuses on a single study and includes sufficient detail for its content to stand alone.
- Most chapters are further supplemented with one or more short, boxed essays entitled *Focus on Research.* Some of these essays describe seminal studies that provided a new perspective on an important question. Others describe how basic research has solved everyday problems relating to health or the environment. Another set introduces model research organisms—such as *Escherichia coli, Drosophila, Arabidopsis, Caenorhabditis,* and *Anolis*—and explains why they have been selected as subjects for in-depth analysis.

Three types of specially designed Research Figures provide more detailed information about how biologists formulate and test specific hypotheses by gathering and interpreting data. In the second edition we have included one or more Research Figures in practically every chapter (see the list on the endpapers at the back of the book).

- *Research Method* figures provide examples of important techniques, such as gel electrophoresis, the use of radioisotopes, and cladistic analysis. Each *Research Method* figure leads a student through the technique's purpose and protocol and describes how scientists interpret the data it generates.
- *Observational Research* figures describe specific studies in which biologists have tested hypotheses by comparing systems under varying natural circumstances.
- *Experimental Research* figures describe specific studies in which researchers used both experimental and control treatments—either in the laboratory or in the field—to test hypotheses by manipulating the system they studied.

Integrating spectacular visuals into the narrative . . .

Today's students are accustomed to receiving ideas and information visually, making the illustrations and photographs in a textbook more important than ever before. Our illustration program provides an exceptionally clear supplement to the narrative in a style that is consistent throughout the book. Graphs and anatomical drawings are annotated with interpretative explanations that lead students, step by step, through the major points they convey.

In preparing this edition, we undertook a rigorous review of all the art in the text. The publishing team has made an extraordinary effort to identify the key elements of effective illustrations. In focus groups and surveys, instructors helped us identify the "Key Visual Learning Figures" covering concepts or processes that demand premier visual learning support. Each of these figures has been critiqued by our Art Advisory Board, revised, and revised again to insure its usability and accuracy.

Organizing chapters around important concepts . . .

As authors and college teachers, we know how easily students can get lost within a chapter that spans 15 or more pages. When students request advice about how to approach such a large task, we usually suggest that, after reading each section, they pause and quiz themselves on the material they have just encountered. After completing all of the sections in a chapter, they should quiz themselves again, even more rigorously, on the individual sections and, most important, on how the concepts developed in different sections fit together. To assist these efforts, we have adopted a structure for each chapter that will help students review concepts as they learn them.

- The organization within chapters presents material in digestible chunks, building on students' knowledge and understanding as they acquire it. Each major section covers one broad topic. Each subsection, titled with a declarative sentence that summarizes the main idea of its content, explores a narrower range of material.
- Whenever possible, we include the derivation of unfamiliar terms so that students will see connections between words that share etymological roots. Mastery of the technical language of biology will allow students to discuss ideas and processes precisely. At the same time, we have minimized the use of unnecessary jargon as much as possible.
- Sets of embedded *Study Break* questions follow every major section. These questions encourage students to pause at the end of a section and review what they have learned before going on to the next topic within the chapter. Short answers to these questions appear in an appendix.

Encouraging active learning, critical thinking, and self-assessment of learning outcomes . . .

The second edition of *Biology: The Dynamic Science* includes a new active learning feature, *Think Outside the Book,* which will help students think analytically and critically about the material as they are learning it. *Think Outside the Book* activities have been designed to encourage students to explore the biological world directly or through high-quality electronic resources. Students can engage in these activities either individually or in small groups.

Supplementary materials at the end of each chapter help students review the material they have learned, assess their understanding, and think analytically as they apply the principles developed in the chapter to novel situations. Many of the end-of-chapter questions also serve as good starting points for class discussions or out-of-class assignments.

- *Review Key Concepts* references figures and tables in the chapter, providing a summary of important ideas developed in the chapter. These *Reviews* are much too short to serve as a substitute for reading the chapter. Instead, students may use them as an outline of the material, filling in the details on their own.
- Each chapter also closes with *Test Your Knowledge,* a set of 10 questions that focus on factual material.
- Several open-ended *Discuss the Concepts* questions emphasize key ideas, the interpretation of data, and practical applications of the material.
- To help students hone their critical thinking ability, another question asks students to *Design an Experiment* to test hypotheses that relate to the chapter's main topic.
- To help students develop analytical and quantitative skills, each chapter also includes an *Interpret the Data* feature. The hypothesis and methods of an experimental or observational study are summarized, and some of its results are presented in either graphical or tabular format. Students are asked to interpret these results in the context of the experimental design.
- *Apply Evolutionary Thinking* asks students to interpret a relevant topic in relation to the principles of evolutionary biology.
- The *Express Your Opinion* exercise allows students to weigh both sides of an issue by reading pro/con articles, and then making their opinion known through an online voting process.

Helping students master key concepts throughout the course. . .

As teachers, we know that student effort is an important determinant of student success. Unfortunately, many of us simply cannot spare the time to develop novel learning tools for every concept—or even every chapter—in a large and complex introductory textbook. To help address this problem, we are pleased to offer **Aplia for Biology,** an automatically graded homework management system tailored to this edition. For students, Aplia provides a structure within which they can expand their efforts, master key concepts throughout the course, and increase their success. For faculty, Aplia can help us transform our teaching and raise our productivity by allowing us to require more—and more consistent—effort from students without adding to our work load. By providing students with continuous exposure to key concepts and their applications throughout the course, Aplia allows us to do what we do best—respond to questions, lead discussions, and challenge our students.

We hope you agree that we have developed a clear, fresh, and well-integrated introduction to biology as it is understood by researchers today. Just as important, we hope that our efforts will excite students about the research process and the new discoveries it generates.

New to This Edition

This section highlights the changes we made to enhance the effectiveness of the text. Every chapter has been updated to ensure currency of information. We made organizational changes to more closely link related topics and reflect preferred teaching sequences. New features in the text have been developed to help students actively engage in their study of biology. Enriched media offerings provide students a broad spectrum of learning opportunities.

Organizational Changes

The genetic and molecular regulation of development has been integrated into Chapter 16 (Regulation of Gene Expression), which introduces students to the mechanisms that produce a complex multicellular organism from a single cell. The new section expands their understanding of the scope of gene regulation by showing how sequential regulatory events such as induction, determination, and differentiation drive the formation of a complex organism, and it prepares them for the application of these principles more specifically to plants and animals in Chapters 34 and 48, respectively.

Coverage of viruses is now complete in Chapter 17 (Bacterial and Viral Genetics); brief coverage of viroids and prions is also included in that chapter. Placing the entire coverage of viruses in one chapter makes it easier for instructors to tailor the treatment of viruses in their course.

The Ecology Unit has been reorganized to begin with a large-scale overview of the science of ecology and the biosphere (combining part of the first edition's Chapter 49 with the first edition's Chapter 52). This reorganized Chapter 49 lays out a conceptual framework for the scope of ecology and describes the patterns of distribution of life on Earth. After this orientation, the chapters continue in order with Chapter 50 focusing on population ecology, Chapter 51 treating population interactions and community ecology, and Chapter 52 covering ecosystems.

New Features

We carefully considered all of the existing features and made several improvements to enhance their effectiveness as learning and teaching tools. The *Insights from the Molecular Revolution,* which showcase exciting and interesting research based on molecular techniques, have been rewritten to improve readability and ease of understanding. The topics of these features have been updated to reflect recent discoveries, and references to the original papers have been included. In addition, more than 40 illustrations have been developed to add clarity and stimulate student interest in this feature.

The *Unanswered Questions* essays appearing at the end of every chapter, each prepared by an expert in the field, identify important unresolved issues relating to the chapter topic. The essays now conclude with a *Think Critically* question, which encourages students to think about the unanswered question, ponder possible next steps in the research, or consider the benefits of answering the question.

The *Research Figures* in the text provide more detailed information about how biologists formulate and test specific hypotheses by gathering and interpreting data. We have developed 26 new *Research Figures.* In addition, references to the original papers have been added to each figure so that the research can be explored in more depth, as needed.

The second edition also includes new features that are designed to encourage students to engage with the material and develop the quantitative skills necessary for biological study and investigation. A feature called *Think Outside the Book* helps students think critically about the material as they are learning it. The questions presented in these boxes are related to topics that the student has just studied and are designed to help explore the biological world directly or through high-quality electronic resources. In addition, new *Interpret the Data* exercises are included in the study material at the end of each chapter. These exercises, most of which are drawn from published biological research, help students build their skills in analyzing figures and in reading graphs and tables.

Enhanced Art Program

Helping today's students understand biological processes requires effective visual learning support. In preparation for this edition we undertook a rigorous review of all the art in the text. Focus groups and instructor surveys helped identify key figures that are essential teaching and learning tools. The figures identified in this process were scrutinized by our Art Advisory Board and other content area experts to insure their effectiveness in the classroom. In some cases we developed new views of structures to facilitate a three-dimensional visualization. We also included additional "orientation" diagrams to help students visualize levels of organization and how systems function as a whole.

The effective integration of text and illustration has also been a top priority for the development of the art. We have increased the number of illustrations supported by numbered step-by-step annotations placed directly on the figure. These annotations help students interpret detailed illustrations and develop deeper understanding of complex processes.

Enriched Media

New to the second edition is Aplia for Biology, an automatically graded homework management system tailored to this edition. Aplia courses are customized to fit with each instructor's syllabus and provide automatically graded homework with detailed, immediate feedback on every question. Aplia's interactive tools serve to increase student engagement and understanding.

New, interactive 3-D animations have been developed that help students visualize processes in a more dynamic way. These animations promote in-depth understanding of key biological topics, including cellular respiration, photosynthesis, DNA replication, and evolutionary processes. Embedded assessments in the animations ensure that students have the necessary foundation for understanding these important concepts.

Also new to the second edition are clips from the BBC Motion Gallery. This diverse and robust library of high-quality videos features clips from well-respected scientists and naturalists, including Sir David Attenborough. The clips can be used in conjunction with the text to spark discussion and help students connect the material to their lives outside of the classroom.

We now invite you and your students to preview the many exciting features that will help them think and engage like scientists . . .

THE BIG PICTURE

Each chapter is carefully organized and presented in "digestible chunks" so you can stay focused on the most important concepts. Easy-to-use learning tools point out the topics covered in each chapter, show why they are important, and help you learn the material.

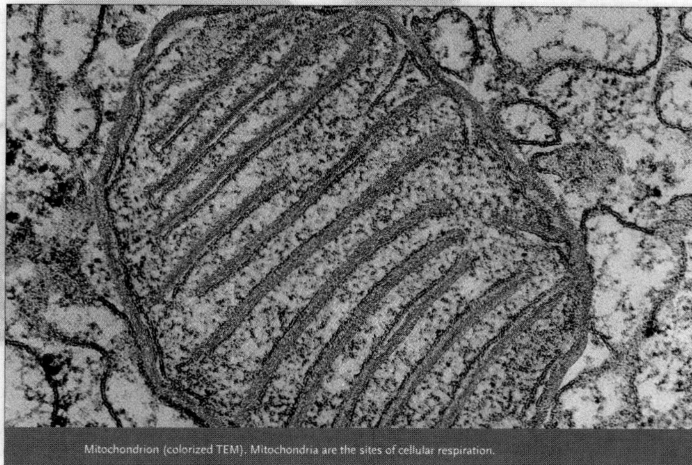

Mitochondrion (colorized TEM). Mitochondria are the sites of cellular respiration.

Dr. Donald Fawcett/Visuals Unlimited, Inc.

8

STUDY OUTLINE

Harvesting Chemical Energy: Cellular Respiration

Why It Matters. . . In the early 1960s, Swedish physician Rolf Luft mulled over some odd symptoms of a patient. The young woman felt weak and too hot all the time (with a body temperature of up to 38.4°C). Even on the coldest winter days she never stopped perspiring, and her skin was always flushed. She was also underweight (40 kg), despite consuming about 3,500 calories per day.

Luft inferred that his patient's symptoms pointed to a metabolic disorder. Her cells were very active, but much of their activity was being dissipated as metabolic heat. He decided to order tests to measure her metabolic rate, the amount of energy her body was expending. The results showed the patient's oxygen consumption was the highest ever recorded—about twice the normal rate!

Luft also examined a tissue sample from the patient's skeletal muscles. Using a microscope, he found that her muscle cells contained many more mitochondria—the ATP-producing organelles of the cell—than are normally present in muscle cells. In addition, her mitochondria were abnormally shaped and their interior was packed to an abnormal degree with cristae, the infoldings of the inner mitochondrial membrane (see Section 5.3). Other studies showed that the mitochondria were engaged in cellular respiration—their prime function—but little ATP was being generated.

The disorder, now called *Luft syndrome,* was the first disorder to be linked directly to a defective mitochondrion. By analogy, someone with this disorder functions like a city with half of its power plants shut down. Skeletal and heart muscles, the brain, and other hardworking body parts with high energy demands are hurt the most by the inability of mitochondria to provide enough energy for metabolic demands. More than 100 mitochondrial disorders are now known.

155

‹ **Study Outline** The Study Outline provides an overview of all the topics and key concepts in the chapter. Each section breaks the material into a manageable amount of information, building on knowledge and understanding as you acquire it.

‹ **Why It Matters** Engaging introductory sections capture the excitement of biology and help you understand why the topic is important and how the material you are about to read fits into the Big Picture.

› **Study Break** Encourages you to pause and think about the key concepts you have just encountered before moving to the next section.

STUDY BREAK 8.2

1. **What are the energy-requiring and energy-releasing steps of glycolysis?**
2. **What is the redox reaction in glycolysis?**
3. **How is ATP synthesized in glycolysis?**
4. **Why is phosphofructokinase a target for inhibition by ATP?**

THINK LIKE A SCIENTIST

Your study of biology focuses not only on *what* scientists now know about the living world, but also *how* they know it. Use these unique features to learn through example how scientists ask scientific questions, pose hypotheses, and test them.

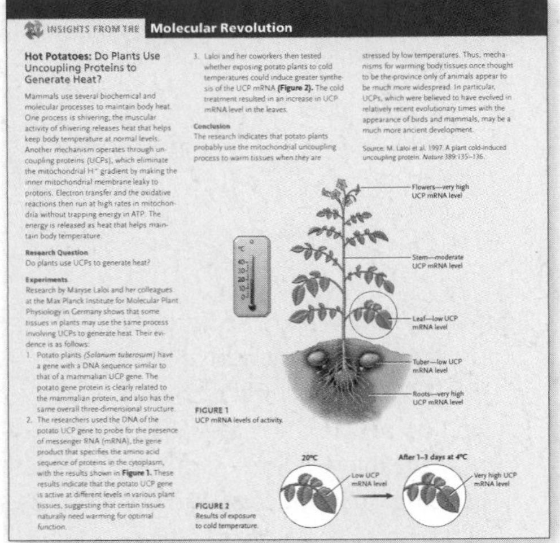

INSIGHTS FROM THE **Molecular Revolution**

Hot Potatoes: Do Plants Use Uncoupling Proteins to Generate Heat?

❮ Insights from the Molecular Revolution Showcases exciting and interesting research relying on molecular techniques; these molecular tools allow researchers to answer questions that they could not even pose 20 or 30 years ago.

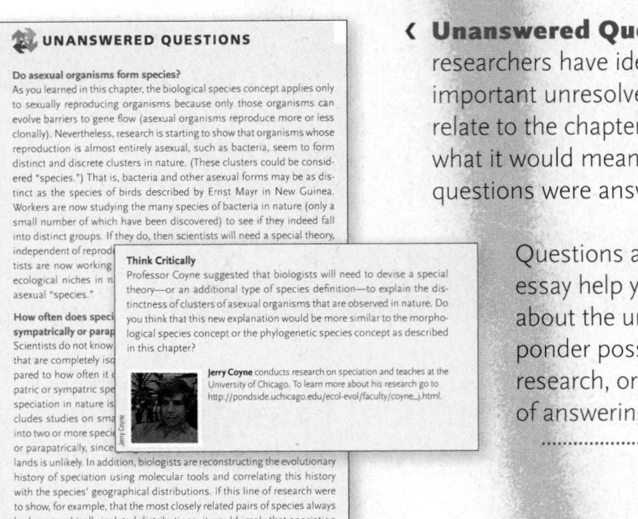

UNANSWERED QUESTIONS

Do asexual organisms form species?

As you learned in this chapter, the biological species concept applies only to sexually reproducing organisms because only those organisms can evolve barriers to gene flow (asexual organisms reproduce more or less clonally). Nevertheless, research is starting to show that organisms whose reproduction is almost entirely asexual, such as bacteria, seem to form distinct and discrete clusters in nature. (These clusters could be considered "species.") That is, bacteria and other asexual forms may be as distinct as the species of birds described by Ernst Mayr in New Guinea. Workers are now studying the many species of bacteria in nature (only a small number of which have been discovered) to see if they indeed fall into distinct groups. If they do, then scientists will need a special theory, independent of reprod... tists are now working... ecological niches in n... asexual "species."

How often does speci... sympatrically or parap...

Scientists do not know... that are completely iso... pared to how often it c... patric or sympatric spe... speciation in nature is... cludes studies on sma... into two or more speci... or parapatrically, since... lands is unlikely. In addition, biologists are reconstructing the evolutionary history of speciation using molecular tools and correlating this history with the species' geographical distributions. If this line of research were to show, for example, that the most closely related pairs of species always had geographically isolated distributions, it would imply that speciation was usually allopatric. These lines of research should eventually answer

Think Critically

Professor Coyne suggested that biologists will need to devise a special theory—or an additional type of species definition—to explain the distinctness of clusters of asexual organisms that are observed in nature. Do you think that this new explanation would be more similar to the morphological species concept or the phylogenetic species concept as described in this chapter?

Jerry Coyne conducts research on speciation and teaches at the University of Chicago. To learn more about his research go to http://pondside.uchicago.edu/ecol-evol/faculty/coyne_j.html.

❮ Unanswered Questions Top researchers have identified one to three important unresolved questions that relate to the chapter topic and talk about what it would mean to the field if the questions were answered.

Questions at the end of each essay help you think critically about the unanswered question, ponder possible next steps in the research, or consider the benefits of answering the question.

❯ Focus on Research Special boxes in most chapters present research topics in more depth.

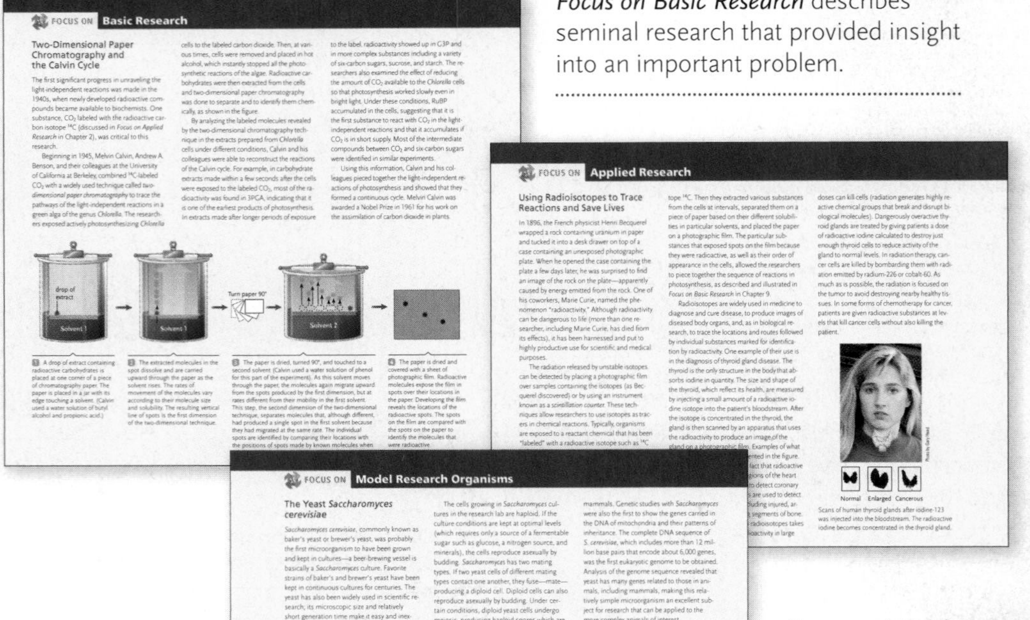

FOCUS ON **Basic Research**

Two-Dimensional Paper Chromatography and the Calvin Cycle

FOCUS ON **Applied Research**

Using Radioisotopes to Trace Reactions and Save Lives

FOCUS ON **Model Research Organisms**

The Yeast *Saccharomyces cerevisiae*

Focus on Basic Research describes seminal research that provided insight into an important problem.

Focus on Applied Research describes how scientific research has solved everyday problems.

Focus on Model Research Organisms explains why researchers use certain organisms as research subjects.

ENGAGE LIKE A SCIENTIST

Be Active. Engage yourself in the process of learning and doing biology.

› **Research Figures** Specially designed figures provide information about how biologists formulate and test specific hypotheses by gathering and interpreting data.

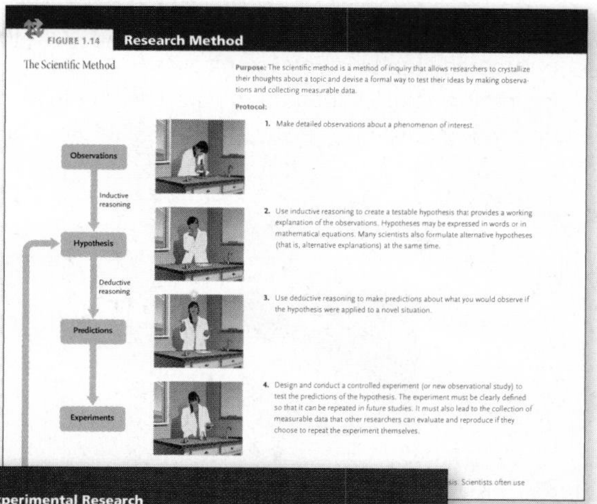

Research Method

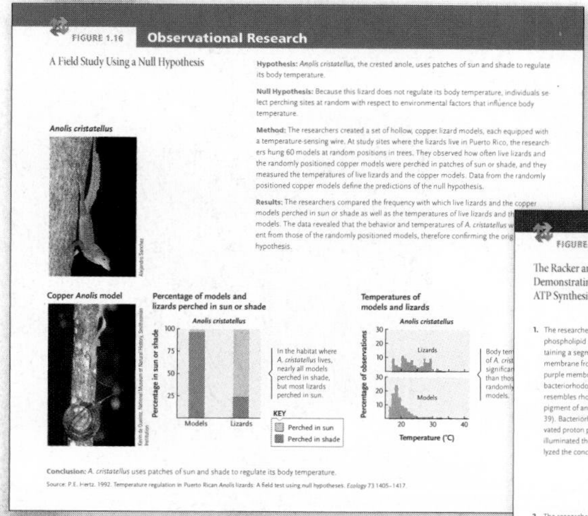

FIGURE 1.14 **Research Method**

The Scientific Method

Observational Research

FIGURE 1.16 **Observational Research**

A Field Study Using a Null Hypothesis

Experimental Research

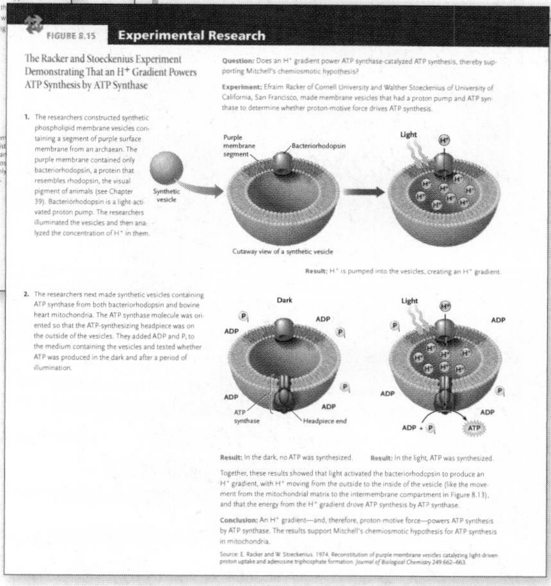

FIGURE 8.15 **Experimental Research**

The Racker and Stoeckenius Experiment Demonstrating That an H^+ Gradient Powers ATP Synthesis by ATP Synthase

› **Interpret the Data** These exercises, most of which are drawn from published biological research, help you build your skills in analyzing figures and in reading graphs and tables.

Interpret the Data

Some cancer treatments target rapidly dividing cells while leaving non-proliferating cells undisturbed. The chemicals 5-fluorouracil (5-FU) and cisplatin (CDDP) are drugs that work in this way. 5-FU inhibits DNA replication, while CDDP binds to DNA causing changes that cannot be corrected by DNA repair enzymes so that programmed cell death (see Chapter 7) is triggered. Researchers suspected that these drugs might be useful in treating human gastric cancer and tested their effectiveness in gastric cancer cells growing in culture. They added 5-FU alone, or 5-FU and CDDP in various timed combinations (schedules) to cultured gastric cancer cells and measured the inhibitory effects of the drugs on cell proliferation compared with untreated cells. The results in the figure to the right shows for each schedule the % cell proliferation, meaning the proliferation of treated cells/proliferation of control, untreated cells × 100%.

1. Which drug schedule was the most effective?
2. How did the drug schedule in the most effective treatment differ from the schedules in all the other treatments?

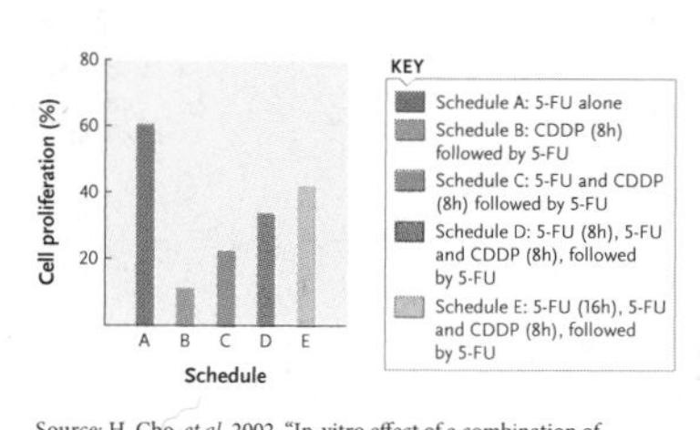

Cell proliferation (%)
80
60
40
20
A
B
C
D
E
Schedule
KEY
Schedule A: 5-FU alone
Schedule B: CDDP (8h) followed by 5-FU
Schedule C: 5-FU and CDDP (8h) followed by 5-FU
Schedule D: 5-FU (8h), 5-FU and CDDP (8h), followed by 5-FU
Schedule E: 5-FU (16h), 5-FU and CDDP (8h), followed by 5-FU

Source: H. Cho, *et al.* 2002. "In-vitro effect of a combination of 5-fluorouracil (5-FU) and cisplatin (CDDP) on human gastric cancer cell lines: timing of cisplatin treatment." *Gastric Cancer* 5:43–46.

› **Think Outside the Book** These individual, team, or Internet research activities encourage active learning and peer discussion. They foster critical thinking skills by engaging you in the topic.

THINK OUTSIDE THE BOOK

Scientists have been working to develop an artificial version of photosynthesis that can be used to produce liquid fuels from CO_2 and H_2O. Collaboratively or individually, find an example of research on artificial photosynthesis and prepare an outline of how the system works or is anticipated to work.

VISUAL LEARNING

Spectacular illustrations—developed with great care—help you visualize biological processes, relationships, and structures.

› Illustrations of complex biological processes are annotated with numbered step-by-step explanations that lead you through all the major points. Orientation diagrams are inset on figures and help you identify the specific biological process being depicted and where the process takes place.

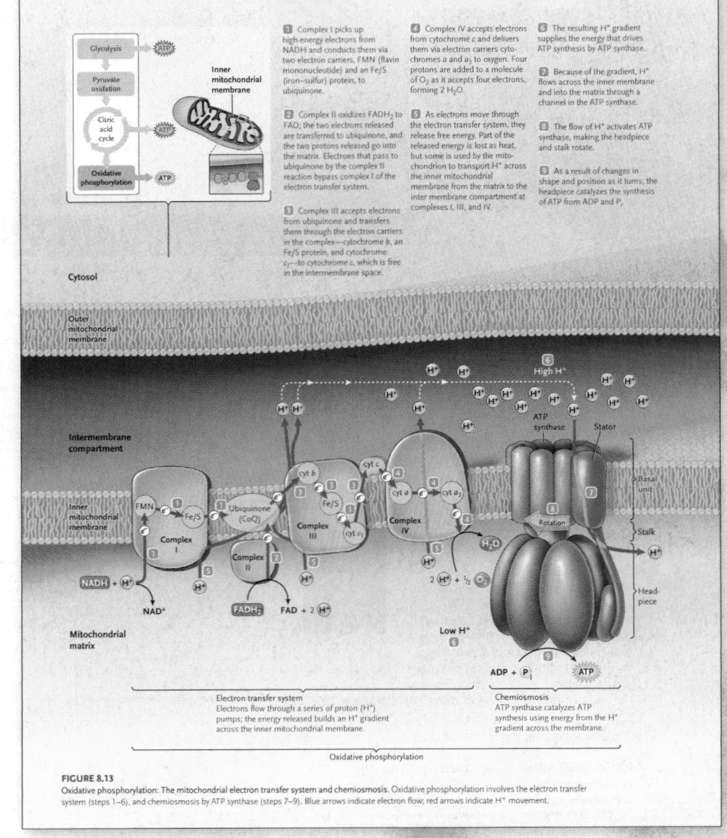
Glycolysis
Pyruvate oxidation
Citric acid cycle
Oxidative phosphorylation
ATP
Inner mitochondrial membrane
Cytosol
Outer mitochondrial membrane
Intermembrane compartment
Inner mitochondrial membrane
Mitochondrial matrix
High H+
Low H+
ATP synthase
Stator
Basal unit
Stalk
Headpiece
Rotation
Complex I
Complex II
Complex III
Complex IV
FMN
Fe/S
Ubiquinone (CoQ)
cyt b
cyt c
cyt a
cyt a_3
NADH + H+
NAD+
$FADH_2$
FAD + 2 H+
2 H+ + ½ O_2
H_2O
ADP + P_i
ATP
Electron transfer system
Electrons flow through a series of proton (H+) pumps; the energy released builds an H+ gradient across the inner mitochondrial membrane.
Chemiosmosis
ATP synthase catalyzes ATP synthesis using energy from the H+ gradient across the membrane.
Oxidative phosphorylation

FIGURE 8.13
Oxidative phosphorylation: The mitochondrial electron transfer system and chemiosmosis. Oxidative phosphorylation involves the electron transfer system (steps 1–6), and chemiosmosis by ATP synthase (steps 7–9). Blue arrows indicate electron flow; red arrows indicate H+ movement.

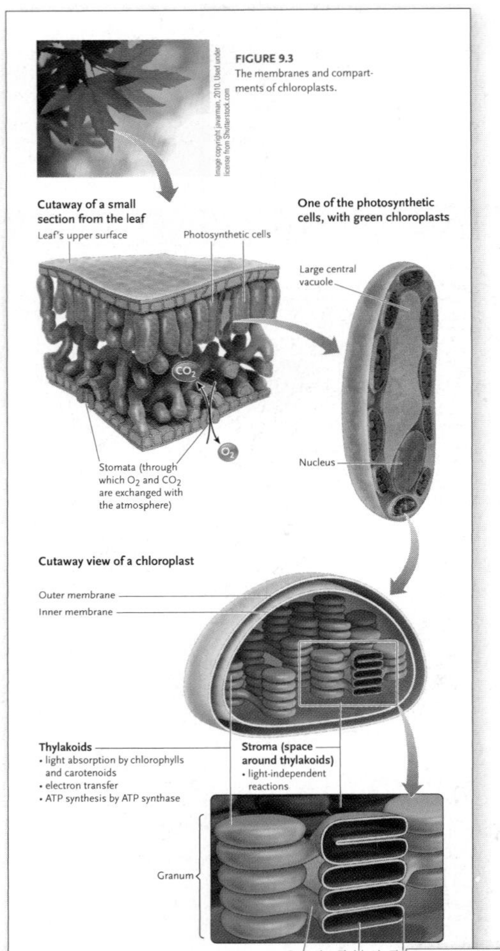
Image copyright jayarman, 2010. Used under license from Shutterstock.com
Cutaway of a small section from the leaf
Leaf's upper surface
Photosynthetic cells
One of the photosynthetic cells, with green chloroplasts
Large central vacuole
CO_2
O_2
Stomata (through which O_2 and CO_2 are exchanged with the atmosphere)
Nucleus
Cutaway view of a chloroplast
Outer membrane
Inner membrane
Thylakoids
• light absorption by chlorophylls and carotenoids
• electron transfer
• ATP synthesis by ATP synthase
Stroma (space around thylakoids)
• light-independent reactions
Granum
Stromal lamella
Thylakoid lumen

FIGURE 9.3
The membranes and compartments of chloroplasts.

‹ From Macro to Micro: Multiple views help you visualize the levels of organization of biological structures and how systems function as a whole.

› Electron micrographs are keyed to selected illustrations to help clarify biological structures.

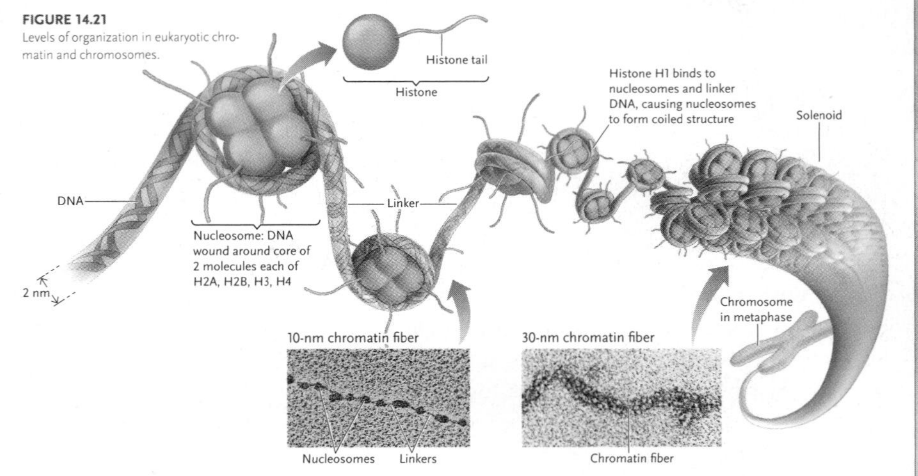
Histone tail
Histone
Histone H1 binds to nucleosomes and linker DNA, causing nucleosomes to form coiled structure
Solenoid
DNA
Linker
Nucleosome: DNA wound around core of 2 molecules each of H2A, H2B, H3, H4
2 nm
10-nm chromatin fiber
30-nm chromatin fiber
Chromosome in metaphase
Nucleosomes
Linkers
Chromatin fiber

FIGURE 14.21
Levels of organization in eukaryotic chromatin and chromosomes.

End-of-chapter material encourages you to review the content, assess your understanding, think analytically, and apply what you have learned to novel situations.

› **Review Key Concepts** This brief review references figures and tables in the chapter and provides an outline summary of important ideas developed in the chapter.

REVIEW KEY CONCEPTS

Go to **CENGAGENOW** at www.cengage.com/login to access quizzing, animations, exercises, articles, and personalized homework help.

8.1 Overview of Cellular Energy Metabolism

- Plants and almost all other organisms obtain energy for cellular activities through cellular respiration, the process of transferring electrons from donor organic molecules to a final acceptor molecule such as oxygen; the energy that is released drives ATP synthesis (Figure 8.1).
- Oxidation–reduction reactions, called redox reactions, partially or completely transfer electrons from donor to acceptor atoms; the donor is oxidized as it releases electrons, and the acceptor is reduced (Figure 8.2).
- Cellular respiration occurs in three stages: (1) In glycolysis, glucose is converted to two molecules of pyruvate through a series of enzyme-catalyzed reactions; (2) in pyruvate oxidation and the citric ... is ... phos-

phorylation, which is comprised of the electron transfer system and chemiosmosis, high-energy electrons produced from the first two stages pass through the transfer system, with much of their energy being used to establish an H^+ gradient across the membrane that drives the synthesis of ATP from ADP and P_i (Figure 8.3).

- In eukaryotes, most of the reactions of cellular respiration occur in mitochondria. In prokaryotes, glycolysis, pyruvate oxidation, and the citric acid cycle occur in the cytosol, while the rest of cellular respiration occurs on the plasma membrane (Figure 8.4).

Animation: The functional zones in mitochondria

8.2 Glycolysis: Splitting the Sugar in Half

- In glycolysis, which occurs in the cytosol, glucose (six carbons) is oxidized into two molecules of pyruvate (three carbons each). Electrons removed in the oxidations are delivered to NAD^+, producing NADH. The reaction sequence produces a net gain of 2 ATP, 2 NADH, and 2 pyruvate molecules for each molecule of glucose oxidized (Figures 8.5 and 8.7).

UNDERSTAND AND APPLY

Test Your Knowledge

1. What is the final acceptor for electrons in cellular respiration?
   a. oxygen
   b. ATP
   c. carbon dioxide
   d. hydrogen
   e. water
2. In glycolysis:
   a. free oxygen is required for the reactions to occur.
   b. ATP is used when glucose and fructose-6-phosphate are phosphorylated, and ATP is synthesized when 3-phosphoglycerate and pyruvate are formed.
   c. the enzymes that move phosphate groups on and off the molecules are uncoupling proteins.
   d. the product with the highest potential energy in the pathway is pyruvate.
   e. the end product of glycolysis moves to the electron transfer system.
3. Which of the following statements about phosphofructokinase is *false*?
   a. It is located and has its main activity in the inner mitochondrial membrane.
   b. It catalyzes a reaction to form a product with the highest potential energy in the pathway.
   c. It can be inactivated by ATP at an inhibitory site on its surface.
   d. It can be activated by ADP at an excitatory site on its surface.
   e. It can cause ADP to form.
4. Which of the following statements is *false*? Imagine that you ingested three chocolate bars just before sitting down to study this chapter. Most likely:
   a. your brain cells are using ATP.
   b. there is no deficit of the initial substrate to begin glycolysis.
   c. the respiratory processes in your brain cells are moving atoms from glycolysis through the citric acid cycle to the electron transfer system.
   d. after a couple of hours, you change position and stretch to rest certain muscle cells, which removes lactate from these muscles.
   e. after 2 hours, your brain cells are oxygen-deficient.
5. If ADP is produced in excess in cellular respiration, this excess ADP will:
   a. bind glucose to turn off glycolysis.
   b. bind glucose-6-phosphate to turn off glycolysis.
   c. bind phosphofructokinase to turn on or keep glycolysis turned on.
   d. cause lactate to form.
   e. increase oxaloacetate binding to increase NAD^+ production.
6. Which of the following statements is *false*? In cellular respiration:
   a. one molecule of glucose can produce about 32 ATP.
   b. oxygen combines directly with glucose to form carbon dioxide.
   c. a series of energy-requiring reactions is coupled to a series of energy-releasing reactions.
   d. NADH and $FADH_2$ allow H^+ to be pumped across the inner mitochondrial membrane.
   e. the electron transfer system occurs in the inner mitochondrial membrane.

‹ **Understand and Apply** End-of-chapter questions focus on factual content in the chapter while encouraging you to apply what you have learned.

Design an Experiment

There are several ways to measure cellular respiration experimentally. For example, CO_2 and O_2 gas sensors measure changes over time in the concentration of carbon dioxide or oxygen, respectively. Design two experiments to test the effects of changing two different variables or conditions (one per experiment) on the respiration of a research organism of your choice.

Design an Experiment challenges your understanding of the chapter and helps deepen your understanding of the scientific method as you consider how to develop and test hypotheses about a situation that relates to a main chapter topic.

Discuss the Concepts

1. Why do you think nucleic acids are not oxidized extensively as a cellular energy source?
2. A hospital patient was regularly found to be intoxicated. He denied that he was drinking alcoholic beverages. The doctors and nurses made a special point to eliminate the possibility that the patient or his friends were smuggling alcohol into his room, but he was still regularly intoxicated. Then, one of the doctors had an idea that turned out to be correct and cured the patient of his intoxication. The idea involved the patient's digestive system and one of the oxidative reactions covered in this chapter. What was the doctor's idea?

Discuss the Concepts enables you to participate in discussions on key questions to build your knowledge and learn from others.

Express Your Opinion

Developing new drugs is costly. There is little incentive for pharmaceutical companies to target ailments that affect relatively few individuals, such as Luft syndrome. Should governments allocate some funds to private companies that search for cures for rare disorders? Go to CengageNOW to investigate both sides of the issue and then vote online.

Express Your Opinion encourages you to weigh both sides of an issue by reading pro/con articles, and then make your opinion known through an online voting process.

Supplements

Student Supplements

APLIA FOR BIOLOGY

Engage in Biology: Aplia homework promotes active learning by:

- leading students through the thought processes of doing science.
- encouraging students to think like scientists.
- helping students understand the scientific procedures that led to scientific results covered in the textbook.
- connecting conceptual figures with doing science.

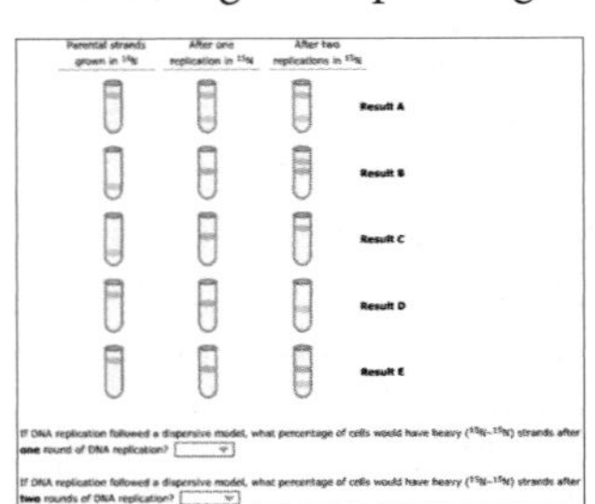

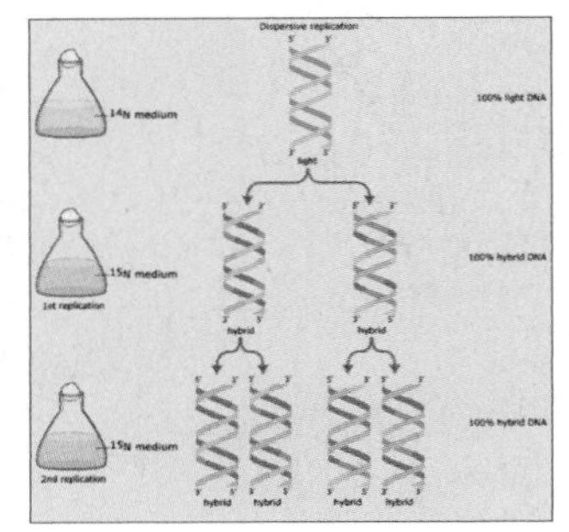

Simple, Elegant Pedagogy: Aplia uses clear, straightforward language to clarify challenging concepts. Aplia homework problems guide students through the process of learning important details and provide explanations of complex processes and systems. To further enhance student understanding, rich, clear visuals are included in the homework.

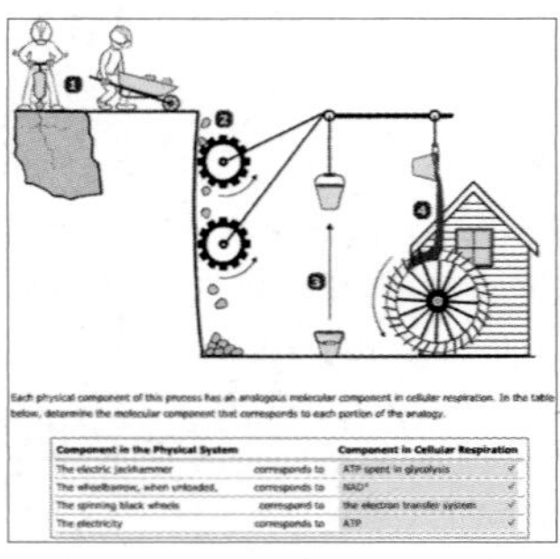

Real-World Connections: To help students better grasp the material, Aplia takes the most difficult and important biological concepts and relates them to the concrete and familiar. In this way, key biological concepts such as cell respiration are broken down to help students truly understand them.

Analyze Research Data: Illustrations and concepts within the Aplia homework are closely tied to the art in the textbook to strengthen connections between homework and the concepts students are mastering throughout the course. Students learn to analyze scientific data and interact with visuals present in the textbook and in Aplia.

Interpreting Figures: The Aplia homework integrates the textbook, the art, and the media to give students a comprehensive, visual, and interactive experience. By breaking processes down visually, Aplia helps students step through complex processes to better understand them. Within the homework, students are then asked to apply their understanding of the process to further solidify their learning.

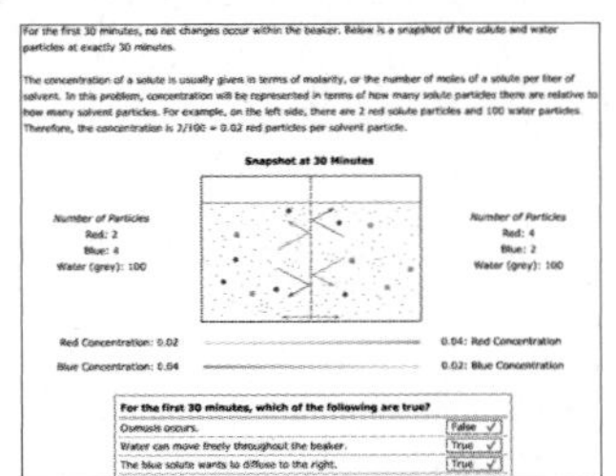

Straightforward language, a simple and intuitive user interface, and online material that reinforces the textbook all combine to make Aplia a solution that makes students think, and helps students learn.

CENGAGENOW: CengageNOW saves time for both students and professors. The Personalized Study Plan for students tests their knowledge and then guides them to focus on topics that need more of their attention. Students also can access our many high-quality animations and videos through CengageNOW. Prebuilt homework and testing capabilities for professors, as well as the ability to create your own exams in a snap, make this a versatile tool. The instructor gradebook allows you to assign study and homework options, track the progress of individual students, review automatically graded assignments, and see how students are spending most of their time.

COURSEMATE: Cengage Learning's Biology CourseMate brings course concepts to life with interactive learning, study, and exam preparation tools that support the printed textbook.

STUDENT STUDY GUIDE (0-538-49366-6): A student study tool that includes key terms, labeling exercises, self-quizzes, review questions, and critical thinking exercises to help with retention and better understanding.

AUDIO STUDY TOOLS (0-538-48896-4): These useful study aids are a great way to preview or review key concepts in the book as well as key terms and clarifications of common misconceptions.

STUDY CARD (0-538-49371-2): An at-a-glance study tool. The study card includes art and brief descriptions of major topics that focus on key points of the chapter.

PROBLEM-BASED GUIDE TO GENETICS (0-495-38468-2): A guide to learning genetics, with plenty of solved and practice problems so students can learn by doing.

ESSENTIAL STUDY SKILLS FOR SCIENCE STUDENTS (0-534-37595-2): This practical book provides tips for developing better study habits, getting more out of lectures, and how to prepare for tests.

SPANISH GLOSSARY: The glossary of biology terms helps Spanish-speaking students better understand the terminology.

Instructor Supplements

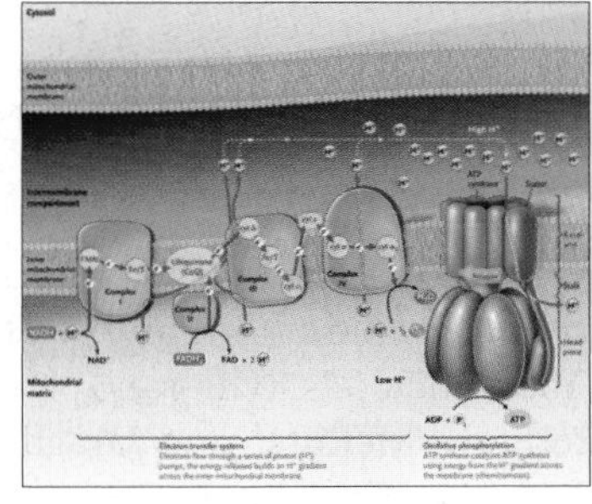

POWERLECTURE: Provides all your lecture resources on a single DVD. Power Lecture allows you to edit the text and art slides to fit your needs. Included are all the art and photos from the book plus bonus photos for you to use. All the art is available with removable labels, and key pieces of art have been "stepped" so they can be presented one segment at a time. With one click you can launch animations and videos related to the chapter, without leaving PowerPoint.

New 3-D animations help bring important concepts to life with topics such as DNA replication, mitosis, and photosynthesis.

BROOKS/COLE VIDEO LIBRARY (FEATURING BBC MOTION GALLERY VIDEO CLIPS): The Brooks/Cole Video Library contains over 40 high-quality videos that can be used alongside the text. A wide range of video topics offers professors a great tool to engage students and help them connect the material to their lives outside of the classroom.

APLIA FOR BIOLOGY: More *targeted* effort from students with Aplia's powerful and flexible tools.

Aplia is an online interactive homework solution that improves learning by increasing student effort and engagement. Aplia courses are customized to fit with each instructor's syllabus and provide automatically graded homework with detailed, immediate feedback on every question. Aplia's interactive tools also serve to increase student engagement and understanding. The Aplia assignments match the language, style, and structure of the textbook, allowing your students to apply what they learn in the text directly to their homework.

Real-time data on student and class performance are available in the Aplia Gradebook. Aplia Gradebook Analytics show how each student is doing relative to peers in the class. This allows instructors to quickly and easily:

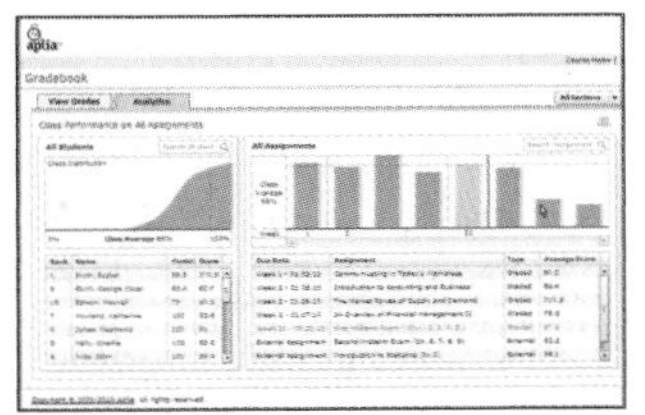

- review progress student-by-student, and even drill down to the homework and question level.
- use graphs to monitor student, and class, performance.
- determine where your students are struggling and where they may need more review.

INSTRUCTOR MANUAL: A great planning tool that includes chapter outlines, objectives, suggestions for presenting the material, classroom and laboratory enrichment ideas, possible answers to critical thinking questions, and more. Also included in Microsoft Word format on the PowerLecture DVD.

TEST BANK: Test items that are ranked according to difficulty and Bloom's Taxonomy. Questions include multiple-choice (organized by chapter heading), matching, classification, selecting the exception, and labeling exercises. Also included in Microsoft Word format on the PowerLecture DVD.

RESOURCE INTEGRATION GUIDE: A chapter-by-chapter guide to help you use the book's resources effectively. Please ask your local sales representative for more information.

EXAMVIEW: Create, deliver, and customize tests and study guides, both print and online, in minutes with this easy-to-use assessment and tutorial system.

Course Management Systems

- Aplia: This full course management system can be used either independently or in conjunction with other course management systems such as Blackboard and WebCT.
- CengageNOW: An easy way to assign homework and track the results while giving students an efficient way to study.
- WebTutors for Blackboard and WebCT: Now all your favorite media, quizzing, and other online assets are available at your fingertips in the easy-to-use cartridge, with the classroom system of your choice.
- CourseMate: Cengage Learning's text-specific website, *Biology CourseMate* brings course concepts to life with interactive learning, study, and exam preparation tools that support the printed textbook.

Lab Options

VIRTUAL BIOLOGY LABS (VBL) 4.0: This online tool includes 14 labs with 130 activities that give students virtual experience gathering data and performing experiments through engaging simulations. Students can change parameters to see what happens in each simulation, generate their own data, and write up results. Each experiment includes a general introduction followed by a series of interactive laboratory exercises. Labs can be assigned in a learning management system, and each exercise includes its own quiz questions that students can submit electronically or print out and submit.

SIGNATURE LABS: Whether you prefer to select the labs you need from our extensive collection, co-develop our content to match your lab exactly, or create and publish your own experiments, we will help you create a lab manual that is a perfect fit for your lab.

MAJORS BIOLOGY LAB MANUAL (0-495-11505-3): This lab manual includes 30 laboratory exercises, many of which incorporate inquiry-based experiments. Students are provided with exciting, relevant activities and experiments that allow them to explore some of the rapidly developing areas of biological science. Each laboratory exercise includes objectives, an introduction, materials lists, procedures, and pre- and post-lab questions.

Acknowledgments

Revising a text from its first edition to its second is a gigantic project, and the helpful assistance of many people enabled us to accomplish the task in a timely manner.

Michelle Julet and Yolanda Cossio have provided the essential support and encouragement, and "cracked the whip" when necessary, enabling us to bring the project to fruition.

Our Developmental Editors have served as "midwives" to this book. They have compiled, interpreted, and sometimes deconstructed reviewer comments; their analyses and insights have helped us tighten the narrative and maintain a steady course. Our hats are off to Mary Arbogast, who has worked patiently on this project since its conception. Shelley Parlante has provided very helpful guidance as the manuscript matured. Suzannah Alexander helped to organize our art development program and kept it on track; she also offered helpful suggestions on many chapters. Jake Warde coordinated the efforts of the numerous contributors of Unanswered Questions essays, as well as a host of reviewers; Jake also helped us organize the end-of-chapter material.

We are grateful to Elizabeth Momb for coordinating the print supplements and our Editorial Assistant Brandy Radoias for managing all our reviewer information.

We offer many thanks to Lauren Oliveira and Shelley Ryan, who supervised our partnership with our technology authors and media advisory board. Their collective efforts allow us to create a set of tools that support students in learning and instructors in teaching.

We thank the Aplia for Biology team, Qinzi Ji, Andy Marinkovich, and Peg Knight for building a learning solution that is truly integrated with our text.

We appreciate the help of the production staff led by Teresa Trego and Dan Fitzgerald at Graphic World. We thank our Creative Director Rob Hugel and Art Director John Walker.

The outstanding art program is the result of the collaborative talent, hard work, and dedication of a select group of people. The meticulous styling and planning of the program are credited to Steve McEntee and Dragonfly Media Group, led by Mike Demaray. The DMG group created hundreds of complex, vibrant art pieces. Steve's role was crucial in overseeing the development and consistency of the art program; he was also the illustrator for the unique—and brilliantly rendered—Research figures. We would like to thank Cecie Starr for the use of selected art pieces.

We also wish to acknowledge Tom Ziolkowski, our Marketing Manager, whose expertise ensured that all of you would know about this new book.

Peter Russell thanks Stephen Arch of Reed College for his expert input, valuable discussions, and advice during the revision of the Unit Six chapters on Animal Structure and Function. Paul Hertz thanks Hilary Callahan, John Glendinning, and Brian Morton of Barnard College for their generous advice on many phases of this project, and Eric Dinerstein of the World Wildlife Fund for his contributions to the discussion of Conservation Biology. Paul especially thanks Jamie Rauchman for extraordinary patience and endless support as this book was written (and rewritten, and rewritten again) as well as his thousands of past students at Barnard College, who have taught him at least as much as he has taught them. Beverly McMillan thanks John A. Musick, Acuff Professor Emeritus at the College of William and Mary—and an award-winning teacher and mentor to at least two generations of college students—for patient and thoughtful discussions about effective ways to present the often complex subject matter of biological science.

We would also like to thank our advisors and contributors:

SECOND EDITION REVIEWERS AND CONTRIBUTORS

Tracey M. Anderson, *University of Minnesota Morris*
Stephen Arch, *Reed College*
Mitchell F. Balish, *Miami University*
Timothy J. Baroni, *State University of New York at Cortland*
Erwin A. Bautista, *University of California, Davis*
Michael C. Bell, *Richland College*
William L. Bischoff, *The University of Toledo*
Catherine Black, *Idaho State University*
Andrew R. Blaustein, *Oregon State University*
Robert S. Boyd, *Auburn University*
Arthur L. Buikema, *Virginia Polytechnic Institute and State University*
Carolyn J. W. Bunde, *Idaho State University*
Domenic Castignetti, *Loyola University Chicago–Lake Shore*
Peter Chen, *College of DuPage*
James W. Clack, *Indiana University–Purdue University Indianapolis*
Karen Curto, *University of Pittsburgh*
Rebekka Darner, *University of Florida*
Eric Dinerstein, *World Wildlife Fund*
Nick Downey, *University of Wisconsin–LaCrosse*
Kathryn A. Durham, *Luzerne County Community College*
Jamin Eisenbach, *Eastern Michigan University*
Kathleen Engelmann, *University of Bridgeport*
Helene Engler, *Science Consultant and Lecturer*
Jose Luis Ergemy, *Northwest Vista College*
Frederick B. Essig, *University of South Florida*
Brent E. Ewers, *University of Wyoming*
Mark A. Farmer, *University of Georgia*
Michael B. Ferrari, *University of Missouri–Kansas City*
David H. A. Fitch, *New York University*
Anne M. Galbraith, *University of Wisconsin–LaCrosse*
E. Eileen Gardner, *William Paterson University*
David W. Garton, *Georgia Institute of Technology*
John R. Geiser, *Western Michigan University*
Florence Gleason, *University of Minnesota Twin Cities*
Erich Grotewold, *Ohio State University*
R. James Hickey, *Miami University*

Kelly Hogan, *University of North Carolina*
Eric Jellen, *Brigham Young University*
Walter S. Judd, *University of Florida*
David Kiewlich, *Science Consultant and Research Biologist*
Scott L. Kight, *Montclair State University*
Richard Knapp, *University of Houston*
David Kooyman, *Brigham Young University*
Olga Ruiz Kopp, *Utah Valley State University*
Shannon Lee, *California State University, Northridge*
Charly Mallery, *University of Miami*
Paul Manos, *Duke University*
Patricia Matthews, *Grand Valley State University*
Jacqueline S. McLaughlin, *Penn State University–Lehigh Valley*
Jennifer Metzler, *Ball State University*
Melissa Michael, *University of Illinois at Urbana-Champaign*
Jeanne M. Mitchell, *Truman State University*
Roderick Morgan, *Grand Valley State University*
Jacalyn S. Newman, *University of Pittsburgh*
Todd W. Osmundson, *University of California, Berkeley*
Kathryn Perez, *University of Wisconsin–LaCrosse*
Vinnie Peters, *Indiana University–Purdue University Fort Wayne*
Steve Vincent Pollock, *Louisiana State University*
Elena Pravosudova, *University of Nevada, Reno*
Jason M. Rauceo, *John Jay College of Criminal Justice*
Michael Reagan, *College of Saint Benedict and Saint John's University*
Melissa Murray Reedy, *University of Illinois at Urbana-Champaign*
Laurel Roberts, *University of Pittsburgh*
Scott D. Russell, *University of Oklahoma*
Stephen G. Saupe, *College of Saint Benedict and Saint John's University*
Nancy N. Shontz, *Grand Valley State University*
Jennifer L. Siemantel, *Cedar Valley College*
Richard Stalter, *College of St. Benedict and St. John's University*
Sonja Stampfler, *Kellogg Community College*
Brian Stout, *Northwest Vista College*
Gregory W. Stunz, *Texas A&M University*
Mark T. Sugalski, *Southern Polytechnic State University*
Salvatore Tavormina, *Austin Community College*
Ken Thomas, *Northern Essex Community College*
Terry M. Trier, *Grand Valley State University*
William Velhagen, *New York University*
Alexander Wait, *Missouri State University*
R. Douglas Watson, *The University of Alabama at Birmingham*
Cindy Wedig, *The University of Texas–Pan American*
Michael N. Weintraub, *The University of Toledo*
Sue Simon Westendorf, *Ohio University*
Ward Wheeler, *American Museum of Natural History, Division of Invertebrate Zoology*
Elizabeth Willott, *The University of Arizona*
Yunde Zhao, *University of California, San Diego*
Heping Zhou, *Seton Hall University*

ART ADVISORY BOARD

Robert S. Boyd, *Auburn University*
Patricia J. S. Colberg, *University of Wyoming*
Jose Egremy, *Northwest Vista College*
Michael B. Ferrari, *University of Missouri–Kansas City*
Tim Gerber, *University of Wisconsin–LaCrosse*
Stephen G. Saupe, *College of Saint Benedict and Saint John's University*
Brian Stout, *Northwest Vista College*
Mark Sugalski, *Southern Polytechnic State University*
Terry Trier, *Grand Valley State University*
R. Douglas Watson, *The University of Alabama at Birmingham*
Cindy Wedig, *The University of Texas–Pan American*

ACCURACY CHECKERS

Asim Bej, *University of Alabama at Birmingham*
Anne Bergey, *Truman State University*
Domenic Castignetti, *Loyola University Chicago–Lake Shore*
Robin Cooper, *University of Kentucky*
Eric Dinerstein, *World Wildlife Fund*
Frederick B. Essig, *University of South Florida*
Michael B. Ferrari, *University of Missouri–Kansas City*
Anne Galbraith, *University of Wisconsin–LaCrosse*
Scott Gleeson, *University of Kentucky*
Erich Grotewold, *Ohio State University*
Donna Koslowsky, *Michigan State University*
Michael Meighan, *University of California, Berkeley*
Todd Osmundson, *University of California, Berkeley*
Kathryn Perez, *University of Wisconsin–LaCrosse*
Jason Rauceo, *John Jay College of Criminal Justice*
Ann Rushing, *Baylor University*
Scott Russell, *The University of Oklahoma*
Mark Sheridan, *North Dakota State University*
Tom Stidham, *Texas A&M University*
Gregory W. Stunz, *Texas A&M University*
R. Douglas Watson, *The University of Alabama at Birmingham*
Michael Weintraub, *The University of Toledo*
Yunde Zhao, *University of California, San Diego*

SUPPLEMENTS AUTHORS

David Asch, *Youngstown State University*
Carolyn Bunde, *Idaho State University*
Frederick B. Essig, *University of South Florida*
Brent Ewers, *University of Wyoming*
Anne Galbraith, *University of Wisconsin–LaCrosse*
Kathleen Hecht, *Nassau Community College*
William Kroll, *Loyola University Chicago–Lake Shore*
Todd Osmundson, *University of California, Berkeley*
Debra Pires, *University of California, Los Angeles*
Elena Pravosudova, *University of Reno, Nevada*
Jeff Roth-Vinson, *Cottage Grove High School*
Mark Sheridan, *North Dakota State University*
Gary Shin, *California State University, Long Beach*
Michael Silva, *El Paso Community College*
Catherine Anne Ueckert, *Northern Arizona University*
Jyoti Wagle, *Houston Community College, Central College*
Alexander Wait, *Missouri State University*

APLIA FOR BIOLOGY REVIEWERS AND CLASS TESTERS

Thomas Abbott, *University of Connecticut*
John Bell, *Brigham Young University*
Anne Bergey, *Truman State University*
Carolyn Bunde, *Idaho State University*
Jung H. Choi, *Georgia Institute of Technology*
Tim W. Christensen, *East Carolina University*
Patricia J. S. Colberg, *University of Wyoming*
Robin Cooper, *University of Kentucky*
Karen Curto, *University of Pittsburgh*
Joe Demasi, *Massachusetts College of Pharmacy and Health Science*
Nicholas Downey, *University of Wisconsin–LaCrosse*
Lisa Elfring, *University of Arizona*
Kathleen Engelmann, *University of Bridgeport*
Monika Espinasa, *State University of New York at Ulster*
Michael Ferrari, *University of Missouri–Kansas City*
David Fitch, *New York University*
Paul Fitzgerald, *Northern Virginia Community College*
Steven Francoeur, *Eastern Michigan University*
Ed Himelblau, *California Polytechnic State University–San Luis Obispo*
Justin Hoffman, *McNeese State University*
Ashok Jain, *Albany State University*
Susan Jorstad, *University of Arizona*
Christopher Kirkhoff, *McNeese State University*
Richard Knapp, *University of Houston*
Nathan Lents, *John Jay College*
Janet Loxterman, *Idaho State University*
Susan McRae, *East Carolina University*
Brad Mehrtens, *University of Illinois at Urbana-Champaign*
Jennifer Metzler, *Ball State University*
Bruce Mobarry, *University of Idaho*
Jennifer Moon, *The University of Texas at Austin*
Robert Osuna, *State University of New York at Albany*
Matt Palmer, *Columbia University*
Roger Persell, *Hunter College*
Michael Reagan, *College of Saint Benedict and Saint John's University*
Ann Rushing, *Baylor University*
Jeanne Serb, *Iowa State University*
Leah Sheridan, *University of Northern Colorado*
Mark Sheridan, *North Dakota State University*
Nancy N. Shontz, *Grand Valley State University*
Linda Stabler, *University of Central Oklahoma*
Mark Staves, *Grand Valley State University*
Eric Strauss, *University of Wisconsin–LaCrosse*
David Tam, *University of North Texas*
Rebecca Thomas, *Montgomery College*
David H. Townson, *University of New Hampshire*
David Vleck, *Iowa State University*
Miryam Wahrman, *William Paterson University*
Alexander Wait, *Missouri State University*
Johanna Weiss, *Northern Virginia Community College*
Lisa Williams, *Northern Virginia Community College*
Marilyn Yoder, *University of Missouri–Kansas City*

MEDIA REVIEWERS AND CONTRIBUTORS

David Asch, *Youngstown State University*
Gerald Bergtrom, *University of Wisconsin–Milwaukee*
Scott Bowling, *Auburn University*
Joi Braxton-Sanders, *Northwest Vista College*
Albia Dugger, *Miami-Dade College*
Natalie Dussourd, *Illinois State University*
Bert Ely, *University of South Carolina*
Helene Engler, *Science Writer*
Daria Hekmat-Scafe, *Stanford University*
Jutta Heller, *Loyola University Chicago–Lake Shore*
Kelly Howe, *University of New Mexico*
Judy Kaufman, *Monroe Community College*
David Kiewlich, *Research Biologist*
William Kroll, *Loyola University Chicago–Lake Shore*
Michael Silva, *El Paso Community College*
Mark Sturtevant, *Oakland University*
Mark Sugalski, *Southern Polytechnic State University*
Salvatore Tavormina, *Austin Community College*
Neal Voelz, *St. Cloud State University*
Camille Wagner, *San Jacinto College*
Suzanne Wakim, *Butte Community College*
Martin Zhan, *Thomas Nelson Community College*

FIRST EDITION REVIEWERS AND CONTRIBUTORS

Heather Addy, *The University of Calgary*
Adrienne Alaie-Petrillo, *Hunter College–CUNY*
Richard Allison, *Michigan State University*
Terry Allison, *The University of Texas–Pan American*
Deborah Anderson, *Saint Norbert College*
Robert C. Anderson, *Idaho State University*
Andrew Andres, *University of Nevada, Las Vegas*
Steven M. Aquilani, *Delaware County Community College*
Jonathan W. Armbruster, *Auburn University*
Peter Armstrong, *University of California, Davis*
John N. Aronson, *The University of Arizona*
Joe Arruda, *Pittsburgh State University*
Karl Aufderheide, *Texas A&M University*
Charles Baer, *University of Florida*
Gary I. Baird, *Brigham Young University*
Aimee Bakken, *University of Washington*
Marica Bakovic, *University of Guelph*
Michael Baranski, *Catawba College*
Michael Barbour, *University of California, Davis*
Edward M. Barrows, *Georgetown University*
Anton Baudoin, *Virginia Polytechnic Institute and State University*
Penelope H. Bauer, *Colorado State University*
Kevin Beach, *The University of Tampa*
Mike Beach, *Southern Polytechnic State University*
Ruth Beattie, *University of Kentucky*
Robert Beckmann, *North Carolina State University*
Jane Beiswenger, *University of Wyoming*
Andrew Bendall, *University of Guelph*
Catherine Black, *Idaho State University*
Andrew Blaustein, *Oregon State University*
Anthony H. Bledsoe, *University of Pittsburgh*
Harriette Howard-Lee Block, *Prairie View A&M University*
Dennis Bogyo, *Valdosta State University*
David Bohr, *University of Michigan*
Emily Boone, *University of Richmond*
Hessel Bouma III, *Calvin College*
Nancy Boury, *Iowa State University*
Scott Bowling, *Auburn University*
Laurie Bradley, *Hudson Valley Community College*
William Bradshaw, *Brigham Young University*
J. D. Brammer, *North Dakota State University*
G. L. Brengelmann, *University of Washington*
Randy Brewton, *University of Tennessee–Knoxville*

Bob Brick, *Blinn College–Bryan*
Mirjana Brockett, *Georgia Institute of Technology*
William Bromer, *University of Saint Francis*
William Randy Brooks, *Florida Atlantic University–Boca Raton*
Mark Browning, *Purdue University*
Gary Brusca, *Humboldt State University*
Alan H. Brush, *University of Connecticut*
Arthur L. Buikema, Jr., *Virginia Polytechnic Institute and State University*
Carolyn Bunde, *Idaho State University*
E. Robert Burns, *University of Arkansas for Medical Sciences*
Ruth Buskirk, *The University of Texas at Austin*
David Byres, *Florida Community College at Jacksonville*
Christopher S. Campbell, *The University of Maine*
Angelo Capparella, Illinois State University
Marcella D. Carabelli, *Broward Community College–North*
Jeffrey Carmichael, *University of North Dakota*
Bruce Carroll, *North Harris Montgomery Community College*
Robert Carroll, *East Carolina University*
Patrick Carter, *Washington State University*
Christine Case, *Skyline College*
Domenic Castignetti, *Loyola University Chicago–Lake Shore*
Jung H. Choi, *Georgia Institute of Technology*
Kent Christensen, *University of Michigan Medical School*
John Cogan, *Ohio State University*
Linda T. Collins, *University of Tennessee–Chattanooga*
Lewis Coons, *University of Memphis*
Joe Cowles, *Virginia Polytechnic Institute and State University*
George W. Cox, *San Diego State University*
David Crews, *The University of Texas at Austin*
Paul V. Cupp, Jr., *Eastern Kentucky University*
Karen Curto, *University of Pittsburgh*
Anne M. Cusic, *The University of Alabama at Birmingham*
David Dalton, *Reed College*
Frank Damiani, *Monmouth University*
Peter J. Davies, *Cornell University*
Fred Delcomyn, *University of Illinois at Urbana-Champaign*
Jerome Dempsey, *University of Wisconsin–Madison*
Philias Denette, *Delgado Community College–City Park*
Nancy G. Dengler, *University of Toronto*
Jonathan J. Dennis, *University of Alberta*
Daniel DerVartanian, *University of Georgia*
Donald Deters, *Bowling Green State University*
Kathryn Dickson, *California State University, Fullerton*
Kevin Dixon, *University of Illinois at Urbana-Champaign*
Gordon Patrick Duffie, *Loyola University Chicago–Lake Shore*
Charles Duggins, *University of South Carolina*
Carolyn S. Dunn, *University of North Carolina–Wilmington*
Roland R. Dute, *Auburn University*
Melinda Dwinell, *Medical College of Wisconsin*
Gerald Eck, *University of Washington*
Gordon Edlin, *University of Hawaii*
William Eickmeier, *Vanderbilt University*
Ingeborg Eley, *Hudson Valley Community College*
Paul R. Elliott, *Florida State University*
John A. Endler, *University of Exeter*
Brent Ewers, *University of Wyoming*
Daniel J. Fairbanks, *Brigham Young University*
Piotr G. Fajer, *Florida State University*
Richard H. Falk, *University of California, Davis*
Ibrahim Farah, *Jackson State University*
Jacqueline Fern, *Lane Community College*
Daniel P. Fitzsimons, *University of Wisconsin–Madison*
Daniel Flisser, *Camden County College*
R. G. Foster, *University of Virginia*
Dan Friderici, *Michigan State University*
J. W. Froehlich, *The University of New Mexico*
Paul Garcia, *Houston Community College–Southwest*
Umadevi Garimella, *University of Central Arkansas*
Robert P. George, *University of Wyoming*
Stephen George, *Amherst College*
John Giannini, *St. Olaf College*
Joseph Glass, *Camden County College*
John Glendinning, *Barnard College*
Elizabeth Godrick, *Boston University*
Judith Goodenough, *University of Massachusetts Amherst*
H. Maurice Goodman, *University of Massachusetts Medical School*
Bruce Grant, *College of William and Mary*
Becky Green-Marroquin, *Los Angeles Valley College*
Christopher Gregg, *Louisiana State University*
Katharine B. Gregg, *West Virginia Wesleyan College*
John Griffin, *College of William and Mary*
Samuel Hammer, *Boston University*
Aslam Hassan, *University of Illinois at Urbana-Champaign*
Albert Herrera, *University of Southern California*
Wilford M. Hess, *Brigham Young University*
Martinez J. Hewlett, *The University of Arizona*
Christopher Higgins, *Tarleton State University*
Phyllis C. Hirsch, *East Los Angeles College*
Carl Hoagstrom, *Ohio Northern University*
Stanton F. Hoegerman, *College of William and Mary*
Ronald W. Hoham, *Colgate University*
Margaret Hollyday, *Bryn Mawr College*
John E. Hoover, *Millersville University*
Howard Hosick, *Washington State University*
William Irby, *Georgia Southern University*
John Ivy, *Texas A&M University*
Alice Jacklet, *University at Albany, State University of New York*
John D. Jackson, *North Hennepin Community College*
Jennifer Jeffery, *Wharton County Junior College*
John Jenkin, *Blinn College–Bryan*
Leonard R. Johnson, *The University of Tennessee College of Medicine*
Walter Judd, *University of Florida*
Prem S. Kahlon, *Tennessee State University*
Thomas C. Kane, *University of Cincinnati*
Peter Kareiva, *University of Washington*
Gordon I. Kaye, *Albany Medical College*
Greg Keller, *Eastern New Mexico University–Roswell*
Stephen Kelso, *University of Illinois at Chicago*
Bryce Kendrick, *University of Waterloo*
Bretton Kent, *University of Maryland*
Jack L. Keyes, *Linfield College Portland Campus*
John Kimball, *Tufts University*
Hillar Klandorf, *West Virginia University*
Michael Klymkowsky, *University of Colorado at Boulder*
Loren Knapp, *University of South Carolina*
Ana Koshy, *Houston Community College–Northwest*
Kari Beth Krieger, *University of Wisconsin–Green Bay*
David T. Krohne, *Wabash College*
William Kroll, *Loyola University Chicago–Lake Shore*
Josepha Kurdziel, *University of Michigan*
Allen Kurta, *Eastern Michigan University*
Howard Kutchai, *University of Virginia*
Paul K. Lago, *The University of Mississippi*
John Lammert, *Gustavus Adolphus College*
William L'Amoreaux, *College of Staten Island–CUNY*
Brian Larkins, *The University of Arizona*
William E. Lassiter, *University of North Carolina–Chapel Hill*
Shannon Lee, *California State University, Northridge*
Lissa Leege, *Georgia Southern University*
Matthew Levy, *Case Western Reserve University*

Harvey Liftin, *Broward Community College–Central*
Tom Lonergan, *University of New Orleans*
Lynn Mahaffy, *University of Delaware*
Alan Mann, *University of Pennsylvania*
Kathleen Marrs, *Indiana University–Purdue University Indianapolis*
Robert Martinez, *Quinnipiac University*
Joyce B. Maxwell, *California State University, Northridge*
Jeffrey D. May, *Marshall University*
Geri Mayer, *Florida Atlantic University*
Jerry W. McClure, *Miami University*
Andrew G. McCubbin, *Washington State University*
Mark McGinley, *Texas Tech University*
F. M. Anne McNabb, *Virginia Polytechnic Institute and State University*
Mark Meade, *Jacksonville State University*
Bradley Mehrtens, *University of Illinois at Urbana-Champaign*
Michael Meighan, *University of California, Berkeley*
Catherine Merovich, *West Virginia University*
Richard Merritt, *Houston Community College*
Ralph Meyer, *University of Cincinnati*
Melissa Michael, *University of Illinois at Urbana-Champaign*
James E. "Jim" Mickle, *North Carolina State University*
Hector C. Miranda, Jr., *Texas Southern University*
Jasleen Mishra, *Houston Community College–Southwest*
David Mohrman, *University of Minnesota Medical School Duluth*
John M. Moore, *Taylor University*
Roderick M. Morgan, *Grand Valley State University*
David Morton, *Frostburg State University*
Alexander Motten, *Duke University*
Alan Muchlinski, *California State University, Los Angeles*
Michael Muller, *University of Illinois at Chicago*
Richard Murphy, *University of Virginia*
Darrel L. Murray, *University of Illinois at Chicago*
Allan Nelson, *Tarleton State University*
David H. Nelson, *University of South Alabama*
Jacalyn Newman, *University of Pittsburgh*
David O. Norris, *The University of Colorado*
Bette Nybakken, *Hartnell College*
Tom Oeltmann, *Vanderbilt University*
Diana Oliveras, *The University of Colorado at Boulder*
Alexander E. Olvido, *Virginia State University*
Karen Otto, *The University of Tampa*
William W. Parson, *University of Washington School of Medicine*
James F. Payne, *The University of Memphis*
Craig Peebles, *University of Pittsburgh*
Joe Pelliccia, *Bates College*
Susan Petro, *Ramapo College of New Jersey*
Debra Pires, *University of California, Los Angeles*
Thomas Pitzer, *Florida International University*
Roberta Pollock, *Occidental College*
Jerry Purcell, *San Antonio College*
Kim Raun, *Wharton County Junior College*
Tara Reed, *University of Wisconsin–Green Bay*
Lynn Robbins, *Missouri State University*
Carolyn Roberson, *Roane State Community College*
Laurel Roberts, *University of Pittsburgh*
Kenneth Robinson, *Purdue University*
Frank A. Romano, *Jacksonville State University*
Michael R. Rose, *University of California, Irvine*
Michael S. Rosenzweig, *Virginia Polytechnic Institute and State University*
Linda S. Ross, *Ohio University*
Ann Rushing, *Baylor University*
Linda Sabatino, *Suffolk Community College*
Tyson Sacco, *Cornell University*
Peter Sakaris, *Southern Polytechnic State University*
Frank B. Salisbury, *Utah State University*
Mark F. Sanders, *University of California, Davis*
Andrew Scala, *Dutchess Community College*
John Schiefelbein, *University of Michigan*
Deemah Schirf, *The University of Texas at San Antonio*
Kathryn J. Schneider, *Hudson Valley Community College*
Jurgen Schnermann, *University of Michigan Medical School of Medicine*
Thomas W. Schoener, *University California, Davis*
Brian Shea, *Northwestern University*
Mark Sheridan, *North Dakota State University*
Dennis Shevlin, *The College of New Jersey*
Richard Showman, *University of South Carolina*
Bill Simcik, *Lone Star College–Tomball*
Robert Simons, *University of California, Los Angeles*
Roger Sloboda, *Dartmouth College*
Jerry W. Smith, *St. Petersburg College*
Nancy Solomon, *Miami University*
Bruce Stallsmith, *The University of Alabama in Huntsville*
Karl Sternberg, *Western New England College*
Pat Steubing, *University of Nevada, Las Vegas*
Karen Steudel, *University of Wisconsin–Madison*
Richard D. Storey, *The Colorado College*
Michael A. Sulzinski, *The University of Scranton*
Marshall Sundberg, *Emporia State University*
David Tam, *University of North Texas*
David Tauck, *Santa Clara University*
Jeffrey Taylor, *Slippery Rock University of Pennsylvania*
Franklyn Te, *Miami Dade College*
Roger E. Thibault, *Bowling Green State University*
Megan Thomas, *University of Nevada, Las Vegas*
Patrick Thorpe, *Grand Valley State University*
Ian Tizard, *Texas A&M University*
Robert Turner, *Western Oregon University*
Joe Vanable, *Purdue University*
Linda H. Vick, *North Park University*
J. Robert Waaland, *University of Washington*
Douglas Walker, *Wharton County Junior College*
James Bruce Walsh, *The University of Arizona*
Fred Wasserman, *Boston University*
Edward Weiss, *Christopher Newport University*
Mark Weiss, *Wayne State University*
Adrian M. Wenner, *University of California, Santa Barbara*
Adrienne Williams, *University of California, Irvine*
Mary Wise, *Northern Virginia Community College*
Charles R. Wyttenbach, *The University of Kansas*
Robert Yost, *Indiana University–Purdue University Indianapolis*
Xinsheng Zhu, *University of Wisconsin–Madison*
Adrienne Zihlman, *University of California, Santa Cruz*

Unanswered Questions Contributors

Chapter 2
Peter J. Russell, *Reed College*

Chapter 3
Michael S. Brown and Joseph L. Goldstein, *University of Texas Southwestern Medical School*

Chapter 4
Peter J. Russell, *Reed College*

Chapter 5
Matthew Welch, *University of California, Berkeley*

Chapter 6
Peter Agre, *Johns Hopkins Malaria Research Institute*

Chapter 7
Jeffrey Blaustein, *University of Massachusetts Amherst*

Chapter 8
Gail A. Breen, *University of Texas at Dallas*

Chapter 9
David Kramer, *Washington State University*

Chapter 10
Raymond Deshaies, *California Institute of Technology*

Chapter 11
Peter J. Russell, *Reed College*

Chapter 12
Nicholas Katsanis, *Duke University*

Chapter 13
Peter J. Russell, *Reed College*

Chapter 14
Janis Shampay, *Reed College*

Chapter 15
Harry Noller, *University of California, Santa Cruz*

Chapter 16
Mark A. Kay, *Stanford School of Medicine*

Chapter 17
Gerald Baron, *Rocky Mountain Laboratories*

Chapter 18
Larisa H. Cavallari, *University of Illinois at Chicago College of Pharmacy*

Chapter 19
Douglas J. Futuyma, *Stony Brook University*

Chapter 20
Mohamed Noor, *Duke University*

Chapter 21
Jerry Coyne, *University of Chicago*

Chapter 22
Elena M. Kramer, *Harvard University*

Chapter 23
Richard Glor, *University of Rochester*

Chapter 24
Andrew Pohorille, *National Aeronautics and Space Administration (NASA)*

Chapter 25
Rachel Y. Samson and Stephen D. Bell, *Oxford University*

Chapter 26
Geoff McFadden, *University of Melbourne*

Chapter 27
Amy Litt, *The New York Botanical Garden*

Chapter 28
Todd Osmundson, *University of California, Berkeley*

Chapter 29
William S. Irby, *Georgia Southern University*

Chapter 30
Marvalee H. Wake, *University of California, Berkeley*

Chapter 31
Jennifer Fletcher, *University of California, Berkeley*

Chapter 32
Beverly McMillan

Chapter 33
Michael Weintraub, *University of Toledo*

Chapter 34
Ravi Palanivelu, *University of Arizona*

Chapter 35
Christopher A. Cullis, *Case Western Reserve University*

Chapter 36
R. Daniel Rudic, *Medical College of Georgia*

Chapter 37
Paul S. Katz, *Georgia State University*

Chapter 38
Josh Dubnau, *Cold Spring Harbor Laboratory*

Chapter 39
Rona Delay, *University of Vermont*

Chapter 40
Peter J. Russell, *Reed College*

Chapter 41
Buel (Dan) Rodgers, *Washington State University*

Chapter 42
Russell Doolittle, *University of California at San Diego*

Chapter 43
Peter J. Russell, *Reed College*

Chapter 44
Ralph Fregosi, *University of Arizona*

Chapter 45
Mark Sheridan, *North Dakota State University*

Chapter 46
Martin Pollak, *Harvard Medical School*

Chapter 47
David Miller, *University of Illinois, Urbana-Champaign*

Chapter 48
Laura Carruth, *Georgia State University*

Chapter 49
Camille Parmesan, *University of Texas at Austin*

Chapter 50
David Reznick, *University of California, Riverside*

Chapter 51
Anurag Agrawal, *Cornell University*

Chapter 52
Kevin Griffin, *Lamont-Doherty Earth Observatory of Columbia University*

Chapter 53
Eric Dinerstein, *World Wildlife Fund*

Chapter 54
Gene E. Robinson, *University of Illinois at Urbana-Champaign*

Chapter 55
Michael J. Ryan, *University of Texas at Austin*

Contents

Unit One Molecules and Cells

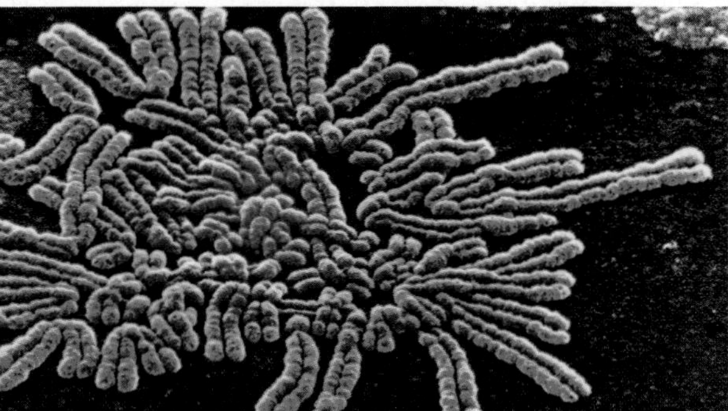

Unit Two Genetics

Unit Three Evolutionary Biology

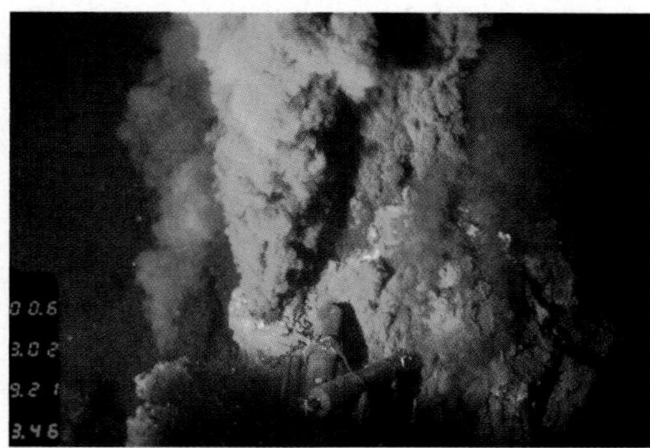
0 0.6
3.0 2
9.2 1
3.4 6

Unit Four Biodiversity

Unit Five Plant Structure and Function

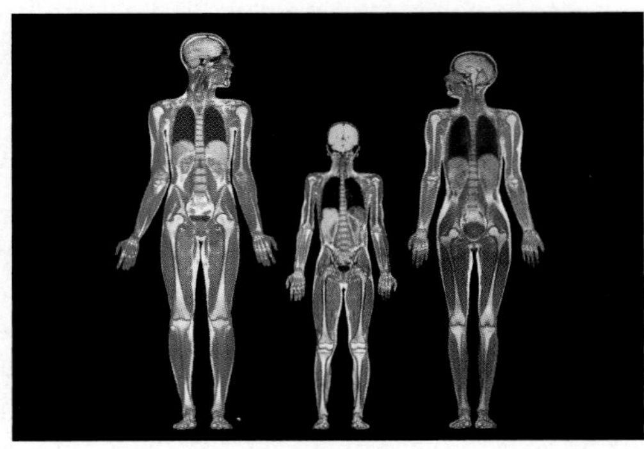

Unit Six Animal Structure and Function

Unit Seven Ecology and Behavior

Earth, a planet teeming with life, is seen here in a satellite photograph.

NASA Goddard Space Flight Center

1

STUDY OUTLINE

Introduction to Biological Concepts and Research

Why It Matters. . . Life abounds in almost every nook and cranny on Earth. A lion creeps through the brush of an African plain, ready to spring at a zebra. The leaves of a sunflower in Kansas turn slowly through the day, keeping their surfaces fully exposed to the sun's light. Fungi and bacteria in the soil of a Canadian forest obtain nutrients by decomposing dead organisms. A child plays in a park in Madrid, laughing happily as his dog chases a tennis ball. In one room of a nearby hospital, a mother hears the first cry of her newborn baby; in another room, an elderly man sighs away his last breath. All over the world, countless organisms are born, live, and die every second of every day. How did life originate, how does it persist, and how is it changing? Biology, the science of life, provides scientific answers to these questions.

What *is* life? Offhandedly, you might say that although you cannot define it, you know it when you see it. The question has no simple answer, because life has been unfolding for billions of years, ever since nonliving materials assembled into the first organized, living cells. Clearly, any list of criteria for the living state only hints at the meaning of "life." Deeper scientific insight requires a wide-ranging examination of the characteristics of life, which is what this book is all about.

Over the next semester or two, you will encounter examples of how organisms are constructed, how they function, where they live, and what they do. The examples provide evidence in support of concepts that will greatly enhance your appreciation and understanding of the living world, including its fundamental unity and striking diversity. This chapter provides a brief overview of these basic concepts. It also describes some of the ways in which biologists conduct research, the process by which they observe nature, formulate explanations of their observations, and test their ideas. <

Image copyright Mishella, 2010. Used under license from Shutterstock.com

FIGURE 1.1
Living organisms and inanimate objects. Living organisms, such as this lizard *(Iguana iguana)*, have characteristics that are fundamentally different from those of inanimate objects, like the rock on which it is sitting.

1.1 What Is Life? Characteristics of Living Organisms

Picture a lizard on a rock, slowly turning its head to follow the movements of another lizard nearby **(Figure 1.1)**. You know that the lizard is alive and that the rock is not. At the atomic and molecular levels, however, the differences between them blur. Lizards, rocks, and all other matter are composed of atoms and molecules, which behave according to the same physical laws. Nevertheless, living organisms share a set of characteristics that collectively set them apart from nonliving matter.

The differences between a lizard and a rock depend not only on the kinds of atoms and molecules present, but also on their organization and their interactions. Individual organisms are at the middle of a hierarchy that ranges from the atoms and molecules within their bodies to the assemblages of organisms that occupy Earth's environments. Within every individual, certain biological molecules contain instructions for building other molecules, which, in turn, are assembled into complex structures. Living organisms must gather energy and materials from their surroundings to build new biological molecules, grow in size, maintain and repair their parts, and produce offspring. They must also respond to environmental changes by altering their chemistry and activity in ways that allow them to survive. Finally, the structure and function of living organisms often change from one generation to the next.

Life on Earth Exists at Several Levels of Organization, Each with Its Own Emergent Properties

The organization of life extends through several levels of a hierarchy **(Figure 1.2)**. Complex biological molecules exist at the lowest level of organization, but by themselves, these molecules are not alive. The properties of life do not appear until they are arranged into cells. A **cell** is an organized chemical system that includes many specialized molecules surrounded by a membrane. A cell is

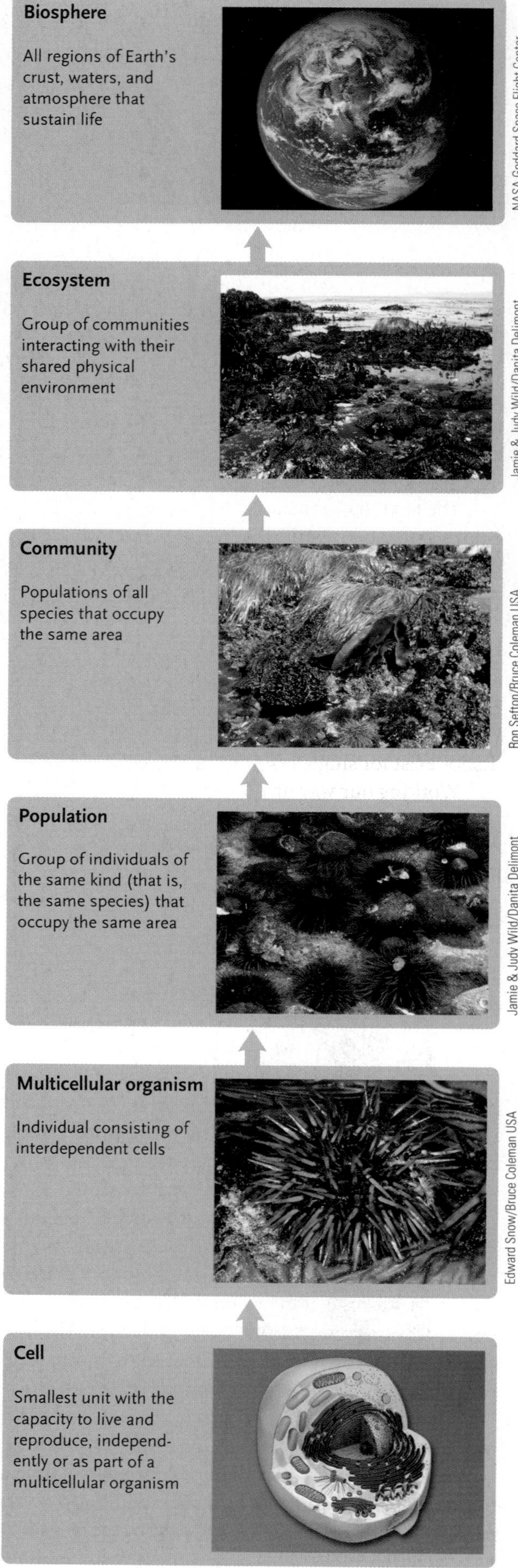
Biosphere
All regions of Earth's crust, waters, and atmosphere that sustain life
NASA Goddard Space Flight Center
Ecosystem
Group of communities interacting with their shared physical environment
Jamie & Judy Wild/Danita Delimont
Community
Populations of all species that occupy the same area
Ron Sefton/Bruce Coleman USA
Population
Group of individuals of the same kind (that is, the same species) that occupy the same area
Jamie & Judy Wild/Danita Delimont
Multicellular organism
Individual consisting of interdependent cells
Edward Snow/Bruce Coleman USA
Cell
Smallest unit with the capacity to live and reproduce, independently or as part of a multicellular organism

FIGURE 1.2
The hierarchy of life. Each level in the hierarchy of life exhibits emergent properties that do not exist at lower levels. The middle four photos depict a rocky intertidal zone on the coast of Washington State.

the lowest level of biological organization that can survive and reproduce—as long as it has access to a usable energy source, the necessary raw materials, and appropriate environmental conditions. However, a cell is alive only as long as it is organized as a cell; if broken into its component parts, a cell is no longer alive even if the parts themselves are unchanged. Characteristics that depend on the level of organization of matter, but do not exist at lower levels of organization, are called **emergent properties.** Life is thus an emergent property of the organization of matter into cells.

Many single cells, such as bacteria and protozoans, exist as **unicellular organisms.** By contrast, plants and animals are **multicellular organisms.** Their cells live in tightly coordinated groups and are so interdependent that they cannot survive on their own. Human cells cannot live by themselves in nature because they must be bathed in body fluids and supported by the activities of other cells. Like individual cells, multicellular organisms have emergent properties that their individual components lack; for example, humans can learn biology.

The next, more inclusive level of organization is the **population,** a group of organisms of the same kind that live together in the same place. The humans who occupy the island of Tahiti and a group of sea urchins living together on the coast of Washington State are examples of populations. Like multicellular organisms, populations have emergent properties that do not exist at lower levels of organization. For example, a population has characteristics such as its birth or death rate—that is, the number of individual organisms who are born or die over a period of time—that do not exist for single cells or individual organisms.

Working our way up the biological hierarchy, all the populations of different organisms that live in the same place form a **community.** The algae, snails, sea urchins, and other organisms that live along the coast of Washington State, taken together, make up a community. The next higher level, the **ecosystem,** includes the community *and* the nonliving environmental factors with which it interacts. For example, a coastal ecosystem comprises a community of living organisms, as well as rocks, air, seawater, minerals, and sunlight. The highest level, the **biosphere,** encompasses all the ecosystems of Earth's waters, crust, and atmosphere. Communities, ecosystems, and the biosphere also have emergent properties. For example, communities can be described in terms of their *diversity*—the number and types of different populations they contain—and their *stability*—the degree to which the populations within the community remain the same through time.

Living Organisms Contain Chemical Instructions That Govern Their Structure and Function

The most fundamental and important molecule that distinguishes living organisms from nonliving matter is **deoxyribonucleic acid (DNA; Figure 1.3).** DNA is a large, double-stranded, helical molecule that contains instructions for assembling a living organism from simpler molecules. We recognize bacteria, trees, fishes, and humans as different because differences in their DNA produce differences in their appearance and function. (Some nonliving systems, notably certain viruses, also contain DNA, but biologists do not consider viruses to be alive because they cannot reproduce independently of the organisms they infect.)

DNA functions similarly in all living organisms. As you will discover in Chapters 14 and 15, the instructions in DNA are copied into molecules of a related substance, **ribonucleic acid (RNA),** which then directs the synthesis (production) of different protein molecules **(Figure 1.4). Proteins** carry out most of the activities of life, including the synthesis of all other biological molecules. This pathway is preserved from generation to generation by the ability of DNA to copy itself so that offspring receive the same basic molecular instructions as their parents.

FIGURE 1.3
Deoxyribonucleic acid (DNA). A computer-generated model of DNA illustrates that it is made up of two strands twisted into a double helix.

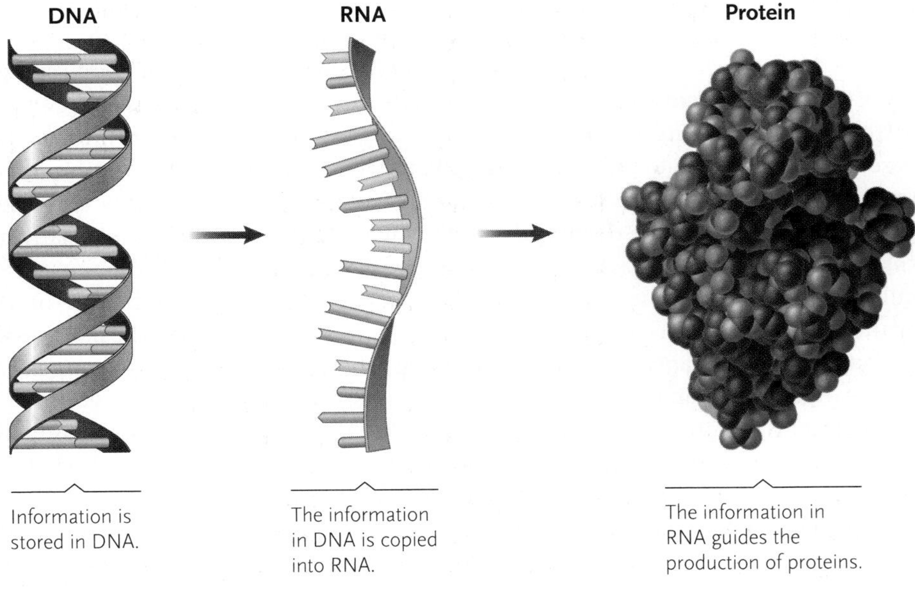
DNA
RNA
Protein
Information is stored in DNA.
The information in DNA is copied into RNA.
The information in RNA guides the production of proteins.

FIGURE 1.4
The pathway of information flow in living organisms. Information stored in DNA is copied into RNA, which then directs the construction of protein molecules. The protein shown here is lysozyme.

Living Organisms Engage in Metabolic Activities

Metabolism, described in Chapters 8 and 9, is another key property of living cells and organisms. **Metabolism** describes the ability of a cell or organism to extract energy from its surroundings and use that energy to maintain itself, grow, and reproduce. As a part of metabolism, cells carry out chemical reactions that assemble, alter, and disassemble molecules **(Figure 1.5).** For example, a growing sunflower plant carries out **photosynthesis,** in which the electromagnetic energy in sunlight is absorbed and converted into chemical energy. The cells of the plant store some chemical energy in sugar and starch molecules, and they use the rest to manufacture other biological molecules from simple raw materials obtained from the environment.

Sunflowers concentrate some of their energy reserves in seeds from which more sunflower plants may grow. The chemical energy stored in the seeds also supports other organisms, such as insects, birds, and humans, that eat them. Most organisms, including sunflower plants, tap stored chemical energy through another metabolic process, **cellular respiration.** In cellular respiration complex biological molecules are broken down with oxygen, releasing some of their energy content for cellular activities.

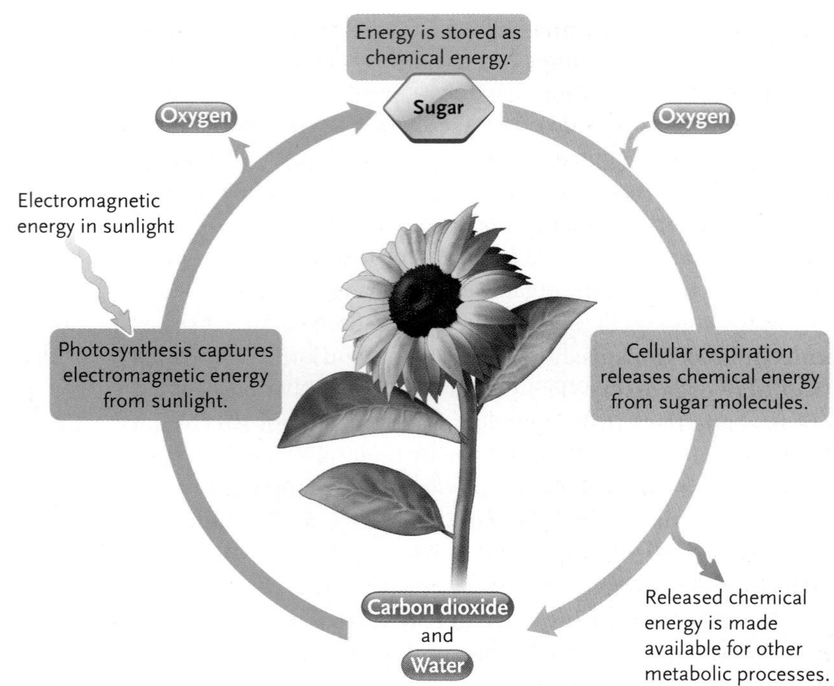
Energy is stored as chemical energy.
Sugar
Oxygen
Oxygen
Electromagnetic energy in sunlight
Photosynthesis captures electromagnetic energy from sunlight.
Cellular respiration releases chemical energy from sugar molecules.
Carbon dioxide
and
Water
Released chemical energy is made available for other metabolic processes.

FIGURE 1.5
Metabolic activities. Photosynthesis converts the electromagnetic energy in sunlight into chemical energy, which is stored in sugars and starches built from carbon dioxide and water; oxygen is released as a by-product of the reaction. Cellular respiration uses oxygen to break down sugar molecules, releasing their chemical energy and making it available for other metabolic processes.

Energy Flows and Matter Cycles through Living Organisms

With few exceptions, energy from sunlight supports life on Earth. Plants and other photosynthetic organisms absorb energy from sunlight and convert it into chemical energy. They use this chemical energy to assemble complex molecules, such as sugars, from simple raw materials, such as water and carbon dioxide. As such, photosynthetic organisms are the **primary producers** of the food on which all other organisms rely **(Figure 1.6)**. By contrast, animals are **consumers:** directly or indirectly, they feed on the complex molecules manufactured by plants. For example, zebras tap directly into the molecules of plants when they eat grass, and lions tap into it indirectly when they eat zebras. Certain bacteria and fungi are **decomposers:** they feed on the remains of dead organisms, breaking down complex biological molecules into simpler raw materials, which may then be recycled by the producers.

As you will see in Chapter 52, some of the energy that photosynthetic organisms trap from sunlight *flows* within and between populations, communities, and ecosystems. But because the transfer of energy from one organism to another is not 100% efficient, some of that energy is lost as heat. Although some animals can use this form of energy to maintain body temperature, it cannot sustain other life processes. By contrast, matter—nutrients such as carbon and nitrogen—*cycles* between living organisms and the nonliving components of the biosphere, to be used again and again (see Figure 1.6).

Living Organisms Compensate for Changes in the External Environment

All objects, whether living or nonliving, respond to changes in the environment; for example, a rock warms up on a sunny day and cools at night. But only living organisms have the capacity to detect environmental changes and *compensate* for them through controlled responses. They do so by means of diverse and varied *receptors*—molecules or larger structures, located on individual cells and body surfaces, that can detect changes in external and internal conditions. When stimulated, the receptors trigger reactions that produce a compensating response.

For example, your internal body temperature remains reasonably constant, even though the environment in which you live is usually either cooler or warmer than you are. Your body compensates for these environmental variations and maintains its internal temperature at about 37° Celsius (C). When the environmental temperature drops significantly, receptors in your skin detect the change and transmit that information to your brain. Your brain may send a signal to your muscles, causing you to shiver, thereby releasing heat that keeps your body temperature from dropping below its optimal level. When the environmental temperature rises significantly, glands in your skin secrete sweat, which evaporates, cooling the skin and its underlying blood supply. The cooled blood circulates internally and keeps your body temperature from rising above 37°C. People also compensate behaviorally by dressing warmly on a cold winter day or jumping into a swimming pool in the heat of summer. Keeping your internal temperature within a narrow range is one example of **homeostasis**—a steady internal condition maintained by responses that compensate for changes in the external environment. As described in Units 5 and 6, all organisms have mechanisms that maintain homeostasis in relation to temperature, blood chemistry, and other important factors.

Living Organisms Reproduce and Many Undergo Development

Humans and all other organisms are part of an unbroken chain of life that began billions of years ago. This chain continues today through **reproduction,** the process by which parents produce offspring. Offspring generally resemble their parents because the parents pass copies of their DNA—with all the accompanying instructions for virtually every life process—to their offspring. The transmission of DNA (that is, genetic information) from one generation to the next is called **inheritance.** For example, the eggs produced by storks hatch into little storks, not into pelicans, because they inherited stork DNA, which is different from pelican DNA.

Multicellular organisms also undergo a process of **development,** a series of programmed changes encoded in DNA, through which a fertilized egg divides into many cells that ultimately are transformed into an adult, which is itself capable of reproduction. As an example, consider the development of a moth **(Figure 1.7).** This insect begins its life as a tiny egg that contains all the instructions necessary for its development into an adult moth. Following these

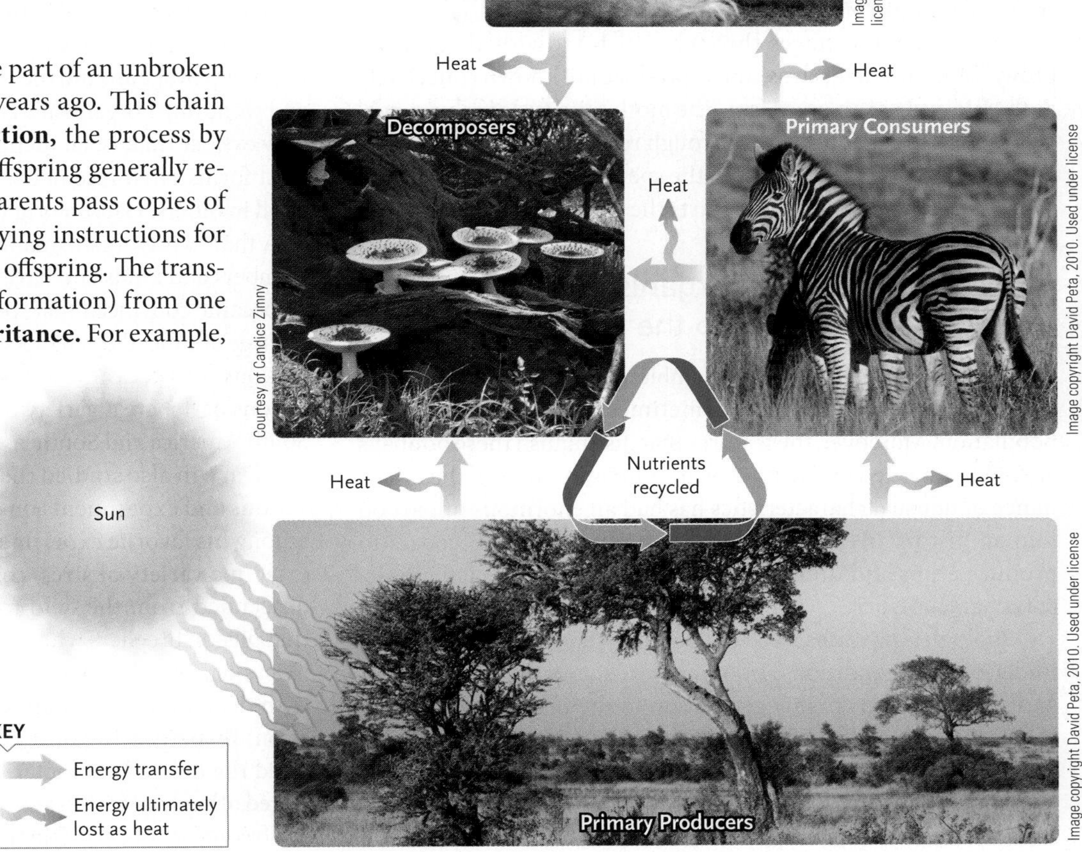
Secondary Consumers
Heat
Heat
Decomposers
Primary Consumers
Heat
Nutrients recycled
Heat
Heat
Sun
Primary Producers
KEY
Energy transfer
Energy ultimately lost as heat
Image copyright Nick Biemans, 2010. Used under license from Shutterstock.com
Courtesy of Candice Zimny
Image copyright David Peta, 2010. Used under license from Shutterstock.com
Image copyright David Peta, 2010. Used under license from Shutterstock.com

FIGURE 1.6
Energy flow and nutrient recycling. In most ecosystems, energy flows from the sun to producers to consumers to decomposers. On the African savanna, the sun provides energy to grasses (producers); zebras (primary consumers) then feed on the grasses before being eaten by lions (secondary consumers); and fungi (decomposers) absorb nutrients and energy from the digestive wastes of animals and from the remains of dead animals and plants. All of the energy that enters an ecosystem is ultimately lost from the system as heat. Nutrients move through the same pathways, but they are conserved and recycled.

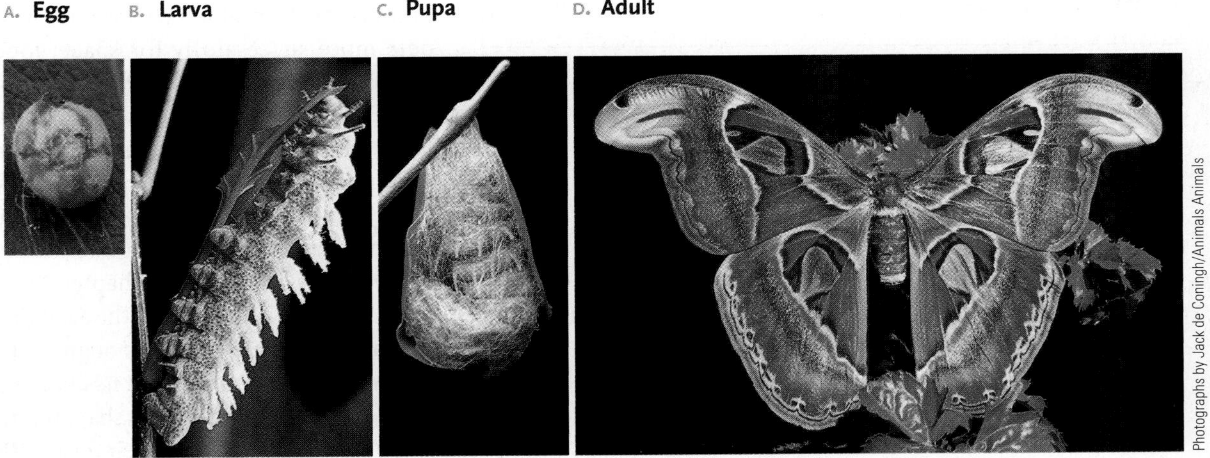
A. Egg
B. Larva
C. Pupa
D. Adult
Photographs by Jack de Coninghs/Animals Animals

FIGURE 1.7
Life cycle of an atlas moth *(Attacus atlas)*.

instructions, the egg first hatches into a caterpillar, a larval form adapted for feeding and rapid growth. The caterpillar increases in size until internal chemical signals indicate that it is time to spin a cocoon and become a pupa. Inside its cocoon, the pupa undergoes profound developmental changes that remodel its body completely. Some cells die; others multiply and become organized in different patterns. When these transformations are complete, the adult moth emerges from the cocoon. It is equipped with structures and behaviors, quite different from those of the caterpillar, that enable it to reproduce.

The sequential stages through which individuals develop, grow, maintain themselves, and reproduce are known collectively as the **life cycle** of an organism. The moth's life cycle includes egg, larva, pupa, and adult stages. Through reproduction, adult moths continue the cycle by producing the sperm and eggs that unite to form the fertilized egg, which starts the next generation.

Populations of Living Organisms Change from One Generation to the Next

Although offspring generally resemble their parents, individuals with unusual characteristics sometimes suddenly appear in a population. Moreover, the features that distinguish these oddballs are often inherited by their offspring. Our awareness of the inheritance of unusual characteristics has had an enormous impact on human history because it allows plant and animal breeders to produce crops and domesticated animals with especially desirable characteristics.

Biologists have observed that similar changes also take place under natural conditions. In other words, populations of all organisms change from one generation to the next, because some individuals experience changes in their DNA and they pass those modified instructions along to their offspring. We introduce this fundamental process, **biological evolution,** in the next section and explore it further in Unit 3.

STUDY BREAK 1.1 <

1. **List the major levels in the hierarchy of life, and identify one emergent property of each level.**
2. **What do living organisms do with the energy they collect from the external environment?**
3. **What is a life cycle?**

1.2 Biological Evolution

All research in biology—ranging from analyses of the precise structure of biological molecules to energy flow through the biosphere—is undertaken with the knowledge that biological evolution has shaped life on Earth. Our understanding of the evolutionary process reveals several truths about the living world: (1) all populations change through time, (2) all organisms are descended from a common ancestor that lived in the distant past, and (3) evolution has produced the spectacular diversity of life that we see around us. Evolution is the unifying theme that links all the subfields of the biological sciences, and it provides cohesion to our treatment of the many topics discussed in this book.

Darwin and Wallace Explained How Populations of Organisms Change through Time

How do evolutionary changes take place? One important mechanism was first explained in the mid-nineteenth century by two British naturalists, Charles Darwin and Alfred Russel Wallace. On a five-year voyage around the world, Darwin observed many "strange and wondrous" organisms. He also found fossils of species that are now extinct (that is, all members of the species are dead). The extinct forms often resembled living species in some traits but differed in others. Darwin originally believed in special creation—the idea that living organisms were placed on Earth in their present numbers and kinds and have not changed since their creation. But he became convinced that species do not remain constant with the passage of time: they change from one form into another over generations. Wallace came to the same conclusion through his observations of the great variety of plants and animals in the jungles of South America and Southeast Asia.

Darwin also studied the process of evolution through observations and experiments on domesticated animals. Pigeons were among his favorite experimental subjects. Domesticated pigeons exist in a variety of sizes, colors, and shapes, but all of them are descended from the wild rock dove **(Figure 1.8)**. Darwin noted that pigeon breeders who wished to promote a certain characteristic, such as elaborately curled tail feathers, selected individuals with the most curl in their feathers as parents for the next generation. By permitting only these birds to mate, the breeders fostered the desired characteristic and gradually eliminated or reduced other traits. The same practice is still used today to increase the frequency of desirable traits in tomatoes, dogs, and other domesticated plants and animals. Darwin called this practice **artificial selection.** He termed the equivalent process that occurs in nature **natural selection.**

In 1858, Darwin and Wallace formally summarized their observations and conclusions explaining biological evolution. (1) Most organisms can produce numerous offspring, but environmental factors limit the number that actually survive and reproduce. (2) Heritable variations allow some individuals to compete more successfully for space, food, and mates. (3) These successful individuals somehow pass the favorable characteristics to their offspring. (4) As a result, the favorable traits become more common in the next generation, and less successful traits become less common. This process of natural selection results in evolutionary change. Today, evolutionary biologists recognize that natural selection is just one of several potent evolutionary processes, as described in Chapter 20.

Over many generations, the evolutionary changes in a population may become extensive enough to produce a population of organisms that is distinct from its ancestors. Nevertheless, parental and descendant species often share many characteristics, allowing researchers to understand their relationships and reconstruct their

Wild rock dove

FIGURE 1.8
Artificial selection. Using artificial selection, pigeon breeders have produced more than 300 varieties of domesticated pigeons from ancestral wild rock doves *(Columba livia)*.

Photographs courtesy Derrell Fowler, Tecumseh, Oklahoma

shared evolutionary history, as described below and in Chapter 23. Starting with the first organized cells, this aspect of evolutionary change has contributed to the diversity of life that exists today.

Darwin and Wallace described evolutionary change largely in terms of how natural selection changes the commonness or rarity of particular variations over time. Their intellectual achievement was remarkable for its time. Although Darwin and Wallace understood the central importance of variability among organisms to the process of evolution, they could not explain how new variations arose or how they were passed to the next generation.

Mutations in DNA Are the Raw Materials That Allow Evolutionary Change

Today, we know that both the origin and the inheritance of new variations arise from the structure and variability of DNA, which is organized into functional units called **genes.** Each gene contains the code for (that is, the instructions for building) a protein molecule or one of its parts. Proteins are the molecules that establish the structures and perform important biological functions within organisms.

Variability among individuals—the raw material molded by evolutionary processes—arises ultimately through **mutations,** random changes in the structure, number, or arrangement of DNA molecules. Mutations in the DNA of reproductive cells may change the instructions for the development of offspring that the reproductive cells produce. Many mutations are of neutral value to individuals bearing them, and some turn out to be harmful. On rare occasions, however, a mutation is beneficial under the prevailing environmental conditions. Beneficial mutations increase the likelihood that individuals carrying the mutation will survive and reproduce. Thus, through the persistence and spread of beneficial mutations among individuals and their descendants, the genetic makeup of a population will change from one generation to the next.

Adaptations Enable Organisms to Survive and Reproduce in the Environments Where They Live

Favorable mutations may produce **adaptations,** characteristics that help an organism survive longer or reproduce more under a particular set of environmental conditions. To understand how organisms benefit from adaptations, consider an example from the recent literature on *cryptic coloration* (camouflage) in animals. Many animals have skin, scales, feathers, or fur that matches the color and appearance of the background in their environment, enabling them to blend into their surroundings. Camouflage makes it harder for predators to identify and then catch them—an obvious advantage to survival. Animals that are not camouflaged are often just sitting ducks.

The rock pocket mouse *(Chaetodipus intermedius)*, which lives in the deserts of the southwestern United States, is mostly nocturnal (that is, active at night). At most desert localities, the rocks are pale brown, and rock pocket mice have sandy-colored fur on their backs. However, at several sites, the rocks—remnants of lava flows from now-extinct volcanoes—are black; here, the rock pocket mice have black fur on their backs. Thus, like the sandy-colored mice in other areas, they are camouflaged in their habitats, the types of areas in which they live **(Figure 1.9).**

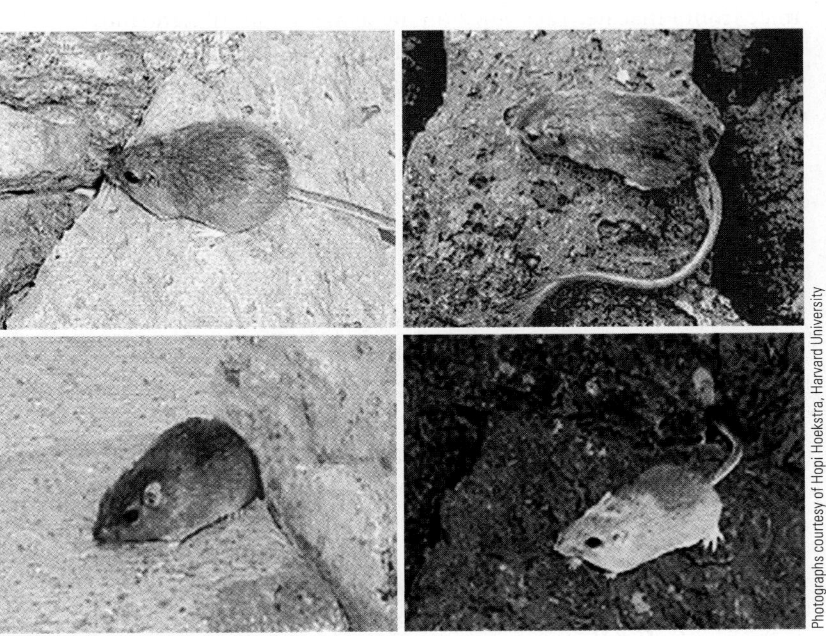
Photographs courtesy of Hopi Hoekstra, Harvard University

FIGURE 1.9
Camouflage in rock pocket mice *(Chaetodipus intermedius)*. Sandy-colored mice are well camouflaged on pale rocks, and black mice are well camouflaged on dark rocks (top); but mice with fur that does not match their backgrounds (bottom) are easy to see.

Camouflage appears to be important to these mice because owls, which locate prey using their exceptionally keen eyesight, frequently eat nocturnal desert mice.

Examples of cryptic coloration are well documented in the scientific literature, and biologists generally interpret them as adaptations that reduce the likelihood of being captured by a predator. Nevertheless, few researchers have identified the precise genetic mutations that produced these adaptations. Michael W. Nachman, Hopi E. Hoekstra, and their colleagues at the University of Arizona tackled this problem in a study of the genetic and evolutionary basis for the color difference between rock pocket mice that live on light and dark backgrounds. In an article published in 2003, they reported the results of an analysis of mice sampled at six sites in southern Arizona and New Mexico. In two regions (Pinacate, AZ, and Armendaris, NM), both light and dark rocks were present, allowing the researchers to compare mice that lived on differently colored backgrounds. Two other sites had only light rocks and sandy-colored mice.

Nachman and his colleagues found that nearly all of the mice they captured on dark rocks had dark fur and that nearly all of the mice they captured on light rocks had light fur **(Figure 1.10)**. The researchers then studied the structure of *Mc1r,* a gene known to influence fur color in laboratory mice; random mutations in this gene can produce fur colors ranging from light to dark in any population of mice, regardless of the habitat it occupies. The 17 black mice from Pinacate all shared certain mutations in their *Mc1r* gene, which established four specific changes in the structure of the Mc1r protein. However, none of the 12 sandy-colored mice from Pinacate carried these mutations. The exact match between the presence of the mutations and the color of the mouse strongly suggests that these mutations in the *Mc1r* gene are responsible for the dark fur in the mice from Pinacate. These data on the distributions of light and dark mice coupled with analyses of their DNA suggest that the color difference is the product of specific mutations that were favored by natural selection. In other words, natural selection *conserved* random mutations that produced black fur in mice that live on black rocks.

Nachman's team then analyzed the *Mc1r* gene in the dark and light mice from Armendaris and in the light mice at two intermediate sites. Because the mice in these regions also closely matched the color of their environments, the researchers expected to find the *Mc1r* mutations in the dark mice but not in the light mice. However, none of the mice from Armendaris shared any of the mutations that contributed to the dark color of mice from Pinacate. Apparently, mutations in some other gene or genes, which the researchers have not yet identified, are responsible for the camouflaging black coloration of mice that live on black rocks in Armendaris.

The example of an adaptation provided by the rock pocket mice illustrates the observation that genetic differences often develop between populations. Sometimes these differences become so great that the organisms develop different appearances and adopt different ways of life. If they become different enough, biologists may regard them as distinct types, as described in Chapter 21. Over immense spans of time, evolutionary processes have produced many types of organisms, which constitute the diversity of life on Earth. In the next section, we survey this diversity and consider how it is studied.

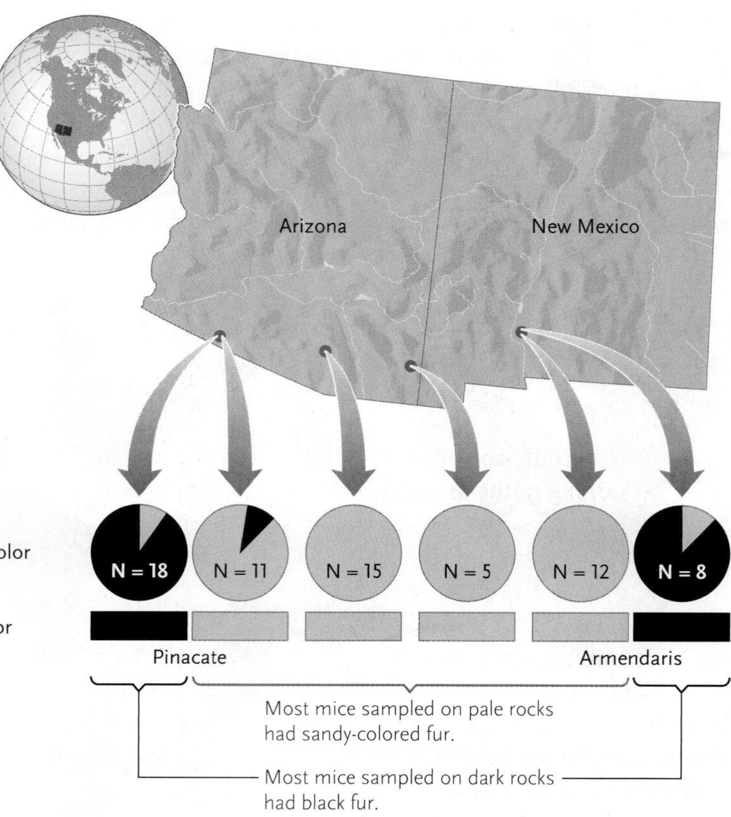
Arizona
New Mexico
Mouse color
N = 18
N = 11
N = 15
N = 5
N = 12
N = 8
Rock color
Pinacate
Armendaris
Most mice sampled on pale rocks had sandy-colored fur.
Most mice sampled on dark rocks had black fur.

FIGURE 1.10
Distributions of rock pocket mice with light and dark fur. At six sites in Arizona and New Mexico, mouse fur color closely matched the colors of the backgrounds where they lived. The pie charts below the map show the proportion of mice with sandy-colored or black fur. N indicates the number of mice sampled at each site. The bars beneath the pie charts indicate the rock color.

STUDY BREAK 1.2 <

1. **What is the difference between artificial selection and natural selection?**
2. **How do random changes in the structure of DNA affect the characteristics of organisms?**
3. **What is the usefulness of being camouflaged in natural environments?**

1.3 Biodiversity and the Tree of Life

Millions of different kinds of organisms live on Earth today, and many millions more existed in the past and became extinct. This mind-boggling biodiversity, the product of evolution, represents the many ways in which the common elements of life have combined to survive and reproduce. To make sense of the past and present diversity of life on Earth, biologists analyze the evolutionary relationships of these organisms and use classification systems to keep track of them. As described in Chapter 23, the task is daunting, and there is no clear consensus on the numbers and kinds of divisions and categories to use. Moreover, our understanding of evolutionary relationships is constantly changing as researchers develop new analytical techniques and learn more about extinct and living organisms.

Researchers Traditionally Defined Species and Grouped Them into Successively More Inclusive Hierarchical Categories

Biologists generally consider the species to be the most fundamental grouping in the diversity of life. As described in Chapter 21, a **species** is a group of populations in which the individuals are so similar in structure, biochemistry, and behavior that they can successfully interbreed. Biologists recognize a **genus** (plural, *genera*) as a group of similar species that share recent common ancestry. Species in the same genus usually also share many characteristics. For example, a group of closely related animals that have large bodies, four stocky legs, long snouts, shaggy hair, nonretractable claws, and short tails are classified together in the genus *Ursus*, commonly known as bears.

Each species is assigned a two-part **scientific name:** the first part identifies the genus to which it belongs, and the second part designates a particular species within that genus. In the genus *Ursus*, for example, *Ursus americanus* is the scientific name of the American black bear; *Ursus maritimus*, the polar bear, and *Ursus arctos*, the brown bear, are two other species in the same genus. Scientific names are always written in italics, and only the genus name is capitalized. After its first mention in a discussion, the genus name is frequently abbreviated to its first letter, as in *U. americanus.*

In a traditional classification, described further in Chapter 23, biologists first identified species and then grouped them into successively more inclusive categories **(Figure 1.11A)**: related genera are placed in the same **family,** related families in the same **order,** and related orders in the same **class.** Related classes are grouped into a **phylum** (plural, *phyla*), and related phyla are assigned to a **kingdom.** In recent years, biologists have added the **domain** as the most inclusive group.

Today Biologists Identify the Trunks, Branches, and Twigs on the Tree of Life

For hundreds of years, biologists classified biodiversity within the hierarchical scheme described above, mostly using structural similarities and differences as clues to evolutionary relationships. With the development of new techniques late in the twentieth century, biologists began to use the precise structure of DNA and other biological molecules to trace the evolutionary pathways through which biodiversity evolved. This new approach allows the comparison of species as different as bread molds and humans because all living organisms share the same genetic code. It also provides so much data that biologists are now able to construct very detailed **phylogenetic trees** (*phylon* = race; *genetikos* = origin)—illustrations of the evolutionary pathways through which species and more inclusive groups appeared—for all organisms **(Figure 1.11B).** Phylogenetic trees are like family genealogies spanning the millions of years that evolution has been occurring.

Phylogenetic trees contain more information than simple hierarchical classifications do because the trees illustrate which ancestors gave rise to which descendants as well as when those evolutionary events occurred. Each fork between trunks, branches, and twigs on the phylogenetic tree represents an evolutionary event in which one ancestral species gave rise to two descendant species. Over time, descendant species gave rise to their own descendants, producing the great diversity of life. In the phylogenetic trees you will encounter in Units 3 and 4 of this book, time is represented vertically, with forks closer to the base of the tree representing evolutionary events in the distant past and those near the top representing more recent events.

In many ways, information in a phylogenetic tree parallels the traditional hierarchical classification, because organisms on the same branch share the common ancestor that is represented at the base of their branch. If the base of a branch that includes two species is near the bottom of the tree, biologists would judge the species to be only distant relatives because their ancestries separated very long ago. By contrast, if the base of the branch containing two species is close to the top of the tree, we would describe the species as being close relatives. Major branches on the tree are therefore roughly equivalent to kingdoms and phyla; progressively smaller branches represent classes, orders, families, and genera. The twigs represent species or the individual populations they comprise.

Since 1994, with substantial support from the National Science Foundation, biologists have collaborated on the Tree of Life web project to share and disseminate their discoveries about how all organisms on Earth are related. The "Tree of Life" has been reconstructed from data on the genetics, structure, metabolic processes, and behavior of living organisms as well as data gathered from the fossils of extinct species. It is constantly updated and revised as scientists accumulate new data.

Three Domains and Several Kingdoms Form the Major Trunks and Branches on the Tree of Life

Biologists distinguish three domains—Bacteria, Archaea, and Eukarya—each of which is a group of organisms with characteristics that set it apart as a major trunk on the Tree of Life. Species in two of the three domains, Bacteria and Archaea, are described as **prokaryotes** (*pro* = before; *karyon* = nucleus). Their DNA is suspended inside the cell without being separated from other cellular components **(Figure 1.12A).** By contrast, the domain Eukarya comprises organisms that are described as **eukaryotes** (*eu* = typical) because their DNA is enclosed in a nucleus, a separate structure within the cells **(Figure 1.12B).** The nucleus and other specialized internal compartments of eukaryotic cells are called **organelles** ("little organs").

THE DOMAIN BACTERIA The Domain Bacteria **(Figure 1.13A)** comprises unicellular organisms (bacteria) that are generally visible

A. Traditional hierarchical classification for animals

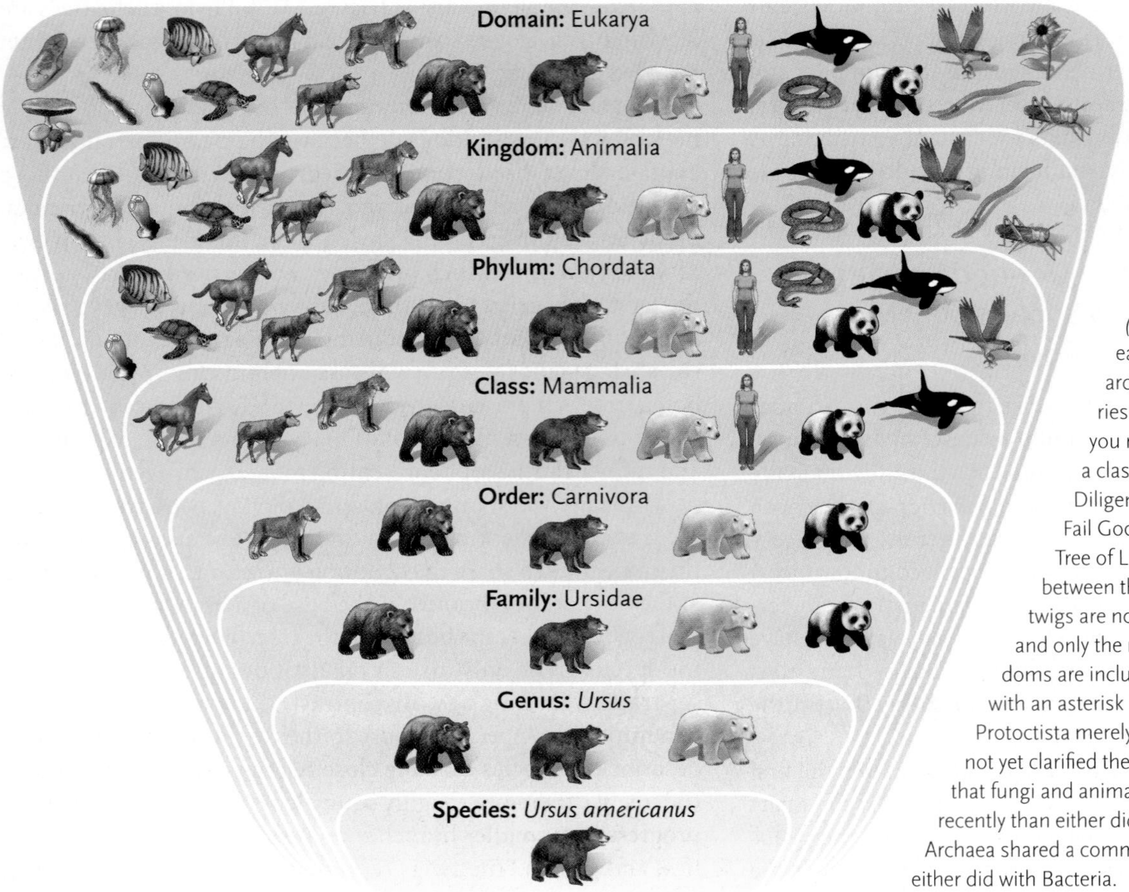

Domain: Eukarya

Kingdom: Animalia

Phylum: Chordata

Class: Mammalia

Order: Carnivora

Family: Ursidae

Genus: *Ursus*

Species: *Ursus americanus*

FIGURE 1.11
Hierarchical classification and the Tree of Life. **(A)** The classification of the American black bear *(Ursus americanus)* illustrates how each species fits into a nested hierarchy of ever-more inclusive categories. The following sentence can help you remember the order of categories in a classification, from *Domain* to *Species:* Diligent Kindly Professors Cannot Often Fail Good Students. **(B)** An overview of the Tree of Life illustrates the relationships between the three domains. Branches and twigs are not included for Bacteria or Archaea, and only the main branches indicating the kingdoms are included for Eukarya. Branches marked with an asterisk (*) are grouped in the Kingdom Protoctista merely as a convenience; biologists have not yet clarified their evolutionary relationships. Note that fungi and animals shared a common ancestor more recently than either did with plants and that Eukarya and Archaea shared a common ancestor more recently than either did with Bacteria.

B. Large scale view of the Tree of Life

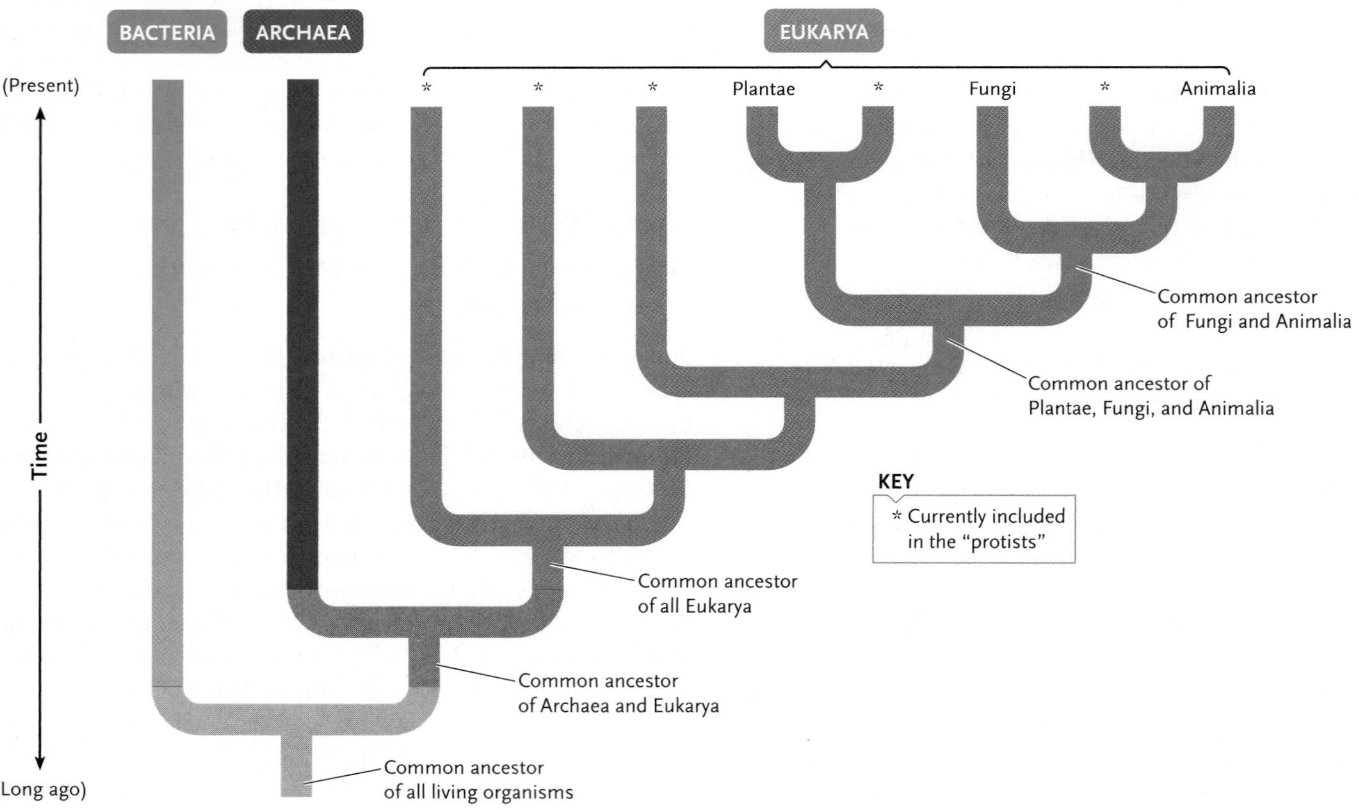
BACTERIA
ARCHAEA
EUKARYA
(Present)
Time
(Long ago)
*
*
*
Plantae
*
Fungi
*
Animalia
Common ancestor of Fungi and Animalia
Common ancestor of Plantae, Fungi, and Animalia
KEY
* Currently included in the "protists"
Common ancestor of all Eukarya
Common ancestor of Archaea and Eukarya
Common ancestor of all living organisms

A. ***Escherichia coli*, a prokaryote**

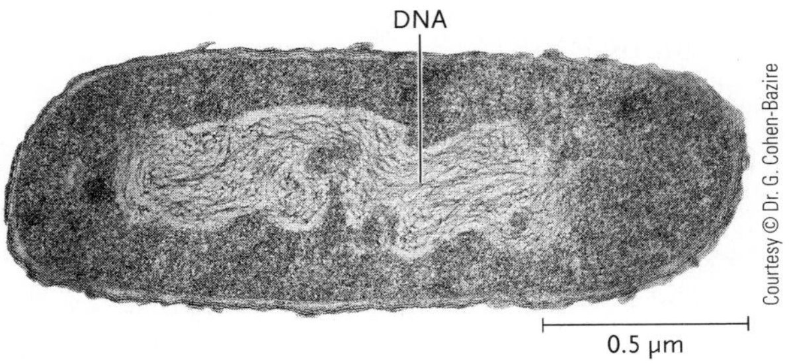
DNA
0.5 μm
Courtesy © Dr. G. Cohen-Bazire

B. ***Paramecium aurelia*, a eukaryote**

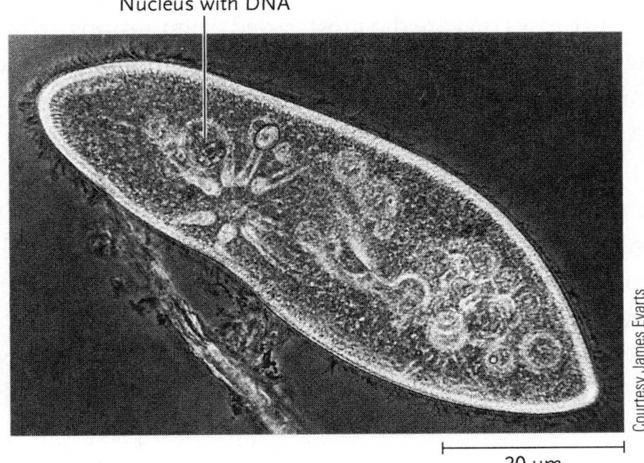
Nucleus with DNA
20 μm
Courtesy James Evarts

FIGURE 1.12
Prokaryotic and eukaryotic cells. **(A)** *Escherichia coli,* a prokaryote, lacks the complex internal structures apparent in **(B)** *Paramecium aurelia,* a eukaryote.

A. **Domain Bacteria**

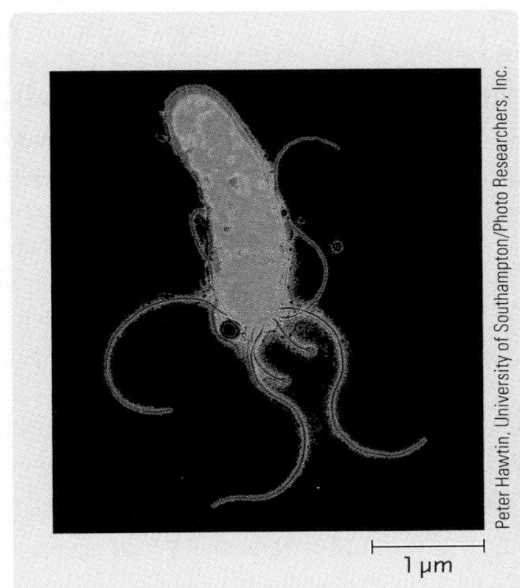
1 μm
Peter Hawtin, University of Southampton/Photo Researchers, Inc.

C. **Domain Eukarya**

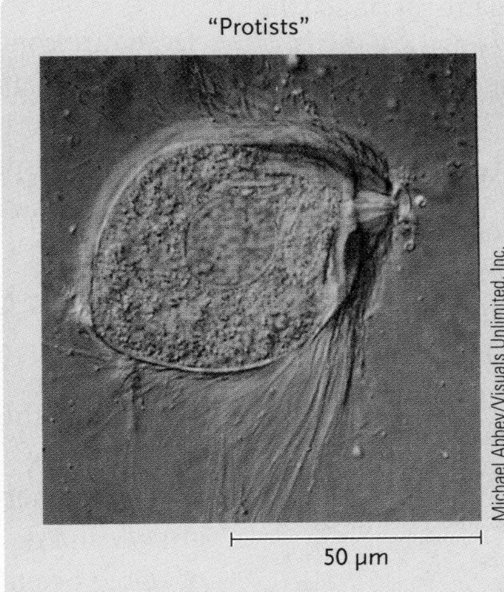
"Protists"
50 μm
Michael Abbey/Visuals Unlimited, Inc.

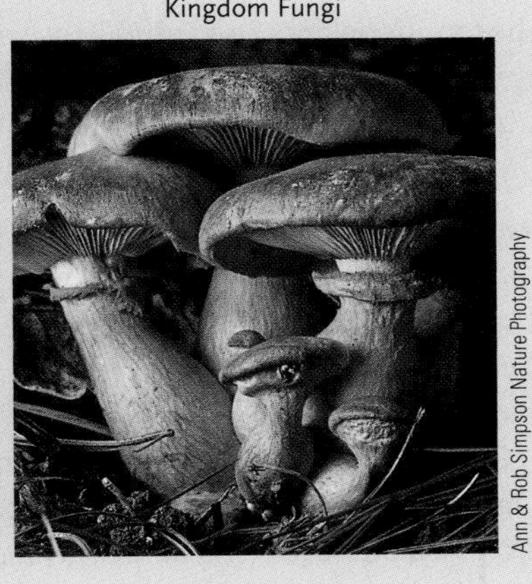
Kingdom Fungi
Ann & Rob Simpson Nature Photography

B. **Domain Archaea**

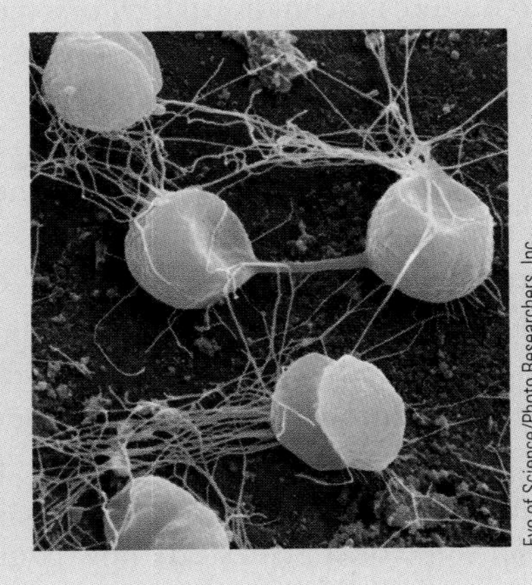
Eye of Science/Photo Researchers, Inc.

Kingdom Plantae
Image copyright Nickolay Stanev, 2010.
Used under license from Shutterstock.com

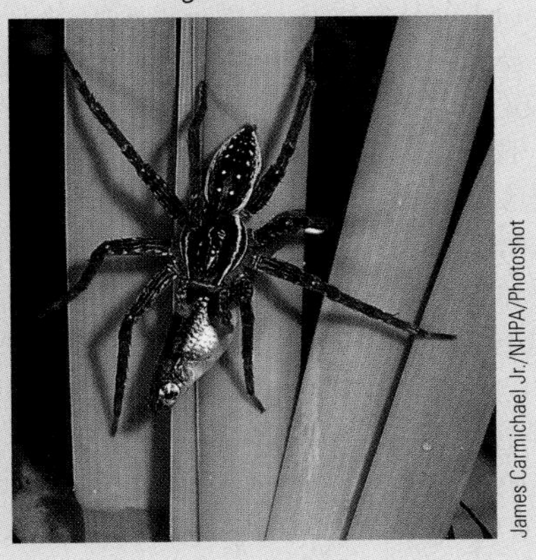
Kingdom Animalia
James Carmichael Jr./NHPA/Photoshot

FIGURE 1.13
Three domains of life. **(A)** This member of the Domain Bacteria *(Helicobacter pylori)* causes ulcers in the digestive systems of humans. **(B)** These organisms from the Domain Archaea *(Pyrococcus furiosus*) live in hot ocean sediments near an active volcano. **(C)** The Domain Eukarya includes the "protists" and three kingdoms in this book. The "protists" are represented by a trichomonad (*Trichonympha* species) that lives in the gut of a termite. Coast redwoods *(Sequoia sempervirens)* are among the largest members of the Kingdom Plantae; the picture shows the large trunk of an older tree in the foreground. The Kingdom Fungi includes the big laughing mushroom (*Gymnopilus* species), which lives on the forest floor. Members of the Kingdom Animalia are consumers, as illustrated by the fishing spider (*Dolomedes* species), which is feasting on a minnow it has captured.

only under the microscope. These prokaryotes live as producers, consumers, or decomposers almost everywhere on Earth, utilizing metabolic processes that are the most varied of any group of organisms. They share with the archaeans a relatively simple cellular organization of internal structures and DNA, but bacteria have some unique structural molecules and mechanisms of photosynthesis.

THE DOMAIN ARCHAEA Similar to bacteria, species in the Domain Archaea **(Figure 1.13B)**, known archaeans, are unicellular, microscopic organisms that live as producers or decomposers. However, many archaeans inhabit extreme environments—hot springs, extremely salty ponds, or habitats with little or no oxygen—that other organisms cannot tolerate. They have some distinctive structural molecules and a primitive form of photosynthesis that is unique to their domain. Although archaeans are prokaryotic, they have some molecular and biochemical characteristics that are typical of eukaryotes, including features of DNA and RNA organization and processes of protein synthesis.

THE DOMAIN EUKARYA All the remaining organisms on Earth, including the familiar plants and animals, are members of the Domain Eukarya **(Figure 1.13C).** The organisms of this domain, all eukaryotic in cell structure, are currently divided into the "protists" and three kingdoms: Plantae, Fungi, and Animalia.

1. **The "Protists"** The "protists" are a large grouping of diverse single-celled and multicellular eukaryotic species. They do not constitute a kingdom, because they do not share a unique common ancestry (see asterisks in Figure 1.11B). Protozoans, which are primarily unicellular, and algae, which range from single-celled, microscopic species to large, multicellular seaweeds, are the most familiar "protists." Protozoans live as consumers and decomposers, but almost all algae are photosynthetic producers.
2. **The Kingdom Plantae** Members of the Kingdom Plantae are multicellular organisms that, with few exceptions, carry out photosynthesis; they therefore function as producers in ecosystems. Except for the reproductive cells (pollen) and seeds of some species, plants do not move from place to place. The kingdom includes the familiar flowering plants, conifers, and mosses.
3. **The Kingdom Fungi** The Kingdom Fungi includes a highly varied group of unicellular and multicellular species, among them the yeasts and molds. Most fungi live as decomposers by breaking down and then absorbing biological molecules from dead organisms. Fungi do not carry out photosynthesis.
4. **The Kingdom Animalia** Members of the Kingdom Animalia are multicellular organisms that live as consumers by ingesting organisms from all three domains. One of the distinguishing features of animals is their ability to move actively from one place to another during some stage of their life cycles. The kingdom encompasses a great range of organisms, including groups as varied as sponges, worms, insects, fishes, amphibians, reptiles, birds, and mammals.

Now that we have introduced the characteristics of living organisms, basic concepts of evolution, and biological diversity, we turn our attention to the ways in which biologists examine the living world to make new discoveries and gain new insights about life on Earth.

STUDY BREAK 1.3

1. **What is a major difference between prokaryotic and eukaryotic organisms?**
2. **In which domain and kingdom are humans classified?**

THINK OUTSIDE THE BOOK

Learn more about the Tree of Life web project by visiting this web site: http://tolweb.org/tree/.

1.4 Biological Research

The entire content of this book—every observation, experimental result, and generality—is the product of **biological research,** the collective effort of countless individuals who have worked to understand every aspect of the living world. This section describes how biologists working today pose questions and find answers to them.

Biologists Confront the Unknown by Conducting Basic and Applied Research

As you read this book, you may at first be uncomfortable discovering how many fundamental questions have not yet been answered. How and where did life begin? How exactly do genes govern the growth and development of an organism? What triggers the signs of aging? Scientists embrace these "unknowns" as opportunities to apply creative thinking to important problems. To show you how exciting it can be to venture into unknown territory, most chapters close with a discussion of Unanswered Questions. Although the concepts and facts that you will learn are profoundly interesting, you will discover that unanswered questions are even more exciting. In many cases, we do not know how you and other scientists of your generation will answer these questions.

Research science is often broken down into two complementary activities—basic research and applied research—that constantly inform one another. Biologists who conduct **basic research** often seek explanations about natural phenomena to satisfy their own curiosity and to advance our collective knowledge. Sometimes, they may not have a specific practical goal in mind. For example, some biologists study how lizards control their body temperatures in different environments. At other times, basic research is inspired by specific practical concerns. For example, understanding how certain bacteria attack the cells of larger organisms might someday prove useful for the development of a new antibiotic (that is, a bacteria-killing agent). Many chapters in this book include a *Focus on Basic Research,* which describes particularly elegant or insightful basic research that advanced our knowledge.

Other scientists conduct **applied research,** with the goal of solving specific practical problems. For example, biomedical scientists conduct applied research to develop new drugs and to learn how illnesses spread from animals to humans or through human populations. Similarly, agricultural scientists try to develop varieties of important crop plants that are more productive and more pest-resistant than the varieties currently in use. Examples of applied research are presented throughout this book, some of them described in detail as a *Focus on Applied Research.*

The Scientific Method Helps Researchers Crystallize and Test Their Ideas

People have been adding to our knowledge of biology ever since our distant ancestors first thought about gathering food or hunting game. However, beginning about 500 years ago in Europe, inquisitive people began to understand that direct observation is the most reliable and productive way to study natural phenomena. By the nineteenth century, researchers were using the **scientific method,** an investigative approach to acquiring knowledge in which scientists make observations about the natural world, develop working explanations about what they observe, and then test those explanations by collecting more information.

Grade school teachers often describe the scientific method as a stepwise, linear procedure for observing and explaining the world around us **(Figure 1.14).** But because scientists usually think faster than they can work, they often undertake the different steps simultaneously. Application of the scientific method requires both curiosity and skepticism: successful scientists question the current state of our knowledge and challenge old concepts with new ideas and new observations. Scientists like to be shown *why* an idea is correct, rather than simply being told that it is: explanations of natural phenomena must be backed up by objective evidence rooted in observation and measurement.

Most importantly, scientists share their ideas and results through the publication of their work. Publications typically include careful descriptions of the methods employed and details of the results obtained so that other researchers can repeat and verify their findings at a later time.

Biologists Conduct Research by Collecting Observational and Experimental Data

Biologists generally use one of two complementary approaches—*descriptive science* and *experimental science*—or a combination of the two to advance our knowledge. In many cases, they collect **observational data,** basic information on biological structures or the details of biological processes. This approach, which is sometimes called descriptive science, provides information about systems that have not yet been well studied. For example, biologists are now collecting observational data about the precise chemical structure of the DNA in different species of organisms for the purpose of understanding their evolutionary relationships.

In other cases, researchers collect **experimental data,** information that describes the result of a careful manipulation of the system under study. This approach, which is known as experimental science, often answers questions about why or how systems work as they do. For example, a biologist who wonders whether a particular snail species influences the distribution of algae on a rocky shoreline might remove the snail from some enclosed patches of shoreline and examine whether the distribution of algae changes as a result. Similarly, a geneticist who wants to understand the role of a particular gene in the functioning of an organism might make mutations in the gene and examine the consequences.

Researchers Often Test Hypotheses with Controlled Experiments

Research on a previously unexplored system usually starts with basic observations (step 1 in Figure 1.14). Once the facts have been carefully observed and described, scientists may develop a **hypothesis** to explain them (step 2). In a 2008 report, the National Academy of Sciences defined a hypothesis as "a tentative explanation for an observation, phenomenon, or scientific problem that can be tested by further investigation." And whenever scientists create a hypothesis, they simultaneously define—either explicitly or implicitly—a **null hypothesis,** a statement of what they would see if the hypothesis being tested is not correct.

Many scientists structure their hypotheses with one crucial requirement: it must be *falsifiable* by experimentation or further observation. In other words, scientists must describe an idea in such a way that, if it is wrong, they will be able to demonstrate that it is wrong. The principle of falsifiability helps scientists define testable, focused hypotheses. Hypotheses that are testable and falsifiable fall within the realm of science, whereas those that cannot be falsified—although possibly valid and true—do not fall within the realm of science.

Hypotheses generally explain the relationship between **variables,** environmental factors that may differ among places or organismal characteristics that may differ among individuals. Thus, hypotheses yield testable **predictions** (step 3), statements about what the researcher expects to happen to one variable if another variable changes. Scientists then test their hypotheses and prediction with experimental or observational tests that generate relevant data (step 4). And if data from just one study refute a scientific hypothesis (that is, demonstrate that its predictions are incorrect), the scientist must modify the hypothesis and test it again or abandon it altogether. However, no amount of data can *prove* beyond a doubt that a hypothesis is correct; there may always be a contradictory example somewhere on Earth, and it is impossible to test every imaginable example. That is why scientists say that positive results *are consistent with, support,* or *confirm* a hypothesis.

To make these ideas more concrete, consider a simple example of hypothesis development and testing. Say that a friend gives you a plant that she grew on her windowsill. Under her loving care, the plant always flowered. You place the plant on your windowsill and water it regularly, but the plant never blooms. You know that your friend always gave fertilizer to the plant—your observation—and you wonder whether fertilizing the plant would make it flower. In other words, you create a hypothesis with a specific prediction: "This type of plant will flower if it receives fertilizer." This is a good scientific hypothesis because it is falsifiable. To test the hypothesis,

The Scientific Method

Purpose: The scientific method is a method of inquiry that allows researchers to crystallize their thoughts about a topic and devise a formal way to test their ideas by making observations and collecting measurable data.

Protocol:

Observations

Inductive reasoning

Hypothesis

Deductive reasoning

Predictions

Experiments

1. Make detailed observations about a phenomenon of interest.

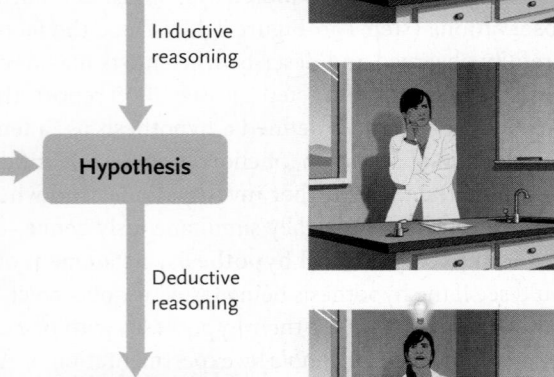

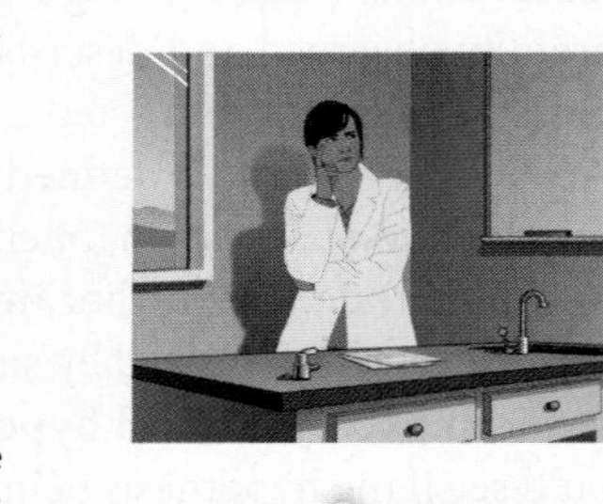

2. Use inductive reasoning to create a testable hypothesis that provides a working explanation of the observations. Hypotheses may be expressed in words or in mathematical equations. Many scientists also formulate alternative hypotheses (that is, alternative explanations) at the same time.

3. Use deductive reasoning to make predictions about what you would observe if the hypothesis were applied to a novel situation.

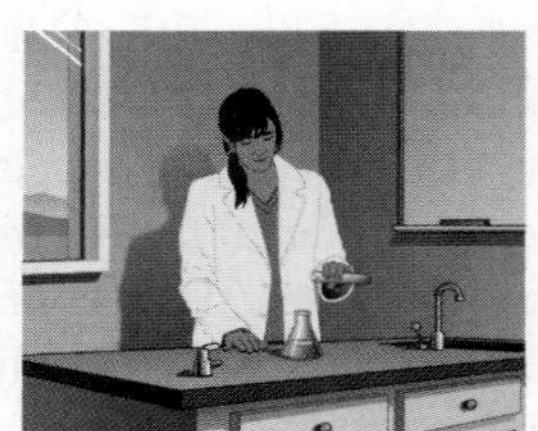

4. Design and conduct a controlled experiment (or new observational study) to test the predictions of the hypothesis. The experiment must be clearly defined so that it can be repeated in future studies. It must also lead to the collection of measurable data that other researchers can evaluate and reproduce if they choose to repeat the experiment themselves.

Interpreting the Results: Compare the results of the experiment or new observations with those predicted by the hypothesis. Scientists often use formal statistical tests to determine whether the results match the predictions of the hypothesis.

If the results do not match the predictions, the hypothesis is refuted, and it must be rejected or revised.

If the statistical tests suggest that the prediction was correct, the hypothesis is confirmed—until new data refute it in the future.

NOBEL
PRIZE

you would simply give the plant fertilizer. If it blooms, the data—the fact that it flowers—confirm your hypothesis. If it does not bloom, the data force you to reject or revise your hypothesis.

One problem with this experiment is that the hypothesis does not address other possible reasons that the plant did not flower. Maybe it received too little water. Maybe it did not get enough sunlight. Maybe your windowsill was too cold. All of these explanations could be the basis of **alternative hypotheses,** which a conscientious scientist always considers when designing experiments. You could easily test any of these hypotheses by

FIGURE 1.15 Experimental Research

Hypothetical Experiment Illustrating the Use of Control Treatment and Replicates

Question: Your friend fertilizes a plant that she grows on her windowsill, and it flowers. After she gives you the plant, you put it on your windowsill, but you do not give it any fertilizer and it does not flower. Will giving the plant fertilizer induce it to flower?

Add fertilizer

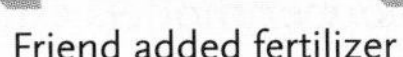

Friend added fertilizer

You did not add fertilizer

Experiment: Establish six replicates of an experimental treatment (identical plants grown with fertilizer) and six replicates of a control treatment (identical plants grown without fertilizer).

Experimental Treatment

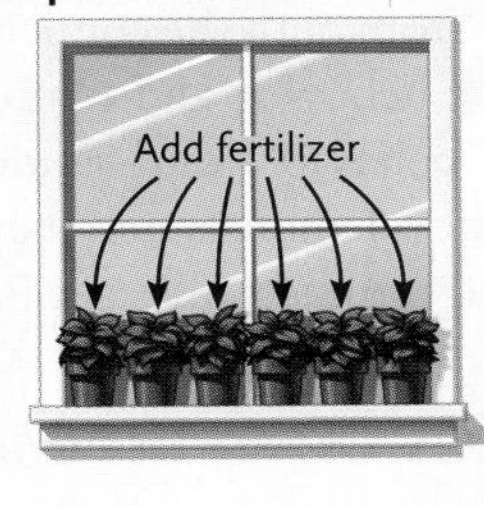
Add fertilizer

Control Treatment

No fertilizer

Possible Result 1: Neither experimental nor control plants flower.

Experimentals | Controls

Conclusion: Fertilizer alone does not cause the plants to flower. Consider alternative hypotheses and conduct additional experiments, each testing a different experimental treatment, such as the amount of water or sunlight the plant receives or the temperature to which it is exposed.

Possible Result 2: Plants in the experimental group flower, but plants in the control group do not.

Experimentals

Controls

Conclusion: The application of fertilizer induces flowering in this type of plant, confirming your original hypothesis. Pat yourself on the back and apply to graduate school in plant biology.

providing more water, more hours of sunlight, or warmer temperatures to the plant.

But even if you provide each of these necessities in turn, your efforts will not definitively confirm or refute your hypothesis unless you introduce a control treatment. The **control,** as it is often called, represents a null hypothesis; it tells us what we would see in the absence of the experimental manipulation. For example, your experiment would need to compare plants that received fertilizer (the experimental treatment) with plants grown without fertilizer (the control treatment). The presence or absence of fertilizer is the **experimental variable,** and in a controlled experiment, everything except the experimental variable—the flower pots, the soil, the amount of water, and exposure to sunlight—is exactly the same, or as close to exactly the same as possible. Thus, if your experiment is well controlled **(Figure 1.15),** any difference in flowering pattern observed between plants that receive the experimental treatment (fertilizer) and those that receive the control treatment (no fertilizer) can be attributed to the experimental variable. If the plants that receive fertilizer did not flower more than the control plants, you would reject your initial hypothesis. The elements of a typical experimental approach, as well as our hypothetical experiment, are summarized in Figure 1.15. Figures that present observational and experimental research using this basic format are provided throughout this book.

Notice that in the preceding discussion we discussed plants (plural) that received fertilizer and plants that did not. Nearly all experiments in biology include **replicates,** multiple subjects that receive either the same experimental treatment or the same control treatment. Scientists use replicates in experiments because individuals typically vary in genetic makeup, size, health, or other characteristics—and because accidents may disrupt a few replicates. By exposing multiple subjects to both treatments, we can use a statistical test to compare the average result of the experimental treatment with the average result of the control treatment, giving us more confidence in the overall findings. Thus, in the fertilizer experiment we described, we might expose six or more individual plants to each treatment and compare the results obtained for the experimental group with those obtained for the control group. We would also try to ensure that the individuals included in the experiment were as similar as possible. For example, we might specify that they all must be the same age or size.

When Controlled Experiments Are Unfeasible, Researchers Employ Null Hypotheses to Evaluate Observational Data

In some fields of biology, especially ecology and evolution, the systems under study may be too large or complex for experimental manipulation. In such cases, biologists can use a null hypothesis to evaluate observational data. For example, Paul E. Hertz of Barnard College studies temperature regulation in lizards. As in many other animals, a lizard's body temperature can vary substantially as environmental temperatures change. Research on many lizard species has demonstrated that they often compensate for fluctuations in environmental temperature—that is, maintain thermal homeostasis—by perching in the sun to warm up or in the shade when they feel hot.

Hertz hypothesized that the crested anole, *Anolis cristatellus,* a lizard species in Puerto Rico, regulates its body temperature by perching in patches of sun when environmental temperatures are low. To test this hypothesis, Hertz needed to determine what he would see if lizards were *not* trying to control their body temperatures. In other words, he needed to know the predictions of a null hypothesis that states: "Lizards do not regulate their body temperature, and they select perching sites at random with respect to factors that influence body temperature" **(Figure 1.16).** Of course, it would be impossible to force a natural population of lizards to perch in places that define the null hypothesis. Instead, he and his students created a population of artificial lizards, copper models that served as lizard-sized, lizard-shaped, and lizard-colored thermometers. Each hollow copper model was equipped with a built-in temperature-sensing wire that can be plugged into an electronic thermometer. After constructing the copper models, Hertz and his students verified that the models reached the same internal temperatures as live lizards under various laboratory conditions. They then traveled to Puerto Rico and hung 60 models at randomly selected positions in the habitats where this lizard species lives.

How did the copper models allow Hertz and his students to interpret their data? Because the researchers placed these inanimate objects at random positions in the lizards' habitats, the percentages of models observed in sun and in shade provided a measure of how sunny or shady a particular habitat was. In other words, the copper models established the null hypothesis about the percentage of lizards that would perch in sunlit spots just by chance. Similarly, the temperatures of the models provided a null hypothesis about what the temperatures of lizards would be if they perched at random in their habitats. Hertz and his students gathered data on the use of sunny perching places and temperatures from both the copper models and live lizards. By comparing the behavior and temperatures of live lizards with the random "behavior" and random temperatures of the copper models, they demonstrated that *A. cristatellus* did, in fact, regulate its body temperature (see Figure 1.16).

Biologists Often Use Model Organisms to Study Fundamental Biological Processes

Certain species or groups of organisms have become favorite subjects for laboratory and field studies because their characteristics make them relatively easy subjects of research. In most cases, biologists began working with these **model organisms** because they have rapid development, short life cycles, and small adult size. Thus, researchers can rear and house large numbers of them in the laboratory. Also, as fuller portraits of their genetics and other aspects of their biology emerge, their appeal as research subjects grows because biologists have a better understanding of the biological context within which specific processes occur.

Because many forms of life share similar molecules, structures, and processes, research on these small and often simple organisms provides insights into biological processes that operate in larger and more complex organisms. For example, early analyses of inheritance in a fruit fly *(Drosophila melanogaster)* established our basic understanding of genetics in all eukaryotic organisms. Research in the mid-twentieth century with the bacterium *Escherichia coli* demonstrated the mechanisms that control whether the information in any particular gene is used to manufacture a protein molecule. In fact, the body of research with *E. coli* formed the foundation that now allows scientists to make and clone (that is, produce multiples copies of) DNA molecules. Similarly, research on a tiny mustard plant *(Arabidopsis thaliana)* is providing information about the genetic and molecular control of development in all plants, including important agricultural crops. Other model organisms facilitate research in ecology and evolution. For example, *Anolis cristatellus* is just 1 of more than 400 *Anolis* species. The geographic distribution of these species allows researchers to study general processes and

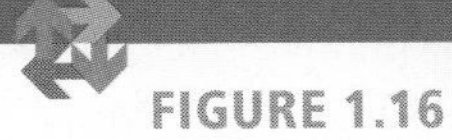

FIGURE 1.16 Observational Research

A Field Study Using a Null Hypothesis

Hypothesis: *Anolis cristatellus,* the crested anole, uses patches of sun and shade to regulate its body temperature.

Null Hypothesis: Because this lizard does not regulate its body temperature, individuals select perching sites at random with respect to environmental factors that influence body temperature.

Method: The researchers created a set of hollow, copper lizard models, each equipped with a temperature-sensing wire. At study sites where the lizards live in Puerto Rico, the researchers hung 60 models at random positions in trees. They observed how often live lizards and the randomly positioned copper models were perched in patches of sun or shade, and they measured the temperatures of live lizards and the copper models. Data from the randomly positioned copper models define the predictions of the null hypothesis.

Results: The researchers compared the frequency with which live lizards and the copper models perched in sun or shade as well as the temperatures of live lizards and the copper models. The data revealed that the behavior and temperatures of *A. cristatellus* were different from those of the randomly positioned models, therefore confirming the original hypothesis.

Anolis cristatellus

Alejandro Sanchez

Copper *Anolis* model

Kevin de Queiroz, National Museum of Natural History, Smithsonian Institution

Percentage of models and lizards perched in sun or shade

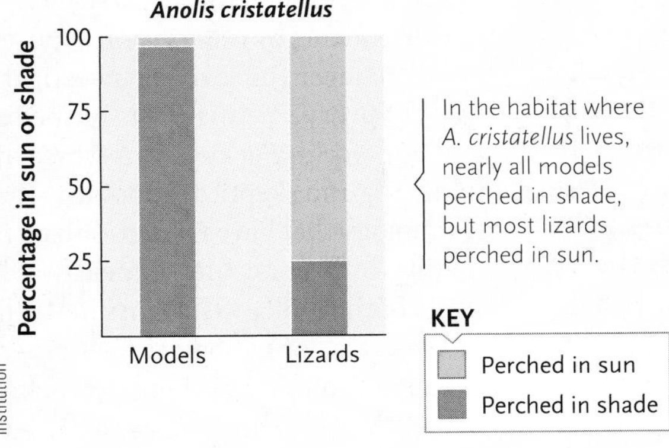

Anolis cristatellus
Percentage in sun or shade
100
75
50
25
Models
Lizards
In the habitat where A. cristatellus lives, nearly all models perched in shade, but most lizards perched in sun.
KEY
Perched in sun
Perched in shade

Temperatures of models and lizards

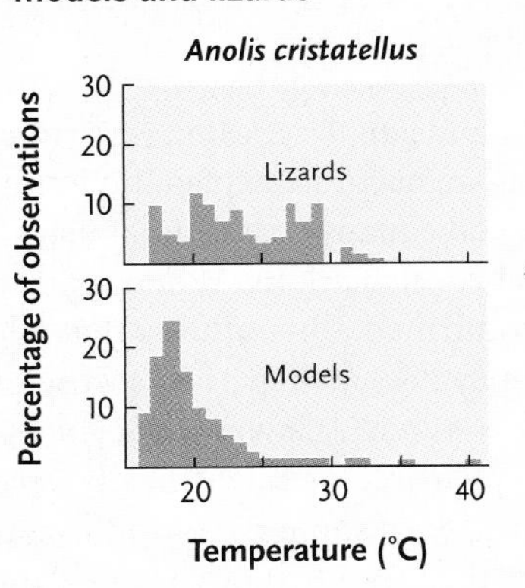

Anolis cristatellus
Percentage of observations
30
20
10
Lizards
Models
20
30
40
Temperature (°C)
Body temperatures of A. cristatellus were significantly higher than those of the randomly placed models.

Conclusion: *A. cristatellus* uses patches of sun and shade to regulate its body temperature.

Source: P.E. Hertz. 1992. Temperature regulation in Puerto Rican *Anolis* lizards: A field test using null hypotheses. *Ecology* 73:1405–1417.

interactions that affect the ecology and evolution of all forms of life. You will read about eight of the organisms most frequently used in research in *Focus on Model Organisms* boxes distributed throughout this book.

Molecular Techniques Have Revolutionized Biological Research

In 1941, George Beadle and Edward Tatum used a simple bread mold *(Neurospora crassa)* as a model organism to demonstrate that genes provide the instructions for constructing certain proteins. Their work represents the beginning of the molecular revolution. In 1953, James Watson and Francis Crick determined the structure of DNA, giving us a molecular vision of what a gene is. In the years since those pivotal discoveries, our understanding of the molecular aspects of life has increased exponentially because many new techniques allow us to study life processes at the molecular level. For example, we can isolate individual genes and study them in detail—even manipulate them—in the test tube. We can modify organisms by replacing or adding genes. We can explore the interactions that each individual protein in the cell has with other proteins. We can identify and characterize each of the genes in an organism to unravel its evolutionary relationships. The list of experimental possibilities is nearly endless.

This molecular revolution has made it possible to answer questions about biological systems that we could not even ask just a few years ago. For example: What specific DNA changes are responsible for genetic diseases? How is development controlled at the molecular level? What genes do humans and chimpanzees share? In particular, the continuing analysis of the structure of DNA in many organisms is fueling a new intensity of scientific enquiry focused on the role of whole genomes (all of the DNA of an organism) in directing biological processes. To give you a sense of the exciting impact of molecular research on all areas of biology, most chapters in this book include a box that focuses on *Insights from the Molecular Revolution.*

Advances in molecular biology have also revolutionized applied research. DNA "fingerprinting" allows forensic scientists to identify individuals who left molecular traces at crime scenes. **Biotechnology,** the manipulation of living organisms to produce useful products, has also revolutionized the pharmaceutical industry. For example, insulin—a protein used to treat the metabolic disorder diabetes—is now routinely produced by bacteria into which the gene coding for this protein has been inserted. Current research on gene therapy and the cloning of stem cells also promises great medical advances in the future.

Scientific Theories Are Grand Ideas That Have Withstood the Test of Time

When a hypothesis stands up to repeated experimental tests, it is gradually accepted as an accurate explanation of natural events. This acceptance may take many years, and it usually involves repeated experimental confirmations. When many different tests have consistently confirmed a hypothesis that addresses many broad questions, it may become regarded as a **scientific theory.** In a 2008 report, the National Academy of Sciences defined a scientific theory as "a plausible or scientifically acceptable, well-substantiated explanation of some aspect of the natural world . . . that applies in a variety of circumstances to explain a specific set of phenomena and predict the characteristics of as yet unobserved phenomena."

Most scientific theories are supported by exhaustive experimentation; thus, scientists usually regard them as established truths that are unlikely to be contradicted by future research. Note that this use of the word *theory* is quite different from its informal meaning in everyday life. In common usage, the word *theory* most often labels an idea as either speculative or downright suspect, as in the expression "It's only a theory." But when scientists talk about theories, they do so with respect for ideas that have withstood the test of many experiments.

Because of the difference between the scientific and common usage of the word *theory,* many people fail to appreciate the extensive evidence that supports most scientific theories. For example, virtually every scientist accepts the theory of evolution as fully supported scientific truth: all species change with time, new species are formed, and older species eventually die off. Although evolutionary biologists debate the details of how evolutionary processes bring about these changes, very few scientists doubt that the theory of evolution is essentially correct. Moreover, *no scientist who has tried to cast doubt on the theory of evolution has ever devised or conducted a study that disproves any part of it.* Unfortunately, the confusion between the scientific and common usage of the word *theory* has led, in part, to endless public debate about supposed faults and inadequacies in the theory of evolution.

Curiosity and the Joy of Discovery Motivate Scientific Research

What drives scientists in their quest for knowledge? The motivations of scientists are as complex as those driving people toward any goal. Intense curiosity about ourselves, our fellow creatures, and the chemical and physical objects of the world and their interactions is a basic ingredient of scientific research. The discovery of information that no one knew before is as exciting to a scientist as finding buried treasure. There is also an element of play in science, a joy in the manipulation of scientific ideas and apparatus, and the chase toward a scientific goal. Biological research also has practical motivations—for example, to cure disease or improve agricultural productivity. In all of this research, one strict requirement of science is honesty—without honesty in the gathering and reporting of results, the work of science is meaningless. Dishonesty is actually rare in science, not least because repetition of experiments by others soon exposes any funny business.

Whatever the level of investigation or the motivation, the work of every scientist adds to the fund of knowledge about us and our world. For better or worse, the scientific method—that inquiring and skeptical approach—has provided knowledge and technology that have revolutionized the world and improved the quality of human life immeasurably. This book presents the fruits of the biologists' labors in the most important and fundamental areas of biological science—cell and molecular biology, genetics, evolution, systematics, physiology, developmental biology, ecology, and behavioral science.

STUDY BREAK 1.4

1. **In your own words, explain the most important requirement of a scientific hypothesis.**
2. **What information did the copper lizard models provide in the study of temperature regulation described earlier?**
3. **Why do biologists often use model organisms in their research?**
4. **How would you respond to a nonscientist who told you that Darwin's ideas about evolution were "just a theory"?**

THINK OUTSIDE THE BOOK

Learn more about the scientific method and the richness of the scientific enterprise by visiting this web site, maintained by researchers at the University of California at Berkeley: http://undsci.berkeley.edu/.

REVIEW KEY CONCEPTS

Go to **CENGAGENOW** at www.cengage.com/login to access quizzing, animations, exercises, articles, and personalized homework help.

1.1 What Is Life? Characteristics of Living Organisms

- Living systems are organized in a hierarchy, each level having its own emergent properties (Figure 1.2): cells, the lowest level of organization that is alive, are organized into unicellular or multicellular organisms; populations are groups of organisms of the same kind that live together in the same area; an ecological community comprises all the populations living in an area, and ecosystems include communities that interact through their shared physical environment; the biosphere includes all of Earth's ecosystems.
- Living organisms have complex structures established by instructions coded in their DNA (Figure 1.3). The information in DNA is copied into RNA, which guides the production of protein molecules (Figure 1.4). Proteins carry out most of the activities of life.
- Living cells and organisms engage in metabolism, obtaining energy and using it to maintain themselves, grow, and reproduce. The two primary metabolic processes are photosynthesis and cellular respiration (Figure 1.5).
- Energy that flows through the hierarchy of life is eventually released as heat. By contrast, matter is recycled within the biosphere (Figure 1.6).
- Cells and organisms use receptors to detect environmental changes and trigger a compensating reaction that allows the organism to survive.
- Organisms reproduce, and their offspring develop into mature, reproductive adults (Figure 1.7).
- Populations of living organisms undergo evolutionary change as generations replace one another over time.

Animation: Life's levels of organization

Animation: One-way energy flow and materials cycling

Animation: Insect development

1.2 Biological Evolution

- The structure, function, and types of organisms in populations change with time. According to the theory of evolution by natural selection, certain characteristics allow some organisms to survive better and reproduce more than others in their population. If the instructions that produce those characteristics are coded in DNA, successful characteristics will become more common in later generations. As a result, the characteristics of the offspring generation will differ from those of the parent generation (Figure 1.8).
- The instructions for many characteristics are coded by segments of DNA called *genes,* which are passed from parents to offspring in reproduction.
- Mutations—changes in the structure, number, or arrangement of DNA molecules—create variability among individuals. Variability is the raw material of natural selection and other processes that cause biological evolution.
- Over many generations, the accumulation of favorable characteristics may produce adaptations, which enable individuals to survive longer or reproduce more (Figures 1.9 and 1.10).
- Over long spans of time, the accumulation of different adaptations and other genetic differences between populations has produced the diversity of life on Earth.

1.3 Biodiversity and the Tree of Life

- Scientists classify organisms in a hierarchy of categories. The species is the most fundamental category, followed by genus, family, order, class, phylum, and kingdom as increasingly inclusive categories (Figure 1.11A). Analyses of DNA structure now allow biologists to construct the Tree of Life, a model of the evolutionary relationships among all known living organisms (Figure 1.11B).
- Most biologists recognize three domains—Bacteria, Archaea, and Eukarya—based on fundamental characteristics of cell structure and molecular analysis. The Bacteria and Archaea each include one kingdom; the Eukarya is divided into the "protists" and three kingdoms: Plantae, Fungi, and Animalia (Figures 1.12 and 1.13).

Animation: Life's diversity

1.4 Biological Research

- Biologists conduct basic research to advance our knowledge of living organisms and applied research to solve practical problems.
- The scientific method allows researchers to crystallize and test their ideas. Scientists develop hypotheses—working explanations about the relationships between variables. Scientific hypotheses must be falsifiable (Figure 1.14).
- Scientists may collect observational data, which describe biological organisms or the details of biological processes, or experimental data, which describe the results of an experimental manipulation.
- A well-designed experiment considers alternative hypotheses and includes control treatments and replicates (Figure 1.15). When experiments are unfeasible, biologists often use null hypotheses, explanations of what they would see if their hypothesis was wrong, to evaluate data (Figure 1.16).
- Model organisms, which are easy to maintain in the laboratory, have been the subject of much research.
- Molecular techniques allow detailed analysis of the DNA of many species and the manipulation of specific genes in the laboratory.
- A scientific theory is a set of broadly applicable hypotheses that have been supported by repeated tests under many conditions and in many different situations. The theory of evolution by natural selection is of central importance to biology because it explains how life evolved through natural processes.

Animation: Sample size and accuracy

Animation: How do scientists use random samples to test hypotheses?

Test Your Knowledge

1. What is the lowest level of biological organization that biologists consider to be alive?
   a. a protein
   b. DNA
   c. a cell
   d. a multicellular organism
   e. a population of organisms
2. Which category falls immediately below "class" in the systematic hierarchy?
   a. species
   b. order
   c. family
   d. genus
   e. phylum
3. Which of the following represents the application of the "scientific method"?
   a. comparing one experimental subject to one control subject
   b. believing an explanation that is too complex to be tested
   c. using controlled experiments to test falsifiable hypotheses
   d. developing one testable hypothesis to explain a natural phenomenon
   e. observing a once-in-a-lifetime event under natural conditions
4. Houseflies develop through a series of programmed stages from egg, to larva, to pupa, to flying adult. This series of stages is called:
   a. artificial selection.
   b. respiration.
   c. homeostasis.
   d. a life cycle.
   e. metabolism.
5. Which structure allows living organisms to detect changes in the environment?
   a. a protein
   b. a receptor
   c. a gene
   d. RNA
   e. a nucleus
6. Which of the following is *not* a component of Darwin's theory as he understood it?
   a. Some individuals in a population survive longer than others.
   b. Some individuals in a population reproduce more than others.
   c. Heritable variations allow some individuals to compete more successfully for resources.
   d. Mutations in genes produce new variations in a population.
   e. Some new variations are passed to the next generation.
7. What role did the copper lizard models play in the field study of temperature regulation?
   a. They attracted live lizards to the study site.
   b. They measured the temperatures of live lizards.
   c. They established null hypotheses about basking behavior and temperatures.
   d. They scared predators away from the study site.
   e. They allowed researchers to practice taking lizard temperatures.
8. Which of the following questions best exemplifies basic research?
   a. How did life begin?
   b. How does alcohol intake affect aging?
   c. How fast does H1N1 flu spread among humans?
   d. How can we reduce hereditary problems in purebred dogs?
   e. How does the consumption of soft drinks promote obesity?
9. When researchers say that a scientific hypothesis must be falsifiable, they mean that:
   a. the hypothesis must be proved correct before it is accepted as truth.
   b. the hypothesis has already withstood many experimental tests.
   c. they have an idea about what will happen to one variable if another variable changes.
   d. appropriate data can prove without question that the hypothesis is correct.
   e. if the hypothesis is wrong, scientists must be able to demonstrate that it is wrong.
10. Which of the following characteristics would *not* qualify an animal as a model organism?
    a. It has rapid development.
    b. It has small adult size.
    c. It has a rapid life cycle.
    d. It has unique genes and unusual cells.
    e. It is easy to raise in the laboratory.

Discuss the Concepts

1. Viruses are infectious agents that contain either DNA or RNA surrounded by a protein coat. They cannot reproduce on their own, but they can take over the cells of the organisms they infect and force those cells to produce more virus particles. Based on the characteristics of living organisms described in this chapter, should viruses be considered living organisms?
2. While walking in the woods, you discover a large rock covered with a gelatinous, sticky substance. What tests could you perform to determine whether the substance is inanimate, alive, or the product of a living organism?
3. Explain why control treatments are a necessary component of well-designed experiments.

Design an Experiment

Design an experiment to test the hypothesis that the color of farmed salmon is produced by pigments in their food.

Interpret the Data

While working in Puerto Rico, Paul E. Hertz and his students studied a second species of lizard, the yellow-chinned anole, *Anolis gundlachi,* using the procedures described in Figure 1.16. Their results for copper models and living lizards are presented below. Based on these data, do you think that *A. gundlachi* regulates its body temperature? Why or why not?

Apply Evolutionary Thinking

When a biologist first tested a new pesticide on a population of insects, she found that only 1% of the insects survived their exposure to the poison. She allowed the survivors to reproduce and discovered that 10% of the offspring survived exposure to the same concentration of pesticide. One generation later, 50% of the insects survived this experimental treatment. What is a likely explanation for the increasing survival rate of these insects over time?

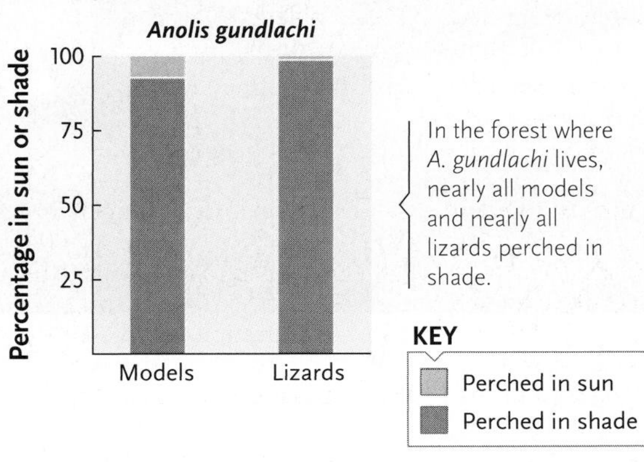
Percentage of models and lizards perched in sun or shade
Anolis gundlachi
Percentage in sun or shade
100
75
50
25
Models
Lizards
In the forest where A. gundlachi lives, nearly all models and nearly all lizards perched in shade.
KEY
Perched in sun
Perched in shade

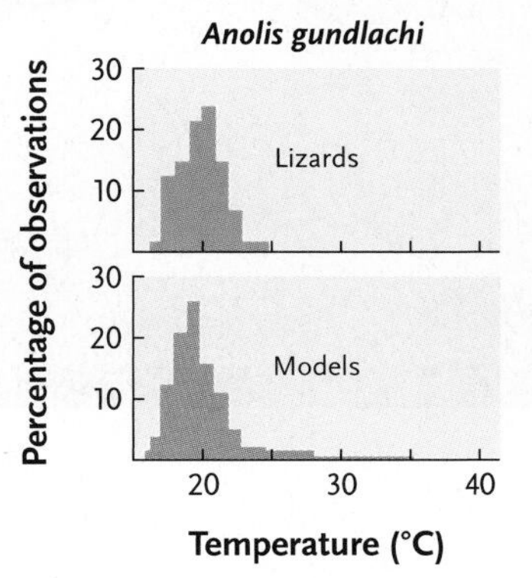
Temperatures of models and lizards
Anolis gundlachi
Percentage of observations
30
20
10
Lizards
Models
20
30
40
Temperature (°C)
Body temperatures of A. gundlachi were not significantly different from those of the randomly placed models.

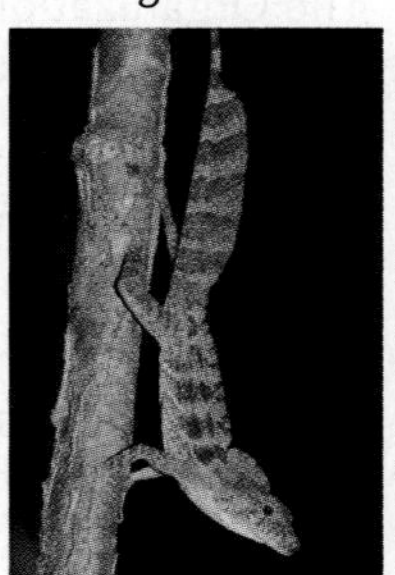
Anolis gundlachi
Kevin de Queiroz, National Museum of Natural History, Smithsonian Institution

2

STUDY OUTLINE

2.1 The Organization of Matter: Elements and Atoms

2.2 Atomic Structure

2.3 Chemical Bonds and Chemical Reactions

2.4 Hydrogen Bonds and the Properties of Water

2.5 Water Ionization and Acids, Bases, and Buffers

Life as we know it would be impossible without water, a small inorganic compound with unique properties.

Life, Chemistry, and Water

Why It Matters. . . We—like all plants, animals, and other organisms—are collections of atoms and molecules linked together by chemical bonds. The chemical nature of life makes it impossible to understand biology without knowledge of basic chemistry and chemical interactions.

For example, the element selenium is a natural ingredient of rocks and soils. In minute amounts it is necessary for the normal growth and survival of humans and many other animals, but in high concentrations selenium is toxic. In 1983, thousands of dead or deformed waterfowl were discovered at the Kesterson National Wildlife Refuge in the San Joaquin Valley of California. The deaths and deformities were traced to high concentrations of selenium in the environment, which had built up over decades as irrigation runoff washed selenium-containing chemicals from the soil into the water of the refuge. Since the problem was identified, engineers have diverted agricultural drainage water from the area, and the Kesterson refuge is now being restored.

The study of selenium and its biological effects suggested possible ways to prevent it from accumulating in the environment. For example, Norman Terry and his coworkers at the University of California, Berkeley designed a large-scale experiment to test natural methods for removing excess selenium from contaminated soils. Terry found that some wetland plants could remove up to 90% of the selenium in wastewater from a gasoline refinery. The plants convert much of the selenium into a relatively nontoxic gas, methyl selenide, which can pass into the atmosphere without harming plants and animals.

To test further the ability of plants to remove selenium, Terry and his coworkers grew wetland plants in 10 experimental plots watered by runoff from agricultural irrigation **(Figure 2.1)**. The researchers measured how much selenium remained in the soil of the plots, how much was incorpo-

Photo by Gary Head

FIGURE 2.1
Researcher Norman Terry in an experimental wetlands plot in Corcoran, California. Terry tested the ability of cattails, bulrushes, and marsh grasses to reduce selenium contamination in water draining from irrigated fields.

rated into plant tissues, and how much escaped into the air as a gas. Terry's results indicate that, before the runoff trickles through to local ponds, the plants in his plots reduce selenium to nontoxic levels, less than two parts per billion. Such applications of chemical and biological knowledge to decontaminate polluted environments are known as **bioremediation.** Bioremediation could help safeguard our food supplies, our health, and the environment.

The selenium example shows the importance of understanding and applying chemistry in biology. However, reactions involving selenium are only a few of the many thousands of chemical reactions that take place inside living organisms. Decades of research have taught us much about these reactions and have confirmed that the same laws of chemistry and physics govern both living and nonliving things. We can therefore apply with confidence information obtained from chemical experiments in the laboratory to the processes inside living organisms. An understanding of the relationship between the structure of chemical substances and their behavior is the first step in learning biology. <

2.1 The Organization of Matter: Elements and Atoms

Selenium is an example of an **element**—a pure substance that cannot be broken down into simpler substances by ordinary chemical or physical techniques. All **matter** of the universe—anything that occupies space and has mass—is composed of elements and combinations of elements. Ninety-two different elements occur naturally on Earth, and more than fifteen artificial elements have been synthesized in the laboratory.

Living Organisms Are Composed of about 25 Key Elements

Four elements—carbon, hydrogen, oxygen, and nitrogen—make up more than 96% of the weight of living organisms. Seven other elements—calcium, phosphorus, potassium, sulfur, sodium, chlorine, and magnesium—contribute most of the remaining 4%. Several other elements occur in organisms in quantities so small (<0.01%) that they are known as **trace elements. Figure 2.2** compares the relative proportions of different elements in a human, a plant, Earth's crust, and seawater, and lists the most important trace elements in a human. The proportions of elements in living organisms, as represented by the human and the plant, differ markedly from those of Earth's crust and seawater; these differences reflect the highly ordered chemical structure of living organisms.

Trace elements are vital for normal biological functions. For example, iodine makes up only about 0.0004% of a human's weight. However, a lack of iodine in the human diet severely impairs the function of the thyroid gland, which produces hormones that regulate metabolism and growth. Symptoms of iodine deficiency include lethargy, apathy, and sensitivity to cold temperatures. Prolonged iodine deficiency causes a *goiter,* a condition in which the thyroid gland enlarges so much that the front

Seawater	
Oxygen	88.3
Hydrogen	11.0
Chlorine	1.9
Sodium	1.1
Magnesium	0.1
Sulfur	0.09
Potassium	0.04
Calcium	0.04
Carbon	0.003
Silicon	0.0029
Nitrogen	0.0015
Strontium	0.0008

Human	
Oxygen	65.0
Carbon	18.5
Hydrogen	9.5
Nitrogen	3.3
Calcium	2.0
Phosphorus	1.1
Potassium	0.35
Sulfur	0.25
Sodium	0.15
Chlorine	0.15
Magnesium	0.05
Iron	0.004
Iodine	0.0004

Pumpkin	
Oxygen	85.0
Hydrogen	10.7
Carbon	3.3
Potassium	0.34
Nitrogen	0.16
Phosphorus	0.05
Calcium	0.02
Magnesium	0.01
Iron	0.008
Sodium	0.001
Zinc	0.0002
Copper	0.0001

Earth's crust	
Oxygen	46.6
Silicon	27.7
Aluminum	8.1
Iron	5.0
Calcium	3.6
Sodium	2.8
Potassium	2.6
Magnesium	2.1
Other elements	1.5

Image copyright Andriano, 2010. Used under license from Shutterstock.com

iStockphoto.com/Hon Lau

FIGURE 2.2
The proportions by mass of different elements in seawater, the human body, a fruit, and Earth's crust. Trace elements in humans include boron, chromium, cobalt, copper, fluorine, iodine, iron, manganese, molybdenum, selenium, tin, vanadium, and zinc, as well as variable traces of other elements.

of the neck swells significantly. Once a common condition, goiter has almost been eliminated by adding iodine to table salt, especially in regions where soils are iodine-deficient.

Elements Are Composed of Atoms, Which Combine to Form Molecules

Elements are composed of individual **atoms**—the smallest units that retain the chemical and physical properties of an element. Any given element has only one type of atom. Several million atoms arranged side by side would be needed to equal the width of the period at the end of this sentence.

Atoms are identified by a standard one- or two-letter symbol. The element carbon is identified by the single letter *C,* which stands for both the carbon atom and the element; iron is identified by the two-letter symbol *Fe* (*ferrum* = iron). **Table 2.1** lists the chemical symbols of these and other atoms common in living organisms.

Atoms combined chemically in fixed numbers and ratios form the **molecules** of living and nonliving matter. For example, the oxygen we breathe is a molecule formed from the chemical combination of two oxygen atoms; a molecule of the carbon dioxide we exhale contains one carbon atom and two oxygen atoms. The name of a molecule is written in chemical shorthand as a **formula,** using the standard symbols for the elements and using subscripts to indicate the number of atoms of each element in the molecule. The subscript is omitted for atoms that occur only once in a molecule. For example, the formula for an oxygen molecule is written as O_2 (two oxygen atoms); for a carbon dioxide molecule the formula is CO_2 (one carbon atom and two oxygen atoms).

TABLE 2.1 Atomic Number and Mass Number of the Most Common Elements in Living Organisms

Element	Symbol	Atomic Number	Mass Number of the Most Common Form
Hydrogen	H	1	1
Carbon	C	6	12
Nitrogen	N	7	14
Oxygen	O	8	16
Sodium	Na	11	23
Magnesium	Mg	12	24
Phosphorus	P	15	31
Sulfur	S	16	32
Chlorine	Cl	17	35
Potassium	K	19	39
Calcium	Ca	20	40
Iron	Fe	26	56
Iodine	I	53	127

Molecules whose component atoms are different (such as carbon dioxide) are called **compounds.** The chemical and physical properties of compounds are typically distinct from those of their atoms or elements. For example, we all know that water is a liquid at room temperature. We also know that water does not burn. However, the individual elements of water—hydrogen and oxygen—are gases at room temperature, and both are highly reactive.

STUDY BREAK 2.1 <

Distinguish between an element and an atom, and between a molecule and a compound.

2.2 Atomic Structure

Each element consists of one type of atom. However, all atoms share the same basic structure **(Figure 2.3).** Each atom consists of an **atomic nucleus,** surrounded by one or more smaller, fast-moving particles called **electrons.** Although the electrons occupy more than 99.99% of the space of an atom, the nucleus makes up more than 99.99% of its total mass.

The Atomic Nucleus Contains Protons and Neutrons

All atomic nuclei contain one or more positively charged particles called **protons.** The number of protons in the nucleus of each kind of atom is referred to as the **atomic number.** This number does not vary and thus specifically identifies the atom. The smallest atom, hydrogen, has a single proton in its nucleus, so its atomic number is 1. The heaviest naturally occurring atom, uranium, has 92 pro-

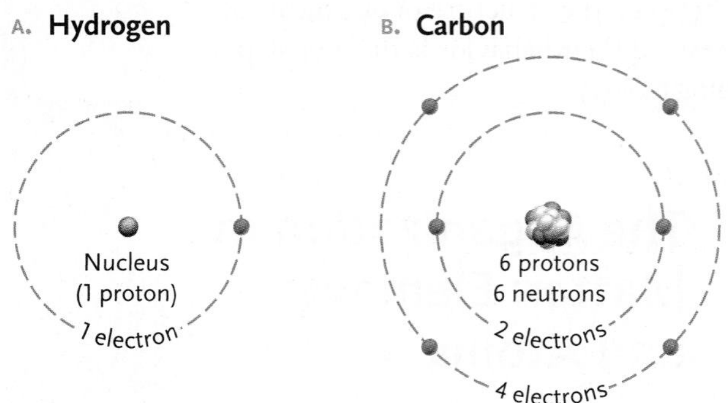
A. Hydrogen
Nucleus
(1 proton)
1 electron
B. Carbon
6 protons
6 neutrons
2 electrons
4 electrons

FIGURE 2.3
Atomic structure. The nucleus of an atom contains one or more protons and, except for the most common form of hydrogen, a similar number of neutrons. Fast-moving electrons, in numbers equal to the protons, surround the nucleus. **(A)** The most common form of hydrogen, the simplest atom, has a single proton in its nucleus and a single electron. **(B)** Carbon, a more complex atom, has a nucleus surrounded by electrons at two levels. The electrons in the outer level follow more complex pathways than shown here.

tons in its nucleus and therefore has an atomic number of 92. Similarly, carbon with six protons, nitrogen with seven protons, and oxygen with eight protons have atomic numbers of 6, 7, and 8, respectively (see Table 2.1).

With one exception, the nuclei of all atoms also contain uncharged particles called **neutrons.** Neutrons occur in variable numbers approximately equal to the number of protons. The lone exception is the most common form of hydrogen, which has a nucleus that contains a single proton. There are two less common forms of hydrogen as well. One form, named deuterium, has one neutron and one proton in its nucleus. The other form, named tritium, has two neutrons and one proton.

Other atoms also have common and less common forms with different numbers of neutrons. For example, the most common form of the carbon atom has six protons and six neutrons in its nucleus, but about 1% of carbon atoms have six protons and seven neutrons in their nuclei and an even smaller percentage has six protons and eight neutrons.

The distinct forms of the atoms of an element, all with the same number of protons but different numbers of neutrons, are called **isotopes (Figure 2.4).** The various isotopes of an atom differ in mass and other physical characteristics, but all have essentially the same chemical properties. Therefore, organisms can use any hydrogen or carbon isotope, for example, without a change in their chemical reactions.

A neutron and a proton have almost the same mass, about 1.66×10^{-24} grams (g). This mass is defined as a standard unit, the **dalton** (Da), named after John Dalton, a nineteenth-century English scientist who contributed to the development of atomic theory. Atoms are assigned a **mass number** based on the total number of protons and neutrons in the atomic nucleus (see Table 2.1). Electrons are ignored in determinations of atomic mass because the mass of an electron, at only 1/1,800 of the mass of a proton or neutron, does not contribute significantly to the mass of an atom. Thus, the mass number of the hydrogen isotope with one proton in its nucleus is 1, and its mass is 1 dalton. The mass number of the hydrogen isotope deuterium is 2, and the mass number of tritium is 3. The carbon isotope with six protons and six neutrons in its nucleus has a mass number of 12; the isotope with six protons and seven neutrons has a mass number of 13, and the isotope with six protons and eight neutrons has a mass number of 14 (see Figure 2.4). These carbon mass numbers are written as ^{12}C, ^{13}C, and ^{14}C, or carbon-12, carbon-13, and carbon-14, respectively. However, all carbon isotopes have the same atomic number of 6, because this number reflects only the number of protons in the nucleus.

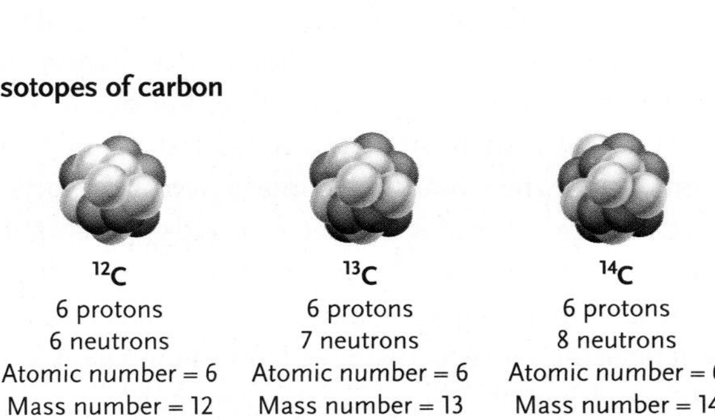
Isotopes of hydrogen
¹H
1 proton
Atomic number = 1
Mass number = 1
²H (deuterium)
1 proton
1 neutron
Atomic number = 1
Mass number = 2
³H (tritium)
1 proton
2 neutrons
Atomic number = 1
Mass number = 3
Isotopes of carbon
¹²C
6 protons
6 neutrons
Atomic number = 6
Mass number = 12
¹³C
6 protons
7 neutrons
Atomic number = 6
Mass number = 13
¹⁴C
6 protons
8 neutrons
Atomic number = 6
Mass number = 14

FIGURE 2.4
The atomic nuclei of hydrogen and carbon isotopes. Note that isotopes of an atom have the same atomic number but different mass numbers.

You might wonder about the meaning of *mass* as compared to *weight.* **Mass** is the amount of matter in an object, whereas **weight** is a measure of the pull of gravity on an object. Mass is constant, but the weight of an object may vary because of differences in gravity. For example, the mass of a piece of lead is the same on Earth and in outer space, but the same piece of lead that weighs 1 kilogram (kg) on Earth is weightless in an orbiting spacecraft, even though its mass remains the same. However, as long as an object is on Earth's surface, its mass and weight are equivalent. Thus, we can weigh an object in the laboratory and be assured that its weight accurately reflects its mass.

The Nuclei of Some Atoms Are Unstable and Tend to Break Down to Form Simpler Atoms

The nuclei of some isotopes are unstable and break down, or *decay,* giving off particles of matter and energy that can be detected as **radioactivity.** The decay transforms the unstable, radioactive isotope—called a **radioisotope**—into an atom of another element. The decay continues at a steady, clocklike rate, with a constant proportion of the radioisotope breaking down at any instant. The rate of decay is not affected by chemical reactions or environmental conditions such as temperature or pressure. For example, the carbon isotope ^{14}C is unstable and undergoes radioactive decay in which one of its neutrons splits into a proton and an electron. The electron is ejected from the nucleus, but the proton is retained, giving a new total of seven protons and seven neutrons, which is characteristic of the most common form of nitrogen. Thus, the decay transforms the carbon atom into an atom of nitrogen.

Because unstable isotopes decay at a clocklike rate, they can be used to estimate the age of organic material, rocks, or fossils that contain them. These techniques have been vital in dating animal remains and tracing evolutionary lineages, as described in Chapter 22. Isotopes are also used in biological research as **tracers** to label molecules so that they can be tracked as they pass through biochemical reactions. Radioactive isotopes of carbon (^{14}C), phosphorus (^{32}P), and sulfur (^{35}S) can be traced easily by their radioactivity. A number of stable, nonradioactive isotopes, such as ^{15}N (called heavy nitrogen), can be detected by their mass differences and have also proved valuable as tracers in biological experiments. *Focus on Applied Research* describes some applications of radioisotopes in research and medicine.

Using Radioisotopes to Trace Reactions and Save Lives

In 1896, the French physicist Henri Becquerel wrapped a rock containing uranium in paper and tucked it into a desk drawer on top of a case containing an unexposed photographic plate. When he opened the case containing the plate a few days later, he was surprised to find an image of the rock on the plate—apparently caused by energy emitted from the rock. One of his coworkers, Marie Curie, named the phenomenon "radioactivity." Although radioactivity can be dangerous to life (more than one researcher, including Marie Curie, has died from its effects), it has been harnessed and put to highly productive use for scientific and medical purposes.

The radiation released by unstable isotopes can be detected by placing a photographic film over samples containing the isotopes (as Becquerel discovered) or by using an instrument known as a *scintillation counter.* These techniques allow researchers to use isotopes as tracers in chemical reactions. Typically, organisms are exposed to a reactant chemical that has been "labeled" with a radioactive isotope such as ^{14}C or ^{3}H. After being exposed to the tracer, the chemical products in which the isotope appears, and their sequence of appearance, can be detected by their radioactivity and identified.

For example, algae and plants use carbon dioxide (CO_2) as a raw material in photosynthesis. To trace the reactions of photosynthesis, Melvin Calvin and his coworkers grew algal cells in a medium with CO_2 that contained the radioisotope ^{14}C. Then they extracted various substances from the cells at intervals, separated them on a piece of paper based on their different solubilities in particular solvents, and placed the paper on a photographic film. The particular substances that exposed spots on the film because they were radioactive, as well as their order of appearance in the cells, allowed the researchers to piece together the sequence of reactions in photosynthesis, as described and illustrated in *Focus on Basic Research* in Chapter 9.

Radioisotopes are widely used in medicine to diagnose and cure disease, to produce images of diseased body organs, and, as in biological research, to trace the locations and routes followed by individual substances marked for identification by radioactivity. One example of their use is in the diagnosis of thyroid gland disease. The thyroid is the only structure in the body that absorbs iodine in quantity. The size and shape of the thyroid, which reflect its health, are measured by injecting a small amount of a radioactive iodine isotope into the patient's bloodstream. After the isotope is concentrated in the thyroid, the gland is then scanned by an apparatus that uses the radioactivity to produce an image of the gland on a photographic film. Examples of what the scans may show are presented in the figure. Another application uses the fact that radioactive thallium is not taken up by regions of the heart muscle with poor circulation to detect coronary artery disease. Other isotopes are used to detect bone injuries and defects, including injured, arthritic, or abnormally growing segments of bone.

Treatment of disease with radioisotopes takes advantage of the fact that radioactivity in large doses can kill cells (radiation generates highly reactive chemical groups that break and disrupt biological molecules). Dangerously overactive thyroid glands are treated by giving patients a dose of radioactive iodine calculated to destroy just enough thyroid cells to reduce activity of the gland to normal levels. In radiation therapy, cancer cells are killed by bombarding them with radiation emitted by radium-226 or cobalt-60. As much as is possible, the radiation is focused on the tumor to avoid destroying nearby healthy tissues. In some forms of chemotherapy for cancer, patients are given radioactive substances at levels that kill cancer cells without also killing the patient.

Photo by Gary Head

Normal Enlarged Cancerous

Scans of human thyroid glands after iodine-123 was injected into the bloodstream. The radioactive iodine becomes concentrated in the thyroid gland.

The Electrons of an Atom Occupy Orbitals around the Nucleus

In an atom, the number of electrons surrounding the nucleus is equal to the number of protons in the nucleus. An electron carries a negative charge that is exactly equal and opposite to the positive charge of a proton. The equality of numbers of electrons and protons in an atom balances the positive and negative charges and makes the total structure of an atom electrically neutral.

An atom is often drawn with electrons following a pathway around the nucleus similar to planets orbiting a sun. The reality is different. The speed of electrons in motion around the nucleus approaches the speed of light. At any instant, an electron may be in any location with respect to its nucleus, from the immediate vicinity of the nucleus to practically infinite space. An electron moves so fast that it almost occupies all the locations at the same time; however, it passes through some locations much more frequently than others. The locations where an electron occurs most frequently around the atomic nucleus define a path called an orbital. An **orbital** is essentially the region of space where the electron "lives" most of the time. Although either one or two electrons may occupy an orbital, the most stable and balanced condition occurs when an orbital contains a pair of electrons.

Electrons are maintained in their orbitals by a combination of attraction to the positively charged nucleus and mutual repulsion because of their negative charge. The orbitals take different shapes depending on their distance from the nucleus and their degree of repulsion by electrons in other orbitals.

Under certain conditions, electrons may pass from one orbital to another within an atom, enter orbitals shared by two or more atoms, or pass completely from orbitals in one atom to orbitals in another. As discussed later in this chapter, the ability of electrons to move from one orbital to another underlies the chemical reactions that combine atoms into molecules.

Orbitals Occur in Discrete Layers around an Atomic Nucleus

Within an atom, electrons are found in regions of space called **energy levels,** or more simply, **shells.** Within each energy level, electrons are grouped into orbitals. The lowest energy level of an atom, the one nearest the nucleus, may be occupied by a maximum of two electrons in a single orbital **(Figure 2.5A).** This orbital, which has a spherical shape, is called the 1*s* orbital. (The "1" signifies that the orbital is in the energy level closest to the nucleus, and the "*s*" signifies the shape of the orbital, in this case, spherically symmetric around the nucleus.) Hydrogen has one electron in this orbital, and helium has two.

Atoms with atomic numbers between 3 (lithium) and 10 (neon) have two energy levels, with two electrons in the 1*s* orbital and one to eight electrons in orbitals at the next highest energy level. The electrons at this second energy level occupy one spherical orbital, called the 2*s* orbital **(Figure 2.5B),** and as many as three orbitals that are pushed into a dumbbell shape by repulsions between electrons, called 2*p* orbitals **(Figure 2.5C).** The orbitals for neon are shown in **Figure 2.5D.**

Larger atoms have more energy levels. The third energy level, which may contain as many as 18 electrons in 9 orbitals, includes the atoms from sodium (11 electrons) to argon (18 electrons). (**Figure 2.6** shows the 18 elements that have electrons in the lowest

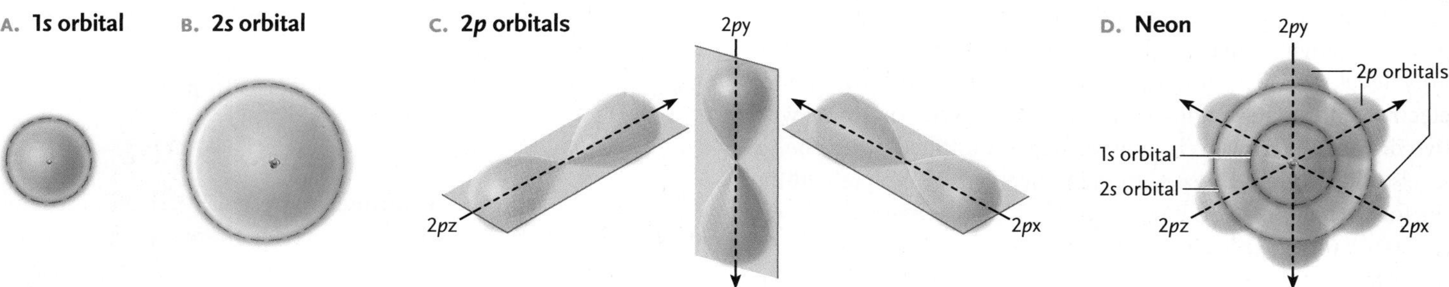

A. 1s orbital
B. 2s orbital
C. 2p orbitals
2py
2pz
2px
D. Neon
2py
2p orbitals
1s orbital
2s orbital
2pz
2px

FIGURE 2.5
Electron orbitals. **(A)** The single 1*s* orbital of hydrogen and helium approximates a sphere centered on the nucleus. **(B)** The 2*s* orbital. **(C)** The 2*p* orbitals lie in the three planes x, y, and z, each at right angles to the others. **(D)** In atoms with two energy levels, such as neon, the lowest energy level is occupied by a single 1*s* orbital as in hydrogen and helium. The second, higher energy level is occupied by a maximum of four orbitals—a spherical 2*s* orbital and three dumbbell-shaped 2*p* orbitals.

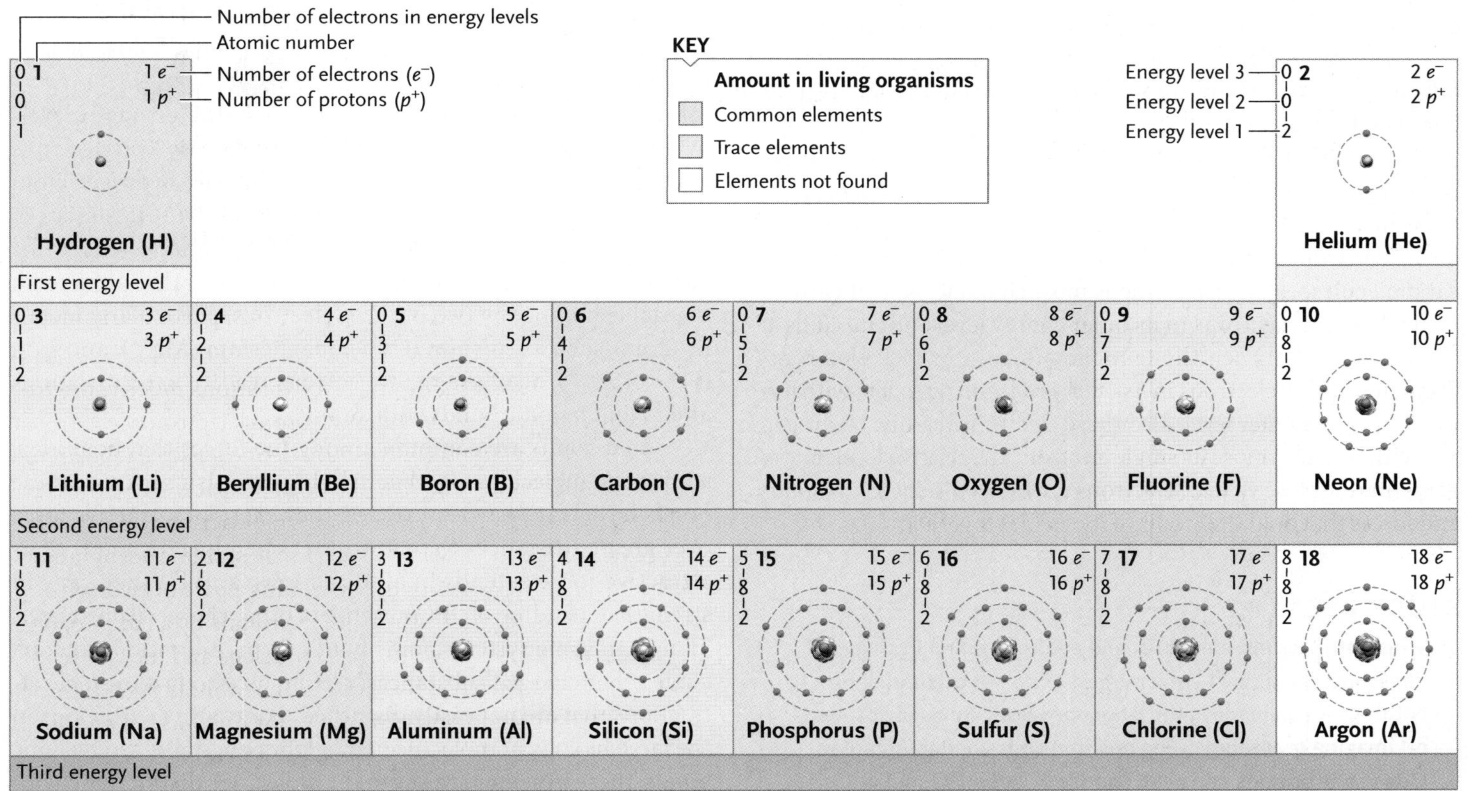

Number of electrons in energy levels
Atomic number
Number of electrons (e^-)
Number of protons (p^+)
KEY
Amount in living organisms
Common elements
Trace elements
Elements not found
Energy level 3
Energy level 2
Energy level 1
Hydrogen (H)
Helium (He)
First energy level
Lithium (Li)
Beryllium (Be)
Boron (B)
Carbon (C)
Nitrogen (N)
Oxygen (O)
Fluorine (F)
Neon (Ne)
Second energy level
Sodium (Na)
Magnesium (Mg)
Aluminum (Al)
Silicon (Si)
Phosphorus (P)
Sulfur (S)
Chlorine (Cl)
Argon (Ar)
Third energy level

FIGURE 2.6
The atoms with electrons distributed in one, two, or three energy levels. The atomic number of each element (shown in boldface in each panel) is equivalent to the number of protons in its nucleus.

three energy levels only.) The fourth energy level may contain as many as 32 electrons in 16 orbitals. In all cases, the total number of electrons in the orbitals is matched by the number of protons in the nucleus. However, whatever the size of an atom, the outermost energy level typically contains one to eight electrons occupying a maximum of four orbitals.

The Number of Electrons in the Outermost Energy Level of an Atom Determines Its Chemical Activity

The electrons in an atom's outermost energy level are known as **valence electrons** (*valentia* = power or capacity). Atoms in which the outermost energy level is not completely filled with electrons tend to be chemically reactive; those with a completely filled outermost energy level are nonreactive, or inert. For example, hydrogen has a single, unpaired electron in its outermost and only energy level, and it is highly reactive; helium has two valence electrons filling its single orbital, and it is inert. For atoms with two or more energy levels, only those with unfilled outer energy levels are reactive. Those with eight electrons completely filling the four orbitals of the outer energy level, such as neon and argon, are stable and chemically unreactive (see Figure 2.6).

Atoms with outer energy levels that contain electrons near the stable numbers tend to gain or lose electrons to reach the stable configuration. For example, sodium has two electrons in its first energy level, eight in the second, and one in the third and outermost level (see Figure 2.6). The outermost electron is readily lost to another atom, giving the sodium atom a stable second energy level (now the outermost level) with eight electrons. Chlorine, with seven electrons in its outermost energy level, tends to take up an electron from another atom to attain the stable number of eight electrons.

Atoms that differ from the stable configuration by more than one or two electrons tend to attain stability by *sharing* electrons in joint orbitals with other atoms rather than by gaining or losing electrons completely. Among the atoms that form biological molecules, electron sharing is most characteristic of carbon, which has four electrons in its outer energy level and thus falls at the midpoint between the tendency to gain or lose electrons. Oxygen, with six electrons in its outer level, and nitrogen, with five electrons in its outer level, also share electrons readily. Hydrogen may either share or lose its single electron. The relative tendency to gain, share, or lose valence electrons underlies the chemical bonds and forces that hold the atoms of molecules together.

STUDY BREAK 2.2

1. **Where are protons, electrons, and neutrons found in an atom?**
2. **The isotopes carbon-11 and oxygen-15 do not occur in nature, but they can be made in the laboratory. Both are used in a medical imaging procedure called positron emission tomography. How many protons and neutrons are in carbon-11 and in oxygen-15?**
3. **What determines the chemical reactivity of an atom?**

2.3 Chemical Bonds and Chemical Reactions

Atoms of inert elements, such as helium, neon, and argon, occur naturally in uncombined forms, but atoms of reactive elements tend to combine into molecules by forming **chemical bonds.** The four most important chemical linkages in biological molecules are *ionic bonds, covalent bonds, hydrogen bonds,* and *van der Waals forces.* Chemical reactions occur when atoms or molecules interact to form new chemical bonds or break old ones.

Ionic Bonds Are Multidirectional and Vary in Strength

Ionic bonds result from electrical attractions between atoms that gain or lose valence electrons completely. A sodium atom (Na) readily loses a single electron to achieve a stable outer energy level, and chlorine (Cl) readily gains an electron:

$$Na^{\cdot} + \cdot\ddot{\underset{\cdot\cdot}{Cl}}: \rightarrow Na^{+} :\ddot{\underset{\cdot\cdot}{Cl}}:^{-}$$

(The dots in the preceding formula represent the electrons in the outermost energy level.) After the transfer, the sodium atom, now with 11 protons and 10 electrons, carries a single positive charge. The chlorine atom, now with 17 protons and 18 electrons, carries a single negative charge. In this charged condition, the sodium and chlorine atoms are called **ions** instead of atoms and are written as Na^+ and Cl^- **(Figure 2.7).** A positively charged ion such as Na^+ is called a **cation,** and a negatively charged ion such as Cl^- is called an **anion.** The difference in charge between cations and anions creates an attraction—the ionic bond—that holds the ions together in solid NaCl (sodium chloride).

Many other atoms that differ from stable outer energy levels by one electron, including hydrogen, can gain or lose electrons completely to form ions and ionic bonds. When a hydrogen atom loses its single electron to form a hydrogen ion (H^+), it consists of only a proton and is often simply called a proton to reflect this fact. A number of atoms with outer energy levels that differ from the stable number by two or three electrons, particularly metallic atoms such as calcium (Ca^{2+}), magnesium (Mg^{2+}), and iron (Fe^{2+} or Fe^{3+}), also lose their electrons readily to form cations and to join in ionic bonds with anions.

Ionic bonds are common among the forces that hold ions, atoms, and molecules together in living organisms because these bonds have three key features: (1) they exert an attractive force over greater distances than any other chemical bond; (2) their attractive force extends in all directions; and (3) they vary in strength depending on the presence of other charged substances. That is, in some systems, ionic bonds form in locations that exclude other charged substances, setting up strong and stable attractions that are not easily disturbed. For example, iron ions in the large biological molecule hemoglobin are stabilized by ionic bonds; these iron ions are key to the molecule's distinctive chemical properties. In other systems, particularly at molecular surfaces exposed to water molecules, ionic bonds are relatively weak, allowing ionic attractions to be established or broken quickly.

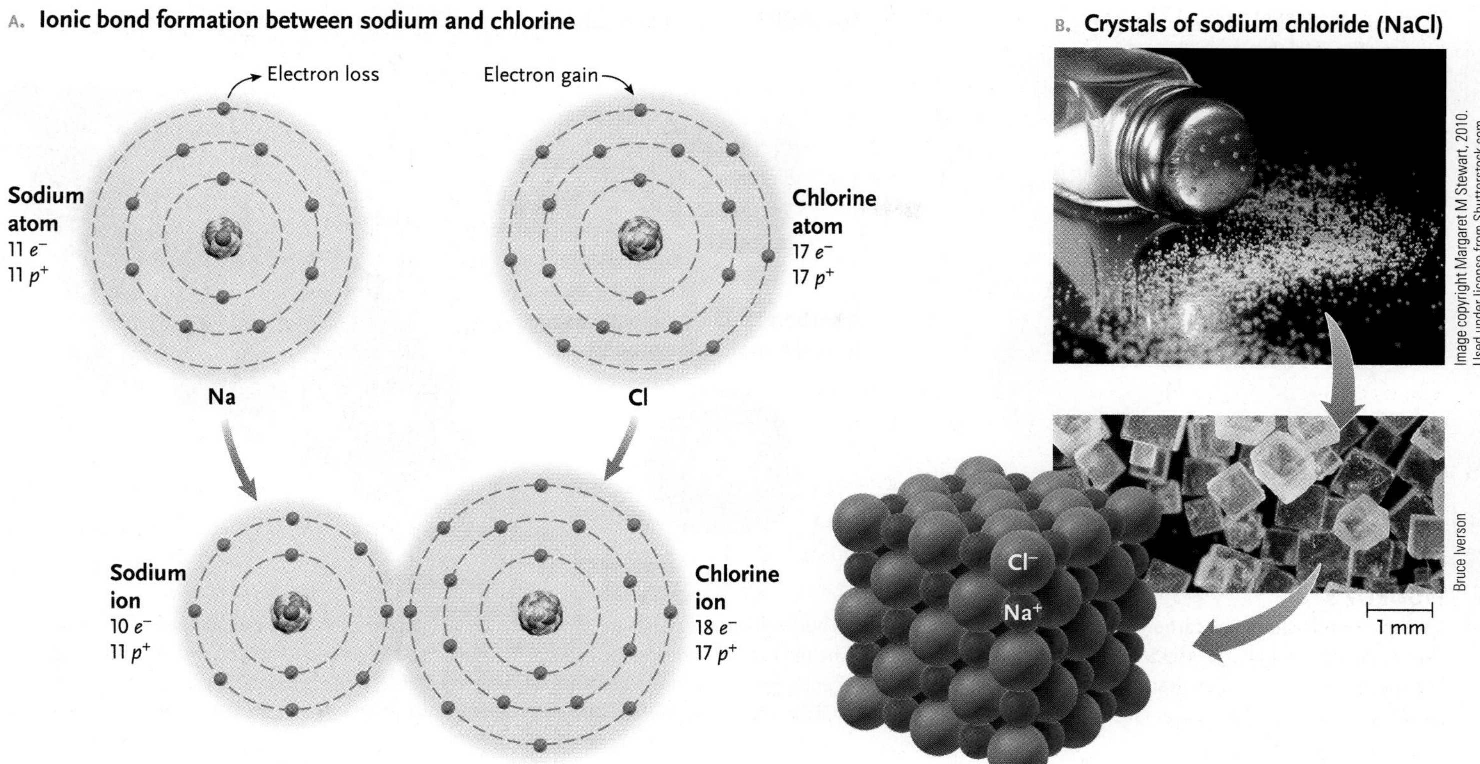
A. Ionic bond formation between sodium and chlorine
Electron loss
Electron gain
Sodium atom
11 e⁻
11 p⁺
Chlorine atom
17 e⁻
17 p⁺
Na
Cl
Sodium ion
10 e⁻
11 p⁺
Chlorine ion
18 e⁻
17 p⁺
Na⁺
Cl⁻
B. Crystals of sodium chloride (NaCl)
Image copyright Margaret M Stewart, 2010. Used under license from Shutterstock.com
Bruce Iverson
Cl⁻
Na⁺
1 mm

FIGURE 2.7
Formation of an ionic bond. **(A)** Sodium, with one electron in its outermost energy level, readily loses that electron to attain a stable state in which its second energy level, with eight electrons, becomes the outer level. Chlorine, with seven electrons in its outer energy level, readily gains an electron to attain the stable number of eight. The transfer creates the ions Na^+ and Cl^-. **(B)** The combination forms sodium chloride (NaCl), common table salt.

For example, as part of their activity in speeding biological reactions, many enzymatic proteins bind and release molecules by forming and breaking relatively weak ionic bonds.

Covalent Bonds Are Formed by Electrons in Shared Orbitals

Covalent bonds form when atoms share a pair of valence electrons rather than gaining or losing them. For example, if two hydrogen atoms collide, the single electron of each atom may join in a new, combined two-electron orbital that surrounds both nuclei. The two electrons fill the orbital; thus, the hydrogen atoms tend to remain linked stably together in the form of molecular hydrogen, H_2. The linkage formed by the shared orbital is a covalent bond.

In molecular diagrams, a covalent bond is designated by a pair of dots or a single line that represents a pair of shared electrons. For example, in H_2, the covalent bond that holds the molecule together is represented as H:H or H—H.

Unlike ionic bonds, which extend their attractive force in all directions, the shared orbitals that form covalent bonds extend between atoms at discrete angles and directions, giving covalently bound molecules distinct, three-dimensional forms. For biological molecules such as proteins, which are held together primarily by covalent bonds, the three-dimensional form imparted by these bonds is critical to their functions.

An example of a molecule that contains covalent bonds is methane, CH_4, the main component of natural gas. Carbon, with four unpaired outer electrons, typically forms four covalent bonds to complete its outermost energy level, here with four atoms of hydrogen. The four covalent bonds are fixed at an angle of 109.5° from each other, forming a tetrahedron **(Figure 2.8A, B)**. The tetrahedral arrangement of the bonds allows carbon "building blocks" **(Figure 2.8C)** to link to each other in both branched and unbranched chains and rings **(Figure 2.8D)**. Such structures form the backbones of an almost unlimited variety of molecules. Carbon can also form double bonds, in which atoms share two pairs of electrons, and triple bonds, in which atoms share three pairs of electrons.

Oxygen, hydrogen, nitrogen, and sulfur also share electrons readily to form covalent linkages, and they commonly combine with carbon in biological molecules. In these linkages with carbon, oxygen typically forms two covalent bonds; hydrogen, one; nitrogen, three; and sulfur, two.

Unequal Electron Sharing Results in Polarity

Electronegativity is the measure of an atom's attraction for the electrons it shares in a chemical bond with another atom. The more electronegative an atom is, the more strongly it attracts shared electrons. Among atoms, electronegativity increases as the number of protons in the nucleus increases and as the distance of electrons from the nucleus increases.

Although all covalent bonds involve the sharing of valence electrons, they differ widely in the degree of sharing. Depending

A. **Shared orbitals of methane (CH_4)**

B. **Space-filling model of methane**

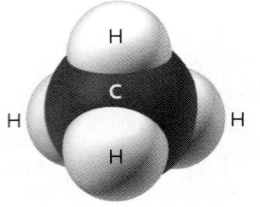

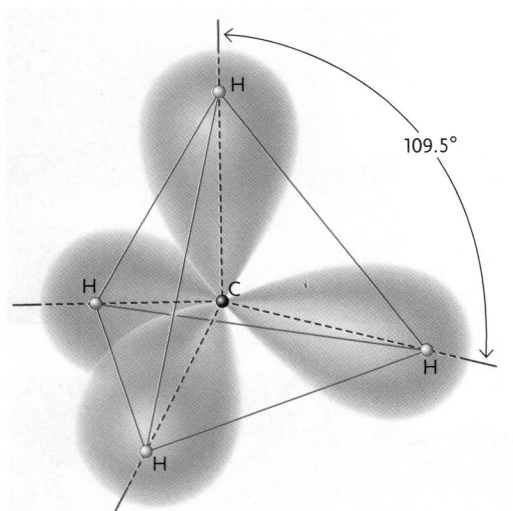

C. **A carbon "building block" used to make molecular models**

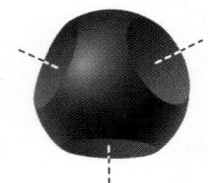

D. **Cholesterol**

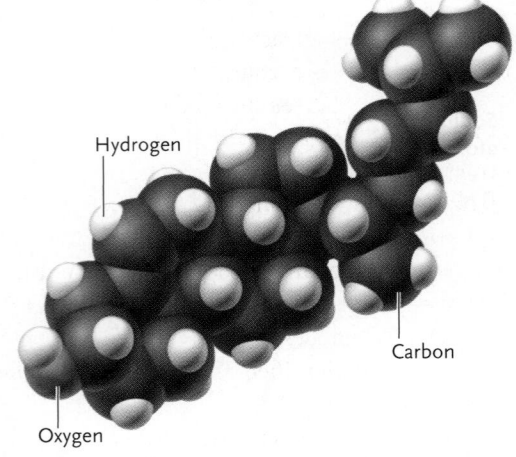

FIGURE 2.8
Covalent bonds shared by carbon. (A) The four covalent bonds of carbon in methane (CH_4) are shown as shared orbitals. The bonds extend outward from the carbon nucleus at angles of 109.5° from each other (dashed lines). The red lines connecting the hydrogen nuclei form a regular tetrahedron with four faces. **(B)** In the space-filling model of methane, the diameter of the sphere representing an atom shows the approximate limit of its electron orbitals. **(C)** A tetrahedral carbon "building block." One of the four faces of the block is not visible. **(D)** Carbon atoms assembled into rings and chains forming a complex molecule.

on the difference in electronegativity between the bonded atoms, covalent bonds are classified as **nonpolar covalent bonds** or **polar covalent bonds.** In a nonpolar covalent bond, electrons are shared equally, whereas in a polar covalent bond, they are shared unequally. When electron sharing is unequal, as in polar covalent bonds, the atom that attracts the electrons more strongly carries a partial negative charge, δ^- ("delta minus"), and the atom deprived of electrons carries a partial positive charge, δ^+ ("delta plus"). The atoms carrying partial charges may give the whole molecule partially positive and negative ends; in other words, the molecule is *polar,* hence the name given to the bond.

Nonpolar covalent bonds are characteristic of molecules that contain atoms of one element, such as elemental hydrogen (H_2) and oxygen (O_2), although there are some exceptions. Polar covalent bonds are characteristic of molecules that contain atoms of different elements.

For example, in water, an oxygen atom forms polar covalent bonds with two hydrogen atoms. Because the oxygen nucleus with its eight protons attracts electrons much more strongly than the hydrogen nuclei do, the bonds are strongly polar **(Figure 2.9).** In addition, the water molecule is asymmetric, with the oxygen atom located on one side and the hydrogen atoms on the other. This arrangement gives the entire molecule an unequal charge distribution, with the hydrogen end partially positive and the oxygen end partially negative, and makes water molecules strongly polar. In fact, water is the primary biological example of a polar molecule. The polar nature of water underlies its ability to adhere to ions and weaken their attractions.

Oxygen, nitrogen, and sulfur, which all share electrons unequally with hydrogen, are located asymmetrically in many biological molecules. Therefore, the presence of —OH, —NH, or —SH groups tends to make regions in biological molecules containing them polar.

Although carbon and hydrogen share electrons somewhat unequally, these atoms tend to be arranged symmetrically in biological molecules. Thus, regions that contain only carbon–hydrogen chains are typically nonpolar. For example, the C—H bonds in methane are located symmetrically around the carbon atom (see Figure 2.8), so their partial charges cancel each other and the molecule as a whole is nonpolar.

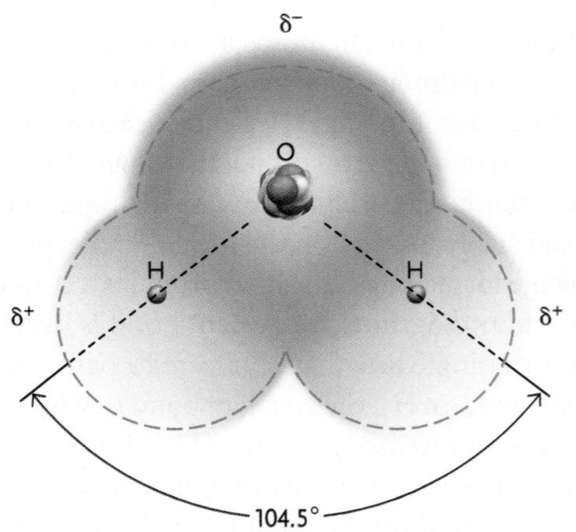

δ−
O
H
H
δ+
δ+
104.5°

FIGURE 2.9
Polarity in the water molecule, created by unequal electron sharing between the two hydrogen atoms and the oxygen atom and the asymmetric shape of the molecule. The unequal electron sharing gives the hydrogen end of the molecule a partial positive charge, δ^+ ("delta plus"), and the oxygen end of the molecule a partial negative charge, δ^- ("delta minus"). Regions of deepest color indicate the most frequent locations of the shared electrons. The orbitals occupied by the electrons are more complex than the spherical forms shown here.

Polar Molecules Tend to Associate with Each Other and Exclude Nonpolar Molecules

Polar molecules attract and align themselves with other polar molecules and with charged ions and molecules. These **polar associations** create environments that tend to exclude nonpolar molecules. When present in quantity, the excluded nonpolar molecules tend to clump together in arrangements called **nonpolar associations;** these nonpolar associations reduce the surface area exposed to the surrounding polar environment. Polar molecules that associate readily with water are identified as **hydrophilic** (*hydro* = water; *philic* = having an affinity or preference for). Nonpolar substances that are excluded by water and other polar molecules are identified as **hydrophobic** (*phobic* = having an aversion to or fear of).

Polar and nonpolar associations can be demonstrated with an apparatus no more complex than a bottle containing water and vegetable oil. If the bottle has been placed at rest for some time, the nonpolar oil and polar water form separate layers, with the oil on top. If you shake the bottle, the oil becomes suspended as spherical droplets in the water; the harder you shake, the smaller the oil droplets become (the spherical form of the oil droplets exposes the least surface area per unit volume to the watery polar surroundings). If you place the bottle at rest, the oil and water quickly separate again into distinct polar and nonpolar layers.

Hydrogen Bonds Also Involve Unequal Electron Sharing

When hydrogen atoms are made partially positive by sharing electrons unequally with oxygen, nitrogen, or sulfur, they may be attracted to nearby oxygen, nitrogen, or sulfur atoms made partially negative by unequal electron sharing in a different covalent bond. This attractive force, the **hydrogen bond,** is illustrated by a dotted line in structural diagrams of molecules **(Figure 2.10A)**. Hydrogen bonds may be intramolecular (between atoms in the same molecule) or intermolecular (between atoms in different molecules).

Individual hydrogen bonds are weak compared with ionic and covalent bonds. However, large biological molecules may offer many opportunities for hydrogen bonding, both within and between molecules. When numerous, hydrogen bonds are collectively strong and lend stability to the three-dimensional structure of molecules such as proteins **(Figure 2.10B).** Hydrogen bonds between water molecules are responsible for many of the properties that make water uniquely important to life (see Section 2.4 for a more detailed discussion).

The weak attractive force of hydrogen bonds makes them much easier to break than covalent and ionic bonds, particularly when elevated temperature increases the movements of molecules. Hydrogen bonds begin to break extensively as temperatures rise above 45°C and become practically nonexistent at 100°C. The disruption of hydrogen bonds by heat—for instance, the bonds in proteins—is one of the primary reasons most organisms cannot survive temperatures much greater than 45°C. Thermophilic (temperature-loving) organisms, which live at temperatures higher than 45°C, some at 120°C or more, have different molecules from those of organisms that live at lower temperatures. The proteins in thermophiles are stabilized at high temperatures by van der Waals forces and other noncovalent interactions.

A. Representation of hydrogen bond

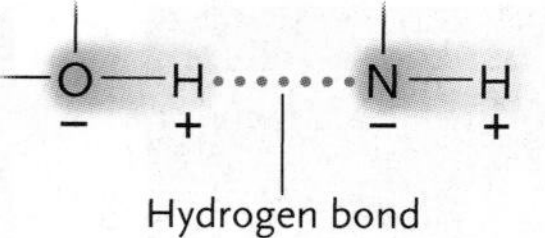
—O—H·······N—H
− + − +
Hydrogen bond

B. Stabilizing effect of hydrogen bonds

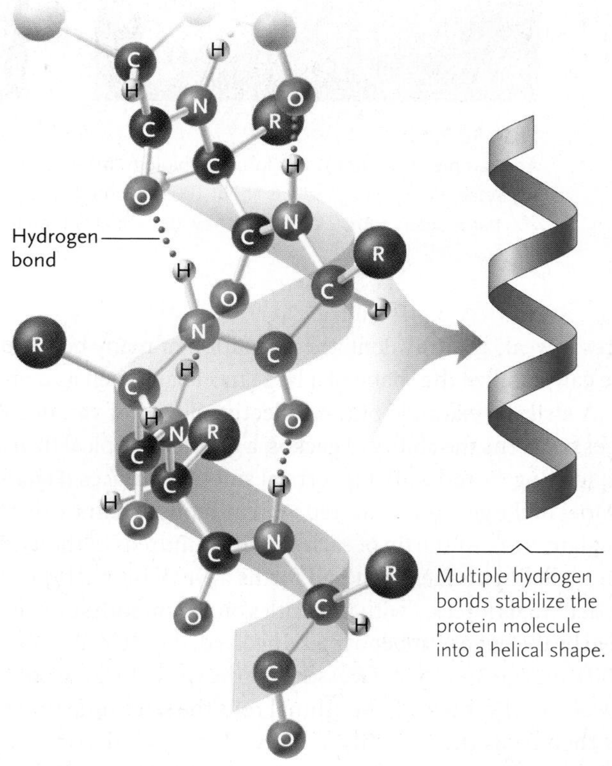
Hydrogen bond
Multiple hydrogen bonds stabilize the protein molecule into a helical shape.

FIGURE 2.10

Hydrogen bonds. **(A)** A hydrogen bond (dotted line) between the hydrogen of an —OH group and a nearby nitrogen atom, which also shares electrons unequally with another hydrogen. Regions of deepest blue indicate the most likely locations of electrons. **(B)** Multiple hydrogen bonds stabilize the backbone chain of a protein molecule into a spiral called the alpha helix. The spheres labeled *R* represent chemical groups of different kinds.

Van der Waals Forces Are Weak Attractions over Very Short Distances

Van der Waals forces are even weaker than hydrogen bonds. These forces develop over very short distances between nonpolar molecules or regions of molecules when, through their constant motion, electrons accumulate by chance in one part of a molecule or another. This process leads to zones of positive and negative charge, making the molecule polar. If they are oriented in the right way, the polar parts of the molecules are attracted electrically to one another and cause the molecules to stick together briefly. Although an individual bond formed with van der Waals

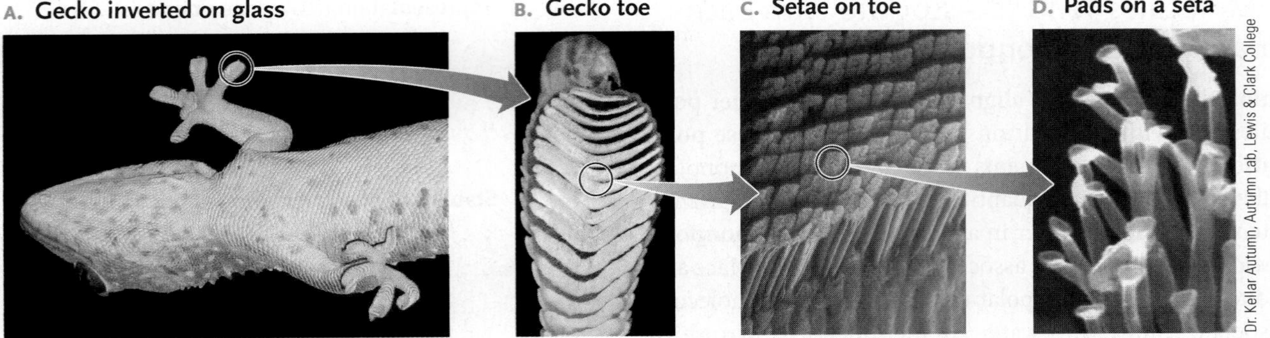
A. Gecko inverted on glass
B. Gecko toe
C. Setae on toe
D. Pads on a seta
Dr. Kellar Autumn, Autumn Lab, Lewis & Clark College

FIGURE 2.11
An example of van der Waals forces in biology. **(A)** A Tokay gecko *(Gekko gecko)* climbing while inverted on a glass plate. **(B)** Gecko toe. **(C)** Setae on a toe. Each seta is about 100 micrometers (μm; 0.004 inch) long. **(D)** Pads (spatulae) on an individual seta. Each pad is about 200 nanometers (nm; 0.000008 inch) wide—smaller than the wavelength of visible light.

forces is weak and transient, the formation of many bonds of this type can stabilize the shape of a large molecule, such as a protein.

A striking example of the collective power of van der Waals forces concerns the ability of geckos, a group of tropical lizard species, to cling to and walk up vertical smooth surfaces **(Figure 2.11).** The toes of the gecko are covered with millions of hairs, called *setae* (singular, *seta*). At the tip of each seta are hundreds of thousands of pads, called *spatulae*. Each pad forms a weak interaction—using van der Waals forces—with molecules on the smooth surface. Magnified by the huge number of pads involved, the attractive forces are 1,000 times greater than necessary for the gecko to hang on a vertical wall. To climb a wall, the animal rolls the setae onto the surface and then peels them off like a piece of tape. Understanding the gecko's remarkable ability to climb has led to the development of gecko tape, a super adhesive prototype tape capable of holding a 3-kg weight with a 1-cm^2 (centimeter squared) piece.

Bonds Form and Break in Chemical Reactions

In chemical reactions, atoms or molecules interact to form new chemical bonds or break old ones. As a result of bond formation or breakage, atoms are added to or removed from molecules, or the linkages of atoms in molecules are rearranged. When any of these alterations occur, molecules change from one type to another, usually with different chemical and physical properties. In biological systems, chemical reactions are accelerated by molecules called *enzymes* (which are discussed in more detail in Chapter 4).

The atoms or molecules entering a chemical reaction are called the **reactants,** and those leaving a reaction are the **products.** A chemical reaction is written with an arrow showing the direction of the reaction; reactants are placed to the left of the arrow, and products are placed to the right. Both reactants and products are usually written in chemical shorthand as formulas.

For example, the overall reaction of photosynthesis, in which carbon dioxide and water are combined to produce sugars and oxygen (see Chapter 9), is written as follows:

$$\underset{\text{carbon dioxide}}{6\,CO_2} + \underset{\text{water}}{6\,H_2O} \rightarrow \underset{\text{a sugar}}{C_6H_{12}O_6} + \underset{\text{molecular oxygen}}{6\,O_2}$$

The number in front of each formula indicates the number of molecules of that type among the reactants and products (the number 1 is not written). Notice that there are as many atoms of each element to the left of the arrow as there are to the right, even though the products are different from the reactants. This balance reflects the fact that in such reactions, atoms may be rearranged but not created or destroyed. Chemical reactions written in balanced form are known as **chemical equations.**

With the information about chemical bonds and reactions provided thus far, you are now ready to examine the effects of chemical structure and bonding, particularly hydrogen bonding, in the production of the unusual properties of water, the most important substance to life on Earth.

STUDY BREAK 2.3 <

1. **How does an ionic bond form?**
2. **How does a covalent bond form?**
3. **What is electronegativity, and how does it relate to nonpolar covalent bonds and polar covalent bonds?**
4. **What is a chemical reaction?**

2.4 Hydrogen Bonds and the Properties of Water

All living organisms contain water, and many kinds of organisms live directly in water. Even those that live in dry environments contain water in all their structures—different organisms range from 50% to more than 95% water by weight. The water inside organisms is crucial for life: it is required for many important biochemical reactions and plays major roles in maintaining the shape and organization of cells and tissues. Key properties of water molecules that make them so important to life include:

- Hydrogen bonds between water molecules produce a *water lattice.* (A lattice is a cross-linked structure.) The lattice arrangement makes water denser than ice and gives water several other properties that make it a highly suitable medium for the molecules and reactions of life. For example, *water absorbs or releases relatively large amounts of energy as heat without undergoing extreme changes in temperature.* This property stabilizes both living organisms and their environments. Also, the water lattice has an unusually high internal *cohesion* (resistance of water molecules to separate). This property plays an important role, for example, in water transport from the roots to the leaves of plants. Further, water molecules at surfaces facing air are even more resistant to separation, producing the force called *surface tension* which, for example, allows droplets of water to form and small insects to walk on water.

- The polarity of water molecules in the hydrogen-bond lattice contributes to the formation of distinct polar and nonpolar environments that are critical to the organization of cells.
- Water is a *solvent* for charged or polar molecules, meaning that it is a solution in which such molecules can dissolve. Water's solvent properties made life possible because it enabled the chemical reactions needed for life to evolve. The chemical reactions in all organisms take place in aqueous solutions.
- Water molecules separate into ions. Those ions are important for maintaining an environment within cells that is optimal for the chemical reactions that occur there.

A Lattice of Hydrogen Bonds Gives Water Several Unusual, Life-Sustaining Properties

Hydrogen bonds form readily between water molecules in both liquid water and ice. In liquid water, each water molecule establishes an average of 3.4 hydrogen bonds with its neighbors, forming an arrangement known as the **water lattice (Figure 2.12A).** In liquid water, the hydrogen bonds that hold the lattice together constantly break and reform, allowing the water molecules to break loose from the lattice, slip past one another, and reform the lattice in new positions.

THE DIFFERING DENSITIES OF WATER AND ICE In ice, the water lattice is a rigid, crystalline structure in which each water molecule forms four hydrogen bonds with neighboring molecules **(Figure 2.12B).** The rigid **ice lattice** spaces the water molecules farther apart than the water lattice. Because of this greater spacing, water has the unusual property of being about 10% less dense when solid than when liquid. (Almost all other substances are denser in solid form than in liquid form.) Hence, ice cubes are a little larger than the water volume poured into the ice tray, and water filling a closed glass vessel will expand, breaking the vessel when the water freezes. At atmospheric pressure, water reaches its greatest density at a temperature of 4°C, while it is still a liquid.

Because it is less dense than liquid water, ice forms at the surface of a body of water and remains floating at the surface. The ice creates an insulating layer that helps keep the water below from freezing. If ice were denser than liquid water, it would sink to the bottom as it freezes, continually exposing liquid water at the surface to freezing temperatures. Under those conditions, most bodies of water would freeze entirely solid, making life difficult or impossible for aquatic plants and animals.

THE BOILING POINT AND TEMPERATURE-STABILIZING EFFECTS OF WATER The hydrogen-bond lattice of liquid water retards the escape of individual water molecules as the water is heated. As a result, relatively high temperatures and the addition of considerable heat are required to break enough hydrogen bonds

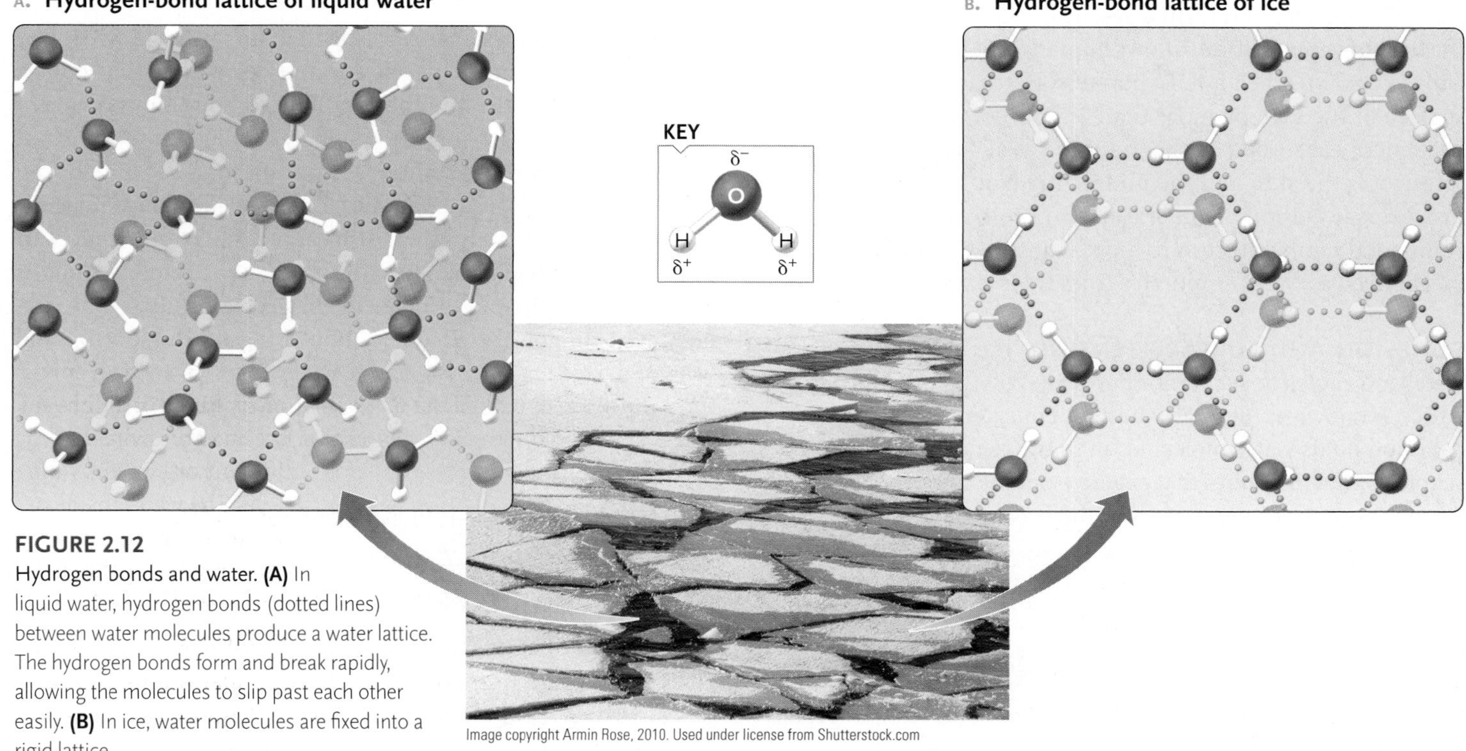

A. Hydrogen-bond lattice of liquid water
B. Hydrogen-bond lattice of ice
KEY
δ⁻
O
H
H
δ⁺
δ⁺

FIGURE 2.12
Hydrogen bonds and water. **(A)** In liquid water, hydrogen bonds (dotted lines) between water molecules produce a water lattice. The hydrogen bonds form and break rapidly, allowing the molecules to slip past each other easily. **(B)** In ice, water molecules are fixed into a rigid lattice.

to make water boil. Its high boiling point maintains water as a liquid over the wide temperature range of 0°C to 100°C. Similar molecules that do not form an extended hydrogen-bond lattice, such as H_2S (hydrogen sulfide), have much lower boiling points and are gases rather than liquids at room temperature. The properties of these related substances indicate that, without its hydrogen-bond lattice, water would boil at −81°C. If this were the case, most of the water on Earth would be in gaseous form and life as described in this book could not exist.

As a result of water's stabilizing hydrogen-bond lattice, it also has a relatively high **specific heat**—that is, the amount of energy as heat required to increase the temperature of a given quantity of water. As heat energy flows into water, much of it is absorbed in the breakage of hydrogen bonds. As a result, the temperature of water, reflected in the average motion of its molecules, increases relatively slowly as heat energy is added. For example, a given amount of heat energy increases the temperature of water by only half as much as that of an equal quantity of ethyl alcohol. High specific heat allows water to absorb or release relatively large quantities of heat energy without undergoing extreme changes in temperature; this gives it a moderating and stabilizing effect on both living organisms and their environments.

The specific heat of water is measured in **calories.** This unit, used both in the sciences and in dieting, is the amount of heat energy required to raise 1 g of water by 1°C (technically, from 14.5°C to 15.5°C at one atmosphere of pressure). This amount of heat is known as a "small" calorie and is written with a small *c*. The unit most familiar to dieters, equal to 1,000 small calories, is written with a capital *C* as a **Calorie;** the same 1,000-calorie unit is known scientifically as a **kilocalorie (kcal).** A 300-Calorie candy bar therefore really contains 300,000 calories.

A large amount of heat, 586 calories per gram, must be added to give water molecules enough energy of motion to break loose from liquid water and form a gas. This required heat, known as the **heat of vaporization,** allows humans and many other organisms to cool off when hot. In humans, water is released onto the surface of the skin by more than 2.5 million sweat glands; the heat energy absorbed by the water in sweat as the sweat evaporates cools the skin and the underlying blood vessels. The heat loss helps keep body temperature from increasing when environmental temperatures are high. Plants use a similar cooling mechanism as water evaporates from their leaves.

COHESION AND SURFACE TENSION The high resistance of water molecules to separation, provided by the hydrogen-bond lattice, is known as internal **cohesion.** For example, in land plants, cohesion holds water molecules in unbroken columns in microscopic conducting tubes that extend from the roots to the highest leaves. As water evaporates from the leaves, water molecules in the columns, held together by cohesion, move upward through the tubes to replace the lost water. This movement raises water from roots to the tops of the tallest trees (see discussion in Chapter 32). Maintenance of the long columns of water in the tubes is aided by **adhesion,** in which molecules "stick" to the walls of the tubes by forming hydrogen bonds with charged and polar groups in molecules that form the walls of the tubes.

A. **Creation of surface tension by unbalanced hydrogen bonding**

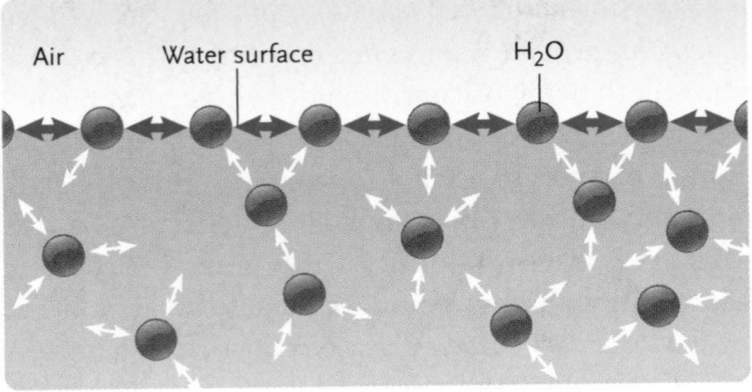
Air
Water surface
H_2O

B. **Spider supported by water's surface tension**

iStockphoto.com/Alasdair Thomson

FIGURE 2.13
Surface tension in water. **(A)** Unbalanced hydrogen bonding places water molecules under lateral tension where a water surface faces the air. **(B)** A raft spider *(Dolomedes fimbriatus)* is supported by the surface tension of water.

Water molecules at surfaces facing air can form hydrogen bonds with water molecules beside and below them but not on the sides that face the air. This unbalanced bonding produces a force that places the surface water molecules under tension, making them more resistant to separation than the underlying water molecules **(Figure 2.13A).** The force, called **surface tension,** is strong enough to allow small insects such as raft spiders to walk on water **(Figure 2.13B).** Similarly, the surface tension of water will support a sewing needle placed carefully on the surface, even though the needle is about 10 times denser than the water. Surface tension also causes water to form water droplets; the surface tension pulls the water in around itself to produce the smallest possible area, which is a spherical bead or droplet.

The Polarity of Water Molecules in the Hydrogen-Bond Lattice Contributes to Polar and Nonpolar Environments in and around Cells

The polarity of water molecules in the hydrogen-bond lattice gives water other properties that make it unique and ideal as a life-sustaining medium. In liquid water, the lattice resists invasion by other molecules unless the invading molecule also contains polar or charged regions that can form competing attractions with water molecules. If present, the competing attractions open the water lattice, creating a cavity into which the polar or charged molecule can move. By contrast, nonpolar molecules are unable to disturb the water lattice. The lattice thus excludes nonpolar substances, forcing them to form the nonpolar associations that expose the least surface area to the surrounding water—such as the spherical droplets of oil that form when oil and water are mixed together and shaken.

The distinct polar and nonpolar environments created by water are critical to the organization of cells. For example, biological membranes, which form boundaries around and inside cells, consist of lipid molecules with dual polarity: one end of each molecule is polar, and the other end is nonpolar. (Lipids are described in more detail in Chapter 3.) The membranes are surrounded on both sides by strongly polar water molecules. Exclusion by the water molecules forces the lipid molecules to associate into a double layer, a **bilayer,** in which only the polar ends of the surface molecules are exposed to the water **(Figure 2.14).** The nonpolar ends of the molecules associate in the interior of the bilayer, where they are not exposed to the water. Exclusion of their nonpolar regions by water is all that holds membranes together.

The membrane at the surface of cells prevents the watery solution inside the cell from mixing directly with the watery solution outside the cell. By doing so, the surface membrane, kept intact by nonpolar exclusion by water, maintains the internal environment and organization necessary for cellular life.

The Small Size and Polarity of Its Molecules Makes Water a Good Solvent

Because water molecules are small and strongly polar, they readily penetrate or coat the surfaces of other polar and charged molecules and ions. The surface coat, called a **hydration layer,** reduces the attraction between the molecules or ions and promotes their separation and entry into a **solution,** where they are suspended individually, surrounded by water molecules. Once in solution, the hydration layer prevents the polar molecules or ions from reassociating. In such a solution, water is called the **solvent,** and the molecules of a substance dissolved in water are called the **solute.**

For example, when a teaspoon of table salt is added to water, water molecules quickly form hydration layers around the Na^+ and Cl^- ions in the salt crystals, reducing the attraction between the ions so much that they separate from the crystal and enter the surrounding water lattice as individual ions **(Figure 2.15).** If the water evaporates, the hydration layer is eliminated, exposing the strong positive and negative charges of the ions. The opposite charges attract and reestablish the ionic bonds that hold the ions in salt crystals. As the last of the water evaporates, all of the Na^+ and Cl^- ions reassociate, reestablishing the solid, crystalline form.

In the cell, chemical reactions depend on solutes dissolved in aqueous solutions. To understand these reactions, you need to know

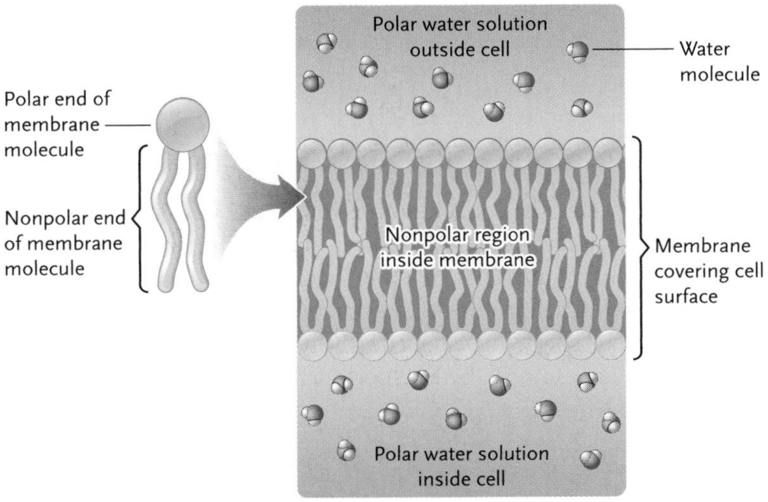

Polar water solution outside cell
Water molecule
Polar end of membrane molecule
Nonpolar end of membrane molecule
Nonpolar region inside membrane
Membrane covering cell surface
Polar water solution inside cell

FIGURE 2.14
Formation of the membrane covering the cell surface by lipid molecules. Exclusion by polar water molecules forces the nonpolar ends of lipid molecules to associate into the bilayer that forms the membrane.

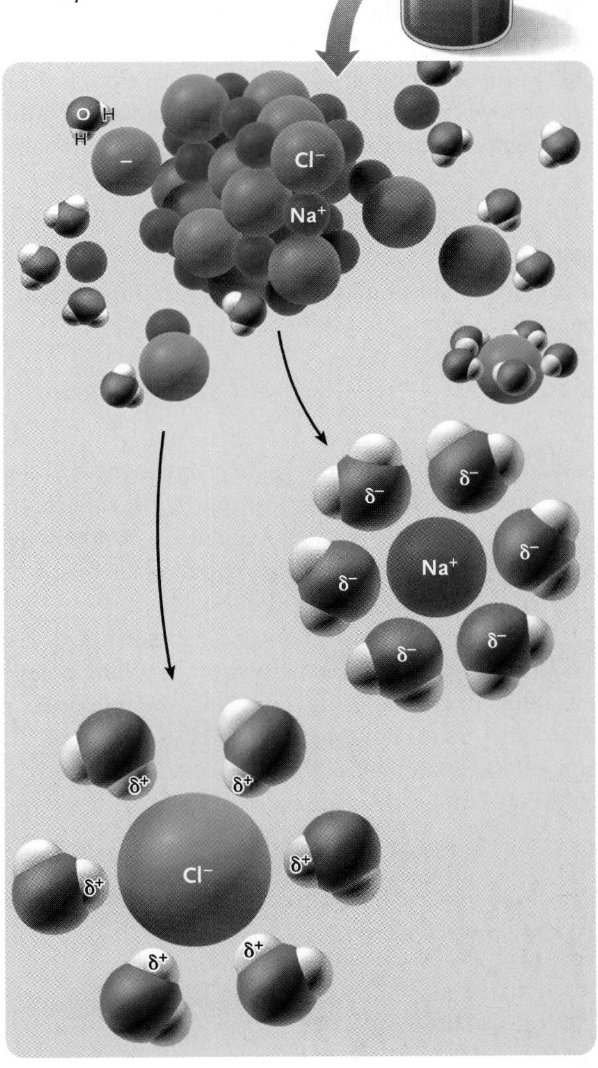
Salt
O
H
H
Cl⁻
Na⁺
δ⁻
Na⁺
δ⁺
Cl⁻

FIGURE 2.15
Water molecules forming a hydration layer around Na^+ and Cl^- ions, which promotes their separation and entry into solution.

the number of atoms and molecules involved. **Concentration** is the number of molecules or ions of a substance in a unit volume of space, such as a milliliter (mL) or liter (L). The number of molecules or ions in a unit volume cannot be counted directly but can be calculated indirectly by using the mass number of atoms as the starting point. The same method is used to prepare a solution with a known number of molecules per unit volume.

The mass number of an atom is equivalent to the number of protons and neutrons in its nucleus. From the mass number, and the fact that neutrons and protons are approximately the same weight (that is, 1.66×10^{-24} g), you can calculate the weight of an atom of any substance. For an atom of the most common form of carbon, with 6 protons and 6 neutrons in its nucleus, the total weight is calculated as follows:

$$12 \times (1.66 \times 10^{-24}\text{ g}) = 1.992 \times 10^{-23}\text{ g}$$

For an oxygen atom, with 8 protons and 8 neutrons in its nucleus, the total weight is calculated as follows:

$$16 \times (1.66 \times 10^{-24}\text{ g}) = 2.656 \times 10^{-23}\text{ g}$$

Dividing the total weight of a sample of an element by the weight of a single atom gives the number of atoms in the sample. Suppose you have a carbon sample that weighs 12 g—a weight in grams equal to the atom's mass number. (A weight in grams equal to the mass number is known as an **atomic weight** of an element.) Dividing 12 g by the weight of one carbon atom gives the following result:

$$\frac{12}{(1.992 \times 10^{-23}\text{ g})} = 6.022 \times 10^{23}\text{ atoms}$$

If you divide the atomic weight of oxygen (16 g) by the weight of one oxygen atom, you get the same result:

$$\frac{16}{(2.656 \times 10^{-23}\text{ g})} = 6.022 \times 10^{23}\text{ atoms}$$

In fact, dividing the atomic weight of any element by the weight of an atom of that element always produces the same number: 6.022×10^{23}. This number is called **Avogadro's number** after Amedeo Avogadro, the nineteenth-century Italian chemist who first discovered the relationship.

The same relationship holds for molecules. The **molecular weight** of any molecule is the sum of the atomic weights of all of the atoms in the molecule. For NaCl, the total mass number is $23 + 35 = 58$ (a sodium atom has 11 protons and 12 neutrons, and a chlorine atom has 17 protons and 18 neutrons). The weight of an NaCl molecule is therefore:

$$58 \times (1.66 \times 10^{-24}\text{ g}) = 9.628 \times 10^{-23}\text{ g}$$

Dividing a molecular weight of NaCl (58 g) by the weight of a single NaCl molecule gives:

$$\frac{58}{(9.628 \times 10^{-23}\text{ g})} = 6.022 \times 10^{23}\text{ molecules}$$

When concentrations are described, the atomic weight of an element or the molecular weight of a compound—the amount that contains 6.022×10^{23} atoms or molecules—is known as a **mole** (abbreviated **mol**). More strictly, chemists define the mole as the amount of a substance that contains as many atoms or molecules as there are atoms in exactly 12 g of carbon-12. As we saw above, the number of atoms in 12 g of carbon-12 is 6.022×10^{23}. The number of moles of a substance dissolved in 1 L of solution is known as the **molarity** (abbreviated ***M***) of the solution. This relationship is highly useful in chemistry and biology because we know that two solutions having the same volume and molarity but composed of different substances will contain the same number of molecules of the substances.

Water has still other properties that contribute to its ability to sustain life, the most important being that its molecules separate into ions. These ions help maintain an environment inside living organisms that promotes the chemical reactions of life.

STUDY BREAK 2.4

1. **How do hydrogen bonds between water molecules contribute to the properties of water?**
2. **How do a solute, a solvent, and a solution differ?**

2.5 Water Ionization and Acids, Bases, and Buffers

The most critical property of water that is unrelated to its hydrogen-bond lattice is its ability to separate, or **dissociate,** to produce positively charged *hydrogen ions* (H^+, or protons) and *hydroxide ions* (OH^-):

$$H_2O \rightleftharpoons H^+ + OH^-$$

(The double arrow means that the reaction is **reversible**—that is, depending on conditions, it may go from left to right or from right to left.) The proportion of water molecules that dissociates to release protons and hydroxide ions is small. However, because of the dissociation, water always contains some H^+ and OH^- ions.

Substances Act as Acids or Bases by Altering the Concentrations of H^+ and OH^- Ions in Water

In pure water, the concentrations of H^+ and OH^- ions are equal. However, adding other substances may alter the relative concentrations of H^+ and OH^-, making them unequal. Some substances, called **acids,** are proton donors that release H^+ (and anions) when they are dissolved in water, effectively increasing the H^+ concentration. For example, hydrochloric acid (HCl) dissociates into H^+ and Cl^- when dissolved in water:

$$HCl \rightleftharpoons H^+ + Cl^-$$

Other substances, called **bases,** are proton acceptors that reduce the H^+ concentration of a solution. Most bases dissociate in water into a hydroxide ion (OH^-) and a cation. The hydroxide ion can act as a base by accepting a proton (H^+) to produce water. For example, sodium hydroxide (NaOH) separates into Na^+ and OH^- ions when dissolved in water:

$$NaOH \rightarrow Na^+ + OH^-$$

The excess OH^- combines with H^+ to produce water:

$$OH^- + H^+ \rightarrow H_2O$$

thereby reducing the H^+ concentration.

Other bases do not dissociate to produce hydroxide ions directly. For example, ammonia (NH_3), a poisonous gas, acts as a base when dissolved in water by directly accepting a proton from water to produce an ammonium ion and releasing a hydroxide ion:

$$NH_3 + H_2O \rightarrow NH_4 + OH^-$$

The concentration of H^+ ions in a water solution, as compared with the concentration of OH^- ions, determines the **acidity** of the solution. Scientists measure acidity using a numerical scale from 0 to 14, called the **pH scale.** Because the number of H^+ ions in solution increases exponentially as the acidity increases, the scale is based on logarithms of this number to make the values manageable:

$$pH = -\log_{10}[H^+]$$

In this formula, the brackets indicate concentration in moles per liter of the substance within them. The negative of the logarithm is used to give a positive number for the pH value. For example, in a water solution that is *neutral*—neither acidic nor basic—the concentration of *both* H^+ and OH^- ions is 1×10^{-7} *M* (0.0000001 *M*), with the product of the two ion concentrations being constant in an aqueous solution at 25°C and given by $[H^+][OH^-] = (1 \times 10^{-7}) \times (1 \times 10^{-7}) = 1 \times 10^{-14}$. For the water solution, the $\log_{10}$ of 1×10^{-7} is -7. The negative of the logarithm -7 is 7. Thus, a neutral water solution with an H^+ concentration of 1×10^{-7} *M* has a pH of 7. *Acidic* solutions have pH values less than 7, with pH 0 being the value for the highly acidic 1 *M* hydrochloric acid (HCl); *basic* solutions have pH values greater than 7, with pH 14 being the value for the highly basic 1 *M* sodium hydroxide (NaOH) (basic solutions are also called *alkaline* solutions). Each whole number on the pH scale represents a value 10 times greater or less than the next number. Thus, a solution with a pH of 4 is 10 times more acidic than one with a pH of 5, and a solution with a pH of 6 is 100 times more acidic than a solution with a pH of 8. (The pH of many familiar solutions is shown in **Figure 2.16.**)

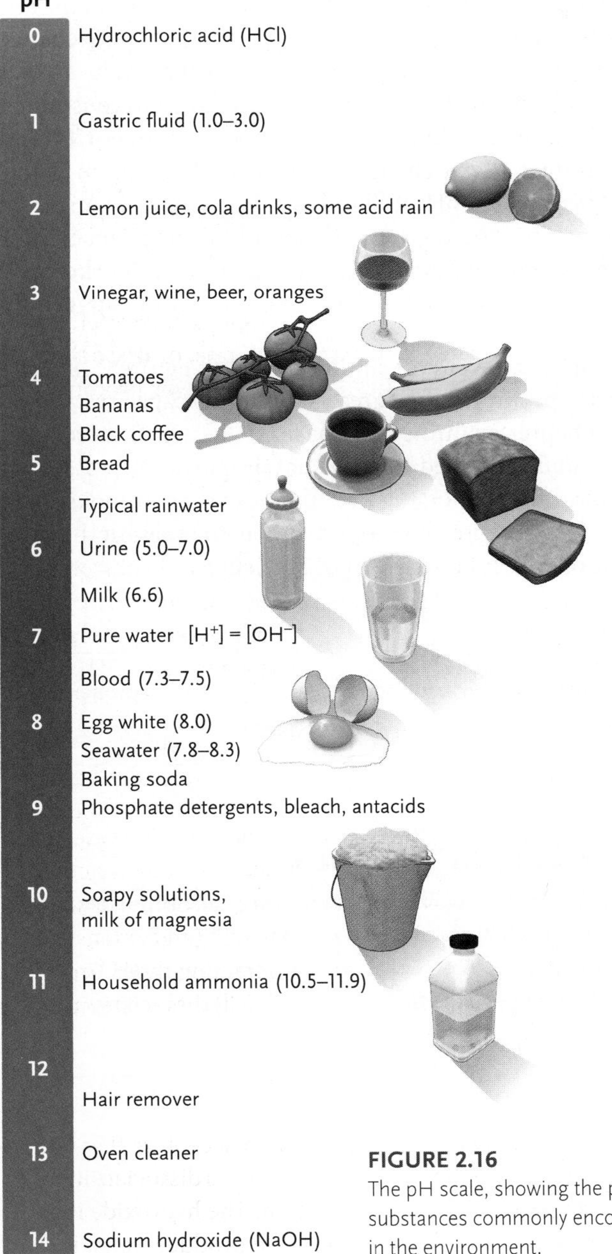
pH
0 Hydrochloric acid (HCl)
1 Gastric fluid (1.0–3.0)
2 Lemon juice, cola drinks, some acid rain
3 Vinegar, wine, beer, oranges
4 Tomatoes
Bananas
Black coffee
5 Bread
Typical rainwater
6 Urine (5.0–7.0)
Milk (6.6)
7 Pure water [H+] = [OH−]
Blood (7.3–7.5)
8 Egg white (8.0)
Seawater (7.8–8.3)
Baking soda
9 Phosphate detergents, bleach, antacids
10 Soapy solutions, milk of magnesia
11 Household ammonia (10.5–11.9)
12
Hair remover
13 Oven cleaner
14 Sodium hydroxide (NaOH)

FIGURE 2.16
The pH scale, showing the pH of substances commonly encountered in the environment.

Acidity is important to cells because even small changes, on the order of 0.1 or even 0.01 pH unit, can drastically affect biological reactions. In large part, this effect reflects changes in the structure of proteins that occur when the water solution surrounding the proteins has too few or too many hydrogen ions. Consequently, all living organisms have elaborate systems that control their internal acidity by regulating H^+ concentration near the neutral value of pH 7.

Acidity is also important to the environment in which we live. Where the air is unpolluted, rainwater is only slightly acidic. However, in regions where certain pollutants are released into the air in large quantities by industry and automobile exhaust, the polluting chemicals combine with atmospheric water to produce "acid rain" with a pH as low as 3, about the same pH as that of vinegar. Acid rain can sicken and kill wildlife such as fishes and birds, as well as plants and trees (**Figure 2.17;** see also discussion in Chapter 53). Humans are also affected; acid rain and acidified water vapor in the air can contribute to human respiratory diseases such as bronchitis and asthma.

Buffers Help Keep pH under Control

Living organisms control the internal pH of their cells with **buffers,** substances that compensate for pH changes by absorbing or releasing H^+. When H^+ ions are released in excess by biological reactions, buffers combine with them and remove them from the solution; if the concentration of H^+ decreases greatly, buffers re-

FIGURE 2.17
Forest affected by acid rain and other forms of air pollution in the Great Smoky Mountains National Park. The trees are susceptible to drought, disease, and insect pests.

lease additional H^+ to restore the balance. Most buffers are weak acids or bases, or combinations of these substances that dissociate reversibly in water solutions to release or absorb H^+ or OH^-. (Weak acids, such as acetic acid, or weak bases, such as ammonia, are substances that release relatively few H^+ or OH^- ions in a water solution. Strong acids or bases are substances that dissociate extensively in a water solution. HCl is a strong acid; NaOH is a strong base.)

The buffering mechanism that maintains blood pH near neutral values is a primary example. In humans and many other animals, blood pH is buffered by a *carbonic acid–bicarbonate buffer system*. In water solutions, carbonic acid (H_2CO_3), which is a weak acid, dissociates readily into bicarbonate ions (HCO_3^-) and H^+:

$$H_2CO_3 \rightleftharpoons HCO_3^- + H^+$$

The reaction is reversible. If H^+ is present in excess, the reaction is pushed to the left—the excess H^+ ions combine with bicarbonate ions to form H_2CO_3. If the H^+ concentration declines below normal levels, the reaction is pushed to the right—H_2CO_3 dissociates into HCO_3^- and H^+, restoring the H^+ concentration. **Figure 2.18** graphs the pH of a solution (such as blood) as its relative proportions of H_2CO_3 and HCO_3^- change. The colored zone of the graph indicates the range of greatest *buffering capacity,* that is, the range in which changes in the relative proportions of H_2CO_3 and HCO_3^- produce *little* change in the pH of the solution. Note, however, that at pH values lower than 5.1 and higher than 7.1, the slope of the curve is much greater. Here, changes in the relative proportions of H_2CO_3 and HCO_3^- produce a *large* change in the pH of the solution. Interestingly, the normal pH of blood is 7.4 which, as Figure 2.18 shows, is outside the region of greatest buffering capacity for this buffer system. In the body, other mechanisms help keep the blood pH relatively constant. (More on blood pH appears in Section 44.4.)

All buffers have curves similar to that of Figure 2.18. Each buffer has a specific range of greatest buffering capacity.

This chapter examined the basic structure of atoms and molecules and discussed the unusual properties of water that make it ideal for supporting life. The next chapter looks more closely at the structure and properties of carbon and at the great multitude of molecules based on this element.

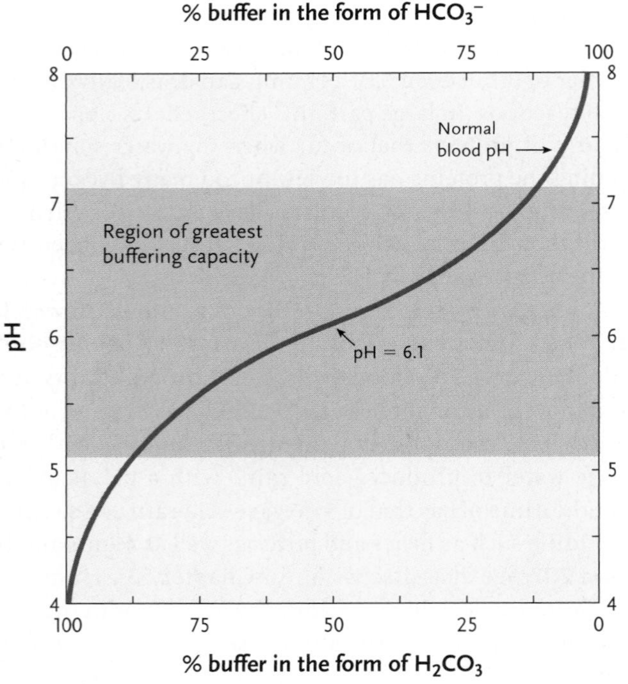
% buffer in the form of HCO_3^-
0
25
50
75
100
8
7
6
5
4
Normal blood pH
Region of greatest buffering capacity
pH = 6.1
pH
100
75
50
25
0
% buffer in the form of H_2CO_3

FIGURE 2.18
Properties of the carbonic acid–bicarbonate buffer system. The colored zone is the pH range of greatest buffering capacity for this buffer system.

STUDY BREAK 2.5

1. **Distinguish between acids and bases. What are their properties?**
2. **Why are buffers important for living organisms?**

THINK OUTSIDE THE BOOK

The H_2CO_3–HCO_3^- buffering system is only one of the mechanisms by which the human body maintains blood pH at a relatively constant level. Collaboratively, or on your own, research what happens to the body when blood pH becomes abnormal.

UNANSWERED QUESTIONS

Can arsenic be removed from water by bioremediation?

In the Why It Matters section, we learned that bioremediation of selenium in wastewater is possible using plants. Research is showing that bioremediation can also be used to remove other toxic chemicals in the environment, including perchlorate and arsenic. For example, arsenic-contaminated soils and sediments are the major sources of arsenic contamination in surface water and groundwater, which leads to contamination of foods. In some parts of the world, the drinking water is contaminated. Arsenic poses serious health risks to humans and other animals; for example, some cancers have been correlated with high levels of arsenic. Arsenic contamination is a worldwide concern, with arsenic levels in the environment in some parts of the world being tens of thousands of times higher than the maximum contaminant level set in the United States.

One research group at University College London, led by Joanne Santini, is exploring whether bacteria can be used for arsenic bioremediation in contaminated wastewater on mining sites and groundwater in Bangladesh and West Bengal, India. Their approach has been to study 13 rare bacteria isolated from gold mines, which are typical places to find arsenic. Arsenic is present in water in two toxic forms; one of these is easy and safe to get rid of, but the other is not. Santini's group has identified a bacterium that can "eat" the difficult-to-get-rid-of form of arsenic and convert it to the easy-to-get-rid-of form. Potentially, this bacterium could be used in bioremediation of arsenic in contaminated locations.

Think Critically

Referring to what you have learned about characteristics of living systems in Chapter 1, list some questions you would like to ask a biologist about this bacterium that "eats" arsenic, a semi-metal.

Peter J. Russell

Peter J. Russell

REVIEW KEY CONCEPTS

Go to **CENGAGENOW** at www.cengage.com/login to access quizzing, animations, exercises, articles, and personalized homework help.

2.1 The Organization of Matter: Elements and Atoms

- Matter is anything that occupies space and has mass. Matter is composed of elements, each consisting of atoms of the same kind.
- Atoms combine chemically in fixed numbers and ratios to form the molecules of living and nonliving matter. Compounds are molecules in which the component atoms are different.

2.2 Atomic Structure

- Atoms consist of an atomic nucleus that contains protons and neutrons surrounded by one or more electrons traveling in orbitals. Each orbital can hold a maximum of two electrons (Figure 2.3).
- All atoms of an element have the same number of protons, but the number of neutrons is variable. The number of protons in an atom is designated by its atomic number; the number of protons plus neutrons is designated by the mass number (Figure 2.4 and Table 2.1).
- Isotopes are atoms of an element with differing numbers of neutrons. The isotopes of an atom differ in physical but not chemical properties (Figure 2.4).
- Electrons surround an atomic nucleus in orbitals occupying energy levels that increase in discrete steps (Figures 2.5 and 2.6).
- The chemical activities of atoms are determined largely by the number of electrons in the outermost energy level. Atoms that have the outermost level filled with electrons are nonreactive, whereas atoms in which that level is not completely filled with electrons are reactive. Atoms tend to lose, gain, or share electrons to fill the outermost energy level.

Animation: Electron arrangements in atoms

Animation: Isotopes of hydrogen

Animation: The shell model of electron distribution

Practice: Predicting the number of bonds of elements

2.3 Chemical Bonds and Chemical Reactions

- An ionic bond forms between atoms that gain or lose electrons in the outermost energy level completely, that is, between a positively charged cation and a negatively charged anion (Figure 2.7).
- A covalent bond is established by a pair of electrons shared between two atoms. If the electrons are shared equally, the covalent bond is nonpolar (Figure 2.8).
- If electrons are shared unequally in a covalent bond, the atoms carry partial positive and negative charges and the bond is polar (Figure 2.9).
- Polar molecules tend to associate with other polar molecules and to exclude nonpolar molecules. Polar molecules that associate readily with water are hydrophilic; nonpolar molecules excluded by water are hydrophobic.
- A hydrogen bond is a weak attraction between a hydrogen atom made partially positive by unequal electron sharing and another atom—usually oxygen, nitrogen, or sulfur—made partially negative by unequal electron sharing (Figure 2.10).
- Van der Waals forces, bonds even weaker than hydrogen bonds, can form when natural changes in the electron density of molecules produce regions of positive and negative charge, which cause the molecules to stick together briefly.
- Chemical reactions occur when molecules form or break chemical bonds. The atoms or molecules entering into a chemical reaction are the reactants, and those leaving a reaction are the products.

Animation: How atoms bond

2.4 Hydrogen Bonds and the Properties of Water

- The hydrogen-bond lattice gives water unusual properties that are vital to living organisms, including high specific heat, boiling point, cohesion, and surface tension (Figures 2.12 and 2.13).
- The polarity of the water molecules in the hydrogen-bond lattice makes it difficult for nonpolar substances to penetrate the lattice. The distinct polar and nonpolar environments created by water are critical to the organization of cells (Figure 2.14).
- The polar properties of water allow it to form a hydration layer over the surfaces of polar and charged biological molecules, particularly proteins. Many chemical reactions depend on the special molecular conditions created by the hydration layer (Figure 2.15).
- The polarity of water allows ions and polar molecules to dissolve readily in water, making it a good solvent.

Animation: Structure of water

Animation: Spheres of hydration

2.5 Water Ionization and Acids, Bases, and Buffers

- Acids are substances that increase the H^+ concentration by releasing additional H^+ as they dissolve in water; bases are substances that decrease the H^+ concentration by gathering H^+ or releasing OH^- as they dissolve.
- The relative concentrations of H^+ and OH^- in a water solution determine the acidity of the solution, which is expressed quantitatively as pH on a number scale ranging from 0 to 14. Neutral solutions, in which the concentrations of H^+ and OH^- are equal, have a pH of 7. Solutions with pH less than 7 have H^+ in excess and are acidic; solutions with pH greater than 7 have OH^- in excess and are basic or alkaline (Figure 2.16).
- The pH of living cells is regulated by buffers, which absorb or release H^+ to compensate for changes in H^+ concentration (Figure 2.18).

Animation: The pH scale

UNDERSTAND AND APPLY

Test Your Knowledge

1. Which of the following statements about the mass number of an atom is *incorrect*?
   a. It has a unit defined as a dalton.
   b. On Earth, it equals the atomic weight.
   c. Unlike the atomic weight of an atom, it does not change when gravitational forces change.
   d. It equals the number of electrons in an atom.
   e. It is the sum of the protons and neutrons in the atomic nucleus.
2. Oxygen (O) is a(n) ______, while the oxygen we breathe (O_2) is a(n) _____, and the carbon dioxide we exhale is a(n) _______.
   a. compound; molecule; element
   b. atom; compound; element
   c. element; atom; molecule
   d. atom; element; molecule
   e. element; molecule; compound
3. The chemical activity of an atom:
   a. depends on the electrons in the outermost energy level.
   b. is increased when the outermost energy level is filled with electrons.
   c. depends on its 1*s* but not its 2*s* or 2*p* orbitals.
   d. is increased when valence electrons completely fill the outer orbitals.
   e. of oxygen prevents it from sharing its electrons with other atoms.
4. When electrons are shared equally between atoms, they form:
   a. a polar covalent bond.
   b. a nonpolar covalent bond.
   c. an ionic bond.
   d. a hydrogen bond.
   e. a van der Waals force.
5. Which of the following is *not* a property of water?
   a. It has a low boiling point compared with other molecules.
   b. It has a high heat of vaporization.
   c. Its molecules resist separation, a property called cohesion.
   d. It has the property of adhesion, the ability to stick to charged and polar groups in molecules.
   e. It can form hydrogen bonds to molecules below but not above its surface.
6. Which of the following is *not* a hydrophilic body fluid?
   a. blood
   b. sweat
   c. tears
   d. oil
   e. saliva
7. The water lattice:
   a. is formed from hydrophobic bonds.
   b. causes ice to be denser than water.
   c. reduces water's ability as a solvent.
   d. excludes polar substances.
   e. contributes to polar and nonpolar spaces around cells.
8. A hydrogen bond is:
   a. a strong attraction between hydrogen and another atom.
   b. a bond between a hydrogen atom already covalently bound to one atom and made partially negative by unequal electron sharing with another atom.
   c. a bond between a hydrogen atom already covalently bound to one atom and made partially positive by unequal electron sharing with another atom.
   d. weaker than van der Waals forces.
   e. exemplified by the two hydrogens covalently bound to oxygen in the water molecule.
9. If the water in a pond has a pH of 5, the hydroxide concentration would be
   a. $10^{-5}\ M$.
   b. $10^{-10}\ M$.
   c. $10^{5}\ M$.
   d. $10^{9}\ M$.
   e. $10^{-9}\ M$.
10. Because of a sudden hormonal imbalance, a patient's blood was tested and shown to have a pH of 7.5. What does this pH value mean?
    a. This is more acidic than normal blood.
    b. It represents a weak alkaline fluid.
    c. This is caused by a release of large amounts of hydrogen ions into the system.
    d. The reaction $H_2CO_3 \rightarrow HCO_3^- + H^+$ is pushed to the left.
    e. This is probably caused by excess CO_2 in the blood.

Discuss the Concepts

1. Detergents allow particles of oil to mix with water. From the information presented in this chapter, how do you think detergents work?
2. What would living conditions be like on Earth if ice were denser than liquid water?
3. You place a metal pan full of water on the stove and turn on the heat. After a few minutes, the handle is too hot to touch but the water is only warm. How do you explain this observation?
4. You are studying a chemical reaction accelerated by an enzyme. H^+ forms during the reaction, but the enzyme's activity is lost at low pH. What could you include in the reaction mix to keep the enzyme's activity at high levels? Explain how your suggestion might solve the problem.

Design an Experiment

You know that adding NaOH to HCl results in the formation of common table salt, NaCl. You have a 0.5 *M* HCl solution. What weight of NaOH would you need to add to convert all of the HCl to NaCl? (**Note:** Chemical reactions have the potential to be dangerous. Please do not attempt to perform this reaction.)

Interpret the Data

The pH of human stomach acid ranges from 1.0–3.0 whereas a healthy esophagus has a pH of approximately 7.0. In gastroesophageal reflux disorder (GERD), often called acid reflux, stomach acid flows backward from the stomach into the esophagus. Repeated episodes in which esophageal pH goes below 4.0, considered clinical acid reflux, can result in bleeding ulcers and damage to the esophageal lining. The data in the figure, from a patient with GERD, show esophageal pH during a sleeping reflux event.

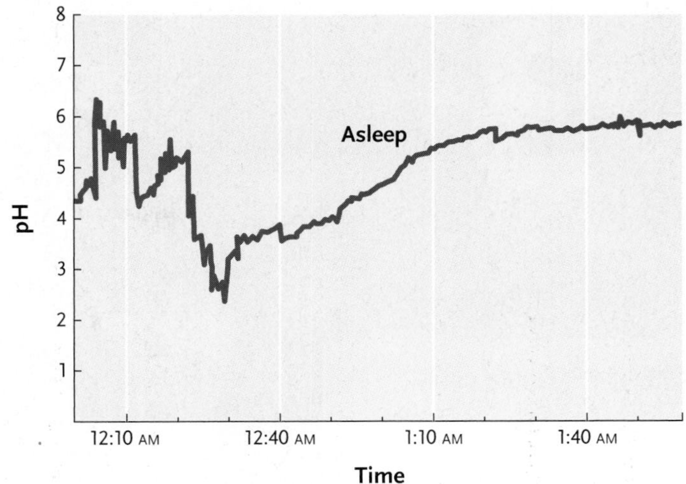
pH
8
7
6
5
4
3
2
1
Asleep
12:10 AM
12:40 AM
1:10 AM
1:40 AM
Time

1. How many minutes does it take to go from the peak of the reflux event (when pH is most acidic) to when the reflux event is over?
2. What is the molar concentration of H^+ and OH^- ions (a) during sleep after the reflux event, and (b) during the peak of the reflux event? Be sure to include the correct concentration units in your answer.
3. What is the change in concentration of H^+ and OH^- during the peak of the reflux attack compared to the clinical value of acid reflux? Be sure to include the correct concentration units in your answer.

Source: T. Demeester et al. 1976. Patterns of gastroesophageal reflux in health and disease. *Annals of Surgery* 184:459–469.

Apply Evolutionary Concepts

What properties of water made the evolution of life possible?

3

STUDY OUTLINE

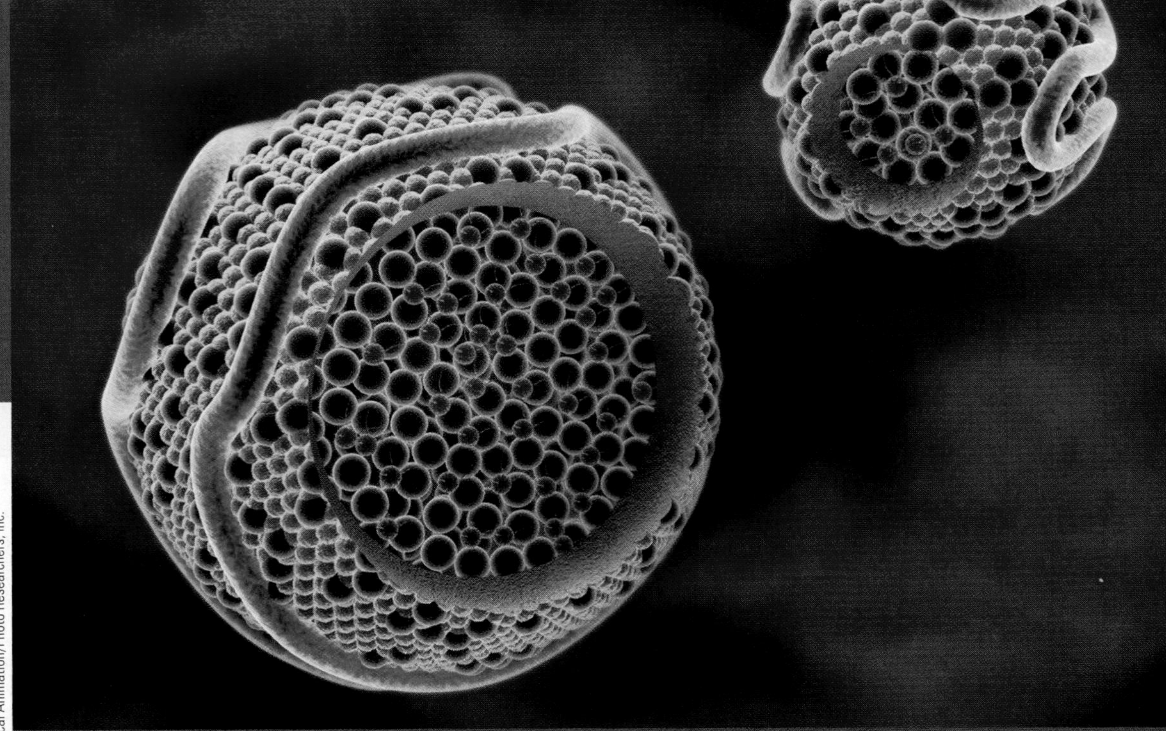

Hybrid Medical Animation/Photo Researchers, Inc.

The lipoproteins HDL and LDL, cholesterol-transporting molecules composed of protein and lipid units, which are found in the bloodstream (computer illustration).

Biological Molecules: The Carbon Compounds of Life

Why It Matters. . . High in the mountains of the Pacific Northwest, vast forests of coniferous trees have survived another cold winter **(Figure 3.1).** With the arrival of spring, rising temperatures and water from melting snow stimulate renewed growth. Carbon dioxide (CO_2) from the air enters the needlelike leaves of the trees through microscopic pores. Using energy from sunlight, the trees combine the water and carbon dioxide into sugars and other carbon-based compounds through the process known as photosynthesis. The lives of plants, and almost all other organisms, depend directly or indirectly on the products of photosynthesis.

The amount of CO_2 in the atmosphere is critical to photosynthesis. Researchers have been studying the atmospheric concentration of CO_2 since the early 1950s. Among other things, they found that CO_2 concentration shifts with the seasons. It declines during spring and summer, when plants and other photosynthetic organisms withdraw large amounts of the gas from the air and convert it into sugars and other complex carbon compounds. It increases during autumn and winter, when global photosynthesis decreases and decomposers that release the gas as a metabolic by-product increase. Great quantities of CO_2 are also added to the atmosphere by forest fires and by the burning of coal, oil, gasoline, and other fossil fuels in automobiles, aircraft, trains, power plants, and other industries. The resulting increase in atmospheric CO_2 contributes to global warming, which may have profound effects on life in years to come.

The importance of atmospheric CO_2 to food production and world climate are just two examples of how carbon and its compounds are fundamental to the entire living world, from the structures and activities of single cells to physical effects that take place on a global scale. Carbon

iStockphoto.com/Mark Hatfield

FIGURE 3.1
Conifers around Mount Rainier in Washington State. As is true of all other organisms, the structure, activities, and survival of these trees start with the carbon atom and its diverse molecular partners in organic compounds.

compounds form the structures of living organisms and take part in all biological reactions. They also serve as sources of energy for living organisms and as an energy resource for much of the world's industry—for example, coal and oil are the fossil remains of long-dead organisms. This chapter outlines the structures and functions of biological carbon compounds. <

A. Two-carbon hydrocarbons with single, double, and triple bonding

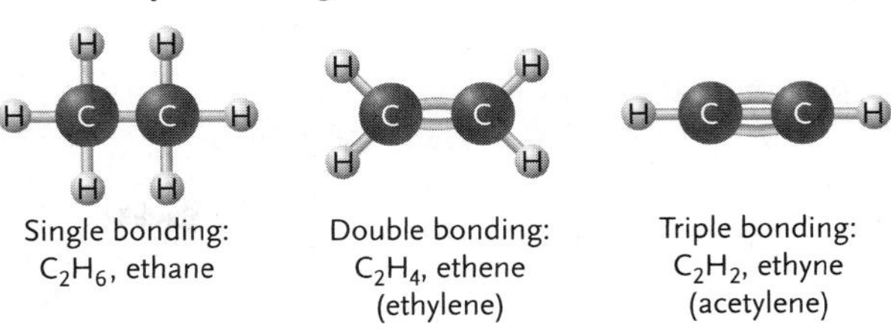
Single bonding:
C_2H_6, ethane
Double bonding:
C_2H_4, ethene
(ethylene)
Triple bonding:
C_2H_2, ethyne
(acetylene)

B. Linear and branched hydrocarbon chains

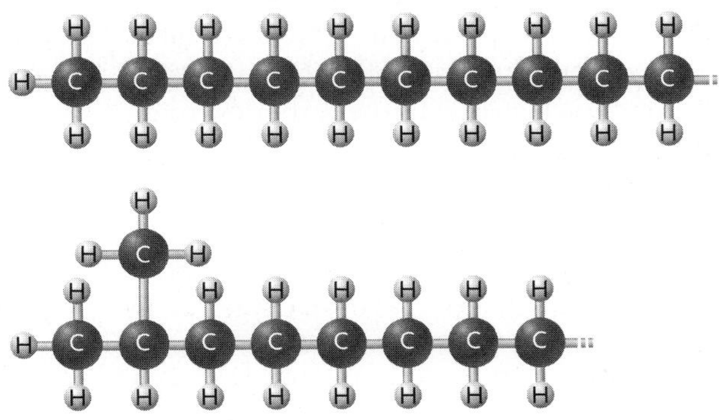

C. Hydrocarbon ring, in this case with double bonds

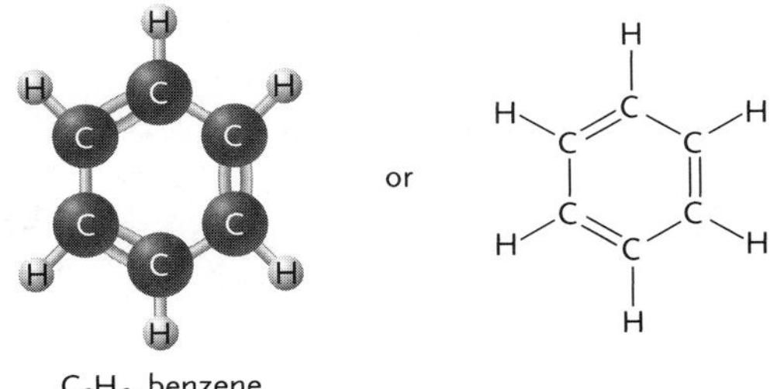
or
C_6H_6, benzene

FIGURE 3.2
Examples of hydrocarbon structures.

3.1 Formation and Modification of Biological Molecules

The bonding properties of carbon enable it to form an astounding variety of chain and ring structures that are the backbones of all biological molecules. In fact, the wide variety of carbon-based molecules has been responsible for the wide diversity of organisms that have evolved.

Carbon Chains and Rings Form the Backbones of All Biological Molecules

Collectively, molecules based on carbon are known as **organic molecules.** All other substances—those without carbon atoms in their structures—are **inorganic molecules.** A few of the smallest carbon-containing molecules that occur in the environment as minerals or atmospheric gases, such as CO_2, are also considered inorganic molecules. Outside of water, the four major classes of organic molecules—*carbohydrates, lipids, proteins,* and *nucleic acids*—form almost the entire substance of living organisms. They are discussed in turn in subsequent sections in this chapter.

In organic molecules, carbon atoms bond covalently to each other and to other atoms (chiefly hydrogen, oxygen, nitrogen, and sulfur), in molecules that range in size from a few atoms to thousands, or even millions of atoms. Molecules consisting of carbon linked only to hydrogen atoms are called **hydrocarbons** (*hydro-* refers to hydrogen, not to water).

As discussed in Section 2.3, carbon has four unpaired outer electrons that it readily shares to complete its outermost energy level, forming four covalent bonds. The simplest hydrocarbon, CH_4 (methane), consists of a single carbon atom bonded to four hydrogen atoms (see Figure 2.8A and B). More complex hydrocarbons involve two or more carbon atoms arranged in a linear unbranched chain, a linear branched chain, or a structure with one or more rings. The number of bonds between neighboring carbon atoms diversifies the structures. A triple bond can only occur in a two-carbon hydrocarbon, but single and double bonds are found in both linear and ring hydrocarbons **(Figure 3.2).** All in all, there is almost no limit to the number of different hydrocarbon structures that carbon and hydrogen can form.

Functional Groups Confer Specific Properties to Biological Molecules

Carbohydrates, lipids, proteins, and nucleic acids contain particular small, reactive groups of atoms called **functional groups.** Each of those groups has specific chemical properties, which are then also found in the larger molecules containing them. Thus, the number and arrangement of functional groups in a larger molecule give that molecule its particular properties.

Functional groups can participate in biological reactions. The functional groups that enter most frequently into biological reactions are the *hydroxyl, carbonyl, carboxyl, amino, phosphate,* and *sulfhydryl* groups **(Table 3.1).** The functional groups (boxed in blue in the table) are linked by covalent bonds to other atoms in biological molecules, usually carbon atoms. The symbol *R* is often used to represent the chain of carbon atoms. A double bond, such as that in the carbonyl group, indicates that two pairs of electrons are shared between the carbon and oxygen atoms.

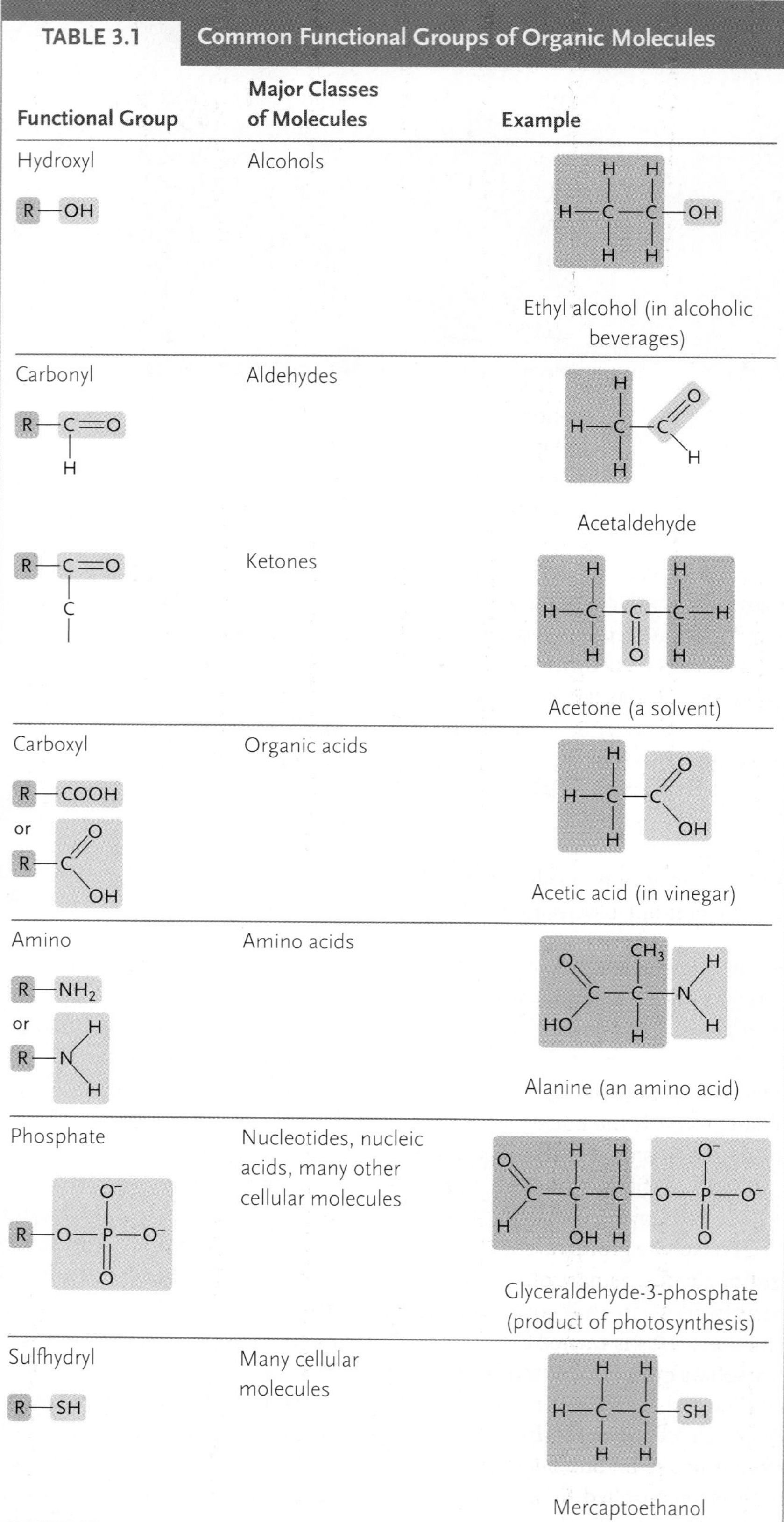

TABLE 3.1 Common Functional Groups of Organic Molecules

Functional Group	Major Classes of Molecules	Example
Hydroxyl R—OH	Alcohols	Ethyl alcohol (in alcoholic beverages)
Carbonyl R—C=O (with H)	Aldehydes	Acetaldehyde
R—C=O (with C)	Ketones	Acetone (a solvent)
Carboxyl R—COOH or R—C(=O)OH	Organic acids	Acetic acid (in vinegar)
Amino $R—NH_2$ or R—N(H)H	Amino acids	Alanine (an amino acid)
Phosphate $R—O—P(O^-)(=O)—O^-$	Nucleotides, nucleic acids, many other cellular molecules	Glyceraldehyde-3-phosphate (product of photosynthesis)
Sulfhydryl R—SH	Many cellular molecules	Mercaptoethanol

HYDROXYL GROUP A **hydroxyl group** (—OH) is a functional group that consists of an oxygen atom linked to a hydrogen atom on one side; in a molecule it is linked to an R group on the other side (see Table 3.1). Hydroxyl groups are polar, and confer polarity on the parts of the molecules that contain them (see Section 2.3 for a discussion of polarity).

The hydroxyl group is a key component of **alcohols,** such as ethyl alcohol. Ethyl alcohol (ethanol) is the alcohol found in beer, wine, and spirits, and it is also used to precipitate DNA from solutions in molecular biology experiments. The hydroxyl group enables an alcohol to form linkages to other organic molecules through dehydration synthesis reactions (see Figure 3.5A).

CARBONYL GROUP A **carbonyl group** (=C=O) consists of an oxygen atom linked to a carbon atom by a double bond (see Table 3.1). The oxygen atom of a carbonyl group is highly reactive, especially with substances that act as bases (see Section 2.5 for a discussion of acids and bases).

Carbonyl groups are the reactive parts of aldehydes and ketones, molecules that act as major building blocks of carbohydrates and that also take part in the reactions supplying energy for cellular activities. In an **aldehyde,** the carbonyl group is linked—along with a hydrogen atom—to a carbon atom at the end of a carbon chain, along with a hydrogen atom, as in acetaldehyde (see Table 3.1). In a **ketone,** the carbonyl group is linked to a carbon atom in the interior of a carbon chain, as in acetone (see Table 3.1).

CARBOXYL GROUP A carbonyl group and a hydroxyl group combine to form a **carboxyl group** (—COOH), the characteristic functional group of **organic acids** (also called *carboxylic acids*); an example is acetic acid (see Table 3.1). The carboxyl group gives organic molecules acidic properties because its —OH group readily releases the hydrogen as a proton (H^+) in water solutions (see Section 2.5):

$$R—C(=O)—OH \rightleftharpoons R—C(=O)—O^- + H^+$$

Many organic acids, such as citric acid and acetic acid, are central components of energy-generating reactions in living organisms.

AMINO GROUP The **amino group** (—NH_2) consists of a nitrogen atom bonded on one side to two hydrogen atoms; in a molecule it is linked to an R group on the other side, as in the amino acid alanine (see Table 3.1) and all other amino acids. It readily acts as an organic base by accepting H^+ (a proton) in water solutions:

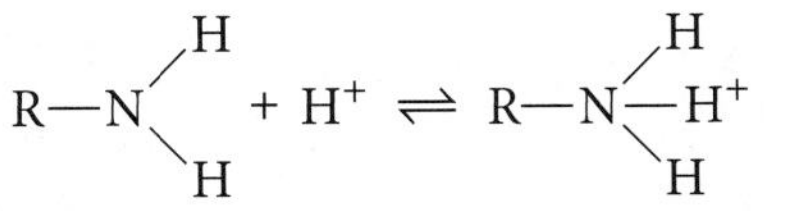

$$\mathrm{R{-}N}\begin{matrix}\diagup \mathrm{H}\\ \diagdown \mathrm{H}\end{matrix} + \mathrm{H^+} \rightleftharpoons \mathrm{R{-}N}\begin{matrix}\diagup \mathrm{H}\\ -\mathrm{H^+}\\ \diagdown \mathrm{H}\end{matrix}$$

PHOSPHATE GROUP The **phosphate group** (—OPO_3^{2-}) consists of a central phosphorus atom held in four linkages. Two of the linkages bind —OH groups to the central phosphorus atom; a third linkage, formed by a double bond, binds an oxygen atom to the central phosphorus atom. The remaining bond links the phosphate group to an oxygen atom, which, in turn, binds to an R group (see Table 3.1). An example is glyceraldehyde-3-phosphate, a product of photosynthesis.

Phosphate groups give molecules that contain them the ability to react as weak acids because one or both —OH groups readily release their hydrogens as H^+:

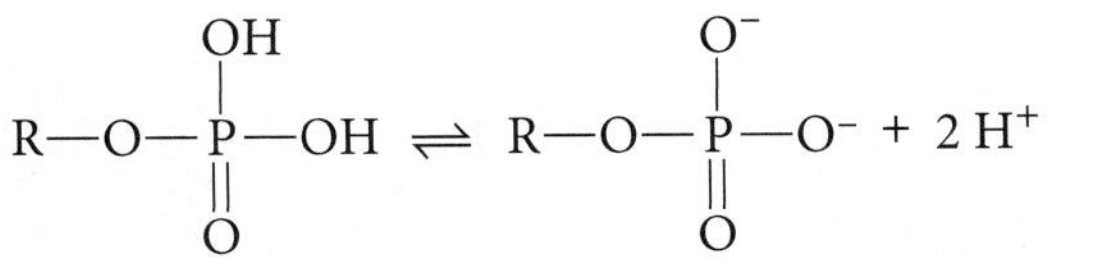

$$\mathrm{R{-}O{-}}\overset{\mathrm{OH}}{\underset{\mathrm{O}}{\overset{|}{\underset{||}{\mathrm{P}}}}}\mathrm{{-}OH} \rightleftharpoons \mathrm{R{-}O{-}}\overset{\mathrm{O^-}}{\underset{\mathrm{O}}{\overset{|}{\underset{||}{\mathrm{P}}}}}\mathrm{{-}O^-} + 2\,\mathrm{H^+}$$

A phosphate group can also form a chemical bridge that links two organic building blocks into a larger structure:

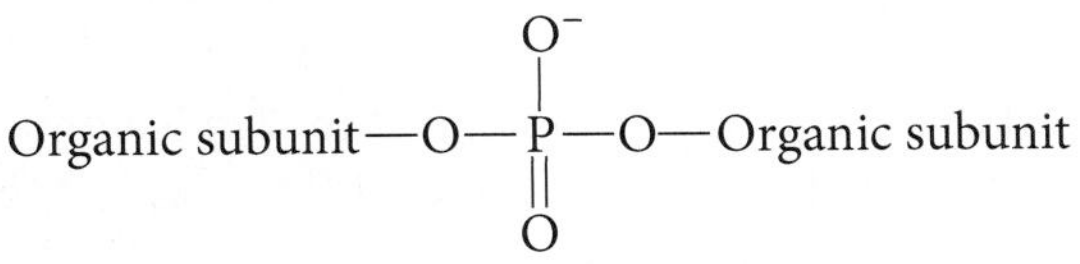

$$\text{Organic subunit}\mathrm{{-}O{-}}\overset{\mathrm{O^-}}{\underset{\mathrm{O}}{\overset{|}{\underset{||}{\mathrm{P}}}}}\mathrm{{-}O{-}}\text{Organic subunit}$$

Among the large biological molecules linked together by phosphate groups is the nucleic acid DNA, the genetic material of all living organisms. When connecting organic subunits, a phosphate group still has one —OH group that can dissociate, releasing H^+ and leaving O^- as part of the new molecule.

Phosphate groups are also added to or removed from biological molecules as part of reactions that conserve or release energy. In addition, they control biological activity—the activity of many proteins is turned on or off by the addition or removal of phosphate groups.

SULFHYDRYL GROUP In the **sulfhydryl group** (—SH), a sulfur atom is linked on one side to a hydrogen atom; in a molecule, the other side is linked to an R group, as in mercaptoethanol (see Table 3.1). The sulfhydryl group is easily converted into a covalent linkage, in which it loses its hydrogen atom as it binds. In many of these linking reactions, two sulfhydryl groups interact to form a **disulfide linkage** (—S—S—):

$$\mathrm{R{-}SH + HS{-}R \rightarrow R{-}\underbrace{S{-}S}_{\text{disulfide linkage}}{-}R + 2\,H^+ + 2\ electrons}$$

In many proteins, the disulfide linkage forms a sort of molecular fastener that holds proteins in their folded form or links protein subunits into larger structures (see Figure 3.18).

Isomers Have the Same Chemical Formula but Different Molecular Structures

Often, one or more of the carbon atoms in an organic molecule links to four different atoms or functional groups. Carbons linked in this way are called *asymmetric* carbons; they have important effects on the structure of the molecule because they can take either of two fixed positions with respect to other carbons in a carbon chain. For example, the middle carbon of the three-carbon sugar glyceraldehyde is asymmetric because it shares electrons in covalent bonds with four different atoms or groups: —H, —OH, —CHO, and —CH_2OH **(Figure 3.3)**. The —H and —OH groups can take either of two positions, with the —OH extending to either the left or right of the carbon chain relative to the —CHO and —CH_2OH groups.

The two forms of glyceraldehyde have the same chemical formula, $C_3H_6O_3$. The difference between the two forms is similar to the difference between your two hands. Although both hands have four fingers and a thumb, they are not identical; rather, they are mirror images of each other. That is, when you hold your right hand in front of a mirror, the reflection looks like your left hand and vice versa.

Two or more molecules with the same chemical formula but different molecular structures are called **isomers.** Isomers that are mirror images of each other, like the two forms of glyceraldehyde, are called **enantiomers,** or **optical isomers.** One of the enantiomers—the one in which the hydroxyl group extends to the left—is called the L-form (*laevus* = left). The other enantiomer, in which the —OH extends to the right, is called the D-form (*dexter* = right) (see Figure 3.3). The difference between L- and D-enantiomers is critical to biological function. Typically, one of the two forms enters much more readily into cellular reactions; just as your left hand does not fit readily into a right-hand glove, enzymes (proteins that accelerate chemical reactions in living organisms) fit best to one of the two forms of an enantiomer. For example, most of the enzymes that catalyze the biochemical reactions involving sugars recognize the D-form, making this form much more common among cellular carbohydrates than L-forms. Many other kinds of biological molecules besides sugars form enantiomers; an example is the amino acids. Most of the enzymes that catalyze the biochemical reactions involving amino acids recognize the L-form.

Another form of isomerism is found in sugars, as well as in other molecules. **Structural isomers** are two molecules with the same chemical formula but atoms that are arranged in different ways. The sugars glucose and fructose are examples of structural isomers **(Figure 3.4)**.

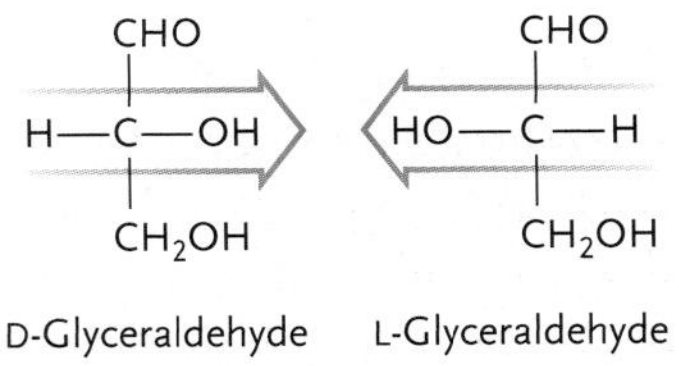

FIGURE 3.3
Enantiomers—mirror-image alternative forms—of the sugar glyceraldehyde.

CHO
H—C—OH
CH_2OH
D-Glyceraldehyde

CHO
HO—C—H
CH_2OH
L-Glyceraldehyde

A. **Glucose (an aldehyde)**

B. **Fructose (a ketone)**

FIGURE 3.4
Glucose and fructose, structural isomers of a six-carbon sugar. (A) In glucose, the aldehyde isomer, the carbonyl group (shaded region) is located at the end of the carbon chain. **(B)** In fructose, the ketone isomer, the carbonyl group is located inside the carbon chain. For convenience, the carbons of the sugars are numbered, with 1 being the carbon at the end nearest the carbonyl group.

A Water Molecule Is Added or Removed in Many Reactions Involving Functional Groups

In many of the reactions that involve functional groups, the components of a water molecule, —H and —OH, are removed from or added to the groups as they interact. When the components of a water molecule are *removed* during a reaction (usually as part of the assembly of a larger molecule from smaller subunits), the reaction is called a **dehydration synthesis reaction** or **condensation reaction (Figure 3.5A)**. For example, this type of reaction occurs when individual sugar molecules combine to form a starch molecule. In **hydrolysis,** the reverse reaction, the components of a water molecule are *added* to functional groups as molecules are broken into smaller subunits **(Figure 3.5B)**. For example, the breakdown of a protein molecule into individual amino acids occurs by hydrolysis in the digestive processes of animals.

Of the functional groups, hydroxyl groups readily enter dehydration synthesis reactions and are formed as part of hydrolysis reactions. Carboxyl groups readily enter into dehydration synthesis reactions, giving up hydroxyl groups as organic molecules combine into larger assemblies. Amino groups also readily enter dehydration synthesis reactions, releasing H^+ as it links subunits into larger molecules. For example, the joining of amino acids in the synthesis of proteins involves a dehydration reaction involving the carboxyl group of one amino acid and the amino group of another amino acid (see Figure 3.19).

Many Carbohydrates, Lipids, Proteins, and Nucleic Acids Are Macromolecules

Carbohydrates, lipids, proteins, and nucleic acids are large *polymers* (*poly* = many; *mer* = unit). A polymer is a molecule assembled from subunit molecules called monomers into a chain by covalent bonds. The process of assembly of a polymer from monomers is called polymerization. The polymerization reactions are dehydration synthesis reactions. The opposite reactions—breakdown of polymers into monomers—occur by hydrolysis.

A. Dehydration synthesis reactions remove the components of a water molecule as new covalent bonds join subunits into a larger molecule.

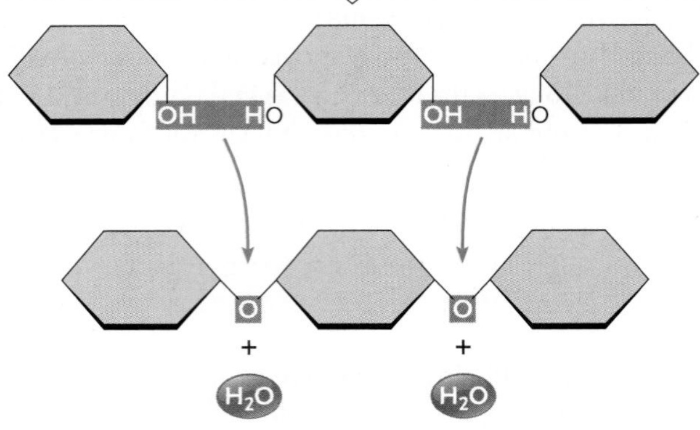
OH
HO
OH
HO
O
O
+
+
H_2O
H_2O

B. Hydrolysis adds the components of a water molecule as covalent bonds are broken, splitting a molecule into smaller subunits.

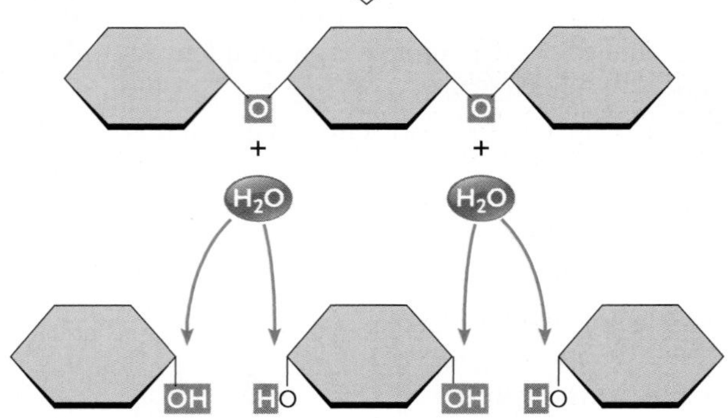
O
O
+
+
H_2O
H_2O
OH
HO
OH
HO

FIGURE 3.5
Dehydration synthesis and hydrolysis reactions.

Each type of polymeric biological molecule contains one type of monomer. The monomers may be identical, or they may have chemical variations, depending on the molecule. The variations among monomer structures are responsible for the highly diverse and varied biological molecules found in living organisms. For instance, proteins are polymers consisting of amino acid monomers. There are twenty different amino acids, each with identical amino and carboxyl functional groups that enable them to undergo polymerization (see Figure 3.17), but each also with a different R group. The proteins assembled from these amino acids vary in number and organization in the polymer chains, resulting in a huge variety of proteins with different structures and, therefore, functions.

Somewhat arbitrarily, a single polymer molecule with a mass of 1,000 daltons (Da) or more is called a macromolecule (*macro* = large). By that criterion, many representatives of carbohydrates, proteins, and nucleic acids are macromolecules. Lipids are not large enough to be classed as macromolecules. In a number of instances, macromolecules interact to form even larger functional molecular structures in cells. For example, the ribosome, the cellular structure that plays the central role in the polymerization of amino acids into a protein chain, consists of several RNA macromolecules and many protein macromolecules.

STUDY BREAK 3.1

1. What is the difference between hydrocarbons and other organic molecules?
2. What is the maximum number of bonds that a carbon atom can form?
3. Do carboxyl groups, amino groups, and phosphate groups act as acids or bases?
4. What is the difference between a dehydration synthesis reaction (condensation reaction) and hydrolysis?

3.2 Carbohydrates

Carbohydrates, the most abundant organic molecules in the world, serve many functions. Together with fats, they act as the major fuel substances providing chemical energy for cellular activities. The carbohydrate sucrose, common table sugar, is consumed in large quantities as an energy source in the human diet. Energy-providing carbohydrates are stored in plant cells as **starch** and in animal cells as **glycogen,** both consisting of long chains of repeating carbohydrate subunits linked end to end. Chains of carbohydrate subunits also form many structural molecules, such as **cellulose,** one of the primary constituents of plant cell walls.

Carbohydrates contain only carbon, hydrogen, and oxygen atoms, in an approximate ratio of 1 carbon:2 hydrogens:1 oxygen (CH_2O). The names of many carbohydrates end in *-ose.* The smallest carbohydrates, the **monosaccharides** (*mono* = one; *saccharum* = sugar), contain three to seven carbon atoms. For example, the monosaccharide glucose consists of a chain of six carbons and has the molecular formula $C_6H_{12}O_6$. Two monosaccharides polymerize to form a *disaccharide* such as sucrose, which is common table sugar. Carbohydrate polymers with more than 10 linked monosaccharide monomers are called **polysaccharides**. Starch, glycogen, and cellulose are common polysaccharides.

Monosaccharides Are the Structural Units of Carbohydrates

Carbohydrates occur either as monosaccharides or as polymers of monosaccharide units linked together. Monosaccharides are soluble in water, and most have a distinctly sweet taste. Of the monosaccharides, those that contain three carbons *(trioses),* five carbons *(pentoses),* and six carbons *(hexoses)* are most common in living organisms **(Figure 3.6).**

All monosaccharides can occur in the linear form shown in Figure 3.6. In this form, each carbon atom in the chain except one has both an —H and an —OH group attached to it. The remaining carbon is part of a carbonyl group, which may be located at the end of the carbon chain in the aldehyde position, or inside the chain in the ketone position, resulting in structural isomers (see Figure 3.4).

Monosaccharides with five or more carbons can fold back on themselves to assume a ring form. Folding into a ring occurs through a reaction between two functional groups in the same

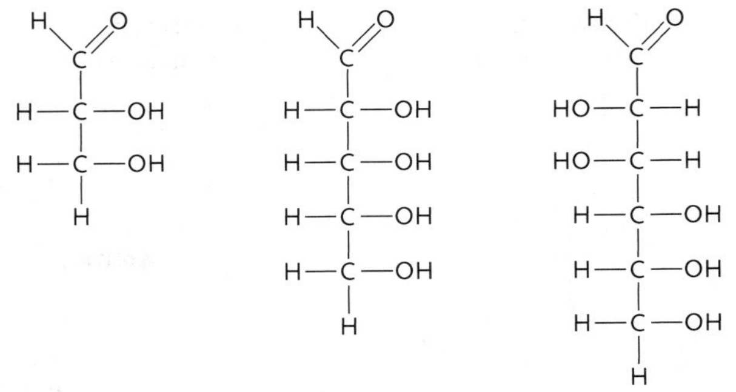

Glyceraldehyde (3 carbons; a triose) | **Ribose** (5 carbons; a pentose) | **Mannose** (6 carbons; a hexose)

FIGURE 3.6

Some representative monosaccharides. The triose, glyceraldehyde, takes part in energy-yielding reactions and photosynthesis. The pentose, ribose, is a component of RNA and of molecules that carry energy. The hexose, mannose, is a fuel substance and a component of glycolipids and glycoproteins.

monosaccharide, as occurs in glucose **(Figure 3.7).** The ring form of most five- and six-carbon sugars is much more common in cells than the linear form.

In the ring form of many five- or six-carbon monosaccharides, including glucose, the carbon at the 1 position of the ring is asymmetric because its four bonds link to different groups of atoms. This asymmetry allows monosaccharides such as glucose to exist as two different enantiomers. The glucose enantiomer with an —OH group pointing below the plane of the ring is known as *alpha-glucose,* or *α-glucose;* the enantiomer with an —OH group pointing above the plane of the ring is known as *beta-glucose,* or *β-glucose* (see Figure 3.7B). Other five- and six-carbon monosaccharide rings have similar α- and β-configurations.

The α- and β-rings of monosaccharides can give the polysaccharides assembled from them vastly different chemical properties. For example, starches, which are assembled from α-glucose units, are biologically reactive polysaccharides easily digested by animals; cellulose, which is assembled from β-glucose units, is relatively unreactive and, for most animals, completely indigestible.

Two Monosaccharides Link to Form a Disaccharide

Disaccharides typically are assembled from two monosaccharides covalently joined by a dehydration synthesis reaction. For example, the disaccharide maltose is formed by the linkage of two α-glucose molecules **(Figure 3.8A)** with oxygen as a bridge between the number 1 carbon of the first glucose unit and the 4 carbon of the second glucose unit. Bonds of this type, which commonly link monosaccharides into chains, are known as **glycosidic bonds.** A glycosidic bond between a 1 carbon and a 4 carbon is written in chemical shorthand as a 1→4 linkage; 1→2, 1→3, and 1→6 linkages are also common in carbohydrate chains. The linkages are designated as α or β depending on the orientation of the —OH group at the 1 carbon that forms the

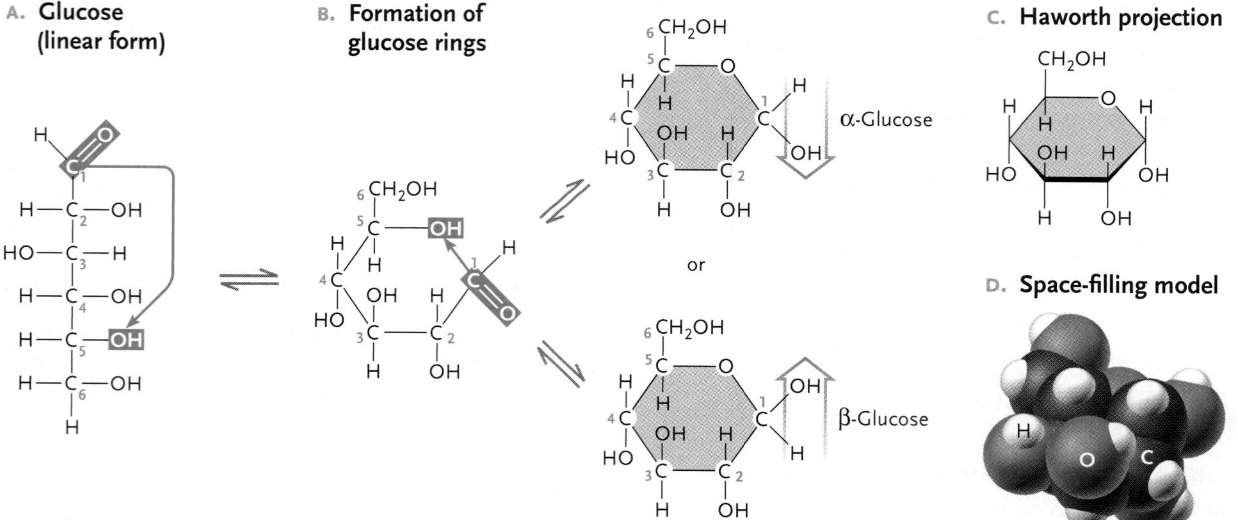

A. Glucose (linear form)
B. Formation of glucose rings
α-Glucose
or
β-Glucose
C. Haworth projection
D. Space-filling model

FIGURE 3.7
Ring formation by glucose. **(A)** Glucose in linear form. **(B)** The ring form of glucose is produced by a reaction between the aldehyde group at the 1 carbon and the hydroxyl group at the 5 carbon. The reaction produces two alternate glucose enantiomers, α- and β-glucose. If the ring is considered to lie in the plane of the page, the —OH group points below the page in α-glucose and upward from the page in β-glucose. For simplicity, the group at the 6 carbon is shown as CH_2OH in this and later diagrams. **(C)** A commonly used, simplified representation of the glucose ring, in which the Cs designating carbons of the ring are omitted. The thicker lines along one side indicate that the ring lies in a flat plane with the thickest edge closest to the viewer. **(D)** A space-filling model of glucose, showing the volumes occupied by the atoms. Carbon atoms are black, oxygen atoms are red, and hydrogen atoms are white.

bond. In maltose, the —OH group is in the α position. Therefore, the link between the two glucose subunits of maltose is written as an α (1→4) linkage.

Maltose, sucrose, and lactose are common disaccharides. Maltose is present in germinating seeds and is a major sugar used in the brewing industry. Sucrose, which contains a glucose and a fructose unit **(Figure 3.8B)**, is transported to and from different parts of leafy plants. It is probably the most plentiful sugar in nature. Table sugar is made by extracting and crystallizing sucrose from plants, such as sugar cane and sugar beets. Lactose, assembled from a glucose and a galactose unit **(Figure 3.8C)**, is the primary sugar of milk.

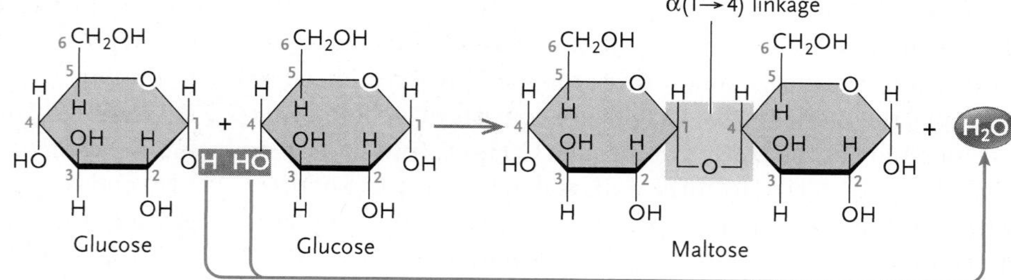

A. Formation of maltose
α(1→4) linkage
Glucose
Glucose
Maltose

B. Sucrose

α(1→2) linkage

Glucose unit

Fructose unit

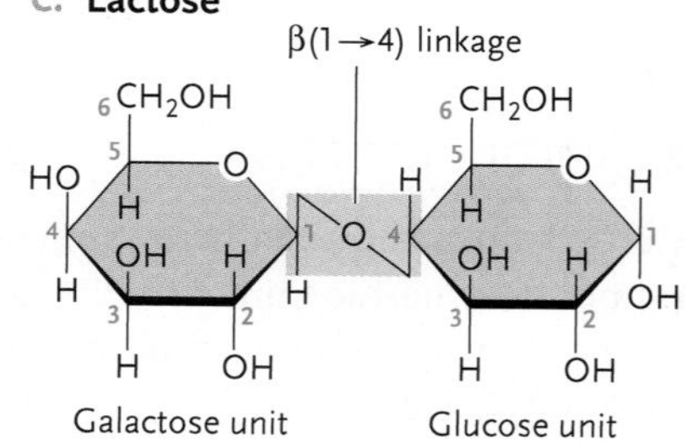

C. Lactose
β(1→4) linkage
Galactose unit
Glucose unit

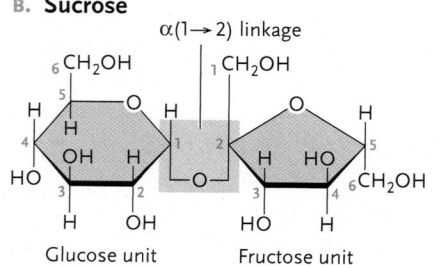

FIGURE 3.8
Disaccharides. **(A)** Combination of two glucose molecules by a dehydration synthesis reaction to form the disaccharide maltose. The components of a water molecule (in blue) are removed from the monosaccharides as they join. **(B)** Sucrose, assembled from glucose and fructose. **(C)** Lactose, assembled from galactose and glucose.

Monosaccharides Link in Longer Chains to Form Polysaccharides

Polysaccharides are the macromolecules formed by polymerization of monosaccharide monomers through dehydration synthesis reactions. The most common polysaccharides—the plant starches, glycogen, and cellulose—are polymers of hundreds or thousands of glucose units. Other polysaccharides are built up from a variety of different sugar units. Polysaccharides may be linear, unbranched molecules, or they may contain one or more branches in which side chains of sugar units attach to a main chain.

Figure 3.9 shows four common polysaccharides. Plant starches include both linear, unbranched forms such as amylose (Figure 3.9A) and branched forms such as amylopectin. Glycogen (Figure 3.9B), a more highly branched polysaccharide than amylopectin, can be assembled or disassembled readily to take

A. Amylose, formed from α-glucose units joined end to end in $\alpha(1 \rightarrow 4)$ linkages. The coiled structures are induced by the bond angles in the α-linkages.

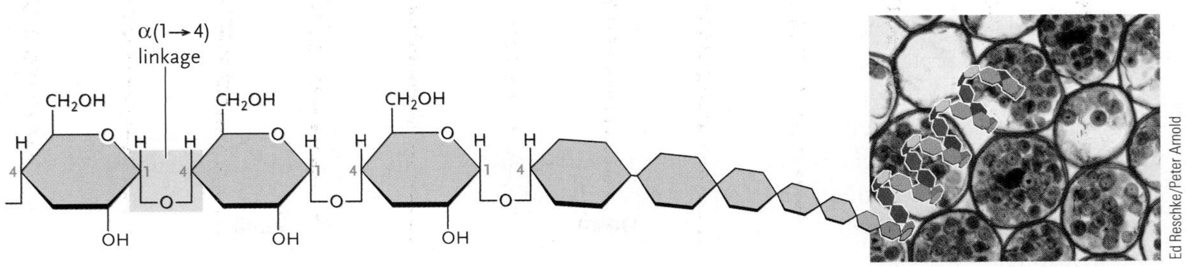
α(1→4) linkage
CH₂OH
OH
Amylose grains (purple) in plant root tissue
Ed Reschke/Peter Arnold

B. Glycogen, formed from glucose units joined in chains by $\alpha(1 \rightarrow 4)$ linkages; side branches are linked to the chains by $\alpha(1 \rightarrow 6)$ linkages (boxed in blue).

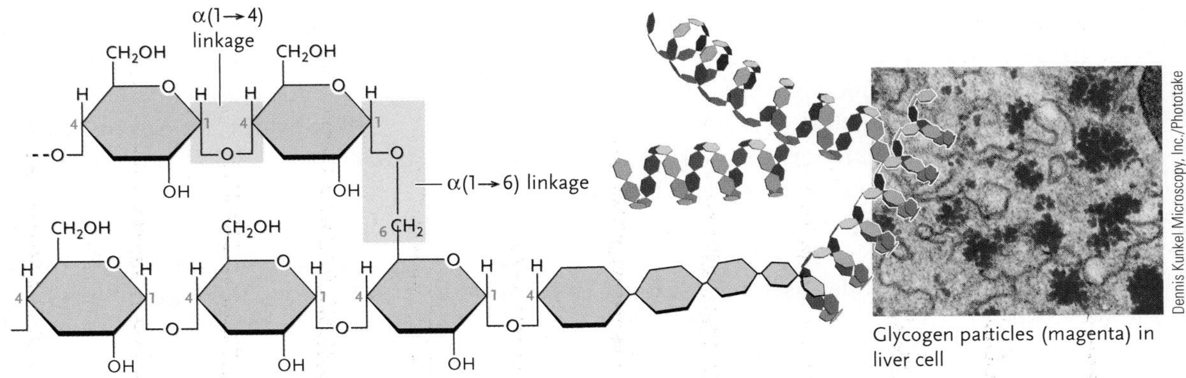
α(1→4) linkage
α(1→6) linkage
CH₂OH
OH
Glycogen particles (magenta) in liver cell
Dennis Kunkel Microscopy, Inc./Phototake

C. Cellulose, formed from glucose units joined end to end by $\beta(1 \rightarrow 4)$ linkages. Hundreds to thousands of cellulose chains line up side by side, in an arrangement reinforced by hydrogen bonds between the chains, to form cellulose microfibrils in plant cells.

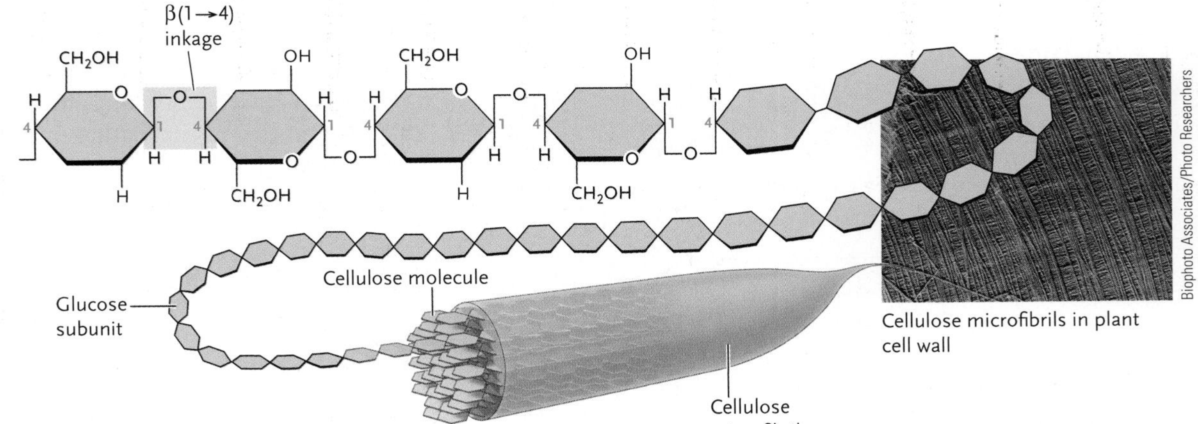
β(1→4) inkage
CH₂OH
OH
Cellulose molecule
Glucose subunit
Cellulose microfibril
Cellulose microfibrils in plant cell wall
Biophoto Associates/Photo Researchers

D. Chitin, formed from $\beta(1 \rightarrow 4)$ linkages joining glucose units modified by the addition of nitrogen-containing groups. The external body armor of the tick is reinforced by chitin fibers.

β(1→4) linkage

Image copyright Carolina K. Smith, M.D., 2010. Used under license from Shutterstock.com

FIGURE 3.9

Four common polysaccharides: **(A)** amylose, a plant starch; **(B)** glycogen, found in animal tissues; **(C)** cellulose, the primary fiber in plant cell walls; and **(D)** chitin, a reinforcing fiber in the external skeleton of arthropods and the cell walls of some fungi.

up or release glucose; it is stored in large quantities in the liver and muscle tissues of many animals.

Cellulose (Figure 3.9C), probably the most abundant carbohydrate on Earth, is an unbranched polysaccharide assembled from glucose monomers bound together by β-linkages. It is the primary structural fiber of plant cell walls; in this role, cellulose has been likened to the steel rods in reinforced concrete. Its tough fibers enable the cell walls of plants to withstand enormous weight and stress. Fabrics such as cotton and linen are made from cellulose fibers extracted from plant cell walls. Animals such as mollusks, crustaceans, and insects synthesize an enzyme that digests the cellulose they eat. In ruminant mammals, such as cows, microorganisms in the digestive tract break down cellulose. Cellulose passes unchanged through the human digestive tract as indigestible fibrous matter. Many nutritionists maintain that the bulk provided by cellulose fibers helps maintain healthy digestive function.

Chitin (Figure 3.9D), another tough and resilient polysaccharide, is assembled from glucose units modified by the addition of nitrogen-containing groups. Similar to the subunits of cellulose, the modified glucose monomers of chitin are held together by β-linkages. Chitin is the main structural fiber in the external skeletons and other hard body parts of arthropods such as insects, crabs, and spiders. It is also a structural material in the cell walls of fungi such as mushrooms and yeasts. Unlike cellulose, chitin is digested by enzymes that are widespread among microorganisms, plants, and many animals. In plants and animals, including humans and other mammals, chitin-digesting enzymes occur primarily as part of defenses against fungal infections. However, humans cannot digest chitin as a food source.

Polysaccharides also occur on the surfaces of cells, particularly in animals. These surface polysaccharides are attached to both the protein and lipid molecules in membranes. They help hold the cells of animals together and serve as recognition sites between cells.

STUDY BREAK 3.2 <

What is the difference between a monosaccharide, a disaccharide, and a polysaccharide? Give examples of each.

3.3 Lipids

Lipids are a diverse group of water-insoluble, primarily nonpolar biological molecules composed mostly of hydrocarbons. Some are large molecules, but they are not large enough to be considered macromolecules. As a result of their nonpolar character, lipids typically dissolve much more readily in nonpolar solvents, such as acetone and chloroform, than in water, the polar solvent of living organisms. Their insolubility in water underlies their ability to form cell membranes, the thin molecular films that create boundaries between and within cells.

In addition to forming membranes, some lipids are stored and used in cells as an energy source. Other lipids serve as hormones that regulate cellular activities. Three types of lipid molecules—*neutral lipids, phospholipids,* and *steroids*—occur most commonly in living organisms.

Neutral Lipids Are Familiar as Fats and Oils

Neutral lipids, commonly found in cells as energy-storage molecules, are called "neutral" because at cellular pH they have no charged groups; they are therefore nonpolar. There are two types of neutral lipids: **oils** and **fats.** Oils are liquid at biological temperatures, and fats are semisolid. Generally, neutral lipids are insoluble in water. Almost all neutral lipids contain a three-carbon backbone chain formed from glycerol, an alcohol, with each of the three carbons linked to a side chain consisting of a *fatty acid.*

FATTY ACIDS A **fatty acid** contains a single hydrocarbon chain with a carboxyl group (—COOH) at one end **(Figure 3.10).** The carboxyl group gives the fatty acid its acidic properties. The fatty acids in living organisms contain four or more carbons in their hydrocarbon chain, with the most common forms having even-numbered chains of 14 to 22 carbons. Only the shortest fatty acid chains are water-soluble. As chain length increases, fatty acids become progressively less water-soluble and more oily.

If the hydrocarbon chain of a fatty acid binds the maximum possible number of hydrogen atoms, so that only single bonds link the carbon atoms, the fatty acid is said to be **saturated** with hydrogen atoms (as in stearic acid in Figure 3.10A). If one or more double bonds link the carbons (see Figure 3.10B, arrow), reducing the number of hydrogen atoms bound, the fatty acid is **unsaturated.** Fatty acids with one double bond are **monounsaturated;** those with more than one double bond are **polyunsaturated.**

Unsaturated fatty acid chains tend to bend or "kink" at a double bond (see Figures 3.10B and 3.14C). The kink makes the chains more disordered and thus more fluid at biological temperatures. Consequently, unsaturated fatty acids—and lipids that contain them—melt at lower temperatures than saturated fatty acids of the same length, and they generally have oily rather than fatty characteristics.

A. **Stearic acid,** $CH_3(CH_2)_{16}COOH$

B. **Oleic acid,** $CH_3(CH_2)_7CH{=}CH(CH_2)_7COOH$

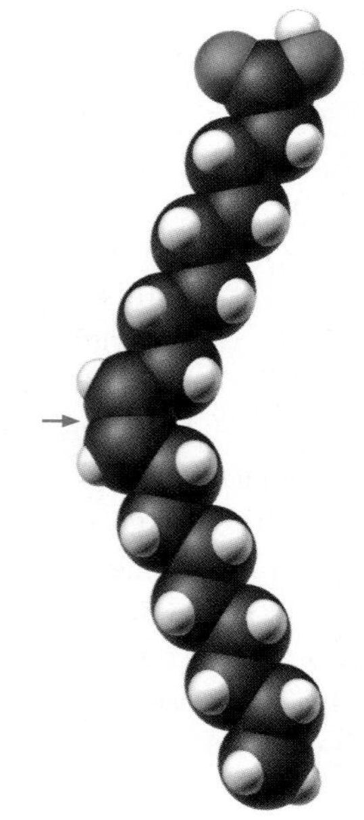

FIGURE 3.10
Fatty acids, one of two components of a neutral lipid. **(A)** Stearic acid, a saturated fatty acid. **(B)** Oleic acid, an unsaturated fatty acid. An arrow marks the "kink" introduced by the double bond.

A. **Formation of a triglyceride**

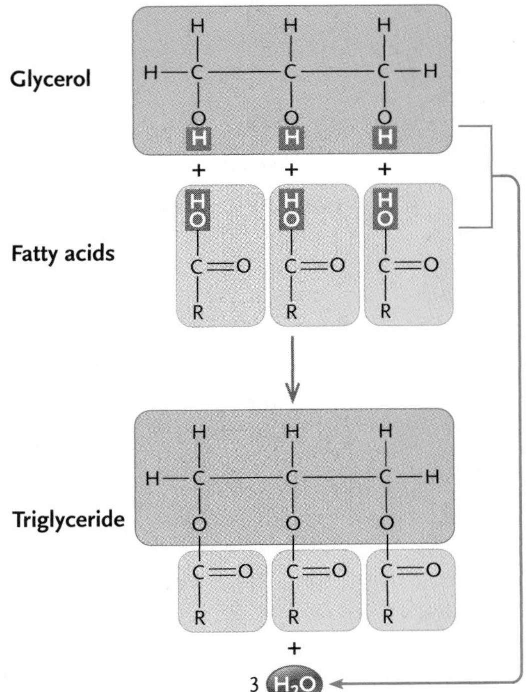
Glycerol
Fatty acids
Triglyceride
3 H2O

B. **Glyceryl palmitate**

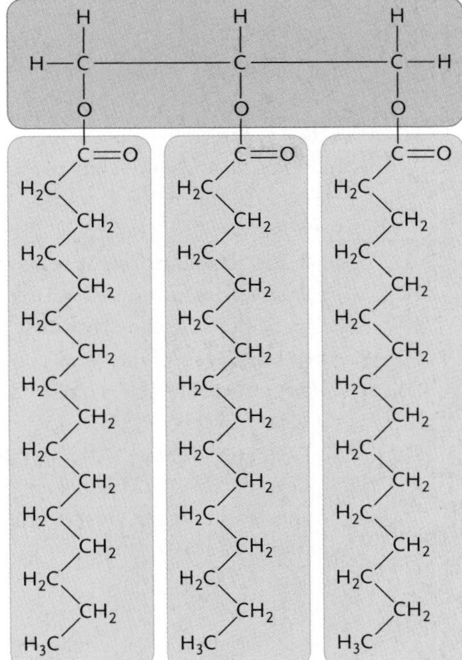

C. **Triglyceride model**

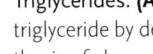

FIGURE 3.11
Triglycerides. **(A)** Formation of a triglyceride by dehydration synthesis of glycerol with three fatty acids. The R groups represent the hydrocarbon chains of the fatty acids. The components of a water molecule (in blue) are removed from the glycerol and fatty acids in each of the three bonds formed. **(B)** Chemical structure and **(C)** space-filling model of glyceryl palmitate, a triglyceride.

In foods, saturated fatty acids are usually found in solid animal fats, such as butter, whereas unsaturated fatty acids are usually found in vegetable oils, such as liquid canola oil. Nonetheless, both solid animal fats and liquid vegetable oils contain some saturated and some unsaturated fatty acids.

GLYCEROL AND TRIGLYCERIDE FORMATION The glycerol unit that forms the backbone of neutral lipids has three —OH groups at which fatty acids may link **(Figure 3.11A)**. In its free state, glycerol is a polar, water-soluble, sweet-tasting alcohol. If a fatty acid binds by a dehydration synthesis reaction at each of glycerol's three —OH bearing sites, the polar groups are eliminated, producing a nonpolar compound known as a **triglyceride** (see Figure 3.11). Most lipids stored as an energy reserve in living systems are triglycerides.

The fatty acids linked to glycerol may be different or the same. Different organisms usually have distinctive combinations of fatty acids in their triglycerides. As with individual fatty acids, triglycerides generally become less fluid as the length of their fatty acid chains increases; those with shorter chains remain liquid as oils at biological temperatures, and those with longer chains solidify as fats. The degree of saturation of the fatty acid chains also affects the fluidity of triglycerides—the more saturated, the less fluid the triglyceride. Plant oils are converted commercially to fats by *hydrogenation*—that is, adding hydrogen atoms to increase the degree of saturation, as in the conversion of vegetable oils to margarines and shortening.

Triglycerides are used widely as stored energy in animals. Gram for gram, they yield more than twice as much energy as carbohydrates do. Therefore, fats are an excellent source of energy in the diet. Storing the equivalent amount of energy as carbohydrates rather than fats would add more than 100 pounds to the weight of an average man or woman. A layer of fatty tissue just under the skin also serves as an insulating blanket in humans, other mammals, and birds. Triglycerides secreted from special glands in waterfowl and other birds help make their feathers water repellent (as in the penguins shown in **Figure 3.12**).

Unsaturated fats are considered healthier than saturated fats in the human diet. Saturated fats have been implicated in the development of atherosclerosis (see *Focus on Applied Research*), a disease in which arteries, particularly those serving the heart, become clogged with fatty deposits.

WAXES Fatty acids may also combine with long-chain alcohols or hydrocarbon structures to form **waxes,** which are harder and less greasy than fats. Insoluble in water, waxy coatings help keep skin, hair, or feathers of animals protected, lubricated, and pliable.

FIGURE 3.12
Penguins of the Antarctic, one of several animals that have a thick, insulating layer of fatty tissue that contains triglycerides under the skin. Penguins also use their face and bill to spread oil, secreted by a gland near their tail, over their feathers. The oily coating keeps their feathers watertight and dry.

iStockphoto.com/Keith Szafranski

FOCUS ON Applied Research

Fats, Cholesterol, and Coronary Artery Disease

Butter! Bacon and eggs! Ice cream! Cheesecake! Possibly you think of such foods as irresistible, off limits, or both. After all, who doesn't know about animal fats, cholesterol, and hardening of the arteries? Hardening of the arteries, or *atherosclerosis,* is a condition in which deposits of lipid and fibrous material called plaque build up in the walls of arteries, the vessels that supply oxygenated blood to body tissues. Plaque reduces the internal diameter of the arteries, restricting or even completely blocking the flow of blood. Blockage of the coronary arteries that supply oxygenated blood to the heart muscle (see figure) can severely impair heart function, a condition called coronary heart disease. In extreme cases, it can lead to destruction of heart muscle tissue, as occurs in a heart attack (myocardial infarction).

Your body requires a certain amount of cholesterol, but the liver normally makes enough to meet this demand. Additional cholesterol is made from fats taken in as food. Cholesterol is found in the blood bound to low-density lipoprotein (LDL) and high-density lipoprotein (HDL). LDL cholesterol is considered "bad" because clinical studies have shown a positive correlation between its level in the blood and the risk for coronary heart disease. LDL cholesterol contributes to plaque formation as atherosclerosis proceeds. In contrast, HDL cholesterol is "good" because clinical studies have shown that high levels of this form appear to provide some protection against coronary heart disease. Simplifying, HDL cholesterol removes excess cholesterol from plaques in arteries, thereby reducing plaque buildup. The cholesterol that has been removed is transported by the HDL cholesterol to the liver where it is broken down.

Fats in food affect cholesterol levels in the blood. Diets high in saturated fats raise LDL cholesterol levels, but levels of HDL cholesterol appear not to be affected by such a diet. Foods of animal origin typically contain saturated fats, and foods of plant origin typically contain unsaturated fats.

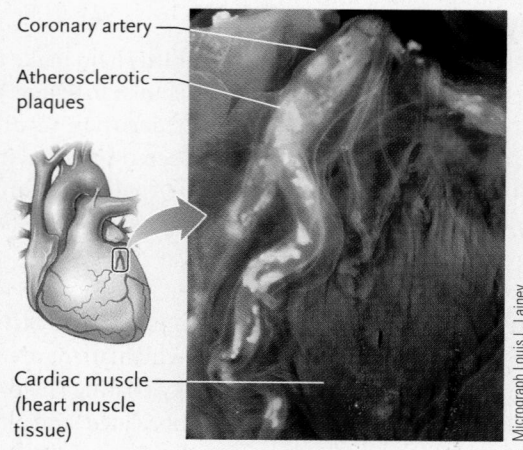

Coronary artery

Atherosclerotic plaques

Cardiac muscle (heart muscle tissue)

Micrograph Louis L. Lainey

Atherosclerotic plaques (bright areas) in the coronary arteries of a patient with heart disease.

In the food industry, unsaturated vegetable oils are often processed to solidify the fats. The process, partial hydrogenation, adds hydrogen atoms to unsaturated sites, eliminating many double bonds and generating substances known as *trans* fatty acids (or *trans* fats). Usually the hydrogen atoms at a double bond are positioned on the same side of the carbon chain, producing a *cis* (Latin, "on the same side") fatty acid:

```
 H   H
 |   |
—C═C—
```

but in a *trans* (Latin, "across") fatty acid, the hydrogen atoms are on different sides of the chain at some double bonds:

```
  H
  |
—C═C—
     |
     H
```

Trans fatty acids are found in many vegetable shortenings, some margarines, cookies, cakes, doughnuts, and other foods made with or fried in partially hydrogenated fats.

Research from human feeding studies has shown that *trans* fatty acids raise LDL cholesterol levels nearly as much as saturated fatty acids do. More seriously, intake of *trans* fatty acids at levels found in a typical U.S. diet also appears to reduce HDL cholesterol levels. In addition, clinical studies have demonstrated a positive correlation between the intake of *trans* fatty acids and the occurrence of coronary heart disease. A regulation to add the *trans* fatty acid content to nutritional labels went into effect in the United States in January 2006. A number of federal and state agencies and cities have or are considering legislation to ban *trans* fatty acids in food. For instance, in 2008, California became the first state to ban *trans* fatty acids in restaurants, following a number of cities around the country.

One of the many unanswered questions about dietary cholesterol is why some populations can consume large quantities of fatty foods of the "wrong" kind yet rarely develop atherosclerosis. For example, atherosclerosis was once virtually nonexistent in Inuits, whose diet in their native culture contained more than 90% animal fat; however, atherosclerosis developed in that same population when they adopted a "civilized" diet and lifestyle. In France, the incidence of atherosclerosis is relatively low even though cheese and other dairy products are diet staples. Of course, the French say that wine keeps them healthy!

In humans, earwax lubricates the outer ear canal and protects the eardrum. Honeybees use a wax secreted by glands in their abdomen to construct the comb in which larvae are raised and honey is stored **(Figure 3.13A)**.

Many plants secrete waxes that form a protective exterior layer, which greatly reduces water loss from the plants and resists invasion by infective agents such as bacteria and viruses. This waxy covering gives cherries, apples, and many other fruits their shiny appearance **(Figure 3.13B)**.

Phospholipids Provide the Framework of Biological Membranes

Phosphate-containing lipids called **phospholipids** are the primary lipids of cell membranes. In the most common phospholipids, glycerol forms the backbone of the molecule as in triglycerides, but only two of its binding sites are linked to fatty acids **(Figure 3.14)**. The third site is linked to a polar phosphate group, which binds to yet another polar unit. The end of the

A. **Comb constructed by wax secretion from honeybee abdominal glands**

B. **Waxy cuticle of cherries**

FIGURE 3.13
Waxy structures in nature.

molecule containing the fatty acids is nonpolar and hydrophobic, and the end with the phosphate group is polar and hydrophilic.

In polar environments, such as a water solution, phospholipids assume arrangements in which only their polar ends are exposed to the water; their nonpolar ends collect together in a region that excludes water. One of these arrangements, the *bilayer,* is the structural basis of membranes, the organizing boundaries of all living cells (see Figure 2.14). In a bilayer, formed by a film of phospholipids just two molecules thick, the phospholipid molecules are aligned so that the polar groups face the surrounding water molecules at the surfaces of the bilayer. The hydrocarbon chains of the phospholipids are packed together in the interior of the bilayer, where they form a nonpolar, hydrophobic region that excludes water. The bilayer remains stable because, if disturbed, the hydrophobic, nonpolar hydrocarbon chains of the phospholipids become exposed to the surrounding watery solution, and the molecule returns to its normal bilayer arrangement.

Steroids Contribute to Membrane Structure and Work as Hormones

Steroids are a group of lipids with structures based on a framework of four carbon rings **(Figure 3.15A)**. Small differences in the side groups attached to the rings distinguish one steroid from another. The most abundant steroids, the **sterols,** have a

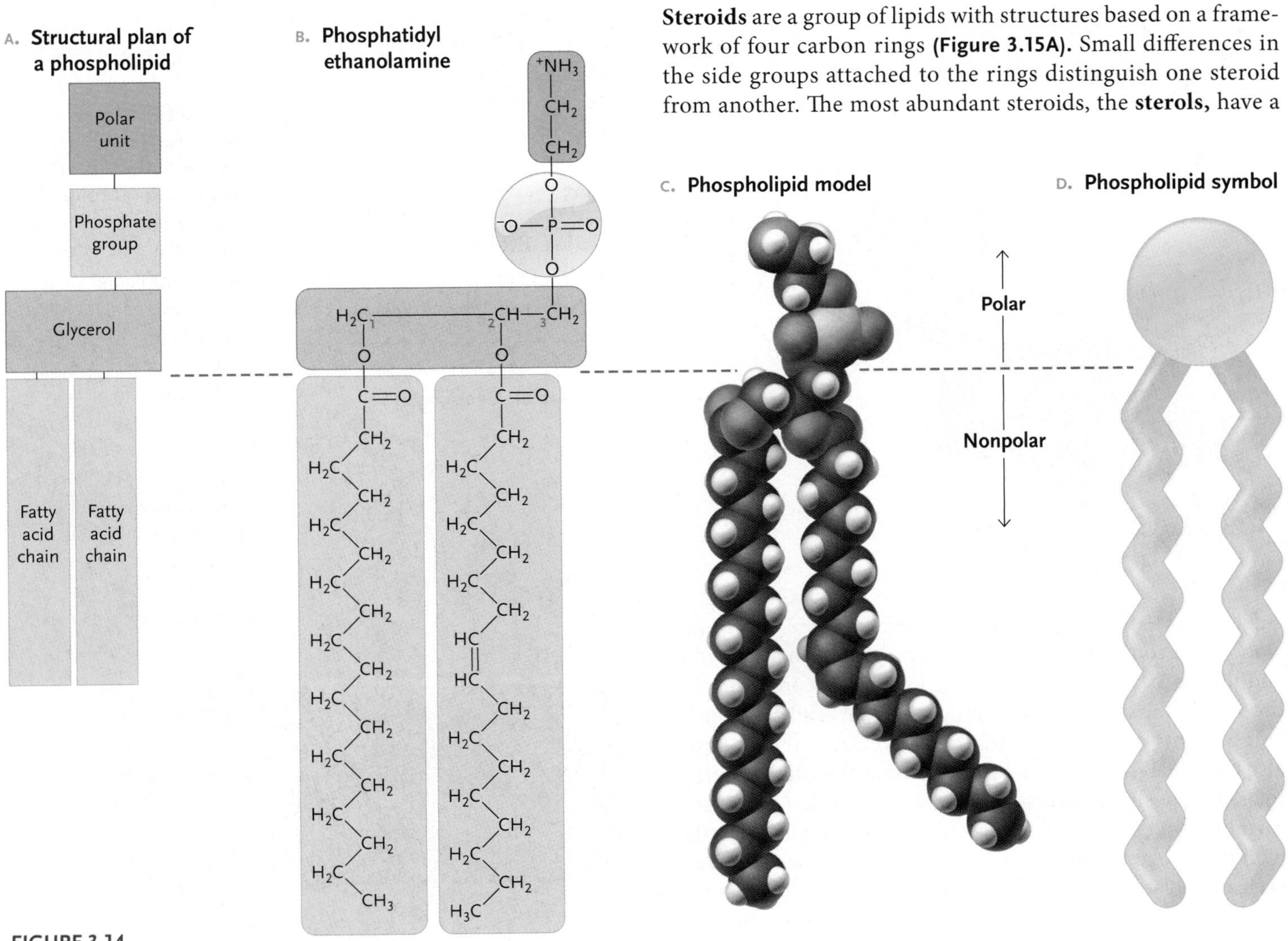

A. **Structural plan of a phospholipid**

Polar unit

Phosphate group

Glycerol

Fatty acid chain

Fatty acid chain

B. **Phosphatidyl ethanolamine**

C. **Phospholipid model**

Polar

Nonpolar

D. **Phospholipid symbol**

FIGURE 3.14
Phospholipid structure. (A) The arrangement of components in phospholipids. **(B)** Phosphatidyl ethanolamine, a common membrane phospholipid. **(C)** Space-filling model of phosphatidyl ethanolamine. The kink in the fatty acid chain on the right reflects a double bond at this position. **(D)** Diagram widely used to depict a phospholipid molecule in cell membrane diagrams. The sphere represents the polar end of the molecule, and the zigzag lines represent the nonpolar fatty acid chains. (The kink in the fatty acid chain is not depicted.)

A. Arrangement of carbon rings in a steroid

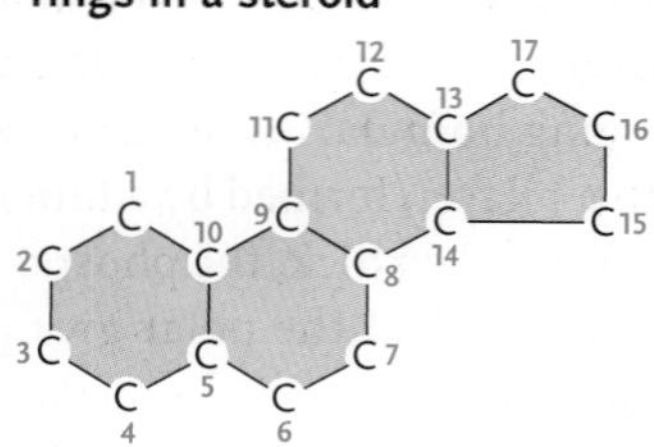
12
17
C
13
C
11C
C
C16
1
C
10
9C
C
C15
2C
C
C8
14
3C
C
C7
C
5
C
4
6

B. Cholesterol, a sterol

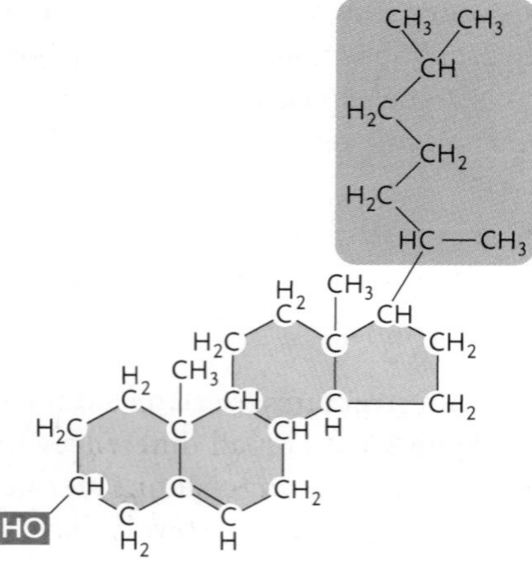
CH3 CH3
CH
H2C
CH2
H2C
HC—CH3
H2 CH3
C
CH
H2C
C
CH2
H2 CH3
C
CH
C
CH2
H2C
C
CH H
CH
C
CH2
HO
C
H2
C
H

C. Cholesterol model

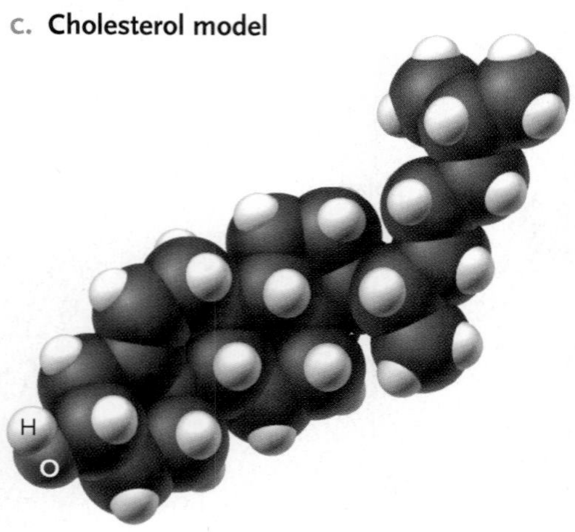
H
O

FIGURE 3.15
Steroids. **(A)** Typical arrangement of four carbon rings in a steroid molecule. **(B)** A sterol, cholesterol. Sterols have a hydrocarbon side chain linked to the ring structure at one end and a single —OH group at the other end (boxed in red). The —OH group makes its end of a sterol slightly polar. The rest of the molecule is nonpolar. **(C)** A space-filling model of cholesterol.

single polar —OH group linked to one end of the ring framework and a complex, nonpolar hydrocarbon chain at the other end **(Figure 3.15B)**. Although sterols are almost completely hydrophobic, the single hydroxyl group gives one end of the molecule a slightly polar, hydrophilic character. As a result, sterols also have dual solubility properties and, like phospholipids, tend to assume positions that satisfy these properties. In biological membranes, they line up beside the phospholipid molecules with their polar —OH group facing the membrane surface and their nonpolar ends buried in the nonpolar membrane interior.

Cholesterol (see **Figure 3.15B, C**) is an important component of the boundary membrane surrounding animal cells; similar sterols, called **phytosterols,** occur in plant cell membranes. Deposits derived from cholesterol also collect inside arteries in atherosclerosis (see *Focus on Applied Research*).

Other steroids, the *steroid hormones,* are important regulatory molecules in animals; they control development, behavior, and many internal biochemical processes. The sex hormones that control differentiation of the sexes and sexual behavior are primary examples of steroid hormones **(Figure 3.16)**. Small differences in the functional groups of steroid hormones have vastly different effects in animals. For instance, the two key differences between the estrogen estradiol, the primary female sex hormone, and the androgen testosterone, the male sex hormone, are that estradiol has an —OH in the position where testosterone has an =O, and testosterone has a methyl group (—CH_3) that is absent from estradiol.

Bodybuilders and other athletes sometimes use hormone-like steroids (anabolic-androgenic steroids) to increase their muscle mass (see *Focus on Basic Research* in Chapter 40). Unfortunately, these substances also produce numerous side effects, including elevated cholesterol, elevated blood pressure, and acne. Other steroids occur as poisons in the venoms of toads and other animals.

Several other lipid types have structures unrelated to triglycerides, phospholipids, or steroids. Among these are *chlorophylls* and *carotenoids,* pigments that absorb light and participate in its conversion to chemical energy in plants (see Chapter 9). Lipid groups also combine with carbohydrates to form *glycolipids* and with proteins to form *lipoproteins.* Both glycolipids and lipoproteins form parts of cell membranes, where they perform vital structural and functional roles.

STUDY BREAK 3.3 <

What are the three most common lipids in living organisms? Distinguish between their structures.

A. Estradiol, an estrogen

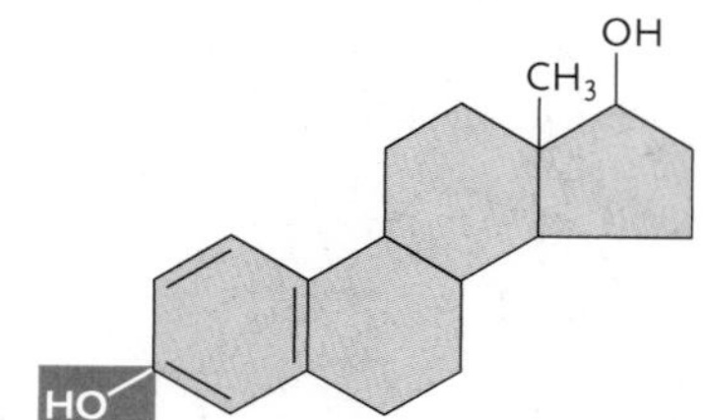
OH
CH3
HO

B. Testosterone

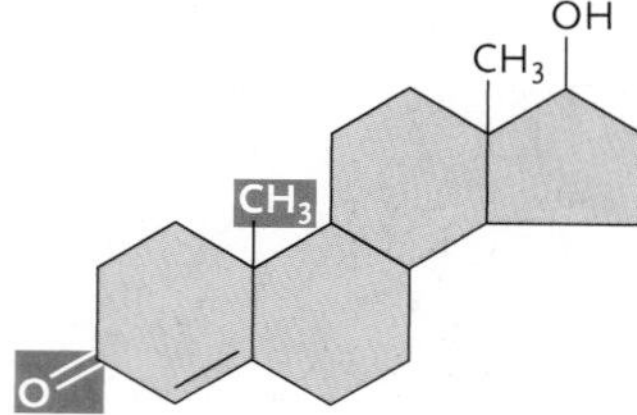
OH
CH3
CH3
O

C.

Female wood duck
Male wood duck

Tim Davis/Photo Researchers, Inc.

FIGURE 3.16
Steroid sex hormones and their effects. The female sex hormone, estradiol **(A)**, and the male sex hormone, testosterone **(B)**, differ only in substitution of an —OH group for an oxygen and the absence of one methyl group (—CH_3) in the estrogen. Although small, these differences greatly alter sexual structures and behavior in animals, such as humans, and the wood ducks *(Aix sponsa)* shown in **(C)**.

3.4 Proteins

Proteins perform many vital functions in living organisms **(Table 3.2)**. Some provide structural support for cells; others, called **enzymes,** increase the rate of cellular reactions; still others impart movement to cells and cellular structures. Proteins also transport substances across biological membranes, serve as recognition and receptor molecules at cell surfaces, or regulate the activity of other proteins and DNA. Some proteins work as hormones or defend against foreign substances, such as infectious microorganisms. Many toxins and venoms are also based on proteins.

All of the protein molecules that carry out these and other functions are fundamentally similar in structure. All are macromolecules—polymers consisting of one or more unbranched chains of monomers called amino acids. An **amino acid** is a molecule that contains both an amino and a carboxyl group. Although the most common proteins contain 50 to 1,000 amino acids, some proteins found in nature have as few as 3 or as many as 50,000 amino acid units. Proteins range in shape from globular or spherical forms to elongated fibers, and they vary from soluble to completely insoluble in water solutions. Some proteins have single functions, whereas others have multiple functions.

Cells Assemble 20 Kinds of Amino Acids into Proteins by Forming Peptide Bonds

The cells of all organisms use 20 different amino acids as the initial building blocks of proteins **(Figure 3.17)**. Of these 20 amino acids, 19 have the same structural plan—a central carbon atom is attached to an amino group (NH_2), a carboxyl group (—COOH), and a hydrogen atom:

$$\begin{array}{c} \text{R} \\ | \\ \text{H}_2\text{N—C—COOH} \\ | \\ \text{H} \end{array}$$

TABLE 3.2 Major Protein Functions

Protein Type	Function	Examples
Structural proteins	Support	Microtubule and microfilament proteins form supporting fibers inside cells; collagen and other proteins surround and support animal cells; cell wall proteins support plant cells.
Enzymatic proteins	Increase the rate of biological reactions	Among thousands of examples, DNA polymerase increases the rate of duplication of DNA molecules; RuBP (ribulose 1,5-bisphosphate) carboxylase/oxygenase increases the rates of the first synthetic reactions of photosynthesis; the digestive enzymes lipases and proteases increase the rate of breakdown of fats and proteins, respectively.
Membrane transport proteins	Speed up movement of substances across biological membranes	Ion transporters move ions such as Na^+, K^+, and Ca^{2+} across membranes; glucose transporters move glucose into cells; aquaporins allow water molecules to move across membranes.
Motile proteins	Produce cellular movements	Myosin acts on microfilaments (called thin filaments in muscle) to produce muscle movements; dynein acts on microtubules to produce the whipping movements of sperm tails, flagella, and cilia (the last two are whiplike appendages on the surfaces of many eukaryotic cells); kinesin acts on microtubules of the cytoskeleton (the three-dimensional scaffolding of eukaryotic cells responsible for cellular movement, cell division, and the organization of organelles).
Regulatory proteins	Promote or inhibit the activity of other cellular molecules	Nuclear regulatory proteins turn genes on or off to control the activity of DNA; protein kinases add phosphate groups to other proteins to modify their activity.
Receptor proteins	Bind molecules at cell surface or within cell; some trigger internal cellular responses	Hormone receptors bind hormones at the cell surface or within cells and trigger cellular responses; cellular adhesion molecules help hold cells together by binding molecules on other cells; LDL receptors bind cholesterol-containing particles to cell surfaces.
Hormones	Carry regulatory signals between cells	Insulin regulates sugar levels in the bloodstream; growth hormone regulates cellular growth and division.
Antibodies	Defend against invading molecules and organisms	Antibodies recognize, bind, and help eliminate essentially any protein of infecting bacteria and viruses, and many other types of molecules, both natural and artificial.
Storage proteins	Hold amino acids and other substances in stored form	Ovalbumin is a storage protein of eggs; apolipoproteins hold cholesterol in stored form for transport through the bloodstream.
Venoms and toxins	Interfere with competing organisms	Ricin is a castor-bean protein that stops protein synthesis; bungarotoxin is a snake venom that causes muscle paralysis.

FIGURE 3.17
The 20 amino acids used by cells to make proteins. The side group of each amino acid is boxed in brown. The amino acids are shown in the ionic forms in which they are found at the pH within the cell; the amino group becomes $—NH_3^+$, and the carboxyl group becomes $—COO^-$. Three-letter and one-letter abbreviations commonly used for the amino acids appear below each diagram. All amino acids assembled into proteins are in the L-form, one of two possible enantiomers.

Nonpolar amino acids

Alanine Ala A
Valine Val V
Leucine Leu L
Isoleucine Ile I
Glycine Gly G
Cysteine Cys C
Phenylalanine Phe F
Tryptophan Trp W
Methionine Met M
Proline Pro P

Uncharged polar amino acids

Serine Ser S
Threonine Thr T
Tyrosine Tyr Y
Asparagine Asn N
Glutamine Gln Q

Negatively charged (acidic) polar amino acids

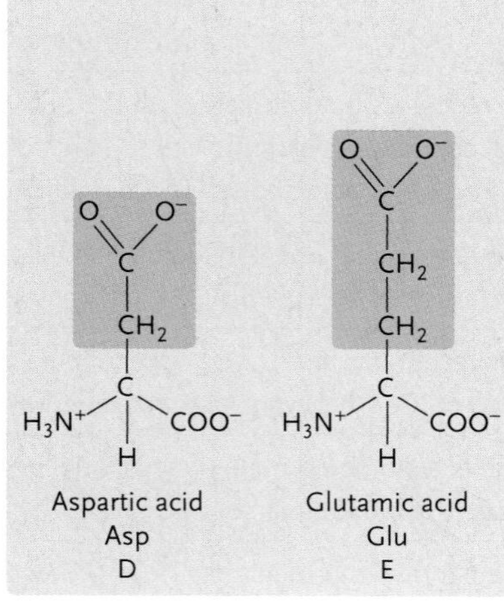
Aspartic acid
Asp
D
Glutamic acid
Glu
E

Positively charged (basic) polar amino acids

Lysine Lys K
Arginine Arg R
Histidine His H

The remaining bond of the central carbon is linked to 1 of 19 different side groups represented by the *R* (see shaded regions in Figure 3.17); its usage for amino acids refers to a range from a single hydrogen atom to complex carbon-containing chains or rings. The remaining amino acid, proline, differs slightly in that it has a ring structure that includes the central carbon atom; the central carbon bonds to a —COOH group on one side and to an =NH (imino) group that forms part of the ring at the other side (see Figure 3.17). (Strictly speaking, proline is an imino acid.) Although they are called acids, all 20 of the amino acids can act as either acids or bases—depending on cellular conditions, the amino group can produce a basic reaction by accepting H^+, or the carboxyl group can produce an acidic reaction by releasing H^+.

Differences in the side groups give the amino acids their individual properties. Some side groups are polar, and some are nonpolar; among the polar side groups, some carry a positive or negative charge and some act as acids or bases (see Figure 3.17). Many of the side groups contain reactive functional groups, such as $—NH_2$, —OH, —COOH, or —SH, which may interact with atoms located elsewhere in the same protein or with molecules and ions outside the protein.

The sulfhydryl group (—SH) in the amino acid cysteine is particularly important in protein structure. The sulfhydryl groups in the side groups of two cysteines located in different regions of the same protein, or in different proteins, can react to produce disulfide linkages (—S—S—). The linkages fasten amino acid chains together **(Figure 3.18)** and help hold proteins in their three-dimensional shape.

Overall, the varied properties and functions of proteins depend on the types and locations of the different amino acid side groups in their structures. The variations in the number and types of amino acids mean that the total number of possible proteins is extremely large.

Covalent bonds link amino acids into the chains of subunits that make proteins. The link, a **peptide bond,** is formed by a dehydration synthesis reaction between the —NH_2 group of one amino acid and the —COOH group of a second **(Figure 3.19).** An amino acid chain always has an —NH_2 group (—NH_3^+ in ionized form) at one end, called the **N-terminal end,** and a —COOH group (—COO^- in ionized form) at the other end, called the **C-terminal end.** In cells, amino acids link only to the —COOH end of another amino acid.

The chain of amino acids formed by sequential peptide bonds, that is, a **polypeptide,** is only part of the complex structure of proteins. Once assembled, an amino acid chain may fold in various patterns, and more than one chain may combine to form a finished protein, adding to the structural and functional variability of proteins.

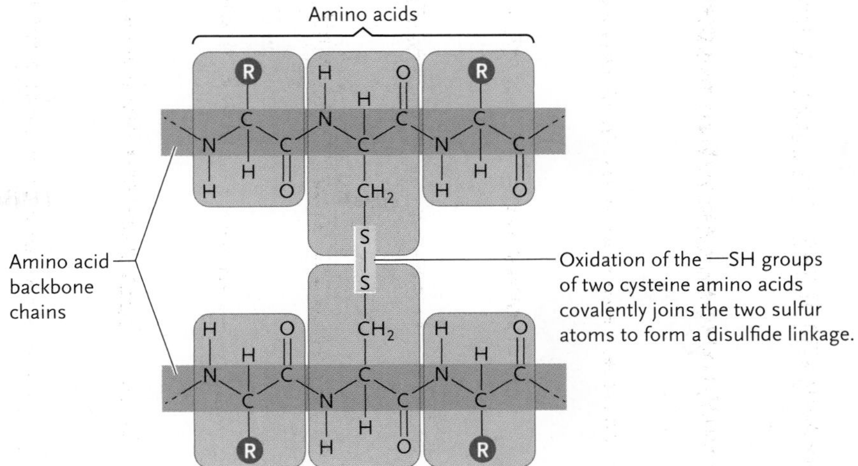

Amino acids
Amino acid backbone chains
Oxidation of the —SH groups of two cysteine amino acids covalently joins the two sulfur atoms to form a disulfide linkage.

FIGURE 3.18
A disulfide linkage between two amino acid chains or two regions of the same chain. The linkage is formed by a reaction between the sulfhydryl groups (—SH) of cysteines. The circled *R*s indicate the side groups of other amino acids in the chains. Figure 3.21 shows disulfide linkages in a real protein, and Figure 3.24 shows how disulfide linkages help maintain a protein's conformation.

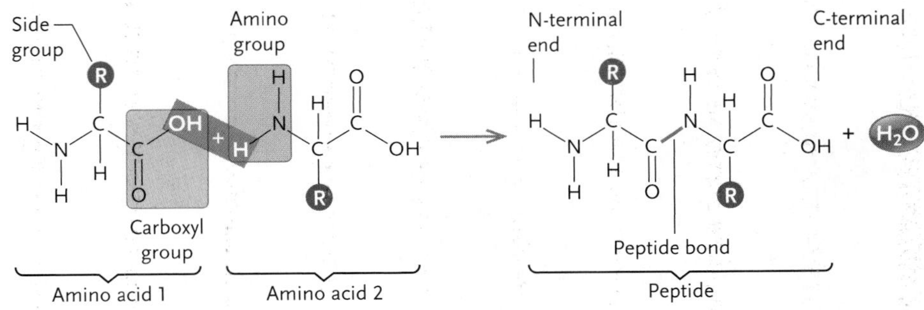

Side group
Amino group
Carboxyl group
Amino acid 1
Amino acid 2
N-terminal end
C-terminal end
Peptide bond
Peptide

FIGURE 3.19
A peptide bond formed by reaction of the carboxyl group of one amino acid with the amino group of a second amino acid. The reaction is a typical dehydration synthesis reaction.

Proteins Have as Many as Four Levels of Structure

Proteins potentially have four levels of structure, with each level imparting different characteristics and degrees of structural complexity to the molecule **(Figure 3.20). Primary structure** is the particular and unique sequence of amino acids forming a polypeptide; **secondary structure** is produced by the twists and turns of the amino acid chain. **Tertiary structure** is the folding of the amino acid chain, with its secondary structures, into the overall three-dimensional shape of a protein. All proteins have primary, secondary, and tertiary structures. **Quaternary structure,** when present, refers to the arrangement of polypeptide chains in a protein that is formed from more than one chain.

Primary Structure Is the Fundamental Determinant of Protein Form and Function

The primary structure of a protein—the sequence in which amino acids are linked—underlies the other, higher levels of structure. Changing even a single amino acid of the primary structure alters the secondary, tertiary, and quaternary structures to at least some degree and, by so doing, can alter or even destroy the biological functions of a protein. For example, substitution of a single amino acid in the blood protein hemoglobin produces an altered form responsible for sickle-cell disease (see Chapter 15); a number of other blood disorders are caused by single amino acid substitutions in other parts of the protein.

Because primary structure is so fundamentally important, many years of intensive research have been devoted to determining the amino acid sequence of proteins. Initial success came in 1953, when the English biochemist Frederick Sanger deduced the amino acid sequence of insulin, a protein-based hormone, using samples obtained from cows **(Figure 3.21).** Now, the amino acid sequences of literally thousands of proteins have been determined, and more are constantly being added to the list. Knowledge of the primary structure of proteins often allows their three-dimensional structure and functions to be predicted and reveals relationships among proteins.

--- Ser | Glu | Gly | Asp | Trp | Gln | Leu | His ---

A. Primary structure: the sequence of amino acids in a protein

B. Secondary structure: regions of alpha helix, beta strand, or random coil in a polypeptide chain

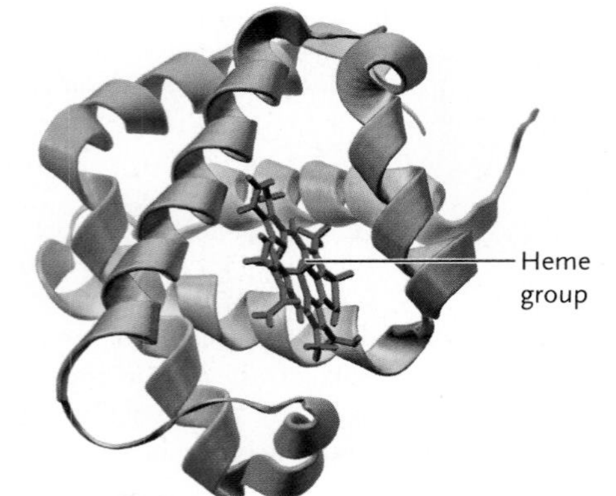

Heme group

C. Tertiary structure: overall three-dimensional folding of a polypeptide chain

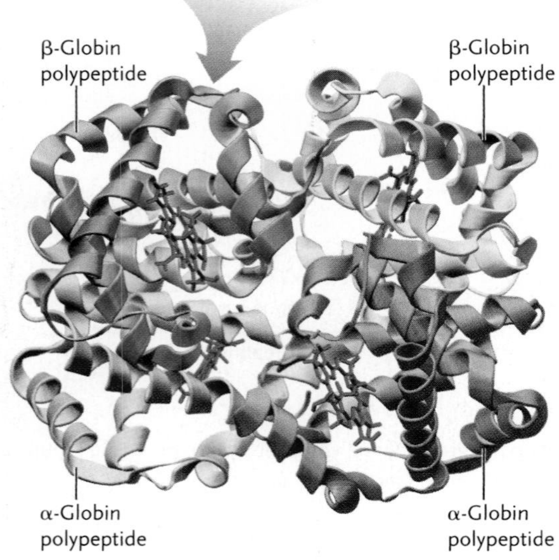

β-Globin polypeptide
β-Globin polypeptide
α-Globin polypeptide
α-Globin polypeptide

D. Quaternary structure: the arrangement of polypeptide chains in a protein that contains more than one chain

FIGURE 3.20
The four levels of protein structure. The protein shown in **(C)** is one of the subunits of a hemoglobin molecule; the heme group (in red) is an iron-containing group that binds oxygen. **(D)** A complete hemoglobin molecule.

Twists and Other Arrangements of the Amino Acid Chain Form the Secondary Structure of a Protein

The amino acid chain of a protein, rather than being stretched out in linear form, is folded into arrangements that form the protein's secondary structure. Two highly regular secondary structures, the *alpha helix* and the *beta strand,* are particularly stable and make an amino acid chain resistant to bending. A third, less regular arrangement, the *random coil* or *loop,* provides flexible regions that allow sections of amino acid chains containing them to bend. Most proteins have segments of all three arrangements.

THE ALPHA HELIX In the **alpha (α) helix,** first identified by Linus Pauling and Robert Corey at the California Institute of Technology in 1951, the backbone of the amino acid chain is twisted into a regular, right-hand spiral **(Figure 3.22).** The amino acid side groups extend outward from the twisted backbone. The structure is stabilized by regularly spaced hydrogen bonds (see dotted lines in Figure 3.22) between atoms in the backbone.

Most proteins contain segments of α helix, which are rigid and rodlike, in at least some regions. Globular proteins usually contain several short α-helical segments that run in different directions, connected by segments of random coil. Fibrous proteins, such as the collagens, a major component of tendons, bone, and other extracellular structures in animals, typically contain one or more α-helical segments that run the length of the molecule, with few or no bendable regions of random coil.

THE BETA STRAND Pauling and Corey were also the first to identify the beta (β) strand as a major secondary protein structure. In a β strand, the amino acid chain zigzags in a flat plane rather than twisting into a coil.

In many proteins, β strands are aligned side by side in the same or opposite directions to form a structure known as a **beta (β) sheet (Figure 3.23).** Hydrogen bonds between adjacent β strands stabilize the sheet, making it a highly rigid structure. Beta sheets may lie in a flat plane or may twist into propeller- or barrel-like structures.

Beta strands and sheets occur in many proteins, usually in combination with α-helical segments. One notable exception is

FIGURE 3.21
The primary structure of the peptide hormone insulin, which consists of two polypeptide chains connected by disulfide linkages. (Bovine insulin is shown.)

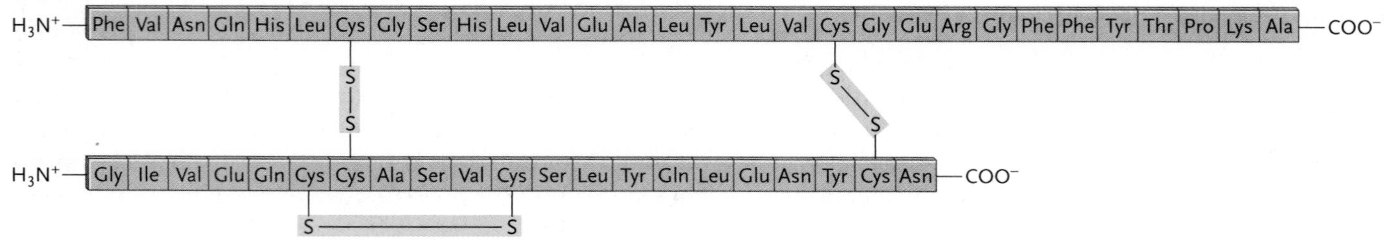

H_3N^+— Phe Val Asn Gln His Leu Cys Gly Ser His Leu Val Glu Ala Leu Tyr Leu Val Cys Gly Glu Arg Gly Phe Phe Tyr Thr Pro Lys Ala —COO^-
S—S
S—S
H_3N^+— Gly Ile Val Glu Gln Cys Cys Ala Ser Val Cys Ser Leu Tyr Gln Leu Glu Asn Tyr Cys Asn —COO^-
S—S

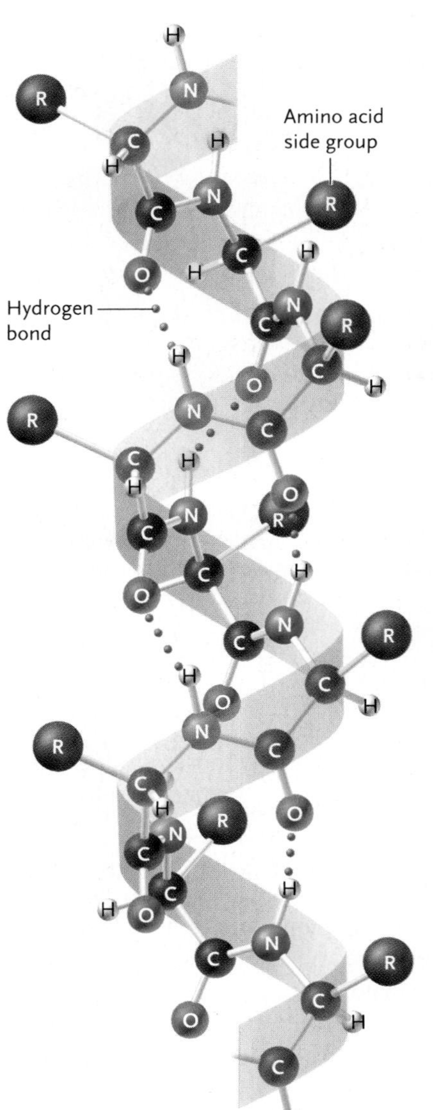
A. Ball-and-stick model of α helix
Amino acid side group
Hydrogen bond

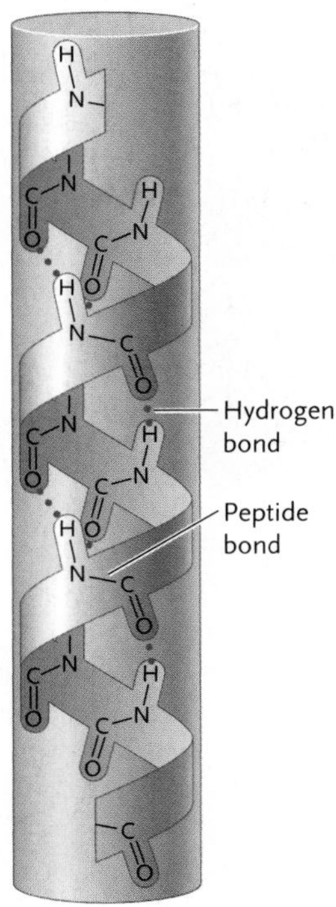
B. Cylinder representation of α helix
Hydrogen bond
Peptide bond

FIGURE 3.22
The α helix, a type of secondary structure in proteins. **(A)** A model of the α helix showing atoms as spheres and covalent bonds as rods. The backbone of the amino acid chain is held in a spiral by hydrogen bonds formed at regular intervals. **(B)** The cylinder often is used to depict an α helix in protein diagrams, with peptide and hydrogen bonds also shown.

FIGURE 3.23
A β sheet formed by side-by-side alignment of two β strands, a type of secondary structure in proteins. The β strands are held together stably by hydrogen bonds. In this sheet, the β strands run in opposite directions, as shown by the arrows, which point in the direction of the C-terminal end of each polypeptide chain. Strands may also run in the same direction in a β sheet. Arrows alone often are used to represent β strands in protein diagrams.

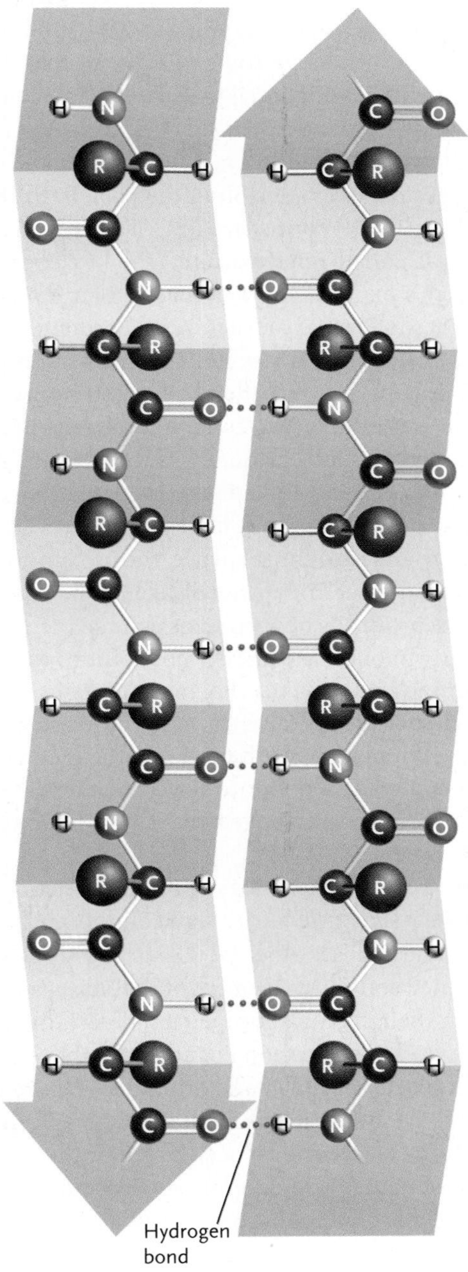
Hydrogen bond

in the silk protein secreted by silk worms, which contains only β sheets. This exceptionally stable structure, reinforced by an extensive network of hydrogen bonds, underlies the unusually high tensile strength of silk fibers.

THE RANDOM COIL In a **random coil** the amino acid chain has an irregularly folded arrangement. The amino acid proline is often present in random-coil structures. Its ring form does not fit into an α helix or β sheet, and it has no sites available for formation of stabilizing hydrogen bonds.

Segments of random coil provide flexible sites that allow α-helical or β-strand segments to bend or fold back on themselves. The fold-back loops of random coils often occur at the surfaces of proteins, at points where they link segments of α helix or β strand located deeper in the protein. Segments of random coil also commonly act as "hinges" that allow major parts of proteins to move with respect to one another.

The Tertiary Structure of a Protein Is Its Overall Three-Dimensional Conformation

The tertiary structure of a protein is its overall three-dimensional shape, or **conformation (Figure 3.24).** The contents of α-helical, β-strand, and random-coil segments, together with the number and position of disulfide linkages and hydrogen bonds, play the major roles in folding each protein into its tertiary structure. Attractions between positively and negatively charged side groups and polar or nonpolar associations also contribute to the tertiary structure.

The first insight into how a protein assumes its tertiary structure came from a classic experiment published by Christian Anfinsen and Edgar Haber of the National Institutes of Health in 1962 **(Figure 3.25)**. The researchers studied ribonuclease, an enzyme that hydrolyzes RNA. When they treated the enzyme chemically to break the disulfide linkages holding the protein in its functional state, the protein unfolded and had no enzyme activity. Unfolding a protein from its active conformation so that it loses its structure and function is called **denaturation.** When they removed the denaturing chemicals, the ribonuclease slowly regained full activity because the disulfide linkages reformed, enabling the protein to reassume its functional conformation. The reversal of denaturation is called **renaturation.** The key conclusion from Anfinsen's experiment was that the amino acid sequence specifies the tertiary structure of a protein. Christian Anfinsen received a Nobel Prize in 1972 for this work.

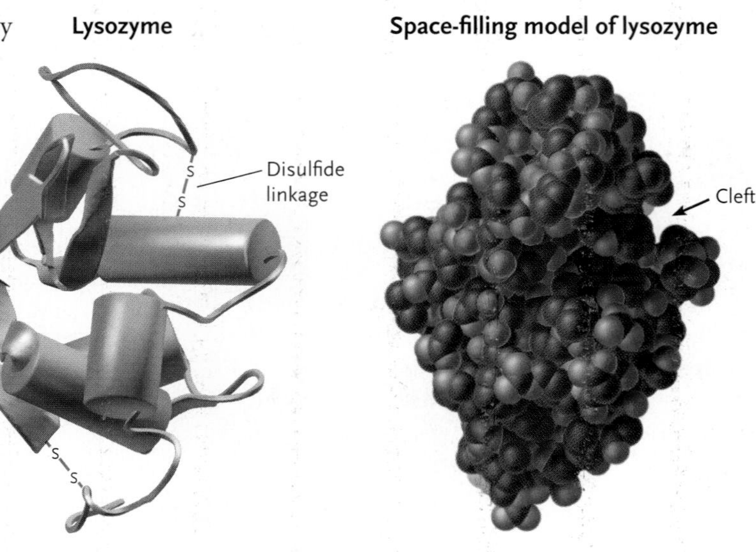
Lysozyme
Space-filling model of lysozyme
S
S
Disulfide linkage
S
S
Cleft

FIGURE 3.24
Tertiary structure of the protein lysozyme, with α helices shown as cylinders, β strands as arrows, and random coils as ropes. Lysozyme is an enzyme found in nasal mucus, tears, and other body secretions; it destroys the cell walls of bacteria by breaking down molecules in the wall. Disulfide bonds are shown in red. A space-filling model of lysozyme is shown for comparison.

How do proteins fold into their tertiary structure in the cell? This question is the subject of contemporary research. Results indicate that proteins fold gradually as they are assembled—as successive amino acids are linked into the primary structure, the chain folds into increasingly complex structures. As the final amino acids are added to the sequence, the protein completes its folding into final three-dimensional form. One nagging question about this process is how proteins assume their correct tertiary structure among the different possibilities that may exist for a given amino acid sequence. For many proteins, "guide" proteins called **chaperone proteins** or **chaperonins** solve this problem; they bind temporarily with newly synthesized proteins, directing their conformation toward the correct tertiary structure and inhibiting incorrect arrangements as the new proteins fold **(Figure 3.26)**.

Tertiary structure determines a protein's function. That is, a protein's tertiary structure buries some amino acid side groups in its interior and exposes others at the surface. The distribution and three-dimensional arrangement of the side groups, in combination with their chemical properties, determine the overall chemical activity of the protein. For example, the tertiary structure of the antibacterial enzyme lysozyme (see Figure 3.24) has a cleft that binds a polysaccharide found in bacterial cell walls; hydrolysis of the polysaccharide is accelerated by the enzyme.

Tertiary structure also determines the solubility of a protein. Water-soluble proteins have mostly polar or charged amino acid side groups exposed at their surfaces, whereas nonpolar side groups are clustered in the interior. Proteins embedded in nonpolar membranes are arranged in patterns similar to phospholipids, with their polar (hydrophilic) segments facing the surrounding watery solution and their nonpolar surfaces embedded in the nonpolar (hydrophobic) membrane interior. These dual-solubility proteins perform many important functions in membranes, such as transporting ions and molecules into and out of cells.

The tertiary structure of most proteins is flexible, allowing them to undergo limited alterations in three-dimensional shape known as **conformational changes.** These changes contribute to the function of many proteins, particularly those working as enzymes, in cellular movements or in the transport of substances across cell membranes.

As we learned earlier, chemical treatment can denature a protein in the test tube (see Figure 3.25). Excessive heat can also break the hydrogen bonds holding a protein in its natural conformation, causing it to denature and lose its biological activity. Denaturation is one of the major reasons few living organisms can tolerate temperatures greater than 45°C. Extreme changes in pH, which alter the charge of amino acid side groups and weaken or destroy ionic bonds, can also cause protein denaturation.

For some proteins, denaturation is permanent. A familiar example of a permanently denatured protein is a cooked egg white. In its natural form, the egg white protein albumin dissolves in water to form a clear solution. The heat of cooking denatures it permanently into an insoluble, whitish mass. For other proteins such as ribonuclease (mentioned previously), denaturation is reversible; the proteins can renature and return to their functional form if the temperature or pH returns to normal values.

Multiple Polypeptide Chains Form Quaternary Structure

Some complex proteins, such as hemoglobin and antibody molecules, have quaternary structure—that is, the presence and arrangement of two or more polypeptide chains (see Figure 3.20D). The same bonds and forces that fold single polypeptide chains into tertiary structures, including hydrogen bonds, polar and nonpolar attractions, and disulfide linkages, also hold the multiple polypeptide chains together. During the assembly of multichain proteins, chaperonins also promote correct association of the individual amino acid chains and inhibit incorrect formations.

FIGURE 3.25 **Experimental Research**

Anfinsen's Experiment Demonstrating That the Amino Acid Sequence of a Protein Specifies Its Tertiary Structure

Question: What is the relationship between the amino acid sequence of a protein and its conformation?

Experiment: Anfinsen and Haber studied the 124-amino acid enzyme ribonuclease in the test tube. They knew that the native (functional) enzyme has four disulfide linkages between amino acids 26 and 84, 40 and 95, 58 and 110, and 65 and 72 (see figure below). They treated the active enzyme with a mixture of urea and β-mercaptoethanol, which breaks disulfide linkages. They then removed the two chemicals and left the enzyme solution in air.

Result: The chemical treatment broke the four disulfide linkages, which caused the protein to denature and lose its enzyme activity. After the chemicals had been removed and the enzyme solution exposed to air, Anfinsen made the crucial observation that the protein slowly regained enzyme activity.

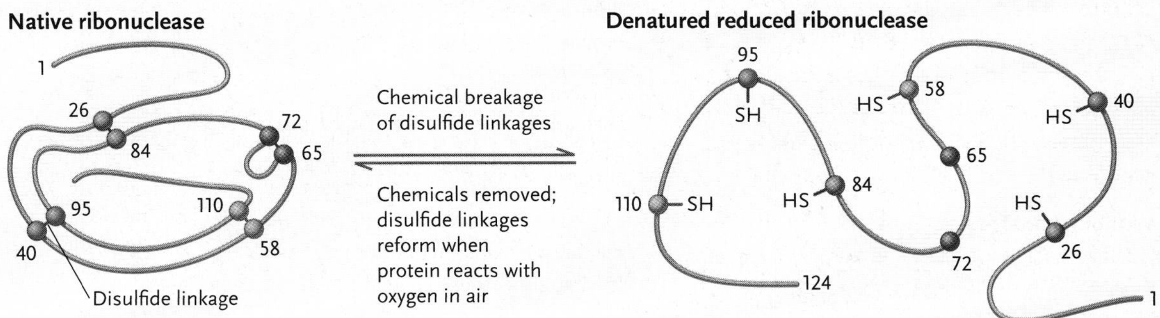
Native ribonuclease
1
26
84
72
65
95
110
40
58
Disulfide linkage
Chemical breakage of disulfide linkages
Chemicals removed; disulfide linkages reform when protein reacts with oxygen in air
Denatured reduced ribonuclease
95
SH
HS
58
HS
40
65
110
SH
HS
84
HS
72
26
124
1

He realized that oxygen from the air had reacted with the —SH groups of the denatured enzyme causing disulfide linkages to reform, and that the enzyme had spontaneously refolded into its native, active conformation. All physical and chemical properties of the refolded enzyme the researchers measured were the same as those of the native enzyme.

Conclusion: Anfinsen concluded that the information for determining the three-dimensional shape of ribonuclease is in its amino acid sequence.

Source: E. Haber and C. Anfinsen. 1962. Side-chain interactions governing the pairing of half-cystine residues in ribonuclease. *Journal of Biological Chemistry* 237:1839–1844.

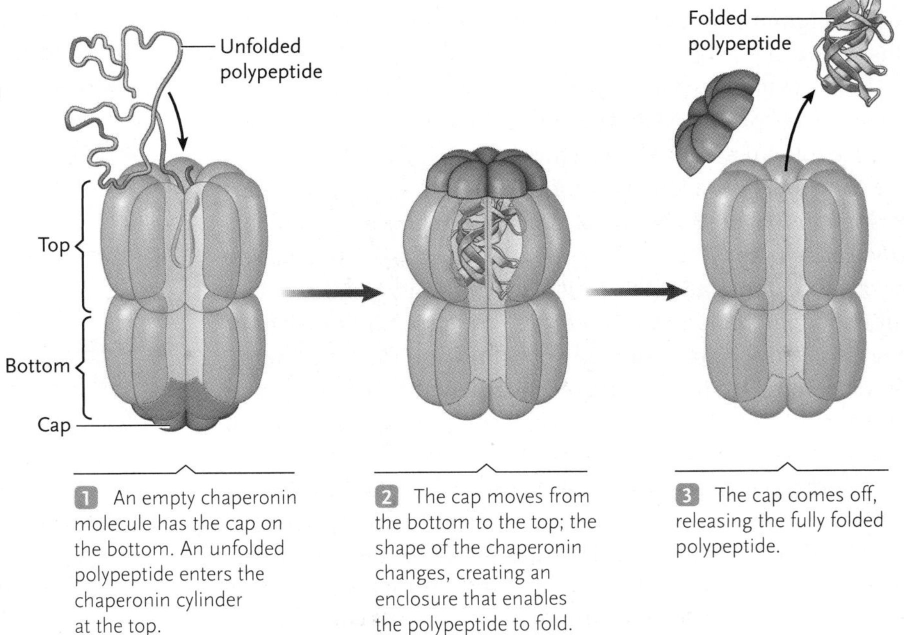
Unfolded polypeptide
Top
Bottom
Cap
Folded polypeptide
1 An empty chaperonin molecule has the cap on the bottom. An unfolded polypeptide enters the chaperonin cylinder at the top.
2 The cap moves from the bottom to the top; the shape of the chaperonin changes, creating an enclosure that enables the polypeptide to fold.
3 The cap comes off, releasing the fully folded polypeptide.

FIGURE 3.26
Role of a chaperonin in folding a polypeptide. The three parts of the chaperonin are the top and bottom, which form a cylinder, and the cap.

INSIGHTS FROM THE Molecular Revolution

A Big Bang in Protein Structure Evolution: How did the domain organization in proteins evolve?

Many proteins have distinct, large structural regions called domains (see Figure 3.27). In multifunctional proteins, different domains are responsible for the various functions. Further, proteins that have similar functions will have similar domains responsible for those functions. In other words, protein domains may be considered as units of structure and function that have combined to produce a wide variety of complex domain arrangements.

Research Question

How did the domain organization in proteins evolve?

Method of Analysis

Gustavo Caetano-Anollés and Minglei Wang at the University of Illinois at Urbana-Champaign used a bioinformatics approach to answer the question. Bioinformatics is a fusion of biology with mathematics and computer science used to manage and analyze biological data. Typically, bioinformatics analysis involves DNA sequences and/or protein sequences in comprehensive databases. The researchers analyzed domain structure and organization in proteins encoded in hundreds of fully sequenced genomes. From the data they reconstructed phylogenetic trees of protein structures. These phylogenetic trees laid out a rough timeline for, and details of, the evolution of domain organization in proteins.

Conclusion

The researchers discovered that, before the emergence of the three taxonomic domains—Bacteria, Archaea, and Eukarya (see Chapter 1)—most proteins contained only single domains that performed multiple tasks. With time, these protein domains began to combine with one another becoming more specialized. After a long period of gradual evolution, there was an explosion ("big bang") of change in which protein domains combined with each other or split apart to produce a wide range of novel domains, each typically more specialized than its ancestors. The explosion in the evolution of domain organization is of particular interest because it coincided with the rapidly increasing diversity of the three taxonomic domains. Thereafter, the protein domains diverged markedly from each other as the diversification of life continued. Diversification of protein domains was most extensive in the Eukarya, allowing eukaryotic organisms to exhibit functions of their proteins not possible in other organisms.

Source: M. Wang and G. Caetano-Anollés. 2009. The evolutionary mechanics of domain organization in proteomes and the rise of modularity in the protein world. *Structure* 17:66–78.

Combinations of Secondary, Tertiary, and Quaternary Structure Form Functional Domains in Many Proteins

In many proteins, folding of the amino acid chain (or chains) produces distinct, large structural subdivisions called **domains (Figure 3.27A).** Often, one domain of a protein is connected to another by a segment of random coil. The hinge formed by the flexible random coil allows domains to move with respect to one another. Hinged domains of this type are typical of proteins that produce motion and also occur in many enzymes. *Insights from the Molecular Revolution* discusses how the domain organization in proteins might have evolved.

Many proteins have multiple functions. For instance, the sperm surface protein SPAM1 (sperm adhesion molecule 1) plays multiple roles in mammalian fertilization. In proteins with multiple functions, individual functions are often located in different domains **(Figure 3.27B),** meaning that domains are functional as well as structural subdivisions. Different proteins often share one or more domains with particular functions. For example, a type of domain that releases energy to power biological reactions appears in similar form in many enzymes and motile proteins. The appearance of similar domains in different proteins suggests that the proteins may have evolved through a mechanism that mixes existing domains into new combinations.

The three-dimensional arrangement of amino acid chains within and between domains also produces highly specialized regions called **motifs.** Several types of motifs, each with a specialized function, occur in proteins. Examples are presented in Section 16.2.

Proteins Combine with Units Derived from Other Classes of Biological Molecules

We have already mentioned the linkage of proteins to lipids to form lipoproteins. Proteins also link with carbohydrates to form *glycoproteins,* which function as enzymes, antibodies, recognition and receptor molecules at the cell surface, and parts of extracellular supports such as collagen. In fact, most of the known proteins located at the cell surface or in the spaces between cells are glycoproteins. Linkage of proteins to nucleic acids produces *nucleoproteins,* which form such vital structures as *chromosomes,* the structures that organize DNA inside cells, and ribosomes, which carry out the process of protein synthesis in the cell.

This section has demonstrated the importance of amino acid sequence to the structure and function of proteins and highlighted the great variability in proteins produced by differences in their amino acid sequence. The next section considers the nucleic acids, which store and transmit the information required to arrange amino acids into particular sequences in proteins.

STUDY BREAK 3.4

1. **What gives amino acids their individual properties?**
2. **What is a peptide bond, and what type of reaction forms it?**
3. **What are functional domains of proteins, and how are they formed?**

THINK OUTSIDE THE BOOK

Collaboratively or on your own, investigate whether chaperonins are the same in bacteria, archaeans, and eukaryotes.

A. **Two domains in an enzyme that assembles DNA molecules**

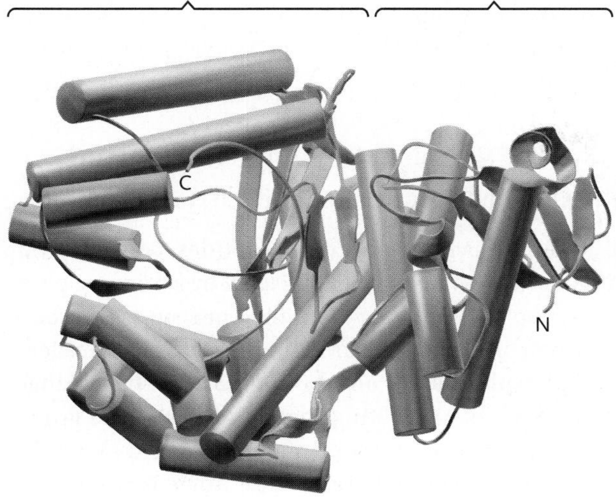
Domain a
Domain b
C
N

B. **The same protein, showing the domain surfaces**

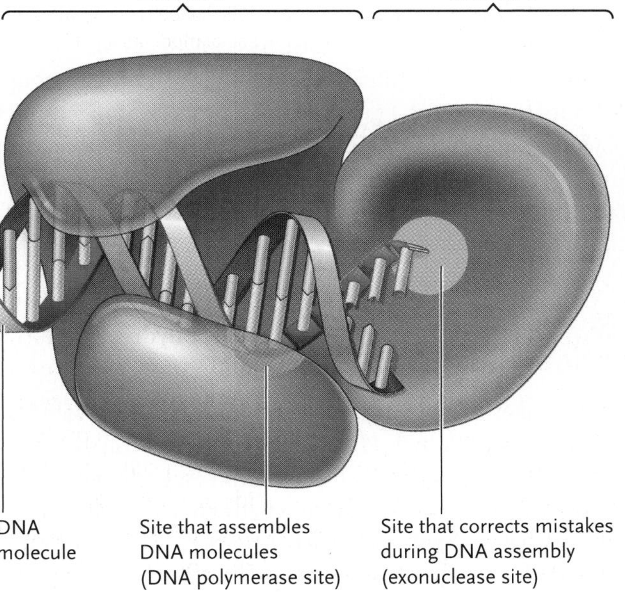
Domain a
Domain b
DNA molecule
Site that assembles DNA molecules (DNA polymerase site)
Site that corrects mistakes during DNA assembly (exonuclease site)

FIGURE 3.27

Domains in proteins. (A) Two domains in part of an enzyme that assembles DNA molecules in the bacterium *Escherichia coli*. The α helices are shown as cylinders, the β strands as arrows, and the random coils as ropes. "N" indicates the N-terminal end of the protein, and "C" indicates the C-terminal end. **(B)** The same view of the protein as in **(A)**, showing the domain surfaces and functional sites.

3.5 Nucleotides and Nucleic Acids

Nucleic acids are another class of macromolecules, in this case, long polymers assembled from repeating monomers called *nucleotides*. Two types of nucleic acids exist: DNA and RNA. **DNA (deoxyribonucleic acid)** stores the hereditary information responsible for inherited traits in all eukaryotes and prokaryotes and in a large group of viruses. **RNA (ribonucleic acid)** is the hereditary molecule of another large group of viruses; in all organisms, one major type of RNA carries the instructions for assembling proteins from DNA to the sites where the proteins are made inside cells. Another major type of RNA forms part of ribosomes, the structural units that assemble proteins, and a third major type of RNA brings amino acids to the ribosomes for their assembly into proteins (see Chapter 15).

Nucleotides Consist of a Nitrogenous Base, a Five-Carbon Sugar, and One or More Phosphate Groups

A **nucleotide,** the monomer of nucleic acids, consists of three parts linked together by covalent bonds: (1) a **nitrogenous base** (a nitrogen-containing molecule that accepts protons), formed from rings of carbon and nitrogen atoms; (2) a five-carbon, ring-shaped sugar; and (3) one to three phosphate groups **(Figure 3.28).** The two types of nitrogenous bases are **pyrimidines,** with one carbon-nitrogen ring, and **purines,** with two carbon–nitrogen rings **(Figure 3.29).** Three pyrimidine bases—uracil (U), thymine (T), and cytosine (C)—and two purine bases—adenine (A) and guanine (G)—form parts of nucleic acids in cells.

In nucleotides, the nitrogenous bases link covalently to either **deoxyribose** in DNA or **ribose** in RNA. Nucleotides containing deoxyribose are called **deoxyribonucleotides** and nucleotides containing ribose are called **ribonucleotides.** Deoxyribose and ribose are five-carbon sugars. The carbons of these two sugars are numbered with a prime symbol—1′, 2′, 3′, 4′, and 5′—to distinguish them from the carbons and nitrogens in the nitrogenous bases, which are numbered without primes (see Figure 3.28). The two sugars differ only in the chemical group bound to the 2′ carbon (boxed in red in Figure 3.28B): deoxyribose has an —H at this position and ribose has an —OH group. The prefix *deoxy-* in deoxyribose indicates that oxygen is absent at this position in the DNA sugar. In individual, unlinked nucleotides, a chain of one, two, or three phosphate groups bonds to the ribose or deoxyribose sugar at the 5′ carbon; nucleotides are called monophosphates, diphosphates, or triphosphates according to the length of this phosphate chain.

A structure containing only a nitrogenous base and a five-carbon sugar is called a **nucleoside** (see Figure 3.28B). Thus, nucleotides are *nucleoside phosphates.* For example, the nucleoside containing adenine and ribose is called *adenosine.* Adding one phosphate group to this structure produces *adenosine monophosphate (AMP),* adding two phosphate groups produces *adenosine diphosphate (ADP),* and adding three produces *adenosine triphosphate (ATP).* The corresponding adenine–deoxyribose complexes are named *deoxyadenosine monophosphate (dAMP), deoxyadenosine diphosphate (dADP),* and *deoxyadenosine triphosphate (dATP).* The lowercase *d* in the abbreviations indicates that the nucleoside contains the deoxyribose form of the sugar. Equivalent names and abbreviations are used for the other nucleotides (see Figure 3.28B). Whether a nucleotide is a monophosphate, diphosphate, or triphosphate has fundamentally important effects on its activities.

Nucleotides perform many functions in cells in addition to serving as the building blocks of nucleic acids. Two ribose-containing nucleotides in particular, adenosine triphosphate (ATP) and guanosine triphosphate (GTP), are the primary molecules that transport chemical energy from one reaction system to another; the same nucleotides regulate and adjust cellular activity. Molecules derived from nucleotides play important roles in biochemical reactions by delivering reactants or electrons from one system to another.

Nucleic Acids DNA and RNA Are the Informational Molecules of All Organisms

DNA and RNA consist of chains of nucleotides, *polynucleotide chains,* with one nucleotide linked to the next by a bridging phosphate group between the 5′ carbon of one sugar and the 3′ carbon of the next sugar in line; this linkage is called a **phosphodiester bond (Figure 3.30).** This arrangement of alternating sugar and phosphate groups forms the backbone of a nucleic acid chain. The nitrogenous bases of the nucleotides project from this backbone.

Each nucleotide of a DNA chain contains deoxyribose and one of the four bases A, T, G, or C. Each nucleotide of an RNA chain contains ribose and one of the four bases A, U, G, or C. Thymine and uracil differ only in a single functional group: in T a methyl (—CH_3) group is linked to the ring, but in U it is replaced by a hydrogen (see Figure 3.29). The differences in sugar and pyrimidine bases between DNA and RNA account for important differences in the structure and functions of these nucleic acids inside cells.

DNA Molecules in Cells Consist of Two Nucleotide Chains Wound Together

In cells, DNA takes the form of a **double helix,** first discovered by James D. Watson and Francis H. C. Crick in 1953, in collaboration with Maurice Wilkins and Rosalind Franklin (see Chapter 14 for details of their discovery). The double helix they described consists of two nucleotide chains wrapped around each other in a spiral that resembles a twisted ladder **(Figure 3.31).** The sides of the ladder are the sugar–phosphate backbones of the two chains, which twist around each other in a right-handed direction to form the double spiral. The rungs of the ladder are the nitrogenous bases, which extend inward from the sugars toward the cen-

A. Overall structural plan of a nucleotide

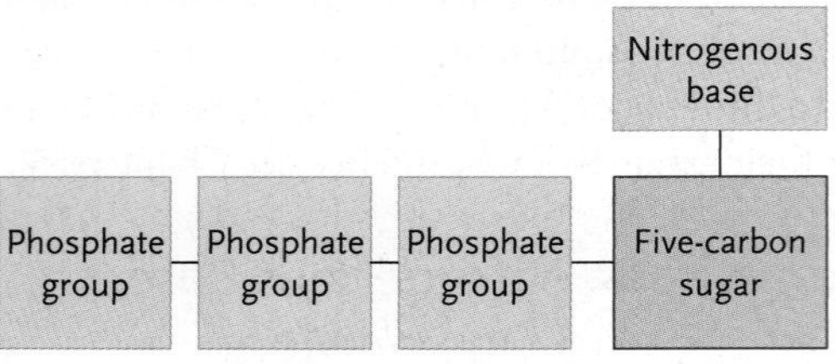
Nitrogenous base
Phosphate group
Phosphate group
Phosphate group
Five-carbon sugar

B. Chemical structures of nucleotides

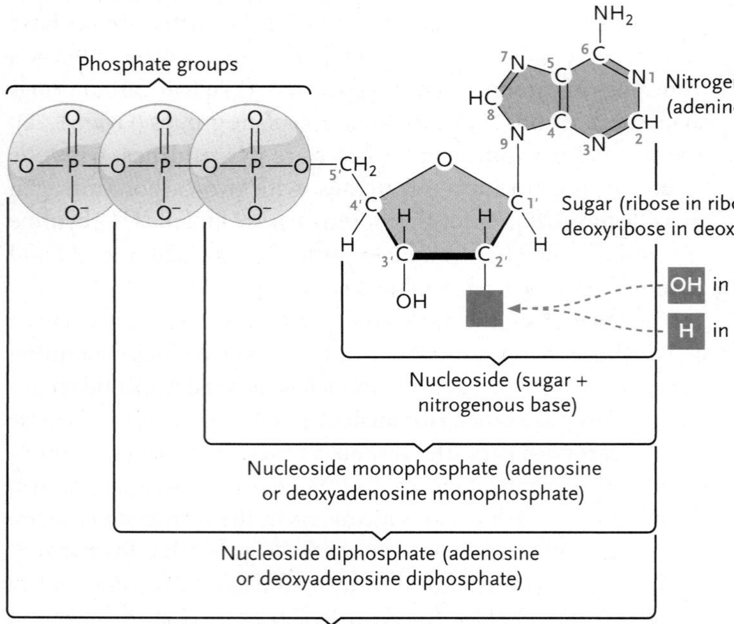
Phosphate groups
Nitrogenous base (adenine shown)
Sugar (ribose in ribonucleotides; deoxyribose in deoxyribonucleotides)
OH in ribose
H in deoxyribose
Nucleoside (sugar + nitrogenous base)
Nucleoside monophosphate (adenosine or deoxyadenosine monophosphate)
Nucleoside diphosphate (adenosine or deoxyadenosine diphosphate)
Nucleoside triphosphate (adenosine or deoxyadenosine triphosphate)

Other nucleotides:

Containing guanine:	Guanosine or deoxyguanosine monophosphate, diphosphate, or triphosphate
Containing cytosine:	Cytidine or deoxycytidine monophosphate, diphosphate, or triphosphate
Containing thymine:	Thymidine monophosphate, diphosphate, or triphosphate
Containing uracil:	Uridine monophosphate, diphosphate, or triphosphate

FIGURE 3.28
Nucleotide structure.

Pyrimidines

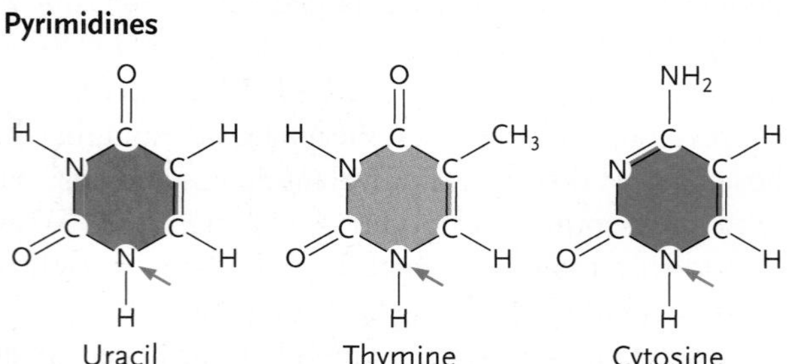
Uracil
Thymine
Cytosine

Purines

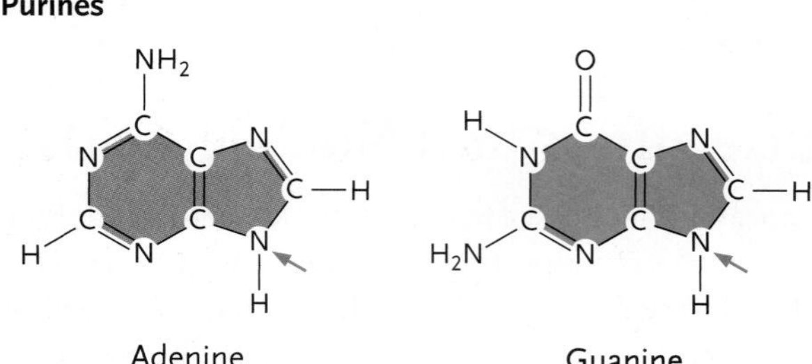
Adenine
Guanine

FIGURE 3.29
Pyrimidine and purine bases of nucleotides and nucleic acids. Red arrows indicate where the bases link to ribose or deoxyribose sugars in the formation of nucleotides.

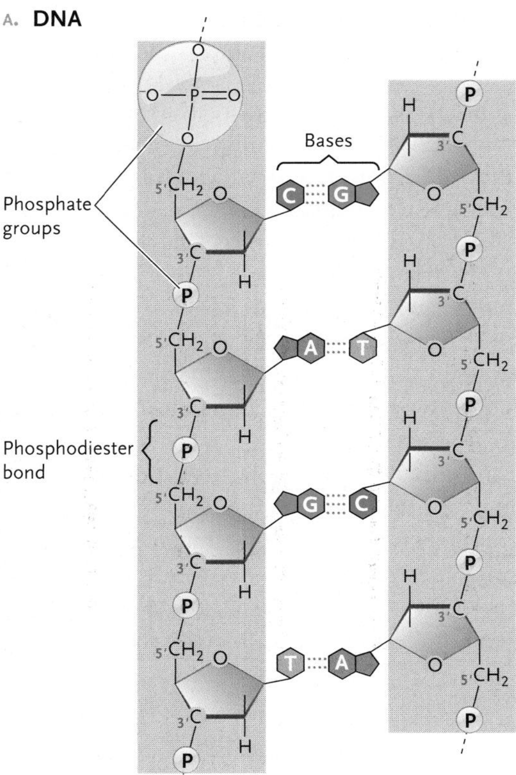
A. DNA
Phosphate groups
Bases
Phosphodiester bond

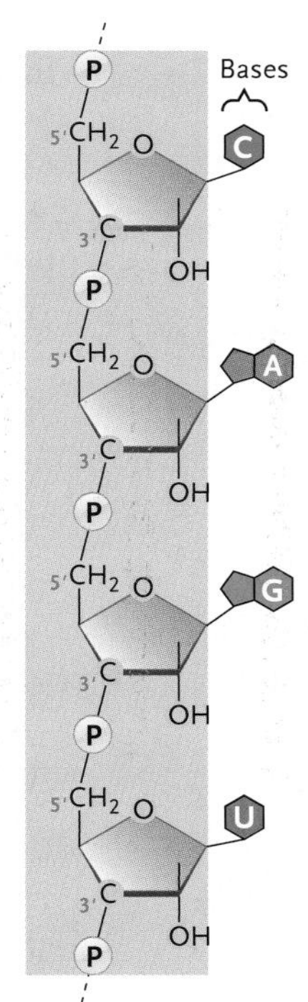
B. RNA
Bases

FIGURE 3.30
Linkage of nucleotides to form the nucleic acids DNA and RNA. P is a phosphate group (see Figure 3.28). **(A)** In DNA, the bases adenine (A), thymine (T), cytosine (C), or guanine (G) may be bound at the base positions. The lilac zones are the sugar (deoxyribose)–phosphate backbones of the two polynucleotide strands of the DNA molecule. Dotted lines designate hydrogen bonds. **(B)** In RNA, A, G, C, or uracil (U) may be bound at the base positions. The light green zone is the sugar (ribose)–phosphate backbone of the polynucleotide strand of the RNA molecule.

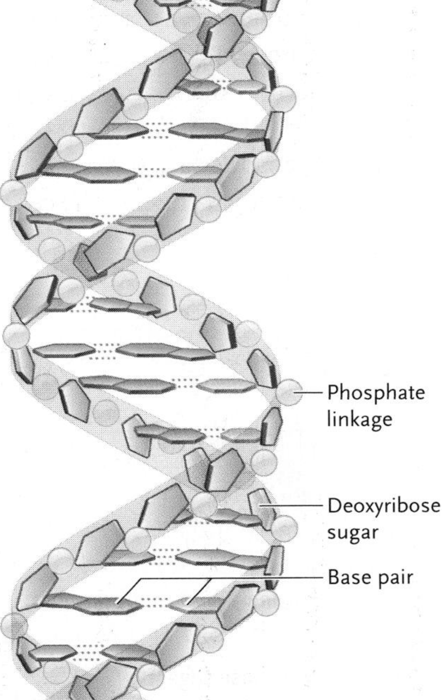
A. DNA double helix, showing arrangement of sugars, phosphate groups, and bases
Phosphate linkage
Deoxyribose sugar
Base pair

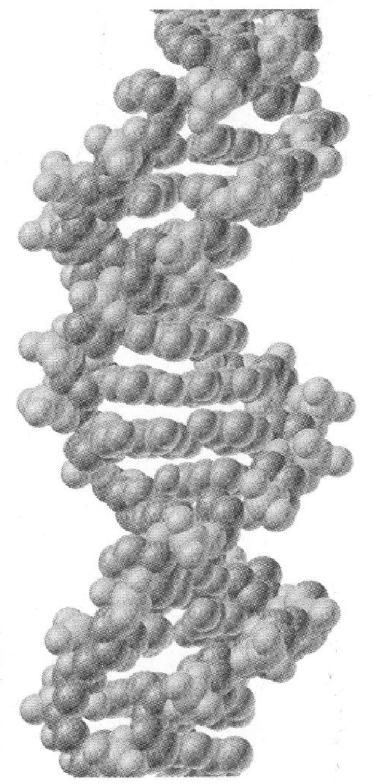
B. Space-filling model of DNA double helix

FIGURE 3.31
The DNA double helix. **(A)** Arrangement of sugars, phosphate groups, and bases in the DNA double helix. The dotted lines between the bases designate hydrogen bonds. **(B)** Space-filling model of the DNA double helix. The paired bases, which lie in flat planes, are seen edge-on in this view.

ter of the helix. Each rung consists of a pair of nitrogenous bases held in a flat plane roughly perpendicular to the long axis of the helix. The two nucleotide chains of a DNA double helix are held together primarily by hydrogen bonds between the base pairs. Slightly more than 10 base pairs are packed into each turn of the double helix. A DNA double-helix molecule is also referred to as double-stranded DNA.

The space separating the sugar–phosphate backbones of a DNA double helix is just wide enough to accommodate a base pair that consists of one purine and one pyrimidine. Purine–purine base pairs are too wide and pyrimidine–pyrimidine pairs are too narrow to fit this space exactly. More specifically, of the possible purine–pyrimidine pairs, only two combinations, adenine with thymine, and guanine with cytosine, can form stable hydrogen bonds so that the base pair fits precisely within the double helix **(Figure 3.32)**. An adenine–thymine (A–T) pair forms two stabilizing hydrogen bonds; a guanine–cytosine (G–C) pair forms three.

As Watson and Crick pointed out in the initial report of their discovery, the formation of A–T and G–C pairs allows the sequence of one nucleotide chain to determine the sequence of its partner in the double helix. Thus, wherever a T occurs on one chain of a DNA double helix, an A occurs opposite it on the other chain; wherever a C occurs on one chain, a G occurs on the other side (see Figure 3.30). That is, the nucleotide sequence of one chain is said to be *complementary* to the nucleotide sequence of the other chain. The complementary nature of the two chains underlies the processes when DNA molecules are copied—replicated—to pass hereditary information from parents to offspring and when RNA copies are made of DNA molecules to transmit information within cells. In DNA replication, one nucleotide chain is used as a **template** for the assembly of a complementary chain according to the A–T and G–C base-pairing rules **(Figure 3.33)**.

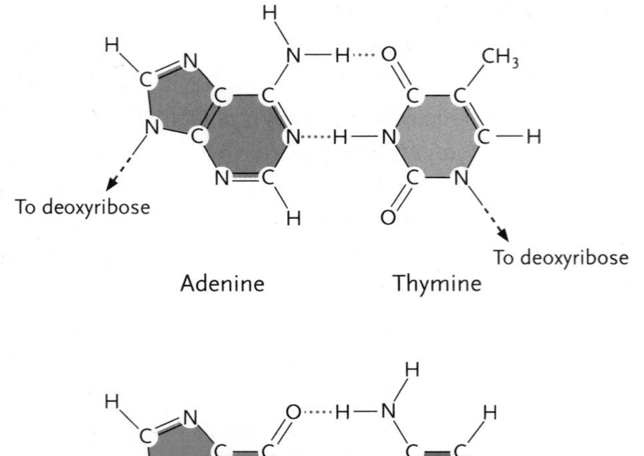
To deoxyribose
To deoxyribose
Adenine
Thymine

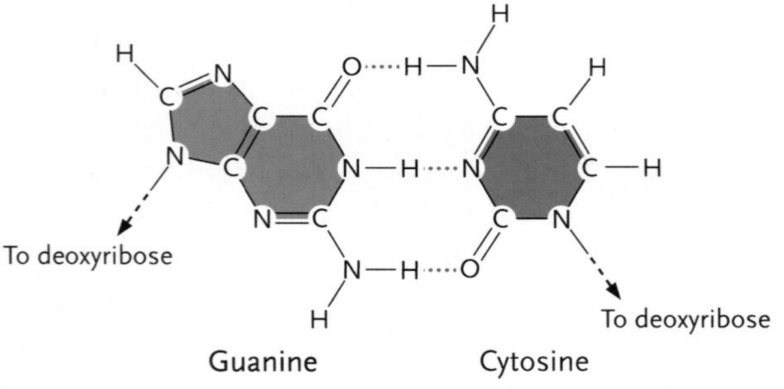
To deoxyribose
To deoxyribose
Guanine
Cytosine

FIGURE 3.32
The DNA base pairs A–T (adenine–thymine) and G–C (guanine–cytosine), as seen from one end of a DNA molecule. Dotted lines between the bases designate hydrogen bonds.

RNA Molecules Are Usually Single Nucleotide Chains

In contrast to DNA, RNA molecules exist largely as single, rather than double, polynucleotide chains in living cells. That is, RNA is typically single-stranded. However, RNA molecules can fold and twist back on themselves to form double-helical regions. The patterns of these fold-back double helices are as vital to RNA function as the folding of amino acid chains is to protein function. "Hybrid" double helices, which consist of an RNA chain paired with a DNA chain, are formed temporarily when RNA copies are made of DNA chains. In the RNA–RNA or hybrid RNA–DNA helices, U in RNA takes over the pairing functions of T, forming A–U rather than A–T base pairs.

The description of nucleic acid molecules in this section, with the discussions of carbohydrates, lipids, and proteins in earlier sections, completes our survey of the major classes of organic molecules found in living organisms. The next chapter discusses the functions of molecules in one of these classes, the enzymatic proteins, and the relationships of energy changes to the biological reactions speeded by enzymes.

STUDY BREAK 3.5

1. **What is the monomer of a nucleic acid macromolecule?**
2. **What are the chemical differences between DNA and RNA?**

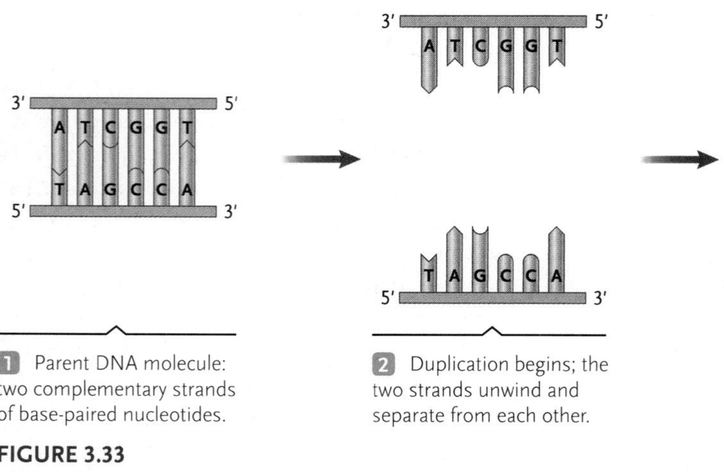
3'
5'
5'
3'

1 Parent DNA molecule: two complementary strands of base-paired nucleotides.

2 Duplication begins; the two strands unwind and separate from each other.

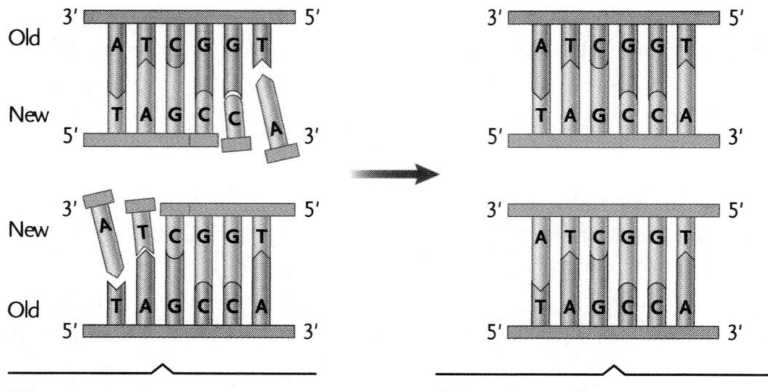
Old
New
New
Old
3'
5'

3 Each "old" strand serves as a template for addition of bases according to the A–T and G–C base-pairing rules.

4 Bases positioned on each old strand are joined together into a "new" strand. Each half-old, half-new DNA molecule is an exact duplicate of the parent molecule.

FIGURE 3.33
How complementary base pairing allows DNA molecules to be replicated precisely.

UNANSWERED QUESTIONS

How are the enzymes that synthesize fatty acids and cholesterol regulated in the cells?

The cells of humans and animals contain thousands of different lipids that form the membranes delimiting cells and subcellular organelles. Lipids are also converted into signaling molecules or hormones that regulate many metabolic processes. Despite their ubiquitous functions, we know very little about how the body controls the level of each lipid substance to assure a concentration that is adequate for function but guards against an over-accumulation. Over-accumulation can produce disorders ranging from brain degeneration to heart attacks.

One area of recent progress revolves about the mechanism that controls the body's production of two of the most abundant lipids, cholesterol and fatty acids. Much of this work has been carried out in a laboratory that is led jointly by Joseph L. Goldstein and myself at the University of Texas Southwestern Medical School in Dallas, Texas. Several years ago our group discovered Sterol Regulatory Element Binding Proteins (SREBPs), which are members of a class of regulatory proteins called transcription factors, that bind to specific sequences in DNA and activate transcription of nearby genes. (Transcription is the first step in the process of gene expression, in which the DNA sequence of a gene is copied into an RNA molecule. This RNA sequence later directs assembly of the protein's amino acid sequence in translation.) The SREBPs selectively activate several dozen genes, including all of the ones necessary for the production of cholesterol or its uptake from plasma.

Unlike most other transcription factors, SREBPs are synthesized initially on membranes of the endoplasmic reticulum (ER). To reach the nucleus, the SREBPs must be transported from the ER to the cell's Golgi complex where they are cleaved by proteases to release a soluble fragment that enters the nucleus and activates transcription. Transport of SREBPs from ER to Golgi is accomplished by an escort protein called Scap, which binds SREBPs immediately after their synthesis.

Why are SREBPs attached to ER membranes, and why do cells go through this elaborate transport process to release the active fragments? The reason is simple: it allows the synthesis of membrane cholesterol to be regulated by the concentration of cholesterol in the ER membrane. When cholesterol builds up in ER membranes, the cholesterol binds to Scap, and this blocks the transport of SREBPs to the Golgi. The SREBPs are no longer processed by the Golgi proteases, and the amount of soluble nuclear fragment declines. Synthesis of cholesterol is diminished, and it remains low until the cholesterol content of ER membranes falls, upon which the SREBPs are again transported to the Golgi where they activate the genes required for cholesterol synthesis.

The elucidation of the SREBP processing pathway is a step forward, but many questions remain unanswered. Many of these center on the precise way in which Scap senses the cholesterol level in membranes. Also, we must begin to understand the mechanisms that control the other constituents of cell membranes, primarily phospholipids. Only then will we understand how our cells manage to create membranes with the precise chemical and physical properties that allow them to perform their vital functions.

Thinking Critically

Defects in SREBP regulation contribute to common diseases, ranging from heart attacks to obesity and diabetes. How might better understanding of the SREBP regulatory mechanism contribute to treatment of diabetes?

Dr. Michael S. Brown

Dr. Joseph L. Goldstein

Dr. Michael S. Brown and **Dr. Joseph L. Goldstein,** both of the University of Texas Southwestern Medical School, discovered the low-density lipoprotein (LDL) receptor. They shared many awards for this work, including the Nobel Prize for Medicine or Physiology. Dr. Brown is Paul J. Thomas Professor of Molecular Genetics and Director of the Jonsson Center for Molecular Genetics; Dr. Goldstein is chairman of the Department of Molecular Genetics. To learn more about their research, go to http://www4.utsouthwestern.edu/moleculargenetics/pages/brown/lab.html.

REVIEW KEY CONCEPTS

Go to **CENGAGENOW** at www.cengage.com/login to access quizzing, animations, exercises, articles, and personalized homework help.

3.1 Formation and Modification of Biological Molecules

- Carbon atoms readily share electrons, allowing each carbon atom to form four covalent bonds with other carbon atoms or atoms of other elements. The resulting extensive chain and ring structures form the backbones of diverse organic compounds (Figure 3.2).
- The structure and behavior of organic molecules, as well as their linkage into larger units, depend on the chemical properties of functional groups. Particular combinations of functional groups determine whether an organic molecule is an alcohol, aldehyde, ketone, or acid (Table 3.1).
- Isomers have the same chemical formula but different molecular structures. Enantiomers, also called optical isomers, are isomers that are mirror images of each other (Figure 3.3). Structural isomers are two molecules with atoms arranged in different ways (Figure 3.4).
- In a dehydration synthesis reaction, the components of a water molecule are removed as subunits assemble. In hydrolysis, the components of a water molecule are added as subunits are broken apart (Figure 3.5).
- Carbohydrates, lipids, proteins, and nucleic acids are large polymers of monomer subunits. Polymerization reactions assemble monomers into polymers. Polymers larger than 1,000 daltons in mass are considered to be macromolecules.

Animation: Functional groups

Animation: Dehydration synthesis and hydrolysis

3.2 Carbohydrates

- Carbohydrates are molecules in which carbon, hydrogen, and oxygen occur in the approximate ratio 1:2:1.
- Monosaccharides are carbohydrate subunits that contain three to seven carbons (Figures 3.5–3.7).
- Monosaccharides have D and L enantiomers. Typically, one of the two forms is used in cellular reactions because it has a molecular shape that can be recognized by the enzyme accelerating the reaction, whereas the other form does not.
- Two monosaccharides join to form a disaccharide; greater numbers form polysaccharides (Figures 3.8 and 3.9).

Animation: Structure of starch and cellulose

3.3 Lipids

- Lipids are hydrocarbon-based, water-insoluble, nonpolar molecules. Biological lipids include neutral lipids, phospholipids, and steroids.
- Neutral lipids, which are primarily energy-storing molecules, have a glycerol backbone and three fatty acid chains (Figures 3.10 and 3.11).
- Phospholipids are similar to neutral lipids except that a phosphate group and a polar organic unit substitute for one of the fatty acids (Figure 3.13). In polar environments (such as a water solution), phospholipids orient with their polar end facing the water and their nonpolar ends clustered in a region that excludes water. This orientation underlies the formation of bilayers, the structural framework of biological membranes.

- Steroids, which consist of four carbon rings carrying primarily nonpolar groups, function chiefly as components of membranes and as hormones in animals (Figures 3.15 and 3.16).
- Lipids link with carbohydrates to form glycolipids and with proteins to form lipoproteins, both of which play important roles in cell membranes.

Animation: Structure of a phospholipid

3.4 Proteins

- Proteins are assembled from 20 different amino acids. Amino acids have a central carbon to which is attached an amino group, a carboxyl group, a hydrogen atom, and a side group that differs in each amino acid (Figure 3.17).
- Peptide bonds between the amino group of one amino acid and the carboxyl group of another amino acid link amino acids into chains (Figure 3.19).
- A protein may have four levels of structure. Its primary structure is the linear sequence of amino acids in a polypeptide chain; secondary structure is the arrangement of the amino acid chain into α helices, β strands and sheets, or random coils; tertiary structure is the protein's overall conformation. Quaternary structure is the number and arrangement of polypeptide chains in a protein (Figures 3.20–3.24).
- In many proteins, combinations of secondary, tertiary, and quaternary structure form functional domains (Figure 3.27).
- Proteins combine with lipids to produce lipoproteins, with carbohydrates to produce glycoproteins, and with nucleic acids to form nucleoproteins.

Animation: Structure of an amino acid

Animation: Peptide bond formation

Animation: The primary and secondary structure of proteins

Animation: Secondary and tertiary structure

Animation: Globin and hemoglobin structure

3.5 Nucleotides and Nucleic Acids

- A nucleotide consists of a nitrogenous base, a five-carbon sugar, and one to three phosphate groups (Figures 3.28 and 3.29).
- Nucleotides are linked into nucleic acid chains by covalent bonds between their sugar and phosphate groups. Alternating sugar and phosphate groups form the backbone of a nucleic acid chain (Figure 3.30).
- There are two nucleic acids: DNA and RNA. DNA contains nucleotides with the nitrogenous bases adenine (A), thymine (T), guanine (G), or cytosine (C) linked to the sugar deoxyribose; RNA contains nucleotides with the nitrogenous bases adenine, uracil (U), guanine, or cytosine linked to the sugar ribose (Figures 3.28–3.30).
- In a DNA double helix, two nucleotide chains wind around each other like a twisted ladder, with the sugar–phosphate backbones of the two chains forming the sides of the ladder and the nitrogenous bases forming the rungs of the ladder (Figure 3.31).
- A–T and G–C base pairs mean that the sequences of the two nucleotide chains of a DNA double helix are complements of each other. Complementary pairing underlies the processes that replicate DNA and copy RNA from DNA (Figures 3.32 and 3.33).

UNDERSTAND AND APPLY

Test Your Knowledge

1. Which functional group has a double bond and forms organic acids?
   a. carboxyl
   b. amino
   c. hydroxyl
   d. carbonyl
   e. sulfhydryl
2. Which of the following characteristics is *not* common to carbohydrates, lipids, and proteins?
   a. They are composed of a carbon backbone with functional groups attached.
   b. Monomers of these molecules undergo dehydration synthesis to form polymers.
   c. Their polymers are broken apart by hydrolysis.
   d. The backbones of the polymers are primarily polar molecules.
   e. The molecules are held together by covalent bonding.
3. Cellulose is to carbohydrate as:
   a. amino acid is to protein.
   b. lipid is to fat.
   c. collagen is to protein.
   d. nucleic acid is to DNA.
   e. nucleic acid is to RNA.
4. Maltose, sucrose, and lactose differ from one another:
   a. because not all contain glucose.
   b. because not all of them exist in ring form.
   c. in the number of carbons in the sugar.
   d. in the number of hexose monomers involved.
   e. by the linkage of the monomers.
5. Lipids that are liquid at room temperature:
   a. are fats.
   b. contain more hydrogen atoms than lipids that are solids at room temperature.
   c. if polyunsaturated, contain several double bonds in their fatty acid chains.
   d. lack glycerol.
   e. are not stored in cells as triglycerides.
6. Which of the following statements about steroids is *false*?
   a. They are classified as lipids because, like lipids, they are nonpolar.
   b. They can act as regulatory molecules in animals.
   c. They are composed of four carbon rings.
   d. They are highly soluble in water.
   e. Their most abundant form is as sterols.
7. The term *secondary structure* refers to a protein's:
   a. sequence of amino acids.
   b. structure that results from local interactions between different amino acids in the chain.
   c. interactions with a second protein chain.
   d. interaction with a chaperonin.
   e. interactions with carbohydrates.
8. The first and major effect in denaturation of proteins is that:
   a. peptide bonds break.
   b. α helices unwind.
   c. β sheet structures unfold.
   d. tertiary structure is changed.
   e. quaternary structures disassemble.

9. In living systems:
   a. proteins rarely combine with other macromolecules.
   b. enzymes are always proteins.
   c. proteins are composed of 24 amino acids.
   d. chaperonins inhibit protein movement.
   e. a protein domain refers to the place in the cell where proteins are synthesized and function.
10. RNA differs from DNA because:
   a. RNA may contain the pyrimidine uracil, and DNA does not.
   b. RNA is always single-stranded when functioning, and DNA is always double-stranded.
   c. the pentose sugar in RNA has one less O atom than the pentose sugar in DNA.
   d. RNA is more stable and is broken down by enzymes less easily than DNA.
   e. RNA is a much larger molecule than DNA.

Discuss the Concepts

1. Identify the following structures as a carbohydrate, fatty acid, amino acid, or polypeptide:
   a. $H_3N^+—C(R)(H)—COO^-$ (The *R* indicates an organic group.)
   b. $C_6H_{12}O_6$
   c. $(glycine)_{20}$
   d. $CH_3(CH_2)_{16}COOH$
2. Lipoproteins are relatively large, spherical clumps of protein and lipid molecules that circulate in the blood of mammals. They are like suitcases that move cholesterol, fatty acid remnants, triglycerides, and phospholipids from one place to another in the body. Given what you know about the insolubility of lipids in water, which of the three kinds of lipids would you predict to be on the outside of a lipoprotein clump, bathed in the fluid portion of blood?
3. The shapes of a protein's domains often give clues to its functions. For example, protein HLA (human leukocyte antigen) is a type of recognition protein on the outer surface of all vertebrate body cells. Certain cells of the immune system use HLAs to distinguish self (the body's own cells) from nonself (invading cells). Each HLA protein has a jawlike region that can bind to molecular parts of an invader. It thus alerts the immune system that the body has been invaded. Speculate on what might happen if a mutation makes the jawlike region misfold.
4. Explain how polar and nonpolar groups are important in the structure and functions of lipids, proteins, and nucleic acids.

Design an Experiment

A clerk in a health food store tells you that natural vitamin C extracted from rose hips is better for you than synthetic vitamin C. Given your understanding of the structure of organic molecules, how would you respond? Design an experiment to test whether the rose hips and synthetic vitamin C preparations differ in their effects.

Interpret the Data

Cholesterol does not dissolve in blood. It is carried through the bloodstream by lipoproteins, as described in *Focus on Applied Research.* Low-density lipoprotein (LDL) carries cholesterol to body tissues, such as artery walls, where it can form health-endangering deposits. LDL is often called "bad" cholesterol. High-density lipoprotein (HDL) carries cholesterol away from tissues to the liver for disposal; it is often called "good" cholesterol.

In 1990, Ronald Mensink and Martijn Katan tested the effects of different dietary fats on blood lipoprotein levels. They placed 59 men and women on a diet in which 10% of their daily energy intake consisted of *cis* fatty acids, *trans* fatty acids, or saturated fats. (All subjects were tested on each of the diets.) Blood LDL and HDL levels of the subjects were measured after 3 weeks on the diet. The table displays the averaged results, shown in mg/dL (milligrams per deciliter) of blood.

Effect of Diet on Lipoprotein Levels (mg/dL)

	cis fatty acids	*trans* fatty acids	Saturated fats	Optimal level
LDL	103	117	121	<100
HDL	55	48	55	>40
LDL-to-HDL ratio	1.87	2.43	2.2	<2

1. In which group was the level of LDL ("bad" cholesterol) highest and in which group was the level of HDL ("good" cholesterol) lowest?
2. An elevated risk of heart disease has been correlated with increasing LDL-to-HDL ratios. Which group had the highest LDL-to-HDL ratio?
3. Rank the three diets from best to worst according to their potential effect on cardiovascular health.

Source: R. P. Mensink and M. B. Katan. 1990. Effect of dietary trans fatty acids on high-density and low-density lipoprotein cholesterol levels in healthy subjects. *New England Journal of Medicine* 323:439–445.

Apply Evolutionary Thinking

How do you think the primary structure (amino acid) sequence of proteins could inform us about the evolutionary relationships of proteins?

Express Your Opinion

Scientists have discovered vast reservoirs of methane (natural gas, an important fossil fuel) under sediments covering the seafloor. It occurs in a highly unstable form that can cause immense explosions if the temperature rises or the pressure falls slightly. Should we work toward developing these vast undersea methane deposits as an energy source, given that the environmental costs and risks to life are unknown? Go to academic.cengage.com/login to investigate both sides of the issue and then vote.

4

STUDY OUTLINE

Fly caught by a leaf of the English sundew *(Drosera anglica)*. Enzymes secreted by the hairs on the leaf digest trapped insects, provid nutrients to the plant.

Energy, Enzymes, and Biological Reactions

Why It Matters. . . Earth is a cold place, at least when it comes to chemical reactions. Life cannot survive at the high temperatures routinely used in most laboratories and industrial plants for chemical synthesis. Instead, life relies on substances called catalysts, which speed up the rates of reaction without the need for an increase in temperature. The acceleration of a reaction by a catalyst is called catalysis. Most of the catalysts in biological systems are proteins called enzymes.

How good are enzymes at increasing the rate of a reaction? Richard Wolfenden and his colleagues at the University of North Carolina experimentally measured the rates for a range of uncatalyzed and enzyme-catalyzed biological reactions. They found the greatest difference between the uncatalyzed rate and the enzyme-catalyzed rate for a reaction that removes a phosphate group from a molecule.

In the cell, a group of enzymes called phosphatases catalyze the removal of phosphate groups from a number of molecules, including proteins. The reversible addition and removal of a phosphate group from particular proteins is a central mechanism of intracellular communication in almost all cells. In a cell using a phosphatase enzyme, the phosphate removal reaction takes approximately 10 milliseconds (msec). Wolfenden's research group calculated that in an aqueous environment such as a cell, without an enzyme, the phosphate removal reaction would take over 1 trillion (10^{12}) years to occur. This exceeds the current estimate for the age of the universe! By contrast, the enzyme-catalyzed reaction is 21 orders of magnitude (10^{21}) faster.

For most reactions, the difference between the uncatalyzed rate and the enzyme-catalyzed rate is many millions of times. Because life requires temperatures that are relatively low (below 100°C), without enzymes to speed up the rates of reaction, life as we know would not have evolved.

Enzymes are key players in **metabolism**—the biochemical modification and use of organic molecules and energy to support the activities of life. Metabolism, which occurs only in living organisms, comprises thousands of biochemical reactions that accomplish the special activities we associate with life, such as growth, reproduction, movement, and the ability to respond to stimuli.

Understanding how biological reactions occur and how enzymes work requires knowledge of the basic laws of chemistry and physics. All reactions, whether they occur inside living organisms or in the outside, inanimate world, obey the same chemical and physical laws that operate everywhere in the universe. These fundamental laws are our starting point for an exploration of the nature of energy and how cells use it to conduct their activities. <

4.1 Energy, Life, and the Laws of Thermodynamics

Life, like all chemical and physical activities, is an energy-driven process. Energy cannot be measured or weighed directly. We can detect it only through its effects on matter, including its ability to move objects against opposing forces, such as friction, gravity, or pressure, or to push chemical reactions toward completion. Therefore, energy is most conveniently defined as *the capacity to do work.* Even when you are asleep, cells of your muscles, brain, and other parts of your body are at work and using energy.

Energy Exists in Different Forms and States

Energy can exist in many different forms, including heat, chemical, electrical, mechanical, and radiant energy. Visible light, infrared and ultraviolet light, gamma rays, and X-rays are all types of radiant energy. Although the forms of energy are different, energy can be converted readily from one form to another. For example, chemical energy is transformed into electrical energy in a flashlight battery, and electrical energy is transformed into light and heat energy in the flashlight bulb. In green plants, the radiant energy of sunlight is transformed into chemical energy in the form of complex sugars and other organic molecules.

All forms of energy can exist in one of two states: kinetic and potential. **Kinetic energy** (*kinetikos* = putting in motion) is the energy of an object because it is in motion. Examples of everyday objects that possess kinetic energy are waves in the ocean, a hit baseball, and a falling rock. Examples from the natural world are electricity (which is a flow of electrons), photons of light, and heat. The movement present in kinetic energy is useful because it can perform work by making other objects move. **Potential energy** is stored energy, that is, the energy an object has because of its location or chemical structure. A boulder on the top of a hill has potential energy because of its position in Earth's gravitational field. Chemical energy, nuclear energy, gravitational energy, and stored mechanical energy are forms of potential energy.

Potential energy can be converted to kinetic energy and vice versa. Consider a roller coaster. When the train descends from its maximum height, its potential energy is converted into kinetic energy, and the coaster accelerates. On the next hill, kinetic energy is converted back to potential energy and the coaster slows. In the ups and downs of the ride, the coaster's potential energy and kinetic energy interchange continuously but, importantly, the sum of potential energy and kinetic energy remains constant.

The Laws of Thermodynamics Describe the Energy Flow in Natural Systems

The study of energy and its transformations is called **thermodynamics.** Quantitative research by chemists and physicists in the nineteenth century regarding energy flow between systems and the surroundings led to the formulation of two fundamental laws of thermodynamics that apply equally to living cells and to stars and galaxies. These laws allow us to predict whether reactions of any kind, including biological reactions, can occur. That is, if particular groups of molecules are placed together, are they likely to react chemically and change into different groups of molecules? The laws also give us the information needed to trace energy flows in biological reactions: They allow us to estimate the amount of energy released or required as a reaction proceeds.

When discussing thermodynamics, scientists refer to a *system,* which is the object under study. A system is whatever we define it to be—a single molecule, a cell, a planet. Everything outside a system is its *surroundings.* The universe, in this context, is the total of the system and the surroundings. There are three types of systems: isolated, closed, and open. An *isolated system* does not exchange matter or energy with its surroundings **(Figure 4.1A).** A perfectly insulated Thermos flask is an example of an isolated system. A *closed system* can exchange energy but not matter with its surroundings **(Figure 4.1B).** Earth can be considered a closed system. It takes in a great amount of energy from the sun and releases heat, but essentially no matter is exchanged between Earth and the rest of the universe (barring the odd space probe, of course). An *open system* can exchange both energy and matter with its surroundings **(Figure 4.1C).** All living organisms are open systems.

The First Law of Thermodynamics Addresses the Energy Content of Systems and Their Surroundings

The **first law of thermodynamics** states that *energy can be transformed from one form to another or transferred from one place to another, but it cannot be created or destroyed.* That is, in any process that involves an energy change, the *total amount of energy in a system and its surroundings remains constant.* This law is also called the *principle of conservation of energy.*

If energy can be neither created nor destroyed, what is the ultimate source of the energy we and other living organisms use? For almost all organisms, the ultimate source is the sun **(Figure 4.2).** Plants capture the kinetic energy of light radiating from the sun by absorbing it and converting it to the chemical potential energy of complex organic molecules—primarily sugars, starches, and lipids. These substances are used as fuels by the plants themselves, by animals that feed on plants, and by organisms (such as fungi and bacteria) that break down the bodies of dead organisms. The chemical potential energy stored in sug-

FIGURE 4.1 Isolated, closed, and open systems in thermodynamics.

A. **Isolated system: does not exchange matter or energy with its surroundings**

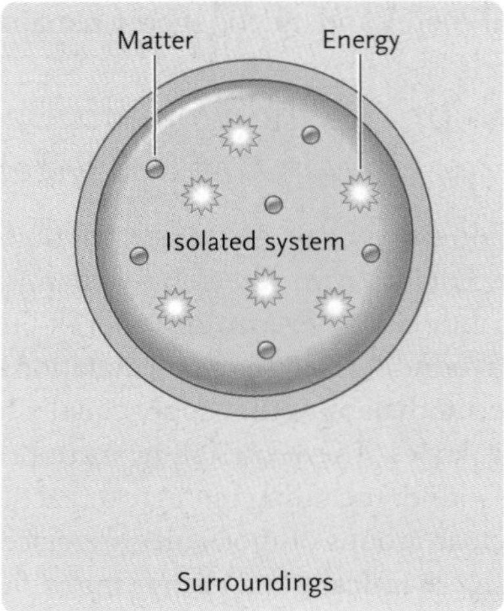
Matter
Energy
Isolated system
Surroundings

B. **Closed system: exchanges energy with its surroundings**

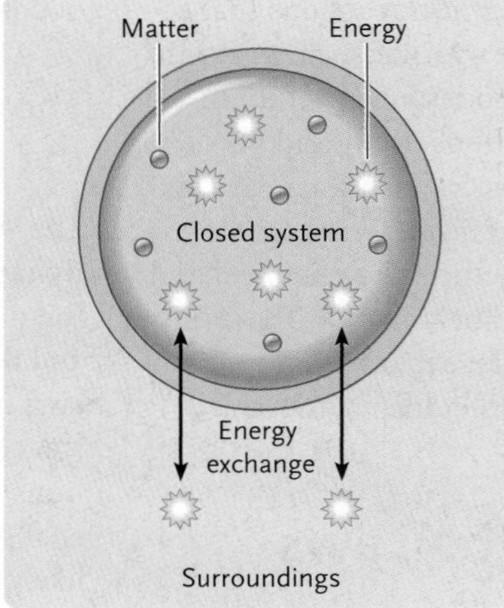
Matter
Energy
Closed system
Energy exchange
Surroundings

C. **Open system: exchanges both energy and matter with its surroundings**

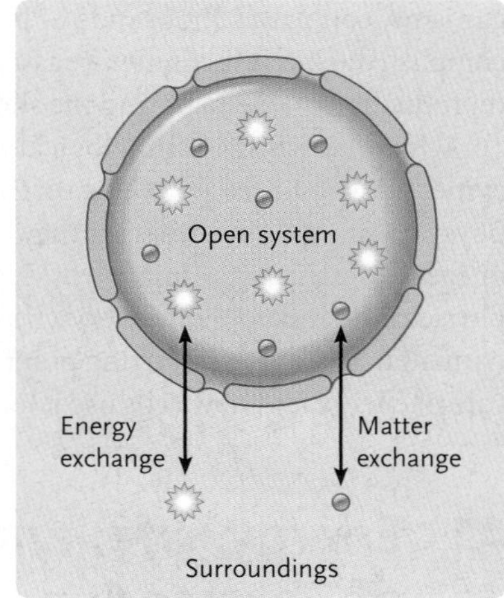
Open system
Energy exchange
Matter exchange
Surroundings

ars and other organic molecules is used for growth, reproduction, and other work of living organisms.

Eventually, most of the solar energy absorbed by green plants is converted into heat energy as the activities of life take place. Heat (a form of kinetic energy) is largely unusable by living organisms; as a result, most of the heat released by the reactions of living organisms radiates to their surroundings, and then from Earth into space.

How does the principle of conservation of energy apply to biochemical reactions? Molecules have both kinetic and potential energy. Kinetic energy for molecules above absolute zero (−273°C) is reflected in the constant motion of the molecules, whereas potential energy for molecules is the energy contained in the arrangement of atoms and chemical bonds. The energy content of reacting systems provides part of the information required to predict the likelihood and direction of chemical reactions. Usually, the energy content of the reactants in a chemical reaction (see Section 2.3) is larger than the energy content of the products. Thus, reactions usually progress to a state in which the products have *minimum energy content*. When this is the case, the difference in energy content between reactants and products in the reacting system is released to the surroundings.

FIGURE 4.2 Energy flow from the sun to photosynthetic organisms (here, colonies of the green alga *Volvox*), which capture the kinetic radiant energy of sunlight and convert it to potential chemical energy in the form of complex organic molecules.

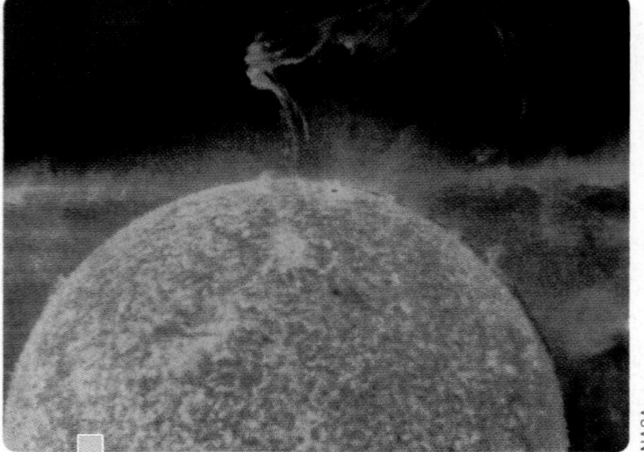
NASA

Radiant energy lost from the sun

Light energy absorbed by photosynthetic organisms and converted into potential chemical energy

Manfred Kage/Peter Arnold, Inc.

Heat energy lost from photosynthetic organisms

The Second Law of Thermodynamics Considers Changes in the Degree of Order in Reacting Systems

The second law of thermodynamics explains why, as any energy change occurs, the objects (matter) involved in the change typically become more disordered. (Your room and the kitchen at home are probably the best examples of this phenomenon.) You know from experience that it takes energy to straighten out (decrease) the disorder (as when you clean up your room).

The **second law of thermodynamics** states this tendency toward disorder formally, in terms of a system and its surroundings: In any process in which a system changes from an initial to a final state, *the total disorder of a system and its surroundings always increases.* In thermodynamics, disorder is called **entropy.** If the system and its surroundings are defined as the entire universe, the second law means that as changes occur anywhere in the universe, the total disorder or entropy of the universe constantly increases. As the first law of thermodynamics asserts, however, the total energy in the universe does not change.

At first glance, living organisms—which are open systems—appear to violate the second law of thermodynamics. As a fertilized egg develops into an adult animal, it becomes more highly ordered (decreases its entropy) as it synthesizes organic molecules from less complex substances. However, the entropy of the whole system—the surroundings as well as the organism—must be considered as growth proceeds. For the fertilized egg—the initial state—its surroundings include all the carbohydrates, fats, and other complex organic molecules the animal will use as it develops into an adult. For the adult—the final state—the surroundings include the animal's waste products (water, carbon dioxide, and many relatively simple organic molecules), which are collectively much less complex than the organic molecules used as fuels. When the total reactants, including all the nutrients, and the total products, including all the waste materials, are tallied, the total change satisfies both laws of thermodynamics—the total energy content remains constant, and the entropy of the system and its surroundings increases.

STUDY BREAK 4.1

1. **What are kinetic energy and potential energy?**
2. **In thermodynamics, what is meant by an isolated system, a closed system, and an open system?**

4.2 Free Energy and Spontaneous Reactions

Applying the first and second laws of thermodynamics together allows us to predict whether any particular chemical or physical reaction will occur without an input of energy. Such reactions are called **spontaneous reactions** in thermodynamics. In this usage, the word *spontaneous* means only that a reaction will occur—it does not describe the rate of a reaction. Spontaneous reactions may proceed very slowly, such as the formation of rust on a nail, or very quickly, such as a match bursting into flame.

Energy Content and Entropy Contribute to Making a Reaction Spontaneous

Two factors related to the first and second laws of thermodynamics must be taken into account to determine whether a reaction is spontaneous: the change in energy content of a system, and its change in entropy.

1. *Reactions tend to be spontaneous if the products have less potential energy than the reactants.* The potential energy in a system is its **enthalpy** (*en* = to put into; *-thalpein* = to heat), symbolized by H. Reactions that release energy are termed **exothermic**—the products have less potential energy than the reactants. Reactions that absorb energy are termed **endothermic**—the products have more potential energy than the reactants.

   When natural gas burns, methane reacts spontaneously with oxygen producing carbon dioxide and water. This reaction is exothermic, producing a large quantity of heat, as the products have less potential energy than the reactants. In a system composed of a glass of ice cubes in water melting at room temperature, the system absorbs energy from its surroundings and the water has greater potential energy than the ice. The process of ice melting is endothermic, yet it is spontaneous. Clearly, some other factor besides potential energy is involved.
2. *Reactions tend to be spontaneous when the products are less ordered (more random) than the reactants.* Reactions tend to occur spontaneously if the entropy of the products is greater than the entropy of the reactants.

   Consider again the glass of ice in water. It is an increase in entropy that makes the melting of the ice a spontaneous process at room temperature. Molecules of ice are far more ordered (possess lower entropy) than molecules of water moving around randomly (see Section 2.4).

The Change in Free Energy Indicates Whether a Reaction Is Spontaneous

Recall from the second law of thermodynamics that energy transformations are not 100% efficient; some of the energy is lost as an increase in entropy. How much energy is available? The portion of a system's energy that is available to do work is called **free energy,** which is symbolized as G in recognition of the physicist Josiah Willard Gibbs, who developed the concept. In living organisms, free energy accomplishes the chemical and physical work involved in activities such as the synthesis of molecules, movement, and reproduction.

The change in free energy, ΔG ($\Delta G = G_{\text{final state}} - G_{\text{initial state}}$; Δ, pronounced delta, means change in), can be calculated for any chemical reaction from the formula

$$\Delta G = \Delta H - T\Delta S$$

where ΔH is the change in enthalpy, T is the absolute temperature in degrees Kelvin (K, where K = °C + 273.16), and ΔS is the change in entropy.

For a reaction to be spontaneous, ΔG must be negative. As the above formula tells us, both the entropy and the enthalpy of a reaction can influence the overall ΔG. In some processes, such as the combustion of methane, the large loss of potential energy, negative enthalpy (ΔH), dominates in making a reaction spontaneous. In other reactions, such as the melting of ice at room temperature, a decrease in order (ΔS increases) dominates. Once we know what the ΔG is for a reaction, we can determine if the reaction will proceed spontaneously.

Spontaneous Reactions Typically Reach an Equilibrium Point Rather than Going to Completion

In many spontaneous biological reactions, the reactants may not convert completely to products even though the reactions have a negative ΔG. Instead, the reactions run in the direction of completion (toward reactants or toward products) until they reach the equilibrium point, a state of balance between the opposing factors pushing the reaction in either direction. At the equilibrium point, both reactants and products are present and the reactions typically are reversible.

Consider as an example the chemical reaction in which glucose-1-phosphate is converted into glucose-6-phosphate, starting with 0.02 M glucose-1-phosphate as the reactant **(Figure 4.3).** The reaction proceeds spontaneously until there is 0.019 M glucose-6-phosphate (product) and 0.001 M of glucose-1-phosphate (reactant) in the solution. In fact, regardless of the amounts of each you start with, the reaction reaches a point at which there is 95% glucose-6-phosphate and 5% glucose-1-phosphate. This is the point of *chemical equilibrium:* The reaction does not stop, but the rate of the forward reaction equals the rate of the reverse reaction. As a system moves toward equilibrium, its free energy becomes progressively lower and reaches its lowest point when the system achieves equilibrium ($\Delta G = 0$). You can think of a reaction as an energy valley with the equilibrium point being at the bottom. To move away from equilibrium requires free energy and thus will not be spontaneous.

The point of equilibrium of a reaction is related to its ΔG. The more negative the ΔG, the further toward completion the reaction will move before equilibrium is established. If the reaction shown in Figure 4.3 had a positive ΔG, the reaction would run in reverse toward glucose-1-phosphate. Many reactions have a ΔG that is near zero and are thus readily **reversible** by adjusting the concentration of products and reactants slightly. Reversible reactions are written with a double arrow:

$$\underset{\text{reactants}}{A + B} \rightleftharpoons \underset{\text{products}}{C + D}$$

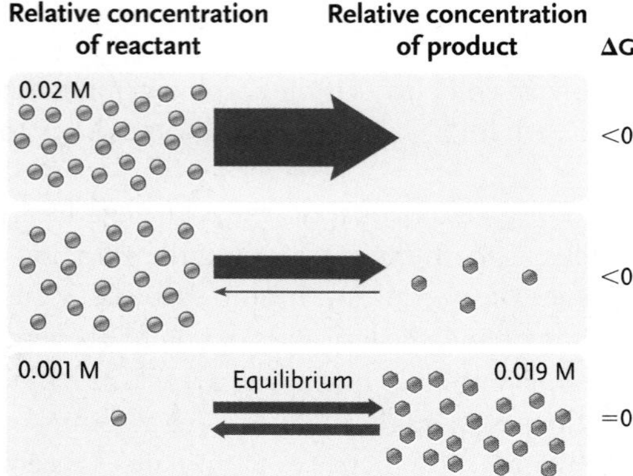
Relative concentration of reactant
Relative concentration of product
ΔG
0.02 M
<0
<0
0.001 M
Equilibrium
0.019 M
=0

FIGURE 4.3
The equilibrium point of a reaction. No matter what quantities of glucose-1-phosphate and glucose-6-phosphate are dissolved in water, when equilibrium is reached, there is 95% glucose-6-phosphate (product) and 5% glucose-1-phosphate (reactant). At equilibrium, the number of reactant molecules being converted to product molecules equals the number of product molecules being converted back to reactant molecules. The reaction at the equilibrium point is reversible; it may be made to run to the right (forward) by adding more reactants, or to the left (reverse) by adding more products.

The reaction in Figure 4.3 is an isolated system and, over time, equilibrium is reached, with ΔG becoming zero. However, many individual reactions in living organisms never reach an equilibrium point because living systems are open; thus, the supply of reactants is constant and, as products are formed, they do not accumulate but become the reactants of another reaction. In fact, the ΔG of life is always negative as organisms constantly take in energy-rich molecules (or light, if they are photosynthetic) and use them to do work. Organisms reach equilibrium, with $\Delta G = 0$, only when they die.

Metabolic Pathways Consist of Exergonic and Endergonic Reactions

Based on the free energy of reactants and products, every reaction can be placed into one of two groups. An **exergonic reaction** (*ergon* = work) **(Figure 4.4A)** is one in that releases free energy—the ΔG is negative because the products contain less free energy than the reactants. In an **endergonic reaction (Figure 4.4B),** the products contain more free energy than the reactants; therefore, ΔG is positive. The reactants involved in endergonic reactions need to gain free energy from the surroundings to form the products of the reaction.

In metabolism, individual reactions tend to be part of a *metabolic pathway,* a series of reactions in which the products of one reaction are used immediately as the reactants for the next reaction in the series. In one type of metabolic pathway, called a **catabolic pathway** (*cata* = downward, as in the sense of a rock releasing energy as it rolls down a hill), energy is released by the breakdown of complex molecules to simpler compounds. (An individual reaction from which energy is released is called a **catabolic reaction.**) An example of a catabolic pathway is cellular respiration, the topic of Chapter 8, in which energy is extracted from the breakdown of food such as glucose. By contrast, in an **anabolic pathway** (*ana* = upward, as in the sense of using energy to push a rock up a hill), energy is used to build complicated molecules from simpler ones; these pathways are often called

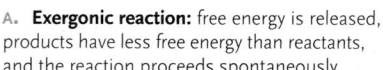

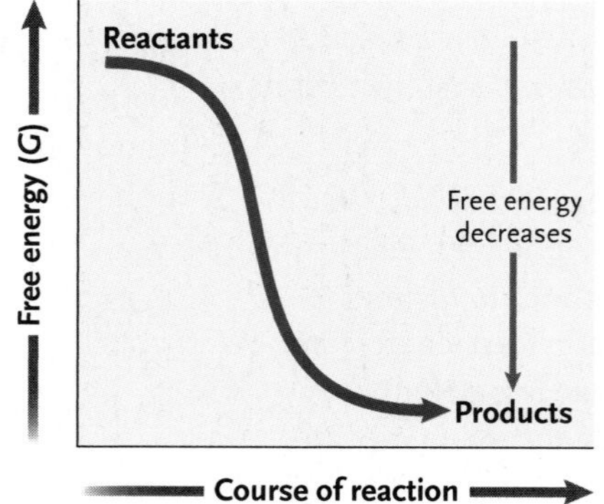
A. **Exergonic reaction:** free energy is released, products have less free energy than reactants, and the reaction proceeds spontaneously
Reactants
Free energy decreases
Products
Free energy (G)
Course of reaction

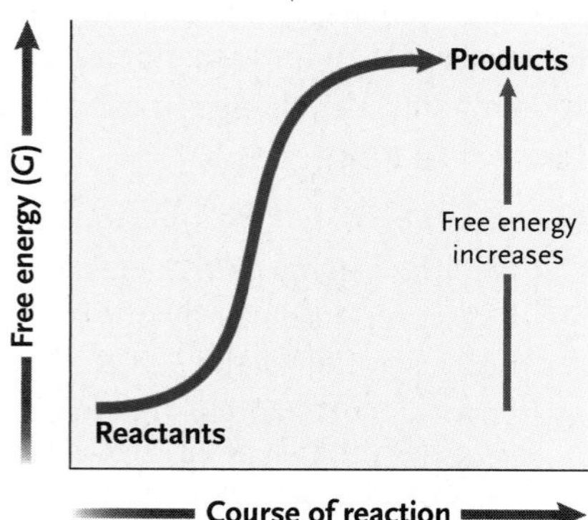
B. **Endergonic reaction:** free energy is gained, products have more free energy than reactants, and the reaction is not spontaneous
Products
Free energy increases
Reactants
Free energy (G)
Course of reaction

FIGURE 4.4
Exergonic (A) and endergonic (B) reactions. An endergonic reaction proceeds only if energy is supplied by an exergonic reaction.

biosynthetic pathways. (An individual reaction that requires energy input is called an **anabolic reaction,** or a **biosynthetic reaction.**) Examples of anabolic pathways include photosynthesis, the topic of Chapter 9, as well as the synthesis of macromolecules such as proteins and nucleic acids.

The overall ΔG of a catabolic pathway is negative, whereas the overall ΔG of an anabolic pathway is positive. Any one pathway consists of a number of individual reactions, each of which may have a positive or negative ΔG. However, when you sum the individual reaction ΔG values for a catabolic pathway, the overall free energy is negative, and for an anabolic reaction, the overall free energy is positive.

STUDY BREAK 4.2 <

1. **What two factors must be taken into account to determine if a reaction will proceed spontaneously?**
2. **What is the relation between ΔG and the concentrations of reactants and products at the equilibrium point of a reaction?**
3. **Distinguish between exergonic and endergonic reactions, and between catabolic and anabolic reactions. How are the two categories of reactions related?**

4.3 Adenosine Triphosphate (ATP): The Energy Currency of the Cell

Many reactions within cells involve the assembly of complex molecules from more simple components. As we learned in the previous section, these reactions have a positive ΔG and are called endergonic; they may be part of both catabolic and anabolic pathways. How the cell supplies energy to drive these endergonic reactions is highly conserved among all forms of life and involves the nucleotide adenosine triphosphate (ATP).

ATP Hydrolysis Releases Free Energy

ATP is the best example of a molecule that contains large amounts of free energy because of its high-energy phosphate bonds. ATP consists of the five-carbon sugar ribose linked to the nitrogenous base adenine and a chain of three phosphate groups **(Figure 4.5).** Much of the potential energy of ATP is associated with the arrangement of the three phosphate groups. As shown in **Figure 4.5A,** each phosphate group is closely associated with the others and their negative charges repel each other strongly, making the bonding arrangement unstable. Removal of one or two of the three phosphate groups is a spontaneous reaction that relieves the repulsion and releases large amounts of free energy.

The breakdown of ATP is a hydrolysis reaction **(Figure 4.5B)** and results in the formation of adenosine diphosphate (ADP) and a molecule of inorganic phosphate (P_i).

$$ATP + H_2O \rightarrow ADP + P_i$$

$$\Delta G = -7.3 \text{ kcal/mol}$$

A. Chemical structure of AMP, ADP, and ATP

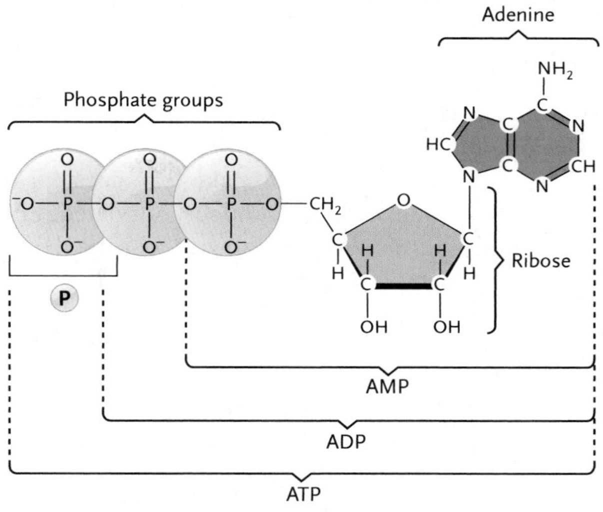
Adenine
Phosphate groups
Ribose
P
AMP
ADP
ATP

With one phosphate group, the molecule is known as AMP; with two phosphates, the molecule is called ADP. Each added phosphate packs additional potential chemical energy into the molecular structure.

B. Hydrolysis reaction removing a phosphate group from ATP

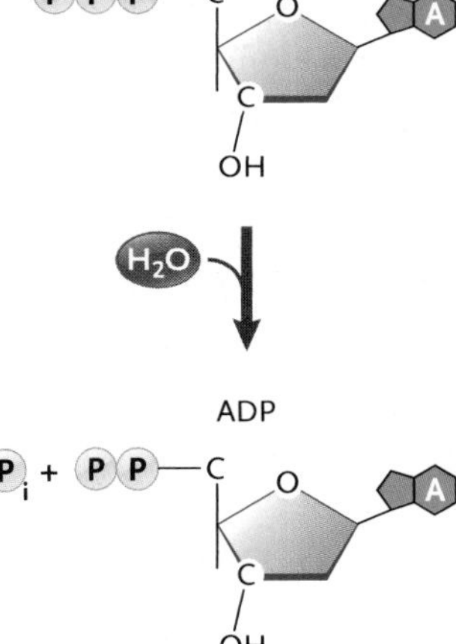
ATP
H_2O
ADP
P_i +

FIGURE 4.5 ATP, the primary molecule that couples energy-requiring reactions to energy-releasing reactions in living organisms. (P_i is the symbol used in this book for inorganic phosphate.)

ADP can be further hydrolyzed to adenosine monophosphate (AMP); however, this releases somewhat less free energy than the hydrolysis of ATP.

Phosphate Groups from ATP Hydrolysis Couple Reactions

The hydrolysis of ATP in water in a test tube releases free energy that simply warms up the water. Within cells, heat production by the isolated hydrolysis of ATP is rare, but does occur during shivering in muscle tissue to maintain body heat. If most ATP were hydrolyzed in this way, it would be difficult for the cell to trap the heat generated and use it to do work. Further, significant heat generation could kill the cell. Given these two points, how do living cells couple the hydrolysis of ATP to an endergonic reaction so that energy is not simply wasted as heat?

In a process called **energy coupling,** ATP comes into close contact with a reactant molecule involved in an endergonic reaction and, when the ATP is hydrolyzed, the terminal phosphate group is transferred to the reactant molecule. The addition of a phosphate group to a molecule is called **phosphorylation,** and the modified molecule is said to have been *phosphorylated*. Phosphorylation makes a molecule less stable (more reactive) than when it is unphosphorylated. Energy coupling requires the action of an enzyme to bring the ATP and reactant molecule into close association. The enzyme has a specific site on it that binds both the ATP and the reactant molecule, allowing for transfer of the phosphate group.

An example of energy coupling that is very common in most cells is the reaction in which ammonia (NH_3) is added to glutamic acid, an amino acid with one amino group, to produce glutamine, an amino acid with two amino groups **(Figure 4.6A):**

$$\text{glutamic acid} + NH_3 \rightarrow \text{glutamine} + H_2O$$
$$\Delta G = +3.4 \text{ kcal/mol}$$

The positive value for ΔG indicates that the reaction cannot proceed spontaneously. Both glutamic acid and glutamine are used in the assembly of proteins. Glutamine is a donor of nitrogen for other reactions in the cell.

How do cells carry out this reaction? As shown in **Figure 4.6B,** the endergonic reaction proceeds by using energy released by ATP hydrolysis. As a first step, the phosphate group removed from ATP is transferred to glutamic acid, forming glutamyl phosphate:

$$\text{glutamic acid} + \text{ATP} \rightarrow \text{glutamyl phosphate} + \text{ADP}$$

The ΔG for this reaction is negative, making the reaction spontaneous, but much less free energy is released than in the hydrolysis of ATP to ADP + P_i. In the second step, glutamyl phosphate reacts with NH_3:

$$\text{glutamyl phosphate} + NH_3 \rightarrow \text{glutamine} + P_i$$

This second reaction also has a negative value for ΔG and is spontaneous.

Even though the reaction proceeds in two steps, it is usually written for convenience as one reaction, with a combined negative value for ΔG:

$$\text{glutamic acid} + NH_3 + \text{ATP} \rightarrow \text{glutamine} + \text{ADP} + P_i$$
$$\Delta G = -3.9 \text{ kcal/mol}$$

Because ΔG is negative, the coupled reaction is spontaneous and releases energy. The difference between -3.9 kcal/mol and the -7.3 kcal/mol released by hydrolyzing ATP to ADP + P_i (that is, $+3.4$ kcal/mol) represents potential chemical energy transferred to the glutamine molecules produced by the reaction. In effect, the coupling system works by joining an exergonic reaction—ATP hydrolysis—to the endergonic biosynthesis reaction, producing an overall reaction that is exergonic. All the endergonic reactions of living organisms, including those of growth, reproduction, movement, and response to stimuli, are made possible by coupling reactions in this way.

A. **Without ATP, reaction is not spontaneous because ΔG is positive**

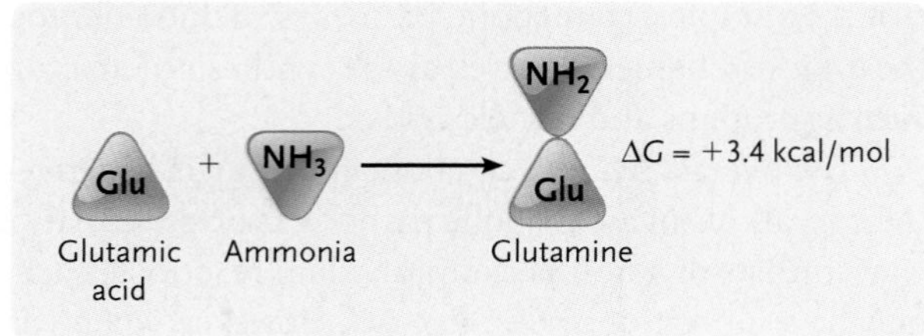
Glu
+
NH_3
NH_2
Glu
$\Delta G = +3.4$ kcal/mol
Glutamic acid
Ammonia
Glutamine

B. **With ATP hydrolysis, reaction is spontaneous because overall ΔG is negative**

1 Free energy of the terminal phosphate is transferred to glutamic acid.

2 Complex reacts spontaneously with ammonia to produce glutamine.

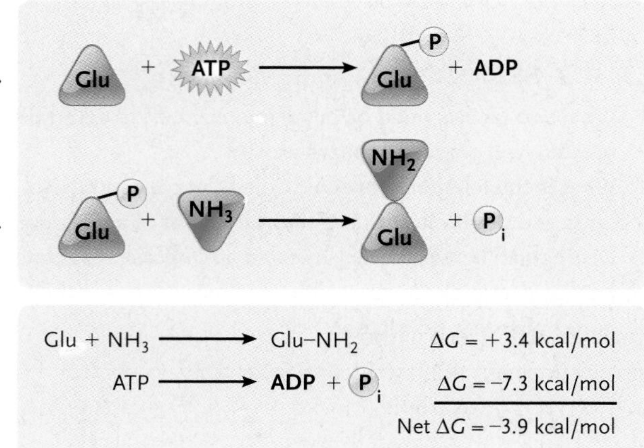
Glu
+
ATP
P
Glu
+ ADP
P
Glu
+
NH_3
NH_2
Glu
+ P_i

Glu + NH_3 → Glu–NH_2 $\Delta G = +3.4$ kcal/mol

ATP → ADP + P_i $\Delta G = -7.3$ kcal/mol

Net $\Delta G = -3.9$ kcal/mol

FIGURE 4.6
Energy coupling using ATP in the synthesis of glutamine from glutamic acid and ammonia.

Cells also Couple Reactions to Regenerate ATP

We have just seen how the hydrolysis of ATP to ADP and P_i is an exergonic reaction that can be coupled to make otherwise endergonic reactions proceed spontaneously. These coupling reactions occur continuously in living cells, consuming a tremendous amount of ATP. How do cells generate that ATP? ATP is a renewable resource that is synthesized by recombining ADP and P_i. Although ATP hydrolysis is an exergonic reaction, ATP synthesis from ADP and P_i is an energy-requiring, endergonic reaction. The energy for ATP synthesis comes from the exergonic breakdown of complex molecules that contain an abundance of free energy: carbohydrates, proteins, and fats. Essentially, what we are referring to is food.

The continual hydrolysis and resynthesis of ATP is called the **ATP/ADP cycle (Figure 4.7).** Approximately 10 million ATP molecules are hydrolyzed and resynthesized each second in a typical cell, illustrating that this cycle operates at an astonishing rate. In fact, if ATP were not regenerated from ADP and P_i, the average human would use an estimated 75 kg of ATP per day.

STUDY BREAK 4.3 <

1. **How does the structure of the ATP molecule store and release energy?**
2. **How are coupled reactions important to cell function? How is ATP involved in coupled reactions?**

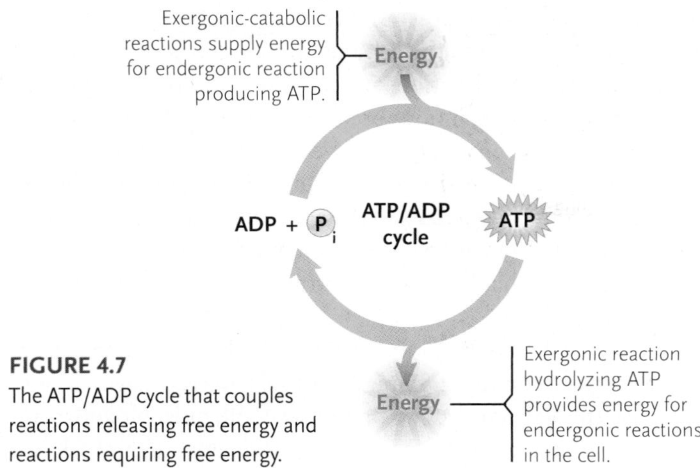
Exergonic-catabolic reactions supply energy for endergonic reaction producing ATP.
Energy
ADP + P i
ATP/ADP cycle
ATP
Energy
Exergonic reaction hydrolyzing ATP provides energy for endergonic reactions in the cell.

FIGURE 4.7
The ATP/ADP cycle that couples reactions releasing free energy and reactions requiring free energy.

A. Activation energy barrier in the oxidation of glucose

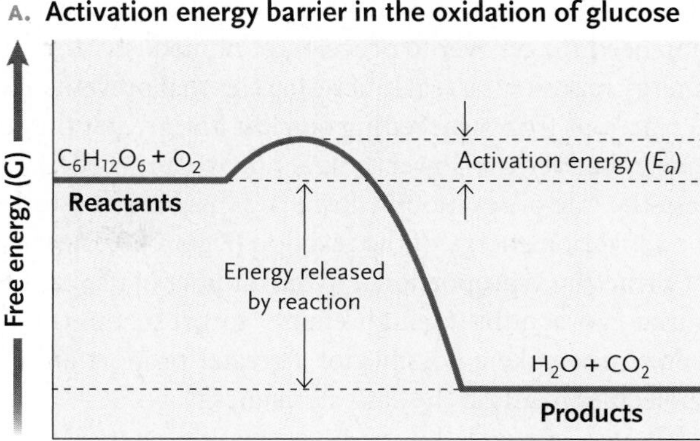
Free energy (G)
$C_6H_{12}O_6 + O_2$
Reactants
Activation energy (E_a)
Energy released by reaction
$H_2O + CO_2$
Products
Direction of reaction

B. "Activation energy" barrier in the movement of a rock downhill

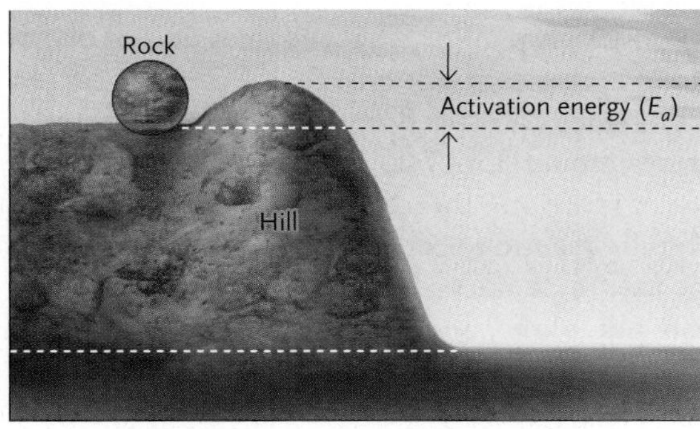
Rock
Activation energy (E_a)
Hill

FIGURE 4.8
Activation energy (E_a).

4.4 Role of Enzymes in Biological Reactions

The laws of thermodynamics are useful because they can tell us if a process will occur without an input of energy, that is, spontaneously. However, the laws do not tell us anything about the rate of a reaction. For example, even though the breakdown of sucrose into glucose and fructose is a spontaneous reaction with a ΔG of -7 kcal/mol, a solution of sucrose will sit for years without any detectable glucose or fructose forming. That is, *spontaneous reactions do not necessarily proceed rapidly.* In this section, we discuss how the speed of a reaction can be altered by enzymes.

Activation Energy Represents a Kinetic Barrier for a Reaction

In our example above, what prevents sucrose from being converted rapidly into glucose and fructose? Chemical reactions require bonds to break and new bonds to form. For bonds to be broken, they must be strained or otherwise made less stable so that breakage actually can occur. To get reacting molecules into a less stable (more unstable) state requires a small input of energy. Thus, even though a reaction is spontaneous (negative ΔG), the reaction will not start unless a relatively small boost of energy is added **(Figure 4.8A)**. This initial energy investment required to start a reaction is called the **activation energy,** symbolized E_a. Molecules that gain the necessary activation energy occupy what is called the transition state, where bonds are unstable and are ready to be broken.

A rock resting in a depression at the top of a hill provides a physical example of activation energy **(Figure 4.8B)**. The rock will not roll downhill spontaneously, even though its position represents considerable potential energy and the total "reaction"—the downward movement of the rock—is spontaneous and releases free energy. In this example, the activation energy is the effort required to raise the rock over the rim of the depression and start its downhill roll.

What provides the activation energy in chemical reactions? The molecules taking part in chemical reactions are in constant motion at temperatures above absolute zero. Periodically, reacting molecules gain enough energy to reach the transition state. For a solution of sucrose, the number of molecules that reach the transition state at any one time is very small. However, if a significant number of reactant molecules reach the transition state, the free energy that is released may be enough to get the remaining reactants to the transition state. For example, in the chemistry lab, heat commonly provides the energy needed for reactant molecules to get to the transition state and, therefore, to speed up the rate of a reaction. In biology, however, using heat to speed up a reaction is problematic for two reasons: (1) High temperatures destroy the structural components of cells, particularly proteins; and (2) an increase in temperature would speed up all possible chemical reactions in a cell, not just the specific reactions that are part of metabolism.

Enzymes Accelerate Reactions by Reducing the Activation Energy

How can you increase the rate of a reaction without raising the temperature? The answer is that you can use a **catalyst,** a chemical agent that accelerates the rate of a reaction without itself being changed by the reaction. The process of accelerating a reaction with a catalyst is called **catalysis,** and we say that the chemical agent responsible *catalyzes* the reaction. The most common biological catalysts are proteins called **enzymes** (*enzym* = in yeast).

Recall that the activation energy represents a hurdle that a reaction needs to get over to proceed spontaneously. This activation energy represents a real kinetic barrier that prevents spontaneous reactions from proceeding quickly. The greater the activation energy barrier, the slower the reaction will proceed. Enzymes increase the rate of reaction by lowering this barrier—by lowering the activation energy of the reaction **(Figure 4.9).** Because the rate of a reaction is proportional to the number of reactant molecules that can acquire enough energy to get to the transition state, enzymes make it possible for a greater proportion of reactant molecules to attain the activation energy.

Although enzymes lower the activation energy of a reaction, as shown in Figure 4.9, they do not alter the change in free energy (ΔG) of the reaction. The free energy values of the reactants and products are the same; the only difference is the path the reaction takes.

Let us be clear about what enzymes do and do not do with regard to biological reactions. By lowering the activation energy, enzymes DO speed up the rate of spontaneous (exergonic) reactions. However, enzymes DO NOT supply free energy to a reaction. Therefore, enzymes CANNOT make an endergonic reaction proceed spontaneously. ATP hydrolysis can be used to make an endergonic reaction proceed spontaneously but, alone, an enzyme cannot. Lastly, enzymes DO NOT change the ΔG of a reaction.

Cells have thousands of different enzymes. They vary from relatively small molecules, with single polypeptide chains containing as few as 100 amino acids, to large complexes that include many polypeptide chains totaling thousands of amino acids. It is the three-dimensional structure of a protein—its *conformation*—that determines the protein's function. Conformation is determined by the number of polypeptides in the protein and the amino acid sequence of each polypeptide. (The three-dimensional structure of proteins is described in Section 3.4.) Each enzyme has a specific protein structure that has evolved to catalyze a specific reaction.

Different enzymes are found in all areas of the cell, from the aqueous cell solution to the cell membranes. Other enzymes are released to catalyze reactions outside the cell. For example, enzymes that catalyze reactions breaking down food molecules are released from cells into the digestive cavity in all animals.

The majority of enzymes have names ending in *-ase*. The rest of the name typically relates to the substrate of the enzyme or to the type of reaction with which the enzyme is associated. For example, enzymes that break down proteins are called *proteinases* or *proteases*.

Enzymes are not the only biological molecules capable of accelerating reaction rates. Some RNA molecules (see Section 4.6) also have this capacity. As we do in this book, most biologists reserve the term *enzyme* for protein molecules that can accelerate reaction rates and call the RNA molecules with this capacity *ribozymes*.

Enzymes Combine with Reactants and Are Released Unchanged

In enzymatic reactions, an enzyme combines briefly with reacting molecules and is released unchanged when the reaction is complete. For example, the enzyme in **Figure 4.10,** hexokinase, catalyzes the reaction:

$$\text{glucose} + \text{ATP} \xrightarrow{\text{hexokinase}} \text{glucose-6-phosphate} + \text{ADP}$$

Writing the enzyme name (here, hexokinase) above the reaction arrow indicates that it is required but not involved as a reactant or a product. The reactant that an enzyme acts on is called the **substrate,** or substrates if the enzyme binds two or more molecules. In this reaction, the substrates are glucose and ATP.

Each type of enzyme catalyzes the reaction of a single type of substrate molecule or a group of closely related molecules. This **enzyme specificity** explains why a typical cell needs about 4,000 different enzymes to function properly. Notice in Figure 4.10 that the enzyme is much larger than the substrate. Moreover, the substrate interacts with only a very small region of the

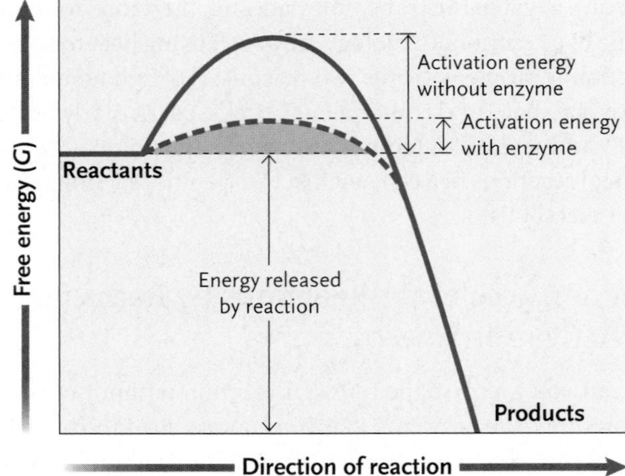
Activation energy without enzyme
Activation energy with enzyme
Reactants
Free energy (G)
Energy released by reaction
Products
Direction of reaction

FIGURE 4.9
Effect of enzymes in reducing the activation energy *(E_a)*. The reduction allows biological reactions to proceed rapidly at the relatively low temperatures that can be tolerated by living organisms.

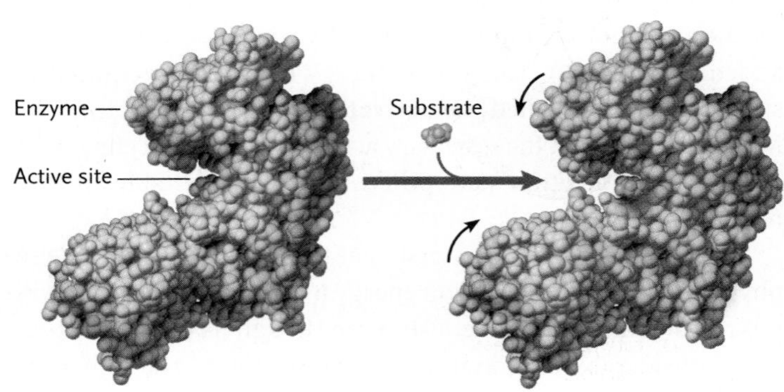
Enzyme
Active site
Substrate

FIGURE 4.10
Combination of an enzyme, hexokinase (in blue), with its substrate, glucose (in yellow). Hexokinase catalyzes the phosphorylation of glucose to form glucose-6-phosphate. The phosphate group that is transferred to glucose is not shown. Note how the enzyme undergoes a conformational change, closing the active site more tightly as it binds the substrate.

enzyme called the **active site,** the place on the enzyme where catalysis occurs. The active site is usually a pocket or groove that is formed when the enzyme protein folds into its functional three-dimensional shape.

When the substrate binds initially at the active site of the enzyme, both the enzyme and substrate molecules are distorted, which stabilizes the substrate molecule in the transition state and makes its chemical bonds ready for reaction. Figure 4.10 illustrates the change in enzyme conformation on substrate binding, a phenomenon called *induced fit.*

Once an enzyme–substrate complex is formed, catalysis occurs, with the enzyme converting the substrate into one or more products. Because enzymes are released unchanged after a reaction, enzyme molecules cycle repeatedly through reactions, combining with reactants and releasing products **(Figure 4.11).** Depending on the enzyme, the rate at which reactants are bound and catalyzed and at which products are released varies from 100 times to 10 million times per second. These astoundingly high rates of catalysis mean that a small number of enzyme molecules can catalyze large numbers of reactions.

Some enzymes consist of polypeptide chains only. However, many enzymes require a **cofactor,** a nonprotein group that binds precisely to the enzyme, for catalytic activity. The role of a cofactor in an enzyme's catalytic activity varies with the cofactor and the enzyme. Some cofactors are metallic ions, including iron, copper, magnesium, zinc, and manganese. Other cofactors are small organic molecules called **coenzymes,** which are often derived from vitamins. Some coenzymes bind loosely to enzymes; others—called *prosthetic groups*—bind tightly.

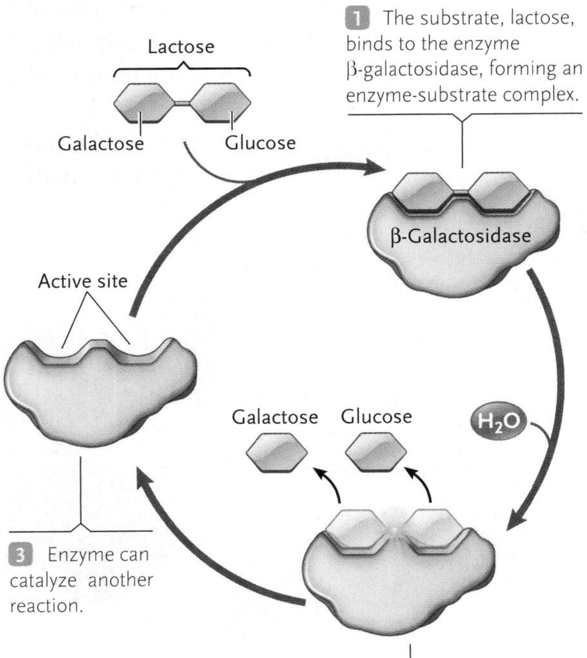

1 The substrate, lactose, binds to the enzyme β-galactosidase, forming an enzyme-substrate complex.

2 β-Galactosidase catalyzes the breakage of the bond between the two sugars of lactose, and the products are released.

3 Enzyme can catalyze another reaction.

FIGURE 4.11
The catalytic cycle of enzymes.
Shown is the enzyme β-galactosidase, which cleaves the sugar lactose to produce glucose and galactose.

Enzymes Reduce the Activation Energy by Stabilizing the Transition State

A central question of enzyme function is how do enzymes reduce the activation energy of a reaction? Recall that substrate molecules need to be in the transition state for catalysis to occur. Enzymes function by stabilizing the transition state, doing so through three major mechanisms:

1. Enzymes *bring the reacting molecules together.* Reacting molecules can assume the transition state only when they collide. Binding of reactant molecules to an enzyme's active site brings them close together in the correct orientation for catalysis to take place.
2. Enzymes *expose the reactant molecules to altered charge environments that promote catalysis.* In some systems, the active site of the enzyme contains ionic groups whose positive or negative charges alter the substrate in a way that favors catalysis.
3. Enzymes *change the shape of the substrate molecules.* As mentioned previously, the active site can distort substrate molecules into a conformation that mimics the transition state.

Regardless of the mechanism, the binding of the substrate molecule(s) to the active site results in stabilization of the substrate in the transition state conformation. Substrate molecules do attain the transition state in the absence of an enzyme, but that is a rare event. The inclusion of an enzyme enables many more molecules to reach the transition state more rapidly. Fundamentally, this is why an enzyme speeds up the rate of a reaction.

STUDY BREAK 4.4

1. **How do enzymes increase the rates of the reaction they catalyze?**
2. **Can enzymes alter the ΔG of a reaction?**

THINK OUTSIDE THE BOOK

Use the Internet to determine which enzymes are involved in meat tenderizers, the production of soft cheeses such as brie and camembert, and the production of yogurt. For each enzyme you identify, give the organism that makes it, as well as the reaction catalyzed and how that reaction relates to its use in food.

4.5 Conditions and Factors That Affect Enzyme Activity

Several conditions can alter enzyme activity, including changes in the concentration of substrate and other molecules that can bind to enzymes. In addition, a number of control mechanisms modify enzyme activity, thereby adjusting reaction rates to meet a cell's requirements for chemical products. Changes in temperature and pH can also have a significant effect on enzyme activity.

Enzyme and Substrate Concentrations Influence the Rate of Catalysis

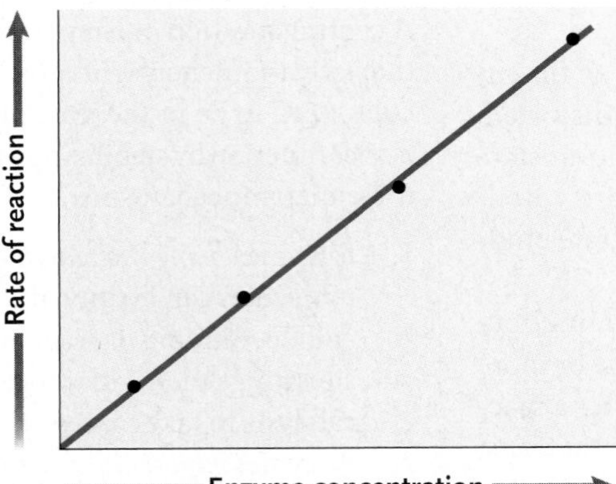
A. Rate of reaction as function of enzyme concentration (substrate at high concentration)
Rate of reaction
Enzyme concentration

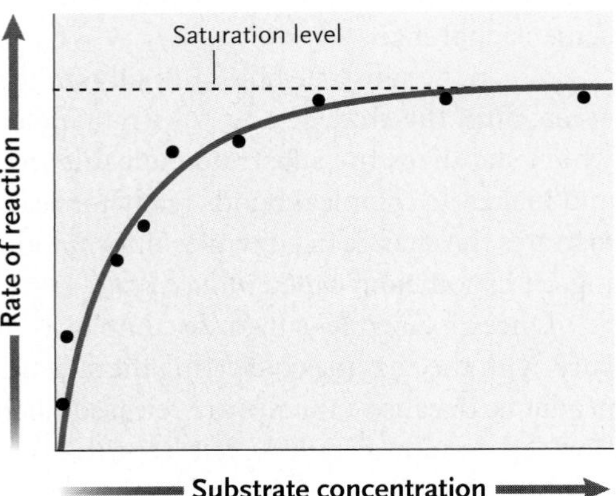
B. Rate of reaction as function of substrate concentration (enzyme amount constant)
Saturation level
Rate of reaction
Substrate concentration

FIGURE 4.12
Effect of increasing enzyme concentration or substrate concentration on the rate of an enzyme-catalyzed reaction.

Biochemists use a wide range of approaches to study enzymes. These include molecular tools to study the structure and regulation of the gene encoding the enzyme and sophisticated computer algorithms to model the three-dimensional structure of the enzyme itself. The most fundamental and central approach has been to determine the rate of an enzyme-catalyzed reaction and how it changes in response to altering certain experimental parameters. Typically this requires isolating the enzyme from a cell, incubating it in an appropriate buffered solution, and supplying the reaction mixture with substrate. With these constituents, a researcher can then determine the rate of catalysis, usually by measuring the rate at which the product of the reaction is formed.

As shown in **Figure 4.12A,** in the presence of excess substrate (that is, at high concentrations), the rate of catalysis is proportional to the amount of enzyme. As enzyme concentration increases, the rate of product formation increases. In this system, the rate of the reaction is limited by the amount of enzyme in the reaction mixture. What happens to the reaction rate if we keep the concentration of enzyme constant at some intermediate level and change the concentration of substrate from low to high? As shown in **Figure 4.12B,** at very low concentrations, substrate molecules collide so infrequently with enzyme molecules that the reaction proceeds slowly. As the substrate concentration increases, the reaction rate initially increases as enzyme and substrate molecules collide more frequently. But, as the enzyme molecules approach the maximum rate at which they can combine with reactants and release products, increasing substrate concentration has a smaller and smaller effect, and the rate of reaction eventually levels off. When the enzymes are cycling as rapidly as possible, further increases in substrate concentration have no effect on the reaction rate. At this point, the enzymes are said to be **saturated** with the substrate (the saturation level is shown by a horizontal dashed line in Figure 4.12B).

Enzyme Inhibitors Have Characteristic Effects on Enzyme Activity

The rate at which an enzyme can catalyze a reaction can be lowered by *enzyme inhibitors,* nonsubstrate molecules that bind to an enzyme and decrease its activity. Some inhibitors work by binding to the active site of an enzyme, whereas other inhibitors bind to critical sites located elsewhere in the structure of an enzyme.

Inhibitors that bind to the active site have shapes resembling the normal substrate closely enough to fit into and occupy the site, thereby blocking access for the normal substrate and slowing the reaction rate. If the concentration of the inhibitor is high enough, the reaction may stop completely. Inhibition of this type is called **competitive inhibition** because the inhibitor competes with the normal substrate for binding to the active site **(Figure 4.13A).** Competitive inhibitors are useful in enzyme research because their structure helps identify the region of a normal substrate that binds to an enzyme.

Specific molecules inhibit enzyme activity without competing with substrate molecules at the active site **(Figure 4.13B).** Instead, they bind to an enzyme at a location other than the active site. This binding, called **noncompetitive inhibition,** often results in a change to the conformation of the enzyme that reduces the ability of the active site to bind substrate efficiently.

Inhibitors differ with respect to how strongly they bind to enzymes. In *reversible inhibition,* the binding of an inhibitor to

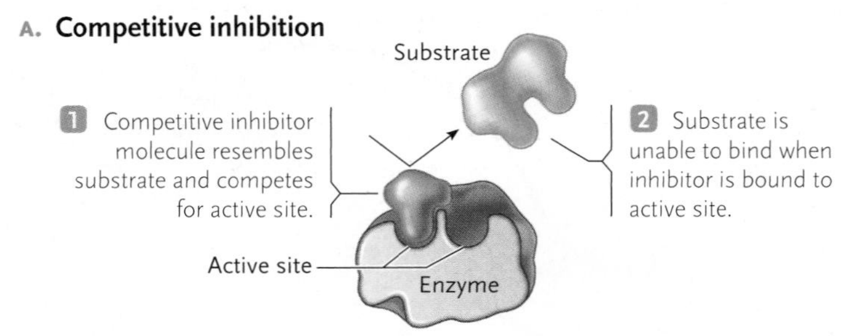
A. Competitive inhibition
Substrate
1 Competitive inhibitor molecule resembles substrate and competes for active site.
2 Substrate is unable to bind when inhibitor is bound to active site.
Active site
Enzyme

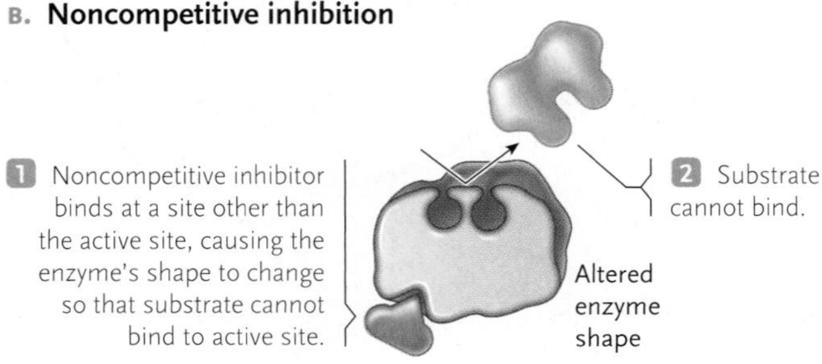
B. Noncompetitive inhibition
1 Noncompetitive inhibitor binds at a site other than the active site, causing the enzyme's shape to change so that substrate cannot bind to active site.
2 Substrate cannot bind.
Altered enzyme shape

FIGURE 4.13
How competitive **(A)** and noncompetitive **(B)** inhibitors reduce enzyme activity.

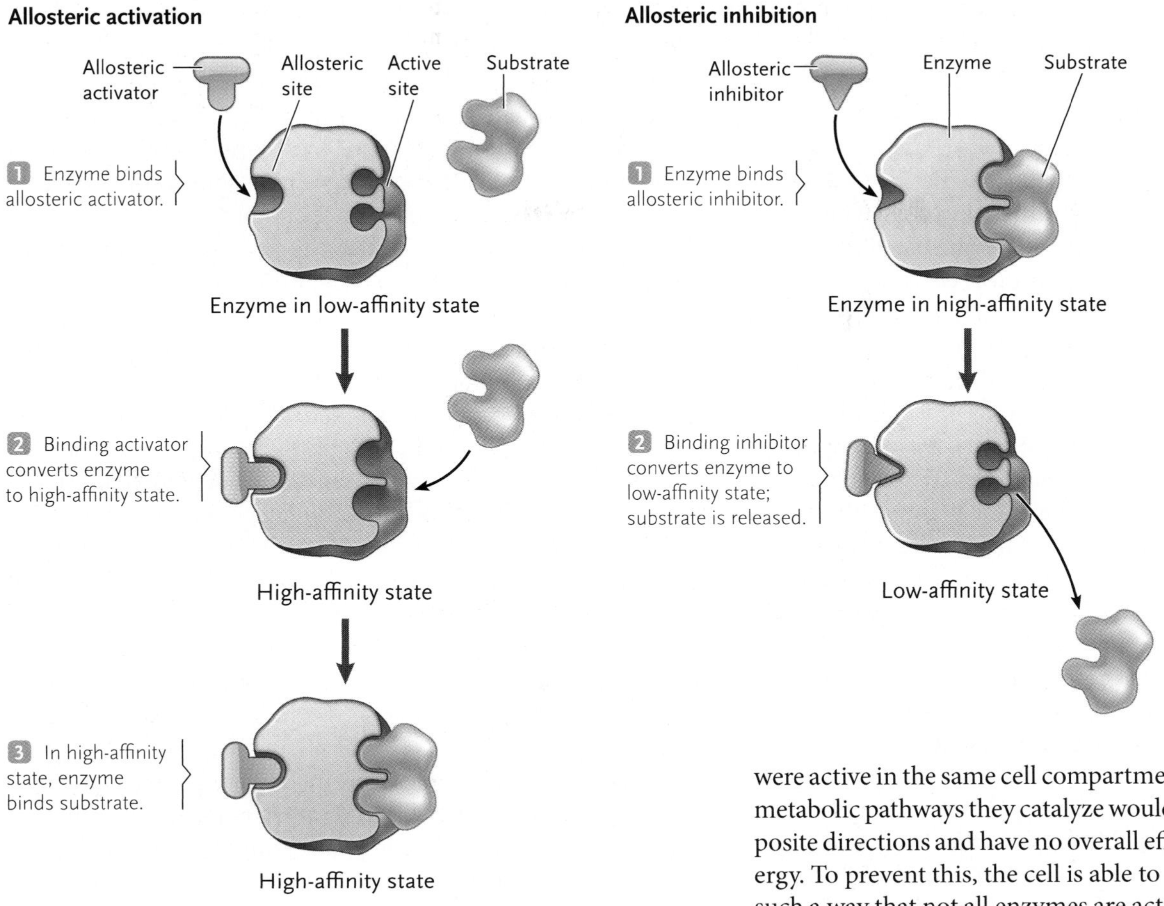

Allosteric activation
Allosteric activator
Allosteric site
Active site
Substrate
1 Enzyme binds allosteric activator.
Enzyme in low-affinity state
2 Binding activator converts enzyme to high-affinity state.
High-affinity state
3 In high-affinity state, enzyme binds substrate.
High-affinity state
Allosteric inhibition
Allosteric inhibitor
Enzyme
Substrate
1 Enzyme binds allosteric inhibitor.
Enzyme in high-affinity state
2 Binding inhibitor converts enzyme to low-affinity state; substrate is released.
Low-affinity state

FIGURE 4.14 Allosteric regulation.

an enzyme is weak, and when the inhibitor releases, enzyme activity returns to normal. By contrast, some inhibitors bind so strongly to an enzyme through the formation of covalent bonds that the enzyme is completely disabled—this is *irreversible inhibition*. Not surprisingly, many irreversible inhibitors that act on critical enzymes are toxic to the cell. They include a wide variety of drugs and pesticides. Cyanide is a potent poison because it binds strongly to and inhibits cytochrome oxidase, the enzyme that catalyzes the last step of electron transfer in cellular respiration (see Chapter 8). In addition, many antibiotics are toxins that inhibit enzyme activity in bacteria. Irreversible inhibition can only be overcome if the cell synthesizes more of the enzyme.

Cells Adjust Enzyme Activity to Meet Their Needs for Reaction Products

Cells adjust the activity of many enzymes upward or downward to meet their needs for reaction products. Several mechanisms are used in this regulation, including competitive and noncompetitive inhibition, a form of noncompetitive control called *allosteric regulation,* and covalent modification of enzyme structure by the addition or removal of chemical groups.

REGULATION BY INHIBITORS Reversible inhibition of enzyme activity serves as an important mechanism of metabolism regulation. A typical cell contains thousands of enzymes, and for many enzymes that synthesize a specific molecule, usually another enzyme exists that catalyzes its breakdown. If both of these enzymes were active in the same cell compartment at the same time, the two metabolic pathways they catalyze would run simultaneously in opposite directions and have no overall effect other than using up energy. To prevent this, the cell is able to regulate enzyme activity in such a way that not all enzymes are active at the same time.

Many enzymes are regulated by natural inhibitors that work either competitively or noncompetitively. Typically, the combination of these inhibitors and the enzyme is fully reversible. If the concentration of the inhibitor increases, it combines with the enzymes in greater numbers, thereby interfering with enzyme activity and decreasing the rate of the reaction. If the concentration of the inhibitor decreases, its combination with enzymes decreases proportionately and the rate of the reaction increases. Control by the inhibitors changes enzyme activity precisely to meet the needs of the cell for the products of the reaction catalyzed by the enzyme.

In the mechanism of **allosteric regulation** (*allo* = different; *stereo* = shape), enzyme activity is controlled by the reversible binding of a regulatory molecule to the **allosteric site,** a location on the enzyme outside the active site. The mechanism may either increase or decrease enzyme activity. Because allosteric inhibitors work by binding to sites separate from the active site, their action is noncompetitive.

Enzymes controlled by allosteric regulation typically have two alternate conformations controlled from the allosteric site. In one conformation, called the *high-affinity state* (the active form), the enzyme binds strongly to its substrate; in the other conformation, *the low-affinity state* (the inactive form), the enzyme binds the substrate weakly or not at all. Binding with regulatory substances may induce either state: Binding an **allosteric activator** converts it from the low- to high-affinity state and therefore increases enzyme activity, and binding an **allosteric inhibitor** converts an allosteric enzyme from the high- to low-affinity state and therefore decreases enzyme activity **(Figure 4.14).**

Frequently, allosteric inhibitors are a product of the metabolic pathway they regulate. If the product accumulates in excess, its effect as an inhibitor automatically slows or stops the enzymatic reaction producing it, typically by inhibiting the enzyme that catalyzes the first reaction of the pathway. If the product becomes too scarce, the inhibition is reduced and its production increases. Regulation of this type, in which the product of a reaction acts as a regulator of the reaction, is termed **feedback inhibition** (also called **end-product inhibition**). Feedback inhibition prevents cellular resources from being wasted in the synthesis of molecules made at intermediate steps of the pathway.

The biochemical pathway that makes the amino acid isoleucine from threonine is an excellent example of feedback inhibition. The pathway proceeds in five steps, each catalyzed by an enzyme **(Figure 4.15).** The end product of the pathway, isoleucine, is an allosteric inhibitor of the first enzyme of the pathway, threonine deaminase. If the cell makes more isoleucine than it needs, isoleucine binds reversibly with threonine deaminase at the allosteric site, converting the enzyme to the low-affinity state and inhibiting its ability to combine with threonine, the substrate for the first reaction in the pathway. If isoleucine levels drop too low, the allosteric site of threonine deaminase is vacated, the enzyme is converted to the high-affinity state, and isoleucine production increases.

REGULATION BY CHEMICAL MODIFICATION Many key enzymes are regulated by chemical linkage to other substances, typically ions, functional groups such as phosphate or methyl groups, or units derived from nucleotides. The regulatory substances induce folding changes in the enzyme that increase or decrease its activity.

For example, chemical modification by the addition or removal of phosphate groups is a highly significant mechanism of cellular regulation that is used by all organisms from bacteria to humans. Typically, regulatory phosphate groups derived from ATP or other nucleotides are added to the regulated enzymes by other enzymes known as protein kinases. The addition of a phosphate group—*phosphorylation*—either increases or decreases enzyme activity or activates or deactivates the enzyme, depending on the particular enzyme and where the phosphate group is added to the enzyme.

Removal of phosphate groups—*dephosphorylation*—reverses the effects of phosphorylation. Dephosphorylation is carried out by a different group of enzymes called *protein phosphatases.* The balance between phosphorylation and dephosphorylation of the enzymes modified by the protein kinases and protein phosphatases closely regulates cellular activity, often as a part of the response to external signal molecules (see Chapter 7).

pH and Temperature Are Key Factors Affecting Enzyme Activity

The activity of most enzymes is strongly altered by changes in pH and temperature. Characteristically, enzymes reach maximal activity within a narrow range of pH or temperature; at levels outside this range, enzyme activity decreases. These effects produce a typically peaked curve when enzyme activity is plotted, with the peak where pH or temperature produces maximal activity.

EFFECTS OF pH CHANGES Typically, each enzyme has an optimal pH where it operates at peak efficiency in speeding the rate of its biochemical reaction **(Figure 4.16).** On either side of this pH optimum, the rate of the catalyzed reaction decreases. The effects become more extreme at pH values farther from the optimum,

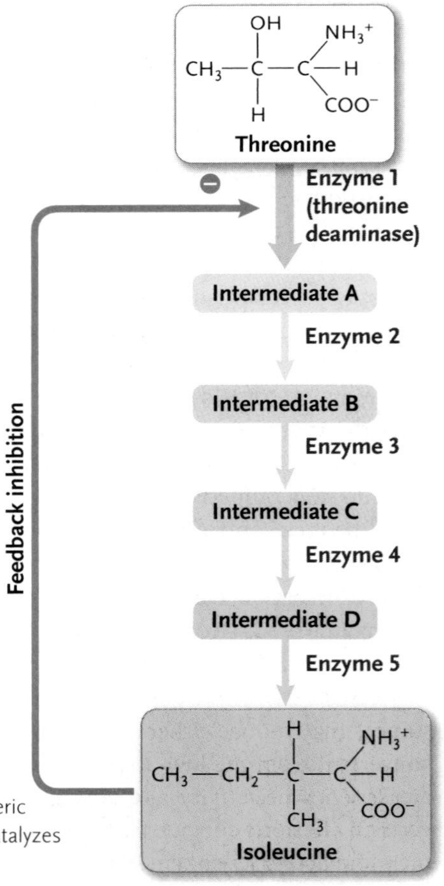

$CH_3-CH(OH)-C(H)(NH_3^+)(COO^-)$
Threonine
Enzyme 1 (threonine deaminase)
Intermediate A
Enzyme 2
Intermediate B
Enzyme 3
Intermediate C
Enzyme 4
Intermediate D
Enzyme 5
$CH_3-CH_2-CH(CH_3)-C(H)(NH_3^+)(COO^-)$
Isoleucine
Feedback inhibition

FIGURE 4.15
Feedback inhibition in the pathway that produces isoleucine from threonine.
If the product of the pathway, isoleucine, accumulates in excess, it slows or stops the pathway by acting as an allosteric inhibitor of the enzyme that catalyzes the first step in the pathway.

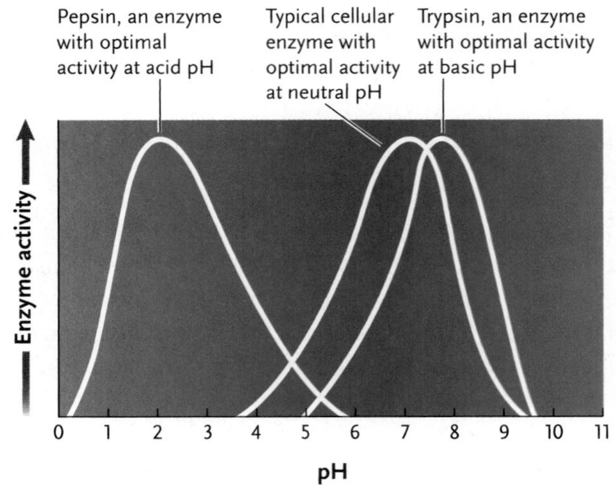
Pepsin, an enzyme with optimal activity at acid pH
Typical cellular enzyme with optimal activity at neutral pH
Trypsin, an enzyme with optimal activity at basic pH
Enzyme activity
0 1 2 3 4 5 6 7 8 9 10 11
pH

FIGURE 4.16
Effects of pH on enzyme activity. An enzyme typically has an optimal pH at which it is most active; at pH values above or below the optimum, the rate of enzyme activity drops off. At extreme pH values, the rate drops to zero.

until the rate drops to zero. An enzyme's dependence on pH typically is caused by ionizable amino acids—a change in pH from the optimal value alters the charges of those amino acids, which modifies the conformation of the protein and eventually causes denaturation of the enzyme.

Most enzymes have an optimum of about pH 7, near the pH of the cellular contents. Enzymes that are secreted from cells may have pH optima farther from neutrality. For example, pepsin, a protein-digesting enzyme secreted into the stomach, has an optimum of pH 1.5, close to the acidity of stomach contents. Similarly, trypsin, also a protein-digesting enzyme, has an optimum of about pH 8, allowing it to function well in the somewhat alkaline contents of the intestine where it is secreted.

EFFECTS OF TEMPERATURE CHANGES The effects of temperature changes on enzyme activity reflect two distinct processes. First, temperature has a general effect on chemical reactions of all kinds. As the temperature rises, the rate of chemical reactions typically increases. This effect reflects increases in the kinetic motion of all molecules, with more frequent and stronger collisions as the temperature rises. Second, temperature has an effect on all proteins, including enzymes. As the temperature rises, the kinetic motions of the amino acid chains of an enzyme increase, along with the strength and frequency of collisions between enzymes and surrounding molecules. At some point, these disturbances become strong enough to denature the enzyme: The hydrogen bonds and other forces that maintain its three-dimensional structure break, making the enzyme unfold and lose its function.

The two effects of temperature act in opposition to each other to produce characteristic changes in the rate of enzymatic catalysis **(Figure 4.17)**. In the range of 0° to about 40°C, the reaction rate doubles for every 10°C increase in temperature. Above 40°C, the increasing kinetic motion begins to unfold the enzyme, reducing the rate of increase in enzyme activity. At some point, as temperature continues to rise, the unfolding causes the reaction rate to level off at a peak. Further increases cause such extensive unfolding that the reaction rate decreases rapidly to zero. For most enzymes, the peak in activity lies between 40° and 50°C; the drop-off becomes steep at 55°C and falls to zero at about 60°C. Thus, the rate of an enzyme-catalyzed reaction peaks at a temperature at which kinetic motion is greatest but no significant unfolding of the enzyme has occurred.

Although most enzymes have a temperature optimum between 40° and 50°C, some have activity peaks below or above this range. For example, the enzymes of maize (corn) pollen function best near 30°C and undergo steep reductions in activity above 32°C. As a result, environmental temperatures above 32°C can seriously inhibit the growth of corn crops. Many animals living in frigid regions have enzymes with much lower temperature optima than average. For example, the enzymes of arctic snow fleas are most active at −10°C. At the other extreme are the enzymes of archaeans that live in hot springs, which are so resistant to denaturation that they remain active at temperatures of 85°C or more.

A. **Effect of temperature on enzyme activity**

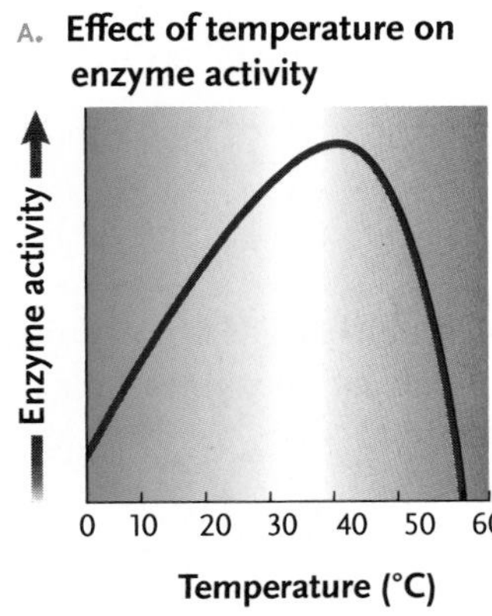
Enzyme activity
0
10
20
30
40
50
60
Temperature (°C)

B. **Visible effect of a heat-sensitive enzyme**

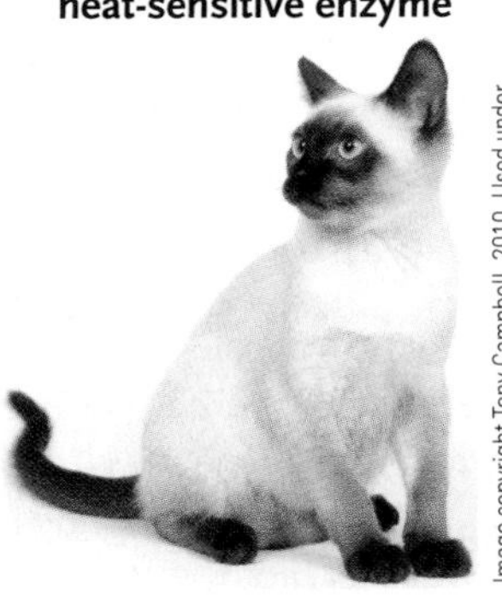

FIGURE 4.17
Effect of temperature on enzyme activity. **(A)** As the temperature rises, the rate of the catalyzed reaction increases proportionally until the temperature reaches the point at which the enzyme begins to denature. The rate drops off steeply as denaturation progresses and becomes complete. **(B)** Visible effects of environmental temperature on enzyme activity in Siamese cats. The fur on the extremities—ears, nose, paws, and tail—contains more dark brown pigment (melanin) than the rest of the body. A heat-sensitive enzyme controlling melanin production is denatured in warmer body regions, so dark pigment is not produced and fur color is lighter.

STUDY BREAK 4.5 <

1. **Why do enzyme-catalyzed reactions reach a saturation level when substrate concentration is increased?**
2. **Distinguish between competitive and noncompetitive inhibition.**
3. **Why will the activity of an enzyme eventually decrease to zero as the temperature rises?**

4.6 RNA-Based Biological Catalysts: Ribozymes

In 1981, biochemist Thomas R. Cech of the University of Colorado at Boulder discovered a group of RNA molecules that appeared to be capable of accelerating the rate of certain biological reactions without being changed by the reactions. This discovery was a great surprise to the scientific community. Further work demonstrated that these RNA-based catalysts, now called **ribozymes,** are part of the biochemical machinery of all cells. Cech and another scientist, Yale University biochemist Sidney Altman, received the Nobel Prize in 1989 for their research establishing that ribozymes are essential cellular catalysts.

Most of the known ribozymes speed the cutting and splicing reactions that remove surplus segments from RNA molecules as part of their conversion into finished form. Some have other functions, however. For example, Harry F. Noller and his coworkers at the University of California, Santa Cruz, found that ribosomes, the cell structures that assemble amino acids into proteins, can still link amino acids together even if their proteins are removed. After the proteins are extracted, only RNA molecules are left in the ribosomes, indicating that a ribozyme catalyzes this central reaction of protein synthesis. After Noller's dis-

INSIGHTS FROM THE Molecular Revolution

Ribozymes: Can RNA catalyze peptide bond formation in protein synthesis?

Harry Noller's experiment showed that if proteins were removed from ribosomes, the remaining ribosomal RNA (rRNA) molecules could still catalyze the central reaction of protein synthesis, linkage of amino acids into chains via peptide bonds. However, his work did not eliminate the possibility that undetectable small amounts of ribosomal proteins in the preparations might be catalyzing peptide bond formation.

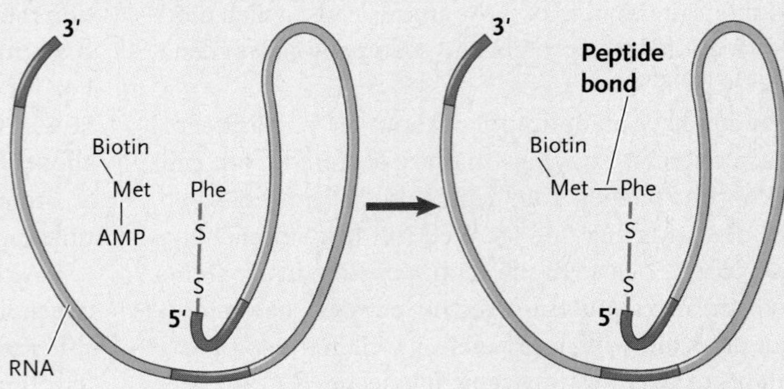
3′
Biotin
Met
AMP
Phe
S
S
5′
RNA
3′
Peptide bond
Biotin
Met — Phe
S
S
5′

Research Question

Can RNA catalyze peptide bond formation in protein synthesis?

Experiment

Research by Biliang Zhang and Thomas R. Cech of the University of Colorado at Boulder showed that rRNA synthesized artificially and, therefore, never exposed to ribosomal proteins, could catalyze the formation of peptide bonds between amino acids (see Figure 3.19).

1. The researchers first synthesized an extremely large pool of RNA molecules in the test tube. Part of the RNA sequence was the same in every molecule and part differed randomly from molecule to molecule, but all were the same length. The investigators linked the amino acid phenylalanine (Phe) to the 5′ end of each RNA molecule by a disulfide (—S—S—) bond (see figure, left).
2. To the pool of RNAs, they added the amino acid methionine (Met) linked to the nucleotide AMP (see Figure 3.28). (In the cell, single amino acids linked to AMP are used in the pathway that makes proteins.) The methionine–AMP was "tagged" by linking the small molecule biotin to it (see figure, left).
3. They allowed the molecules to react, hypothesizing that some of the RNA molecules would have the right sequence to act as a ribozyme and form a peptide bond between the methionine and the phenylalanine (see figure, right).

   To determine if peptide bonds had formed, Zhang and Cech poured the reaction mixture through a column packed with chemical beads that bind to biotin. Such binding trapped any RNA molecules that were able to catalyze the joining of the two amino acids, whereas unreactive RNA molecules flowed out the bottom of the column. The RNA molecules with the biotin tag were then washed from the column and separated from the linked amino acids by adding a reagent that breaks disulfide bonds.
4. The ribozyme RNA molecules washed from the column were analyzed and refined. Eventually, the researchers obtained ribozymes that catalyzed peptide bond formation at rates 100,000 times faster than the same reaction occurring spontaneously without a catalyst.

Conclusion

Zhang and Cech's experiments confirmed a feature of ribozyme activity that is critical to the role proposed for these RNA-based catalysts in the primitive RNA world—their ability to catalyze formation of the fundamental linkage tying amino acids together in proteins. Thus, during the evolution of life, proteins could have been made first in quantity by RNA, with no requirement for either DNA or enzymatic proteins.

Source: B. Zhang and T. R. Cech. 1997. Peptide bond formation by in vitro selected ribozymes. *Nature* 390:96–100.

covery, Cech and his colleague, Biliang Zhang, confirmed that ribozymes can actually catalyze this reaction (see *Insights from the Molecular Revolution* for an outline of Cech and Zhang's experiment).

Ribozymes provide a possible solution to a long-standing "chicken-or-egg" paradox about the evolution of life: Did proteins or nucleic acids come first in evolution? It is difficult to understand how DNA could exist without the enzymatic proteins required for its duplication. At the same time, it is difficult to understand how enzymes could exist without nucleic acids, which contain the information required to make them. Ribozymes offer a way around this dilemma because they could have acted as *both* enzymes and informational molecules when cellular life first appeared. The earliest forms of life therefore might have inhabited an "RNA world" in which neither DNA nor proteins played critical roles (see Chapter 24). If so, ribozymes—the most recently discovered biological catalysts—may have existed for the longest time.

This chapter concludes our survey of the chemical underpinnings of biology. In the next chapter, we survey the structure of cells, the fundamental units into which biological molecules are organized and where molecules interact to produce the characteristics of life.

STUDY BREAK 4.6

What is a ribozyme, and how does it fit the definition of an enzyme?

UNANSWERED QUESTIONS

Many biological processes rely on enzymes to catalyze key reactions. A complete understanding of those processes requires knowledge about the structure and function of the enzymes involved. Much research continues to be done to elucidate enzyme structure and function.

Ribozymes are catalytic RNA molecules. Various types of ribozymes exist, each type differing in its three-dimensional structure and mechanism of catalysis.

How does ribozyme structure relate to function, and how might ribozymes be used as therapeutic agents?

Researcher John M. Burke at the University of Vermont and his group are studying hairpin ribozymes and hammerhead ribozymes, which are catalytically active once they fold into those two shapes (the hammerhead shape is similar to that of the head of a hammerhead shark). Their research has four directions: determining the molecular structure of ribozymes, characterizing RNA conformational changes during catalysis, elucidating the mechanisms of catalysis, and exploring ways to use ribozymes as therapeutic agents.

For example, Burke's group has shown that the hairpin ribozyme undergoes a dramatic conformational change when the substrate binds to the active site. Furthermore, they have engineered hairpin ribozymes that can inhibit viral replication in mammalian cells. The particular viruses targeted have RNA genomes and include HIV-1 (the causative agent of AIDS) and hepatitis B virus. To achieve their goal, the researchers had to identify appropriate target sites within the viral RNA molecules and to express the engineered ribozymes efficiently within the cell. Current research focuses on optimizing the inhibition of viral replication by the ribozymes, determining the mechanism of antiviral activity, and extending this technology to develop therapeutic approaches for significant infectious diseases such as AIDS and hepatitis B.

Think Critically

Given what you know about protein folding and the significance of conformational changes in proteins, what are some potential experimental variables that might be used *in situ* to study RNA-based biological catalysts such as hairpin ribozymes?

Peter J. Russell

Peter J. Russell

REVIEW KEY CONCEPTS

Go to **CENGAGENOW** at www.cengage.com/login to access quizzing, animations, exercises, articles, and personalized homework help.

4.1 Energy, Life, and the Laws of Thermodynamics

- Energy is the capacity to do work. Kinetic energy is the energy of motion; potential energy is energy stored in an object because of its location or chemical structure. Energy may be readily converted between potential and kinetic states.
- Thermodynamics is the study of energy flow between a system and its surroundings during chemical and physical reactions. A system that does not exchange energy or matter with its surroundings is an isolated system. A system that exchanges energy but not matter with its surroundings is a closed system. A system that exchanges both energy and matter with its surroundings is an open system (Figure 4.1).
- The first law of thermodynamics states that the total amount of energy in a system and its surroundings remains constant. The second law states that in any process involving a spontaneous (possible) change from an initial to a final state, the total entropy (disorder) of the system and its surroundings always increases.

4.2 Free Energy and Spontaneous Reactions

- A spontaneous reaction is one that will occur without the input of energy from the surroundings. A spontaneous reaction releases free energy—energy that is available to do work.
- The free energy equation, $\Delta G = \Delta H - T\Delta S$, states that the free energy change, ΔG, is influenced by two factors: The change in enthalpy (potential energy in a system) and the change in entropy of the system as a reaction goes to completion.
- Factors that oppose the completion of spontaneous reactions, such as the relative concentrations of reactants and products, produce an equilibrium point at which reactants are converted to products and products are converted back to reactants, at equal rates (Figure 4.3).
- Organisms reach equilibrium ($\Delta G = 0$) only when they die.
- Reactions with a negative ΔG are spontaneous; they release free energy and are known as exergonic reactions. Reactions with a positive ΔG require free energy and are known as endergonic reactions (Figure 4.4).
- Metabolism is the biochemical modification and use of energy in the synthesis and breakdown of organic molecules. A catabolic reaction releases the potential energy of a molecule in breaking it down to a simpler molecule (ΔG is negative). An anabolic (biosynthetic) reaction uses energy to convert a simple molecule to a more complex molecule (ΔG is positive). Typically, individual reactions operate in metabolic pathways. Individual reactions in a particular pathway can be catabolic or anabolic; it is the sum of the reactions that makes the pathway catabolic or anabolic.

Animation: Chemical equilibrium

4.3 Adenosine Triphosphate (ATP): The Energy Currency of the Cell

- The hydrolysis of ATP releases free energy that can be used as a source of energy for the cell (Figure 4.5).
- A cell can couple the exergonic reaction of ATP hydrolysis to make an otherwise endergonic (anabolic) reaction proceed spontaneously. These coupling reactions require enzymes (Figure 4.6).
- The ATP used in coupling reactions is replenished by reactions that link ATP synthesis to catabolic reactions. ATP thus cycles between reactions that release free energy and reactions that require free energy (Figure 4.7).

Animation: Structure of ATP

Animation: Active transport

4.4 Role of Enzymes in Biological Reactions

- What prevents many exergonic reactions from proceeding rapidly is that they need to overcome an energy barrier (the activation energy, E_a) to get to the transition state (Figure 4.8).
- Enzymes are catalysts that greatly speed the rate at which spontaneous reactions occur because they lower the activation energy (Figure 4.9).
- Enzymes usually are specific: they catalyze reactions of only a single type of molecule or a group of closely related molecules (Figure 4.10).
- Catalysis occurs at the active site, which is the site where the enzyme binds to the substrate (reactant molecule). After combining briefly with the substrate, the enzyme is released unchanged when the reaction is complete (Figure 4.11).
- Many enzymes require a cofactor, a nonprotein group that binds to the enzyme, for catalytic activity. Some cofactors are ions; others are small organic molecules called coenzymes. Some coenzymes bind loosely to enzymes whereas others, called prosthetic groups, bind tightly.
- Enzymes reduce the activation energy by inducing the transition state of the reaction, from which the reaction can move easily in the direction of either products or reactants.
- Three major mechanisms contribute to enzymatic catalysis by reducing the activation energy: (1) Enzymes bring reacting molecules together; (2) Enzymes expose reactant molecules to altered charge environments that promote catalysis; and (3) Enzymes change the shape of a substrate molecule.

Animation: Activation energy

Animation: How catalase works

Animation: Enzymes and their role in lowering activation energy

4.5 Conditions and Factors That Affect Enzyme Activity

- When substrate is abundant, the rate of a reaction is proportional to the amount of enzyme. At a fixed enzyme concentration, the rate of a reaction increases with substrate concentration until the enzyme becomes saturated with reactants. At that point, further increases in substrate concentration do not increase the rate of the reaction (Figure 4.12).
- Many cellular enzymes are regulated by nonsubstrate molecules called inhibitors. Competitive inhibitors interfere with reaction rates by combining with the active site of an enzyme; noncompetitive inhibitors combine with sites elsewhere on the enzyme (Figure 4.13).
- Allosteric regulation resembles noncompetitive inhibition except that regulatory molecules may either increase or decrease enzyme activity. Allosteric regulation often carries out feedback inhibition, in which a product of an enzyme-catalyzed pathway acts as an allosteric inhibitor of the first enzyme in the pathway (Figures 4.14 and 4.15).
- Many key enzymes are regulated by chemical modification, by substances such as ions and certain functional groups. The modifications change enzyme conformation resulting in increased or decreased activity.
- Typically, an enzyme has optimal activity at a certain pH and a certain temperature; at pH and temperature values above and below the optimum, the reaction rate falls off (Figures 4.16 and 4.17).

Animation: Allosteric activation

Animation: Allosteric inhibition

Interaction: Feedback inhibition

Interaction: Enzymes and temperature

4.6 RNA-Based Biological Catalysts: Ribozymes

- RNA-based catalysts called ribozymes speed some types of biological reactions; these include cutting and splicing reactions in which surplus segments are removed from RNA molecules and linking reactions that combine amino acids into polypeptide chains.

UNDERSTAND AND APPLY

Test Your Knowledge

1. The capacity to do work best defines:
   a. a metabolic pathway.
   b. entropy.
   c. kinetic or potential energy.
   d. a chemical equilibrium.
   e. thermodynamics.
2. The assembly of proteins from amino acids is best described as:
   a. a conversion of kinetic energy to potential energy reaction.
   b. an entropy reaction.
   c. a catabolic reaction.
   d. an anabolic reaction.
   e. an energy free reaction.
3. When two glucose molecules react to form maltose:
   a. the reaction represents a negative ΔG.
   b. free energy had to be available to allow the reaction to proceed.
   c. the reaction is exothermic.
   d. it supports the second law of thermodynamics, which states there is tendency of the universe toward disorder.
   e. the resulting product has less potential energy than the reactants.
4. When glucose reacts with ATP to form glucose-6-phosphate:
   a. the synthesis of glucose-6-phosphate is exergonic.
   b. ADP is at a higher energy level than ATP.
   c. glucose-6-phosphate is at a higher energy level than glucose.
   d. because ATP donates a phosphate to glucose, this is not a coupled reaction.
   e. the reaction is spontaneous.
5. In the following graph:

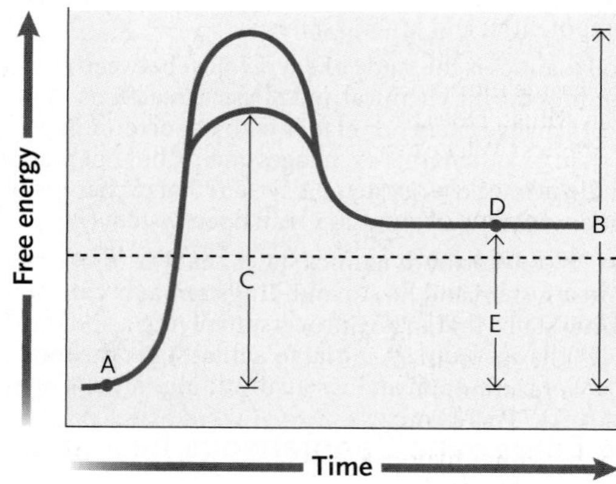

Free energy
Time
A
B
C
D
E

   a. A represents the product.
   b. B represents the energy of activation when enzymes are present.
   c. C is the free energy difference between A and D.
   d. C is the energy of activation without enzymes.
   e. E is the difference in free energy between the reactant and the products.

6. Which of the following methods is *not* used by enzymes to increase the rate of reactions?
   a. covalent bonding with the substrate at their active site
   b. bringing reacting molecules into close proximity
   c. orienting reactants into positions to favor transition states
   d. changing charges on reactants to hasten their reactivity
   e. increasing fit of enzyme and substrate that reduces the energy of activation
7. In an enzymatic reaction:
   a. the enzyme leaves the reaction chemically unchanged.
   b. if the enzyme molecules approach maximal rate, and the substrate is continually increased, the rate of the reaction does not reach saturation.
   c. in the stomach, enzymes would have an optimal activity at a neutral pH.
   d. increasing temperature above the optimal value slows the reaction rate.
   e. the least important level of organization for an enzyme is its tertiary structure.
8. Which of the following statements about the allosteric site is true?
   a. The allosteric site is a second active site on a substrate in a metabolic pathway.
   b. The allosteric site on an enzyme can allow the product of a metabolic pathway to inhibit that enzyme and stop the pathway.
   c. When the allosteric site of an enzyme is occupied, the reaction is irreversible and the enzyme cannot react again.
   d. An allosteric activator prevents binding at the active site.
   e. An enzyme that possesses allosteric sites does not possess an active site.
9. Which of the following statements about inhibition is true?
   a. Allosteric inhibitors and allosteric activators are competitive for a given enzyme.
   b. If an inhibitor binds the active site, it is considered noncompetitive.
   c. If an inhibitor binds to a site other than the active site, this is competitive inhibition.
   d. A noncompetitive inhibitor is believed to change the shape of the enzyme, making its active site inoperable.
   e. Competitive inhibition is usually not reversible.
10. Which of the following statements is *incorrect*?
   a. Ribozymes can link amino acids to form protein.
   b. Ribozymes can act as enzymes.
   c. Ribozymes can act as informational molecules.
   d. Ribozymes are suggested as the first molecules of life.
   e. Ribozymes are proteins.

Discuss the Concepts

1. Trees become more complex as they develop spontaneously from seeds to adults. Does this process violate the second law of thermodynamics? Why or why not?
2. Trace the flow of energy through your body. What products increase the entropy of you and your surroundings?
3. You have found a molecular substance that accelerates the rate of a particular reaction. What kind of information would you need to demonstrate that this molecular substance is an enzyme?
4. The addition or removal of phosphate groups from ATP is a fully reversible reaction. In what way does this reversibility facilitate the use of ATP as a coupling agent for cellular reactions?
5. Researchers once hypothesized that an enzyme and its substrate fit together like a lock and key but that the products do not fit the enzyme. Examine this idea with respect to reversible reactions.

Design an Experiment

Succinate dehydrogenase is part of the cellular biochemical machinery for breaking down sugars, fatty acids, and amino acids into carbon dioxide and water, with the capture of their chemical energy as ATP. Suppose you are measuring the activity of this enzyme extracted from cells in test-tube reactions. You find that the rate of the reaction converting succinate to fumarate catalyzed by succinate dehydrogenase is inhibited by the addition of malonate to the reaction mixture. Design an experiment that will tell you whether malonate is acting as a competitive or a noncompetitive inhibitor.

Interpret the Data

The postsynaptic density 95 (PSD-95) protein plays a key role in mammalian nervous system responses by concentrating and organizing receptors on the nerve cells. Future drugs that change these receptors could change learning and memory at the cellular level. A key part of the interaction between PSD-95 and the proposed drugs is the thermodynamics of binding between them.

The thermodynamics of binding between PSD-95 and two different small polypeptides (similar to those in the proposed drugs) is shown in the graphs. Graph A shows data for a parent polypeptide, and graph B shows the data for a mutated polypeptide (with some altered amino acids) of the same length. On the *x* axis is the ratio (in moles) between the amount of bound polypeptide and PSD-95. When the molar ratio is very near zero on the *x* axis in each graph, all of the PSD-95 molecules available are bound by the polypeptides. The molar ratio can be manipulated by injecting small amounts of polypeptide into the solution, and the change in energy content of the injectant (ΔH measured in kcal/mole) is measured with each accumulating injection.

A. **Parent polypeptide**

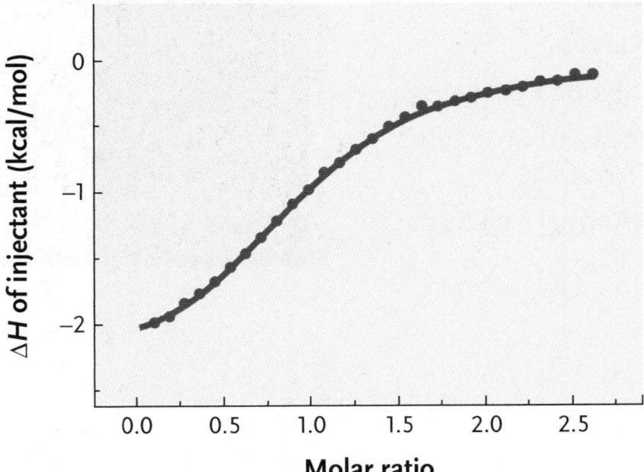

0
−1
−2
ΔH of injectant (kcal/mol)
0.0
0.5
1.0
1.5
2.0
2.5
Molar ratio

B. **Mutated polypeptide**

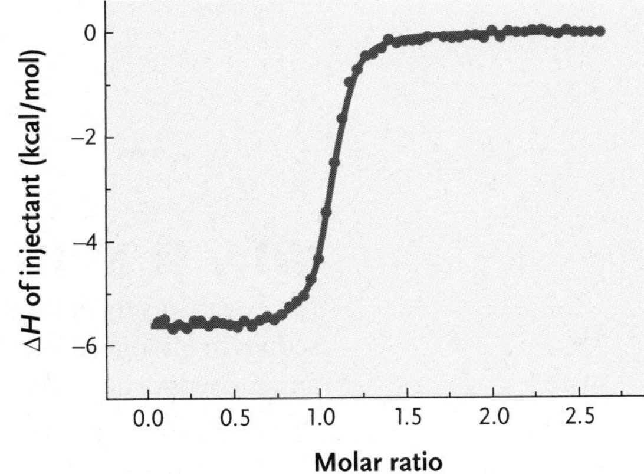

0
−2
−4
−6
ΔH of injectant (kcal/mol)
0.0
0.5
1.0
1.5
2.0
2.5
Molar ratio

1. If $T\Delta S$ for the parent polypeptide (graph A) is 4.0 kcal/mol and for the mutated polypeptide (graph B) is 3.5 kcal/mol, what is the ΔG for each polypeptide binding to the protein?
2. Using the ΔG values calculated in question 1, state whether each polypeptide binding is endergonic or exergonic, spontaneous or not spontaneous, and which would yield more free energy on binding to PSD-95.
3. If the mutated polypeptide represented a drug that needed to compete with the parent polypeptide in a nerve cell, would this mutated polypeptide be very effective?

Source: From D. Saro et al. 2007. A thermodynamic ligand binding study of the third PDZ domain (PDZ3) from the mammalian neuronal protein PSD-95. *Biochemistry* 46:6340–6352.

Apply Evolutionary Thinking

If RNA appeared first in evolution, establishing an RNA world, which do you think would evolve next: DNA or proteins? Why?

5

STUDY OUTLINE

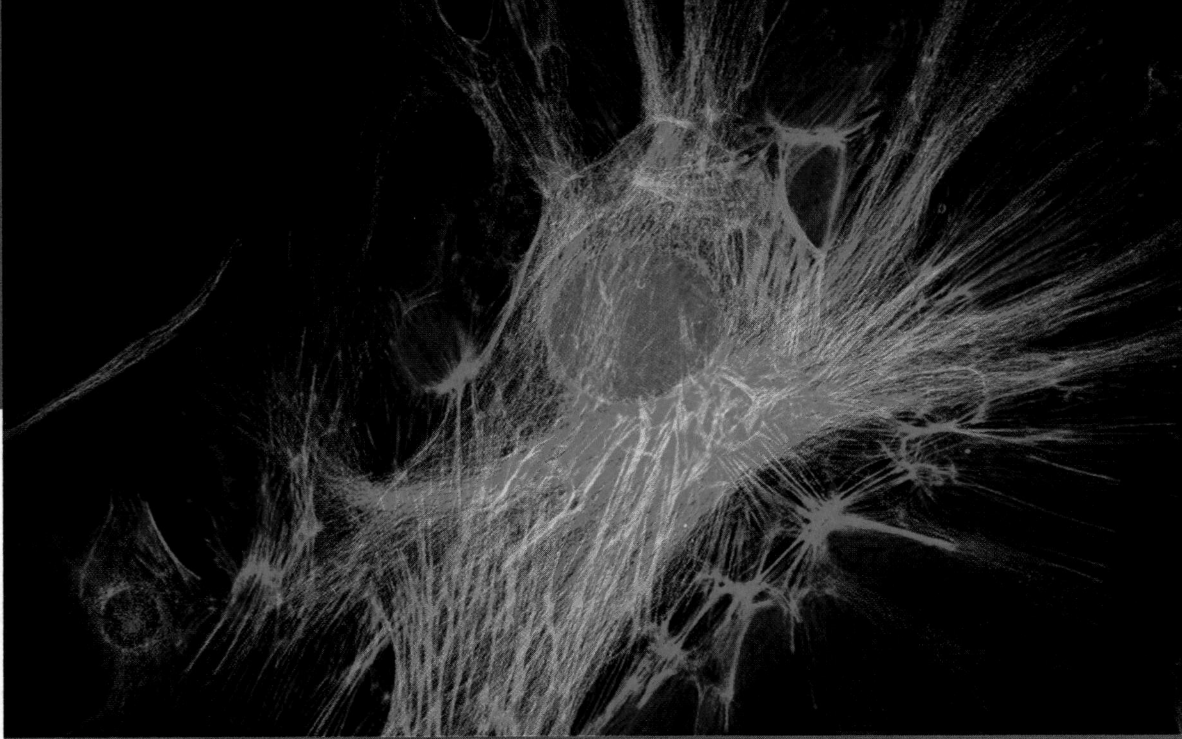

David Becker/SPL/Photo Researchers, Inc.

Cells fluorescently labeled to visualize their internal structure (confocal light micrograph). Cell nuclei are shown in blue and parts of t cytoskeleton in red and green.

The Cell: An Overview

Why It Matters. . . In the mid-1600s, Robert Hooke, Curator of Instruments for the Royal Society of England, was at the forefront of studies applying the newly invented light microscopes to biological materials. When Hooke looked at thinly sliced cork from a mature tree through a microscope, he observed tiny compartments **(Figure 5.1A).** He gave them the Latin name *cellulae,* meaning "small rooms"—hence, the origin of the biological term *cell.* Hooke was actually looking at the walls of dead cells, which is what cork consists of.

Reports of cells also came from other sources. By the late 1600s, Anton van Leeuwenhoek **(Figure 5.1B),** a Dutch shopkeeper, observed "many very little animalcules, very prettily a-moving," using a single-lens microscope of his own construction. Leeuwenhoek discovered and described diverse protists, sperm cells, and even bacteria, organisms so small that they would not be seen by others for another two centuries.

In the 1820s, improvements in microscopes brought cells into sharper focus. Robert Brown, an English botanist, noticed a discrete, spherical body inside some cells; he called it a *nucleus.* In 1838, a German botanist, Matthias Schleiden, speculated that the nucleus had something to do with the development of a cell. The following year, the zoologist Theodor Schwann of Germany expanded Schleiden's idea to propose that all animals and plants consist of cells that contain a nucleus. He also proposed that even when a cell forms part of a larger organism, it has an individual life of its own. However, an important question remained: Where do cells come from? A decade later, the German physiologist Rudolf Virchow answered this question. From his studies of cell growth and reproduction, Virchow proposed that cells arise only from preexisting cells by a process of division.

A. **Hooke's microscope**

National Library of Medicine
Armed Forces Institute of Pathology

B. **Leeuwenhoek and microscope**

Bettmann/CORBIS

FIGURE 5.1
Investigations leading to the first descriptions of cells. (A) The cork cells drawn by Robert Hooke and the compound microscope he used to examine them. **(B)** Anton van Leeuwenhoek holding his microscope, which consisted of a single, small sphere of glass fixed in a holder. He viewed objects by holding them close to one side of the glass sphere and looking at them through the other side.

Thus, by the middle of the nineteenth century, microscopic observations had yielded three profound generalizations, which together constitute what is now known as the **cell theory:**

1. All organisms are composed of one or more cells.
2. The cell is the basic structural and functional unit of all living organisms.
3. Cells arise only from the division of preexisting cells.

These tenets were fundamental to the development of biological science.

This chapter provides an overview of our current understanding of the structure and functions of cells, emphasizing both the similarities among all cells and some of the most basic differences among cells of various organisms. The variations in cells that help make particular groups of organisms distinctive are discussed in later chapters. This chapter also introduces some of the modern microscopes that enable us to learn more about cell structure. <

5.1 Basic Features of Cell Structure and Function

As the basic structural and functional units of all living organisms, cells carry out the essential processes of life. They contain highly organized systems of molecules, including the nucleic acids DNA and RNA, which carry hereditary information and direct the manufacture of cellular molecules. Cells use chemical molecules or light as energy sources for their activities. Cells also respond to changes in their external environment by altering their internal reactions. Further, cells duplicate and pass on their hereditary information as part of cellular reproduction. All these activities occur in cells that, in most cases, are invisible to the naked eye.

Some types of organisms, including almost all bacteria and archaeans, some protists, such as amoebas, and some fungi, such as yeasts, are unicellular. Each of these cells is a functionally independent organism capable of carrying out all activities necessary for its life. In more complex multicellular organisms, including plants and animals, the activities of life are divided among varying numbers of specialized cells. However, individual cells of multicellular organisms are potentially capable of surviving by themselves if placed in a chemical medium that can sustain them.

If cells are broken open, the property of life is lost: They are unable to grow, reproduce, or respond to outside stimuli in a coordinated, potentially independent fashion. This fact confirms the second tenet of the cell theory: Life as we know it does not exist in units more simple than individual cells. *Viruses,* which consist only of a nucleic acid molecule surrounded by a protein coat, cannot carry out most of the activities of life. Their only capacity is to infect living cells and direct them to make more virus particles of the same kind. (Viruses are discussed in Chapter 17.)

Cells Are Small and Are Visualized Using a Microscope

Cells assume a wide variety of forms in different prokaryotes and eukaryotes **(Figure 5.2)**. Individual cells range in size from tiny bacteria to an egg yolk, a single cell that can be several centimeters in diameter. Yet, all cells are organized according to the same basic plan, and all have structures that perform similar activities.

Most cells are too small to be seen by the unaided eye: Humans cannot see objects smaller than about 0.1 mm in diameter. The smallest bacteria have diameters of about 0.5 μm (a micrometer is 1,000 times smaller than a millimeter). The cells of multicellular animals range from about 5 to 30 μm in diameter. Your red blood cells are 7 to 8 μm across—a string of 2,500 of these cells is needed to span the width of your thumbnail. Plant cells range from about 10 μm to a few hundred micrometers in diameter. (**Figure 5.3** explains the units of measurement used in biology to study molecules and cells.)

To see cells and the structures within them we use **microscopy,** a technique for producing visible images of objects, biological or otherwise, that are too small to be seen by the human eye **(Figure 5.4)**. The instrument of microscopy is the **microscope.** The two common types of microscopes are **light microscopes,** which use light to illuminate the specimen (the object being viewed), and **electron microscopes,** which use electrons to illuminate the specimen. Different types of microscopes give different magnification and resolution of the specimen. Just as for a camera or a pair of binoculars, **magnification** is the ratio of the

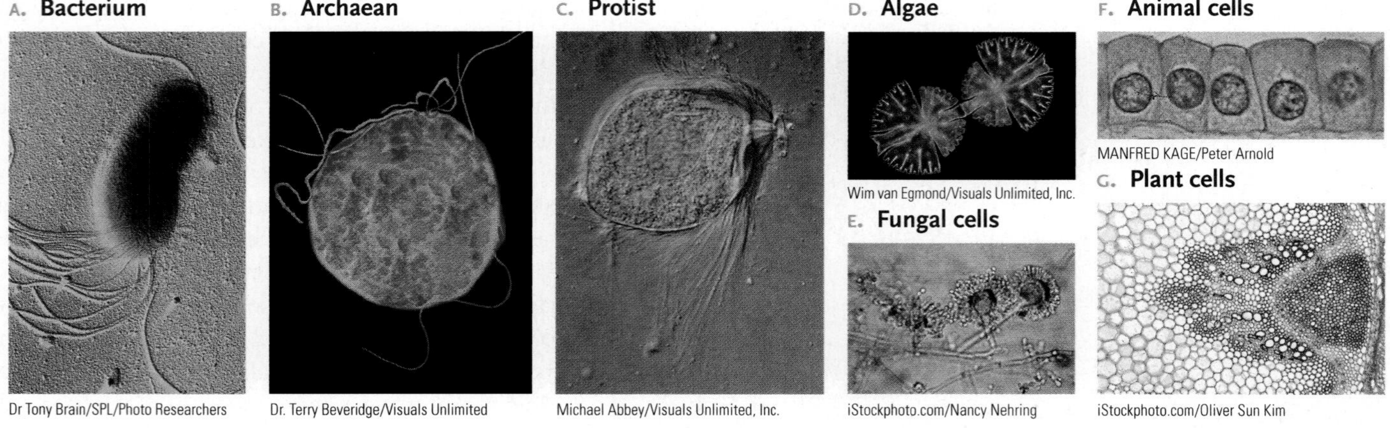
A. Bacterium
B. Archaean
C. Protist
D. Algae
F. Animal cells
MANFRED KAGE/Peter Arnold
G. Plant cells
Wim van Egmond/Visuals Unlimited, Inc.
E. Fungal cells
Dr Tony Brain/SPL/Photo Researchers
Dr. Terry Beveridge/Visuals Unlimited
Michael Abbey/Visuals Unlimited, Inc.
iStockphoto.com/Nancy Nehring
iStockphoto.com/Oliver Sun Kim

FIGURE 5.2
Examples of the varied kinds of cells: **(A)** and **(B)** are prokaryotes, the others are eukaryotes. **(A)** A bacterial cell with flagella, *Pseudomonas fluorescens.* **(B)** An archaean, the extremophile *Sulfolobus acidocaldarius.* **(C)** *Trichonympha,* a protist that lives in a termite's gut. **(D)** Two cells of *Micrasterias,* an algal protist. **(E)** Fungal cells of the bread mold *Aspergillus.* **(F)** Cells of a surface layer in the human kidney. **(G)** Cells in the stem of a sunflower, *Helianthus annuus.*

object as viewed to its real size, usually given as something like 1,200×. **Resolution** is the minimum distance two points in the specimen can be separated and still be seen as two points. Resolution depends primarily on the wavelength of light or electrons used to illuminate the specimen; the shorter the wavelength, the better the resolution. Hence, electron microscopes have higher resolution than light microscopes. Biologists choose the type of microscopy technique based on what they need to see in the specimen; selected examples are shown in Figure 5.4.

1 centimeter (cm) = 1/100 meter or 0.4 inch
3 cm – Chicken egg (the "yolk")

1 millimeter (mm) = 1/1,000 meter
1 mm – Frog egg, fish egg

1 micrometer (μm) = 1/1,000,000 meter
100 μm – Human egg
10–100 – Typical plant cell
5–30 – Typical animal cell
2–10 – Chloroplast
1–5 – Mitochondrion
5 – *Anabaena* (cyanobacterium)
1 – *Escherichia coli*

1 nanometer (nm) = 1/1,000,000,000 meter
100 nm – Large virus (HIV, influenza virus)
25 – Ribosome
7–10 – Cell membrane (thickness)
2 – DNA double helix (diameter)
0.1 – Hydrogen atom

Unaided human eye
Light microscopes
Electron microscopes

1 meter = 10^2 cm = 10^3 mm = 10^6 μm = 10^9 nm

FIGURE 5.3
Units of measure and the ranges in which they are used in the study of molecules and cells. The vertical scale in each box is logarithmic.

Why are most cells so small? The answer depends partly on the change in the surface area-to-volume ratio of an object as its size increases **(Figure 5.5).** For example, doubling the diameter of a cell increases its volume by eight times but increases its surface area by only four times. The significance of this relationship is that the volume of a cell determines the amount of chemical activity that can take place within it, whereas the surface area determines the amount of substances that can be exchanged between the inside of the cell and the outside environment. Nutrients must constantly enter cells, and wastes must constantly leave; however, past a certain point, increasing the diameter of a cell gives a surface area that is insufficient to maintain an adequate nutrient–waste exchange for its entire volume.

Some cells increase their ability to exchange materials with their surroundings by flattening or by developing surface folds or extensions that increase their surface area. For example, human intestinal cells have closely packed, fingerlike extensions that increase their surface area, which greatly enhances their ability to absorb digested food molecules.

Cells Have a DNA-Containing Central Region That Is Surrounded by Cytoplasm

All cells are bounded by the **plasma membrane,** a bilayer made of lipids with embedded protein molecules **(Figure 5.6).** The lipid bilayer is a hydrophobic barrier to the passage of water-soluble substances, but selected water-soluble substances can penetrate cell membranes through transport protein channels. The selective movement of ions and water-soluble molecules through the transport proteins maintains the specialized internal ionic and molecular environments required for cellular life. (Membrane structure and functions are discussed further in Chapter 6.)

Light and Electron Microscopy

Purpose: In biology, microscopy is used to view organisms, cells, and structures within cells in their natural state or after being treated (stained) so that specific structures can be seen more clearly. All of the photographs of cells and cell structures in this book were made using microscopy.

Protocol: A light microscope uses a beam of light to illuminate the specimen and forms a magnified image of the specimen with glass lenses. An electron microscope uses a beam of electrons to illuminate the specimen and forms an image with magnetic fields. Electron microscopy provides higher resolution and higher magnification than light microscopy.

Light microscopy
Micrographs are of the protist *Paramecium*.

Electron microscopy
Micrographs are of the green alga *Scenedesmus*.

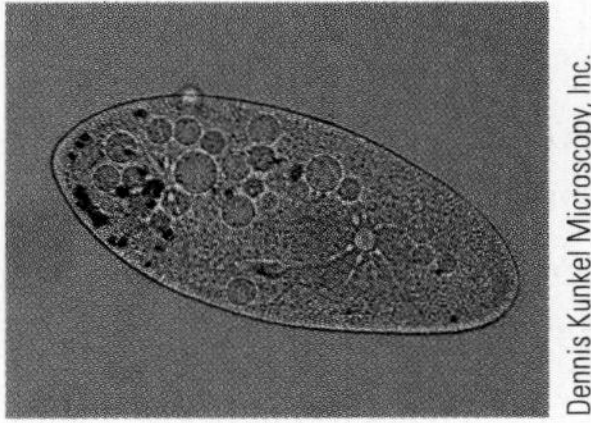

Dennis Kunkel Microscopy, Inc.

Bright field microscopy: Light passes directly through the specimen. Many cell structures have insufficient contrast to be discerned. Staining with a dye is used to enhance contrast in a specimen, as shown here, but this treatment usually fixes and kills the cells.

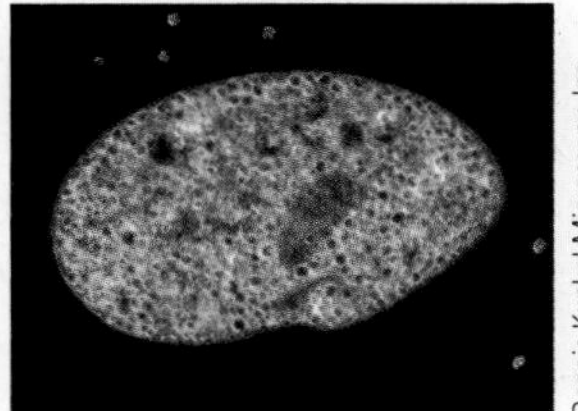

Dennis Kunkel Microscopy, Inc.

Dark field microscopy: Light illuminates the specimen at an angle, and only light scattered by the specimen reaches the viewing lens of the microscope. This gives a bright image of the cell against a black background.

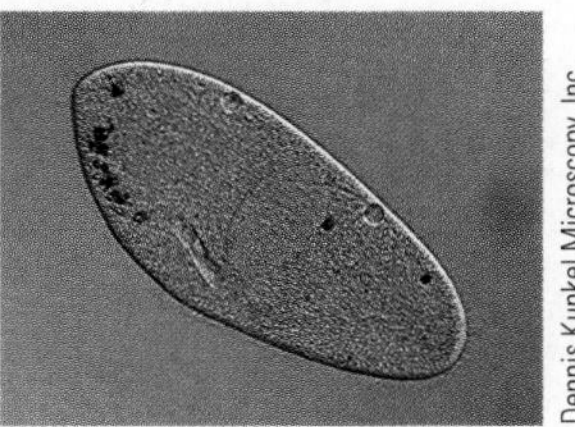
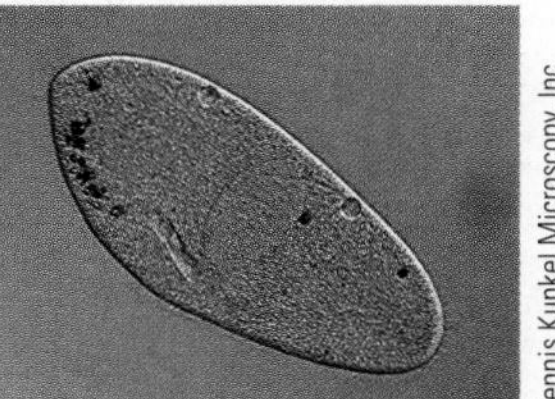

Dennis Kunkel Microscopy, Inc.

Phase-contrast microscopy: Differences in refraction (the way light is bent) caused by variations in the density of the specimen are visualized as differences in contrast. Otherwise invisible structures are revealed with this technique, and living cells in action can be photographed or filmed.

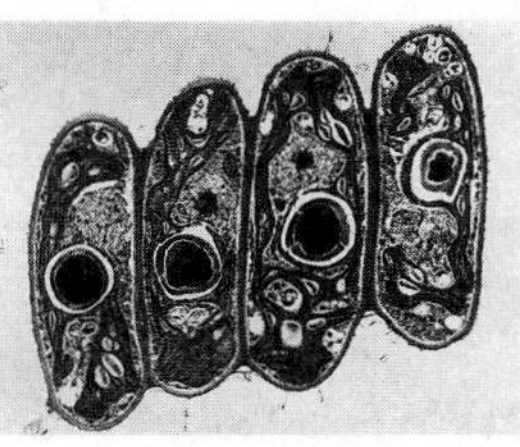

Jeremy Pickett-Heaps, University of Colorado

Transmission electron microscopy (TEM): A beam of electrons is focused on a thin section of a specimen in a vacuum. Electrons that pass through form the image; structures that scatter electrons appear dark. TEM is used primarily to examine structures within cells. Various staining and fixing methods are used to highlight structures of interest.

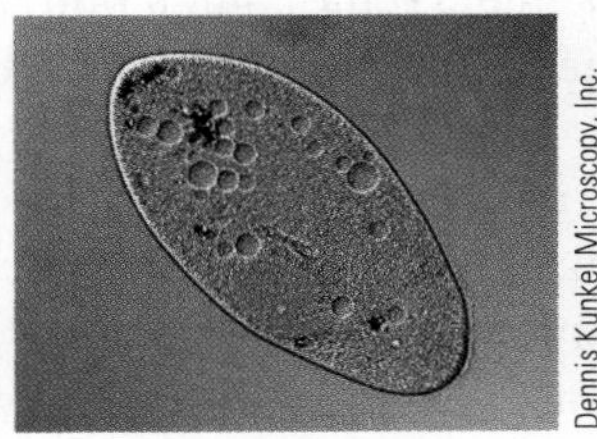

Dennis Kunkel Microscopy, Inc.

Nomarski (differential interference contrast): Similar to phase-contrast microscopy, special lenses enhance differences in density, giving a cell a 3D appearance.

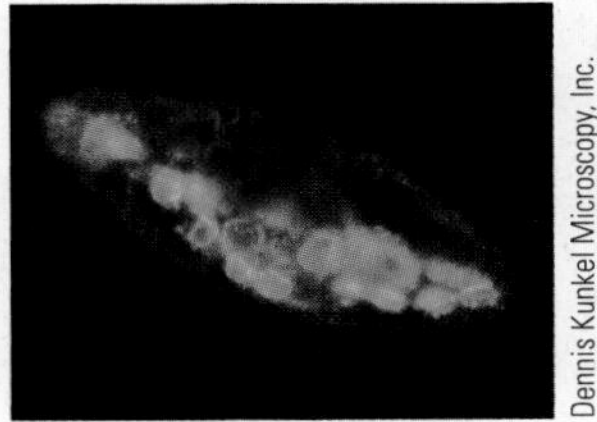

Dennis Kunkel Microscopy, Inc.

Fluorescence microscopy: Different structures or molecules in cells are stained with specific fluorescent dyes. The stained structures or molecules fluoresce when the microscope illuminates them with ultraviolet light, and their locations are seen by viewing the emitted visible light.

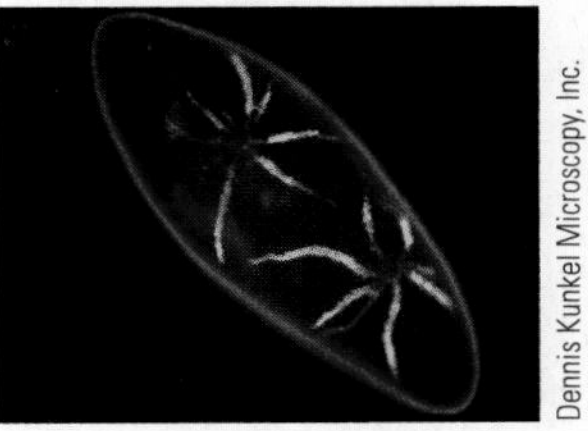

Dennis Kunkel Microscopy, Inc.

Confocal laser scanning microscopy: Lasers scan across a fluorescently stained specimen, and a computer focuses the light to show a single plane through the cell. This provides a sharper 3D image than other light microscopy techniques.

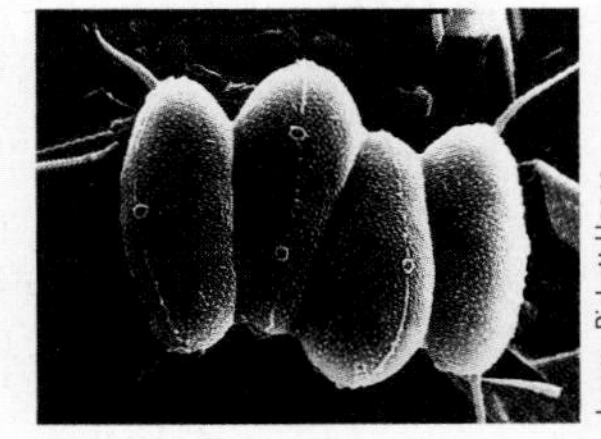

Jeremy Pickett-Heaps, University of Colorado

Scanning electron microscopy (SEM): A beam of electrons is scanned across a whole cell or organism, and the electrons excited on the specimen surface are converted to a 3D-appearing image.

Interpreting the Results: Different techniques of light and electron microscopy produce images that reveal different structures or functions of the specimen. A micrograph is a photograph of an image formed by a microscope.

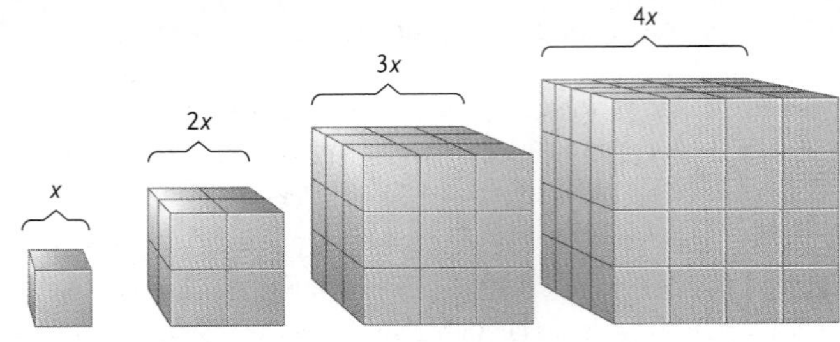

x
2x
3x
4x

Total surface area	$6x^2$	$6(2x)^2 = 24x^2$	$6(3x)^2 = 54x^2$	$6(4x)^2 = 96x^2$
Total volume	x^3	$(2x)^3 = 8x^3$	$(3x)^3 = 27x^3$	$(4x)^3 = 64x^3$
Surface area/ volume ratio	6:1	3:1	2:1	1.5:1

FIGURE 5.5
Relationship between surface area and volume. The surface area of an object increases as a square of the linear dimension, whereas the volume increases as a cube of that dimension.

A central region of all cells contains DNA molecules, which store hereditary information. The hereditary information is organized in the form of *genes*—segments of DNA that code for individual proteins. The central region also contains proteins that help maintain the DNA structure and enzymes that duplicate DNA and copy its information into RNA.

All the parts of the cell between the plasma membrane and the central region comprise the **cytoplasm.** The cytoplasm contains the *organelles,* the *cytosol,* and the *cytoskeleton.* The **organelles** ("little organs") are small, organized structures important for cell function. The **cytosol** is an aqueous (water) solution containing ions and various organic molecules. The **cytoskeleton** is a protein-based framework of filamentous structures that, among other things, helps maintain proper cell shape and plays key roles in cell division and chromosome segregation from cell generation to cell generation. The cytoskeleton was once thought to be specific to eukaryotes, but recent research has shown that all major eukaryotic cytoskeletal proteins have functional equivalents in prokaryotes.

Many of the cell's vital activities occur in the cytoplasm, including the synthesis and assembly of most of the molecules required for growth and reproduction (except those made in the central region) and the conversion of chemical and light energy into forms that can be used by cells. The cytoplasm also conducts stimulatory signals from the outside into the cell interior and carries out chemical reactions that respond to these signals.

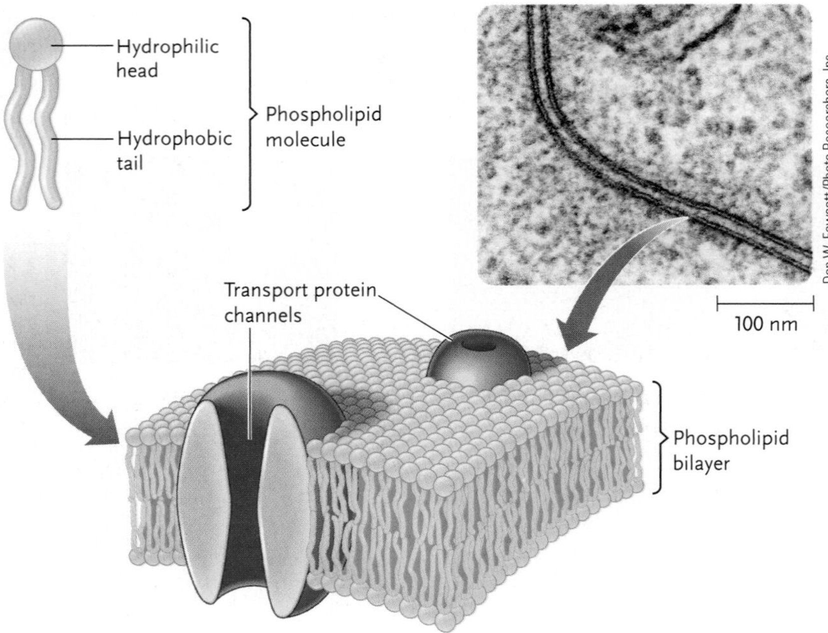

Hydrophilic head
Hydrophobic tail
Phospholipid molecule
Transport protein channels
100 nm
Phospholipid bilayer
Don W. Fawcett/Photo Researchers, Inc.

FIGURE 5.6
The plasma membrane, which forms the outer limit of a cell's cytoplasm. The plasma membrane consists of a phospholipid bilayer, an arrangement of phospholipids two molecules thick, which provides the framework of all biological membranes. Water-soluble substances cannot pass through the phospholipid part of the membrane. Instead, they pass through protein channels in the membrane; two proteins that transport substances across the membrane are shown. Other types of proteins are also associated with the plasma membrane. *(Inset)* Electron micrograph showing the plasma membranes of two adjacent animal cells.

Cells Occur in Prokaryotic and Eukaryotic Forms, Each with Distinctive Structures and Organization

Organisms fall into two fundamental groups, prokaryotes and eukaryotes, based on the organization of their cells. **Prokaryotes** (*pro* = before; *karyon* = nucleus) make up two domains of organisms, the Bacteria and the Archaea. The DNA-containing central region of prokaryotic cells, the **nucleoid,** has no boundary membrane separating it from the cytoplasm. Many species of bacteria contain few if any internal membranes, but a number of other bacterial species contain extensive internal membranes.

The **eukaryotes** (*eu* = true) make up the domain Eukarya, which includes all the remaining organisms. The DNA-containing central region of eukaryotic cells, a true **nucleus,** is separated by membranes from the surrounding cytoplasm. The cytoplasm of eukaryotic cells typically contains extensive membrane systems that form organelles with their own distinct environments and specialized functions. As in prokaryotes, a plasma membrane surrounds eukaryotic cells as the outer limit of the cytoplasm.

The remainder of this chapter surveys the components of prokaryotic and eukaryotic cells in more detail.

STUDY BREAK 5.1

What is the plasma membrane, and what are its main functions?

5.2 Prokaryotic Cells

Most prokaryotic cells are relatively small, usually not much more than a few micrometers in length and a micrometer or less in diameter. A typical human cell is about ten times larger in diameter and over 8,000 times larger in volume than an average prokaryotic cell.

The three shapes most common among prokaryotes are spherical, rodlike, and spiral. *Escherichia coli (E. coli),* a normal inhabitant of the mammalian intestine that has been studied extensively as a model organism in genetics, molecular biology, and genomics research, is rodlike in shape. **Figure 5.7** shows an EM and diagram of *E. coli* to illustrate the basic features of prokaryotic cell structure. More detail about prokaryotic cell structure and function, as well as about the diversity of prokaryotic organisms, is presented in Chapter 25.

The genetic material of prokaryotes is located in the nucleoid; in an electron microscope, that region of the cell is seen to contain a highly folded mass of DNA (see Figure 5.7). For most species, the DNA is a single, circular molecule that unfolds when released from the cell. This DNA molecule is the **prokaryotic chromosome.** (Chapter 17 discusses the genetics of prokaryotes.)

Individual genes in the DNA molecule encode the information required to make proteins. This information is copied into a type of RNA molecule called *messenger RNA (mRNA).* Small, roughly spherical particles in the cytoplasm, the **ribosomes,** use the information in the mRNA to assemble amino acids into proteins. A prokaryotic ribosome consists of a large and a small subunit, each formed from a combination of *ribosomal RNA* (rRNA) and protein molecules. Each prokaryotic ribosome contains three types of rRNA molecules, which are also copied from the DNA, and more than 50 proteins.

In almost all prokaryotes, the plasma membrane is surrounded by a rigid external layer of material, the **cell wall,** which ranges in thickness from 15 to 100 nm or more (a nanometer is one-billionth of a meter). The cell wall provides rigidity to prokaryotic cells and, with the capsule, protects the cell from physical damage. In many prokaryotic cells, the wall is coated with an external layer of polysaccharides called the **glycocalyx** (a "sugar coating" from *glykys* = sweet; *calyx* = cup or vessel). When the glycocalyx is diffuse and loosely associated with the cells, it is a **slime layer;** when it is gelatinous and more firmly attached to cells, it is a **capsule.** The glycocalyx helps protect prokaryotic cells from physical damage and desiccation, and may enable a cell to attach to a surface, such as other prokaryotic cells (as in forming a colony), eukaryotic cells (as in *Streptococcus pneumoniae* attaching to lung cells), or nonliving substrate (such as a rock).

The plasma membrane itself performs several vital functions in prokaryotes. Besides transporting materials into and out of the cells, it contains most of the molecular systems that metabolize food molecules into the chemical energy of ATP. In photosynthetic prokaryotes, the molecules that absorb light energy and convert it to the chemical energy of ATP are also associated with the plasma membrane or with internal, saclike membranes derived from the plasma membrane.

Many prokaryotic species contain few if any internal membranes; in such cells, most cellular functions occur either on the plasma membrane or in the cytoplasm. But some prokaryotes have more extensive internal membrane structures. For example, photosynthetic bacteria and archaeans have complex layers of intracellular membranes formed by invaginations of the plasma membrane on which photosynthesis takes place. And members of the bacterial phylum Planctomycetes have complex internal membranes that form distinct compartments.

As mentioned earlier, prokaryotic cells have filamentous cytoskeletal structures with functions similar to those in eukaryotes. Prokaryotic cytoskeletons play important roles in creating and maintaining the proper shape of cells, in cell division and, for certain prokaryotes, in determining polarity of the cells.

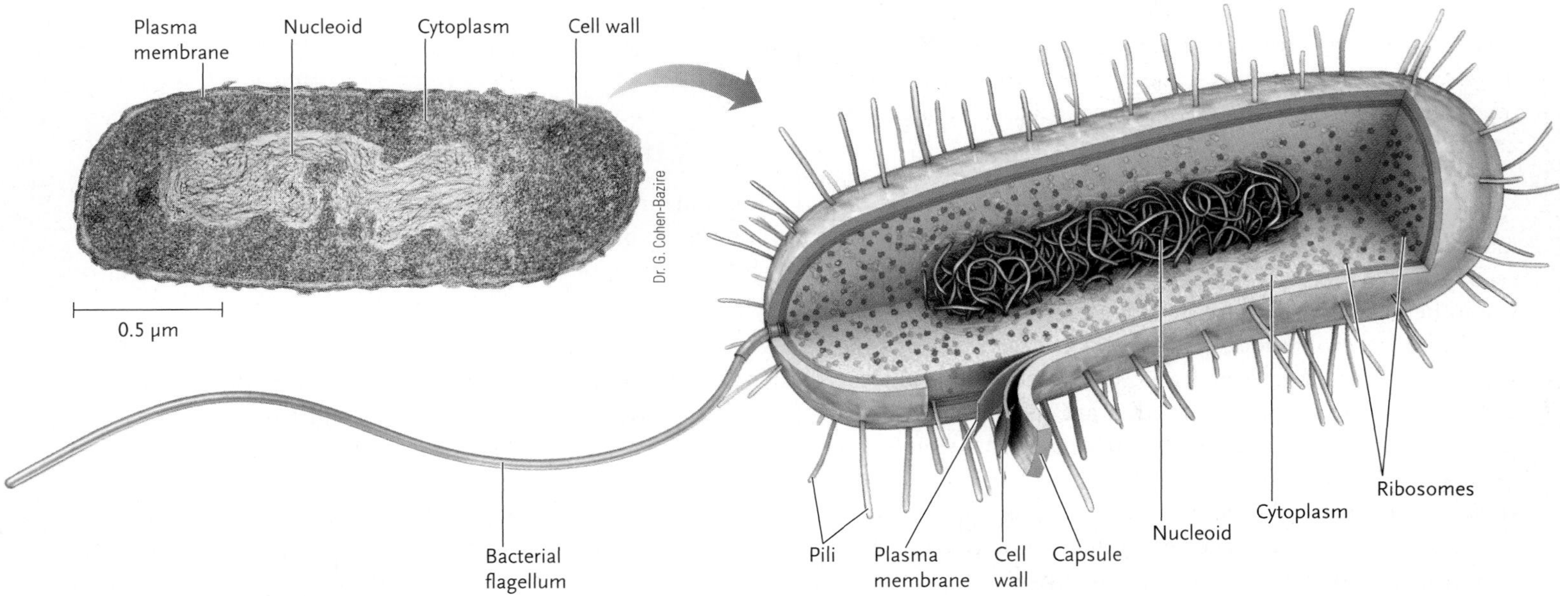
Plasma membrane
Nucleoid
Cytoplasm
Cell wall
Dr. G. Cohen-Bazire
0.5 µm
Bacterial flagellum
Pili
Plasma membrane
Cell wall
Capsule
Nucleoid
Cytoplasm
Ribosomes

FIGURE 5.7
Prokaryotic cell structure. An electron micrograph (left) and a diagram (right) of the bacterium *Escherichia coli.* The pili extending from the cell wall attach bacterial cells to other cells of the same species or to eukaryotic cells as a part of infection. A typical *E. coli* has four flagella.

INSIGHTS FROM THE Molecular Revolution

An Old Kingdom in a New Domain: Do archaeans define a distinct domain of life?

Many archaeans live in extreme environments that can be tolerated by no other organisms, suggesting that they might belong in a distinct domain of life. For example, *Methanococcus jannaschii* was first found in an oceanic hot water vent at a depth of more than 2,600 m (8,500 feet). It can live at temperatures as high as 94°C, which is almost the temperature of boiling water, and can tolerate pressures as high as 200 times the pressure of air at sea level!

Research Question: Are archaeans a distinct domain of life?

Experiment: In 1996, Carol J. Bult, Carl R. Woese, J. Craig Venter, and 37 other scientists at the Institute for Genomic Research published the complete genomic DNA sequence of *Methanococcus jannaschii.* (The sequence was obtained using techniques outlined in Chapter 18.) Using computer algorithms, the scientists compared the sequence with the already known genome sequences of several bacteria and of brewer's yeast *(Saccharomyces cerevisiae),* the first eukaryote to be sequenced completely.

Results: The researchers found genes coding for 1,738 proteins in the *Methanococcus* genome. Of these, only 38% were related to genes coding for known proteins in either bacteria or eukaryotes. The remaining 62% have no known relatives in organisms of those two groups.

Some features of *Methanococcus* DNA are typically prokaryotic. Its single, circular chromosome is in a nucleoid, which is not bounded by a membrane. Its protein-coding genes are organized into functional groups called *operons,* each having several genes copied as a unit into a single mRNA molecule (see discussion in Section 16.1). By contrast, each protein-coding gene in eukaryotes is copied into a separate mRNA molecule. Some of the proteins encoded in *Methanococcus* DNA, including enzymes active in energy metabolism, membrane transport, and cell division, are similar to those of bacteria. Other proteins encoded in the *Methanococcus* DNA are similar to those of eukaryotes, including enzymes and other proteins that carry out DNA replication and the copying of genes into mRNA.

Conclusion: *Methanococcus* has a majority of genes that are unique, some that are typically bacterial, and some that are typically eukaryotic. This finding supports the proposal, first advanced by Woese, that *Methanococcus* and its archaean relatives are a separate domain of life, the Archaea, with the Bacteria and the Eukarya as the other domains. Together, Bacteria and Archaea are the prokaryotes. Woese's three-domain system is used in this book.

Source: C. J. Bult et al. 1996. Complete genome sequence of the methanogenic archaeon, *Methanococcus jannaschii. Science* 273:1058–1073.

Many bacteria and archaeans can move through liquids and across wet surfaces. Most commonly they do so using long, threadlike protein fibers called **flagella** (singular, *flagellum,* meaning whip), which extend from the cell surface (see Figure 5.2A). The **bacterial flagellum,** which is helically shaped, rotates in a socket in the plasma membrane and cell wall to push the cell through a liquid medium (see Chapter 25). In *E. coli,* for instance, rotating bundles of flagella propel the bacterium. Archaeal flagella function similarly to bacterial flagella, but the two types differ significantly in their structures and mechanisms of action. Both types of prokaryotic flagella are also fundamentally different from the much larger and more complex flagella of eukaryotic cells, which are described in Section 5.3.

Some bacteria and archaeans have hairlike shafts of protein called **pili** (singular, *pilus*) extending from their cell walls. The main function of pili is attaching the cell to surfaces or other cells. A special type of pilus, the *sex pilus,* attaches one bacterium to another during mating (see Chapter 17).

Although prokaryotic cells appear relatively simple, their simplicity is deceptive. Most can use a variety of substances as energy and carbon sources, and they are able to synthesize almost all of their required organic molecules from simple inorganic raw materials. In many respects, prokaryotes are more versatile biochemically than eukaryotes. Their small size and metabolic versatility are reflected in their abundance; prokaryotes vastly outnumber all other types of organisms and live successfully in almost all regions of Earth's surface.

The two domains of prokaryotes, the Bacteria and the Archaea, share many biochemical and molecular features. However, the archaeans also share some features with eukaryotes and have other characteristics that are unique to their group. *Insights from the Molecular Revolution* describes the discovery of features that support the classification of the Archaea as a separate domain.

STUDY BREAK 5.2

Where in a prokaryotic cell is DNA found? How is that DNA organized?

5.3 Eukaryotic Cells

The domain of the eukaryotes, Eukarya, is divided into four major groups: the protists, fungi, animals, and plants. The rest of the chapter focuses on the cell components that are common to all or large groups of eukaryotic organisms.

Eukaryotic Cells Have a True Nucleus and Cytoplasmic Organelles Enclosed within a Plasma Membrane

The cells of all eukaryotes have a true nucleus enclosed by membranes. The cytoplasm surrounding the nucleus contains a remarkable system of membranous organelles, each specialized to carry out one or more major functions of energy metabolism and molecular synthesis, storage, and transport. The cytosol, the cytoplasmic solution surrounding the organelles, participates in en-

Cell Fractionation

Purpose: Cell fractionation partitions cells into fractions containing a single cell component, such as mitochondria or ribosomes. Once isolated, the cell component can be disassembled by the same general techniques to analyze its structure and function.

Protocol:

1. Break open intact cells by sonication (high-frequency sound waves), grinding in fine glass beads, or exposure to detergents that disrupt plasma membranes.

2. Use sequential centrifugations at increasing speeds to separate and purify cell structures. The spinning centrifuge drives cellular structures to the bottom of tube at a rate that depends on their shape and density. With each centrifugation, the largest and densest components are isolated and concentrated into a pellet; the remaining solution, the supernatant, is drawn off and can be centrifuged again at higher speed.

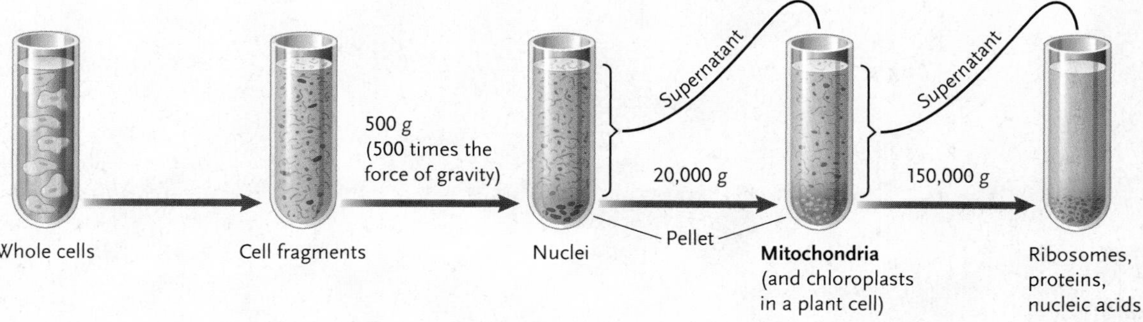
Whole cells
Cell fragments
500 g
(500 times the force of gravity)
Nuclei
Supernatant
20,000 g
Pellet
Mitochondria
(and chloroplasts in a plant cell)
Supernatant
150,000 g
Ribosomes, proteins, nucleic acids

3. Resuspend the pellet containing the isolated cell components and subfractionate using the same general techniques to examine the components of organelles.

Interpreting the Results: Many of the cell or organelle subfractions generated by cell fractionation retain their biological activity, making them useful in studies of various cellular processes. For example, mitochondrial subfractions were used to work out the structure and function of the electron transfer system. Cell fractionation is still used to determine the cellular location of a protein or biological reaction, such as whether it is free in the cytosol or associated with a membrane.

ergy metabolism and molecular synthesis and performs specialized functions in support and motility. Researchers have discovered much about the structures and functions of various cellular organelles by using the research method of cell fractionation to isolate them and purify them **(Figure 5.8).**

The eukaryotic plasma membrane carries out various functions through several types of embedded proteins. Some of these proteins form channels through the plasma membrane that transport substances into and out of the cell. Other proteins in the plasma membrane act as receptors; they recognize and bind specific signal molecules in the cellular environment and trigger internal responses. In some eukaryotes, particularly animals, plasma membrane proteins recognize and adhere to molecules on the surfaces of other cells. Yet other plasma membrane proteins are important markers in the immune system, labeling cells as "self," that is, belonging to the organism. Therefore, the immune system can identify cells without those markers as being foreign, most likely *pathogens* (disease-causing organisms or viruses).

A supportive cell wall surrounds the plasma membrane of fungal, plant, and many protist cells. Because the cell wall lies outside the plasma membrane, it is an *extracellular* structure (*extra* = outside). Although animal cells do not have cell walls, they also form extracellular material with supportive and other functions.

Figure 5.9 presents a diagram of a representative animal cell and **Figure 5.10** presents a diagram of a representative plant cell to show where the nucleus, cytoplasmic organelles, and other structures are located. The following sections discuss the structure and function of eukaryotic cell parts in more detail, beginning with the nucleus.

The Eukaryotic Nucleus Contains Much More DNA Than the Prokaryotic Nucleoid

The nucleus (see Figures 5.9 and 5.10) is separated from the cytoplasm by the **nuclear envelope,** which consists of two membranes, one layered just inside the other and separated by a narrow space **(Figure 5.11).** A network of protein filaments called *lamins* lines and reinforces the inner surface of the nuclear envelope in animal cells. Lamins are a type of intermediate filament (see later in this section). Unrelated proteins line the inner surface of the nuclear envelope in protists, fungi, and plants.

Embedded in the nuclear envelope are many hundreds of nuclear pore complexes. A **nuclear pore complex** is a large, octagonally symmetrical, cylindrical structure formed of many

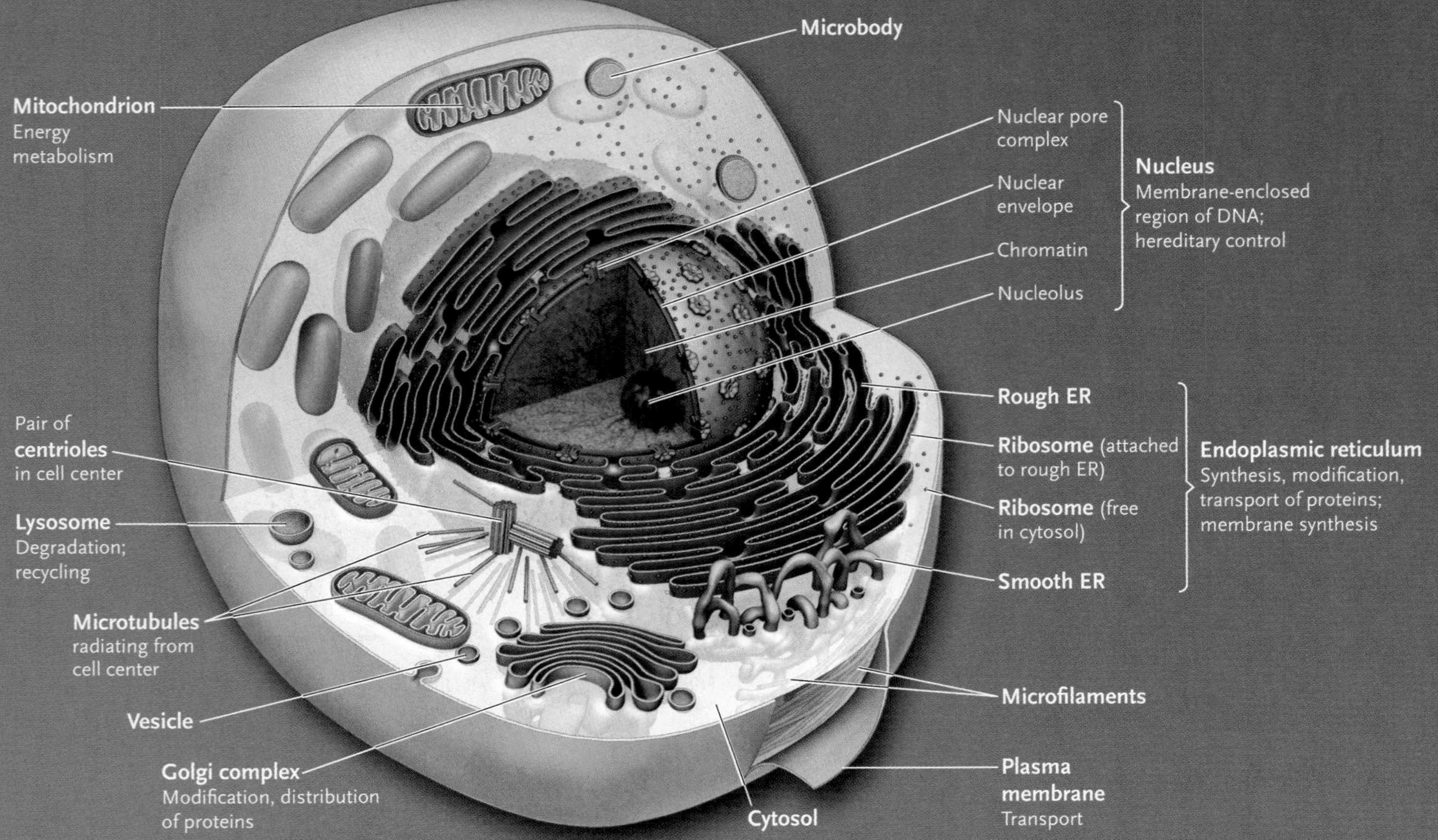

Microbody

Mitochondrion
Energy metabolism

Nuclear pore complex
Nuclear envelope
Chromatin
Nucleolus

Nucleus
Membrane-enclosed region of DNA; hereditary control

Rough ER
Ribosome (attached to rough ER)
Ribosome (free in cytosol)
Smooth ER

Endoplasmic reticulum
Synthesis, modification, transport of proteins; membrane synthesis

Pair of **centrioles** in cell center

Lysosome
Degradation; recycling

Microtubules
radiating from cell center

Vesicle

Golgi complex
Modification, distribution of proteins

Cytosol

Microfilaments

Plasma membrane
Transport

FIGURE 5.9
Diagram of an animal cell, highlighting the major organelles and their primary locations.

Cytosol

Mitochondrion
Energy metabolism

Golgi complex

Vesicle

Central vacuole
Cell growth, support, storage

Tonoplast
(central vacuole membrane)

Chloroplast
Photosynthesis; some starch storage

Microtubules
(components of cytoskeleton)

Cell wall
Protection; structural support

Plasma membrane
Transport

Nuclear pore complex
Nuclear envelope
Chromatin
Nucleolus

Nucleus
Membrane-enclosed region of DNA; hereditary control

Plasmodesmata

Rough ER
Ribosome (attached to rough ER)
Ribosome (free in cytosol)
Smooth ER

Endoplasmic reticulum
Synthesis, modification, transport of proteins; membrane synthesis

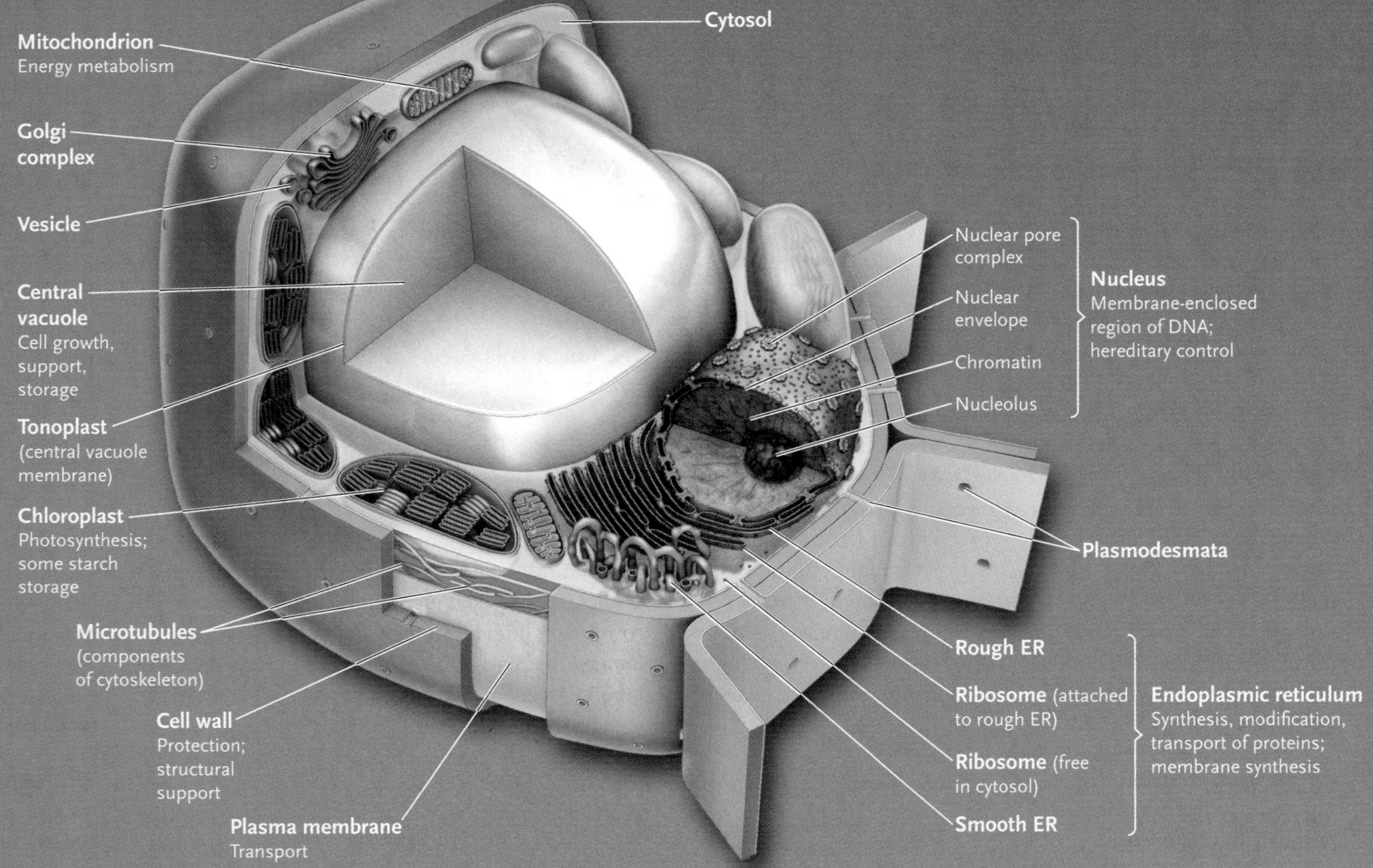

FIGURE 5.10
Diagram of a plant cell, highlighting the major organelles and their primary locations.

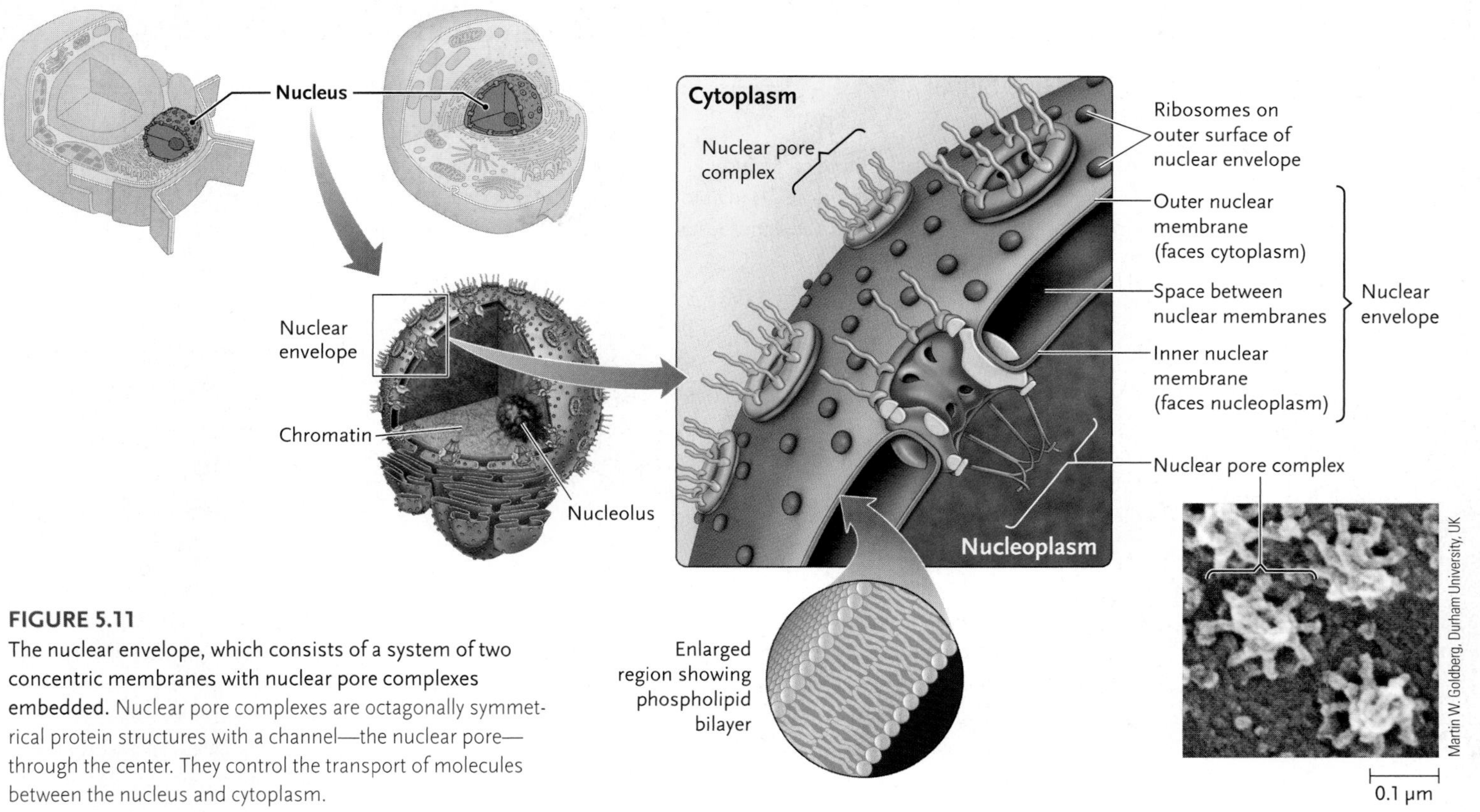
Nucleus
Nuclear envelope
Chromatin
Nucleolus
Cytoplasm
Nuclear pore complex
Ribosomes on outer surface of nuclear envelope
Outer nuclear membrane (faces cytoplasm)
Space between nuclear membranes
Inner nuclear membrane (faces nucleoplasm)
Nuclear envelope
Nuclear pore complex
Nucleoplasm
Enlarged region showing phospholipid bilayer
0.1 μm
Martin W. Goldberg, Durham University, UK

FIGURE 5.11
The nuclear envelope, which consists of a system of two concentric membranes with nuclear pore complexes embedded. Nuclear pore complexes are octagonally symmetrical protein structures with a channel—the nuclear pore—through the center. They control the transport of molecules between the nucleus and cytoplasm.

types of proteins, called *nucleoporins*. Probably the largest protein complex in the cell, it exchanges components between the nucleus and cytoplasm and prevents the transport of material not meant to cross the nuclear membrane. A channel through the nuclear pore complex—a nuclear pore—is the path for the assisted exchange of large molecules such as proteins and RNA molecules with the cytoplasm, whereas small molecules simply pass through unassisted. A protein or RNA molecule (called the *cargo*) associates with a transport protein acting as a chaperone to shuttle the cargo through the pore.

Some proteins—for instance, the enzymes for replicating and repairing DNA—must be imported into the nucleus to carry out their functions. Proteins to be imported into the nucleus are distinguished from those that function in the cytosol by the presence of a special, short amino acid sequence called a **nuclear localization signal.** A specific protein in the cytosol recognizes and binds to the signal and moves the protein containing it to the nuclear pore complex where it is then transported through the pore into the nucleus. **Figure 5.12** shows how researchers discovered the nuclear localization signal.

The liquid or semiliquid substance within the nucleus is called the **nucleoplasm.** Most of the space inside the nucleus is filled with **chromatin,** a combination of DNA and proteins. By contrast with most prokaryotes, most of the hereditary information of a eukaryote is distributed among several to many linear DNA molecules in the nucleus. Each individual DNA molecule with its associated proteins is a **eukaryotic chromosome.** The terms *chromatin* and *chromosome* are similar but have distinct meanings. *Chromatin* refers to any collection of eukaryotic DNA molecules with their associated proteins. *Chromosome* refers to one complete DNA molecule with its associated proteins.

Eukaryotic nuclei contain much more DNA than do prokaryotic nucleoids. For example, the entire complement of 46 chromosomes in the nucleus of a human cell has a total DNA length of about 2 meters (m), compared with about 1,500 mm in prokaryotic cells with the most DNA. Some eukaryotic cells contain even more DNA; for example, a single frog or salamander nucleus, although of microscopic diameter, is packed with about 10 m of DNA!

A eukaryotic nucleus also contains one or more **nucleoli** (singular, *nucleolus*), which look like irregular masses of small fibers and granules (see Figures 5.9 and 5.10). These structures form around the genes coding for the rRNA molecules of ribosomes. Within the nucleolus, the information in rRNA genes is copied into rRNA molecules, which combine with proteins to form ribosomal subunits. The ribosomal subunits then leave the nucleoli and exit the nucleus through the nuclear pore complexes to enter the cytoplasm, where they join on mRNAs to form complete ribosomes.

The genes for most of the proteins that the organism can make are found within the chromatin, as are the genes for specialized RNA molecules such as rRNA molecules. Expression of these genes is carefully controlled as required for the function of each cell. (The other proteins in the cell are specified by DNA in the mitochondria and chloroplasts.)

FIGURE 5.12 **Experimental Research**

Discovery of the Nuclear Localization Signal

Question: How are proteins that are imported into the nucleus identified by the import machinery?

Experiment: Alan Smith and his colleagues at the National Institute for Medical Research, Mill Hill, London, studied a viral protein that normally is found in the nucleus after the virus infects a cell. They mutated the 708-amino-acid–long protein, changing one or more specific amino acids in the protein or deleting segments of the protein, and determined whether the alterations affected the location of the viral protein in rodent or monkey cells in culture.

Results: The researchers obtained the following results:

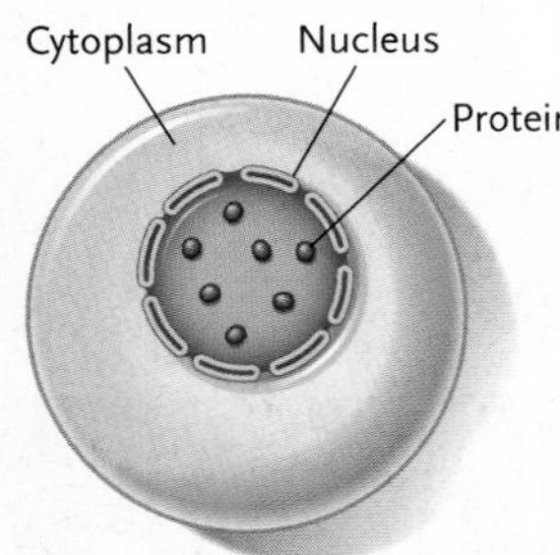
Cytoplasm
Nucleus
Protein

Normal protein: Localized to nucleus.

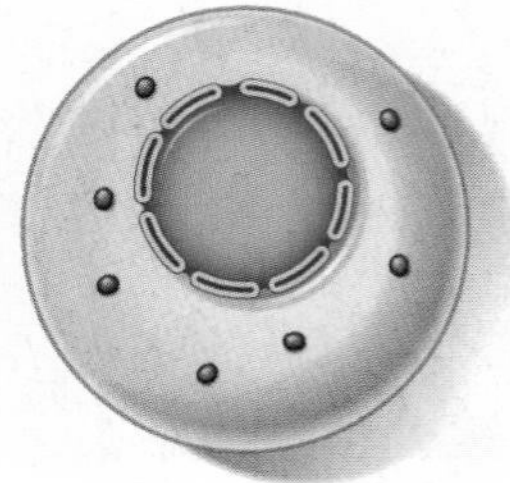

The amino acid at position 128 (thought to be important for the protein to bind to DNA) was mutated from lysine to threonine: Mutated protein localized in cytoplasm. The researchers interpreted this result to mean that the mutated amino acid was important for localizing the protein to the nucleus. Mutating other amino acids in the same region of the protein impaired import of the protein into the nucleus, but did not abolish it.

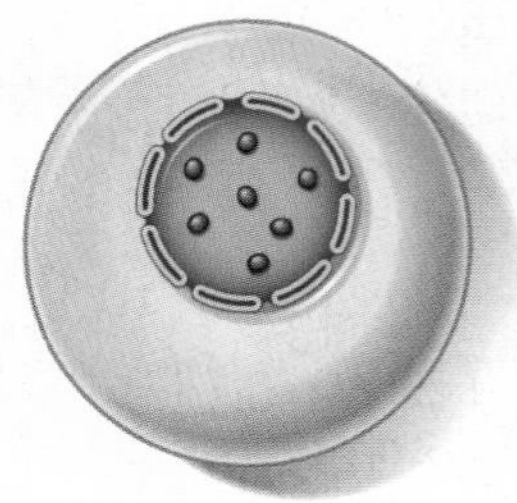

Deleting amino acids 1–126 (or any part of that region) or 136–708 (or any part of that region): Protein localized to nucleus, meaning that amino acids in those regions are not important for nuclear localization.

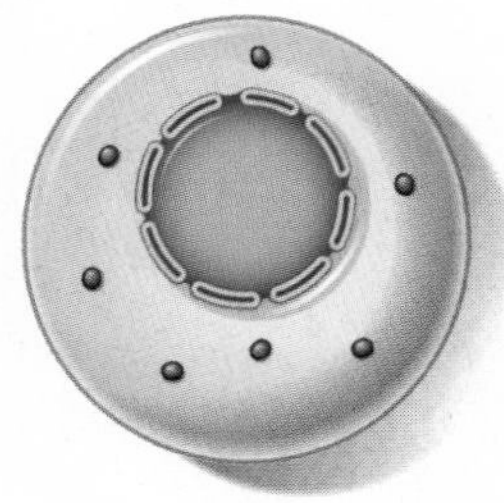

Deleting amino acids 127–133: Protein localized to cytoplasm, meaning that this amino acid sequence is necessary for nuclear localization of the viral protein. Other deletions involving parts of this region gave the same result.

Conclusion: By mutating the viral protein sequence, the researchers identified a seven-amino-acid segment of the protein, amino acids 127–133, that is necessary for localization of the protein to the nucleus. In follow-up experiments, they added this amino acid sequence to a cellular enzyme protein normally found only in the cytoplasm and determined that the modified protein localized to the nucleus. Therefore, the seven-amino-acid sequence is a nuclear localization signal. Continuing research has shown that this sequence is only the first example of similar sequences in other nuclear proteins. Thus, the identification of nuclear localization signals was a key step toward understanding the import of proteins into the nucleus.

Sources: D. Kalderon, W. D. Richardson, A. F. Markham, and A. E. Smith. 1984. Sequence requirements for nuclear location of simian virus 40 large-T antigen. *Nature* 311:33–38; D. Kalderon, B. L. Roberts, W. D. Richardson, and A. E. Smith, 1984. A short amino acid sequence able to specify nuclear location. *Cell* 39:499–509.

Eukaryotic Ribosomes Are Either Free in the Cytosol or Attached to Membranes

Like prokaryotic ribosomes, a eukaryotic ribosome consists of a large and a small subunit **(Figure 5.13)**. However, the structures of bacterial, archaeal and eukaryotic ribosomes, although similar, are not identical. In general, eukaryotic ribosomes are larger than either bacterial or archaeal ribosomes; they contain four types of rRNA molecules and more than 80 proteins. Their function is identical to that of prokaryotic ribosomes: They use the information in mRNA to assemble amino acids into proteins.

Some eukaryotic ribosomes are freely suspended in the cytosol; others are attached to membranes. Proteins made on free ribosomes in the cytosol may remain in the cytosol, pass through the nuclear pores into the nucleus, or become parts of mitochondria, chloroplasts, the cytoskeleton, or other cytoplasmic structures. Proteins that enter the nucleus become part of chromatin, line the nuclear envelope (the lamins), or remain in solution in the nucleoplasm.

Many ribosomes are attached to membranes. Some ribosomes are attached to the nuclear envelope, but most are attached to a network of membranes in the cytosol called the *endoplasmic reticulum* (ER) (described in more detail next). The proteins made on ribosomes attached to the ER follow a special path to other organelles within the cell.

An Endomembrane System Divides the Cytoplasm into Functional and Structural Compartments

Eukaryotic cells are characterized by an **endomembrane system** (*endo* = within), a collection of interrelated internal membranous sacs that divide the cell into functional and structural compartments. The endomembrane system has a number of functions, including the synthesis and modification of proteins and their transport into membranes and organelles or to the outside of the cell, the synthesis of lipids, and the detoxification of some toxins. The membranes of the system are connected either directly in the physical sense or indirectly by **vesicles,** which are small membrane-bound compartments that transfer substances between parts of the system.

The components of the endomembrane system include the nuclear envelope, endoplasmic reticulum, Golgi complex, lysosomes, vesicles, and plasma membrane. The plasma membrane and the nuclear envelope are discussed earlier in this chapter. The functions of the other organelles are described in the following sections.

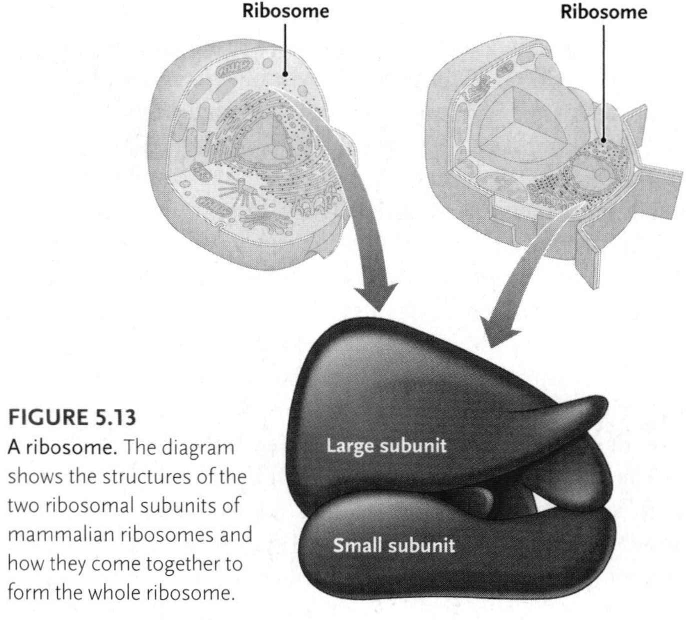
Ribosome
Ribosome
Large subunit
Small subunit

FIGURE 5.13
A ribosome. The diagram shows the structures of the two ribosomal subunits of mammalian ribosomes and how they come together to form the whole ribosome.

ENDOPLASMIC RETICULUM The **endoplasmic reticulum** (ER) is an extensive interconnected network (*reticulum* = little net) of membranous channels and vesicles called **cisternae** (singular, *cisterna*). Each cisterna is formed by a single membrane that surrounds an enclosed space called the **ER lumen (Figure 5.14).** The ER occurs in two forms: rough ER and smooth ER, each with specialized structure and function.

The **rough ER** (see Figure 5.14A) gets its name from the many ribosomes that stud its outer surface. The proteins made on ribosomes attached to the ER enter the ER lumen, where they fold into their final form. Chemical modifications of these proteins, such as addition of carbohydrate groups

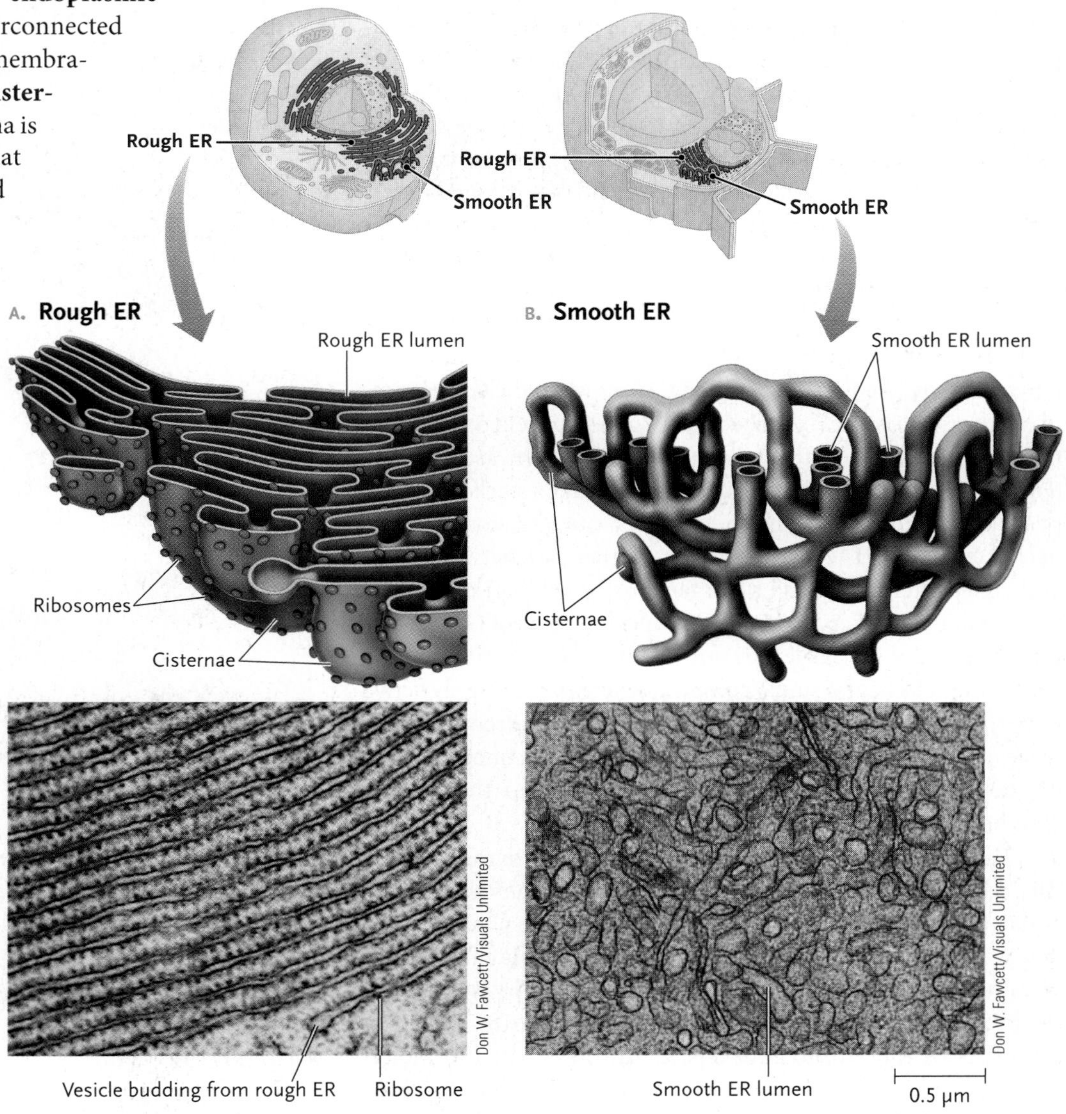

Rough ER
Rough ER
Smooth ER
Smooth ER

A. **Rough ER**
Rough ER lumen
Ribosomes
Cisternae
Don W. Fawcett/Visuals Unlimited
Vesicle budding from rough ER
Ribosome

B. **Smooth ER**
Smooth ER lumen
Cisternae
Don W. Fawcett/Visuals Unlimited
Smooth ER lumen
0.5 μm

FIGURE 5.14
The endoplasmic reticulum. (A) Rough ER, showing the ribosomes that stud the membrane surfaces facing the cytoplasm. Proteins synthesized on these ribosomes enter the lumen of the rough ER where they are modified chemically and then begin their path to their final destinations in the cell. **(B)** Smooth ER membranes. Among their functions are the synthesis of lipids for cell membranes, and enzymatic conversion of certain toxic molecules to safer molecules.

to produce glycoproteins, occur in the lumen. The proteins are then delivered to other regions of the cell within small vesicles that pinch off from the ER, travel through the cytosol, and join with the organelle that performs the next steps in their modification and distribution. For most of the proteins made on the rough ER, the next destination is the Golgi complex, which packages and sorts them for delivery to their final destinations.

The outer membrane of the nuclear envelope is closely related in structure and function to the rough ER, to which it is connected. This membrane is also a "rough" membrane, studded with ribosomes attached to the surface facing the cytoplasm. The proteins made on these ribosomes enter the space between the two nuclear envelope membranes. From there, the proteins can move into the ER and on to other cellular locations.

The **smooth ER** (see Figure 5.14B) is so called because its membranes have no ribosomes attached to their surfaces. The smooth ER has various functions in the cytoplasm, including synthesis of lipids that become part of cell membranes. In some cells, such as those of the liver, smooth ER membranes contain enzymes that convert drugs, poisons, and toxic by-products of cellular metabolism into substances that can be tolerated or more easily removed from the body.

The rough and smooth ER membranes are often connected, making the entire ER system a continuous network of interconnected channels in the cytoplasm. The relative proportions of rough and smooth ER reflect cellular activities in protein and lipid synthesis. Cells that are highly active in making proteins to be released outside the cell, such as pancreatic cells that make digestive enzymes, are packed with rough ER but have relatively little smooth ER. By contrast, cells that primarily synthesize lipids or break down toxic substances are packed with smooth ER but contain little rough ER.

GOLGI COMPLEX Camillo Golgi, a late-nineteenth-century Italian neuroscientist and Nobel laureate, discovered the **Golgi complex.** The Golgi complex consists of a stack of flattened, membranous sacs (without attached ribosomes) known as cisternae **(Figure 5.15).** In most cells, the complex looks like a stack of cupped pancakes, and like pancakes, they are separate sacs, not interconnected as the ER cisternae are. Typically there are between three and eight cisternae, but some organisms have Golgi complexes with several tens of cisternae. The number and size of Golgi complexes can vary with cell type and the metabolic activity of the cell. Some cells have a single complex, whereas cells highly active in secreting proteins from the cell can have hundreds of complexes. Golgi complexes are usually located near concentrations of rough ER membranes, between the ER and the plasma membrane.

The Golgi complex receives proteins that were made in the ER and transported to the complex in vesicles. When the vesicles contact the *cis* face of the complex (which faces the nucleus), they fuse with the Golgi membrane and release their contents directly into the cisternal (see Figure 5.15). Within the Golgi complex, the proteins are chemically modified, for example, by removing segments of the amino acid chain, adding small functional groups, or adding lipid or carbohydrate units. The modified proteins are transported within the Golgi to the *trans* face of the complex (which faces the plasma membrane), where they are sorted into vesicles that bud off from the margins of the Golgi (see Figure 5.15). The content of a vesicle is kept separate from the cytosol by the vesicle membrane. Three quite different models have been proposed for how proteins move through the Golgi complex. The mechanism is a subject of active current research.

The Golgi complex regulates the movement of several types of proteins. Some are secreted from the cell, others become embedded in the plasma membrane as integral membrane proteins, and yet others are placed in lysosomes. The modifications of the proteins within the Golgi complex include adding "zip codes" to

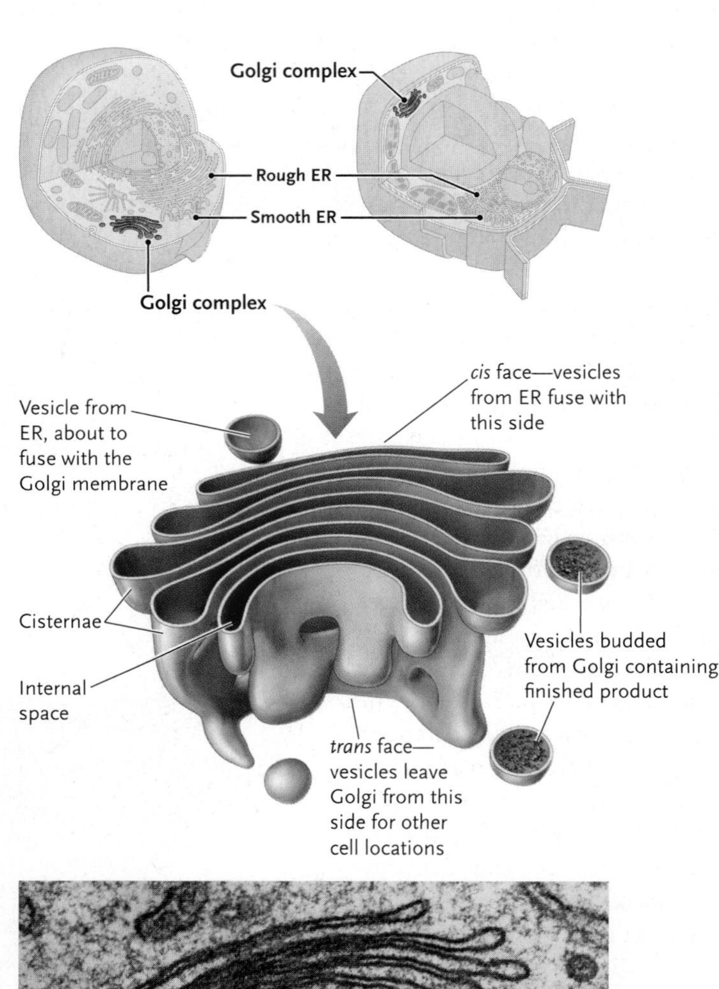
Golgi complex
Rough ER
Smooth ER
Golgi complex
cis face—vesicles from ER fuse with this side
Vesicle from ER, about to fuse with the Golgi membrane
Cisternae
Internal space
Vesicles budded from Golgi containing finished product
trans face—vesicles leave Golgi from this side for other cell locations

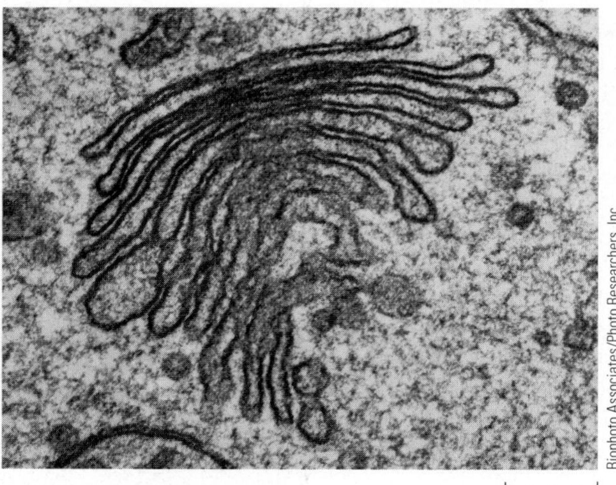
0.25 μm
Biophoto Associates/Photo Researchers, Inc.

FIGURE 5.15
The Golgi complex.

the proteins, which tags them for sorting to their final destinations. For instance, proteins secreted from the cell are transported to the plasma membrane in **secretory vesicles,** which release their contents to the exterior by **exocytosis (Figure 5.16A).** In this process, a secretory vesicle fuses with the plasma membrane and spills the vesicle contents to the outside. The contents of secretory vesicles vary, including signaling molecules such as hormones and neurotransmitters (see Chapter 7), waste products or toxic substances, and enzymes (such as from cells lining the intestine). The membrane of a vesicle that fuses with the plasma membrane becomes part of the plasma membrane. In fact, this process is used to expand the surface of the cell during cell growth.

Vesicles also may form by the reverse process, called endocytosis, which brings molecules into the cell from the exterior **(Figure 5.16B).** In this process, the plasma membrane forms a pocket, which bulges inward and pinches off into the cytoplasm as an **endocytic vesicle.** Once in the cytoplasm, endocytic vesicles, which contain segments of the plasma membrane as well as proteins and other molecules, are carried to the Golgi complex or to other destinations such as lysosomes in animal cells. The substances carried to the Golgi complex are sorted and placed into vesicles for routing to other locations, which may include lysosomes. Those routed to lysosomes are digested into molecular subunits that may be recycled as building blocks for the biological molecules of the cell. Exocytosis and endocytosis are discussed in more detail in Chapter 6.

A. **Exocytosis:** A secretory vesicle fuses with the plasma membrane, releasing the vesicle contents to the cell exterior. The vesicle membrane becomes part of the plasma membrane.

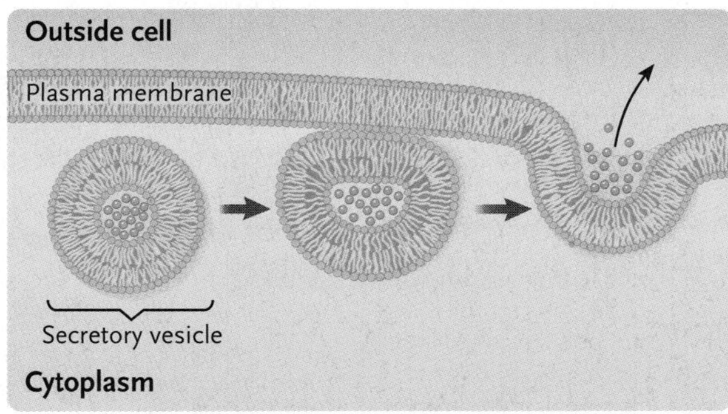
Outside cell
Plasma membrane
Secretory vesicle
Cytoplasm

B. **Endocytosis:** Materials from the cell exterior are enclosed in a segment of the plasma membrane that pockets inward and pinches off as an endocytic vesicle.

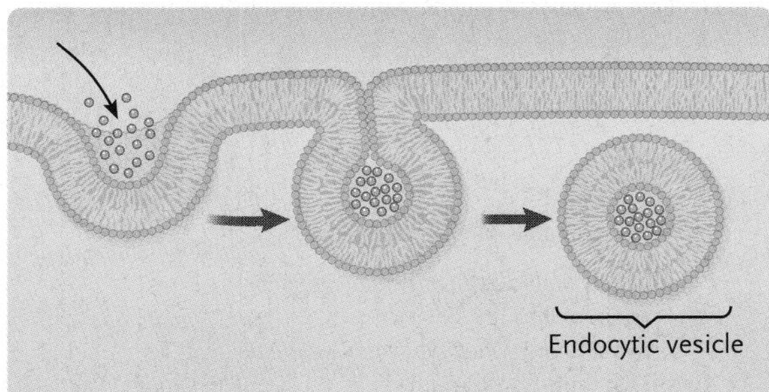
Endocytic vesicle

FIGURE 5.16
Exocytosis and endocytosis.

LYSOSOMES Lysosomes (*lys* = breakdown; *some* = body) are small, membrane-bound vesicles that contain more than 30 hydrolytic enzymes for the digestion of many complex molecules, including proteins, lipids, nucleic acids, and polysaccharides **(Figure 5.17).** The cell recycles the subunits of these molecules. Lysosomes are found in animals, but not in plants. The functions of lysosomes in plants are carried out by the central vacuole (see Section 5.4). Depending on the contents they are digesting, lysosomes assume a variety of sizes and shapes instead of a uniform structure as is characteristic of other organelles. Most commonly, lysosomes are small (0.1–0.5 μm in diameter) oval or spherical bodies. A human cell contains about 300 lysosomes.

Lysosomes are formed by budding from the Golgi complex. Their hydrolytic enzymes are synthesized in the rough ER, modified in the lumen of the ER to identify them as being bound for a lysosome, transported to the Golgi complex in a vesicle, and then packaged in the budding lysosome.

The pH within lysosomes is acidic (pH ~5) and is significantly lower than the pH of the cytosol (pH ~7.2). The hydrolytic enzymes in the lysosomes function optimally at the acidic pH within the organelle, but they do not function well at the pH of the cytosol; this difference reduces the risk to the viability of the cell should the enzymes be released from the vesicle.

Lysosomal enzymes can digest several types of materials. They digest food molecules entering the cell by endocytosis when an endocytic vesicle fuses with a lysosome. In a process called *autophagy,* they digest organelles that are not functioning correctly. A membrane surrounds the defective organelle, forming a large vesicle that fuses with one or more lysosomes; the organelle then is degraded by the hydrolytic enzymes. They also play a role in **phagocytosis,** a process in which some types of cells engulf bacteria or other cellular debris to break them down. These cells

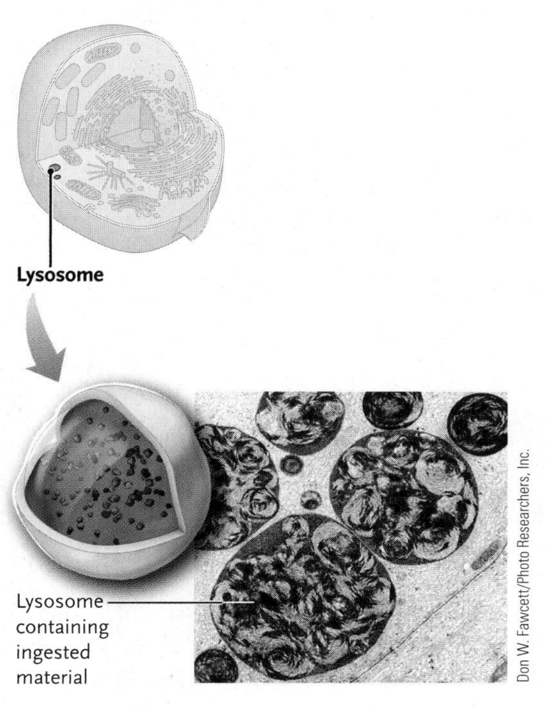
Lysosome
Lysosome containing ingested material
Don W. Fawcett/Photo Researchers, Inc.

FIGURE 5.17
A lysosome.

include the white blood cells known as *phagocytes,* which play an important role in the immune system (see Chapter 43). Phagocytosis produces a large vesicle that contains the engulfed materials until lysosomes fuse with the vesicle and release the hydrolytic enzymes necessary for degrading them.

In certain human genetic diseases known as *lysosomal storage diseases,* one of the hydrolytic enzymes normally found in the lysosome is absent. As a result, the substrate of that enzyme accumulates in the lysosomes, and this accumulation eventually interferes with normal cellular activities. An example is Tay–Sachs disease, which is a fatal disease of the central nervous system caused by the failure to synthesize the enzyme needed for hydrolysis of fatty acid derivatives found in brain and nerve cells.

SUMMARY In summary, the endomembrane system is a major traffic network for proteins and other substances within the cell. The Golgi complex in particular is a key distribution station for membranes and proteins **(Figure 5.18).** From the Golgi complex, lipids and proteins may move to storage or secretory vesicles, and from the secretory vesicles, they may move to the cell exterior by exocytosis. Membranes and proteins may also move between the nuclear envelope and the endomembrane system. Proteins and other materials that enter cells by endocytosis also enter the endomembrane system to travel to the Golgi complex for sorting and distribution to other locations.

Mitochondria Are the Organelles in Which Cellular Respiration Occurs

Mitochondria (singular, *mitochondrion*) are the membrane-bound organelles in which cellular respiration occurs. *Cellular respiration* is the process by which energy-rich molecules such as sugars, fats, and other fuels are broken down to water and carbon dioxide by mitochondrial reactions, with the release of energy. Much of the energy released by the breakdown is captured in ATP. In fact, mitochondria generate most of the ATP of the cell. Mitochondria require oxygen for cellular respiration—when you breathe, you are taking in oxygen primarily for your mitochondrial reactions (see Chapter 8).

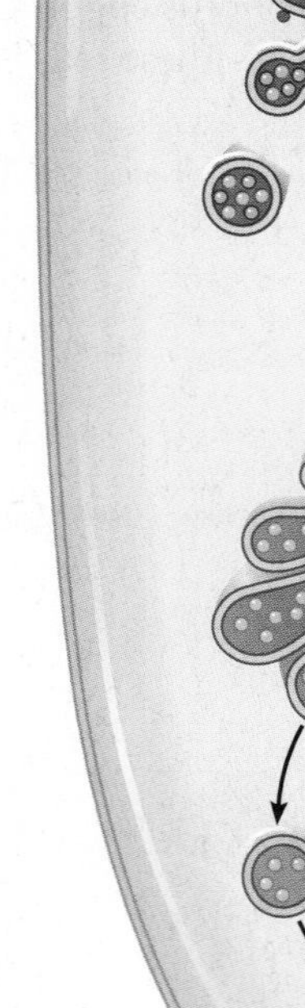
Instructions for building proteins leave the nucleus and enter the cytoplasm.
Proteins (green and yellow) are assembled from amino acids by ribosomes attached to the ER or free in the cytosol.
Nucleus
Ribosomes
Rough ER
Vesicles
Golgi complex
Secretory vesicles
Lysosomes
Damaged organelle
Endocytic vesicle

1 Proteins made by ER ribosomes enter ER membranes or the space inside ER cisternae. Chemical modification of some proteins begins. Membrane lipids are also made in the ER.

2 Vesicles bud from the ER membrane and then transport unfinished proteins and lipids to the Golgi complex.

3 Protein and lipid modification is completed in the Golgi complex, and products are sorted into vesicles that bud from the complex.

4 Secretory vesicles budding from the Golgi membranes transport finished products to the plasma membrane. The products are released by exocytosis. Other vesicles remain in storage in the cytoplasm.

5 Lysosomes budding from the Golgi membranes contain hydrolytic enzymes that digest damaged organelles or the contents of endocytic vesicles that fuse with them. Endocytic vesicles form at the plasma membrane and move into the cytoplasm.

FIGURE 5.18
Vesicle traffic in the cytoplasm. The ER and Golgi complex are part of the endomembrane system, which releases proteins and other substances to the cell exterior and gathers materials from outside the cell.

Mitochondria are enclosed by two membranes **(Figure 5.19).** The **outer mitochondrial membrane** is smooth and covers the outside of the organelle. The surface area of the **inner mitochondrial membrane** is expanded by folds called **cristae** (singular, *crista*). Both membranes surround the innermost compartment of the mitochondrion, called the **mitochondrial matrix.** The ATP-generating reactions of mitochondria occur in the cristae and matrix.

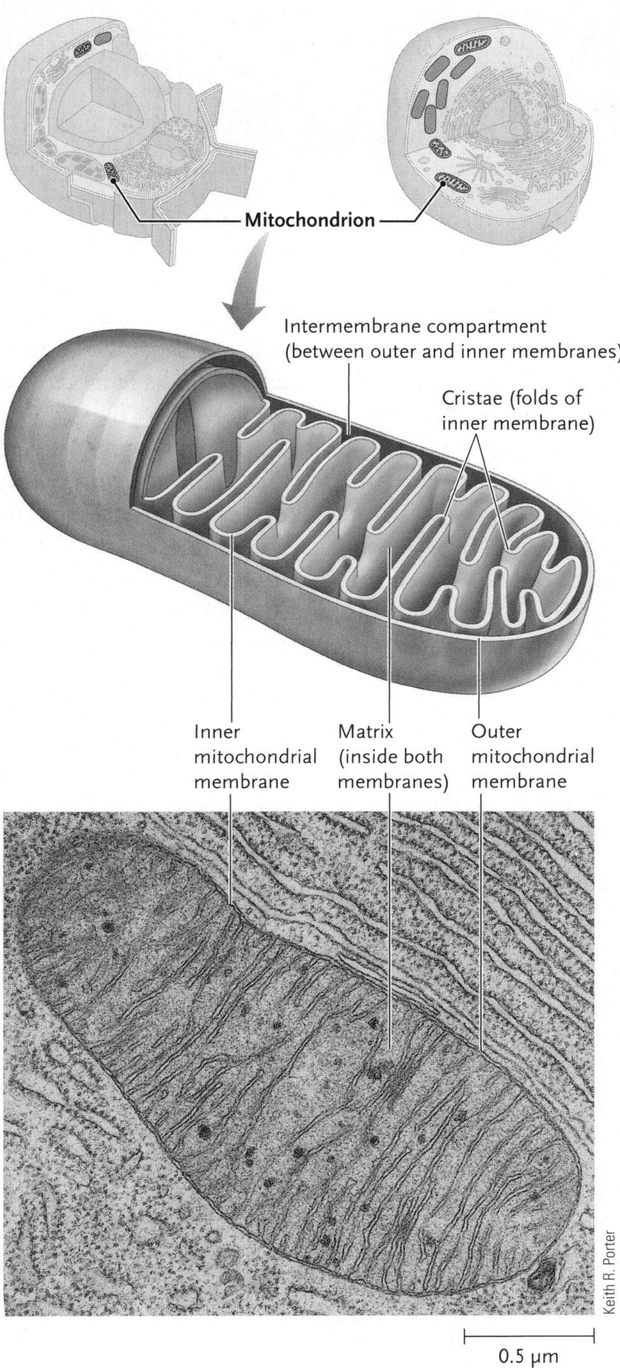

Mitochondrion
Intermembrane compartment (between outer and inner membranes)
Cristae (folds of inner membrane)
Inner mitochondrial membrane
Matrix (inside both membranes)
Outer mitochondrial membrane
Keith R. Porter
0.5 μm

FIGURE 5.19
Mitochondria. The electron micrograph shows a mitochondrion from bat pancreas, surrounded by cytoplasm containing rough ER. Cristae extend into the interior of the mitochondrion as folds from the inner mitochondrial membrane. The darkly stained granules inside the mitochondrion are probably lipid deposits.

The mitochondrial matrix also contains DNA and ribosomes that resemble the equivalent structures in bacteria. These and other similarities suggest that mitochondria originated from ancient bacteria that became permanent residents of the cytoplasm during the evolution of eukaryotic cells (see Chapter 24 for further discussion).

Microbodies Carry Out Vital Reactions That Link Metabolic Pathways

Microbodies are small, relatively simple membrane-bound organelles found in various forms in essentially all eukaryotic cells. They consist of a single boundary membrane that encloses a collection of enzymes and other proteins **(Figure 5.20).** Recent research has shown that the ER is involved in microbody production. Proteins and phospholipids are continuously imported into microbodies. The phospholipids are used for new membrane synthesis, leading to growth of the microbody. Division of a microbody then produces new microbodies.

Microbodies have various functions that are often specific to an organism or cell type. Commonly, they contain enzymes that conduct preparatory or intermediate reactions linking major biochemical pathways. For example, the series of reactions that allows cells to use fats as an energy source begins in microbodies and continues in mitochondria. Beginning or intermediate steps in the breakdown of some amino acids and alcohols also take place in microbodies, including about half of the ethyl alcohol that humans consume. Many types of microbodies produce as a by-product the toxic substance hydrogen peroxide (H_2O_2), which is broken down into water and oxygen by the enzyme *catalase.* Microbodies with this reaction are often termed **peroxisomes.**

Microbodies in plants convert oils or fats to sugars that can be used directly for energy-releasing reactions in mitochondria or for reactions that require sugars as chemical building blocks. These microbody reactions are particularly important in plant embryos that develop from oily seeds, such as those of the peanut

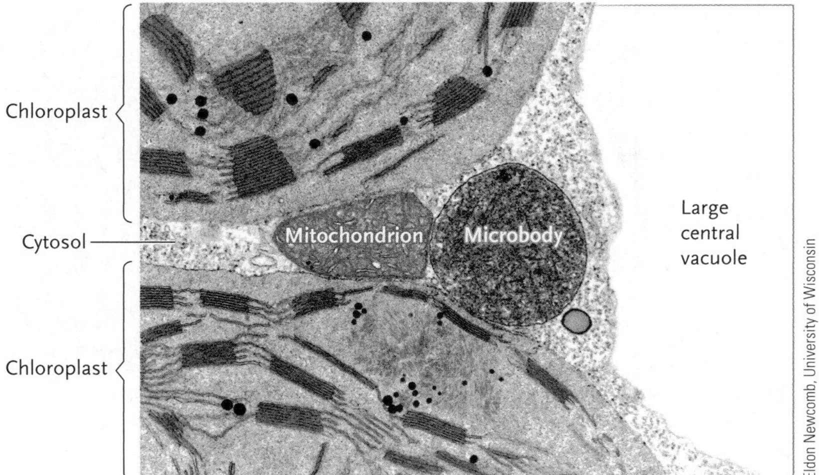

Chloroplast
Cytosol
Chloroplast
Mitochondrion
Microbody
Large central vacuole
Eldon Newcomb, University of Wisconsin

FIGURE 5.20
A microbody in the cytoplasm of a tobacco leaf cell. The EM has been colorized to make the structures easier to identify.

A. **Microtubules**

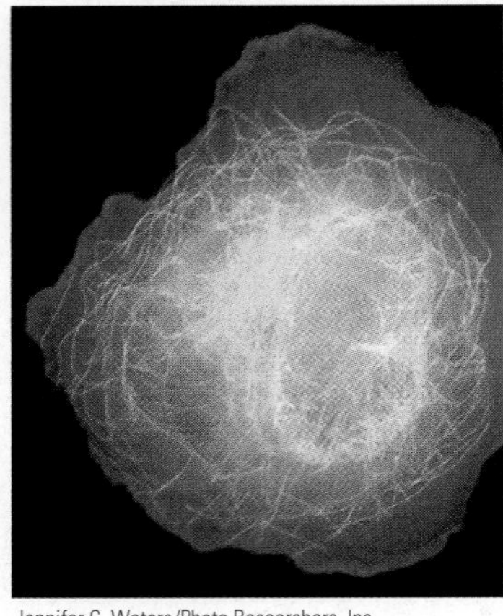

Jennifer C. Waters/Photo Researchers, Inc.

B. **Intermediate filaments**

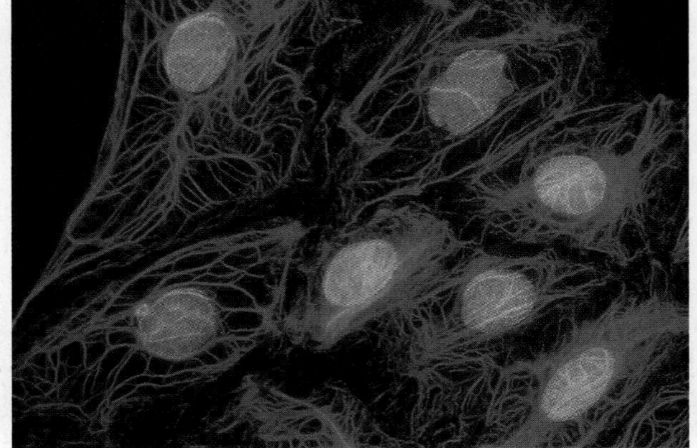

Courtesy of Mary Osborn

C. **Microfilaments**

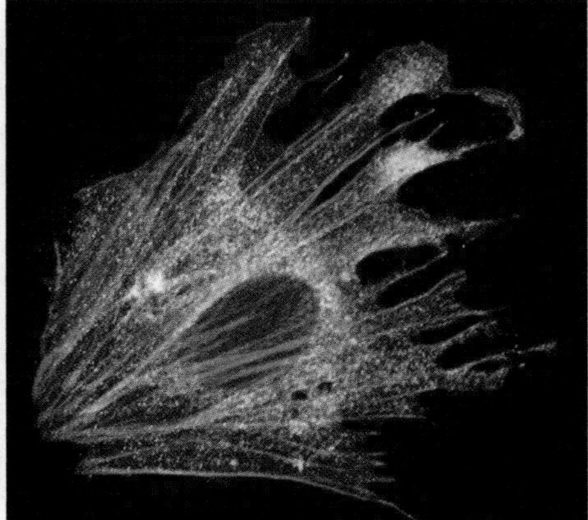

Courtesy Dr. Vincenzo Cerulli, Lab of Developmental Biology, The Whittier Inst. for Diabetes, Univ. of Cal.-San Diego, La Jolla, CA

FIGURE 5.21
Cytoskeletons of eukaryotic cells, as seen in cells stained for light microscopy. **(A)** Microtubules (yellow) and microfilaments (red) in a pancreatic cell. **(B)** Intermediate filaments assembled from keratin proteins in cells of the kangaroo rat. The nucleus is stained blue in these cells. **(C)** Microfilaments (red) in a migrating mammalian cell.

or soybean. Depending on the particular reaction pathways they carry out, plant microbodies are called peroxisomes, *glyoxysomes,* or *glycosomes.*

The Cytoskeleton Supports and Moves Cell Structures

The characteristic shape and internal organization of each type of cell is maintained in part by its cytoskeleton, the interconnected system of protein fibers and tubes that extends throughout the cytoplasm. The cytoskeleton also reinforces the plasma membrane and functions in movement, both of structures within the cell and of the cell as a whole. It is most highly developed in animal cells, in which it fills and supports the cytoplasm from the plasma membrane to the nuclear envelope **(Figure 5.21)**. Although cytoskeletal structures are also present in plant cells, the fibers and tubes of the system are less prominent; much of cellular support in plants is provided by the cell wall and a large central vacuole (described in Section 5.4).

The cytoskeleton of animal cells contains structural elements of three major types: *microtubules, intermediate filaments,* and *microfilaments.* Plant cytoskeletons likewise contain the same three structural elements. Microtubules are the largest cytoskeletal elements, and microfilaments are the smallest. Each cytoskeletal element is assembled from proteins—microtubules from *tubulins,* intermediate filaments from a large and varied group of *intermediate filament proteins,* and microfilaments from *actins* **(Figure 5.22)**. The keratins of animal hair, nails, and claws contain a common form of intermediate filament proteins known as *cytokeratins.* For example, human hair

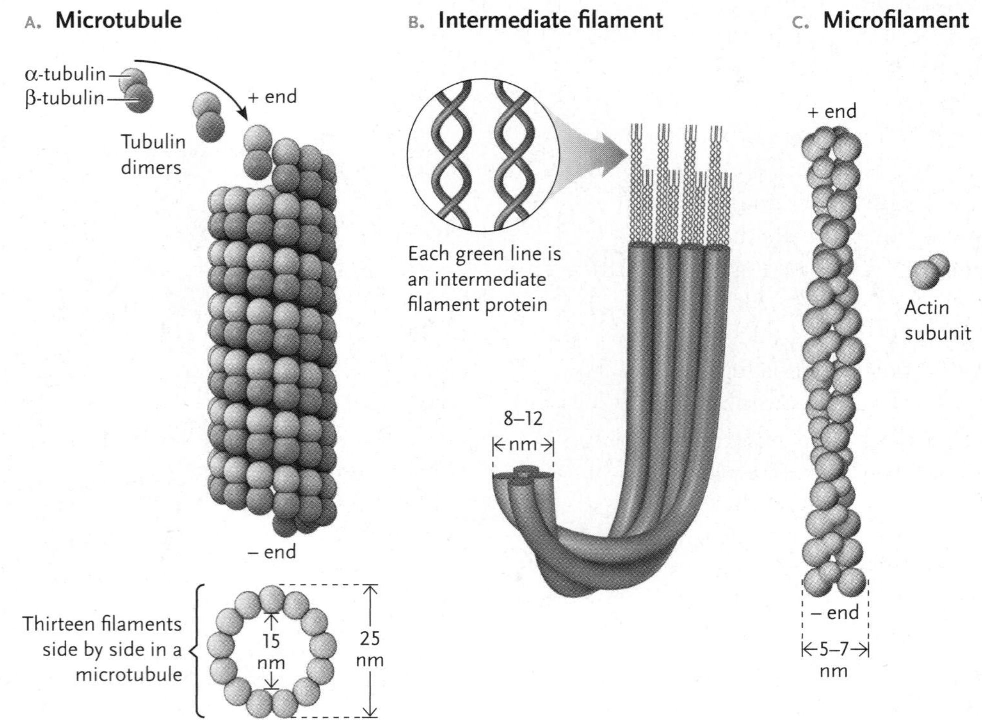

A. Microtubule
α-tubulin
β-tubulin
+ end
Tubulin dimers
– end
Thirteen filaments side by side in a microtubule
15 nm
25 nm
B. Intermediate filament
Each green line is an intermediate filament protein
8–12 nm
C. Microfilament
+ end
Actin subunit
– end
5–7 nm

FIGURE 5.22
The major components of the cytoskeleton. **(A)** A microtubule, assembled from dimers of α- and β-tubulin proteins. **(B)** An intermediate filament. Eight protein chains wind together to form each subunit shown as a green cylinder. **(C)** A microfilament, assembled from two linear polymers of actin proteins, wound around each other into a helical spiral.

consists of thick bundles of cytokeratin fibers extruded from hair follicle cells. The lamins that line the inner surface of the nuclear envelope in animal cells are also assembled from intermediate filament proteins.

Microtubules (Figure 5.22A) are microscopic tubes with an outer diameter of about 25 nm and an inner diameter of about 15 nm; they function much like the tubes used by human engineers to construct supportive structures. Microtubules vary widely in length from less than 200 nm to several micrometers. The wall of the microtubule consists of 13 protein filaments arranged side by side. A filament is a linear polymer of tubulin dimers, each dimer consisting of one α-tubulin and one β-tubulin subunit bound noncovalently together. The dimers are organized head-to-tail in each filament, giving the microtubule a polarity, meaning that the two ends are different. One end, called the + (plus) end, has α-tubulin subunits at the ends of the filaments; the other end, called the − (minus) end, has β-tubulin subunits at the ends of the filaments. Microtubules are dynamic structures, changing their lengths as required by their functions. This is seen readily in animal cells that are changing shape. Microtubules change length by the addition or removal of tubulin dimers; this occurs asymmetrically, with dimers adding or detaching more rapidly at the + end than at the − end. The lengths of microtubules are tightly regulated in the cell.

Many of the cytoskeletal microtubules in animal cells are formed and radiate outward from a site near the nucleus termed the **cell center** or **centrosome** (see Figure 5.9). At its midpoint are two short, barrel-shaped structures also formed from microtubules called the **centrioles** (see Figure 5.26). Often, intermediate filaments also extend from the cell center, apparently held in the same radiating pattern by linkage to microtubules. Microtubules that radiate from the cell center anchor the ER, Golgi complex, lysosomes, secretory vesicles, and at least some mitochondria in position. The microtubules also provide tracks along which vesicles move from the cell interior to the plasma membrane and in the reverse direction. The intermediate filaments probably add support to the microtubule arrays.

Microtubules play other key roles, for instance, in separating and moving chromosomes during cell division, determining the orientation for growth of the new cell wall during plant cell division, maintaining the shape of animal cells, and moving animal cells themselves. Animal cell movements are generated by "motor" proteins that push or pull against microtubules or microfilaments, much as our muscles produce body movements by acting on bones of the skeleton. One end of a motor protein is firmly fixed to a cell structure such as a vesicle or to a microtubule or microfilament. The other end has reactive groups that "walk" along another microtubule or microfilament by making an attachment, forcefully swiveling a short distance, and then releasing **(Figure 5.23)**. ATP supplies the energy for the walking movements. The motor proteins that walk along microfilaments are called *myosins,* and the ones that walk along microtubules are called *dyneins* and *kinesins*. Some cell movements, such as the whipping motions of sperm tails, depend entirely on microtubules and their motor proteins.

FIGURE 5.23 The microtubule motor protein kinesin. **(A)** Structure of the end of a kinesin molecule that "walks" along a microtubule, with α-helical segments shown as spirals and β strands as flat ribbons. **(B)** How a kinesin molecule walks along the surface of a molecule by alternately attaching and releasing its "feet."

A. **"Walking" end of a kinesin molecule**

Connects to cell structure such as a vesicle

One "foot" of motor protein

B. How a kinesin molecule "walks"

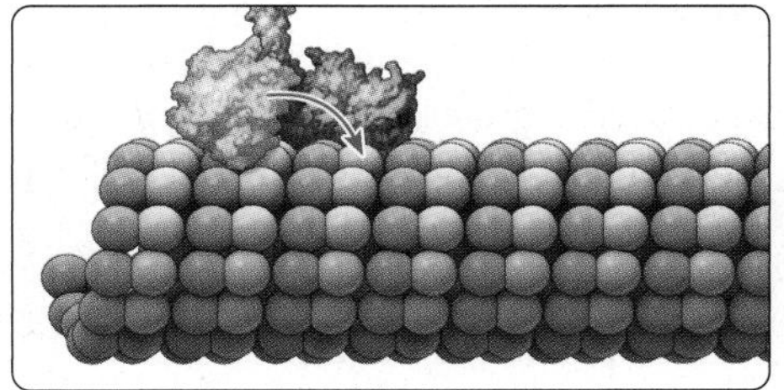

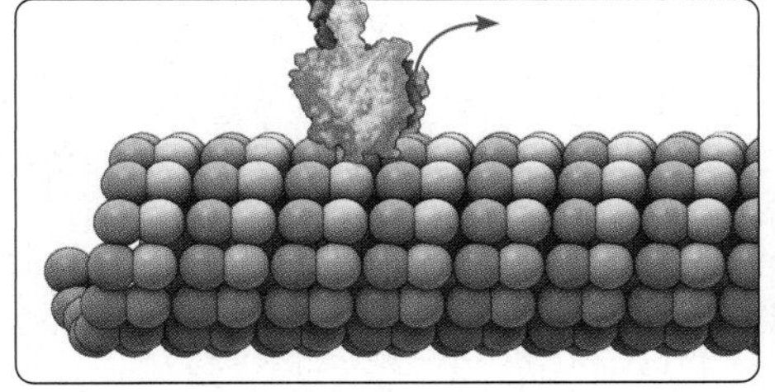

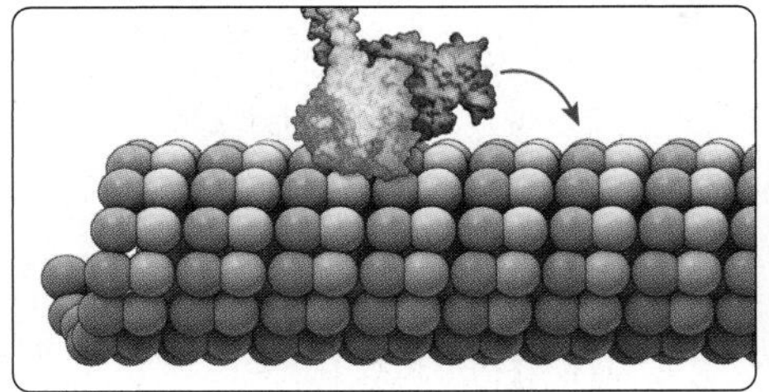

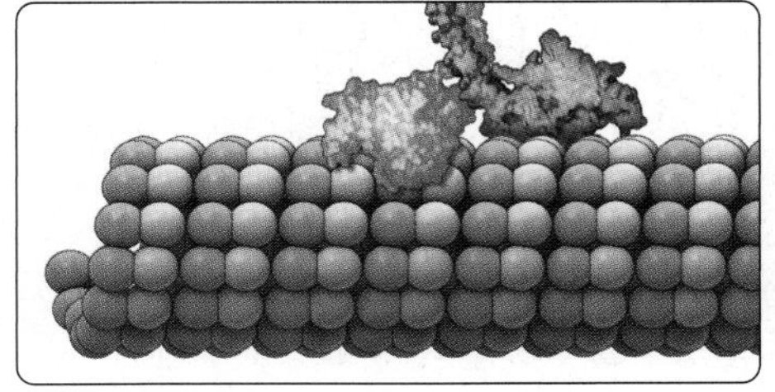

Intermediate filaments (Figure 5.22B) are fibers with diameters of about 8 to 12 nm. ("Intermediate" signifies, in fact, that these filaments are intermediate in size between microtubules and microfilaments.) These fibers occur singly, in parallel bundles, and in interlinked networks, either alone or in combination with microtubules, microfilaments, or both. Intermediate filaments are only found in multicellular organisms. Moreover, whereas microtubules and microfilaments are the same in all tissues, intermediate filaments are tissue-specific in their protein composition. Despite the molecular diversity of intermediate filaments, however, they all play similar roles in the cell, providing structural support in many cells and tissues. For example, the nucleus in epithelial cells is held within the cell by a basketlike network of intermediate filaments made of keratins.

Microfilaments (Figure 5.22C) are thin protein fibers 5 to 7 nm in diameter that consist of two polymers of actin subunits wound around each other in a long helical spiral. The actin subunits are asymmetrical in shape, and they are all oriented in the same way in the polymer chains of a microfilament. Thus, as for microtubules, microfilaments have a polarity; the two ends are designated + (plus) and − (minus). And as for microtubules, growth and disassembly occur more rapidly at the + end than at the − end.

Microfilaments occur in almost all eukaryotic cells and are involved in many processes, including a number of structural and locomotor functions. Microfilaments are best known as one of the two components of the contractile elements in muscle fibers of vertebrates (the roles of myosin and microfilaments in muscle contraction are discussed in Chapter 41). Microfilaments are involved in the actively flowing motion of cytoplasm called *cytoplasmic streaming,* which can transport nutrients, proteins, and organelles in both animal and plant cells, and which is responsible for amoeboid movement. When animal cells divide, microfilaments are responsible for dividing the cytoplasm (see Chapter 10 for further discussion).

Flagella Propel Cells, and Cilia Move Materials over the Cell Surface

Flagella and *cilia* (singular, *cilium*) are elongated, slender, motile structures that extend from the cell surface. They are identical in structure except that cilia are usually shorter than flagella and occur on cells in greater numbers. Whiplike or oarlike movements of a flagellum propel a cell through a watery medium, and cilia move fluids over the cell surface.

A bundle of microtubules extends from the base to the tip of a flagellum or cilium **(Figure 5.24).** In the bundle, a circle of nine double microtubules surrounds a central pair of single microtubules, forming what is known as the 9 + 2 complex. Dynein motor proteins slide the microtubules of the 9 + 2 complex over each other to produce the movements of a flagellum or cilium **(Figure 5.25).**

Flagella and cilia arise from the centrioles. These barrel-shaped structures contain a bundle of microtubules similar to the 9 + 2 complex, except that the central pair of microtubules is missing and the outer circle is formed from a ring of nine triple rather than double microtubules (compare Figure 5.24 and **Figure 5.26**). During the formation of a flagellum or cilium, a centri-

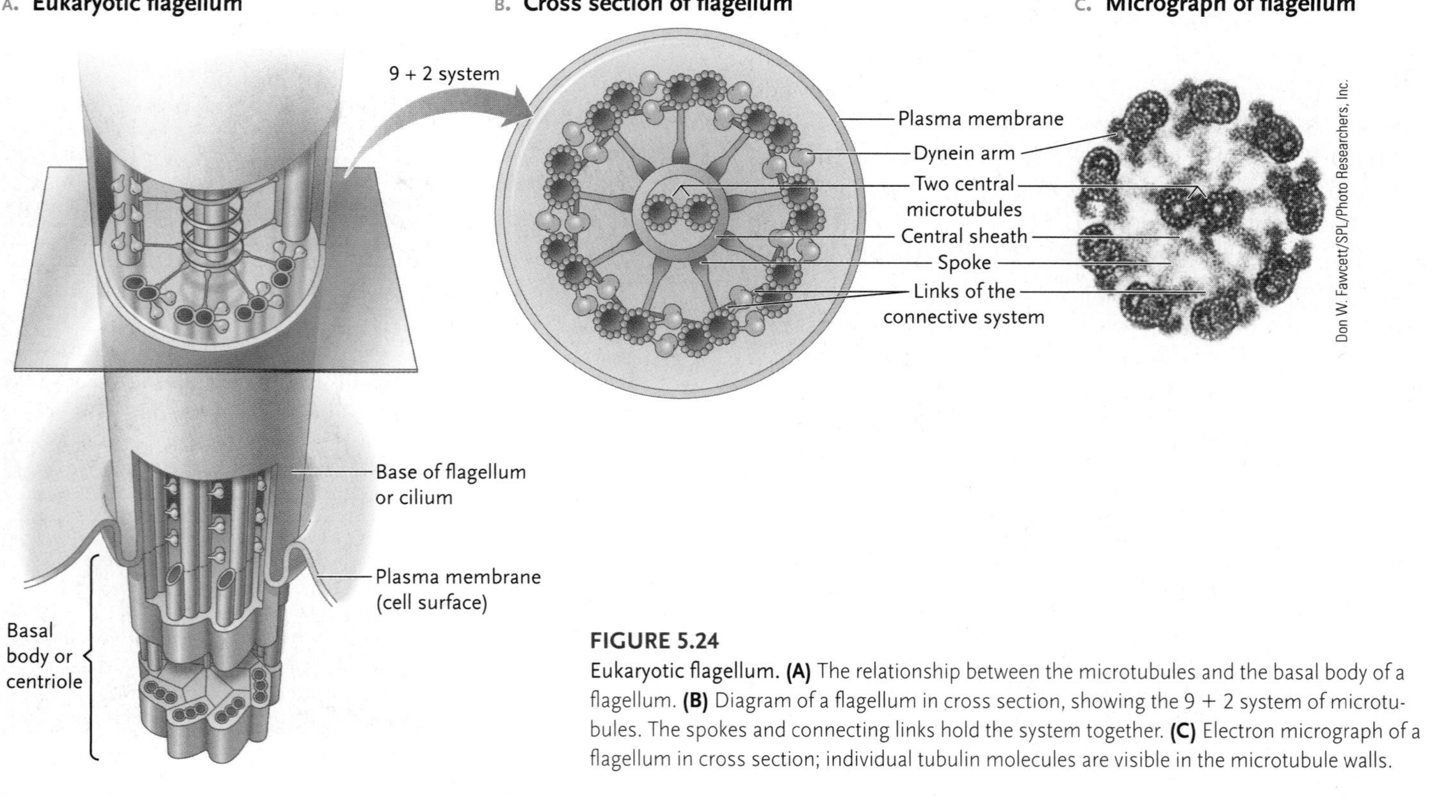
A. Eukaryotic flagellum
B. Cross section of flagellum
C. Micrograph of flagellum
9 + 2 system
Plasma membrane
Dynein arm
Two central microtubules
Central sheath
Spoke
Links of the connective system
Base of flagellum or cilium
Plasma membrane (cell surface)
Basal body or centriole
Don W. Fawcett/SPL/Photo Researchers, Inc.

FIGURE 5.24
Eukaryotic flagellum. **(A)** The relationship between the microtubules and the basal body of a flagellum. **(B)** Diagram of a flagellum in cross section, showing the 9 + 2 system of microtubules. The spokes and connecting links hold the system together. **(C)** Electron micrograph of a flagellum in cross section; individual tubulin molecules are visible in the microtubule walls.

Flagella: Flagella beat in smooth, S-shaped waves that travel from base to tip.

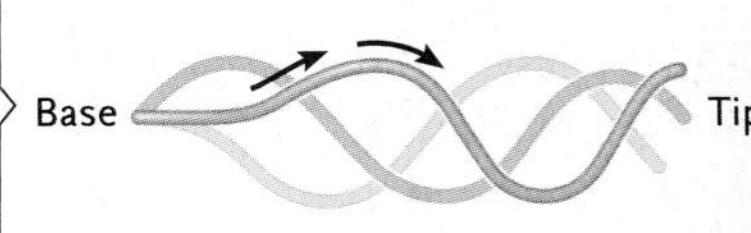

Base
Tip

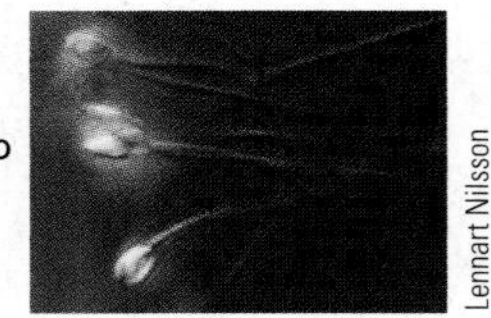

Lennart Nilsson

Cilia: Cilia beat in an oarlike power stroke (dark orange) followed by a recovery stroke (light orange).

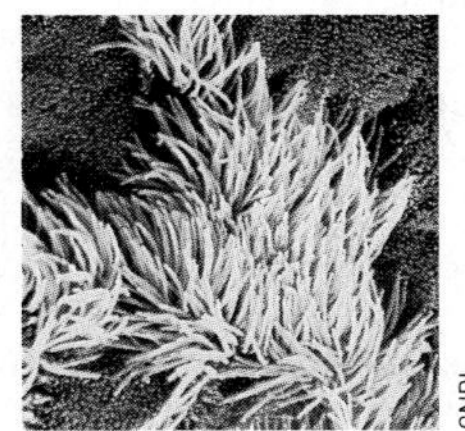

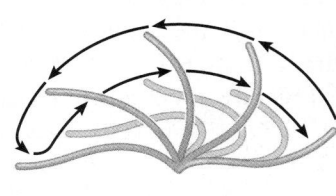

CNRI

Waving and bending mechanism: The waves and bends are produced by dynein motor proteins, which slide the microtubule doublets over each other. An examination of the tip of a bent cilium or flagellum shows that the doublets extend farther toward the tip on the side toward the bend, confirming that the doublets actually slide as the shaft of the cilium or flagellum bends.

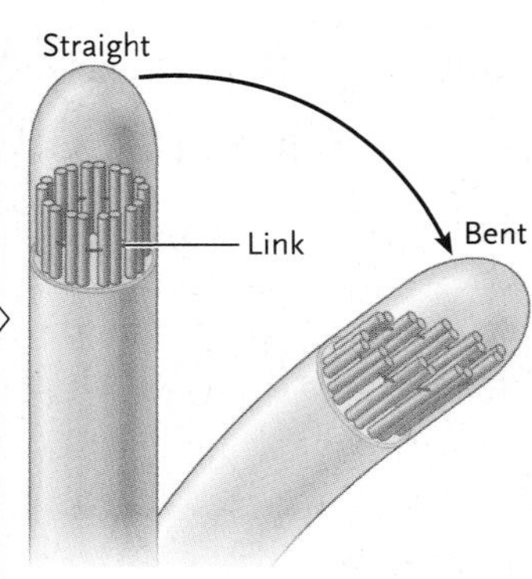

Straight
Link
Bent

FIGURE 5.25
Flagellar and ciliary beating patterns. The micrographs show a few human sperm, each with a flagellum (top), and cilia from the lining of an airway in the lungs (bottom).

ole moves to a position just under the plasma membrane. Then two of the three microtubules of each triplet grow outward from one end of the centriole to form the ring of nine double microtubules. The two central microtubules of the 9 + 2 complex also grow from the end of the centriole, but without direct connection to any centriole microtubules. The centriole remains at the innermost end of a flagellum or cilium when its development is complete as the **basal body** of the structure (see Figure 5.24).

Cilia and flagella are found in protozoa and algae, and many types of animal cells have flagella—the tail of a sperm cell is a flagellum—as do the reproductive cells of some plants. In humans, cilia cover the surfaces of cells lining cavities or tubes in some parts of the body. For example, cilia on cells lining the ventricles (cavities) of the brain circulate fluid through the brain, and cilia in the oviducts conduct eggs from the ovaries to the uterus. Cilia covering cells that line the air passages of the lungs sweep out mucus containing bacteria, dust particles, and other contaminants.

Although the purpose of the eukaryotic flagellum is the same as that of prokaryotic flagella, the genes that encode the components of the flagellar apparatus of cells of Bacteria, Archaea, and Eukarya are different in each case. Thus, as mentioned earlier in the chapter, the three types of flagella are analogous, not homologous, structures, and they must have evolved independently.

With a few exceptions, the cell structures described so far in this chapter occur in all eukaryotic cells. The major exception is lysosomes, which appear to be restricted to animal cells. The next section describes three additional structures that are characteristic of plant cells.

STUDY BREAK 5.3

1. **Where in a eukaryotic cell is DNA found? How is that DNA organized?**
2. **What is the nucleolus, and what is its function?**
3. **Explain the structure and function of the endomembrane system.**
4. **What is the structure and function of a mitochondrion?**
5. **What is the structure and function of the cytoskeleton?**

THINK OUTSIDE THE BOOK

On your own or collaboratively, explore the Internet and the research literature to develop an outline of the molecular steps that a protein with a nuclear localization signal follows for nuclear import (that is, being transported through a nuclear pore complex).

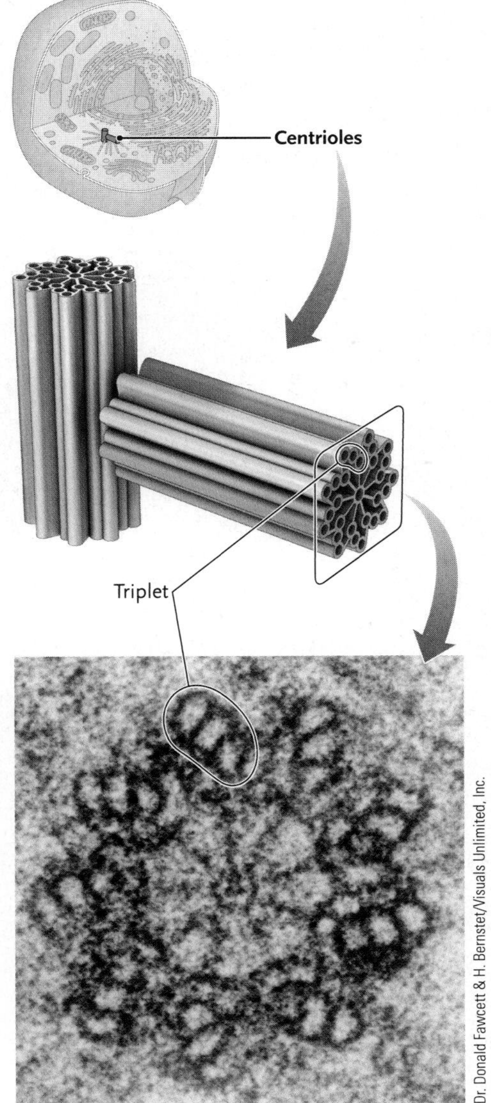

Centrioles
Triplet
Dr. Donald Fawcett & H. Bernstet/Visuals Unlimited, Inc.

FIGURE 5.26
Centrioles. The two centrioles of the pair at the cell center usually lie at right angles to each other as shown. The electron micrograph shows a centriole from a mouse cell in cross section. A centriole gives rise to the 9 + 2 system of a flagellum and persists as the basal body at the inner end of the flagellum.

5.4 Specialized Structures of Plant Cells

Chloroplasts, a large and highly specialized central vacuole, and cell walls give plant cells their distinctive characteristics, but these structures also occur in some other eukaryotes—chloroplasts in algal protists and cell walls in algal protists and fungi.

Chloroplasts Are Biochemical Factories Powered by Sunlight

Chloroplasts (*chloro* = yellow–green), the sites of photosynthesis in plant cells, are members of a family of plant organelles known collectively as **plastids.** Other members of the family include amyloplasts and chromoplasts. **Amyloplasts** (*amylo* = starch) are colorless plastids that store starch, a product of photosynthesis. They occur in great numbers in the roots or tubers of some plants, such as the potato. **Chromoplasts** (*chromo* = color) contain red and yellow pigments and are responsible for the colors of ripening fruits or autumn leaves. All plastids contain DNA genomes and molecular machinery for gene expression and the synthesis of proteins on ribosomes. Some of the proteins within plastids are encoded by their genomes; others are encoded by nuclear genes and are imported into the organelles.

Chloroplasts, like mitochondria, are usually lens- or disc-shaped and are surrounded by a smooth **outer boundary membrane** and an **inner boundary membrane,** which lies just inside the outer membrane **(Figure 5.27).** These two boundary membranes completely enclose an inner compartment, the stroma. Within the stroma is a third membrane system that consists of flattened, closed sacs called thylakoids. In higher plants, the thylakoids are stacked, one on top of another, forming structures called **grana** (singular, *granum*).

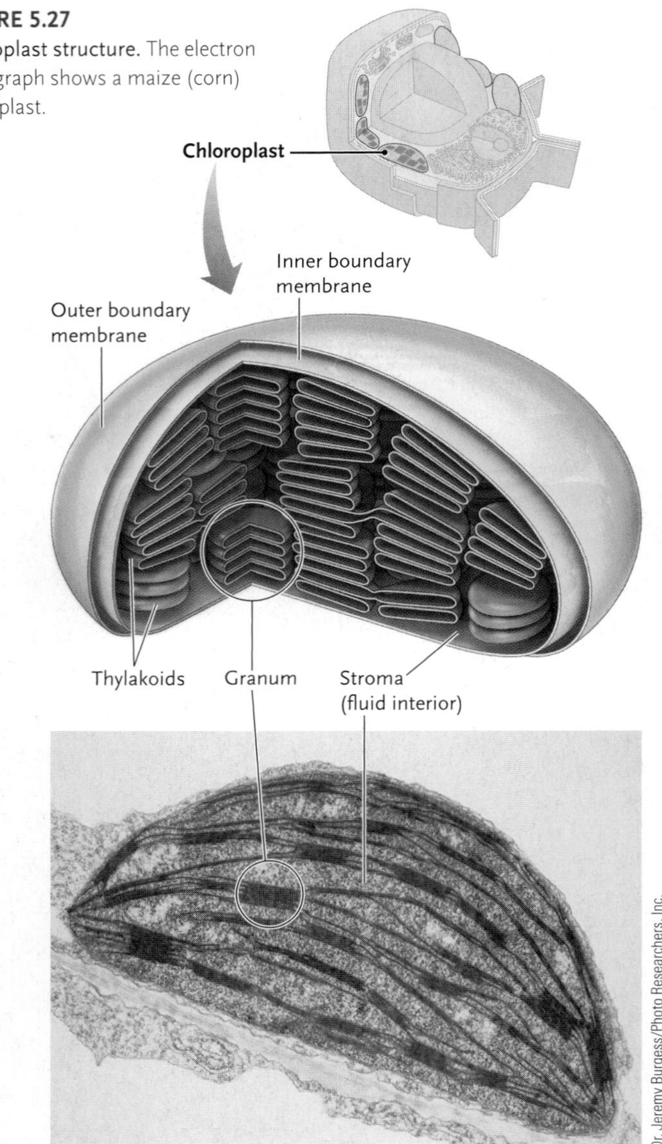
Chloroplast
Inner boundary membrane
Outer boundary membrane
Thylakoids
Granum
Stroma (fluid interior)
1.0 µm
Dr. Jeremy Burgess/Photo Researchers, Inc.

FIGURE 5.27
Chloroplast structure. The electron micrograph shows a maize (corn) chloroplast.

The thylakoid membranes contain molecules that absorb light energy and convert it to chemical energy in photosynthesis. The primary molecule absorbing light is *chlorophyll,* a green pigment that is present in all chloroplasts. The chemical energy is used by enzymes in the stroma to make carbohydrates and other complex organic molecules from water, carbon dioxide, and other simple inorganic precursors. The organic molecules produced in chloroplasts, or from biochemical building blocks made in chloroplasts, are the ultimate food source for most organisms. (The physical and biochemical reactions of chloroplasts are described in Chapter 9.)

The chloroplast stroma contains DNA and ribosomes that resemble those of certain photosynthetic bacteria. Because of these similarities, chloroplasts, like mitochondria, are believed to have originated from ancient prokaryotes that became permanent residents of the eukaryotic cells ancestral to the plant lineage (see Chapter 24 for further discussion).

Central Vacuoles Have Diverse Roles in Storage, Structural Support, and Cell Growth

Central vacuoles (see Figure 5.10) are large vesicles identified as distinct organelles of plant cells because they perform specialized functions unique to plants. In a mature plant cell, 90% or more of the cell's volume may be occupied by one or more large central vacuoles. The remainder of the cytoplasm and the nucleus of these cells are restricted to a narrow zone between the central vacuole and the plasma membrane. The pressure within the central vacuole supports the cells.

The membrane that surrounds the central vacuole, the **tonoplast,** contains transport proteins that move substances into and out of the central vacuole. As plant cells mature, they grow primarily by increases in the pressure and volume of the central vacuole.

Central vacuoles conduct other vital functions. They store salts, organic acids, sugars, storage proteins, pigments, and, in some cells, waste products. Pigments concentrated in the vacuoles produce the colors of many flowers. Enzymes capable of breaking down biological molecules are present in some central vacuoles, giving them some of the properties of lysosomes. Molecules that provide chemical defenses against pathogenic organisms also occur in the central vacuoles of some plants.

Cell Walls Support and Protect Plant Cells

The cell walls of plants are extracellular structures because they are located outside the plasma membrane **(Figure 5.28)**. Cell walls provide support to individual cells, contain the pressure produced in the central vacuole, and protect cells against invading bacteria and fungi.

Cell walls consist of cellulose fibers (see Figure 3.9C), which give tensile strength to the walls, embedded in a network of highly branched carbohydrates. The initial cell wall laid down by a plant cell, the **primary cell wall,** is relatively soft and flexible. As the cell grows and matures, the primary wall expands and additional layers of cellulose fibers and branched carbohydrates are laid down between the primary wall and the plasma membrane. The added wall layer, which is more rigid and may become many times thicker than the primary wall, is the **secondary cell wall.** In woody plants and trees, secondary cell walls are reinforced by *lignin,* a hard, highly resistant substance assembled from complex alcohols, surrounding the cellulose fibers. Lignin-impregnated cell walls are actually stronger than reinforced concrete by weight; hence, trees can grow to substantial size, and the wood of trees is used extensively in human cultures to make many structures and objects, including houses, tables, and chairs.

The walls of adjacent cells are held together by a layer of gel-like polysaccharides called the **middle lamella,** which acts as an intercellular glue (see Figure 5.28). The polysaccharide material of the middle lamella, called *pectin,* is extracted from some plants and used to thicken jams and jellies.

Both primary and secondary cell walls are perforated by minute channels, the **plasmodesmata** (singular, *plasmodesma;* see Figure 5.28). A typical plant cell has between 1,000 and 100,000 plasmodesmata connecting it to abutting cells. These cytosol-filled channels are lined by plasma membranes, so that connected cells essentially all have one continuous surface membrane. Most plasmodesmata also contain a narrow tubelike structure derived from the smooth endoplasmic reticulum of the connected cells. Plasmodesmata allow ions and small molecules to move directly from one cell to another through the connecting cytosol, without having to penetrate the plasma membranes or cell walls. Proteins and nucleic acids move through some plasmodesmata using energy-dependent processes.

Cell walls also surround the cells of fungi and algal protists. Carbohydrate molecules form the major framework of cell walls in most of these organisms, as they do in plants. In some, the wall fibers contain chitin (see Figure 3.9D) instead of cellulose. Details of cell wall structure in the algal protists and fungi, as well as in different subgroups of the plants, are presented in later chapters devoted to these organisms.

As noted earlier, animal cells do not form rigid, external, layered structures equivalent to the walls of plant cells. However, most animal cells secrete extracellular material and have other structures at the cell surface that play vital roles in the support

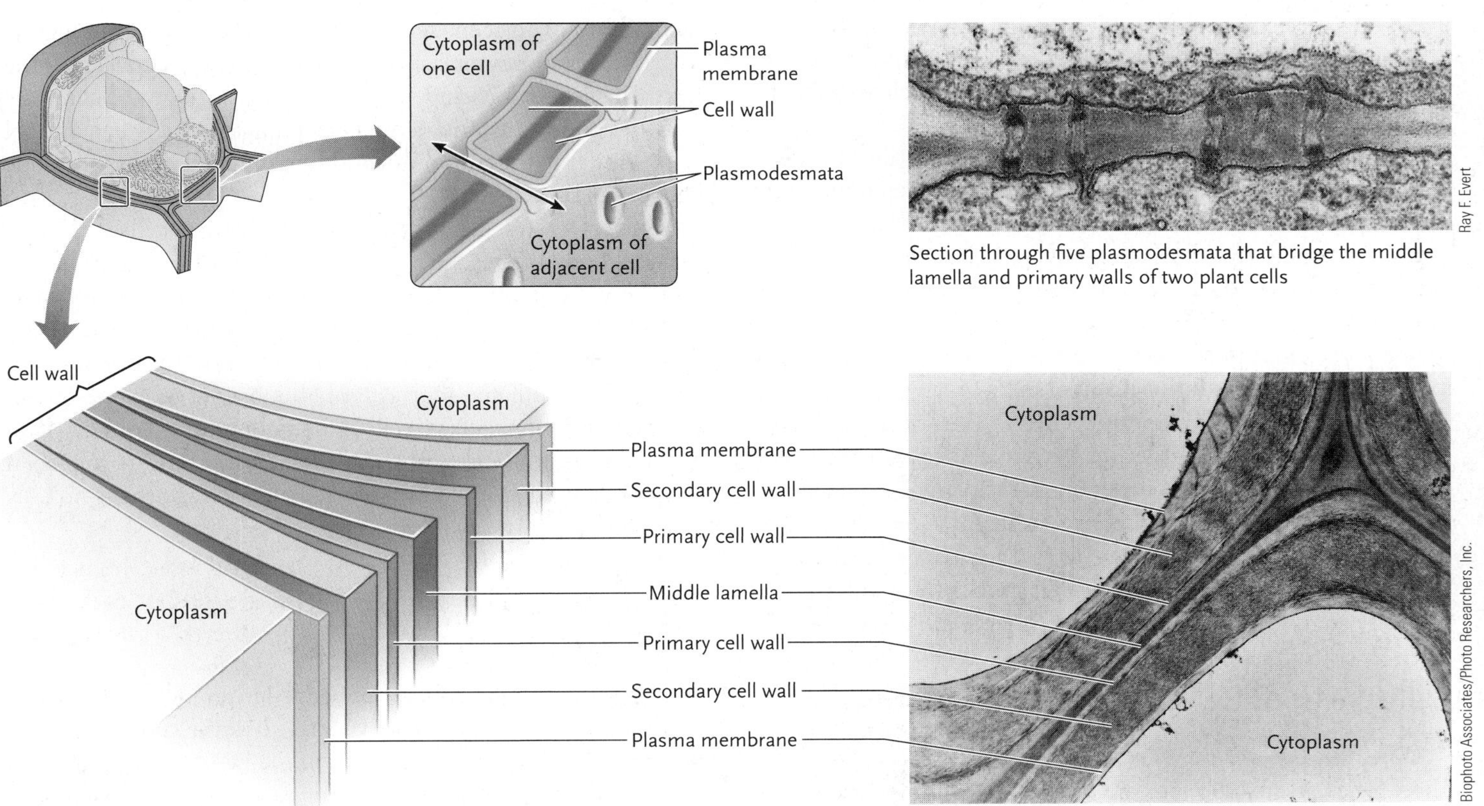
Cytoplasm of one cell
Plasma membrane
Cell wall
Plasmodesmata
Cytoplasm of adjacent cell
Section through five plasmodesmata that bridge the middle lamella and primary walls of two plant cells
Ray F. Evert
Cell wall
Cytoplasm
Cytoplasm
Plasma membrane
Secondary cell wall
Primary cell wall
Middle lamella
Primary cell wall
Secondary cell wall
Plasma membrane
Cytoplasm
Cytoplasm
Biophoto Associates/Photo Researchers, Inc.

FIGURE 5.28
Cell wall structure in plants. The upper right diagram and electron micrograph show plasmodesmata, which form openings in the cell wall that directly connect the cytoplasm of adjacent cells. The lower diagram and electron micrograph show the successive layers in the cell wall between two plant cells that have laid down secondary wall material.

and organization of animal body structures. The next section describes these and other surface structures of animal cells.

STUDY BREAK 5.4 <

1. What is the structure and function of a chloroplast?
2. What is the function of the central vacuole in plants?

5.5 The Animal Cell Surface

Animal cells have specialized structures that help hold cells together, produce avenues of communication between cells, and organize body structures. Molecular systems that perform these functions are organized at three levels: individual **cell adhesion molecules** bind cells together, more complex **cell junctions** seal the spaces between cells and provide direct communication between cells, and the **extracellular matrix (ECM)** supports and protects cells and provides mechanical linkages, such as those between muscles and bone.

Cell Adhesion Molecules Organize Animal Cells into Tissues and Organs

Cell adhesion molecules are glycoproteins embedded in the plasma membrane. They help maintain body form and structure in animals ranging from sponges to the most complex invertebrates and vertebrates. Rather than acting as a generalized intercellular glue, cell adhesion molecules bind to specific molecules on other cells. Most cells in solid body tissues are held together by many different cell adhesion molecules.

Cell adhesion molecules make initial connections between cells early in embryonic development, but then attachments are broken and remade as individual cells or tissues change position in the developing embryo. As an embryo develops into an adult, the connections become permanent and are reinforced by cell junctions. Cancer cells typically lose these adhesions, allowing them to break loose from their original locations, migrate to new locations, and form additional tumors.

Some bacteria and viruses—such as the virus that causes the common cold—target cell adhesion molecules as attachment sites during infection. Cell adhesion molecules are also partially responsible for the ability of cells to recognize one another as being part of the same individual or foreign. For example, rejection of organ transplants in mammals results from an immune response triggered by the foreign cell-surface molecules.

Cell Junctions Reinforce Cell Adhesions and Provide Avenues of Communication

Three types of cell junctions are common in animal tissues **(Figure 5.29)**. **Anchoring junctions** form buttonlike spots, or belts, that run entirely around cells, "welding" adjacent cells together. For some anchoring junctions known as **desmosomes,** intermediate filaments anchor the junction in the underlying cytoplasm; in other anchoring junctions known as **adherens junctions,** microfilaments are the anchoring cytoskeletal component. Anchoring junctions are most common in tissues that are subject to stretching, shear, or other mechanical forces—for example, heart muscle, skin, and the cell layers that cover organs or line body cavities and ducts.

Tight junctions, as the name indicates, are regions of tight connections between membranes of adjacent cells (see Figure 5.29). The connection is so tight that it can keep particles as small as ions from moving between the cells in the layers.

Tight junctions seal the spaces between cells in the cell layers that cover internal organs and the outer surface of the body, or the layers that line internal cavities and ducts. For example, tight junctions between cells that line the stomach, intestine, and bladder keep the contents of these body cavities from leaking into surrounding tissues.

A tight junction is formed by direct fusion of proteins on the outer surfaces of the two plasma membranes of adjacent cells. Strands of the tight junction proteins form a complex network that gives the appearance of stitch work holding the cells together. Within a tight junction, the plasma membrane is not joined continuously; instead, there are regions of intercellular space. Nonetheless, the network of junction proteins is sufficient to make the tight cell connections characteristic of these junctions.

Gap junctions open direct channels that allow ions and small molecules to pass directly from one cell to another (see Figure 5.29). Hollow protein cylinders embedded in the plasma membranes of adjacent cells line up and form a sort of pipeline that connects the cytoplasm of one cell with the cytoplasm of the next. The flow of ions and small molecules through the channels provides almost instantaneous communication between animal cells, similar to the communication that plasmodesmata provide between plant cells.

In vertebrates, gap junctions occur between cells within almost all body tissues, but not between cells of different tissues. These junctions are particularly important in heart muscle tissues and in the smooth muscle tissues that form the uterus, where their pathways of communication allow cells of the organ to operate as a coordinated unit. Although most nerve tissues do not have gap junctions, nerve cells in dental pulp are connected by gap junctions; they are responsible for the discomfort you feel if your teeth are disturbed or damaged, or when a dentist pokes a probe into a cavity.

The Extracellular Matrix Organizes the Cell Exterior

Many types of animal cells are embedded in an ECM that consists of proteins and polysaccharides secreted by the cells themselves **(Figure 5.30)**. The primary function of the ECM is protection and support. The ECM forms the mass of skin, bones, and tendons; it also forms many highly specialized extracellular structures such as the cornea of the eye and filtering networks in the kidney. The ECM also affects cell division, adhesion, motility, and embryonic development, and it takes part in reactions to wounds and disease.

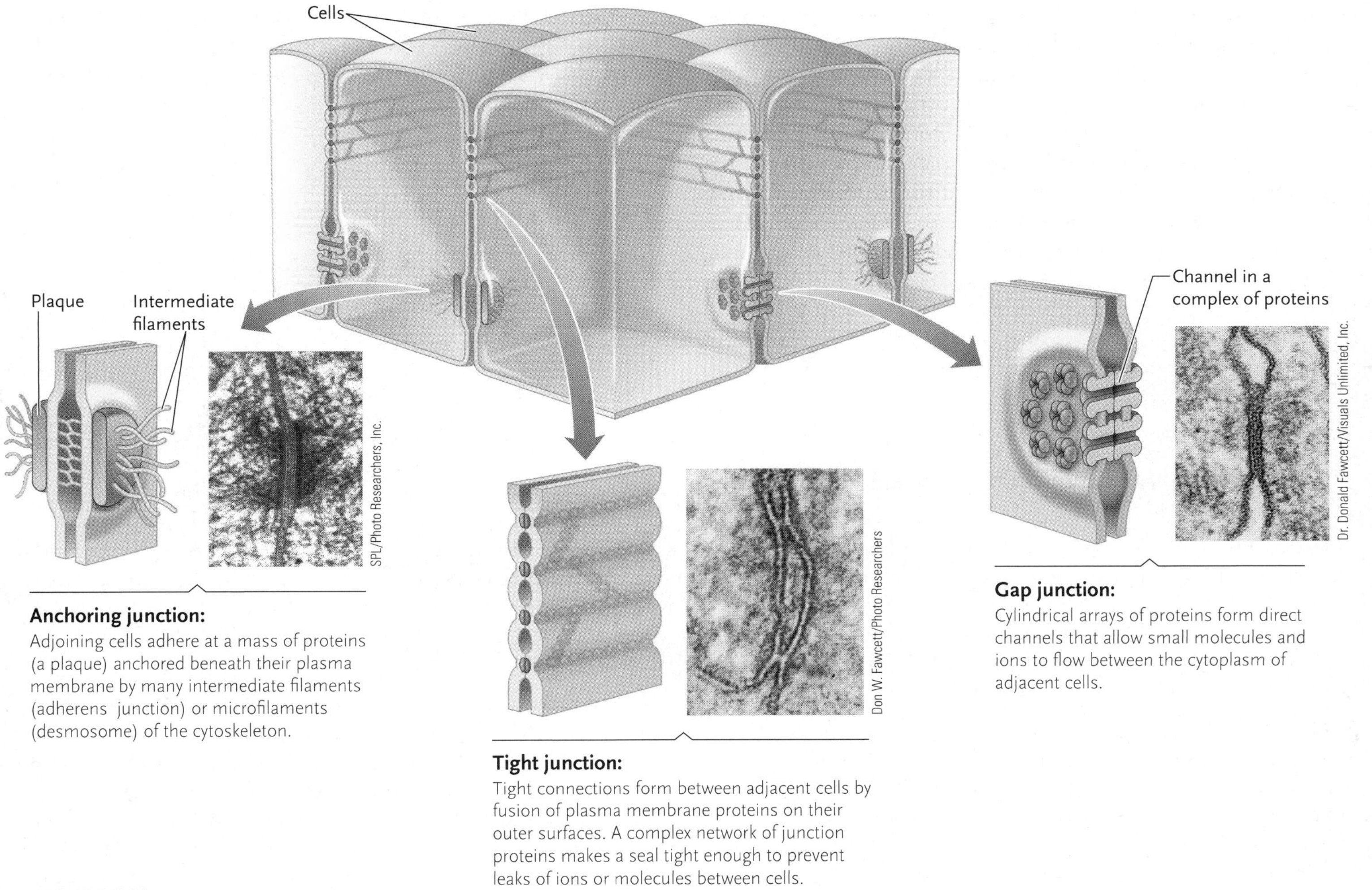

Cells

Plaque

Intermediate filaments

SPL/Photo Researchers, Inc.

Anchoring junction:
Adjoining cells adhere at a mass of proteins (a plaque) anchored beneath their plasma membrane by many intermediate filaments (adherens junction) or microfilaments (desmosome) of the cytoskeleton.

Don W. Fawcett/Photo Researchers

Tight junction:
Tight connections form between adjacent cells by fusion of plasma membrane proteins on their outer surfaces. A complex network of junction proteins makes a seal tight enough to prevent leaks of ions or molecules between cells.

Channel in a complex of proteins

Dr. Donald Fawcett/Visuals Unlimited, Inc.

Gap junction:
Cylindrical arrays of proteins form direct channels that allow small molecules and ions to flow between the cytoplasm of adjacent cells.

FIGURE 5.29
Anchoring junctions, tight junctions, and gap junctions, which connect cells in animal tissues. Anchoring junctions reinforce the cell-to-cell connections made by cell adhesion molecules, tight junctions seal the spaces between cells, and gap junctions create direct channels of communication between animal cells.

Glycoproteins are the main component of the ECM. In most animals, the most abundant ECM glycoprotein is *collagen,* which forms fibers with great tensile strength and elasticity. In vertebrates, the collagens of tendons, cartilage, and bone are the most abundant proteins of the body, making up about half of the total body protein by weight. (Collagens and their roles in body structures are described in further detail in Chapter 36.)

The consistency of the matrix, which may range from soft and jellylike to hard and elastic, depends on a network of proteoglycans that surrounds the collagen fibers. *Proteoglycans* are glycoproteins that consist of small proteins noncovalently attached to long polysaccharide molecules. Matrix consistency depends on the number of interlinks in this network, which determines how much water can be trapped in it. For example, cartilage, which contains a high proportion of interlinked glycoproteins, is relatively soft. Tendons, which are almost pure collagen, are tough and elastic. In bone, the glycoprotein network that sur-

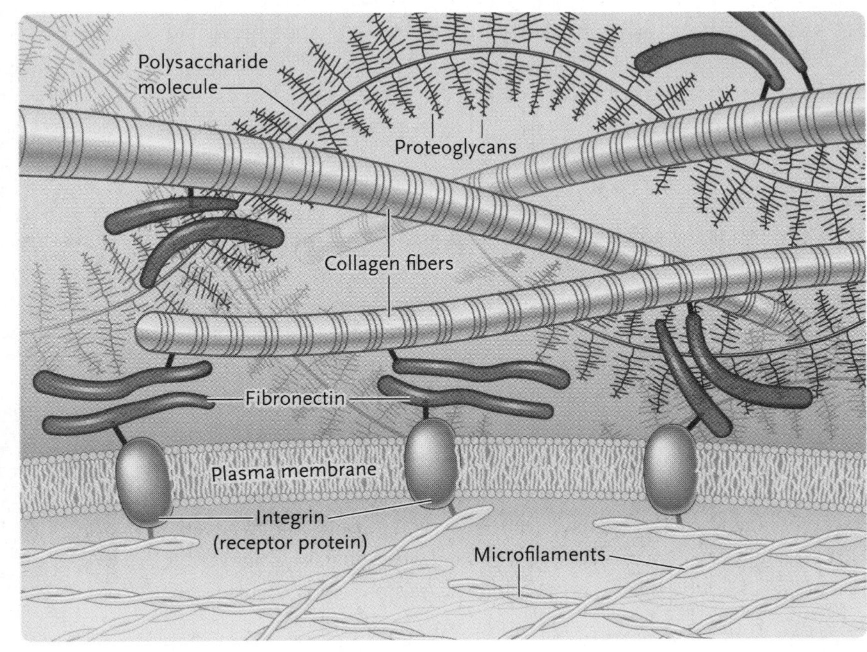
Polysaccharide molecule
Proteoglycans
Collagen fibers
Fibronectin
Plasma membrane
Integrin (receptor protein)
Microfilaments

FIGURE 5.30
Components of the extracellular matrix in an animal cell.

rounds collagen fibers is impregnated with mineral crystals, producing a dense and hard—but still elastic—structure that is about as strong as fiberglass or reinforced concrete.

Yet another class of glycoproteins is *fibronectins,* which aid in organizing the ECM and help cells attach to it. Fibronectins bind to receptor proteins called *integrins* that span the plasma membrane. On the cytoplasmic side of the plasma membrane, the integrins bind to microfilaments of the cytoskeleton. Integrins integrate changes outside and inside the cell by communicating changes in the ECM to the cytoskeleton.

Having laid the groundwork for cell structure and function in this chapter, we next take up further details of individual cell structures, beginning with the roles of cell membranes in transport in the next chapter.

STUDY BREAK 5.5

1. **Distinguish between anchoring junctions, tight junctions, and gap junctions.**
2. **What is the structure and function of the extracellular matrix?**

UNANSWERED QUESTIONS

The field of cell biology seeks to understand the properties and behaviors of cells, including their growth and division, shape and movement, subcellular organization and transport systems, and interactions and communication with each other and the environment. Although cell biology research has dramatically enhanced our basic understanding of cells, many fundamental questions remain to be answered.

How do eukaryotic cells balance assembly and disassembly pathways to maintain a complex cellular and subcellular structure?

Eukaryotic cells have a high degree of organizational complexity in the form of numerous subcellular organelles and structures. Maintenance of this complexity depends on a delicate balance between the pathways that promote the formation and disassembly of each structure. How is this balance achieved? At the whole cell level, researchers such as Paul Nurse at Rockefeller University have worked to understand how the balance between growth and division is coordinated to control cell size. At the subcellular level, work from Peter Walter's lab at the University of California, San Francisco aims to understand how organelle biogenesis and disassembly are balanced to determine organelle abundance, and my own lab studies how the assembly and disassembly of cytoskeletal polymers are balanced to coordinate cell movement and division. As with all questions in cell biology, answers will come from a combination of experiments on cells and experiments aimed at reconstructing subcellular processes outside of the cell.

How intricate is the subcellular organization of prokaryotic cells?

Despite a historical focus on eukaryotic cells, cell biologists are becoming increasingly interested in prokaryotic cells (Bacteria and Archaea), which evolved more than a billion years before eukaryotes and hold clues to the evolutionary origins of basic cellular properties. Although these smaller cells were once thought to be simple in their organization and behaviors, recent advances from Lucy Shapiro's lab at Stanford University and others indicate that prokaryotic cells exhibit a high degree of spatial organization. Moreover, work from Harold Erickson's lab at Duke University and others has shown that prokaryotes have components that contribute to subcellular architecture, such as a cytoskeleton, that were initially thought to be unique to eukaryotic cells. To what extent are components of prokaryotic cells spatially organized, how is their organization achieved, and how does this organization contribute to cell growth and division? Because prokaryotes are small, answering these questions will rely in part on advances in microscopy technologies. The answers will shed light on the evolutionary origins of cellular architecture.

How do the complex properties of cells arise from the functions and interactions of their molecular components?

Over the past few decades, inspired by advances in molecular biology and genetics, most cell biologists have pursued a reductionist approach by determining the detailed role of individual genes and proteins in cell structure and behavior. This approach has been very successful, and together with recent advances in genome sequencing technology, it has enabled cell biologists to catalog many molecules that perform key roles in the cell. In the past few years, cell biologists, including Marc Kirschner at Harvard University and others, have embraced a more holistic approach called systems biology, which is focused on understanding how the complex properties of cells and subcellular systems arise from the properties and interactions of their individual parts. Systems cell biologists work with mathematicians and computer scientists to develop models of cellular processes that can be refined by experimental validation. Which approach, reductionist or holistic, represents the future of cell biology? Both do, as each relies on the other to generate insights and hypotheses that move the field forward.

Think Critically

Even the simplest cells are structurally and functionally complex, so how could such complexity arise through the process of evolution? After thinking about this, read Chapter 24 for some potential answers.

Matthew Welch

Matthew Welch is a professor of Molecular and Cell Biology at the University of California, Berkeley. His research interests include cytoskeleton dynamics and microbial pathogenesis. Learn more about his work at http://mcb.berkeley.edu/labs/welch.

REVIEW KEY CONCEPTS

Go to **CENGAGENOW** at www.cengage.com/login to access quizzing, animations, exercises, articles, and personalized homework help.

5.1 Basic Features of Cell Structure and Function

- According to the cell theory: (1) all living organisms are composed of cells; (2) cells are the structural and functional units of life; and (3) cells arise only from the division of preexisting cells.
- Cells of all kinds are divided internally into a central region containing the genetic material and the cytoplasm, which consists of the cytosol, the cytoskeleton, and organelles and is bounded by the plasma membrane.
- The plasma membrane is a lipid bilayer in which transport proteins are embedded (Figure 5.6).
- In the cytoplasm, proteins are made, most of the other molecules required for growth and reproduction are assembled, and energy absorbed from the surroundings is converted into energy usable by the cell.

Animation: Overview of cells

Animation: Surface-to-volume ratio

Animation: Cell membranes

5.2 Prokaryotic Cells

- Prokaryotic cells are surrounded by a plasma membrane and, in most groups, are enclosed by a cell wall. The genetic material, typically a single, circular DNA molecule, is located in the nucleoid. The cytoplasm contains masses of ribosomes (Figure 5.7).

Animation: Typical prokaryotic cell

5.3 Eukaryotic Cells

- Eukaryotic cells have a true nucleus, which is separated from the cytoplasm by the nuclear envelope perforated by nuclear pores. A plasma membrane forms the outer boundary of the cell. Other membrane systems enclose specialized compartments as organelles in the cytoplasm (Figures 5.9 and 5.10).
- The eukaryotic nucleus contains chromatin, a combination of DNA and proteins. A specialized segment of the chromatin forms the nucleolus, where ribosomal RNA molecules are made and combined with ribosomal proteins to make ribosomes. The nuclear envelope contains nuclear pore complexes with pores that allow passive or assisted transport of molecules between the nucleus and the cytoplasm. Proteins destined for the nucleus contain a short amino acid sequence called a nuclear localization signal (Figures 5.11 and 5.12).
- Eukaryotic cytoplasm contains ribosomes (Figure 5.13), an endomembrane system, mitochondria, microbodies, the cytoskeleton, and some organelles specific to certain organisms. The endomembrane system includes the nuclear envelope, ER, Golgi complex, lysosomes, vesicles, and plasma membrane.
- The endoplasmic reticulum (ER) occurs in two forms, as rough and smooth ER. The ribosome-studded rough ER makes proteins that become part of cell membranes or are released from the cell. Smooth ER synthesizes lipids and breaks down toxic substances (Figure 5.14).
- The Golgi complex chemically modifies proteins made in the rough ER and sorts finished proteins to be secreted from the cell, embedded in the plasma membrane, or included in lysosomes (Figures 5.15, 5.16, and 5.18).
- Lysosomes, specialized vesicles that contain hydrolytic enzymes, digest complex molecules such as food molecules that enter the cell by endocytosis, cellular organelles that are no longer functioning correctly, and engulfed bacteria and cell debris (Figure 5.17).
- Mitochondria carry out cellular respiration, the conversion of fuel molecules into the energy of ATP (Figure 5.19).
- Microbodies conduct the initial steps in fat breakdown and other reactions that link major biochemical pathways in the cytoplasm (Figure 5.20).
- The cytoskeleton is a supportive structure built from microtubules, intermediate filaments, and microfilaments. Motor proteins walking along microtubules and microfilaments produce most movements of animal cells (Figures 5.21–5.23).
- Motor protein-controlled sliding of microtubules generates the movements of flagella and cilia. Flagella and cilia arise from centrioles (Figures 5.24–5.26).

Animation: Common eukaryotic organelles

Animation: Nuclear envelope

Animation: The endomembrane system

Practice: Structure of a mitochondrion

Animation: Cytoskeletal components

Animation: Motor proteins

Animation: Flagellar structure

5.4 Specialized Structures of Plant Cells

- Plant cells contain all the eukaryotic structures found in animal cells except for lysosomes. They also contain three structures not found in animal cells: chloroplasts, a central vacuole, and a cell wall (Figure 5.10).
- Chloroplasts contain pigments and molecular systems that absorb light energy and convert it to chemical energy. The chemical energy is used inside the chloroplasts to assemble carbohydrates and other organic molecules from simple inorganic raw materials (Figure 5.27).
- The large central vacuole, which consists of a tonoplast enclosing an inner space, develops pressure that supports plant cells, accounts for much of cellular growth by enlarging as cells mature, and serves as a storage site for substances including waste materials (Figure 5.10).
- A cellulose cell wall surrounds plant cells, providing support and protection. Plant cell walls are perforated by plasmodesmata, channels that provide direct pathways of communication between the cytoplasm of adjacent cells (Figure 5.28).

Practice: Structure of a chloroplast

Animation: Plant cell walls

5.5 The Animal Cell Surface

- Animal cells have specialized surface molecules and structures that function in cell adhesion, communication, and support.
- Cell adhesion molecules bind to specific molecules on other cells. The adhesions organize and hold together cells of the same type in body tissues.
- Cell adhesions are reinforced by various junctions. Anchoring junctions hold cells together. Tight junctions seal together the plasma membranes of adjacent cells, preventing ions and molecules from moving between the cells. Gap junctions open direct channels between the cytoplasm of adjacent cells (Figure 5.29).
- The extracellular matrix, formed from collagen proteins embedded in a matrix of branched glycoproteins, functions primarily in cell and body protection and support but also affects cell division, motility, embryonic development, and wound healing (Figure 5.30).

Animation: Animal cell junctions

UNDERSTAND AND APPLY

Test Your Knowledge

1. You are examining a cell from a crime scene using an electron microscope. It contains ribosomes, DNA, a plasma membrane, a cell wall, and mitochondria. What type of cell is it?
   a. lung cell
   b. plant cell
   c. prokaryotic cell
   d. cell from the surface of a human fingernail
   e. sperm cell
2. A prokaryote converts food energy into the chemical energy of ATP on/in its:
   a. chromosome.
   b. flagella.
   c. ribosomes.
   d. cell wall.
   e. plasma membrane.
3. Eukaryotic and prokaryotic ribosomes are similar in that:
   a. both contain a small subunit, but only eukaryotes contain a large subunit.
   b. both contain the same number of proteins.
   c. both use mRNA to assemble amino acids into proteins.
   d. both contain the same number of types of rRNA.
   e. both produce proteins that can pass through pores into the nucleus.
4. Which of the following structures does *not* require an immediate source of energy to function?
   a. central vacuoles
   b. cilia
   c. microtubules
   d. microfilaments
   e. microbodies
5. Which of the following structures is *not* used in eukaryotic protein manufacture and secretion?
   a. ribosome
   b. lysosome
   c. rough ER
   d. secretory vesicle
   e. Golgi complex
6. Which of the following are glycoproteins whose function is affected by the common cold virus?
   a. plasmodesmata
   b. desmosomes
   c. cell adhesion molecules
   d. flagella
   e. cilia
7. An electron micrograph shows that a cell has extensive amounts of rough ER throughout. One can deduce from this that the cell is:
   a. synthesizing and metabolizing carbohydrates.
   b. synthesizing and secreting proteins.
   c. synthesizing ATP.
   d. contracting.
   e. resting metabolically.
8. Which of the following contributes to the sealed lining of the digestive tract to keep food inside it?
   a. a central vacuole that stores proteins
   b. tight junctions formed by direct fusion of proteins
   c. gap junctions that communicate between cells of the stomach lining and its muscular wall
   d. desmosomes forming buttonlike spots or a belt to keep cells joined together
   e. plasmodesmata that help cells communicate their activities
9. Which of the following statements about proteins is correct?
   a. Proteins are transported to the rough ER for use within the cell.
   b. Lipids and carbohydrates are added to proteins by the Golgi complex.
   c. Proteins are transported directly into the cytosol for secretion from the cell.
   d. Proteins that are to be stored by the cell are moved to the rough ER.
   e. Proteins are synthesized in vesicles.
10. Which of the following is *not* a component of the cytoskeleton?
    a. microtubules
    b. actins
    c. microfilaments
    d. cilia
    e. cytokeratins

Discuss the Concepts

1. Many compound microscopes have a filter that eliminates all wavelengths except that of blue light, thereby allowing only blue light to pass through the microscope. Use the spectrum of visible light (see Figure 9.4) to explain why the filter improves the resolution of light microscopes.
2. Explain why aliens invading Earth are not likely to be giant cells the size of humans.
3. An electron micrograph of a cell shows the cytoplasm packed with rough ER membranes, a Golgi complex, and mitochondria. What activities might this cell concentrate on? Why would large numbers of mitochondria be required for these activities?
4. Assuming that mitochondria evolved from bacteria that entered cells by endocytosis, what are the likely origins of the outer and inner mitochondrial membranes?
5. Researchers have noticed that some men who were sterile because their sperm cells were unable to move also had chronic infections of the respiratory tract. What might be the connection between these two symptoms?

Design an Experiment

The unicellular alga *Chlamydomonas reinhardtii* has two flagella assembled from tubulin proteins. If a researcher changes the pH from approximately neutral (their normal growing condition) to pH 4.5, *Chlamydomonas* cells spontaneously lose their flagella. After the cells are returned to neutral pH, they regrow the flagella—a process called reflagellation. Assuming that you have deflagellated *Chlamydomonas* cells, devise experiments to answer the following questions:

1. Do new tubulin proteins need to be made for reflagellation to occur, or is there a reservoir of proteins in the cell?
2. Is the production of new mRNA for the tubulin proteins necessary for reflagellation?
3. What is the optimal pH for reflagellation?

Interpret the Data

Investigators studying protein changes during aging examined enzyme activity in cells extracted from the nematode worm *Caenorhabditis elegans.* The cell extracts were treated to conserve enzyme activity, although the investigators noted that some proteins were broken down by the extraction procedure. The extracts were centrifuged, and seven fractions were collected in sequence to isolate the location of activity by protease enzymes called cathepsins. Examine the activity profiles in the figure below. In which fraction and, hence, in which eukaryotic cellular structure are these enzymes most active?

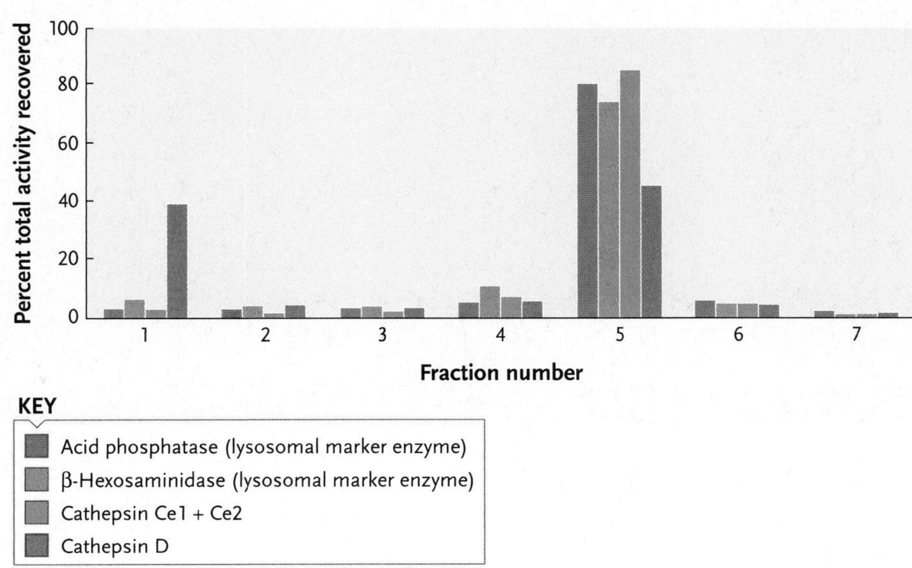
Percent total activity recovered
100
80
60
40
20
0
1
2
3
4
5
6
7
Fraction number
KEY
Acid phosphatase (lysosomal marker enzyme)
β-Hexosaminidase (lysosomal marker enzyme)
Cathepsin Ce1 + Ce2
Cathepsin D

Distribution of enzyme activity in fractions from centrifugation of an organelle pellet. The fractions are numbered 1–7 from the top to the bottom of the centrifuge tube. Fraction 1 contains cytosolic contents and is the supernatant, and fraction 7 contains cellular debris and membrane fragments.

Source: G. J. Sarkis et al. 1988. Decline in protease activities with age in the nematode *Caenorhabiditis elegans. Mechanisms of Ageing and Development* 45:191–201.

Apply Evolutionary Thinking

What aspects of cell structure suggest that prokaryotes and eukaryotes share a common ancestor in their evolutionary history?

Express Your Opinion

Researchers are modifying prokaryotes to identify what it takes to "be alive." They are creating "new" organisms by removing genes from living cells, one at a time. What are the potential advantages or bioethical pitfalls of this kind of research? Go to academic.cengage.com/login to investigate both sides of the issue and then vote.

6

STUDY OUTLINE

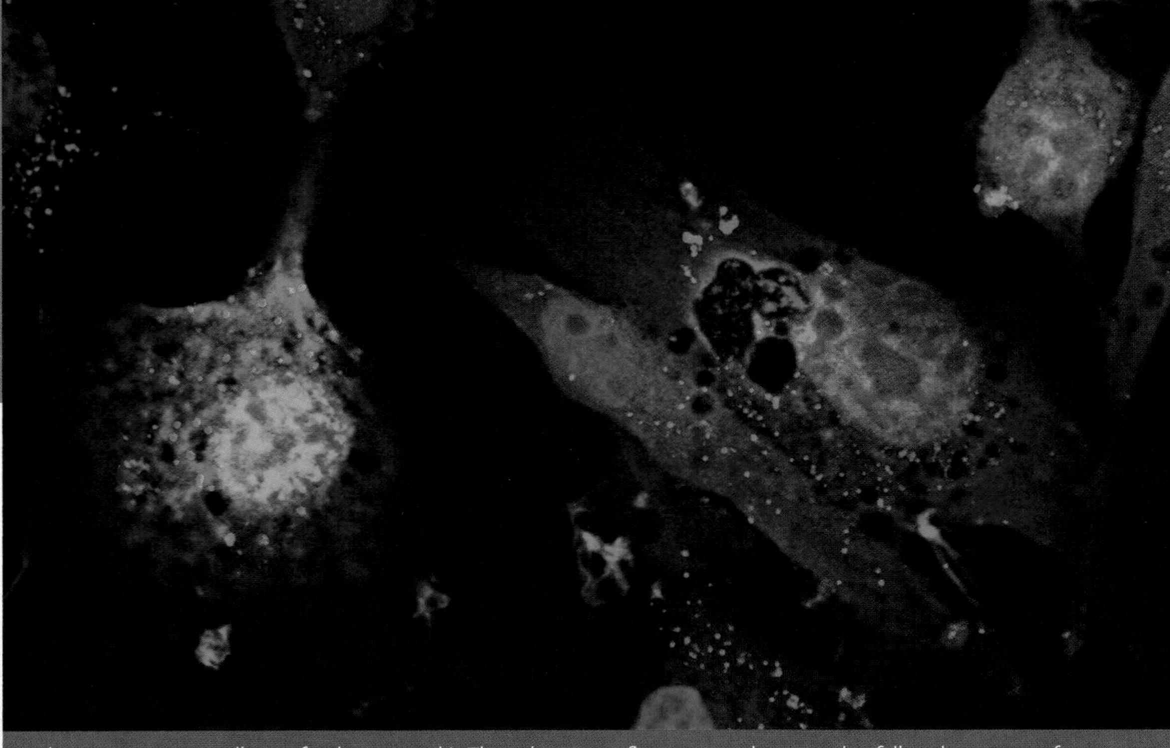

Alex Gray, Wellcome Images

Endocytosis in cancer cells (confocal micrograph). The red spots are fluorescent spheres used to follow the process of endocytosis; some of the spheres have been taken up by cells.

Membranes and Transport

Why It Matters. . . Cystic fibrosis (CF) is one of the most common genetic diseases. It affects approximately 1 in 4,000 children born in the United States. People with CF experience a progressive impairment of lung and gastrointestinal function. Although the treatment of CF patients is slowly improving, their average lifespan remains under 40 years. CF is caused by mutations in a single gene that codes for a protein so vital that a change in a single amino acid has life-altering consequences. What is this protein, and what role does it play in the body?

To maintain their internal environment, the cells of all organisms must constantly exchange molecules and ions with the fluid environment that surrounds them. What makes this possible is the **plasma membrane,** the thin layer of lipids and proteins that separates a cell from its surroundings. Some of the proteins—called transport proteins—move particular ions and molecules, including water, in a directed way across the membrane. Different transport proteins move different molecules or ions. One such transport protein is cystic fibrosis transmembrane conductance regulator, or CFTR, which is found in the plasma membrane of epithelial cells **(Figure 6.1).** Epithelial cells are organized into sheetlike layers that form coverings and linings which are typically exposed to water, air, or fluids within the body. Epithelial cells line the passageways and ducts of the lungs, liver, pancreas, intestines, reproductive system, and skin. CFTR pumps chloride ions out of those cells, and water follows the ions producing a thin, watery film over the epithelial tissue surface. Mucus can slide easily over the tissue because of the moist surface.

Particular mutations in the CFTR gene result in a CFTR molecule that transports chloride ions poorly or not at all. The most common mutation is a small deletion in the gene that removes one amino acid from the CFTR protein. If an individual has one normal and one mutant copy of the gene, there are enough working CFTR molecules for normal chloride transport. However, if

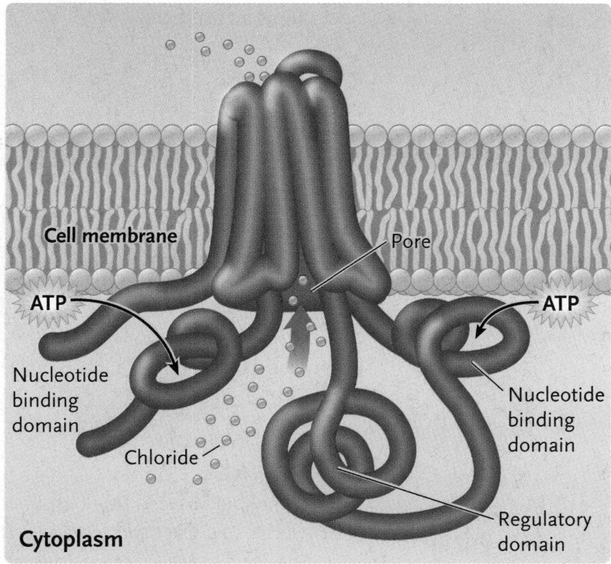
Cell membrane
Pore
ATP
ATP
Nucleotide
binding
domain
Nucleotide
binding
domain
Chloride
Regulatory
domain
Cytoplasm

FIGURE 6.1
Molecular model of cystic fibrosis transmembrane conductance regulator (CFTR), a chloride ion transport protein embedded in the plasma membrane. The nucleotide binding domains bind ATP which provides energy for ion transport by the protein.

an individual has two mutant copies of the gene, all CFTR molecules are mutant, and chloride transport is defective—the individual has CF. Not enough chloride ions are transported out of the cell, so not enough water leaves either. As a result, mucus sticks to the drier epithelial tissue, building up into a thick mass. In the respiratory tract, the abnormally thick mucus means much diminished protection from bacteria, which leads to infections. Chronic lung infections produce progressive damage to the respiratory tract; at some point, lung function becomes insufficient to support life and the CF patient dies.

For a person to have CF, one mutant copy of the gene must be inherited from each parent. The frequency of CF individuals in the United States varies among groups—1/2,500–1/3,500 for Caucasian Americans, 1/17,000 for African Americans, and 1/31,000 for Asian Americans. Most treatments center on managing the symptoms, and include pounding the back or chest to dislodge mucus, and antibiotic or other pharmaceutical treatments to control infections. Research is ongoing to develop more effective treatments. One interesting research approach is to develop pharmaceuticals that target the defective CFTR itself. Basic research has located the disease-causing mutations in the CFTR molecule, so researchers may be able to design treatments that change the transport protein to a more functional form.

The plasma membrane, in which the CFTR molecule is located, forms the outer boundary of every living cell, and encloses the intracellular contents. The plasma membrane also regulates the passage of molecules and ions between the inside and outside of the cell, helping to determine the cell's composition. Within eukaryotic cells, membranes surrounding internal organelles play similar roles, creating environments that differ from the surrounding cytosol.

The structure and function of biological membranes are the focus of this chapter. We first consider the structure of membranes and then examine how membranes selectively transport substances in and out of cells and organelles. Other roles of membranes, including recognition of molecules on other cells, adherence to other cells or extracellular materials, and reception of molecular signals such as hormones, are the subjects of Chapters 7, 40, and 43 in this book. <

6.1 Membrane Structure and Function

A watery fluid medium—or aqueous solution—bathes both surfaces of all biological membranes. The membranes are also fluid, but they are kept separate from their surroundings by the properties of the lipid and protein molecules from which they are formed.

Biological Membranes Contain Both Lipid and Protein Molecules

Biological membranes consist of lipids and proteins assembled into a thin film. The proportions of lipid and protein molecules in membranes vary, depending on the functions of the membranes in the cells.

MEMBRANE LIPIDS *Phospholipids* and *sterols* are the two major types of lipids in membranes (see Section 3.3). Phospholipids have a polar (electrically charged) end containing a phosphate group linked to one of several alcohols or amino acids, and a nonpolar (uncharged) end containing two nonpolar fatty-acid tails **(Figure 6.2A)**. The polar end is hydrophilic—it "prefers" being in an aqueous environment—and the nonpolar end is hydrophobic—it "prefers" being in an environment from which water is excluded. In other words, phospholipids have dual solubility properties.

In an aqueous medium, phospholipid molecules satisfy their dual solubility characteristics by assembling into a **bilayer**—a layer two molecules thick **(Figure 6.2B)**. In a bilayer, the polar ends of the phospholipid molecules are located at the surfaces, where they face the aqueous media. The nonpolar fatty-acid chains arrange themselves end to end in the membrane interior, in a nonpolar region that excludes water.

At low temperatures, the phospholipid bilayer freezes into a semisolid, gel-like state **(Figure 6.2C)**. When a phospholipid bilayer sheet is shaken in water, it breaks and spontaneously forms small *vesicles* **(Figure 6.2D)**. Vesicles consist of a spherical shell of phospholipid bilayer enclosing a small droplet of water.

Membrane sterols also have dual solubility characteristics. As explained in Section 3.4, these molecules have nonpolar carbon rings with a nonpolar side chain at one end and a single polar

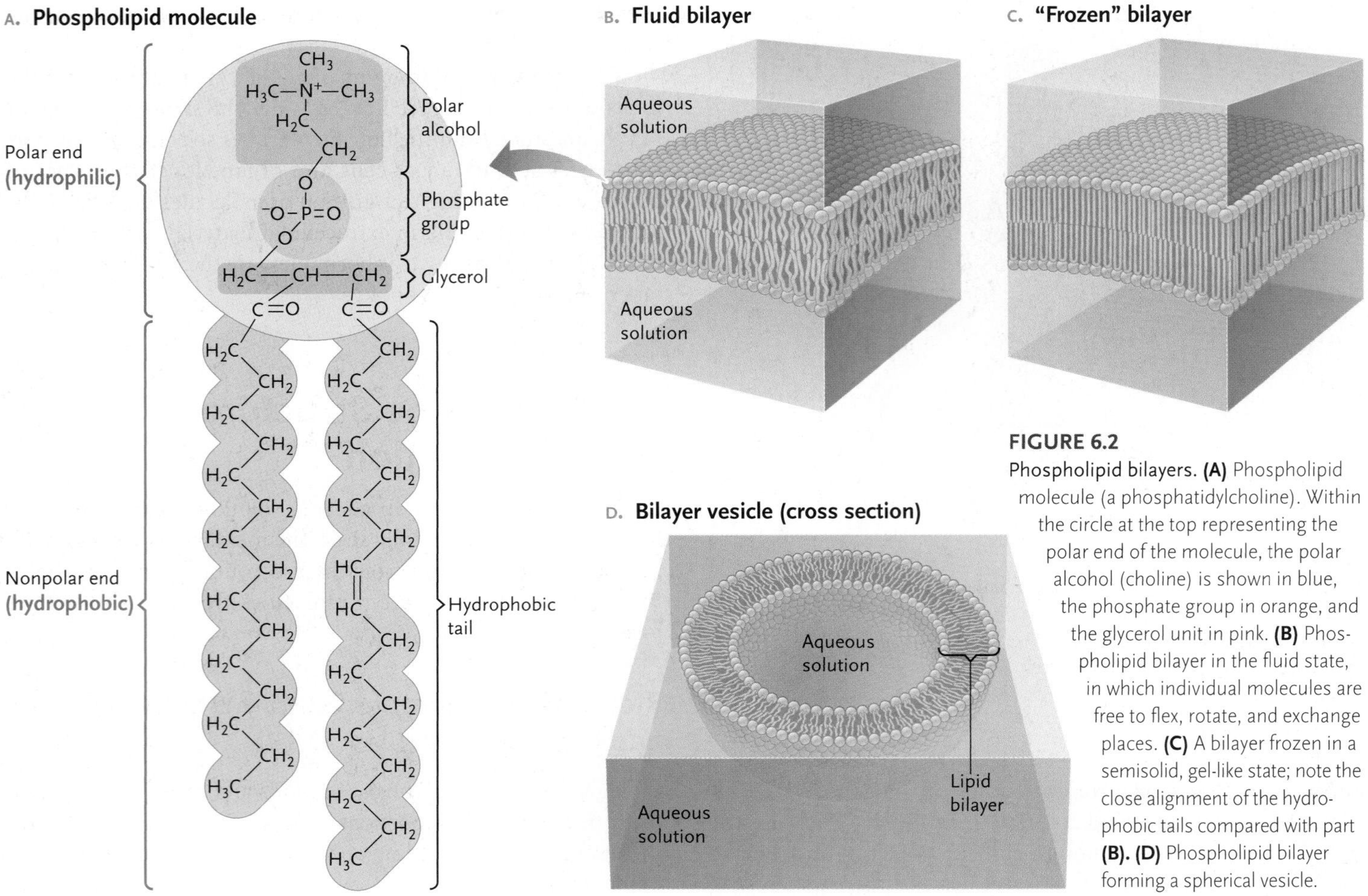

FIGURE 6.2
Phospholipid bilayers. (A) Phospholipid molecule (a phosphatidylcholine). Within the circle at the top representing the polar end of the molecule, the polar alcohol (choline) is shown in blue, the phosphate group in orange, and the glycerol unit in pink. **(B)** Phospholipid bilayer in the fluid state, in which individual molecules are free to flex, rotate, and exchange places. **(C)** A bilayer frozen in a semisolid, gel-like state; note the close alignment of the hydrophobic tails compared with part **(B). (D)** Phospholipid bilayer forming a spherical vesicle.

group (an —OH group) at the other end. In biological membranes, sterols pack into membranes alongside the phospholipid hydrocarbon chains, with only the polar end extending into the polar membrane surface **(Figure 6.3).** The predominant sterol of animal cell membranes is **cholesterol,** which is important for keeping the membranes fluid. A variety of sterols, called *phytosterols,* is found in plants.

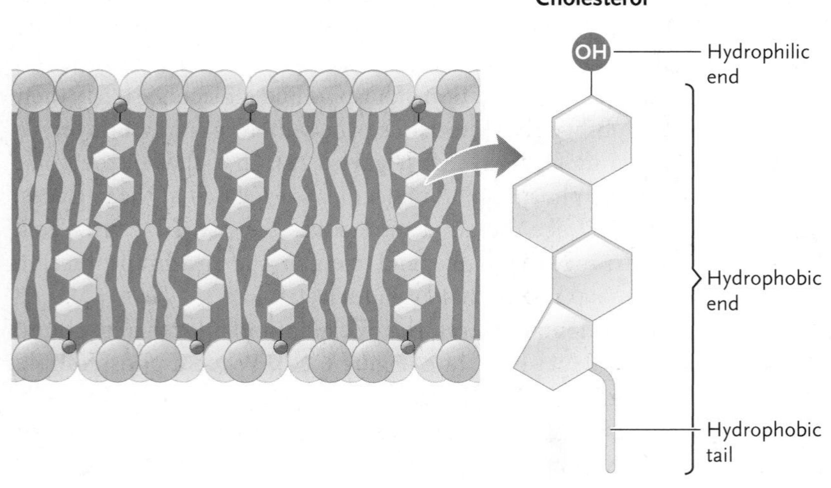

Cholesterol
OH
Hydrophilic end
Hydrophobic end
Hydrophobic tail

FIGURE 6.3
The position taken by cholesterol in bilayers. The hydrophilic —OH group at one end of the molecule extends into the polar regions of the bilayer; the ring structure extends into the nonpolar membrane interior.

MEMBRANE PROTEINS Membrane proteins also have hydrophilic and hydrophobic regions that give them dual solubility properties. The hydrophobic regions of membrane proteins are formed by segments of the amino acid chain with hydrophobic side groups. These hydrophobic segments are often wound into α helices, which span the membrane bilayer **(Figure 6.4).** (Protein structural elements are described in Section 3.4.) The hydrophobic segments are connected by loops of hydrophilic amino acids that extend into the polar regions at the membrane surfaces (for example, see Figure 6.4).

Each type of membrane has a characteristic group of proteins that is responsible for its specialized functions. **Transport proteins** form channels that allow selected polar molecules and ions to pass across a membrane. **Recognition proteins** in the plasma membrane identify a cell as part of the same individual or as foreign. **Receptor proteins** recognize and bind molecules from other cells that act as chemical signals, such as the peptide hormone insulin in animals. **Cell adhesion proteins** bind cells together by recognizing and binding receptors or chemical groups on other cells or on the extracellular matrix. Still other proteins are enzymes that speed chemical reactions carried out by membranes.

MEMBRANE GLYCOLIPIDS AND GLYCOPROTEINS **Glycolipids** are another type of lipid found in membranes **(Figure 6.5)**. As their name suggests, glycolipids are lipid molecules with carbohydrate groups attached. These molecules are found in the part of the membrane that faces the outside of the cell. The carbohydrate groups may form a linear or a branched chain. Carbohydrates are also attached to some of the proteins in the exterior-facing lipid layer; producing **glycoproteins** (see Figure 6.5). In many animal cells, such as intestinal epithelial cells, the carbohydrate groups of the cell surface glycolipids and glycoproteins form a surface coat called the **glycocalyx** ("sugar coat": *glykys* = sweet; *calyx* = cup or vessel).

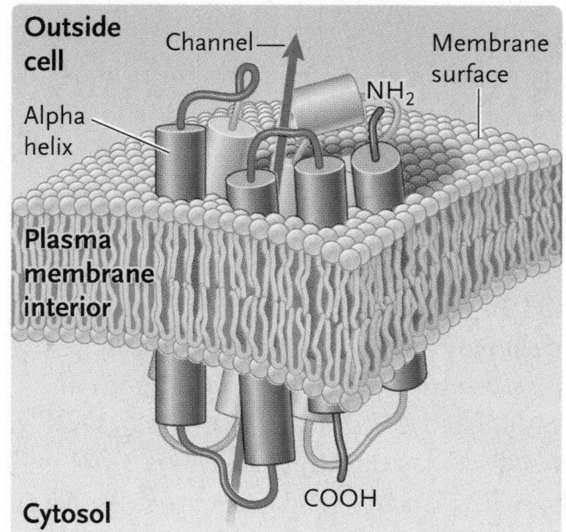

A. Typical membrane protein
Outside cell
Channel
Membrane surface
NH_2
Alpha helix
Plasma membrane interior
COOH
Cytosol

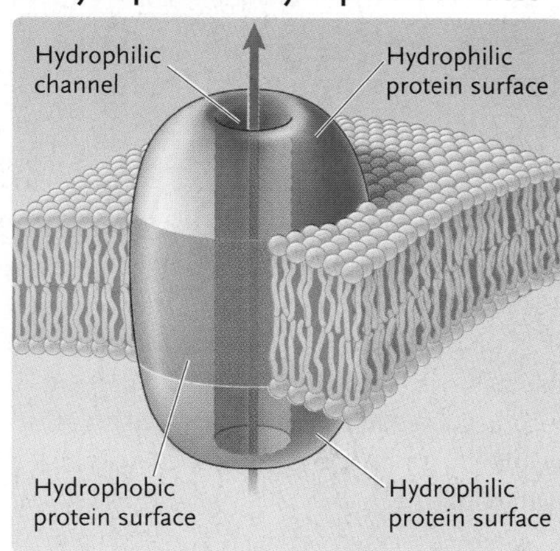

B. Hydrophilic and hydrophobic surfaces
Hydrophilic channel
Hydrophilic protein surface
Hydrophobic protein surface
Hydrophilic protein surface

FIGURE 6.4

Structure of membrane proteins. **(A)** Typical membrane protein, bacteriorhodopsin, showing the membrane-spanning alpha-helical segments (blue cylinders), connected by flexible loops of the amino acid chain at the membrane surfaces. **(B)** The same protein as in **(A)** in a diagram that shows hydrophilic (blue) and hydrophobic (orange) surfaces and the membrane-spanning channel created by this protein. Bacteriorhodopsin absorbs light energy in plasma membranes of photosynthetic archaeans.

The Fluid Mosaic Model Explains Membrane Structure

Membrane structure has been and continues to be a subject for research. The current view of membrane structure is based on the fluid mosaic model advanced by S. Jonathan Singer and Garth L. Nicolson at the University of California, San Diego, in 1972 (see Figure 6.5). The **fluid mosaic model** proposes that the membrane consists of a fluid phospholipid bilayer in which proteins are embedded and float freely. This model revolutionized how scientists think about membrane structure and function.

The "fluid" part of the fluid mosaic model refers to the phospholipid molecules, which vibrate, flex back and forth, spin around their long axis, move sideways, and exchange places

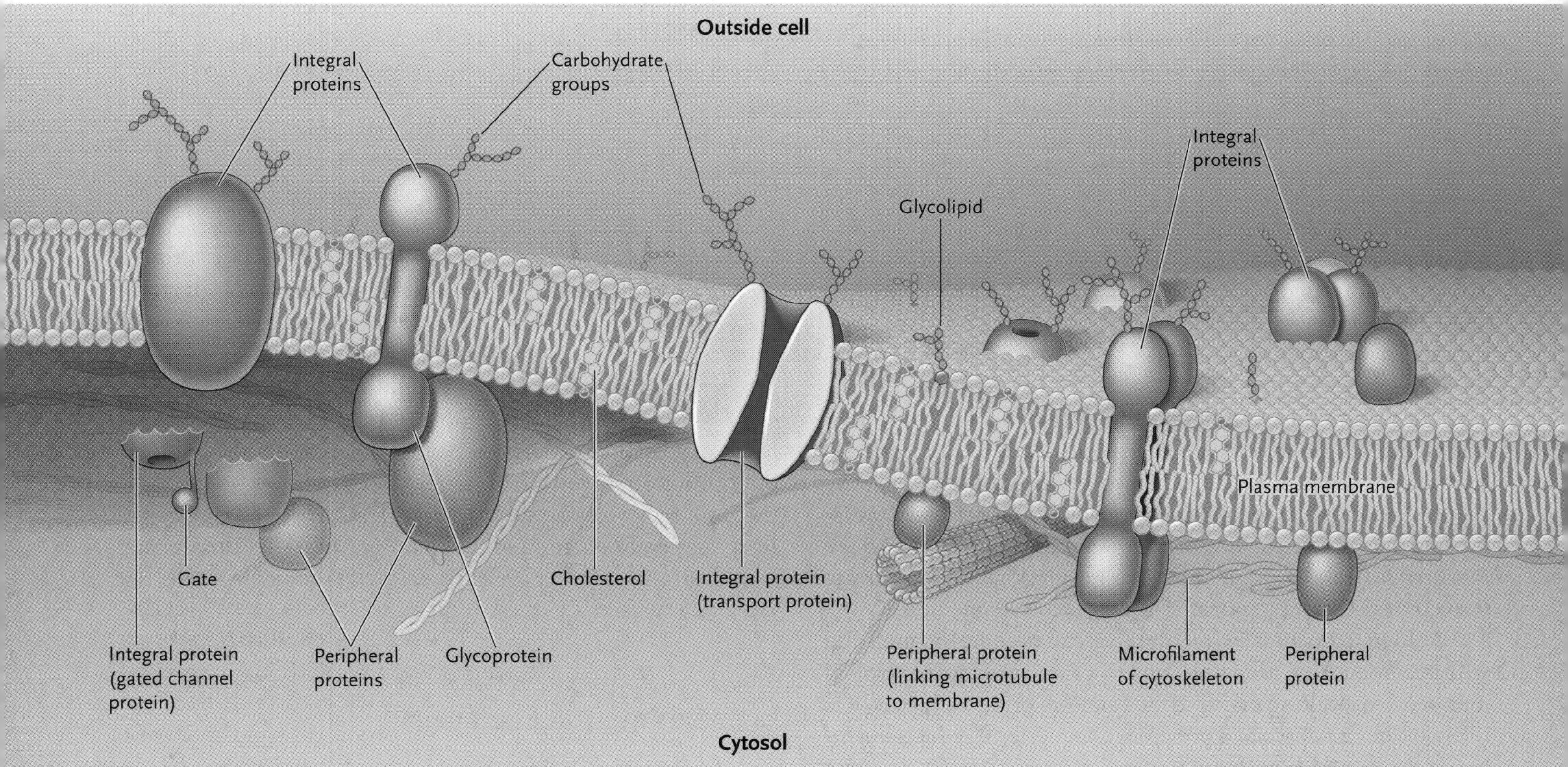

Outside cell
Integral proteins
Carbohydrate groups
Integral proteins
Glycolipid
Plasma membrane
Gate
Cholesterol
Integral protein (transport protein)
Integral protein (gated channel protein)
Peripheral proteins
Glycoprotein
Peripheral protein (linking microtubule to membrane)
Microfilament of cytoskeleton
Peripheral protein
Cytosol

FIGURE 6.5

Membrane structure according to the fluid mosaic model, in which integral membrane proteins are suspended individually in a fluid bilayer. Peripheral proteins are attached to integral proteins or membrane lipids mostly on the cytoplasmic side of the membrane (shown only on the inner surface in the figure). In the plasma membrane, carbohydrate groups of membrane glycoproteins and glycolipids face the cell exterior.

FOCUS ON **Basic Research**

Keeping Membranes Fluid at Cold Temperatures

The fluid state of biological membranes is critical to membrane function and, therefore, vital to cellular life. When membranes freeze, researchers have shown that the phospholipids form a semisolid gel in which they are unable to move (see Figure 6.2C), and proteins become locked in place. Freezing can kill cells by impeding vital membrane functions such as transport.

Many eukaryotic organisms, including algae, higher plants, protozoa, and animals, adapt to colder temperatures by changing membrane lipids. Experiments comparing membrane composition have shown that, in animals with body temperatures that fluctuate with environmental temperature, such as fish, amphibians, and reptiles, both the proportion of double bonds in membrane phospholipids and the cholesterol content are increased at lower temperatures. How do these changes affect membrane fluidity? Double bonds in unsaturated fatty acids introduce "kinks" in their hydrocarbon chain (see Figure 3.14); the kinks help bilayers stay fluid at lower temperatures by interfering with packing of the hydrocarbons. Cholesterol depresses the freezing point by interfering with close packing of membrane phospholipids.

All of these membrane changes also occur in mammals that enter hibernation in cold climates. When mammals enter hibernation, their body temperature may fall to as low as 5°C without freezing their membranes. The resistance to freezing allows the nerve cells of a hibernating mammal to remain active so that the animal can maintain basic body functions and respond, although sluggishly, to external stimuli. In active, nonhibernating mammals, membranes freeze into the gel state at about 15°C.

---

within the same bilayer half. Only rarely does a phospholipid flip-flop between the two layers. Phospholipids exchange places within a layer millions of times a second, making the phospholipid molecules in the membrane highly dynamic. Membrane fluidity is critical to the functions of membrane proteins and allows membranes to accommodate, for example, cell growth, motility, and surface stresses.

Membranes remain fluid at a relatively wide range of temperatures. Low temperatures can be detrimental to membrane structure, and therefore membrane function, because at a sufficiently low temperature the phospholipid molecules become closely packed and the membrane becomes a nonfluid gel. A common modification that helps keep membranes fluid at low temperatures is increasing the proportion of unsaturated fatty acid chains in membrane phospholipids. The double bonds in the unsaturated fatty acid chains produce physical kinks in the chain that interfere with the packing of the phospholipids at low temperatures, thereby reducing the temperature at which the bilayer becomes a nonfluid gel. In a paradoxical way, cholesterol also helps protect against the adverse effects of low temperatures. Cholesterol in the membrane decreases membrane fluidity at moderate to high temperatures because the rigid cholesterol rings keep the membrane phospholipids from packing tightly together (see Figure 6.3). However, when temperatures drop, the high concentrations found in eukaryotic membranes actually help slow the transition of the membrane to the nonfluid gel state by disrupting the ordered packing of phospholipids. (See *Focus on Basic Research* for a description of other strategies that organisms use to keep their membranes from freezing at low temperatures.)

At high temperatures, membranes can become too fluid and will become leaky, allowing ions to cross in an uncontrolled manner. This leaking disrupts the function of the cell; thus, it is likely to die. As described previously, cholesterol reduces membrane fluidity at high temperatures, thereby providing some protection.

The "mosaic" part of the fluid mosaic model refers to the membrane proteins, most of which float individually in the fluid lipid bilayer, like icebergs in the sea. Membrane proteins are larger than membrane lipids, and those that move do so much more slowly than do lipids. A number of membrane proteins are attached to the cytoskeleton. These proteins either are immobile or move in a directed fashion, such as along cytoskeletal filaments.

Membrane proteins are oriented across the membrane so that particular functional groups and active sites face either the inside or the outside membrane surface. The inside and outside halves of the bilayer also contain different mixtures of phospholipids. These differences make biological membranes *asymmetric* and give their inside and outside surfaces different functions.

Proteins that are embedded in the phospholipid bilayer are termed **integral proteins** (see Figure 6.5). Essentially all transport, receptor, recognition, and cell adhesion proteins that give membranes their specific functions are integral membrane proteins.

Other proteins, called **peripheral proteins** (see Figure 6.5), are held to membrane surfaces by noncovalent bonds—hydrogen bonds and ionic bonds—formed with the polar parts of integral membrane proteins or membrane lipids. Most peripheral proteins are on the cytoplasmic side of the membrane. Some peripheral proteins are parts of the cytoskeleton, such as microtubules, microfilaments, or intermediate filaments, or proteins that link the cytoskeleton together. These structures hold some integral membrane proteins in place. For example, this anchoring constrains many types of receptors to the sides of cells that face body surfaces, cavities, or tubes.

The Fluid Mosaic Model Is Fully Supported by Experimental Evidence

The novel ideas of a fluid membrane and a flexible mosaic arrangement of proteins and lipids challenged an accepted model in which a relatively rigid, stable membrane was coated on both sides with proteins arranged like jam on bread. Researchers tested

FIGURE 6.6 **Experimental Research**

The Frye–Edidin Experiment Demonstrating That the Phospholipid Bilayer Is Fluid

Question: Do membrane proteins move in the phospholipid bilayer?

Experiment: Frye and Edidin labeled membrane proteins on cultured human and mouse cells with fluorescent dyes, red for human proteins and green for mouse proteins. Human and mouse cells were then fused and the pattern of fluorescence was followed under a microscope.

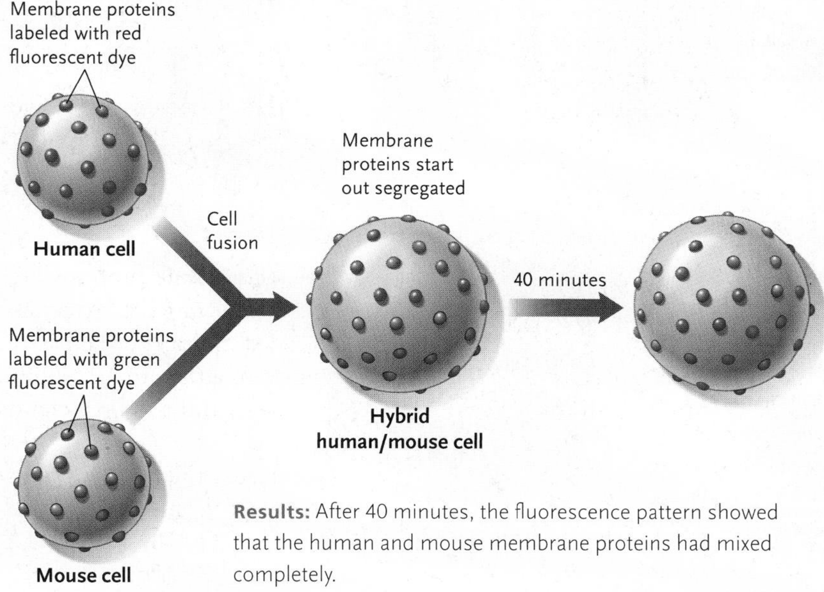

Membrane proteins labeled with red fluorescent dye
Human cell
Cell fusion
Membrane proteins start out segregated
40 minutes
Membrane proteins labeled with green fluorescent dye
Hybrid human/mouse cell
Mouse cell

Results: After 40 minutes, the fluorescence pattern showed that the human and mouse membrane proteins had mixed completely.

Conclusion: The rapid mixing of membrane proteins in the hybrid human/mouse cells showed that membrane proteins move in the phospholipid bilayer, indicating that the membrane is fluid.

Source: L. D. Frye and M. Edidin. 1970. The rapid intermixing of cell surface antigens after formation of mouse–human heterokaryons. *Journal of Cell Science* 7:319–335.

the new model with intensive research. The experimental evidence from that research completely supports every major hypothesis of the model: That membrane lipids are arranged in a bilayer; that the bilayer is fluid; that proteins are suspended individually in the bilayer; and that the arrangement of both membrane lipids and proteins is asymmetric.

EVIDENCE THAT MEMBRANES ARE FLUID In 1970, L. David Frye and Michael A. Edidin grew human cells and mouse cells separately in tissue culture. Then they added antibodies that bound to either human or mouse membrane proteins **(Figure 6.6).** The anti-human antibodies were attached to dye molecules that fluoresce red under ultraviolet light, and the anti-mouse antibodies to molecules that fluoresce green. The researchers then fused the two cells. The red and green tagged proteins started out in separate halves of the fused cell, but over a period of 40 minutes they gradually intermixed, demonstrating that the proteins were floating freely in a fluid membrane.

Based on the measured rates at which molecules mix in biological membranes, the membrane bilayer appears to be about as fluid as light machine oil, such as the lubricants you might use around the house to oil a door hinge, the wheels of a skateboard, or a bicycle.

EVIDENCE FOR MEMBRANE ASYMMETRY AND INDIVIDUAL SUSPENSION OF PROTEINS An experiment that used membranes prepared for electron microscopy by the **freeze-fracture technique** confirmed that the membrane is a bilayer with proteins suspended in it individually and that the arrangement of membrane lipids and proteins is asymmetric **(Figure 6.7).** In this technique, experimenters freeze a block of cells rapidly by dipping it in, for example, liquid nitrogen. Then they fracture the block by giving it a blow from the sharp edge of a microscopic knife. Often, the fracture splits bilayers into inner and outer halves, exposing the hydrophobic membrane interior. In the electron microscope, the split membranes appear as smooth layers in which individual particles the size of proteins are embedded (see Figure 6.7C).

Recent research has shown that the asymmetry of lipids and proteins in membranes is not the result of free mixing of the components. Rather, both lipids and proteins tend to form ordered arrangements in membranes, and their movements may be restricted. Indeed, many cellular processes require a specific structural organization of membrane lipids and proteins. Cell communication, the topic of the next chapter, is an example of such a process.

Membranes Have Several Functions

Membranes perform a diverse array of functions. As you will see, often the membrane protein defines the function of a membrane or membrane segment.

- Membranes define the boundaries of cells and, in eukaryotes, the boundaries of compartments, for example, the nucleus, mitochondria, and chloroplasts.
- Membranes are permeability barriers, permitting regulated control of the contents of cells compared with the extracellular environments.
- Some membranes have enzyme activities that are the properties of the membrane proteins. For instance, the particular enzymes found in or on their membranes define the specific properties of many eukaryotic organelles, such as the endoplasmic reticulum, Golgi complex, lysosome, mitochondrion, and chloroplast (see Sections 5.3 and 5.4). Enzymes on mitochondrial and chloroplast internal membranes, for example, play essential roles in converting the chemical energy in energy-rich nutrients to ATP in cellular respiration, and the conversion of light energy to the chemical energy of ATP in photosynthesis (see Sections 5.3 and 5.4, and Chapters 8 and 9).

Freeze Fracture

Purpose: Quick-frozen cells are fractured to split apart lipid bilayers for analysis of the membrane interior.

Protocol:

1. The specimen is frozen quickly in liquid nitrogen and then fractured by a sharp blow by a knife edge.

2. The fracture may travel over membrane surfaces as it passes through the specimen, or it may split membrane bilayers into inner and outer halves as shown here.

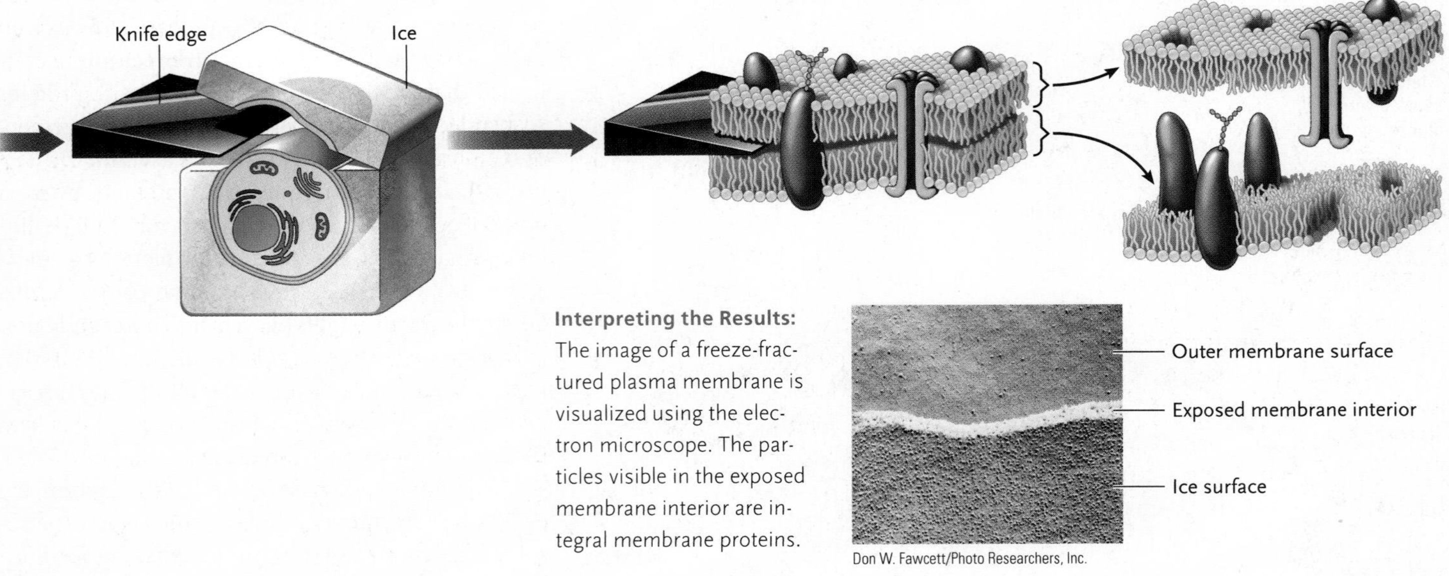

Knife edge
Ice
Outer membrane surface
Exposed membrane interior
Ice surface

Interpreting the Results: The image of a freeze-fractured plasma membrane is visualized using the electron microscope. The particles visible in the exposed membrane interior are integral membrane proteins.

Don W. Fawcett/Photo Researchers, Inc.

- Membrane-spanning channel proteins form channels that selectively transport specific ions or water through the membrane. Proteins that form channels for ions may either allow the ions to pass freely, or may regulate their passage. For example, specific channel proteins control the passage of individual ions such as Na^+, K^+, Ca^{2+}, and Cl^- (see next section).
- Membrane-spanning carrier proteins bind to specific substances and transport them across the membrane. An example is the carrier protein for transporting glucose into cells (see next section).
- Some membrane proteins serve as receptors that recognize and bind specific molecules in the extracellular environment. Depending on the system, binding serves as the first step in bringing the substance into the cell, or it activates the receptor and triggers a series of molecular events within the cell that leads to a cellular response. For instance, some hormones bind to receptors in the plasma membrane and cause the cell to change its gene activity (discussed in Chapter 7).
- Membranes have electrical properties as a result of an uneven distribution of particular ions inside and outside of the cell. These electrical properties can serve as a mechanism of signal conduction when a cell receives an electrical, chemical, or mechanical stimulus. For instance, neurons (nerve cells) and muscle cells conduct electrical signals by using the electrical properties of membranes (discussed in Chapter 37).
- Some membrane proteins facilitate cell adhesion and cell-to-cell communication. For example, in gap junctions particular plasma membrane proteins of adjacent cells line up and form pipelines between the two cells (see Section 5.5 and Figure 5.29).

In the remainder of the chapter we focus on the functions of membranes in transporting substances into and out of the cell.

STUDY BREAK 6.1

1. **Describe the fluid mosaic model for membrane structure.**
2. **Give two examples each of integral proteins and peripheral proteins.**

6.2 Functions of Membranes in Transport: Passive Transport

Transport is the controlled movement of ions and molecules from one side of a membrane to the other. Membrane proteins are the molecules responsible for transport. The movement is typically *directional;* that is, some ions and molecules consistently move into cells, whereas others move out of cells. Transport is also *specific;* that is, only certain ions and molecules move directionally across membranes. Transport is critical to the ionic and molecular organization of cells, and with it, the maintenance of cellular life.

Transport occurs by two mechanisms. The first mechanism, **passive transport,** depends on concentration differences on the two sides of a membrane (concentration = number of molecules or ions per unit volume). In this mechanism, ions and molecules move across the membrane from the side with the higher concentration to the side with the lower concentration (that is, *with* the gradient). The difference in concentration provides the energy for this form of transport.

The second mechanism, **active transport,** moves ions or molecules *against* the concentration gradient; that is, from the side with the lower concentration to the side with the higher concentration. Active transport uses energy directly or indirectly obtained by breaking down ATP. The properties of passive and active transport are compared in **Table 6.1.**

TABLE 6.1 Characteristics of Transport Mechanisms

Characteristic	Passive Transport: Simple Diffusion	Passive Transport: Facilitated Diffusion	Active Transport
Membrane component responsible for transport	Lipids, water	Proteins, water	Proteins
Binding of transported substance	No	Yes	Yes
Energy source	Concentration gradients	Concentration gradients	ATP hydrolysis or concentration gradients
Direction of transport	With gradient of transported substance	With gradient of transported substance	Against gradient of transported substance
Specificity for molecules or molecular classes	Nonspecific	Specific	Specific
Saturation at high concentrations of transported molecules	No	Yes	Yes

Passive Transport Is Based on Diffusion

Passive transport is a form of **diffusion,** the net movement of ions or molecules from a region of higher concentration to a region of lower concentration. If you add a drop of food dye to a container of clear water, the dye molecules, and therefore the color, will spread or *diffuse* from their initial center of high concentration until they are distributed evenly. At this point, the water has an even color. Diffusion depends on the constant motion of ions or molecules at temperatures above absolute zero (−273°C). The constant motion gradually mixes the dye molecules and water molecules until they are distributed uniformly.

The concentration difference that drives diffusion, a **concentration gradient,** is a form of potential energy. At the initial state, when molecules are more concentrated in one region of a solution, as when a dye is dropped into one side of a container of water, the molecules are highly organized and at a state of minimum entropy. In the final state, when they are distributed evenly throughout the solution, they are less organized and at a state of maximum entropy. As the distribution proceeds to the state of maximum disorder, the molecules release free energy that can accomplish work (see Sections 4.1 and 4.2 for a discussion of entropy and free energy).

Diffusion involves a *net* movement of molecules or ions. Molecules and ions actually move in all directions at all times in a solution. But when molecules or ions exist in a concentration gradient, more of them move from the area of higher concentration to areas of lower concentration than in the opposite direction. Even after their concentration is the same in all regions, there is still constant movement of molecules or ions from one space to another, but there is no net change in concentration on either side. This condition is an example of a *dynamic equilibrium.*

Substances Move Passively through Membranes by Simple or Facilitated Diffusion

Hydrophobic (nonpolar) molecules are able to dissolve in the lipid bilayer of a membrane and move through it freely. By contrast, hydrophilic molecules such as ions and polar molecules are impeded in their movement through the membrane by the hydrophobic core; thus, their passage is slow. Charged atoms and molecules are mostly blocked from moving through the membrane because of the hydrophobic core. Membranes that affect diffusion in this way are said to be **selectively permeable.**

TRANSPORT BY SIMPLE DIFFUSION A few small substances diffuse through the lipid part of a biological membrane. With one major exception—water—these substances are nonpolar inorganic gases such as O_2, N_2, and CO_2 and nonpolar organic molecules such as steroid hormones. This type of transport, which depends solely on molecular size and lipid solubility, is **simple diffusion** (see Table 6.1).

Water is a strongly polar molecule. Nevertheless, water molecules are small enough to slip through momentary spaces created between the hydrocarbon tails of phospholipid molecules as they flex and move in a fluid bilayer. This type of water movement across the membrane is relatively slow.

TRANSPORT BY FACILITATED DIFFUSION Many polar and charged molecules such as water, amino acids, sugars, and ions diffuse across membranes with the help of transport proteins, a mechanism termed **facilitated diffusion.** The transport proteins enable polar and charged molecules to avoid interaction with the hydrophobic lipid bilayer (see Table 6.1).

Facilitated diffusion is specific: The membrane proteins involved transport certain polar and charged molecules, but not others. Facilitated diffusion is also dependent on concentration gradients: Proteins aid the transport of polar and charged molecules through membranes, but a favorable concentration gradient provides the energy for transport. Transport stops if the gradient falls to zero.

Two Groups of Transport Proteins Carry Out Facilitated Diffusion

The proteins that carry out facilitated diffusion are integral membrane proteins that extend entirely through the membrane. There are two types of transport proteins involved in facilitated diffusion. One type, **channel proteins,** forms hydrophilic channels in the membrane through which water and ions can pass **(Figure 6.8A).** The channel "facilitates" the diffusion of molecules through the membrane by providing an avenue. For example, facilitated diffusion of water through membranes occurs through specialized water channels called **aquaporins** (see Figure 6.8A). A billion molecules of water per second can move through an aquaporin channel. How the molecules move is fascinating. Each water molecule is severed from its hydrogen-bonded neighbors as it is handed off to a succession of hydrogen-bonding sites on the aquaporin protein in the channel. Peter Agre at Johns Hopkins University in Baltimore, Maryland, received a Nobel Prize in 2003 for his discovery of aquaporins.

Other channel proteins facilitate the transport of ions such as sodium (Na^+), potassium (K^+), calcium (Ca^{2+}), and chlorine (Cl^-). Most of these ion transporters, which occur in all eukaryotes, are **gated channels;** that is, they switch between open, closed, or intermediate states. The gates may open or close in response to changes in voltage across the membrane, for instance, or by binding signal molecules. The opening or closing involves changes in the protein's three-dimensional shape. In animals, voltage-gated ion channels are used in nerve conduction and the control of muscle contraction (see Chapters 37 and 41); **Figure 6.8B** shows a voltage-gated K^+ channel.

Gated ion channels perform functions that are vital to survival, as illustrated by the effects of hereditary defects in the channels. For example, as you learned in *Why It Matters,* the lethal genetic disease *cystic fibrosis* (CF) results from a fault in a Cl^- channel.

Carrier proteins are the second type of transport proteins; they also form passageways through the lipid bilayer **(Figure 6.8C).** Carrier proteins each bind a specific single solute, such as glucose or an amino acid, and transport it across the lipid bilayer (glucose is also transported by active transport, as described in the next section). Because a single solute is transferred in this carrier-mediated fashion, the transfer is called *uniport transport.* In performing the transport step, the carrier protein undergoes conformational changes that progressively move the solute-binding site from one side of the membrane to the other, thereby transporting the solute. This property distinguishes carrier protein function from channel protein function.

Facilitated diffusion by carrier proteins can become *saturated* when there are too few transport proteins to handle all the solute molecules. For example, if glucose is added at higher and higher concentrations to the solution that surrounds an animal cell, the rate at which it passes through the membrane at first increases proportionately with the increase in concentration. However, at some point, as the glucose concentration is increased still further, the increase in the rate of transport slows. Eventually, further increases in concentration cause no additional rise in the rate of transport—the transport mechanism is saturated. By contrast, saturation does not occur for simple diffusion.

Because the proteins that perform facilitated diffusion are specific, cells can control the kinds of molecules and ions that pass through their membranes by regulating the types of transport proteins in their membranes. As a result, each type of cellular membrane, and each type of cell, has its own group of transport proteins and passes a characteristic group of substances by facilitated diffusion. The kinds of transport proteins present in a cell ultimately depend on the activity of genes in the cell nucleus.

STUDY BREAK 6.2 <

1. **What is the difference between passive and active transport?**
2. **What is the difference between simple and facilitated diffusion?**

6.3 Passive Water Transport and Osmosis

As discussed earlier, water can follow concentration gradients and diffuse passively across membranes in response. It diffuses both directly through the membrane and through aquaporins. The passive transport of water, called **osmosis,** occurs constantly in living cells. Inward or outward movement of water by osmosis develops forces that can cause cells to swell and burst or shrink and shrivel up. Much of the energy budget of many cell types, particularly in animals, is spent counteracting the inward or outward movement of water by osmosis.

Osmosis Can Be Demonstrated in a Purely Physical System

The apparatus shown in **Figure 6.9A** is a favorite laboratory demonstration of osmosis. It consists of an inverted thistle tube (so named because its shape resembles a thistle flower) tightly sealed at its lower end by a sheet of cellophane. The tube is filled with a solution of glucose molecules in water and is suspended in a beaker of distilled water. The cellophane film acts as a selectively permeable membrane because its pores are large enough to admit water molecules but not glucose. At the start of the experiment, the position of the tube is set so the level of the liquid in the tube is at the same level as the distilled water in the beaker. Almost immediately, the level of the solution in the tube begins to rise, eventually reaching a maximum height above the liquid in the beaker.

The liquid rises in the tube because water moves by osmosis from the beaker into the thistle tube. The movement occurs passively, in response to a concentration gradient in which the water

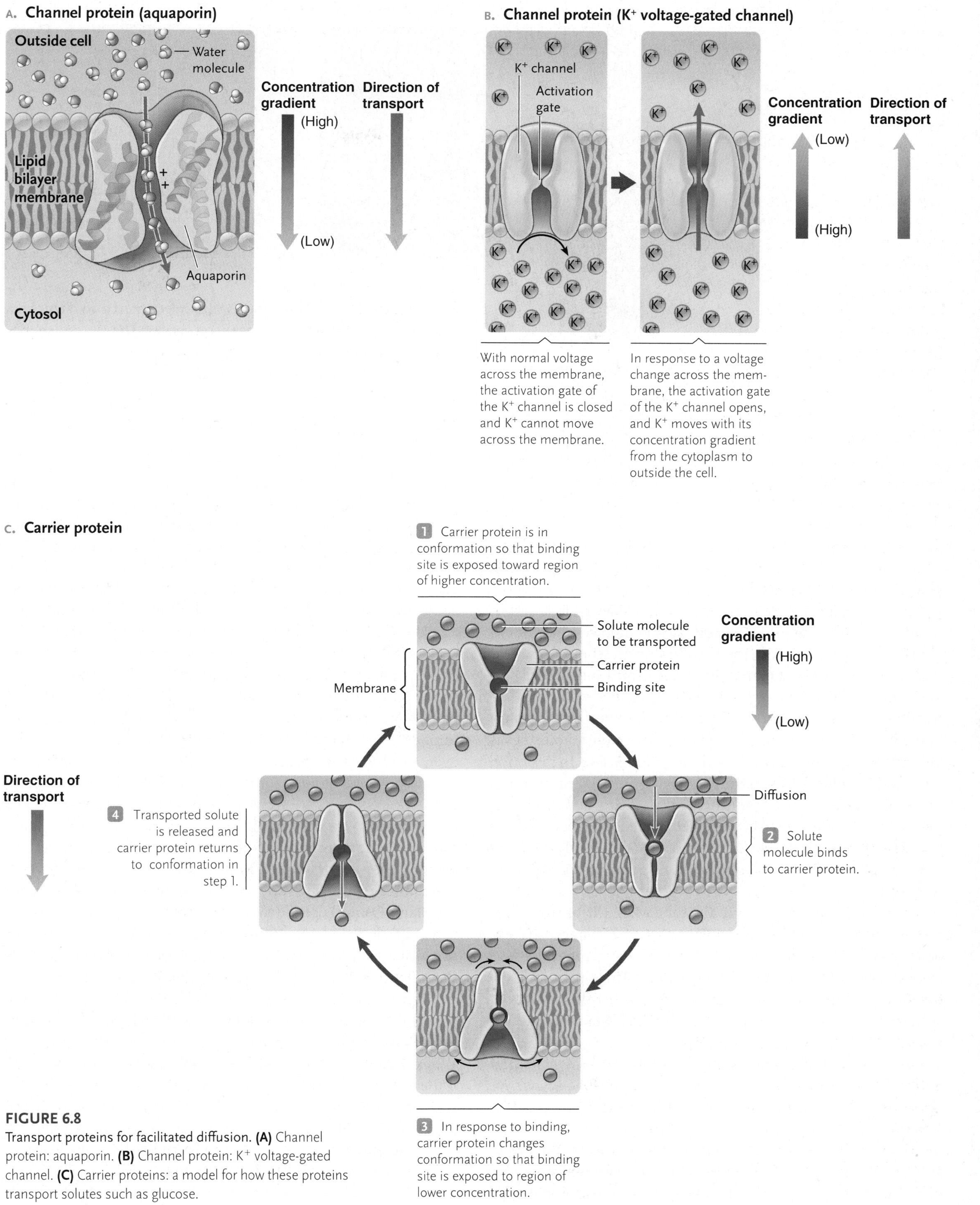

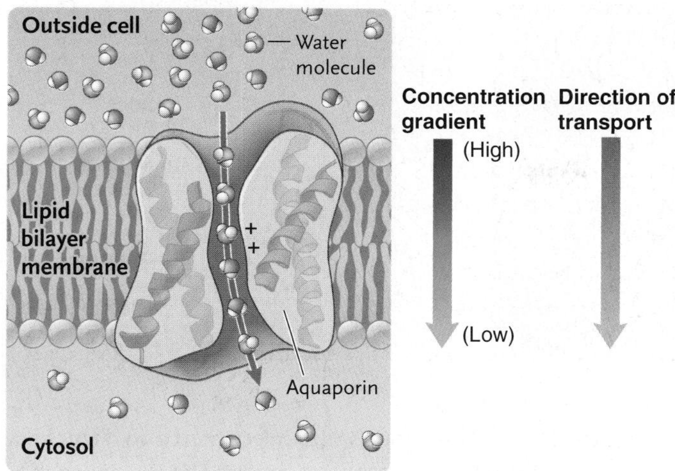
A. Channel protein (aquaporin)
Outside cell
Water molecule
Lipid bilayer membrane
Aquaporin
Cytosol
Concentration gradient
(High)
(Low)
Direction of transport

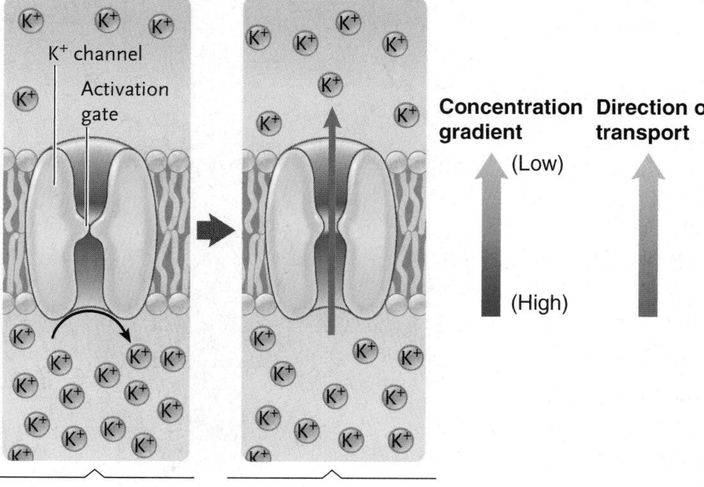
B. Channel protein (K^+ voltage-gated channel)
K^+ channel
Activation gate
Concentration gradient
(Low)
(High)
Direction of transport
With normal voltage across the membrane, the activation gate of the K^+ channel is closed and K^+ cannot move across the membrane.
In response to a voltage change across the membrane, the activation gate of the K^+ channel opens, and K^+ moves with its concentration gradient from the cytoplasm to outside the cell.

C. Carrier protein

1 Carrier protein is in conformation so that binding site is exposed toward region of higher concentration.

Solute molecule to be transported

Carrier protein

Binding site

Membrane

Concentration gradient

(High)

(Low)

Diffusion

2 Solute molecule binds to carrier protein.

3 In response to binding, carrier protein changes conformation so that binding site is exposed to region of lower concentration.

4 Transported solute is released and carrier protein returns to conformation in step 1.

Direction of transport

FIGURE 6.8
Transport proteins for facilitated diffusion. (A) Channel protein: aquaporin. **(B)** Channel protein: K^+ voltage-gated channel. **(C)** Carrier proteins: a model for how these proteins transport solutes such as glucose.

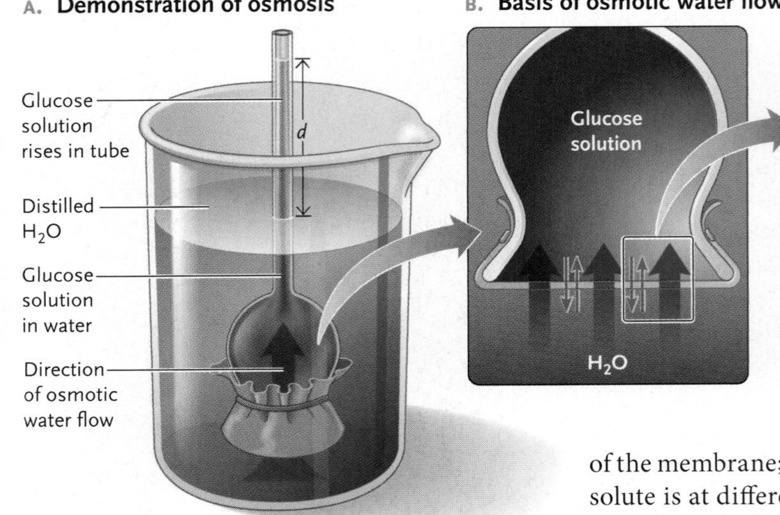
A. Demonstration of osmosis
Glucose solution rises in tube
d
Distilled H_2O
Glucose solution in water
Direction of osmotic water flow
B. Basis of osmotic water flow
Glucose solution
H_2O
Region of lower free water concentration
Glucose molecule
Selectively permeable membrane
Water molecule
Region of higher free water concentration

FIGURE 6.9 Osmosis. **(A)** An apparatus demonstrating osmosis. The fluid in the tube rises because of the osmotic flow of water through the cellophane membrane, which is permeable to water but not to glucose molecules. Osmotic flow continues until the weight of the water in column *d* develops enough pressure to counterbalance the movement of water molecules into the tube. **(B)** The basis of osmotic water flow. The pure water solution on the left is separated from the glucose solution on the right by a membrane permeable to water but not to glucose. The free water concentration on the glucose side is lower than on the water-only side because water molecules are associated with the glucose molecules. That is, water molecules are in greater concentration on the bottom than on the top. Although water molecules move in both directions across the membrane (small red arrows), there is a net upward movement of water (blue arrows), with the water's concentration gradient.

molecules are more concentrated in the beaker than inside the thistle tube. The basis for the gradient is shown in **Figure 6.9B.** The glucose molecules are more concentrated on one side of the selectively permeable membrane. On this side, association of water molecules with those solute molecules reduces the amount of water available to cross the membrane. Thus, although initially there is an equal apparent water concentration on each side of the membrane, there is a difference in the *free water* concentration—that is, the water available to move across the membrane. Specifically, the concentration of free water molecules is lower on the glucose side than on the pure water side. In response, more water molecules from the pure water side will hit the pores in the membrane than from the solute side, producing a net movement of water from the pure water side to the glucose solution side. Osmosis is the net diffusion of water molecules through a selectively permeable membrane in response to a gradient of this type.

The solution stops rising in the tube when the pressure created by the weight of the raised solution exactly balances the tendency of water molecules to move from the beaker into the tube in response to the concentration gradient. This pressure is the **osmotic pressure** of the solution in the tube. At this point, the system is in a state of dynamic equilibrium and no further net movement of water molecules occurs.

A formal definition for osmosis is *the net movement of water molecules across a selectively permeable membrane by passive diffusion, from a solution of lesser solute concentration to a solution of greater solute concentration* (the *solute* is the substance dissolved in water). For osmosis to occur, the selectively permeable membrane must allow water molecules, but not molecules of the solute, to pass. Pure water does not need to be on one side of the membrane; osmotic water movement also occurs if a solute is at different concentrations on the two sides. Because osmosis occurs in response to a concentration gradient, it releases free energy and can accomplish work.

The Free Energy Released by Osmosis May Work for or against Cellular Life

Osmosis occurs in cells because they contain a solution of nonpenetrating solutes—proteins and other molecules that are retained in the cytoplasm by a membrane impermeable to them but freely permeable to water. The resulting osmotic movement of water is used as an energy source for some of the activities of life. However, it can also create a disturbance that cells must counteract by expending energy. If the solution surrounding a cell contains nonpenetrating solutes at lower concentrations than in the cell, the solution is said to be **hypotonic** to the cell (*hypo* = under or below; *tonos* = tension or tone). When a cell is in a hypotonic solution, water enters by osmosis and the cell tends to swell **(Figure 6.10A).** Animal cells, for instance red blood cells, in a hypotonic solution may actually swell to the point of bursting. However, in most plant cells, strong walls prevent the cells from bursting in a hypotonic solution. In most land plants, the cells at the surfaces of roots are surrounded by almost pure water, which is hypotonic to the cells and tissues of the root. As a result, water flows from the surrounding soil into the root cells by osmosis. The osmotic pressure developed by the inward flow contributes part of the force required to raise water from the roots to the leaves of the plant. Osmosis also drives water into cells of the stems and leaves of plants. The resulting osmotic pressure, called **turgor pressure,** pushes the cells tightly against their walls and supports the softer tissues against the force of gravity.

If the solution that surrounds a cell contains nonpenetrating solutes at higher concentrations than in the cell, the outside solution is said to be **hypertonic** to the cells (*hyper* = over or above). When a cell is in a hypertonic solution, water leaves by osmosis. If the outward osmotic movement exceeds the capacity of cells to replace the lost water, both animal and plant cells will shrink **(Figure 6.10B)**. In plants, the shrinkage and loss of internal osmotic pressure under these conditions causes stems and leaves to wilt. In extreme cases, plant cells shrink so much that they retract from their walls, a condition known as **plasmolysis.**

FIGURE 6.10

Tonicity and osmotic water movement. The diagrams show what happens when a cellophane bag filled with a 2 M sucrose solution is placed in a **(A)** hypotonic, **(B)** hypertonic, or **(C)** isotonic solution. They also show the corresponding effects of these three types of solutions on animal and plant cells.

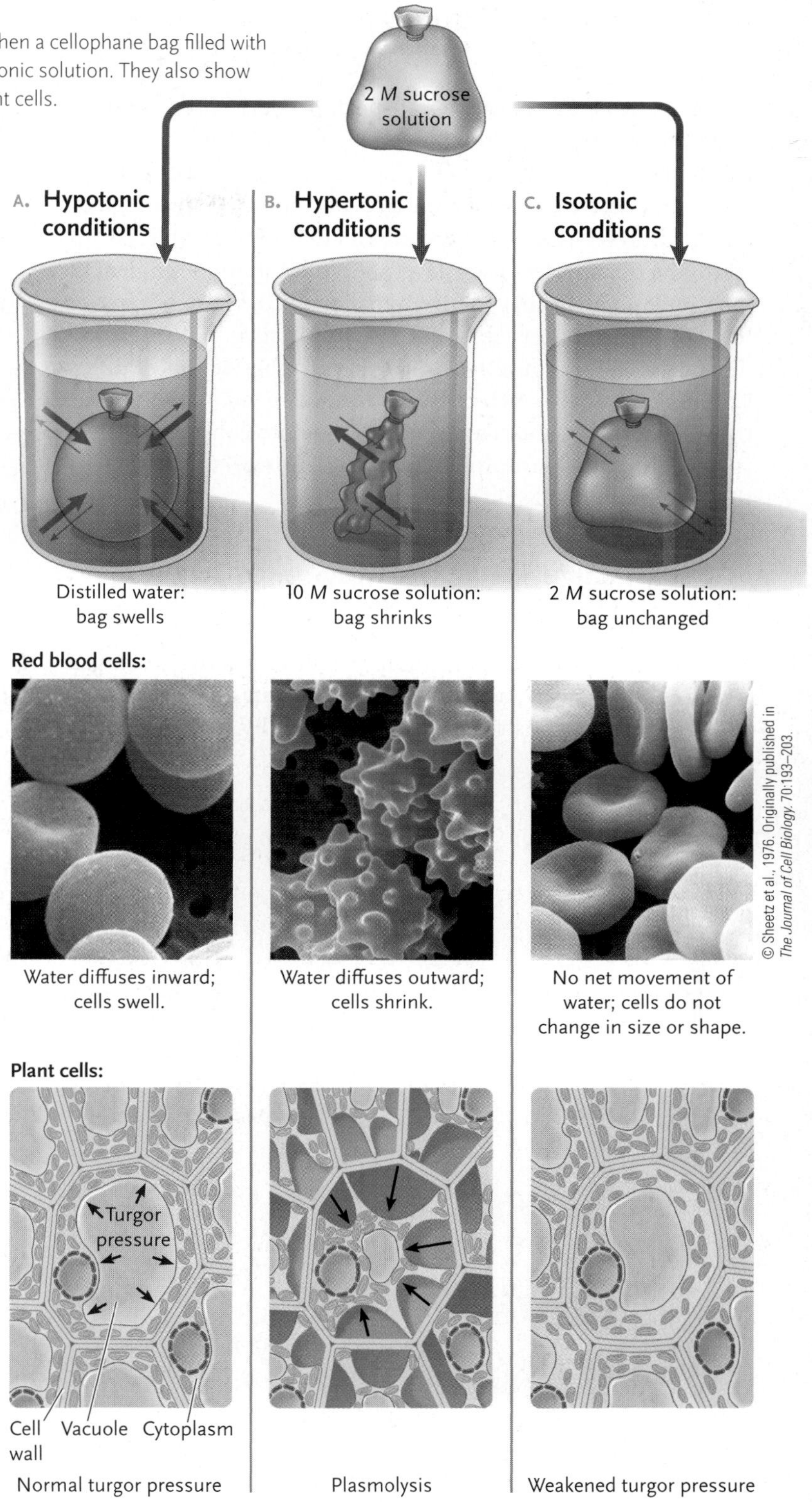

2 *M* sucrose solution
A. Hypotonic conditions
B. Hypertonic conditions
C. Isotonic conditions
Distilled water: bag swells
10 *M* sucrose solution: bag shrinks
2 *M* sucrose solution: bag unchanged
Red blood cells:
Water diffuses inward; cells swell.
Water diffuses outward; cells shrink.
No net movement of water; cells do not change in size or shape.
Plant cells:
Turgor pressure
Cell wall
Vacuole
Cytoplasm
Normal turgor pressure
Plasmolysis
Weakened turgor pressure

In animals, ions, proteins, and other molecules are concentrated in extracellular fluids, as well as inside cells, so that the concentration of water inside and outside cells is usually equal or **isotonic** (*iso* = the same; see **Figure 6.10C**). To keep fluids on either side of the plasma membrane isotonic, animal cells must constantly use energy to pump Na^+ from inside to outside by active transport (see Section 6.4); otherwise, water would move inward by osmosis and cause the cells to burst. For animal cells, an isotonic solution is usually optimal, whereas for plant cells, an isotonic solution results in some loss of turgor. The mechanisms by which plants and animals balance their water content by regulating osmosis are discussed in Chapters 32 and 46.

Passive transport, driven by concentration gradients, accounts for much of the movement of water, ions, and many types of molecules into or out of cells. In addition, all cells transport some ions and molecules against their concentration gradients by active transport (see the next section).

STUDY BREAK 6.3 <

1. **What conditions are required for osmosis to occur?**
2. **Explain the effect of a hypertonic solution that surrounds animal cells.**

6.4 Active Transport

Facilitated diffusion accelerates the movement of substances across cellular membranes, but it is limited to transport with a concentration gradient. To transport substances across a membrane in the opposite direction, that is, against a concentration gradient, requires **active transport,** a process that requires energy input.

The three main functions of active transport in cells and organelles are: (1) Uptake of essential nutrients from the fluid surrounding cells even when their concentrations are lower than in cells; (2) removal of secretory or waste materials from cells or organelles even when the concentration of those materials is higher outside the cells or organelles; and (3) maintenance of essentially constant intracellular concentrations of H^+, Na^+, K^+, and Ca^{2+}.

Because ions are charged molecules, active transport of ions may contribute to voltage—an electrical potential difference—across the plasma membrane, called a **membrane potential.** The unequal distribution of ions across the membrane created by passive transport also contributes to the voltage. Neurons and muscle cells use the membrane potential in a specialized way. That is, in response to electrical, chemical, mechanical, and certain other types of stimuli, their membrane potential changes rapidly and transiently. In nerve cells, for example, this type of transport is the basis for transmission of a nerve impulse.

In short, active transport usually establishes differences in solute concentrations or of voltage across membranes that are important for cell or organelle function. By contrast, passive diffusion and facilitated diffusion act mostly to move substances across membranes in the direction toward equalizing their concentrations on each side.

Active Transport Requires a Direct or Indirect Input of Energy Derived from ATP Hydrolysis

There are two kinds of active transport: primary and secondary. In **primary active transport,** the same protein that transports a substance also hydrolyzes ATP to power the transport directly. In **secondary active transport,** the transport is indirectly driven by ATP hydrolysis. That is, the proteins do not break down ATP; instead, the transporters use a favorable concentration gradient of ions, built up by primary active transport, as their energy source for active transport of a different ion or molecule.

Other features of active transport resemble facilitated diffusion (listed in Table 6.1). Both processes depend on membrane transport proteins, both are specific, and both can be saturated. In both, the transport proteins are carrier proteins that change their conformation as they function.

Primary Active Transport Moves Positively Charged Ions across Membranes

The primary active transport pumps all move positively charged ions—H^+, Ca^{2+}, Na^+, and K^+—across membranes **(Figure 6.11).** The gradients of positive ions established by primary active transport pumps underlie functions that are absolutely essential for cellular life. For example, **H^+ pumps** (also called **proton pumps**) move hydrogen ions across membranes, temporarily binding a phosphate group removed from ATP during the pumping cycle. Proton pumps have various functions. For example, in prokaryotes, plants, and fungi, proton pumps in the plasma membrane generate membrane potential. Proton pumps in lysosomes of animals and vacuoles of plants and fungi keep the pH within the organelle low, serving to activate the enzymes contained within them.

The **Ca^{2+} pump** (or **calcium pump**) is distributed widely among eukaryotes. It pushes Ca^{2+} from the cytoplasm to the cell exterior, and also from the cytosol into the vesicles of the endoplasmic reticulum (ER). As a result, Ca^{2+} concentration is typically high outside cells and inside ER (endoplasmic reticulum) vesicles and low in the cytoplasmic solution. This Ca^{2+} gradient is used universally among eukaryotes as a regulatory control of cellular activities as diverse as secretion, microtubule assembly, and muscle contraction. For the process of muscle contraction in animals, calcium pumps release stored Ca^{2+} from vesicles inside a muscle fiber (a single cell in a muscle) which initiates a series of steps leading to contraction of the fiber. In plants, a calcium pump is involved in pollen growth and fertilization.

FIGURE 6.11
Model for how a primary active transport pump operates.

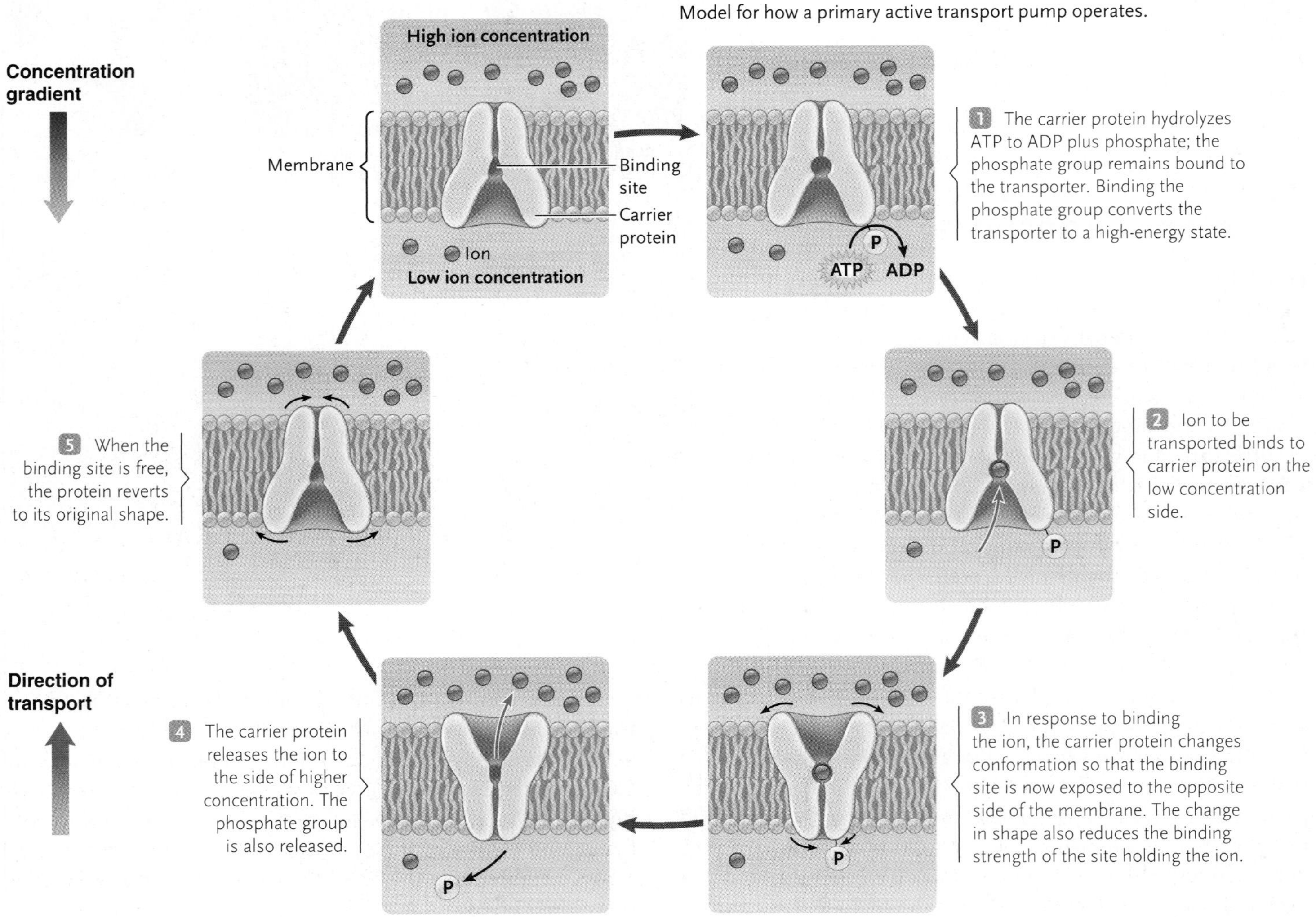

1 The carrier protein hydrolyzes ATP to ADP plus phosphate; the phosphate group remains bound to the transporter. Binding the phosphate group converts the transporter to a high-energy state.

2 Ion to be transported binds to carrier protein on the low concentration side.

3 In response to binding the ion, the carrier protein changes conformation so that the binding site is now exposed to the opposite side of the membrane. The change in shape also reduces the binding strength of the site holding the ion.

4 The carrier protein releases the ion to the side of higher concentration. The phosphate group is also released.

5 When the binding site is free, the protein reverts to its original shape.

FIGURE 6.12

The Na^+/K^+ pump, an active transport protein in the plasma membrane. Energy from the protein's hydrolysis of ATP transports Na^+ out of the cell and K^+ into the cell, each against its concentration gradient. The pump moves three Na^+ out and two K^+ in for each ATP molecule hydrolyzed.

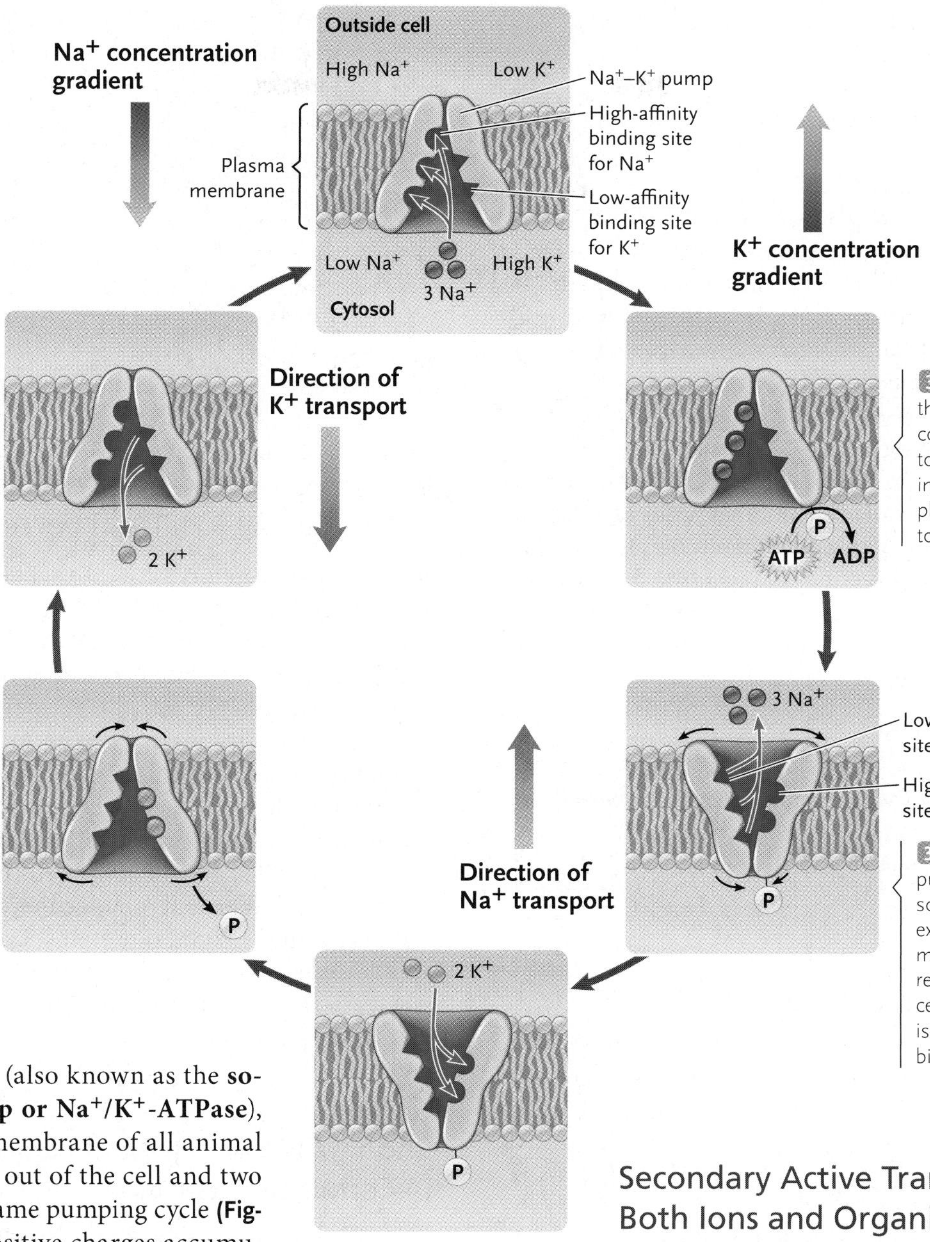

1 Pump has 3 high-affinity sites for Na^+ and 2 low-affinity sites for K^+ when exposed to the cytosol.

2 When 3 Na^+ from the cytosol (where Na^+ concentration is low) bind to pump, it splits ATP into ADP plus phosphate; phosphate group binds to pump.

3 Phosphorylation causes pump to change conformation so that Na^+ binding sites are exposed to opposite side of membrane and 3 Na^+ are released to the fluid outside the cell (where Na^+ concentration is high) as affinity of Na^+ binding sites greatly decreases.

4 Change in shape also exposes pump's binding sites for K^+ to the fluid outside the cell and greatly increases affinity of K^+ sites

5 When 2 K^+ from the fluid outside the cell (where K^+ concentration is low) bind to pump, it releases phosphate group. Dephosphorylation causes pump to revert to its original conformation.

6 Two K^+ are released to the cytosol (where K^+ concentration is high) as affinity of K^+ binding sites markedly decreases during change in shape. At the same time, affinity of Na^+ binding sites greatly increases, returning process to step 1.

The **Na^+/K^+ pump** (also known as the **sodium–potassium pump or Na^+/K^+-ATPase**), located in the plasma membrane of all animal cells, pushes three Na^+ out of the cell and two K^+ into the cell in the same pumping cycle **(Figure 6.12).** As a result, positive charges accumulate in excess outside the membrane, and the inside of the cell becomes negatively charged with respect to the outside. This creates a membrane potential measuring from about −50 to −200 millivolts (mV; 1 millivolt = 1/1,000th of a volt), with the minus sign indicating that the charge inside the cell is negative versus the outside. That is, there is both a concentration difference (of the ions) and an electrical charge difference on the two sides of the membrane, constituting what is called an **electrochemical gradient.** Electrochemical gradients store energy that is used for other transport mechanisms. For instance, an electrochemical gradient across a nerve cell membrane drives the movement of ions involved in nerve impulse transmission (see Chapter 37).

Secondary Active Transport Moves Both Ions and Organic Molecules across Membranes

Secondary active transport pumps use the concentration gradient of an ion established by a primary pump as their energy source. For example, the driving force for most secondary active transport in animal cells is the high outside/low inside Na^+ gradient created by the Na^+/K^+ pump. Also, in secondary active transport, the transfer of the solute across the membrane always occurs coupled with transfer of the ion that supplies the driving force.

Secondary active transport occurs by two mechanisms known as *symport* and *antiport* **(Figure 6.13).** In **symport** (also called **cotransport**), the solute moves through the membrane channel in the same direction as the driving ion. Sugars, such as glucose, and amino acids are examples of molecules actively transported into

FIGURE 6.13
Secondary active transport, in which a concentration gradient of an ion is used as the energy source for active transport of a solute. **(A)** In symport, the transported solute moves in the same direction as the gradient of the driving ion. **(B)** In antiport, the transported solute moves in the direction opposite from the gradient of the driving ion.

A. **Symport**

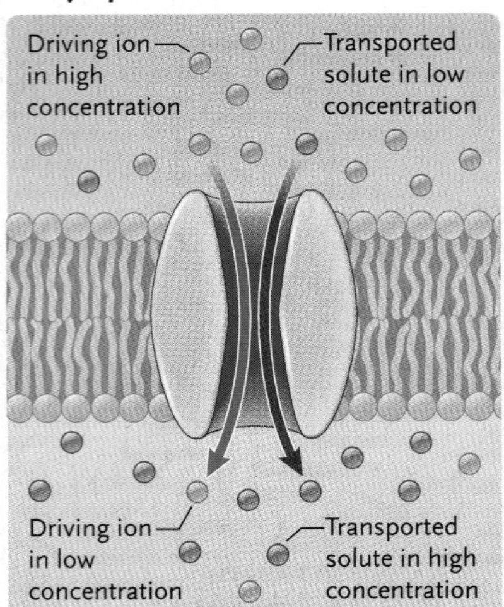
Driving ion in high concentration
Transported solute in low concentration
Driving ion in low concentration
Transported solute in high concentration

B. **Antiport**

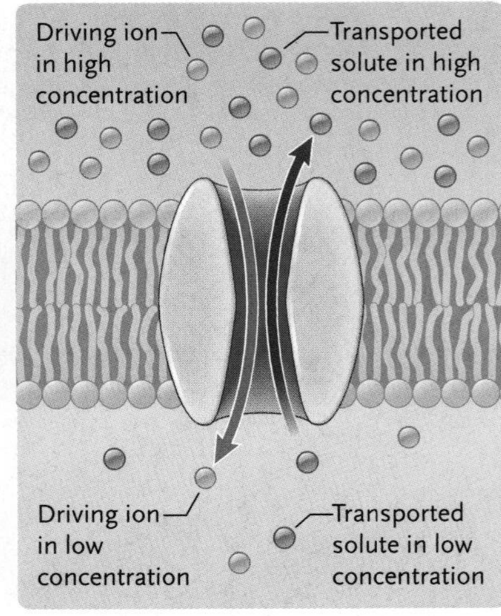
Driving ion in high concentration
Transported solute in high concentration
Driving ion in low concentration
Transported solute in low concentration

cells by symport. In **antiport** (also known as **exchange diffusion**), the driving ion moves through the membrane channel in one direction, providing the energy for the active transport of another molecule through the membrane in the opposite direction. In many cases, ions are exchanged by antiport. For example, in many tissues, a Na^+/Ca^{2+} exchanger is used to remove cytosolic calcium from cells, exchanging one Ca^{2+} for three Na^+.

Active and passive transport move ions and smaller hydrophilic molecules across cellular membranes. Cells can also move much larger molecules or aggregates of molecules from inside to outside, or in the reverse direction, by including them in the inward or outward vesicle traffic of the cell. The mechanisms that carry out this movement—exocytosis and endocytosis—are discussed in the next section.

STUDY BREAK 6.4 <

1. **What is active transport? What is the difference between primary and secondary active transport?**
2. **How is a membrane potential generated?**

>

THINK OUTSIDE THE BOOK

Collaboratively or on your own, draw a diagram showing how ions and solutes exchange in a symport system and an antiport system in the intestinal tract.

6.5 Exocytosis and Endocytosis

The largest molecules transported through cellular membranes by passive and active transport are in the size range of amino acids or monosaccharides such as glucose. Eukaryotic cells import and export larger molecules by exocytosis and endocytosis (introduced in Section 5.3). The export of materials by exocytosis primarily carries secretory proteins and some waste materials from the cytoplasm to the cell exterior. Import by endocytosis may carry proteins, larger aggregates of molecules, or even whole cells from the outside into the cytoplasm. Exocytosis and endocytosis also contribute to the back-and-forth flow of membranes between the endomembrane system and the plasma membrane. Both exocytosis and endocytosis require energy; thus, both processes stop if the ability of a cell to make ATP is inhibited.

Exocytosis Releases Molecules to the Outside of the Cell by Means of Secretory Vesicles

In exocytosis, secretory vesicles originated by budding from the Golgi complex move through the cytoplasm and contact the plasma membrane **(Figure 6.14A)**. The vesicle membrane fuses with the plasma membrane, releasing the contents of the vesicle to the cell exterior.

All eukaryotic cells secrete materials to the outside through exocytosis. For example, in animals, glandular cells secrete peptide hormones or milk proteins, and cells that line the digestive tract secrete mucus and digestive enzymes. Plant cells, fungal cells, and bacterial cells use exocytosis to secrete proteins and other macromolecules associated with the cell wall, including enzymes, proteins, and carbohydrates. Fungi and bacteria also secrete enzymes by exocytosis to digest nutrients in their environments. Finally, all organisms use exocytosis to place integral membrane proteins in the plasma membrane.

Endocytosis Brings Materials into Cells in Endocytic Vesicles

In endocytosis, proteins and other substances are trapped in pitlike depressions that bulge inward from the plasma membrane. The depression then pinches off as an endocytic vesicle. Endocytosis occurs in most eukaryotic cells by one of two distinct but related pathways. The simplest of these mechanisms, **bulk-phase endocytosis** (sometimes called **pinocytosis,** meaning "cell drinking"), takes in a drop of the aqueous fluid surrounding the cell—called the *extracellular fluid* (ECF)—together with any molecules that happen to be in solution in the water **(Figure 6.14B)**. In fact, the primary function of pinocytosis is the absorption of extracellular fluid. The process is nonspecific; that is it takes in any solutes present in the fluid because the membrane lacks surface receptors for specific molecules.

In the second endocytic pathway, **receptor-mediated endocytosis,** the target molecules to be taken in are bound to re-

A. **Exocytosis: vesicle joins plasma membrane, releases contents**

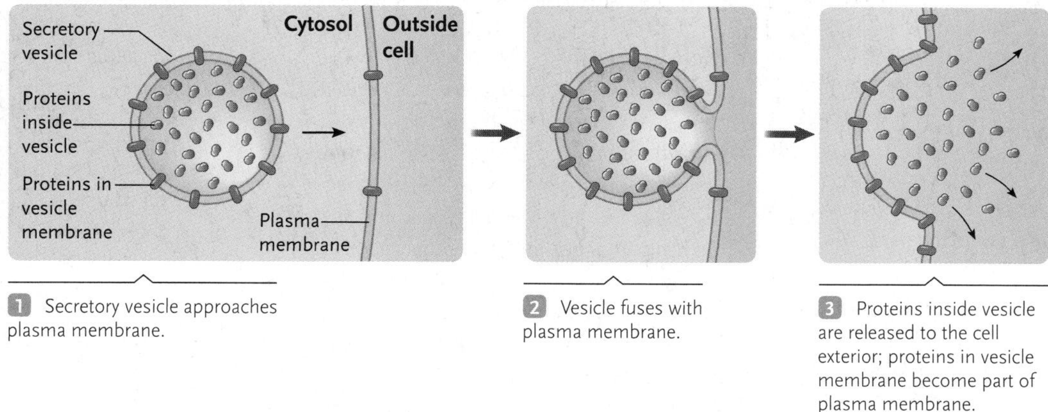
Secretory vesicle
Proteins inside vesicle
Proteins in vesicle membrane
Cytosol
Outside cell
Plasma membrane
1 Secretory vesicle approaches plasma membrane.
2 Vesicle fuses with plasma membrane.
3 Proteins inside vesicle are released to the cell exterior; proteins in vesicle membrane become part of plasma membrane.

B. **Bulk-phase endocytosis (pinocytosis): vesicle imports water and other substances from outside cell**

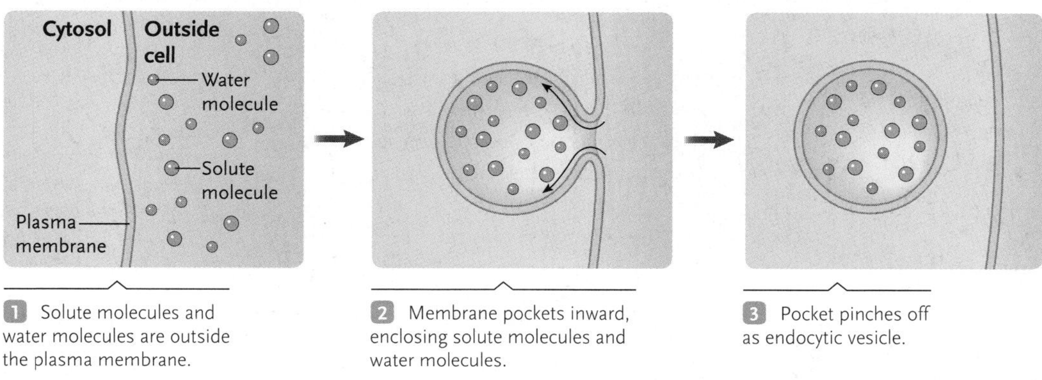
Cytosol
Outside cell
Water molecule
Solute molecule
Plasma membrane
1 Solute molecules and water molecules are outside the plasma membrane.
2 Membrane pockets inward, enclosing solute molecules and water molecules.
3 Pocket pinches off as endocytic vesicle.

C. **Receptor-mediated endocytosis: vesicle imports specific molecules**

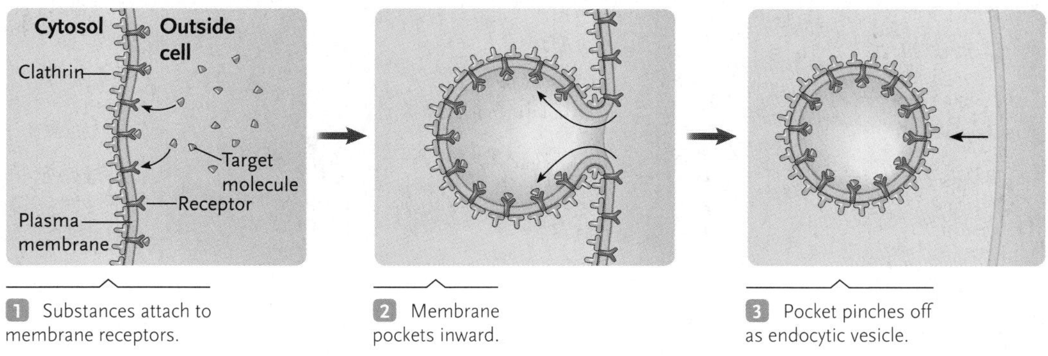
Cytosol
Outside cell
Clathrin
Target molecule
Receptor
Plasma membrane
1 Substances attach to membrane receptors.
2 Membrane pockets inward.
3 Pocket pinches off as endocytic vesicle.

D. **Micrographs of stages of receptor-mediated endocytosis shown in C**

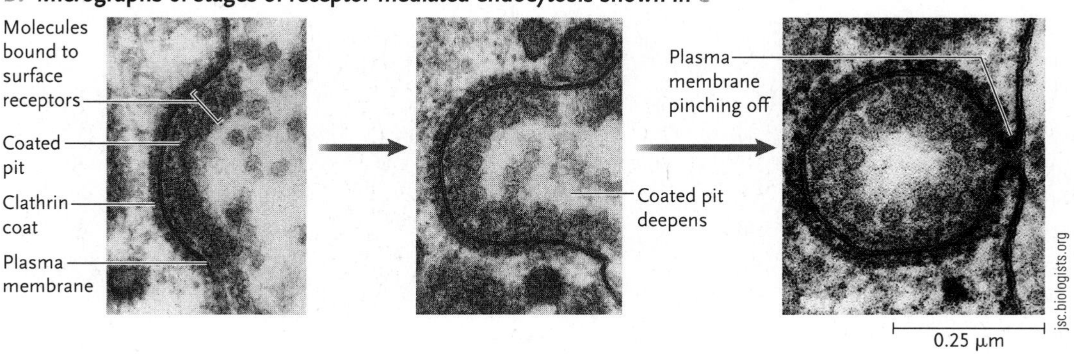
Molecules bound to surface receptors
Coated pit
Clathrin coat
Plasma membrane
Coated pit deepens
Plasma membrane pinching off
0.25 μm
jsc.biologists.org

FIGURE 6.14
Exocytosis and endocytosis.

Research Serendipity: The discovery of receptor-mediated endocytosis

When they were still in training, two physicians, Michael Brown and Joseph Goldstein, were called to treat two young sisters, 6 and 8 years old. The children were dying of recurrent heart attacks brought on by extremely high cholesterol levels in their blood. How could such young children develop a condition that usually appears only in late middle or old age? Brown and Goldstein went on to do groundbreaking research seeking an answer to this puzzling question.

As you have seen, cholesterol is essential for keeping cell membranes fluid. The blood transports cholesterol to cells that need it, but blood cholesterol can be dangerous. It can cause atherosclerosis (thickening of the arteries as a result of the build up of fatty materials such as cholesterol), which can be lethal. To lessen this danger, the body packages cholesterol with proteins to form lipoproteins such as LDL (low density lipoprotein).

Patients with familial hypercholesteremia (FH), have a higher than normal concentration of LDL in their blood and often experience atherosclerosis and heart attacks early in life. Individuals with one copy of the mutated gene responsible for the disease have about twice the normal level of LDL, and begin to have heart attacks at 30–40 years old. Individuals like the two sisters, with two copies of the mutated gene, have 6-to-10 fold higher than normal levels of LDL in their blood, and they often have heart attacks in childhood.

Research Question

What is the molecular basis of familial hypercholesteremia?

Experiments

Brown and Goldstein of the University of Texas Southwestern Medical School, Dallas, Texas, began their studies on the molecular basis of familial hypercholesteremia in 1972. Two key experiments were as follows:

1. The researchers cultured human cells (skin fibroblasts) from normal individuals and found that radioactively labeled LDL bound strongly to the cells. They interpreted this result to mean that the cells had specific surface receptors for LDL. Once bound to the receptor, the LDL was taken into the cell and degraded by lysosomes to release the cholesterol. In fibroblasts from patients with two mutant forms of the FH gene, very little radioactively labeled LDL bound to the cells (see figure). This experiment provided evidence that FH occurs because patients either do not make the receptor, or have defective receptor molecules. As a result, LDL accumulates in the blood because it cannot be taken into the cells. Further research has confirmed that an abnormality in the gene for this receptor is responsible for FH.
2. How does the receptor–LDL complex enter cells? The two researchers, with Richard G. W. Anderson, used the same fibroblast system to seek and answer. To help them visualize the receptor–LDL complex entering the cell, they bound ferritin to the LDL. Ferritin is a protein that binds iron which is electron dense, making it possible to see the LDL as black dots under the electron microscope. When they incubated fibroblast cells from normal individuals with ferritin-labeled LDL, the investigators were surprised to see that the ferritin collected at short segments of the plasma membrane which appeared to be indented and coated on both sides by "fuzzy material." These regions we now know to be clathrin coated pits, and they are the cellular entry site for the receptor–LDL complex.

Conclusion

In a major step toward answering the question of what causes FH, Goldstein and Brown had discovered that LDL enters cells using a specific receptor on the cell surface, which is absent or reduced in FH patients. Serendipitously, they had also discovered the answer to broader question: How do cells take in specific molecules that cannot pass through the membrane? They had discovered receptor-mediated endocytosis (see Figure 6.14C and D). The researchers received the Nobel Prize for their discovery in 1986.

Sources: R. G. W. Anderson, J. L. Goldstein, and M. S. Brown. 1976. Localization of low density lipoprotein receptors on plasma membrane of normal human fibroblasts and their absence in cells from a familial hypercholesteremia homozygote. *Proceedings of the National Academy of Sciences USA* 73:2434–2438; R. G. W. Anderson, J. L. Goldstein, and M. S. Brown. 1977. A mutation that impairs the ability of lipoprotein receptors to localize in coated pits on the cell surface of human fibroblasts. *Nature* 270:695–699; M. S. Brown and J. L. Goldstein. 1974. Familial hypercholesteremia: Defective binding of lipoproteins to cultured fibroblasts associated with impaired regulation of 3-hydroxy-3-methylglutaryl coenzyme A reductase activity. *Proceedings of the National Academy of Sciences USA* 71:788–792.

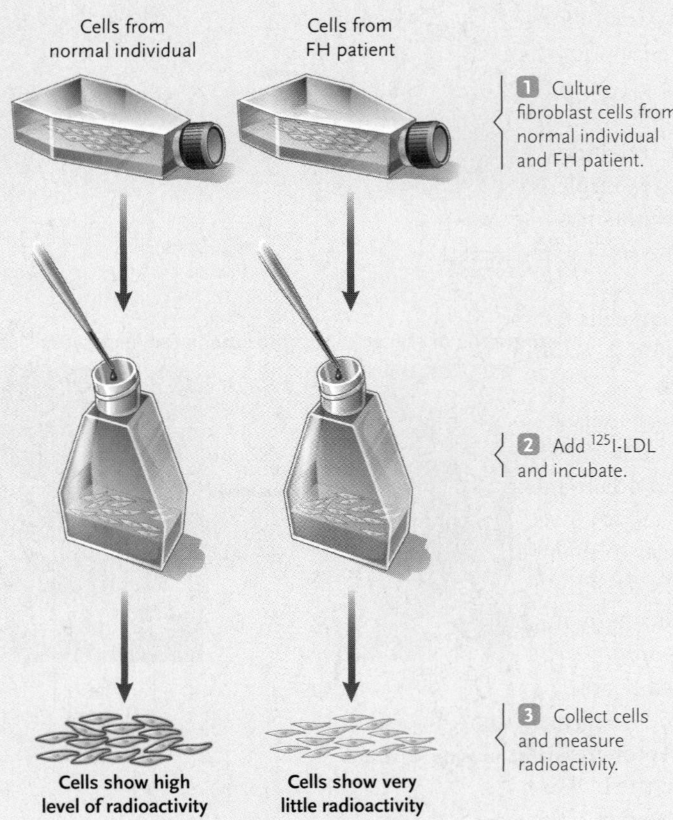
Cells from normal individual
Cells from FH patient
1 Culture fibroblast cells from normal individual and FH patient.
2 Add ^{125}I-LDL and incubate.
3 Collect cells and measure radioactivity.
Cells show high level of radioactivity
Cells show very little radioactivity

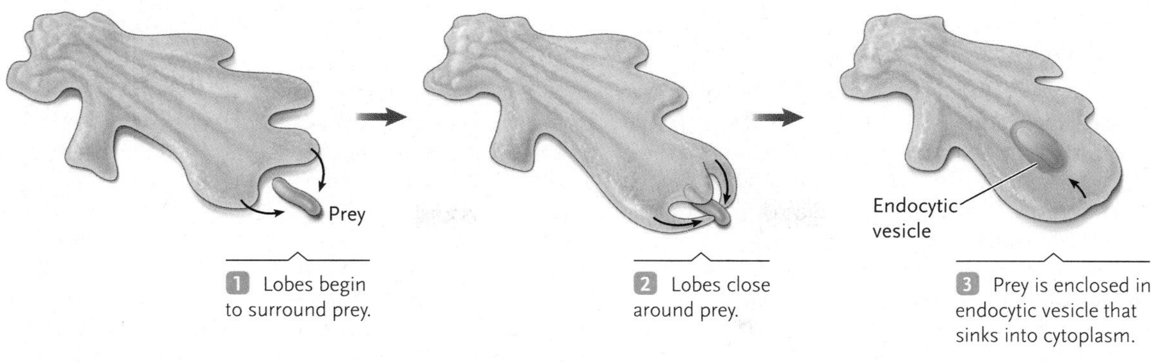
Prey
Endocytic vesicle
1 Lobes begin to surround prey.
2 Lobes close around prey.
3 Prey is enclosed in endocytic vesicle that sinks into cytoplasm.

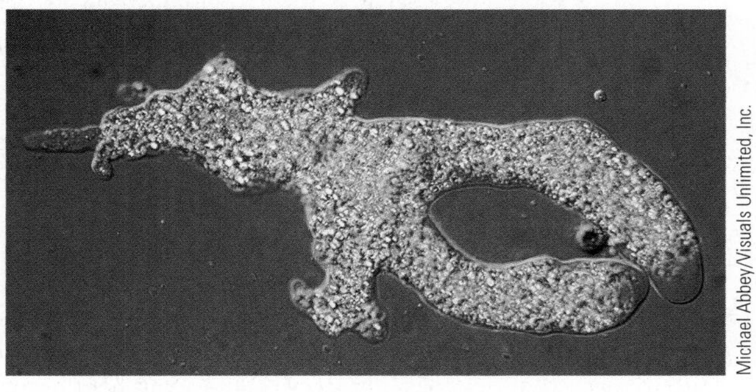
Michael Abbey/Visuals Unlimited, Inc.

FIGURE 6.15
Phagocytosis, in which lobes of the cytoplasm extend outward and surround a cell targeted as prey. The micrograph shows the protist *Chaos carolinense* engulfing a single-celled alga *(Pandorina)* by phagocytosis (corresponding to step 2 in the diagram); white blood cells called phagocytes carry out a similar process in mammals.

ceptor proteins on the outer cell surface **(Figure 6.14C, D).** The receptors, which are integral proteins of the plasma membrane, recognize and bind only certain molecules in the solution that surrounds the cell, which makes this type of endocytosis highly specific. After binding their target molecules, the receptors collect into a depression in the plasma membrane called a **coated pit** because a network of proteins (called **clathrin**) coat and reinforce the cytoplasmic side. With the target molecules attached, the pits deepen and pinch free of the plasma membrane to form endocytic vesicles. Once in the cytoplasm, an endocytic vesicle rapidly loses its clathrin coat and may fuse with a lysosome. The enzymes within the lysosome then digest the contents of the vesicle, breaking them down into smaller molecules useful to the cell. These molecular products—for example, amino acids and monosaccharides—enter the cytoplasm by crossing the vesicle membrane via transport proteins. The membrane proteins are recycled to the plasma membrane.

Mammalian cells take in many substances by receptor-mediated endocytosis, including peptide hormones such as insulin, growth factors, enzymes, antibodies, blood proteins, iron, and vitamin B_{12}. Some viruses exploit the receptor-mediated exocytosis mechanism to enter cells. For instance, HIV, the virus that causes AIDS, binds to membrane receptors that function normally to internalize a needed molecule. The receptors that bind these substances to the plasma membrane are present in thousands to hundreds of thousands of copies. For example, a mammalian cell plasma membrane has about 20,000 receptors for *low-density lipoprotein (LDL).* LDL, a complex of lipids and proteins, is the way cholesterol moves through the bloodstream. When LDL binds to its receptor on the membrane, it is taken into the cell by receptor-mediated endocytosis; then, by the steps described earlier, the LDL is broken down within the cell and cholesterol is released into the cytoplasm. *Insights from the Molecular Revolution* describes the discovery of receptor-mediated endocytosis.

Some specialized cells, such as certain white blood cells *(phagocytes)* in the bloodstream, or protists such as *Amoeba proteus,* can take in large aggregates of molecules, cell parts, or even whole cells by a process related to receptor-mediated endocytosis. The process, called **phagocytosis** (meaning "cell eating"), begins when surface receptors bind molecules on the substances to be taken in **(Figure 6.15).** Cytoplasmic lobes then extend, surround, and engulf the materials, forming a pit that pinches off and sinks into the cytoplasm as a large endocytic vesicle. Enzymes then digest the materials as in receptor-mediated endocytosis, and the cell permanently sequesters any remaining residues into storage vesicles or expels them by exocytosis as wastes.

Working together, exocytosis and endocytosis constantly cycle membrane segments between the internal cytoplasm and the cell surface. The balance of the two mechanisms maintains the surface area of the plasma membrane at controlled levels.

Thus, through the combined mechanisms of passive transport, active transport, exocytosis, and endocytosis, cells maintain their internal concentrations of ions and molecules and exchange larger molecules such as proteins with their surroundings. The next chapter explores cell communication through intercellular chemical messengers. Many of these messengers act through binding to specific proteins embedded in the plasma membrane.

STUDY BREAK 6.5 <

1. **What is the mechanism of exocytosis?**
2. **What is the difference between bulk-phase endocytosis and receptor-mediated endocytosis?**

UNANSWERED QUESTIONS

How do aquaporin channels function?

The discovery of water channels along with the structural and mechanistic studies of ion channels fundamentally changed the scientific understanding of how biological fluids cross cell membranes. By serendipity, we discovered the protein referred to as AQP1 (aquaporin-1) in human red blood cells, and isolated, purified, and cloned the protein before its function was identified. AQP1 is now known to permit movement of water across cell membranes by osmosis. Cell biological determinations established the sites in humans where AQP1 is expressed, thereby predicting its physiological significance and also predicting several diseases states. For example, in the kidney, AQP1 allows the reabsorption of water from primary urine; in capillaries AQP1 facilitates the reabsorption of tissue edema (swelling because of fluid accumulation); in the area of the brain where cerebrospinal fluid is synthesized, AQP1 permits secretion of that fluid; in the eye AQP1 provides for secretion and reabsorption of aqueous humor (the watery fluid between the cornea and the iris). Humans lacking the Colton Blood Group Antigen carry mutations in the gene encoding AQP1. These genetic defects cause the absence of AQP1 and are characterized by inability to concentrate urine despite prolonged thirsting.

The discovery of AQP1 quickly led to the discovery of several other mammalian aquaporins. In humans, AQP2 resides in the final segment of the kidney where the neurohormone vasopressin (also known as antidiuretic hormone) regulates water reabsorption by controlling exocytosis and endocytosis of vesicles containing AQP2 channels. Genetic defects in AQP2 result in severe nephrogenic diabetes insipidus—a disease where children must drink gallons of fluid every day, because they make large volumes of dilute urine. Secondary defects in the biosynthesis of AQP2 are found in milder clinical disorders. One example is bedwetting by small children. Fortunately the children soon outgrow the problem when they recover AQP2 expression. AQP4 exists in multiple sites, including the perivascular membranes of astroglial cells in brain. This protein has been linked to epileptic seizures as well as brain edema after injury. AQP5 is present in secretory glands and is responsible for sweat, tears, and saliva. Other aquaporins are necessary for lens homeostasis; defects in AQP0 are linked to congenital cataracts.

Other members of the aquaporin family were found to be permeated by water plus glycerol—hence, "aquaglyceroporins." In human biology, the aquaglyceroporin AQP3 facilitates glycerol uptake by basal cells in the epidermis (the outermost layer of the skin). Most skin moisturizers consist mainly of glycerol, and the beauty industry now markets products reputed to increase this pathway. AQP3 in red blood cells allows glycerol uptake and is important during the growth of malarial parasites within those cells. The parasite uses its aquaglyceroporins to take up glycerol from the blood cell, and uses the glycerol to synthesize new membranes. An important question is whether the aquaglyceroporins can be exploited as a means of preventing or treating malaria. Aquaglyceroporin AQP7 resides in fat cells where it facilitates the export of glycerol released from triglycerides during fasting. AQP9 in liver permits glycerol uptake by hepatocytes (the chief functional cells of the liver) where glycerol is converted to glucose.

The structures of the aquaporins and aquaglyceroporins are highly related. The narrowest span of the pore allows water to move rapidly in single file. All larger molecules, including hydrated ions, are blocked by their greater diameter. Fixed positively charged residues in the pore serve as barriers to movement of protons, thereby restricting passage of protons. Aquaglyceroporins have a slightly larger pore size with hydrophobic pore-lining residues permitting passage of glycerol.

Thus, the mysterious process of transcellular water movement occurs through molecular water channels. Together these proteins form a "plumbing system" for cells. Research on the aquaporin family is ongoing to understand their structure and their known cellular and physiological functions in detail. Where defects in aquaporins are implicated in clinical disorders, that understanding will fuel new research into treatments or cures. Research is also being done to identify other cellular and physiological roles of individual aquaporins, as well as to search for new members of the family in humans and in other organisms.

Thinking Critically

Why do you think it is functional that the water channels in the cell membrane formed by aquaporins restrict the passage of protons?

Dr. Peter Agre

Dr. Peter Agre is director of the Johns Hopkins Malaria Research Institute. In 2003 he and Roderick McKinnon of Rockefeller University were awarded the Nobel Prize in Chemistry "for discoveries concerning channels in cell membranes." In 2009 Dr. Agre became president of the American Association for the Advancement of Science. To learn more about his research, visit http://faculty.jhsph.edu/people/director/Peter_Agre.

REVIEW KEY CONCEPTS

Go to **CENGAGENOW** at www.cengage.com/login to access quizzing, animations, exercises, articles, and personalized homework help.

6.1 Membrane Structure and Function

- Both membrane phospholipids and membrane proteins have hydrophobic and hydrophilic regions, giving them dual solubility properties.
- Membranes are based on a fluid phospholipid bilayer, with the polar regions of the phospholipids at the surfaces of the bilayer and their nonpolar tails in the interior (Figures 6.2–6.5).
- Membrane proteins are suspended individually in the bilayer, with their hydrophilic regions at the membrane surfaces and their hydrophobic regions in the interior (Figures 6.4 and 6.5).
- The lipid bilayer forms the structural framework of membranes and is a barrier to the passage of most water-soluble molecules.
- Proteins embedded in the phospholipid bilayer perform most membrane functions, including transport of selected hydrophilic substances, recognition, signal reception, cell adhesion, and metabolism.
- Integral membrane proteins are embedded deeply in the bilayer, whereas peripheral membrane proteins associate with membrane surfaces (Figure 6.5).
- Membranes are asymmetric—different arrangements of membrane lipids and proteins occur in the two bilayer halves.
- Membranes have diverse functions, including defining the boundaries of cells and of internal compartments, acting as permeability

barriers, and facilitating electric signal conduction. Membrane proteins also show diverse activities, acting as enzymes, channel proteins, carrier proteins, receptors, and cell adhesion molecules.

Animation: Lipid bilayer organization

Animation: Cell membranes

6.2 Functions of Membranes in Transport: Passive Transport

- Passive transport depends on diffusion, the net movement of molecules from a region of higher concentration to a region of lower concentration. It does not require cells to expend energy (Table 6.1).
- Simple diffusion is the passive transport of substances across the lipid portion of cellular membranes. It proceeds most rapidly for small molecules that are soluble in lipids (Table 6.1).
- Facilitated diffusion is the diffusion of polar and charged molecules through a membrane aided by transport proteins in the membrane. It follows concentration gradients, is specific for certain substances, and becomes saturated at high concentrations of the transported substance (Figure 6.8 and Table 6.1).
- Most proteins that carry out facilitated diffusion of ions are controlled by "gates" that open or close their transport channels (Figure 6.8).

Interaction: Selective permeability

Animation: Passive transport

6.3 Passive Water Transport and Osmosis

- Osmosis is the net diffusion of water molecules across a selectively permeable membrane in response to differences in the concentration of solute molecules (Figure 6.9). Water moves from hypotonic (lower solute concentrations) to hypertonic solutions (higher solute concentrations). When the solutions on each side are isotonic, net osmotic movement of water ceases (Figure 6.10).

Animation: Solute concentration and osmosis

Interaction: Tonicity and water movement

6.4 Active Transport

- Active transport moves substances against their concentration gradients and requires cells to expend energy. It depends on membrane proteins, is specific for certain substances, and becomes saturated at high concentrations of the transported substance (Table 6.1).
- Active transport proteins are either primary transport pumps, which directly use ATP for energy, or secondary transport pumps, which use favorable concentration gradients of positively charged ions, created by primary transport pumps, as their energy source (Figures 6.11–6.12).
- Secondary active transport may occur by symport, in which the transported substance moves in the same direction as the concentration gradient which provides energy, or by antiport, in which the transported substance moves in the direction opposite to the concentration gradient which provides energy (Figure 6.13).

Animation: Active transport

6.5 Exocytosis and Endocytosis

- Large molecules and particles are moved out of and into cells by exocytosis and endocytosis. The mechanisms allow substances to leave and enter cells without directly passing through the plasma membrane (Figure 6.14).
- In exocytosis, a vesicle carrying secreted materials contacts and fuses with the plasma membrane on its cytoplasmic side. The fused vesicle membrane releases the vesicle contents to the cell exterior (Figure 6.14A).
- In endocytosis, materials on the cell exterior are enclosed in a segment of the plasma membrane that pockets inward and pinches off on the cytoplasmic side as an endocytic vesicle. The two forms of endocytosis are bulk-phase (pinocytosis) and receptor-mediated endocytosis. Most of the materials that enter cells are digested into molecular subunits small enough to be transported across the vesicle membranes (Figures 6.14B–D).

Animation: Phagocytosis

UNDERSTAND AND APPLY

Test Your Knowledge

1. In the fluid mosaic model:
   a. plasma membrane proteins orient their hydrophilic sides toward the internal bilayer.
   b. phospholipids often flip-flop between the inner and outer layers.
   c. the mosaic refers to proteins attached to the underlying cytoskeleton.
   d. the fluid refers to the phospholipid bilayer.
   e. the mosaic refers to the symmetry of the internal membrane proteins and sterols.
2. Which of the following statements is false? Proteins in the plasma membrane can:
   a. transport ions.
   b. transport chloride ions when there are two mutant copies of the cystic fibrosis transmembrane conductance regulator gene.
   c. recognize self versus foreign molecules.
   d. allow adhesion between the same tissue cells or cells of different tissues.
   e. combine with lipids or sugars to form complex macromolecules.
3. The freeze-fracture technique demonstrated:
   a. that the plasma membrane is a bilayer with individual proteins suspended in it.
   b. that the plasma membrane is fluid.
   c. that the arrangement of membrane lipids and proteins is symmetric.
   d. that proteins are bound to the cytoplasmic side but not embedded in the lipid bilayer.
   e. the direction of movement of solutes through the membrane.
4. In the following diagram, assume that the setup was left unattended. Which of the following statements is correct?

Selectively permeable membrane
Inside a cell
Outside fluids
Solvent 95%
Solute 5%
Solvent 98%
Solute 2%

   a. The relation of the cell to its environment is isotonic.
   b. The cell is in a hypertonic environment.
   c. The net flow of solvent is into the cell.
   d. The cell will soon shrink.
   e. Diffusion can occur here but not osmosis.

5. Which of the following statements is true for the diagram in question 4?
   a. The net movement of solutes is into the cell.
   b. There is no concentration gradient.
   c. There is a potential for plasmolysis.
   d. The solvent will move against its concentration gradient.
   e. If this were a plant cell, turgor pressure would be maintained.
6. Using the principle of diffusion, a dialysis machine removes waste solutes from a patient's blood. Imagine blood runs through a cylinder wherein diffusion can occur across an artificial selectively permeable membrane to a saline solution on the other side. Which of the following statements is correct?
   a. Solutes move from lower to higher concentration.
   b. The concentration gradient is lower in the patient's blood than in the saline solution wash.
   c. The solutes are transported through a symport in the blood cell membrane.
   d. The saline solution has a lower concentration gradient of solute than the blood.
   e. The waste solutes are actively transported from the blood.
7. A characteristic of carrier molecules in a primary active transport pump is that:
   a. they cannot transport a substance and also hydrolyze ATP.
   b. they retain their same shape as they perform different roles.
   c. their primary role is to move negatively charged ions across membranes.
   d. they move Na^+ into a neural cell and K^+ out of the same cell.
   e. they act to establish an electrochemical gradient.
8. A driving ion moving through a membrane channel in one direction gives energy to actively transport another molecule in the opposite direction. What is this process called?
   a. facilitated diffusion
   b. exchange diffusion
   c. symport transport
   d. primary active transport pump
   e. cotransport
9. Phagocytosis illustrates which phenomenon?
   a. receptor-mediated endocytosis
   b. bulk-phase endocytosis
   c. exocytosis
   d. pinocytosis
   e. cotransport
10. Place in order the following events of receptor-mediated endocytosis.
    (1) Clathrin coat disappears.
    (2) Receptors collect in a coated pit covered with clathrin on the cytoplasmic side.
    (3) Receptors recognize and bind specific molecules.
    (4) Endocytic vesicle may fuse with lysosome whereas receptors are recycled to the cell surface.
    (5) Pits deepen and pinch free of plasma membrane to form endocytic vesicles.
    a. 4, 1, 2, 5, 3
    b. 2, 1, 3, 5, 4
    c. 3, 2, 5, 1, 4
    d. 4, 1, 5, 2, 3
    e. 3, 1, 2, 4, 5

Discuss the Concepts

1. The bacterium *Vibrio cholerae* causes cholera, a disease characterized by severe diarrhea that may cause infected people to lose up to 20 L of fluid in a day. The bacterium enters the body when someone drinks contaminated water. It adheres to the intestinal lining, where it causes cells of the lining to release sodium and chloride ions. Explain how this release is related to the massive fluid loss.
2. Irrigation is widely used in dryer areas of the United States to support agriculture. In those regions, the water evaporates and leaves behind deposits of salt. What problems might these salt deposits cause for plants?
3. In hospitals, solutions of glucose with a concentration of 0.3 *M* can be introduced directly into the bloodstream of patients without tissue damage by osmotic water movement. The same is true of NaCl solutions, but these must be adjusted to 0.15 *M* to be introduced without damage. Explain why one solution is introduced at 0.3 *M* and the other at 0.15 *M*.

Design an Experiment

Design an experiment to determine the concentration of NaCl (table salt) in water that is isotonic to potato cells. Use only the following materials: a knife, small cookie cutters, and a balance.

Interpret the Data

Some cancer cells are insensitive to typical chemotherapy. Research into the mechanisms underlying this insensitivity uncovered an ability by these cells to "pump" the treatment drug out of the cell against its concentration gradient. Additional drugs have been developed that inhibit the pump, thus trapping the chemotherapeutic agent inside to promote cancer cell destruction.

The graphs below show what happens when two types of cells are treated with a ^3H-labeled anti-cancer drug, paclitaxel.

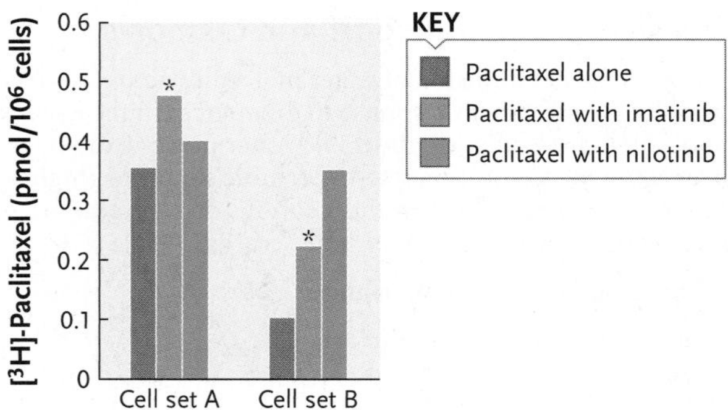
[3H]-Paclitaxel (pmol/10^6 cells)
0.6
0.5
0.4
0.3
0.2
0.1
0
*
*
Cell set A
Cell set B
KEY
Paclitaxel alone
Paclitaxel with imatinib
Paclitaxel with nilotinib

1. Which set of cells (A or B) would be described as resistant to the cancer treatment? Explain your answer. What type of transport are the resistant cells using?
2. Two additional drugs, imatinib and nilotinib, are evaluated for their ability to overcome the cancer cells' ability to "pump out" the chemotherapeutic agent. An * indicates a statistically significant difference from the cells receiving paclitaxel alone. Do the additional drugs seem to be effective in overcoming the pump? Which set of graphs (A or B) best supports your answer? Explain your answer.

Source: Adapted from T. Shen et al. 2009. Imatinib and nilotinib reverse multidrug resistance in cancer cells by inhibiting the efflux activity of the MRP7 (ABCC10). *PLoS ONE* 4(10):e7520. doi:10.1371/journal.pone.0007520.

Apply Evolutionary Thinking

What evidence would convince you that membranes and active transport mechanisms evolved from an ancestor common to both prokaryotes and eukaryotes?

Express Your Opinion

The ability to detect mutant genes that cause severe disorders raises bioethical questions. Should we encourage the mass screening of prospective parents for mutant genes that cause cystic fibrosis? Should society encourage women to give birth only if their child will not develop severe medical problems? Go to academic.cengage.com/login to investigate both sides of the issue and then vote.

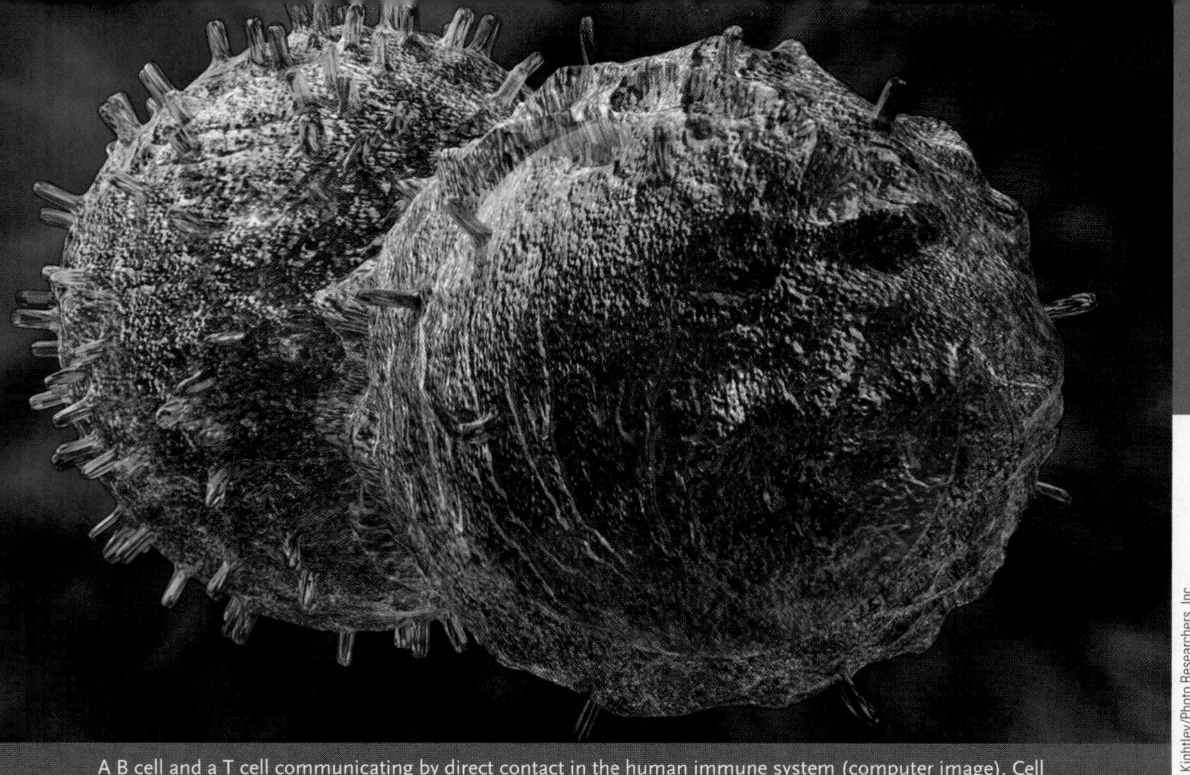
A B cell and a T cell communicating by direct contact in the human immune system (computer image). Cell communication coordinates the cellular defense against disease.

Russell Kightley/Photo Researchers, Inc.

7

Cell Communication

STUDY OUTLINE

Why It Matters. . . Hundreds of aircraft approach and leave airports traveling at various speeds, altitudes, and directions. How are all these aircraft kept separate, and routed to and from their airports safely and efficiently? The answer lies in a highly organized system of controllers, signals, and receivers. As the aircraft arrive and depart, they follow directions issued by air traffic controllers. Although thousands of different messages are traveling through the airspace, each pilot has a radio receiver tuned to a frequency specific for only that aircraft. The flow of directing signals, followed individually by each aircraft in the vicinity, keeps the traffic unscrambled and moving safely.

The principle of the air control system is nothing new. An equivalent system of signals and tuned receivers evolved hundreds of millions of years ago, as one of the developments that made multicellular life possible. Within a multicellular organism, the activities of individual cells are directed by molecular signals such as hormones that are released by certain controlling cells. Although the controlling cells release many signals, each receiving cell—the target cell—has receptors that are "tuned" to recognize only one or a few of the many signal molecules that circulate in its vicinity; other signals pass by without effect because the cell has no receptors for them.

When a target cell binds a signal molecule, it modifies its internal activities in accordance with the signal, coordinating its functions with the activities of other cells of the organism. The responses of the target cell may include changes in gene activity, protein synthesis, transport of molecules across the plasma membrane, metabolic reactions, secretion, movement, division, or even "suicide"—that is, the programmed death of the receiving cell. As part of its response, a cell may itself become a controller by releasing signal molecules that modify the activity of other cell types. A signal molecule triggers a cellular pathway that results in a response in the target cell. The series of steps from signal molecule to response is a *signaling pathway.* The total network of signaling pathways allows multicellular organisms to grow, develop, reproduce, and compensate for environmental changes in an internally coordinated fashion. Maintaining the internal environment within a narrow tolerable range is *homeostasis.* The system of communication between cells through signaling pathways is called **cell signaling.** Research in cell signaling is a highly important field of biology, motivated by the desire to understand the growth, development, and function of organisms.

This chapter describes the major pathways that form parts of the cell communication system based on both surface and internal receptors, including the links that tie the different response pathways into fully integrated networks. (Communication pathways based on neurons—nerve cells—in animals are discussed in Chapter 37.) <

7.1 Cell Communication: An Overview

Cell communication is essential to orchestrate the activities of cells in multicellular organisms, and also takes place among single-celled organisms. In this chapter we focus on the principles of cell communication in animals, and in subsequent chapters you will see how similar principles apply to plants, fungi, and even bacteria and archaea.

Cell Communication in Animals May Involve Nearby or Distant Cells

Communication is critical for the function and survival of cells that compose a multicellular animal. For example, the ability of cells to communicate with one another in a regulated way is responsible for the controlled growth and development of an animal, as well as the integrated activities of its tissues and organs.

Cells communicate with one another in three ways:

1. **By direct contact.** In communication by direct contact, adjacent cells have direct channels linking their cytoplasms. In this rapid means of communication, small molecules and ions exchange directly between the two cytoplasms. In animal cells, the direct channels of communication are *gap junctions,* the specialized connections between the cytoplasms of adjacent cells (see Section 5.5 for a discussion of gap junctions). The main role of gap junctions is to synchronize metabolic activities or electrical signals between cells in a tissue. For example, gap junctions play a key role in spreading electrical signals from one cell to the next in cardiac muscle. In plant cells, the direct channels of communication are plasmodesmata (see discussion in Section 5.4). Small molecules moving between adjacent cells in plants include certain plant hormones that regulate growth (see Chapter 35). In this way, plant hormones are distributed to other cells.

   Cells can also communicate through a process called *cell–cell recognition.* In this process, animal cells with particular membrane-bound cell-surface molecules dock with one another, initiating communication between the cells. For example, cell–cell recognition of this kind activates particular cells in a mammal's immune system in order to mount an immune response (see the figure at the start of this chapter, and Figures 43.5 and 43.10).
2. **By local signaling.** In local signaling, a cell releases a signal molecule that diffuses through the aqueous fluid surrounding and between the cells and causes a response in nearby target cells. Here the effect of cell signaling is local, so the signal molecule is called a *local regulator.* The process is called *paracrine regulation* (see Figure 40.1). In some cases the local regulator acts on the same cell that produces it; this is called *autocrine regulation* (see Figure 40.1). For example, many of the growth factors that regulate cell division are local regulators that act in both a paracrine and autocrine fashion.
3. **By long-distance signaling.** In this form of communication, a controlling cell secretes a long-distance signaling molecule called a **hormone** (*hormaein* = to excite) which produces a response in target cells that may be far from the controlling cell. This method is the most common means of cell communication. Hormones are found in both animals and plants. In animals, hormones secreted by controlling cells enter the circulatory system where they travel to target cells elsewhere in the body. For example, in response to stress, cells of a mammal's adrenal glands (located on top of the kidneys)—the controlling cells—secrete the hormone epinephrine (also known as *adrenaline*), into the bloodstream. Epinephrine acts on target cells to increase the amount of glucose in the blood. In plants, most hormones travel to target cells by moving through cells rather than by moving through vessels. Some plant hormones are gases which diffuse through the air to the target tissues. The actions of plant hormones are discussed in Chapter 35.

Cell communication by long-distance signaling is the focus of this chapter, and we will use the epinephrine example to illustrate the principles involved. In the 1950s, Earl Sutherland and his research team at Case Western Reserve University, Cleveland, Ohio, began investigating this cell communication system. Sutherland wanted to understand how the hormone epinephrine activates the enzyme, *glycogen phosphorylase,* which, in the liver, catalyzes the breakdown of glycogen—a polymer of glucose molecules—into glucose molecules which are then released into the bloodstream. That is, the phosphorylase is inactive in the absence of epinephrine, but active in its presence. The overall effect of this response to epinephrine secretion is to supply energy to the major muscles responsible for locomotion—the body is now ready for physical activity or to handle stress.

Sutherland's key experiments are shown in **Figure 7.1.** He demonstrated that enzyme activation did not involve epinephrine directly but required an unknown (at the time) cellular substance. Sutherland called the hormone the *first messenger* in the system and the unknown cellular substance the *second messenger.* He proposed the following chain of reactions: epinephrine (the first messenger) stimulates the membrane fraction of cell to produce a second messenger molecule, which activates the glycogen phosphorylase for conversion of glycogen to glucose. Later in the chapter we return to this topic and describe the nature and functions of the second messenger molecule. Sutherland was awarded a Nobel Prize in 1971 for his discoveries concerning the mechanisms of the action of hormones.

FIGURE 7.1 **Experimental Research**

Sutherland's Experiments Discovering a Second Messenger Molecule

Question: How does epinephrine activate phosphorylase to break down glycogen into glucose in the liver?

Experiment: Sutherland had shown in one experiment that a homogenate (disrupted cells, consisting of cytoplasm, membranes, and other cell components) would activate glycogen phosphorylase if incubated with epinephrine, ATP, and magnesium ions. Sutherland then set out to learn more about the activation mechanism. In this second experiment, he prepared a liver cell homogenate and then centrifuged it, separating the cytoplasm from membranes and other cell debris.

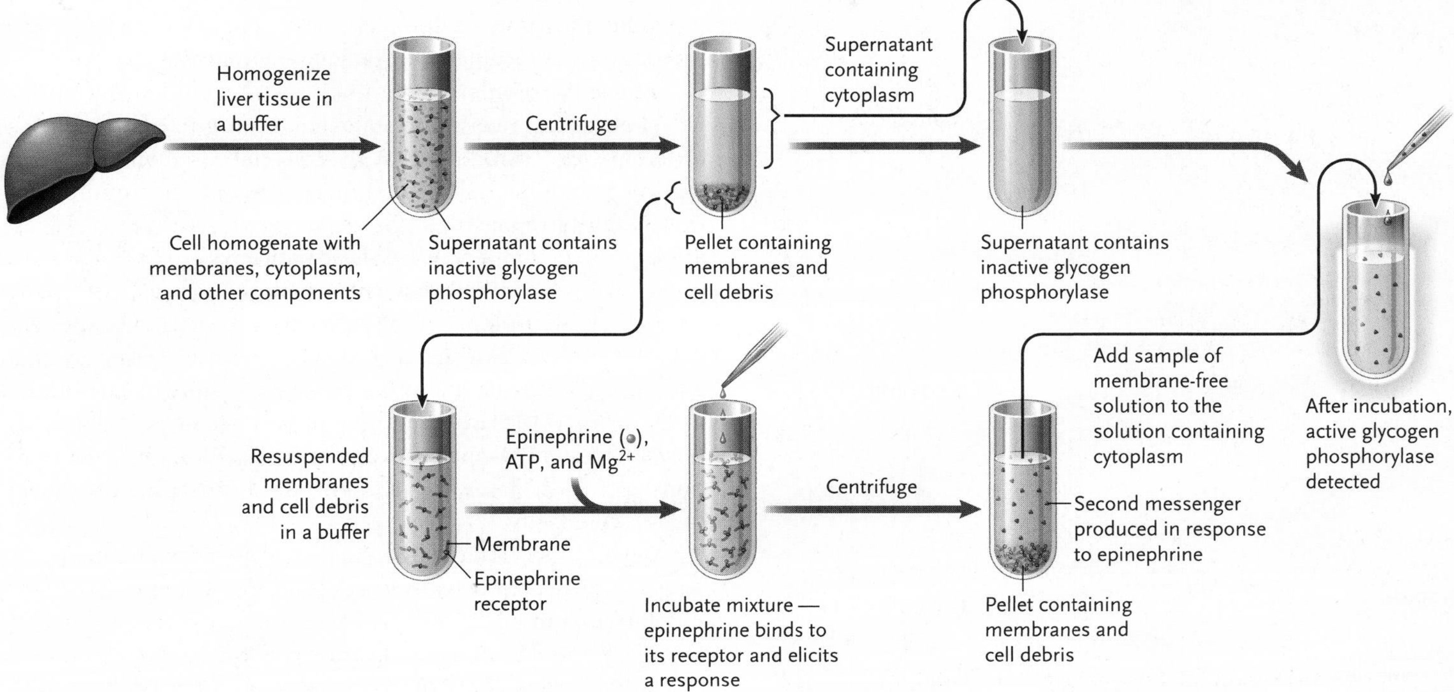

Homogenize liver tissue in a buffer
Centrifuge
Supernatant containing cytoplasm
Cell homogenate with membranes, cytoplasm, and other components
Supernatant contains inactive glycogen phosphorylase
Pellet containing membranes and cell debris
Supernatant contains inactive glycogen phosphorylase
Add sample of membrane-free solution to the solution containing cytoplasm
After incubation, active glycogen phosphorylase detected
Resuspended membranes and cell debris in a buffer
Epinephrine (●), ATP, and Mg2+
Membrane
Epinephrine receptor
Centrifuge
Second messenger produced in response to epinephrine
Incubate mixture — epinephrine binds to its receptor and elicits a response
Pellet containing membranes and cell debris

The cytoplasm was moved to a new tube, and the pellet with the membranes and cell debris was resuspended in a buffer. Neither the cytoplasm nor the membrane fractions had active glycogen phosphorylase. Next, he added epinephrine, ATP, and magnesium ions to the resuspended membranes and incubated the mixture. Centrifuging the mixture pelleted the membranes to the bottom of the tube. A sample of the supernatant (the membrane-free solution above the pellet) was added to the solution containing cytoplasm and the mixture was incubated.

Result: Active glycogen phosphorylase was detected in the mixture.

Conclusion: Sutherland had shown that the response to the hormone epinephrine—the activation of glycogen phosphorylase—does not involve epinephrine directly, but requires another cellular factor. He named the factor the *second messenger,* with the hormone itself being the *first messenger.*

Source: T. W. Rall, E. W. Sutherland, and J. Berthet. 1957. The relationship of epinephrine and glucagon to liver phosphorylase. IV. Effect of epinephrine and glucagon on the reactivation of phosphorylase in liver homogenates. *Journal of Biological Chemistry* 224:463–475.

Sutherland's discovery was critical to understanding the mechanism of action of epinephrine and, in fact, of many other hormones. His work also illustrates how this type of long-distance cell signaling operates: a controlling cell releases a signal molecule that causes a response (affects the function) in target cells. Target cells process the signal in the following three sequential steps **(Figure 7.2)**:

1. **Reception** Reception is the binding of a signal molecule with a specific receptor of target cells. Target cells have receptors that are specific for the signal molecule, which distinguishes them from cells that do not respond to the signal molecule. The signals themselves may be polar (charged, hydrophilic) molecules or nonpolar (hydrophobic) molecules, and their receptors are shaped to recognize and bind them specifically. Receptors for polar signal molecules are embedded in the plasma membrane with a binding site for the signal molecule on the cell surface (see Figure 7.2). Epinephrine, the first messenger in Sutherland's research, is a peptide hormone, a polar molecule that is recognized by a surface receptor embedded in the plasma membrane of target cells. Receptors for nonpolar molecules are located within the cell (described later in the chapter; see Figure 7.13). In this case, the nonpolar signal molecule passes freely through the

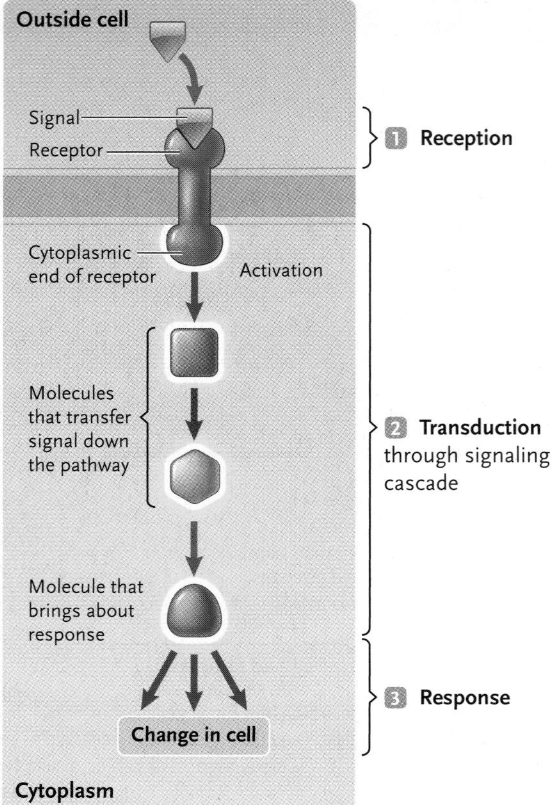
Outside cell
Signal
Receptor
1 Reception
Cytoplasmic end of receptor
Activation
Molecules that transfer signal down the pathway
2 Transduction through signaling cascade
Molecule that brings about response
3 Response
Change in cell
Cytoplasm

FIGURE 7.2
The three stages of signal transduction: **reception**, **transduction**, and **response** (shown for a signal transduction system using a surface receptor).

plasma membrane and interacts with its receptor within the cell. Steroid hormones such as testosterone and estrogen are examples of nonpolar signal molecules.

2. **Transduction** Transduction is the process of changing the signal into the form necessary to cause the cellular response (see Figure 7.2). The initial signal binds to and activates the receptor, changing it to a form that initiates transduction. Transduction typically involves a cascade of reactions that include several different molecules, referred to as a *signaling cascade.* For example, with respect to Sutherland's work, after epinephrine binds to its surface receptor, the signal is transmitted through the plasma membrane into the cell to another protein which, in turn, causes the production of numerous small second messenger molecules. As we shall see later, both proteins and second messengers can be part of the signaling cascade that results in triggering a cellular response.
3. **Response** In the third and last stage, the transduced signal causes a specific cellular response. That response depends on the signal and the receptors on the target cell. In Sutherland's work, the response was the activation of the enzyme glycogen phosphorylase. The active enzyme catalyzes the conversion of stored glycogen to glucose, which is the response to the signal delivered by epinephrine.

The sequence of reception, transduction, and response is common to all the signaling systems we will encounter in this chapter, although they vary greatly in detail.

Cell Communication Is Evolutionarily Ancient

A number of cell communication properties are ancient, evolutionarily speaking. That is, mechanisms for one cell to signal to another cell and elicit a response most likely existed in unicellular organisms prior to the evolution of multicellularity. For instance, research with present-day bacteria has shown that a number of species alter their patterns of gene expression in response to changes in population density. In this process of *quorum sensing,* bacteria release signal molecules in increasing concentration as cell density increases. The molecules are sensed by the cells in the population, and each cell then responds to adapt to the changing environment.

In the unicellular eukaryote, yeast, sexual mating begins when one cell secretes a hormone that is recognized by a cell of a different "sex," signaling that the two cells are compatible for mating. In multicellular eukaryotes complex cell communication pathways coordinate the activities of multiple cell types. Some protein components of these pathways are found in both prokaryotes and eukaryotes, indicating they are evolutionarily ancient. Other proteins in the pathways appeared only after eukaryotes evolved. For instance, protein kinases—enzymes that add phosphate groups to other proteins to control their activity—are one of the largest families of proteins in eukaryotes, yet they are absent in prokaryotes. Scientists believe that the evolution of protein kinases was an important step in the development of multicellularity.

Beyond individual components, entire cell signaling pathways are conserved between organisms. For example, one pathway for cell growth control is conserved between the fruit fly, *Drosophila,* and humans, indicating that the pathway is at least 800 million years old. In short, the principles of cell communication are similar in unicellular and multicellular organisms, and some components are shared between them, but there is no single evolutionary root for the pathways involved. Many of the examples in this chapter focus on the systems working in animals, particularly in mammals, from which most of our knowledge of cell communication has been developed. (The plant communication and control systems are described in more detail in Chapter 35.) This discussion begins with a few fundamental principles that underlie the often complex networks of cell communication.

STUDY BREAK 7.1 <

What accounts for the specificity of a cellular response to a signal molecule?

7.2 Cell Communication Systems with Surface Receptors

Cell communication systems using surface receptors have three components: (1) the extracellular signal molecules released by controlling cells; (2) the surface receptors on target cells that receive the signals; and (3) the internal response pathways triggered when receptors bind a signal.

Peptide Hormones and Neurotransmitters Are Extracellular Signal Molecules Recognized by Surface Receptors in Animals

Surface receptors in mammals and other vertebrates recognize and bind two major types of extracellular signal molecules: *peptide hormones* and *neurotransmitters.* These signal molecules are polar, water-soluble molecules that are released by controlling cells and enter the fluids surrounding and between cells, and then into the blood circulation (in animals with a circulatory system).

PEPTIDE HORMONES Peptide hormones are small proteins with a few to more than 200 amino acids. As a group, they affect all body systems. For example, they regulate sugar levels in blood, pigmentation, and ovulation. A special class of peptide hormones, the *growth factors,* affects cell growth, division, and differentiation.

Cells that release peptide hormones are called gland cells. They may form part of distinct, individual organs such as the thyroid or pituitary gland, or they may be distributed among the cells of organs with other functions, such as the stomach and intestines, heart, brain, liver, and kidneys in humans and other mammals. For example, gland cells scattered through the lining of the human stomach and small intestine secrete peptide hormones that regulate digestive functions. (Peptide hormones and growth factors are discussed in further detail in Chapter 40.)

NEUROTRANSMITTERS Neurotransmitters are molecules released by neurons that trigger activity in other neurons or other cells in the body; they include small peptides, individual amino acids or their derivatives, and other chemical substances. Some neurotransmitters affect only one or a few cells in the immediate vicinity of the neuron that releases the signal molecule, whereas others are released into the body circulation and act essentially as hormones, affecting many types of tissues. (Neurotransmitters are discussed in further detail in Chapter 37.)

Once signal molecules are released into the body's circulation, they remain for only a certain time. They are either broken down at a steady rate by enzymes in their target cells or in organs such as the liver, or they are excreted by the kidneys. The removal process ensures that the signal molecules are active only as long as controlling cells are secreting them.

Surface Receptors Are Integral Membrane Glycoproteins

The surface receptors that recognize and bind signal molecules are all glycoproteins—proteins with attached carbohydrate chains (see Section 3.4). They are integral membrane proteins that extend entirely through the plasma membrane **(Figure 7.3A).** Typically the signal-binding site of the receptor is the part of the protein that extends from the outer membrane surface, and which is folded in a way that closely fits the signal molecule. The fit, similar to the fit of an enzyme to its substrate, is specific, so a particular receptor binds only one type of signal molecule or a closely related group of signal molecules.

A signal molecule brings about specific changes in cells to which it binds. When a signal molecule binds to a surface receptor, the molecular structure of that receptor changes so that it transmits the signal through the plasma membrane, activating the cytoplasmic end of the receptor. The activated receptor then initiates the first step in a cascade of molecular events—the signaling cascade—that triggers the cellular response **(Figure 7.3B).**

Animal cells typically have hundreds to thousands of surface receptors that represent many receptor types. Receptors for a specific peptide hormone may number from 500 to as many as 100,000 or more per cell. Different cell types contain distinct combinations of receptors, allowing them to react individually to the hormones and growth factors circulating in the extracellular fluids. The combination of surface receptors on particular cell types is not fixed but rather changes as cells develop. Changes also occur as normal cells are transformed into cancer cells.

The Signaling Molecule Bound by a Surface Receptor Triggers Response Pathways within the Cell

Signal transduction pathways triggered by surface receptors are common to all animal cells. At least parts of the pathways are also found in protists, fungi, and plants. In all cases, binding of a signal molecule to a surface receptor is sufficient to trigger the cellular response—the signal molecule does not enter the cell. For example, experiments have shown that: (1) a signal molecule produces no response if it is injected directly into the cytoplasm; and (2) unrelated molecules that mimic the structure of the normal extracellular signal molecule can trigger or block a full cellular response as long as they can bind to the recognition site of the receptor. In fact, many

A. Surface receptor

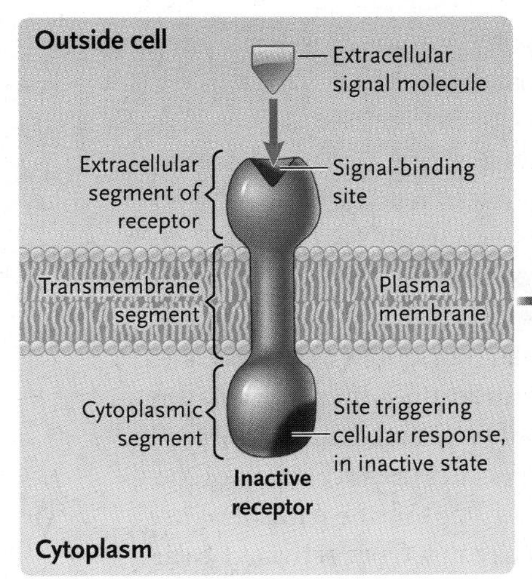
Outside cell
Extracellular signal molecule
Extracellular segment of receptor
Signal-binding site
Transmembrane segment
Plasma membrane
Cytoplasmic segment
Site triggering cellular response, in inactive state
Inactive receptor
Cytoplasm

A surface receptor has an extracellular segment with a site that recognizes and binds a particular signal molecule.

B. Activation of receptor by binding of a specific signal molecule

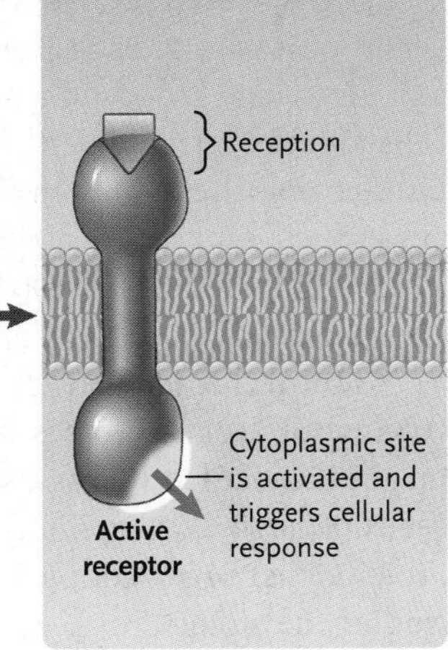
Reception
Active receptor
Cytoplasmic site is activated and triggers cellular response

When the signal molecule is bound, a conformational change is transmitted through the trans-membrane segment that activates a site on the cytoplasmic segment of receptor. The activation triggers a reaction pathway that results in the cellular response.

FIGURE 7.3
The mechanism by which a surface receptor responds when it binds a signal molecule.

medical conditions are treated with drugs that are signal molecule mimics. For example, the drug famotidine, sold under the name Pepcid AC, decreases stomach acid release by selectively binding and inhibiting a type of receptor on stomach cells. Propranolol, a beta-blocker drug, inhibits β-adrenergic receptors involved in contractions of certain muscles, and is used to reduce the strength of cardiac contractions and to reduce blood pressure. Isoproterenol binds to the same receptors but activates them, and is used to stimulate the heart or, as an inhaled aerosol to treat asthma.

Another typical characteristic of signal transduction involving surface receptors is that the signal is relayed inside the cell by **protein kinases,** enzymes that transfer a phosphate group from ATP to one or more sites on particular proteins (see Section 4.5). As shown in **Figure 7.4,** protein kinases often act in a chain catalyzing a series of phosphorylation reactions called a *phosphorylation cascade,* to pass along a signal. The first kinase catalyzes phosphorylation of the second, which then becomes active and phosphorylates the third kinase, which then becomes active, and so on. The last protein in the cascade is the *target protein.* Phosphorylation of a target protein stimulates or inhibits its activity depending on the particular protein. This change in activity brings about the cellular response. For example, phosphorylating a target protein that regulates whether a set of genes are turned on or off could cause cells to start or stop producing the proteins the genes encode. The change in this set of proteins then causes a response related to the functions of those proteins.

The effects of protein kinases in the signal transduction pathways are balanced or reversed by another group of enzymes called **protein phosphatases,** which remove phosphate groups from target proteins. Unlike the protein kinases, which are active only when a surface receptor binds a signal molecule, most of the protein phosphatases are continuously active in cells. By continually removing phosphate groups from target proteins, the protein phosphatases quickly shut off a signal transduction pathway if its signal molecule is no longer bound at the cell surface.

Two scientists, Edwin Krebs and Edmond Fischer at the University of Washington, Seattle, first discovered that protein kinases add phosphate groups to control the activities of key proteins in cells and obtained evidence showing that protein phosphatases reverse these phosphorylations. Krebs and Fischer, who began their experiments in the 1950s, received a Nobel Prize in 1992 for their discoveries concerning reversible protein phosphorylation.

A third characteristic of signal transduction pathways involving surface receptors is **amplification**—an increase in the magnitude of each step as a signal transduction pathway proceeds **(Figure 7.5).** Amplification occurs because many of the proteins that carry out individual steps in the pathways, including the protein kinases, are enzymes. Once activated, each enzyme can activate hundreds of proteins including other enzymes that enter the next step in the pathway. Generally, the more enzyme-catalyzed steps in a response pathway, the greater the amplification. As a result, just a few extracellular signal molecules binding to their receptors can produce a full internal response. For similar reasons, amplification also occurs for signal transduction pathways that involve internal receptors.

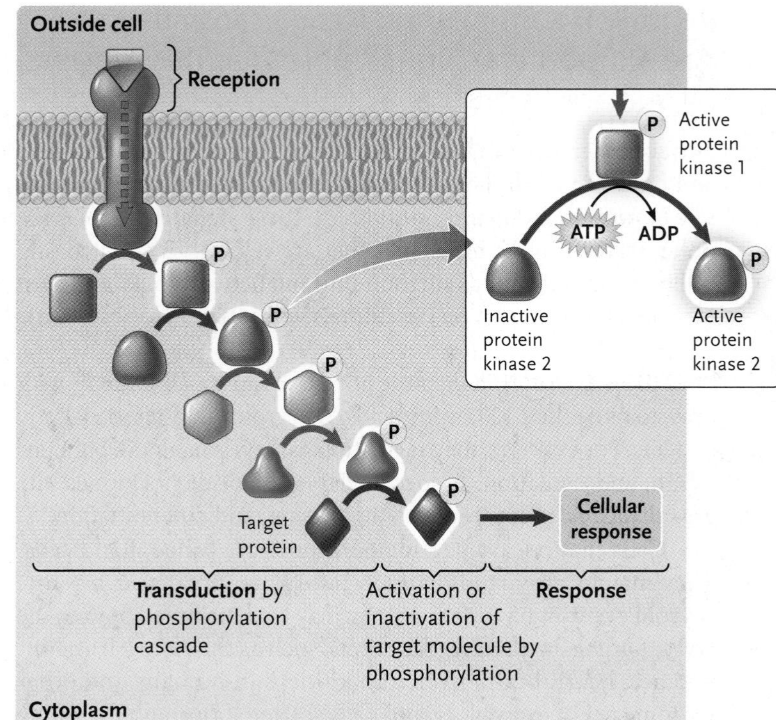

Outside cell
Reception
P
Active protein kinase 1
ATP
ADP
Inactive protein kinase 2
Active protein kinase 2
Target protein
Cellular response
Transduction by phosphorylation cascade
Activation or inactivation of target molecule by phosphorylation
Response
Cytoplasm

FIGURE 7.4
Phosphorylation, a key reaction in many signaling pathways.

As signal transduction runs its course, the receptors and their bound signal molecules are removed from the cell surface by endocytosis (see Section 6.5). Both the receptor and its bound signal molecule may be degraded in lysosomes after entering the cell. Alternatively, the receptors may be separated from the signal molecules and recycled to the cell surface, whereas the signal molecules are degraded. Thus, surface receptors participate in an extremely lively cellular "conversation" with moment-to-moment shifts in the information.

FIGURE 7.5
Amplification in signal transduction.

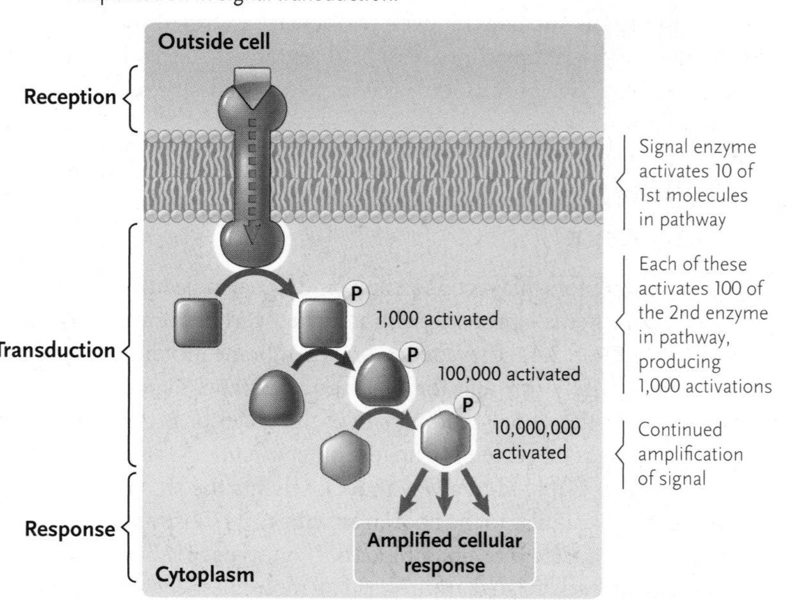

Outside cell
Reception
Transduction
Response
Signal enzyme activates 10 of 1st molecules in pathway
Each of these activates 100 of the 2nd enzyme in pathway, producing 1,000 activations
Continued amplification of signal
P
1,000 activated
100,000 activated
10,000,000 activated
Amplified cellular response
Cytoplasm

Next, you will see how the three hallmarks of surface receptor pathways (surface receptor, kinase cascade, amplification) play out in two large surface receptor families: the receptor tyrosine kinases and the G-protein–coupled receptors.

STUDY BREAK 7.2 <

1. What are protein kinases, and how are they involved in signal transduction pathways?
2. How is amplification accomplished in a signal transduction pathway?

7.3 Surface Receptors with Built-In Protein Kinase Activity: Receptor Tyrosine Kinases

One major type of surface receptors, the **receptor tyrosine kinases,** have their own protein kinase activity on the cytoplasmic end of the protein. Binding of a signal molecule to this type of receptor turns on the receptor's built-in protein kinase which leads to activation of the receptor. The activated receptor then initiates a signaling cascade which results in a cellular response.

For this type of receptor, initiation of transduction occurs when two receptor molecules each bind a signal molecule in the reception step, move together in the membrane, and assemble into a pair called a *dimer* **(Figure 7.6A).** The protein kinases of each receptor monomer are activated by dimer formation and they phosphorylate the partner monomer in the dimer, a process called *autophosphorylation* **(Figure 7.6B).** The phosphorylation is of tyrosine amino acids, which gives this type of receptors their name. The multiple phosphorylations activate many different sites on the dimer. When a signaling protein binds to an activated site, it initiates a transduction pathway leading to a cellular response **(Figure 7.6C).** Since different receptor tyrosine kinases bind different combinations of signaling proteins, the receptors initiate different responses.

Receptor tyrosine kinases are found in all multicellular animals, but not in plants or fungi. Fifty eight genes in the human genome encode receptor tyrosine kinases. In mammals, more generally, receptor tyrosine kinases fall into about 20 different families, all related to one another in structure and amino acid sequence. The cellular responses triggered by receptor tyrosine

A. Signal molecules bind and two receptors assemble into a dimer

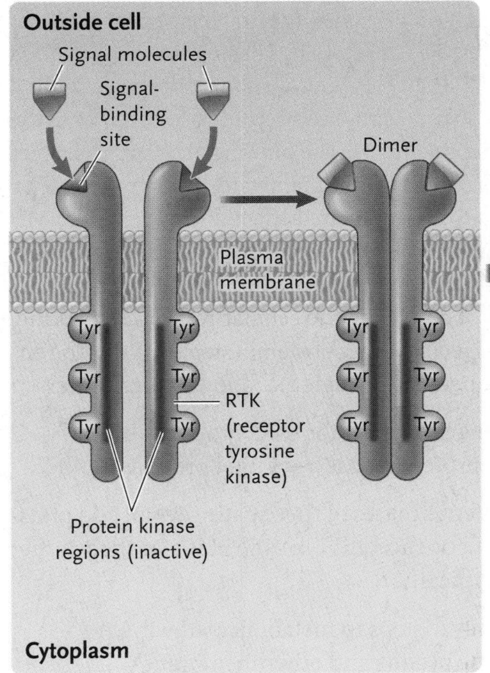
Outside cell
Signal molecules
Signal-binding site
Dimer
Plasma membrane
Tyr
RTK (receptor tyrosine kinase)
Protein kinase regions (inactive)
Cytoplasm

When no signal molecules are bound, the receptors are distributed singly in the plasma membrane and their protein kinase sites on the cytoplasmic segment are inactive. Binding a signal molecule causes the receptors to assemble into dimers.

B. Activation of protein kinases and autophosphorylation of the receptor

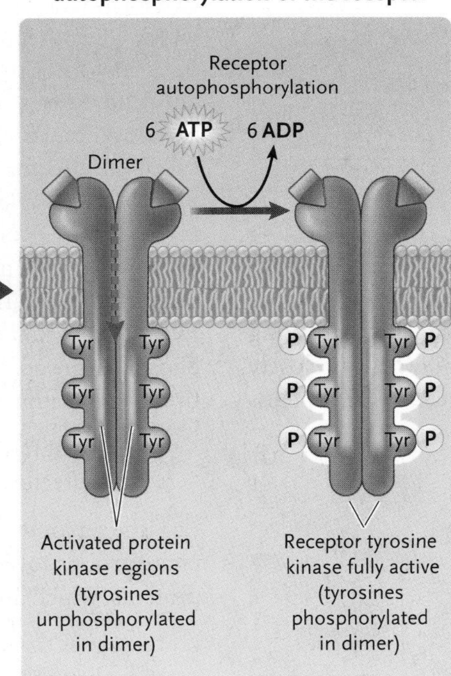
Receptor autophosphorylation
6 ATP
6 ADP
Dimer
Tyr
P
Activated protein kinase regions (tyrosines unphosphorylated in dimer)
Receptor tyrosine kinase fully active (tyrosines phosphorylated in dimer)

Conformational changes induced by the binding and pairing activate the protein kinases on cytoplasmic segments of the receptor monomers, which leads to phosphorylation of tyrosine amino acids in each monomer.

C. Transduction and cellular responses

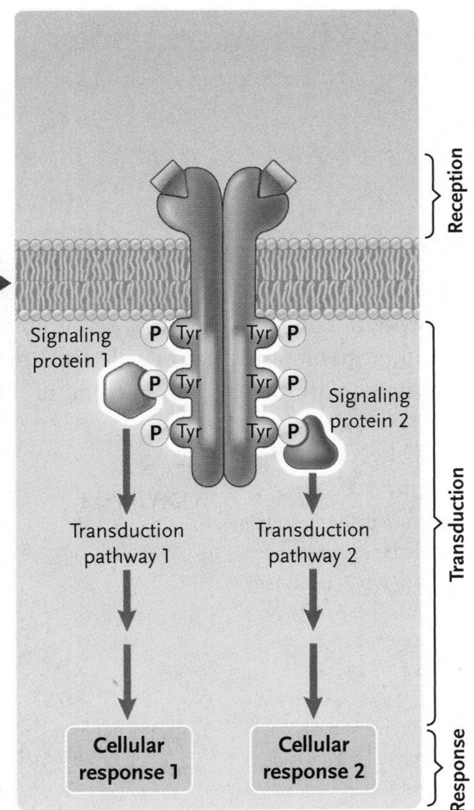
Reception
Signaling protein 1
Signaling protein 2
Tyr
P
Transduction pathway 1
Transduction pathway 2
Transduction
Cellular response 1
Cellular response 2
Response

Signaling proteins bind to the activated receptor and become activated. Each signaling protein initiates a transduction pathway that produces a cellular response.

FIGURE 7.6
The action of a receptor tyrosine kinase, a receptor type with built-in protein kinase activity. Binding of a signal molecule causes receptor monomers to form a dimer which becomes active by autophosphorylation, in this case by protein kinase regions of each monomer phosphorylating particular tyrosines on the other monomer. Signaling proteins bind to the activated receptor and become activated, each initiating a transduction pathway which produces a cellular response.

kinases are among the most important processes of animal cells. For example, the receptor tyrosine kinases binding the peptide hormone *insulin,* a regulator of carbohydrate metabolism, triggers diverse cellular responses, including effects on glucose uptake, the rates of many metabolic reactions, and cell growth and division. (The insulin receptor is exceptional because it is permanently in a tetrameric {four-monomer} form.) Other receptor tyrosine kinases bind growth factors, including *epidermal growth factor, platelet-derived growth factor,* and *nerve growth factor,* which are all important peptide hormones that regulate cell growth and division in higher animals.

Hereditary defects in the insulin receptor are responsible for some forms of *diabetes,* a disease in which glucose accumulates in the blood because it cannot be absorbed in sufficient quantity by body cells. The cells with faulty receptors do not respond to insulin's signal to add glucose receptors to take up glucose. (The role of insulin in glucose metabolism and diabetes is discussed further in Chapter 40.)

STUDY BREAK 7.3

1. **How does a receptor tyrosine kinase become activated?**
2. **Once fully activated, how does a receptor tyrosine kinase bring about a cellular response?**

THINK OUTSIDE THE BOOK

Using a specific example, outline how an alteration in a receptor tyrosine kinase can contribute to the development of a cancer.

7.4 G-Protein–Coupled Receptors

A second large family of surface receptors, known as the **G-protein–coupled receptors,** respond to a signal by activating an inner membrane protein called a G protein, which is closely associated with the cytoplasmic end of the receptor. G proteins are so named because they bind the guanine nucleotides GDP (guanosine diphosphate) and GTP (guanosine triphosphate). G-protein–coupled receptors are found in animals (both multicellular and unicellular forms), plants, fungi, and certain protists. Researchers have identified thousands of different G-protein–coupled receptors in mammals, including thousands involved in recognizing and binding odor molecules as part of the mammalian sense of smell, light-activated receptors in the eye, and many receptors for hormones and neurotransmitters. Almost all of the receptors of this group are large glycoproteins built up from a single polypeptide chain anchored in the plasma membrane by seven segments of the amino acid chain that zigzag back and forth across the membrane seven times **(Figure 7.7).**

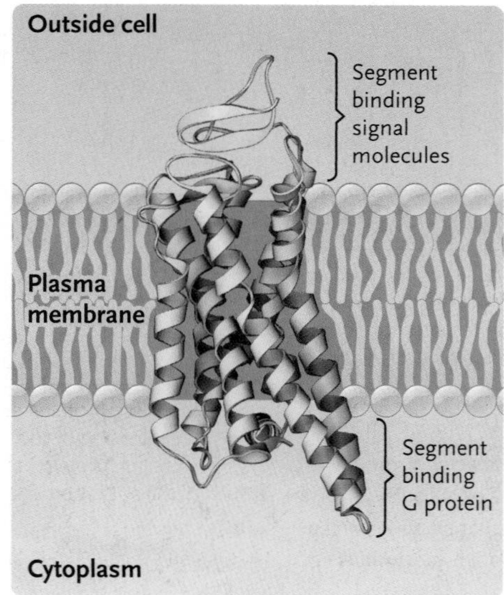
Outside cell
Segment binding signal molecules
Plasma membrane
Segment binding G protein
Cytoplasm

FIGURE 7.7
Structure of the G-protein–coupled receptors, which activate separate protein kinases. These receptors have seven transmembrane α-helical segments that zigzag across the plasma membrane. Binding of a signal molecule at the cell surface, by inducing changes in the positions of some of the helices, activates the cytoplasmic end of the receptor.

Unlike receptor tyrosine kinases, G-protein–coupled receptors lack built-in protein kinase activity.

G Proteins Are Key Molecular Switches in Second-Messenger Pathways

In signal transduction pathways controlled by G-protein–coupled receptors, the extracellular signal molecule is called the **first messenger.** The binding of the first messenger to the receptor activates it (**Figure 7.8,** step 1). Coupled to the receptor is a G protein, which is called a molecular switch because it switches between an inactive form with GDP bound to it (step 2), and an active form in which GDP is replaced by GTP. When the first messenger activates the receptor, it activates the G protein by causing it to release GDP and bind GTP (step 3). The GTP-bound subunit of the G-protein breaks off and binds to a plasma membrane-associated enzyme called the **effector** (steps 4 and 5). The binding activates the effector and then the G-protein subunit inactivates itself by hydrolyzing GTP to GDP (step 6). The activated effector now generates one or more internal, nonprotein signal molecules called **second messengers** (step 7). The second messengers directly or indirectly activate protein kinases, which elicit the cellular response by adding phosphate groups to specific target proteins (step 8). Thus, the entire control pathway operates through the following sequence:

first messenger → receptor → G proteins →
effector → protein kinases → target proteins

The separate protein kinases of these pathways all add phosphate groups to serine or threonine amino acids in their target proteins, which are typically:

- enzymes that catalyze steps in metabolic pathways.
- ion channels in the plasma and other membranes.
- regulatory proteins that control gene activity and cell division.

The pathway from first messengers to target proteins is common to all G-protein–coupled receptors.

As long as a G-protein–coupled receptor is bound to a first messenger, the receptor keeps the G protein active. The activated G protein, in turn, keeps the effector active in generating second messengers. If the first messenger is released from the receptor, or if the receptor is taken into the cell by endocytosis, GTP is hydrolyzed to GDP, which inactivates the G protein. As a result, the effector becomes inactive, turning "off" the response pathway.

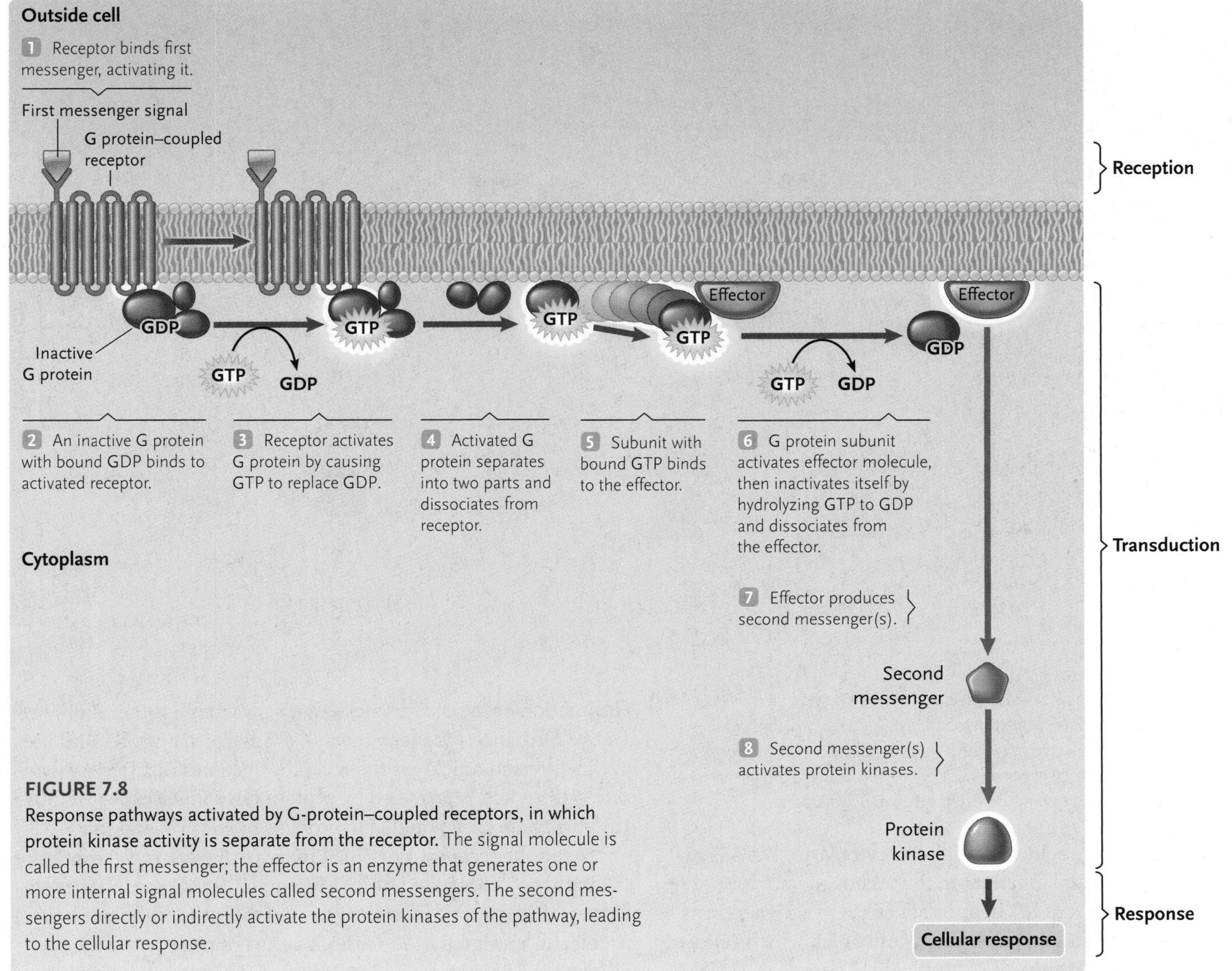
Outside cell
1 Receptor binds first messenger, activating it.
First messenger signal
G protein–coupled receptor
Reception
Effector
Effector
GDP
GTP
GTP
GTP
GDP
Inactive G protein
GTP
GDP
GTP
GDP
2 An inactive G protein with bound GDP binds to activated receptor.
3 Receptor activates G protein by causing GTP to replace GDP.
4 Activated G protein separates into two parts and dissociates from receptor.
5 Subunit with bound GTP binds to the effector.
6 G protein subunit activates effector molecule, then inactivates itself by hydrolyzing GTP to GDP and dissociates from the effector.
Transduction
Cytoplasm
7 Effector produces second messenger(s).
Second messenger
8 Second messenger(s) activates protein kinases.
Protein kinase
Response
Cellular response

FIGURE 7.8

Response pathways activated by G-protein–coupled receptors, in which protein kinase activity is separate from the receptor. The signal molecule is called the first messenger; the effector is an enzyme that generates one or more internal signal molecules called second messengers. The second messengers directly or indirectly activate the protein kinases of the pathway, leading to the cellular response.

Cells can make a variety of G proteins, with each type activating a different cellular response. Alfred G. Gilman at the University of Virginia, Charlottesville, and Martin Rodbell at the National Institutes of Health, Bethesda, Maryland, received a Nobel Prize in 1994 for their discovery of G proteins and the role of these proteins in signal transduction in cells.

The importance of G proteins to cellular metabolism is underscored by the fact that they are targets of toxins released by some infecting bacteria. The cholera toxin produced by *Vibrio cholerae,* the pertussis toxin (which causes whooping cough)produced by *Bordetella pertussis,* and a toxin produced by a disease-causing strain of *Escherichia coli* are all enzymes that modify the G proteins, making them continuously active and keeping their response pathways turned "on" at high levels. For example, the cholera toxin prevents a G protein from hydrolyzing GTP, keeping the G protein switched on and the pathway in a permanently active state. Among other effects, the pathway opens ion channels in intestinal cells, causing severe diarrhea through a massive release of salt and water from the body into the intestinal tract. Unless the resulting dehydration of the body is relieved, death can result quickly. The *E. coli* toxin, which has similar but milder effects, is the cause of many cases of traveler's diarrhea.

Two Major G-Protein–Coupled Receptor–Response Pathways Involve Different Second Messengers

Activated G proteins bring about a cellular response through two major receptor–response pathways in which different effectors generate different second messengers. One pathway involves the second messenger **cyclic AMP (cAMP**—cyclic 3′,5′-adenosine monophosphate), a relatively small, water-soluble molecule derived from ATP **(Figure 7.9).** The effector that produces cAMP is the enzyme *adenylyl cyclase,* which converts ATP to cAMP **(Figure 7.10).** cAMP diffuses through the cytoplasm and activates protein kinases that add phosphate groups to target proteins. The other pathway involves two second messengers: **inositol triphosphate (IP_3)** and **diacylglycerol (DAG).** The effector of this pathway, an enzyme called *phospholipase C,* produces both of these second messengers by breaking down a membrane phospholipid **(Figure 7.11).** IP_3 is a small, water-soluble

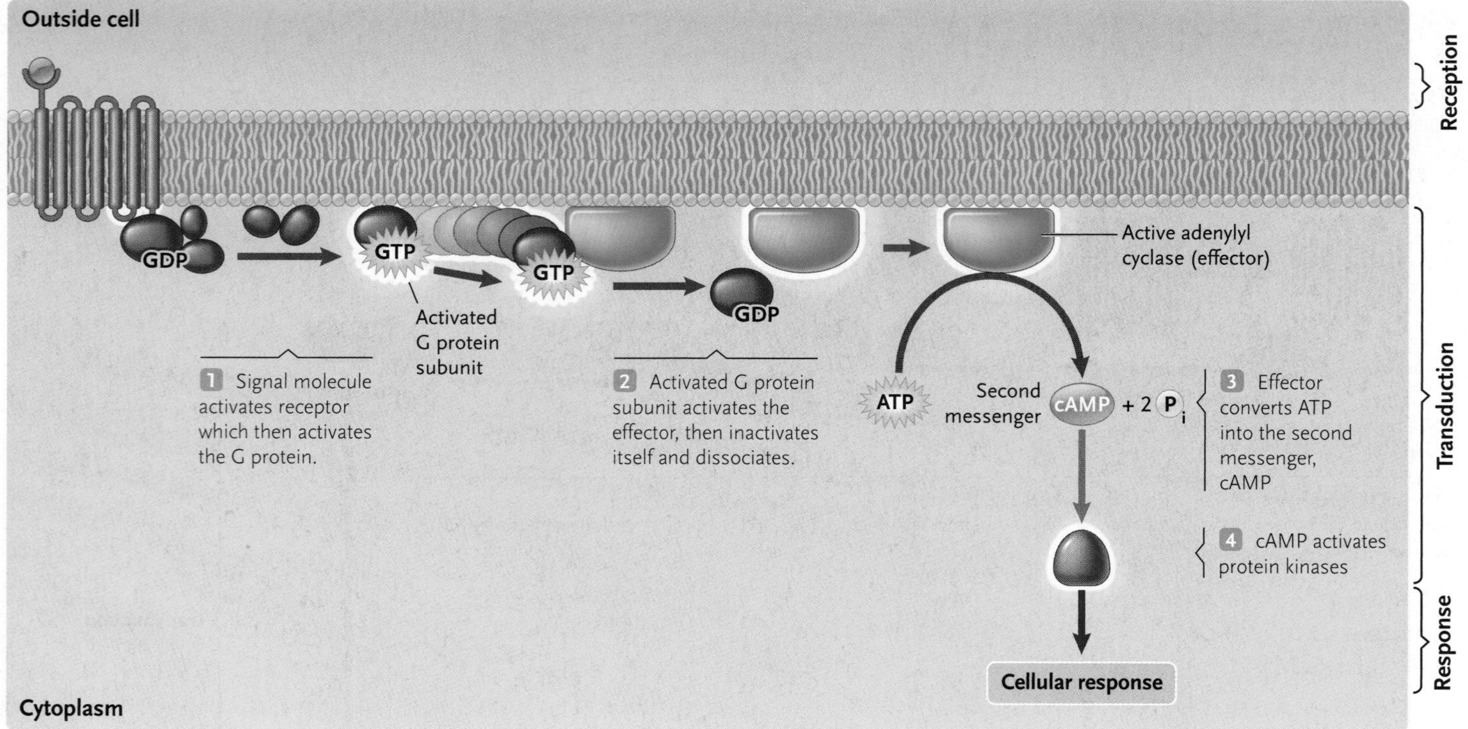
Outside cell
GDP
GTP
GTP
GDP
Activated G protein subunit
Active adenylyl cyclase (effector)
1 Signal molecule activates receptor which then activates the G protein.
2 Activated G protein subunit activates the effector, then inactivates itself and dissociates.
ATP
Second messenger
cAMP + 2 P_i
3 Effector converts ATP into the second messenger, cAMP
4 cAMP activates protein kinases
Cellular response
Cytoplasm
Reception
Transduction
Response

FIGURE 7.9
The operation of cAMP receptor–response pathways. Reception occurs as shown in Figure 7.8, leading to activation of adenylyl cyclase, which produces the second messenger of the pathway, cAMP. cAMP activates one or more cAMP-dependent protein kinases for a phosphorylation cascade which, at its last step, activates a target protein that causes the cellular response.

molecule that diffuses rapidly through the cytoplasm. DAG is hydrophobic; it remains and functions in the plasma membrane.

The primary effect of IP_3 in animal cells is to activate transport proteins in the endoplasmic reticulum (ER), which release Ca^{2+} stored in the ER into the cytoplasm. The released Ca^{2+}, either alone or in combination with DAG, activates a protein kinase cascade that brings about the cellular effect. Techniques designed to detect Ca^{2+} release inside cells are among the most important tools of researchers studying cell signaling (see *Focus on Basic Research* for a description of two of these techniques.)

Both major G-protein–coupled receptor–response pathways are balanced by reactions that constantly eliminate their second messengers. For example, cAMP is quickly converted to AMP (5′-adenosine monophosphate) by *phosphodiesterase,* an enzyme that is continuously active in the cytoplasm (see Figure 7.10). The rapid elimination of the second messengers provides another highly effective off switch for the pathways, ensuring that protein kinases are inactivated quickly if the receptor becomes inactive. Still another off switch is provided by protein phosphatases that remove the phosphate groups added to proteins by the protein kinases.

As in the receptor tyrosine kinase pathways, the activities of the pathways controlled by cAMP and IP_3/DAG second messengers are also stopped by endocytosis of receptors and their bound extracellular signals. As with all cell signaling pathways, cells vary in their response to cAMP or IP_3/DAG pathways depending on the type of G-protein–coupled receptors on the cell surface and the kinds of protein kinases present in the cytoplasm.

The cAMP second messenger pathway is found in animals and fungi. In plants, cAMP may be involved in germination and in some plant defensive responses (see Chapter 35), although the pathways are not well understood. The IP_3/DAG second messenger pathway is universally distributed among eukaryotic organisms, including both vertebrate and invertebrate animals, fungi, and plants. In plants, IP_3 releases Ca^{2+} primarily from the large central vacuole rather than from the ER.

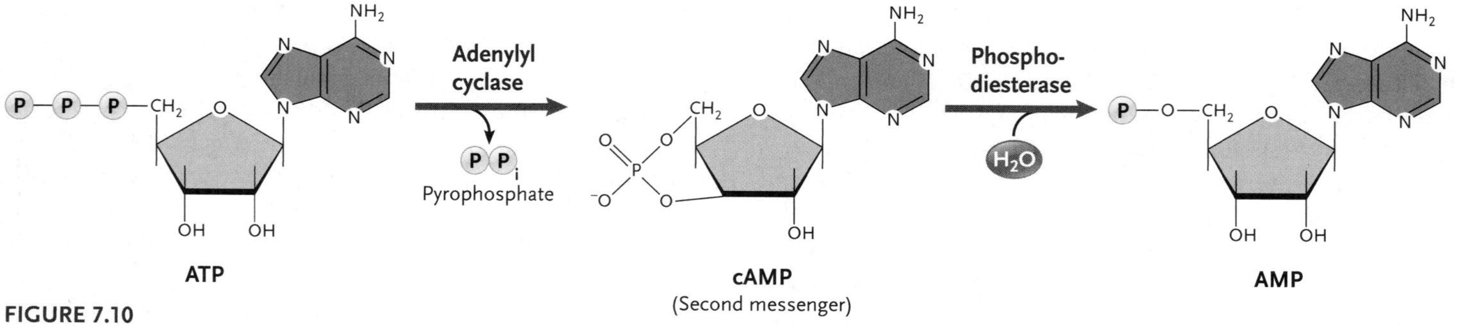

ATP — Adenylyl cyclase → (Pyrophosphate, PP_i) — cAMP (Second messenger) — Phosphodiesterase (H_2O) → AMP

FIGURE 7.10
cAMP. The second messenger, cAMP, is made from ATP by adenylyl cyclase and is broken down to AMP by phosphodiesterase.

FOCUS ON **Basic Research**

Detecting Calcium Release in Cells

Because calcium ions are used as a control element in all eukaryotic cells, it was important to develop techniques for detecting Ca^{2+} when it is released into the cytosol. One of the most interesting techniques uses substances that release a burst of light when they bind the ion. One of these substances is *aequorin,* a protein produced by jellyfish, ctenophores, and many other luminescent organisms. Aequorin is injected into the cytoplasm of cells using microscopic needles, and it releases light when IP_3 opens Ca^{2+} channels in the ER, causing an increase in cytosolic Ca^{2+} concentration.

Artificially made, water-soluble molecules called *fura-2* and *quin-2* are also used as indicators of Ca^{2+} release. These molecules fluoresce (emit light) when exposed to ultraviolet (UV) light. They emit fluorescence at different wavelengths of UV light depending on whether the molecules are bound to or free of Ca^{2+}. Therefore, the amount of Ca^{2+} released into the cytosol can be quantified by measuring the amount of fluorescence at each of the two wavelengths. Rather than injecting fura-2 or quin-2 into cells, investigators combine them with a hydrophobic organic molecule that allows them to pass directly through the plasma membrane. After fura-2 or quin-2 are inside the cell, cellular enzymes remove the added organic group, releasing the Ca^{2+} indicators into the cytosol.

In a typical experiment designed to follow steps in the IP_3/DAG pathway, an investigator might want to know whether a given hormone triggers the pathway in a group of cells. The investigator first adds aequorin or quin-2 to the cells, and then the hormone. If the cells emit a bright flash of light, it is a good indication that the hormone triggers the IP_3/DAG pathway.

Experiments using these methods and others have revealed the many cellular processes controlled by Ca^{2+} concentration inside cells, including cellular response pathways, cell movements, assembly and disassembly of the cytoskeleton, secretion, and endocytosis.

SPECIFIC EXAMPLES OF CYCLIC AMP PATHWAYS Many peptide hormones act as first messengers for cAMP pathways in mammals and other vertebrates. The receptors that bind these hormones control such varied cellular responses as the uptake and oxidation of glucose, glycogen breakdown or synthesis, ion transport, the transport of amino acids into cells, and cell division.

For example, a cAMP pathway is involved in regulating the level of glucose, the fundamental fuel of cells. When the level of blood glucose falls too low in mammals, cells in the pancreas release the peptide hormone glucagon. Glucagon triggers a cAMP receptor–response pathway (see Figure 7.9) in liver cells, which stimulates them to break down glycogen into glucose units that pass

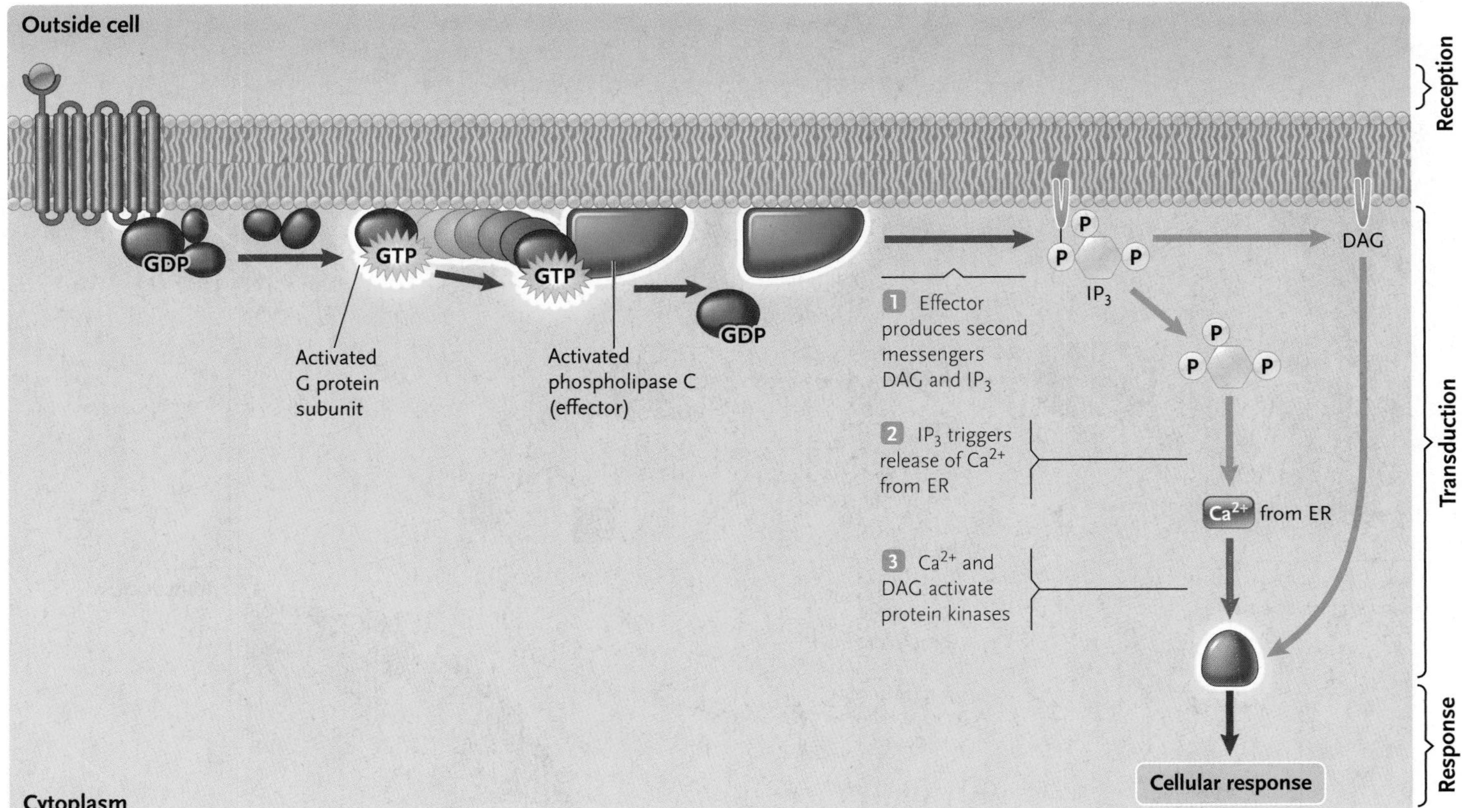

Outside cell
Reception
GDP
GTP
GTP
GDP
Activated
G protein
subunit
Activated
phospholipase C
(effector)
P
P
P
IP3
DAG
1 Effector produces second messengers DAG and IP3
P
P
P
2 IP3 triggers release of Ca2+ from ER
Ca2+ from ER
3 Ca2+ and DAG activate protein kinases
Transduction
Cellular response
Response
Cytoplasm

FIGURE 7.11
The operation of IP3/DAG receptor–response pathways. Two second messengers, IP_3 and DAG, are produced by the pathway. IP_3 opens Ca^{2+} channels in ER membranes, releasing the ion into the cytoplasm. The Ca^{2+}, with DAG in some cases, directly or indirectly activates the protein kinases of the pathway, which add phosphate groups to target proteins to initiate the cellular response.

from the liver cells into the bloodstream. The other enzyme involved in glucose regulation is *glycogen synthase,* which adds glucose units to glycogen when the level of blood glucose is excessive.

SPECIFIC EXAMPLES OF IP_3/DAG PATHWAYS The IP_3/DAG-response–pathways are also activated by a large number of peptide hormones (including growth factors) and neurotransmitters, leading to responses as varied as sugar and ion transport, glucose oxidation, cell growth and division, and movements such as smooth muscle contraction.

Among the mammalian hormones that activate the pathways are vasopressin, angiotensin, and norepinephrine. Vasopressin, also known as antidiuretic hormone, helps the body conserve water by reducing the output of urine. Angiotensin helps maintain blood volume and pressure (see Chapter 46). Norepinephrine (also known as noradrenaline), together with epinephrine, brings about the fight-or-flight response in threatening or stressful situations (see Chapter 40).

Many growth factors operate through IP_3/DAG pathways. Defects in the receptors or other parts of the pathways that lead to higher-than-normal levels of DAG in response to growth factors are often associated with the progression of some forms of cancer. This is because DAG, in turn, causes an overactivity of the protein kinases responsible for stimulating cell growth and division. Also, plant substances in a group called *phorbol esters* resemble DAG so closely that they can promote cancer in animals by activating the same protein kinases.

IP_3/DAG pathways have also been linked to mental disease, particularly *bipolar disorder* (previously called *manic depression*), in which patients experience periodic changes in mood. Lithium has been used for many years as a therapeutic agent for bipolar disorder. Recent research has shown that lithium reduces the activity of IP_3/DAG pathways that release neurotransmitters, among them are some neurotransmitters that take part in brain function. Lithium also relieves cluster headaches and premenstrual tension, suggesting that IP_3/DAG pathways may be linked to these conditions as well.

In plants, IP_3/DAG pathways control responses to conditions such as water loss and changes in light intensity or salinity. Plant hormones—relatively small, nonprotein molecules such as the *cytokinins* (derivatives of the nucleotide base adenine)—act as first messengers activating some of the IP_3/DAG pathways of these organisms.

Some Signaling Pathways Combine a Receptor Tyrosine Kinase with the G Protein Ras

Some pathways important in gene regulation link certain receptor tyrosine kinases to a specific type of G protein called Ras. When the receptor tyrosine kinase receives a signal (**Figure 7.12,**

FIGURE 7.12
The pathway from receptor tyrosine kinases to gene regulation, including the G protein, Ras, and MAP kinase.

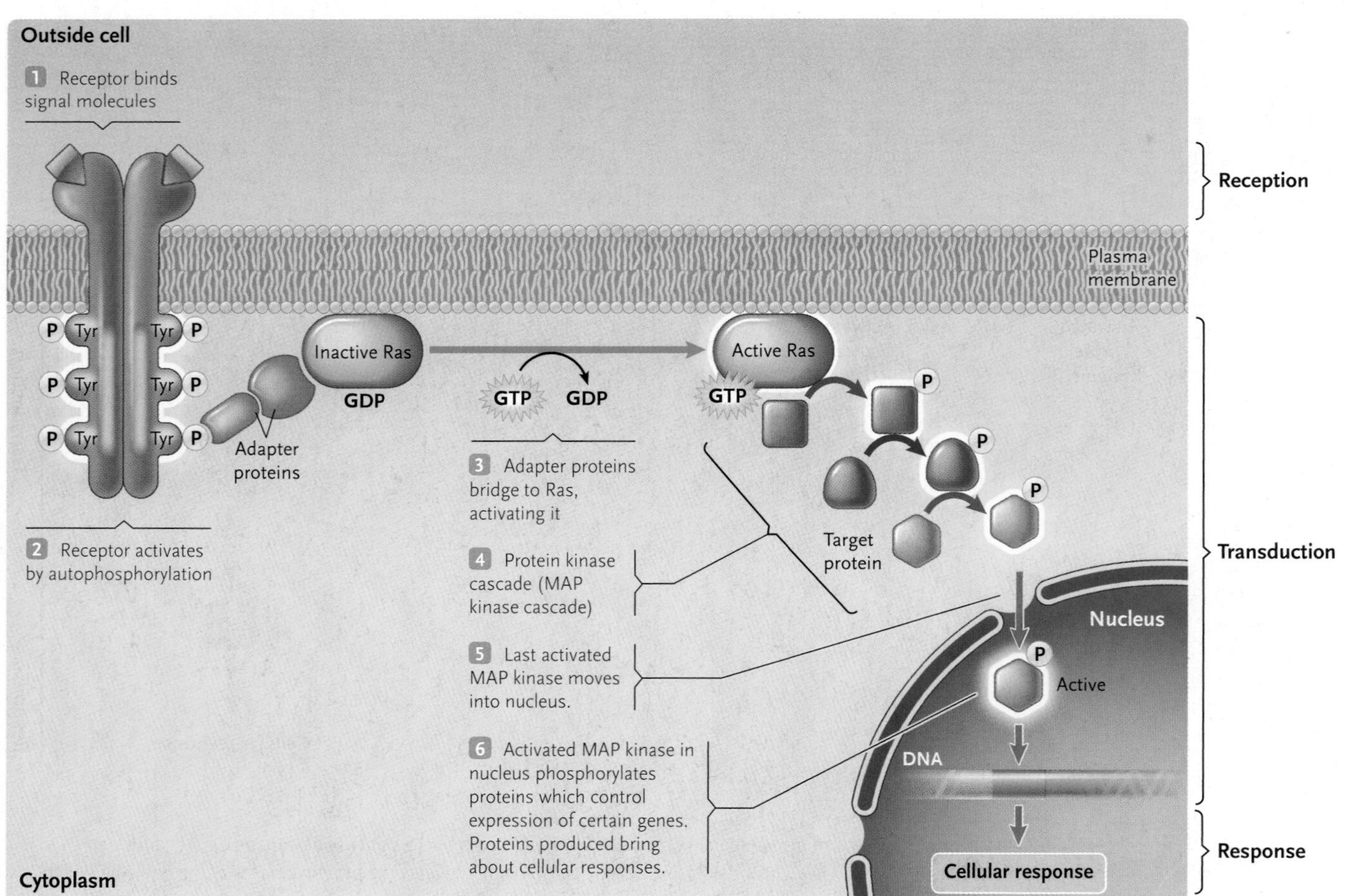
Outside cell
1 Receptor binds signal molecules
Reception
Plasma membrane
P Tyr
Tyr P
Inactive Ras
GDP
Active Ras
GTP
GTP
GDP
P
Adapter proteins
3 Adapter proteins bridge to Ras, activating it
2 Receptor activates by autophosphorylation
4 Protein kinase cascade (MAP kinase cascade)
Target protein
Transduction
Nucleus
5 Last activated MAP kinase moves into nucleus.
Active
6 Activated MAP kinase in nucleus phosphorylates proteins which control expression of certain genes. Proteins produced bring about cellular responses.
DNA
Cellular response
Response
Cytoplasm

Virus Infections and Cell Signaling Pathways: Does influenza virus propagation involve a cellular MAP kinase cascade?

Epidemic outbreaks of influenza occur almost every year, even in developed countries. Because flu virus strains differ in virulence, the death rate varies with each epidemic. Moreover, some influenza epidemics become pandemics, epidemics of new strains that spread through many parts of the world. One particularly virulent pandemic, the so-called Spanish Flu of 1918–1919, infected an estimated 500 million people; 50–100 million people died, most of whom were between the ages of 20 and 40 years old. The cause of the 1918–1919 influenza pandemic was a viral strain called influenzavirus A/H1N1. An influenza pandemic which began in 2009 involved a new strain of influenzavirus A/H1N1 and was not as deadly as the 1918 flu virus, although the exact molecular properties responsible for the differences in virulence between the two strains are not completely understood. The deadly potential of flu viruses has spurred research into antiviral drugs.

What is influenza virus? An influenza virus consists of an RNA genome (the genomes of all living organisms are DNA), that is divided into eight segments, each of which contains one to two genes. The RNA genome is within a protein coat. The RNA genome and protein coat together form the virus particle. New genomes can be assembled when two influenza viruses simultaneously infect a cell. That is, if a cell is infected by two different strains of the virus, the genome segments are replicated to produce many copies. New virus particles will then receive various combinations of genome segments derived from the two parent viruses. These new viruses may be more or less virulent than the parental viruses depending on the segments they contain. The genes may also mutate, and this can also affect the virulence of a strain.

Like all viruses, influenza viruses have a limited number of genes and, therefore, must manipulate host cell functions in order to propagate. When influenza virus infects a cell, it elicits responses in the cell to combat the infection. To be successful in its infection, the virus must overcome these antiviral activities. Various viruses support their propagation by activating MAP kinase cascades in cells they infect. The prototypic MAP kinase signaling pathway, called the Raf/MEK/ERK cascade, has a major role in regulating cell growth and proliferation.

Research Question

Does influenza A virus propagation involve the Raf/MEK/ERK MAP kinase cascade?

Experiments

Stephan Pleschka of the University of Geissen, Germany, Stephan Ludwig of the University of Würzburg, Germany, and other researchers performed two important experiments with mammalian cells in culture to answer the question.

1. First, the researchers examined how influenza A virus infection of mammalian cells in culture affected the activity of the Raf/MEK/ERK MAP kinase cascade when the chemical U0126 was either present or absent. U0126 specifically inhibits MEK, the middle kinase in the Raf/MEK/ERK MAP kinase cascade (see figure).

   The researchers assayed for ERK activity as an indicator of whether the pathway had been activated. In the absence of U0126, they detected ERK activity in influenza A virus-infected cells, showing that the pathway was activated. In the presence of U0126, they detected very little ERK activity, which suggests that U0126 had inhibited MEK so ERK was not activated. Overall, the results of this experiment show that influenza A virus activates the MAP kinase cascade.

2. The researchers then did a similar experiment, this time measuring the propagation of influenza A viruses in the cells in the absence or presence of U0126.This assay measured the production of progeny influenza viruses released from infected cells. The results indicated that U0126 reduces the production of progeny viruses by about 80% compared with cultures not treated with the inhibitor.

Conclusion

Experiment 1 showed that influenza A virus activates the Raf/MEK/ERK MAP kinase cascade. Experiment 2 showed that influenza A virus propagation requires virus-induced signaling through that kinase cascade pathway. More recent experiments have shown that specific inhibition of the Raf/MEK/ERK signaling cascade causes a marked impairment in the growth of all influenza A viruses tested. The targeting of signaling pathways that are essential for virus propagation has become the basis for clinical research with the goal of developing effective antiviral drugs.

Sources: S. Pleschka et al. 2001. Influenza virus propagation is impaired by inhibition of the Raf/MEK/ERK signalling cascade. *Nature Cell Biology* 3:301–305; S. Ludwig. 2009. Targeting cell signalling pathways to fight the flu: towards a paradigm change in anti-influenza therapy. *Journal of Antimicrobial Chemotherapy* 64:1–4.

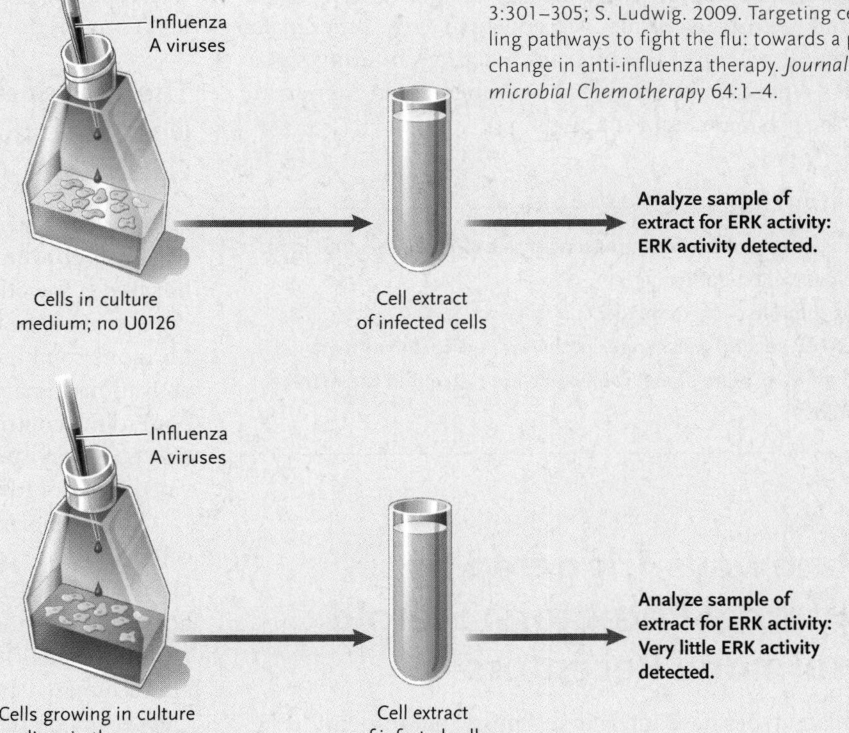
Influenza A viruses
Analyze sample of extract for ERK activity: ERK activity detected.
Cells in culture medium; no U0126
Cell extract of infected cells
Influenza A viruses
Analyze sample of extract for ERK activity: Very little ERK activity detected.
Cells growing in culture medium in the presence of inhibitor U0126
Cell extract of infected cells

The researchers infected mammalian cells growing in culture with influenza A virus and incubated them for a few hours in either the absence (top) or presence (bottom) of chemical U0126.

The scientists broke open cells, centrifuged to remove cell debris, and collected the cell extract.

Samples of the extract were analyzed for the presence of ERK, the last kinase in the cascade.

step 1), it activates by autophosphorylation (step 2). Adapter proteins then bind to the phosphorylated receptor and bridge to Ras, stimulating the activation of Ras (step 3). Like other G proteins, Ras is activated by binding GTP. The activated Ras sets in motion a phosphorylation cascade that involves a series of three enzymes known as *mitogen-activated protein kinases* (MAP kinases; step 4). The last MAP kinase in the cascade, when activated, enters the nucleus (step 5) and phosphorylates other proteins, which then change the expression of certain genes, particularly activating those involved in cell division (step 6). (A *mitogen* is a substance that controls cell division, hence the name of the kinases.) Changes in gene expression can have far-reaching effects on the cell, such as determining whether a cell divides or how frequently it divides. The Ras proteins are of major interest to investigators because of their role in linking receptor tyrosine kinases to gene regulation, as well as their major roles in the development of many types of cancer when their function is altered.

Both the Ras proteins and the MAP kinases are widely distributed among eukaryotes. Ras has been detected in eukaryotic organisms ranging from yeasts to humans and higher plants. Similarly, MAP kinases have been identified in eukaryotes as diverse as yeasts, roundworms, insects, humans, and plants. *Insights from the Molecular Revolution* presents evidence that influenza virus uses a MAP kinase cascade to aid its propagation in infected cells.

In this section, we have surveyed major response pathways linked to surface receptors that bind peptide hormones, growth factors, and neurotransmitters. We now turn to the other major type of signal receptor: the internal receptors binding signal molecules—primarily steroid hormones—that penetrate through the plasma membrane.

STUDY BREAK 7.4 <

1. **What is the role of the first messenger in a G-protein–coupled receptor-controlled pathway?**
2. **What is the role of the effector?**
3. **For a cAMP second-messenger pathway, how is the pathway turned off if no more signal molecules are present in the extracellular fluids?**

7.5 Pathways Triggered by Internal Receptors: Steroid Hormone Receptors

Cells of many types have internal receptors that respond to signals arriving from the cell exterior. Unlike the signal molecules that bind to surface receptors, these signals, primarily steroid hormones, penetrate through the plasma membrane to trigger response pathways inside the cells. The internal receptors, called **steroid hormone receptors,** are typically control proteins that turn on (sometimes off) specific genes when they are activated by binding a signal molecule.

Steroid Hormones Have Widely Different Effects That Depend on Relatively Small Chemical Differences

Steroid hormones are relatively small, nonpolar molecules derived from cholesterol, with a chemical structure based on four carbon rings (see Figure 3.15). Steroid hormones combine with hydrophilic carrier proteins that mask their hydrophobic groups and hold them in solution in the blood and extracellular fluids. When a steroid hormone–carrier protein complex collides with the surface of a cell, the hormone is released and penetrates directly through the nonpolar part of the plasma membrane. On the cytoplasmic side, the hormone binds to its internal receptor.

The various steroid hormones differ only in the side groups attached to their carbon rings. Although the differences are small, they are responsible for highly distinctive effects. For example, the male and female sex hormones of mammals, testosterone and estrogen, respectively, which are responsible for many of the structural and behavioral differences between male and female mammals, differ only in minor substitutions in side groups at two positions (see Figure 3.16). The differences cause the hormones to be recognized by different receptors, which activate specific group of genes leading to development of individuals as males or females.

The Response of a Cell to Steroid Hormones Depends on Its Internal Receptors and the Genes They Activate

Steroid hormone receptors are proteins with two major domains **(Figure 7.13).** One domain recognizes and binds a specific steroid hormone. The other domain interacts with the regions of target genes that control the expression of those genes. When a steroid hormone combines with the hormone-binding domain, the gene activation domain changes shape, thus enabling the complex to bind to the control regions of the target genes that the hormone affects. For most steroid hormone receptors, binding of the activated receptor to a gene control region activates that gene.

Steroid hormones, like peptide hormones, are released by cells in one part of an organism and are carried by the organism's circulation to other cells. Whether a cell responds to a steroid hormone depends on whether it has a receptor for the hormone within the cell. The type of response depends on the genes that are recognized and turned on (or off) by an activated receptor. Depending on the receptor type and the particular genes it recognizes, even the same steroid hormone can have highly varied effects on different cells. (The effects of steroid hormones are described in more detail in Chapter 40.)

Taken together, the various types of receptor tyrosine kinases, G-protein–coupled receptors, and steroid hormone recep-

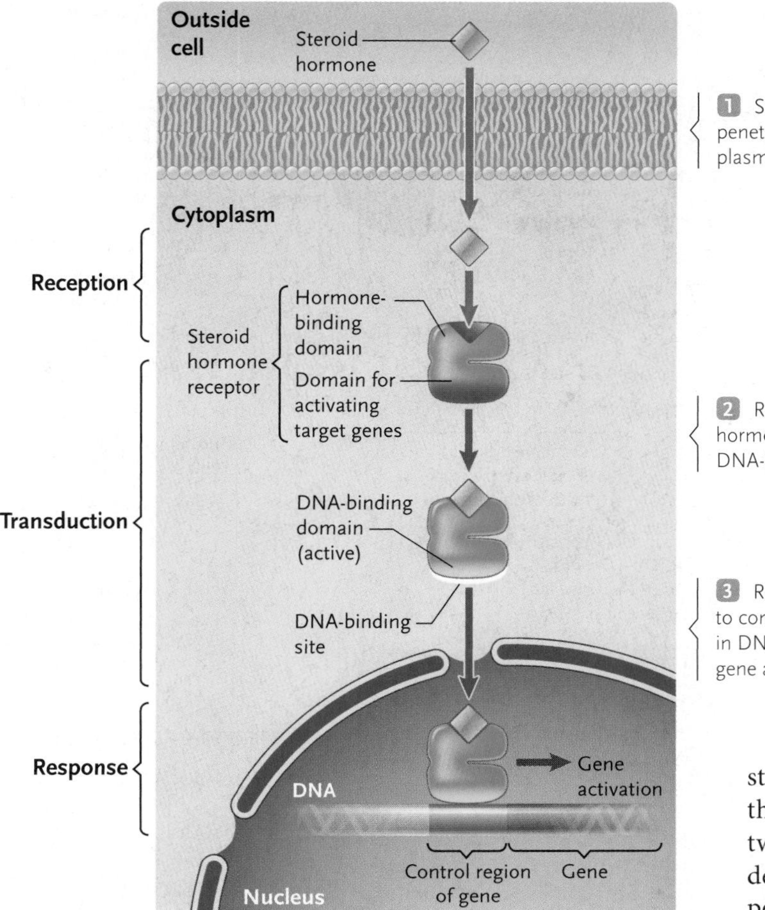

Outside cell
Steroid hormone
Cytoplasm
Reception
Transduction
Response
Steroid hormone receptor
Hormone-binding domain
Domain for activating target genes
DNA-binding domain (active)
DNA-binding site
DNA
Nucleus
Gene activation
Control region of gene
Gene
1 Steroid hormone penetrates through plasma membrane
2 Receptor binds hormone, activating DNA-binding site
3 Receptor binds to control sequence in DNA, leading to gene activation

FIGURE 7.13
Pathway of gene activation by steroid hormone receptors.

tors, prime cells to respond to a stream of specific signals that continuously fine-tune their function. How are the signals integrated within the cell and organism to produce harmony rather than chaos? The next section shows how the various signal pathways are integrated into a coordinated response.

STUDY BREAK 7.5

1. **What distinguishes a steroid receptor from a receptor tyrosine kinase receptor or a G-protein–coupled receptor?**
2. **By what means does a specific steroid hormone result in a specific cellular response?**

7.6 Integration of Cell Communication Pathways

Cells are under the continual influence of many simultaneous signal molecules. The cell signaling pathways may operate independently, or communicate with one another to integrate their responses to cellular signals coming from different controlling cells. The interpathway interaction is called **cross-talk;** a conceptual example that involves two second-messenger pathways is shown in **Figure 7.14.** For example, a protein kinase in one pathway might phosphorylate a site on a target protein in another signal transduction pathway, activating or inhibiting that protein, depending on the site of the phosphorylation. The cross-talk can be extensive, resulting in a complex network of interactions between cell communication pathways.

Cross-talk often leads to modifications of the cellular responses controlled by the pathways. Such modifications fine-tune the effects of combinations of signal molecules binding to the receptors of a cell. For example, cross-talk between second-messenger pathways is involved in particular types of olfactory (smell) signal transduction in rats, and probably in many animals including humans. The two pathways involved are activated upon stimulation with distinct odors. One pathway involves cAMP as the second messenger, and the other involves IP_3. However, the two olfactory second-messenger pathways do not work independently; rather, they operate in an antagonistic way. That is, experimentally blocking key enzymes of one signal transduction cascade inhibits that pathway, while simultaneously augmenting the activity of the other pathway. The cross-talk may be a way to refine an animal's olfactory sensory perception by helping discriminate different odor molecules more effectively.

Direct channels of communication may also be involved in a cross-talk network. For example, gap junctions between the cytoplasms of adjacent cells admit ions and small molecules, including the Ca^{2+}, cAMP, and IP_3 second messengers released by the receptor–response pathways. (Gap junctions are discussed in further detail in Section 5.5.) Thus, one cell that receives a signal through its surface receptors can transmit the signal to other cells in the same tissue via the connecting gap junctions, thereby coordinating the functions of those cells. For instance, cardiac muscle cells are connected by gap junctions, and the Ca^{2+} flow regulates coordinated muscle fiber contractions.

The entire system integrating cellular response mechanisms, tied together by many avenues of cross-talk between individual pathways, creates a sensitively balanced control mechanism that regulates and coordinates the activities of individual cells into the working unit of the organism.

STUDY BREAK 7.6

What cell communication pathways might be integrated in a cross-talk network?

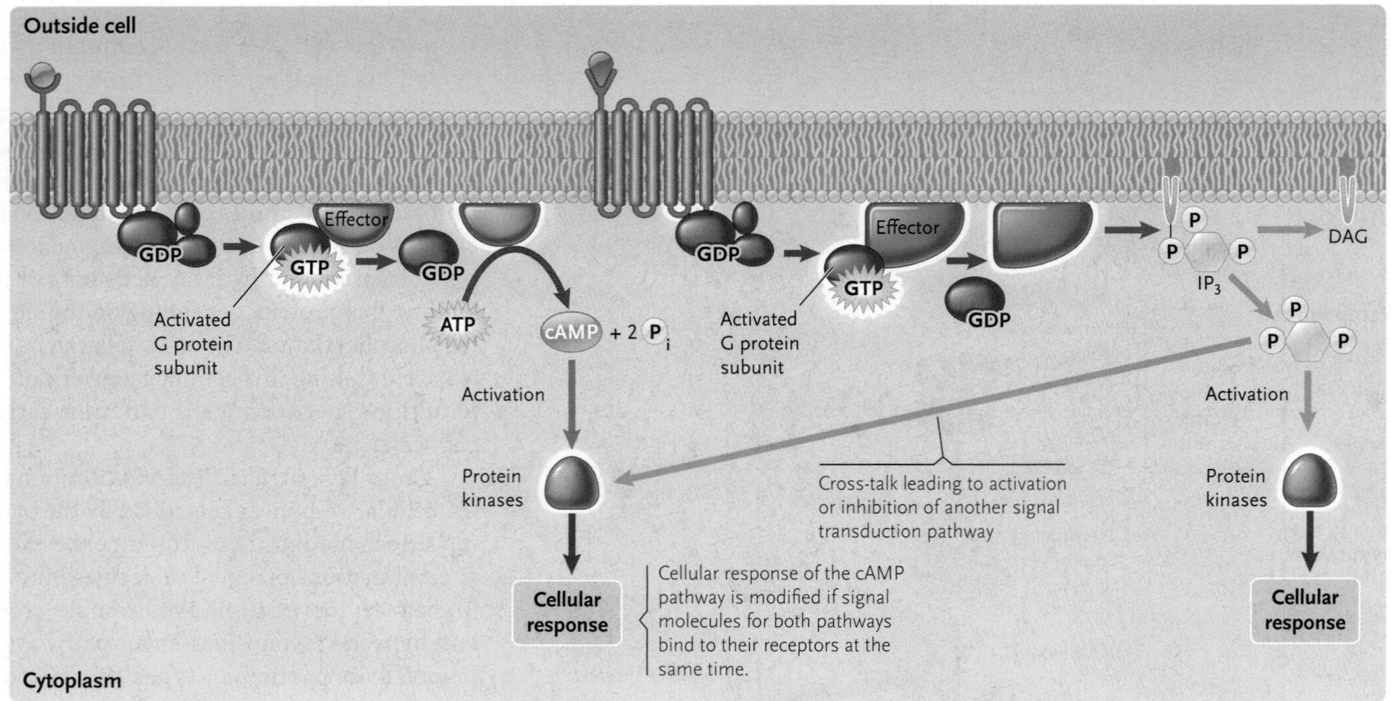

Outside cell
Effector
GDP
GTP
GDP
Activated G protein subunit
ATP
cAMP + 2 P i
Activation
Protein kinases
Cellular response
Effector
GDP
GTP
GDP
Activated G protein subunit
P
P
P
IP3
DAG
P
P
P
Activation
Protein kinases
Cellular response
Cross-talk leading to activation or inhibition of another signal transduction pathway
Cellular response of the cAMP pathway is modified if signal molecules for both pathways bind to their receptors at the same time.
Cytoplasm

FIGURE 7.14
Cross-talk, the interaction between cell communication pathways to integrate the responses to signal molecules.

UNANSWERED QUESTIONS

How does cross-talk between signaling pathways influence a behavior?
As you learned in this chapter, cell signaling by the sex hormones of animals is responsible for many of the structural and behavioral characteristics of males and females. Another type of cell signaling, called neural signaling, is also important in reproductive behavior. Currently, many laboratories are actively investigating possible cross-talk between these two major types of cell signaling. Specifically, researchers are investigating the cellular processes by which steroid hormones that are involved in mammalian reproductive behavior act on neurons. During the estrous cycle of many animals, including female rats, guinea pigs, hamsters and mice, the ovarian hormones estradiol and progesterone regulate the expression of reproductive behaviors via cellular processes, including binding to steroid hormone receptors in neurons that are involved in the behaviors. But do neural signals also activate steroid hormone receptors?

The steroid hormone receptor model presented in this chapter is that an intracellular steroid hormone receptor becomes activated when the steroid hormone binds to it. This mechanism, which involves both estrogen receptors (for estradiol) and progestin receptors (for progesterone) in the brain, is consistent with a great deal of the research on the cellular mechanisms of hormonal regulation of reproductive behaviors. However, research based on work of Shaila Mani and collaborators has now shown that the regulation of reproductive behaviors involves cross-talk between neurotransmitter signaling pathways and steroid hormone receptors. Although steroid hormones bind directly to the steroid hormone receptors, neurotransmitters, including dopamine, acting *via* second messenger pathways, can also activate steroid hormone receptors in the absence of a hormone. In addition, my research group at the University of Massachusetts, Amherst, has shown that when a male rat attempts to mate with a female rat, the mating stimulation somehow activates the female's neural progestin receptors, presumably by a process that involves the release of particular neurotransmitters onto neurons containing the receptors. In fact, although it had always been thought that progesterone is required to facilitate the full expression of sexual behaviors in female rats, stimulation by the male can substitute for progesterone. Mating stimulation induces reproductive behavior similar to that induced by the secretion of steroid hormones. That is, how a male behaves toward a female alters neurotransmitter release in her brain, presumably then activating steroid hormone receptors in some neurons. This activation results in neuronal changes, many of which are the same as those caused by the hormone secretions from the female's ovaries.

How does this hormone-independent steroid hormone receptor activation occur? In which neurons would you expect these events occur, and what characteristics would you expect of the neurons (for example, inputs and outputs)? What might regulate the process? The results of experiments designed to answer these questions will give valuable insights into the mechanisms of steroid hormone action in the brain.

Think Critically

1. Why do you think the male's behavior can substitute for the hormone in facilitating sexual behavior?
2. Progestin receptor regulation of sexual behavior is a useful model for studying the interaction between neurotransmitters and steroid hormone receptors. Would you expect this cross-talk mechanism to be limited to sexual behavior, or do you think it might come into play with other hormone-regulated behaviors?

Jeffrey Blaustein

Jeffrey Blaustein is a professor in the Neuroscience and Behavior Program and is a member of the Center for Neuroendocrine Studies at the University of Massachusetts Amherst. His research interests are in the many ways in which the environment can influence hormonal processes in the brain resulting in changes in behavior. In recent years, the interest of his group has expanded to the influences of stress around the time of puberty on response to ovarian hormones in adulthood. To learn more about the work of his research group, go to http://www.umass.edu/cns/blaustein.

REVIEW KEY CONCEPTS

Go to **CENGAGENOW** at www.cengage.com/login to access quizzing, animations, exercises, articles, and personalized homework help.

7.1 Cell Communication: An Overview

- Cells communicate with one another by direct contact, local signaling, and long distance signaling.
- In long distance signaling, a controlling cell releases a signal molecule that causes a response of target cells. Target cells process the signal in three steps: reception, transduction, and response. This process is called signal transduction (Figure 7.2).
- Some cell communication properties are evolutionarily ancient. In some cases, entire cell signaling pathways are conserved between distantly related organisms.

7.2 Cell Communication Systems with Surface Receptors

- Cell communication systems based on surface receptors have three components: (1) extracellular signal molecules; (2) surface receptors that receive the signals; and (3) internal response pathways triggered when receptors bind a signal.
- The systems based on surface receptors respond to peptide hormones and neurotransmitters. Peptide hormones are small proteins. They include growth factors, which affect cell growth, division, and differentiation. Neurotransmitters include small peptides, individual amino acids or their derivatives, and other chemical substances.
- Surface receptors are integral membrane proteins that extend through the plasma membrane. Binding a signal molecule induces a molecular change in the receptor that activates its cytoplasmic end (Figure 7.3).
- Many cellular response pathways operate by activating protein kinases, which add phosphate groups that stimulate or inhibit the activities of the target proteins, bringing about the cellular response (Figure 7.4). Protein phosphatases that remove phosphate groups from target proteins reverse the response. In addition, receptors are removed by endocytosis when signal transduction has run its course.
- Each step of a response pathway catalyzed by an enzyme is amplified, because each enzyme can activate hundreds or thousands of proteins that enter the next step in the pathway. Through amplification, a few signal molecules can bring about a full cellular response (Figure 7.5).

Animation: Signal transduction

7.3 Surface Receptors with Built-In Protein Kinase Activity: Receptor Tyrosine Kinases

- When receptor tyrosine kinases bind a signal molecule, the protein kinase site is activated and adds phosphate groups to tyrosines in the receptor itself activating those sites. When a signaling protein binds to an activated site, it initiates a transduction pathway leading to a cellular response. The binding of different combinations of signaling proteins to different tyrosine kinases produces different responses (Figure 7.6).

7.4 G-Protein–Coupled Receptors

- In the pathways activated by G-protein–coupled receptors, binding of the extracellular signal molecule (the first messenger) activates a site on the cytoplasmic end of the receptor (Figure 7.7). The activated receptor turns on a G protein, which acts as a molecular switch. The G protein is active when it is bound to GTP and inactive when it is bound to GDP (Figure 7.8).
- The active G protein switches on the effector, an enzyme that generates small internal signal molecules called second messengers. The second messengers activate the protein kinases of the pathway (Figure 7.8).
- In one of the two major pathways triggered by G-protein–coupled receptors, the effector, adenylyl cyclase, generates cAMP as second messenger. cAMP activates specific protein kinases (Figures 7.9 and 7.10).
- In the other major pathway, the activated effector, phospholipase C, generates two second messengers, IP_3 and DAG. IP_3 activates transport proteins in the ER, which release stored Ca^{2+}. The Ca^{2+} alone, or with DAG, activates specific protein kinases that phosphorylate their target proteins (Figure 7.11).
- Both the cAMP and IP_3/DAG pathways are balanced by reactions that constantly eliminate their second messengers. Both pathways are also stopped by protein phosphatases that remove phosphate groups from target proteins and by endocytosis of receptors and their bound signals.
- Mutated systems can turn on the pathways permanently, contributing to the progression of some forms of cancer.
- Some pathways important in gene regulation link certain receptor tyrosine kinases to a specific G protein called Ras. When the receptor binds a signal molecule, it phosphorylates itself, and adapter proteins then bind, bridging to Ras, activating it. Activated Ras turns on the MAP kinase cascade. The last MAP kinase in the cascade phosphorylates target proteins in the nucleus, which turn on specific genes (Figure 7.12). Many of those genes control cell division.

Practice: Response pathways activated by G-protein–coupled receptors

7.5 Pathways Triggered by Internal Receptors: Steroid Hormone Receptors

- Steroid hormones penetrate through the plasma membrane to bind to receptors within the cell, activating the receptors. The internal receptors are regulatory proteins that turn on specific genes, producing the cellular response (Figure 7.13).
- Steroid hormone receptors have a domain that recognizes and binds a specific steroid hormone and a domain that interacts with the controlling regions of target genes. A cell responds to a steroid hormone only if it has an internal receptor for the hormone; and the type of response depends on the genes that are turned on by an activated receptor (Figure 7.13).

Animation: Pathway of gene activation by steroid hormone receptors

7.6 Integration of Cell Communication Pathways

- In cross-talk, cell signaling pathways communicate with one another to integrate responses to cellular signals. Cross-talk may result in a complex network of interactions between cell communication pathways. Cross-talk often modifies the cellular responses controlled by the pathways, fine-tuning the effects of combinations of signal molecules binding to a cell (Figure 7.14).
- In animals, inputs from other cellular response systems, including cell adhesion molecules and molecules arriving through gap junctions, also can be involved in the cross-talk network.

Animation: Animal cell junctions

UNDERSTAND AND APPLY

Test Your Knowledge

1. In signal transduction, which of the following is *not* a target protein?
   a. proteins that regulate gene activity
   b. hormones that activate the receptor
   c. enzymes of pathways
   d. transport proteins
   e. enzymes of cell reactions
2. Which of the following could *not* elicit a signal transduction response?
   a. a signal molecule injected directly into the cytoplasm
   b. a virus mimicking a normal signal molecule
   c. a peptide hormone
   d. a steroid hormone
   e. a neurotransmitter

3. A cell that responds to a signal molecule is distinguished from a cell that does not respond by the fact that it has:
   a. a cell adhesion molecule.
   b. cAMP.
   c. a first messenger molecule.
   d. a receptor.
   e. a protein kinase.
4. The mechanism to activate an immune cell to make an antibody involves signal transduction using tyrosine kinases. Place in order the following series of steps to activate this function.
   (1) The activated receptor phosphorylates cytoplasmic proteins.
   (2) Conformational change occurs in the receptor tyrosine kinase.
   (3) Cytoplasmic protein crosses the nuclear membrane to activate genes.
   (4) An immune hormone signals the immune cell.
   (5) Activation of protein kinase site(s) adds phosphates to the receptor to activate it.
   a. 2, 1, 4, 3, 5
   b. 5, 3, 4, 2, 1
   c. 4, 1, 5, 2, 3
   d. 4, 2, 5, 1, 3
   e. 2, 5, 3, 4, 1
5. Which of the following describes the ability of enzymes, involving few surface receptors, to activate thousands of molecules in a stepwise pathway?
   a. autophosphorylation
   b. second-messenger enhancement
   c. amplification
   d. ion channel regulation
   e. G protein turn-on
6. Which of the following is *incorrect* about pathways activated by G-protein–coupled receptors?
   a. The extracellular signal is the first messenger.
   b. When activated, plasma membrane-bound G protein can switch on an effector.
   c. Second messengers enter the nucleus.
   d. ATP converts to cAMP to activate protein kinases.
   e. Protein kinases phosphorylate molecules to change cellular activity.
7. Which of the following would *not* inhibit signal transduction?
   a. Phosphate groups are removed from proteins.
   b. Endocytosis acts on receptors and their bound signals.
   c. Receptors and signals separate.
   d. Receptors and bound signals enter lysosomes.
   e. Autophosphorylation targets the cytoplasmic portion of the receptor.
8. An internal receptor binds both a signal molecule and controlling region of a gene. What type of receptor is it?
   a. protein
   b. steroid
   c. IP_3/DAG
   d. receptor tyrosine kinase
   e. switch protein
9. Place in order the following steps for the normal activity of a Ras protein.
   (1) Ras turns on the MAP kinase cascade.
   (2) Adaptor proteins connect phosphorylated tyrosine on a receptor to Ras.
   (3) GTP activates Ras by binding to it, displacing GDP.
   (4) The last MAP kinase in the cascade phosphorylates proteins in the nucleus that activate genes.
   (5) Receptor tyrosine kinase binds a signal molecule and is activated.
   a. 1, 2, 3, 4, 5
   b. 2, 3, 5, 1, 4
   c. 5, 2, 3, 1, 4
   d. 2, 3, 1, 5, 4
   e. 4, 1, 5, 3, 2
10. Cross-talk is exemplified by all of the following except:
   a. a protein kinase in one pathway that phosphorylates a site on a target protein in another signal transduction pathway.
   b. modifications of cellular responses controlled by pathways.
   c. two second-messenger pathways interacting.
   d. olfactory sensory perception.
   e. signal transduction pathways controlled by G-protein-coupled receptors.

Discuss the Concepts

1. Describe the possible ways in which a G-protein–coupled receptor pathway could become defective and not trigger any cellular responses.
2. Is providing extra insulin an effective cure for an individual who has diabetes that is caused by a hereditary defect in the insulin receptor? Why or why not?
3. There are molecules called GTP analogs that resemble GTP so closely that they can be bound by G proteins. However, they cannot be hydrolyzed by cellular GTPases. What differences in effect would you expect if you inject GTP or a nonhydrolyzable GTP analog into a liver cell that responds to glucagon?
4. Why do you suppose cells evolved internal response mechanisms using switching molecules that bind GTP instead of ATP?

Design an Experiment

How would you set up an experiment to determine whether a hormone receptor is located on the cell surface or inside the cell?

Interpret the Data

In most individuals with cystic fibrosis, the 508th amino acid of the CFTR protein (a phenylalanine) is missing. A CFTR protein with this change is synthesized correctly, and it can transport ions correctly, but it never reaches the plasma membrane to do its job.

Sergei Bannykh and his coworkers developed a procedure to measure the relative amounts of the CFTR protein localized in different regions of the cell. They compared the pattern of CFTR distribution in normal cells with the pattern in CFTR-mutated cells. A summary of their results is shown in the accompanying figure.

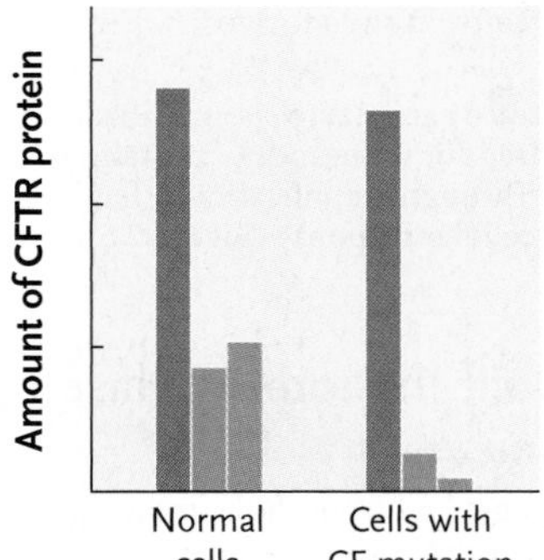

Amount of CFTR protein
Normal cells
Cells with CF mutation

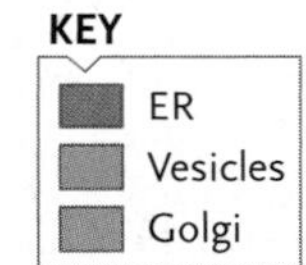

KEY
ER
Vesicles
Golgi

Comparison of the amounts of CFTR protein associated with endoplasmic reticulum (blue), vesicles traveling from ER to Golgi (green), and Golgi bodies (orange). The patterns of CFTR distribution in normal cells, and the cells with the most common cystic fibrosis mutation, were compared.

1. Which organelle contains the least amount of CFTR protein in normal cells? In CF cells? Which contains the most?
2. In which organelle is the amount of CFTR protein in CF cells closest to the amount in normal cells?
3. Where is the mutated CFTR protein getting held up?

Apply Evolutionary Thinking

Based on their distributions among different groups of organisms, which signaling pathway is the oldest?

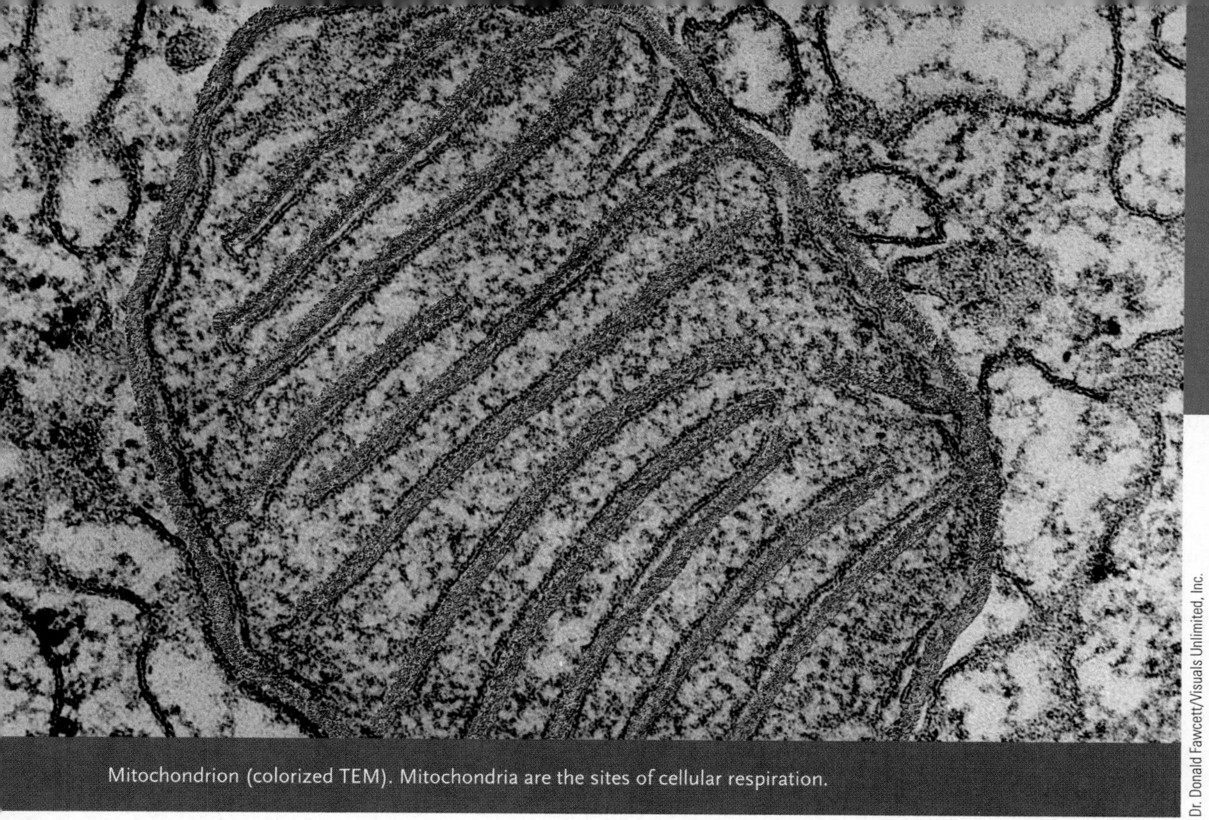

Mitochondrion (colorized TEM). Mitochondria are the sites of cellular respiration.

Dr. Donald Fawcett/Visuals Unlimited, Inc.

8

STUDY OUTLINE

Harvesting Chemical Energy: Cellular Respiration

Why It Matters. . . In the early 1960s, Swedish physician Rolf Luft mulled over some odd symptoms of a patient. The young woman felt weak and too hot all the time (with a body temperature of up to 38.4°C). Even on the coldest winter days she never stopped perspiring, and her skin was always flushed. She was also underweight (40 kg), despite consuming about 3,500 calories per day.

Luft inferred that his patient's symptoms pointed to a metabolic disorder. Her cells were very active, but much of their activity was being dissipated as metabolic heat. He decided to order tests to measure her metabolic rate, the amount of energy her body was expending. The results showed the patient's oxygen consumption was the highest ever recorded—about twice the normal rate!

Luft also examined a tissue sample from the patient's skeletal muscles. Using a microscope, he found that her muscle cells contained many more mitochondria—the ATP-producing organelles of the cell—than are normally present in muscle cells. In addition, her mitochondria were abnormally shaped and their interior was packed to an abnormal degree with cristae, the infoldings of the inner mitochondrial membrane (see Section 5.3). Other studies showed that the mitochondria were engaged in cellular respiration—their prime function—but little ATP was being generated.

The disorder, now called *Luft syndrome,* was the first disorder to be linked directly to a defective mitochondrion. By analogy, someone with this disorder functions like a city with half of its power plants shut down. Skeletal and heart muscles, the brain, and other hardworking body parts with high energy demands are hurt the most by the inability of mitochondria to provide enough energy for metabolic demands. More than 100 mitochondrial disorders are now known.

Defective mitochondria also contribute to many age-related problems, including type 1 diabetes, atherosclerosis, amyotrophic lateral sclerosis (ALS, also called Lou Gehrig disease), as well as Parkinson, Alzheimer, and Huntington diseases.

Clearly, human health depends on mitochondria that are structurally sound and that function properly. More broadly, every animal, plant, fungus, and most protists depend on mitochondria that are functioning correctly to grow and survive.

In mitochondria, ATP forms as part of the reactions of cellular respiration. **Cellular respiration** is the collection of metabolic reactions within cells that breaks down food molecules to produce energy in the form of ATP. ATP fuels nearly all the reactions that keep cells, and organisms, metabolically active. In eukaryotes and many prokaryotes, oxygen is a reactant in the ATP-producing process. This form of cellular respiration is **aerobic respiration** (*aero* = air, *bios* = life). In some prokaryotes, a molecule other than oxygen, such as sulfate or nitrate, is used in the ATP-producing process. This form of cellular respiration is **anaerobic respiration** (*an* = without). The discussion of the reactions of cellular respiration in this chapter concerns aerobic respiration except where noted.

The primary source of the food molecules broken down in cellular respiration is *photosynthesis,* which is described in Chapter 9. Photosynthesis captures energy from light by splitting water molecules, and hydrogen from the water is combined with carbon dioxide to synthesize carbohydrates and other organic molecules. A major by-product of photosynthesis is oxygen, a molecule needed for cellular respiration. Photosynthesis occurs in most plants, many protists, and some prokaryotes.

Together, cellular respiration and photosynthesis are the major biological steps of the carbon cycle, which is the global circulation of carbon atoms. The carbon cycle is described in Section 52.3. Atmospheric carbon in the carbon cycle is mostly in the form of CO_2, which is a product of cellular respiration. <

8.1 Overview of Cellular Energy Metabolism

Electron-rich food molecules synthesized by plants are used by the plants themselves, by animals, and by other eukaryotes. The electrons are removed from fuel substances, such as sugars, and donated to other molecules, such as oxygen, that act as electron acceptors. In the process, some of the energy of the electrons is released and used to drive the synthesis of ATP. ATP provides energy for most of the energy-consuming activities in the cell. Thus, life and its systems are driven by a cycle of electron flow that is powered by light in photosynthesis and oxidation in cellular respiration **(Figure 8.1).**

Coupled Oxidation and Reduction Reactions Produce the Flow of Electrons for Energy Metabolism

The removal of electrons (e^-) from a substance is termed an **oxidation,** and the substance from which the electrons are removed—called the *electron donor*—is said to be **oxidized.** The addition of electrons to a substance is termed a **reduction,** and the substance that receives the electrons—called the *electron acceptor*—is said to be **reduced.** A simple mnemonic to remember the direction of electron transfer is OIL RIG: Oxidation Is Loss (of electrons), Reduction Is Gain (of electrons). The term *oxidation* was originally used to describe the reaction that occurs when fuel substances are burned in air, in which oxygen directly accepts electrons removed from the fuels. However, although the term oxidation suggests that oxygen is involved in electron removal, most cellular oxidations occur without the direct participation of oxygen. The term *reduction* refers to the decrease in positive electrical charge that occurs when electrons, which are negatively charged, are added to a substance. Although the term reduction suggests that the energy level of molecules is decreased when they accept electrons, molecules typically gain energy from added electrons.

FIGURE 8.1
The flow of energy from sunlight to ATP. **(A)** Photosynthesis occurs in plants, many protists, and some prokaryotes; **(B)** cellular respiration (aerobic respiration) occurs in all eukaryotes, including plants, and in some prokaryotes.

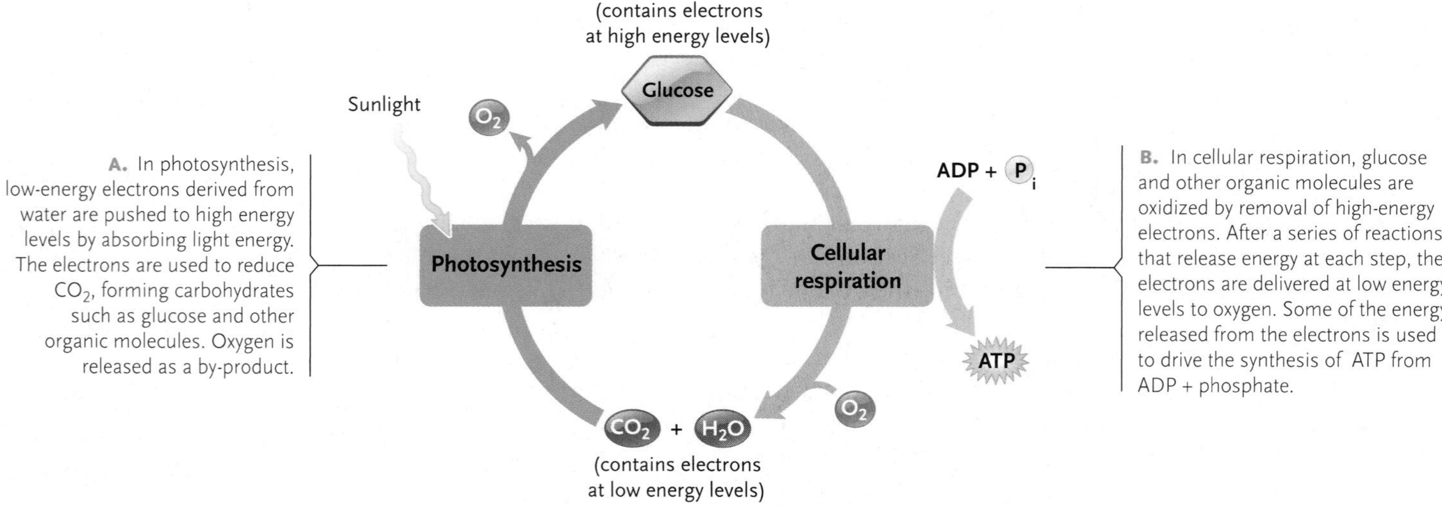
(contains electrons
at high energy levels)
Glucose
Sunlight
O_2
A. In photosynthesis, low-energy electrons derived from water are pushed to high energy levels by absorbing light energy. The electrons are used to reduce CO_2, forming carbohydrates such as glucose and other organic molecules. Oxygen is released as a by-product.
Photosynthesis
Cellular respiration
ADP + P_i
ATP
B. In cellular respiration, glucose and other organic molecules are oxidized by removal of high-energy electrons. After a series of reactions that release energy at each step, the electrons are delivered at low energy levels to oxygen. Some of the energy released from the electrons is used to drive the synthesis of ATP from ADP + phosphate.
O_2
CO_2 + H_2O
(contains electrons
at low energy levels)

Oxidation and reduction *invariably* are coupled reactions that remove electrons from a donor molecule and simultaneously add them to an acceptor molecule. In such coupled oxidation–reduction reactions, also called **redox reactions,** electrons release some of their energy as they pass from a donor molecule to an acceptor molecule. This free energy (see Section 4.1) is available for cellular work, such as ATP synthesis. ATP is the primary agent that couples exergonic and endergonic reactions in the cell to facilitate the synthesis of complex molecules (see Section 4.2). The harnessing of energy into a useful form such as ATP when electrons move from a high-energy state to a low-energy state is analogous to what happens when water moves through turbines at a dam and produces electricity.

Frequently, protons (hydrogen atoms that have been stripped of electrons, symbolized as H^+) are also removed from a molecule during oxidation. (Recall from Chapter 2 that a hydrogen atom, H, consists of a proton and an electron: $H = H^+ + e^-$.) The molecules that accept electrons may also combine with protons, as oxygen does when it is reduced to form water.

The gain or loss of an electron in a redox reaction is not always complete. That is, depending on the redox reaction, electrons are transferred completely from one atom to another, or alternatively, the degree of electron sharing in covalent bonds changes. The condition of electron sharing is said to involve a relative loss or gain of electrons; most redox reactions in the electron transfer system discussed later in the chapter are of this type. The redox reaction between methane and oxygen (the burning of natural gas in air) that produces carbon dioxide and water illustrates a change in the degree of electron sharing **(Figure 8.2).** The dots in the figure indicate the positions of the electrons involved in the covalent bonds of the reactants and products.

Compare the reactant methane with the product carbon dioxide. In methane, the covalent electrons are shared essentially equally between bonded C and H atoms because C and H are almost equally electronegative. In carbon dioxide, electrons are closer to the O atoms than to the C atom in the C=O bonds because O atoms are highly electronegative. This means that the C atom has partially "lost" its shared electrons in the reaction. In short, methane has been oxidized. Now compare the oxygen reactant with the product water. In the oxygen molecule, the two O atoms share their electrons equally. The oxygen reacts with the hydrogen from methane, producing water in which the electrons are closer to the O atom than to the H atoms. This means that each O atom has partially "gained" electrons; in short, oxygen has been reduced.

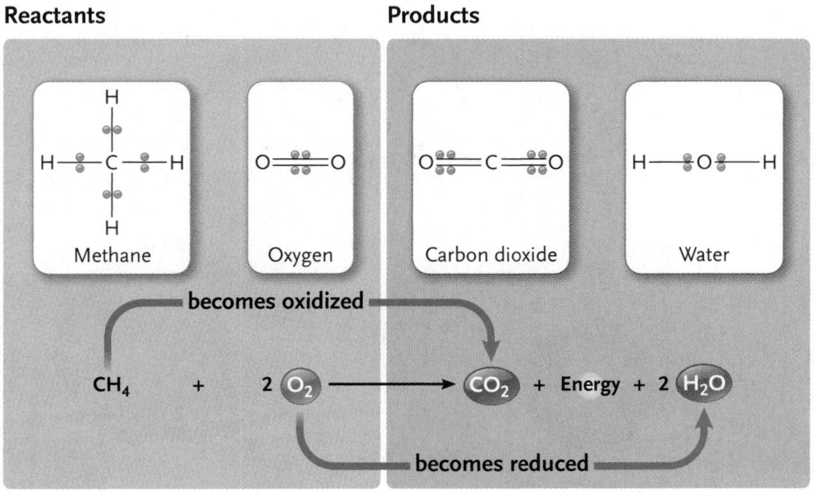

The movement of electrons away from an atom requires energy. The more electronegative an atom is, the greater the force that holds the electrons to that atom and therefore the greater the energy required to remove an electron from the atom. The changes in electron positions in a redox reaction consequently change the amount of chemical energy in the reactants and products. In our example of methane burning in oxygen, electrons are held more tightly in the product molecules (by being closer to the highly electronegative O atoms) than in the reactant molecules. Therefore, in this redox reaction, the potential energy of the reactants has dropped and chemical energy that can be used for cellular work is released (see Section 4.1).

Electrons Flow from Fuel Substances to Final Electron Acceptors

The energy of the electrons removed during cellular oxidations originates in the reactions of photosynthesis (see Figure 8.1A). During photosynthesis, electrons derived from water are pushed to very high energy levels using energy from the absorption of light. These high-energy electrons, together with H^+ from water, are combined with carbon dioxide to form sugar molecules and then are removed by the oxidative reactions that release energy for cellular activities (see Figure 8.1B). As electrons pass to acceptor molecules, they lose much of their energy. Some of this energy drives the synthesis of ATP from ADP and P_i (a phosphate group from an inorganic source) (see Section 4.2).

The total amount of energy obtained from electrons flowing through cellular oxidative pathways depends on the difference between their high energy level in fuel substances and the lower energy level in the molecule that acts as the *final acceptor* for electrons, that is, the last molecule reduced in cellular pathways. The lower the energy level in the final acceptor, the greater the yield of energy for cellular activities. Oxygen is the final acceptor in the most efficient and highly developed form of cellular oxidation: cellular respiration (see Figure 8.1B). The very low energy level of the electrons added to oxygen allows a maximum output of energy for ATP synthesis. As part of the final reduction, oxygen combines with protons and electrons to form water.

FIGURE 8.2
Relative loss and gain of electrons in a redox reaction, the burning of methane (natural gas) in oxygen. Compare the positions of the electrons in the covalent bonds of reactants and products. In this redox reaction, methane is oxidized and oxygen is reduced.

In Cellular Respiration, Cells Make ATP by Oxidative Phosphorylation

Cellular respiration includes both the reactions that transfer electrons from organic molecules to oxygen and the reactions that make ATP. These reactions are often written in a summary form that uses glucose ($C_6H_{12}O_6$) as the initial reactant:

$$C_6H_{12}O_6 + 6\ O_2 + 32\ ADP + 32\ P_i \rightarrow 6\ H_2O + 6\ CO_2 + 32\ ATP$$

In this overall reaction, electrons and protons are transferred from glucose to oxygen, forming water, and the carbons left after this transfer are released as carbon dioxide. ATP synthesis by the addition of P_i to ADP is the key, and final, step of this reaction. As discussed in Section 4.2, phosphorylation is the term for a reaction that adds a phosphate group to a substance such as ADP. How we derive the 32 ATP molecules is explained later in this chapter.

The entire process of cellular respiration can be divided into three stages **(Figure 8.3)**:

1. In **glycolysis,** enzymes break a molecule of glucose (contains six carbon atoms) into two molecules of pyruvate (an organic compound with a backbone of three carbon atoms). Some ATP is synthesized by **substrate-level phosphorylation,** an enzyme-catalyzed reaction that transfers a phosphate group from a substrate to ADP.
2. In **pyruvate oxidation,** enzymes convert the three-carbon pyruvate into a two-carbon acetyl group which enters the **citric acid cycle** where it is completely oxidized to carbon dioxide. Some ATP is synthesized during the citric acid cycle.
3. In **oxidative phosphorylation,** high-energy electrons produced from stages 1 and 2 are first delivered to oxygen by a sequence of electron carriers in the **electron transfer system.** Free energy released by the electron flow then generates an H^+ gradient in a process called **chemiosmosis.** The enzyme **ATP synthase** uses the H^+ gradient as the energy source to make ATP.

In eukaryotes, most of the reactions of cellular respiration occur in various regions of the mitochondrion **(Figure 8.4)**; only glycolysis is located in the cytosol. Pyruvate oxidation and the citric acid cycle take place in the mitochondrial matrix. The inner mitochondrial membrane houses the electron transfer system and the ATP synthase enzymes. Transport proteins, concentrated primarily in the inner membrane, control the substances that enter and leave mitochondria.

Researchers determined the locations of these reactions by studying mitochondria that had been isolated from broken-open cells by **cell fractionation**—a technique that divides the cell contents into fractions containing a single type of organ-

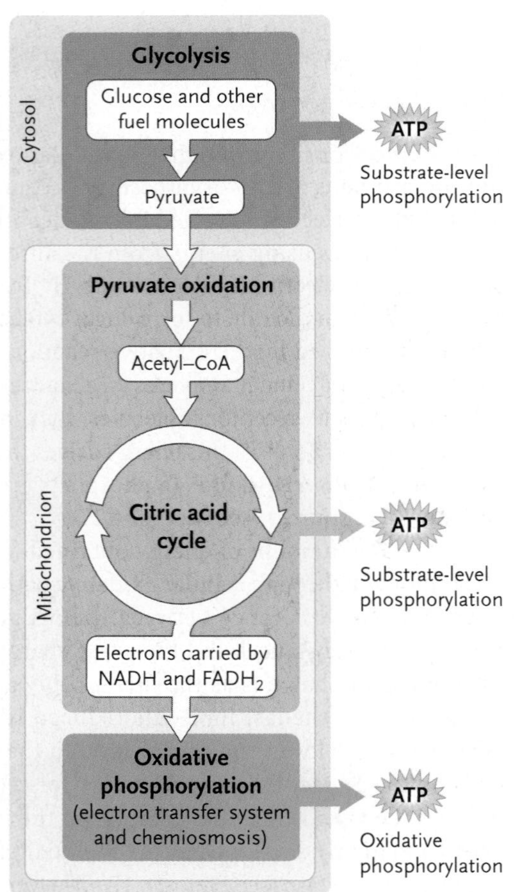

Cytosol
Glycolysis
Glucose and other fuel molecules
ATP
Substrate-level phosphorylation
Pyruvate
Mitochondrion
Pyruvate oxidation
Acetyl–CoA
Citric acid cycle
ATP
Substrate-level phosphorylation
Electrons carried by NADH and FADH$_2$
Oxidative phosphorylation (electron transfer system and chemiosmosis)
ATP
Oxidative phosphorylation

FIGURE 8.3
The three stages of cellular respiration: (1) glycolysis; (2) pyruvate oxidation and the citric acid cycle; and (3) oxidative phosphorylation which includes the electron transfer system and chemiosmosis.

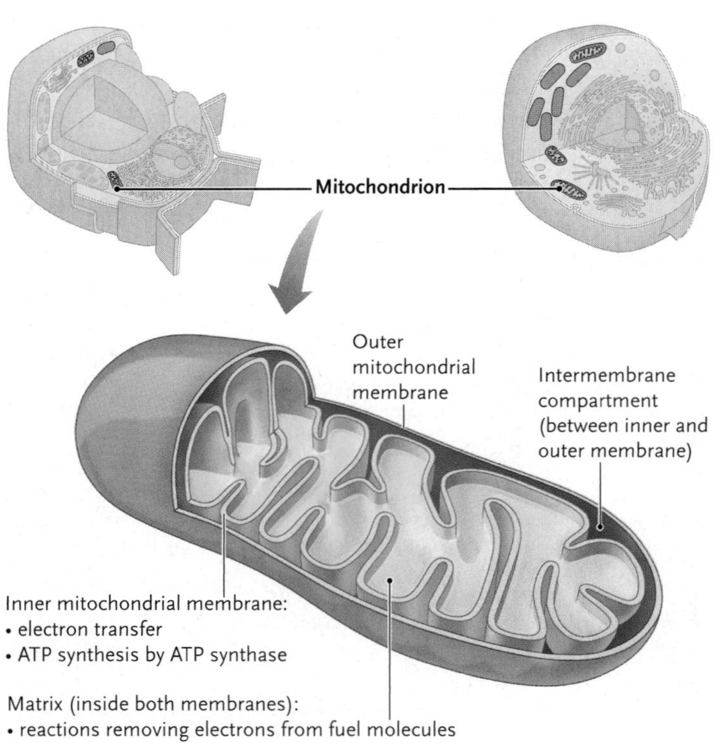

Mitochondrion
Outer mitochondrial membrane
Intermembrane compartment (between inner and outer membrane)
Inner mitochondrial membrane:
• electron transfer
• ATP synthesis by ATP synthase
Matrix (inside both membranes):
• reactions removing electrons from fuel molecules (pyruvate oxidation, citric acid cycle)

FIGURE 8.4
Membranes and compartments of a mitochondrion. Label lines that end in a dot indicate a compartment enclosed by the membranes.

elle, such as mitochondria or chloroplasts, or other structure, such as ribosomes (see Figure 5.8). The collected mitochondria were, in turn, fractionated into different subfractions using experimental treatments. For example, the outer and inner mitochondrial membranes react differently to particular detergents, allowing each membrane, as well as the solutions in the matrix and intermembrane compartment, to be purified individually. Each subfraction was then analyzed in detail to identify the locations of the individual reactions of cellular respiration.

In prokaryotes, glycolysis, pyruvate oxidation, and the citric acid cycle all take place in the cytosol. The other reactions of cellular respiration occur on the plasma membrane.

The next three sections examine the three stages of cellular respiration in turn.

STUDY BREAK 8.1

1. **Distinguish between oxidation and reduction.**
2. **Distinguish between cellular respiration and oxidative phosphorylation.**

THINK OUTSIDE THE BOOK

Collaboratively or individually, find techniques for cell fractionation and prepare an outline of the steps needed to isolate and purify mitochondria from an animal cell.

8.2 Glycolysis: Splitting the Sugar in Half

Glycolysis, the first series of oxidative reactions that remove electrons from cellular fuel molecules, takes place in the cytosol of all organisms. In glycolysis (*glykys* = sweet, *lysis* = breakdown), sugars such as glucose are partially oxidized and broken down into smaller molecules, and a relatively small amount of ATP is produced. Glycolysis is also known as the Embden–Meyerhof pathway in honor of Gustav Embden and Otto Meyerhof, two German physiological chemists who (separately) made the most important contributions to determining the sequence of reactions in the pathway. Meyerhof received a Nobel Prize in 1922 for his work.

Glycolysis starts with the six-carbon sugar glucose and produces two molecules of the three-carbon organic substance *pyruvate* or *pyruvic acid* in 10 sequential enzyme-catalyzed reactions. (The *-ate* suffix indicates the ionized form of an organic acid such as pyruvate, in which the carboxyl group —COOH dissociates to —COO^- + H^+ which is usual under cellular conditions.) Pyruvate still contains many electrons that can be removed by oxidation, and it is the primary fuel substance for the second stage of cellular respiration.

The Reactions of Glycolysis Include Energy-Requiring and Energy-Releasing Steps

The initial steps of glycolysis (marked by the red arrow in **Figure 8.5**) are energy-requiring reactions—2 ATP are hydrolyzed; they convert glucose into a phosphorylated derivative (not shown in the figure). In the subsequent energy-releasing part of glycolysis (see the blue arrow in Figure 8.5), electrons are removed from the phosphorylated derivatives of glucose and 4 ATP are produced, giving a net gain of 2 ATP. Two molecules of pyruvate are generated in the final reaction of the pathway.

The electrons removed from fuel molecules in glycolysis are accepted by the electron carrier molecule *nicotinamide adenine dinucleotide* **(Figure 8.6).** The oxidized form of this electron carrier is NAD^+; the reduced form, NADH, carries a pair of electrons and a proton removed from fuel molecules. Nicotinamide

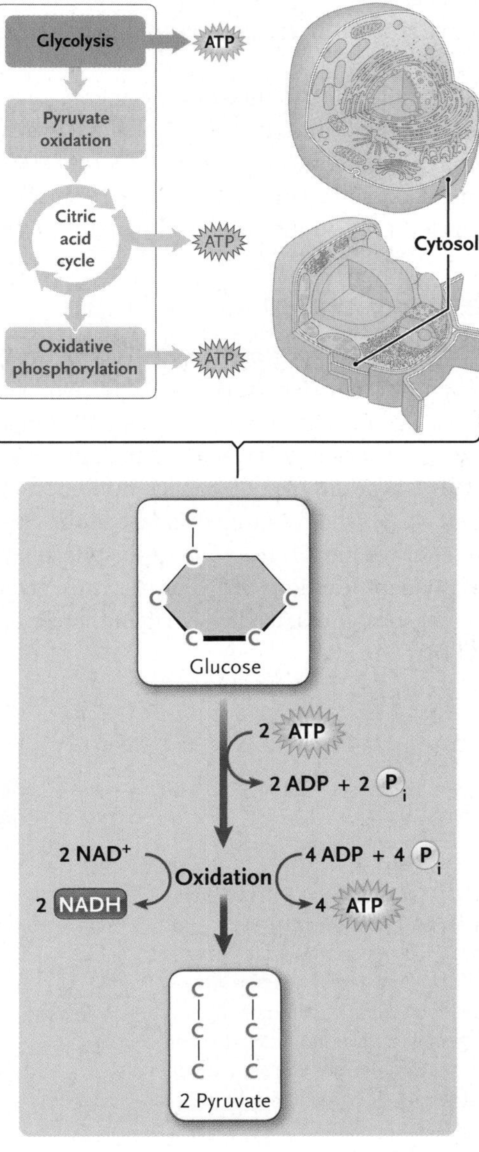
Glycolysis
ATP
Pyruvate oxidation
Citric acid cycle
ATP
Oxidative phosphorylation
ATP
Cytosol
C
C
C
C
C
C
Glucose
2 ATP
2 ADP + 2 Pi
2 NAD+
Oxidation
4 ADP + 4 Pi
2 NADH
4 ATP
C C
C C
C C
2 Pyruvate

FIGURE 8.5
Overall reactions of glycolysis. Glycolysis splits glucose (six carbons) into pyruvate (three carbons) and yields ATP and NADH.

adenine dinucleotide is one of several nucleotide-based carriers that move electrons in biological redox reactions (nucleotides are discussed in Section 3.6).

The reactions of glycolysis are shown in **Figure 8.7.** The one redox reaction in glycolysis, which occurs in reaction 6, removes two electrons and two protons from the three-carbon substance *glyceraldehyde-3-phosphate* (G3P). G3P is one of the phosphorylated glucose derivatives mentioned previously. Both electrons and one proton are picked up by NAD^+ to form NADH (see Figure 8.6). The other proton is released into the cytosol.

For each molecule of glucose that enters the pathway (see Figure 8.7), reactions 1–5 generate 2 molecules of G3P using 2 ATP, and reactions 6–10 convert the 2 molecules of G3P to 2 molecules of pyruvate, producing 4 ATP and 2 NADH. The net reactants and products of glycolysis are:

$$\text{glucose} + 2\ \text{ADP} + 2\ P_i + 2\ \text{NAD}^+ \rightarrow 2\ \text{pyruvate} + 2\ \text{NADH} + 2\ \text{H}^+ + 2\ \text{ATP}$$

The total of six carbon atoms in the two molecules of pyruvate is the same as in a glucose molecule; no carbons are released as CO_2 by glycolysis.

Each ATP molecule produced in the energy-releasing steps of glycolysis—steps 7 and 10 (see Figure 8.7)—results from *substrate-level phosphorylation* **(Figure 8.8).** In step 7 the molecule donating a phosphate group to ADP is 1,3-bisphosphoglycerate, and in step 10 the donor is phosphoenolpyruvate (PEP).

Glycolysis Is Regulated at Key Points

The rate of sugar oxidation by glycolysis is closely regulated by several mechanisms to match the cell's need for ATP. For example, if excess ATP is present in the cytosol it binds to *phosphofructokinase,* the enzyme that catalyzes reaction 3 in Figure 8.7, inhibiting its action. This is an example of feedback inhibition (see Section 4.5 and Figure 4.18). The resulting decrease in the concentration of the product of reaction 3, fructose-1,6-bisphosphate, slows or stops the subsequent reactions of glycolysis. Thus, glycolysis does not oxidize fuel substances needlessly when there is an adequate supply of ATP.

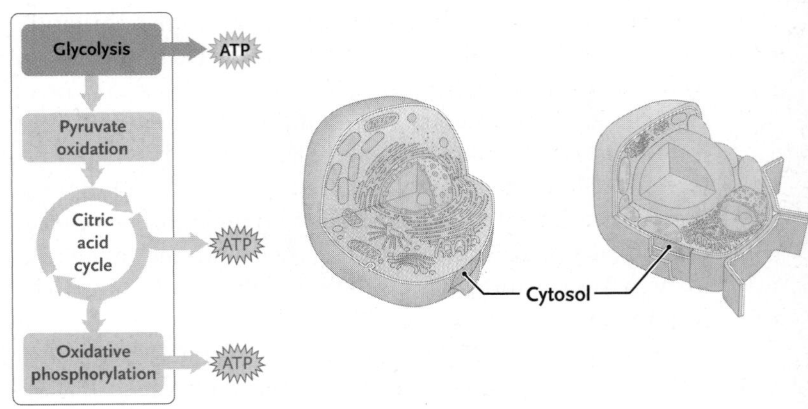
Glycolysis
ATP
Pyruvate oxidation
Citric acid cycle
ATP
Oxidative phosphorylation
ATP
Cytosol

FIGURE 8.7
Reactions of glycolysis, which occur in the cytosol. Because two molecules of G3P are produced in reaction 5, all the reactions from 6 to 10 are doubled (not shown). The names of the enzymes that catalyze each reaction are in rust.

If energy-requiring activities then take place in the cell, ATP concentration in the cytosol decreases, and ADP concentration increases. As a result, ATP is released from phosphofructokinase, relieving inhibition of the enzyme. In addition, ADP activates the enzyme. Therefore, the rates of glycolysis and ATP production increase proportionately as cellular activities convert ATP to ADP.

NADH also inhibits phosphofructokinase. This inhibition slows glycolysis if excess NADH is present, such as when oxidative phosphorylation has been slowed by limited oxygen supplies. The systems that regulate phosphofructokinase and other enzymes of glycolysis closely balance the rate of the pathway to produce adequate supplies of ATP and NADH without oxidizing excess quantities of glucose and other sugars.

Our discussion of the oxidative reactions that supply electrons now moves from the cytosol to mitochondria, the site of pyruvate oxidation and the citric acid cycle. These reactions complete the breakdown of fuel substances into carbon dioxide and provide most of the electrons that drive electron transfer and ATP synthesis.

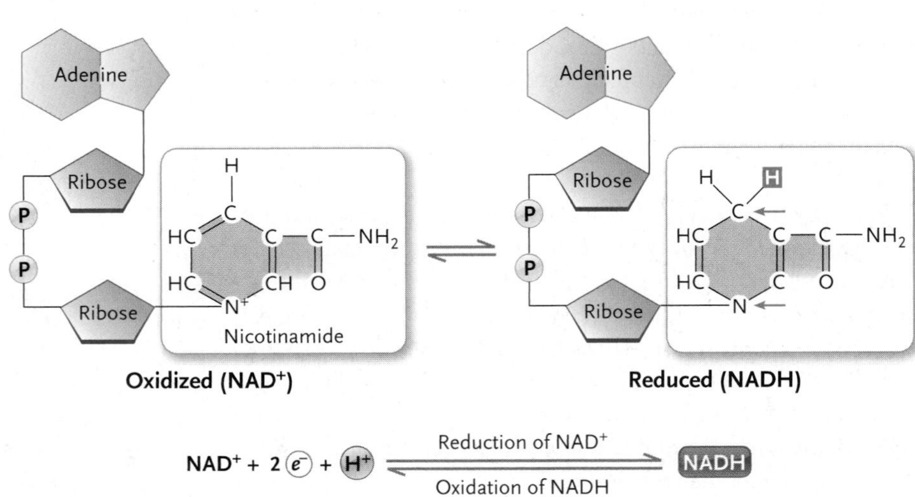
Adenine
Ribose
P
P
Ribose
H
C
HC
C
C
NH2
HC
CH
O
N+
Nicotinamide
Oxidized (NAD+)
Adenine
Ribose
P
P
Ribose
H
H
C
HC
C
C
NH2
HC
C
O
N
Reduced (NADH)
NAD+ + 2 e + H+
Reduction of NAD+
Oxidation of NADH
NADH

FIGURE 8.6
Electron carrier NAD^+. As the carrier is reduced to NADH, an electron is added at each of the two positions marked by a red arrow; a proton is also added at the position boxed in red. The nitrogenous base (blue) that adds and releases electrons and protons is nicotinamide, which is derived from the vitamin niacin (nicotinic acid).

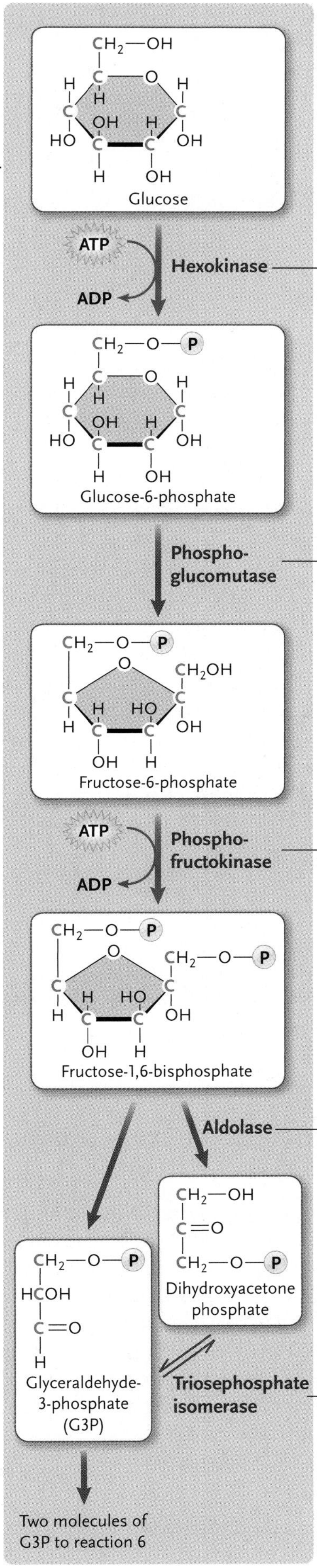

Glucose

ATP
Hexokinase
ADP

1 Glucose receives a phosphate group from ATP, producing glucose-6-phosphate. (phosphorylation reaction)

Glucose-6-phosphate

Phospho-glucomutase

2 Glucose-6-phosphate is rearranged into its isomer, fructose-6-phosphate. (isomerization reaction)

Fructose-6-phosphate

ATP
Phospho-fructokinase
ADP

3 Another phosphate group derived from ATP is attached to fructose-6-phosphate, producing fructose-1,6-bisphosphate. (phosphorylation reaction)

Fructose-1,6-bisphosphate

Aldolase

4 Fructose-1,6-bisphosphate is split into glyceraldehyde-3-phosphate (G3P) and dihydroxyacetone phosphate (DAP). (hydrolysis reaction)

Glyceraldehyde-3-phosphate (G3P)

Dihydroxyacetone phosphate

Triosephosphate isomerase

5 The DAP produced in reaction 4 is converted into G3P, giving a total of two of these molecules per molecule of glucose. (isomerization reaction)

Two molecules of G3P to reaction 6

Continued from reaction 5

G3P
(2 molecules)

NAD^+
Triosephosphate dehydrogenase
NADH
H^+
P_i

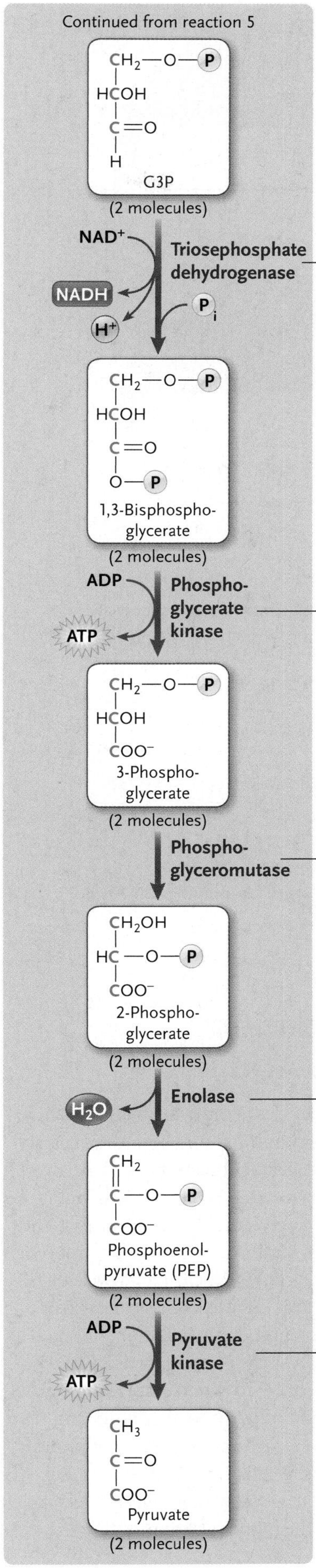

6 Two electrons and two protons are removed from G3P. Some of the energy released in this reaction is trapped by the addition of an inorganic phosphate group from the cytosol (not derived from ATP). The electrons are accepted by NAD^+, along with one of the protons. The other proton is released to the cytosol. (redox reaction)

1,3-Bisphosphoglycerate
(2 molecules)

ADP
Phosphoglycerate kinase
ATP

7 One of the two phosphate groups of 1,3-bisphosphoglycerate is transferred to ADP to produce ATP. (substrate-level phosphorylation reaction)

3-Phosphoglycerate
(2 molecules)

Phosphoglyceromutase

8 3-Phosphoglycerate is rearranged, shifting the phosphate group from the 3 carbon to the 2 carbon to produce 2-phosphoglycerate. (mutase reaction—shifting of a chemical group to another within same molecule)

2-Phosphoglycerate
(2 molecules)

Enolase
H_2O

9 Electrons are removed from one part of 2-phosphoglycerate and delivered to another part of the molecule. Most of the energy lost by the electrons is retained in the product, phosphoenolpyruvate. (redox reaction)

Phosphoenolpyruvate (PEP)
(2 molecules)

ADP
Pyruvate kinase
ATP

10 The remaining phosphate group is removed from phosphoenolpyruvate and transferred to ADP. The reaction forms ATP and the final product of glycolysis, pyruvate. (substrate-level phosphorylation reaction)

Pyruvate
(2 molecules)

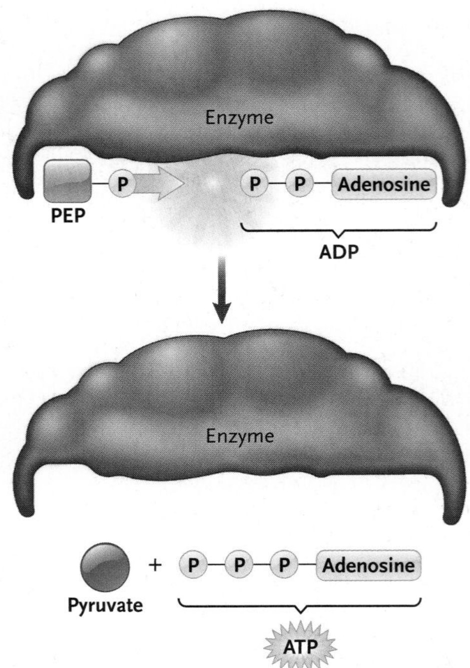
Enzyme
P
P
P
Adenosine
PEP
ADP
Enzyme
P
P
P
Adenosine
Pyruvate
ATP

FIGURE 8.8
Mechanism that synthesizes ATP by substrate-level phosphorylation. A phosphate group is transferred from a high-energy donor directly to ADP, forming ATP.

STUDY BREAK 8.2

1. **What are the energy-requiring and energy-releasing steps of glycolysis?**
2. **What is the redox reaction in glycolysis?**
3. **How is ATP synthesized in glycolysis?**
4. **Why is phosphofructokinase a target for inhibition by ATP?**

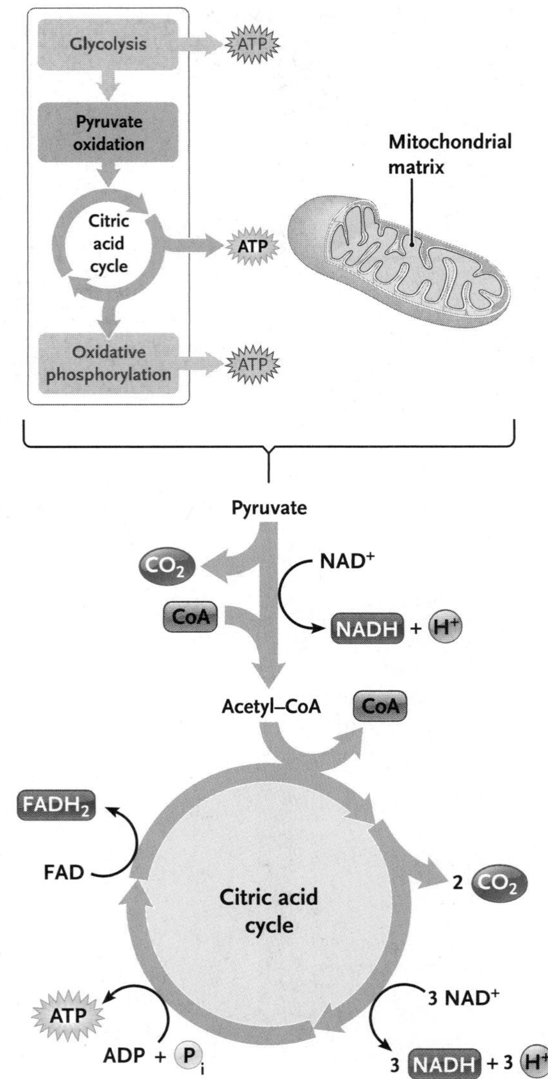
Glycolysis
ATP
Pyruvate oxidation
Mitochondrial matrix
Citric acid cycle
ATP
Oxidative phosphorylation
ATP
Pyruvate
CO_2
NAD^+
CoA
NADH + H^+
Acetyl–CoA
CoA
$FADH_2$
FAD
Citric acid cycle
2 CO_2
ATP
3 NAD^+
ADP + P_i
3 NADH + 3 H^+

FIGURE 8.9
Overall reactions of pyruvate oxidation and the citric acid cycle. Each turn of the cycle oxidizes an acetyl group of acetyl–CoA to 2 CO_2. Acetyl–CoA, NAD^+, FAD, and ADP enter the cycle; CoA, NADH, $FADH_2$, ATP, and CO_2 are released as products.

8.3 Pyruvate Oxidation and the Citric Acid Cycle

Glycolysis produces pyruvate molecules in the cytosol, and an active transport mechanism moves them into the mitochondrial matrix where pyruvate oxidation and the citric acid cycle proceed. An overview of these two processes is presented in **Figure 8.9.** Oxidation of pyruvate generates CO_2, acetyl–coenzyme A (acetyl–CoA), and NADH in a redox reaction involving NAD^+. The acetyl group of acetyl–CoA enters the **citric acid cycle.** As the citric acid cycle turns, every available electron carried into the cycle from pyruvate oxidation is transferred to NAD^+ or to another nucleotide-based molecule, *flavin adenine dinucleotide* (FAD; the reduced form is $FADH_2$). With each turn of the cycle, substrate-level phosphorylation produces 1 ATP. The combined action of pyruvate oxidation and the citric acid cycle oxidizes the three-carbon products of glycolysis completely to carbon dioxide. The 3 NADH and 1 $FADH_2$ produced for each acetyl–CoA during this stage carry high-energy electrons to the electron transfer system in the mitochondrion.

Pyruvate Oxidation Produces the Two-Carbon Fuel of the Citric Acid Cycle

In pyruvate oxidation (also called **pyruvic acid oxidation**), a multienzyme complex removes the $—COO^-$ from pyruvate as CO_2 and then oxidizes the remaining two-carbon fragment of pyruvate to an acetyl group ($CH_3CO—$) **(Figure 8.10).** Two electrons and two protons are released by these reactions; the electrons and one proton are accepted by NAD^+, reducing it to NADH, and the other proton is released as free H^+. The acetyl group is transferred to the nucleotide-based carrier *coenzyme A* (CoA). As acetyl–CoA, it carries acetyl groups to the citric acid cycle.

In summary, the pyruvate oxidation reaction is:

$$\text{pyruvate} + \text{CoA} + NAD^+ \rightarrow \text{acetyl–CoA} + \text{NADH} + H^+ + CO_2$$

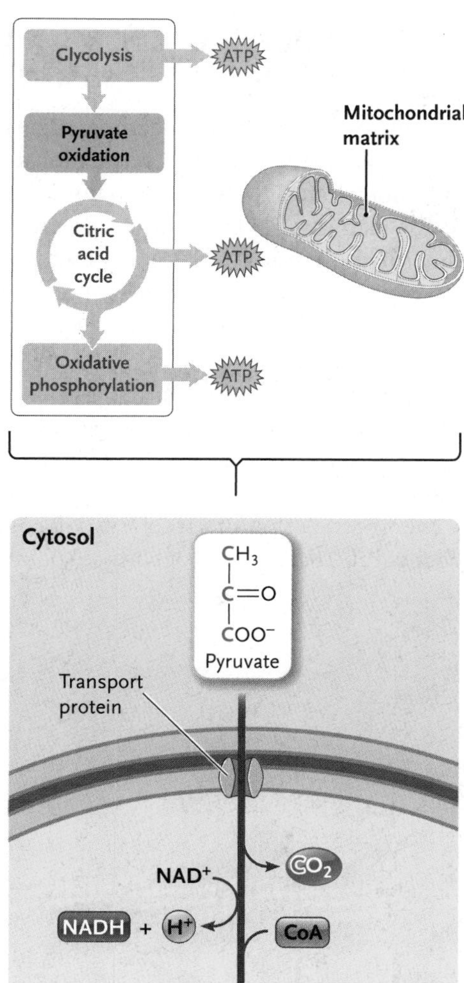
Glycolysis
ATP
Pyruvate oxidation
Mitochondrial matrix
Citric acid cycle
ATP
Oxidative phosphorylation
ATP
Cytosol
CH3
C=O
COO−
Pyruvate
Transport protein
NAD+
CO2
NADH + H+
CoA
Acetyl group
CH3
C=O
CoA
Acetyl–CoA
Mitochondrial matrix

FIGURE 8.10

Reactions of pyruvate oxidation. Pyruvate (three carbons) is oxidized to an acetyl group (two carbons), which is carried to the citric acid cycle by CoA. The third carbon is released as CO_2. NAD^+ accepts two electrons and one proton removed in the oxidation. The acetyl group carried from the reaction by CoA is the fuel for the citric acid cycle.

Because each glucose molecule that enters glycolysis produces two molecules of pyruvate, all the reactants and products in this equation are doubled when pyruvate oxidation is considered a continuation of glycolysis.

The Citric Acid Cycle Oxidizes Acetyl Groups Completely to CO_2

The reactions of the citric acid cycle **(Figure 8.11)** oxidize acetyl groups completely to CO_2 and synthesize some ATP molecules. The citric acid cycle gets its name from citrate, the product of the first reaction of the cycle. It is also called the **tricarboxylic acid cycle** or the **Krebs cycle,** named after Hans Krebs, a German-born scientist who worked out the majority of the reactions in the cycle in research he conducted in England beginning in 1932. How did Krebs discover the citric acid cycle? Using minced pigeon breast muscle (a tissue used in flight that has a very high rate of cellular respiration) he demonstrated that the four-carbon organic acids succinate, fumarate, malate, and oxaloacetate stimulated the consumption of oxygen by the tissue. He also showed that the oxidation of pyruvate by the muscle tissue is stimulated by the six-carbon organic acids citrate and isocitrate, and the five-carbon acid α-ketoglutarate. While several other researchers pieced together segments of the reaction series, Krebs was the first to reason that the organic acids were linked into a cycle of reactions rather than a linear series. Krebs was awarded a Nobel Prize in 1953 for his elucidation of the citric acid cycle.

The citric acid cycle has eight reactions, each catalyzed by a specific enzyme. All of the enzymes are located in the mitochondrial matrix except the enzyme for reaction 6, succinate dehydrogenase, which is bound to the inner mitochondrial membrane on the matrix side. In a complete turn of the cycle, one two-carbon acetyl unit is consumed and two molecules of CO_2 are released (at reactions 3 and 4), thereby completing the conversion of all the C atoms originally in glucose to CO_2. The CoA molecule that carried the acetyl group to the cycle is released and participates again in pyruvate oxidation to pick up another acetyl group. Electron pairs are removed at each of four oxidations in the cycle (reactions 3, 4, 6, and 8). Three of the oxidations use NAD^+ as the electron acceptor, producing 3 NADH, and one uses FAD, producing 1 $FADH_2$. Substrate-level phosphorylation generates 1 ATP as part of reaction 5. Therefore, the net reactants and products of one turn of the citric acid cycle are:

$$1\ \text{acetyl–CoA} + 3\ NAD^+ + 1\ FAD + 1\ ADP + 1\ P_i + 2\ H_2O \rightarrow 2\ CO_2 + 3\ NADH + 1\ FADH_2 + 1\ ATP + 3\ H^+ + 1\ CoA$$

Because one molecule of glucose is converted to two molecules of pyruvate by glycolysis and each molecule of pyruvate is converted to one acetyl group, all the reactants and products in this equation are doubled when the citric acid cycle is considered a continuation of glycolysis and pyruvate oxidation.

Most of the energy released by the four oxidations of the cycle is associated with the high-energy electrons carried by the 3 NADH and 1 $FADH_2$. These high-energy electrons enter the electron transfer system, where their energy is used to make most of the ATP produced in cellular respiration.

Like glycolysis, the citric acid cycle is regulated at several steps to match its rate to the cell's requirements for ATP. For example, the enzyme that catalyzes the first reaction of the citric acid cycle, *citrate synthase,* is inhibited by elevated ATP concentrations. The inhibitions automatically slow or stop the cycle when ATP production exceeds the demands of the cell and, by doing so, conserve cellular fuels.

FIGURE 8.11

Reactions of the citric acid cycle. Acetyl–CoA, NAD^+, FAD, and ADP enter the cycle; CoA, NADH, $FADH_2$, ATP, and CO_2 are released as products. The CoA released in reaction 1 can cycle back for another turn of pyruvate oxidation. Enzyme names are in rust.

Carbohydrates, Fats, and Proteins Can Function as Electron Sources for Oxidative Pathways

In addition to glucose and other six-carbon sugars, reactions leading from glycolysis through pyruvate oxidation also oxidize a wide range of carbohydrates, lipids, and proteins. These molecules enter the reaction pathways at various points. **Figure 8.12** summarizes the cellular pathways involved. It shows the central role of CoA in funneling acetyl groups from different pathways into the citric acid cycle and of the mitochondrion as the site where most of these groups are oxidized.

Carbohydrates such as sucrose and other disaccharides are easily broken into monosaccharides such as glucose and fructose that enter glycolysis at early steps. Starch (see Figure 3.7A) is hydrolyzed by digestive enzymes into individual glucose molecules, which enter the first reaction of glycolysis. Glycogen, a more complex carbohydrate that consists of glucose subunits (see Fig-

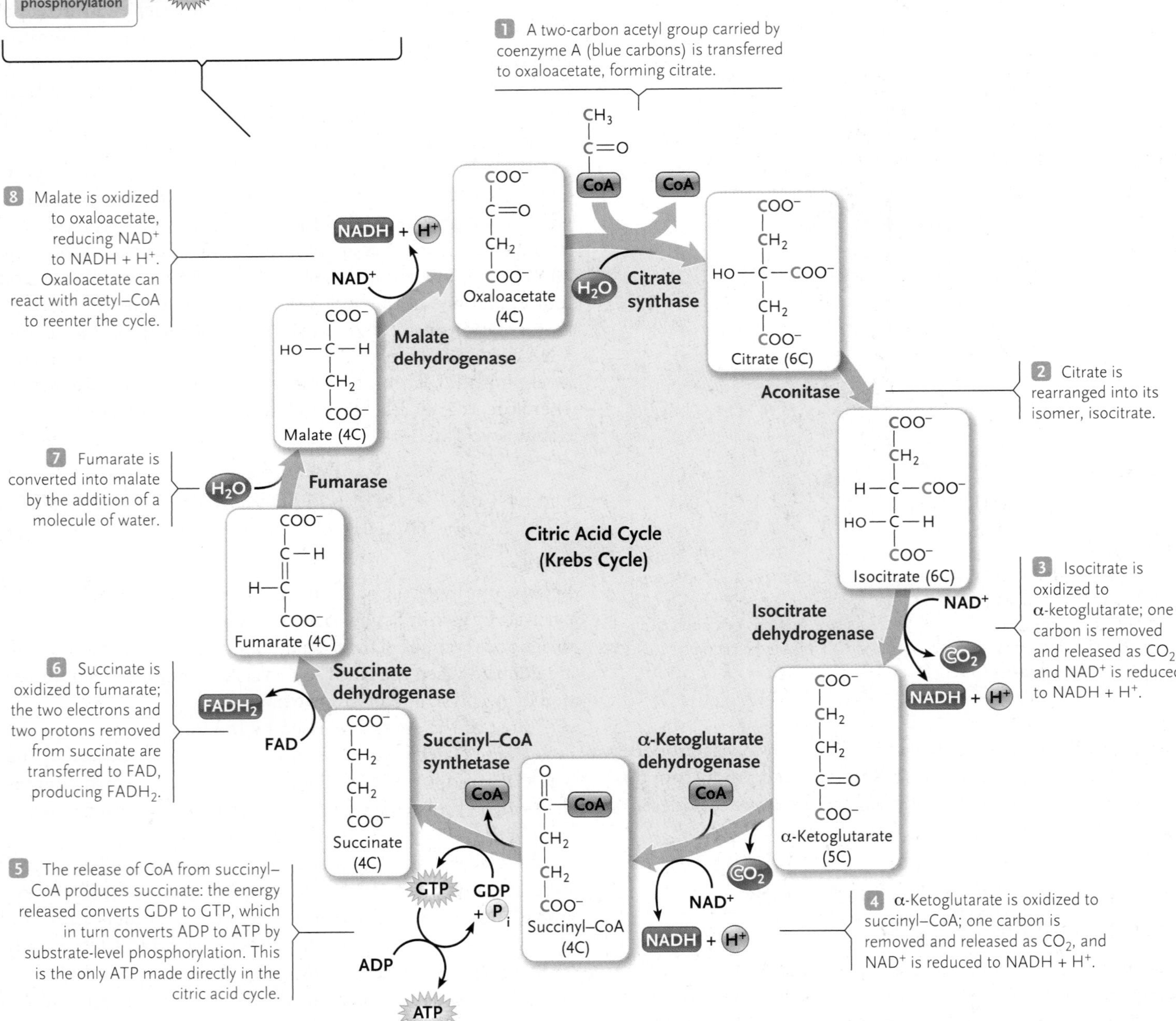

Glycolysis
ATP
Pyruvate oxidation
Citric acid cycle
ATP
Oxidative phosphorylation
ATP
Mitochondrial matrix
1 A two-carbon acetyl group carried by coenzyme A (blue carbons) is transferred to oxaloacetate, forming citrate.
CH3
C=O
CoA
CoA
COO⁻
C=O
CH2
COO⁻
Oxaloacetate (4C)
H2O
Citrate synthase
COO⁻
CH2
HO—C—COO⁻
CH2
COO⁻
Citrate (6C)
Aconitase
2 Citrate is rearranged into its isomer, isocitrate.
COO⁻
CH2
H—C—COO⁻
HO—C—H
COO⁻
Isocitrate (6C)
NAD+
Isocitrate dehydrogenase
CO2
NADH + H+
3 Isocitrate is oxidized to α-ketoglutarate; one carbon is removed and released as CO2, and NAD+ is reduced to NADH + H+.
COO⁻
CH2
CH2
C=O
COO⁻
α-Ketoglutarate (5C)
α-Ketoglutarate dehydrogenase
CoA
CO2
NAD+
NADH + H+
4 α-Ketoglutarate is oxidized to succinyl–CoA; one carbon is removed and released as CO2, and NAD+ is reduced to NADH + H+.
O
C—CoA
CH2
CH2
COO⁻
Succinyl–CoA (4C)
Succinyl–CoA synthetase
CoA
GTP
GDP + Pi
ADP
ATP
5 The release of CoA from succinyl–CoA produces succinate: the energy released converts GDP to GTP, which in turn converts ADP to ATP by substrate-level phosphorylation. This is the only ATP made directly in the citric acid cycle.
COO⁻
CH2
CH2
COO⁻
Succinate (4C)
Succinate dehydrogenase
FADH2
FAD
6 Succinate is oxidized to fumarate; the two electrons and two protons removed from succinate are transferred to FAD, producing FADH2.
COO⁻
C—H
H—C
COO⁻
Fumarate (4C)
Fumarase
H2O
7 Fumarate is converted into malate by the addition of a molecule of water.
COO⁻
HO—C—H
CH2
COO⁻
Malate (4C)
Malate dehydrogenase
NAD+
NADH + H+
8 Malate is oxidized to oxaloacetate, reducing NAD+ to NADH + H+. Oxaloacetate can react with acetyl–CoA to reenter the cycle.
Citric Acid Cycle (Krebs Cycle)

ure 3.7B), is broken down and converted by enzymes into glucose-6-phosphate, which enters glycolysis at reaction 2 of Figure 8.7.

Among the fats, triglycerides (see Figure 3.9) are major sources of electrons for ATP synthesis. Before entering the oxidative reactions, the triglycerides are hydrolyzed into glycerol and individual fatty acids. The glycerol is converted to G3P and enters glycolysis at reaction 6 of Figure 8.7, in the ATP-producing portion of the pathway. The fatty acids—and many other types of lipids—are split into two-carbon fragments, which enter the citric acid cycle as acetyl–CoA. The energy released by the oxidation of fats, by weight, is comparatively high—a little more than twice the energy of oxidation of carbohydrates or proteins. This fact explains why fats are an excellent source of energy in the diet.

Proteins are hydrolyzed to amino acids before oxidation. During oxidation the amino group is removed, and the remainder of the molecule enters the pathway of carbohydrate oxidation as either pyruvate, acetyl units carried by CoA, or intermediates of the citric acid cycle. For example, the amino acid alanine is converted into pyruvate, leucine is converted into acetyl units, and phenylalanine is converted into fumarate. Fumarate enters the citric acid cycle at reaction 7 of Figure 8.11.

STUDY BREAK 8.3 <

Summarize the fate of pyruvate molecules produced by glycolysis.

8.4 Oxidative Phosphorylation: The Electron Transfer System and Chemiosmosis

From the standpoint of ATP synthesis, the most significant products of glycolysis, pyruvate oxidation, and the citric acid cycle are the many high-energy electrons removed from fuel molecules and picked up by the carrier molecules NAD^+ or FAD as a result of redox reactions. These electrons are released by the carriers into the electron transfer system of mitochondria (in eukaryotes).

The **mitochondrial electron transfer system** consists of a series of electron carriers that alternately pick up and release electrons, and ultimately transfer them to their final acceptor, oxygen. As the electrons flow through the system they release free energy which is used to build a gradient of H^+ across the inner mitochondrial membrane. The gradient goes from a high concentration of H^+ in the intermembrane compartment to a low concentration of H^+ in the matrix. The H^+ gradient supplies the energy that drives ATP synthesis by mitochondrial ATP synthase.

Ralph Pleasant

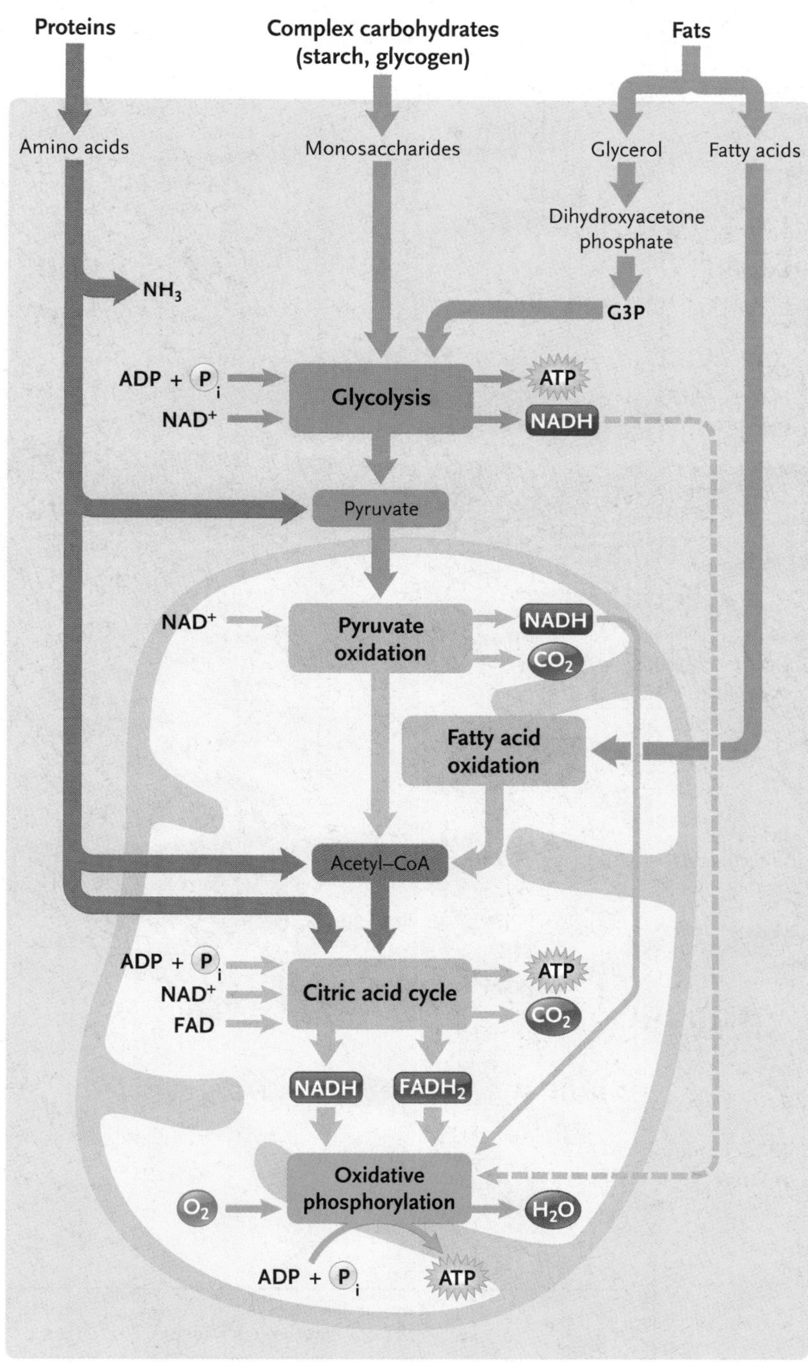
Proteins
Complex carbohydrates (starch, glycogen)
Fats
Amino acids
Monosaccharides
Glycerol
Fatty acids
Dihydroxyacetone phosphate
G3P
NH3
ADP + Pi
NAD+
Glycolysis
ATP
NADH
Pyruvate
NAD+
Pyruvate oxidation
NADH
CO2
Fatty acid oxidation
Acetyl–CoA
ADP + Pi
NAD+
FAD
Citric acid cycle
ATP
CO2
NADH
FADH2
O2
Oxidative phosphorylation
H2O
ADP + Pi
ATP

FIGURE 8.12
Major pathways that oxidize carbohydrates, fats, and proteins. Reactions that occur in the cytosol are shown against a tan background; reactions that occur in mitochondria are shown inside the organelle. CoA funnels the products of many oxidative pathways into the citric acid cycle.

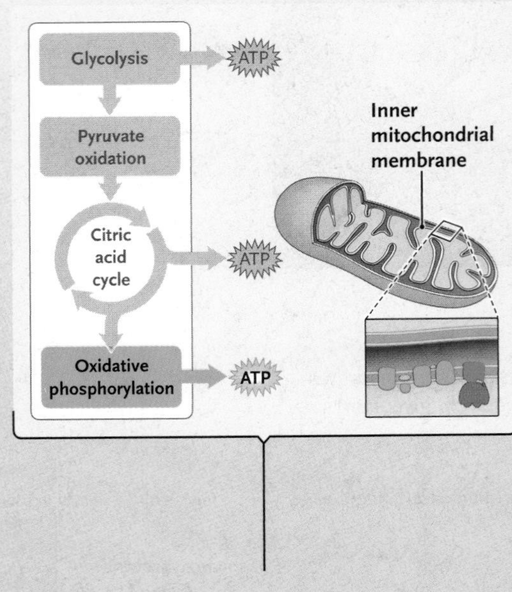

Glycolysis
ATP
Pyruvate oxidation
Citric acid cycle
ATP
Oxidative phosphorylation
ATP
Inner mitochondrial membrane

1 Complex I picks up high-energy electrons from NADH and conducts them via two electron carriers, FMN (flavin mononucleotide) and an Fe/S (iron–sulfur) protein, to ubiquinone.

2 Complex II oxidizes $FADH_2$ to FAD; the two electrons released are transferred to ubiquinone, and the two protons released go into the matrix. Electrons that pass to ubiquinone by the complex II reaction bypass complex I of the electron transfer system.

3 Complex III accepts electrons from ubiquinone and transfers them through the electron carriers in the complex—cytochrome *b*, an Fe/S protein, and cytochrome c_1—to cytochrome *c*, which is free in the intermembrane space.

4 Complex IV accepts electrons from cytochrome *c* and delivers them via electron carriers cytochromes *a* and a_3 to oxygen. Four protons are added to a molecule of O_2 as it accepts four electrons, forming 2 H_2O.

5 As electrons move through the electron transfer system, they release free energy. Part of the released energy is lost as heat, but some is used by the mitochondrion to transport H^+ across the inner mitochondrial membrane from the matrix to the inter membrane compartment at complexes I, III, and IV.

6 The resulting H^+ gradient supplies the energy that drives ATP synthesis by ATP synthase.

7 Because of the gradient, H^+ flows across the inner membrane and into the matrix through a channel in the ATP synthase.

8 The flow of H^+ activates ATP synthase, making the headpiece and stalk rotate.

9 As a result of changes in shape and position as it turns, the headpiece catalyzes the synthesis of ATP from ADP and P_i.

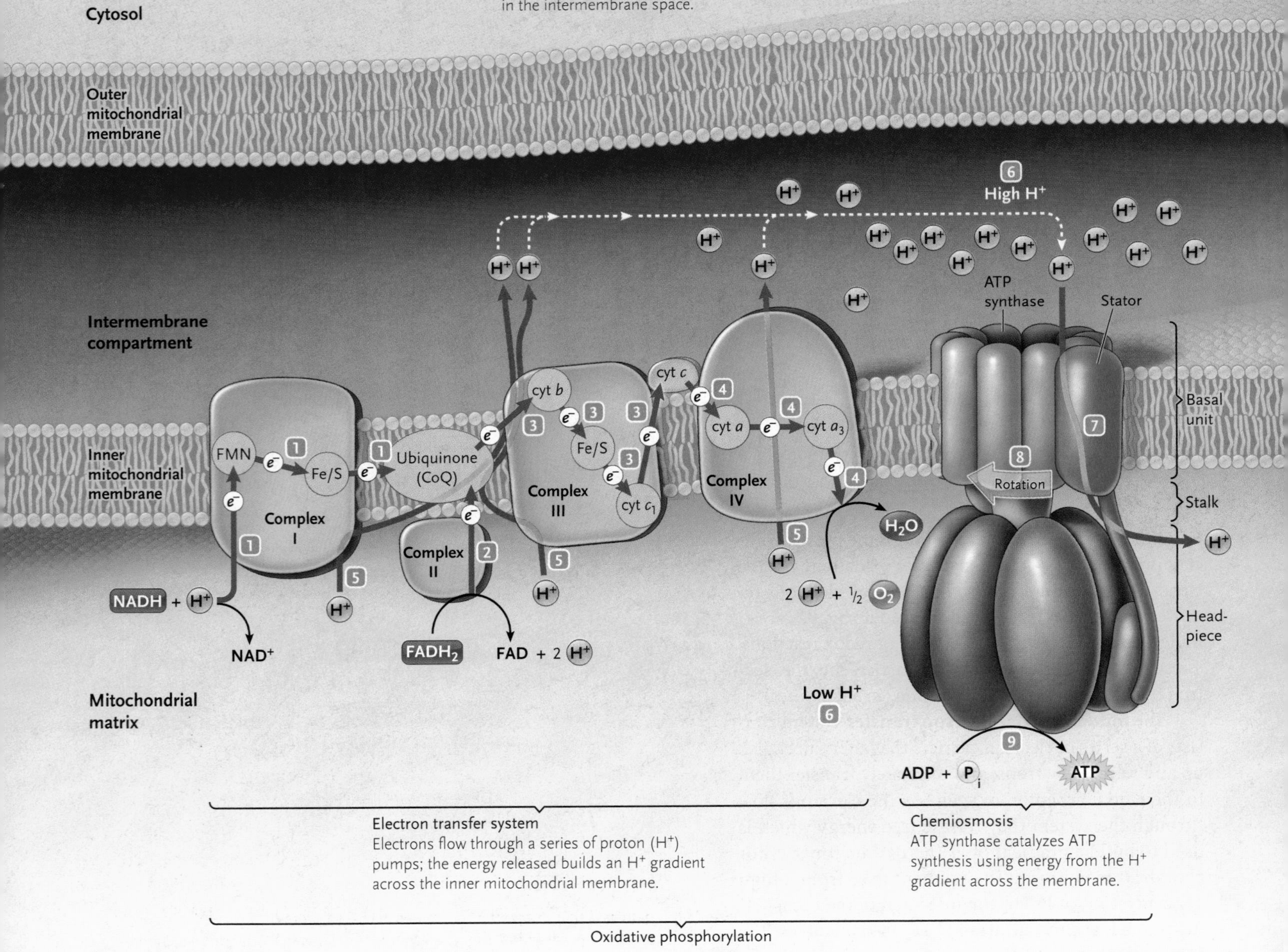

Cytosol

Outer mitochondrial membrane

Intermembrane compartment

Inner mitochondrial membrane

Mitochondrial matrix

Complex I · FMN · Fe/S · Ubiquinone (CoQ) · Complex II · Complex III · cyt *b* · Fe/S · cyt c_1 · cyt *c* · Complex IV · cyt *a* · cyt a_3 · ATP synthase · Stator · Basal unit · Stalk · Headpiece · Rotation

NADH + H^+ → NAD^+

$FADH_2$ → FAD + 2 H^+

2 H^+ + ½ O_2 → H_2O

High H^+ · Low H^+

ADP + P_i → ATP

Electron transfer system
Electrons flow through a series of proton (H^+) pumps; the energy released builds an H^+ gradient across the inner mitochondrial membrane.

Chemiosmosis
ATP synthase catalyzes ATP synthesis using energy from the H^+ gradient across the membrane.

Oxidative phosphorylation

FIGURE 8.13
Oxidative phosphorylation: The mitochondrial electron transfer system and chemiosmosis. Oxidative phosphorylation involves the electron transfer system (steps 1–6), and chemiosmosis by ATP synthase (steps 7–9). Blue arrows indicate electron flow; red arrows indicate H^+ movement.

In the Electron Transfer System, Electrons Flow through Protein Complexes in the Inner Mitochondrial Membrane

The mitochondrial electron transfer system includes three major protein complexes, which serve as electron carriers **(Figure 8.13).** These protein complexes, numbered I, III, and IV, are integral membrane proteins located in the inner mitochondrial membrane. In addition, a smaller complex, complex II, is bound to the inner mitochondrial membrane on the matrix side. Associated with the system are two small, highly mobile electron carriers, *cytochrome c* and *ubiquinone* (also known as coenzyme Q, or CoQ), which shuttle electrons between the major complexes.

Cytochromes are proteins with a heme prosthetic group that contains an iron atom. (Heme is an iron-containing group that binds oxygen; see Figure 3.20. A prosthetic group is a cofactor that binds tightly to a protein or enzyme; see Section 4.4.) The iron atom accepts and donates electrons. One cytochrome in particular, cytochrome *c,* was important historically in determining evolutionary relationships. Because the amino acid sequence of cytochrome *c* has been conserved by evolution, comparisons of cytochrome *c* amino acid sequences from a wide variety of eukaryotic species contributed to the construction of phylogenetic trees (see Figure 23.10). Present-day phylogeny studies typically use DNA or RNA sequences.

Electrons flow through the major protein complexes as shown in Figure 8.13, steps 1–4. Note that complex II, a succinate dehydrogenase complex, catalyzes two reactions. One is reaction 6 of the citric acid cycle, the conversion of succinate to fumarate (see Figure 8.11). In that reaction, FAD accepts two protons and two electrons and is reduced to $FADH_2$. The other reaction is shown in Figure 8.13, step 2.

The poison cyanide does its deadly work by blocking the transfer of electrons from complex IV to oxygen. The gas carbon monoxide inhibits complex IV activity, leading to abnormalities in mitochondrial function. In this way, the carbon monoxide in tobacco smoke contributes to the development of diseases associated with smoking.

What is the driving force for electron flow through the protein complexes of the electron transfer system? **Figure 8.14** shows that the individual electron carriers of the system are organized specifically from high to low free energy. Any single component has a higher affinity for electrons than the preceding carrier in the series has. Overall, molecules such as NADH contain an abundance of free energy and can be oxidized readily, whereas O_2, the terminal electron acceptor of the series, can be reduced easily. As a consequence of this organization, electron movement through the system is spontaneous, releasing free energy.

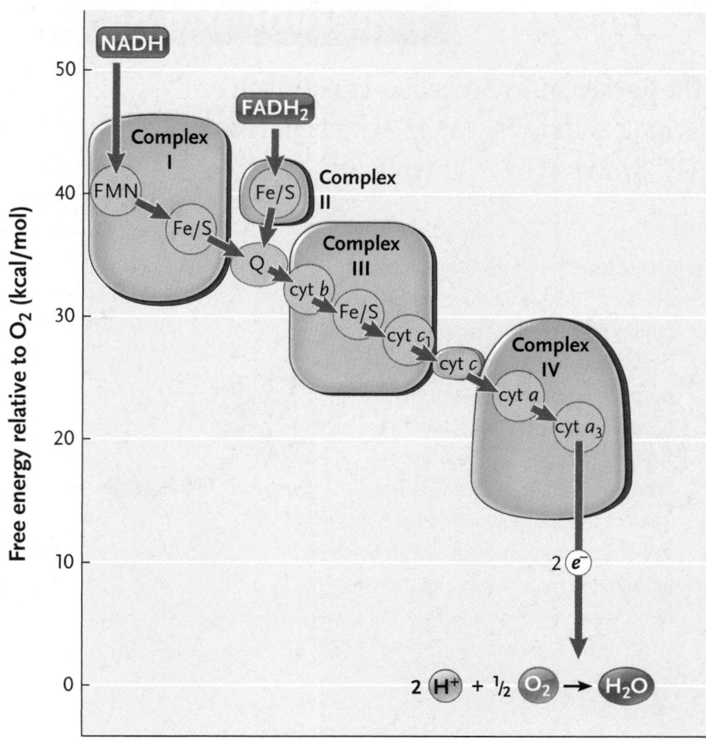
NADH
FADH₂
Complex I
FMN
Fe/S
Fe/S
Complex II
Q
Complex III
cyt b
Fe/S
cyt c₁
cyt c
Complex IV
cyt a
cyt a₃
2 e⁻
2 H⁺ + ½ O₂ → H₂O
50
40
30
20
10
0
Free energy relative to O₂ (kcal/mol)
Direction of electron flow

FIGURE 8.14
Organization of the mitochondrial electron transfer system from high to low free energy. Electrons flow spontaneously from one molecule to the next in the series.

Ubiquinone and the Three Major Electron Transfer Complexes Pump H^+ across the Inner Mitochondrial Membrane

Using energy from electron flow, ubiquinone and the proteins of complexes I, III, and IV pump (actively transport) H^+ (protons) from the matrix to the intermembrane compartment of the mitochondrion (see Figure 8.13, step 5). The result is an H^+ gradient with a high concentration of H^+ in the intermembrane compartment and a low concentration of H^+ in the matrix (step 6). Because protons carry a positive charge, the asymmetric distribution of protons generates an electrical and chemical gradient across the inner mitochondrial membrane, with the intermembrane compartment more positively charged than the matrix. The combination of a proton gradient and voltage gradient across the membrane produces stored energy known as the **proton-motive force.** This force contributes energy for ATP synthesis, as well as for the cotransport of substances to and from mitochondria (see Section 6.4).

Chemiosmosis Powers ATP Synthesis by an H^+ Gradient

Within the mitochondrion ATP is synthesized by ATP synthase, an enzyme embedded in the inner mitochondrial membrane. In 1961, British scientist Peter Mitchell of Glynn Research Laboratories proposed that mitochondrial electron transfer produces an H^+ gradient and that the gradient powers ATP synthesis by ATP synthase. He called this pioneering model the **chemiosmotic hypothesis;** the process is commonly called *chemiosmosis* (see Figure 8.13). At the time, this hypothesis was a radical proposal because most researchers thought that the energy of electron transfer was stored as a high-energy chemical intermediate. No such in-

FIGURE 8.15 **Experimental Research**

The Racker and Stoeckenius Experiment Demonstrating That an H^+ Gradient Powers ATP Synthesis by ATP Synthase

Question: Does an H^+ gradient power ATP synthase-catalyzed ATP synthesis, thereby supporting Mitchell's chemiosmotic hypothesis?

Experiment: Efraim Racker of Cornell University and Walther Stoeckenius of University of California, San Francisco, made membrane vesicles that had a proton pump and ATP synthase to determine whether proton-motive force drives ATP synthesis.

1. The researchers constructed synthetic phospholipid membrane vesicles containing a segment of purple surface membrane from an archaean. The purple membrane contained only bacteriorhodopsin, a protein that resembles rhodopsin, the visual pigment of animals (see Chapter 39). Bacteriorhodopsin is a light-activated proton pump. The researchers illuminated the vesicles and then analyzed the concentration of H^+ in them.

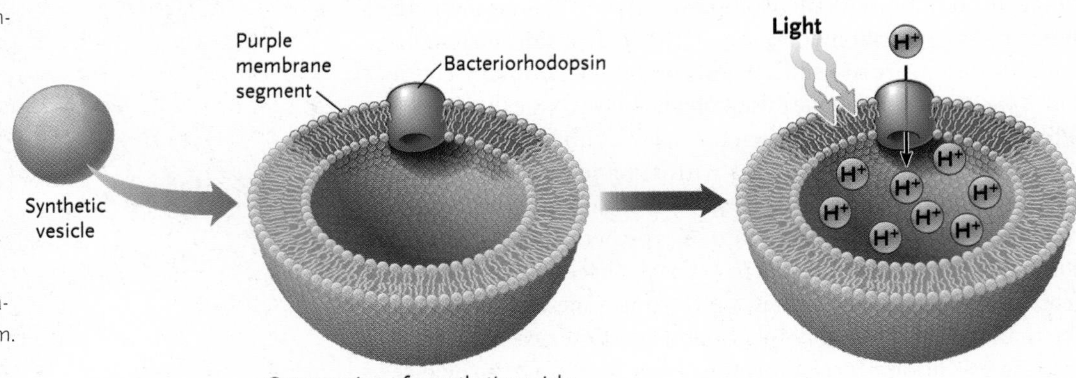

Synthetic vesicle
Purple membrane segment
Bacteriorhodopsin
Light
H+
Cutaway view of a synthetic vesicle

Result: H^+ is pumped into the vesicles, creating an H^+ gradient.

2. The researchers next made synthetic vesicles containing ATP synthase from both bacteriorhodopsin and bovine heart mitochondria. The ATP synthase molecule was oriented so that the ATP-synthesizing headpiece was on the outside of the vesicles. They added ADP and P_i to the medium containing the vesicles and tested whether ATP was produced in the dark and after a period of illumination.

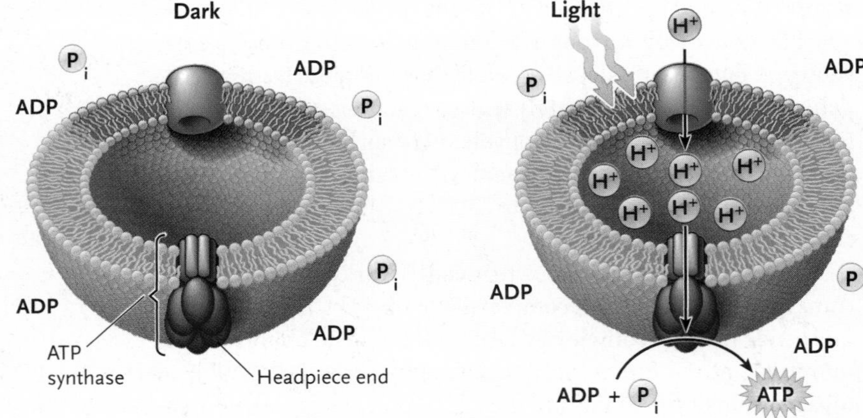

Dark
Light
H+
ADP
Pi
ATP synthase
Headpiece end
ADP + Pi
ATP

Result: In the dark, no ATP was synthesized. **Result:** In the light, ATP was synthesized.

Together, these results showed that light activated the bacteriorhodopsin to produce an H^+ gradient, with H^+ moving from the outside to the inside of the vesicle (like the movement from the mitochondrial matrix to the intermembrane compartment in Figure 8.13), and that the energy from the H^+ gradient drove ATP synthesis by ATP synthase.

Conclusion: An H^+ gradient—and, therefore, proton-motive force—powers ATP synthesis by ATP synthase. The results support Mitchell's chemiosmotic hypothesis for ATP synthesis in mitochondria.

Source: E. Racker and W. Stoeckenius. 1974. Reconstitution of purple membrane vesicles catalyzing light-driven proton uptake and adenosine triphosphate formation. *Journal of Biological Chemistry* 249:662–663.

termediate was ever found, and eventually, Mitchell's hypothesis was supported by the results of many experiments, one of which is described in **Figure 8.15.** Mitchell received a Nobel Prize in 1978 for his model and supporting research.

How does ATP synthase use the H^+ gradient to power ATP synthesis in chemiosmosis? ATP synthase consists of a *basal unit* embedded in the inner mitochondrial membrane connected by a *stalk* to a *headpiece* located in the matrix. A peripheral stalk called a *stator* bridges the basal unit and headpiece (see Figure 8.13). ATP synthase functions like an active transport ion pump. In Chapter 6, we described active transport pumps that use the energy created by hydrolysis of ATP to ADP and P_i to transport ions across membranes against their concentration gradients (see Figure 6.11). However, if the concentration of an ion is very high on the side toward which it is normally transported, the pump runs in reverse—that is, the ion is transported backward through the pump, and the pump adds phosphate to ADP to generate ATP. That is how ATP synthase operates in mitochondrial membranes. Proton-motive force moves protons in the intermembrane space through the channel in the enzyme's basal unit, down their concentration gradient, and into the matrix (see Figure 8.13, step 7). The flow of H^+ powers ATP synthesis by the headpiece; this phosphorylation reaction is chemiosmosis (steps 8 and 9). ATP synthase occurs in similar form and works in the same way in mitochondria, chloroplasts, and prokaryotes capable of oxidative phosphorylation.

Many details of the chemiosmotic mechanism are still being investigated. Paul D. Boyer of the University of California at Los Angeles, one of the major contributors to this area of research, proposed the novel idea that passage of protons through the channel of the basal unit makes the stalk and headpiece spin like a top just as the flow of water makes a waterwheel turn. The turning motion cycles each of three catalytic sites on the headpiece through sequential conformational changes that pick up ADP and phosphate, combine them, and release the ATP product. Another researcher, John E. Walker of the Laboratory of Molecular Biology in Cambridge, United Kingdom, used X-ray diffraction to create a three-dimensional picture of ATP synthase that clearly verified Boyer's model by showing the head in different rotational positions as ATP synthesis proceeds. Boyer and Walker jointly received a Nobel Prize in 1997 for their research into the mechanisms by which ATP synthase makes ATP.

Thirty-Two ATP Molecules Are Produced for Each Molecule of Glucose Completely Oxidized to CO_2 and H_2O

How many ATP molecules are produced as electrons flow through the mitochondrial electron transfer system? The most recent research indicates that approximately 2.5 ATP are synthesized as a pair of electrons released by NADH travels through the entire electron transfer pathway to oxygen. The shorter pathway, followed by an electron pair released from $FADH_2$ by complex II to oxygen, synthesizes about 1.5 ATP. (Some accounts of ATP production round these numbers to 3 and 2 molecules of ATP, respectively.)

These numbers allow us to estimate the total amount of ATP that would be produced by the complete oxidation of glucose to CO_2 and H_2O if the entire H^+ gradient produced by electron transfer is used for ATP synthesis **(Figure 8.16)**. During glycolysis, substrate-level phosphorylation produces 2 ATP. Glycolysis also produces 2 NADH, which leads to 5 ATP (see earlier discussion). In pyruvate oxidation 2 NADH are produced from the two molecules of pyruvate which again lead to 5 ATP. In summary, glycolysis and pyruvate oxidation together yield 2 ATP, 4 NADH, and 2 CO_2 and, in the end, are responsible for 12 of the ATP produced by oxidation of glucose.

The subsequent citric acid cycle turns twice for each molecule of glucose that enters glycolysis, yielding a total of 2 ATP produced by substrate-level phosphorylation, as well as 6 NADH, 2 $FADH_2$, and 4 CO_2. The 6 NADH lead to 15 ATP, and the 2 $FADH_2$ lead to 3 ATP, for a total of 20 ATP from the citric acid cycle. With the ATP from glycolysis and pyruvate oxidation, the total yield is 32 ATP from each molecule of glucose oxidized to carbon dioxide and water.

The combination of glycolysis, pyruvate oxidation, and the citric acid cycle has the following summary reaction:

$$\text{glucose} + 4\,\text{ADP} + 4\,\text{P}_i + 10\,\text{NAD}^+ + 2\,\text{FAD} \rightarrow 4\,\text{ATP} + 10\,\text{NADH} + 10\,\text{H}^+ + 2\,\text{FADH}_2 + 6\,\text{CO}_2$$

The total of 32 ATP assumes that the two pairs of electrons carried by the 2 NADH reduced in glycolysis each drive the synthesis of 2.5 ATP when traversing the mitochondrial electron transfer system. However, because NADH cannot penetrate the mitochon-

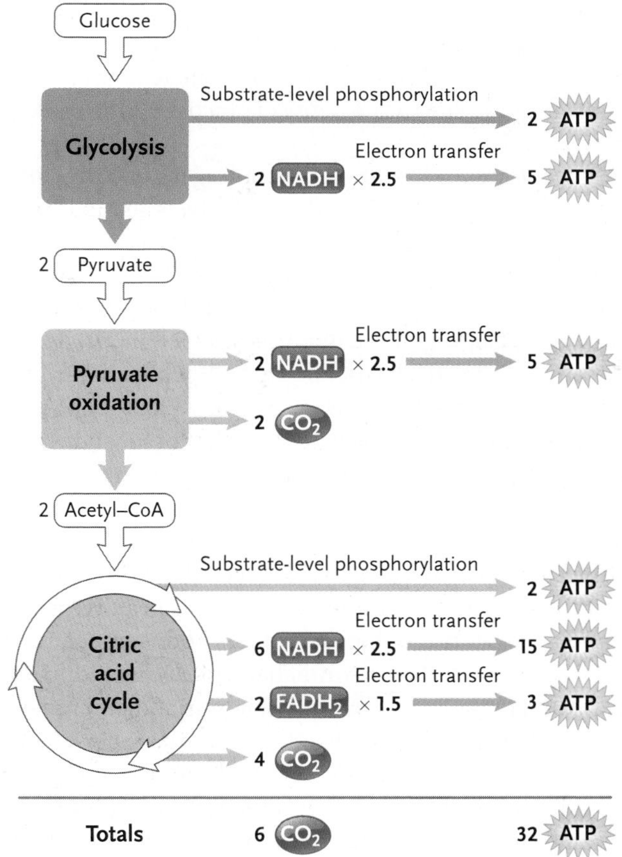
Glucose
Substrate-level phosphorylation
Glycolysis
2 ATP
Electron transfer
2 NADH × 2.5
5 ATP
2 Pyruvate
Pyruvate oxidation
Electron transfer
2 NADH × 2.5
5 ATP
2 CO2
2 Acetyl–CoA
Substrate-level phosphorylation
2 ATP
Citric acid cycle
Electron transfer
6 NADH × 2.5
15 ATP
Electron transfer
2 FADH2 × 1.5
3 ATP
4 CO2
Totals
6 CO2
32 ATP

FIGURE 8.16
Summary of ATP production from the complete oxidation of a molecule of glucose. The total of 32 ATP assumes that electrons carried from glycolysis by NADH are transferred to NAD^+ inside mitochondria. If the electrons from glycolysis are instead transferred to FAD inside mitochondria, total production will be 30 ATP.

drial membranes, its electrons are transferred into the mitochondrion by one of two shuttle systems. The more efficient shuttle mechanism transfers the electrons to NAD^+ as the acceptor inside mitochondria. These electron pairs, when passed through the electron transfer system, result in the synthesis of 2.5 ATP each, producing the grand total of 32 ATP. The less efficient shuttle transfers the electrons to FAD as the acceptor inside mitochondria. These electron pairs, when passed through the electron transfer system, result in the synthesis of only 1.5 ATP each and produce a grand total of 30 ATP instead of 32.

Which shuttle system predominates depends on the particular species and the cell types involved. For example, heart, liver, and kidney cells in mammals use the more efficient shuttle system; skeletal muscle and brain cells use the less efficient shuttle system. Regardless, the numbers of ATP produced are idealized, because mitochondria also use the H^+ gradient to drive cotransport; any of the energy in the gradient used for this activity would reduce ATP production proportionately.

Cellular Respiration Conserves More Than 30% of the Chemical Energy of Glucose in ATP

The process of cellular respiration is not 100% efficient; it does not convert all the chemical energy of glucose to ATP. By using the estimate of 32 ATP produced for each molecule of glucose oxidized under ideal conditions, we can estimate the overall efficiency of cellular glucose oxidation. Efficiency refers to the percentage of the chemical energy of glucose conserved as ATP energy.

Under standard conditions, including neutral pH (pH = 7) and a temperature of 25°C, the hydrolysis of ATP to ADP yields about 7.0 kilocalories per mole (kcal/mol). Assuming that complete glucose oxidation produces 32 ATP, the total energy conserved in ATP production would be about 224 kcal/mol. By contrast, if glucose is simply burned in air, it releases 686 kcal/mol. On this basis, the efficiency of cellular glucose oxidation would be about 33% ($224/686 \times 100 =$ about 33%). This value is considerably better than that of most devices designed by human engineers—for example, the engine of an automobile extracts only about 25% of the energy in the fuel it burns.

The chemical energy released by cellular oxidations that is not captured in ATP synthesis is released as heat. In mammals and birds, this source of heat maintains body temperature at a constant level. In certain mammalian tissues, including *brown fat* (see Chapter 46), the inner mitochondrial membranes contain *uncoupling proteins* (UCPs) that make the inner mitochondrial membrane "leaky" to H^+. As a result, electron transfer runs without building an H^+ gradient or synthesizing ATP and releases all the energy extracted from the electrons as heat. Brown fat with UCPs occurs in significant quantities in hibernating mammals and in very young mammals, including human infants. (*Insights from the Molecular Revolution* describes research showing that some plants also use UCPs in mitochondrial membranes to heat tissues.)

STUDY BREAK 8.4

1. **What distinguishes the four complexes of the mitochondrial electron transfer system?**
2. **Explain how the proton pumps of complexes I, III, and IV relate to ATP synthesis.**

THINK OUTSIDE THE BOOK

A few human genetic diseases result from mutations that affect mitochondrial function. Collaboratively or individually, find an example of such a disease and research how the genetic mutation disrupts mitochondrial function and leads to the disease symptoms.

8.5 Fermentation

When oxygen is plentiful, electrons carried by the 2 NADH produced by glycolysis are passed to the electron transfer system inside mitochondria, and the released energy drives ATP synthesis. When oxygen is absent or limited, the electrons may be used in fermentation. In **fermentation,** electrons carried by NADH are transferred to an organic acceptor molecule rather than to the electron transfer system. This transfer converts the NADH to NAD^+, which is required to accept electrons in reaction 6 of glycolysis (see Figure 8.7). As a result, glycolysis continues to supply ATP by substrate-level phosphorylation.

Two types of fermentation reactions exist: lactate fermentation and alcoholic fermentation **(Figure 8.17). Lactate fermentation** converts pyruvate into lactate (Figure 8.17A). This reaction occurs in many bacteria, in some plant tissues, and in certain animal tissues such as skeletal muscle cells. When vigorous contraction of muscle cells calls for more oxygen than the circulation can supply, lactate fermentation takes place. For example, lactate accumulates in the leg muscles of a sprinter during a 100-meter race. The lactate temporarily stores electrons, and when the oxygen content of the muscle cells returns to normal levels the reverse of the reaction in Figure 8.17A regenerates pyruvate and NADH. The pyruvate can then be used in the second stage of cellular respiration, and the NADH contributes its electron pair to the electron transfer system. Lactate is also the fermentation product of some bacteria; the sour taste of buttermilk, yogurt, and dill pickles is a sign of their activity.

Alcoholic fermentation (Figure 8.17B) occurs in some plant tissues, in certain invertebrates and protists, in certain bacteria, and in some single-celled fungi such as yeasts. In this reaction, pyruvate is converted into ethyl alcohol (which has two carbons) and carbon dioxide in a two-step series that also converts NADH into NAD^+. Alcoholic fermentation by yeasts has widespread commercial applications. Bakers use the yeast *Saccharomyces cerevisiae* to make bread dough rise. They mix the yeast with a small amount of sugar and blend the mixture into the dough,

INSIGHTS FROM THE Molecular Revolution

Hot Potatoes: Do plants use uncoupling proteins to generate heat?

Mammals use several biochemical and molecular processes to maintain body heat. One process is shivering; the muscular activity of shivering releases heat that helps keep body temperature at normal levels. Another mechanism operates through uncoupling proteins (UCPs), which eliminate the mitochondrial H^+ gradient by making the inner mitochondrial membrane leaky to protons. Electron transfer and the oxidative reactions then run at high rates in mitochondria without trapping energy in ATP. The energy is released as heat that helps maintain body temperature.

Research Question

Do plants use UCPs to generate heat?

Experiments

Research by Maryse Laloi and her colleagues at the Max Planck Institute for Molecular Plant Physiology in Germany shows that some tissues in plants may use the same process involving UCPs to generate heat. Their evidence is as follows:

1. Potato plants *(Solanum tuberosum)* have a gene with a DNA sequence similar to that of a mammalian UCP gene. The potato gene protein is clearly related to the mammalian protein, and also has the same overall three-dimensional structure.
2. The researchers used the DNA of the potato UCP gene to probe for the presence of messenger RNA (mRNA), the gene product that specifies the amino acid sequence of proteins in the cytoplasm, with the results shown in **Figure 1.** These results indicate that the potato UCP gene is active at different levels in various plant tissues, suggesting that certain tissues naturally need warming for optimal function.
3. Laloi and her coworkers then tested whether exposing potato plants to cold temperatures could induce greater synthesis of the UCP mRNA **(Figure 2).** The cold treatment resulted in an increase in UCP mRNA level in the leaves.

Conclusion

The research indicates that potato plants probably use the mitochondrial uncoupling process to warm tissues when they are stressed by low temperatures. Thus, mechanisms for warming body tissues once thought to be the province only of animals appear to be much more widespread. In particular, UCPs, which were believed to have evolved in relatively recent evolutionary times with the appearance of birds and mammals, may be a much more ancient development.

Source: M. Laloi et al. 1997. A plant cold-induced uncoupling protein. *Nature* 389:135–136.

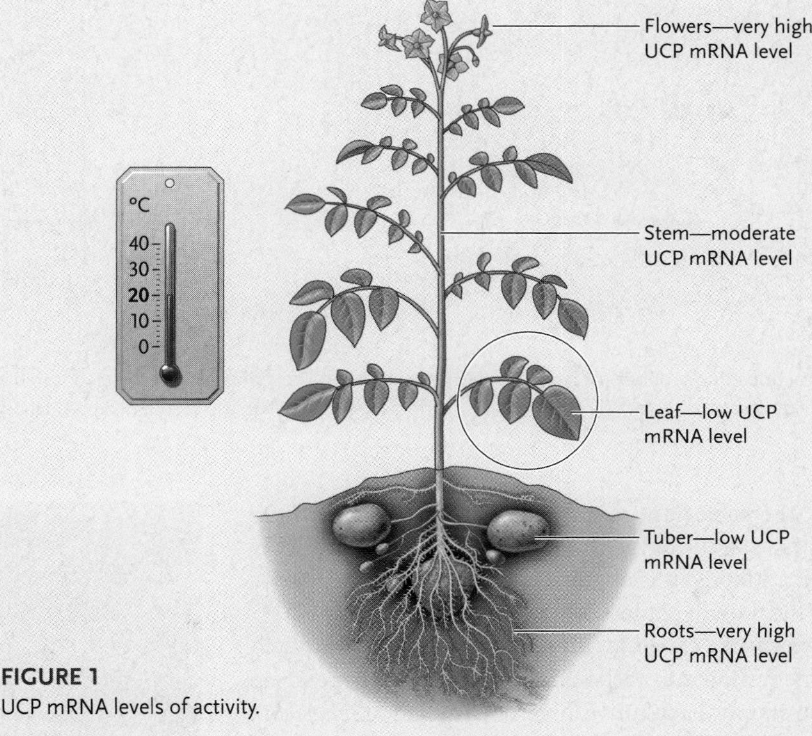

°C
40
30
20
10
0
Flowers—very high UCP mRNA level
Stem—moderate UCP mRNA level
Leaf—low UCP mRNA level
Tuber—low UCP mRNA level
Roots—very high UCP mRNA level

FIGURE 1
UCP mRNA levels of activity.

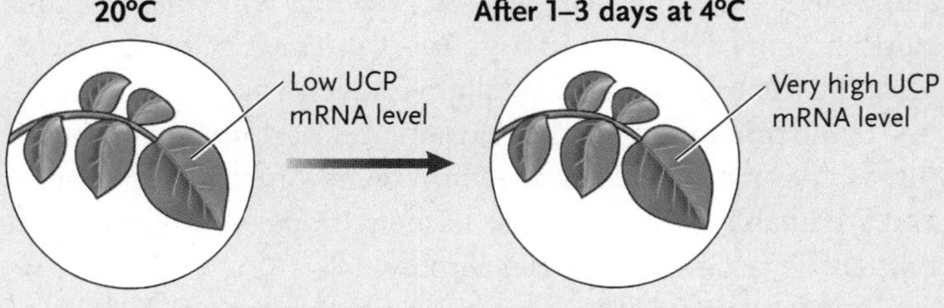

20°C
After 1–3 days at 4°C
Low UCP mRNA level
Very high UCP mRNA level

FIGURE 2
Results of exposure to cold temperature.

where oxygen levels are low. As the yeast cells convert the sugar into ethyl alcohol and carbon dioxide, the gaseous CO_2 expands and creates bubbles that cause the dough to rise. Oven heat evaporates the alcohol and causes further expansion of the bubbles, producing a light-textured product. Alcoholic fermentation is also the mainstay of beer and wine production. Fruits are a natural home to wild yeasts **(Figure 8.18)**; for example, winemakers rely on a mixture of wild and cultivated yeasts to produce wine. Alcoholic fermentation also occurs naturally in the environment; for example, overripe or rotting fruit frequently will start to ferment, and birds that eat the fruit may become too drunk to fly.

Fermentation is a lifestyle for some organisms. In bacteria and fungi that lack the enzymes and factors to carry out oxidative phosphorylation, fermentation is the only source of ATP. These organisms are called **strict anaerobes.** In general, strict anaerobes require an oxygen-free environment; they cannot utilize oxygen as a final electron acceptor. Among these organisms are the bacteria that cause botulism, tetanus, and some other serious diseases.

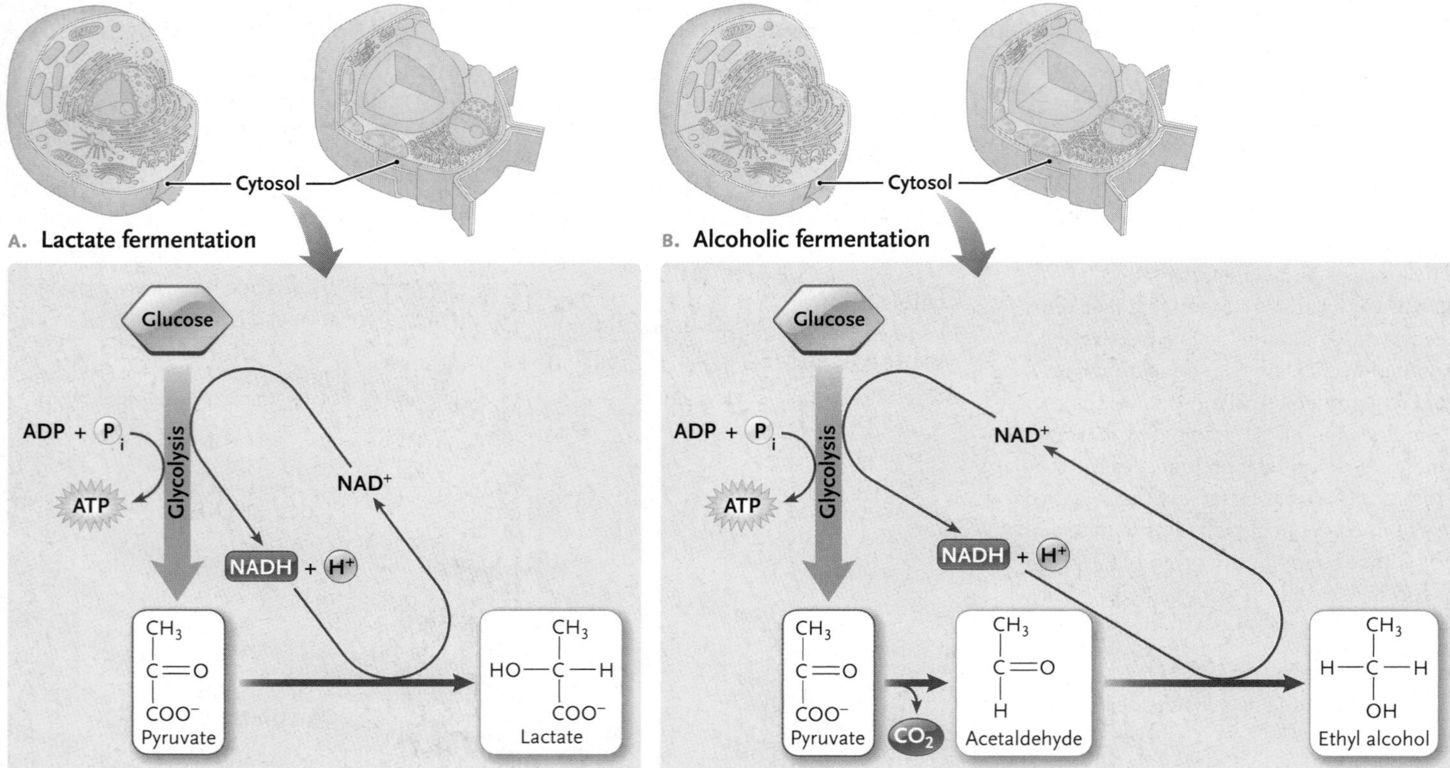

Cytosol
A. Lactate fermentation
Glucose
Glycolysis
ADP + P i
ATP
NADH + H+
NAD+
CH3 C=O COO− Pyruvate
CH3 HO—C—H COO− Lactate
Cytosol
B. Alcoholic fermentation
Glucose
Glycolysis
ADP + P i
ATP
NAD+
NADH + H+
CH3 C=O COO− Pyruvate
CO2
CH3 C=O H Acetaldehyde
CH3 H—C—H OH Ethyl alcohol

FIGURE 8.17

Fermentation reactions that produce (A) lactate and (B) ethyl alcohol. The fermentations, which occur in the cytosol, convert NADH to NAD^+, allowing the electron carrier to cycle back to glycolysis. This process keeps glycolysis running, with continued production of ATP.

For example, the bacterium that causes botulism *(Clostridium botulinum)* thrives in the oxygen-free environment of improperly sterilized canned foods. It is absence of oxygen in canned foods that prevents the growth of most other microorganisms.

Other organisms, called **facultative anaerobes,** can switch between fermentation and full oxidative pathways, depending on the oxygen supply. Facultative anaerobes include *Escherichia coli,* a bacterium that inhabits the digestive tract of humans; the *Lactobacillus* bacteria used to produce buttermilk and yogurt; and *S. cerevisiae,* the yeast used in brewing, wine making, and baking. Many cell types in higher organisms, including vertebrate muscle cells, are also facultative anaerobes.

Some prokaryotic and eukaryotic cells are **strict aerobes**—that is, they have an absolute requirement for oxygen to survive and are unable to live solely by fermentations. Vertebrate brain cells are key examples of strict aerobes.

Fermentation differs from anaerobic respiration, the form of cellular respiration used by some prokaryotes for ATP production (see *Why It Matters* for this chapter). In fermentation the electrons carried by NADH are transferred to an organic acceptor molecule, while in anaerobic respiration the electrons are transferred to an electron transfer system. However, in contrast to aerobic respiration, in anaerobic respiration the electrons are transferred through the electron transfer system to a molecule other than oxygen, such as sulfate.

This chapter traced the flow of high-energy electrons from fuel molecules to ATP. As part of the process, the fuels are broken into molecules of carbon dioxide and water. The next chapter shows how photosynthetic organisms use these inorganic raw materials to produce organic molecules in a process that pushes the electrons back to high energy levels by absorbing the energy of sunlight.

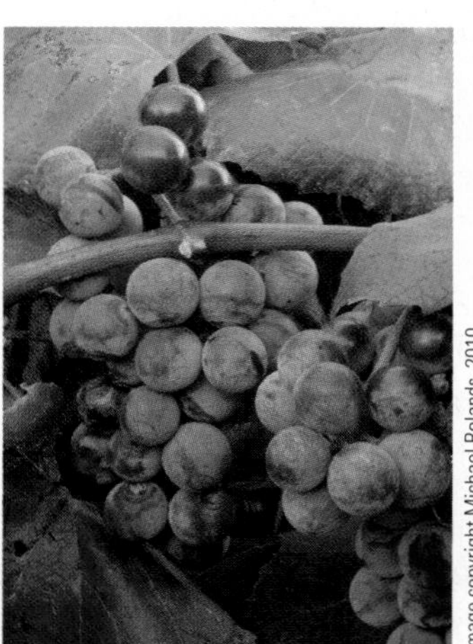

FIGURE 8.18

Alcoholic fermentation in nature. Wild yeast cells are visible as a dustlike coating on grapes.

STUDY BREAK 8.5 <

What is fermentation, and when does it occur? What are the two types of fermentation?

UNANSWERED QUESTIONS

Glycolysis and energy metabolism are crucial for the normal functioning of an animal. Research of many kinds is being conducted in this area, such as characterizing the molecular components of the reactions in detail and determining how they are regulated. The goal is to generate comprehensive models of cellular respiration and its regulation. Following is a specific example of ongoing research related to human disease caused by defects in cellular respiration.

How do mitochondrial proteins change in patients with Alzheimer disease?

Alzheimer disease (AD) is an age-dependent, irreversible, neurodegenerative disorder in humans. Symptoms include a progressive deterioration of cognitive functions and, in particular, a significant loss of memory. Neuropathologically, AD is characterized by the presence of extracellular amyloid plaques, intracellular neurofibrillary tangles, and synaptic and neuronal loss. Reduced brain metabolism occurs early in the onset of AD. One of the mechanisms for this physiological change appears to be damage to or reduction of key mitochondrial components, including enzymes of the citric acid cycle and the oxidative phosphorylation system. However, the complete scope of mitochondrial protein changes has not been established, nor have detailed comparisons been made of mitochondrial protein changes among AD patients.

To begin to address these questions, research is being carried out in my laboratory at The University of Texas at Dallas to analyze the complete set of mitochondrial proteins in healthy versus AD brains both qualitatively and quantitatively. The results will show the changes that occur in mitochondrial protein synthesis in the two tissues. This approach is called quantitative comparative proteomic profiling. (Proteomics is the complete characterization of the proteins present in a cell, cell compartment, tissue, organ, or organism; see Chapter 18. The set of proteins identified in such a study is called the proteome.) We are using a transgenic mouse model of AD in the first stage of this research; that is, the mice have been genetically engineered with altered genes so that they have AD. (The generation of transgenic organisms is described in Chapter 18.)

The results of our experiments have demonstrated that regulation of the amounts of many mitochondrial proteins is altered in the brains of transgenic AD mice. These dysregulated mitochondrial proteins participate in many different metabolic functions, including the citric acid cycle, oxidative phosphorylation, pyruvate metabolism, fatty acid oxidation, ketone body metabolism, metabolite transport, oxidative stress, mitochondrial protein synthesis, mitochondrial protein import, and cell growth and apoptosis (programmed cell death; see Chapter 43). Interestingly, both down-regulated and up-regulated mitochondrial proteins were identified in AD brains. We have also determined that this altered regulation of mitochondrial protein synthesis occurs early in AD before the development of significant plaque and tangle pathologies. Future experiments will be directed towards examining changes in the mitochondrial proteome in the brains of human AD patients. Ultimately, the results of our experiments may lead to the development of treatments that can slow or halt the progression of AD in humans.

Think Critically

What significance do you attribute to the statement in this essay that the altered regulation of mitochondrial protein synthesis occurs early in AD before the development of significant plaque and tangle pathologies? Develop a scientific hypothesis based on this statement.

Gail A. Breen

Gail A. Breen is an associate professor in the Department of Molecular and Cell Biology at The University of Texas at Dallas. Her current research focuses on mitochondrial biogenesis and the role of mitochondria in neurodegenerative diseases, such as Alzheimer disease. To learn more about Dr. Breen's research go to http://www.utdallas.edu/biology/faculty/breen.html.

REVIEW KEY CONCEPTS

Go to **CENGAGENOW** at www.cengage.com/login to access quizzing, animations, exercises, articles, and personalized homework help.

8.1 Overview of Cellular Energy Metabolism

- Plants and almost all other organisms obtain energy for cellular activities through cellular respiration, the process of transferring electrons from donor organic molecules to a final acceptor molecule such as oxygen; the energy that is released drives ATP synthesis (Figure 8.1).
- Oxidation–reduction reactions, called redox reactions, partially or completely transfer electrons from donor to acceptor atoms; the donor is oxidized as it releases electrons, and the acceptor is reduced (Figure 8.2).
- Cellular respiration occurs in three stages: (1) In glycolysis, glucose is converted to two molecules of pyruvate through a series of enzyme-catalyzed reactions; (2) in pyruvate oxidation and the citric acid cycle, pyruvate is converted to an acetyl compound that is oxidized completely to carbon dioxide; and (3) in oxidative phosphorylation, which is comprised of the electron transfer system and chemiosmosis, high-energy electrons produced from the first two stages pass through the transfer system, with much of their energy being used to establish an H^+ gradient across the membrane that drives the synthesis of ATP from ADP and P_i (Figure 8.3).
- In eukaryotes, most of the reactions of cellular respiration occur in mitochondria. In prokaryotes, glycolysis, pyruvate oxidation, and the citric acid cycle occur in the cytosol, while the rest of cellular respiration occurs on the plasma membrane (Figure 8.4).

Animation: The functional zones in mitochondria

8.2 Glycolysis: Splitting the Sugar in Half

- In glycolysis, which occurs in the cytosol, glucose (six carbons) is oxidized into two molecules of pyruvate (three carbons each). Electrons removed in the oxidations are delivered to NAD^+, producing NADH. The reaction sequence produces a net gain of 2 ATP, 2 NADH, and 2 pyruvate molecules for each molecule of glucose oxidized (Figures 8.5 and 8.7).

- ATP molecules produced in the energy-releasing steps of glycolysis result from substrate-level phosphorylation, an enzyme-catalyzed reaction that transfers a phosphate group from a substrate to ADP (Figure 8.8).

Animation: The overall reactions of glycolysis

Practice: Recreating the reactions of glycolysis

8.3 Pyruvate Oxidation and the Citric Acid Cycle

- In pyruvate oxidation, which occurs inside mitochondria, 1 pyruvate (three carbons) is oxidized to 1 acetyl group (two carbons) and 1 carbon dioxide (CO_2). Electrons removed in the oxidation are accepted by 1 NAD^+ to produce 1 NADH. The acetyl group is transferred to coenzyme A, which carries it to the citric acid cycle (Figure 8.10).
- In the citric acid cycle, acetyl groups are oxidized completely to CO_2. Electrons removed in the oxidations are accepted by NAD^+ or FAD, and substrate-level phosphorylation produces ATP. For each acetyl group oxidized by the cycle, 2 CO_2, 1 ATP, 3 NADH, and 1 $FADH_2$ are produced (Figure 8.11).

Animation: Pyruvate oxidation and the citric acid cycle

Animation: Major pathways oxidizing carbohydrates, fats, and proteins

8.4 Oxidative Phosphorylation: The Electron Transfer System and Chemiosmosis

- Electrons are passed from NADH and $FADH_2$ to the electron transfer system, which consists of four protein complexes and two smaller shuttle carriers. As the electrons flow from one carrier to the next through the system, some of their energy is used by the complexes to pump protons across the inner mitochondrial membrane (Figures 8.13 and 8.14).
- Ubiquinone and the three major protein complexes (I, III, and IV) pump H^+ from the matrix to the intermembrane compartment, generating an H^+ gradient with a high concentration in the intermembrane compartment and a low concentration in the matrix (Figure 8.13).
- The H^+ gradient produced by the electron transfer system is used by ATP synthase as an energy source for synthesis of ATP from ADP and P_i. The ATP synthase is embedded in the inner mitochondrial membrane together with the electron transfer system (Figure 8.13).
- An estimated 2.5 ATP are synthesized as each electron pair travels from NADH to oxygen through the mitochondrial electron transfer system; about 1.5 ATP are synthesized as each electron pair travels through the system from $FADH_2$ to oxygen. Using these totals gives an efficiency of more than 30% for the utilization of energy released by glucose oxidation if the H^+ gradient is used only for ATP production (Figure 8.16).

Animation: The mitochondrial electron transfer system and oxidative phosphorylation

8.5 Fermentation

- Lactate fermentation and alcoholic fermentation are reaction pathways that deliver electrons carried from glycolysis by NADH to organic acceptor molecules, thereby converting NADH back to NAD^+. The NAD^+ is required to accept electrons generated by glycolysis, allowing glycolysis to supply ATP by substrate-level phosphorylation (Figure 8.17).

Animation: The fermentation reactions

UNDERSTAND AND APPLY

Test Your Knowledge

1. What is the final acceptor for electrons in cellular respiration?
   a. oxygen
   b. ATP
   c. carbon dioxide
   d. hydrogen
   e. water
2. In glycolysis:
   a. free oxygen is required for the reactions to occur.
   b. ATP is used when glucose and fructose-6-phosphate are phosphorylated, and ATP is synthesized when 3-phosphoglycerate and pyruvate are formed.
   c. the enzymes that move phosphate groups on and off the molecules are uncoupling proteins.
   d. the product with the highest potential energy in the pathway is pyruvate.
   e. the end product of glycolysis moves to the electron transfer system.
3. Which of the following statements about phosphofructokinase is *false*?
   a. It is located and has its main activity in the inner mitochondrial membrane.
   b. It catalyzes a reaction to form a product with the highest potential energy in the pathway.
   c. It can be inactivated by ATP at an inhibitory site on its surface.
   d. It can be activated by ADP at an excitatory site on its surface.
   e. It can cause ADP to form.
4. Which of the following statements is *false*? Imagine that you ingested three chocolate bars just before sitting down to study this chapter. Most likely:
   a. your brain cells are using ATP.
   b. there is no deficit of the initial substrate to begin glycolysis.
   c. the respiratory processes in your brain cells are moving atoms from glycolysis through the citric acid cycle to the electron transfer system.
   d. after a couple of hours, you change position and stretch to rest certain muscle cells, which removes lactate from these muscles.
   e. after 2 hours, your brain cells are oxygen-deficient.
5. If ADP is produced in excess in cellular respiration, this excess ADP will:
   a. bind glucose to turn off glycolysis.
   b. bind glucose-6-phosphate to turn off glycolysis.
   c. bind phosphofructokinase to turn on or keep glycolysis turned on.
   d. cause lactate to form.
   e. increase oxaloacetate binding to increase NAD^+ production.
6. Which of the following statements is *false*? In cellular respiration:
   a. one molecule of glucose can produce about 32 ATP.
   b. oxygen combines directly with glucose to form carbon dioxide.
   c. a series of energy-requiring reactions is coupled to a series of energy-releasing reactions.
   d. NADH and $FADH_2$ allow H^+ to be pumped across the inner mitochondrial membrane.
   e. the electron transfer system occurs in the inner mitochondrial membrane.

7. You are reading this text while breathing in oxygen and breathing out carbon dioxide. The carbon dioxide arises from:
   a. glucose in glycolysis.
   b. NAD^+ redox reactions in the mitochondrial matrix.
   c. NADH redox reactions on the inner mitochondrial membrane.
   d. $FADH_2$ in the electron transfer system.
   e. the oxidation of pyruvate, isocitrate, and α-ketoglutarate in the citric acid cycle.
8. In the citric acid cycle:
   a. NADH and H^+ are produced when α-ketoglutarate is both produced and metabolized.
   b. ATP is produced by oxidative phosphorylation.
   c. to progress from a four-carbon molecule to a six-carbon molecule, CO_2 enters the cycle.
   d. $FADH_2$ is formed when succinate is converted to oxaloacetate.
   e. the cycle "turns" once for each molecule of glucose metabolized.
9. For each NADH produced from the citric acid cycle, about how many ATP are formed?
   a. 38
   b. 36
   c. 32
   d. 2.5
   e. 2.0
10. In the 1950s, a diet pill that had the effect of "poisoning" ATP synthase was tried. The person taking it could not use glucose and "lost weight"—and ultimately his or her life. Today, we know that the immediate effect of poisoning ATP synthase is:
    a. ATP would not be made in the electron transfer system.
    b. H^+ movement across the inner mitochondrial membrane would increase.
    c. more than 32 ATP could be produced from a molecule of glucose.
    d. ADP would be united with phosphate more readily in the mitochondria.
    e. ATP would react with oxygen.

Discuss the Concepts

1. Why do you think nucleic acids are not oxidized extensively as a cellular energy source?
2. A hospital patient was regularly found to be intoxicated. He denied that he was drinking alcoholic beverages. The doctors and nurses made a special point to eliminate the possibility that the patient or his friends were smuggling alcohol into his room, but he was still regularly intoxicated. Then, one of the doctors had an idea that turned out to be correct and cured the patient of his intoxication. The idea involved the patient's digestive system and one of the oxidative reactions covered in this chapter. What was the doctor's idea?

Design an Experiment

There are several ways to measure cellular respiration experimentally. For example, CO_2 and O_2 gas sensors measure changes over time in the concentration of carbon dioxide or oxygen, respectively. Design two experiments to test the effects of changing two different variables or conditions (one per experiment) on the respiration of a research organism of your choice.

Interpret the Data

As CO_2 concentrations increase in the atmosphere, biologists continue to explore the role of respiration from plants as a small, but potentially important contribution beyond fossil fuel combustion. The data in the table were collected from the leaf of a sagebrush plant from a semiarid ecosystem in Wyoming, enclosed in a chamber that measures the rate of CO_2 exchange. The respiration rate is the amount of CO_2 in micromoles lost by the leaf per square meter per second, which results in the negative numbers. The temperature values are from the leaves as they are heated or cooled during the measurements.

Observation	Temperature (°C)	Respiration Rate (μmol/m²/s)
1	25	−2.0
2	30	−2.7
3	35	−4.1
4	40	−5.8
5	20	−1.3
6	15	−1.0
7	10	−0.7

1. Make a graph of the data, with temperature on the x axis and respiration rate on the y axis.
2. The Q10 value of respiration is the increase in respiration, expressed as the ratio of the higher rate to the lower rate, with a 10°C change in temperature. What is the approximate (whole number) Q10 of respiration for this sagebrush leaf?
3. What describes the relationship between temperature and respiration, a line or a curve? Does the Q10 that you calculated in 2 suggest a line or a curve?
4. How might the predicted increase in temperature due to elevated CO_2 concentrations impact respiration of sagebrush leaves? Do you think the data presented here are all that is needed to predict this impact?

Source: Data based on unpublished research by Brent Ewers, University of Wyoming.

Apply Evolutionary Thinking

Which of the two phosphorylation mechanisms, oxidative phosphorylation or substrate-level phosphorylation, is likely to have appeared first in evolution? Why?

Express Your Opinion

Developing new drugs is costly. There is little incentive for pharmaceutical companies to target ailments that affect relatively few individuals, such as Luft syndrome. Should governments allocate some funds to private companies that search for cures for rare disorders? Go to CengageNOW to investigate both sides of the issue and then vote online.

9

STUDY OUTLINE

Chloroplasts in the leaf of the pea plant *Pisum sativum* (colorized TEM). The light-dependent reactions of photosynthesis take place within the thylakoids of the chloroplasts (thylakoid membranes are shown in yellow).

Photosynthesis

Why It Matters. . . By the late 1880s, scientists realized that green algae and plants use light as a source of energy to make organic molecules. This conversion of light energy to chemical energy in the form of sugar and other organic molecules is called **photosynthesis.** The scientists also knew that these organisms release oxygen as part of their photosynthetic reactions. Among these scientists was a German botanist, Theodor Engelmann, who was curious about the particular colors of light used in photosynthesis. Was green light the most effective in promoting photosynthesis or were other colors more effective?

Engelmann used only a light microscope and a glass prism to find the answer to this question. Yet his experiment stands today as a classic one, both for the fundamental importance of his conclusion and for the simple but elegant methods he used to obtain it. Engelmann placed a strand of a filamentous green alga on a glass microscope slide along with water containing aerobic bacteria (bacteria that require oxygen to survive). He adjusted the prism so that it split a beam of visible light into its separate colors, which spread like a rainbow across the strand **(Figure 9.1).** After a short time, he noticed that the

Light

Bacteria

Strand of a filamentous green alga

Number of bacteria

Wavelength

FIGURE 9.1
Engelmann's 1882 experiment revealing the action spectrum of light used in photosynthesis by a filamentous green alga. Using a glass prism, Engelmann broke up a beam of light into a spectrum of colors, which were cast across a microscope slide with a strand of the alga in water containing aerobic bacteria. The bacteria clustered along the algal strand in the regions where oxygen was released in greatest quantity—the regions in which photosynthesis proceeded at the greatest rate. Those regions corresponded to the colors (wavelengths) of light being absorbed most effectively by the alga—violet, blue, and red.

bacteria had formed large clusters in the blue and violet light at one end of the strand, and in the red light at the other end. Very few bacteria were found in the green light. Evidently, violet, blue, and red light caused the most oxygen to be released, and Engelmann concluded that these colors of light—rather than green—were used most effectively in photosynthesis.

Engelmann used the distribution of bacteria in the light to construct a curve called an *action spectrum* for the wavelengths of light that fell on the alga; it showed the relative effect of each color of light on photosynthesis (black curve in Figure 9.1). Engelmann's results were so accurate that an action spectrum obtained with modern equipment fits closely with his bacterial distribution. However, his results were controversial for some 60 years, until instruments that directly measure the effects of specific wavelengths of light on photosynthesis became available.

Scientists now know that photosynthetic organisms, which include plants, some protists (the algae), and some archaeans and bacteria, absorb the radiant energy of sunlight and convert it into chemical energy. The organisms use the chemical energy to convert simple inorganic raw materials—water, carbon dioxide (CO_2) from the air, and inorganic minerals from the soil—into complex organic molecules. Photosynthesis is still not completely understood, so it remains a subject of active research today.

This chapter begins with an overview of the photosynthetic reactions. We then examine light and light absorption and the reactions that use absorbed energy to make organic molecules from inorganic substances. This chapter focuses primarily on photosynthesis in plants and green algae; other eukaryotic photosynthesizers have individual variations on the process (see Chapter 26). Prokaryotic photosynthesis is described in Chapter 25. <

9.1 Photosynthesis: An Overview

Plants and other photosynthetic organisms are the *primary producers* of Earth; they convert the energy of sunlight into chemical energy and use this chemical energy to assemble simple inorganic raw materials into complex organic molecules. Primary producers use some of the organic molecules they make as an energy source for their own activities. But they also serve—directly or indirectly—as a food source for *consumers,* the animals that live by eating plants or other animals. Eventually, the bodies of both primary producers and consumers provide chemical energy for bacteria, fungi, and other *decomposers.*

Photosynthesizers and other organisms that make all of their required organic molecules from CO_2 and other inorganic sources such as water are called **autotrophs** (*autos* = self, *trophos* = feeding). Autotrophs that use light as the energy source to make organic molecules by photosynthesis are called **photoautotrophs.** Consumers and decomposers, which need a source of organic molecules to survive, are called **heterotrophs** (*hetero* = different).

As the pathway of energy flows from the sun through plants (primary producers) and animals to decomposers, the organic molecules made by photosynthesis are broken down into inorganic molecules again, and the chemical energy captured in photosynthesis is released as heat energy. Because the reactions capturing light energy are the first step in this pathway, photosynthesis is the vital link between the energy of sunlight and the vast majority of living organisms.

Electrons Play a Primary Role in Photosynthesis

Photosynthesis proceeds in two stages, each involving multiple reactions **(Figure 9.2).** In the first stage, the **light-dependent reactions,** the energy of sunlight is absorbed and converted into chemical energy in the form of two substances: ATP and NADPH. ATP is the main energy source for plant cells, and NADPH (nicotinamide adenine dinucleotide phosphate) carries electrons that are pushed to high energy levels by absorbed light. In the second stage of photosynthesis, the **light-independent reactions** (also called the *Calvin cycle*), these electrons are used as a source of energy to convert inorganic CO_2 to an organic form. The conversion process, called **CO_2 fixation,** is a reduction reaction, in which electrons are added to CO_2. As part of the reduction, protons are also added to CO_2 (reduction and oxidation are discussed in Section 8.1). With the added electrons and protons (H^+), CO_2 is converted to a carbohydrate that contains carbon, hydrogen, and oxygen atoms in the ratio 1 C : 2 H : 1 O.

$$CO_2 + H^+ + e^- \rightarrow (CH_2O)_n$$

Carbohydrate units are often symbolized as $(CH_2O)_n$, with the "*n*" indicating that different carbohydrates are formed from different multiples of the carbohydrate unit.

In plants, algae, and one group of photosynthetic bacteria (the cyanobacteria), the source of electrons and protons for CO_2 fixation is water (H_2O), the most abundant substance on Earth. Oxygen (O_2) generated from the splitting of the water

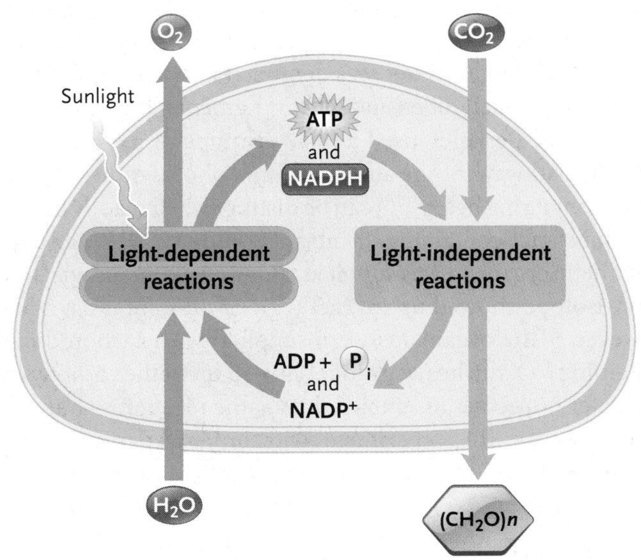

O2
CO2
Sunlight
ATP
and
NADPH
Light-dependent reactions
Light-independent reactions
ADP + P i
and
NADP+
H2O
(CH2O)n

FIGURE 9.2
The light-dependent and light-independent reactions of photosynthesis, and their interlinking reactants and products. Both series of reactions occur in the chloroplasts of plants and algae.

molecule is released into the environment as a by-product of photosynthesis:

$$2\,H_2O \rightarrow 4\,H^+ + 4\,e^- + O_2$$

Thus plants, algae, and cyanobacteria use three resources that are readily available—sunlight, water, and CO_2—to produce almost all the organic matter on Earth, and to supply the oxygen of our atmosphere.

In organisms that are able to split water, the two reactions shown above are combined and multiplied by six to produce a six-carbon carbohydrate such as glucose:

$$6\,CO_2 + 12\,H_2O \rightarrow C_6H_{12}O_6 + 6\,O_2 + 6\,H_2O$$

Note that water appears on both sides of the equation; it is both consumed as a reactant and generated as a product in photosynthesis.

Although glucose is the major product of photosynthesis, other monosaccharides, disaccharides, polysaccharides, lipids, and amino acids are also produced indirectly. In fact, all the organic molecules of plants are assembled as direct or indirect products of photosynthesis.

The relationships between the light-dependent and light-independent reactions are summarized in Figure 9.2. Notice that the ATP and NADPH produced by the light-dependent reactions, along with CO_2, are the reactants of the light-independent reactions. The ADP, inorganic phosphate (P_i), and $NADP^+$ produced by the light-independent reactions, along with H_2O, are the reactants for the light-dependent reactions. The light-dependent and light-independent reactions thus form a cycle in which the net inputs are H_2O and CO_2, and the net outputs are organic molecules and O_2.

Oxygen Released by Photosynthesis Derives from the Splitting of Water

Early investigators thought that the O_2 released by photosynthesis came from the CO_2 entering the process. The fact that it comes from the splitting of water was demonstrated experimentally in 1941 when Samuel Ruben and Martin Kamen of the University of California at Berkeley used a heavy isotope of oxygen, ^{18}O, to trace the pathways of the atoms through photosynthesis. A substance containing heavy ^{18}O can be distinguished readily from the same substance containing the normal isotope, ^{16}O. When a photosynthetic organism was supplied with water containing ^{18}O, the heavy isotope showed up in the O_2 given off in photosynthesis. However, if the organisms were supplied with carbon dioxide containing ^{18}O, the heavy isotope showed up in the carbohydrate and water molecules assembled during the reactions—but not in the oxygen gas. This experiment, and similar experiments using different isotopes, revealed where each atom of the reactants end up in products:

Reactants: $12\,H_2O$ $6\,CO_2$

Products: $6\,O_2$ $C_6H_{12}O_6$ $6\,H_2O$

The water-splitting reaction probably developed even before oxygen-consuming organisms appeared, evolving first in photosynthetic bacteria that resembled present-day cyanobacteria. The oxygen released by the reaction profoundly changed the atmosphere. It allowed for aerobic respiration in which oxygen serves as the final acceptor for electrons removed in cellular oxidations. The existence of all animals depends on the oxygen provided by the water-splitting reaction of photosynthesis.

In Eukaryotes, Photosynthesis Takes Place in Chloroplasts

In eukaryotes, the photosynthetic reactions take place in the chloroplasts of plants and algae; in cyanobacteria, the reactions are distributed between the plasma membrane and the cytosol.

Chloroplasts from individual algal and plant groups differ in structural details. The chloroplasts of plants and green algae are formed from three membranes that enclose three compartments inside the organelles **(Figure 9.3)**. (Chloroplast structure is also described in Section 5.4.) An *outer membrane* covers the entire surface of the organelle. An *inner membrane* lies just inside the outer membrane. Between the outer and inner membranes is an *intermembrane compartment.* The fluid within the compartment formed by the inner membrane is the *stroma.* Within the stroma is the third membrane system, the *thylakoid membranes,* which form flattened, closed sacs called *thylakoids.* The space enclosed by a thylakoid is called the *thylakoid lumen.*

In green algae and higher plants, thylakoids are arranged into stacks called *grana* (singular, *granum;* shown in Figure 9.3). The grana are interconnected by flattened, tubular membranes called *stromal lamellae.* The stromal lamellae probably link the thylakoid lumens into a single continuous space within the stroma.

The thylakoid membranes and stromal lamellae house the molecules that carry out the light-dependent reactions of photosynthesis, which include the pigments, electron transfer carriers, and ATP synthase enzymes for ATP production. The light-independent reactions are concentrated in the stroma.

In higher plants, the CO_2 required for photosynthesis diffuses to cells containing chloroplasts after entering the plant through *stomata* (singular, *stoma*), small pores in the surface of the leaves (particularly the undersurface) and stems. (Stomata are described in Section 27.1, and are shown in Figures 9.16 and 27.3.) The O_2 produced in photosynthesis diffuses from the cells and exits through the stomata, as does the H_2O. The water and minerals required for photosynthesis are absorbed by the roots and transported to cells containing chloroplasts through tubular conducting cells. The organic products of photosynthesis are distributed to all parts of the plant by other conducting cells (see Chapter 32).

STUDY BREAK 9.1

1. **What are the two stages of photosynthesis?**
2. **In which organelle does photosynthesis take place in plants? Where in that organelle are the two stages of photosynthesis carried out?**

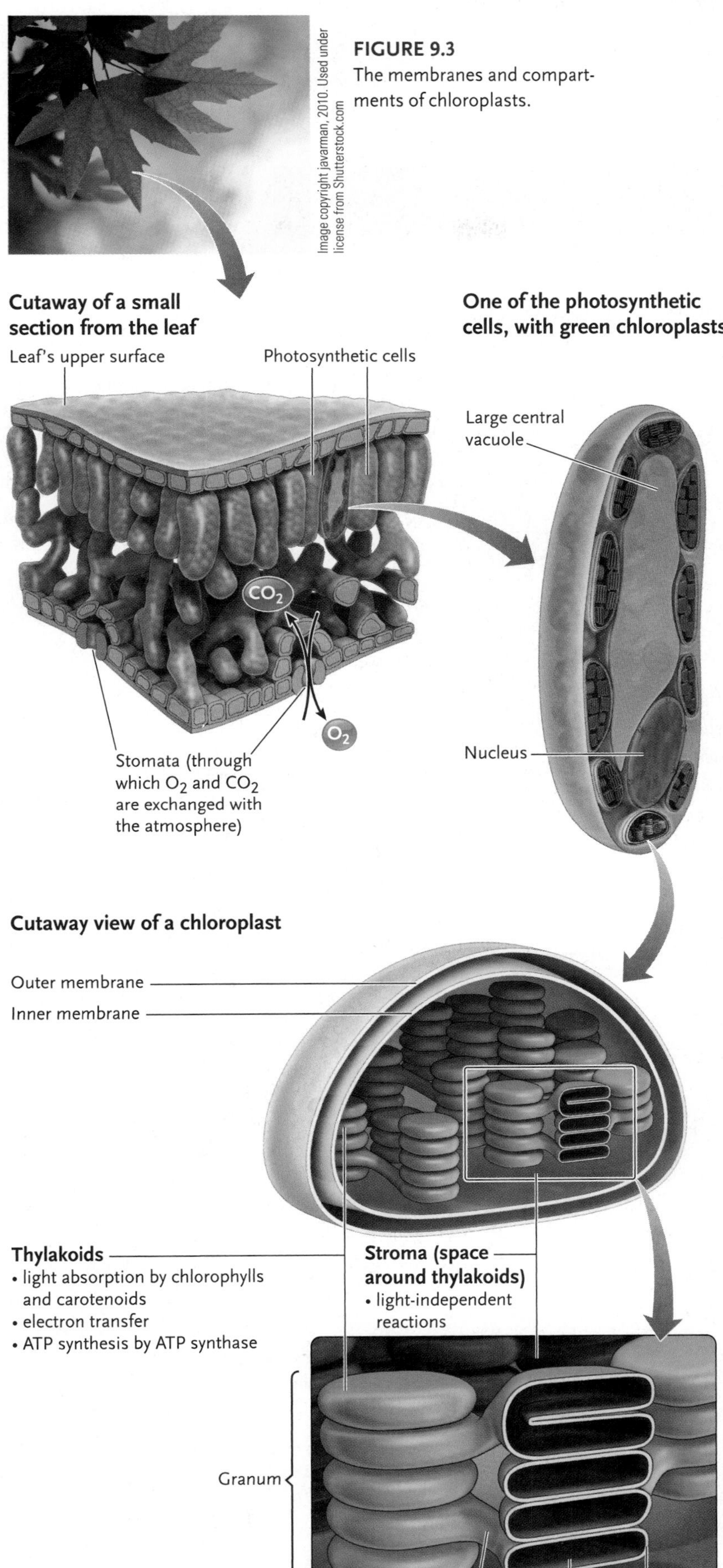
Image copyright javarman, 2010. Used under license from Shutterstock.com
Cutaway of a small section from the leaf
Leaf's upper surface
Photosynthetic cells
CO_2
O_2
Stomata (through which O_2 and CO_2 are exchanged with the atmosphere)
One of the photosynthetic cells, with green chloroplasts
Large central vacuole
Nucleus
Cutaway view of a chloroplast
Outer membrane
Inner membrane
Thylakoids
• light absorption by chlorophylls and carotenoids
• electron transfer
• ATP synthesis by ATP synthase
Stroma (space around thylakoids)
• light-independent reactions
Granum
Stromal lamella
Thylakoid lumen
Thylakoid membrane

FIGURE 9.3
The membranes and compartments of chloroplasts.

> **THINK OUTSIDE THE BOOK**
>
> **Scientists have been working to develop an artificial version of photosynthesis that can be used to produce liquid fuels from CO_2 and H_2O. Collaboratively or individually, find an example of research on artificial photosynthesis and prepare an outline of how the system works or is anticipated to work.**

9.2 The Light-Dependent Reactions of Photosynthesis

In this section we discuss the light-dependent reactions (also referred to more simply as the light reactions), in which light energy is converted to chemical energy. The light-dependent reactions involve two main processes: (1) light absorption; and (2) synthesis of NADPH and ATP. We will describe each of these processes in turn. To keep the bigger picture in perspective, you may find it useful to refer periodically to the summary of photosynthesis shown in Figure 9.2.

Electrons in Pigment Molecules Absorb Light Energy in Photosynthesis

The first process in photosynthesis is light absorption. What is light? Visible light is a form of radiant energy. It makes up a small part of the **electromagnetic spectrum (Figure 9.4),** which ranges from radio waves to gamma rays. The various forms of electromagnetic radiation differ in *wavelength*—the horizontal distance between the crests of successive waves. Radio waves have wavelengths in the range of 10 meters to hundreds of kilometers, and gamma rays have wavelengths in the range of one hundredth to one millionth of a nanometer. The average wavelength of radiowaves for an FM radio station, for example, is 3 m. Generally, the shorter the wavelength, the greater the energy of the radiation.

The radiation humans detect as visible light has wavelengths between about 700 nm, seen as red light, and 400 nm, seen as blue light. We see the entire spectrum of wavelengths from 700 to 400 nm combined together as white light. The energy of light interacts with matter in elementary particles called *photons*. Each photon is a discrete unit that contains a fixed amount of energy. That energy is inversely proportional to its wavelength: the shorter the wavelength, the greater the energy of a photon.

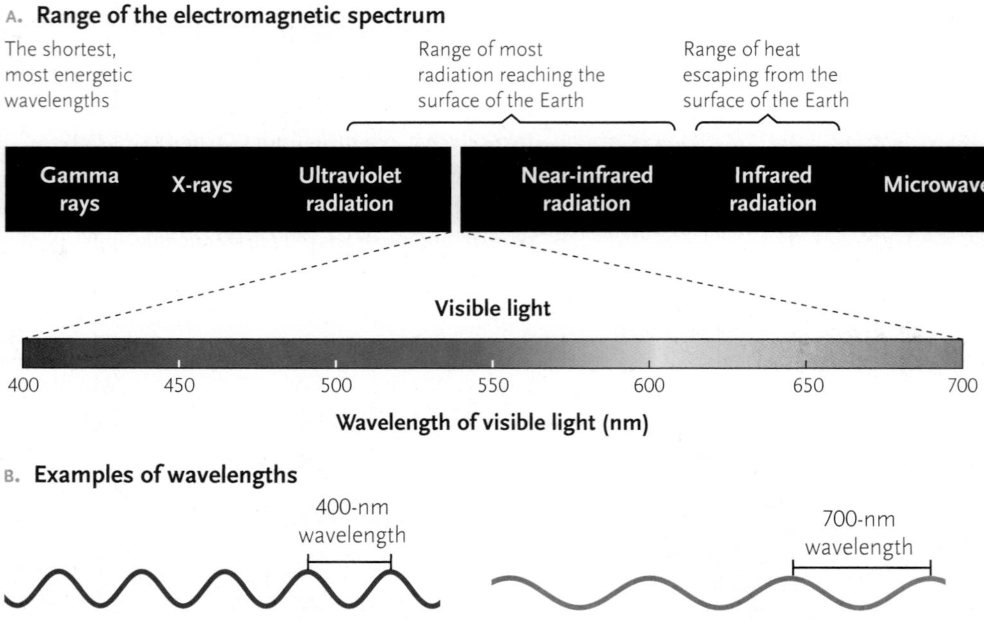

A. **Range of the electromagnetic spectrum**

The shortest, most energetic wavelengths

Range of most radiation reaching the surface of the Earth

Range of heat escaping from the surface of the Earth

The longest, lowest-energy wavelengths

Gamma rays | X-rays | Ultraviolet radiation | Near-infrared radiation | Infrared radiation | Microwaves | Radio waves

Visible light

400 450 500 550 600 650 700

Wavelength of visible light (nm)

B. **Examples of wavelengths**

400-nm wavelength

700-nm wavelength

FIGURE 9.4

The electromagnetic spectrum. **(A)** The electromagnetic spectrum ranges from gamma rays to radio waves; visible light and the wavelengths used for photosynthesis occupy only a narrow band of the spectrum. **(B)** Examples of wavelengths, showing the difference between the longest and shortest wavelengths of visible light.

In photosynthesis, light is absorbed by molecules of green pigments called **chlorophylls** (*chloros* = yellow-green, *phyllon* = leaf) and yellow-orange pigments called **carotenoids** (*carota* = carrot). Pigment molecules such as chlorophyll appear colored to an observer because they absorb the energy of visible light at certain wavelengths and transmit or reflect other wavelengths. The color of a pigment is produced by the transmitted or reflected light. Plants look green because chlorophyll absorbs blue and red light most strongly, and transmits or reflects most of the wavelengths in between; we see the reflected light as green. This green light, as demonstrated by Engelmann's experiment described earlier in *Why It Matters,* is the combination of wavelengths that are *not used* by the plants in photosynthesis.

Light is absorbed in a pigment molecule by excitable electrons occupying certain energy levels in the atoms of the pigments (see Section 2.2 for the discussion on energy levels). When not absorbing light, these electrons are at a relatively low energy level known as the *ground state.* If an electron in the pigment absorbs the energy of a photon, it jumps to a higher energy level that is farther from the atomic nucleus. This condition of the electron is called the *excited state.* The difference in energy level between the ground state and the excited state is equivalent to the energy of the photon of light that was absorbed.

One of three events then occurs, depending on both the pigment that is absorbing the light and other molecules in the vicinity of the pigment **(Figure 9.5):**

- The excited electron from the pigment molecule returns to its ground state, releasing its energy either as heat, or as an emission of light of a longer wavelength than the absorbed light, a process called *fluorescence.*

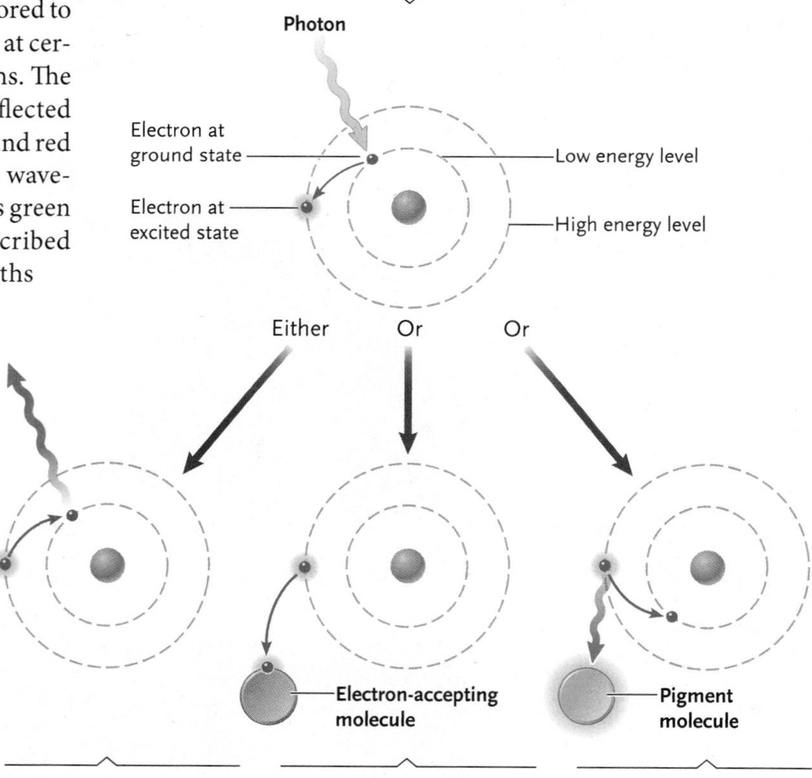
Photon is absorbed by an excitable electron that moves from a relatively low energy level to a higher energy level.
Photon
Electron at ground state
Electron at excited state
Low energy level
High energy level
Either
Or
Or
Electron-accepting molecule
Pigment molecule
The electron returns to its ground state by emitting a less energetic photon (fluorescence) or releasing energy as heat.
The high-energy electron is accepted by an electron-accepting molecule, the primary acceptor.
The electron returns to its ground state, and the energy released transfers to a neighboring pigment molecule, a process called inductive resonance.

FIGURE 9.5

Alternative effects of light absorbed by a pigment molecule.

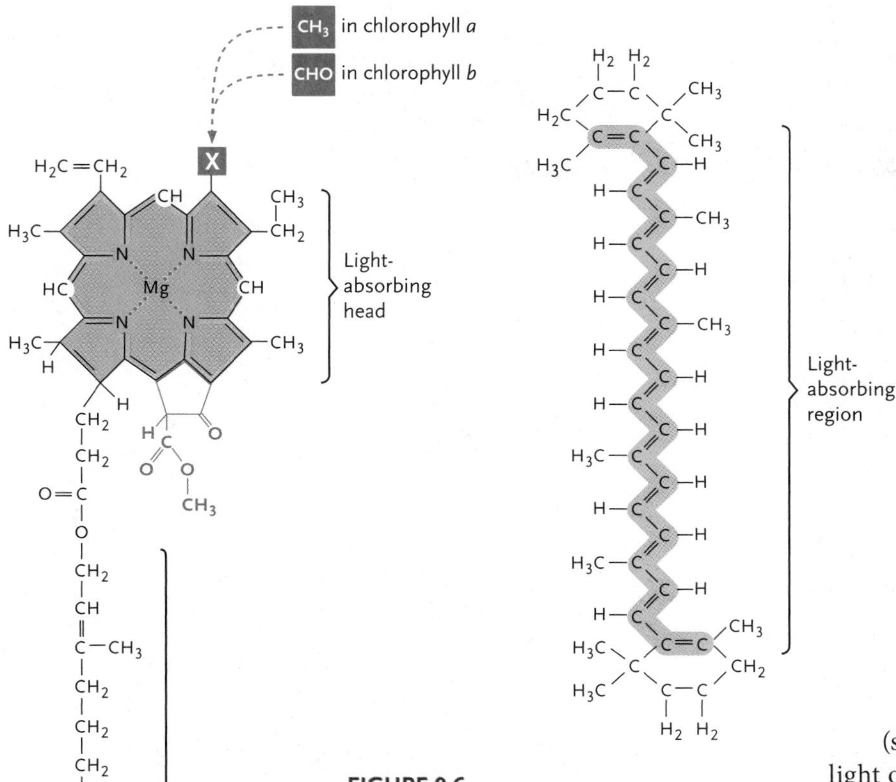

A. Chlorophyll structure
B. Carotenoid structure
CH3 in chlorophyll a
CHO in chlorophyll b
X
Mg
Light-absorbing head
Light-absorbing region
Hydrophobic side chain

FIGURE 9.6
Pigment molecules used in photosynthesis. **(A)** Chlorophylls *a* and *b*, which differ only in the side group attached at the X. Light-absorbing electrons are distributed among the bonds shaded in orange. The chlorophylls are similar in structure to the cytochromes, which occur in both the chloroplast and mitochondrial electron transfer systems. **(B)** In carotenoids, the light-absorbing electrons are distributed in a series of alternating double and single bonds in the backbone of these pigments.

- The excited electron is transferred from the pigment molecule to a nearby electron-accepting molecule called a *primary acceptor*.
- The energy of the excited electron, but not the electron itself, is transferred to a neighboring pigment molecule in a process called *inductive resonance*. This transfer excites the second molecule, while the first molecule returns to its ground state. Very little energy is lost in this energy transfer.

The Chlorophyll and Carotenoid Pigments Cooperate in Light Absorption

Chlorophylls are the major photosynthetic pigments in plants, green algae, and cyanobacteria. They absorb photons and transfer excited electrons to stable orbitals in primary acceptor molecules. In the transfer, the chlorophyll is oxidized because it loses an electron, and the primary acceptor is reduced because it gains an electron. Closely related molecules, the *bacteriochlorophylls*, carry out the same functions in other photosynthetic bacteria. Carotenoids are pigments that absorb light energy and pass it on by inductive resonance to the chlorophylls.

Chlorophylls and carotenoids are bound to proteins that are embedded in photosynthetic membranes. In plants and green algae they are located in the thylakoid membranes of chloroplasts; in photosynthetic bacteria they are located in the cell membrane.

Molecules of the chlorophyll family **(Figure 9.6A)** have a carbon ring structure, to which is attached a long, hydrophobic side chain. A magnesium atom is bound at the center of the ring structure. The most important kind of chlorophyll is chlorophyll *a*, which is found in plants, green algae, and cyanobacteria. A second kind, chlorophyll *b*, is found only in plants and green algae. Chlorophyll *a* and chlorophyll *b* differ only in one side group that is attached to a carbon of the ring structure (shown in Figure 9.6A).

A chlorophyll molecule contains a network of electrons capable of absorbing light (shaded in orange in Figure 9.6A). The amount of light of different wavelengths that is absorbed by a pigment is called an **absorption spectrum;** it is usually shown as a graph in which the height of the curve at any wavelength indicates the amount of light absorbed. **Figure 9.7A** shows the absorption spectra for chlorophylls *a* and *b*.

The carotenoids are built on a long backbone that typically contains 40 carbon atoms **(Figure 9.6B).** Carotenoids expand the range of wavelengths used for photosynthesis because they absorb different wavelengths that chlorophyll does not absorb. Carotenoids transmit or reflect other wavelengths in combinations that appear yellow, orange, red, or brown, depending on the type of carotenoid. The carotenoids contribute to the red, orange, and yellow colors of vegetables and fruits and to the brilliant colors of autumn leaves, in which the green color is lost when the chlorophylls break down.

The light absorbed by the carotenoids and chlorophylls, acting in combination, drives the reactions of photosynthesis. Plotting the effectiveness of light of each wavelength in driving photosynthesis produces a graph called the **action spectrum** of photosynthesis (**Figure 9.7B** shows the action spectrum of higher plants). The action spectrum is usually determined by measuring the amount of O_2 released by photosynthesis carried out at different wavelengths of visible light, as Engelmann did indirectly in his experiment described in *Why It Matters* (compare Figures 9.1 and 9.7B).

In all eukaryotic photosynthesizers, a specialized chlorophyll *a* molecule passes excited electrons to stable orbitals in the primary acceptor. Other chlorophyll molecules, along with carotenoids, act as *accessory pigments* that pass their energy to chlorophyll *a*. Light energy that is absorbed by the entire collection of chlorophyll and carotenoid molecules in chloroplasts is passed by inductive resonance to the specialized chlorophyll *a* molecules that are directly involved in transforming light into chemical energy.

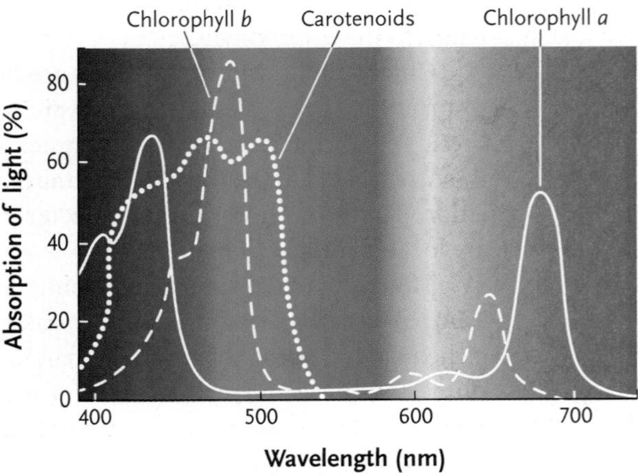
A. The absorption spectra of chlorophylls a and b and carotenoids
Chlorophyll b
Carotenoids
Chlorophyll a
Absorption of light (%)
80
60
40
20
0
400
500
600
700
Wavelength (nm)

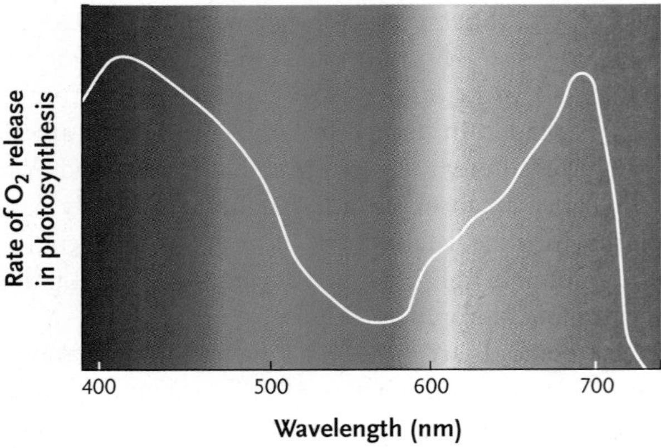
B. The action spectrum in higher plants, representing the combined effects of chlorophylls and carotenoids
Rate of O2 release in photosynthesis
400
500
600
700
Wavelength (nm)

FIGURE 9.7
The absorption spectra of the photosynthetic pigments **(A)** and the action spectrum of photosynthesis **(B)** in higher plants. The absorption spectra in **(A)** were made from pigments that were extracted from cells and purified.

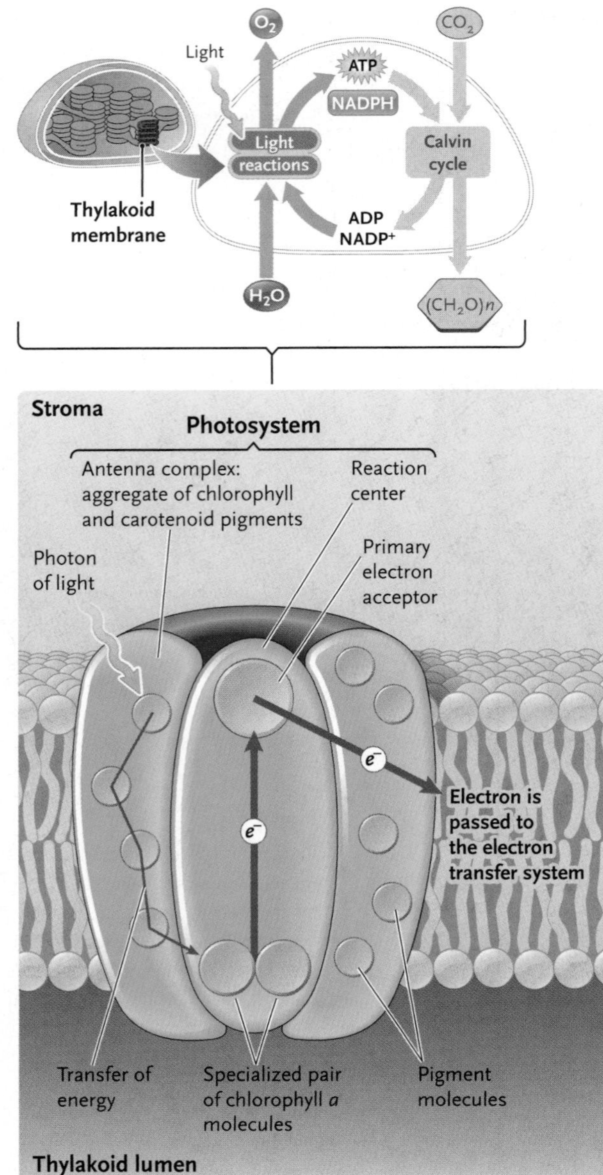
O2
CO2
Light
ATP
NADPH
Light reactions
Calvin cycle
Thylakoid membrane
ADP
NADP+
H2O
(CH2O)n
Stroma
Photosystem
Antenna complex: aggregate of chlorophyll and carotenoid pigments
Reaction center
Photon of light
Primary electron acceptor
e−
e−
Electron is passed to the electron transfer system
Transfer of energy
Specialized pair of chlorophyll a molecules
Pigment molecules
Thylakoid lumen

FIGURE 9.8
Major components of a photosystem: a group of pigments forming an antenna complex (light-harvesting complex) and a reaction center. Light energy absorbed anywhere in the antenna complex is conducted by inductive resonance to a pair of specialized chlorophyll *a* molecules in the reaction center. The absorbed light is converted to chemical energy when an excited electron from the chlorophyll *a* is transferred to a stable orbital in a primary acceptor, also in the reaction center. High-energy electrons are passed out of the photosystem to the electron transfer system. The blue arrows show the path of energy flow.

The Photosynthetic Pigments Are Organized into Photosystems in Chloroplasts

The light-absorbing pigments are organized with proteins and other molecules into large complexes called **photosystems (Figure 9.8),** which are embedded in thylakoid membranes and stromal lamellae. The photosystems are the sites at which light is absorbed and converted into chemical energy.

Plants, green algae, and cyanobacteria have two types of these complexes, called **photosystems I** and **II,** which carry out different parts of the light-dependent reactions. Photosystem II, in addition, is closely linked to a group of enzymes that carries out the initial reaction of splitting water into electrons, protons, and oxygen.

Each photosystem consists of two closely associated components: an **antenna complex** (also called a *light-harvesting complex*), and a **reaction center.** The antenna complex contains an aggregate of many chlorophyll pigments and a number of carotenoid pigments. The chlorophyll molecules are anchored in the complex by being bound to specific membrane proteins. In this form, they are efficiently arranged to optimize the capture of light energy.

The reaction center contains a pair of specialized chlorophyll *a* molecules complexed with proteins. The pair of specialized chlorophyll *a* molecules at the reaction center of photosystem I is called *P700* (P = pigment) because it absorbs light optimally at a wavelength of 700 nm. The reaction center of photosystem II contains a different pair of specialized chlorophyll *a* molecules, *P680,* which absorbs light optimally at a wavelength of 680 nm. P700 and P680 are structurally identical to other chlorophyll *a* molecules; their specific light absorption patterns result from interactions with particular proteins in the photosystems.

Light energy in the form of photons is absorbed by the pigment molecules of the antenna complex. This absorbed light energy reaches P700 and P680 in the reaction center by inductive resonance. On arrival, the energy is captured quickly in the form of an excited electron that is passed to a stable orbital in a primary acceptor molecule. That electron is passed to the electron transfer system, which carries electrons away from the primary acceptor. Some components of the electron transfer system are located within the photosystems and other components are separate.

Electron Flow from Water through Photosystem I to $NADP^+$ Leads to the Synthesis of NADPH and ATP

In the second main process of the light-dependent reactions, the electrons obtained from the splitting of water (two electrons per molecule of water; see Section 9.1) are used for the synthesis of NADPH and ATP. These electrons, which were pushed to higher levels by light energy, pass through an electron transfer system consisting of a series of electron carriers that are alternately reduced and oxidized as they pick up and release electrons in sequence. The electron carriers are embedded in a thylakoid membrane in eukaryotes and in the plasma membrane in prokaryotes.

As in all electron transfer systems, the electron carriers of the photosynthetic system consist of nonprotein organic groups that pick up and release the electrons traveling through the system. The carriers include the same types that act in mitochondrial electron transfer—cytochromes, quinones, and iron–sulfur centers (discussed in Section 8.4). Most of the carriers are organized with proteins into larger complexes, which are distributed among the thylakoid membranes and stromal lamellae of chloroplasts.

The electron carriers of photosynthesis are arranged in a chain **(Figure 9.9).** This chain-like structure was first deduced by Robert Hill and Fay Bendall of Cambridge University. Electrons from water first flow through photosystem II, becoming excited

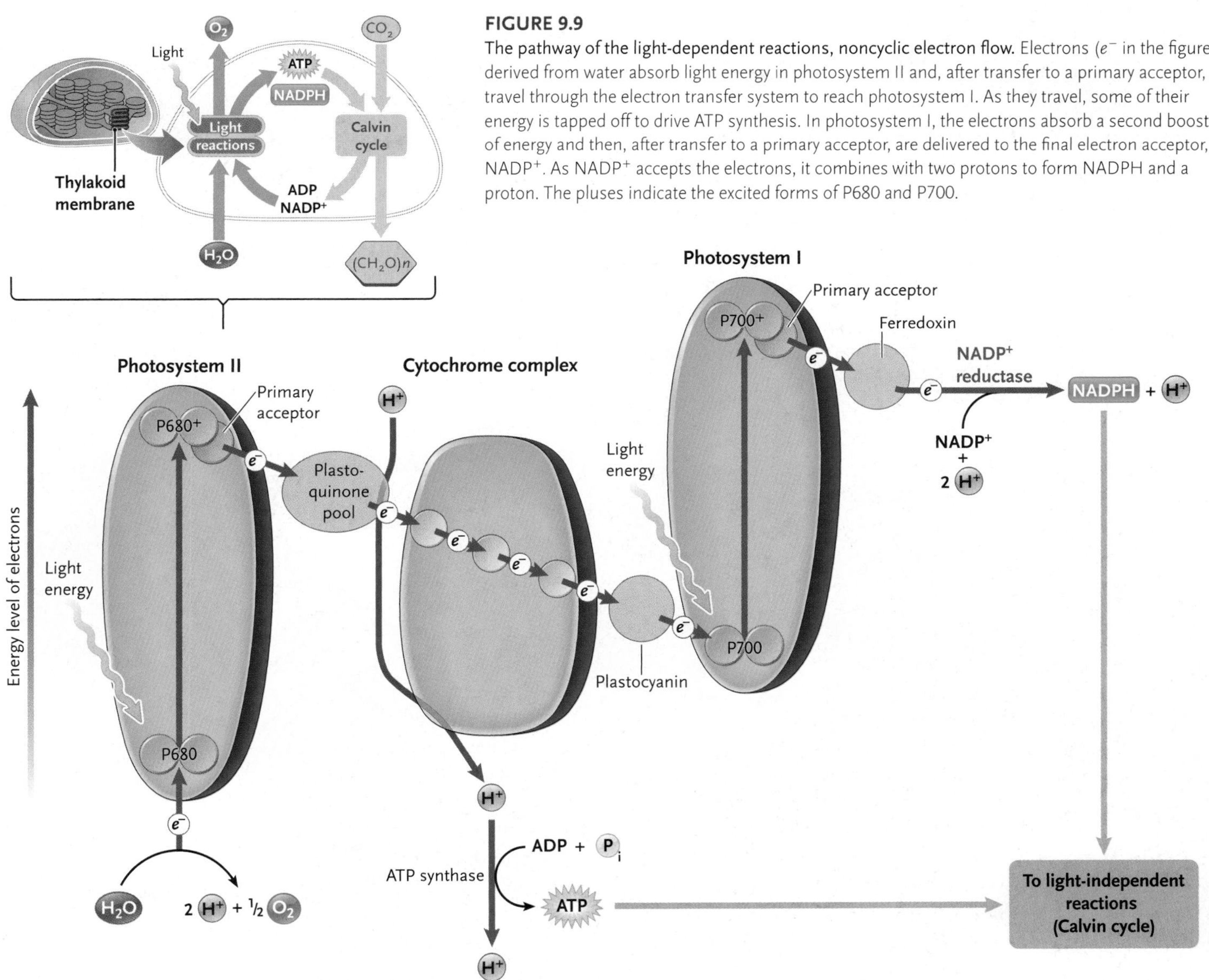

Light
O_2
ATP
NADPH
CO_2
Light reactions
Calvin cycle
Thylakoid membrane
ADP
$NADP^+$
H_2O
$(CH_2O)n$
Photosystem II
Primary acceptor
P680+
Cytochrome complex
Plasto-quinone pool
Energy level of electrons
Light energy
P680
$2H^+ + \frac{1}{2}O_2$
ADP + P_i
ATP synthase
ATP
Photosystem I
Primary acceptor
P700+
Ferredoxin
$NADP^+$ reductase
NADPH + H^+
$NADP^+$ + 2 H^+
Light energy
P700
Plastocyanin
To light-independent reactions (Calvin cycle)

FIGURE 9.9
The pathway of the light-dependent reactions, noncyclic electron flow. Electrons (e^- in the figure) derived from water absorb light energy in photosystem II and, after transfer to a primary acceptor, travel through the electron transfer system to reach photosystem I. As they travel, some of their energy is tapped off to drive ATP synthesis. In photosystem I, the electrons absorb a second boost of energy and then, after transfer to a primary acceptor, are delivered to the final electron acceptor, $NADP^+$. As $NADP^+$ accepts the electrons, it combines with two protons to form NADPH and a proton. The pluses indicate the excited forms of P680 and P700.

FIGURE 9.10
The chloroplast electron transfer system (steps 1-9) and chemiosmosis (steps 10-12), illustrating the synthesis of NADPH and ATP by the noncyclic electron flow pathway. The components of the electron transfer system and chemiosmosis are located in the chloroplast thylakoid membrane.

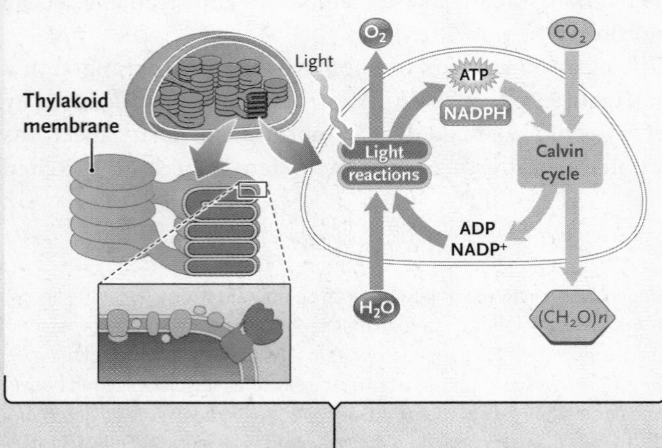

Thylakoid membrane
Light
O_2
CO_2
ATP
NADPH
Light reactions
Calvin cycle
ADP
$NADP^+$
H_2O
$(CH_2O)n$

1 Electrons from the water-splitting reaction system are accepted one at a time by a P680 chlorophyll *a* in the reaction center of photosystem II. As P680 accepts the electrons, they are raised to the excited state, using energy passed to the reaction center from the light-absorbing pigment molecules in the antenna complex. The excited electrons are immediately transferred to the primary acceptor of photosystem II, a modified form of chlorophyll *a*.

2 The electrons flow through a short chain of carriers within photosystem II and then transfer to the mobile carrier plastoquinone. The plastoquinones form a "pool" of molecules within the thylakoid membranes.

3 Plastoquinones pass the electrons to the cytochrome complex. As it accepts and releases electrons, the cytochrome complex pumps H^+ from the stroma into the thylakoid lumen. Those protons drive ATP synthesis (see step 8).

4 Electrons now pass to the mobile carrier *plastocyanin*, which shuttles electrons between the cytochrome complex and photosystem I.

5 Electrons flow to a P700 chlorophyll *a* in the reaction center of photosystem I, where they are excited to high energy levels again by absorbing more light energy. The excited electrons are transferred from P700 to the primary acceptor of photosystem I, a modified chlorophyll *a*.

6 After passage through carriers within photosystem I, the electrons are transferred to the iron-sulfur protein ferredoxin, which acts as a mobile carrier.

7 The ferredoxin transfers the electrons, still at very high energy levels, to $NADP^+$, the final acceptor of the noncyclic pathway. $NADP^+$ is reduced to NADPH by $NADP^+$ reductase. Electron transfer is now complete.

8 Proton pumping by the plastoquinones and the cytochrome complex, as described in step 3, creates a concentration gradient of H^+ with the high concentration within the thylakoid lumen and the low concentration in the stroma.

9 The H^+ gradient supplies the energy that drives ATP synthesis by ATP synthase.

10 Due to the gradient, H^+ ions flow across the inner membrane into the matrix through a channel in the ATP synthase.

11 The flow of H^+ activates ATP synthase, making the headpiece and stalk rotate.

12 As a result of changes in shape and position as it turns, the headpiece catalyzes the synthesis of ATP from ADP and P_i.

Stroma
Electron transfer system
Chemiosmosis
Photosystem II
Cytochrome complex
Photosystem I
Low H^+
To light-independent reactions (Calvin cycle)
Light energy
Antenna complex
Primary acceptor
Pigment molecules
Ferredoxin
$2\ H^+ + NADP^+$
H^+ + NADPH
ATP
ADP + P_i
$NADP^+$ reductase
P680
P700
Plastoquinone
Water-splitting complex
Plastocyanin
Stator
Rotation
ATP synthase
H_2O
$2\ H^+ + \frac{1}{2}\ O_2$
High H^+
Thylakoid lumen
Thylakoid membrane

to a higher energy level in P680 through absorbed light energy. The electrons then flow "downhill" in energy level through an electron transfer system connecting photosystems II and I. (Note: Photosystem I is so named because it was discovered first; the systems were given their numbers before their order of use in the pathway was worked out.)

The electron transfer system consists of a pool of molecules of the electron carrier plastoquinone, a cytochrome complex, and the mobile carrier protein plastocyanin. As electrons pass through the system they release free energy at each transfer from a donor to an acceptor molecule. Some of this free energy is used to create a gradient of H^+ across the membrane. The gradient provides the energy source for ATP synthesis, just as it does in mitochondria.

The electrons then pass to photosystem I, where they are excited a second time in P700 through absorbed light energy. The high-energy electrons enter a short electron transfer system that leads to the final acceptor of the chloroplast system, $NADP^+$. The enzyme $NADP^+$ reductase reduces $NADP^+$ to NADPH by using two electrons and two protons from the surrounding water solution, and by releasing one proton. This pathway is frequently called **noncyclic electron flow** because electrons travel in a one-way direction from H_2O to $NADP^+$; it is sometimes called the *Z scheme* because of the zigzag-like changes in electron energy level (shown by the blue arrows in Figure 9.9).

Figure 9.10 shows how the electron transfer and ATP synthesis systems for the light-dependent reactions are organized in the thylakoid membrane and lays out the noncyclic electron pathway for NADPH and ATP synthesis. The electron transfer system involves steps 1–9, and the chemiosmotic synthesis of ATP involves steps 10–12.

The flow of electrons through the electron transfer system leads to the generation of an H^+ gradient across the inner membrane (steps 1–8). That gradient is enhanced by the addition of two protons to the lumen for each water molecule split, and by the removal of one proton from the stroma for each NADPH molecule synthesized. Because protons carry a positive charge, an electrical gradient forms across the thylakoid membrane, with the lumen more positively charged than the stroma. The combination of a proton gradient and a voltage gradient across the membrane produces stored energy known as the *proton-motive force* (also discussed in Section 8.4), which contributes energy for ATP synthesis by ATP synthase. Just as for the mitochondrial ATP synthase, the chloroplast ATP synthase is embedded in the same membranes as the electron transfer system. Protons flow through a membrane channel from the thylakoid lumen to the stroma along their concentration gradient (step 10). Free energy is released as H^+ moves through the channel; it powers synthesis of ATP from ADP and P_i by the ATP synthase (steps 11–12). This process of using an H^+ gradient to power ATP synthesis—*chemiosmosis*—is the same as that used for ATP synthesis in mitochondria (see Section 8.4).

The overall yield of the noncyclic electron flow pathway is one molecule of NADPH and one molecule of ATP for each pair of electrons produced from the splitting of water. The synthesis of ATP coupled to the transfer of electrons energized by photons of light is called **photophosphorylation.** This process is analogous to oxidative phosphorylation in mitochondria (see Section 8.4), except that in chloroplasts light provides the energy for establishing the proton gradient.

Comparing the noncyclic pathway with the mitochondrial electron transfer system (shown in Figure 8.13) reveals that the pathway from the plastoquinones through plastocyanin in chloroplasts is essentially the same as the pathway from the ubiquinones through cytochrome *c* in mitochondria. The similarities between the two pathways indicate that the electron transfer system is a very ancient evolutionary development that became adapted to both photosynthesis and oxidative phosphorylation.

The elements of the noncyclic pathway are not located in fixed, organized assemblies as Figure 9.10 might suggest. Instead, photosystem II is located almost exclusively in thylakoid membranes, in regions where one thylakoid membrane is fused to the next in the stacks of grana; photosystem I is located primarily in stromal lamellae. Other components of the electron transfer system are distributed among both thylakoids and stromal lamellae.

Electrons Can Also Drive ATP Synthesis by Flowing Cyclically around Photosystem I

Noncyclic electron flow produces ATP and NADPH + H^+. In some cases, however, photosystem I works independently of photosystem II in a circular process called **cyclic electron flow (Figure 9.11).** In this process, electrons pass through the cytochrome complex and plastocyanin to the P700 chlorophyll *a* in the reaction center of photosystem I where they are excited by light energy. The electrons then flow from photosystem I to the mobile carrier ferredoxin, but rather than being used for $NADP^+$ reduction by $NADP^+$ reductase, they flow back to P700 in the cytochrome complex. The electrons again pass to plastocyanin and on to photosystem I where they receive another energy boost from light energy, and so the cycle continues. Each time electrons flow around the cycle, more H^+ is pumped across the thylakoid membranes, driving ATP synthesis in the way already described. The net result of cyclic electron flow is that light energy is converted into the chemical energy of ATP *without* the production of NADPH.

Cyclic electron flow was observed in higher plant chloroplasts over 50 years ago. However, researchers still debate whether it is a real physiological process. Recent experimental results, including studies of mutants that appear to lack the cyclic electron flow pathway, support the hypothesis that the pathway plays an important role in the responses of plants to stress. That is, because cyclic electron flow produces ATP but no NADPH, the plants avoid risky overproduction of reducing power when stressed. Excess reducing power can be damaging because of the potential for some of the extra electrons "leaking out" of the system to form reactive oxygen species such as superoxide. Reactive oxygen species are toxic and can lead to the oxidative destruction of cells.

Photosynthesis occurs also in some groups of bacteria: cyanobacteria, purple sulfur bacteria, and green sulfur bacteria. In those organisms the photosynthetic electron transfer system components are embedded in membranes, but there are no chloroplasts like those of plants. Cyanobacteria, the only prokaryotes that produce oxygen by photosynthesis, contain both photosystems II and I and normally carry out the light-dependent reac-

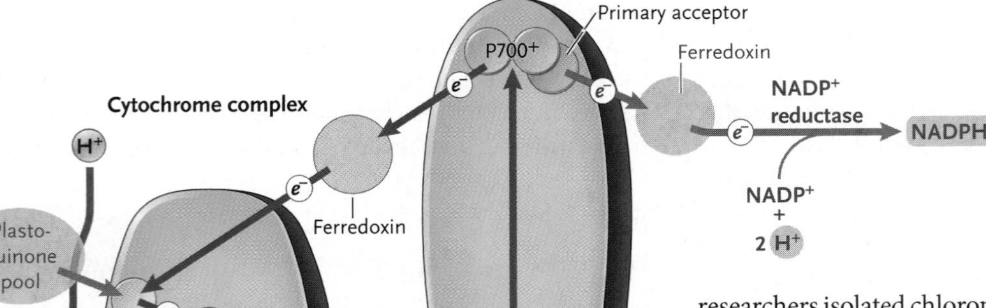

Light
ATP
NADPH
Light reactions
Calvin cycle
Thylakoid membrane
ADP
NADP⁺
H₂O
(CH₂O)n
Photosystem I
Primary acceptor
P700⁺
Ferredoxin
NADP⁺ reductase
NADPH + H⁺
NADP⁺ + 2 H⁺
Cytochrome complex
H⁺
Plasto-quinone pool
Ferredoxin
Plastocyanin
P700
ADP + Pi
ATP synthase
ATP
To light-independent reactions (Calvin cycle)

FIGURE 9.11

Cyclic electron flow around photosystem I. Electrons move in a circular pathway from ferredoxin back to the cytochrome complex, then to plastocyanin, through photosystem I, and back to ferredoxin again. The cycle pumps additional H^+ each time electrons flow through the cytochrome complex. The H^+ drive ATP synthesis as described for the noncyclic flow pathway.

tions using the noncyclic electron flow pathway. They are also capable of using the cyclic electron flow pathway involving photosystem I. The purple sulfur bacteria have a single photosystem structurally related to photosystem II of plants, and the green sulfur bacteria have a single photosystem structurally related to photosystem I of plants. For both of these groups, the light-dependent reactions take place only by cyclic electron flow, and no oxygen is produced by those pathways.

Experiments with Chloroplasts Helped Confirm the Synthesis of ATP by Chemiosmosis

Our present understanding of the connection between electron transfer and ATP synthesis was first proposed for mitochondria in Mitchell's chemiosmotic hypothesis (discussed in Section 8.4). Several experiments have shown that the same mechanism operates in chloroplasts. One of these experiments is shown in **Figure 9.12.** The researchers isolated chloroplasts from cells and treated them to create an H^+ gradient between the thylakoid lumen and the stroma. In the dark, ATP was made in the stroma, indicating that the gradient, and not light-generated electron transfer, powers ATP synthesis.

Our description of photosynthesis to this point shows how the light-dependent reactions generate NADPH and ATP, which provide the reducing power and chemical energy required to produce organic molecules from CO_2. The next section follows NADPH and ATP through the light-independent reactions and shows how the organic molecules are produced.

STUDY BREAK 9.2

1. **What is the difference in function between the chlorophyll *a* molecules in the antenna complexes and the chlorophyll *a* molecules in the reaction centers of the photosystems?**
2. **How is NADPH made in the noncyclic electron flow pathway?**
3. **What is the difference between the noncyclic electron flow pathway and the cyclic electron flow pathway?**

9.3 The Light-Independent Reactions of Photosynthesis

NADPH has the same primary role in all eukaryotes—to deliver high-energy electrons to synthetic reactions that require a reduction. In photosynthesis, the reaction requiring a reduction, the fixation of CO_2, takes place in the second stage of photosynthesis (the light-independent reactions). The electrons carried from the light-dependent reactions by NADPH retain much of the energy absorbed from sunlight. These electrons provide the reducing power required to fix CO_2 into carbohydrates and other organic molecules in the light-independent reactions. The ATP generated in the light-dependent reactions supplies additional energy for

FIGURE 9.12 **Experimental Research**

Demonstration That an H^+ Gradient Drives ATP Synthesis in Chloroplasts

Question: Does chemiosmosis power ATP synthesis by a proton gradient in chloroplasts?

Experiment: Andre T. Jagendorf and Ernest Uribe of Johns Hopkins University placed chloroplasts extracted from cells in darkness, thereby eliminating light absorption and electron transfer as a source of energy for photosynthesis. They then added an acid to the solution, which increased the acidity (lowered the pH) inside the stroma and thylakoids to pH 4. This allowed H^+ to penetrate inside the chloroplasts, including into the thylakoid lumen.

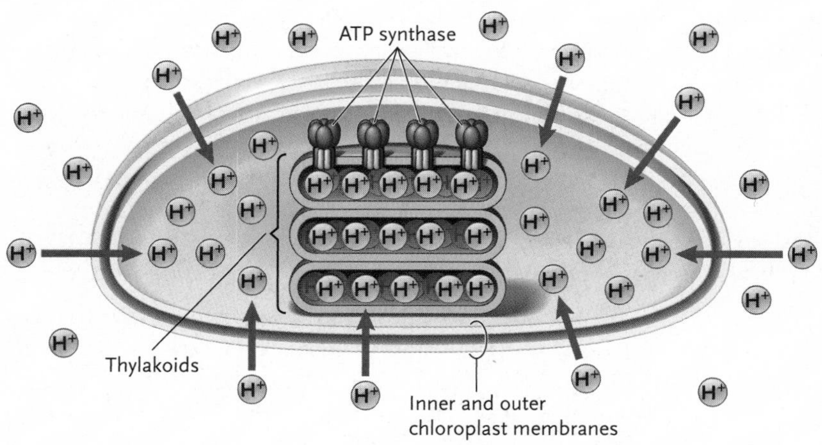

ATP synthase
Thylakoids
Inner and outer chloroplast membranes

The chloroplasts, still in darkness, were then moved to a basic solution that was pH 8. This created an H^+ gradient, high inside the thylakoid lumen and low in the stroma. In response to the gradient, H^+ moved from the thylakoid lumen to the stroma.

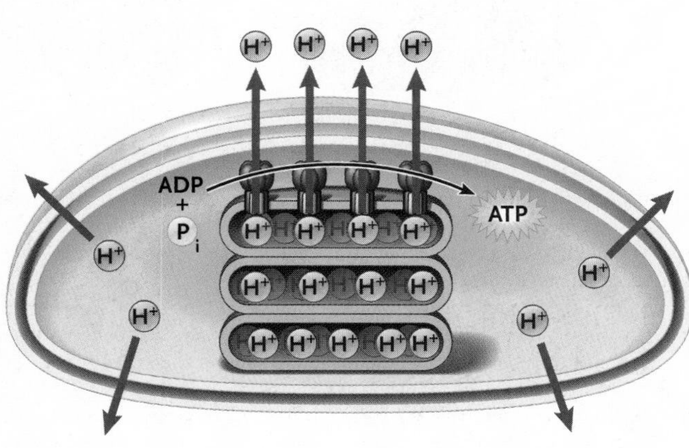

ADP
+
P
i
ATP

Results: ATP was synthesized by the chloroplast.

Conclusion: Because chloroplasts in darkness cannot use electron transfer as an energy source, chemiosmosis must power ATP synthesis by an H^+ gradient.

Source: A. T. Jagendorf and E. Uribe. 1966. ATP formation caused by acid–base transition of spinach chloroplasts. *Proceedings of the National Academy of Sciences USA* 55:170–177.

the light-independent reactions. The reactions using NADPH and ATP to fix CO_2 occur in a circuit known as the **Calvin cycle,** named for its discoverer, Melvin Calvin. *Focus on Basic Research* describes the experiments Calvin and his colleagues used to elucidate the light-independent reactions.

The Calvin Cycle Uses NADPH, ATP, and CO_2 to Generate Carbohydrates

The light-independent reactions of the Calvin cycle use CO_2, ATP, and NADPH as inputs. As products, the cycle releases ADP; $NADP^+$; the three-carbon carbohydrate molecule glyceraldehyde-3-phosphate (G3P), already familiar as part of glycolysis; and inorganic phosphate (outlined in **Figure 9.13A**). In plants, the Calvin cycle takes place entirely in the chloroplast stroma. The Calvin cycle also occurs in most photosynthetic prokaryotes, where it takes place in the cytoplasm. We focus here on plants.

Figure 9.13A focuses primarily on tracking the carbon atoms through the cycle. In phase 1 of the cycle, *carbon fixation,* a carbon atom from CO_2 is added to ribulose 1,5-bisphosphate (RuBP), a five-carbon sugar, to produce two three-carbon molecules of 3-phosphoglycerate (3PGA). Because the product of the carbon fixation reaction is a three-carbon molecule, the Calvin cycle is also called the **C_3 pathway,** and plants that initially fix carbon in this way are termed **C_3** plants. Most plants are of this kind.

In phase 2, *reduction,* reactions using NADPH and ATP from the light-dependent reactions convert 3PGA into G3P, another three-carbon molecule. After several rounds of the Calvin cycle, two molecules of G3P leave the cycle and are used to form the products of the cycle: the six-carbon sugar glucose and other organic compounds. In phase 3, *regeneration,* some G3P molecules are used to produce the five-carbon RuBP with the help of energy from ATP. The cycle then begins again.

Now let us consider the reactions in more detail. **Figure 9.13B** shows the chemical structures, reactions, and enzymes of the cycle. The key reaction of the cycle is the first, carbon fixation, in which CO_2 combines with RuBP, forming a transient six-carbon molecule that is cleaved to form 3PGA. This reaction, which fixes CO_2 into organic form, is catalyzed by the *carboxylase* activity of the key enzyme of the Calvin cycle, **RuBP carboxylase/oxygenase** (abbreviated as **rubisco**). In the next two reactions (reactions 2 and 3 in Figure 9.13B, shown in two parallel paths because two molecules of 3PGA are being processed), the three-carbon molecules are raised in energy level by the addition of phosphate groups transferred from ATP and electrons from NADPH (the ATP and NADPH are products of the light-dependent reactions). The G3P generated by reaction 3 is the carbohydrate product of the Calvin cycle.

Most of the G3P produced by the reactions is used to regenerate the RuBP. The G3P enters a complex series of reactions (reaction series 4 in Figure 9.13B) that yields the five-carbon sugar ribulose 5-phosphate. Then, in the final reaction of the cycle (re-

FOCUS ON **Basic Research**

Two-Dimensional Paper Chromatography and the Calvin Cycle

The first significant progress in unraveling the light-independent reactions was made in the 1940s, when newly developed radioactive compounds became available to biochemists. One substance, CO_2 labeled with the radioactive carbon isotope ^{14}C (discussed in *Focus on Applied Research* in Chapter 2), was critical to this research.

Beginning in 1945, Melvin Calvin, Andrew A. Benson, and their colleagues at the University of California at Berkeley, combined ^{14}C-labeled CO_2 with a widely used technique called *two-dimensional paper chromatography* to trace the pathways of the light-independent reactions in a green alga of the genus *Chlorella*. The researchers exposed actively photosynthesizing *Chlorella* cells to the labeled carbon dioxide. Then, at various times, cells were removed and placed in hot alcohol, which instantly stopped all the photosynthetic reactions of the algae. Radioactive carbohydrates were then extracted from the cells and two-dimensional paper chromatography was done to separate and to identify them chemically, as shown in the figure.

By analyzing the labeled molecules revealed by the two-dimensional chromatography technique in the extracts prepared from *Chlorella* cells under different conditions, Calvin and his colleagues were able to reconstruct the reactions of the Calvin cycle. For example, in carbohydrate extracts made within a few seconds after the cells were exposed to the labeled CO_2, most of the radioactivity was found in 3PGA, indicating that it is one of the earliest products of photosynthesis. In extracts made after longer periods of exposure to the label, radioactivity showed up in G3P and in more complex substances including a variety of six-carbon sugars, sucrose, and starch. The researchers also examined the effect of reducing the amount of CO_2 available to the *Chlorella* cells so that photosynthesis worked slowly even in bright light. Under these conditions, RuBP accumulated in the cells, suggesting that it is the first substance to react with CO_2 in the light-independent reactions and that it accumulates if CO_2 is in short supply. Most of the intermediate compounds between CO_2 and six-carbon sugars were identified in similar experiments.

Using this information, Calvin and his colleagues pieced together the light-independent reactions of photosynthesis and showed that they formed a continuous cycle. Melvin Calvin was awarded a Nobel Prize in 1961 for his work on the assimilation of carbon dioxide in plants.

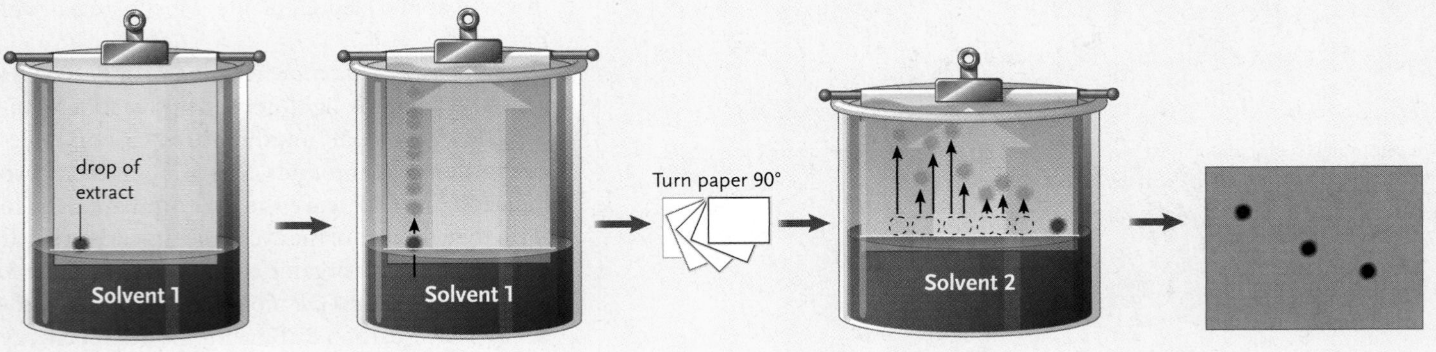
drop of extract
Solvent 1
Solvent 1
Turn paper 90°
Solvent 2

1 A drop of extract containing radioactive carbohydrates is placed at one corner of a piece of chromatography paper. The paper is placed in a jar with its edge touching a solvent. (Calvin used a water solution of butyl alcohol and propionic acid.)

2 The extracted molecules in the spot dissolve and are carried upward through the paper as the solvent rises. The rates of movement of the molecules vary according to their molecule size and solubility. The resulting vertical line of spots is the first dimension of the two-dimensional technique.

3 The paper is dried, turned 90°, and touched to a second solvent (Calvin used a water solution of phenol for this part of the experiment). As this solvent moves through the paper, the molecules again migrate upward from the spots produced by the first dimension, but at rates different from their mobility in the first solvent. This step, the second dimension of the two-dimensional technique, separates molecules that, although different, had produced a single spot in the first solvent because they had migrated at the same rate. The individual spots are identified by comparing their locations with the positions of spots made by known molecules when the "knowns" are run through the same procedure.

4 The paper is dried and covered with a sheet of photographic film. Radioactive molecules expose the film in spots over their locations in the paper. Developing the film reveals the locations of the radioactive spots. The spots on the film are compared with the spots on the paper to identify the molecules that were radioactive.

action 5), a phosphate group is transferred from ATP to regenerate the RuBP used in the first reaction, and the cycle is ready to turn again.

Three Turns of the Calvin Cycle Are Needed to Make One Net G3P Molecule

What we have just described are the events for *one* turn of the Calvin cycle. For each turn, one molecule of CO_2 is converted into one reduced carbon—essentially one (CH_2O) unit of carbohydrate. However, it takes three turns of the cycle to actually produce something the cell can use—one extra G3P that is released as a net product (see Figure 9.13B, lower right). The net G3P released serves as the primary building block for reactions producing glucose and many other organic molecules in chloroplasts. Providing enough (CH_2O) units to make a six-carbon carbohydrate such as glucose requires six turns of the cycle.

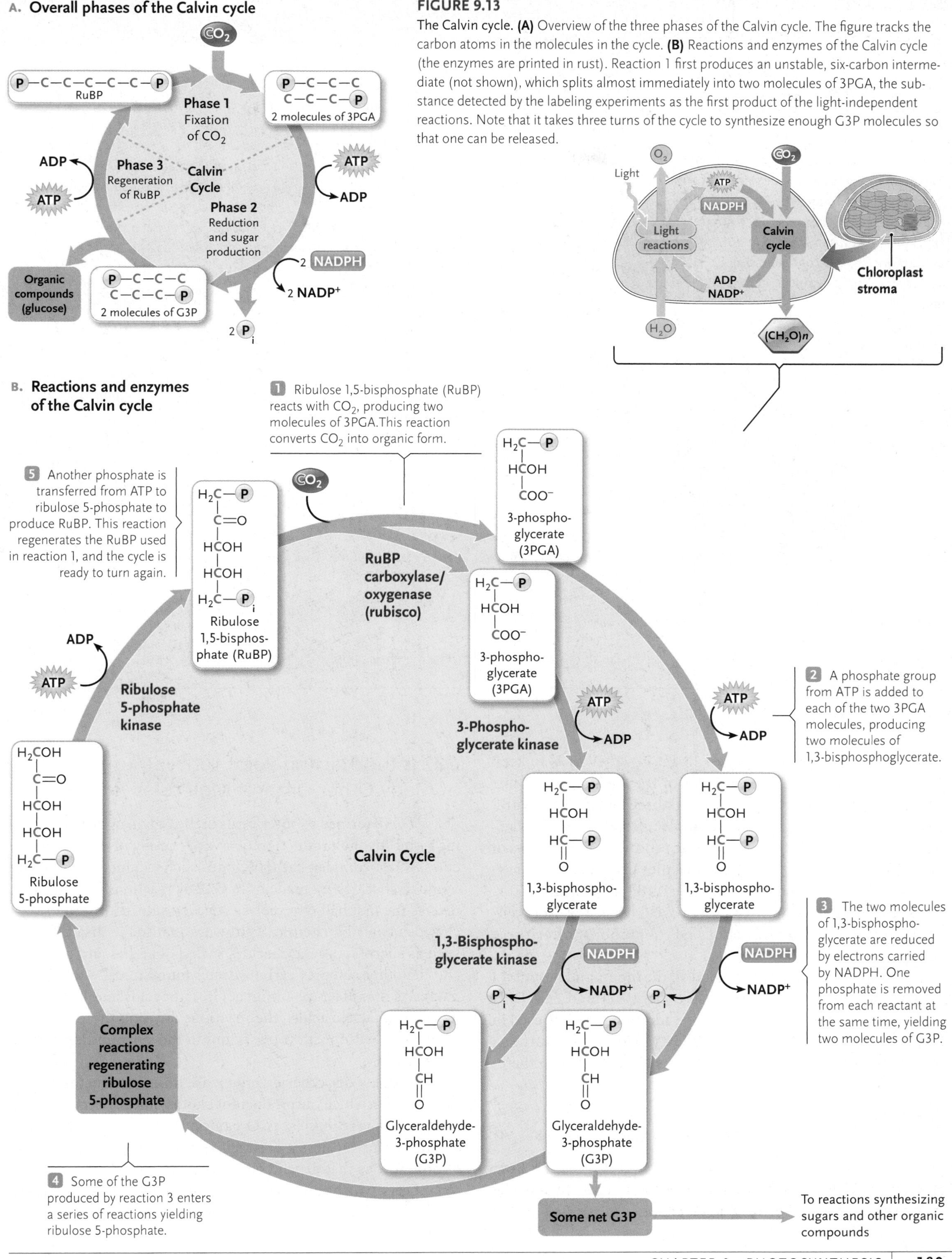
A. Overall phases of the Calvin cycle
CO_2
P—C—C—C—C—C—P
RuBP
Phase 1
Fixation of CO_2
P—C—C—C
C—C—C—P
2 molecules of 3PGA
ADP
ATP
Phase 3
Regeneration of RuBP
Calvin Cycle
ATP
ADP
Phase 2
Reduction and sugar production
2 NADPH
2 $NADP^+$
Organic compounds (glucose)
P—C—C—C
C—C—C—P
2 molecules of G3P
2 P_i
O_2
CO_2
Light
ATP
NADPH
Light reactions
Calvin cycle
ADP
$NADP^+$
Chloroplast stroma
H_2O
$(CH_2O)n$
B. Reactions and enzymes of the Calvin cycle
1 Ribulose 1,5-bisphosphate (RuBP) reacts with CO_2, producing two molecules of 3PGA.This reaction converts CO_2 into organic form.
5 Another phosphate is transferred from ATP to ribulose 5-phosphate to produce RuBP. This reaction regenerates the RuBP used in reaction 1, and the cycle is ready to turn again.
CO_2
Ribulose 1,5-bisphos-phate (RuBP)
3-phospho-glycerate (3PGA)
RuBP carboxylase/ oxygenase (rubisco)
3-phospho-glycerate (3PGA)
ADP
ATP
Ribulose 5-phosphate kinase
ATP
ADP
3-Phospho-glycerate kinase
ATP
ADP
2 A phosphate group from ATP is added to each of the two 3PGA molecules, producing two molecules of 1,3-bisphosphoglycerate.
Ribulose 5-phosphate
Calvin Cycle
1,3-bisphospho-glycerate
1,3-bisphospho-glycerate
3 The two molecules of 1,3-bisphospho-glycerate are reduced by electrons carried by NADPH. One phosphate is removed from each reactant at the same time, yielding two molecules of G3P.
1,3-Bisphospho-glycerate kinase
NADPH
$NADP^+$
NADPH
$NADP^+$
P_i
P_i
Complex reactions regenerating ribulose 5-phosphate
Glyceraldehyde-3-phosphate (G3P)
Glyceraldehyde-3-phosphate (G3P)
4 Some of the G3P produced by reaction 3 enters a series of reactions yielding ribulose 5-phosphate.
Some net G3P
To reactions synthesizing sugars and other organic compounds

FIGURE 9.13

The Calvin cycle. **(A)** Overview of the three phases of the Calvin cycle. The figure tracks the carbon atoms in the molecules in the cycle. **(B)** Reactions and enzymes of the Calvin cycle (the enzymes are printed in rust). Reaction 1 first produces an unstable, six-carbon intermediate (not shown), which splits almost immediately into two molecules of 3PGA, the substance detected by the labeling experiments as the first product of the light-independent reactions. Note that it takes three turns of the cycle to synthesize enough G3P molecules so that one can be released.

INSIGHTS FROM THE Molecular Revolution

Small but Pushy: What is the function of the small subunit of rubisco?

Rubisco is a 16-subunit protein with eight copies of a large polypeptide and eight copies of a small polypeptide (see Figure). We noted that all the active sites of rubisco appear to be on the large polypeptide subunit of the enzyme. Even so, 99% of the enzyme's catalytic activity is lost if the small subunit is removed.

Research Question

What is the function of the small subunit of rubisco?

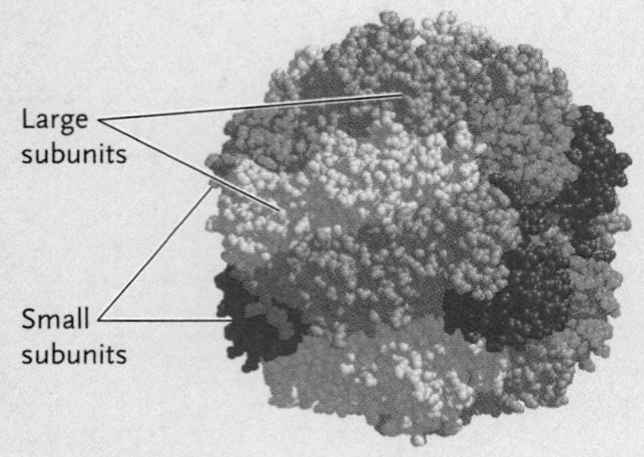
Large subunits
Small subunits

A model for the structure of rubisco in chloroplasts of higher plants. The large subunits are shown in gray and white, and the small subunits in blue and orange.

Experiment

Betsy A. Read and F. Robert Tabita of The Ohio State University, at Columbus, Ohio, set out to answer this question using molecular techniques. They hypothesized that the structure of a specific region of the small subunit was critical to its function. To test this hypothesis, the investigators used DNA cloning techniques (see Chapter 18) to produce five versions of the small subunit, each with a different amino acid substituted for the normal one at five different positions in the protein. Then, they examined the effects of the substitutions on enzyme activity:

Results

1. One of the modified small subunits, which had glutamine substituted for arginine at position 88 in the amino acid sequence, was unable to assemble with the large subunit to form a complete enzyme complex. Thus, the arginine at position 88 is essential for normal enzyme assembly.
2. The four remaining modified versions of the small subunit assembled normally with large subunits, and all recognized and bound their substrates for the initial reaction of the Calvin cycle (see Figure 9.13) as ably as the normal enzyme. Therefore, these four modifications of the small subunit had no effect on the specificity of the enzyme.
3. The investigators next checked the rates at which the enzymes with the four modified small subunits catalyzed CO_2 fixation. Three of the four modified enzymes ran the first reaction of the Calvin cycle at only 35% of the rate of the normal enzyme. The most active worked only about half as fast as the normal enzyme.

Conclusion

The small subunit has a very significant effect on rubisco's rate of catalysis. The effect is critically important when considered in the context of the comparatively slow reaction rate of the normal enzyme. The enzyme's multiple form—eight copies of each subunit, massed together, all doing the same thing—and the very large amount of the enzyme packed into leaves compensate for the slow rate. Evidently, the small subunit evolved as yet another way to compensate for the enzyme's slow action, by pushing the large subunit to do its job faster.

Source: B. A. Read and F. R. Tabita. 1992. Amino acid substitutions in the small subunit of ribulose-1,5-bisphosphate carboxylase/oxygenase that influence catalytic activity of the holoenzyme. *Biochemistry* 31:519–525.

Let's work through Figure 9.13 again—the key is to keep track of the carbons. In three turns of the Calvin cycle, three CO_2 (3 carbons) are incorporated into three molecules of RuBP (15 carbons), which produce six molecules of 3PGA (18 carbons). Three turns of the Calvin cycle produce six molecules of G3P (18 carbons). Of these six molecules of G3P, five (15 carbons) are used to regenerate the three RuBP molecules (15 carbons) required for three turns of the cycle. Thus, the cycle generates one surplus molecule of G3P (3 carbons) after three turns.

This approach allows us to total all the inputs and outputs of the Calvin cycle. For each turn of the cycle, 2 ATP and 2 NADPH are used in reactions 2 and 3, and one additional ATP is used in reaction 5, for a total of 3 ATP and 2 NADPH for each turn. As net reactants and products, one complete turn of the cycle therefore includes the following reactions:

$$CO_2 + 2\,NADPH + 3\,ATP \rightarrow (CH_2O) + 2\,NADP^+ + 3\,ADP + 3\,P_i$$

For three turns of the cycle, 9 ATP and 6 NADPH are used. Both ATP and NADPH are regenerated from ADP and $NADP^+$, respectively, by the light-dependent reactions.

G3P Is the Starting Point for Synthesis of Many Other Organic Molecules

The net G3P formed by three turns of the Calvin cycle is the starting point for the production of a wide variety of organic molecules. More complex carbohydrates such as glucose and other monosaccharides are made from G3P by reactions that, in effect, reverse the first half of glycolysis. Once produced, the monosaccharides enter biochemical pathways that make disaccharides such as sucrose, polysaccharides such as starches and cellulose, and the other complex carbohydrates found in cell walls. Other pathways manufacture amino acids, fatty acids and lipids, proteins, and nucleic acids. The reactions forming these products occur both within chloroplasts and in the surrounding cytosol and nucleus.

Sucrose, a disaccharide of glucose linked to fructose, is the main form in which the products of photosynthesis circulate from cell to cell in vascular plants. Organic nutrients are stored in most vascular plants as sucrose or starch, or as a combination of the two in proportions that depend on the plant species. Sugar cane and sugar beets, which contain stored sucrose in high concentrations, are the main sources of the sucrose we use as table sugar.

Rubisco Is the Key Enzyme of the World's Food Economy

Rubisco, the enzyme that catalyzes the first reaction of the Calvin cycle, is unique to photosynthetic organisms. By catalyzing CO_2 fixation, it provides the source of organic molecules for most of the world's organisms—the enzyme converts about 100 billion tons of CO_2 into carbohydrates annually. There are so many rubisco molecules in chloroplasts that the enzyme may make up 50% or more of the total protein of plant leaves. As such, it is also the world's most abundant protein, estimated to total some 40 million tons worldwide, equivalent to about 6 kg for every human.

Rubisco has essentially the same overall structure in almost all photosynthetic organisms: eight copies each of a large and a small polypeptide, joined together in a 16-subunit structure. The large subunit contains all of the known active sites where substrates, including CO_2 and RuBP, can bind. (Active sites of enzymes are described in Section 4.4.) Although the small subunit has no active sites, it is still essential for efficient operation of the enzyme. *Insights from the Molecular Revolution* describes experiments to determine the molecular functions of the small subunit.

Rubisco is also the key regulatory site of the Calvin cycle. The enzyme is stimulated by both NADPH and ATP; as long as these substances are available from the light-dependent reactions, the enzyme is active and the light-independent reactions proceed. During the daytime, when sunlight powers the light-dependent reactions, the abundant NADPH and ATP supplies keep the Calvin cycle running. In darkness, when NADPH and ATP become unavailable, the enzyme is inhibited and the Calvin cycle slows or stops. Similar controls based on the availability of ATP and NADPH also regulate the enzymes that catalyze other reactions of the Calvin cycle, including reactions 2 and 3 in Figure 9.13B.

STUDY BREAK 9.3

1. **What is the reaction that rubisco catalyzes? Why is rubisco the key enzyme for producing the world's food, and how is it the key regulatory site of the Calvin cycle?**
2. **How many molecules of carbon dioxide must enter the Calvin cycle for a plant to produce a sugar containing 12 carbon atoms? How many ATP and NADPH molecules would be required to make that molecule?**

9.4 Photorespiration and Alternative Processes of Carbon Fixation

In this section we consider *photorespiration* in plants, a mechanism with some features similar to aerobic respiration that uses O_2 as the first step in a pathway to generate CO_2. We also describe two alternative processes of carbon fixation, the C_4 pathway and the CAM pathway.

Photorespiration Produces Carbon Dioxide That Is Used by the Calvin Cycle

The major photosynthetic organ of a plant is the leaf. Because of its large surface area, you might think there would be high water loss by evaporation through the leaf. However, the surface of the leaf is covered by a waxy *cuticle* that prevents water loss. But, the cuticle also prevents the rapid diffusion of gases, such as CO_2, into the leaf. The high rates of gas exchange—CO_2 in and O_2 out—between the air and the cells within the leaf occur through the stomata of the leaves and stem. The plant regulates the size of stomata from fully closed to fully open so as to balance the demands for gas exchange with the need to minimize water loss. As you may suspect, plants that are adapted to hot, dry climates are faced with a constant dilemma: their stomata must be open to let in CO_2 for the Calvin cycle, but to conserve water the stomata should remain closed. A similar situation occurs for nonadapted plants when environmental conditions become hot and dry. With the stomata closed, photosynthesis rapidly consumes the CO_2 in the leaf and produces O_2, which accumulates in the chloroplasts.

Rubisco, RuBP carboxylase/oxygenase, is so named because the enzyme has an active site to which either CO_2 or O_2 can bind. When CO_2 binds, rubisco acts as a carboxylase; when O_2 binds, the enzyme acts as an oxygenase **(Figure 9.14).** Recall from Chapter 4 that a molecule which can compete with the normal substrate for the active site of an enzyme is a competitive inhibitor for that enzyme. In this case, O_2 acts as a competitive inhibitor

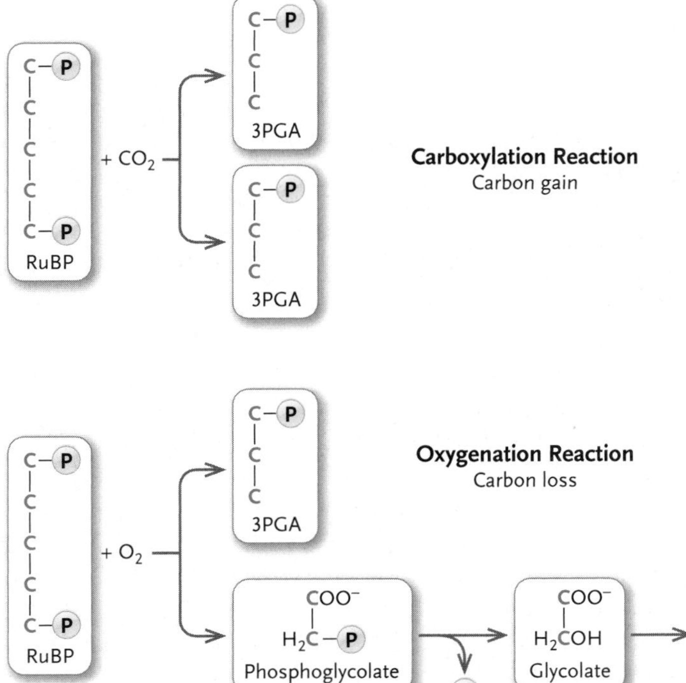

RuBP
+ CO2
3PGA
3PGA
Carboxylation Reaction
Carbon gain
RuBP
+ O2
3PGA
Phosphoglycolate
Glycolate
Pi
To reactions releasing CO2
Oxygenation Reaction
Carbon loss

FIGURE 9.14
Photorespiration, an alternative pathway for rubisco in which, in the presence of oxygen, the oxygenase activity of the enzyme produces glycolate. Glycolate, a toxic product, is eliminated by reactions that convert carbon back to inorganic form as CO_2.

for rubisco because it can compete with CO_2 for binding to the enzyme. When O_2 binds to rubisco, the oxygenase activity of the enzyme uses the O_2 in a reaction to convert RuBP to one molecule of 3PGA and one molecule of a two-carbon substance, phosphoglycolate (see Figure 9.14). No carbon is fixed during this reaction, in contrast to the carboxylase reaction, in which rubisco uses CO_2. Further, there is no carbon gain as is seen for the carboxylase reaction (see Figure 9.13); here there are five carbons in and five carbons out. But photoautotrophs cannot use phosphoglycolate. In the process of breaking it down to salvage the carbon, the toxic compound glycolate is produced. The glycolate is eliminated by oxidation in reaction pathways that result in the release of CO_2. Thus, whereas the carboxylation reaction of rubisco leads to carbon gain, the oxygenation reaction actually results in the cell losing carbon. But, in order to grow, all organisms must gain carbon.

The entire process from the oxygenase reaction of rubisco to the release of CO_2 is called **photorespiration.** The term was coined because the process occurs in the presence of light and, like aerobic respiration, it requires oxygen, and produces carbon dioxide and water. However, unlike aerobic respiration, photorespiration does not generate ATP. Photorespiration also reduces the efficiency of photosynthesis because it removes some of the intermediates of the Calvin cycle. For a plant with high photorespiration rates at elevated temperatures, as much as 50% of the carbon fixed by the Calvin cycle may be lost because only one three-carbon 3PGA molecule is produced instead of two. Unfortunately, many economically important crop plants become seriously impaired by high photorespiration rates when temperatures rise; among them are the C_3 plants rice, barley, wheat, soybeans, tomatoes, and potatoes.

Alternative Processes of Carbon Fixation Minimize Photorespiration

Many plant species that live in hot, dry environments have evolved alternative processes of carbon fixation that minimize photorespiration and, therefore, its negative effect on photosynthesis. C_4 plants use the *C_4 pathway* to first fix CO_2 into a four-carbon molecule, *oxaloacetate,* while CAM plants use the *CAM pathway* at night to fix carbon in the form of oxaloacetate. These special pathways precede the C_3 pathway (Calvin cycle) rather than replacing it.

THE C_4 PATHWAY AND C_4 PLANTS In the **C_4 pathway,** CO_2 initially is fixed in the form of oxaloacetate **(Figure 9.15).** The four-carbon oxaloacetate is formed by combining CO_2 with a three-

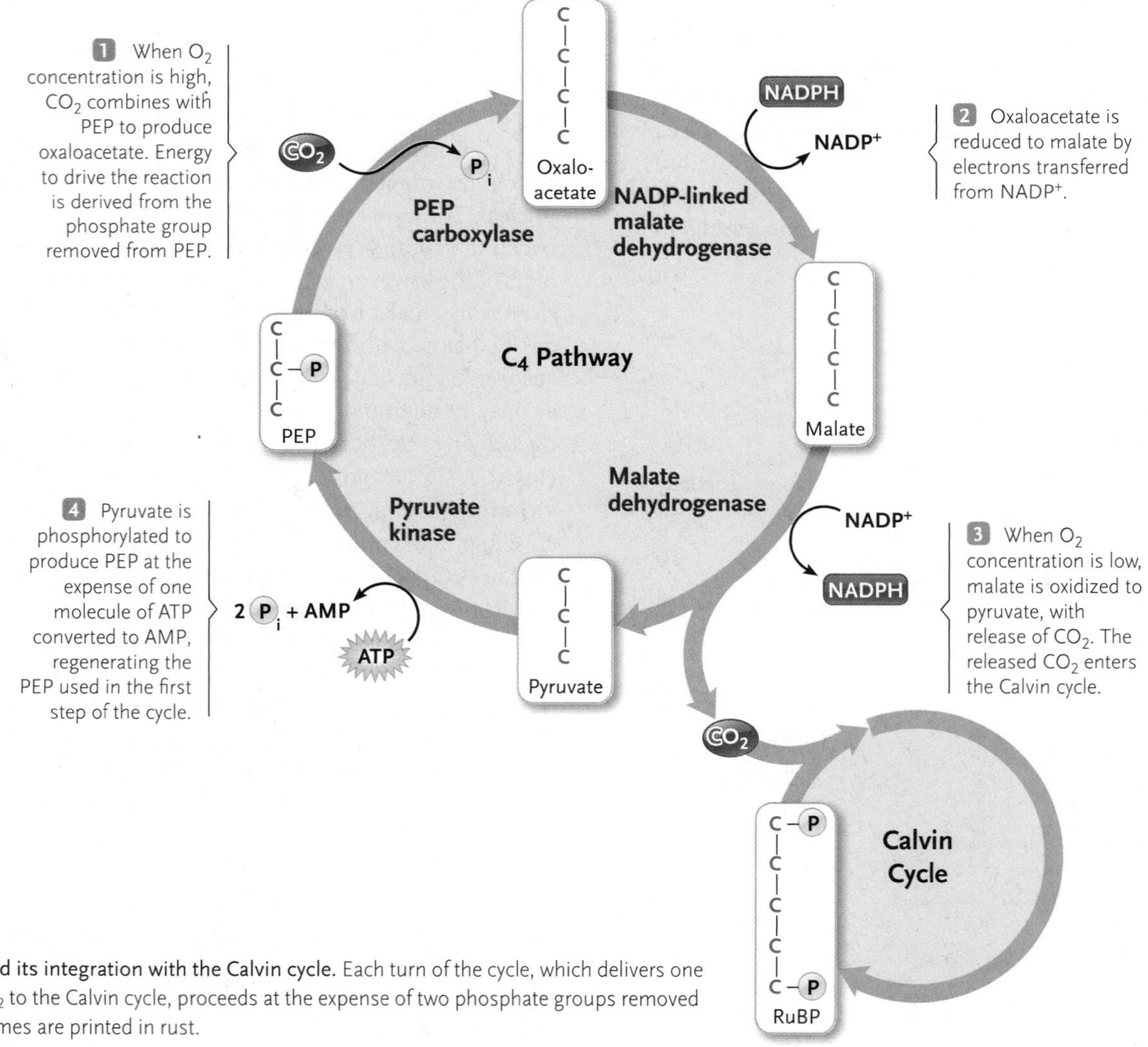
1 When O2 concentration is high, CO2 combines with PEP to produce oxaloacetate. Energy to drive the reaction is derived from the phosphate group removed from PEP.
CO2
Pi
Oxalo-acetate
PEP carboxylase
NADPH
NADP+
NADP-linked malate dehydrogenase
2 Oxaloacetate is reduced to malate by electrons transferred from NADP+.
PEP
C4 Pathway
Malate
Malate dehydrogenase
NADP+
NADPH
3 When O2 concentration is low, malate is oxidized to pyruvate, with release of CO2. The released CO2 enters the Calvin cycle.
Pyruvate kinase
4 Pyruvate is phosphorylated to produce PEP at the expense of one molecule of ATP converted to AMP, regenerating the PEP used in the first step of the cycle.
2 Pi + AMP
ATP
Pyruvate
CO2
Calvin Cycle
RuBP

FIGURE 9.15
The C4 cycle and its integration with the Calvin cycle. Each turn of the cycle, which delivers one molecule of CO_2 to the Calvin cycle, proceeds at the expense of two phosphate groups removed from ATP. Enzymes are printed in rust.

carbon molecule, *phosphoenolpyruvate (PEP)*. The C_4 pathway gets its name because its first product is a four-carbon molecule rather than a three-carbon molecule, as in the Calvin (C_3) cycle. The carbon fixation reaction producing oxaloacetate is catalyzed by *PEP carboxylase*. This enzyme has a much greater affinity for CO_2 than rubisco does and, unlike rubisco, it has no oxygenase activity.

As the C_4 pathway continues, the oxaloacetate is next reduced to *malate* by electrons transferred from NADPH. Malate is then oxidized to pyruvate, releasing CO_2 which is used in the rubisco-catalyzed first step of the Calvin cycle. Pyruvate is converted back into PEP in a reaction that consumes ATP. The oxygenation reaction of rubisco is inhibited by the C_4 pathway because the conversion of malate to pyruvate generates CO_2, resulting in much higher concentrations of that molecule at the rubisco enzyme. As a result, photorespiration is extremely low or negligible in C_4 plants.

At least 19 families of flowering plants are C_4 plants. They include crabgrass and several crops important agriculturally, including corn and sugarcane. C_4 plants have a characteristic leaf anatomy **(Figure 9.16A)**: photosynthetic *mesophyll cells* are tightly associated with specialized, chloroplast-rich *bundle sheath cells*, which encircle the veins of the leaf. In a C_3 plant leaf **(Figure 9.16B)**, the bundle sheath cells are not photosynthetic.

In C_3 plants, the Calvin cycle takes place in the mesophyll cells, the only cells that are photosynthetic. The CO_2 used for the reactions diffuses in through the stomata. In C_4 plants, however, the C_4 pathway takes place in the mesophyll cells, while the Calvin cycle occurs in the bundle sheath cells **(Figure 9.17A)**. That is, carbon fixation and the Calvin cycle take place in different cells. More specifically, reviewing the C_4 pathway in Figure 9.15, carbon fixation to oxaloacetate and conversion of that four-carbon molecule to malate occurs in the mesophyll cells. The malate then diffuses to the bundle sheath cells where it is broken down into pyruvate and CO_2, which enters the Calvin cycle in those cells. That is, the C_4 pathway generates the CO_2 used by the Calvin cycle inside the leaf, in the bundle sheath. The pyruvate diffuses back to the mesophyll cells where the remainder of the C_4 pathway occurs.

If the C_4 pathway is so effective at reducing photorespiration, why is the pathway not used by all plants? Note in Figure 9.15 that the C_4 pathway has an additional energy requirement: for each turn of the C_4 pathway cycle, one ATP molecule is hydrolyzed to regenerate PEP from pyruvate. This adds an energy requirement of six ATP molecules for each G3P produced by the Calvin cycle. However, as mentioned earlier, photorespiration potentially can decrease carbon fixation efficiency by as much as 50% in hot environments, so the additional ATP requirement is worthwhile. Moreover, hot environments typically receive a lot of sunshine; thus the additional ATP requirement can be met easily by increasing the output of the light-dependent reactions.

C_4 plants also perform better where it is dry. Because PEP carboxylase has a very high affinity only for CO_2, C_4 plants are more efficient at fixing CO_2 than are C_3 plants. Therefore, they do not have to keep their stomata open for as long as a C_3 plant does under the same conditions. Because this reduces water loss, C_4 plants are much better suited to arid conditions.

A. C_4 plant leaf cross section

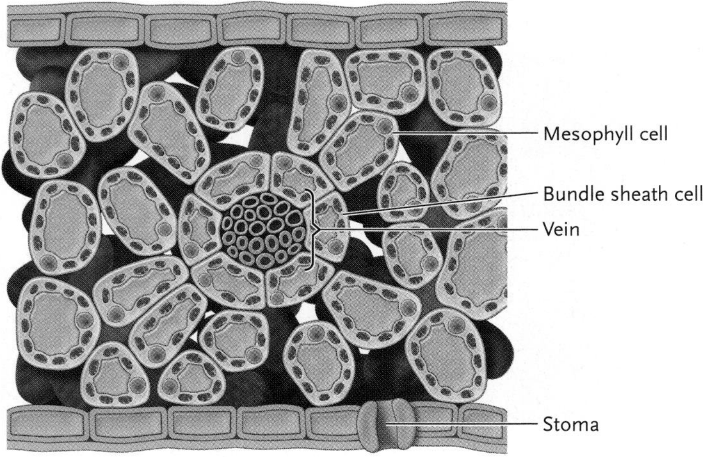
Mesophyll cell
Bundle sheath cell
Vein
Stoma

B. C_3 plant leaf cross section

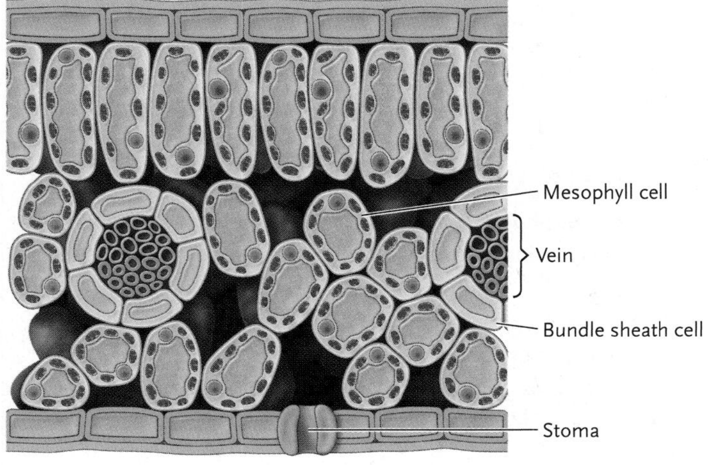
Mesophyll cell
Vein
Bundle sheath cell
Stoma

FIGURE 9.16
Leaf anatomy in **(A)** C_4 plants and **(B)** C_3 plants.

THE CAM PATHWAY AND CAM PLANTS Exactly as with the C_4 pathway in C_4 plants, in the **CAM pathway** CO_2 is initially fixed to oxaloacetate in a reaction catalyzed by PEP carboxylase **(Figure 9.17B)**. The CO_2 produced by the oxidation of malate is used in the rubisco-catalyzed first step of the Calvin cycle. But, whereas carbon fixation and the Calvin cycle run in different cell types in C_4 plants, in CAM plants carbon fixation and the Calvin cycle both occur in mesophyll cells but run at different times, initial carbon fixation at night and the Calvin cycle during the day.

The acronym CAM stands for **crassulacean acid metabolism;** the name comes from the Crassulaceae family (stonecrops) where the pathway was first described. CAM plants live in very dry environments and have several evolutionary adaptations that enable them to survive the arid conditions. Besides succulents (water-storing plants) such as the stonecrops, CAM plants include some members of at least 25 plant families, including the cactus family (see photo in Figure 9.16C), the lily family, and the orchid family.

A. C_4 pathway in C_4 plants

CO_2
Mesophyll cell
PEP
Oxaloacetate
C_4 pathway
Pyruvate
Malate
Pyruvate
Malate
CO_2
RuBP
Calvin cycle
3PGA
Sugar
Bundle sheath cell

CO_2 is incorporated into malate in mesophyll cells.

Malate enters bundle sheath cells, where CO_2 is released for Calvin cycle.

B. CAM pathway in CAM plants

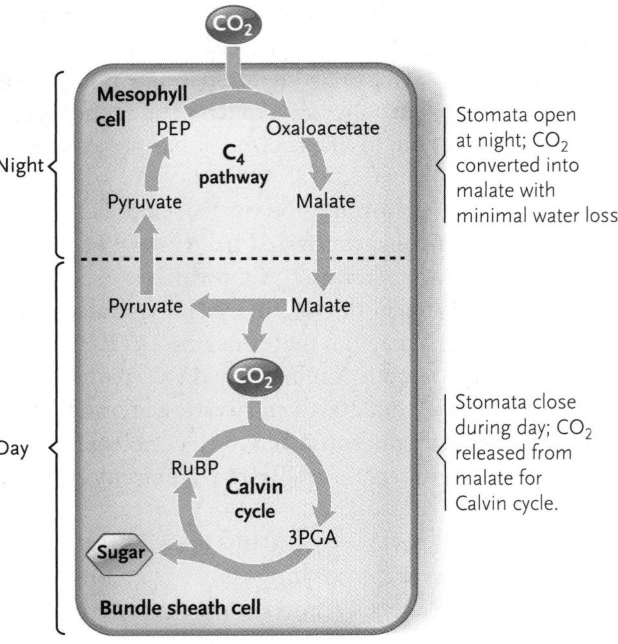

CO_2
Night
Mesophyll cell
PEP
Oxaloacetate
C_4 pathway
Pyruvate
Malate
Pyruvate
Malate
CO_2
Day
RuBP
Calvin cycle
3PGA
Sugar
Bundle sheath cell
Stomata open at night; CO_2 converted into malate with minimal water loss.
Stomata close during day; CO_2 released from malate for Calvin cycle.

Zea mays (corn)

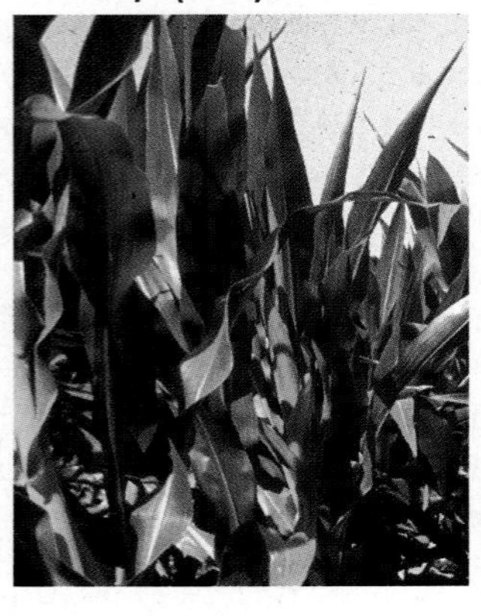

Opuntia basilaris (beavertail cactus)

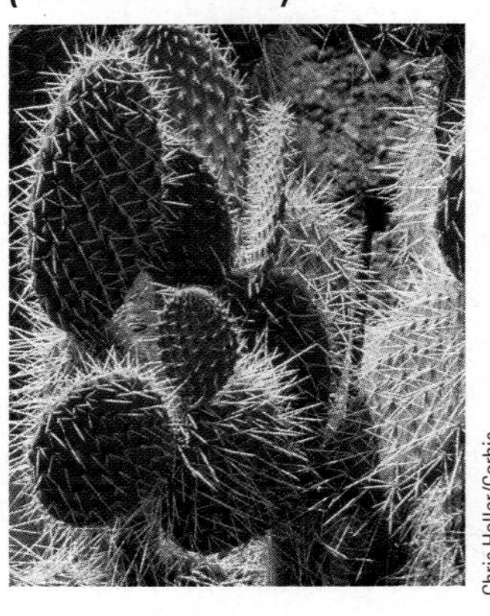

FIGURE 9.17
Two alternative processes of carbon fixation to minimize photorespiration. In each case, carbon fixation produces the four-carbon oxaloacetate, which is processed to generate CO_2 that feeds into the Calvin (C_3) cycle. **(A)** In C_4 plants, carbon fixation and the Calvin cycle occur in different cell types: carbon fixation by the C_4 pathway takes place in mesophyll cells, while the Calvin cycle takes place in bundle sheath cells. **(B)** In CAM plants, carbon fixation and the Calvin cycle occur at different times in mesophyll cells: carbon fixation by the C_4 pathway takes place at night, while the Calvin cycle takes place during the day.

CAM plants live in regions that are hot and dry during the day and cool at night. The stomata on their fleshy leaves or stems open only at night to minimize water loss. When they open, O_2 generated by photosynthesis is released, and CO_2 enters. The CO_2 is fixed into malate in the mesophyll cells by the CAM pathway; malate accumulates throughout the night and is stored in large cell vacuoles. During the day, the stomata close to reduce water loss in the hot conditions; this also cuts off the exchange of gases with the atmosphere. Malate now moves from the vacuole to the chloroplasts, where its oxidation to pyruvate generates CO_2 that is used by the Calvin cycle.

STUDY BREAK 9.4

1. **When does photorespiration occur? What are the reactions of photorespiration, and what are the energetic consequences of the process?**
2. **What is the C_4 pathway, and how does it enable C_4 plants to circumvent photorespiration?**
3. **How are carbon fixation and the Calvin cycle different in C_4 plants and CAM plants?**

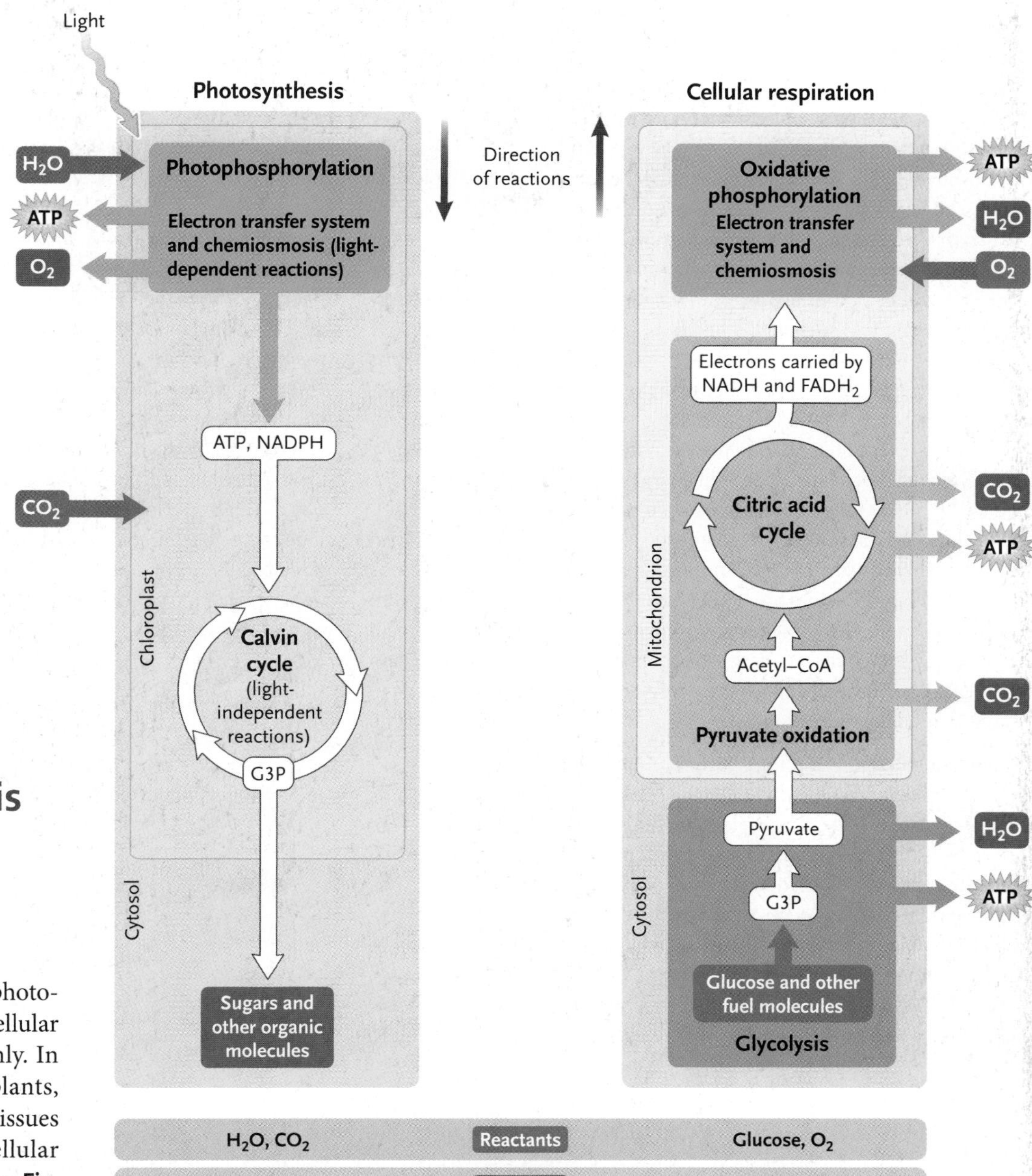
Light
Photosynthesis
Cellular respiration
H_2O
ATP
O_2
Photophosphorylation
Electron transfer system and chemiosmosis (light-dependent reactions)
Direction of reactions
Oxidative phosphorylation
Electron transfer system and chemiosmosis
ATP
H_2O
O_2
Electrons carried by NADH and $FADH_2$
ATP, NADPH
CO_2
Citric acid cycle
CO_2
ATP
Chloroplast
Calvin cycle (light-independent reactions)
G3P
Mitochondrion
Acetyl–CoA
CO_2
Pyruvate oxidation
Pyruvate
H_2O
G3P
ATP
Cytosol
Cytosol
Sugars and other organic molecules
Glucose and other fuel molecules
Glycolysis
H_2O, CO_2
Reactants
Glucose, O_2
O_2, Sugars
Products
H_2O, CO_2

FIGURE 9.18
Schematic diagrams of the process of photosynthesis (left) and cellular respiration (right). Cellular respiration is shown upside down with respect to the direction of reactions to help illustrate the similarities of the process with photosynthesis.

9.5 Photosynthesis and Cellular Respiration Compared

A popular misconception is that photosynthesis occurs in plants, and cellular respiration occurs in animals only. In fact, both processes occur in plants, with photosynthesis confined to tissues containing chloroplasts and cellular respiration taking place in all cells. **Figure 9.18** presents side-by-side schematics of photosynthesis and cellular respiration to highlight their similarities and points of connection. Note that their overall reactions are basically the reverse of each other. That is, the reactants of photosynthesis—CO_2 and H_2O—are the products of cellular respiration, and the reactants of cellular respiration—glucose and O_2—are the products of photosynthesis. Both processes have key phosphorylation reactions involving an electron transfer system—photophosphorylation in photosynthesis and oxidative phosphorylation in cellular respiration—followed by the chemiosmotic synthesis of ATP. Further, G3P is found in the pathways of both processes. In photosynthesis, it is a product of the Calvin cycle and is used for the synthesis of sugars and other organic fuel molecules. In cellular respiration, it is an intermediate generated in glycolysis in the conversion of glucose to pyruvate. Thus, G3P is used by anabolic pathways when it is generated by photosynthesis, and it is a product of a catabolic pathway in cellular respiration.

In this chapter, you have seen how photosynthesis supplies the organic molecules used as fuels by almost all the organisms of the world. It is a story of electron flow: electrons, pushed to high energy levels by the absorption of light energy, are added to CO_2, which is fixed into carbohydrates and other fuel molecules. The high-energy electrons are then removed from the fuel molecules by the oxidative reactions of cellular respiration, which use the released energy to power the activities of life. Among the most significant of these activities are cell growth and division, the subjects of the next chapter.

STUDY BREAK 9.5 <

How are the reactants and products of photosynthesis and cellular respiration related?

UNANSWERED QUESTIONS

Why is photosynthesis so inefficient, and what (if anything) can we do about it?

Photosynthesis is considered by many to be the most important biological process on Earth. It is the process that captures essentially all energy for our ecosystem. Photosynthesis involves the highest energy processes of life; it is the process where (by far) most of the energy in our ecosystem is input. All other biological processes are exergonic (they lose the energy captured by photosynthesis) and thus all other processes involve less energy than photosynthesis. It is also the process where, by far, the most energy in our ecosystem is *lost*. Because the intermediates of photosynthesis are so reactive (and therefore dangerous), photosynthetic organisms must prevent their buildup by dissipating "excess" light energy. As a result, a large part of absorbed light energy is "purposely" dumped (dissipated) to prevent the buildup of reactive oxygen species that can damage the plant. At full sunlight, regulatory dissipation can dump more than 75% of absorbed light energy! Consequently, typical agricultural crops store only about 0.1% to 0.5% of their absorbed solar energy in the form of biomass. Interestingly, some plants and many green algae can store up to 10 times as much of this energy. Research is being done to understand why they are so efficient, and whether crop or biofuels plants can be engineered to produce more biomass.

How is the efficiency of photosynthesis regulated?

My laboratory at Washington State University is interested in how plants set their photosynthetic "strategies" and thus how they control light capture and electron and proton transfer reactions to balance their competing needs for efficient photosynthesis while avoiding toxic side reactions. As you have learned, energy conversion by the chloroplast involves the capture of light energy and the channeling of that energy through an electron transfer system with the eventual synthesis of NADPH and ATP. At high concentrations, many of the intermediates produced in this energy conversion can potentially destroy the photosynthetic apparatus, a phenomenon called photoinhibition. To prevent such damage, the efficiency of some of the photosystem components is decreased in a process called photoprotection. High temperatures lower the efficiency of photosynthesis, however. Evidence from a range of studies indicates that the balance between protection against photoinhibition and photosynthetic efficiency is important in enabling plants to acclimate to environmental changes, but it also limits their productivity. We have found that ATP synthase acts as a major regulator of photosynthesis, sensing the metabolic status of the chloroplast, and in response, regulating the proton-motive force by restricting proton flow out of the lumen. In turn, the proton-motive force controls the photoprotective dissipation of light energy and electron transfer. Among the major unanswered questions are: what controls the ATP synthase, what specifically does it sense, and how does this mechanism differ in high- and low-productivity plants? Answering these questions will illuminate how photosynthesis determines plant growth and survival. In addition, the technology developed as part of the research may lead to applications in plant breeding, particularly for improving energy storage by biofuels and food crops.

Think Critically

Find some specific examples of plants with higher and lower efficiency photosynthesis. Why do you think some plants have higher photosynthetic rates than others? What are some potential selective advantages of lower rates of photosynthesis?

David Kramer

David Kramer is a professor and fellow at the Institute of Biological Chemistry (IBC) at Washington State University. Kramer's research group aims to understand how photosynthesis works, what specific reactions limit its productivity, and how these limitations can be overcome to redirect more output towards readily useable biofuels. Learn more about his research at http://ibc.wsu.edu/research/kramer/index.htm.

REVIEW KEY CONCEPTS

Go to **CENGAGENOW** at www.cengage.com/login to access quizzing, animations, exercises, articles, and personalized homework help.

9.1 Photosynthesis: An Overview

- In photosynthesis, plants, algae, and photosynthetic prokaryotes use the energy of sunlight to drive synthesis of organic molecules from simple inorganic raw materials. The organic molecules are used by the photosynthesizers themselves as fuels; they also form the primary energy source for heterotrophs.
- The two overall stages of photosynthesis are the light-dependent and light-independent reactions. In eukaryotes, both stages take place inside chloroplasts (Figures 9.2 and 9.3).
- Photosynthesizers use light energy to push electrons to elevated energy levels. In eukaryotes and many prokaryotes, water is split as the source of the electrons for this process, and oxygen is released to the environment as a by-product.
- The high-energy electrons provide an indirect energy source for ATP synthesis and also for CO_2 fixation, in which CO_2 is fixed into organic substances by the addition of both electrons and protons.

Animation: Photosynthesis overview

Animation: Sites of photosynthesis

9.2 The Light-Dependent Reactions of Photosynthesis

- In the light-dependent reactions of photosynthesis, light energy is converted to chemical energy when electrons, excited by absorption of light in a pigment molecule, are passed from the pigment to a stable orbital in a primary acceptor molecule (Figure 9.5).
- Chlorophylls and carotenoids, the photon-absorbing pigments in eukaryotes and cyanobacteria, together absorb light energy at a range of wavelengths, enabling a wide spectrum of light to be used in photosynthesis (Figures 9.6 and 9.7).

- In organisms that split water as their electron source, the pigments are organized with proteins into two photosystems. Specialized forms of chlorophyll *a* pass excited electrons to primary acceptor molecules in the photosystems (Figure 9.8).
- Electrons obtained from splitting water are used for the synthesis of NADPH and ATP. In the noncyclic electron flow pathway, electrons first flow through photosystem II, becoming excited there to a higher energy level, and then pass through an electron transfer system to photosystem I releasing energy that is used to create an H^+ gradient across the membrane. The gradient is used by ATP synthase to drive synthesis of ATP. The net products of the light-dependent reactions are ATP, NADPH, and oxygen (Figures 9.9 and 9.10).
- Electrons can also flow cyclically around photosystem I and the electron transfer system, building the H^+ concentration and allowing extra ATP to be produced, but no NADPH (Figure 9.11).

Interaction: Wavelengths of light

Animation: Noncyclic pathway of electron flow

9.3 The Light-Independent Reactions of Photosynthesis

- In the light-independent reactions of photosynthesis, CO_2 is reduced and converted into organic substances by the addition of electrons and hydrogen carried by the NADPH produced in the light-dependent reactions. ATP, also derived from the light-dependent reactions, provides additional energy. The key enzyme of the light-independent reactions is rubisco (RuBP carboxylase/oxygenase), which catalyzes the reaction that combines CO_2 into organic compounds (Figure 9.13).
- In the process, NADPH is oxidized to $NADP^+$, and ATP is hydrolyzed to ADP and phosphate. These products of the light-independent reactions cycle back as inputs to the light-dependent reactions.
- The Calvin cycle produces surplus molecules of G3P, which are the starting point for synthesis of glucose, sucrose, starches, and other organic molecules. The light-independent reactions take place in the chloroplast stroma in eukaryotes and in the cytoplasm of photosynthetic prokaryotes.

Animation: Calvin cycle

9.4 Photorespiration and Alternative Processes of Carbon Fixation

- When oxygen concentrations are high relative to CO_2 concentrations, rubisco acts as an oxygenase, catalyzing the combination of RuBP with O_2 rather than CO_2 and forming toxic products that cannot be used in photosynthesis. The toxic products are eliminated by reactions that release carbon as CO_2, greatly reducing the efficiency of photosynthesis. The entire process is called photorespiration because it uses oxygen and releases CO_2 (Figure 9.14).
- Some plants use alternative processes of carbon fixation that minimize photorespiration. C_4 plants use the C_4 pathway to first fix CO_2 into the four-carbon oxaloacetate in mesophyll cells (the site of the Calvin cycle in C_3 plants), and then produce CO_2 for the Calvin cycle in bundle sheath cells. CAM plants use the CAM pathway also to first fix CO_2 into oxaloacetate and then to generate CO_2 for the Calvin cycle. Here both the carbon fixation and the Calvin cycle occur in mesophyll cells, but they are separated by time; initial carbon fixation occurs at night and the Calvin cycle occurs during the day (Figures 9.15 and 9.16).

Animation: C_3-C_4 comparison

9.5 Photosynthesis and Cellular Respiration Compared

- Photosynthesis occurs in the cells of plants that contain chloroplasts, whereas cellular respiration occurs in all cells. The overall reactions of the two processes are essentially the reverse of each other, with the reactants of one being the products of the other.

UNDERSTAND AND APPLY

Test Your Knowledge

1. An organism exists for long periods by using only CO_2 and H_2O. It could be classified as a(n):
   a. herbivore.
   b. carnivore.
   c. decomposer.
   d. autotroph.
   e. heterotroph.
2. During the light-dependent reactions:
   a. CO_2 is fixed.
   b. NADPH and ATP are synthesized using electrons derived from splitting water.
   c. glucose is synthesized.
   d. water is split and the electrons generated are used for glucose synthesis.
   e. photosystem I is unlinked from photosystem II.
3. Which of the following is a correct step in the light-dependent reactions of the Z system?
   a. Light is absorbed at P700, and electrons flow through a pathway to $NADP^+$, the final acceptor of the noncyclic pathway.
   b. Electrons flow from photosystem II to water.
   c. $NADP^+$ is oxidized to NADPH as it accepts electrons.
   d. Water is degraded to activate P680.
   e. Electrons pass through a thylakoid membrane to create energy to pump H^+ through the cytochrome complex.
4. The light-dependent reactions of photosynthesis resemble aerobic respiration as both:
   a. synthesize NADPH.
   b. synthesize NADH.
   c. require electron transfer systems to synthesize ATP.
   d. require oxygen as the final electron acceptor.
   e. have the same initial energy source.
5. The molecules that link the light-dependent and light-independent reactions are:
   a. ADP and H_2O.
   b. RuBP and CO_2.
   c. cytochromes and water.
   d. G3P and RuBP.
   e. ATP and NADPH.
6. You bite into a spinach leaf. Which one of the following is true?
   a. You are getting 50% of the protein in the leaf in the form of ribulose 1,5-bisphosphate carboxylase.
   b. The major pigment you are ingesting is a carotenoid.
   c. The water in the leaf is a product of the light-independent reactions.

d. Any energy from the leaf you can use directly is in the form of ATP.
e. The spinach most likely was grown in an area with a low CO_2 concentration.

7. Animal metabolism and plant metabolism are related in that:
   a. plants carry out photosynthesis and animals carry out respiration.
   b. G3P is found in the metabolic pathways of both animals and plants.
   c. G3P is used by catabolic pathways when it is generated by photosynthesis, and it is a product of an anabolic pathway in cellular respiration.
   d. light drives electron excitation.
   e. the reactants of photosynthesis drive cellular respiration in animals.
8. Which of the following statements about the C_4 cycle is *incorrect*?
   a. CO_2 initially combines with PEP.
   b. PEP carboxylase catalyzes a reaction to produce oxaloacetate.
   c. Oxaloacetate transfers electrons from NADPH and is reduced to malate.
   d. Less ATP is used to run the C_4 cycle than the C_3 cycle.
   e. The cycle runs when O_2 concentration is high.
9. In one turn of the Calvin cycle, one molecule of CO_2 generates:
   a. 6 ATP.
   b. 6 NADH.
   c. 6 ATP and 6 NADPH.
   d. one (CH_2O) unit of carbohydrate.
   e. one molecule of glucose.
10. The oxygen released by photosynthesis comes from:
   a. CO_2.
   b. H_2O.
   c. light.
   d. NADPH.
   e. electrons.

Discuss the Concepts

1. Suppose a garden in your neighborhood is filled with red, white, and blue petunias. Explain the floral colors in terms of which wavelengths of light are absorbed and reflected by the petals.
2. About 200 years ago, Jan Baptista van Helmont tried to determine the source of raw materials for plant growth. To do so, he planted a young tree weighing 5 pounds in a barrel filled with 200 pounds of soil. He watered the tree regularly. After 5 years, he again weighed the tree and the soil. At that time the tree weighed 169 pounds, 3 ounces, and the soil weighed 199 pounds, 14 ounces. Because the tree's weight had increased so much, and the soil's weight had remained about the same, he concluded that the tree gained weight as a result of the water he had added to the barrel. Analyze his conclusion in terms of the information you have learned from this chapter.
3. Like other accessory pigments, the carotenoids extend the range of wavelengths absorbed in photosynthesis. They also protect plants from a potentially lethal process known as *photooxidation.* This process begins when excitation energy in chlorophylls drives the conversion of oxygen into free radicals, substances that can damage organic compounds and kill cells. When plants that cannot produce carotenoids are grown in light, they bleach white and die. Given this observation, what molecules in the plants are likely to be destroyed by photooxidation?
4. What molecules would you have to provide a plant, theoretically speaking, for it to make glucose in the dark?

Design an Experiment

Space travelers of the future land on a planet in a distant galaxy, where they find populations of a carbon-based life form. The beings on this planet are of a vibrantly purple color. The travelers suspect that the beings secure the energy necessary for survival by a process similar to photosynthesis on Earth. How might they go about testing this conclusion?

Interpret the Data

Photosynthesis directly opposes respiration in determining how plants influence atmospheric CO_2 concentrations. When a leaf is in the light, both photosynthesis and respiration are occurring simultaneously. The data in the table were collected from the leaf of a sagebrush plant that was enclosed in a chamber that measures the rate of CO_2 exchange. The same leaf was used to collect the data in *Interpret the Data* in Chapter 8. Respiration is shown as a negative and photosynthesis as a positive rate of CO_2 exchange. The net photosynthesis rate is the amount of CO_2 (in micromoles per square meter per second) assimilated by the leaf while respiration is occurring; a positive value indicates more photosynthesis is occurring than respiration. The light exposed to the leaf is quantified as the number of photons in the 400 to 700 nm wavelength, the Photosynthetic Photon Flux Density (PPFD); 2,000 $\mu mol/m^2/s$ is equivalent to the amount of light occurring at midday in full sun.

Observation	Photosynthetic Photon Flux Density (PPFD) ($\mu mol/m^2/s$)	Net Photosynthesis ($\mu mol/m^2/s$)
1	2,000	9.1
2	1,500	8.4
3	1,250	8.2
4	1,000	7.4
5	750	6.3
6	500	4.8
7	250	2.2
8	0	−2.0

1. Why is net photosynthesis negative when PPFD is zero? Looking at the respiration data from Chapter 8 *Interpret the Data,* at what temperature do you think these data were collected?
2. Make a graph of the data, with PPFD on the x axis and net photosynthesis rate on the y axis.
3. Is the relationship between PPFD and net photosynthesis best described as linear or curved? Why do you think the data have this type of relationship?
4. How might the predicted increase in temperature due to elevated CO_2 concentrations in the atmosphere impact net photosynthesis? Hint: combine the data from this question with the data from the table in *Interpret the Data* in Chapter 8. Do you think the data presented in both of these questions are sufficient to explain this impact?

Source: Data based on unpublished research by Brent Ewers, University of Wyoming.

Apply Evolutionary Thinking

If global warming raises the temperature of our climate significantly, will C_3 plants or C_4 plants be favored by natural selection? How will global warming change the geographical distributions of plants?

Express Your Opinion

The oxygen in Earth's atmosphere is a sure indicator that photosynthetic organisms flourish here. New technologies will allow astronomers in search of life to measure the oxygen content of the atmosphere of planets too far away for us to visit. Should public funds be used to continue this research? Go to academic.cengage.com/login to investigate both sides of the issue and then vote.

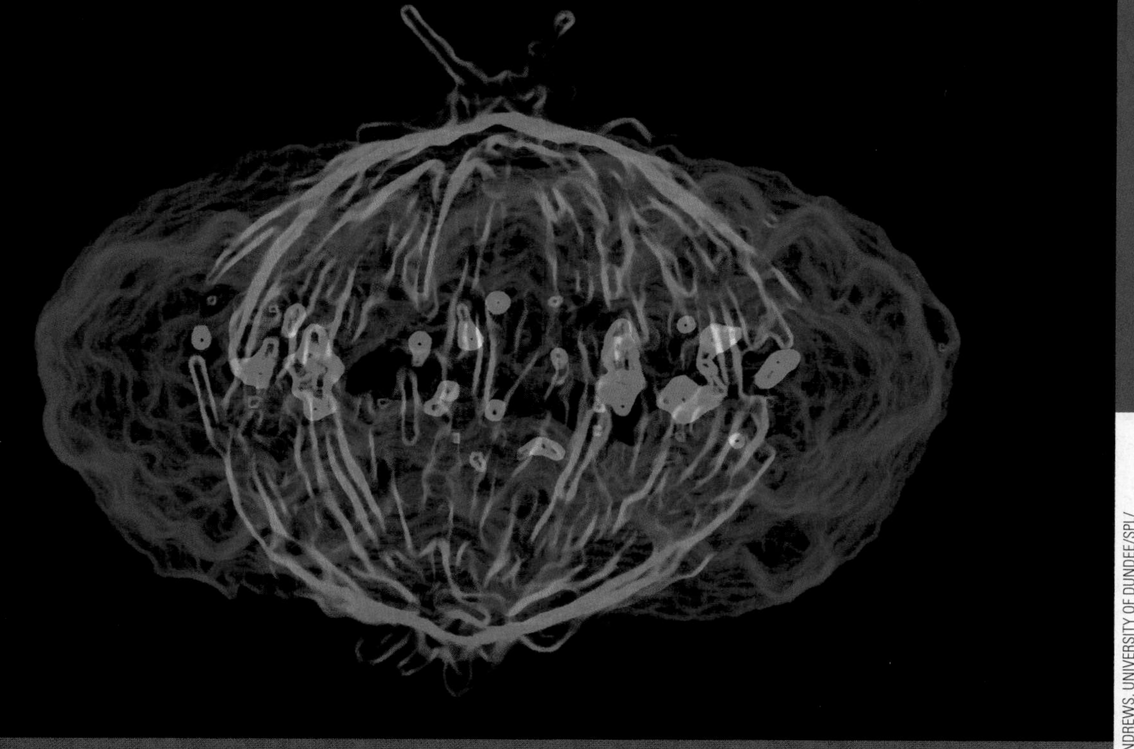
A cell in mitosis (fluorescence micrograph). The spindle (red) is separating copies of the cell's chromosomes (green) prior to cell division.

DR PAUL ANDREWS, UNIVERSITY OF DUNDEE/SPL/ Photo Researchers

10

STUDY OUTLINE

Cell Division and Mitosis

Why It Matters. . . As the rainy season recedes in Northern India, rice paddies and other flooded areas begin to dry out. The resulting shallow, seasonal pools have provided an environment of slow-moving warm water for zebrafish *(Danio rerio)* to spawn **(Figure 10.1).** Over the past few months, many millions of cell divisions have changed the single-celled fertilized eggs into the complex multicellular tissues and organs of these small, boldly striped fish. Most cells in the adult fish have stopped dividing and are dedicated to particular functions.

Moving into the fast-running streams that feed the Ganges River, the young zebrafish often encounter larger predators such as knifefish *(Notopterus notopterus).* Imagine that a knifefish attacks a zebrafish, tearing off part of a fin before its prey escapes. Within a week, the entire zebrafish fin will regenerate—skin, nerves, muscles, bones, and related tissues. The regeneration occurs because cells that were not growing and dividing are suddenly stimulated to grow and divide in a highly regulated way.

As a model organism for vertebrate development, the zebrafish has provided a popular tool for researchers to identify the stages of regeneration at the molecular level. In the first step of regeneration in injured zebrafish, existing skin cells migrate to close the wound and prevent bleeding. Then, cells just under the new skin transform into "regeneration cells" that form a temporary tissue called a blastema. The blastema cells begin to grow and divide. The large numbers of daughter cells produced are capable of maturing into new nerve, blood vessel, muscle and bone cells in response to signal proteins that are produced by the skin. Once the regenerated fin has reached its normal size and shape, the new cells stop growing and dividing.

Since multicellular organisms are made almost entirely of cells and their products, understanding organismal development and structure at its most fundamental level by studying the regulation

WILDLIFE/Peter Arnold Inc.

FIGURE 10.1
Zebrafish *(Danio rerio).*

of cell division is necessary. Which conditions stimulate cells to divide? Which make them stop? This chapter focuses on how cell division occurs, and how that process is regulated. <

10.1 The Cycle of Cell Growth and Division: An Overview

A prokaryotic cell such as *Escherichia coli* or a single-celled eukaryotic microorganism such as baker's yeast *(Saccharomyces cerevisiae)*, will grow and divide as long as environmental conditions allow. But, a cell inside the cheek of a horse or in a yellow rose petal may neither grow nor divide. That is, in multicellular eukaryotes, cell division is under strict control to develop and maintain a mature body consisting of different subpopulations of cells. Most mature cells in multicellular organisms divide infrequently, if at all. However, the tissues of animals, plants and other multicellular organisms also contain small populations of cells that are always actively dividing. The new cells are required, for instance, for growth (new leaves, hair), and replacement of cells lost to wear (blood cells, cells lining the intestine) and tear (wound repair, virus infections).

Before dividing, most cells enter a period of growth in which they synthesize proteins, lipids, and carbohydrates and, during one particular stage, replicate their nuclear DNA. After this growth period, the nuclei divide and, usually, **cytokinesis** (*cyto* = cell; *kinesis* = movement)—the division of the cytoplasm—follows, partitioning nuclei into daughter cells. Each daughter nucleus contains one copy of the replicated parent DNA. This sequence of events—the period of growth followed by nuclear division and cytokinesis—is known as the **cell cycle.** The nuclear division part of the cell cycle is **mitosis.**

The Products of Mitosis Are Genetic Duplicates of the Dividing Cell

Mitosis serves the purpose of dividing the replicated DNA equally and precisely, generating daughter cells which are exact genetic copies of the parent cell. This is accomplished by three interrelated systems: 1) A master program of molecular checks and balances that ensures an orderly and timely progression through the cell cycle; 2) Within the overall regulation of the cell cycle, a process of DNA synthesis which replicates each DNA chromosome into two copies with almost perfect fidelity (see Chapter 14); and 3) A structural and mechanical web of interwoven "cables" and "motors" of the mitotic cytoskeleton that separates the DNA copies precisely into the daughter cells.

However, at a particular stage of the life cycle of sexually reproducing organisms, a cell division process called *meiosis* produces some cells that are genetically different from the parent cells. **Meiosis** produces daughter nuclei that are different in that they have one half the number of chromosomes the parental nucleus had. Also, the mechanisms involved in producing the daughter nuclei produce arrangements of genes on chromosomes that are different from those in the parent cell (see Chapter 11). The cells that are the products of meiosis may function as gametes in animals (fusing with other gametes to make a zygote) and as spores in plants and many fungi (dividing by mitosis).

This chapter concentrates on the mechanical and regulatory aspects of mitosis in eukaryotes and cell division in prokaryotes; meiosis and its role in eukaryotic sexual reproduction are addressed in Chapter 11. We begin our discussion with **chromosomes,** the nuclear units of genetic information divided and distributed by mitotic cell division.

Chromosomes Are the Genetic Units That are Divided by Mitosis

In all eukaryotes, the hereditary information within the nucleus is distributed among individual, linear DNA molecules. These DNA molecules are combined with proteins that stabilize the DNA molecules, assist in packaging DNA during cell division, and influence the expression of individual genes. In a cell, each *chromosome* (*chroma* = color; *soma* = body; **Figure 10.2**) is composed of one of these linear DNA molecules along with its associated proteins.

Most eukaryotes have two copies of each type of chromosome in their nuclei, so their chromosome complement is said to be **diploid,** or $2n$. For example, human body cells have 23 pairs of different chromosomes, for a diploid number of 46 chromosomes ($2n = 46$). The two chromosomes of each pair in a diploid cell are called **homologous chromosomes**—they have the same genes, in the same order in the DNA of the chromosomes. One of the pair derives from the maternal parent and the other derives from the paternal parent. Other eukaryotes, mostly microorganisms, have only one copy of each type of chromosome in their nuclei, so their chromosome complement is said to be **haploid,** or n. For example, the orange bread mold *Neurospora crassa* is haploid ($n = 7$). Baker's yeast *(S. cerevisiae)*, a model organism that helped illuminate regulatory aspects of the cell cycle, is an example of an organism that can grow as haploid cells ($n = 16$) or as diploid cells ($2n = 32$). Still others, such as many plant species, have three, four, or even more complete sets of chromosomes in each cell. The number of chromosome sets is called the **ploidy** of a cell or species.

Before a cell divides in mitosis, duplication of each chromosome produces two identical copies of each chromosome called **sister chromatids.** Duplication of a chromosome involves replicating the DNA molecule it contains, plus duplicating the set of

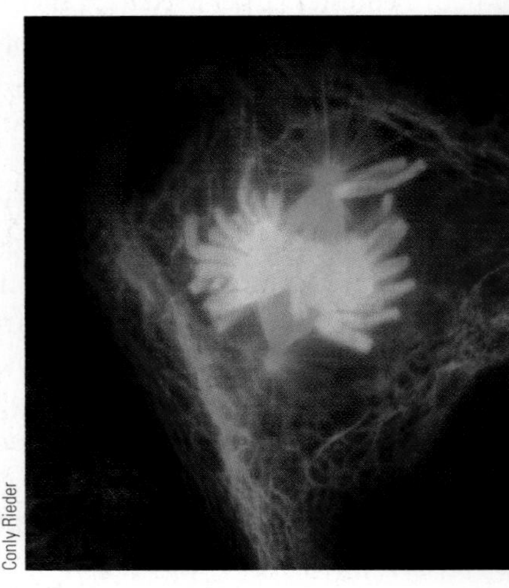
Conly Rieder

FIGURE 10.2
Eukaryotic chromosomes (blue) in a dividing animal cell. At this stage of division, the chromosomes have duplicated and are highly compacted compared to their extended state between cell divisions. The chromosomes are attached to the spindle (green), a structure formed from microtubules, which serves to partition the two sets of chromosomes to the two daughter cells as cell division proceeds.

proteins associated with the DNA. Newly formed sister chromatids are held together until mitosis separates them, placing one in each of the two daughter nuclei. *As a result of this precise division, each daughter nucleus receives exactly the same number and types of chromosomes, and contains the same genetic information, as the parent cell that entered the division.* The equal distribution of daughter chromosomes into each of the two daughter cells that result from cell division is called **chromosome segregation.**

The accuracy of chromosome replication and segregation in the mitotic cell cycle creates a group of genetically identical cells—**clones** of the original cell. Since all the diverse cells types of a complex multicellular organism arose by mitosis from a single zygote, they all contain the same genetic information. Forensic scientists rely on this feature of organisms when, for instance, they match the genetic profile of a small amount of tissue (e.g., cells in dog saliva recovered from a bite victim) with that of a blood sample from the suspected animal. In the laboratory, cells may be grown in **cell cultures,** living cells grown in laboratory vessels. Many types of prokaryotes and eukaryotes can be grown in this way. Cells of each type of organism are cultured in a growth medium optimized for the organism.

STUDY BREAK 10.1

1. **What are the three interrelated systems that contribute to the eukaryotic cell cycle?**
2. **What is the general composition of a eukaryotic chromosome?**
3. **For mitosis, compare the DNA content of daughter cells with that of the parent cell.**

10.2 The Mitotic Cell Cycle

Growth and division of both diploid and haploid cells occurs in the mitotic cell cycle.

Interphase Extends from the End of One Mitosis to the Beginning of the Next Mitosis

If we set the formation of a new daughter cell as the beginning of the mitotic cell cycle, then the first and longest stage is **interphase** **(Figure 10.3).** Three phases of the cell cycle comprise interphase: 1) **G_1 phase,** in which the cell grows; 2) **S phase,** in which DNA replicates and chromosomal proteins are duplicated; and 3) **G_2 phase,** in which cell growth continues and the cell prepares for mitosis (the fourth phase of the cell cycle; also called *M phase*) and cytokinesis.

Usually, G_1 is the only phase of the cell cycle that varies in length. The other phases are typically uniform in length within a species. Thus, whether cells divide rapidly or slowly, depends primarily on the length of G_1. Once DNA replication begins, most mammalian cells take about 10 to 12 hours to proceed through the S phase, about 4 to 6 hours to go through G_2, and about 1 hour or less to complete mitosis.

G_1 is also the stage in which many cell types stop dividing. Cells in a state of division arrest enter a shunt from G_1 called the **G_0 phase.** In some cases, a cell in G_0 may start dividing again by reentering G_1. Some cells never resume the cell cycle; for example, most cells of the human nervous system stop dividing once they are fully mature.

Internal regulatory controls trigger each phase of the cell cycle, ensuring that the processes of one phase are completed successfully before the next phase can begin. Various internal mechanisms also regulate the overall number of cycles that a cell goes through. These internal controls may be subject to various "external" influences such as other cells or viruses, as well as signal molecules, including hormones, growth factors, and death signals.

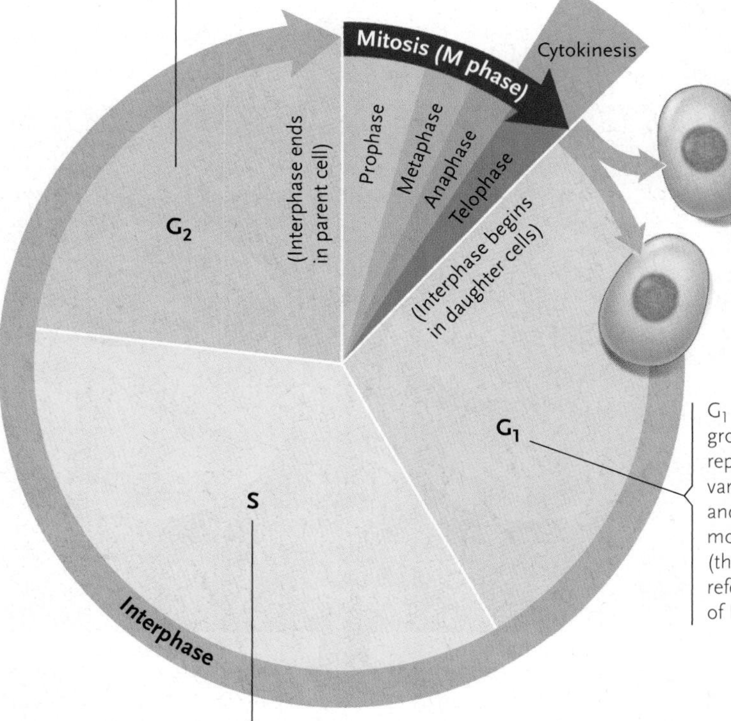
G_2 refers to the second gap in which there is no DNA synthesis. During G_2, the cell continues to synthesize RNAs and proteins, including those for mitosis, and it continues to grow. The end of G_2 marks the end of interphase; mitosis then begins.
Mitosis (M phase)
Cytokinesis
Prophase
Metaphase
Anaphase
Telophase
(Interphase ends in parent cell)
(Interphase begins in daughter cells)
G_2
G_1
S
Interphase
G_1 phase is a period of growth before the DNA replicates. The cell makes various RNAs, proteins, and other types of cellular molecules but not DNA (the G in G_1 stands for *gap*, referring to the absence of DNA synthesis).
If the cell is going to divide, DNA replication begins. During S phase, the cell duplicates each chromosome, including both the DNA and the chromosomal proteins, and it also continues synthesis of other cellular molecules.

FIGURE 10.3
The cell cycle. The length of G_1 varies, but for a given cell type, the timing of S, G_2, and mitosis is usually relatively uniform. Cytokinesis (segment at 2 o'clock) usually begins while mitosis is in progress and reaches completion as mitosis ends. Cells in a state of division arrest enter a shunt from G_1 called G_0 (not shown).

After Interphase, Mitosis Proceeds in Five Stages

Once it begins, mitosis proceeds continuously, without significant pauses or breaks. However, for convenience in study, biologists separate mitosis into five sequential stages: **prophase** (*pro* = before), **prometaphase** (*meta* = between), **metaphase, anaphase** (*ana* = back), and **telophase** (*telo* = end). Mitosis in an animal cell and a plant cell is shown in **Figures 10.4** and **10.5,** respectively. The entire process takes from 1 to 4 hours in most eukaryotes.

PROPHASE During **prophase,** the greatly extended chromosomes that were replicated during interphase begin to *condense* into compact, rodlike structures (see chromatin packaging in Chapter 14). Each diploid human cell, although only about 40–50 μm in diameter, contains *2 meters* of DNA distributed among 23 pairs of chromosomes. Condensation during prophase packs these long DNA molecules into units small enough to be divided successfully during mitosis. As they condense, the chromosomes appear as thin threads under the light microscope. The word *mitosis* (*mitos* = thread) is derived from this threadlike appearance.

While condensation is in progress, the nucleolus becomes smaller and eventually disappears in most species. The disappearance reflects a shutdown of all types of RNA synthesis, including the ribosomal RNA made in the nucleolus.

In the cytoplasm, the mitotic **spindle** (**Figure 10.6**; see also Figure 10.11), the structure that actually separates chromatids, begins to form between the two centrosomes as they start migrating toward the opposite ends of the cell, where they will form the **spindle poles.** The spindle develops as two bundles of microtubules that radiate from the two spindle poles.

FIGURE 10.4
The stages of mitosis. Light micrographs show mitosis in an animal cell (whitefish embryo). Diagrams show mitosis in an animal cell with two pairs of chromosomes. As you study these diagrams, consider that each diploid human cell, although only 40–50 μm in diameter, contains 2 meters of DNA distributed among 23 pairs of chromosomes.

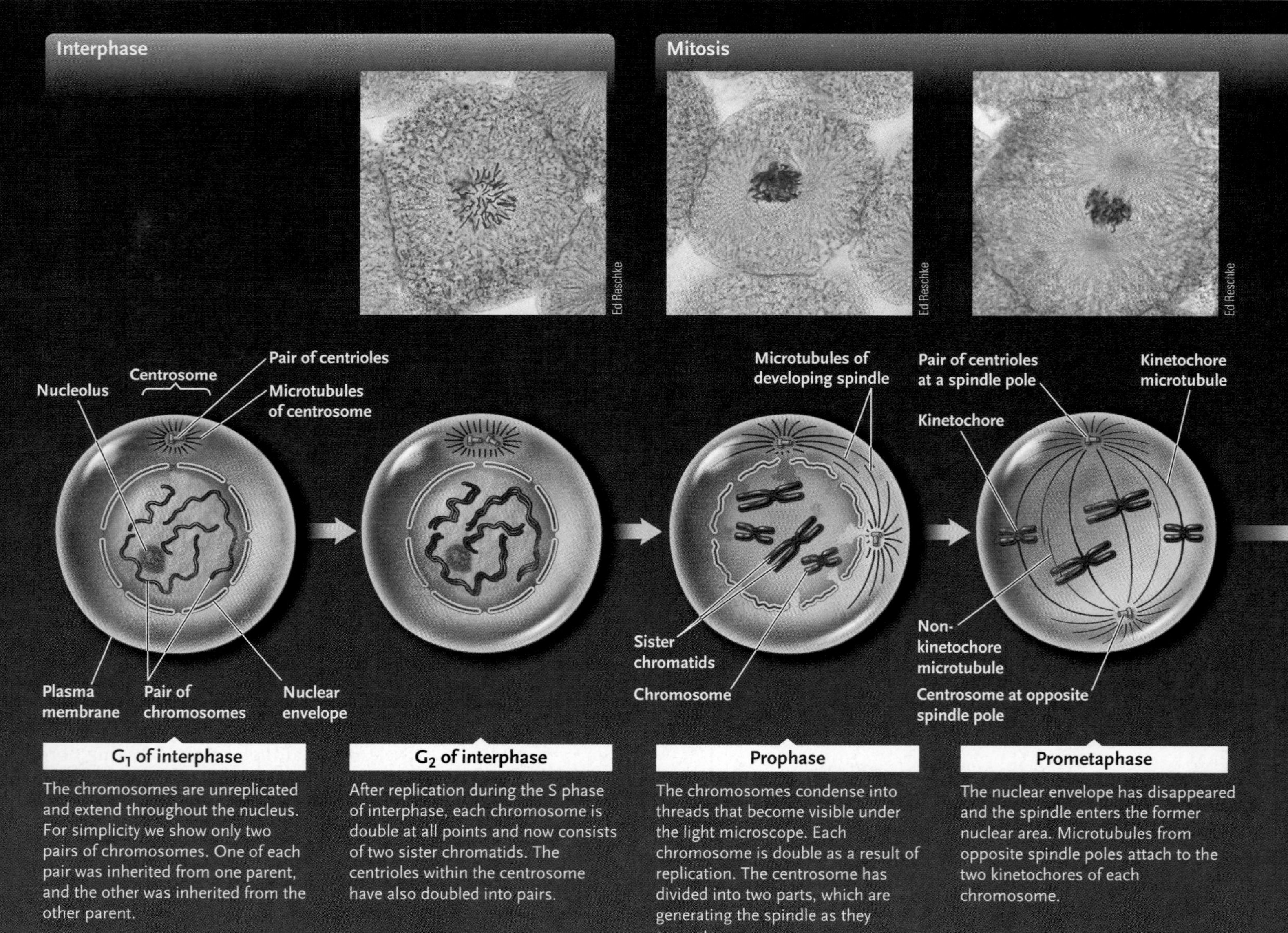

Interphase
Mitosis
Ed Reschke
Ed Reschke
Ed Reschke
Nucleolus
Centrosome
Pair of centrioles
Microtubules of centrosome
Plasma membrane
Pair of chromosomes
Nuclear envelope
Microtubules of developing spindle
Sister chromatids
Chromosome
Pair of centrioles at a spindle pole
Kinetochore
Kinetochore microtubule
Non-kinetochore microtubule
Centrosome at opposite spindle pole
G_1 of interphase
The chromosomes are unreplicated and extend throughout the nucleus. For simplicity we show only two pairs of chromosomes. One of each pair was inherited from one parent, and the other was inherited from the other parent.
G_2 of interphase
After replication during the S phase of interphase, each chromosome is double at all points and now consists of two sister chromatids. The centrioles within the centrosome have also doubled into pairs.
Prophase
The chromosomes condense into threads that become visible under the light microscope. Each chromosome is double as a result of replication. The centrosome has divided into two parts, which are generating the spindle as they separate.
Prometaphase
The nuclear envelope has disappeared and the spindle enters the former nuclear area. Microtubules from opposite spindle poles attach to the two kinetochores of each chromosome.

PROMETAPHASE At the end of prophase, the nuclear envelope breaks down, heralding the beginning of **prometaphase.** Bundles of spindle microtubules grow from centrosomes at the *opposite spindle poles* toward the center of the cell. Some of the developing spindle enters the former nuclear area.

Each chromosome is still a double structure made up of two identical sister chromatids held together only at their **centromeres.** By this time, a complex of several proteins, a **kinetochore,** has formed on each chromatid at the **centromere,** a region located at a particular position in a given chromosome. The centromere is often narrower than the rest of the chromosome. *Kinetochore microtubules* bind to the kinetochores. These connections determine the outcome of mitosis, because they attach the sister chromatids of each chromosome to microtubules that lead to the opposite spindle poles (see Figure 10.6). Microtubules that do not attach to kinetochores—the *nonkinetochore microtubules*—overlap those from the opposite spindle pole.

METAPHASE During **metaphase,** the spindle reaches its final form and the spindle microtubules move the chromosomes into alignment at the spindle midpoint, also called the *metaphase plate.* The chromosomes complete their condensation in this stage. The pattern of condensation gives each chromosome a characteristic shape, determined by the location of the centromere and the length and thickness of the chromatid arms.

Only when the chromosomes are all assembled at the spindle midpoint, with the two sister chromatids of each one attached to microtubules leading to opposite spindle poles, can metaphase give way to actual separation of chromatids in anaphase.

Although chromosomes at metaphase are generally thought of—and are often drawn—as "X" shapes, few chromosomes actually take on that shape. Most have centromeres near one end or the other rather than in the middle, as an X would imply, and they are also too loosely packaged to take on a defined shape.

The complete set of metaphase chromosomes, arranged according to size and shape, forms the **karyotype** of a given species.

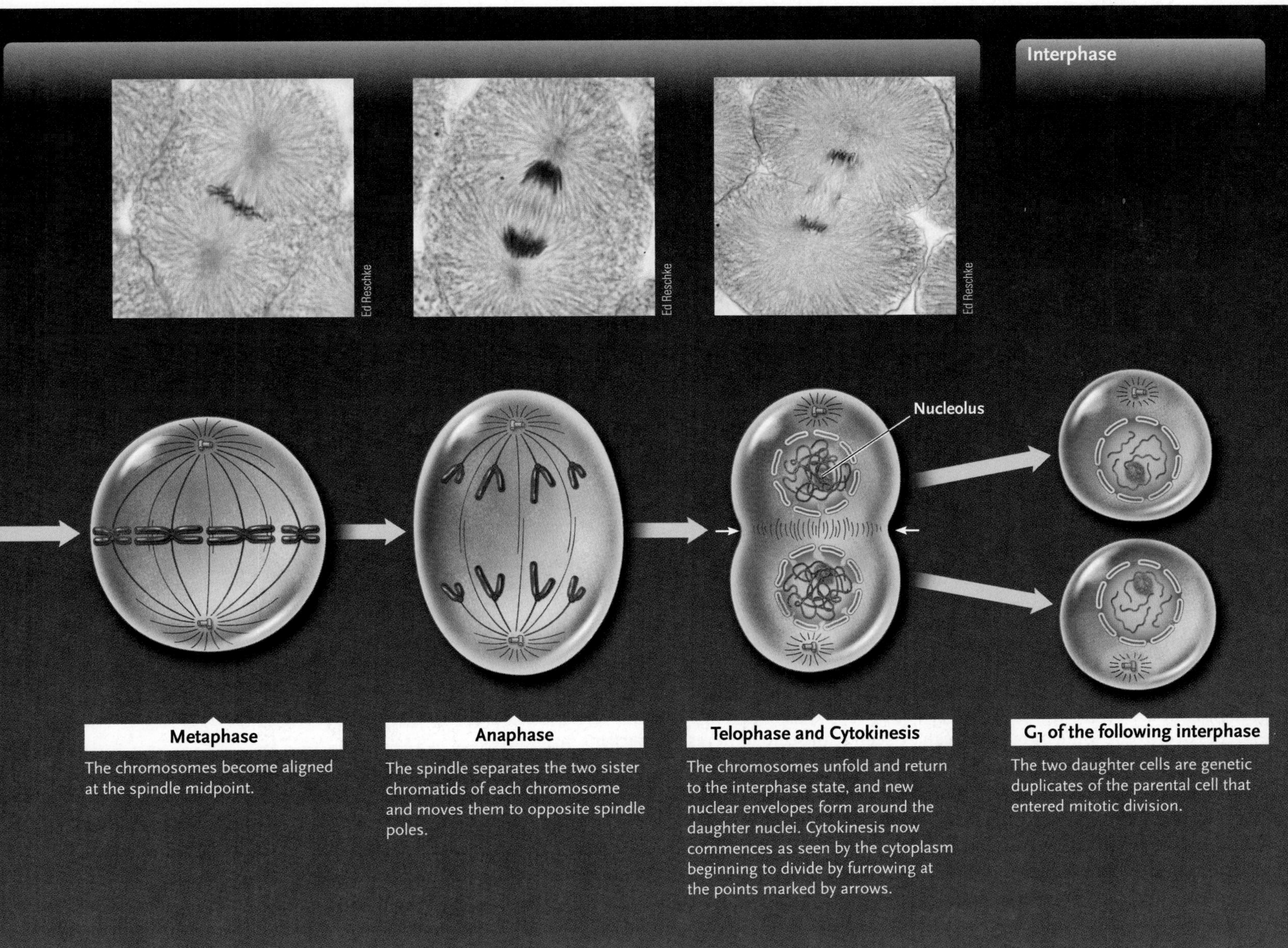

Interphase
Ed Reschke
Ed Reschke
Ed Reschke
Nucleolus
Metaphase
The chromosomes become aligned at the spindle midpoint.
Anaphase
The spindle separates the two sister chromatids of each chromosome and moves them to opposite spindle poles.
Telophase and Cytokinesis
The chromosomes unfold and return to the interphase state, and new nuclear envelopes form around the daughter nuclei. Cytokinesis now commences as seen by the cytoplasm beginning to divide by furrowing at the points marked by arrows.
G_1 of the following interphase
The two daughter cells are genetic duplicates of the parental cell that entered mitotic division.

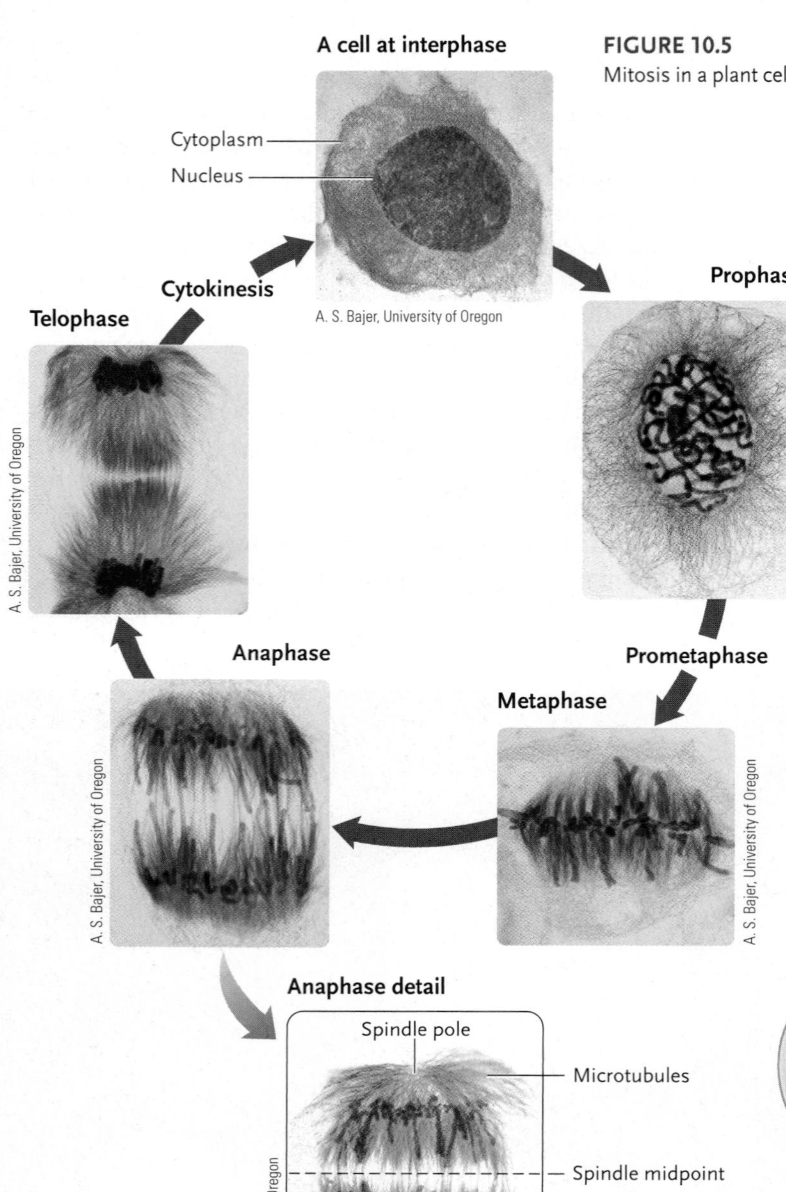
A cell at interphase
Cytoplasm
Nucleus
A. S. Bajer, University of Oregon
Prophase
A. S. Bajer, University of Oregon
Prometaphase
Metaphase
A. S. Bajer, University of Oregon
Anaphase
A. S. Bajer, University of Oregon
Telophase
Cytokinesis
A. S. Bajer, University of Oregon
Anaphase detail
Spindle pole
Microtubules
Spindle midpoint
Chromosomes
Spindle pole
A. S. Bajer, University of Oregon

FIGURE 10.5
Mitosis in a plant cell. The chromosomes are stained blue; the spindle microtubules are stained red.

In many cases, the karyotype is so distinctive that a species can be identified from this characteristic alone. **Figure 10.7** shows how human chromosomes are prepared for analysis as a karyotype.

ANAPHASE During **anaphase,** the spindle separates sister chromatids and pulls them to opposite spindle poles. The first signs of chromosome movement can be seen at the centromeres, where tension developed by the spindle pulls the kinetochores toward opposite poles. The movement continues until the separated chromatids, now called *daughter chromosomes,* have reached the two poles. At this point, chromosome segregation has been completed.

TELOPHASE During **telophase,** the spindle disassembles and the chromosomes at each spindle pole decondense, returning to the extended state typical of interphase. As decondensation proceeds, the nucleolus reappears, RNA transcription resumes, and a new nuclear envelope forms around the chromosomes at each pole producing the two daughter nuclei. At this point, nuclear division is complete, and the cell has two nuclei.

Mitosis produces two daughter nuclei, each with identical genomes to the parental cell. Cytokinesis, the division of the cytoplasm, typically follows the nuclear division stage of mitosis, and produces two daughter cells each with one of the two daughter nuclei. Cytokinesis proceeds by different pathways in the various kingdoms of eukaryotic organisms. In animals, protists, and many fungi, a groove, the **furrow,** girdles the cell and gradually deepens until it cuts the cytoplasm into two parts **(Figure 10.8).** In plants, a new cell wall, called the **cell plate,** forms between the daughter nuclei and grows laterally until it divides the cytoplasm in two **(Figure 10.9).** In both cases, the plane of cytoplasmic division is determined by the layer of microtubules that persist at the former spindle midpoint.

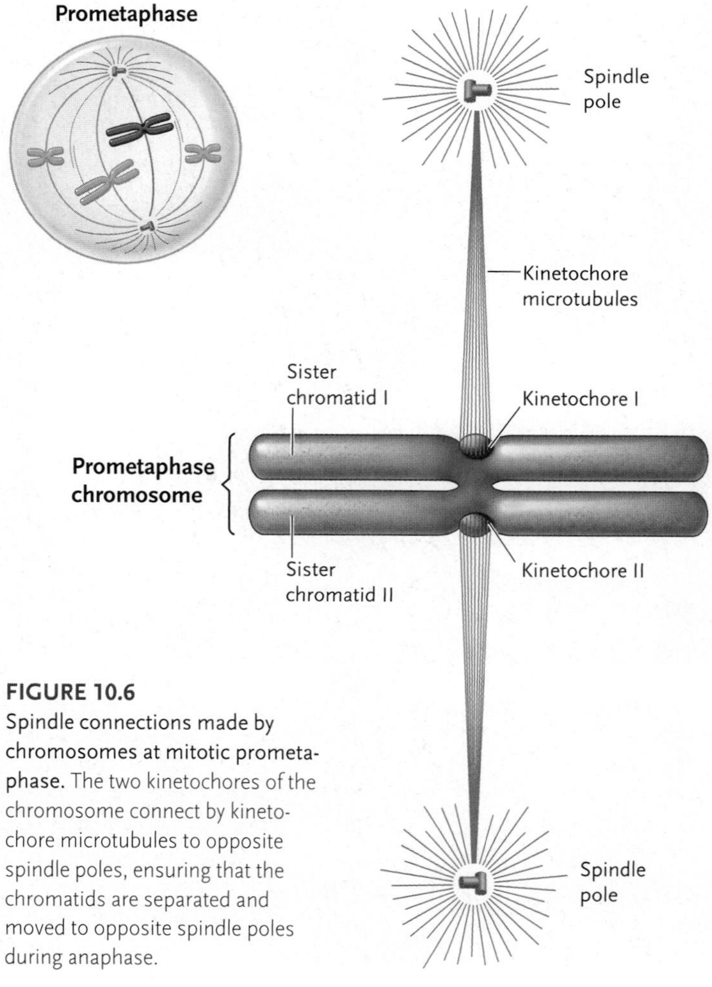
Prometaphase
Spindle pole
Kinetochore microtubules
Sister chromatid I
Kinetochore I
Prometaphase chromosome
Sister chromatid II
Kinetochore II
Spindle pole

FIGURE 10.6
Spindle connections made by chromosomes at mitotic prometaphase. The two kinetochores of the chromosome connect by kinetochore microtubules to opposite spindle poles, ensuring that the chromatids are separated and moved to opposite spindle poles during anaphase.

Preparing a Human Karyotype

Purpose: A karyotype is a display of chromosomes of an organism arranged in pairs. A normal karyotype has a characteristic appearance for each species. Examination of the karyotype of the chromosomes from a particular individual indicates whether the individual has a normal set of chromosomes or whether there are abnormalities in number or appearance of individual chromosomes. A normal karyotype can be used to indicate the species.

Protocol:

1. Add sample to culture medium that has stimulator for growth and division of cells (white blood cells in the case of blood). Incubate at 37°C. Add colchicine, which causes spindle to dissemble, to arrest mitosis at metaphase.

2. Stain the cells so that the chromosomes are distinguished. Some stains produce chromosome-specific banding patterns, as shown in the photograph below.

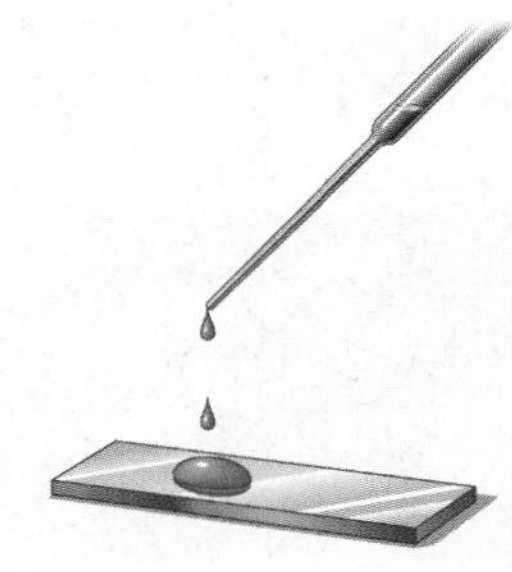

3. View the stained cells under a microscope equipped with a digital imaging system and take a digital photograph. A computer processes the photograph to arrange the chromosomes in pairs and number them according to size and shape.

SIU/Peter Arnold

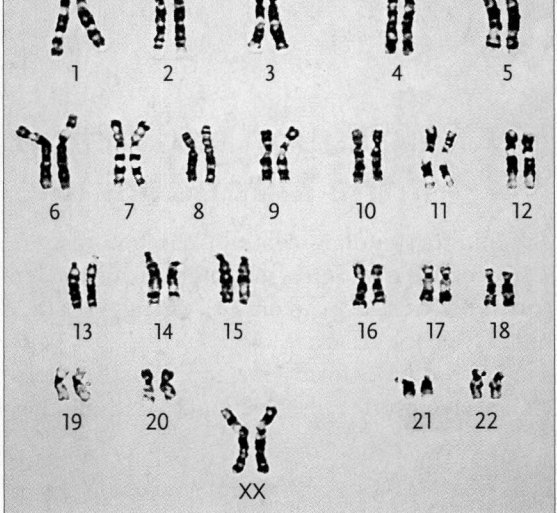

Pair of homologous chromosomes

Pair of sister chromatids closely aligned side-by-side

1 2 3 4 5 6 7 8 9 10 11 12 13 14 15 16 17 18 19 20 21 22 XX

Leonard Lessin/Peter Arnold

Interpreting the Results: The karyotype is evaluated with respect to the scientific question being asked. For example, it may identify a particular species, or it may indicate whether or not the chromosome set of a human (fetus, child, or adult) is normal or aberrant.

The Mitotic Cell Cycle Is Significant for Both Development and Reproduction

The mitotic cycle of interphase, nuclear division, and cytokinesis accounts for the growth of multicellular eukaryotes from single initial cells, such as a fertilized egg, to fully developed adults. Mitosis also serves as a method of reproduction called **vegetative** or **asexual reproduction,** which occurs in many kinds of plants and protists and in some animals. In asexual reproduction, daughter cells produced by mitotic cell division are released from the parent and grow separately by further mitosis into complete individuals. For example, asexual reproduction occurs when a single-celled protist such as an amoeba divides by mitosis to produce two separate individuals, or when a leaf cutting is used to generate an entire new plant.

STUDY BREAK 10.2 <

1. **In what order do the stages of mitosis occur?**
2. **What is the importance of centromeres to mitosis?**
3. **Colchicine, an alkaloid extracted from plants, prevents the formation of spindle microtubules. What would happen if a cell enters mitosis when colchicine is present?**

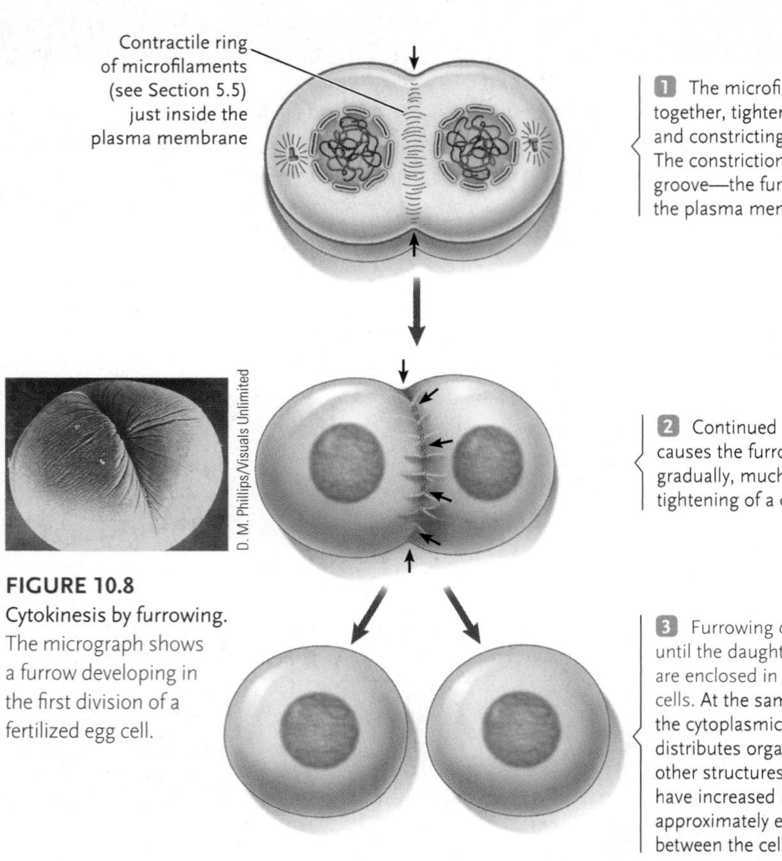
Contractile ring of microfilaments (see Section 5.5) just inside the plasma membrane
D. M. Phillips/Visuals Unlimited
1 The microfilaments slide together, tightening the ring and constricting the cell. The constriction forms a groove—the furrow—in the plasma membrane.
2 Continued constriction causes the furrow to deepen gradually, much like the tightening of a drawstring.
3 Furrowing continues until the daughter nuclei are enclosed in separate cells. At the same time, the cytoplasmic division distributes organelles and other structures (which also have increased in number) approximately equally between the cells.

FIGURE 10.8
Cytokinesis by furrowing. The micrograph shows a furrow developing in the first division of a fertilized egg cell.

10.3 Formation and Action of the Mitotic Spindle

The mitotic spindle is central to both mitosis and cytokinesis. The spindle is made up of microtubules and their motor proteins, and its activities depend on their changing patterns of organization during the cell cycle.

Microtubules form a major part of the interphase cytoskeleton of eukaryotic cells. (Section 5.3 outlines the patterns of microtubule organization in the cytoskeleton.) As mitosis approaches, the microtubules disassemble from their interphase arrangement and reorganize into the spindle, which grows until it fills almost the entire cell. This reorganization follows one of two pathways in different organisms, depending on the presence or absence of a *centrosome* during interphase. However, once organized, the basic function of the spindle is the same, regardless of whether a centrosome is present or not.

Animals and Plants Form Spindles in Different Ways

Figure 10.10 shows the **centrosome,** a site near the nucleus from which microtubules radiate outward in all directions, and its role in spindle formation. The centrosome is the main **microtubule organizing center (MTOC)** of animal cells and many protists. The centrosome contains a pair of **centrioles,** usually arranged at right angles to each other. Although centrioles originally appeared to be important in the construction of the mitotic spindle, they can be removed experimentally with no ill effect. The primary function of centrioles is actually to generate the microtubules for flagella or cilia, the whiplike extensions of certain eukaryotic cells (see Section 5.5).

When the nuclear envelope breaks down at the end of prophase, the spindle (see Figure 10.10, step 4) moves into the region formerly occupied by the nucleus and continues growing until it fills the cytoplasm. The microtubules that extend from the centrosomes also grow in length and extent, producing radiating arrays that appear starlike under the light microscope. Initially

FIGURE 10.9
Cytokinesis by cell plate formation.

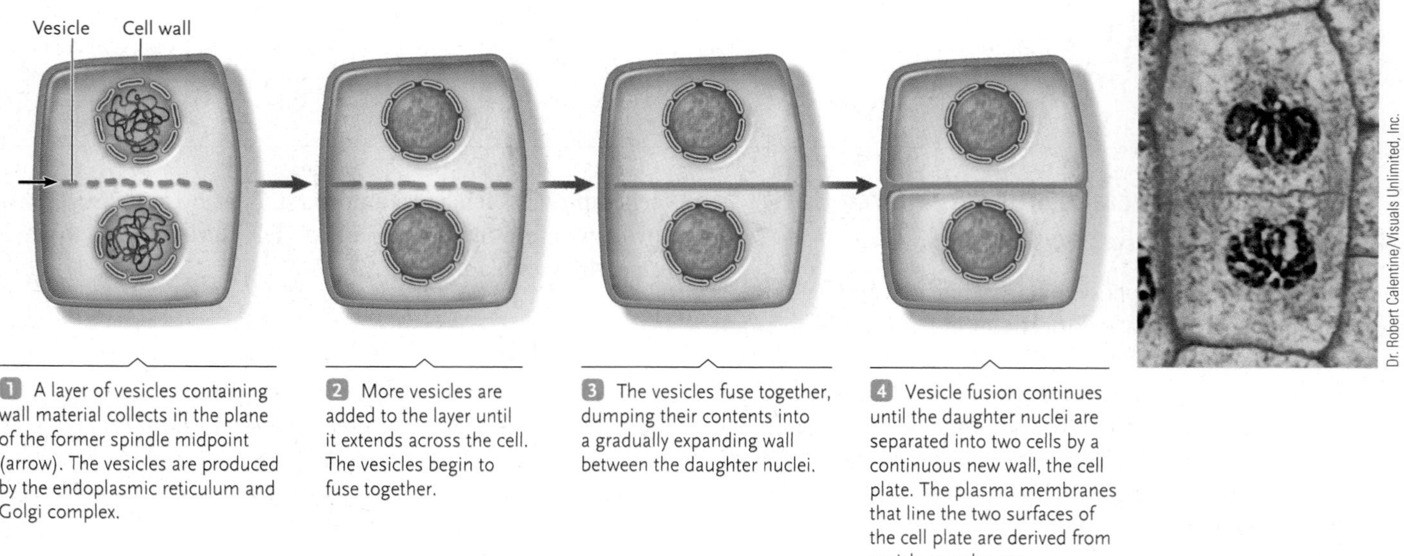
Vesicle
Cell wall
Dr. Robert Calentine/Visuals Unlimited, Inc.
1 A layer of vesicles containing wall material collects in the plane of the former spindle midpoint (arrow). The vesicles are produced by the endoplasmic reticulum and Golgi complex.
2 More vesicles are added to the layer until it extends across the cell. The vesicles begin to fuse together.
3 The vesicles fuse together, dumping their contents into a gradually expanding wall between the daughter nuclei.
4 Vesicle fusion continues until the daughter nuclei are separated into two cells by a continuous new wall, the cell plate. The plasma membranes that line the two surfaces of the cell plate are derived from vesicle membranes.

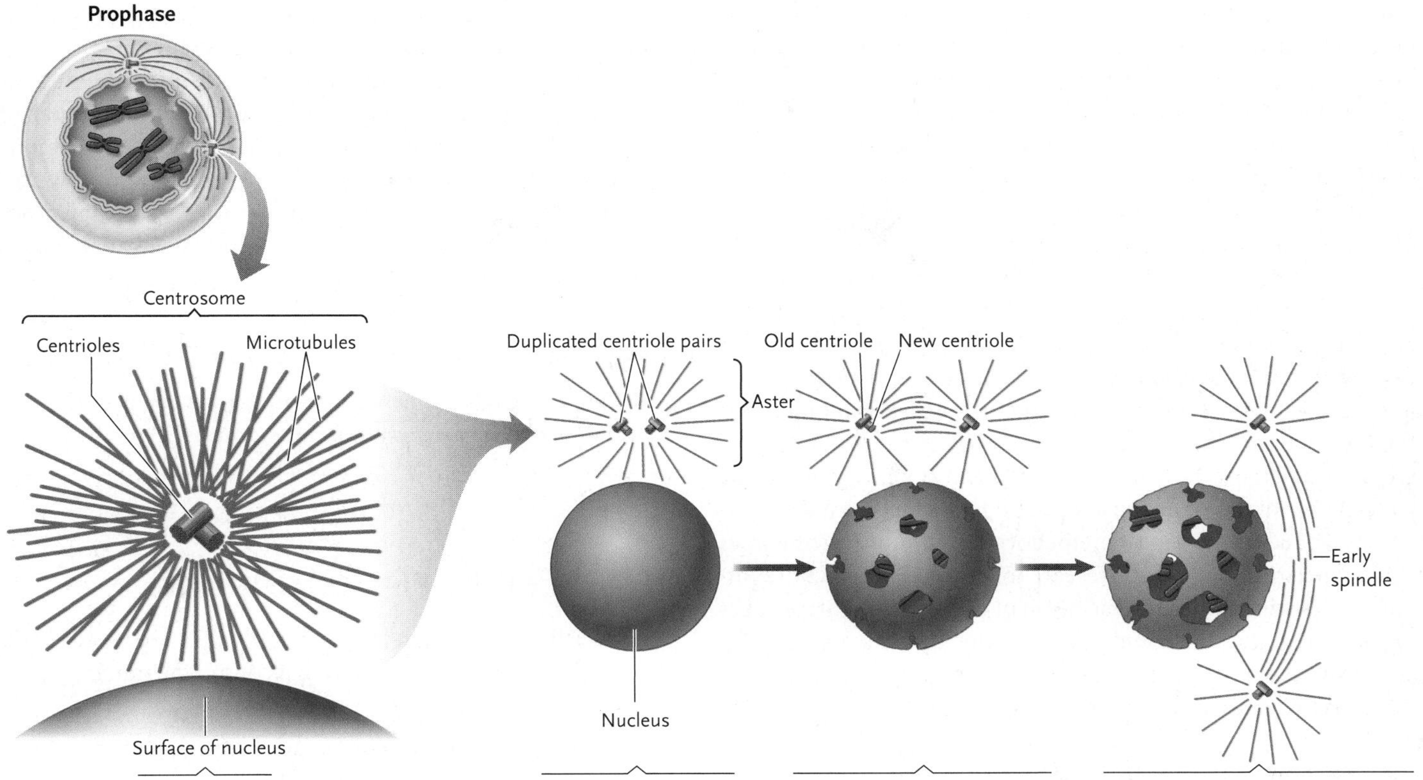

Prophase

Centrosome

Centrioles

Microtubules

Surface of nucleus

1 Centrosome at interphase.

Duplicated centriole pairs

Aster

Nucleus

2 The original pair of centrioles duplicate during the S phase of the cell cycle, producing two pairs of centrioles.

Old centriole

New centriole

3 As prophase begins, the centrosome separates into two parts, each containing one "old" and one "new" centriole—one centriole of the original pair and its copy.

Early spindle

4 The duplicated centrosomes, containing the centrioles, continue to separate until they reach opposite sides of the nucleus. The microtubules between them lengthen and increase in number. By late prophase the early spindle is complete, consisting of the separated centrosomes and a large mass of microtubules between them.

FIGURE 10.10
The centrosome and its role in spindle formation.

named by early microscopists, **asters** (*aster* = star) are the centrosomes at the spindle tips, which form the poles of the spindle. By dividing the duplicated centrioles, the spindle ensures that, when the cytoplasm divides during cytokinesis, the daughter cells each receive a pair of centrioles.

Angiosperms (flowering plants) and most gymnosperms, such as conifers, lack centrosomes and centrioles. In these organisms, the spindle forms from microtubules that assemble in all directions from multiple MTOCs surrounding the entire nucleus (see prophase in Figure 10.5). Then, when the nuclear envelope breaks down at the end of prophase, the spindle moves into the former nuclear region.

Mitotic Spindles May Move Chromosomes by a Combination of Two Mechanisms

When fully formed at metaphase, the spindle may contain from hundreds to many thousands of microtubules, depending on the species **(Figure 10.11)**. In almost all eukaryotes, these microtubules are divided into two groups. *Kinetochore microtubules* connect the chromosomes to the spindle poles **(Figure 10.12A)**. *Nonkineto-*

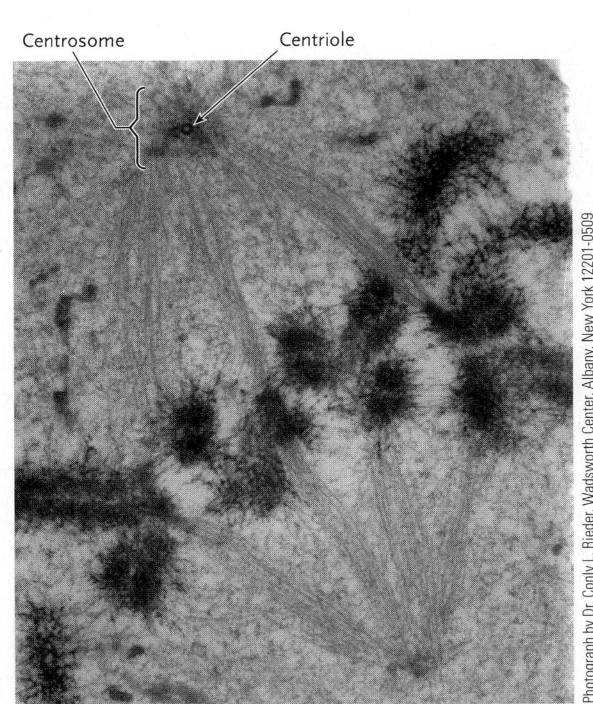
Centrosome
Centriole
Photograph by Dr. Conly L. Rieder, Wadsworth Center, Albany, New York 12201-0509

FIGURE 10.11
A fully developed spindle in a mammalian cell. Only microtubules connected to chromosomes have been caught in the plane of this section. One of the centrioles is visible in cross section in the centrosome at the top of the micrograph.

chore microtubules extend between the spindle poles without connecting to chromosomes; at the spindle midpoint, these microtubules from one pole overlap with microtubules from the opposite pole **(Figure 10.12B)**. The separation of the chromosomes at anaphase appears to result from a combination of separate but coordinated movements produced by the two types of microtubules.

The exact mechanism by which chromosomes move is still uncertain. At one time, scientists thought that microtubules pulled the chromosomes toward the poles of dividing cells. However, recent data suggest that chromosomes "walk" themselves to the poles along stationary microtubules, using motor proteins in their kinetochores **(Figure 10.13)**. The tubulin subunits of the kinetochore microtubules disassemble as the kinetochores pass along them; thus, the microtubules become shorter as the movement progresses (see Figure 10.12A). The movement is similar to pulling yourself, hand over hand, up a rope as it falls apart behind you.

Evidence supporting kinetochore-based movement comes from experiments in which researchers tagged kinetochore microtubules with a microscopic beam of ultraviolet light, producing bleached sites that could be seen in the light microscope **(Figure 10.14)**. As the chromosomes were moved to the spindle poles, the bleached sites stayed in the same place. This result showed that the kinetochore microtubules do not move much with respect to the poles during the anaphase movement.

In nonkinetochore microtubule-based movement, the entire spindle is lengthened, pushing the poles farther apart (see Figure 10.12B). The pushing movement is produced by microtubules sliding over one another in the zone of overlap, powered by proteins acting as microtubule motors. In many species, the nonkinetochore microtubules also push the poles apart by growing in length as they slide along.

Metaphase

Anaphase

A. The kinetochore microtubules connected to the chromosomes become shorter, lessening the distance from the chromosomes to the poles.

B. Sliding of the nonkinetochore microtubules in the zone of overlap at the spindle midpoint pushes poles farther apart and increases the total length of the spindle.

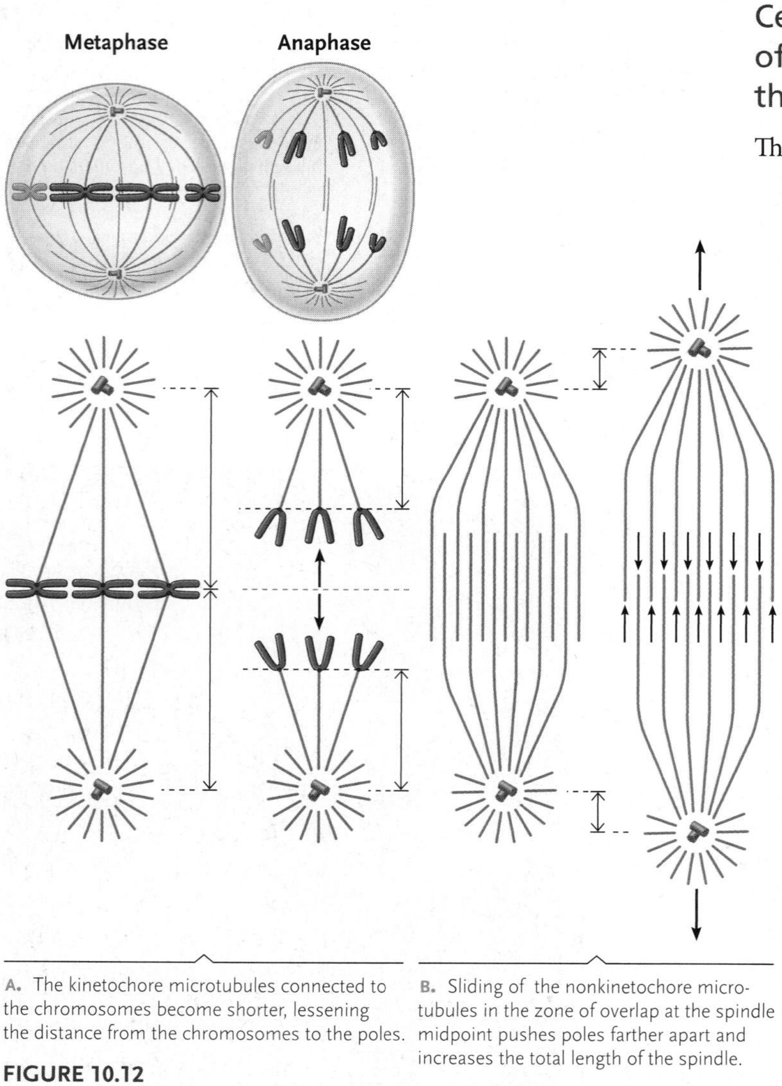

FIGURE 10.12
The two microtubule-based movements of the anaphase spindle.

STUDY BREAK 10.3

1. **How does spindle formation differ in animals and plants?**
2. **How do mitotic spindles move chromosomes?**

10.4 Cell Cycle Regulation

In this section we discuss experimental evidence for (and the operation of) regulatory mechanisms that control the mitotic cell cycle.

Cell Fusion Experiments and Studies of Yeast Mutants Identified Molecules that Control the Cell Cycle

The first insights into how the cell cycle is regulated came from experiments by Robert T. Johnson and Potu N. Rao at the University of Colorado Medical Center, Denver, published in 1970. They fused human HeLa cells (a type of cancer cell that can be grown in cell culture) that were in different

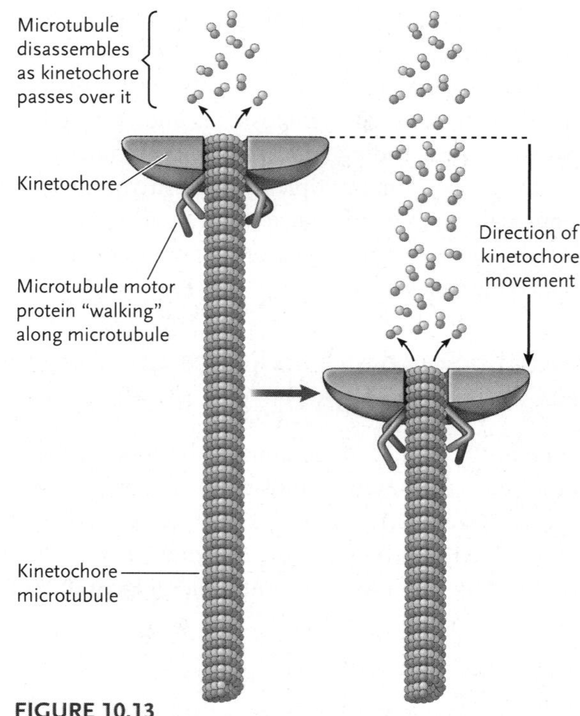
Microtubule disassembles as kinetochore passes over it
Kinetochore
Microtubule motor protein "walking" along microtubule
Direction of kinetochore movement
Kinetochore microtubule

FIGURE 10.13
Microtubule motor proteins "walking" the kinetochore of a chromosome along a microtubule.

FIGURE 10.14 **Experimental Research**

Movement of Chromosomes during Anaphase of Mitosis

Question: How do chromosomes move during anaphase of mitosis?

Experiment: One hypothesis for how chromosomes move during anaphase of mitosis was that the kinetochore microtubules moved, pulling chromosomes to the poles. An alternative hypothesis was that chromosomes move by sliding over or along kinetochore microtubules. To test the hypotheses, G. J. Gorbsky and his colleagues made regions of the kinetochore microtubules visibly distinct.

1. Kinetochore microtubules were combined with a dye molecule that bleaches when it is exposed to light.

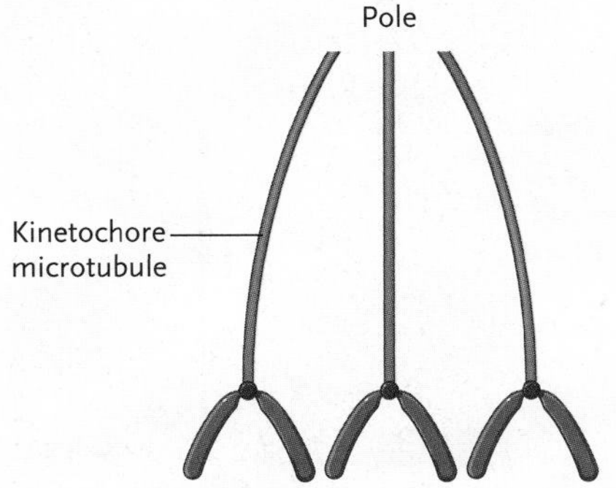

Pole
Kinetochore microtubule

2. The region of the spindle between the kinetochores and the poles was exposed to a microscopic beam of light that bleached a narrow stripe across the microtubules. The bleached region could be seen with a light microscope and analyzed as anaphase proceeded.

Results: The bleached region remained at the same distance from the pole as the chromosomes moved toward the pole.

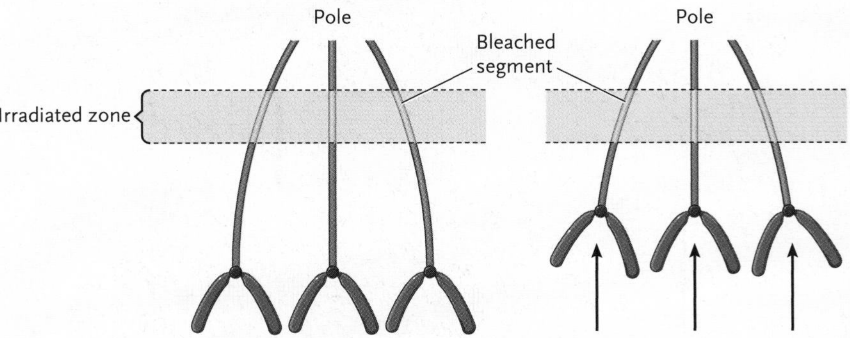

Pole
Bleached segment
Irradiated zone
Pole

Conclusion: The results support the hypothesis that chromosomes move by sliding over or along kinetochore microtubules.

Source: G. J. Gorbsky, P. J. Sammak, and G. G. Borisy. 1987. Chromosomes moved poleward in anaphase along stationary microtubules that coordinately disassemble from their kinetochore microtubules. *Journal of Cell Biology* 104:9–18.

stages of the cell cycle and determined whether one nucleus could influence the other **(Figure 10.15).** Their results suggested that specific molecules in the cytoplasm cause the progression of cells from G_1 to S, and from G_2 into M.

Some key research using baker's yeast, *S. cerevisiae,* helped to identify these cell cycle control molecules and contributed to our general understanding of how the cell cycle is regulated. (*Focus on Model Research Organisms* describes yeast and its role in research in more detail.) In particular, Leland Hartwell of the Fred Hutchinson Cancer Center, Seattle, investigated yeast mutants that become stuck at some point in the cell cycle, but only when they are cultured at a high temperature. By growing the mutants initially at the standard temperature, and then shifting the cells to the higher temperature, Hartwell was able to use time-lapse photomicroscopy to see if and when growth and division were affected. In this way he isolated many *c*ell *d*ivision *c*ycle, or *cdc* mutants. By examining the mutants he could identify the stage in the cell cycle where each cycle was blocked by noting whether nuclei had divided, chromosomes had condensed, the mitotic spindle had formed, cytokinesis had occurred, and so on. Using this approach, Hartwell identified many genes involved in yeast's cell cycle and hypothesized where in the cycle their protein products operated. As might be expected, some of the genes encoded proteins involved in DNA replication, but a number of others were shown to function in cell cycle regulation. Hartwell received a Nobel Prize in 2001 for his discovery.

Paul Nurse of the Imperial Cancer Research Fund, London, carried out similar research with the fission yeast, *Schizosaccharomyces pombe,* a species that divides by fission rather than budding. He identified a gene called *cdc2* which encodes a protein needed for the cell to progress from G_2 to M. Nurse also made the breakthrough discovery that all eukaryotic cells studied have counterparts of the yeast *cdc2* gene. The protein product of *cdc2* is a protein kinase, an enzyme that catalyzes the phosphorylation of a target protein. (Recall from Section 7.2 that phosphorylation of proteins by protein kinases can activate or inactivate proteins.) That discovery was pivotal in determining how cell cycle regulation occurs. Paul Nurse received a Nobel Prize in 2001 for his discovery.

FIGURE 10.15 **Experimental Research**

Demonstrating the Existence of Molecules Controlling the Cell Cycle by Cell Fusion

Question: Do molecules in the cytoplasm direct the progression through the cell cycle?

Experiment: Johnson and Rao fused human HeLa cells at different stages of the cell cycle. Cell fusion produces a single cell with two separate nuclei. The researchers allowed the fused cells to grow and determined whether one nucleus influenced the other in terms of progression through the cell cycle.

1. Fusion of cell in S phase with cell in G_1 phase.

2. Fusion of cell in M (mitosis) with cell in any other stage.

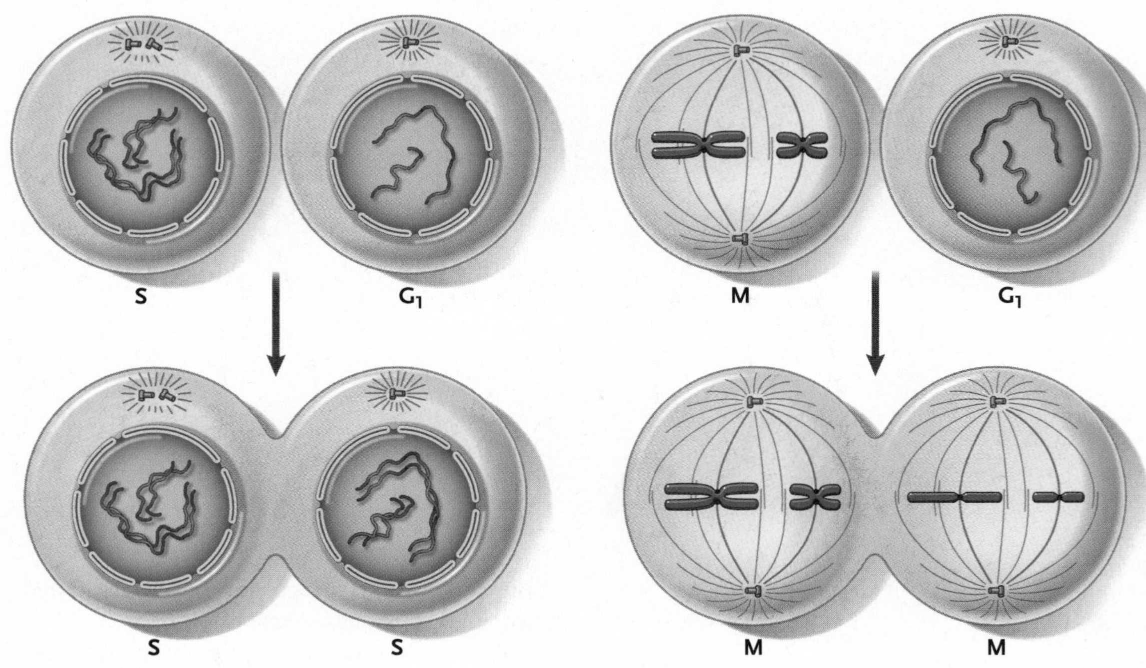
S
G1
M
G1
S
S
M
M

Result: DNA synthesis quickly began in the original G_1 nucleus. Normally, the G_1 nucleus would not have initiated DNA synthesis until it reached S phase itself, which could have been several hours later. The result suggested that one or more molecules that activate S phase are present in the cytoplasm of S phase cells.

Result: Regardless of the phase of the cell, the nucleus of the cell with which the M phase cell was fused immediately began the early stages of mitosis. This included condensation of the chromosomes, spindle formation, and breaking down of the nuclear envelope. For a cell in G_1 (shown in the diagram), the condensed chromosomes that appear have not replicated.

Conclusion: Taken together, the results showed that specific molecules in the cytoplasm direct the progression of cells from G_1 to S, and from G_2 to M in the cell cycle.

Source: R. T. Johnson and P. N. Rao. 1970. Mammalian cell fusion: induction of premature chromosome condensation in interphase nuclei. *Nature* 226:717–722.

The Cell Cycle Can Be Arrested at Specific Checkpoints

A cell has internal controls that monitor its progression through the cell cycle. As part of the internal controls, the cell cycle has three key **checkpoints** to prevent critical phases from beginning until the previous phases are completed correctly **(Figure 10.16)**. The G_1/S checkpoint near the end of the G_1 phase is the main point in the cell cycle at which a cell decides whether to divide or not. Once it passes this checkpoint, the cell is committed to continue the cell cycle, from DNA replication in S to cell division in M. The cell cycle arrests (the cell stops proceeding through the cell cycle) at the G_1/S checkpoint if the DNA is damaged by radiation or chemicals. If the DNA damage is repaired, the cycle starts again. Cell cycle arrest also occurs at this checkpoint if the cell is nutritionally deficient, because then the cell has not reached a sufficient size to proceed to cell division. The G_1/S checkpoint is also the primary point at which cells "read" extracellular signals for cell growth and division. Therefore, if a growth factor required for stimulating cell growth is absent, the cells will arrest at this check-

FOCUS ON **Model Research Organisms**

The Yeast *Saccharomyces cerevisiae*

Saccharomyces cerevisiae, commonly known as baker's yeast or brewer's yeast, was probably the first microorganism to have been grown and kept in cultures—a beer-brewing vessel is basically a *Saccharomyces* culture. Favorite strains of baker's and brewer's yeast have been kept in continuous cultures for centuries. The yeast has also been widely used in scientific research; its microscopic size and relatively short generation time make it easy and inexpensive to culture in large numbers in the laboratory.

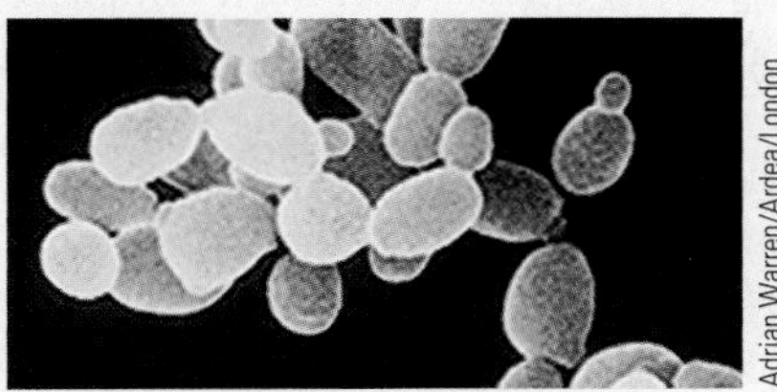
Adrian Warren/Ardea/London

The cells growing in *Saccharomyces* cultures in the research lab are haploid. If the culture conditions are kept at optimal levels (which requires only a source of a fermentable sugar such as glucose, a nitrogen source, and minerals), the cells reproduce asexually by budding. *Saccharomyces* has two mating types. If two yeast cells of different mating types contact one another, they fuse—mate—producing a diploid cell. Diploid cells can also reproduce asexually by budding. Under certain conditions, diploid yeast cells undergo meiosis, producing haploid spores which are resistant to harsh conditions in the environment. Yeast spores germinate into haploid cells, which then reproduce asexually.

Genetic studies with *Saccharomyces* led to the discovery of some of the genes that control the eukaryotic cell cycle. Many of these genes, after their first discovery in yeast cells, were found to have counterparts in animals and plants. Defective versions of the genes often contribute to the development of cancer in mammals. Genetic studies with *Saccharomyces* were also the first to show the genes carried in the DNA of mitochondria and their patterns of inheritance. The complete DNA sequence of *S. cerevisiae,* which includes more than 12 million base pairs that encode about 6,000 genes, was the first eukaryotic genome to be obtained. Analysis of the genome sequence revealed that yeast has many genes related to those in animals, including mammals, making this relatively simple microorganism an excellent subject for research that can be applied to the more complex animals of interest.

The genes of yeast can also be manipulated easily using genetic engineering techniques. This has made it possible for researchers to alter essentially any of the yeast genes experimentally to test their functions and to introduce genes or DNA samples from other organisms for testing or cloning. *Saccharomyces* has been so important to genetic studies in eukaryotes that it is often called the eukaryotic *E. coli.*

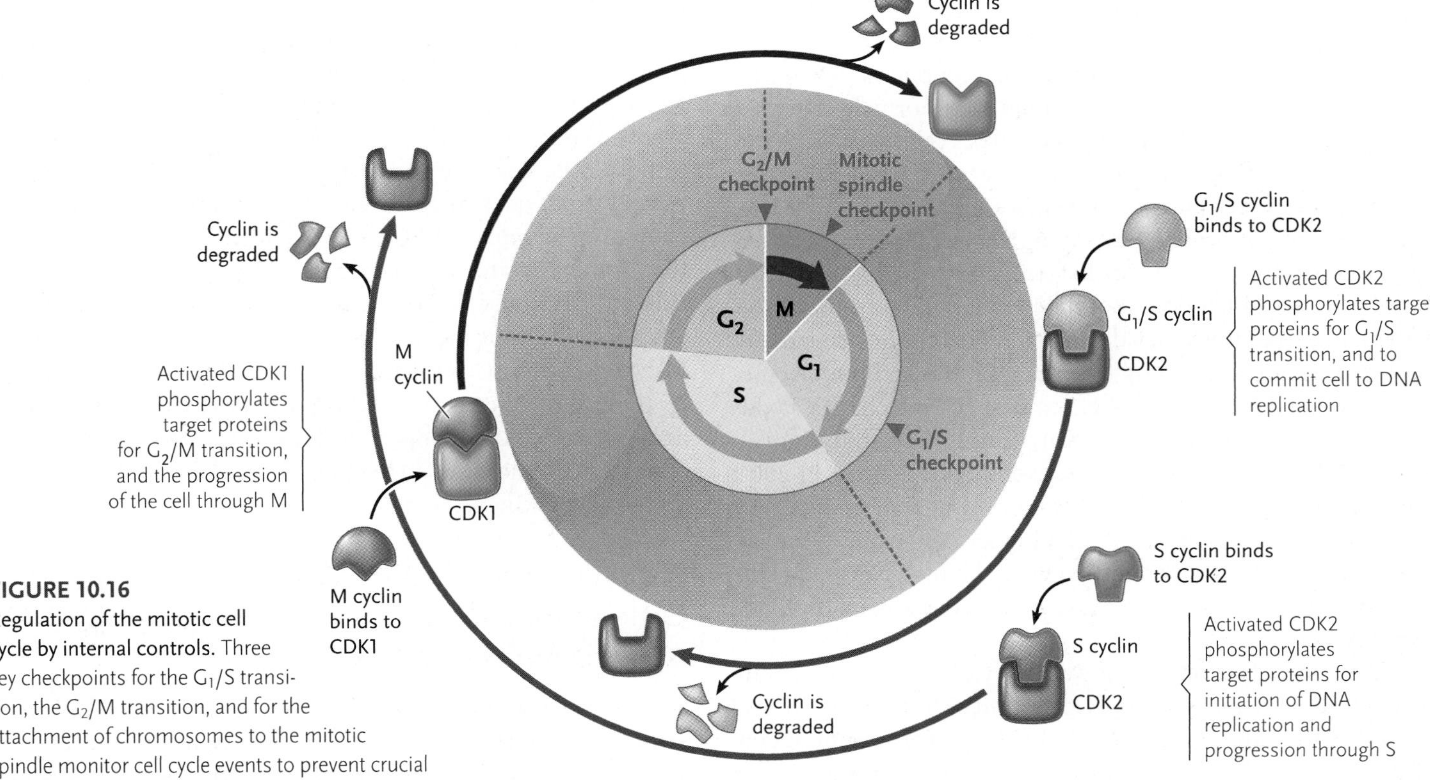
Cyclin is degraded
G_2/M checkpoint
Mitotic spindle checkpoint
G_1/S cyclin binds to CDK2
G_1/S cyclin
CDK2
Activated CDK2 phosphorylates target proteins for G_1/S transition, and to commit cell to DNA replication
Cyclin is degraded
M cyclin
G_2
M
G_1
S
Activated CDK1 phosphorylates target proteins for G_2/M transition, and the progression of the cell through M
CDK1
G_1/S checkpoint
M cyclin binds to CDK1
S cyclin binds to CDK2
S cyclin
CDK2
Activated CDK2 phosphorylates target proteins for initiation of DNA replication and progression through S
Cyclin is degraded

FIGURE 10.16
Regulation of the mitotic cell cycle by internal controls. Three key checkpoints for the G_1/S transition, the G_2/M transition, and for the attachment of chromosomes to the mitotic spindle monitor cell cycle events to prevent crucial phases of the cell cycle from starting until previous phases are completed correctly. Complexes of cyclins and cyclin-dependent kinases (Cdks) regulate the progression of the cell through the cell cycle. The three cyclin–Cdks present in all eukaryotes are shown. The Cdks are present throughout the cell cycle, but they are active only when complexed with a cyclin (shown by the broad arrows in the figure). Each cyclin is synthesized and degraded in a regulated way so that it is present only for a particular phase of the cell cycle. During that phase the Cdk to which it is bound phosphorylates and, thereby, regulates the activity of target proteins in the cell that are involved in initiating or regulating key events of the cell cycle.

point. (Extracellular signals and their effects on the cell cycle are discussed in more detail later.)

The G_2/M checkpoint is at the junction between the G_2 and M phases. Passage through this checkpoint commits a cell to mitosis. Cells arrest at the G_2/M checkpoint if DNA was not replicated accurately in S, or (as for the G_1/M checkpoint) if the DNA has been damaged by radiation or chemicals. Accurate DNA replication is essential for producing genetically identical daughter cells, highlighting the importance of this checkpoint.

The mitotic spindle checkpoint is within the M phase before metaphase. The checkpoint assesses whether chromosomes are attached properly to the mitotic spindle so that they are aligned correctly at the metaphase plate. The checkpoint is essential for production of genetically identical daughter cells, which depends upon separation of daughter chromosomes in anaphase which, in turn, depends upon the correct alignment of the chromosomes on the spindle in metaphase. Once the cell begins anaphase, it is irreversibly committed to completing M, underlining the importance of the mitotic spindle checkpoint.

The control systems that operate at the checkpoints are signals to stop; basically, they are brakes. This becomes evident when a checkpoint is inactivated by mutation or chemical treatment. The consequence of inactivation of the checkpoints is that the cell cycle proceeds, even if DNA is damaged, DNA replication is incomplete, or the spindle did not assemble completely.

Cyclins and Cyclin-Dependent Kinases Are the Internal Controls That Directly Regulate Cell Division

The internal control system that acts at checkpoints serves to ensure that only cells moving through the cycle normally progress through the cycle. The control system causes cells to arrest in the cycle should proceeding through the cycle be harmful. The direct regulation of the cell cycle itself involves an internal control system consisting of proteins called **cyclins,** and enzymes called **cyclin-dependent kinases (Cdks)** (see Figure 10.16). A Cdk is a *protein kinase,* which phosphorylates and thereby regulates the activity of target proteins. Cdk enzymes are "cyclin-dependent" because they are active only when bound to a cyclin molecule. Cyclins are named because their concentrations change as the cell cycle progresses. R. Timothy Hunt, Imperial Cancer Research Fund, London, UK, received a Nobel Prize in 2001 for discovering cyclins. The basic control of the cell cycle by Cdks and cyclins is the same in all eukaryotes, but there are differences in the number and types of the molecules. We will focus on cell cycle regulation in vertebrates to explain how these proteins work.

The concentrations of the various Cdks remain constant throughout the cell cycle, while the concentrations of cyclins change as they are synthesized and degraded at specific stages of the cell cycle. Thus, a specific Cdk becomes active when the cell synthesizes the cyclin that binds to it and remains active until the cyclin is degraded. The cyclin–Cdk complex influences the cell cycle because, while active, Cdk phosphorylates particular target proteins. The phosphorylation regulates the activities of those proteins, which play roles in initiating or regulating key events of the cell cycle, and keep the cycle operating in an orderly way. Those key events are DNA replication, mitosis, and cytokinesis. A succession of cyclin–Cdk complexes, each of which has specific regulatory effects, ensures that these stages follow in sequence somewhat like a clock passing through the sequence of hours. Regulation of the activity of cyclin–Cdk complexes is integrated with the regulatory events at the key cell cycle checkpoints to ensure that daughter cells with damaged DNA or abnormal DNA amount are not generated.

Three classes of cyclins, each named for the stage of the cell cycle at which they bind and activate Cdks, operate in all eukaryotes (see Figure 10.16):

1. G_1/S cyclin binds to Cdk2 near the end of G_1 forming a complex required for the cell to make the transition from G_1 to S, and to commit the cell to DNA replication.
2. S cyclin binds to Cdk2 in the S phase forming a complex required for the initiation of DNA replication and the progression of the cell through S.
3. M cyclin binds to Cdk1 in G_2 forming a complex required for the transition from G_2 and M, and the progression of the cell through mitosis. (Cdk1 is the product of the fission yeast *cdc2* gene described earlier.)

In most cells, a fourth class of cyclins, G_1 cyclin, binds to Cdk4 and Cdk6 before step 1 (the G_1/S transition) to form two cyclin–Cdk complexes. These complexes are needed to move the cell through the G_1 checkpoint stimulating it then to proceed from G_1 to S.

The M cyclin–Cdk1 complex is also called **M phase-promoting factor (MPF).** In addition to initiating mitosis, the M cyclin–Cdk1 complex (MPF) also orchestrates some of its key events. When all chromosomes are correctly attached to the mitotic spindle near the end of metaphase, the M cyclin–Cdk1 complex activates another enzyme complex, the **anaphase-promoting complex (APC).** Activated APC degrades an inhibitor of anaphase, and this leads to the separation of sister chromatids and the onset of daughter chromosome separation in anaphase. Later in anaphase, APC directs the degradation of the M cyclin, causing Cdk1 to lose its activity. The loss of Cdk1 activity then allows the separated chromosomes to become extended again, the nuclear envelope to reform around the two clusters of daughter chromosomes in telophase, and the cytoplasm then to divide in cytokinesis.

External Controls Coordinate the Mitotic Cell Cycle of Individual Cells with the Overall Activities of the Organism

The internal controls that regulate the cell cycle are modified by signal molecules that originate from outside the dividing cells. In animals, these signal molecules include the peptide hormones and similar proteins called *growth factors.*

The hormones and growth factors act on the cell by the reception–transduction–response pattern that applies to cell communication in general (see Chapter 7). The external factors bind to receptors at the cell surface, which respond by triggering reactions inside the cell. These reactions often include steps that add inhibiting or stimulating phosphate groups to the cyclin/Cdk complexes, particularly to the Cdks. The reactions triggered by the activated receptor may also directly affect the same proteins regulated by the cyclin–Cdk complexes. The overall effect is to speed, slow, or stop the progress of cell division, depending on the particular hormone or growth factor and the internal pathway that is stimulated. Some growth factors are even able to break the arrest of cells that have shunted into the G_0 stage and return them to active division.

Cell-surface receptors in animals also recognize contact with other cells or with molecules of the extracellular matrix (see Section 5.5). The contact triggers internal reaction pathways that inhibit division by arresting the cell cycle, usually in the G_1 phase. The response, called **contact inhibition,** stabilizes cell growth in fully developed organs and tissues. As long as the cells of most tissues are in contact with one another or the extracellular matrix, they are shunted into the G_0 phase and prevented from dividing. If the contacts are broken, the freed cells often enter rounds of division.

Contact inhibition is easily observed in cultured mammalian cells grown on a glass or plastic surface. In such cultures, division proceeds until all the cells are in contact with their neighbors in a continuous, unbroken, single layer. At this point, division stops. If a researcher then scrapes some of the cells from the surface, cells at the edges of the "wound" are released from inhibition and divide until they form a continuous layer and all the cells are again in contact with their neighbors.

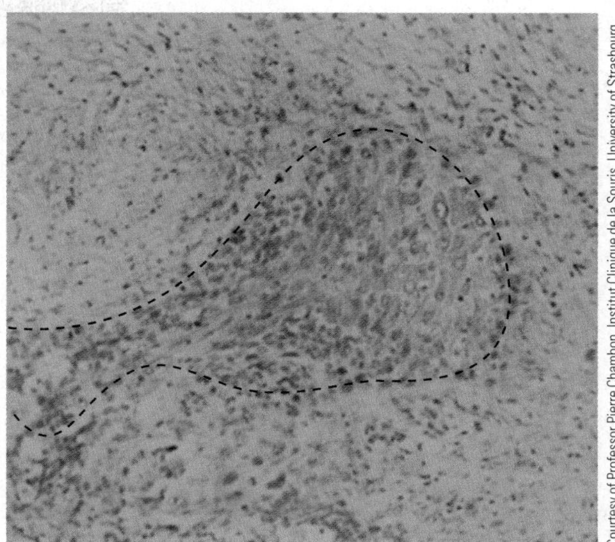

FIGURE 10.17
A mass of tumor cells (dashed line) embedded in normal tissue.

Cell Cycle Controls Are Lost in Cancer

Cancer occurs when cells lose the normal controls that determine when and how often they will divide. Cancer cells divide continuously and uncontrollably, producing a rapidly growing mass called a *tumor* **(Figure 10.17).** Cancer cells also typically lose their adhesions to other cells and often become actively mobile. As a result, in a process called *metastasis,* they break loose from an original tumor, spread throughout the body, and grow into new tumors in other body regions. Metastasis is promoted by changes that block contact inhibition and alter the cell-surface molecules that link cells together or to the extracellular matrix.

Growing tumors damage surrounding normal tissues by compressing them and interfering with blood supply and nerve function. Tumors may also break through barriers such as the outer skin, internal cell layers, or the gut wall. The breakthroughs cause bleeding, open the body to infection by microorganisms, and destroy the separation of body compartments necessary for normal functioning. Both compression and breakthroughs can cause pain that, in advanced cases, may become extreme. As tumors increase in mass, the actively growing and dividing cancer cells may deprive normal cells of their required nutrients, leading to generally impaired body functions, muscular weakness, fatigue, and weight loss.

Cancer cells typically have a number of mutated genes of different types. The altered functions of those mutated genes in some way promote uncontrolled cell division or metastasis. In their normal (non-mutated) form, many of these genes code for components of the cyclin/Cdk system that regulates cell division; others encode proteins that regulate gene expression, form cell surface receptors, or make up elements of the systems controlled by the receptors. The mutated form of the genes, called **oncogenes** (*oncos* = bulk or mass), encode altered versions of these products.

For example, a mutation in a gene that codes for a surface receptor might result in a protein that is constantly active, even without binding an extracellular signal molecule. As a result, the internal reaction pathways triggered by the receptor, which induce cell division, are continually stimulated. Another mutation, this time in a cyclin gene, could decrease the cyclin–Cdk binding that triggers DNA replication and the rest of the cell cycle. (*Insights from the Molecular Revolution* describes an experiment testing the effects of a viral system that induces cancer by overriding normal controls of the cyclin/CDK system.) Cancer, oncogenes, and the alterations that convert normal genes to oncogenes are discussed in further detail in Chapter 16.

The overview of the mitotic cell cycle and its regulation presented in this chapter only hints at the complexity of cell growth and division. The likelihood of any given cell dividing is determined by the interplay of various internal signals in the context of external cues from the environment. If a cell is destined to divide, then the problem of accurately replicating and partitioning its DNA requires a highly regulated, intricately interrelated series of mechanisms. It is a challenging operation even for male Australian Jack Jumper ants *(Myrmecia pilosula),* which have only two chromosomes ($2n = 2$); and think of the challenges faced by the fern species, *Ophioglossum pycnostichum,* which has 1,260 chromosomes in each cell!

INSIGHTS FROM THE **Molecular Revolution**

Herpesviruses and Uncontrolled Cell Division: How does herpesvirus 8 transform normal cells into cancer cells?

Almost all of us harbor one or more herpesviruses as more or less permanent residents in our cells. Fortunately, most of the herpesviruses are relatively benign—one group is responsible for the bothersome but nonlethal oral and genital ulcers known commonly as cold sores or "herpes." But another virus, *herpesvirus 8* **(Figure 1),** is a DNA tumor virus that causes two kinds of cancer: Kaposi's sarcoma and lymphomas of the body cavity. Kaposi's sarcoma is one of the most common cancers that develops in AIDS patients, and is the fourth most common cancer caused by infection worldwide.

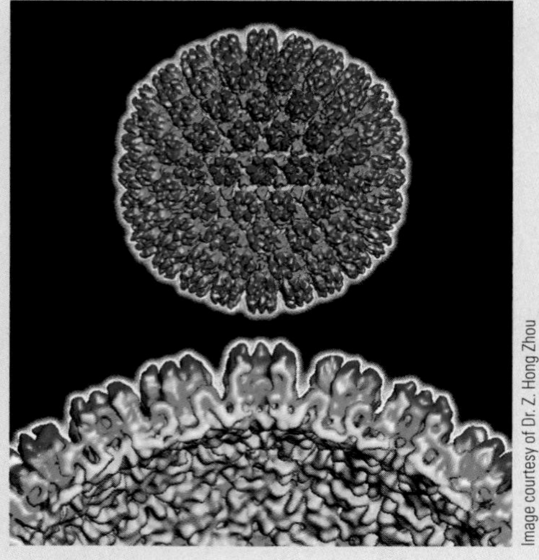
Image courtesy of Dr. Z. Hong Zhou

FIGURE 1
Human herpesvirus 8

Cancers are characterized by uncontrolled cell division. In normal cells, G_1 cyclin combined with either Cdk4 or Cdk6 contributes to the G_1/S transition and thus stimulates cell division. One way that normal cells control cell division is to use regulatory proteins, such as p16, p21, and p27, that bind to and inhibit G_1 cyclin–Cdk complexes, thereby preventing the cells from becoming transformed into cancer cells. (*p* stands for protein; the number indicates the molecular weight in thousands of Daltons). The proteins that have this regulatory ability are called *tumor suppressor proteins.*

Research Question

How does herpesvirus 8 cause the uncontrolled cell division characteristic of malignant tumors?

Experiments

To answer this question, investigators in London and at the Friedrich-Alexander University in Germany examined the effects of herpesvirus 8 on the primary transition point that leads to cell division, the change from G_1 to S. The investigators focused on how the virus might interfere with regulatory mechanisms that control the rate of cell division.

The investigators knew that the DNA of herpesvirus 8 encodes a protein that acts as a cyclin. Could this viral cyclin, *K cyclin,* be the means by which the herpesvirus bypasses normal controls and triggers the rapid cell division characteristic of cancer? To answer this question, researchers first inserted the DNA coding for K cyclin into a benign virus. When they infected cultured human cells with this virus, the virus produced K cyclin, which bound to the human Cdk6. These K cyclin–Cdk6 complexes stimulated the initiation of the S phase much faster than do the cell's normal G_1 cyclin–Cdk6 complexes. In addition, the tumor suppressor proteins that normally regulate cell division by binding to G_1 cyclin–Cdk6 complexes were unable to bind to the K cyclin–Cdk6 complexes, resulting in uncontrolled division.

Conclusion

Herpesvirus 8 has evolved a mechanism that overrides normal cellular controls and triggers cell division. At some point in its evolution, the virus may have picked up a copy of a cyclin gene, which through mutation and selection became the K cyclin that is unaffected by the inhibitors.

Source: C. Swanton et al. Herpes viral cyclin/Cdk6 complexes evade inhibition by CDK inhibitor proteins. 1997. *Nature* 390:184–187.

STUDY BREAK 10.4

1. **Why is a Cdk not active throughout the entire cell cycle?**
2. **How do cyclin–Cdk complexes typically trigger transitions in the cell cycle?**
3. **What is an oncogene? How might an oncogene affect the cell cycle?**
4. **What is metastasis?**

THINK OUTSIDE THE BOOK

Collaboratively or on your own, summarize an experiment demonstrating that higher-than-normal levels of cyclin E (a G_1/S cyclin) can initiate breast cancer in humans.

10.5 Cell Division in Prokaryotes

Prokaryotes undergo a cycle of cytoplasmic growth, DNA replication, and cell division, producing two daughter cells from an original parent cell. The entire mechanism of prokaryotic cell division is called **binary fission**—that is, splitting or dividing into two parts. Although binary fission is regulated, the small size of prokaryotic cells makes it particularly difficult to discover just how the chromosomes move. Although actin-like proteins are present in bacteria, their role in chromosome segregation remains unclear.

Replication Occupies Most of the Cell Cycle in Rapidly Dividing Prokaryotic Cells

All prokaryotes use DNA as their genetic material. The vast majority of prokaryotes have a single, circular DNA molecule known as the **prokaryotic chromosome,** more specifically the **bacterial chromosome** for bacteria, and the **archaeal chromosome** for archaeans. When prokaryotic cells divide at the maximum rate, DNA replication occupies most of the period between cytoplasmic divisions. As soon as replication is complete, the cytoplasm divides to complete the cell cycle. For example, in *E. coli* cells, which are capable of dividing every 20 minutes, DNA replication occupies 19 minutes of the entire 20-minute division cycle.

Replicated Chromosomes Are Distributed Actively to the Halves of the Prokaryotic Cell

In the 1960s, François Jacob of The Pasteur Institute, Paris, France, proposed a model for the segregation of bacterial chromosomes to the daughter cells in which the two chromosomes attach to the plasma membrane near the middle of the cell and separate as a new plasma membrane is added between the two sites during cell elongation. The essence of this model is that chromosome separation is passive. However, current research indicates that bacterial chromosomes rapidly separate in an active way that is linked to DNA replication events and that is independent of cell elongation. The new model is shown in **Figure 10.18.**

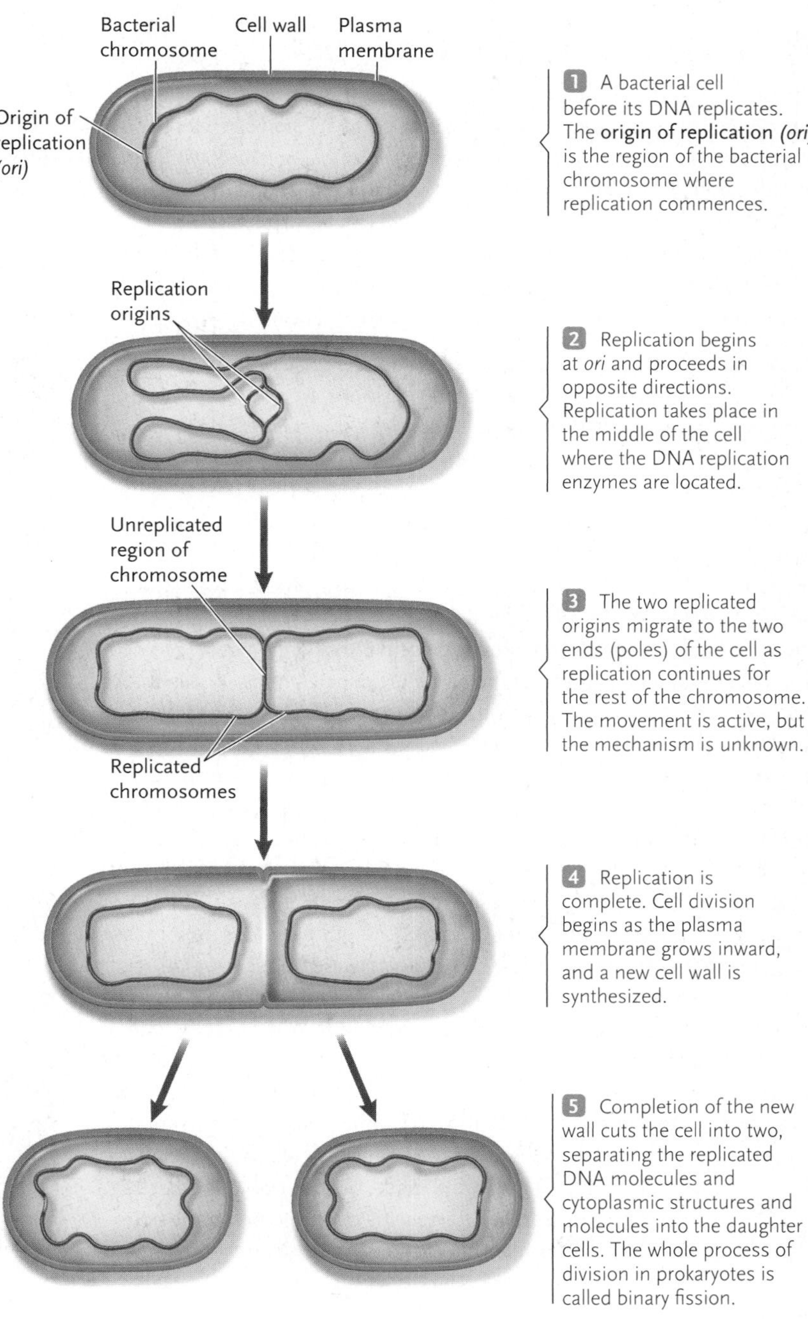
Bacterial chromosome
Cell wall
Plasma membrane
Origin of replication (ori)
1 A bacterial cell before its DNA replicates. The origin of replication (ori) is the region of the bacterial chromosome where replication commences.
Replication origins
2 Replication begins at ori and proceeds in opposite directions. Replication takes place in the middle of the cell where the DNA replication enzymes are located.
Unreplicated region of chromosome
Replicated chromosomes
3 The two replicated origins migrate to the two ends (poles) of the cell as replication continues for the rest of the chromosome. The movement is active, but the mechanism is unknown.
4 Replication is complete. Cell division begins as the plasma membrane grows inward, and a new cell wall is synthesized.
5 Completion of the new wall cuts the cell into two, separating the replicated DNA molecules and cytoplasmic structures and molecules into the daughter cells. The whole process of division in prokaryotes is called binary fission.

FIGURE 10.18
Model for the segregation of replicated bacterial chromosomes to daughter cells.

Mitosis Evolved from Binary Fission

The prokaryotic mechanism works effectively because most prokaryotic cells have only a single chromosome. Thus, if a daughter cell receives at least one copy of the chromosome, its genetic information is complete. By contrast, in most cases the genetic information of eukaryotes is divided among several chromosomes, with each chromosome containing a much greater length of DNA than a bacterial chromosome. If a daughter cell fails to receive a copy of even one chromosome, the effects may be lethal. The evolution of mitosis solved the mechanical problems associated with distributing long DNA molecules without breakage. Mitosis provided the level of precision required to ensure that each daughter cell receives a complete complement of the chromosomes the parent cell had.

Scientists believe that the ancestral division process was binary fission and that mitosis evolved from that process. Variations in the mitotic apparatus in modern-day organisms illuminate possible intermediates in this evolutionary pathway. For example, in many primitive eukaryotes, such as dinoflagellates (a type of single-celled protist), the nuclear envelope remains intact during mitosis, and the chromosomes bind to the inner membrane of the nuclear membrane. When the nucleus divides, the chromosomes are segregated.

A more advanced form of the mitotic apparatus is seen in yeasts and diatoms (diatoms are another type of single-celled protist). In these organisms, the mitotic spindle forms and chromosomes segregate to daughter nuclei without the disassembly and reassembly of the nuclear envelope. Currently, scientists think that the types of mitosis seen in yeasts and diatoms, as well as in animals and higher plants, evolved separately from a common ancestral type.

Mitotic cell division, the subject of this chapter, produces two cells that have the same genetic information as the parental cell entering division. In the next chapter, you will learn about meiosis, a specialized form of cell division that produces gametes, which have half the number of chromosomes as that present in diploid cells.

STUDY BREAK 10.5 <

1. **How do prokaryotes divide?**
2. **What processes involved in eukaryotic cell division are absent from prokaryotic cell division?**

UNANSWERED QUESTIONS

Disrupted or defective control of cell growth and division can lead to diseases such as cancer. Complex, interacting molecular networks within the cell fine tune the division of each cell in both unicellular and multicellular organisms. Identifying the genes and proteins involved in these networks is crucial both for a complete understanding of cell growth and division, and for developing models for diseases caused by cell cycle defects. Many researchers worldwide are working in this area of research.

How are transitions between phases of the cell cycle regulated?

Research in many labs has shown that transitions are important control points for progression through the cell cycle. If a cell in G_1 phase has damaged DNA, for instance, the cell pauses to repair the DNA before entering S phase, to ensure that any mutations are not passed on to progeny cells. My lab at Caltech is interested in how cells execute two transitions in the cell division program: the transition from G_1 phase to S phase and the transition from mitosis to G_1 phase. Most of our studies are focused on the budding yeast *Saccharomyces cerevisiae,* which is used to bake bread and brew beer, because this organism is easy to manipulate and uses many of the same proteins as human cells do to carry out cell division. The G_1-to-S and mitosis-to-G_1 transitions involve turnover—breakdown—of proteins that serve to maintain cells in the pre-transition state. In the G_1 phase, the cyclin-dependent kinase (CDK) inhibitor Sic1 blocks the activity of the S phase cyclin–CDK complexes that promote DNA replication, and thereby delays the onset of S phase. Enzymes rapidly eliminate Sic1 at the end of the G_1 phase, thereby unmasking the activity of the S phase cyclin–CDK, which causes the rapid initiation of chromosome duplication. Although we now know much about how this transition works, a key unanswered question that is of interest to my laboratory is how do the enzymes that eliminate Sic1 do their job? For example, we are working hard to unravel the precise mechanism of action of the ubiquitin ligase enzyme that attaches ubiquitin to Sic1, which serves as a signal that activates Sic1 turnover. (Ubiquitin is a small protein added to proteins to designate those proteins for destruction; see Chapter 16. Ubiquitination of a particular protein is catalyzed by one of a family of ubiquitin ligases in the cell.) Another key question is how the timing of Sic1 turnover is controlled. It may be possible to address this question at the level of single cells using the time-lapse methods being developed in the laboratory of Dr. Fred Cross at Rockefeller University.

Similar to the G_1-to-S transition, the exit from mitosis and return to G_1 is also governed by the controlled elimination of a specific protein. In this case, the cell must degrade the mitotic cyclin that holds cells in mitosis by keeping mitotic cyclin–CDK complexes in an active state. Degradation of the mitotic cyclin is mediated by the ubiquitin ligase known as Anaphase-Promoting Complex/Cyclosome (APC/C), which is switched on immediately prior to the elimination of the cyclin. We know that activation of APC/C in yeast requires the protein phosphatase Cdc14, which is normally stored in the nucleolus, separated from APC/C. (Recall from Section 4.5 that a phosphatase is an enzyme that removes a phosphate from a substrate. Here, a regulatory phosphate that inhibits function of a protein is removed by the enzyme.) Late in mitosis, Cdc14 is abruptly released from the nucleolus to activate APC/C and turnover of mitotic cyclin. A key unanswered question is what is the underlying molecular mechanism that governs the appropriately timed release of Cdc14 from the nucleolus?

From our studies on both the G_1-to-S and mitosis-to-G_1 transitions, it is apparent that many of the key players have already been discovered. However, we still do not know how all of the proteins are organized and communicate with each other to bring about the transition in the cell division program at the right time, and we also do not know how the proteins operate as nano-machines to carry out a specific biological task.

Think Critically

Why would researchers choose *Saccharomyces cerevisiae* as a model organism to address mitotic cell cycle defects over *Escherichia coli* or mammalian cells? How similar would you expect *Saccharomyces* ubiquitin to be to human ubiquitin?

Raymond Deshaies

Raymond Deshaies is professor of biology at CalTech and an investigator of Howard Hughes Medical Institute. The focus of his lab is investigation of the cellular machinery that mediates protein degradation by the ubiquitin-proteasome system, and how this machinery regulates cell division. Learn more about his work at: http://www.its.caltech.edu/~rjdlab.

REVIEW KEY CONCEPTS

Go to **CENGAGENOW** at www.cengage.com/login to access quizzing, animations, exercises, articles, and personalized homework help.

10.1 The Cycle of Cell Growth and Division: An Overview

- In mitotic cell division, DNA replication is followed by the equal separation—that is, segregation—of the replicated DNA molecules and their delivery to daughter cells. The process ensures that the two cell products of a division end up with the same genetic information as the parent cell entering division.
- Mitosis is the basis for growth and maintenance of body mass in multicelled eukaryotes, and for the reproduction of many single-celled eukaryotes.
- The DNA of eukaryotic cells is divided among individual, linear chromosomes located in the cell nucleus.
- DNA replication and duplication of chromosomal proteins converts each chromosome into two exact copies known as sister chromatids.

10.2 The Mitotic Cell Cycle

- Mitosis and interphase constitute the mitotic cell cycle. Mitosis occurs in five stages. In prophase (stage 1), the chromosomes condense into short rods and the spindle forms in the cytoplasm (Figures 10.3 and 10.4).
- In prometaphase (stage 2), the nuclear envelope breaks down, the spindle enters the former nuclear area, and the sister chromatids of each chromosome make connections to opposite spindle poles.

Each chromatid has a kinetochore that attaches to spindle microtubules (Figures 10.3, 10.4, and 10.6).

- In metaphase (stage 3), the spindle is fully formed and the chromosomes, moved by the spindle microtubules, become aligned at the metaphase plate (Figures 10.3 and 10.4).
- In anaphase (stage 4), the spindle separates the sister chromatids and moves them to opposite spindle poles. At this point, chromosome segregation is complete (Figures 10.3 and 10.4).
- In telophase (stage 5), the chromosomes decondense and return to the extended state typical of interphase and a new nuclear envelope forms around the chromosomes (Figures 10.3 and 10.4).
- Cytokinesis, the division of the cytoplasm, completes cell division by producing two daughter cells, each containing a daughter nucleus produced by mitosis (Figures 10.3 and 10.4).
- Cytokinesis in animal cells proceeds by furrowing, in which a band of microfilaments just under the plasma membrane contracts, gradually separating the cytoplasm into two parts (Figure 10.8).
- In plant cytokinesis, cell wall material is deposited along the plane of the former spindle midpoint; the deposition continues until a continuous new wall, the cell plate, separates the daughter cells (Figure 10.9).

Animation: The cell cycle

Animation: Mitosis step-by-step

Animation: Cytoplasmic division

10.3 Formation and Action of the Mitotic Spindle

- In animal cells, the centrosome divides and the two parts move apart. As they do so, the microtubules of the spindle form between them. In plant cells, which do not have a centrosome, the spindle microtubules assemble around the nucleus (Figure 10.10).
- In the spindle, kinetochore microtubules run from the poles to the kinetochores of the chromosomes, and nonkinetochore microtubules run from the poles to a zone of overlap at the spindle midpoint without connecting to the chromosomes (Figure 10.12).
- During anaphase, the kinetochores move along the kinetochore microtubules, pulling the chromosomes to the poles. The nonkinetochore microtubules slide over each other, pushing the poles farther apart (Figures 10.12 and 10.13).

Animation: Mechanisms for chromosome movement

10.4 Cell Cycle Regulation

- The internal controls that monitor progression through the cell cycle include checkpoints at key points to ensure that critical phases do not commence before previous phases are completed correctly (Figure 10.16).
- The internal control system that directly regulates cell division involves complexes of a cyclin and a cyclin-dependent protein kinase (Cdk). Cdk is activated when combined with a cyclin and then phosphorylates target proteins, regulating their activities. The altered target proteins then initiate or regulate key events of the cell cycle. Four classes of cyclins—G_1, G_1/S, S, and M—are distinguished by the stage of the cell cycle at which they activate Cdks (Figure 10.16).
- External controls are based primarily on surface receptors that recognize and bind signals such as peptide hormones and growth factors, surface groups on other cells, or molecules of the extracellular matrix. The binding triggers internal reactions that speed, slow, or stop cell division.
- In cancer, control of cell division is lost, and cells divide continuously and uncontrollably, forming a rapidly growing mass of cells that interferes with body functions. Cancer cells can also break loose from their original tumor (metastasize) to form additional tumors in other parts of the body.

Animation: Cancer and metastasis

10.5 Cell Division in Prokaryotes

- Replication begins at the origin of replication of the bacterial chromosome in reactions catalyzed by enzymes located in the middle of the cell. Once the origin of replication is duplicated, the two origins migrate to the two ends of the cells. Division of the cytoplasm then occurs through a partition of cell wall material that grows inward until the cell is separated into two parts (Figure 10.17).

Animation: Prokaryotic fission

UNDERSTAND AND APPLY

Test Your Knowledge

1. During the cell cycle, the DNA mass of a cell:
   a. decreases during G_1.
   b. decreases during metaphase.
   c. increases during the S phase.
   d. increases during G_2.
   e. decreases during interphase.
2. A tumor suppressor protein, p21, inhibits Cdk1. The earliest effect of p21 on the cell cycle would be to stop the cell cycle at:
   a. early G_1.
   b. late G_1.
   c. the S phase.
   d. G_2.
   e. the mitotic prophase.
3. A major difference between hereditary information in eukaryotes and prokaryotes is:
   a. in prokaryotes, the hereditary information is distributed among individual, linear DNA molecules in the nucleus.
   b. in eukaryotes, the hereditary information is encoded in a single, circular DNA molecule.
   c. in prokaryotes, the hereditary information is usually distributed among multiple circular DNA molecules in the cytoplasm.
   d. in eukaryotes, the hereditary information is distributed among individual, linear DNA molecules in the cytoplasm.
   e. in eukaryotes, the hereditary information is distributed among individual, linear DNA molecules in the nucleus.
4. The major microtubule organizing center of the animal cell is:
   a. chromosomes, composed of chromatids.
   b. the centrosome, composed of centrioles.
   c. the chromatin, composed of chromatids.
   d. chromosomes, composed of centromere.
   e. centrioles, composed of centrosome.
5. The chromatids separate into chromosomes:
   a. during prophase.
   b. going from prophase to metaphase.
   c. going from anaphase to telophase.
   d. going from metaphase to anaphase.
   e. going from telophase to interphase.

6. Which of the following statements about mitosis is *incorrect*?
   a. Microtubules from the spindle poles attach to the kinetochores on the chromosomes.
   b. In anaphase, the spindle separates sister chromatids and pulls them apart.
   c. In metaphase, spindle microtubules align the chromosomes at the spindle midpoint.
   d. Cytokinesis describes the movement of chromosomes.
   e. Both the animal cell furrow and the plant cell plate form at their former spindle midpoint.
7. Mitomycin C is an anticancer drug that stops cell division by inserting itself into the strands of DNA and binding them together. This action is thought to have its major effect at:
   a. late G_1, early S phases.
   b. late G_2.
   c. prophase.
   d. metaphase.
   e. anaphase.
8. Which of the following statements about cell cycle regulators is *incorrect*?
   a. The concentrations of cyclins change throughout the cell cycle.
   b. Cyclins are present in all stages of the cell cycle except S.
   c. Cyclin–Cdk complexes phosphorylate target proteins.
   d. Cdks combine with cyclin to move the cycle into mitosis.
   e. During anaphase of mitosis, cyclin is degraded, allowing mitosis to end.
9. All of the following are characteristic of cancer cells *except*:
   a. less cytoplasmic volume than normal cells.
   b. an absence of cyclin.
   c. loss of adhesion to other cells.
   d. loss of control of cell division.
   e. loss of normal control of G_1/S phase transition.
10. In bacteria:
   a. several chromosomes undergo mitosis.
   b. binary fission produces four daughter cells.
   c. replication begins at the origin *(ori)*, and the DNA strands separate.
   d. the plasma membrane plays an important role in separating the duplicated chromosomes into the two daughter cells.
   e. the daughter cells receive different genetic information from the parent cell.

Discuss the Concepts

1. You have a means of measuring the amount of DNA in a single cell. You first measure the amount of DNA during G_1. At what points during the remainder of the cell cycle would you expect the amount of DNA per cell to change?
2. A cell has 38 chromosomes. After mitosis and cell division, one daughter cell has 39 chromosomes and the other has 37. What might have caused these abnormal chromosome numbers? What effects do you suppose this might have on cell function? Why?
3. Paclitaxel (Taxol), a substance isolated from Pacific yew *(Taxus brevifolia)*, is effective in the treatment of breast and ovarian cancers. It works by stabilizing microtubules, thereby preventing them from disassembling. Why would this activity slow or stop the growth of cancer cells?
4. A cell has 24 chromosomes at G_1 of interphase. How many chromosomes would you expect it to have at G_2 of interphase? At metaphase of mitosis? At telophase of mitosis?

Design an Experiment

Many chemicals in the food we eat have the potential to affect the growth of cancer cells. Chocolate, for example, contains a number of flavonoid compounds, which act as natural antioxidants. Design an experiment to determine whether any of the flavonoids in chocolate inhibit the cell cycle of breast cancer cells that are growing in culture.

Interpret the Data

Biologists have long been interested in the effects of radiation on cells. In one experiment, researchers examined the effect of radium on mitosis of chick embryo cells growing in culture. A population of experimental cells was examined under the microscope for the number of cells in telophase (as a measure of mitosis occurring) before, during, and after exposure to radium. The results are shown in the following figure.

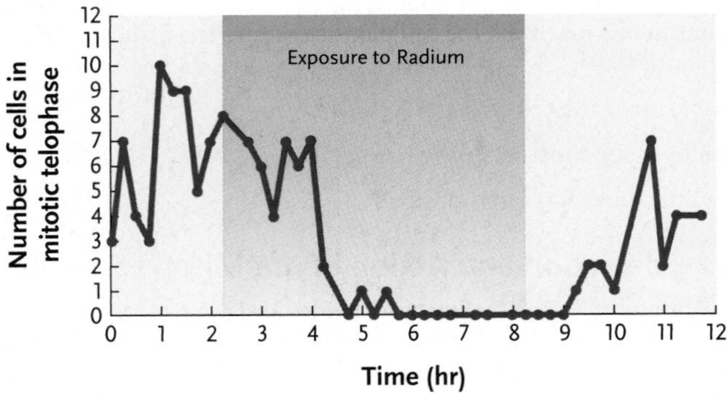
Exposure to Radium
Number of cells in mitotic telophase
0
1
2
3
4
5
6
7
8
9
10
11
12
Time (hr)
0
1
2
3
4
5
6
7
8
9
10
11
12

1. What is the effect of radium exposure on mitosis?
2. Was the effect of radium exposure permanent?

Source: R. G. Canti and M. Donaldson. 1926. The effect of radium on mitosis *in vitro. Proceedings of the Royal Society of London, Series B, Containing Papers of a Biological Character* 100:413–419.

Apply Evolutionary Thinking

The genes and proteins involved in cell cycle regulation in prokaryotes and eukaryotes are very different. However, both types of organisms use similar molecular regulatory reactions to coordinate DNA synthesis with cell division. What does this observation mean from an evolutionary perspective?

Express Your Opinion

It is illegal to sell your organs, but you can sell your cells, including eggs, sperm, and blood cells. HeLa cells are still being sold all over the world by cell culture firms. HeLa cells derive from tumor cells from a patient, Henrietta Lacks, who died from cancer in the early 1950s. The cells were then sold without the family of Henrietta Lacks knowing about it. Should the family of Henrietta Lacks share in the profits? Go to academic .cengage.com/login to investigate both sides of the issue and then vote.

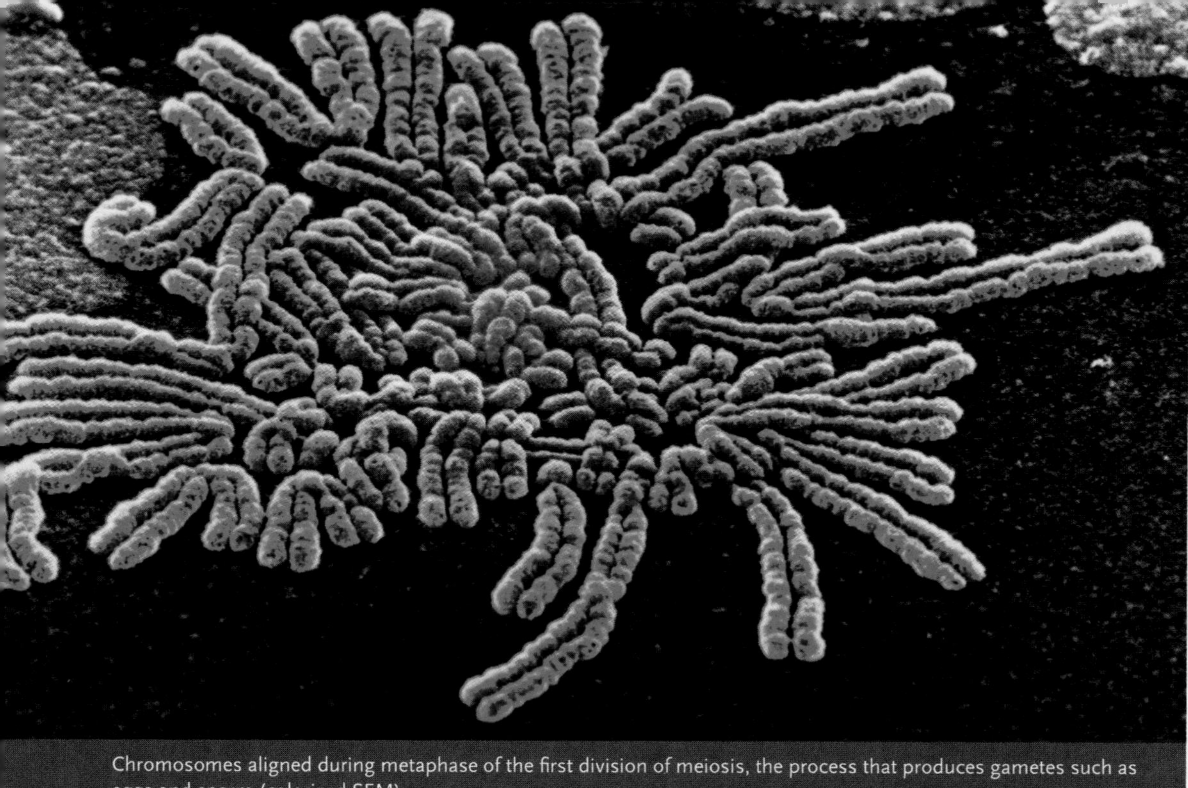

Chromosomes aligned during metaphase of the first division of meiosis, the process that produces gametes such as eggs and sperm (colorized SEM).

Adrian T. Sumner/Science Photo Library/Photo Researchers, Inc.

11

STUDY OUTLINE

Meiosis: The Cellular Basis of Sexual Reproduction

Why It Matters. . . A couple clearly shows mutual interest. First, he caresses her with one arm, then another—then another, another, and another. She reciprocates. This interaction goes on for hours; a hug here, a squeeze there. At the climactic moment, the male reaches deftly under his mantle and removes a packet of sperm, which he inserts under the mantle of the female. For every one of his sperm that successfully performs its function, a fertilized egg can develop into a new octopus.

For the octopus, sex is an occasional event, preceded by a courtship ritual that involves intermingled tentacles. For another marine animal, the slipper limpet, sex is a lifelong group activity. Slipper limpets are relatives of snails. Like many other animals, a slipper limpet passes through a free-living immature stage before it becomes a sexually mature adult. When the time comes for an immature limpet to transform into an adult, it settles onto a rock or other firm surface. If the limpet settles by itself, it develops into a female. If instead it settles on top of a female, it develops into a male. If another slipper limpet settles down on that male, it, too, becomes a male. Adult slipper limpets almost always live in such piles, with the one on the bottom always being a female. All the male limpets continually contribute sperm that fertilize eggs shed by the female. If the one female dies, the surviving male at the bottom of the pile changes into a female and reproduction continues.

These octopuses and slipper limpets are engaged in forms of **sexual reproduction,** the production of offspring through union of male and female **gametes**—for example, eggs and sperm cells in animals. Sexual reproduction depends on **meiosis,** a specialized process of cell division that, in animals, produces gametes. Meiosis reduces the number of chromosomes, producing gametes with half the number of chromosomes present in the **somatic cells** (body cells) of a species. The derivation of the word *meiosis* (*meioun* = to diminish) reflects this reduction. At **fertilization,** the nuclei of an egg and sperm cell fuse, producing a cell called the **zygote,** in which the chromosome number typical of the species is restored. Without the halving of chromosome number by the meiotic divisions, fertilization would double the number of chromosomes in each subsequent generation.

Both meiosis and fertilization also mix genetic information into new combinations; thus, none of the offspring of a mating pair is likely to be genetically identical. By contrast, asexual reproduction generates genetically identical offspring because they are the products of mitotic divisions (asexual reproduction is discussed in Chapter 10). Sexual reproduction generates the variability that is the basis of most inherited differences among individual sexually reproducing organisms. This variability is a source of raw material for the process of evolution.

The halving of the chromosome number and mixing of genetic information into new combinations—both by meiosis—and the restoration of the chromosome number by fertilization, are the biological foundations of sexual reproduction. Intermingled tentacles in octopuses, communal sex among limpets, and the courting and mating rituals of humans are nothing more or less than variations of the means for accomplishing fertilization. <

11.1 The Mechanisms of Meiosis

Meiosis occurs only in eukaryotes that reproduce sexually and only in organisms that are at least diploid—that is, organisms that have at least two representatives of each chromosome. In humans, and other animals, meiosis takes place in the primary reproductive organs, the **gonads.** Meiosis in mature gonads of the male, the **testes,** produces **spermatozoa (sperm),** the gametes of the male. Meiosis in mature gonads of the female, the **ovaries,** produces **ova (eggs),** the gametes of the female. The cellular mechanisms of gamete formation—**gametogenesis**—are described in Chapter 47.

Meiosis Is Based on the Interactions and Distribution of Homologous Chromosome Pairs

To follow the steps of meiosis, you must understand clearly the significance of the chromosome pairs in diploid organisms. As discussed in Section 10.1, the two representatives of each chromosome in a diploid cell constitute a *homologous pair*—they have the same genes, arranged in the same order in the DNA of the chromosomes. One chromosome of each homologous pair, the **paternal chromosome,** comes from the male parent of the organism, and the other chromosome, the **maternal chromosome,** comes from its female parent.

Although the two chromosomes of a homologous pair contain the same genes, arranged in the same order, the versions of each gene, called **alleles,** present in the members of the pair may be the same or different. For a gene that encodes a protein, the different alleles might encode distinct versions of the same protein, which have different structures, molecular properties, or both, or perhaps an allele will not encode a protein at all.

For example, humans normally have 46 chromosomes in their diploid cells, which make up 23 homologous pairs (see Figure 10.7). One chromosome in each pair comes from the mother and the other from the father. However, each individual (except for identical twins, identical triplets, and so forth) has a unique combination of the alleles in the two chromosomes of each homologous pair. The distinct set of alleles, arising from the mixing mechanisms of meiosis and fertilization, gives each individual his or her unique combination of inherited traits, including such attributes as height, hair and eye color, susceptibility to certain diseases, and even aspects of personality and intelligence.

Meiosis separates the homologous pairs, thereby reducing the diploid or $2n$ number of chromosomes to the **haploid** or n number **(Figure 11.1).** Each gamete produced by meiosis receives only one member of each homologous pair. For example, a human egg produced in an ovary or sperm cell produced in a testis contains 23 chromosomes, one of each pair. When the egg and sperm combine in sexual reproduction to produce the

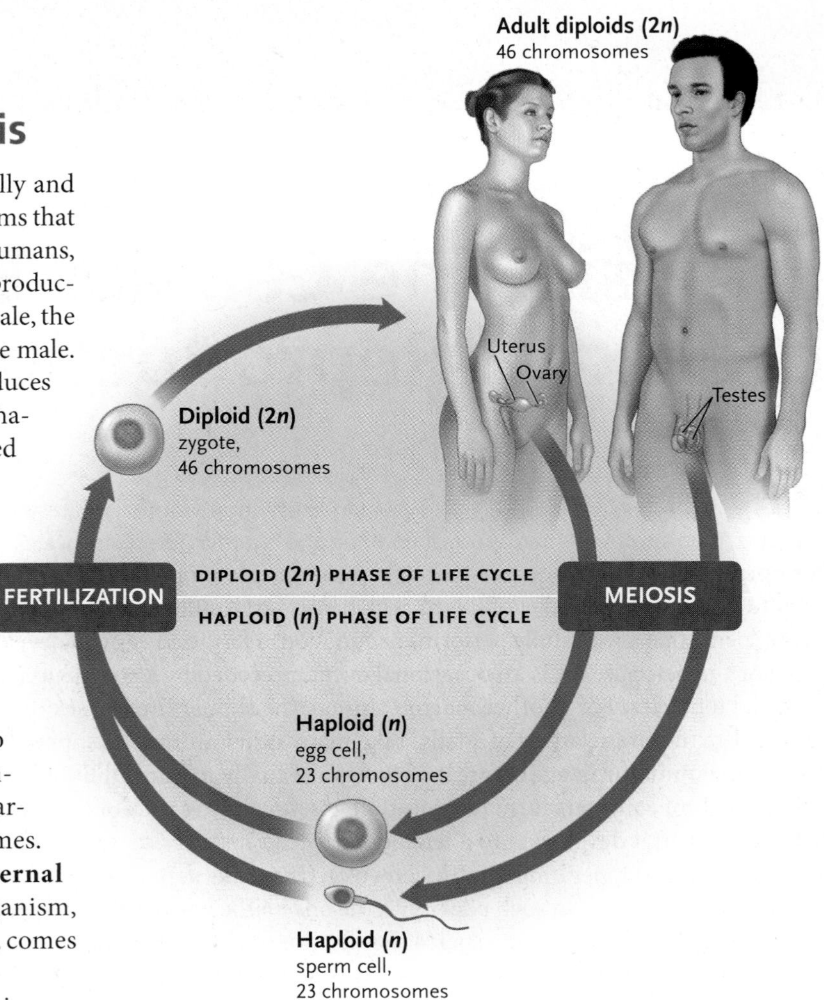
Adult diploids (2n)
46 chromosomes
Uterus
Ovary
Testes
Diploid (2n)
zygote,
46 chromosomes
FERTILIZATION
DIPLOID (2n) PHASE OF LIFE CYCLE
HAPLOID (n) PHASE OF LIFE CYCLE
MEIOSIS
Haploid (n)
egg cell,
23 chromosomes
Haploid (n)
sperm cell,
23 chromosomes

FIGURE 11.1
The cycle of meiosis and fertilization in animals, with humans as an example. Meiosis in animals produces gametes, spermatozoa (sperm) in the testes of the male, and ova (eggs) in the ovaries of the female. Meiosis reduces the chromosome number from the diploid level of two representatives of each chromosome to the haploid level of one representative of each chromosome. Fertilization restores the chromosome number to the diploid level.

FIGURE 11.2
Production of four haploid nuclei by the two meiotic divisions. (For simplicity, just one pair of homologous chromosomes is followed through the divisions.)

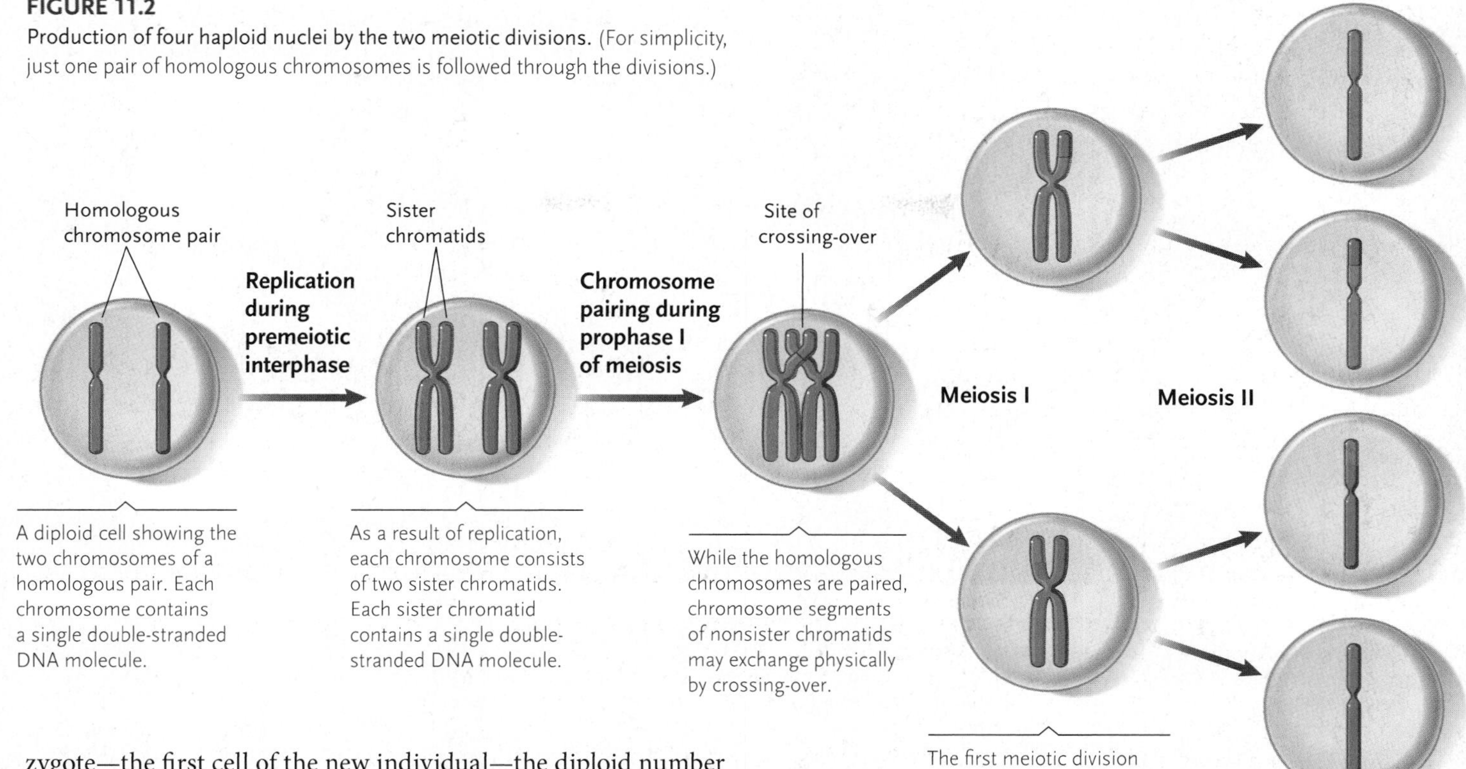
Homologous chromosome pair
Replication during premeiotic interphase
Sister chromatids
Chromosome pairing during prophase I of meiosis
Site of crossing-over
Meiosis I
Meiosis II
A diploid cell showing the two chromosomes of a homologous pair. Each chromosome contains a single double-stranded DNA molecule.
As a result of replication, each chromosome consists of two sister chromatids. Each sister chromatid contains a single double-stranded DNA molecule.
While the homologous chromosomes are paired, chromosome segments of nonsister chromatids may exchange physically by crossing-over.
The first meiotic division separates the homologs, placing one in each of the two cells resulting from the division. These products have half the diploid number of chromosomes, but each chromosome still consists of two chromatids. Each of those chromatids contains a single double-stranded DNA molecule.
The second meiotic division separates the sister chromatids, and one of each of the now-daughter chromosomes is placed in each cell resulting from the division. Each daughter chromosome contains a single double-stranded DNA molecule.

zygote—the first cell of the new individual—the diploid number of 46 chromosomes (23 pairs) is regenerated. The processes of DNA replication and mitotic cell division ensure that this diploid number is maintained in the body cells as the zygote develops.

The Meiotic Cell Cycle Produces Four Genetically Different Daughter Cells with Half the Parental Number of Chromosomes

Meiosis is a two-part (**meiosis I** and **meiosis II**) process of cell division in sexually reproducing organisms. In meiosis, the duplicated chromosomes in the parental cell are distributed to four daughter cells, each of which, therefore, has half the number of chromosomes as does the parental cell **(Figure 11.2)**. By contrast, in mitosis, each chromosome duplication is followed by a division. Consequently, the chromosome number remains constant from one cell generation to the next.

The meiotic cell cycle begins with a premeiotic interphase in which DNA replicates and the chromosomal proteins are duplicated. (This interphase passes through G_1, S, and G_2 stages as does a premitotic interphase.) As in a premitotic interphase, the two resulting copies are the identical *sister chromatids* of each chromosome. Each sister chromatid contains a single double-stranded DNA molecule. Following premeiotic interphase, cells enter the two meiotic divisions, meiosis I and meiosis II. Homologous chromosomes pair during meiosis I, and nonsister chromatids undergo a physical exchange of chromosome segments in a process called *crossing-over.* The homologous chromosomes subsequently separate as the cell continues through the first division. Completion of meiosis I produces two cells, each with half the diploid number of chromosomes, with each chromosome still consisting of two chromatids. Again, each of these chromatids contains a single double-stranded DNA molecule. During the second meiotic division, meiosis II, the sister chromatids separate—and are now called daughter chromosomes—then segregate into different cells. Each daughter chromosome contains a single double-stranded DNA molecule. A total of four cells, each with the haploid number of chromosomes, is the result of the two meiotic divisions.

Figure 11.3 illustrates the two meiotic divisions. For convenience, biologists separate each meiotic division into the same key stages as mitosis: *prophase, prometaphase, metaphase, anaphase,* and *telophase.* The stages are identified as belonging to the two divisions, meiosis I and meiosis II, by a *I* or *II,* as in *prophase I* and *prophase II.* A brief interphase called **interkinesis** separates the two meiotic divisions, *but no DNA replication occurs during interkinesis.*

PROPHASE I At the beginning of prophase I, the replicated chromosomes, each consisting of two sister chromatids, begin to fold and condense into threadlike structures in the nucleus (**Figure 11.3,** step 1). The two chromosomes of each homologous pair then come

First meiotic division

Prophase I

Plasma membrane

Duplicated centrioles

Nuclear envelope

Tetrad

Crossover

Kinetochore microtubule

Condensation of chromosomes

1 At the beginning of prophase I the chromosomes begin to condense into threadlike structures. Each consists of two sister chromatids, as a result of DNA replication during premeiotic interphase. The chromosomes of two homologous pairs, one long and one short, are shown.

Synapsis

2 Homologous chromosomes come together and pair. The fully paired homologs are called tetrads because each consists of four chromatids.

Crossing-Over

3 While they are paired, the chromatids of homologous chromosomes exchange segments by crossing-over. The enlarged circle shows a site of crossing-over, a crossover (arrow).

Prometaphase I

4 In prometaphase I, the nuclear envelope breaks down, and the spindle moves into the former nuclear area. Kinetochore microtubules connect to the chromosomes—kinetochore microtubules from one pole attach to both sister kinetochores of one duplicated chromosome, and kinetochore microtubules from the other pole attach to both sister kinetochores of the other duplicated chromosome.

Second meiotic division

Prophase II

8 The chromosomes condense and a spindle forms.

Prometaphase II

9 The nuclear envelope breaks down, the spindle enters the former nuclear area, and kinetochore microtubules from the opposite spindle poles attach to the kinetochores of each chromosome.

FIGURE 11.3

The meiotic divisions. Two homologous pairs of chromosomes are shown. (Micrographs with thanks to the John Innes Foundation Trustees.)

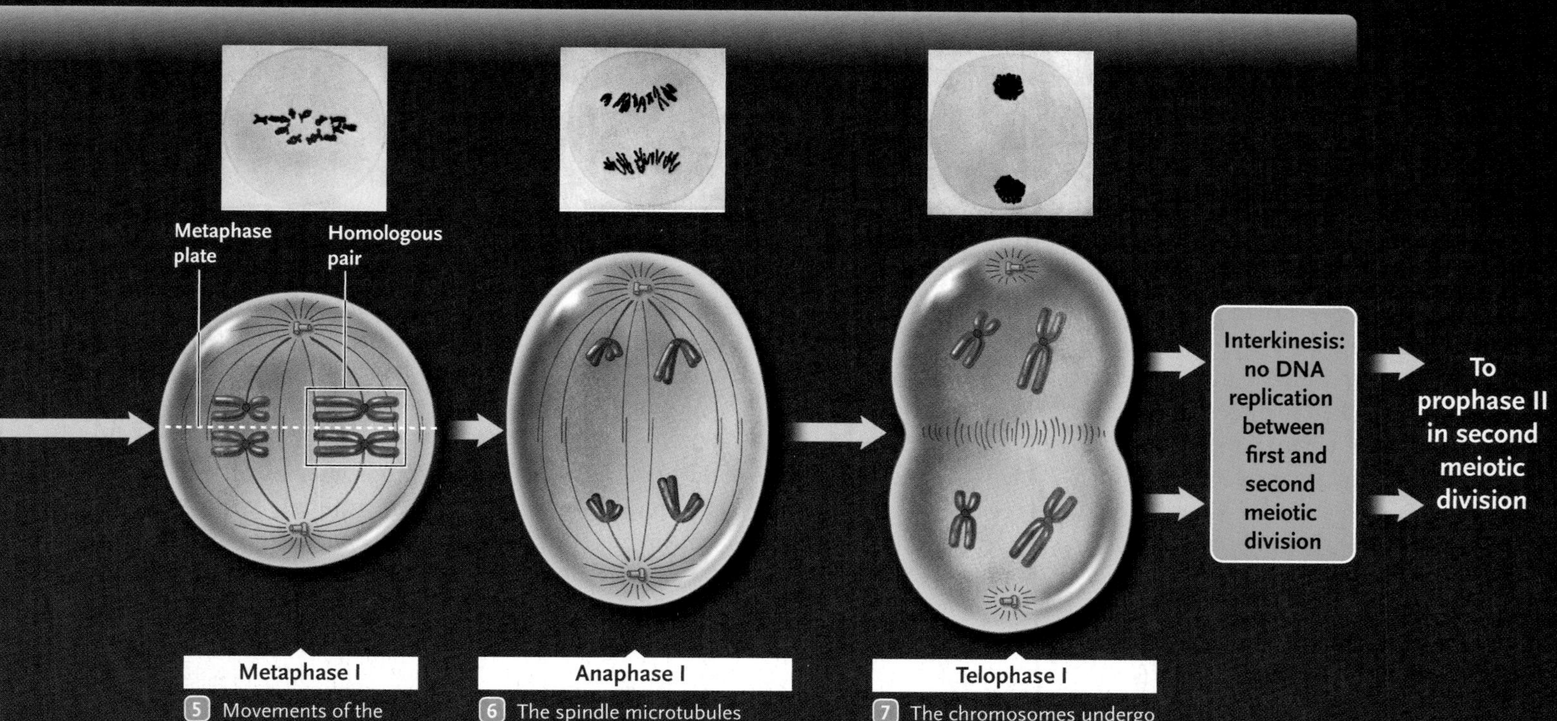

Metaphase plate
Homologous pair
Interkinesis: no DNA replication between first and second meiotic division
To prophase II in second meiotic division
Metaphase I
Anaphase I
Telophase I

5 Movements of the spindle microtubules align the tetrads in the equatorial plane–metaphase plate–between the two spindle poles.

6 The spindle microtubules separate the two chromosomes of each homologous pair and move them to opposite spindle poles. The poles now contain the haploid number of chromosomes. However, each chromosome at the poles still contains two chromatids.

7 The chromosomes undergo little or no change except for limited decondensation or unfolding in some species. The spindle of the first meiotic division disassembles.

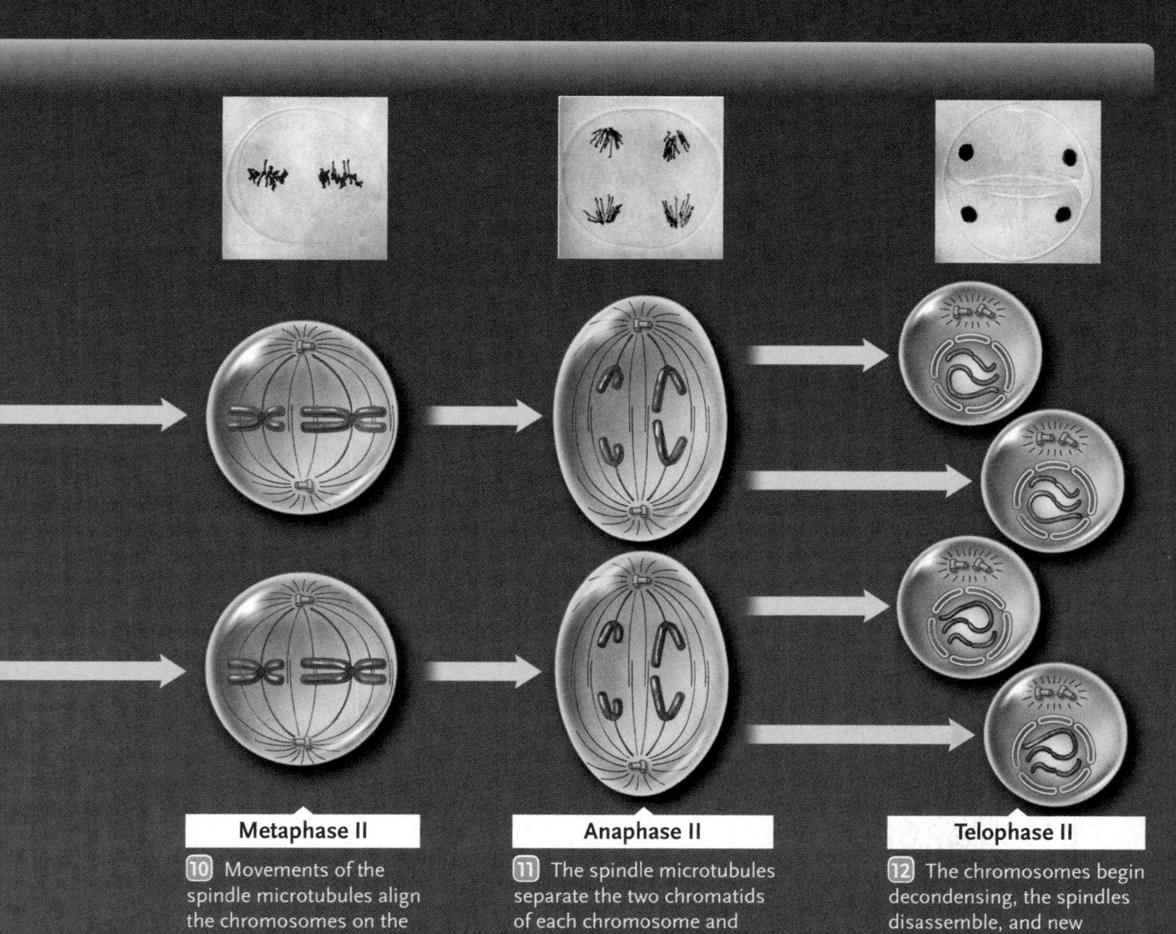

Metaphase II
Anaphase II
Telophase II

10 Movements of the spindle microtubules align the chromosomes on the metaphase plate.

11 The spindle microtubules separate the two chromatids of each chromosome and deliver them to opposite spindle poles.

12 The chromosomes begin decondensing, the spindles disassemble, and new nuclear envelopes form. Cytokinesis typically follows.

together and line up side-by-side in a zipperlike way in a process called **pairing** or **synapsis** (step 2). The fully paired homologs are called **tetrads,** referring to the fact that each homologous pair consists of four chromatids. No equivalent of chromosome pairing exists in mitosis.

While they are paired, the chromatids of homologous chromosomes physically exchange segments (step 3). The physical exchange process is called **crossing-over.** In crossing-over, enzymes break and rejoin DNA molecules of chromatids with great precision. The sites where crossing-over has occurred become visible under the light microscope when the chromosomes condense and thicken further as prophase I proceeds. The sites, called **crossovers** or **chiasmata** (singular, *chiasma* = crosspiece), clearly show that two of the four chromatids have exchanged segments.

As prophase I finishes, a spindle forms in the cytoplasm by the same basic mechanisms described in Section 10.3.

PROMETAPHASE I In prometaphase I, the nuclear envelope breaks down and the spindle enters the former nuclear area (Figure 11.3, step 4). The two chromosomes of each pair attach to kinetochore microtubules leading to opposite spindle poles. That is, both sister chromatids of one homolog attach to microtubules leading to one spindle pole, whereas both sister chromatids of the other homolog attach to microtubules leading to the opposite pole.

METAPHASE I AND ANAPHASE I In metaphase I, movements of the spindle microtubules align the tetrads on the equatorial plane—the *metaphase plate*—between the two spindle poles (Figure 11.3, step 5). Then, the two chromosomes of each homologous pair separate and move to opposite spindle poles as the spindle microtubules disassemble during anaphase I (step 6). The movement segregates homologous pairs, delivering one-half the diploid number of chromosomes to each pole of the spindle. However, all the chromosomes at the poles are still double structures composed of two sister chromatids.

TELOPHASE I AND INTERKINESIS Telophase I is a brief, transitory stage in which there is little or no change in the chromosomes (Figure 11.3, step 7). New nuclear envelopes form in some species but not in others. Telophase I is followed by an interkinesis in which the single spindle of the first meiotic division disassembles and the microtubules reassemble into two new spindles for the second division.

PROPHASE II, PROMETAPHASE II, AND METAPHASE II The second meiotic division, meiosis II, is similar to a mitotic division. During prophase II, the chromosomes condense and a spindle forms (Figure 11.3, step 8). During prometaphase II, the nuclear envelope breaks down, the spindle enters the former nuclear area, and spindle microtubules leading to opposite spindle poles attach to the two kinetochores of each chromosome (step 9). At metaphase II, movements of the spindle microtubules have aligned the chromosomes on the metaphase plate (step 10).

ANAPHASE II AND TELOPHASE II Anaphase II begins as the spindles separate the two chromatids of each chromosome and pull them to opposite spindle poles (Figure 11.3, step 11). At the completion of anaphase II, the chromatids—now called chromosomes—have been segregated to the two poles. During telophase II, the chromatids decondense to the extended interphase state, the spindles disassemble, and new nuclear envelopes form around the masses of chromatin (step 12). The result is four haploid cells, each with a nucleus containing half the number of chromosomes present in a G_1 nucleus of the same species.

CHROMOSOME SEGREGATION FAILURE Rarely, chromosome segregation fails. That is, both chromosomes of a homologous pair may connect to a kinetochore microtubule from the same spindle pole in meiosis I. In the resulting **nondisjunction,** as it is called, the spindle fails to separate the homologous chromosomes. As a result, one pole receives both chromosomes of the homologous pair, whereas the other pole has no copies of that chromosome. Alternatively, both chromatids of a sister-chromatid pair may connect to a kinetochore microtubule from the same spindle pole in meiosis II. Nondisjunction in this case produces a similar result, that is, one pole receives both sister chromatids (as daughter chromosomes), whereas the other pole receives no copies of that chromosome. Zygotes that receive an extra chromosome because of nondisjunction have three copies of one chromosome instead of two. In humans, most zygotes of this kind do not result in live births. One exception is Down syndrome, which results from three copies of chromosome 21 instead of the normal two copies. Chapter 13 discusses nondisjunction as well as Down syndrome in more detail.

SEX CHROMOSOMES IN MEIOSIS In many eukaryotes, including most animals, one or more pairs of chromosomes, called the **sex chromosomes,** are different in male and female individuals of the same species. For example, in humans, the cells of females contain a pair of sex chromosomes called the *XX* pair (the sex chromosomes are visible in Figure 10.7). Male humans contain a pair of sex chromosomes that consist of one X chromosome and a smaller chromosome called the *Y chromosome.* Developing into a male, in fact, is directly determined by the presence of the Y chromosome because of a gene it contains. In the absence of a Y chromosome, as in an XX individual, a female is produced.

The two X chromosomes in females are fully homologous. In mammals, the X and Y chromosomes in males are homologous through a short region. This means that an X chromosome from the mother is able to pair up with either an X or a Y from the father and follow the same pathways through the meiotic divisions as the other chromosome pairs.

As a result of meiosis, a gamete formed in a female (an egg) may receive either member of the X pair. A gamete formed in a male (a sperm) receives either an X or a Y chromosome. (*Insights from the Molecular Revolution* describes experiments showing how a key molecular signal controls whether a gamete will become an egg or a sperm in mammals.)

Molecular Revolution

Meiosis and Mammalian Gamete Formation: What determines whether an egg or a sperm will form?

The sex of a mammal is determined genetically. That is, the presence of a Y chromosome and the specific male-determining gene it contains, directs the development of a male. In the absence of a Y chromosome, the default development of a female occurs. The gonads contain germ cells—cells that eventually turn into eggs or sperm. Interestingly, germ cells in embryos can become eggs or sperm, regardless of the genetic sex of the individual. Whether a germ cell develops into an egg or a sperm depends on the time at which meiosis begins. In females, meiosis commences in the fetus before birth and, as a result, eggs are produced in the ovaries. In males, meiosis begins after birth and sperm is produced in the testes. The working model was that germ cells in fetuses are programmed to enter meiosis and produce eggs unless prevented from doing so by a meiosis-inhibiting factor, in which case sperm is produced. Until recently, that factor had not been identified.

Research Question

What molecular signal determines germ cell fate in mammals?

Experiment

The research group of Peter Koopman at the Institute for Molecular Bioscience, University of Queensland, Brisbane, Australia, set out to answer the question using mouse as the model mammal. The researchers used molecular tools to search for genes expressed in a sex-related fashion during gonad formation. They deemed one gene, *Cyp26b1*, to be a strong candidate for the meiosis-inhibiting factor gene. *Cyp26b1* codes for an enzyme that breaks down retinoic acid, which is a natural derivative of vitamin A. Retinoic acid regulates the development of many organ systems. Relevant here is the fact that retinoic acid causes germ cells in female embryos to begin meiosis.

The researchers showed that the *Cyp26b1* gene is expressed in embryonic mouse testes but not in embryonic ovaries. This result argued that the delay in meiosis in males was caused by the *Cyp26b1*-encoded enzyme degrading the retinoic acid. Koopman's research group also studied *Cyp26b1*-knockout male mice, that is, mice in which the gene had been deleted. Germ cells in these mice enter meiosis much earlier. The interpretation is that, in the absence of the enzyme to degrade retinoic acid, meiosis was stimulated to occur at the same time as it does in the females. These two lines of evidence support the hypothesis that the product of the *Cyp26b1* gene is the meiosis-inhibiting factor in males.

Conclusion

The timing of the initiation of meiosis in mice (and probably in other mammals) is determined by the regulation of retinoic acid presence by the *Cyp26b1*-encoded enzyme. In females, the gene is not active in embryonic ovaries, so the retinoic acid present in the fetus stimulates the initiation of meiosis prior to birth leading to the production of eggs. In males, the gene is active in embryonic testes, and the resulting enzyme degrades the retinoic acid. As a result, meiosis is delayed, initiating after birth, leading to the production of sperm.

Although potentially answering a very important biological question, the study does not provide all of the answers for how germ cell fate is regulated. Other questions remain, including how does early meiosis favor egg formation over sperm formation.

Source: J. Bowles et al. 2006. Retinoid signaling determines germ cell fate in mice. *Science* 312:596–600.

The sequence of steps in the two meiotic divisions accomplishes the major outcomes of meiosis: the generation of genetic variability and the reduction of chromosome number. (**Figure 11.4** reviews the two meiotic divisions and compares them with the single division of mitosis.)

STUDY BREAK 11.1

1. **How does the outcome of meiosis differ from that of mitosis?**
2. **What is recombination, and in what stage of meiosis does it occur?**
3. **Which of the two meiotic divisions is similar to a mitotic division?**

11.2 Mechanisms That Generate Genetic Variability

The generation of genetic variability is a prime evolutionary advantage of sexual reproduction. The variability increases the chance that at least some offspring will be successful in surviving and reproducing in changing environments.

The variability produced by sexual reproduction is apparent all around us, particularly in the human population. Except for identical twins (or identical triplets, identical quadruplets, and so forth), no two humans look alike, act alike, or have identical biochemical and physiological characteristics, even if they are members of the same immediate family. Other species that reproduce sexually show equivalent variability arising from meiosis.

During meiosis and fertilization, genetic variability arises from three sources:

1. Genetic recombination
2. The differing combinations of maternal and paternal chromosomes segregated to the poles during anaphase I
3. The particular sets of male and female gametes that unite in fertilization.

The three mechanisms, working together, produce so much total variability that no two gametes produced by the same or different individuals and no two zygotes produced by union of the gametes are likely to have the same genetic makeup. Each of these sources of variability is discussed in further detail in the following sections.

FIGURE 11.4
Comparison of key steps in meiosis and mitosis. Both diagrams use an animal cell as an example. Maternal chromosomes are shown in red; paternal chromosomes are shown in blue.

Mitosis

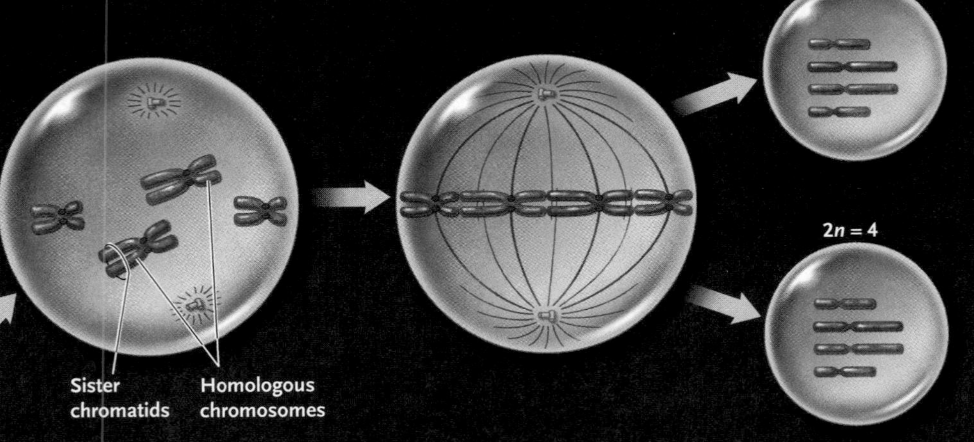

2n = 4
2n = 4
Sister chromatids
Homologous chromosomes

Chromosome duplication

Parental cell
2n = 4

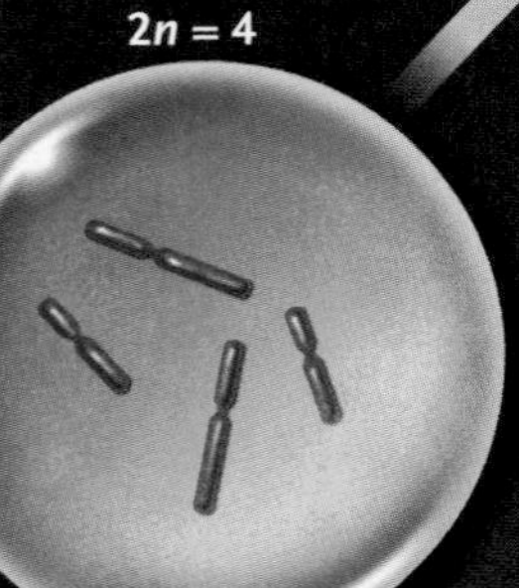

Prophase

Homologous chromosomes, each consisting of two sister chromatids, remain separate.

Metaphase

Homologous chromosomes attach individually to the spindle and align on the metaphase plate.

Anaphase/telophase

Sister chromatids separate in anaphase, becoming daughter chromosomes. In telophase, two diploid nuclei form; cytokinesis produces two diploid cells.

Chromosome duplication

Homologous chromosomes
Crossover (site of crossing-over)
Sister chromatids

n = 2
n = 2
n = 2
n = 2

Prophase I

Homologous chromosomes, each consisting of two sister chromatids, pair and crossing-over takes place.

Metaphase I

Homologous chromosome pairs attach to the spindle and align at the metaphase plate.

Anaphase I/telophase I

Homologous chromosomes separate in anaphase I. After telophase I, each cell has one copy of each homologous chromosome pair, each consisting of two sister chromatids attached at the centromere.

Metaphase II

Homologous chromosomes align on the metaphase plate in metaphase II.

Anaphase II/telophase II

Sister chromatids separate in anaphase II, becoming daughter chromosomes. In telophase II four haploid nuclei form; cytokinesis produces four haploid cells.

Meiosis II

Variability Generated by Genetic Recombination Depends on Chromosome Pairing and Crossing-Over Events between Homologous Chromatids

You learned earlier that crossing-over, the physical exchange of chromosome segments at corresponding positions along homologous chromosomes, occurs during prophase I of meiosis. Crossing-over is facilitated by the tight association of homologous chromosomes after they pair in a protein framework called the **synaptonemal complex (Figure 11.5).** When the exchange is complete toward the end of prophase I, the synaptonemal complex disassembles and disappears.

Sister chromatids of one of a homologous pair of chromosomes

Synaptonemal complex

Sister chromatids of the other of a homologous pair of chromosomes

Courtesy Diter von Wettstein

FIGURE 11.5
The synaptonemal complex as seen in a meiotic cell of the fungus *Neotiella*.

If there are genetic differences between the homologs, crossing-over can produce new allele combinations in a chromatid. Consider two genes on a homologous pair of chromosomes. The alleles for the two genes are *A* and *B* on the chromosome from one parent, and *a* and *b* on the chromosome from the other parent. After chromosome duplication the homologous chromosomes, each consisting of two sister chromatids, will pair in prophase I (**Figure 11.6,** step 1). Crossing-over exchanges segments of nonsister chromatids, producing new combinations of alleles (step 2). Any single crossing-over event involves only two of the four chromatids. In our example, crossing-over occurred between the two genes, exchanging chromosome segments to produce two nonparental arrangements of alleles on two of the chromatids. Specifically, one sister chromatid of the left homologous chromosome now has an *a* allele and a *B* allele, and one sister chromatid of the right homologous chromosome now has an *A* allele and a *b* allele.

Homologous chromosomes then separate in the first meiotic division (Figure 11.6, step 3) and, after the second meiotic division four nuclei are produced, each receiving one of the four chromatids. Two nuclei receive unchanged chromatids (now called chromosomes) with parental combinations of alleles, and two receive chromosomes that have new combinations of alleles caused by crossing-over. Geneticists call the chromosomes with parental combinations of alleles **parental chromosomes,** and chromosomes with new, nonparental combinations of alleles **recombinant chromosomes.** Therefore, crossing-over is a

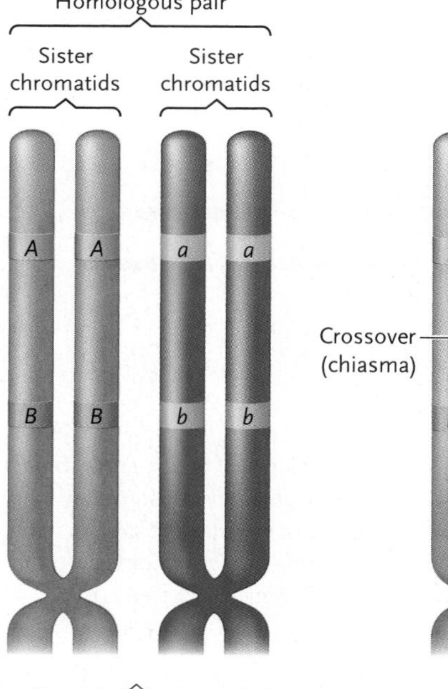
Homologous pair
Sister chromatids
Sister chromatids
A
A
a
a
B
B
b
b

1 Homologous chromosomes pair. Each sister chromatid consists of a single double-stranded DNA molecule.

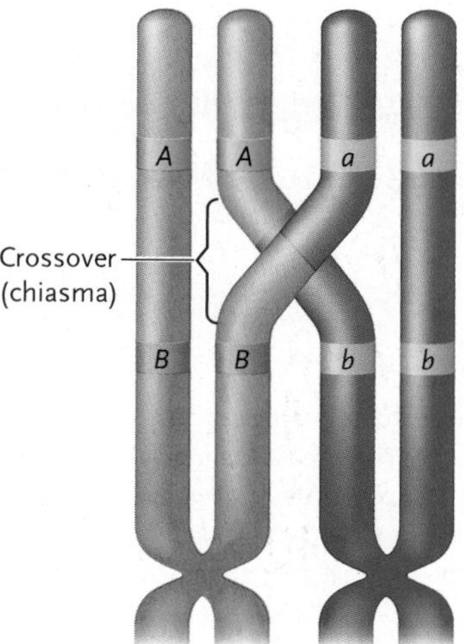
A
A
a
a
Crossover (chiasma)
B
B
b
b

2 Crossing-over occurs between homologous chromatids, resulting in the exchange of segments.

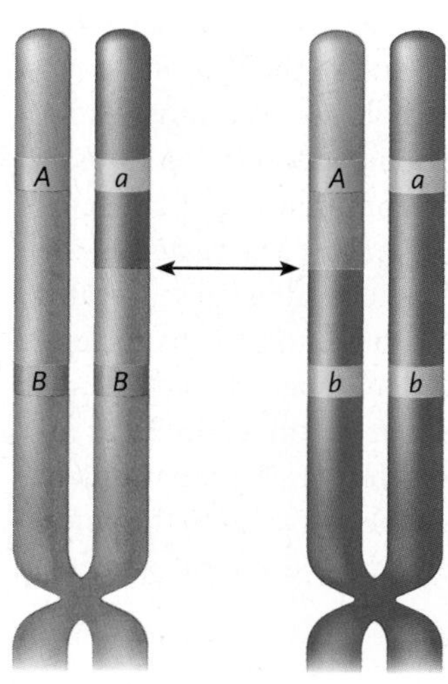
A
a
A
a
B
B
b
b

3 Homologous chromosomes separate at first meiotic division.

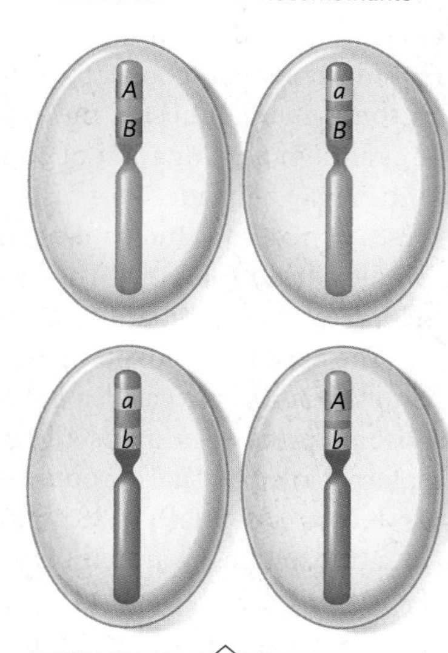
Parentals
Recombinants
A
B
a
B
a
b
A
b

4 After meiosis II, two daughter chromosomes have parental combinations of alleles, *A–B* and *a–b*, and the other two daughter chromosomes are recombinants for the alleles, *A–b* and *a–B*.

FIGURE 11.6
Generation of genetic recombinants by crossing-over. The letters indicate two alleles, *A* and *a*, of one gene, and two alleles, *B* and *b*, of another gene. The parents have these alleles in the combinations *A–B* and *a–b;* as a result of crossing-over, two of the chromosomes produced from this meiosis—the recombinants—have the new combinations *a–B* and *A–b*.

mechanism for **genetic recombination;** that is, it produces **genetic recombinants,** also called more simply **recombinants.**

Crossing-over takes place largely at random, at almost any position along the chromosome arms, between any two of the four chromatids of a homologous pair. One or more additional crossing-over events may occur in the same chromosome pair and involve the same or different chromatids exchanging segments in the first event. In most species, crossing-over occurs at two or three sites in each set of paired chromosomes.

Random Segregation of Maternal and Paternal Chromosomes Is the Second Major Source of Genetic Variability in Meiosis

Recall that a diploid individual inherits one set of chromosomes from the mother and one from the father. The maternal and paternal members of a chromosome pair are *homologous chromosomes.* Before meiosis, each homologous chromosome duplicates to produce a pair of sister chromatids that remain connected at the centromere until they are separated in meiosis II and become daughter chromosomes.

Random segregation of maternal and paternal chromosomes accounts for the second major source of genetic variability in meiosis. Recall that prometaphase I is the stage of meiosis in which the homologous pairs of chromosomes attach to the spindle poles. The maternal and paternal chromosomes of each homologous pair typically carry different alleles of many of the genes on that chromosome. For each homologous pair, one chromosome (at this stage consisting of a pair of sister chromatids) makes spindle connections leading to one pole and the other chromosome (also a pair of sister chromatids) connects to the opposite pole. In making these connections, all the maternal chromosomes may connect to one pole and all the paternal chromosomes may connect to the opposite pole. Or, as is most likely, any random combination of connections between these possibilities may be made. As a result, any combination of chromosomes of maternal and paternal origin may be segregated to the spindle poles **(Figure 11.7).** The second meiotic division segregates these random combinations of chromosomes to gamete nuclei.

The number of possible combinations depends on the number of chromosome pairs in a species. For example, the 23 chromosome pairs of humans allow 2^{23} different combinations of maternal and paternal chromosomes to be delivered to the poles, producing potentially 8,388,608 genetically different gametes from this source of variability alone.

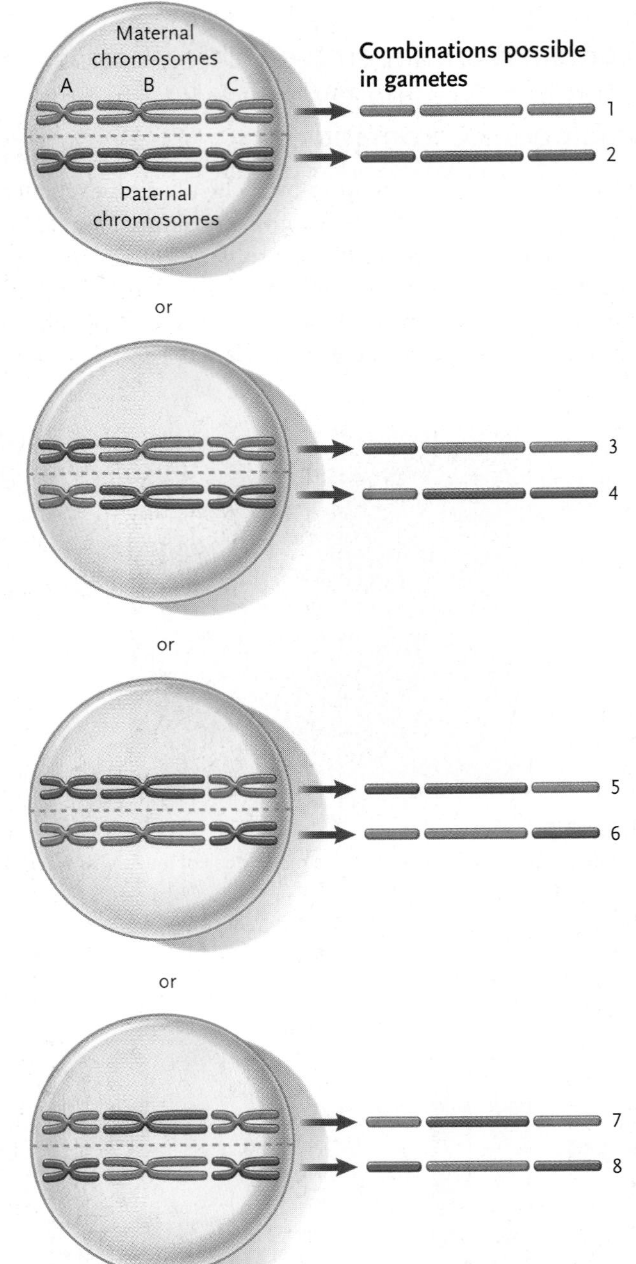
Maternal chromosomes
A
B
C
Paternal chromosomes
Combinations possible in gametes
1
2
or
3
4
or
5
6
or
7
8

FIGURE 11.7
Possible outcomes of the random spindle connections of three pairs of chromosomes at metaphase I of meiosis. The three types of chromosomes are labeled A, B, and C. Maternal chromosomes are red; paternal chromosomes are blue. There are four possible patterns of connections, giving eight possible combinations of maternal and paternal chromosomes in gametes (labeled 1–8).

Random Joining of Male and Female Gametes in Fertilization Adds Additional Variability

The male and female gametes produced by meiosis are genetically diverse. Which two gametes join in fertilization is a matter of chance. This chance union of gametes amplifies the variability of sexual reproduction. Considering just the variability from random separation of homologous chromosomes and that from fertilization, the possibility that two children of the same parents could receive the same combination of maternal and paternal chromosomes is 1 chance out of $(2^{23})^2$ or 1 in 70,368,744,000,000 ($\sim$70 trillion), a number that far exceeds the number of people in the entire human population. The further variability introduced by recombination makes it practically impossible for humans and most other sexually reproducing organisms to produce genetically identical gametes or offspring. The only exception is identical twins (or identical triplets, identical quadruplets, and so forth), which arise not from the combination of identical gametes during fertilization but from mitotic division of a single fertilized egg into separate cells that give rise to genetically identical individuals.

STUDY BREAK 11.2

1. **What are the three ways in which sexual reproduction generates genetic variability?**
2. **Consider an animal with six pairs of chromosomes; one set of six chromosomes is from this animal's male parent, and the homologous set of six chromosomes is from this animal's female parent. When this animal produces gametes, what proportion of these gametes will have chromosomes, all of which originate from the animal's female parent?**

THINK OUTSIDE THE BOOK

Mutants of several organisms have been identified that affect meiosis. Individually or collaboratively, use the Internet or research literature to find two examples of mutants that affect meiosis and outline: 1) how meiosis differs in the mutants compared with normal cells; and 2) what we have learned about the molecular mechanisms of meiosis by studying the mutants.

11.3 The Time and Place of Meiosis in Organismal Life Cycles

The time and place at which meiosis occurs follows one of three major patterns in the life cycles of eukaryotes **(Figure 11.8)**. The differences reflect the portions of the life cycle spent in the haploid and diploid phases and whether mitotic divisions intervene between meiosis and the formation of gametes.

In Animals, the Diploid Phase Is Dominant, the Haploid Phase Is Reduced, and Meiosis Is Followed Directly by Gamete Formation

Animals follow the pattern (see Figure 11.8A) in which the diploid phase dominates the life cycle, the haploid phase is reduced, and meiosis is followed directly by gamete formation. In male animals, each of the four nuclei produced by meiosis is enclosed in a separate cell by cytoplasmic divisions, and each of the four cells differentiates into a functional sperm cell. In female animals, only one of the four nuclei becomes functional as an egg cell nucleus.

Fertilization restores the diploid phase of the life cycle. Thus, animals are haploids only as sperm or eggs, and no mitotic divisions occur during the haploid phase of the life cycle.

A. **Animal life cycles**

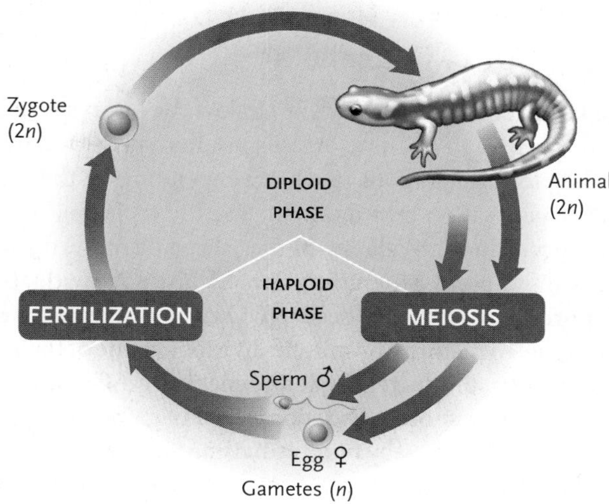
Zygote
(2n)
DIPLOID
PHASE
Animal
(2n)
HAPLOID
PHASE
FERTILIZATION
MEIOSIS
Sperm ♂
Egg ♀
Gametes (n)

B. **All plants and some fungi and algae (fern shown; relative length of the two phases varies widely in plants)**

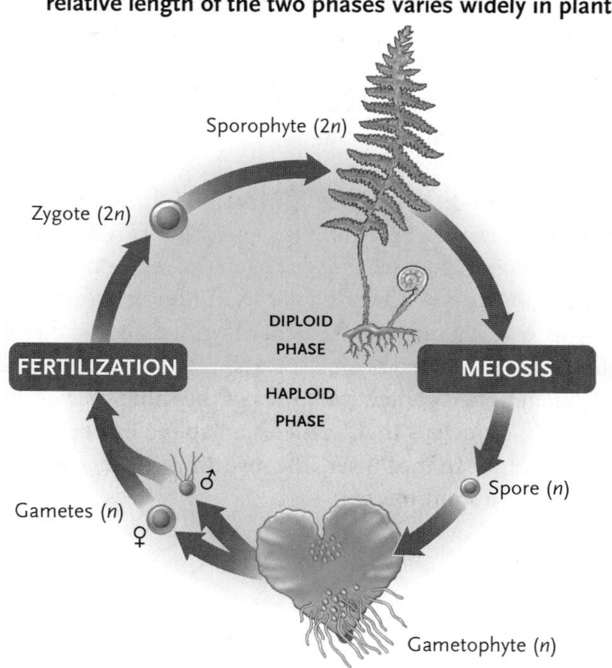
Sporophyte (2n)
Zygote (2n)
DIPLOID
PHASE
FERTILIZATION
MEIOSIS
HAPLOID
PHASE
♂
Spore (n)
Gametes (n)
♀
Gametophyte (n)

C. **Other fungi and algae**

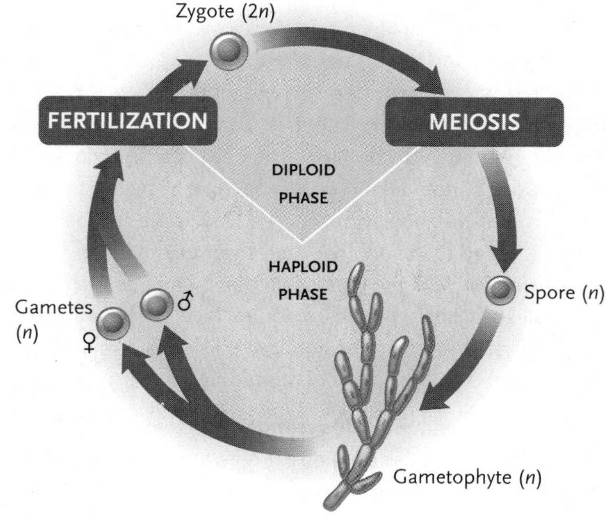
Zygote (2n)
FERTILIZATION
MEIOSIS
DIPLOID
PHASE
HAPLOID
PHASE
Gametes
(n)
♂
♀
Spore (n)
Gametophyte (n)

FIGURE 11.8
Variations in the time and place of meiosis in eukaryotes. The diploid phase of the life cycles is shaded in green; the haploid phase is shaded in yellow. n = haploid number of chromosomes; $2n$ = diploid number.

In All Plants and Some Fungi, Generations Alternate between Haploid and Diploid Phases That Are Both Multicellular

All plants and some algae and fungi follow the life cycle pattern shown in Figure 11.8B. These organisms alternate between haploid and diploid generations in which, depending on the organism, either generation may dominate the life cycle, and mitotic divisions occur in both phases. In these organisms, fertilization produces the diploid generation, in which the individuals are called **sporophytes** (*spora* = seed; *phyta* = plant). After the sporophytes grow to maturity by mitotic divisions, some of their cells undergo meiosis, producing haploid, genetically different, reproductive cells called **spores.** The spores are not gametes; they germinate and grow directly by mitotic divisions into a generation of haploid individuals called **gametophytes** (*gameta* = gamete). At maturity, the nuclei of some cells in gametophytes develop into egg or sperm nuclei. All the egg or sperm nuclei produced by a particular gametophyte are genetically identical because they arise through mitosis; meiosis does not occur in gametophytes. Fusion of a haploid egg and sperm nucleus produces a diploid zygote nucleus that divides by mitosis to produce the diploid sporophyte generation again.

In many plants, including most bushes, shrubs, trees, and flowers, the diploid sporophyte generation is the most visible part of the plant. The gametophyte generation is reduced to an almost microscopic stage that develops in the reproductive parts of the sporophytes—in flowering plants, in the structures of the flower. The female gametophyte remains in the flower; the male gametophyte is released from flowers as microscopic pollen grains. When pollen contacts a flower of the same species, it releases a haploid nucleus that fertilizes a haploid egg cell of a female gametophyte in the flower. The resulting cell reproduces by mitosis to form a sporophyte.

In Some Fungi and Other Organisms, the Haploid Phase Is Dominant and the Diploid Phase Is Reduced to a Single Cell

The life cycle of some fungi and algae follows the third life cycle pattern (see Figure 11.8C). In these organisms, the diploid phase is limited to a single cell, the zygote, produced by fertilization. Immediately after fertilization, the diploid zygote undergoes meiosis to produce the haploid phase. Mitotic divisions occur only in the haploid phase.

During fertilization, two haploid gametes, often designated as plus (+) or minus (−) because they are similar in structure, fuse to form a diploid nucleus. This nucleus immediately enters meiosis, producing four haploid cells. These cells develop directly or after one or more mitotic divisions into haploid spores. These spores germinate to produce haploid individuals, the gametophytes, which grow or increase in number by mitotic divisions. Eventually positive and negative gametes are formed in these individuals by differentiation of some of the cells produced by the mitotic divisions. Because the gametes are produced by mitosis, all the gametes of an individual are genetically identical.

In this chapter, we have seen that meiosis has three outcomes that are vital to sexual reproduction. Meiosis reduces the chromosomes to the haploid number so that the chromosome number does not double at fertilization. Through recombination and random separation of maternal and paternal chromosomes, meiosis produces genetic variability; further variability is provided by the random combination of gametes in fertilization. The next chapter shows how the outcomes of meiosis and fertilization underlie the inheritance of traits in sexually reproducing organisms.

STUDY BREAK 11.3 <

How does the place of meiosis differ in the life cycles of animals and most plants?

UNANSWERED QUESTIONS

The overall mechanism and outcomes of meiosis have been known for a long time, since the turn of the twentieth century. However, despite the fundamental importance of meiosis in sexual reproduction, the biochemical, genetic, and molecular mechanisms of meiosis are poorly understood. For example, how do homologous chromosomes recognize their appropriate pairing partners? How do they become aligned in a configuration that allows the formation of crossovers? How is the number of crossover events regulated to ensure that each chromosome pair will have a crossover? Developing a deeper understanding of the molecular mechanisms that regulate meiosis is highly important in human biology, because mis-segregation of chromosomes during meiosis I is a major cause of birth defects and the leading cause of miscarriages.

What are the genetic and molecular controls of meiosis?

Many unanswered questions surround the regulatory controls that switch cells from mitosis and that control the many individual steps of the division process. Both the yeast *Saccharomyces cerevisiae* and the fruit fly, *Drosophila melanogaster,* have been used widely in this research. Researchers have discovered a large number of mutant genes—well over a hundred in the two species combined—with effects in meiosis. Many of the genes identified are related to genes in humans. Many questions surround the control and operation of these genes and the identity and functions of the proteins they encode. For instance, Michael Lichten at the National Cancer Institute, National Institutes of Health, is using yeast in a project to describe the molecular steps of meiotic recombination from start to finish, including the chromosomal structure changes that occur when homologous chromosomes pair and recombine. For example, one of Lichten's research projects involves using genomics scale techniques to study where in the genome meiotic recombination occurs, and how that is related to chromosomal structure during meiosis. That is, he is using the molecular properties of recombination intermediates to isolate all of the recombination intermediates that form in cells undergoing meiosis. By determining what DNA sequences these intermediates contain, Lichten can map their location in the genome, and determine how changes in chromosome structure affect their distribution across chromosomes.

Think Critically

One key to understanding mis-segregation is studying the genes that code for proteins involved in chromosomes attaching to microtubules. Which protein structure on chromosomes would you think is of great interest to researchers and why?

Peter J. Russell

Peter J. Russell

REVIEW KEY CONCEPTS

Go to **CENGAGENOW** at www.cengage.com/login to access quizzing, animations, exercises, articles, and personalized homework help.

11.1 The Mechanisms of Meiosis

- The major cellular processes that underlie sexual reproduction are the halving of chromosome number by meiosis and restoration of the number by fertilization. Meiosis and fertilization also produce new combinations of genetic information (Figure 11.1).
- Meiosis occurs only in eukaryotes that reproduce sexually and only in organisms that are at least diploid—that is, organisms that have at least two representatives of each chromosome.
- DNA replicates and the chromosomal proteins are duplicated during the premeiotic interphase, producing two copies, the sister chromatids, of each chromosome.
- During prophase I of the first meiotic division (meiosis I), the replicated chromosomes condense and come together and pair. While they are paired, the exchange of segments of chromatids of homologous chromosomes occurs by crossing-over. While these events are in progress, the spindle forms in the cytoplasm. Crossing-over events become visible later as crossovers (chiasmata) (Figures 11.2 and 11.3).
- During prometaphase I, the nuclear envelope breaks down, the spindle enters the former nuclear area, and kinetochore microtubules leading to opposite spindle poles attach to one kinetochore of each pair of sister chromatids of homologous chromosomes (Figure 11.3).
- At metaphase I, spindle microtubule movements have aligned the tetrads on the metaphase plate, the equatorial plane between the two spindle poles. The connections of kinetochore microtubules to opposite poles ensure that the homologous pairs separate and move to opposite spindle poles during anaphase I, reducing the chromosome number to the haploid value. Each chromosome at the poles still contains two chromatids.
- Telophase I and interkinesis are brief and transitory stages; no DNA replication occurs during interkinesis. During these stages, the single spindle of the first meiotic division disassembles.
- During prophase II, the chromosomes condense and a spindle forms. During prometaphase II, the nuclear envelope breaks down, the spindle enters the former nuclear area, and spindle microtubules leading to opposite spindle poles attach to the two kinetochores of each chromosome. At metaphase II, the chromosomes become aligned on the metaphase plate. The connections of kinetochore microtubules to opposite spindle poles ensure that during anaphase II, the chromatids of each chromosome are separated and segregate to those opposite spindle poles.
- During telophase II, the chromosomes decondense to their extended interphase state, the spindles disassemble, and new nuclear envelopes form. The result is four haploid cells, each containing half the number of chromosomes present in a G_1 nucleus of the same species.

Animation: Gamete-producing organs

Animation: Meiosis step-by-step

Animation: Meiosis I and II

Animation: Crossing-over

Animation: Comparing mitosis and meiosis

11.2 Mechanisms That Generate Genetic Variability

- Recombination is the first source of the genetic variability produced by meiosis (Figures 11.5 and 11.6). Chromatids generate new combinations of alleles—recombinants—by physically exchanging segments in crossing-over. The exchange process involves precise breakage and joining of DNA molecules. It is catalyzed by enzymes and occurs while the homologous chromosomes are held together tightly by the synaptonemal complex.
- The crossovers visible between the chromosomes at late prophase I reflect the exchange of chromatid segments that occurred during the molecular steps of genetic recombination.
- The random segregation of homologous chromosomes is the second source of genetic variability produced by meiosis. The homologous pairs separate at anaphase I of meiosis, segregating random combinations of maternal and paternal chromosomes to the spindle poles (Figure 11.7).
- Random joining of male and female gametes in fertilization is the third source of genetic variability.

Animation: Random alignment

11.3 The Time and Place of Meiosis in Organismal Life Cycles

- The time and place of meiosis follow one of three major pathways in the life cycles of eukaryotes, which reflect the portions of the life cycle spent in the haploid and diploid phases and whether mitotic divisions intervene between meiosis and the formation of gametes (Figure 11.8).
- In animals, the diploid phase dominates the life cycle; mitotic divisions occur only in this phase. Meiosis in the diploid phase gives rise to products that develop directly into egg and sperm cells without undergoing mitosis (Figure 11.8A).

- In all plants and some fungi, the life cycle alternates between haploid and diploid generations that both grow by mitotic divisions. Fertilization produces the diploid sporophyte generation; after growth by mitotic divisions, some cells of the sporophyte undergo meiosis and produce haploid spores. The spores germinate and grow by mitotic divisions into the gametophyte generation. After growth of the gametophyte, cells develop directly into egg or sperm nuclei, which fuse in fertilization to produce the diploid sporophyte generation again (Figure 11.8B).
- In some fungi and protists, meiosis occurs immediately after fertilization, producing a haploid phase, which dominates the life cycle; mitosis occurs only in the haploid phase. At some point in the life cycle, haploid cells differentiate directly into gametes, which fuse together as pairs to produce the brief diploid phase (Figure 11.8C).

Animation: Generalized life cycles

UNDERSTAND AND APPLY

Test Your Knowledge

1. The diploid number of this individual is 6.

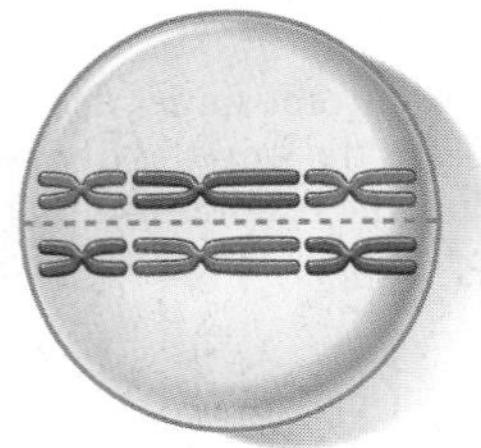

   This figure represents:
   a. mitotic metaphase.
   b. meiotic metaphase I.
   c. meiotic metaphase II.
   d. a gamete.
   e. six nonhomologous chromosomes.
2. Which of the following is *not* associated with sperm production?
   a. daughter cells identical to the parent cell
   b. variety in resulting cells
   c. chromosome number halved in resulting cells
   d. four daughter cells arising from one parent cell
   e. 23 chromosomes in the human sperm
3. Chiasmata:
   a. form during metaphase II of meiosis.
   b. occur between two nonhomologous chromosomes.
   c. represent chromosomes independently assorting.
   d. are sites of DNA exchange between homologous chromatids.
   e. ensure the resulting cells are identical to the parent cell.
4. If $2n$ is four, the number of possible combinations in the resulting gametes is:
   a. 1.
   b. 2.
   c. 4.
   d. 8.
   e. 16.
5. In meiosis:
   a. homologous chromosomes pair up at prophase II.
   b. chromosomes separate from their homologous partners at anaphase I.
   c. the centromeres split at anaphase I.
   d. a female gamete has two X chromosomes.
   e. reduction of chromosome number occurs in meiosis II.
6. The DNA content in a diploid cell in G_2 is X. If that cell goes into meiosis at its metaphase II, the DNA content would be:
   a. 0.1X.
   b. 0.5X.
   c. X.
   d. 2X.
   e. 4X.
7. Metaphase in mitosis is similar to what stage in meiosis?
   a. prophase I
   b. prophase II
   c. metaphase I
   d. metaphase II
   e. crossing-over
8. In the human sperm:
   a. there must be one chromosome of each type, except for the sex chromosomes, where either an X or a Y chromosome is present.
   b. a chromosome must be represented from each parent.
   c. there must be an unequal mixture of chromosomes from both parents.
   d. there must be representation of chromosomes from only one parent.
   e. there is the possibility of 246 different combinations of maternal and paternal chromosomes.
9. In plants, the adult diploid individuals are called:
   a. spores.
   b. sporophytes.
   c. gametes.
   d. gametophytes.
   e. zygotes.
10. Which of the following sequences of events describes the general life cycle of an animal?
   a. 2N→meiosis→2N→fertilization→1N
   b. 1N→meiosis→2N→fertilization→1N
   c. 2N→meiosis→1N→fertilization→2N
   d. 2N→mitosis→1N→fertilization→2N
   e. 2N→mitosis→1N→fertilization→1N

Discuss the Concepts

1. You have a technique that allows you to measure the amount of DNA in a cell nucleus. You establish the amount of DNA in a sperm cell of an organism as your baseline. Which multiple of this amount would you expect to find in a nucleus of this organism at G_2 of premeiotic interphase? At telophase I of meiosis? During interkinesis? At telophase II of meiosis?
2. One of the human chromosome pairs carries a gene that influences eye color. In an individual human, one chromosome of this pair has an allele of this gene that contributes to the formation of blue eyes. The other chromosome of the pair has an allele that contributes to brown eye color (other genes also influence eye color in humans). After meiosis in the cells of this individual, what fraction of the nuclei will carry the allele that contributes to blue eyes? To brown eyes?
3. Mutations are changes in DNA sequence that can create new alleles. In which cells of an individual, somatic or meiotic cells, would mutations be of greatest significance to that individual? What about to the species to which the individual belongs?

Design an Experiment

Design experiments to determine whether a new pesticide on the market adversely affects egg production and fertilization in frogs.

Interpret the Data

Aphids are bizarre. A wingless female aphid can generate up to one hundred almost-identical female offspring who themselves reproduce similarly, thereby creating a clone of offspring similar to the original female. In these cases, no eggs are laid, and the young are born as juveniles. However, if day length (photoperiod) and temperature change significantly (e.g., in fall in temperate climates), the wingless female can switch and produce a combination of winged males, winged females that reproduce without laying eggs, or egg-laying females with wings. Depending on circumstances and aphid species, different proportions of each of these are produced.

The table represents some data from an early study looking at the effect of temperature and day length (or night length) on the number of males produced by individual females of the species *Megoura viciae.*

			Number of males per mother	
Temperature	**Photoperiod (hr light/24 hr)**	**Number of mothers**	**Range**	**Mean**
25°C	12	9	0–0	0
	16	9	0–0	0
20°C	12	27	5–21	13.7
	16	44	0–21	8.0
15°C	12	37	3–17	11.6
	16	25	1–20	10.4
11°C	12	19	0–5	2.5
	16	32	0–9	2.9

1. At the temperatures used females produce about 100 offspring each, typically a mix of males and females. Assuming this number of offspring, what was the highest and lowest percentage of male offspring per female parent at 20°C? What temperature range appears to be optimal for male production? Which temperature(s) correspond(s) with the production of no males? With less than 10% males?
2. The effect of photoperiod on male production was also studied. Various day lengths were simulated in the laboratory using artificial light. Citing examples from the data, what effect does photoperiod have on male production? How does it compare to the effect of temperature?

Source: A. D. Lees. 1959. The role of photoperiod and temperature in the determination of parthenogenetic and sexual forms in the aphid *Megoura viciae* Buckton—I: The influence of these factors on apterous virginoparae and their progeny. *Journal of Insect Physiology* 3:92–117.

Apply Evolutionary Thinking

Explain aspects of the processes of mitosis and meiosis that would lead you to conclude that they are evolutionarily related processes. Do you think that mitosis evolved from meiosis, or did the opposite occur? Explain your conclusion.

Express Your Opinion

Japanese researchers have successfully created a "fatherless" mouse that contains the genetic material from the eggs of two females. The mouse is healthy and fully fertile. Do you think researchers should be allowed to try the same process with human eggs? Go to academic.cengage.com/login to investigate both sides of the issue and then vote.

12

STUDY OUTLINE

Biosphoto/Gunther Michel/Peter Arnold

Rabbits, showing genetic variation in coat color.

Mendel, Genes, and Inheritance

Why It Matters. . . Parties and champagne were among the last things on Ernest Irons' mind on New Year's Eve, 1904. Irons, a medical intern, was examining a blood specimen from a new patient and was sketching what he saw through his microscope—peculiarly elongated red blood cells **(Figure 12.1)**. He and his supervisor, James Herrick, had never seen anything like them. The shape of the cells was reminiscent of a sickle, a cutting tool with a crescent-shaped blade.

A. **A normal red blood cell**

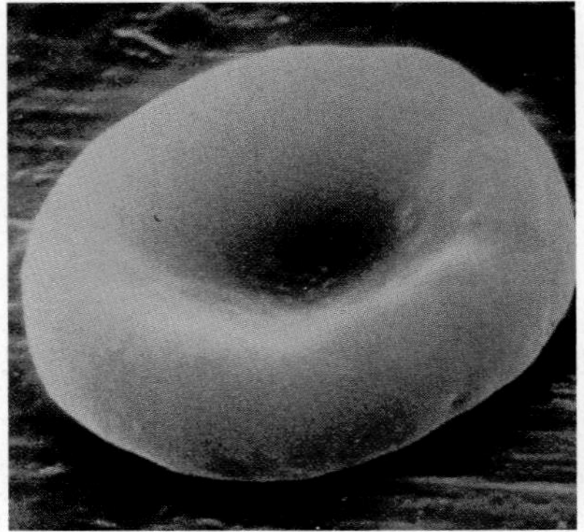

Stanley Flegler/Visuals Unlimited

B. **A sickled red blood cell**

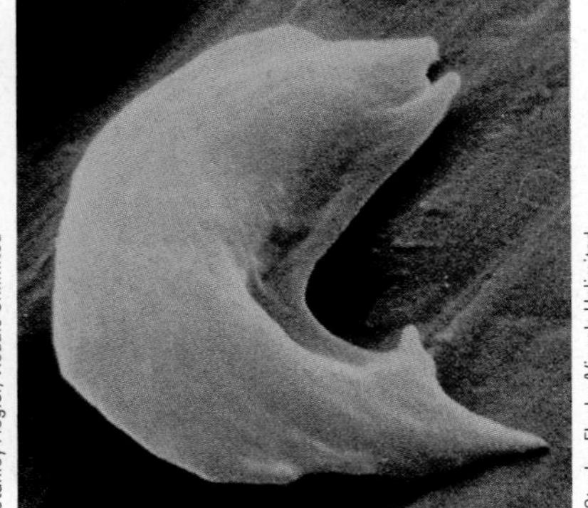

Stanley Flegler/Visuals Unlimited

FIGURE 12.1
Red blood cell shape in sickle-cell disease.

The patient had complained of weakness, dizziness, shortness of breath, and pain. His father and two sisters had died from mysterious ailments that had damaged their lungs or kidneys. Did those deceased family members also have sickle-shaped red cells in their blood? Was there a connection between the abnormal cells and the ailments? How did the cells become sickled?

The medical problems that baffled Irons and Herrick killed their patient when he was only 32 years old. The patient's symptoms were characteristic of a genetic disorder now called *sickle-cell disease*. This disease develops when a person has received two copies of a gene (one from each parent) that codes for an altered sub-

FIGURE 12.2 Gregor Mendel (1822–1884), the founder of modern genetics.

M. Hofer/National Library of Medicine

unit of hemoglobin, the oxygen-transporting protein in red blood cells. When oxygen supplies are low, the altered hemoglobin forms long, fibrous, crystal-like structures that push red blood cells into the sickle shape. The altered protein differs from the normal protein by just a single amino acid.

The sickled red blood cells are too elongated and inflexible to pass through the capillaries, the smallest vessels in the circulatory system. As a result, the cells block the capillaries. The surrounding tissues become starved for oxygen and saturated with metabolic wastes, causing the symptoms experienced by Irons and Herrick's patient. The problem worsens as oxygen concentration falls in tissues and more red blood cells are pushed into the sickled form. (You will learn more about sickle-cell disease in this chapter and in Chapter 13.)

Researchers have studied sickle-cell disease in great detail at both the molecular and the clinical levels. You may find it curious, though, that our understanding of sickle-cell disease—and all other heritable traits—actually began with studies of pea plants in a monastery garden.

Fifty years before Ernest Irons sketched sickled red blood cells, a scholarly monk named Gregor Mendel **(Figure 12.2)** used garden peas to study patterns of inheritance. To test his hypotheses about inheritance, Mendel bred generation after generation of pea plants and carefully observed the patterns by which parents transmit traits to their offspring. Through his experiments and observations, Mendel discovered the fundamental rules that govern inheritance. His discoveries and conclusions founded the science of genetics and still have the power to explain many of the puzzling and sometimes devastating aspects of inheritance. <

12.1 The Beginnings of Genetics: Mendel's Garden Peas

Until about 1900, scientists and the general public believed in the **blending theory of inheritance,** which suggested that hereditary traits blend evenly in offspring through mixing of the parents' blood, much like the effect of mixing coffee and cream. Even today, many people assume that parental characteristics such as skin color, body size, and facial features blend evenly in their offspring, with the traits of the children appearing about halfway between those of their parents. Yet if blending takes place, why don't extremes, such as very tall and very short individuals, gradually disappear over generations as repeated blending takes place? Also, why do children with blue eyes keep turning up among the offspring of brown-eyed parents?

Gregor Mendel's experiments with garden peas, performed in the 1860s, provided the first answers to these questions and many more. Mendel was an Augustinian monk who lived in a monastery in Brünn, now part of the Czech Republic. But he had an unusual education for a monk in the mid-nineteenth century. He had studied mathematics, chemistry, zoology, and botany at the University of Vienna under some of the foremost scientists of his day. He had also been reared on a farm and was well aware of agricultural principles and their application. He kept abreast of breeding experiments published in scientific journals. Mendel also won several awards for developing improved varieties of fruits and vegetables.

In his work with peas, Mendel studied a variety of *characters.* A **character** is a specific heritable attribute or property of an organism. It may be a visible attribute, or it may be one that can only be detected by biochemical or molecular analysis. The characters Mendel studied included seed shape, seed color, and flower color. Mendel studied plants that had alternative forms of these characters, known as **character differences** or **traits.** For example, for the flower color character, the two traits studied—the alternative forms—were purple and white. Mendel established that characters are passed to offspring in the form of discrete hereditary factors, which now are known as genes. Mendel observed that, rather than blending evenly, many parental traits appear unchanged in offspring, whereas others disappear in one generation to reappear unchanged in the next. Although Mendel did not know it, the inheritance patterns he observed are the result of the segregation of chromosomes, on which the genes are located, to gametes in meiosis (see Chapter 11). Mendel's methods illustrate, perhaps as well as any experiments in the history of science, how rigorous scientific work is conducted: through observation, making hypotheses, and testing the hypotheses with experiments. And, although others had studied inheritance patterns before him, Mendel's most important innovation was his quantitative approach to science, specifically his rigor and statistical analysis in an era when qualitative, descriptive science was the accepted practice.

Mendel Chose True-Breeding Garden Peas for His Experiments

Mendel chose the garden pea *(Pisum sativum)* for his genetics experiments because the plant could be grown easily in the monastery garden, without elaborate equipment. **Figure 12.3** shows the structure of a pea flower. Normally, pea plants **self-fertilize** (also known as **self-pollinate,** or more simply, *self*): sperm nuclei in

Making a Genetic Cross between Two Pea Plants

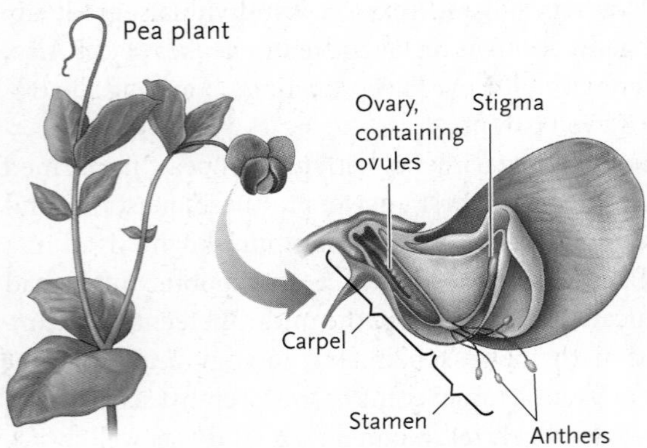
Pea plant
Ovary, containing ovules
Stigma
Carpel
Stamen
Anthers

Purpose: Mendel used the garden pea, *Pisum sativum,* for his genetic experiments. The goal of the experiments was to test various hypotheses about the patterns of inheritance by cross-breeding plants with easily observable characters, such as flower color and seed shape. He could then analyze whether the characters he observed and counted in the offspring supported the predictions made by a particular hypothesis.

In cross-breeding, the sperm and the egg must come from different plants. However, this type of flowering plant has both male and female structures within the same flower and is capable of self-fertilization, also called "selfing." The figure to the left shows a pea flower sectioned to show the location of the reproductive structures. A stamen is a male structure; it ends with an anther where pollen—the male gametes—are produced. The carpal is a female structure; its lower end contains the ovary which contains ovules in which eggs—female gametes—are produced and in which fertilization takes place. At the upper end of the carpal is the stigma, the structure where pollen lands and begins the fertilization process. Fertilization in the ovules leads to the development of seeds. (Details of plant fertilization are presented in Chapter 34).

The figure below shows how Mandel designed his experiments to eliminate selfing so he could be sure that only cross-breeding took place.

Protocol:

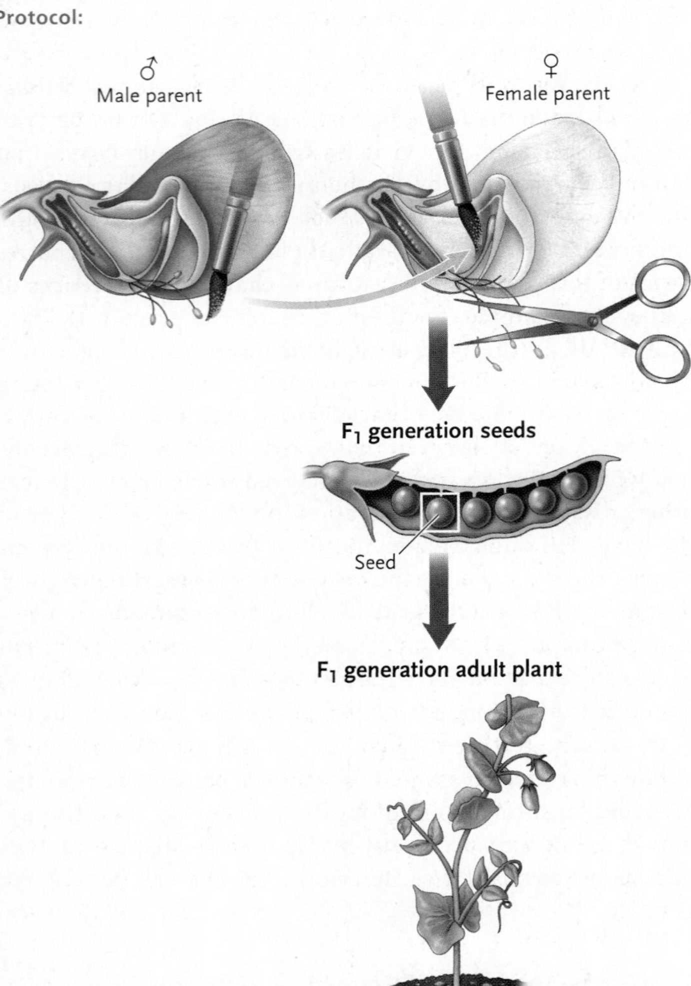
♂
Male parent
♀
Female parent
F_1 generation seeds
Seed
F_1 generation adult plant

1. Cut off the anthers from one of the parents (the white-flowered plant) to prevent self-fertilization. Dust pollen from the male parent (the purple-flowered plant) onto the style of the white flower (the female parent). This results in cross-fertilization, the fertilization of one plant with pollen from another.

2. The cross-fertilized plant produces seeds. Seeds may be scored for seed traits, such as smooth vs. wrinkled shape.

3. Seeds are grown into adult plants. Plants may be scored for adult traits, such as purple vs. white flower color.

pollen produced by anthers fertilize egg cells housed in the ovule of the same flower. However, for his experiments, Mendel prevented self-fertilization simply by cutting off the anthers (see Figure 12.3). Pollen to fertilize these flowers had to come from a different plant. This technique is called **cross-fertilization** or **cross-pollination,** or more simply and more generally, a *cross.* Using this method Mendel tested the effects of mating pea plants of different parental types.

To begin his experiments, Mendel chose pea plants that were known to be **true-breeding** (also called *pure-breeding*); that is, when self-fertilized, or more simply, *selfed,* they passed traits without change from one generation to the next.

Mendel First Worked with Crosses of Plants Differing in One Character

Flower color was among the seven characters Mendel selected for study; one true-breeding variety of peas had purple flowers, and the other true-breeding variety had white flowers (see Figure 12.3). Would these traits blend evenly if plants with purple flowers were crossed with plants with white flowers?

To answer this question, Mendel fertilized a white flower with pollen from a purple flower as shown in Figure 12.3. In this cross, the white-flowered plant is the female parent and purple-flowered plant is the male parent. A simple way geneticists present crosses like this is:

purple ♂ × white ♀
where the "×" stands for "cross."

Mendel also carried out a **reciprocal cross,** that is, a cross in which the two parents were switched. For this particular study, he used a purple-flowered plant as the female parent and a white-flowered plant as the male parent, performing a cross-pollination experiment in the opposite direction from that shown in Figure 12.3. The simple representation for this cross is:

white ♂ × purple ♀

Seeds were the result of the crosses; each seed contains a zygote, or embryo, that develops into a new pea plant. We call the plants used in an initial cross the parental or **P generation.** We call the first generation of offspring from a cross the **F_1 generation** (F stands for *filial; filius* = son).

From the purple ♂ × white ♀ cross, the plants that grew from the F_1 seeds all formed purple flowers, as if the trait for white flowers had disappeared. The flowers showed no evidence of blending. Exactly the same result was obtained for the reciprocal cross of white ♂ × purple ♀, showing that the sex of the parent did not affect the segregation of the traits.

Mendel then allowed purple-flowered F_1 plants to self, producing seeds that represented the **F_2 generation.** When he planted the F_2 seeds produced by this cross, plants with the white-flowered trait reappeared along with purple-flowered plants. Mendel counted 705 plants with purple flowers and 224 with white flowers, in a ratio that he noted was close to 3 purple:1 white.

Mendel made similar crosses that involved six other characters with pairs of traits; for example, the character of seed color has the traits yellow and green. In all cases, he observed a uniform F_1 generation, in which only one of the two traits was present **(Figure 12.4).** In the F_2 generation, the missing trait reappeared, and both traits were present among the offspring. Moreover, Mendel noted that, for each character, the ratio of the two traits was also close to 3:1.

Single-Character Crosses Led Mendel to Propose the Principle of Segregation

Using his knowledge of mathematics, Mendel developed a set of hypotheses to explain the results of his crosses of plants differing in one character. His first hypothesis was: *The adult plants carry a pair of factors that govern the inheritance of each trait.* He correctly deduced that for each trait, an organism inherits one factor from each parent.

In modern terminology, Mendel's factors are *genes,* which are located on chromosomes; the different versions of a gene, producing different traits of a character, are **alleles** of the gene (see Section 11.1). Although Mendel did not use the terms *genes* and *alleles,* we use them in this chapter in our description of Mendel's work. Thus, there are two alleles of the gene that governs flower color in garden peas: one allele for purple flower color and another allele for white flower color. Organisms with two copies of each gene are known as diploids (see Section 11.1); the two alleles of a gene in a diploid individual may be identical or different.

How can we explain why one of the traits, such as white flowers, disappears in the F_1 generation and then reappears in the F_2 generation? Mendel deduced that the trait that "disappeared" in the F_1 generation was actually present but was masked by the "stronger" allele. Mendel called the masking effect **dominance.** Accordingly, Mendel's second hypothesis stated: *If an individual's pair of genes consists of different alleles, one allele is* ***dominant*** *over the other, which is* ***recessive.*** When a dominant allele of a gene is paired with a recessive allele of that gene, the dominant allele is expressed. By contrast, a recessive allele is expressed only when two copies of that allele are present. For example, the allele for purple flowers is dominant and the allele for white flowers is recessive.

As a third hypothesis, Mendel proposed: *The pairs of alleles that control a character* ***segregate*** *(separate) as gametes are formed; half the gametes carry one allele, and the other half carry the other allele.* This hypothesis is now known as Mendel's **Principle of Segregation.** During fertilization, fusion of the haploid maternal and paternal gametes produces a diploid nucleus called the *zygote nucleus.* The zygote nucleus receives one allele for the character from the male gamete and one allele for the same character from the female gamete, reuniting the pairs.

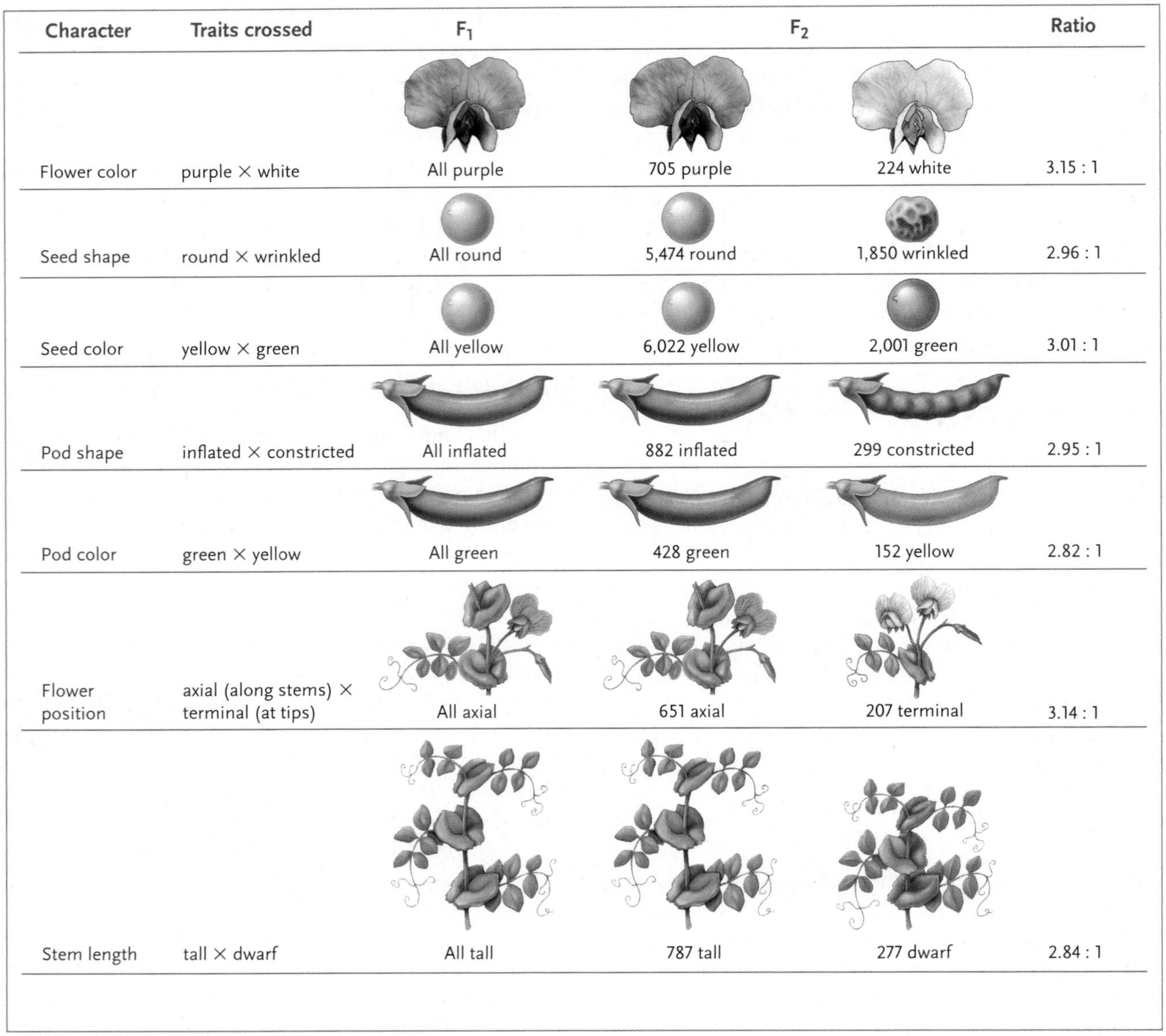

Character	Traits crossed	F_1	F_2		Ratio
Flower color	purple × white	All purple	705 purple	224 white	3.15 : 1
Seed shape	round × wrinkled	All round	5,474 round	1,850 wrinkled	2.96 : 1
Seed color	yellow × green	All yellow	6,022 yellow	2,001 green	3.01 : 1
Pod shape	inflated × constricted	All inflated	882 inflated	299 constricted	2.95 : 1
Pod color	green × yellow	All green	428 green	152 yellow	2.82 : 1
Flower position	axial (along stems) × terminal (at tips)	All axial	651 axial	207 terminal	3.14 : 1
Stem length	tall × dwarf	All tall	787 tall	277 dwarf	2.84 : 1

FIGURE 12.4
Mendel's crosses with seven characters in peas, including his results and the calculated ratios of offspring.

Figure 12.5 illustrates the principle of segregation in Mendel's flower color cross. Some important genetic terms are used in the figure:

- **Homozygote** (*homo* = same): A true-breeding individual with both alleles of a gene the same. A homozygote produces only one type of gamete: It contains one copy of that allele.
- **Homozygous:** An individual that is a homozygote is said to be homozygous for the particular allele of the gene.
- **Heterozygote** (*hetero* = different): An individual with two different alleles of a gene. A heterozygote produces two types of gametes: one type has a copy of one allele, the other type has a copy of the other, different, allele.
- **Heterozygous:** An individual that is a heterozygote is said to be heterozygous for the pair of different alleles of a gene.
- **Monohybrid** (*mono* = one; *hybrid* = an offspring of parents with different traits): An F_1 heterozygote produced from a cross that involves a single character.
- **Monohybrid cross:** A cross between two individuals that are each heterozygous for the same pair of alleles.

Mendel's hypotheses explain how individuals may differ genetically but still look the same. The *PP* and *Pp* plants, although genetically different, both have purple flowers (see Figure 12.5). In modern terminology, **genotype** refers to the *genetic constitution of an organism in terms of genes and alleles,* and **phenotype** (Greek *phainein* = to show) refers to its *appearance.* In this case, the two different *genotypes PP* and *Pp* produce the same purple-flower *phenotype.*

How does the genotype relate to the phenotype for the flower-color trait? Conceptually, the *P* allele of the gene encodes a product that is needed for the synthesis of the purple pigment in the flower. Therefore, a *PP* plant has purple flowers. The *p* allele is a mutant form of the gene; the product of this allele is inactive or mostly inactive and therefore, in *pp* plants, purple pigment cannot be made. In the absence of purple pigment, the flower is

FIGURE 12.5 **Experimental Research**

The Principle of Segregation in Mendel's Crosses Studying the Inheritance of Flower Color in Garden Peas

Question: How is flower color in garden peas inherited?

Experiment: Mendel crossed a true-breeding purple-flowered plant with a true-breeding white-flowered plant and analyzed the progeny through the F_1 and F_2 generations. We explain this cross here in modern terms.

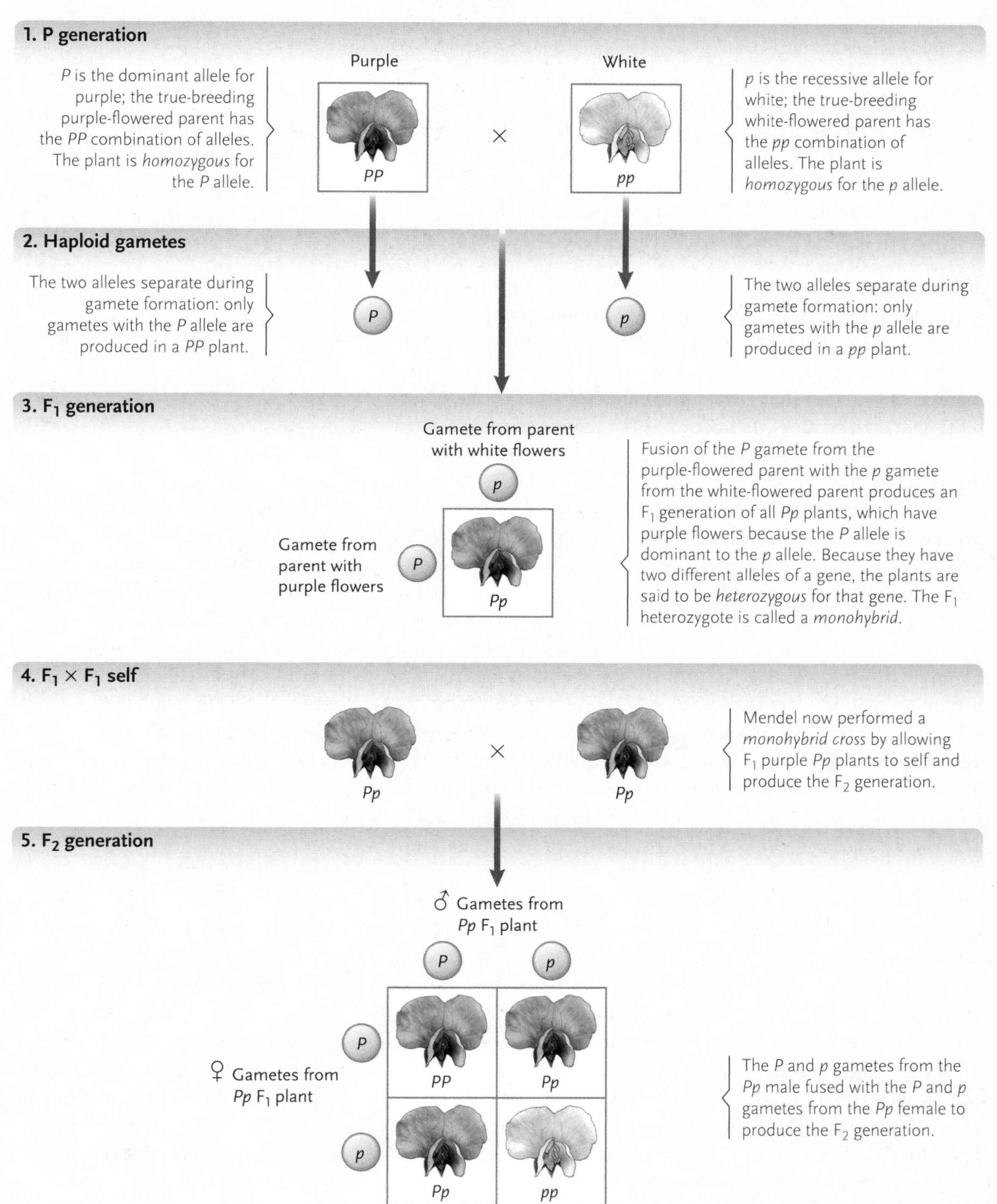

1. P generation
Purple
White
PP
pp
P is the dominant allele for purple; the true-breeding purple-flowered parent has the *PP* combination of alleles. The plant is *homozygous* for the *P* allele.
p is the recessive allele for white; the true-breeding white-flowered parent has the *pp* combination of alleles. The plant is *homozygous* for the *p* allele.
2. Haploid gametes
The two alleles separate during gamete formation: only gametes with the *P* allele are produced in a *PP* plant.
The two alleles separate during gamete formation: only gametes with the *p* allele are produced in a *pp* plant.
3. F_1 generation
Gamete from parent with white flowers
Gamete from parent with purple flowers
Pp
Fusion of the *P* gamete from the purple-flowered parent with the *p* gamete from the white-flowered parent produces an F_1 generation of all *Pp* plants, which have purple flowers because the *P* allele is dominant to the *p* allele. Because they have two different alleles of a gene, the plants are said to be *heterozygous* for that gene. The F_1 heterozygote is called a *monohybrid*.
4. $F_1 \times F_1$ self
Pp
Pp
Mendel now performed a *monohybrid cross* by allowing F_1 purple *Pp* plants to self and produce the F_2 generation.
5. F_2 generation
♂ Gametes from *Pp* F_1 plant
♀ Gametes from *Pp* F_1 plant
PP
Pp
Pp
pp
The *P* and *p* gametes from the *Pp* male fused with the *P* and *p* gametes from the *Pp* female to produce the F_2 generation.

Results: Mendel's selfing of the F_1 purple-flowered plants produced an F_2 generation consisting of 3/4 purple-flowered and 1/4 white-flowered plants. White flowers were inherited as a recessive trait, disappearing in the F_1 and reappearing in the F_2.

Conclusion: The results supported Mendel's Principle of Segregation hypothesis that the pairs of alleles that control a character segregate as gametes are formed, with half of the gametes carrying one allele, and the other half carrying the other allele

white. The *Pp* heterozygote is purple, rather than intermediate between purple and white, because the amount and activity of the product of the *P* allele is sufficient to enable a normal amount of purple pigment to be synthesized.

Thus, the results of Mendel's crosses support his three hypotheses:

1. The genes that govern genetic characters occur in pairs in individuals.
2. If different alleles are present in an individual's pair of genes, one allele is dominant over the other.
3. The two alleles of a gene segregate and enter gametes singly.

Mendel Could Predict Both Classes and Proportions of Offspring from His Hypotheses

Mendel could predict both which traits would appear in the offspring of a cross and their proportions. To understand how Mendel's hypotheses allowed him to predict the proportions of offspring resulting from a genetic cross, let us review the mathematical rules that govern **probability**—that is, the possibility that an outcome will occur if it is a matter of chance, as in the random fertilization of an egg by a sperm cell that contains one allele or another.

In the mathematics of probability, we predict the likelihood of an outcome on a scale of 0 to 1. A certain outcome has a probability of 1, and an impossible outcome has a probability of 0. If two different outcomes are equally likely, as in getting heads or tails in flipping a coin, we determine the probability of one of the outcomes by dividing that outcome by the total number of possible outcomes. The probability of obtaining a head in flipping a coin is 1 divided by 2, or 1/2. The probability of obtaining a tail is also 1 divided by 2, or 1/2. The probabilities of all the possible outcomes, when added together, must equal 1. Thus, a coin flip has only two possible outcomes, heads or tails, each with a probability of 1/2; the sum of these probabilities is: $1/2 + 1/2 = 1$.

THE PRODUCT RULE IN PROBABILITY What is the chance of flipping two heads in succession? Because the outcome of one flip has no effect on the next one, the two successive flips are independent. When two or more events are independent, we calculate the probability that they will occur in succession using the **product rule**—we multiply their individual probabilities. That is, the probability that events A and B *both* will occur equals the probability of event A *multiplied* by the probability of event B. For example, the probability of getting heads on the first flip is 1/2; the probability of heads on the second flip is also 1/2 **(Figure 12.6).** Because the events are independent, the probability of getting two heads in a row is $1/2 \times 1/2 = 1/4$. Applying the same principles, the probability of getting two tails is also $1/2 \times 1/2 = 1/4$ (see Figure 12.6). Similarly, because the sex of one child has no effect on the sex of the next child in a family, the probability of having four girls in a row is the product of their individual probabilities (very close to 1/2 for each birth): $1/2 \times 1/2 \times 1/2 \times 1/2 = 1/16$.

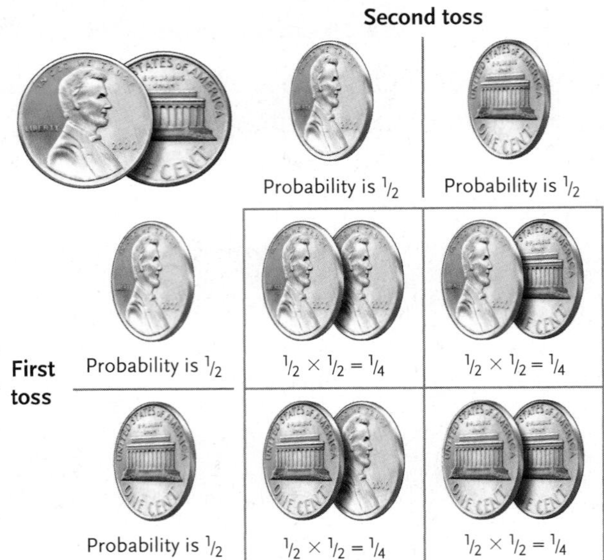

Second toss
Probability is 1/2
Probability is 1/2
First toss
Probability is 1/2
1/2 × 1/2 = 1/4
1/2 × 1/2 = 1/4
Probability is 1/2
1/2 × 1/2 = 1/4
1/2 × 1/2 = 1/4

FIGURE 12.6
Rules of probability. For each coin toss, the probability of a head is 1/2; the probability of a tail is also 1/2. Because the outcome of the first toss is independent of the outcome of the second, the combined probabilities of the outcomes of successive tosses are calculated by multiplying their individual probabilities according to the product rule.

THE SUM RULE IN PROBABILITY We apply a different relationship, the **sum rule,** when there are two or more different ways of obtaining the same outcome. Returning to the coin toss example, we can determine the probability of getting a head and a tail in two tosses. We could toss the coin twice and get a head, then a tail. The probability that this will occur is 1/2 for the head × 1/2 for the tail = 1/4 (see Figure 12.6). However, we could toss the coin twice and get first a tail, then a head. The probability that this will occur is 1/2 for the tail × 1/2 for the head, which also = 1/4 (see Figure 12.6). Therefore, for the probability of tossing a head and a tail, we sum the individual probabilities to get the final probability: here, $1/4 + 1/4 = 1/2$.

PROBABILITY IN MENDEL'S CROSSES The same rules of probability just discussed apply to Mendel's crosses. For example, **Figure 12.7** shows the rules of probability applied to calculating the proportion of *PP, Pp,* and *pp* F_2 plants from a cross of two F_1 *Pp* purple-flowered plants.

What if we want to know the probability of obtaining purple flowers in the cross *Pp* × *Pp*? In this case, the rule of addition applies, because there are two ways to get purple flowers: genotypes *PP* and *Pp*. Adding the individual probabilities of these combinations shown in Figure 12.7, 1/4 *PP* + 1/2 *Pp,* gives a total of 3/4, indicating that three-fourths of the F_2 offspring are expected to have purple flowers.

What is shown in Figure 12.7 is the **Punnett square** method for determining the genotypes of offspring and their expected proportions. The method was named for its originator, Reginald Punnett, a British geneticist who worked in the early part of the twentieth century. To use the Punnett square, write the probability of obtaining gametes with each type of allele from one parent at the top of the diagram and write the chance of obtaining each type

FIGURE 12.7
Punnett square method for predicting offspring and their ratios in genetic crosses. The example is the $F_1 \times F_1$ self of purple-flowered plants from Figure 12.5 to produce the F_2 generation. Each cell shows the genotype and proportion of one type of F_2 plant.

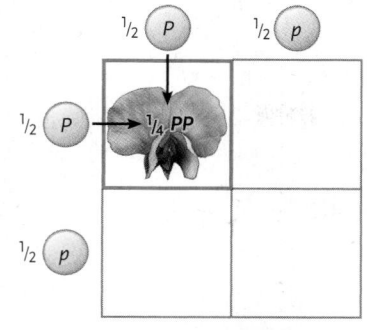

Gametes from F_1 purple *Pp* plant

Gametes from F_1 purple *Pp* plant

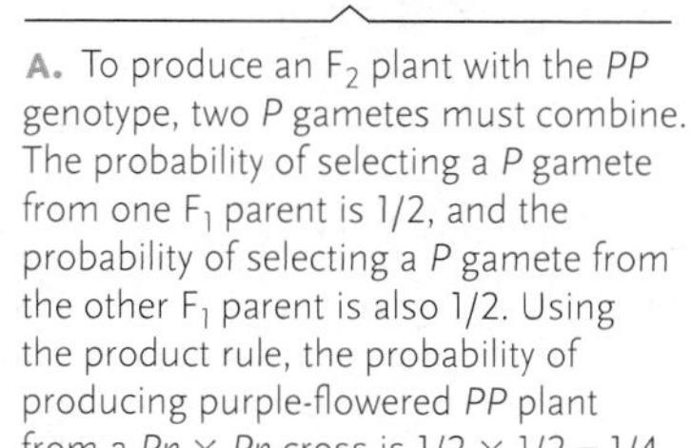

1/2 P
1/2 p
1/2 P
1/2 p
1/4 PP

A. To produce an F_2 plant with the *PP* genotype, two *P* gametes must combine. The probability of selecting a *P* gamete from one F_1 parent is 1/2, and the probability of selecting a *P* gamete from the other F_1 parent is also 1/2. Using the product rule, the probability of producing purple-flowered *PP* plant from a $Pp \times Pp$ cross is $1/2 \times 1/2 = 1/4$.

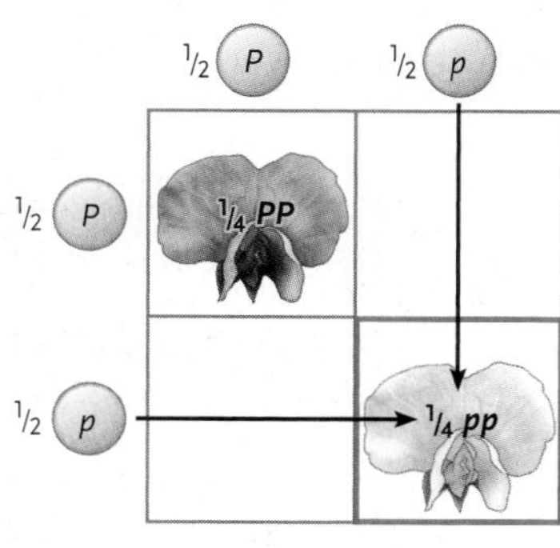

1/2 P
1/2 p
1/2 P
1/2 p
1/4 PP
1/4 pp

B. To produce an F_2 plant with the *pp* genotype, two *p* gametes must combine. The probability of selecting a *p* gamete from one F_1 parent is 1/2, and the probability of selecting a *p* gamete from the other F_1 parent is also 1/2. Using the product rule, the probability of producing white-flowered *pp* plant from a $Pp \times Pp$ cross is $1/2 \times 1/2 = 1/4$.

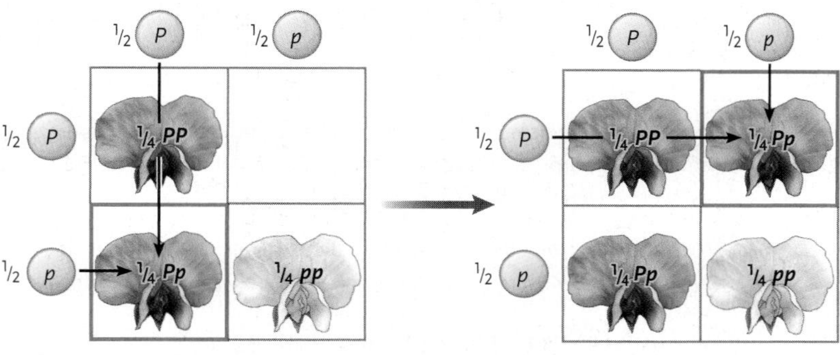

1/2 P
1/2 p
1/2 P
1/2 p
1/4 PP
1/4 Pp
1/4 pp
1/2 P
1/2 p
1/2 P
1/2 p
1/4 PP
1/4 Pp
1/4 Pp
1/4 pp

C. To produce an F_2 plant with the *Pp* genotype, a *P* gamete must combine with a *p* gamete. The cross $Pp \times Pp$ can produce *Pp* offspring in two different ways: (1) a *P* gamete from the first parent can combine with a *p* gamete from the second parent; or (2) a *p* gamete from the first parent can combine with a *P* gamete from the second parent. We apply the sum rule to obtain the combined probability: each of the ways to get *Pp* has an individual probability of 1/4, so the probability of *Pp*, purple-flowered offspring is $1/4 + 1/4 = 1/2$.

of allele from the other parent on the left side. Then fill in the cells by combining the alleles from the top and from the left and multiply their individual probabilities.

Mendel Used a Testcross to Check the Validity of His Hypotheses

Mendel realized that he could assess the validity of his hypotheses by determining whether they could be used successfully to *predict* the outcome of a cross of a different type than he had tried so far. Accordingly, he crossed an F_1 plant with purple flowers, assumed to have the heterozygous genotype *Pp,* with a true-breeding white-flowered plant, with the homozygous genotype *pp* **(Figure 12.8, Experiment 1).** There are two expected classes of offspring, *Pp* and *pp,* both with a probability of 1/2. Thus, the phenotypes of the offspring are expected to be 1 purple-flowered : 1 white-flowered. Mendel's actual results closely approach that ratio. Mendel also made the same type of cross with all the other traits used in his study, including those traits affecting seed shape, seed color, and plant height, and found the same 1 : 1 ratio.

A cross between an individual with the dominant phenotype and a homozygous recessive individual, such as the one just described, is called a **testcross.** Geneticists use a testcross to determine whether an individual with a dominant trait is a heterozygote or a homozygote, because their phenotypes are identical. If the offspring of the testcross are of two types, with half displaying the dominant trait and half the recessive trait, then the individual in question must be a heterozygote (see Figure 12.8, Experiment 1). If all the offspring display the dominant trait, the individual in question must be a homozygote. For example, the cross $PP \times pp$ gives all *Pp* progeny, which show the dominant purple flower phenotype **(Figure 12.8, Experiment 2).**

Obviously, the testcross method cannot be used for humans. However, it can be used in reverse, by noting the traits present in families over several generations and working backward to deduce whether a parent must have been a homozygote or a heterozygote (see also Chapter 13).

Mendel Tested the Independence of Different Genes in Crosses

Mendel next asked what happens in crosses when more than one character is involved. Would the alleles of the genes controlling the different characters be inherited independently, or would they interact to alter their expected proportions in offspring?

To answer these questions, Mendel crossed parental strains of peas that had differences in two of the characters he was studying: seed shape and seed color. His single-character crosses had shown each was controlled by a pair of alleles of one gene. For seed shape, round is dominant to wrinkled: the homozygous *RR* or heterozygous *Rr* genotypes produce round seeds and the homozygous *rr* genotype produces wrinkled seeds. For seed color, yellow is dominant to green: the *YY* or *Yy* genotypes produce yellow seeds; the homozygous *yy* genotype produces green seeds. **Figure 12.9** shows the cross of a true-breeding plant with round and yellow seeds *(RR YY)* with a true-breeding plant with wrinkled and green seeds *(rr yy)* through to the F_2 generation.

New genetic terms are used in the figure:

Dihybrid (*di* = two): An F_1 that is produced from a cross that involves two characters and is heterozygous for each of the pairs of alleles of the two genes involved. An example would

FIGURE 12.8 **Experimental Research**

Testing the Predicted Outcomes of Genetic Crosses

Question: How can it be determined whether a plant with the dominant phenotype is a heterozygote or a homozygote?

Experiment 1: Mendel crossed an F_1 plant with purple flowers, predicted to have a *Pp* genotype, with a true-breeding white-flowered plant and analyzed the flower color phenotypes in the offspring.

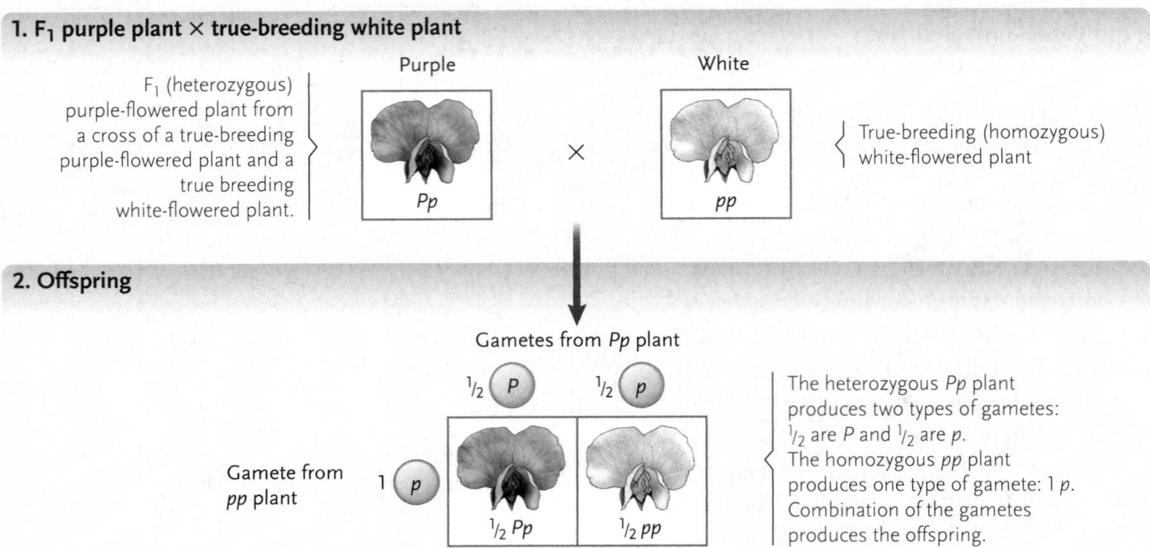

1. F1 purple plant × true-breeding white plant
Purple
White
F1 (heterozygous) purple-flowered plant from a cross of a true-breeding purple-flowered plant and a true breeding white-flowered plant.
Pp
×
pp
True-breeding (homozygous) white-flowered plant
2. Offspring
Gametes from Pp plant
1/2 P
1/2 p
Gamete from pp plant
1 p
1/2 Pp
1/2 pp
The heterozygous Pp plant produces two types of gametes: 1/2 are P and 1/2 are p. The homozygous pp plant produces one type of gamete: 1 p. Combination of the gametes produces the offspring.

Results: The predicted outcome of crossing a purple-flowered heterozygote with a true-breeding white-flowered plant is a ratio of 1 *Pp* purple-flowered : 1 *pp* white-flowered plants. Mendel observed 85 purple-flowered and 81 white-flowered plants, results that were close to those predicted.

Experiment 2: Mendel crossed a true-breeding plant with purple flowers, predicted to have a *PP* genotype, with a true-breeding white-flowered plant and analyzed the flower color phenotypes in the offspring.

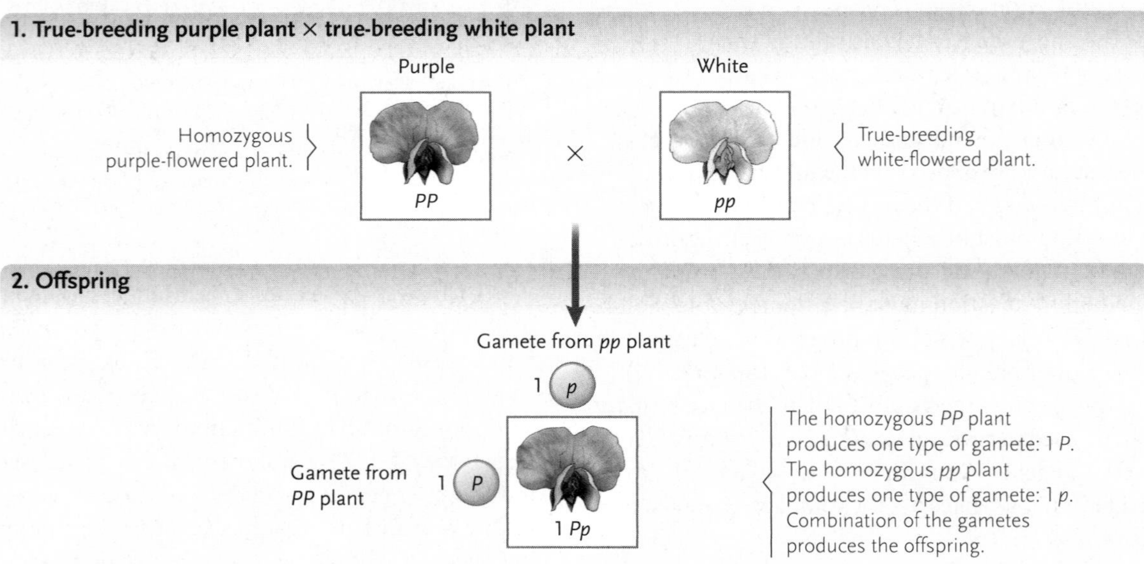

1. True-breeding purple plant × true-breeding white plant
Purple
White
Homozygous purple-flowered plant.
PP
×
pp
True-breeding white-flowered plant.
2. Offspring
Gamete from pp plant
1 p
Gamete from PP plant
1 P
1 Pp
The homozygous PP plant produces one type of gamete: 1 P. The homozygous pp plant produces one type of gamete: 1 p. Combination of the gametes produces the offspring.

Results: The outcome of crossing a purple-flowered homozygote with a true-breeding white-flowered plant is all *Pp* plants, which have purple flowers.

Conclusion: The outcome of a cross between (here) a plant with a dominant phenotype and a plant with a recessive phenotype—a testcross—gives a different result depending on whether the plant with the dominant phenotype is a homozygote or a heterozygote. Therefore, a testcross is a useful way to determine the genotype for an individual with a dominant phenotype.

FIGURE 12.9 **Experimental Research**

The Principle of Independent Assortment

Question: Do alleles of genes for two different characters in garden peas assort independently in a cross?

Experiment: Mendel crossed a true-breeding plant with round and yellow seeds with a true-breeding plant with wrinkled and green seeds and analyzed the progeny through the F_1 and F_2 generations. We explain this cross here in modern terms.

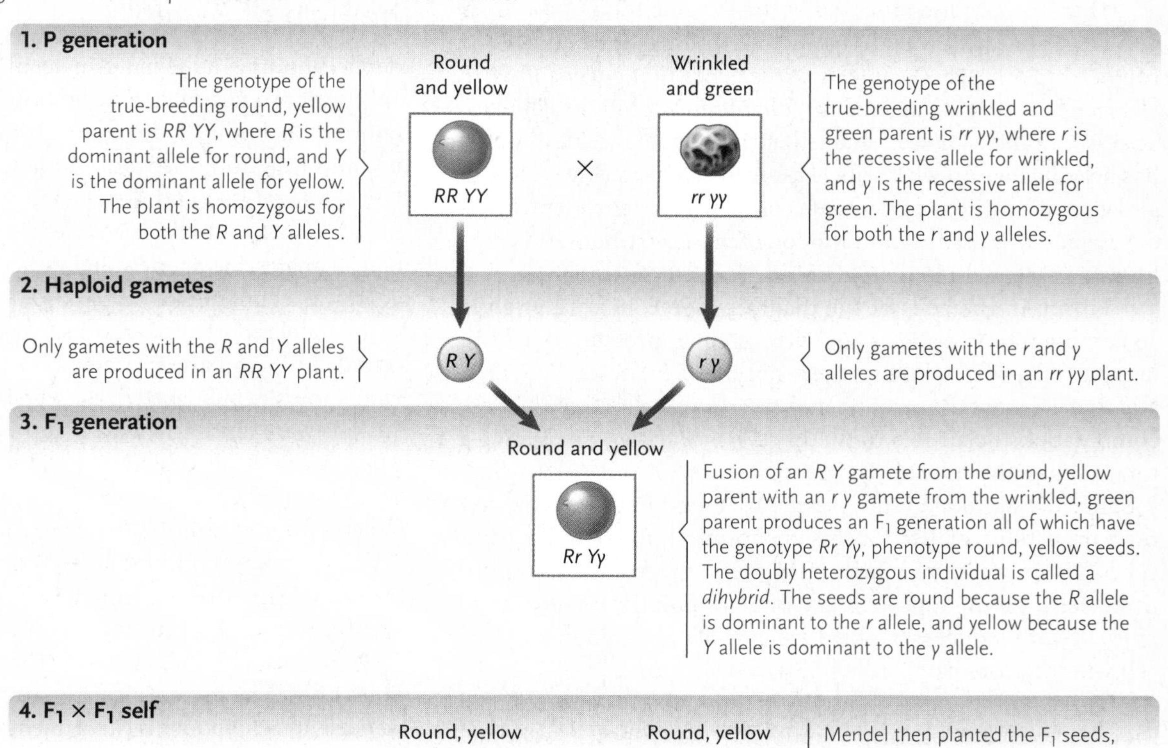

1. P generation
Round and yellow
Wrinkled and green
The genotype of the true-breeding round, yellow parent is *RR YY*, where *R* is the dominant allele for round, and *Y* is the dominant allele for yellow. The plant is homozygous for both the *R* and *Y* alleles.
RR YY
×
rr yy
The genotype of the true-breeding wrinkled and green parent is *rr yy*, where *r* is the recessive allele for wrinkled, and *y* is the recessive allele for green. The plant is homozygous for both the *r* and *y* alleles.
2. Haploid gametes
Only gametes with the *R* and *Y* alleles are produced in an *RR YY* plant.
R Y
r y
Only gametes with the *r* and *y* alleles are produced in an *rr yy* plant.
3. F_1 generation
Round and yellow
Rr Yy
Fusion of an *R Y* gamete from the round, yellow parent with an *r y* gamete from the wrinkled, green parent produces an F_1 generation all of which have the genotype *Rr Yy*, phenotype round, yellow seeds. The doubly heterozygous individual is called a *dihybrid*. The seeds are round because the *R* allele is dominant to the *r* allele, and yellow because the *Y* allele is dominant to the *y* allele.

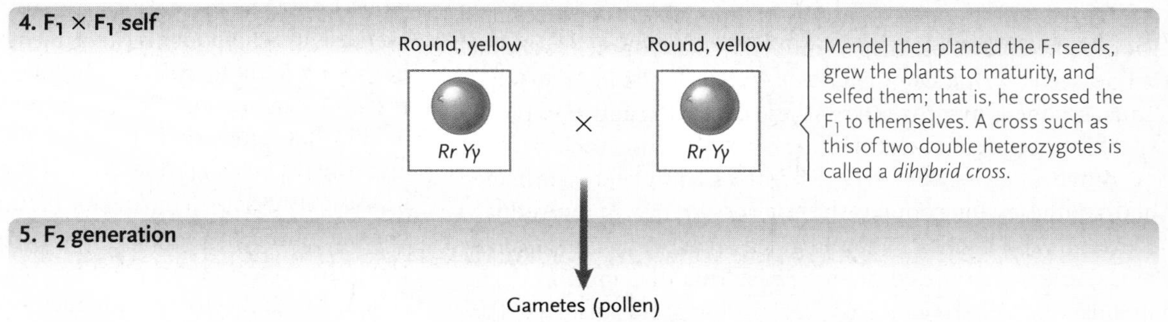

4. $F_1 \times F_1$ self
Round, yellow
Round, yellow
Rr Yy
×
Rr Yy
Mendel then planted the F_1 seeds, grew the plants to maturity, and selfed them; that is, he crossed the F_1 to themselves. A cross such as this of two double heterozygotes is called a *dihybrid cross*.

5. F_2 generation

Gametes (pollen)

Gametes (eggs)	1/4 *R Y*	1/4 *R y*	1/4 *r Y*	1/4 *r y*
1/4 *R Y*	1/16 *RR YY*	1/16 *RR Yy*	1/16 *Rr YY*	1/16 *Rr Yy*
1/4 *R y*	1/16 *RR Yy*	1/16 *RR yy*	1/16 *Rr Yy*	1/16 *Rr yy*
1/4 *r Y*	1/16 *Rr YY*	1/16 *Rr Yy*	1/16 *rr YY*	1/16 *rr Yy*
1/4 *r y*	1/16 *Rr Yy*	1/16 *Rr yy*	1/16 *rr Yy*	1/16 *rr yy*

If the alleles that control seed shape and seed color assort independently, each F_1 plant grown from the seeds would produce four types of gametes: the *R* allele for seed shape would go to a gamete with either the *Y* or *y* allele for seed color, and similarly, the *r* allele would go to a gamete with either the *Y* or *y* allele. Thus, independent assortment of genes from the *Rr Yy* parents is expected to produce four types of gametes with equal probability: 1/4 *R Y*, 1/4 *R y*, 1/4 *r Y*, and 1/4 *r y*. Random fusion of the four different male gametes with the four different female gametes produces the F_2 generation.

Results: Filling in the cells of the Punnett square gives 16 combinations, all with an equal probability of 1 in every 16 offspring if the alleles of the genes for the two characters assort independently. The 16 combinations resulting from independent assortment yield an expected phenotypic ratio in the F_2 of 9 round yellow : 3 round green : 3 wrinkled yellow : 1 wrinkled green. Mendel's selfing of the F_1 *Rr Yy* round–yellow plants produced an F_2 generation consisting of 315 round yellow : 108 round green : 101 wrinkled yellow : 32 wrinkled green, which is close to a 9 : 3 : 3 : 1 ratio (3 : 1 for round : wrinkled, and 3 : 1 for yellow : green).

Conclusion: The results indicate that the alleles of the genes for the two characters assort independently during the formation of gametes.

be *Aa Bb,* where genes *A* and *B* control different characters, and the upper and lower case letters for the two genes are dominant and recessive alleles, respectively.

Dihybrid cross: A cross between two individuals that are each heterozygous for the pairs of alleles of two genes.

This 9:3:3:1 ratio observed in the F_2 generation of the cross was consistent with Mendel's previous findings if he added one further hypothesis: *The alleles of the genes that govern the two characters assort independently during formation of gametes.* That is, the allele for seed shape that the gamete receives (*R* or *r*) has no influence on which allele for seed color it receives (*Y* or *y*) and vice versa. The two events are completely independent. Mendel termed this assumption **independent assortment;** it is now known as Mendel's **Principle of Independent Assortment.**

Filling in the cells of the diagram (see Figure 12.9) gives 16 combinations of alleles, all with an equal probability of 1 in every 16 offspring. Of these, the genotypes *RR YY, RR Yy, Rr YY,* and *Rr Yy* all have the same phenotype: round yellow seeds. These combinations occur in 9 of the 16 cells in the diagram, giving a total probability of 9/16. The genotypes *rr YY* and *rr Yy,* which produce the wrinkled yellow seeds, are found in three cells, giving a probability of 3/16 for this phenotype. Similarly, the genotypes *RR yy* and *Rr yy,* which yield round green seeds, occur in three cells, giving a probability of 3/16. Finally, the genotype *rr yy,* which produces wrinkled green seeds, is found in only one cell and therefore has a probability of 1/16.

The actual results obtained by Mendel closely approximated the expected 9:3:3:1 ratio of round yellow:round green:wrinkled yellow seeds:wrinkled green seeds, as developed in Figure 12.9. Thus, Mendel's first three hypotheses, with the added hypothesis of independent assortment, explain the observed results of his dihybrid cross. Mendel's testcrosses completely confirmed his hypotheses; for example, the testcross *Rr Yy* × *rr yy* produced 55 round yellow seeds, 51 round green seeds, 49 wrinkled yellow seeds, and 53 wrinkled green seeds. This distribution corresponds well with the expected 1:1:1:1 ratio in the offspring. (Try to set up a Punnett square for this cross and predict the expected classes of offspring and their frequencies.)

What is the molecular basis for the seed shape character differences? The normal *R* allele encodes an enzyme, starch-branching enzyme 1 (SBEI), which is required to produce a branched form of starch called amylopectin. The *r* allele is a mutant form of the gene, resulting in an inactive form of the enzyme and, therefore, no amylopectin production. Round seeds (*RR* and *Rr* genotypes) contain amylopectin, but wrinkled seeds (*rr* genotype) do not. The presence or absence of amylopectin is responsible for the seed shape. When *RR* or *Rr* seeds dry as they mature, they lose water and shrink uniformly, staying round. When *rr* seeds dry, they lose water irregularly, giving a wrinkled appearance.

Mendel's first three hypotheses provided a coherent explanation of the pattern of inheritance for alternate traits of the same character, such as purple and white for flower color. His fourth hypothesis, independent assortment, addressed the inheritance of traits for different characters, such as seed shape, seed color, and flower color, and showed that, instead of being inherited together, the traits of different characters were distributed independently to offspring.

Mendel's Research Founded the Field of Genetics

Mendel's findings anticipated in detail the patterns by which genes and chromosomes determine inheritance. Yet, when Mendel first reported his findings, during the nineteenth century, the structure and function of chromosomes and the patterns by which they are separated and distributed to gametes were unknown; meiosis remained to be discovered. In addition, his use of mathematical analysis was a new and radical departure from the usual biological techniques of his day.

Mendel reported his results to a small group of fellow intellectuals in Brünn and presented his results in 1866 in a natural history journal published in the city. But, Mendel's scientific conclusions were not immediately appreciated. His article received little notice outside of Brünn, and those who read it were unable to appreciate the significance of his findings. His work was overlooked until the early 1900s, when three investigators—Hugo de Vries in Holland, Carl Correns in Germany, and Erich von Tschermak in Austria—independently performed a series of breeding experiments similar to Mendel's and reached the same conclusions. These investigators, in searching through previously published scientific articles, discovered to their surprise Mendel's article about his experiments conducted 34 years earlier. Each gave credit to Mendel's discoveries, and the quality and far-reaching implications of his work were at last realized. Mendel died in 1884, 16 years before the rediscovery of his experiments and conclusions, and thus he never received the recognition that he so richly deserved during his lifetime. Mendel was also unable to relate the behavior of his "factors" (genes) to cell structures because the critical information he required was not obtained until later, through the discovery of meiosis during the 1890s.

Sutton's Chromosome Theory of Inheritance Related Mendel's Genes to Chromosomes

By the time Mendel's results were rediscovered in the early 1900s, critical information from studies of meiosis was available. It was not long before Walter Sutton, a genetics graduate student at Columbia University in New York, recognized the similarities between the inheritance of the genes discovered by Mendel and the behavior of chromosomes in meiosis and fertilization **(Figure 12.10).**

In a historic article published in 1903, Sutton drew all the necessary parallels between genes and chromosomes:

- Chromosomes occur in pairs in sexually reproducing, diploid organisms, as do the alleles of each gene.
- The chromosomes of each pair are separated and delivered singly to gametes, as are the alleles of a gene.
- The separation of any pair of chromosomes in meiosis and gamete formation is independent of the separation of other

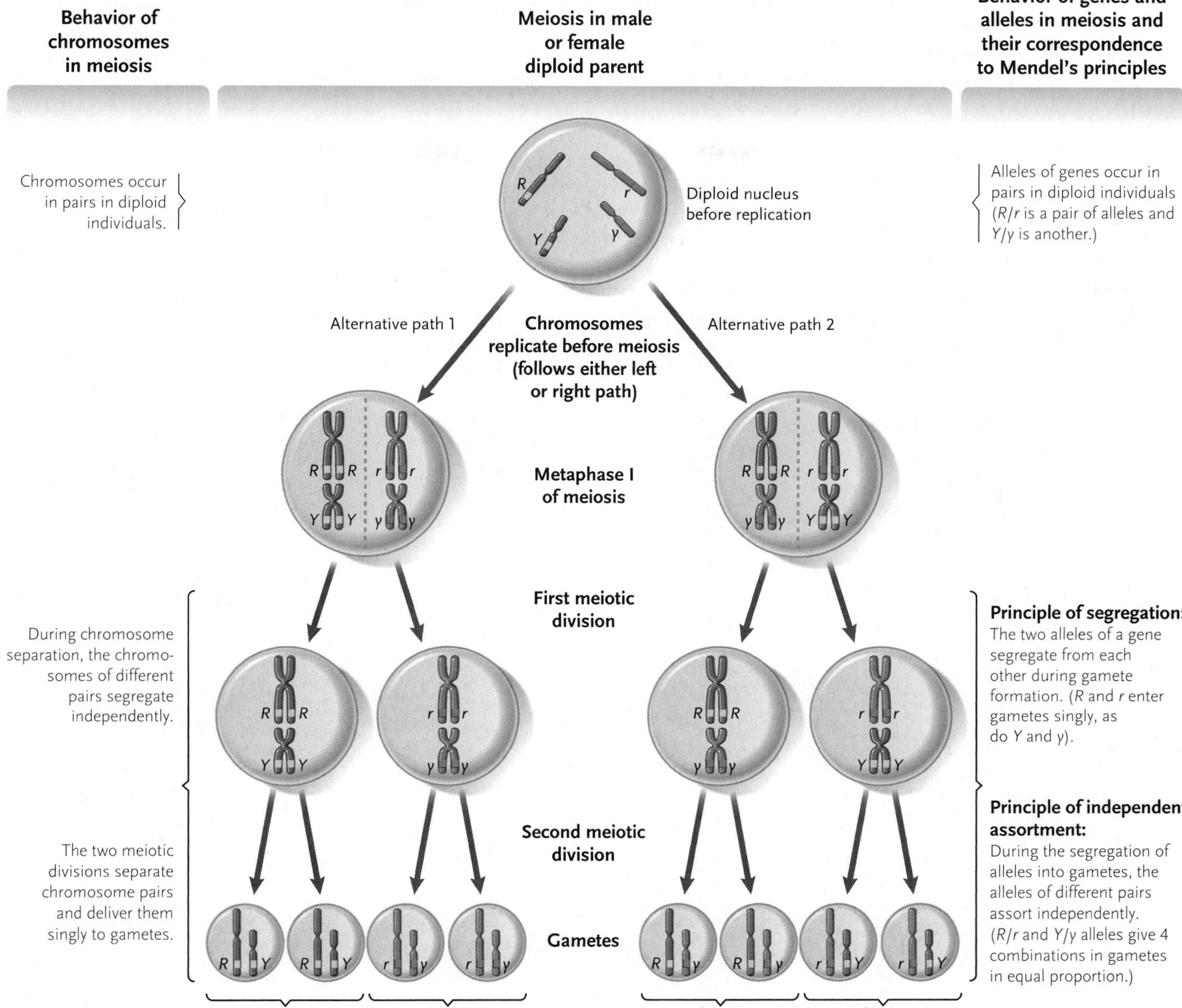

Behavior of chromosomes in meiosis
Meiosis in male or female diploid parent
Behavior of genes and alleles in meiosis and their correspondence to Mendel's principles
Chromosomes occur in pairs in diploid individuals.
Diploid nucleus before replication
Alleles of genes occur in pairs in diploid individuals (R/r is a pair of alleles and Y/y is another.)
Alternative path 1
Chromosomes replicate before meiosis (follows either left or right path)
Alternative path 2
Metaphase I of meiosis
First meiotic division
During chromosome separation, the chromosomes of different pairs segregate independently.
Principle of segregation: The two alleles of a gene segregate from each other during gamete formation. (R and r enter gametes singly, as do Y and y).
Second meiotic division
The two meiotic divisions separate chromosome pairs and deliver them singly to gametes.
Principle of independent assortment: During the segregation of alleles into gametes, the alleles of different pairs assort independently. (R/r and Y/y alleles give 4 combinations in gametes in equal proportion.)
Gametes
1/4 R Y
1/4 r y
1/4 R y
1/4 r Y

FIGURE 12.10
The parallels between the behavior of chromosomes and genes and alleles in meiosis. The gametes show four different combinations of alleles produced by independent segregation of chromosome pairs.

pairs (see Figure 12.10), as in the independent assortment of the alleles of different genes in Mendel's dihybrid crosses.

- One member of each chromosome pair is derived in fertilization from the male parent, and the other member is derived from the female parent, in an exact parallel with the two alleles of a gene.

From this total coincidence in behavior, Sutton correctly concluded that genes and their alleles are carried on the chromosomes, a conclusion known today as the **chromosome theory of inheritance.**

The exact parallel between the principles set forth by Mendel, and the behavior of chromosomes and genes during meiosis, is shown in Figure 12.10 for an *Rr Yy* diploid. For a dihybrid cross of *Rr Yy* × *Rr Yy,* when the gametes produced as in Figure 12.10 use randomly, the progeny will show a phenotypic ratio of 9:3:3:1. This mechanism explains the same ratio of gametes and progeny as the *Rr Yy* × *Ry Yy* cross in Figure 12.9.

The particular site on a chromosome at which a gene is located is called the **locus** (plural, *loci*) of the gene. The locus is a particular DNA sequence that encodes (typically) a protein responsible for the phenotype controlled by the gene. A locus for a

INSIGHTS FROM THE Molecular Revolution

Mendel's Dwarf Pea Plants: How does a gene defect produce dwarfing?

One of the seven characters Mendel studied was stem length. The stem length gene *Le* controls the length of the stem between the leaf branches of the pea plant. Plants homozygous or heterozygous for the dominant *Le* allele have a normal stem length, resulting in tall plants, whereas plants homozygous for the recessive allele *le* have a much reduced stem length, resulting in dwarf plants.

Research Question

What is the function of Mendel's *Le* gene in controlling stem length?

Experiment

Two independent research teams worked out the molecular basis for stem length in garden peas. The investigators, including Diane Lester and her colleagues at the University of Tasmania in Australia, and David Martin and his coworkers at Oregon State University, and Peter Hedden at the University of Bristol, England, were interested in learning the molecular differences between the *Le* and *le* alleles of the stem length gene.

Lester's team discovered that the gene codes for an enzyme that carries out a preliminary step in the synthesis of the plant hormone gibberellin, which, among other effects, causes the stems of plants to elongate. Martin's group cloned the gene, determined its complete DNA sequence, and analyzed its function. (Cloning techniques and DNA sequencing are described in Sections 18.1 and 18.3.)

Results

The sequence showed that the *Le* and *le* alleles of the gene encode two versions of the enzyme that catalyzes gibberellin synthesis, which differ by only a single amino acid (see Figure). Lester's group found that the faulty enzyme encoded by the *le* allele carries out its step (addition of a hydroxyl group to a precursor) much more slowly than the enzyme encoded by the normal *Le* allele. As a result, plants with the *le* allele have only about 5% as much gibberellin in their stems as *Le* plants. The reduced gibberellin levels limit stem elongation, resulting in dwarf plants.

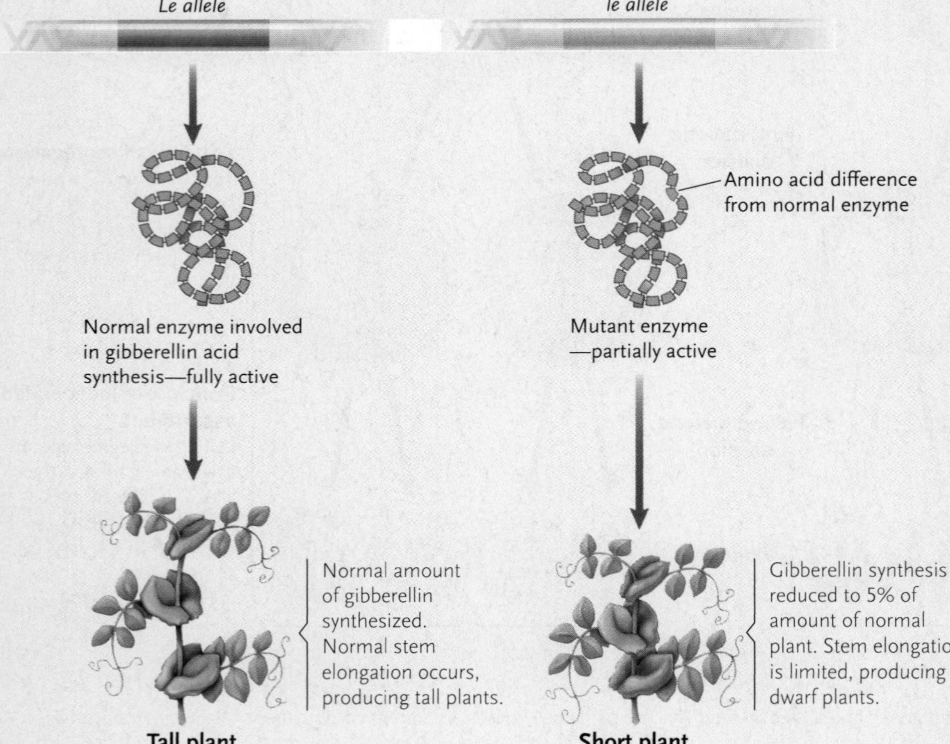

Le allele
le allele
Amino acid difference from normal enzyme
Normal enzyme involved in gibberellin acid synthesis—fully active
Mutant enzyme —partially active
Normal amount of gibberellin synthesized. Normal stem elongation occurs, producing tall plants.
Gibberellin synthesis reduced to 5% of amount of normal plant. Stem elongation is limited, producing dwarf plants.
Tall plant
Short plant

Conclusion

The methods of molecular biology allowed contemporary researchers to study a gene first studied genetically in the mid-nineteenth century. The findings leave little doubt that the gene codes for a plant hormone responsible for causing plant stems to elongate. Moreover, the recessive allele of the gene has a less active hormone due to a change in a single amino acid, and this change leads to the dwarf phenotype Mendel observed in his monastery garden.

Sources: D. R. Lester, J. J. Ross, P. J. Davies, and J. B. Reid. 1997. Mendel's stem length gene *(Le)* encodes a gibberellin 3ß-hydroxylase. *Plant Cell* 9:1435–1443; D. N. Martin, W. M. Proebsting, and P. Hedden. 1997. Mendel's dwarfing gene: cDNAs from the *Le* alleles and function of the expressed proteins. *Proceedings of the National Academy of Sciences USA* 94:8907–8911.

gene with two alleles, *A* and *a,* on a homologous pair of chromosomes is shown in **Figure 12.11.** At the molecular level, different alleles consist of differences in the DNA sequence of a gene, which may result in functional differences in the protein encoded by the gene. These differences are detected as distinct phenotypes in the offspring of a cross. *Insights from the Molecular Revolution* describes a molecular study that uncovered the mechanisms that control height in pea plants, one of the seven characteristics originally examined by Mendel.

All the genetics research conducted since the early 1900s has confirmed Mendel's basic hypotheses about inheritance. This research has shown that Mendel's conclusions apply to all types of organisms, from yeast and fruit flies to humans, and has led to the rapidly growing field of human genetics. In humans, a number of easily seen traits show inheritance patterns that follow Mendelian principles **(Figure 12.12);** for example, albinism, the lack of normal skin color, is recessive to normal skin color, and normally separated fingers are recessive to fingers with webs between them. Similarly, achondroplasia, the most frequent form of short-limb dwarfism, is a dominant trait that involves abnormal bone growth. Many human disorders that cannot be seen easily also show simple inheritance patterns. For instance, cystic fibrosis, in which a defect in the membrane transport of chloride ions leads to pulmonary and digestive dysfunctions and eventually death, is a recessive trait.

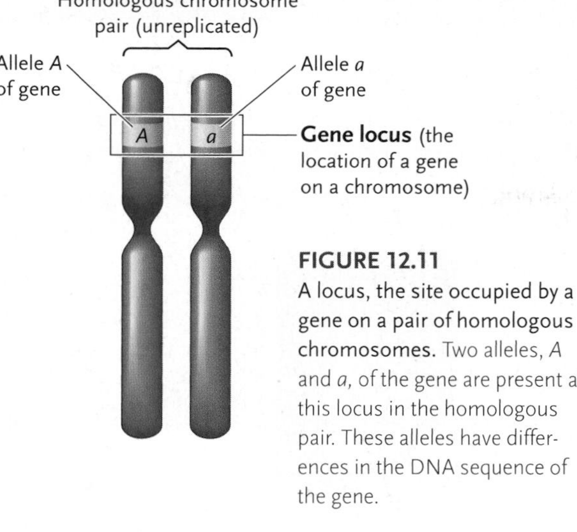
Homologous chromosome pair (unreplicated)
Allele A of gene
Allele a of gene
A
a
Gene locus (the location of a gene on a chromosome)

FIGURE 12.11
A locus, the site occupied by a gene on a pair of homologous chromosomes. Two alleles, *A* and *a*, of the gene are present at this locus in the homologous pair. These alleles have differences in the DNA sequence of the gene.

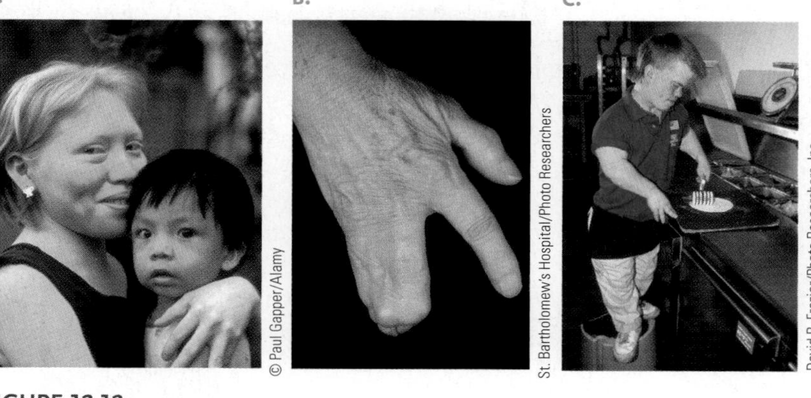
A.
B.
C.
© Paul Gapper/Alamy
St. Bartholomew's Hospital/Photo Researchers
David R. Frazier/Photo Researchers, Inc.

FIGURE 12.12
Human traits showing inheritance patterns that follow Mendelian principles. **(A)** Lack of normal skin color (albinism), a recessive trait. **(B)** Webbed fingers, a dominant trait. **(C)** Achondroplasia, or short-limbed dwarfism, a dominant trait.

The post-Mendel research has demonstrated additional patterns of inheritance (see the next section) that were not anticipated by Mendel and, in some circumstances, require modifications or additions to his hypotheses.

STUDY BREAK 12.1

1. **Two pairs of traits are segregating in a cross. Two parents produce 156 progeny that fall into 4 phenotypes. The numbers of offspring in the 4 phenotypes are 89, 31, 28, and 8. What are the genotypes of the two parents?**
2. **If instead, the four phenotypes in question 1 occur in approximately equal numbers, what are the genotypes of the parents? What is this kind of cross called?**

12.2 Later Modifications and Additions to Mendel's Hypotheses

The rediscovery of Mendel's research in the early 1900s produced an immediate burst of interest in genetics, and the research that followed greatly expanded our understanding of genes and their inheritance. That research fully supported Mendel's hypotheses, but also revealed many variations on the basic principles he had outlined. The following sections discuss some of these so-called extensions of Mendel's fundamental principles.

In Incomplete Dominance, Dominant Alleles Do Not Completely Mask Recessive Alleles

When one allele of a gene is not completely dominant over another allele of the same gene, it is said to show **incomplete dominance.** With incomplete dominance, the phenotype of the heterozygote is somewhere between the phenotypes of individuals that are homozygous for either of the alleles. Flower color in snapdragons shows incomplete dominance. One gene controls the flower color character, with one allele for red and another allele for white. Because one allele is not completely dominant to the other in incomplete dominance, we use a different genetic symbolism. That is, we use an italic letter that relates to the character, with superscripts for the different alleles. In this case, C signifies flower color and the superscript R is for red, and the superscript W is for white. That is, C^R is the allele for red color and C^W is the allele for white color. We use these symbols in **Figure 12.13,** which follows a cross of a true-breeding red-flowered snapdragon with a true-breeding, white-flowered snapdragon through to the F_2 generation. The key differences from a cross in which complete dominance is the case are: (1) The F_1 phenotype is intermediate between the phenotypes of the two parents; and (2) The phenotypes of F_2 individuals are directly determined by the different genotypes of those individuals, giving a 1 : 2 : 1 ratio rather than a 3 : 1 ratio as is characteristic for complete dominance.

We can explain the flower colors as follows: The C^R allele encodes an enzyme that produces a red pigment, but two alleles (C^RC^R) are necessary to produce enough of the active form of the enzyme to produce fully red flowers. The enzyme is completely inactive in C^WC^W plants, which produce colorless flowers that appear white because of the scattering of light by cell walls and other structures. With their single C^R allele, the C^RC^W heterozygotes of the F_1 generation can produce only enough pigment to give the flowers a pink color. When pink C^RC^W F_1 plants are crossed, the fully red and white colors reappear, together with the pink color, in exactly the same ratio as the ratio of genotypes produced from a cross of two heterozygotes in Mendel's experiments (for example, see Figure 12.7).

Some human disorders show incomplete dominance. For example, sickle-cell disease (see *Why It Matters*) is characterized by an alteration in the hemoglobin molecule that changes the shape of red blood cells when oxygen levels are low. An individual with sickle-cell disease is homozygous for a recessive allele that encodes a defective form of one of the polypeptides of the hemoglobin molecule. Individuals heterozygous for that recessive allele and the normal allele have a condition known as *sickle-cell trait,* which is a milder form of the disease because the individuals produce some normal polypeptides from the normal allele.

FIGURE 12.13 **Experimental Research**

Experiment Showing Incomplete Dominance of a Trait

Question: How is flower color in snapdragons inherited?

Experiment: Cross a true-breeding red-flowered snapdragon with a true-breeding white-flowered snapdragon and analyze the progeny through the F_1 and F_2 generations.

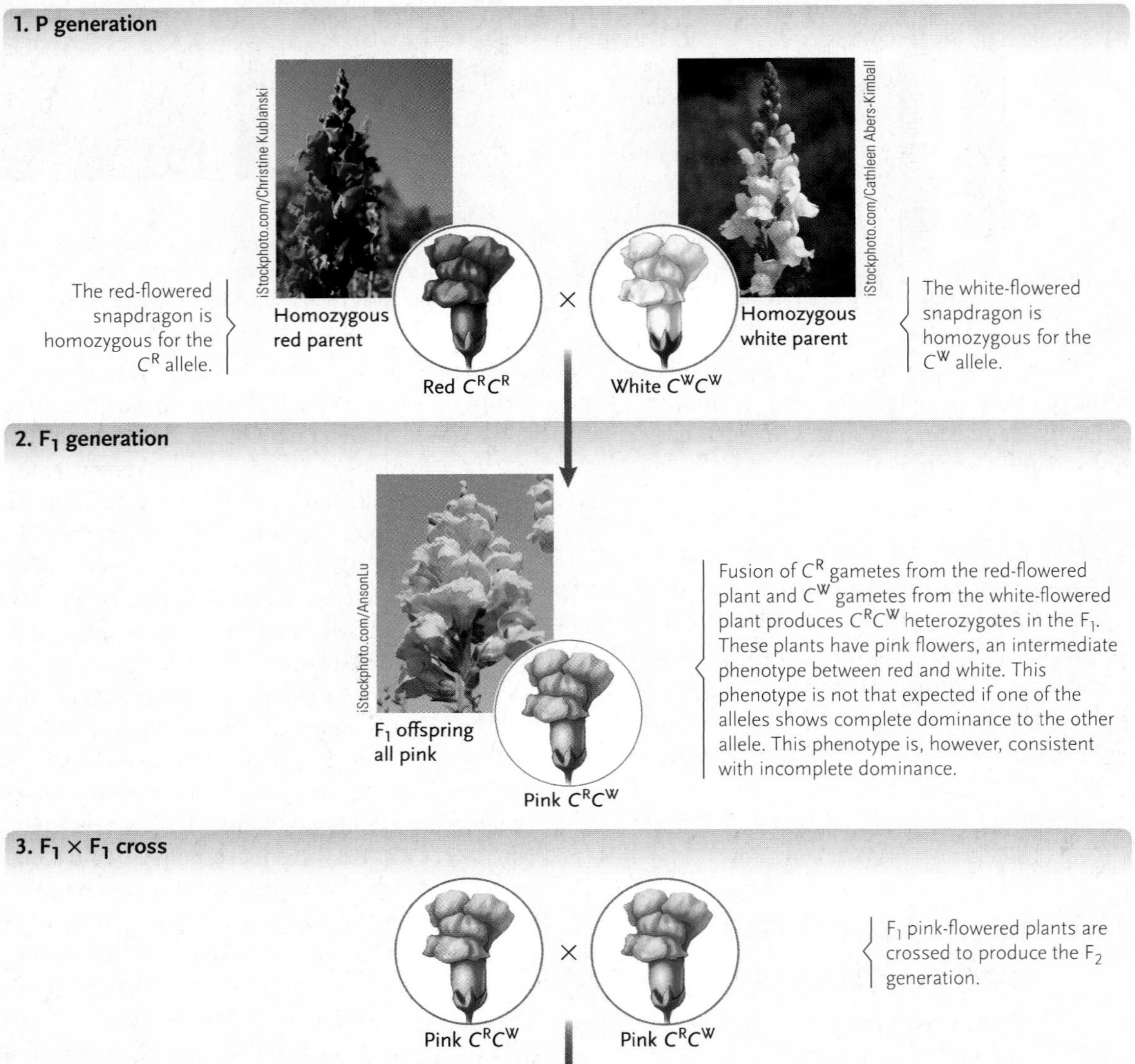

1. P generation

iStockphoto.com/Christine Kublanski

iStockphoto.com/Cathleen Abers-Kimball

The red-flowered snapdragon is homozygous for the C^R allele.

Homozygous red parent

Homozygous white parent

Red C^RC^R × White C^WC^W

The white-flowered snapdragon is homozygous for the C^W allele.

2. F_1 generation

iStockphoto.com/AnsonLu

F_1 offspring all pink

Pink C^RC^W

Fusion of C^R gametes from the red-flowered plant and C^W gametes from the white-flowered plant produces C^RC^W heterozygotes in the F_1. These plants have pink flowers, an intermediate phenotype between red and white. This phenotype is not that expected if one of the alleles shows complete dominance to the other allele. This phenotype is, however, consistent with incomplete dominance.

3. $F_1 \times F_1$ cross

Pink C^RC^W × Pink C^RC^W

F_1 pink-flowered plants are crossed to produce the F_2 generation.

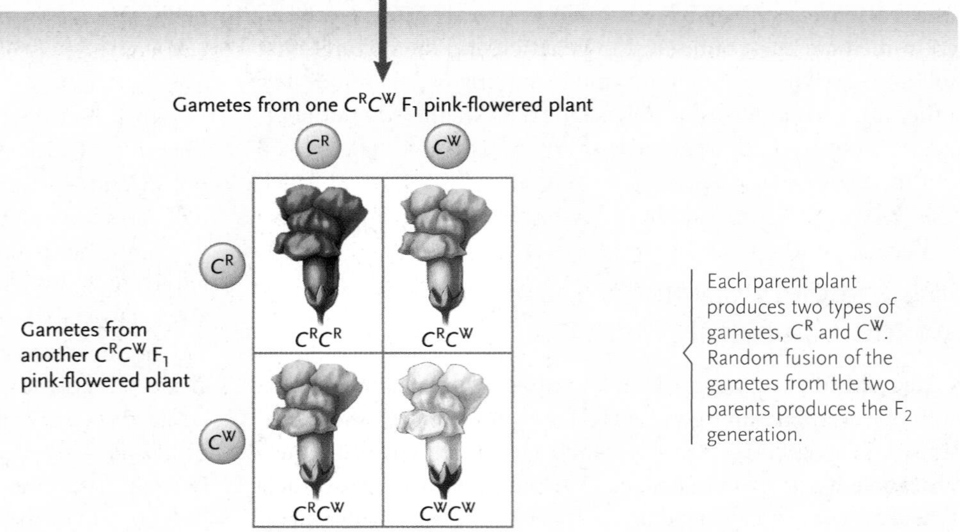

4. F_2 generation

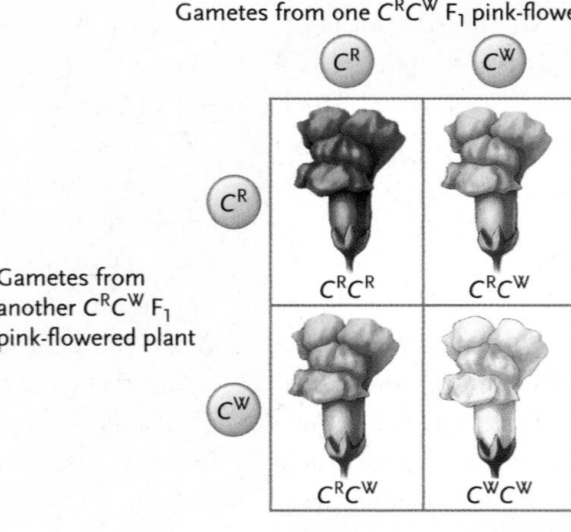

Gametes from one C^RC^W F1 pink-flowered plant
C^R
C^W
C^R
C^W
C^RC^R
C^RC^W
C^RC^W
C^WC^W
Gametes from another C^RC^W F1 pink-flowered plant

Each parent plant produces two types of gametes, C^R and C^W. Random fusion of the gametes from the two parents produces the F_2 generation.

Results: The F_2 generation shows a phenotypic ratio of 1 red : 2 pink : 1 white. Each phenotype results from a distinct genotype, C^RC^R for red flowers, C^RC^W for pink flowers, and C^WC^W for white flowers. The 1 : 2 : 1 phenotypic ratio is consistent with incomplete dominance.

Conclusion: In incomplete dominance, each genotype has a distinct phenotype. From a cross of two heterozygotes, the outcome is a phenotypic ratio of 1 : 2 : 1 rather than the 3 : 1 ratio characteristic of complete dominance.

In Codominance, the Effects of Different Alleles Are Equally Detectable in Heterozygotes

Codominance occurs when alleles have approximately equal effects in individuals, making the alleles equally detectable in heterozygotes. The inheritance of the human MN blood group presents an example of codominance. The L^M and L^N alleles of the MN blood group gene that control this character encode different forms of a glycoprotein molecule located on the surface of red blood cells. If the genotype is L^ML^M, only the M form of the glycoprotein is present and the blood type is M; if it is L^NL^N, only the N form is present and the blood type is N. In heterozygotes with the L^ML^N genotype, both glycoprotein types are present and can be detected, producing the blood type MN. Because each genotype has a different phenotype, the inheritance pattern for the MN blood group alleles is generally the same as for incompletely dominant alleles. The MN blood types do not affect blood transfusions and have relatively little medical importance.

In Multiple Alleles, More Than Two Alleles of a Gene Are Present in a Population

One of Mendel's major and most fundamental assumptions was that alleles (his factors) occur in pairs in individuals; in the pairs, the alleles may be the same or different. After the rediscovery of Mendel's principles, it soon became apparent that although alleles do indeed occur in pairs in individuals, **multiple alleles** (more than two different alleles of a gene) may be present if all the individuals of a population are considered. For example, for a gene *B*, there could be the normal allele, *B*, and several alleles with alterations in this gene, for example, b_1, b_2, b_3, and so on. Some individuals in a population may have the *B* and b_1 alleles of a gene; others, the b_2 and b_3 alleles; still others, the b_3 and b_5 alleles; and so on, for all possible combinations. That is, although any one individual can have only two alleles of the gene, there are more than two alleles in the population as a whole. Genes may certainly occur in many more than the four alleles of the example; for instance, one of the genes that plays a part in the acceptance or rejection of organ transplants in humans has more than 200 different alleles.

The multiple alleles of a gene each contain differences at one or more points in their DNA sequences **(Figure 12.14)**, which cause detectable alterations in the structure and function of proteins encoded by the alleles. Multiple alleles present no real difficulty in genetic analysis because each diploid individual still has only two of the alleles, allowing gametes to be predicted and traced through crosses by the usual methods.

HUMAN ABO BLOOD GROUP The human *ABO* blood group provides an example of multiple alleles, in a system that also exhibits both dominance and codominance. The ABO blood group was discovered in 1901 by Karl Landsteiner, an Austrian biochemist who was investigating the sometimes fatal outcome of attempts to transfer blood from one person to another. Landsteiner found that only certain combinations of four blood types, designated A, B, AB, and O, can be mixed safely in transfusions **(Table 12.1)**.

B allele 5'... ATGCAGATACCGATTACAGACCATAGG... 3'
3'... TACGTCTATGGCTAATGTCTGGTATCC... 5'

b_1 allele 5'... ATGCAGAGACCGATTACAGACCATAGG... 3'
3'... TACGTCTCTGGCTAATGTCTGGTATCC... 5'

b_2 allele 5'... ATGCAGATACCGACTACAGACCATAGG... 3'
3'... TACGTCTATGGCTGATGTCTGGTATCC... 5'

b_3 allele 5'... ATGCAGATACCGATTACAGTCCATAGG... 3'
3'... TACGTCTATGGCTAATGTCAGGTATCC... 5'

FIGURE 12.14
Multiple alleles. Multiple alleles consist of differences in the DNA sequence of a gene at one or more points, which result in detectable differences in the structure of the protein encoded by the gene. The differences shown here are single base-pair changes. The *B* allele is the normal allele, which encodes a protein with normal function. The three *b* alleles each have alterations of the normal protein-coding DNA sequence that may adversely affect the function of that protein.

Landsteiner determined that, in the wrong combinations, red blood cells from one blood type are agglutinated (clumped) by an agent in the serum of another type (the serum is the fluid in which the blood cells are suspended). The clumping was later found to depend on the action of an antibody in the blood serum. (Antibodies, protein molecules that interact with specific substances called antigens, are discussed in Chapter 43.)

The antigens responsible for the blood types of the ABO blood group are the carbohydrate parts of glycoproteins located on the surfaces of red blood cells (unrelated to the glycoprotein carbohydrates responsible for the blood types of the MN blood group). For example, people with type A blood have *antigen A* on their red blood cells, and anti-B antibodies in their blood. If a person with type A blood receives a transfusion of type B blood, their anti-B antibodies will cause the blood to clump. Table 12.1 shows how the four blood types of the human ABO blood group determine compatibility in transfusions.

The four blood types—A, B, AB, and O—are produced by different combinations of multiple (three) alleles of a single gene *I* designated I^A, I^B, and *i* **(Figure 12.15).** I^A and I^B are codominant alleles that are each dominant to the recessive *i* allele.

TABLE 12.1 Blood Types of the Human ABO Blood Group

Blood Type	Antigens	Antibodies	Blood Types Accepted in a Transfusion
A	A	Anti-B	A or O
B	B	Anti-A	B or O
AB	A and B	None	A, B, AB, or O
O	None	Anti-A, anti-B	O

FIGURE 12.15
Inheritance of the blood types of the human ABO blood group.

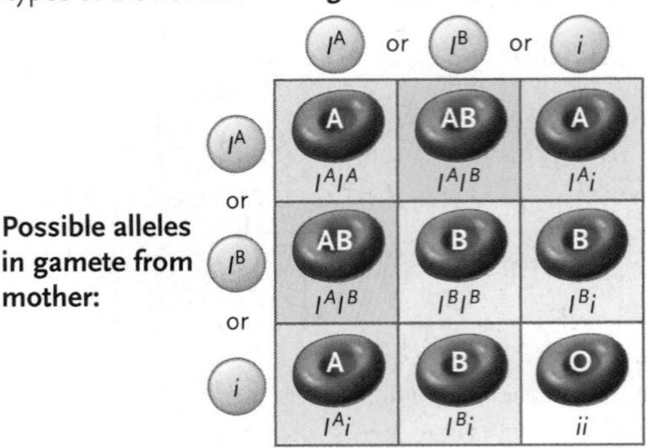
Possible alleles in gametes from father:
I^A or I^B or i
Possible alleles in gamete from mother:
I^A or I^B or i
A
I^AI^A
AB
I^AI^B
A
I^Ai
AB
I^AI^B
B
I^BI^B
B
I^Bi
A
I^Ai
B
I^Bi
O
ii

In Epistasis, Genes Interact, with the Activity of One Gene Influencing the Activity of Another Gene

The genetic characters discussed so far in this chapter, such as flower color, seed shape, and the blood types of the ABO group, are all produced by the alleles of single genes, with each gene functioning on its own. This is not the case for every gene. In **epistasis** (*epi* = on or over; *stasis* = standing or stopping), genes interact, with one or more alleles of a gene at one locus inhibiting or masking the effects of one or more alleles of a gene at a different locus. The result of epistasis is that some expected phenotypes do not appear among offspring.

FIGURE 12.16
An example of epistasis: the inheritance of coat color in Labrador retrievers.

A. **Black labrador**

B. **Chocolate brown labrador**

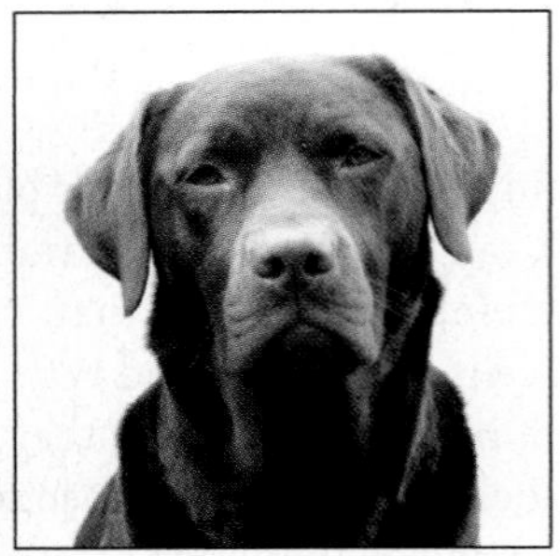

C. **Yellow labrador**

D. **Black × yellow labrador cross**

Homozygous parents:

F_1 puppies:

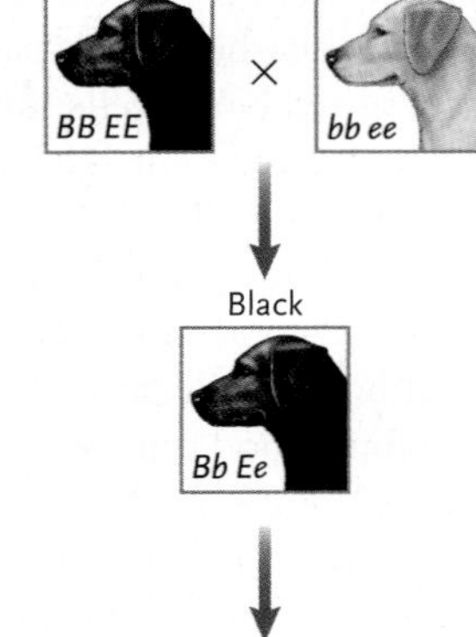
Black
Yellow
BB EE
×
bb ee
Black
Bb Ee

F_2 offspring from cross of two F_1 *Bb Ee* dogs:

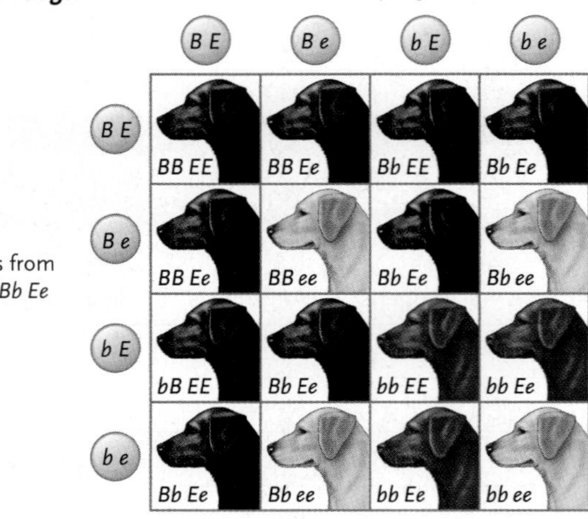
Gametes from one Bb Ee F1 dog:
B E
B e
b E
b e
Gametes from another Bb Ee F1 dog:
B E
B e
b E
b e
BB EE
BB Ee
Bb EE
Bb Ee
BB Ee
BB ee
Bb Ee
Bb ee
bB EE
Bb Ee
bb EE
bb Ee
Bb Ee
Bb ee
bb Ee
bb ee

F_2 phenotypic ratio is 9 black : 3 chocolate : 4 yellow

Labrador retrievers (Labs) may have black, chocolate brown, or yellow fur **(Figure 12.16A–C).** The different colors result from variations in the amount and distribution in hairs of a brownish black pigment called melanin. One gene, coding for an enzyme involved in melanin production, determines how much melanin is produced. The dominant *B* allele of this gene produces black fur color in *BB* or *Bb* Labs; less pigment is produced in *bb* dogs, which are chocolate brown. Another gene at a different locus determines whether the black or chocolate color appears at all, by controlling the deposition of pigment in hairs. A dominant allele *E* of this second gene permits pigment deposition, so that the black color in *BB* or *Bb* individuals, or the chocolate color in *bb* individuals, actually appears in the fur. Pigment deposition is almost completely blocked in homozygous recessive *ee* individuals, so the fur lacks melanin and has a yellow color whether the genotype for the *B* gene is *BB, Bb,* or *bb.* Thus, the *E* gene is epistatic to the *B* gene (that is, *E* and *B* interact).

Epistasis by the *E* gene eliminates some of the expected classes from crosses among Labs **(Figure 12.16D).** Rather than two separate classes, as would be expected from a dihybrid cross without epistasis, the *BB ee, Bb ee, bB ee,* and *bb ee* genotypes produce a single yellow phenotype. Therefore if we cross a true-breeding black Labrador with a true-breeding yellow Labrador of genotype *bb ee,* the F_1 puppies are *Bb Ee* black heterozygotes (see Figure 12.16). F_2 progeny produced by crossing F_1 dogs have the distribution: 9/16 black, 3/16 chocolate, and 4/16 yellow because of epistasis. That is, the ratio is 9:3:4 instead of the expected 9:3:3:1 ratio. Many other dihybrid crosses that involve epistatic interactions produce distributions that differ from the expected 9:3:3:1 ratio.

In human biology, researchers believe that gene interactions and epistasis are common. Current thinking is that epistasis is an important factor in determining an individual's susceptibility to common human diseases. That is, the different degrees of susceptibility are the result of different gene interactions in the individuals. A specific example is insulin resistance, a disorder in which muscle, fat, and liver cells do not use insulin correctly, with the result that glucose and insulin levels become high in the blood. This disorder is believed to be determined by several genes often interacting with one another.

In Polygenic Inheritance, a Character Is Controlled by the Common Effects of Several Genes

Some characters follow a pattern of inheritance in which there is a more or less even gradation of types, forming a continuous distribution, rather than "on" or "off" (discontinuous) effects such as the production of purple or white flowers in pea plants. For example, in the human population, people range from short to tall, in a continuous distribution of gradations in height between limits of about 4 and 7 feet. Typically, a continuous distribution of this type is the result of **polygenic inheritance,** in which several to many different genes contribute to the *same* character. Other characters that undertake a similar continuous distribution include skin color and body weight in humans, ear length in corn, seed color in wheat, and color spotting in mice. These characters are also known as **quantitative traits.** The individual genes that contribute to a quantitative trait are known as **quantitative trait loci** or QTLs.

Polygenic inheritance can be detected by defining classes of a variation, such as human body height of 60 inches in one class, 61 inches in the next class, 62 inches in the next class, and so on **(Figure 12.17)**. The number of individuals in each class is then plotted as a graph. If the plot produces a bell-shaped curve, with fewer individuals at the extremes and the greatest numbers clustered around the midpoint, it is a good indication that the trait is quantitative.

Polygenic inheritance is often modified by the environment. For example, height in humans is not the result of genetics alone. Poor nutrition during infancy and childhood is one environmental factor that can limit growth and prevent individuals from reaching the height expected from genetic inheritance; good nutrition can have the opposite effect. Thus, the average young adult in Japan today is several inches taller than the average adult in the 1930s, when nutrition was poorer. Similarly, individuals who live in cloudy, northern or southern climates usually have lighter skin color than individuals with the same genotype who live in sunny climates.

At first glance, the effects of polygenic inheritance might appear to support the idea that characteristics of parents are blended in their offspring. Commonly, people believe that the children in a family with one tall and one short parent will be of intermediate height. Although the children of such parents are most likely to be of intermediate height, careful genetic analysis of many such families shows that their offspring actually range over a continuum from short to tall, forming a typical bell-shaped curve. Careful analysis of the inheritance of skin color produces the same result: Although the skin color of children is most often intermediate between that of the parents, a typical bell-shaped distribution is obtained in which some

A. **Students at Brigham Young University, arranged according to height**

Y
BYU

Dan Fairbanks/Brigham Young University

B. **Actual distribution of individuals in the photo according to height**

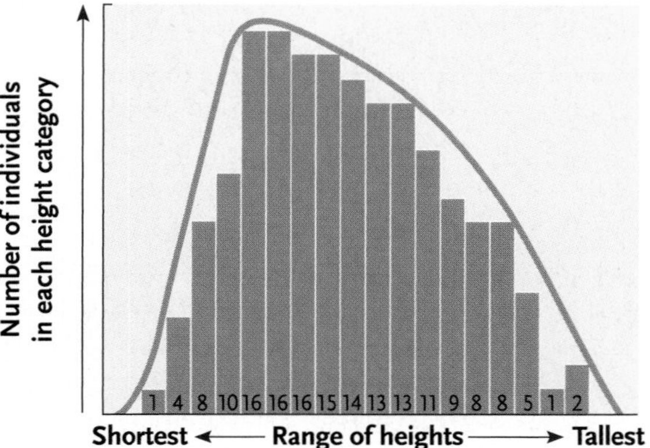
Number of individuals in each height category
1 4 8 10 16 16 16 15 14 13 13 11 9 8 8 5 1 2
Shortest ← Range of heights → Tallest

C. **Idealized bell-shaped curve for a population that displays continuous variation in a trait**

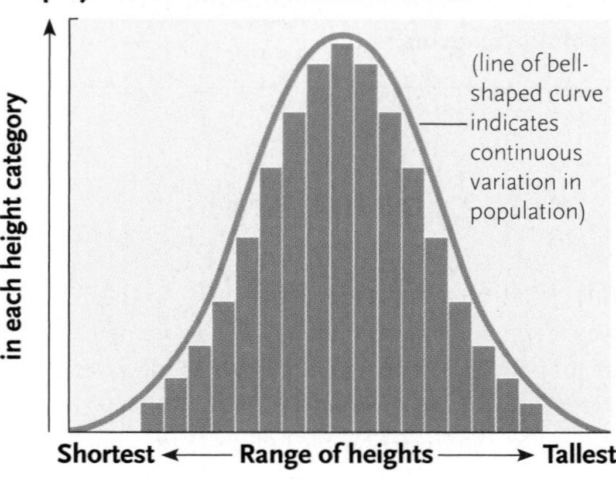
Number of individuals in each height category
(line of bell-shaped curve indicates continuous variation in population)
Shortest ← Range of heights → Tallest

If the sample in the photo included more individuals, the distribution would more closely approach this ideal.

FIGURE 12.17
Continuous variation in height due to polygenic inheritance.

children at the extremes are lighter or darker than either parent. Thus, genetic analysis does not support the idea of blending or even mixing of parental traits in polygenic characteristics such as body size or skin color.

In Pleiotropy, Two or More Characters Are Affected by a Single Gene

In **pleiotropy**, single genes affect more than one character of an organism. For example, sickle-cell disease (see earlier discussion) is caused by a recessive allele of a single gene that affects hemoglobin structure and function. However, the altered hemoglobin, the primary phenotypic change of the sickle-cell mutation, leads to blood vessel blockage, which can damage many tissues and organs in the body and affect many body functions, producing such wide-ranging symptoms as fatigue, abdominal pain, heart failure, paralysis, and pneumonia **(Figure 12.18)**. Physicians recognize these wide-ranging pleiotropic effects as symptoms of sickle-cell disease.

The next chapter describes additional patterns of inheritance that were not anticipated by Mendel, including the effects of recombination during meiosis. These additional patterns also extend, rather than contradict, Mendel's fundamental principles.

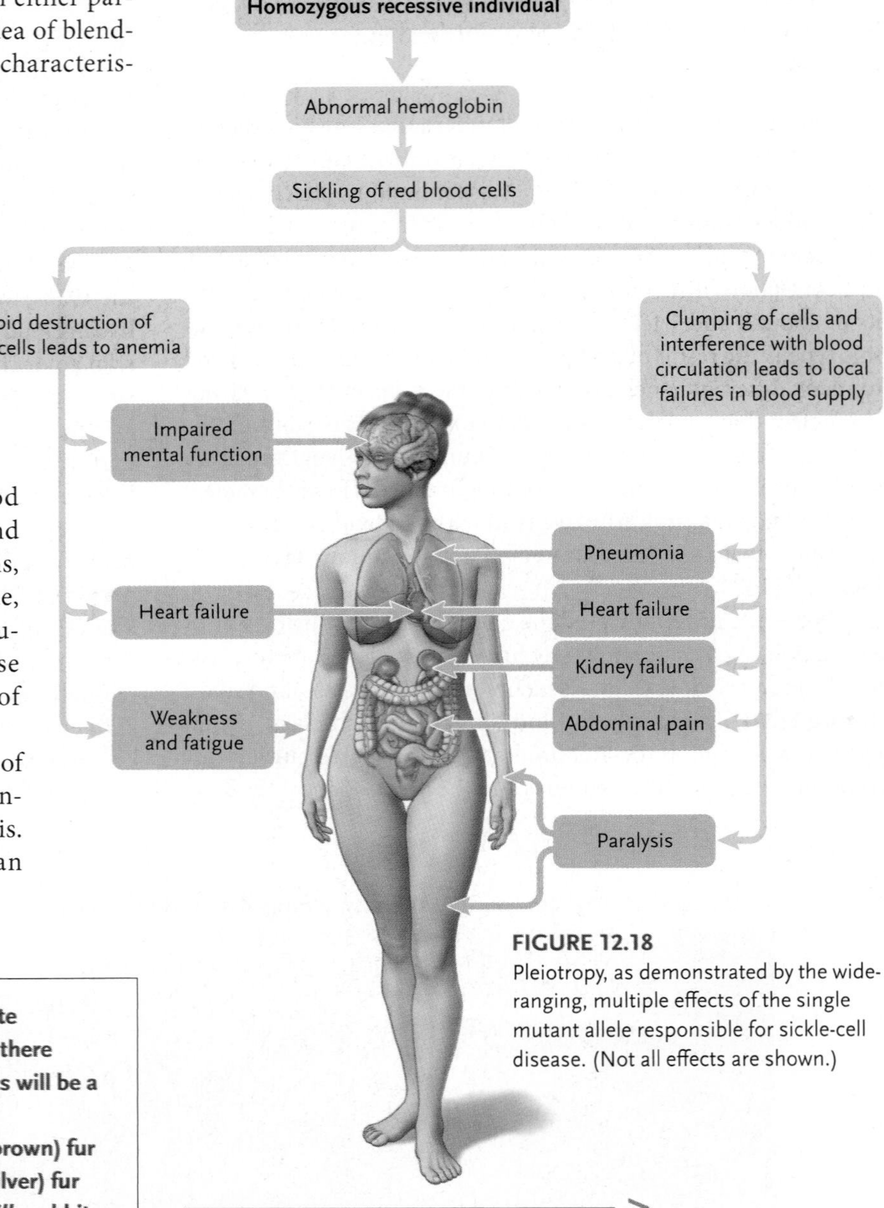
Homozygous recessive individual
Abnormal hemoglobin
Sickling of red blood cells
Rapid destruction of sickle cells leads to anemia
Clumping of cells and interference with blood circulation leads to local failures in blood supply
Impaired mental function
Heart failure
Weakness and fatigue
Pneumonia
Heart failure
Kidney failure
Abdominal pain
Paralysis

FIGURE 12.18
Pleiotropy, as demonstrated by the wide-ranging, multiple effects of the single mutant allele responsible for sickle-cell disease. (Not all effects are shown.)

STUDY BREAK 12.2

1. **Palomino horses have a golden coat color, with a white mane and tail. Palominos do not breed true. Instead, there is a 50% chance that a foal with two Palomino parents will be a Palomino. What is the explanation?**
2. **A true-breeding rabbit with *agouti* (mottled, grayish brown) fur crossed with a true-breeding rabbit with *chinchilla* (silver) fur produces all agouti offspring. A true-breeding *chinchilla* rabbit crossed with a true-breeding *Himalayan* rabbit (white fur with pigmented nose, ears, tail, and legs) produces all *chinchilla* offspring. A true-breeding *Himalayan* rabbit crossed with a true-breeding *albino* rabbit produces all *Himalayan* offspring. Explain the inheritance of the fur colors.**

THINK OUTSIDE THE BOOK

Individually or collaboratively explore the Internet or research literature to outline what is known about the genetics of human eye color, and about how certain individuals can have irises of different colors.

UNANSWERED QUESTIONS

What is next for the genetics of human disease?

Technological and conceptual advances have propelled our understanding of the genetic basis of human disease to unprecedented levels of refinement. Comparing the *status quo* to a short decade ago, we now have detailed genetic and genomic information for humans, several primates, and a host of other organisms across all phyla, and our progress is set to accelerate. We have also witnessed the rapid identification of mutations that cause rare diseases, and we have recently discovered over 1,000 genomic segments that are associated with susceptibility to complex traits such as type I and type II diabetes, obesity, high cholesterol, and others.

Despite these advances, the most fundamental questions in genetics, raised a century ago, have yet to be fully addressed. Most fundamental of all is the question of how the genotype relates to the phenotype. Once a genetic disorder has been diagnosed, we can often confirm the diagnosis by mutation analysis. However, our ability to use such analysis to predict whether or not a patient will develop a particular genetic disorder is still extremely limited. This is in part because we still evaluate the effects of genetic and genomic variations in isolation. We have not yet developed the tools to understand the effects of such variations in the context of an entire genome (as well as the environment). Examples from a wide variety of single-gene disorders have shown us that individuals can be genotypically affected but clinically normal. Likewise, the typical clinical experience is that patients with the same mutation can vary greatly in the presence and severity of particular symptoms.

One of the challenges ahead of us is to understand the effect of mutations in the context of an individual's total genomic variation; that is, the sum total of likely pathogenic mutations in the genome. Coupled to that goal, the availability of the total genome sequence poses both an opportunity and a problem. Given that each human has several thousand variants predicted to affect gene/protein function, we need to solve the problems of how to assign the effect of genetic variation for any particular phenotype and how to predict the frequency with which each allele results in disease symptoms.

An overlapping question is the century-old debate over the relative contribution of rare and common alleles to genetic disease (the so-called common-allele common disease versus rare-allele common disease argument). The premise that complex traits such as schizophrenia, hypertension, and diabetes must be caused by common mutations in the general population (otherwise they would not be so common) has also proven true of some disorders, such as age-related macular degeneration, the most frequent cause of blindness in the elderly. However, exhaustive analysis of large cohorts has also shown that common alleles can account for only a modest fraction of the genetic risk for any given complex trait and that most of that risk maps to regions of the genome that are not directly involved in coding protein sequences. For example, a common allele in a noncoding region of a gene called *Fatso (FTO)* that shows exceptionally significant association with obesity and diabetes accounts for only about 3 kg of increased weight. The questions ahead of us are therefore: (1) where are the genetic risk alleles that account for the majority of complex traits (the so-called genetic "dark matter"); (2) why is there a dearth of coding sequence changes in complex traits; and (3) how do common, probably mild, alleles (alleles that have small effects) interact with possibly rare alleles having strong effects to magnify the risk of a complex trait? For example, work in our laboratory on Bardet–Biedl syndrome, a genetic disorder characterized by obesity and learning defects, and related disorders has shown that some alleles that essentially abolish protein function (and therefore have a strong effect) can interact with mutations in other genes that have a modest effect on protein function (mild alleles). The interaction between the two can significantly enhance the severity of the clinical phenotype, but until we are able to model the effect of these mutations, genetic analysis alone is insufficient to detect such disorders. It is likely that a combination of extensive sequencing, dense genotyping, and a new generation of computational and biological tools will be required to address these challenges.

Think Critically

Given significant resources, what would you choose to do as a means of understanding variation in the human genome and its involvement in disease: sequence 1,000 humans or sequence 1,000 other species? (Hint: One would give you access to a lot of human variation, the other would help you with evolutionary arguments.)

Nicholas Katsanis

Nicholas Katsanis is professor of Cell Biology and director of the Center for Human Disease Modeling at Duke University. His research interests focus on the genetic basis of Bardet–Biedl sydnrome, where his laboratory is engaged in the identification of causative genes. The Katsanis lab also pursues questions centered on the signaling roles of vertebrate cilia, the translation of signaling pathway defects on the causality of ciliary disorders, and the dissection of second-site modification phenomena as a consequence of genetic load in a functional system. To learn more, go to http://www.cellbio.duke.edu/faculty/research/Katsanis.html.

REVIEW KEY CONCEPTS

Go to **CENGAGENOW** at www.cengage.com/login to access quizzing, animations, exercises, articles, and personalized homework help.

12.1 The Beginnings of Genetics: Mendel's Garden Peas

- Mendel was successful in his research because of his good choice of experimental organism, which had clearly defined characters, such as flower color or seed shape, and because he analyzed his results quantitatively (Figures 12.3 and 12.4).
- Mendel showed that traits are passed from parents to offspring as hereditary factors (now called genes and alleles) in predictable ratios and combinations, disproving the notion of blended inheritance (Figure 12.5).
- Mendel realized that his results with crosses that involve single characters (monohybrid crosses) could be explained if three hypotheses were true: (1) The genes that govern genetic characters occur in pairs in individuals; (2) If different alleles of a gene are present in the pair of an individual, one allele is dominant over the other; and (3) The two alleles of a gene segregate and enter gametes singly (Figures 12.5 and 12.7).
- Mendel confirmed his hypotheses by a testcross between an F_1 heterozygote with a homozygous recessive parent. This type of testcross is still used to determine whether an individual is homozygous or heterozygous for a dominant allele (Figure 12.8).
- To explain the results of his crosses with individuals showing differences in two characters—dihybrid crosses—Mendel added a fourth hypothesis: The alleles of the genes that govern the two characters segregate independently during formation of gametes (Figure 12.9).
- Walter Sutton was the first person to note the similarities between the inheritance of genes and the behavior of chromosomes in meiosis and fertilization. These parallels made it obvious that genes and alleles are carried on the chromosomes. Sutton's parallels are called the chromosome theory of inheritance (Figure 12.10).
- A locus is the site occupied by a gene on a chromosome (Figure 12.11).

Animation: Crossing garden pea plants

Animation: Monohybrid cross

Animation: F_2 ratios interaction

Practice: Testcross

Animation: Dihybrid cross

Animation: Review of crossing over

Animation: Genetic terms

12.2 Later Modifications and Additions to Mendel's Hypotheses

- In incomplete dominance, some or all alleles of a gene are neither completely dominant nor recessive. In such cases, the phenotype of heterozygotes with different alleles of the gene can be distinguished from that of either homozygote (Figure 12.13).
- In codominance, different alleles of a gene have approximately equal effects in heterozygotes, also allowing heterozygotes to be distinguished from either homozygote.
- Many genes may have multiple alleles if all the individuals in a population are taken into account. However, any diploid individual in a population has only two alleles of these genes, which are inherited and passed on according to Mendel's principles (Figures 12.14 and 12.15).
- In epistasis, genes interact, with one or more alleles of one locus inhibiting or masking the effects of one or more alleles at a different locus. The result is that some expected phenotypes do not appear among offspring (Figure 12.16).
- In polygenic inheritance, genes at several to many different loci interact to control the same character, producing a more or less continuous variation in the character from one extreme to another. Plotting the distribution of such characters among individuals typically produces a bell-shaped curve (Figure 12.17).
- In pleiotropy, one gene affects more than one character of an organism (Figure 12.18).

Animation: Comb shape in chickens

Animation: Incomplete dominance

Animation: Codominance: ABO blood types

Animation: Coat color in Labrador retrievers

Interaction: Continuous variation in height

Animation: Pleiotropic effects of Marfan syndrome

Animation: Coat color in the Himalayan rabbit

UNDERSTAND AND APPLY

Test Your Knowledge

1. The dominant *C* allele of a gene that controls color in corn produces kernels with color; plants homozygous for a recessive *c* allele of this gene have colorless or white kernels. What kinds of gametes, and in what proportions, would be produced by the plants in the following crosses? What seed color, and in what proportions, would be expected in the offspring of the crosses?
   a. *CC* × *Cc*
   b. *Cc* × *Cc*
   c. *Cc* × *cc*
2. In peas, the allele *T* produces tall plants and the allele *t* produces dwarf plants. The *T* allele is dominant to *t*. If a tall plant is crossed with a dwarf, the offspring are distributed about equally between tall and dwarf plants. What are the genotypes of the parents?
3. The ability of humans to taste the bitter chemical phenylthiocarbamide (PTC) is a genetic trait. People with at least one copy of the normal, dominant allele of the *PTC* gene can taste PTC; those who are homozygous for a mutant, recessive allele cannot taste it. Could two parents able to taste PTC have a nontaster child? Could nontaster parents have a child able to taste PTC? A pair of taster parents, both of whom had one parent able to taste PTC and one nontaster parent, are expecting their first child. What is the chance that the child will be able to taste PTC? Unable to taste PTC? Suppose the first child is a nontaster. What is the chance that their second child will also be unable to taste PTC?
4. One gene has the alleles *A* and *a;* another gene has the alleles *B* and *b.* For each of the following genotypes, what types of gametes will be produced, and in what proportions, if the two gene pairs assort independently?
   a. *AA BB* b. *Aa bb*
   c. *Aa BB* d. *Aa Bb*
5. What genotypes, and in what frequencies, will be present in the offspring from the following matings?
   a. *AA BB* × *aa BB* b. *Aa BB* × *AA Bb*
   c. *Aa Bb* × *aa bb* d. *Aa Bb* × *Aa Bb*
6. In addition to the two genes in problem 4, assume you now study a third independently assorting gene that has the alleles *C* and *c.* For each of the following genotypes, indicate what types of gametes will be produced:
   a. *AA BB CC* b. *Aa BB cc*
   c. *Aa BB Cc* d. *Aa Bb Cc*
7. A man is homozygous dominant for alleles at 10 different genes that assort independently. How many genotypically different types of sperm cells can he produce? A woman is homozygous recessive for the alleles of 8 of these 10 genes, but she is heterozygous for the other 2 genes. How many genotypically different types of eggs can she produce? What hypothesis can you suggest to describe the relationship between the number of different possible gametes and the number of heterozygous and homozygous genes that are present?
8. In guinea pigs, an allele for rough fur *(R)* is dominant over an allele for smooth fur *(r);* an allele for black coat *(B)* is dominant over that for white *(b*). You have an animal with rough, black fur. What cross would you use to determine whether the animal is homozygous for these traits? What phenotype would you expect in the offspring if the animal is homozygous?
9. You cross a lima bean plant from a variety that breeds true for green pods with another lima bean from a variety that breeds true for yellow pods. You note that all the F_1 plants have green pods. These green-pod F_1 plants, when crossed, yield 675 plants with green pods and 217 with yellow pods. How many genes probably control pod color in this experiment? Give the alleles letter designations. Which is dominant?
10. Some recessive alleles have such a detrimental effect that they are lethal when present in both chromosomes of a pair. Homozygous recessives cannot survive and die at some point during embryonic development. Suppose that the allele *r* is lethal in the homozygous *rr* condition. What genotypic ratios would you expect among the living offspring of the following crosses?
    a. *RR* × *Rr*
    b. *Rr* × *Rr*
11. In garden peas, the genotypes *GG* or *Gg* produce green pods and *gg* produces yellow pods; *TT* or *Tt* plants are tall and *tt* plants are dwarfed; *RR* or *Rr* produce round seeds and *rr* produces wrinkled seeds. If a plant of a true-breeding, tall variety with green pods and round seeds is crossed with a plant of a true-breeding, dwarf variety with yellow pods and wrinkled seeds, what phenotypes are expected, and in what ratios, in the F_1 generation? What phenotypes, and in what ratios, are expected if F_1 individuals are crossed?
12. In chickens, feathered legs are produced by a dominant allele *F.* Another allele *f* of the same gene produces featherless legs. The

dominant allele *P* of a gene at a different locus produces pea combs; a recessive allele *p* of this gene causes single combs. A breeder makes the following crosses with birds 1, 2, 3, and 4; all parents have both feathered legs and pea combs:

Cross	Offspring
1 × 2	all feathered, pea comb
1 × 3	3/4 feathered; 1/4 featherless, all pea comb
1 × 4	9/16 feathered, pea comb; 3/16 featherless, pea comb; 3/16 feathered, single comb; 1/16 featherless, single comb

What are the genotypes of the four birds?

13. A mixup in a hospital ward caused a mother with O and MN blood types to think that a baby given to her really belonged to someone else. Tests in the hospital showed that the doubting mother was able to taste PTC (see problem 3). The baby given to her had O and MN blood types and had no reaction when the bitter PTC chemical was placed on its tongue. The mother had four other children with the following blood types and tasting abilities for PTC:
    a. Type A and MN blood, taster
    b. Type B and N blood, nontaster
    c. Type A and M blood, taster
    d. Type A and N blood, taster

    Without knowing the father's blood types and tasting ability, can you determine whether the child is really hers? (Assume that all her children have the same father.)
14. In cats, the genotype *AA* produces tabby fur color; *Aa* is also a tabby, and *aa* is black. Another gene at a different locus is epistatic to the gene for fur color. When present in its dominant *W* form (*WW* or *Ww*), this gene blocks the formation of fur color and all the offspring are white; *ww* individuals develop normal fur color. What fur colors, and in what proportions, would you expect from the cross *Aa Ww* × *Aa Ww*?
15. Having malformed hands with shortened fingers is a dominant trait controlled by a single gene; people who are homozygous for the recessive allele have normal hands and fingers. Having woolly hair is a dominant trait controlled by a different gene; homozygous recessive individuals have normal, nonwoolly hair. Suppose a woman with normal hands and nonwoolly hair marries a man who has malformed hands and woolly hair. Their first child has normal hands and nonwoolly hair. What are the genotypes of the mother, the father, and the child? If this couple has a second child, what is the probability that it will have normal hands and woolly hair?

Discuss the Concepts

1. Explain how individuals of an organism that are phenotypically alike can produce different ratios of progeny phenotypes.
2. ABO blood type tests can be used to exclude paternity. Suppose a defendant who is the alleged father of a child takes a blood type test and the results do not exclude him as the father. Do the results indicate that he is the father? What arguments could a lawyer make based on the test results to exclude the defendant from being the father? (Assume the tests were performed correctly.)

Design an Experiment

Imagine that you are a breeder of Labrador retriever dogs. Labs can be black, chocolate brown, or yellow. Suppose that a yellow Lab is donated to you and you need to know its genotype. You have a range of dogs with known genotypes. What cross would you make to determine the genotype of the donated dog? Explain how the resulting puppies show you the Lab's genotype.

Interpret the Data

Half of the world's population eats rice at least twice a day. Much of this rice is grown in flooded conditions, and different strains of rice are tolerant (survive) or intolerant (die) under these conditions. Rice breeders used genetic crosses to test whether tolerance to flooding is a dominant trait. Researchers used three true-breeding flood-tolerant strains, FR143, BKNFR, and Kurk, and two true-breeding flood-intolerant strains, IR42 and NB, in the crosses. Results were obtained from three sets of crosses and are reported in the table below:

1. F_2 results: Intolerant and tolerant strains were crossed and the resulting F_1 were interbred to produce the F_2.
2. Results of cross of F_1 to intolerant parent: F_1 plants were crossed with the intolerant parent of the cross.
3. Results of cross of F_1 to tolerant parent: F_1 plants were crossed with the tolerant parent of the cross.

Progeny Analyzed from Intolerant × Tolerant Cross	Number of Plants		
	Alive	**Dead**	**Total**
1. F_2 results of cross:			
IR42 × FR13A	187	77	264
IR42 × BKNFR	192	73	265
NB × Kurk	142	52	195
2. Results of cross of F_1 to intolerant parent:			
(F_1 of IR42 × FR13A) × IR42	14	17	31
(F_1 of IR42 × BKNFR) × IR42	15	10	25
(F_1 of NB × Kurk) × NB	21	35	56
3. Results of cross of F_1 to tolerant parent:			
(F_1 of IR42 × FR13A) × FR13A	31	0	31
(F_1 of IR42 × BKNFR) × BKNFR	28	0	28
(F_1 of NB × Kurk) × Kurk	40	0	40

Do the data support the hypothesis that the tolerance trait is dominant? Justify your conclusion by explaining the results from each of the three sets of crosses in terms of genotypes and phenotypic ratios.

Source: T. Setter et al. 1997. Physiology and genetics of submergence tolerance in rice. *Annals of Botany* 79:67–77.

Apply Evolutionary Thinking

How could an epistatic interaction shelter a harmful allele from the action of natural selection?

13

STUDY OUTLINE

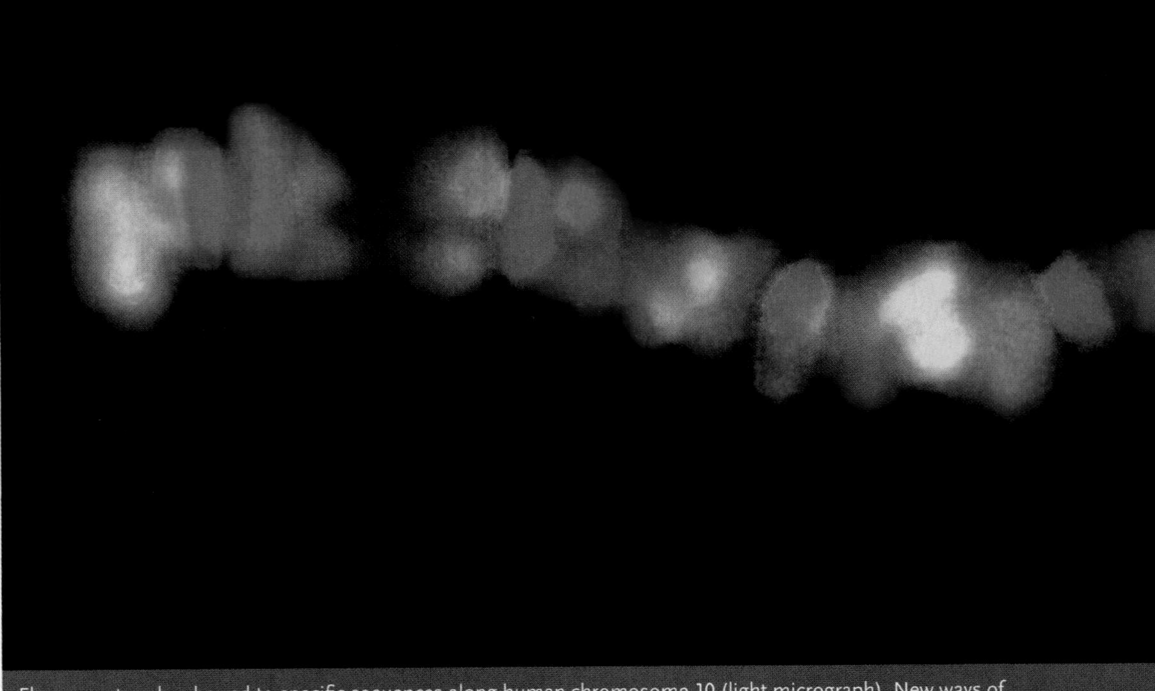

Fluorescent probes bound to specific sequences along human chromosome 10 (light micrograph). New ways of mapping chromosome structure yield insights into the inheritance of normal and abnormal traits.

Genes, Chromosomes, and Human Genetics

Why It Matters. . . Imagine being 10 years old and trapped in a body that each day becomes more shriveled, frail, and old. You are just tall enough to peer over the top of the kitchen counter, and you weigh less than 35 pounds. Already you are bald, and you probably have only a few more years to live. But if you are like Mickey Hayes or Fransie Geringer **(Figure 13.1)**, you still have not lost your courage or your childlike curiosity about life. Like them, you still play, laugh, and celebrate birthdays.

Progeria, the premature aging that afflicts Mickey and Fransie, is caused by a genetic error that is present in only 1 of every 4–8 million human births. The error is in the gene for a lamin A, one of the lamin proteins that reinforces the inner surface of the nuclear envelope in animal cells. In some way as yet not understood, the defective lamin A makes the nucleus unstable, leading to the premature aging and reduced life expectancy characteristic of progeria.

Progeria affects both sexes equally, and all races and ethnic groups. Usually, symptoms begin to appear between 18 and 24 months of age. The rate of body growth declines to abnormally low levels. Skin becomes thinner, muscles become flaccid, and limb bones start to degenerate. Children with progeria die from a stroke or heart attack brought on by hardening of the arteries, a condition typical of advanced age. Death occurs at an average age of 13, with a range of about 8–21 years.

The plight of Mickey and Fransie provides a telling and tragic example of the dramatic effects that gene defects can have on living organisms. We are the products of our genes, and the characteristics of each individual, from humans to pine trees to protozoa, depend on the combination of

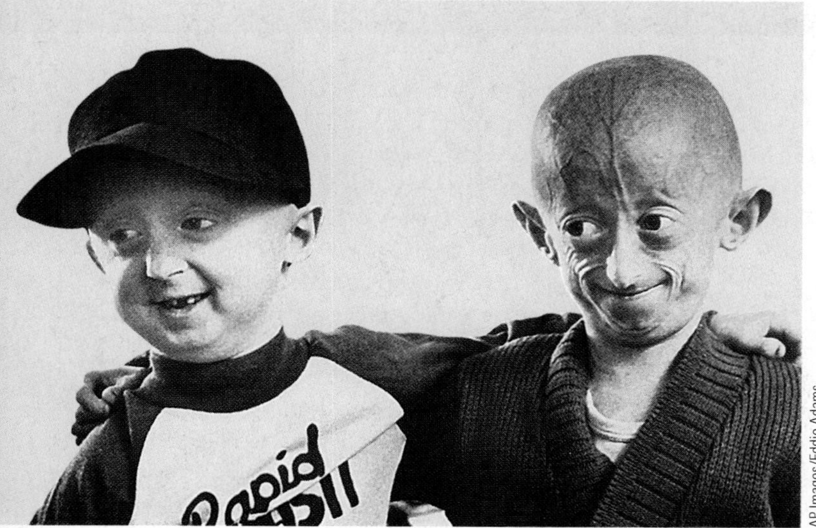

AP Images/Eddie Adams

FIGURE 13.1
Two boys, both younger than 10, who have progeria, a genetic disorder characterized by accelerated aging and extremely reduced life expectancy.

genes, alleles, and chromosomes inherited from its parents, as well as on environmental effects. This chapter delves deeply into genes and the role of chromosomes in inheritance. <

13.1 Genetic Linkage and Recombination

In his historic experiments, Gregor Mendel found that each of the seven genes he studied assorted independently of the others in the formation of gametes. If Mendel had extended his study to additional characters, he would have found exceptions to this principle. This should not be surprising, because an organism has far more genes than chromosomes. Chromosomes contain many genes, with each gene at a particular locus. Genes located on different chromosomes assort independently in gamete formation because the two chromosomes behave independently of one another during meiosis. Genes located on the same chromosome may be inherited together in genetic crosses—that is, *not* assort independently—because the chromosome is inherited as a single physical entity in meiosis. Genes on the same chromosome are known as **linked genes,** and the phenomenon is called **linkage.**

The Principles of Linkage and Recombination Were Determined with *Drosophila*

In the early part of the twentieth century, Thomas Hunt Morgan and his coworkers at Columbia University were using the fruit fly, *Drosophila melanogaster,* as a model organism to investigate Mendel's principles in animals. (*Focus on Model Research Organisms* describes the development and use of *Drosophila* in research.) In 1911, Morgan crossed a true-breeding fruit fly with normal red eyes and normal wing length, genotype $pr^+pr^+\ vg^+vg^+$, with a true-breeding fly with the recessive traits of purple eyes and vestigial (that is, short and crumpled) wings, genotype *prpr vgvg,* to analyze the segregation of the two traits.

This gene symbolism is new to us. Morgan devised this symbolism, and it is commonly used, much more so than the *A/a* system we have used until now. In this system, the superscript plus (+) symbol associated with a letter or letters indicates a wild-type—normal—allele of a gene. Typically, but not always, a wild-type allele is the most common allele found in a population. In most instances, the wild-type allele is dominant to mutant alleles, but there are exceptions. The letters for the gene are based on the phenotype of the organism that expresses the *mutant* allele, for example, *pr* for *purple* eyes. Thus, we refer to the gene as the *purple* or *pr* gene; the dominant wild-type allele of the gene, pr^+, gives the wild-type red eye color.

Figure 13.2 steps through Morgan's cross of the two parents and his testcross of F_1 flies. Based on Mendel's principle of independent assortment (see Section 12.1), there should be four classes of phenotypes in the testcross offspring, in a 1:1:1:1 ratio of red eyes, normal wings:purple, vestigial:red, vestigial:purple, normal. But Morgan did not observe this result (step 4); instead, of the 2,839 progeny flies, 1,339 were red, normal and 1,195 were purple, vestigial. These phenotypes are identical to the two original P generation flies and are called **parental** phenotypes. The remaining progeny flies consisted of 151 red, vestigial and 154 purple, normal. These phenotypes have different combinations of traits from those of the P generation flies and are called **recombinant** phenotypes. If the genes had shown independent assortment, there would have been 25% of each of the four classes, or 50% parental and 50% recombinant phenotypes. In numbers, there would have been 710 (approximately) of each of the 4 classes.

How could the low frequency of recombinant phenotypes be explained? Morgan hypothesized that the two genes are linked genetically—physically associated on the same chromosome. That is, *pr* and *vg* are linked genes. He further hypothesized that the behavior of these linked genes in the testcross is explained by what he called *chromosome recombination,* a process in which two homologous chromosomes exchange segments with each other by crossing-over during meiosis (see Figure 11.6). Furthermore, he proposed that the frequency of this recombination is a function of the distance between linked genes. The nearer two genes are, the greater the chance they will be inherited together (resulting in parental phenotypes) and the lower the chance that recombinant phenotypes will be produced. The farther apart two genes are, the lower the chance that they will be inherited together and the greater the chance that recombinant phenotypes will be produced. These brilliant and far-reaching hypotheses were typical of Morgan, who founded genetics research in the United States, developed *Drosophila* as a research organism, and made discoveries that were almost as significant to the development of genetics as those of Mendel.

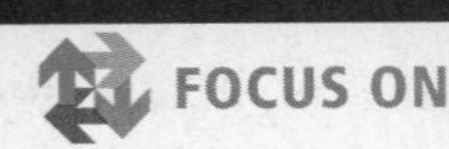 FOCUS ON **Model Research Organisms**

The Marvelous Fruit Fly, *Drosophila melanogaster*

Herman Eisenbeiss/ Photo Researchers, Inc.

The unobtrusive little fruit fly that appears seemingly from nowhere when rotting fruit or a fermented beverage is around is one of the mainstays of genetic research. It was first described in 1830 by C. F. Fallén, who named it *Drosophila,* meaning "dew lover." The species identifier became *melanogaster,* which means "black belly."

The great geneticist Thomas Hunt Morgan began to culture *D. melanogaster* in 1909 in the famous "Fly Room" at Columbia University. Many important discoveries in genetics were made in the Fly Room, including sex-linked genes and sex linkage and the first chromosome map. The subsequent development of methods to induce mutations in *Drosophila* led, through studies of the mutants produced, to many other discoveries that collectively established or confirmed essentially all the major principles and conclusions of eukaryotic genetics.

One reason for the success of *D. melanogaster* as a subject for genetics research is the ease of culturing it. It is grown usually at 25°C in small milk bottles stopped with a cotton or plastic foam wad and filled about one-third of the way with a fermenting medium that contains water, corn meal, agar, molasses, and yeast. The several hundred eggs laid by each adult female hatch rapidly and progress through larval and pupal stages to produce adult flies in about 10 days, which are ready to breed within 10 to 12 hours.

Males and females can be identified easily with the unaided eye. Many types of mutations produce morphologic differences, such as changes in eye color, wing shape, or the numbers and shapes of bristles, which can be seen with the unaided eye or under a low-power binocular microscope. The salivary gland cells of the fly larvae also have giant chromosomes. The chromosomes are so large that differences can be observed directly with the light microscope.

The availability of a wide range of mutants, comprehensive linkage maps of each of its chromosomes, and the ability to manipulate genes readily by molecular techniques made the fruit fly one of the model organisms for genome sequencing in the Human Genome Project. The sequencing of *Drosophila*'s genome was completed in 2001; there are approximately 14,000 protein-coding genes in its 165 million base-pair genome. (A database of the *Drosophila* genome is available at http://flybase.bio.indiana.edu). Importantly, the relationship between fruit fly and human genes is close, to the point that many human disease genes have counterparts in the fruit fly genome. This similarity enables the fly genes to be studied as models of human disease genes in efforts to understand better the functions of those genes and how alterations in them lead to disease.

Drosophila has also become established as an excellent experimental model for neurobiology studies (the investigation of the structure and function of the nervous system). Using molecular, genetic, electrophysiological, imaging, and developmental techniques, scientists are investigating, for example, neural development, and analyzing behavior. Again, findings are helping us understand the human nervous system also.

The analysis of fruit fly embryonic development has also contributed significantly to the understanding of development in humans. For example, experiments on mutants that affect fly development have provided insights into the genetic basis of many human birth defects. Recognition of the importance of research with *Drosophila* developmental genetics to our understanding of development in general, came in the form of the award of the Nobel Prize in 1995 to three scientists who pioneered the fruit fly work: Edward Lewis of the California Institute of Technology, Christiane Nusslein-Volhard of the Max Planck Institute for Developmental Biology in Tübingen, Germany, and Eric Wieschaus of Princeton University.

We show how Morgan's thinking applies to the purple-vestigial cross in **Figure 13.3,** which presents the alleles of the genes with cartoons of the chromosomes themselves. This figure allows us to follow pictorially the consequences of crossing-over during meiosis in the production of gametes, and then the fusion of parental and recombinant gametes with the gamete from the male testcross parent. We can see that the parental or recombinant phenotypes of the offspring directly reflect the genotypes of the gametes of the dihybrid parent. Because linked genes are involved, the number of offspring in the testcross progeny with parental phenotypes exceeds the number with recombinant phenotypes. As we learned in Chapter 11, **genetic recombination** is the process by which the combinations of alleles for different genes in two parental individuals become shuffled into new combinations in offspring individuals as we are seeing here.

To determine the distance between the two genes on the chromosome, we calculate the **recombination frequency,** the percentage of testcross progeny that are recombinants. For this testcross, the recombination frequency is 10.7% (see Figure 13.3).

Recombination Frequency Can Be Used to Map Chromosomes

The recombination frequency of 10.7% for the *pr* and *vg* genes of *Drosophila* means that 10.7% of the gametes originating from the $pr^+pr\ vg^+vg$ parent contained recombined chromosomes. That recombination frequency is characteristic for those two genes. In other crosses that involve linked genes, Morgan found that the recombination frequency was characteristic of the two genes involved, varying from less than 1% to 50% (see the next section).

From these observations, Alfred Sturtevant, then an undergraduate at Columbia University working with Morgan, realized that the variations in recombination frequencies could be used as a means of mapping genes on chromosomes. Sturtevant himself later recalled his lightbulb moment:

> I suddenly realized that the variations in the strength of linkage already attributed by Morgan to difference in the spatial separation of the gene offered the possibility of determining sequence in the linear dimensions of a chromosome. I went home and spent most of the night (to the neglect of my undergraduate homework) in producing the first chromosome map.

FIGURE 13.2 **Experimental Research**

Evidence for Gene Linkage

Question: Do the purple-eye and vestigial-wing genes of *Drosophila* assort independently?

Experiment: Morgan crossed true-breeding wild-type flies with red eyes and normal wings with true-breeding purple-eyed, vestigial-winged flies. He then testcrossed the F_1 flies, which were wild type in phenotype, and analyzed the distribution of phenotypes in the progeny.

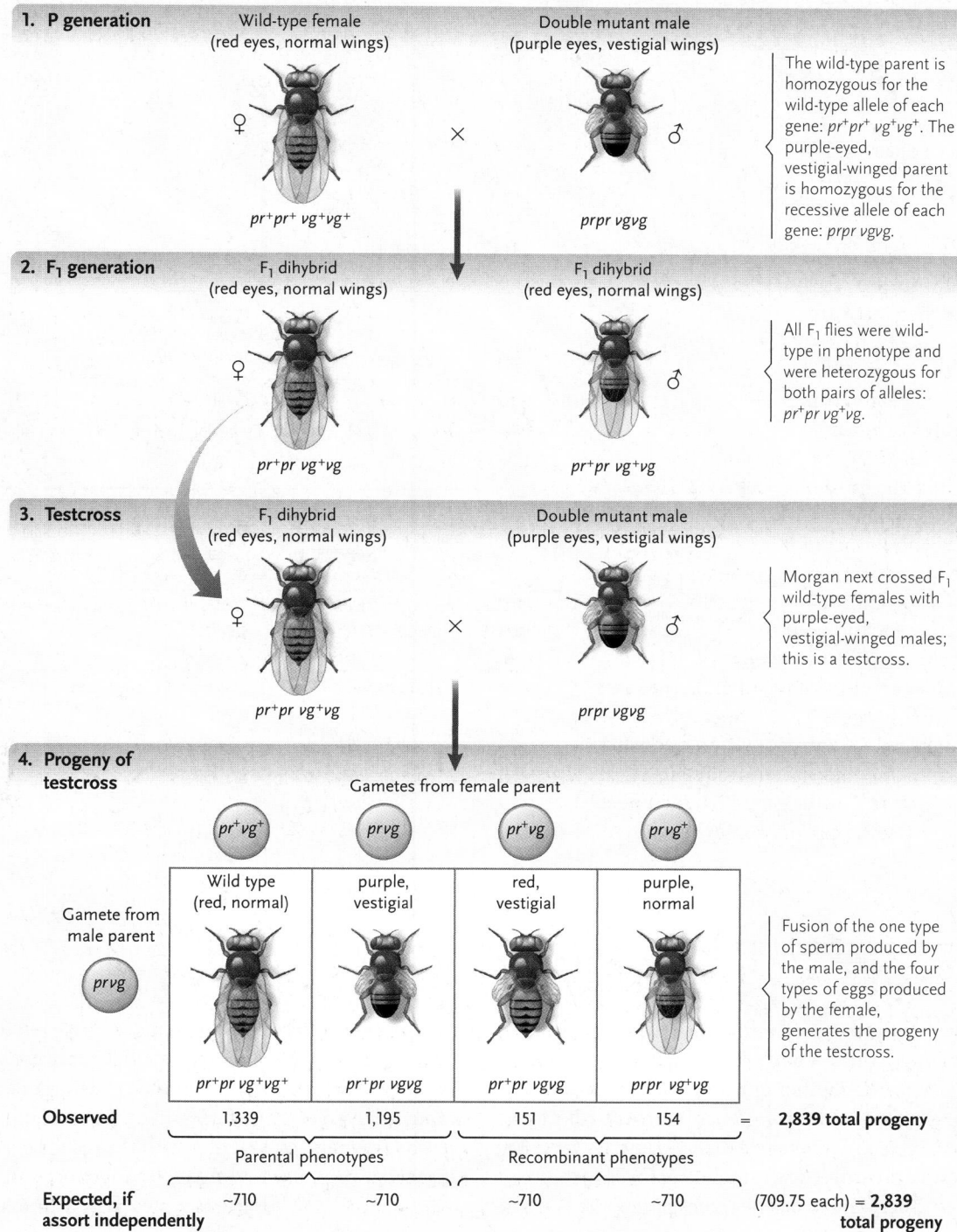

1. P generation

Wild-type female (red eyes, normal wings) ♀ $pr^+pr^+\ vg^+vg^+$ × Double mutant male (purple eyes, vestigial wings) ♂ $prpr\ vgvg$

The wild-type parent is homozygous for the wild-type allele of each gene: $pr^+pr^+\ vg^+vg^+$. The purple-eyed, vestigial-winged parent is homozygous for the recessive allele of each gene: $prpr\ vgvg$.

2. F_1 generation

F_1 dihybrid (red eyes, normal wings) ♀ $pr^+pr\ vg^+vg$; F_1 dihybrid (red eyes, normal wings) ♂ $pr^+pr\ vg^+vg$

All F_1 flies were wild-type in phenotype and were heterozygous for both pairs of alleles: $pr^+pr\ vg^+vg$.

3. Testcross

F_1 dihybrid (red eyes, normal wings) ♀ $pr^+pr\ vg^+vg$ × Double mutant male (purple eyes, vestigial wings) ♂ $prpr\ vgvg$

Morgan next crossed F_1 wild-type females with purple-eyed, vestigial-winged males; this is a testcross.

4. Progeny of testcross

Gametes from female parent

Gamete from male parent: $pr\ vg$	pr^+vg^+	$pr\ vg$	pr^+vg	$pr\ vg^+$
Phenotype	Wild type (red, normal)	purple, vestigial	red, vestigial	purple, normal
Genotype	$pr^+pr\ vg^+vg^+$	$pr^+pr\ vgvg$	$pr^+pr\ vgvg$	$prpr\ vg^+vg$
Observed	1,339	1,195	151	154
	Parental phenotypes		Recombinant phenotypes	
Expected, if assort independently	~710	~710	~710	~710

Observed = **2,839 total progeny**

Expected: (709.75 each) = **2,839 total progeny**

Fusion of the one type of sperm produced by the male, and the four types of eggs produced by the female, generates the progeny of the testcross.

Results: 2,534 of the testcross progeny flies had parental phenotypes, wild-type and purple, vestigial, whereas 305 of the progeny had recombinant phenotypes of red, vestigial and purple, normal. If the genes assorted independently, the expectation is a 1 : 1 : 1 : 1 ratio for testcross progeny: approximately 1,420 of both parental and recombinant progeny.

Conclusion: The purple-eye and vestigial-wing genes do not assort independently. The simplest alternative hypothesis is that the two genes are linked on the same chromosome. The small number of recombinant phenotypes is explained by crossing over.

Source: C. B. Bridges. 1919. The genetics of purple eye color in *Drosophila*. *The Journal of Experimental Zoology* 28:265–305.

FIGURE 13.3
Recombination between the purple-eye gene and the vestigial-wing gene, resulting from crossing-over between homologous chromosomes. The testcross of Figure 13.2 is redrawn here showing the two linked genes on chromosomes. Chromosomes or chromosome segments with wild-type alleles are red, whereas chromosomes or segments with mutant alleles are blue. The parental phenotypes in the testcross progeny are generated by segregation of the parental chromosomes, whereas the recombinant phenotypes are generated by crossing-over between the two linked genes.

Testcross parents

F_1 dihybrid (red eyes, normal wings): pr^+ vg^+ / pr vg

♀ × ♂

Double mutant (purple eyes, vestigial wings): pr vg / pr vg

Meiosis in the F_1 female pr^+ pr vg^+ vg dihybrid parent produces four types of gametes. Two parental gametes, pr^+ vg^+ and pr vg, are generated by chromosome segregation with no crossing-over between the genes. Two recombinant gametes, pr^+ vg and pr vg^+, result from crossing-over between the homologous chromatids when they are paired in prophase I of meiosis (see Figure 11.4).

Meiosis in the male testcross parent produces one type of gamete with genotype pr vg.

Female gametes (eggs)

Parental: pr^+ vg^+; pr vg — Recombinant: pr^+ vg; pr vg^+

Male gamete (sperm): pr vg

Testcross progeny are produced by pairing of the parental and recombinant gametes of the female with a pr vg gamete of the male.

	pr^+ vg^+ / pr vg	pr vg / pr vg	pr^+ vg / pr vg	pr vg^+ / pr vg
Phenotype	Wild-type (red, normal)	purple, vestigial	red, vestigial	purple, normal
Number	1,339	1,195	151	154
Class	Parental	Parental	Recombinant	Recombinant

Counting the phenotypic classes in the testcross progeny gives:

Parentals = 2,534
Recombinants = 305
Total progeny = 2,839

The percentage of the progeny that are recombinants, the recombination frequency = $\frac{305 \text{ recombinants}}{2{,}839 \text{ total progeny}} \times 100 = 10.7\%$

Sturtevant's revelation was that the recombination frequency observed between any two linked genes reflects the distance between them on their chromosome. The greater this distance, the greater the chance that a crossover can form between the genes and the greater the recombination frequency.

Therefore, recombination frequencies can be used to make a **linkage map** of a chromosome showing the relative locations of genes. For example, assume that the three genes *a, b,* and *c* are carried together on the same chromosome. Crosses reveal a 9.6% recombination frequency for *a* and *b*, an 8% recombination frequency for *a* and *c*, and a 2% recombination frequency for *b* and *c*. These frequencies allow the genes to be arranged in only one sequence on the chromosomes as follows:

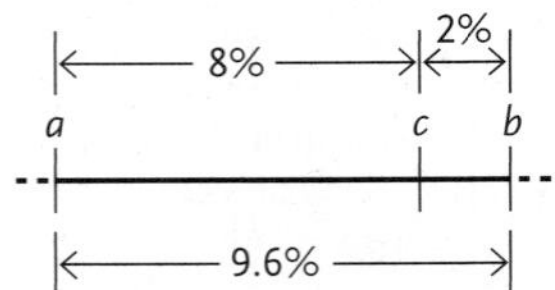
8%
2%
a
c
b
9.6%

You will note that the *a–b* recombination frequency does not exactly equal the sum of the *a–c* and *c–b* recombination frequencies. This is because genes farther apart on a chromosome are more likely to have more than one crossover occur between them. Whereas a single crossover between two genes gives recombinants, a double crossover (two single crossovers occurring in the same meiosis) between two genes gives parentals. You can see this simply by drawing single and double crossovers between two genes on a piece of paper. In our example, double crossovers that occur between *a* and *b* have slightly decreased the recombination frequency between these two genes.

Using this method, Sturtevant created the first linkage map showing the arrangement of six genes on the *Drosophila* X chromosome. (A partial linkage map of a *Drosophila* chromosome is shown in **Figure 13.4.**)

Since the time of Morgan, many *Drosophila* genes and those of other eukaryotic organisms widely used for genetic research, including *Neurospora* (a fungus), yeast, maize (corn), and the mouse, have been mapped using the same approach. Recombination frequencies, together with the results of other techniques, have also been used to create linkage maps of the locations of genes in the DNA of prokaryotes such as the human intestinal bacterium *Escherichia coli* (see Chapter 17).

The unit of a linkage map, called a **map unit** (abbreviated mu), is equivalent to a recombination frequency of 1%. The map unit is also called the **centimorgan** (cM) in honor of Morgan's discoveries of linkage and recombination. Map units are not absolute physical distances in micrometers or nanometers; rather, they are *relative,* showing the positions of genes with respect to each other. One of the reasons that the units are

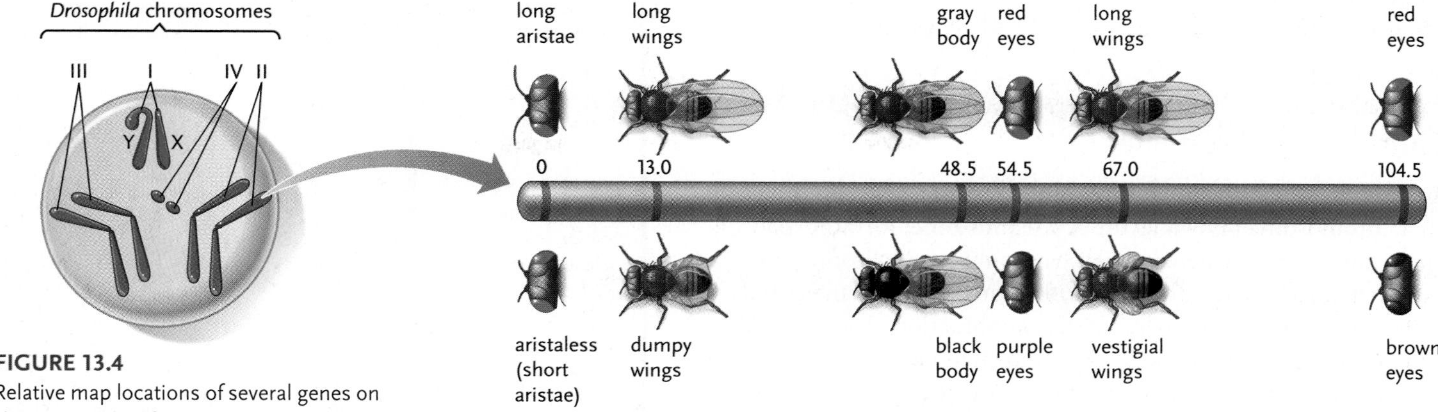
Drosophila chromosomes
III
I
IV
II
Y
X
Wild-type phenotypes
long aristae
long wings
gray body
red eyes
long wings
red eyes
0
13.0
48.5
54.5
67.0
104.5
aristaless (short aristae)
dumpy wings
black body
purple eyes
vestigial wings
brown eyes
Mutant phenotypes

FIGURE 13.4
Relative map locations of several genes on chromosome 2 of *Drosophila*, as determined by recombination frequencies. For each gene, the diagram shows the normal or "wild-type" phenotype on the top and the mutant phenotype on the bottom. Mutant alleles at two different locations alter wing structure, one producing the dumpy wing and the other the vestigial wing phenotypes; the normal allele at these locations results in normal long-wing structure. Mutant alleles at two different locations also alter eye color.

relative and not absolute distances is that the frequency of crossing-over varies to some extent from one position to another on chromosomes.

In recent years, the linkage maps of a number of species have been supplemented by DNA sequencing of whole genomes, which shows the precise physical locations of genes in the chromosomes.

Widely Separated Linked Genes Assort Independently

Genes can be so widely separated on a chromosome that recombination is likely to occur at some point between them in every cell undergoing meiosis. When this is the case, no linkage is detected and the genes assort independently. In other words, even though the alleles of the genes are carried on the same chromosome, the approximate 1 : 1 : 1 : 1 ratio of phenotypes is seen in the offspring of a dihybrid × double mutant testcross. That is, 50% of the progeny are parentals and 50% are recombinants. Linkage between such widely separated genes can still be detected, however, by testing their linkage to one or more genes that lie between them. For example, the genes *a* and *c* in **Figure 13.5** are located so far apart that they assort independently and show no linkage. However, crosses show that *a* and *b* are 23 map units apart (recombination frequency of 23%), and crosses that show *b* and *c* are 34 map units apart. Therefore, *a* and *c* must also be linked and carried on the same chromosome at $23 + 34 = 57$ map units apart. We could not see a recombination frequency of 57% in testcross progeny because the maximum frequency of recombinants is 50%.

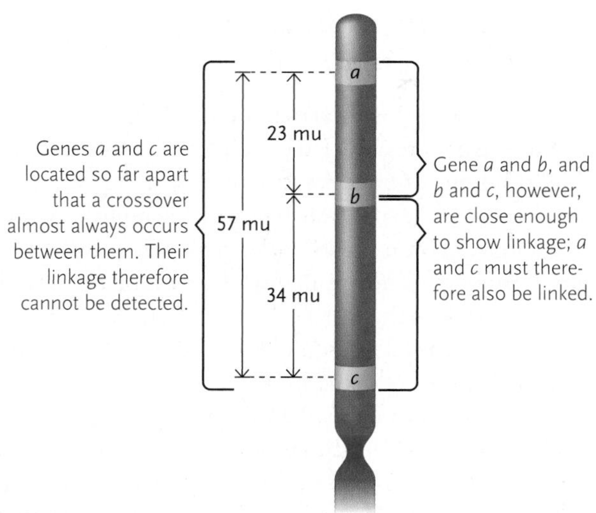
a
b
c
23 mu
57 mu
34 mu
Genes a and c are located so far apart that a crossover almost always occurs between them. Their linkage therefore cannot be detected.
Gene a and b, and b and c, however, are close enough to show linkage; a and c must therefore also be linked.

FIGURE 13.5
Genes far apart on the same chromosome. Genes *a* and *c* are far apart and will not show linkage, suggesting they are on different chromosomes. However, linkage between such genes can be established by noting their linkage to another gene or genes located between them—in this case, gene *b*.

We now know that some of the genes Mendel studied assort independently even though they are on the same chromosome. For example, the genes for flower color and seed color are located on the same chromosome, but they are so far apart that the frequent recombination between them makes them appear to be unlinked.

STUDY BREAK 13.1

You want to determine whether genes *a* and *b* are linked. What cross would you use and why? How would this cross tell you if they are linked?

13.2 Sex-Linked Genes

In many organisms, one or more pairs of chromosomes are different in males and females (see Section 11.1). Genes located on these chromosomes, the *sex chromosomes,* are called **sex-linked genes;** they are inherited differently in males and females. (Note that the word *linked* in *sex-linked gene* means that the gene is on a sex chromosome, whereas the use of the term *linked* when considering two or more genes means that the genes are on the same chromosome, not necessarily a sex chromosome.) Chromosomes other than the sex chromosomes are called **autosomes;** genes on these chromosomes have the same patterns of inheritance in both sexes. In humans, chromosomes 1 to 22 are the autosomes.

Females Are XX and Males Are XY in Both Humans and Fruit Flies

In most species with sex chromosomes, females have two copies of a chromosome known as the **X chromosome,** forming a homologous XX pair, whereas males have only one X chromosome. Another chromosome, the **Y chromosome,** occurs in males but not in females, giving males an XY combination. The XX–XY human chromosome complement is shown in Figure 10.7.

Because an XX female produces only one type of gamete with respect to the sex chromosomes, she is called the **homogametic sex.** That is, each normal gamete produced by an XX female carries an X chromosome. Because the XY male produces two types of gametes with respect to the sex chromosomes, one with an X and one with a Y, the male is called the **heterogametic sex.** That is, half the gametes produced by an XY male carry an X chromosome and half carry a Y. When a sperm cell carrying an X chromosome fertilizes an X-bearing egg cell, the new individual develops into an XX female. Conversely, when a sperm cell carrying a Y chromosome fertilizes an X-bearing egg cell, the combination produces an XY male **(Figure 13.6).** The Punnett square shows that fertilization is expected to produce females and males with an equal frequency of 1/2. This expectation is closely matched in human and *Drosophila* populations.

Other sex chromosome arrangements occur, as in some insects with XX females and XO males (the O means there is no Y chromosome). In birds, butterflies, and some reptiles, the situation is reversed: Males are the homogametic sex with a homologous pair of sex chromosomes (termed ZZ instead of XX), and females are the heterogametic sex with ZW sex chromosomes, equivalent to an XY combination. Researchers have compared the genes on sex chromosomes and have determined that the X and Y chromosomes of mammals are quite different from the Z and W chromosomes of birds. That is, mammalian X and Y chromosome genes typically are on bird autosomes, whereas bird Z and W chromosome genes are on mammalian autosomes. The interpretation is that mammalian and bird sex chromosomes have evolved from different autosomal pairs.

In bees and wasps, and certain other arthropods, sex is determined not by sex chromosomes but whether the individual is haploid or diploid. Essentially this means that sex depends on the number of sets of chromosomes. In this system, an individual produced by fusion of an egg and a sperm is diploid and develops into a female, whereas an unfertilized egg, which is haploid, develops into a male.

A number of eukaryotic microorganisms do not have sex chromosomes but have a "sex" system specified by simple alleles of a gene. For example, budding yeast, *Saccharomyces cerevisiae* (see *Focus on Model Research Organisms,* p. 211), is a haploid eukaryote with two *mating types* or sexes, designated **a** and α. The mating types are identical in appearance, but matings will only occur between two individuals of opposite type.

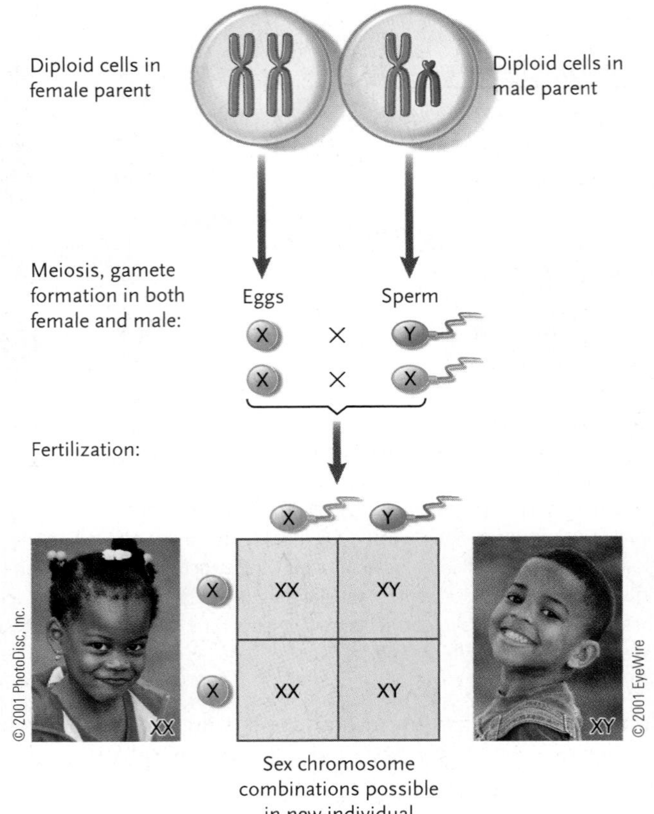
Diploid cells in female parent
Diploid cells in male parent
Meiosis, gamete formation in both female and male:
Eggs
Sperm
X × Y
X × X
Fertilization:
X Y
X XX XY
X XX XY
XX
XY
© 2001 PhotoDisc, Inc.
© 2001 EyeWire
Sex chromosome combinations possible in new individual

FIGURE 13.6
Sex chromosomes and the chromosomal basis of sex determination in humans. Females have two X chromosomes and produce gametes (eggs), all of which have the X sex chromosome. Males have one X and one Y chromosome and produce gametes, half with an X and half with a Y chromosome. Males transmit their Y chromosome to their sons, but not to their daughters. Males receive their X chromosome only from their mother.

Human Sex Determination Depends on the Y Chromosome

The human X chromosome carries about 2,350 genes. Although some of these genes are associated with sexual traits, such as differing distributions of body fat in males and females, most are concerned with nonsexual traits such as the ability to perceive color, metabolize certain sugars, or form blood clots when tissues are injured. Human sex determination depends on the Y chromosome, which contains the *SRY* gene (for *s*ex-determining *r*egion of the *Y*) that switches development toward maleness at an early point in embryonic development.

For the first month or so of embryonic development in humans and other mammals, the rudimentary structures that give rise to reproductive organs and tissues are the same in XX or XY embryos. After 6 to 8 weeks, the *SRY* gene becomes active in XY embryos, producing a protein that regulates the expression of other genes, thereby stimulating part of these structures to develop as testes. As a part of stimulation by hormones secreted in the developing testes and elsewhere, tissues degenerate that would otherwise develop into female structures such as the vagina and oviducts. The remaining structures develop into the penis and scrotum. In XX embryos, which do not have a copy of the *SRY* gene, development proceeds toward female reproductive structures. The rudimentary male structures degenerate in XX embryos because the hormones released by the developing testes in XY embryos are not present. Further details of the *SRY* gene and its role in human sex determination are presented in Chapter 48 (specifically, see *Insights from the Molecular Revolution* in that chapter).

Sex-Linked Genes Were First Discovered in *Drosophila*

The different sets of sex chromosomes in males and females affect the inheritance of the alleles on these chromosomes in a distinct pattern known as *sex linkage.* Two features of the XX–XY arrangement cause sex linkage. One is that alleles carried on the X chromosome occur in two copies in females but in only one copy in males. The second feature is that alleles carried on the Y chromosome are present in males but not females; those alleles do not correspond to alleles on the X chromosome.

Morgan discovered sex-linked genes and their pattern of sex linkage in 1910. It started when he found a male fly in his stocks with white eyes instead of the normal red eyes **(Figure 13.7).** How is the white-eye gene inherited? Morgan first crossed a true-breeding female with red eyes with the white-eyed male, and then interbred the F_1 generation to produce the F_2 generation. The F_1 flies of both sexes had red eyes, indicating that the white-eye trait is recessive. The F_2 flies showed a phenotypic ratio of 3 red-eyed:1 white-eyed. But, Morgan observed that, unexpectedly, the phenotypic ratio was not the same in males and females: All F_2 female flies had red eyes, whereas 1/2 the F_2 males had red eyes and the other 1/2 had white eyes.

A. Normal, red wild-type eye color

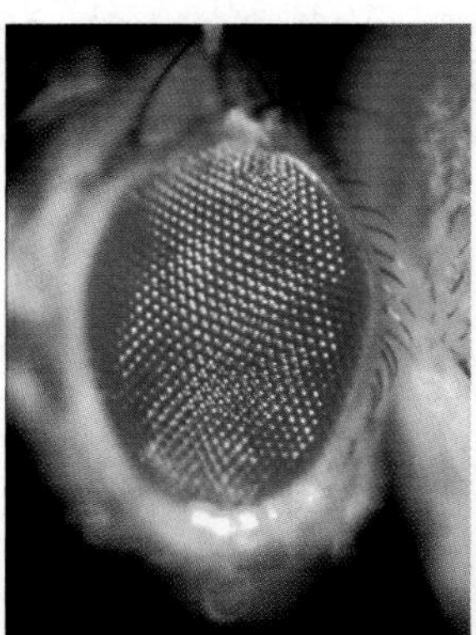
Carolina Biological Supply/Visuals Unlimited

B. Mutant white eye color caused by recessive allele of a sex-linked gene on the X chromosome

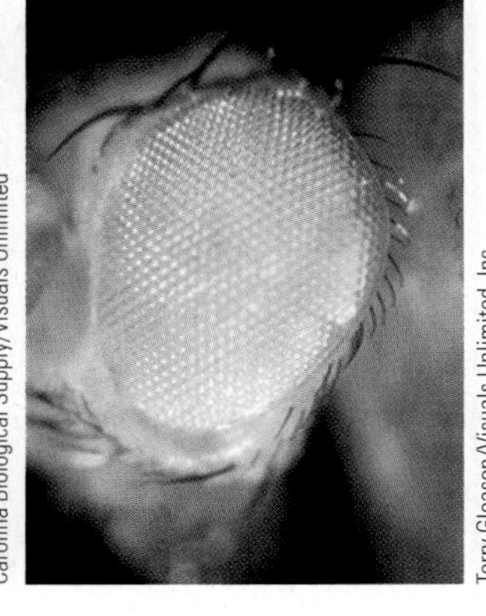
Terry Gleason/Visuals Unlimited, Inc.

FIGURE 13.7
Eye color phenotypes in *Drosophila.* **(A)** Normal, red wild-type eye color. **(B)** Mutant white eye color caused by a recessive allele of a sex-linked gene carried on the X chromosome.

Morgan hypothesized that the alleles segregating in the cross were of a gene located on the X chromosome and that the unexpected distribution of phenotypes in males and females in the F_2 generation could be accounted for by the inheritance pattern of X and Y chromosomes. A gene on a sex chromosome is a *sex-linked gene,* as we discussed earlier. A gene on the X chromosome more precisely is called an **X-linked gene;** the pattern of inheritance of an X-linked gene is called **X-linked inheritance.** To designate X-linked genes and their alleles, we use a symbolism using the upper-case letter X for the X chromosome and superscript letter(s) for the gene. In this case, the white mutant allele is X^w, a white-eyed female is X^wX^w and a white-eyed male is X^wY. The wild-type allele of the white gene is X^{w^+}; X^{w^+} is recessive to this allele.

We can follow the alleles in Morgan's cross of a true-breeding red-eyed female ($X^{w^+}X^{w^+}$) with the white-eyed male (X^wY) through to the F_2 generation in **Figure 13.8A.** The transmission of the white-eye allele shown in this cross—from a male parent to a female offspring ("child") to a male "grandchild" is called **criss-cross inheritance.**

Morgan also performed a **reciprocal cross,** meaning that he switched the phenotypes of the parents. The reciprocal cross here was a true-breeding white-eyed female (X^wX^w) with a red-eyed male ($X^{w^+}Y$); we can follow the alleles in this cross through to the F_2 generation in **Figure 13.8B.**

The results of the reciprocal crosses differed markedly in both the F_1 and F_2 generations. Morgan's experiments had shown that there is a distinctive pattern in the phenotypic ratios for reciprocal crosses in which the gene involved is on the X chromosome. A key indicator of X-linked inheritance of a recessive trait is when all male offspring of a cross between a true-breeding mutant female and a wild-type male have the mutant phenotype. As we have seen, this occurs because a male receives his X chromosome from his female parent.

X-Linked Genes in Humans Are Inherited as They Are in *Drosophila*

For obvious reasons, experimental genetic crosses cannot be conducted with humans. However, a similar analysis can be made by interviewing and testing living members of a family and reconstructing the genotypes and phenotypes of past generations from family records. The results are summarized in a chart called a **pedigree,** which shows all parents and offspring for as many generations as possible, the sex of individuals in the different generations, and the presence or absence of the trait of interest. Females are designated by a circle and males by a square; a solid circle or square indicates the presence of the trait.

In humans, as in fruit flies, X-linked recessive traits appear more frequently among males than females because males need to receive only one copy of the allele on the X chromosome inherited from their mothers to develop the trait. Females must receive two copies of the recessive allele, one from each parent, to develop the trait. Two examples of human X-linked traits are red–green color blindness, a recessive trait in which the affected individual is unable to distinguish between the colors red and

FIGURE 13.8 **Experimental Research**

Evidence for Sex-Linked Genes

Question: How is the white-eye gene of *Drosophila* inherited?

Experiment: Morgan crossed a white-eyed male *Drosophila* with a true-breeding female with red eyes and then interbred the F_1 flies to produce the F_2 generation. He also performed the reciprocal cross in which the phenotypes were switched in the parental flies—true-breeding white-eyed female × red-eyed male.

A. True-breeding red-eyed female × white-eyed male

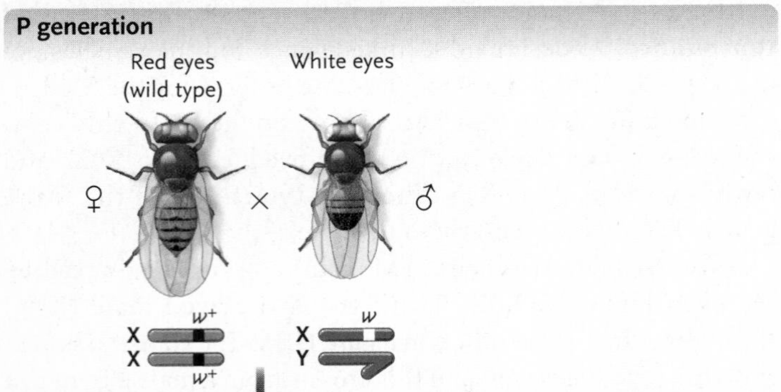

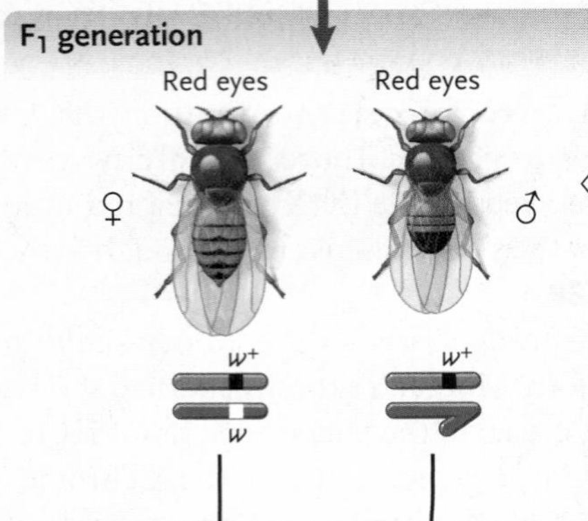

P generation

Red eyes (wild type) ♀ × White eyes ♂

$X^{w^+}X^{w^+}$ × $X^{w}Y$

F_1 generation

Red eyes ♀, Red eyes ♂

All F_1 flies have red eyes, indicating that the white-eye trait is recessive. The F_1 females inherit one X from each parent; their genotype is $X^{w^+}X^{w}$, and their phenotype is red eyes because the X^{w^+} allele is dominant. The F_1 males inherit their X chromosome from their mothers; their genotype is $X^{w^+}Y$, and their phenotype is red eyes.

F_2 generation

Sperm: w^+, Y

Eggs: w^+, w

All red-eyed females | $^1/_2$ red-eyed, $^1/_2$ white-eyed males

$^3/_4$ red eyes : $^1/_4$ white eyes

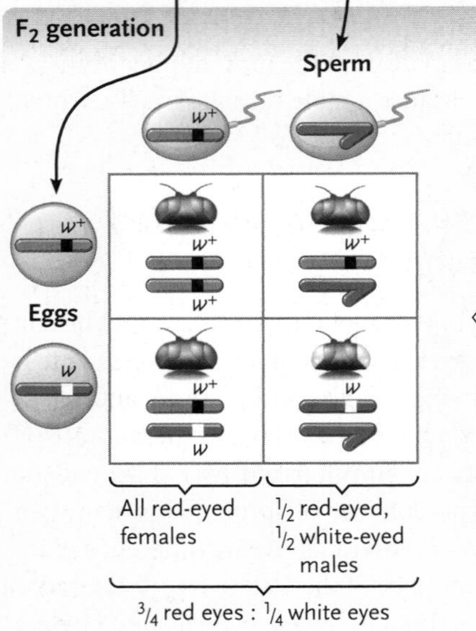

The F_2 females receive an X^{w^+} allele from the F_1 father and either an X^{w^+} or X^{w} allele from the F_1 mother; both these genotypes result in red eyes. The F_2 males inherit their one X chromosome from the F_1 mother whose genotype is $X^{w^+}X^{w}$. Therefore, F_2 males are half $X^{w^+}Y$ (red eyes) and half $X^{w}Y$ (white eyes). Females and males together show a phenotypic ratio of 3 red : 1 white.

B. White-eyed female × red-eyed male

P generation

White eyes ♀ × Red eyes ♂

$X^{w}X^{w}$ × $X^{w^+}Y$

F_1 generation

Red eyes ♀, White eyes ♂

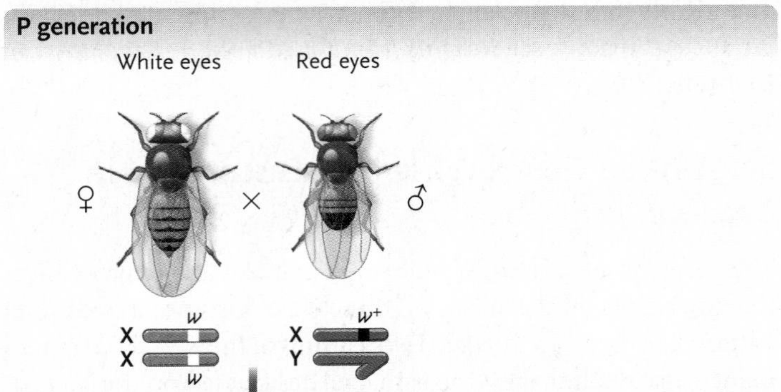

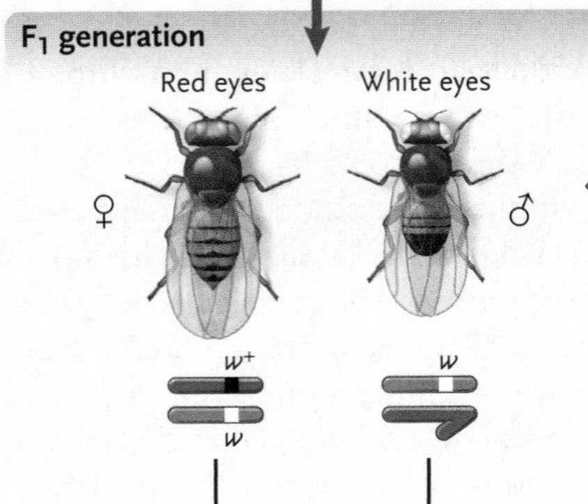

The F_1 females have red eyes: they are heterozygous $X^{w^+}X^{w}$. Their X^{w^+} allele comes from the mother, and their X^{w} allele from the father. The F_1 males all have white eyes because they received the X^{w}-bearing chromosome from the mother; their genotype is $X^{w}Y$. This result is distinctly different from that of the reciprocal cross.

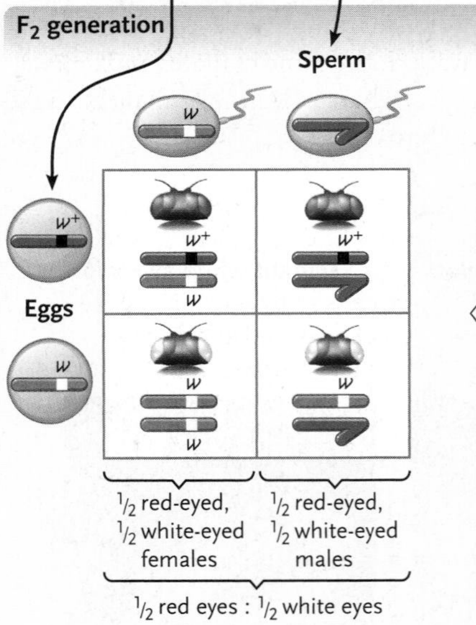

F_2 generation

Sperm: w, Y

Eggs: w^+, w

$^1/_2$ red-eyed, $^1/_2$ white-eyed females | $^1/_2$ red-eyed, $^1/_2$ white-eyed males

$^1/_2$ red eyes : $^1/_2$ white eyes

In this reciprocal cross, the F_2 females show a phenotypic ratio of 1 red ($X^{w^+}X^{w}$) : 1 white ($X^{w}X^{w}$), and the F_2 males also show a phenotypic ratio of 1 red ($X^{w^+}Y$) : 1 white ($X^{w}Y$). Combined, the F_2 progeny show a phenotypic ratio of 1 red : 1 white.

Results: Differences were seen in both the F_1 and F_2 generations for the red ♀ × white ♂ and white ♀ × red ♂ reciprocal crosses.

Conclusion: The segregation pattern for the white-eye trait showed that the white-eye gene is a sex-linked gene located on the X chromosome.

green because of a defect in light-sensing cells in the retina, and hemophilia, a recessive trait in which affected individuals have a defect in blood clotting.

Hemophiliacs—people with hemophilia—are "bleeders"; that is, if they are injured, they bleed uncontrollably because a protein required for forming blood clots is not produced in functional form. Males are bleeders if they receive an X chromosome that carries the recessive allele, X^h. The disease also develops in females with the recessive allele on both of their X chromosomes, genotype X^hX^h—a rare combination. Although affected persons, with luck and good care, can reach maturity, their lives are tightly circumscribed by the necessity to avoid injury of any kind. Even internal bleeding from slight bruises can be fatal. The disease, which affects about 1 in 7,000 males, can be treated by injection of the required clotting molecules.

Hemophilia has had effects reaching far beyond individuals who inherit the disease. The most famous cases occurred in the royal families of Europe descended from Queen Victoria of England **(Figure 13.9).** The disease was not recorded in Queen Victoria's ancestors, so the recessive allele for the trait probably appeared as a spontaneous mutation in the queen or one of her parents. Queen Victoria was heterozygous for the recessive hemophilia allele ($X^{h^+}X^h$); that is, she was a **carrier,** meaning that she carried the mutant allele and could pass it on to her offspring but she did not have symptoms of the disease. A carrier is indicated in a pedigree by a male or female symbol with a central dot.

Note in Queen Victoria's pedigree in Figure 13.9 that Leopold, Duke of Albany, had hemophilia, as did his grandson, Rupert, Viscount Trematon. The trait alternates from generation to generation in males because a father does not pass his X chromosome to his sons; the X chromosome received by a male always comes from his mother. The appearance of a trait in the males of alternate generations therefore indicates that the allele under study is recessive and carried on the X chromosome.

At one time, 18 of Queen Victoria's 69 descendants were affected males or female carriers. Because so many sons of European royalty were affected, the trait influenced the course of history. In Russia, Crown Prince Alexis (highlighted in Figure 13.9) was one of Victoria's hemophiliac descendants. His affliction drew together his parents, Czar Nicholas II and Czarina Alexandra (a granddaughter of Victoria and a carrier), and the hypnotic monk Rasputin, who manipulated the family to his advantage by convincing them that only he could control the boy's bleeding. The situation helped trigger the Russian Revolution of 1917, which ended the Russian monarchy and led to the establishment of a Communist government in the former Soviet Union, a significant event in twentieth century history.

Hemophilia affected only sons in the royal lines but could have affected daughters if a hemophiliac son had married a carrier female. Because the disease is rare in the human population as a whole, the chance of such a mating is so low that only a few unions of this type have been recorded.

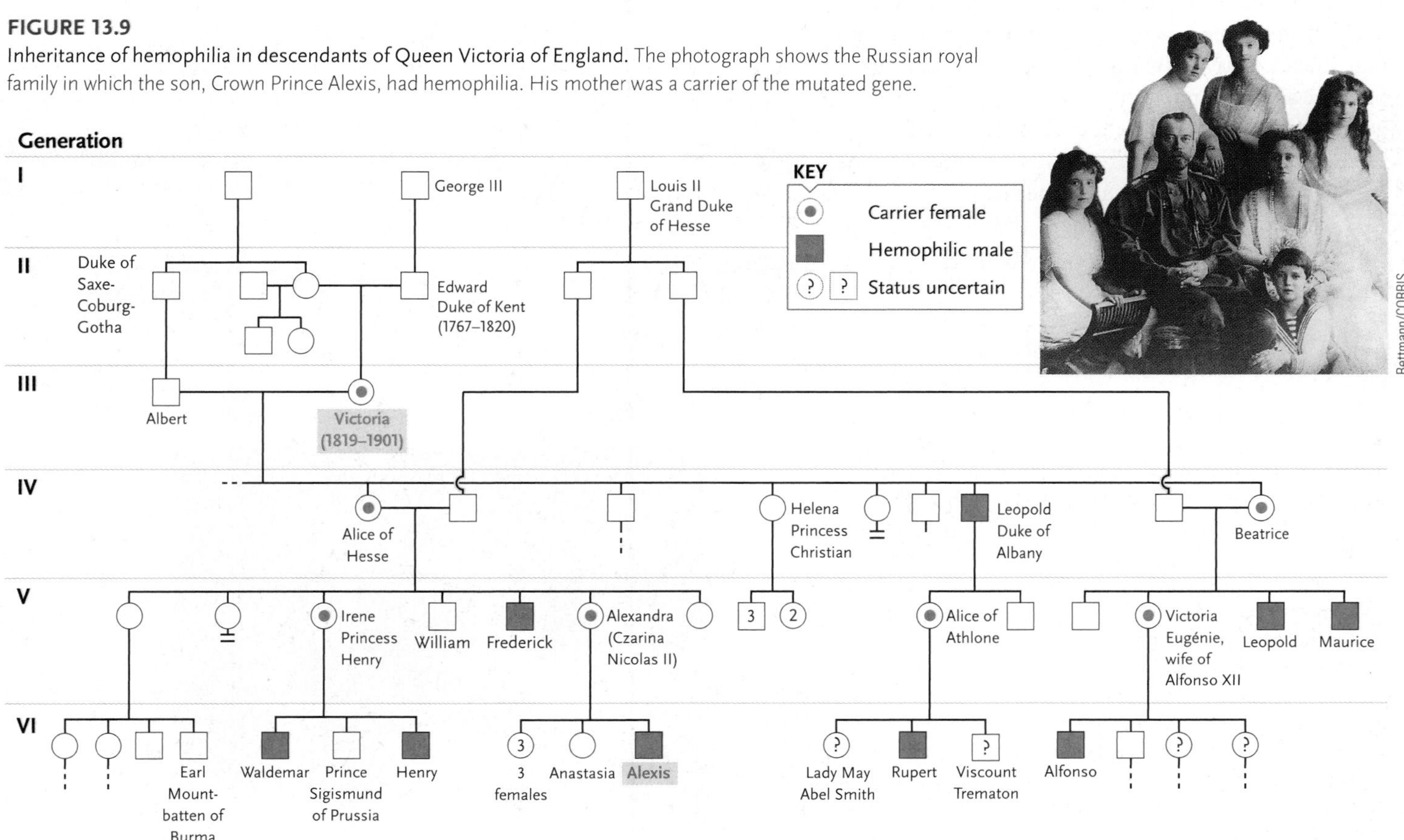

Generation
I
George III
Louis II
Grand Duke
of Hesse
KEY
Carrier female
Hemophilic male
Status uncertain
II
Duke of
Saxe-
Coburg-
Gotha
Edward
Duke of Kent
(1767–1820)
III
Albert
Victoria
(1819–1901)
IV
Alice of
Hesse
Helena
Princess
Christian
Leopold
Duke of
Albany
Beatrice
V
Irene
Princess
Henry
William
Frederick
Alexandra
(Czarina
Nicolas II)
Alice of
Athlone
Victoria
Eugénie,
wife of
Alfonso XII
Leopold
Maurice
VI
Earl
Mount-
batten of
Burma
Waldemar
Prince
Sigismund
of Prussia
Henry
3
females
Anastasia
Alexis
Lady May
Abel Smith
Rupert
Viscount
Trematon
Alfonso
Bettmann/CORBIS

FIGURE 13.9

Inheritance of hemophilia in descendants of Queen Victoria of England. The photograph shows the Russian royal family in which the son, Crown Prince Alexis, had hemophilia. His mother was a carrier of the mutated gene.

Inactivation of One X Chromosome Evens out Gene Effects in Mammalian Females

Although mammalian females have twice as many X chromosomes as males, the effects of most genes carried on the X chromosome in females is equalized in the male and female offspring of placental mammals by a **dosage compensation mechanism** that inactivates one of the two X chromosomes in most body cells of female mammals.

As a result of the equalizing mechanism, the activity of most genes carried on the X chromosome is essentially the same in males and females. The inactivation occurs by a condensation process that folds and packs the chromatin of one of the two X chromosomes into a tightly coiled state similar to the condensed state of chromosomes during cell division. The inactive, condensed X chromosome can be seen at one side of the nucleus in cells of females as a dense mass of chromatin called the **Barr body.**

The inactivation occurs during early embryonic development. Which of the two X chromosomes becomes inactive in a particular embryonic cell line is a random event. But once one of the X chromosomes is inactivated in a cell, that same X is inactivated in all descendants of the cell. Thus, within one female, one of the X chromosomes is active in particular cells and inactive in others and vice versa.

If the two X chromosomes carry different alleles of a gene, one allele will be active in cell lines in which one X chromosome is active, and the other allele will be active in cell lines in which the other X chromosome is active. For many sex-linked alleles, such as the recessive allele that causes hemophilia, random inactivation of either X chromosome has little overall whole-body effect in heterozygous females because the dominant allele is active in enough of the critical cells to produce a normal phenotype. However, for some genes, the inactivation of either X chromosome in heterozygotes produces recognizably different effects in distinct regions of the body.

For example, the orange and black patches of fur in calico cats result from inactivation of one of the two X chromosomes in regions of the skin of heterozygous females **(Figure 13.10).** Males, which get only one of the two alleles, normally have either black or orange fur.

An X-linked trait in humans has a similar, but less visible phenotype. Called anhidrotic ectodermal dysplasia, the trait is characterized by the absence of sweat glands. Females heterozygous for the mutation that causes the trait may have a patchy distribution of skin areas with and without the glands.

As we have seen, the discovery of genetic linkage, recombination, and sex-linked genes led to the elaboration and expansion of Mendel's principles of inheritance. Next, we examine what happens when patterns of inheritance are modified by changes in the chromosomes.

STUDY BREAK 13.2

You have a true-breeding strain of miniature-winged fruit flies, where this wing trait is recessive to the normal long wings. How would you show whether the miniature wing trait is sex-linked or autosomal?

THINK OUTSIDE THE BOOK

While *Drosophila* has X and Y chromosomes with XX flies being female and XY males being Y, this species does not have an *SRY* gene. Individually or collaboratively, explore the Internet or research papers to determine how sex determination occurs in *Drosophila*.

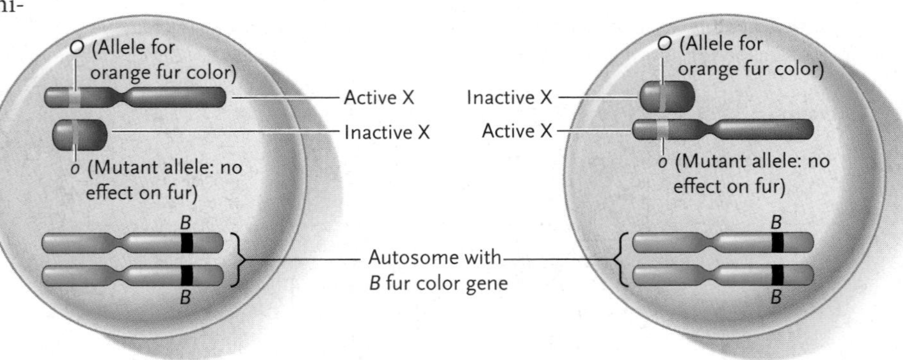
O (Allele for orange fur color)
Active X
Inactive X
o (Mutant allele: no effect on fur)
B
B
Autosome with B fur color gene
Inactive X
Active X
O (Allele for orange fur color)
o (Mutant allele: no effect on fur)
B
B
Orange fur
Black fur

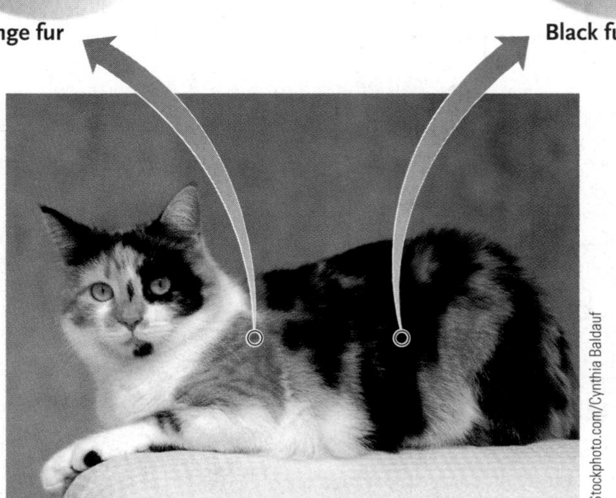
iStockphoto.com/Cynthia Baldauf

FIGURE 13.10
A female cat with the calico color pattern in which patches of orange and black fur are produced by random inactivation of one of the two X chromosomes. Two genes control the black and orange colors: The *O* allele on the X chromosome is for orange fur color, and the mutated *o* allele has no effect on color. The *B* gene on an autosome is for black fur color. A calico cat has the genotype *Oo BB* or *Oo Bb*; the former genotype is illustrated in the figure. An orange patch results when the X chromosome carrying the mutant *o* allele is inactivated. In this case, the *O* gene masks the expression of the *B* gene and orange fur is produced. (This example is of epistasis; see Section 12.2.) A black patch results when the X chromosome carrying the *O* allele is inactivated. In this case, the mutant *o* allele cannot mask *B* gene expression and black fur results. The white patches result from interactions with a different, autosomal gene that entirely blocks pigment deposition in the fur. In tortoiseshell cats, the same orange–black patching occurs as in calico cats but the gene for the white patching is not active.

13.3 Chromosomal Mutations That Affect Inheritance

Chromosomal mutations are variations from the normal condition in chromosome structure or chromosome number. Changes in chromosome structure occur when the DNA breaks, which can be generated by agents such as radiation or certain chemicals or by enzymes encoded in some infecting viruses. The broken chromosome fragments may be lost, or they may reattach to the same or different chromosomes. Chromosomal structure changes may have genetic consequences if alleles are eliminated, mixed in new combinations, duplicated, or placed in new locations by the alterations in cell lines that lead to the formation of gametes.

Genetic changes may also occur through changes in chromosome number, including addition or loss of one or more chromosomes or even entire sets of chromosomes. Both forms of chromosomal mutations, changes in chromosome structure and changes in chromosome number, can be a source of disease and disability, as well as a source of variability during evolution.

Deletions, Duplications, Translocations, and Inversions Are the Most Common Chromosomal Mutations Affecting Chromosome Structure

Chromosomal mutations after breakages occur in four major forms **(Figure 13.11)**:

- A **deletion** occurs if a broken segment is lost from a chromosome.
- A **duplication** occurs if a segment is broken from one chromosome and inserted into its homolog. In the receiving homolog, the alleles in the inserted fragment are added to the ones already there.
- A **translocation** occurs if a broken segment is attached to a different, nonhomologous chromosome.
- An **inversion** occurs if a broken segment reattaches to the same chromosome from which it was lost, but in reversed orientation, so that the order of genes is reversed.

To be inherited, chromosomal alterations must occur or be included in cells of the germ line leading to development of eggs or sperm.

DELETIONS AND DUPLICATIONS A deletion (see Figure 13.11A) may cause severe problems if the missing segment contains genes that are essential for normal development or cellular functions. For example, an individual heterozygous for a deletion of part of human chromosome 5 typically has severe mental retardation, a variety of physical abnormalities and a malformed larynx. The cries of an affected infant sound more like a meow than a human cry, hence the name of the disorder, *cri-du-chat* (meaning "cry of the cat").

A duplication (see Figure 13.11B) may have effects that vary from harmful to beneficial, depending on the genes and alleles contained in the duplicated region. Although most duplications are likely to be detrimental, some have been important sources of evolutionary change. That is, because there are duplicate genes, one copy can mutate into new forms without seriously affecting the basic functions of the organism. For example, mammals have genes that encode several types of hemoglobin that are not present in vertebrates, such as sharks, which evolved earlier; the additional hemoglobin genes of mammals are believed to have appeared through duplications, followed by mutations in the duplicates that created new and beneficial forms of hemoglobin as further evolution took place. Duplications sometimes arise if crossing-over occurs unequally during meiosis, so that a segment is deleted from one chromosome of a homologous pair and inserted in the other.

TRANSLOCATIONS AND INVERSIONS In a translocation, a segment breaks from one chromosome and attaches to another, nonhomologous chromosome. In many cases, a translocation is reciprocal, meaning that two nonhomologous chromosomes exchange segments (see Figure 13.11C). Reciprocal translocations resemble genetic recombination, except that the two chromosomes involved in the exchange do not contain the same genes.

Many cancers have chromosomal mutations, and the most common type of chromosomal mutation involved is a translocation. For example, 90% of patients with chronic myelogenous leukemia (CML) have a chromosomal mutation called the Philadelphia chromosome. CML is a type of cancer of the blood involving the uncontrolled division of stem cells for white blood cells. The Philadelphia chromosome arises from a reciprocal translocation event involving chromosomes 9 and 22 **(Figure 13.12)**. The movement of the segment of chromosome 9 to chromosome 22 fuses together on the resulting

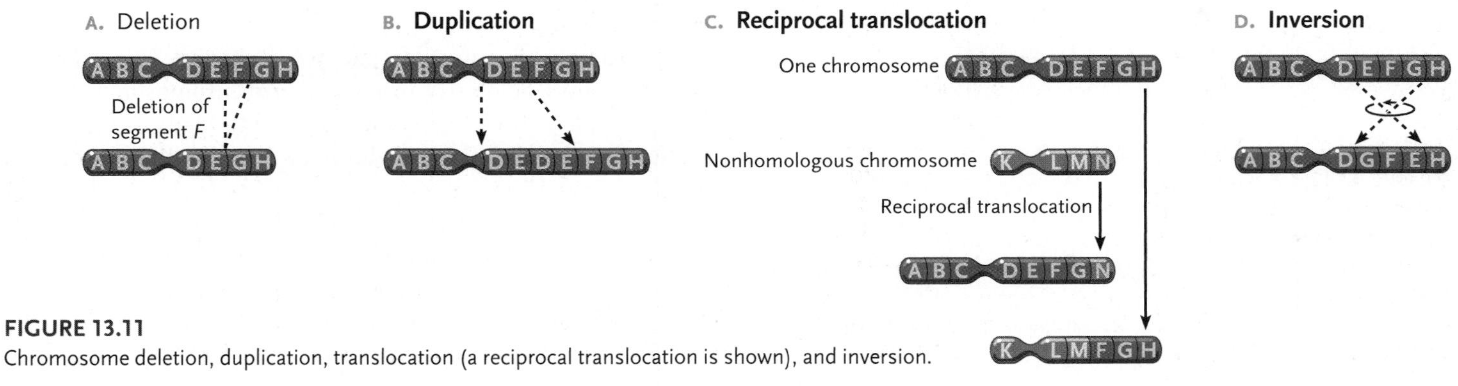
A. Deletion
ABC DEFGH
Deletion of segment F
ABC DEGH
B. Duplication
ABC DEFGH
ABC DEDEFGH
C. Reciprocal translocation
One chromosome
ABC DEFGH
Nonhomologous chromosome
K LMN
Reciprocal translocation
ABC DEFGN
K LMFGH
D. Inversion
ABC DEFGH
ABC DGFEH

FIGURE 13.11
Chromosome deletion, duplication, translocation (a reciprocal translocation is shown), and inversion.

Philadelphia chromosome, the *ABL* gene from 9 with the *BCR* gene on 22. The *ABL* gene is one of many genes that control cell growth and division. (Cell division control is described in Chapter 11.) Its product is a tyrosine kinase, an enzyme that adds phosphate to tyrosine amino acids in target proteins. We learned about tyrosine kinases in Chapter 7. In its new location on the Philadelphia chromosome, the *ABL* gene becomes much more active because of its fusion with the *BCR* gene and much more than normal of its tyrosine kinase product is made. As a result of its overactivity, normal cell cycle control breaks down and the cells are stimulated to growth and divide uncontrollably, becoming cancer cells. The drug Gleevec® is used to treat CML patients. It works by inhibiting the tyrosine kinase enzyme so that the body stops, or at least reduces, the production of too many white blood cells.

In an inversion, a chromosome segment breaks and then reattaches to the same chromosome, but in reverse order (see Figure 13.11D). Inversions have essentially the same effects as translocations—genes may be broken internally by the inversion, with loss of function, or they may be transferred intact to a new location within the same chromosome, producing effects that range from beneficial to harmful.

Inversions and translocations have been important factors in the evolution of plants and some animals, including insects and primates. For example, five of the chromosome pairs of humans show evidence of translocations and inversions that are not present in one of our nearest primate relatives, gorillas, and therefore must have occurred after the gorilla and human evolutionary lineages split.

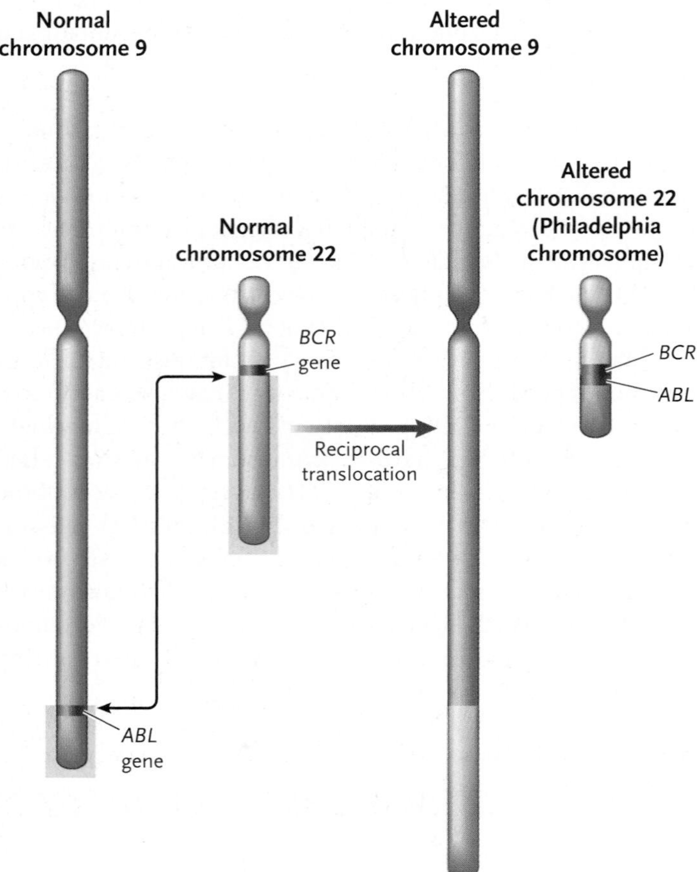
Normal chromosome 9
Altered chromosome 9
Altered chromosome 22 (Philadelphia chromosome)
Normal chromosome 22
BCR gene
BCR
ABL
Reciprocal translocation
ABL gene

FIGURE 13.12
Translocation found in many patients with a form of blood cancer called chronic myelogenous leukemia (CML). A reciprocal translocation involving chromosomes 9 and 22 produce a short chromosome named the Philadelphia. On this chromosome the chromosome 9 *ABL* gene has become fused to the chromosome 22 *BCR*. The resulting overactivity of the *ABL* gene, which normally helps control cell division, causes the cell to convert to a cancer cell.

Some Chromosome Mutations Involve Changes in the Number of Entire Chromosomes

At times, whole, single chromosomes are lost or gained from cells entering or undergoing meiosis, resulting in a change of chromosome number. Most often, these changes occur through **nondisjunction**—the failure of homologous pairs to separate during the first meiotic division or of chromatids to separate during the second meiotic division **(Figure 13.13)**. As a result, gametes are produced that lack one or more chromosomes or contain extra copies of the chromosomes. Fertilization by these gametes produces an individual with extra or missing chromosomes. Such individuals are called **aneuploids,** whereas individuals with a normal set of chromosomes are called **euploids.**

Changes in chromosome number can also occur through duplication or loss of entire sets, meaning individuals may receive fewer or more than the normal number of the entire haploid complement of chromosomes. Individuals with one set of chromosomes instead of the normal two are **monoploids;** individuals with more than the normal number of chromosomes are called **polyploids.** *Triploids* have three copies of each chromosome instead of two; *tetraploids* have four copies of each chromosome; *hexaploids* have six copies of each chromosome. Multiples higher than hexaploids also occur.

ANEUPLOIDS The effects of addition or loss of whole chromosomes vary depending on the chromosome and the species. In animals, aneuploidy of autosomes usually produces debilitating or lethal developmental abnormalities. In humans addition or loss of an autosomal chromosome causes embryos to develop so abnormally that generally they are aborted naturally. For reasons that are not understood, aneuploidy is as much as 10 times more frequent in humans than in other mammals. Of human embryos that have been miscarried and examined, about 70% are aneuploids.

In some cases, autosomal aneuploids survive. This is the case with humans who receive an extra copy of chromosome 21—one of the smallest chromosomes **(Figure 13.14A).** Many of these individuals survive until young adulthood. The condition produced by the extra chromosome, called *Down syndrome* or *trisomy 21* (for "three chromosome 21s"), is characterized by short stature and moderate to severe mental retardation. About 40% of individuals with Down syndrome have heart defects, and skeletal development is slower than normal. Most do not mature sexually and remain sterile. However, with attentive care and special

A. **Nondisjunction during first meiotic division**

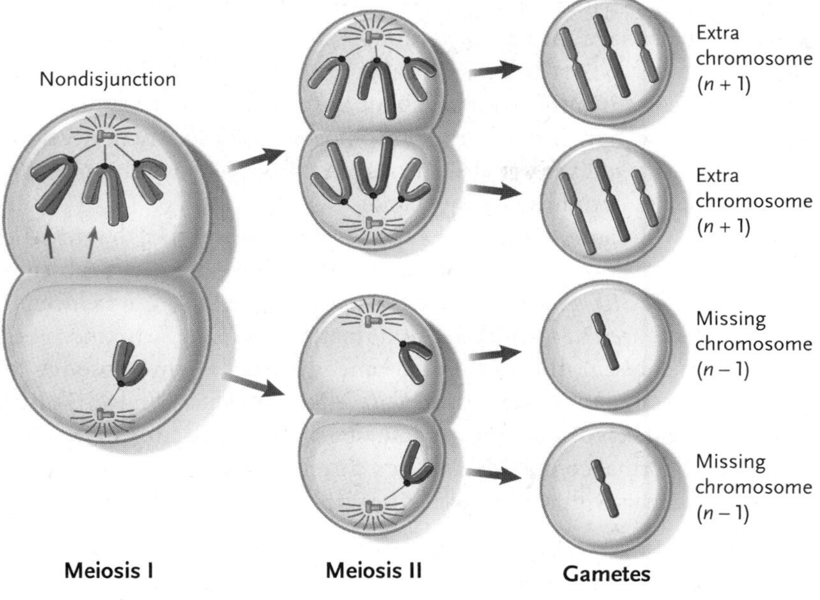
Nondisjunction
Extra chromosome (n + 1)
Extra chromosome (n + 1)
Missing chromosome (n – 1)
Missing chromosome (n – 1)
Meiosis I
Meiosis II
Gametes

Nondisjunction during the first meiotic division causes both chromosomes of one pair to be delivered to the same pole of the spindle. The nondisjunction produces two gametes with an extra chromosome and two with a missing chromosome.

B. **Nondisjunction during second meiotic division**

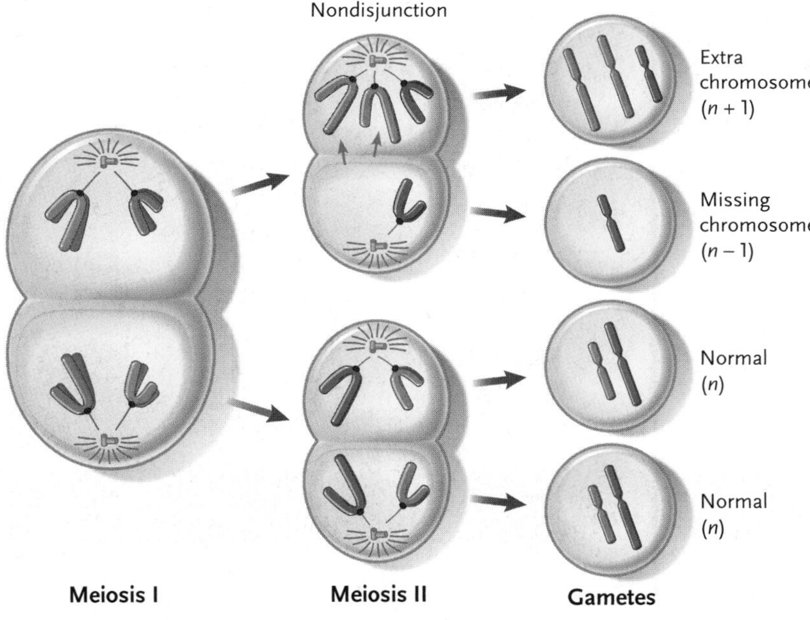
Nondisjunction
Extra chromosome (n + 1)
Missing chromosome (n – 1)
Normal (n)
Normal (n)
Meiosis I
Meiosis II
Gametes

Nondisjunction during the second meiotic division produces two normal gametes, one gamete with an extra chromosome and one gamete with a missing chromosome.

FIGURE 13.13
Nondisjunction during (A) the first meiotic division and (B) the second meiotic division.

A. **The chromosomes of a human female with Down syndrome showing three copies of chromosome 21 (circled in red).**

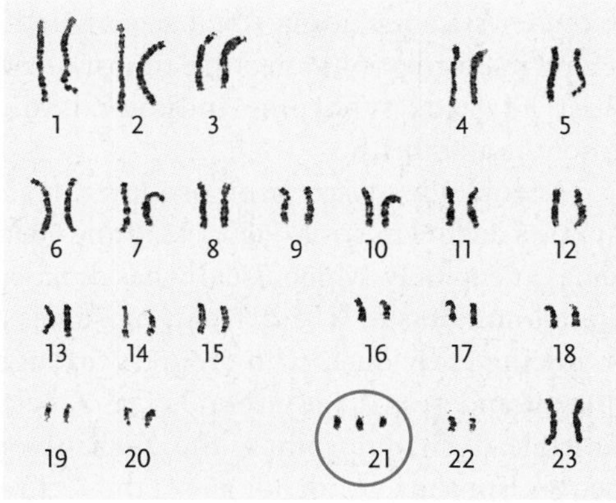
1
2
3
4
5
6
7
8
9
10
11
12
13
14
15
16
17
18
19
20
21
22
23

B. **The increase in the incidence of Down syndrome with increasing age of the mother, from a study conducted in Victoria, Australia, between 1942 and 1957.**

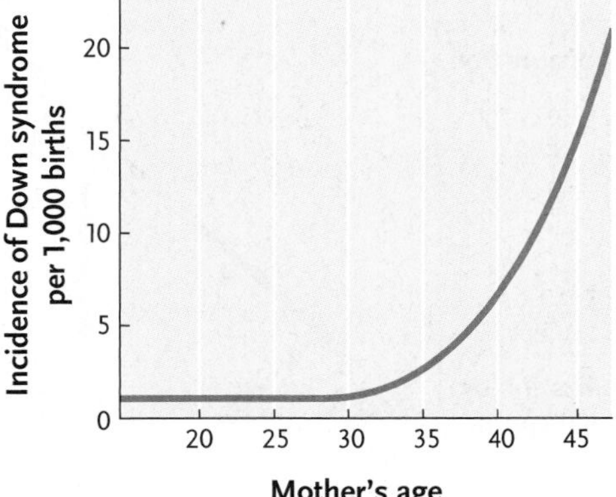
Incidence of Down syndrome per 1,000 births
0
5
10
15
20
20
25
30
35
40
45
Mother's age

C. **Person with Down syndrome.**

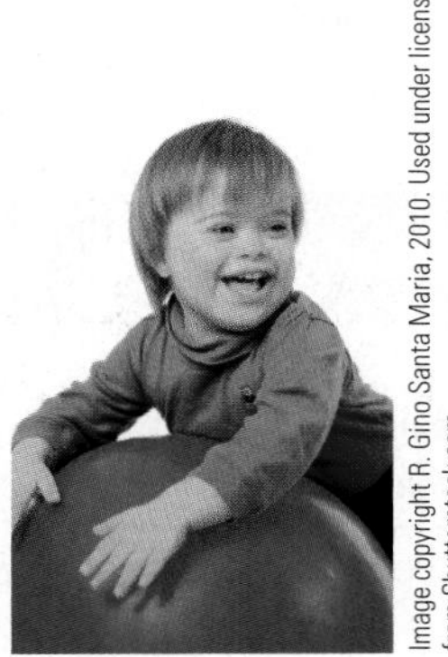

FIGURE 13.14
Down syndrome.

training, individuals with Down syndrome can participate with reasonable success in many activities.

Down syndrome arises from nondisjunction of chromosome 21 during the meiotic divisions, primarily in women (about 5% of nondisjunctions that lead to Down syndrome occur in men). The nondisjunction occurs more frequently as women age, increasing the chance that a child may be born with the syndrome **(Figure 13.14B).** In all, 1 in every 1,000 children is born with Down syndrome in the United States, making it one of the most common serious human genetic defects.

Trisomy of chromosomes 13 and 18 also is seen in live births of humans. Trisomy-13 produces Patau syndrome, with characteristics including cleft lip and palate, small eyes, extra fingers and toes, mental and developmental retardation, and cardiac anomalies. Most Patau syndrome individuals die before the age of 3 months. Trisomy-18 produces Edwards syndrome, with characteristics including small size at birth and multiple congenital malformations affecting almost every organ of the body. Most Edwards syndrome individuals die within the first 6 months after birth.

Aneuploidy of sex chromosomes can also arise by nondisjunction during meiosis (**Figure 13.15** and **Table 13.1**). Unlike autosomal aneuploidy, which usually has drastic effects on survival, altered numbers of X and Y chromosomes are often tolerated, producing individuals who progress through embryonic development and grow to adulthood. This is because, in the case of multiple X chromosomes, the X-chromosome inactivation mechanism converts all but one of the X chromosomes to a Barr body, so the dosage of active X-chromosome genes is the same as in normal XX females and XY males. Thus, XO (Turner syndrome) females have no Barr bodies, XXY (Klinefelter syndrome) males have one Barr body, and XXX (triple-X syndrome) females have two Barr bodies (see Table 13.1). However, X chromosomes are not inactivated until about 15 to 16 days after fertilization. Expression of the extra X chromosome genes early in development results in any deleterious effects associated with a particular sex chromosome aneuploidy.

Because sexual development in humans is pushed toward male or female reproductive organs primarily by the presence or absence of the *SRY* gene on the Y chromosome, people with a Y chromosome are externally malelike, no matter how many X chromosomes are present. If no Y chromosome is present, X chromosomes in various numbers give rise to femalelike individuals. (Table 13.1 lists the effects of some alterations in sex chromosome number.) Similar abnormal combinations of sex chromosomes also occur in other animals with varying effects on viability.

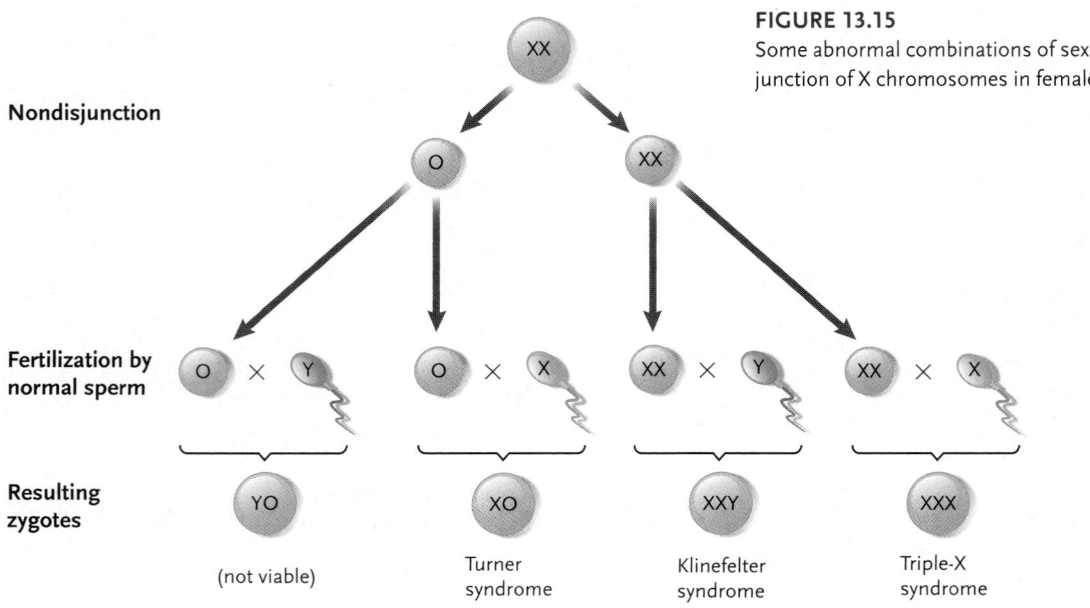
Nondisjunction
XX
O
XX
Fertilization by normal sperm
O × Y
O × X
XX × Y
XX × X
Resulting zygotes
YO
XO
XXY
XXX
(not viable)
Turner syndrome
Klinefelter syndrome
Triple-X syndrome

FIGURE 13.15
Some abnormal combinations of sex chromosomes resulting from nondisjunction of X chromosomes in females.

TABLE 13.1 Effects of Unusual Combinations of Sex Chromosomes in Humans

Combination of Sex Chromosomes	Approximate Frequency	Barr Bodies	Effects
XO	1 in 5,000 births	0	Turner syndrome: females with underdeveloped ovaries; sterile; intelligence and external genitalia are normal; typically, individuals are short in stature with underdeveloped breasts
XXY	1 in 2,000 births	1	Klinefelter syndrome: male external genitalia with very small and underdeveloped testes; sterile; intelligence usually normal; sparse body hair and some development of the breasts; similar characteristics in XXXY and XXXXY individuals
XYY	1 in 1,000 births	0	XYY syndrome: apparently normal males but often taller than average
XXX	1 in 1,000 births	2	Triple-X syndrome: apparently normal female with normal or slightly retarded mental function

MONOPLOIDS A monoploid has only one set of chromosomes instead of the normal two. Monoploidy is lethal in most animal species, but is tolerated more in plants. Certain animal species produce monoploid organisms as a normal part of their life cycle. For instance, as mentioned earlier, some male wasps, ants, and bees are monoploid individuals which have developed from unfertilized eggs.

POLYPLOIDS Polyploidy often originates from failure of the spindle to function normally during mitosis in cell lines leading to germ-line cells. In these divisions, the spindle fails to separate the duplicated chromosomes, which are incorporated into a single nucleus with twice the usual number of chromosomes. Eventually, meiosis takes place and produces gametes with two copies of each chromosome instead of one. Fusion of one such gamete with a normal haploid gamete produces a triploid, and fusion of two such gametes produces a tetraploid. Polyploidy may also occur when a single egg is fertilized by more than one sperm. For example, fertilization by two sperm produces a triploid with three sets of chromosomes. Such an event is rare in animals because of mechanisms that operate during fertilization of an egg with a sperm to prevent the subsequent fusion of other sperm with that egg (discussed more in Chapter 47).

The effects of polyploidy vary widely between plants and animals. In plants, polyploids are often hardier and more successful in growth and reproduction than the diploid plants from which they were derived. As a result, polyploidy is common and has been an important source of variability in plant evolution. About half of all flowering plant species are polyploids, including important crop plants such as wheat and other cereals, cotton, strawberries, and bananas. Commercial wheat, for instance, is hexaploid, and cultivated bananas are triploid.

By contrast, among animals, polyploidy is uncommon because it usually has lethal effects during embryonic development. For example, in humans, all but about 1% of polyploids die before birth, and the few who are born die within a month. The lethality is probably caused by disturbance of animal developmental pathways, which are typically much more complex than those of plants.

We now turn to a description of the effects of altered alleles on human health and development.

STUDY BREAK 13.3

What mechanisms are responsible for (a) duplication of a chromosome segment, (b) generation of a Down syndrome individual, (c) a chromosome translocation, and (d) polyploidy?

THINK OUTSIDE THE BOOK

Diagram the various ways a normal XX female and normal XY male could produce an XXY zygote.

13.4 Human Genetics and Genetic Counseling

We have already noted a number of human genetic traits and conditions caused by mutant alleles or chromosomal alterations. All these traits are of interest as examples of patterns of inheritance that amplify and extend Mendel's basic principles. Those with harmful effects are also important because of their impact on human life and society.

In Autosomal Recessive Inheritance, Heterozygotes Are Carriers and Homozygous Recessives Are Affected by the Trait

Sickle-cell disease, cystic fibrosis, and phenylketonuria are examples of human traits caused by recessive alleles on autosomes. Many other human genetic traits follow a similar pattern of inheritance. These traits are passed on according to the pattern known as **autosomal recessive inheritance,** in which individuals who are homozygous for the dominant allele are free of symptoms and are not carriers; heterozygotes are usually symptom-free but are carriers. People who are homozygous for the recessive allele show the trait.

In sickle-cell disease, the amino acid change in hemoglobin causes red blood cells to assume a sickle shape (see Figure 12.1, the Chapter 12 *Why It Matters,* and Section 12.2). The problems the sickled red blood cells have in passing through capillaries cause the serious symptoms of sickle-cell disease. Between 10% and 15% of African Americans in the United States are carriers for this disease—they have sickle-cell trait (see Section 12.2). Although carriers make enough normal hemoglobin through the activity of the dominant allele to be essentially unaffected, the mutant, sickle-cell form of the hemoglobin molecule is also present in their red blood cells. Carriers can be identified by a molecular test for the mutant hemoglobin. In countries where malaria is common, including several countries in Africa, carriers are less susceptible to contraction of the disease, which helps explain the increased proportions of the recessive allele among races that originated in malarial areas.

Cystic fibrosis (CF), one of the most common genetic disorders among persons of Northern European descent, is another autosomal recessive trait **(Figure 13.16)** (see Chapter 6 *Why It Matters*). About 1 in every 25 people from this line of descent is an unaffected carrier with one copy of the recessive allele. Approximately 1 in 4,000 children born in the U.S. have CF. They have a mutated form of the transport protein called cystic fibrosis transmembrane conductance regulator (CFTR: see Figure 6.1). CFTR is embedded in the plasma membrane of epithelial cells, such as those of the passageways and ducts of the lungs, pancreas, and digestive tract. The mutant CFTR is deficient in the transport of Cl^- (chloride ions) out of the cells into the extracellular fluids. This alteration in chloride transport causes thick, sticky mucus to collect in airways of the lungs, in the ducts of glands such as the pancreas, and in the digestive tract. The accumulated mucus impairs body functions and, in the lungs, promotes pneumonia and other

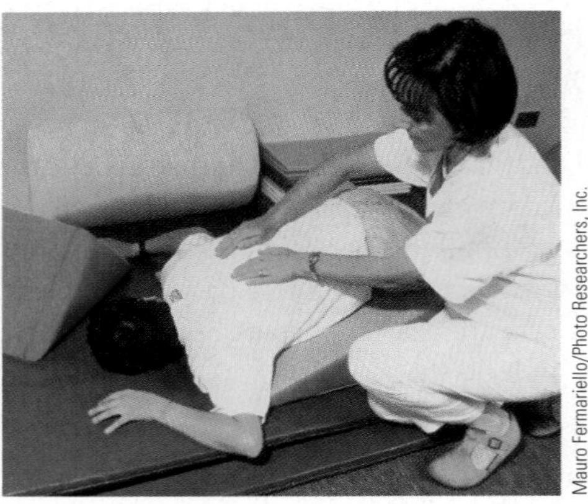
Mauro Fermariello/Photo Researchers, Inc.

FIGURE 13.16
A child affected by cystic fibrosis. Daily chest thumps, back thumps, and repositioning dislodge thick mucus that collects in airways to the lungs.

infections. With current management procedures, the life expectancy for a person with cystic fibrosis is about 40 years.

Another autosomal recessive disease, *phenylketonuria* (PKU), appears in about 1 of every 15,000 births. Affected individuals cannot produce an enzyme that converts the amino acid phenylalanine to another amino acid, tyrosine. As a result, phenylalanine builds up in the blood and is converted in the body into other products, including phenylpyruvate. Elevations in both phenylalanine and phenylpyruvate damage brain tissue and can lead to mental retardation. If diagnosed early enough, an affected infant can be placed on a phenylalanine-restricted diet, which can prevent the PKU symptoms. The diet must maintain a level of phenylalanine in the blood high enough to allow normal development of the nervous system, but be low enough to prevent mental retardation. Treatment must begin within the first one or two months after birth, or the brain will be damaged and treatment will be ineffective. By U.S. law, all infants born in the country are tested for PKU.

You may have seen warnings on certain foods and drinks for phenylketonuriacs (individuals with PKU) not to use them. This is because they contain the artificial sweetener aspartame (trade name NutraSweet®). Aspartame is a small molecule consisting of the amino acids aspartic acid and phenylalanine joined together. Aspartame binds to taste receptors in the mouth signaling that the substance is sweet. However, it has essentially no calories. Once ingested, aspartame is broken down to its component amino acids. Large amounts of phenylalanine released in this way could be harmful to an individual with PKU, hence the warnings.

In Autosomal Dominant Inheritance, Only Homozygous Recessives Are Unaffected

Some human traits follow a pattern of **autosomal dominant inheritance.** In this case, the allele that causes the trait is dominant, and people who are either homozygous or heterozygous for the dominant allele are affected. Individuals homozygous for the recessive normal allele are unaffected.

Achondroplasia (see Figure 12.12), a type of dwarfing that occurs in about 1 in 25,000 births worldwide, is caused by an autosomal dominant allele of a gene on chromosome 4. Of individuals with the dominant allele, only heterozygotes survive embryonic development; homozygous dominants are usually stillborn. When limb bones develop in heterozygous children, cartilage formation is defective, leading to disproportionately short arms and legs. They also have a relatively large head, but the trunk and torso are of normal size. Affected adults are usually not much more than 4 feet tall. Achondroplastic dwarfs are of normal intelligence, are fertile, and can have children. The gene responsible for this trait has been identified and is described in *Insights from the Molecular Revolution.* Progeria, the condition described in the *Why It Matters* of this chapter, is also an autosomal dominant trait.

X-Linked Recessive Traits Affect Males More than Females

Red–green color blindness and hemophilia have already been presented as examples of human traits that demonstrate **X-linked recessive inheritance,** that is, traits resulting from inheritance of recessive alleles carried on the X chromosome. Another X-linked recessive human disease trait is Duchenne muscular dystrophy (DMD) **(Figure 13.17).** In affected individuals, muscle tissue begins to degenerate late in childhood; by the onset of puberty, most individuals with this disease are unable to walk. Muscular weakness progresses, with later involvement of the heart muscle; the average life expectancy for individuals with DMD is 25 years. The gene that causes DMD has been identified. The normal form of the gene encodes the protein dystrophin, which anchors a particular glycoprotein complex in the plasma membrane of a muscle fiber to the cytoskeleton in the cytosol. In patients with DMD, the mutation in the dystrophin gene results in a nonfunctional protein. As a result, the plasma membrane of muscle fibers is susceptible to tearing during contraction, which leads to muscle destruction. (*Insights from the Molecular Revolution* in Chapter 41 discusses molecular research that could lead to a cure for the disease.)

EXIT
REGISTER TO VOTE!
© Sacramento Bee/ZUMA Press

FIGURE 13.17
Individual with Duchenne muscular dystrophy (DMD), an X-linked recessive trait.

Achondroplasia: What is the gene defect that is responsible for the trait?

Achondroplasia (also called achondroplastic dwarfing: see Figure 12.12C) is an autosomal dominant trait that is the most common form of short-limb dwarfism in humans. Over 80% of the cases of achondroplasia result from a new mutation; that is, the individual is born to parents each of normal height. Research has shown that the new gene mutation occurs during sperm formation in the male parent and is inherited from that parent.

Using genetic analysis, researchers found that the gene responsible for achondroplastic dwarfing is on chromosome 4. Mapping to the same region of the chromosome is a gene encoding a receptor that binds *fibroblast growth factor (FGF)*, a protein that stimulates a wide range of cells to grow and divide. The receptor is a protein embedded in the plasma membrane. (Receptors are described in Chapter 7; this particular receptor is a receptor tyrosine kinase.).The many receptors that bind this hormone form a family called the *fibroblast growth factor receptors (FGFRs)*. The gene on chromosome 4, *FGFR3*, is active in chondrocytes—cells that form cartilage and bone.

Research Question

Is achondroplasia caused by a mutation in the *FGFR3* gene? In other words, are the achondroplasia and *FGFR3* genes one and the same?

Experiments

To answer the questions, Arnold Munnich and his colleagues at the Hospital of Children's Diseases in Paris, France, and John Wasmuth and his colleagues at the University of California, Irvine independently analyzed the DNA sequence of the *FGFR3* gene in people with achondroplasia.

Results

Both research groups identified mutations in the *FGFR3* gene in DNA from all individuals they examined with achondroplasia, and no such mutations in DNA from normal people. Surprisingly, both groups found that every mutation in the gene affected the same nucleotide pair. In all but one case a guanine–cytosine (G–C) base pair had mutated to an adenine–thymine (A–T) base pair; in the other case the G–C base pair had mutated to a C–G base pair. Both mutations result in the substitution of arginine for glycine in the amino acid sequence of the encoded protein. These amino acids have very different chemical properties. The substitution occurs in a segment of the protein that extends across the membrane, connecting the growth factor-binding site outside the cell with a site inside the cell that triggers the internal response.

Conclusions

The correlations between mutations in the *FGFR3* gene and achondroplasia supported the hypothesis that a mutation in the gene for the fibroblast growth factor receptor 3 protein is responsible for achondroplastic dwarfing.

How does the single amino acid substitution cause dwarfing? In individuals homozygous for the normal *FGFR3* gene, the growth factor receptor has an inhibitory effect on bone growth, and participates in pathways to regulate bone growth. In this role, the receptor is sometimes active, other times not and, thus, the appropriate amount of chondrocyte proliferation occurs for normal development (figure). In heterozygotes with the mutation in *FGFR3*, the receptor is active all the time, resulting in a long-term negative effect on bone growth and, hence, the dwarfing symptoms of achondroplasia (see figure).

The conclusions have been confirmed with a mouse model of achondroplasia. David Givol and colleagues at the Weizmann Institute of Science, Rehovot, Israel, and the Agricultural Research Organization, Bet Dagan, Israel, introduced the A–T to G–C mutation studied in the human *FGFR3* gene into one copy of the mouse *FGFR3* gene, creating genetically engineered mice heterozygous for the mutation. These mice have small size and other similar characteristics of achondroplasia. The experiments showed that the dominant mutant allele of *FGFR3* leads to inhibition of chondrocyte proliferation, producing the symptoms of the disease.

Source: F. Rousseau et al. 1994. Mutations in the gene encoding fibroblast growth factor receptor-3 in achondroplasia. *Nature* 371:252–254; R. Shiang et al. 1994. Mutations in the transmembrane domain of FGFR3 cause the most common genetic form of dwarfism, achondroplasia. *Cell* 78:335–342; Y. Wang et al. 1999. A mouse model for achondroplasia produced by targeting fibroblast growth factor receptor 3. *Proceedings of the National Academy of Sciences USA* 96:4455–4460.

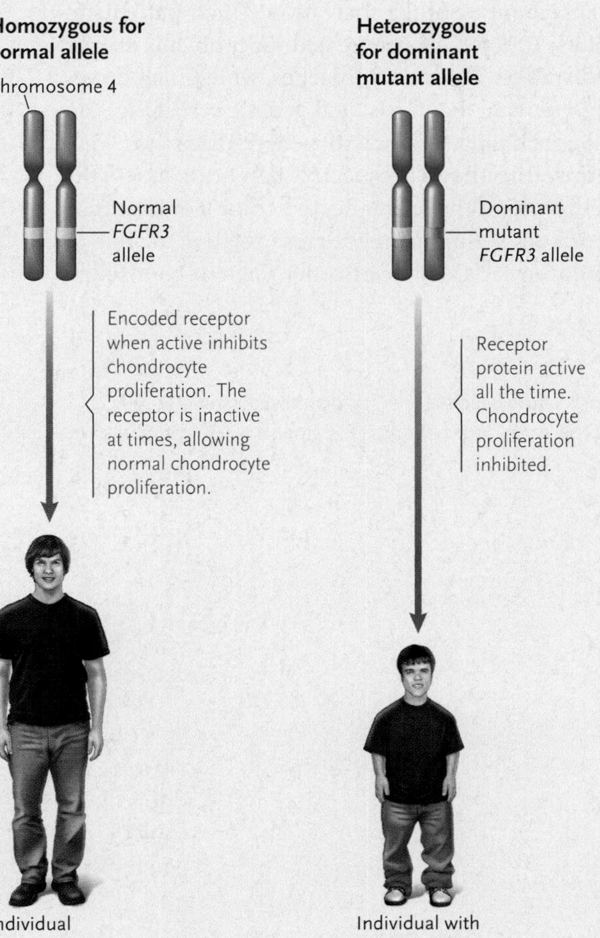

Homozygous for normal allele
Chromosome 4
Normal FGFR3 allele
Encoded receptor when active inhibits chondrocyte proliferation. The receptor is inactive at times, allowing normal chondrocyte proliferation.
Individual with normal size develops
Heterozygous for dominant mutant allele
Dominant mutant FGFR3 allele
Receptor protein active all the time. Chondrocyte proliferation inhibited.
Individual with achondroplasia develops

Molecular basis for achondroplasia.

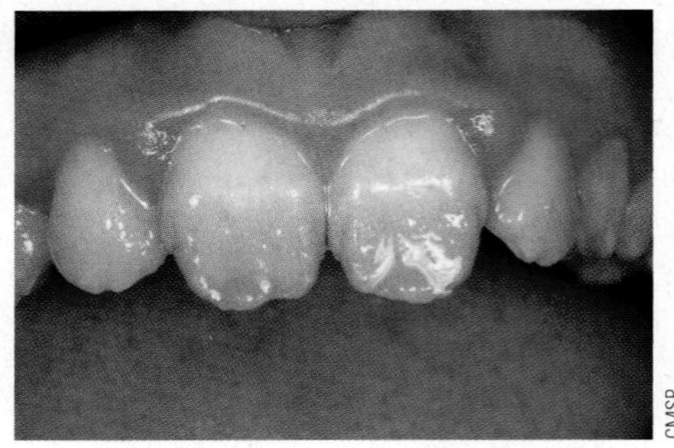
CMSP

FIGURE 13.18 Teeth showing hereditary faulty enamel and dental discoloration, an X-linked dominant trait.

X-Linked Dominant Traits are Rare

Only a few X-linked dominant traits have been identified in humans. One example is hereditary faulty enamel and dental discoloration, technical name *hereditary enamel hypoplasia* **(Figure 13.18).** Another is a severe bleeding anomaly called constitutional thrombopathy.

Human Genetic Disorders Can Be Predicted, and Many Can Be Treated

Of all newborns, between 1% and 3% have mutant alleles that encode defective forms of proteins required for normal functions. Possibly 1% have pronounced difficulties resulting from a chromosomal rearrangement or other aberration. Of all patients in children's hospitals, 10% to 25% are treated for problems arising from inherited disorders. Several approaches, which include genetic counseling, prenatal diagnosis, and genetic screening, can reduce the number of children born with genetic diseases.

Genetic counseling allows prospective parents to assess the possibility that they might have an affected child. For example, parents may seek counseling if they, a close relative, or one of their existing children has a genetic disorder. Genetic counseling begins with identification of parental genotypes through family pedigrees or direct testing for an altered protein or DNA sequence. With this information in hand, counselors can often predict the chances of having a child with the trait in question. Couples can then make an informed decision about whether to have a child.

Genetic counseling is often combined with techniques of **prenatal diagnosis,** in which cells derived from a developing embryo or its surrounding tissues or fluids are tested for the presence of mutant alleles (by DNA testing or biochemical analysis) or for chromosomal mutations. In **amniocentesis,** cells are obtained from the amniotic fluid—the watery fluid surrounding the embryo or fetus in the mother's uterus **(Figure 13.19).** In **chorionic villus sampling,** cells are obtained from portions of the placenta that develop from tissues of the embryo. More than 100 genetic disorders can now be detected by these tests. If prenatal diagnosis detects a serious genetic defect, the prospective parents can reach an informed decision about whether to continue the pregnancy, including religious and moral considerations, as well as genetic and medical advice.

Once a child is born, inherited disorders are identified by **genetic screening,** in which biochemical or molecular tests for disorders are routinely applied to children and adults or to newborn infants in hospitals. The tests can detect inherited disorders early enough to start any available preventive measures before symptoms develop. We have noted that most hospitals in the United States now test all newborns for PKU, making it possible to use dietary restrictions to prevent symptoms of the disorder from developing. As a result, it is becoming less common to see individuals debilitated by PKU.

In addition to the characters and traits described so far in this chapter, some patterns of inheritance depend on genes located not in the cell nucleus, but in mitochondria or chloroplasts in the cytoplasm, as discussed in the following section.

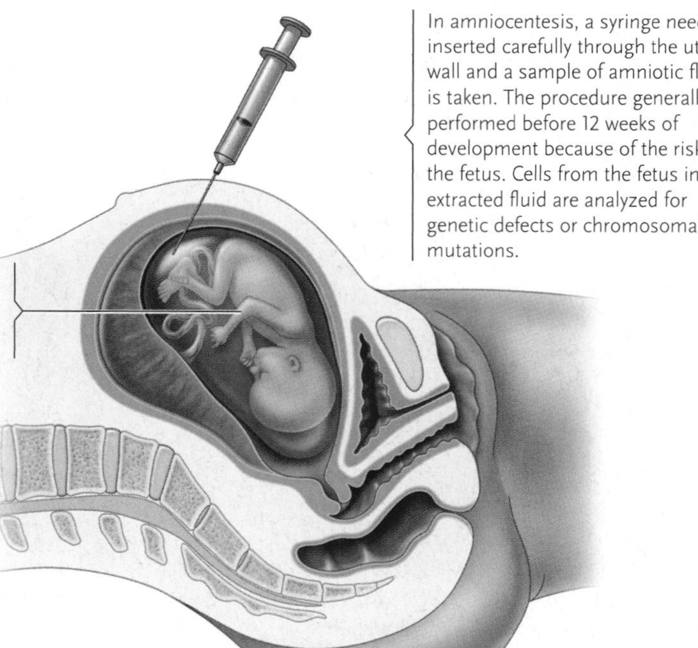
In amniocentesis, a syringe needle is inserted carefully through the uterine wall and a sample of amniotic fluid is taken. The procedure generally is performed before 12 weeks of development because of the risk to the fetus. Cells from the fetus in the extracted fluid are analyzed for genetic defects or chromosomal mutations.
Embryo and fetus develops surrounded by amniotic fluid to cushion it against shock

FIGURE 13.19 Amniocentesis, a procedure used for prenatal diagnosis of genetic defects. The procedure is complicated and costly and, therefore, it is used primarily in high-risk cases.

STUDY BREAK 13.4

1. A man has Simpson syndrome, an addiction to a certain television show. His wife does not have this syndrome. This couple has four children, two boys and two girls. One of the boys and one of the girls has this syndrome; the other children are normal. Can Simpson syndrome be an autosomal recessive trait? A sex-linked recessive trait?
2. In another family, a female child has wiggly ears, whereas her brother does not. Both parents are normal. Can the wiggly ear trait be an autosomal recessive trait? A sex-linked recessive trait?

13.5 Non-Mendelian Patterns of Inheritance

We consider two examples of patterns of inheritance in this section that do not follow the principles of Mendelian inheritance we have developed in this chapter and the previous one. In **cytoplasmic inheritance,** the pattern of inheritance follows that of genes in the genomes of mitochondria or chloroplasts. In **genomic imprinting,** the expression of a nuclear gene is based on whether an individual organism inherits the gene from the male or female parent.

Cytoplasmic Inheritance Follows the Pattern of Inheritance of Mitochondria or Chloroplasts

As noted in Chapter 5, not all DNA is contained in the nucleus; both chloroplasts and mitochondria also contain DNA. Like nuclear genes, chloroplast and mitochondrial genes are subject to mutation. However, the inheritance pattern of these mutant genes—called **cytoplasmic inheritance**—is fundamentally different from that of mutant genes in the nucleus. First, these genes do not segregate by meiosis, so the ratios of mutated and parental genes typical of Mendelian segregation are absent. Second, the genes usually show uniparental inheritance from generation to generation. In **uniparental inheritance,** all progeny (both males and females) have the phenotype of only one of the parents. For most multicellular eukaryotes, offspring inherit only the mother's phenotype, a phenomenon called **maternal inheritance.** In sexual reproduction, both the male and female gamete provide nuclear DNA, but the female provides most of the cytoplasm in the fertilized cell. Maternal inheritance occurs because a zygote receives most of its cytoplasm, including mitochondria and (in plants) chloroplasts, from the female parent and little from the male parent.

FIGURE 13.20
A four-o'clock *(Mirabilis)* plant with a variegated (patchy) distribution of green and white segments.

The first example of cytoplasmic inheritance of a mutant trait was found in 1909 by the German scientist Carl Correns, one of the geneticists who rediscovered Mendel's principles. Correns made his discovery through his genetic studies of a plant, *Mirabilis* (the four-o'clock), using mutant plants that had a variegated pattern of green and white **(Figure 13.20).** In the white segments, chloroplasts are colorless instead of green. Correns fertilized flowers in a green region of the plant with pollen from a white region and vice versa. He discovered that the phenotype of the progeny seedlings showed maternal inheritance. That is, it was always that of the female segment the pollen fertilized: white for the white ♀ × green ♂ cross, and green for the green ♀ × white ♂ cross.

Many examples of maternal inheritance of mutant traits involving the chloroplast are now known. In each case, the trait results from a mutation of one of the genes in the chloroplast genome. Maternal inheritance of mutant traits involving the mitochondria have also been characterized in many eukaryotic species, including plants, animals, protists, and fungi. Similar to the mutant traits of chloroplasts, each mutant trait results from an alteration of a gene in the mitochondrial genome.

In humans, several inherited diseases have been traced to mutations in mitochondrial genes. Recall that the mitochondrion plays a critical role in synthesizing ATP, the energy source for many cellular reactions. Several maternally inherited diseases in humans involve mutations in mitochondrial genes that encode components of the ATP-generating system of the organelle. The resulting mitochondrial defects are especially destructive to the organ systems most dependent on mitochondrial reactions for energy: the central nervous system, skeletal and cardiac muscle, the liver, and the kidneys.

For example, *Leber's hereditary optic neuropathy* (LHON) is a maternally inherited human disease which affects midlife adults, and is characterized by complete or partial blindness caused by optic nerve degeneration. Mutations in any one of the genes for eight electron transfer system proteins (see Chapter 8) all lead to LHON. The electron transfer system is responsible for ATP synthesis by oxidative phosphorylation (see Chapter 8). Death of the optic nerve is a common result of defects in oxidative phosphorylation which, in LHON, are caused by the inhibition of the electron transfer system.

Another example of a maternally inherited human disease is *myoclonic epilepsy and ragged-red fiber (MERRF) disease.* The "ragged-red fibers" in the name refers to the abnormal appearance of tissues under the microscope. Symptoms of the disease include jerking spasms of the limbs or the whole body, a defect in coordinating movement, and the accumulation of lactic acid in the blood. MERRF disease is caused by a mutation in a transfer RNA (tRNA) gene in the mitochondrial genome. Transfer RNA molecules play important roles in protein synthesis. The mutated tRNA adversely affects protein synthesis in the mitochondria which, in some way, causes the various phenotypes of the disease.

In Genomic Imprinting, the Allele Inherited from One of the Parents Is Expressed Whereas the Other Allele Is Silent

Throughout our discussions of Mendelian inheritance we have assumed that a particular allele has the same effect in an individual whether it was inherited from the mother or father. For the vast majority of genes, the assumption is correct. However, in mammals, researchers have identified 30 or so genes whose effects do in fact depend on whether an allele is inherited from the mother or the father. For some of these genes, only the egg-derived or sperm-derived allele of the gene is expressed. For some genes, only the paternal, sperm derived, allele is expressed; for others, only the maternal, egg derived, allele is expressed. The phenomenon in which the expression of an allele of a gene depends on the parent that contributed it is called **genomic imprinting.** The silent allele—the inherited allele that is not expressed—is called the *imprinted allele.*

The first imprinted gene identified was *Igf2* in the mouse. *Igf2* encodes insulin-like growth factor 2, a protein that stimulates cell growth and division. The growth factor is needed for early embryos to develop normally. Researchers studying mice heterozygous for a deletion of the entire *Igf2* gene from the genome observed that, if mice inherited the mutated chromosome from the father they were small, but if they had inherited the mutated chromosome from the mother they were normal size **(Figure 13.21A).** It appeared that only the paternally inherited gene had an effect on size: The maternally inherited gene, whether normal or not, had no effect. The scientists reasoned that, in normal mice homozygous for *Igf2,* the active form of the gene is the copy on the paternal chromosome, whereas the maternal copy of the gene is imprinted (silent) **(Figure 13.21B).** If a heterozygote inherits the deletion from the father, that copy of the gene is inactive. Even if the maternal copy is normal, it is not expressed because it is imprinted. As a result, normal development does not occur and the adult mouse produced is small.

The mechanism of genomic imprinting involves the modification of the DNA in the region that controls the expression of an allele by the addition of methyl ($—CH_3$) groups to cytosine (C) nucleotides. The methylation of the control region of a gene prevents it from being expressed. (You will learn more about the regulation of gene expression by methylation of DNA in Chapter 16.) Genomic imprinting occurs in the germ cells that develop into gametes. In those germ cells, the allele destined to be inactive in the new embryo after fertilization is methylated. That is, in the production of sperm, alleles for paternally imprinted genes are methylated, and in the production of eggs, alleles for maternally imprinted genes are methylated. That methylated (silenced) state of the gene is passed on cell generation to cell generation as the cells grow and divide to produce the somatic (body) cells of the organism.

Inherited imprints must first be erased in the germ cells before new imprinting occurs in the production of gametes. Consider a gene that is maternally imprinted, for example. In the adult, the maternal chromosome has an imprinted allele of that gene whereas the paternal chromosome has an active allele. When that adult produces gametes, it needs to imprint all alleles in a way appropriate to its sex—for example, if it is male, it must erase the imprint from the maternally inherited gene. In the diploid cells that go through meiosis to produce the gametes, all imprints are first erased providing a clean slate for individuals to imprint alleles appropriately to their gender.

We must be clear that, although we are talking about the expression of alleles inherited from one or the other parent,

A. Phenotypes of mice heterozygous for a deletion of gene *Igf2*.

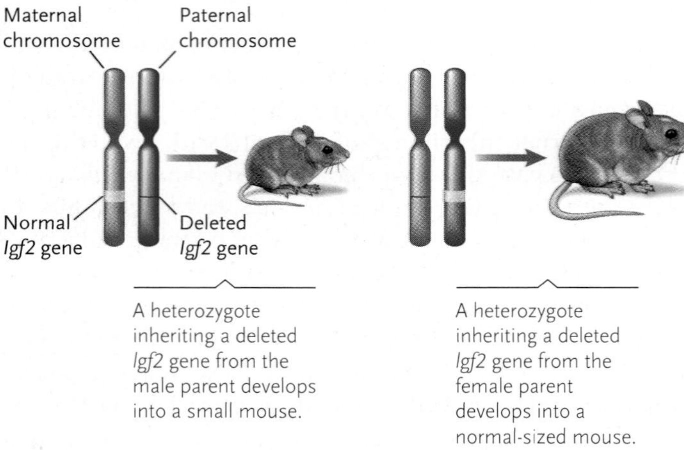
Maternal chromosome
Paternal chromosome
Normal *Igf2* gene
Deleted *Igf2* gene
A heterozygote inheriting a deleted *Igf2* gene from the male parent develops into a small mouse.
A heterozygote inheriting a deleted *Igf2* gene from the female parent develops into a normal-sized mouse.

B. Phenotype of mice homozygous for the normal allele of *Igf2*.

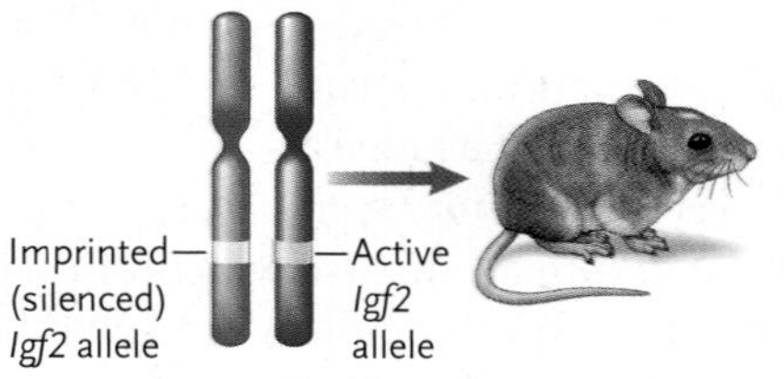
Imprinted (silenced) *Igf2* allele
Active *Igf2* allele
In a mouse heterozygous for the normal allele of *Igf2*, the parental allele is active, and the maternal allele is imprinted (silenced). As long as a normal allele is inherited from the male parent, the mouse develops into a normal-sized adult.

FIGURE 13.21
Imprinting of the mouse *Igf2* (insulin-like growth factor 2) gene.

genomic imprinting is a completely different phenomenon than sex linkage. For sex-linked genes, alleles inherited from the mother or father are both expressed. The phenotypic ratios are different than those for autosomal genes, but that is because of the difference in sex chromosome composition in males and females. And, in fact, most imprinted genes known are autosomal genes.

In this chapter, you have learned about genes and the role of chromosomes in inheritance. In the next chapter, you will learn about the molecular structure and function of the genetic material and about the molecular mechanism by which DNA is replicated.

STUDY BREAK 13.5 <

What key feature or features would suggest to you that a mutant trait shows cytoplasmic inheritance?

UNANSWERED QUESTIONS

Does recombination protect against cancer?

In this chapter, we learned that genetic recombination during meiosis in germ-line cells is a mechanism for the exchange of genetic information between chromosomes. Researchers have discovered that genetic recombination also occurs in somatic cells and is a vital mechanism for the repair of damaged or broken chromosomes. If recombination is defective, unrepaired chromosome breaks or gene translocations can have serious consequences to the cell and even lead to cancer. For instance, the *BRCA1* and *BRCA2* genes, which predispose patients to breast cancer, have recombination and repair defects that lead to genome instability. Researchers are using tissue culture cells derived from patients with these genes to find out exactly what goes wrong when recombination is inadequate and how recombination may act to maintain genome stability and provide protection against cancer and other diseases.

Think Critically

Does the discovery of recombination defects in somatic cells potentially change research on germ-line cells with respect to cancers such as breast cancer? Are somatic cell defects heritable?

Peter J. Russell

Peter J. Russell

REVIEW KEY CONCEPTS

Go to **CENGAGENOW** at www.cengage.com/login to access quizzing, animations, exercises, articles, and personalized homework help.

13.1 Genetic Linkage and Recombination

- Genes, consisting of sequences of nucleotides in DNA, are arranged linearly in chromosomes.
- Genes carried on the same chromosome are linked together in their transmission from parent to offspring. Linked genes are inherited in patterns similar to those of single genes, except for changes in the linkage caused by recombination (Figures 13.2–13.3).
- In genetic recombination, alleles linked on the same chromosome are mixed into new combinations by exchange of segments between the chromosomes of a homologous pair. The exchanges occur by the process of crossing-over while homologous chromosomes are paired during prophase I of meiosis.
- The amount of recombination between any two genes located on the same chromosome pair reflects the distance between them on the chromosome. The greater this distance, the greater the chance that chromatids will exchange segments at points between the genes and the greater the recombination frequency.
- The relationship between separation and recombination frequencies is used to produce chromosome maps in which genes are assigned relative locations with respect to each other (Figure 13.4).

13.2 Sex-Linked Genes

- Sex linkage is a pattern of inheritance produced by genes carried on sex chromosomes: chromosomes that differ in males and females. In humans and fruit flies, which have XX females and XY males, most sex-linked genes are carried on the X chromosome.
- Because males have only one X chromosome, they need to receive only one copy of a recessive allele from their mothers to develop the trait. Females must receive two copies of the recessive allele, one from each parent, to develop the trait (Figures 13.6–13.8).
- In mammals, inactivation of one of the two X chromosomes in cells of the female makes the dosage of X-linked genes the same in males and females (Figure 13.10).

Animation: Human sex determination

Animation: Morgan's reciprocal crosses

13.3 Chromosomal Mutations That Affect Inheritance

- Inheritance is influenced by processes that delete, duplicate, or invert segments within chromosomes, or translocate segments between chromosomes (Figures 13.11-13.12).

- Chromosomes also change in number by addition or removal of individual chromosomes or entire sets. Changes in single chromosomes usually occur through nondisjunction, in which homologous pairs fail to separate during meiosis I, or sister chromatids fail to separate during meiosis II. As a result, one set of gametes receives an extra copy of a chromosome and the other set is deprived of the chromosome (Figures 13.13–13.15).
- Monoploids have only one set of chromosomes; monoploidy is lethal in most species. Polyploids have one or more extra copies of the entire chromosome set. Polyploids usually arise when the spindle fails to function during mitosis in cell lines leading to gamete formation, producing gametes that contain double the number of chromosomes typical for the species.

Animation: Karyotype preparation

Animation: Deletion

Animation: Duplication

Animation: Inversion

Animation: Translocation

13.4 Human Genetics and Genetic Counseling

- Three modes of inheritance are most significant in human heredity: autosomal recessive, autosomal dominant, and X-linked recessive inheritance. X-linked dominant traits are rare.
- In autosomal recessive inheritance, males or females carry a recessive allele on an autosome. Heterozygotes are carriers that are usually unaffected, but homozygous individuals show symptoms of the trait (Figure 13.16).
- In autosomal dominant inheritance, a dominant gene is carried on an autosome. Individuals that are homozygous or heterozygous for the trait show symptoms of the trait; homozygous recessives are normal.
- In X-linked recessive inheritance, a recessive allele for the trait is carried on the X chromosome. Male individuals with the recessive allele on their X chromosome or female individuals with the recessive allele on both X chromosomes show symptoms of the trait. Heterozygous females are carriers but usually show no symptoms of the trait (Figure 13.17).
- In X-linked dominant inheritance, a dominant allele for the trait is carried on the X chromosome. The trait is expressed in males and females who receive such an X chromosome (Figure 13.18).
- Genetic counseling, based on identification of parental genotypes by constructing family pedigrees and prenatal diagnosis, allow prospective parents to reach an informed decision about whether to have a child or continue a pregnancy (Figure 13.19).

Animation: Autosomal recessive inheritance

Animation: Autosomal dominant inheritance

Animation: Pedigree diagrams

Animation: X-linked inheritance

Animation: Amniocentesis

13.5 Non-Mendelian Patterns of Inheritance

- Cytoplasmic inheritance depends on genes carried on DNA in mitochondria or chloroplasts. Cytoplasmic inheritance follows the maternal line: it parallels the inheritance of the cytoplasm in fertilization, in which most or all of the cytoplasm of the zygote originates from the egg cell (Figure 13.20).
- Genomic imprinting is a phenomenon in which the expression of an allele of a gene is determined by the parent that contributed it. In some cases, the allele inherited from the father is expressed; in others, the allele from the mother is expressed. Commonly, the silencing of the other allele is the result of methylation of the region adjacent to the gene that is responsible for controlling the expression of that gene (Figure 13.21).

UNDERSTAND AND APPLY

Test Your Knowledge

1. In humans, red–green color blindness is an X-linked recessive trait. If a man with normal vision and a color-blind woman have a son, what is the chance that the son will be color-blind? What is the chance that a daughter will be color-blind?
2. The following pedigree shows the pattern of inheritance of red–green color blindness in a family. Females are shown as circles and males as squares; the squares or circles of individuals affected by the trait are filled in black.

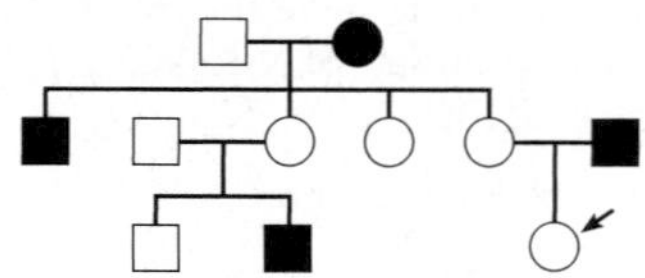

What is the chance that a son of the third-generation female indicated by the arrow will be color-blind if the father is a normal man? If the father is color-blind?

3. Individuals affected by a condition known as polydactyly have extra fingers or toes. The following pedigree shows the pattern of inheritance of this trait in one family:

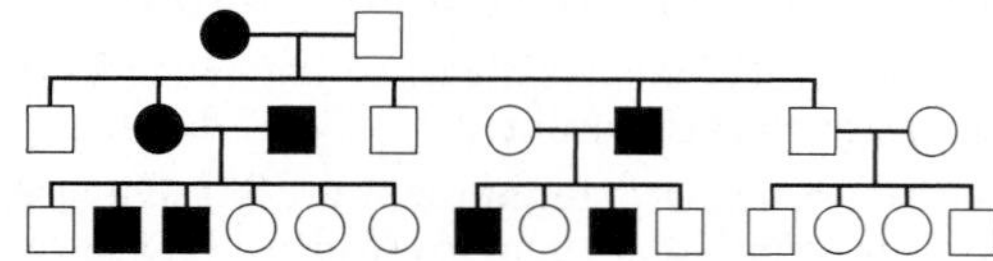

From the pedigree, can you tell if polydactyly comes from a dominant or recessive allele? Is the trait sex-linked? As far as you can determine, what is the genotype of each person in the pedigree with respect to the trait?

4. A number of genes carried on the same chromosome are tested and show the following crossover frequencies. What is their sequence in the map of the chromosome?

Genes	Crossover Frequencies between Them
C and A	7%
B and D	3%
B and A	4%
C and D	6%
C and B	3%

5. In *Drosophila,* two genes, one for body color and one for eye color, are carried on the same chromosome. The wild-type gray body color is dominant to black body color, and wild-type red eyes are dominant to purple eyes. You make a cross between a fly with gray body and red eyes and a fly with black body and purple eyes. Among the offspring, about half have gray bodies and red eyes and half have black bodies and purple eyes. A small percentage have (a) black bodies and red eyes or (b) gray bodies and purple eyes. What alleles are carried together on the chromosomes in each of the flies used in the cross? What alleles are carried together on the chromosomes of the F_1 flies with black bodies and red eyes, and those with gray bodies and purple eyes?
6. Another gene in *Drosophila* determines wing length. The dominant wild-type allele of this gene produces long wings; a recessive allele produces vestigial (short) wings. A female that is true-breeding for red eyes and long wings is mated with a male that has purple eyes and vestigial wings. F_1 females are then crossed with purple-eyed, vestigial-winged males. From this second cross, a total of 600 offspring are obtained with the following combinations of traits:

   252 with red eyes and long wings
   276 with purple eyes and vestigial wings
   42 with red eyes and vestigial wings
   30 with purple eyes and long wings

   Are the genes linked, unlinked, or sex-linked? If they are linked, how many map units separate them on the chromosome?

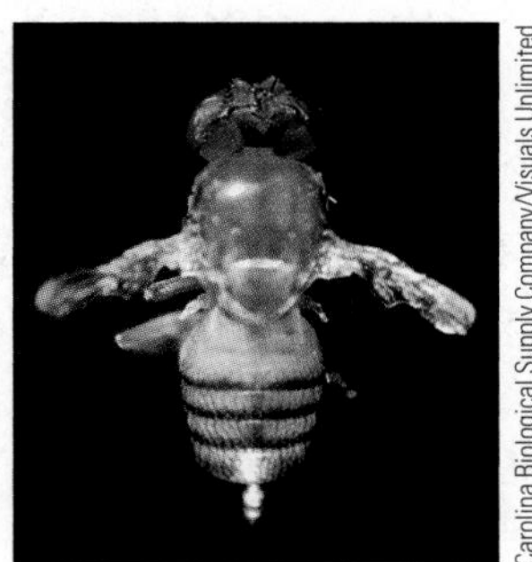
Carolina Biological Supply Company/Visuals Unlimited

Drosophilia with vestigial wings

7. In *Drosophila,* white eyes is an X-linked recessive trait. A white-eyed female is crossed with a male with normal red eyes. An F_1 female from this cross is mated with her father, and an F_1 male is mated with his mother. What will be the phenotypic ratios for eye color in the two sexes of the offspring of the last two crosses?
8. You conduct a cross in *Drosophila* that produces only half as many male as female offspring. What might you suspect as a cause?

Discuss the Concepts

1. Can a linkage map be made for a haploid organism that reproduces sexually?
2. Crossing-over does not occur between any pair of homologous chromosomes during meiosis in male *Drosophila.* From what you have learned about meiosis and crossing-over, propose one hypothesis for why this might be the case.
3. Even though X inactivation occurs in XXY (Klinefelter syndrome) humans, they do not have the same phenotype as normal XY males. Similarly, even though X inactivation occurs in XX individuals, they do not have the same phenotype as XO (Turner syndrome) humans. Why might this be the case?
4. All mammals have evolved from a common ancestor. However, the chromosome number varies among mammals. By what mechanism might this have occurred?

Design an Experiment

Assume that genes *a, b, c, d, e,* and *f* are linked. Explain how you would construct a linkage map that shows the order of these six genes and the map units between them.

Interpret the Data

Exposure to tobacco, by chewing, smoking cigars, cigarettes or pipes and passive exposure to smoke has been linked to cancer of the mouth and throat. Chemical compounds released during tobacco use form covalent bonds with the DNA in the oral cavity cells to generate structures called "adducts." These oral cavity cells replicate frequently and the presence of DNA adducts is suspected to contribute to mutations, chromosomal alterations, and hence cellular defects passed onto offspring cells. Below is a graph showing the incidence (attomol = 10^{-18} mol) of adduct formation in samples of *healthy* tissue from nonsmokers, smokers of various frequencies, and ex-smokers, all of whom required surgery to remove oral cancerous growths.

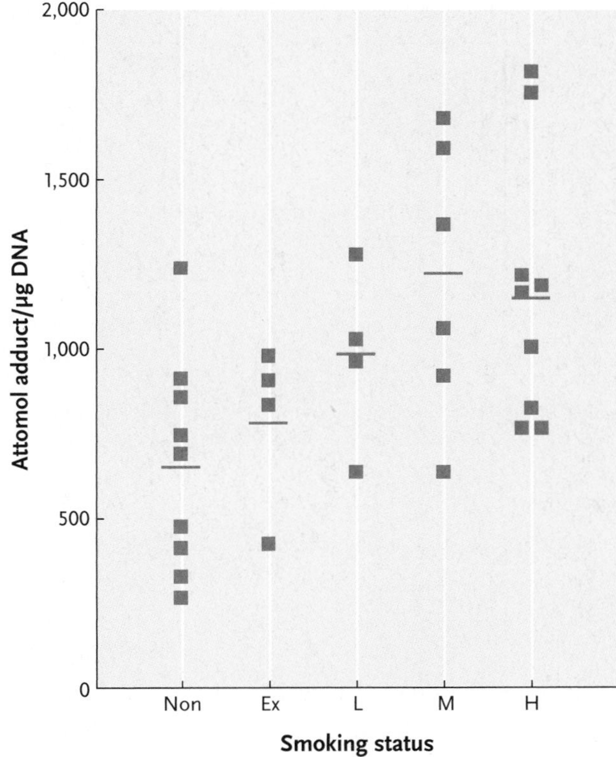
2,000
1,500
1,000
500
0
Attomol adduct/µg DNA
Non
Ex
L
M
H
Smoking status

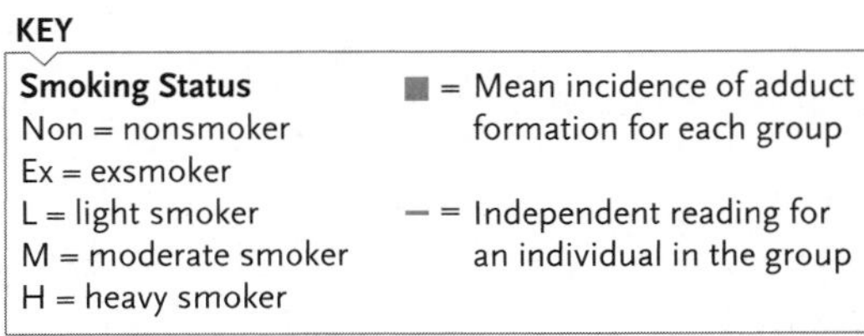
KEY
Smoking Status
Non = nonsmoker
Ex = exsmoker
L = light smoker
M = moderate smoker
H = heavy smoker
■ = Mean incidence of adduct formation for each group
— = Independent reading for an individual in the group

1. On the *y* axis, why is it necessary to express the data in terms of the quantity of adduct per µg of DNA, rather than simply noting the total quantity of adduct?
2. What do these data suggest about the relation between smoking and the formation of adducts?

3. What impact does smoking cessation have on the frequency of adduct formation?
4. Which data point(s) could suggest a relation between the potential to develop oral cancer and passive exposure to tobacco smoke?

Source: N. J. Jones, A. D. McGregor, and R. Waters. 1993. Detection of DNA adducts in human oral tissue: Correlation of adduct levels with tobacco smoking and differential enhancement of adducts using the butanol extraction and nuclease P1 versions of ^{32}P postlabeling. *Cancer Research* 53:1522–1528.

Apply Evolutionary Thinking

How would the effects of natural selection differ on alleles that cause diseases fatal in childhood (such as progeria) and those that cause diseases that shorten life expectancy to 40 or 50 years (such as cystic fibrosis)?

Express Your Opinion

Advances in genetics have led to our ability to detect mutant genes that cause medical disorders in human embryos and fetuses. Should society encourage women to give birth only if their child will not develop severe medical problems? How severe? Go to academic.cengage.com/login to investigate both sides of the issue and then vote.

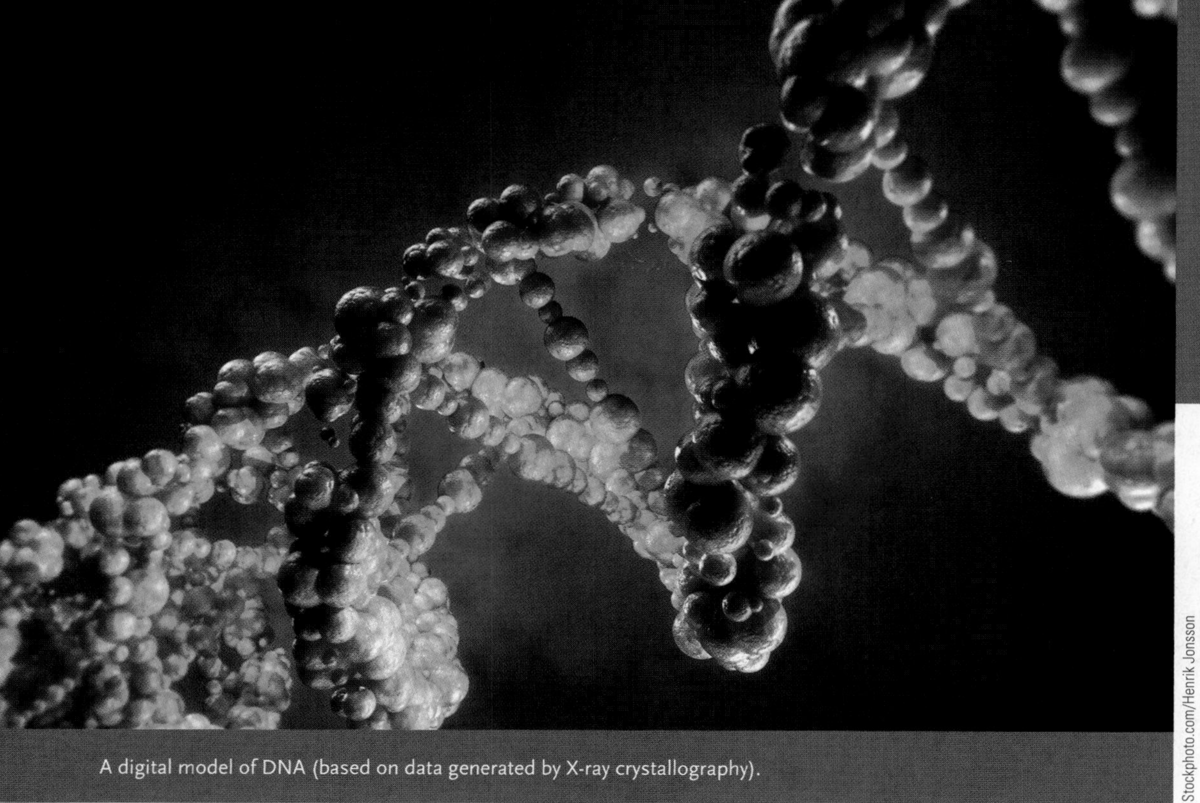

A digital model of DNA (based on data generated by X-ray crystallography).

iStockphoto.com/Henrik Jonsson

14

STUDY OUTLINE

DNA Structure, Replication, and Organization

Why It Matters. . . In the spring of 1868, Johann Friedrich Miescher, a Swiss physician and physiological chemist, was collecting pus from discarded bandages. One may have wondered why Miescher would be interested in collecting bloody bandages, but his intentions were purely scientific: Miescher wanted to study the chemical composition of the nuclei of the white blood cells found in the pus. He used the white blood cells found in pus because much of their volume is occupied by the nucleus. From the nuclei of these cells, Miescher extracted large quantities of an acidic substance with a high phosphorus content. He called the unusual substance "nuclein." His discovery is at the root of the development of our molecular understanding of life: nuclein is now known by its modern name, **deoxyribonucleic acid,** or **DNA,** the molecule that is the genetic material of all living organisms.

At the time of Miescher's discovery, scientists knew nothing about the molecular basis of heredity and very little about genetics. Although Mendel had already published the results of his genetic experiments with garden peas, the significance of his findings was not widely known or appreciated. It was not known which chemical substance in cells actually carries the instructions for reproducing parental traits in offspring. Not until 1952, more than 80 years after Miescher's discovery, did scientists fully recognize that the hereditary molecule was DNA.

After DNA was established as the hereditary molecule, the focus of research changed to the three-dimensional structure of DNA. Among the scientists striving to work out the structure were James D. Watson, a young American postdoctoral student at Cambridge University in England,

A. Barrington Brown/Photo Researchers, Inc.

FIGURE 14.1
James D. Watson and Francis H. C. Crick demonstrating their 1953 model for DNA structure, which revolutionized the biological sciences.

and the British scientist Francis H. C. Crick, then a graduate student at Cambridge University. Using chemical and physical information about DNA, in particular Rosalind Franklin's analysis of the arrangement of atoms in DNA, the two investigators assembled molecular models from pieces of cardboard and bits of wire. Eventually they constructed a model for DNA that fit all the known data **(Figure 14.1).** Their discovery was of momentous importance in biology. The model immediately made it apparent how genetic information is stored and how it could be replicated faithfully. Unquestionably, the discovery launched a molecular revolution within biology, making it possible for the first time to relate the genetic traits of living organisms to a universal molecular code present in the DNA of every cell. In addition, Watson and Crick's discovery opened the way for numerous advances in fields such as medicine, forensics, pharmacology, and agriculture, and eventually gave rise to the current rapid growth of the biotechnology industry. <

14.1 Establishing DNA as the Hereditary Molecule

In the first half of the twentieth century, many scientists believed that proteins were the most likely candidates for the hereditary molecules because they appeared to offer greater opportunities for information coding than did nucleic acids. That is, proteins contain 20 different amino acids, whereas nucleic acids have only four different nitrogenous bases available for coding. Other scientists believed that nucleic acids were the hereditary molecules. In this section, we describe the experiments showing that DNA, and not protein, is the genetic material.

Experiments Began When Griffith Found a Substance That Could Genetically Transform Pneumonia Bacteria

In 1928, Frederick Griffith, a British medical officer, observed an interesting phenomenon in his experiments with the bacterium *Streptococcus pneumoniae,* the species that causes a severe form of pneumonia in mammals. He studied two strains of the bacterium. The smooth strain—*S*—has a polysaccharide capsule surrounding each cell and forms colonies that appear smooth and shiny when grown on a culture plate. The *S* strain is virulent (highly infective, or pathogenic), causing pneumonia and killing the mice in a day or two. The rough strain—*R*—does not have a polysaccharide capsule and forms colonies with a rough, nonshiny appearance. The *R* strain is nonvirulent (not infective, or nonpathogenic), meaning that it does not affect mice.

Griffith's experiments are shown in **Figure 14.2.** His critical experimental result was that the mice died if he injected them with a mixture of living *R* bacteria and heat-killed *S* bacteria. He was able to isolate living *S* bacteria with polysaccharide capsules from the infected mice. In some way, living *R* bacteria had acquired the ability to make the polysaccharide capsule from the dead *S* bacteria, and they had changed—transformed—into virulent *S* cells. The smooth, virulence trait was stably inherited by the descendants of the transformed bacteria. Griffith called the conversion of *R* bacteria to *S* bacteria *transformation* and called the agent responsible the *transforming principle.* What was the nature of the molecule responsible for the transformation? The most likely candidates were proteins or nucleic acids.

Avery and His Coworkers Identified DNA as the Molecule That Transforms Nonvirulent *Streptococcus* Bacteria to the Virulent Form

In the 1940s, Oswald Avery, a physician and medical researcher at the Hospital at Rockefeller Institute for Medical Research, and his coworkers Colin MacLeod and Maclyn McCarty performed an experiment designed to identify the chemical nature of the transforming principle that can change *R Streptococcus* bacteria into the *S* virulent form. Rather than working with mice, they reproduced the transformation using bacteria growing in culture tubes. They used heat to kill virulent *S* bacteria and then treated the macromolecules extracted from the *S* bacterial cells in turn with enzymes that break down each of the three main candidate molecules for the hereditary material—protein; DNA; and the other nucleic acid, RNA. When they destroyed proteins or RNA, the researchers saw no effect; the extract of *S* bacteria still transformed *R* bacteria into virulent *S* bacteria—the cells had polysaccharide capsules and produced smooth colonies on culture plates. When they destroyed DNA, however, no transformation occurred and no smooth colonies were seen on culture plates.

In 1944, Avery and his colleagues published their discovery that the transforming principle was DNA. At that time, many biologists were convinced that the genetic material was protein. So, although their findings were clearly revolutionary, Avery and his colleagues presented their conclusions in the paper cautiously,

FIGURE 14.2 Experimental Research

Griffith's Experiment with Virulent and Nonvirulent Strains of *Streptococcus pneumoniae*

Question: What is the nature of the genetic material?

Experiment: Frederick Griffith studied the conversion of a nonvirulent (noninfective) *R* form of the bacterium *Streptococcus pneumoniae* to a virulent (infective) *S* form. The *S* form has a capsule surrounding the cell, giving colonies of it on a laboratory dish a smooth, shiny appearance. The *R* form has no capsule, so the colonies have a rough, nonshiny appearance. Griffith injected the bacteria into mice and determined how the mice were infected.

1. Mice injected with live *S* cells (control to show effect of *S* cells)

Result: Mice die. Live *S* cells in their blood; shows that *S* cells are virulent.

2. Mice injected with live *R* cells (control to show effect of *R* cells)

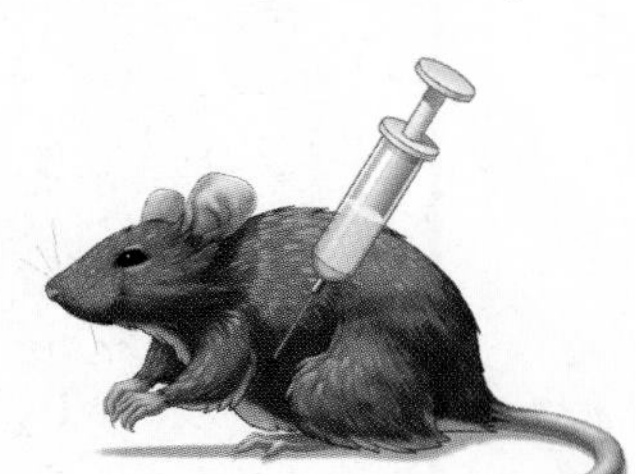

Result: Mice live. No live *R* cells in their blood; shows that *R* cells are nonvirulent. Evidently the capsule is responsible for virulence of the *S* strain.

3. Mice injected with heat-killed *S* cells (control to show effect of dead *S* cells)

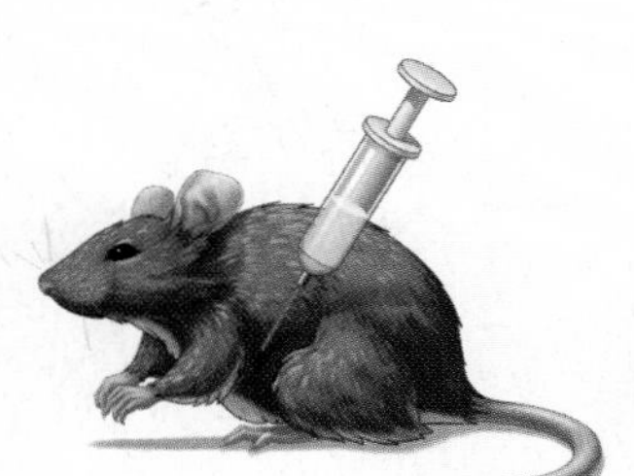

Result: Mice live. No live *S* cells in their blood; shows that live *S* cells are necessary to be virulent to mice.

4. Mice injected with heat-killed *S* cells plus live *R* cells

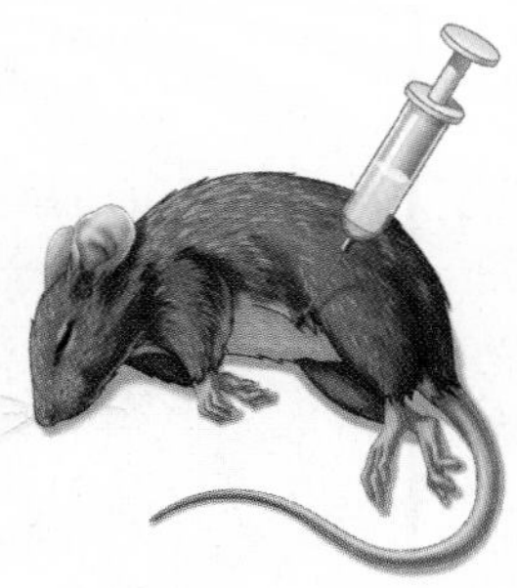

Result: Mice die. Live *S* cells in their blood; shows that living *R* cells can be converted to virulent *S* cells with some factor from dead *S* cells.

Conclusion: Griffith concluded that some molecules released when *S* cells were killed could change living nonvirulent *R* cells genetically to the virulent *S* form. He called the molecule the *transforming principle* and the process of genetic change *transformation.*

Source: F. Griffith. 1928. The significance of pneumococcal types. *Journal of Hygiene (London)* 27:113–159.

offering several interpretations of their results. Some biologists accepted their results almost immediately. However, those who believed that the genetic material was protein argued that it was possible that not all protein was destroyed by the enzyme treatments and, as contaminants in their DNA transformation reaction, those remaining proteins were responsible for the transformation.

Hershey and Chase Found the Final Experimental Evidence Establishing DNA as the Hereditary Molecule

A final series of experiments conducted in 1952 by bacteriologist Alfred D. Hershey and his laboratory assistant Martha Chase at the Cold Spring Harbor Laboratory removed any remaining doubts that DNA is the hereditary molecule. Hershey and Chase studied the infection of the bacterium *Escherichia coli* by bacteriophage T2. *E. coli* is a bacterium normally found in the intestines of mammals. **Bacteriophages** (or simply **phages;** see Chapter 17) are viruses that infect bacteria. A **virus** is an infectious agent that is made of either DNA or RNA surrounded by a protein coat. Viruses can reproduce only in a host cell. When a virus infects a cell, it can use the cell's resources to produce more virus particles.

The phage life cycle begins when a phage attaches to the surface of a bacterium and infects it. Phages such as T2 quickly stop the infected cell from producing its own molecules and instead use the cell's resources for making progeny phages. After about 100 to 200 phages are assembled inside the bacterial cell, a viral enzyme breaks down the cell wall, killing the cell and releasing the new phages. The whole life cycle takes approximately 90 minutes.

The T2 phage that Hershey and Chase studied consists of only a core of DNA surrounded by proteins. Therefore, one of these molecules must be the genetic material that enters the bacterial cell and directs the infective cycle within. But which one? Hershey and Chase's definitive experiment to answer that question is presented in **Figure 14.3.** Their experimental approach was to label the DNA or protein radioactively and then to use the label as a tag to follow the molecule through the phage life cycle. Their results showed that labeled DNA, but not labeled protein, entered the cell and appeared in progeny phages. Unequivocally they had shown that the genetic material of phages is DNA, not protein.

When considered together, the experiments of Griffith, Avery and his coworkers, and Hershey and Chase established that DNA, not proteins, carries genetic information. The research also established the term *transformation,* which is still used in molecular biology. **Transformation** is the alteration of a cell's hereditary type by the uptake of DNA released by the breakdown of another cell, as in the Griffith and Avery experiments. Having identified DNA as the hereditary molecule, scientists turned next to determining its structure.

FIGURE 14.3 **Experimental Research**

The Hershey and Chase Experiment Demonstrating That DNA Is the Hereditary Molecule

Question: Is DNA or protein the genetic material?

Experiment: Hershey and Chase performed a definitive experiment to show whether DNA or protein is the genetic material. They used phage T2 for their experiment; it consists only of DNA and protein.

1. They infected *E. coli* growing in the presence of radioactive ^{32}P or ^{35}S with phage T2. The progeny phages were either labeled in their protein with ^{35}S (top), or in their DNA with ^{32}P (bottom).
2. Separate cultures of *E. coli* were infected with the radioactively labeled phages.
3. After a short period of time to allow the genetic material to enter the bacterial cell, the bacteria were mixed in a blender. The blending sheared from the cell surface the phage coats that did not enter the bacteria. The components were analyzed for radioactivity.
4. Progeny phages analyzed for radioactivity.

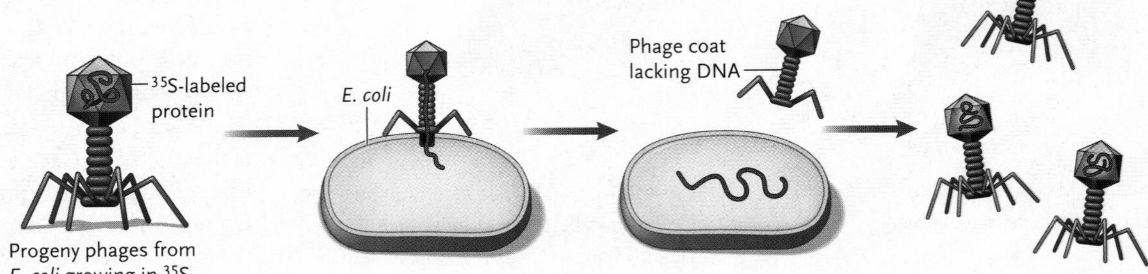

^{35}S-labeled protein
Progeny phages from *E. coli* growing in ^{35}S
E. coli
Phage coat lacking DNA

Result: No radioactivity within cell; ^{35}S in phage coat

Result: No radioactivity in progeny phages

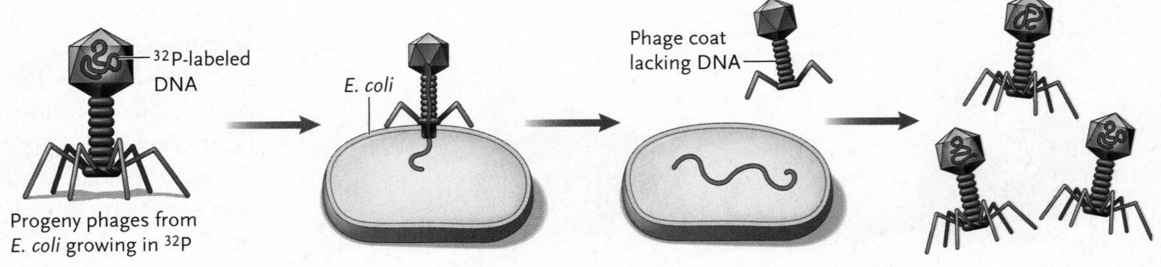

^{32}P-labeled DNA
Progeny phages from *E. coli* growing in ^{32}P
E. coli
Phage coat lacking DNA

Result: ^{32}P within cell; not in phage coat

Result: ^{32}P in progeny phages

Conclusion: ^{32}P, the isotope used to label DNA, was found within phage-infected cells and in progeny phages, indicating that DNA is the genetic material. ^{35}S, the radioisotope used to label proteins, was found in phage coats after infection, but was not found in the infected cell or in progeny phages, showing that protein is not the genetic material.

Source: A. D. Hershey and M. Chase. 1952. Independent functions of viral protein and nucleic acid in growth of bacteriophage. *Journal of General Physiology* 36:39–56.

STUDY BREAK 14.1 <

Imagine that ^{35}S labeled *both* protein and DNA, whereas ^{32}P labeled only DNA. How would Hershey and Chase's results have been different?

14.2 DNA Structure

The experiments that established DNA as the hereditary molecule were followed by a highly competitive scientific race to discover the structure of DNA. The race ended in 1953, when Watson and Crick elucidated the structure of DNA, ushering in a new era of molecular biology.

Watson and Crick Brought Together Information from Several Sources to Work Out DNA Structure

Before Watson and Crick began their research, other investigators had established that DNA contains four different nucleotides. Nucleotides were first described in Section 3.5, and are shown in detail in Figure 3.28; recall that the nucleotide in DNA is called a deoxyribonucleotide because the sugar it contains is deoxyribose. As shown in **Figure 14.4,** each deoxyribonucleotide consists of the five-carbon sugar *deoxyribose* (carbon atoms on deoxyribose are numbered with primes from 1′ to 5′), a phosphate group, and one of the four *nitrogenous bases*—adenine (A), guanine (G), thymine (T), or cytosine (C). (The chemical structures of the nitrogenous bases—nitrogen-containing molecules with the property of a base—are also shown in Figure 3.29.) Two of the bases, **adenine** and **guanine,** are *purines,* nitrogenous bases built from a pair of fused rings of carbon and nitrogen atoms. The other two bases, **thymine** and **cytosine,** are *pyrimidines,* built from a single carbon–nitrogen ring. Erwin Chargaff, a biochemist, measured the amounts of nitrogenous bases in DNA and discovered that they occur in definite ratios. He observed that the amount of purines equals the amount of pyrimidines, but more specifically, the amount of adenine equals the amount of thymine, and the amount of guanine equals the amount of cytosine; these relationships are known as *Chargaff's rules.*

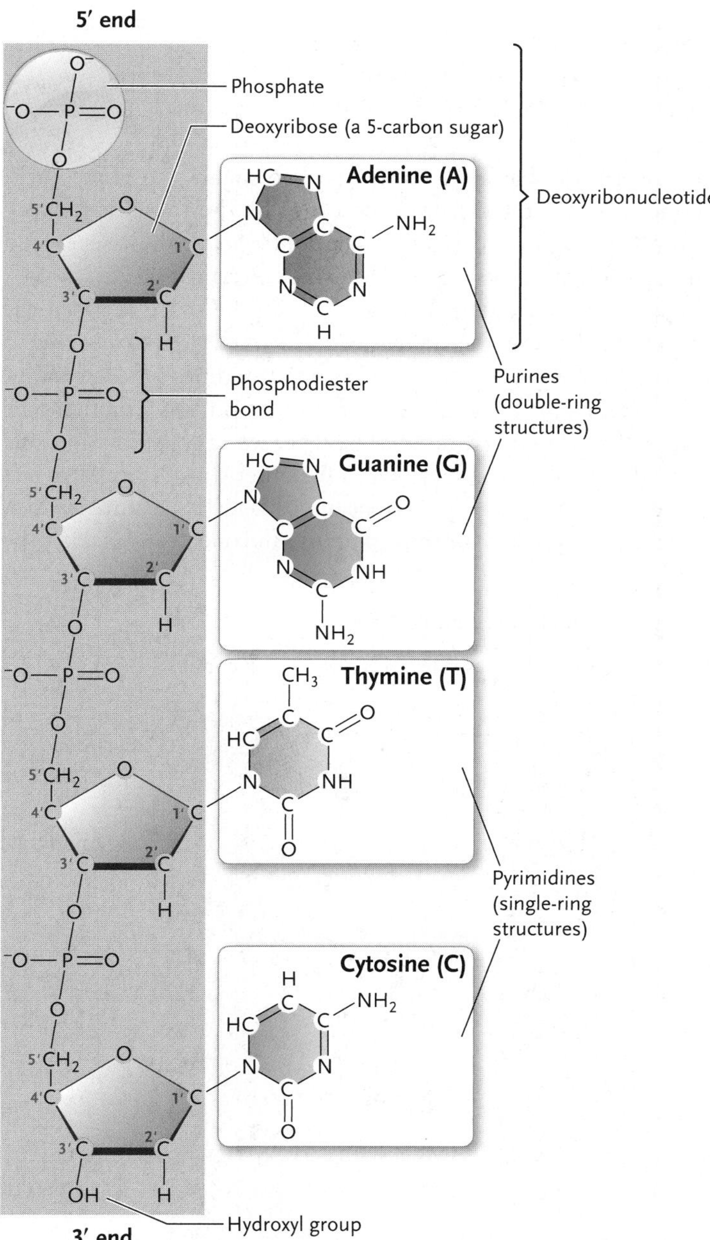

5′ end
Phosphate
Deoxyribose (a 5-carbon sugar)
Adenine (A)
Deoxyribonucleotide
Phosphodiester bond
Purines (double-ring structures)
Guanine (G)
Thymine (T)
Pyrimidines (single-ring structures)
Cytosine (C)
Hydroxyl group
3′ end

FIGURE 14.4
The four deoxyribonucleotide subunits of DNA, linked into a polynucleotide chain. The sugar–phosphate backbone of the chain is highlighted in gray. The connection between adjacent deoxyribose sugars is a phosphodiester bond. The polynucleotide chain has polarity; at one end, the 5′ end, a phosphate group is bound to the 5′ carbon of a deoxyribose sugar, whereas at the other end, the 3′ end, a hydroxyl group is bound to the 3′ carbon of a deoxyribose sugar.

Researchers had also determined that DNA contains nucleotides joined to form a *polynucleotide chain* (see Figure 14.4). In a DNA polynucleotide chain, the deoxyribose sugars are linked by phosphate groups in an alternating sugar–phosphate–sugar–phosphate pattern, forming a **sugar–phosphate backbone** (highlighted in gray in Figure 14.4). Each phosphate group is a "bridge" between the 3′ carbon of one sugar and the 5′ carbon of the next sugar; the entire linkage, including the bridging phosphate group, is called a *phosphodiester bond,* also shown in Figure 14.4.

The polynucleotide chain of DNA has polarity—directionality. That is, the two ends of the chain are not the same: at one end, a phosphate group is bound to the 5′ carbon of a deoxyribose sugar, whereas at the other end, a hydroxyl group is bonded to the 3′ carbon of a deoxyribose sugar (see Figure 14.4). Consequently, the two ends are called the **5′ end** and **3′ end,** respectively.

Those were the known facts when Watson and Crick began their collaboration in the early 1950s. However, the number of polynucleotide chains in a DNA molecule and the manner in which they fold or twist in DNA were unknown. Watson and Crick themselves did not conduct experiments to study the structure of DNA; instead, they used the research data of others for their analysis, relying heavily on data gathered by physicist Maurice H. F. Wilkins and his research associate Rosalind Franklin **(Figure 14.5A),** at King's College, London. These researchers were using X-ray diffraction to study the structure of DNA **(Figure 14.5B).** In **X-ray diffraction,** an X-ray beam is directed at a molecule in the form of a regular solid, ideally in the form of a crystal. Within the crystal, regularly arranged rows and banks of atoms bend and reflect the X-rays into smaller beams that exit the crystal at definite angles determined by the arrangement of atoms in the crystal. If a photographic film is placed behind the crystal, the exiting beams produce a pattern of exposed spots. From that pattern, researchers can deduce the positions of the atoms in the crystal.

Wilkins and Franklin did not have DNA crystals with which to work, but they were able to obtain X-ray diffraction patterns from DNA molecules that had been pulled out into a fiber **(Figure 14.5C).** The patterns indicated that the DNA molecules within the fiber were cylindrical and about 2 nm in diameter. Separations between the spots showed that major patterns of atoms repeat at intervals of 0.34 and 3.4 nm within the DNA. Franklin interpreted an X-shaped distribution of spots in the diffraction pattern (see dashed lines in Figure 14.5C) to mean that DNA has a helical structure.

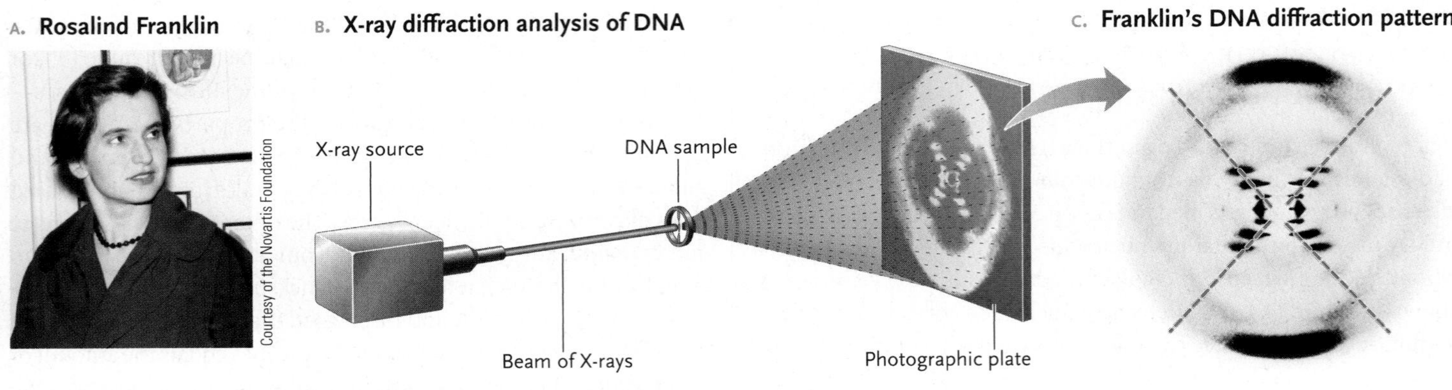
A. Rosalind Franklin
B. X-ray diffraction analysis of DNA
C. Franklin's DNA diffraction pattern
X-ray source
DNA sample
Beam of X-rays
Photographic plate
Courtesy of the Novartis Foundation
A. Barrington Brown/Photo Researchers, Inc.

FIGURE 14.5
X-ray diffraction analysis of DNA. **(A)** Rosalind Franklin. **(B)** The X-ray diffraction method to study DNA. **(C)** The diffraction pattern Rosalind Franklin obtained. The X-shaped pattern of spots (dashed lines) was correctly interpreted by Franklin to indicate that DNA has a helical structure similar to a spiral staircase.

The New Model Proposed That Two Polynucleotide Chains Wind into a DNA Double Helix

Watson and Crick constructed scale models of the four DNA nucleotides and fitted them together in different ways until they arrived at an arrangement that satisfied both Wilkins' and Franklin's X-ray data and Chargaff's chemical analysis. Watson and Crick's trials led them to the **double-helix model** for DNA, a double-stranded model for DNA structure in which two polynucleotide chains twist around each other in a right-handed way, like a double-spiral staircase **(Figure 14.6)**.

In the double-helix model the two sugar–phosphate backbones are separated from each other by the same distance—2 nm—throughout the length of the double helix. The bases extend into and fill this central space. A purine and a pyrimidine, if paired together, are exactly wide enough to fill the space between the backbone chains in the double helix. However, two purines are too wide to fit the space exactly, and two pyrimidines are too narrow. The purine–pyrimidine base pairs in Watson and Crick's model are A–T and G–C pairs. That is, wherever an A occurs in one strand, a T must be opposite it in the other strand; wherever a G occurs in one strand, a C must be opposite it. This feature of DNA is called **complementary base pairing,** and one strand is said to be

A. **Space filling model overlaid with sugar–phosphate backbones**

5′ end 3′ end

5′ end 3′ end

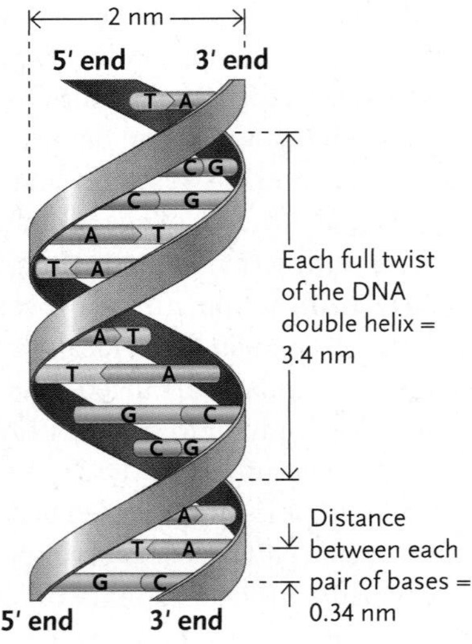
B. Schematic drawing
2 nm
5′ end
3′ end
Each full twist of the DNA double helix = 3.4 nm
Distance between each pair of bases = 0.34 nm
5′ end
3′ end

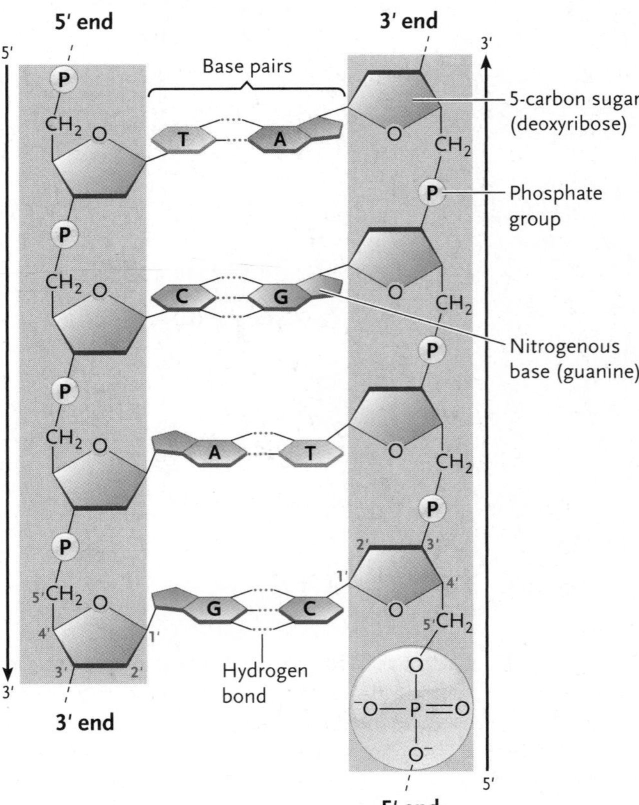
C. Chemical structure drawing
5′ end
3′ end
Base pairs
5-carbon sugar (deoxyribose)
Phosphate group
Nitrogenous base (guanine)
Hydrogen bond
3′ end
5′ end

FIGURE 14.6
DNA double helix. **(A)** Space-filling model. **(B)** Schematic drawing. **(C)** Chemical structure drawing. Arrows and labeling of the ends show that the two polynucleotide chains of the double helix are antiparallel—that is, they have opposite polarity in that they run in opposite directions. In the space-filling model at the top, the spaces occupied by atoms are indicated by spheres. There are 10 base pairs per turn of the helix; only 8 base pairs are visible because the other 2 are obscured where the backbones pass over each other.

complementary to the other. Complementary base pairing involving A–T and G–C base pairs fits Chargaff's rules. The base pairs, which fit together like pieces of a jigsaw puzzle, are stabilized by hydrogen bonds—two between A and T and three between G and C (see Figures 14.6 and 3.32; hydrogen bonds are discussed in Section 2.3. Note that A cannot pair with C, and G cannot pair with T because of the hydrogen bonding requirements.) The hydrogen bonds between the paired bases, repeated along the double helix, hold the two strands together in the helix.

The base pairs lie in flat planes almost perpendicular to the long axis of the DNA molecule. In this state, each base pair occupies a length of 0.34 nm along the long axis of the double helix (see Figure 14.6). This spacing accounts for the repeating 0.34 nm pattern noted in the X-ray diffraction patterns. The larger 3.4 nm repeating pattern was interpreted to mean that each full turn of the double helix takes up 3.4 nm along the length of the molecule and therefore 10 base pairs are packed into a full turn.

Watson and Crick also realized that the two strands of a double helix fit together in a stable chemical way only if they are **antiparallel,** that is, only if they run in opposite directions (see Figure 14.6, arrows). In other words, the *3′ end* of one strand is opposite the *5′ end* of the other strand. This antiparallel arrangement is highly significant for the process of replication, which is discussed in the next section.

As hereditary material, DNA must faithfully store and transmit genetic information for the entire life cycle of an organism. Watson and Crick recognized that this information is coded into the DNA by the particular sequence of the four nucleotides. Although only four different kinds of nucleotides exist, combining them in groups allows an essentially infinite number of different sequences to be "written," just as the 26 letters of the alphabet can be combined in groups to create a virtually unlimited number of words. Chapter 15 shows how the four nucleotides form sequences of three nucleotides that form enough "words" to "spell out" the structure of any conceivable protein.

Watson and Crick announced their model for DNA structure in a brief but monumental paper published in the journal *Nature* in 1953. Watson and Crick shared a Nobel Prize with Wilkins in 1962 for their discovery of the molecular structure of DNA. Rosalind Franklin might have been a candidate for a Nobel Prize had she not died of cancer at age 38 in 1958. (The Nobel Prize is given only to living investigators.) Watson and Crick's discovery of the structure of DNA opened the way to molecular studies of genetics and heredity, leading to our modern understanding of gene structure and action at the molecular level.

STUDY BREAK 14.2 <

1. **Which bases in DNA are purines? Which are pyrimidines?**
2. **What bonds form between complementary base pairs? Between a base and the deoxyribose sugar?**
3. **Which features of the DNA molecule did Watson and Crick describe?**
4. **The percentage of A in a double-stranded DNA molecule is 20. What is the percentage of C in that DNA molecule?**

14.3 DNA Replication

Once they had discovered the structure of DNA, Watson and Crick realized immediately that complementary base pairing between the two strands could explain how DNA replicates **(Figure 14.7).** They imagined that, for replication, the hydrogen bonds between the two strands break, and the two strands unwind and separate. Each strand then acts as a template for the synthesis of its partner strand. When replication is complete, there are two double helices, each of which has one strand derived from the parental DNA molecule base paired with a newly synthesized strand. Most importantly, each of the two new double helices has base-pair sequence identical to that of the parental DNA molecule.

The model of replication Watson and Crick proposed is termed **semiconservative replication (Figure 14.8A).** Other scien-

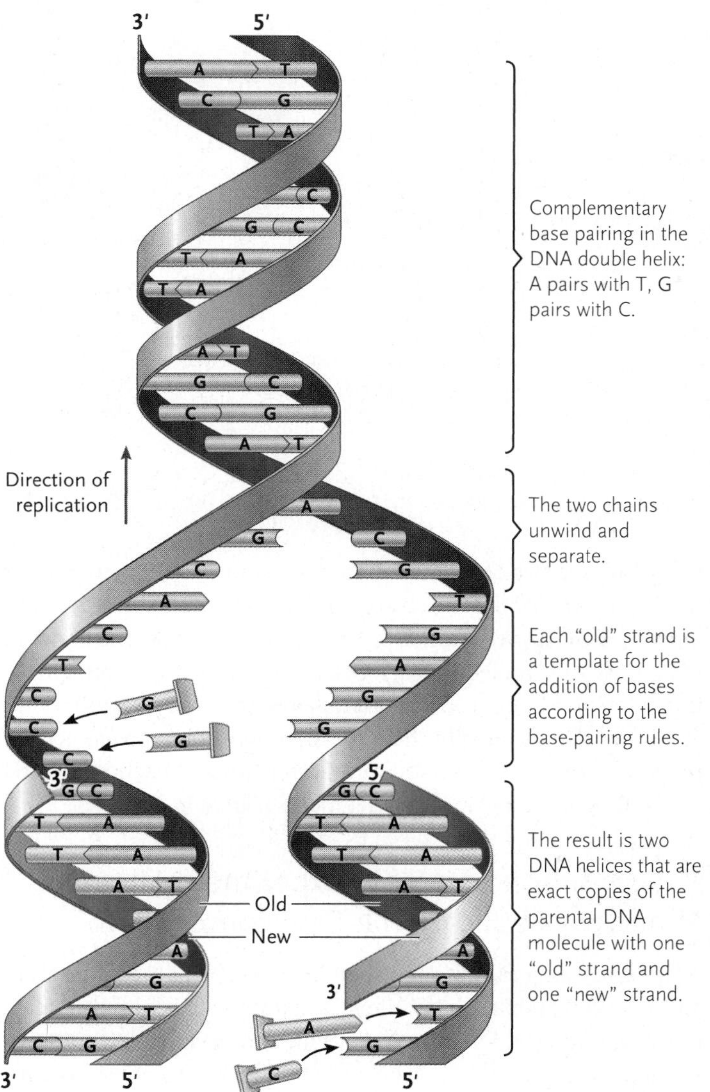

3′
5′
Complementary base pairing in the DNA double helix: A pairs with T, G pairs with C.
Direction of replication
The two chains unwind and separate.
Each "old" strand is a template for the addition of bases according to the base-pairing rules.
The result is two DNA helices that are exact copies of the parental DNA molecule with one "old" strand and one "new" strand.
Old
New
3′
5′
3′
5′
5′

FIGURE 14.7
Watson and Crick's model for DNA replication. The original DNA molecule is shown in gray. A new polynucleotide chain (red) is assembled on each original chain as the two chains unwind. The template and complementary copy chains remain wound together when replication is complete, producing molecules that are half old and half new. The model is known as the semiconservative model for DNA replication.

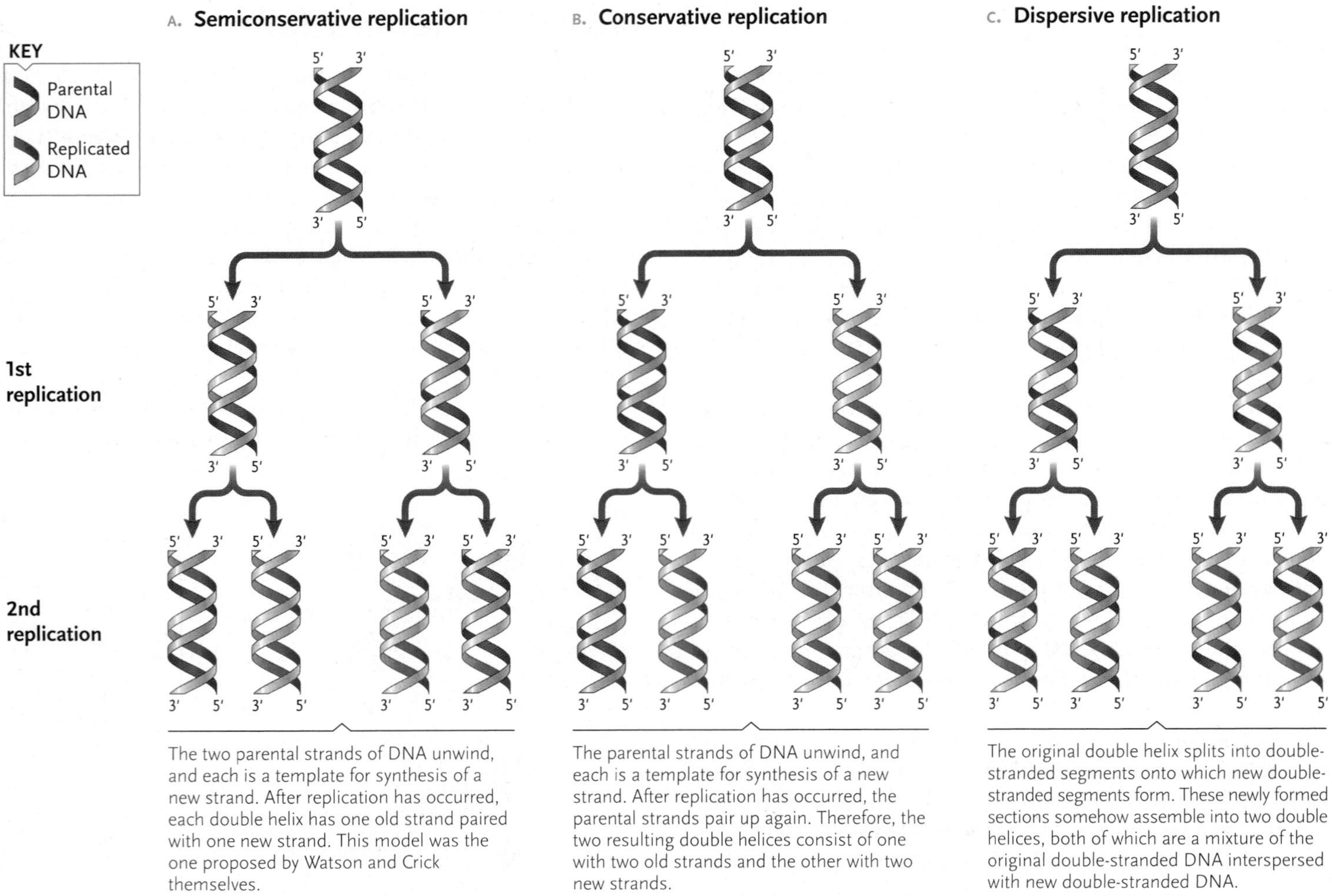

A. **Semiconservative replication**

B. **Conservative replication**

C. **Dispersive replication**

KEY

Parental DNA

Replicated DNA

1st replication

2nd replication

The two parental strands of DNA unwind, and each is a template for synthesis of a new strand. After replication has occurred, each double helix has one old strand paired with one new strand. This model was the one proposed by Watson and Crick themselves.

The parental strands of DNA unwind, and each is a template for synthesis of a new strand. After replication has occurred, the parental strands pair up again. Therefore, the two resulting double helices consist of one with two old strands and the other with two new strands.

The original double helix splits into double-stranded segments onto which new double-stranded segments form. These newly formed sections somehow assemble into two double helices, both of which are a mixture of the original double-stranded DNA interspersed with new double-stranded DNA.

FIGURE 14.8
Semiconservative **(A)**, conservative **(B)**, and dispersive **(C)** models for DNA replication.

tists proposed two other models for replication. In the *conservative replication model,* the two strands of the original molecule serve as templates for the two strands of a new DNA molecule, then rewind into an all "old" molecule **(Figure 14.8B).** After the two complementary copies separate from their templates, they wind together into an all "new" molecule. In the *dispersive replication model,* neither parental strand is conserved and both chains of each replicated molecule contain old and new segments **(Figure 14.8C).**

The Meselson and Stahl Experiment Showed That DNA Replication Is Semiconservative

A definitive experiment published in 1958 by Matthew Meselson and Franklin Stahl of the California Institute of Technology demonstrated that DNA replication is semiconservative **(Figure 14.9).** In their experiment, Meselson and Stahl had to be able to distinguish parental DNA strands from newly synthesized DNA. To do this they used a nonradioactive "heavy" nitrogen isotope to tag the parental DNA strands. The heavy isotope, ^{15}N, has one more neutron in its nucleus than the normal ^{14}N isotope. Molecules containing ^{15}N are measurably heavier (denser) than molecules of the same type containing ^{14}N. DNA molecules with different densities were distinguished by a special type of centrifugation.

DNA Polymerases Are the Primary Enzymes of DNA Replication

During replication, complementary polynucleotide chains are assembled from individual deoxyribonucleotides by enzymes known as DNA polymerases. More than one kind of DNA polymerase is required for DNA replication in bacteria, archaea, and eukaryotes. *Deoxyribonucleoside triphosphates* are the substrates for the polymerization reaction catalyzed by DNA polymerases **(Figure 14.10).** A nucleoside triphosphate is a nitrogenous base linked to a sugar, which is linked, in turn, to a chain of three phosphate groups (see Figure 3.28). You have encountered a nucleoside triphosphate before, namely the ATP produced in cellular respiration (see Chapter 8). In that case, the sugar is ribose, making ATP a ribonucleoside triphosphate. The deoxyribonucleoside triphosphates used in DNA replication have the sugar *deoxyribose* rather than the sugar *ribose.* Because four different bases are found in DNA—adenine (A), guanine (G), cytosine (C), and thymine (T)—four different deoxyribonucleoside triphosphates are used for DNA replication. In keeping with the ATP naming convention, the deoxyribonucleoside triphosphates for DNA replication are given the short names dATP, dGTP, dCTP, and dTTP, where the "d" stands for "deoxyribose."

FIGURE 14.9 **Experimental Research**

The Meselson and Stahl Experiment Demonstrating the Semiconservative Model for DNA Replication to Be Correct

Question: Does DNA replicate semiconservatively?

Experiment: Matthew Meselson and Franklin Stahl proved that the semiconservative model of DNA replication is correct and that the conservative and dispersive models are incorrect.

1. Bacteria grown in ^{15}N (heavy) medium. The heavy isotope is incorporated into the bases of DNA, resulting in all the DNA being heavy, that is, labeled with ^{15}N.

2. Bacteria transferred to ^{14}N (light) medium and allowed to grow and divide for several generations. All new DNA is light.

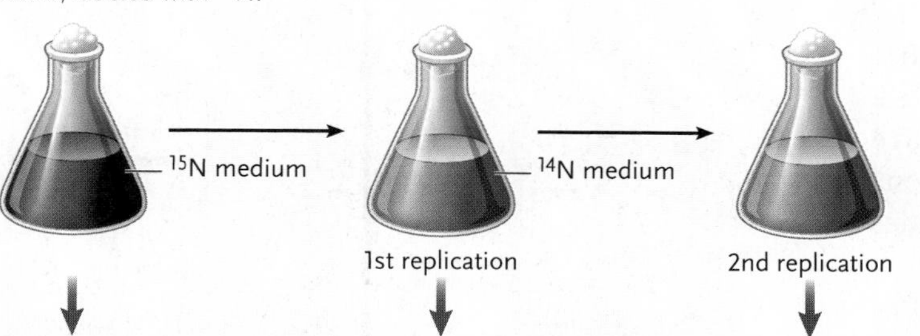

^{15}N medium
^{14}N medium
1st replication
2nd replication

3. DNA extracted from bacteria cultured in ^{15}N medium and after each generation in ^{14}N medium.

4. DNA mixed with cesium chloride (CsCl) and centrifuged at very high speed for about 48 hours.

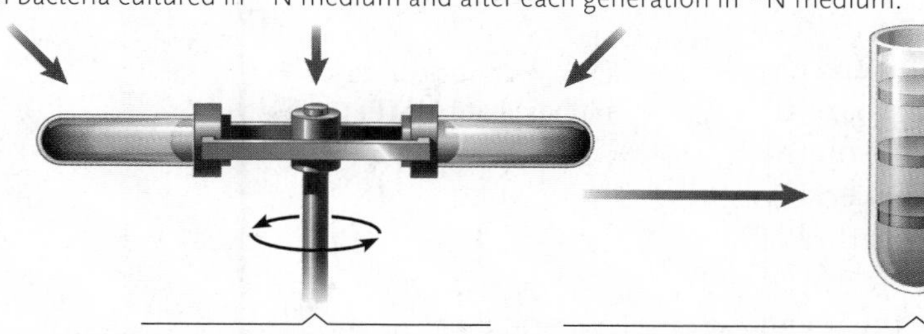

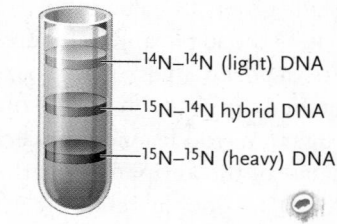

^{14}N–^{14}N (light) DNA
^{15}N–^{14}N hybrid DNA
^{15}N–^{15}N (heavy) DNA

CsCl forms a density gradient during centrifugation, with the highest density at the bottom of the tube.

DNA molecules move to positions where their density equals that of the CsCl solution and form bands. The densest DNA ends up closest to the bottom of the tube. Shown are the positions of differently labeled DNA molecules. Experimentally the bands are detected by absorbance of UV light.

Results: Meselson and Stahl obtained the following results:

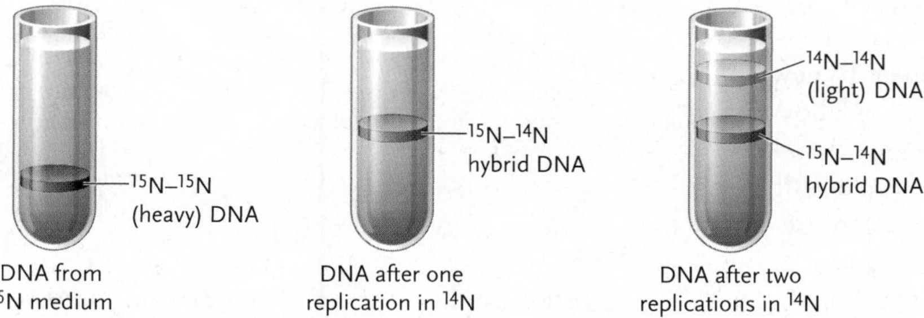

^{15}N–^{15}N (heavy) DNA
^{15}N–^{14}N hybrid DNA
^{14}N–^{14}N (light) DNA
^{15}N–^{14}N hybrid DNA
DNA from ^{15}N medium
DNA after one replication in ^{14}N
DNA after two replications in ^{14}N

Conclusion: The predicted DNA banding patterns for the three DNA replications were:

	^{15}N medium	One replication in ^{14}N	Two replications in ^{14}N	
Semiconservative				√ **Matches results**
Conservative				X **Does not match results**
Dispersive				X **Does not match results**

The results support the semiconservative model.

Source: M. Meselson and F. W. Stahl. 1958. The replication of DNA in *Escherichia coli*. *Proceedings of the National Academy of Sciences USA* 44:671–682.

FIGURE 14.10

Reaction assembling a complementary DNA chain in the 5′→3′ direction on a template DNA strand, showing the phosphodiester bond formed when the DNA polymerase enzyme adds each deoxyribonucleotide to the chain.

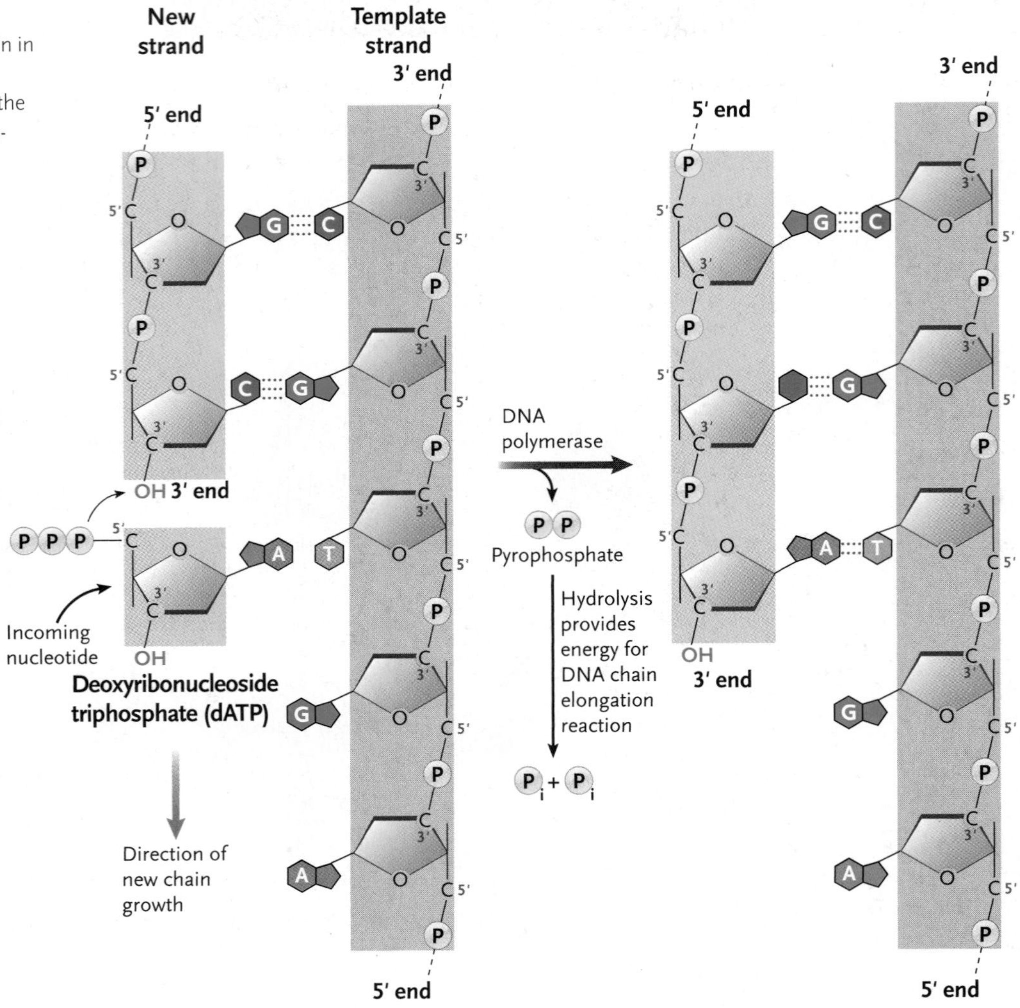

New strand
Template strand
3′ end
5′ end
OH 3′ end
Incoming nucleotide
Deoxyribonucleoside triphosphate (dATP)
Direction of new chain growth
5′ end
DNA polymerase
Pyrophosphate
Hydrolysis provides energy for DNA chain elongation reaction
3′ end
5′ end
OH
3′ end
5′ end

1 DNA polymerase forms a complementary base pair between a deoxyribonucleoside triphosphate with an A base (dATP) from the surrounding solution with the next, T, nucleotide of the template strand.

2 DNA polymerase catalyzes the formation of a phosphodiester bond involving the 3′–OH group at the end of the new chain and the innermost of the three phosphate groups of the dATP. The other two phosphates are released as a pyrophosphate molecule. The new chain has been lengthened by one nucleotide. The process continues, with DNA polymerase adding complementary nucleotides one by one to the growing DNA chain.

Figure 14.10 presents a section of a DNA polynucleotide chain being replicated, and shows how DNA polymerase catalyzes the assembly of a new DNA strand that is complementary to the template strand. To understand Figure 14.10, remember that the carbons in the deoxyriboses of nucleotides are numbered with primes. Each DNA strand has two distinct ends: the 5′ end has an exposed phosphate group attached to the 5′ carbon of the sugar, and the 3′ end has an exposed hydroxyl group attached to the 3′ carbon of the sugar. As you learned earlier, because of the antiparallel nature of the DNA double helix, the 5′ end of one strand is opposite the 3′ end of the other.

DNA polymerase can add a nucleotide *only to the* 3′ end of an existing nucleotide chain. As a new DNA strand is assembled, a 3′–OH group is always exposed at its "newest" end; the "oldest" end of the new chain has an exposed 5′ triphosphate. DNA polymerases are therefore said to assemble nucleotide chains in the 5′→3′ direction. Because of the antiparallel nature of DNA, the template strand is "read" in the 3′→5′ direction for this new synthesis.

DNA polymerases of bacteria, archaea, and eukaryotes all consist of several polypeptide subunits arranged to form different domains (see Figure 3.27, and Section 3.4). The polymerases share a shape that is said to resemble a partially-closed human right hand in which the template DNA lies over the "palm" in a groove formed by the "fingers" and "thumb" **(Figure 14.11A).** The palm domain is evolutionarily related among the polymerases of bacteria, archaea, and eukaryotes, while the finger and thumb domains are different sequences in each of those three types of organisms. The template strand does not pass through the tunnel formed by the thumb and finger domains, however. Instead, the template strand and the 3′–OH of the new strand meet at the active site for the polymerization reaction of DNA synthesis, located in the palm domain. A nucleotide is added to the new strand when an incoming dNTP enters the active site carrying a base complementary to the template strand base positioned in the active site. By moving along the template strand, one nucleotide at a time, DNA polymerase extends the new DNA strand as we saw in Figure 14.10.

Figure 14.11B shows the representation of DNA polymerase used in the following DNA replication figures, and it also shows a *sliding DNA clamp.* The **sliding DNA clamp** is a protein that encircles the DNA and binds to the rear of the DNA polymerase in terms of the enzyme's forward movement during replication. The function of the sliding DNA clamp is to tether the DNA polymerase to the template strand. Tethering the DNA polymerase makes replication more efficient because without it, the enzyme will detach from the template after only a few tens of polymerizations. But, with the clamp, many tens of thousands of polymerizations occur before the enzyme detaches. Overall, the rate of DNA synthesis is much faster because of the sliding DNA clamp.

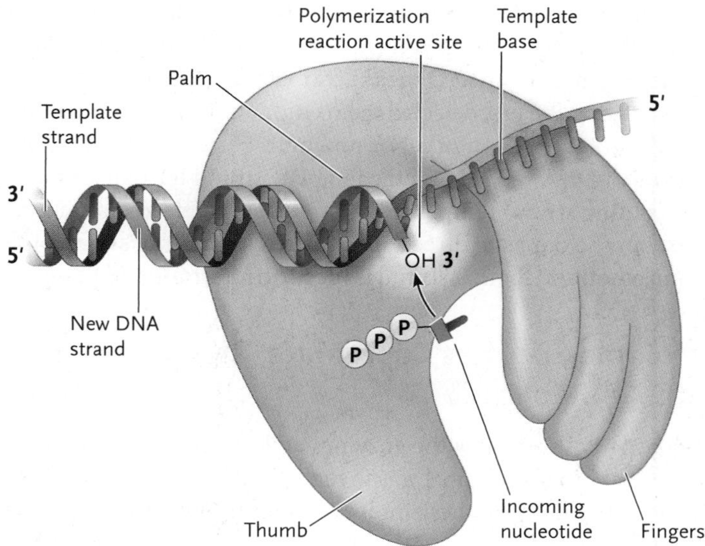
A. Bacterial DNA polymerase
Polymerization reaction active site
Template base
Palm
Template strand
5′
3′
5′
OH 3′
New DNA strand
P
P
P
Incoming nucleotide
Thumb
Fingers

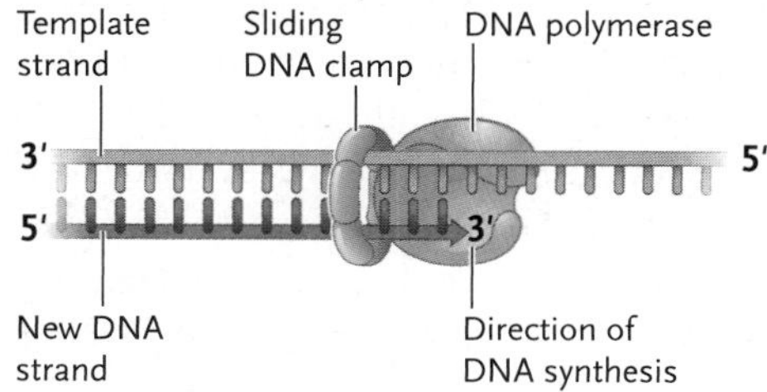
B. How a DNA polymerase and sliding clamp is shown in the book
Template strand
Sliding DNA clamp
DNA polymerase
3′
5′
5′
3′
New DNA strand
Direction of DNA synthesis

FIGURE 14.11

DNA polymerase structure. (A) Stylized drawing of a bacterial DNA polymerase. The enzyme viewed from the side resembles a human right hand. The polymerization reaction site lies on the palm. When the incoming nucleotide is added, the thumb and fingers close over the site to facilitate the reaction. **(B)** How DNA polymerase is shown in subsequent figures of DNA replication. The figure also shows a sliding DNA clamp tethering the DNA polymerase to the template strand.

In sum, the key molecular events of DNA replication are as follows:

1. The two strands of the DNA molecule unwind for replication to occur.
2. DNA polymerase can add nucleotides only to an existing chain.
3. The overall direction of new synthesis is in the 5′→3′ direction, which is a direction antiparallel to that of the template strand.
4. Nucleotides enter into a newly synthesized chain according to the A–T and G–C complementary base-pairing rules.

The following sections describe how enzymes and other proteins conduct these molecular events. Our focus is on the well-characterized replication system of *E. coli*. Replication in archaea and eukaryotes is highly similar, although there are differences in the replication machinery. The replication machinery of archaea is strikingly similar to that of eukaryotes and is clearly different from that of bacteria.

Helicases Unwind DNA for New DNA Synthesis and Other Proteins Stabilize the DNA at the Replication Fork

In semiconservative replication, the two strands of the parental DNA molecule unwind and separate to expose the template strands for new DNA synthesis **(Figure 14.12)**. Unwinding of the DNA for replication occurs at a small, specific region in the bacterial chromosome known as an **origin of replication (*ori*)**. Specific proteins recognize an *ori* and recruit **DNA helicase,** which unwinds the DNA strands. The unwinding produces a Y-shaped structure called a **replication fork,** which consists of the two unwound template strands transitioning to double-helical DNA. **Single-stranded binding proteins (SSBs)** coat the exposed single-stranded DNA segments, stabilizing the DNA and keeping the two strands from pairing (see Figure 14.12). The SSBs are displaced as the replication enzymes make the new polynucleotide chain on the template strands.

For circular chromosomes, such as the genomes of most bacteria, unwinding the DNA will eventually cause the still-wound DNA ahead of the unwinding to become highly twisted. You can visualize this phenomenon with some string. Take two equal lengths of string and twist them around each other. Now tie the two ends of each string together. You have created a model of a circular DNA double helix. Pick anywhere in the circle and pull apart the two pieces of string. The more you pull, the more the region where the two strings are still together becomes highly twisted. In the cell,

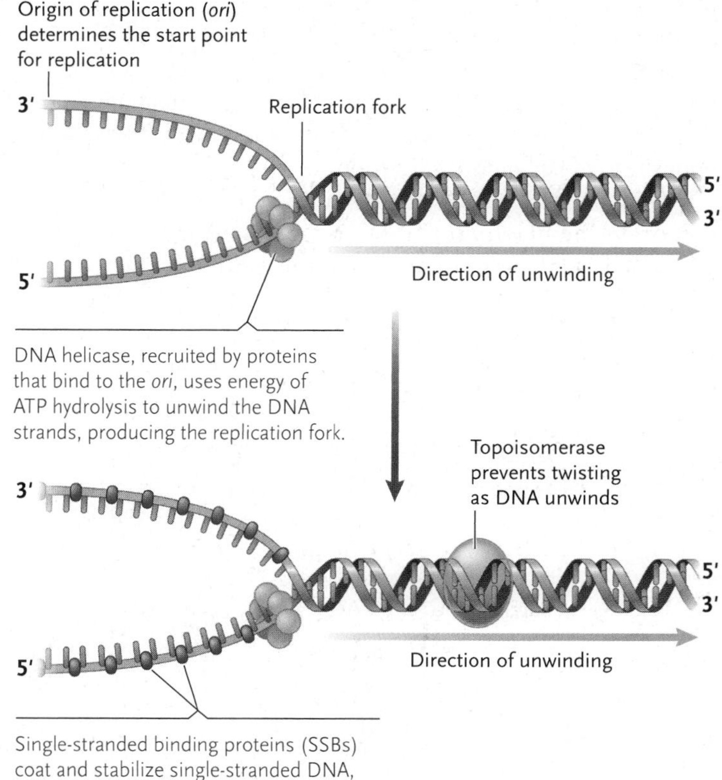
Origin of replication (*ori*) determines the start point for replication
3′
Replication fork
5′
3′
5′
Direction of unwinding
DNA helicase, recruited by proteins that bind to the *ori*, uses energy of ATP hydrolysis to unwind the DNA strands, producing the replication fork.
3′
Topoisomerase prevents twisting as DNA unwinds
5′
3′
5′
Direction of unwinding
Single-stranded binding proteins (SSBs) coat and stabilize single-stranded DNA, preventing the two strands from reforming double-stranded DNA.

FIGURE 14.12

The roles of DNA helicase, single-stranded binding proteins (SSBs), and topoisomerase in DNA replication.

the twisting of DNA during replication is prevented by **topoisomerase,** which cuts the DNA ahead of the replication fork, turns the DNA on one side of the break in the opposite direction of the twisting force, and rejoins the two strands again (see Figure 14.12).

RNA Primers Provide the Starting Point for DNA Polymerase to Begin Synthesizing a New DNA Chain

DNA polymerases can add nucleotides only to the 3′ end of an existing strand. How can a new strand begin when there is no existing strand in place? The answer lies in a short chain a few nucleotides long called a **primer** that is made of RNA instead of DNA **(Figure 14.13).** The primer is synthesized by the enzyme **primase.** Primase then leaves the template, and DNA polymerase takes over, extending the RNA primer with DNA nucleotides as it synthesizes the new DNA chain. RNA primers are removed and replaced with DNA later in replication.

One New DNA Strand Is Synthesized Continuously; the Other, Discontinuously

DNA polymerases synthesize a new DNA strand on a template strand in the 5′→3′ direction. Because the two strands of a DNA molecule are antiparallel, only one of the template strands runs in a direction that allows DNA polymerase to make a 5′→3′ complementary copy in the direction of unwinding. That is, on this template strand—top strand in **Figure 14.14**—the new DNA strand is synthesized continuously in the direction of unwinding of the double helix. However, the other template strand—bottom strand in Figure 14.14—runs in the opposite direction; this means DNA polymerase has to copy it in the direction opposite to the unwinding direction.

How is the new DNA strand made in the opposite direction to the unwinding? The polymerases make this strand in short lengths that are synthesized in the direction opposite to that of DNA unwinding (see Figure 14.14). The short lengths produced by this **discontinuous replication** are then covalently linked into a continuous polynucleotide chain. The short lengths are called **Okazaki fragments,** after Reiji Okazaki, the scientist who first detected them. The new DNA strand synthesized in the direction of DNA unwinding is called the **leading strand** of DNA replication; the template strand for that strand is the **leading strand template.** The strand synthesized discontinuously in the opposite direction is called the **lagging strand;** the template strand for that strand is the **lagging strand template.**

Multiple Enzymes Coordinate Their Activities in DNA Replication

Figure 14.15 shows how the enzymes and proteins we have introduced act in a coordinated way to replicate DNA. Primase initiates all new strands by synthesizing an RNA primer. **DNA polymerase III,** the main polymerase, extends the primer by adding DNA nucleotides. For the lagging strand, **DNA polymerase I** removes the RNA primer at the 5′ end of the previous newly synthesized Okazaki fragment, replacing the RNA nucleotides one by one with DNA nucleotides. RNA nucleotide removal uses the 5′→3′ exonuclease activity of the enzyme. (An exonuclease removes nucleotides from the end of a molecule.) DNA synthesis uses the 5′→3′ polymerization activity. DNA ligase (*ligare* = to tie) seals the nick left between the two fragments. The replication process continues in the same way until the entire DNA molecule is copied. **Table 14.1** summarizes the activities of the major enzymes replicating DNA.

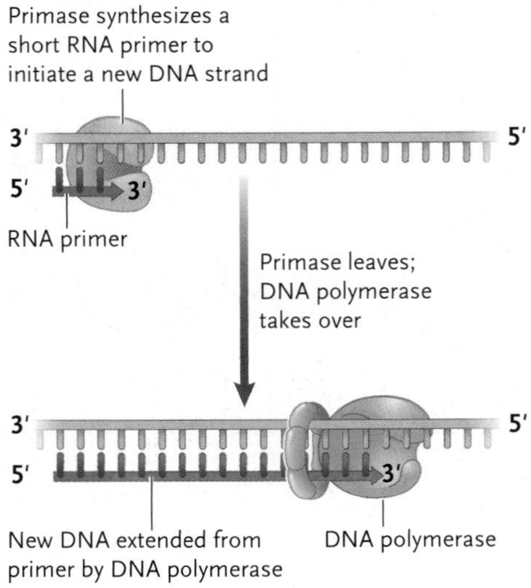

Primase synthesizes a short RNA primer to initiate a new DNA strand
3′ 5′
5′ 3′
RNA primer
Primase leaves; DNA polymerase takes over
3′ 5′
5′ 3′
New DNA extended from primer by DNA polymerase
DNA polymerase

FIGURE 14.13
Initiation of a new DNA strand by synthesis of a short RNA primer by primase, and the extension of the primer as DNA by DNA polymerase.

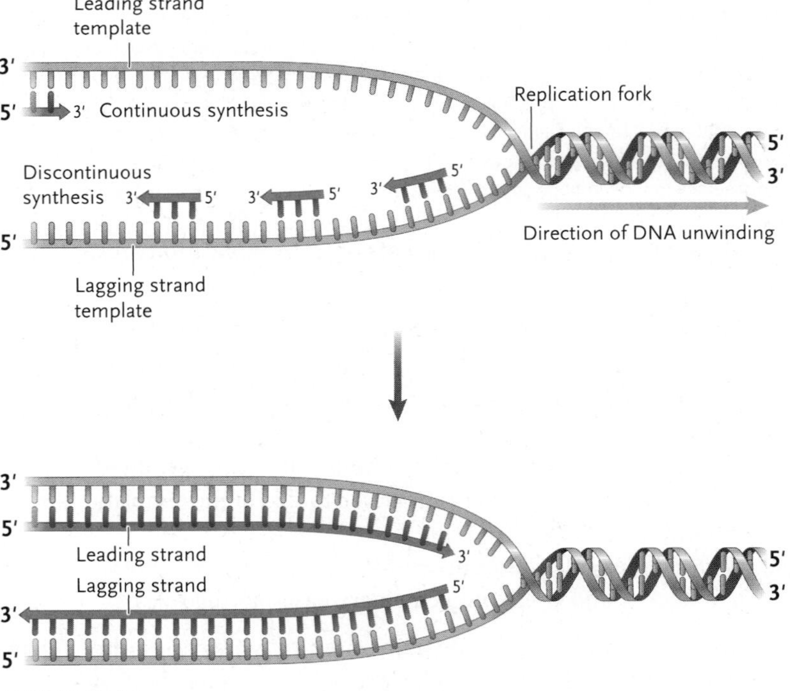
Leading strand template
3′
5′
3′ Continuous synthesis
Replication fork
5′
3′
Discontinuous synthesis
3′ 5′
3′ 5′
3′ 5′
5′
Direction of DNA unwinding
Lagging strand template
3′
5′
Leading strand
Lagging strand
3′
5′
5′
3′
3′
5′

FIGURE 14.14
Replication of antiparallel template strands at a replication fork. Synthesis of the new DNA strand on the top template strand is continuous. Synthesis on the new DNA strand on the bottom template strand is discontinuous—short lengths of DNA are made which are then joined into a continuous chain. The overall effect is synthesis of both strands in the direction of replication fork movement.

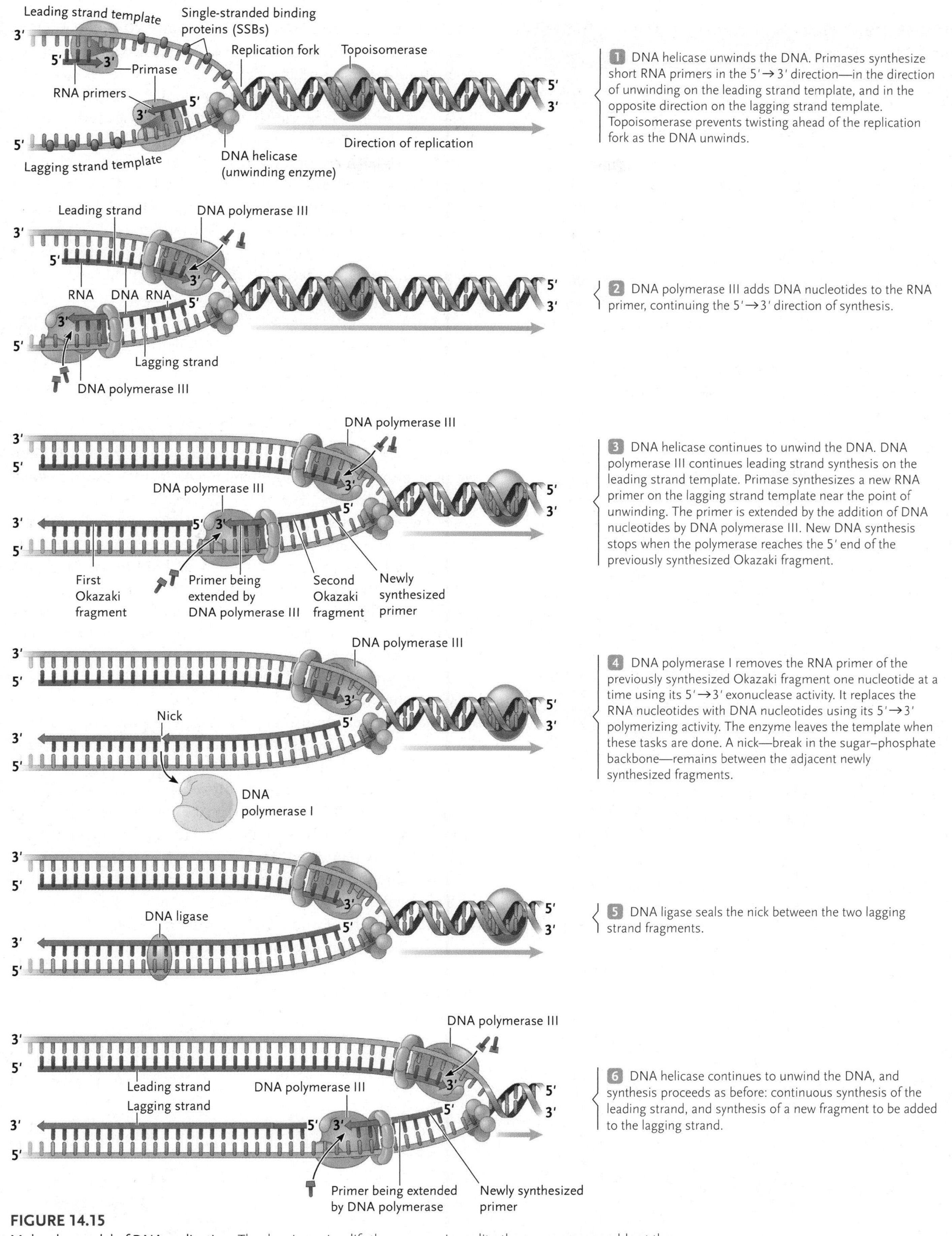

Leading strand template
Single-stranded binding proteins (SSBs)
Replication fork
Topoisomerase
Primase
RNA primers
Direction of replication
Lagging strand template
DNA helicase (unwinding enzyme)
1 DNA helicase unwinds the DNA. Primases synthesize short RNA primers in the 5′→3′ direction—in the direction of unwinding on the leading strand template, and in the opposite direction on the lagging strand template. Topoisomerase prevents twisting ahead of the replication fork as the DNA unwinds.
Leading strand
DNA polymerase III
RNA
DNA
RNA
Lagging strand
DNA polymerase III
2 DNA polymerase III adds DNA nucleotides to the RNA primer, continuing the 5′→3′ direction of synthesis.
DNA polymerase III
DNA polymerase III
First Okazaki fragment
Primer being extended by DNA polymerase III
Second Okazaki fragment
Newly synthesized primer
3 DNA helicase continues to unwind the DNA. DNA polymerase III continues leading strand synthesis on the leading strand template. Primase synthesizes a new RNA primer on the lagging strand template near the point of unwinding. The primer is extended by the addition of DNA nucleotides by DNA polymerase III. New DNA synthesis stops when the polymerase reaches the 5′ end of the previously synthesized Okazaki fragment.
DNA polymerase III
Nick
DNA polymerase I
4 DNA polymerase I removes the RNA primer of the previously synthesized Okazaki fragment one nucleotide at a time using its 5′→3′ exonuclease activity. It replaces the RNA nucleotides with DNA nucleotides using its 5′→3′ polymerizing activity. The enzyme leaves the template when these tasks are done. A nick—break in the sugar–phosphate backbone—remains between the adjacent newly synthesized fragments.
DNA ligase
5 DNA ligase seals the nick between the two lagging strand fragments.
DNA polymerase III
Leading strand
Lagging strand
DNA polymerase III
Primer being extended by DNA polymerase
Newly synthesized primer
6 DNA helicase continues to unwind the DNA, and synthesis proceeds as before: continuous synthesis of the leading strand, and synthesis of a new fragment to be added to the lagging strand.

FIGURE 14.15
Molecular model of DNA replication. The drawings simplify the process. In reality, the enzymes assemble at the fork, replicating both strands from that position as the template strands fold and pass through the assembly.

TABLE 14.1 Major Enzymes of DNA Replication

Enzyme	Symbol	Function
Helicase		Unwinds DNA helix
Single-stranded binding proteins		Stabilize single-stranded DNA and prevent the two strands at the replication fork from reforming double-stranded DNA
Topoisomerase		Avoids twisting of the DNA ahead of replication fork (in circular DNA) by cutting the DNA, turning the DNA on one side of the break in the direction opposite to that of the twisting force, and rejoining the two strands again
Primase		Synthesizes RNA primer in the 5′→3′ direction to initiate a new DNA strand
DNA polymerase III		Main replication enzyme in *E. coli*. Extends the RNA primer by adding DNA nucleotides to it.
DNA polymerase I		*E. coli* enzyme that uses its 5′→3′ exonuclease activity to remove the RNA of the previously synthesized Okazaki fragment, and uses its 5′→3′ polymerization activity to replace the RNA nucleotides with DNA nucleotides.
Sliding DNA clamp		Tethers DNA polymerase III to the DNA template, making replication more efficient.
DNA ligase		Seals nick left between adjacent fragments after RNA primers replaced with DNA

Replication advances at a rate of about 500–1,000 nucleotides per second in *E. coli* and other bacteria, and at a rate of about 50–100 per second in eukaryotes. The entire process is so rapid that the RNA primers and nicks left by discontinuous synthesis persist for only seconds or fractions of a second. A short distance behind the fork, the new DNA chains are fully continuous and wound into complete DNA double helices. Each helix consists of one "old" and one "new" polynucleotide chain.

Researchers identified the enzymes that replicate DNA through experiments with a variety of bacteria and eukaryotes and with viruses that infect both types of cells. Experiments with the bacterium *E. coli* have provided the most complete information about DNA replication, particularly in the laboratory of Arthur Kornberg at Stanford University. Kornberg received a Nobel Prize in 1959 for his discovery of the mechanism for DNA synthesis.

Multiple Replication Origins Enable Rapid Replication of Large Chromosomes

Unwinding at an *ori* within a DNA molecule actually produces two replication forks: two Ys joined together at their tops to form a **replication bubble.** Typically, each of the replication forks moves away from the *ori* as DNA replication proceeds with the events at each fork mirroring those in the other **(Figure 14.16).**

For small circular genomes, such as those found in *E. coli,* and in many bacteria and archaea, there is a single *ori.* Eukaryotic genomes, by contrast, are distributed among several linear chromosomes, each of which can be very long. The average human chromosome, for instance, is about 25 times longer than the *E. coli* chromosome. Nonetheless, replication of long, eukaryotic chromosomes is relatively rapid—sometimes faster than the *E. coli* chromosome—because there are many, sometimes hundreds of origins of replications along eukaryotic chromosomes. Replication initiates at each origin, forming a replication bubble at each **(Figure 14.17).** Movement of the two forks in opposite directions from each origin extends the replication bubbles until the forks eventually meet along the chromosomes to produce fully replicated DNA molecules.

Normally, a replication origin is activated only once during the S phase of a eukaryotic cell cycle, so no portion of the DNA is replicated more than once. *Insights from the Molecular Revolution* describes the production of an abnormal number of copies of a segment of DNA that underlies a common cause of mental retardation in humans.

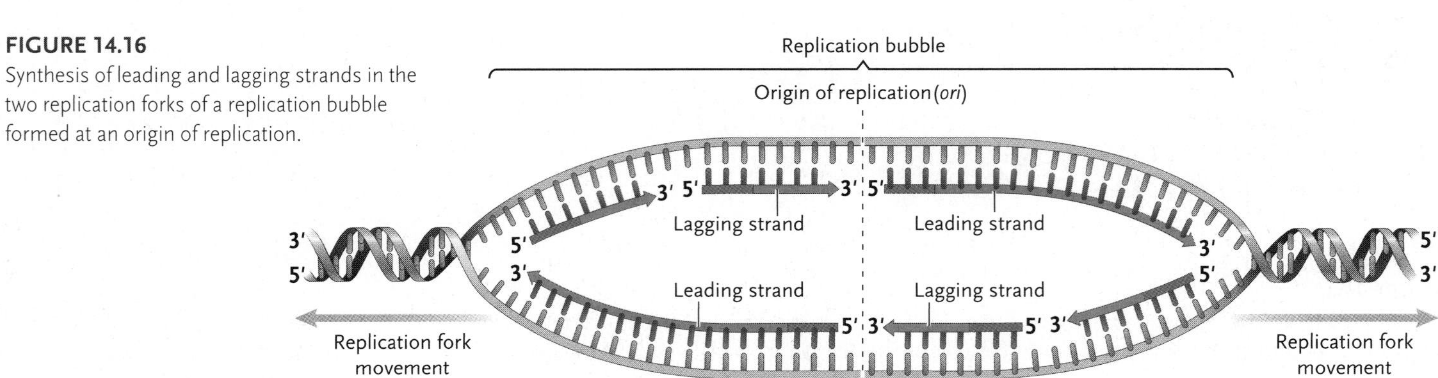

Replication bubble
Origin of replication (*ori*)
3′ 5′
3′ 5′
Lagging strand
Leading strand
Leading strand
Lagging strand
5′ 3′
5′ 3′
3′
5′
5′
3′
3′
5′
5′
3′
Replication fork movement
Replication fork movement

FIGURE 14.16
Synthesis of leading and lagging strands in the two replication forks of a replication bubble formed at an origin of replication.

A Fragile Connection between DNA Replication and Mental Retardation: What is the molecular basis for fragile X syndrome?

The second most common cause of mental retardation in humans after Down syndrome results from breaks that occur in a narrow, constricted region near one end of the X chromosome (see **Figure 1**). Because the region breaks easily when cultured cells divide, the associated disabilities are called *fragile X syndrome.* In addition to mental retardation, affected individuals may have an unusually long face and protruding ears.

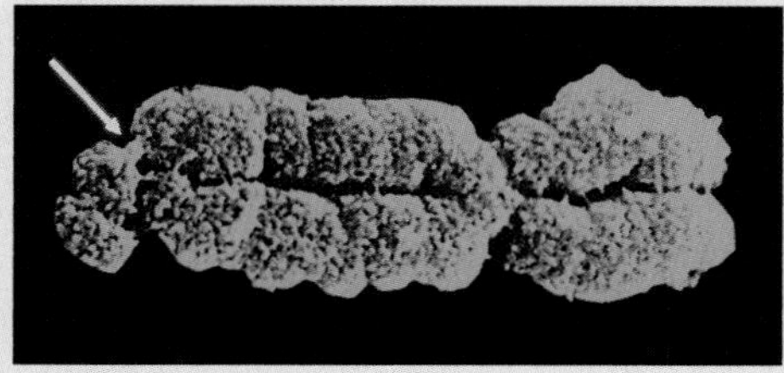

C. J. Harrison

FIGURE 1
The constricted region *(arrow)* in the human X chromosome associated with fragile X syndrome. The chromosome is double because it has been duplicated in preparation for cell division.

Like most other X-linked traits, the disorder affects males more frequently than females; about 1 in 4,000 males and 1 in 6,000–8,000 females worldwide have fragile X syndrome. However, the syndrome has an unusual inheritance pattern. The disease can be passed from an unaffected grandfather through his unaffected daughter to his grandchildren, in whom abnormal X chromosomes and the symptoms of the disease are seen.

Research Question: What is the molecular basis for fragile X syndrome?

Experiments: Geneticists were baffled by this unusual pattern of inheritance until a partial explanation was supplied by the application of molecular techniques in the laboratories of Grant R. Sutherland of Adelaide Children's Hospital in Australia and others. The investigators examined DNA from individuals with fragile X syndrome using "probes"—short, artificially synthesized DNA sequences that are complementary to, and can pair with, DNA sequences that are of interest. (The technique used—Southern blot analysis—is described in Chapter 18.) They found that probes containing C and G nucleotides in high proportions paired most strongly with DNA in the fragile X region. Sequencing of DNA that paired with those probes showed that the region contains a few-to-many repeats of the three-nucleotide sequence 5′-CGG-3′, the number of copies varying with the individual. The fragility of the X chromosome correlated with higher numbers of the repeated sequence.

Conclusion: The researchers concluded that the molecular basis for the fragility of the X chromosome in fragile X syndrome is an abnormally high number of the three-nucleotide repeated sequence. The normal allele of the gene contains between 6 and 54 copies of the repeat; no symptoms are associated with this allele. When the copies exceed 54 in number, the allele is considered abnormal. Individuals with 55–200 copies are said to have a premutation allele because the gene is abnormal, yet there are few or no symptoms of fragile X syndrome. Individuals with more than 200 copies (sometimes thousands of copies) have the fully mutant allele; all males with this allele have all the symptoms of fragile X syndrome while females have somewhat milder symptoms.

But what accounts for this unusual inheritance pattern? Let's look more closely at the grandfather-daughter-grandson inheritance pattern we mentioned in the beginning **(Figure 2).**

1. The grandfather usually has no symptoms, but has a premutation allele near the beginning of the *FMR1* gene **(Figure 2A).**
2. The grandfather passes to his daughter a copy of his X chromosome with additional CGG repeats produced during DNA replication. Each repeat is, itself, a premutation allele **(Figure 2B).**
3. If the daughter passes on her paternal X chromosome to her son, there is a high likelihood that the three-nucleotide sequence region will expand further to produce a fully mutant allele. The *FMR1*-encoded protein is not produced, control of neuron connection strength is altered and, in ways that are not understood, mental retardation results **(Figure 2C).**

If the mother passed the premutation X chromosome to her daughter, the daughter would likely have the premutation allele, perhaps with some increase in the number of repeats. In a few instances, the daughter will experience expansion of the repeats to produce the full mutation and, therefore, have fragile X syndrome with the milder symptoms characteristic of females.

Source: E. J. Kremer, et al. 1991. Mapping of DNA instability at the fragile X to a trinucleotide repeat sequence p(CCG)*n*. *Science* 252:1711–1714.

A. Grandfather's X chromosome

Premutation allele:
55–200 repeats of CGG

FMR1 Gene
5′ . . . CGG CGG CGG CGG CGG . . . 3′
3′ . . . GCC GCC GCC GCC GCC . . . 5′

Grandfather's X chromosome has premutation allele

FMR1 protein produced

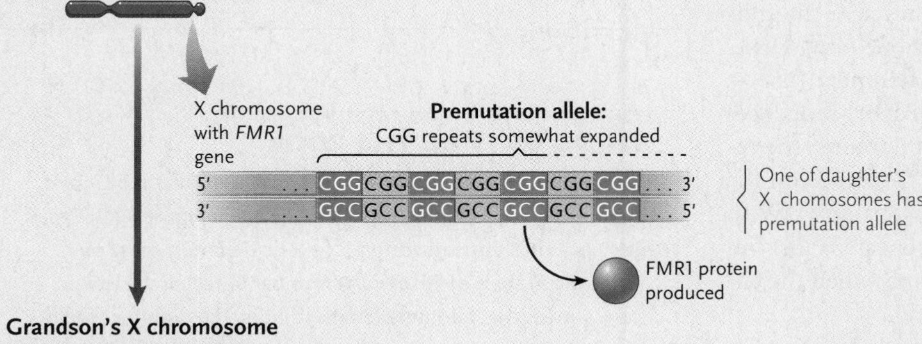
B. Daughter's paternal X chromosome
X chromosome with FMR1 gene
Premutation allele:
CGG repeats somewhat expanded
5′ . . . CGG CGG CGG CGG CGG CGG CGG . . . 3′
3′ . . . GCC GCC GCC GCC GCC GCC GCC . . . 5′
One of daughter's X chomosomes has premutation allele
FMR1 protein produced

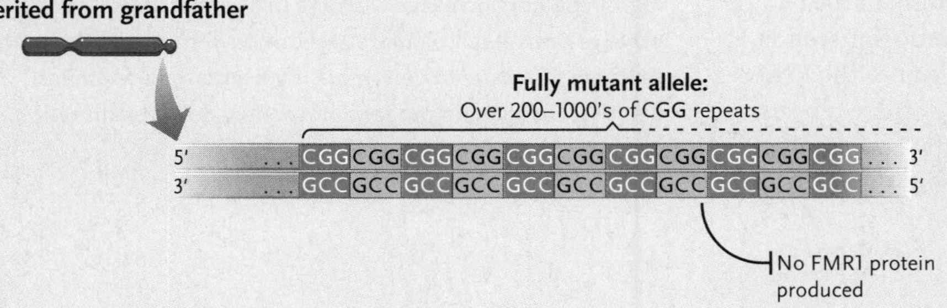
C. Grandson's X chromosome inherited from grandfather
Fully mutant allele:
Over 200–1000's of CGG repeats
5′ . . . CGG CGG CGG CGG CGG CGG CGG CGG CGG CGG CGG . . . 3′
3′ . . . GCC GCC GCC GCC GCC GCC GCC GCC GCC GCC GCC . . . 5′
A grandson who inherits this allele develops full mutation
No FMR1 protein produced

FIGURE 2
Inheritance pattern in fragile X syndrome.

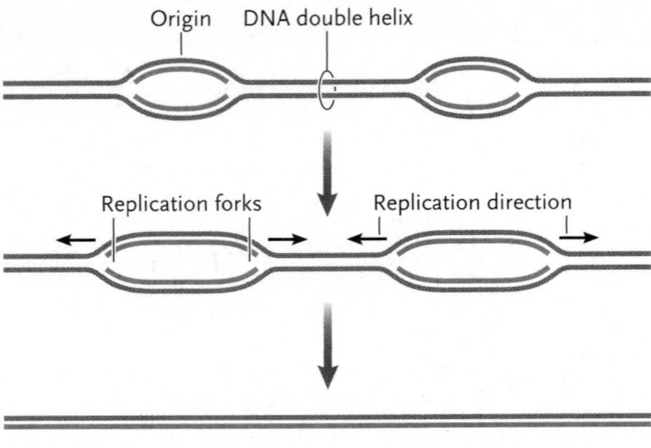
Origin
DNA double helix
Replication forks
Replication direction

FIGURE 14.17
Replication from multiple origins in the linear chromosomes of eukaryotes.

Telomerases Solve a Special Replication Problem at the Ends of Linear DNA Molecules in Eukaryotes

The RNA primer synthesized in DNA replication (see Figures 14.13 and 14.15) produces a problem for replicating the linear chromosomes of eukaryotes. Think about the end of a linear DNA molecule. New DNA synthesis on the 3′→5′ template strand starts with an RNA primer. That primer will subsequently be removed, leaving a gap at the 5′ end of the new DNA strand **(Figure 14.18).** But, because there is no existing nucleotide chain that can be used, DNA polymerase cannot fill in the gap with DNA nucleotides. In a similar way, a gap is produced at the 5′ end of the new strand made starting at the other end of the chromosome. When these new, now shortened DNA strands are used as a template for the next round of DNA replication, the new chromosome will be shorter. Indeed, when most somatic cells go through the cell cycle, the chromosomes shorten with each division. Deletion of genes by such shortening can eventually have lethal consequences for the cell.

In most eukaryotic chromosomes the genes near the ends of chromosomes are protected by a buffer of noncoding DNA. The region of noncoding DNA is called the **telomere** (*telo* = end, *mere* = segment). A telomere consists of short sequences repeated hundreds to thousands of times. In humans, the repeated sequence, the *telomere repeat,* is 5′-TTAGGG-3′ on the template strand (the top strand in Figure 14.18, step 1). With each replication, a fraction of the telomere repeats is lost but the genes are unaffected. The buffering fails only when the entire telomere is lost.

The enzyme **telomerase** can stop the shortening of the telomeres by adding telomere repeats to the chromosome ends. Telomerase consists of proteins and an RNA molecule. The RNA of telomerase is the template for the addition of telomere repeats. Telomerase binds to the DNA template strand by complementary base pairing between the telomerase RNA and the DNA. It then adds telomere repeats to the DNA using the RNA as a template (see Figure 14.18). Now when the top strand is used as a template and the RNA primer is removed, there will be a single-stranded region at the end of the chromosome as before. However, the chromosome has not shortened because of the extra telomere repeats added by the telomerase.

In most multicellular organisms, telomerase is active only in the rapidly dividing cells of the early embryo, and in germ cells (the precursors to sperm and eggs). However, telomerase is inactive in somatic cells, meaning telomeres shorten when such cells divide. As a result, somatic cells are capable of only a certain number of mitotic divisions before they stop dividing and die.

Telomerase explains how cancer cells can divide indefinitely and not be limited to a certain number of divisions as a result of telomere shortening. For many cancers, as normal cells develop into cancer cells, their telomerases are reactivated, preserving chromosome length during the rapid divisions characteristic of cancer. A positive side of this discovery is that it may lead to an effective cancer treatment, if a means can be found to switch off the telomerases in tumor cells. The chromosomes in the rapidly dividing cancer cells would then eventually shorten to the length at which they break down, leading to cell death and elimination of the tumor.

Elizabeth Blackburn, Carol Greider, and Jack Szostak were awarded a Nobel Prize in 2009 for their discovery of how chromosomes are protected by telomeres and the enzyme telomerase.

STUDY BREAK 14.3

1. **What is the importance of complementary base pairing to DNA replication?**
2. **Why is a primer needed for DNA replication? How is the primer made?**
3. **DNA polymerase III and DNA polymerase I are used in DNA replication in *E. coli*. What are their roles?**
4. **Why are telomeres important?**

THINK OUTSIDE THE BOOK

We introduced progeria in *Why It Matters* of Chapter 13, and pictured two individuals with the disease in Figure 13.1. Progeria is a rare premature aging genetic disorder; most patients die in their early teens. A number of research studies have shown that telomere length (that is, the number of telomere repeats) decreases with age in humans (and in many other organisms). Collaboratively, or on your own, explore the research literature to determine if a decrease in telomere length is involved in progeria. If so, how does the mutation in progeria cause the decrease?

FIGURE 14.18
Addition of telomere repeats to the 3′ end of a eukaryotic linear chromosome by telomerase.

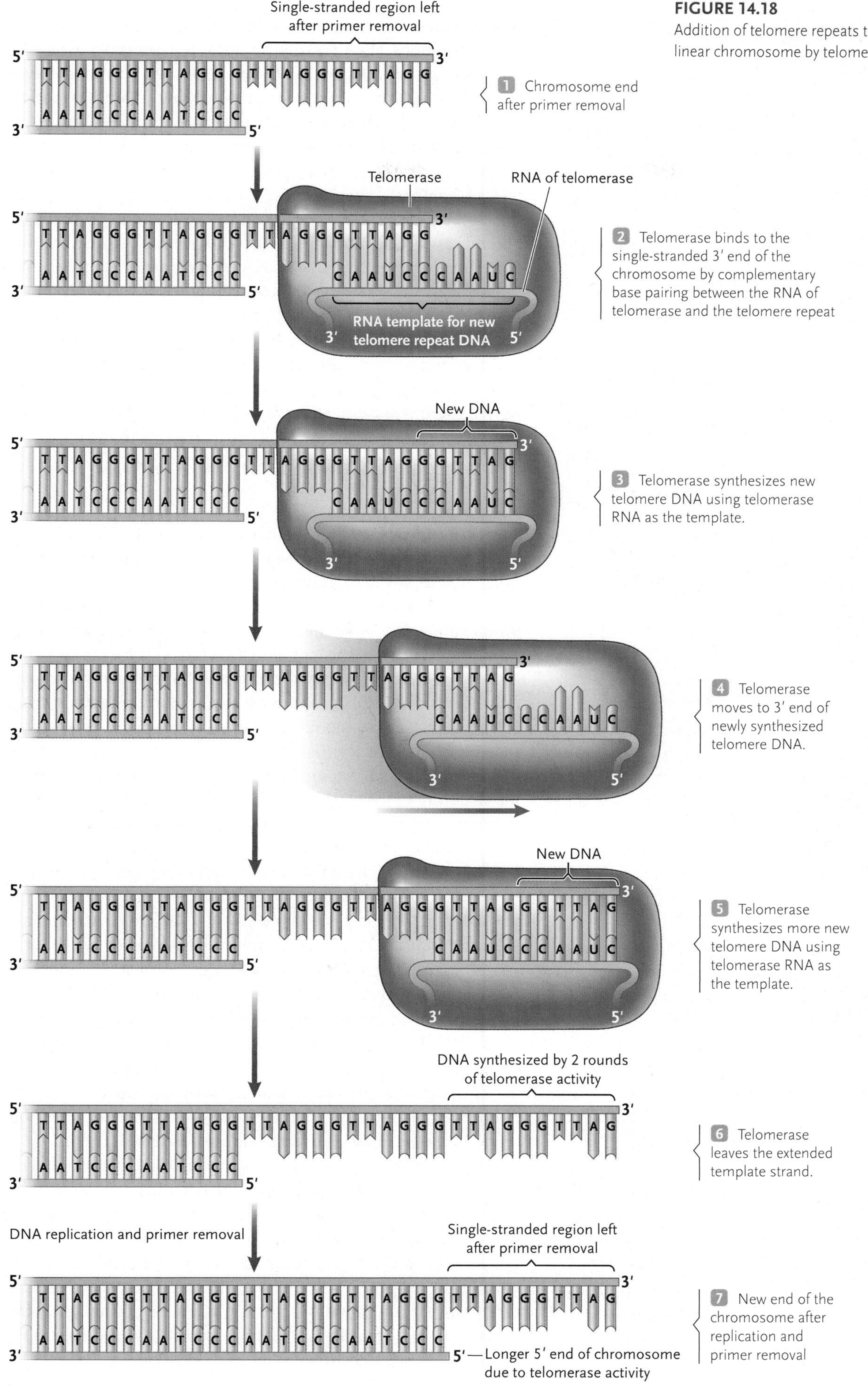
Single-stranded region left after primer removal
1 Chromosome end after primer removal
Telomerase
RNA of telomerase
RNA template for new telomere repeat DNA
2 Telomerase binds to the single-stranded 3′ end of the chromosome by complementary base pairing between the RNA of telomerase and the telomere repeat
New DNA
3 Telomerase synthesizes new telomere DNA using telomerase RNA as the template.
4 Telomerase moves to 3′ end of newly synthesized telomere DNA.
New DNA
5 Telomerase synthesizes more new telomere DNA using telomerase RNA as the template.
DNA synthesized by 2 rounds of telomerase activity
6 Telomerase leaves the extended template strand.
DNA replication and primer removal
Single-stranded region left after primer removal
5′—Longer 5′ end of chromosome due to telomerase activity
7 New end of the chromosome after replication and primer removal

14.4 Mechanisms That Correct Replication Errors

DNA polymerases make very few errors as they assemble new nucleotide chains. Most of the mistakes that do occur, called **base-pair mismatches,** are corrected, either by a proofreading mechanism carried out during replication by the DNA polymerases themselves or by a DNA repair mechanism that corrects mismatched base pairs after replication is complete.

Proofreading Depends on the Ability of DNA Polymerases to Reverse and Remove Mismatched Bases

The **proofreading mechanism,** first proposed in 1972 by Arthur Kornberg and Douglas L. Brutlag of Stanford University, depends on the ability of DNA polymerases to back up and remove mispaired nucleotides from a DNA strand. For most of the polymerization reactions, DNA polymerase adds the correct nucleotide to the growing chain (**Figure 14.19,** step 1). But the polymerase also has a small window of time for proofreading. If a newly added nucleotide is mismatched (step 2), the DNA polymerase reverses, using a built-in 3′→5′ exonuclease activity to remove the newly added incorrect nucleotide (step 3). The enzyme then resumes forward synthesis, now inserting the correct nucleotide (step 4).

Several experiments have confirmed that the major DNA polymerases of replication can actually proofread their work. For example, when the *E. coli* DNA polymerase III is fully functional, its overall error rate is astonishingly low with only about 1 mispair surviving in the DNA for every 1 million nucleotides polymerized in the test tube. If the proofreading activity of the enzyme is experimentally inhibited, the error rate increases to about 1 mistake for every 1,000 to 10,000 nucleotides polymerized. Experiments with eukaryotes have yielded similar results.

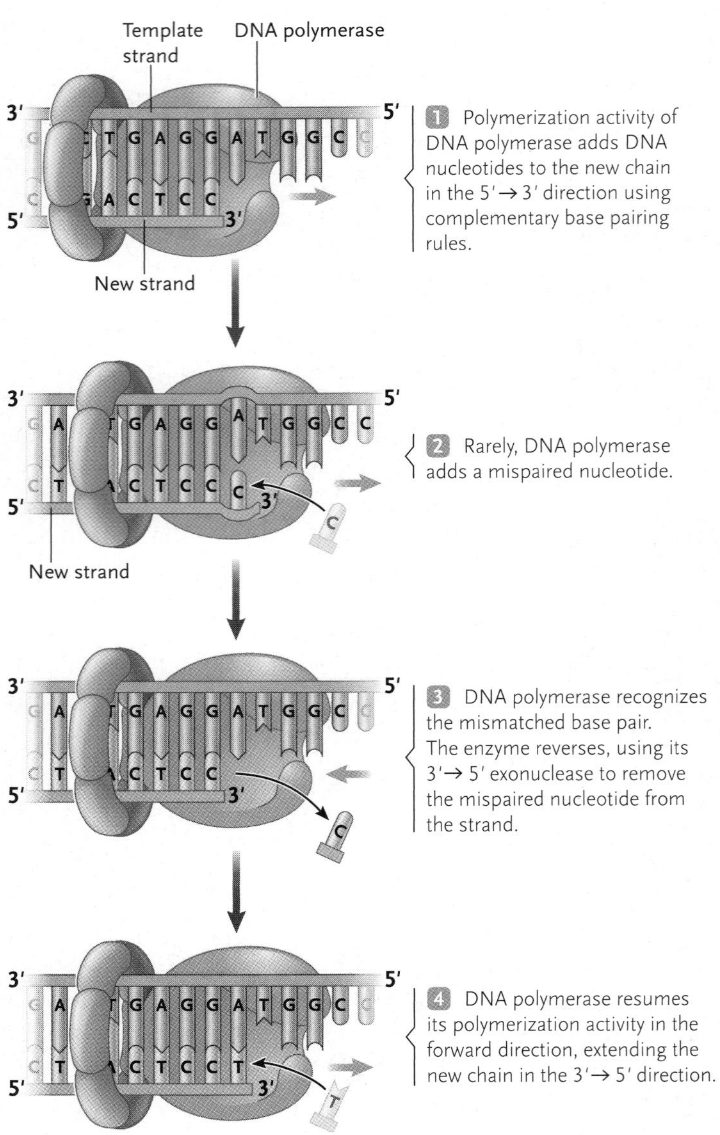
Template strand
DNA polymerase
New strand
3′
5′
1 Polymerization activity of DNA polymerase adds DNA nucleotides to the new chain in the 5′→3′ direction using complementary base pairing rules.
2 Rarely, DNA polymerase adds a mispaired nucleotide.
3 DNA polymerase recognizes the mismatched base pair. The enzyme reverses, using its 3′→5′ exonuclease to remove the mispaired nucleotide from the strand.
4 DNA polymerase resumes its polymerization activity in the forward direction, extending the new chain in the 3′→5′ direction.

FIGURE 14.19
Proofreading by a DNA polymerase.

DNA Repair Corrects Errors That Escape Proofreading

Any base-pair mismatches that remain after proofreading face still another round of correction by **DNA repair mechanisms.** These **mismatch repair** mechanisms increase the accuracy of DNA replication well beyond the one-in-a million errors that persist after proofreading. The mechanisms operate similarly in all organisms.

As noted earlier, the "correct" A–T and G–C base pairs fit together like pieces of a jigsaw puzzle, and their dimensions separate the sugar–phosphate backbone chains by a constant distance. Mispaired bases are too large or small to maintain the correct separation, and they cannot form the hydrogen bonds characteristic of the normal base pairs. As a result, base mismatches distort the structure of the DNA helix. These distortions provide recognition sites for the enzymes catalyzing mismatch repair.

The repair enzymes detect the mispaired base, cut the new DNA strand on each side of the mismatch, and remove a portion of the chain **(Figure 14.20).** DNA polymerase fills in the gap with new DNA. The repair is completed by DNA ligase, which seals the nucleotide chain into a continuous DNA molecule.

Similar repair systems also detect and correct alterations in DNA caused by the damaging effects of chemicals and radiation, including the ultraviolet light in sunlight. Some idea of the importance of the repair mechanisms comes from the unfortunate plight of individuals with *xeroderma pigmentosum,* a hereditary disorder in which the repair mechanism is faulty. Because of the effects of unrepaired alterations in their DNA, skin cancer can develop quickly in these individuals if they are exposed to sunlight.

Very few replication errors remain in DNA after proofreading and DNA repair. The errors that persist, although extremely rare, are a primary source of **mutations,** differences in DNA sequence that appear and remain in the replicated copies. When a mutation occurs in a gene, it can alter the property of the protein encoded by the gene, which, in turn, may alter how the organism functions. Hence, mutations are highly important to the evolutionary process because they are the ultimate source of the variability in offspring acted on by natural selection.

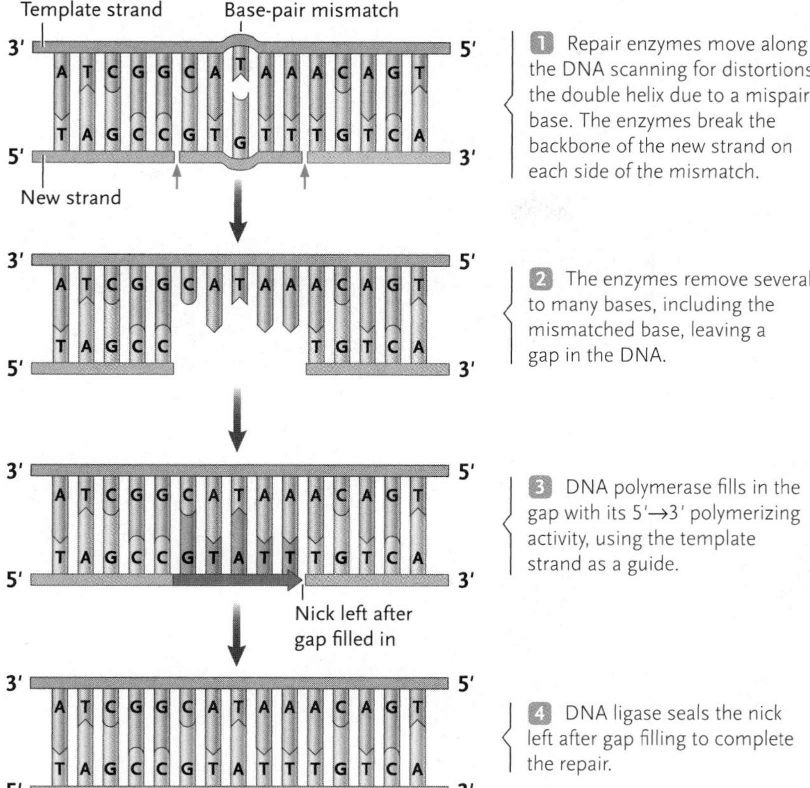
Template strand
Base-pair mismatch
New strand
1 Repair enzymes move along the DNA scanning for distortions in the double helix due to a mispaired base. The enzymes break the backbone of the new strand on each side of the mismatch.
2 The enzymes remove several to many bases, including the mismatched base, leaving a gap in the DNA.
3 DNA polymerase fills in the gap with its 5'→3' polymerizing activity, using the template strand as a guide.
Nick left after gap filled in
4 DNA ligase seals the nick left after gap filling to complete the repair.

FIGURE 14.20
Repair of mismatched bases in replicated DNA.

We now turn from DNA replication and error correction to the arrangements of DNA in eukaryotic and prokaryotic cells. These arrangements organize superstructures that fit the long DNA molecules into the microscopic dimensions of cells and also contribute to the regulation of DNA activity.

STUDY BREAK 14.4 <

Why is a proofreading mechanism important for DNA replication, and what are the mechanisms that correct errors?

14.5 DNA Organization in Eukaryotes and Prokaryotes

Enzymatic proteins are the essential catalysts of every step in DNA replication. In addition, numerous proteins of other types organize the DNA in both eukaryotes and prokaryotes and control its function.

In eukaryotes, two major types of proteins, the histone and nonhistone proteins, are associated with DNA structure and regulation in the nucleus. These proteins are known collectively as the **chromosomal proteins** of eukaryotes. The complex of DNA and its associated proteins, termed **chromatin,** is the structural building block of a chromosome.

By comparison, the single chromosome of a prokaryotic cell is more simply organized and has fewer associated proteins. However, prokaryotic DNA is still associated with two classes of proteins with functions similar to those of the eukaryotic histones and nonhistones: one class that organizes the DNA structurally and one that regulates gene activity. We begin this section with the major DNA-associated proteins of eukaryotes.

Histones Pack Eukaryotic DNA at Successive Levels of Organization

The **histones** are a class of small, positively charged (basic) proteins that are complexed with DNA in the chromosomes of eukaryotes. (Most other cellular proteins are larger and are neutral or negatively charged.) The histones bind to DNA by an attraction between their positive charges and the negatively charged phosphate groups of the DNA.

Five types of histones exist in most eukaryotic cells: H1, H2A, H2B, H3, and H4. The amino acid sequences of these proteins are highly similar among eukaryotes, suggesting that they perform the same functions in all eukaryotic organisms.

One function of histones is to pack DNA molecules into the narrow confines of the cell nucleus. For example, each human cell nucleus contains 2 meters of DNA. Combination with the histones compacts this length so much that it fits into nuclei that are only about 10 mm in diameter. Another function is the regulation of DNA activity.

HISTONES AND DNA PACKING The histones pack DNA at several levels of chromatin structure. In the most fundamental structure, called a **nucleosome,** two molecules each of H2A, H2B, H3, and H4 combine to form a beadlike, eight-protein **nucleosome core particle** around which DNA winds for almost two turns **(Figure 14.21).** A short segment of DNA, the **linker,** extends between one nucleosome and the next. Under the electron microscope, this structure looks like beads on a string. The diameter of the beads (the nucleosomes) gives this structure its name—the **10-nm chromatin fiber** (see Figure 14.21).

Each nucleosome and linker includes about 200 base pairs of DNA. Nucleosomes compact DNA by a factor of about 7; that is, a length of DNA becomes about 7 times shorter when it is wrapped into nucleosomes.

HISTONES AND CHROMATIN FIBERS The fifth histone, H1, brings about the next level of chromatin packing. One H1 molecule binds both to the nucleosome (at the point where the DNA enters and leaves the core particle) and to the linker DNA. This binding causes the nucleosomes to package into a coiled structure 30 nm in diameter, called the **30-nm chromatin fiber.** One possible model for the 30-nm fiber is the **solenoid model,** with the nucleosomes spiraling helically with about six nucleosomes per turn (see Figure 14.21).

The arrangement of DNA in nucleosomes and the 30-nm fiber compacts the DNA and probably also protects it from chemical and mechanical damage. In the test tube, DNA wound

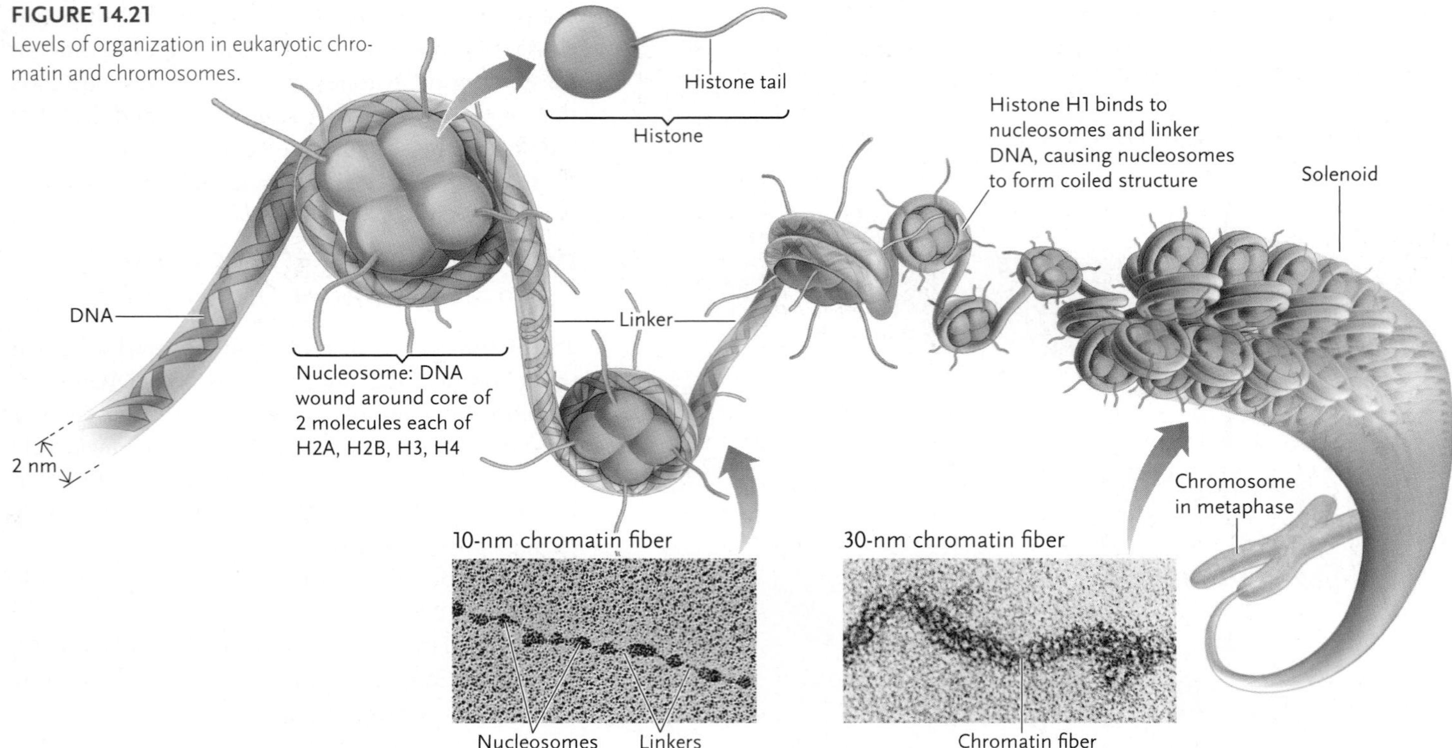
Histone tail
Histone
Histone H1 binds to nucleosomes and linker DNA, causing nucleosomes to form coiled structure
Solenoid
DNA
Linker
Nucleosome: DNA wound around core of 2 molecules each of H2A, H2B, H3, H4
2 nm
Chromosome in metaphase
10-nm chromatin fiber
30-nm chromatin fiber
Nucleosomes
Linkers
Chromatin fiber

FIGURE 14.21
Levels of organization in eukaryotic chromatin and chromosomes.

into nucleosomes and chromatin fibers is much more resistant to attack by deoxyribonuclease (a DNA-digesting enzyme) than when it is not bound to histone proteins. It is also less accessible to the proteins and enzymes required for gene expression. Therefore, the association of the DNA with histones must loosen in order for a gene to become active (see Section 16.2).

PACKING OF EUKARYOTIC CHROMOSOMES AT STILL HIGHER LEVELS: EUCHROMATIN AND HETEROCHROMATIN In interphase nuclei, chromatin fibers are loosely packed in some regions and densely packed in others. The loosely packed regions are known as euchromatin (*eu* = true, regular, or typical), and the densely packed regions are called heterochromatin (*hetero* = different). Chromatin fibers also fold and pack into the thick, rodlike chromosomes visible during mitosis and meiosis. Experiments indicate that links formed between H1 histone molecules contribute to the packing of chromatin fibers, both into heterochromatin and into the chromosomes visible during nuclear division (see discussion in Section 10.2). However, the exact mechanism for the more complex folding and packing is not known.

Several experiments indicate that some regions of heterochromatin include genes that have been turned off and placed in a compact storage form. For example, recall the process of X-chromosome inactivation in mammalian females (see Section 13.2). As one of the two X chromosomes becomes inactive in cells early in development, it packs down into a block of heterochromatin called the *Barr body,* which is large enough to see under the light microscope. These findings support the idea that, in addition to organizing nuclear DNA, histones play a role in regulating gene activity.

Many Nonhistone Proteins Have Key Roles in the Regulation of Gene Expression in Eukaryotes

Nonhistone proteins are loosely defined as all the proteins associated with DNA that are not histones. Nonhistones vary widely in structure; most are negatively charged or neutral, but some are positively charged. They range in size from polypeptides smaller than histones to some of the largest cellular proteins.

Many nonhistone proteins help control the expression of individual genes. (The regulation of gene expression is the subject of Chapter 16.) For example, expression of a gene requires that the enzymes and proteins for that process be able to access the gene in the chromatin. If a gene is packed into heterochromatin it is unavailable for activation. If the gene is in the form of the more-extended euchromatin it is more accessible. Many nonhistone proteins affect gene accessibility by modifying histones to change how the histones associate with DNA in chromatin, either loosening or tightening the association. Other nonhistone proteins are regulatory proteins that activate or repress the expression of a gene. Yet others are components of the enzyme–protein complexes that are needed for the expression of any gene.

DNA Is Organized More Simply in Bacteria Than in Eukaryotes

Several features of DNA organization in bacteria differ fundamentally from eukaryotic DNA organization. In contrast to the linear DNA in eukaryotes, the primary DNA molecule of most bacteria is circular, with only one copy per cell. In parallel with eukaryotic terminology, the DNA molecule is called a **bacterial**

chromosome. The chromosome of the best-known bacterium, *E. coli,* includes about 1,460 mm of DNA, which is equivalent to 4.6 million base pairs. There are exceptions: some bacteria have two or more different chromosomes in the cell, and some bacterial chromosomes are linear.

Replication begins from a single origin in the DNA circle, forming two forks that travel around the circle in opposite directions. Eventually, the forks meet at the opposite side from the origin to complete replication **(Figure 14.22).**

Inside bacterial cells, the DNA circle is packed and folded into an irregularly shaped mass called the **nucleoid** (shown in Figure 5.7). The DNA of the nucleoid is suspended directly in the cytoplasm with no surrounding membrane.

Many bacterial cells also contain other DNA molecules, called **plasmids,** in addition to the main chromosome of the nucleoid. Most plasmids are circular, although some are linear. Plasmids have replication origins and are duplicated and distributed to daughter cells together with the bacterial chromosome during cell division.

Although bacterial DNA is not organized into nucleosomes, certain positively charged proteins do combine with bacterial DNA. Some of these proteins help organize the DNA into loops, thereby providing some compaction of the molecule. Bacterial DNA also combines with many types of genetic regulatory proteins that have functions similar to those of the nonhistone proteins of eukaryotes (see Chapter 16).

With this description of bacterial DNA organization, our survey of DNA structure and its replication and organization is complete. The next chapter revisits the same structures and discusses how they function in the expression of information encoded in the DNA.

STUDY BREAK 14.5

1. **What is the structure of the nucleosome?**
2. **What is the role of histone H1 in eukaryotic chromosome structure?**

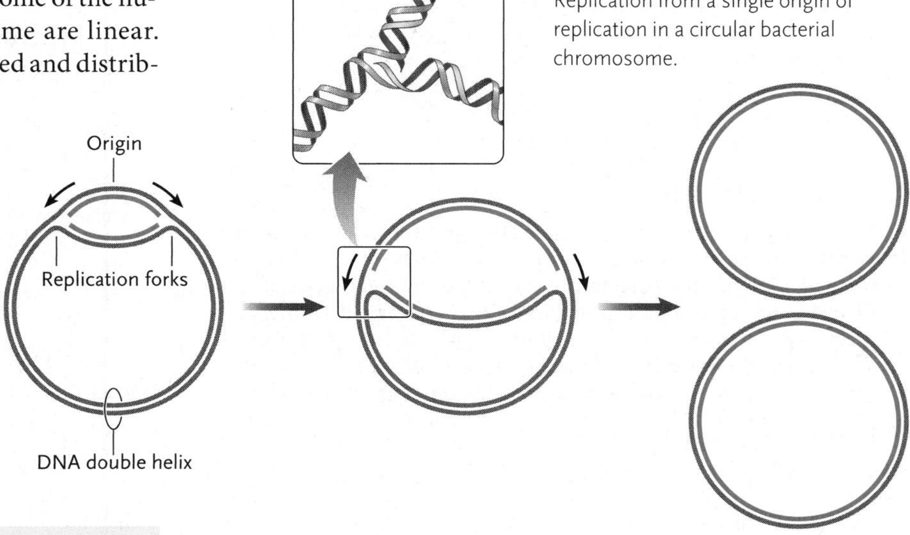
Origin
Replication forks
DNA double helix

FIGURE 14.22
Replication from a single origin of replication in a circular bacterial chromosome.

UNANSWERED QUESTIONS

Does size matter?

In this chapter, you learned that the addition of DNA onto telomeres by the enzyme telomerase can counteract the shortening of chromosomes which is predicted by the end replication problem. In humans, telomerase is active in cells destined to become sperm and egg, and in stem cells (cells capable of differentiating into almost any adult cell type and tissue) and other highly proliferative cells. It is also active in most cancers and immortalized cell lines (cells that can grow and divide indefinitely in a culture dish). The presence of telomerase does not result in ever-growing telomeres, however. Cancer cells maintain a specific average telomere length, which is often shorter than the average telomere length for normal somatic cells. Furthermore, some species express high amounts of telomerase activity in all tissues, yet maintain species-specific, and average telomere lengths.

How do cells measure the length of the telomeric DNA tract?

Several proteins that assemble at the telomere have been identified, including two that bind specifically to the telomeric repeat sequence, and are thought to regulate telomere length. Exactly how this is accomplished is an active area of research. The function of these proteins can be tested by manipulating the protein in question with contemporary genetic approaches and observing effects on the length of the telomeric DNA. For example, if a protein's role is to prevent telomerase access to the telomere, preventing the protein from being synthesized should result in telomere elongation; if it functions to recruit telomerase, telomeres should shorten. Proteins that bind to the double-stranded telomeric DNA sequences can affect access of telomerase to the single-stranded 3' end. Biochemical experiments and studies of the 3D structure of these proteins are being used to dissect how information about the length of the double-stranded telomere region is communicated to the very terminus where telomerase acts.

Shortened telomeres, whether due to cell division in the absence of telomerase or experimental manipulations of telomere proteins, resemble broken chromosomes and can prevent the cell from progressing through the cell cycle. When normal human cells lacking telomerase are cultured in a dish, they undergo only a finite number of cell divisions. What kind of evidence would rigorously link this behavior specifically to shortened telomeres? A more complicated question is whether short telomeres are involved in organismal aging. Model organisms lacking the telomerase gene are an important tool to address this question, as are humans who have rare defects in telomerase function.

Think Critically

From what you learned in the chapter and in this essay, develop a brief statement that helps explain why researchers consider telomerase an important potential target in the fight against cancer.

Janis Shampay

Janis Shampay is professor of biology at Reed College. Her research interests include the regulation of telomere metabolism and conservation of telomere protein function in non-mammalian model systems. Learn more about her work at http://academic.reed.edu/biology/professors/jshampay/index.html.

REVIEW KEY CONCEPTS

Go to **CENGAGENOW** at www.cengage.com/login to access quizzing, animations, exercises, articles, and personalized homework help.

14.1 Establishing DNA as the Hereditary Molecule

- Griffith found that a substance derived from killed virulent *Streptococcus pneumoniae* bacteria could transform nonvirulent living *S. pneumoniae* bacteria to the virulent type (Figure 14.2).
- Avery and his coworkers showed that DNA, and not protein or RNA, was the molecule responsible for transforming *S. pneumoniae* bacteria into the virulent form.
- Hershey and Chase showed that the DNA of a phage, not the protein, enters bacterial cells to direct the life cycle of the virus. Taken together, the experiments of Griffith, Avery and his coworkers, and Hershey and Chase established that DNA is the hereditary molecule (Figure 14.3).

Animation: Griffith's experiment

Animation: The Hershey and Chase experiment

14.2 DNA Structure

- Watson and Crick discovered that a DNA molecule consists of two polynucleotide chains twisted around each other into a right-handed double helix. Each nucleotide of the chains consists of deoxyribose, a phosphate group, and either adenine, thymine, guanine, or cytosine. The deoxyribose sugars are linked by phosphate groups to form an alternating sugar–phosphate backbone. The two strands are held together by adenine–thymine (A–T) and guanine–cytosine (G–C) base pairs. Each full turn of the double helix involves 10 base pairs (Figures 14.4 and 14.6).
- The two strands of the DNA double helix are antiparallel.

Animation: The nucleotides of DNA

Animation: The DNA double helix

Practice: Constructing DNA

14.3 DNA Replication

- DNA is duplicated by semiconservative replication, in which the two strands of a parental DNA molecule unwind and each serves as a template for the synthesis of a complementary copy (Figure 14.7–14.9).
- DNA replication is catalyzed by several enzymes. Helicase unwinds the DNA; primase synthesizes an RNA primer used as a starting point for nucleotide assembly by DNA polymerases. DNA polymerases assemble nucleotides into a chain one at a time, in a sequence complementary to the sequence of bases in the template strand. After a DNA polymerase removes the primers and fills in the resulting gaps, DNA ligase closes the remaining single-strand nicks (Figures 14.10–14.13 and 14.15).
- As the DNA helix unwinds, only one template strand runs in a direction allowing the new DNA strand to be made continuously in the direction of unwinding. The other template strand is copied in short lengths that run in the direction opposite to unwinding. The short lengths produced by this discontinuous replication are then linked into a continuous strand (Figures 14.14 and 14.15).
- DNA synthesis begins at sites that act as replication origins and proceeds from the origins as two replication forks moving in opposite directions (Figures 14.16 and 14.17).
- The ends of eukaryotic chromosomes consist of telomeres, short sequences repeated hundreds to thousands of times. These repeats provide a buffer against chromosome shortening during replication. Although most somatic cells show this chromosome shortening, some cell types do not because they have a telomerase enzyme that adds telomere repeats to the chromosome ends (Figure 14.18).

Animation: Overview of DNA replication and base pairing

Animation: DNA replication in detail

14.4 Mechanisms That Correct Replication Errors

- In proofreading, the DNA polymerase reverses and removes the most recently added base if it is mispaired as a result of a replication error. The enzyme then resumes DNA synthesis in the forward direction (Figure 14.19).
- In DNA mismatch repair, enzymes recognize distorted regions caused by mispaired base pairs and remove a section of DNA that includes the mispaired base from the newly synthesized nucleotide chain. A DNA polymerase then resynthesizes the section correctly, using the original template chain as a guide (Figure 14.20).

14.5 DNA Organization in Eukaryotes and Prokaryotes

- Eukaryotic chromosomes consist of DNA complexed with histone and nonhistone proteins.
- In eukaryotic chromosomes, DNA is wrapped around a core consisting of two molecules each of histones H2A, H2B, H3, and H4 to produce a nucleosome. Linker DNA connects adjacent nucleosomes. The chromosome structure in this form is the 10-nm chromatin fiber. The binding of histone H1 causes the nucleosomes to package into a coiled structure called the 30-nm chromatin fiber (Figure 14.21).
- Chromatin is distributed between euchromatin, a loosely packed region in which genes are active in RNA transcription, and heterochromatin, densely packed masses in which genes, if present, are inactive. Chromatin also folds and packs to form thick, rodlike chromosomes during nuclear division.
- Nonhistone proteins help control the expression of individual genes.
- The bacterial chromosome is a closed, circular molecule of DNA; it is packed into the nucleoid region of the cell. Replication begins from a single origin and proceeds in both directions. Many bacteria also contain plasmids, which replicate independently of the host chromosome (Figure 14.22).
- Bacterial DNA is organized into loops through interaction with proteins. Other proteins similar to eukaryotic nonhistones regulate gene activity in prokaryotes.

Animation: Chromosome structural organization

UNDERSTAND AND APPLY

Test Your Knowledge

1. Working on the Amazon River, a biologist isolated DNA from two unknown organisms, P and Q. He discovered that the adenine content of P was 15% and the cytosine content of Q was 42%. This means that:
   a. the amount of guanine in P is 15%.
   b. the amount of guanine and cytosine combined in P is 70%.
   c. the amount of adenine in Q is 42%.
   d. the amount of thymine in Q is 21%.
   e. it takes more energy to unwind the DNA of P than the DNA of Q.
2. The Hershey and Chase experiment showed that phage:
   a. ^{35}S entered bacterial cells.
   b. ^{32}P remained outside of bacterial cells.
   c. protein entered bacterial cells.
   d. DNA entered bacterial cells.
   e. DNA mutated in bacterial cells.
3. Pyrimidines built from a single carbon ring are:
   a. cytosine and thymine.
   b. adenine, cytosine, and guanine.
   c. adenine and thymine.
   d. cytosine and guanine.
   e. adenine and guanine.
4. Which of the following statements about DNA replication is *false*?
   a. Synthesis of the new DNA strand is from 3′ to 5′.
   b. Synthesis of the new DNA strand is from 5′ to 3′.
   c. DNA unwinds, primase adds RNA primer, and DNA polymerases synthesize the new strand and remove the RNA primer.
   d. Many initiation points exist in each eukaryotic chromosome.
   e. Okazaki fragments are synthesized in the opposite direction from the direction in which the replication fork moves.
5. Which of the following statements about DNA is *false*?
   a. Phosphate is linked to the 5′ and 3′ carbons of adjacent deoxyribose molecules.
   b. DNA is bidirectional in its synthesis.
   c. Each side of the helix is antiparallel to the other.
   d. The binding of adenine to thymine is through three hydrogen bonds.
   e. Avery identified DNA as the transforming factor in crosses between smooth and rough bacteria.
6. In the Meselson and Stahl experiment, the DNA in the parental generation was all $^{15}N^{15}N$, and after one round of replication, the DNA was all $^{15}N^{14}N$. What DNAs were seen after three rounds of replication, and in what ratio were they found?
   a. one $^{15}N^{14}N$: one ^{14}N:^{14}N
   b. one $^{15}N^{14}N$: two ^{14}N:^{14}N
   c. one $^{15}N^{14}N$: three ^{14}N:^{14}N
   d. one $^{15}N^{14}N$: four ^{14}N:^{14}N
   e. one $^{15}N^{14}N$: seven ^{14}N:^{14}N
7. During replication, DNA is synthesized in a 5′→3′ direction. This implies that:
   a. the template is read in a 5′→3′ direction.
   b. successive nucleotides are added to the 3′-OH end of the newly forming chain.
   c. because both strands are replicated nearly simultaneously, replication must be continuous on both.
   d. ligase unwinds DNA in a 5′→3′ direction.
   e. primase acts on the 3′ end of the replicating strand.
8. Telomerase:
   a. is active in many cancer cells.
   b. is more active in adult than embryonic cells.
   c. complexes with the ribosome to form telomeres.
   d. acts on unique genes called telomeres.
   e. shortens the ends of chromosomes.
9. Mismatch repair is the ability:
   a. to seal Okazaki fragments with ligase into a continual DNA strand.
   b. of primase to remove the RNA primer and replace it with the correct DNA.
   c. of some enzymes to sense the insertion of an incorrect nucleotide, remove it, and use a DNA polymerase to insert the correct one.
   d. to correct mispaired chromosomes in prophase I of meiosis.
   e. to remove worn-out DNA by telomerase and replace it with newly synthesized nucleotides.
10. Which of the following does not accurately characterize bacterial DNA?
    a. The bacterial chromosome is a closed, circular molecule of DNA.
    b. Bacterial DNA is organized into nucleosomes.
    c. The primary DNA molecule of most bacteria is circular.
    d. DNA is organized more simply in bacteria than in eukaryotes.
    e. Positively charged proteins can combine with bacterial DNA.

Discuss the Concepts

1. Chargaff's data suggested that adenine pairs with thymine and guanine pairs with cytosine. What other data available to Watson and Crick suggested that adenine–guanine and cytosine–thymine pairs normally do not form?
2. Eukaryotic chromosomes can be labeled by exposing cells to radioactive thymidine during the S phase of interphase. If cells are exposed to radioactive thymidine during the S phase, would you expect both or only one of the sister chromatids of a duplicated chromosome to be labeled at metaphase of the following mitosis (see Section 10.2)?
3. If the cells in question 2 finish division and then enter another round of DNA replication in a medium that has been washed free of radioactive label, would you expect both or only one of the sister chromatids of a duplicated chromosome to be labeled at metaphase of the following mitosis?
4. During replication, an error uncorrected by proofreading or mismatch repair produces a DNA molecule with a base mismatch at the indicated position:

   ```
   AATTCCGACTCCTATGG
   TTAAGGTTGAGGATACC
          ↑
   ```

   The mismatch results in a mutation. This DNA molecule is received by one of the two daughter cells produced by mitosis. In the next round of replication and division, the mutation appears in only one of the two daughter cells. Develop a hypothesis to explain this observation.
5. Strains of bacteria that are resistant to an antibiotic sometimes appear spontaneously among other bacteria of the same type that are killed by the antibiotic. In view of the information in this chapter about DNA replication, what might account for the appearance of this resistance?

Design an Experiment

Design an experiment using radioactive isotopes to show that the process of bacterial transformation involves DNA and not protein.

Interpret the Data

Some cancer treatments target rapidly dividing cells while leaving non-proliferating cells undisturbed. The chemicals 5-fluorouracil (5-FU) and cisplatin (CDDP) are drugs that work in this way. 5-FU inhibits DNA replication, while CDDP binds to DNA causing changes that cannot be corrected by DNA repair enzymes so that programmed cell death (see Chapter 7) is triggered. Researchers suspected that these drugs might be useful in treating human gastric cancer and tested their effectiveness in gastric cancer cells growing in culture. They added 5-FU alone, or 5-FU and CDDP in various timed combinations (schedules) to cultured gastric cancer cells and measured the inhibitory effects of the drugs on cell proliferation compared with untreated cells. The results in the figure to the right shows for each schedule the % cell proliferation, meaning the proliferation of treated cells/proliferation of control, untreated cells × 100%.

1. Which drug schedule was the most effective?
2. How did the drug schedule in the most effective treatment differ from the schedules in all the other treatments?

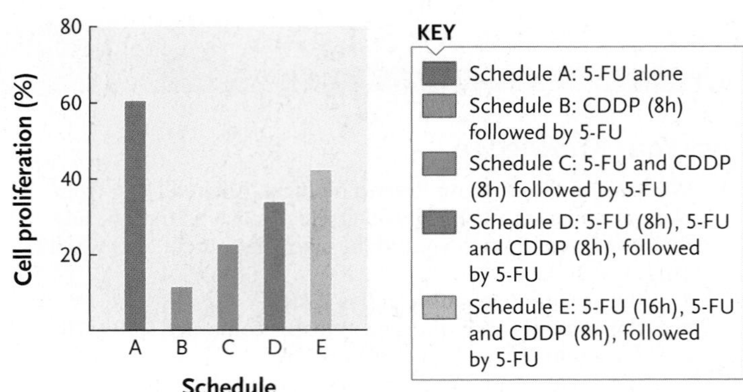
Cell proliferation (%)
80
60
40
20
A
B
C
D
E
Schedule
KEY
Schedule A: 5-FU alone
Schedule B: CDDP (8h) followed by 5-FU
Schedule C: 5-FU and CDDP (8h) followed by 5-FU
Schedule D: 5-FU (8h), 5-FU and CDDP (8h), followed by 5-FU
Schedule E: 5-FU (16h), 5-FU and CDDP (8h), followed by 5-FU

Source: H. Cho, *et al.* 2002. "In-vitro effect of a combination of 5-fluorouracil (5-FU) and cisplatin (CDDP) on human gastric cancer cell lines: timing of cisplatin treatment." *Gastric Cancer* 5:43–46.

Apply Evolutionary Thinking

The amino acid sequences of the DNA polymerases found in bacteria show little similarity to those of the DNA polymerases found in eukaryotes and in archaea. By contrast, the amino acid sequences of the DNA polymerases of eukaryotes and archaea show a high degree of similarity. Interpret these observations from an evolutionary point of view.

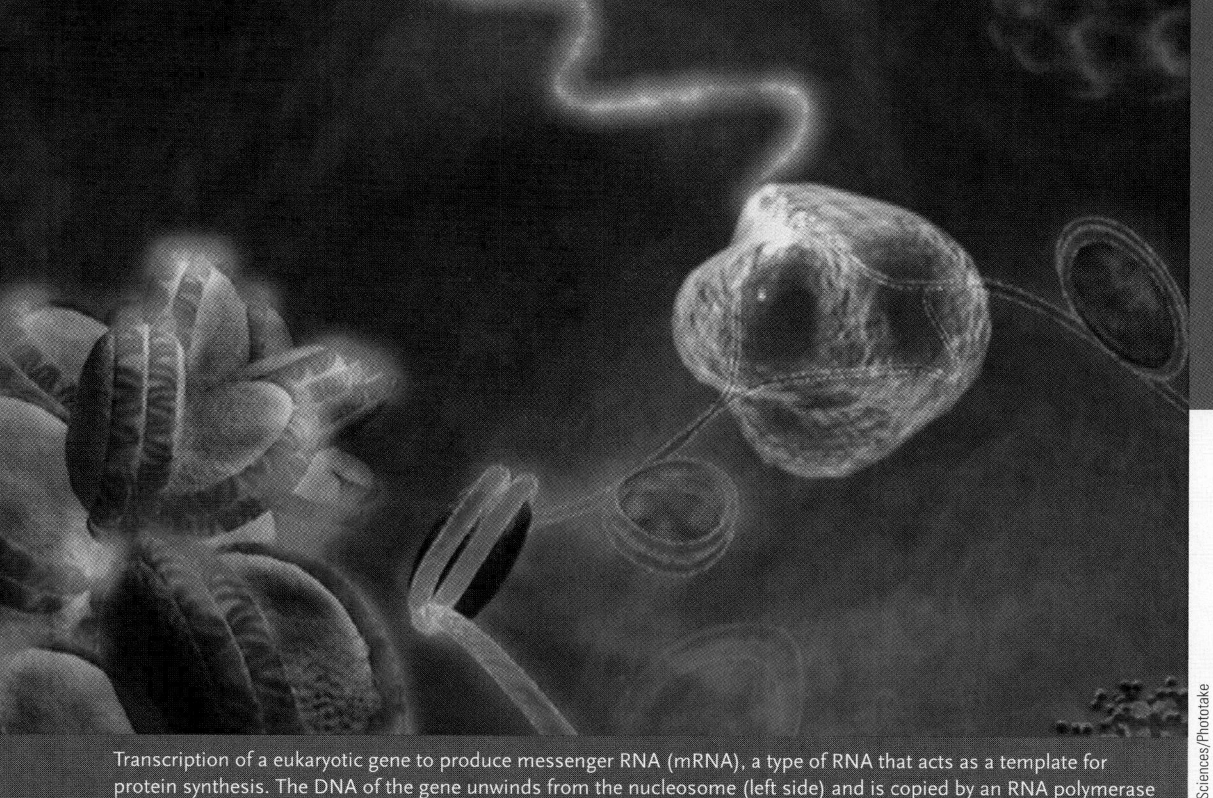

Transcription of a eukaryotic gene to produce messenger RNA (mRNA), a type of RNA that acts as a template for protein synthesis. The DNA of the gene unwinds from the nucleosome (left side) and is copied by an RNA polymerase (center) into mRNA (exiting the top).

LookatSciences/Phototake

15

STUDY OUTLINE

From DNA to Protein

Why It Matters. . . The marine mussel *Mytilus* **(Figure 15.1)** lives in one of the most demanding environments on Earth—it clings permanently to rocks pounded by surf day in and day out, constantly in danger of being dashed to pieces or torn loose by foraging predators. The mussel is remarkably resistant to disturbance, however; if you try to pry one loose you will find how difficult it is to tear the tough, flexible fibers that hold it fast, or even to cut them with a knife.

The fibers holding mussels to the rocks are a complex of proteins secreted by the muscular foot of the animal. The proteins, which include *keratin* (an intermediate filament protein; discussed in Section 5.3) form a tough, adhesive material called *byssus*.

Byssus is a premier underwater adhesive. It interests biochemists, adhesive manufacturers, dentists, and surgeons looking for better ways to hold repaired body parts together. Genetic engineers have inserted segments of mussel DNA into yeast cells, which reproduce in large numbers and serve as "factories" translating the mussel genes into byssus and other proteins. With byssus produced in this way, investigators are learning how to use or imitate the mussel glue for human needs. This work, like the mussel's own byssus-building, starts with one of life's universal truths: *every protein is assembled on ribosomes according to instructions that are copied from DNA.*

In this chapter we trace the reactions by which proteins are made, beginning with the instructions encoded in DNA and leading through RNA to the sequence of amino acids in a protein. Many enzymes and other proteins are players as well as products in this story, as are several kinds of RNA and the cell's protein-making molecular machines, the ribosomes. The same basic

FIGURE 15.1
The marine mussel *Mytilus*.

© WildPictures/Alamy

FIGURE 15.2 Experimental Research

Relationship between Genes and Enzymes

Question: Are enzymes specified by genes?

Experiment: Beadle and Tatum isolated auxotrophic mutants of the orange bread mold *Neurospora crassa*. Auxotrophic mutants require a nutritional supplement added to MM (minimal medium) to grow. They analyzed arginine auxotrophs—*arg* mutants—to determine the relationship between genes and enzymes. The wild type grows on MM, whereas *arg* mutants cannot. All *arg* mutants grow if arginine is added to the medium. Beadle and Tatum tested whether the *arg* mutants could also grow on MM supplemented with ornithine, citrulline, or argininosuccinate, three compounds known to be involved in the synthesis of arginine.

Results:

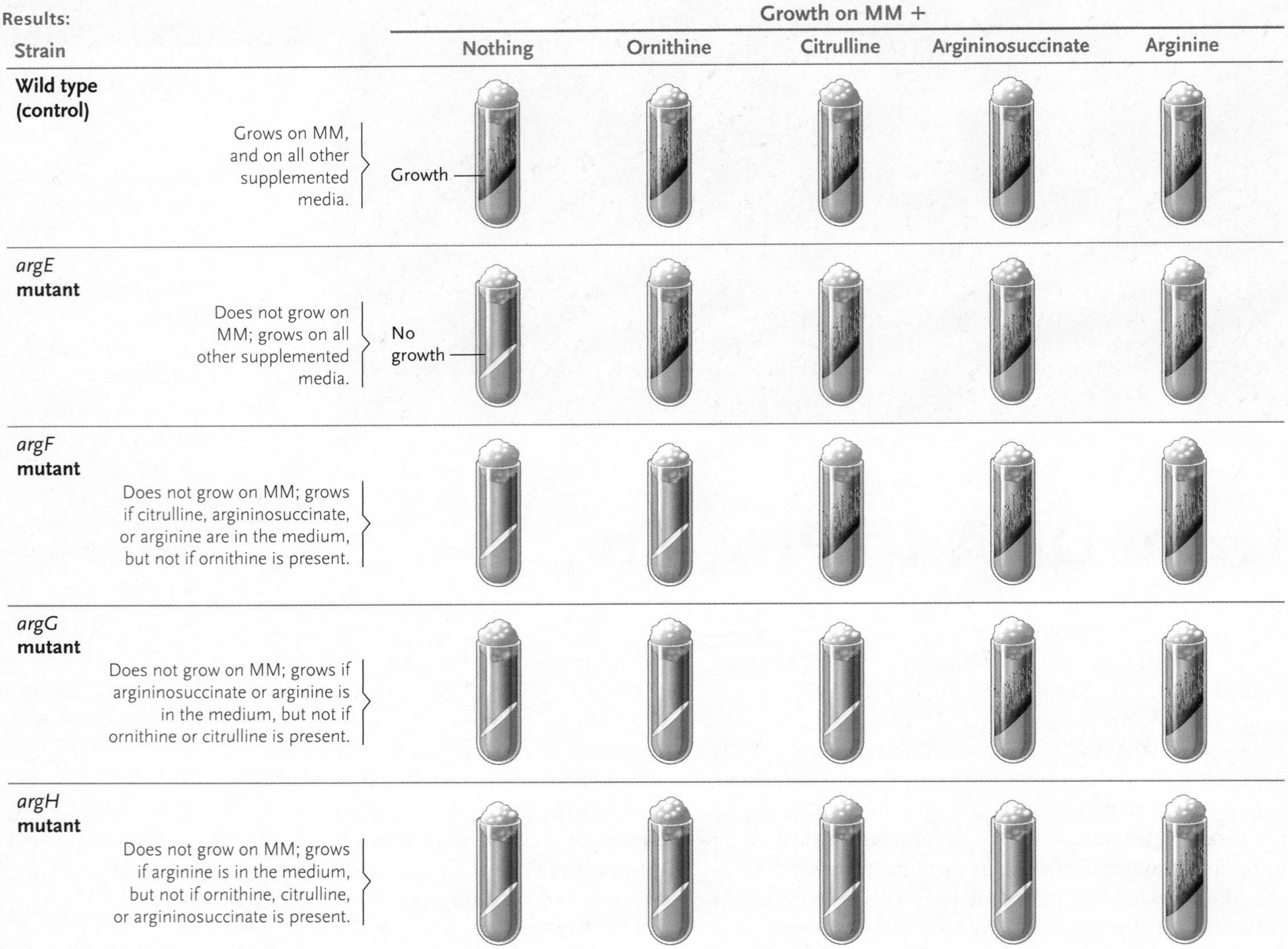

Growth on MM +

Nothing | Ornithine | Citrulline | Argininosuccinate | Arginine

Strain	
Wild type (control)	Grows on MM, and on all other supplemented media.
argE **mutant**	Does not grow on MM; grows on all other supplemented media.
argF **mutant**	Does not grow on MM; grows if citrulline, argininosuccinate, or arginine are in the medium, but not if ornithine is present.
argG **mutant**	Does not grow on MM; grows if argininosuccinate or arginine is in the medium, but not if ornithine or citrulline is present.
argH **mutant**	Does not grow on MM; grows if arginine is in the medium, but not if ornithine, citrulline, or argininosuccinate is present.

Growth

No growth

steps produce the proteins of all organisms. Our discussion begins with an overview of the entire process, starting with DNA and ending with a finished protein. <

15.1 The Connection between DNA, RNA, and Protein

You have learned that genes encode—specify the amino acid sequence of—proteins. In this section you will learn how that connection was discovered. This section also presents an overview of the molecular steps from gene to protein: transcription and translation.

Proteins Are Specified by Genes

How do scientists know that genes code for proteins? Two key pieces of research involving defects in metabolism proved this connection unequivocally. The first began in 1896 with Archibald Garrod, an English physician. He studied *alkaptonuria,* a human disease that does little harm but is easily detected: The patient's urine turns black in air. Garrod and an English geneticist, William Bateson, studied families of patients with the disease and con-

Conclusion: Each of the *arg* mutants shows a different pattern of growth on the supplemented MM. Beadle and Tatum deduced that the biosynthesis of arginine occurs in a number of steps, with each step controlled by a gene that encodes the enzyme for the step.

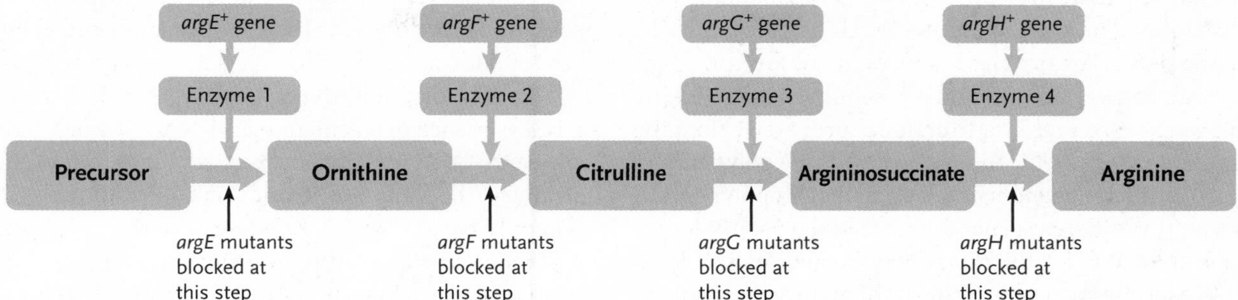

argE+ gene
argF+ gene
argG+ gene
argH+ gene
Enzyme 1
Enzyme 2
Enzyme 3
Enzyme 4
Precursor
Ornithine
Citrulline
Argininosuccinate
Arginine
argE mutants blocked at this step
argF mutants blocked at this step
argG mutants blocked at this step
argH mutants blocked at this step

The logic is as follows, working from the end of the pathway back to its beginning:

- The *argH* mutant grows on MM + arginine, but not on MM + any of the other three compounds; this means that the mutant is blocked at the last step in the pathway that produces arginine.
- The *argG* mutant grows on MM + arginine or argininosuccinate, but not on MM + any of the other supplements; this means that *argG* is blocked in the pathway before argininosuccinate is made.
- Similarly, the *argF* mutant's growth pattern shows that it is blocked in the pathway before citrulline is made, and the *argE* mutant's growth pattern shows that it is blocked in the pathway before ornithine is made.

In this way Beadle and Tatum worked out the genetic control of the arginine biosynthesis pathway. With this and similar studies of other types of auxotrophs, they showed that there is a direct relationship between genes and enzymes.

Source: G. W. Beadle and E. L. Tatum. 1942. Genetic control of biochemical reactions in *Neurospora*. *Proceedings of the National Academy of Sciences USA* 27:499–506.

cluded that it is an inherited trait. Garrod also found that people with alkaptonuria excrete a particular chemical in their urine. It is this chemical that turns black in air. Garrod deduced that normal people can metabolize the chemical, whereas people with alkaptonuria cannot. In 1908 Garrod concluded that the disease was an *inborn error of metabolism.* He did not know it at the time, but alkaptonuria results from a change in a gene that encodes an enzyme that metabolizes a key chemical. The altered gene causes a defect in the function of the enzyme, which leads to the disease phenotype. Garrod's work was the first evidence of a specific relationship between genes and metabolism.

In the second piece of research, George Beadle and Edward Tatum, working at Stanford University in the 1940s with the orange bread mold *Neurospora crassa,* obtained results showing a direct relationship between genes and enzymes. Beadle and Tatum chose *Neurospora* for their work because it is a haploid fungus, with simple nutritional needs. That is, wild-type *Neurospora*—the form of the mold found in nature—grows readily on a minimal medium (MM) consisting of a number of inorganic salts, sucrose, and vitamins. The researchers reasoned that the fungus uses the simple chemicals in MM to synthesize all of the more complex molecules needed for growth and reproduction, including amino acids for proteins and nucleotides for DNA and RNA.

Beadle and Tatum exposed spores of wild-type *Neurospora* to X-rays. An X-ray is a *mutagen,* an agent that causes mutations. They found that some of the treated spores would not germinate and grow on MM unless they supplemented the medium with additional nutrients, such as amino acids or vitamins. Mutant strains that require a nutrient supplement in the MM to grow are called **auxotrophs** (*auxo* = increased; *troph* = eater) or *nutritional mutants.* Beadle and Tatum hypothesized that each auxotrophic strain had a defect in a gene that codes for an enzyme needed to synthesize a particular nutrient. The wild-type strain could make the nutrient for itself from raw materials in the MM, but the mutant strain could grow only if the researchers supplied the nutrient. By testing each mutant strain on MM with a single added nutrient, they discovered what specific nutrient the strain needed to grow and, therefore, generally what gene defect it had. For example, a mutant that requires the addition of the amino acid arginine to grow has a defect in a gene for an enzyme involved in the synthesis of arginine. Such arginine auxotrophs are known as *arg* mutants.

The synthesis of arginine in the cell from raw materials is a multistep "assembly-line" process with a different enzyme catalyzing each step. Beadle and Tatum studied four *arg* mutants—*argE, argF, argG,* and *argH*—to determine the metabolic defect each had; that is, where in the arginine synthesis pathway each was blocked. Their experimental approach was to test whether the *arg* mutants, all of which could grow on MM + arginine but not on MM, could also grow on MM supplemented with compounds known to be involved in arginine synthesis **(Figure 15.2).** Their analysis of *arg* auxotrophs and of auxotrophs of other kinds demonstrated a direct relationship between genes and enzymes, which they put forward as the **one gene–one enzyme hypothesis.** Their experiment was a keystone in the development of molecular biology. As a result of their work, they were awarded a Nobel Prize in 1958.

As you learned in Chapter 3, enzymes are just one form of proteins, the amino acid-containing macromolecules that carry out many vital functions in living organisms. A functional protein consists of one or more subunits, called *polypeptides.* The protein hemoglobin, for instance, is made up of four polypeptides, two each of an α subunit and a β subunit. Hemoglobin's ability to transport oxygen is a functional property belonging only to the complete protein, and not to any of the polypeptides individually. A different gene encodes each distinct polypeptide, meaning that two different genes are needed to specify the hemoglobin protein: one for the α polypeptide and one for the β polypeptide. Since some proteins consist of more than one polypeptide, and not all proteins are enzymes, Beadle and Tatum's hypothesis was updated to the **one gene–one polypeptide hypothesis.** It is important to keep in mind the distinction between a protein, the functional molecule, and a polypeptide, the molecule specified by a gene, as we discuss transcription and translation in the rest of this chapter.

The Pathway from Gene to Polypeptide Involves Transcription and Translation

The pathway from gene to polypeptide has two major steps, *transcription* and *translation.* **Transcription** is the mechanism by which the information encoded in DNA is made into a complementary RNA copy. It is called transcription because the information in one nucleic acid type is transferred to another nucleic acid type. **Translation** is the use of the information encoded in the RNA to assemble amino acids into a polypeptide. It is called translation because the information in a nucleic acid, in the form of nucleotides, is converted into a different kind of molecule—amino acids. In 1956, Francis Crick gave the name **central dogma** to the flow of information from DNA → RNA → protein.

In transcription, the enzyme RNA polymerase copies the DNA sequence of a gene into an RNA sequence. The process is similar to DNA replication, except that only one of the two DNA strands—the **template strand**—is copied into an RNA strand, and only part of the DNA sequence of the genome is copied in any cell at any given time. A gene encoding a polypeptide is a **protein-coding gene,** and the RNA transcribed from it is called **messenger RNA (mRNA).**

Some genes do not encode a polypeptide. Instead, they encode various molecules that play roles in transcription and translation, and in some other processes in the cell.

In translation, an mRNA associates with a *ribosome,* a particle on which amino acids are linked into polypeptide chains. As the ribosome moves along the mRNA, the amino acids specified by the mRNA are joined one by one to form the polypeptide encoded by the gene.

Transcription and translation occur in all organisms. Both processes are similar in prokaryotes and eukaryotes but there are differences **(Figure 15.3).** One key difference is that in eukaryotes, transcription in the nucleus produces a precursor-mRNA

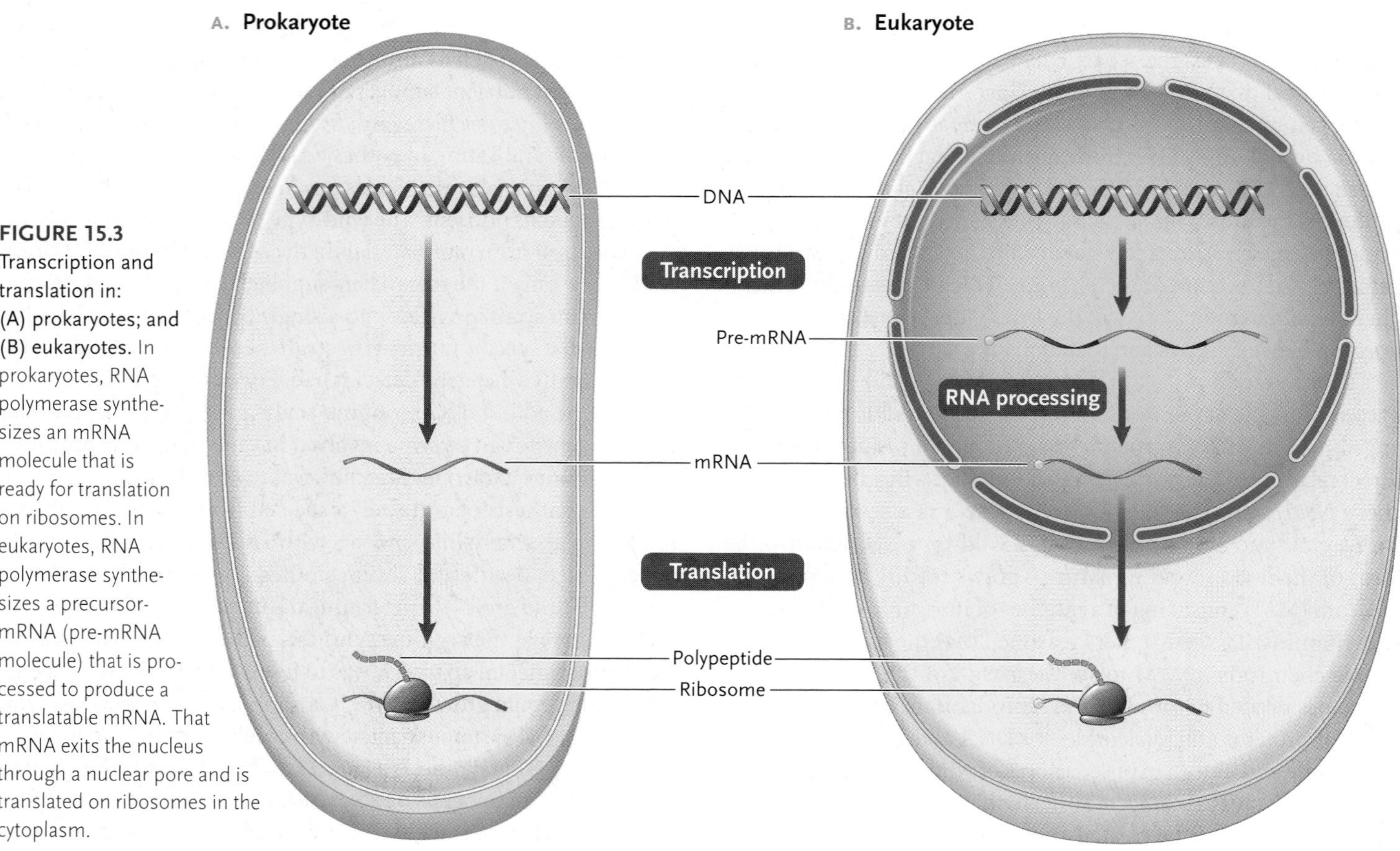

A. Prokaryote
B. Eukaryote
DNA
Transcription
Pre-mRNA
RNA processing
mRNA
Translation
Polypeptide
Ribosome

FIGURE 15.3 Transcription and translation in: (A) prokaryotes; and (B) eukaryotes. In prokaryotes, RNA polymerase synthesizes an mRNA molecule that is ready for translation on ribosomes. In eukaryotes, RNA polymerase synthesizes a precursor-mRNA (pre-mRNA molecule) that is processed to produce a translatable mRNA. That mRNA exits the nucleus through a nuclear pore and is translated on ribosomes in the cytoplasm.

(pre-mRNA) that must be altered to generate the functional mRNA. Specifically, each end of the pre-mRNA is modified, and then extra segments within its sequence are removed by RNA processing. The result is the functional mRNA that exits the nucleus and is translated in the cytoplasm. In prokaryotes, transcription in the cytoplasm produces a functional mRNA directly, with no modifications.

The Genetic Code Is Written in Three-Letter Words Using a Four-Letter Alphabet

Conceptually, the transcription of DNA into RNA is straightforward. The DNA "alphabet" consists of the four letters A, T, G, and C, representing the four bases of DNA nucleotides: adenine, thymine, guanine, and cytosine. The RNA "alphabet" consists of the four letters A, U, G, and C, representing the four RNA bases: adenine, uracil, guanine, and cytosine. In other words, the nucleic acids share three of the four bases but differ in the other one; T in DNA is equivalent to U in RNA. But although there are four RNA bases, there are 20 amino acids. How is nucleotide information in an mRNA translated into the amino acid sequence of a polypeptide?

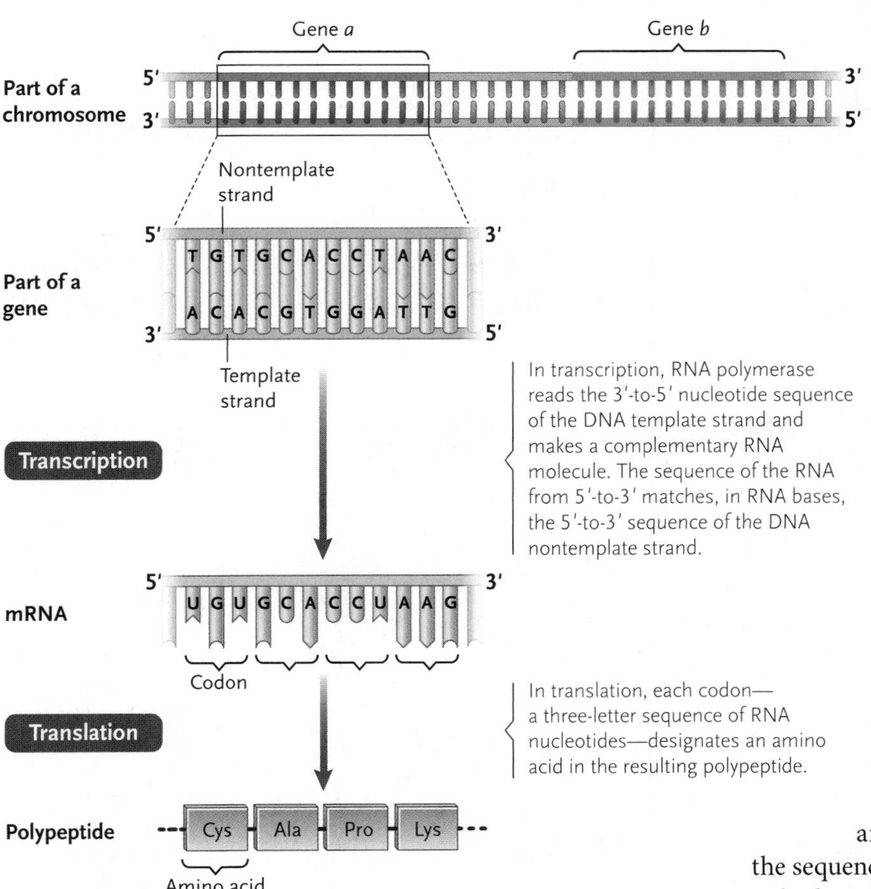

Gene *a*
Gene *b*
Part of a chromosome
5'
3'
3'
5'
Nontemplate strand
5'
3'
T G T G C A C C T A A C
Part of a gene
A C A C G T G G A T T G
3'
5'
Template strand
Transcription
In transcription, RNA polymerase reads the 3'-to-5' nucleotide sequence of the DNA template strand and makes a complementary RNA molecule. The sequence of the RNA from 5'-to-3' matches, in RNA bases, the 5'-to-3' sequence of the DNA nontemplate strand.
5'
3'
mRNA
U G U G C A C C U A A G
Codon
Translation
In translation, each codon—a three-letter sequence of RNA nucleotides—designates an amino acid in the resulting polypeptide.
Polypeptide
Cys
Ala
Pro
Lys
Amino acid

KEY

Cys = cysteine	Pro = proline
Ala = alanine	Lys = lysine

FIGURE 15.4
Relationship between a gene, codons in an mRNA, and the amino acid sequence of a polypeptide.

BREAKING THE GENETIC CODE The nucleotide information that specifies the amino acid sequence of a polypeptide is called the **genetic code.** Scientists realized that the four bases in an mRNA (A, U, G, and C) would have to be used in combinations of at least three to provide the capacity to code for 20 different amino acids. One- and two-letter words were eliminated because if the code used one-letter words, only four different amino acids could be specified (that is, 4^1); if two-letter words were used, only 16 different amino acids could be specified (that is, 4^2). But if the code used three-letter words, 64 different amino acids could be specified (that is, 4^3), more than enough to specify 20 amino acids. Experimental research showed that the genetic code is a three-letter code; each three-letter word (triplet) of the code is called a **codon. Figure 15.4** illustrates the relationship between codons in a gene, codons in an mRNA, and the amino acid sequence of a polypeptide. The three-letter codons in DNA are first transcribed into complementary three-letter RNA codons. The process is similar to DNA replication except that in mRNA, the complement to adenine (A) in the template strand is uracil (U) instead of thymine (T) as in DNA replication.

How do the RNA codons correspond to the amino acids? The identity of most of the codons was established in 1964 by Marshall Nirenberg and Philip Leder of the National Institutes of Health (NIH). These researchers found that short, artificial mRNAs of codon length—three nucleotides—could bind to ribosomes in a test tube and cause a single transfer RNA (tRNA), with its linked amino acid, to bind to the ribosome. (As you will learn in Section 15.4, tRNAs are a special class of RNA molecules that bring amino acids to the ribosome for assembly into the polypeptide chain.) Nirenberg and Leder then made 64 of the short mRNAs, each consisting of a different, single codon. They added the mRNAs, one at a time, to a test tube containing ribosomes and all the different tRNAs, each linked to its own amino acid. The idea was that each single-codon mRNA would link to the tRNA in the mixture that carried the amino acid corresponding to the codon. The experiment worked for 50 of the 64 codons, allowing those codons to be assigned to amino acids definitively.

Another approach, carried out in 1966 by H. Gobind Khorana and his coworkers at Massachusetts Institute of Technology, used long, artificial mRNA molecules containing only one nucleotide repeated continuously, or different nucleotides in repeating patterns. The researchers added each artificial mRNA to ribosomes in a test tube, and analyzed the sequence of amino acids in the polypeptide chain made by the ribosomes. For example, an artificial mRNA containing only uracil nucleotides in the sequence UUUUUU. . . resulted in a polypeptide containing only the amino acid phenylalanine: UUU must be the codon for phenylalanine. Khorana's approach, combined with the results of Nirenberg and Leder's experiments, identified the coding assignments of all the codons. Nirenberg and Khorana received a Nobel Prize in 1968 for their research in solving the nucleic acid code.

FEATURES OF THE GENETIC CODE **Figure 15.5** shows the genetic code of the 64 possible codons. By convention, scientists write the codons in the 5′→3′ direction, as they appear in mRNAs, in which U substitutes for the T of DNA. Of the 64 codons, 61 specify amino acids. These are known as **sense codons.** One of these codons, AUG, specifying the amino acid methionine, is the first codon read in an mRNA in translation in both prokaryotes and eukaryotes. In that position, AUG is called a **start codon** or **initiator codon.** The three codons that do not specify amino acids—UAA, UAG, and UGA—are **stop codons** (also called **nonsense codons** and **termination codons**) that act as "periods" indicating the end of a polypeptide-encoding "sentence." When a ribosome reaches one of the stop codons, polypeptide synthesis stops and the new polypeptide chain is released from the ribosome.

Only two amino acids, methionine and tryptophan, are specified by a single codon. All the rest are represented by more than one codon, some by as many as six. In other words, there are many *synonyms* in the nucleic acid code, a feature known as **degeneracy** (also called *redundancy*). For example, UGU and UGC both specify cysteine, and CCU, CCC, CCA, and CCG all specify proline.

The genetic code is also **commaless;** that is, the words of the nucleic acid code are sequential, with no indicators such as commas or spaces to mark the end of one codon and the beginning of the next. The code can be read correctly only by starting at the right place—at the first base of the first three-letter codon at the beginning of a coded message—and reading three nucleotides at a time from this beginning codon. In other words, there is only one correct *reading frame* for each mRNA. By analogy, if you read the message SADMOMHASMOPCUTOFFBOYTOT three letters at a time, starting with the first letter of the first "codon," you would find that a mother reluctantly had her small child's hair cut. However, if you start incorrectly at the second letter of the first codon, you read the gibberish message ADM OMH ASM OPC UTO FFB OYT OT.

The code is **universal.** With a few exceptions, the same codons specify the same amino acids in all living organisms, and also in viruses. In other words, the eukaryotic translation machinery can read a prokaryotic mRNA to make the same polypeptide as in the prokaryote, and vice versa. The universality of the nucleic acid code indicates that it was established in its present form very early in the evolution of life and has remained virtually unchanged since then. (The evolution of life and the genetic code are discussed further in Chapter 24.) Minor exceptions to the universality of the genetic code have been found in a few organisms including a yeast, some protozoans, a prokaryote, and in the genetic systems of mitochondria and chloroplasts.

STUDY BREAK 15.1

1. **On the basis of their work with auxotrophic mutants of the fungus *Neurospora crassa*, Beadle and Tatum proposed the one gene–one enzyme hypothesis. Why is it now known as the one gene–one polypeptide hypothesis?**
2. **If the codon were five bases long, how many different codons would exist in the genetic code?**

THINK OUTSIDE THE BOOK

The section you have just read states that the genetic code is not completely universal. Use the Internet or research literature to determine what variants of the genetic code exist and in which organisms they occur.

15.2 Transcription: DNA-Directed RNA Synthesis

An organism's genome contains a large number of genes. For example, the human genome sequence has about 20,000 protein-coding genes. Transcription is the process of transferring the in-

Second base of codon

First base of codon	U	C	A	G	Third base of codon
U	UUU, UUC: Phe UUA, UUG: Leu	UCU, UCC, UCA, UCG: Ser	UAU, UAC: Tyr UAA UAG	UGU, UGC: Cys UGA UGG: Trp	U C A G
C	CUU, CUC, CUA, CUG: Leu	CCU, CCC, CCA, CCG: Pro	CAU, CAC: His CAA, CAG: Gln	CGU, CGC, CGA, CGG: Arg	U C A G
A	AUU, AUC, AUA: Ile AUG: Met	ACU, ACC, ACA, ACG: Thr	AAU, AAC: Asn AAA, AAG: Lys	AGU, AGC: Ser AGA, AGG: Arg	U C A G
G	GUU, GUC, GUA, GUG: Val	GCU, GCC, GCA, GCG: Ala	GAU, GAC: Asp GAA, GAG: Glu	GGU, GGC, GGA, GGG: Gly	U C A G

KEY

- Ala = alanine
- Arg = arginine
- Asn = asparagine
- Asp = aspartic acid
- Cys = cysteine
- Gln = glutamine
- Glu = glutamic acid
- Gly = glycine
- His = histidine
- Ile = isoleucine
- Leu = leucine
- Lys = lysine
- Met = methionine
- Phe = phenylalanine
- Pro = proline
- Ser = serine
- Thr = threonine
- Trp = tryptophan
- Tyr = tyrosine
- Val = valine

FIGURE 15.5
The genetic code, written in the form in which the codons appear in mRNA. The AUG initiator codon, which codes for methionine, is shown in green; the three terminator codons are boxed in red.

formation coded in the DNA sequences of particular genes to complementary RNA copies. Some of those genes are protein-coding genes that encode mRNAs that are translated; others are non-protein-coding genes that encode RNAs that are not translated, such as ribosomal RNAs (rRNAs) and transfer RNAs (tRNAs). Much of our initial understanding of transcription came from experimental research done with the model organism, *Escherichia coli.* We now know that transcription is generally similar, but not identical, in prokaryotes and eukaryotes. Throughout this section, we will point out the important differences between bacterial and eukaryotic processes.

RNA Transcription Is Similar to DNA Replication

The process of RNA transcription is similar to DNA replication. Like DNA, RNA is made in the 5′→3′ direction using the 3′→5′ DNA strand as template. Thus, we refer to the beginning of the RNA strand as the *5′ end,* and the other end as the *3′ end.* However, there are some important differences between replication and transcription. In transcription:

- Only one of the two DNA nucleotide strands acts as a template for synthesis of a complementary copy, instead of both as in replication.
- Only a relatively small part of a DNA molecule—the sequence encoding a single gene—serves as a template, rather than all of both strands as in DNA replication.
- **RNA polymerases** catalyze the assembly of RNA nucleotides into an RNA strand, rather than the DNA polymerases that catalyze the assembly of DNA nucleotides into a DNA strand in replication. RNA polymerases require a specific DNA sequence to direct them where to begin transcription, but they do not require a primer as DNA polymerases do.
- The RNA molecules resulting from transcription are single polynucleotide chains, not double ones as in DNA replication.
- Where adenine appears in the DNA template chain, a uracil is matched to it in the RNA transcript instead of thymine as in DNA replication (see Figure 15.4).

Transcription Proceeds in Three Stages

Figure 15.6 illustrates the general structure of a eukaryotic protein-coding gene and shows how it is transcribed. The gene consists of two main parts, a **promoter,** which is a control sequence for transcription, and a **transcription unit,** the section of the gene that is copied into an RNA molecule. Transcription takes place in three stages. The first stage is initiation, in which the molecular machinery that carries out transcription assembles at the promoter and begins synthesizing an RNA copy of the gene. The second stage is elongation, in which the RNA polymerase moves along the gene extending the RNA chain; and the third stage is termination, in which transcription ends and the RNA molecule—the transcript—and the RNA polymerase are released from the DNA template. Roger Kornberg of Stanford University received a Nobel Prize in 2006 for describing the molecular structure of the eukaryotic transcription apparatus and how it acts in transcription.

Similarities and differences in transcription of eukaryotic and bacterial protein-coding genes are as follows:

- Gene organization is the same, although the specific sequences in the promoter where the transcription apparatus assembles differ.
- In eukaryotes, RNA polymerase II, the enzyme that transcribes protein-coding genes, cannot bind directly to DNA; it is recruited to the promoter once proteins called **transcription factors** have first bound. In bacteria, RNA polymerase binds directly to DNA; it is directed to the promoter by a protein factor that is then released once transcription begins.
- Elongation is essentially identical in the two types of organisms.
- In bacteria, specific DNA sequences called **terminators** signal the end of transcription of the gene. A specific protein binds to the terminator, triggering the termination of transcription and the release of the RNA and RNA polymerase from the template. Eukaryotic DNA has no equivalent sequences. Instead, the 3′ end of the mRNA is specified by a very different process, which is discussed in the next section.

Once an RNA polymerase molecule has started transcription and progressed past the beginning of a gene, another molecule of RNA polymerase may start transcribing as soon as there is room at the promoter. In most genes this process continues until there are many RNA polymerase molecules spaced closely along a gene, each making an RNA transcript.

Transcription of Non-Protein-Coding Genes Occurs in a Similar Way

Non-protein-coding genes include those for tRNAs and rRNAs (ribosomal RNAs, the RNA components of ribosomes). In eukaryotes, whereas RNA polymerase II transcribes protein-coding genes, RNA polymerase III transcribes tRNA genes and the gene for one of the four rRNAs, and RNA polymerase I transcribes the genes for the three other rRNAs. The promoters for these non-protein-coding genes are different from those of protein-coding genes, being specialized for the assembly of the transcription machinery that involves the correct RNA polymerase type. In bacteria, only a single RNA polymerase type exists, and it transcribes all types of genes. The promoters for bacterial non-protein-coding genes are essentially the same as those of protein-coding genes.

STUDY BREAK 15.2 <

1. **For the DNA template below, what would be the sequence of an RNA transcribed from it?**

   3′-CAAATTGGCTTATTACCGGATG-5′
2. **What is the role of the promoter in transcription?**

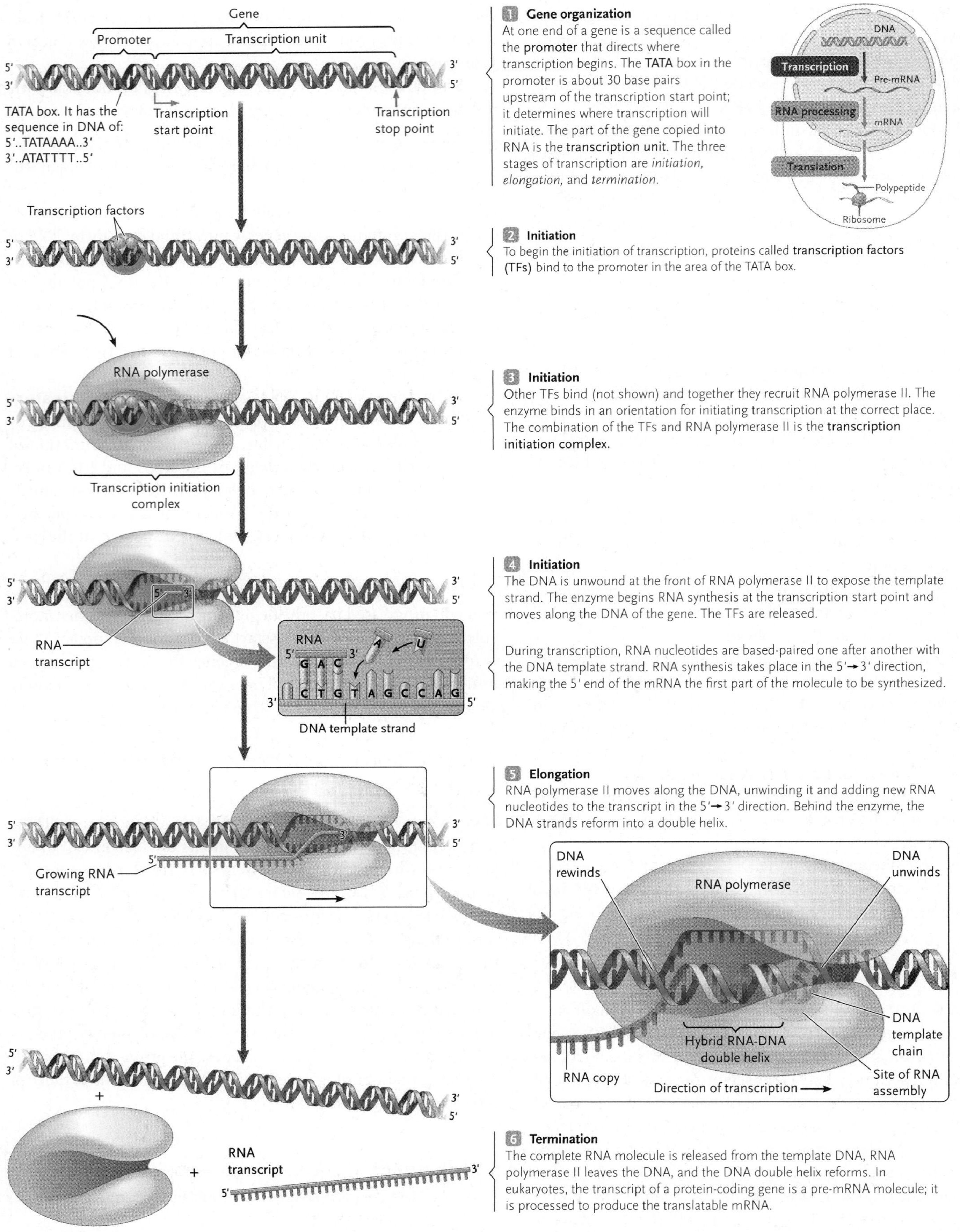
Gene
Promoter
Transcription unit
5'
3'
3'
5'
TATA box. It has the sequence in DNA of:
5'..TATAAAA..3'
3'..ATATTTT..5'
Transcription start point
Transcription stop point
Transcription factors
RNA polymerase
Transcription initiation complex
RNA transcript
RNA
5'
3'
G A C
A
U
C T G T A G C C A G
DNA template strand
Growing RNA transcript
RNA transcript
DNA
Transcription
Pre-mRNA
RNA processing
mRNA
Translation
Polypeptide
Ribosome
DNA rewinds
DNA unwinds
RNA polymerase
Hybrid RNA-DNA double helix
DNA template chain
RNA copy
Direction of transcription
Site of RNA assembly

1 Gene organization
At one end of a gene is a sequence called the **promoter** that directs where transcription begins. The **TATA** box in the promoter is about 30 base pairs upstream of the transcription start point; it determines where transcription will initiate. The part of the gene copied into RNA is the **transcription unit.** The three stages of transcription are *initiation, elongation,* and *termination.*

2 Initiation
To begin the initiation of transcription, proteins called **transcription factors (TFs)** bind to the promoter in the area of the TATA box.

3 Initiation
Other TFs bind (not shown) and together they recruit RNA polymerase II. The enzyme binds in an orientation for initiating transcription at the correct place. The combination of the TFs and RNA polymerase II is the **transcription initiation complex.**

4 Initiation
The DNA is unwound at the front of RNA polymerase II to expose the template strand. The enzyme begins RNA synthesis at the transcription start point and moves along the DNA of the gene. The TFs are released.

During transcription, RNA nucleotides are based-paired one after another with the DNA template strand. RNA synthesis takes place in the 5'→3' direction, making the 5' end of the mRNA the first part of the molecule to be synthesized.

5 Elongation
RNA polymerase II moves along the DNA, unwinding it and adding new RNA nucleotides to the transcript in the 5'→3' direction. Behind the enzyme, the DNA strands reform into a double helix.

6 Termination
The complete RNA molecule is released from the template DNA, RNA polymerase II leaves the DNA, and the DNA double helix reforms. In eukaryotes, the transcript of a protein-coding gene is a pre-mRNA molecule; it is processed to produce the translatable mRNA.

FIGURE 15.6
Transcription of a eukaryotic protein-coding gene. Transcription has three stages: initiation, elongation, and termination. RNA polymerase moves along the gene, separating the two DNA strands to allow RNA synthesis in the 5'→3' direction using the 3'→5' DNA strand as template.

15.3 Production of mRNAs in Eukaryotes

Both prokaryotic and eukaryotic mRNAs contain regions that code for proteins, along with noncoding regions that play key roles in the process of protein synthesis. In prokaryotic mRNAs, the coding region is flanked by untranslated ends, the **5′ untranslated region (5′ UTR)** and the **3′ untranslated region (3′ UTR).** The same elements are present in eukaryotic mRNAs along with additional noncoding elements. The synthesis of mRNA in eukaryotes is the focus of this section.

Eukaryotic Protein-Coding Genes Are Transcribed into Precursor-mRNAs That Are Modified in the Nucleus

A eukaryotic protein-coding gene is typically transcribed into a **precursor-mRNA (pre-mRNA)** that must be processed in the nucleus to produce the translatable mRNA (**Figure 15.7;** and see Figure 15.3). The mRNA exits the nucleus and is translated in the cytoplasm.

MODIFICATIONS OF PRE-mRNA AND mRNA ENDS At the 5′ end of the pre-mRNA is the **5′ cap,** consisting of a guanine-containing nucleotide that is reversed so that its 3′-OH group faces the beginning rather than the end of the molecule. A *capping enzyme* adds the 5′ cap to the pre-mRNA soon after RNA polymerase II begins transcription. The cap, which is connected to the rest of the chain by three phosphate groups, remains when pre-mRNA is processed to mRNA. The cap is the site where ribosomes attach to mRNAs at the start of translation.

Transcription of a eukaryotic protein-coding gene is terminated differently from that of a prokaryotic gene. The eukaryotic gene has no terminator sequence that signals RNA polymerase to stop transcription. Instead, near the 3′ end of the gene is a sequence called a **polyadenylation signal** that is transcribed into the pre-mRNA. Proteins bind to this *sequence,* and cleave the pre-mRNA just downstream of that point. Then, the enzyme **poly(A) polymerase** adds a chain of 50 to 250 adenine nucleotides, one nucleotide at a time, to that 3′ end of the pre-mRNA. This string of A nucleotides, called the **poly(A) tail,** enables the mRNA produced from the pre-mRNA to be translated efficiently, and protects it from attack by RNA-digesting enzymes in the cytoplasm. If the poly(A) tail of an mRNA is removed experimentally, the mRNA is quickly degraded inside cells.

SEQUENCES INTERRUPTING THE PROTEIN-CODING SEQUENCE The transcription unit of a eukaryotic protein-coding gene—the RNA-coding sequence—also contains one or more non–protein-coding sequences called **introns** that interrupt the protein-coding sequence (shown in Figure 15.7). The introns are transcribed into pre-mRNAs, but removed from pre-mRNAs during processing in the nucleus, so that the coded messages in the finished mRNAs are read continuously, without interruptions. The amino acid-coding sequences that are retained in finished mRNAs are called **exons.**

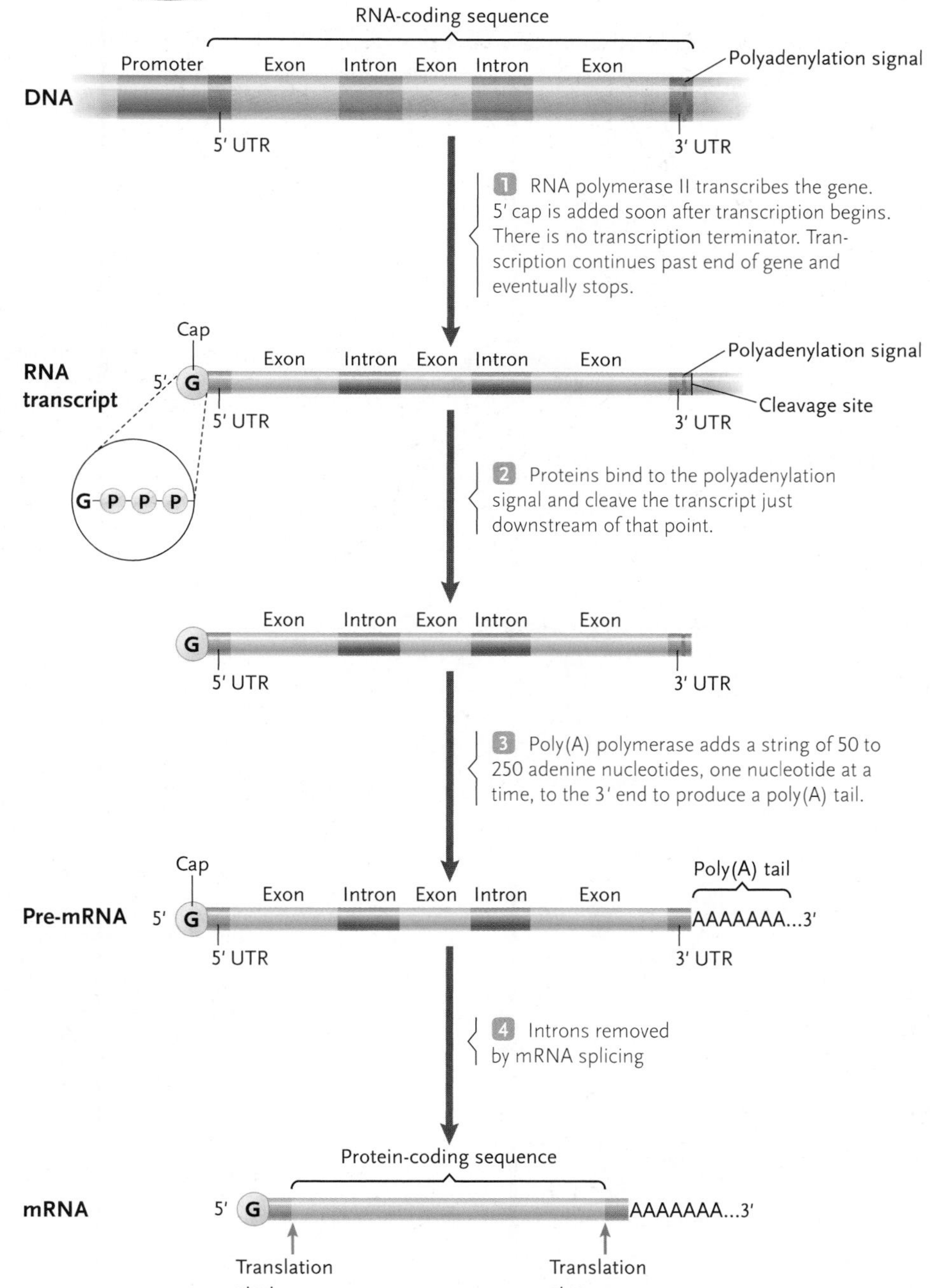

RNA-coding sequence
Promoter
Exon
Intron
Exon
Intron
Exon
Polyadenylation signal
DNA
5′ UTR
3′ UTR
1 RNA polymerase II transcribes the gene. 5′ cap is added soon after transcription begins. There is no transcription terminator. Transcription continues past end of gene and eventually stops.
Cap
RNA transcript
5′
G
G-P-P-P
Cleavage site
2 Proteins bind to the polyadenylation signal and cleave the transcript just downstream of that point.
3 Poly(A) polymerase adds a string of 50 to 250 adenine nucleotides, one nucleotide at a time, to the 3′ end to produce a poly(A) tail.
Poly(A) tail
Pre-mRNA
AAAAAAA...3′
4 Introns removed by mRNA splicing
Protein-coding sequence
mRNA
Translation start
Translation stop

FIGURE 15.7
Relationship between a eukaryotic protein-coding gene, the pre-mRNA transcribed from it, and the mRNA processed from the pre-mRNA.

Introns were discovered by several methods, including direct comparisons between the nucleotide sequences of mature mRNAs and either pre-mRNAs or the genes encoding them. The majority of known eukaryotic genes contain at least one intron; some contain more than 60. The original discoverers of introns, Richard Roberts of New England Biolabs and Phillip Sharp of Massachusetts Institute of Technology, received a Nobel Prize in 1993 for their findings.

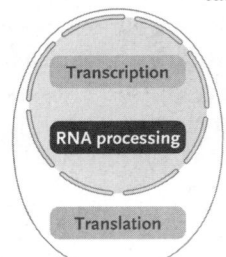

Transcription
RNA processing
Translation

FIGURE 15.8
mRNA splicing—the removal from pre-mRNA of introns and joining of exons in the spliceosome.

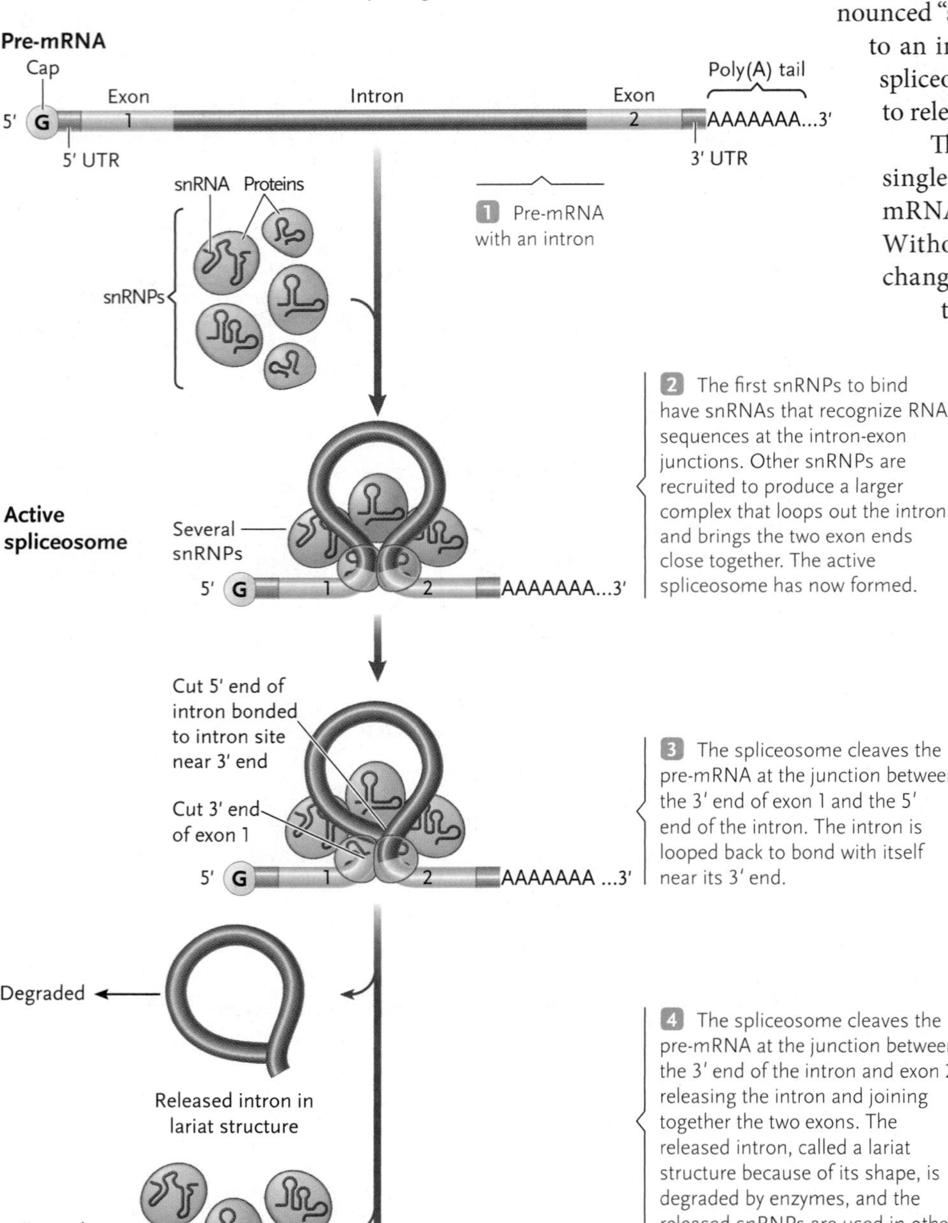

Pre-mRNA
Cap
Exon 1
Intron
Exon 2
Poly(A) tail
5′ G
AAAAAAA...3′
5′ UTR
3′ UTR
snRNA
Proteins
snRNPs
1 Pre-mRNA with an intron
2 The first snRNPs to bind have snRNAs that recognize RNA sequences at the intron-exon junctions. Other snRNPs are recruited to produce a larger complex that loops out the intron and brings the two exon ends close together. The active spliceosome has now formed.
Active spliceosome
Several snRNPs
Cut 5′ end of intron bonded to intron site near 3′ end
Cut 3′ end of exon 1
3 The spliceosome cleaves the pre-mRNA at the junction between the 3′ end of exon 1 and the 5′ end of the intron. The intron is looped back to bond with itself near its 3′ end.
Degraded
Released intron in lariat structure
4 The spliceosome cleaves the pre-mRNA at the junction between the 3′ end of the intron and exon 2, releasing the intron and joining together the two exons. The released intron, called a lariat structure because of its shape, is degraded by enzymes, and the released snRNPs are used in other mRNA splicing reactions.
Reused
Released snRNPs

Introns Are Removed During Pre-mRNA Processing to Produce the Translatable mRNA

A process called **mRNA splicing,** which occurs in the nucleus, removes introns from pre-mRNAs and joins exons together. As an illustration of mRNA splicing, **Figure 15.8** shows the processing of a pre-mRNA with a single intron to produce a mature mRNA. mRNA splicing takes place in a **spliceosome,** a complex formed between the pre-mRNA and a handful of **small ribonucleoprotein particles.** A ribonucleoprotein particle is a complex of RNA and proteins. The small ribonucleoprotein particles involved in mRNA splicing are located in the nucleus; each consists of a relatively short *small nuclear RNA* (snRNA) bound to a number of proteins. The particles are therefore known as snRNPs, pronounced "snurps." The snRNPs bind in a particular order to an intron in the pre-mRNA and form the active spliceosome. The spliceosome cleaves the pre-mRNA to release the intron, and joins the flanking exons.

The cutting and splicing are so exact that not a single base of an intron is retained in the finished mRNA, nor is a single base removed from the exons. Without this precision, removing introns would change the reading frame of the coding portion of the mRNA, producing gibberish from the point of a mistake onward.

Introns Contribute to Protein Variability

Introns require elaborate cellular machinery to remove them during pre-mRNA processing. Why are they present in mRNA-encoding genes? Among a number of possibilities, introns may provide advantages by increasing the coding capacity of existing genes through a process called *alternative splicing* and by generating new proteins through a process called *exon shuffling.*

ALTERNATIVE SPLICING Many pre-mRNAs are processed by reactions that join exons in different combinations to produce different mRNAs from a single gene. The mechanism, called **alternative splicing,** greatly increases the number and variety of proteins encoded in the cell nucleus without increasing the size of the genome. For example, geneticists estimate that three-quarters of all human pre-mRNAs are subjected to alternative splicing. In each case the different mRNAs produced from the "parent" pre-mRNA are translated to produce a family of related proteins with various combinations of amino acid sequences derived from the

FIGURE 15.9
Alternative splicing of the α-tropomyosin pre-mRNA to distinct mRNA forms found in smooth muscle and skeletal muscle. All of the introns are removed in both mRNA splicing pathways, but exons 3, 10, and 11 are also removed to produce the smooth muscle mRNA, and exons 2 and 12 are also removed to produce the skeletal muscle mRNA.

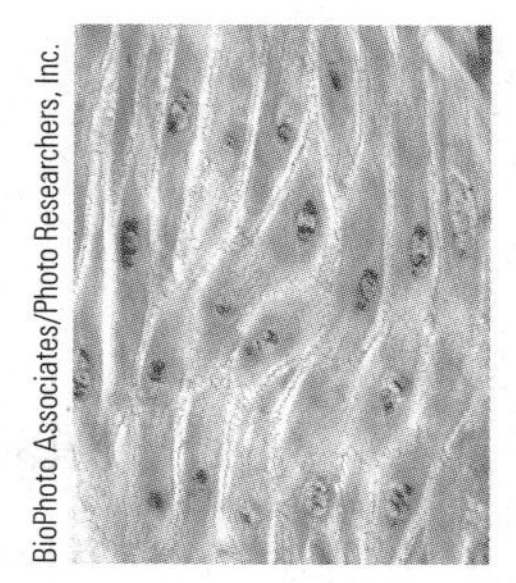

BioPhoto Associates/Photo Researchers, Inc.

Smooth muscle
Found in walls of tubes and cavities of the body, including blood vessels, the stomach and intestine, the bladder, and the uterus. Contraction of smooth muscles typically produce a squeezing motion.

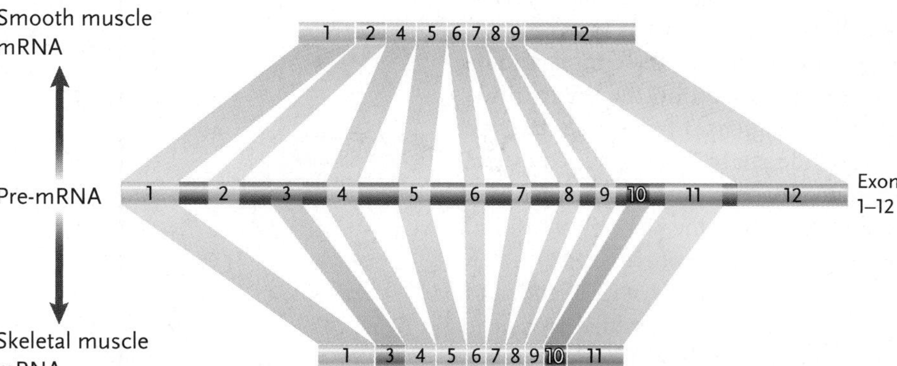
Smooth muscle mRNA
1 2 4 5 6 7 8 9 12
Pre-mRNA
1 2 3 4 5 6 7 8 9 10 11 12
Exons 1–12
Skeletal muscle mRNA
1 3 4 5 6 7 8 9 10 11

Skeletal muscle
Most muscles of this type are attached by tendons to the skeleton. Their function is locomotion and movement of body parts. The human body has more than 600 skeletal muscles, ranging in size from the small muscles that move the eyeballs, to the large muscles that move the legs.

Ed Reschke

exons. Each protein in the family, then, will vary to a degree in its function. Alternative splicing helps us understand why humans have only about 20,000 genes. As a result of the alternative splicing process, the number of proteins produced far exceeds the number of genes, and it is proteins that mostly direct an organism's functions.

As an example, the pre-mRNA transcript of the mammalian α-tropomyosin gene is alternatively spliced in various ways in different tissues—smooth muscle (for example, muscles of the intestine and bladder), skeletal muscle (for example, biceps, glutes), fibroblast (connective tissue cell that makes collagen), liver, and brain—to produce different forms of the α-tropomyosin protein. **Figure 15.9** shows the alternative splicing of the α-tropomyosin pre-mRNA to the mRNAs found in smooth muscle and striated muscle. Exons 2 and 12 are exclusive to the smooth muscle mRNA, whereas exons 3, 10, and 11 are exclusive to the striated muscle mRNA.

The polypeptides made from the two mRNAs have some identical stretches of amino acids, along with others that differ. As we learned in Section 3.4, the primary structure of a protein—its amino acid structure—directs the folding of the chain into its three-dimensional shape. Therefore, the two forms of α-tropomyosin fold into related, but different shapes. In its role in muscle contraction in smooth muscles and striated muscles, α-tropomyosin interacts with other proteins. The interactions depend upon the specific structural form of the α-tropomyosin and, as you might expect, the two forms participate in different types of muscle action; typically smooth muscles perform squeezing actions in blood vessels and internal organs, whereas skeletal muscles pull on the bones of the skeleton to move body parts.

Alternative splicing forces us to reconsider the one gene–one polypeptide hypothesis introduced earlier in the chapter. We must now accept the fact that for some genes at least, one gene may specify a number of polypeptides each of which has a related function.

EXON SHUFFLING Another advantage provided by introns may come from the fact that intron–exon junctions often fall at points dividing major functional regions in encoded proteins, as they do in the genes for antibody proteins, hemoglobin blood proteins, and the peptide hormone insulin. The functional divisions may have allowed new proteins to evolve by **exon shuffling,** a process by which existing protein regions or domains, already selected for their functions by the evolutionary process, are mixed into novel combinations to create new proteins. Evolution of new proteins by this mechanism would produce changes much more quickly and efficiently than by alterations in individual amino acids at random points. The process resembles automobile design, in which new models are produced by combining proven parts and substructures of previous models, rather than starting with an entirely new design each year.

STUDY BREAK 15.3

1. **What are the similarities and differences between pre-mRNAs and mRNAs?**
2. **What is the role of snRNPs in mRNA splicing?**

THINK OUTSIDE THE BOOK

Explore the Internet or research literature to find a molecular model that explains how an exon and its flanking introns is removed by alternative splicing.

15.4 Translation: mRNA-Directed Polypeptide Synthesis

Translation is the reading of an mRNA to assemble amino acids into a polypeptide. In prokaryotes, translation takes place throughout the cell, whereas in eukaryotes it takes place mostly in the cytoplasm, although, as you will see, a few specialized genes are transcribed and translated in mitochondria and chloroplasts.

Figure 15.10 summarizes the translation process. For prokaryotes, the mRNA produced by transcription is immediately available for translation. For eukaryotes, the mRNA produced by splicing of the pre-mRNA first exits the nucleus, and then is translated in the cytoplasm. In translation, the mRNA associates with a ribosome, and tRNAs, another type of RNA, bring amino acids to the complex to be joined one by one into the polypeptide chain. The sequence of amino acids in the polypeptide chain is determined by the sequence of codons in the mRNA, whereas the ribosome is simply a facilitator of the translation process. The mRNA is read from the 5′ end to the 3′ end, and the polypeptide is assembled from the N-terminal end to the C-terminal end.

In this section we start by discussing the key players in the process, the tRNAs and ribosomes, and then walk through the translation process from a start codon to a stop codon.

FIGURE 15.10
An overview of translation, in which ribosomes assemble amino acids into a polypeptide chain. The figure shows a ribosome in the process of translation. A tRNA molecule with an amino acid bound to it is entering the ribosome on the right. The anticodon on the tRNA will pair with the codon in the mRNA. Its amino acid will then be added to the growing polypeptide that is currently attached to the tRNA in the middle of the ribosome. As it assembles a polypeptide chain, the ribosome moves from one codon to the next along the mRNA in the 5′→3′ direction.

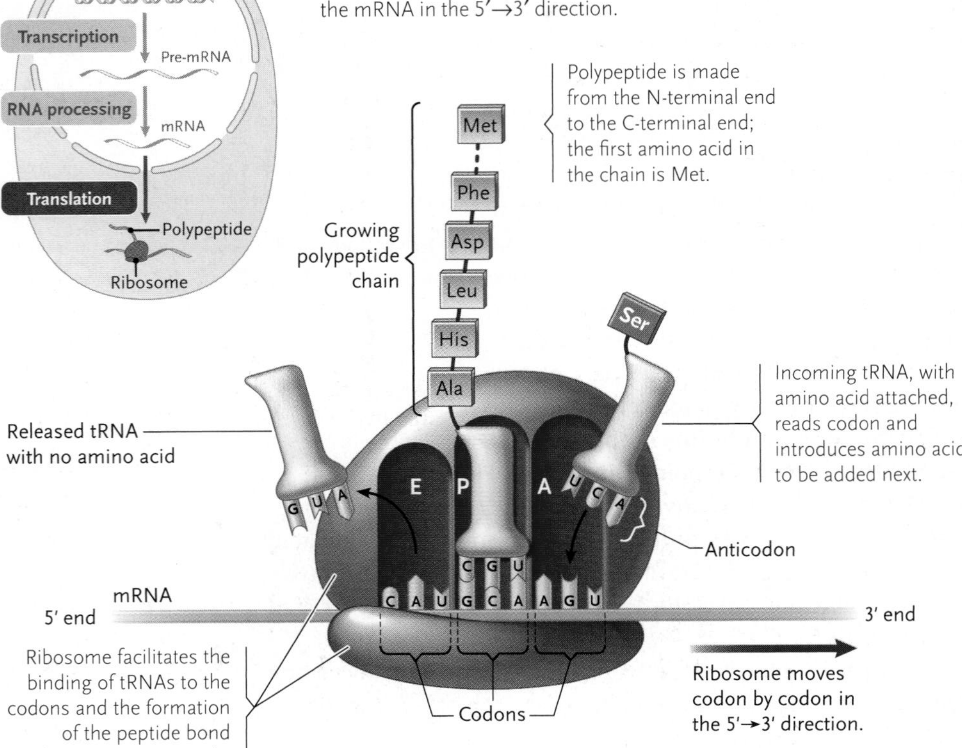
DNA
Transcription
Pre-mRNA
RNA processing
mRNA
Translation
Polypeptide
Ribosome
Polypeptide is made from the N-terminal end to the C-terminal end; the first amino acid in the chain is Met.
Met
Phe
Asp
Leu
His
Ala
Ser
Growing polypeptide chain
Released tRNA with no amino acid
Incoming tRNA, with amino acid attached, reads codon and introduces amino acid to be added next.
G U A
E
P
A
U C A
Anticodon
C G U
C A U G C A A G U
mRNA
5′ end
3′ end
Ribosome facilitates the binding of tRNAs to the codons and the formation of the peptide bond between amino acids.
Codons
Ribosome moves codon by codon in the 5′→3′ direction.

tRNAs Are Small, Highly Specialized RNAs That Bring Amino Acids to the Ribosome

A **transfer RNA (tRNA)** brings an amino acid to the ribosome for addition to the polypeptide chain.

tRNA STRUCTURE tRNAs are small RNAs, about 75 to 90 nucleotides long (mRNAs are typically hundreds of nucleotides long), with a highly distinctive structure that accomplishes their role in translation **(Figure 15.11)**. All tRNAs can wind into four double-helical segments, forming in two dimensions what is known as the *cloverleaf* pattern. At the tip of one of the double-helical segments is the **anticodon,** the three-nucleotide segment that base pairs with a codon in mRNAs. Opposite the anticodon, at the other end of the cloverleaf, is a double-helical segment that links to the amino acid corresponding to the anticodon. For example, a tRNA that base pairs with the codon 5′-AGU-3′ has serine (Ser) linked to it (see Figure 15.10). The anticodon of the tRNA that pairs with this codon is 3′-UCA-5′. (The anticodon and codon pair in an antiparallel manner, as do the strands in DNA. We will write anticodons in the 3′→5′ direction to make it easy to see how they pair with codons.)

The tRNA cloverleaf folds in three dimensions into the structure shown in Figure 15.11B—it is generally referred to as an upside-down L. The anticodon and the segment binding the amino acid are located at the opposite tips of the structure.

We learned earlier that 61 of the 64 codons of the genetic code specify an amino acid. Does this mean that 61 different tRNAs read the sense codons? The answer is no. Francis Crick's **wobble hypothesis** states that the complete set of 61 sense codons can be read by fewer than 61 distinct tRNAs because of particular pairing properties of the bases in the anticodons. That is, the pairing of the anticodon with the first two nucleotides of the codon is always precise, but the anticodon has more flexibility in pairing with the third nucleotide of the codon. In many cases the same tRNA anticodon can read codons that have either U or C in the third position; for example, a tRNA carrying phenylalanine can read both codons 5′-UUU-3′ and 5′-UUC-3′. Similarly the same tRNA anticodon can read two codons that have A or G in the third position; for example, a tRNA carrying glutamine can read both codons 5′-CAA-3′ and 5′-CAG-3′. The special inosine purine in the alanine tRNA shown in Figure 15.11A allows even more extensive wobble by allowing the tRNA to pair with codons that have either U, C, or A in the third position.

ADDITION OF AMINO ACIDS TO THEIR CORRESPONDING tRNAS The correct amino acid must be present on a tRNA if translation is to be accurate. The process of adding an amino acid to a tRNA is called **aminoacylation** (literally, the addition of an amino acid) or **charging** (because the process adds free energy as

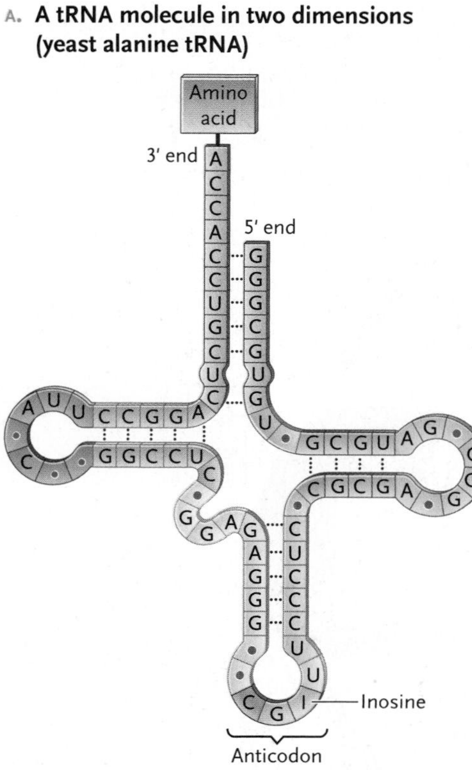

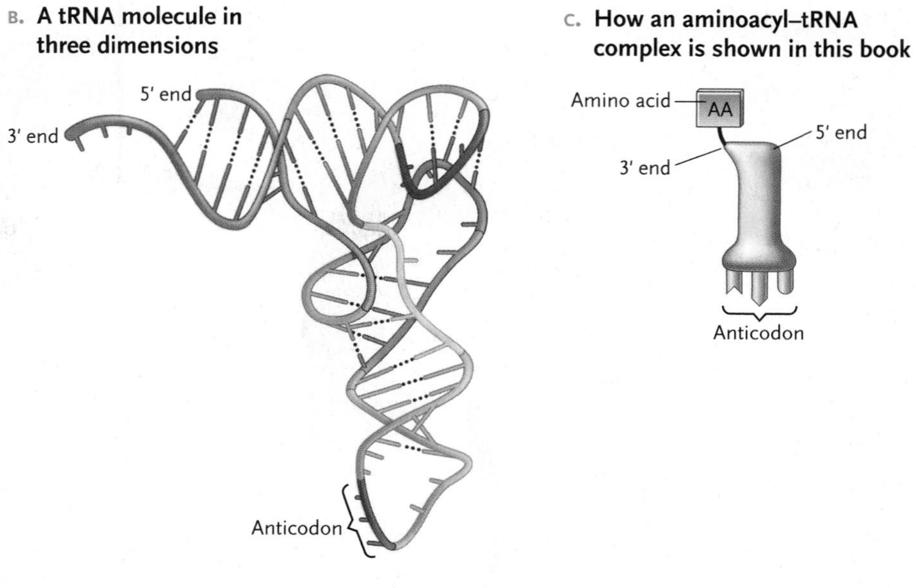

A. A tRNA molecule in two dimensions (yeast alanine tRNA)

B. A tRNA molecule in three dimensions

C. How an aminoacyl–tRNA complex is shown in this book

FIGURE 15.11
tRNA structure. The red dots show sites where bases are chemically modified into other forms; chemical modification of certain bases is typical of tRNAs. This tRNA has the purine inosine (I) in the anticodon which has relatively loose base pairing ability, allowing this single tRNA to pair with each of three alanine codons 5′-GCU-3′, 5′-GCC-3′, and 5′-GCA-3′. This tRNA also has the unusual base pair G–U. Unusual base pairs, allowed by the greater flexibility of short RNA chains, are common in tRNAs.

the amino acid–tRNA combinations are formed). The finished product of charging, a tRNA linked to its "correct" amino acid, is called an **aminoacyl–tRNA.** Twenty different enzymes called **aminoacyl–tRNA synthetases**—one synthetase for each of the 20 amino acids—catalyze aminoacylation in the four steps shown in **Figure 15.12.**

With the tRNAs attached to their corresponding amino acids, our attention moves to the ribosome, where the amino acids are removed from their tRNAs and linked into polypeptide chains.

Ribosomes Are rRNA–Protein Complexes That Work as Automated Protein Assembly Machines

Ribosomes are ribonucleoprotein particles that carry out protein synthesis by translating mRNA into chains of amino acids. A ribosome reads the codons on an mRNA and joins the appropriate amino acids to make a polypeptide chain.

In prokaryotes, ribosomes carry out their assembly functions throughout the cell. In eukaryotes, ribosomes function in the cytoplasm, either suspended freely in the cytoplasmic solution, or attached to the membranes of the endoplasmic reticulum (ER), the system of tubular or flattened sacs in the cytoplasm (discussed in Section 5.3).

A finished ribosome is made up of two parts of dissimilar size, called the *large* and *small ribosomal subunits* **(Figure 15.13).** Each subunit is a combination of **ribosomal RNA (rRNA)** and ribosomal proteins.

Prokaryotic and eukaryotic ribosomes are similar in function, and quite similar in structure. However, certain differences in their molecular structure, particularly in the ribosomal proteins, give them distinguishable properties. For example, the antibiotics streptomycin and erythromycin are effective antibacterial agents because they inhibit the function of the bacterial ribosome, but not the eukaryotic ribosome. Streptomycin and erythromycin affect translation activities in the small and large ribosomal subunit, respectively.

In translation, the mRNA follows a bent path through a groove in the ribosome. The ribosome also has binding sites where tRNAs interact with the mRNA (see Figure 15.13 and refer also to Figure 15.10). The **A site** (aminoacyl site) is where the incoming aminoacyl–tRNA carrying the next amino acid to be added to the polypeptide chain binds to the RNA. The **P site** (peptidyl site) is where the tRNA carrying the growing polypeptide chain is bound. The **E site** (exit site) is where a tRNA, now without an attached amino acid, binds before exiting the ribosome. You will learn more about these functional sites as we discuss the stages of translation.

Translation Initiation Brings the Ribosomal Subunits, an mRNA, and the First Aminoacyl–tRNA Together

Translation is similar in bacteria and eukaryotes. We will present translation from a eukaryotic perspective, and indicate how it differs in bacteria.

There are three major stages of translation: *initiation, elongation,* and *termination.* During initiation the translation components assemble on the start codon of the mRNA. In elongation the assembled complex reads the string of codons in the mRNA one at a time while joining the specified amino acids

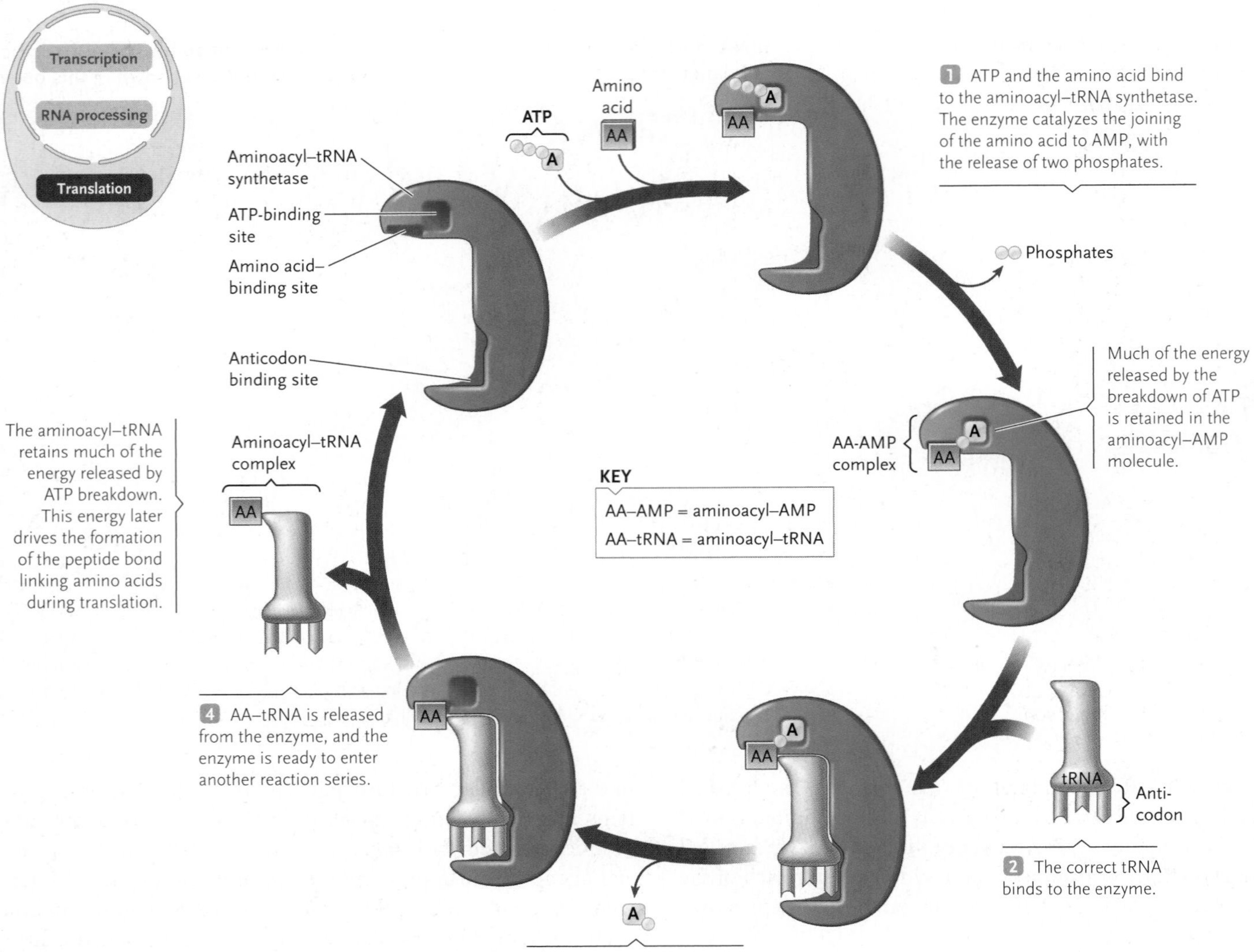

Transcription
RNA processing
Translation

Aminoacyl–tRNA synthetase
ATP-binding site
Amino acid–binding site
Anticodon binding site

ATP
Amino acid
AA
A

1 ATP and the amino acid bind to the aminoacyl–tRNA synthetase. The enzyme catalyzes the joining of the amino acid to AMP, with the release of two phosphates.

Phosphates

Much of the energy released by the breakdown of ATP is retained in the aminoacyl–AMP molecule.

AA-AMP complex

KEY

AA–AMP = aminoacyl–AMP
AA–tRNA = aminoacyl–tRNA

tRNA
Anti-codon

2 The correct tRNA binds to the enzyme.

3 The enzyme transfers the amino acid from AA–AMP to the tRNA, forming AA–tRNA. AMP is released.

4 AA–tRNA is released from the enzyme, and the enzyme is ready to enter another reaction series.

Aminoacyl–tRNA complex

The aminoacyl–tRNA retains much of the energy released by ATP breakdown. This energy later drives the formation of the peptide bond linking amino acids during translation.

FIGURE 15.12
Aminoacylation (aka charging): the addition of an amino acid to a tRNA.

FIGURE 15.13
Ribosome structure. **(A)** Computer model of a ribosome in the process of translation. **(B)** The ribosome as we will show it during translation.
(A: Michael W. Davidson/Molecular Expressions, Florida State Research Foundation.)

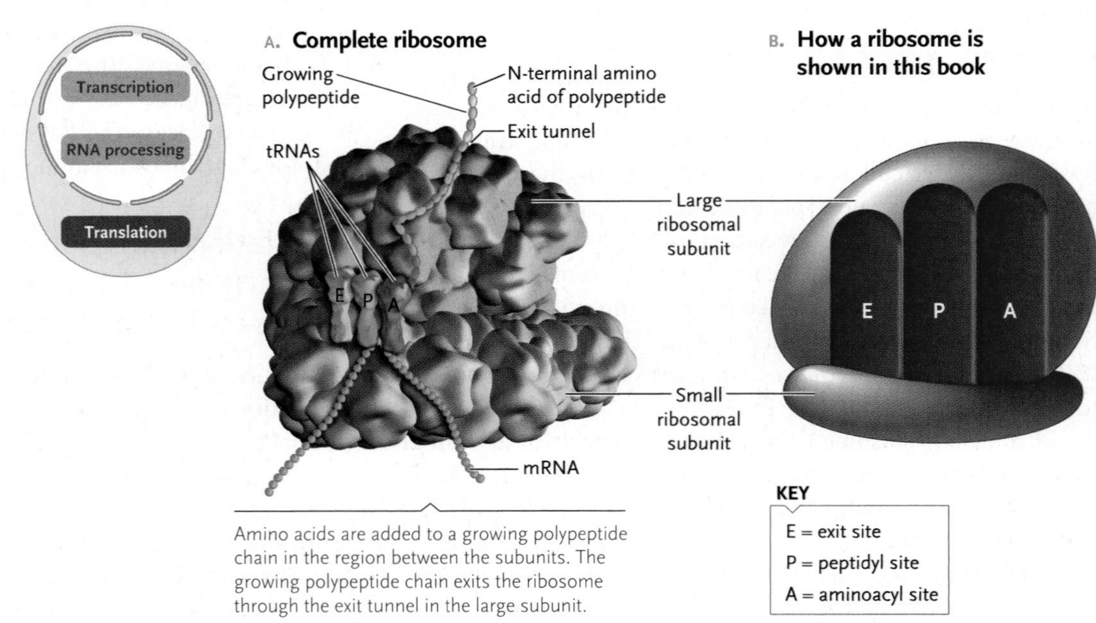

Transcription
RNA processing
Translation
A. Complete ribosome
Growing polypeptide
N-terminal amino acid of polypeptide
Exit tunnel
tRNAs
E
P
A
Large ribosomal subunit
Small ribosomal subunit
mRNA
Amino acids are added to a growing polypeptide chain in the region between the subunits. The growing polypeptide chain exits the ribosome through the exit tunnel in the large subunit.
B. How a ribosome is shown in this book
E
P
A
KEY
E = exit site
P = peptidyl site
A = aminoacyl site

into the polypeptide. Termination completes the translation process when the complex disassembles after the last amino acid of the polypeptide specified by the mRNA has been added to the polypeptide.

TRANSLATION INITIATION **Figure 15.14** illustrates the steps of translation initiation in eukaryotes. Each initiation step is aided by proteins called **initiation factors (IFs).** The IFs are released when initiation is complete in step 3.

In bacteria, translation initiation is similar in using a special initiator Met–tRNA, GTP, and IFs, but the way in which the ribosome assembles at the start codon is different. Rather than scanning from the 5′ end of the mRNA, the small ribosomal subunit, the initiator Met–tRNA, GTP, and IFs bind directly to the region of the mRNA with the AUG start codon. A **ribosome binding site**—a short, specific RNA sequence—just upstream of the start codon directs the small ribosomal subunit in this initiation step. The large ribosomal subunit then binds to the small subunit to complete the ribosome. GTP hydrolysis then releases the IFs.

After the initiator tRNA pairs with the AUG initiator codon, the subsequent stages of translation simply read the codons one at a time on the mRNA. The initiator tRNA–AUG pairing thus establishes the correct **reading frame**—the series of codons for the polypeptide encoded by the mRNA.

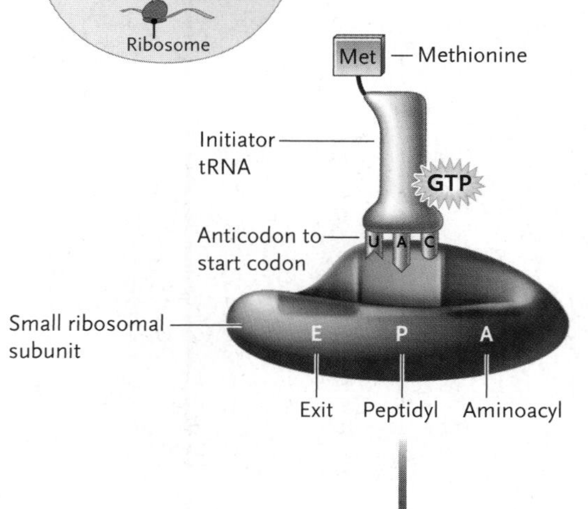
DNA
Transcription
Pre-mRNA
RNA processing
mRNA
Translation
Polypeptide
Ribosome

Polypeptide Chains Grow during the Elongation Stage of Translation

The central reactions of translation take place in the elongation stage, which adds amino acids one at a time to a growing polypeptide chain. The individual steps of elongation depend on the binding properties of the P, A, and E sites of the ribosome.

The P site, with one exception, can bind only to a **peptidyl–tRNA,** that is, a tRNA linked to a growing polypeptide chain containing two or more amino acids. The exception is the initiator tRNA, which is recognized by the P site as a peptidyl–tRNA even though it carries only a single amino acid, methionine. The A site can bind only to an aminoacyl–tRNA. The tRNA that previously was in the P site binds to the E site and then leaves the ribosome.

Figure 15.15 shows the elongation cycle of translation. The cycle begins at the point when an initiator tRNA with its attached methionine is bound to the P site, and the A site is empty (top of figure). The first step in each round of the cycle is the

FIGURE 15.14

Translation initiation in eukaryotes. Protein initiation factors (IFs) participate in the event but, for simplicity, they are not shown in the figure. The IFs are released when the large ribosomal subunit binds and GTP is hydrolyzed.

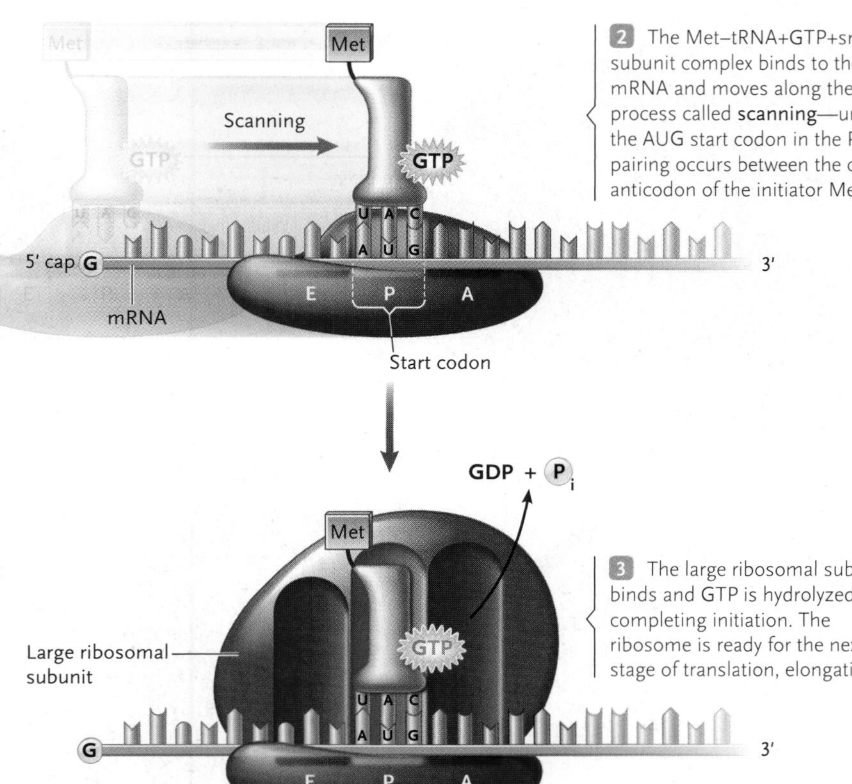
Met
Methionine
Initiator tRNA
GTP
Anticodon to start codon
U A C
Small ribosomal subunit
E P A
Exit
Peptidyl
Aminoacyl
1 A specialized methionine–tRNA is used as an initiator tRNA in translation. The initiator Met–tRNA has an anticodon 3′-UAC-5′ for the AUG start codon. The initiator Met–tRNA with GTP bound to it binds to the small ribosomal subunit and forms a complex.
Scanning
5′ cap G
A U G
3′
mRNA
Start codon
2 The Met–tRNA+GTP+small ribosomal subunit complex binds to the 5′ cap of the mRNA and moves along the mRNA—a process called scanning—until it reaches the AUG start codon in the P site. Base pairing occurs between the codon and the anticodon of the initiator Met–tRNA.
GDP + P i
Large ribosomal subunit
3 The large ribosomal subunit binds and GTP is hydrolyzed, completing initiation. The ribosome is ready for the next stage of translation, elongation.

binding of the appropriate aminoacyl–tRNA to the codon in the A site of the ribosome (step 1). This binding is facilitated by a protein **elongation factor (EF)** that is bound to the aminoacyl–tRNA and that is released once the tRNA binds to the codon. Another EF is used when the ribosome translocates along the mRNA to the next codon (step 3). Each EF is released after its job is completed.

In elongation, a peptide bond is formed between the C-terminal end of the growing polypeptide on the P site tRNA and the amino acid on the A site tRNA (step 2). **Peptidyl transferase** catalyzes this reaction. Researchers were surprised to discover that this enzyme is not a protein but a part of an rRNA of the large ribosomal subunit. An RNA molecule that catalyzes a reaction like a protein enzyme does is called a *catalytic RNA* or a **ribozyme** (*ribo*nucleic acid en*zyme*). (*Insights from the Molecular Revolution* describes the experimental evidence showing that peptidyl transferase is a ribozyme.)

Elongation is highly similar in prokaryotes and eukaryotes, with no significant conceptual differences. The elongation cycle turns at the rate of about one to three times per second in eukaryotes and 15 to 20 times per second in bacteria. Once it is long enough, the growing polypeptide chain extends from the ribosome through the exit tunnel (see Figure 15.13) as elongation continues.

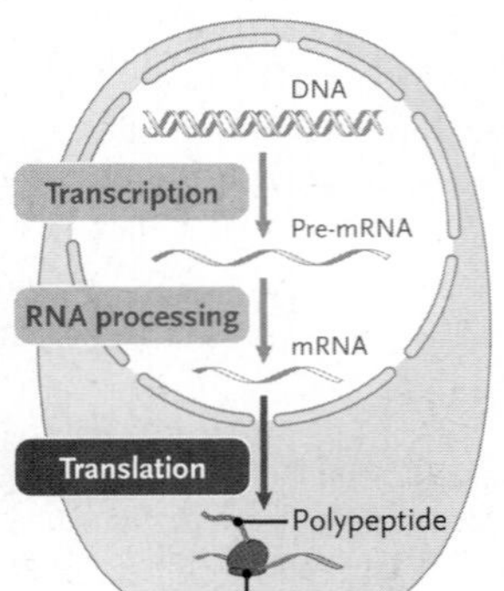

DNA
Transcription
Pre-mRNA
RNA processing
mRNA
Translation
Polypeptide
Ribosome

FIGURE 15.15
Translation elongation. A protein elongation factor (EF) complexes with the aminoacyl–tRNA to bring it to the ribosome, and another EF is needed for ribosome translocation. For simplicity, the EFs are not shown in the figure.

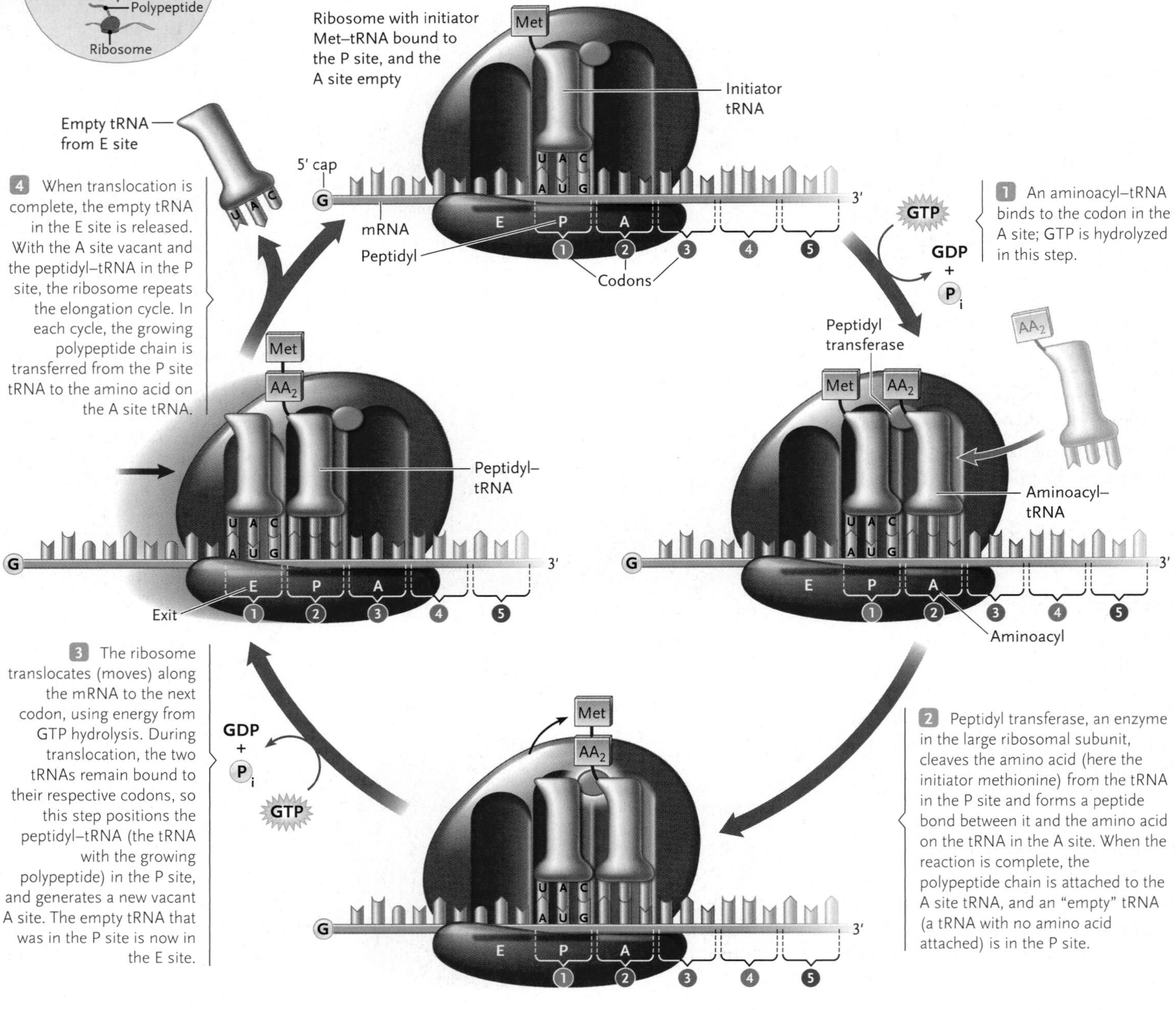

Ribosome with initiator Met–tRNA bound to the P site, and the A site empty

Met
Initiator tRNA
5′ cap
G
U A C
A U G
3′
mRNA
Peptidyl
E P A
1 2 3 4 5
Codons

1 An aminoacyl–tRNA binds to the codon in the A site; GTP is hydrolyzed in this step.

GTP
GDP + P_i
Peptidyl transferase
AA_2
Met
Aminoacyl–tRNA
Aminoacyl

2 Peptidyl transferase, an enzyme in the large ribosomal subunit, cleaves the amino acid (here the initiator methionine) from the tRNA in the P site and forms a peptide bond between it and the amino acid on the tRNA in the A site. When the reaction is complete, the polypeptide chain is attached to the A site tRNA, and an "empty" tRNA (a tRNA with no amino acid attached) is in the P site.

Met
AA_2
U A C
A U G
G
3′
E P A
1 2 3 4 5

3 The ribosome translocates (moves) along the mRNA to the next codon, using energy from GTP hydrolysis. During translocation, the two tRNAs remain bound to their respective codons, so this step positions the peptidyl–tRNA (the tRNA with the growing polypeptide) in the P site, and generates a new vacant A site. The empty tRNA that was in the P site is now in the E site.

GDP + P_i
GTP
Met
AA_2
Peptidyl–tRNA
Exit
E P A
1 2 3 4 5

4 When translocation is complete, the empty tRNA in the E site is released. With the A site vacant and the peptidyl–tRNA in the P site, the ribosome repeats the elongation cycle. In each cycle, the growing polypeptide chain is transferred from the P site tRNA to the amino acid on the A site tRNA.

Empty tRNA from E site

INSIGHTS FROM THE Molecular Revolution

Peptidyl Transferase: Protein or RNA?

Ribosomes are responsible for polypeptide synthesis in the process of translation. Ribosomes of both prokaryotes and eukaryotes consist of two subunits, each with one or more ribosomal RNA (rRNA) molecules and many ribosomal proteins. A key event in polypeptide synthesis is the formation of a peptide bond when the amino acid or growing polypeptide on the tRNA in the P site of the ribosome is transferred to the amino acid on the tRNA in the A site. Peptide bond formation is catalyzed by a peptidyl transferase enzyme in the large subunit. In the large 50S subunit of bacteria, regions of the 23S rRNA and several ribosomal proteins are in the peptidyl transferase area, indicating that they may participate in its enzymatic activity.

Research Question

Is the peptidyl transferase activity of the bacterial ribosome a function of a ribosomal protein or an RNA molecule?

Experiment

To answer this question, Harry F. Noller and his coworkers at the University of California, Santa Cruz set up a *fragment reaction,* a simplified peptidyl transferase reaction in a test tube, and tested the effects of removing or degrading the ribosomal proteins or 23S rRNA on peptide bond formation.

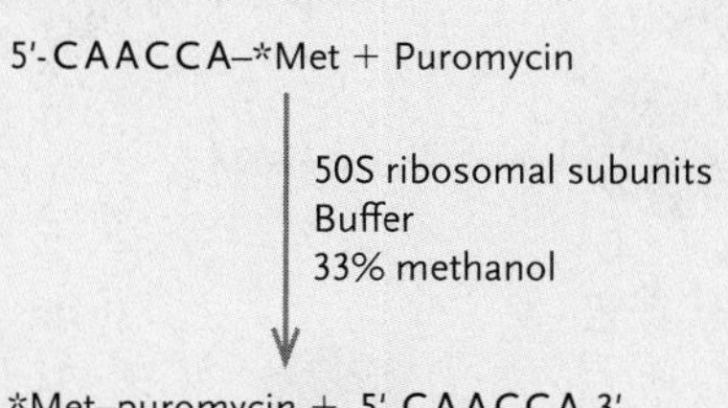

Fragment Reaction
5'-CAACCA–*Met + Puromycin
50S ribosomal subunits
Buffer
33% methanol
*Met–puromycin + 5'-CAACCA-3'

The figure outlines the fragment reaction; its contents and operation are as follows:

- 5′-CAACCA–*Met: A short RNA segment with radioactively labeled Met (methionine) attached; it mimics the initiator tRNA.
- Puromycin: An antibiotic that causes premature polypeptide chain termination during translation in bacteria. It enters the A site of the ribosome, mimicking a tRNA. The amino acid or growing polypeptide chain in the P site is transferred to the antibiotic and further polypeptide synthesis is blocked.
- 50S ribosomal subunits: Subunits from *Thermus aquaticus,* a heat-tolerant bacterium. No 30S subunits, tRNAs, GTP, or protein factors were added.
- Buffer and methanol: To enable the reaction to proceed in vitro.

If peptidyl transferase is active in the reaction, the radioactive Met transfers to the puromycin to form Met–puromycin. Biochemical separation techniques distinguished the Met–puromycin from the Met; the amount of radioactivity in the product is a measure of peptidyl transferase activity.

Results

The treatments of the reaction used, and the results obtained were:

Treatment	Result	Interpretation
None	Met–puromycin forms	Peptidyl transferase activity is present in 50S subunits
Sodium dodecyl sulfate (SDS) or phenol to remove proteins from subunits	Met–puromycin forms	Peptidyl transferase activity remains after proteins removed from 50S subunits
Proteinase K to digest proteins of subunits	Met–puromycin forms	Peptidyl transferase activity remains after proteins of 50S subunits digested away
Ribonuclease to degrade RNA of subunits	No Met–puromycin forms	Peptidyl transferase activity is lost when RNA of subunits is degraded

Conclusion

The researchers concluded that the peptidyl transferase activity of the 50S ribosomal subunit is a function of the 23S rRNA component of that subunit, rather than of ribosomal proteins. That is, the peptidyl transferase activity was resistant to treatments that removed or degraded proteins, but it was sensitive to treatment with ribonuclease that degrades RNA. The investigators allowed that their protein removal techniques may have left fragments of ribosomal proteins that potentially may be part of the peptidyl transferase region of the 50S subunit. They argued that definitive proof that the 23S rRNA alone is responsible for peptidyl transferase activity in translation would require demonstration that a completely protein-free preparation of the rRNA can carry out the peptidyl transferase reaction in vitro. To date, however, peptidyl transferase activity by rRNA alone has not been demonstrated.

Source: H. F. Noller, V. Hoffarth, and L. Zimniak. 1992. Unusual resistance of peptidyl transferase to protein extraction procedures. *Science* 256:1416–1419.

Termination Releases a Completed Polypeptide from the Ribosome

Translation termination takes place when the A site of a ribosome arrives at one of the stop codons on the mRNA, UAG, UAA, or UGA **(Figure 15.16).** The stop codon is read by a protein **release factor** (**RF;** also called a **termination factor**). Termination is highly similar in prokaryotes and eukaryotes.

Multiple Ribosomes Simultaneously Translate a Single mRNA

Once the first ribosome has begun translation, another one can assemble with an initiator tRNA as soon as there is room on the mRNA. Ribosomes continue to attach as translation continues and become spaced along the mRNA like beads on a string. The entire structure of an mRNA molecule and the multiple ribosomes

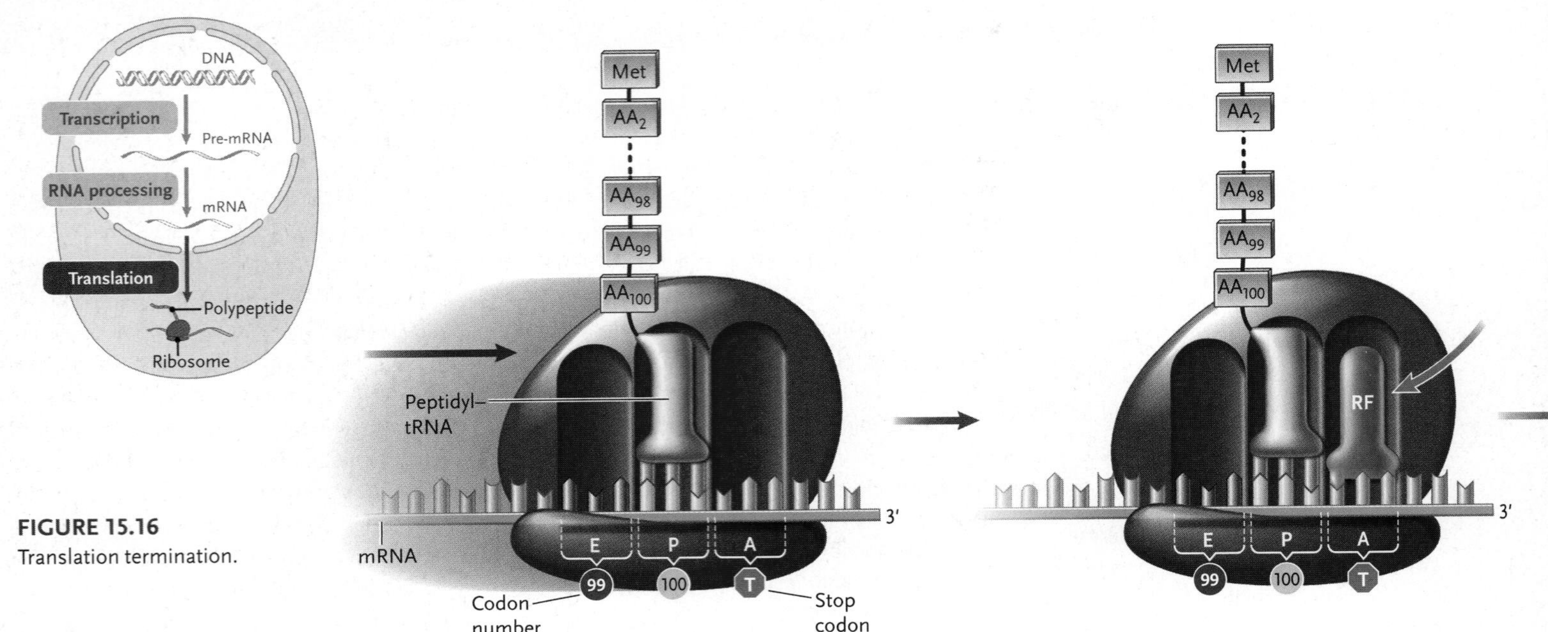
DNA
Transcription
Pre-mRNA
RNA processing
mRNA
Translation
Polypeptide
Ribosome
Met
AA_2
AA_{98}
AA_{99}
AA_{100}
Peptidyl-
tRNA
mRNA
E
P
A
99
100
T
3'
Codon
number
Stop
codon
RF

FIGURE 15.16
Translation termination.

1 The ribosome reaches a stop codon, UAG, UAA, or UGA.

2 No tRNA has an anticodon that can pair with a stop codon. Instead, a release factor (RF) binds to the stop codon in the A site. The shape of the release factor mimics that of a tRNA, including regions that read the stop codons.

attached to it is known as a **polysome** (a contraction of *polyribosome;* **Figure 15.17**). The multiple ribosomes greatly increase the overall rate of polypeptide synthesis from a single mRNA. The total number of ribosomes in a polysome depends on the length of the coding region of its mRNA molecule, ranging from a minimum of one or two ribosomes on the smallest mRNAs to as many as 100 on the longest mRNAs.

In prokaryotes, because of the absence of a nuclear envelope, transcription and translation typically are coupled. That is, as soon as the 5′ end of a new mRNA emerges from the RNA polymerase, ribosomal subunits attach and initiate translation **(Figure 15.18).** In essence the polysome forms while the mRNA is still being made. By the time the mRNA is completely transcribed, it is covered with ribosomes from end to end, each assembling a copy of the encoded polypeptide.

Newly Synthesized Polypeptides Are Processed and Folded into Finished Form

Most eukaryotic proteins are in an inactive, unfinished form when ribosomes release them. Processing reactions that convert the new proteins into finished form include the removal of amino acids from the ends or interior of the polypeptide chain and the addition of larger organic groups, including carbohydrate or lipid structures.

Proteins fold into their final three-dimensional shapes as the processing reactions take place. For many proteins, helper proteins called **chaperones** or **chaperonins** assist the folding process by combining with the folding protein, promoting "correct" three-dimensional structures, and inhibiting incorrect ones (see Section 3.4 and Figure 3.26).

In some cases the same initial polypeptide may be processed by alternative pathways that produce different mature polypep-

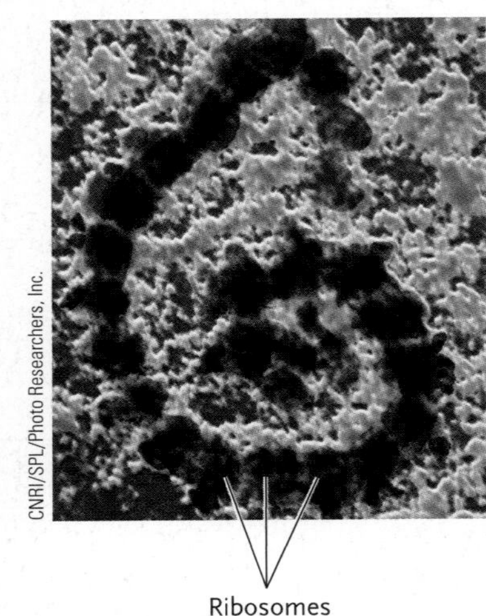
CNRI/SPL/Photo Researchers, Inc.
Ribosomes

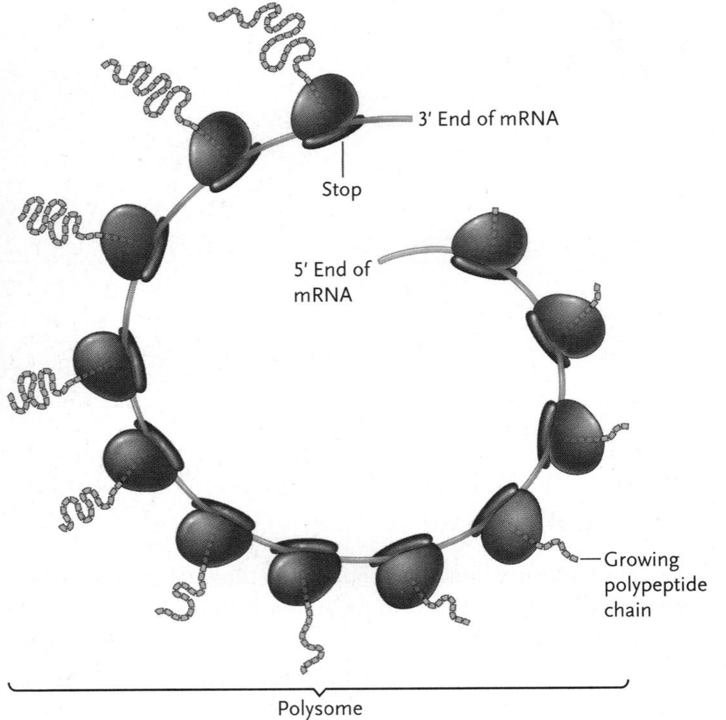
3′ End of mRNA
Stop
5′ End of
mRNA
Growing
polypeptide
chain
Polysome

FIGURE 15.17
Polysomes, consisting of a series of ribosomes reading the same mRNA.

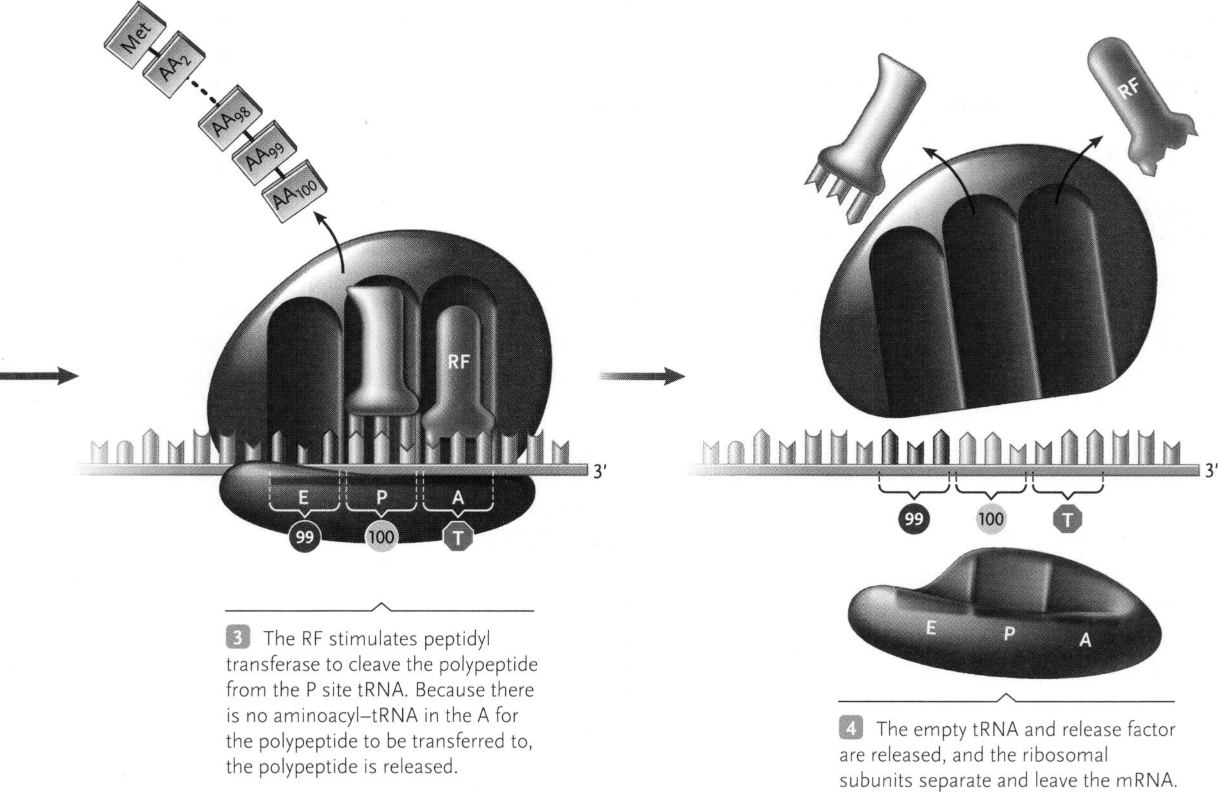
Met
AA2
AA98
AA99
AA100
RF
3′
E
P
A
99
100
T
RF
3′
99
100
T
E
P
A
3 The RF stimulates peptidyl transferase to cleave the polypeptide from the P site tRNA. Because there is no aminoacyl–tRNA in the A for the polypeptide to be transferred to, the polypeptide is released.
4 The empty tRNA and release factor are released, and the ribosomal subunits separate and leave the mRNA.

tides, usually by removing different, long stretches of amino acids from the interior of the polypeptide chain. Alternative processing of a pre-mRNA is another mechanism that increases the number of polypeptides encoded by a single gene.

Other proteins are processed into an initial, inactive form that is later activated at a particular time or location by removal of a covering segment of the amino acid chain. The digestive enzyme pepsin, for example, is made by processing reactions within cells lining the stomach into an inactive form called *pepsinogen.* When the cells secrete pepsinogen into the stomach, the high acidity of that organ triggers removal of a segment of amino acids from one end of the protein's amino acid chain; the removal converts the enzyme into the active form in which it rapidly breaks proteins in food particles into shorter pieces. The initial production of the protein as inactive pepsinogen protects the cells that make it from having their proteins degraded by the enzyme.

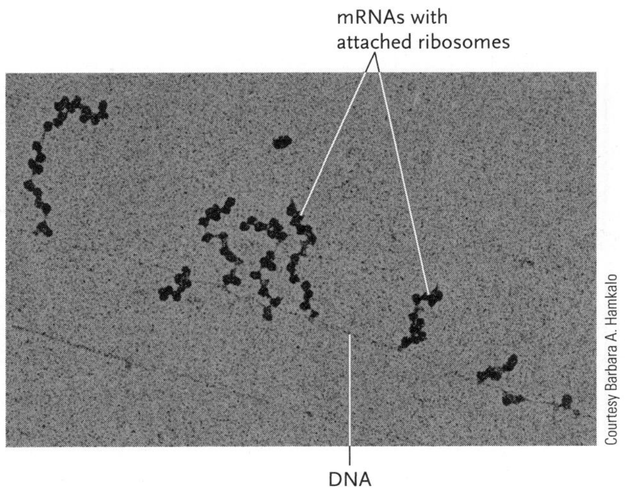
mRNAs with attached ribosomes
DNA
Courtesy Barbara A. Hamkalo

FIGURE 15.18
Simultaneous transcription and translation in progress in an electron microscope preparation extracted from *E. coli.* ×57,000.

Finished Proteins Are Sorted to the Cellular Locations Where They Function

In a eukaryotic cell, every protein that is made must be sorted to the compartment where it performs a necessary function. Without a sorting system, cells would wind up as a jumble of proteins floating about in the cytoplasm, with none of the spatial organization that makes cellular life possible.

With respect to how proteins are sorted, we can consider three types of compartments: (1) the cytosol; (2) the endomembrane system, which includes the endoplasmic reticulum (ER), Golgi complex, lysosomes, secretory vesicles, the nuclear envelope, and the plasma membrane (see Section 5.3); and (3) other membrane-bound organelles distinct from the endomembrane system, including mitochondria, chloroplasts, microbodies (for example, peroxisomes), and within the nucleus (see Section 5.3).

PROTEIN SORTING TO THE CYTOSOL Proteins that function in the cytosol are synthesized on *free ribosomes* in the cytosol. The polypeptides are simply released from the ribosomes once translation is completed. Examples of proteins that function in the cytoplasm are microtubule proteins, and the enzymes that carry out glycolysis (see Section 8.2).

PROTEIN SORTING TO THE ENDOMEMBRANE SYSTEM The endomembrane system is a major traffic network for proteins. Polypeptides that sort to the endomembrane system begin their synthesis on free ribosomes in the cytosol. Specific to these polypeptides is a short segment of amino acids called a **signal sequence** (also called a **signal peptide**) near their N-terminal ends. As **Figure 15.19** shows, the signal sequence initiates a series of steps that result in the polypeptide entering the lumen of the rough ER. This mechanism is called **cotranslational import** because import of the polypeptide into the ER occurs simultaneously with translation of the mRNA encoding the polypeptide. The signal sequence was discovered in 1975 by Günter Blobel, B. Dobberstein, and colleagues at Rockefeller University, when they observed that proteins sorted through the endomembrane system initially contain extra amino acids at their N-terminal ends. Blobel received a Nobel Prize in 1999 for his work with the mechanism of sorting proteins in cells.

Once in the lumen of the rough ER, the proteins fold into their final form. They also have or attain a tag—a "zip code" if you will—that targets each protein for sorting to its final destination. Depending on the protein and its destination, the tag may be an amino acid sequence already in the protein, or a functional group or short sugar chain added in the lumen. Some proteins remain in the ER, whereas others are transported to the Golgi complex where they may be modified further. From the Golgi complex, proteins are packaged into vesicles, which may deliver them to lysosomes, secrete them from the cell (digestive enzymes, for example), or deposit them in the plasma membrane (cell surface receptors, for instance). Vesicle traffic in the cytoplasm involving the rough ER and Golgi complex is illustrated in Figure 5.15.

PROTEIN SORTING TO MITOCHONDRIA, CHLOROPLASTS, MICROBODIES, AND NUCLEUS Proteins are sorted to mitochondria, chloroplasts, microbodies, and the nucleus after they have been made on free ribosomes in the cytosol. This mechanism of sorting is called **posttranslational import.** Proteins destined for the mitochondria, chloroplasts, and microbodies have short amino acid sequences called **transit sequences** at their N-terminal ends that target them to the appropriate organelle. The protein is taken up into the correct organelle by interactions between its transit sequences and organelle-specific transport complexes in the membrane of the appropriate organelle. A transit peptidase enzyme within the organelle then removes the transit sequence. Proteins sorted to the nucleus, such as the enzymes for DNA replication, and RNA transcription, have short amino acid sequences called **nuclear localization signals.** A cytosolic protein binds to the signal and moves it to the nuclear pore complex (see Section 5.3 and Figures 5.11 and 5.12) where it is then transported into the nucleus through the pore. For these proteins, the nuclear localization signal remains because the proteins need to enter the nucleus each time the nuclear envelope breaks down and reforms during the cell division cycle.

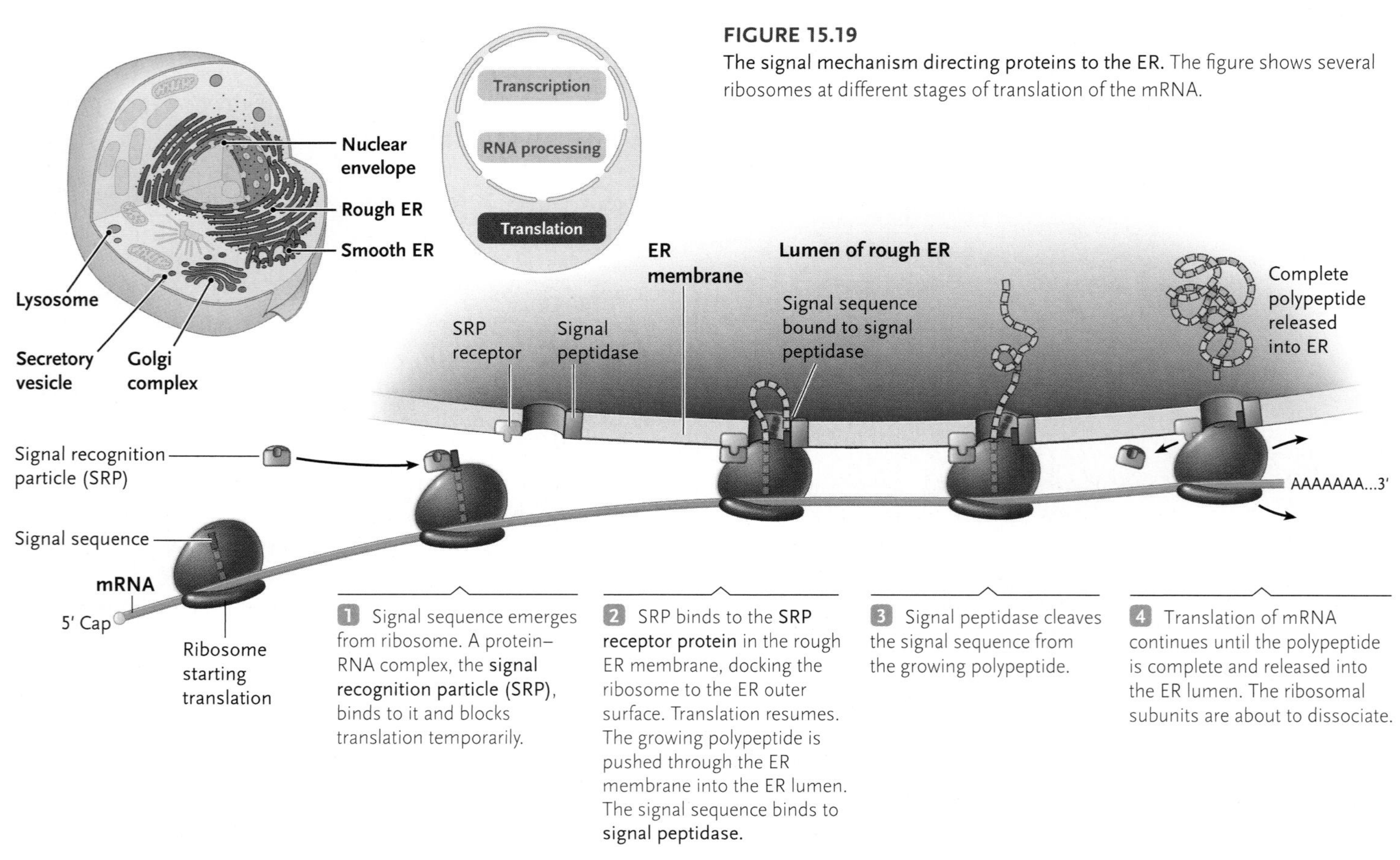

Transcription
RNA processing
Translation
Nuclear envelope
Rough ER
Smooth ER
Lysosome
Secretory vesicle
Golgi complex
ER membrane
Lumen of rough ER
SRP receptor
Signal peptidase
Signal sequence bound to signal peptidase
Complete polypeptide released into ER
Signal recognition particle (SRP)
AAAAAAA...3'
Signal sequence
mRNA
5′ Cap
Ribosome starting translation
1 Signal sequence emerges from ribosome. A protein–RNA complex, the signal recognition particle (SRP), binds to it and blocks translation temporarily.
2 SRP binds to the SRP receptor protein in the rough ER membrane, docking the ribosome to the ER outer surface. Translation resumes. The growing polypeptide is pushed through the ER membrane into the ER lumen. The signal sequence binds to signal peptidase.
3 Signal peptidase cleaves the signal sequence from the growing polypeptide.
4 Translation of mRNA continues until the polypeptide is complete and released into the ER lumen. The ribosomal subunits are about to dissociate.

FIGURE 15.19
The signal mechanism directing proteins to the ER. The figure shows several ribosomes at different stages of translation of the mRNA.

The same basic system of sorting signals distributes proteins in prokaryotic cells, indicating that this mechanism probably evolved with the first cells. In prokaryotes, signals similar to the ER-directing signals of eukaryotes direct newly synthesized bacterial proteins to the plasma membrane (bacteria do not have ER membranes); further information built into the proteins keeps them in the plasma membrane or allows them to enter the cell wall or to be secreted outside the cell. Proteins without sorting signals remain in the cytoplasmic solution.

STUDY BREAK 15.4 <

1. **How does translation initiation occur in eukaryotes versus prokaryotes?**
2. **Distinguish between the P, A, and E sites of the ribosome.**
3. **How are proteins directed to different compartments of a eukaryotic cell?**

15.5 Genetic Changes That Affect Protein Structure and Function

We learned in Chapters 12 and 13 that a mutant allele of a gene can alter the phenotype controlled by the gene. In this section we discuss two types of genetic change and how they can alter protein structure and function and, therefore, produce an altered phenotype. One type is mutation of a base pair in the DNA, that is, a change from one base pair to another. The other is when certain genetic elements known as transposable elements (TEs) move from one location to another in the genome.

Base-Pair Mutations Can Alter Protein Structure and Function

Mutations, in general, are changes to the genetic material. Base-pair substitution mutations are particular mutations involving changes to individual base pairs in the genetic material. If a base-pair substitution mutation occurs in the protein-coding portion of a gene, it can change a base in a codon in the mRNA and thereby affect the structure and function of the encoded protein. More generally, mutations affecting the functions of genes are known as gene mutations.

Consider a theoretical stretch of normal (unmutated) DNA encoding a string of amino acids in a polypeptide **(Figure 15.20A).** Four types

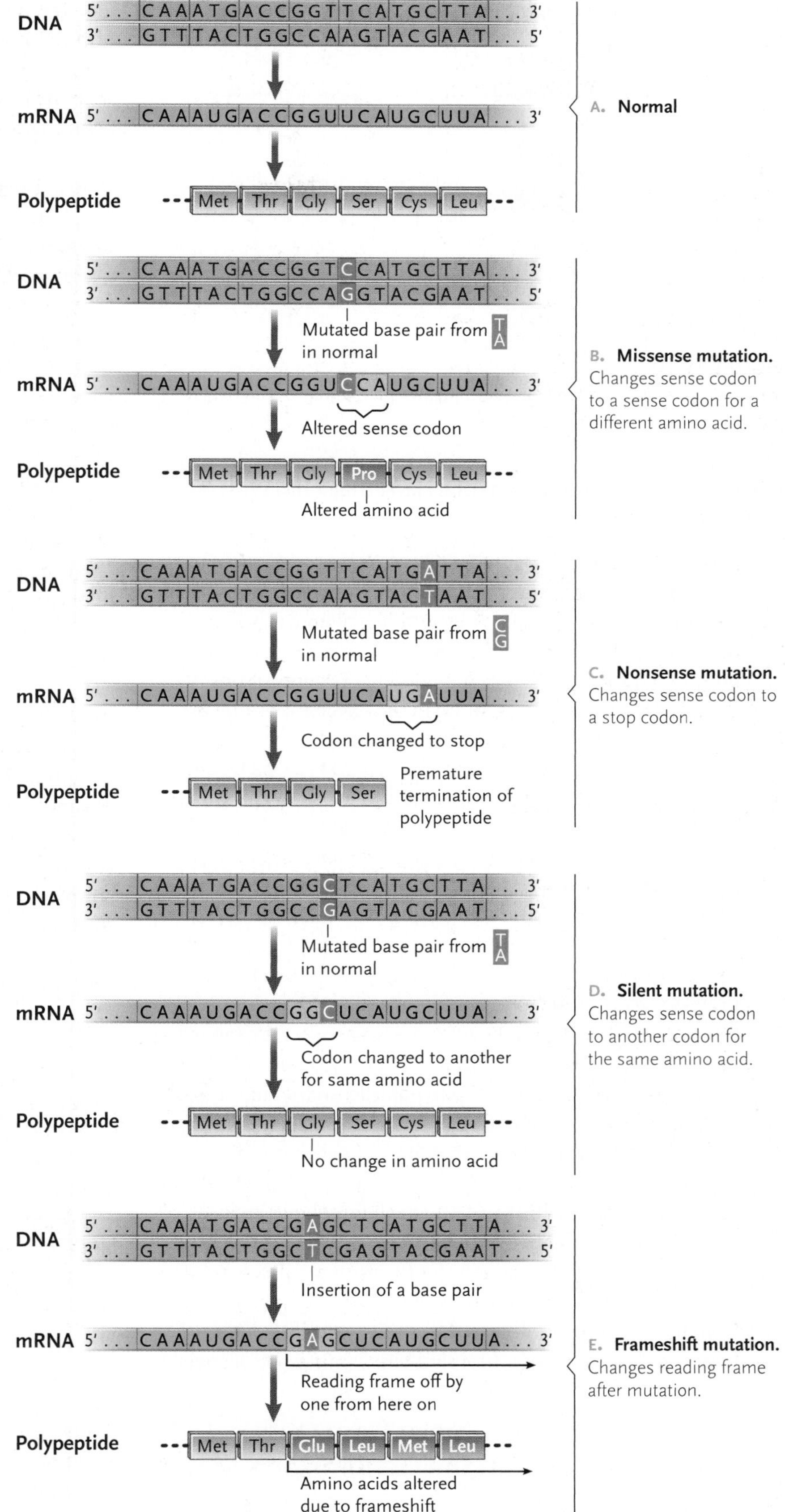
DNA
5' . . . CAAATGACCGGTTCATGCTTA . . . 3'
3' . . . GTTTACTGGCCAAGTACGAAT . . . 5'
mRNA 5' . . . CAAAUGACCGGUUCAUGCUUA . . . 3'
Polypeptide
Met Thr Gly Ser Cys Leu
A. Normal
DNA
5' . . . CAAATGACCGGTCCATGCTTA . . . 3'
3' . . . GTTTACTGGCCAGGTACGAAT . . . 5'
Mutated base pair from T A in normal
mRNA 5' . . . CAAAUGACCGGUCCAUGCUUA . . . 3'
Altered sense codon
Polypeptide
Met Thr Gly Pro Cys Leu
Altered amino acid
B. Missense mutation.
Changes sense codon to a sense codon for a different amino acid.
DNA
5' . . . CAAATGACCGGTTCATGATTA . . . 3'
3' . . . GTTTACTGGCCAAGTACTAAT . . . 5'
Mutated base pair from C G in normal
mRNA 5' . . . CAAAUGACCGGUUCAUGAUUA . . . 3'
Codon changed to stop
Polypeptide
Met Thr Gly Ser
Premature termination of polypeptide
C. Nonsense mutation.
Changes sense codon to a stop codon.
DNA
5' . . . CAAATGACCGGCTCATGCTTA . . . 3'
3' . . . GTTTACTGGCCGAGTACGAAT . . . 5'
Mutated base pair from T A in normal
mRNA 5' . . . CAAAUGACCGGCUCAUGCUUA . . . 3'
Codon changed to another for same amino acid
Polypeptide
Met Thr Gly Ser Cys Leu
No change in amino acid
D. Silent mutation.
Changes sense codon to another codon for the same amino acid.
DNA
5' . . . CAAATGACCGAGCTCATGCTTA . . . 3'
3' . . . GTTTACTGGCTCGAGTACGAAT . . . 5'
Insertion of a base pair
mRNA 5' . . . CAAAUGACCGAGCUCAUGCUUA . . . 3'
Reading frame off by one from here on
Polypeptide
Met Thr Glu Leu Met Leu
Amino acids altered due to frameshift
E. Frameshift mutation.
Changes reading frame after mutation.

FIGURE 15.20
Effects of base-pair mutations in protein-coding genes on the amino acid sequence of the encoded polypeptide.

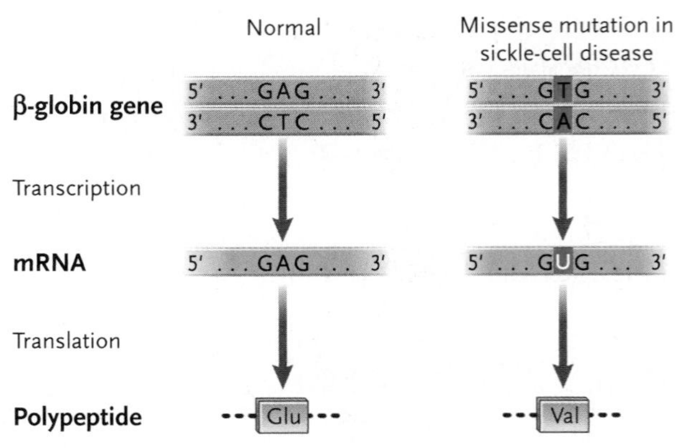
Normal
Missense mutation in sickle-cell disease
β-globin gene
5' . . . GAG . . . 3'
3' . . . CTC . . . 5'
5' . . . GTG . . . 3'
3' . . . CAC . . . 5'
Transcription
mRNA
5' . . . GAG . . . 3'
5' . . . GUG . . . 3'
Translation
Polypeptide
Glu
Val

FIGURE 15.21
Missense mutation in a gene for one of the two polypeptides of hemoglobin that is the cause of sickle-cell disease.

of base-pair substitution mutations affecting a protein-coding gene are:

1. **Missense mutation (Figure 15.20B):** A sense codon is changed to a different sense codon that specifies a different amino acid. Whether the function of a polypeptide is altered significantly depends on the amino acid change that occurs. Individuals homozygous for a missense mutation in the gene for one of the two polypeptide types found in the oxygen-carrying protein hemoglobin **(Figure 15.21)** have the genetic disease sickle-cell disease, described in Chapter 12 (pp. 234–235; 247; 252). Many other human genetic diseases are caused by missense mutations, including albinism, hemophilia, and achondroplasia (see Chapter 13 *Insights from the Molecular Revolution*).
2. **Nonsense mutation (Figure 15.20C):** A sense codon is changed to a nonsense (stop) codon. Translation of an mRNA containing a nonsense mutation results in a shorter-than-normal polypeptide and, in many cases, this polypeptide will be only partially functional at best.
3. **Silent mutation (Figure 15.20D):** A sense codon is changed to a different sense codon but that codon specifies the same amino acid as in the normal polypeptide, so the function of the polypeptide is unchanged.
4. **Frameshift mutation (Figure 15.20E):** A single base pair deletion or insertion in the coding region of a gene alters the reading frame of the resulting mRNA (the figure shows an insertion). After the point of the mutation, the ribosome reads codons that are not the same as for the normal mRNA, producing a different amino acid sequence in the polypeptide from then on. The resulting polypeptide typically is nonfunctional because of the significantly altered amino acid sequence.

Transposable Elements Move from One Location to Another in the Genome and May Affect Gene Function

All organisms contain particular segments of DNA that can move from one place to another within a cell's genome. The movable sequences are called *transposable genetic elements,* or more simply, **transposable elements (TEs).**

The movement of TEs, called **transposition,** involves a type of genetic recombination process. However, the location in the DNA where the TE moves to—the **target site**—is not homologous with the TE. In this respect, transposition differs from genetic recombination in meiosis in eukaryotes, and in the processes that produce recombinants in bacteria (see Chapter 17), which involve crossing-over between homologous DNA molecules.

Transposition of a TE occurs at a low frequency. Depending on the TE, transposition occurs in one of two ways: (1) a cut-and-paste process, in which the TE leaves its original location and transposes to a new location **(Figure 15.22A);** or (2) a copy-and-paste process, in which a copy of a TE transposes to a new location, leaving the original TE behind **(Figure 15.22B).** For most TEs, transposition starts with contact between the TE and the target site. This also means that TEs do not exist free of the DNA in which they are integrated.

TEs are important because of the genetic changes they cause. For example, they produce mutations by transposing into and thereby disrupting the coding sequences of genes, knocking out their functions. If they transpose instead into regulatory sequences of genes, they may increase or decrease the level of gene expression of those genes. As such, TEs are an important source of genetic variability.

BACTERIAL TRANSPOSABLE ELEMENTS Bacterial TEs were discovered in the 1960s. They move from site to site within the bacterial chromosome, between the bacterial chromosome and plasmids, and between plasmids.

The two major types of bacterial TEs are *insertion sequences (IS)* and *transposons* **(Figure 15.23). Insertion sequences** are the simplest TEs. They are relatively small and contain only the gene for **transposase,** an enzyme that catalyzes the reactions inserting or removing the TE from the DNA (see Figure 15.23A). At the two ends of an IS is a short *inverted repeat* sequence—the same DNA sequence running in opposite directions (shown by directional arrows in the figure). The inverted repeat sequences enable the transposase enzyme to identify the ends of the TE when it catalyzes transposition.

The second type of bacterial TE, called a **transposon,** has an inverted repeat sequence at each end enclosing a central region with one or more genes (Figure 15.23B). In a number of bacterial transposons, the inverted repeat sequences are insertion sequences, one of which provides the transposase for movement of the element. Bacterial transposons without IS ends have short inverted repeat end sequences, and a transposase gene is within the central region. Additional gene(s) in the central region of both types of transposons typically are for antibiotic resistance; they originated from the main bacterial DNA circle or from plasmids. The additional genes are carried along as the TEs move from place to place within and between species.

Many antibiotics, such as penicillin, erythromycin, tetracycline, ampicillin, and streptomycin, that were once successful in curing bacterial infections have lost much of their effectiveness because of resistance genes carried in transposons. Movements of the transposons, particularly to plasmids that have been transferred between bacteria within the same species and between

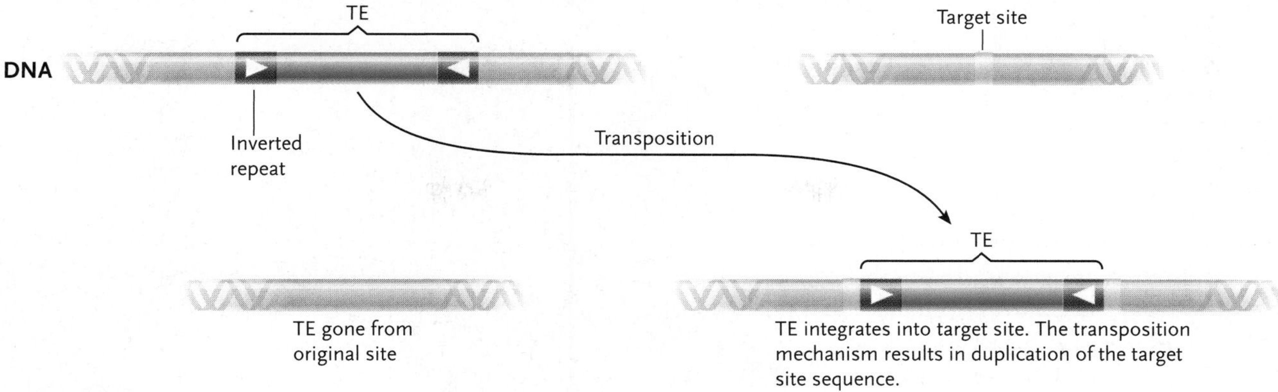
A. Cut-and-paste transposition. The TE leaves one location in the DNA and moves to a new location.
TE
Target site
DNA
Inverted repeat
Transposition
TE
TE gone from original site
TE integrates into target site. The transposition mechanism results in duplication of the target site sequence.

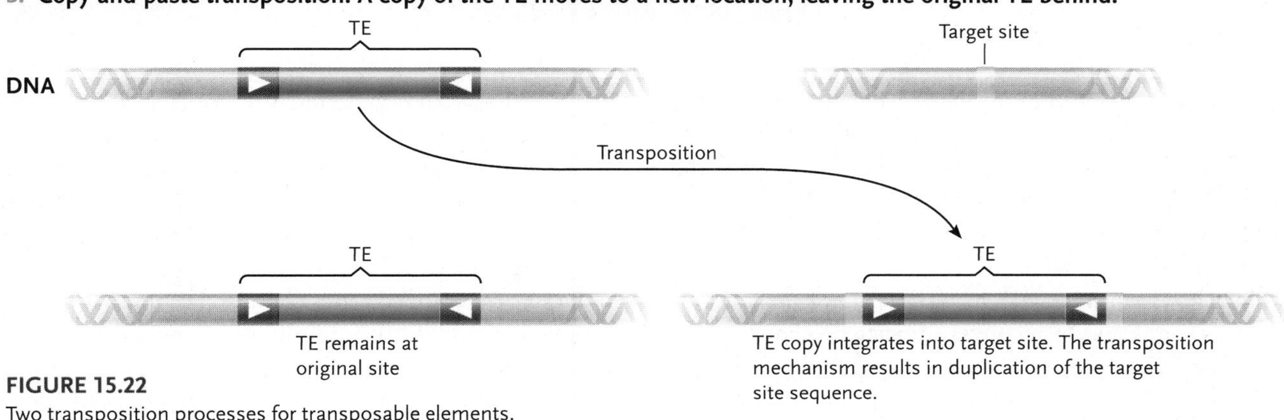
B. Copy-and-paste transposition. A copy of the TE moves to a new location, leaving the original TE behind.
TE
Target site
DNA
Transposition
TE
TE
TE remains at original site
TE copy integrates into target site. The transposition mechanism results in duplication of the target site sequence.

FIGURE 15.22
Two transposition processes for transposable elements.

different species, greatly increase the spread of genes providing antibiotic resistance. Resistance genes have made many bacterial diseases difficult or impossible to treat with standard antibiotics. (Chapter 25 discusses bacterial antibiotic resistance further.)

TRANSPOSABLE ELEMENTS IN EUKARYOTES TEs actually were first discovered in a eukaryote, maize (corn), in the 1940s by Barbara McClintock, a geneticist working at the Cold Spring Harbor Laboratory in New York. McClintock noted that some mutations affecting kernel and leaf color appeared and disappeared rapidly under certain conditions. Mapping the alleles by linkage studies produced a surprising result—the map positions changed frequently, indicating that the alleles could move from place to place in the corn chromosomes. Some of the movements were so frequent that changes in their effects could be noticed at different times in a single developing kernel **(Figure 15.24).**

When McClintock first reported her results, her findings were regarded as an isolated curiosity, possibly applying only to corn. This was because the then-prevailing opinion among geneticists was that genes are fixed in the chromosomes and do not move to other locations. Her conclusions were widely accepted only after TEs were detected and characterized in bacteria in the 1960s. By the 1970s, further examples of TEs were discovered in other eukaryotes, including yeast and mammals. We now know that TEs are probably universally distributed among both prokaryotes and eukaryotes. McClintock was awarded a Nobel Prize in 1983 for her discovery of mobile genetic elements.

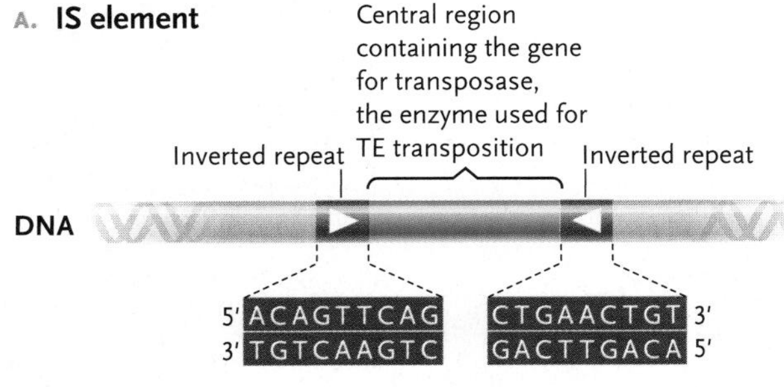
A. IS element
Central region containing the gene for transposase, the enzyme used for TE transposition
Inverted repeat
Inverted repeat
DNA
5′ ACAGTTCAG
3′ TGTCAAGTC
CTGAACTGT 3′
GACTTGACA 5′

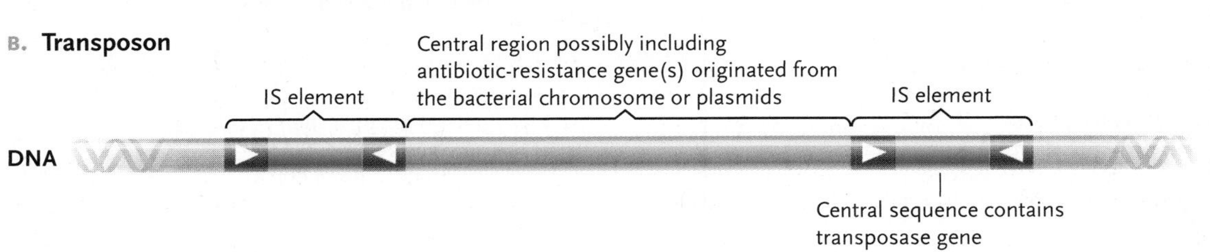
B. Transposon
Central region possibly including antibiotic-resistance gene(s) originated from the bacterial chromosome or plasmids
IS element
IS element
DNA
Central sequence contains transposase gene

FIGURE 15.23
Types of bacterial transposable elements. **(A)** Insertion sequence. **(B)** Transposon.

FIGURE 15.24
Barbara McClintock and corn kernels showing different color patterns because of the movement of transposable elements. As TEs move into or out of genes controlling pigment production in developing kernels, the ability of cells and their descendants to produce the dark pigment is destroyed or restored. The result is random patterns of pigmented and colorless (yellow) segments in individual kernels.

Nik Kleinberg

Eukaryotic TEs fall into two major classes, *transposons* and *retrotransposons,* distinguished by the way the TE sequence moves from place to place in the DNA. Researchers detect both classes of eukaryotic TEs through DNA sequencing or through their effects on genes at or near their sites of insertion. Unlike prokaryotes, eukaryotes have no TEs resembling insertion sequences.

Eukaryotic transposons are similar to bacterial transposons in their general structure and in the way they transpose. A gene for transposase is in the central region of the transposon, and most have inverted repeat sequences at their ends. Depending on the transposon, transposition is by the cut-and-paste or copy-and-paste mechanism (see Figure 15.22).

Members of the other class of eukaryotic TEs, the **retrotransposons,** transpose by a copy-and-paste mechanism but, unlike the other TEs we have discussed, their transposition occurs via an intermediate RNA copy of the TE **(Figure 15.25).** Some retrotransposons are bounded by sequences that are directly repeated rather than in inverted form; others have no repeated sequences at their ends.

Cellular genes may become incorporated into the central region of either a transposon or a retrotransposon and travel with it as it moves to a new location. The trapped genes may become continuously active through the effects of regulatory sequences in the TE. The trapped genes may also become abnormally active if moved in a TE to the vicinity of a regulatory region or promoter of an intensely transcribed cellular gene. Certain forms of cancer have been linked to the TE-instigated abnormal activation of genes that regulate cell division.

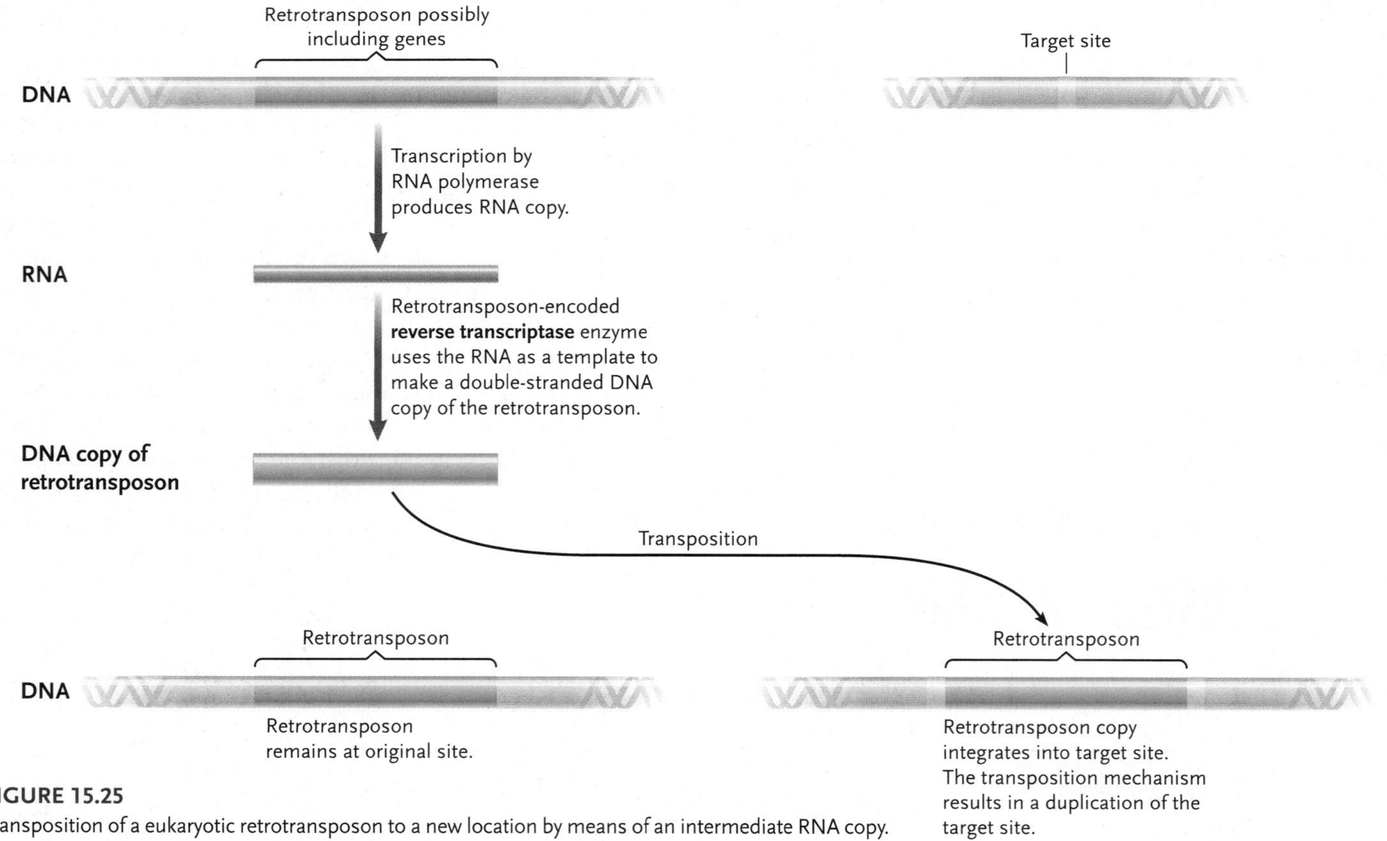
Retrotransposon possibly including genes
DNA
Target site
Transcription by RNA polymerase produces RNA copy.
RNA
Retrotransposon-encoded **reverse transcriptase** enzyme uses the RNA as a template to make a double-stranded DNA copy of the retrotransposon.
DNA copy of retrotransposon
Transposition
Retrotransposon
DNA
Retrotransposon
Retrotransposon remains at original site.
Retrotransposon copy integrates into target site. The transposition mechanism results in a duplication of the target site.

FIGURE 15.25
Transposition of a eukaryotic retrotransposon to a new location by means of an intermediate RNA copy.

Once TEs are inserted in the chromosomes, they become more or less permanent residents, duplicated and passed on during cell division along with the rest of the DNA. TEs inserted into the DNA of reproductive cells that produce gametes may be inherited, thereby becoming a permanent part of the genetic material of a species.

Long-standing TEs are subject to mutation along with other sequences in the genome. Such mutations may accumulate in a TE, gradually altering it into a nonmobile, residual sequence in the DNA. The genomes of many eukaryotes, including humans, contain many nonfunctional TEs likely created in this way.

In this chapter we have learned how gene expression occurs by the processes of transcription and translation. We also learned how gene expression may be changed by mutation or by the actions of transposable elements. In the next chapter, you will see how organisms and cells exert control over how their genes are expressed.

STUDY BREAK 15.5 <

1. **How does a missense mutation differ from a silent mutation?**
2. **How do genetic recombination and TE transposition differ?**
3. **In what ways are bacterial IS elements and transposons alike?**
4. **How do eukaryotic transposons and retrotransposons differ?**

UNANSWERED QUESTIONS

How does the ribosome work?

When they were first discovered, ribosomes were thought to be passive "workbenches" on which proteins were synthesized, presumably by enzymes. But in the late 1960s, it was found that the ribosome itself catalyzes formation of peptide bonds. Binding and lining up the mRNA and the correct tRNAs were other things that appeared to be jobs carried out by the ribosome. But the ribosome is a huge macromolecular complex, containing more than 50 ribosomal proteins and large ribosomal RNAs containing many thousands of nucleotides.

In my laboratory, we wanted to identify the parts of the ribosome that carried out these important functions, as a step toward understanding *how* they manage to do it. Following the common wisdom that it is the proteins that carry out the biological functions of the cell, we assumed that ribosomal proteins were responsible for ribosome functions, but which ones? We decided to use *chemical modification* to approach this question: Our plan was to inactivate the functions of the ribosome by attacking the ribosome with chemical reagents that were known to modify the reactive groups in proteins; we would then figure out which proteins were inactivated by rescuing the inactivated proteins with individual active ones, using *in vitro* reconstitution of a functional ribosome as the measure for a successful result. To our great surprise, it was very difficult to inactivate the ribosome using reagents that react with proteins. This disappointed me, because I was eager to use my skills as a protein chemist to understand how the ribosome works. As a sort of control experiment, we tried using a reagent that attacks RNA, and for the first time we observed rapid inactivation! Furthermore, ribosome activity could be rescued by reconstituting with active RNA, proving that inactivation of the ribosome was caused by modification of the ribosomal RNA. In 1972 we published a paper proposing that ribosomal RNA plays a functional role in protein synthesis, and followed it with evidence from many other kinds of studies in the following years, but no one was ready to believe such a "crackpot idea" (as one prominent scientist was heard to describe our findings).

Twenty years after our original proposal, I was frustrated that people were still resistant to the idea, although less so because of the discovery of ribozymes (catalytic RNAs) in the 1980s. So in 1992 we carried out another set of experiments in which we subjected ribosomes to relatively brutal extraction procedures that are used to remove proteins from complexes with RNA and DNA (see *Insights from the Molecular Revolution* for this chapter). We washed the ribosomes with a strong detergent, treated them with a protease that is known to chew up proteins into small fragments and vigorously extracted them with phenol, which is known to separate protein from RNA. At the end of this ordeal, the ribosomes were completely active in catalyzing formation of peptide bonds! Although we had not completely removed all of the protein from the ribosome, the idea that the biological activity of ribosomes might be based on its RNA, rather than its proteins, now became accepted widely by researchers. It was finally demonstrated convincingly by high-resolution crystallography of the ribosome that the site of peptide bond formation contains no protein, only RNA.

This experience provides insight into how science works; the scientific community is resistant to changes in fundamental paradigms ("Only proteins can carry out biological functions"), and acceptance of new ideas often depends on proposing them at "the right time." Also typical of science, the new insight leaves many questions still unanswered. Even after 50 years of research by hundreds of investigators, we are still trying to figure out the ribosome's fundamental mechanisms of action. And an even more daunting question is, how did the ribosome evolve?

Think Critically

1. Ribosomes contain both RNA and protein. Because ribosomes are responsible for making proteins, how can you explain the evolution of the first ribosome?
2. If you think you can answer that question, how about the following: The chance that you could make a functional protein from mRNA containing a random RNA sequence is essentially zero. Why, then, would you evolve a ribosome in the first place?

Harry Noller

Harry Noller is Robert L. Sinsheimer Professor of Molecular Biology at UC, Santa Cruz, where he studies the structure and function of the ribosome, using such diverse approaches as biochemistry, molecular genetics, fluorescence spectroscopy, and X-ray crystallography. To learn more about his research, go to http://rna.ucsc.edu/rnacenter/noller_lab.html.

REVIEW KEY CONCEPTS

Go to **CENGAGENOW** at www.cengage.com/login to access quizzing, animations, exercises, articles, and personalized homework help.

15.1 The Connection between DNA, RNA, and Protein

- In their genetic experiments with *Neurospora crassa,* Beadle and Tatum found a direct correspondence between gene mutations and alterations of enzymes. Their one gene–one enzyme hypothesis was updated as the one gene–one polypeptide hypothesis (Figure 15.2).
- The pathway from genes to proteins involves transcription then translation. In transcription, a sequence of nucleotides in DNA is copied into a complementary sequence in an RNA molecule. In translation, the sequence of nucleotides in an mRNA molecule specifies an amino acid sequence in a polypeptide (Figure 15.3).
- The genetic code is a triplet code. AUG at the beginning of a coded message establishes a reading frame for reading the codons three nucleotides at a time until a stop codon is reached. The code is degenerate: most of the amino acids are specified by more than one codon (Figures 15.4 and 15.5).
- The genetic code is essentially universal.

Animation: Uracil–thymine comparison

Animation: Protein synthesis summary

Practice: The major differences between prokaryotic and eukaryotic protein synthesis

15.2 Transcription: DNA-Directed RNA Synthesis

- Transcription, the process by which information coded in DNA is transferred to a complementary RNA copy, begins when an RNA polymerase binds to a promoter sequence in the DNA and starts synthesizing an RNA molecule. The enzyme then adds RNA nucleotides in sequence according to the DNA template. At the end of the transcribed sequence, the enzyme and the completed RNA transcript release from the DNA template.
- Transcription occurs in three stages: initiation, elongation, and termination. Each stage is similar in eukaryotes and bacteria. For transcription of eukaryotic protein-coding genes, transcription factors bind to the promoter and recruit RNA polymerase II which then begins transcription of the mRNA. Elongation of the mRNA occurs as the polymerase reads the coding sequence of the gene. The 3′ end of the mRNA is specified differently in prokaryotes and eukaryotes; termination of transcription in bacteria is determined by a terminator sequence in the gene (Figure 15.6).
- Transcription of non-protein-coding genes occurs in a similar way to transcription of protein-coding genes. In eukaryotes, special RNA polymerases are used to transcribe tRNA and rRNA genes, whereas in bacteria the same RNA polymerase that transcribes protein-coding genes transcribes those genes.

Animation: Gene transcription details

15.3 Production of mRNAs in Eukaryotes

- A gene encoding an mRNA molecule includes the promoter, which is recognized by the regulatory proteins and transcription factors that promote DNA unwinding and the initiation of transcription by an RNA polymerase. Transcription in eukaryotes produces a pre-mRNA molecule that consists of a 5′ cap, the 5′ untranslated region, interspersed exons (amino acid-coding segments) and introns, the 3′ untranslated region, and the 3′ poly(A) tail. The 5′ cap and 3′ poly(A) tail are not encoded in the DNA. The 5′ cap is added by a capping enzyme, and the 3′ poly(A) tail is added by poly(A) polymerase once a 3′ end of the pre-mRNA is generated by cleavage downstream of a polyadenylation signal in the transcript (Figure 15.7).
- Introns in pre-mRNAs are removed to produce functional mRNAs by splicing. snRNPs bind to the introns, loop them out of the pre-mRNA, clip the intron at each exon boundary, and join the adjacent exons together (Figure 15.8).
- Many pre-mRNAs are subjected to alternative splicing, a process that joins exons in different combinations to produce different mRNAs encoded by the same gene. Translation of each mRNA produced in this way generates a protein with different function (Figure 15.9).

Animation: Pre-mRNA transcript processing

15.4 Translation: mRNA-Directed Polypeptide Synthesis

- Translation is the assembly of amino acids into polypeptides. Translation occurs on ribosomes. The P, A, and E sites of the ribosome are used for the stepwise addition of amino acids to the polypeptide as directed by the mRNA (Figures 15.10 and 15.13).
- Amino acids are brought to the ribosome attached to specific tRNAs. Amino acids are linked to their corresponding tRNAs by aminoacyl–tRNA synthetases. By matching amino acids with tRNAs, the reactions also provide the ultimate basis for the accuracy of translation (Figures 15.11 and 15.12).
- Translation proceeds through the stages of initiation, elongation, and termination. In initiation, a ribosome assembles with an mRNA molecule and an initiator methionine-tRNA. In elongation, amino acids linked to tRNAs add one at a time to the growing polypeptide chain. In termination, the new polypeptide is released from the ribosome and the ribosomal subunits separate from the mRNA (Figures 15.14–15.16).
- Multiple ribosomes translate the same mRNA simultaneously, forming a polysome (Figure 15.17 and Figure 15.18).
- After they are synthesized on ribosomes, polypeptides are converted into finished form by processing reactions, which include removal of one or more amino acids from the protein chains, addition of organic groups, and folding guided by chaperones.
- Finished proteins are sorted to the cellular locations where they function. Proteins that function in the cytosol are synthesized on free ribosomes. Proteins are sorted to the endomembrane system by cotranslational import and proteins are sorted to the mitochondria, chloroplast, microbodies, and nucleus by posttranslational import. In each of these two cases, specific amino acid sequences in the polypeptides direct them to their destinations (Figure 15.19).

Animation: Translation

Animation: Structure of a ribosome

15.5 Genetic Changes That Affect Protein Structure and Function

- Base-pair substitution mutations alter the mRNA and can lead to changes in the amino acid sequence of the encoded polypeptide. A missense mutation changes one sense codon to one that specifies a different amino acid, a nonsense mutation changes a sense codon to a stop codon, and a silent mutation changes one sense codon to another sense codon that specifies the same amino acid. A base-pair insertion or deletion is a frameshift mutation that alters the reading frame beyond the point of the mutation, leading to a different amino acid sequence from then on in the polypeptide (Figures 15.20 and 15.21).

- Both prokaryotes and eukaryotes contain TEs (transposable elements)—DNA sequences that can move from place to place in the DNA. The TEs may move from one location in the DNA to another, or generate duplicated copies that insert in new locations while leaving the "parent" copy in its original location (Figure 15.22).
- Genes of the host cell DNA may become incorporated into a TE and may be carried with it to a new location. There, the genes may become abnormally active when placed near sequences that control the activity of genes within the TE, or near the control elements of active host genes.
- Bacterial TEs occur as insertion sequences and transposons. Both contain a gene for the transposase enzyme needed for transposition. The transposon may also contain genes, such as for antibiotic resistance, which originate in host DNA (Figure 15.23).
- Eukaryotic TEs occur as transposons, which release from one location in the DNA and insert at a different site, or as retrotransposons, which move by making an RNA copy, which is then replicated into a DNA copy that is inserted at a new location. The original copy remains at the original location (Figures 15.24–15.25).
- TE-instigated abnormal activation of genes regulating cell division has been linked to the development of some forms of cancer in humans and other complex animals.

Animation: Base-pair substitution

Animation: Frameshift mutation

UNDERSTAND AND APPLY

Test Your Knowledge

1. Which statement about the following pathway is false?

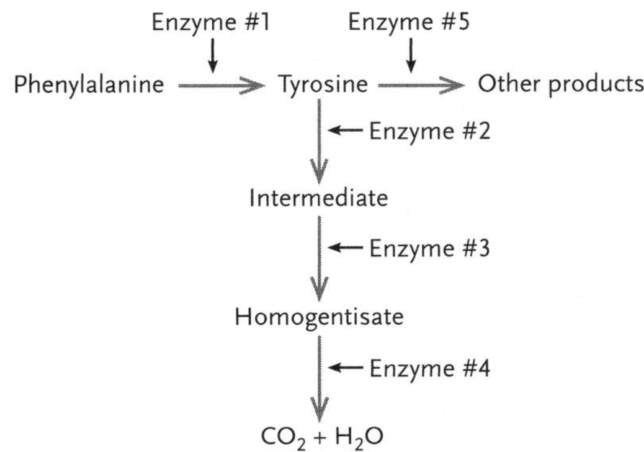
Enzyme #1
Enzyme #5
Phenylalanine
Tyrosine
Other products
Enzyme #2
Intermediate
Enzyme #3
Homogentisate
Enzyme #4
$CO_2 + H_2O$

   a. A mutation for enzyme #1 causes phenylalanine to build up.
   b. A mutation for enzyme #2 prevents tyrosine from being synthesized.
   c. A mutation at enzyme #3 prevents homogentistate from being synthesized.
   d. A mutation for enzyme #2 could hide a mutation in enzyme #4.
   e. Each step in a pathway such as this is catalyzed by an enzyme, which is coded by a gene.

2. Eukaryotic mRNA:
   a. uses snRNPs to cut out introns and seal together translatable exons.
   b. uses a spliceosome mechanism made of DNA to recognize consensus regions to cut and splice.
   c. has a guanine cap on its 3′ end and a poly(A) tail on its 5′ end.
   d. is composed of adenine, thymine, guanine, and cytosine.
   e. codes the guanine cap and poly(A) tail from the DNA template.

3. A segment of a strand of DNA has a base sequence of 5′-GCATTAGAC-3′. What would be the sequence of an RNA molecule complementary to that sequence?
   a. 5′-GUCTAATGC-3′
   b. 5′-GCAUUAGAC-3′
   c. 5′-CGTAATCTG-3′
   d. 5′-GUCUAAUGC-3′
   e. 5′-CGUAAUCUG-3′

4. Which of the following statements about the initiation phase of translation is false?
   a. An initiation factor allows 5′ mRNA to attach to the small ribosomal subunit.
   b. Initiation factors complex with GTP to help Met–tRNA and AUG pair.
   c. mRNA attaches first to the small ribosomal subunit.
   d. GTP is synthesized.
   e. 3′-UAC-5′ on the tRNA binds 5′-AUG-3′ on mRNA.

5. Which of the following statements about aminoacylation is false?
   a. It precedes translation.
   b. It occurs in the ribosome.
   c. It requires ATP to bind an aminoacyl–tRNA synthetase.
   d. It joins the correct amino acid to a specific tRNA based on the tRNA's anticodon.
   e. It uses three binding sites on aminoacyl–tRNA synthetase.

6. Which of the following statements is false?
   a. GTP is an energy source during various stages of translation.
   b. In the ribosome, peptidyl transferase catalyses peptide bond formation between amino acids.
   c. When the mRNA code UAA reaches the ribosome, there is no tRNA to bind to it.
   d. A long polypeptide is cut off the tRNA in the A site so its Met amino acid links to the amino acid in the P site.
   e. Forty-two amino acids of a protein are encoded by 126 nucleotides of the mRNA.

7. Which item binds to SRP receptor and to the signal sequence to guide a newly synthesized protein to be secreted to its proper "channel"?
   a. ribosome
   b. signal recognition particle
   c. endoplasmic reticulum
   d. signal peptidase
   e. receptor protein

8. A part of an mRNA molecule with the sequence 5′-UGC GCA-3′ is being translated by a ribosome. The following activated tRNA molecules are available. Two of them can correctly bind the mRNA so that a dipeptide can form.

tRNA Anticodon	Amino Acid
3′-GGC-5′	Proline
3′-CGU-5′	Alanine
3′-UGC-5′	Threonine
3′-CCG-5′	Glycine
3′-ACG-5′	Cysteine
3′-CGG-5′	Alanine

   a. cysteine–alanine
   b. proline–cysteine
   c. glycine–cysteine
   d. alanine–alanine
   e. threonine–glycine

9. A missense mutation cannot be:
   a. the code for the sickle-cell gene.
   b. caused by a frameshift.
   c. the deletion of a base in a coding sequence.
   d. the addition of two bases in a coding sequence.
   e. the same as a silent mutation.
10. Which of the following is *not* correct about transposable elements?
   a. They can be recognized by their ends of inverted transposable elements.
   b. They have an internal portion that can be transcribed.
   c. They encode a transposase enzyme.
   d. They have no harmful effects on cell function.
   e. They move by a cut-and-paste or copy-and-paste mechanism.

Discuss the Concepts

1. Which do you think are more important to the accuracy by which amino acids are linked into proteins: nucleic acids or enzymatic proteins? Why?
2. A mutation occurs that alters an anticodon in a tRNA from 3′-AAU-5′ to 3′-AUU-5′. What effect will this mutation have on protein synthesis?
3. The normal form of a gene contains the nucleotide sequence:

   5′-ATGCCCGCCTTTGCTACTTGGTAG-3′
   3′-TACGGGCGGAAACGATGAACCATC-5′

   When this gene is transcribed, the result is the following mRNA molecule:

   5′-AUGCCCGCCUUUGCUACUUGGUAG-3′

   In a mutated form of the gene, two extra base pairs (underlined) are inserted:

   5′-ATGCCCGCCTAATTGCTACTTGGTAG-3′
   3′-TACGGGCGGATTAACGATGAACCATC-5′

   What effect will this particular mutation have on the structure of the protein encoded in the gene?
4. A geneticist is attempting to isolate mutations in the genes for four enzymes acting in a metabolic pathway in the bacterium *Escherichia coli*. The end product *E* of the pathway is absolutely essential for life:

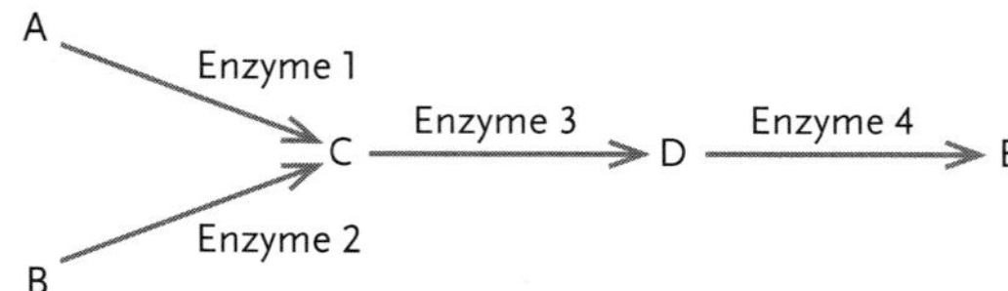
A
Enzyme 1
C
Enzyme 3
D
Enzyme 4
E
Enzyme 2
B

   The geneticist has been able to isolate mutations in the genes for enzymes 1 and 2, but not for enzymes 3 and 4. Develop a hypothesis to explain why.
5. Experimental systems have been developed in which transposable elements can be induced to move under the control of a researcher. Following the induced transposition of a yeast TE element, two mutants were identified with altered activities of enzyme X. One of the mutants lacked enzyme activity completely, whereas the other had five times as much enzyme activity as normal cells did. Both mutants were found to have the TE inserted into the gene for enzyme X. Propose hypotheses for how the two different mutant phenotypes were produced.

Design an Experiment

How could you show experimentally that the genetic code is universal; namely, that it is the same in bacteria as it is in eukaryotes such as fungi, plants, and animals?

Interpret the Data

The figure below shows amino acid changes that occurred at a particular position in a polypeptide as a result of mutations in the gene encoding the polypeptide. The amino acids connected by a line are specified by codons that differ in a single base. Using the genetic code shown in Figure 15.5, deduce the codons that specify the amino acids.

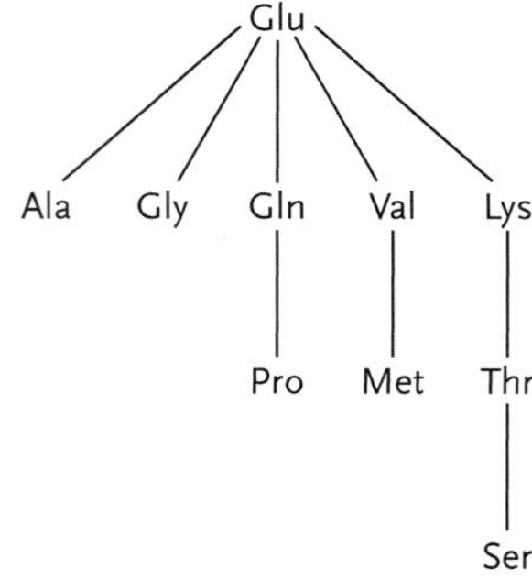
Glu
Ala
Gly
Gln
Val
Lys
Pro
Met
Thr
Ser

Apply Evolutionary Thinking

How might the process of alternative splicing and exon shuffling affect the rate at which new proteins evolve?

Express Your Opinion

Ricin, a molecule that inactivates ribosomes, is difficult to disperse through the air and is unlikely to be used in a large-scale terrorist attack. However, ricin powder did turn up in a Senate office building. Scientists are working to develop a vaccine against ricin. If mass immunizations were to be offered, would you sign up to be vaccinated? Go to academic.cengage.com/login to investigate both sides of the issue and then vote.

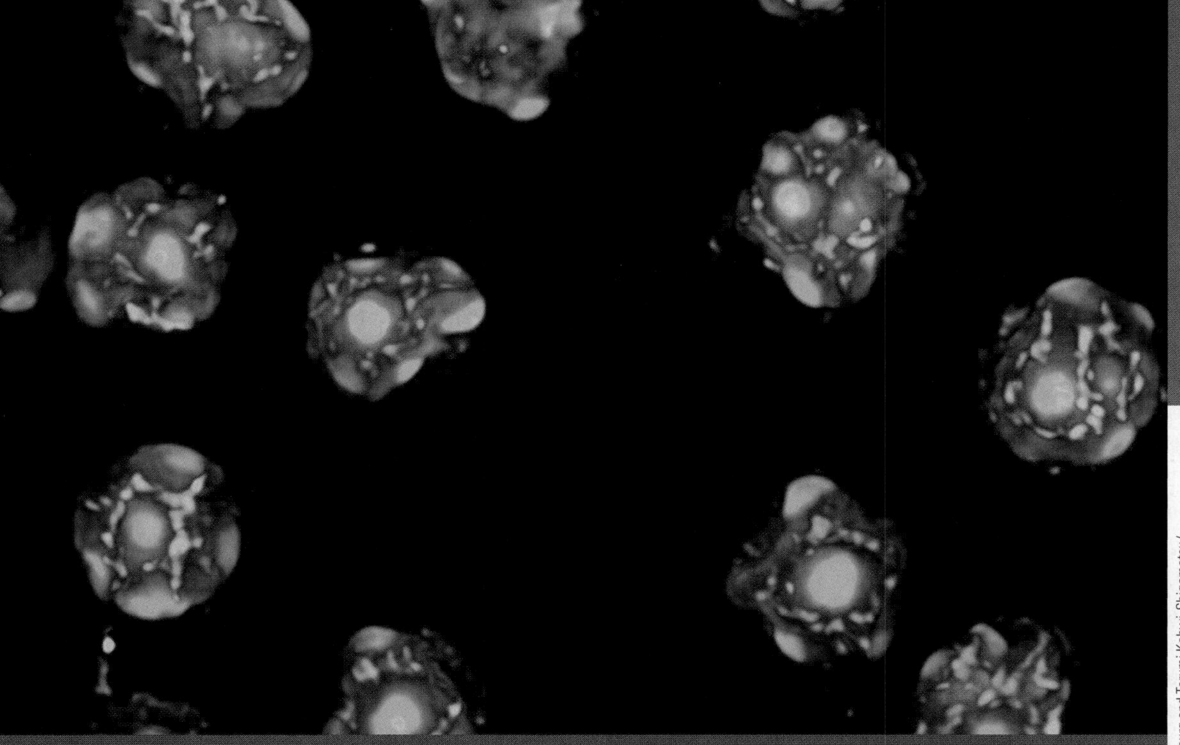

Chromatin remodeling proteins (gold) binding to chromatin (blue). Chromatin remodeling, a change in chromosome structure in the region of a gene, is a key step in the activation of genes in eukaryotes.

Abby Dernburg and Terumi Kohwi-Shigematsu/Lawrence Berkeley National Laboratory

16

STUDY OUTLINE

Regulation of Gene Expression

Why It Matters. . . A human egg cell is almost completely metabolically inactive when it is released from the ovary. It remains quiescent as it begins its journey down the oviduct leading from the ovary to the uterus, carried along by movements of cilia lining the walls of the tube **(Figure 16.1)**. It is here, in the oviduct, that egg and sperm cells meet and embryonic development begins. Within seconds after the cells unite, the fertilized egg breaks its quiescent state and begins a series of divisions that continues as it moves through the fallopian tube and enters the uterus. Subsequent divisions produce specialized cells that *differentiate* into the distinct types tailored for specific functions in the body, such as muscle cells and cells of the nervous system.

All the nucleated cells in both the developing embryo and the adult retain the same set of genes. The structural and functional differences in the cell types are determined not by the presence or absence of certain genes but rather through differences in patterns of *gene expression* that result in cell type-specific sets of proteins. Some genes, known as housekeeping genes, are expressed in almost all cell types; other genes are expressed or not expressed depending on the cell type. Each differentiated cell is characterized by genes that are active in only that cell type. For example, all mammalian cells carry the genes for hemoglobin, but these genes are active only in cells that give rise to red blood cells. Specific regulatory events activate the hemoglobin genes only in the precursor cells to red blood cells and not in other cell types.

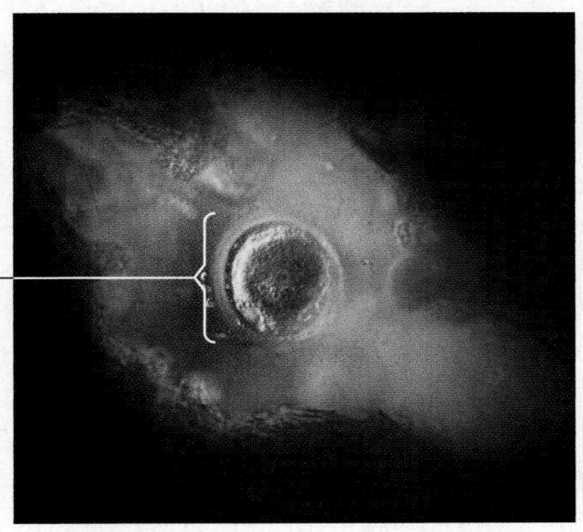
Egg

Lennart Nilsson/Bonnier Fakta

FIGURE 16.1
A human egg at the time of its release from the ovary. The outer layer appearing light blue in color is a coat of polysaccharides and glycoproteins that surrounds the egg. Within the egg, genes and regulatory proteins are poised to enter the pathways initiating embryonic development.

The fundamental mechanisms that control gene expression are common to all multicellular eukaryotes. Even single-celled eukaryotes and prokaryotes have systems that regulate gene expression. With few exceptions, however, prokaryotic systems are limited almost exclusively to short-term responses to environmental changes; eukaryotic cells exhibit both short-term responses and long-term differentiation.

The processes that directly control gene expression are known collectively as **transcriptional regulation.** Transcriptional regulation, the fundamental level of control, determines which genes are transcribed into mRNA. Additional controls fine-tune regulation by affecting the processing of mRNA *(posttranscriptional regulation),* its translation into proteins *(translational regulation),* and the life span and activity of the proteins themselves *(posttranslational regulation).*

These levels of regulation ultimately affect more than proteins, because among the proteins are enzymes that determine the types and kinds of all other molecules made in the developing cell. So, effectively, these regulatory mechanisms tailor the production of all cellular molecules. The entire spectrum of controls constitutes an exquisitely sensitive mechanism regulating when, where, and what kinds and numbers of cellular molecules are produced.

In this chapter we examine the mechanisms of transcriptional regulation and its fine-tuning by additional controls at the posttranscriptional, translational, and posttranslational levels. Our discussion begins with bacterial systems, where researchers first discovered a mechanism for transcriptional regulation, and then moves to eukaryotic systems where regulation of gene activity is more complex. We then discuss the mechanisms by which genes regulate embryonic development, and finally how cancers develop when regulatory controls are lost. <

16.1 Regulation of Gene Expression in Prokaryotes

Transcription and translation are closely regulated in prokaryotes in ways that reflect prokaryotic life histories. Prokaryotes are relatively simple, single-celled organisms with generations that come and go in a matter of minutes. Rather than the complex patterns of long-term cell differentiation and development typical of multicellular eukaryotes, prokaryotic cells typically undergo rapid and reversible alterations in biochemical pathways that allow them to adapt quickly to changes in their environment.

The human intestinal bacterium *Escherichia coli,* for example, can catabolize a number of sugars and other molecules to provide carbon and energy for the cell. One of those sugars is lactose (milk sugar). Lactose is not essential for *E. coli* growth but, when lactose is present, the bacterium makes enzymes for catabolizing the sugar. In the absence of lactose, however, it does not make those enzymes. The versatile and responsive control system allows the bacterium to use the nutrients available in the surrounding medium with maximum efficiency.

The Operon Is the Unit of Transcription in Prokaryotes

When the environment in which a bacterium lives changes, some metabolic processes are stopped and others are started. Typically, this involves turning off the genes for the metabolic processes not needed and turning on the genes for the new metabolic processes. Each metabolic process involves a few to many genes, and the regulation of those genes must be coordinated. For example, three genes encode enzymes for the catabolism of lactose by *E. coli.* In the absence of lactose, the three genes are not transcribed, while in the presence of lactose, the genes are transcribed.

In 1961, François Jacob and Jacques Monod of the Pasteur Institute in Paris proposed the *operon model* for the control of the expression of genes for lactose catabolism in *E. coli.* Subsequently, research has shown that the *operon model* is applicable to the regulation of expression of many genes in prokaryotes and their viruses. Jacob and Monod received the Nobel Prize in 1965 for their discovery of bacterial operons and their regulation by repressors.

An **operon** is a cluster of prokaryotic genes and associated **regulatory sequences.** Regulatory sequences are DNA sequences involved in the regulation of a gene or genes. **Regulatory proteins** bind to these sequences to control the transcription of the gene or genes. One of those regulatory DNA sequences is the **promoter,** which is the site to which RNA polymerase binds to begin transcription. Operons contain several genes and all are transcribed using the single promoter to direct RNA polymerase to start transcription. The result is a single mRNA which contains the amino acid-coding sequences for each of the proteins encoded by the operon's genes. The cluster of genes transcribed into a single mRNA is called a **transcription unit.** Ribosomes translate the mRNA, synthesizing each protein encoded in the mRNA. Typically, the proteins encoded by an operon are involved in the same function, such as enzymes acting in sequence in a biochemical pathway.

The other regulatory DNA sequence in the operon is the **operator,** a short segment to which a regulatory protein binds. The regulatory protein is encoded by a gene that is separate from the operon the protein controls. Some operons are controlled by a regulatory protein termed a **repressor,** which, when active, prevents the operon genes from being expressed. Other operons are controlled by a regulatory protein termed an **activator,** which, when active, turns on the expression of the genes.

Many operons are controlled by more than one regulatory mechanism, and a number of the repressors or activators control more than one operon. The result is a complex network of superimposed controls that provides total regulation of transcription and allows almost instantaneous responses to changing environmental conditions.

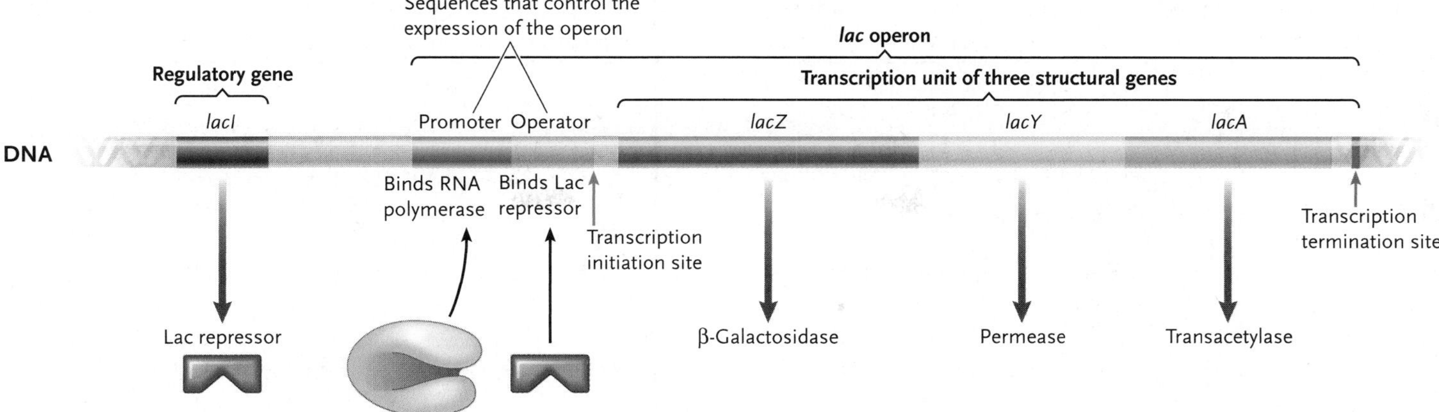
Sequences that control the expression of the operon
lac operon
Regulatory gene
Transcription unit of three structural genes
lacI
Promoter
Operator
lacZ
lacY
lacA
DNA
Binds RNA polymerase
Binds Lac repressor
Transcription initiation site
Transcription termination site
Lac repressor
β-Galactosidase
Permease
Transacetylase

FIGURE 16.2
The *E. coli lac* operon. The *lacZ, lacY,* and *lacA* genes encode the enzymes taking part in lactose metabolism. The separate regulatory gene, *lacI,* encodes the Lac repressor, which plays a pivotal role in the control of the operon. The promoter binds RNA polymerase, and the operator binds activated Lac repressor. The transcription unit, which extends from the transcription initiation site to the transcription termination site, contains the structural genes.

The *lac* Operon for Lactose Metabolism Is Transcribed When an Inducer Inactivates a Repressor

Jacob and Monod researched the genetic control of lactose metabolism in *E. coli.* Lactose is a sugar that, when catabolized, provides energy for the cell. Jacob and Monod used genetic and biochemical approaches to study the genetic control of lactose metabolism in *E. coli.* Their genetic studies showed that the enzyme products of three genes, *lacZ, lacY,* and *lacA,* are involved in lactose catabolism **(Figure 16.2).** The three genes are adjacent to one another on the chromosome in the order *Z-Y-A.* A single promoter is upstream of *lacZ,* and the genes are transcribed as a unit into a single mRNA starting with the *lacZ* gene. The *lacZ* gene encodes the enzyme β-galactosidase, which catalyzes the hydrolysis of the disaccharide sugar, lactose, into the monosaccharide sugars, glucose and galactose. These sugars are then catabolized by other enzymes, producing energy for the cell. The *lacY* gene encodes a permease enzyme that actively transports lactose into the cell. The *lacA* gene encodes a transacetylase enzyme; its function is unclear.

Jacob and Monod called the cluster of genes and adjacent sequences that control their expression the *lac operon* (see Figure 16.2). They coined the name *operon* from a key DNA sequence they discovered between the promoter and the *lacZ* gene that, through binding a protein, regulates transcription of the operon—the *operator.*

The two investigators found that the *lac* operon was a negatively regulated system controlled by a regulatory protein they termed the **Lac repressor.** The Lac repressor is encoded by the regulatory gene *lacI,* which is nearby but separate from the *lac* operon, and is synthesized in active form (see Figure 16.2). In general, we use the term **structural gene** for a gene that encodes a protein that has a function other than gene regulation; examples are the three *lac* operon enzymes. We use the term **regulatory gene** for a gene that encodes a protein that regulates the expression of structural genes.

Figure 16.3A shows how the Lac repressor inhibits transcription when lactose is absent from the medium. When lactose is added to the medium, expression of the *lac* operon increases about 100-fold **(Figure 16.3B).** The molecular switch for this change is set when some lactose molecules are converted by β-galactosidase in the cell to *allolactose,* an isomer of lactose. Allolactose is the **inducer** for the *lac* operon, so-called because it causes—induces—the transcription of the operon's structural genes. Allolactose works by inactivating the Lac repressor (see Figure 16.3B). The *lac* operon is called an **inducible operon** because an inducer molecule increases its expression.

When the lactose is used up from the medium, the regulatory system again switches the *lac* operon off. That is, the absence of lactose means that there are no allolactose inducer molecules to inactivate the repressor so the now-active repressor binds to the operator, blocking transcription of the structural genes. The controls are aided by the fact that bacterial mRNAs are very short-lived, about 3 minutes on the average. This quick turnover permits the cytoplasm to be cleared quickly of the mRNAs transcribed from an operon. The enzymes themselves also have short lifetimes, and are quickly degraded.

Transcription of the *lac* Operon Is Also Controlled by a Positive Regulatory System

A *positive gene regulation* system also regulates the *lac* operon. This system ensures that the *lac* operon is transcribed efficiently if lactose is provided as an energy source, but not if glucose is present in addition to lactose. This is because glucose is a more efficient source of energy than lactose. Glucose can be used directly in the glycolysis pathway to produce energy for the cell (see Chapter 8). Lactose, on the other hand, must first be converted into glucose and galactose, and the galactose then converted into glucose. These conversions require energy from the cell. Thus the cell gains more net energy by catabolizing

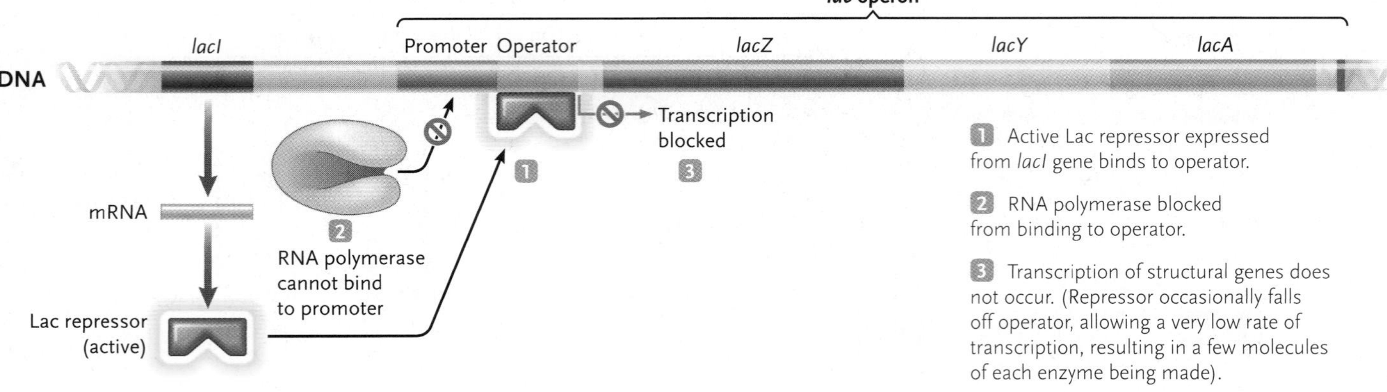
A. Lactose absent from medium: structural genes not transcribed
lac operon
lacI
Promoter Operator
lacZ
lacY
lacA
DNA
mRNA
Lac repressor (active)
RNA polymerase cannot bind to promoter
Transcription blocked
1 Active Lac repressor expressed from lacI gene binds to operator.
2 RNA polymerase blocked from binding to operator.
3 Transcription of structural genes does not occur. (Repressor occasionally falls off operator, allowing a very low rate of transcription, resulting in a few molecules of each enzyme being made).

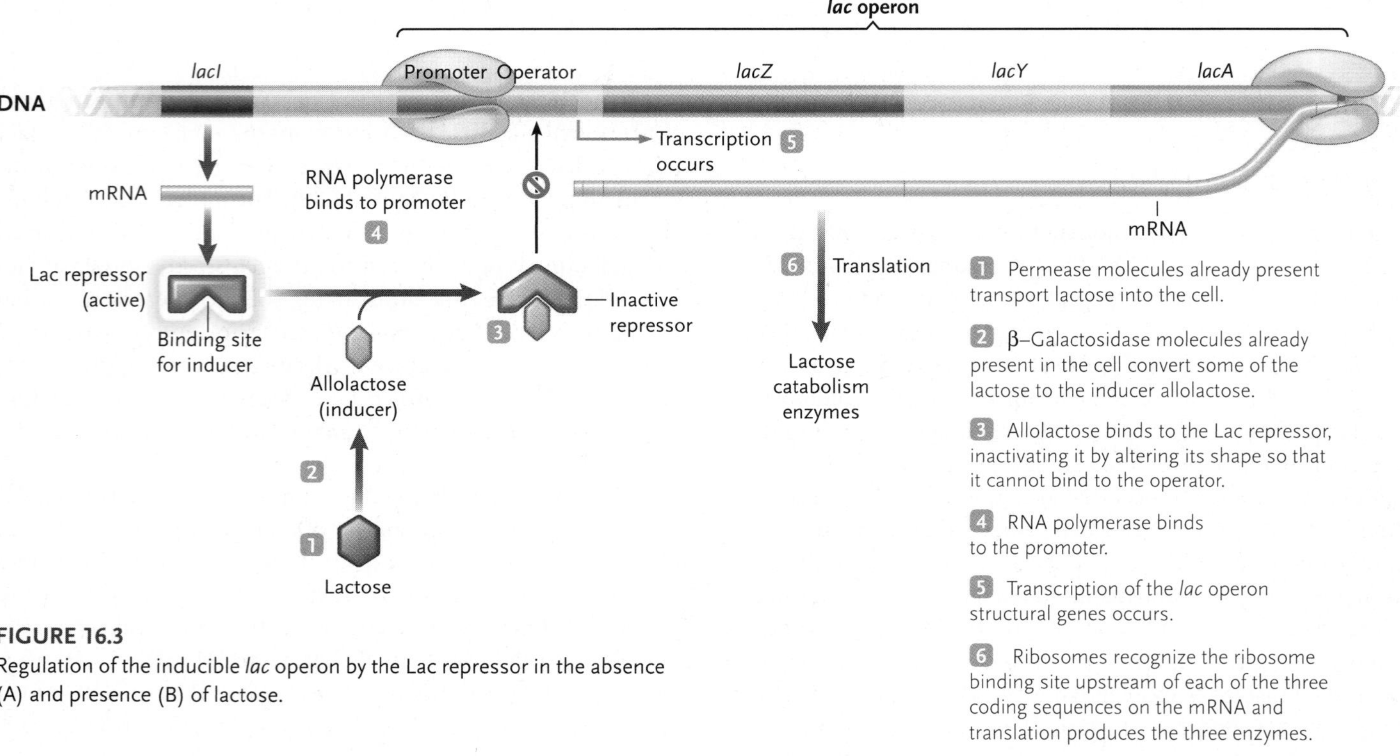
B. Lactose present in medium: structural genes transcribed
lac operon
lacI
Promoter Operator
lacZ
lacY
lacA
DNA
mRNA
RNA polymerase binds to promoter
Transcription occurs
mRNA
Lac repressor (active)
Binding site for inducer
Inactive repressor
Allolactose (inducer)
Lactose
Translation
Lactose catabolism enzymes
1 Permease molecules already present transport lactose into the cell.
2 β–Galactosidase molecules already present in the cell convert some of the lactose to the inducer allolactose.
3 Allolactose binds to the Lac repressor, inactivating it by altering its shape so that it cannot bind to the operator.
4 RNA polymerase binds to the promoter.
5 Transcription of the lac operon structural genes occurs.
6 Ribosomes recognize the ribosome binding site upstream of each of the three coding sequences on the mRNA and translation produces the three enzymes.

FIGURE 16.3
Regulation of the inducible *lac* operon by the Lac repressor in the absence (A) and presence (B) of lactose.

glucose than by catabolizing lactose, or for that matter, any other sugar.

Figure 16.4 shows the positive gene regulation system under the two conditions of lactose present + glucose low or absent (efficient transcription of *lac* operon genes), and lactose present + glucose present (very low level of transcription of *lac* operon genes). In essence, we are adding to the model shown Figure 16.3B. The key regulatory molecule involved in positive gene regulation of the *lac* operon is **CAP (catabolite activator protein).** CAP is an **activator,** a regulatory protein that stimulates gene expression. It is synthesized in *inactive* form. When activated by cAMP (cyclic AMP, a nucleotide that plays a role in regulating cellular processes in both prokaryotes in eukaryotes; see Section 7.4), CAP binds to the **CAP site** in the promoter for the *lac* operon and enables RNA polymerase to bind efficiently and transcribe the operon's genes. If glucose is absent, cAMP levels are high, resulting in active CAP, whereas if glucose is present, cAMP levels are low, resulting in inactive CAP.

The same positive gene regulation system using CAP and cAMP regulates a large number of other operons that control the catabolism of many sugars. In each case, the system functions so that glucose, if it is present in the growth medium, is catabolized first.

Transcription of the *trp* Operon Genes for Tryptophan Biosynthesis Is Repressed When Tryptophan Activates a Repressor

Tryptophan is an amino acid that is used in the synthesis of proteins. As such, tryptophan is a critical component for cell growth and survival since protein synthesis is the main activity that the cell performs. Therefore, if tryptophan is absent from

A. Lactose present and glucose low or absent: structural genes expressed at high levels

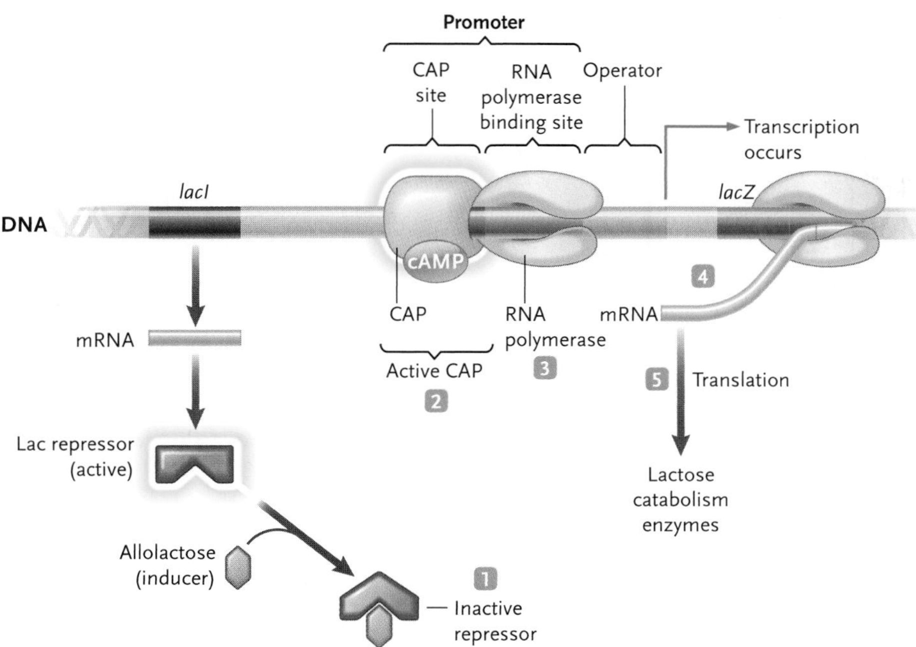
Promoter
CAP site
RNA polymerase binding site
Operator
Transcription occurs
lacI
lacZ
DNA
cAMP
4
CAP
RNA polymerase
mRNA
mRNA
Active CAP
2
3
5
Translation
Lac repressor (active)
Lactose catabolism enzymes
Allolactose (inducer)
1
Inactive repressor

1 Lactose converted to the inducer, allolactose, which inactivates Lac repressor.

2 Active adenylyl cyclase synthesizes cAMP to high levels. cAMP binds to activator CAP, activating it. Activated CAP binds to CAP site in the promoter.

3 RNA polymerase binds efficiently to the promoter.

4 Genes of operon transcribed to high levels.

5 Translation produces high amounts of enzymes.

B. Lactose present and glucose present: structural genes expressed at very low levels

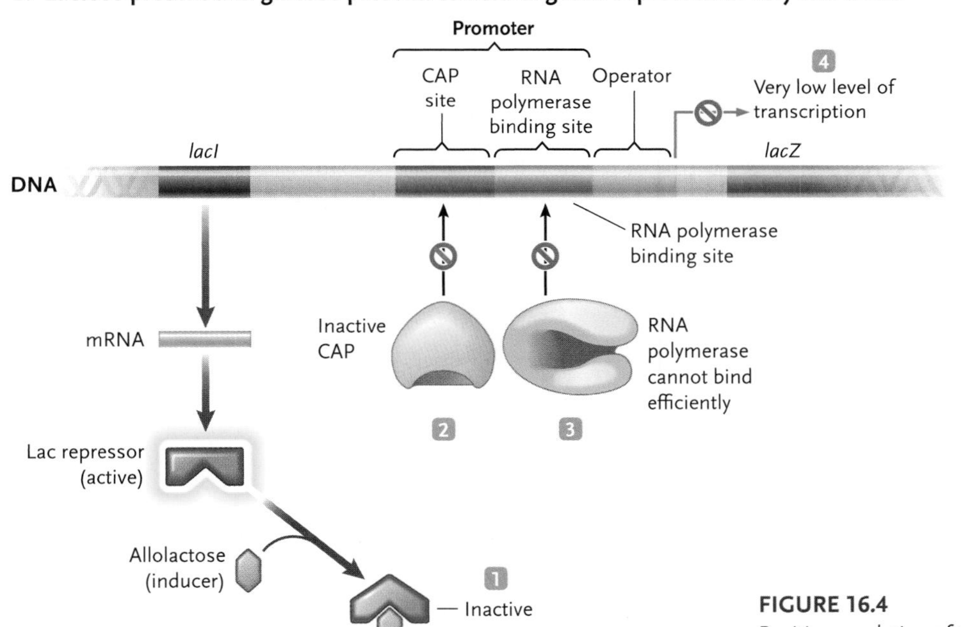
Promoter
CAP site
RNA polymerase binding site
Operator
4
Very low level of transcription
lacI
lacZ
DNA
RNA polymerase binding site
mRNA
Inactive CAP
RNA polymerase cannot bind efficiently
2
3
Lac repressor (active)
Allolactose (inducer)
1
Inactive repressor

1 Lactose converted to the inducer, allolactose, which inactivates Lac repressor.

2 Catabolism of incoming glucose leads to inactivation of adenylyl cyclase, which causes the amount of cAMP in the cell to drop to a level too low to activate CAP. Inactive CAP cannot bind to the CAP site.

3 RNA polymerase is unable to bind to the promoter efficiently.

4 Transcription occurs at a very low level: Because the Lac repressor is not present to block RNA polymerase from binding to the promoter, the level of transcription is higher than when lactose is absent, but far lower than when lactose is present and glucose is absent.

FIGURE 16.4
Positive regulation of the *lac* operon by the CAP activator.

the growth medium, *E. coli* must synthesize tryptophan so that it can make its proteins. If tryptophan is present in the medium, then the cell will use that source of the amino acid rather than synthesizing its own.

Tryptophan biosynthesis also involves an operon, the *trp* operon **(Figure 16.5).** The five structural genes in this operon, *trpE–trpA,* encode the enzymes for the steps in the tryptophan biosynthesis pathway. Upstream of the *trpE* gene are the operon's promoter and operator sequences. Expression of the *trp* operon is controlled by the Trp repressor, a regulatory protein encoded by the regulatory gene, *trpR,* which is located elsewhere in the genome (not nearby as was the case for the repressor gene for the *lac* operon). In contrast to the Lac repressor, the Trp repressor is synthesized in an inactive form in which it cannot bind to the operator. The inactive state of the repressor leads to the default state of the operon, the expression of the *trp* operon structural genes (see Figure 16.5A). The repressor is activated when tryptophan levels are high, such as when tryptophan is present in the medium (see Figure 16.5B). Because the presence of tryptophan represses the expression of the tryptophan biosynthesis genes, this operon is an example of a **repressible operon.** Here, tryptophan acts as a **corepressor,** a regula-

A. Tryptophan absent from medium: tryptophan must be made by the cell—structural genes transcribed

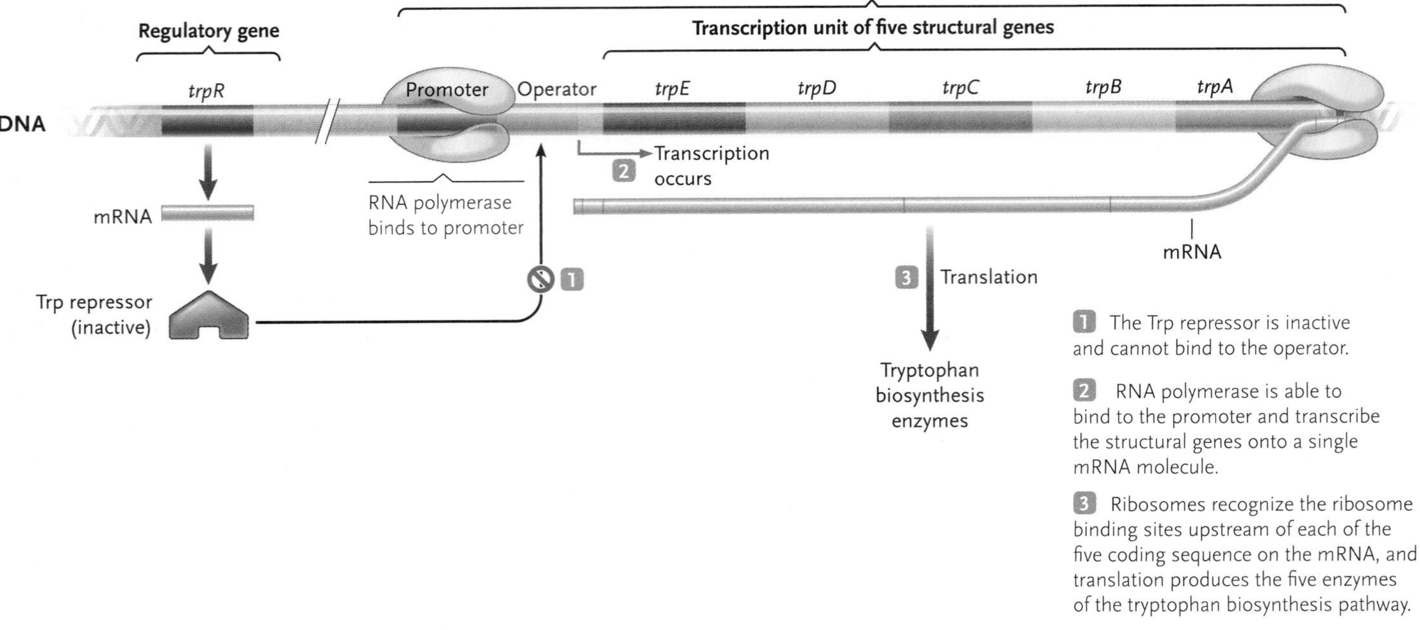

trp operon
Regulatory gene
Transcription unit of five structural genes
trpR
Promoter
Operator
trpE
trpD
trpC
trpB
trpA
DNA
2 Transcription occurs
mRNA
RNA polymerase binds to promoter
1
3 Translation
mRNA
Trp repressor (inactive)
Tryptophan biosynthesis enzymes
1 The Trp repressor is inactive and cannot bind to the operator.
2 RNA polymerase is able to bind to the promoter and transcribe the structural genes onto a single mRNA molecule.
3 Ribosomes recognize the ribosome binding sites upstream of each of the five coding sequence on the mRNA, and translation produces the five enzymes of the tryptophan biosynthesis pathway.

B. Tryptophan present in medium: cell uses tryptophan in medium rather than synthesizing it—structural genes not transcribed

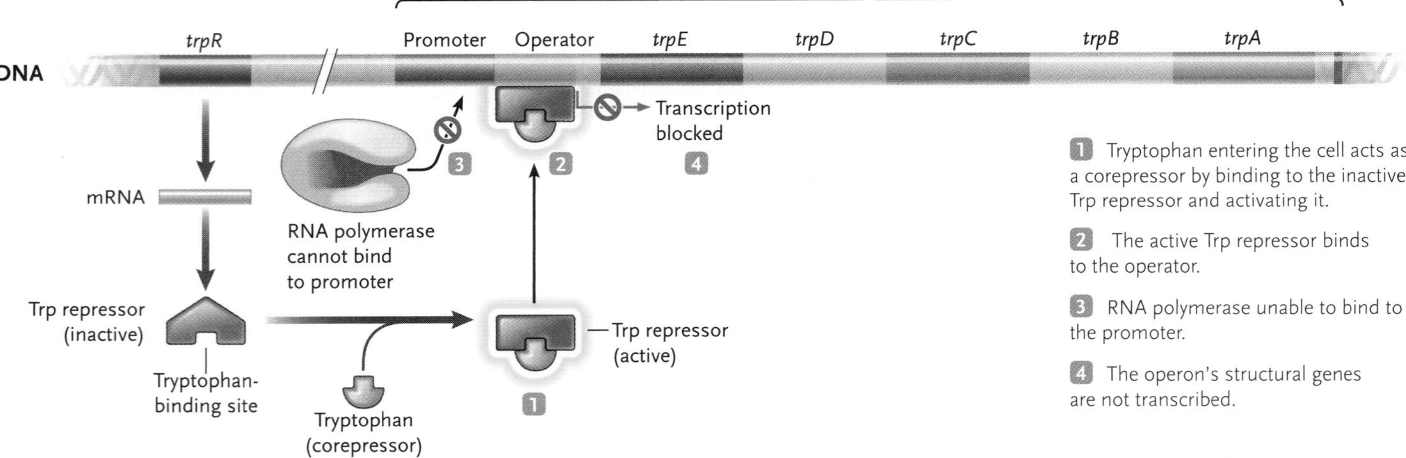

trp operon
trpR
Promoter
Operator
trpE
trpD
trpC
trpB
trpA
DNA
Transcription blocked
3
2
4
mRNA
RNA polymerase cannot bind to promoter
Trp repressor (inactive)
Trp repressor (active)
Tryptophan-binding site
Tryptophan (corepressor)
1
1 Tryptophan entering the cell acts as a corepressor by binding to the inactive Trp repressor and activating it.
2 The active Trp repressor binds to the operator.
3 RNA polymerase unable to bind to the promoter.
4 The operon's structural genes are not transcribed.

FIGURE 16.5
Regulation of the repressible *trp* operon by the Trp repressor in the absence (A) and presence (B) of tryptophan.

tory molecule that activates the repressor to turn off expression of the operon.

To compare and contrast the two operons we have discussed:

- *lac* operon: the repressor is synthesized in an active form. The inducer allolactose inactivates the repressor. The structural genes are then transcribed.
- *trp* operon: the repressor is synthesized in an inactive form so the structural genes are transcribed. The corepressor (tryptophan from the growth medium) activates the repressor. Active repressor blocks transcription of the operon.

Inducible and repressible operons illustrate two types of *negative gene regulation* because both are regulated by a repressor that turns off gene expression when it is in active form. Genes are expressed only when the repressor is in inactive form.

In sum, regulation of gene expression in prokaryotes occurs primarily at the transcription level. There are also some examples of regulation at the translation level. For example, some proteins can bind to the mRNAs that produce them and modulate their translation. This serves as a feedback mechanism to fine-tune the amounts of the proteins in the cell. In the next section we discuss the regulation of gene expression in eukaryotes. You will see that regulation occurs at several points between the gene and the protein, and that regulatory mechanisms of eukaryotes are more complex than those in prokaryotes.

STUDY BREAK 16.1

1. Suppose the *lacI* gene is mutated so that the Lac repressor is not made. How does this mutation affect the regulation of the *lac* operon?
2. Answer the equivalent question for the *trp* operon: How would a mutation that prevents the Trp repressor from being made affect the regulation of the *trp* operon?

THINK OUTSIDE THE BOOK

In addition to the mechanism described, transcription of the *trp* operon is also regulated by another regulatory mechanism known as attenuation. Individually or collaboratively, use the Internet or the research literature to develop an outline of regulation of the *trp* operon by attenuation.

16.2 Regulation of Transcription in Eukaryotes

As you just learned, gene expression in prokaryotes is commonly regulated at the transcription level with genes organized in functional units called operons. The molecular mechanisms of operon regulation provide a simple means of coordinating synthesis of proteins that have related functions. In eukaryotes, the coordinated synthesis of proteins that have related functions also occurs, but the genes involved are usually scattered around the genomes; that is, they are not organized into operons. Nonetheless, like operons, individual eukaryotic genes also consist of protein-coding sequences and adjacent regulatory sequences.

There are two general categories of eukaryotic gene regulation. Short-term regulation involves regulatory events in which gene sets are quickly turned on or off in response to changes in environmental or physiological conditions in the cell's or organism's environment. This type of regulation is most similar to prokaryotic gene regulation. Long-term gene regulation involves regulatory events required for an organism to develop and differentiate. Long-term gene regulation occurs in multicellular eukaryotes and not in simpler, unicellular eukaryotes. The mechanisms we discuss in this and the next section are applicable to both short-term and long-term regulation. The specific molecules and genes involved in short-term and long-term regulation are different and, of course, so is the outcome to the cell or organism.

In Eukaryotes, Regulation of Gene Expression Occurs at Several Levels

The regulation of gene expression is more complicated in eukaryotes than in prokaryotes because eukaryotic cells are more complex, because nuclear DNA is organized with histones into chromatin, and because multicellular eukaryotes produce large numbers and types of cells. Further, the eukaryotic nuclear envelope separates the processes of transcription and translation, whereas in prokaryotes translation can start on an mRNA that is still being synthesized. Consequently, gene expression in eukaryotes is regulated at more levels. That is, there is transcriptional regulation, posttranscriptional regulation, translational regulation, and posttranslational regulation **(Figure 16.6).** The most important of these is transcriptional regulation.

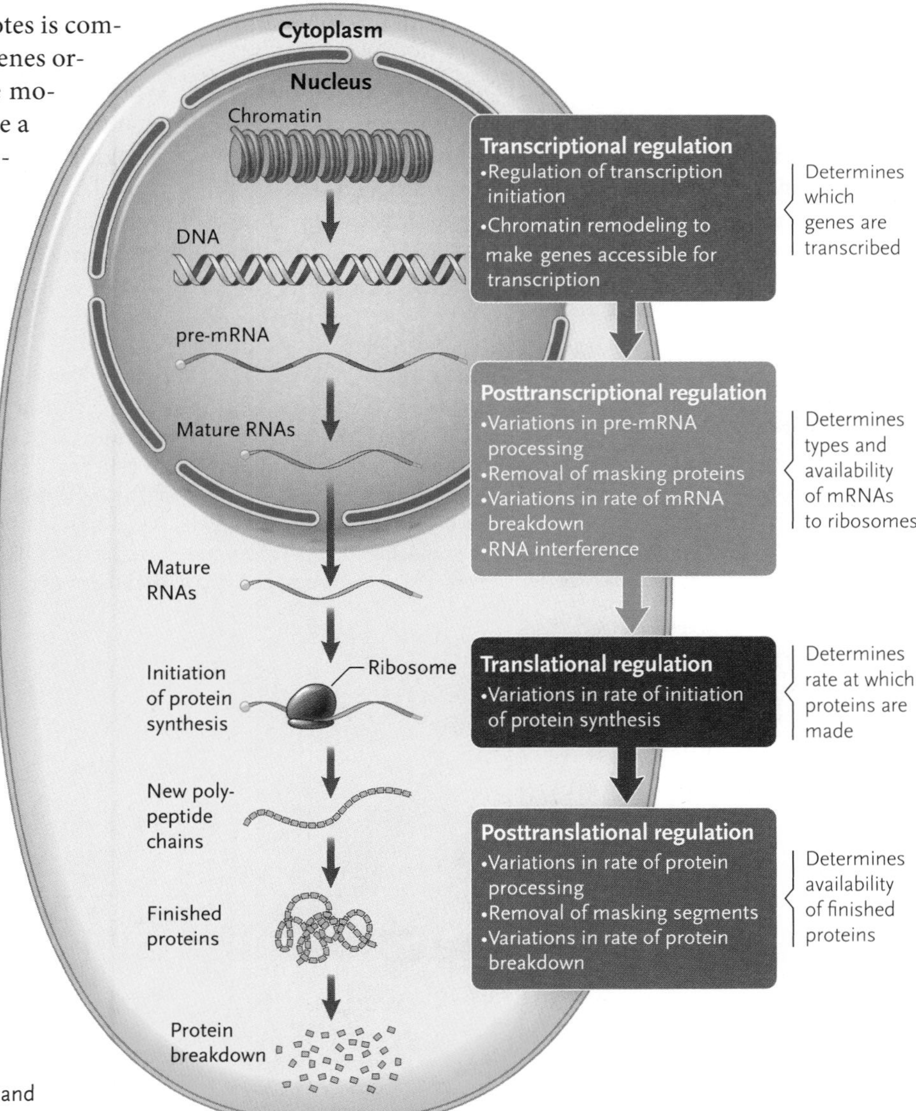

Cytoplasm
Nucleus
Chromatin
DNA
pre-mRNA
Mature RNAs
Mature RNAs
Initiation of protein synthesis
Ribosome
New polypeptide chains
Finished proteins
Protein breakdown
Transcriptional regulation
•Regulation of transcription initiation
•Chromatin remodeling to make genes accessible for transcription
Determines which genes are transcribed
Posttranscriptional regulation
•Variations in pre-mRNA processing
•Removal of masking proteins
•Variations in rate of mRNA breakdown
•RNA interference
Determines types and availability of mRNAs to ribosomes
Translational regulation
•Variations in rate of initiation of protein synthesis
Determines rate at which proteins are made
Posttranslational regulation
•Variations in rate of protein processing
•Removal of masking segments
•Variations in rate of protein breakdown
Determines availability of finished proteins

FIGURE 16.6
Steps in transcriptional, posttranscriptional, translational, and posttranslational regulation of gene expression in eukaryotes.

Regulation of Transcription Initiation Involves the Effects of Proteins Binding to a Gene's Promoter and Regulatory Sites

Transcription initiation is the most important level at which regulation of gene expression takes place.

ORGANIZATION OF A EUKARYOTIC PROTEIN-CODING GENE **Figure 16.7** shows a eukaryotic gene, emphasizing the regulatory sites involved in its expression. Immediately upstream of the transcription unit is the promoter. The promoter in the figure contains a TATA box, a sequence about 25 bp upstream of the start point for transcription that, as we will shortly see, plays an important role in transcription initiation in many promoters. The TATA box has the 7-bp consensus sequence 5′-TATAAAA-3′ / 3′-ATATTTT-5′. Promoters without TATA boxes have other sequence elements that play a similar role. In the following discussions, we describe transcription initiation involving a TATA box-containing promoter.

RNA polymerase II itself cannot recognize the promoter sequence. Instead, proteins called **transcription factors** recognize and bind to the TATA box and then recruit the polymerase. Once the RNA polymerase II–transcription factor complex forms, the polymerase unwinds the DNA and transcription begins. Adjacent to the promoter, further upstream, is the **promoter proximal region,** which contains regulatory sequences called **promoter proximal elements.** Regulatory proteins that bind to promoter proximal elements may stimulate or inhibit the rate of transcription initiation. More distant from the beginning of the gene is the **enhancer.** Regulatory proteins binding to regulatory sequences within an enhancer also stimulate or inhibit the rate of transcription initiation. Next we see more specifically how these regulatory sequences are involved in transcription initiation.

ACTIVATION OF TRANSCRIPTION To initiate transcription, proteins called **general transcription factors** (also called *basal transcription factors*) bind to the promoter in the area of the TATA box **(Figure 16.8).** These factors recruit the enzyme RNA polymerase II, which alone cannot bind to the promoter, and orient the enzyme to start transcription at the correct place. The combination of general transcription factors with RNA polymerase II is the **transcription initiation complex.** On its own, this complex brings about only a low rate of transcription initiation, which leads to just a few mRNA transcripts.

Activators are regulatory proteins that play a role in a positive regulatory system which controls the expression of one or more genes. Activators that bind to the promoter proximal elements interact directly with the general transcription factors at the promoter to stimulate transcription initiation, so many more transcripts are synthesized in a given time. Housekeeping genes—genes that are expressed in all cell types for basic cellular functions such as glucose metabolism—have promoter proximal elements that are recognized by activators present in all cell types. By contrast, genes expressed only in particular cell types or at particular times have promoter proximal elements that are recognized by activators found only in those cell types, or at those times when transcription of these genes needs to be activated. To turn this around, the particular set of activators present within a cell at a given time is responsible for determining which genes in that cell are expressed to a significant level.

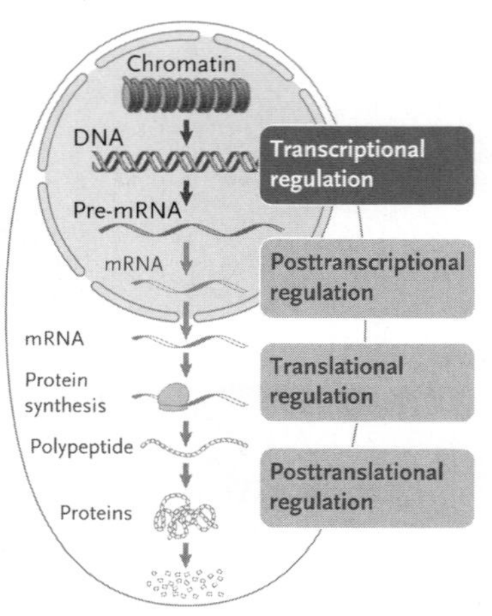

Chromatin
DNA
Transcriptional regulation
Pre-mRNA
mRNA
Posttranscriptional regulation
mRNA
Protein synthesis
Translational regulation
Polypeptide
Posttranslational regulation
Proteins

FIGURE 16.7
Organization of a eukaryotic gene. The transcription unit is the segment that is transcribed into the pre-mRNA; it contains the 5′ UTR (untranslated region), exons, introns, and 3′ UTR. Immediately upstream of the transcription unit is the promoter, which often contains the TATA box. Adjacent to the promoter and further upstream of the transcription unit is the promoter proximal region, which contains regulatory sequences called promoter proximal elements. More distant from the gene is the enhancer, which contains regulatory sequences that control the rate of transcription of the gene. Transcription of the gene produces a pre-mRNA molecule with a 5′ cap and 3′ poly(A) tail; processing of the pre-mRNA to remove introns generates the functional mRNA (see Chapter 15).

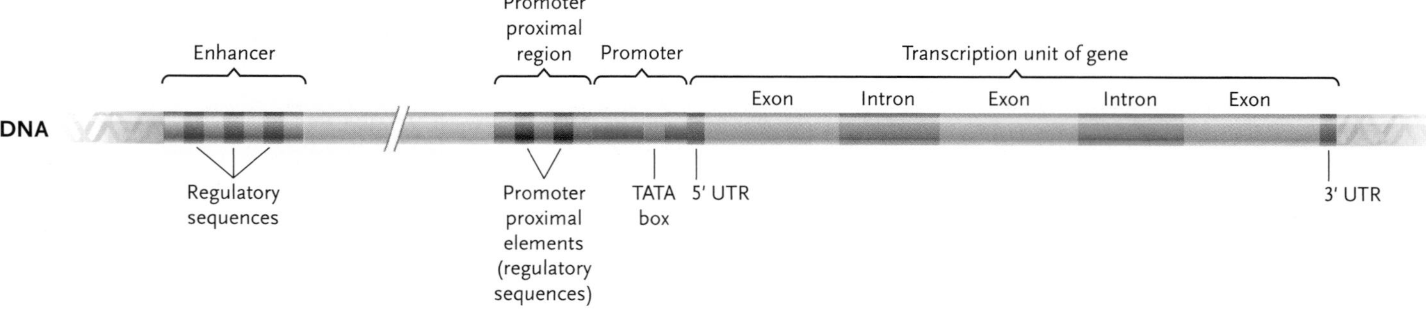

Enhancer
Promoter proximal region
Promoter
Transcription unit of gene
Exon
Intron
Exon
Intron
Exon
DNA
Regulatory sequences
Promoter proximal elements (regulatory sequences)
TATA box
5′ UTR
3′ UTR

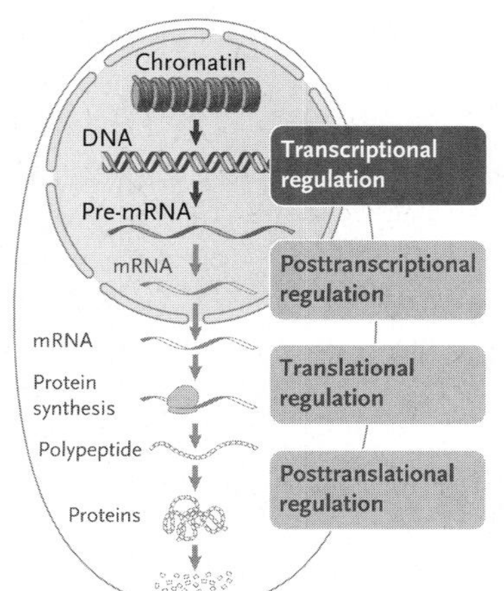
Chromatin
DNA
Transcriptional regulation
Pre-mRNA
mRNA
Posttranscriptional regulation
mRNA
Protein synthesis
Translational regulation
Polypeptide
Posttranslational regulation
Proteins

FIGURE 16.8
Formation of the transcription complex on the promoter of a protein-coding gene by the combination of general transcription factors with RNA polymerase. The general transcription factors are needed for RNA polymerase to bind and initiate transcription at the correct place.

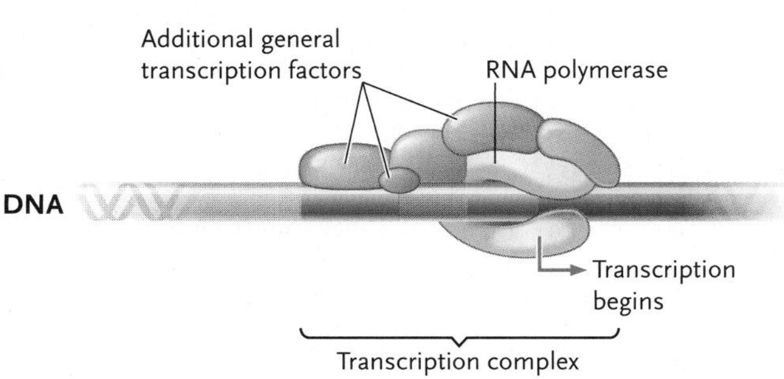
Initial general transcription factor
DNA
TATA box
Site where transcription starts
Promoter
Additional general transcription factors
RNA polymerase
DNA
Transcription begins

Transcription complex

1 The first general transcription factor recognizes and binds to the TATA box of a protein-coding gene's promoter.

2 Additional general transcription factors and then RNA polymerase adds to the complex. A general transcription factor unwinds the promoter DNA, and then transcription begins.

The DNA binding and activation functions of activators are properties of two distinct domains in the proteins. (Protein domains were introduced in Section 3.4.) The three-dimensional arrangement of amino acid chains within and between domains also produces highly specialized regions called **motifs.** Several types of motifs, each with a specialized function, are found in proteins, including motifs which insert into the DNA double helix. Motifs found in the DNA-binding domains of regulatory proteins, such as activators, include the helix-turn-helix, zinc finger, and leucine zipper **(Figure 16.9).**

Activators binding at the enhancer greatly increase transcription rates **(Figure 16.10).** The enhancers of different genes have different sets of regulatory sequences, which bind particular activators. A **coactivator** (also called a *mediator*), a large multiprotein complex, forms a bridge between the activators at the enhancer and the proteins at the promoter and promoter proximal region, causing the DNA to form a loop. The interactions between the activators at the enhancer, the coactivator, the proteins at the promoter, and the RNA polymerase greatly stimulate transcription up to its maximal rate.

REPRESSION OF TRANSCRIPTION In some genes, repressors oppose the effect of activators, thereby blocking or reducing the rate of transcription. The final rate of transcription then depends upon the "battle" between the activation signal and the repression signal.

FIGURE 16.9
Three DNA-binding motifs found in activators and other regulatory proteins.

A. **Helix-turn-helix**

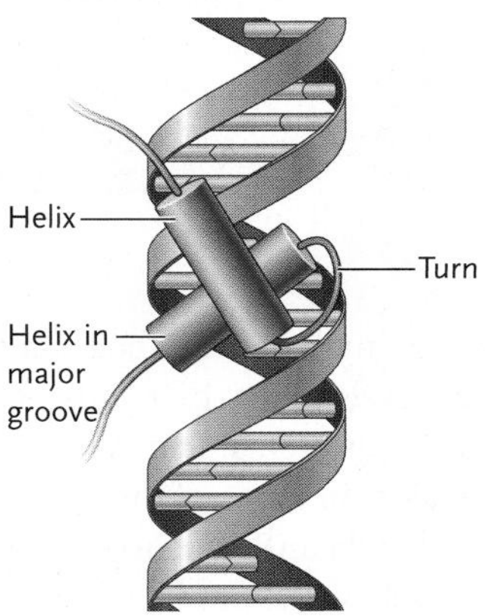
Helix
Turn
Helix in major groove

A helix-turn-helix motif part of a protein bound to DNA. One of the α-helices binds to base pairs in the major groove of the DNA. A looped region of the protein—the turn—connects to a second α-helix that helps hold the first helix in place.

B. **Zinc finger**

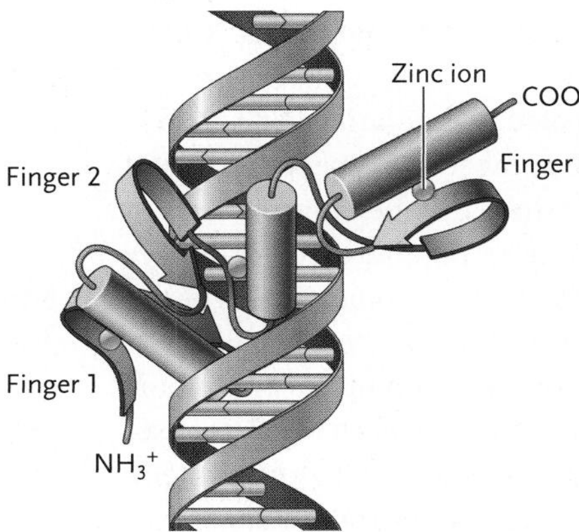
Zinc ion
COO⁻
Finger 3
Finger 2
Finger 1
NH₃⁺

Zinc fingers motifs are parts of proteins named for their resemblance to fingers projecting from a protein, and the presence of a bound zinc atom. Zinc fingers bind to specific base pairs in the grooves of DNA.

C. **Leucine zipper**

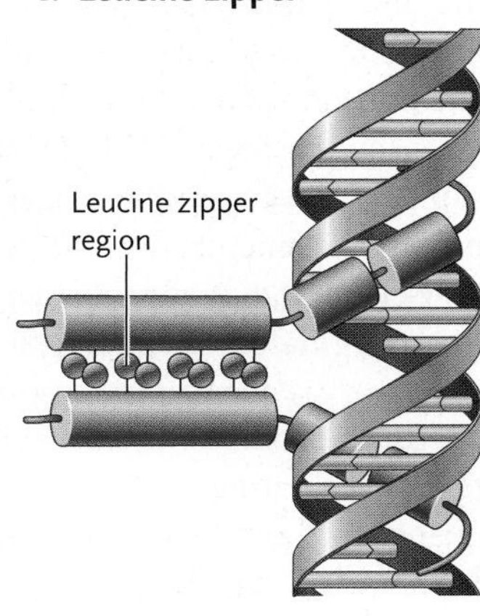
Leucine zipper region

Leucine zipper proteins are dimers, with each monomer consisting of α-helical segments. Hydrophobic interactions between leucine residues within the leucine zipper motif hold the two monomers together. Other α-helices bind to DNA base pairs in the major groove.

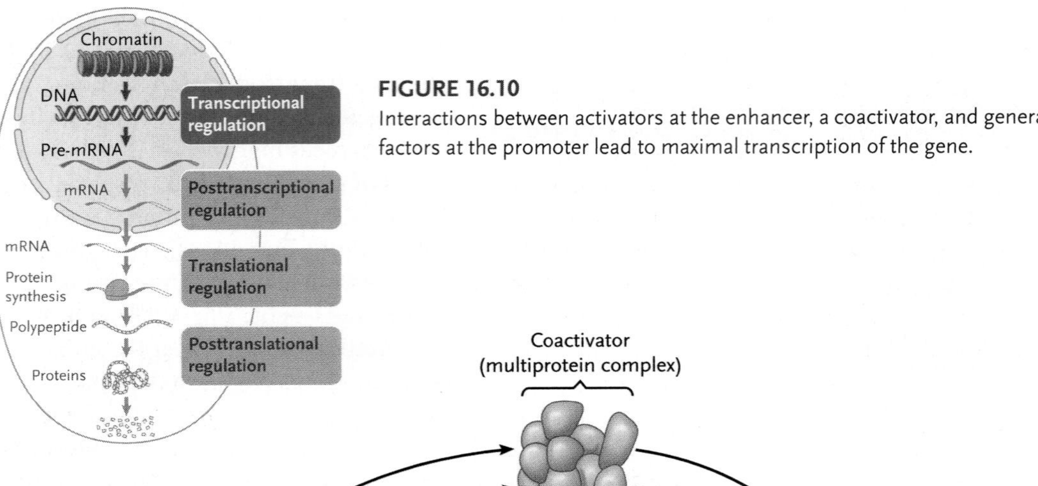
Chromatin
DNA
Transcriptional regulation
Pre-mRNA
mRNA
Posttranscriptional regulation
mRNA
Protein synthesis
Translational regulation
Polypeptide
Posttranslational regulation
Proteins
Coactivator (multiprotein complex)
Activator
Enhancer
Activator
Transcription initiation site
Gene
Promoter proximal region
Promoter
Interaction between activators at the enhancer, coactivator, and proteins at the promoter and promoter proximal region
DNA loop
Transcription initiation site
Gene
Maximal transcription

FIGURE 16.10
Interactions between activators at the enhancer, a coactivator, and general transcription factors at the promoter lead to maximal transcription of the gene.

Different kinds of repressors in eukaryotes work in different ways. Some repressors bind to the same regulatory sequence to which activators bind (often in the enhancer), thereby preventing activators from binding to that site. Other repressors bind to their own specific site in the DNA near where the activator binds and interact with the activator so that it cannot interact with the coactivator. Yet other repressors bind to specific sites in the DNA and recruit **corepressors,** multiprotein complexes analogous to coactivators but which are negative regulators, inhibiting transcription initiation.

COMBINATORIAL GENE REGULATION Let us review the key elements of transcription regulation for a protein-coding gene. General transcription factors bind to certain promoter sequences such as the TATA box and recruit RNA polymerase II; this results in a basal level of transcription. Specific activators bind to promoter proximal elements and stimulate the rate of transcription initiation. Activators also bind to the enhancer to greatly stimulate transcription of the gene.

How are these events coordinated in regulating gene expression? Any given gene has a specific number and types of promoter proximal elements. In some genes there may be only one regulatory element, but genes under complex regulatory control have many regulatory elements. Similarly, the number and types of regulatory sequences in the enhancer is specific for each gene.

Both promoter proximal regions and enhancers are important in regulating the transcription of a gene. Each regulatory sequence in those two regions binds a specific regulatory protein. Since some regulatory proteins are activators and others are repressors, the overall effect of regulatory sequences on transcription depends on the particular proteins that bind to them. If activators bind to both the regulatory sequences in the promoter proximal region and to the enhancer, transcription is activated maximally, meaning a high rate of transcription and therefore the production of a high level of the mRNA encoded by the gene. But, if a repressor binds to the enhancer and an activator binds to the promoter proximal element, the amount of gene expression depends upon the relative effects of those two regulatory proteins. For example, if the repressor is strong, gene expression, in terms of the rate of transcription and the consequent level of the mRNA encoded by the gene, will be reduced.

A relatively small number of regulatory proteins (activators and repressors) control transcription of all protein-coding genes. By combining a few regulatory proteins in particular ways, the transcription of a wide array of genes can be controlled. The process is called **combinatorial gene regulation.** Consider a theoretical example of two genes, each with activators already bound to the respective promoter proximal elements **(Figure 16.11).** Maximal transcription of gene *A* requires activators 2, 5, 7, and 8 binding to their regulatory sequences in the enhancer, whereas maximal transcription of gene *B* requires activators 1, 5, 8, and 11 binding to its enhancer. Looked at another way, both genes require activators 5 and 8 combined with other different activators for full activation.

This operating principle solves a basic dilemma in gene regulation—if each gene were regulated by a single, distinct protein, the number of genes encoding regulatory proteins would have to equal the number of genes to be regulated. Regulating the regulators would require another set of genes of equal number, and so on until the coding capacity of any chromosome set, no matter how large, would be exhausted. But because different genes require different combinations of regulatory proteins, the number of genes encoding regulatory proteins can be much lower than the number of genes the regulatory proteins control.

FIGURE 16.11

Combinatorial gene regulation. A relatively small number of regulatory proteins control transcription of all protein-coding genes. Different combinations of activators bind to enhancer regulatory sequences to control the rate of transcription of each gene.

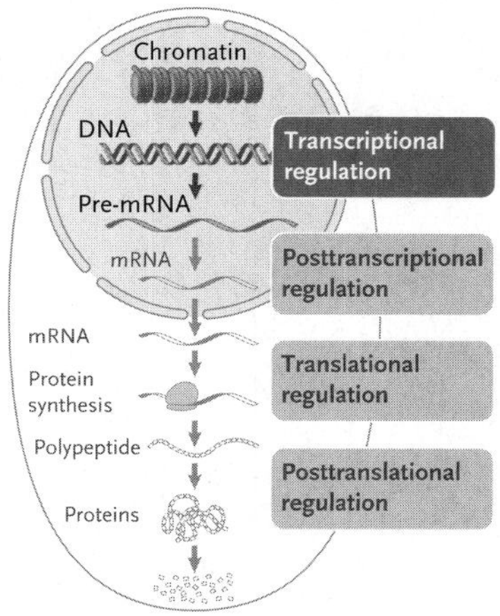
Chromatin
DNA
Transcriptional regulation
Pre-mRNA
mRNA
Posttranscriptional regulation
mRNA
Protein synthesis
Translational regulation
Polypeptide
Posttranslational regulation
Proteins

A. A unique combination of activators controls gene *A*.

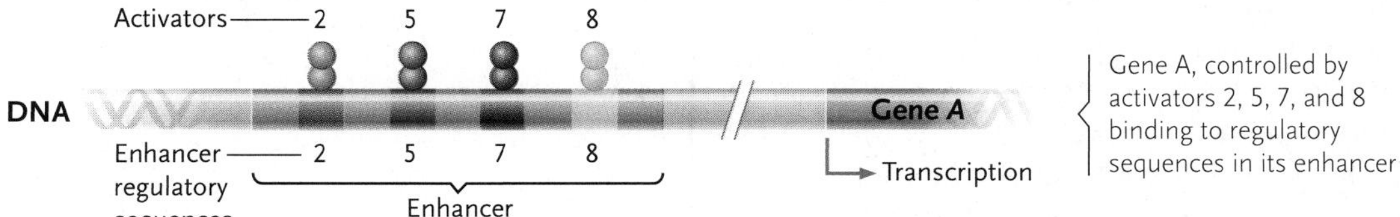
Activators
2
5
7
8
DNA
Gene A
Enhancer regulatory sequences
2
5
7
8
Enhancer
Transcription
Gene A, controlled by activators 2, 5, 7, and 8 binding to regulatory sequences in its enhancer

B. A different combination of activators controls gene *B*.

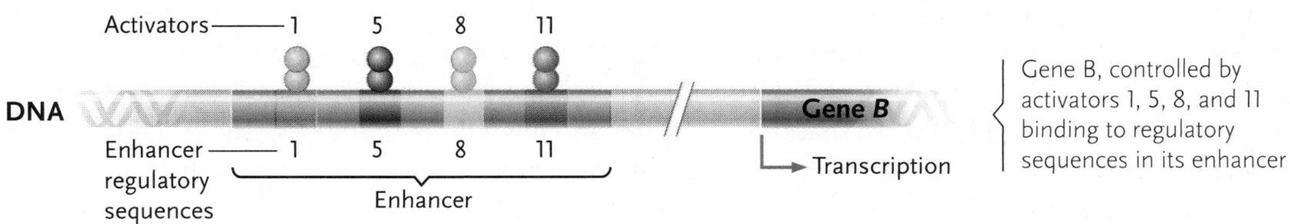
Activators
1
5
8
11
DNA
Gene B
Enhancer regulatory sequences
1
5
8
11
Enhancer
Transcription
Gene B, controlled by activators 1, 5, 8, and 11 binding to regulatory sequences in its enhancer

COORDINATED REGULATION OF TRANSCRIPTION OF GENES WITH RELATED FUNCTIONS In the discussion of prokaryotic operons, you learned that genes with related function are often clustered *and* they are transcribed from one promoter onto a single mRNA. That mRNA is translated to produce the several proteins encoded by the genes. There are no operons in eukaryotes, yet the transcription of genes with related function is coordinately controlled. How is this accomplished?

The answer is that all genes that are coordinately regulated have the same regulatory sequences associated with them. Therefore, with one signal, the transcription of all of the genes can be controlled simultaneously. Consider the control of gene expression by steroid hormones in mammals. A **hormone** is a molecule produced by one tissue and transported via the bloodstream to a target tissue or tissues to alter physiological activity. A **steroid** is a type of lipid derived from cholesterol (see Section 3.3). Examples of steroid hormones are testosterone and glucocorticoid. Testosterone regulates the expression of a large number of genes associated with the maintenance of primary and secondary male characteristics. Glucocorticoid, among other actions, regulates the expression of genes involved in the maintenance of the concentration of glucose and other fuel molecules in the blood. Figure 16.12 illustrates how a steroid hormone, when it enters a cell, activates gene transcription.

A steroid hormone acts on specific target tissues in the body because only cells in those tissues have **steroid hormone receptors** in their cytoplasm that recognize and bind the hormone (see Section 7.5). The steroid hormone moves through the plasma membrane into the cytoplasm and the receptor binds to it **(Figure 16.12)**. The hormone–receptor complex then enters the nucleus and binds to specific regulatory sequences that are adjacent to the genes whose expression is controlled by the hormone. This binding activates transcription of those genes, and proteins encoded by the genes are synthesized rapidly.

A single steroid hormone can regulate many different genes because all of the genes have an identical DNA sequence—a **steroid hormone response element**—to which the hormone–receptor complex binds. For example, all genes controlled by glucocorticoid have a glucocorticoid response element associated with them. Therefore, the release of glucocorticoid into the bloodstream coordinately activates the transcription of genes with that response element.

Methylation of DNA Can Control Gene Transcription

In **DNA methylation,** enzymes add a methyl group (—CH_3) to cytosine bases in the DNA. Methylated cytosines in promoter regions can regulate transcription through a process called **silencing,** in which transcription of genes controlled by those promoters is turned off. This is an example of **epigenetics,** a phenomenon in which a change in gene expression does not involve a change in the DNA sequence of the gene or of the genome.

Silencing by methylation is not universal among eukaryotes, but is common among vertebrates. For example, genes encoding the blood protein hemoglobin are methylated and inactive in most vertebrate body cells. In the cell lines giving rise to red blood cells, enzymes remove the methyl groups from the hemoglobin genes, which are then transcribed.

DNA methylation in some cases silences large blocks of genes, or even chromosomes. Recall from Section 13.2 that a dosage compensation mechanism inactivates one of the two X chromosomes in most body cells of female placental mammals, including humans. In X chromosome inactivation—another example of an epigenetic phenomenon—one of the two X chromosomes packs tightly into a mass known as a Barr body, in which most genes of the X chromosome are turned off. The inactivation occurs during embryonic development, and which X chromosome is inactivated in a particular embryonic cell line is a random event. As part of X chromosome inactivation, cytosines in the DNA become methylated.

DNA methylation underlies **genomic imprinting,** an epigenetic phenomenon in which the expression of an allele is determined by the parent that contributed it (see Section 13.5). In genomic imprinting, methylation permanently silences transcription of either the inherited maternal or paternal allele of a particular gene. The methylation occurs during gametogenesis in a parent. An inherited methylated allele, the *imprinted allele,* is not expressed—it is silenced. The expression of the gene involved therefore depends upon expression of the nonimprinted allele inherited from the other parent. The methylation of the parental allele is maintained as the DNA is replicated, so that the silenced allele remains inactive in progeny cells. Figure 13.22 shows an example of genomic imprinting.

FIGURE 16.12

Steroid hormone regulation of gene expression. A steroid hormone enters the cell and forms a complex in the cytoplasm with a steroid hormone receptor that is specific to the hormone. Steroid hormone-receptor complexes migrate to the nucleus, bind to the steroid hormone response element next to each gene they control (one such gene is shown in the figure), and affect transcription of those genes.

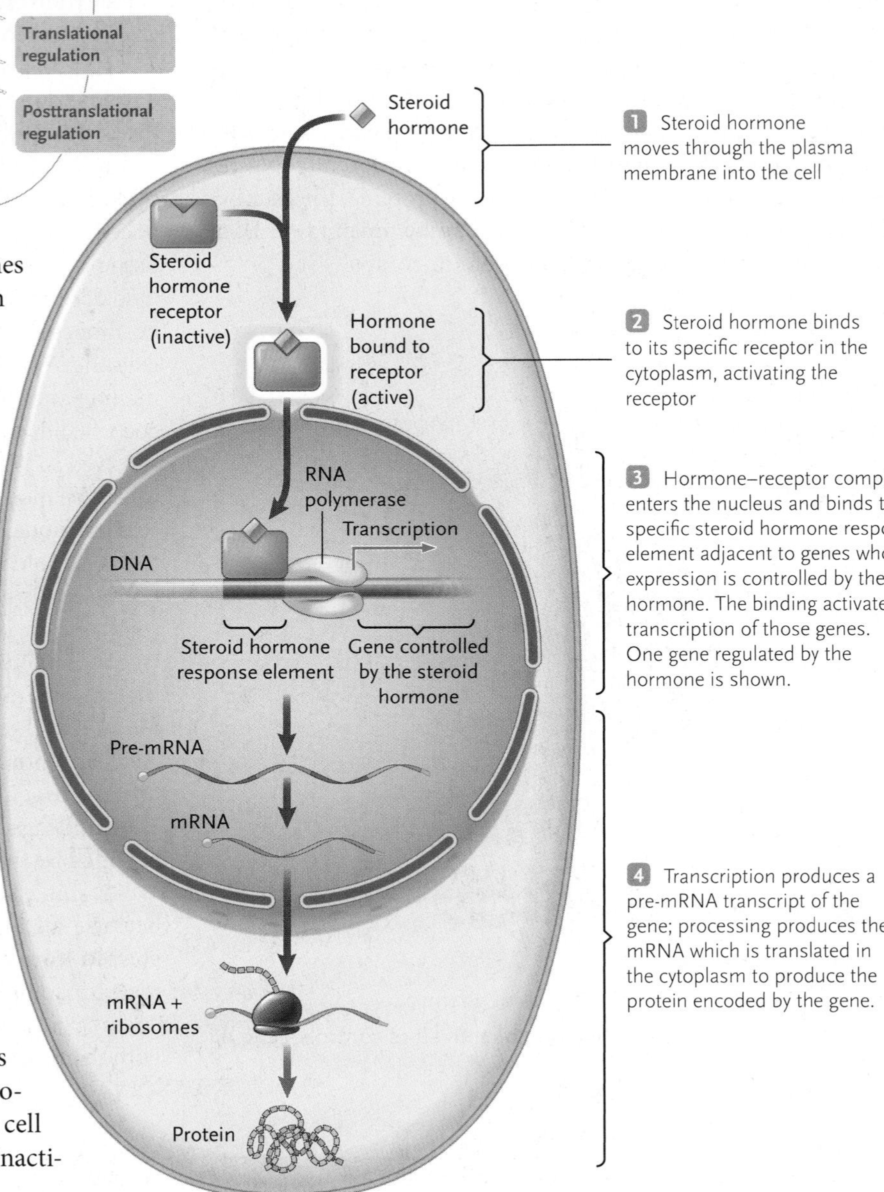

Chromatin
DNA
Transcriptional regulation
Pre-mRNA
mRNA
Posttranscriptional regulation
mRNA
Protein synthesis
Translational regulation
Polypeptide
Proteins
Posttranslational regulation
Steroid hormone
1 Steroid hormone moves through the plasma membrane into the cell
Steroid hormone receptor (inactive)
Hormone bound to receptor (active)
2 Steroid hormone binds to its specific receptor in the cytoplasm, activating the receptor
RNA polymerase
Transcription
DNA
Steroid hormone response element
Gene controlled by the steroid hormone
3 Hormone–receptor complex enters the nucleus and binds to a specific steroid hormone response element adjacent to genes whose expression is controlled by the hormone. The binding activates transcription of those genes. One gene regulated by the hormone is shown.
Pre-mRNA
mRNA
mRNA + ribosomes
Protein
4 Transcription produces a pre-mRNA transcript of the gene; processing produces the mRNA which is translated in the cytoplasm to produce the protein encoded by the gene.

Chromatin Structure Plays an Important Role in Whether a Gene Is Active or Inactive

Eukaryotic DNA is organized into chromatin by combination with histone proteins (discussed in Section 14.5). Recall that DNA is wrapped around a core of two molecules each of histones H2A, H2B, H3, and H4, forming the nucleosome (see Figure 14.21). The negative charge of DNA and the positive charges of the histone proteins naturally attract each other in nucleosomes and contribute to the structure's stability. Higher levels of chromatin organization occur when histone H1 links adjacent nucleosomes.

Genes in regions of the DNA that are tightly wound around histones in chromatin are inactive, because their promoters are not accessible to the proteins that initiate transcription. For a eukaryotic gene to be activated, the chromatin structure must be altered in the vicinity of the promoter to provide access to the general transcription factors for transcription initiation. The process of changing chromatin structure is called **chromatin remodeling.** In one type of chromatin remodeling, an activator binds to a regulatory sequence upstream of the gene's promoter and recruits a **nucleosome remodeling complex.** The multiprotein complex uses the energy of ATP hydrolysis to slide the nucleosome along the DNA to expose the promoter, or to restructure the nucleosome without moving it to allow transcription factors to bind **(Figure 16.13).** In a second type of chromatin remodeling, an activator binds to a regulatory sequence upstream of the gene's promoter, and recruits an enzyme that acetylates (adds acetyl groups: $CH_3CO—$) lysine amino acids in the tails of histones in the nucleosome, where the promoter is located. Acetylation removes the positively charged amino group of the lysine, which makes the histone less attractive to the negatively charged DNA. As a result, the histones loosen their association with DNA, and the promoter becomes accessible. This type of remodeling is reversed by deacetylation enzymes that remove the acetyl groups from the histones. Many activators use both of these chromatin remodeling mechanisms to regulate gene activity.

The tails of histones can also be modified by the covalent addition of methyl groups or phosphate groups to affect chromatin structure and gene expression. For example, histone methylation is associated with gene inactivation. That is, methylation of histones is a property of condensed regions of chromatin, including heterochromatin, where genes are inactive. Unmethylated histones is a property of chromatin where genes are active.

Once mRNAs are transcribed from active genes, further regulation occurs at each of the major steps in the pathway from genes to proteins: during pre-mRNA processing and the movement of finished mRNAs to the cytoplasm (posttranscriptional regulation), during protein synthesis (translational regulation), and after translation is complete (posttranslational regulation). The next section takes up the regulatory mechanisms operating at each of these steps.

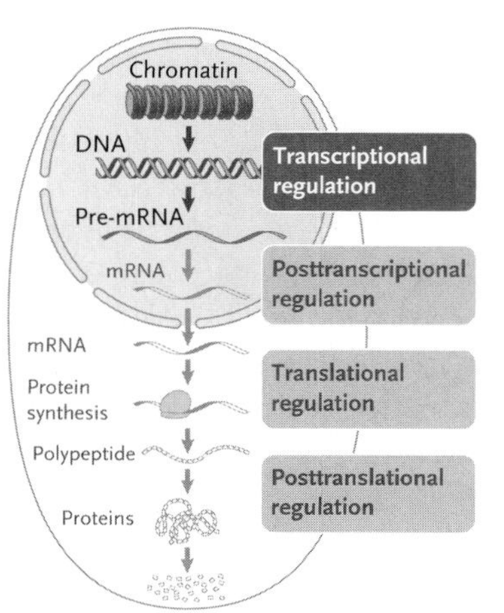
Chromatin
DNA
Transcriptional regulation
Pre-mRNA
mRNA
Posttranscriptional regulation
mRNA
Protein synthesis
Translational regulation
Polypeptide
Proteins
Posttranslational regulation

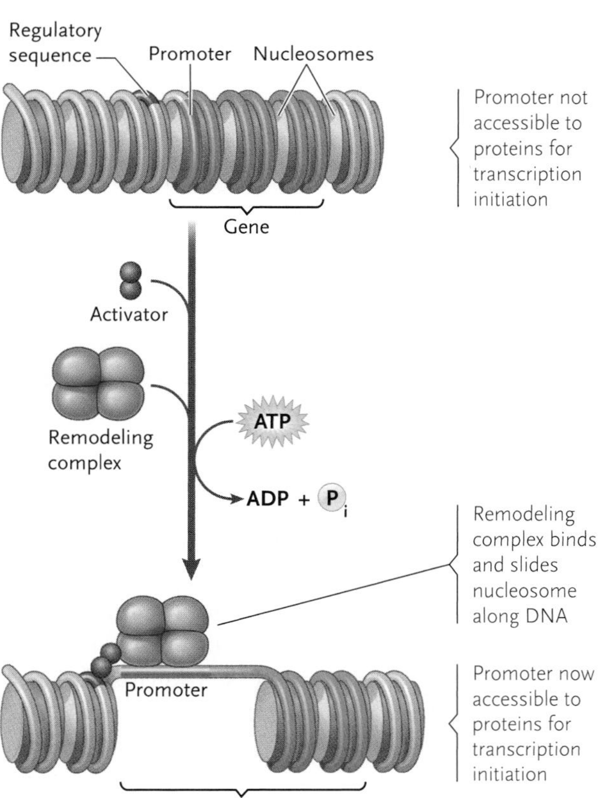
Regulatory sequence
Promoter
Nucleosomes
Promoter not accessible to proteins for transcription initiation
Gene
Activator
Remodeling complex
ATP
ADP + P$_i$
Remodeling complex binds and slides nucleosome along DNA
Promoter
Promoter now accessible to proteins for transcription initiation
Gene

FIGURE 16.13
Exposing a gene's promoter by chromatin remodeling.

STUDY BREAK 16.2

1. **What are the roles of general transcription factors, activators, and coactivators in transcription of a protein-coding gene?**
2. **What is the role of histones in gene expression? How does acetylation of the histones affect gene expression?**

16.3 Posttranscriptional, Translational, and Posttranslational Regulation

Transcriptional regulation determines which genes are copied into mRNAs. This basic level of regulation is fine-tuned by posttranscriptional, translational, and posttranslational controls, the subjects of this section (refer again to Figure 16.6).

Posttranscriptional Regulation Controls mRNA Availability

Posttranscriptional regulation regulates translation by controlling the availability of mRNAs to ribosomes. The controls work by several mechanisms, including changes in pre-mRNA processing and the rate at which mRNAs are degraded.

VARIATIONS IN PRE-mRNA PROCESSING In Chapter 15 we noted that mRNAs are transcribed initially as pre-mRNA molecules. These pre-mRNAs are processed to produce the finished mRNAs, which are then available for protein synthesis. Variations in pre-mRNA processing can regulate *which* proteins are made in cells. As described in Section 15.3, pre-mRNAs can be processed by *alternative splicing,* which produces different mRNAs from the same pre-mRNA by removing different combinations of exons (the amino acid-coding segments) along with the introns (the noncoding spacers). The resulting mRNAs are translated to produce a family of related proteins with various combinations of amino acid sequences derived from the exons. Alternative splicing itself is under regulatory control. Regulatory proteins specific to the type of cell control which exons are removed from pre-mRNA molecules by binding to regulatory sequences within those molecules. The outcome of alternative splicing is that related, but structurally different proteins within a family are synthesized in different cell types or tissues, and have functions that are keyed to the activities of those cell types and tissues. Perhaps three-quarters of human genes are alternatively spliced at the pre-mRNA level.

VARIATIONS IN THE RATE OF THE mRNA BREAKDOWN The rate at which eukaryotic mRNAs break down can also be controlled posttranscriptionally. Regulatory molecules, such as a steroid hormone, directly or indirectly affect the mRNA breakdown steps, either slowing or increasing the rate of those steps. For example, in the mammary gland of the rat, the mRNA for casein (a milk protein) has a half-life of about 5 hours (meaning that it takes 5 hours for half of the mRNA present at a given time to break down). The half-life of casein mRNA changes to about 92 hours if the peptide hormone prolactin is present. Prolactin is synthesized in the brain and in other tissues, including the breast. The most important effect of prolactin is to stimulate the mammary glands to produce milk. During milk production, a large amount of casein must be synthesized, and this is accomplished in part by radically decreasing the rate of breakdown of the casein mRNA.

Nucleotide sequences in the 5′ UTR (untranslated region at the 5′ end of an mRNA preceding the start codon; see Section 15.3) appear also to be important in determining mRNA half-life. If the 5′ UTR is transferred experimentally from one mRNA to another, the half-life of the receiving mRNA becomes the same as that of the donor mRNA. The controlling sequences in the 5′ UTR of an mRNA might be recognized by proteins that regulate its stability.

REGULATION OF GENE EXPRESSION IN EUKARYOTES BY SMALL RNAs Until relatively recently, the commonly accepted view was that regulation of gene expression in prokaryotes and eukaryotes involved only protein-based mechanisms. However, in 1998, Andrew Fire of Stanford University School of Medicine and Craig Mello of University of Massachusetts Medical School showed that RNA silenced the expression of a particular gene in the nematode worm, *Caenorhabditis elegans.* They called the phenomenon **RNA interference (RNAi).** Their discovery revolutionized the way scientists thought about and studied gene regulation in eukaryotes. They now understand that post-transcriptional regulation may be carried out not only by regulatory proteins, but also by non-coding single-stranded RNAs which can bind to mRNAs and affect their translation. We now know that RNAi is widespread among eukaryotes. Fire and Mello received a Nobel Prize in 2006 for their discovery of RNA interference.

Two major groups of small regulatory RNAs are involved in RNAi, **microRNAs (miRNAs)** and **short interfering RNAs (siRNAs).** The transcription of an miRNA gene and the processing of the transcript to produce the functional miRNA molecule is shown in **Figure 16.14.** The miRNA, in a protein complex called the **miRNA-induced silencing complex (miRISC)** binds to sequences in the 3′ UTRs of target mRNAs. If the miRNA and mRNA pair imperfectly, the double-stranded segment formed between the miRNA and the mRNA blocks ribosomes from translating the mRNA (shown in Figure 16.14). In this case, the target mRNA is not destroyed, but its expression is silenced. If the miRNA and mRNA pair perfectly, an enzyme in the protein complex cleaves the target mRNA where the miRNA is bound to it, destroying the mRNA and silencing its expression. RNAi by imperfect pairing and translation inhibition is the most common mechanism in animals. RNAi by perfect pairing and RNA degradation is the most common mechanism in plants.

As of September 2009 almost 11,000 miRNAs have been identified among the organisms examined. MicroRNA genes have been found in all multicellular eukaryotes that have been examined, and also in some unicellular ones. MicroRNAs play central roles in controlling gene expression in a variety of cellular, physiological, and developmental processes in animals and plants. In animals, for example, miRNAs help regulate specific developmental timing events, gene expression in neurons, brain development, and stem cell division.

The other major type of small regulatory RNAs is the **small interfering RNA (siRNA).** Whereas miRNA is produced from RNA that is encoded in the cell's genome, siRNA is produced from double-stranded RNA that is *not* encoded by nuclear genes. For example, the life cycle and replication of many viruses with RNA genomes involves a double-stranded RNA stage. Cells attacked by such a virus can defend themselves using siRNA which they produce from the virus' own RNA. The viral double-stranded RNA enters the cell's RNAi process in a way very similar to that described for miRNAs; double-stranded RNA is cut by Dicer (see Figure 16.14) into short double-stranded RNA molecules, and then a protein complex binds to the molecules and degrades one of the RNA strands to produce single-stranded siRNA. The protein complex is similar to one that acts on the double-stranded RNA precursors of miRNAs. The siRNA with the protein complex in this case is the **siRNA-induced silencing complex (siRISC).** In the RNAi process, the siRNA in the siRISC acts like miRNA in the miRISC—single-stranded RNAs complementary to the siRNA are targeted and, in this case, the target RNA is cleaved and the pieces are then degraded. In our viral example, the targeted RNAs would be viral mRNAs for proteins the virus uses to replicate itself, or a single-stranded RNA that is the viral genome itself, or that is produced from the viral genome during replication.

The expression of any gene can be knocked down to low levels or knocked out completely in experiments involving RNAi with siRNA. To silence a gene, researchers introduce into the cell a double-stranded RNA that can be processed by Dicer and the protein complex into an siRNA complementary to the mRNA tran-

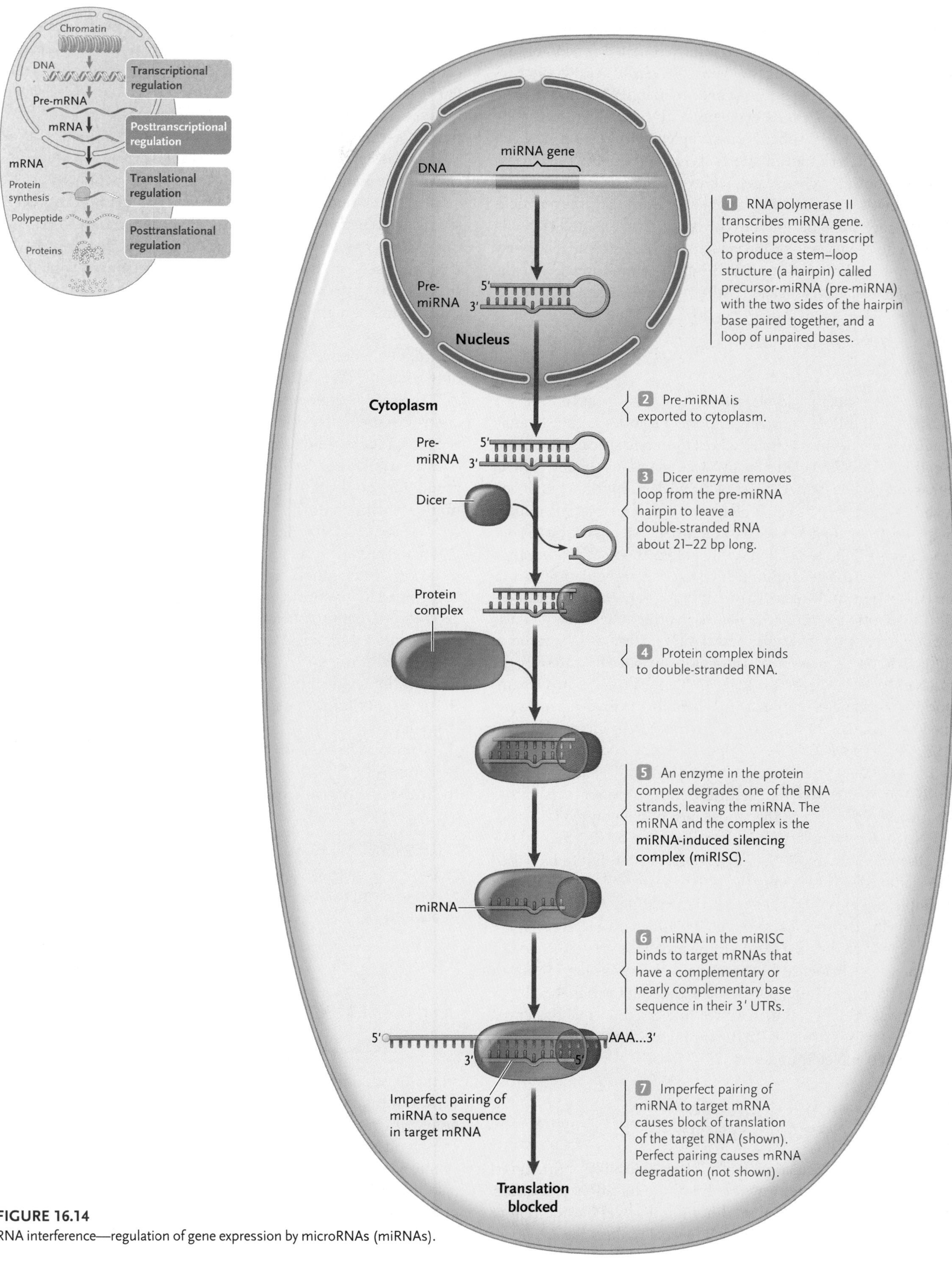

Chromatin
DNA
Transcriptional regulation
Pre-mRNA
mRNA
Posttranscriptional regulation
mRNA
Protein synthesis
Translational regulation
Polypeptide
Posttranslational regulation
Proteins
miRNA gene
DNA
Pre-miRNA
5'
3'
Nucleus
1 RNA polymerase II transcribes miRNA gene. Proteins process transcript to produce a stem–loop structure (a hairpin) called precursor-miRNA (pre-miRNA) with the two sides of the hairpin base paired together, and a loop of unpaired bases.
Cytoplasm
2 Pre-miRNA is exported to cytoplasm.
Pre-miRNA
Dicer
3 Dicer enzyme removes loop from the pre-miRNA hairpin to leave a double-stranded RNA about 21–22 bp long.
Protein complex
4 Protein complex binds to double-stranded RNA.
5 An enzyme in the protein complex degrades one of the RNA strands, leaving the miRNA. The miRNA and the complex is the miRNA-induced silencing complex (miRISC).
miRNA
6 miRNA in the miRISC binds to target mRNAs that have a complementary or nearly complementary base sequence in their 3' UTRs.
5'
AAA...3'
3'
5'
Imperfect pairing of miRNA to sequence in target mRNA
7 Imperfect pairing of miRNA to target mRNA causes block of translation of the target RNA (shown). Perfect pairing causes mRNA degradation (not shown).
Translation blocked

FIGURE 16.14
RNA interference—regulation of gene expression by microRNAs (miRNAs).

scribed from that gene. Knocking down or knocking out the function of a gene is equivalent to creating a mutated version of that gene, but without changing the gene's DNA sequence. Researchers are using this experimental approach to identify the functions of genes whose presence has been detected by sequencing complete genomes, but whose function is completely unknown. After an siRNA specific to a gene of interest is introduced into the cell, researchers look for a change in phenotype, such as properties relating to growth or metabolism. If such a change is seen, the researchers now have some insight into the gene's function, and they can investigate the gene with more focus. RNAi using siRNAs may have some applications in medicine also, perhaps to regulate the expression of genes associated with particular human diseases.

Translational Regulation Controls the Rate of Protein Synthesis

At the next regulatory level, translational regulation controls the rate at which mRNAs are used in protein synthesis. Translational regulation occurs in essentially all cell types and species. For example, translational regulation is involved in cell cycle control in all eukaryotes and in many processes during development in multicellular eukaryotes, such as red blood cell differentiation in animals. Significantly, many viruses exploit translational regulation to control their infection of cells and to shut off the host cell's own genes.

Consider the general role of translational regulation in animal development. During early development of most animals, little transcription occurs. The changes in protein synthesis patterns seen in developing cell types and tissues instead are the result of the activation, repression, or degradation of maternal mRNAs, the mRNAs that were in the mother's egg before fertilization. One important mechanism for translational regulation involves adjusting the length of the poly(A) tail of the mRNA. (Recall from Section 15.3 that the poly(A) tail—a string of adenine-containing nucleotides—is added to the 3′ end of pre-mRNA and is retained on the mRNA produced from the pre-mRNA after introns are removed.) That is, enzymes can change the length of the poly(A) tail on an mRNA in the cytoplasm in either direction: by shortening it or lengthening it. Increases in poly(A) tail length result in increased translation; decreases in length result in decreased translation. For example, during embryogenesis (the formation of the embryo) of the fruit fly, *Drosophila,* key proteins are synthesized when the poly(A) tails on the mRNAs for those proteins are lengthened in a regulated way. Evidence for this came from experiments in which poly(A) tail lengthening was blocked; the result was that embryogenesis was inhibited. But, while researchers know that the length of poly(A) tails is regulated in the cytoplasm, how this process occurs is not completely understood.

Posttranslational Regulation Controls the Availability of Functional Proteins

Posttranslational regulation controls the availability of functional proteins mainly in three ways: chemical modification, processing, and degradation. Chemical modification involves the addition or removal of chemical groups, which reversibly alters the activity of the protein. For example, you saw in Section 7.2 how the addition of phosphate groups to proteins involved in signal transduction pathways either stimulates or inhibits the activity of those proteins. Further, in Section 10.4 you learned how the addition of phosphate groups to target proteins plays a crucial role in regulating how a cell progresses through the cell division cycle. And, in Section 16.2 you saw how acetylation of histones altered the properties of the nucleosome, loosening its association with DNA in chromatin.

In processing, a protein is synthesized as an inactive precursor that is activated under regulatory control by removal of a segment. For example, you saw in Section 15.4 that the digestive enzyme pepsin is synthesized as pepsinogen, an inactive precursor that activates by removal of a segment of amino acids. Similarly, the glucose-regulating hormone insulin is synthesized as a precursor called proinsulin; processing of the precursor removes a central segment but leaves the insulin molecule, which consists of two polypeptide chains linked by disulfide bridges.

The rate of degradation of proteins is also under regulatory control. Some proteins in eukaryotic cells last for the lifetime of the individual, while others persist only for minutes. Proteins with relatively short cellular lives include many of the proteins regulating transcription. Typically, these short-lived proteins are marked for breakdown by enzymes that attach a "doom tag" consisting of a small protein called *ubiquitin.* The protein is given this name because it is indeed ubiquitous—present in almost the same form in essentially all eukaryotes. A ubiquinated protein is recognized by the **proteasome,** a large cytoplasmic complex of several different proteins, where degradation of the protein occurs (**Figure 16.15**). Aaron Ciechanover and Avram Hershko, both of the Israel Institute of Technology, Haifa, Israel, and Irwin Rose of the University of California, Irvine, received a Nobel Prize in 2004 for the discovery of ubiquitin-mediated protein degradation.

Control of protein breakdown is the last of the opportunities for control of gene expression. In the next section we will see how the regulation of gene expression is involved in development.

STUDY BREAK 16.3 <

1. **How does a microRNA silence gene expression?**
2. **If the poly(A) tail on a mRNA was removed, what would likely be the effect on the translation of that mRNA?**

16.4 Genetic and Molecular Regulation of Development

In the development of multicellular eukaryotes, a fertilized egg divides into many cells that are ultimately transformed into an adult, which is itself capable of reproduction. The process is orchestrated by a series of programmed changes encoded in the genome, although the development of a multicellular organism is also influenced to some extent by the environment. The programmed changes do not alter the genome itself, which remains identical in virtually all of the cells of the developing organism.

(Thus, the processes involved in development are all examples of epigenetic phenomena.) Instead, controlled changes in gene expression direct sequential developments which are appropriate both in time—certain stages must unfold before others—and in place—new structures must arise in the correct location. An understanding of how genes are regulated therefore aids researchers in their genetic analysis of development. Developmental geneticists are very interested in identifying and characterizing the genes involved in development, and understanding how the products of the genes regulate and bring about the elaborate events that occur.

One productive research approach has been to isolate mutants that affect developmental processes. Researchers can then identify the genes involved, molecularly clone the genes (see Chapter 18), and analyze them in detail to build models for the molecular functions of the gene products in development. A number of model organisms are used for these studies because of the relative ease with which mutants can be made and studied, and the ease of performing molecular analyses. These organisms include the fruit fly *(Drosophila melanogaster)* (see *Focus on Model Research Organisms* in Chapter 13) and *Caenorhabditis elegans* (a nematode worm) (see *Focus on Model Research Organisms* in Chapter 29) among invertebrates, and the zebrafish *(Danio rerio)* (see Figure 10.1), and the house mouse *(Mus musculus)* (see *Focus on Model Research Organisms* in Chapter 43) among vertebrates.

In this section we discuss some of the genetic and molecular mechanisms that regulate development in animals to illustrate the principles involved. In Chapter 48 you will learn about the cellular and morphological events that occur as an adult animal develops from a fertilized egg. The gene regulation of development in plants is described in Chapter 34.

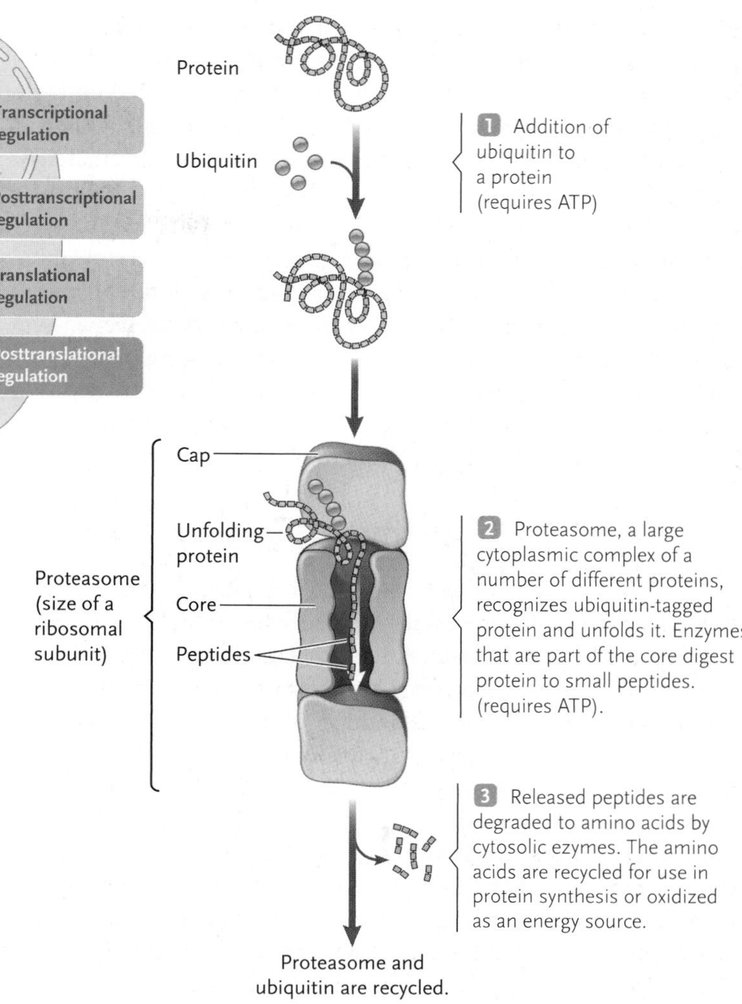

Protein
Ubiquitin
1 Addition of ubiquitin to a protein (requires ATP)
Cap
Unfolding protein
Core
Peptides
Proteasome (size of a ribosomal subunit)
2 Proteasome, a large cytoplasmic complex of a number of different proteins, recognizes ubiquitin-tagged protein and unfolds it. Enzymes that are part of the core digest protein to small peptides. (requires ATP).
3 Released peptides are degraded to amino acids by cytosolic ezymes. The amino acids are recycled for use in protein synthesis or oxidized as an energy source.
Proteasome and ubiquitin are recycled.

FIGURE 16.15
Protein degradation by ubiquitin addition and enzymatic digestion within a proteasome.

Development in Animals Is Accomplished by Several Genetically Regulated Mechanisms

Fertilization of an egg by a sperm cell produces a zygote. Development begins at this point. Development in all animals is accomplished by a number of mechanisms that are under genetic control but are influenced to some extent by the environment. The mechanisms include mitotic cell divisions, movements of cells, *induction, determination,* and *differentiation.*

Mitotic divisions produce the cells that are the subjects of the gene regulatory processes that create the various cell types in the adults. Cell movements during development are part of the program to create specific tissues and organs.

Induction is the process in which one group of cells (the inducer cells) causes or influences another nearby group of cells (the responder cells) to follow a particular developmental pathway.

Determination is the process by which the developmental fate of a cell is set. Prior to determination, a cell is **totipotent** (has the potential to become any cell type of the adult) but, after determination, the cell commits to becoming a particular cell type. Induction is the major process responsible for determination.

Differentiation, which follows determination, involves the establishment of a cell-specific developmental program in cells. Differentiation results in cell types with clearly defined structures and functions; those features derive from specific patterns of gene expression in cells.

Developmental Information Is Located in both the Nucleus and Cytoplasm of the Fertilized Egg

The information that directs the initiation of development is stored in two locations in the zygote. Part of the information is stored in the zygote nucleus, in the DNA derived from the egg and sperm nuclei. This information directs development as individual genes are activated or turned off in a highly ordered manner. The rest of the information is stored in the egg cytoplasm, in the form of messenger RNA (mRNA) and protein molecules.

Because the fertilizing sperm contributes essentially no cytoplasm, nearly all the cytoplasmic information of the fertilized egg is maternal in origin. The mRNA and proteins stored in the egg cytoplasm are known as **cytoplasmic determinants.** When the zygote divides, the cytoplasmic determinants are distributed asymmetrically to the daughter cells. As a result, the two daughter cells resulting from division differ in the signals they have, and this leads to different patterns of gene expression. Cytoplasmic determinants direct the first stages of animal development, in the period before genes of the zygote become active. Depending on the animal group, the control of early development by cytoplasmic determinants may be limited to the first few divisions of the zygote, as in mammals, or it may last until the actual tissues of the embryo are formed, as in most invertebrates.

Induction Is the Major Process Responsible for Determination

Induction is the major process responsible for determination, which is the process by which the developmental fate of a cell is set. Induction is a highly selective process; only certain responder cells can respond to the signal from the inducer cells. Many experiments have shown that induction occurs through the interaction of signal molecules of inducer cells with surface receptors on the responding cells. In some cases, a signal molecule released by the inducer cell interacts with a plasma membrane-embedded receptor on the surface of a responder cell. The binding activates the receptor, triggering a signal transduction pathway within the cell which produces a developmental change, often involving a change in gene activity (the cellular response to the signal). (Surface receptors and their associated signal transduction pathways are discussed in Sections 7.3 and 7.4.) In other cases, induction occurs by direct cell-to-cell contact involving interaction between a membrane-embedded protein on the inducer cell and a membrane-embedded receptor protein on the surface of the responder cell. The receptor is activated by the interaction, and in the same way as for the diffusible signal molecule example, the cell undergoes a developmental change.

Differentiation Produces Specialized Cells without Loss of Genes

Differentiation is the process by which cells that have been committed to a particular developmental fate by the determination process now develop into specialized cell types with distinct structures and functions. As part of differentiation, cells produce molecules characteristic of the specific types. For example, in the lens cells of the eye, 80% to 90% of the total protein synthesized is crystallin, the protein responsible for the transparency of the lens.

Research into differentiation confirmed that, as cells specialize, the DNA in the genome remains constant, matching that of the original zygote. In other words, differentiation generally occurs as a result of differential gene activity, and not by a process in which DNA is lost, so that each type of differentiated cell retains only those genes required for that cell type. (There are just a few examples in particular organisms where gene loss occurs during differentiation.)

The most compelling evidence showing that the DNA in the genome remains constant during differentiation has come from experiments in which animals and plants have been cloned. Animal clones have been made by taking an egg produced by one animal, and then replacing its nucleus with a nucleus taken from a somatic cell of a different adult animal, showing that the adult nucleus is still totipotent. Different species of frogs and many types of mammals have been cloned in this way. The first mammal cloned was Dolly, a sheep (the cloning experiment is described in Chapter 18, and shown in Figure 18.13). Similarly, plants can be cloned from single cells isolated from a mature plant.

Genes Control Determination and Differentiation

The end result of determination is differentiation, which produces cell types of particular kinds, such as skin cells or nerve cells. Both determination and differentiation involve specific, regulated changes in gene expression.

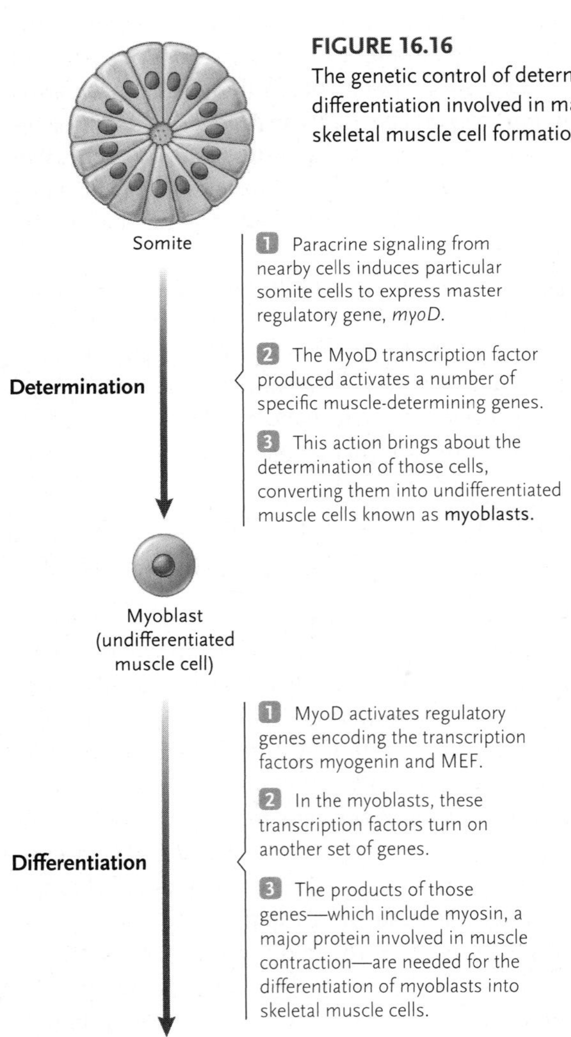
Somite
Determination
1 Paracrine signaling from nearby cells induces particular somite cells to express master regulatory gene, *myoD*.
2 The MyoD transcription factor produced activates a number of specific muscle-determining genes.
3 This action brings about the determination of those cells, converting them into undifferentiated muscle cells known as **myoblasts**.
Myoblast (undifferentiated muscle cell)
Differentiation
1 MyoD activates regulatory genes encoding the transcription factors myogenin and MEF.
2 In the myoblasts, these transcription factors turn on another set of genes.
3 The products of those genes—which include myosin, a major protein involved in muscle contraction—are needed for the differentiation of myoblasts into skeletal muscle cells.
Skeletal muscle cell

FIGURE 16.16
The genetic control of determination and differentiation involved in mammalian skeletal muscle cell formation.

One well-studied example of the genetic control of determination and differentiation is the production of skeletal muscle cells from somites in mammals **(Figure 16.16)**. Somites give rise to the vertebral column, the ribs, the repeating sets of skeletal muscles associated with the ribs and vertebral column, and the skeletal muscles of the limbs. Under the control of the master regulatory gene, *myoD*, particular cells of a somite differentiate into skeletal muscle cells. The product of *myoD* is the transcription factor MyoD. In this context, "transcription factor" means a regulatory protein that binds to regulatory sequences associated with a gene or genes to stimulate or inhibit the expression of those genes.

Generally speaking, the molecular mechanisms involved in determination and differentiation depend on master regulatory genes that encode transcription factors, which are regulatory proteins controlling the expression of other genes. Expression of the master regulatory genes is controlled by induction in most cases. The genes controlled by the products of the master regulatory genes may themselves be regulatory genes encoding transcription factors that, in turn, regulate the expression of other sets of genes.

Genes Regulate Pattern Formation during Development

As a part of the signals guiding differentiation, cells receive positional information that tells them where they are in the embryo. The positional information is vital to **pattern formation:** the arrangement of organs and body structures in their proper three-dimensional relationships. Positional information is laid down primarily in the form of concentration gradients of regulatory molecules produced under genetic control. In most cases, gradients of several different regulatory molecules interact to tell a cell, or a cell nucleus, where it is in the embryo. Below, we describe in brief the results of studies of the genetic control of pattern formation during the development of the fruit fly, *Drosophila melanogaster.* The developmental principles discovered from these studies apply to many other animal species, including humans.

THE LIFE CYCLE OF *DROSOPHILA* The production of an adult fruit fly from a fertilized egg occurs in a sequence of genetically controlled development events. The *Drosophila* life cycle is shown in **Figure 16.17.** As is typical of most insects, the life cycle proceeds from fertilized egg to embryo within the egg. The embryo then hatches from the egg and the life cycle proceeds through three larval stages to a pupal stage and then to the adult stage. The stages of development from fertilized egg to hatching collectively are called **embryogenesis.** As illustrated by the color usage in Figure 16.17, the segments of the embryo can be mapped to the segments of the adult fly. The development of vertebrates, including mammals, occurs quite differently, as you will learn in Chapter 48.

GENETIC ANALYSIS OF *DROSOPHILA* DEVELOPMENT The study of developmental mutants by a large number of researchers has given us important information about *Drosophila* development. Three researchers performed key, pioneering research with

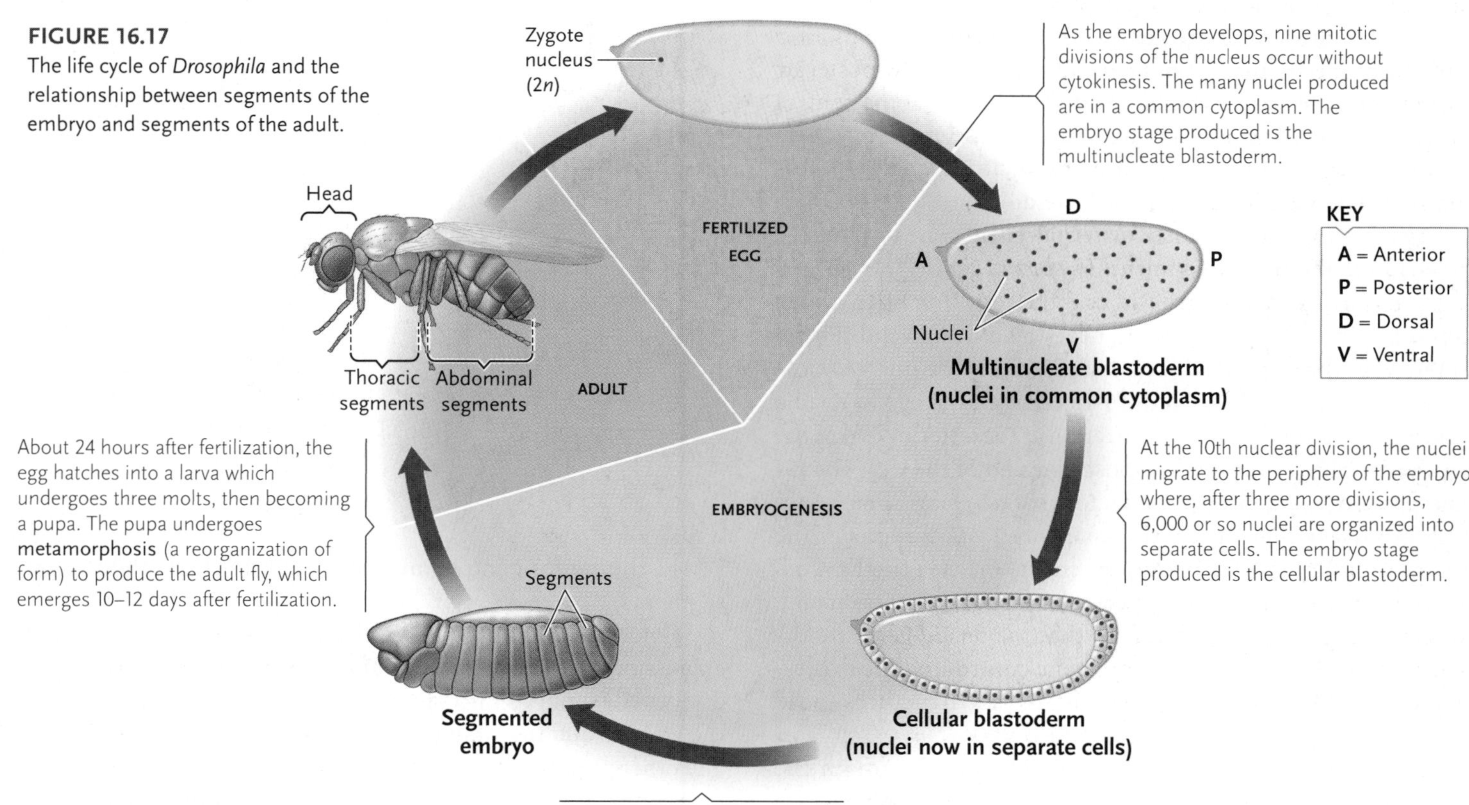

Zygote nucleus (2n)
As the embryo develops, nine mitotic divisions of the nucleus occur without cytokinesis. The many nuclei produced are in a common cytoplasm. The embryo stage produced is the multinucleate blastoderm.
FERTILIZED EGG
D
A
P
V
Nuclei
Multinucleate blastoderm (nuclei in common cytoplasm)
KEY
A = Anterior
P = Posterior
D = Dorsal
V = Ventral
Head
Thoracic segments
Abdominal segments
ADULT
About 24 hours after fertilization, the egg hatches into a larva which undergoes three molts, then becoming a pupa. The pupa undergoes **metamorphosis** (a reorganization of form) to produce the adult fly, which emerges 10–12 days after fertilization.
EMBRYOGENESIS
At the 10th nuclear division, the nuclei migrate to the periphery of the embryo where, after three more divisions, 6,000 or so nuclei are organized into separate cells. The embryo stage produced is the cellular blastoderm.
Segments
Segmented embryo
Cellular blastoderm (nuclei now in separate cells)
The cellular blastoderm develops into a segmented embryo. At this point, 10 hours have passed since the egg was fertilized.

FIGURE 16.17
The life cycle of *Drosophila* and the relationship between segments of the embryo and segments of the adult.

developmental mutants: Edward B. Lewis of the California Institute of Technology, Christiane Nüsslein-Volhard of the Max Planck Institute for Developmental Biology in Tübingen, Germany, and Eric Wieschaus of Princeton University. The three shared a Nobel Prize in 1995 "for their discoveries concerning the genetic control of early embryonic development."

Nüsslein-Volhard and Wieschaus studied early embryogenesis. They searched for *every* gene required for early pattern formation in the embryo. They did this by looking for recessive *embryonic lethal* mutations. These mutations, when homozygous, result in the death of the embryo during development. By examining the stage of development at which an embryo died, and how development was disrupted, they gained insights into the role of the particular genes in embryogenesis.

Lewis studied mutants that changed the fates of cells in particular regions in the embryo, producing structures in the adult that normally were produced by other regions. His work was the foundation of research identifying master regulatory genes that control the development of body regions in a wide range of organisms.

MATERNAL-EFFECT GENES AND SEGMENTATION GENES FOR ESTABLISHING THE BODY PLAN IN THE EMBRYO A number of genes control the establishment of the embryo's body plan; that is, how the organism is laid out in its anterior-to-posterior, ventral-to-dorsal, and side-to-side axes. These genes code for regulatory proteins which regulate the expression of other genes. There are two classes: *maternal-effect genes,* and *segmentation genes* that work sequentially **(Figure 16.18).**

Many **maternal-effect genes** are expressed by the mother during oogenesis. These genes control the anterior-to-posterior polarity of the egg and, therefore, of the embryo. Some control the formation of the anterior structures of the embryo, others control the formation of the posterior structures, and yet others control the formation of the terminal end.

The *bicoid* gene is the key maternal-effect gene responsible for head and thorax development. The *bicoid* gene is transcribed in the mother during oogenesis, and the resulting mRNAs are deposited in the egg, localizing near the anterior end **(Figure 16.19).** After the egg is fertilized, translation of the mRNAs produces Bicoid protein, which diffuses through the egg to form a gradient with its highest concentration at the anterior end of the egg, and fading to none at the posterior end of the egg. The Bicoid protein is a transcription factor that activates some genes and represses others along the anterior–posterior axis of the embryo. Embryos with mutations in the *bicoid* gene have no thoracic structures, but have posterior structures at each end. Researchers concluded, therefore, that the *bicoid* gene in normal embryos is a master regulator gene controlling the expression of genes for the development of anterior structures (head and thorax).

A number of other maternal-effect genes, through the activities of their products in gradients in the embryo, are also involved in axis formation. The *nanos* gene, for instance, is the key maternal-effect gene for the posterior structures.

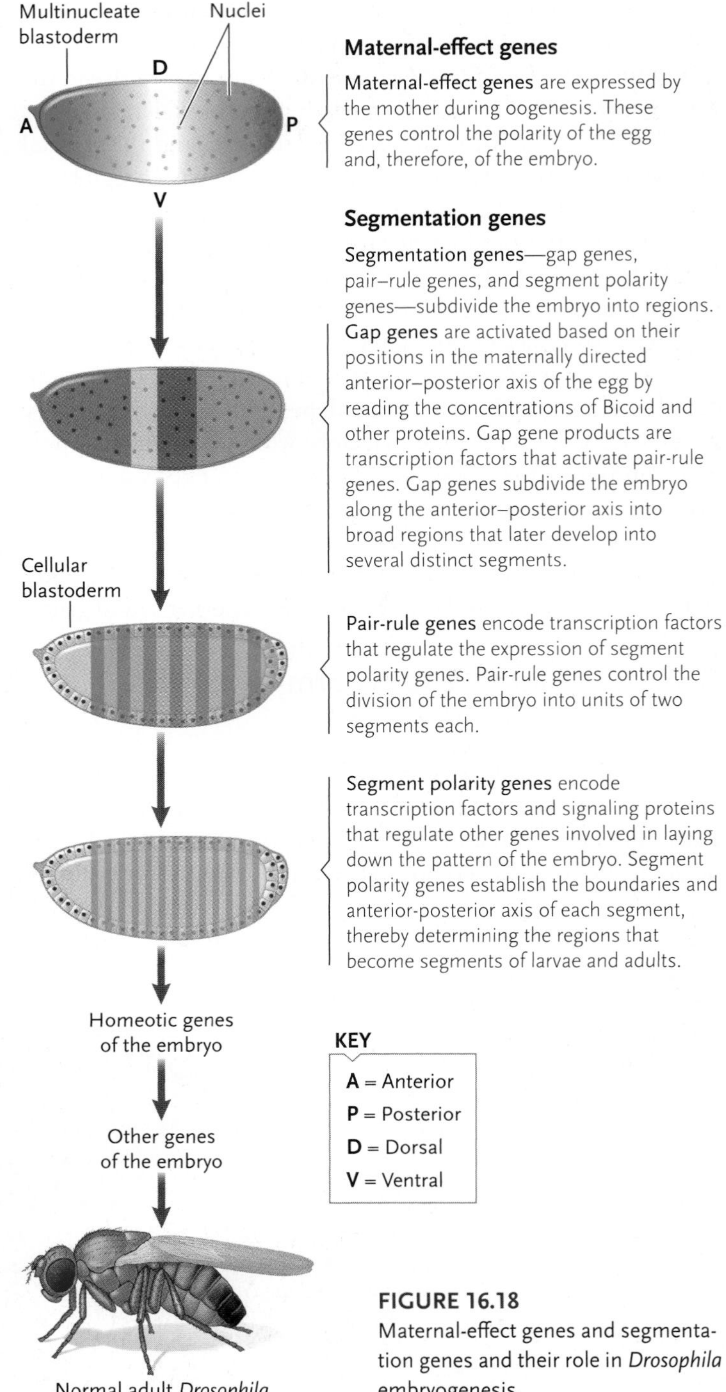

Multinucleate blastoderm
Nuclei
D
A
P
V
Maternal-effect genes
Maternal-effect genes are expressed by the mother during oogenesis. These genes control the polarity of the egg and, therefore, of the embryo.
Segmentation genes
Segmentation genes—gap genes, pair–rule genes, and segment polarity genes—subdivide the embryo into regions.
Gap genes are activated based on their positions in the maternally directed anterior–posterior axis of the egg by reading the concentrations of Bicoid and other proteins. Gap gene products are transcription factors that activate pair-rule genes. Gap genes subdivide the embryo along the anterior–posterior axis into broad regions that later develop into several distinct segments.
Cellular blastoderm
Pair-rule genes encode transcription factors that regulate the expression of segment polarity genes. Pair-rule genes control the division of the embryo into units of two segments each.
Segment polarity genes encode transcription factors and signaling proteins that regulate other genes involved in laying down the pattern of the embryo. Segment polarity genes establish the boundaries and anterior-posterior axis of each segment, thereby determining the regions that become segments of larvae and adults.
Homeotic genes of the embryo
Other genes of the embryo
Normal adult Drosophila
KEY
A = Anterior
P = Posterior
D = Dorsal
V = Ventral

FIGURE 16.18 Maternal-effect genes and segmentation genes and their role in *Drosophila* embryogenesis.

When the *nanos* gene is mutated, embryos lack abdominal segments.

Once the anterior–posterior axis of the embryo is set, the expression of at least 24 **segmentation genes** progressively subdivides the embryo into regions, determining the segments of the embryo and the adult (see Figure 16.18). Gradients of Bicoid and other proteins encoded by maternal-effect genes regulate expression of the embryo's segmentation genes differentially. That is, each segmentation gene is expressed at a particular time and in a particular location during embryogenesis. Three classes of

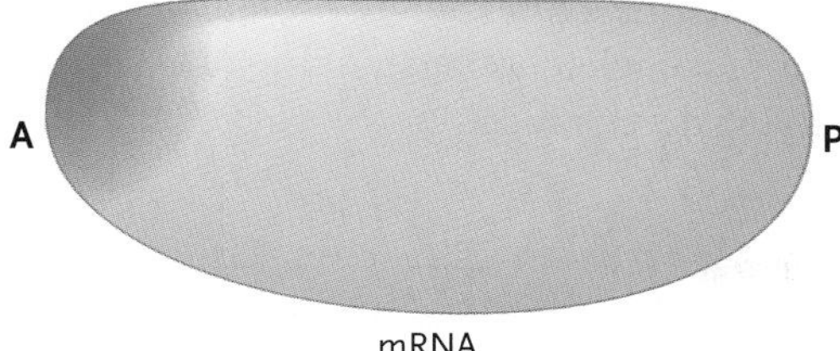
Distribution of maternal *bicoid* mRNA (green)
A
P
mRNA

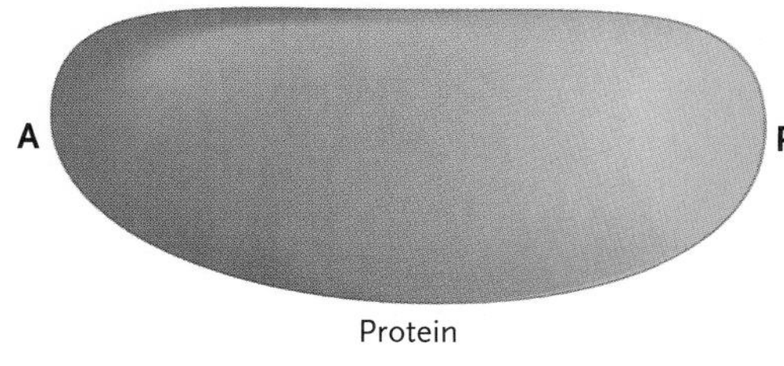
Distribution of Bicoid protein (blue)
A
P
Protein

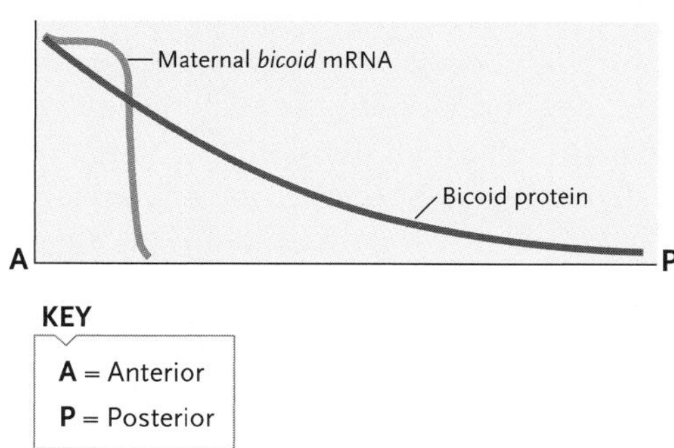
Maternal *bicoid* mRNA
Bicoid protein
A
P
KEY
A = Anterior
P = Posterior

FIGURE 16.19
Gradients of *bicoid* mRNA and Bicoid protein in the *Drosophila* egg.

segmentation genes act sequentially; their activities are regulated in a cascade of gene activations:

- **Gap genes,** through their activation of the next genes in the regulatory cascade, control the subdivision of the embryo along the anterior–posterior axis into several broad regions. Mutations in gap genes result in the loss of one or more body segments in the embryo.
- **Pair-rule genes,** through their activation of the next genes in the regulatory cascade, control the division of the embryo into units of two segments each. Mutations in pair-rule genes delete every other segment of the embryo.
- **Segment polarity genes** set the boundaries and anterior–posterior axis of each segment in the embryo. Mutations in segment polarity genes produce segments in which one part is missing and the other part is duplicated as a mirror image.

HOMEOTIC GENES FOR SPECIFYING THE DEVELOPMENTAL FATE OF EACH SEGMENT Once the segmentation pattern has been set, **homeotic** (structure-determining) **genes** of the embryo specify what each segment will become after metamorphosis. In normal flies, homeotic genes are master regulatory genes that control the development of structures such as eyes, antennae, legs, and wings on particular segments (see Figure 16.17). Researchers discovered the role of homeotic genes from the study of mutations in these genes; such mutations alter the developmental fate of a segment in the embryo in a major way. For example, in flies with a mutation in the *Antennapedia* gene, legs develop in place of antennae **(Figure 16.20).**

How do homeotic genes regulate development? Homeotic genes encode transcription factors that regulate expression of genes responsible for the development of adult structures. Each homeotic gene has a common region called the **homeobox** that is key to its function. A homeobox corresponds to an amino acid section of the encoded transcription factor called the **homeodomain.** The homeodomain of each protein binds to regulatory sequences of the genes whose transcription it regulates.

Homeobox-containing genes are called *Hox* genes. There are eight *Hox* genes in *Drosophila* and, interestingly, they are organized along a chromosome in the same order as they are expressed along the anterior–posterior body axis **(Figure 16.21).**

The discovery of *Hox* genes in *Drosophila* led to a search for equivalent genes in other organisms. The result of that search has shown that *Hox* genes are present in all major animal phyla. In each case, the genes control the development of the segments/regions of the body and are arranged in order in the genome. (This is a good example of how the results of studies in a model research organism are broadly applicable.) The homeobox sequences in the *Hox* genes are highly conserved, indicating a common function in the wide range of animals in which they are found. For example, the homeobox sequences of mammals are the same or very similar to those of the fruit fly (see Figure 16.21).

Homeotic genes are also found in plants. For example, many homeotic mutations that affect flower development have been identified and analyzed in *Arabidopsis* (see Section 34.5).

Normal

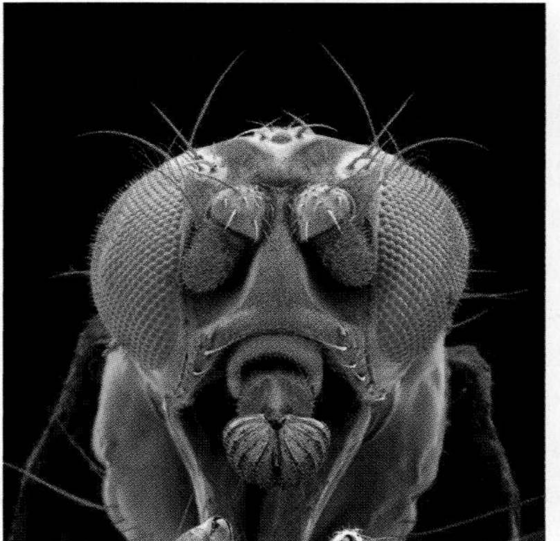

Antennapedia mutant

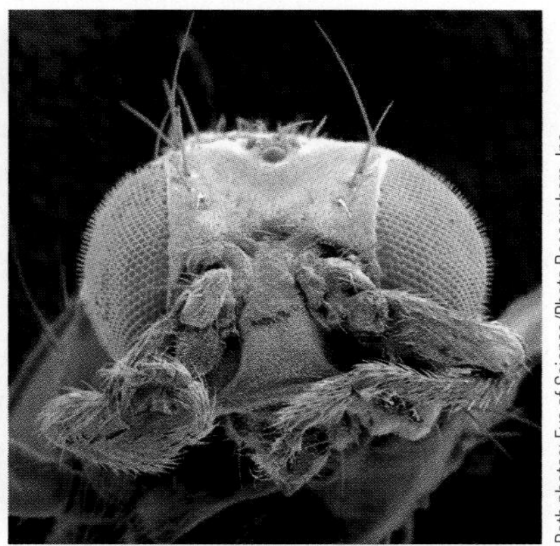

Both photos: Eye of Science/Photo Researchers, Inc.

FIGURE 16.20
Antennapedia, a homeotic mutant of *Drosophila,* in which legs develop in place of antennae.

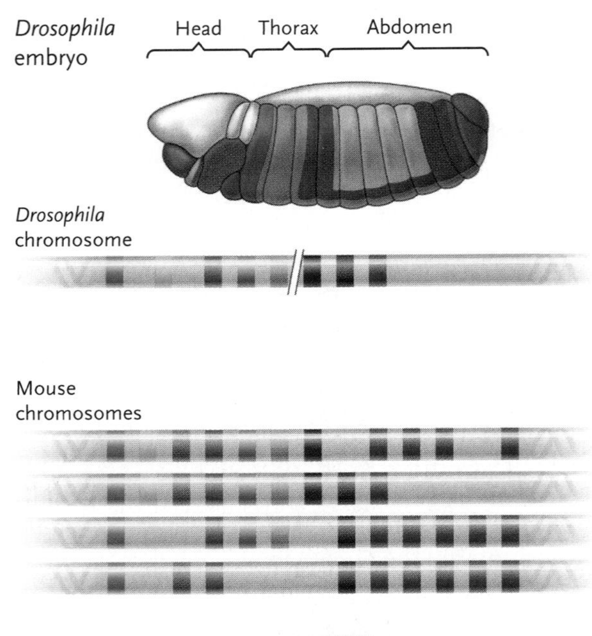
Drosophila embryo
Head
Thorax
Abdomen
Drosophila chromosome
Mouse chromosomes
Mouse embryo

FIGURE 16.21
The *Hox* genes of the fruit fly and the corresponding regions of the embryo they affect. The mouse has four sets of *Hox* genes on four different chromosomes. Their relationship to the fruit fly genes is shown by the colors.

MicroRNAs Play Critical Roles in Development

Earlier in the chapter you learned how microRNAs (miRNAs) can regulate gene expression at the translational level. Studies of miRNA gene mutants, and of organisms with defective miRNA synthesis, have shown that miRNAs have critical roles in development and differentiation, including embryogenesis, the formation of organs, and the development of the germline (the lineage of cells from which gametes are produced). For instance, in *Drosophila,* miRNAs are required for development of both somatic tissues and the germline, and in the maintenance of stem cells in the germline. (Stem cells are cells capable of differentiating into almost any adult cell type.) In zebrafish, miRNAs are essential for development; for example, a knockout of Dicer results in a developmental arrest at 7-to-10 days post-fertilization. miRNAs are also involved in brain formation, somitogenesis (generation of somites), and heart development. In mouse (and, by extrapolation, other mammals), miRNAs are essential for development; for example, a knockout of Dicer dies at 7.5 days of gestation. Dicer is also required for embryonic stem cell differentiation *in vitro* and, therefore, probably *in vivo* also.

In short, regulatory proteins play critical roles in development and differentiation, however they are but one player, and not the only player. Researchers must now discover how miRNAs regulate the expression of protein-coding mRNAs to develop a more complete understanding of the regulatory circuits underlying development.

STUDY BREAK 16.4

1. **What are determination and differentiation and, in general, how are they controlled?**
2. **How do the segmentation genes and homeotic genes of *Drosophila* differ in function?**

16.5 The Genetics of Cancer

In Chapter 10 you learned that the cell division cycle in all eukaryotes is carefully regulated by genes (see Section 10.4 and Figure 10.16). For normal cells, the relationship between gene products that stimulate cell division and gene products that inhibit cell division govern whether the cell remains in a nondividing state or whether it grows and divides. Only when the balance shifts towards stimulatory signals does the cell grow and divide.

Occasionally, differentiated cells of complex multicellular organisms deviate from their normal genetic program and begin to grow and divide, giving rise to tissue masses called *tumors* (see Figure 10.17). Those cells have lost their normal regulatory controls and have reverted partially or completely to an embryonic developmental state, in a process called *dedifferentiation.* If the altered cells stay together in a single mass, the tumor is *benign.* Benign tumors usually are invasive, but not life threatening, and their surgical removal generally results in a complete cure.

If the cells of a tumor invade and disrupt surrounding tissues, the tumor is *malignant* and is called a cancer. Sometimes, cells from malignant tumors break off and move through the blood system or lymphatic system, forming new tumors at other locations in the body. The spreading of a malignant tumor is called *metastasis* (meaning "change of state"). Malignant tumors can result in debilitation and death in various ways, including damage to critical organs, metabolic problems, hemorrhage, and secondary malignancies. In some cases, malignant tumors can be eliminated from the body by surgery or be destroyed by chemicals *(chemotherapy)* or radiation.

Cancers Are Genetic Diseases

Experimental evidence of various kinds shows that cancers are genetic diseases:

1. Particular cancers can have a high incidence in some human families. Cancers that run in families are known as **familial (hereditary) cancers.** Cancers that do not appear to be inherited are known as **sporadic (nonheredi-**

 INSIGHTS FROM THE **Molecular Revolution**

A Viral Tax on Transcriptional Regulation: How does human T-cell leukemia virus cause cancer?

The human *T-cell leukemia virus (HTLV)* causes a form of cancer by triggering rapid, uncontrolled division of white blood cells. It does so by speeding up a pathway that triggers division at normal rates in uninfected cells. The pathway is a G-protein coupled receptor-response pathway involving cyclic AMP (cAMP) as a second messenger (see Section 7.4).

Normally, the pathway is triggered by cAMP released when an infection takes place. In the pathway, a specific activator called CREB (CRE-binding protein) is activated by phosphorylation and binds to CRE (cAMP-response element) in the enhancers for particular regulatory genes that control cell division. CREB binding turns on the genes and leads to the rapid division of white blood cells.

HTLV takes advantage of the pathway by means of a short sequence 5′-CGTCA-3′ in its DNA that mimics part of the human CRE enhancer sequence 5′-TGACGTCA-3′. When the host cell's CREB is activated, it binds strongly to the enhancer sequence in the virus, turns on the viral genes, and leads to reproduction of the virus. Unfortunately for someone infected with HTLV, CREB in an infected cell also binds more strongly to the enhancers of cell division genes, leading to the uncontrolled division of white cells and, hence, to leukemia.

Research Question

In the test tube, CREB binds only weakly to the viral mimic of the CRE enhancer, so how does HTLV cause the effects it has on cell division?

Experiment

An answer to the question was provided by Susanne Wagner and Michael R. Green at the University of Massachusetts Medical School in Worcester. They found that HTLV uses one of its own encoded proteins, called *Tax,* to get around the problem of weak binding. When Tax is present, CREB binds strongly to the viral mimic of the CRE enhancer. Tax also greatly increases the ability of CREB to bind to the cell's normal CRE enhancer, leading to rapid and uncontrolled growth of infected white cells and to leukemia.

How does the Tax protein accomplish this feat? The CREB activator interacts with DNA as a dimer (a pair of molecules that together form the functional form). Wagner and Green found that the Tax protein greatly increases the ability of CREB to form dimers and thus to bind to the viral mimic of the CRE enhancer. Evidently, Tax acts as a sort of molecular "safety pin" that holds the dimer together with either the viral enhancer mimic or the cell's CRE enhancer.

Conclusion

The Tax protein compensates for the imperfection of the viral sequence that mimics the CRE enhancer sequence by greatly increasing the ability of CREB to form dimers and bind to it.

Source: S. Wagner and M. R. Green. 1993. HTLV-1 Tax protein stimulation of DNA binding of bZIP proteins by enhancing dimerization. *Science* 262:395–399.

---

tary) cancers. Familial cancers are less frequent than sporadic cancers.

2. Descendants of cancer cells are all cancer cells. In fact, it is the cloned descendants of a cancer cell that forms a tumor.
3. The incidence of cancers increases upon exposure to mutagens, agents that cause mutations in DNA. Particular chemicals and certain kinds of radiation are effective mutagens.
4. Particular chromosomal mutations are associated with specific forms of cancer (see Section 13.3 and Figure 13.13). In these cases, chromosomal breakage affects the expression of genes associated with the regulation of cell division.
5. Some viruses can induce cancer. They do so by the expression of viral genes in the host which disrupts normal cell cycle control.

Basically, all the characteristics of cancer cells that have been mentioned—dedifferentiation, uncontrolled division, and metastasis—reflect changes in gene activity.

Three Main Classes of Genes Are Implicated in Cancer

Three major classes of genes are altered frequently in cancers: *proto-oncogenes, tumor suppressor genes,* and microRNA (miRNA) genes.

PROTO-ONCOGENES Proto-oncogenes (*onkos* = bulk or mass) are genes in normal cells that encode various kinds of proteins which stimulate cell division. Examples are growth factors, receptors on target cells that are activated by growth factors (see Chapter 7), components of cellular signal transduction pathways triggered by cell division stimulatory signals (see Chapter 7), and transcription factors that regulate the expression of the structural genes for progression through the cell cycle. In normal cells, the products of proto-oncogenes are balanced by inhibitory proteins encoded by tumor suppressor genes. In cancer cells, the proto-oncogenes are altered to become **oncogenes,** genes that stimulate the cell to progress to the cancerous state. Various alterations can convert a proto-oncogene into an oncogene, and only one allele needs to be altered for the cellular changes to occur:

- A mutation in a gene's promoter or other control sequences that results in the gene becoming abnormally active.
- A mutation in the coding segment of the gene may produce an altered form of the encoded protein that is abnormally active.
- Translocation, a process in which a segment of a chromosome breaks off and attaches to a different chromosome (discussed in Section 13.3), may move a gene that controls cell division to a new location near the promoter or enhancer

sequence of a highly active gene, making the cell division gene overactive (see Figure 13.13).

- Infecting viruses may introduce genes to regions in the chromosomes where the expression of the genes disrupts cell cycle control or alters regulatory proteins to turn genes on. (*Insights from the Molecular Revolution* describes a virus that causes a blood cancer by altering a transcription factor.)

TUMOR SUPPRESSOR GENES **Tumor suppressor genes** are genes in normal cells that encode proteins which inhibit cell division. The best known tumor suppressor gene is *TP53,* so called because its encoded protein, p53, has a molecular weight of 53,000 daltons. Among other activities, normal p53 stops cell division by combining with and inhibiting cyclin-dependent protein kinases that trigger the cell's transition from the G_1 phase to the S phase of the cell cycle (discussed in Section 10.4). This activity is particularly important if the cell has sustained DNA damage. If such a cell undergoes DNA replication and divides, the damage is propagated to progeny cells. Damage means mutations, and mutations can mean cancer. However, cancer may be avoided if p53's action to block the cell from entering S gives the cell time to repair the damage, or, if the damage cannot be repaired, to trigger the cell to undergo programmed cell death (apoptosis: see Chapter 43). If the *TP53* gene is mutated so that the p53 protein is not produced or is produced in an inactive form, the cyclin-dependent protein kinases are continually active in triggering cell division. As a result, any mutations that occur can be passed on to progeny cells. Inactive *TP53* genes are found in at least 50 percent of all cancers. In general, mutations of tumor-suppressor genes contribute to the onset of cancer because the mutations result in a decrease in or a loss of the inhibitory action of the cell cycle controlling proteins they encode.

Both alleles of a tumor suppressor gene must be inactivated for inhibitory activity to be lost in cancer cells. **Figure 16.22** illustrates inactivation of the tumor suppressor gene *BRCA1* (*br*east *ca*ncer *1*) in sporadic and familial forms of breast cancer. Inactivating both alleles of *BRCA1* is not by itself sufficient for the development of breast cancer, but is one of the gene changes typically involved. Since sporadic breast cancer requires the mutational inactivation of two normal alleles of *BRCA1,* this form of the disease typically occurs later in life than the familial form. For familial breast cancer and other familial cancers we use the term *predisposition* for the cancer. This term relates to the inactivation mechanism just described. That is, an individual is predisposed to develop a particular cancer if they inherit one mutant allele of an associated tumor suppressor disease because then a mutation inactivating the other allele is all that is needed to lose the growth inhibitory properties of the tumor suppressor gene's product.

miRNA GENES You learned about the role of microRNAs (miRNAs) in regulating expression of target mRNAs. In human cancers, many miRNA genes show altered, cancer-specific expression patterns. Studying these miRNA genes has given scientists insights into the normal activities of their encoded miRNAs in cell cycle control. Some miRNAs regulate the expression of mRNAs that are the transcripts of tumor suppressor genes. If these miRNAs are overexpressed because of altera-

FIGURE 16.22
Mutational inactivation of tumor suppressor gene alleles in sporadic (A) and familial (B) cancers as exemplified by the *BRCA1* gene associated with breast cancer.

A. Sporadic breast cancer. Two independent mutations of the *BRCA1* tumor suppressor.

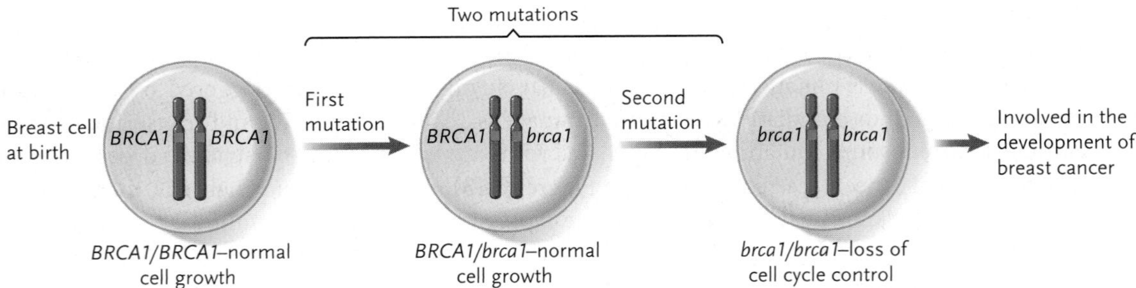
Two mutations
Breast cell at birth
BRCA1
BRCA1
First mutation
BRCA1
brca1
Second mutation
brca1
brca1
Involved in the development of breast cancer
BRCA1/BRCA1–normal cell growth
BRCA1/brca1–normal cell growth
brca1/brca1–loss of cell cycle control

B. Familial breast cancer. An individual has a predisposition for breast cancer because of inheriting one mutated *brca1* allele; mutation of the other normal *BRCA1* allele then occurs.

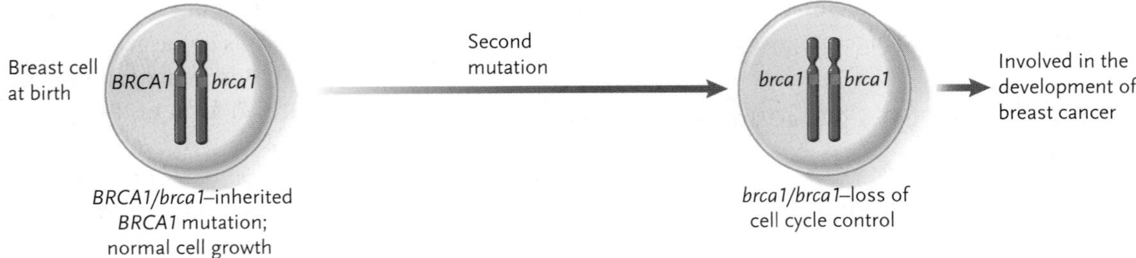
Breast cell at birth
BRCA1
brca1
Second mutation
brca1
brca1
Involved in the development of breast cancer
BRCA1/brca1–inherited BRCA1 mutation; normal cell growth
brca1/brca1–loss of cell cycle control

tions of the genes encoding them, expression of the target mRNAs can be completely blocked, thereby removing or decreasing inhibitory signals for cell proliferation. Other miRNAs regulate the translation of mRNAs that are transcripts of particular proto-oncogenes. If these miRNA genes are inactivated, or expression of these genes is markedly reduced, expression of the proto-oncogenes is higher than normal and cell proliferation is stimulated.

Cancer Develops Gradually by Multiple Steps

Cancer rarely develops by alteration of a single proto-oncogene to an oncogene, or inactivation of a single tumor-suppressor gene. Rather, in almost all cancers, successive alterations in several to many genes gradually accumulate to transform normal cells to cancer cells. This gradual mechanism is called the *multistep progression of cancer.* One example of the steps that can occur, in this case for a form of colorectal cancer, is shown in **Figure 16.23.**

The ravages of cancer, probably more than any other example, bring home the critical extent to which humans and all other multicellular organisms depend on the mechanisms controlling gene expression to develop and live normally. In a sense, the most amazing thing about these control mechanisms is that, in spite of their complexity, they operate without failures throughout most of the lives of all eukaryotes.

In the next chapter, you will learn about the molecular genetics of bacteria and their phages, and about DNA sequences in prokaryotic and eukaryotic genomes that have the ability to move to different chromosomal locations.

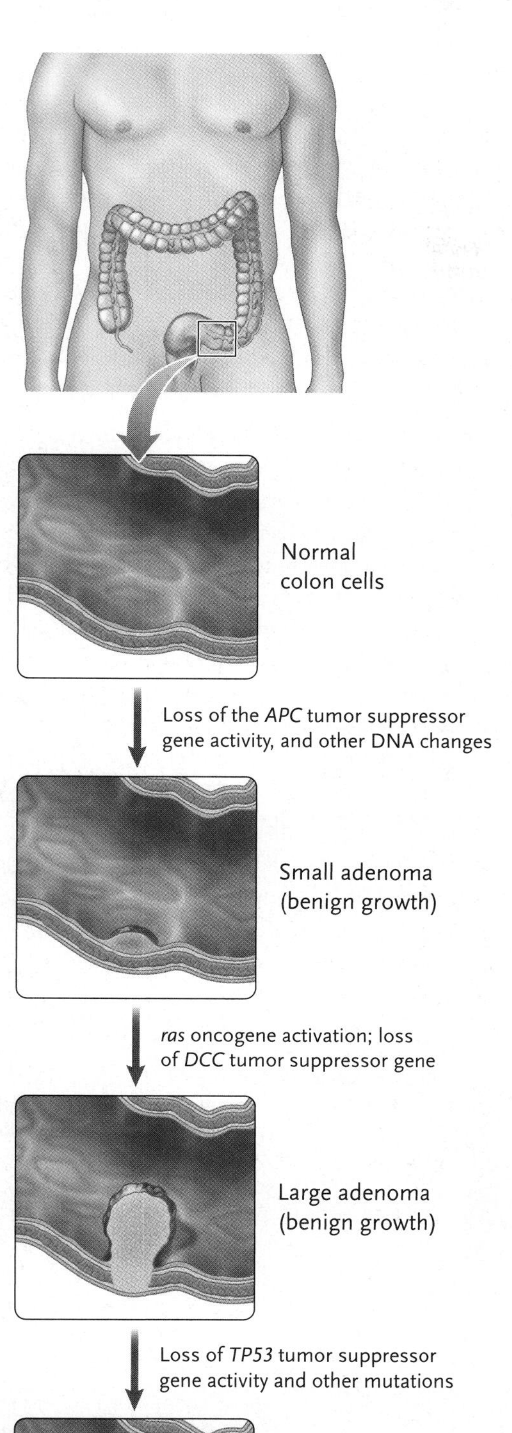
Normal
colon cells
Loss of the *APC* tumor suppressor
gene activity, and other DNA changes
Small adenoma
(benign growth)
ras oncogene activation; loss
of *DCC* tumor suppressor gene
Large adenoma
(benign growth)
Loss of *TP53* tumor suppressor
gene activity and other mutations
Carcinoma (malignant
tumor with metastasis)

FIGURE 16.23
A multistep model for the development of a type of colorectal cancer.

STUDY BREAK 16.5

1. **What is the normal function of a tumor-suppressor gene? How do mutations in tumor-suppressor genes contribute to the onset of cancer?**
2. **What is the normal function of a proto-oncogene? How can mutations in proto-oncogenes contribute to the onset of cancer?**
3. **How can changes in expression of miRNA genes contribute to the onset of cancer?**

THINK OUTSIDE THE BOOK

Individually or collaboratively use the Internet and/or research papers to find two examples of how an miRNA potentially may be an effective therapeutic molecule to treat cancer. For the examples, outline the natures of the cancers and the state of the research on the particular miRNAs under investigation.

UNANSWERED QUESTIONS

Can RNA interference silence disease?

RNA interference (RNAi) is the process in which a single-stranded RNA such as small interfering RNA (siRNA) of about 21 nucleotides associated with a protein complex (RISC) inhibits gene expression by associating with an mRNA containing a totally or partially complementary sequence. The siRNA inhibits the mRNA by site-specific cleavage, enhanced degradation, or inducing a block to its translation. RNAi can be induced experimentally in cells by delivering either a double-stranded siRNA that is converted within the cell to mature siRNA, or a DNA transcriptional template, which is transcribed to produce short-hairpin RNAs that are then converted into the mature siRNAs. Each of these approaches has advantages and disadvantages. Physicians and scientists are working together to develop RNAi for many clinical applications. Let us consider some examples of RNAi therapies that are in the works.

What are examples of diseases that are amenable to RNAi based therapies?

Anti-viral treatment Viruses are essentially exogenous genes that enter an organism and through the infection process can cause disease. Can RNAi treat viral infections? Our laboratory has had a long-standing interest in using RNAi to inhibit virus genes in order to short-circuit hepatitis virus infection. After we were able to establish that RNAi worked in whole mammals, we were successful in knocking down both hepatitis virus C gene sequences and hepatitis virus B replication in mice. Currently, scientists are using RNAi approaches to inactivate all types of viruses by directly attacking the viral genes and/or key cell proteins (e.g. viral receptor) required for viral infection and spread. A number of different treatments for viral infections are in various phases of testing in animals and humans.

Macular degeneration treatment Macular degeneration, the leading cause of blindness among those age 55 and older in the United States, is characterized by an overabundance of blood vessel growth in the retina of the eye due to the upregulation of the VEGF (vascular endothelial growth factor) pathway. These blood vessels leak, leading to clouded and often complete loss of vision. Researchers are investigating whether RNAi could be an effective therapy for such diseases by targeting VEGF or its receptor. While there was early enthusiasm, some of the medical improvement may have been due to a side-effect, due to the siRNA eliciting the non-specific release of proteins which indirectly interfere with the VEGF pathway. Such side-effects can be reduced by modification of the siRNA and/or its delivery method.

Anti-cancer treatment RNAi can also be used to turn down various genes that are mis-regulated in cancer. Again, promising results in animals have prompted early clinical studies. RNAi therapies in combination with more standard therapies will also likely be tried.

How long will it take before there are approved RNAi drugs?

Scientists believe that it may take some years because clinical trials proceed at a slower rate than preclinical animal studies. Moreover, clinical scientists will remind us that a mouse is not a man, and that while success in humans is likely there will likely be technical barriers that need to worked out before the full benefits of this new therapeutic will be realized. Interestingly, scientists are finding all sorts of new classes of small and very large noncoding RNAs. In fact, scientists have found that over 90% of the human genome is transcribed into RNA. What are the functions of these RNAs? The biological importance of these non-coding RNAs is an intense area of fundamental research. While discovery is ongoing, the next intriguing question is whether these RNAs can be harnessed or manipulated in a manner similar to RNAi for therapeutic benefit.

Think Critically

Researchers proceed cautiously with trials of RNAi disease therapies in animals before jumping into human trials. If you were working in Dr. Kay's lab, what would be your strategy to test for dangerous side-effects of the drugs?

Mark A. Kay

Mark A. Kay is the Dennis Farrey Family Professor in the Departments of Pediatrics and Genetics at the Stanford University School of Medicine. One major focus of his work is the role of small RNAs in mammalian gene regulation and their manipulation to produce new therapeutics. To learn more about his research, go to http://kaylab.stanford.edu.

REVIEW KEY CONCEPTS

Go to **CENGAGENOW** at www.cengage.com/login to access quizzing, animations, exercises, articles, and personalized homework help.

16.1 Regulation of Gene Expression in Prokaryotes

- Transcriptional control in prokaryotes involves short-term changes that turn specific genes on or off in response to changes in environmental conditions. The changes in gene activity are controlled by regulatory proteins that recognize operators of operons (Figure 16.2).
- Regulatory proteins may be repressors, which slow the rate of transcription of operons, or activators, which increase the rate of transcription.
- Some repressors are made in an active form, in which they bind to the operator of an operon and inhibit its transcription. Combination with an inducer blocks the activity of the repressor and allows the operon to be transcribed (Figure 16.3).
- Other repressors are made in an inactive form, in which they are unable to inhibit transcription of an operon unless they combine with a corepressor (Figure 16.4).
- Activators typically are made in inactive form, in which they cannot bind to their binding site next to an operon. Combining with another molecule, often a nucleotide, converts the activator into the form in which it binds with its binding site and recruits RNA polymerase, thereby stimulating transcription of the operon (Figure 16.5).

Animation: The *lac* operon

Animation: Negative control of the *lac* operon

16.2 Regulation of Transcription in Eukaryotes

- Operons are not found in eukaryotes. Instead, genes that encode proteins with related functions typically are scattered through the genome, while being regulated in a coordinated manner.
- Two general types of gene regulation occur in eukaryotes. Short-term regulation involves relatively rapid changes in gene expression in response to changes in environmental or physiological conditions. Long-term regulation involves changes in gene expression that are associated with the development and differentiation of an organism.
- Gene expression in eukaryotes is regulated at the transcriptional level (where most regulation occurs) and at posttranscriptional, translational, and posttranslational levels (Figure 16.6).
- Transcriptionally active genes have a looser chromatin structure than transcriptionally inactive genes. The change in chromatin structure that accompanies the activation of transcription of a gene involves chromatin remodeling—specific histone modifications—particularly in the region of a gene's promoter (Figure 16.7).
- Regulation of transcription initiation involves proteins binding to a gene's promoter and regulatory sites. At the promoter, general transcription factors bind and recruit RNA polymerase II, giving a very low level of transcription. Activator proteins bind to promoter proximal elements and increase the rate of transcription. Other activators bind to the enhancer and, through interaction with a coactivator, which also binds to the proteins at the promoter, greatly stimulate the rate of transcription (Figures 16.8–16.10).
- The overall control of transcription of a gene depends on the particular regulatory proteins that bind to promoter proximal elements and enhancers. The regulatory proteins are cell-type specific and may be activators or repressors. This gene regulation is achieved by a relatively low number of regulatory proteins, acting in various combinations (Figure 16.11).
- The coordinate expression of genes with related functions is achieved by each of the related genes having the same regulatory sequences associated with them.
- Sections of chromosomes or whole chromosomes can be inactivated by DNA methylation, a phenomenon called silencing. DNA methylation is also involved in genomic imprinting, in which transcription of either the inherited maternal or paternal allele of a gene is inhibited permanently.

Animation: Controls of eukaryotic gene expression

Animation: X-chromosome inactivation

16.3 Posttranscriptional, Translational, and Posttranslational Regulation

- Posttranscriptional, translational, and posttranslational controls operate primarily to regulate the quantities of proteins synthesized in cells (Figure 16.6).
- Posttranscriptional controls regulate pre-mRNA processing, mRNA availability for translation, and the rate at which mRNAs are degraded. In alternative splicing, different mRNAs are derived from the same pre-mRNA. In another process, small single-stranded RNAs complexed with proteins bind to mRNAs that have complementary sequences, and either the mRNA is cleaved or translation is blocked (Figure 16.13).
- Translational regulation controls the rate at which mRNAs are used by ribosomes in protein synthesis.
- Posttranslational controls regulate the availability of functional proteins. Mechanisms of regulation include the alteration of protein activity by chemical modification, protein activation by processing of inactive precursors, and affecting the rate of degradation of a protein.

16.4 Genetic and Molecular Regulation of Development

- Development proceeds as a result of cell division, cell movements, induction, determination, and differentiation.
- Developmental information is stored in both the nucleus and cytoplasm of the fertilized egg. The mRNA and protein molecules that direct the first stages of development are the cytoplasmic determinants.
- In determination, the developmental fate of a cell is set. The major process responsible for determination is induction, which results from the effects of signaling molecules of inducing cells on responding cells.
- In differentiation, cells change from embryonic form to specialized types with distinct structures and functions. Differentiation produces specialized cells without loss of genes from the genome.
- Determination and differentiation both involve regulated changes in gene expression (Figure 16.16).
- Pattern formation derives from the positions of cells in the embryo. Typically, positional information is detected by the cells in the form of concentration gradients of regulatory molecules encoded by genes (Figures 16.17–16.19).
- *Hox* genes are evolutionarily conserved regulatory genes that control the development of the segments or regions of the body (Figures 16.20 and 16.21).

16.5 The Genetics of Cancer

- In cancer, cells partially or completely dedifferentiate, divide rapidly and uncontrollably, and may break loose to form additional tumors in other parts of the body.
- Proto-oncogenes, tumor suppressor genes, and miRNA genes typically are altered in cancer cells. Proto-oncogenes encode proteins that stimulate cell division. Their altered forms, oncogenes, are abnormally active. Tumor suppressor genes in their normal form encode proteins that inhibit cell division. Mutated forms of these genes lose this inhibitory activity (Figure 16.22). MicroRNA genes control the activity of mRNA transcripts of particular tumor suppressor genes and proto-oncogenes. Alteration of activity of such an miRNA gene can lead to a lower than normal activity of tumor suppressor gene products or a higher than normal activity of proto-oncogene products depending on the target of the miRNA. In either case, cell proliferation can be stimulated.
- Most cancers develop by multistep progression involving the successive alteration of several to many genes (Figure 16.23).

UNDERSTAND AND APPLY

Test Your Knowledge

1. The control of the delivery of finished mRNAs to the cytoplasm is an example of:
   a. translational regulation.
   b. posttranslational regulation.
   c. transcriptional regulation.
   d. posttranscriptional regulation.
   e. deoxyribonucleic regulation.

2. For the *E. coli lac* operon, when lactose is present:
   a. and glucose is absent, cAMP binds and activates catabolic activator protein (CAP).
   b. and glucose is absent, the level of cAMP decreases.
   c. activated CAP binds the repressor protein to remove it from the operator gene.
   d. the cell prefers lactose over glucose.
   e. RNA polymerase cannot bind to the promoter.
3. For the *trp* operon:
   a. tryptophan is an inducer.
   b. when tryptophan binds to the Trp repressor, transcription is blocked.
   c. Trp repressor is synthesized in an active form.
   d. low levels of tryptophan bind to the *trp* operator and block transcription of the tryptophan biosynthesis genes.
   e. high levels of tryptophan activate RNA polymerase and induce transcription.
4. Transcriptional regulation is important because it:
   a. is the final and most important step in gene regulation.
   b. determines the availability of finished proteins.
   c. determines which genes are expressed.
   d. determines the rate at which proteins are made.
   e. removes masking proteins that block initiation of transcription.
5. Chromatin remodeling activates gene expression when it:
   a. allows proteins initiating transcription to disengage from the promoter.
   b. winds genes tightly around histones.
   c. deacetylates histones.
   d. inserts nucleosomes into chromatin.
   e. recruits a protein complex that displaces a nucleosome, thereby exposing the promoter.
6. Which statement about activation of transcription is *not* correct?
   a. A transcription factor binds to the promoter in the area of the TATA box.
   b. A coactivator forms a bridge between the promoter and the gene to be transcribed.
   c. Transcription factors bind the promoter and RNA polymerase.
   d. Activators bind to the enhancer region on DNA.
   e. RNA is transcribed downstream from the promoter region.
7. Normal ears in a certain mammal are perky; mutants have droopy ears. In males of these mammals, the gene encoding perky ears is transcribed only from the female parent. This is because the gene from the male parent is silenced by methylation. If the maternal gene is mutated:
   a. male offspring have droopy ears.
   b. offspring have perky ears.
   c. male offspring have one droopy ear and one perky ear.
   d. the genetic mechanism is called alternative splicing.
   e. this is an example of posttranscriptional regulation.
8. Which of the following statements does *not* describe microRNA?
   a. MicroRNA is encoded by non-protein-coding genes.
   b. MicroRNA has a precursor that is folded and then cut by a Dicer enzyme.
   c. MicroRNA is an example of a molecule that induces RNA interference or gene silencing.
   d. MicroRNA is synthesized *in vitro* but probably not *in vivo.*
   e. MicroRNA has a similar function to that of small interfering RNAs.
9. In mammals, the nose is located at the anterior end of the embryo, and the heart at the center of the embryo. These positions are the result of activation of:
   a. *Hox* genes that are arranged along a number of chromosomes in the same order as they are expressed along the anterior–posterior body axis.
   b. maternal-effect genes, after somites differentiate into muscle.
   c. *Hox* genes that are scattered randomly among different chromosomes.
   d. a transcription factor called the homeobox.
   e. a homeodomain that binds ribosomes.
10. Which of the following is not a characteristic of cancer cells?
   a. proto-oncogenes altered to become oncogenes
   b. the mutation of a suppressor gene that results in normal genes becoming inactive
   c. the mutation of the *TP53* gene
   d. multistep progression
   e. amplification of growth factors and growth factor receptors

Discuss the Concepts

1. In a mutant strain of *E. coli,* the CAP protein is unable to combine with its target region of the *lac* operon. How would you expect the mutation to affect transcription when cells of this strain are subjected to the following conditions?

   lactose and glucose are both available

   lactose is available but glucose is not

   both lactose and glucose are unavailable
2. Duchenne muscular dystrophy, an inherited genetic disorder, affects boys almost exclusively. Early in childhood, muscle tissue begins to break down in affected individuals, who typically die in their teens or early twenties as a result of respiratory failure. Muscle samples from women who carry the mutation reveal some regions of degenerating muscle tissue adjacent to other regions that are normal. Develop a hypothesis explaining these observations.
3. Eukaryotic transcription is generally controlled by the binding of regulatory proteins to DNA sequences rather than by the modification of RNA polymerases. Develop a hypothesis explaining why this is so.

Design an Experiment

Design an experiment using rats as the model organism to test the hypothesis that human chorionic gonadotrophin (hCG), a hormone produced during pregnancy, leads to a significant protection against breast cancer.

Interpret the Data

Investigating a correlation between specific cancer-causing mutations and risk of mortality in humans is challenging, in part because each cancer patient is given the best treatment available at the time. There are no "untreated control" cancer patients, and the idea of what treatments are optimal changes quickly as new drugs become available and new discoveries are made.

The table on the next page shows the results of a study in which 442 women who had been diagnosed with breast cancer were checked for mutations of the *BRCA1* tumor suppressor gene and of a second tumor suppressor gene involved with breast cancer, *BRCA2.* The treatments and progress of the women were followed over several years. All of the women in the study had at least two affected close relatives, so their risk of developing breast cancer due to an inherited factor was estimated to be greater than that of the general population.

1. According to this study, what is the woman's risk of dying of cancer if two of her close relatives have breast cancer?
2. What is her risk of dying of cancer if she carries a mutated *BRCA1* gene?
3. Is a *BRCA1* or *BRCA2* mutation more dangerous in breast cancer cases?
4. What other data would you have to see in order to make a conclusion about the effectiveness of preventive surgeries?

BRCA Mutations in Women Diagnosed with Breast Cancer[1]	*BRCA1* Mutations	*BRCA2* Mutations	No *BRCA* Mutations	Total
Total number of patients	89	35	318	442.0
Average age at diagnosis	43.9	46.2	50.4	
Preventive mastectomy[2]	6	3	14	23
Preventive oophorectomy[2]	38	7	22	67
Number of deaths	16	1	21	38
Percent died	18.0	2.8	6.9	8.6

[1]Results from a 2007 study investigating *BRCA* mutations in women diagnosed with breast cancer. All women in the study had a family history of breast cancer.
[2]Some of the women underwent preventive mastectomy (removal of the noncancerous breast) during their course of treatment. Others had preventive oophorectomy (surgical removal of the ovaries).

Apply Evolutionary Thinking

Fruit flies homozygous for a mutation in the tumor suppressor gene *HIPPO* develop tumors in every organ. Expression of the human gene *MST2* in flies homozygous for *HIPPO* show greatly reduced or no tumors. What does this result suggest about the evolution of tumor suppressor genes in animals?

Express Your Opinion

Some females at high risk of developing breast cancer opt for prophylactic mastectomy, the surgical removal of one or both breasts even before cancer develops. Many of them would never have developed cancer. Should the surgery be restricted to cancer treatment? Go to academic.cengage.com/login to investigate both sides of the issue and then vote.

17

STUDY OUTLINE

Hybrid Medical Animation/Photo Researchers, Inc.

Escherichia coli, a model research organism for several types of biological studies, including bacterial genetics (computer rendering).

Bacterial and Viral Genetics

Why It Matters. . . In 1885, a German pediatrician, Theodor Escherich, identified a bacterium that caused severe diarrhea in infants. He named it *Bacterium coli.* Researchers were surprised to discover, however, that *Bacterium coli* is also present in healthy infants, and is a normal inhabitant of the human intestine. It was also discovered that only certain strains of *B. coli* cause human diseases. Further, researchers in the twentieth century found that if bacteria of different strains were mixed together, the organisms produced some progeny with a mixture of traits from more than one strain—evidence that bacteria could undergo genetic recombination.

As an organism that is readily available and easy to grow, this intestinal bacterium has been of central interest to scientists since its first discovery. Renamed *Escherichia coli* in honor of Escherich, the bacterium brings several distinct advantages to scientific investigation. It can be grown quickly in huge numbers in nutrient solutions that are simple to prepare. And, it can be infected with a group of viruses called **bacteriophages** (**phages** for short) that have been as valuable to scientists as *E. coli* has because phages can be grown by the billions in cultures of the host bacterium. The rapid generation times and numerous offspring of *E. coli* and its phages make them especially valuable to geneticists. Geneticists have used them to analyze genetic crosses and their outcomes much more quickly than they can with eukaryotes. Researchers can also detect genetic events that occur only once within millions of offspring. The characteristics of these rare events helped scientists to work out the structure, activity, and recombination of genes at the molecular level.

Because *E. coli* can be cultured in completely defined chemical media—solutions in which the identity and amount of each chemical is known—it is particularly useful for biochemical investigations. What goes in, what comes out, and what biochemical changes occur inside the cells while growing in a particular medium can be detected and closely followed in normal and mutant bacteria, and

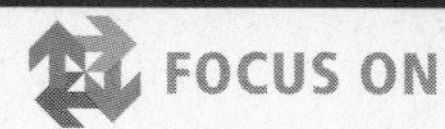 FOCUS ON **Model Research Organisms**

Escherichia coli

We probably know more about *E. coli* than any other organism. For example, microbiologists have deciphered the complete DNA sequence of the genome of a standard laboratory strain of *E. coli*, including the sequence of its approximately 4,400 genes. However the functions of about one third of these genes are still unidentified.

E. coli got its start in laboratory research because of the ease with which it can be grown in culture. *E. coli* cells divide about every 20 minutes under optimal conditions, producing a clone of 1 billion cells in a matter of hours, in only 10 mL of culture medium. The same amount of medium could accommodate as many as 10 billion cells before the growth rate of *E. coli* begins to slow. Different strains of *E. coli* can be grown with minimal equipment, requiring little more than culture vessels in an incubator that is held at 37°C.

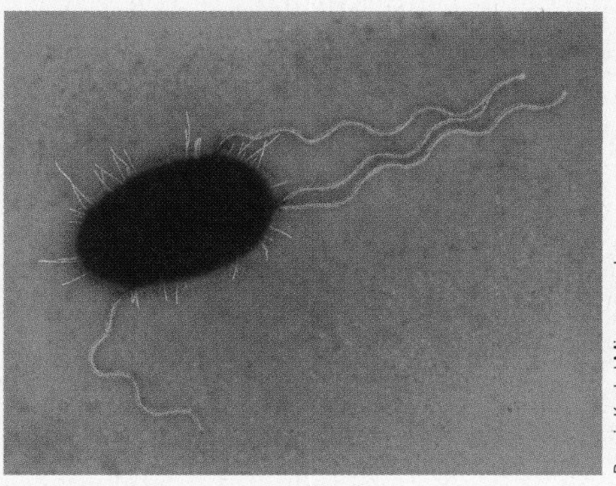
Dennis Kunkel Microscopy, Inc.

Early on, however, the major advantage of *E. coli* for research could be summed up in a single word: sex. When Joshua Lederberg and Edward Tatum discovered that *E. coli* could conjugate, with genetic material passing from one bacterium to the other, they and other scientists realized they could carry out genetic crosses with the bacterium, producing genetic recombinants that could indicate the relative positions of genes on the chromosome. Knowing these relative positions, researchers were able to generate a genetic map of the *E. coli* chromosome. The map showed that genes with related functions are clustered together, a fact that had significant implications for the regulation of expression of those genes. For example, François Jacob and Jacques Monod's work with the genes for lactose metabolism led to the pioneering operon model (described in Section 16.1). In their work, at the Pasteur Institute in Paris, they used conjugation to map the genes and generated partial diploids to help understand the details of the regulation of transcription of those genes.

The development of *E. coli* as a model organism for studying gene organization and the regulation of gene expression led to the field of molecular genetics. The study of naturally occurring plasmids in *E. coli* and of enzymes that cut DNA at specific sequences eventually resulted in techniques for combining DNA from different sources, such as inserting a gene from an organism into a plasmid. Today *E. coli* is used for amplifying (cloning) plasmids that contain inserted genes or other sequences.

In essence, the biotechnology industry has its foundation in molecular genetics studies of *E. coli*, and large-scale *E. coli* cultures are widely used as "factories" for production of desired proteins. For example, the human insulin hormone, required for treatment of certain forms of diabetes, is produced by *E. coli* factories. (Chapter 18 explains more about cloning and other types of DNA manipulation.)

Laboratory strains of *E. coli* are harmless to humans. Similarly, the natural *E. coli* cells in the colon of humans and other mammals are usually harmless. There are pathogenic strains of *E. coli*, though; sometimes they make the news when humans who eat food that contains a pathogenic strain develop disease symptoms, notably colitis (inflammation of the colon) and bloody diarrhea. The genomes of several pathogenic *E. coli* strains have been sequenced; each has more genes than either the lab strain or the strain that normally lives in the human colon. The extra genes of the pathogenic strains include the genes that make the bacteria pathogenic.

---

in bacteria infected by phages. These biochemical studies have added immeasurably to the definition of genes and their activities, and have identified many biochemical pathways and the enzymes catalyzing them. (*Focus on Model Research Organisms* tells more about *E. coli's* advantages as a model laboratory organism.)

Biologists successfully applied the same techniques used with bacteria and their phages to eukaryotes such as *Neurospora* and *Aspergillus,* fungi with short generation times that can also be grown and analyzed biochemically in large numbers. Molecular studies are now easy to carry out with a wide variety of eukaryotes, including the yeast *Saccharomyces cerevisiae,* the fruit fly *Drosophila melanogaster,* and the plant *Arabidopsis thaliana.* The results of this research showed that the molecular characteristics of genes discovered in prokaryotes apply to eukaryotes as well.

This chapter outlines the basic findings of molecular genetics in bacteria and their viruses. It also describes more broadly the structure and properties of viruses, viroids, and prions. We begin our discussion with genetic recombination mechanisms in bacteria. <

17.1 Gene Transfer and Genetic Recombination in Bacteria

In the first half of the twentieth century, foundational genetic experiments with eukaryotes revealed the processes of genetic recombination during sexual reproduction, which led to the construction of genetic maps of chromosomes for a number of organisms (see Chapters 12 and 13). Bacteria became the subject of genetics research in the middle of the twentieth century. A key early question was whether gene transfer and genetic recombination can occur in bacteria even though these organisms do not undergo meiosis. For particular bacteria, the answer to the question was yes—genes can be transferred from one bacterium to another by several different mechanisms, and the newly introduced DNA can recombine with DNA already present. Such genetic recombination performs the same function as it does in eukaryotes: it generates genetic variability through the exchange of alleles between homologous regions of DNA molecules from two different individuals.

FIGURE 17.1 Experimental Research

Genetic Recombination in Bacteria

Question: Does genetic recombination occur in bacteria?

Experiment: Lederberg and Tatum tested whether genetic recombination occurred between two mutant strains of *E. coli*. Mutant strain 1 required biotin and methionine to grow, but not leucine, threonine, or thiamine—its genotype was bio^- met^- leu^+ thr^+ thi^+. Mutant strain 2 required leucine, threonine, and thiamine to grow, but not biotin or methionine—its genotype was bio^+ met^+ leu^- thr^- thi^-.

Lederberg and Tatum mixed together large numbers of the two mutant strains and plated them on minimal medium, which lacked any of the nutrients the strains needed for growth. As controls, they also plated large numbers of the two mutant strains individually on minimal medium.

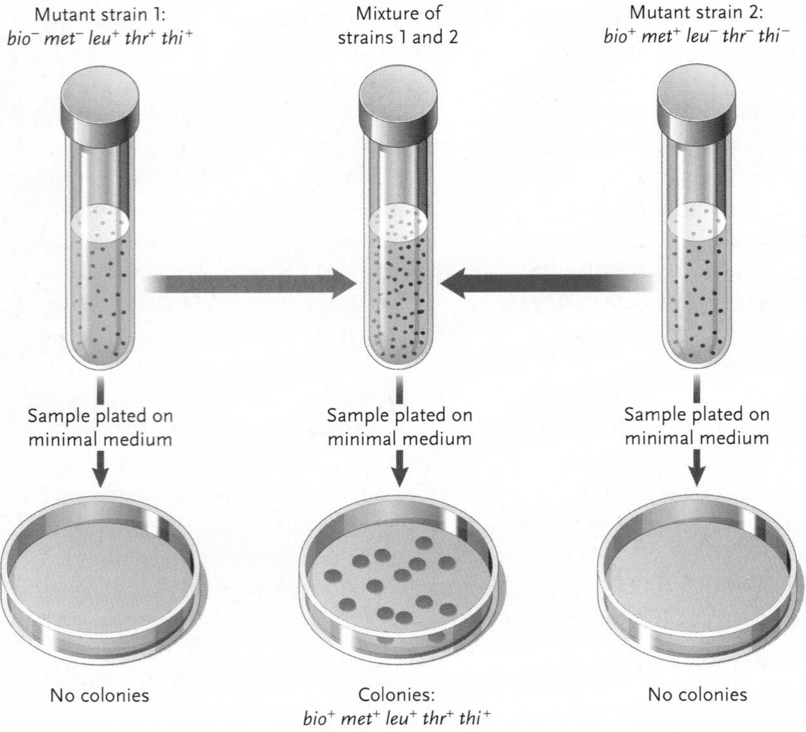

Mutant strain 1:
bio^- met^- leu^+ thr^+ thi^+
Mixture of strains 1 and 2
Mutant strain 2:
bio^+ met^+ leu^- thr^- thi^-
Sample plated on minimal medium
Sample plated on minimal medium
Sample plated on minimal medium
No colonies
Colonies:
bio^+ met^+ leu^+ thr^+ thi^+
No colonies

Results: No colonies grew on the control plates, indicating that the mutant alleles in the two strains did not mutate back to normal alleles, which would have produced growth on minimal medium. However, many colonies grew on plates spread with a mixture of mutant strain 1 and mutant strain 2.

Conclusion: To grow on minimal medium, the bacteria must have the genotype bio^+ met^+ leu^+ thr^+ thi^+. Lederberg and Tatum reasoned that the colonies on the plate must have resulted from genetic recombination between mutant strains 1 and 2.

Source: J. Lederberg and E. Tatum. 1946. Gene recombination in *Escherichia coli*. *Nature* 158:558.

E. coli is one of the bacteria in which genetic recombination occurs. By the 1940s geneticists knew that *E. coli* and many other bacteria could be grown in a **minimal medium** containing water, an organic carbon source such as glucose, and a selection of inorganic salts—including one that provides nitrogen—such as ammonium chloride. The growth medium can be in liquid form or in the form of a gel that is made by adding agar to the liquid medium. (Agar is a polysaccharide material, indigestible by most bacteria, that is extracted from algae.)

Since it is not practical to study a single bacterium for most experiments, researchers soon developed techniques for starting bacterial cultures from a single cell, generating cultures with a large number of genetically identical cells. Cultures of this type are called **clones.** To start bacterial clones, the scientist spreads a drop of a bacterial culture over a sterile agar gel in a culture dish. The culture has been diluted enough to ensure that cells will be widely separated on the agar surface. Each individual cell divides many times to produce a separate colony that is a clone of the initial cell. Cells can be removed from a clone and introduced into liquid cultures or spread on agar and grown in essentially any quantity.

Genetic Recombination Occurs in *E. coli*

In 1946, Joshua Lederberg and Edward L. Tatum of Yale University set out to determine if genetic recombination occurs in bacteria, using *E. coli* as their experimental organism. In essence, they tested whether bacteria had a genetic recombination process. As a first step, they induced genetic mutations in *E. coli* bacteria by exposing the cells to mutagens such as X-rays or ultraviolet light. After the exposure, Lederberg and Tatum found that some bacteria had become *auxotrophs,* mutant strains that could not grow on minimal medium (see Section 15.1). One mutant strain could grow only if the vitamin biotin and the amino acid methionine were added to the culture medium; genes that encoded the enzymes required to make these substances had mutated in this strain. A second mutant strain did not need biotin or methionine in its growth medium, but could grow only if the amino acids leucine and threonine were added along with the vitamin thiamine. These two genetic strains of *E. coli* are represented in genetic shorthand as shown:

Strain 1: bio^- met^- leu^+ thr^+ thi^+

Strain 2: bio^+ met^+ leu^- thr^- thi^-

In this shorthand, *bio* refers to the gene that governs a cell's ability to synthesize biotin from inorganic precursors. The designation bio^+ indicates that the allele is normal; bio^- represents the mutant allele, which produces cells that cannot make biotin for themselves. Similarly, met^+ and met^-, leu^+ and leu^-, thr^+ and thr^-, and thi^+ and thi^- are the respective normal and mutant alleles for methionine, leucine, threonine, and thiamine synthesis.

Lederberg and Tatum mixed about 100 million cells of the two mutant strains together and placed them on a minimal medium **(Figure 17.1).** None of the cells were expected to be able to grow on the minimal medium unless some form of recombination between DNA molecules from the two parental types produced the new combination with normal alleles for each of the five genes:

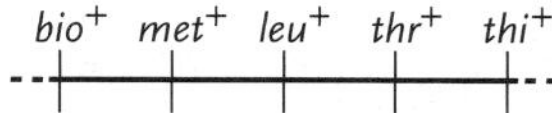
bio^+ met^+ leu^+ thr^+ thi^+

Several hundred colonies grew on the minimal medium, indicating that genetic recombination had actually taken place in the bacteria. Lederberg and Tatum eliminated the possibility that the colonies arose from chance mutations back to normal alleles by placing hundreds of millions of cells of strains 1 or 2 separately on the surface of a minimal medium: no colonies grew on these control plates.

Bacterial Conjugation Brings DNA of Two Cells Together, Allowing Genetic Recombination to Occur

Lederberg and Tatum's results led to a major question: How did DNA molecules with different alleles get together to undergo genetic recombination? Recombination in eukaryotes occurs in diploid cells undergoing meiosis by an exchange of segments between the chromatids of homologous chromosome pairs (discussed in Section 13.1). Bacteria typically have a single, circular chromosome—they are haploid organisms (see Section 14.5). Rather than fusing together to produce the prokaryotic equivalent of a diploid zygote, bacterial cells *conjugate:* they contact each other, initially becoming connected by a long tubular structure called a *sex pilus* **(Figure 17.2A),** and then forming a cytoplasmic bridge that connects two cells **(Figure 17.2B).** During **conjugation,** a copy of part of the DNA of one cell, the *donor* (the bristly cell in Figure 17.2A), moves through the cytoplasmic bridge into the other cell, the *recipient.* Once donor DNA enters the recipient, it pairs with the homologous region of the recipient cell's DNA, and genetic recombination can occur.

THE F FACTOR AND CONJUGATION The donor bacterial cell in a pairing initiates conjugation. The ability to conjugate depends on the presence within a donor cell of a plasmid called the **F factor** (F = fertility). Plasmids are small circles of DNA that occur in bacteria in addition to the main circular chromosomal DNA molecule **(Figure 17.3).** Plasmids contain several to many genes and a replication origin that permits them to be duplicated and passed on during bacterial division. Donor cells in conjugation are called F^+ cells because they contain the F factor. They are able to conjugate (mate) with recipient cells but not with other donor cells. Recipient cells, which lack the F factor, are called **F^- cells.**

A. **Attachment by sex pilus**

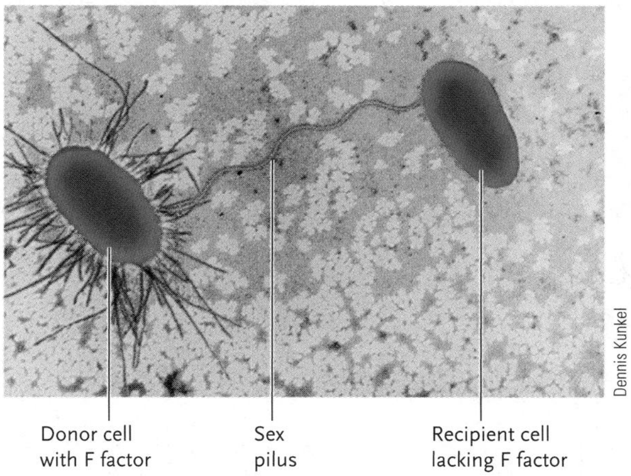
Dennis Kunkel
Donor cell with F factor
Sex pilus
Recipient cell lacking F factor

B. **Cytoplasmic bridge formed**

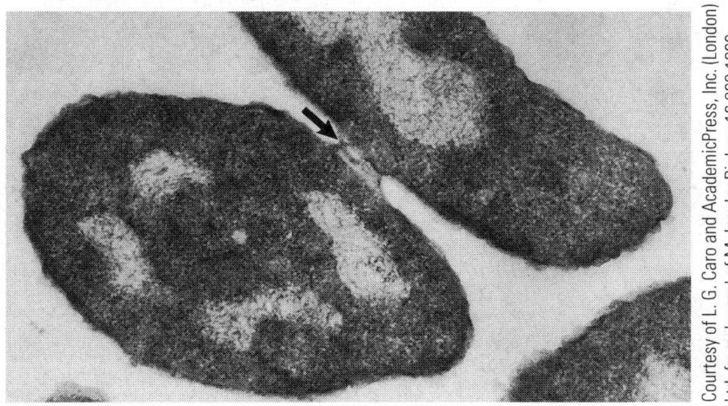
Courtesy of L. G. Caro and AcademicPress, Inc. (London) Ltd. from Journal of Molecular Biology 16.269.1966

FIGURE 17.2
Conjugating *E. coli* cells. **(A)** Initial attachment of two cells by the sex pilus. **(B)** A cytoplasmic bridge (arrow) has formed between the cells, through which DNA moves from one cell to the other.

A. **Bacterial DNA released from cell**

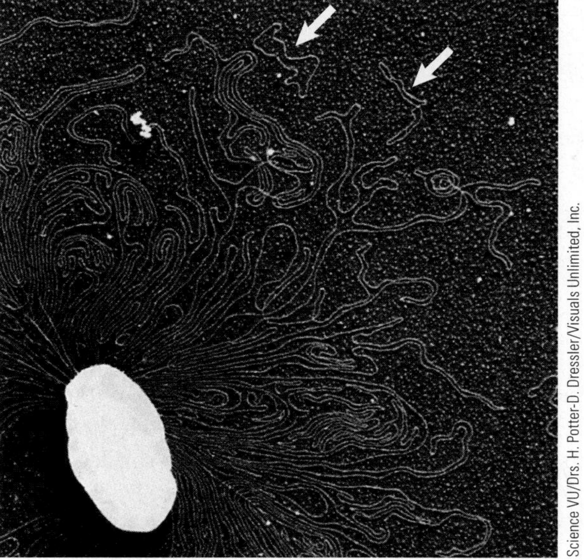
Science VU/Drs. H. Potter-D. Dressler/Visuals Unlimited, Inc.

B. **Plasmid**

Professor Stanley Cohen/Science Photo Library/ Photo Researchers, Inc.

FIGURE 17.3
Electron micrographs of DNA released from a disrupted bacterial cell. **(A)** Plasmids (arrows) near the mass of chromosomal DNA. **(B)** A single plasmid at higher magnification (colorized).

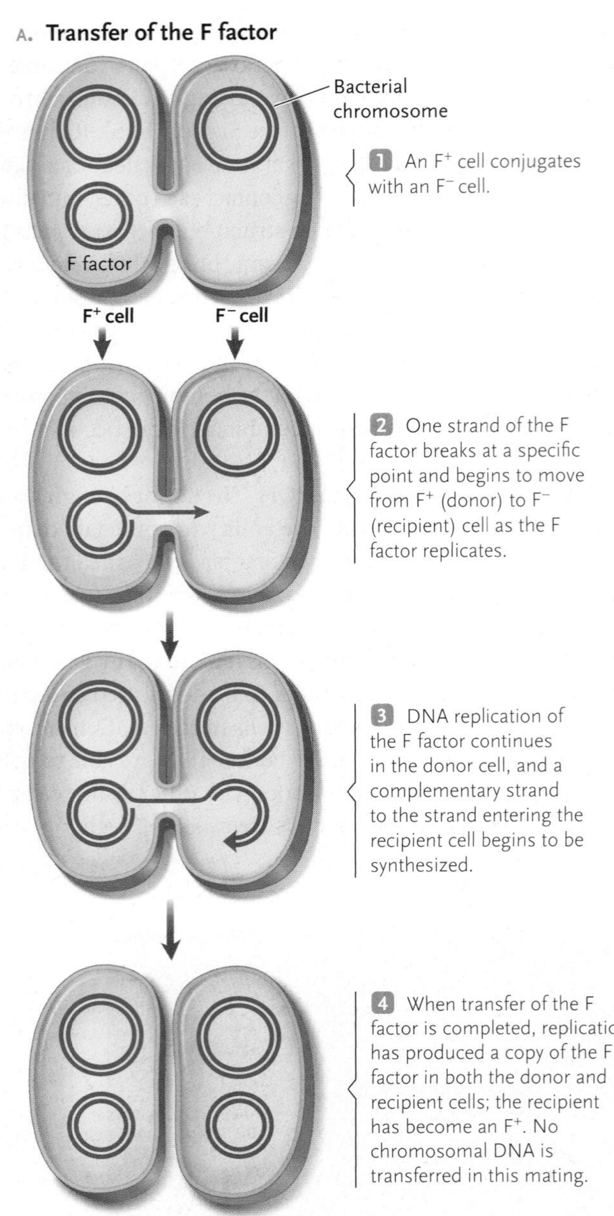
A. Transfer of the F factor
Bacterial chromosome
F factor
F⁺ cell
F⁻ cell
1 An F⁺ cell conjugates with an F⁻ cell.
2 One strand of the F factor breaks at a specific point and begins to move from F⁺ (donor) to F⁻ (recipient) cell as the F factor replicates.
3 DNA replication of the F factor continues in the donor cell, and a complementary strand to the strand entering the recipient cell begins to be synthesized.
4 When transfer of the F factor is completed, replication has produced a copy of the F factor in both the donor and recipient cells; the recipient has become an F⁺. No chromosomal DNA is transferred in this mating.
F⁺ cell
F⁺ cell
B. Transfer of bacterial genes
Bacterial chromosome
F factor
F⁺ cell
1 F factor integrates into the E. coli chromosome in a single crossover event producing an Hfr cell.
Bacterial chromosome
2 An Hfr cell and an F⁻ cell conjugate. These two cells differ in alleles: the Hfr is a⁺ b⁺ c⁺ d⁺, and the F⁻ cell is a⁻ b⁻ c⁻ d⁻.
Hfr cell
F⁻ cell
3 As with the F⁺ × F⁻ conjugation, one strand of the F factor breaks at a specific point and begins to move from the Hfr (donor) to F⁻ (recipient) cell as replication takes place.
4 In the F⁻ cell the entering single-stranded F factor segment and the attached chromosomal DNA are replicated by synthesis of the complementary DNA strand. Recombination occurs between the entering donor chromosomal DNA and the recipient's chromosome.
5 Here, as a result of recombination two cross overs produce a b⁺ recombinant. When the conjugating pair breaks apart, the linear piece of donor DNA is degraded and all descendants of the recipient will be b⁺. The recipient remains F⁻ because not all the F factor has been transferred.
Hfr chromosome (part of F factor, followed by bacterial genes)
Conjugation bridge breaks, resulting in two cells. F⁻ is a b⁺ recombinant.

FIGURE 17.4
Transfer of genetic material during conjugation between *E. coli* cells. **(A)** Transfer of the F factor during conjugation between F^+ and F^- cells. **(B)** Transfer of bacterial genes and production of recombinants during conjugation between Hfr and F^- cells.

The F factor carries about 20 or so genes. Several of the genes encode proteins of the **sex pilus,** also called the **F pilus** (plural, *pili*). The sex pilus is a long, tubular structure on the cell surface that allows an F^+ donor cell to attach to an F^- recipient (see Figure 17.2A). Once attached, the cells form a cytoplasmic bridge and conjugate (**Figure 17.4A,** step 1; see also Figure 17.2B). During conjugation, the F plasmid replicates using a special type of DNA replication. When the two strands of the plasmid DNA separate during replication, one of the strands is transferred from the F^+ cell through the cytoplasmic bridge to the F^- cell (Figure 17.4A, step 2). In the recipient cell, synthesis of the complementary strand to the entering DNA strand occurs (Figure 17.4A, step 3). When the entire F factor strand has entered and its complementary strand has been synthesized, the F factor circularizes into a complete F factor, changing the cell to F^+ (Figure 17.4A, step 4). No chromosomal DNA is transferred between cells in this process, however, so no genetic recombination results from $F^+ \times F^-$ conjugation.

Hfr CELLS AND GENETIC RECOMBINATION How does genetic recombination of bacterial genes occur as a result of conjugation if no chromosomal DNA transfers when an F factor is transferred in conjugation? The answer is that in some F^+ cells the F factor integrates into the bacterial chromosome by crossing-over (**Figure 17.4B,** step 1). This produces a donor that can transfer genes on the bacterial chromosome to a recipient. These special donor cells are known as **Hfr cells** (Hfr = high frequency recombination). Because the F factor genes are still active when the plasmid is integrated into the bacterial chromosome, an Hfr cell can conjugate with an F^- cell. In an Hfr × F^- conjugation where the two cell types differ in alleles (Figure 17.4B, step 2), DNA replication begins in the middle of the integrated F factor, and a segment of the F factor moves through the conjugation bridge into the recipient (Figure 17.4B, step 3), bringing the chromosomal DNA behind it (Figure 17.4B, step 4). Synthesis of the DNA strand that is complementary to the entering DNA from the donor occurs in the recipient cell.

Before the conjugating pair breaks apart, some genes of the donor cell follow the F segment into the recipient cell. The recipient therefore becomes a **partial diploid** for the donor chromosomal DNA segment that goes through the conjugation bridge. For our example, the recipient cell in Figure 17.4B, step 4, has become $a^+ b^+/a^- b^-$. The recipient's DNA and the homologous DNA fragment from the donor can pair and recombine. Genetic recombination occurs by a double crossover event exchanging donor gene(s) with recipient gene(s) using essentially the same mechanisms as in eukaryotes (discussed in Section 13.1)—Figure 17.4B, step 5, shows the generation of an b^+ recombinant. In other pairs in the mating population, the a^+ gene could recombine with the homologous recipient gene, or both a^+ and b^+ genes could recombine. The genetic recombinants observed in Lederberg and Tatum's experiment (see Figure 17.1) were produced in this same general way. Since conjugation usually breaks off long before the second part of the F plasmid has been transferred (it would be the last DNA piece transferred), the recipient cell remains F^-.

Recombinants produced during conjugation can be detected only if the alleles of the genes in the DNA transferred from the donor differ from those in the recipient's chromosome. Following recombination, the bacterial DNA replicates and the cell divides normally, producing a cell line with the new gene combination. Any remnants of the DNA fragment that originally entered the cell are degraded as division proceeds and do not contribute further to genetic recombination or cell heredity.

MAPPING GENES BY CONJUGATION Genetic recombination by conjugation was discovered by two scientists, François Jacob (who also proposed the operon model for the regulation of gene expression in bacteria; see Section 16.1) and Elie L. Wollman, at the Pasteur Institute in Paris. They began their experiments by conjugating Hfr and F^- cells that differed in a number of alleles. At regular intervals after conjugation commenced, they removed some of the cells and agitated them in a blender to break apart attached cells. They then cultured the separated cells and analyzed them for recombinants. They found that the longer they allowed cells to conjugate before separation, the greater the number of donor genes that entered the recipient and produced recombinants. From this result, Jacob and Wollman concluded that during conjugation, the Hfr cell slowly injects a copy of its DNA into the F^- cell. Full transfer of an entire DNA molecule to an F^- cell would take about 90 to 100 minutes. In nature, however, the entire DNA molecule is rarely transferred because the cytoplasmic bridge between conjugating cells is fragile and easily broken by random molecular motions before transfer is complete.

The pattern of gene transfer from Hfr to F^- cells was used to map the *E. coli* chromosome. The F factor integrates into one of a few possible fixed positions around the circular *E. coli* DNA. As a result, the genes of the bacterial DNA follow the F factor segment into the recipient cell in a definite order, with the gene immediately behind the F factor segment entering first and the next genes following. In the theoretical example shown in Figure 17.4B, donor genes will enter in the order a^+–b^+–c^+–d^+. By breaking off conjugation at gradually increasing times, investigators allowed longer and longer pieces of DNA to enter the recipient cell, carrying more and more genes from the donor cell (detected by the appearance of recombinants). By noting the order and time at which genes were transferred, investigators were able to map and assign the relative positions of most genes in the *E. coli* chromosome. The resulting genetic map has distances between genes in units of minutes. To this day, the genetic map of *E. coli* shows map distances as minutes, reflecting the mapping of genes by conjugation.

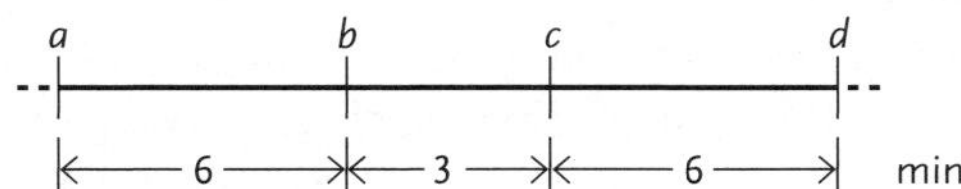

a
b
c
d
6
3
6
min

The genetic maps from *E. coli* conjugation experiments indicate that the genes are arranged in a circle, reflecting the circular form of the *E. coli* chromosome. More recently, direct sequencing of the *E. coli* genome has confirmed the results obtained by genetic mapping.

In addition to the F plasmid, bacteria also contain other types of plasmids. **R plasmids,** for example, contain genes that provide resistance to unfavorable conditions, such as exposure to antibiotics. The competitive advantage provided by the genes in some plasmids may account for the wide distribution of plasmids of all kinds in prokaryotic cells.

In Transformation, DNA Taken Up by a Bacterium Is the Source of Genetic Recombination

The discovery of conjugation and genetic recombination in *E. coli* showed that genetic recombination is not restricted to eukaryotes. DNA can transfer from one bacterial cell to another by two additional mechanisms, *transformation* and *transduction.* Like conjugation, these mechanisms transfer DNA in one direction, and create partial diploids in which recombination can occur between alleles in the homologous DNA regions.

In **transformation,** bacteria take up pieces of DNA that are released as other cells disintegrate. Frederick Griffith, a medical officer in the British Ministry of Health, London, discovered this mechanism in 1928, when he found that a noninfective form of the bacterium *Streptococcus pneumoniae,* unable to cause pneumonia in mice, could be transformed to the infective form if it was exposed to heat-killed cells of an infective strain (see Figure 14.2). The key difference between the strains is a polysaccharide capsule around the infective strain that is absent in the noninfective strain because of a genetic difference between the strains. In 1944, Oswald Avery and his colleagues at New York University found that the substance capable of transforming noninfective bacteria to the infective form was DNA.

Subsequently, geneticists established that in these transformation experiments the linear DNA fragments taken up from disrupted infective cells recombine with the chromosomal DNA of the noninfective cells by double crossovers, in much the same way as genetic recombination takes place in conjugation. Recombination introduces the normal allele for capsule formation into the DNA of the noninfective cells; expression of that normal allele generates a capsule around the cell and its descendants, making them infective.

Approximately 1% of bacterial species can take up DNA from the surrounding medium by natural mechanisms. Transformation in those species is known as *natural transformation.* Such bacteria typically have a DNA-binding protein on the outer surface of the cell wall. When DNA from the cell's surroundings binds to the protein, a deoxyribonuclease enzyme breaks the DNA into short pieces that pass through the cell wall and plasma membrane into the cytosol. The entering DNA can then recombine with the recipient cell's chromosome if it contains homologous regions.

E. coli cells do not normally take up DNA from their surroundings. However, they can be induced to take up DNA by *artificial transformation.* One transformation technique is to expose *E. coli* cells to calcium ions and the DNA of interest, incubate them on ice, and then give them a quick heat shock. This treatment alters the plasma membrane so that DNA can penetrate and enter. The entering DNA undergoes recombination if it contains regions that are homologous to part of the chromosomal DNA.

Another technique for artificial transformation, called *electroporation,* exposes cells briefly to rapid pulses of an electrical current. The electrical shock alters the plasma membrane so that DNA can enter. The method works well with many bacterial species that are unable to take up DNA on their own, and also with many types of eukaryotic cells.

Artificial transformation is often used to insert plasmids containing DNA sequences of interest into *E. coli* cells as a part of DNA cloning techniques. After the cells are transformed, clones of the cells are grown in large numbers to increase the quantity of the inserted DNA to the amounts necessary for sequencing or genetic engineering. (DNA cloning and genetic engineering are discussed further in Chapter 18.)

In Transduction, DNA Introduced into a Bacterium by Phage Infection Is the Source of Genetic Recombination

In **transduction,** DNA is transferred to recipient bacterial cells by an infecting phage (see Section 14.1). When new phages assemble in an infected bacterial cell, they sometimes incorporate fragments of the host cell DNA instead of all or some of the viral DNA. After the phages are released from the host cell, they may attach to another cell and inject—transduce—the bacterial DNA (and the viral DNA if it is present) into that cell. The introduction of this DNA, as in conjugation and transformation, makes the recipient cell a partial diploid and creates the potential for recombination to take place. And, because the phage carrying the donor bacterium's DNA does not have a complete phage genome, it cannot kill the recipient cell, allowing recombinant colonies to form. The DNA-containing capacity of a phage is limited, so the amount of donor DNA that can be transferred to a recipient by this mechanism is far less than is the case for conjugation. Joshua Lederberg and his graduate student, Norton Zinder, then at the University of Wisconsin–Madison, discovered transduction in 1952 in experiments with the bacterium *Salmonella typhimurium* and phage P22. Lederberg received a Nobel Prize in 1958 for his discovery of conjugation and transduction in bacteria.

Replica Plating Allows Genetic Recombinants to Be Identified and Counted

How do researchers identify and count genetic recombinants in conjugation, transformation, or transduction experiments? Joshua Lederberg and Esther Lederberg developed a now widely applied technique for doing this called **replica plating (Figure 17.5)**. In replica plating, researchers begin with bacterial colonies grown on a plate with a **complete medium,** that is, a medium containing a full complement of nutrients, including amino acids and other chemicals that normal strains can make for themselves. This master plate is pressed gently onto sterile velveteen, which transfers some of each colony to the velveteen in the same pattern as the colonies on the plate. The velveteen is then pressed onto new plates containing a minimal growth medium—the replica plates—thereby transferring some cells from each original colony to those plates. The composition of the minimal medium on the replica plates is adjusted to promote the growth of colonies with particular characteristics. The transfer inoculates each new plate with a "replica" of the original set of colonies on the starting plate, and the replica plates are incubated to allow new colonies to grow.

Figure 17.5 shows the identification of auxotrophic mutants of *E. coli* by replica plating. An investigator can determine the mutations a strain carries by comparing the original plate to the replica plates to identify missing colonies on the minimal-medium plate. Normal cells will grow in a minimal medium, but auxotrophic mutants will not, because they are unable to make one or more of the missing substances. Thus, the investigator takes for further study the colonies from the original plate that correspond to the missing colonies on the minimal-medium plates.

FIGURE 17.5 **Research Method**

Replica Plating

Purpose: Replica plating is used to identify different strains of bacteria with respect to their growth requirements in a heterogeneous mixture of strains.

Protocol:

1. Press sterile velveteen gently onto the master plate of solid growth medium with bacterial colonies on it. Some of each colony transfers to the velveteen in the same pattern as the colonies on the plate. In the example, a mixture of colonies of normal and auxotrophic strains is on a plate of complete medium.
2. Press the velveteen gently onto a sterile replica plate to transfer some of each strain. In the example, the replica plate contains minimal medium. Incubate to allow colonies to grow, and compare the pattern of colonies on the replica plate with that on the master plate.

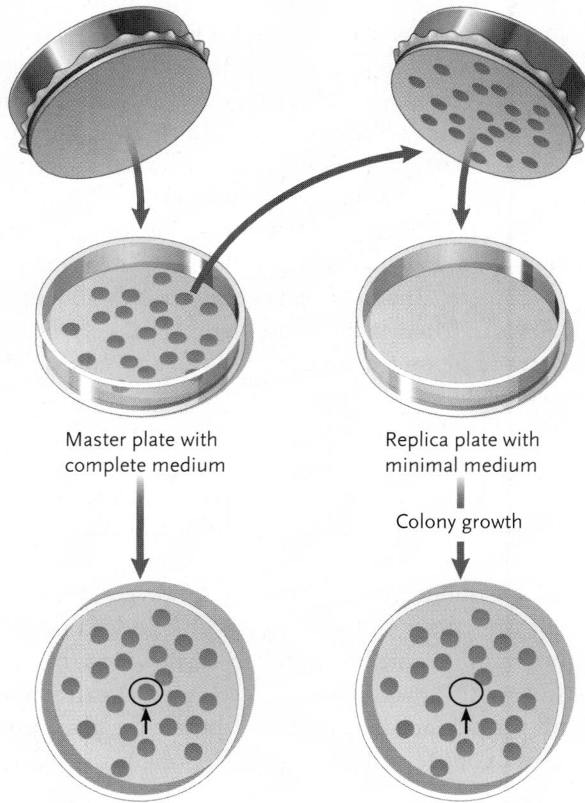

Master plate with complete medium
Replica plate with minimal medium
Colony growth

Interpreting the Results: A colony present on the master plate but not on the replica plate indicates that the strain requires some substance missing from the minimal medium in order to grow. In other words, the strain is an auxotroph. In actual experiments, the compositions of the master plate and replica plate media are chosen to be appropriate for the goals of the experiment.

Source: J. Lederberg and E. M. Lederberg. 1952. Replica plating and indirect selection of bacterial mutants. *Journal of Bacteriology* 63:399–406.

In an actual experiment, the compositions of the media are appropriate for the goals of the experiment. For example, to identify a met^+ recombinant in a conjugation experiment, the starting plate contains methionine and the colonies are replica plated to a plate lacking methionine. Comparison of the colony patterns on the two plates identifies met^+ recombinants because they grow on the plate lacking methionine, whereas met^- parentals do not.

Conjugation, Transformation, and Transduction Are Mechanisms for Horizontal Gene Transfer

Our examples of gene inheritance patterns in earlier chapters have all involved the movement of genetic material from generation to generation, whether that is from cell to cell or from parent to offspring. This is movement of genetic material by descent and, at its root, it is a feature of sexual reproduction. By contrast, **horizontal gene transfer** is the movement of genetic material between organisms other than by descent. Conjugation, transformation, and transduction are three major mechanisms by which horizontal gene transfer occurs in bacterial species and seemingly in some archaeal species.

In conjugation, genetic material is transferred horizontally from donor to recipient under the control of a plasmid that facilitates cell–cell contact, and by the formation of a bridge through which DNA is transferred. Relatively long segments of DNA can be transferred in this process. The particular donor segment of DNA transferred depends on where the plasmid responsible for conjugation is integrated into the donor bacterium's chromosome. Our examples involved conjugation between donor and recipient cells of the same species, but the process can occur between bacteria that are not closely related.

In transformation, a DNA molecule in the environment surrounding a bacterial cell is taken up by that cell. In natural populations of bacteria, the bacteria taking up DNA typically are those capable of natural transformation. Any piece of DNA can be taken up in transformation, but usually only relatively short DNA molecules are involved. Griffith's experiment with *Streptococcus* was the first laboratory demonstration of horizontal gene transfer.

In transduction, a phage transfers bacterial DNA from a donor to a recipient cell. The amount of DNA that can be transferred in this way is relatively small because it is limited by the capacity of the phage head in which the DNA must be packed. Horizontal gene transfer by transduction generally is limited to closely related bacteria because the phage carrying the donor DNA has to bind to specific receptors on a recipient bacterium in order to inject the genetic material. Unless the bacteria are closely related, the necessary receptor will be absent.

In our examples of conjugation, transformation, and transduction in the same species, genetic recombination between donor and recipient DNA occurs because the sequences match, allowing crossing-over to occur. When different species are involved, the transferred DNA may have little sequence similarity to the recipient's DNA. However, if some stretches of sequences match sufficiently, crossing-over can lead to the integration of a segment of the transferred donor DNA. In this way, a recipient species can acquire new genes. If those genes confer some type of selective advantage to the recipient, then the altered genome will be retained and potentially the old genome will be lost in the population. In this way, horizontal gene transfer has resulted in the evolution of genomes.

Now that the genomes of many bacterial and archaeal species have been sequenced completely, geneticists can compare whole genomes to see how much horizontal gene transfer has occurred. Such analysis shows that horizontal gene transfer has involved a significant fraction of prokaryotic genes, making it a major process in the evolution of prokaryotic genomes. For instance, 20% of the *E. coli* genome is thought to have derived from horizontal gene transfer. The genes acquired in this way have enabled *E. coli* to adapt to the various environments in which it lives, and have contributed to its effectiveness as a pathogen. As an example, the genome of *E. coli* strain O157:H7, the cause of a number of food-related deaths, has 1,387 genes not present in the genome of the standard lab strain of *E. coli*. Some of these extra genes, which represent about 25% of the organism's total number of genes, encode proteins responsible for the pathogenicity of the bacterium. All of the extra genes were acquired by horizontal gene transfer from other bacterial species.

Compared with prokaryotes, far fewer eukaryotic genomes are available to analyze for horizontal gene transfer. Nonetheless, such analysis reveals that horizontal gene transfer also appears to have had an important role in eukaryotic genome evolution. Too little information is available to give us a complete understanding of the phenomenon in eukaryotes, however. For instance, we know little about the mechanism of horizontal gene transfer in eukaryotes, or the factors that enhance or discourage it. The most common transfers seen are from prokaryotes, although the amount of prokaryotic DNA found in eukaryotes varies widely. Most of these transfers are to the nuclear genomes, with only a few to organelle genomes. Eukaryote–eukaryote transfers of nuclear genes are also known, as are eukaryote–prokaryote transfers. Both of these types of transfer are rarer than prokaryote–prokaryote transfers, with eukaryote–prokaryote being extremely rare.

In the next section we turn to viruses. Whereas the phages mentioned earlier in this section infect only bacteria, viruses infect all living organisms. Many viruses have become important subjects for research into, among other things, the molecular nature of recombination and the genetic control of viral infection.

STUDY BREAK 17.1

1. **What are the properties of F^+, F^-, and Hfr cells of *E. coli*?**
2. **How is horizontal gene transfer in bacteria distinctive from gene segregation in sexually reproducing organisms?**

17.2 Viruses and Viral Genetics

As agents of transduction, the phages that infect bacteria are important tools in research on bacterial genetics. The same viruses are also important for studying *viral* recombination and genetics. Viruses can undergo genetic recombination when the DNA of two viruses, carrying different alleles of one or more genes, infect a single cell. Using phages, researchers study viral genetics with the same molecular and biochemical techniques used to investigate their bacterial hosts. In this section we examine the structure and infectious properties of viruses.

Viruses in the Free Form Consist of a Nucleic Acid Core Surrounded by a Protein Coat

A **virus** (Latin for poison) is a biological particle that can infect the cells of a living organism. Viral infections usually have detrimental effects on their hosts. The study of viruses is called *virology*, and researchers studying viruses are known as *virologists*.

All viruses consist of a DNA or RNA genome surrounded by a protein coat—a layer of proteins—called the **capsid.** The complete viral particle is also called a **virion.** Most, but not all, viruses are significantly smaller than bacteria. Among the few exceptions are viruses that are 300–500 nm long and 120–300 nm in diameter. The smallest bacteria are 200–300 nm long.

Are viruses living or nonliving? Many scientists consider viruses to be on the border of living and nonliving. That is, although they share some properties with living things, viruses are missing several important characteristics. They cannot reproduce independently, but only within a host cell. They are not made up of cells; they are merely pieces of DNA or RNA surrounded by a protein capsid and in mammals an additional membranous envelope. They also do not grow, develop, or generate metabolic energy. They do, however, share their chemical constituents with living things, and indeed these constituents possess the same biological order; however, this order typically stops at the macromolecule level. Viruses can adapt very readily over time since they routinely mutate the glycoproteins in their capsids or their envelopes, which provide the lock and key fit for entry into their host cells. In summary, viruses can not be defined as being alive at the cellular level; they do, however, share enough characteristics with organisms that they can not be defined as non-living either, like a rock.

Viral Structure Is Reduced to the Minimum Necessary to Transmit Nucleic Acid Molecules from One Host Cell to Another

The genetic material of all organisms is double-stranded DNA, but the nucleic acid genome of a virus may be double-stranded DNA, singled-stranded DNA, double-stranded RNA, or single-stranded RNA, depending on the viral type. Moreover, the genome of some viruses consists of one nucleic acid molecule, while the genomes of others are distributed among two or more nucleic acid molecules. For example, the genome of a herpes virus consists of a single molecule of double-stranded DNA, while the genome of influenza virus is distributed among eight (influenza A and B strains) or seven (influenza C strains) segments of single-stranded RNA.

The simplest viruses contain only a few genes, whereas those of the most complex viruses may contain a hundred or more. All viruses have genes encoding the proteins of their capsid. For some viruses the capsid is assembled from protein molecules of a single type. More complex viruses have capsids assembled from several different proteins, including the recognition proteins that bind to host cells. All viruses also have a gene or genes encoding the proteins that make them pathogenic to their hosts. The particles of some viruses also contain the DNA or RNA polymerase enzymes required for viral genome replication and/or an enzyme that attacks cell walls or membranes.

Most viruses take one of two basic structural forms, helical or polyhedral. In **helical viruses** the capsid proteins assemble in a rodlike spiral around the genome **(Figure 17.6A)**. A number of viruses that infect plant cells are helical. In **polyhedral viruses** the capsid proteins form triangular units that form an icosahedral structure **(Figure 17.6B)**. The polyhedral viruses include forms that infect animals, plants, and bacteria. In some polyhedral viruses, protein spikes that provide host cell recognition extend from the corners where the facets fit together. Some viruses, the **enveloped viruses,** are covered by a surface membrane derived from the plasma membrane of their host cells; both enveloped helical and enveloped polyhedral viruses are known **(Figure 17.6C)**. For example, **HIV** (for *H*uman *I*mmunodeficiency *V*irus), the virus that causes AIDS, is an enveloped polyhedral virus. Protein spikes extend through the membrane, giving the particle its recognition and adhesion functions.

A number of bacteriophages with DNA genomes, such as T2 (see Section 14.1), have a **tail** attached at one side of a polyhedral **head,** forming what is known as a **complex virus (Figure 17.6D)**. The genome is packed into the head; the tail is made up of proteins forming a collar, sheath, baseplate, and tail fibers. The tail has recognition proteins at its tip and, once attached to a host cell, functions as a sort of syringe that injects the DNA genome into the cell.

Viruses of Different Kinds Infect All Living Cells

Although they are considered to be nonliving material, viruses are classified by the International Committee on Taxonomy of Viruses into orders, families, genera, and species using several criteria, including size and structure, type and number of nucleic acid molecules, method of replication of the nucleic acid molecules inside host cells, host range, and infective cycle. More than 4,000 species of viruses have been classified into more than 80 families according to these criteria.

One or more kinds of viruses probably infect all living organisms. Usually a virus infects only a single species or a few closely related species. A virus may even infect only one organ system, or a single tissue or cell type in its host. However, some viruses are able to infect unrelated species, either naturally or after mutating. For example, some humans have contracted bird flu from being infected with the natural avian flu virus as a result of contact with virus-infected birds. Of the viral families, 21 include viruses that cause human diseases. Viruses also cause diseases of wild and domestic animals; plant viruses cause annual losses of millions of tons of crops, especially cereals, potatoes, sugar beets, and sugar cane. **Table 17.1** lists some virus families that infect animals, and **Table 17.2** lists some virus families that infect plants.

The effects of viruses on the organisms they infect range from undetectable, through merely bothersome, to seriously debilitating or lethal diseases. For instance, some viral infections of humans, such as those causing cold sores, chicken pox, and the common cold, are usually little more than a nuisance to healthy adults. Others, including AIDS, encephalitis, yellow fever, and

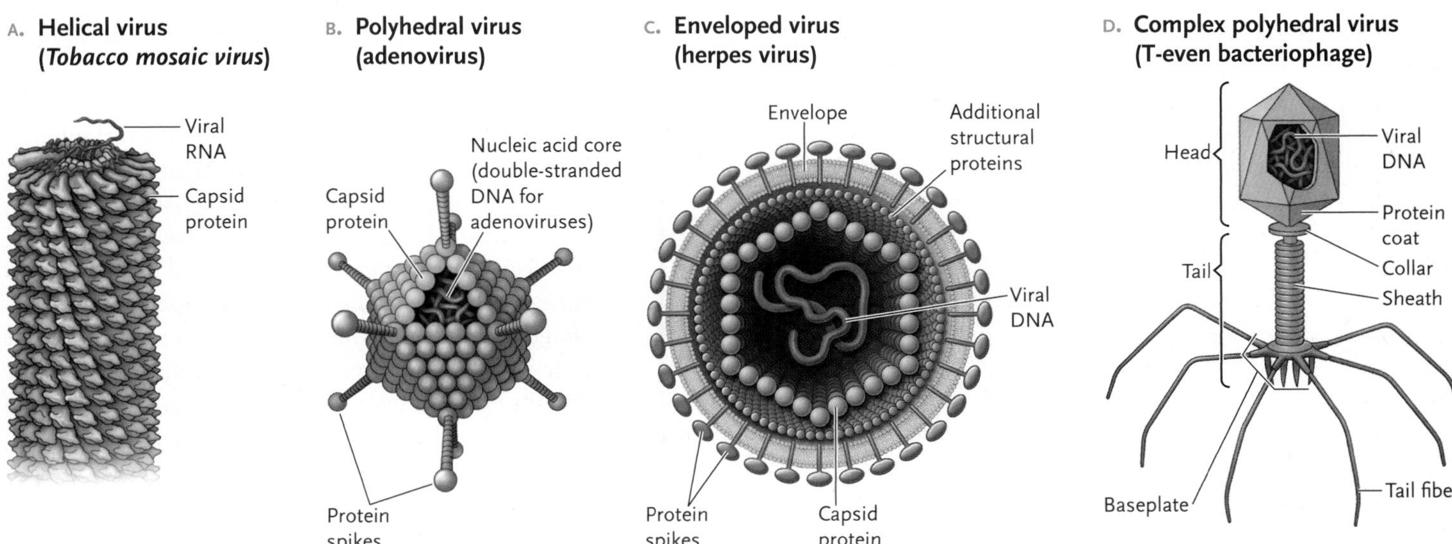
A. Helical virus (Tobacco mosaic virus)
Viral RNA
Capsid protein
B. Polyhedral virus (adenovirus)
Nucleic acid core (double-stranded DNA for adenoviruses)
Capsid protein
Protein spikes
C. Enveloped virus (herpes virus)
Envelope
Additional structural proteins
Viral DNA
Protein spikes
Capsid protein
D. Complex polyhedral virus (T-even bacteriophage)
Head
Viral DNA
Protein coat
Tail
Collar
Sheath
Baseplate
Tail fiber

FIGURE 17.6

Basic viral structures. The tobacco mosaic virus in **(A)** assembles from more than 2,000 identical protein subunits. Protein spikes contain recognition proteins that allow the viral particle to bind to the surface of a host cell.

TABLE 17.1 Examples of Animal Viruses

Virus Family	Envelope	Nucleic Acid	Diseases and Infections in Humans and Other Animals
Adenoviridae	No	dsDNA	Respiratory infections, tumors
Baculoviridae	Yes	dsDNA	Gastrointestinal infections in arthropods
Filoviridae	No	ssRNA	Ebola hemorrhagic fever, Marburg hemorrhagic fever
Flaviviridae	Yes	ssRNA	Yellow fever, dengue, hepatitis C, West Nile fever
Hepadnaviridae	Yes	dsDNA	Hepatitis B
Herpesviridae	Yes	dsDNA	
Herpes simplex I			Oral herpes, cold sores
Herpes simplex II			Genital herpes
Varicella-zoster			Chicken pox, shingles
Orthomyxoviridae	Yes	ssRNA	Influenza
Papillomaviridae	No	dsDNA	Benign and malignant warts
Paramyxoviridae	Yes	ssRNA	Measles, mumps, pneumonia
Parvoviridae	No	ssDNA	Fifth disease (humans), canine parvovirus disease (dogs)
Picornaviridae	No	ssRNA	
Enteroviruses			Polio, hemorrhagic eye disease, gastroenteritis
Rhinoviruses			Common cold
Hepatitis A virus			Hepatitis A
Aphthovirus			Foot-and-mouth disease (livestock)
Poxviridae	Yes	dsDNA	Smallpox, cowpox
Retroviridae	Yes	ssRNA	
FeLV			Feline leukemia
HTLV I, II			T-cell leukemia
HIV			AIDS
Rhabdoviridae	Yes	ssRNA	Rabies, other animal diseases

ss = single-stranded; ds = double-stranded.

TABLE 17.2 Examples of Plant Viruses

Virus Family	Nucleic Acid	Examples of Viruses (the name indicates a plant infected and its symptoms)
Bromoviridae	ssRNA	*Brome mosaic virus, Broad bean mottle virus*
Caulimoviridae	dsDNA	*Cauliflower mosaic virus, Blueberry red ringspot virus*
Geminiviridae	ssDNA	*Maize streak virus, Beet curly top virus*
Luteoviridae	ssRNA	*Barley yellow dwarf virus, Soybean dwarf virus*
Rhabdoviridae	ssRNA	*Broccoli necrotic yellow virus, Potato yellow dwarf virus*
Tobamoviruses	ssRNA	*Tobacco mosaic virus, Pepper mold mottle virus*
Tombusviridae	ssRNA	*Tomato bushy stunt virus, Carnation Italian ringspot virus*

ss = single-stranded; ds = double-stranded.

smallpox, are and have been among the most severe and deadly human diseases.

While most viruses have detrimental effects, some may be considered beneficial. One of the primary reasons why bacteria do not completely overrun the planet is that they are destroyed in incredibly huge numbers by bacteriophages. Viruses also provide a natural means to control some insect pests.

Viruses Infect Bacterial, Animal, and Plant Cells by Similar Pathways

Free viruses move by random molecular motions until they contact the surface of a host cell. For infection to occur, the virus or the viral genome must then enter the cell. Inside the cell, typically the viral genes are expressed, leading to replication of the viral genome and assembly of progeny viruses. The viruses are then released from the host cell, a process that often ruptures the host cell, killing it.

INFECTION OF BACTERIAL CELLS Bacteriophages vary as to whether they have a DNA or an RNA genome, and whether that nucleic acid is double-stranded or single-stranded. They also differ in how they infect and kill hosts cells; **virulent bacteriophages** kill their host cells during each cycle of infection, and **temperate bacteriophages** may enter an inactive phase in which the host cell replicates and passes on the bacteriophage DNA for generations before the phage becomes active and kills the host. Bacteriophages of both types that infect *E. coli* are widely used in genetic research, and we will use examples of each type to illustrate their life cycles.

Virulent Bacteriophages. Among the virulent bacteriophages infecting *E. coli,* the **T-even bacteriophages** T2, T4, and T6 have been most valuable in genetic studies (see Figure 17.6D). The coat

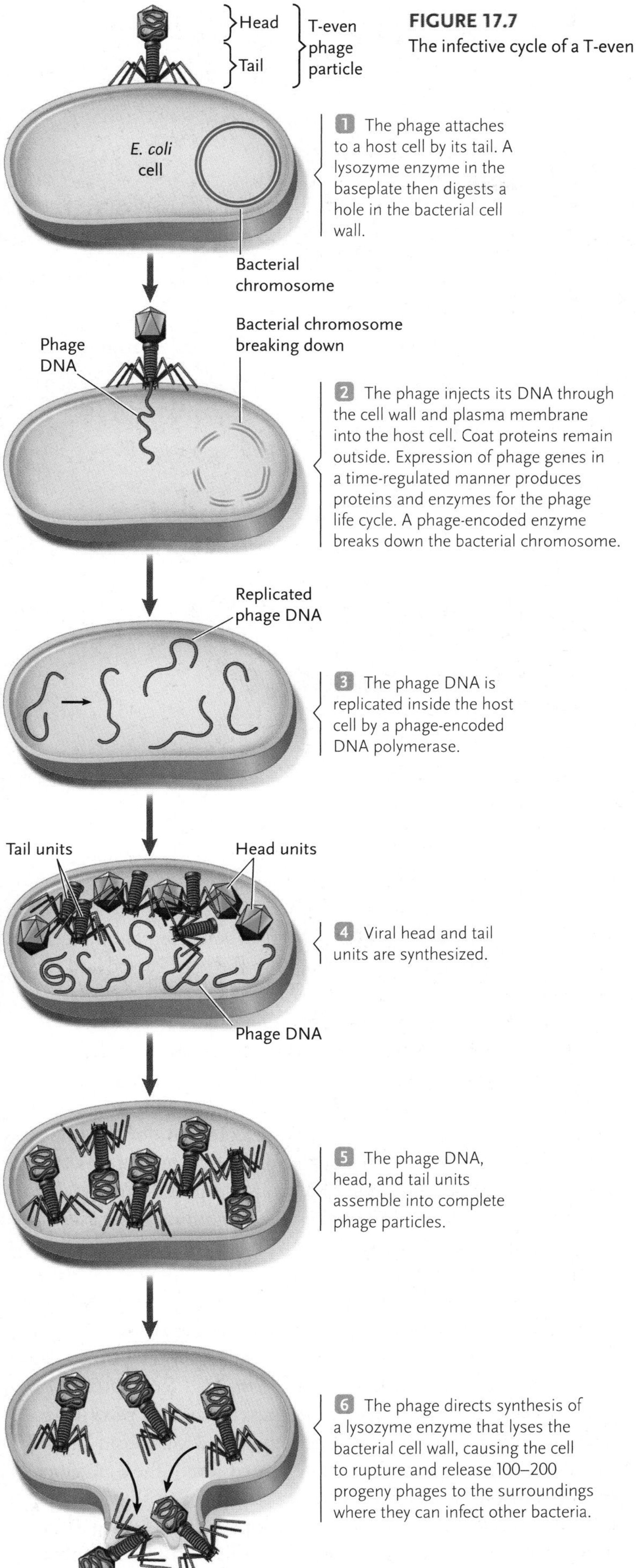

FIGURE 17.7
The infective cycle of a T-even bacteriophage, an example of a virulent phage.

Head
Tail
T-even phage particle
E. coli cell
Bacterial chromosome
1 The phage attaches to a host cell by its tail. A lysozyme enzyme in the baseplate then digests a hole in the bacterial cell wall.
Phage DNA
Bacterial chromosome breaking down
2 The phage injects its DNA through the cell wall and plasma membrane into the host cell. Coat proteins remain outside. Expression of phage genes in a time-regulated manner produces proteins and enzymes for the phage life cycle. A phage-encoded enzyme breaks down the bacterial chromosome.
Replicated phage DNA
3 The phage DNA is replicated inside the host cell by a phage-encoded DNA polymerase.
Tail units
Head units
Phage DNA
4 Viral head and tail units are synthesized.
5 The phage DNA, head, and tail units assemble into complete phage particles.
6 The phage directs synthesis of a lysozyme enzyme that lyses the bacterial cell wall, causing the cell to rupture and release 100–200 progeny phages to the surroundings where they can infect other bacteria.

of these phages is divided into a *head* and a *tail.* Packed into the head is a single linear molecule of double-stranded DNA. The tail, assembled from several different proteins, has recognition proteins at its tip that can bind to the surface of the host cell.

When a virulent DNA bacteriophage such as phage T2 infects *E. coli,* it enters the **lytic cycle (Figure 17.7),** in which the host cell is killed in each cycle of infection. The phage attaches to a host bacterial cell and injects its DNA genome into the cell; there, expression of phage genes directs the phage life cycle, leading to the production of progeny phages, which are released from the cell when it breaks open, or lyses. Those phages can now infect other bacteria.

For some virulent phages (although not T-even phages), fragments of the host DNA may be included in the heads as the viral particles assemble, providing the basis for transduction of bacterial genes during the next cycle of infection. Because genes are randomly incorporated from essentially any DNA fragments, gene transfer by this mechanism is termed **generalized transduction.**

Temperate Bacteriophages. Temperate bacteriophages alternate between a lytic cycle and a **lysogenic cycle,** in which the viral DNA inserts into the host cell DNA and production of new viral particles is delayed. During the lysogenic cycle, the integrated viral DNA, known as the **prophage,** remains partially or completely inactive, but is replicated and passed on with the host DNA to all descendants of the infected cell. In response to certain environmental signals, the prophage loops out of the chromosome and the lytic cycle of the phage proceeds.

A much-researched temperate phage that infects *E. coli* is lambda (λ). Phage λ infects *E. coli* in much the same way as the T-even phages do. The phage injects its linear DNA chromosome into the bacterium (**Figure 17.8,** step 1). Once inside, the linear chromosome forms a circle, and then follows one of two paths. Viral gene products act as molecular switches to govern which path is followed at the time of infection.

One path is the lytic cycle, which is like the lytic cycles of virulent phages. The lytic cycle starts with steps 1 and 2 (infection), then goes directly to steps 7 through 9 (production and release of progeny virus), and back to step 1.

The second and more common path, the lysogenic cycle, begins when the circular λ chromosome integrates into the host cell's DNA by crossing-over, becoming a prophage (Figure 17.8, step 3). The DNA of a temperate phage typically inserts at one or possibly a few specific sites in the bacterial chromosome through the action of a phage-encoded enzyme that recognizes certain sequences in the host DNA. In the case of λ, there is one integration site in the *E. coli* chromosome. Once integrated, the λ genes are mostly inactive and, therefore, no phage components are made. As a consequence, the λ prophage does not affect its host cell and its descendants. In the integrated state, the viral DNA is replicated and passed on in division along with the host cell DNA (steps 4 and 5).

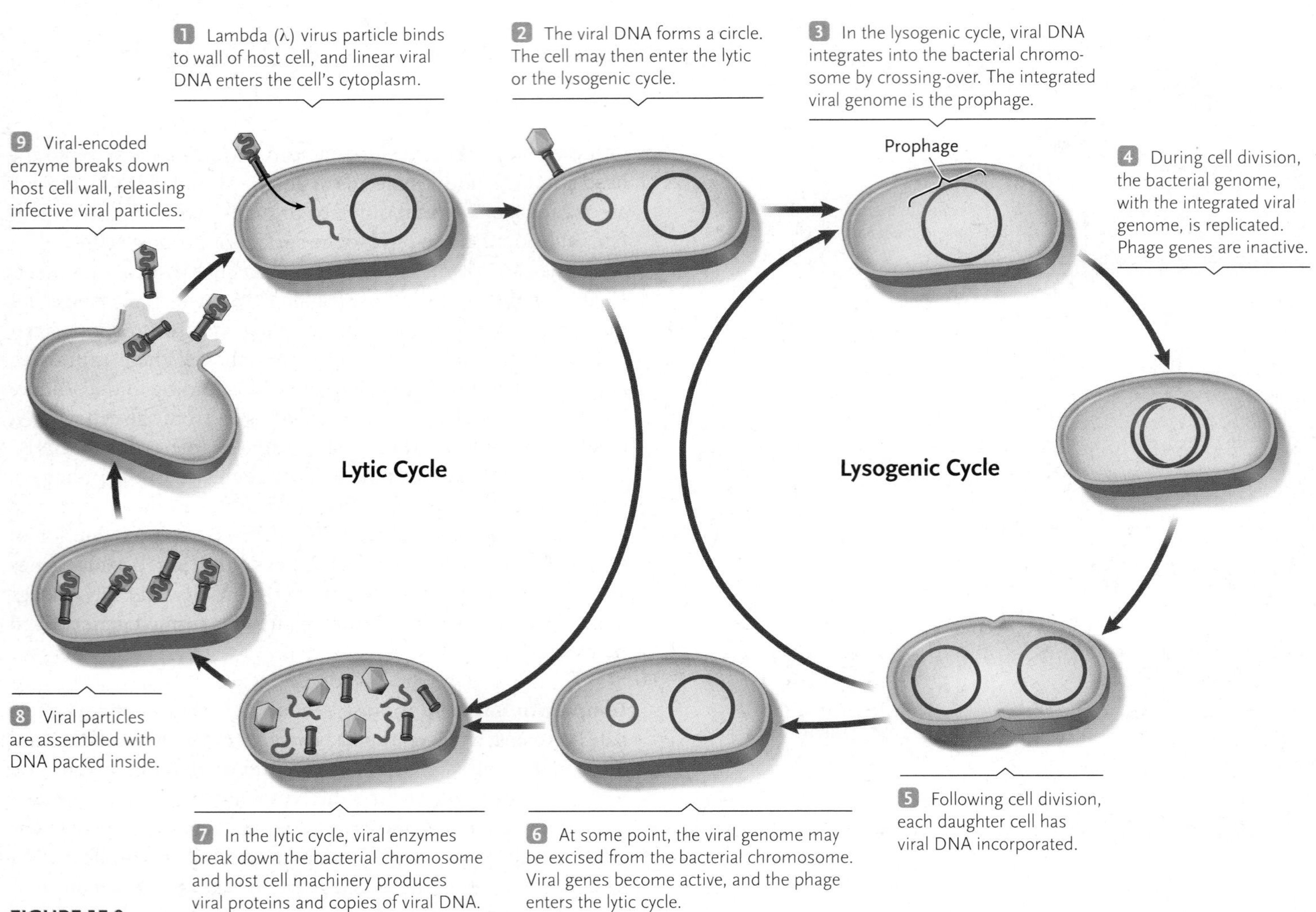
1 Lambda (λ) virus particle binds to wall of host cell, and linear viral DNA enters the cell's cytoplasm.
2 The viral DNA forms a circle. The cell may then enter the lytic or the lysogenic cycle.
3 In the lysogenic cycle, viral DNA integrates into the bacterial chromosome by crossing-over. The integrated viral genome is the prophage.
Prophage
4 During cell division, the bacterial genome, with the integrated viral genome, is replicated. Phage genes are inactive.
9 Viral-encoded enzyme breaks down host cell wall, releasing infective viral particles.
Lytic Cycle
Lysogenic Cycle
8 Viral particles are assembled with DNA packed inside.
7 In the lytic cycle, viral enzymes break down the bacterial chromosome and host cell machinery produces viral proteins and copies of viral DNA.
6 At some point, the viral genome may be excised from the bacterial chromosome. Viral genes become active, and the phage enters the lytic cycle.
5 Following cell division, each daughter cell has viral DNA incorporated.

FIGURE 17.8
The infective cycle of lambda (λ), an example of a temperate phage, which can go through the lytic cycle or the lysogenic cycle.

In response to certain environmental signals, such as UV irradiation, the λ phage in an infected cell becomes active and enters the lytic cycle. Genes that were inactive in the prophage are now transcribed. Among the first viral proteins synthesized in response to the environmental signal are enzymes that excise the λ genome from the host chromosome (step 6). Excision occurs by a crossing-over process that reverses the integration step. The result is a circular λ chromosome. Replication of that chromosome produces many copies of linear λ chromosomes. Expression of genes on those chromosomes generates coat proteins, which assemble with the chromosomes to produce the viral particles (steps 7 and 8). This active stage culminates in rupture of the host cell with the release of infective viral particles (step 9), and the beginning of a new cycle (step 1).

At times, excision of the λ chromosome from the *E. coli* DNA is not precise, resulting in the inclusion of one or more host cell genes. These genes are replicated with the viral DNA and packed into the coats, and may be carried to a new host cell in the next cycle of infection. Because of the mechanism involved, only genes that are adjacent to the integration site(s) of a temperate phage can be cut out with the viral DNA, included in phage particles during the lytic stage, and undergo transduction. Accordingly, this mechanism of gene transfer is termed **specialized transduction.**

INFECTION OF ANIMAL CELLS Viruses infecting animal cells follow a similar pattern except that both the viral capsid and the genome—which is DNA or RNA—enter a host cell.

As for bacteriophages, infection of animal cells requires interaction between specific proteins on the surface of the viral particle and specific cell surface proteins—receptors. The specificity of this interaction is responsible for the specific host range of each animal virus. It is important to note that, for the infection process, animal viruses exploit cell surface proteins that have specific cellular functions. For example, poliovirus attaches to a receptor for intracellular adhesion, and influenza viruses attach to the acetylcholine receptor, which is used in the transmission of neuronal impulses.

Viruses without an envelope, such as adenovirus (DNA genome) and poliovirus (RNA genome), bind by their recognition proteins to the plasma membrane and are then taken into the host cell by receptor-mediated endocytosis (see Section 6.5). For some enveloped viruses, such as herpesviruses and poxviruses (DNA genome), and HIV (RNA genome), the viral capsid and its contained genome enter the host cell by fusion of their envelope with the host cell plasma membrane. In this case, the envelope does not enter the cell. For other enveloped viruses, such as orthomyxoviruses (RNA genome; for example, influenza virus)

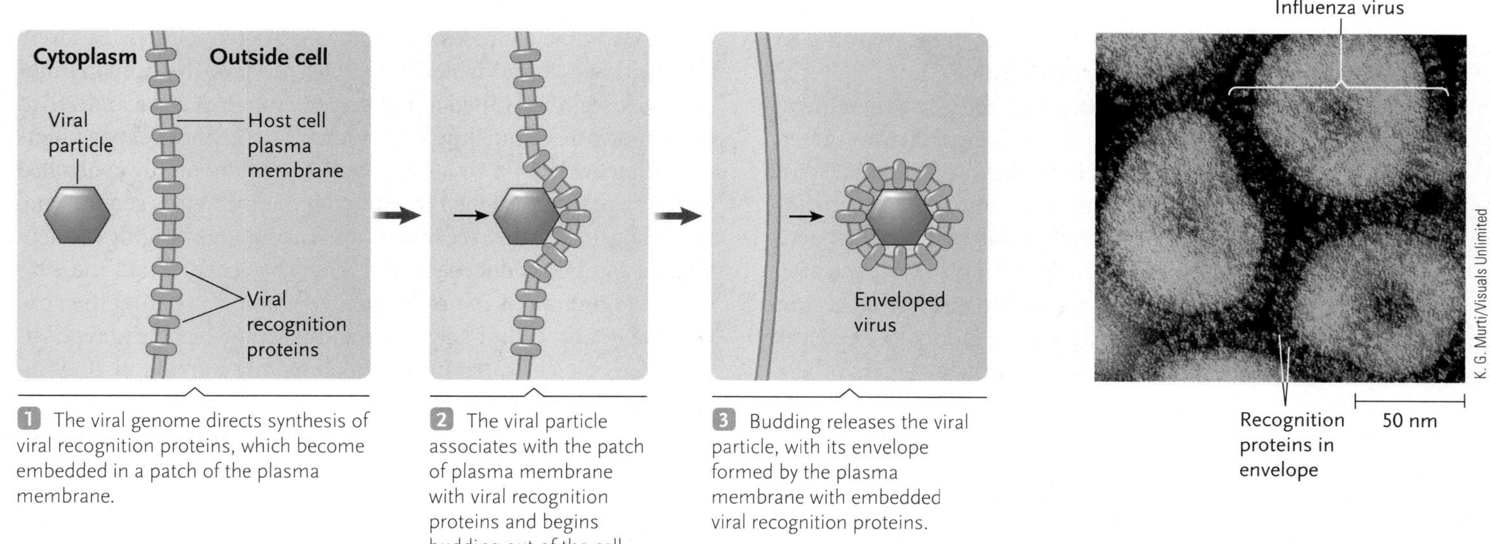
Cytoplasm
Outside cell
Viral particle
Host cell plasma membrane
Viral recognition proteins
Enveloped virus
Influenza virus
Recognition proteins in envelope
50 nm
K. G. Murti/Visuals Unlimited

1 The viral genome directs synthesis of viral recognition proteins, which become embedded in a patch of the plasma membrane.

2 The viral particle associates with the patch of plasma membrane with viral recognition proteins and begins budding out of the cell.

3 Budding releases the viral particle, with its envelope formed by the plasma membrane with embedded viral recognition proteins.

FIGURE 17.9
How enveloped viruses acquire their envelope. The micrograph shows the influenza virus with its envelope. Note the recognition proteins studding the envelope.

and rhabdoviruses (RNA genome), the entire virus, including the envelope, enters the cell by endocytosis and the viral capsid with its contained genome then is released into the cytoplasm.

Once inside the host cell, the viral capsid is disassembled to release the genome. The genome then directs the synthesis of additional viral particles by basically the same pathways as bacterial viruses. Newly completed viruses that do not acquire an envelope are released by rupture of the cell's plasma membrane, which lyses and kills the cell. In contrast, most enveloped viruses receive their envelope as they pass through the plasma membrane, usually without breaking the membrane **(Figure 17.9).** This pattern of viral release typically does not injure the host cell.

Some animal viruses enter a **latent phase** in which the virus remains in the cell in inactive form: the viral nucleic acid is present in the cytoplasm or nuclear DNA, but no complete viral particles or viral release can be detected. (The latent phase is similar to the lysogenic cycle that is part of the life cycle of some bacteriophages.) At some point, the latent phase may end as the viral DNA is replicated in quantity, capsid proteins are made, and completed viral particles are released from the cell. The herpesviruses that cause oral and genital ulcers in humans remain in a latent phase in the cytoplasm of some body cells for the life of the individual. At times, particularly during periods of metabolic stress, the virus becomes active in some cells, directing viral replication and causing ulcers to form as cells break down during viral release.

INFECTION OF PLANT CELLS Plant viruses may be rodlike or polyhedral; although most include RNA as their nucleic acid, some contain DNA. None of the known plant viruses has an envelope. Plant viruses enter cells through mechanical injuries to leaves and stems or through transmission from plant to plant by biting and feeding insects such as leafhoppers and aphids, by nematode worms, and by pollen during fertilization. Plant viruses can also be transmitted from generation to generation in seeds. Once inside a cell, plant viruses replicate in the same patterns as animal viruses. Within plants, viral particles pass from infected to healthy cells through plasmodesmata, the openings in cell walls that directly connect plant cells (see Figure 5.28), and through the vascular system.

Plant viruses are generally named and classified by the type of plant they infect and their most visible effects. *Tomato bushy stunt virus,* for example, causes dwarfing and overgrowth of leaves and stems of tomato plants, and *Tobacco mosaic virus* causes a mosaic-like pattern of spots on leaves of tobacco plants. Most species of crop plants can be infected by at least one destructive virus.

Tobacco mosaic virus (see Figure 17.6A) was the first virus to be isolated, crystallized, disassembled, and reassembled in the test tube, and the first viral structure to be established in full molecular detail.

Viral Infections Are Typically Difficult to Treat

The vast majority of animal viral infections are asymptomatic; pathogenesis—the causation of disease—is of no value to the virus. However, there are many types of pathogenic viruses, and they employ a variety of ways to cause disease. In some instances, release of progeny viruses from the cell leads to massive cell death, destroying vital tissues such as nervous tissue or white or red blood cells, or causing lesions such as ulcers in skin and mucous membranes. Some other viruses release cellular molecules when infected cells break down, which can induce fever and inflammation. Yet other viruses alter gene function when they insert into the host cell DNA, leading to cancer and other abnormalities.

Viral infections are unaffected by the antibiotics and other treatments used for bacterial infections. As a result, many viral infections are allowed to run their course, with treatment limited to relieving the symptoms while the immune defenses of the patient attack the virus. Some viruses, however, cause serious and sometimes deadly symptoms upon infection and, consequently, researchers have spent considerable effort to develop antiviral drugs to treat them. Many of these drugs target a stage of the viral life cycle. For example, amantadine inhibits entry of hepatitis B and hepatitis C virus into cells; acyclovir (an analog of nucleosides; *analog* means it is chemically similar) inhibits replication of the genomes of herpesviruses; and zanamivir inhibits release of influenza virus particles from cells.

The influenza viruses (see Figure 17.9) illustrate the difficulties inherent in treating viral diseases. Influenza viruses infect a number of vertebrates, including birds (avian flu), pigs (swine flu), and humans and other mammals. In humans, influenza viruses are transmitted from person to person primarily by aerosols and droplets, and enter the host through the respiratory tract. The virus enters a host cell when the viral protein spike binds to a cell surface receptor and induces endocytosis. Release of progeny viruses lyses and kills the cell, which triggers inflammatory responses. The presence of viruses in the body also stimulates the immune system to make antibodies targeted at the virus. (Antibodies are highly specific protein molecules produced by the immune system that recognize and bind to foreign proteins originating from a pathogen; see Chapter 43.)

Virologists classify influenza viruses into three types (related species, if you will), influenzavirus A, influenzavirus B, and influenzavirus C. Strains of each type can infect humans; the A and B types are highly contagious. Yearly epidemic outbreaks of flu are routine, even in developed countries, resulting in many deaths. However, each virus strain that causes an epidemic has a particular virulence because of differences in the molecular properties of the viruses, so the death rate varies with each epidemic. Moreover, influenza viruses have the potential to cause pandemics, meaning the spread of a new strain through many human populations in a country, region, or even worldwide. Some pandemics are relatively benign, while others result in high mortality. For example, the so-called Spanish Flu of 1918–1919 was caused by a particularly virulent strain of the influenza virus; 50–100 million people died worldwide after an estimated 500 million people, one third of the human population at the time, were infected. Most who died were aged between 20 and 40 years.

The influenza virus type that caused the 1918–1919 influenza pandemic was influenzavirus A/H1N1. The "H" and "N" numbers refer to the subtypes of two glycoproteins found in the envelope of the virus. The "H" glycoprotein is hemagglutinin (the protein spike that binds to cell surface receptors), and the "N" glycoprotein is neuraminidase. The virus type was determined experimentally in 2005, when researchers led by Jeffrey Taubenberger at the U.S. Armed Forces Institute of Pathology reconstructed the genome of the virus and produced infectious, pathogenic viruses in the laboratory. The team worked mainly with tissue from a 1918 flu victim found in permafrost in Alaska. Using modern DNA technology (see Chapter 18), they pieced together the sequences of the virus's eleven genes and characterized their protein products. They also transformed clones of the genes into animal cells and were able to produce complete viruses. These reconstructed 1918 viruses were about 50 times more virulent than most modern-day human influenza viruses; they killed a higher percentage of mice and killed them much more quickly, for instance. (All of these experiments were done with appropriate approval and under highly controlled experimental conditions.) By studying the 1918 virus genome and its pathogenicity, the researchers are learning how highly virulent viruses can be produced. So far, they have learned that the 1918 virus had mutations in polymerase genes for replicating the viral genome in host cells, likely making this strain capable of replicating more efficiently. An influenza pandemic starting in 2009 involving a new strain of influenzavirus A/H1N1 has not been as deadly as the 1918 flu virus (at least 18,000 people have died worldwide as of May 2010), although the exact molecular properties responsible for their differences in virulence are not completely understood.

The influenza type A and B viruses have many unusual features that tend to keep them a step ahead of efforts to counteract their infections. One is the genome of the virus, which consists of eight segments of RNA. When two different influenza viruses infect the same individual, the pieces can assemble in random combinations derived from either parent virus. The new combinations can change the proteins of the capsid, making the virus unrecognizable to antibodies developed against either parent virus. The invisibility to antibodies means that new virus strains can infect and produce flu symptoms in people who have already had the flu or who have had flu shots that stimulate the formation of antibodies effective only against the earlier strains of the virus. Random mutations in the RNA genome of the virus add to the variations in the capsid proteins that make previously formed antibodies ineffective, and also can alter the virulence of the virus.

Retroviruses Are Viruses That Replicate Their RNA Genomes via a DNA Intermediate

Most viruses with RNA genomes directly replicate those genomes to produce progeny RNA genomes. However, a **retrovirus** is an enveloped virus with an RNA genome that replicates via a DNA intermediate. In that respect, retroviruses resemble transposable elements known as retrotransposons (see Section 15.5). When a retrovirus infects a host cell, a *reverse transcriptase* enzyme carried in the viral particle (and encoded by the viral genome) is released and copies the single-stranded RNA genome into a double-stranded DNA copy. The viral DNA is then inserted into the host DNA, where it is replicated and passed to progeny cells during cell division. Similar to the prophage of temperate bacteriophages, the inserted viral DNA is

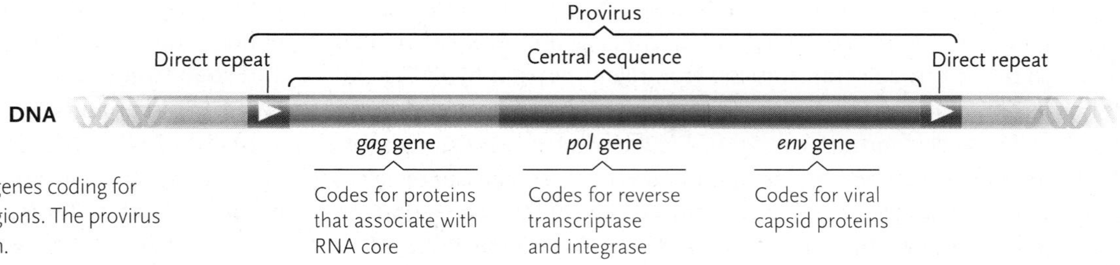

FIGURE 17.10
A mammalian retrovirus in the provirus form in which it is inserted into chromosomal DNA. The direct repeats at either end contain sequences capable of acting as enhancer, promoter, and termination signals for transcription. The central sequence contains genes coding for proteins, concentrated in the *gag*, *pol*, and *env* regions. The provirus of HIV, the virus that causes AIDS, takes this form.

Provirus
Direct repeat
Central sequence
Direct repeat
DNA
gag gene
pol gene
env gene
Codes for proteins that associate with RNA core
Codes for reverse transcriptase and integrase
Codes for viral capsid proteins

INSIGHTS FROM THE Molecular Revolution

Reversing the Central Dogma: How do RNA tumor viruses replicate their genomes?

RNA tumor viruses, which represent some of the viruses with RNA genomes, convert the animal cells they infect to tumor cells. In 1964, Howard Temin of the University of Wisconsin–Madison hypothesized that the mechanism by which RNA tumor viruses convert normal cells to tumor cells involves an enzyme encoded by the virus that copies the RNA genome into DNA. He proposed that the provirus DNA integrated into the DNA of a host cell chromosome, enabling the viral genetic material to persist as the tumor cell reproduced. At the time, most virologists were highly doubtful that such an enzyme existed because it went against the central dogma of molecular biology, namely that DNA is transcribed to produce RNA, and RNA is translated into protein.

Research Question

How do RNA tumor viruses replicate their genomes?

Experiment

David Baltimore of MIT searched for the hypothesized enzyme using reaction mixtures that contained virions of Rauscher murine leukemia virus (R-MuLV, an RNA tumor virus that infects mice), and the four precursors for DNA, one of which was radioactive. Only if the enzyme was present could the precursors be polymerized into a DNA molecule. To detect DNA synthesis, he took samples of the reaction mixtures and added an acid to precipitate macromolecules. Then, he measured the radioactivity in the precipitated, so-called acid-insoluble material.

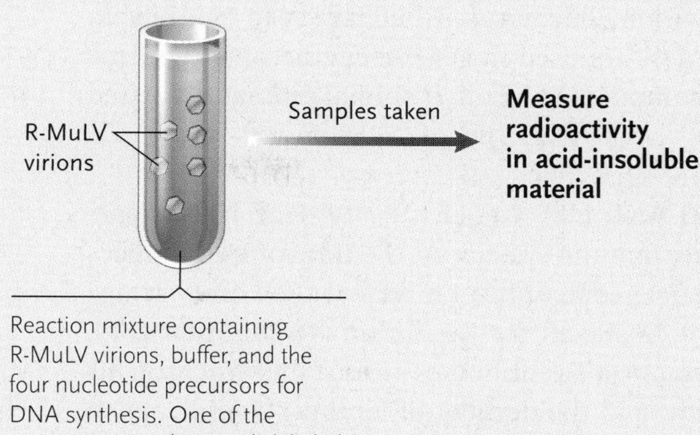
R-MuLV virions
Samples taken
Measure radioactivity in acid-insoluble material
Reaction mixture containing R-MuLV virions, buffer, and the four nucleotide precursors for DNA synthesis. One of the percursors is radioactively labeled.

Results

1. Radioactivity was detected in the acid-insoluble material when a radioactive DNA precursor was used in the reaction. This result showed that the virions were capable of catalyzing the synthesis of DNA.
2. No radioactivity was detected in acid-insoluble material when a radioactive DNA precursor was used in the reaction if the virions were first treated with RNase. RNase is an enzyme that degrades RNA. This result showed that the template for DNA synthesis was RNA.
3. No radioactivity was detected in acid-insoluble material when the four RNA precursors, one of which was radioactive, were used in the reaction instead of DNA precursors. This result showed that the virions were not capable of catalyzing the synthesis of RNA.

Conclusion

Baltimore concluded that R-MuLV virions contain an enzyme that can synthesize DNA using an RNA template. In parallel research, he demonstrated that the same enzyme was present in virions of another RNA tumor virus, Rous sarcoma virus, a result confirmed independently by Howard Temin. In commenting on the work, the editors of *Nature* dubbed the enzyme reverse transcriptase because the synthesis direction was the opposite of that for transcription, namely RNA to DNA rather than DNA to RNA. The name stuck. RNA tumor viruses with this enzyme then became known as retroviruses. Baltimore and Temin received a Nobel Prize in 1975 "for their discoveries concerning the interaction between tumor viruses and the genetic material of the cell."

Source: D. Baltimore. 1970. Viral RNA-dependent DNA polymerase: RNA–dependent DNA polymerase in virions of RNA tumour viruses. *Nature* 226:1209–1211.

known as a **provirus (Figure 17.10).** *Insights from the Molecular Revolution* describes the discovery of reverse transcriptase.

PROPERTIES OF RETROVIRUSES Retroviruses are found in a wide range of organisms, with most so far identified in vertebrates. You, as well as most other humans and mammals, probably contain from one to as many as 100 or more retroviruses in your genome as proviruses. Many of these retroviruses never produce viral particles. However, they may sometimes cause genetic disturbances of various kinds, including alterations of gene activity or DNA rearrangements such as deletions and translocations, some of which may be harmful to the host. HIV, the causal agent of acquired immune deficiency syndrome (AIDS), does produce viral particles. (More discussion of AIDS is in Chapter 43.)

Some retroviruses, such as *avian sarcoma virus,* have been linked to cancer (in this case in chickens). Many of the cancer-causing retroviruses have picked up a host gene that triggers the entry of cells into uncontrolled DNA replication and cell division. When included in a retrovirus, the host gene comes under the influence of the highly active retroviral promoter, which makes the gene continually active and leads to the uncontrolled cell division which is characteristic of cancer. In other words, the host gene has become an oncogene, a gene that promotes the development of cancer by stimulating cell division (see Section 16.5). Usually, the host gene replaces one or more retrovirus genes, making the virus unable to produce viral particles.

Retroviruses also may activate genes related to cell division by moving them to the vicinity of an active host cell promoter or enhancer, or by delivering an enhancer or active promoter to the vicinity of a host cell gene. In either case, the result may be uncontrolled cell division. Retroviruses that have been modified to make them harmless are used in genetic engineering to introduce genes into mammalian and other animal cells (as discussed in Chapter 18).

THE PROPERTIES AND LIFE CYCLE OF HIV HIV infects particular cells of the immune system. At the time of initial infection with HIV, many people suffer a mild fever and other symptoms that may be mistaken for the flu or the common cold. The symptoms disappear as antibodies against the viral proteins appear in the body, and the number of viral particles drops in the bloodstream. However, the genome of the virus is still present, integrated into the DNA of the host cells as a provirus, and the virus steadily spreads to infect other immune system cells. An infected person may remain apparently healthy for years, yet can transmit the virus to others. Both the transmitter and recipient of the virus may be unaware that the disease is present, making it difficult to control the spread of HIV infections. Ultimately, many important cells of the immune system are destroyed, wiping out the body's immune response and making the HIV-infected person susceptible to other infections. Steady debilitation may then occur, sometimes resulting in death.

Like all retroviruses, HIV is an enveloped virus with two copies of a single-stranded RNA genome contained within a capsid **(Figure 17.11)**. When HIV first infects a cell, a *gp120* glycoprotein of the capsid attaches the virus to a particular type of receptor in the cell's plasma membrane. Then, another viral protein triggers fusion of the viral envelope with the host cell's plasma membrane, releasing the virus into the cell **(Figure 17.12)**. Once inside, the viral *reverse transcriptase* that entered the cell in the viral capsid makes a double-stranded DNA copy of the viral RNA genome. Another viral enzyme, *integrase*, then splices the viral DNA into the host cell's nuclear DNA producing the provirus. Once it is part of the host cell DNA, the provirus becomes dormant and is replicated and passed on as the cell divides. As part of the host cell DNA, the virus is effectively hidden in the host cell and protected from attack by the immune system.

When the provirus ceases to be dormant (for example when the host cell in which it is located is stimulated to grow and divide), the viral DNA is copied into new viral RNA molecules, and into mRNAs that direct the host cell's ribosomes to make viral proteins. The viral RNAs are added to the viral proteins to make infective HIV particles, which are released from the host cell by budding. The viral particles may infect more body cells, or another person. Transmission to another person occurs when an infected person's body fluids, especially blood or semen, enter the blood or tissue fluids of another person's body.

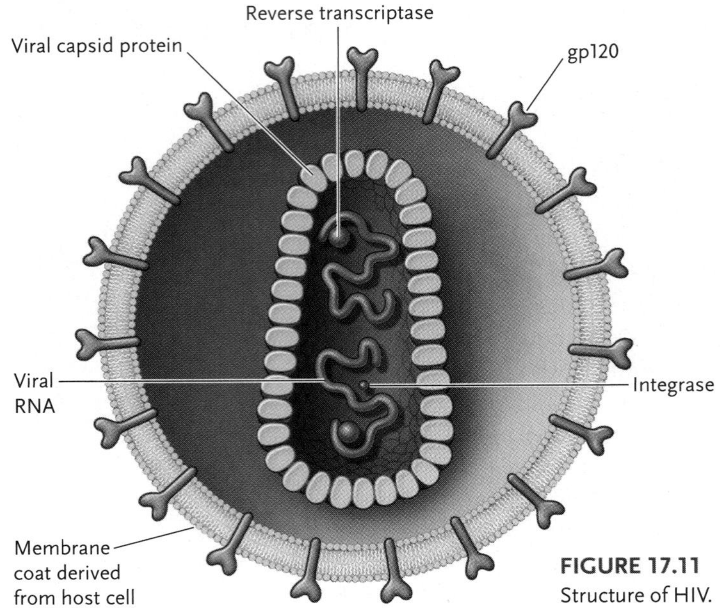
Reverse transcriptase
Viral capsid protein
gp120
Viral RNA
Integrase
Membrane coat derived from host cell

FIGURE 17.11
Structure of HIV.

Viruses May Have Evolved from Fragments of Cellular DNA or RNA

Where did viruses come from? Because viruses can duplicate only by infecting a host cell, they probably evolved after cells appeared. They may represent fragments of DNA molecules that once formed part of the genetic material of living cells, or an

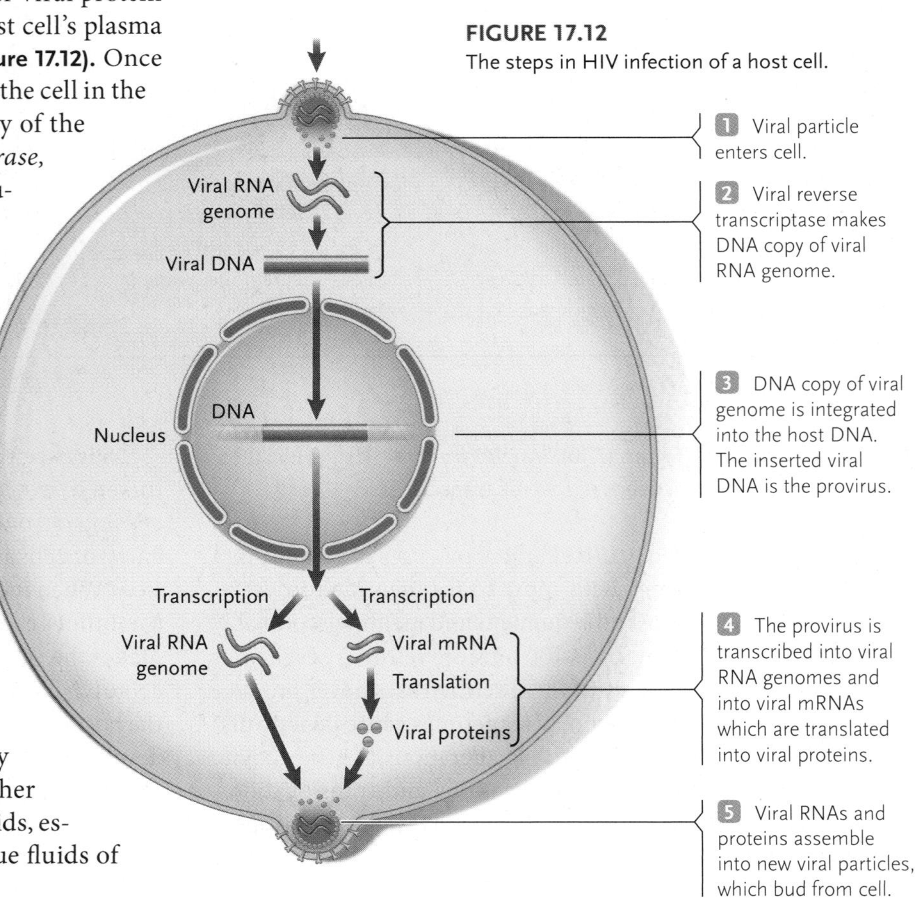
Viral RNA genome
Viral DNA
Nucleus
DNA
Transcription
Transcription
Viral RNA genome
Viral mRNA
Translation
Viral proteins
1 Viral particle enters cell.
2 Viral reverse transcriptase makes DNA copy of viral RNA genome.
3 DNA copy of viral genome is integrated into the host DNA. The inserted viral DNA is the provirus.
4 The provirus is transcribed into viral RNA genomes and into viral mRNAs which are translated into viral proteins.
5 Viral RNAs and proteins assemble into new viral particles, which bud from cell.

FIGURE 17.12
The steps in HIV infection of a host cell.

RNA copy of such a fragment. In some way, the fragments became surrounded by a protective layer of protein with recognition functions and escaped from their parent cells. As viruses evolved, the information encoded in the core of the virus became reduced to a set of directions for producing more viral particles of the same kind.

STUDY BREAK 17.2

1. **What is the difference between a virulent phage and a temperate phage?**
2. **How does viral infection of an animal cell and a plant cell differ?**
3. **How are retroviruses distinctive among RNA viruses with respect to replication of their genome?**

THINK OUTSIDE THE BOOK

Use the Internet or research literature to determine which gene(s) of the influenza virus genome is(are) responsible for the pathogenicity of the virus, and to outline what is known about their molecular functions.

17.3 Viroids and Prions, Infectious Agents Lacking Protein Coats

Although viruses are so chemically simple that most scientists do not consider them to be living organisms, even simpler and smaller infective agents exist. Two infectious agents that are simpler than viruses are *viroids* and *prions*. Both of these agents lack capsids.

Viroids, first discovered in 1971, are plant pathogens that consist solely of single-stranded, circular RNA without a protein coat. Infection by viroids can rapidly destroy entire fields of citrus, potatoes, tomatoes, coconut palms, and other crop plants.

About 30 types of viroids are known. Their RNA genomes range from 246–401 nucleotides and in no case do they encode a protein. The genomes are about 10-fold smaller than the smallest known viral RNA genome. Depending on the viroid, the genomes replicate and accumulate either in the nucleus or in the chloroplast of the host cell. Replication is catalyzed by host nuclear or chloroplast RNA polymerases.

Somehow the tiny RNA genomes of viroids contain sufficient information to infect host plants and to induce the host to replicate the genomes. As a consequence of the replication, the viroids cause specific diseases. The manner in which viroids cause disease remains ill-defined. In fact, researchers believe that there is more than one mechanism. Some recent research has defined one pathway to disease in which viroid RNA activates a protein kinase (an enzyme that adds phosphate groups to proteins; see Sections 4.5 and 7.2) in plants. This process leads to a reduction in protein synthesis and protein activity, and disease symptoms result.

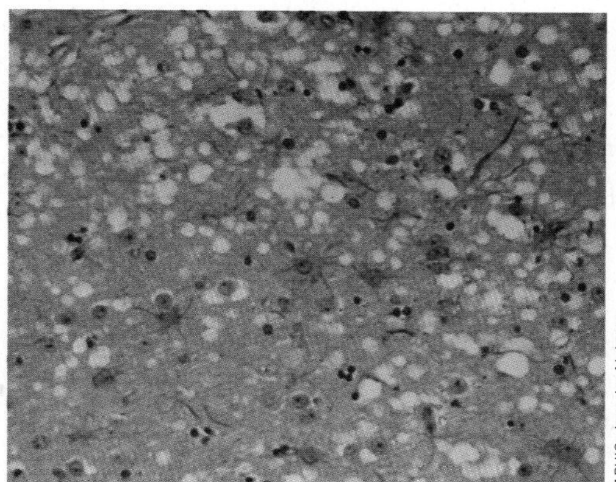
APHIS photo by Al Jenny

FIGURE 17.13
Bovine spongiform encephalopathy (BSE). The light-colored patches in this section from a brain damaged by BSE are areas where tissue has been destroyed.

Prions, named in 1982 by Stanley Prusiner of the University of California, San Francisco for *pro*teinaceous infect*ion,* are the only known infectious agents that do not include a nucleic acid molecule.

Prions have been identified as the causal agents of certain diseases that degenerate the nervous system in mammals. One of these diseases is *scrapie,* a brain disease that causes sheep to rub against fences, rocks, or trees until they scrape off most of their wool. Another prion-based disease is bovine spongiform encephalopathy (BSE), also called *mad cow disease.* The disease produces spongy holes and deposits of proteinaceous material in brain tissue (**Figure 17.13**). In 1996, 150,000 cattle in Great Britain died from an outbreak of BSE, which was traced to cattle feed containing ground-up tissues of sheep that had died of scrapie. Humans are subject to a fatal prion infection called *Creutzfeldt-Jakob disease (CJD).* The symptoms of CJD include rapid mental deterioration, loss of vision and speech, and paralysis; autopsies show spongy holes and deposits in brain tissue similar to those of cattle with BSE. Classic CJD occurs as a result of the spontaneous transformation of normal proteins into prion proteins. Fewer than 300 cases a year occur in the United States. Variant CJD is a form of the disease caused by eating meat or meat products containing nervous system tissue from cattle with BSE. Another prion-based disease of humans, *kuru,* is originally thought to have spread in a tribe in New Guinea when relatives ritualistically ate a deceased individual, including the brain, as a way (in their view) of returning that person's life force to the tribe.

STUDY BREAK 17.3

What distinguishes viroids and prions from viruses?

UNANSWERED QUESTIONS

How do prion proteins move within the brain?

The brain-wasting diseases caused by prions are not well understood, despite much research, and the mechanisms by which prions spread between animals are still under investigation. In scrapie, a prion disease of sheep, and chronic wasting disease (CWD), which affects deer and elk, the infected animals shed prions into the environment. Remarkably, infectivity can persist in the soil for years. Under natural circumstances, prions enter the host at a peripheral site (for example, through ingestion of contaminated material), invade the central nervous system (CNS), and ultimately spread within the CNS and throughout the brain. These events require that prions infect new cells and spread between distant locations in the body. Scientists have shown that prion propagation in the CNS follows the same circuits that nerves use to communicate with each other.

To understand mechanisms of prion infection, scientists have investigated how prion protein aggregates move through the nervous system. Research led by Byron Caughey and myself at the Rocky Mountain Laboratories (RML) in Montana, and Marco Prado at the University of Minas Gerais, Brazil, followed prion aggregates as they invaded mouse brain cells growing in tissue culture. Using prion preparations labeled with a red fluorescent dye, we were able to track the fate of the prion particles in living cells using fluorescence microscopy. One exciting observation was that the prion protein aggregates moved through the wirelike projections of the nerve cells to points of contact with other cells. In rare cases, there seemed to be transfer between cells.

Our team proposed that newly described cell surface structures, called tunneling nanotubes (TNTs), might mediate the intercellular transfer of prion aggregates. TNTs are delicate intercellular projections that can transport proteins between cells in carrier vesicles similar to those containing internalized prion aggregates. Using methodologies developed by the RML/Brazil team, another group later reported that prion aggregates can be transported between cells through TNTs.

Initiation of infection requires that infectious prion aggregates interact with the normal prion protein in host cells which induces its conversion to the disease-associated form. However, we were unable to observe these interactions by microscopy because the normal prion protein lacked a fluorescent tag. To permit simultaneous visualization of the normal and disease-associated aggregates of prion proteins, my group at RML developed a technique to label the normal prion protein in live cells with a green fluorescent dye. Importantly, we found that the green normal prion protein could convert to the aggregated prion disease-associated form. This technique, called IDEAL-labeling, provides the long-sought tools for visualizing the entire prion infection process in living cells by tracking green and red fluorescent prion proteins together. The advances in understanding how prions move through the nervous system were heralded as a significant step toward developing therapies to stop the spread of brain-wasting diseases by blocking the pathways by which prions invade cells, replicate, move within the cell, and invade adjacent cells.

There are several important outstanding questions in the prion field. Some of these include the following:

What are the most relevant mechanisms involved in the spread of prions in animals? Do they involve TNTs or perhaps transport vesicles containing prions that are secreted from infected cells?

How does prion propagation cause the death of nerve cells? Does this involve corruption of the function of the normal prion protein during its conversion to the aggregated form?

Think Critically

One direction of research on prions is to determine if development of the infectious form involves corruption of the function of the normal prion protein. What might be some approaches to further understanding of these processes?

Gerald Baron

Gerald Baron is an investigator and head of the TSE/Prion cell biology section at the Rocky Mountain Laboratories (RML) in Montana. RML is part of the National Institute of Allergy and Infectious Diseases within the National Institutes of Health. To learn more about Dr. Baron's research go to http://www3.niaid.nih.gov/labs/aboutlabs/lpvd/TSEPrionCellBiologySection/.

This research was supported in part by the Intramural Research Program of the NIH, NIAID.

REVIEW KEY CONCEPTS

Go to **CENGAGENOW** at www.cengage.com/login to access quizzing, animations, exercises, articles, and personalized homework help.

17.1 Gene Transfer and Genetic Recombination in Bacteria

- The rapid generation times and numerous offspring of bacteria and viruses make it possible to trace genetic crosses and their outcomes much more quickly than in eukaryotes. These characteristics make it possible to detect rare genetic events. The results of these crosses show that recombination may occur within the boundaries of a gene as well as between genes.
- Recombination occurs in both bacteria and eukaryotes by exchange of segments between homologous DNA molecules. In bacteria, the DNA of the bacterial chromosome may recombine with DNA brought into the cell from outside.
- Three primary mechanisms bring DNA into bacterial cells from the outside: conjugation, transformation, and transduction.
- In conjugation, two bacterial cells form a cytoplasmic bridge, and part or all of the DNA of one cell moves into the other through the bridge. The donated DNA can then recombine with homologous sequences of the recipient cell's DNA (Figures 17.1, 17.2, and 17.4).
- *E. coli* bacteria that are able to act as DNA donors in conjugation have an F plasmid, making them F^+; recipients have no F plasmid and are F^-. In Hfr strains of *E. coli,* the F plasmid is integrated into the main chromosome. As a result, genes of the main chromosome are often transferred into F^- cells along with a portion of the F plasmid DNA. Researchers have mapped genes on the *E. coli* chromosome by noting the order in which they are transferred from Hfr to F^- cells during conjugation (Figure 17.4).

- In transformation, intact cells take up pieces of DNA released from cells that have disintegrated. The entering DNA fragments can recombine with the recipient cell's chromosomal DNA.
- In transduction, bacterial DNA is transferred from one cell to another by an infecting phage.
- Conjugation, transformation, and transduction do not move genes from generation to generation (that is, by descent) but move genes between organisms by horizontal gene transfer.

Practice: Distinguishing between the three major processes: conjugation, transformation, and transduction

Animation: Prokaryotic conjugation

17.2 Viruses and Viral Genetics

- A virus is a biological particle that can infect the cells of a living organism. A free viral particle consists of a nucleic acid core, either double-stranded or single-stranded DNA or double-stranded or single-stranded RNA, surrounded by a protein coat (the capsid). Some animal viruses also have a surrounding envelope derived from the host cell's plasma membrane (Figure 17.6). Many viruses have recognition proteins on the virus surface that enable the virus to attach to host cells.
- One or more kinds of viruses likely infect all living organisms. Viruses show specificity with respect to the host(s) they infect; typically, a particular virus infects only a single species or a few closely related species.
- The cycle of viral infection begins when the nucleic acid molecule of a virus is introduced into a host cell. The virus then directs the cellular machinery to assemble viral capsid proteins with the new viral genomes into new viral particles.
- Virulent phages kill a host bacterial cell during a lytic cycle by releasing an enzyme that ruptures the plasma membrane and cell wall and releases the new viral particles (Figure 17.7).
- Temperate phages do not always kill their host bacterial cell. They may enter the lytic cycle, in which the phage DNA becomes active, exits the host DNA, and begins replication, or a lysogenic cycle. In the lysogenic cycle, the phage's DNA is integrated into the host cell's DNA and may remain for many generations. At some point, the phage may enter the lytic cycle and begin replication. After production of viral capsids, the DNA is assembled into new progeny phages, which are released as the cell ruptures (Figure 17.8).
- During a cycle of viral infection with particular phages, one or more fragments of host cell DNA may be incorporated into phage particles. As an infected cell breaks down, it releases the viral particles containing host cell DNA. These phages, which form the basis of bacterial transduction, may infect a second cell and introduce the bacterial DNA segment into the new host, where it may recombine with the host DNA.
- The infection of animal cells by viruses depends on interaction between specific proteins on the viral particle surface and specific cell surface proteins. Viruses without envelopes enter the cell by receptor-mediated endocytosis. Enveloped viruses infect cells by endocytosis, or by release of the genome-containing capsid into the cell following fusion of the envelope with the plasma membrane.
- Progeny of nonenveloped animal viruses typically are released from a cell by rupture of the cell's plasma membrane, which lyses and kills the cell. Most enveloped viruses bud from the cell in a process that surrounds the protein capsid–nucleic acid genome with an envelope derived from the cell's plasma membrane and containing embedded viral encoded recognition proteins (Figure 17.9).
- Viruses are unaffected by antibiotics and most other treatment methods; hence, infections caused by them are difficult to treat.
- Retroviruses are viruses with single-stranded RNA genomes that are replicated via a DNA intermediate that integrates into a nuclear chromosome to produce a provirus (Figures 17.10–17.12).

Animation: Body plans of viruses

Animation: Bacteriophage multiplication cycles

17.3 Viroids and Prions, Infectious Agents Lacking Protein Coats

- Viroids, which infect crop plants, consist only of a very small, single-stranded RNA molecule. Prions, which cause brain diseases in some animals, are infectious proteins with no associated nucleic acid. Prions are misfolded versions of normal cellular proteins that can induce other normal proteins to misfold (Figure 17.13).

UNDERSTAND AND APPLY

Test Your Knowledge

1. When studying the differences in the genes of bacteria, researchers:
   a. do not grow bacteria on a minimal medium as the medium lacks needed nutrients.
   b. use a bacterial clone, which is a group of cells from different bacteria of varying genetic makeup.
   c. use bacteria diploid for their full genome because they can grow on minimal medium.
   d. can study only one genetic trait in a single recombinant event.
   e. can measure the passage of genes between cells during conjugation, transduction, and transformation.
2. If crossovers occurred between two bacterial genomes as shown in the figure, the result would be:

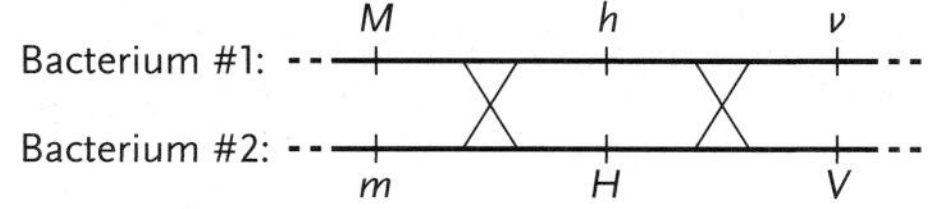

M
h
v
Bacterium #1:
Bacterium #2:
m
H
V

   a. *MHv* and *mhV.*
   b. *MHV* and *mhv.*
   c. *Mhv* and *mHv.*
   d. *MHV* and *mhV.*
   e. *mhv* and *MhV.*
3. In conjugation, when a bacterial F factor is transferred:
   a. the donor cell becomes F^-.
   b. the recipient cell becomes F^+.
   c. the recipient cell becomes F^-.
   d. the donor cell turns into a recipient cell.
   e. chromosomal DNA is transferred in the mating.
4. Which of the following is *not* correct for bacterial conjugation?
   a. Both Hfr and F^+ bacteria have the ability to code for a sex pilus.
   b. After an F^- cell has conjugated with an F^+, its plasmid holds the F^+ factor.
   c. The recipient cell is Hfr following conjugation.
   d. In an Hfr $\times$ F^- conjugation, DNA of the main chromosome moves to a recipient cell.
   e. Genes on the F factor encode proteins of the sex pilus.

5. Which of the following is *not* correct for bacterial transformation?
   a. Artificial transformation is used in cloning procedures.
   b. Avery was able to transform live noninfective bacteria with DNA from dead infective bacteria.
   c. The cell wall and plasma membrane must be penetrated for transformation to proceed.
   d. A virus is required for the process.
   e. Electroporation is a form of artificial transformation used to introduce DNA into cells.
6. Transduction:
   a. may allow recombination of newly introduced DNA with host cell DNA.
   b. is the movement of DNA from one bacterial cell to another by means of a plasmid.
   c. can cause the DNA of the donor to change but not the DNA of the recipient.
   d. is the movement of viral DNA, but not bacterial DNA, into a recipient bacterium.
   e. requires a physical contact between two bacteria.
7. Viruses:
   a. have a protein core.
   b. have a nucleic acid coat.
   c. that infect and kill bacteria during a cell cycle are called virulent bacteriophages.
   d. were probably the first forms of life on Earth.
   e. if they are temperate bacteriophages, kill host cells immediately.
8. When a virus enters the lysogenic stage:
   a. the viral DNA is replicated outside the host cell.
   b. it enters the host cell and kills it immediately.
   c. it enters the host cell, picks up host DNA, and leaves the cell unharmed.
   d. it sits on the host cell plasma membrane with which it covers itself and then leaves the cell.
   e. the viral DNA integrates into the host genome.
9. An infectious material is isolated from a nerve cell. It contains protein with amino acid sequences identical to the host protein but no nucleic acids. It belongs to the group:
   a. prions.
   b. Archaea.
   c. toxin producers.
   d. viroids.
   e. sporeformers.
10. Which is *not* correct about retroviruses?
   a. They are RNA viruses.
   b. They are believed to be the source of retrotransposons.
   c. They encode an enzyme for their insertion into host cell DNA.
   d. They encode single-stranded viral DNA from viral RNA.
   e. They encode a reverse transcriptase enzyme for RNA to DNA synthesis.

Discuss the Concepts

1. You set up an experiment like the one carried out by Lederberg and Tatum, mixing millions of *E. coli* of two strains with the following genetic constitutions:

   Strain 1: bio^- met^- thr^+ leu^+

   Strain 2: bio^+ met^+ thr^- leu^-

   Among the bacteria obtained after mixing, you find some cells that do not require threonine, leucine, or biotin to grow, but still need methionine. How might you explain this result?
2. As a control for their experiments with bacterial recombination, Lederberg and Tatum placed cells of either "parental" strain 1 or 2 on the surface of a minimal medium. If you set up this control and a few scattered colonies showed up, what might you propose as an explanation? How could you test your explanation?
3. What rules would you suggest to prevent the spread of mad cow disease (BSE)?

Design an Experiment

You have a culture of Hfr *E. coli* cells that cannot make biotin for themselves. To this culture you add some wild-type *E. coli* cells that have been heat killed, and then subject the culture to electroporation. After the addition, you find some cells that can grow on minimal medium. How could you establish whether the wild-type bio^+ allele was inserted in a plasmid or the chromosomal DNA of the Hfr cells?

Interpret the Data

As indicated in the chapter, the F factor integrates into one of several positions around the circular *E. coli* chromosome, producing in each case a different Hfr strain. The various Hfr strains can be used to map the locations of genes around the bacterial chromosome. The following table presents the results of conjugation experiments with four different Hfr strains. For each Hfr strain the order of transfer of the bacterial genes to the recipient is given. From the data, construct a map to show the arrangement of the genes on the circular *E. coli* chromosome.

	Order of gene transfer					
Hfr strain	**First**					**Last**
1	*e*	*i*	*f*	*u*	*j*	*b*
2	*w*	*k*	*c*	*r*	*b*	*j*
3	*r*	*c*	*k*	*w*	*h*	*p*
4	*w*	*h*	*p*	*y*	*e*	*i*

Apply Evolutionary Thinking

Are viruses evolutionarily derived from complex organisms that can reproduce themselves, or are they remnants of precellular "life"? Argue your case.

Protein microarrays, a key tool in proteomics, the study of the complete set of proteins that can be expressed by an organism's genome. Each colored dot is a protein, with a specific color for each protein being studied.

Alfred Pasieka/Photo Researchers, Inc.

18

STUDY OUTLINE

DNA Technologies and Genomics

Why It Matters. . . J. Craig Venter and his research colleagues have been pioneers in genome sequencing. Among other achievements, Venter's group determined the genome sequence of the archaean *Methanococcus jannaschii* in 1996; their analysis affirmed that Archaea and Bacteria are separate domains. *M. jannaschii* was isolated from a hot, deep sea vent in the Pacific Ocean. Venter's current research derives from working with the DNA of that organism. He recognized that microbes play important roles in marine ecosystems, yet we know very little about them. Therefore, Venter's goal is to evaluate the microbial diversity in the world's oceans using the molecular tools and techniques developed to sequence genomes.

On their first expedition to achieve this goal, in 2003, researchers from the J. Craig Venter Institute (JCVI) set sail to Bermuda's Sargasso Sea. About every 200 nautical miles, they collected a 200 L sample and passed it through a series of filters to capture bacteria and archaeans, and then viruses. The filtered samples were frozen and shipped back to Venter's institute where scientists extracted genomic DNA from the samples, sequenced it, and stored the sequences in computers. Computer algorithms were used to reassemble the DNA information for genes and large sections of genomes from the diverse microbial communities in the samples. The results showed unexpectedly high microbial diversity in a region of the ocean known to be relatively nutrient poor, and therefore predicted to have low microbial diversity. In just this one small study, sequencing analysis revealed more than 1.2 million new genes and inferred the presence of 1,800 bacterial species.

Venter has continued his marine expeditions. In a 2009–2010 expedition, for example, his research group collected samples from the Baltic, Black, and Mediterranean seas, which are the largest seas iso-

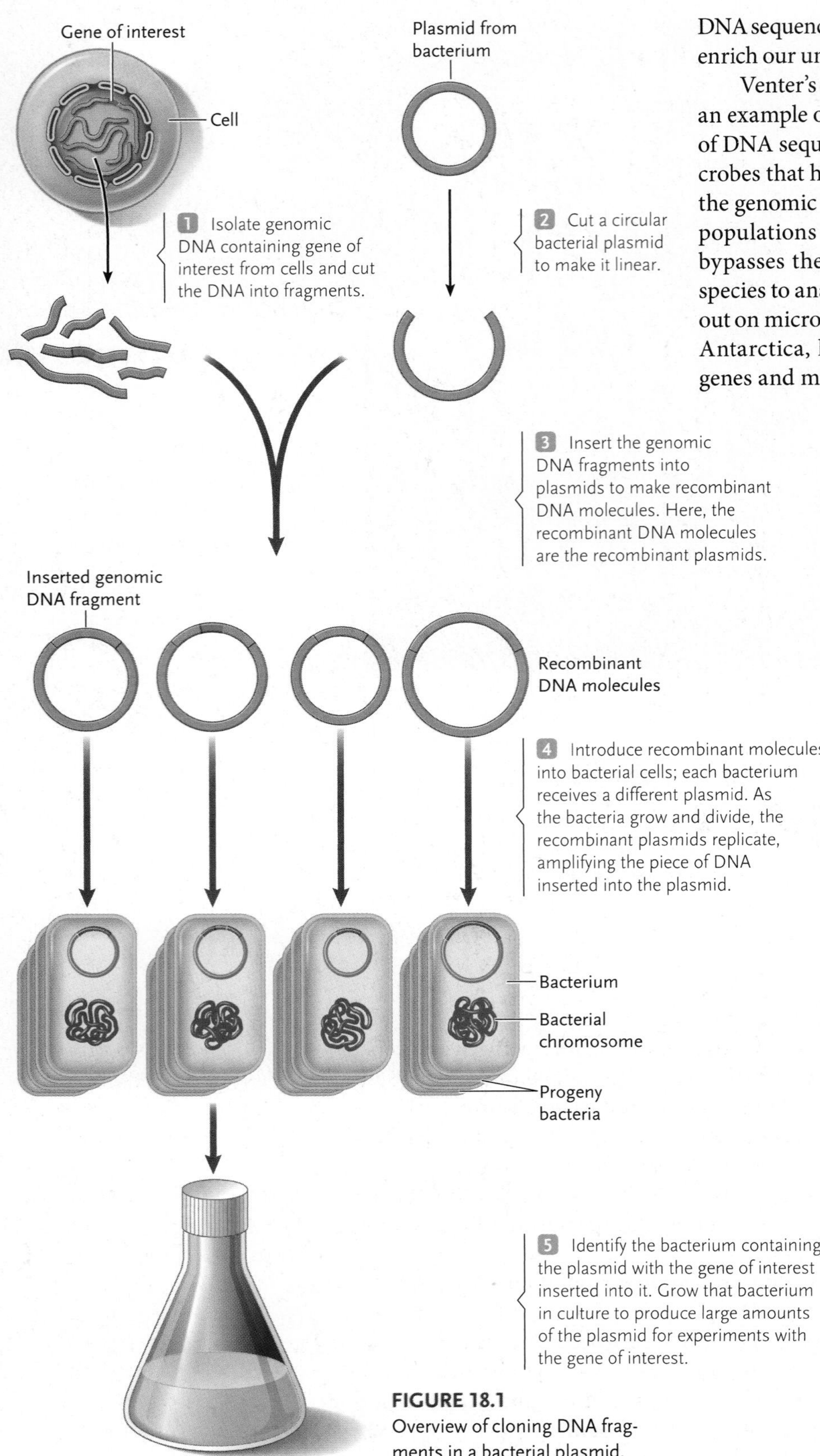

Gene of interest
Cell
Plasmid from bacterium
1 Isolate genomic DNA containing gene of interest from cells and cut the DNA into fragments.
2 Cut a circular bacterial plasmid to make it linear.
3 Insert the genomic DNA fragments into plasmids to make recombinant DNA molecules. Here, the recombinant DNA molecules are the recombinant plasmids.
Inserted genomic DNA fragment
Recombinant DNA molecules
4 Introduce recombinant molecules into bacterial cells; each bacterium receives a different plasmid. As the bacteria grow and divide, the recombinant plasmids replicate, amplifying the piece of DNA inserted into the plasmid.
Bacterium
Bacterial chromosome
Progeny bacteria
5 Identify the bacterium containing the plasmid with the gene of interest inserted into it. Grow that bacterium in culture to produce large amounts of the plasmid for experiments with the gene of interest.

FIGURE 18.1
Overview of cloning DNA fragments in a bacterial plasmid.

lated from major oceans and potentially contain unique microbial communities. Venter's group has so far discovered about 20 million new genes, and thousands of new protein families. Deciphering all of the information to learn about the organisms from which the DNA sequences are derived is a monumental task, but it will greatly enrich our understanding of microbial life.

Venter's collection of data on genetic diversity in the ocean is an example of the emerging field of **metagenomics;** the analysis of DNA sequences of the genomes in entire communities of microbes that have been isolated from the environment, it extends the genomic analysis of a single species to the analysis of mixed populations of microbes. Importantly, metagenomic analysis bypasses the need to isolate and culture individual microbial species to analyze them. Metagenomic studies have been carried out on microbial communities from various locations, including Antarctica, hot springs, and the human intestine. Many new genes and microbial species have been identified. Metagenomic analysis also has utility for comparing conditions, for instance, a polluted ecosystem versus a pristine one, or establishing the effects of pharmaceuticals on human intestinal microbes.

Novel genes identified in metagenomic studies potentially may code for enzymes that will be useful, for instance, in pharmaceutical development, food production and processing, and bioremediation. The techniques used to isolate, purify, analyze, and manipulate DNA sequences for such purposes are known as **DNA technologies.** Scientists use DNA technologies both for basic research into the biology of organisms and for applied research. The use of DNA technologies to alter genes for practical purposes is called **genetic engineering.**

Genetic engineering is part of a broad area known as **biotechnology,** which is any technique applied to biological systems or living organisms to make or modify products or processes for a specific purpose. Thus, biotechnology includes manipulations that do not involve DNA technologies, such as the use of yeast to brew beer and bake bread and the use of bacteria to make yogurt and cheese.

In this chapter, we focus on how biologists isolate genes and manipulate them for basic and applied research. You will learn about the basic DNA technologies and their applications to research in biology, to genetic engineering, and to the analysis of genomes. You will also learn about some of the risks and controversies surrounding genetic engineering and about some of the scientific, social, and ethical questions related to its application.

We begin our discussion with a description of methods used to obtain genes in large quantities, an essential step for their analysis or manipulation. <

18.1 DNA Cloning

Remember from Chapter 17 that a *clone* is a line of genetically identical cells or individuals derived from a single ancestor. DNA cloning is a method for producing many copies of a piece of DNA, such as a gene of interest, defined as a gene that a researcher wants to study or manipulate. Scientists clone DNA for many reasons. For example, a researcher might be interested in how a particular human gene functions. Each human cell contains only two copies of most genes, amounting to a very small fraction of the total amount of DNA in a diploid cell. Therefore, in its natural state in the genome, the gene is extremely difficult to study. However, through DNA cloning, a researcher can produce a large amount of the gene for scientific experimentation.

Cloned genes are used in basic research to find out about their biological functions. For example, researchers can determine the DNA sequence of a cloned gene, giving them the ultimate information about its structure. Also, by manipulating the gene and inducing mutations in it, they can gain information about its function, and about how its expression is regulated. Cloned genes can be expressed in bacteria or other microorganisms, and the proteins encoded by the cloned genes can be produced in quantity and then purified. Those proteins can be used in basic research or, in the case of genes encoding proteins of pharmaceutical or clinical importance, they can be used in applied research.

An overview of one common method for cloning a gene of interest from a genome is shown in **Figure 18.1.** The method uses bacteria (commonly, *Escherichia coli*) and plasmids, small circular DNA molecules that replicate separately from the bacterial chromosome (see Section 17.1). The researcher extracts genomic DNA containing a gene of interest from cells and cuts the genomic DNA into fragments. The fragments are inserted into plasmids producing *recombinant DNA molecules.* **Recombinant DNA** is DNA from two or more different sources that have been joined together to form a single molecule, such as a plasmid. The recombinant plasmids are introduced into bacteria. Each bacterium receives a different plasmid and, as the bacterium continues to grow and divide, the plasmid replicates. Through the replication of the plasmid, amplification of the piece of DNA inserted into the plasmid occurs. The final step is to identify the bacteria containing the plasmid with the gene of interest and then isolate the plasmid for further study of the gene.

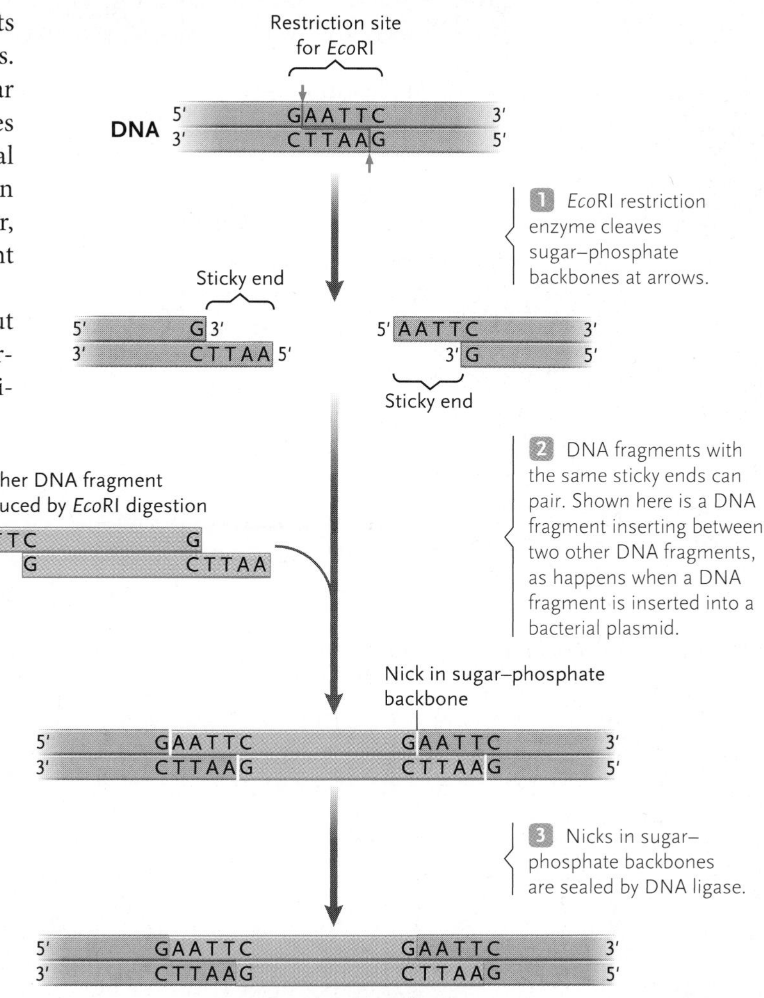

Restriction site for *Eco*RI
DNA
5' GAATTC 3'
3' CTTAAG 5'
1 *Eco*RI restriction enzyme cleaves sugar–phosphate backbones at arrows.
Sticky end
5' G 3'
3' CTTAA 5'
5' AATTC 3'
3' G 5'
Sticky end
2 DNA fragments with the same sticky ends can pair. Shown here is a DNA fragment inserting between two other DNA fragments, as happens when a DNA fragment is inserted into a bacterial plasmid.
Another DNA fragment produced by *Eco*RI digestion
AATTC G
G CTTAA
Nick in sugar–phosphate backbone
5' GAATTC GAATTC 3'
3' CTTAAG CTTAAG 5'
3 Nicks in sugar–phosphate backbones are sealed by DNA ligase.
5' GAATTC GAATTC 3'
3' CTTAAG CTTAAG 5'
Recombinant DNA molecule

FIGURE 18.2
The restriction site for the restriction enzyme *Eco*RI, and the generation of a recombinant DNA molecule by complementary base pairing of DNA fragments which are produced by digestion with the same restriction enzyme.

Bacterial Enzymes Called Restriction Endonucleases Are the Basis of DNA Cloning

The key to DNA cloning is the joining of two DNA molecules from different sources, such as a genomic DNA fragment and a bacterial plasmid (see Figure 18.1). Joining of DNA is made possible, in part, by bacterial enzymes called **restriction endonucleases** (also called **restriction enzymes**), which were discovered in the late 1960s. Restriction enzymes recognize *restriction sites,* specific DNA sequences that are typically 4–8 base pairs (bp) long, and cut the DNA at specific locations within those sites. The DNA fragments produced by a restriction enzyme are known as **restriction fragments.**

The "restriction" in the name of the enzymes refers to their normal role inside bacteria, in which the enzymes defend against viral attack by breaking down (restricting) the DNA molecules of infecting viruses. The bacterium protects the restriction sites in its own DNA from cutting by modifying bases in those sites enzymatically, thereby blocking the action of its restriction enzyme.

Hundreds of different restriction enzymes have been identified, each one cutting DNA at a specific restriction site. As illustrated by the restriction site of *Eco*RI **(Figure 18.2),** most restriction sites are symmetrical in that the sequence of nucleotides read in the 5′→3′ direction on one strand is same as the sequence read in the 5′→3′ direction on the complementary strand. The restriction enzymes most used in cloning—such as *Eco*RI—cleave the sugar–phosphate backbones of DNA to produce DNA fragments with single-stranded ends (step 1). These ends are called **sticky ends** because the short single-stranded regions can form hydrogen bonds with complementary sticky ends on any other DNA molecules cut with the same enzyme (step 2). The pairings leave nicks in the sugar–phosphate backbones of the DNA strands that are sealed by *DNA ligase,* an enzyme that has the same function in DNA replication (step 3; see Section 14.3). The result is a recombinant DNA molecule.

Cloning a Gene of Interest in a Plasmid Cloning Vector

Purpose: Cloning a gene produces many copies of a gene of interest that can be used, for example, to determine the DNA sequence of the gene, to manipulate the gene in basic research experiments, to understand its function, and to produce the protein encoded by the gene.

Protocol:

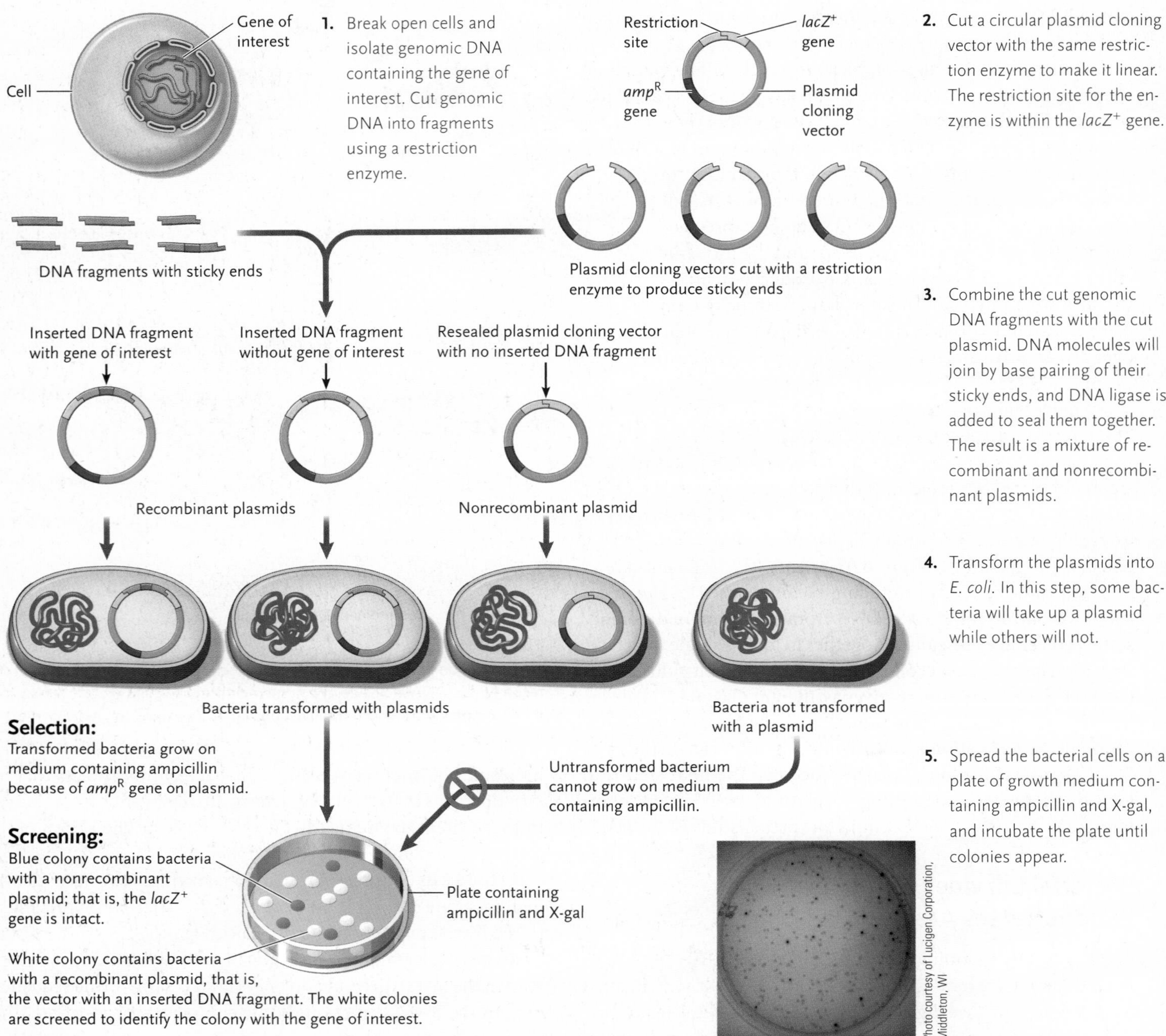

Gene of interest
Cell
1. Break open cells and isolate genomic DNA containing the gene of interest. Cut genomic DNA into fragments using a restriction enzyme.
Restriction site
lacZ+ gene
ampR gene
Plasmid cloning vector
2. Cut a circular plasmid cloning vector with the same restriction enzyme to make it linear. The restriction site for the enzyme is within the lacZ+ gene.
DNA fragments with sticky ends
Plasmid cloning vectors cut with a restriction enzyme to produce sticky ends
3. Combine the cut genomic DNA fragments with the cut plasmid. DNA molecules will join by base pairing of their sticky ends, and DNA ligase is added to seal them together. The result is a mixture of recombinant and nonrecombinant plasmids.
Inserted DNA fragment with gene of interest
Inserted DNA fragment without gene of interest
Resealed plasmid cloning vector with no inserted DNA fragment
Recombinant plasmids
Nonrecombinant plasmid
4. Transform the plasmids into E. coli. In this step, some bacteria will take up a plasmid while others will not.
Bacteria transformed with plasmids
Bacteria not transformed with a plasmid
Selection:
Transformed bacteria grow on medium containing ampicillin because of ampR gene on plasmid.
Untransformed bacterium cannot grow on medium containing ampicillin.
5. Spread the bacterial cells on a plate of growth medium containing ampicillin and X-gal, and incubate the plate until colonies appear.
Screening:
Blue colony contains bacteria with a nonrecombinant plasmid; that is, the lacZ+ gene is intact.
Plate containing ampicillin and X-gal
White colony contains bacteria with a recombinant plasmid, that is, the vector with an inserted DNA fragment. The white colonies are screened to identify the colony with the gene of interest.
Photo courtesy of Lucigen Corporation, Middleton, WI

Interpreting the Results: All of the colonies on the plate contain plasmids because the bacteria that form the colonies are resistant to the ampicillin present in the growth medium. Blue-white screening distinguishes bacterial colonies with nonrecombinant plasmids from those with recombinant plasmids. Blue colonies have nonrecombinant plasmids. These plasmids have intact *lacZ*$^+$ genes and produce β-galactosidase, which changes X-gal to a blue product. White colonies have recombinant plasmids. These plasmids have DNA fragments inserted into the *lacZ*$^+$ gene, so they do not produce β-galactosidase. As a result, they cannot convert X-gal to the blue product and the colonies are white. Among the white colonies is a colony containing the plasmid with the gene of interest. Further screening is done to identify that particular white colony (see Figure 18.4). Once identified, the colony is cultured to produce large quantities of the recombinant plasmid for analysis or manipulation of the gene.

Bacterial Plasmids Illustrate the Use of Restriction Enzymes in Cloning

The bacterial plasmids used for cloning are examples of **cloning vectors**—DNA molecules into which a DNA fragment can be inserted to form a recombinant DNA molecule for the purpose of cloning. Bacterial plasmid cloning vectors are derivatives of plasmids naturally found in bacteria, and have been engineered to have special features. Commonly, plasmid cloning vectors contain two genes that are useful in the final steps of a cloning experiment for sorting bacteria that have recombinant plasmids from those that do not: (1) The amp^R gene encodes an enzyme that breaks down the antibiotic ampicillin. When the plasmid is introduced into an *E. coli* cell and the amp^R gene is expressed, the bacterium becomes resistant to ampicillin. (2) The $lacZ^+$ gene encodes β-galactosidase (recall the *lac* operon from Section 16.1), which hydrolyzes the sugar lactose, as well as a number of synthetic substrates. A cluster of restriction sites is located within the $lacZ^+$ gene, but does not alter the gene's function. Each of the restriction sites within the cluster is unique to the plasmid. For a given cloning experiment, one or two of these restriction sites is chosen.

CLONING A GENE OF INTEREST **Figure 18.3** expands on the overview of Figure 18.1 to show the steps used to clone a gene of interest using a plasmid cloning vector and restriction enzymes. Genomic DNA isolated from the organism in which the gene is found is cut with a restriction enzyme, and a plasmid cloning vector is cut within the $lacZ^+$ gene with the same restriction enzyme (steps 1 and 2). Mixing the DNA fragments and cut plasmid together with DNA ligase produces various joined molecules as the sticky ends pair and the enzyme seals them together. Some of these molecules are recombinant plasmids consisting of, in each case, a DNA fragment inserted into the plasmid cloning vector. Other molecules are nonrecombinant plasmids resulting from the cut plasmid having resealed into a circle without an inserted fragment (step 3). In addition, ligase joins together pieces of genomic DNA with no plasmid involved. Only the recombinant plasmids are important in cloning the gene of interest; the other two undesired molecules are sorted out in later steps.

Next, the DNA molecules are *transformed*—introduced—into ampicillin-sensitive, $lacZ^-$ *E. coli* (which cannot make β-galactosidase), and the transformed bacteria are spread on a plate of agar growth medium containing ampicillin and the β-galactosidase synthetic substrate X-gal (steps 4 and 5). (Section 17.1 describes techniques for the transformation of DNA into bacteria.) Only bacteria with a plasmid can grow and form colonies because expression of the plasmid's amp^R gene makes the bacteria resistant to ampicillin. Within each cell of a colony, the plasmids have replicated until a hundred or so are present.

The X-gal in the medium distinguishes between bacteria that have been transformed with recombinant plasmids and nonrecombinant plasmids by *blue-white screening* (Figure 18.3, Interpreting the Results). Blue colonies contain nonrecombinant plasmids, whereas white colonies contain recombinant plasmids. The white colonies are examined to find the one containing a recombinant plasmid with the gene of interest.

Three researchers, Paul Berg, Stanley N. Cohen, and Herbert Boyer, in 1973 pioneered the development of DNA cloning techniques using restriction enzymes and bacterial plasmids. Berg received a Nobel Prize in 1980 for his research, which helped push DNA technology to the forefront of biological investigations.

IDENTIFYING A CLONE CONTAINING THE GENE OF INTEREST How is a clone containing the gene of interest identified amongst the population of clones? The gene of interest has a unique DNA sequence, which is the basis for one commonly used identification technique. In this technique, called **DNA hybridization,** the gene of interest is identified in the set of clones when it base-pairs with a short, single-stranded complementary DNA or RNA molecule called a *nucleic acid probe* **(Figure 18.4)**. The probe is labeled with a radioactive or a nonradioactive tag, so investigators can detect it. In our example, if we know the sequence of part of the gene of interest, we can use that information to design and synthesize a nucleic acid probe. Or, we can take advantage of DNA sequence similarities of evolutionarily related organisms. For instance, we could make a probe for a human gene based on the sequence of an equivalent cloned mouse gene. Once a colony containing plasmids with the gene of interest has been identified, that colony can be used to produce large quantities of the cloned gene.

DNA Libraries Contain Collections of Cloned DNA Fragments

As you have seen, the starting point for cloning a gene of interest is a large set of plasmid clones carrying fragments representing all of the DNA of an organism's genome. A collection of clones that contains a copy of every DNA sequence in a genome is called a **genomic library.** A genomic library can be made using plasmid cloning vectors or any other kind of cloning vector. The number of clones in a genomic library increases with the size of the genome. For example, a yeast genomic library of plasmid clones consists of hundreds of plasmids, whereas a human genomic library of plasmid clones consists of thousands of plasmids.

A genomic library is a resource containing the entire DNA of an organism cut into pieces. Similar to a book library, where you can search through the same set of books on various occasions to find different passages of interest, researchers can search through the same genomic library on various occasions to find and isolate different genes or other DNA sequences.

Researchers also commonly use another kind of DNA library that is made by starting with mRNA molecules isolated from a cell. To convert single-stranded mRNA to double-stranded DNA for cloning (RNA cannot be cloned) first they use the enzyme *reverse transcriptase* (isolated from retroviruses; see Section 17.2) to make a single-stranded DNA that is complementary to the mRNA. Then they degrade the mRNA strand with an enzyme, and use DNA polymerase to make a second DNA strand that is complementary to the first. The result is a double-stranded **complementary DNA (cDNA).** After adding restriction sites to each end, they insert the cDNA into a cloning vector as described for the genomic library. The entire collection of cloned cDNAs made from the mRNAs isolated from a cell is a **cDNA library.**

DNA Hybridization to Identify a DNA Sequence of Interest

Purpose: Hybridization with a specific DNA probe allows researchers to detect a specific DNA sequence, such as a gene, within a population of DNA molecules. Here, DNA hybridization is used to screen a collection of bacterial colonies to identify those containing a recombinant plasmid with a gene of interest.

Protocol:

1. Prepare master plates of white colonies detected in the blue-white screening step of Figure 18.3. These colonies contain bacteria with recombinant plasmids. Hundreds or thousands of colonies can be screened for the gene of interest by using many master plates.

2. Lay a special filter paper on the plate to pick up some cells from each colony. This produces a replica of the colony pattern on the filter.

3. Treat the filter to break open the cells and to denature the released DNA into single strands. The single-stranded DNA sticks to the filter in the same position as the colony from which it was derived.

4. Add a labeled single-stranded DNA probe (DNA or RNA) for the gene of interest and incubate. The label can be radioactive or nonradioactive. If a recombinant plasmid's inserted DNA fragment is complementary to the probe, the two will hybridize, that is, form base pairs. Wash off excess labeled probe.

5. Detect the hybridization event by looking for the labeled tag on the probe. If the probe was radioactively labeled, place the filter against photographic film. The decaying radioactive compound exposes the film, giving a dark spot when the film is developed. Correlate the position of any dark sport on the film to the original colony pattern on the master plate. Isolate the colony and use it to produce large quantities of the gene of interest.

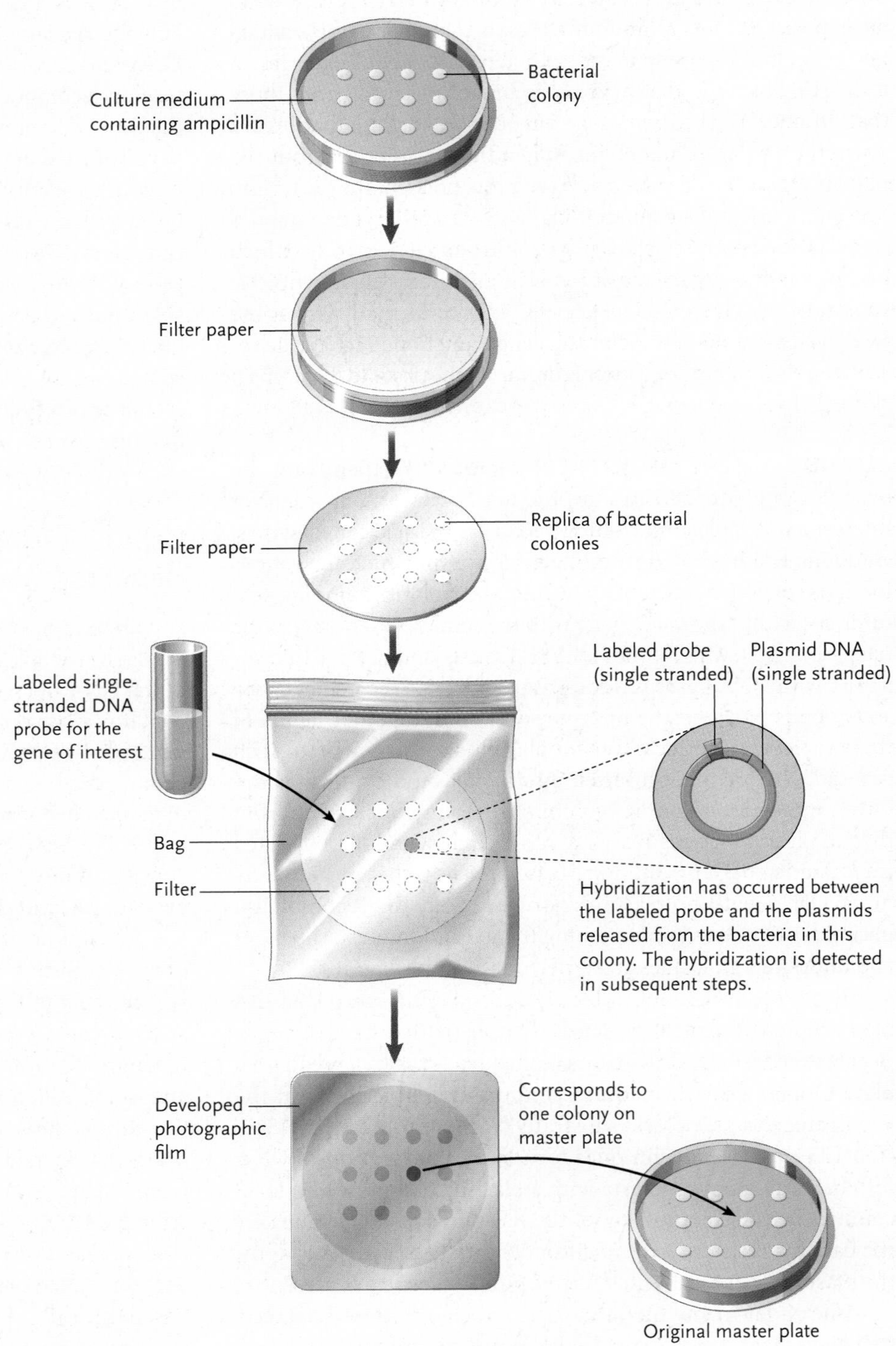

Culture medium containing ampicillin
Bacterial colony
Filter paper
Filter paper
Replica of bacterial colonies
Labeled single-stranded DNA probe for the gene of interest
Bag
Filter
Labeled probe (single stranded)
Plasmid DNA (single stranded)
Hybridization has occurred between the labeled probe and the plasmids released from the bacteria in this colony. The hybridization is detected in subsequent steps.
Developed photographic film
Corresponds to one colony on master plate
Original master plate

Interpreting the Results: DNA hybridization with a labeled probe enables a researcher to identify a sequence of interest. If the probe is for a particular gene, it allows the specific identification of a colony containing bacteria with recombinant plasmids carrying that gene. The specificity of the method depends directly on the probe used. The same collection of bacterial clones can be used again and again to search for recombinant plasmids carrying different genes or different plasmids of interest simply by changing the probe used in the experiment.

Not all genes are active in every cell. Therefore, a cDNA library is limited in that it includes copies of only the genes that were active in the cells used as the starting point for creation of the library. This limitation can be an advantage, however, in identifying genes active in one cell type and not another. cDNA libraries are useful, therefore, for providing clues to the changes in gene activity that are responsible for cell differentiation and specialization. An ingenious method for comparing the cDNA libraries produced by different cell types—the DNA chip—is described later in this chapter.

cDNA libraries provide a critical advantage to genetic engineers who wish to insert eukaryotic genes into bacteria, particularly when the bacteria are to be used as "factories" for making the protein encoded by the gene. Eukaryotic genes typically contain many *introns,* spacer sequences that interrupt the amino acid-coding sequence of a gene (see Section 15.3). Because bacterial genes do not contain introns, bacteria are not equipped to remove introns from transcripts of eukaryotic genes. However, the cDNA copy of a eukaryotic mRNA already has the introns removed, so bacteria can transcribe and translate it accurately to synthesize the encoded eukaryotic protein.

The Polymerase Chain Reaction (PCR) Amplifies DNA in Vitro

Producing multiple DNA copies by cloning in a living cell requires a series of techniques and considerable time. A much more rapid process, **polymerase chain reaction (PCR),** produces an extremely large number of copies of a specific DNA sequence from a DNA mixture without having to clone the sequence in a host organism. The process is called *amplification* because it increases the amount of DNA to the point where it can be analyzed or manipulated easily. Developed in 1983 by Kary B. Mullis and Fred Faloona at Cetus Corporation (Emeryville, CA), PCR has become one of the most important tools in modern molecular biology, having a wide range of applications in all areas of biology. Mullis received a Nobel Prize in 1993 for his role in the development of PCR.

How PCR is performed is shown in **Figure 18.5.** PCR is essentially DNA replication, but a special case of replication in which a DNA polymerase replicates just a portion of a DNA molecule rather than the whole molecule. PCR takes advantage of a characteristic common to all DNA polymerases, namely, that these enzymes add nucleotides only to the end of an existing chain called the *primer* (see Section 14.3). For replication to take place, a primer therefore must be in place, base-paired to the template chain at which replication is to begin. By cycling 20 to 30 times through a series of steps, PCR amplifies the target sequence, producing millions of copies.

Since the primers used in PCR are designed to bracket only the sequence of interest, the cycles replicate only this sequence from a mixture of essentially any DNA molecules. Thus PCR not only finds the "needle in the haystack" among all the sequences in a mixture, but also makes millions of copies of the "needle"—the DNA sequence of interest. Usually, no further purification of the amplified sequence is necessary.

The characteristics of PCR allow extremely small DNA samples to be amplified to concentrations high enough for analysis. PCR is used, for example, to produce enough DNA for analysis from the root of a single human hair, or from a small amount of blood, semen, or saliva, such as the traces left at the scene of a crime. It is also used to extract and multiply DNA sequences from skeletal remains; ancient sources such as mammoths, Neanderthals, and Egyptian mummies; and, in rare cases, from amber-entombed fossils, fossil bones, and fossil plant remains.

A successful outcome of PCR is shown by analyzing a sample of the amplified DNA using **agarose gel electrophoresis** to see if the copies are the same length as the target DNA sequence **(Figure 18.6).** Gel electrophoresis is a technique by which DNA, RNA, or protein molecules are separated in a gel that is subjected to an electric field. The type of gel and the conditions used vary with the experiment, but in each case the gel functions as a molecular sieve to separate the macromolecules based on their size, electrical charge, or other properties. For separating large DNA molecules, such as those typically produced by PCR or by cutting with a restriction enzyme, a gel made of agarose, a natural molecule isolated from seaweed, is used because of its large pore size.

For PCR experiments, the size of the amplified DNA is determined by comparing the position of the DNA band with the positions of bands of a DNA ladder, that is, DNA fragments of known size that are separated on the gel at the same time. If the sample fragment size matches the predicted size for the target DNA fragment, PCR is deemed successful. In some cases, such as DNA from ancient sources, a size prediction may not be possible; in this case, agarose gel electrophoresis analysis simply indicates whether there was DNA in the sample that could be amplified.

The advantages of PCR have made it the technique of choice for researchers, law enforcement agencies, and forensic specialists whose primary interest is in the amplification of specific DNA fragments up to a practical maximum of a few thousand base pairs. Cloning remains the technique of choice for amplification of longer fragments. The major limitation of PCR relates to the primers. In order to design a primer for PCR, the researcher must first have sequence information about the target DNA. By contrast, cloning can be used to amplify DNA of unknown sequence.

STUDY BREAK 18.1 <

1. **What features do restriction enzymes have in common? How do they differ?**
2. **Plasmid cloning vectors are one type of cloning vector that can be used with *E. coli* as a host organism. What features of a plasmid cloning vector make it useful for constructing and cloning recombinant DNA molecules?**
3. **What is a cDNA library, and from what cellular material is it derived? How does a cDNA library differ from a genomic library?**
4. **What information and materials are needed to amplify a region of DNA using PCR?**

The Polymerase Chain Reaction (PCR)

Purpose: To amplify—produce large numbers of copies of—a target DNA sequence in the test tube without cloning.

Protocol: A polymerase chain reaction mixture has four key elements: (1) the DNA with the target sequence to be amplified; (2) a pair of DNA primers, one complementary to one end of the target sequence and the other complementary to the other end of the target sequence; (3) the four nucleoside triphosphate precursors for DNA synthesis (dATP, dTTP, dGTP, and dCTP); and (4) DNA polymerase. Since PCR uses high temperatures that would break down normal DNA polymerases, a heat-stable DNA polymerase is used. Heat-stable polymerases are isolated from microorganisms that grow in a high-temperature area such as a thermal pool or near a deep-sea vent.

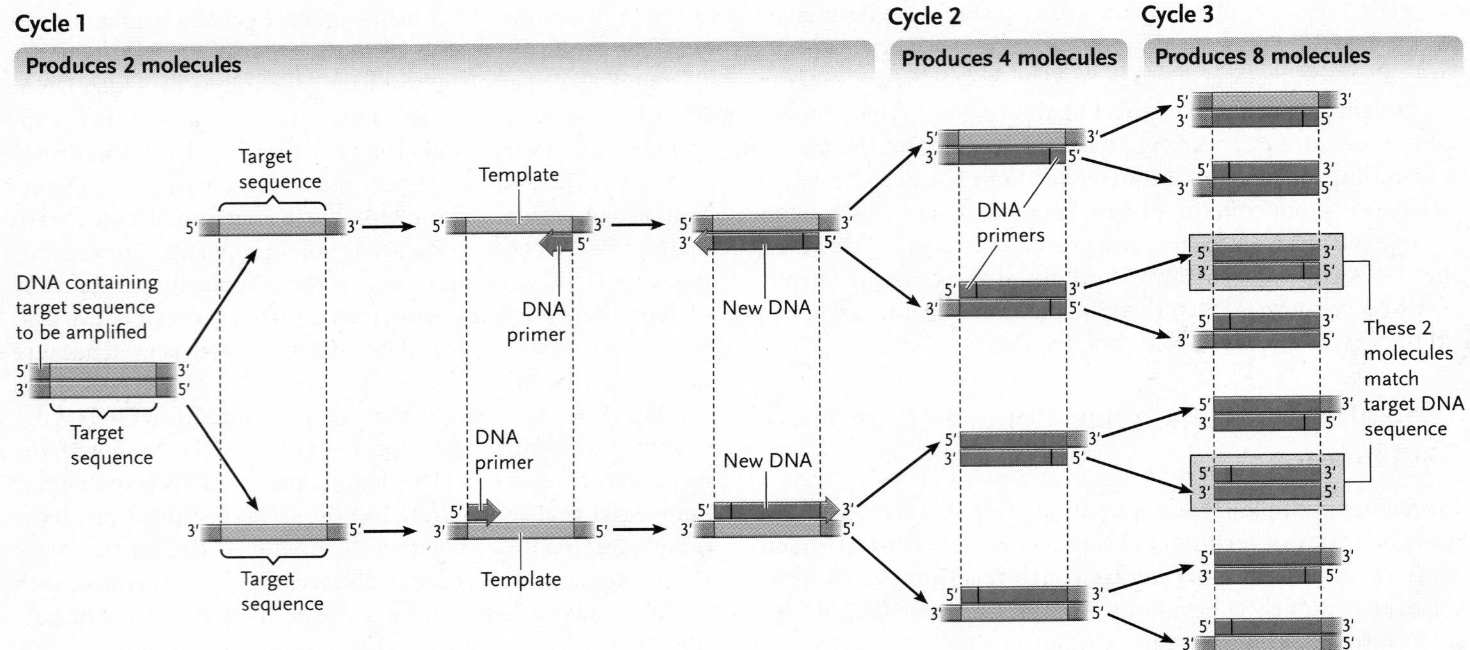

Cycle 1
Produces 2 molecules
Cycle 2
Produces 4 molecules
Cycle 3
Produces 8 molecules
DNA containing target sequence to be amplified
Target sequence
Target sequence
Template
DNA primer
DNA primer
Template
Target sequence
New DNA
New DNA
DNA primers
These 2 molecules match target DNA sequence
5'
3'

1. Denaturation: Heat DNA containing target sequence to 95°C to denature it to single strands.
2. Annealing: Cool the mixture to 55–65°C (depending on the primers) to allow the two primers to anneal their complementary sequences at the two ends of the target sequence.
3. Extension: Heat to 72°C, the optimal temperature for DNA polymerase to extend the primers, using the four nucleoside triphosphate precursors to make complementary copies of the two template strands. This completes cycle 1 of PCR; the end result is two molecules.
4. Repeat the same steps of denaturation, annealing of primers, and extension in cycle 2, producing a total of four molecules.
5. Repeat the same steps in cycle 3, producing a total of eight molecules. Two of the eight match the exact length of the target DNA sequence (highlighted in yellow).

Interpreting the Results: After three cycles, PCR produces a pair of molecules matching the target sequence. Subsequent cycles amplify these molecules to the point where they outnumber all other molecules in the reaction by many orders of magnitude.

18.2 Applications of DNA Technologies

The ability to clone pieces of DNA—genes, especially—and to amplify specific segments of DNA by PCR revolutionized biology. These and other DNA technologies are now used for research in all areas of biology, including cloning genes to determine their structure, function, and regulation of expression; manipulating genes to determine how their products function in cellular or developmental processes; and identifying differences in DNA sequences among individuals in ecological studies. The same DNA technologies also have practical applications, including medical and forensic detection, modification of animals and plants, and the manufacture of commercial products. In this section, case studies provide examples of how the techniques are used to answer questions and solve problems.

Separation of DNA Fragments by Agarose Gel Electrophoresis

Purpose: Gel electrophoresis separates DNA molecules, RNA molecules, or proteins according to their sizes, electrical charges, or other properties through a gel in an electric field. Different gel types and conditions are used for different molecules and types of applications. A common gel for separating large DNA fragments is made of agarose.

Protocol:

1. Prepare a gel consisting of a thin slab of agarose and place it in a gel box in between two electrodes. The gel has wells for placing the DNA samples to be analyzed. Add buffer to cover the gels.

2. Load DNA sample solutions, such as PCR products, into wells of the gel, alongside a well loaded with marker DNA fragments of known sizes. (The DNA samples, as well as the marker DNA sample, have a dye added to help see the liquid when loading the wells. The dye migrates during electrophoresis, enabling the progress of electrophoresis to be followed.)

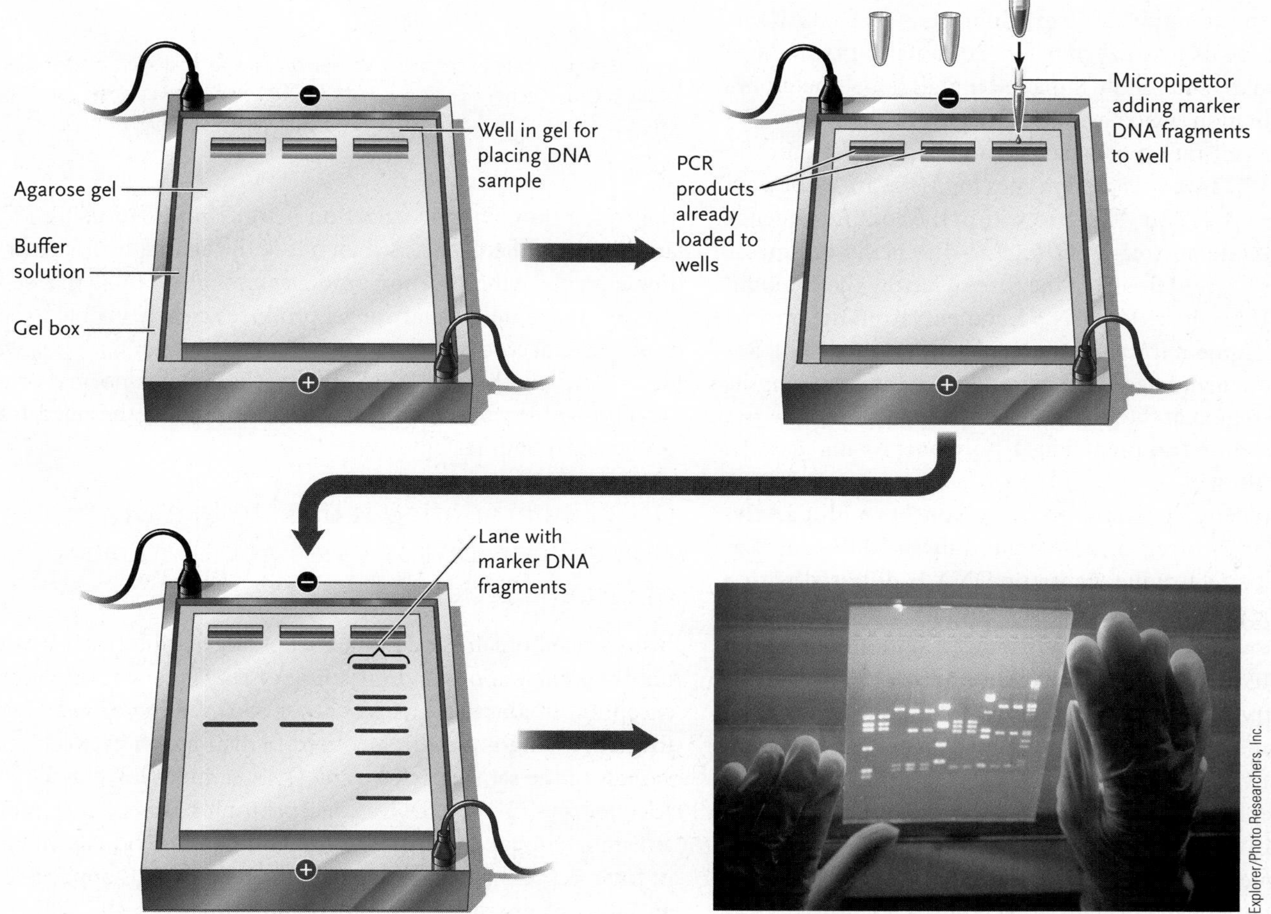

Well in gel for placing DNA sample
Agarose gel
Buffer solution
Gel box
PCR products already loaded to wells
Micropipettor adding marker DNA fragments to well
Lane with marker DNA fragments
Explorer/Photo Researchers, Inc.

3. Apply an electric current to the gel; DNA fragments are negatively charged, so they migrate to the positive pole. Shorter DNA fragments migrate faster than longer DNA fragments. At the completion of separation, DNA fragments of the same length have formed bands in the gel. At this point, the bands are invisible.

4. Stain the gel with a dye that binds to DNA. The dye fluoresces under UV light, enabling the DNA bands to be seen and photographed. An actual gel showing separated DNA bands stained and visualized this way is shown.

Interpreting the Results: Agarose gel electrophoresis separates DNA fragments according to their length. The lengths of the DNA fragments being analyzed are determined by measuring their migration distances and comparing those distances to a calibration curve of the migration distances of the marker bands, which have known lengths. For PCR, agarose gel electrophoresis shows whether DNA of the correct length was amplified. For restriction enzyme digests, this technique shows whether fragments are produced as expected.

DNA Technologies Are Used in Molecular Testing for Many Human Genetic Diseases

Many human genetic diseases are caused by defects in enzymes or other proteins that result from mutations at the DNA level. Once scientists have identified the specific mutations responsible for human genetic diseases, they can often use DNA technologies to develop molecular tests for those diseases. One example is sickle-cell disease (see *Why It Matters* in Chapter 12 and Sections 12.2 and 13.4). People with this disease are homozygous for a DNA mutation that affects hemoglobin, the oxygen-carrying molecule of the blood. Hemoglobin consists of two copies each of the α-globin and β-globin polypeptides. The mutation, which is in the β-globin gene, alters one amino acid in the polypeptide. As a consequence, the function of hemoglobin is significantly impaired in individuals homozygous for the mutation (who have sickle-cell anemia), and mildly impaired in individuals heterozygous for the mutation (who have sickle-cell trait).

The sickle-cell mutation changes a restriction site in the DNA **(Figure 18.7).** Three restriction sites for *Mst*II are associated with the normal β-globin gene, two within the coding sequence of the gene and one upstream of the gene. The sickle-cell mutation eliminates the middle site of the three. Cutting the β-globin gene with *Mst*II produces two DNA fragments from the normal gene and one fragment from the mutated gene (Figure 18.7). Restriction enzyme-generated DNA fragments of different lengths from the same region of the genome, such as in this example, are known as **restriction fragment length polymorphisms (RFLPs,** pronounced "riff-lips").

RFLPs typically are analyzed using **Southern blot analysis** (named after its inventor, researcher Edwin Southern) **(Figure 18.8).** In this technique, genomic DNA is digested with a restriction enzyme, and the DNA fragments are separated using agarose gel electrophoresis. The fragments are then transferred—blotted—to a filter paper, and a labeled probe is used to identify a DNA sequence of interest from among the many thousands of fragments on the filter paper. As shown in Figure 18.8 for the sickle-cell disease example, the DNA bands visualized after hybridization indicate the allele or alleles present in each individual.

A technique called **northern blot analysis** is very similar to Southern blot analysis but is used to study RNA rather than DNA. (The name is a play on words on Southern's name.) In northern blot analysis, RNA molecules extracted from cells or a tissue are separated by size using electrophoresis and then transferred to a filter paper. Hybridization with a labeled probe reveals the RNAs that were present, both their size and concentration. Northern blot analysis is useful, for example, for determining the size of an mRNA encoded by a gene or for quantifying gene expression in different tissues or at different stages of development.

Restriction enzyme digestion and Southern blot analysis may be used to test for a number of human genetic diseases, including phenylketonuria and Duchenne muscular dystrophy. In some cases, restriction enzyme digestion is combined with PCR for a quicker, easier analysis. The gene or region of the gene with the restriction enzyme variation is first amplified using PCR, and the amplified DNA is then cut with the diagnostic restriction enzyme. Amplification produces enough DNA so that separation by size on an agarose gel produces clearly visible bands, positioned according to fragment length. Researchers can then determine whether the fragment lengths match a normal or abnormal RFLP pattern. This method eliminates the need for a probe or for Southern blotting.

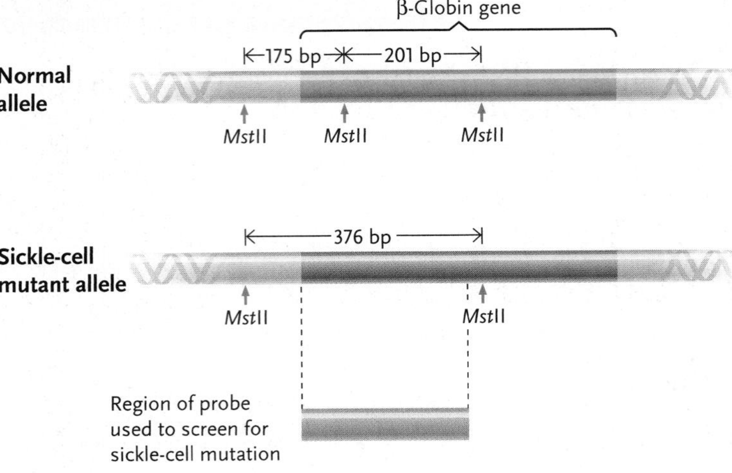

β-Globin gene
Normal allele
175 bp
201 bp
MstII
MstII
MstII
Sickle-cell mutant allele
376 bp
MstII
MstII
Region of probe used to screen for sickle-cell mutation

FIGURE 18.7
Restriction site differences between the normal and sickle-cell mutant alleles of the β-globin gene. The figure shows a DNA segment that can be used as a probe to identify these alleles in subsequent analysis (see Figure 18.8).

DNA Fingerprinting Is Used to Identify Human Individuals as well as Individuals of Other Species

Just as each human has a unique set of fingerprints, each also has unique combinations and variations of DNA sequences (with the exception of identical twins) known as *DNA fingerprints.* **DNA fingerprinting** is a technique used to distinguish between individuals of the same species using DNA samples. Invented by Sir Alec Jeffreys in 1984, DNA fingerprinting has become a mainstream technique for distinguishing human individuals, notably in forensics and paternity testing. The technique is applicable to all kinds of organisms. We focus on humans in the following discussion.

DNA FINGERPRINTING PRINCIPLES In DNA fingerprinting, scientists use molecular techniques, most typically PCR, to analyze DNA variations at various loci in the genome. In the United States, 13 loci in noncoding regions of the genome are the standards for PCR analysis. Each locus is an example of a **short tandem repeat (STR)** sequence (also called a **microsatellite**), meaning that it has a short sequence of DNA repeated in series, with each repeat 2–6 bp. Each locus has a different repeated sequence, and the number of repeats varies among individuals in a population. For example, one STR locus has the sequence AGAT repeated between eight and twenty times. As a further source of variation, a given individual is either homozygous or heterozy-

Southern Blot Analysis

Purpose: The Southern blot technique allows researchers to identify DNA fragments of interest after separating DNA fragments on a gel. One application is to compare different samples of genomic DNA cut with a restriction enzyme to detect specific restriction fragment length polymorphisms. Here the technique is used to distinguish between individuals with sickle-cell disease, individuals with sickle-cell trait, and normal individuals.

Protocol:

1. Isolate genomic DNA and digest with a restriction enzyme. Here, genomic DNA is isolated from three individuals: A, sickle-cell disease (homozygous for the sickle-cell mutant allele); B, normal (homozygous for the normal allele); and C, sickle-cell trait (heterozygous for sickle-cell mutant allele). Digest the DNA with *Mst*II.

2. Separate the DNA fragments by agarose gel electrophoresis. The thousands of differently sized DNA fragments produce a smear of DNA down the length of each lane in the gel, which can be seen after staining the DNA. (Gel electrophoresis and gel staining are shown in Figure 18.6.)

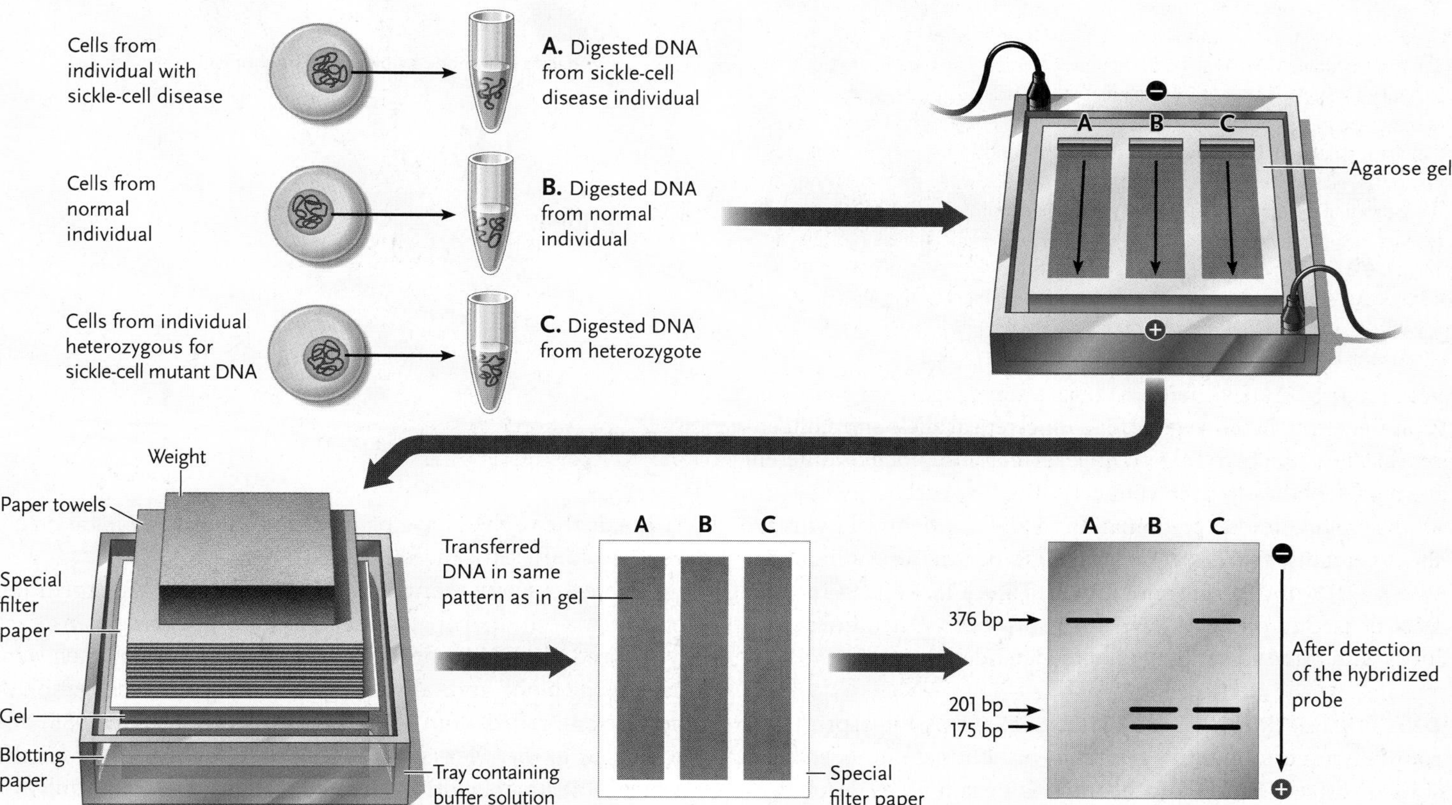

Cells from individual with sickle-cell disease
A. Digested DNA from sickle-cell disease individual
Cells from normal individual
B. Digested DNA from normal individual
Cells from individual heterozygous for sickle-cell mutant DNA
C. Digested DNA from heterozygote
A
B
C
Agarose gel
Weight
Paper towels
Special filter paper
Gel
Blotting paper
Tray containing buffer solution
Transferred DNA in same pattern as in gel
A
B
C
Special filter paper
A
B
C
376 bp
201 bp
175 bp
After detection of the hybridized probe

3. Hybridization with a labeled DNA probe to identify DNA fragments of interest cannot be done directly with an agarose gel. Edward Southern devised a method to transfer the DNA fragments from a gel to a special filter paper. First, treat the gel with a solution to denature the DNA into single strands. Next, place the gel on a piece of blotting paper with ends of the paper in the buffer solution and place the special filter paper on top of the gel. Capillary action wicks the buffer solution in the tray up the blotting paper, through the gel and special filter paper, and into the weighted stack of paper towels on top of the gel. The movement of the solution transfers—blots—the single-stranded DNA fragments to the filter paper, where they stick. The pattern of DNA fragments is the same as it was in the gel.

4. To focus on a particular region of the genome, use DNA hybridization with a labeled probe. That is, incubate a labeled, single-stranded probe with the filter and, after washing off excess probe, detect hybridization of the probe with DNA fragments on the filter. For a radioactive probe, place the filter against photographic film, which, after development will show a band or bands where the probe hybridized. In this experiment, the probe is a cloned piece of DNA from the area shown in Figure 18.7 (the β-globin gene), that can bind to all three of the *Mst*II fragments of interest.

Interpreting the Results: The hybridization result indicates that the probe has identified a very specific DNA fragment or fragments in the digested genomic DNA. The RFLPs for the β-globin gene can be seen in Figure 18.7. DNA from the sickle-cell disease individual cut with *Mst*II results in a single band of 376 bp detected by the probe, while DNA from the normal individual results in two bands, of 201 bp and 175 bp. DNA from a sickle-cell trait heterozygote results in three bands, of 376 bp (from the sickle-cell mutant allele), and 201 bp and 175 bp (both from the normal allele). This type of analysis in general is useful for distinguishing normal and mutant alleles of genes where the mutation involved alters a restriction site.

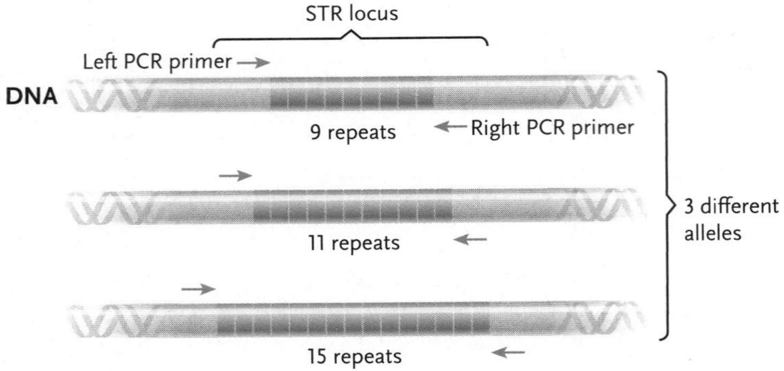
A. Alleles at an STR locus
STR locus
Left PCR primer
DNA
9 repeats
Right PCR primer
11 repeats
15 repeats
3 different alleles

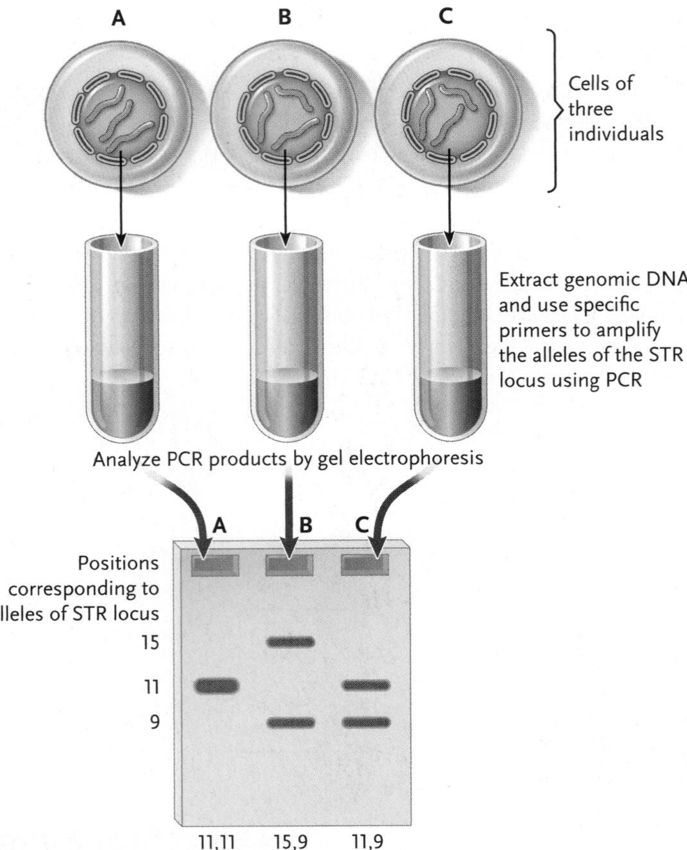
B. DNA fingerprint analysis of the STR locus by PCR
A
B
C
Cells of three individuals
Extract genomic DNA and use specific primers to amplify the alleles of the STR locus using PCR
Analyze PCR products by gel electrophoresis
Positions corresponding to alleles of STR locus
15
11
9
11,11
15,9
11,9

FIGURE 18.9
Using PCR to obtain a DNA fingerprint for an STR locus. **(A)** Three alleles of the STR locus with 9, 11, and 15 copies of the tandemly repeated sequence. The arrows indicate where left and right PCR primers can bind to amplify the STR locus. **(B)** DNA fingerprint analysis of the STR locus by PCR. The number of bands on the gel and the sizes of the DNA in the bands show the STR alleles that were amplified. One band indicates that the individual was homozygous for an STR allele with a particular number of repeats, while two bands indicates the individual is heterozygous for two STR alleles with different numbers of repeats. Here, individual A is homozygous for an 11-repeat allele (designated 11,11), B is heterozygous for a 15-repeat allele and a 9-repeat allele (15,9), and C is heterozygous for the 11-repeat allele and the 9-repeat allele (11,9).

gous for an STR allele; perhaps you are homozygous for an eleven-repeat allele, or heterozygous for a nine-repeat allele and a fifteen-repeat allele. Likely your DNA fingerprint for this locus is different from most of the others in your class. Because each individual has an essentially unique combination of alleles (identical twins are the exception), analysis of multiple STR loci can discriminate between DNA of different individuals. **Figure 18.9** illustrates how PCR is used to obtain a DNA fingerprint for a theoretical STR locus with three alleles of 9, 11, and 15 tandem repeats.

DNA FINGERPRINTING IN FORENSICS DNA fingerprints are routinely used to identify criminals or eliminate innocent persons as suspects in legal proceedings. For example, a DNA fingerprint prepared from a hair found at the scene of a crime, or from a semen sample, might be compared with the DNA fingerprint of a suspect to link the suspect with the crime. Or, a DNA fingerprint of blood found on a suspect's clothing or possessions might be compared with the DNA fingerprint of a victim. Typically, the evidence is presented in terms of the probability that the particular DNA sample could have come from a random individual. Hence the media report probability values, such as one in several million, or in several billion, that a person other than the accused could have left his or her DNA at the crime scene.

Although courts initially met with legal challenges to the admissibility of DNA fingerprints, experience has shown that they are highly dependable as a line of evidence if DNA samples are collected and prepared with care, and if a sufficient number of polymorphic loci are examined. There is always concern, though, about the possibility of contamination of the sample with DNA from another source during the path from the crime scene to the forensic lab for analysis. Moreover, in some cases criminals themselves have planted fake DNA samples at crime scenes to confuse the investigation.

There are many examples of the use of DNA fingerprinting to identify a criminal or rule out a suspect. For example, in a case in England, the DNA fingerprints of more than 5,000 men were made using blood and saliva samples during an investigation of the rape and murder of two teenage girls in 1983 and 1986. The results led to the release of a man wrongly imprisoned for the crimes—the first person exonerated by DNA fingerprinting evidence. But, no DNA fingerprints matched the DNA at the crime scene, posing a great dilemma for the police. Then a woman overheard a colleague at work saying that he had given his samples in the name of his friend, Colin Pitchfork. Pitchfork was arrested, and his DNA fingerprint subsequently was shown to match the DNA at the crime scene. In 1988 he was convicted of the murders, the first conviction on the basis of DNA evidence. And the application of DNA fingerprinting techniques to stored forensic samples has also led to the release of a number of persons wrongly convicted for rape or murder.

DNA FINGERPRINTING IN TESTING PATERNITY AND ESTABLISHING ANCESTRY DNA fingerprints are also widely used as evidence of paternity because parents and their children share common alleles in their DNA fingerprints. That is, each child receives one allele of each locus from one parent and the other allele from the other parent. A comparison of DNA fingerprints for a number of loci can prove almost infallibly whether a child has been fathered or mothered by a given person. DNA fingerprints have also been used for other investigations, such as confirming that remains discovered in a remote region of Russia were actually those of Czar Nicholas II and members of his family, murdered in 1918 during the Russian revolution.

DNA fingerprinting is also widely used in studies of other organisms, including other animals, plants, and bacteria. Examples include testing for pathogenic *E. coli* in food sources such as hamburger meat, investigating cases of wildlife poaching, detecting genetically modified organisms among living organisms or in food, and comparing the DNA of ancient organisms with present-day descendants.

Genetic Engineering Uses DNA Technologies to Alter the Genes of a Cell or Organism

We have seen the many ways scientists use DNA technologies to ask, and answer, questions that were once completely inaccessible. Genetic engineering goes beyond gathering information; it is the use of DNA technologies to modify genes of a cell or organism. The goals of genetic engineering include using prokaryotes, fungi, animals, and plants as factories for the production of proteins needed in medicine and scientific research; correcting hereditary disorders; and improving animals and crop plants of agricultural importance. In many of these areas genetic engineering has already been highly successful. The successes and potential benefits of genetic engineering, however, are tempered by ethical and social concerns about its use, along with the fear that the methods may produce toxic or damaging foods, or release dangerous and uncontrollable organisms to the environment.

Genetic engineering uses DNA technologies of the kind discussed already in this chapter. DNA—perhaps a modified gene—is introduced into target cells of an organism where that gene is then expressed. Organisms that have undergone a gene transfer are called **transgenic,** meaning that they have been modified to contain genetic information—the *transgene*—from an external source.

The following sections discuss applications of the genetic engineering of bacteria, animals, and plants, and assess the major controversies arising from these projects.

GENETIC ENGINEERING OF BACTERIA TO PRODUCE PROTEINS Transgenic bacteria have been made, for example, to synthesize proteins for medical applications, break down toxic wastes such as oil spills, produce industrial chemicals such as alcohols, and process minerals. *E. coli* is the organism of choice for many of these applications of DNA technologies.

Using *E. coli* to make a protein from a foreign source is conceptually straightforward **(Figure 18.10).** First, the gene for the protein is cloned from the appropriate organism. Then the gene is inserted into an **expression vector** which, in addition to the usual features of a cloning vector, contains the regulatory sequences that allow transcription and translation of the gene. For a bacterial expression vector, this means having a promoter and a transcription terminator that are recognized by the *E. coli* transcriptional machinery, and having the ribosome binding site

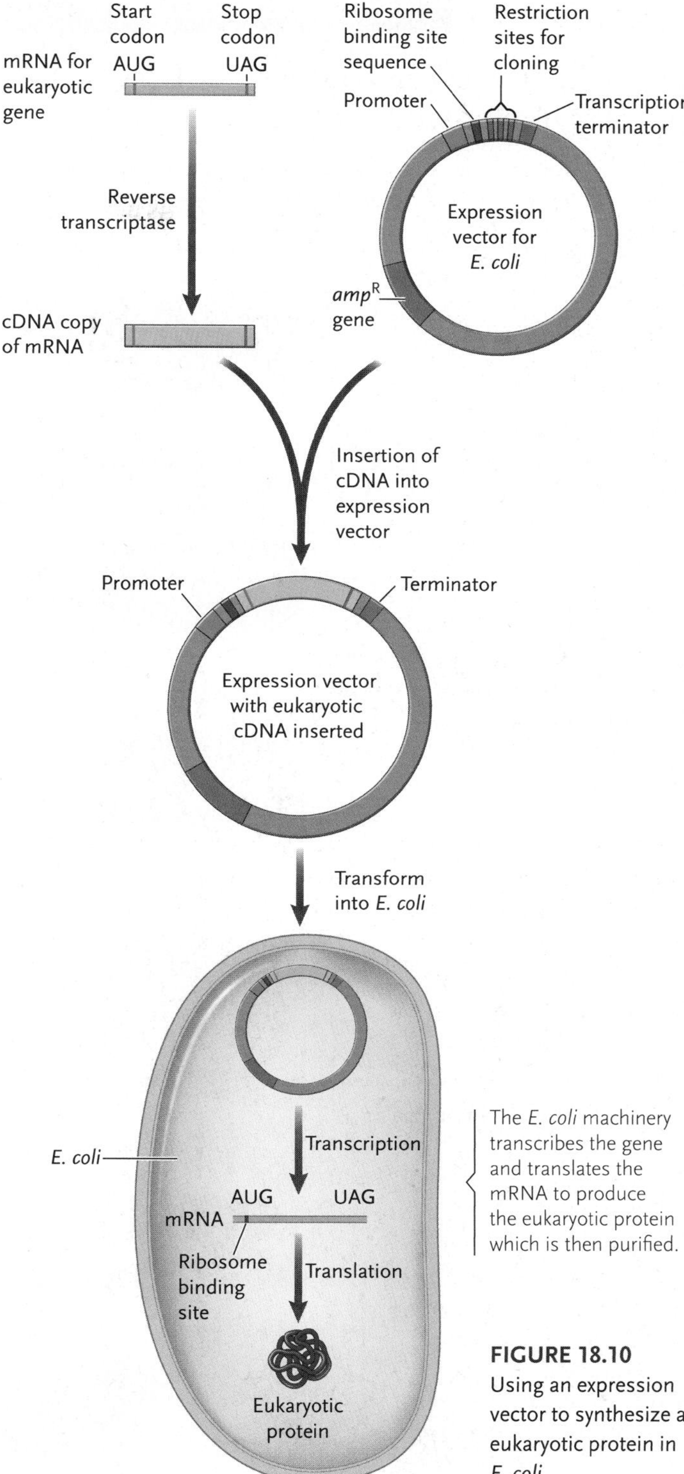
Start codon AUG
Stop codon UAG
mRNA for eukaryotic gene
Reverse transcriptase
cDNA copy of mRNA
Ribosome binding site sequence
Restriction sites for cloning
Promoter
Transcription terminator
Expression vector for E. coli
amp^R gene
Insertion of cDNA into expression vector
Promoter
Terminator
Expression vector with eukaryotic cDNA inserted
Transform into E. coli
E. coli
Transcription
AUG
UAG
mRNA
Ribosome binding site
Translation
Eukaryotic protein
The E. coli machinery transcribes the gene and translates the mRNA to produce the eukaryotic protein which is then purified.

FIGURE 18.10 Using an expression vector to synthesize a eukaryotic protein in *E. coli.*

needed for the bacterial ribosome to recognize the start codon of the transgene (see Section 15.4). The regulatory sequences flank the cluster of restriction sites of the expression vector that are used for cloning so that the inserted gene is correctly placed for transcription and translation when the recombinant plasmid is transformed into *E. coli.*

As we mentioned earlier while discussing DNA libraries, a cDNA copy of a eukaryotic gene is used when we want to use bacteria to express the protein encoded by the gene (see Figure 18.10). This is because the gene itself typically contains introns, which

FIGURE 18.11 **Research Method**

Introduction of Genes into Mouse Embryos Using Embryonic Germ-Line Cells

Purpose: To make a transgenic animal that can transmit the transgene to offspring. The embryonic germ-line cells that receive the transgene develop into the reproductive cells of the animal.

Protocol:

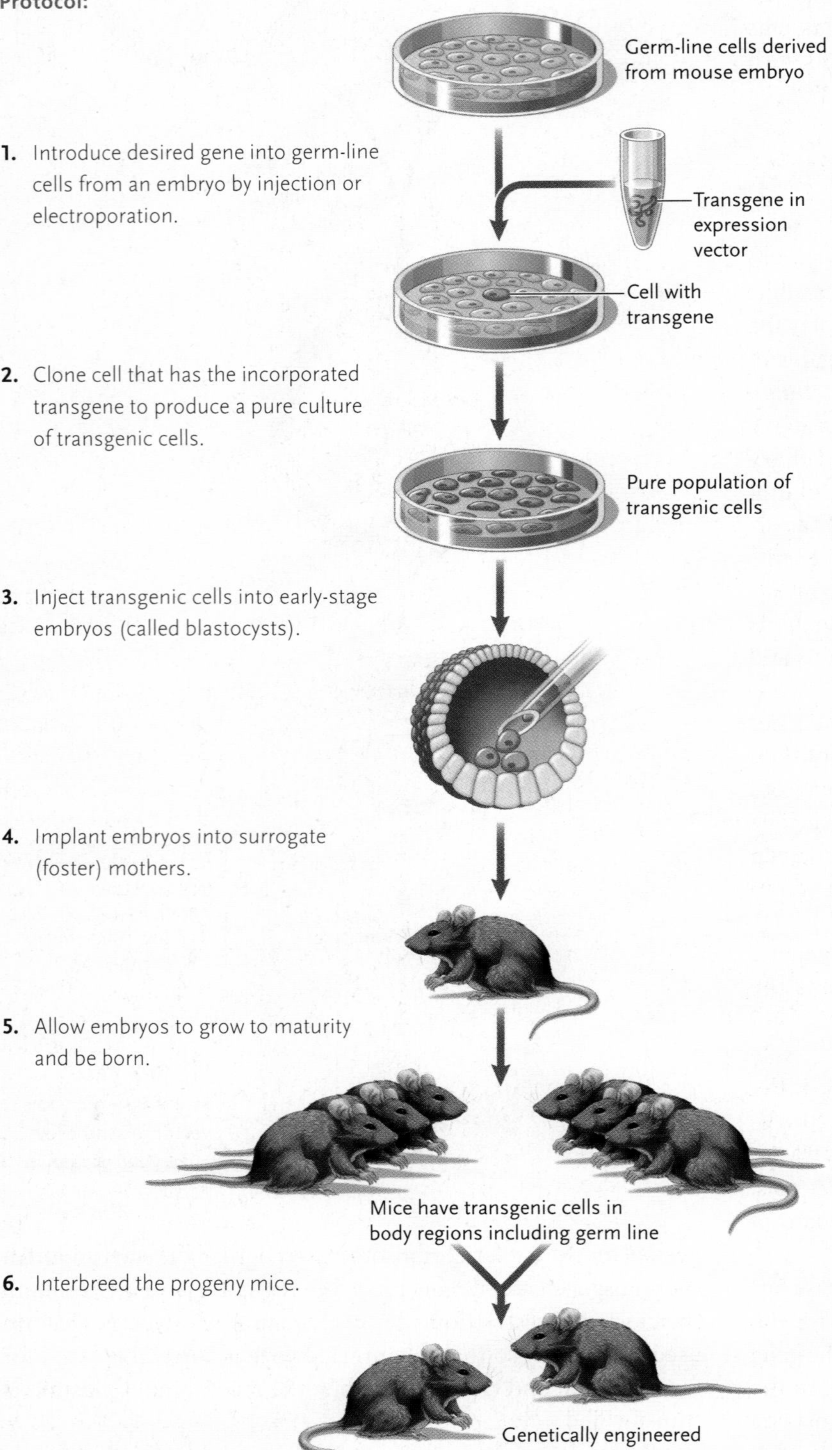

Germ-line cells derived from mouse embryo
1. Introduce desired gene into germ-line cells from an embryo by injection or electroporation.
Transgene in expression vector
Cell with transgene
2. Clone cell that has the incorporated transgene to produce a pure culture of transgenic cells.
Pure population of transgenic cells
3. Inject transgenic cells into early-stage embryos (called blastocysts).
4. Implant embryos into surrogate (foster) mothers.
5. Allow embryos to grow to maturity and be born.
Mice have transgenic cells in body regions including germ line
6. Interbreed the progeny mice.
Genetically engineered offspring—all cells transgenic

Interpreting the Results: The result of the breeding is some offspring in which all cells are transgenic—a genetically engineered animal has been produced.

bacteria cannot remove when they transcribe a eukaryotic gene. However, the eukaryotic mRNA that is copied by reverse transcriptase to synthesize cDNA has had its introns removed; thus, when that cDNA is expressed in bacteria it can be transcribed and translated to make the encoded eukaryotic protein. The protein is either extracted from the bacterial cells and purified or, if the protein is secreted, it is purified from the culture medium.

Expression vectors are available for a number of organisms. They vary in the regulatory sequences they contain and the selectable marker they carry so that the host organism transformed with the vector carrying a gene of interest can be detected and the host can express that gene.

For example, *E. coli* bacteria have been genetically engineered to make the human hormone insulin; the commercial product is called Humulin. Insulin is required by persons with some forms of diabetes. Humulin is a perfect copy of the human insulin hormone. Many other proteins, including human growth hormone to treat human growth disorders, tissue plasminogen activator to dissolve blood clots that cause heart attacks, and a vaccine against foot-and-mouth disease of cattle and other cloven-hoofed animals (a highly contagious, and sometimes fatal viral disease), have been developed for commercial production in bacteria using similar methods.

A concern is that genetically engineered bacteria may be released accidentally into the environment where possible adverse effects of the organisms are currently unknown. Scientists minimize the danger of accidental release by growing the bacteria in laboratories that follow appropriate biosafety protocols. In addition, the bacterial strains typically used are genetic mutants that cannot survive outside the growth media used in the laboratory.

GENETIC ENGINEERING OF ANIMALS Many animals, including fruit flies, fish, mice, pigs, sheep, goats, and cows, have been altered successfully by genetic engineering. Typically the organism has been altered to express a transgene, which means using an expression vector with the appropriate molecular features for gene expression in that organism. There are many purposes for these alterations, including basic research, correcting genetic disorders in humans and other mammals, and producing pharmaceutically important proteins.

Genetic Engineering Methods for Animals. Several methods are used to introduce a gene of interest into animal cells. The gene may be introduced into **germ-line cells,** which develop into sperm or eggs and thus enable the introduced gene to be passed from generation to generation. Or, the gene may be introduced into **somatic** (body) **cells,** differentiated cells that are not part of lines producing sperm or eggs, in which case the gene is not transmitted from generation to generation.

Germ-line cells of embryos are often used as targets for introducing genes, particularly in mammals other than humans **(Figure 18.11).** The treated cells are then cultured in quantity and reintroduced into early embryos. If the technique is successful, some of the introduced cells become founders of cell lines that develop into eggs or sperm with the desired genetic information integrated into their DNA. Individuals produced by crosses using the engineered eggs and sperm then contain the introduced sequences in all of their cells. Several genes have been introduced into the germ lines of mice by this approach, resulting in permanent, heritable changes in the engineered individuals.

A related technique involves introducing desired genes into **stem cells,** which are cells capable of undergoing many divisions in an unspecialized, undifferentiated state, but which also have the ability to differentiate into specialized cell types. In mammals, *embryonic stem cells* are found in a mass of cells inside an early-stage embryo (the blastocyst) (see Figure 18.11, step 3) and can differentiate into all of the tissue types of the embryo, whereas *adult stem cells* function to replace specialized cells in various tissues and organs. In mice and other non-human mammals, transgenes are introduced into embryonic stem cells, which are then injected into early-stage embryos as in Figure 18.11 (step 3). The stem cells then differentiate into a variety of tissues along with cells of the embryo itself, including sperm and egg cells. Males and females are then bred, leading to offspring that are either homozygotes, containing two copies of the introduced gene, or heterozygotes, containing one introduced gene and one gene that was native to the embryo receiving the engineered stem cells.

Introduction of genes into stem cells has been performed mostly in mice. One of the highly useful results is the production of a "knockout mouse," a homozygous recessive that receives two copies of a gene altered to a nonfunctional state, and thus has no functional copies. The effect of the missing gene on the knockout mouse is a clue to the normal function of the gene. In some cases, knockout mice are used to model human genetic diseases.

For introducing genes into somatic cells, typically somatic cells are removed from the body, cultured, and then transformed with an expression vector containing the transgene. The modified cells are then reintroduced into the body where the transgene functions. Because germ cells and their products are not involved, the transgene remains in the engineered individual and is not passed to offspring.

Gene Therapy: Correcting Genetic Disorders. The path to **gene therapy**—correcting genetic disorders—in humans began with experiments using mice. In 1982, Richard Palmiter at the University of Washington, Ralph Brinster of the University of Pennsylvania, and their colleagues injected a growth hormone gene from rats into fertilized mouse eggs and implanted the eggs into a surrogate mother. She gave birth to some normal-sized mouse pups that grew faster than normal and became about twice the size of their normal litter mates. These *giant mice* **(Figure 18.12)** attracted media attention from around the world.

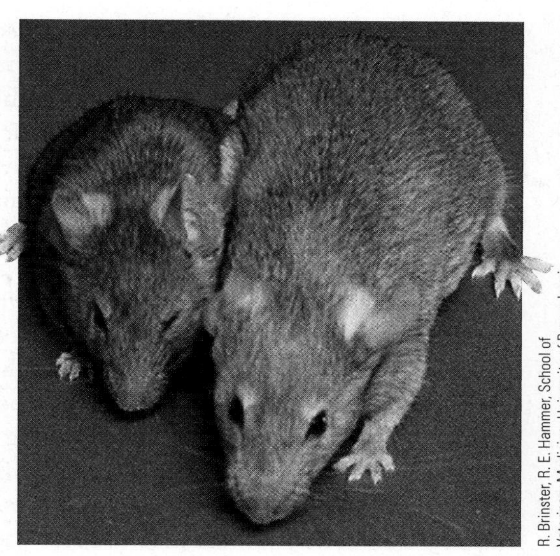

FIGURE 18.12 A genetically engineered giant mouse (right) produced by the introduction of a rat growth hormone gene into the animal. A mouse of normal size is on the left.

R. Brinster, R. E. Hammer, School of Veterinary Medicine, University of Pennsylvania

Palmiter and Brinster next attempted to cure a genetic disorder by gene therapy. In this experiment, they introduced a normal copy of the growth hormone gene into fertilized eggs taken from mutant dwarf mice with a genetic growth hormone deficiency and implanted them into a surrogate mother. The transgenic mouse pups grew to slightly larger than normal, demonstrating that the genetic defect in those mice had been corrected.

This sort of experiment, in which a gene is introduced into germ-line cells of an animal to correct a genetic disorder, is **germ-line gene therapy.** For ethical reasons, germ-line gene therapy is not permitted with humans. Instead, humans are treated with **somatic gene therapy,** in which genes are introduced into somatic cells (as described in the previous section).

The first successful use of somatic gene therapy with a human subject who had a genetic disorder was carried out in the 1990s by W. French Anderson and his colleagues at the National Institutes of Health (NIH). The subject was a young girl with *adenosine deaminase deficiency (ADA).* Without the adenosine deaminase enzyme, white blood cells cannot mature (see Chapter 43). In the absence of normally functioning white blood cells, the body's immune response is so deficient that most children with ADA die of infections before reaching puberty. The researchers successfully introduced a functional ADA gene into mature white blood cells isolated from the patient. Those cells were reintroduced into the girl, and expression of the ADA gene provided a temporary cure for her ADA deficiency. The cure was not permanent because mature white blood cells, produced by differentiation of stem cells in the bone marrow, are nondividing cells with a finite life time. Therefore, the somatic gene therapy procedure has to be repeated every few months. The subject of this example still receives periodic gene therapy to maintain the necessary levels of the ADA enzyme in her blood. In addition, she receives direct doses of the normal enzyme.

Successful somatic gene therapy has also been achieved for sickle-cell disease. In December 1998, a 13-year-old boy's bone marrow cells were replaced with stem cells from the umbilical cord of an unrelated infant. The hope was that the stem cells would produce healthy bone marrow cells, the source of blood cells. The procedure worked, and the patient has been declared cured of the disease. Stem cell therapy is also the basis of the bone marrow transplant, which is used to treat leukemia and some other types of cancer, as well as certain blood diseases.

Despite enormous efforts, human somatic gene therapy has not been the panacea people expected. Relatively little progress has been made since the first gene therapy clinical trial for ADA deficiency, and, in fact, there have been major setbacks. In 1999, for example, a teenage patient in a somatic gene therapy trial died as a result of a severe immune response to the viral vector being used to introduce a normal gene to correct his genetic deficiency. Furthermore, some children in gene therapy trials using retrovirus vectors to introduce genes into blood stem cells have developed a leukemia-like condition. Research and clinical trials continue as scientists try to circumvent the difficulties.

Turning Domestic Animals into Protein Factories. Another successful application of genetic engineering turns animals into pharmaceutical factories for the production of proteins required to treat human diseases or other medical conditions. Most of these *pharming* projects, as they are called, engineer the animals to produce the desired proteins in milk, making the isolation and purification of the proteins easy, as well as harmless to the animals.

One of the first successful applications of this approach was carried out with sheep engineered to produce a protein required for normal blood clotting in humans. The protein, called a *clotting factor,* is deficient in persons with one form of hemophilia. These people require frequent injections of the factor to avoid bleeding to death from even minor injuries. Using DNA cloning techniques, researchers joined the gene encoding the normal form of the clotting factor to the promoter sequences of the β-lactoglobulin gene, which encodes a protein secreted in milk, and introduced it into fertilized eggs. Those cells were implanted into a surrogate mother, and the transgenic sheep born were allowed to mature. The β-lactoglobulin promoter controlling the clotting factor gene became activated in mammary gland cells of females, resulting in the production of clotting factor. The clotting factor was then secreted into the milk.

Other similar projects are under development to produce particular proteins in transgenic mammals. These include a protein to treat cystic fibrosis, collagen to correct scars and wrinkles, human milk proteins to be added to infant formulas, and normal hemoglobin for use as an additive to blood transfusions.

Producing Animal Clones. Making transgenic mammals is expensive and inefficient. And, because only one copy of the transgene typically becomes incorporated into the treated cell, not all progeny of a transgenic animal inherit that gene. Scientists reasoned that an alternative to breeding a valuable transgenic mammal to produce progeny with the transgene would be to clone the mammal. Each clone would be identical to the original, including the expression of the transgene. That this is possible was shown when two Scottish scientists, Ian Wilmut of the Roslin Institute and Keith Campbell of PPL Therapeutics, Roslin, Scotland, announced in 1997 that they had successfully cloned a sheep using a single somatic cell taken from an adult sheep **(Figure 18.13)**—Dolly, the first cloned mammal.

After the successful cloning experiment that produced Dolly, many other mammals have been cloned, including mice, goats, pigs, monkeys, rabbits, dogs, a male calf appropriately named Gene, and a domestic cat called *CC* (for *Copy Cat*).

Cloning farm animals has been so successful that several commercial enterprises now provide cloned copies of champion animals. One example is a clone of an American Holstein cow, Zita, who was the U.S. national champion milk producer for many years. Animal breeders estimate that there are now more than 100 cloned animals on American farms, and breeders plan to produce entire herds if government approval is granted.

The cloning of domestic animals has its drawbacks. Many cloning attempts fail, leading to the death of the transplanted embryos. Cloned animals often suffer from conditions such as birth defects and poor lung development. Genes may be lost during the cloning process or may be expressed abnormally in the cloned animal. For example, molecular studies have shown that the expression of perhaps hundreds of genes in the genomes of clones is regulated abnormally.

GENETIC ENGINEERING OF PLANTS Genetic engineering of plants has led to increased resistance to pests and disease; greater tolerance to heat, drought, and salinity; larger crop yields; faster growth; and resistance to herbicides. Another aim is to produce seeds with higher levels of amino acids. The essential amino acid lysine, for example, is present only in limited quantities in cereal grains such as wheat, rice, oats, barley, and corn; the seeds of legumes such as beans, peas, lentils, soybeans, and peanuts are deficient in the essential amino acids methionine or cysteine. Increasing the amounts of the deficient amino acids in plant seeds by genetic engineering would greatly improve the diet of domestic animals and human populations that rely on seeds as a primary food source. Efforts are also under way to increase the vitamin and mineral content of crop plants.

Other possibilities for plant genetic engineering include plant pharming to produce pharmaceutical products. Plants are ideal for this purpose, because they are primary producers at the bottom rung of the food chain and can be grown in huge numbers with maximum conservation of the sun's energy that is captured in photosynthesis.

Some plants, such as *Arabidopsis thaliana* (thale cress), tobacco, potato, cabbage, and carrot, have special advantages for genetic engineering because individual cells can be removed from an adult, altered by the introduction of a desired gene, and then grown in cultures into a multicellular mass of cloned cells called a *callus*. Subsequently, roots, stems, and leaves develop in the callus, forming a young plant that can then be cultivated by the usual methods. Each cell in the plant contains the introduced gene. The gametes produced by the transgenic plants can then be used in crosses to produce offspring, some of which will have the transgene, as in the similar experiments with animals.

Methods Used to Insert Genes into Plants. Genes are inserted into plant cells by several techniques. A commonly used method takes advantage of a natural process that causes crown gall disease, which is characterized by bulbous, irregular

FIGURE 18.13 **Experimental Research**

The First Cloning of a Mammal

Question: Does the nucleus of an adult mammal contain all the genetic information to specify a new organism? In other words, can mammals be cloned starting with adult cells?

Experiment: Ian Wilmut, Keith Campbell, and their colleagues fused a mammary gland cell from an adult sheep with an unfertilized egg cell from which the nucleus had been removed and tested whether that fused cell could produce a lamb.

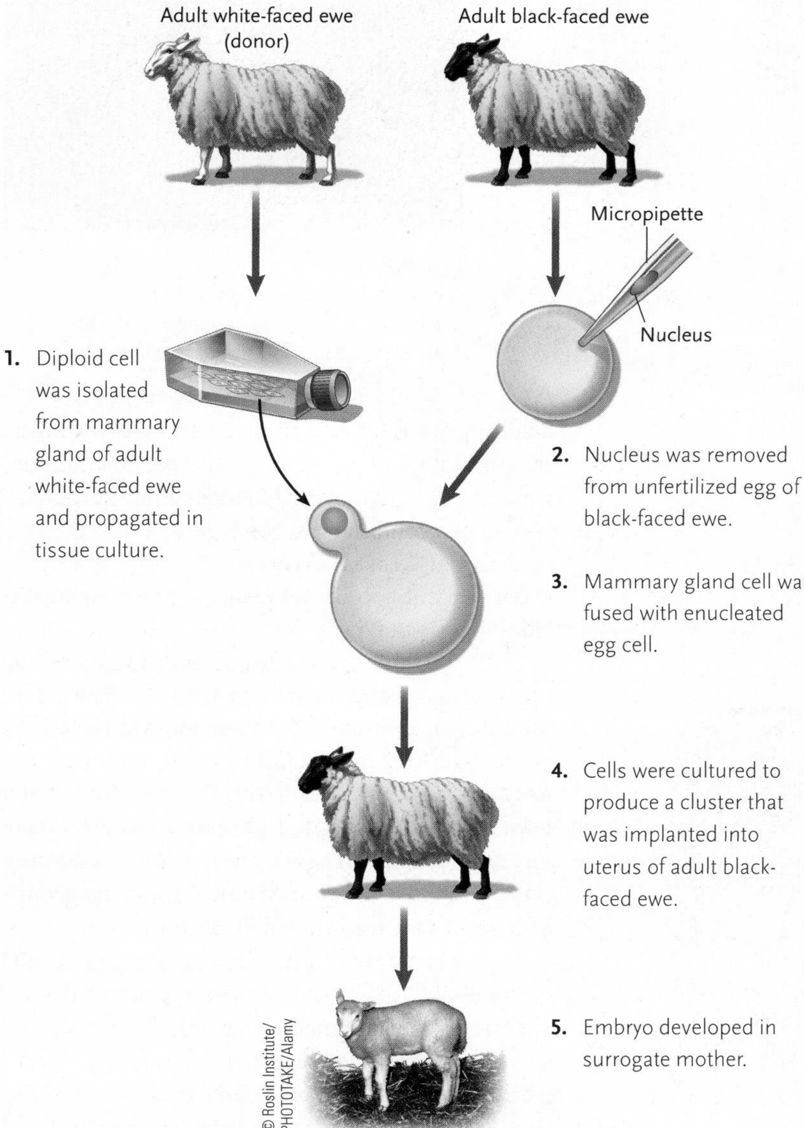

Adult white-faced ewe (donor)
Adult black-faced ewe
Micropipette
Nucleus
1. Diploid cell was isolated from mammary gland of adult white-faced ewe and propagated in tissue culture.
2. Nucleus was removed from unfertilized egg of black-faced ewe.
3. Mammary gland cell was fused with enucleated egg cell.
4. Cells were cultured to produce a cluster that was implanted into uterus of adult black-faced ewe.
5. Embryo developed in surrogate mother.
© Roslin Institute/ PHOTOTAKE/Alamy

Result: Dolly was born and grew normally. She was white-faced—a clone of the donor ewe. DNA fingerprinting using STR loci showed her DNA matched that of the donor ewe and not either the ewe who donated the egg, or the ewe who was the surrogate mother.

Conclusion: An adult nucleus of a mammal contains all the genetic material necessary to direct the development of a normal new organism, a clone of the original. Dolly was the first cloned mammal. The success rate for Wilmut and Campbell's experiment was very low—Dolly represented less than 0.4% of the fused cells they made—but its significance was huge.

Source: I. Wilmut et al. 1997. Viable offspring derived from fetal and adult mammalian cells. *Nature* 385:810–813.

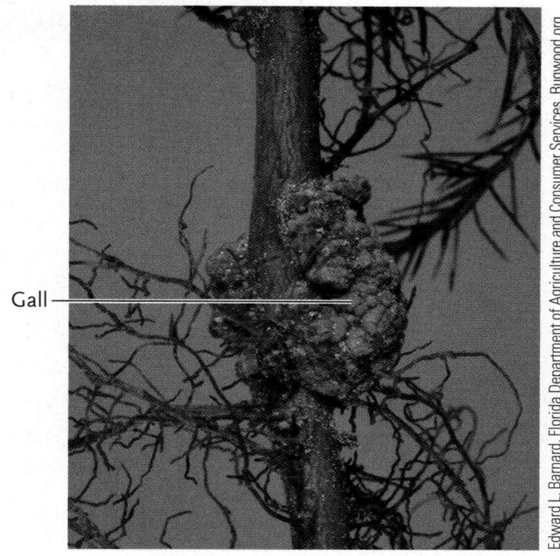

Gall
Edward L. Barnard, Florida Department of Agriculture and Consumer Services, Bugwood.org

FIGURE 18.14
A crown gall tumor on the trunk of a California pepper tree. The tumor, stimulated by genes introduced from the bacterium *Agrobacterium tumefaciens,* is the bulbous, irregular growth extending from the trunk.

growths—tumors, essentially—that can develop at wound sites on the trunks and limbs of deciduous trees **(Figure 18.14).** Crown gall disease is caused by the bacterium *Agrobacterium tumefaciens.* This bacterium contains a large, circular plasmid called the **Ti (tumor-inducing) plasmid (Figure 18.15).** The interaction between the bacterium and a plant cell it infects stimulates the excision of a segment of the Ti plasmid called *T DNA* (for transforming DNA), which then integrates into the plant cell's genome. Genes on the T DNA are then expressed; the products stimulate the transformed cell to grow and divide, producing a tumor. The tumors provide essential nutrients for the bacterium. The Ti plasmid is used as a vector for making transgenic plants, in much the same way as bacterial plasmids are used as vectors to introduce genes into bacteria.

Successful Plant Genetic Engineering Projects. An early visual demonstration of the successful use of genetic engineering techniques to produce a transgenic plant is the glowing tobacco plant **(Figure 18.16).** The transgenic plant contained the gene for luciferase, the enzyme that makes fireflies glow, in a plant expression vector. When the plant was soaked in the substrate for the enzyme, it became luminescent.

The most widespread application of genetic engineering of plants involves the production of transgenic crops. Thousands of such crops have been developed and field tested, and many have

FIGURE 18.15 **Research Method**

Using the Ti Plasmid of *Agrobacterium tumefaciens* to Produce Transgenic Plants

Purpose: To make transgenic plants. This technique is one way to introduce a transgene into a plant for genetic engineering purposes.

Protocol:

1. Isolate the Ti plasmid from *Agrobacterium tumefaciens*. The plasmid contains a segment called T DNA (T 5 transforming), which induces tumors in plants.

T DNA
Agrobacterium tumefaciens
Bacterial chromosome

2. Digest the Ti plasmid with a restriction enzyme that cuts within the T DNA. Mix with a gene of interest on a DNA fragment that was produced by digesting with the same enzyme. Use DNA ligase to join the two DNA molecules together to produce a recombinant plasmid.

Restriction site
T DNA
Ti plasmid
DNA fragment with gene of interest

3. Transform the recombinant Ti plasmid into a disarmed *A. tumefaciens* that cannot induce tumors, and use the transformed bacterium to infect cells in plant fragments in a test tube. In infected cells, the T DNA with the inserted gene of interest excises from the Ti plasmid and integrates into the plant cell genome.

Recombinant plasmid

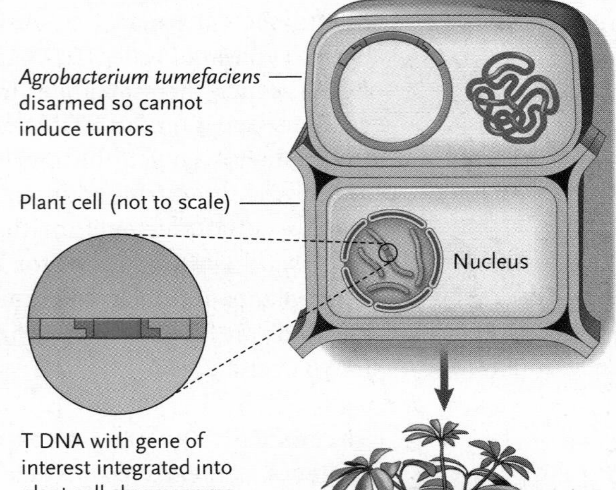
Agrobacterium tumefaciens disarmed so cannot induce tumors
Plant cell (not to scale)
Nucleus
T DNA with gene of interest integrated into plant cell chromosome

Regenerated transgenic plant

4. Culture the transgenic plant fragments to regenerate whole plants.

Interpreting the Results: The plant has been genetically engineered to contain a new gene. The transgenic plant will express a new trait based on that gene, perhaps resistance to an herbicide or production of an insect toxin, according to the goal of the experiment.

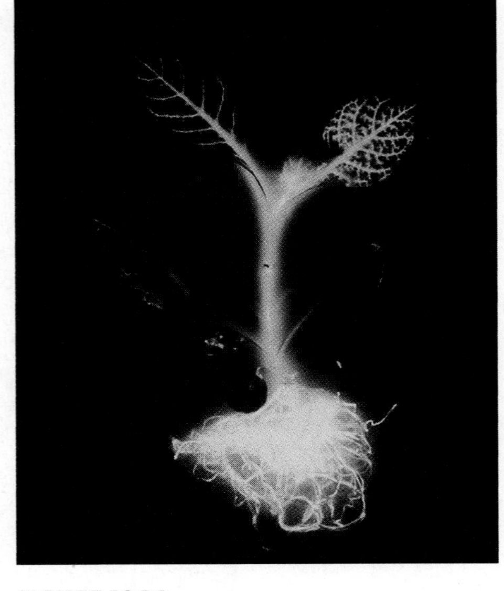

Science VU/Keith Wood/Visuals Unlimited, Inc.

FIGURE 18.16
A genetically engineered tobacco plant, made capable of luminescence by the introduction of a firefly gene coding for the enzyme luciferase.

been approved for commercial use. If you examine the processed, plant-based foods sold at a national supermarket chain, you will likely find that at least two-thirds contain transgenic plants.

In many cases, plants are modified to make them resistant to insect pests, viruses, or herbicides. Crops that have been modified for insect resistance include corn, cotton, and potatoes. The most common approach to making plants resistant to insects is to introduce the gene from the bacterium *Bacillus thuringiensis* that encodes the *Bt* toxin, a natural pesticide. This toxin has been used in powder form to kill insects in agriculture for many years, and now transgenic plants making their own *Bt* toxin are resistant to specific groups of insects that feed on them. Millions of acres of crop plants planted in the United States, amounting to about 70% of the nation's agricultural acreage, are now *Bt*-engineered varieties.

Virus infections cause enormous crop losses worldwide. Transgenic crops that are virus-resistant would be highly valuable to the agricultural community. There is some promise in this area. By

Regular rice
Genetically engineered golden rice containing β-carotene

Dr. Jorge Mayer, Golden Rice Project

FIGURE 18.17
Rice genetically engineered to contain β-carotene.

some unknown process, transgenic plants expressing certain viral proteins become resistant to infections by whole viruses that contain those same proteins. Two virus-resistant genetically modified crops made so far are papaya and squash.

Several crops have also been engineered to become resistant to herbicides. For example, *glyphosate* (commonly known by its brand name, Roundup) is a potent herbicide that is widely used in weed control. The herbicide works by inhibiting a particular enzyme in the chloroplast. Unfortunately, it also kills crops. But transgenic crops have been made in which a bacterial form of the chloroplast enzyme was added to the plants. The bacteria-derived enzyme is not affected by Roundup, and farmers who use these herbicide-resistant crops can spray fields of crops to kill weeds without killing the crops. Now most of the corn, soybean, and cotton plants grown in the United States and many other countries are genetically engineered, glyphosate-resistant ("Roundup-ready") varieties.

Crop plants are also being engineered to alter their nutritional qualities. For example, a strain of rice plants has been produced with seeds rich in β-carotene, a precursor of vitamin A **(Figure 18.17).** The new rice, which is given a yellow or golden color by its carotene content, may provide improved nutrition for the billions of people that depend on rice as a diet staple. In particular, the rice may help improve the nutrition of children younger than five years of age in southeast Asia, 70% of whom suffer from impaired vision because of vitamin A deficiency. *Insights from the Molecular Revolution* describes an experiment in which rice plants were genetically engineered to develop resistance to a damaging bacterial blight.

Plant pharming is also an active area in both university research labs and biotechnology companies. Plant pharming involves the engineering of transgenic plants to produce medically valuable products. The approach is one described earlier: the gene for the product is cloned into an expression vector, in this case one with a promoter active in plants, and the recombinant DNA molecule is introduced into plants. Products under development include vaccines for various bacterial and viral diseases, protease inhibitors to treat or prevent virus infections, collagen to treat scars and wrinkles, and the protein aprotinin, which is used to reduce bleeding and clotting during heart surgery.

In contrast to animal genetic engineering, genetically altered plants have been widely developed and appear to have become mainstays of agriculture. But, as the next section discusses, both animal and plant genetic engineering have not proceeded without concerns.

DNA Technologies and Genetic Engineering Are a Subject of Public Concern

When recombinant DNA technology was developed in the early 1970s, researchers quickly recognized that in addition to the anticipated benefits, there might be deleterious outcomes. One key concern at the time was that a bacterium carrying a recombinant DNA molecule might escape into the environment, transfer the recombinant molecule to other bacteria, and produce new, potentially harmful, strains. To address these concerns, the U.S. scientists who developed the technology drew up comprehensive safety guidelines for recombinant DNA research in the United States. Since that time countless thousands of experiments involving recombinant DNA molecules have been done in laboratories around the world. Those experiments have shown that recombinant DNA manipulations can be done safely. Over time, therefore, the recombinant DNA guidelines have become more relaxed. Nonetheless, stringent regulations still exist for certain areas of recombinant DNA research that pose significant risk, such as cloning genes from highly pathogenic bacteria or viruses, or gene therapy experiments.

Guidelines for genetic engineering also extend to research in several areas that have been the subject of public concern and debate. While the public is concerned little about genetically engineered microorganisms, for example those cleaning up oil spills and hazardous chemicals, it is concerned about possible problems with **genetically modified organisms (GMOs)** used as food. A GMO is a transgenic organism; the majority of GMOs are crop plants. Issues of concern include the safety of GMO-containing food, and the possible adverse effects of the GMOs on the environment, such as by interbreeding with natural species or by harming beneficial insect species. For example, could introduced genes providing herbicide or insect resistance move from crop plants into related weed species through cross pollination, producing "super weeds" that might be difficult or impossible to control? And *Bt*-expressing corn was originally thought to have adverse effects on monarch butterflies who fed on the pollen. The most recent of a series of independent studies investigating this possibility has indicated that the risk to the butterflies from *Bt* toxin is extremely low.

More broadly, different countries have reacted to GMOs in different ways. In the United States, transgenic crops are widely planted and harvested. Before commercialization, such GMOs are evaluated for potential risk by appropriate government regulatory agencies, including the NIH, Food and Drug Administration (FDA), Department of Agriculture, and Environmental Protection Agency (EPA). Usually, the opposition to GMOs has come from particular activist and consumer groups.

Political opposition to GMOs has been greater in Europe, dampening the use of transgenic crop plants in the fields and GMOs in food. In 1999, the European Union (EU) imposed a 6-year moratorium on all GMOs, leading to a bitter dispute with the United States, Canada, and Argentina, the leading growers of transgenic crops. More recently, the EU has decided that using genetic engineering in agriculture and food production is permissible provided the GMO or food containing it is safe for humans, animals, and the environment. All use of GMOs in the field or in food requires authorization following a careful review process.

On a global level, an international agreement, the **Cartagena Protocol on Biosafety,** "promotes biosafety by establishing practical rules and procedures for the safe transfer [between countries], handling and use of GMOs." Separate procedures have been set up for GMOs that are to be introduced into the environment and those that are to be used as food or feed or for processing. To date, 157 countries have ratified the Protocol. (As of May 2010, the United States is not one of them.)

INSIGHTS FROM THE **Molecular Revolution**

Rice Blight: Engineering rice for resistance to the disease.

Rice is common in the diet of the entire human population; for one-third of humanity, more than 2 billion people, it is the primary nutrient source. Worldwide, some 146 million hectares are planted with rice, producing 560 million tons of the grain annually.

This major human staple is threatened by a rice blight caused by the bacterium *Xanthomonas oryzae*. Rice plants infected by the bacterium turn yellow and wilt (see figure); in many Asian and African rice fields, as much as half of the crop is lost to the bacterial blight. Although some wild forms of rice, which are not usable as crop plants, have a natural resistance to the *Xanthomonas* blight, none of the cultivated varieties is resistant. Attempts to develop resistant strains by crossbreeding crop rice with wild varieties have produced only one resistant type that, unfortunately, is essentially unusable as a food source.

Research Goal

To develop a blight-resistant rice strain using a genetic engineering approach.

Experiments

Pamela Ronald and her colleagues at the University of California, Davis used genetic engineering techniques to produce a blight-resistant rice strain. They were aided by the results of genetic experiments in which a gene, *Xa21*, was identified that confers resistance to *Xanthomonas* in a wild rice.

Ronald and her coworkers cloned the *Xa21* gene following these steps:

1. Using traditional genetic mapping crosses, they found that *Xa21* is located near several known genes, including one that was so close that it rarely recombined with *Xa21* in the crosses.
2. The DNA of the known gene was used as a probe to find nearby sequences in a genomic library of the resistant wild rice genome. The probe paired with 16 DNA chromosome fragments in the library.
3. To determine whether any of the fragments included the *Xa21* gene, the researchers introduced each fragment individually into rice crop plants that were susceptible to *Xanthomonas* infection. Out of 1,500 plants that incorporated the fragments, 50 plants, all containing the same 9,600-bp fragment, proved to be resistant to the blight. These plants were the first rice crop plants to be engineered successfully to resist *Xanthomonas* infections.
4. The 9,600-base-pair fragment was then digested with restriction enzymes into subfragments that were cloned and tested individually for their ability to confer resistance. This technique allowed the researchers to identify and isolate the specific part of the fragment containing the resistance gene.

Conclusion

Sequencing the *Xa21* gene revealed that it encodes a typical plasma membrane receptor protein, which in some way triggers an internal cellular response on exposure to *Xanthomonas*, or its molecular components. The response alters cell structure or biochemistry to inhibit growth of the bacterium.

Ronald and her coworkers then introduced the *Xa21* gene successfully into three varieties of rice that are widely grown as crops in Asia and Africa. Those varieties became resistant to the bacterial blight, thereby showing that expression of the *Xa21* gene is sufficient to confer resistance.

Normal rice plants (left) and rice plants infected by the blight bacterium (right).

Source: W.-Y. Song, et al. 1995. A receptor kinase-like protein encoded by the rice disease resistance gene, *Xa21*. *Science* 270:1804–1806.

In sum, the use of DNA technologies in biotechnology has the potential for tremendous benefits to humankind. Such experimentation is not without risk, and so for each experiment, researchers must assess that risk and make a judgment about whether to proceed and, if so, how to do so safely. Furthermore, agreed-upon guidelines and protocols should ensure a level of biosafety that is acceptable to researchers, consumers, politicians, and governments.

We now turn to the analysis of whole genomes.

STUDY BREAK 18.2

1. **What are the principles of DNA fingerprinting?**
2. **What is a transgenic organism?**
3. **What is the difference between using germ-line cells and somatic cells for gene therapy?**

> **THINK OUTSIDE THE BOOK**
>
> Individually or collaboratively, investigate how it can be determined experimentally that processed foods that contain plant material contain transgenic plants. List the steps you would take to investigate whether a can of corn from a supermarket contains genetically modified corn.

18.3 Genome Analysis

The development of DNA technologies for analyzing genes and gene expression revolutionized experimental biology. DNA sequencing techniques (described in this section) have made it possible to analyze the sequences of cloned genes and genes amplified by PCR. Having the complete sequence of a gene aids researchers tremendously in unraveling how the gene functions. But a gene is only part of a genome. Researchers want to know about the organization of genes in a complete genome, and how genes work together in networks to control life. Of particular interest, of course, is the human genome. The complete sequencing of the approximately 3 billion base-pair human genome—the Human Genome Project (HGP)—began in 1990. The task was completed in 2003 by an international consortium of researchers and by a private company, Celera Genomics. As part of the official HGP, for purposes of comparison the genomes of several important model organisms commonly used in genetic studies were sequenced: *E. coli* (representing prokaryotes), the yeast *Saccharomyces cerevisiae* (representing single-celled eukaryotes), *Drosophila melanogaster* and *Caenorhabditis elegans* (the fruit fly and a nematode worm, respectively, representing multicellular animals of moderate genome complexity), and *Mus musculus* (the mouse, representing a mammal of genome complexity comparable to that of humans). In addition, the sequences of the genomes of many organisms not listed here, including plants, have been completed or are in progress at this time. What researchers are learning from analyzing complete genomes is of enormous importance to our understanding of biology and the evolution of organisms.

DNA Sequencing Techniques Are Based on DNA Replication

DNA sequencing is the key technology for genome sequencing projects. It is also used to determine the sequence of individual genes that have been cloned or amplified by PCR. DNA sequencing was first developed in the late 1970s by Allan M. Maxam, a graduate student, and his mentor, Walter Gilbert of Harvard University. Within a few years Frederick Sanger, of Cambridge University, designed the method that is most commonly used today. Gilbert and Sanger were awarded a Nobel Prize in 1980.

The Sanger method is based on the properties of nucleotides known as *dideoxyribonucleotides,* therefore, the method is also called *dideoxy sequencing* **(Figure 18.18).** Dideoxyribonucleotides have a single —H bound to the 3′ carbon of the deoxyribose sugar instead of the —OH normally at this position in deoxyribonucleotides. DNA polymerases, the replication enzymes, recognize the dideoxyribonucleotides and place them in the DNA just as they do normal deoxyribonucleotides. However, because a dideoxyribonucleotide has no 3′–OH group available for addition of the next base, replication of a nucleotide chain stops when one of these nucleotides is added to a growing nucleotide chain. (Remember from Section 14.3 that a 3′–OH group must be present at the growing end of a nucleotide chain for the next nucleotide to be added during DNA replication.)

In a dideoxy sequencing reaction, researchers use a mixture of dideoxyribonucleotides and normal nucleotides, so that chain termination will occur randomly at each position where a particular nucleotide appears in the population of DNA molecules being replicated. Each chain-termination event generates a newly synthesized DNA strand that ends with the dideoxyribonucleotide; hence, for this particular strand, the base at the 3′ end is known and, because of base-pairing rules, the base on the template strand being sequenced can be deduced. Once they know the base at the end of each terminated DNA strand, researchers can work out the complete sequence of the template DNA strand.

The dideoxy sequencing method can be used with any pure piece of DNA, such as a cloned DNA fragment, or a fragment amplified by PCR. An unambiguous sequence of about 500 to 750 nucleotides can be obtained from each sequencing experiment.

Genome Analysis Consists of Three Main Areas

The following are the three main areas of genome analysis:

1. Genome sequence determination and annotation, which means obtaining the sequences of complete genomes and analyzing the sequences to locate putative genes and other functionally important sequences within the genome.
2. **Functional genomics,** the study of the functions of genes and other parts of the genome. In relation to the genes, this includes developing an understanding of how their expression is regulated, the proteins they encode, and the role of those proteins in the organism's metabolic processes.
3. **Comparative genomics,** the comparison of entire genomes (or extensive portions of them) to understand evolutionary relationships and the basic biological similarities and differences among species.

Genome Sequence Determination and Annotation Involves Obtaining and Analyzing the Sequences of Complete Genomes

The first genome sequence reported, that of the bacterium *Haemophilus influenzae,* was determined using the whole-genome shotgun method **(Figure 18.19).** Developed by J. Craig Venter and his associates at Celera Genomics, in this method, the entire genome is broken into thousands to millions of random, overlapping fragments. Each fragment is cloned and sequenced. The genome sequence is then assembled by computer on the basis of the sequence overlaps between fragments. Originally thought

Dideoxy (Sanger) Method for Sequencing DNA

Purpose: Obtain the sequence of a piece of DNA, such as in gene sequencing or genome sequencing. The method is shown here with an automated sequencing system.

Protocol:

1. A dideoxy sequencing reaction has the following components: the fragment of DNA to be sequenced (denatured to single strands); a DNA primer that will bind to the 3′ end of the sequence to be determined; a mixture of the four deoxyribonucleotide precursors for DNA synthesis; and a mixture of the four dideoxyribonucleotide (dd) precursors, each labeled with a different fluorescent molecule, and DNA polymerase to catalyze the DNA synthesis reaction.

2. The new DNA strand is synthesized in the 5′→3′ direction starting at the 3′ end of the primer. New synthesis continues until a dideoxyribonucleotide is incorporated into the DNA instead of a normal deoxyribonucleotide. For a large population of template DNA strands, the dideoxy sequencing reaction produces a series of new strands, with lengths from one on up. At the 3′ end of each new strand is the labeled dideoxyribonucleotide that terminated the synthesis.

3. The labeled strands produced by the reaction are separated by gel electrophoresis. The principle of separation is the same as for agarose gel electrophoresis described in Figure 18.6. But here it is necessary to discriminate between DNA strands that differ in length by one nucleotide, which agarose gels cannot do. Therefore, a gel made of polyacrylamide is prepared in a capillary tube for separating the DNA fragments. As the bands of DNA fragments move near the bottom of the tube, a laser beam shining through the gel excites the fluorescent labels on each DNA fragment. The fluorescence is registered by a detector, with the wavelength of the fluorescence indicating which of the four dideoxyribonucleotides is at the end of the fragment in each case.

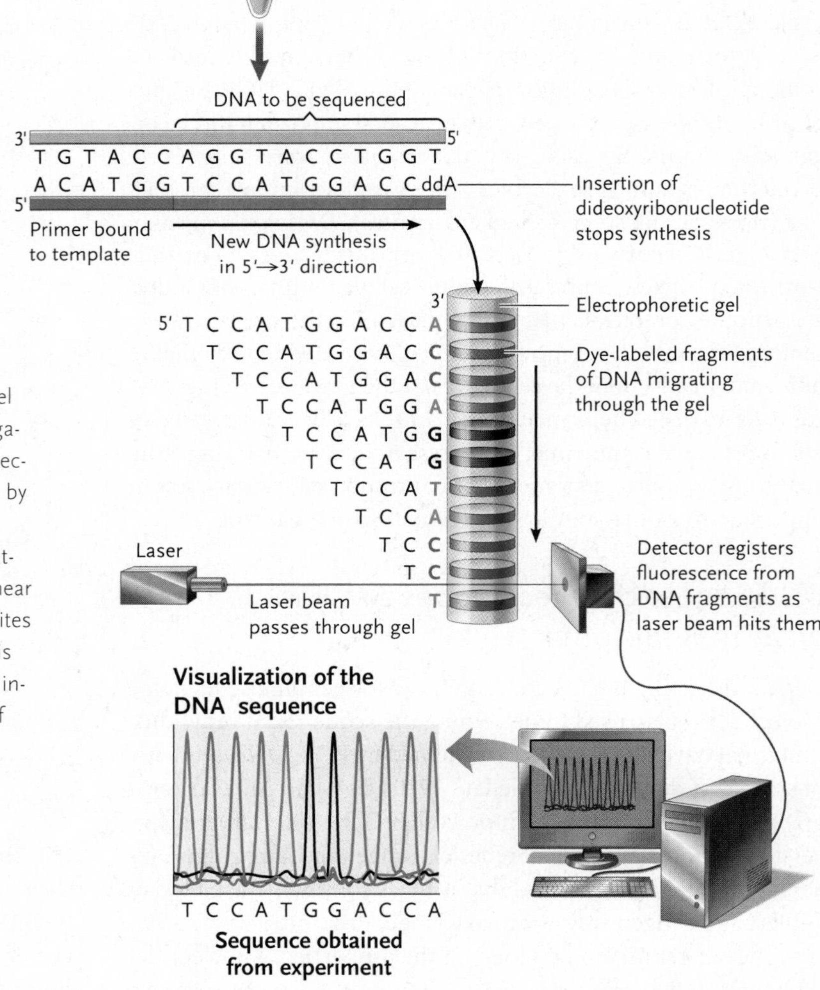

DNA strand to be sequenced 3′ 5′
T G T A C C A G G T A C C T G G T
A C A T G G
DNA primer 5′ 3′
DNA polymerase
P P P O A
dATP
OH H
P P P O A
ddATP
H H
Deoxyribonucleotide precursors
dATP
dTTP
dGTP
dCTP
ddATP
ddTTP
ddGTP
ddCTP
Dideoxyribonucleotide (dd) precursors (fluorescently labeled)
DNA to be sequenced
3′ 5′
T G T A C C A G G T A C C T G G T
A C A T G G T C C A T G G A C C ddA
5′
Insertion of dideoxyribonucleotide stops synthesis
Primer bound to template
New DNA synthesis in 5′→3′ direction
3′
5′ T C C A T G G A C C A
T C C A T G G A C C
T C C A T G G A C
T C C A T G G A
T C C A T G G
T C C A T G
T C C A T
T C C A
T C C
T C
T
Electrophoretic gel
Dye-labeled fragments of DNA migrating through the gel
Laser
Laser beam passes through gel
Detector registers fluorescence from DNA fragments as laser beam hits them
Visualization of the DNA sequence
T C C A T G G A C C A
Sequence obtained from experiment

Interpreting the Results: The data from the laser system are sent to a computer that interprets which of the four possible fluorescent labels is at the end of each DNA strand. The results show the colors of the labels as the DNA bands passed the detector. They may be seen on the computer screen or in printouts. The sequence of the newly synthesized DNA, which is complementary to the template strand, is read from left (5′) to right (3′). (The sequence shown here begins after the primer.)

FIGURE 18.19 Research Method

Whole-Genome Shotgun Sequencing

Purpose: Obtain the complete sequence of the genome of an organism.

Protocol:

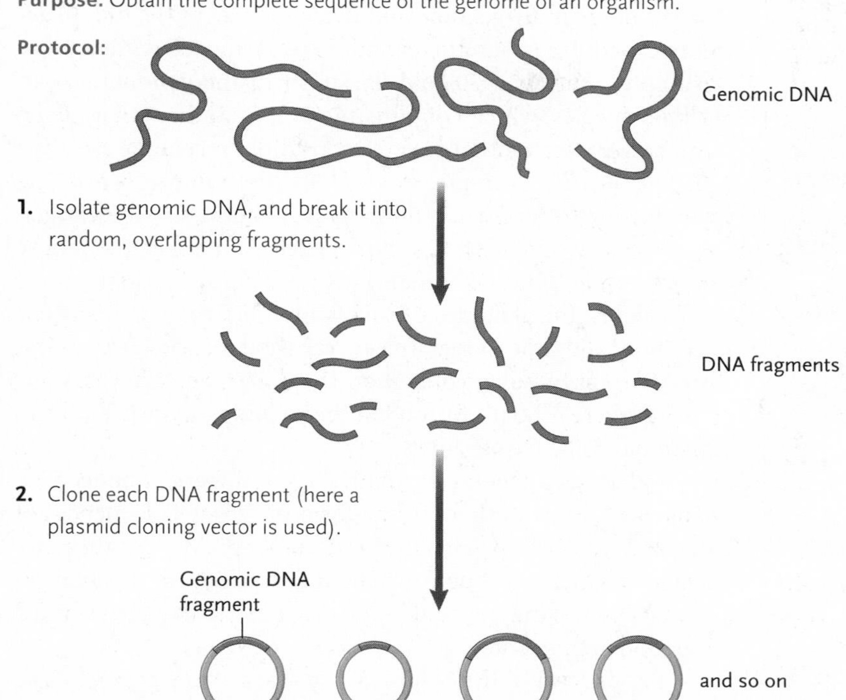

Genomic DNA
DNA fragments
Genomic DNA fragment
and so on
Plasmid cloning vector

1. Isolate genomic DNA, and break it into random, overlapping fragments.

2. Clone each DNA fragment (here a plasmid cloning vector is used).

3. Sequence the genomic DNA fragment in each clone.

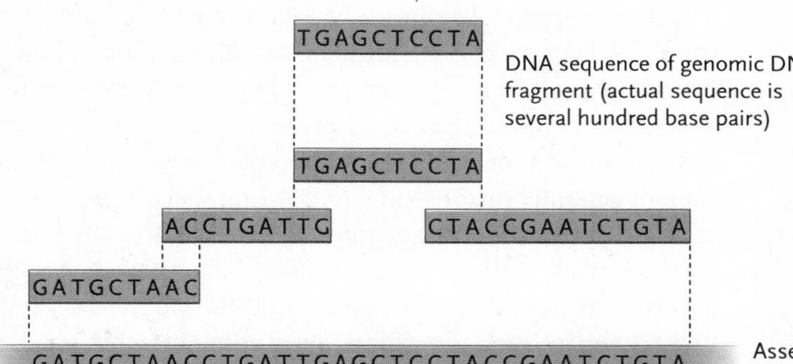

TGAGCTCCTA
DNA sequence of genomic DNA fragment (actual sequence is several hundred base pairs)
TGAGCTCCTA
ACCTGATTG
CTACCGAATCTGTA
GATGCTAAC
GATGCTAACCTGATTGAGCTCCTACCGAATCTGTA
Assembled sequence

4. Enter the DNA sequences of the fragments into a computer, and use the computer to assemble overlapping sequences into the continuous sequence of each chromosome of the organism. This technique is analogous to taking 10 copies of a book that has been torn randomly into smaller sets of a few pages each and, by matching overlapping pages of the leaflets, assembling a complete copy of the book with the pages in the correct order.

Interpreting the Results: The method generates the complete sequence of the genome of an organism.

to be inapplicable to the large genomes of eukaryotes, improvements in sequencing technologies and in the computer algorithms have now made it the method of choice for sequencing genomes of all organisms.

By 2009, the genomes of a large number of viruses and hundreds of organisms had been sequenced, and those of more species are continually being added to the list of fully sequenced genomes. Among those already sequenced are cytomegaloviruses, bacteria including *E. coli,* various archaean species, and eukaryotes including the budding yeast *Saccharomyces cerevisiae,* the protist *Plasmodium falciparum* (a malarial parasite), the roundworm *Caenorhabditis elegans,* the plants *Arabidopsis thaliana* and rice, the fruit fly *Drosophila melanogaster,* the chicken, the mouse, the rat, the dog, the chimpanzee, and the human.

Once the complete sequence of a genome has been determined, the next step is *annotation,* the identification of genes and other sequences of importance. Researchers in the rapidly growing field of **bioinformatics,** which fuses biology with mathematics and computer science, apply sophisticated computer algorithms in this endeavor. Bioinformatics also is used to predict the structure and function of gene products and to postulate evolutionary relationships of sequences, which are issues for functional and comparative genomicists, respectively.

Protein-coding genes are of particular interest in genome analysis. Using computer analysis, researchers identify possible protein-coding genes by searching for **open reading frames (ORFs),** that is, a start codon (ATG, at the DNA level) separated by a multiple of three nucleotides from one of the stop codons (TAG, TAA, or TGA at the DNA level). This process is easy for prokaryotic genomes, because the genes have no introns. In eukaryotic protein-coding genes, which typically have introns, more sophisticated algorithms are used to try to identify the junctions between exons and introns in scanning for open reading frames.

With many genomes sequenced, researchers can compare the genomes to learn about genome sizes and the number of protein-coding genes. **Table 18.1** gives examples of such data for Bacteria, Archaea, and Eukarya. We can make some general conclusions about the genomes presented. Members of the domain Bacteria have genomes that vary widely in size. *Carsonella ruddii,* a symbiotic bacterium living in the guts of certain insects, has the smallest known cellular genome of any organism (viral genomes are significantly smaller), and has 182 genes. That gene number is, at the moment, the minimum number of genes known to be required for life. The largest known bacterial genome is more than 80 times larger than that of *C. ruddii.* Archaean genomes also vary widely in size, the largest known being that of *Methanosarcina acetivorans,* which lives in oxygen-depleted environments such as oil wells and deep sea

TABLE 18.1 Genome Sizes and Estimated Number of Genes for Selected Members of Domains Bacteria, Archaea, and Eukarya

Domain and Organism	Genome Size (Mb)	Number of Genes
Bacteria		
Carsonella ruddii	0.16	182
Mycoplasma genitalium	0.58	523
Escherichia coli	4.6	4,200
Rhizobium radiobacter	5.7	5,482
Archaea		
Thermoplasma acidophilum	1.56	1,509
Methanosarcina acetivorans	5.75	4,662
Eukarya		
Protozoa		
Tetrahymena thermophila (a ciliated protist)	220	>20,000
Fungi		
Saccharomyces cerevisiae (a budding yeast)	12	~6,000
Neurospora crassa (orange bread mold)	40	~10,100
Plants		
Arabidopsis thaliana (thale cress)	125	25,900
Oryza sativa (rice)	430	~56,000
Invertebrates		
Caenorhabditis elegans (a nematode worm)	100	20,443
Drosophila melanogaster (fruit fly)	180	13,700
Vertebrates		
Takifugu rubripes (pufferfish)	393	>31,000
Mus musculus (mouse)	2,700	~22,000
Homo sapiens (human)	2,900	~20,000

vents. For members of both domain Bacteria and Archaea, genes are densely packed in the genomes, with little space between. Thus, larger genomes of organisms in these two domains tend to reflect increased gene number.

Members of the domain Eukarya vary markedly in form and complexity, and their genomes also show great differences in size. For example, yeast has a very small genome that is about 0.4% the size of the human genome, yet humans have only a little more than three times the number of genes compared with yeast. There are no rules relating organism complexity and genome size. For instance, the fruit fly and the locust have similar physiological complexity, but the locust genome is 50 times the size of the fruit fly genome and twice the size of the mouse genome. Within a genus, there is not necessarily a consistency. For example, there is a 50-fold variation in the genome size of *Allium* species, which contains the onions and relatives. Even among vertebrates, there is great variation in genome size. The pufferfish, for example, has a 393-Mb genome, while the genomes of the mouse and humans are about seven times larger. And yet, the pufferfish has more genes than either the mouse or the human. But the human genome is not the largest among eukaryotes; the genomes of some amphibians and some ferns are about 200 times larger. In general, though, genes are packed less densely in eukaryotes than they are in prokaryotes, although there is no uniformity in the packing, as the pufferfish–mammal comparison shows.

Eukaryotic genomes contain large numbers of noncoding sequences, most of them in the form of repeated sequences of nucleotides of various lengths and numbers. Most of these sequences, which make up from about 25% to 50% of the total genomic DNA in different eukaryotic species, have no determined function at this time.

Let us learn a little more about the human genome. The human genome sequence consists of almost 3 *billion* base pairs. Before genomics, researchers had predicted that human cells had as many as 100,000 different protein-coding genes. The best current estimate is about 20,000 such genes. However, although the number of protein-coding genes is unexpectedly small, the total number of different proteins produced in humans is much greater, and probably approaches the 100,000 figure originally proposed for genes. The additional proteins arise through such processes as alternative splicing during mRNA processing (see Section 15.3) and differences in protein processing.

All the protein-coding sequences occupy less than 2% of the human genome. Introns—the noncoding spacers in genes—occupy another 24% of the genome. The rest of the DNA, almost three-quarters of the genome, occupies the spaces between genes. Some of this intergenic DNA is functional and includes regulatory sequences such as promoters and enhancers, but much of it, more than 50% of the total genome, consists of repeated sequences that have no known function.

There are bioethics issues concerning the human genome. To address them, the U.S. Department of Energy and the NIH have funded studies of the ethical, legal, and social issues surrounding the availability of genetic information from human genome research. The following questions are some that are being looked at. Who should have access to personal genetic information, and how should it be used? To what extent should genetic information be private and confidential? How will genetic tests be evaluated and regulated? How can people be informed sufficiently about the genetic information from genomic analysis so that they can make informed personal medical choices? Does a set of genes predispose a person's behavior, and can the person control that behavior?

Functional Genomics Focuses on the Functions of Genes and Other Parts of the Genome

The complete genome sequence for an organism is basically a very long string of A, T, G, and C letters, which means little without further analysis. Discovering the functions of genes and other parts of the genome is one important goal of the analysis. Most of this functional genomics research is focused on the genes because they control the functions of cells and, therefore, of organisms. Functional genomics relies on laboratory experiments by molecular biologists as well as computer analysis by bioinformaticists.

ASSIGNING GENE FUNCTION BY SEQUENCE SIMILARITY Computer analysis of a genome sequence will reveal its putative genes. These days there are databases with an enormous amount of sequence information for both DNA and amino acids; the latter is typically inferred from gene sequences. For a newly sequenced genome, researchers use those databases to assign functions to the putative genes identified in the initial computer analysis. They look for sequence matches in *sequence similarity searches,* in which an input sequence is compared with all sequences in a database. Such searches can be done using an Internet browser to access the computer programs. For example, to use the BLAST (Basic Local Alignment Search Tool) program at the National Center for Biotechnology Information, a user pastes the putative gene DNA sequence, or the sequence of the protein it encodes, into a browser window and sets the program to work. The BLAST program searches the databases of known sequences and returns the best matches, indicating the degree to which the entered sequence is similar to the sequences found in the databases. The matches are listed in order, from the closest match to the least likely match.

Sequence similarity searching can assign probable functions to genes in a newly obtained genome sequence because homology—descent from a common ancestor—reflects evolutionary relationships. That is, the DNA sequences of two genes from different organisms will be similar if they are homologous genes with a common ancestor. Differences between the genes will have resulted from mutations that occurred over evolutionary time. For example, if a gene from a newly synthesized genome has close sequence similarity to genes that are known to be RNA polymerases in other organisms, then it is highly likely the gene encodes an RNA polymerase. This information is useful to researchers who wish to study the function of the new gene.

Despite the extensive databases we now have, there are still many putative genes with unknown functions. Indeed, it is a surprising characteristic of each new genome sequence that the functions of many of its genes are previously unknown. Naturally it is a challenging task to determine their functions because there are no clues to begin the investigation.

ASSIGNING GENE FUNCTION EXPERIMENTALLY To prove unequivocally that a gene identified in genome sequencing has a particular function requires experimental analysis. One important approach to assigning gene function experimentally is to knock out or knock down the function of a gene and determine what phenotypic process changes. The altered phenotype informs the researcher about the function of the normal gene. Major projects have been done, or are being done, to systematically knock out or knock down the function of each gene in the genomes of several organisms, including yeast, the fruit fly, and the nematode worm *C. elegans.*

The two main ways used in these experiments are *gene knockouts* and RNA interference (RNAi). A gene knockout uses molecular techniques to disrupt the gene in the chromosome. RNA interference, as discussed in Chapter 16, knocks down the expression of a gene at the translation level. When used experimentally, a small regulatory RNA is transcribed from an expression plasmid introduced into the cell. The sequence of that regulatory RNA can form complementary base pairs with the mRNA of the gene of interest. The base pairing triggers the RNA interference molecular mechanisms, which knocks down the expression of the gene by causing degradation of that gene's mRNA by blocking its translation.

RNA interference has been used to knock down gene expression of each of the approximately 20,000 genes of the nematode worm *C. elegans* one by one. It is not practical to look for any and all changes in phenotype caused by each knock down, so researchers usually screen for changes in particular phenotypes and correlate those changes to the specific genes affected. For example, screening for changes in fat metabolism reveals the genes involved in that process, and so on. In this general way, RNA interference has been useful in assigning functions for some genes that had not been characterized by sequence similarity searching or other experimental approaches.

STUDYING DIFFERENTIAL GENE ACTIVITY IN ENTIRE GENOMES WITH DNA MICROARRAYS As a part of functional genomics research, investigators are interested in comparing which genes are active in different cell types of humans and other organisms, and in tracking the changes in total gene activity in the same cell types as development progresses or as conditions change. In some cases, the researcher wants to know whether or not particular genes are being expressed, and in other cases how the level of expression varies in different circumstances. This research has been revolutionized by a technique using **DNA microarrays.** The microarrays are also called **DNA chips** because the techniques used to "print" the arrays resemble those used to lay out electronic circuits on a computer chip. The surface of a DNA microarray is divided into a microscopic grid of about 60,000 spaces. On each space of the grid, a computerized system deposits a microscopic spot containing about 10,000,000 copies of a DNA probe that is about 20 nucleotides long.

Studies of gene activity using DNA microarrays involve comparing gene expression under a defined experimental condition with expression under a reference (control) condition. DNA microarrays can be used to answer basic biological questions, such as: How does gene expression change when a cell goes from a resting state (reference condition) to a dividing state (experimental condition); that is, how is gene expression different in different stages of development? DNA microarrays can also be used to address many questions of medical significance, such as: How are genes differentially expressed in normal cells and cells of various cancers? In

DNA Microarray Analysis of Gene Expression Levels

Purpose: DNA microarrays can be used in various experiments, including comparing the levels of gene expression in two different tissues, as illustrated here. The power of the technique is that the entire set of genes in a genome can be analyzed simultaneously.

Protocol:

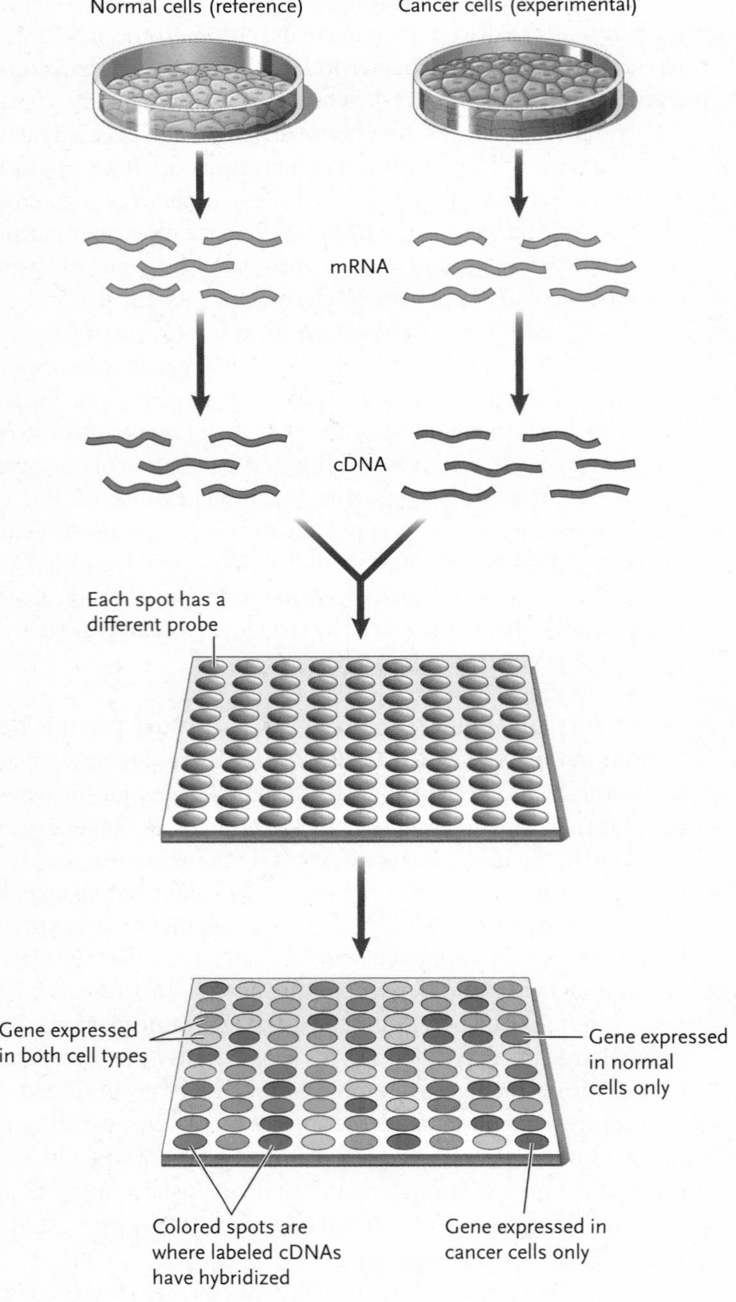
Normal cells (reference)
Cancer cells (experimental)
mRNA
cDNA
Each spot has a different probe
Gene expressed in both cell types
Gene expressed in normal cells only
Colored spots are where labeled cDNAs have hybridized
Gene expressed in cancer cells only

1. Isolate mRNAs from a control cell type (here, normal human cells) and an experimental cell type (here, human cancer cells).

2. Prepare cDNA libraries from each mRNA sample. For the normal cell (control) library use nucleotides with a green fluorescent label, and for the cancer cell (experimental) library use nucleotides with a red fluorescent label.

3. Denature the cDNAs to single strands, mix them, and pump them across the surface of a DNA microarray containing a set of single-stranded probes representing every protein-coding gene in the human genome. The probes are spotted on the surface, with each spot containing a probe for a different gene. Allow the labeled cDNAs to hybridize with the gene probes on the surface of the chip, and then wash excess cDNAs off.

4. Locate and quantify the fluorescence of the labels on the hybridized cDNAs with a laser detection system.

Actual DNA microarray result

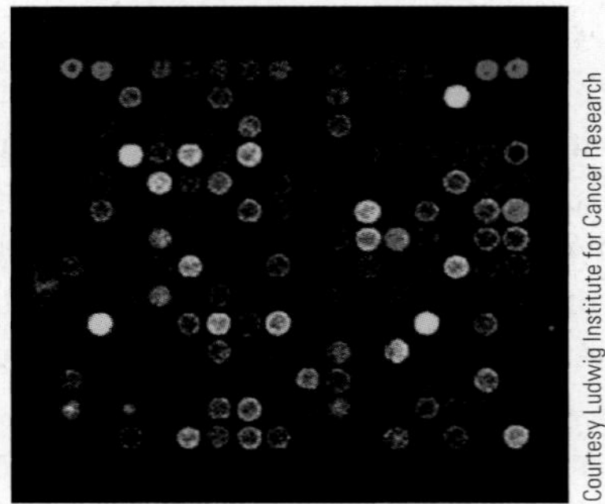

Courtesy Ludwig Institute for Cancer Research

Interpreting the Results: The colored spots on the microarray indicate where the labeled cDNAs have bound to the gene probes attached to the chip and, therefore, which genes were active in normal and/or cancer cells. Moreover, we can quantify the gene expression in the two cell types by the color detected. A purely green spot indicates the gene was active in the normal cell, but not in the cancer cell. A purely red spot indicates the gene was active in the cancer cell, but not in the normal cell. A yellow spot indicates the gene was equally active in the two cell types, and other colors tell us the relative levels of gene expression in the two cell types. For this particular experiment, we would be able to see how many genes have altered expression in the cancer cells, and exactly how their expression was changed.

these experiments, investigators might focus on which genes are active and inactive under the two conditions, or on how the levels of expression of genes change under the two conditions.

Figure 18.20 shows how a DNA microarray is used to compare gene expression in normal cells and cancer cells in humans. mRNAs are isolated from each cell type, and cDNAs are made from them, incorporating different fluorescent labels; green for one cDNA, red for the other. The two cDNAs are mixed and added to the DNA chip, where they hybridize with any complementary probes. A laser locates and quantifies the green and red fluorescence, enabling a researcher to see which genes are expressed in the cells and, for those that are expressed, to quantify differences in gene expression between the two cell types (see *Interpreting the Results* in Figure 18.20). The results can help researchers understand how the cancer develops and progresses.

DNA microarrays are also used to screen individuals for particular mutations. To detect mutations, the probes spotted onto the chip include probes for the normal sequence of the genes of interest along with probes for sequences of all known mutations. A fluorescent spot at a site on the chip printed with a probe for a given mutation immediately shows the presence of the mutation in the individual. Such a test is currently used to screen patients for whether they carry any one of a number of mutations of the *breast cancer 1 (BRCA1)* gene, known to be associated with the possible development of breast cancer.

In Comparative Genomics, Whole or Large Parts of Genomes Are Compared to Study Basic Biological Differences between Species and Evolutionary Relationships

In comparative genomics studies, genomes or sections of genomes from two or more species, strains, or individuals are analyzed to determine similarities and differences between sequences. Depending on the study, the focus may be on gene sequences, nongene sequences, or both. Various types of research questions can be asked in comparative genomics studies, including the following:

1. What are the evolutionary relationships between two or more genomes? Using DNA or amino acid sequence comparisons to determine evolutionary relationships among organisms is **molecular phylogenetics,** a topic discussed more in Chapter 23. Historically, amino acid sequences were the first to be used in such comparisons, and then DNA sequences. A phylogeny—the evolutionary history of a group of organisms—potentially can be inaccurate if it is based on a small set of sequences, such as those of a gene or genes. However, genome sequences contain much more information. Researchers comparing genome sequences are learning a great deal about how genomes relate to each other, including which sequences are conserved, as well as about how genomes have evolved. For example, in a three-way comparison of genome sequences, investigators have shown that modern humans share approximately 95% of their genes with chimpanzees (our nearest relative), but perhaps as much as 99.5% with Neanderthals **(Figure 18.21).** Complete genome sequence analysis was also crucial in demonstrating the evolutionary relationships and distinctions between members of the domains Bacteria, Archaea, and Eukarya, in particular confirming the archaeans as being a distinct lineage (see *Why It Matters*).

   A revelation of genome sequence comparisons is the degree to which different organisms, some of them widely separated in evolutionary origins, contain similar genes. For example, even though the yeast *Saccharomyces cerevisiae* is separated from our species by millions of years of evolutionary history, about 2,300 of its approximately 6,000 genes are related to those of mammals, including many genes that control progress through the cell cycle. The similarities are so close that the yeast and human versions of many genes can be interchanged with little or no effect on cell functions in either organism.
2. What are the functions of human genes? For ethical reasons, direct experimentation with humans is not possible. Comparative genomics provides a way to explore the functions of human genes by identifying homologous genes in nonhuman organisms. By studying the gene homolog in another organism, researchers can gain insight into the normal function of the gene in humans and how its function may be altered. Naturally, there is a lot of interest in studying genes that cause human disease.
3. What genes make us human? Researchers have compared the chimpanzee genome with the human genome to identify genes that potentially make us human. Chimpanzees and humans last shared a common ancestor about 6 million years ago. The investigators found a number of genes unique to the human genome known as *human accelerated regions.* One of these genes encodes an RNA that is not translated into protein. The gene is expressed in a region of the brain that undergoes a developmental change in humans but not in chimpanzees. Exactly what the RNA does is the subject of active research.
4. What organisms or viruses are in a sample? *Why It Matters* for this chapter discussed a metagenomics study of marine microbes. Metagenomic studies are comparative genomics studies because the similarities and differences between the genomic sequences obtained indicate the microbial diversity and potentially the array of specific sequences in the sample. Metagenomic studies can be descriptive, or they can be function based. For example, in function-based metagenomic analyses, researchers analyze the DNA sequences in an envi-

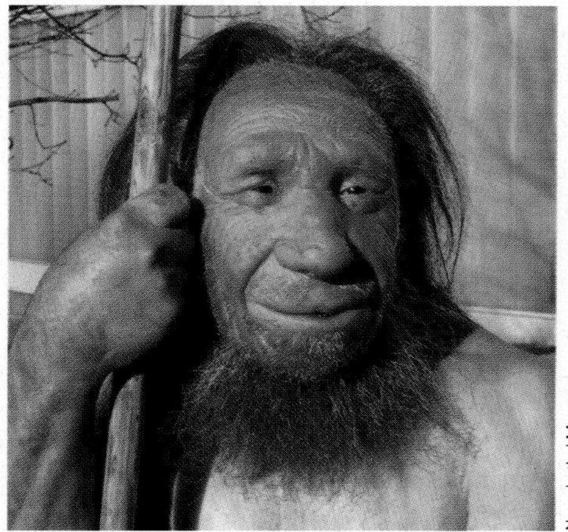

FIGURE 18.21 A Neanderthal (a reconstruction)

Neanderthal Museum

ronmental sample for genes with specific functions. New antibiotics have been discovered using this approach.

Studying the Array of Expressed Proteins Is the Next Level of Study of Biological Systems

Genome research also includes analysis of the proteins encoded by a genome, for proteins are largely responsible for cell function and, therefore, for all the functions of an organism. The term **proteome** has been coined to refer to the complete set of proteins that can be expressed by an organism's genome. A *cellular proteome* is a subset of those proteins, the collection of proteins found in a particular cell type under a particular set of environmental conditions.

The study of the proteome is the field of **proteomics.** The number of possible proteins encoded by the genome is larger than the number of protein-coding genes in the genome, at least in eukaryotes. In eukaryotes, alternative splicing of gene transcripts and variation in protein processing means that expression of a gene may yield more than one protein product. Therefore, proteomics is a more challenging area of research than is genomics.

Proteomics has the following two major immediate goals: (1) to determine the number and structure of proteins in the proteome; and (2) to determine the functional interactions between the proteins. The interactions between proteins are particularly important because they help us understand how proteins work together to determine the phenotype of the cell. For instance, if a particular set of interacting proteins characterized a lung tumor cell, then drugs could be developed that specifically target the interactions.

What are the tools of proteomics? For many years it has been possible to separate and identify proteins by gel electrophoresis (using polyacrylamide to make the gels, the same material used for separating DNA fragments in DNA sequencing) or mass spectrometry. However, to study an entire cellular proteome, many more proteins must be analyzed simultaneously than is possible with either of those techniques. A big step in that direction is the development of **protein microarrays (protein chips)** (see chapter opener photograph), which are similar in concept to DNA microarrays. One type of protein microarray involves binding antibodies prepared against different proteins to different locations on the protein chip. An antibody for a foreign substance such as a protein is generated by the immune system of an animal that has been injected with that substance. The antibody is then isolated from the animal's blood and can be used to bind specifically to the protein in experiments. Proteins are isolated from cells, labeled, and then pumped over the surface of the protein microarray. Each labeled protein binds to the antibody for that protein. After washing off excess proteins, the protein microarray is analyzed much the same way DNA microarrays are, to determine where the proteins bound and to quantify that binding. With this technique, a researcher can quantify proteins in different cell types and in different tissues. Researchers can also compare proteins under different conditions, such as during differentiation, or with and without a particular disease condition, or with and without a particular drug treatment. In the future, we can expect protein arrays to become routine for studying cellular proteomes.

Systems Biology Studies the Interactions among All Components of an Organism

Traditional biology research focuses on identifying and studying the functions of individual genes, proteins, and cells. Although such research has provided an enormous body of knowledge—this textbook being an example—it provides only a limited insight into how a whole organism functions at the cellular and molecular levels. For instance, studying separately the individual components of a bicycle does not tell you what the whole bicycle is or what it does.

Systems biology seeks to overcome the limitations of the approaches of traditional biology by studying the organism as a whole to unravel the integrated and interacting network of genes, proteins, and biochemical reactions responsible for life. Systems biologists work from the premise that those interactions are responsible for an organism's form and function. Present-day research in systems biology has been stimulated by the development of techniques for genomic and proteomic analysis and by the data from those analyses.

Systems biologists use genomics and proteomics techniques, such as those discussed in this section, along with information from other sources. They typically obtain very complex data and use sophisticated quantitative analysis to generate models for the interactions within an organism.

Systems biologists study organisms of many kinds. Some focus on humans and have the ambitious goal of transforming the practice of medicine. Their vision is to define the interactions between all the components that affect the health of an individual human. It may then be possible to predict more accurately than is currently possible whether a person will develop particular diseases, and to personalize treatments for those diseases.

STUDY BREAK 18.3 <

1. **What are the three main areas of genome analysis?**
2. **What is the principle behind whole-genome shotgun sequencing of genomes?**
3. **How are possible protein-coding genes identified in a genome sequence of a bacterium? Of a mammal?**
4. **How can gene function be assigned by a sequence similarity search?**
5. **How would you determine how a steroid hormone affects gene expression in human tissue culture cells?**

>

THINK OUTSIDE THE BOOK

Earlier in the chapter we mentioned the fact that cloned animals may have many genes whose expression is abnormal compared to gene expression in a noncloned animal. Individually or collaboratively, outline the steps you would take experimentally to determine, on a genome-wide scale, if genes are abnormally expressed in a cloned mammal. Your answer should include how the experiment reveals both qualitative and quantitative differences in gene expression.

UNANSWERED QUESTIONS

It is well known that individuals respond differently to medications. For example, a blood pressure lowering medication may effectively lower blood pressure in one patient, have no effect on blood pressure in another patient, and cause intolerable adverse effects in a third patient. Factors such as age, weight, or liver and kidney function may influence response to medication. However, it remains difficult to predict how effective or safe a drug will be for a particular patient based on these factors alone. In clinical medicine, a trial-and-error approach is often used for drug therapy—several drugs or drug doses may be tried before achieving optimal response with acceptable tolerability.

Genes encoding proteins involved in the metabolism and transport of a drug through the body and for proteins at the drug target site may greatly influence drug response. Genes involved in disease progression or phenotype may also influence how well a drug works. How can molecular approaches be used to personalize drug therapy? The answer to that question may well come from the relatively new field of pharmacogenomics. Pharmacogenomics is the study of how genes affect individual responses to drug therapy. Its goal is to choose the most appropriate drug, drug dose, and treatment duration for a particular patient based on genetic information.

How can pharmacogenomics be used to optimize therapy for blood clots?

Prescribed to prevent the formation of blood clots in the circulatory system, the drug warfarin ranks among the top 50 most commonly prescribed drugs in the United States. Because it inhibits the blood from clotting, warfarin also increases the chance of bleeding. The risk for bleeding increases when the warfarin dose is too high, while the risk for clotting increases when the dose is too low. The dose of warfarin needed to prevent clotting without significantly increasing the chance for bleeding is highly variable among patients, with doses ranging by as much as 20-fold. Age, body size, other diseases the person may have, and medications the person may be taking, all influence the dose of warfarin a person needs. However, considering age, body size, and clinical factors alone is not enough to choose the right warfarin dose for a given patient. Genes encoding for proteins involved in warfarin metabolism and proteins at the warfarin target site are now known to influence warfarin dose requirements. Our group is working with other research groups from around the world to study variations in a person's genome that affect the dose of warfarin that he or she needs. The goal is to develop an algorithm, or equation, for dosing warfarin that includes both genetic and clinical factors.

Will genotype-guided warfarin dosing become routine clinical practice?

The United States Food and Drug Administration recently revised the guidelines for prescribing warfarin, to recommend lower doses for individuals with certain genotypes. However, the labeling falls short of recommending routine genotyping before warfarin is prescribed. Before genotyping to determine warfarin dose becomes routine, the next step will be to compare warfarin dosing based on both genetic and clinical information with traditional dosing based on clinical information alone. This effort is currently underway. If genotype-guided dosing improves the accuracy of choosing a warfarin dose and consequently, decreases a person's risk for clotting or bleeding, it will likely become a routine part of clinical practice.

Think Critically

What are the ethical implications of pharmacogenomics if the genes in question are associated with disease risk as well as drug response? What are the ethical implications of pharmacogenomics if a patient possesses a genotype predictive of poor drug response and there are no alternative treatments to the drug?

Larisa H. Cavallari

Larisa H. Cavallari is an Associate Professor in the Department of Pharmacy Practice at the University of Illinois at Chicago. She is involved in both clinical and basic science research focusing on genetic contributions to cardiovascular drug response. To learn more about Dr. Cavallari's laboratory and research, go to http://pmpr.pharm.uic.edu/pharmacogenomics/index.htm.

REVIEW KEY CONCEPTS

Go to **CENGAGENOW** at www.cengage.com/login to access quizzing, animations, exercises, articles, and personalized homework help.

18.1 DNA Cloning

- Producing multiple copies of genes by cloning is a common first step for studying the structure and function of genes, or for manipulating genes. Cloning involves cutting genomic DNA and a cloning vector with the same restriction enzyme, joining the fragments to produce recombinant plasmids, and introducing those plasmids into a living cell such as a bacterium, where replication of the plasmid takes place (Figures 18.1–18.3).
- A clone containing a gene of interest may be identified among a population of clones by using DNA hybridization with a labeled nucleic acid probe (Figure 18.4).
- A genomic library is a collection of clones that contains a copy of every DNA sequence in the genome. A cDNA (complementary DNA) library is the entire collection of cloned cDNAs made from the mRNAs isolated from a cell. A cDNA library contains only sequences from the genes that are active in the cell when the mRNAs are isolated.
- PCR amplifies a specific target sequence in DNA, such as a gene, which is defined by a pair of primers. PCR increases DNA quantities by successive cycles of denaturing the template DNA, annealing the primers, and extending the primers in a DNA synthesis reaction catalyzed by DNA polymerase. With each cycle of PCR, the amount of DNA doubles (Figure 18.5).

Animation: Formation of recombinant DNA

Animation: Base pairing of DNA fragments

Animation: Restriction enzymes

Animation: How to make cDNA

Animation: Use of a radioactive probe

Animation: Polymerase chain reaction (PCR)

Animation: Automated DNA sequencing

18.2 Applications of DNA Technologies

- Recombinant DNA and PCR techniques are used in DNA molecular testing for human genetic disease mutations. One approach exploits restriction site differences between normal and mutant alleles of a gene that create restriction fragment length polymorphisms (RFLPs) which are detectable by DNA hybridization with a labeled nucleic acid probe (Figures 18.7 and 18.8).
- Human DNA fingerprints are produced from a number of loci in the genome characterized by short, tandemly repeated sequences that vary in number in all individuals (except identical twins). To produce a DNA fingerprint, the PCR is used to amplify the region of genomic DNA for each locus, and the lengths of the PCR products indicate the alleles an individual has for the repeated sequences at each locus. DNA fingerprints are widely used to establish paternity, ancestry, or criminal guilt (Figure 18.9).
- Genetic engineering is the introduction of new genes or genetic information to alter the genetic makeup of humans, other animals, plants, and microorganisms such as bacteria and yeast. Genetic engineering primarily aims to correct hereditary defects, improve domestic animals and crop plants, and provide proteins for medicine, research, and other applications (Figures 18.10–18.12 and 18.15).
- Genetic engineering has enormous potential for research and applications in medicine, agriculture, and industry. Potential risks include unintended damage to living organisms or to the environment.

Animation: DNA fingerprinting

Animation: Gene transfer using a Ti plasmid

Animation: Transferring genes into plants

18.3 Genome Analysis

- Sequencing a genome involves a replication reaction with a DNA template, a DNA primer, the four normal deoxyribonucleotides, and a mixture of four dideoxyribonucleotides, each labeled with a different fluorescent tag, and DNA polymerase. Replication stops at any place in the sequence in which a dideoxyribonucleotide is substituted for the normal deoxyribonucleotide. The lengths of the terminated DNA chains and the label on them indicate the overall sequence of the DNA chain being sequenced (Figure 18.18).
- Genome analysis consists of three main areas: genome sequence determination followed by annotation to locate putative genes and other functionally important sequences; functional genomics, the study of the function of genes and other parts of the genome; and comparative genomics, the comparison of entire genomes, or large parts of genomes, to understand evolutionary relationships and biological similarities and differences among species.
- The whole-genome shotgun method of sequencing a genome involves breaking up the entire genome into random, overlapping fragments, cloning each fragment, determining the sequence of the fragment in each clone, and using computer algorithms to assemble overlapping sequences into the sequence of the complete genome (Figure 18.19).
- Once a gene is sequenced, the sequence of the protein encoded in a prokaryotic gene can be deduced by reading the coding portion of the gene three nucleotides at a time, starting at the AUG codon that indicates the beginning of a coding sequence. More sophisticated algorithms are used for eukaryotes because of the possible presence of introns in protein-coding genes.
- Complete genome sequences have been obtained for many viruses, a large number of prokaryotes, and many eukaryotes, including the human. The sequences have revealed that all eukaryotes share related gene sequences. They have also revealed a significant proportion of genes whose functions are not presently known.
- The functions of putative genes identified in analysis of genome sequences can be determined experimentally by sequence similarity searches.
- Having the complete genome of an organism makes it possible to study the expression of all of the genes in the genome simultaneously, including comparing gene expression in two different cell types. The DNA microarray (or DNA chip) is typically used for the comparison; this technique can provide information about which genes are active in the two cell types as well as relative levels of expression of those genes (Figure 18.20).
- Comparative genomics studies provide information about biological issues such as the evolutionary relationships between two or more genomes, the functions of human genes, and what organisms or viruses are in a sample.
- Proteomics is the study of the complete set of proteins in an organism or in a particular cell type. Protein numbers, protein structure, and protein interactions are all topics of proteomics.
- Systems biology combines data derived from genomics, proteomics, and other sources of information. Using sophisticated quantitative analysis, it seeks to model the total array of interactions responsible for an organism's form and function.

UNDERSTAND AND APPLY

Test Your Knowledge

1. A complementary DNA library (cDNA) and a genomic library are similar in that both:
   a. use bacteria to make eukaryotic proteins.
   b. provide information on whether genes are active.
   c. contain all of the DNA of an organism cut into pieces.
   d. clone mRNA.
   e. depend on cloning in a living cell to produce multiple copies of the DNA of interest.
2. Restriction endonucleases, ligases, plasmids, *E. coli*, electrophoretic gels, and a bacterial gene resistant to an antibiotic are all required for:
   a. dideoxyribonucleotide analysis.
   b. PCR.
   c. DNA cloning.
   d. DNA fingerprinting.
   e. DNA sequencing.
3. Polymerase chain reaction (PCR) differs from other techniques discussed in this chapter by its use of:
   a. primers.
   b. DNA.
   c. RNA.
   d. DNA polymerase tolerant to high temperatures.
   e. the four nucleoside triphosphates.
4. Restriction fragment length polymorphisms (RFLPs):
   a. are produced by reaction with restriction endonucleases and are detected by Southern blot analysis.
   b. are of the same length for mutant and normal β-globin alleles.
   c. determine the sequence of bases in a DNA fragment.
   d. have in their middle short fragments of DNA that are palindromic.
   e. are used as vectors.

5. DNA fingerprinting, which is often used in forensics, paternity testing, and for establishing ancestry:
   a. compares one stretch of the same DNA between two or more people.
   b. measures different lengths of DNA from many repeating noncoding regions.
   c. requires the largest DNA lengths to run the greatest distance on a gel.
   d. requires amplification after the gels are run.
   e. can easily differentiate DNA between identical twins.
6. Dolly, a sheep, was an example of reproductive (germ-line) cloning. Required to perform this process was:
   a. implantation of uterine cells from one strain into the mammary gland of another.
   b. fusion of the mammary cell from one strain with an enucleated egg of another strain.
   c. fusion of an egg from one strain with the egg of a different strain.
   d. fusion of an embryonic diploid cell with an adult haploid cell.
   e. fusion of two nucleated mammary cells from two different strains.
7. Which of the following is *not* true of somatic cell gene therapy?
   a. White blood cells can be used.
   b. Somatic cells are cultured, and the desired DNA is introduced into them.
   c. Cells with the introduced DNA are returned to the body.
   d. The technique is still very experimental.
   e. The inserted genes are passed to the offspring.
8. The sequence of the human genome:
   a. was obtained by sequencing overlapping DNA fragments.
   b. revealed far more genes than expected.
   c. revealed 3 trillion base pairs.
   d. used techniques not applicable to mapping other species.
   e. revealed 250,000 protein-coding genes.
9. Which of the following is true about genome size?
   a. Bacteria have genomes that vary widely in size.
   b. The human genome is the largest among eukaryotes.
   c. Organisms with large genomes are always more complex than organisms with small genomes.
   d. As genome size increases in a lineage, the number of genes also always increases.
   e. The smallest known cellular genome is found in a species of Archaea.
10. The human genome is characterized by all of the following except:
   a. introns occupy 24% of the genome in humans.
   b. the protein-coding sequences occupy less than 2% of the human genome.
   c. more than 50% of the genome consists of repeated sequences.
   d. the genome sequence is comprised of approximately 30 million base pairs.
   e. human cells have about 20,000 different protein-coding genes.

Discuss the Concepts

1. Do you think that genetic engineering is worth the risk? Who do you think should decide whether genetic engineering experiments and projects should be carried out: scientists, judges, or politicians?
2. Do you think that human germ-line cells should be modified by genetic engineering to cure birth defects? To increase intelligence or beauty?
3. Write a paragraph supporting genetic engineering, and one paragraph arguing against it. Which argument carries more weight, in your opinion?
4. What should juries know to be able to interpret DNA evidence? Why might juries sometimes ignore DNA evidence?
5. A forensic scientist obtained a small DNA sample from a crime scene. In order to examine the sample, he increased its quantity by cycling the sample through the polymerase chain reaction. He estimated that there were 50,000 copies of the DNA in his original sample. Derive a simple formula and calculate the number of copies he will have after 15 cycles of the PCR.
6. A market puts out a bin of tomatoes that have outstanding color, flavor, and texture. A sign posted above them identifies them as genetically engineered produce. Most shoppers pick unmodified tomatoes in an adjacent bin, which are pale, mealy, and nearly tasteless. Which tomatoes would you pick? Why?

Design an Experiment

Suppose a biotechnology company has developed a GMO, a transgenic plant that expresses *Bt* toxin. The company sells its seeds to a farmer under the condition that the farmer may plant the seed, but not collect seed from the plants that grow and use it to produce crops in the subsequent season. The seeds are expensive, and the farmer buys seeds from the company only once. How could the company show experimentally that the farmer has violated the agreement and is using seeds collected from the first crop to grow the next crop?

Interpret the Data

You learned in the chapter that an STR locus is a locus where alleles differ in the number of copies of a short, tandemly repeated DNA sequence. PCR is used to determine the number of alleles present, as shown by the size of the DNA fragment amplified. In the figure below are the results of PCR analysis for STR alleles at a locus where the repeat unit length is 9 bp, and alleles are known which have 5–11 copies of the repeat. Given the STR alleles present in the adults, state whether each of the four juveniles could or could not be an offspring of those two adults. Explain your answers.

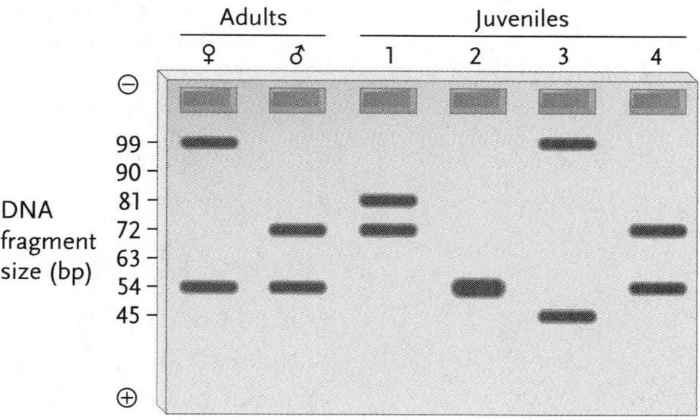

Adults
Juveniles
♀
♂
1
2
3
4
⊖
99
90
81
72
63
54
45
DNA fragment size (bp)
⊕

Apply Evolutionary Thinking

Search for the words "comparative genomics" on the Internet to answer this question: How can complete genome sequences provide more accurate information about the evolution of species than can the sequences of one or a few genes?

Express Your Opinion

Nutritional labeling is required on all packaged food in the United States, but genetically modified food products may be sold without labeling. Should food distributors be required to label all products made from genetically modified plants or livestock? Go to academic.cengage.com/login to investigate both sides of the issue and then vote.

19

STUDY OUTLINE

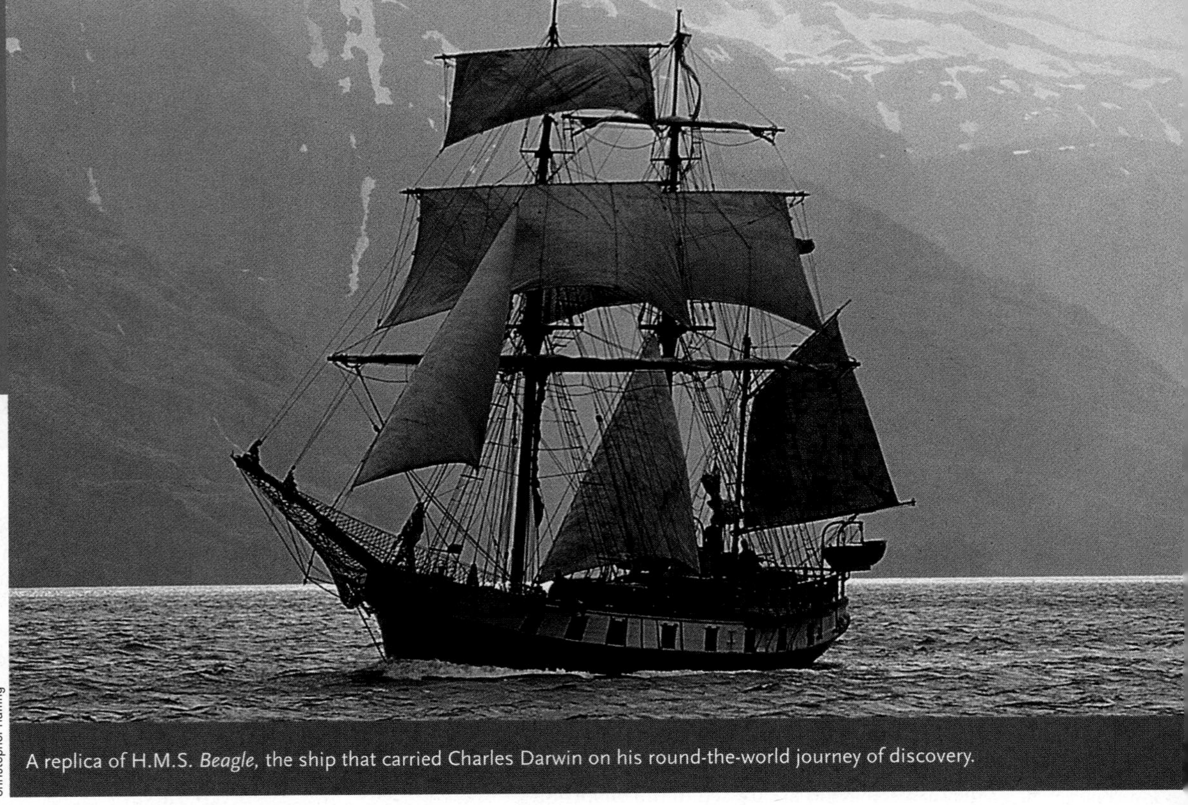

Christopher Ralling

A replica of H.M.S. *Beagle*, the ship that carried Charles Darwin on his round-the-world journey of discovery.

Development of Evolutionary Thought

Why It Matters. . . On June 18, 1858, Charles Darwin received the shock of his life. Alfred Russel Wallace, a young naturalist working in the Asian tropics, solicited Darwin's opinion of a manuscript he had written about how species change through time. Darwin quickly realized that Wallace had independently described a mechanism for biological evolution that was nearly identical to the one he had been studying for more than 20 years, but had not yet described in print.

Like researchers today, scientists in the nineteenth century had to publish their work quickly to establish the "priority" on which their scientific reputations were made. Darwin's friend and colleague, the geologist Charles Lyell, had encouraged him to publish a preliminary essay on evolution two years before Wallace's letter arrived. But Darwin procrastinated, and because Wallace was the first to prepare his work for publication, Darwin feared that history would credit the younger man with these new ideas. Despite his anxiety, Darwin forwarded Wallace's manuscript to Lyell, who passed it along to the botanist Joseph Hooker. Lyell and Hooker engineered a solution that gave credit to both men **(Figure 19.1)**. On July 1, 1858, papers by Darwin and Wallace were presented to the Linnaean Society of London, a prestigious scientific organization.

Darwin worked feverishly after this harrowing experience, and his now-famous book, *On the Origin of Species by Means of*

Charles Darwin

English Heritage. NMR/Bridgeman Art Library

Alfred Russel Wallace

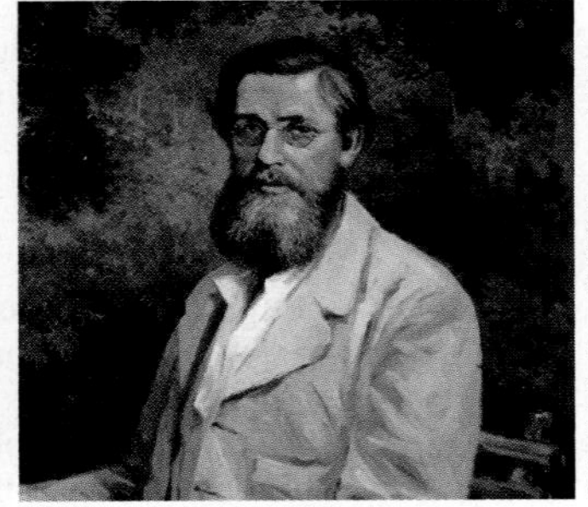

© English Heritage. NMR/Bridgeman Art Library

FIGURE 19.1
Pioneers of evolutionary theory. Charles Darwin (1809–1882) and Alfred Russel Wallace (1823–1913) independently discovered the mechanism of natural selection.

Natural Selection, was published on November 24, 1859. The first printing of 1,250 copies sold out in one day. Today, we honor Darwin for developing the seminal idea about how biological evolution occurs and for accumulating and documenting vast quantities of evidence over decades of study.

In *The Origin,* Darwin proposed that natural mechanisms produce and transform the diversity of life on Earth. His concept of evolution still forms the unifying intellectual paradigm within which all biological research is undertaken. Even when researchers do not address explicitly evolutionary questions, their observations, theories, hypotheses, and experiments are formulated with the implicit knowledge that all forms of life are related and have evolved from ancestral forms.

Biological evolution occurs in populations when specific *processes* cause the genomes of organisms to differ from those of their ancestors. These genetic changes, and the phenotypic modifications they cause, are the *products* of evolution. By studying the products of evolution, biologists strive to understand the processes that cause evolutionary change.

The theory of evolution is so widely accepted that most people cannot think about the biological world in any other way. But the biological changes implied by Darwin's ideas and by modern evolutionary theory had not been included in earlier European worldviews. <

19.1 Recognition of Evolutionary Change

The historical development of evolutionary theory is a fascinating tale of scientists struggling to reconcile evidence of change with a prevailing philosophy that change was impossible in a perfectly created universe.

Europeans Integrated Ideas from Ancient Greek Philosophy into Christian Doctrine

The Greek philosopher Aristotle (384–322 B.C.) was a keen observer of nature, and he is generally considered to have been the first student of **natural history,** the branch of biology that examines the form and variety of organisms in their natural environments. Aristotle believed that both inanimate objects and living species had fixed characteristics. Careful study of their differences and similarities enabled him to create a ladder-like classification of nature, from the simplest to the most complex forms: minerals ranked below plants, plants below animals, animals below humans, and humans below the gods of the spiritual realm.

By the fourteenth century, Europeans had merged Aristotle's classification system with the biblical account of creation: each organism had been specially created by God, species could never change or become extinct, and new species could never arise. Biological research became dominated by **natural theology,** which sought to name and catalog all of God's creation. Careful study of each species would identify its position and purpose in the *Scala Naturae,* or Great Chain of Being, as Aristotle's ladder of life was called. In the eighteenth century, the Swedish botanist Carolus Linnaeus (1707–1778), who developed the science of **taxonomy,** the branch of biology that classifies organisms (see Chapter 23), undertook this important work *ad majorem Dei gloriam* ("for the greater glory of God").

Scholars also used a literal interpretation of scripture to date the time of creation precisely. By tabulating the human generations described in the Bible, they determined that the creation had occurred around 4000 B.C., making Earth a bit less than 6,000 years old—hardly old enough for much change to have taken place.

Scientists Slowly Became Aware of Change in the Natural World

Modern science came of age in the fifteenth through eighteenth centuries. The English philosopher and statesman Sir Francis Bacon (1561–1626) established the importance of observation, experimentation, and inductive reasoning. Other scientists, notably Nicolaus Copernicus (1473–1543), Galileo Galilei (1564–1642), René Descartes (1596–1650), and Sir Isaac Newton (1643–1727), proposed mechanistic theories to explain physical events. In addition, three new disciplines—biogeography, comparative morphology, and geology—promoted a growing awareness of change.

QUESTIONS ABOUT BIOGEOGRAPHY As long as naturalists encountered organisms only from Europe and surrounding lands, the task of understanding the *Scala Naturae* was manageable. But global explorations in the fifteenth through seventeenth centuries provided naturalists with thousands of unknown plants and animals from Asia, Africa, the Pacific Islands, and the Americas. Although some were similar to European species, others were new and very strange.

Studies of the world distribution of plants and animals, now called **biogeography,** raised puzzling questions. Was there no limit to the number of species created by God? Where did all these species fit in the *Scala Naturae*? If they had all been created in the Garden of Eden, why did some species have limited geographical distributions, whereas others were widespread? And why were some species found in Africa or Asia different from those found in Europe, whereas other species from far-flung places were similar to each other **(Figure 19.2)**?

QUESTIONS ABOUT COMPARATIVE MORPHOLOGY When biologists began to compare the **morphology** (anatomical structure) of organisms, they discovered interesting similarities and differences. For example, the front legs of pigs, the flippers of dolphins, and the wings of bats differ markedly in size, shape, and function **(Figure 19.3).** But these appendages have similar locations in the animals' bodies; all are constructed of bones, muscles, and skin; and all develop similarly in the animals' embryos. If these limbs were specially created for different means of locomotion, naturalists wondered, why didn't the Creator use different materials and structures for walking, swimming, and flying?

FIGURE 19.2
Large, flightless birds. Three large bird species with greatly reduced wings occupy similar habitats in geographically separated regions.

African ostrich
(Struthio camelus)

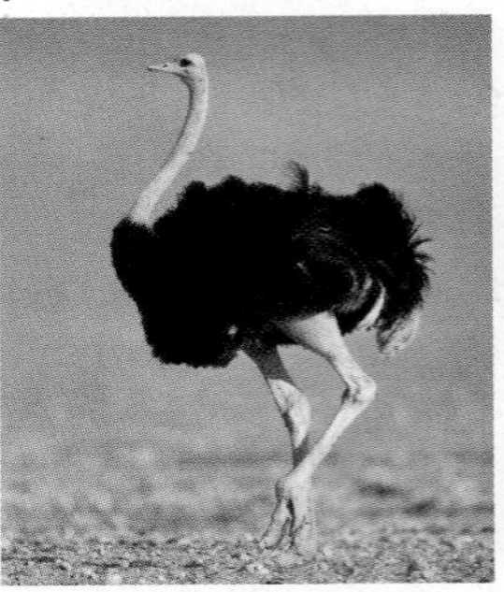

Image copyright Johan Swanepoel, 2010. Used under license from Shutterstock.com

South American rhea
(Rhea americana)

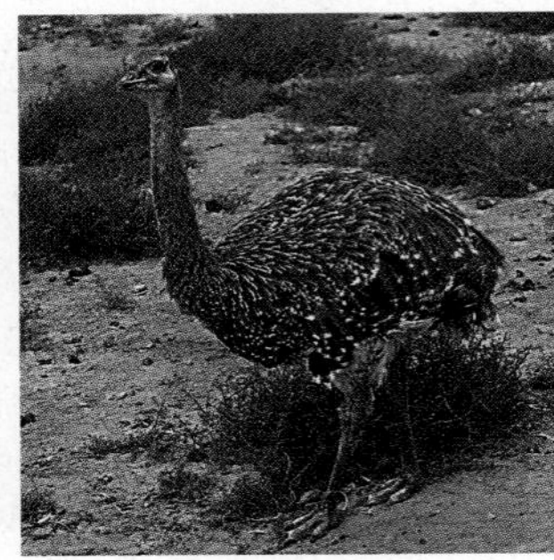

Kenneth W. Fink/Photo Researchers, Inc.

Australian emu
(Dromaius novaehollandiae)

Image copyright S. Cooper Digital, 2010. Used under license from Shutterstock.com

Natural theologians countered this argument by stating that the body plans were perfect, and there was no need to invent a new plan for every species. But a French scientist, George-Louis Leclerc (1707–1788), le Comte (Count) de Buffon, was still puzzled by the existence of body parts with no apparent function. For example, he noted that the feet of pigs and some other mammals have two toes that never touch the ground (pig digits 2 and 5 in Figure 19.3). If each species was anatomically perfect for its particular way of life, Buffon asked, why did useless structures exist?

Buffon proposed that some animals must have *changed* since their creation; he suggested that **vestigial structures,** the useless parts we observe today, must have functioned in ancestral organisms. Buffon offered no explanation of how functional structures became vestigial, but he clearly recognized that some species were "conceived by Nature and produced by Time."

FIGURE 19.3
Mammalian forelimbs and locomotion. Pigs use their legs to walk or run, dolphins use their flippers to swim, and bats use their wings to fly. Homologous (equivalent) bones are pictured in the same color, and digits (fingers) are numbered; pigs have lost the first digit over evolutionary time. (Limbs are not drawn to scale.)

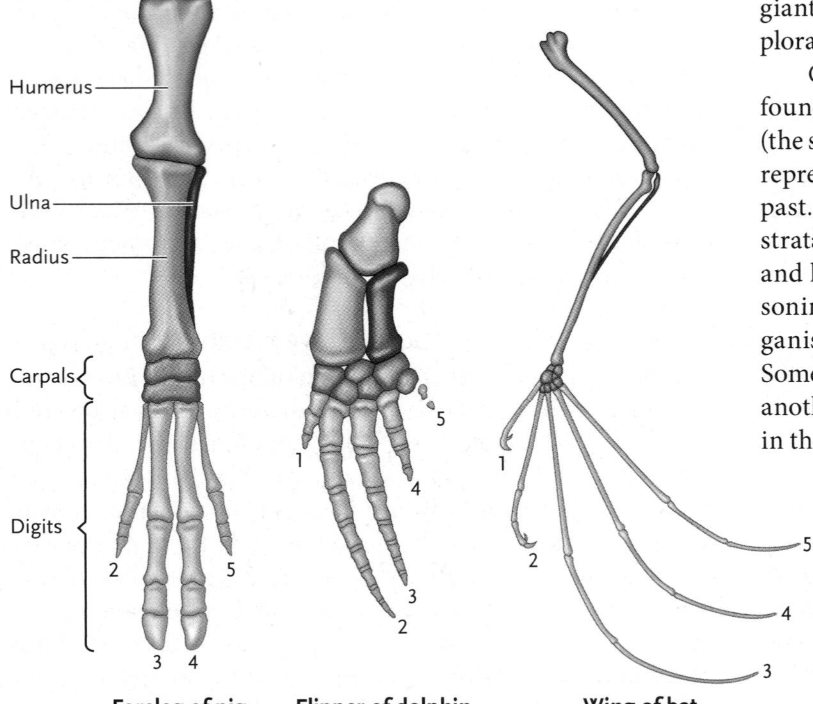
Humerus
Ulna
Radius
Carpals
Digits
2
5
3
4
1
4
5
3
2
1
2
5
4
3
Foreleg of pig
Flipper of dolphin
Wing of bat

QUESTIONS ABOUT FOSSILS By the mid-eighteenth century, geologists were mapping the **stratification,** or horizontal layering, of sedimentary rocks beneath the soil surface (see Figure 22.2). Different layers held different kinds of **fossils** (*fossilis* = dug up). Relatively small and simple fossils appeared in the deepest layers. Fossils in the layers above them were more complex. Those in the uppermost layers often resembled living organisms. Moreover, fossils found in any particular layer were often similar, even if they were collected from geographically separated sites. What were these fossils, and why did they vary more from one layer of rock to another than from one geographical region to another?

Some scientists suggested that fossils were the remains of extinct organisms, but natural theology did not allow extinction. Thomas Jefferson, the third president of the United States and an amateur fossil hunter, thought that fossils were the remains of species that were now extremely rare; he believed that nature could not have "permitted any one race of her animals to become extinct" or "formed any link in her great works so weak [as] to be broken." He even asked Lewis and Clark to keep an eye out for giant ground sloths, now known to be extinct, during their exploration of the Pacific Northwest.

Georges Cuvier (1769–1832), a French zoologist and a founder of comparative morphology, as well as **paleobiology** (the study of ancient organisms), realized that the layers of fossils represented organisms that had lived at successive times in the past. He suggested that the abrupt changes between geological strata marked dramatic shifts in ancient environments. Cuvier and his followers developed the theory of **catastrophism,** reasoning that each layer of fossils represented the remains of organisms that had died in a local catastrophe, such as a flood. Somewhat different species then recolonized the area, and when another catastrophe struck, they formed a different set of fossils in the next, higher layer of rock.

Lamarck Developed an Early Theory of Biological Evolution

A contemporary of Cuvier and a student of Buffon, Jean Baptiste de Lamarck (1744–1829) proposed the first comprehensive theory of biological evolution that was based on specific mechanisms. He proposed that a meta-

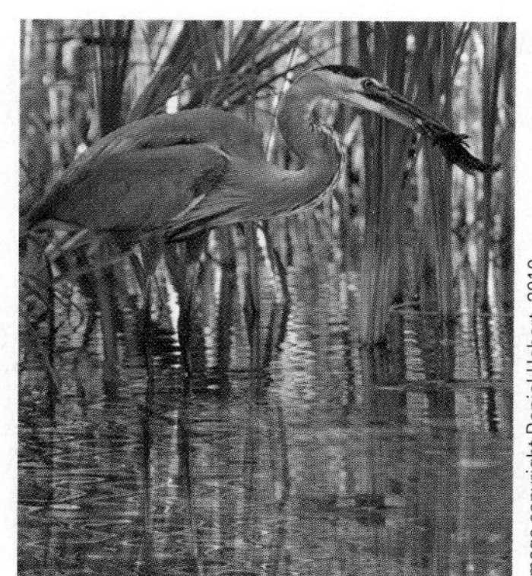

FIGURE 19.4
A great blue heron *(Ardea herodias)*. Like many other wading birds, herons have long, stilt-like legs. Lamarck hypothesized that as wading birds stretched their legs to keep their bodies dry while feeding, successive generations of their offspring would have progressively longer legs.

physical "perfecting principle" caused organisms to become better suited to their environments. Simple organisms evolved into more complex ones, moving up the ladder of life, and microscopic organisms were replaced at the bottom of the ladder by spontaneous generation.

Lamarck theorized that two mechanisms fostered evolutionary change. According to his *principle of use and disuse,* body parts grow in proportion to how much they are used, as anyone who "pumps iron" well knows. Conversely, structures that are not often used get weaker and shrink, such as the muscles of an arm immobilized in a cast. According to his second principle, the *inheritance of acquired characteristics,* changes that an animal acquires during its lifetime are inherited by its offspring. Thus, Lamarck argued that long-legged wading birds, such as herons **(Figure 19.4),** are descended from short-legged ancestors that stretched their legs to stay dry while feeding in shallow water. Their offspring inherited slightly longer legs, and after many generations, their legs became extremely long.

Today, we know that Lamarck's proposed mechanisms do not cause evolutionary change. Although muscles do grow larger through continued use, most structures do not respond in the way Lamarck predicted. Moreover, structural changes acquired during an organism's lifetime are not inherited by the next generation. Even in his own day, Lamarck's ideas were not widely accepted.

Despite the shortcomings of his theory, Lamarck made four tremendously important contributions to the development of an evolutionary worldview. First, he proposed that all species change through time. Second, he recognized that new characteristics are passed from one generation to the next. Third, he suggested that organisms change in response to their environments. And fourth, he hypothesized the existence of specific mechanisms that caused evolutionary change. All four of these ideas became cornerstones of Darwin's evolutionary theory, although he proposed a different mechanism, natural selection, as the cause of evolutionary change. Perhaps Lamarck's most important contribution was that he fostered discussion. By the mid-nineteenth century, most educated Europeans were talking about evolutionary change, whether they believed in it or not.

Geologists Recognized That Earth Had Changed over Time

In 1795, the Scottish geologist James Hutton (1726–1797) argued that slow and continuous physical processes, *acting over long periods of time,* produced Earth's major geological features; for example, the movement of water in a river slowly erodes the land and deposits sediments near the mouth of the river. Given enough time, erosion creates deep canyons, and sedimentation creates thick topsoil on flood plains. Hutton's **gradualism,** the view that Earth changed *slowly* over its history, contrasted sharply with Cuvier's catastrophism.

The English geologist Charles Lyell (1797–1875) championed and extended Hutton's ideas in an influential series of books, *Principles of Geology.* Lyell argued that the geological processes that sculpted Earth's surface over long periods of time—such as volcanic eruptions, earthquakes, erosion, and the formation and movement of glaciers—are exactly the same as the processes observed today. This concept, called **uniformitarianism,** undermined any remaining notions of an unchanging Earth. Also, because geological processes proceed very slowly, it must have taken millions of years, not just a few thousand, to mold the landscape into its current configuration.

STUDY BREAK 19.1

1. **Why did the existence of vestigial structures make Buffon question the idea that living systems never changed?**
2. **What were Lamarck's contributions to an evolutionary worldview?**
3. **How do the concepts of gradualism and uniformitarianism in geology undermine the belief that Earth is only about 6,000 years old?**

19.2 Darwin's Journeys

In 1831, in the midst of this intellectual ferment, young Charles Darwin wondered what to do with his life. Raised in a wealthy English household, he had always collected shells and studied the habits of insects and birds; he preferred hunting and fishing to classical studies. Despite his lackluster performance as a student, Darwin was expected to continue the family tradition of practicing medicine. But after two years, he abandoned his medical studies at the University of Edinburgh. Instead, he followed his interest in natural history despite the objections of his father, who reputedly told him, "You care for nothing but shooting, dogs, and rat-catching and you will be a disgrace to yourself and all of your family."

At the suggestion of his father, Darwin studied for a career as a clergyman, earning a degree from Cambridge University. There, he found a mentor in the Reverend John Henslow, a leading botanist, who arranged for Darwin to travel as the captain's dining companion aboard H.M.S. *Beagle,* a naval surveying ship. Darwin thus embarked on a sea voyage and an intellectual journey that altered the foundations of modern thought.

Darwin Saw the World on the Voyage of the *Beagle*

The *Beagle* sailed westward to map the coastline of South America and then circumnavigated the globe **(Figure 19.5)**. When the ship's naturalist quit his post mid journey, Darwin replaced him in an unofficial capacity. For nearly five years Darwin toured the world, and because he suffered from seasickness, he seized every chance to go ashore. He collected plants and animals in Brazilian rain forests and fossils in Patagonia. He hiked the grasslands of the Pampas and climbed the Andes in Chile. Armed with Henslow's parting gift, the first volume of Lyell's *Principles of Geology*, Darwin was primed to apply gradualism and uniformitarianism to the living world.

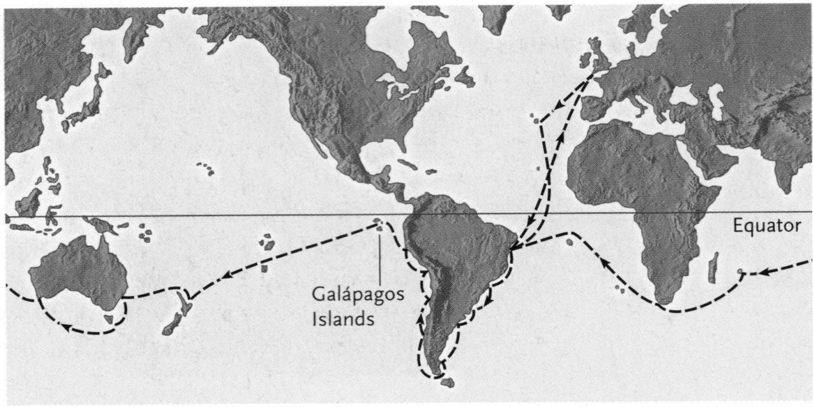
Equator
Galápagos Islands

FIGURE 19.5
Darwin's voyage. H.M.S. *Beagle* circumnavigated the globe between 1831 and 1836.

WHAT DARWIN SAW When he began his travels, Darwin had no clue that biological evolution had produced the mind-boggling variety of species that he would encounter. Three broad sets of observations later helped him unravel the mystery of evolutionary change.

First, while exploring along the coast of Argentina, Darwin discovered fossils that often resembled organisms inhabiting the same region today. For example, despite an enormous size difference, living armadillos and fossilized glyptodonts had similar body armor, but they were unlike any other species known to science **(Figure 19.6)**. If both species had been created at the same time and both were found in South America, why didn't glyptodonts live alongside armadillos? Darwin later wondered whether armadillos might be living relatives of the now-extinct glyptodonts.

Second, Darwin observed that the animals he encountered in different South American habitats clearly resembled each other but differed from species that occupied similar habitats in Europe. For example, he noted that nutria *(Myocastor coypus)*, a semiaquatic rodent native to South America, bore a closer resemblance to rodent species from the mountains or grasslands of that continent than it did to the European beaver *(Castor fiber)*, another semiaquatic rodent **(Figure 19.7)**. Why did animals from markedly different South American environments resemble

FIGURE 19.6
Ancestors and descendants. Darwin hypothesized that even though an extinct glyptodont (top) probably weighed 300 to 400 times as much as a living nine-banded armadillo *(Dasypus novemcinctus)*, their obvious resemblance suggested that they are related.

South American nutria *(Myocastor coypus)*

European beaver *(Castor fiber)*

FIGURE 19.7
Morphologic differences in species from different continents. Darwin noted that the South American nutria and the European beavers differ in appearance, even though both species are semiaquatic rodents that feed on vegetation. Notice that nutria have long, round tails, whereas beavers have short, flat tails.

A. The Galápagos **B. Galápagos tortoise *(Geochelone elephantopus)*** **C. Marine iguana *(Amblyrhynchus cristatus)***

Image copyright rebvt, 2010. Used under license from Shutterstock.com

iStockphoto.com/Sebastien Cote

FIGURE 19.8
The Galápagos Islands and some of their unusual animal inhabitants. **(A)** Volcanic eruptions created the Galápagos archipelago (located 1,000 km west of Ecuador) between 3 and 5 million years ago. **(B)** The islands were named for the giant tortoises found there (in Spanish, *galápa* means tortoise); this tortoise is native to Isla Santa Cruz. **(C)** Marine iguanas dive into the Pacific Ocean to feed on algae.

each other, and why were animals that lived in similar environments on separate continents different? Darwin later understood that animals in South America resembled each other because they had inherited their similarities from a common ancestor.

Third, Darwin observed fascinating patterns in the distributions of species on the Galápagos Islands **(Figure 19.8).** There he found strange and wonderful creatures, including giant tortoises and lizards that dove into the sea to feed on algae. Darwin quickly noted that the animals on different islands varied slightly in form. Indeed, experienced sailors could easily identify a tortoise's island of origin by the shape of its shell. Moreover, many species resembled those on the distant South American mainland. Why did so many different organisms occupy one small island cluster, and why did these species resemble others from the nearest continent? Darwin later hypothesized that the plants and animals of the Galápagos Islands were descended from South American ancestors, and that each species had changed after being isolated on a particular island.

DARWIN'S REFLECTIONS AFTER HIS VOYAGE The *Beagle* returned to England in 1836, and Darwin began his first notebook on the "transmutation of species" the following year. He realized that changes in species over time provided the only plausible explanation for his observations.

A diverse group of finches from the Galápagos Islands **(Figure 19.9)** provided the single greatest spark for Darwin's work. He had noticed great variability in the shapes of their bills, but he

A. Warbler finch *(Certhidea olivacea)* **B. Common cactus-finch *(Geospiza scandens)*** **C. Large ground-finch *(Geospiza magnirostris)*** **D. Woodpecker finch *(Camarhynchus pallidus)***

Dr. P. Evans/Bruce Coleman

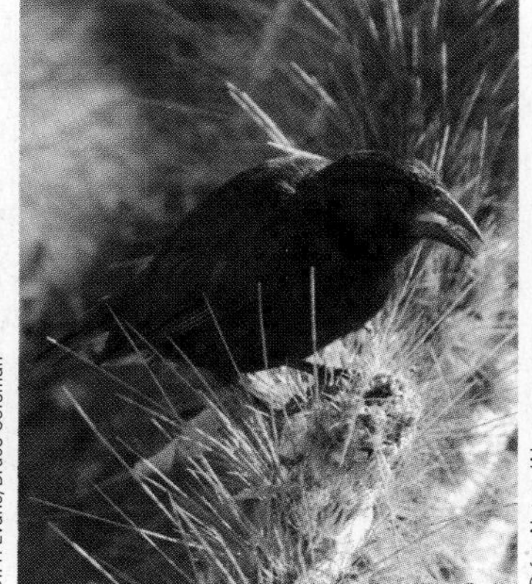
© fotoNatura/Alamy

Mark Moffett/Minden Pictures

Alan Root/Bruce Coleman Ltd.

FIGURE 19.9
Bill shape and food habits. The 13 finch species that inhabit the Galápagos Islands are descended from a common ancestor, a seed-eating ground finch that migrated to the islands from South America. **(A)** The warbler finch uses its slender bill to probe for insects in vegetation. **(B)** The common cactus-finch has a medium-sized bill suitable for eating cactus flowers and fruit. **(C)** The large ground-finch uses its thick, strong bill to crush cactus seeds. **(D)** The woodpecker finch uses its bill to hammer at bark and to hold cactus spines, with which it probes for wood-boring insects, such as termites.

had incorrectly assumed that birds on different islands belonged to the same species. Thus, he had not recorded the island where he had captured each specimen. Luckily, the *Beagle's* captain, Robert Fitzroy, had more thoroughly documented his own collection, allowing Darwin to study the relationships and geographical distributions of a dozen species. As Darwin reviewed the data, he began to focus on two aspects of a general problem. Why were the finches on a particular island slightly different from those on nearby islands, and how did all these different species arise?

Darwin Used Common Knowledge and Several Inferences to Develop His Theory

With a substantial inheritance, and burdened by chronic illness, Darwin led a reclusive life as he embarked on an intellectual journey every bit as exciting as his voyage on the *Beagle* (see *Focus on Basic Research*). His lifetime goal was to accumulate evidence of evolutionary change and identify the mechanism that caused it.

SELECTIVE BREEDING AND HEREDITY Having grown up in the country, Darwin was well aware that "like begets like"; that is, offspring frequently resemble their parents. Plant breeders and animal breeders had applied this basic truth of inheritance for thousands of years. By selectively breeding individuals with desired characteristics, they enhanced those traits in future generations.

Farmers use selective breeding to improve domesticated plants and animals. If one cow produces more milk than any other, the farmer selectively breeds her (rather than others), hoping that her offspring will also be good milk producers. Although the mechanism of heredity was not yet understood, this principle had been applied countless times to produce bigger beets, plumper pigs, and prize-winning pigeons (see Figure 1.8). Darwin was well aware of this process, which he called **artificial selection,** but he puzzled over how it could operate in nature.

THE STRUGGLE FOR EXISTENCE Darwin had a revelation about how selective breeding could occur naturally when he read the famous publication by Thomas Malthus, *Essay on the Principles of Population*. Malthus, an English clergyman and economist, was worried about the fate of the nation's poor. England's population was growing much faster than its agricultural capacity, and with individuals competing for limited food resources, some would inevitably starve.

Darwin applied Malthus's argument to organisms in nature. Species typically produce many more offspring than are needed to replace the parent generation, yet the world is not overrun with sunflowers, tortoises, or bears. Darwin even calculated that, if its reproduction went unchecked, a single pair of elephants, the slowest breeding animal known, would leave roughly 19 million descendants after only 750 years. Happily for us (and all other species that might get underfoot), the world is not so crowded with elephants. Instead, some members of every population survive and reproduce, whereas others die without reproducing.

DARWIN'S INFERENCES Darwin's discovery of a mechanism for evolutionary change required him to infer the nature of a process that no one had envisioned, much less documented **(Figure 19.10)**. First, individuals within populations vary in size, form, color, behavior, and other characteristics. Second, many of these variations are hereditary. What if variations in hereditary traits enabled some individuals to survive and reproduce more than others? Organisms with such advantageous traits would leave many offspring, whereas those that lacked such traits would die leaving few, if any, descendants. Thus, advantageous hereditary traits would become more common in the next generation. If the next generation was subjected to the same process of selection, the traits would be even more common in the third generation. Because this process is analogous to artificial selection, Darwin called it **natural selection.**

As an evolutionary mechanism, natural selection favors **adaptive traits,** genetically based characteristics that make organisms more likely to survive and reproduce. And by favoring individuals that are well adapted to the environments in which they live, natural selection causes species to change through time. As shown in Figure 19.9, each species of Galápagos finch has a distinctive bill. Variations in bill size and shape make some birds better adapted for crushing seeds and others for capturing insects. Imagine an island where large seeds were the only food available; individuals with a stout bill would be more likely to survive and reproduce than would birds with slender bills. These favored individuals would pass the genes that produce stout bills to their descendants, and after many generations, their bills might resemble those of *Geospiza magnirostris* (see Figure 19.9C).

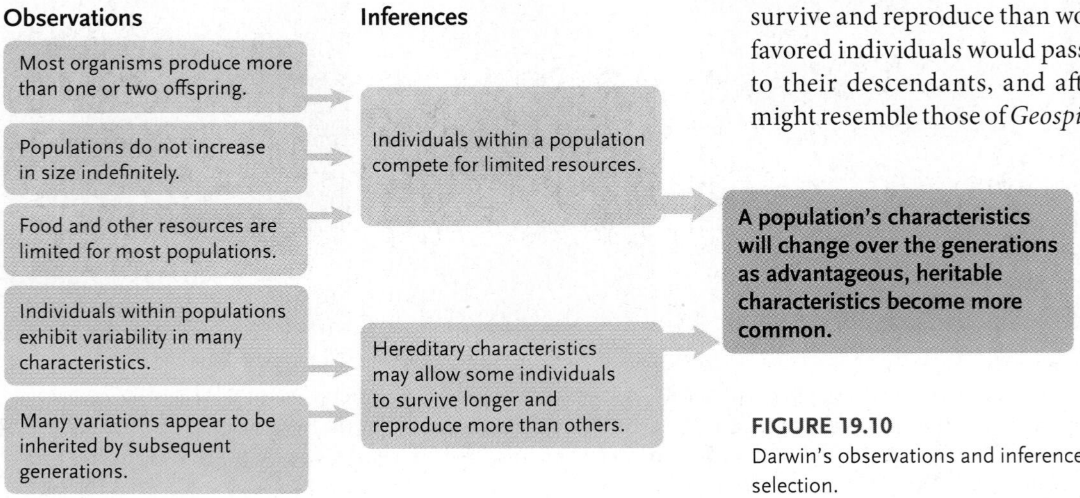

Observations
Inferences
Most organisms produce more than one or two offspring.
Populations do not increase in size indefinitely.
Food and other resources are limited for most populations.
Individuals within populations exhibit variability in many characteristics.
Many variations appear to be inherited by subsequent generations.
Individuals within a population compete for limited resources.
Hereditary characteristics may allow some individuals to survive longer and reproduce more than others.
A population's characteristics will change over the generations as advantageous, heritable characteristics become more common.

FIGURE 19.10
Darwin's observations and inferences about evolution by means of natural selection.

FOCUS ON Basic Research

Charles Darwin's Life as a Scientist

Darwin's observations during the voyage of H.M.S. *Beagle* convinced him that species change through time, and that natural processes produced Earth's biodiversity. He spent the rest of his life gathering data to test his ideas and unravel the workings of natural selection.

Shortly after the *Beagle* returned to England in 1836, Darwin began his first notebook on the "transmutation of species." But he put his study of evolution aside while he wrote up the geological and biological research that he had undertaken during the voyage. This task took him 10 years to complete—twice as long as the journey itself. The results of these efforts were numerous articles and several books, including the now famous *Journal of the Voyage of the Beagle,* published in 1839.

After preparing a sketch of his ideas about evolution in 1844, Darwin continued to write up his observations from the voyage. But he had trouble classifying one species of barnacle, a small marine invertebrate, which he had collected in Chile. For the next eight years he studied barnacles, examining more than 10,000 specimens and revising the entire classification of these animals. His colleagues saw this study as a strange diversion from his work on evolution, but Darwin's detailed examination of barnacle anatomy sharpened his observational skills and provided a test case in which he could apply his ideas about descent with modification to a large and diverse group of organisms. He published four volumes about barnacles in 1854.

While studying barnacles, Darwin continued to think about "the species question." He kept notebooks about variation in plants and animals, focusing on variation that was amplified by selective breeding. He was a tireless collector of facts, which he sought from every possible source. He badgered dog breeders, horse farmers, and horticulturists with long lists of questions about their work. His enthusiasm was infectious, and workers throughout the world supplied him with data and specimens. Darwin was also an eager and skilled experimentalist, and he took up pigeon breeding, marveling at the huge variety of morphological traits that he and other breeders could produce. In the late 1850s, a communication from Alfred Russel Wallace (see *Why It Matters*) forced him to finally complete *The Origin,* which revolutionized the study of biology.

Even after *The Origin* was published, Darwin continued to gather facts and write about evolution, working almost to the day he died in 1882 at age 74. He published a detailed analysis of how earthworms improve the soil *(The Formation of Vegetable Mould through the Action of Worms)* and wrote books on several botanical topics, among them plants that eat animals *(Insectivorous Plants),* and the tendency of plants to grow toward sunlight *(The Power of Movement in Plants).* Darwin's work always had an evolutionary focus, however, and he produced several revisions of *The Origin,* as well as books on artificial selection *(Variation of Animals and Plants under Domestication),* human ancestry *(The Descent of Man),* and animal behavior *(The Expression of the Emotions in Men and Animals).*

William Perlman/Corbis

Darwin's study. Darwin undertook most of his life's work in this room at Down House. He hesitated to discard old papers and specimens, believing that he would find a use for them as soon as they were carried away in the trash.

Natural selection also changes nonmorphologic characteristics of populations; for example, insect populations that are exposed to insecticides develop resistance to these toxic chemicals over time (see Figure 19.12).

Darwin realized that natural selection could also account for striking differences between populations and, given enough time, for the production of new species. For example, suppose that small insects were the only food available to finches on a different island. Birds with long thin bills might be favored by natural selection, and the population of finches might eventually possess a bill shaped like that of *Certhidea olivacea* (see Figure 19.9A). If we apply parallel reasoning to the many characteristics that affect survival and reproduction, natural selection would cause the populations to become more different over time, a process called **evolutionary divergence.**

Darwin's Theory Revolutionized the Way We Think about the Living World

It would be hard to overestimate the impact of Darwin's theory on Western thought. In *The Origin,* Darwin proposed a logical mechanism for evolutionary change and provided enough supporting evidence to convince the educated public.

Darwin argued that all the organisms that have ever lived arose through **descent with modification,** the evolutionary alteration and diversification of ancestral species. He envi-

Present
Time
Origin of life

FIGURE 19.11
The Tree of Life. Darwin envisioned the history of life as a tree. Branching points represent the origins of new lineages; branches that do not reach the top represent extinct groups.

sioned this pattern of descent as a tree growing through time **(Figure 19.11).** The base of the trunk represents the ancestor of all organisms. Branching points above it represent the evolutionary divergence of ancestors into their descendants. Each limb represents a body plan suitable for a particular way of life; smaller branches represent more narrowly defined groups of organisms; and the uppermost twigs represent living species. Biologists still apply this analogy today when analyzing the Tree of Life (see Figure 1.11B and Chapter 23).

Darwin proposed natural selection as the mechanism that drives evolutionary change. In fact, most of *The Origin* was an explanation of how natural selection acted on the variability within groups of organisms, preserving advantageous traits and eliminating disadvantageous ones.

Four characteristics distinguish Darwin's theory from earlier explanations of biological diversity and adaptive traits:

1. Darwin provided purely physical, rather than spiritual, explanations for the origins of biological diversity.
2. Darwin recognized that evolutionary change occurs in groups of organisms, rather than in individuals: some members of a group survive and reproduce more successfully than others.
3. Darwin described evolution as a multistage process: variations arise within groups, natural selection eliminates unsuccessful variations, and the next generation inherits successful variations.
4. Like Lamarck, Darwin understood that evolution occurs because some organisms function better than others *in a particular environment.*

What is most amazing about Darwin's intellectual achievement is that he knew nothing about genetics (see Chapter 12). Thus, he had no clear idea of how variation arose or how it was passed from one generation to the next.

Evolution was a popular topic in Victorian England, and Darwin's theory was both praised and ridiculed. Although he had not speculated about the evolution of humans in *The Origin,* many readers were quick to extrapolate Darwin's ideas to our own species. Needless to say, certain influential Victorians were not amused by the suggestion that humans and apes share a common ancestry.

Nevertheless, Darwin's painstaking logic and careful documentation convinced most readers that evolution really does take place. Thomas Huxley, so staunch an advocate that he was known as "Darwin's bulldog," summed up the reaction of many when he quipped that the theory was so obvious, once articulated, that he was surprised he had not thought of it himself. Darwin's vision of common ancestry quickly became the intellectual framework for nearly all biological research. Many readers, however, did not readily accept the mechanism of natural selection. The major stumbling block was that Darwin had not provided any plausible theory of heredity.

STUDY BREAK 19.2

1. **What observations that Darwin made on his round-the-world voyage influenced his later thoughts about evolution?**
2. **How did Darwin's understanding of artificial selection enable him to envision the process of natural selection?**
3. **What were the four great intellectual triumphs of Darwin's theory?**

THINK OUTSIDE THE BOOK

Find a copy of *On the Origin of Species by Means of Natural Selection* in a library or bookstore or online, and read a chapter or two. How did Darwin's careful observation of organisms allow him to draw conclusions? Do you think that straightforward observation still has a role in twenty-first–century science?

19.3 Evolutionary Biology since Darwin

Although Gregor Mendel published his work on genetics in 1866, it was not well known in England until 1900. At that time, scientists perceived a fundamental conflict between Darwin's and Mendel's theories. One problem was that Darwin had used complex characteristics, such as the structure of bird bills, to illustrate how natural selection worked. We now know that several genes often control such traits. By contrast, Mendel had studied simpler characteristics, such as the height of pea plants (see Chapter 12). A single gene often controls simple traits, which is one reason Mendel could interpret his experimental results so clearly. Biologists had a hard time applying Mendel's straightforward experimental results to Darwin's complex examples.

A second problem arose because Darwin believed that biological evolution occurred gradually over many generations. However, early twentieth-century geneticists, focusing on simple traits such as those Mendel had studied, sometimes observed very rapid and dramatic changes in certain characteristics. A widely accepted theory, *mutationism,* suggested that evolution occurred in spurts, induced by the chance appearance of "hopeful monsters," rather than by gradual change.

The Modern Synthesis Created a Unified Theory of Evolution

In the early twentieth century, Thomas Hunt Morgan of Columbia University discerned that genes are carried on chromosomes. His experiments, described in Chapter 13, enabled geneticists and mathematicians to forge a critical link between Darwin's and Mendel's ideas. The new discipline, **population genetics,** recognized the importance of genetic variation as the raw material of evolution. Population geneticists constructed mathematical models, which applied equally well to simple and complex traits, to predict how natural selection and other processes influence a population's genetics.

In the 1930s and 1940s, a unified theory of evolution, the **modern synthesis,** interpreted data from biogeography, comparative morphology, comparative embryology, paleontology, and taxonomy within an evolutionary framework. The authors of the modern synthesis focused on evolutionary change within populations, and although they considered natural selection the primary mechanism of evolution, they acknowledged the importance of other processes (see Chapter 20). Proponents of the modern synthesis also embraced Darwin's idea of gradualism and deemphasized the significance of mutations that changed traits suddenly and dramatically.

The modern synthesis also tried to link the two levels of evolutionary change that Darwin had identified: microevolution and macroevolution. **Microevolution** describes the small-scale genetic changes that populations undergo, often in response to shifting environmental circumstances; a small evolutionary shift in the size of the bill of a finch species is an example of microevolution. **Macroevolution** describes larger-scale evolutionary changes observed in species and more inclusive groups. According to the modern synthesis, macroevolution results from the gradual accumulation of microevolutionary changes; researchers have recently begun to unravel the genetic mechanisms that establish a relationship between these two levels of evolutionary change (see Chapter 22).

Research in Many Fields Has Provided Evidence of Evolutionary Change

Since the emergence of the modern synthesis, scientists have assembled a huge and compelling body of evidence from many biological disciplines indicating that biological evolution is a fact of life on Earth.

ADAPTATION BY NATURAL SELECTION Biologists interpret the products of natural selection as evolutionary adaptations. For example, the wings of birds, which have been modified by evolutionary processes over millions of years, have an obvious function that helps these animals survive and reproduce. Sometimes, however, natural selection operates on a short time scale, as illustrated by the development of pesticide resistance in insects. When we first use a new pesticide, a low concentration often kills a large percentage of the pests. However, just by chance, a few insects may have genetic characteristics that confer resistance to the poison. These individuals survive and produce offspring, many of which inherit the resistance. As a result, a given concentration of the poison kills a smaller percentage of insects in the next generation, and over time, the entire population may become highly resistant **(Figure 19.12).**

THE FOSSIL RECORD Because evolution results from the modification of existing species, Darwin's theory proposes that all species that have ever lived are genetically related. The fossil record documents such continuity in morphological characteristics, providing clear evidence of ongoing change in **biological lineages,** evolutionary sequences of ancestral organisms and their descendants (see Chapter 22). For example, the evolution of modern birds can be traced from a dinosaur ancestor through fossils such as *Archaeopteryx lithographica* **(Figure 19.13).** This species, discovered only two years after *The Origin* was published, resembled both dinosaurs and birds. Like the small carnivorous dinosaur *Dromaeosaurus, Archaeopteryx* walked on its hind legs and had teeth, claws on its forelimbs, and a long, bony tail. Like modern birds, it had hollow bones, an enlarged sternum, and feathers that covered its body.

HISTORICAL BIOGEOGRAPHY Analyses of **historical biogeography,** the study of the geographical distributions of plants and animals in relation to their evolutionary history, are generally consistent with Darwin's theory of evolution. Species on oceanic islands often closely resemble species on the nearest mainland, suggesting that the island and mainland species share a common ancestry. Moreover, species on a continental land mass are clearly related to one another and are often distinct from those on other continents. For example, monkeys in South America have long, prehensile tails and broad noses, traits that they inherited from a shared South American ancestor. By contrast, monkeys in Africa and Asia evolved from a different common ancestor in the Old World, and their shorter tails and narrower noses distinguish them from their American cousins.

COMPARATIVE MORPHOLOGY Other evidence of evolution comes from **comparative morphology,** analyses of the structure of living and extinct organisms. Such analyses are based on the comparison of **homologous traits,** characteristics that are similar in two species because they inherited the genetic basis of the trait from their common ancestor. For example, the forelimbs of all four-legged vertebrates are homologous because they evolved from a common ancestor with a forelimb composed of the same component parts (see Figure 19.3, which shows homologous

FIGURE 19.12 Experimental Research

How Exposure to Insecticide Fosters the Evolution of Insecticide Resistance

Question: Does exposure to insecticide foster the evolution of insecticide resistance in insect populations?

Experiment: Researchers studied samples of wild mosquitoes *(Anopheles culicifacies)* captured at a small village in India, where public health officials frequently sprayed the insecticide dichloro-diphenyl-trichloroethane (DDT) to control these pests. For each test, the researchers exposed samples of mosquitoes to a 4% concentration of DDT for 1 hour and then measured the percentage that died during the next 24 hours. Tests were repeated 12 months and 16 months after the first experiment.

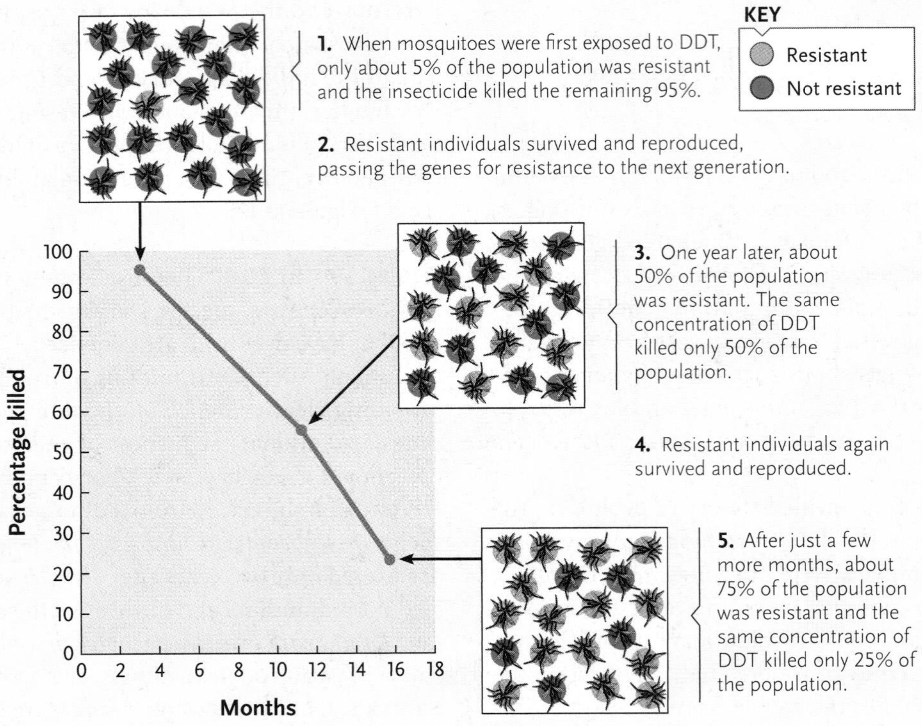
KEY
Resistant
Not resistant
1. When mosquitoes were first exposed to DDT, only about 5% of the population was resistant and the insecticide killed the remaining 95%.
2. Resistant individuals survived and reproduced, passing the genes for resistance to the next generation.
3. One year later, about 50% of the population was resistant. The same concentration of DDT killed only 50% of the population.
4. Resistant individuals again survived and reproduced.
5. After just a few more months, about 75% of the population was resistant and the same concentration of DDT killed only 25% of the population.
Percentage killed
100
90
80
70
60
50
40
30
20
10
0
0 2 4 6 8 10 12 14 16 18
Months

Results: Over the course of the experiment, smaller and smaller percentages of the mosquitoes died after their exposure to the test concentration of the insecticide.

Conclusion: The indiscriminate use of DDT established natural selection favoring DDT-resistant individuals. Exposure to DDT therefore fostered the evolution of an adaptive resistance to DDT in the mosquito population.

Source: A.M. Shalaby. 1968. Susceptibility studies on *Anopheles culicifacies* with DDT and Dieldrin in Gujarat state, India. *Journal of Economic Entomology* 61:533–541.

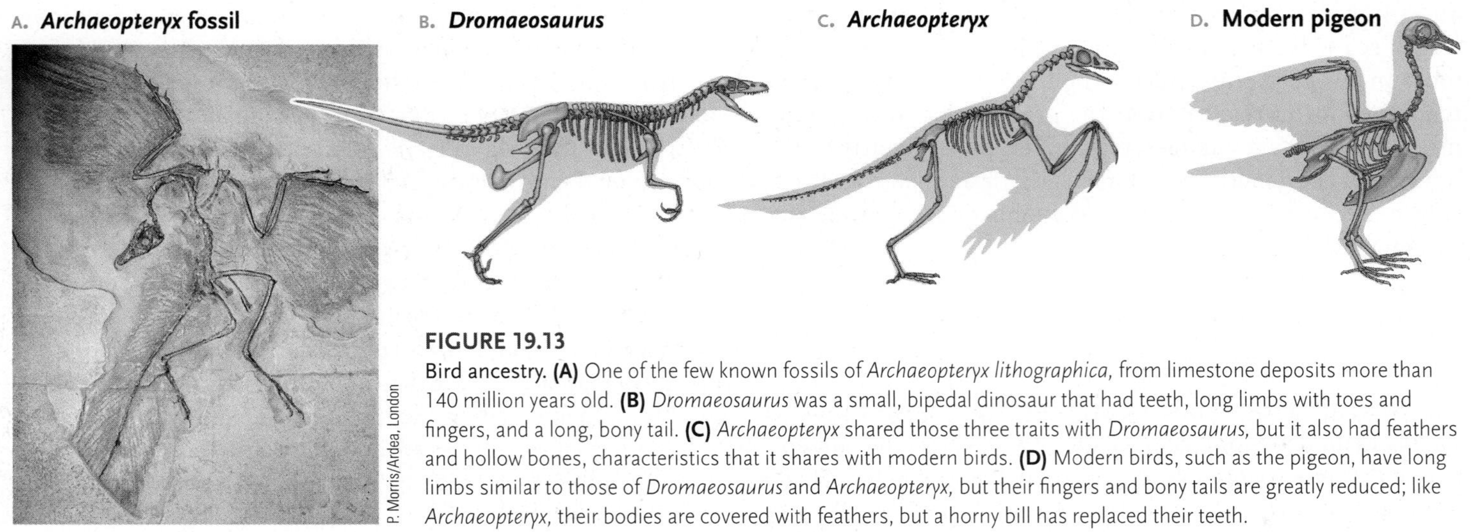
A. Archaeopteryx fossil
B. Dromaeosaurus
C. Archaeopteryx
D. Modern pigeon
P. Morris/Ardea, London

FIGURE 19.13
Bird ancestry. **(A)** One of the few known fossils of *Archaeopteryx lithographica,* from limestone deposits more than 140 million years old. **(B)** *Dromaeosaurus* was a small, bipedal dinosaur that had teeth, long limbs with toes and fingers, and a long, bony tail. **(C)** *Archaeopteryx* shared those three traits with *Dromaeosaurus,* but it also had feathers and hollow bones, characteristics that it shares with modern birds. **(D)** Modern birds, such as the pigeon, have long limbs similar to those of *Dromaeosaurus* and *Archaeopteryx,* but their fingers and bony tails are greatly reduced; like *Archaeopteryx,* their bodies are covered with feathers, but a horny bill has replaced their teeth.

A. **Monitor lizard (*Varanus* species)**

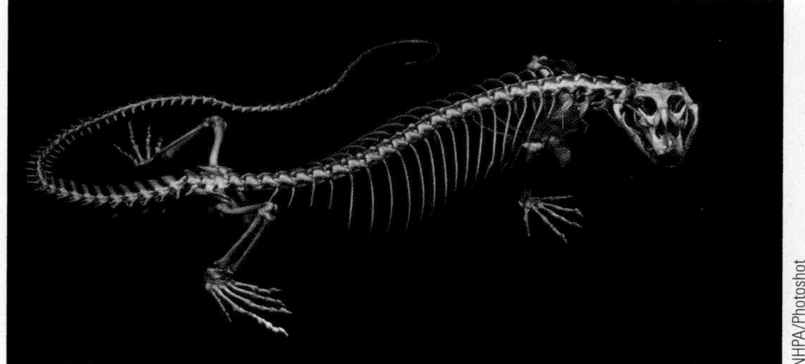
NHPA/Photoshot

B. **Grass snake (*Natrix* species)**

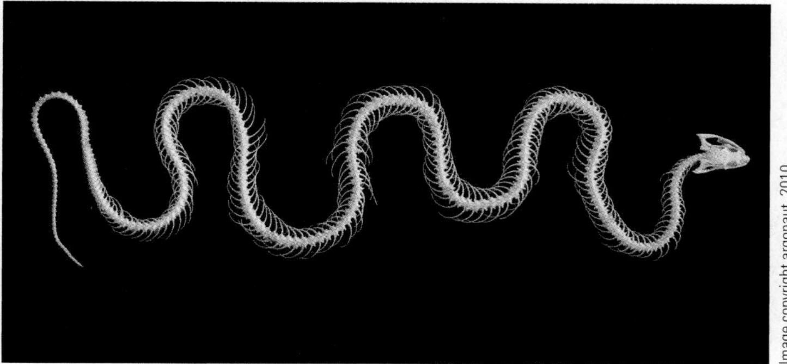
Image copyright argonaut, 2010.
Used under license from Shutterstock.com

C. **Ball python (*Python regius*)**

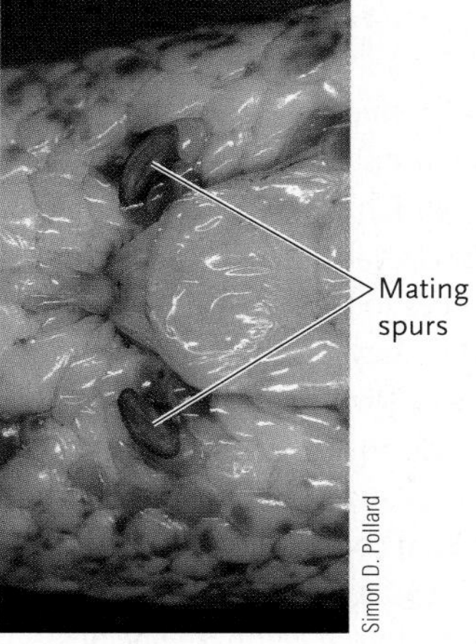
Mating
spurs
Simon D. Pollard

D. **Chick embryo**

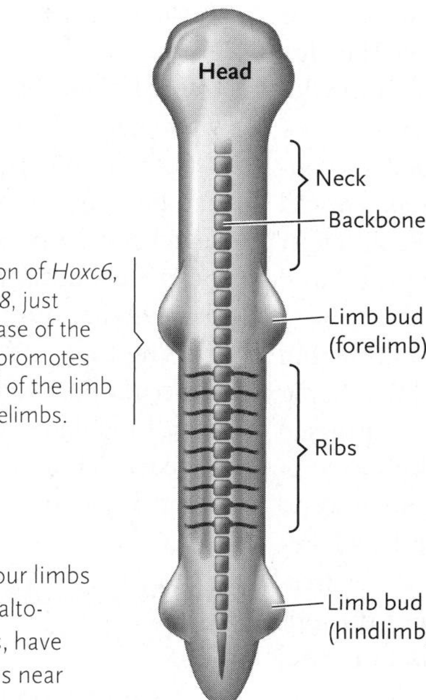
Head
Neck
Backbone
Limb bud
(forelimb)
Ribs
Limb bud
(hindlimb)

The expression of *Hoxc6*, but not *Hoxc8*, just behind the base of the chick's neck promotes development of the limb buds and forelimbs.

E. **Python embryo**

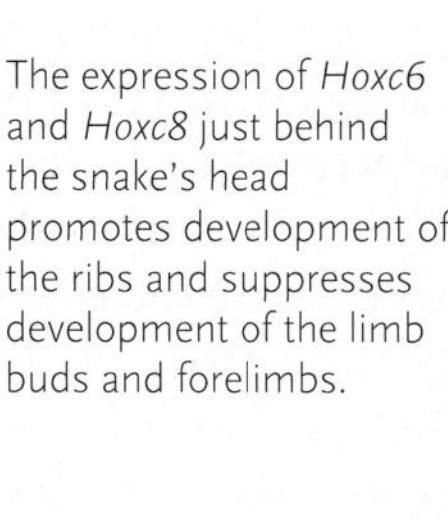
The expression of *Hoxc6* and *Hoxc8* just behind the snake's head promotes development of the ribs and suppresses development of the limb buds and forelimbs.

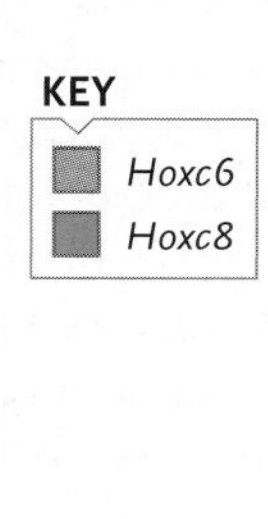
KEY
Hoxc6
Hoxc8

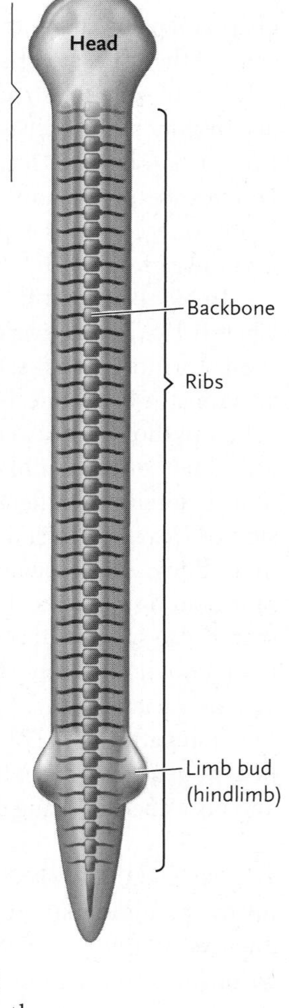
Head
Backbone
Ribs
Limb bud
(hindlimb)

FIGURE 19.14
Genetics of limb loss in snakes. **(A)** Most lizards have four limbs attached to their backbones; **(B)** most snakes lack limbs altogether. **(C)** Some primitive snakes, like boas and pythons, have vestigial hindlimbs, visible as a pair of claw-like structures near the base of the tail. **(D)** Molecular analyses reveal that the expression of *Hoxc6*, but not *Hoxc8*, at the base of the neck in a chick embryo causes forelimbs to develop nearby. **(E)** In pythons, both *Hoxc6* and *Hoxc8* are expressed alongside the backbone, beginning just behind the head. The expression of these two genes appears to suppress the development of limbs and promote the development of ribs at that location.

bones in the same color). Even though the shapes of the bones are different in pigs, dolphins, and bats, similarities in the three limbs are apparent. The differences in structural details arose over evolutionary time, allowing pigs to walk, dolphins to swim, and bats to fly. The arms of humans and the wings of birds are constructed of comparable elements, suggesting that they, too, share a common ancestor with the three species that are illustrated.

Molecular Techniques Extend the Achievements of the Modern Synthesis

Molecular techniques provide biologists with powerful tools for exploring all aspects of life—and evolutionary biology is no exception. In this section, we describe two examples that illustrate how molecular techniques have furthered our understanding of evolution since the modern synthesis was articulated.

HOW SNAKES LOST THEIR LIMBS

From Darwin's time until the mid-twentieth century, biologists tried to discern evolutionary patterns in animals by comparing their embryos and patterns of development **(Figure 19.14)**. The early embryos of related species are often strikingly similar, but morphological differences appear as the embryos grow and develop their adult forms (see Chapter 48). For example, the early embryos of most four-legged vertebrates (such as lizards, mammals, and birds) develop "limb buds" from which the legs or wings typical of each species later grow. Forelimbs and their supporting structures grow at the base of the neck, just in front of the ribcage, and hindlimbs grow right behind the ribcage (Figure 19.14A). Similarities in the limb buds and the positions of the limbs in these animals provide evidence of their descent from a shared ancestor. Differences in their adult structures are caused by additional genetic instructions that have evolved over time.

However, most snakes—which are very closely related to lizards—show no traces of limbs or necks; their ribcages are positioned right behind their heads (Figure 19.14B). The fossil record shows us that snakes evolved from four-legged ancestors in stages: early snakes had small hindlimbs, and the most recently evolved snakes have no limbs at all. Only the most ancient living snakes, pythons and boas, have any traces of limbs—vestigial hindlimbs, which appear as a pair of tiny claw-like structures near the base of the tail (Figure 19.14C). Observational studies of their embryos reveal that most living snakes never develop limb buds; by contrast, pythons and boas develop hind limb buds, which grow only slightly as the animal develops.

How did pythons, and presumably all other snakes, lose their necks and forelimbs? Although many genes control the differences in the adult forelimbs among species (see Figure 19.3), research reported in the 1990s revealed that two regulatory genes, *Hoxc6* and *Hoxc8,* determine whether forelimbs or ribs grow at a particular site along an animal's backbone. The *Hox* genes either activate or suppress other genes that direct the development of these structures. Forelimbs—but not ribs—grow just in front of the tissues where only *Hoxc6* is expressed. By contrast, ribs—but not forelimbs—grow where both *Hoxc6* and *Hoxc8* are expressed. In chickens and other vertebrates with four limbs, only *Hoxc6* is expressed just in front of the ribcage, and forelimbs grow nearby (Figure 19.14D).

In 1999, Martin J. Cohn of the University of Reading and Cheryll Tickle of University College London reported that in pythons (primitive snakes), both *Hoxc6* and *Hoxc8* are expressed all along the backbone, beginning at the base of the skull; as a result, a python's ribcage develops right behind its head, and no limb buds or forelimbs develop (Figure 19.14E). Thus, snakes have no forelimbs or necks because a mutation causes the expression of *Hoxc8* to extend into a more forward region of the animal's body. All descendants of the ancestor with that original mutation now lack necks and forelimbs. Cohn and Tickle's research also suggests that the second stage in snake evolution, the reduction or complete absence of hindlimbs, is caused by other genetic variations that appeared some time after the altered expression pattern of *Hoxc8*. Thus, molecular research has identified the genetic changes that caused snakes to lose their forelimbs and necks before losing their hindlimbs.

THE WOOLLY MAMMOTH'S CLOSEST LIVING RELATIVE Evolutionary biologists now routinely compare the genetic sequences of different species to determine their evolutionary relationships (see Chapter 23). The process is straightforward as long as researchers can collect small tissue samples from which they extract DNA. For many years, paleobiologists could not use these techniques to determine the evolutionary relationships between extinct organisms and those living today because they could not obtain the necessary samples from fossilized material. Sequencing the DNA of extinct organisms was the "holy grail" of paleobiology.

Recent advances in molecular techniques now allow researchers to sequence DNA preserved in some fossils of extinct organisms. In a study published in 2006, Hendrick Poinar of McMaster University and colleagues at other institutions sequenced 13 million base pairs of mitochondrial and nuclear DNA extracted from the jawbone of a woolly mammoth *(Mammuthus primigenius),* a species that has been extinct for at least 4,000 years **(Figure 19.15).** The mammoth, which died 27,000 years ago, had been preserved in a frigid Siberian ice cave. When the researchers compared its DNA sequences to those from a living African elephant *(Loxodonta africana),* they discovered that more than 98% of the sequence was identical in the two species, confirming their close evolutionary relationship. *Insights from the Molecular Revolution* describes another study, published the same year, in which several European researchers analyzed the complete mitochondrial genome sequence of a woolly mammoth and compared it to the sequences from two living elephant species.

Mauricio Anton

FIGURE 19.15
Woolly mammoths.

Some People Misinterpret the Theory of Evolution

The theory of evolution has always been a contentious subject because it challenges deeply held traditional views of how living organisms originated. Many of Darwin's contemporaries were dismayed by the suggestion that all organisms share a common ancestry. Some people even misinterpreted this assertion as "humans evolved from chimpanzees or gorillas." But the theory of evolution makes no such claims. Instead, it suggests that humans and apes are descended from an apelike common ancestor (see Section 30.13). In other words, an ancient population of organisms left descendants, which now include the living species of apes, as well as our own species. Moreover, the theory recognizes that evolution is an ongoing process: humans and apes have been evolving up until this very moment and will continue to evolve for as long as their descendants persist.

Early in the twentieth century, some scientists embraced the notion of **orthogenesis,** or progressive, goal-oriented evolution. This idea, derived from the *Scala Naturae,* suggests that evolution produces new species with the goal of improvement. We now know that evolution proceeds as an ongoing process of dynamic adjustment, not toward any fixed goal. Natural selection preserves the genes of organisms that function well in particular environments, but it cannot predict future environmental change. Imagine a population of plants with genes that affect how well they function under wet versus dry conditions. After a

Asian Elephant or African Elephant: Which is the woolly mammoth's closest living relative?

Recent technical innovations now allow biologists to extract DNA from some fossils of extinct organisms, providing answers to long-puzzling questions. Based on morphological evidence, paleobiologists suspected that woolly mammoths *(Mammuthus primigenius)* were more closely related to living Asian elephants *(Elephas maximus)* than to living African elephants *(Loxodonta africana)*. When they recovered samples of mammoth DNA, researchers knew they could settle the matter.

Research Question

Were woolly mammoths genetically more similar to Asian elephants than to African elephants?

Experiment

Svante Pääbo and his colleagues at the Max Planck Institute for Evolutionary Anthropology, Leipzig, Germany, and collaborating researchers at three other institutions in England, Germany, and the United States, used molecular techniques—specifically, PCR amplification, cloning, and sequencing of mitochondrial DNA—to answer the question (see Chapter 18).

1. They extracted DNA from a mammoth bone, estimated to be between 11,900 and 13,400 years old, from Berelek, Yakutia (in Siberia).
2. Based on known mitochondrial DNA sequences from the African and Asian elephant, they designed and synthesized 46 pairs of PCR primers. Assuming homology with the elephant sequences, these primer pairs were predicted to amplify the entire circular mitochondrial genome of the mammoth (yellow circle) as overlapping DNA fragments. All of the primer pairs were used in a single PCR, and the amplified DNA fragments produced were cloned and sequenced. Based on their overlaps, the sequences could be arranged in a circle (blue and red), which showed that the researchers had succeeded in amplifying the entire mitochondrial genome.
3. Using computer algorithms, this mitochondrial DNA was compared with the mitochondrial genomes of Asian and African elephants.

Conclusion

The mammoth DNA sequences were more similar to those from the Asian elephant than those from the African elephant. Thus, Asian elephants are the woolly mammoth's closest living relatives. This conclusion makes biogeographic sense because woolly mammoths and Asian elephants occupied the same land mass (that is, Asia), whereas African elephants live on a different continent.

Source: J. Krause et al. 2006. Multiplex amplification of the mammoth mitochondrial genome and the evolution of the Elephantidae. *Nature* 439:724–727.

DNA extracted

Mammoth bone

DNA

Mammoth mitochondrial genome sequence assembled

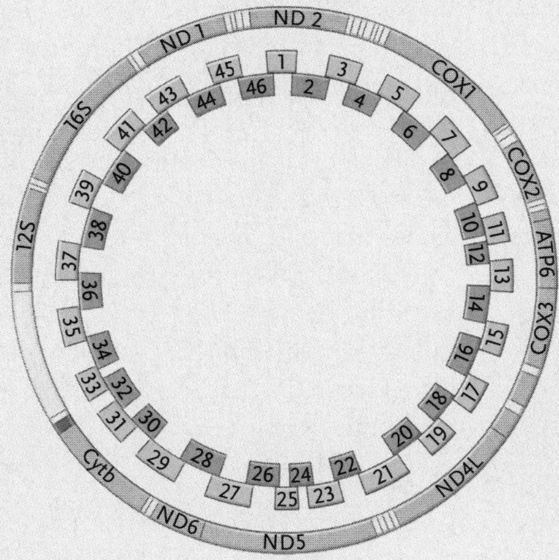
ND 1
ND 2
COX1
COX2
ATP6
COX3
ND4L
ND5
ND6
Cytb
12S
16S

Mitochondrial DNA compared

Woolly mammoth

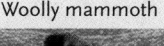

Mauricio Anton

African elephant

iStockphoto.com/Alan Lucas

Asian elephant

five-year drought, the population would include mostly dry-adapted plants. If a series of wet years follows the drought, these plants will be poorly adapted to the altered conditions. The process that favored drought-adapted plants operated under the prevailing dry conditions, not in anticipation of how conditions might change in the future.

Evolution is the core theory of modern biology because its explanatory power touches on every aspect of the living world. The application of molecular techniques to the study of evolutionary biology has greatly enhanced our knowledge. In the remaining chapters of this unit, you will discover how contemporary evolutionary theory explains changes at every level of biological organization, from adaptive modifications within populations (see Chapter 20), to the development of new species (see Chapter 21), major patterns in the history of life (see Chapter 22), and unraveling the evolutionary history of all organisms on Earth (see Chapter 23).

STUDY BREAK 19.3 <

1. **What two problems slowed the acceptance of Darwin's theory among scientists?**
2. **What is the difference between microevolution and macroevolution?**
3. **What types of data provide evidence that evolution has adapted organisms to their environments and promoted the diversification of species?**

THINK OUTSIDE THE BOOK >

Find examples from popular publications or advertisements for consumer products that misrepresent the theory of biological evolution. Explain how the theory is misrepresented.

UNANSWERED QUESTIONS

What determines whether a species adapts to a changing environment or becomes extinct?

Natural selection has produced marvelous adaptations in every species on Earth, and we know that evolutionary adaptation to certain environmental changes has allowed many species to persist. But we also know that more than 99% of the species that have ever lived became extinct, evidently because they failed to adapt to changes in climate, natural competitors or enemies, or other environmental factors. But what kinds of genetic variation are required for adaptation, and what kinds of characteristics must evolve to allow survival? This is a critical question today, because human activities are changing environments so rapidly and drastically that many species face the threat of extinction. Can aquatic species adapt to various kinds of water pollution? Can animals and plants that lived in prairies adapt to different habitats now that most prairies have been destroyed?

Is adaptation by natural selection responsible for most of the genetic differences between species?

New genetic variations sometimes become more common within populations or species because the proteins for which they code are advantageous and preserved by natural selection. But biologists who study molecular evolution have discovered that a large part of the genome in most organisms (about 98% of the human genome, for example) does not code for proteins and therefore appears to have no function. If this observation is generally correct, why do the noncoding parts of genomes exist? Are evolutionary changes in noncoding regions and the differences in noncoding sequences among species adaptive? For example, only about 1% of the DNA base pairs differ between human and chimpanzee genomes—but this amounts to about 34 million base-pair differences altogether, at least 60,000 of which alter the amino acid sequences of proteins. How can we determine which of these differences are adaptive and which differences underlie the unique characteristics of humans?

How do pathways of embryonic development evolve?

The characteristics of adult organisms are the product of developmental events, starting with the fertilized egg, that include growth in size, changes in the shape of various body parts, and the differentiation of cell types. These processes are largely controlled by genes, with input from the environment. Although biologists are beginning to learn how the genetic foundations of developmental processes evolve, many questions remain. For example, how do genetic changes induce differences in the branching patterns of antlers among species of deer, or differences in the length of the tails of monkeys and apes (including humans), or differences in the number and size of scales among species of lizards? We know that the proteins forming the lens of the eye are actually enzymes that play different roles in other cells, and that they have been "recruited" to form the lens, but what mechanisms induce them to assume this new role? And why do different enzymes form the lens in eyes of birds and mammals? Evolutionary developmental biology, which is discussed in Chapter 22, is one of the most active, exciting fields in biology at this time.

Think Critically

As our production of carbon dioxide increases Earth's average temperature faster than ever before, will Arctic species be able to adapt to changes in climate?

Douglas J. Futuyma

Douglas J. Futuyma is Distinguished Professor of evolution and ecology at Stony Brook University. His research interests focus on speciation and the evolution of ecological interactions among species, and in particular on insect–plant interactions. Learn more about Dr. Futuyma's work at http://life.bio.sunysb.edu/ee/people.htm.

REVIEW KEY CONCEPTS

Go to **CENGAGENOW** at www.cengage.com/login to access quizzing, animations, exercises, articles, and personalized homework help.

19.1 Recognition of Evolutionary Change

- Ancient Greek philosophers classified the natural world, ranking inanimate objects and living organisms from simple to complex.
- Natural theologians believed that all species were specially created and perfectly adapted; that existing species do not change or become extinct; and that new species do not arise. Studies in biogeography, comparative morphology, and paleontology led scientists to question whether species change through time (Figures 19.2 and 19.3).
- Lamarck proposed that species evolved into more complex forms that functioned better in their environments. He hypothesized that structures in an organism changed when they were used, and that those changes were inherited by the organism's offspring (Figure 19.4). Experiments have refuted Lamarck's proposed mechanisms.
- Two geologists, Hutton and Lyell, recognized that major features on Earth were created by the long-term action of the very slow geological processes that scientists observe today. Their insights suggested that Earth was much older than natural theologians had supposed.

19.2 Darwin's Journeys

- Darwin's observations during his voyage on the *Beagle* provided much of the data and inspiration for the development of his theory of evolution (Figures 19.5–19.8).
- Darwin based the theory of evolution by means of natural selection on three inferences: (1) individuals within a population compete for limited resources, (2) hereditary characteristics allow some individuals to survive longer and reproduce more than others, and (3) a population's characteristics change over time as advantageous heritable characteristics become more common (Figure 19.10).
- Darwin also proposed that the accumulation of differences fostered by natural selection could cause populations to diverge over time. Such evolutionary divergence can lead to the production of new species, which can, in turn, give rise to new evolutionary lineages (Figures 19.9 and 19.11).

Animation: The Galápagos Islands

Animation: Finches of the Galápagos Islands

19.3 Evolutionary Biology since Darwin

- Scientists working in population genetics developed theories of evolutionary change by integrating Darwin's ideas with Mendel's research on genetics.
- In the 1930s and 1940s, the modern synthesis provided a unified view of evolution that drew on studies from many biological disciplines. It emphasized evolution within populations, the central role of variation in the evolutionary process, and the gradualism of evolutionary change.
- Studies of adaptation, the fossil record, historical biogeography, and comparative morphology provide compelling evidence of evolutionary change (Figures 19.12–19.13).
- Molecular techniques have extended the achievements of the modern synthesis, allowing precise analysis of the genetic basis of evolutionary change (Figure 19.14) and the genetic relatedness of living and extinct organisms (Figure 19.15).

UNDERSTAND AND APPLY

Test Your Knowledge

1. Which of the following statements about evolutionary studies is *not* true?
   a. Biologists study the products of evolution to understand the processes causing it.
   b. Biologists design molecular experiments to examine evolutionary processes operating over short time periods.
   c. Biologists study the inheritance of characteristics that a parent acquired during its lifetime.
   d. Biologists study variation in homologous structures among related organisms.
   e. Biologists examine why a huge variety of species may inhabit a small island cluster.
2. Which of the following ideas is *not* included in Darwin's theory?
   a. All organisms that have ever existed arose through evolutionary modifications of ancestral species.
   b. The great variety of species alive today resulted from the diversification of ancestral species.
   c. Natural selection drives some evolutionary change.
   d. Natural selection preserves advantageous traits.
   e. Natural selection eliminates adaptive traits.
3. The "father" of taxonomy is:
   a. Charles Darwin.
   b. Charles Lyell.
   c. Alfred Wallace.
   d. Carolus Linnaeus.
   e. Jean Baptiste de Lamarck.
4. The wings of birds, the legs of pigs, and the flippers of dolphins provide an example of:
   a. vestigial structures.
   b. homologous structures.
   c. acquired characteristics.
   d. artificial selection.
   e. uniformitarianism.
5. Which of the following statements is *not* compatible with Darwin's theory?
   a. All organisms have arisen by descent with modification.
   b. Evolution has altered and diversified ancestral species.
   c. Evolution occurs in individuals rather than in groups.
   d. Natural selection eliminates unsuccessful variations.
   e. Evolution occurs because some individuals function better than others in a particular environment.
6. Which of the following does *not* contribute to the study of evolution?
   a. population genetics
   b. inheritance of acquired characteristics
   c. the fossil record
   d. DNA sequencing
   e. comparative morphology

7. Which of the following could be an example of microevolution?
   a. a slight change in a bird population's color due to a small genetic change in the population
   b. large differences between fossils found near the ground surface and those found in deep rock layers
   c. the sudden disappearance of an entire genus
   d. the direct evolutionary link between living primates and humans
   e. a flood that drowns all members of a population
8. Which of the following ideas proposed by Lamarck was *not* included in Darwin's theory?
   a. Organisms change in response to their environments.
   b. Changes that an organism acquires during its lifetime are passed to its offspring.
   c. All species change with time.
   d. New variations may be passed from one generation to the next.
   e. Specific mechanisms cause evolutionary change.
9. Medical advances now allow many people who suffer from genetic diseases to survive and reproduce. These advances:
   a. refute Darwin's theory.
   b. support Lamarck's theory.
   c. disprove descent with modification.
   d. reduce the effects of natural selection.
   e. eliminate adaptive traits.
10. The belief that evolution is progressive or goal-oriented is called:
   a. gradualism.
   b. uniformitarianism.
   c. taxonomy.
   d. orthogenesis.
   e. the modern synthesis.

Discuss the Concepts

1. Explain why the characteristics we see in living organisms adapt them to the environments in which their ancestors lived rather than to the environments in which they live today.
2. Imagine a population of mice that includes both brown and black individuals. They live in a habitat with brown soil, where predatory hawks can see black mice more easily than they can see brown ones. Design a study that would allow you to determine whether the brown mice are better adapted to this environment than the black mice.

Design an Experiment

Design an experiment to test Lamarck's hypothesis that characteristics acquired during an organism's lifetime are inherited by their offspring. (You may wish to review the components of a well-designed experiment in Chapter 1 before formulating your answer.) Can you think of examples of acquired characteristics that are *not* inherited by offspring?

Interpret the Data

For centuries, animal breeders have used artificial selection to increase the speed of racehorses. The results of such efforts are illustrated in a graph that identifies how quickly winning horses ran the $1\frac{1}{4}$ mile track at the Kentucky Derby, a famous race held annually at Churchill Downs, Kentucky. (The time it takes a horse to run a track of fixed length is inversely proportional to the horse's speed.) Based on the data in the graph, has artificial selection increased the speed of horses consistently between 1900 and 2000? Do you think that artificial selection can increase the speed of racehorses in the future?

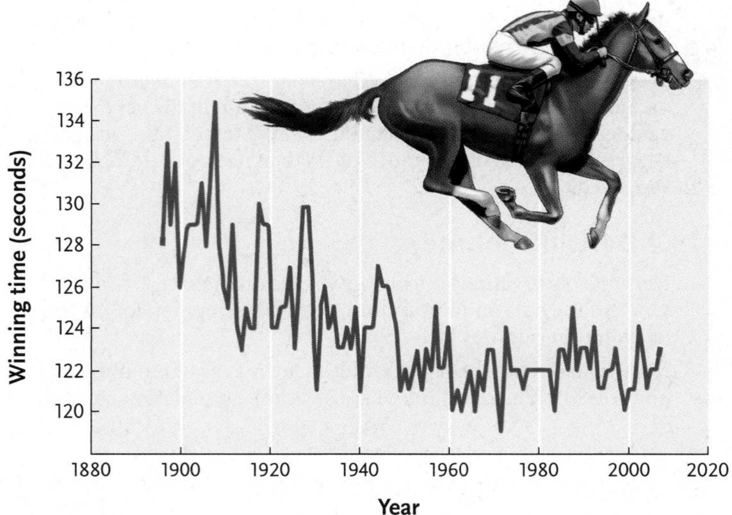
Winning time (seconds)
136
134
132
130
128
126
124
122
120
1880
1900
1920
1940
1960
1980
2000
2020
Year
11

Apply Evolutionary Thinking

Identify three discoveries or inventions that have changed how humans are affected by natural selection. Describe in detail how each discovery influences survival or reproduction in our species.

Phenotypic variation. The frog *Dendrobates pumilio* exhibits dramatic color variation in populations that inhabit the Bocas del Toro Islands, Panama.

Mark Moffett/Minden Pictures

20

STUDY OUTLINE

Microevolution: Genetic Changes within Populations

Why It Matters. . . On November 28, 1942, at the height of American involvement in World War II, a disastrous fire killed more than 400 people in Boston's Cocoanut Grove nightclub. Many more would have died later but for a new experimental drug, penicillin. A product of *Penicillium* mold, penicillin fought the usually fatal infections of *Staphylococcus aureus,* a bacterium that enters the body through damaged skin. Penicillin was the first antibiotic drug based on a naturally occurring substance that kills bacteria.

Until the disaster at the Cocoanut Grove, the production and use of penicillin had been a closely guarded military secret. But after its public debut, the pharmaceutical industry hailed penicillin as a wonder drug, promoting its use for the treatment of the many diseases caused by infectious microorganisms. Penicillin became widely available as an over-the-counter remedy, and Americans dosed themselves with it, hoping to cure all sorts of ills **(Figure 20.1)**. But in 1945, Sir Alexander Fleming, the scientist who discovered penicillin, predicted that some bacteria could survive low doses, and that the offspring of those germs would be more resistant to its effects. In 1946—just 4 years after penicillin's use in Boston—14% of the *Staphylococcus* strains isolated from patients in a London hospital were resistant. By 1950, more than half the strains were resistant.

Scientists have discovered numerous antibiotics since the 1940s, and many strains of bacteria have developed resistance to these drugs. In fact, according to the Centers for Disease Control and Prevention, between 30,000 and 40,000 Americans die each year from infections caused by antibiotic-resistant bacteria.

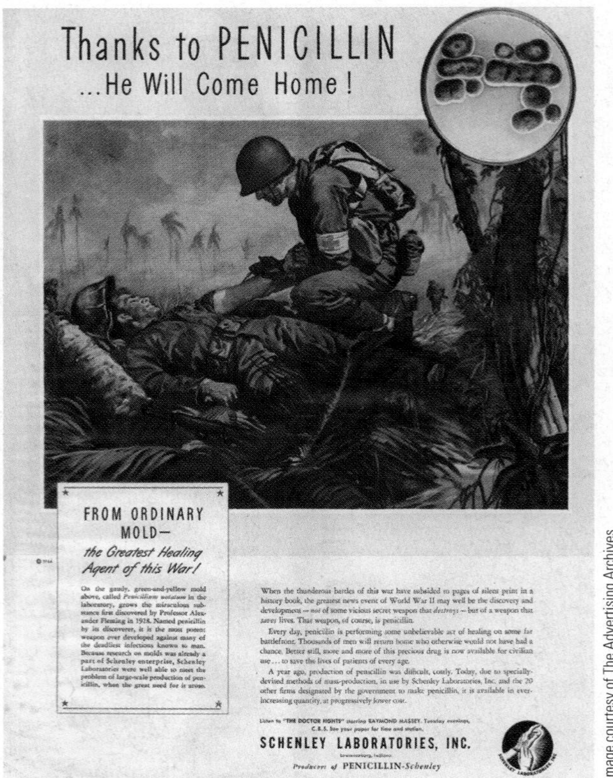
Thanks to PENICILLIN
...He Will Come Home!
FROM ORDINARY MOLD—
the Greatest Healing Agent of this War!
SCHENLEY LABORATORIES, INC.
Producers of PENICILLIN-Schenley

Image courtesy of The Advertising Archives

FIGURE 20.1
Selling penicillin. This ad, from a 1944 issue of *Life* magazine, credits penicillin with saving the lives of wounded soldiers.

How do bacteria become resistant to antibiotics? The genomes of bacteria—like those of all other organisms—vary among individuals, and some bacteria have genetic traits that allow them to withstand attack by antibiotics. When we administer antibiotics to an infected patient, we create an environment favoring bacteria that are even slightly resistant to the drug. The surviving bacteria reproduce, and resistant microorganisms—along with the genes that confer antibiotic resistance—become more common in later generations. In other words, bacterial strains adapt to the presence of antibiotics in their environment through the evolutionary process of selection. Our use of antibiotics is comparable to artificial selection by plant and animal breeders (see Chapter 19), but when we use antibiotics, we inadvertently select for the success of the organisms we are trying to eradicate.

The evolution of antibiotic resistance in bacteria is an example of **microevolution,** which is a heritable change in the genetics of a population. A **population** of organisms includes all the individuals of a single species that live together in the same place and time. Today, when scientists study microevolution, they analyze variation—the differences between individuals—in natural populations and determine how and why these variations are inherited. Darwin recognized the importance of heritable variation within populations; he also realized that natural selection can change the pattern of variation in a population from one generation to the next. Scientists have since learned that microevolutionary change results from several processes, not just natural selection, and that sometimes these processes counteract each other.

In this chapter, we first examine the extensive variation that exists within natural populations. We then take a detailed look at the most important processes that alter genetic variation within populations, causing microevolutionary change. Finally, we consider how microevolution can fine-tune the functioning of populations within their environments. <

20.1 Variation in Natural Populations

In some species, individuals vary dramatically in appearance, but in most species, the members of a population look pretty much alike **(Figure 20.2)**. However, even those that look alike, such as the *Cerion* snails on the right in Figure 20.2, are not identical. With a scale and ruler, you could detect differences in their weight as well as in the length and diameter of their shells. With suitable techniques, you could also document variations in their individual biochemistry, physiology, internal anatomy, and behavior. All of these are examples of **phenotypic variation,** differences in appearance or function that are passed from generation to generation.

Evolutionary Biologists Describe and Quantify Phenotypic Variation

Darwin's theory recognized the importance of heritable phenotypic variation, and today, microevolutionary studies often begin by assessing phenotypic variation within populations. Most characters exhibit **quantitative variation:** individuals differ in small, incremental ways. If you weighed everyone in your biology class, for example, you would see that weight varies almost continuously from your lightest to your heaviest classmate. Humans also exhibit quantitative variation in the length of their

A. **European garden snails *(Cepaea nemoralis)***

George Bernard/Photo Researchers, Inc.

B. **Bahaman land snails *(Cerion christophei)***

Timothy A. Pearce, Ph.D./Section of Mollusks/ Carnegie Museum of Natural History. Photograph by Mindy McNaugher

FIGURE 20.2
Phenotypic variation. **(A)** Shells of the European garden snail from a population in Scotland vary considerably in appearance. **(B)** By contrast, shells of Bahaman land snails from a population in the Bahamas look very similar.

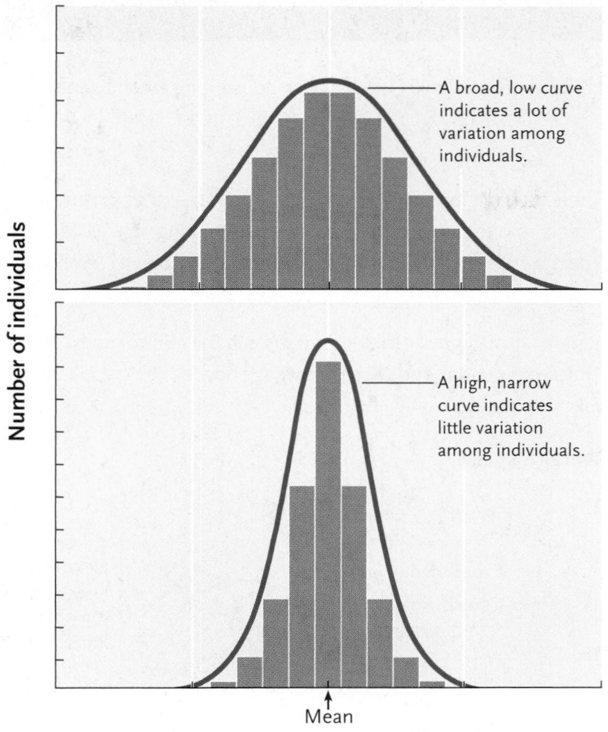
A broad, low curve indicates a lot of variation among individuals.
A high, narrow curve indicates little variation among individuals.
Number of individuals
Mean
Measurement or value of trait

FIGURE 20.3
Quantitative variation. Many traits vary continuously among members of a population, and a bar graph of the data often approximates a bell-shaped curve. The mean defines the average value of the trait in the population, and the width of the curve is proportional to the variability among individuals.

toes, the number of hairs on their heads, and their height, as discussed in Chapter 12.

We usually display data on quantitative variation in a bar graph or, if the sample is large enough, as a curve **(Figure 20.3).** The width of the curve is proportional to the variability—the amount of variation—among individuals, and the *mean* describes the average value of the character. As you will see shortly, natural selection often changes the mean value of a character or its variability within populations.

Other characters, like those Mendel studied (see Section 12.1), exhibit **qualitative variation:** they exist in two or more discrete states, and intermediate forms are often absent. Snow geese, for example, have *either* blue *or* white feathers **(Figure 20.4).** The existence of discrete variants of a character is called a **polymorphism** (*poly* = many, *morphos* = form); we describe such traits as *polymorphic.* The *Cepaea nemoralis* snail shells in Figure 20.2A are polymorphic in background color, number of stripes, and color of stripes. Biochemical polymorphisms, like the human A, B, AB, and O blood types (described in Section 12.2), are also common.

We describe phenotypic polymorphisms quantitatively by calculating the percentage or *frequency* of each trait. For example, if you counted 123 blue snow geese and 369 white ones in a population of 492 geese, the frequency of the blue phenotype would be 123/492 or 0.25, and the frequency of the white phenotype would be 369/492 or 0.75.

Arthur Morris/VIREO

FIGURE 20.4
Qualitative variation. Individual snow geese *(Chen caerulescens)* are either blue or white. Although both colors are present in many populations, geese tend to associate with others of the same color.

Phenotypic Variation Can Have Genetic and Environmental Causes

Phenotypic variation within populations may be caused by genetic differences between individuals, by differences in the environmental factors that individuals experience, or by an interaction between genetics and the environment. As a result, genetic and phenotypic variations may not be perfectly correlated. Under some circumstances, organisms with different genotypes exhibit the same phenotype. For example, the black coloration of some rock pocket mice from Arizona is caused by certain mutations in the *Mc1r* gene (see Section 1.2); but black rock pocket mice from New Mexico do not share those mutations—that is, they have different genotypes—even though they exhibit the same phenotype. On the other hand, organisms with the same genotype sometimes exhibit different phenotypes. For example, the acidity of soil influences flower color in some plants **(Figure 20.5).**

Knowing whether phenotypic variation is caused by genetic differences, environmental factors, or an interaction of the two is

iStockphoto.com/mcswin

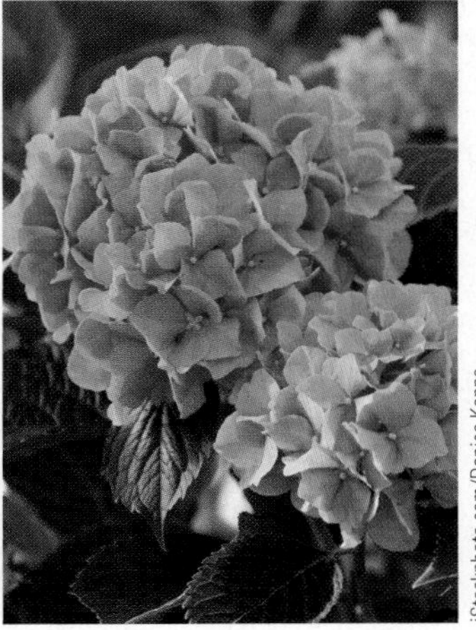
iStockphoto.com/Denise Kappa

FIGURE 20.5
Environmental effects on phenotype. Soil acidity affects the expression of the gene controlling flower color in the common garden plant *Hydrangea macrophylla.* When grown in acid soil, it produces deep blue flowers. In neutral or alkaline soil, its flowers are bright pink.

FIGURE 20.6 Experimental Research

Using Artificial Selection to Demonstrate That Activity Level in Mice Has a Genetic Basis

Question: Do observed differences in activity level among house mice have a genetic basis?

Experiment: Swallow, Carter, and Garland knew that a phenotypic character responds to artificial selection only if it has a genetic, rather than an environmental, basis. In an experiment with house mice *(Mus domesticus)*, they selected for the phenotypic character of increased wheel-running activity. In four experimental lines, they bred those mice that ran the most. Four other lines, in which breeders were selected at random with respect to activity level, served as controls.

Results: After 10 generations of artificial selection, mice in the experimental lines ran longer distances and ran faster than mice in the control lines. Thus, artificial selection on wheel-running activity in house mice increased **(A)** the distance that mice run per day and **(B)** their average speed. The data illustrate responses of females in four experimental lines and four control lines. Males showed similar responses.

A. **Distance run**

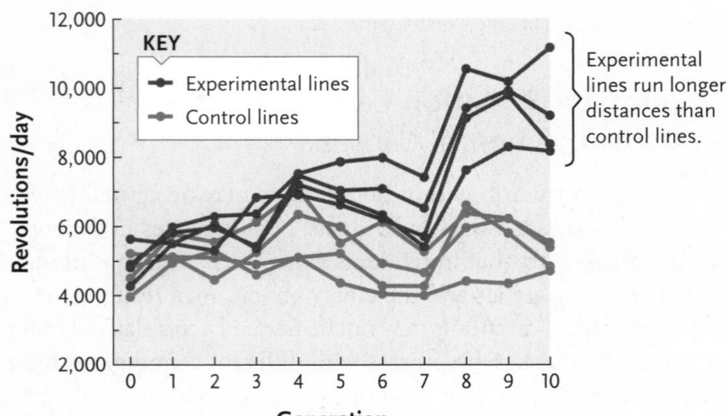

12,000
10,000
8,000
6,000
4,000
2,000
Revolutions/day
KEY
Experimental lines
Control lines
Experimental lines run longer distances than control lines.
0 1 2 3 4 5 6 7 8 9 10
Generation

B. **Average speed**

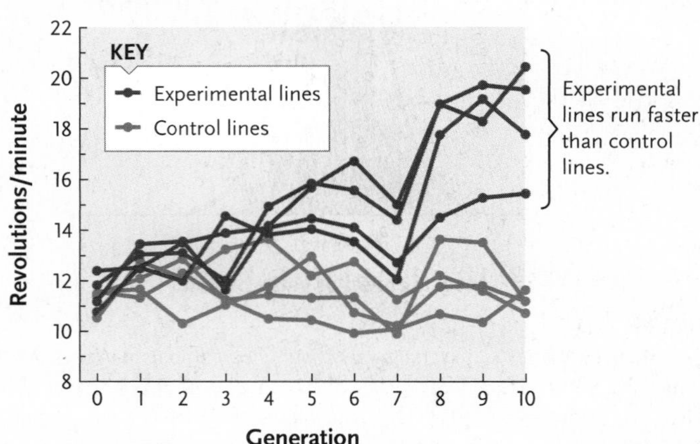

22
20
18
16
14
12
10
8
Revolutions/minute
KEY
Experimental lines
Control lines
Experimental lines run faster than control lines.
0 1 2 3 4 5 6 7 8 9 10
Generation

Conclusion: Because the two measures of activity level responded to artificial selection, researchers concluded that variation in this behavioral character has a genetic basis.

Source: J. G. Swallow, P. A. Carter, and T. Garland, Jr. 1998. Artificial selection for increased wheel-running behavior in house mice. *Behavior Genetics* 28:227–237.

important because *only genetically based variation is subject to evolutionary change.* Moreover, knowing the causes of phenotypic variation has important practical applications. Suppose, for example, that one field of wheat produced more grain than another. If a difference in the availability of nutrients or water caused the difference in yield, a farmer might choose to fertilize or irrigate the less productive field. But if the difference in productivity resulted from genetic differences between plants in the two fields, a farmer might plant only the more productive genotype. Because environmental factors can influence the expression of genes, an organism's phenotype is frequently the product of an interaction between its genotype and its environment. In our hypothetical example, the farmer may maximize yield by fertilizing and irrigating the more productive genotype of wheat.

How can we determine whether phenotypic variation is caused by environmental factors or by genetic differences? We can test for an environmental cause experimentally by changing one environmental variable and measuring the effects on genetically similar subjects. You can try this yourself by growing some cuttings from an ivy plant in shade and other cuttings from the same plant in full sun. Although they all have the same genotype, the cuttings grown in sun will produce smaller leaves and shorter stems.

Breeding experiments can demonstrate the genetic basis of phenotypic variation. For example, Mendel inferred the genetic basis of qualitative traits, such as flower color in peas, by crossing plants with different phenotypes. Moreover, traits that vary quantitatively will respond to artificial selection only if the variation has some genetic basis. For example, researchers observed that individual house mice *(Mus domesticus)* differ in activity levels, as measured by how much they use an exercise wheel and how fast they run. John G. Swallow, Patrick A. Carter, and Theodore Garland, Jr., then at the University of Wisconsin, Madison, used artificial selection to produce lines of mice that exhibit increased wheel-running behavior, demonstrating that the observed differences in these two aspects of activity level have a genetic basis **(Figure 20.6).**

Breeding experiments are not always practical, however, particularly for organisms with long generation times. Ethical concerns also render these techniques unthinkable for humans.

Instead, researchers sometimes study the inheritance of particular traits by analyzing genealogical pedigrees, as discussed in Section 13.2, but this approach often provides poor results for analyses of complex traits.

Several Processes Generate Genetic Variation

Genetic variation, the raw material molded by microevolutionary processes, has two potential sources: the production of new alleles and the rearrangement of existing alleles. Most new alleles probably arise from small-scale mutations in DNA (described later in this chapter). The rearrangement of existing alleles into new combinations can result from larger scale changes in chromosome structure or number and from several forms of genetic recombination, including crossing over between homologous chromosomes during meiosis, independent assortment of nonhomologous chromosomes during meiosis, and random fertilizations between genetically different sperm and eggs.

The shuffling of *existing* alleles into new combinations can produce an extraordinary number of novel genotypes in the next generation. By one estimate, more than 10^{600} combinations of alleles are possible in human gametes, yet there are fewer than 10^{10} humans alive today. So unless you have an identical twin, it is extremely unlikely that another person with your genotype has ever lived or ever will.

Populations Often Contain Substantial Genetic Variation

How much genetic variation actually exists within populations? In the 1960s, evolutionary biologists began to use gel electrophoresis (see Figure 18.6) to identify biochemical polymorphisms in diverse organisms. This technique separates two or more forms of a given protein if they differ significantly in shape, mass, or net electrical charge. The identification of a protein polymorphism allows researchers to infer genetic variation at the locus coding for that protein.

Researchers discovered much more genetic variation than anyone had imagined. For example, nearly half the loci surveyed in many populations of plants and invertebrates are polymorphic. Moreover, gel electrophoresis actually underestimates genetic variation because it doesn't detect different amino acid substitutions if the proteins for which they code migrate at the same rate.

Advances in molecular biology now allow scientists to survey genetic variation directly, and researchers have accumulated an astounding knowledge of the structure of DNA and its nucleotide sequences. In general, studies of chromosomal and mitochondrial DNA suggest that every locus exhibits some variability in its nucleotide sequence. The variability is apparent in comparisons of individuals from a single population, populations of one species, and related species. However, some variations detected in the protein-coding regions of DNA may not affect phenotypes because, as explained in Section 15.1, they do not change the amino acid sequences of the proteins for which the genes code.

STUDY BREAK 20.1

1. **If a population of skunks includes some individuals with stripes and others with spots, would you describe the variation as quantitative or qualitative?**
2. **In the experiment on house mice described in Figure 20.6, how did researchers demonstrate that variations in activity level had a genetic basis?**
3. **What factors contribute to phenotypic variation in a population?**

THINK OUTSIDE THE BOOK

Go to a local park, field, or woodland and choose a population of plants or animals to observe. Make a list of the characteristics that show phenotypic variation in this population, describe their patterns of variation, and identify whether each characteristic exhibits quantitative or qualitative variation.

20.2 Population Genetics

To predict how certain factors may influence genetic variation, population geneticists first describe the genetic structure of a population. They then create hypotheses, formalized in mathematical models, to describe how evolutionary processes may change the genetic structure under specified conditions. Finally, researchers test the predictions of these models to evaluate the ideas about evolution that are embodied within them.

All Populations Have a Genetic Structure

Populations are made up of individuals, each with its own genotype. In diploid organisms, which have pairs of homologous chromosomes, an individual's genotype includes two copies of every gene. The sum of all gene copies at all gene loci in all individuals is called the population's **gene pool.**

To describe the structure of a gene pool, scientists first identify the genotypes in a representative sample and calculate **genotype frequencies,** the percentages of individuals possessing each genotype. Knowing that each diploid organism has two copies of each gene (either two copies of the same allele or two different alleles), a scientist can then calculate **allele frequencies,** the relative abundances of the different alleles. For a gene with two alleles, scientists use the symbol p to identify the frequency of one allele, and q the frequency of the other.

The calculation of genotype and allele frequencies for the two alleles at the gene locus governing flower color in snapdragons (genus *Antirrhinum*) is straightforward **(Table 20.1).** This locus is easy to study because it exhibits incomplete dominance (see Section 12.2). Individuals that are homozygous for the C^R allele (C^RC^R) have red flowers; those homozygous for the C^W allele (C^WC^W) have white flowers; and heterozygotes (C^RC^W) have pink flowers. Genotype frequencies represent how the C^R and C^W alleles are distributed among individuals. In this example, examination of the plants reveals that 45% of individuals have the

TABLE 20.1 Calculation of Genotype Frequencies and Allele Frequencies for the Snapdragon Flower Color Locus

Because each diploid individual has two alleles at each gene locus, the entire sample of 1,000 individuals has a total of 2,000 alleles at the C locus.

Flower Color Phenotype	Genotype	Number of Individuals	Genotype Frequency[1]	Total Number of C^R Alleles[2]	Total Number of C^W Alleles[2]
Red	C^RC^R	450	450/1,000 = 0.45	2 × 450 = 900	0 × 450 = 0
Pink	C^RC^W	500	500/1,000 = 0.50	1 × 500 = 500	1 × 500 = 500
White	C^WC^W	50	50/1,000 = 0.05	0 × 50 = 0	2 × 50 = 100
	Total	1,000	0.45 + 0.50 + 0.05 = 1.0	1,400	600

Calculate allele frequencies using the total of 1,400 + 600 = 2,000 alleles in the sample:

$$p = \text{frequency of } C^R \text{ allele} = 1{,}400/2{,}000 = 0.7$$
$$q = \text{frequency of } C^W \text{ allele} = 600/2{,}000 = 0.3$$
$$p + q = 0.7 + 0.3 = 1.0$$

[1]Genotype frequency = the number of individuals possessing a particular genotype divided by the total number of individuals in the sample.
[2]Total number of C^R or C^W alleles = the number of C^R or C^W alleles present in one individual with a particular genotype multiplied by the number of individuals with that genotype.

C^RC^R genotype, 50% have the heterozygous C^RC^W genotype, and the remaining 5% have the C^WC^W genotype. Allele frequencies represent the commonness or rarity of each allele in the gene pool. As calculated in the table, 70% of the alleles in the population are C^R and 30% are C^W. Remember that for a gene locus with two alleles, there are three genotype frequencies, but only two allele frequencies (p and q). The sum of the three genotype frequencies must equal 1; so must the sum of the two allele frequencies.

The Hardy–Weinberg Principle Is a Null Model That Defines How Evolution Does Not Occur

When designing experiments, scientists often use control treatments to evaluate the effect of a particular factor: the control tells us what we would see if the experimental treatment had no effect. As you may recall from the hypothetical example presented in Chapter 1 (see Figure 1.15), to determine whether fertilizer has an effect on plant growth, you must compare the growth of fertilized plants (the experimental treatment) to the growth of plants that received no fertilizer (the control treatment). However, in studies that use observational rather than experimental data, there is often no suitable control. In such cases, investigators develop conceptual models, called **null models,** which predict what they would see if a particular factor had no effect. Null models serve as theoretical reference points against which observations can be evaluated.

Early in the twentieth century, geneticists were puzzled by the persistence of recessive traits because they assumed that natural selection replaced recessive or rare alleles with dominant or common ones. An English mathematician, G. H. Hardy, and a German physician, Wilhelm Weinberg, tackled this problem independently in 1908. Their analysis, now known as the **Hardy–Weinberg principle,** specifies the conditions under which a population of diploid organisms achieves **genetic equilibrium,** the point at which neither allele frequencies nor genotype frequencies change in succeeding generations. Their work also showed that dominant alleles need not replace recessive ones, and that the shuffling of genes in sexual reproduction does not in itself cause the gene pool to change.

The Hardy–Weinberg principle describes how genotype frequencies are established in sexually reproducing organisms. According to this mathematical model, genetic equilibrium is possible only if *all* of the following conditions are met:

1. No mutations are occurring.
2. The population is closed to migration from other populations.
3. The population is infinite in size.
4. All genotypes in the population survive and reproduce equally well.
5. Individuals in the population mate randomly with respect to genotypes.

If the conditions of the model are met, the allele frequencies of the population will never change, and the genotype frequencies will stop changing after one generation. In short, under these restrictive conditions, microevolution will *not* occur (see *Focus on Basic Research* on p. 438). The Hardy–Weinberg principle is thus a null model that serves as a reference point for evaluating the circumstances under which evolution *may* occur.

If a population's genotype frequencies do not match the predictions of this model or if its allele frequencies change over time, microevolution may be occurring. Determining which of the model's conditions are *not* met is a first step in understanding how and why the gene pool is changing. Natural populations never fully meet all five requirements simultaneously, but they often come pretty close.

STUDY BREAK 20.2 <

1. What is the difference between the genotype frequencies and the allele frequencies in a population?
2. Why is the Hardy–Weinberg principle considered a null model of evolution?
3. If the conditions of the Hardy–Weinberg principle are met, when will genotype frequencies stop changing?

20.3 The Agents of Microevolution

A population's allele frequencies will change over time if conditions of the Hardy–Weinberg model are violated. The processes that foster microevolutionary change—mutation, gene flow, genetic drift, natural selection, and nonrandom mating—are summarized in **Table 20.2.**

Mutations Create New Genetic Variations

A **mutation** is a spontaneous and heritable change in DNA. Mutations are rare events; during any particular breeding season, between one gamete in 100,000 and one in 1 million will include a new mutation at a particular gene locus. New mutations are so infrequent, in fact, that they exert little or no immediate effect on allele frequencies in most populations. But over evolutionary time scales, their numbers are significant—mutations have been accumulating in biological lineages for billions of years. And because it is a mechanism through which entirely new genetic variations arise, *mutation is a major source of heritable variation.*

For most animals, only mutations in the germ line (the cell lineage that produces gametes) are heritable; mutations in other cell lineages have no direct effect on the next generation. In plants, however, mutations may occur in meristem cells, which eventually produce flowers as well as nonreproductive structures (see Chapter 31); in such cases, a mutation may be passed to the next generation and ultimately influence the gene pool.

Deleterious mutations alter an individual's structure, function, or behavior in harmful ways. In mammals, for example, a protein called collagen is an essential component of most extracellular structures. Several simple mutations in humans cause forms of Ehlers–Danlos syndrome, a disruption of collagen synthesis that may result in loose skin, weak joints, or sudden death from the rupture of major blood vessels, the colon, or the uterus.

Lethal mutations can cause great harm to organisms carrying them. If a lethal allele is dominant, both homozygous and heterozygous carriers will, by definition, die from its effects; if recessive, it kills only homozygous recessive individuals. Any lethal mutation that causes an individual to die before it reproduces is eliminated from the population.

Neutral mutations are neither harmful nor helpful. Recall from Section 15.1 that in the construction of a polypeptide chain, a particular amino acid can be specified by several different codons. As a result, some DNA sequence changes—especially certain changes at the third nucleotide of the codon—do not alter the amino acid sequence. Not surprisingly, mutations at the third position appear to persist longer in populations than those at the first two positions. Other mutations may change an organism's phenotype without influencing its survival and reproduction. A neutral mutation might even be beneficial later if the environment changes.

Sometimes a change in DNA produces an *advantageous mutation,* which confers some benefit on an individual that carries it. However slight the advantage, natural selection may preserve the new allele and even increase its frequency over time. Once the mutation has been passed to a new generation, other agents of microevolution determine its long-term fate.

TABLE 20.2 Agents of Microevolutionary Change

Agent	Definition	Effect on Genetic Variation
Mutation	Heritable change in DNA	Introduces new genetic variation into population; does not change allele frequencies quickly
Gene flow	Change in allele frequencies as individuals join a population and reproduce	May introduce genetic variation from another population
Genetic drift	Random changes in allele frequencies caused by chance events	Reduces genetic variation, especially in small populations; can eliminate alleles
Natural selection	Differential survivorship or reproduction of individuals with different genotypes	One allele can replace another or allelic variation can be preserved
Nonrandom mating	Choice of mates based on their phenotypes and genotypes	Does not directly affect allele frequencies, but usually prevents genetic equilibrium

Gene Flow Introduces Novel Genetic Variants into Populations

Organisms or their gametes (for example, pollen) sometimes move from one population to another. If the immigrants reproduce, they may introduce novel alleles into a population, shifting its allele and genotype frequencies away from the values predicted by the Hardy–Weinberg model. This phenomenon, called **gene flow,** violates the Hardy–Weinberg requirement that populations must be closed to migration.

Gene flow is common in some animal species. For example, young male baboons typically move from one local population to another after experiencing aggressive behavior by older males. And many marine invertebrates disperse long distances as larvae carried by ocean currents.

Using the Hardy–Weinberg Principle

To see how the Hardy–Weinberg principle can be applied, we will analyze the snapdragon flower color locus, using the hypothetical population of 1,000 plants described in Table 20.1. This locus includes two alleles—C^R (with its frequency designated as p) and C^W (with its frequency designated as q)—and three genotypes—homozygous C^RC^R, heterozygous C^RC^W, and homozygous C^WC^W. Table 20.1 lists the number of plants with each genotype: 450 have red flowers (C^RC^R), 500 have pink flowers (C^RC^W), and 50 have white flowers (C^WC^W). It also shows the calculation of both the genotype frequencies ($C^RC^R = 0.45$, $C^RC^W = 0.50$, and $C^WC^W = 0.05$) and the allele frequencies ($p = 0.7$ and $q = 0.3$) for the population.

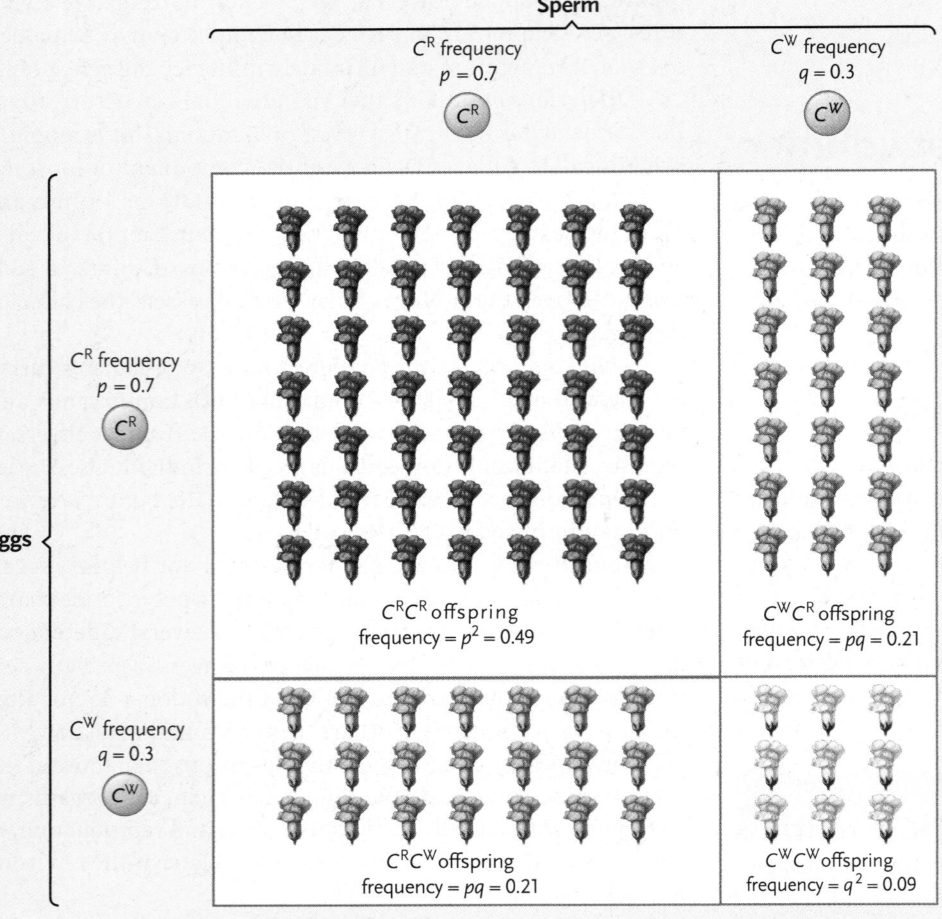

Sperm
C^R frequency $p = 0.7$
C^R
C^W frequency $q = 0.3$
C^W
Eggs
C^R frequency $p = 0.7$
C^R
C^W frequency $q = 0.3$
C^W
C^RC^R offspring frequency = $p^2 = 0.49$
C^WC^R offspring frequency = $pq = 0.21$
C^RC^W offspring frequency = $pq = 0.21$
C^WC^W offspring frequency = $q^2 = 0.09$

Let's assume for simplicity that each individual produces only two gametes and that both gametes contribute to the production of offspring. This assumption is unrealistic, of course, but it meets the Hardy–Weinberg requirement that all individuals in the population contribute equally to the next generation. In each parent, the two alleles segregate and end up in different gametes:

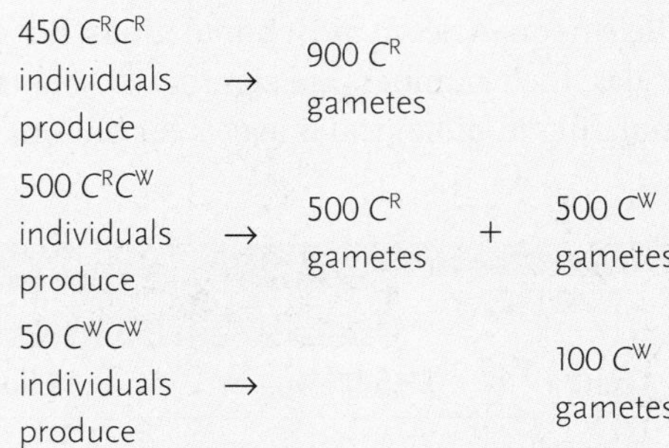

450 C^RC^R individuals produce	→	900 C^R gametes		
500 C^RC^W individuals produce	→	500 C^R gametes	+	500 C^W gametes
50 C^WC^W individuals produce	→			100 C^W gametes

David Neal Parks

W. Carter Johnson

FIGURE 20.7
Gene flow. Blue jays *(Cyanocitta cristata)* serve as agents of gene flow for oaks (genus *Quercus*) when they carry acorns from one oak population to another. An uneaten acorn may germinate and contribute to the gene pool of the population into which it was carried.

Dispersal agents, such as pollen-carrying wind or seed-carrying animals, are responsible for gene flow in most plant populations. For example, blue jays foster gene flow among populations of oaks by carrying acorns from nut-bearing trees to their winter caches, which may be as much as a mile away **(Figure 20.7)**. Transported acorns that go uneaten may germinate and contribute to the gene pool of a neighboring oak population.

Documenting gene flow among populations is not always easy, particularly if it occurs infrequently. Researchers can use phenotypic or genetic markers to identify immigrants in a population, but they must also demonstrate that immigrants reproduced, thereby contributing to the gene pool of their adopted population. In the San Francisco Bay area, for example, Bay checkerspot butterflies *(Euphydryas editha bayensis)* rarely move from one population to another because they are poor fliers (see Figure 50.19). When adult females do change populations, it is often late in the breeding season, and their offspring have virtually no chance of finding enough food to mature. Thus, many immigrant females do not foster gene flow

You can readily see that 1,400 of the 2,000 total gametes carry the C^R allele and 600 carry the C^W allele. The frequency of C^R gametes is 1,400/2,000 or 0.7, which is equal to p; the frequency of C^W gametes is 600/2,000 or 0.3, which is equal to q. Thus, the allele frequencies in the gametes are exactly the same as the allele frequencies in the parent generation—it could not be otherwise because each gamete carries one allele at each locus.

Now assume that these gametes, both sperm and eggs, encounter each other at random. In other words, individuals reproduce without regard to the genotype of a potential mate. We can visualize the process of random mating in the mating table on the left.

We can also describe the consequences of random mating—$(p + q)$ sperm fertilizing $(p + q)$ eggs—with an equation that predicts the genotype frequencies in the offspring generation:

$$(p + q) \times (p + q) = p^2 + 2pq + q^2$$

If the population is at genetic equilibrium for this locus, p^2 is the predicted frequency of the C^RC^R genotype, $2pq$ the predicted frequency of the C^RC^W genotype, and q^2 the predicted frequency of the C^WC^W genotype. Using the gamete frequencies determined above, we can calculate the predicted genotype frequencies in the next generation:

frequency of C^RC^R $= p^2 = (0.7 \times 0.7) = 0.49$

frequency of C^RC^W $= 2pq = 2(0.7 \times 0.3) = 0.42$

frequency of C^WC^W $= q^2 = (0.3 \times 0.3) = 0.09$

Notice that the predicted genotype frequencies in the offspring generation have changed from those in the parent generation: the frequency of heterozygous individuals has decreased, and the frequencies of both types of homozygous individuals have increased. This result occurred because the starting population was *not already* in equilibrium at this gene locus. In other words, the distribution of parent genotypes did not conform to the predicted $p^2 + 2pq + q^2$ distribution.

The 2,000 gametes in our hypothetical population produced 1,000 offspring. Using the genotype frequencies we just calculated, we can predict how many offspring will carry each genotype:

490 red (C^RC^R)

420 pink (C^RC^W)

90 white (C^WC^W)

In a real study, we would examine the offspring to see how well their numbers match these predictions.

What about the allele frequencies in the offspring? The Hardy–Weinberg principle predicts that they did not change. Let's calculate them and see. Using the method shown in Table 20.1 and the prime symbol (′) to indicate offspring allele frequencies:

$$p' = ([2 \times 490] + 420)/2{,}000 = 1{,}400/2{,}000 = 0.7$$

$$q' = ([2 \times 90] + 420)/2{,}000 = 600/2{,}000 = 0.3$$

You can see from this calculation that the allele frequencies did not change from one generation to the next, even though the alleles were rearranged to produce different proportions of the three genotypes. Thus, the population is now at genetic equilibrium for the flower color locus; neither the genotype frequencies nor the allele frequencies will change in succeeding generations as long as the population meets the conditions specified in the Hardy–Weinberg model.

To verify this, you can calculate the allele frequencies of the gametes for this offspring generation and predict the genotype frequencies and allele frequencies for a third generation. You could continue calculating until you ran out of either paper or patience, but these frequencies will not change.

Researchers use calculations like these to determine whether an actual population is near its predicted genetic equilibrium for one or more gene loci. When they discover that a population is not at equilibrium, they infer that microevolution is occurring and can investigate the factors that might be responsible.

because they do not contribute to the gene pool of the population they join.

The evolutionary importance of gene flow depends on the degree of genetic differentiation between populations and the rate of gene flow between them. If two gene pools are very different, a little gene flow may increase genetic variability within the population that receives immigrants, and it will make the two populations more similar. But if populations are already genetically similar, even lots of gene flow will have little effect.

Genetic Drift Reduces Genetic Variability within Populations

Chance events sometimes cause the allele frequencies in a population to change unpredictably. This phenomenon, known as **genetic drift,** has especially dramatic effects on small populations, which clearly violate the Hardy–Weinberg assumption of infinite population size.

A simple analogy clarifies why genetic drift is more pronounced in small populations than in large ones. When individuals reproduce, male and female gametes often pair up randomly, as though the allele in any particular sperm or ovum was determined by a coin toss. Imagine that "heads" specifies the R allele and that "tails" specifies the r allele. If the two alleles are equally common (that is, their frequencies, p and q, are both equal to 0.5), heads should be as likely an outcome as tails. But if you toss the coin 20 or 30 times to simulate random mating in a small population, you won't often see a 50:50 ratio of heads and tails. Sometimes heads will predominate and sometimes tails will—just by chance. Tossing the coin 500 times to simulate random mating in a somewhat larger population is more likely to produce a 50:50 ratio of heads and tails. And if you tossed the coin 5,000 times, you would get even closer to a 50:50 ratio.

Chance deviations from expected results—which cause genetic drift—occur whenever organisms engage in sexual reproduction, simply because their population sizes are not infinitely large. But genetic drift is particularly common in small populations because only a few individuals contribute to the gene pool and because any given allele is present in very few individuals.

Genetic drift generally leads to the loss of alleles and reduced genetic variability; it therefore causes allele and genotype frequencies to differ from those predicted by the Hardy–Weinberg model. Two general circumstances, population bottlenecks and founder effects, often foster genetic drift.

POPULATION BOTTLENECKS On occasion, a stressful factor such as disease, starvation, or drought kills a large proportion of the individuals in a population, producing a **population bottleneck,** a dramatic reduction in population size. This cause of genetic drift greatly reduces genetic variation even if the population numbers later rebound **(Figure 20.8)**.

In the late nineteenth century, for example, hunters nearly wiped out northern elephant seals *(Mirounga angustirostris)* along the Pacific coast of North America. Since the 1880s, when the species received protected status, the population has increased to more than 30,000, all descended from a group of about 20 survivors. Today the population exhibits no variation in 24 proteins studied by gel electrophoresis. This low level of genetic variation, which is unique among seal species, is consistent with the hypothesis that genetic drift eliminated many alleles when the population experienced the bottleneck.

FOUNDER EFFECT When a few individuals colonize a distant locality and start a new population, they carry only a small sample of the parent population's genetic variation. By chance, some alleles may be totally missing from the new population, whereas other alleles that were rare "back home" might occur at relatively high frequencies. This change in the gene pool is called the **founder effect.**

The human medical literature provides some of the best-documented examples of the founder effect. The Old Order Amish, an essentially closed religious community in Lancaster County, Pennsylvania, have an exceptionally high incidence of Ellis–van Creveld syndrome, a genetic disorder caused by a recessive allele. In the homozygous state, the allele produces dwarfism, shortened limbs, and polydactyly (extra fingers). Genetic analysis suggests that, although this syndrome affects less than 1% of the Amish in Lancaster County, as many as 13% may be heterozygous carriers of the allele. All of the individuals exhibiting the syndrome are descended from one couple who helped found the community in the mid-1700s.

CONSERVATION IMPLICATIONS Genetic drift has important implications for conservation biology. By definition, endangered species experience severe population bottlenecks, which result in the loss of genetic variability. Moreover, the small number of individuals available for captive breeding programs may not fully represent a species' genetic diversity. Without such variation, no matter how large a population may become in the future, it will be less resistant to diseases or less able to cope with environmental change.

For example, scientists believe that an environmental catastrophe produced a bottleneck in the African cheetah *(Acinonyx jubatus)* population 10,000 years ago. Cheetahs today are remarkably uniform in genetic make-up. Their populations are highly susceptible to diseases; males also have a high proportion of sperm cell abnormalities and a reduced reproductive capacity. Thus, limited genetic variation, as well as small numbers, threatens populations of endangered species. *Insights from the Molecular Revolution* describes techniques used to determine whether hunting has had the same effect on humpback whales.

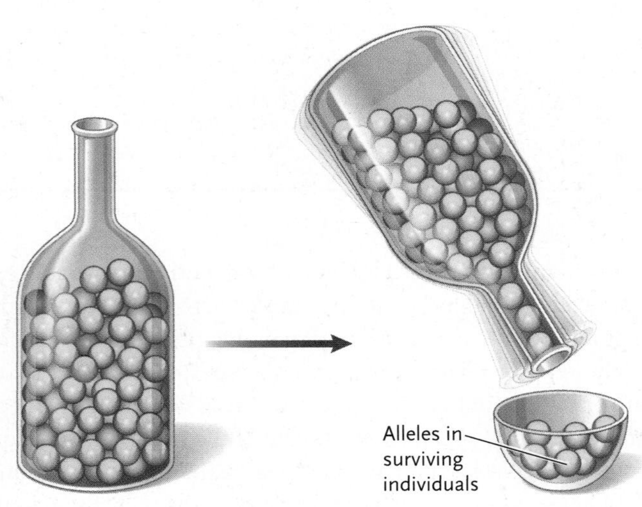
Alleles in surviving individuals

The gene pool of the original population, represented by a bottle filled with colored marbles, includes a locus with three alleles. Two of the alleles, represented by blue and green marbles, occur at high frequency; the third allele, represented by red marbles, occurs at low frequency.

If an environmental event randomly kills a large number of individuals in the population, the drastic reduction in population size is described as a population bottleneck. The process is analogous to shaking only a few of the marbles—the survivors—through the neck of the bottle. As a consequence of chance events associated with population bottlenecks, surviving individuals may not have the same allele frequencies as the original population. Rare alleles are inevitably lost.

FIGURE 20.8
Population bottlenecks, genetic drift, and the loss of genetic variability.

Natural Selection Shapes Genetic Variability by Favoring Some Traits over Others

The Hardy–Weinberg model requires all genotypes in a population to survive and reproduce equally well. But as you know from reading Section 19.2, heritable traits enable some individuals to survive better and reproduce more than others. **Natural selection** is the process by which such traits become more common in subsequent generations. Thus, natural selection violates a requirement of the Hardy–Weinberg equilibrium and causes allele and genotype frequencies to differ from those predicted by the model.

Although natural selection can change allele frequencies, *it is the phenotype of an individual organism, rather than any particular allele, that is successful or not.* When individuals survive and reproduce, their alleles—both favorable and unfavorable—are passed to the next generation. Of course, an organism with harmful or lethal dominant alleles will probably die before reproducing, and all the alleles it carries will share that unhappy fate, even those that are advantageous.

INSIGHTS FROM THE Molecular Revolution

Back from the Brink: Did humpback whales lose genetic variability when they squeezed through a population bottleneck?

Hunted commercially for more than 250 years as a source of meat and oil, humpback whales (*Megaptera novaeangliae*) experienced a disastrous population decline—from about 125,000 to 5,000 individuals—over that time. After an international agreement limited whale hunting in 1966, the number of humpbacks has rebounded strongly; about 80,000 individuals now form three distinct populations in the North Atlantic, North Pacific, and Southern oceans. Yet, conservation biologists worried that the population bottleneck the humpbacks experienced may have reduced their genetic variability. Such a loss could have adverse effects on the population's reproductive capacity, resistance to disease, and ability to survive unfavorable environmental changes.

Research Question

Do living populations of humpback whales exhibit reduced genetic variability as a result of the population bottleneck they experienced before 1966?

Experiment

A large group of researchers working in Hawaii, the continental United States, Australia, South Africa, Canada, Mexico, and the Dominican Republic used molecular techniques to measure the genetic variability in surviving humpback populations. They measured the variability in mitochondrial DNA (mtDNA) because it is small, easily extracted and identified, and almost all of its variability comes from chance mutations that occur at a steady rate (rather than from genetic recombination; see Section 13.5). Except for variations produced by mutation that occurred after the population bottleneck—which can be estimated from the mutation rate and subtracted from the total variation detected—the variability in mtDNA should reflect the level present in the population at the time of the bottleneck.

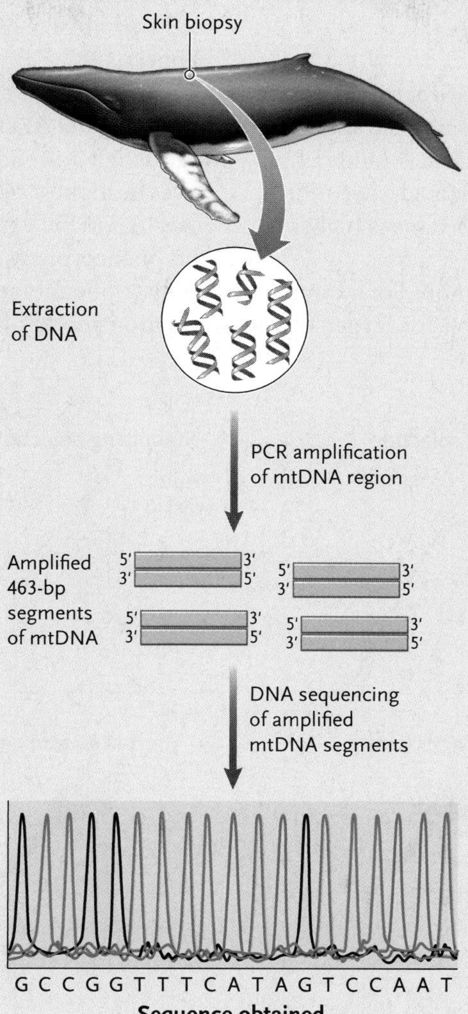

Skin biopsy
Extraction of DNA
PCR amplification of mtDNA region
Amplified 463-bp segments of mtDNA
5′
3′
DNA sequencing of amplified mtDNA segments
G C C G G T T T C A T A G T C C A A T
Sequence obtained

Using biopsy darts, the researchers obtained skin samples from 90 humpback whales distributed among the three populations. They extracted DNA from each sample and used the polymerase chain reaction (PCR; see Figure 18.5) to amplify a 463-base pair segment of mtDNA that includes most of the variable nucleotide positions. They then determined the DNA sequences of the PCR products. (DNA sequencing is described in Section 18.3.)

Results

The researchers were surprised to find that the mtDNA sequence variation was relatively high in most of their sample, between 76% and 82% of the average variation found in all animal species studied to date. However, a subpopulation of the North Pacific population living near Hawaii showed no variability at all in the mtDNA segment examined. One possible hypothesis for this result is that the Hawaiian subpopulation originated recently, perhaps during the twentieth century. Whaling records from the nineteenth century list no sightings or catches of humpbacks around Hawaii, providing indirect support for this hypothesis. Moreover, the native Hawaiian people have no legends or words describing whales of the humpback type (baleen whales). Perhaps the subpopulation was started by a few whales with the same genetic make-up, an example of the founder effect.

Conclusion

With the exception of this Hawaiian subpopulation, humpback whales appear to have retained genetic variability comparable to the levels seen in other animals. The whales may have retained their genetic variability, despite their near extinction, because they have a potential life span of about 50 years. Thus, some individuals still alive today survived the period of commercial hunting, providing a reservoir of variability from the old populations. These results suggest that the hunting ban came just in time to prevent a significant loss of genetic variability in humpback whales.

Source: C. S. Baker et al. 1993. Abundant mitochondrial DNA variation and world-wide population structure in humpback whales. *Proceedings of the National Academy of Sciences USA* 90:8239–8243.

To evaluate reproductive success, evolutionary biologists consider **relative fitness,** the number of surviving offspring that an individual produces compared with the number left by others in the population. Thus, a particular allele will increase in frequency in the next generation if individuals carrying that allele leave *more* offspring than individuals carrying other alleles. Differences in the *relative* success of individuals are the essence of natural selection.

Natural selection tests fitness differences at nearly every stage of the life cycle. One plant may be fitter than others in the population because its seeds survive colder conditions, because the arrangement of its leaves captures sunlight more efficiently, or because its flowers are more attractive to pollinators. However, natural selection exerts little or no effect on traits that appear during an individual's postreproductive life. For example, Huntington disease, a dominant-allele disorder that first strikes humans after the age of 40, is not subject to strong selection. Carriers of the disease-causing allele can reproduce before the onset of the condition, passing it to the next generation.

Biologists measure the effects of natural selection on phenotypic variation by recording changes in the mean and variability of characters over time (see Figure 20.3). Three modes of natural selection have been identified: directional selection, stabilizing selection, and disruptive selection **(Figure 20.9).**

DIRECTIONAL SELECTION Traits undergo **directional selection** when individuals near one end of the phenotypic spectrum have the highest relative fitness. Directional selection shifts a trait away from the existing mean and toward the favored extreme (see Figure 20.9A). After selection, the trait's mean value is higher or lower than before.

Directional selection is extremely common. For example, predatory fish promote directional selection for larger body size in guppies when they selectively feed on the smallest individuals in a guppy population (see *Focus on Basic Research* in Chapter 50). And most cases of artificial selection, including the experiment on the activity levels of house mice, are directional, aimed at increasing or decreasing specific phenotypic traits. Humans routinely use directional selection to produce domestic animals and crops with desired characteristics, such as the small size of chihuahuas and the intense "bite" of chili peppers.

STABILIZING SELECTION Traits undergo **stabilizing selection** when individuals expressing intermediate phenotypes have the highest relative fitness (see Figure 20.9B). By eliminating phenotypic extremes, stabilizing selection reduces genetic and phenotypic variation and increases the frequency of intermediate phenotypes. Stabilizing selection is probably the most common mode of natural selection, affecting many familiar

FIGURE 20.9
Three modes of natural selection. This hypothetical example uses tail length of birds as the quantitative trait subject to selection. The yellow shading in the top graphs indicates phenotypes that natural selection does *not* favor. Notice that the area under each curve is constant because each curve presents the frequencies of all phenotypes in the population. When stabilizing selection **(B)** reduces variability in the trait, the curve becomes higher and narrower.

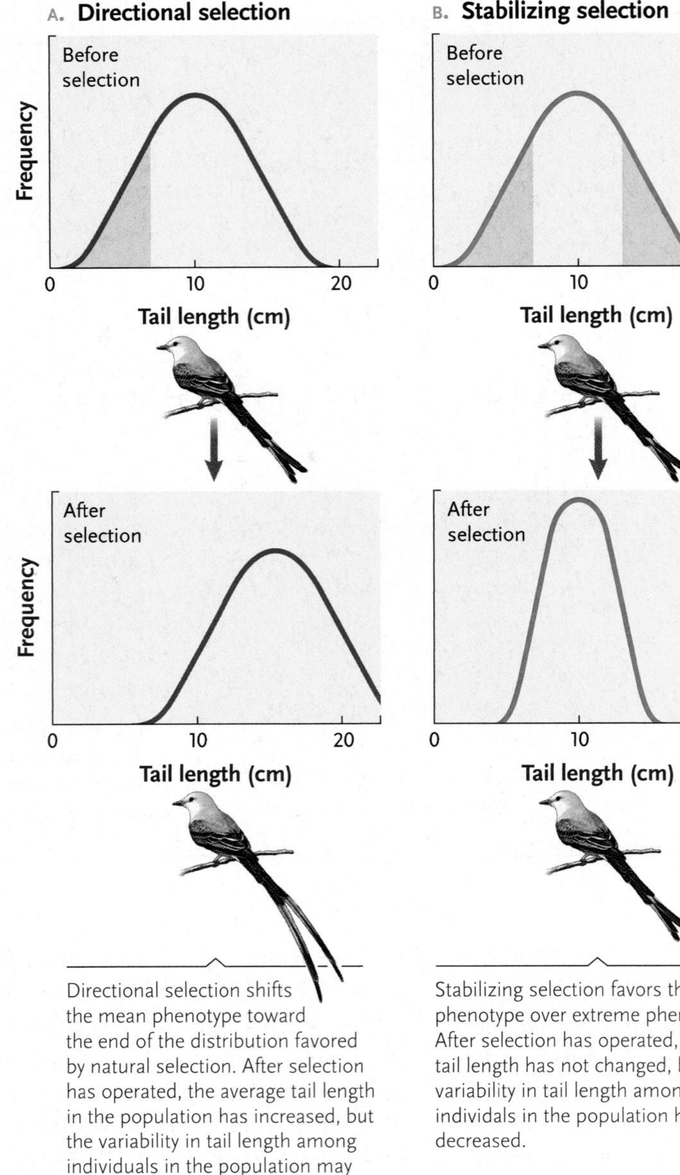
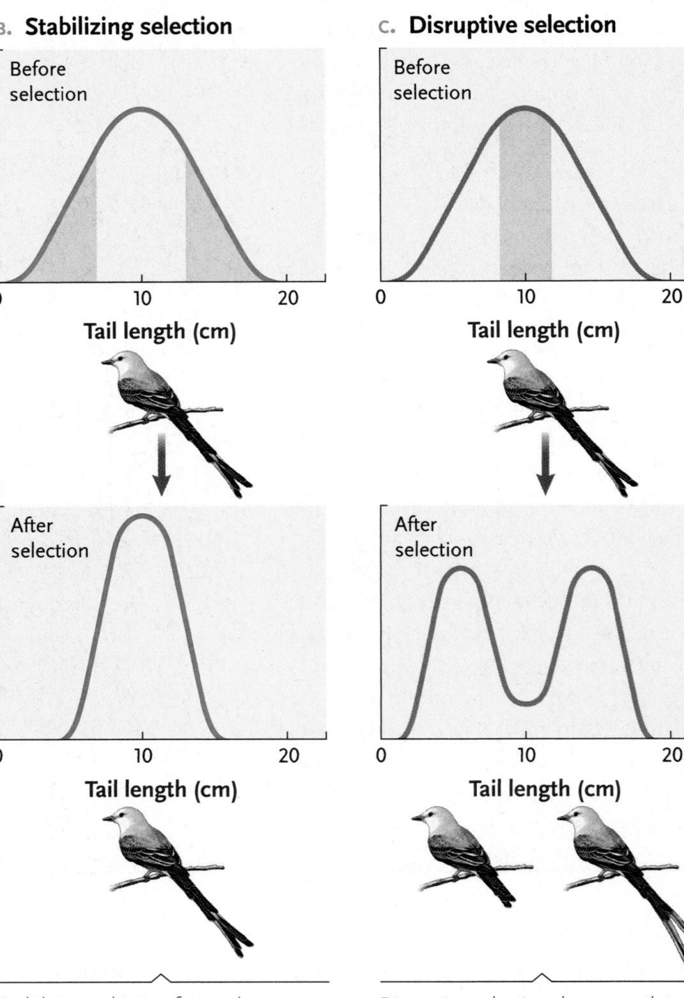

A. Directional selection
B. Stabilizing selection
C. Disruptive selection
Before selection
After selection
Frequency
0
10
20
Tail length (cm)

Directional selection shifts the mean phenotype toward the end of the distribution favored by natural selection. After selection has operated, the average tail length in the population has increased, but the variability in tail length among individuals in the population may be unchanged.

Stabilizing selection favors the mean phenotype over extreme phenotypes. After selection has operated, the mean tail length has not changed, but the variability in tail length among individuals in the population has decreased.

Disruptive selection decreases the frequency of the mean phenotype, but increases the frequencies of phenotypes at both extremes of the distribution. After selection has operated, the mean tail length may be unchanged, but the variability in tail length among individuals has increased. Disruptive selection can produce a polymorphism.

FIGURE 20.10 Observational Research

Do Humans Experience Stabilizing Selection?

Hypothesis: Human birth weight has been adjusted by natural selection.

Null Hypothesis: Natural selection has not affected human birth weight.

Method: Geneticists Luigi Cavalli-Sforza and Sir Walter Bodmer of Stanford University collected data on the variability in human birth weight, a character exhibiting quantitative variation, and on the mortality rates of babies born at different weights. They then searched for a relationship between birth weight and mortality rate by plotting both data sets on the same graph. A lack of correlation between birth weight and mortality rate would support the null hypothesis.

Results: When birth weight (the bar graph) and mortality rate (the curve) are plotted together, the mean birth weight is seen to be very close to the optimum birth weight (the weight at which mortality is lowest). The two data sets also show that few babies are born at the very low weights and very high weights associated with high mortality.

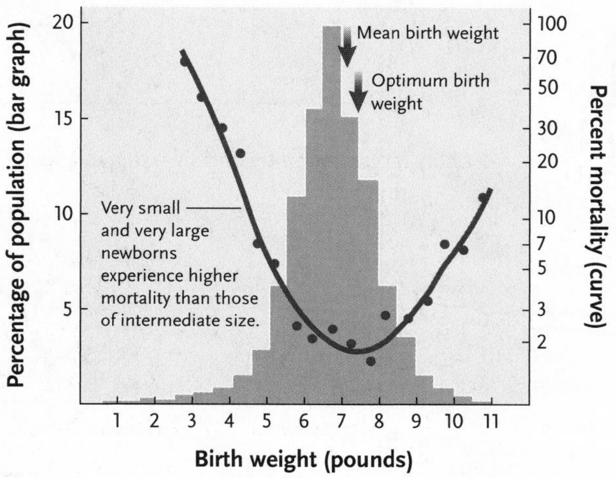

Percentage of population (bar graph)
20
15
10
5
Mean birth weight
Optimum birth weight
Very small and very large newborns experience higher mortality than those of intermediate size.
100
70
50
30
20
10
7
5
3
2
Percent mortality (curve)
1 2 3 4 5 6 7 8 9 10 11
Birth weight (pounds)

Conclusion: The shapes and positions of the birth weight bar graph and the mortality rate curve suggest that stabilizing selection has adjusted human birth weight to an average of 7 to 8 pounds.

Source: L. L. Cavalli-Sforza and W. F. Bodmer. 1971. *The Genetics of Human Populations*. Freeman, San Francisco.

traits. For example, very small and very large human newborns are less likely to survive than those born at an intermediate weight **(Figure 20.10).**

Warren G. Abrahamson and Arthur E. Weis of Bucknell University have shown that opposing forces of directional selection can sometimes produce an overall pattern of stabilizing selection **(Figure 20.11).**

The gallmaking fly *(Eurosta solidaginis)* is a small insect that feeds on the tall goldenrod plant *(Solidago altissima)*. When a fly larva hatches from its egg, it bores into a goldenrod stem, and the plant responds by producing a spherical growth deformity called a gall. The larva feeds on plant tissues inside the gall. Galls vary dramatically in size; genetic experiments indicate that gall size is a heritable trait of the fly, although plant genotype also has an effect.

Fly larvae inside galls are subjected to two opposing patterns of directional selection. On one hand, a tiny wasp *(Eurytoma gigantea)* parasitizes gallmaking flies by laying eggs in fly larvae inside their galls. After hatching, the young wasps feed on the fly larvae, killing them in the process. However, adult wasps are so small that they cannot easily penetrate the thick walls of a large gall; they generally lay eggs in fly larvae occupying small galls. Thus, wasps establish directional selection favoring flies that produce large galls, which are less likely to be parasitized. On the other hand, several bird species open galls to feed on mature fly larvae; these predators preferentially open large galls, fostering directional selection in favor of small galls.

In about one-third of the populations surveyed in central Pennsylvania, wasps and birds attacked galls with equal frequency, and flies producing galls of intermediate size had the highest survival rate. The smallest and largest galls—as well as the genetic predisposition to make very small or very large galls—were eliminated from the population.

DISRUPTIVE SELECTION Traits undergo **disruptive selection** when extreme phenotypes have higher relative fitness than intermediate phenotypes (see Figure 20.9C). Thus, alleles producing extreme phenotypes become more common, promoting polymorphism. Under natural conditions, disruptive selection is much less common than directional selection and stabilizing selection.

Peter and Rosemary Grant of Princeton University, the world's experts on the ecology and evolution of the Galápagos finches, have analyzed a likely case of disruptive selection on the size and shape of the bill in a population of cactus finches *(Geospiza conirostris)* on the island of Genovesa. During normal weather cycles the finches feed on ripe cactus fruits, seeds, and exposed insects. During drought years, when food is scarce, they also search for insects by stripping bark from the branches of bushes and trees.

During the long drought of 1977, about 70% of the cactus finches on Genovesa died; the survivors exhibited unusually high variability in their bills **(Figure 20.12).** The researchers suggested that this morphological variability allowed birds to specialize on particular foods. Birds that stripped bark from branches to look for insects had particularly deep bills, and birds that opened cactus fruits to feed on the fleshy interior had especially long bills. Thus, birds with extreme bill phenotypes appeared to feed more efficiently on specific resources, establishing disruptive selection on the size and shape of their bills. The selection may be particularly strong when drought limits the variety and overall availability of food. However, intermediate bill morphologies may be favored during nondrought years when insects and small seeds are abundant.

FIGURE 20.11 Observational Research

How Opposing Forces of Directional Selection Produce Stabilizing Selection

Hypothesis: The size of galls made by larvae of the gallmaking fly *(Eurosta solidaginis)* is governed by conflicting selection pressures established by parasitic wasps and predatory birds.

Prediction: If parasitic wasps and predatory birds exert selection for different sizes of galls made by gallmaking fly larvae, fly populations that suffer from wasp parasitism and bird predation will make galls of intermediate size.

Method: Abrahamson and his colleagues surveyed galls made by the larvae of the gallmaking fly in Pennsylvania. They measured the diameters of the galls they encountered, and, for those galls in which the larvae had died, they determined whether they had been killed by **(A)** a parasitic wasp or **(B)** a predatory bird, such as the downy woodpecker.

A. ***Eurytoma gigantea*, a parasitic wasp**

Forrest W. Buchanan/Visuals Unlimited

B. ***Picoides pubescens*, a predatory bird**

Gregory K. Scott/Photo Researchers, Inc.

C. **Fly larvae parasitized by wasps versus gall size**

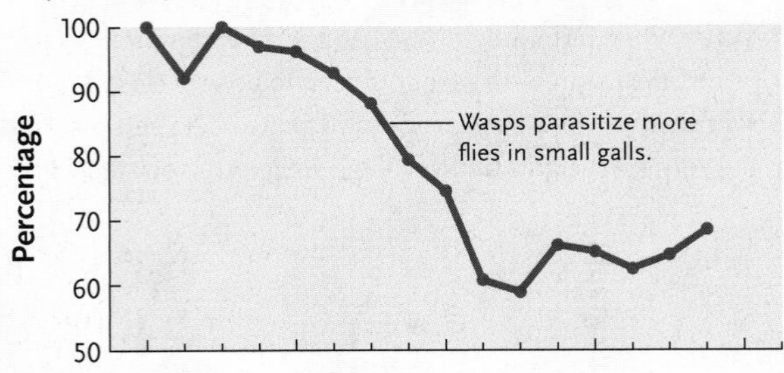
Percentage
100
90
80
70
60
50
Wasps parasitize more flies in small galls.

D. **Fly larvae eaten by birds versus gall size**

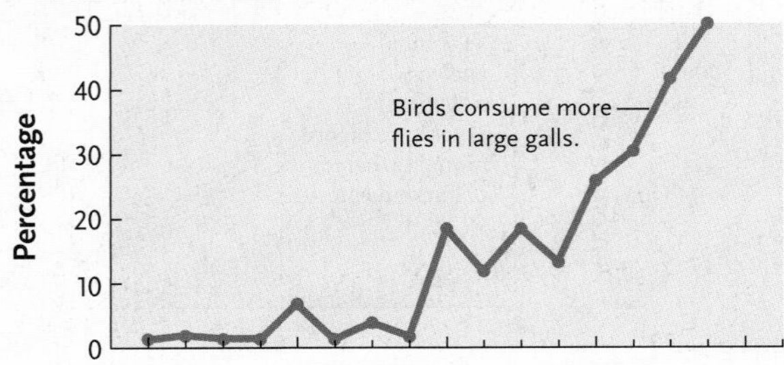
Percentage
50
40
30
20
10
0
Birds consume more flies in large galls.

E. **Fly larvae alive or killed versus gall size**

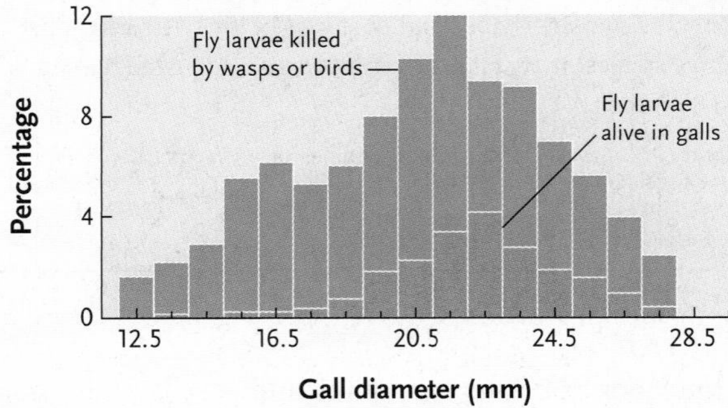
Percentage
12
8
4
0
Fly larvae killed by wasps or birds
Fly larvae alive in galls
12.5
16.5
20.5
24.5
28.5
Gall diameter (mm)

Results: Tiny wasps are more likely to parasitize gallmaking fly larvae inside small galls **(C),** fostering directional selection in favor of large galls. By contrast, birds usually feed on fly larvae inside large galls **(D),** fostering directional selection in favor of small galls. These opposing patterns of directional selection create stabilizing selection for the size of galls that the fly larvae make **(E).**

Conclusion: Because wasps preferentially parasitize fly larvae in small galls and birds preferentially eat fly larvae in large galls, the opposing forces of directional selection establish an overall pattern of stabilizing selection in favor of medium-sized galls.

Source: A. E. Weis and W. G. Abrahamson. 1986. Evolution of host-plant manipulation by gall makers: Ecological and genetic factors in the *Solidago–Eurosta* system. *The American Naturalist* 127:681–695.

Geospiza conirostris

© Krystyna Szulecka/Alamy

Birds with long bills open cactus fruits to feed on the fleshy pulp.

Birds with intermediate bills may be favored during nondrought years when many types of food are available.

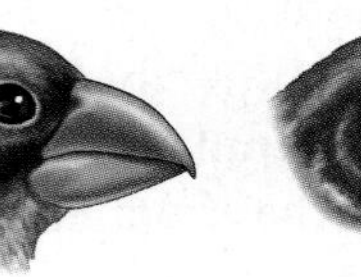

Birds with deep bills strip bark from trees to locate insects.

FIGURE 20.12 Disruptive selection. Cactus finches *(Geospiza conirostris)* on Genovesa exhibit extreme variability in the size and shape of their bills.

FIGURE 20.13 Experimental Research

Sexual Selection in Action

Question: Is the long tail of the male long-tailed widowbird *(Euplectes progne)* the product of intrasexual selection, intersexual selection, or both?

Experiment: Andersson counted the number of females that associated with individual male widowbirds in the grasslands of Kenya. He then shortened the tails of some males by cutting the feathers, lengthened the tails of others by gluing feather extensions to their tails, and left a third group essentially unaltered as a control. One month later, he again counted the number of females associating with each male and compared the results from the three groups.

Results: Males with experimentally lengthened tails attracted more than twice as many mates as males in the control group, and males with experimentally shortened tails attracted fewer. Andersson observed no differences in the ability of altered males and control group males to maintain their display areas.

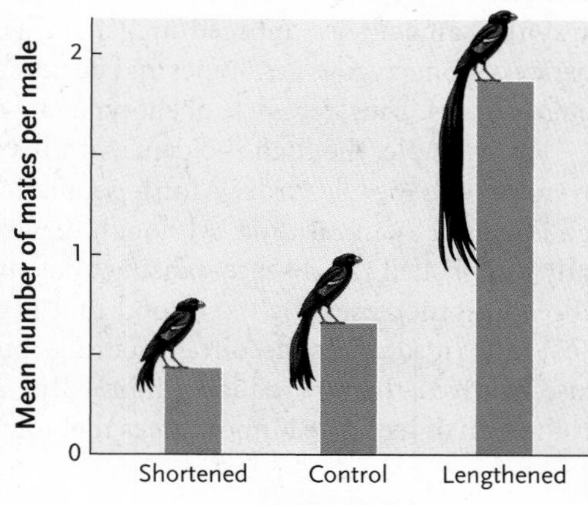
Mean number of mates per male
2
1
0
Shortened
Control
Lengthened
Tail treatment

Conclusion: Female widowbirds clearly prefer males with experimentally lengthened tails to those with normal tails or experimentally shortened tails. Tail length had no obvious effect on the interactions between males. Thus, the long tail of male widowbirds is the product of intersexual selection.

Source: M. Andersson. 1982. Female choice selects for extreme tail length in a widowbird. *Nature* 299:818–820.

Sexual Selection Often Exaggerates Showy Structures in Males

Darwin hypothesized that a special process, which he called **sexual selection,** has fostered the evolution of showy structures—such as brightly colored feathers, long tails, or impressive antlers—as well as elaborate courtship behavior in the males of many animal species. Sexual selection encompasses two related processes. As the result of *intersexual selection* (that is, selection based on the interactions between males and females), males produce these otherwise useless structures simply because females find them irresistibly attractive. Under *intrasexual selection* (that is, selection based on the interactions between members of the same sex), males use their large body size, antlers, or tusks to intimidate, injure, or kill rival males. In many species, sexual selection is the most probable cause of **sexual dimorphism,** differences in the size or appearance of males and females.

Like directional selection, sexual selection pushes phenotypes toward one extreme. But the products of sexual selection are sometimes bizarre—such as the ridiculously long tail feathers of male African widowbirds. How could evolutionary processes favor such elaborate structures, which are costly to produce? Research by Malte Andersson of the University of Gothenburg, Sweden, suggests that the males' long tail feathers are a product of intersexual selection because females are more strongly attracted to males with long tails than to males with short tails, while tail length had no effect on a male's ability to compete with other males for space in the habitat **(Figure 20.13).** Behavioral aspects of sexual selection are described further in Chapter 55.

Nonrandom Mating Can Influence Genotype Frequencies

The Hardy–Weinberg model requires individuals to select mates randomly with respect to their genotypes. This requirement is, in fact, often met; humans, for example, generally marry one another in total ignorance of their genotypes for digestive enzymes or blood types.

Nevertheless, many organisms mate nonrandomly, selecting a mate with a particular phenotype and underlying genotype. Snow geese, for example, usually select mates of their own color, and a tall woman is more likely to marry a tall man than a short man. If no one phenotype is preferred by all potential mates, nonrandom mating does not change allele frequencies by establishing selection for one phenotype over another. But because individuals with similar genetically based phenotypes mate with each other, the next generation will contain fewer heterozygous offspring—and more homozygous offspring—than the Hardy–Weinberg model predicts.

Inbreeding is a special form of nonrandom mating in which genetically related individuals mate with each other. Self-fertilization in plants (see Chapter 34) and a few animals (see Chapter 47) is an extreme example of inbreeding because offspring are produced from the gametes of a single parent. However, other organisms that live in small, relatively closed populations often mate with related individuals. Because relatives often carry the same alleles, inbreeding generally increases the frequency of homozygous genotypes and decreases the frequency of heterozygotes. Thus, recessive phenotypes are often expressed.

For example, the high incidence of Ellis–van Creveld syndrome among the Old Order Amish population, mentioned earlier, is caused by inbreeding. Although the founder effect originally established the disease-causing allele in this population, inbreeding increases the likelihood that it will be expressed. Most human societies discourage matings between genetically close relatives, thereby reducing inbreeding and the inevitable production of recessive homozygotes that inbreeding causes.

STUDY BREAK 20.3

1. **Which agents of microevolution tend to increase genetic variation within populations, and which ones tend to decrease it?**
2. **Which mode of natural selection increases the representation of the average phenotype in a population?**
3. **In what way is sexual selection like directional selection?**

THINK OUTSIDE THE BOOK

Search the Internet for examples of diseases or unusual phenotypes caused by inbreeding in specific human populations. Have human geneticists discovered the molecular basis of these conditions?

20.4 Maintaining Genetic and Phenotypic Variation

Evolutionary biologists continue to discover extraordinary amounts of genetic and phenotypic variation in most natural populations. How can so much variation persist in the face of stabilizing selection and genetic drift?

Diploidy Can Hide Recessive Alleles from the Action of Natural Selection

The diploid condition reduces the effectiveness of natural selection in eliminating harmful recessive alleles from a population. Although such alleles are disadvantageous in the homozygous state, they may have little or no effect on heterozygotes. Thus, recessive alleles can be protected from natural selection by the phenotypic expression of the dominant allele.

In most cases, the masking of recessive alleles in heterozygotes makes it almost impossible to eliminate them completely through selective breeding. Experimentally, we can prevent homozygous recessive organisms from mating. But, as the frequency of a recessive allele decreases, an increasing proportion of its remaining copies is "hidden" in heterozygotes **(Table 20.3)**. Thus, the diploid state preserves recessive alleles at low frequencies, at least in large populations. In small populations, a combination of natural selection and genetic drift can eliminate harmful recessive alleles.

Natural Selection Can Maintain Balanced Polymorphisms

A **balanced polymorphism** is one in which two or more phenotypes are maintained in fairly stable proportions over many generations. Natural selection preserves balanced polymorphisms when heterozygotes have higher relative fitness, when different alleles are favored in different environments, and when the rarity of a phenotype provides an advantage.

HETEROZYGOTE ADVANTAGE A balanced polymorphism can be maintained by **heterozygote advantage,** when heterozygotes for a particular locus have higher relative fitness than either homozygote. The best-documented example of heterozygote ad-

TABLE 20.3 Masking of Recessive Alleles in Diploid Organisms

When a recessive allele is common in a population (top of table), most copies of the allele are present in homozygotes. But when the allele is rare (bottom of table), most copies of it exist in heterozygotes. Thus, rare alleles that are completely recessive are protected from the action of natural selection because they are masked by dominant alleles in heterozygous individuals.

Frequency of Allele *a*	Genotype Frequencies*			% of Allele *a* Copies in	
	AA	*Aa*	*aa*	*Aa*	*aa*
0.99	0.0001	0.0198	0.9801	1	99
0.90	0.0100	0.1800	0.8100	10	90
0.75	0.0625	0.3750	0.5625	25	75
0.50	0.2500	0.5000	0.2500	50	50
0.25	0.5625	0.3750	0.0625	75	25
0.10	0.8100	0.1800	0.0100	90	10
0.01	0.9801	0.0198	0.0001	99	1

*Population is assumed to be in genetic equilibrium.

A. **Distribution of *HbS* allele**

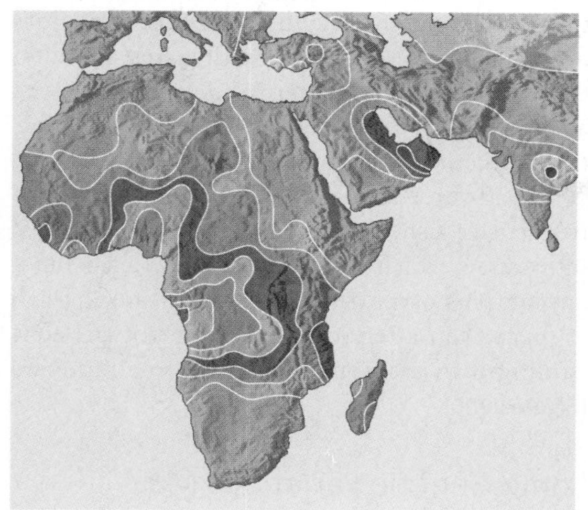

B. **Distribution of malarial parasite**

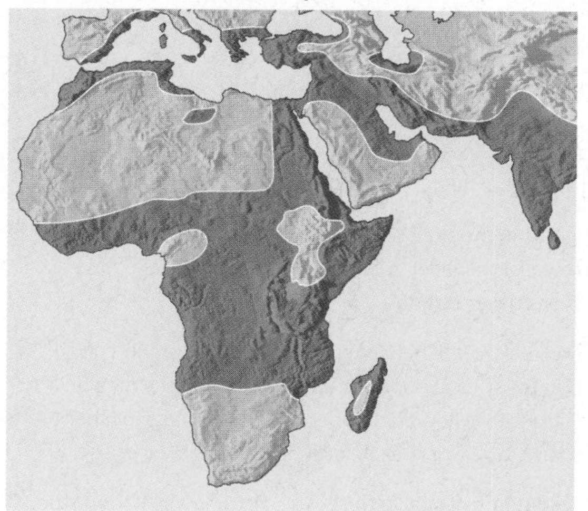

FIGURE 20.14
Heterozygote advantage. The distribution of the *HbS* allele **(A),** which causes sickle-cell disease in homozygotes, roughly matches the distribution of the malarial parasite *Plasmodium falciparum* **(B)** in southern Europe, Africa, the Middle East, and India. Gene flow among human populations has carried the *HbS* allele to some malaria-free regions.

KEY

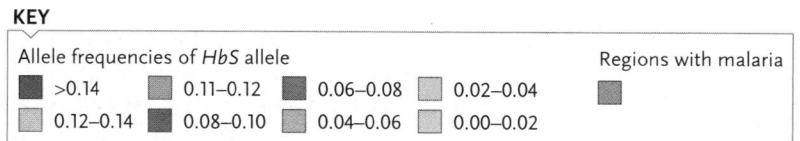

Allele frequencies of *HbS* allele: >0.14; 0.12–0.14; 0.11–0.12; 0.08–0.10; 0.06–0.08; 0.04–0.06; 0.02–0.04; 0.00–0.02

Regions with malaria

vantage is the maintenance of the *HbS* (sickle) allele, which codes for a defective form of hemoglobin in humans. As you learned in Chapter 12, hemoglobin is an oxygen-transporting molecule in red blood cells. The hemoglobin produced by the *HbS* allele differs from normal hemoglobin (coded by the *HbA* allele) by just one amino acid. In *HbS*/*HbS* homozygotes, the faulty hemoglobin forms long fibrous chains under low oxygen conditions, causing red blood cells to assume a sickle shape (as shown in Figure 12.1). Homozygous *HbS*/*HbS* individuals often die of sickle-cell disease before reproducing, yet in tropical and subtropical Africa, *HbS*/*HbA* heterozygotes make up nearly 25% of many populations.

Why is the harmful allele maintained at such high frequency in some populations? It turns out that sickle-cell disease is common in regions where malarial parasites infect red blood cells in humans **(Figure 20.14).** When heterozygous *HbA*/*HbS* individuals contract malaria, their infected red blood cells assume the same sickle shape as those of homozygous *HbS*/*HbS* individuals. The sickled cells lose potassium, killing the parasites, which limits their spread within the infected individual. Heterozygous individuals often survive malaria because the parasites do not multiply quickly inside them, their immune systems can effectively fight the infection, and they retain a large population of uninfected red blood cells. Homozygous *HbA*/*HbA* individuals are also subject to malarial infection, but because their infected cells do not sickle, the parasites multiply rapidly, causing a severe infection with a high mortality rate.

Therefore, *HbA*/*HbS* heterozygotes have greater resistance to malaria and are more likely to survive severe infections in areas where malaria is prevalent. Natural selection preserves the *HbS* allele in these populations because heterozygotes in malaria-prone areas have higher relative fitness than homozygotes for the normal *HbA* allele.

SELECTION IN VARYING ENVIRONMENTS Genetic variability can also be maintained within a population when different alleles are favored in different places or at different times. For example, the shells of European garden snails range in color from nearly white to pink, yellow, or brown, and may be patterned by one to five stripes of varying color (see Figure 20.2A). This polymorphism, which is relatively stable through time, is controlled by several gene loci. The variability in color and striping pattern can be partially explained by selection for camouflage in different habitats.

Predation by song thrushes *(Turdus ericetorum)* is a major agent of selection on the color and pattern of these snails in England. When a thrush finds a snail, it smacks it against a rock to break the shell. The bird eats the snail, but leaves the shell near its "anvil." Researchers used the broken shells near an anvil to compare the phenotypes of captured snails to a random sample of the entire snail population. Their analyses indicated that thrushes are visual predators, usually capturing snails that are easy to find. Thus, well-camouflaged snails survive, and the alleles that specify their phenotypes increase in frequency.

The success of camouflage varies with habitat, however; local subpopulations of the snail, which occupy different habitats, often differ markedly in shell color and pattern. The predators eliminate the most conspicuous individuals in each habitat; thus, natural selection differs from place to place **(Figure 20.15).** In woods where the ground is covered with dead leaves, snails with unstriped pink or brown shells predominate. In hedges and fields, where the vegetation includes thin stems and grass, snails with striped yellow shells are the most common. In populations that span several habitats, selection preserves different alleles in different places, thus maintaining variability in the population as a whole.

FREQUENCY-DEPENDENT SELECTION Sometimes, genetic variability is maintained in a population simply because rare phenotypes—whatever they happen to be—have higher relative fitness than more common phenotypes. The rare phenotype will increase in frequency until it becomes so common

FIGURE 20.15 Observational Research

Habitat Variation in Color and Striping Patterns of European Garden Snails

Hypothesis: Birds and other visual predators are more likely to find snails that are not well camouflaged in their habitats than snails that blend in with their habitats.

Prediction: Genetically based variations in the shell color and striping patterns of the European garden snail *(Cepaea nemoralis)* will differ substantially between local populations of snails that live in different types of vegetation.

Method: British researchers A. J. Cain and P. M. Shepard surveyed the distribution of color and striping patterns of snails in many local populations. They plotted the data on a graph showing the percentage of snails with yellow shells versus the percentage of snails with striped shells, noting the vegetation type where each local population lived.

Results: The shell color and striping patterns of snails living in a particular vegetation type tend to be clustered on the graph, reflecting phenotypic differences that enable the snails to be camouflaged in different habitats. Thus, the alleles that control these characters vary from one local population to another.

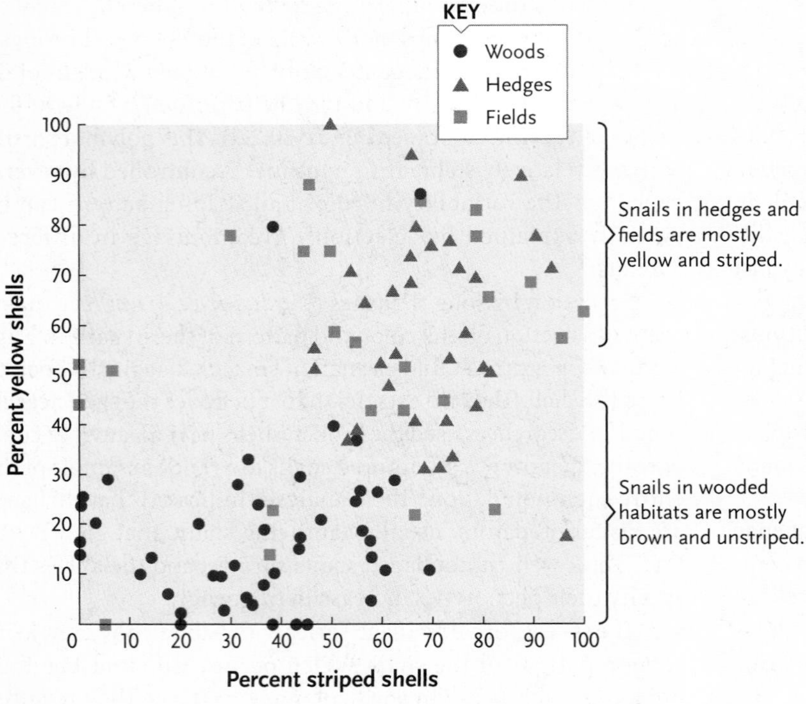
KEY
Woods
Hedges
Fields
Percent yellow shells
Percent striped shells
0
10
20
30
40
50
60
70
80
90
100
Snails in hedges and fields are mostly yellow and striped.
Snails in wooded habitats are mostly brown and unstriped.

Conclusion: Variations in the color and striping patterns on the shells of European garden snails allow most snails to be camouflaged in whatever habitat they occupy. Because these traits are genetically based, the frequencies of the alleles that control them also differ among snails living in different vegetation types. Natural selection therefore favors different alleles in different local populations, maintaining genetic variability in populations that span several vegetation types.

Source: A. J. Cain and P. M. Shepard. 1954. Natural selection in *Cepaea*. *Genetics* 39:89–116.

that it loses its advantage. Such phenomena are examples of **frequency-dependent selection** because the selective advantage enjoyed by a particular phenotype depends on its frequency in the population.

Predator-prey interactions can establish frequency-dependent selection because predators often focus their attention on the most common types of prey (see Chapter 51). For example, the aquatic insects called water boatmen *(Sigara distincta)* occur in three different shades of brown. When all three shades are available at moderate frequencies, fish preferentially feed on the darkest individuals, which are the least camouflaged. But if any one phenotype is very common, fish will learn to focus their attention on that phenotype, consuming it in disproportionately large numbers **(Figure 20.16).**

Some Genetic Variations May Be Selectively Neutral

Many biologists believe that some genetic variations are neither preserved nor eliminated by natural selection. According to the **neutral variation hypothesis,** some of the genetic variation at loci coding for enzymes and other soluble proteins is **selectively neutral.** Even if various alleles code for slightly different amino acid sequences in proteins, the different forms of the proteins may function equally well. In those cases, natural selection would not favor some alleles over others.

Biologists who support the neutral variation hypothesis do not question the role of natural selection in producing complex anatomical structures or useful biochemical traits. They also recognize that selection reduces the frequency of harmful alleles. But they argue that we should not simply assume that every genetic variant that persists in a population has been preserved by natural selection. In practice, it is often very difficult to test the neutral variation hypothesis because the fitness effects of different alleles are often subtle and can vary with small changes in the environment.

The neutral variation hypothesis helps to explain why we see different levels of genetic variation in different populations. It proposes that genetic variation is directly proportional to a population's size and the length of time over which variations have accumulated. Small populations experience fewer mutations than large populations simply because they include fewer replicating genomes. Small populations also lose rare alleles more readily through genetic drift. Thus, small populations should exhibit less genetic variation than large ones, and a population, like the northern elephant seals, that has experienced a recent population bottleneck should exhibit an exceptionally low level of genetic variation. These predictions of the neutral variation hypothesis are generally supported by empirical data.

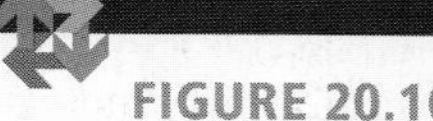

FIGURE 20.16 **Experimental Research**

Demonstration of Frequency-Dependent Selection

Question: How does the frequency of a prey type influence the likelihood that it will be captured by predators?

Experiment: Water boatmen *(Sigara distincta)* occur in three color forms, which vary in the effectiveness of their camouflage. Researchers offered different proportions of the three color forms to predatory fishes in the laboratory and recorded how many of each form were eaten.

Result: When all three phenotypes were available, predatory fishes consumed a disproportionately large number of the most common form, thereby reducing its frequency in the population.

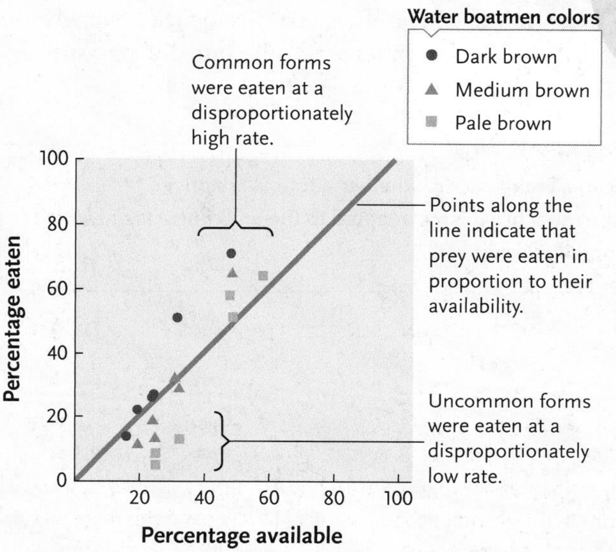

Water boatmen colors
Dark brown
Medium brown
Pale brown
Common forms were eaten at a disproportionately high rate.
Points along the line indicate that prey were eaten in proportion to their availability.
Uncommon forms were eaten at a disproportionately low rate.
Percentage eaten
Percentage available
100
80
60
40
20
0
20
40
60
80
100

Conclusion: Predators tend to feed disproportionately on whatever form of their prey is most abundant, thereby reducing its frequency in the prey population.

Source: B. Clarke. 1962. Balanced polymorphism and the diversity of sympatric species. *Systematics Association Publications* 4:47–70.

STUDY BREAK 20.4

1. **How does the diploid condition protect harmful recessive alleles from natural selection?**
2. **What is a balanced polymorphism?**
3. **Why is the allele that causes sickle-cell disease very rare in human populations that are native to northern Europe?**

20.5 Adaptation and Evolutionary Constraints

Although natural selection preserves alleles that confer high relative fitness on the individuals that carry them, researchers are cautious about interpreting the benefits that particular traits may provide.

Scientists Construct Hypotheses about the Evolution of Adaptive Traits

An **adaptive trait** is any product of natural selection that increases the relative fitness of an organism in its environment. **Adaptation** is the accumulation of adaptive traits over time, and this book describes many examples. The change in the oxygen-binding capacity of hemoglobin in response to carbon dioxide concentration, the water-retaining structures and special photosynthetic pathways of desert plants, and the warning coloration of poisonous animals can all be interpreted as adaptive traits.

In fact, we can concoct an adaptive explanation for almost any characteristic we observe in nature. But such explanations are just fanciful stories unless they are framed as testable hypotheses about the relative fitness of different phenotypes and genotypes. Unfortunately, evolutionary biologists cannot always conduct straightforward experiments because they sometimes study traits that do not vary much within a population or species. In such cases, they may compare variations of a trait in closely related species living in different environments. For example, one can test how the traits of desert plants are adaptive by comparing them to traits in related species from moister habitats.

When biologists try to unravel how and why a particular characteristic evolved, they must also remember that a trait they observe today may have had a different function in the past. For example, the structure of the shoulder joint in birds allows them to move their wings first upward and backward and then downward and forward during flapping flight. But analyses of the fossil record reveal that this adaptation, which is essential for flight, did not originate in birds: some predatory nonflying dinosaurs, including the ancestors of birds, had a similarly constructed shoulder joint. Researchers hypothesize that these fast-running predators may have struck at prey with a flapping motion similar to that used by modern birds. Thus, the structure of the shoulder may have evolved first as an adaptation for capturing prey, and only later proved useful for flapping flight. This hypothesis—however plausible it may be—cannot be tested by direct experimentation because the nonflying ancestors of birds have been extinct for millions of years. Instead, evolutionary biologists must use anatomical studies of birds and their ancestors as well as theoretical models about the mechanics of movement to challenge and refine the hypothesis.

Finally, although evolution has produced all the characteristics of organisms, not all are necessarily adaptive. Some traits may be the products of chance events and genetic drift. Others are produced by alleles that were selected for unrelated reasons (see Section 12.2). And still other characteristics result from the action of basic physical laws. For example, the seeds of many plants fall to the ground when they mature, reflecting the inevitable effect of gravity.

Several Factors Constrain Adaptive Evolution

When we analyze the structure and functions of an organism, we often marvel at how well adapted it is to its environment and mode of life. However, the adaptive traits of most organisms are compromises produced by competing selection pressures. Sea turtles, for example, must lay their eggs on beaches because their embryos cannot acquire oxygen under water. Although flippers allow females to crawl to nesting sites on beaches, they are not ideally suited for terrestrial locomotion. Their structure reflects their primary function, swimming.

Moreover, no organism can be perfectly adapted to its environment because environments change over time. When selection occurs in a population, it preserves alleles that are successful under the prevailing environmental conditions. Thus, each generation is adapted to the environmental conditions under which its parents lived. If the environment changes from one generation to the next, adaptation will always lag behind.

Another constraint on the evolution of adaptive traits is historical. Natural selection is not an engineer that designs new organisms from scratch. Instead, it acts on new mutations and existing genetic variation. Because new mutations are fairly rare, natural selection works primarily with alleles that have been present for many generations. Thus, adaptive changes in the morphology of an organism are almost inevitably based on small modifications of existing structures. The bipedal (two-footed) posture of humans, for example, evolved from the quadrupedal (four-footed) posture of our ancestors. Natural selection did not produce an entirely new skeletal design to accompany this radical behavioral shift. Instead, existing characteristics of the spinal column and the musculature of the legs and back were modified, albeit imperfectly, for an upright stance.

The agents of evolution cause microevolutionary changes in the gene pools of populations. In the next chapter, we examine how microevolution in different populations can cause their gene pools to diverge. The extent of genetic divergence is sometimes sufficient to cause the populations to evolve into different species.

STUDY BREAK 20.5 <

1. **How can a biologist test whether a trait is adaptive?**
2. **Why are most organisms adapted to the environments in which their parents lived?**

UNANSWERED QUESTIONS

What are the evolutionary forces affecting molecular variation within populations?

This question may sound like a simple restatement of the entire chapter you have just read, but it is one of the *fundamental* questions in population genetics today—and we have only begun to scratch its surface. The Hardy–Weinberg principle provides a useful null hypothesis, but since we know that evolution happens routinely, that null hypothesis is very frequently rejected. Recent studies have attempted to address this question using theoretical models, extensive DNA sequence data, and detailed measurements of recombination rate.

Recombination generates new variation, and, most importantly, it causes the evolutionary forces acting on some genes to become independent of forces acting on other genes. Let's imagine that genes *A* and *B* are on the same chromosome, as shown in this depiction of chromosomes sampled from different individuals within a population:

Gene *B* has two alleles (*B* and *b*), but they have no phenotypic effect, and natural selection does not act on them. Suppose that a new advantageous allele at gene *A* (designated *a*) arises in one chromosome. If there is no recombination between genes *A* and *B*, then as allele *a* spreads in the population by selection, so too will allele *b*, even though there was no selection directly favoring the *b* allele. This effect of selection on nearby genes is called a *selective sweep*. By contrast, if genes *A* and *B* frequently recombine, then allele *a* may not remain associated with allele *b*. Under frequent recombination, the spread of allele *a* may have little or no effect on gene *B:* sometimes *a* will be associated with *b*, but at other times *a* will be associated with *B*.

In the 1990s, evolutionary geneticists were greatly excited by several studies that identified a strong and positive relationship between the recombination rate between particular genes and the amount of genetic variation within those genes. In other words, genes that experienced a lot of recombination also exhibited a great deal of variability. This relationship is consistent with the hypothesis that natural selection often occurs throughout the genome—new advantageous alleles arise frequently, and the impact of their "sweeps" is proportional to their recombination rates. This relationship between recombination and genetic variation was first documented in *Drosophila* (fruit flies) by Chip Aquadro and his team at Cornell University, but it has since been demonstrated in humans and various plants. Hence, this pattern appears to be very general.

However, our initial interpretation may be too simplistic. Brian Charlesworth, then at the University of Chicago, suggested that the observed pattern may result from the frequent appearance of detrimental mutations that eliminate variation in regions of low recombination—called *background selection*—rather than from sweeps associated with the spread of advantageous alleles. Given that detrimental mutations arise far more frequently than advantageous ones, background selection surely explains some of this general pattern, and perhaps much of it.

An alternative hypothesis that may explain the relationship between recombination rate and genetic variation suggests that recombination rate and the level of genetic variation may be mechanistically connected. The connection may be direct if recombination itself induces mutations, resulting in higher mutation rates in regions of high recombination. Or the connection may be indirect: recombination rate is related to the base composition in specific regions of the genome, and base composition influences mutation rates. In 2006, Chris Spencer and his colleagues at Oxford University examined the impact of recombination rates on patterns of nucleotide variation at a very fine scale across the human genome. They found that recombination rates had very local effects on variation, an observation that is consistent with the hypothesis of a mechanistic connection between recombination and mutation rate; their results are not consistent with explanations involving natural selection.

Although biologists first thought that the observed relationship between recombination rate and genetic variation had solved questions about the evolutionary forces that affect molecular variation, this observation has become a puzzle in and of itself. Many researchers continue to address this question, now using whole-genome sequences and theoretical and empirical tools for estimating recombination rates. Evolutionary biologists know that the "final answer" will be that all of the processes described above contribute to this relationship, but knowing their specific contributions will help us understand how, how much, and what kinds of natural selection shape variation within genomes.

Think Critically

If recombination itself induces mutations, should genes in regions of high recombination have many differences or few differences when the sequences of these genes are compared among closely related species? Does this test suggest a means for evaluating whether recombination rate and genetic variation are connected mechanistically or connected by the action of natural selection?

Mohamed Noor

Mohamed Noor is professor of biology at Duke University. His research interests include speciation and evolutionary genetics, and recombination. To learn more about Dr. Noor's research go to http://www.biology.duke.edu/noor-lab/Noorlab.html.

Dr. Noor was a PhD student with Dr. Jerry Coyne, who contributed the Unanswered Questions for Chapter 21.

REVIEW KEY CONCEPTS

Go to **CENGAGENOW** at www.cengage.com/login to access quizzing, animations, exercises, articles, and personalized homework help.

20.1 Variation in Natural Populations

- Phenotypic traits exhibit either quantitative or qualitative variation within populations (Figures 20.2 and 20.3).
- Genetic variation, environmental factors, or an interaction between the two cause phenotypic variation within populations (Figures 20.4 and 20.5). Only genetically based phenotypic variation is heritable and subject to evolutionary change.
- Genetic variation arises within populations largely through mutation and genetic recombination. Artificial selection experiments and analyses of protein and DNA sequences reveal that most populations include significant genetic variation (Figure 20.6).

20.2 Population Genetics

- All the gene copies in a population comprise its gene pool, which can be described in terms of allele frequencies and genotype frequencies (Table 20.1).
- The Hardy–Weinberg principle of genetic equilibrium is a null model that describes the conditions under which microevolution, a change in allele frequencies through time, will not take place. Microevolution occurs in populations when the restrictive requirements of the model are not met.

Animation: How to find out if a population is evolving

20.3 The Agents of Microevolution

- Several processes cause microevolution in populations (Table 20.2). Mutation introduces completely new genetic variation. Gene flow carries novel genetic variation into a population through the arrival and reproduction of immigrants (Figure 20.7). Genetic drift causes random changes in allele frequencies, especially in small populations (Figure 20.8). Natural selection occurs when the genotypes of some individuals enable them to survive and reproduce more than others.
- Natural selection alters phenotypic variation in three ways (Figure 20.9). Directional selection increases or decreases the mean value of a trait, shifting it toward a phenotypic extreme. Stabilizing selection increases the frequency of the mean phenotype and reduces variability in the trait (Figures 20.10 and 20.11). Disruptive selection increases the frequencies of extreme phenotypes and decreases the frequency of intermediate phenotypes (Figure 20.12).
- Sexual selection promotes the evolution of exaggerated structures and behaviors (Figure 20.13).
- Although nonrandom mating does not change allele frequencies, it can produce more homozygotic and fewer heterozygotic genotypes than the Hardy–Weinberg model predicts.

Animation: Directional selection

Animation: Change in moth population

Animation: Stabilizing selection

Animation: Disruptive selection

Animation: Disruptive selection among African finches

20.4 Maintaining Genetic and Phenotypic Variation

- Diploidy can maintain genetic variation in a population if recessive alleles are not expressed in heterozygotes and are thus hidden from natural selection (Table 20.3).
- Polymorphisms are maintained in populations when heterozygotes have higher relative fitness than both homozygotes (Figure 20.14), when natural selection occurs in variable environments (Figure 20.15), or when the relative fitness of a phenotype varies with its frequency in the population (Figure 20.16).
- The neutral variation hypothesis proposes that many genetic variations are selectively neutral, conferring neither advantages nor disadvantages on the individuals that carry them. It explains why large populations and those that have not experienced a recent population bottleneck exhibit the highest levels of genetic variation.

Animation: Distribution of sickle-cell trait

Animation: Life cycle of *Plasmodium*

20.5 Adaptation and Evolutionary Constraints

- Adaptive traits increase the relative fitness of individuals carrying them. Adaptive explanations of traits must be framed as testable hypotheses.
- Natural selection cannot result in perfectly adapted organisms because most adaptive traits represent compromises among conflicting needs, because most environments change constantly, and because natural selection can affect only existing genetic variation.

Animation: Adaptation to what?

UNDERSTAND AND APPLY

Test Your Knowledge

1. Which of the following represents an example of qualitative phenotypic variation?
   a. the lengths of people's toes
   b. the body sizes of pigeons
   c. human ABO blood types
   d. the birth weights of humans
   e. the number of leaves on oak trees
2. A population of mice is at Hardy–Weinberg equilibrium at a gene locus that controls fur color. The locus has two alleles, *M* and *m*. A genetic analysis of one population reveals that 60% of its gametes carry the *M* allele. What percentage of mice contains both the *M* and *m* alleles?
   a. 60%
   b. 48%
   c. 40%
   d. 36%
   e. 16%
3. If the genotype frequencies in a population are 0.60 *AA*, 0.20 *Aa*, and 0.20 *aa*, and if the requirements of the Hardy–Weinberg principle apply, the genotype frequencies in the offspring generation will be:
   a. 0.60 *AA*, 0.20 *Aa*, 0.20 *aa*.
   b. 0.36 *AA*, 0.60 *Aa*, 0.04 *aa*.
   c. 0.49 *AA*, 0.42 *Aa*, 0.09 *aa*.
   d. 0.70 *AA*, 0.00 *Aa*, 0.30 *aa*.
   e. 0.64 *AA*, 0.32 *Aa*, 0.04 *aa*.
4. The reason spontaneous mutations do not have an immediate effect on allele frequencies in a large population is that:
   a. mutations are random events, and mutations may be either beneficial or harmful.
   b. mutations usually occur in males and have little effect on eggs.
   c. many mutations exert their effects after an organism has stopped reproducing.
   d. mutations are so rare that mutated alleles are greatly outnumbered by nonmutated alleles.
   e. most mutations do not change the amino acid sequence of a protein.
5. The phenomenon in which chance events cause unpredictable changes in allele frequencies is called:
   a. gene flow.
   b. genetic drift.
   c. inbreeding.
   d. balanced polymorphism.
   e. stabilizing selection.
6. An Eastern European immigrant carrying the allele for Tay Sachs disease settled in a small village on the St. Lawrence River. Many generations later, the frequency of the allele in that village is statistically higher than it is in the immigrant's homeland. The high frequency of the allele in the village probably provides an example of:
   a. natural selection.
   b. the concept of relative fitness.
   c. the Hardy–Weinberg genetic equilibrium.
   d. phenotypic variation.
   e. the founder effect.
7. If a storm kills many small sparrows in a population, but only a few medium-sized and large ones, which type of selection is probably operating?
   a. directional selection
   b. stabilizing selection
   c. disruptive selection
   d. intersexual selection
   e. intrasexual selection
8. Which of the following phenomena explains why the allele for sickle-cell hemoglobin is common in some tropical and subtropical areas where the malaria parasite is prevalent?
   a. balanced polymorphism
   b. heterozygote advantage
   c. sexual dimorphism
   d. neutral selection
   e. stabilizing selection
9. The neutral variation hypothesis proposes that:
   a. complex structures in most organisms have not been fostered by natural selection.
   b. most mutations have a strongly harmful effect.
   c. some mutations are not affected by natural selection.
   d. natural selection cannot counteract the action of gene flow.
   e. large populations are subject to stronger natural selection than small populations.
10. Phenotypic characteristics that increase the fitness of individuals are called:
    a. mutations.
    b. founder effects.
    c. heterozygote advantages.
    d. adaptive traits.
    e. polymorphisms.

Discuss the Concepts

1. Most large commercial farms routinely administer antibiotics to farm animals to prevent the rapid spread of diseases through a flock or herd. Explain why you think that this practice is either wise or unwise.
2. Many human diseases are caused by recessive alleles that are not expressed in heterozygotes. Some people think that eugenics—the selective breeding of humans to eliminate undesirable genetic traits—provides a way for us to rid our populations of such harmful alleles. Explain why eugenics cannot eliminate such genetic traits from human populations.
3. Using two types of beans to represent two alleles at the same gene locus, design an exercise to illustrate how population size affects genetic drift.
4. In what ways are the effects of sexual selection, disruptive selection, and nonrandom mating different? How are they similar?

Design an Experiment

Design an experiment to test the hypothesis that the differences in size among adult guppies are determined by the amount of food they eat rather than by genetic factors.

Interpret the Data

Peter and Rosemary Grant of Princeton University have studied the ecology and evolution of finches on the Galápagos Islands since the early 1970s. They have shown that finches with large bills (as measured by bill depth; see illustration) can eat both small seeds and large seeds, but finches with small bills can only eat small seeds. In 1977, a severe drought on the island of Daphne Major reduced seed production by plants. After the birds consumed whatever small seeds they found, only large seeds were still available. The resulting food shortage killed a majority of the medium ground finches *(Geospiza fortis)* on Daphne Major; their population plummeted from 751 in 1976 to just 90 in 1978. The Grants' research also documented a change in the distributions of bill depths in the birds from 1976 to 1978, as illustrated in the graphs to the right. In light of what you now know about the relationship between bill size and food size for these birds, interpret the change illustrated in the graph. What type of natural selection does this example illustrate?

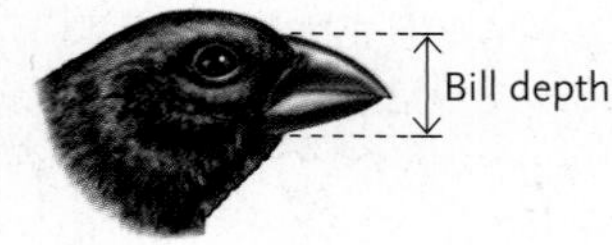
Bill depth

1976 (before drought)
All 751 finches on the island

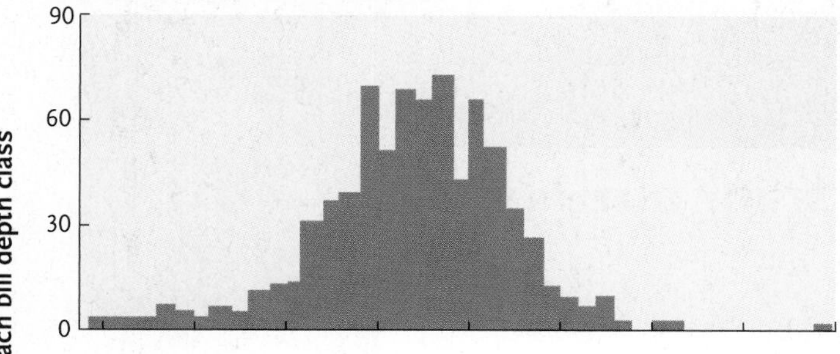
90
60
30
0

1978 (after drought)
The 90 surviving finches on the island

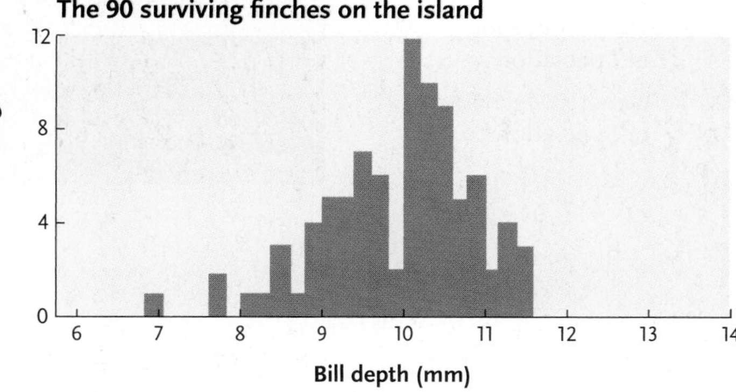
Percentage of finches in each bill depth class
12
8
4
0
6
7
8
9
10
11
12
13
14
Bill depth (mm)

Source: P. R. Grant. 1986. *Ecology and Evolution of Darwin's Finches.* Princeton University Press.

Apply Evolutionary Thinking

Captive breeding programs for endangered species often have access to a limited supply of animals for a breeding stock. As a result, their offspring are at risk of being highly inbred. Why and how might zoological gardens and conservation organizations avoid or minimize inbreeding?

Express Your Opinion

The symptoms of Huntington disease and some other genetically based diseases in humans appear only after the carriers of the disease-causing allele have already reproduced. As a result, they pass the alleles to their offspring and the disease persists in the population. Do you think that all people should be screened for disease-causing alleles and that carriers of such alleles should be discouraged or even prevented from having children? Go to www.cengage.com/login to investigate both sides of the issue and then vote.

21

STUDY OUTLINE

NGS Image Collection

Two closely related bird species, purple martins *(Progne subis)* and tree swallows *(Tachycineta bicolor)* perching together on a branch Crane Creek, Ohio.

Speciation

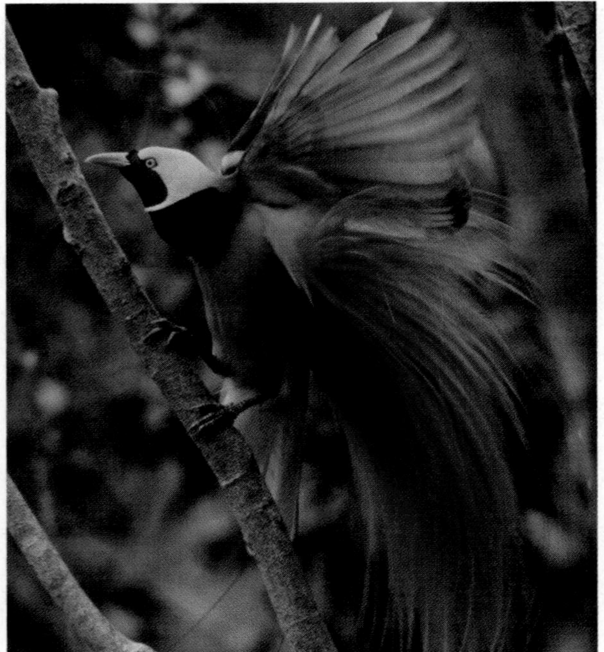

Bruce Beehler

FIGURE 21.1
Birds of paradise. A male Count Raggi's bird of paradise *(Paradisaea raggiana)* tries to attract the attention of a female (not pictured) with his showy plumage and conspicuous display. There are 43 known bird of paradise species, 35 of them found only on the island of New Guinea.

Why It Matters. . . In 1927, nearly 100 years after Darwin boarded the *Beagle,* a young German naturalist named Ernst Mayr embarked on his own journey, to the highlands of New Guinea. He was searching for rare "birds of paradise" **(Figure 21.1).** These birds were known in Europe only through their ornate and colorful feathers, which were used to decorate ladies' hats. On his trek through the remote Arfak Mountains, Mayr identified 137 bird species (including many birds of paradise) based on differences in their size, plumage, color, and other external characteristics.

To Mayr's surprise, the native Papuans—who were untrained in the ways of Western science, but who hunted these birds for food and feathers—had their own names for 136 of the 137 species he had identified. The close match between the two lists confirmed Mayr's belief that the *species* is a fundamental level of organization in nature. Each species has a unique combination of genes underlying its distinctive appearance and habits. Thus, people who observe them closely—whether indigenous hunters or Western scientists—can often distinguish one species from another.

Mayr also discovered some remarkable patterns in the geographical distributions of the bird species in New Guinea. For example, each mountain range he explored was home to some species that lived nowhere else. Closely related species often lived on different mountaintops, separated by deep valleys of unsuitable habitat. In 1942, Mayr published the book *Systematics and the Origin of Species,* in which he described the role of geography in the evolution of new species; the book quickly became a cornerstone of the modern synthesis (which was outlined in Section 19.3).

What mechanisms produce distinct species? As you discovered in Chapter 20, microevolutionary processes alter the pattern and extent of genetic and phenotypic variation within populations. When these processes differ between populations, the populations will diverge, and they may eventually become so different that we recognize them as distinct species. Although Darwin's famous book was titled *On the Origin of Species,* he didn't dwell on the question of *how* new species arise. But the concept of **speciation**—the process of species formation—was implicit in his insight that similar species often share inherited characteristics and a common ancestry. Darwin also recognized that "descent with modification" had generated the amazing diversity of organisms on Earth.

Today evolutionary biologists view speciation as a *process,* a series of events that occur through time. However, they usually study the *products* of speciation, species that are alive today. Because they can rarely witness the process of speciation from start to finish, scientists make inferences about it by studying organisms in various stages of species formation. In this chapter, we consider four major topics: how biologists define and recognize species; how species maintain their genetic identity; how the geographical distributions of organisms influence speciation; and how different genetic mechanisms produce new species. <

Yellow-throated warbler
(*Dendroica dominica*)

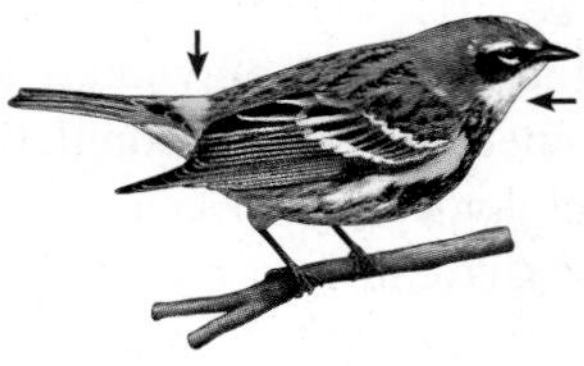

Myrtle warbler
(*Dendroica coronata*)

FIGURE 21.2
Diagnostic characters. Yellow-throated warblers and myrtle warblers can be distinguished by the color of feathers on the throat and rump.

21.1 What Is a Species?

Like the hunters of the Arfak Mountains, most of us recognize the different species that we encounter every day. We can distinguish a cat from a dog and sunflowers from roses. The concept of species is based on our perception that Earth's biological diversity is packaged in discrete, recognizable units, and not as a continuum of forms grading into one another. As evolutionary scientists learn more about the causes of microevolution, they refine our understanding of what a species really is.

The Morphological Species Concept Is a Practical Way to Identify Species

Biologists often describe new species on the basis of visible anatomical characteristics, a process that dates back to Linnaeus' classification of organisms in the eighteenth century (described in Chapter 23). This approach is based on the **morphological species concept,** the idea that all individuals of a species share measurable traits that distinguish them from individuals of other species.

The morphological species concept has many practical applications. For example, paleobiologists use morphological criteria to identify the species of fossilized organisms (see Chapter 22). And because we can observe the external traits of organisms in nature, field guides to plants and animals list diagnostic (that is, distinguishing) physical characters that allow us to recognize them **(Figure 21.2).**

Nevertheless, relying exclusively on morphology to identify species can present problems. Consider the variation in the shells of the European garden snail (*Cepaea nemoralis;* shown earlier, in Figure 20.2). How could anyone imagine that so variable a collection of shells represents just one species of snail? Moreover, morphology does not help us distinguish some closely related species that are nearly identical in appearance. Finally, morphological species definitions tell us little about the evolutionary processes that produce new species.

The Biological Species Concept Is Based on Reproductive Characteristics

The **biological species concept** emphasizes the dynamic nature of species. Ernst Mayr defined biological species as "groups of. . . interbreeding natural populations that are reproductively isolated from [do not produce fertile offspring with] other such groups." The concept is based on reproductive criteria and is easy to apply, at least in principle: if the members of two populations interbreed and produce fertile offspring *under natural conditions,* they belong to the same species; their fertile offspring will, in turn, produce the next generation of that species. If two populations do not interbreed in nature, or fail to produce fertile offspring when they do, they belong to different species.

The biological species concept defines species in terms of population genetics and evolutionary theory. The first half of Mayr's definition notes the genetic *cohesiveness* of species: populations of the same species experience gene flow, which mixes their genetic material. Thus, we can think of a species as one large gene pool, which may be subdivided into local populations.

The second part of the biological species concept emphasizes the genetic *distinctness* of each species. Because populations of different species are reproductively isolated, they cannot exchange genetic information. In fact, the process of speciation is frequently defined as the evolution of reproductive isolation between populations.

The biological species concept also explains why individuals of a species generally look alike: members of the same gene pool share genetic traits that determine their appearance. Individuals of different species generally do not resemble one another as closely because they share fewer genetic characteristics. In practice, biologists often use similarities or differences in morphological traits as convenient markers of genetic similarity or reproductive isolation.

However, the biological species concept does not apply to the many forms of life that reproduce asexually, including most bacteria; some protists, fungi, and plants; and a few animals. In these species, individuals don't interbreed, so it is pointless to ask whether different populations do. Similarly, we cannot use the biological species concept to study extinct organisms, because we have little or no data on their reproductive habits. These species must all be defined using morphological or biochemical criteria. Yet, despite its limitations, the biological species concept currently provides the best evolutionary definition of a sexually reproducing species.

The Phylogenetic Species Concept Focuses on Evolutionary History

Recognizing the limitations of the biological species concept, biologists have developed dozens of other ways to define a species. A widely accepted alternative is the **phylogenetic species concept.** Using both morphological and genetic sequence data, scientists first reconstruct the evolutionary tree for the organisms of interest. They then define a phylogenetic species as a cluster of populations—the tiniest twigs on this part of the Tree of Life—that emerge from the same small branch. Thus, a phylogenetic species comprises populations that share a recent evolutionary history. We will consider this approach for understanding the evolutionary relationships of organisms in Chapter 23.

One advantage of the phylogenetic species concept is that biologists can apply it to any group of organisms, including species that have long been extinct as well as living organisms that reproduce asexually. Proponents of this approach also argue that the morphological and genetic distinctions between organisms on different branches of the Tree of Life reflect the absence of gene flow between them—one of the key requirements of the biological species definition. Nevertheless, because detailed evolutionary histories have been described for relatively few groups of organisms, biologists are not yet able to apply the phylogenetic species concept to all forms of life. As noted in *Unanswered Questions* for Chapter 23, continued research on the details of evolutionary relationships will increase this concept's usefulness in the future.

Many Species Exhibit Substantial Geographical Variation

Populations change in response to shifting environments, and separate populations of a species frequently differ both genetically and phenotypically. Neighboring populations often have shared characteristics because they live in similar environments, exchange individuals, and experience comparable patterns of natural selection. Widely separated populations, by contrast, may live under different conditions and experience different patterns of selection; because gene flow is less likely to occur between distant populations, their gene pools and phenotypes often differ.

When geographically separated populations of a species exhibit dramatic, easily recognized phenotypic variation, biologists may identify them as different **subspecies (Figure 21.3),** which are local variants of a species. Individuals from different subspecies usually interbreed where their geographical distributions meet, and their offspring often exhibit intermediate phenotypes. Biologists sometimes use the word "race" as shorthand for the term "subspecies."

Various patterns of geographical variation have provided great insight into the speciation process. Two of the best-studied patterns are *ring species* and *clinal variation.*

RING SPECIES Some plant and animal species have a ring-shaped geographical distribution that surrounds uninhabitable terrain. Adjacent populations of these so-called **ring species** can exchange genetic material directly, but gene flow between distant populations occurs only through the intermediary populations.

The lungless salamander *Ensatina eschscholtzii,* an example of a ring species, is widely distributed in the coastal mountains and the Sierra Nevada of California, but it cannot survive in the hot, dry Central Valley **(Figure 21.4).** Seven subspecies differ in biochemical traits, color, size, and ecology. Individuals from adjacent subspecies often interbreed where their geographical distributions overlap, and intermediate phenotypes are fairly common. But at the southern end of the Central Valley, adjacent subspecies rarely interbreed. Apparently, they have differentiated to such an extent that they can no longer exchange genetic material directly.

FIGURE 21.3
Subspecies. Five subspecies of rat snake (*Elaphe obsoleta*) in eastern North America differ in color and in the presence or absence of stripes or blotches.

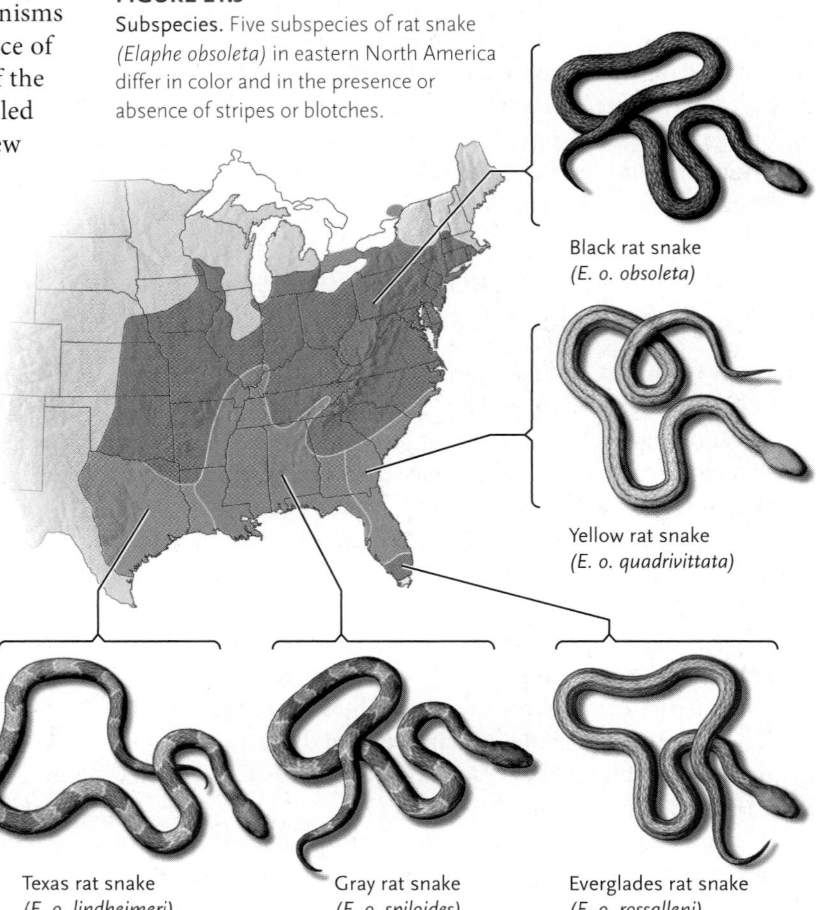
Black rat snake
(E. o. obsoleta)
Yellow rat snake
(E. o. quadrivittata)
Texas rat snake
(E. o. lindheimeri)
Gray rat snake
(E. o. spiloides)
Everglades rat snake
(E. o. rossalleni)

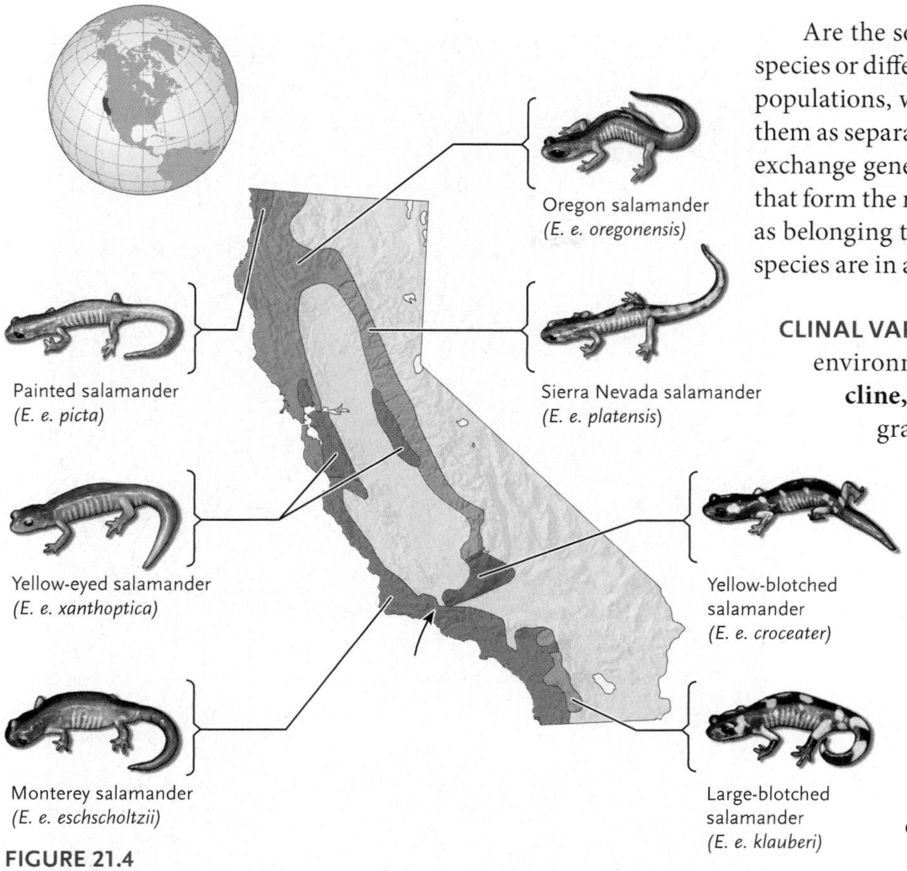

Are the southernmost populations of this salamander subspecies or different species? A biologist who saw *only* the southern populations, which coexist without interbreeding, might define them as separate species. However, they still have the potential to exchange genetic material through the intervening populations that form the ring. Hence, biologists recognize these populations as belonging to the same species. Most likely, the southern subspecies are in an intermediate stage of species formation.

CLINAL VARIATION When a species is distributed over a large, environmentally diverse area, some traits may exhibit a **cline,** a pattern of smooth variation along a geographical gradient. Clinal variation usually results from gene flow between adjacent populations that are each adapting to slightly different conditions. For example, many birds and mammals in the northern hemisphere show clinal variation in body size **(Figure 21.5)** and the relative length of their appendages: in general, populations living in colder environments have larger bodies and shorter appendages, a pattern that is usually interpreted as a mechanism to conserve heat (see Chapter 46). If a cline extends over a large geographical gradient, populations at the opposite ends may be very different.

Despite the geographical variation that many species exhibit, most closely related species are genetically and morphologically different from each other. In the next section, we consider the mechanisms that maintain the genetic distinctness of closely related species by preventing their gene pools from mixing.

Oregon salamander
(E. e. oregonensis)
Painted salamander
(E. e. picta)
Sierra Nevada salamander
(E. e. platensis)
Yellow-eyed salamander
(E. e. xanthoptica)
Yellow-blotched salamander
(E. e. croceater)
Monterey salamander
(E. e. eschscholtzii)
Large-blotched salamander
(E. e. klauberi)

FIGURE 21.4
Ring species. Six of the seven subspecies of the salamander *Ensatina eschscholtzii* are distributed in a ring around California's Central Valley. Subspecies often interbreed where their geographical distributions overlap. However, the two subspecies that nearly close the ring in the south (marked with an arrow), the Monterey salamander and the yellow-blotched salamander, rarely interbreed.

STUDY BREAK 21.1

1. **How do the morphological, biological, and phylogenetic species concepts differ?**
2. **What is clinal variation?**

THINK OUTSIDE THE BOOK

Search the biological literature and the Internet for a definition and application of the ecological species concept. How does the ecological species concept differ from the three species concepts described above? Under what circumstances or for what purpose would the ecological species concept be useful?

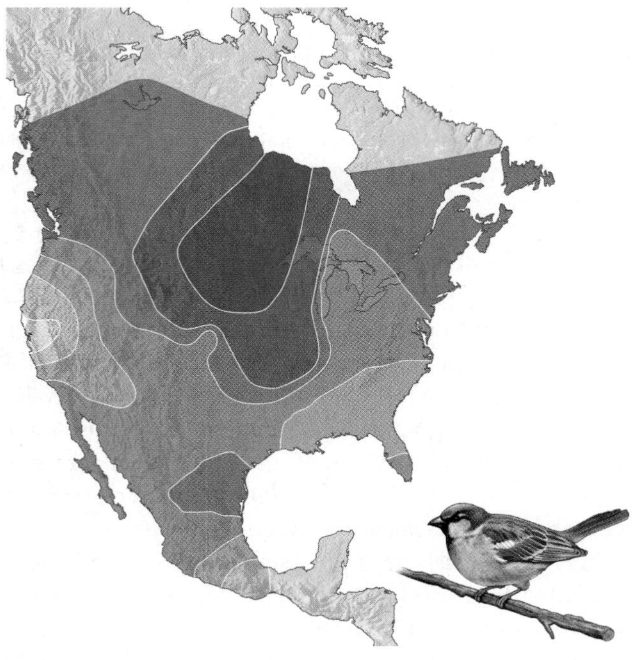

FIGURE 21.5
Clinal variation. House sparrows *(Passer domesticus)* exhibit clinal variation in overall body size, which was summarized from measurements of 16 skeletal features. Darker shading in the map indicates larger size.

21.2 Maintaining Reproductive Isolation

Reproductive isolation is central to the biological species concept. A **reproductive isolating mechanism** is a biological characteristic that prevents the gene pools of two species from mixing. Biologists classify reproductive isolating mechanisms into two catego-

TABLE 21.1	Reproductive Isolating Mechanisms	
Timing Relative to Fertilization	**Mechanism**	**Mode of Action**
Prezygotic ("premating") mechanisms	Ecological isolation	Species live in different habitats
	Temporal isolation	Species breed at different times
	Behavioral isolation	Species cannot communicate
	Mechanical isolation	Species cannot physically mate
	Gametic isolation	Species have non-matching receptors on gametes
Postzygotic ("postmating") mechanisms	Hybrid inviability	Hybrid offspring do not complete development
	Hybrid sterility	Hybrid offspring cannot produce gametes
	Hybrid breakdown	Hybrid offspring have reduced survival or fertility

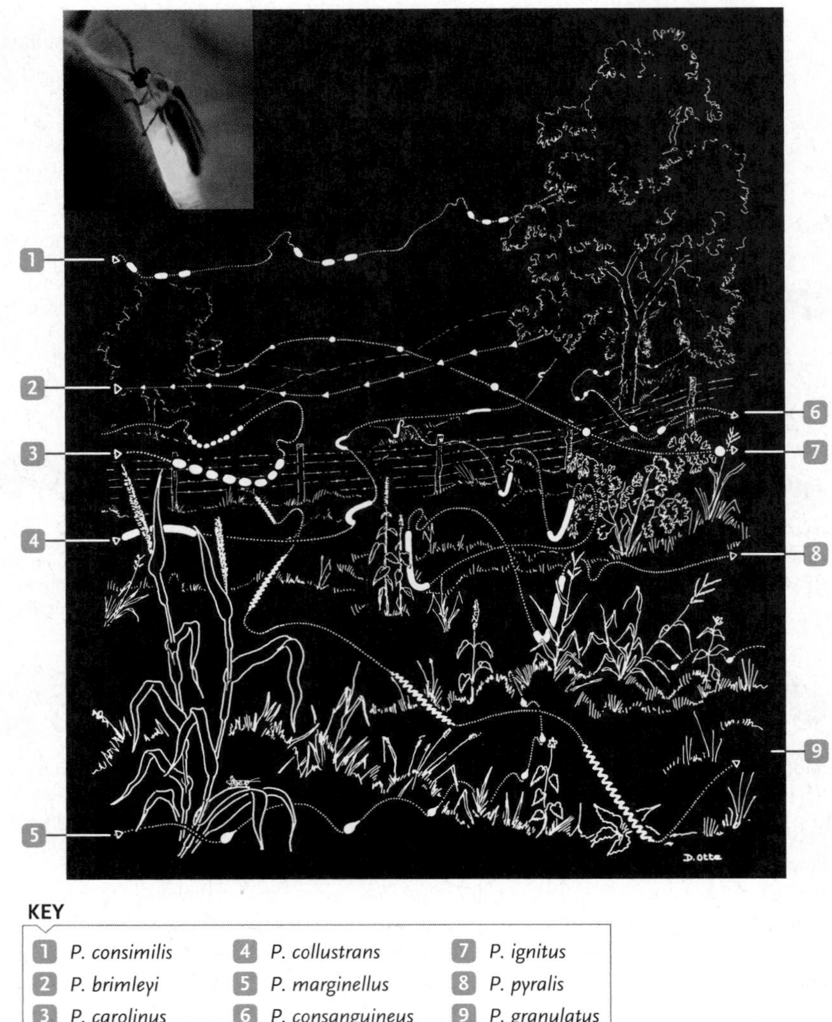
1
2
3
4
5
6
7
8
9
D. Otte
KEY
1 P. consimilis
2 P. brimleyi
3 P. carolinus
4 P. collustrans
5 P. marginellus
6 P. consanguineus
7 P. ignitus
8 P. pyralis
9 P. granulatus

FIGURE 21.6
Behavioral reproductive isolation. Male fireflies use bioluminescent signals to attract potential mates. The different flight paths and flashing patterns of males in nine North American *Photinus* species are represented here. Females respond only to the display given by males of their own species. The inset photo shows *P. pyralis.* (Illustration courtesy of James E. Lloyd. Miscellaneous Publications of the Museum of Zoology of the University of Michigan, 130:1–195, 1966. Photo by Gail Shumway/Getty Images.)

ries (summarized in **Table 21.1**): **prezygotic isolating mechanisms** exert their effects before the production of a zygote, or fertilized egg, and **postzygotic isolating mechanisms** operate after zygote formation. These isolating mechanisms are not mutually exclusive; two or more of them may operate simultaneously.

Prezygotic Isolating Mechanisms Prevent the Production of Hybrid Individuals

Biologists have identified five mechanisms that can prevent interspecific (between species) matings or fertilizations, and thus prevent the production of hybrid (mixed species) offspring. These five prezygotic mechanisms are *ecological, temporal, behavioral, mechanical,* and *gametic isolation.*

Species living in the same geographical region may experience **ecological isolation** if they live in different habitats. For example, lions and tigers were both common in India until the mid-nineteenth century, when hunters virtually exterminated the Asian lions. However, because lions live in open grasslands and tigers in dense forests, the two species did not encounter one another and did not interbreed. Lion-tiger hybrids are sometimes born in captivity, but do not occur under natural conditions.

Species living in the same habitat can experience **temporal isolation** if they mate at different times of day or different times of year. For example, the fruit flies *Drosophila persimilis* and *Drosophila pseudoobscura* overlap extensively in their geographical distributions, but they do not interbreed, in part because *D. persimilis* mates in the morning and *D. pseudoobscura* in the afternoon. Two species of pine in California are reproductively isolated where their geographical distributions overlap: even though both rely on the wind to carry male gametes (pollen grains) to female gametes (ova) in other cones, *Pinus radiata* releases pollen in February and *Pinus muricata* releases pollen in April.

Many animals rely on specific signals, which may differ dramatically between species, to identify the species of a potential mate. **Behavioral isolation** results when the signals used by one species are not recognized by another. For example, female birds rely on the song, color, and displays of males to identify members of their own species. Similarly, female fireflies identify males by their flashing patterns **(Figure 21.6).** These behaviors (collectively called *courtship displays*) are often so complicated that signals sent by one species are like a foreign language that another species simply does not understand.

Mate choice by females and sexual selection (discussed in Section 20.3) generally drive the evolution of mate recognition signals. Females often spend substantial energy in reproduction, and choosing an appropriate mate—that is, a male of her own species—is critically important for the production of successful young. By contrast, a female that mates with a male from a different species is unlikely to leave any surviving offspring at all. Over time, the number of males with recognizable traits, as well as the number of females able to recognize the traits, increases in the population.

Differences in the structure of reproductive organs or other body parts—**mechanical isolation**—may prevent individuals of different species from interbreeding. In particular, many plants have anatomical features that allow only certain pollinators, usually particular bird or insect species, to collect and distribute pollen (see Chapter 27). For example, the flowers and nectar of two native California plants, the monkey-flowers *Mimulus lewisii* and *Mimulus cardinalis* **(Figure 21.7)**, attract different animal pollinators. *Mimulus lewisii* is pollinated by bumblebees. It has shallow pink flowers with broad petals that provide a landing platform for the bees. Bright yellow streaks on the petals serve as "nectar guides," directing bumblebees to the short nectar tube and reproductive parts, which are located among the petals. Bees enter the flowers to drink their concentrated nectar, and they pick up and deliver pollen as they brush against the reproductive parts of the flowers. *Mimulus cardinalis,* by contrast, is pollinated by hummingbirds. It has long red flowers with no yellow streaks, and the reproductive parts extend above the petals. The red color attracts hummingbirds but lies outside the color range detected by bumblebees. The nectar of *M. cardinalis* is more dilute than that of *M. lewisii* but is produced in much greater quantity, making it easier for hummingbirds to ingest. When a hummingbird visits *M. cardinalis* flowers, it pushes its long bill down the nectar tube, and its forehead touches the reproductive parts, picking up and delivering pollen. Recent research has demonstrated that where the two monkey-flower species grow side-by-side, animal pollinators restrict their visits to either one species or the other 98% of the time, providing nearly complete reproductive isolation.

Even when individuals of different species mate, **gametic isolation,** an incompatibility between the sperm of one species and the eggs of another, may prevent fertilization. Many marine invertebrates release gametes into the environment for external fertilization. The sperm and eggs of each species recognize one another's complementary surface proteins (see Chapter 47), but the surface proteins on the gametes of different species don't match. In animals with internal fertilization, sperm of one species may not survive and function within the reproductive tract of another. Interspecific matings between some *Drosophila* species, for example, induce a reaction in the female's reproductive tract that blocks "foreign" sperm from reaching eggs. Parallel physiological incompatibilities between a pollen tube and a stigma prevent interspecific fertilization in some plants.

Mimulus lewisii

iStockphoto.com/Carol Mattsson

Mimulus cardinalis

Photo copyright South12th Photography, 2010. Used under license from Shutterstock.com

FIGURE 21.7
Mechanical reproductive isolation. Because of differences in floral structure, two species of monkey-flower attract different animal pollinators. *Mimulus lewisii* attracts bumblebees and *Mimulus cardinalis* attracts hummingbirds.

Postzygotic Isolating Mechanisms Reduce the Success of Hybrid Individuals

If prezygotic isolating mechanisms between two closely related species are incomplete or ineffective, sperm from one species sometimes fertilizes an egg of the other species. In such cases the two species will be reproductively isolated if their offspring, called interspecific (between species) hybrids, have lower fitness than those produced by intraspecific (within species) matings. Three postzygotic isolating mechanisms—*hybrid inviability, hybrid sterility,* and *hybrid breakdown*—can reduce the fitness of hybrid individuals.

Many genes govern the complex processes that transform a zygote into a mature organism. Hybrid individuals have two sets of developmental instructions, one from each parent species, which may not interact properly for the successful completion of embryonic development. As a result, hybrid organisms frequently die as embryos or at an early age, a phenomenon called **hybrid inviability.** For example, domestic sheep and goats can mate and fertilize one another's ova, but the hybrid embryos always die before coming to term, presumably because the developmental programs of the two parent species are incompatible.

Although some hybrids between closely related species develop into healthy and vigorous adults, they may not produce functional gametes. This **hybrid sterility** often results when the parent species differ in the number or structure of their chromosomes, which cannot pair properly during meiosis. Such hybrids have zero fitness because they leave no descendants. The most familiar example is a mule, the product of mating between a female horse ($2n = 64$) and a male donkey ($2n = 62$). Zebroids, the offspring of matings between horses and zebras, are also usually sterile **(Figure 21.8)**.

Some first-generation hybrids (F_1; see Section 12.1) are healthy and fully fertile. They can breed with other hybrids and with both parental species. However, the second generation (F_2), produced by matings between F_1 hybrids, or between F_1 hybrids and either parental species, may exhibit reduced survival or fertility, a phenomenon known as **hybrid breakdown.** For example, experimental crosses between *Drosophila* species may produce functional hybrids, but their offspring experience a high rate of chromosomal abnormalities and harmful types of genetic recombination. Thus, reproductive isolation is maintained between the species because there is little long-term mixing of their gene pools.

Jen and Des Bartlett/Bruce Coleman USA

FIGURE 21.8
Interspecific hybrids. Horses and zebroids (hybrid offspring of horses and zebras) run in a mixed herd. Zebroids are usually sterile.

STUDY BREAK 21.2

1. **What is the difference between prezygotic and postzygotic isolating mechanisms?**
2. **When a male duck of one species performed a courtship display to a female of another species, she interpreted his behavior as aggressive rather than amorous. What type of reproductive isolating mechanism does this scenario illustrate?**

21.3 The Geography of Speciation

As Ernst Mayr recognized, geography has a huge impact on whether gene pools have the opportunity to mix. Biologists define three modes of speciation, based on the geographical relationship of populations as they become reproductively isolated: *allopatric speciation* (*allo* = different, *patria* = homeland), *parapatric speciation* (*para* = beside), and *sympatric speciation* (*sym* = together).

Allopatric Speciation Occurs between Geographically Separated Populations

Allopatric speciation may take place when a physical barrier subdivides a large population or when a small population becomes separated from a species' main geographical distribution. Probably the most common mode of speciation in large animals, allopatric speciation occurs in two stages. First, two populations become *geographically* separated, preventing gene flow between them. Then, as the populations experience distinct mutations as well as different patterns of natural selection and genetic drift, they may accumulate genetic differences that isolate them *reproductively.*

Geographical separation sometimes occurs when a barrier divides a large population into two or more units **(Figure 21.9)**. For example, hurricanes may create new channels that divide low coastal islands and the populations inhabiting them. Uplifting mountains or landmasses as well as rivers or advancing glaciers can also produce barriers that subdivide populations. The uplift of the Isthmus of Panama, caused by movements of Earth's crust about 5 million years ago, separated a once-continuous shallow sea into the eastern tropical Pacific Ocean and the western tropical Atlantic Ocean. Populations of marine organisms were subdivided by this event, and pairs of closely related species now live on either side of this divide **(Figure 21.10)**.

In other cases, small populations may become isolated at the edge of a species' geographical distribution. Such peripheral populations often differ genetically from the central population because they are adapted to somewhat different environments. Once a small population is isolated, genetic drift and natural selection as well as limited gene flow from the parent population foster further genetic differentiation. In time, the accumulated genetic differences may lead to reproductive isolation.

Populations on oceanic islands represent extreme examples of this phenomenon. The founder effect, an example of genetic drift (see page 439), makes the populations genetically distinct. And on oceanic archipelagos, such as the Galápagos and Hawaiian islands, individuals from one island may

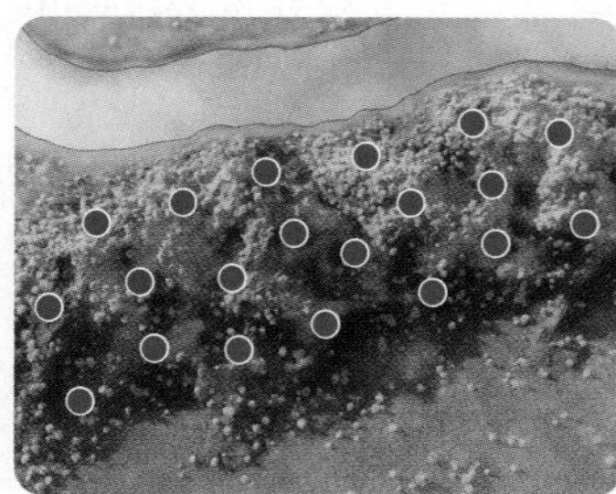

1 At first, a population is distributed over a large geographical area. A river flows along one edge of the population's geographical range.

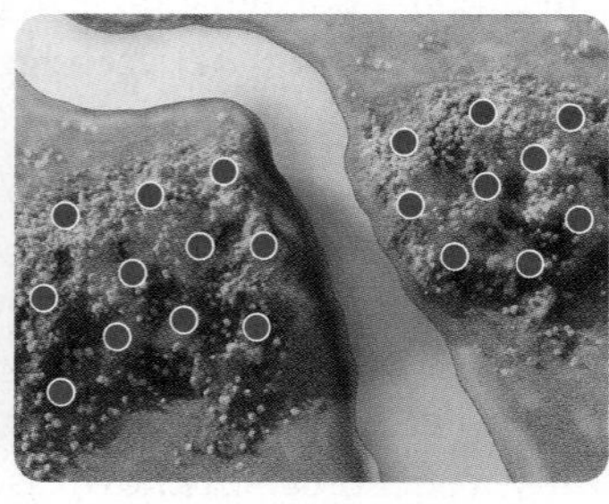

2 A geographical change, such as a change in the river's course, separates the original population, creating a barrier to gene flow.

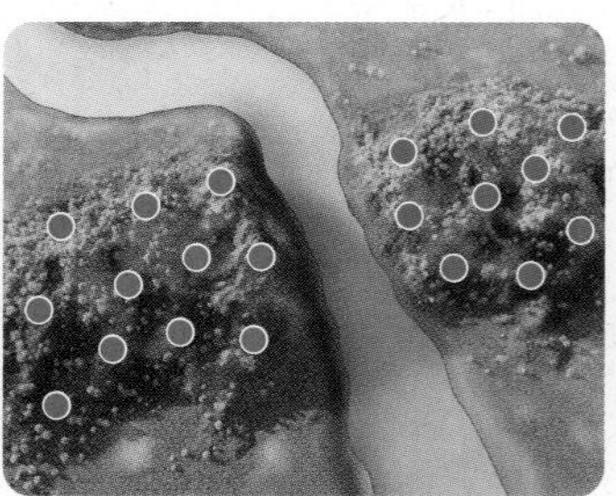

3 In the absence of gene flow, the separated populations evolve independently and diverge into different species.

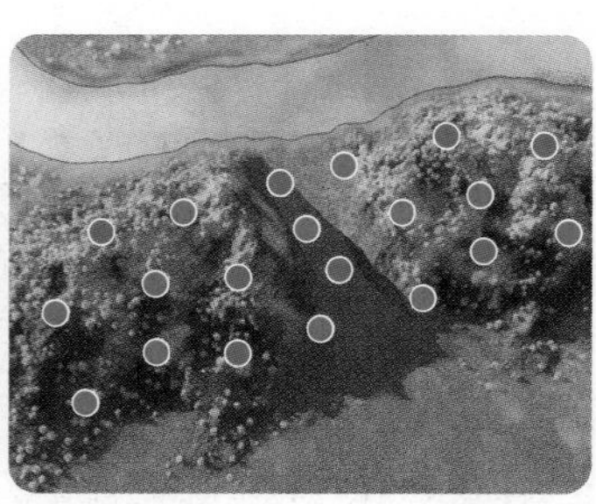

4 When the river later changes course again, allowing individuals of the two species to come into secondary contact, they do not interbreed.

FIGURE 21.9
The model of allopatric speciation and secondary contact.

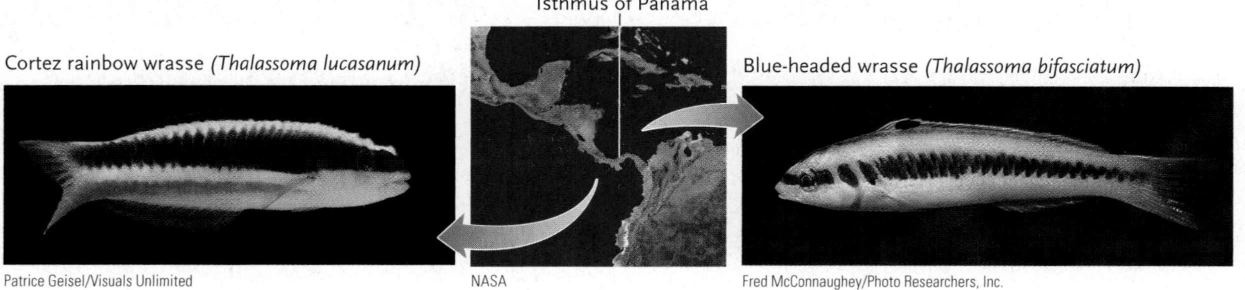
Isthmus of Panama
Cortez rainbow wrasse (Thalassoma lucasanum)
Blue-headed wrasse (Thalassoma bifasciatum)
Patrice Geisel/Visuals Unlimited
NASA
Fred McConnaughey/Photo Researchers, Inc.

FIGURE 21.10
Geographical separation. The uplift of the Isthmus of Panama divided an ancestral wrasse population. The Cortez rainbow wrasse now occupies the eastern Pacific Ocean, and the blue-headed wrasse now occupies the western Atlantic Ocean.

FIGURE 21.11
Evolution of a species cluster on an archipelago. Letters identify four islands in a hypothetical archipelago, and colored dots represent different species. The ancestor of all the species is represented by black dots on the mainland. At the end of the process, islands A and B are each occupied by two species, and islands C and D are each occupied by one species, all of which evolved on the islands.

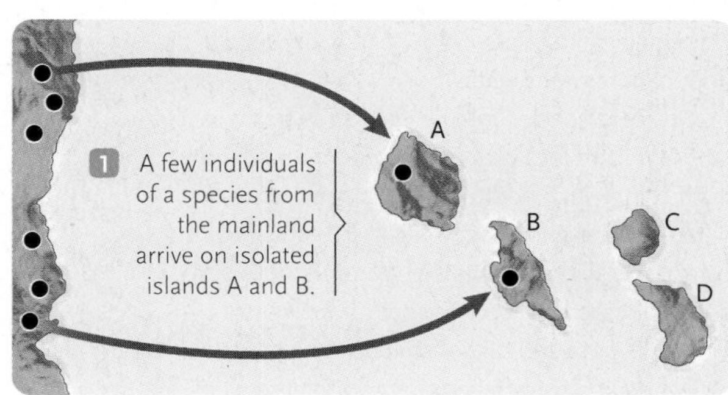
1 A few individuals of a species from the mainland arrive on isolated islands A and B.
A
B
C
D

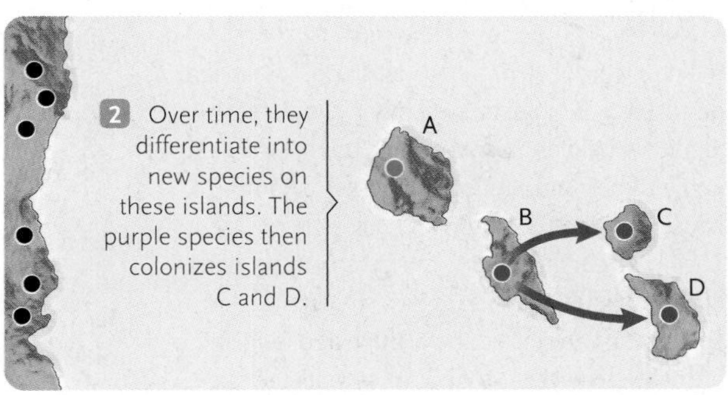
2 Over time, they differentiate into new species on these islands. The purple species then colonizes islands C and D.
A
B
C
D

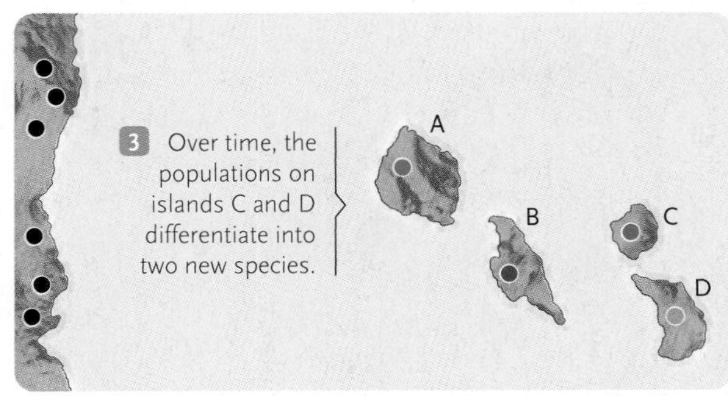
3 Over time, the populations on islands C and D differentiate into two new species.
A
B
C
D

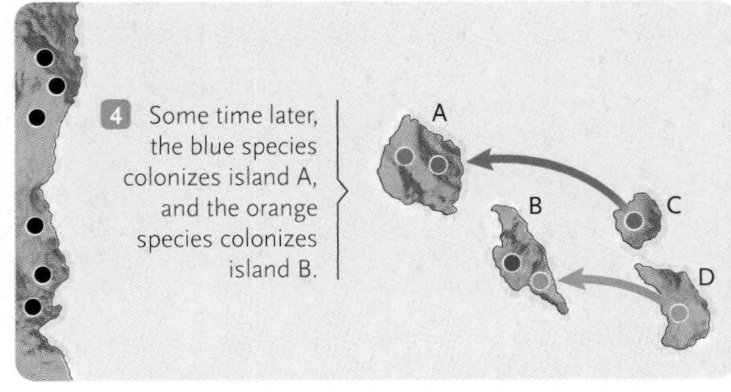
4 Some time later, the blue species colonizes island A, and the orange species colonizes island B.
A
B
C
D

colonize nearby islands, founding populations that differentiate into distinct species. Each island may experience multiple invasions, and the process may be repeated many times within the archipelago, leading to the evolution of a **species cluster,** a group of closely related species recently descended from a common ancestor **(Figure 21.11).** The nearly 800 species of fruit flies on the Hawaiian Islands, described in *Focus on Basic Research,* form several species clusters.

Sometimes, allopatric populations reestablish contact when a geographical barrier is eliminated or breached (see Figure 21.9, step 4). This *secondary contact* provides a test of whether or not the populations have diverged into separate species. If their gene pools did not differentiate much during geographical separation, the populations will interbreed and merge. But if the populations have differentiated enough to be reproductively isolated, they have become separate species.

During the early stages of secondary contact, prezygotic reproductive isolation may be incomplete. Some members of each population may mate with individuals from the other, producing viable, fertile offspring in areas called **hybrid zones.** Although some hybrid zones have persisted for hundreds or thousands of years **(Figure 21.12),** they are generally narrow, and ecological or geographical factors maintain the separation of the gene pools for the majority of individuals in both species.

If postzygotic isolating mechanisms cause hybrid offspring to have lower fitness than those produced within each population, natural selection will promote the evolution of prezygotic isolating mechanisms, favoring individuals that mate only with members of their own population. Recent studies of *Drosophila* suggest that this phenomenon, called **reinforcement,** enhances reproductive isolation that had begun to develop while the populations were geographically separated.

Speciation in Hawaiian Fruit Flies

The islands of the Hawaiian archipelago have been geographically isolated throughout their history, lying at least 3,200 km (approximately 2,000 miles) from the nearest continents or other islands **(Figure 1).** Built by undersea volcanic eruptions over millions of years, they emerged from northwest to southeast: Kauai is at least 5 million years old, and Hawaii, the "Big Island," is less than 1 million years old. Individual islands differ in maximum elevation and include diverse habitats, from sparse, dry vegetation to lush, wet forests.

Resident species must have arrived from distant mainland localities or evolved on the islands from colonizing ancestors. The islands' isolation, different ages, and geographical and ecological complexity allowed repeated interisland colonizations followed by allopatric speciation events. Thus, it is not surprising that species clusters have evolved in several groups of organisms (including flowering plants, insects, and birds).

Nearly 800 species of Hawaiian fruit flies, most of which live on only one island, have been discovered. Biologists used many characters to identify the different species, including external and internal anatomy, cell structure, chromosome structure, ecology, and mating behavior. Their data suggest that the vast majority of native Hawaiian species arose from a single ancestral species that colonized the archipelago long ago, probably from eastern Asia. After repeated speciation events, the fruit flies of the Hawaiian Islands represent more than 25% of all known fruit fly species.

Hampton Carson, of the University of Hawaii, spearheaded studies on the evolutionary relationships of Hawaiian fruit flies. He and his colleagues gathered data on hundreds of fly species—a daunting task. Most species of fruit flies are sexually dimorphic. The females of different species may be similar in appearance, but the males of even closely related species differ in virtually every aspect of their external anatomy: body size, head shape, and the structure of their eyes, antennae, mouthparts, bristles, legs, and wings. Their mating behavior and choice of mating sites also vary dramatically.

Nevertheless, closely related species on different islands occupy comparable habitats and associate with related plant species. Carson suggested that speciation in these flies resulted from the evolution of different genetically determined *mating systems,* the behaviors and morphological characteristics that males display when seeking a mate. The mating systems serve as prezygotic isolating mechanisms.

The 100 or more species of "picture-wing" *Drosophila,* relatively large flies with patterns on their wings, illustrate the evolution of a species cluster. Carson and his colleagues used similarities and differences in the banding patterns on the flies' giant salivary chromosomes (described in *Focus on Model Research Organisms* in Chapter 13), to trace the evolutionary origin of species on the younger islands by identifying their closest relatives on the older islands. Their analysis of 26 species on Hawaii, the youngest island, suggested that flies from the older islands colonized Hawaii at least 19 different times, and each founder population evolved into a new species there. Additional species apparently evolved when lava flows on Hawaii subdivided existing populations.

Among the picture-wing fruit flies, some interspecies matings result in hybrid sterility or hybrid breakdown. But for most species, prezygotic reproductive isolation is maintained by differences in their mating systems. For example, *Drosophila silvestris* and *Drosophila heteroneura,* which produce healthy and fertile hybrids in the laboratory, have similar geographical distributions; however, differences in courtship behavior and in the shape of the males' heads, a characteristic that females use to recognize males of their own species **(Figure 2),** keep these two species reproductively isolated. In nature, they hybridize only in one small geographical area.

The work of Carson and his colleagues suggests that most speciation in Hawaiian *Drosophila* has resulted from founder effects. When a fertile female—or a small group of males and females—moves to a new island, this founding population responds to novel selection pressures in its new environment. Sexual selection then exaggerates distinctive morphological and behavioral characteristics, maintaining the population's reproductive isolation from its new neighbors. The tremendous variety of Hawaiian fruit flies has undoubtedly been produced by repeated colonizations of newer islands by flies from older islands and by the back-colonization of older islands by newly evolved species. Thus, they represent what evolutionary biologists describe as an *adaptive radiation,* a cluster of closely related species that are ecologically different (as described further in Chapter 22).

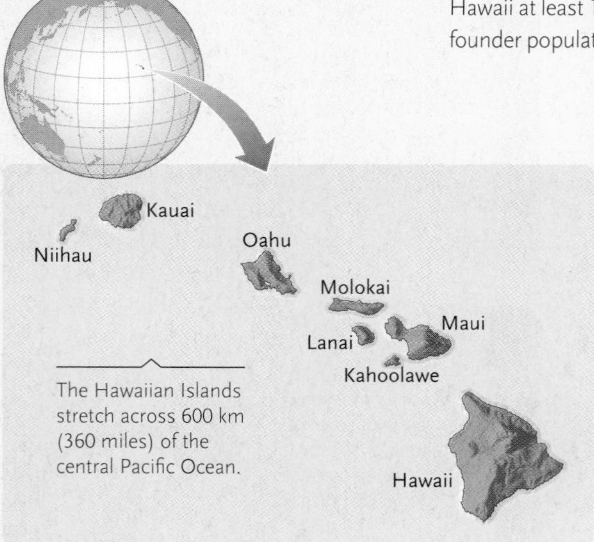
Kauai
Niihau
Oahu
Molokai
Maui
Lanai
Kahoolawe
Hawaii
The Hawaiian Islands stretch across 600 km (360 miles) of the central Pacific Ocean.

FIGURE 1
Geographic isolation of the Hawaiian Islands.

Drosophila heteroneura

Kenneth Y. Kaneshiro, University of Hawaii

Drosophila silvestris

Kenneth Y. Kaneshiro, University of Hawaii

FIGURE 2
Two *Drosophila* species in which the males' head shapes differ.

Bullock's oriole *(Icterus bullockii)*

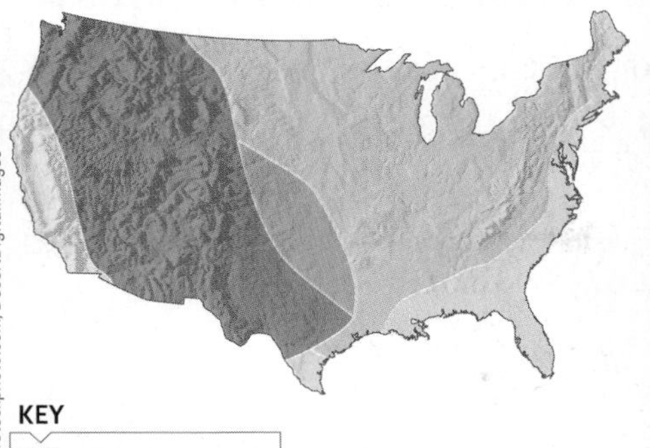
iStockphoto.com/DesertDigitalImages
KEY
Bullock's oriole
Hybrid zone
Baltimore oriole

Baltimore oriole *(Icterus galbula)*

Image copyright Daniel Hebert, 2010. Used under license from ShutterStock.com

FIGURE 21.12
Hybrid zones. Males of the Bullock's oriole and Baltimore oriole differ in color and courtship song. The populations have maintained a hybrid zone for hundreds of years, and once were considered subspecies of the same species. The American Ornithologists' Union recognized them as separate species in 1997. They now hybridize less frequently than they once did, leading some researchers to suggest that their reproductive isolation evolved recently.

Parapatric Speciation May Occur between Adjacent Populations

Sometimes a single species is distributed across a discontinuity in environmental conditions, such as a major change in soil type. Although organisms on one side of the discontinuity may interbreed freely with those on the other side, natural selection may favor different alleles on either side, limiting gene flow. In such cases, **parapatric speciation**—speciation arising between adjacent populations—may occur if hybrid offspring have low relative fitness.

Some strains of bent grass *(Agrostis tenuis),* a common pasture plant in Great Britain, have the physiological ability to grow on mine tailings where the soil is heavily polluted by copper or other metals. Plants of the copper-tolerant strains grow well on polluted soils, but plants of the pasture strain do not. Conversely, copper-tolerant plants don't survive as well as pasture plants on unpolluted soils. These strains often grow within a few meters of each other where polluted and unpolluted soils form an intricate mosaic. Because bent grass is wind-pollinated, pollen is readily transferred from one strain to another. Laboratory tests showed that the strains are fully interfertile. However, copper-tolerant plants flower about one week earlier than nearby pasture plants, which promotes prezygotic (temporal) isolation of the two strains **(Figure 21.13).** If the flowering times become further separated, the two strains may attain complete reproductive isolation and become separate species.

Some biologists argue that the places where parapatric populations of bent grass interbreed are really hybrid zones where allopatric populations have established secondary contact. Unfortunately, there is no way to determine whether the hybridizing populations were parapatric or allopatric in the past. Thus, a thorough evaluation of the parapatric speciation hypothesis must await the development of techniques that enable biologists to distinguish clearly between the products of allopatric and parapatric speciation.

Sympatric Speciation Occurs within One Continuously Distributed Population

In **sympatric speciation,** reproductive isolation evolves between distinct subgroups that arise within one population. Models of sympatric speciation do not require that the populations be either geographically or environmentally separated as their gene pools diverge. We examine below general models of sympatric speciation in animals and plants; the genetic basis of sympatric speciation is one of the topics we consider in the next section.

Insects that feed on just one or two plant species are among the animals most likely to evolve by sympatric speciation. These insects generally carry out most important life cycle activities on or near their "host" plants. Adults mate on the host plant; females lay their eggs on it; and larvae feed on the host plant's tissues, eventually developing into adults, which initiate another round of the life cycle. Host plant choice is genetically determined in many insect species. In others, individuals associate with the host plant species they ate as larvae.

Theoretically, a genetic mutation could suddenly change some insects' choice of host plant. Mutant individuals would shift their life cycle activities to the new host, and then interact primarily with others preferring the same new host, an example of ecological isolation. These individuals would collectively form a separate subpopulation, called a **host race.** Reproductive isolation could evolve between different host races if the individuals of each host race are more likely to mate with members of their own host race than with members of another. Some biologists criticize this model, however, because it assumes that the genes controlling two traits, the insects' host plant choice and their mating preferences, change simultaneously. Moreover, host plant choice is controlled by multiple gene loci in some insect species, and it is clearly influenced by prior experience in others.

The apple maggot *(Rhagoletis pomonella)* is the most thoroughly studied example of possible sympatric speciation in animals **(Figure 21.14).** This fly's natural host plant in eastern North America is the hawthorn (*Crataegus* species), but at least two host races have appeared in little more than 100 years. The larvae of a new host race were first discovered feeding on apples in

FIGURE 21.13 Observational Research

Evidence for Reproductive Isolation in Bent Grass

Question: Do adjacent populations of bent grass living on different soil types exhibit signs of reproductive isolation?

Hypothesis: Adjacent populations of bent grass *(Agrostis tenuis)* flower at slightly different times, which could foster prezygotic reproductive isolation between them.

Prediction: On any given summer day, the flowers in one population will be more mature (thus, being pollinated earlier) than flowers in the adjacent population.

Method: On a late summer day in 1965, Thomas McNeilly and Janis Antonovics of University College of North Wales compared the flowers of bent grass growing on polluted soil at a copper mine with the flowers of plants growing on unpolluted soil in a nearby pasture. A meter-wide stretch of polluted pasture (indicated by cross-hatching in the graph) formed a boundary between the two populations. Researchers assigned a score to every flower, with immature flowers scored as 3 and mature flowers as 4.

Results: On the day that they were surveyed, flowers of the copper-tolerant plants had higher scores, indicating that they were more mature—and thus would complete pollination earlier—than the flowers of the pasture plants.

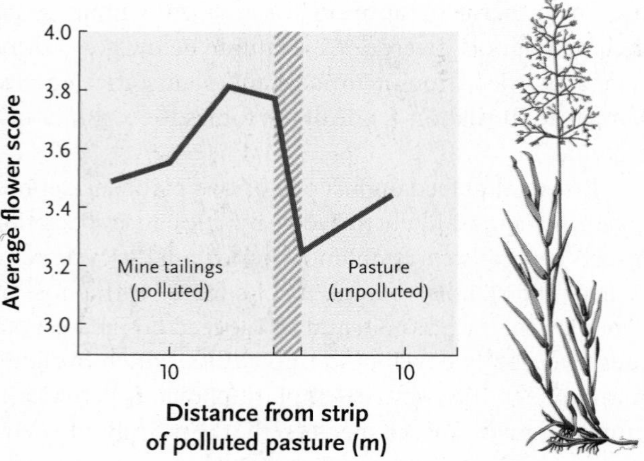

Average flower score
4.0
3.8
3.6
3.4
3.2
3.0
Mine tailings (polluted)
Pasture (unpolluted)
10
10
Distance from strip of polluted pasture (m)

Conclusion: Because adjacent populations of bent grass flower at slightly different times, temporal reproductive isolation may be developing between them.

Source: T. McNeilly and J. Antonovics. 1968. Evolution in closely adjacent plant populations. IV. Barriers to gene flow. *Heredity* 23:205–218.

New York state in the 1860s. In the 1960s, a cherry-feeding host race appeared in Wisconsin.

Recent research has shown that variations at just a few gene loci underlie differences in the feeding preferences of *Rhagoletis* host races; other genetic differences cause the host races to develop at different rates. Moreover, adults of the three races mate during different summer months. Nevertheless, individuals show no particular preference for mates of their own host race, at least under simplified laboratory conditions. Thus, although behavioral isolation has not developed between races, ecological and temporal isolation may separate adults in nature. Researchers are still not certain that the different host races are reproductively isolated under natural conditions.

Sympatric speciation often occurs in plants through a genetic phenomenon, **polyploidy,** in which an individual receives one or more *extra* copies of the entire haploid complement of chromosomes (see Section 13.3). Polyploidy can lead to speciation because these large-scale genetic changes may prevent polyploid individuals from breeding with individuals of the parent species. Nearly half of all flowering plant species are polyploid, including many important crops and ornamental species. The genetic mechanisms that produce polyploid individuals in plant populations are well understood; we describe them in the next section as part of a larger discussion of the genetics of speciation.

STUDY BREAK 21.3 <

1. **What are the two stages required for allopatric speciation?**
2. **What factor appears to promote parapatric speciation in bent grass?**
3. **Why might insects from different host races be unlikely to mate with each other?**

FIGURE 21.14
Sympatric speciation in animals. Male and female apple maggots *(Rhagoletis pomonella)* court on a hawthorn leaf. The female will later lay her eggs on the fruit, and the offspring will feed, mate, and lay their eggs on hawthorns as well.

Jim Smith, Michigan State University

THINK OUTSIDE THE BOOK

Figure 21.12 presents an example of a hybrid zone between two closely related bird species in North America. Search the biological literature and the Internet for three other examples of hybrid zones in animals. How large are the areas of hybridization? What prezygotic isolating mechanisms prevent the species from interbreeding outside the hybrid zones? Is the frequency of hybrid matings increasing or decreasing through time?

21.4 Genetic Mechanisms of Speciation

What genetic changes lead to reproductive isolation between populations, and how do these changes arise? In this section we examine three genetic mechanisms that can lead to reproductive isolation: *genetic divergence* between allopatric populations, *polyploidy* in sympatric populations, and *chromosome alterations,* which occur independently of the geographical distributions of populations.

Genetic Divergence in Allopatric Populations Can Lead to Speciation

In the absence of gene flow, geographically separated populations inevitably accumulate genetic differences. Most postzygotic isolating mechanisms probably develop as accidental by-products of mutation, genetic drift, and natural selection. Note, however, that natural selection cannot promote the evolution of reproductive isolating mechanisms between *allopatric* populations directly: individuals in such populations do not encounter one another and therefore have no opportunity to produce hybrid offspring. And if there are no hybrid offspring, natural selection cannot select against the matings that would have produced them. Nevertheless, natural selection may sometimes foster adaptive changes that create postzygotic reproductive isolation between populations when they later reestablish contact. And, if postzygotic isolating mechanisms reduce the fitness of hybrid offspring, natural selection can promote the evolution of prezygotic isolating mechanisms, the phenomenon described earlier as reinforcement.

How much genetic divergence is necessary for speciation to occur? To understand the genetic basis of speciation in closely related species, researchers first identify the specific causes of reproductive isolation. They then use standard techniques of genetic analysis along with new molecular approaches such as gene mapping and sequencing to analyze the genetic mechanisms that establish reproductive isolation. As explained in *Insights from the Molecular Revolution,* these techniques now allow researchers to determine the minimum number of genes responsible for reproductive isolation in particular pairs of species.

In cases of postzygotic reproductive isolation, mutations in at least a few gene loci establish reproductive isolation. For example, if two common aquarium fishes, swordtails *(Xiphophorus helleri)* and platys *(Xiphophorus maculatus),* mate, two genes induce the development of lethal tumors in their hybrid offspring. When hybrid sterility is the primary cause of reproductive isolation between *Drosophila* species, at least 5 gene loci are responsible. About 55 gene loci contribute to postzygotic reproductive isolation between the European fire-bellied toad *(Bombina bombina)* and the yellow-bellied toad *(Bombina variegata).*

In cases of prezygotic reproductive isolation, some mechanisms have a surprisingly simple genetic basis. For example, a single mutation reverses the direction of coiling in the shells of some snail species. Snails with shells that coil in opposite directions cannot approach each other closely enough to mate, making reproduction between them mechanically impossible.

Many traits that now function as prezygotic isolating mechanisms may originally have evolved in response to sexual selection (described on page 445). This evolutionary process exaggerates showy structures and courtship behaviors in males, the traits that females use to identify appropriate mates. When two species encounter one another on secondary contact, these traits may also prevent interspecific mating. For example, many closely related duck species exhibit dramatic variation in the appearance of males, but not females **(Figure 21.15),** an almost certain sign of sexual selection. Yet these species hybridize readily in captivity, producing offspring that are both viable and fertile. Speciation in these birds probably resulted from geographical isolation and sexual selection without significant genetic divergence: only a few morphological and behavioral characters are responsible for their reproductive isolation. Thus, sometimes the evolution of reproductive isolation may not require much genetic change at all.

Polyploidy Is a Common Mechanism of Sympatric Speciation in Plants

Polyploidy is common among plants, and it may be an important factor in the evolution of some fish, amphibian, and reptile species. Polyploid individuals can arise from chromosome duplications within a single species (autopolyploidy) or through hybridization of different species (allopolyploidy).

AUTOPOLYPLOIDY In **autopolyploidy (Figure 21.16),** a diploid ($2n$) individual may produce, for example, tetraploid ($4n$) offspring, each of which has four complete chromosome sets. Autopolyploidy often results when gametes, through an error in either mitosis or meiosis, spontaneously receive the same number of chromosomes as a somatic cell. Such gametes are called **unreduced gametes** because their chromosome number has not been reduced compared with that of somatic cells.

Mallard ducks *(Anas platyrhynchos)*

Image copyright Mumuism, 2010. Used under license from Shutterstock.com

Pintail ducks *(Anas acuta)*

David Chapman/Alamy

FIGURE 21.15
Sexual selection and prezygotic isolation. In closely related species, such as mallard and pintail ducks, males have much more distinctive coloration than females, a sure sign of sexual selection at work.

INSIGHTS FROM THE **Molecular Revolution**

Monkey-Flower Speciation: How many genes contribute to reproductive isolation?

Reproductive isolation is the primary criterion that biologists use to distinguish species. As noted earlier, the monkey-flower species *Mimulus lewisii* and *Mimulus cardinalis* experience mechanical reproductive isolation in nature because differences in flower structure keep bumblebees or hummingbirds from carrying pollen from one species to the other (see Figure 21.7). However, the two species are easily crossed in the laboratory and produce fertile F_1 hybrids. The F_2 offspring have flowers with various forms intermediate between the two parental types (see Figure), suggesting that several gene loci control the traits separating the species.

Research Question

How many gene loci contribute to the differences in flower traits that foster prezygotic reproductive isolation between the two *Mimulus* species?

Experiment

Because little was known about the genetics of the two monkey-flower species, it was not possible to use a direct genetic analysis to identify and map the flower trait genes to the chromosomes. Instead, H. D. ("Toby") Bradshaw and other researchers at the University of Washington used an indirect molecular approach that identified DNA sequence variations (analogous to those used in DNA fingerprinting; see Chapter 18) at various loci in the genome. Just as morphological and biochemical traits vary within a population, DNA sequences vary at particular sites in the genome; and the different DNA sequence alleles can be distinguished using PCR.

The researchers used a random set of 153 DNA sequence variations. They correlated the segregation of those variations with the segregation of flower traits in 93 plants of the F_2 generation. Some of the DNA sequences segregated closely with a particular flower trait. This result indicated that the particular DNA sequence variation locus was very near the gene for that flower trait on the chromosome. In other words, the flower trait locus was identified indirectly through the close linkage between the DNA variation locus and the flower trait locus. But where are the genes? To answer this question, the researchers used the DNA sequences linked to flower trait loci as probes to find the sites on the chromosomes where they originated. (The use of DNA probes to identify complementary DNA in the genome is described in Figure 18.8.) For any given DNA variation locus, once its position on the chromosomes was determined, the investigators knew that the flower trait locus correlated with it must be nearby.

Mimulus lewisii

F_1 hybrid

Mimulus cardinalis

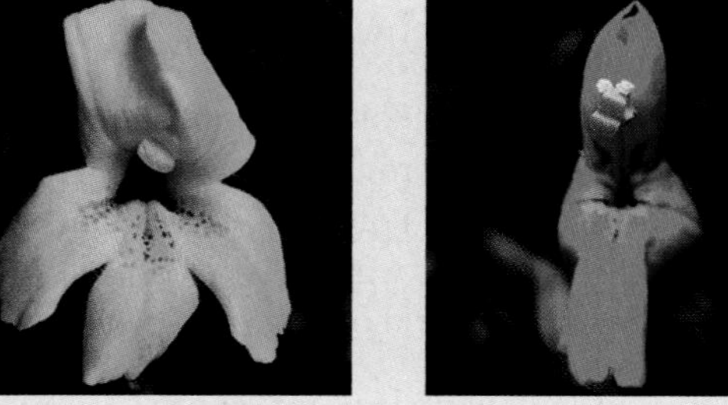

F_2 hybrids

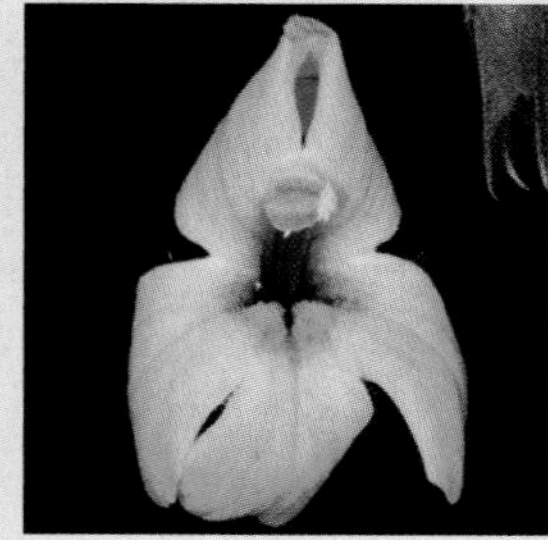

Results

Their results indicate that reproductive isolation of *M. lewisii* and *M. cardinalis* results from differences in eight floral traits: the amount of (1) anthocyanin pigments and (2) carotenoid pigments in petals; (3) flower width; (4) petal width; (5) nectar volume; (6) nectar concentration; and the lengths of the stalks supporting the (7) male and (8) female reproductive parts. Although the investigators could not directly determine the number of genes controlling each trait, the characteristics of the traits, their locations at eight sites on six of the eight chromosomes, and their pattern of inheritance make it most likely that each trait is controlled by a single gene, giving a likely minimum of eight genes.

Conclusion

Mutations in as few as eight genes may have established reproductive isolation between these two species of monkey-flower. Thus, it appears that in some cases surprisingly little genetic change is required for the evolution of a new species. This research was the first in which random differences in DNA sequences were used to answer the fundamental evolutionary question of how much genetic change is needed to produce a new species.

Source: H. D. Bradshaw Jr., S. M. Wilbert, K. G. Otto, and D. W. Schemske. 1995. Genetic mapping of floral traits associated with reproductive isolation in monkeyflowers (*Mimulus*). *Nature* 376:762–765.

Diploid pollen can fertilize the diploid ovules of a self-fertilizing individual, or it may fertilize diploid ovules on another plant with unreduced gametes. The resulting tetraploid offspring can reproduce either by self-pollination or by breeding with other tetraploid individuals. However, a tetraploid plant cannot produce fertile offspring by hybridizing with its diploid parents. The fusion of a diploid gamete with a normal haploid gamete produces a triploid ($3n$) offspring, which is usually sterile because its odd number of chromosomes cannot segregate properly during meiosis. Thus, the tetraploid is reproductively isolated from the original diploid population. Many species of grasses, shrubs, and ornamental plants, including violets, chrysanthemums, and

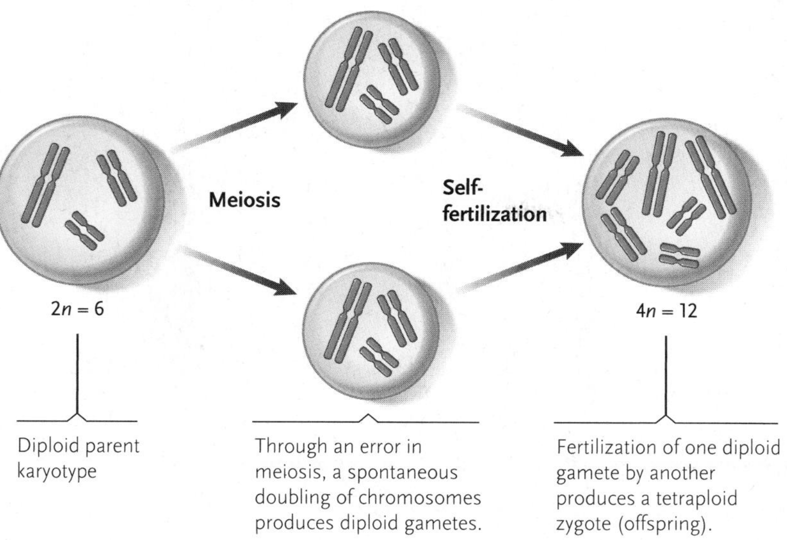
Meiosis
Self-
fertilization
2n = 6
4n = 12
Diploid parent
karyotype
Through an error in
meiosis, a spontaneous
doubling of chromosomes
produces diploid gametes.
Fertilization of one diploid
gamete by another
produces a tetraploid
zygote (offspring).

FIGURE 21.16
Speciation by autopolyploidy in plants. A spontaneous doubling of chromosomes during meiosis produces diploid gametes. If the plant fertilizes itself, a tetraploid zygote will be produced.

nasturtiums, are autopolyploids, having anywhere from 4 to 20 complete chromosome sets.

ALLOPOLYPLOIDY In **allopolyploidy (Figure 21.17),** two closely related species hybridize and subsequently form polyploid offspring. Hybrid offspring are sterile if the two parent species have diverged enough that their chromosomes do not pair properly during meiosis. However, if the hybrid's chromosome number is doubled, the chromosome complement of the gametes is also doubled, producing homologous chromosomes that *can* pair during meiosis. The hybrid can then produce polyploid gametes and, through self-fertilization or fertilization with other doubled hybrids, establish a population of a new polyploid species. Compared with speciation by genetic divergence, speciation by allopolyploidy is extremely rapid, causing a new species to arise in one generation without geographical isolation.

Even when sterile, polyploids are often robust, growing larger than either parent species. For that reason, both autopolyploids and allopolyploids have been important to agriculture. For example, the wheat used to make flour *(Triticum aestivum)* has six sets of chromosomes **(Figure 21.18).** Other polyploid crop plants include plantains (cooking bananas), coffee, cotton, potatoes, sugarcane, and tobacco.

Plant breeders often try to increase the probability of forming an allopolyploid by using chemicals that foster nondisjunction of chromosomes during mitosis. In the first such experiment, undertaken in the 1920s, scientists crossed a radish and a cabbage, hoping to develop a plant with both edible roots and leaves. Instead, the new species, *Raphanobrassica,* combined the least desirable characteristics of each parent, growing a cabbagelike root and radishlike leaves. Recent experiments have been more successful. For example, plant scientists have produced an allopolyploid grain, triticale, that has the disease-resistance of its rye parent and the high productivity of its wheat parent.

Chromosome Alterations Can Foster Speciation

Other changes in chromosome structure or number may also foster speciation. Closely related species often have a substantial number of chromosome differences between them, including in-

FIGURE 21.17
Speciation by hybridization and allopolyploidy in plants. A hybrid mating between two species followed by a doubling of chromosomes during mitosis in gametes of the hybrid can instantly create sets of homologous chromosomes. Self-fertilization can then generate polyploid individuals that are reproductively isolated from both parent species.

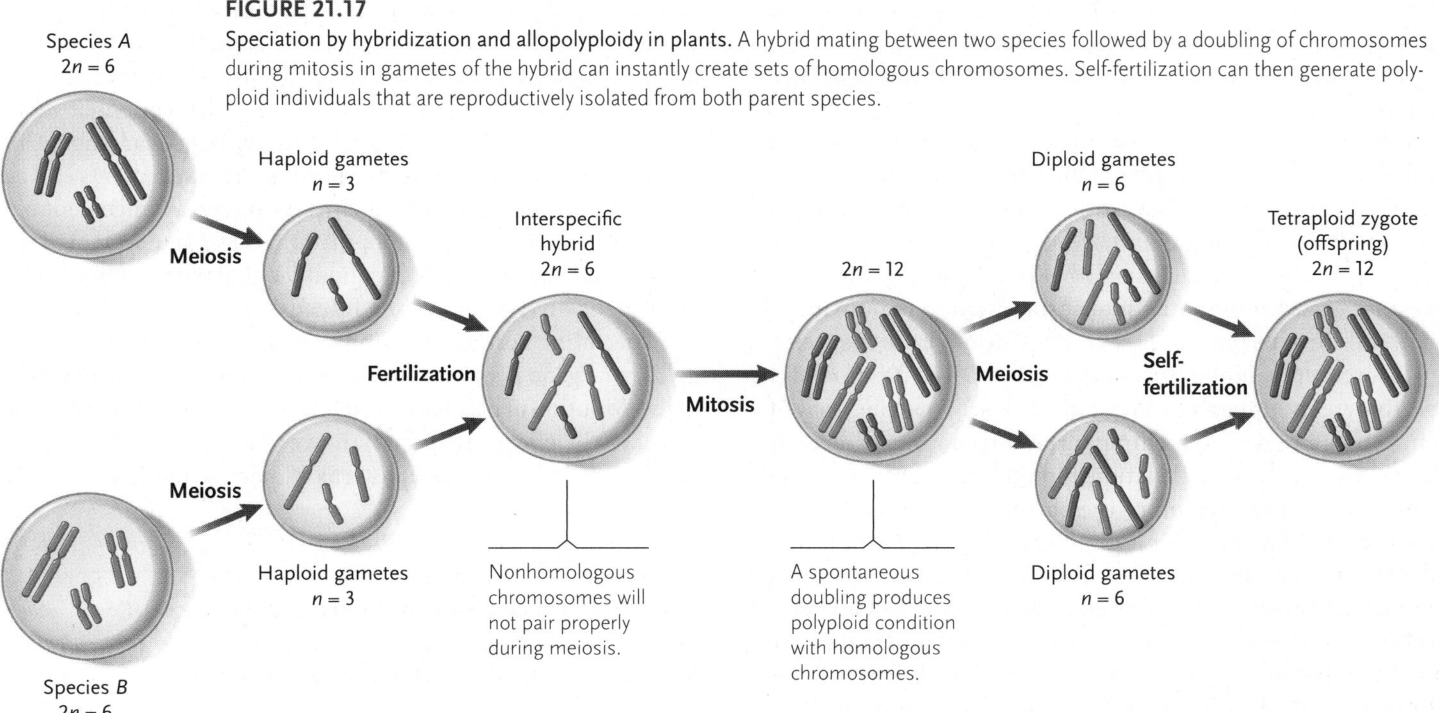
Species *A*
2n = 6
Haploid gametes
n = 3
Meiosis
Interspecific
hybrid
2n = 6
Fertilization
Mitosis
2n = 12
Meiosis
Diploid gametes
n = 6
Self-
fertilization
Tetraploid zygote
(offspring)
2n = 12
Meiosis
Haploid gametes
n = 3
Nonhomologous
chromosomes will
not pair properly
during meiosis.
A spontaneous
doubling produces
polyploid condition
with homologous
chromosomes.
Diploid gametes
n = 6
Species *B*
2n = 6

FIGURE 21.18

The evolution of wheat. Cultivated wheat grains more than 11,000 years old have been found in the Eastern Mediterranean region. Researchers believe that speciation in wheat occurred through hybridization and polyploidy.

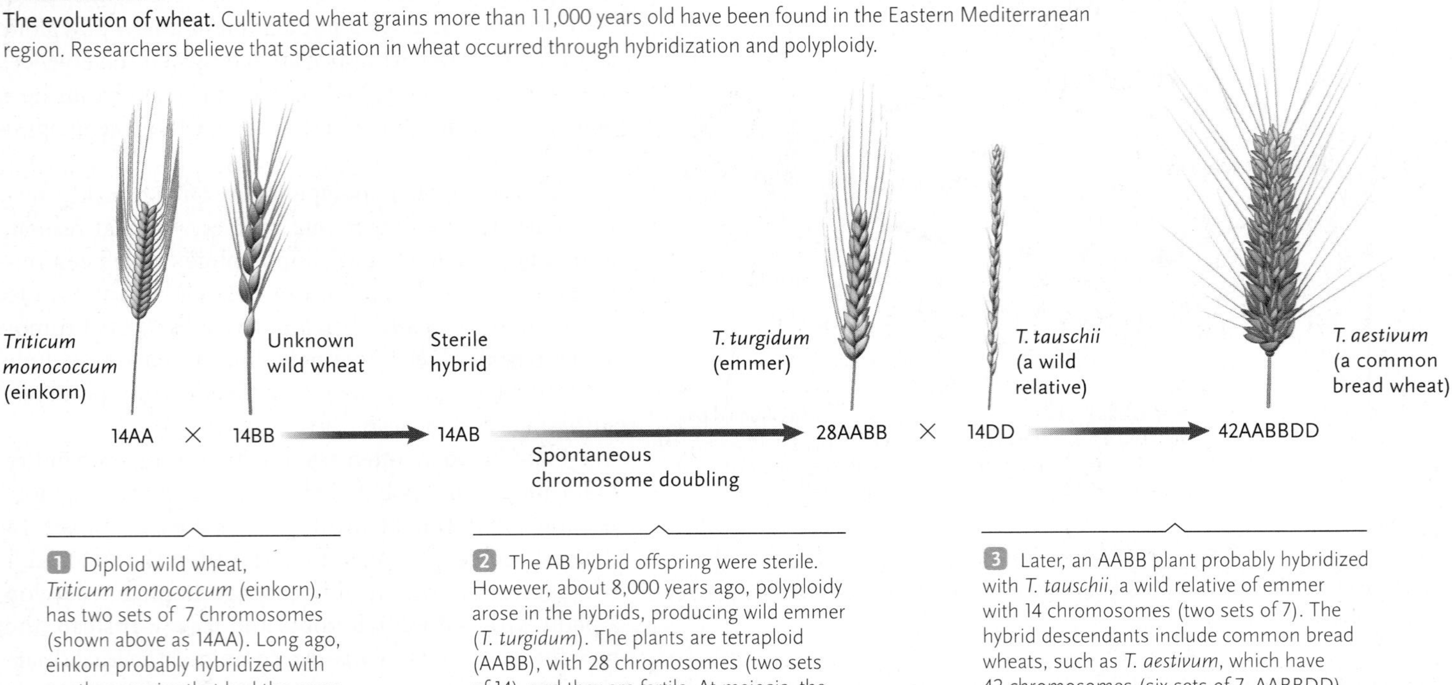

Triticum monococcum (einkorn)
Unknown wild wheat
Sterile hybrid
T. turgidum (emmer)
T. tauschii (a wild relative)
T. aestivum (a common bread wheat)
14AA × 14BB
14AB
Spontaneous chromosome doubling
28AABB × 14DD
42AABBDD

1. Diploid wild wheat, *Triticum monococcum* (einkorn), has two sets of 7 chromosomes (shown above as 14AA). Long ago, einkorn probably hybridized with another species that had the same number of chromosomes (14BB).

2. The AB hybrid offspring were sterile. However, about 8,000 years ago, polyploidy arose in the hybrids, producing wild emmer (*T. turgidum*). The plants are tetraploid (AABB), with 28 chromosomes (two sets of 14), and they are fertile. At meiosis, the A chromosomes pair with each other, and the B chromosomes pair with each other.

3. Later, an AABB plant probably hybridized with *T. tauschii*, a wild relative of emmer with 14 chromosomes (two sets of 7). The hybrid descendants include common bread wheats, such as *T. aestivum*, which have 42 chromosomes (six sets of 7, AABBDD).

versions, translocations, deletions, and duplications (described in Section 13.3). These differences, which may foster postzygotic isolation, can often be identified by comparing the *banding patterns* in stained chromosome preparations from the different species. In all species, banding patterns vary from one chromosome segment to another. When researchers find identical banding patterns in chromosome segments from two or more related species, they know that they are examining comparable portions of the species' genomes. Thus, the banding patterns allow scientists to identify specific chromosome segments and compare their positions in the chromosomes of different species.

The banding patterns of humans and their closest relatives among the apes—chimpanzees, gorillas, and orangutans—reveal that whole sections of chromosomes have been rearranged over evolutionary time **(Figure 21.19)**. For example, humans have a diploid complement of 46 chromosomes, whereas chimpanzees, gorillas, and orangutans have 48 chromosomes. The difference can be traced to the fusion (that is, the joining together) of two ancestral chromosomes into chromosome 2 of humans; the ancestral chromosomes are separate in the other three species.

Moreover, banding patterns suggest that the position of the centromere in human chromosome 2 closely matches that of a centromere in one of the chimpanzee chromosomes, reflecting their close evolutionary relationship. But this centromere falls within an inverted region of the chromosome in gorillas and orangutans, reflecting their evolutionary divergence from chimpanzees and humans. (Recall from Section 13.3 that an inverted chromosome segment has a reversed orientation, so the order of genes on it is reversed relative to the order in a segment that is not inverted.) Nevertheless, humans and chimps differ from each other in centromeric inversions in six other chromosomes.

How might such chromosome rearrangements promote speciation? In a paper published in 2003, Arcadi Navarro of the Universitat Pompeu Fabra in Spain and Nick H. Barton of the University of Edinburgh in Scotland compared the rates of evolution in protein-coding genes that lie within rearranged chromosome segments of humans and chimpanzees to those in genes outside the rearranged segments. They discovered that proteins evolved more than twice as quickly in the rearranged chromosome segments. Navarro and Barton reasoned that because chromosome rearrangements inhibit chromosome pairing and recombination during meiosis, new genetic variations favored by natural selection would be conserved within the rearranged segments. These variations accumulate over time, contributing to genetic divergence between populations with the rearrangement and those without it. Thus, chromosome rearrangements can be a trigger for speciation: once a chromosome rearrangement becomes established within a population, that population will diverge more rapidly from populations lacking the rearrangement. The genetic divergence eventually causes reproductive isolation.

Speciation results from microevolutionary processes that divide one gene pool into two. It is also sometimes the first step in the evolution of new biological lineages. In the next chapter we consider the effects of speciation over vast spans of time as we examine paleobiology and patterns of macroevolution.

STUDY BREAK 21.4 <

1. **How can natural selection promote reproductive isolation in allopatric populations?**
2. **How does polyploidy promote speciation in plants?**

FIGURE 21.19 **Observational Research**

Chromosomal Similarities and Differences among the Great Apes

Question: Does chromosome structure differ between humans and their closest relatives among the apes?

Hypothesis: Large scale chromosome rearrangements contributed to the development of reproductive isolation between species within the evolutionary lineage that includes humans and apes.

Prediction: Chromosome structure differs markedly between humans and their close relatives among the great apes: chimpanzees, gorillas, and orangutans.

Method: Jorge J. Yunis and Om Prakash of the University of Minnesota Medical School used Giemsa stain to visualize the banding patterns on metaphase chromosome preparations from humans, chimpanzees, gorillas, and orangutans. They identified about 1,000 bands that are present in humans and in the three ape species. By matching the banding patterns on the chromosomes, the researchers verified that they were comparing the same segments of the genomes in the four species. They then searched for similarities and differences in the structure of the chromosomes.

Results: Analysis of human chromosome 2 reveals that it was produced by the fusion of two smaller chromosomes that are still present in the other three species. Although the position of the centromere in human chromosome 2 matches that of the centromere in one of the chimpanzee chromosomes, in gorillas and orangutans it falls within an inverted segment of the chromosome.

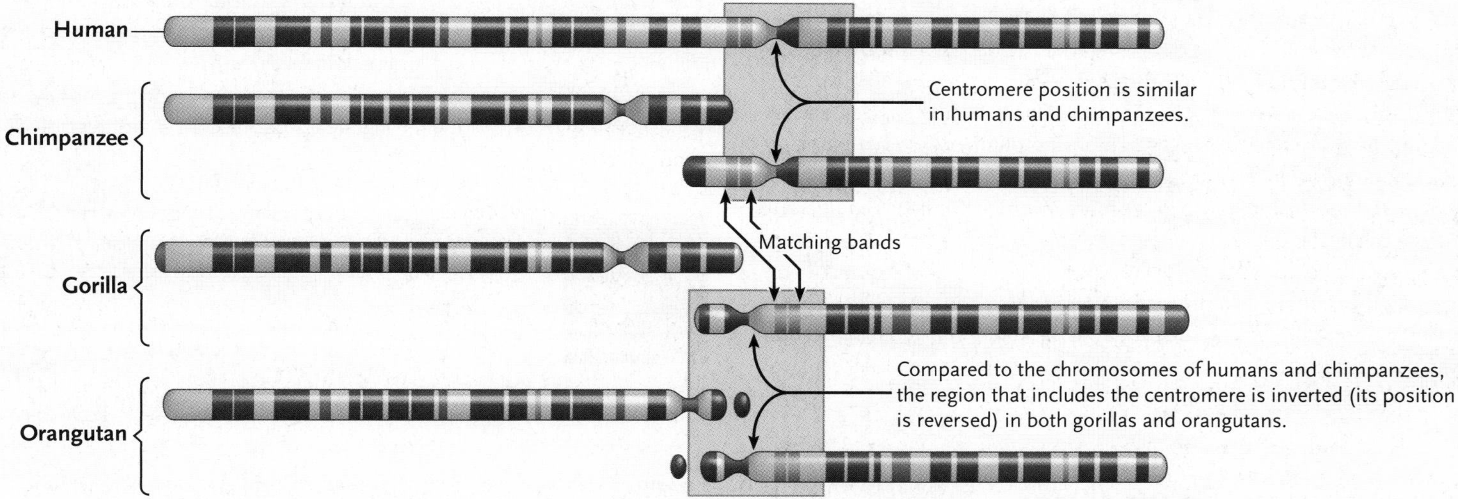
Human
Chimpanzee
Gorilla
Orangutan
Centromere position is similar in humans and chimpanzees.
Matching bands
Compared to the chromosomes of humans and chimpanzees, the region that includes the centromere is inverted (its position is reversed) in both gorillas and orangutans.

Conclusion: Differences in chromosome structure between humans and both gorillas and orangutans are more pronounced than they are between humans and chimpanzees. Structural differences in the chromosomes of these four species may contribute to their reproductive isolation.

Source: J. J. Yunis and O. Prakash. 1982. The origin of man: A chromosomal pictorial legacy. *Science* 215:1525–1530.

UNANSWERED QUESTIONS

Do asexual organisms form species?

As you learned in this chapter, the biological species concept applies only to sexually reproducing organisms because only those organisms can evolve barriers to gene flow (asexual organisms reproduce more or less clonally). Nevertheless, research is starting to show that organisms whose reproduction is almost entirely asexual, such as bacteria, seem to form distinct and discrete clusters in nature. (These clusters could be considered "species.") That is, bacteria and other asexual forms may be as distinct as the species of birds described by Ernst Mayr in New Guinea. Workers are now studying the many species of bacteria in nature (only a small number of which have been discovered) to see if they indeed fall into distinct groups. If they do, then scientists will need a special theory, independent of reproductive isolation, to explain this distinctness. Scientists are now working on theories of whether the existence of discrete ecological niches in nature might explain the possible discreteness of asexual "species."

How often does speciation occur allopatrically versus sympatrically or parapatrically?

Scientists do not know how often speciation occurs between populations that are completely isolated geographically (allopatric speciation), compared to how often it occurs in populations that exchange genes (parapatric or sympatric speciation). The relative frequency of these modes of speciation in nature is an active area of research. The ongoing work includes studies on small isolated islands: if an invading species divides into two or more species in this situation, it probably did so sympatrically or parapatrically, since geographical isolation of populations in small islands is unlikely. In addition, biologists are reconstructing the evolutionary history of speciation using molecular tools and correlating this history with the species' geographical distributions. If this line of research were to show, for example, that the most closely related pairs of species always had geographically isolated distributions, it would imply that speciation was usually allopatric. These lines of research should eventually answer the controversial question of the relative frequency of various forms of speciation.

What are the genetic changes underlying speciation?

Biologists know a great deal about the types of reproductive isolation that prevent gene flow between species, but almost nothing about their genetic underpinnings. Which genes control the difference between flower shape in monkey-flower species? Which genes lead to inviability and sterility of *Drosophila* hybrids? Which genes cause species of ducks to preferentially mate with members of their own species over members of other species? Do the genetic changes that lead to reproductive isolation tend to occur repeatedly at the same gene loci in a group of organisms, or at different gene loci? Do the changes occur mostly in protein-coding regions of genes, or in the noncoding regions that control the production of proteins? Were the changes produced by natural selection or by genetic drift? Biologists are now isolating "speciation genes" and sequencing their DNA. With only a handful of such genes known, and all of these causing hybrid sterility or inviability, there will undoubtedly be a lot to learn about the genetics of speciation in the next decade.

Think Critically

Professor Coyne suggested that biologists will need to devise a special theory—or an additional type of species definition—to explain the distinctness of clusters of asexual organisms that are observed in nature. Do you think that this new explanation would be more similar to the morphological species concept or the phylogenetic species concept as described in this chapter?

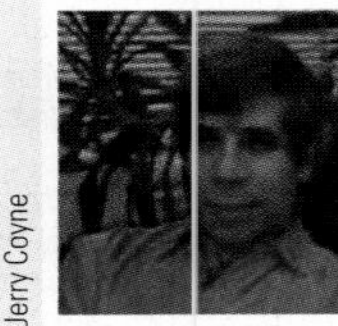
Jerry Coyne

Jerry Coyne conducts research on speciation and teaches at the University of Chicago. To learn more about his research go to http://pondside.uchicago.edu/ecol-evol/faculty/coyne_j.html.

REVIEW KEY CONCEPTS

Go to **CENGAGENOW** at www.cengage.com/login to access quizzing, animations, exercises, articles, and personalized homework help.

21.1 What Is a Species?

- In practice, most biologists describe, identify, and recognize species on the basis of morphological characteristics that serve as indicators of their genetic similarity to or divergence from other species (Figure 21.2).
- The biological species concept defines species as groups of interbreeding populations that are reproductively isolated from populations of other species in nature. A biological species thus represents a gene pool within which genetic material is potentially shared among populations. The biological species concept cannot be applied to organisms that reproduce only asexually, to those that are extinct, or to geographically separated populations.
- The phylogenetic species concept defines a species as a group of populations with a recently shared evolutionary history.
- Most species exhibit geographical variation of phenotypic and genetic traits. When marked geographical variation in phenotypes is discontinuous, biologists sometimes name subspecies (Figure 21.3). In ring species, populations are distributed in a ring around unsuitable habitat (Figure 21.4). Many species exhibit clinal variation of characteristics, which change smoothly over a geographical gradient (Figure 21.5).

Animation: Morphological differences within a species

21.2 Maintaining Reproductive Isolation

- Reproductive isolating mechanisms are characteristics that prevent two species from interbreeding (Table 21.1).
- Prezygotic isolating mechanisms either prevent individuals of different species from mating or prevent fertilization between their gametes. Prezygotic isolation occurs because species live in different habitats, breed at different times, use different courtship behavior (Figure 21.6), or differ anatomically (Figure 21.7). Prezy-

gotic isolation can also result from genetic and physiological incompatibilities between male and female gametes.

- Postzygotic isolating mechanisms reduce the fitness of interspecific hybrids through hybrid inviability, hybrid sterility (Figure 21.8), or hybrid breakdown.

Animation: Reproductive isolating mechanisms

Animation: Temporal isolation among cicadas

21.3 The Geography of Speciation

- The model of allopatric speciation proposes that speciation results from divergent evolution in geographically separated populations (Figures 21.9–21.11). If allopatric populations accumulate enough genetic differences, they will be reproductively isolated upon secondary contact. Nevertheless, some species hybridize over small areas of secondary contact (Figure 21.12).
- The model of parapatric speciation suggests that reproductive isolation can evolve between parts of a population that occupy opposite sides of an environmental discontinuity (Figure 21.13).
- A model of sympatric speciation in insects suggests that reproductive isolation may evolve between host races that rarely contact one another under natural conditions (Figure 21.14). Sympatric speciation commonly occurs in flowering plants by allopolyploidy.

Animation: Models of speciation

Animation: Allopatric speciation on an archipelago

Animation: Sympatric speciation in wheat

21.4 Genetic Mechanisms of Speciation

- Allopatric populations inevitably accumulate genetic differences, some of which contribute to their reproductive isolation. Reproductive isolating mechanisms evolve as by-products of genetic changes that occur during divergence. Prezygotic isolating mechanisms may evolve in populations experiencing secondary contact (Figure 21.15).
- We cannot yet generalize about how many gene loci participate in the process of speciation, but at least several gene loci are usually involved.
- Speciation by polyploidy in flowering plants involves the duplication of an entire chromosome complement through nondisjunction of chromosomes during meiosis or mitosis. Polyploids can arise among the offspring of a single species (autopolyploidy; Figure 21.16) or, more commonly, after hybridization between closely related species (allopolyploidy; Figures 21.17 and 21.18).
- Chromosome alterations can promote speciation by fostering the genetic divergence of, and reproductive isolation between, populations with different numbers of chromosomes or different chromosome structure (Figure 21.19).

UNDERSTAND AND APPLY

Test Your Knowledge

1. The biological species concept defines species on the basis of:
   a. reproductive characteristics.
   b. biochemical characteristics.
   c. morphological characteristics.
   d. behavioral characteristics.
   e. all of the above.
2. Biologists can apply the biological species concept *only* to species that:
   a. reproduce asexually.
   b. lived in the past.
   c. are allopatric to each other.
   d. hybridize in captivity.
   e. reproduce sexually.
3. A characteristic that exhibits smooth changes in populations distributed along a geographical gradient is called a:
   a. ring species.
   b. hybrid.
   c. cline.
   d. hybrid breakdown.
   e. subspecies.
4. If two species of holly (genus *Ilex*) flower during different months, their gene pools may be kept separate by:
   a. mechanical isolation.
   b. ecological isolation.
   c. gametic isolation.
   d. temporal isolation.
   e. behavioral isolation.
5. Prezygotic isolating mechanisms:
   a. reduce the fitness of hybrid offspring.
   b. generally prevent individuals of different species from producing zygotes.
   c. are found only in animals.
   d. are found only in plants.
   e. are observed only in organisms that reproduce asexually.
6. In the model of allopatric speciation, the geographical separation of two populations:
   a. is sufficient for speciation to occur.
   b. occurs only after speciation is complete.
   c. allows gene flow between them.
   d. reduces the relative fitness of hybrid offspring.
   e. inhibits gene flow between them.
7. Adjacent populations that produce hybrid offspring with low relative fitness may be undergoing:
   a. clinal isolation.
   b. parapatric speciation.
   c. allopatric speciation.
   d. sympatric speciation.
   e. geographical isolation.
8. An animal breeder, attempting to cross a llama with an alpaca for finer wool, found that the hybrid offspring rarely lived more than a few weeks. This outcome probably resulted from:
   a. genetic drift.
   b. prezygotic reproductive isolation.
   c. postzygotic reproductive isolation.
   d. sympatric speciation.
   e. polyploidy.

9. Which of the following could be an example of allopolyploidy?
   a. One parent has 8 chromosomes, the other has 10, and their offspring have 36.
   b. Gametes and somatic cells have the same number of chromosomes.
   c. Chromosome number increases by one in a gamete and in the offspring it produces.
   d. Chromosome number decreases by one in a gamete and in the offspring it produces.
   e. Chromosome number in the offspring is exactly half of what it is in the parents.
10. Which of the following genetic characteristics is shared by humans and chimpanzees?
    a. They have the same number of chromosomes.
    b. The position of the centromere on human chromosome 2 matches the position of a centromere on a chimpanzee chromosome.
    c. A fusion of ancestral chromosomes formed chromosome 2.
    d. Centromeres on all of their chromosomes fall within inverted chromosome segments.
    e. all of the above

Discuss the Concepts

1. All domestic dogs are classified as members of the species *Canis familiaris.* But it is hard to imagine how a tiny Chihuahua could breed with a gigantic Great Dane. Do you think that artificial selection for different breeds of dogs will eventually create different dog species?
2. Human populations often differ dramatically in external morphological characteristics. On what basis are all human populations classified as a single species?
3. If intermediate populations in a ring species go extinct, eliminating the possibility of gene flow between populations at the two ends of the ring, would you now identify those remaining populations as full species? Explain your answer.

Design an Experiment

Design an experiment to test whether populations of birds on different islands belong to the same species.

Interpret the Data

David Hillis of Baylor University noted that three closely related species of leopard frog (genus *Rana*) exhibit substantial—but not complete—postzygotic reproductive isolation when crossed in the laboratory. Field surveys of numerous populations in Texas and surrounding states revealed that populations of the three species breed at various times during the year. Data on the breeding schedule of both allopatric and sympatric populations of these species are presented in the figure below. Interpret the data in the figure and explain how they may demonstrate that these frogs experience prezygotic reproductive isolation in nature. What type of prezygotic reproductive isolation do the data suggest?

Apply Evolutionary Concepts

How do human activities (such as destruction of natural habitats, diversion of rivers, and the construction of buildings) influence the chances that new species of plants and animals will evolve in the future? Frame your answer in terms of the geographical and genetic factors that foster speciation.

Express Your Opinion

Often, when a species is at the brink of extinction, some individuals are captured and brought to zoos for captive breeding programs. However, some people say that keeping a species alive in a zoo is a distraction from more meaningful conservation efforts, and captive animals seldom are successfully restored to the wild. Do you support captive breeding of highly endangered species? Go to www.cengage.com/login to investigate both sides of the issue and then vote.

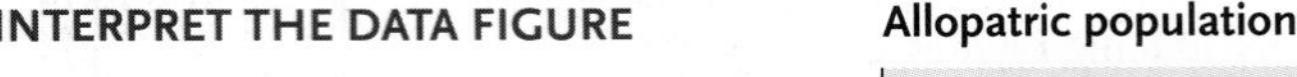
INTERPRET THE DATA FIGURE

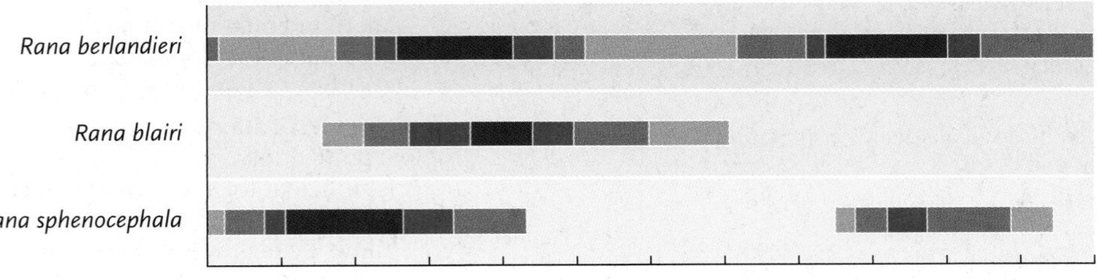
Allopatric populations
Rana berlandieri
Rana blairi
Rana sphenocephala

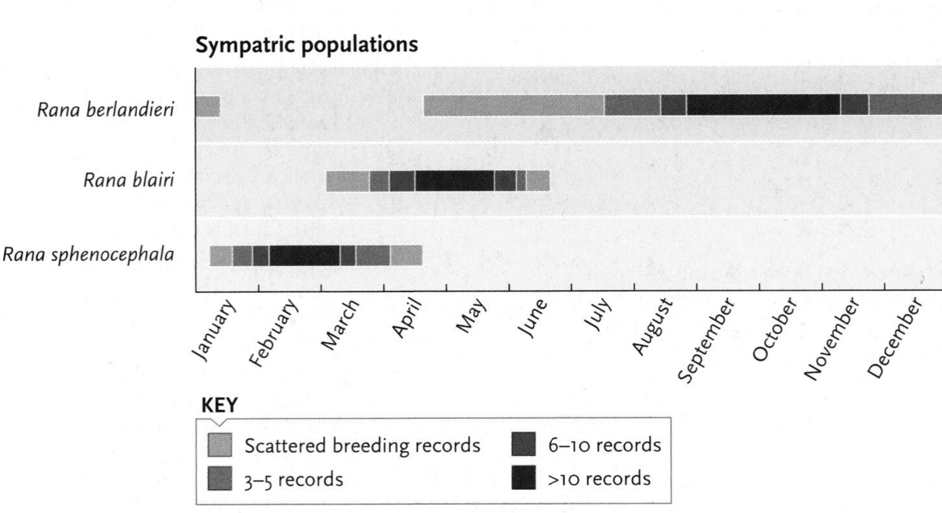
Sympatric populations
Rana berlandieri
Rana blairi
Rana sphenocephala
January
February
March
April
May
June
July
August
September
October
November
December
KEY
Scattered breeding records
3–5 records
6–10 records
>10 records

Source: D. M. Hillis. 1981. Premating isolating mechanisms among three species of the *Rana pipiens* complex in Texas and southern Oklahoma. *Copeia* 1981:312–319.

Fossil of a dragonfly *(Cordulagomphus tuberculatus)* from the Cretaceous period, discovered in Ceara Province, Brazil.

Courtesy Lowcountry Geologic

22

STUDY OUTLINE

Paleobiology and Macroevolution

Why It Matters. . . In April 1842, the anatomist Sir Richard Owen published a report on the fossil reptiles of Britain. His analysis included fragmentary fossils of three species—*Iguanodon, Megalosaurus,* and *Hylaeosaurus*—that another researcher had unearthed in 1818. Recognizing that these gigantic creatures were distinct from lizards and snakes, Owen assigned them to a new group, which he named Dinosauria (*deinos* = terrible; *sauros* = lizards). Although his discoveries were based on limited material, Owen was anxious to bring them to the attention of the broader public. In 1851, he designed a display of "inhabitants of the ancient world" for the Crystal Palace exhibition in London. He and 20 invited guests dined inside one of the exhibits, a hollow reconstruction of *Iguanodon,* which was portrayed in a clumsy, lizard-like pose on its four legs.

Owen's published accounts and Crystal Palace exhibition models led to a dinosaur craze that still grips scientists and the public today. Dinosaur discoveries continued apace, especially in North America, where the most complete fossils are found. In 1858, a nearly intact skeleton of the duck-billed dinosaur *Hadrosaurus foulkii* was uncovered in New Jersey. Analyses of its leg bones showed clearly that it was bipedal (that is, it stood on its hind legs rather than on all fours), a shocking hypothesis at the time. Ten years later, the specimen was exhibited at the Philadelphia Academy of Science, where it is still on display. (Analyses of *Iguanodon* fossils reveal that it, too, was bipedal, making it less desirable as a dining room.)

The *Hadrosaurus* exhibit tripled attendance at the Philadelphia Academy, setting off a race to uncover other complete dinosaur skeletons. In the late 1870s, abundant fossils turned up in Colorado, Wyoming, and Alberta; and scientists and entrepreneurs scoured the region for fossils that would bring fame and fortune. Their quest was not unlike the California gold rush of the mid-nineteenth century, with rivals reputedly hijacking one another's fossils and sometimes coming to blows over their ownership.

As natural history museums exhibited these discoveries, a major question plagued preparators (professionals who clean and reassemble fossils) as they mounted the specimens. What did dinosaurs really look like? The fossilized remains were mostly disarticulated skeletons, which scientists could reassemble. But how did these animals stand when they were alive? How did they hold their heads and tails?

The continued study of dinosaur biology has answered many of these questions. In the 1960s, biologists hypothesized that dinosaurs were very active creatures, like mammals and birds are today. According to this hypothesis, many dinosaurs held their heads up and their tails off the ground; others adopted a bipedal stance. As our knowledge of dinosaurs increased, natural history museums have remounted their specimens and models in more active postures. But one question seemed impossible to answer. Did dinosaurs blend in with their surroundings, or were they brightly colored?

As you will discover later in this chapter, many dinosaurs were feathered, just as birds are today. In 2010, two groups of researchers published analyses of the colors of fossilized feathers from several species of small dinosaurs. Using scanning electron microscopes, the researchers compared the fine structure of melanosomes (pigment-bearing organelles) in the fossilized feathers to those of living birds, allowing them to reconstruct the feather colors of the dinosaurs. The most complete analysis at the time of this writing suggests that *Anchiornis huxleyi,* a dinosaur that lived about 150 million years ago in what is now northern China, had gray or black feathers over most of its body, black and white feathers on its limbs, and a crown of red feathers on its head **(Figure 22.1)**. These animals must surely have used their colorful feathers in social displays.

This description of a truly remarkable discovery opens our chapter on paleobiology and macroevolution, the large-scale changes in morphology and diversity observed over life's 3.8-billion-year history. Macroevolution has occurred over so vast a span of time and space that the evidence for it is fundamentally different from that for microevolution and speciation. In this chapter we consider what paleobiology and evolutionary developmental biology tell us about macroevolutionary patterns. <

FIGURE 22.1
Dinosaurs in living color. By examining the fine structure of pigment-containing organelles inside fossilized feathers, scientists were able to reconstruct the colors of *Anchiornis huxleyi*, a dinosaur that lived about 150 million years ago.

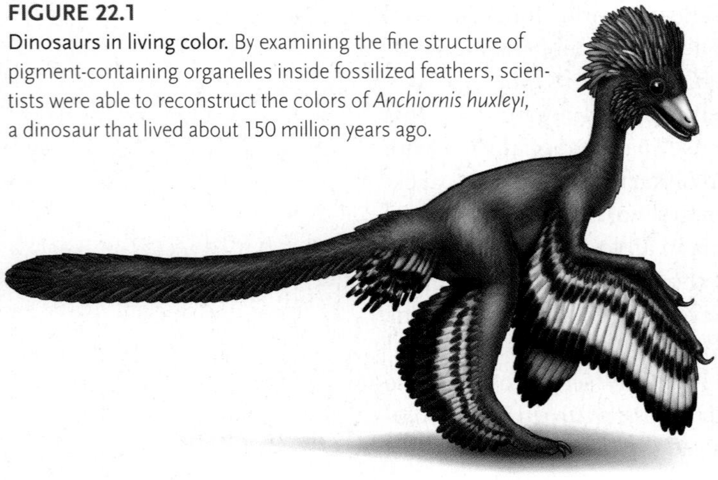

22.1 The Fossil Record

Paleobiologists discover, describe, and name new fossil species and analyze the morphology and ecology of extinct organisms. Because fossils provide physical evidence of life in the past, they are a primary source of data about the evolutionary history of many organisms.

Fossils Form When Organisms Are Buried by Sediments or Preserved in Oxygen-Poor Environments

Most fossils form in sedimentary rocks. Rain and runoff constantly erode the land, carrying fine particles of rock and soil downstream to a swamp, a lake, or the sea. Particles settle to the bottom as sediments, forming successive layers, called *strata,* over millions of years **(Figure 22.2)**. The weight of newer sediments compresses the older layers beneath them into a solid matrix: sand into sandstone and silt or mud into shale. Fossils form within the layers when the remains of organisms are buried in the accumulating sediments. Because sedimentation superimposes new layers over old ones, the lowest strata in a sedimentary rock formation are usually the oldest and the highest layers are the newest.

The process of fossilization is a race against time because the soft remains of organisms are quickly consumed by scavengers or decomposed by microorganisms. Thus, fossils usually preserve the details of hard structures, such as the bones, teeth, and shells of animals and the wood, leaves, and pollen of plants. During fossilization, dissolved minerals replace some parts molecule by molecule, leaving a fossil made of stone **(Figure 22.3A)**; other fossils form as molds, casts, or impressions in material that is later transformed into solid rock **(Figure 22.3B)**.

In some environments, the near absence of oxygen prevents decomposition, and even soft-bodied organisms are preserved. Some insects, plants, and tiny lizards and frogs are embedded in amber, the fossilized resin of coniferous trees **(Figure 22.3C)**. Other organisms are preserved in glacial ice, coal, tar pits, or the highly acidic water of peat bogs **(Figure 22.3D)**. Sometimes organisms are so well preserved that researchers can examine their internal anatomy, cell structure, and even food in their digestive tracts.

The Fossil Record Provides an Incomplete Portrait of Life in the Past

The 300,000 described fossil species represent less than 1% of all the species that have ever lived. Several factors make the fossil record incomplete. First, soft-bodied organisms do not fossilize as readily as species with hard body parts. Moreover, we are unlikely to find fossilized remains of species that were rare and locally distributed. Finally, fossils rarely form in habitats where sediments

A. **Sedimentation**

Highest strata contain the most recent fossils

Lowest strata contain the oldest fossils

B. **Geological strata in the Painted Desert, Arizona**

Nick Greaves/Alamy

FIGURE 22.2
Sedimentation and geological strata. **(A)** Sedimentation deposits successive layers at the bottom of a lake or sea. **(B)** Over millions of years, the upper layers compress those below them into rock. When the rocks are later exposed by uplifting or erosion, the different layers are evident as geological strata.

do not accumulate, such as mountain forests. The most common fossils are those of hard-bodied, widespread, and abundant organisms that lived in swamps or shallow seas, where sedimentation is ongoing.

Although most fossils are composed of stone, they don't last forever. Many are deformed by pressure from overlying rocks or destroyed by geological disturbances like volcanic eruptions and earthquakes. Once they are exposed on Earth's surface, where scientists are most likely to find them, rain and wind cause them to erode. Because the effects of these destructive processes are additive, old fossils are much less common than those formed more recently.

A. **Petrified wood (Araucariaceae)**

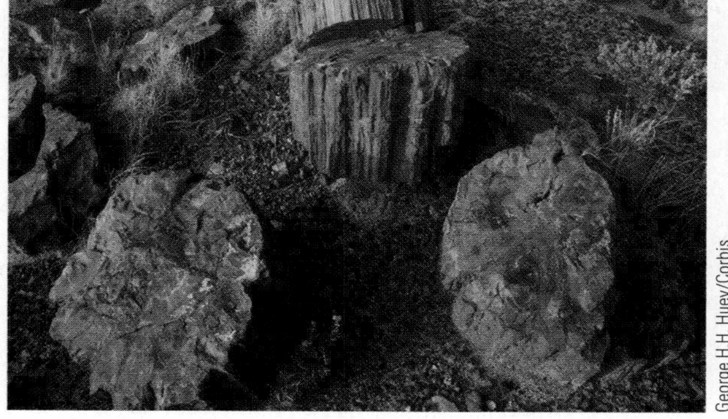

George H.H. Huey/Corbis

B. ***Dickinsonia***

Neville Pledge/South Australian Museum

C. **Mosquitoes in amber**

Image copyright Katrina Brown, 2010. Used under license from Shutterstock.com

D. **Mammoth *(Mammonteus)* in permafrost**

Topham/The Image Works

FIGURE 22.3
Fossils. **(A)** Petrified wood, from Petrified Forest National Park in Arizona, formed when minerals replaced the wood of dead trees, molecule by molecule. These forests lived during the late Triassic period, about 225 million years ago. **(B)** The remains of an invertebrate from the Proterozoic era, 560–555 million years ago, were preserved as an impression in very fine sediments. **(C)** These 10-million-year-old mosquitoes were trapped in the oozing resin of a coniferous tree and are now encased in amber. **(D)** A frozen baby mammoth that lived about 40,000 years ago was discovered embedded in Siberian permafrost in 1977.

Radiometric Dating

Purpose: Radiometric dating allows researchers to estimate the absolute age of a rock sample or fossil.

Protocol:

1. Knowing the approximate age of a rock or fossil, select a radioisotope that has an appropriate half-life. Because different radioisotopes have half-lives ranging from seconds to billions of years, it is usually possible to choose one that brackets the estimated age of the sample under study. For example, if you think that your fossil is more than 10 million years old, you might use uranium-235. The half life of ^{235}U, which decays into the lead isotope ^{207}Pb, is about 700 million years. Or if you think that your fossil is less than 70,000 years old, you might select carbon-14. The half-life of ^{14}C, which decays into the nitrogen isotope ^{14}N, is 5,730 years.

Radioisotopes Commonly Used in Radiometric Dating

Radioisotope (Unstable)		More Stable Breakdown Product	Half-Life (Years)	Useful Range (Years)
Samarium-147	⟶	Neodymium-143	106 billion	>100 million
Rubidium-87	⟶	Strontium-87	48 billion	>10 million
Thorium-232	⟶	Lead-208	14 billion	>10 million
Uranium-238	⟶	Lead-206	4.5 billion	>10 million
Uranium-235	⟶	Lead-207	700 million	>10 million
Potassium-40	⟶	Argon-40	1.25 billion	>100,000
Carbon-14	⟶	Nitrogen-14	5,730	<70,000

2. Prepare a sample of the material and measure the quantities of the parent radioisotope and its more stable breakdown product.

Interpreting the Results: Compare the relative quantities of the parent radioisotope and its breakdown product (or some other stable isotope) to determine what percentage of the original parent radioisotope remains in the sample. Then use a graph of radioactive decay for that isotope to determine how many half-lives have passed since the sample formed.

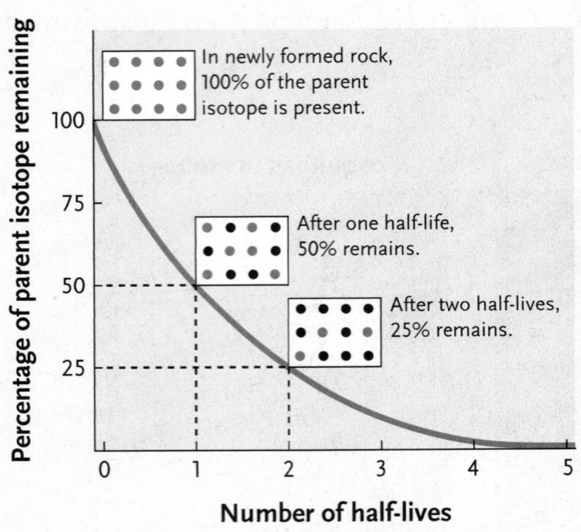
In newly formed rock, 100% of the parent isotope is present.
After one half-life, 50% remains.
After two half-lives, 25% remains.
Percentage of parent isotope remaining
100
75
50
25
0
1
2
3
4
5
Number of half-lives

Knowing the number of half-lives that have passed allows you to estimate the age of the sample.

© 2001 PhotoDisc

A living mollusk absorbed trace amounts of ^{14}C, a rare radioisotope of carbon, and large amounts of ^{12}C, which is the more stable and common isotope of carbon.

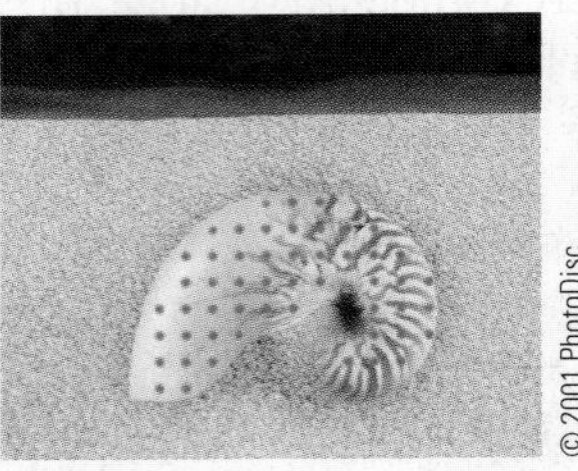
© 2001 PhotoDisc

When the mollusk died, it was buried in sand and fossilized. From the moment of its death, the ratio of ^{14}C to ^{12}C began to decline through radioactive decay. Because the half-life of ^{14}C is 5,730 years, half of the original ^{14}C was eliminated from the fossil in 5,730 years and half of what remained was eliminated in another 5,730 years.

© 2001 PhotoDisc

After the fossil was discovered, a scientist determined that its ^{14}C to ^{12}C ratio was one-eighth of the ^{14}C to ^{12}C ratio in living organisms. Thus, radioactive decay had proceeded for three half-lives—or about 17,000 years—since the mollusk's death.

Scientists Assign Relative and Absolute Dates to Geological Strata and the Fossils They Contain

The sediments found in any one place form distinctive strata (layers) that differ in color, mineral composition, particle size, and thickness (see Figure 22.2B). If they have not been disturbed, the strata are arranged in the order in which they formed, with the youngest layers on top. However, strata are sometimes uplifted, warped, or even inverted by geological processes.

Geologists of the early nineteenth century deduced that the fossils discovered in a particular sedimentary stratum, no matter where on Earth it is found, represent organisms that lived and died at roughly the same time in the past. Because each stratum formed at a specific time, the sequence of fossils in the lowest (oldest) to the highest (newest) strata reveals their *relative ages*. Geologists used the sequence of strata and their distinctive fossil assemblages to establish the geological time scale **(Table 22.1)**.

Although the geological time scale provides a relative dating system for sedimentary strata, it does not tell us how old the rocks and fossils actually are. But many rocks contain radioisotopes, which, from the moment they form, begin to break down into other, more stable elements. The breakdown proceeds at a steady rate that is unaffected by chemical reactions or environmental conditions such as temperature or pressure. Using a technique called **radiometric dating,** scientists can estimate the age of a rock by noting how much of an unstable "parent" isotope has decayed to another form. By measuring the relative amounts of the parent radioisotope and its breakdown products and comparing this ratio with the isotope's **half-life**—the time it takes for half of a given amount of radioisotope to decay—researchers can estimate the *absolute age* of the rock **(Figure 22.4).** Table 22.1 presents these age estimates along with the major geological and evolutionary events of each period.

Radiometric dating works best with volcanic rocks, which form when lava cools and solidifies. But most fossils are found in sedimentary rocks. To date a sedimentary fossil, scientists determine the age of volcanic rocks from the same stratum. Using this method, investigators have linked fossils to deposits that are hundreds of millions of years old.

Fossils that still contain organic matter, such as the remains of bones or wood, can be dated directly by measuring their content of the radioactive carbon isotope ^{14}C, which decays to ^{14}N. Living organisms absorb traces of ^{14}C and large quantities of ^{12}C, a stable carbon isotope, from the environment and incorporate them into biological molecules. As long as an organism is alive, its ^{14}C content remains constant because the ratio of ^{14}C to ^{12}C in its tissues is an equilibrium with the ratio in the environment. But as soon as the organism dies, ^{14}C begins its steady radioactive decay. Scientists use the ratio of ^{14}C to ^{12}C in a fossil to determine its age, as explained in Figure 22.4.

To develop a feeling for geological time, imagine the 4.5-billion-year history of Earth scaled onto a calendar year; each day represents a little over 12 million years **(Figure 22.5).** The planet was formed on January 1. Animal life originated in mid-November, dinosaurs lived between December 14 and December 26, and the genus that includes modern humans appeared during the last 4 hours of December 31.

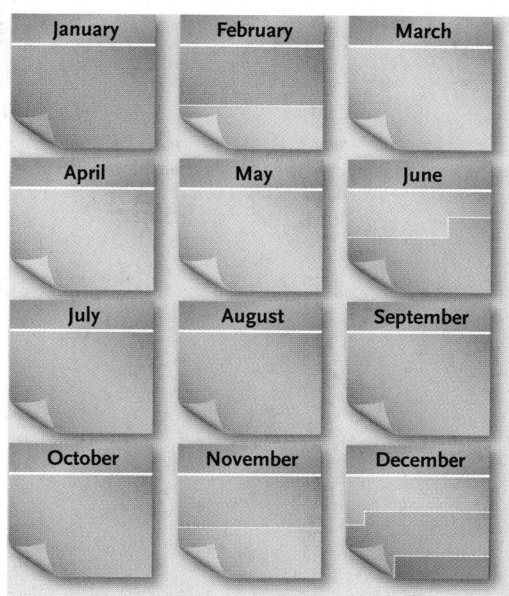
January
February
March
April
May
June
July
August
September
October
November
December

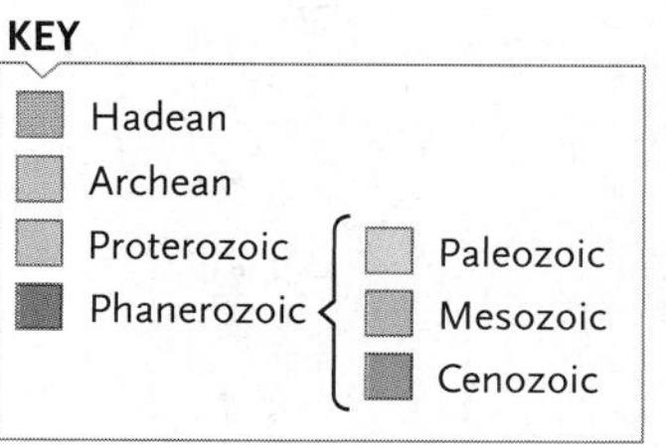
KEY
Hadean
Archean
Proterozoic
Phanerozoic
Paleozoic
Mesozoic
Cenozoic

FIGURE 22.5
A calendar of Earth history. Geological time can be scaled onto a calendar year in which each day is a little more than 12 million years.

Fossils Provide Abundant Information about Life in the Past

Although imperfect, the fossil record provides our only direct information about life in the past. Fossilized skeletons, shells, stems, leaves, and flowers tell us about the size and appearance of ancient animals and plants. The fossil record also allows scientists to see how structures were modified as they became adapted for specialized uses (see Figure 19.3). Moreover, fossils chronicle the proliferation and extinction of evolutionary lineages and provide data on their past geographical distributions.

Recent applications of new techniques allow paleobiologists to make truly remarkable discoveries about extinct organisms. For example, researchers appear to have solved a long-standing puzzle about lambeosaurines, a group of huge duck-billed dinosaurs that lived in the swampy habitats of western North America in the late Cretaceous period. The heads of lambeosaurines sported bony crests with gigantic nasal passages. Paleobiologists had proposed that the crests served as weapons in male combat, adornments that attracted mates, snorkels that facilitated breathing underwater, radiators that cooled the dinosaurs' bodies, structures that enhanced the sense of smell, or resonating chambers that produced honking vocalizations. Based on anatomical analyses, researchers accepted the vocalization hypothesis as the most probable: computerized acoustic models predicted that air flowing through the nasal passages would have produced low frequency sounds (30 to 375 hertz).

But could lambeosaurines hear sounds in that frequency range? In 2009, David Evans of the Royal Ontario Museum in Toronto, Canada, and colleagues in Ohio used computed tomography and 3-D visualization software to scan and reconstruct the interior anatomy of the skulls and brains of several lambeosaurine species **(Figure 22.6).** Their findings indicate that the inner ears of lambeosaurines were attuned to hear low-frequency sounds that matched those predicted by the earlier research. The authors concluded that the elaborate nasal passages in lambeosaurine crests as well as the structure of their inner ears facilitated vocal communication, perhaps between parents and their offspring.

Fossils also provide indirect data about behavior, physiology, and ecology. For example, the fossilized footprints of some dinosaurs suggest that adults surrounded their young when the group moved, possibly to protect them from predators. Complex scrolls of bone in the nasal passages of early mammals suggest that they had a well-developed sense of smell, and fossilized teeth and dung provide data about the diets of extinct animals. The study of fossilized pollen allows paleobiologists to reconstruct

TABLE 22.1 The Geological Time Scale and Major Evolutionary Events

Eon	Era	Period	Epoch	Millions of Years Ago	Major Evolutionary Events
Phanerozoic	Cenozoic	Quaternary	Holocene		
				0.01	
			Pleistocene		Origin of humans; major glaciations
				2.6	
		Neogene	Pliocene		Origin of ape-like human ancestors
				5.3	
			Miocene		Angiosperms and mammals further diversify and dominate terrestrial habitats
				23.0	
		Paleogene	Oligocene		Primates diversify; origin of apes
				33.9	
			Eocene		Angiosperms and insects diversify; modern orders of mammals differentiate
				55.8	
			Paleocene		Grasslands and deciduous woodlands spread; modern birds, mammals, snakes, pollinating insects diversify; continents approach current positions
				65.5	
	Mesozoic	Cretaceous			Angiosperms, insects, marine invertebrates, fishes, dinosaurs diversify; asteroid impact causes mass extinction at end of period, eliminating dinosaurs and many other groups
				145.5	
		Jurassic			Gymnosperms abundant in terrestrial habitats; modern fishes diversify; dinosaurs diversify and dominate terrestrial habitats; frogs, salamanders, lizards, and birds appear; continents continue to separate
				201.6	
		Triassic			Predatory fishes and reptiles dominate oceans; gymnosperms dominate terrestrial habitats; radiation of dinosaurs; early mammals; Pangaea starts to break up; mass extinction at end of period
				251.0	

Eon	Era	Period	Epoch	Millions of Years Ago	Major Evolutionary Events
Phanerozoic	Paleozoic	Permian			Insects and reptiles abundant and diverse in swamp forests; some reptiles colonize oceans; fishes colonize freshwater habitats; continents coalesce into Pangaea, causing glaciation and decline in sea level; mass extinction at end of period eliminates 85% of species
				299.0	
		Carboniferous			Vascular plants form large swamp forests; first flying insects; amphibians diversify; first reptiles appear
				359.0	
		Devonian			Terrestrial vascular plants diversify; fungi, invertebrates, amphibians colonize land; first insects and seed plants; major glaciation at end of period; mass extinction, mostly of marine life
				416.0	
		Silurian			Jawless fishes diversify; first jawed fishes, arthropods, terrestrial vascular plants
				444.0	
		Ordovician			Major radiations of marine invertebrates and jawless fishes; major glaciation at end of period causes mass extinction of marine life
				488.0	
		Cambrian			Appearance of modern animal phyla, including earliest vertebrates (Cambrian explosion); simple marine communities
				542.0	
Proterozoic					High concentration of oxygen in atmosphere; origin of eukaryotic cells; evolution and diversification of "protists," fungi, soft-bodied animals
				2,500	
Archean					Evolution of prokaryotes, including anaerobic and photosynthetic bacteria; oxygen starts to accumulate in atmosphere; origin of aerobic respiration
				3,850	
Hadean					Formation of Earth, including crust, atmosphere, and oceans; origin of life
				4,600	

FIGURE 22.6
Honking dinosaurs. Analysis of the sinuses and braincase of lambeosaurines (*Corythosaurus* species) revealed that the nasal passages in their crests served as resonating chambers for the production of low-frequency sounds that their inner ears could detect.

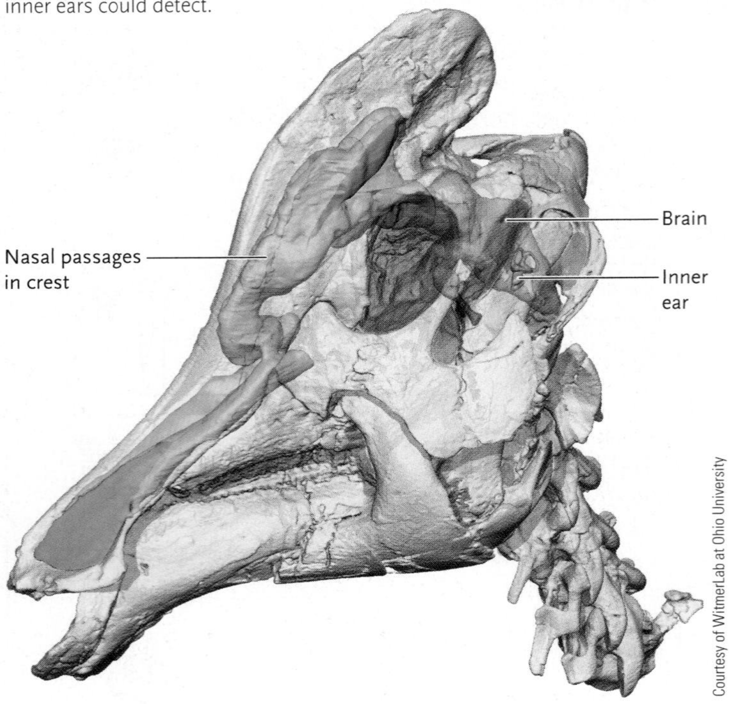

Brain
Nasal passages in crest
Inner ear
Courtesy of WitmerLab at Ohio University

the vegetation and climate of ancient sites. The changing arrays of fossils that document biological evolution partly reflect large-scale shifts in Earth's physical environments, a topic that we explore in the next section.

STUDY BREAK 22.1

1. **What biological materials are the most likely to fossilize?**
2. **Why does the fossil record provide an incomplete portrait of life in the past?**
3. **What sorts of information can paleobiologists discern from the fossil record?**

22.2 Earth History

Organisms have interacted constantly with their environments over geological time. These interactions have triggered fundamental changes in Earth's physical environment as well as in the organisms that occupy them. The development of an oxidizing atmosphere, discussed extensively in Chapter 24, was a landmark effect of living systems on the environment. Photosynthesis by prokaryotic organisms increased the concentration of oxygen in the atmosphere, facilitating the evolution of eukaryotic cells and later allowing animals to invade terrestrial habitats. In this section we focus on long-term shifts in geography and climate—as well as some brief but catastrophic events—that have significantly altered the environments where organisms live.

Continental Drift Has Altered the Configuration of Landmasses and Oceans

Major geological and climatic shifts occur because the planet's crust is constantly in motion. According to the theory of **plate tectonics,** Earth's crust is broken into irregularly shaped plates of

A. **Earth's crustal plates**

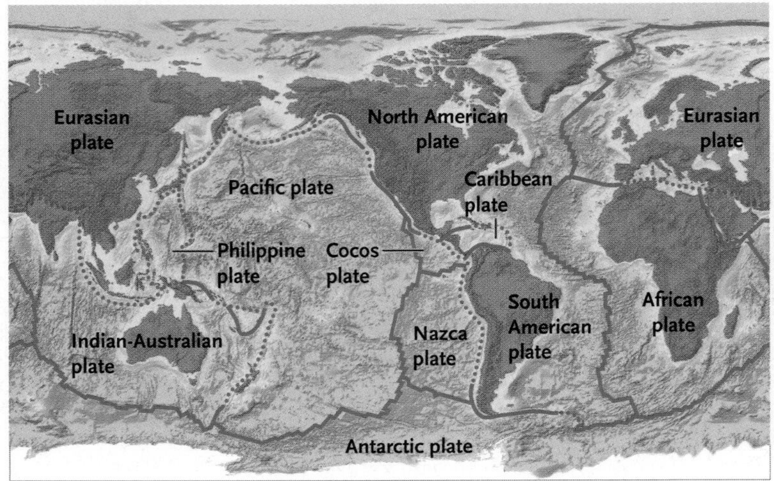

Eurasian plate
North American plate
Eurasian plate
Pacific plate
Caribbean plate
Philippine plate
Cocos plate
South American plate
African plate
Indian-Australian plate
Nazca plate
Antarctic plate
KEY
Oceanic ridge
Oceanic trench

B. **Model of plate tectonics**

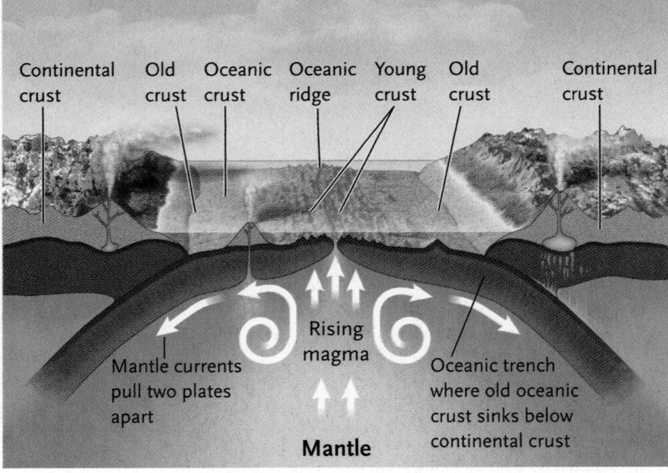

Continental crust
Old crust
Oceanic crust
Oceanic ridge
Young crust
Old crust
Continental crust
Rising magma
Mantle currents pull two plates apart
Oceanic trench where old oceanic crust sinks below continental crust
Mantle

FIGURE 22.7
Plate tectonics. **(A)** Earth's crust is broken into large, rigid plates. New crust is added at oceanic ridges, and old crust is recycled into the mantle at oceanic trenches. **(B)** Oceanic ridges form where pressure in the mantle forces magma (molten rock) through fissures in the sea floor. Mantle currents pull the plates apart on either side of the ridge, forcing the sea floor to move laterally away from the ridge. This phenomenon, seafloor spreading, is widening the Atlantic Ocean about 3 cm per year. Oceanic trenches form where plates collide. The heavier oceanic crust sinks below the lighter continental crust, and it is recycled into the mantle, a process called subduction. The highest mountain ranges (including the Rockies, Himalayas, Alps, and Andes) formed where subduction uplifted continental crust. Earthquakes and volcanoes are common near trenches.

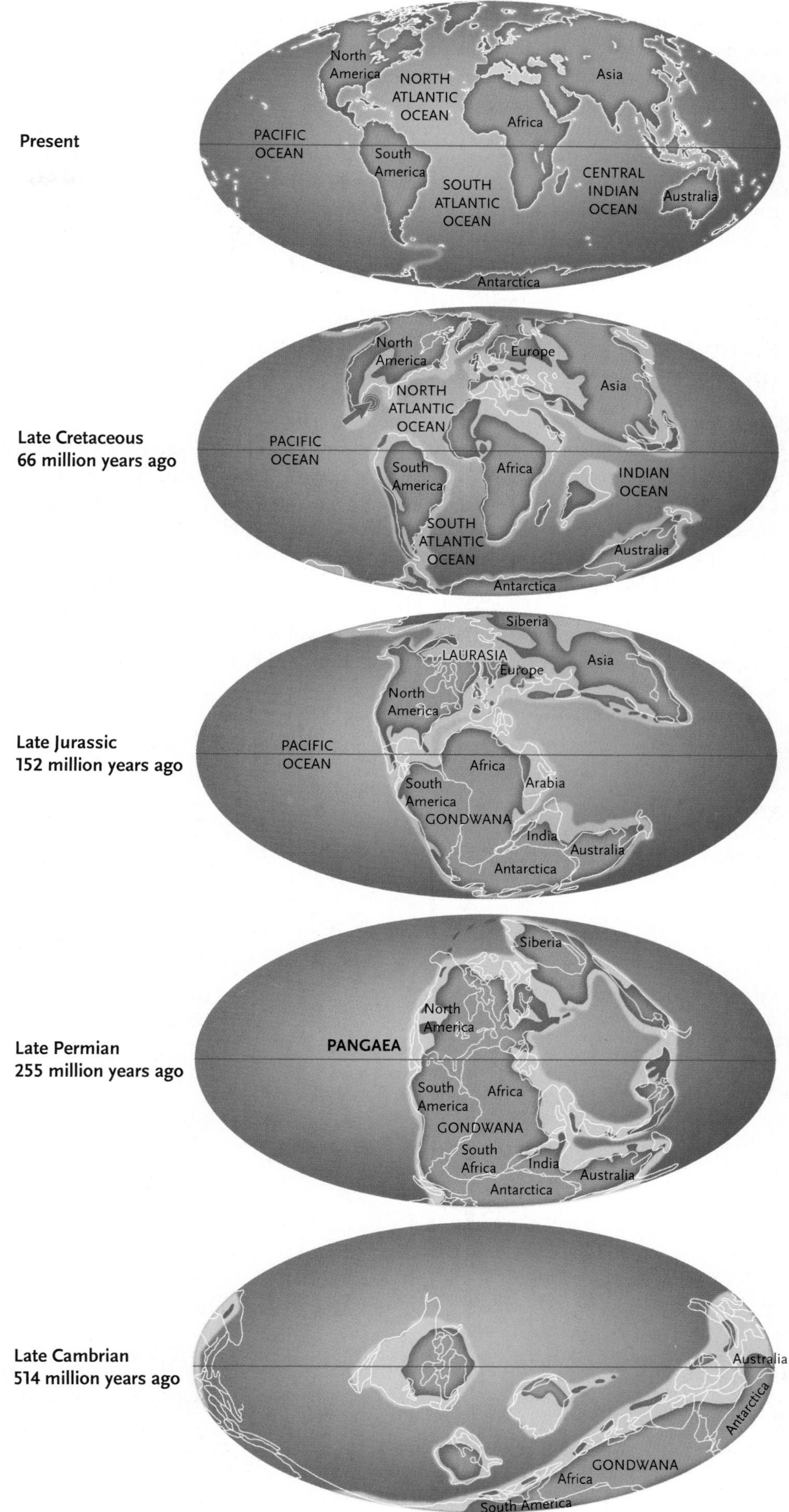
Present
North America
NORTH ATLANTIC OCEAN
Asia
Africa
PACIFIC OCEAN
South America
SOUTH ATLANTIC OCEAN
CENTRAL INDIAN OCEAN
Australia
Antarctica
Late Cretaceous
66 million years ago
North America
Europe
Asia
NORTH ATLANTIC OCEAN
PACIFIC OCEAN
South America
Africa
INDIAN OCEAN
SOUTH ATLANTIC OCEAN
Australia
Antarctica
Late Jurassic
152 million years ago
Siberia
LAURASIA
Europe
Asia
North America
PACIFIC OCEAN
Africa
South America
Arabia
GONDWANA
India
Australia
Antarctica
Late Permian
255 million years ago
Siberia
North America
PANGAEA
South America
Africa
GONDWANA
South Africa
India
Australia
Antarctica
Late Cambrian
514 million years ago
Australia
Antarctica
GONDWANA
Africa
South America

FIGURE 22.8

Continental drift. Earth's many landmasses coalesced during the Permian period, forming the supercontinent Pangaea. About 160 million years ago, Pangaea began to separate into a southern continent, Gondwana, and a northern continent, Laurasia. Then Gondwana began to break apart. The landmasses that would form Africa and India pulled away first, opening the South Atlantic and Indian Oceans. Australia separated from Antarctica about 55 million years ago and slowly drifted northward. South America separated from Antarctica shortly thereafter. Laurasia remained nearly intact until 43 million years ago when North America and Greenland both separated from Europe and Asia. Movement of the continents also changed the shapes and the sizes of the oceans.

rock—roughly 40 km thick—that float on its semisolid mantle **(Figure 22.7).** Currents in the mantle cause the plates—and the continents embedded in them—to move, a phenomenon called **continental drift.**

Continent-sized landmasses have coalesced and broken apart several times in Earth's history. About 250 million years ago, they merged into a single supercontinent called Pangaea; continental drift later separated Pangaea into a northern continent, Laurasia, and a southern continent, Gondwana. Laurasia and Gondwana subsequently broke into the continents we know today **(Figure 22.8).**

The breakup and separation of the continents had a huge impact on the evolution of terrestrial and aquatic organisms. For example, the widespread distribution of many terrestrial groups on Pangaea was a defining characteristic of the Paleozoic era. But continental drift later established large-scale patterns of geographical isolation. Each continent became a separate arena in which organisms evolved independently of those on other landmasses. Thus, the organisms that live on southern continents, the remnants of Gondwana, are very different from those that live on northern continents, which are the remnants of Laurasia. For example, the large flightless birds known as ratites (pictured in Figure 19.2) occur only in Africa, Australia, South America, and on nearby islands; no comparable or closely related species live in Eurasia or North America.

Moreover, the smaller landmasses produced by plate tectonics are surrounded by more extensive coastlines and shallow marine habitats than were the large continents. These environments have always harbored tremendous biodiversity, and as the continents became increasingly separated, the populations living in these shallow seas became geographically isolated from each other. As the continents assumed their present positions at different latitudes, locally distributed organisms diversified and differentiated from those in other regions. Today, shallow marine habitats in the temperate zone are often occupied by giant kelp beds, whereas those in the tropics harbor coral reefs (described further in Chapter 49).

Geological Processes and Unpredictable Events Changed the Environments Where Organisms Lived

Continental drift affected climate, the extent of glaciations, and sea levels, changing the physical environments where organisms live on local, regional, and global scales. Most environmental changes occurred incrementally over millions of years—and millions of generations—but some catastrophes have also had a sudden impact. The combined effects of these changes on biological evolution have been profound.

CLIMATE, GLACIATIONS, AND SEA LEVEL As you will learn in Chapter 49, environmental temperatures vary with latitude: they are lower near the poles than near the equator. As the continents drifted, their latitudinal positions shifted, causing what must have been huge changes in local temperatures. For example, one billion years ago Queensland, Australia, was situated near the North Pole; it then drifted southward for 800 million years, moving to the equator 440 million years ago and to mid-latitudes in the southern hemisphere (40° S latitude) 200 million years ago. It then reversed direction and moved northward to its present position near the equator (12° S).

On a regional level, the sizes of landmasses also have an effect on their climates. Coastal regions generally experience smaller daily and seasonal fluctuations in temperature than interior regions because their proximity to the sea moderates their climate (described further in Chapter 49). Whenever landmasses were joined into large continents, vast expanses of the interior landscape must have experienced frigid winters and hot summers. As the landmasses broke up into smaller continents, a larger fraction of the land was close to the sea and its moderating influence.

On a global scale, the overall climate has shifted from warm and wet to cool and dry several times in Earth's history **(Figure 22.9)**. The changing positions of continents altered the flow of ocean currents, which, along with small changes in Earth's orbit around the sun, contributed to these climatic shifts. At times when Earth's climate was cooler than it is today, the polar ice caps grew larger. And because much of Earth's water was incorporated into these enormous glaciers, rainfall was greatly reduced and sea level fell dramatically. Between the cold spells, Earth's climate was warmer than it is today. Under those conditions, glaciers retreated, rainfall was luxuriant, temperature differences between equatorial and polar regions were reduced, and sea levels rose.

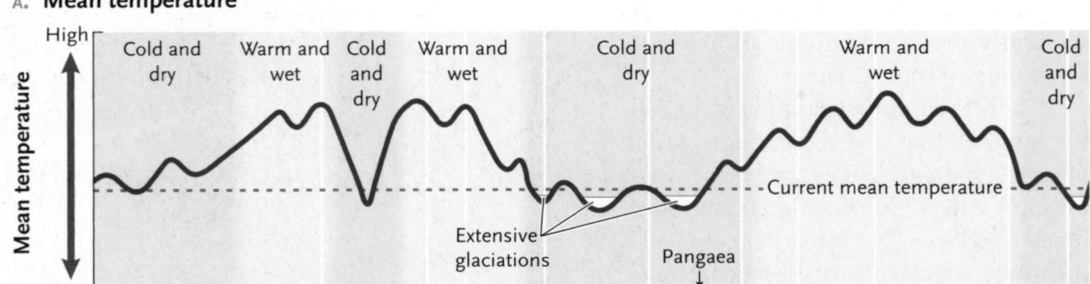
A. Mean temperature
High
Low
Mean temperature
Cold and dry
Warm and wet
Cold and dry
Warm and wet
Cold and dry
Warm and wet
Cold and dry
Current mean temperature
Extensive glaciations
Pangaea
488
444
416
359
299
251
202
145
65
23

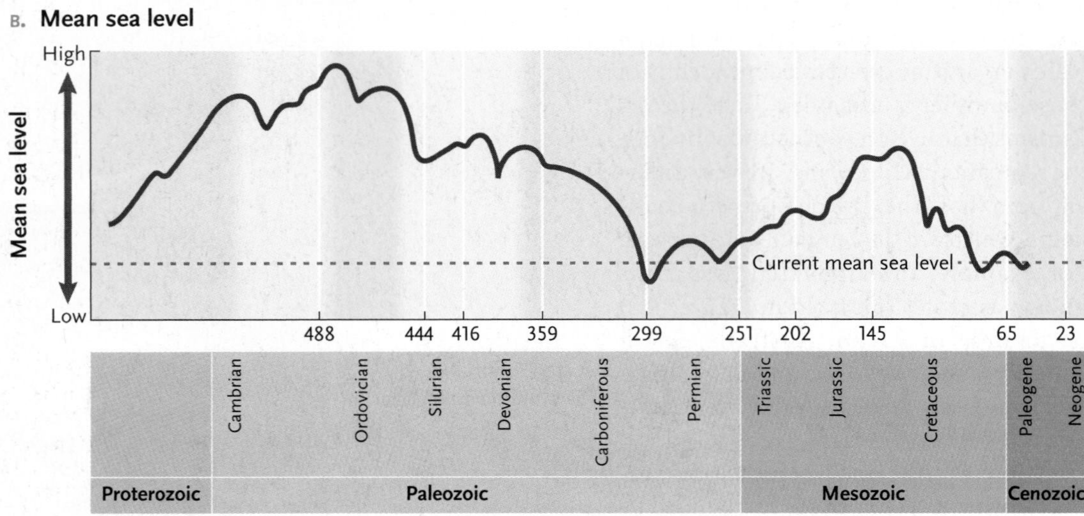
B. Mean sea level
High
Low
Mean sea level
Current mean sea level
488
444
416
359
299
251
202
145
65
23
Cambrian
Ordovician
Silurian
Devonian
Carboniferous
Permian
Triassic
Jurassic
Cretaceous
Paleogene
Neogene
Proterozoic
Paleozoic
Mesozoic
Cenozoic
Millions of years ago

FIGURE 22.9
Variations in climate and sea level. **(A)** Earth's climate has shifted from cold and dry to warm and wet multiple times in Earth's history. Particularly cold periods were accompanied by extensive glaciation. **(B)** As a result of climatic shifts and continental drift, sea level has often risen and fallen. Large drops in sea level eliminated many near-shore, shallow marine environments.

Changing climate patterns and the accompanying rise or fall of sea level influenced the evolution of living systems. Organisms adapted to the altered conditions, shifted their geographical ranges into suitable environments, or became extinct. Declining sea levels took an especially severe toll on organisms living in shallow marine environments. As their habitats disappeared, many marine species ceased to exist. Nevertheless, some organisms always survived, and their descendants flourished and later diversified in the altered environments.

Today we are living in one of the cooler periods of Earth's history. But human activities, including the combustion of fossil fuels and industrial-scale agriculture, are causing an extraordinarily rapid warming of the atmosphere and oceans. Although Earth's climate has experienced repeated temperature shifts, these changes occurred slowly relative to the lifetimes of individual organisms, and many populations, species, and evolutionary lineages adapted to them. But global climate change is occurring much faster today, and scientists have documented its effects: melting glaciers, rising temperatures, and rising sea level. They have also documented changes in the geographical ranges and breeding schedules of many species. We explore the effects of global climate change more fully in Unit Seven.

VOLCANIC ERUPTIONS AND ASTEROID IMPACTS Sudden, unpredictable events have also changed physical environments on Earth. For example, volcanoes are especially active near oceanic trenches where old crust is recycled into Earth's mantle. Volcanic eruptions often spew enormous quantities of ash and gas into the atmosphere, blocking incoming sunlight. Massive eruptions may cause Earth's surface temperature to decrease several degrees for as long as a year. Other types of tectonic activity also release molten rock on Earth's surface, sometimes with the opposite effect on climate, as you will discover in Section 22.4.

Asteroid strikes have also had devastating effects on living systems. In addition to obliterating a gigantic area around the site of impact, the collisions threw massive amounts of material into the atmosphere, blocking sunlight and potentially altering the planet's climate. As we describe next, these cataclysmic events have sometimes caused many forms of life to disappear over relatively short periods of geological time.

STUDY BREAK 22.2 <

1. How did continental drift affect the geographical distributions of organisms?
2. What effect did glaciations have on sea level?

>

THINK OUTSIDE THE BOOK

Search Internet and library resources for information about how continental drift and other major geological phenomena affected the area where you live. At what latitude was your home or university 65 million years ago, 100 million years ago, and 250 million years ago? Was it ever submerged below a shallow sea?

22.3 Historical Biogeography and Convergent Biotas

Continental drift facilitated the diversification of distinct evolutionary lineages in different regions on Earth. In this section we consider the large-scale geographical distributions of organisms.

Historical Biogeography Explains the Broad Geographical Distributions of Organisms

More than a century after Darwin published his observations, the theory of plate tectonics refocused attention on biogeography. Historical biogeographers try to explain how organisms acquired their geographical distributions over evolutionary time.

CONTINUOUS AND DISJUNCT DISTRIBUTIONS Many species have a **continuous distribution:** they live in suitable habitats throughout a geographical area. For example, herring gulls *(Larus argentatus)* live along the coastlines of all northern continents. Continuous distributions, especially in mobile organisms like birds, usually require no special historical explanation.

Other groups exhibit **disjunct distributions,** in which closely related species live in widely separated locations. Two phenomena—dispersal and vicariance—can create disjunct distributions. **Dispersal** is the movement of organisms away from their place of origin; it can produce a disjunct distribution if a new population becomes established on the far side of a geographical barrier. **Vicariance** is the fragmentation of a once-continuous geographical distribution by external factors.

How can researchers determine whether a disjunct distribution is the product of dispersal or vicariance? Analysis of the fossil record for a group of organisms, interpreted in the context of their evolutionary relationships as well as continental drift, can provide an answer. For example, southern beech trees (genus *Nothofagus*) are currently found in Australasia and South America **(Figure 22.10).** (Australasia is the region that includes Australia, New Guinea, New Zealand, and other islands of the South Pacific.) For many years, the genus' distribution only on southern continents and islands has been the classic example of vicariance: the group presumably originated on Gondwana; and its subsequent fragmentation apparently isolated ancestral *Nothofagus* populations on different continents and islands where their descendants live today. The fossil record generally supports that interpretation: researchers have identified fossilized *Nothofagus* pollen, dating from 55 to 34 million years ago, in locations that still formed a nearly contiguous landmass at that time (Figure 22.10C).

However, Lyn G. Cook and Michael D. Crisp of The Australian National University recently undertook a molecular analysis of the relationships and geographical distributions of living *Nothofagus* species. Their research reveals that the genus had a much more complicated biogeographic history. It confirms that the distributions of some *Nothofagus* species in Australia and South America are the products of vicariance: their common ancestor originated on Gondwana *before* the two continents separated from each other, and their descendants now live in both places. However, the analysis

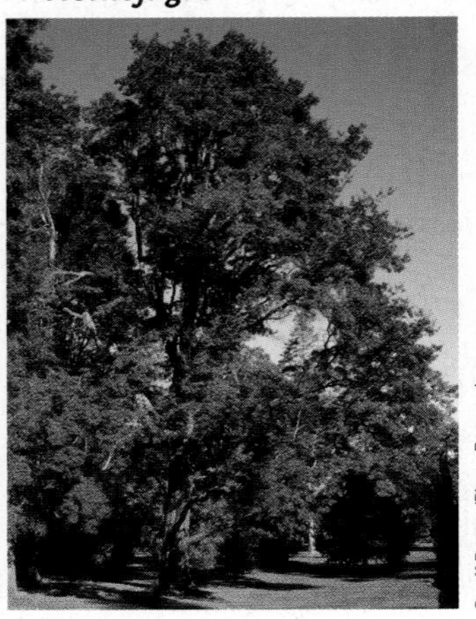
A. Nothofagus
Geoff Bryant/Photo Researchers

FIGURE 22.10
Distribution of southern beech trees *(Nothofagus)*. The genus *Nothofagus* **(A)** exhibits a disjunct distribution **(B);** it includes species living in South America and in Australasia. **(C)** A reconstruction of the positions of Australia and South America in the late Cretaceous reveals that they were parts of a nearly continuous landmass at that time.

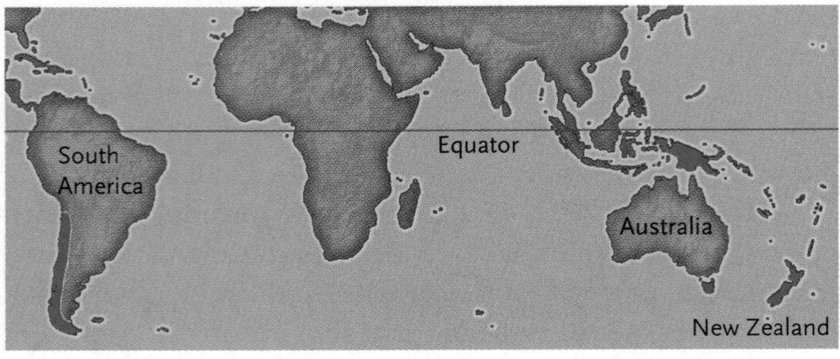
B. Current distribution of Nothofagus species
South America
Equator
Australia
New Zealand

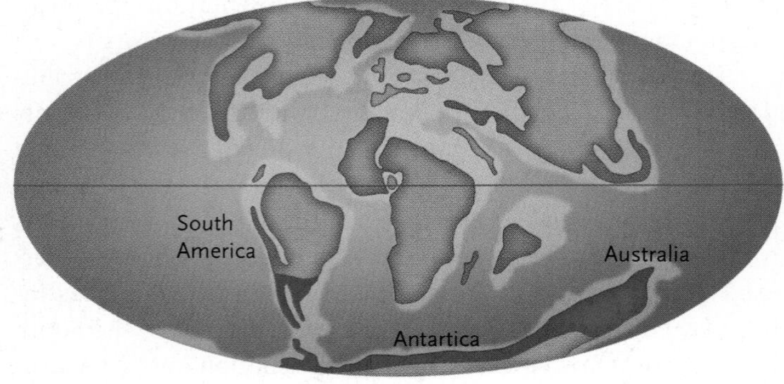
C. Position of southern continents in the late Cretaceous
South America
Australia
Antartica

also suggests that the common ancestor of certain closely related *Nothofagus* species in New Zealand and Australia originated more than 30 million years *after* New Zealand had separated from Australia. Thus, vicariance cannot explain their current distribution. Instead, seeds of these plants must have dispersed across the open ocean from one place to the other. Therefore, both vicariance and dispersal have played important roles in establishing the present geographical distribution of *Nothofagus*.

BIOGEOGRAPHICAL REALMS For species that were widespread in the Mesozoic era, Pangaea's breakup was a powerful vicariant experience. The subsequent geographical isolation of continents fostered the evolution of distinctive regional **biotas** (all organisms living in a region). Alfred Russel Wallace used the biotas to define six **biogeographical realms,** which we still recognize today **(Figure 22.11).**

FIGURE 22.11
Wallace's biogeographical realms. Each realm contains a distinctive biota.

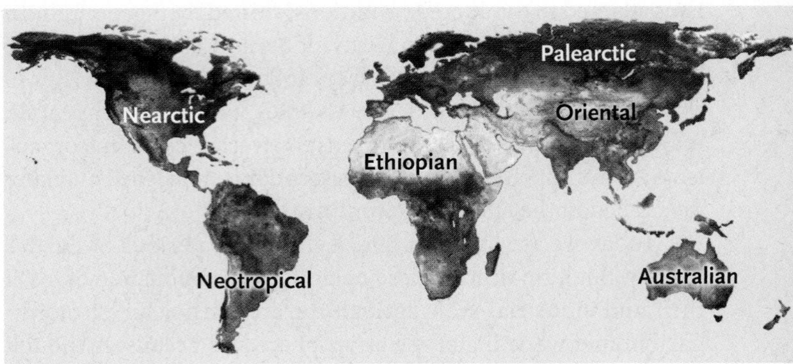
Palearctic
Nearctic
Oriental
Ethiopian
Neotropical
Australian

The Australian and Neotropical realms, which have been geographically isolated since the Mesozoic, contain many **endemic species** (those that occur nowhere else on Earth). The Australian realm, in particular, has had no complete land connection to any other continent for approximately 55 million years. As a result, Australia's mammalian fauna (all the mammals living in the region) is unique, made up almost entirely of endemic marsupials (mammals that give birth after a short gestation period and then carry their young in a pouch). Although Eocene fossils of placental mammals (mammals that do not rear their young in a pouch) have been found in Australia, all of them became extinct. The only native placental mammals that live there now are bats and rodents, which apparently colonized Australia from either Asia or New Guinea after the marsupial fauna had already diversified there.

The biotas of the Nearctic and Palearctic realms are, by contrast, fairly similar. North America and Eurasia were frequently connected by land bridges: eastern North America was attached to Western Europe until the breakup of Laurasia 43 million years ago, and northwestern North America had periodic contact with northeastern Asia over the Bering land bridge during much of the past 60 million years. As a result, many plant and animal species on these continents are closely related.

Evolution Has Produced Convergent Biotas in Widely Separated Regions

Distantly related species living in different biogeographical realms are sometimes very similar in appearance. For example, the overall form of cactuses in the Americas is almost identical to that of spurges in Africa **(Figure 22.12).** But these lineages arose independently long after those continents had separated; thus, cactuses and spurges did not inherit their similarities from a shared ancestor. Their overall resemblance is the product of **convergent evolution,** the evolution of similar adaptations in distantly related organisms that occupy similar environments.

Convergent evolution has also fostered morphological similarities in distantly related animals that feed on similar foods and occupy similar habitats in widely separated geographic ranges. Sometimes, large portions of entire faunas develop convergent morphologies. For example, the marsupial mammals of

A. Cactus (*Echinocereus* species)

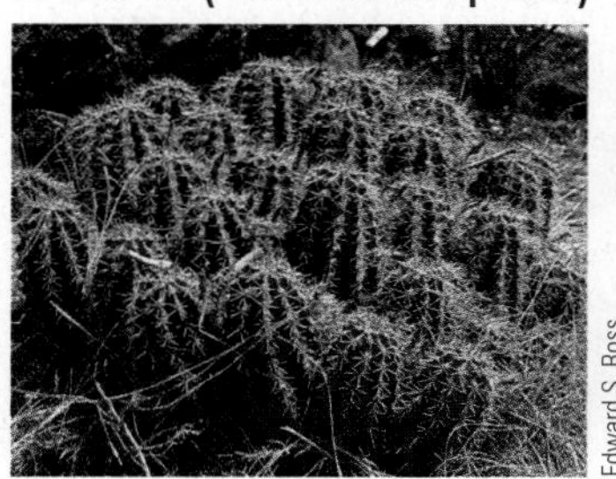
Edward S. Ross

B. Spurge (*Euphorbia* species)

Edward S. Ross

FIGURE 22.12
Convergent evolution in plants. (A) North American cactuses (family Cactaceae) are strikingly similar to **(B)** African spurges (family Euphorbiaceae). Convergent evolution adapted both groups to desert environments with thick, water-storing stems, spiny structures that discourage animals from feeding on them, CAM photosynthesis (see Section 9.4), and stomata that open only at night.

Australia and the placental mammals of North America—groups that arose long after the breakup of Pangaea—include many pairs of morphologically convergent species **(Figure 22.13).** Although the species in each pair differ in most of their anatomical details, their overall appearance and structure are remarkably similar. Biologists interpret these convergent morphologies as evolutionary responses to similar patterns of natural selection in the habitats where these animals live.

STUDY BREAK 22.3

1. **Which type of geographical distribution requires no special explanation?**
2. **Why do distantly related species that live in different biogeographical realms sometimes resemble each other?**

22.4 The History of Biodiversity

As organisms were isolated on different landmasses or in the seas that surrounded them, the number of species living on Earth—its overall **biodiversity**—changed over time as the result of adaptive radiations and extinctions.

A. **Mammals and continental drift**

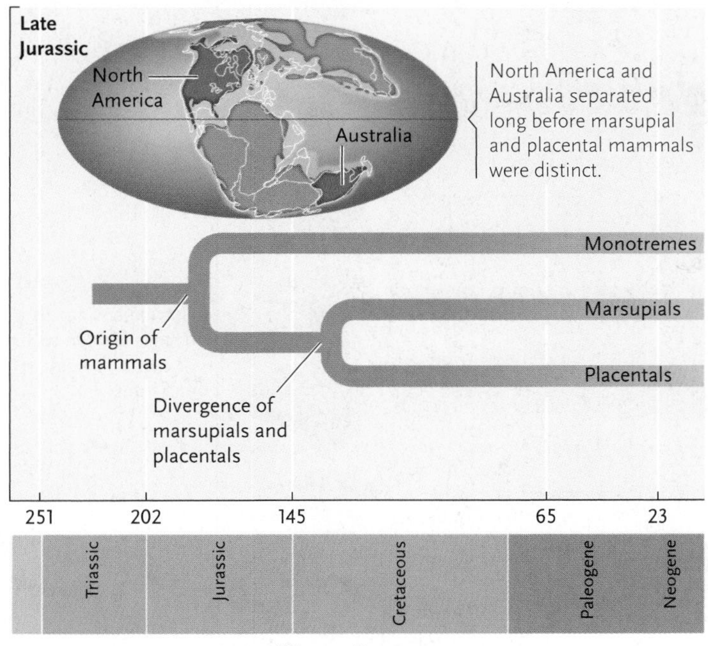
Late Jurassic
North America
Australia
North America and Australia separated long before marsupial and placental mammals were distinct.
Monotremes
Marsupials
Placentals
Origin of mammals
Divergence of marsupials and placentals
251
202
145
65
23
Triassic
Jurassic
Cretaceous
Paleogene
Neogene
Millions of years ago

B. **Convergence of placentals and marsupials**

Flying squirrel (*Glaucomys volans*)

Arboreal gliders

Honey glider (*Petaurus breviceps*)

Gray wolf (*Canis lupus*)

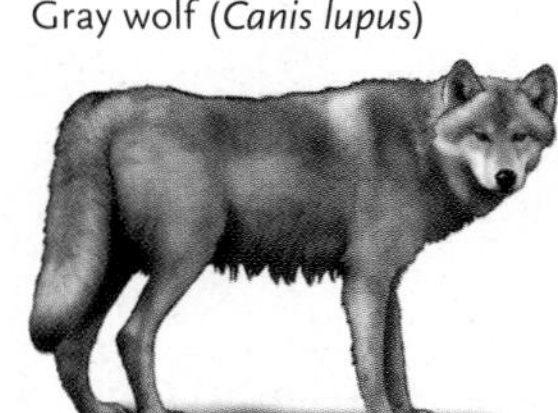

Large predators

Tasmanian wolf (*Thylacinus cynocephalus*)

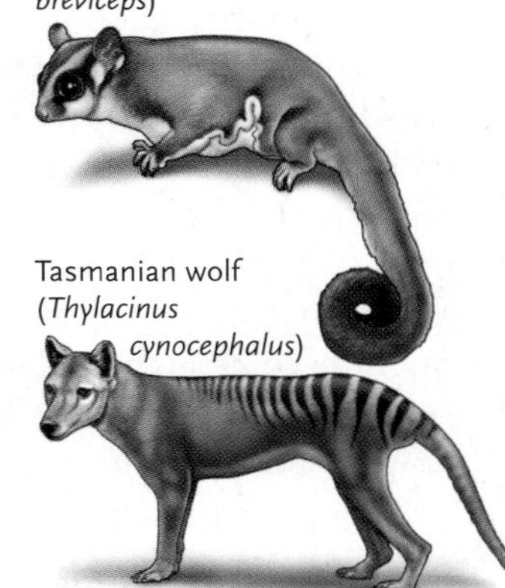

Common mole (*Scalopus aquaticus*)

Burrowers

Marsupial mole (*Notoryctes typhlops*)

House mouse (*Mus musculus*)

Terrestrial seed eaters

Yellow-footed marsupial mouse (*Antechinus flavipes*)

Woodchuck (*Marmota monax*)

Terrestrial herbivores

Wombat (*Vombatus ursinus*)

FIGURE 22.13
Convergent evolution in mammalian faunas. (A) Two distinctive lineages of mammals, the placentals and the marsupials, diverged after Pangaea had already split into northern and southern continents. **(B)** Thus, biologists conclude that convergent evolution, rather than shared ancestry, produced similar body forms in North American placental mammals (on left) and Australian marsupial mammals (on right). The members of each pair differ in their anatomical details, but their overall forms are similar even though they are only distantly related. Biologists interpret these similarities as convergent adaptations in animals that fulfill similar ecological roles on the two continents.

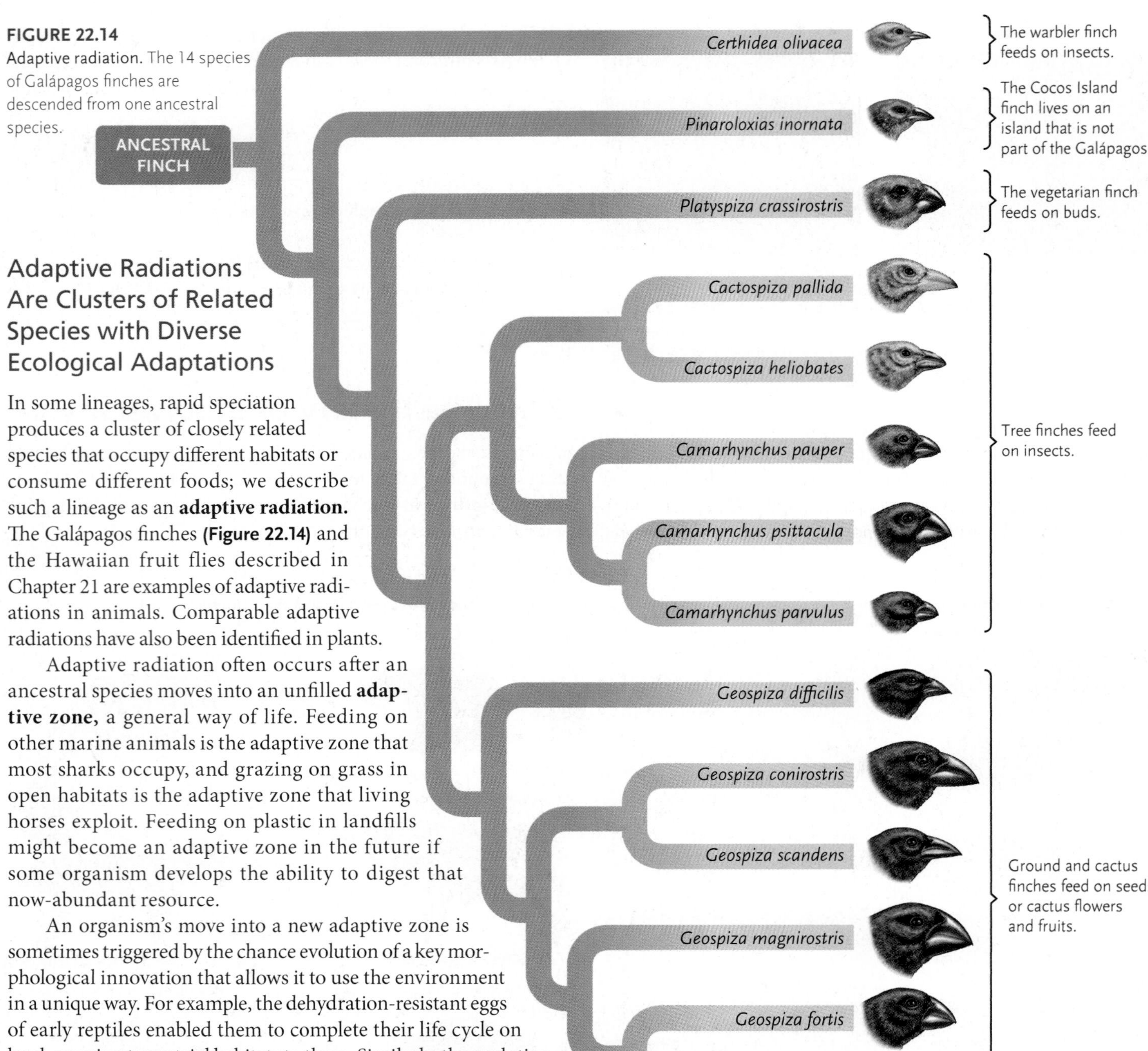
ANCESTRAL FINCH
Certhidea olivacea
Pinaroloxias inornata
Platyspiza crassirostris
Cactospiza pallida
Cactospiza heliobates
Camarhynchus pauper
Camarhynchus psittacula
Camarhynchus parvulus
Geospiza difficilis
Geospiza conirostris
Geospiza scandens
Geospiza magnirostris
Geospiza fortis
Geospiza fuliginosa
The warbler finch feeds on insects.
The Cocos Island finch lives on an island that is not part of the Galápagos.
The vegetarian finch feeds on buds.
Tree finches feed on insects.
Ground and cactus finches feed on seeds or cactus flowers and fruits.

FIGURE 22.14
Adaptive radiation. The 14 species of Galápagos finches are descended from one ancestral species.

Adaptive Radiations Are Clusters of Related Species with Diverse Ecological Adaptations

In some lineages, rapid speciation produces a cluster of closely related species that occupy different habitats or consume different foods; we describe such a lineage as an **adaptive radiation.** The Galápagos finches **(Figure 22.14)** and the Hawaiian fruit flies described in Chapter 21 are examples of adaptive radiations in animals. Comparable adaptive radiations have also been identified in plants.

Adaptive radiation often occurs after an ancestral species moves into an unfilled **adaptive zone,** a general way of life. Feeding on other marine animals is the adaptive zone that most sharks occupy, and grazing on grass in open habitats is the adaptive zone that living horses exploit. Feeding on plastic in landfills might become an adaptive zone in the future if some organism develops the ability to digest that now-abundant resource.

An organism's move into a new adaptive zone is sometimes triggered by the chance evolution of a key morphological innovation that allows it to use the environment in a unique way. For example, the dehydration-resistant eggs of early reptiles enabled them to complete their life cycle on land, opening terrestrial habitats to them. Similarly, the evolution of flowers that attract insect pollinators was a key innovation in the history of flowering plants.

An adaptive radiation may also be triggered after the demise of a successful group. Mammals, for example, were relatively inconspicuous during their first 150 million years on Earth, presumably because dinosaurs dominated terrestrial habitats. But after dinosaurs declined in the late Mesozoic era, mammals underwent an explosive adaptive radiation. Today they are the dominant vertebrates in many terrestrial habitats.

Extinctions Have Been Common in the History of Life

Biodiversity has not always increased through the history of life. The additions to biodiversity caused by adaptive radiations and other speciation events are counteracted by **extinction,** the death of the last individual in a species or the last species in a lineage. Paleobiologists recognize two distinct patterns of extinction in the fossil record, background extinction and mass extinction.

Species and lineages have been going extinct since life first appeared. We should expect species to disappear at some low rate, the **background extinction rate;** as environments change, poorly adapted organisms will not survive and reproduce. In all likelihood, more than 99.9% of the species that have ever lived are now extinct. David Raup of the University of Chicago has suggested that, on average, as many as 10% of species go extinct every million years and that more than 50% go extinct every 100 million years. Thus, the history of life has been characterized by an ongoing turnover of species.

FIGURE 22.15
Mass extinctions. Biodiversity, indicated by the height of the dark blue area in the graph, was temporarily reduced by at least five mass extinctions (arrows) during the history of life. The data presented in this graph record the family-level diversity of marine animals. A family is a group of genera descended from a common ancestor.

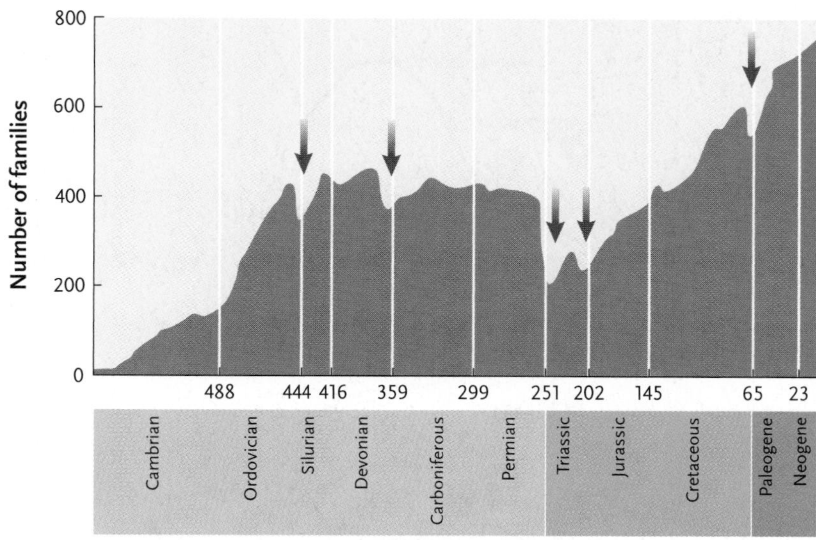

Number of families
800
600
400
200
0
488
444 416
359
299
251 202
145
65
23
Cambrian
Ordovician
Silurian
Devonian
Carboniferous
Permian
Triassic
Jurassic
Cretaceous
Paleogene
Neogene
Millions of years ago

On at least five occasions, however, extinction rates rose well above the background rate. During these **mass extinctions,** large numbers of species and lineages died out over relatively short periods of geological time **(Figure 22.15).** The Permian extinction was the most severe: more than 85% of the species alive at that time—including all trilobites, many insects and amphibians, and the trees of the coal swamp forests—disappeared forever. During the last mass extinction, at the end of the Cretaceous, half the species on Earth, including most dinosaurs, became extinct. A sixth mass extinction, potentially the largest of all, may be occurring now as a result of global climate change and human degradation of the environment (see Unit Seven).

Different factors were responsible for the five mass extinctions. For example, the Ordovician extinction occurred after Gondwana moved toward the South Pole, triggering a glaciation that cooled the world's climate and lowered sea levels, eliminating many shallow marine environments.

Scientists hypothesize that the massive Permian extinction was caused by a chain of events that resulted in severe climate warming. The trigger was apparently a series of ongoing eruptions that released a flood of molten lava hundreds—or even thousands—of meters thick. Once solidified, the lava formed the "Siberian traps," a rock formation that covers 1.6 million km^2 of northeastern Asia. Scientists estimate that the lava flood released so much carbon dioxide, a greenhouse gas (see Section 52.4), into the atmosphere that it may have increased the average global surface temperature by 6°C. The resulting higher ocean temperature melted frozen submarine deposits of methane, which then bubbled into the atmosphere, causing even more climate warming. Michael J. Benton and Richard J. Twitchett of the University of Bristol, United Kingdom, describe the situation as a "runaway greenhouse" in which global temperature increased so much that organisms simply could not adapt to the change. One hundred million years passed before global diversity returned to pre-extinction levels.

Many researchers agree that an asteroid impact was a major cause of the Cretaceous mass extinction. The resulting dust cloud may have blocked the sunlight necessary for photosynthesis, setting up a chain reaction of extinctions that began with microscopic marine organisms. Geological evidence supports this hypothesis. Rocks dating to the end of the Cretaceous period (65 million years ago) contain a highly concentrated layer of iridium, a metal that is rare on Earth but common in asteroids. The impact from an iridium-laden asteroid only 10 km in diameter could have caused an explosion equivalent to that of 100 million megatons of TNT, scattering iridium dust around the world. Geologists have identified the Chicxulub crater, 180 km in diameter, on the edge of Mexico's Yucatán peninsula as the likely site of the impact.

Biodiversity Has Increased Repeatedly over Evolutionary History

Although mass extinctions temporarily reduce biodiversity, they also create evolutionary opportunities. Some species survive because they have highly adaptive traits, large population sizes, or widespread distributions. And some of the surviving species undergo adaptive radiation, filling adaptive zones that mass extinctions made available.

Sometimes, the success of one lineage comes at the expense of another. Although the diversity of terrestrial vascular plants has increased almost continuously since the Devonian period, this trend includes booms and busts in several lineages **(Figure 22.16).** Ferns and conifers recovered rapidly after the Permian extinction, maintaining their diversity until the end of the Mesozoic era. However, angiosperms, which arose and diversified in the early Cretaceous period, may have hastened the decline of these groups by replacing them in many environments.

The superb fossil record left by certain marine animals reveals three major periods of adaptive radiation **(Figure 22.17).** The first occurred during the Cambrian, more than 500 million years ago, when many animal phyla, the major categories of animal life, first appeared. Most of these phyla became extinct, and a second wave of radiations established the dominant Paleozoic fauna during the Ordovician period. A third evolutionary fauna emerged in the Neogene period, right after the great Permian extinction; it produced the immediate ancestors of modern marine animals. The diversity of marine animals has increased consistently since the early Neogene, in large measure because of continental drift. As continents and shallow seas became increasingly isolated, regional biotas diversified independently of one another, increasing worldwide biodiversity.

Historical increases in biodiversity can also be attributed to the evolution of ecological interactions. For example, the number of plant species found *within* fossil assemblages has increased over time, suggesting the evolution of mechanisms that allow more species to coexist. In addition, insects continued to diver-

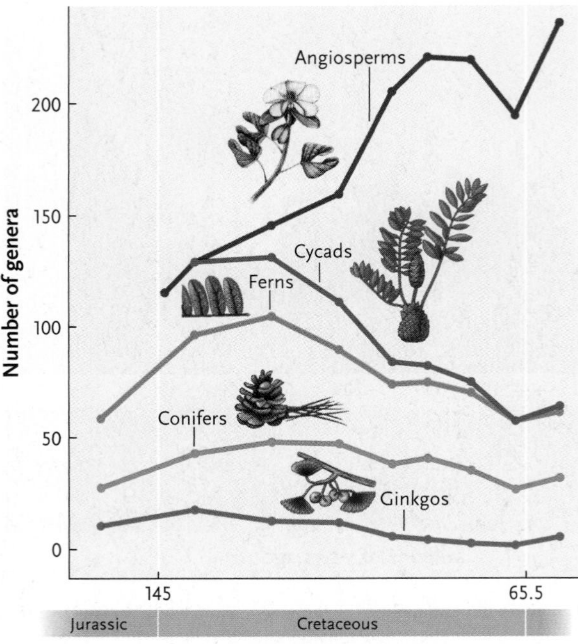

Angiosperms
Cycads
Ferns
Conifers
Ginkgos
Number of genera
200
150
100
50
0
145
65.5
Jurassic
Cretaceous
Millions of years ago

FIGURE 22.16
History of vascular plant diversity. The diversity of angiosperms increased during the Cretaceous period as the diversity of other groups declined.

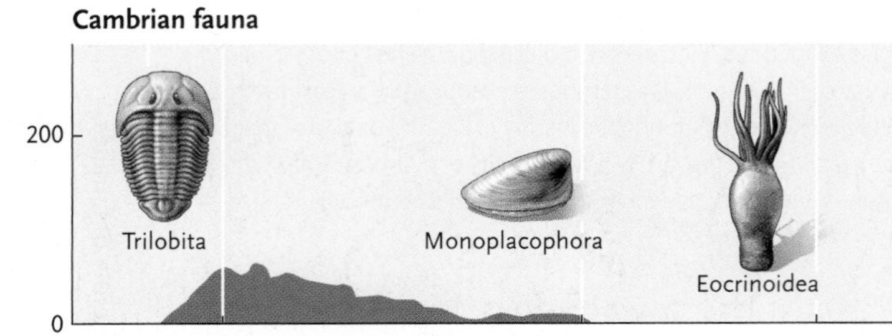

Cambrian fauna
Number of families
200
0
Trilobita
Monoplacophora
Eocrinoidea

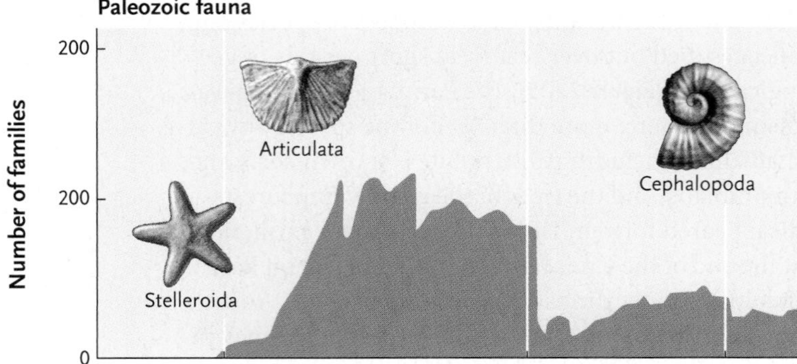

Paleozoic fauna
200
Number of families
200
0
Articulata
Cephalopoda
Stelleroida

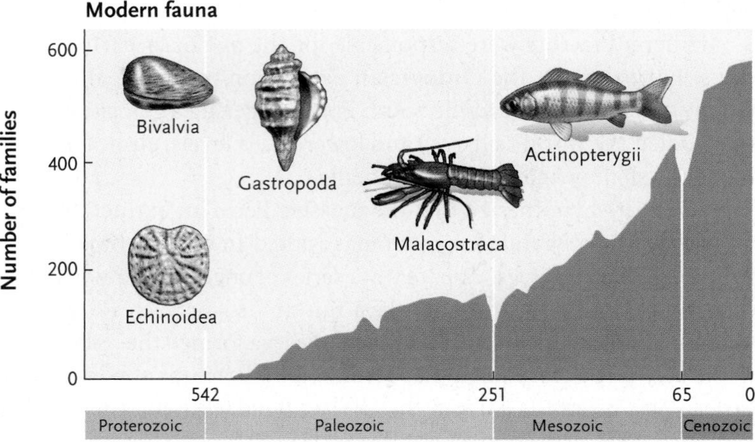

Modern fauna
600
400
200
0
Number of families
Bivalvia
Gastropoda
Actinopterygii
Malacostraca
Echinoidea
542
251
65
0
Proterozoic
Paleozoic
Mesozoic
Cenozoic
Millions of years ago

FIGURE 22.17
History of marine animal diversity. Marine animals have undergone three major radiations. Few remnants of the Cambrian fauna remain alive today.

sify dramatically in the Cretaceous period, possibly because the angiosperms created a new adaptive zone for them. New insect species then provided a novel set of pollinators that may have stimulated the radiation of angiosperms. Such long-term evolutionary interactions between ecologically intertwined lineages have played an important role in structuring ecological communities, which are described more fully in Chapter 51.

STUDY BREAK 22.4

1. **What factors might allow a population of organisms to occupy a new adaptive zone?**
2. **What events apparently triggered the mass extinction at the end of the Permian period?**
3. **When did the first major adaptive radiation of animals occur?**

THINK OUTSIDE THE BOOK

You have learned how the geography of the Galápagos and the Hawaiian Islands fostered adaptive radiations in finches and fruit flies. Search Internet or library sources for examples of adaptive radiations in other types of organisms on these archipelagos. Also look for information about the adaptive radiation of mammals on continental land masses. Do adaptive radiations on archipelagos and on continental landmasses differ in substantive ways?

22.5 Interpreting Evolutionary Lineages

As newly discovered fossils demand the reinterpretation of old hypotheses, biologists constantly refine their ideas about the history of life. In this section we describe how our interpretation of evolutionary lineages is constantly changing.

Modern Horses Are Living Representatives of a Once-Diverse Lineage

The earliest known ancestors of modern horses were first identified by Othniel C. Marsh of Yale University just a year after Darwin published *On the Origin of Species*. These early horses, *Hyracotherium*, stood 25 to 50 cm high and weighed no more than 20 kg. Their toes (four on the front feet and three on the hind) were each capped with a tiny hoof, but the animals walked on soft pads

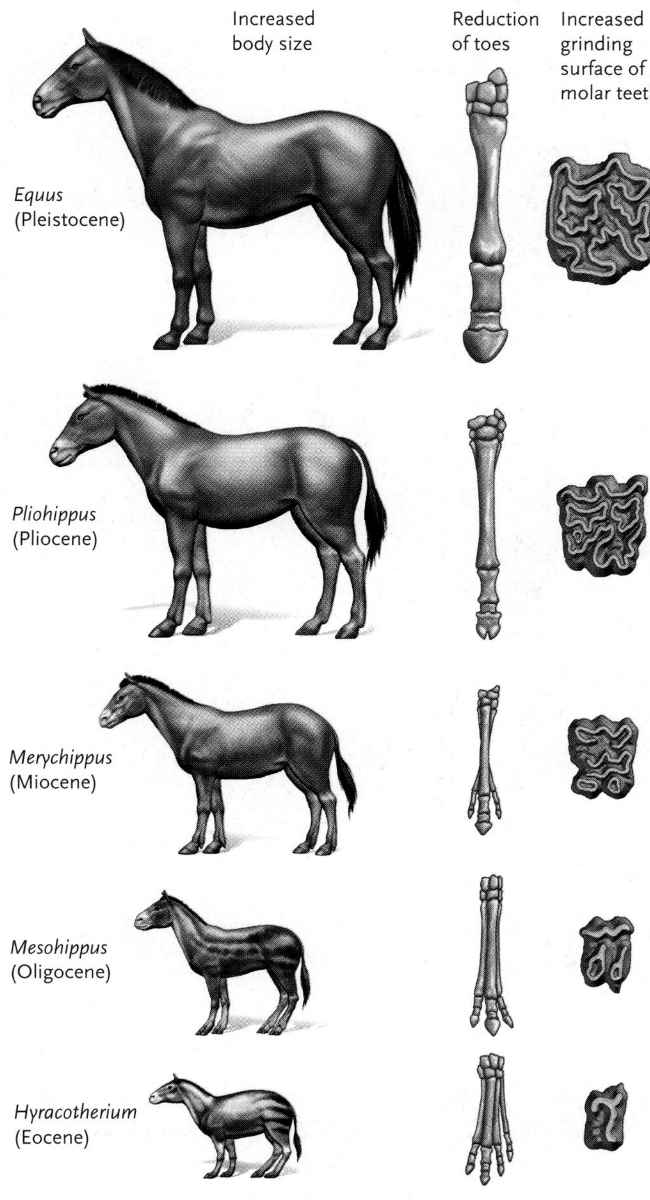
A. Marsh's reconstruction of horse evolution
Increased body size
Reduction of toes
Increased grinding surface of molar teeth
Equus (Pleistocene)
Pliohippus (Pliocene)
Merychippus (Miocene)
Mesohippus (Oligocene)
Hyracotherium (Eocene)

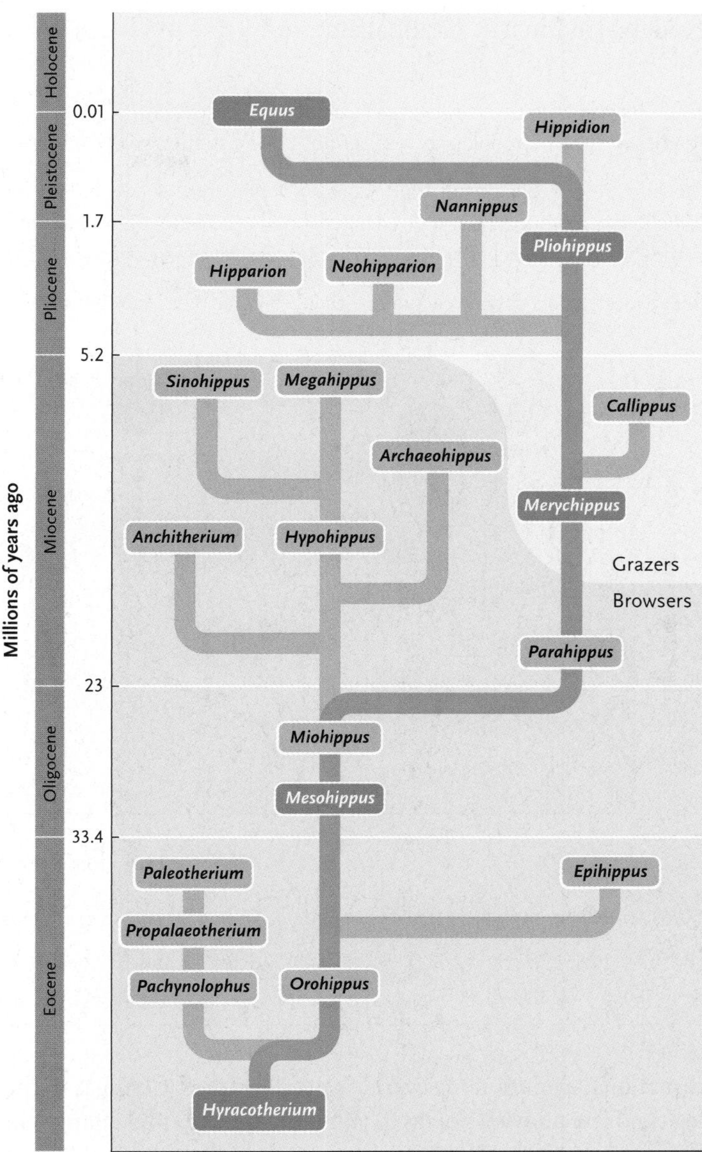
B. Modern reconstruction of horse evolution
Millions of years ago
Holocene
Pleistocene
Pliocene
Miocene
Oligocene
Eocene
0.01
1.7
5.2
23
33.4
Equus
Hippidion
Nannippus
Pliohippus
Hipparion
Neohipparion
Sinohippus
Megahippus
Callippus
Archaeohippus
Merychippus
Anchitherium
Hypohippus
Grazers
Browsers
Parahippus
Miohippus
Mesohippus
Paleotherium
Epihippus
Propalaeotherium
Pachynolophus
Orohippus
Hyracotherium

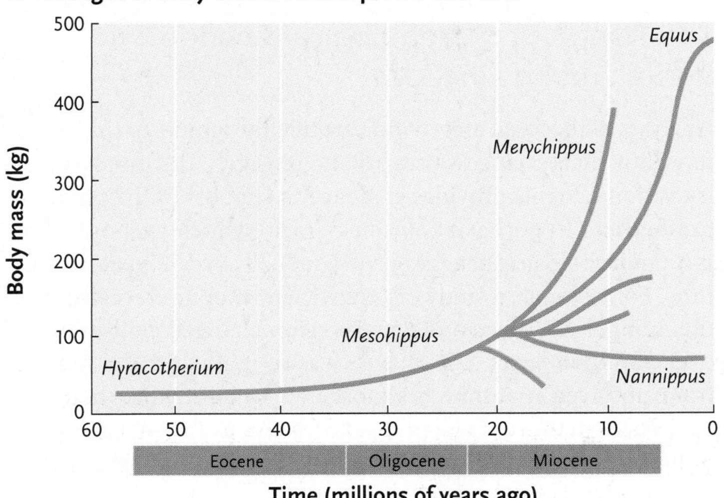
C. Changes in body size of horse species over time
Body mass (kg)
500
400
300
200
100
0
Equus
Merychippus
Mesohippus
Hyracotherium
Nannippus
60
50
40
30
20
10
0
Eocene
Oligocene
Miocene
Time (millions of years ago)

FIGURE 22.18
Evolution of horses. **(A)** Marsh depicted the evolution of horses as a linear pattern of descent characterized by an increase in body size, a reduction in the number of toes, increased fusion of the bones in the lower leg, elongation of the face, and an increase in the size of grinding teeth at the back of the mouth. **(B)** Recent studies revealed that the horse family includes numerous evolutionary branches with variable morphology. The horses in Marsh's analysis are highlighted in green. **(C)** Although many branches of the lineage evolved a larger body size, some remained as small as the earliest horses.

as dogs do today. Their faces were short, their teeth were small, and they browsed on soft leaves in woodland habitats.

In 1879, Marsh published his analysis of 55 million years of horse family history. He described the evolution of this group of mammals as a sequence of stages from the tiny *Hyracotherium* through intermediates represented by *Mesohippus, Merychippus,* and *Pliohippus* to the modern *Equus* **(Figure 22.18A).** (Each of these names refers to a genus, a group of closely related species.) Marsh inferred a pattern of descent characterized by gradual,

FIGURE 22.19 **Observational Research**

Evidence of Phyletic Gradualism

Hypothesis: The phyletic gradualism hypothesis states that most morphological change within evolutionary lineages results from the accumulation of small, incremental changes over long periods of time.

Prediction: The morphology of fossils from a given stratum will be intermediate between those of fossils from the strata immediately below and above it.

Method: Peter R. Sheldon of Trinity College, Dublin, Ireland, counted the number of "ribs" in the tail region of the exoskeletons of approximately 15,000 trilobite fossils from central Wales, United Kingdom. The fossils had formed over a span of about 3 million years during the Ordovician period. Shelton plotted the mean number of ribs found in successive samples of eight lineages.

Results: Sheldon's data reveal gradual changes in the mean number of "ribs" in these animals with no evidence of speciation.

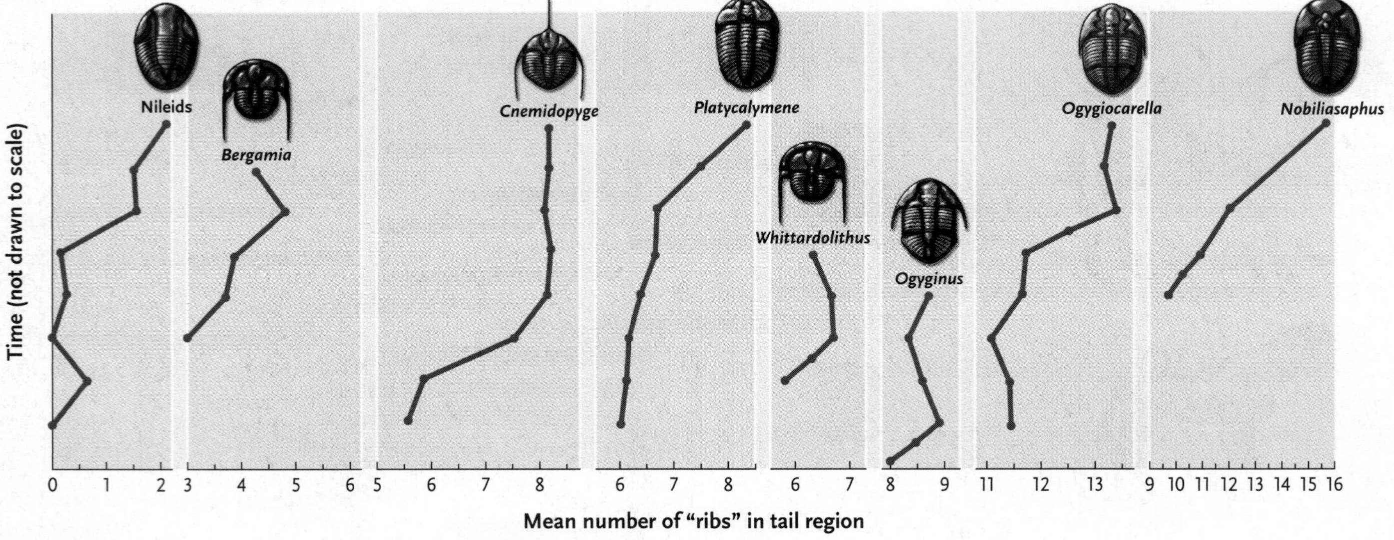
Nileids
Bergamia
Cnemidopyge
Platycalymene
Whittardolithus
Ogyginus
Ogygiocarella
Nobiliasaphus
Time (not drawn to scale)
0 1 2 3 4 5 6 5 6 7 8 6 7 8 6 7 8 9 11 12 13 9 10 11 12 13 14 15 16
Mean number of "ribs" in tail region

Conclusion: Morphological changes in Ordovician trilobites from central Wales are consistent with the predictions of the phyletic gradualism hypothesis.

Source: P. R. Sheldon. 1987. Parallel gradualistic evolution of Ordovician trilobites. *Nature* 330:561–563.

directional evolution in several skeletal features. Changes in the legs and feet allowed horses to run more quickly, and changes in the face and teeth accompanied a switch in diet from soft leaves to tough grasses.

The fossil record for horses is superb, and we now have fossils of more than 100 extinct species from five continents. These data reveal a macroevolutionary history very different from Marsh's interpretation. *Hyracotherium* was not gradually transformed into *Equus* along a linear track. Instead, the evolutionary tree for horses was highly branched **(Figure 22.18B)**, and *Hyracotherium's* descendants differed in size, number of toes, tooth structure, and other traits. Although many branches of this lineage lived in the Miocene and Pliocene epochs, all but one are now extinct. The species of the genus *Equus* living today (horses, donkeys, and zebras) are the surviving tips of that one branch.

When we study extinct organisms, we tend to focus on traits that characterize modern species. Marsh, for example, assumed that the differences between *Hyracotherium* and *Equus* were typical of the changes that characterized the group's evolutionary history. But not all fossil horses were larger **(Figure 22.18C)**, had fewer toes, or were better adapted to feed on grass than their ancestors. And if a branch other than *Equus* had survived, Marsh's description of trends in horse evolution would have been very different. All evolutionary lineages have extinct branches, and any attempt to trace a linear evolutionary path—as Marsh did for horses and many people do for humans—imposes artificial order on an inherently disorderly history.

The Tempo of Morphological Change Varies among Lineages

Analyses of the fossil record indicate that the tempo, or timing, of morphological change varies among lineages. The discovery of transitional fossils provides evidence in support of the **phyletic gradualism hypothesis** (*phylon* = race), which proposes that most morphological change occurs gradually over long periods of time. For example, a study of Ordovician trilobites revealed that the number of "ribs" in their tail region changed continuously over 3 million years. The change was so gradual that a sample from any given stratum was almost always intermediate between samples from the strata just above and below it. The changes in rib number probably evolved without the evolution of new, reproductively isolated species **(Figure 22.19)**.

FIGURE 22.20 Observational Research

Evidence of a Punctuated Pattern of Morphological Change

Hypothesis: The punctuated equilibrium hypothesis states that most morphological change within evolutionary lineages appears during speciation.

Prediction: The fossil record will reveal that most species experienced relatively little morphological change for long periods of time, but that new, morphologically distinctive species arose suddenly.

Method: Cheetham examined numerous fossilized samples of populations of a small marine invertebrate, the ectoproct *Metrarabdotos,* from the Dominican Republic. He measured 46 morphological characters in populations representing 18 species and then used a complex statistical analysis to summarize the extent of morphological difference between populations.

Results: The morphology of most *Metrarabdotos* species changed very little over millions of years, but new species, morphologically very different from their ancestors, often appeared suddenly in the fossil record. Each dot in the graph represents a sample of a population from the fossil record. The horizontal axis reflects the overall morphological difference between samples of one species over time or between samples of ancestral species and descendant species.

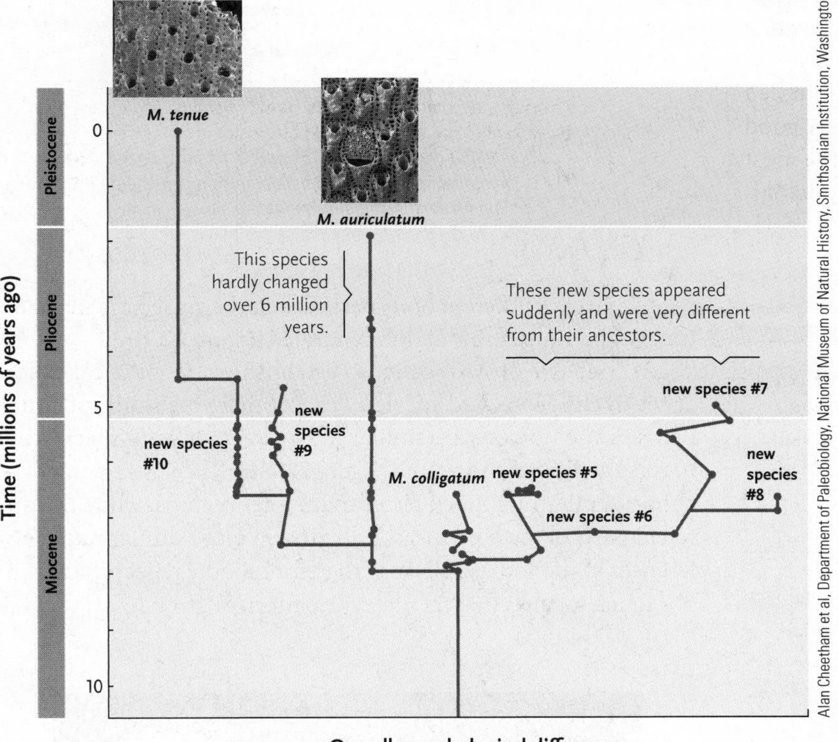

M. tenue
M. auriculatum
Pleistocene
Pliocene
Miocene
Time (millions of years ago)
0
5
10
This species hardly changed over 6 million years.
These new species appeared suddenly and were very different from their ancestors.
new species #7
new species #9
new species #10
M. colligatum
new species #5
new species #6
new species #8
Overall morphological difference

Alan Cheetham et al., Department of Paleobiology, National Museum of Natural History, Smithsonian Institution, Washington, D. C.

Conclusion: The *Metrarabdotos* lineage exhibits a pattern of morphological evolution that is consistent with the predictions of the punctuated equilibrium hypothesis.

Source: A. H. Cheetham. 1987. Tempo of evolution in a Neogene bryozoan: Are trends in single morphologic characters misleading? *Paleobiology* 13:286–296.

But for most groups of organisms the fossil record is not nearly as good as it is for horses or trilobites. In fact, the discovery of intermediate or transitional fossils is fairly rare. Instead, most species appear suddenly in a particular layer, persist for some time with little change, and then disappear from the fossil record. Then another species with different traits suddenly appears in the next higher stratum.

In the early 1970s, Niles Eldredge of the American Museum of Natural History and Stephen Jay Gould of Harvard University published an explanation for the absence of transitional forms, or "missing links." Their **punctuated equilibrium hypothesis** suggested that speciation usually occurs in isolated populations at the edge of a species' geographical distribution. Such populations experience substantial genetic drift and distinctive patterns of natural selection (as described in Section 21.3). According to this hypothesis, morphological variations appear rapidly as new species arise; then, most species exhibit long periods of morphological equilibrium (that is, little change in form), punctuated by brief periods of speciation and rapid morphological evolution. Eldridge and Gould further argued that transitional forms live only for short periods of geological time in small, localized populations—the very conditions that discourage broad representation in the fossil record. Darwin himself used this line of reasoning to explain puzzling gaps in the fossil record: new species appear as fossils only after they become abundant and widespread and begin a period of morphological stasis.

Some evolutionists challenged the hypothesis' definition of rapid morphological change, particularly given our inability to resolve time precisely in the fossil record. To a paleobiologist with a geological perspective, "instantaneous" events occur over tens or hundreds of thousands of years. But to a population geneticist, those time scales may encompass thousands of generations, ample time for gradual microevolutionary change.

Moreover, examples of evolutionary stasis may not be as static as they appear. Alternating periods of directional selection that favor opposite patterns of change could produce the appearance of stasis. For example, if natural selection favored slight increases in body size for 2,000 years and then favored slight decreases for the next 2,000 years, paleobiologists would probably detect no change in body size at all.

Nevertheless, punctuated patterns of evolution have been observed in some groups, such as *Metrarabdotos,* a genus of ectoprocts from the Caribbean Sea. Ectoprocts are small colonial animals that build hard skeletons (see Figure 29.15A), the details of which are well preserved in fossils. Alan Cheetham of the Smithsonian Institution measured the morphological differences between populations of individual *Metrarabdotos* species over time and between ancestral species and their descendants. His results indicate that most species did not change much for millions of years, but new species, which were morphologically different from their ancestors, often appeared suddenly **(Figure 22.20).**

A. **Allometric growth in humans**

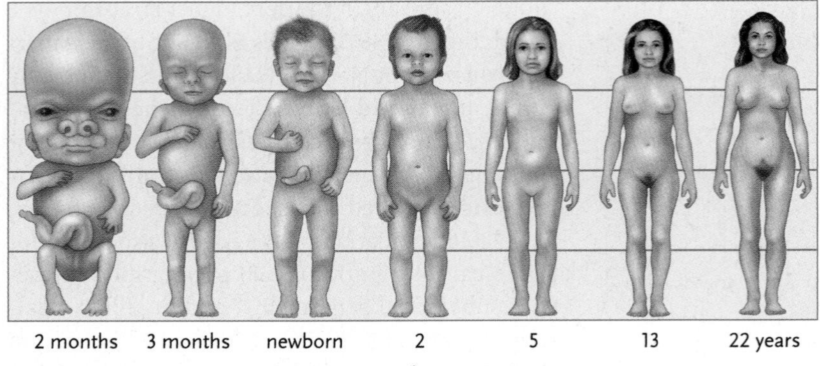
2 months
3 months
newborn
2
5
13
22 years

Humans exhibit allometric growth from prenatal development until adulthood. Our heads grow more slowly than other body parts; our legs grow faster.

B. **Differential growth in the skulls of chimpanzees and humans**

Changes in chimpanzee skull

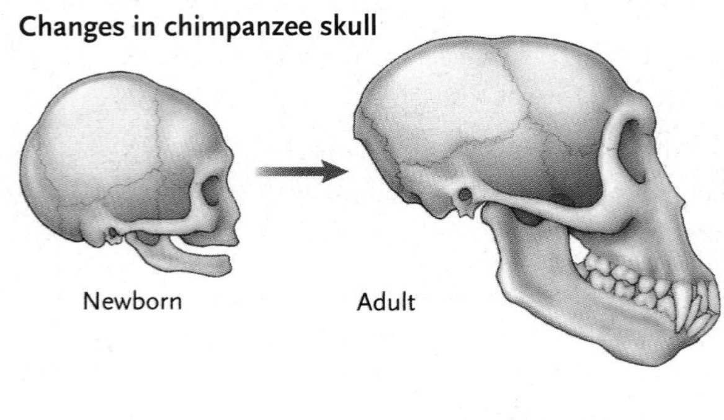
Newborn
Adult

Changes in human skull

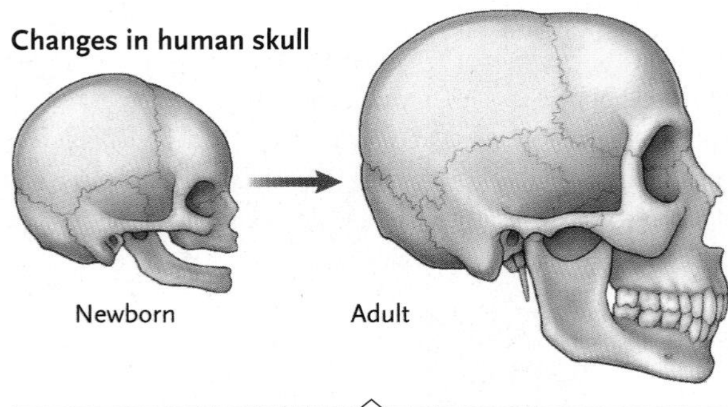
Newborn
Adult

Although the skulls of newborn humans and chimpanzees are remarkably similar, differential patterns of growth make them diverge during development. The jaw of a chimpanzee grows much faster than other regions of the skull. In humans the different parts of the skull grow at more similar rates.

FIGURE 22.21
Examples of allometric growth.

Researchers have found evidence of both gradual and punctuated morphological changes as well as some intermediate patterns. The punctuated equilibrium hypothesis caused a stir when it was first proposed because it challenged long-held assumptions about the tempo of evolution. But its publication rekindled interest in paleobiology and macroevolution, and like any good idea in science, it inspired much new research. Some of the most interesting results have focused on the evolution of morphological novelties, which biologists now analyze with studies in paleontology, embryology, and genetics.

STUDY BREAK 22.5

1. **Did the horse lineage undergo a steady increase in body size over its evolutionary history? Explain your answer.**
2. **What is the difference between the morphological change predicted by the phyletic gradualism hypothesis and the punctuated equilibrium hypothesis?**

22.6 The Evolution of Morphological Novelties

The fossil record documents the appearance of distinctive morphological novelties, such as the limbs of four-legged vertebrates and the appearance of crests and horns on some dinosaurs. In this section we consider explanations for how such novelties may arise.

Differential Growth of Body Parts and Changes in the Timing of Developmental Events Produce Distinctive Structures

The morphology of individuals sometimes changes over time because of **allometric growth** (*allos* = other; *metron* = measure), the differential growth of body parts. In humans, for example, the relative sizes of different body parts change because human heads, torsos, and limbs grow at different rates (**Figure 22.21A**).

Allometric growth can also establish morphological differences in closely related species. For example, the skulls of chimpanzees and humans are similar in newborns of both species, but markedly different in adults (**Figure 22.21B**). The jaw region of the chimp skull grows much faster than other regions, while the different parts of the human skull all grow at more similar rates. Differences in the adult skulls may therefore simply reflect changes in one or a few genes that regulate the pattern of growth.

David Scott/SREL

FIGURE 22.22
Paedomorphosis in salamanders. Some small-mouthed salamanders *(Ambystoma talpoideum)* undergo metamorphosis, losing their gills and developing lungs (left). Others are paedomorphic: they retain juvenile morphological characteristics, such as gills, after attaining sexual maturity (right).

FIGURE 22.23 Observational Research

Paedomorphosis in *Delphinium* Flowers

Hypothesis: The narrow tubular shape of the flowers of *Delphinium nudicaule,* which are pollinated by hummingbirds, is the product of paedomorphosis, the retention of juvenile characteristics in a reproductive adult.

Prediction: The flowers of *D. nudicaule* grow more slowly and mature at an earlier stage of development than those of *Delphinium decorum,* a species with broad, open flowers that are pollinated by bees.

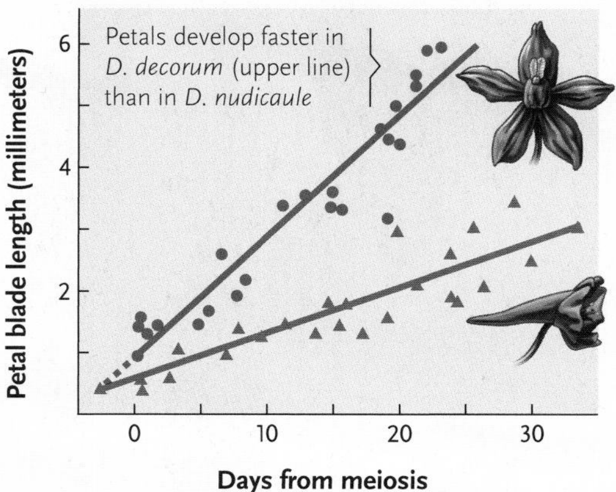

Petals develop faster in *D. decorum* (upper line) than in *D. nudicaule*
Petal blade length (millimeters)
6
4
2
0
10
20
30
Days from meiosis

Method: Edward O. Guerrant of the University of California, Berkeley, measured 42 bud and flower characteristics in *D. nudicaule* and *D. decorum* as their flowers developed and used the number of days since the completion of meiosis in pollen grains as a measure of flower maturity. He then used a complex statistical analysis to compare the characteristics of the buds and flowers of both species.

Results: The mature flowers of *D. nudicaule* resemble the buds of both species more closely than they resemble the flowers of *D. decorum.* Although the time required for maturation of the reproductive structures is similar in the two species, the rate of petal growth (measured as petal blade length) is slower in *D. nudicaule.* As a result, the mature flowers of *D. nudicaule* do not open as widely as those of *D. decorum.* Because of these morphological differences, bees can pollinate flowers of *D. decorum,* but they cannot land on the flowers of *D. nudicaule,* which are instead pollinated by hummingbirds.

D. decorum

Gary Head

D. nudicaule

Gary Head

Conclusion: The narrower and more tubular shape of *D. nudicaule* flowers, which mature at an earlier stage of development than *D. decorum* flowers, is the product of paedomorphosis.

Source: E. O. Guerrant. 1982. Neotenic evolution of *Delphinium nudicaule* (Ranunculaceae): A hummingbird-pollinated larkspur. *Evolution* 36:699–712.

Changes in the timing of developmental events, called **heterochrony** (*heteros* = different; *khronos* = time), also cause the morphology of closely related species to differ. **Paedomorphosis** (*paedo-* = child; *morphos* = form or shape), the development of reproductive capability in an organism with juvenile characteristics, is a common form of heterochrony.

Many salamanders, for example, undergo metamorphosis from an aquatic juvenile into a morphologically distinct terrestrial adult. However, populations of several species are paedomorphic—they grow to adult size and become reproductively mature without changing to the adult form **(Figure 22.22).** The evolutionary change causing these differences may be surprisingly simple. In amphibians, including salamanders, the hormone thyroxine induces metamorphosis (see Chapter 40). Paedomorphosis could result from a mutation that either reduces thyroxine production or limits the responsiveness of some developmental processes to thyroxine concentration.

Changes in developmental rates also influence the morphology of plants **(Figure 22.23).** The flower of a larkspur species, *Delphinium decorum,* includes a ring of petals that guide bees to its nectar tube and structures on which bees can perch. By contrast, *Delphinium nudicaule,* a more recently evolved species, has tight flowers that attract hummingbird pollinators, which can hover in front of the flowers. Slower development in *D. nudicaule* flowers causes the structural difference: a mature flower in the descendant species resembles an unopened (juvenile) flower of the ancestral species.

Morphological Novelties Often Arise as Modifications of Existing Structures

Sometimes a trait that is adaptive in one context is also advantageous under different circumstances. Biologists describe the original version of the trait as an **exaptation** (formerly called a *preadaptation*). Natural selection may then exaggerate the trait or modify it altogether to enhance its new function. Exaptations are just lucky accidents; they never evolve *in anticipation* of future evolutionary needs or benefits.

The forelimbs and feathers of certain dinosaurs provide a good example of an exaptation. Shortly after the first discovery of an *Archaeopteryx* skeleton in 1861, Thomas Henry Huxley, a protégé of Darwin, developed the controversial hypothesis that birds are the living representatives of the theropods, a lineage of bipedal predatory dinosaurs (see Figure 19.13). More than a century passed before John Ostrom of Yale University provided sup-

A. **Fossil of *Microraptor gui***

Feathers on forelimb
Feathers on hindlimb
Xu, et. al, Nature 421, 335-340

B. **Forelimb with feathers**

Xu, et. al, Nature 421, 335-340

C. **Hindlimb with feathers**

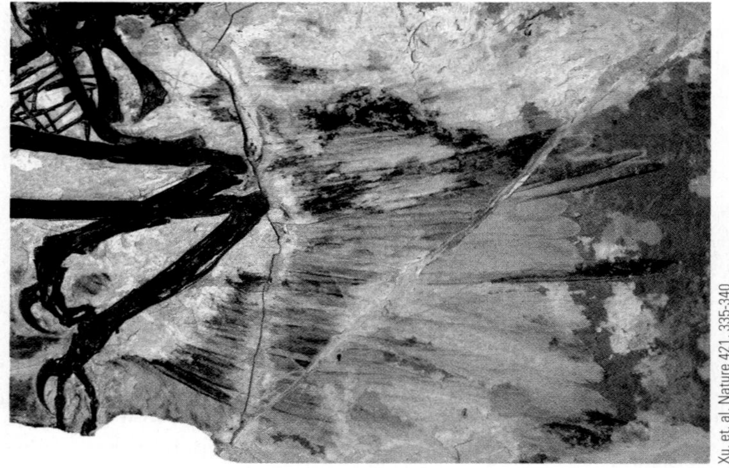
Xu, et. al, Nature 421, 335-340

D. **Reconstruction of *Microraptor gui* with limbs extended for gliding**

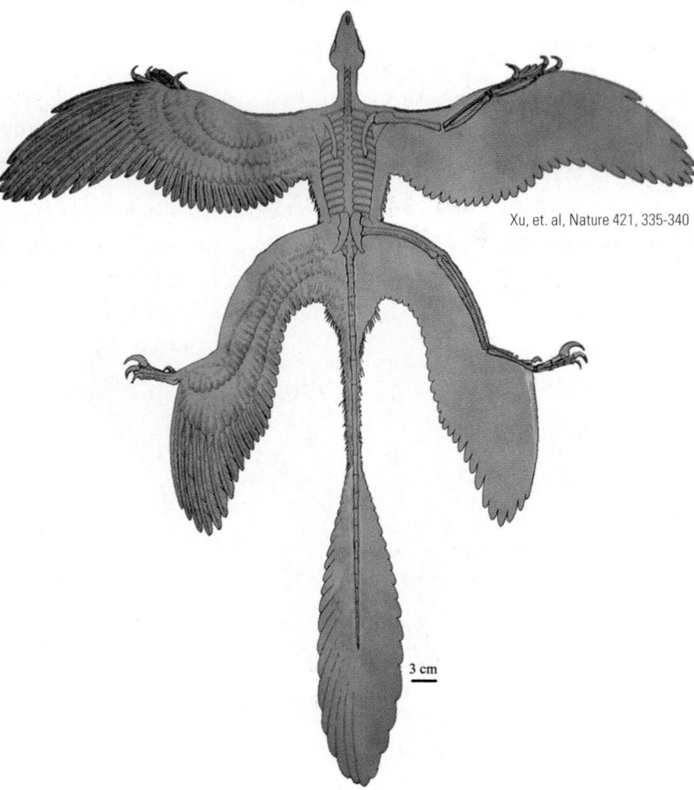
Xu, et. al, Nature 421, 335-340
3 cm

FIGURE 22.24
Feathered dinosaurs. *Microraptor gui* is one of many fossils of feathered dinosaurs that researchers have found in China since the mid-1990s. The reconstruction shows how this species may have looked with its four limbs extended for gliding flight. The scale bar in parts A and B is 6 cm.

port for Huxley's hypothesis; in 1964, he described a previously unknown theropod dinosaur, *Deinonychus antirrhopus,* with a very bird-like skeleton. But *Archaeopteryx* remained a puzzling creature because of its mix of dinosaurian characteristics—teeth, a flat breastbone, a long tail with many vertebrae, and three clawed fingers on each hand—and avian (that is, bird-like) characteristics—a furcula (wishbone), wing-like forelimbs, and feathers. Although *Archaeopteryx* was the only feathered dinosaur-like fossil known, scientists hypothesized that at least some of its dinosaur ancestors must have been feathered as well. Nevertheless, the known fossil record suggested that the transition from dinosaurs to feathered birds was abrupt: feathers were absent in the dinosaur ancestors of birds, but present in what biologists consider the earliest known bird.

Since the early 1990s, however, paleontologists have uncovered a wealth of feathered dinosaur fossils; most have been found in early Cretaceous deposits, dated at 124 million years ago, in northern China. The fossilized dinosaur feathers range from filamentous or downy structures that covered their bodies to long stiff feathers on their forelimbs and tails. One dinosaur species, *Microraptor gui,* had well developed feathers on all four of its limbs **(Figure 22.24).** Research has confirmed that the feathers in one fossil included β-keratin, the protein found in the feathers of modern birds. Based on this anatomical and biochemical evidence, biologists hypothesize that feathers arose as modifications of ancestral scales in the skin of early dinosaurs. Thus, feathers are not unique to birds. They appeared in bird ancestors, perhaps as a form of insulation or as structures used for social communication; birds merely retain this ancestral characteristic.

But flight requires wings as well as feathers. How did wings and flapping flight evolve from the forelimbs of bipedal dinosaurs? Scientists have proposed two hypotheses to explain the origin of flight in birds, both of which view feathered forelimbs as exaptations. According to the "from the ground up" hypothesis, the dinosaur ancestors of birds were bipedal predators that chased prey on the ground. If they jumped at flying insect prey and used their feathered forelimbs as "fly-swatters," natural selection may have enlarged the forelimbs—along with their muscles and feathers—in stages, first to improve jumping ability and later to power flapping flight. The alternative "from the trees down" hypothesis argues that bird ancestors were at least partially arboreal. If they leapt from limb to limb or from tree to tree to chase prey or escape from predators, feathered forelimbs may have allowed them to glide and extend their jumps. Natural selection might then have modified these structures and behaviors to improve gliding ability and eventually allow flapping flight.

Evolutionary Developmental Biology May Explain the Sudden Appearance of Some Morphological Novelties

During much of the twentieth century, the evolutionary biologists who studied morphological novelties in extinct and living organisms worked independently of the scientists studying embryonic development. As a result, evolutionary biologists were unable to construct a coherent picture of the specific developmental mechanisms that contributed to morphological innovations. Since the late 1980s, however, advances in molecular genetics have allowed scientists to explore the genomes of organisms in great detail, fostering a new approach to these studies. **Evolutionary developmental biology**—evo-devo, for short—asks how evolutionary changes in the genes regulating embryonic development can lead to the changes in body shape and form that foster adaptive radiations, increasing biodiversity over geological time.

In the life cycle of a multicellular organism, the many different body parts of the adult develop in a highly controlled sequence of steps that is specified by genetic instructions in the single cell of a fertilized egg. Developmental biologists study how regulatory genes control the development of phenotypes and their variations. As described in Section 16.4, **homeotic genes** (described further for plants in Chapter 34) code for transcription factors that bind regulatory sites on DNA, either activating or repressing the expression of other genes that contribute to an organism's form. In the remainder of this chapter, we describe a few intriguing discoveries about the genetic mechanisms that underlie some macroevolutionary changes in animals. You have already encountered one example when you read about the evolution of limblessness in snakes (Section 19.3).

Most Animals Share the Genetic Tool Kit That Regulates Their Development

Comparisons of genome sequence data reveal that most animals, regardless of their complexity or position in the Tree of Life, share a set of several hundred homeotic genes that control their development. Collectively, these genes have been dubbed the "genetic tool kit," because they govern the basic design of the body plan by controlling the activity of thousands of other genes. Some of the tool-kit genes must be at least 500 million years old, because all living animals inherited them from a common ancestor alive at that time. Not surprisingly, they do not differ much structurally among the animals that possess them, and they generally play the same role in development for all species. For example, as you learned in Section 16.4, genes in the ***Hox*** family control the overall body plan of animals, including where appendages—wings in flies and legs in mice—will develop on the animal's body. Some of the same tool-kit genes are also present in plants, fungi, and prokaryotes, suggesting that those genes may date back to the earliest forms of life.

Another example of a highly conserved and widely distributed tool-kit gene, the *Pax-6* gene, triggers the formation of light-sensing organs as diverse as the eye spots in flatworms, the compound eyes of insects and other arthropods, and the camera eyes of vertebrates (see Chapter 39). Like the *Hox* genes, *Pax-6* also codes for a protein that either activates or represses gene transcription. The proteins coded by *Pax-6* in different animals are so similar that when researchers genetically engineered fruit fly larvae to express the *Pax-6* gene taken from a squid or a mouse, the flies responded by developing eyes. The induced eyes were, however, fruit fly eyes—not squid eyes or mouse eyes. Thus, *Pax-6* triggers activity in the genes that carry the specific instructions for making an eye typical of the species. Apparently, the ancient genetic sequence for *Pax-6*, the master regulatory gene for eye development, has been conserved over the hundreds of millions of years since the common ancestor of squids, fruit flies, and mice lived.

Evolutionary Changes in Developmental Switches May Account for Much Evolutionary Change

If most animals share the same tool-kit genes, how has evolution produced different body plans among species? What makes a squid, a fruit fly, and a mouse different? Researchers in evo-devo have proposed that morphological differences among species arise when mutations alter the effects of developmental regulatory genes. As described in Chapter 16, the developmental programs of animals involve complex networks of many interacting genes. Varying combinations of tool-kit genes may be expressed at different times and in different body regions. According to this hypothesis, the several hundred tool-kit genes encode proteins that work as either activators or repressors in a multitude of possible combinations. Thus, they can generate an unimaginably large number of different gene expression patterns, each with the potential to alter morphology.

Sean B. Carroll of the University of Wisconsin–Madison and the Howard Hughes Medical Institute has described the regulatory sites that transcription factors can bind as *switches*, like those we use to turn lights on or off. When a combination of transcription factors turns on a regulatory switch, a gene further downstream is activated. When transcription factors turn off a regulatory switch, a downstream gene is inactivated.

Although all the cells in an animal contain exactly the same set of genes, the differential expression of genes in different body regions and at different times during embryonic development causes different structures to be made. Allometric growth can result from evolutionary changes in developmental switches that cause certain body parts to grow larger or faster than others. Sim-

Fancy Footwork from Fins to Fingers

Both fishes and tetrapods (four-footed animals) have two pairs of appendages, one anterior and one posterior. The appendages develop similarly in both groups during early embryonic stages. They start out as buds of mesoderm—the middle of the three primary embryonic tissue layers—that thicken by increased cell division. As the buds elongate, cartilage is deposited at localized centers, where bones of the appendages later form.

In fishes, the bones that support the fin develop along a central axis from the base to the tip of the fin. Like fins, tetrapod limbs have a central axis, but they also have distal structures, the digits (that is, the fingers and toes on the hands and feet). In tetrapods, centers of cartilage formation generate both the long bones of the limb and the five digits of the foot.

Groups of homeobox *(Hox)* genes control limb development in all animals with paired anterior and posterior appendages. Researchers have studied the expression of the *HoxD* genes to determine how their activities correlate with appendage development. Expression of these genes is detected by using a nucleic acid probe that can pair with mRNA products of the genes, resulting in the generation of a color. The color is visualized under the microscope in embryos at different stages of development to determine where and when gene expression occurs. For tetrapods, researchers have shown that genes at the 5′ end of the *HoxD* cluster—designated 5′ *HoxD* genes—are expressed in two distinct phases (part A). In the first phase, gene activity is restricted to the posterior half of the limb; this period of activity corresponds to development of the long limb bones. Later, in the second phase, the 5′ *HoxD* genes became active in a band of cells perpendicular to the central axis; the cartilage centers that form the bones of the digits develop in this anterior-posterior band.

In 1995 Paolo Sordino, Frank van der Hoeven, and Denis Duboule at the University of Geneva in Switzerland published a study on *HoxD* expression in the zebrafish *Danio rerio* (a common aquarium fish and a model research organism). They found that only a single phase of 5′ *HoxD* gene activity occurs during zebrafish fin development (part B). Because the second phase of 5′ *HoxD* gene activity is apparently absent in zebrafishes, they concluded that tetrapod digits must be an entirely novel evolutionary structure fostered by the addition of the second phase of 5′ *HoxD* gene activity.

Research Question

Is the pattern of 5′ *HoxD* gene expression seen in zebrafish typical of all fishes? That is, are tetrapods the only vertebrates that have a second phase of 5′ *HoxD* gene expression?

Experiment

A number of research groups have carried out experiments to answer the question. One study, published in 2007 by Renata Freitas, GuangJun Zhang, and Martin Cohn at the University of Florida, Gainesville, used the probing/microscopic visualization technique to determine the pattern of 5′ *HoxD* gene expression in developing catshark *(Scyliorhinus canicula)* fins. Catsharks are included in an evolutionary lineage (Chondrichthyes, fishes with cartilaginous skeletons) that predates the lineage that includes zebrafishes (Actinopterygii, ray-finned fishes). Thus, data on fin development in catsharks may portray a more ancient developmental pattern than the one observed in zebrafishes.

Results

Their results showed that in the early stages of catshark fin development, the 5′ *HoxD* genes are expressed in a phase like that seen during fin development in zebrafish and in the first phase of limb development in tetrapods. Unexpectedly, the researchers observed a *second* phase of expression of two of the 5′ *HoxD* genes along the distal margin of the catshark fin buds (part C). Strikingly, that second phase of expression is similar to the second phase of 5′ *HoxD* gene expression in tetrapod limb development.

Conclusion

The second phase of 5′ *HoxD* gene expression is not uniquely associated with tetrapod limb development, as the earlier zebrafish data had suggested. Indeed, the more recent research suggests that two phases of 5′ *HoxD* gene expression were present in the common ancestor of sharks, ray-finned fishes, and tetrapods. The authors of the later article hypothesized that digits may have arisen if the *HoxD* genes in the distal parts of the limb were expressed for a longer period of development in tetrapods than in catsharks, an example of heterochrony. The loss of the second phase of 5′ *HoxD* gene expression in the zebrafish and their relatives may explain why most ray-finned fishes have a fin skeleton that is much smaller than that of catsharks and their relatives.

A. Tetrapod
B. Zebrafish
C. Catshark
Phase 2 activity
Phase 1 activity
Anterior
Posterior

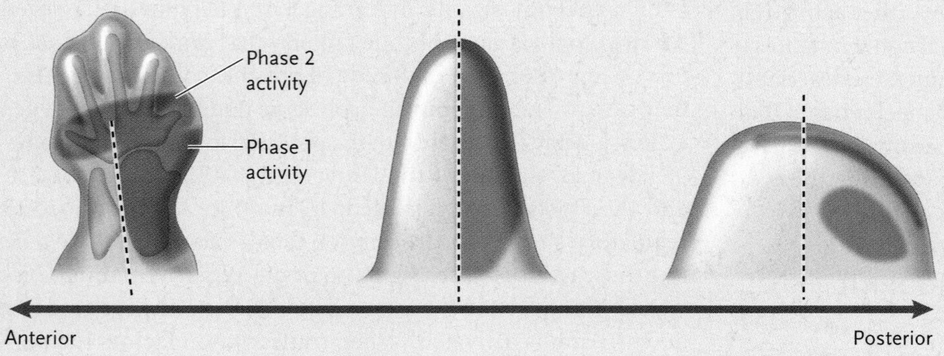

During development of the limb and digits in tetrapods, *HoxD* genes first become active in cells on the posterior side of the limb (blue). Later, these genes are active in a broad band of cells perpendicular to the central axis of the limb (green).

During development of the fin in zebrafish, *HoxD* genes become active in cells only on the posterior side of the developing fin (blue). Zebrafish exhibit no second phase of *HoxD* activity in the developing fin.

During development of the fin in catsharks, *HoxD* genes first become active in cells on the posterior side of the developing fin (blue). Later these genes become active in a very narrow band of cells near the edge of the developing fin (green).

Sources: P. Sordino, F. van der Hoeven, and D. DuBoule. 1995. *Hox* gene expression in teleost fins and the origin of vertebrate digits. *Nature* 375:678–681; R. Freitas, G. Zhang, and M. J. Cohn. 2007. Biphasic *Hoxd* gene expression in shark paired fins reveals an ancient origin of the distal limb domain. *PLoS ONE* 2:e754.

ilarly, heterochrony can be explained as an evolutionary change in the switches that either delays the development of adult characteristics or speeds up the development of reproductive maturity.

If Carroll's hypothesis is correct, morphological novelties arise when evolutionary changes in developmental switches alter the expression patterns of *existing* genes. This view contrasts markedly with the explanation proposed in the modern synthesis (the unified theory of evolution described in Chapter 19), that most morphological novelties arise as mutations that slowly accumulate in the genes that carry blueprints for building particular structures. According to the modern synthesis, the accumulated mutations eventually create *new* genes that specify the creation of new structures.

Although Carroll's hypothesis argues that changes in the genes regulating development cause most morphological change, proponents of evo-devo recognize that mutations in developmental regulatory genes and their effects on morphology are subject to the action of the same microevolutionary processes—natural selection, genetic drift, and gene flow—that influence the frequencies of genotypes and phenotypes in populations. Thus, every morphological change induced by a mutation in a homeotic gene or developmental switch is tested by the success or failure of the individual that carries it.

Numerous studies have shown that changes in the expression of homeotic genes can have dramatic effects on morphology. *Insights from the Molecular Revolution* explains how a change in the number and expression of *Hox* genes changed the fins of fishes into the limbs of four-legged vertebrates.

In another example, researchers have determined how an adaptive morphological change in a small fish, the three-spined stickleback *(Gasterosteus aculeatus),* results from the deactivation of a homeotic gene. The freshwater stickleback populations in North American lakes are the descendants of marine ancestors that colonized the lakes after the retreat of glaciers between 10,000 and 20,000 years ago. Marine sticklebacks have bony armor along their sides and prominent spines; lake-dwelling sticklebacks have greatly reduced armor and, in many populations, lack spines on their pelvic fins **(Figure 22.25).**

Natural selection has apparently fostered these morphological differences in response to the dominant predators in each habitat. In marine environments, long spines prevent some predatory fishes from swallowing sticklebacks. But long spines are a liability in lakes, where voracious dragonfly larvae grab sticklebacks by their spines and devour them; freshwater sticklebacks that lack spines are more likely to escape from their clutches.

In a paper published in 2004, Michael D. Shapiro of the Stanford University School of Medicine and the Howard Hughes Medical Institute and colleagues at several other institutions described how the presence or absence of spines on the pelvic fins of these fishes is governed by the expression of the gene *Pitx1.* Pelvic spines are part of the pelvic fin skeleton, the fishes' equivalent of a hind limb. In fact, *Pitx1* also contributes to the development of hind limbs in four-legged vertebrates and to certain glands and sensory organs in the head.

In long-spined marine sticklebacks, *Pitx1* is expressed in the embryonic buds from which pelvic fins develop, promoting the development of spines. But *Pitx1* is not expressed in the fin buds of the freshwater sticklebacks; hence, pelvic spines do not develop. However, freshwater sticklebacks have not *lost* the *Pitx1* gene; it is still expressed elsewhere in the fishes' bodies. In a

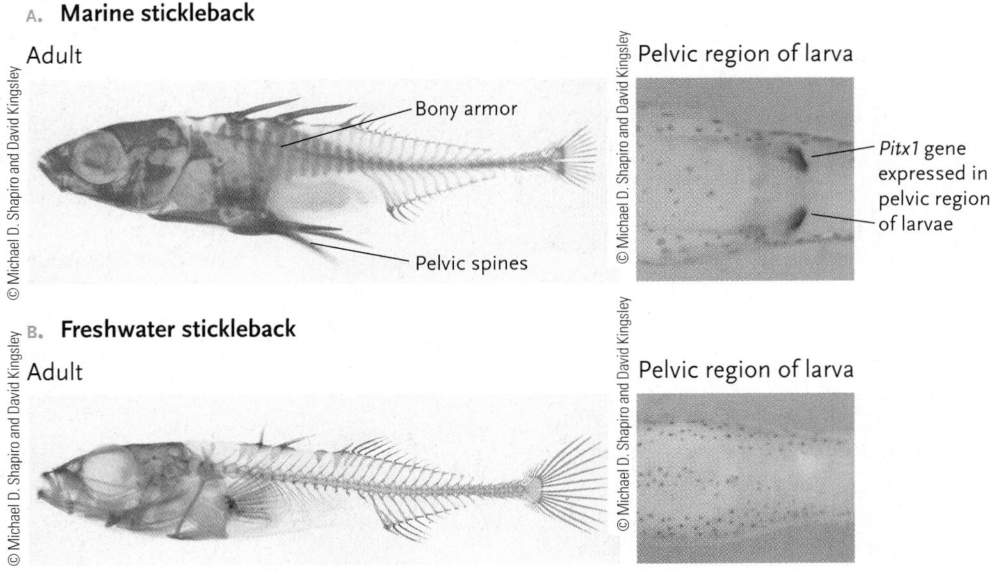
A. Marine stickleback
Adult
Bony armor
Pelvic spines
Pelvic region of larva
Pitx1 gene expressed in pelvic region of larvae
B. Freshwater stickleback
Adult
Pelvic region of larva
© Michael D. Shapiro and David Kingsley
© Michael D. Shapiro and David Kingsley
© Michael D. Shapiro and David Kingsley
© Michael D. Shapiro and David Kingsley

FIGURE 22.25
Sticklebacks with and without pelvic spines. (A) Marine populations of three-spined sticklebacks *(Gasterosteus aculeatus)* have prominent bony plates along their sides and large spines on their dorsal and pelvic fins. The growth of pelvic spines is induced by the expression of the *Pitx1* gene (the purple crescents in the photo on the right) in the pelvic region during embryonic development. **(B)** Many freshwater populations of the same species lack the bony plates and spines. Pelvic spines do not develop in the freshwater sticklebacks because the *Pitx1* gene is not expressed in the pelvic region. The skeletons of these specimens, each about 8 cm long, were dyed bright red.

follow-up paper, published in 2009, the Shapiro team reported that the failure of freshwater sticklebacks to express the *Pitx1* gene in the developing pelvic region results from the deletion of a tissue-specific enhancer of hind limb expression; without this enhancer, the *Pitx1* gene is not expressed, and the fishes do not grow pelvic spines.

In the next chapter, we examine how biologists explore the evolutionary relationships among species and how they organize that information into a useful framework for researchers in every biological discipline. In the following unit, we will revisit evo-devo and examine some recent discoveries about how changes in homeotic genes have diversified body plans in the major evolutionary groups of organisms.

STUDY BREAK 22.6 <

1. **What is the difference between allometry and heterochrony?**
2. **What evidence suggests that many developmental control genes have been conserved by evolution?**
3. **What genetic factor is apparently responsible for the presence or absence of spines in stickleback fish?**

UNANSWERED QUESTIONS

Does morphological evolution always proceed gradually or can it occur in great leaps and bounds?

As you read in this chapter, biologists disagree about whether evolutionary changes in morphology can occur very rapidly. Although biologists have proposed various hypotheses to explain the abrupt changes that we sometimes find in the fossil record, evo-devo studies provide insight into one mechanism for how dramatic changes can arise: the spatial redeployment of homeotic genes. *Homeosis* is defined as the complete replacement of one type of organ with another. In one famous example, a *Hox* gene mutation in *Drosophila* replaces the antennae with legs. If such a mutation were to occur in nature, the organism would probably not have a selective advantage. But what if it did? These kinds of mutant phenotypes first inspired Richard Goldschmidt to develop his idea of the "hopeful monster" early in the twentieth century.

Stephen Jay Gould later revised and updated this idea in the context of his punctuated equilibrium hypothesis about the tempo and mode of evolution. The hypothesis suggests that if, on very rare occasions, truly dramatic morphological changes provide a selective advantage, they may lead to the rapid formation of a new species based on only a few genetic differences. What types of organisms are the most likely to exhibit homeotic change in an evolutionary context? The best candidates are those with highly modular bodies made up of repeating units—like the segments of an insect or the bones in the spine of a vertebrate. Such animals often express different organ identities in different modular units—such as the antennae, claws, and legs of a lobster—and these identities may be redeployed to different positions along the body axis. Plants are among the most modular organisms on Earth. They produce serially repeated structures—a leaf, a bud at the base of the leaf, and a stem—to generate their bodies. Many exciting and promising questions in plant evo-devo relate to how evolutionary homeosis may have generated rapid change in plant morphology.

Has homeosis contributed to the appearance and diversification of the angiosperms?

The sudden appearance of flowering plants, the angiosperms, in the fossil record so puzzled Charles Darwin that he dubbed their evolution an "abominable mystery." How did the gymnosperms, which always bear their male and female reproductive structures separately, give rise to the hermaphroditic (that is, bearing both male and female structures) flower? Our current understanding of the genetics of floral developmental provides a simple solution to this puzzle: the genetic program controlling floral organ identity is homeotic. Thus, it is possible for very simple genetic changes to transform an entirely male set of reproductive organs into a combination of male and female parts. Such models have been outlined by Günter Theissen at the Friedrich-Schiller-Universität Jena in Germany as well as by David Baum and Lena Hileman at the University of Wisconsin–Madison and the University of Kansas, respectively. In addition to fostering the origin of the angiosperms, homeosis may have played a role in the group's diversification. Commonly observed shifts in the morphology of sepals and petals or in the number of stamens (male reproductive structures) are suggestive of homeotic changes. These examples are more suitable for experimental verification than is the question on the origin of the angiosperms because they are much more recent occurrences. Although these hypotheses are very attractive, they remain to be confirmed through molecular genetic analyses.

How have new floral organ identity programs evolved?

The homeotic scenarios described here involve spatial shifts in the expression of preexisting identity programs, such as a stamen developing where there was previously a petal. But what about cases where a whole new type of floral organ appears? There are many examples of flowers that have more than the four most common types of organs (sepals, petals, stamens, and carpels, described in Chapter 34), which suggests that new organ identity programs must have evolved. In such instances, do the new organs evolve through modification of preexisting identity programs, or are entirely new gene pathways recruited? Studies using a new model plant for genetics, *Aquilegia*—commonly known as columbine—suggest that a fifth type of floral organ has evolved through modification of the stamen identity program. Many questions about how this process actually occurred remain unanswered. Was the new program derived through just a few genetic changes, or many? Did it involve changes in regulatory gene function, or only shifts in gene expression? There is much more to learn about how completely new types of floral organs have evolved.

Think Critically

One point of disagreement among evolutionary biologists is whether evolutionary changes primarily occur in gene regulatory regions or in the coding regions. Which of these two types of changes do you think is more likely with homeotic evolution?

David Kramer

Elena M. Kramer is the John L. Loeb Associate Professor of Biology at Harvard University, where she studies the evolution of the genetic mechanisms controlling floral development. To learn more about Dr. Kramer's research go to http://www.oeb.harvard.edu/faculty/kramer/index.htm.

REVIEW KEY CONCEPTS

Go to **CENGAGENOW** at www.cengage.com/login to access quizzing, animations, exercises, articles, and personalized homework help.

22.1 The Fossil Record

- Fossils are the parts of organisms preserved in sedimentary rocks or in oxygen-poor environments (Figures 22.2 and 22.3).
- The fossil record is incomplete because few organisms fossilize completely, because some organisms are more likely to fossilize than others, and because natural processes destroy many fossils.
- Fossils provide a relative dating system, the geological time scale, for the strata in which they occur. Radiometric dating techniques establish the absolute age of rocks and fossils (Figures 22.4 and 22.5, Table 22.1).
- The fossil record provides data on changes in morphology, biogeography, ecology, and the behavior of organisms (Figure 22.6).

Animation: Radioisotope decay

Animation: Radiometric dating

Animation: Geologic time scale

22.2 Earth History

- Earth's crust is composed of plates of solid rock that float on a semisolid mantle (Figure 22.7). New crust is constantly generated and old crust is recycled. Currents in the mantle cause the continents to move over geological time (Figure 22.8). The breakup of the continents profoundly influenced biological evolution by separating populations and creating more shallow marine habitats.
- Continental movements caused variations in climate, the extent of glaciation, and sea levels (Figure 22.9). Asteroid impacts and volcanic eruptions have also influenced Earth's environment, sometimes triggering large-scale extinctions.

Animation: Plate margins

Animation: Geologic forces

22.3 Historical Biogeography and Convergent Biotas

- Disjunct distributions of species can be produced by dispersal and/or vicariance (Figure 22.10). Dispersal results in a disjunct distribution when a new population is established on the far side of a barrier. Vicariance results in a disjunct distribution when external factors such as continental drift fragment the landscape.
- Continental drift has created six major biogeographical realms, each with a characteristic biota (Figure 22.11).
- Convergent evolution produces similar adaptations in distantly related species that live in similar environments (Figures 22.12 and 22.13).

22.4 The History of Biodiversity

- Adaptive radiation produces morphologically diverse species within lineages (Figure 22.14), increasing biodiversity.
- Extinction decreases species diversity. Mass extinctions have occurred at least five times in the history of life (Figure 22.15). Tectonic activity, climate change, and asteroid strikes are probable causes of mass extinctions.
- Biodiversity has increased repeatedly since life first evolved, partly in response to increased geographical separation of the continents and partly because complex interactions evolve among existing species (Figures 22.16 and 22.17).

Animation: Morphological divergence

22.5 Interpreting Evolutionary Lineages

- The horse lineage is complex and highly branched. It includes species of various sizes and diverse morphological adaptations (Figure 22.18).
- The tempo of morphological change varies among lineages. Some exhibit a pattern of gradual change through time (Figure 22.19). Other lineages confirm the predictions of the punctuated equilibrium hypothesis (Figure 22.20).

Animation: Evolutionary tree diagrams

22.6 The Evolution of Morphological Novelties

- Morphological novelties can arise from evolutionary changes in the relative growth rates of body parts (Figure 22.21) or in the timing of developmental events (Figures 22.22 and 22.23).
- Morphological novelties are often modifications of existing structures. An exaptation is a trait that turns out to be useful in a new environmental context even before natural selection refines its form. For example, the feathers and wings of birds evolved from the ancestral feathers and elongated forelimbs of some bipedal dinosaurs (Figure 22.24).
- Evolutionary developmental biology—evo-devo—examines how evolutionary changes in genes that regulate embryonic development can foster changes in body shape and form.
- Most organisms share an ancient tool kit of several hundred genes that regulate the expression of the thousands of genes involved in development. Evolutionary changes in developmental switches may account for many morphological changes. The differential expression of the *Pitx1* gene in sticklebacks determines whether or not a fish grows pelvic spines (Figure 22.25).

UNDERSTAND AND APPLY

Test Your Knowledge

1. The fossil record:
   a. provides direct and indirect evidence about life in the past.
   b. shows that all morphological novelties arise rapidly.
   c. provides abundant data about rare species with local distributions.
   d. is equally good for all organisms that ever lived.
   e. provides no evidence about the physiology or behavior of ancient organisms.
2. The absolute age of a geological stratum is determined by:
   a. the thickness of its rocks.
   b. the particle size in its rocks.
   c. the types of fossils found within it.
   d. its position relative to other layers.
   e. radiometric dating techniques.
3. The observation that fossils of *Premedosaurus* are found only in Argentina and Northern Europe provides an example of:
   a. a continuous distribution.
   b. a disjunct distribution.
   c. species selection.
   d. allometry.
   e. gradualism.

4. The evolutionary history of horses demonstrates that:
   a. modern horses are the direct, lineal descendants of the earliest horses.
   b. the leg bones of modern horses are more complex than those of the earliest horses.
   c. horses have always had specialized teeth that allow them to feed on tough grasses.
   d. horses diversified greatly, but only a few types survived to the present.
   e. the first horses lived in open, grassy habitats.
5. The punctuated equilibrium hypothesis:
   a. recognizes that morphological evolution may occur slowly or quickly.
   b. suggests that major morphological novelties arise gradually.
   c. may help explain why there are so many "missing links" in the fossil record.
   d. suggests that the fossil record is usually complete.
   e. links mass extinctions to the impact of asteroids striking Earth.
6. Biologists believe that the overall similarities between marsupial mammals in Australia and placental mammals in North America were caused by:
   a. plate tectonics.
   b. dispersal.
   c. convergent evolution.
   d. homeotic genes.
   e. punctuated equilibria.
7. The differential growth of body parts is called:
   a. allometry.
   b. paedomorphosis.
   c. heterochrony.
   d. phyletic gradualism.
   e. evo-devo.
8. Exaptations are traits that:
   a. prepare some organisms for future environmental changes.
   b. appear in lineages as a result of an adaptive radiation.
   c. evolve in anticipation of a species' future needs.
   d. are useful in new situations before natural selection alters them for a new function.
   e. occur in animals, but not in plants.
9. Adaptive radiations often follow mass extinctions because:
   a. mass extinctions limit the impact of paedomorphosis.
   b. mass extinctions foster allometry and heterochrony.
   c. mass extinctions decimate all forms of life on Earth.
   d. species that form transitional fossils often survive mass extinctions.
   e. extinctions open adaptive zones that had been previously occupied.
10. Homeotic genes are defined as genes that:
   a. bind directly to a regulatory site on DNA.
   b. code for transcription factors activating or repressing genes that influence an organism's form.
   c. determine whether or not a morphological innovation leads to an adaptive radiation.
   d. have been inherited from an ancient ancestor by nearly all forms of life.
   e. help biologists differentiate between plants and animals.

Discuss the Concepts

1. Many millions of years from now, continental drift may obliterate the Pacific Ocean, pushing North America into physical contact with Asia. What effects might these events have on the organisms living at that time?
2. One possible measure of a species' evolutionary success is the number of descendant species it produces. Should our species, *Homo sapiens,* be considered successful under this definition?
3. Extinctions are common in the history of life. Why are biologists alarmed by the current wave of extinctions caused by human activity?

Design an Experiment

Design a study to determine whether the wings of birds, bats, and insects and their ability to fly are the products of convergent evolution.

Interpret the Data

The graphs presented below provide data on the body sizes (above) and brain sizes (below) of modern humans and three extinct species in the evolutionary lineage that includes humans. How would you describe the relative changes of body size and brain size in this lineage through time?

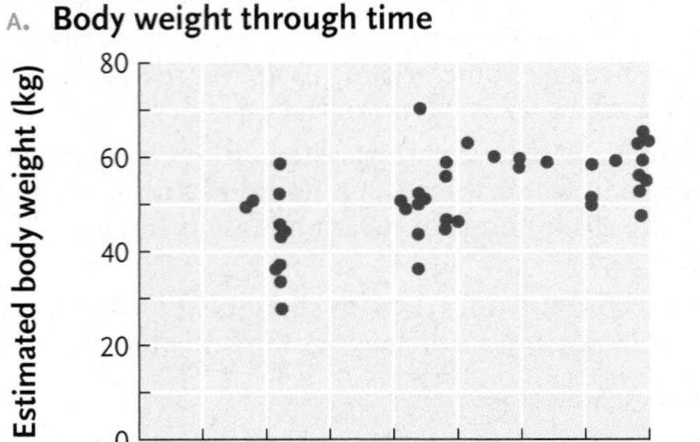
A. Body weight through time
Estimated body weight (kg)
80
60
40
20
0
3.0
2.0
1.0
Present

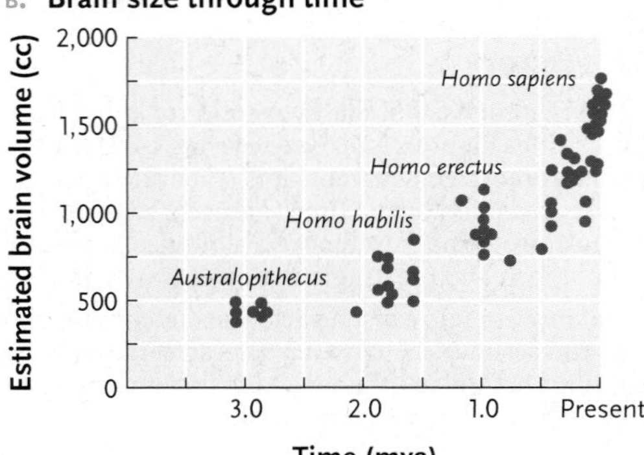
B. Brain size through time
Estimated brain volume (cc)
2,000
1,500
1,000
500
0
Homo sapiens
Homo erectus
Homo habilis
Australopithecus
3.0
2.0
1.0
Present
Time (mya)

Source: D. J. Futuyma. 2009. *Evolution,* 2e. Sinauer Associates, Sunderland, MA (after S. Jones, R. Martin, and D. Pilbeam, eds. 1992. *The Cambridge Encyclopedia of Human Evolution.* Cambridge University Press, Cambridge).

Apply Evolutionary Thinking

The geological evolution of Earth has had an obvious effect on biological evolution. You have read about how the release of oxygen by photosynthetic organisms increased atmospheric oxygen concentration. How are human activities changing the physical environment on Earth? What new selection pressures do these environmental alterations establish?

Express Your Opinion

Scientifically important fossils are sometimes found on privately owned land, creating disputes about who owns the fossils and how they should be used. For example, ownership of "Sue," the largest *Tyrannosaurus* fossil ever discovered, had to be settled in a court of law. Although the fossil was unearthed on a privately owned ranch on a Sioux Indian reservation, the land was held in trust by the U.S. government. The government argued that the fossil was public property, but the court eventually decided that the rancher owned the fossil. He could keep it or dispose of it however he chose. He sold it at auction for more than $8 million, the highest price ever paid for a fossil. A group of corporate sponsors raised the funds to buy the fossil on behalf of the Field Museum in Chicago, where it is now on public display. Do you think that scientifically important specimens should be the property of any one individual, or should they belong to the government, a museum, or some other research institution? Go to www.cengage.com/login to investigate both sides of the issue and then vote.

A new plant species from Idaho. Sacajawea's bitterroot *(Lewisia sacajaweana)* was formally described in 2006. It is named in honor of Sacajawea, the Native American woman who guided Lewis and Clark in their exploration of the Pacific Northwest in the early 1800s.

23

STUDY OUTLINE

Systematics and Phylogenetics: Revealing the Tree of Life

Why It Matters. . . Mention the word "malaria," and people envision the tropics: explorers wander through the jungle in pith helmets and sleep under mosquito netting; clouds of insects hover nearby, ready to infect them with *Plasmodium,* the protistan parasite that causes this disease. You may be surprised to learn, however, that less than 100 years ago, malaria was also a serious threat in the southeastern United States and much of western Europe.

Scientists puzzled over the cause of malaria for thousands of years. Hippocrates, a Greek physician who worked in the fifth century B.C., knew that people who lived near malodorous marshes often suffered from fevers and swollen spleens. Indeed, the name malaria is derived from the Latin for "bad air." By 1900, scientists had established that mosquitoes, *Plasmodium*'s intermediate hosts, transmit the parasite to humans. Mosquitoes breed in standing water, and anyone living nearby is likely to suffer their bites.

Until the 1920s, scientists thought that the mosquito species *Anopheles maculipennis* carried malaria in Europe. But some areas with huge mosquito populations had little human malaria, whereas other areas had relatively few mosquitoes and a high incidence of the disease.

Then, a French researcher reported variation in the mosquitoes, and Dutch scientists identified two forms of the "species," only one of which seemed to carry malaria. The breakthrough came in 1924, when a retired public health inspector in Italy discovered that individual mosqui-

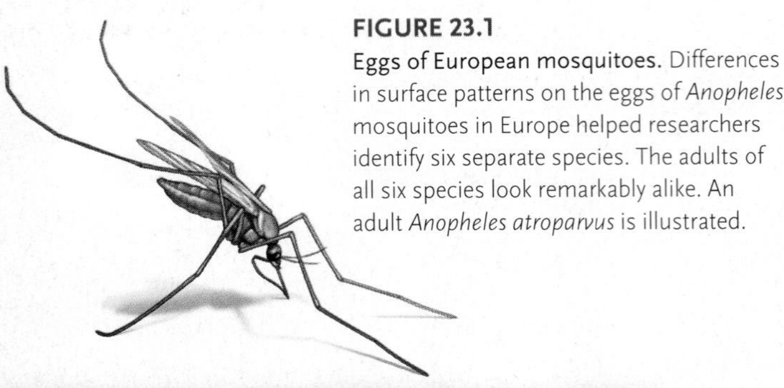

FIGURE 23.1

Eggs of European mosquitoes. Differences in surface patterns on the eggs of *Anopheles* mosquitoes in Europe helped researchers identify six separate species. The adults of all six species look remarkably alike. An adult *Anopheles atroparvus* is illustrated.

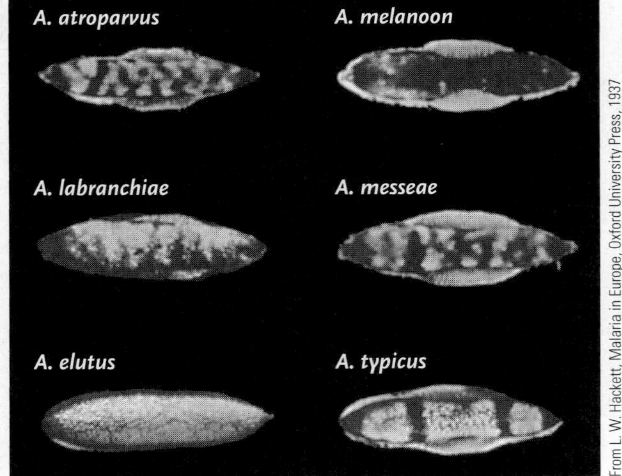
A. atroparvus
A. melanoon
A. labranchiae
A. messeae
A. elutus
A. typicus

From L. W. Hackett, Malaria in Europe, Oxford University Press, 1937

toes—all thought to be the same species—produced eggs with one of six distinctive surface patterns **(Figure 23.1).**

Further research revealed that the name *Anopheles maculipennis* had been applied to six separate mosquito species. Although the adults of these species are almost indistinguishable, their eggs are clearly different. The species are reproductively isolated from each other, and they differ ecologically: some breed in brackish coastal marshes, others in freshwater inland marshes, and still others in slow-moving streams. Only some of these species have a preference for human blood, and researchers eventually determined that only three of them routinely transmit malaria to humans.

These discoveries explained why the geographical distributions of mosquitoes and malaria did not always match. And government agencies could finally fight malaria by eradicating the disease-carrying species. Health workers drained marshes to prevent mosquitoes from breeding. They used insecticides to kill mosquito larvae or introduced fish of the genus *Gambusia,* the mosquitofish, which eats them. These targeted control programs were very successful in the early and middle decades of the twentieth century.

Today, with the increased mobility of humans, agricultural products, and other goods, some mosquito species—as well as many other organisms—are invading habitats made more hospitable by global climate change (discussed further in Unit Seven). Some have expanded their geographical ranges substantially: introduced mosquito species have been discovered on all continents except Antarctica. Mosquitoes carry numerous agents of disease, and biologists are now devising new ways to recognize—and eradicate—the species that pose the greatest threats to human welfare.

The historic eradication of malaria in Europe owed a debt to **systematics,** the branch of biology that studies the diversity of life and its evolutionary relationships. Systematic biologists—systematists for short—identify, describe, name, and classify organisms, organizing their observations within a framework that reflects the organisms' evolutionary relationships. In this chapter we briefly describe the traditional approach to classification. We then focus attention on how systematists working today develop hypotheses about the evolutionary relationships of all the branches, twigs, and leaves on the Tree of Life. <

23.1 Nomenclature and Classification

The Swedish naturalist Carl von Linné (1707–1778), better known by his Latinized name, Carolus Linnaeus, was the first modern practitioner of **taxonomy,** the science that identifies, names, and classifies new species. A professor at the University of Uppsala, Linnaeus sent ill-prepared students around the world to gather specimens, losing perhaps a third of his followers to the rigors of their expeditions. Although he may not have been a commendable student adviser, Linnaeus developed the basic system of naming and classifying organisms that biologists embraced for two centuries.

Linnaeus Developed the System of Binomial Nomenclature

Linnaeus invented the system of **binomial nomenclature,** in which species are assigned a Latinized two-part name, or **binomial.** The first part of the name identifies a **genus** (plural, *genera*), a group of species with similar characteristics. The second part is the **specific epithet,** or species name. When identifying and naming a new species, Linnaeus used the morphological species concept (described in Section 21.1), assigning the same scientific name to individuals that shared anatomical characteristics.

A combination of the generic name and the specific epithet provides a unique name for every species. For example, *Ursus maritimus* is the polar bear and *Ursus arctos* is the brown bear. By convention, the first letter of a generic name is always capitalized; the specific epithet is never capitalized; and the entire binomial is italicized. In addition, the specific epithet is never used without the full or abbreviated generic name preceding it because the same specific epithet is often given to species in different genera. For instance, *Ursus americanus* is the American black bear, *Homarus americanus* is the Atlantic lobster, and *Bufo americanus* is the American toad. If you were to order just "*americanus*" for dinner, you might be dismayed when your plate arrived—unless you have an adventurous palate!

Nonscientists often use different common names to identify a species. For example, *Bothrops asper,* a poisonous snake native to Central and South America, is called *barba amarilla* (meaning "yellow beard") in some places and *cola blanca* (meaning "white tail") in others; biologists have recorded about 50 local names for this species. Adding to the confusion, the same common name is sometimes used for several different species. Binomials, however, allow people everywhere to discuss organisms unambiguously.

Many binomials are descriptive of the organism or its habitat. *Asparagus horridus,* for example, is a spiny plant. Other species, such as the South American bird *Rhea darwinii,* are named for notable biologists. The naming of newly discovered species follows a formal process of publishing a description of the species in a scientific journal. International commissions meet periodically to settle disputes about scientific names.

Linnaeus Devised the Taxonomic Hierarchy to Organize Information about Species

Linnaeus described and named thousands of species on the basis of their morphological similarities and differences. Keeping track of so many species was no easy task, so he devised a **classification,** a conceptual filing system that arranges organisms into ever more inclusive categories. Linnaeus' classification, called the **taxonomic hierarchy,** includes a nested series of formal categories: domain, kingdom, phylum, class, order, family, genus, species, and subspecies. (You may find it useful to review the traditional taxonomic categories, which were described in Section 1.3 and illustrated in Figure 1.11A.) The organisms included within any category of the taxonomic hierarchy compose a **taxon** (plural, *taxa*). Woodpeckers, for example, are a taxon (Picidae) at the family level, and pine trees are a taxon *(Pinus)* at the genus level.

Species that are included in the same taxon at the bottom of the hierarchy (that is, in the same genus or family) generally share many characteristics. By contrast, species that are included in the same taxon only near the top of the hierarchy (that is, the same kingdom or phylum) generally share many fewer characteristics. The hierarchy has been a great convenience for biologists because every taxon is defined by a set of shared characteristics. Thus, when a biologist refers to a member of the family Picidae, all of his or her colleagues understand that the biologist is talking about a medium-sized bird that uses its stout bill to drill holes in tree trunks.

STUDY BREAK 23.1

1. **How does the system of binomial nomenclature minimize ambiguity in the naming and identification of species?**
2. **How does the taxonomic hierarchy help biologists organize information about different species?**

23.2 Phylogenetic Trees

Linnaeus devised the taxonomic hierarchy long before Darwin published his theory of evolution. His goals were to illuminate the details of God's creation and to devise a practical way for naturalists to keep track of their discoveries. But the science of systematics changed in response to Darwin's idea that all organisms in the Tree of Life are descended from a distant common ancestor: systematists began to focus on discovering the evolutionary relationships between groups of organisms.

Systematists Adapted Linnaeus' Approach to a Darwinian Worldview

The taxonomic hierarchy that Linnaeus defined was easily adapted to Darwin's concept of branching evolution, which is itself a hierarchical phenomenon **(Figure 23.2).** As we discussed in Chapters 21 and 22, ancestral species give rise to descendant species through repeated branching of a lineage. Organisms in the same genus generally share a fairly recent common ancestor, whereas those assigned to the same higher taxonomic category, such as a class or phylum, share a common ancestor that lived in the more distant past.

In the second half of the nineteenth century, systematists began to reconstruct the **phylogeny** (that is, the evolutionary history) of organisms. Phylogenies are illustrated in **phylogenetic trees,** which are formal hypotheses that identify likely relationships among species and higher taxonomic groups. And like all hypotheses, they are constantly revised as scientists gather new data.

A Phylogenetic Tree Depicts the Evolutionary History of a Group of Organisms

Contemporary evolutionary biologists construct phylogenetic trees to illustrate the hypothesized evolutionary history of organisms. Researchers tailor the breadth of their analyses to match specific research questions. Thus, some trees might include the evolutionary history of all known organisms, others a small clus-

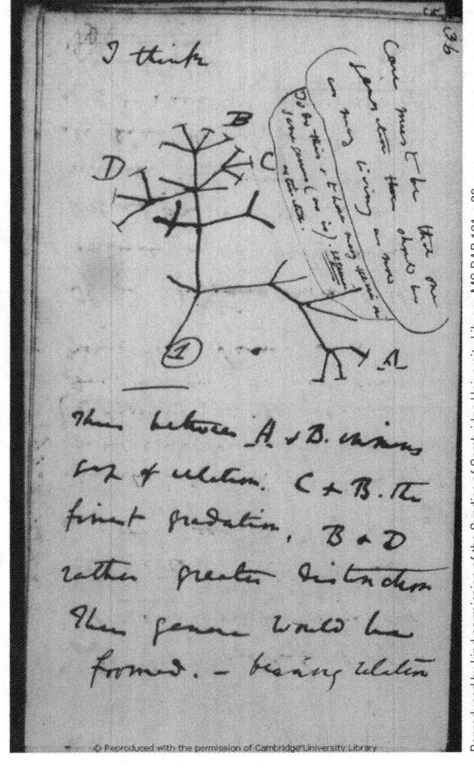
I think
A
B
C
D
1

FIGURE 23.2
Darwin and branching evolution. This entry from his notebook on the "transmutation of species" demonstrates that Charles Darwin first thought about the branching pattern of evolution in 1837, more than 20 years before he published *On the Origin of Species.*

ter of closely related populations within a species, and still others a group somewhere between those extremes. Regardless of how wide a range of organisms is included, all phylogenetic trees share a specific structure and depict key relationships in similar ways **(Figure 23.3).**

For example, phylogenetic trees are usually drawn along an implicit or explicit time line. In this book, phylogenetic trees are generally depicted vertically; the most ancient organisms and evolutionary events are at the bottom of the tree (labeled "long ago"), and the most recent are at the top (labeled "present"). The common ancestor of all species included in the tree is described as the **root** of the tree. In a few cases in this book, trees are presented with the root on the left, with time passing from left to right.

As you know from the discussion in Section 22.5, the tempo of evolution varies within and among lineages. In some cases, evolutionary changes may accumulate slowly in a lineage as the environment shifts over time. This pattern of gradual phyletic change is often described as *anagenesis.* If the changes through time are substantial and the fossil record is incomplete, paleontologists who discover morphologically distinct fossils in different strata may assign them different species names and say that "the ancestral species A evolved into the descendant species B." But the production of such "new" species by anagenesis does not increase biodiversity; rather, it is simply the gradual transformation of one "species" into another as new characteristics accumulated in its populations. Anagenesis is often illustrated by a straight line in a phylogenetic tree.

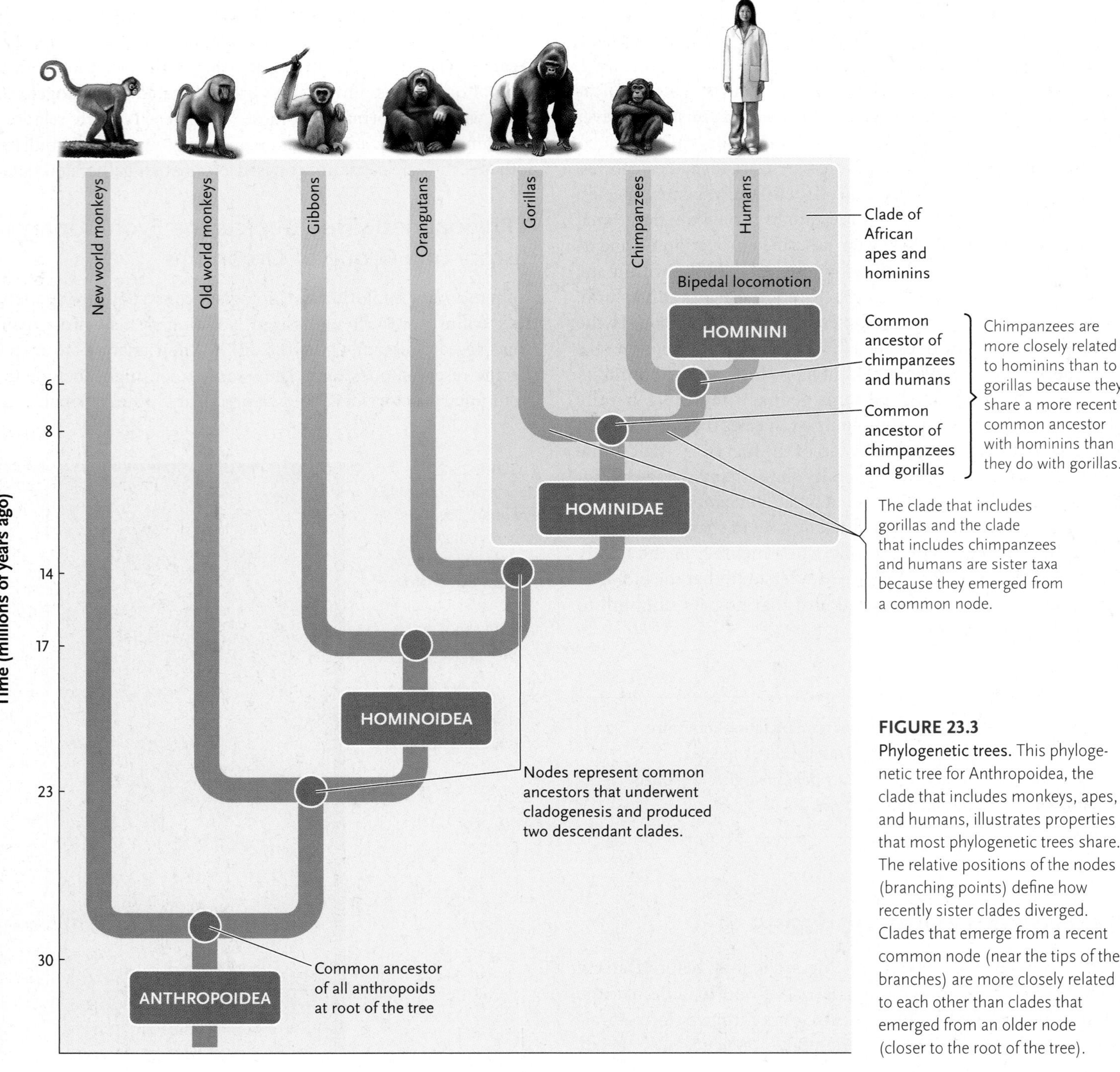

New world monkeys
Old world monkeys
Gibbons
Orangutans
Gorillas
Chimpanzees
Humans
Clade of African apes and hominins
Bipedal locomotion
HOMININI
Common ancestor of chimpanzees and humans
Common ancestor of chimpanzees and gorillas
Chimpanzees are more closely related to hominins than to gorillas because they share a more recent common ancestor with hominins than they do with gorillas.
HOMINIDAE
The clade that includes gorillas and the clade that includes chimpanzees and humans are sister taxa because they emerged from a common node.
HOMINOIDEA
Nodes represent common ancestors that underwent cladogenesis and produced two descendant clades.
ANTHROPOIDEA
Common ancestor of all anthropoids at root of the tree
Time (millions of years ago)
6
8
14
17
23
30

FIGURE 23.3
Phylogenetic trees. This phylogenetic tree for Anthropoidea, the clade that includes monkeys, apes, and humans, illustrates properties that most phylogenetic trees share. The relative positions of the nodes (branching points) define how recently sister clades diverged. Clades that emerge from a recent common node (near the tips of the branches) are more closely related to each other than clades that emerged from an older node (closer to the root of the tree).

In other circumstances, an ancestral species undergoes speciation, producing two descendant species, which may each be morphologically distinct from their common ancestor. This pattern of evolution is described as *cladogenesis,* a process that *does* increase biodiversity. Cladogenesis is depicted by branching points in a phylogenetic tree. Initially, each of the two branches that emerged from a common ancestor represented a new species. But as cladogenesis continued repeatedly through evolutionary time—with branches giving rise to branchlets and branchlets to twigs—each of those new species may have become the common ancestor of its own descendants. Thus, each new species produced by cladogenesis had the potential to become the "root" of the lineage that includes all of its descendants.

Scientists use the term **node** to identify a branching point in a phylogenetic tree; and each branch—with all of its branchlets and twigs—emerging from a node is called a **clade** (*klados* = branch). You can identify a clade on any phylogenetic tree by following a lineage from the node where it first emerges to the tips of its twigs at the very top of the tree. Some clades, such as Aves (birds), include thousands of species, whereas others, such as *Geospiza* (a genus of ground and cactus finches from the Galápagos Islands), include just a few (see Figure 22.14). Like the taxonomic hierarchy, phylogenetic trees have a nested structure: younger and smaller clades, like the genus *Geospiza,* are nested within larger and older clades, such as Aves. Two clades that emerge from the same node are called **sister clades** (or sister taxa) because they are each other's closest relatives; similarly, two species that emerge from the same node near the top of the tree are described as **sister species.**

You should keep certain conventions in mind when you read the phylogenetic trees in this text:

1. If a phylogenetic tree includes an explicit time axis, the positions of the nodes reveal when on the geological time scale two clades originated, and the length of the vertical branch between nodes indexes how long an ancestral group persisted before it diversified. However, in most of the phylogenetic trees presented in this book, the time scale is not precise; often it is not even specified. Thus, the sequence of nodes only indicates the sequence in which clades appeared, and the lengths of the vertical lines contain no specific information about the time since two clades diverged. When reading a phylogenetic tree, you should always remember that if two species or clades emerge from the same node near the top of the tree, they are closely related. If their common node is at the bottom of the tree, they are only distantly related.
2. Horizontal spacing between clades in most of this book's phylogenetic trees does not indicate their degree of difference or their degree of relatedness. However, when the horizontal distance between species or clades is meaningful, as it is in the tree for extinct ectoprocts in Figure 22.20, the horizontal axis label, the figure legend, or the text explain how the distances should be interpreted.
3. Most of the nodes in the phylogenetic trees have two branches protruding from them, reflecting the evolution of two descendant species from one ancestor. However, for some groups of organisms, biologists have not yet discovered the detailed pattern of branching that produced the diversity of clades living today. In these cases, you may see three or even more branches emerging from a node or from a horizontal branch. Biologists describe these nodes as "unresolved." Hopefully, future research will allow us to portray these evolutionary relationships more precisely.
4. In most phylogenetic trees in this book, the clades have been arranged from oldest on the left to youngest on the right. But any clade can be rotated around a node without changing the meaning of the phylogenetic tree **(Figure 23.4).** Similarly, biologists sometimes construct "cladograms" (see Figure 23.4C); although they have a slightly different formal structure, cladograms contain essentially the same information as a phylogenetic tree. When reading both types of diagrams, you should focus primarily on which clades share the more recent common ancestors, indicated by the relative positions of the nodes from which they emerge.

Phylogenetic Trees Allow Biologists to Define Evolutionary Classifications

Evolutionary biologists working today want a classification that mirrors the branching patterns of a group's phylogenetic history. When converting a phylogenetic tree into a classification, they try to identify only **monophyletic taxa** or lineages. A monophyletic taxon comprises one clade—an ancestral species and *all* of its descendants, but no other species **(Figure 23.5).** Monophyletic taxa are defined at every level of the taxonomic hierarchy. For example, biologists consider the Felidae (the traditional family-level taxon that includes all cat species) to be monophyletic: all cat species living on Earth today, from housecats to tigers, are the descendants—and the only living descendants—of a common ancestor that lived about 25 million years ago. Thus, the Felidae is one small, but complete, branch on the Tree of Life. At a much broader scale, the Animalia is a monophyletic taxon (at the traditional kingdom level) that comprises all animals, which are the descendants of one common ancestor. Even if biologists have not yet identified the very first animal in the fossil record, they can infer its existence at the root of the phylogenetic tree for Animalia.

Biologists have not always defined strictly monophyletic taxa. Because of missing data, or as a matter of convenience, they sometimes named taxa that included species from different clades or taxa that included some, but not all, of an ancestral species' descendants. A **polyphyletic taxon** is one that includes organisms from different clades, but not their common ancestor. For example, a taxon that included only birds and bats, two clades of vertebrates that are capable of flight, would be considered polyphyletic because it would not include their last common ancestor, which lived many millions of years before birds and bats first appeared. A **paraphyletic taxon** is one that includes an ancestor and some, *but not all,* of its descendants. For example, people commonly used to define terrestrial dinosaurs and

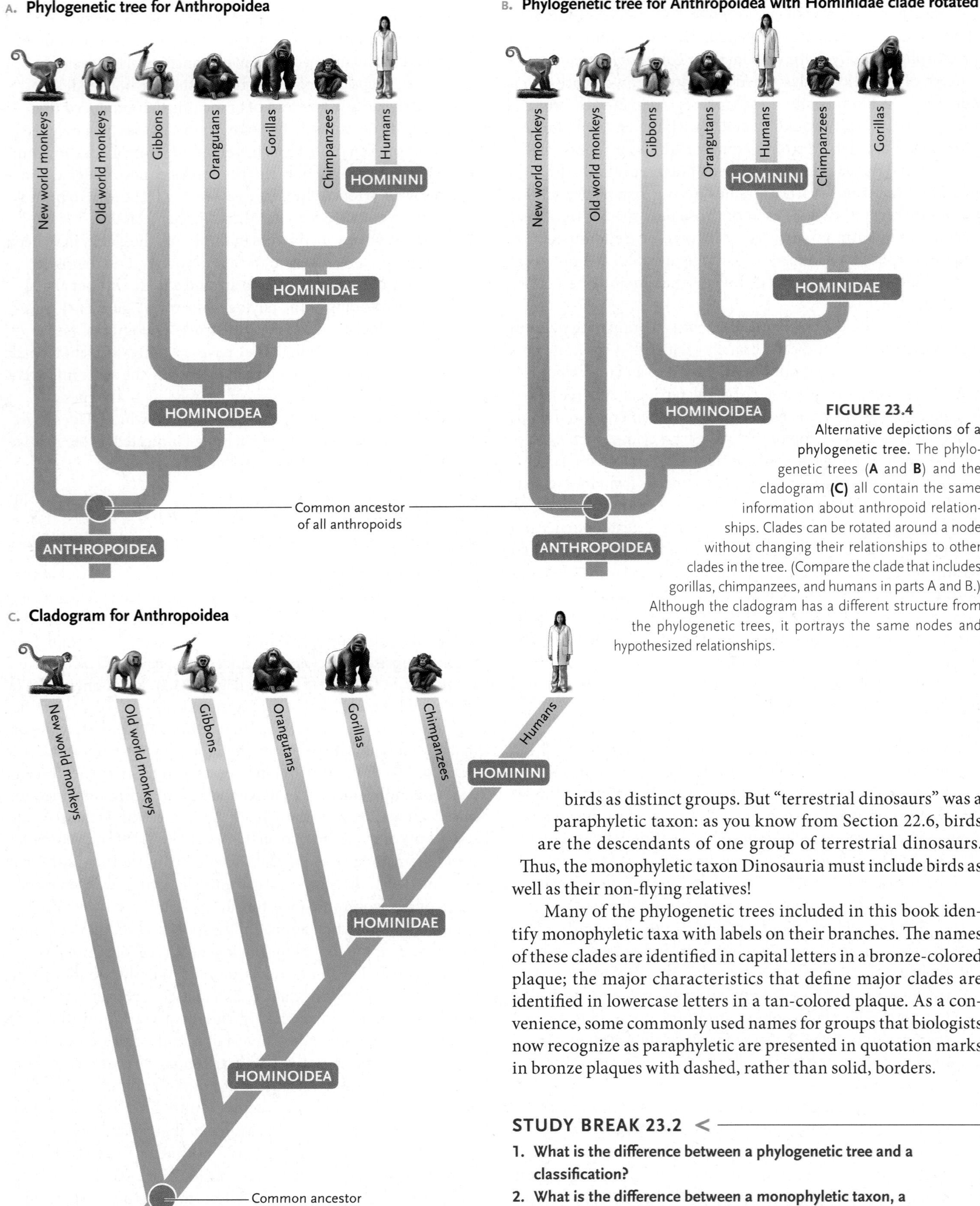
A. Phylogenetic tree for Anthropoidea
New world monkeys
Old world monkeys
Gibbons
Orangutans
Gorillas
Chimpanzees
Humans
HOMININI
HOMINIDAE
HOMINOIDEA
Common ancestor of all anthropoids
ANTHROPOIDEA
B. Phylogenetic tree for Anthropoidea with Hominidae clade rotated
New world monkeys
Old world monkeys
Gibbons
Orangutans
Humans
Chimpanzees
Gorillas
HOMININI
HOMINIDAE
HOMINOIDEA
ANTHROPOIDEA
C. Cladogram for Anthropoidea
New world monkeys
Old world monkeys
Gibbons
Orangutans
Gorillas
Chimpanzees
Humans
HOMININI
HOMINIDAE
HOMINOIDEA
Common ancestor of all anthropoids
ANTHROPOIDEA

FIGURE 23.4
Alternative depictions of a phylogenetic tree. The phylogenetic trees (**A** and **B**) and the cladogram **(C)** all contain the same information about anthropoid relationships. Clades can be rotated around a node without changing their relationships to other clades in the tree. (Compare the clade that includes gorillas, chimpanzees, and humans in parts A and B.) Although the cladogram has a different structure from the phylogenetic trees, it portrays the same nodes and hypothesized relationships.

birds as distinct groups. But "terrestrial dinosaurs" was a paraphyletic taxon: as you know from Section 22.6, birds are the descendants of one group of terrestrial dinosaurs. Thus, the monophyletic taxon Dinosauria must include birds as well as their non-flying relatives!

Many of the phylogenetic trees included in this book identify monophyletic taxa with labels on their branches. The names of these clades are identified in capital letters in a bronze-colored plaque; the major characteristics that define major clades are identified in lowercase letters in a tan-colored plaque. As a convenience, some commonly used names for groups that biologists now recognize as paraphyletic are presented in quotation marks in bronze plaques with dashed, rather than solid, borders.

STUDY BREAK 23.2 <

1. **What is the difference between a phylogenetic tree and a classification?**
2. **What is the difference between a monophyletic taxon, a polyphyletic taxon, and a paraphyletic taxon?**

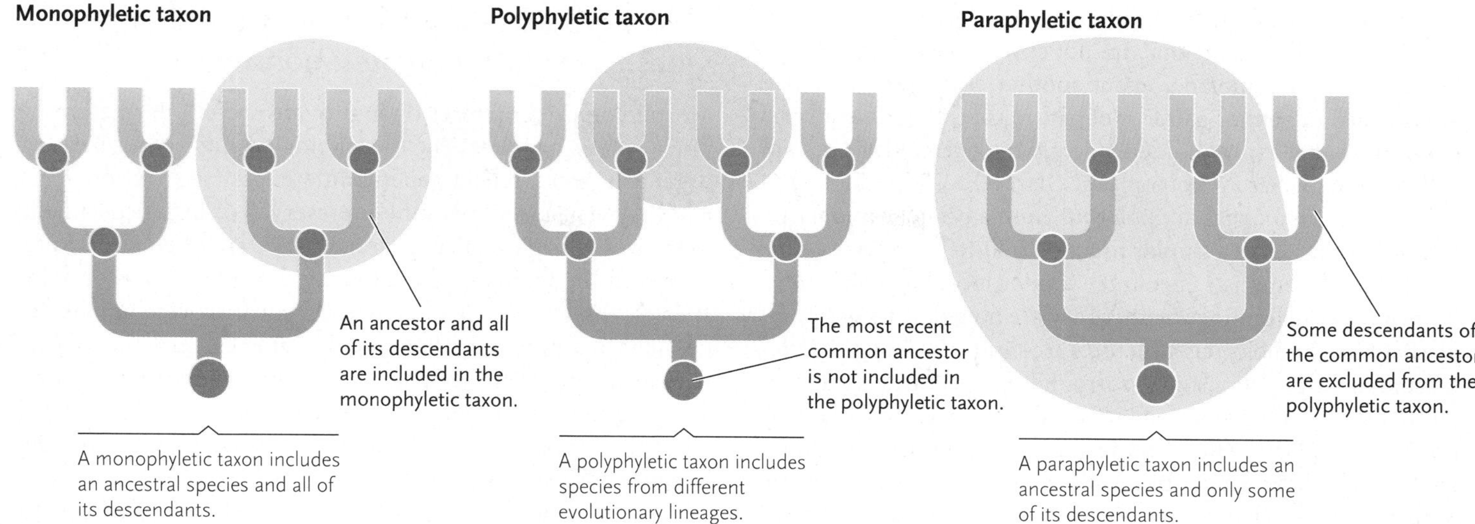
Monophyletic taxon
Polyphyletic taxon
Paraphyletic taxon
An ancestor and all of its descendants are included in the monophyletic taxon.
The most recent common ancestor is not included in the polyphyletic taxon.
Some descendants of the common ancestor are excluded from the polyphyletic taxon.
A monophyletic taxon includes an ancestral species and all of its descendants.
A polyphyletic taxon includes species from different evolutionary lineages.
A paraphyletic taxon includes an ancestral species and only some of its descendants.

FIGURE 23.5
Defining taxa in a classification. Although systematists can identify different groups of species in a phylogenetic tree as a taxon, the taxon highlighted in blue is the only monophyletic group.

23.3 Sources of Data for Phylogenetic Analyses

Linnaeus classified organisms on the basis of their morphological similarities and differences—even though he did not understand how those characteristics arose. For example, he defined birds as a class of oviparous (egg-laying) animals with feathered bodies, two wings, two feet, and a bony beak. No other animals possess all these characteristics, which distinguish birds from "quadrupeds" (his term for mammals), "amphibians" (among which he included reptiles), fishes, insects, and "worms."

FIGURE 23.6
Homologous bones, different structure and function. The hyomandibula, which braced the jaw joint against the skull in early jawed fishes **(A)**, is homologous to the stapes, which transmits sound to the inner ear in the four-legged vertebrates, exemplified here by an early amphibian **(B)**. Both large illustrations show a cross section through the head just behind the jaw joint, as depicted in the small illustrations.

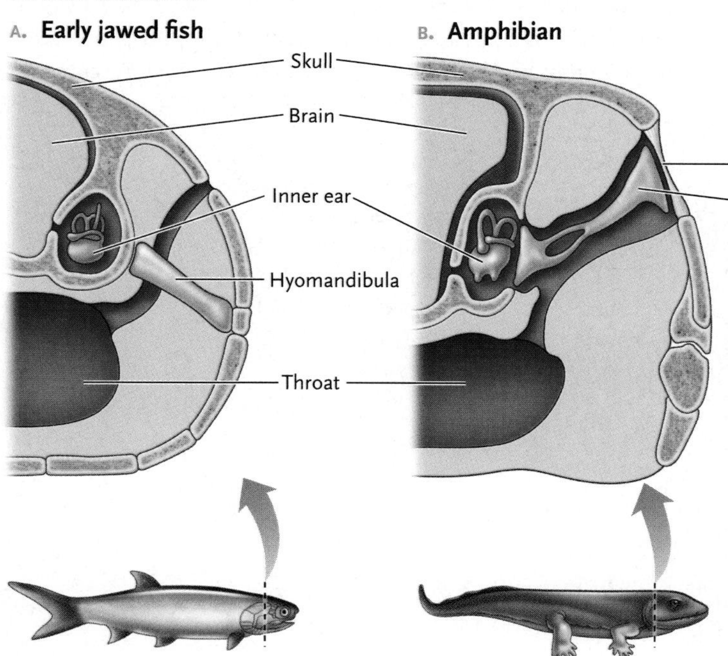
A. Early jawed fish
B. Amphibian
Skull
Brain
Inner ear
Hyomandibula
Throat
Eardrum
Stapes

Mendel's work on the heritable basis of morphological variations provided the scientific rationale for this endeavor: modern systematists infer that morphological differences serve as indicators of underlying genetic differences between species and lineages. Today, with our much deeper understanding of the genetic basis of variation, systematists undertake phylogenetic analyses using a variety of organismal and molecular characters. Indeed, any heritable trait (that is, any trait with a genetic basis) that is intrinsic to the organism can be used in a phylogenetic analysis; phenotypic differences caused by environmental variation (see Section 20.1) are excluded. In this section we first consider a general criterion for evaluating characters and then examine examples of how a few specific types of characters are useful in this effort.

The Analysis of Homologous Characters Sheds Light on Evolutionary Relationships

A basic premise of phylogenetic analyses is that phenotypic similarities between organisms reflect their underlying genetic similarities. As you may recall from Figure 19.3, species that are morphologically similar have often inherited the genetic basis of their resemblance from a common ancestor. Similarities that result from shared ancestry, such as the four limbs of all tetrapod vertebrates, are called **homologies** (or homologous characters). Any trait, from genetic sequences to anatomical structures to mating behaviors, can be described as homologous in two or more species as long as they inherited the trait from their common ancestor.

Even though characters are homologous, they may differ greatly among species, especially if their function has changed over time. For example, the stapes, a bone in the middle ear of tetrapod vertebrates, evolved from—and is therefore homologous to—the hyomandibula, a bone that supported the jaw joint of early fishes. The ancestral function of the bone is retained in some modern fishes, but its structure, position, and function are different in tetrapods **(Figure 23.6)**.

As you know from the discussion of convergent evolution in Section 22.3, organisms that are not closely related sometimes bear a striking resemblance to one another, especially when they are exposed to similar patterns of natural selection. Phenotypic similarities that evolved independently in different lineages are called **homoplasies** (or homoplastic characters). Some biologists use the terms *analogies* or *analogous characters* for homoplastic characters that serve a similar function in different species.

When systematists encounter similar morphological traits, how can they determine whether they are homologous or homoplastic? First, homologous structures are similar in anatomical detail and in their relationship to surrounding structures. For example, the bones in the wings of birds and bats are considered to be homologous **(Figure 23.7)**. They include the same basic structural elements and have similar connections to the rest of the skeleton. Moreover, the fossil record documents that birds and bats inherited the basic skeletal structure of the forelimb from their most recent common ancestor, a tetrapod vertebrate that lived more than 300 million years ago. However, the large flat surfaces of their wings, as well as flying behavior itself, are homoplastic. The wing surfaces are made of different materials—feathers in birds and membranous skin in bats—and their common ancestor, lacking any hint of such structures on its forelimbs, was confined to life on the ground. Thus, for birds and bats, flight and the anatomical structures that produce it are the products of convergent evolution.

Second, in multicellular organisms, homologous characters grow from the same embryonic tissues and in similar ways during development. Systematists have always put great stock in embryological indications of homology on the assumption that evolution has conserved the pattern of embryonic development in related organisms. Indeed, recent discoveries in evolutionary developmental biology (described in Sections 16.4 and 22.6 and explored further in Chapters 29 and 30) have revealed that the genetic controls of developmental pathways are very similar across a wide variety of organisms.

Finally, as we will examine in Section 23.4, the structure of a phylogenetic tree can reveal whether two or more species inherited specific similarities from a common ancestor. If they did, the structures are homologous; if not, they are homoplastic.

Morphological Characters Provide Abundant Clues to Evolutionary Relationships

Morphological structures often provide useful information for phylogenetic analyses. The morphological differences between organisms often reflect genetic differences (see Section 20.1), and they are easy to measure in preserved or living specimens. Moreover, morphological characteristics are often clearly preserved in the fossil record, allowing the comparison of living species with their extinct relatives. As you discovered in the discussions of dinosaur feathers and the inner ear structure of lambeosaurine dinosaurs in Chapter 22, researchers are constantly applying new techniques, such as computed tomography and electron microscopy, to expand our knowledge of the anatomy of extinct organisms.

The morphological traits that are useful in phylogenetic analyses vary from group to group. In flowering plants, the details of flower anatomy may reveal common ancestry. Among vertebrates, the presence or absence of scales, feathers, and fur as well as the structure of the skull help scientists to reconstruct the evolutionary history of major groups. Sometimes systematists use obscure characters of unknown function. But differences in the number of scales on the backs of lizards or in the curvature of a vein in the wings of bees may be good indicators of the genetic differentiation that accompanied or followed speciation—even if we do not know *why* these differences evolved.

Sometimes characteristics found only in the earliest stages of an organism's life cycle can provide evidence of evolutionary relationships. As described in Chapter 30, analyses of the embryos of vertebrates reveal that they are rather closely related to sea cucumbers, sea stars, and sea urchins and even more closely related to a group of nearly shapeless marine invertebrates called sea squirts or tunicates.

As useful as they are, morphological characters alone cannot reveal the details of all evolutionary relationships. For example, some salamander species in North America differ in relatively few morphological features, even when they are genetically, physiologically, and behaviorally distinct. Moreover, researchers

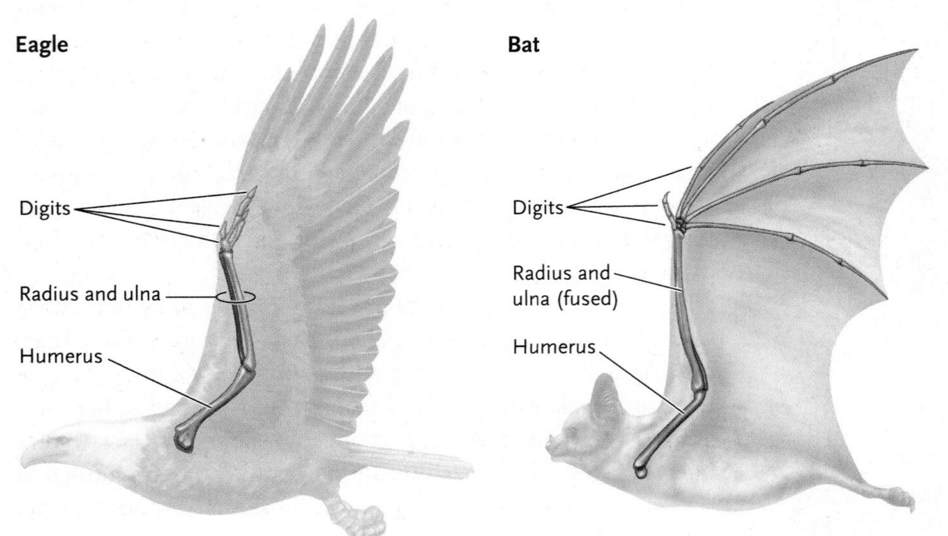
Eagle
Digits
Radius and ulna
Humerus
Bat
Digits
Radius and ulna (fused)
Humerus

FIGURE 23.7
Assessing homology. The wing skeletons of birds and bats are homologous structures with the same basic elements. However, similarities in the flat wing surfaces are homoplastic because the surfaces are composed of different tissues.

cannot easily compare the structures of organisms, such as flatworms and dogs, that share very few morphological traits.

Behavioral Characters Are Useful When Animal Species Are Not Morphologically Distinct

When external morphology cannot be used to differentiate animal species, systematists often examine their behaviors for clues about their relationships. For example, two species of treefrog (*Hyla versicolor* and *Hyla chrysoscelis*) commonly occur together in forests of the central and eastern United States. Both species have bumpy skin and adhesive pads on their toes that enable them to climb vegetation. They also have gray backs, white bellies, yellowish-orange coloration on their thighs, and large white spots below their eyes. The frogs are so similar that even experts cannot easily tell them apart.

How do we know that these frogs represent two species? During the breeding season, males of each species use a distinctive mating call to attract females **(Figure 23.8)**. The difference in calls is a prezygotic reproductive isolating mechanism that prevents females from mating with males of a different species (see Section 21.2). Prezygotic isolating mechanisms are excellent systematic characters because they are often the traits that animals themselves use to recognize members of their own species. The two frog species also differ in chromosome number—*Hyla chrysoscelis* is diploid and *Hyla versicolor* is tetraploid—which is a postzygotic isolating mechanism.

Hyla versicolor
Hyla chrysoscelis
Frequency
Time
Time

FIGURE 23.8
Morphologically similar frog species. The frogs *Hyla versicolor* and *Hyla chrysoscelis* are so similar in appearance that one photo can depict both species. Male mating calls, visualized in sound spectrograms for the two species, are very different. The spectrograms, which depict call frequency on the vertical axis and time on the horizontal axis, show that *H. chrysoscelis* has a faster trill rate.

Molecular Sequences Are Now a Commonly Used Source of Phylogenetic Data

For roughly 200 years, systematists building on Linnaeus' work relied on a variety of organismal traits to analyze evolutionary relationships and classify organisms. Today, most systematists conduct phylogenetic analyses using molecular characters, such as the nucleotide base sequences of DNA and RNA. Because DNA is inherited, shared changes in molecular sequences—insertions, deletions, or substitutions—provide clues to the evolutionary relationships of organisms.

The polymerase chain reaction (PCR) makes it easy for researchers to produce numerous copies of specific segments of DNA for analysis (see Section 18.1); the technique is so effective that it allows scientists to sequence minute quantities of DNA taken from dried or preserved specimens in museums and even from some fossils. Technological advances have automated the sequencing process, and researchers use analytical software to compare new data to known sequences filed in data banks that are accessible over the Internet. Nuclear DNA is frequently used in phylogenetic analyses, and the publication of complete genome sequences for an ever-expanding list of organisms allows researchers to undertake broad comparative studies.

Molecular sequences have certain practical advantages over organismal characters. First, they provide abundant data: every base in a nucleic acid can serve as a separate, independent character for analysis. Moreover, because many genes have been conserved by evolution, molecular sequences can be compared between distantly related organisms that share no organismal characteristics. Molecular characters can also be used to study closely related species with only minor morphological differences. Finally, nucleic acids are not directly affected by the developmental or environmental factors that cause the nongenetic morphological variations described in Section 20.1.

Molecular characters have certain drawbacks, however. For example, only four alternative character states (the four nucleotide bases) exist at each position in a DNA or RNA sequence and only 20 alternative character states (the 20 amino acids) at each position in a protein. (You may want to review Sections 14.2 and 15.1 on the structure of these molecules.)

Because of the limited number of character states, researchers often find it difficult to assess the homology of a nucleotide base substitution that appears at the same position in the DNA of two or more species. For organismal characters, biologists can determine homology by analyzing the characters' embryonic development, details of their function, or their presence in the fossil record. But molecular characters have no embryonic development; biologists still do not understand the functional significance of many molecular differences they discover; and researchers have only recently developed techniques that allow them to sequence DNA found in fossils. Nevertheless, systematists have devised complex statistical tools that allow them to discern whether molecular similarities are likely to be homologous or homoplastic.

Despite their disadvantages, molecular sequences allow researchers to sample the genome directly, and systematists have successfully used sequence data to analyze phylogenetic rela-

INSIGHTS FROM THE Molecular Revolution

Evolution of Whales: SINEs yield signs of the LINEs of evolution

The fossil record reveals that whales evolved from terrestrial mammals into streamlined creatures adapted to life in the sea more than 50 million years ago. Using morphological, fossil, and molecular data, researchers have determined that modern cetaceans—whales, dolphins, and porpoises—are closely related to artiodactyls (even-toed ungulates). Systematists now group cetaceans and artiodactyls in the clade Cetartiodactyla, which, in addition to whales and their relatives, includes camels, pigs, ruminants, and hippopotamuses.

Research Question

Which living cetartiodactyls are the closest relatives of whales?

Experiment

A number of research groups have attempted to answer the question using molecular approaches. Norihiro Okada and his research group at the Tokyo Institute of Technology, Yokohama, Japan, and collaborators in the United States used the patterns of locations of transposable elements (TEs) in genomes to determine the evolutionary relationship between living members of the various cetartiodactyl groups. TEs are sequences that can move to new locations in DNA. The type of TEs the researchers studied are retrotransposons, which move by making RNA copies of themselves; the RNA copies then act as templates for making DNA copies, which insert in new locations (see Section 15.5).

Okada and his collaborators analyzed specific representatives of two classes of retrotransposons, SINEs (for *S*hort *IN*terspersed *E*lements) and LINEs (for *L*ong *IN*terspersed *E*lements), which are known to be inserted into particular chromosomal loci. SINEs and LINEs are valuable tools for phylogenetic studies; their pattern of duplication and movement to new locations represent a derived character that is unique to each clade. Once they have inserted themselves into new locations, they are rarely lost. Thus, if SINEs and LINEs occur at the same sites in the nuclear DNA of several species, those species are likely to be members of the same lineage.

To determine whether a chromosomal locus contained the SINE or LINE associated with it, the researchers isolated DNA, amplified the locus using PCR (see Section 18.1), and then used probes—labeled nucleic acid molecules—that could pair specifically with the SINE or LINE of the locus. A signal from the probing indicated that the SINE or LINE was present; no signal indicated the SINE or LINE was absent.

Results

The investigators analyzed 16 retrotransposons in 10 loci in the DNA from living members of all cetartiodactyl groups. Using the patterns of SINEs and LINEs identified in their phylogenetic analysis, the researchers identified the most parsimonious hypothesis about the evolutionary relationships of the Cetartiodactyla (see **figure**).

Conclusion

As the phylogenetic tree indicates, hippopotamuses are the closest relatives of whales; ruminants (such as cattle and bison) are more distantly related, and pigs and camels even more distantly related. Other phylogenetic analyses based on combined data sets that include morphological characters from extinct and living species as well other molecular characters generally support this conclusion.

Source: M. Nikaido et al. 1999. Phylogenetic relationships among cetartiodactyls based on insertions of short and long interspersed elements: Hippopotamuses are the closest extant relatives of whales. *Proceedings of the National Academy of Sciences USA* 96:10261–10266.

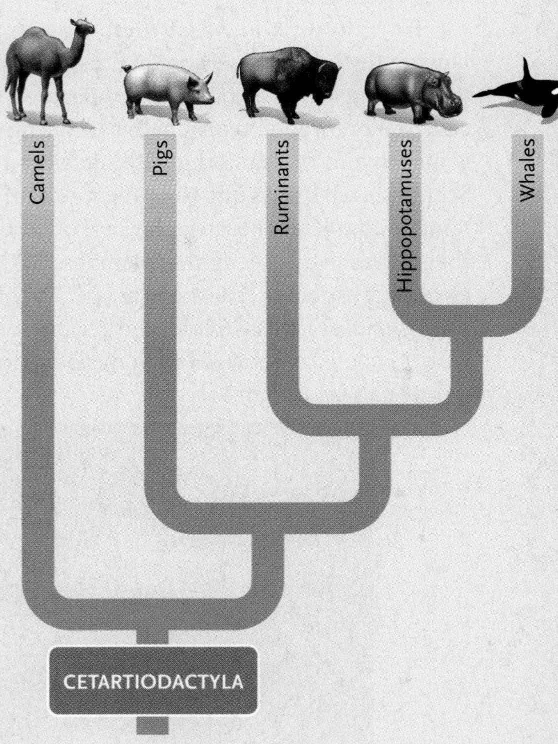
Camels
Pigs
Ruminants
Hippopotamuses
Whales
CETARTIODACTYLA

tionships that organismal characters were unable to resolve. For example, the phylogenetic tree for animals in Figure 29.6 is based on data from several different nucleic acid molecules. *Insights from the Molecular Revolution* describes another example using sequences called transposable elements.

STUDY BREAK 23.3 <

1. **Why do systematists use homologous characters in their phylogenetic analyses?**
2. **Why are morphological traits often helpful in tracing the long-term evolutionary relationships within a group of organisms?**
3. **What are three advantages of using molecular characters in phylogenetic analyses?**

23.4 Traditional Classification and Paraphyletic Groups

For a century after Darwin published his theory of evolution, systematists followed an approach called **traditional systematics.** Researchers constructed phylogenetic trees and classified organisms by assessing the amount of phenotypic divergence between lineages as well as the patterns of branching evolution that had produced them. In other words, they focused on the products of anagenesis (that is, evolutionary change through the accumulation of new characteristics) as well as the products of cladogenesis (that is, the new species and lineages produced through branching evolution). Thus, their classifications did not always strictly reflect the patterns of branching evolution **(Figure 23.9A).**

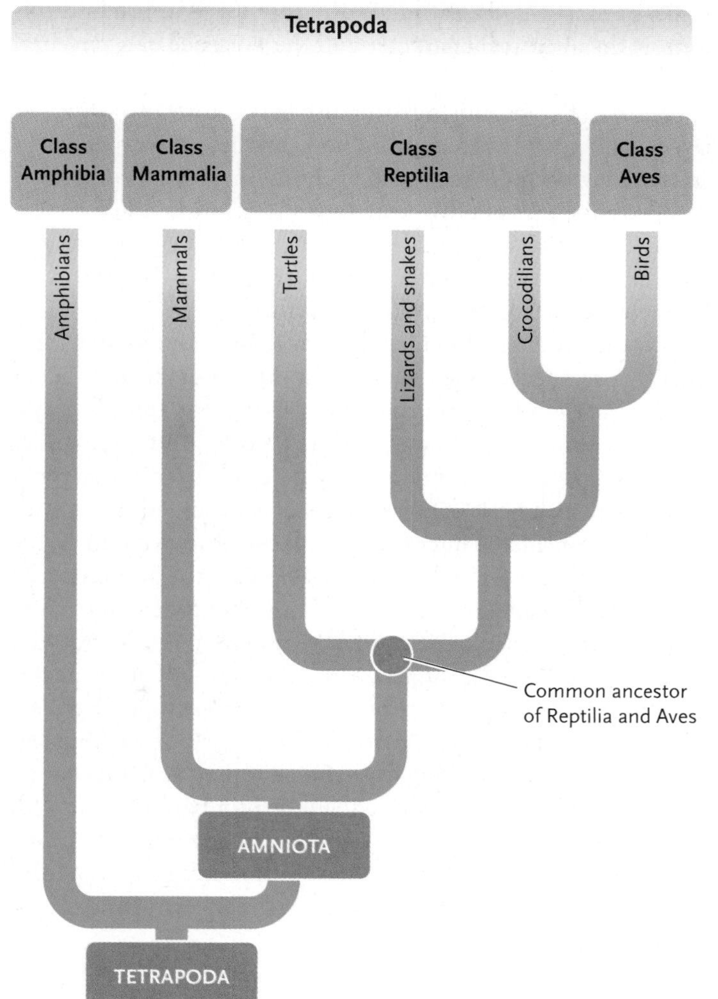
A. Traditional phylogenetic tree with classification
Tetrapoda
Class Amphibia
Class Mammalia
Class Reptilia
Class Aves
Amphibians
Mammals
Turtles
Lizards and snakes
Crocodilians
Birds
Common ancestor of Reptilia and Aves
AMNIOTA
TETRAPODA

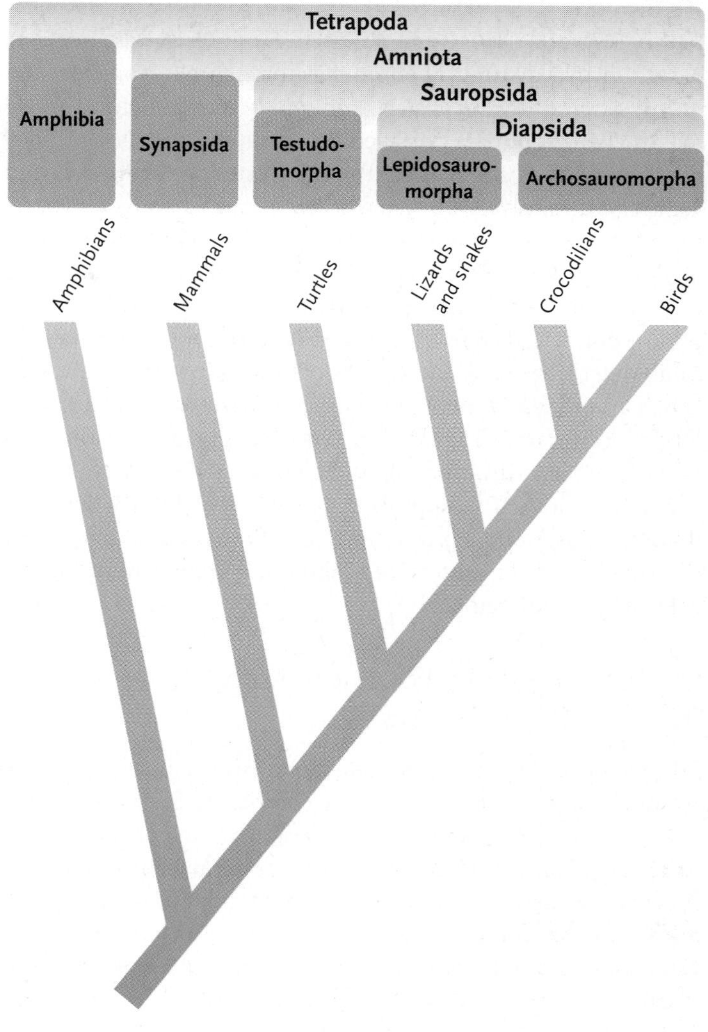
B. Cladogram with classification
Tetrapoda
Amniota
Sauropsida
Diapsida
Amphibia
Synapsida
Testudo-morpha
Lepidosauro-morpha
Archosauromorpha
Amphibians
Mammals
Turtles
Lizards and snakes
Crocodilians
Birds

FIGURE 23.9
Phylogenetic trees and classifications for tetrapod vertebrates. (A) Traditional and **(B)** cladistic approaches produce different phylogenetic trees and classifications. Classifications are presented above the trees. As discussed in Chapter 30, the phylogenetic position of turtles is the subject of much debate. Both parts of this figure follow a traditional view, based on the fossil record, that turtles are not included in the diapsid clade.

For example, the fossil record for tetrapod vertebrates reveals that the amphibian and mammalian lineages each diverged early, followed shortly thereafter by the turtle lineage. (The phylogenetic position of turtles is controversial, as discussed in Chapter 30; here we follow the traditional view based on paleontological evidence.) The remaining organisms then diverged into two groups: one includes living lizards and snakes and another includes living crocodilians and birds. Thus, although crocodilians, with their scaly skin and sprawling posture outwardly resemble lizards, evolutionary biologists have long recognized that crocodilians share a more recent common ancestor with birds.

Even though the phylogenetic tree shows six living clades, the traditional classification recognizes four classes of tetrapod vertebrates: Amphibia, Mammalia, Reptilia, and Aves. These groups are given equal ranking because each represents a distinctive body plan and way of life. The traditionally defined taxon Reptilia is clearly paraphyletic because, even though birds share a common ancestor with reptiles, they are placed in a separate taxon. Traditional systematists justify this definition of Reptilia because it includes morphologically similar animals with close evolutionary relationships. Crocodilians are classified with lizards, snakes, and turtles because they share a distant common ancestry and are covered with dry, scaly skin. Traditional systematists also argued that the key innovations initiating the adaptive radiation of birds—a high metabolic rate, wings, and flight—represent such extreme divergence from the ancestral morphology that birds merit recognition as a separate class.

STUDY BREAK 23.4 <

Why does a classification produced by traditional systematics sometimes include paraphyletic groups?

23.5 The Cladistic Revolution

In the 1950s and 1960s, some researchers criticized classifications based on two distinct phenomena—branching evolution and morphological divergence—as inherently unclear. How can we tell *why* two groups are classified in the same higher taxon? They may have shared a recent common ancestor, as did lizards and snakes. Alternatively, they may have retained some ancestral characteristics after being separated on different branches of a phylogenetic tree, as is the case for lizards and crocodilians.

To avoid such confusion, many systematists followed the philosophical and analytical lead of Willi Hennig, a German entomologist (that is, a scientist who studies insects), who published an influential book, *Phylogenetic Systematics,* in 1950; its English translation appeared in 1966. Hennig and his followers argued that classifications should be based solely on evolutionary relationships. This approach, which is called **cladistics,** produces phylogenetic hypotheses and classifications that reflect only the branching pattern of evolution; it ignores morphological divergence altogether.

Cladistic Analyses Focus on Recently Evolved Character States

Like traditional systematics, cladistics analyzes the evolutionary relationships among organisms by comparing their organismal and, more recently, genetic characteristics. Each **character** can exist in two or more forms, described as **character states.** Evolutionary processes change characters over time from an original, **ancestral character state** to a newer, **derived character state.** Characters that were present in the ancestors of a clade are considered ancestral; those that are new in a descendant are considered derived. For example, ancient fishes, which represent the ancestral vertebrates, had fins; but some of their descendants, the tetrapods, which appeared much later in the fossil record, have limbs. In this example, fins are the ancestral character state, and limbs are the derived character state.

In the jargon of cladistics, a derived character is called an **apomorphy** (*apo* = away from, *morphe* = form), and a derived character found in two or more species is called a **synapomorphy** (*syn* = together). The presence of a synapomorphy (that is, a *shared derived character*) among species provides a clue that they may be members of the same clade. As you learned in Section 22.4, morphological innovations often trigger an adaptive radiation. And once a derived character becomes established in a species, it is likely to be present in that species' descendants. Thus, unless they are lost or replaced by newer characters over evolutionary time, synapomorphies can serve as markers for monophyletic lineages.

Systematists define synapomorphies only when comparing characters among species. Thus, any particular character state is derived *only in relation to* an ancestral character state observed in other organisms—either an older version of the character or its absence. For example, most species of animals lack a vertebral column. However, one animal clade—the vertebrates, including fishes, amphibians, reptiles, birds, and mammals—has that structure. Thus, when systematists compare vertebrates to all other animals, the absence of a vertebral column is the ancestral character and the presence of a vertebral column is derived.

How can systematists distinguish between ancestral and derived characters? In other words, how can they determine the direction in which a character has evolved? The fossil record, if it is detailed enough, can provide unambiguous information. For example, biologists are confident that the presence of a vertebral column is a derived character because fossils of animals that lived before vertebrates lack that structure.

In the absence of evidence from fossils, systematists frequently use a technique called **outgroup comparison** to identify ancestral and derived characters. Using this approach, systematists compare characters in the *ingroup,* the clade under study, to those in an *outgroup,* a closely related species that is not a member of the clade. Character states observed in the outgroup are considered ancestral, and those observed *only* in the ingroup are considered derived. And because the outgroup and the ingroup are phylogenetically related, outgroup comparison allows researchers to hypothesize the root (that is, the common ancestor shared by the outgroup and the ingroup) of the phylogenetic tree. Most modern butterflies, for example, have six walking legs, but species in two families have four walking legs and two small, nonwalking legs. Which is the ancestral character state, and which is derived? Outgroup comparison with other insects, which are not included in the butterfly clade, demonstrates that most insects have six walking legs as adults; this result suggests that six walking legs is ancestral and four is derived **(Figure 23.10).**

Cladistics Uses Synapomorphies to Reconstruct Evolutionary History

Following the cladistic method, biologists construct phylogenetic trees by grouping together species that *share derived characters.* Ancestral characters, because they are shared by the ingroup and the outgroup, do not help to define the ingroup. For example, mammals are a monophyletic lineage—a clade—because they possess a unique set of synapomorphies: hair, mammary glands, and a reduced number of bones in the lower jaw. The ancestral characters found in mammals, such as a vertebral column and four legs, do not distinguish them from other tetrapod vertebrates. Thus, these shared ancestral characters are not useful in defining the mammal clade.

The results of a cladistic analysis are presented in a **cladogram,** a diagram that illustrates the distribution of character states in the organisms being studied and the hypothesized sequence of evolutionary branching that produced them **(Figure 23.9B):** a common ancestor is hypothesized at each node, and every branch portrays a strictly monophyletic group. The synapomorphies that define each clade are sometimes listed on the branches.

Once a researcher identifies a suitable outgroup and the ancestral and derived character states, the manual construction of a cladogram is straightforward **(Figure 23.11).** You will note, when examining Figures 23.9B and 23.11 that the structure of a clado-

FIGURE 23.10
Outgroup comparison. Most adult insects, like **(A)** the caddis fly and **(B)** the orange palm dart butterfly, have six walking legs. This comparison of butterflies with other insects suggests that the four walking legs of **(C)** the monarch butterfly represents the derived character state.

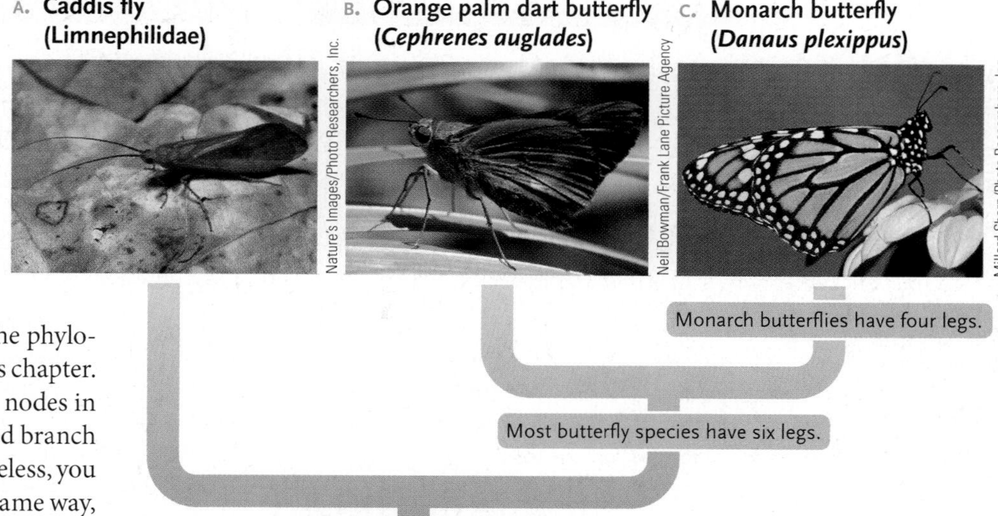
A. Caddis fly (Limnephilidae)
B. Orange palm dart butterfly (Cephrenes auglades)
C. Monarch butterfly (Danaus plexippus)
Nature's Images/Photo Researchers, Inc.
Neil Bowman/Frank Lane Picture Agency
Millard Sharp/Photo Researchers, Inc.
Monarch butterflies have four legs.
Most butterfly species have six legs.
Most insects have six legs.

gram is somewhat different from that of the phylogenetic trees you encountered earlier in this chapter. The major difference between them is that nodes in the cladograms are represented as V-shaped branch points along a diagonal main axis. Nevertheless, you can read the two types of diagrams in the same way, and all of the conventions described above for phylogenetic trees also apply to cladograms.

The use of molecular sequence data in phylogenetic analyses relies on the same logic that underlies analyses based on organismal characters, like those considered above: species included within a clade are expected to exhibit more molecular synapomorphies than do species from different clades. The comparison of sequences in the ingroup to those in the outgroup may allow a researcher to define ancestral and derived character states.

The classifications produced by cladistic analysis often differ radically from those of traditional systematics (Figure 23.9B, compared to Figure 23.9A). Pairs of nested taxa are defined directly from the two-way branching pattern of the cladogram. Thus, the clade Tetrapoda (the traditional amphibians, reptiles, birds, and mammals) is divided into two taxa, the Amphibia (tetrapods that do not have an amnion, as discussed in Section 30.3) and the Amniota (tetrapods that have an amnion). The Amniota is subdivided into two taxa on the basis of skull morphology and other characteristics: Synapsida (mammals) and Sauropsida (turtles, lizards, snakes, crocodilians, and birds). The Sauropsida is further divided into the Testudomorpha (turtles) and the Diapsida (lizards and snakes, crocodilians, and birds). Finally, the Diapsida is subdivided into two more recently evolved taxa—the Lepidosauromorpha (lizards and snakes) and the Archosauromorpha (crocodilians and birds). Thus, a strictly cladistic classification exactly parallels the pattern of branching evolution that produced the organisms included in the classification. These parallels are the essence and strength of the cladistic method.

Most biologists value the evolutionary focus, clear goals, and precise methods of the cladistic approach. In fact, some systematists advocate abandoning the Linnaean hierarchy for classifying and naming organisms. They propose using a strictly phylogenetic system, called **PhyloCode,** that identifies and names clades instead of pigeonholing organisms into traditional taxonomic categories. This approach is further described in *Unanswered Questions* at the end of this chapter. Although traditional systematics has guided many people's understanding of biological diversity, we use a cladistic approach to describe evolutionary lineages and taxa in the Biodiversity unit that follows this chapter.

Systematists Use Several Techniques to Identify an Optimal Cladogram

In practice, most phylogenetic studies are far more complicated than the examples discussed above. Researchers may collect data on hundreds of characters in dozens of species. After scoring the presence or absence of each character in every species, a systematist uses one or more computer programs to generate a set of alternative cladograms. The output of these analyses is often substantial: an analysis of five species can produce 15 possible cladograms; an analysis of 50 species can produce 3×10^{76}!

Faced with such an unimaginably large number of alternative hypotheses, how does a systematist decide which cladogram is the "best" representation of a clade's evolutionary history? This problem is complex, because, when evaluating large data sets, we expect to see some similarities that arise when convergent evolution causes distantly related organisms to evolve similar traits independently; because such traits are not synapomorphies, they are false indicators of relatedness that confound the analysis. We also expect to find some differences between closely related organisms if natural selection or some other microevolutionary process caused a derived trait to be reversed or lost. How can we tell which similarities are synapomorphies and which are homoplasies? Researchers use several approaches to sort through the alternatives, two of which we describe below.

PARSIMONY APPROACH Many systematists adopt a philosophical concept, the **principle of parsimony,** to identify the optimal cladogram. This principle states that the simplest plausible explanation of any phenomenon is the best. If we assume that any complex evolutionary change is an unlikely event, then it is extremely unlikely that the same complex change evolved twice in one lineage. Thus, when the principle is applied to phylogenetic analyses, it suggests that the "best" cladogram is the one that hypothesizes the smallest number of evolutionary changes needed to account for the distribution of character states within a clade; in effect, this approach minimizes the number of homoplasies (that is, the inde-

FIGURE 23.11 **Research Method**

Constructing a Cladogram

Purpose: Systematists construct cladograms to visualize hypothesized evolutionary relationships by grouping together organisms that share derived characters. The derived characters identified in the tan plaques are the synapomorphies that define each clade.

Protocol:

1. *Select the organisms to study.* To demonstrate the method, we develop a cladogram for the nine groups of living vertebrates: lampreys, sharks (and their relatives), bony fishes (and their relatives), amphibians (frogs and salamanders), turtles, lizards (including snakes), crocodilians (including alligators), birds, and mammals. We also include marine animals called lancelets (Chordata, Cephalochordata) as the outgroup. Lancelets are closely related to, but not included within, the vertebrates (see Chapter 30). The inclusion of an outgroup allows biologists to identify ancestral versus derived character states and root the tree.

2. *Choose the characters on which the cladogram will be based.* Our simplified example is based on the presence or absence of 10 characters: (1) vertebral column, (2) jaws, (3) swim bladder or lungs, (4) paired limbs (with one bone connecting each limb to the body), (5) extraembryonic membranes (such as the amnion), (6) mammary glands, (7) dry, scaly skin somewhere on the body, (8) two openings on each side near the back of the skull, (9) one opening on each side of the skull in front of the eye, and (10) feathers.

3. *Score the characters in each group.* Because lancelets serve as the outgroup in this analysis, we consider character states observed in lancelets as ancestral; any deviation from the lancelet pattern is considered derived. Because lancelets lack all of the characters in our analysis, the presence of each character is the derived condition. We tabulate data on the distribution of ancestral (−) and derived (+) characters in all species included in the analysis.

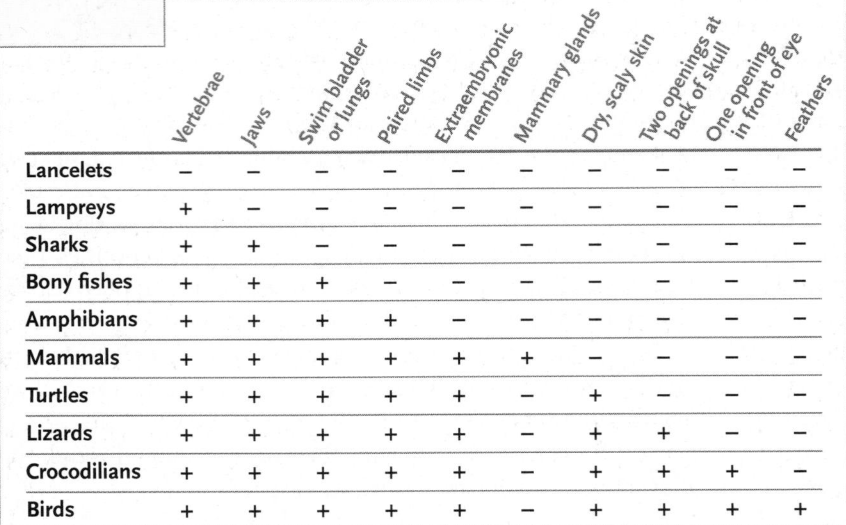

	Vertebrae	Jaws	Swim bladder or lungs	Paired limbs	Extraembryonic membranes	Mammary glands	Dry, scaly skin	Two openings at back of skull	One opening in front of eye	Feathers
Lancelets	−	−	−	−	−	−	−	−	−	−
Lampreys	+	−	−	−	−	−	−	−	−	−
Sharks	+	+	−	−	−	−	−	−	−	−
Bony fishes	+	+	+	−	−	−	−	−	−	−
Amphibians	+	+	+	+	−	−	−	−	−	−
Mammals	+	+	+	+	+	+	−	−	−	−
Turtles	+	+	+	+	+	−	+	−	−	−
Lizards	+	+	+	+	+	−	+	+	−	−
Crocodilians	+	+	+	+	+	−	+	+	+	−
Birds	+	+	+	+	+	−	+	+	+	+

4. *Construct the cladogram from information in the table, grouping organisms that share derived characters.* All groups except lancelets have vertebrae. Thus, we group organisms that share this derived character on the right-hand branch, identifying them as a monophyletic lineage. Lancelets are on their own branch to the left, indicating that they lack vertebrae.

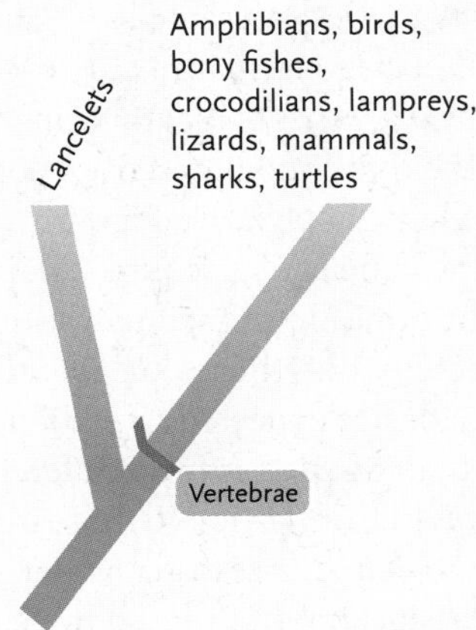

Lancelets
Amphibians, birds, bony fishes, crocodilians, lampreys, lizards, mammals, sharks, turtles
Vertebrae

All of the remaining organisms except lampreys have jaws. (Lancelets also lack jaws, but we have already separated them out, and do not consider them further.) Place all groups with jaws, a derived character, on the right-hand branch. Lampreys are separated out to the left, because they lack jaws. Again, the branch on the right represents a monophyletic lineage.

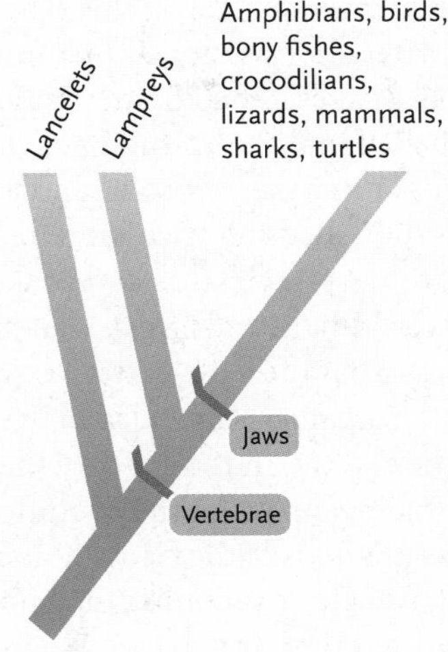

Lancelets
Lampreys
Amphibians, birds, bony fishes, crocodilians, lizards, mammals, sharks, turtles
Jaws
Vertebrae

pendent evolution of similar traits) in the cladogram **(Figure 23.12)**. To apply the principle, computer programs evaluate the number of evolutionary changes hypothesized by each cladogram they generate, and the researcher identifies the one with the fewest hypothesized changes as the most plausible.

The principle of parsimony also allows researchers to identify homologous characters and infer their ancestral and derived states. Once the most parsimonious cladogram is identified, a researcher can visualize the distribution of derived character states and pinpoint when the derived state evolved.

5. *Construct the rest of the cladogram using the same step-by-step procedure to separate the remaining groups.* In our completed cladogram, seven groups share a swim bladder or lungs; six share paired limbs; and five have extraembryonic membranes during development. Some groups are distinguished by the unique presence of a derived character, such as feathers in birds.

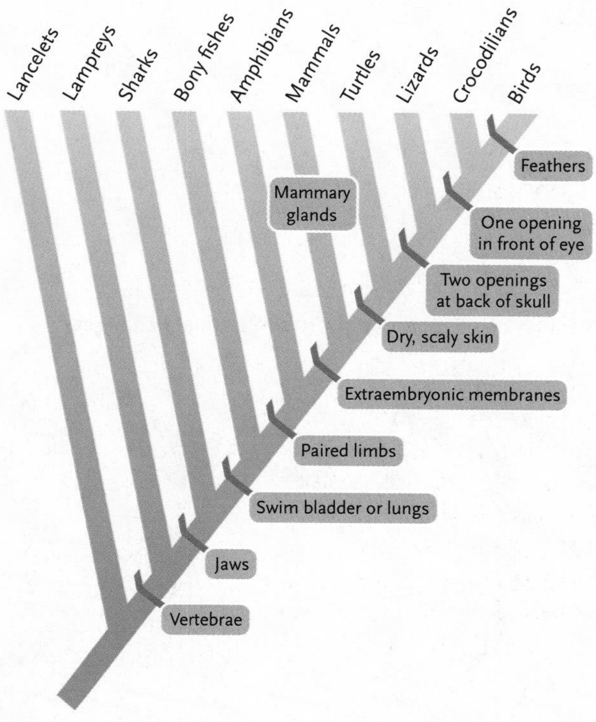
Lancelets
Lampreys
Sharks
Bony fishes
Amphibians
Mammals
Turtles
Lizards
Crocodilians
Birds
Feathers
Mammary glands
One opening in front of eye
Two openings at back of skull
Dry, scaly skin
Extraembryonic membranes
Paired limbs
Swim bladder or lungs
Jaws
Vertebrae

Interpreting the Results: Although cladograms provide information about evolutionary relationships, the common ancestors represented by the branch points are often hypothetical. You can tell from the cladogram, however, that birds are more closely related to lizards than they are to mammals. Follow the branches of the cladogram from birds and lizards back to their node. Next, trace the branches of birds and mammals to their node. You can see that the bird–mammal node is closer to the root of the cladogram than the bird–lizard node is. Nodes that are closer to the bottom of the cladogram indicate a more distant common ancestry than those closer to the top. Note also that this simplified example produces a cladogram that is easy to interpret because the data set includes no homoplastic similarities and no conflicting evidence. Most phylogenetic analyses include many such complications.

STATISTICAL APPROACHES Researchers have faulted the parsimony approach, especially for cladograms based on molecular sequence data. Given that there are only four possible character states at each position in a nucleic acid, identical changes in nucleotides often arise independently. To avoid this problem, systematists develop statistical models of evolutionary change that take into account variations in the evolutionary rates at different nucleotide positions or in different genes or species as well as changes in evolutionary rates over time.

For example, segments of DNA that do not code for proteins are less likely than coding regions to be affected by natural selection. As a result, mutations accumulate faster in noncoding regions, causing them to evolve rapidly. Similarly, because of the degeneracy of the genetic code (described in Section 15.1), mutations in the third codon position do not often influence the amino acid composition of the protein for which a gene codes. As a result, third codon mutations are often selectively neutral (see Section 20.4), and they accumulate more rapidly than do mutations in the first or second positions. Finally, certain nucleotide substitutions are more common than others: transitions (the substitution of a purine for another purine or a pyrimidine for another pyrimidine) occur more frequently than transversions (substitutions between purines and pyrimidines).

Two statistical approaches, maximum likelihood methods and Bayesian methods, allow researchers to compare cladograms by accounting for variation in the rates at which molecular sequences evolve. Although the details of these methods are beyond the scope of an introductory biology text, we can describe their rationale in general terms. In **maximum likelihood methods,** the alternative trees are compared with specific models of evolutionary change, and the cladogram that is most likely to have produced the observed distribution of character states is identified as the best hypothesis. With **Bayesian methods,** researchers evaluate a given cladogram by determining the probability that it is correct given the distribution of character states and the assumptions of the evolutionary model. Although computationally intensive, these approaches have gained favor among systematists because they account for variations in the rate at which different molecular characters evolve; parsimony approaches do not address those issues.

STUDY BREAK 23.5

1. **How does outgroup comparison facilitate the identification of ancestral and derived character states?**
2. **What characteristics are used to group organisms in a cladistic analysis?**
3. **How is the principle of parsimony applied in phylogenetic analyses?**

THINK OUTSIDE THE BOOK

Access the web page for the Tree of Life project at http://www.tolweb.org/tree/. Select a group of animals or plants that is of interest to you, and study the structure of its phylogenetic tree. How many major clades does it include? On the basis of what shared derived characters are those clades defined?

A. **Distribution of character states in six clades of vascular plants**

Characters (possible states)	Ferns (outgroup)	Gnetophytes	Cycads	Ginkgophytes	Conifers	Angiosperms
Archegonium (present or lost)	Present	Lost	Present	Present	Present	Lost
Double fertilization (absent or present)	Absent	Present	Absent	Absent	Absent	Present
Pollen tube growth (haustorium or tube)	Absent	Tube	Haustorium	Haustorium	Tube	Tube
Sperm flagella (present or lost)	Present	Lost	Present	Present	Lost	Lost
Vessels (absent or present)	Absent	Present	Absent	Absent	Absent	Present

B. **Phylogenetic tree hypothesizing five evolutionary changes**

C. **Phylogenetic tree hypothesizing 10 evolutionary changes**

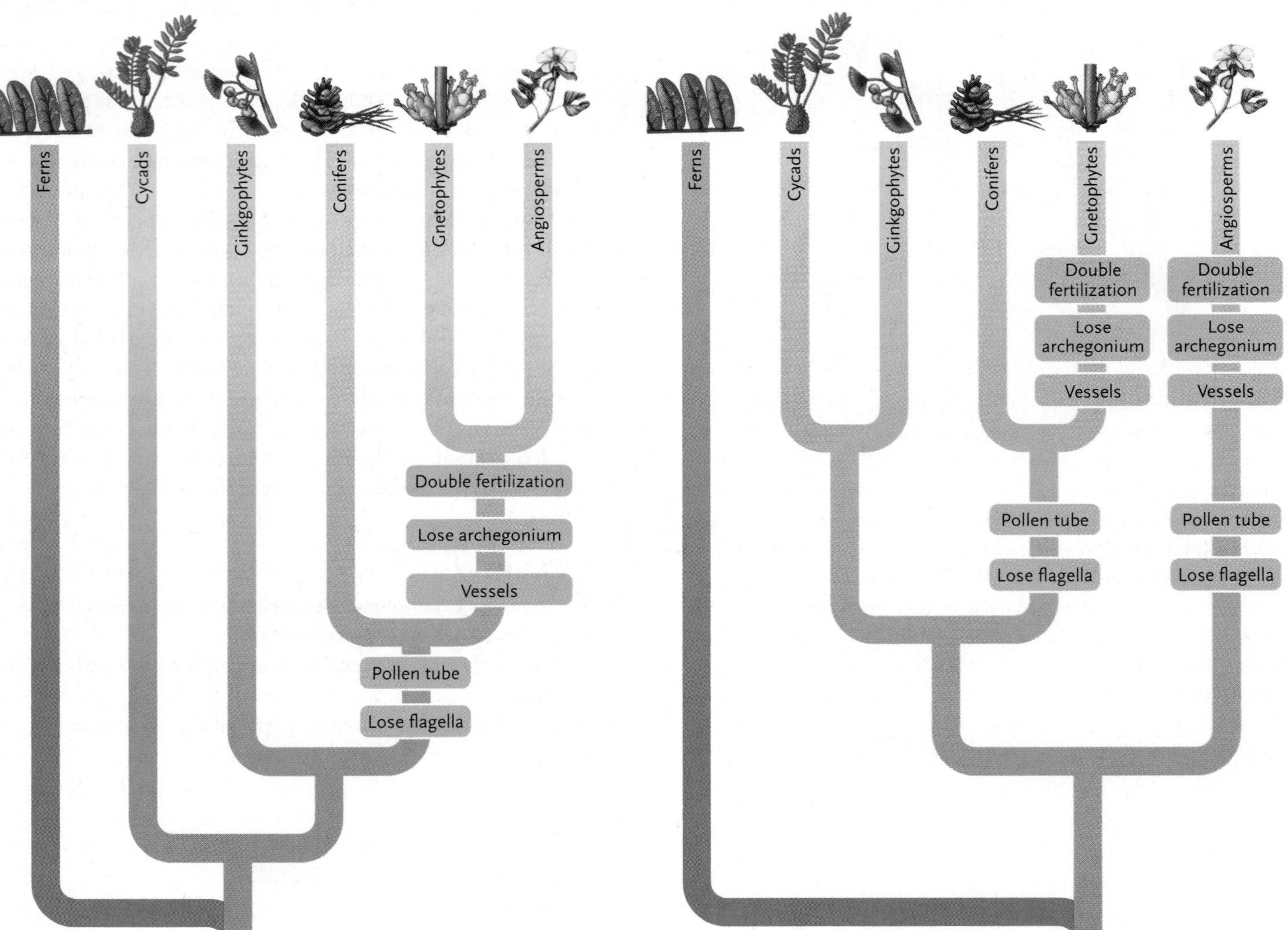
Ferns
Cycads
Ginkgophytes
Conifers
Gnetophytes
Angiosperms
Double fertilization
Lose archegonium
Vessels
Pollen tube
Lose flagella
Ferns
Cycads
Ginkgophytes
Conifers
Gnetophytes
Angiosperms
Double fertilization
Lose archegonium
Vessels
Double fertilization
Lose archegonium
Vessels
Pollen tube
Lose flagella
Pollen tube
Lose flagella

FIGURE 23.12

The principle of parsimony. **(A)** A table presents the distribution of morphological character states in six clades of vascular plants. Ferns represent the outgroup, and all of its character states are considered ancestral. Two alternative phylogenetic trees for the six clades illustrate different hypotheses about their evolutionary relationships. Bars across the branches mark the hypothesized evolution of derived character states in both trees. Based on this set of characters, the phylogenetic tree with five hypothesized evolutionary changes **(B)** is more parsimonious than the tree that hypothesizes 10 evolutionary changes **(C).** In the absence of additional data, a systematist would accept the more parsimonious tree as the best working hypothesis. (The characters and clades of vascular plants are discussed in detail in Chapter 27.)

23.6 Phylogenetic Trees as Research Tools

In addition to providing a wealth of information about the patterns of branching evolution across the entire spectrum of living organisms, phylogenetic trees are useful tools that facilitate research in all areas of biology.

Molecular Clocks Estimate the Time of Evolutionary Divergences

Although many biological molecules have been conserved by evolution, different adaptive changes and neutral mutations accumulate in separate lineages from the moment they first diverge. Because mutations that arise in non-coding regions of DNA do not affect protein structure, they are probably not often eliminated by natural selection. If mutations accumulate in these segments at a reasonably constant rate, differences in their DNA sequences can serve as a **molecular clock,** indexing the time at which two species diverged. Large differences imply divergence in the distant past, whereas small differences suggest a more recent common ancestor.

Because different molecules exhibit individual rates of evolutionary change, every molecule is an independent clock, ticking at its own rate. Researchers study different molecules to track evolutionary divergences that occurred at different times in the past. For example, mitochondrial DNA (mtDNA) evolves relatively quickly; it is useful for dating evolutionary divergences that occurred within the last few million years. Studies of mtDNA have illuminated aspects of the evolutionary history of humans, as described in Section 30.13. By contrast, chloroplast DNA (cpDNA) and genes that encode ribosomal RNA evolve much more slowly, providing information about divergences that date back hundreds of millions of years.

To calibrate molecular clocks, researchers examine the degree of genetic difference between species in relation to their time of divergence estimated from the fossil record. Alternatively, the clock can be calibrated biogeographically with independent data on when volcanic islands first emerged from the sea or when landmasses separated **(Figure 23.13).**

The reliability of molecular clocks depends on the constancy of evolutionary change in the DNA segment analyzed. Some researchers have noted that even DNA segments that are thought to be selectively neutral may show variable rates of evolution. Many factors can influence the rates at which mutations accumulate, and researchers must be cautious when evaluating divergence times estimated with this technique, especially if there are no independent data to corroborate the estimates.

Phylogenetic Trees Allow Biologists to Propose and Test Hypotheses

Accurate phylogenetic trees are essential tools for analyses that biologists describe as the "comparative method." With this approach researchers compare the characteristics of different species to assess the homology of their similarities and infer where on the phylogenetic tree a particular trait appeared. The comparative method is used to study almost any sort of organismal trait, but in this section we will focus on parental care behavior.

As noted above, birds and crocodilians are included within the Archosauria, a clade that also includes non-avian dinosaurs (that is, extinct terrestrial dinosaurs that are not included within the bird clade), pterosaurs (an extinct group of flying vertebrates, not closely related to bats, with wing surfaces formed by skin), and a number of other groups that became extinct in the early Mesozoic era. Crocodilians and birds share certain anatomical characteristics, such as a four-chambered heart and the one-way flow of air through their lungs. They also share behavioral characteristics, including the production of mating calls (songs in birds and roars in crocodilians), nest-building behavior, and parental care of their young. Female crocodilians guard their nests and keep them moist with urine. They also excavate the nest as the young hatch, and then carry them to standing water. Young stay with their mother for about a year, feeding on scraps of food that fall from her mouth.

Did similar parental care behavior evolve independently in birds and crocodilians, or is it truly a synapomorphy? Did most Mesozoic archosaurs care for their young as birds and crocodilians do today? The comparative method seeks answers to these questions by examining the phylogenetic tree for archosaurs **(Figure 23.14).** As you can readily see, crocodilians and birds lie on widely separated branches of the archosaur tree, with pterosaurs and non-avian dinosaurs positioned between them. The most parsimonious inference about the evolution of parental care behavior is that it evolved once in the common ancestor of crocodilians and birds. If that inference is correct, then non-avian dinosaurs and pterosaurs probably also cared for their young in the nest. Indeed, that prediction was confirmed in 1995, when Mark A. Norell of George Washington University and his colleagues discovered a fossil of a non-avian dinosaur *(Oviraptor)* sitting on a nest full of eggs!

Phylogenetic Analyses Help Track the Origin and Spread of Infectious Diseases

Phylogenetic analyses also allow physicians and public health workers to identify the origin of infectious agents and follow their spread through a population. Many pathogenic organisms and viruses mutate as they proliferate, establishing derived character states that are ripe for phylogenetic analysis.

The human immunodeficiency virus (HIV), the agent that causes acquired immunodeficiency syndrome (AIDS) in humans, began to infect large numbers of people in the 1980s. As its devastating effects on humans became apparent, scientists scrambled to discover its origin. Genetic analyses linked it to the lentiviruses, specifically simian immunodeficiency virus (SIV), which infects dozens of monkey species as well as chimpanzees in Africa. Surprisingly, SIV does not cause illness in those animals, perhaps because their populations developed immunity to it after a long period of exposure.

Two distinct strains of HIV infect humans: HIV-1 is common in central Africa, and HIV-2 is common is West Africa. Did these strains evolve within human hosts, or did they exist before the virus was first transmitted to humans? An analysis by Beatrice H. Hahn of the University of Alabama at Birmingham and

FIGURE 23.13 **Observational Research**

Do Molecular Clocks Tick at a Constant Rate?

Hypothesis: Some DNA segments accumulate mutations at a constant rate, allowing researchers to use genetic differences between species as molecular clocks.

Prediction: The genetic differences observed between pairs of species will be proportional to the time since they diverged from their common ancestor.

Method: Robert C. Fleischer of the Smithsonian Institution and colleagues at the Smithsonian and Pennsylvania State University compared gene sequences among pairs of related bird species (honeycreepers) and fruit fly species *(Drosophila)* on different Hawaiian Islands. They used radiometric dating of volcanic rocks to estimate the age of each island and used the estimates of island age to measure the divergence time between pairs of species. This approach assumes that birds and flies founded populations on each new island shortly after the island's emergence from the Pacific Ocean and that newly founded populations began to diverge from ancestral populations immediately. Thus, if a population of *Drosophila* now lives on Oahu, the researchers estimate that it diverged from its ancestor 1.6 million years ago, which is the estimated time of Oahu's origin. The researchers plotted a summary statistic of genetic differences between species pairs versus estimated island age to test their hypothesis about constant rates of evolution in the DNA segments they studied.

Results: The extent of genetic difference in the cytochrome *b* sequences of Hawaiian honeycreepers and in the *Yp1* gene sequences of Hawaiian fruit flies is strongly correlated with the time since the members of each species pair were separated. The results indicate that these DNA sequences have evolved at constant rates.

A. **Hawaiian Islands (with time of origin)**

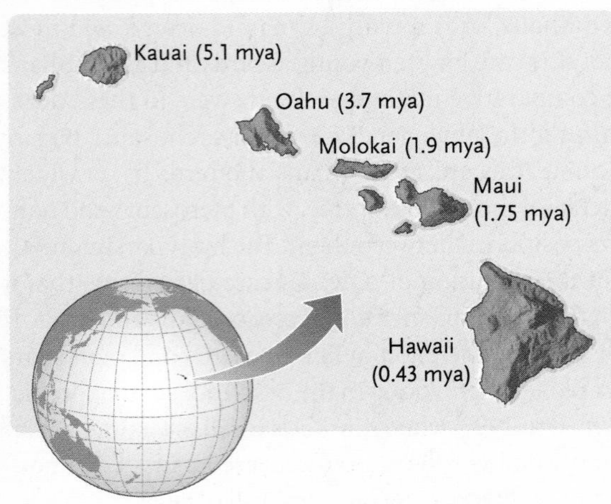

Kauai (5.1 mya)
Oahu (3.7 mya)
Molokai (1.9 mya)
Maui
(1.75 mya)
Hawaii
(0.43 mya)

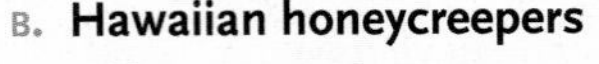
B. **Hawaiian honeycreepers**

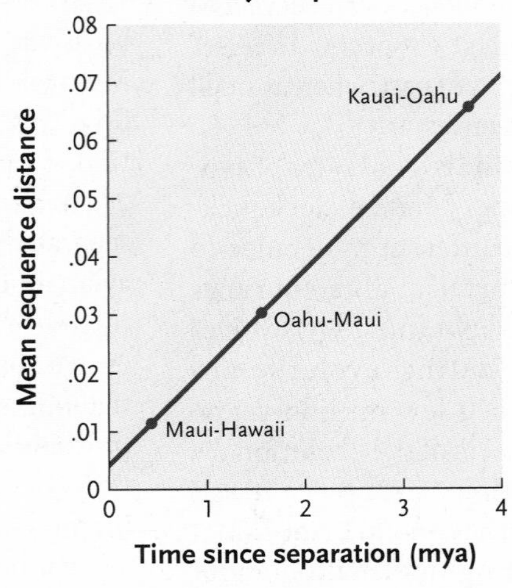

Mean sequence distance
.08
.07
.06
.05
.04
.03
.02
.01
0
Kauai-Oahu
Oahu-Maui
Maui-Hawaii
0
1
2
3
4
Time since separation (mya)

C. **Hawaiian *Drosophila***

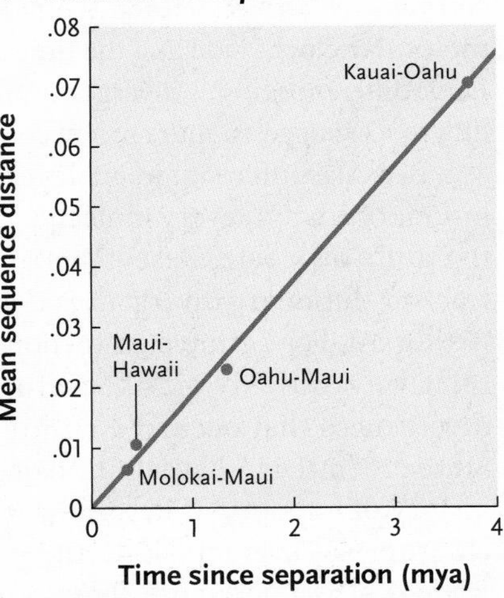

Mean sequence distance
.08
.07
.06
.05
.04
.03
.02
.01
0
Kauai-Oahu
Maui-
Hawaii
Oahu-Maui
Molokai-Maui
0
1
2
3
4
Time since separation (mya)

Conclusion: Because the gene sequences evolve at a constant rate, they can be used as molecular clocks. Each 0.016 units of genetic difference in the cytochrome *b* sequence in Hawaiian honeycreepers and each 0.019 units of genetic difference in the *Yp1* sequence in Hawaiian *Drosophila* indexes 1 million years of independent evolution.

Source: R. C. Fleischer et al. 1998. Evolution on a volcanic conveyor belt: using phylogeographic reconstructions and K–Ar-based ages of the Hawaiian Islands to estimate molecular evolutionary rates. *Molecular Ecology* 7:533–545.

the Howard Hughes Medical Institute and colleagues at other institutions identified three major clades of SIV. The clade that infects chimpanzees includes HIV-1, and one of the clades that infect monkeys includes HIV-2 **(Figure 23.15).** Thus, the two strains of HIV apparently originated in non-human hosts. Scientists suspect that the transmission to humans occurred multiple times when hunters who were butchering bush meat—chimpanzees in central Africa and sooty mangabey monkeys in West Africa—acquired the virus through cuts on their hands.

STUDY BREAK 23.6

1. **What assumption underlies the use of genetic sequence differences between species as a molecular clock?**
2. **Are birds more closely related to non-avian dinosaurs or to crocodilians?**

23.7 Molecular Phylogenetic Analyses

The application of molecular techniques to phylogenetic analyses has allowed systematic biologists to resolve some long-standing evolutionary puzzles. In this section we describe the identification of the most ancestral flowering plant and discoveries about the shape of the entire Tree of Life.

Molecular Phylogenetics Has Identified the Most Ancestral Angiosperm

The origin of angiosperms, the flowering plants, which Darwin described as an "abominable mystery," confounded evolutionary biologists until the end of the twentieth century. Although no one doubted that angiosperms were a monophyletic lineage, morphological analyses of flowers and other structures had produced conflicting hypotheses about their origin and relationships. In 1999, four teams of researchers, analyzing different parts of flowering plant genomes, independently identified *Amborella trichopoda,* a bush native to the South Pacific island of New Caledonia, as a living representative of the lineage that is closest to the root of the angiosperm phylogenetic tree; *Amborella* is therefore considered the most "primitive" angiosperm. But those analyses were based on relatively small numbers of genes, and the statistical support for the phylogenetic position of *Amborella* was weak. Moreover, the analyses did not clearly resolve the phylogenetic positions of several other ancient clades of flowering plants, which are collectively described as the "basal angiosperms" because their clades emerged near the base of the phylogenetic tree.

In 2007, Robert K. Jansen at the University of Texas at Austin and 15 colleagues in the United States and Germany published an ambitious phylogenetic study of 76,583 nucleotide positions in 81 plastid genes from the genomes of 64 plant species. Three species of gymnosperms, which are not flowering plants—*Pinus* (pine), *Ginkgo* (ginkgo), and *Cycas* (cycad)—served as the outgroup. The researchers used their data to generate alternative phylogenetic hypotheses. They evaluated these hypotheses, along with six other

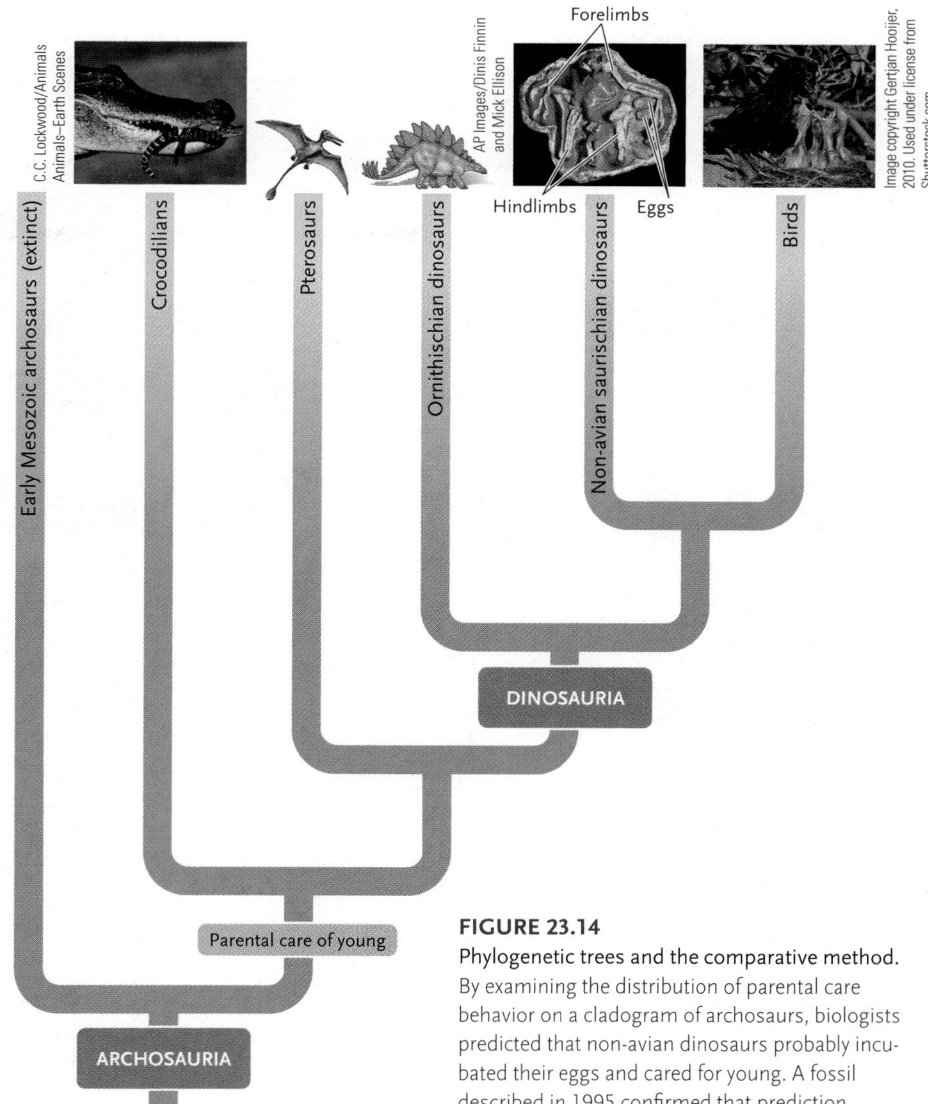

C.C. Lockwood/Animals Animals–Earth Scenes
AP Images/Dinis Firmin and Mick Ellison
Image copyright Gertjan Hooijer, 2010. Used under license from Shutterstock.com
Forelimbs
Hindlimbs
Eggs
Early Mesozoic archosaurs (extinct)
Crocodilians
Pterosaurs
Ornithischian dinosaurs
Non-avian saurischian dinosaurs
Birds
DINOSAURIA
Parental care of young
ARCHOSAURIA

FIGURE 23.14
Phylogenetic trees and the comparative method. By examining the distribution of parental care behavior on a cladogram of archosaurs, biologists predicted that non-avian dinosaurs probably incubated their eggs and cared for young. A fossil described in 1995 confirmed that prediction.

FIGURE 23.15
Phylogenetic trees and public health. A phylogenetic tree for strains of simian immunodeficiency virus (SIV) and human immunodeficiency virus (HIV) suggests that the virus was transmitted to humans independently by chimpanzees and sooty mangabey monkeys.

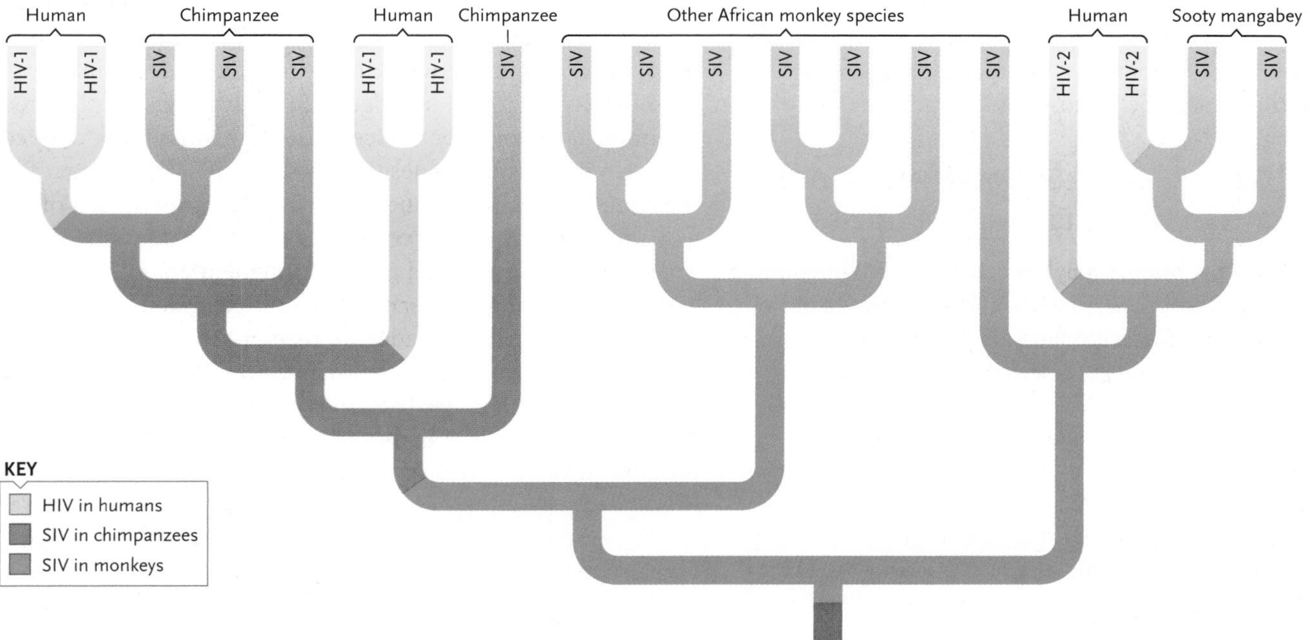

Human
Chimpanzee
Human
Chimpanzee
Other African monkey species
Human
Sooty mangabey
HIV-1
HIV-1
SIV
SIV
SIV
HIV-1
HIV-1
SIV
SIV
SIV
SIV
SIV
SIV
SIV
SIV
HIV-2
HIV-2
SIV
SIV
KEY
HIV in humans
SIV in chimpanzees
SIV in monkeys

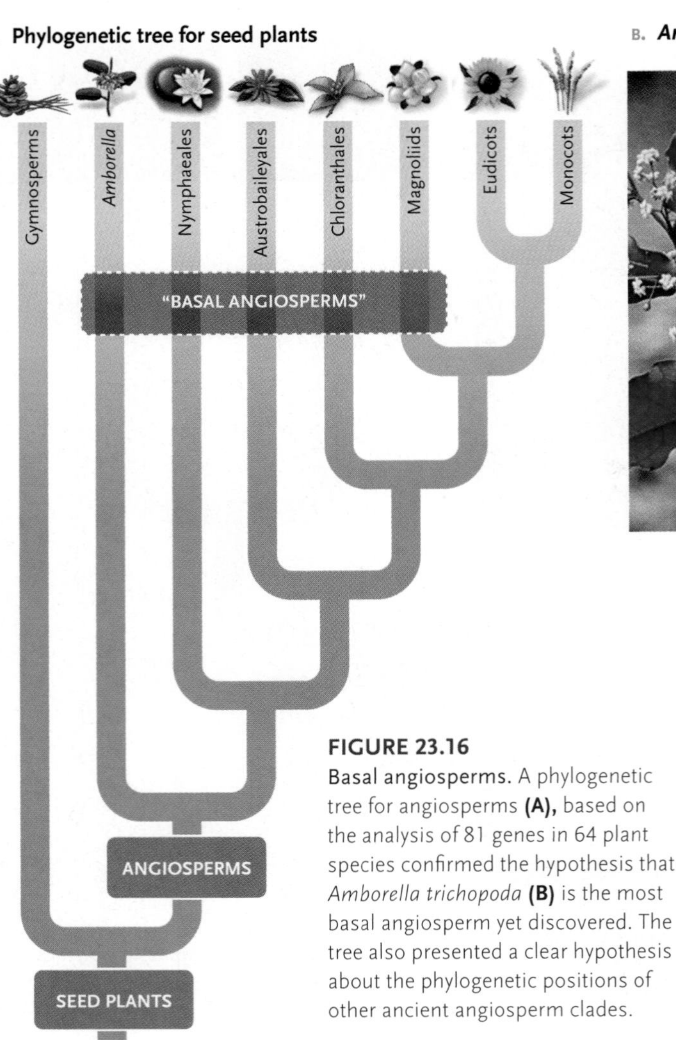
A. Phylogenetic tree for seed plants
Gymnosperms
Amborella
Nymphaeales
Austrobaileyales
Chloranthales
Magnoliids
Eudicots
Monocots
"BASAL ANGIOSPERMS"
ANGIOSPERMS
SEED PLANTS

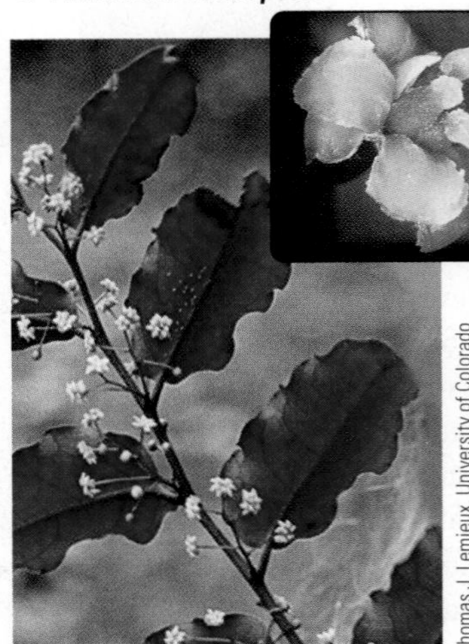
B. Amborella trichopoda
Sandra Floyd, University of Colorado
Thomas J. Lemieux, University of Colorado

FIGURE 23.16
Basal angiosperms. A phylogenetic tree for angiosperms **(A)**, based on the analysis of 81 genes in 64 plant species confirmed the hypothesis that *Amborella trichopoda* **(B)** is the most basal angiosperm yet discovered. The tree also presented a clear hypothesis about the phylogenetic positions of other ancient angiosperm clades.

previously published hypotheses, using parsimony, maximum likelihood, and Bayesian criteria. Their statistical analyses provided the strongest support for a phylogenetic tree that confirmed *Amborella*'s position closest to its root, and clearly indicated the relationships among the other basal angiosperms **(Figure 23.16).** This analysis and other studies like it allow researchers to use a well-supported angiosperm phylogeny to explore the evolution of specific angiosperm characters.

Analyses of Gene Sequences Have Revealed the Branching Pattern of the Entire Tree of Life

On a very grand scale, molecular phylogenetics has revolutionized our view of the entire Tree of Life. The first efforts to create a phylogenetic tree for all forms of life were based on morphological analyses. However, these analyses did not resolve the branches of the Tree containing prokaryotes, which lack significant structural variability for analysis, or the relationships of those branches to eukaryotes.

In the 1960s and early 1970s, biologists organized living systems into five kingdoms. All prokaryotes were grouped into the kingdom Monera. The eukaryotic organisms were grouped into four kingdoms: Fungi, Plantae, Animalia, and "Protista." The kingdom "Protista" was always recognized as a polyphyletic "grab bag" of unicellular or acellular organisms (discussed further in Chapter 26). Unfortunately, phylogenetic analyses based on morphology were unable to sort them into distinct evolutionary lineages.

In the 1970s, biologists realized that molecular phylogenetics provides an alternative approach. They simply needed to identify and analyze molecules that have been conserved by evolution over billions of years. Carl R. Woese, a microbiologist at the University of Illinois at Urbana–Champaign, identified the small subunit of ribosomal RNA (rRNA) as a suitable molecule for analysis. Ribosomes, the structures that translate messenger RNA molecules into proteins (see Section 15.1), are remarkably similar in all forms of life. They are apparently so essential to cellular processes that the genes specifying ribosomal structure exhibit similarities in their nucleotide sequences in organisms as different as bacteria and humans. Thus, it is possible to compare the sequences of these genes in a phylogenetic analysis.

The phylogenetic tree based on rRNA sequences divides living organisms into three primary lineages called domains—Bacteria, Archaea, and Eukarya—which differ in many molecular and cellular characters (**Figure 23.17** and **Table 23.1**). According to this hypothesis, two domains, Bacteria and Archaea, consist of prokaryotic organisms, and one, Eukarya, consists of eukaryotes. Bacteria includes well-known microorganisms. Archaea includes microorganisms that live in physiologically harsh environments, such as hot springs or very salty habitats, as well as less extreme environments. Eukarya includes the familiar animals, plants, and fungi, as well as the many lineages formerly included among the "Protista," which is not a monophyletic group. As the Tree in Figure 23.17 suggests, Archaea and Eukarya are more closely related to each other than either domain is to Bacteria. The next unit of this book is devoted to detailed analyses of the biology and evolutionary relationships between and within these three domains.

Our discussion of phylogenetic analysis has emphasized the importance of direct descent: the transmission of derived traits from ancestors to descendants. But as scientists analyze the complete genome sequences of an ever-growing list of organisms, they are discovering that the three domains in the Tree of Life have not had entirely independent evolutionary histories. Although inheritance from one generation to the next has produced the clades we recognize today, **horizontal gene transfer,** the movement of genetic material from one clade to another, has also been important (discussed further in Chapters 25 and 26). The mechanisms that facilitate horizontal gene transfer include transformation and transduction (discussed in Section 17.1), viral infection, and the incorporation of one organism into another. For example, the genes that regulate metabolism in eukaryotes are more similar to those in Bacteria than to those in Archaea, leading researchers to

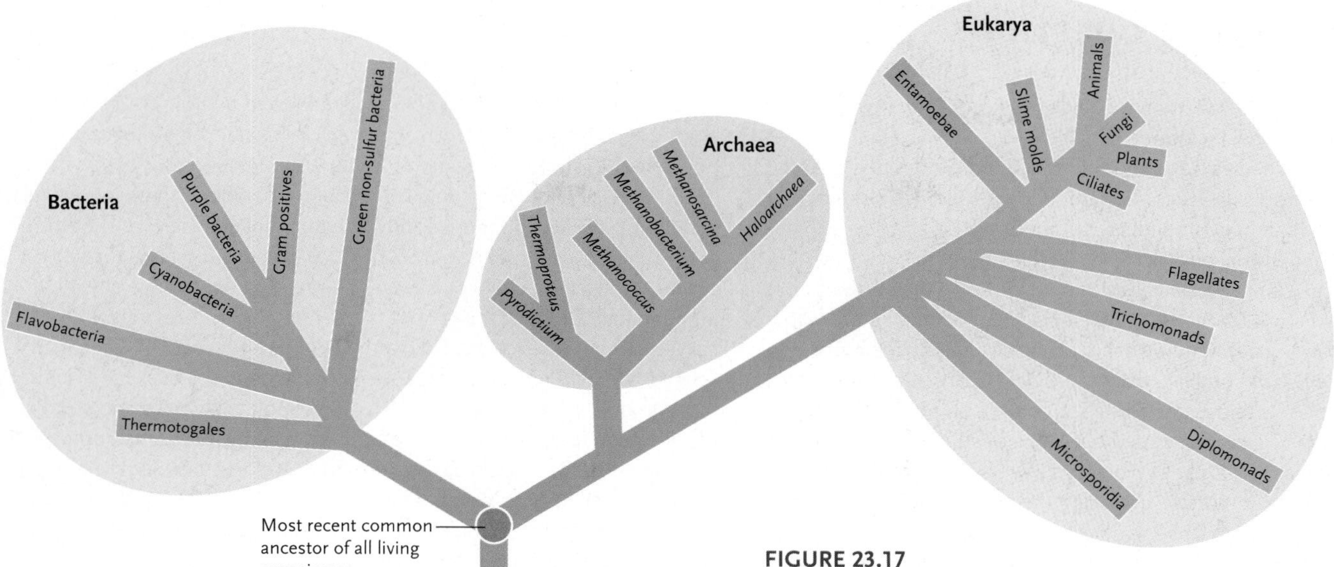
Bacteria
Green non-sulfur bacteria
Gram positives
Purple bacteria
Cyanobacteria
Flavobacteria
Thermotogales
Archaea
Methanosarcina
Methanobacterium
Haloarchaea
Thermoproteus
Methanococcus
Pyrodictium
Eukarya
Entamoebae
Slime molds
Animals
Fungi
Plants
Ciliates
Flagellates
Trichomonads
Microsporidia
Diplomonads
Most recent common ancestor of all living organisms

FIGURE 23.17
Three domains in the Tree of Life. Carl R. Woese's analysis of rRNA sequences suggests that all living organisms can be classified into one of three domains: Bacteria, Archaea, and Eukarya.

hypothesize that the mitochondria in eukaryotic cells may have originated when bacterial genes, or entire bacteria, were transferred horizontally into an ancestral eukaryote. The chloroplasts of plants may well have originated in a similar way **(Figure 23.18).** Thus, a more accurate portrait of the evolutionary history of life on Earth may require network-like connections between the three major trunks on the Tree of Life. Researchers are hard at work analyzing these connections.

TABLE 23.1 Some Differences among the Three Domains

Character	Bacteria	Archaea	Eukarya
Chromosome structure	Circular	Circular	Linear
DNA location	Nucleoid	Nucleoid	Nucleus
Chromosome segregation	Binary fission	Binary fission	Meiosis/mitosis
Introns in genes	Rare	Common	Common
Operons	Present	Present	Absent
Initiator tRNA	Formylmethionine	Methionine	Methionine
Ribosomes	70S	70S	80S
Membrane-enclosed organelles	Absent	Absent	Present
Membrane lipids	Ester-linked	Ether-linked	Ester-linked
Peptidoglycan in cell wall	Present	Absent	Absent
Methanogenesis	Absent	Present	Absent
Temperature tolerance	Up to 90°C	Up to 120°C	Up to 70°C

STUDY BREAK 23.7

1. **Which group of organisms is the sister clade of angiosperms in the phylogenetic tree in Figure 23.16?**
2. **Why was a phylogenetic analysis of prokaryotes based on molecular sequence data more successful than the analysis based on morphological data?**

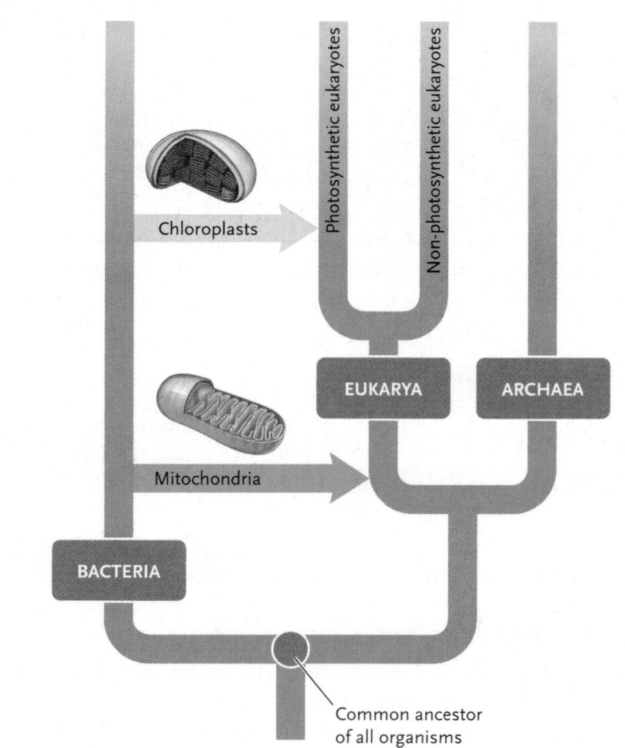
Photosynthetic eukaryotes
Non-photosynthetic eukaryotes
Chloroplasts
EUKARYA
ARCHAEA
Mitochondria
BACTERIA
Common ancestor of all organisms

FIGURE 23.18
Horizontal gene transfer. Molecular analyses suggest that horizontal gene transfer from bacteria brought mitochondria to an ancestral eukaryote. A later transfer brought chloroplasts into the cells of some eukaryotes.

UNANSWERED QUESTIONS

Should we abandon the traditional Linnaean hierarchy in favor of a more evolutionary classification?

Diligent Kindly Professors Cannot Often Fail Good Students—or some equally silly mnemonic device for remembering the Linnaean taxonomic hierarchy—is all that many students recall about systematics. Even if they remember the underlying rank names—is *G* for "group" or "genus"—they often forget that Linnaeus conceived his system of classification more than a century before Darwin articulated his theory of evolution, which revolutionized our understanding of biological diversity. In the more than 150 years since Darwin published *On the Origin of Species,* systematists have sought not only to categorize life's diversity but, more importantly, to understand its origins. The broad relevance of studies in systematics has become increasingly clear as biologists have discovered that systematic principles are as important in tracing the emergence and spread of HIV as they are in distinguishing a duck from a dove.

A century and a half after it was first published, Darwin's theory has an increasingly revolutionary impact. Perhaps the most striking recent example is a call for the complete abandonment of the Linnaean taxonomic hierarchy. Although biologists thought they had reconciled the perspectives of Darwin and Linnaeus, a growing minority of systematists now believe that any effort to catalog and categorize life's diversity must be explicitly phylogenetic and free of the arbitrary ranks that Linnaeus invented. This movement, which has been codified in the PhyloCode initiative, is fueled largely by newly available molecular data, vastly improved phylogenetic methods, and increasingly fast computers. These advances offer the potential to reconstruct accurate and fully resolved phylogenetic trees at a scale never before possible. For the first time, biologists see real progress in accurately reconstructing the entire Tree of Life. Although we are still far from achieving this goal, every day millions of new, phylogenetically informative DNA fragments are being sequenced and analyzed by thousands of computers running around the clock.

Although PhyloCode's synthesis of taxonomy and evolutionary systematics may be long overdue, this attempted coup is not without controversy. For example, some systematists contend that such a radical revision of our taxonomic system will introduce confusion and instability in the naming of species. Even the revolution's adherents recognize that we still face many challenging limitations to the synthesis between taxonomic practice and Darwinian principles. Nowhere is this more evident than in the definition of species.

During Linnaeus' time, species were viewed as immutable natural types created by God. Darwin, however, formulated his theory on the principle that species change over time. Although the truth of this basic hypothesis is no longer a subject of debate, its practical implications for delimiting species boundaries and understanding how new species form are among the most exciting areas of study in modern systematics. Most practicing systematists view species as real (that is, biologically meaningful) categories, but the criteria for recognizing species vary dramatically among systematists working on different types of organisms (plants versus animals, or organisms that reproduce asexually versus those that reproduce sexually). Using new molecular tools and sophisticated genetic experiments, evolutionary biologists are beginning to probe the precise genetic basis of species. Over the past decade a small number of "speciation genes" have been identified; more such discoveries are sure to follow in the coming years. Although many of these studies have been restricted to model research organisms, such as fruit flies, the new tools offered by the fields of genomics and bioinformatics offer the potential to address similar questions in an increasingly broad array of organisms.

Simply put, the systematics of today is not that of your grandparents. Given the enormous challenge involved in categorizing and understanding the origin and evolutionary relationships of millions of species, many additional changes are on the horizon. For the next generation of systematists, however, a better mnemonic to remember may be "Do Keep Probing Charles' *Origin* For Good Systematics."

Think Critically

How easily does the Linnaean system of classification accommodate the phenomenon of horizontal gene transfer? Is a completely phylogenetic view of biodiversity, like PhyloCode, more consistent with the idea that genes have moved horizontally between clades?

Richard Glor

Richard Glor conducts research on the evolution of *Anolis* lizards at the University of Rochester. To learn more about Dr. Glor's research, go to http://www.lacertilia.com.

REVIEW KEY CONCEPTS

Go to **CENGAGENOW** at www.cengage.com/login to access quizzing, animations, exercises, articles, and personalized homework help.

23.1 Nomenclature and Classification

- Linnaeus invented a system of binomial nomenclature in which each species receives a unique two-part name.
- Species are organized into a taxonomic hierarchy, which largely reflects the pattern of branching evolution (Figure 23.2). Species classified in the same lower level taxon have a more recent common ancestor than species classified in the same higher level taxon.

Animation: Classification systems

23.2 Phylogenetic Trees

- Phylogenetic trees are hypotheses that portray the branching pattern of evolution (Figure 23.3). Most phylogenetic trees have an implicit or explicit time line that indicates the relative times for cladogenesis. Branch points are described as nodes, and monophyletic lineages are called clades. Two lineages that share a node are called sister clades. Clades can be rotated at nodes without changing the meaning of the tree (Figure 23.4).
- Evolutionary biologists define monophyletic taxa directly from the structure of a phylogenetic tree (Figure 23.5). Biologists never intentionally define paraphyletic or polyphyletic taxa.

Animation: Evolutionary tree for plants

23.3 Sources of Data for Phylogenetic Analyses

- Homologies are characters that two or more descendant species inherited from their common ancestor (Figure 23.6). Homologous structures are similar in anatomical detail, and they often show a similar pattern of embryonic development. Characters that are similar for any other reason are described as homoplasies (Figure 23.7).
- Systematists often use morphological characters in phylogenetic analyses.
- If morphological characters do not distinguish species, behavioral characters may be useful (Figures 23.8).
- Systematists today use genetic sequence data to reconstruct phylogenetic history.

23.4 Traditional Classification and Paraphyletic Groups

- In traditional systematics, biologists used the similarities and differences between organisms to construct a phylogenetic tree and a classification.
- Traditional classifications sometimes included paraphyletic groups (Figure 23.9A).

23.5 The Cladistic Revolution

- Derived character states can serve as markers of clades.
- Systematists use evidence from the fossil record as well as outgroup comparison to identify which character states are derived and which are ancestral (Figure 23.10).
- Cladistic analyses use synapomorphies (derived character states) to construct phylogenetic hypotheses (Figure 23.9B).
- The manual construction of a simple cladogram is straightforward (Figure 23.11).
- Systematists use the principle of parsimony and statistical techniques to evaluate alternate phylogenetic hypotheses (Figure 23.12).

Animation: Constructing a cladogram

Animation: Interpreting a cladogram

Animation: Current evolutionary tree

23.6 Phylogenetic Trees as Research Tools

- If a nucleic acid accumulates mutations at a reasonably constant rate, it can be used as a molecular clock (Figure 23.13).
- Phylogenetic trees allow biologists to propose and test hypotheses about when certain characters evolved (Figure 23.14).
- Phylogenetic analyses are useful tools in medicine because they allow researchers to track the origin and spread of infectious diseases (Figure 23.15).

23.7 Molecular Phylogenetic Analyses

- Molecular techniques allow systematists to address questions that they could not have posed in the mid-twentieth century.
- Molecular analyses have identified the most basal angiosperm and clarified the phylogenetic relationships of several early clades of flowering plants (Figure 23.16).
- Sequence data from the small subunit ribosomal RNA allowed researchers to recognize the three domains of life (Figure 23.17). Horizontal gene transfer complicates our understanding of the history of biodiversity (Figure 23.18).

Animation: Cytochrome *c* comparison

UNDERSTAND AND APPLY

Test Your Knowledge

1. The evolutionary history of a group of organisms is called its:
   a. classification.
   b. taxonomy.
   c. phylogeny.
   d. domain.
   e. outgroup.
2. In the Linnaean hierarchy, the organisms classified within the same taxonomic category are called:
   a. a phylum.
   b. a taxon.
   c. a genus.
   d. a binomial.
   e. an epithet.
3. When systematists study morphological or behavioral traits to reconstruct the evolutionary history of a group of animals, they assume that:
   a. similarities and differences in phenotypic characters reflect underlying genetic similarities and differences.
   b. the animals use exactly the same traits to identify appropriate mates.
   c. differences in these traits caused speciation in the past.
   d. the adaptive value of these traits can be explained.
   e. variations in these traits are produced by environmental effects during development.
4. Which of the following pairs of structures are homoplastic?
   a. the wing skeleton of a bird and the wing skeleton of a bat
   b. the wing of a bird and the wing of a fly
   c. the eye of a fish and the eye of a human
   d. the bones in the foot of a duck and the bones in the foot of a chicken
   e. the toes on the foot of a lizard and the toes on the foot of a human
5. Which of the following does *not* help systematists determine whether a morphological character state is ancestral or derived?
   a. outgroup comparison
   b. patterns of embryonic development
   c. studies of the fossil record
   d. studies of the character in more related species
   e. dating of the character by molecular clocks
6. In a cladistic analysis, a systematist groups together organisms that share:
   a. derived homologous traits.
   b. derived homoplastic traits.
   c. ancestral homologous traits.
   d. ancestral homoplastic traits.
   e. all of the above.
7. A monophyletic taxon is one that includes:
   a. an ancestor and all of its descendants.
   b. an ancestor and some of its descendants.
   c. organisms from different evolutionary lineages.
   d. an ancestor and those descendants that still resemble it.
   e. organisms that resemble each other because they live in similar environments.
8. Which of the following is *not* an advantage of using molecular characters in a systematic analysis?
   a. Molecular characters provide abundant data.
   b. Systematists can compare molecules among species that are morphologically very similar.
   c. Systematists can compare molecules among species that are morphologically very different.
   d. Nucleotide sequences in DNA are generally not influenced by environmental factors.
   e. Systematists can easily determine whether base substitutions in the DNA of two species are synapomorphies.

9. To construct a cladogram by applying the principles of parsimony to molecular sequence data, one would:
   a. start by making assumptions about variations in the rates at which different DNA segments evolve.
   b. group together organisms that share the largest number of ancestral sequences.
   c. group together organisms that share derived sequences, matching the groups to those defined by morphological characters.
   d. group together organisms that share derived sequences, minimizing the number of hypothesized evolutionary changes.
   e. identify derived sequences by studying the embryology of the organisms.
10. Which of the following underlying assumptions allows differences in a particular molecular sequence to be used as a molecular clock?
   a. The sequence never experiences any mutations.
   b. The sequence codes for a protein.
   c. The sequence accumulates mutations at a reasonably constant rate.
   d. The sequence is part of a mitochondrial gene.
   e. The sequence codes for small subunit ribosomal RNA.

Discuss the Concepts

1. In the past, systematists commonly used the amino acid sequences of proteins and DNA sequences in phylogenetic analyses. Think about the genetic code (Section 15.1), and explain why phylogenetic hypotheses based on DNA sequences may be more accurate than those based on amino acid sequences.
2. Traditional evolutionary systematists identify the Reptilia as one class of vertebrates, even though we know that this taxon is paraphyletic. Describe disadvantages of defining paraphyletic taxa in a classification.
3. The following table provides information about the distribution of ancestral and derived states for six systematic characters (labeled 1 through 6) in five species (labeled A through E). A "d" means that the species has the derived form of the character, and an "a" means that it has the ancestral form. Construct a cladogram for the five species using the principle of parsimony; in other words, assume that each derived character evolved only once in this group of organisms. Mark the branches of the cladogram to show where each character changed from the ancestral to the derived state.

	Character					
Species	**1**	**2**	**3**	**4**	**5**	**6**
A	a	a	a	a	a	a
B	d	a	a	a	a	d
C	d	d	d	a	a	a
D	d	d	d	a	d	a
E	d	d	a	d	a	a

4. Imagine that you are a systematist studying a group of little-known flowering plants. You discover that the phylogenetic tree based on flower morphology differs dramatically from the phylogenetic tree based on DNA sequences. How would you try to resolve the discrepancy? Which tree would you believe is more accurate?
5. Create an imaginary phylogenetic tree for an ancestral species and its 10 descendants. Circle a monophyletic group, a polyphyletic group, and a paraphyletic group on the tree. Explain why the groups you identify match the definitions of the three types of groups.

Design an Experiment

Imagine that you are trying to determine the evolutionary relationships among six groups of animals that look very much alike because they have few measurable morphological characters. What data would you collect to reconstruct their phylogenetic history?

Interpret the Data

The phylogenetic tree for 12 cat species (Felidae) reproduced below was assembled from molecular sequence data. Which species is the domestic cat's closest relative? Which clade is the sister taxon to tigers? Are bobcats more closely related to cougars or to ocelots?

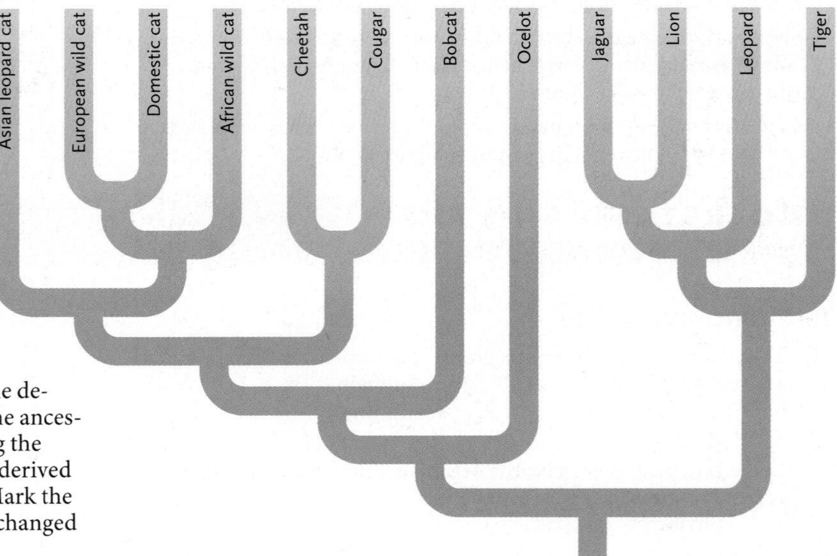

Asian leopard cat
European wild cat
Domestic cat
African wild cat
Cheetah
Cougar
Bobcat
Ocelot
Jaguar
Lion
Leopard
Tiger

Source: Adapted from Warren E. Johnson et al. 2006. The late Miocene radiation of modern Felidae: A genetic assessment. *Science* 311:73–77.

Apply Evolutionary Thinking

Is the construction of parsimonious phylogenetic trees more consistent with the predictions of the phyletic gradualism hypothesis or the punctuated equilibrium hypothesis of evolutionary change? You may want to review material in Section 22.5 before answering this question.

Black smoker hydrothermal vents on the ocean floor. Many scientists support the theory that life developed near hydrothermal vents, where superheated, mineral-rich water is found.

Ralph White/Corbis

24

STUDY OUTLINE

The Origin of Life

Why It Matters. . . In 1927, Belgian priest and astronomer George Lemaître proposed the Big Bang Theory, which is now the dominant scientific theory about the origin of the universe. According to this theory, an incomprehensibly vast explosion about 14 billion years ago produced the matter and energy of our universe. As the universe expanded, slight variations in temperature and density led to an uneven distribution of matter. Gravitational attraction caused gas to form clouds and then collapse into more concentrated collections of matter. Stars, in which nuclear reactions generated the major elements we find in present-day biological molecules, were formed. Stars that were aging before the birth of our solar system, expanded to produce red supergiants, which eventually exploded as supernovae. Such an explosion ejects the contents of a star at near light speed, in effect distributing all the elements of the periodic table across the universe. When our Sun formed, some of the elements coalesced with it and also formed the planets of our solar system. In other words, stardust, the remnants of long-ago stars, is the elemental origin of life on Earth. The origin of cellular life required two other fundamental conditions: a continual source of energy, and a temperature range in which liquid water can form. The continual energy source is, of course, the Sun. The relatively narrow temperature range suitable for life relates to the nature of the Sun, the distance of Earth from the Sun, and the properties of the atmosphere that cause trapping of heat.

Earth is estimated to have formed approximately 4.6 billion years ago. When and how did living cells form from inert materials? Answering these questions is difficult because we must rely on indirect geological evidence, and on experimental data that are open to debate and multiple interpretations.

Figure 24.1 outlines the key events in the origin and early evolution of life, which we will examine in this chapter. The earliest events are uncertain, but probably include the development of **protocells,** primitive cell-like structures that have some of the properties of life and that might have been the

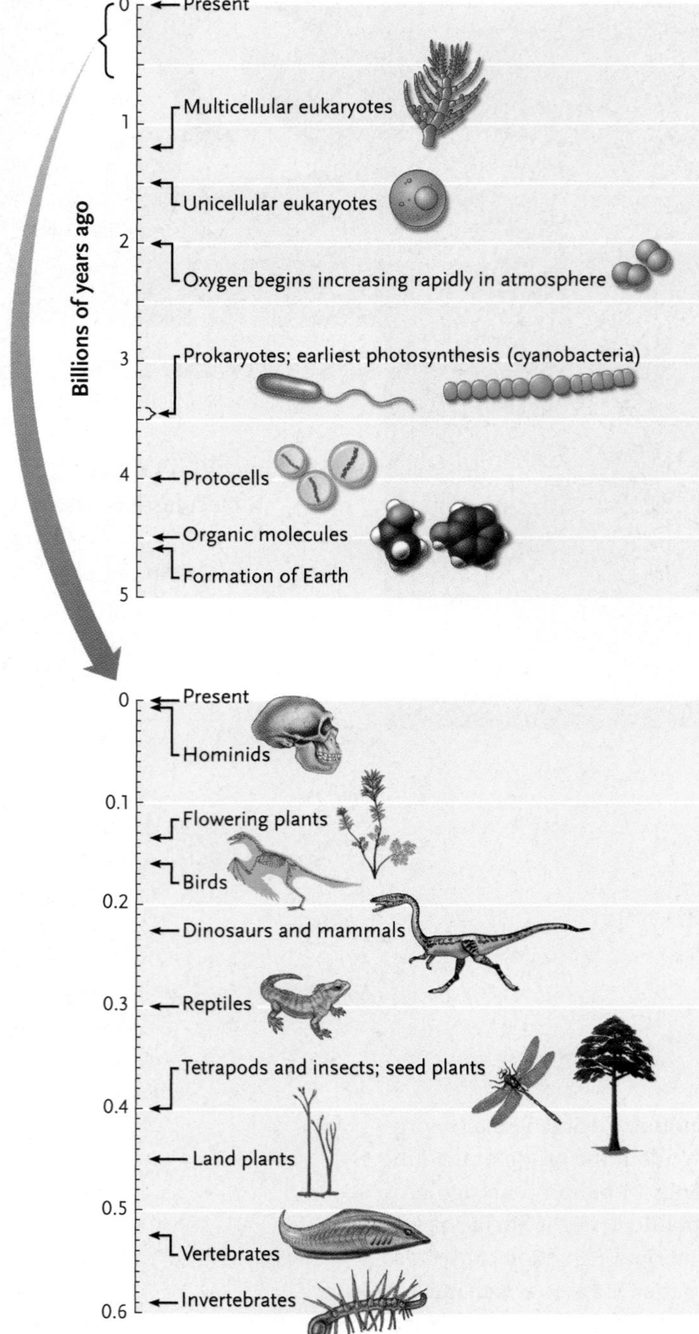
Billions of years ago
0
Present
1
Multicellular eukaryotes
Unicellular eukaryotes
2
Oxygen begins increasing rapidly in atmosphere
3
Prokaryotes; earliest photosynthesis (cyanobacteria)
4
Protocells
Organic molecules
Formation of Earth
5
0
Present
Hominids
0.1
Flowering plants
Birds
0.2
Dinosaurs and mammals
0.3
Reptiles
Tetrapods and insects; seed plants
0.4
Land plants
0.5
Vertebrates
Invertebrates
0.6

FIGURE 24.1
A timeline for the evolution of cellular life, in the context of the evolution of major present-day organisms. The dates for the origin of cellular life derive from geological evidence. Each date is the subject of debate, as discussed in the text. A more complete presentation of the timeline for the evolution of organismal life is in Table 22.1.

precursors of cells. Some geological evidence shows that prokaryotic cells were present in the Archaean eon by 3.5 billion years ago (see Table 22.1 for the geological time scale). The activity of photosynthetic prokaryotes generated an oxygen-enriched atmosphere, a condition probably necessary for the development of the first eukaryotic cells. Fossils of cells that resemble modern eukaryotic cells appeared about 1.5 billion years ago. <

24.1 The Formation of Molecules Necessary for Life

All present-day living cells have the following characteristics: (1) a boundary membrane separating the cell interior from the exterior; (2) one or more nucleic acid coding molecules located in a nuclear region (a nucleus in eukaryotes and a nucleoid region in prokaryotes); (3) a system using the coded information in the nucleic acids to make proteins and, through them, other biological molecules; and (4) a metabolic system providing energy for these activities. The complexity of these systems makes it highly unlikely that living cells appeared suddenly from nonliving matter. Rather, there must have been a transition from nonliving to living matter, although we lack direct evidence for how it occurred.

Most scientists study the origin of life by assuming that it originated from nonliving matter on Earth, through chemical and physical processes no different from those operating today. Hypotheses made under these assumptions are testable to the extent that the chemical and physical processes can be duplicated in the laboratory.

But some scientists consider a possible extraterrestrial origin of life. Analysis of meteorites has shown that they contain some organic molecules that are characteristic of living organisms. Could a living cell or organism, or perhaps a spore, have arrived in such a way? Most scientists believe it is unlikely that a cell or an organism could have survived a long journey in space, even if protected from radiation, or that it could have survived intense heating while traveling through Earth's atmosphere and the actual impact with Earth. At this point the hypothesis that life arrived on Earth by interplanetary transport is not testable, so it cannot be ruled out. Nonetheless, even if a living organism arrived from space and spawned a population on this planet, life would still have had to arise from nonliving matter in a similar way on the organism's home planet.

Conditions on Primordial Earth Led to the Formation of Organic Molecules

The formation of Earth led to a planet with three layers. In its center is the core, which is surrounded by the mantle. Both the core and the mantle are rich in iron. Surrounding the mantle is the relatively thin crust, which consists mainly of silicates. Volcanic activity on early Earth released gases from the crust and mantle such as CO_2, H_2, and N_2, which helped form the first atmosphere. Water vapor emitted from volcanoes formed the first oceans, lakes, and rivers.

Primordial Earth met several basic conditions necessary for life to begin. Although Earth's gravitational pull was strong enough to retain an atmosphere, it was not strong enough to compress the atmospheric gases into liquid form. Earth's distance from the Sun was such that, on average, sunlight warmed the surface enough to keep much of the liquid water from freezing, but not enough to boil the water. *Liquid water is essential for the chemistry of biological systems* (see Chapter 2).

In the primordial atmosphere, any molecular oxygen would have reacted with elements of the crust and atmosphere to form oxides. Spontaneous reactions of hydrogen, nitrogen, and carbon would have produced ammonia (NH_3) and methane (CH_4).

As Earth's surface cooled, natural sources of energy caused chemical bonds to break and reform, leading to the formation of organic molecules at about 4.5 billion years ago. In addition to sunlight and electrical discharges during storms and volcanic activity, radioactivity from atomic decay and heat from volcanoes, geysers, and hydrothermal (hot water) vents in the sea floor, all acted on the primordial atmosphere and crust—as they still do today. As many as a half-billion years may have passed before the concentrations of organic molecules reached levels where their interactions formed more complex organic substances. We now consider the current thinking about how simple molecules were converted into the key molecules of life.

The Oparin–Haldane Hypothesis Initiated Scientific Investigations into the Origin of Life

Scientific efforts to explain the origin of life began with a major hypothesis proposed independently in the 1920s by two investigators, Aleksandr I. Oparin, a Russian plant biochemist at Moscow State University in Russia, and J. B. S. Haldane, a British geneticist and evolutionary biologist at Cambridge University in England. Their hypothesis rested on the critical assumption that Earth's primordial atmosphere was radically different from today's atmosphere. They proposed that, rather than being an oxygen-rich (oxidizing) atmosphere as it is now, the early atmosphere was a strongly reducing atmosphere, due to the concentration of substances such as hydrogen (H_2), methane (CH_4), ammonia (NH_3), and water, which are *fully reduced*—they contain the maximum possible number of electrons and hydrogens (see Section 8.1). They reasoned that the primordial atmosphere would have had an abundance of electrons and hydrogens for reduction reactions, which could create organic molecules from inorganic elements and compounds. Energy to drive the reductions, according to the hypothesis, came from solar energy and other natural sources such as the electrical energy of lightning in atmospheric storms and from volcanic activity.

The absence of oxygen in the primitive atmosphere is essential to the Oparin–Haldane hypothesis. Oxygen can reverse reductions by removing electrons and hydrogens from organic molecules (see Section 8.1). In other words, if oxygen was present, the newly formed molecules would have been broken down quickly by oxidation.

Oparin and Haldane proposed that reductions occurring on the primordial Earth produced great quantities of organic molecules. The molecules accumulated because the two main routes by which such substances break down today, chemical attack by oxygen and decay by microorganisms, could not take place. According to Oparin and Haldane's hypothesis, the organic substances would have become so concentrated that the oceans and other bodies of water resembled a prebiotic soup. Recent geochemical evidence, however, suggests an alternate mechanism for the formation of organic molecules. According to this evidence, early Earth's oceans actually contained some amount of oxidized forms of nitrogen, sulfur, and iron that were generated by UV radiation that penetrated the atmosphere in the absence of an ozone layer. Those oxidized materials could have reacted with reduced molecules of the crust, releasing energy that could have been used to produce some more complex biomolecules.

In their model, Oparin and Haldane assumed that the highly concentrated organic molecules would tend to aggregate in random combinations and that, by chance, some of the combinations were able to carry out one or more primitive reactions characteristic of life, such as increasing in mass by adding new materials. Later, scientists reasoned that these combinations were able to compete successfully against less efficient combinations for space and materials in an organic soup. As a result, they persisted and became more numerous.

Chemistry Simulation Experiments Support the Oparin–Haldane Hypothesis

In 1953, Stanley L. Miller, a graduate student in Harold Urey's laboratory at the University of Chicago, tested the Oparin–Haldane hypothesis by creating a laboratory simulation of conditions Oparin and Haldane believed existed on early Earth (**Figure 24.2**). After one week of running the apparatus, as much as 15% of the carbon was now in the form of organic compounds, including amino acids. The significance of the finding at the time was enormous; amino acids, which are essential to cellular life, could be made under the conditions scientists believed existed on early Earth.

Various other chemicals and combinations of chemicals have been tested in the Miller–Urey apparatus. Produced were all the building blocks of complex biological molecules, including amino acids; fatty acids; the purine and pyrimidine building blocks of nucleic acids; sugars such as glyceraldehyde, ribose, deoxyribose, glucose, and fructose; and phospholipids, which form the lipid bilayers of biological membranes.

The synthesis of complex biological molecules in a reducing atmosphere in the Miller–Urey experiment supported the Oparin–Haldane hypothesis. However, it is only a conjecture that a reducing atmosphere was present at the time key organic molecules were formed on early Earth. Indeed, current thinking is that early Earth's atmosphere was not reducing but that it contained large amounts of oxidants such as CO_2 and N_2 that were released by volcanic activity, as mentioned earlier. In such an oxidizing atmosphere, any organic molecules generated spontaneously in the environment would be oxidized quickly back to inorganic forms by combination with the oxygen in the atmosphere. This is supported experimentally: running the Miller–Urey experiment in the presence of oxygen results in essentially no organic molecules.

FIGURE 24.2 **Experimental Research**

The Miller–Urey Apparatus Demonstrating That Organic Molecules Can Be Synthesized Spontaneously under Conditions Simulating Primordial Earth

Question: Do chemistry simulation experiments support the Oparin–Haldane hypothesis that reduction reactions on primordial Earth produced organic molecules?

Experiment: Stanley Miller set up an experiment to simulate chemical conditions of early Earth (figure). He placed components of a reducing atmosphere—hydrogen, methane, and ammonia—in a closed apparatus. Water vapor was added to the "atmosphere" by boiling water in one part of the apparatus, and it was removed by cooling and condensation in another part. This simulated the cycle of release of water vapor into the atmosphere, and condensation into water. Miller exposed the gases to an energy source in the form of continuously sparking electrodes to simulate lightning.

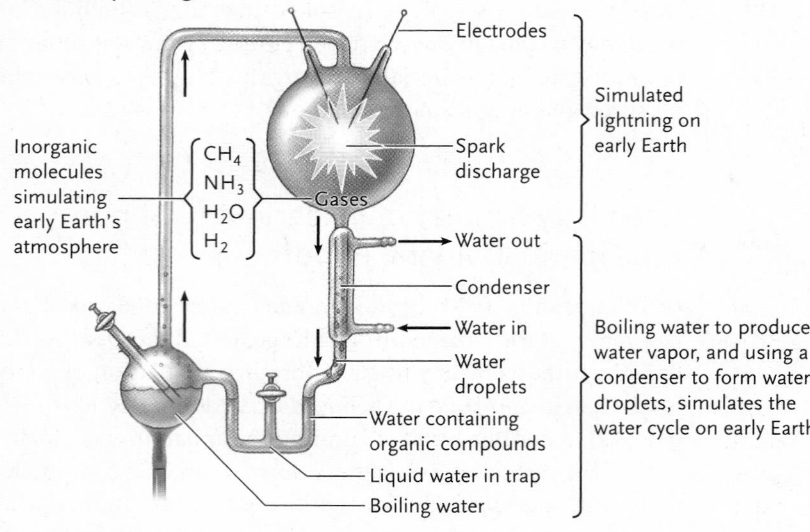
~4.5 billion years ago
Electrodes
Simulated lightning on early Earth
Spark discharge
Inorganic molecules simulating early Earth's atmosphere
CH_4
NH_3
H_2O
H_2
Gases
Water out
Condenser
Water in
Water droplets
Boiling water to produce water vapor, and using a condenser to form water droplets, simulates the water cycle on early Earth.
Water containing organic compounds
Liquid water in trap
Boiling water

Results: After running the apparatus for only a week, Miller found a large assortment of organic compounds in the water, including urea, amino acids, and lactic, formic, and acetic acids—as much as 15% of the carbon was now in the form of organic compounds. Two percent of the carbon was in the form of amino acids, which form easily under sufficiently reducing conditions.

Conclusion: The synthesis of organic molecules, including biological molecules, in a reducing atmosphere supported the Oparin–Haldane hypothesis.

Source: S. Miller. 1953. A production of amino acids under possible primitive Earth conditions. *Science* 117:528–529. (Figure redrawn from an original courtesy of S. L. Miller, Copyright 1955 by the American Chemical Society.)

Scientists Have New Theories about the Sites for the Origin of Organic Molecules

If organic compounds were not generated in a reducing atmosphere, how else could they have arisen? Scientists have developed a number of theories, all of which have the reasonable assumption of the presence of liquid water. Remember that water is essential for the chemistry of biological systems (see Chapter 2). All of the theories have their advocates and critics. It is fair to say that there are controversial issues in each of them. Two of the theories are outlined here.

One current theory for the origin of life, which has significant support among scientists, is that life developed near hydrothermal vents in the sea floor (see photo on first page of the chapter). Many such vents exist in today's oceans, emitting bursts of mineral-rich water superheated to up to 400°C by submarine volcanoes. Scientists exploring hydrothermal vents find complex ecosystems associated with them.

Life might have originated near oceanic hydrothermal vents because reducing conditions existed there along with an abundance of the chemicals that are essential for life. Even now, there are high levels of hydrogen gas, methane, and ammonia around the vents. Indeed, based on simulation experiments, scientists hypothesize that hydrothermal vents could have produced a lot more organic material than that generated in the Miller–Urey experiment. However, if life did evolve near hydrothermal vents, we would expect many present-day hydrothermal-vent life forms to be ancient. This is not the case; in most cases these organisms are closely related to modern organisms that do not live in hydrothermal vent environments. Critics of the hydrothermal-vent origin of life theory also argue that the temperature at the vents is too high to permit the origin of life. The critics argue that at the high temperature found at vents, the organic molecules are too unstable and would be destroyed as soon as they form. Supporters of the theory counter that the necessary organic molecules for life form not at the vent itself, but somewhere in the gradient between the hot water at the vent and the near-freezing water surrounding the vent.

Another theory is that some organic compounds had an extraterrestrial origin. Indeed, many of the compounds made in the Miller–Urey experiment exist in outer space. For example, a meteorite that fell on Murchison, Australia, in 1969 contained more than 90 amino acids, only 19 of which are found on Earth. Since amino acids appear to be able to survive in outer space, they could potentially have been present when Earth was formed. DNA precursors have also been identified in some meteorites. It is possible that other organic compounds arrived by meteor impact, too. Repeated bombardment by meteors occurred during the first period of Earth's existence, the Hadean eon, from 4.6–3.85 billion years ago (see Figure 24.1 and Table 22.1), causing a cycle of vaporization of water in the oceans, which then cooled and recondensed.

STUDY BREAK 24.1

1. **Why is the reducing nature of early Earth's atmosphere key to the Oparin–Haldane theory of the origin of molecules necessary for life?**
2. **How do the theories about the sites for the origin of life differ?**

24.2 The Origin of Cells

No matter where organic molecules originated, they do not qualify as life. In this section, we discuss the key stage in the origin of life—the formation of the first cells.

Protocells Formed with Some of the Properties of Life

How did organic building blocks such as amino acids assemble into macromolecules such as proteins and nucleic acids? To answer this question, researchers have proposed and tested several processes. One process is the concentration of subunits by the evaporation of water. Another is *dehydration synthesis (condensation)*, in which subunits assemble into larger molecules through removal of the elements of a molecule of water (see Section 3.1). Experiments with these processes under simulated conditions showed that both evaporation and condensation reactions can produce polypeptide chains from amino acids, polysaccharides from glucose and other monosaccharides, and nucleotides and nucleic acids from nitrogenous bases, ribose, and phosphates.

Scientists reason that spontaneous condensations and other reactions produced significant quantities of all the major biological molecules over the hundreds of millions of years following the initial formation of Earth. They hypothesize that the accumulation of organic matter set up the conditions necessary for the next stage, the chance assembly of molecules into aggregations that became membrane-bound to form primitive protocells, perhaps by 4.0 billion years ago. Protocells—shown in Figure 24.3—are key to the origin of life, because life depends upon reactions occurring in a controlled and sequestered environment, the cell. Researchers have proposed several mechanisms for the assembly of organic molecules into aggregates, each of which has been successfully duplicated in laboratory experiments simulating primordial conditions. Two of those mechanisms are adsorption into clays and lipid bilayer assembly.

EXPERIMENTAL SUPPORT FOR ADSORPTION INTO CLAYS Could clays have provided an ideal environment for molecular aggregation and interaction on the primitive Earth? Clays consist of very thin layers of minerals separated by layers of water only a few nanometers thick. The layered structure readily adsorbs ions and organic molecules and promotes their interactions, including condensations and other assembly reactions. Clays can also store potential energy, and therefore could have channeled some of the energy into reactions taking place between molecules adsorbed to them.

Several experiments have supported these proposals. For example, Noam Lahav at the Hebrew University of Jerusalem and Sherwood Chang of NASA's Ames Research Center added amino acids to clays and exposed the mixtures to water-content changes and fluctuating temperatures that are likely to occur in a tidal flat. After several cycles of the fluctuating conditions, polypeptides were detected in the clays. Other researchers found that RNA nucleotides linked to phosphate chains could combine into RNA-like molecules in clays. Accumulation of these and other macromolecules in the clays could have provided an environment in which they could react to carry out the first reactions of life.

However, even if molecules became organized in clay and some of the reactions of life commenced, it is not clear how a lipid bilayer membrane could have formed around them. Such a membrane is necessary to organize the molecules into protocells, the presumed precursors of cells. (The biological importance of lipid bilayers and membranes are discussed in Sections 2.4, 3.3, and 6.1.)

EXPERIMENTAL SUPPORT FOR LIPID BILAYER ASSEMBLY In the 1950s, R. J. Goldacre at Chester Beatty Research Institute, London, hypothesized that protocells could have formed starting with lipid bilayers that had assembled spontaneously. In the 1970s, David W. Deamer at the University of California, Davis and other investigators tested this hypothesis, finding that phospholipids and some other types of lipid molecules could form under simulated conditions. The phospholipids self-assembled readily into bilayers when suspended in water (see Section 6.1). Often, the bilayers rounded up into stable, closed vesicles called liposomes, consisting of a continuous-boundary "membrane" surrounding an inner space **(Figure 24.3)**.

Further tests showed that the bilayers formed in these experiments have many properties of living membranes. For ex-

~4.0 billion years ago

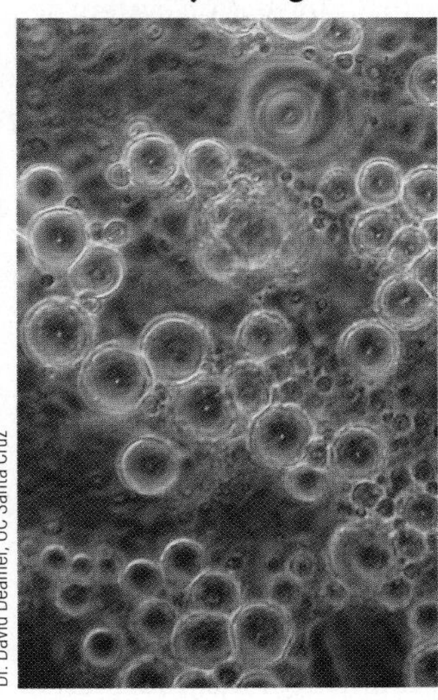
Dr. David Deamer, UC Santa Cruz

Key events in the origin of life

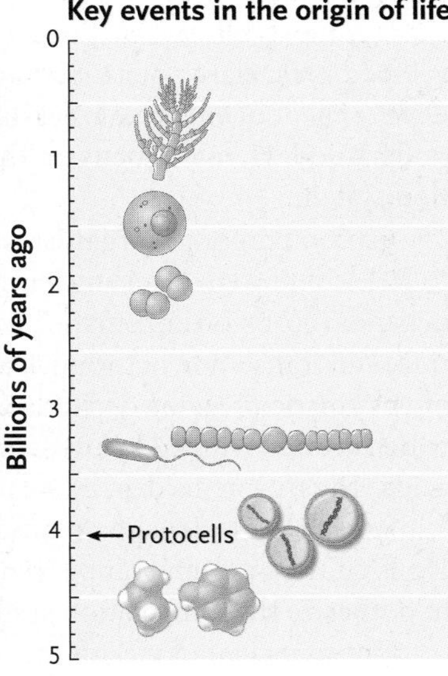
Billions of years ago
0
1
2
3
4
5
Protocells

FIGURE 24.3
Phase micrograph of lipid vesicles assembled from phospholipids. The larger spherical structures consist of hundreds of concentric lipid bilayers surrounding a fluid core. Given enough time, or if mechanically agitated, the multilayered structures disperse to form smaller vesicles bounded by one or a few lipid bilayers. Simpler molecules can also form membranous structures, including ordinary soap (fatty acids) and soap-like molecules extracted from carbonaceous meteorites. The first forms of cellular life probably did not use phospholipids for membranes, because these are synthesized enzymatically. Instead the membranes were more likely to be composed of soap-like molecules that were present in the environment.

ample, they can incorporate proteins onto their surfaces or into the hydrophobic membrane interior and they form vesicles that can trap other substances in the fluid enclosed by the membrane. Potentially, on early Earth, the concentration of organic molecules in such vesicles could have stimulated their expansion in size and fragmentation into smaller vesicles, providing a primitive form of reproduction. These mechanisms of aggregation, as well as others, may have worked separately or together to form protocells.

Researchers continue to experiment on producing protocells in the laboratory as a step toward understanding the transition from the protocell to the living cell. Present-day thinking is that a boundary membrane system must have evolved simultaneously with a genetic information system, rather than one being formed before the other, as some earlier models have proposed. Experimental approaches being used are therefore directed toward generating self-replicating nucleic acid and membrane vesicle systems.

Living Cells May Have Developed from Protocells

Eventually the chemical reactions taking place in the primitive protocells became organized enough to make the transition to living cells. Exactly how this occurred is open to speculation. Two of the critical events necessary for the transition are the development of pathways that captured and harnessed the energy required to drive molecular synthesis, and the development of a system for the storage, replication, and translation of information for protein synthesis. Remember that proteins are the catalysts for most cellular reactions. How the information system developed is crucial to our understanding of the origin of life. The development of energy-harnessing reaction pathways is not well understood. Here we focus on the development of the information system.

In contemporary organisms, information flows from DNA to RNA to protein. This nucleic acid-based information system depends mostly on enzymatic proteins for replication, transcription, and translation of the nucleic acids. However, the specificity of enzymatic proteins depends on their amino acid sequences, which are determined by the sequences of nucleotides in nucleic acids. Thus, proteins depend on nucleic acids for their structure, and nucleic acids depend on proteins to catalyze their activities. How could one have appeared before the other? Scientists hypothesize the information system developed in stages.

The prevalent model for the development of the information system is the **RNA world model.** The RNA world model states that the first genes *and* enzymes were RNA molecules. That is, *ribozymes*—RNA molecules capable of catalyzing biochemical reactions—may have functioned both as informational molecules and as catalysts in protocells, without requiring protein enzymes for catalytic reactions (ribozymes are discussed in Section 4.6). Thus, a self-catalyzed RNA world may have been the first step in the development of an information system.

Ribozymes may have originally developed by the chance assembly of RNA nucleotides taking part in oxidative and other metabolic reactions in protocells (RNA nucleotides such as ATP, NAD, and coenzyme A form important parts of many metabolic pathways, including glycolysis, respiration, and photosynthesis; see Chapters 8 and 9). The RNA molecules then developed the capacity to replicate themselves and other RNA molecules. That is, these RNA molecules acted both as templates—like mRNA—and as catalysts—like ribosomal RNA (see Section 15.4). Then ribozymes could replicate ribozymes, with no need for protein enzymes. Such self-replicating systems may have provided the basis of an RNA-based informational system, and founded the RNA world. *Insights from the Molecular Revolution* describes an experiment in which ribozymes that can replicate RNA were generated in a test tube.

In the RNA world, DNA would have developed as a subsequent step. At first, DNA nucleotides may have been produced by random removal of an oxygen atom from the ribose subunits of the RNA nucleotides. At some point, the DNA nucleotides paired with the RNA informational molecules, and were assembled into complementary copies of the RNA sequences. Some modern day viruses carry out this RNA-to-DNA reaction using the enzyme reverse transcriptase (see Section 18.1). Once the DNA copies were made, selection may have favored DNA as the informational storage molecule because it has greater chemical stability and can be assembled into much longer coding sequences than RNA can. RNA was left to function at intermediate steps between the stored information in DNA and protein synthesis, as it still does today.

As the RNA-based information system evolved, some RNAs may have acted as tRNA-like molecules, linking to amino acids and pairing with the RNA informational molecules. These associations could have led to the assembly of polypeptides with an ordered sequence of amino acids—the development of an RNA genetic code. When DNA took over information storage from RNA, the DNA would have picked up the code information that could then be transferred back to RNA by transcription.

Modern analysis of the ribosome, the organelle responsible for translation of mRNA (see Section 15.4) has shown that the enzymatic activity that catalyzes the formation of a peptide bond between amino acids is a function of one of the RNA molecules of the ribosome. This finding supports the proposal that, in addition to replicating themselves, RNA molecules also generated the first proteins.

Of course, we have no way of knowing exactly how life originated. Sifting through the various models and theories we can perhaps agree that there were some basic steps: (1) the abiotic (nonliving) synthesis of organic molecules such as amino acids; (2) the assembly of complex organic molecules from simple molecules, including protein or RNA or both; and (3) the aggregation of complex organic molecules (likely including those of the information system) inside membrane-bound protocells. Once the information system fully developed in the protocells, and it could replicate, and the protocells

INSIGHTS FROM THE Molecular Revolution

Toward the Evolution of Life in the Lab: The construction of protocell-like vesicles containing an active ribozyme.

Life is characterized by membrane-bound cells that grow and divide, and in which enzymatic reactions take place in a controlled and sequestered environment. We have learned in this chapter so far that living cells may have developed from protocells. And, the discovery of ribozymes led to the proposal that an RNA world was the first step in the evolution of a molecular information system that could store, replicate, and translate the information required for protein synthesis. Constructing a simple protocell in the lab that contains an active ribozyme would provide a system with which to explore the possible steps for the origin and early evolution of life. That is, the simplest cellular system capable of evolution requires a self-replicating informational molecule, and a means to compartmentalize that molecule.

Research Goal

Construct stable protocell-like vesicles capable of growth, and that contain an active ribozyme.

Experiments

Jack Szostak and his research team at the Howard Hughes Medical Institute and Harvard Medical School set out to construct model protocell vesicles. Many chemicals are capable of forming simple vesicles, but the key requirement here is that the vesicle membrane must be stable and capable of growth under conditions that favor ribozyme activity. For example, fatty acids form vesicles that can grow and divide, but ribozymes cannot function in them because they need magnesium ions (Mg^{2+}) for activity, and divalent cations such as Mg^{2+} cause the fatty acids to precipitate.

Results

The researchers found that a mixture of particular, chemically simple amphiphiles (molecules with both hydrophilic and lipophilic properties) form vesicles with the desired protocell-like properties. That is, these vesicles are stable and grow, but, importantly, remain intact in the presence of Mg^{2+}. The vesicles are also permeable to Mg^{2+}, as shown by an experiment in which a ribozyme included in the vesicles became active and self-cleaved when Mg^{2+} was added to the solution.

Conclusion

The researchers successfully constructed simple vesicles containing active ribozymes. They concluded that the stability of the vesicles, and their ability to grow and allow the diffusion into them of Mg^{2+}, are "critical for building model protocells in the laboratory and may have been important for early cellular evolution."

Source: I. A. Chen, K. Salehi-Ashtiani, and J. W. Szostak. 2005. RNA catalysis in model protocell vesicles. *Journal of the American Chemical Society* 127:13213–13219.

could divide, they became true living cells. The advent of living cells marked the beginning of biological evolution, which depends on cells that can reproduce and pass on information to their descendants.

Prokaryotic Cells Were the First Living Cells

The change to biological evolution set the stage for the appearance of all the features of cellular life. One of these features was a nuclear region that contained the DNA of the coding system and the mechanisms replicating the DNA and transcribing it into RNA. Another feature was a cytoplasmic region containing ribosomes and the enzymes required to translate RNA information into sequences of amino acids in proteins. The cytoplasm also contained an oxidative system supplying chemical energy for protein synthesis and assembly of other required molecules. A mechanism of cell division also evolved, allowing replicated DNA to be distributed equally between daughter cells. All these systems were enclosed by a membrane controlling the flow of molecules and ions in and out of the cell. The stages leading to this level may have taken more than a billion years, occupying the period from Earth's formation 4.6 billion years ago to about 3.5 billion years ago.

There is geological evidence of different kinds for early life:

- **Isotope ratios determined by radiometric dating.** Radiometric dating is a technique used to date a rock and is based on the decay of isotopes in the rock (see Section 22.1). Each isotope of an element has a specific decay rate. However, if rock is subjected to biological activity that uses one isotope preferentially over another, then the ratio of the two isotopes will differ from that in inert rock from the same time. Thus, an altered isotope ratio in rock or mineral samples can be interpreted as evidence of biological activity. The earliest rock samples showing a significant alteration of an isotope ratio, interpreted to involve preferential use of $^{12}CO_2$ rather than $^{13}CO_2$ (as is seen in present-day bacteria), date to about 3.7 billion years ago. This may mark the earliest evidence of biochemical reactions by living cells which, by any present theory, would be prokaryotes.
- **Stromatolites.** Fossil stromatolites are layers of carbonate or silicate rock that resemble present-day stromatolites **(Figure 24.4A)**. Geologists think that the fossils formed as layers of phototrophic prokaryotes grew and died, and their remaining forms were filled in by calcium carbonate or silica. ("Phototrophic" refers to an organism that obtains energy from light.) The earliest fossil stromatolites that are accepted by geologists date to about 3.4 billion years ago in the oldest rocks of the Archaean eon. However, the fossil rock is too deformed to allow details of cell structure to be analyzed, and some researchers question whether they are of biological origin.

FIGURE 24.4 Geological evidence for early life. **(A)** Stromatolites exposed at low tide in Western Australia's Shark Bay. These mounds, which consist of mineral deposits made by prokaryotic communities (predominantly cyanobacteria), are about 2,000 years old; they are highly similar in structure to fossil stromatolites that formed 3.4 billion years ago. **(B)** Microfossils—fossil algae from the Gunflint Formation, Ontario, Canada. The earliest microfossils date to about 2.0 billion years ago.

A. ~3.4 billion years ago

Image copyright Monica Johansen, 2010. Used under license from Shutterstock.com

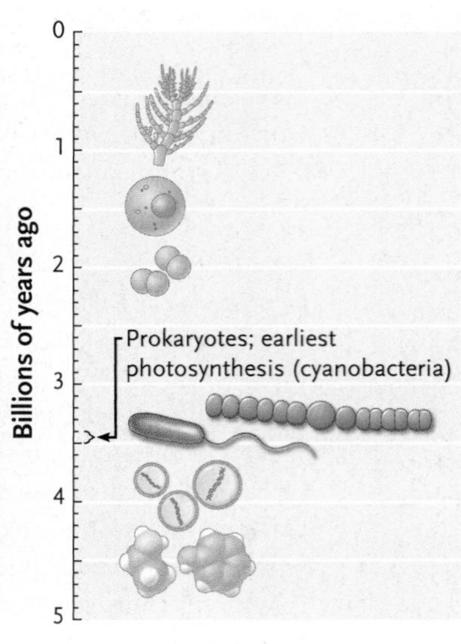

0
1
2
3
4
5
Billions of years ago
Prokaryotes; earliest photosynthesis (cyanobacteria)

B. ~2.0 billion years ago

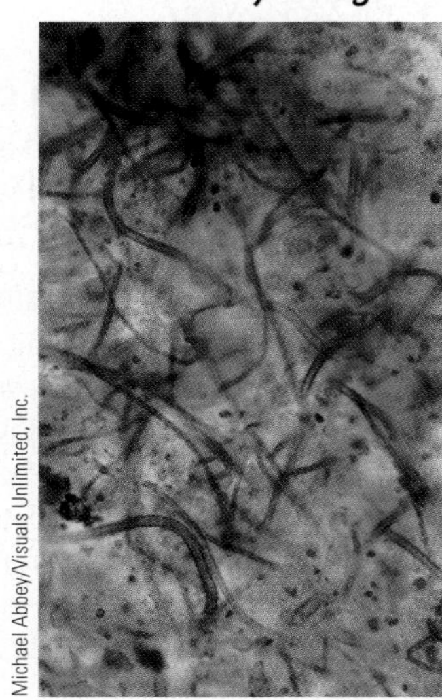

Michael Abbey/Visuals Unlimited, Inc.

- **Biosignatures. Biosignatures** are particular organic molecules in sedimentary rocks that could only have been formed by cellular activity. There are limited examples of biosignature molecules. One is a steroid-like molecule that is found only in the membrane lipids of cyanobacteria, and in no other organism known today. That particular biosignature molecule has been found in rock dated to 2.5 billion years ago, providing evidence for the presence of cyanobacteria by the end of the Archaean eon. Cyanobacteria are photosynthetic organisms. Since photosynthesis is a complex metabolic process, more primitive prokaryotes must have developed at an earlier time.
- **Microfossils.** A **microfossil** is the remains of a cell that has decayed and been filled in by calcium carbonate or silica **(Figure 24.4B).** (Section 22.1 discusses the fossil record and its use in providing information about life in the past.) The earliest microfossils are of filamentous and single-celled prokaryotes, and they date to about 2.0 billion years ago. Those microfossils are considered the most convincing geological evidence of early life. Microfossils from around 1.5 billion years ago include some with larger cells that resemble cells of present-day eukaryotic protists (see Chapter 26).

In sum, prokaryotes probably appeared first about 3.5 billion years ago and eukaryotes first appeared about 1.5 billion years ago, or perhaps earlier.

Subsequent Events Increased the Oxidizing Nature of the Atmosphere

According to Richard E. Dickerson of University of California, Los Angeles and others, the earliest form of photosynthesis evolved about 3.5 billion years ago in the early prokaryotes. This form of photosynthesis probably used electron donors such as hydrogen sulfide (H_2S) that do not release oxygen. However, at some point, an enzymatic system evolved that could use the most abundant molecule of the environment, water (H_2O), as the electron donor for photosynthesis. This reaction split water into protons, electrons, and oxygen, which was released into the atmosphere.

The oxygen released by the water-splitting reaction accumulated in the atmosphere and set the stage for the development of electron transfer systems using oxygen as the final electron acceptor. These transfer systems arose when some cells developed cytochromes that could deliver low-energy electrons to oxygen (see Section 8.4). These cells were able to tap the greatest possible amount of energy from the electrons before releasing them from electron transfer, making the cells highly successful in their environment.

The first water-splitting photosynthesizers confirmed by fossil evidence are the cyanobacteria. As you learned earlier, convincing evidence from biosignatures indicates cyanobacteria were present 2.5 billion years ago. Present-day stromatolites predominantly contain cyanobacteria but, as stated earlier, the types of prokaryotes in fossil stromatolites cannot be identified with any certainty. If the stromatolites do contain cyanobacteria, that would date early water-splitting photosynthesis to about 3.4 billion years ago.

The photosynthesizing activity of cyanobacteria produced a gradual rise in the amount of oxygen in early Earth's atmosphere. That rising oxygen concentration likely killed off many prokaryotic groups to which oxygen is lethal. Around 2 billion years ago, the concentration of oxygen began to increase more rapidly. This abrupt change may correlate with the evolution of eukaryotic cells containing chloroplasts.

These major events established the preconditions for the evolution of eukaryotic cells. The next section traces this evolution, which was pivotal to the later evolution of large-scale multicellularity and the plants, animals, and the other organisms of the domain Eukarya.

STUDY BREAK 24.2 <

Several mechanisms have been proposed for the assembly of organic molecules into protocells. Why is the model involving a lipid bilayer membrane a particularly attractive one?

THINK OUTSIDE THE BOOK

Individually or collaboratively investigate banded iron formations and outline the following in your investigation: (1) what they are; and (2) how researchers consider that they provide evidence for the presence of particular oxygen-producing organisms on early Earth.

24.3 The Origins of Eukaryotic Cells

Present-day eukaryotic cells have several interrelated characteristics that distinguish them from prokaryotes: (1) the separation of DNA and cytoplasm by a nuclear envelope; (2) the presence in the cytoplasm of membrane-bound compartments with specialized metabolic and synthetic functions—mitochondria, chloroplasts, the endoplasmic reticulum (ER), and the Golgi complex, among others; and (3) highly specialized motor (contractile) proteins that move cells and internal cell parts. In this section we discuss how eukaryotes most probably evolved from associations of prokaryotes. As you read, note that there are more good theories for the origin of eukaryotes than there are good data.

The Endosymbiotic Theory Proposes That Mitochondria and Chloroplasts Evolved from Ingested Prokaryotes

The **endosymbiotic theory,** put forward by Lynn Margulis at the University of Massachusetts, Amherst, proposes that the membranous organelles of eukaryotic cells, the mitochondria and chloroplasts, may each have originated from mutualistic (mutually advantageous) relationships between two prokaryotic cells **(Figure 24.5).** The theory proposes that the following events occurred:

1. **Generation of endosymbionts.** Mitochondria began to develop when photosynthetic and nonphotosynthetic prokaryotes coexisted in an oxygen-rich atmosphere. The nonphotosynthetic prokaryotes fed themselves by ingesting organic molecules from their environment. These prokaryotes included both anaerobes, which are unable to use oxygen as the final acceptor for electron transfer, and aerobes, which are fully capable of using oxygen. Only the aerobes could fully exploit the energy stored in organic molecules, but predatory anaerobes could capture that energy by eating aerobic cells. These anaerobic prokaryotes had become efficient predators, and lived by ingesting other cells. Among the ingested cells were some aerobic prokaryotes; instead of being digested, some of them persisted in the cytoplasm of the predators and continued to respire aerobically in their new location. They had become **endosymbionts,** organisms that live symbiotically within a host cell. The cytoplasm of the host anaerobe, formerly limited to the use of organic molecules as final electron acceptors, was now home to an aerobe capable of carrying out the much more efficient transfer of electrons to oxygen.
2. **Formation of the nucleus and other membranous structures.** As a part of the transition to a true eukaryotic cell, the cell also evolved to acquire other membranous structures, the major ones being the nuclear envelope, the endoplasmic reticulum (ER), and the Golgi complex. Endocytosis, the process of infolding of the plasma membrane (see Figure 5.16), is thought to be responsible for the evolution

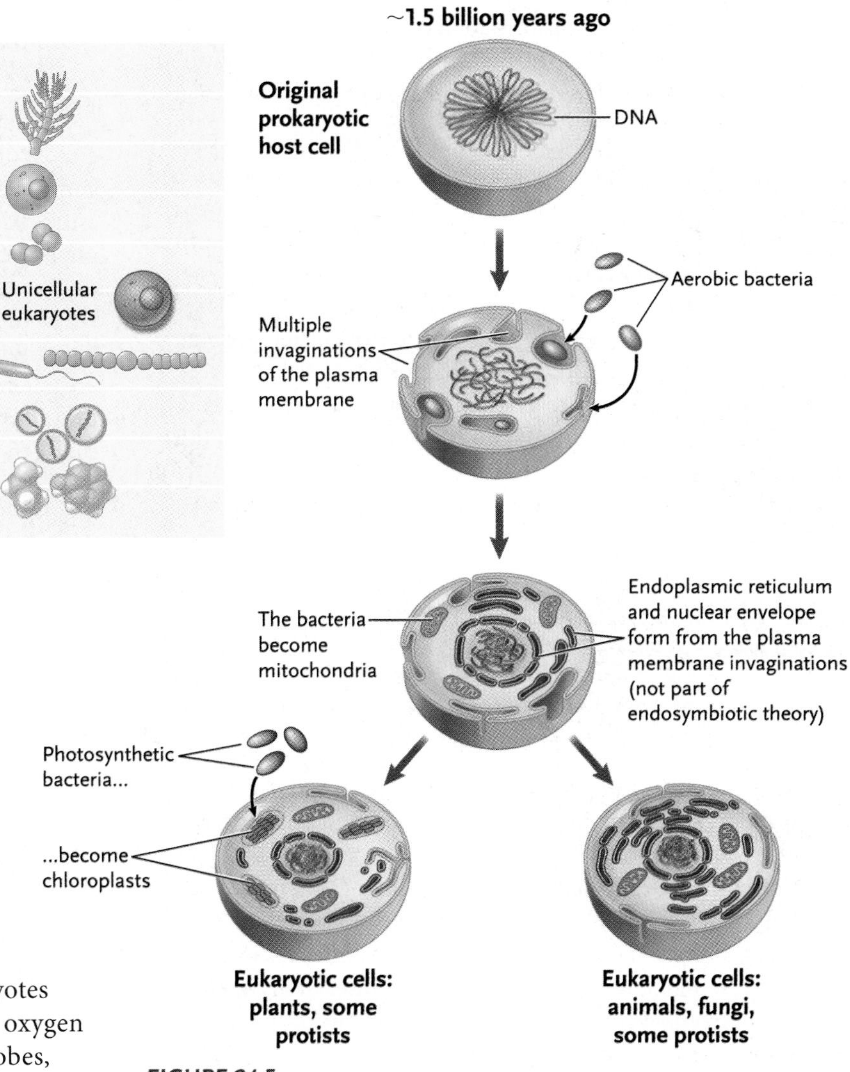
0
1
2
3
4
5
Billions of years ago
Unicellular eukaryotes
~1.5 billion years ago
Original prokaryotic host cell
DNA
Aerobic bacteria
Multiple invaginations of the plasma membrane
The bacteria become mitochondria
Endoplasmic reticulum and nuclear envelope form from the plasma membrane invaginations (not part of endosymbiotic theory)
Photosynthetic bacteria...
...become chloroplasts
Eukaryotic cells: plants, some protists
Eukaryotic cells: animals, fungi, some protists

FIGURE 24.5
The endosymbiotic theory. The mitochondrion is thought to have originated from an aerobic prokaryote that lived as an endosymbiont within an anaerobic prokaryote. The chloroplast is thought to have originated from a photosynthetic prokaryote that became an endosymbiont within an aerobic cell that had mitochondria.

of these structures. Researchers think that, in cell lines leading from prokaryotes to eukaryotes, pockets of the plasma membrane formed during endocytosis may have extended inward and surrounded the nuclear region. Some of these membranes fused around the DNA, forming the nuclear envelope and, hence, the nucleus. The remaining membranes formed vesicles in the cytoplasm that gave rise to the ER and the Golgi complex **(Figure 24.6).**

3. **Transfer of many functions from endosymbiont to the host cell.** Next, many functions duplicated in the aerobic endosymbiont were taken over by the host cell. As part of this transfer of function, most of the genes of the aerobe moved to the cell nucleus and became integrated into the host cell's DNA. At the same time, the host anaerobe became dependent for its survival on the respiratory capacity of the symbiotic aerobe. The ingested aerobe presumably benefited as well, because the host cell brought in large quantities of food molecules to be oxidized. This gradual process of mutual adaptation culminated in transformation of the cytoplasmic aerobe into a mitochondrion. The first eukaryotic cells had appeared, the ancestors of all modern-day eukaryotes.

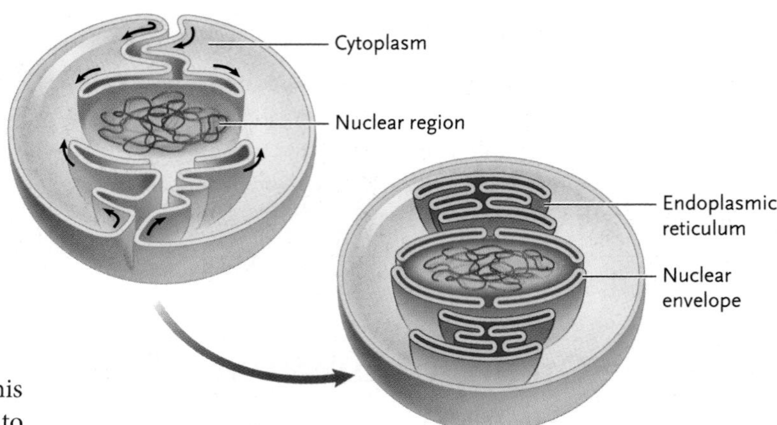
Cytoplasm
Nuclear region
Endoplasmic reticulum
Nuclear envelope

FIGURE 24.6
A hypothetical route for formation of the nuclear envelope and endoplasmic reticulum, through segments of the plasma membrane that were brought into the cytoplasm by endocytosis.

The endosymbiotic theory also proposes that a similar mechanism led to the appearance of the membrane-bound **plastids** (the general term for chloroplasts and related organelles, both photosynthetic and nonphotosynthetic) some time after mitochondria evolved. A plastid originated when an aerobic cell that had a mitochondrion, but was unable to carry out photosynthesis, ingested one or more photosynthetic prokaryotes resembling present-day cyanobacteria (see Figure 24.5). A photosynthetic prokaryote gradually changed into a plastid by evolutionary processes similar to those that produced a mitochondrion. A cell with both a plastid and a mitochondrion founded the cell lines that gave rise to the modern eukaryotic algae and plants.

Several Lines of Evidence Support the Endosymbiotic Theory

There is a lot of evidence to support the endosymbiotic theory, some of which includes the following:

1. Mitochondria and plastids are similar in size to bacteria.
2. Both organelles divide by a process similar to binary fission (see Section 10.5).
3. Both organelles typically contain circular DNA molecules that closely resemble bacterial chromosomes.
4. The genomes of both organelles contain codes for ribosomes that resemble those found in bacteria rather than those found in eukaryotes. The rRNA components of the ribosomes are encoded by organelle genes.
5. Both organelles are surrounded by two or more membranes, the innermost of which has a chemical composition similar to that of a bacterial plasma membrane.
6. Chloroplasts resemble cyanobacteria in their internal structure, including the presence of particular chlorophylls and the existence of thylakoids.

Another line of evidence supports a key assumption of the endosymbiont hypothesis by showing that engulfed cells or organelles can survive in the cytoplasm of the ingesting cell. Among animals, no less than 150 living genera, distributed among 11 phyla, include species that contain eukaryotic algae or cyanobacteria as residents in the cytoplasm of their cells. For example, larvae of the marine slug *Elysia* initially contain no chloroplasts, but after they begin feeding on algae, chloroplasts from the algal cells are taken up into the cells lining the gut. When the larvae develop into adult snails, the chloroplasts continue to carry out photosynthesis in their new location and produce carbohydrates that are used by the snails. Some of the chloroplast proteins synthesized during this time are encoded by genes in the slug's nuclear genome. The uptake of functional chloroplasts has also been observed among the protists (see Chapter 26); **Figure 24.7** shows a protist with chloroplasts that closely resemble cyanobacteria.

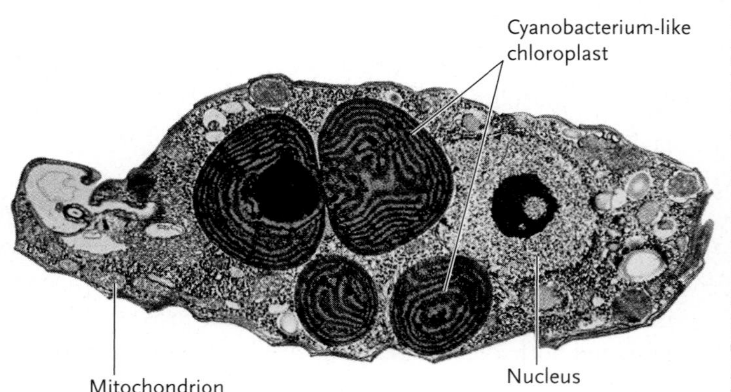
Cyanobacterium-like chloroplast
Mitochondrion
Nucleus

Robert Trench, Professor Emeritus, University of British Columbia

FIGURE 24.7
Cyanophora paradoxa, a protist with chloroplasts that closely resemble cyanobacteria without cell walls.

How long did it take for evolutionary mechanisms to produce fully eukaryotic cells? The oldest known convincing microfossil eukaryotes are 1.5 billion years old. If prokaryotic cells first evolved some 3.5 billion years ago, it took up to 2 billion years for eukaryotic cells to evolve from prokaryotes (see Figure 24.1). If so, this long interval probably reflects the complexity of the adaptations leading from prokaryotic to eukaryotic cells. Of course, it is possible that eukaryotic cells evolved more quickly, and we have yet to find the supporting evidence.

Eukaryotic Cells May Have Evolved from a Common Ancestral Line Shared with Archaeans

The system of classification that has gained acceptance among biologists, and the one used in this book, groups all living organisms into three domains. One domain, Eukarya, contains the eukaryotes. The second domain, Bacteria, includes one of two groups of prokaryotes, the bacteria, which consists of both photosynthesizing and nonphotosynthesizing species. The third domain, Archaea, contains the other group of prokaryotes, the archaeans, many of which inhabit extreme environments, including highly saline environments and hot springs.

There is little question that the three domains originated from a common ancestral cell line. However, the events leading from this common ancestry to the three domains of life remain unclear. The most difficult questions surround the role of the archaeans in both bacterial and eukaryotic evolution.

Archaeans have some features that are typical of bacteria, including a circular DNA genome in a nucleoid region without a surrounding nuclear envelope, and no membrane-bound organelles in the cytoplasm. However, the archaeans also have some features that are typically eukaryotic, including the presence of introns (see Section 15.3) in some of their genes. The archaeans also have some characteristics that are unique to their domain, including features of gene and rRNA sequences, and features of cell wall and plasma membrane structure that are found nowhere else among living organisms. The characteristics shared by archaeans and eukaryotes suggest that their roots may lie in a common ancestral line that split off from the line leading to bacteria (see Figure 23.17). At some point, this ancestral line split into the lines leading to Domain Archaea and Domain Eukarya.

Multicellular Eukaryotes Probably Evolved in Colonies of Cells

The first eukaryotes were unicellular. They are the ancestors of the present-day diversity of unicellular eukaryotes. Multicellular eukaryotes evolved from unicellular eukaryotes and then diverged to produce multiple eukaryotic lineages. Molecular clock analysis indicates the first multicellular eukaryote likely arose between 800 and 1,000 million years ago, while the first fossil records (of small algae) are from 600 to 800 million years ago.

According to the prevalent theory, multicellular eukaryotes arose by the aggregation of cells of the same species into a colony. The ability to act in a coordinated way probably increased the capacity of colonies to adapt to changes in the environment. Subsequently, differentiation of cells into various specialized cell types with distinct functions produced organisms with a wider range of capabilities and adaptability. Cell differentiation in a colony would have required cell signals that affected gene expression. That is, because each cell in the colony has the same genome, the development of specific functions (phenotypes) would require intracellular signals that would change the program of gene regulation. Over time, as genomes evolved, the division of function among cells led to the evolution of the tissues and organ systems of complex eukaryotes.

Multicellularity evolved several times in early eukaryotes, producing a number of lineages of algae as well as the ancestors of present day fungi, plants, and animals.

Life May Have Been the Inevitable Consequence of the Physical Conditions of the Primitive Earth

The events outlined in this chapter, leading from Earth's origin to the appearance of eukaryotic cells, may seem improbable. But, as scientist and author George Wald of Harvard University put it, given the total time span of these events, more than 3.5 billion years, "the impossible becomes possible, the possible probable, and the probable virtually certain. One has only to wait; time itself performs the miracles."

Some researchers go a step further and maintain that the evolution of life on our planet was an inevitable outcome of the initial physical and chemical conditions established by Earth's origin, among them a reducing atmosphere (at least in some locations), a size that generates moderate gravitational forces, and a distance from the Sun that results in average surface temperatures between the freezing and boiling points of water. Given the same conditions and sufficient time, according to these scientists, it is inevitable that life has evolved or is evolving now on other planets in the universe.

The chapters that follow in this unit trace the course of evolution and its products after eukaryotic cells were added to the prokaryotes already living on Earth. Among prokaryotes, evolution established two major groups, the bacteria and archaea; among eukaryotes, further evolution established the protists, fungi, plants, and animals. The survey of life's diversity begins in the next chapter with a description of present-day bacteria and archaeans.

STUDY BREAK 24.3 <

What are the key points of the theory of endosymbiont origins for mitochondria and chloroplasts?

UNANSWERED QUESTIONS

What was the first polymer of life?

As discussed in this chapter, many researchers favor RNA for this role. This is because it both self-replicates and can catalyze chemical reactions, and is neatly connected with the contemporary life based on nucleic acids and proteins. There are, however, several problems with this hypothesis. One of them is that synthesis of RNA and its building blocks, nucleotides, is quite difficult under primordial conditions. To circumvent this difficulty, several researchers proposed that other genetic polymers that are simpler to synthesize might have preceded RNA. For example, Albert Eschenmoser from the Swiss Federal Institute of Technology (ETH) in Zurich replaced the difficult-to-synthesize ribose portion of nucleotides by the six-membered sugar pyranose, and Peter Nielsen from the University of Copenhagen synthesized a polymer with a peptide-like backbone. These polymers are stable and capable of self-replication. Another proposal is that the initial complement of nucleic acid bases was different than A, U, G, and C. This proposal is motivated by the poor stability of cytosine in water. Although we have no evidence that polymers alternative to contemporary nucleic acids were present on the early Earth, it is important to realize that such a possibility exists and might be used by life elsewhere.

The protein-first hypothesis is currently less popular, even though these polymers are excellent catalysts of chemical reactions, and their building blocks, amino acids, existed on prebiotic Earth and are found in relatively large quantities in meteorites. This is because there is no known mechanism for proteins to self-replicate. Some researchers speculate that a limited replication of proteins is possible. An alternative hypothesis is that replication of individual polymers was not necessary at the origin of life and, instead, the reproduction of protein functions in a population was initially sufficient. Currently, neither view has much experimental support, but as we learn more about the structure and functions of small proteins major surprises might be in store.

Think Critically

If RNA molecules were the first molecules capable of catalyzing biochemical reactions, why were they replaced by proteins?

Andrew Pohorille

Andrew Pohorille heads the NASA Center for Computational Astrobiology and Fundamental Biology at NASA's Ames Research Center. He is also professor of Chemistry and Pharmaceutical Chemistry at the University of California, San Francisco. For his work on the origin of life he was awarded the 2002 NASA Exceptional Scientific Achievement Medal.

REVIEW KEY CONCEPTS

Go to **CENGAGENOW** at www.cengage.com/login to access quizzing, animations, exercises, articles, and personalized homework help.

24.1 The Formation of Molecules Necessary for Life

- Living cells are characterized by a boundary membrane, one or more nucleic acid coding molecules in a nuclear region, a system for using the coded information to make proteins, and a metabolic system providing energy for those activities.
- Oparin and Haldane independently hypothesized that life arose *de novo* under the conditions they thought prevailed on the primitive Earth, including a reducing atmosphere that lacked oxygen. Reduction reactions, fueled by natural energy acting on the primitive atmosphere, produced organic molecules. Random aggregations of these molecules were able to carry out primitive reactions which were characteristic of life and that gradually became more complex until life appeared. Chemistry simulation experiments support the hypothesis that organic molecules would form under these conditions (Figure 24.2).
- Present thinking is that early Earth's atmosphere was not reducing, but in fact contained significant amounts of oxidants. This has caused skepticism about Oparin and Haldane's hypothesis. One new theory proposes that life developed near hydrothermal vents in the sea floor.

Animation: Miller's reaction chamber experiment

Animation: Milestones in the history of life

24.2 The Origin of Cells

- Organic molecules produced in early Earth's environment by chance formed aggregates that became membrane-bound in protocells, primitive cell-like structures with some of the properties of life. Protocells may have been the precursors of cells (Figure 24.3).
- Next, living cells may have developed from protocells by the development of several critical components, in particular a system based on nucleic acids that could store and pass on the information required to make proteins.
- Subsequently, fully cellular life evolved, with a nuclear region containing DNA and the mechanisms for copying its information into RNA messages; a cytoplasmic region containing systems for utilizing energy and systems for translating RNA messages into proteins; a membrane separating the cell from its surroundings; and a reproductive system duplicating the informational molecules and dividing them among daughter cells.
- The first living cells were prokaryotes. Eventually, some early cells developed the capacity to carry out photosynthesis using water as an electron donor; the oxygen produced as a by-product accumulated and the oxidizing character of Earth's atmosphere increased. From this time on organic molecules produced in the environment were quickly broken down by oxidation, and life could arise only from preexisting life, as in today's world.

24.3 The Origins of Eukaryotic Cells

- According to the endosymbiotic theory, mitochondria evolved from ingested prokaryotes that were capable of using oxygen as final electron acceptors; chloroplasts evolved from ingested cyanobacteria (Figure 24.5).
- Eukaryotic structures such as the ER, Golgi complex, and nuclear envelope appeared through infoldings of the plasma membrane as a part of endocytosis (Figure 24.6).
- Multicellular eukaryotes probably evolved by differentiation of cells of the same species that had aggregated into colonies. Multicellularity evolved several times, producing lineages of several algae and ancestors of fungi, plants, and animals.

Animation: Eukaryotic evolution

UNDERSTAND AND APPLY

Test Your Knowledge

1. Earth was formed ____ years ago, whereas the oldest known living cell formed about ____ years ago.
   a. 400×10^3; 3.6×10^6.
   b. 4.6×10^9; 1.0×10^9.
   c. 3.8×10^9; 4.6×10^7.
   d. 4.6×10^9; 3.5×10^9.
   e. 2.0×10^9; 600×10^6.
2. Which of the following is *not* a characteristic of all living organisms?
   a. They replicate genetic information and convert the information into proteins.
   b. They pass genetic information between generations.
   c. They have an internal metabolic system that provides energy for biological activity.
   d. They use external energy to drive internal reactions that require energy.
   e. They use mitochondria for synthesis of ATP.
3. Evidence for the appearance of the following is separated by the greatest amount of time:
   a. nonlife: prokaryotes.
   b. photosynthetic prokaryotes: first eukaryotes.
   c. non-photosynthetic bacteria: photosynthetic bacteria.
   d. unicellular eukaryotes: multicellular eukaryotes.
   e. first prokaryotic cells: first eukaryotic cells.
4. According to the Oparin–Haldane hypothesis, the atmosphere when life began was believed to be composed primarily of the following substances that contain the maximum number of electrons:
   a. H_2O, N_2, and CO_2.
   b. H_2, H_2O, NH_3, and CH_4.
   c. H_2O, N_2, O_2, and CO_2.
   d. O_2 and no H_2.
   e. H_2 only.
5. The Miller–Urey experiment:
   a. was based on the understanding that the atmosphere was oxidizing.
   b. was able to synthesize amino acids and macromolecules from reduced gases.
   c. did not require much energy or a continuous energy source to keep synthesizing.
   d. did not require water to produce organic molecules.
   e. used free oxygen as a reactant.
6. An unknown organism was found in a park. It was unicellular, had no nuclear membrane around its DNA, and contained no mitochondria and no chloroplasts. It is most likely a:
   a. eukaryote.
   b. vertebrate or plant.
   c. bacteria or archaean.
   d. plant or fungi.
   e. virus.
7. Hydrothermal vents are theorized as sources for the origin of life because:
   a. the temperature of the water around them supports most life.
   b. most organic molecules undergo dehydration synthesis at high temperatures.
   c. the amino acids degrade in the colder water soon after synthesis.
   d. water is needed by living things.
   e. reducing conditions and abundant chemicals exist.
8. For protocells to transition to living cells, which of the following probably had to occur:
   a. DNA had to code for cell membranes.
   b. proteins had to be aggregated by RNA.
   c. ribozymes had to adsorb onto clays.
   d. H_2O, which is essential for the chemistry of all biological systems, had to become available to cells via metabolism and dehydration synthesis.
   e. informational molecules, such as ribozymes, had to evolve.
9. As part of the evolution of eukaryotic cell, endocytosis, the process of infolding of the plasma membrane, led to the formation of:
   a. chromosomes.
   b. the cell wall.
   c. ribosomes.
   d. the nuclear envelope.
   e. microtubules.
10. Which of the following is *not* part of the evidence supporting the endosymbiotic theory? Both mitochondria and plastids:
    a. are each the size of many bacterial cells.
    b. have inner membranes with a chemical composition similar to that of a bacterial plasma membrane.
    c. have ribosomes that are similar to bacterial ribosomes.
    d. contain circular DNA.
    e. have DNA similar to nuclear DNA.

Discuss the Concepts

1. What evidence supports the idea that life originated through inanimate chemical processes?
2. Explain, in terms of hydrophilic and hydrophobic interactions, how protocells might have formed in water from aggregations of lipids, proteins, and nucleic acids.
3. What conditions would likely be necessary on a planet located elsewhere in the universe for life to originate in a way similar to that which occurred on Earth?
4. Most scientists agree that life on Earth can arise only from preexisting life, but also that life could have originated spontaneously on the primordial Earth. Can you reconcile these seemingly contradictory statements?

Design an Experiment

Suppose you discover a hot spring-fed pool on a remote mountain that has never before been explored by humans. In the pool you find a cellular life form that appears to be prokaryotic. What experiments would you do to distinguish between the alternative hypotheses that this organism evolved on Earth from ancestral prokaryotes, or is descended from a life form that arrived at that location in a meteorite?

Interpret the Data

Studies of ancient rocks and fossils can provide insight into many of the events in Earth's history, including changes in its atmosphere that have taken place over geologic time. The figure on the next page shows how asteroid impacts and the composition of the atmosphere have changed over time. Use this figure and information in the chapter to answer the following questions.

1. Which occurred first, a decline in asteroid impacts or a rise in the atmospheric level of oxygen?
2. How do modern levels of carbon dioxide and oxygen compare to those at the time when the first cells arose?

3. Which is now more abundant, oxygen or carbon dioxide?

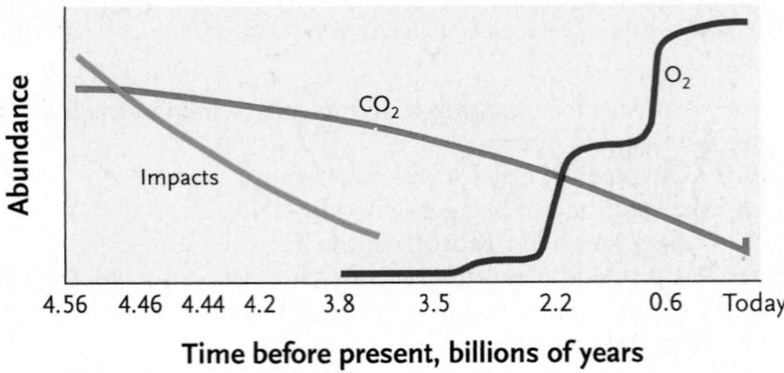
Abundance
CO2
O2
Impacts
4.56 4.46 4.44 4.2 3.8 3.5 2.2 0.6 Today
Time before present, billions of years

Changes over geologic time in asteroid impacts (green), atmospheric carbon dioxide concentration (pink), and oxygen concentration (blue).

Apply Evolutionary Thinking

In the evolution unit, you learned how changes in the environment can foster evolutionary changes in biological systems. How have changing biological systems influenced the evolution of changes in Earth's physical environment?

Express Your Opinion

Private companies make millions of dollars selling an enzyme that was first isolated from cells discovered in Yellowstone National Park. Should the federal government let private companies bioprospect within the boundaries of national parks as long as it shares in the profits from any discoveries? Go to www.cengage.com/login to investigate both sides of the issue and then vote.

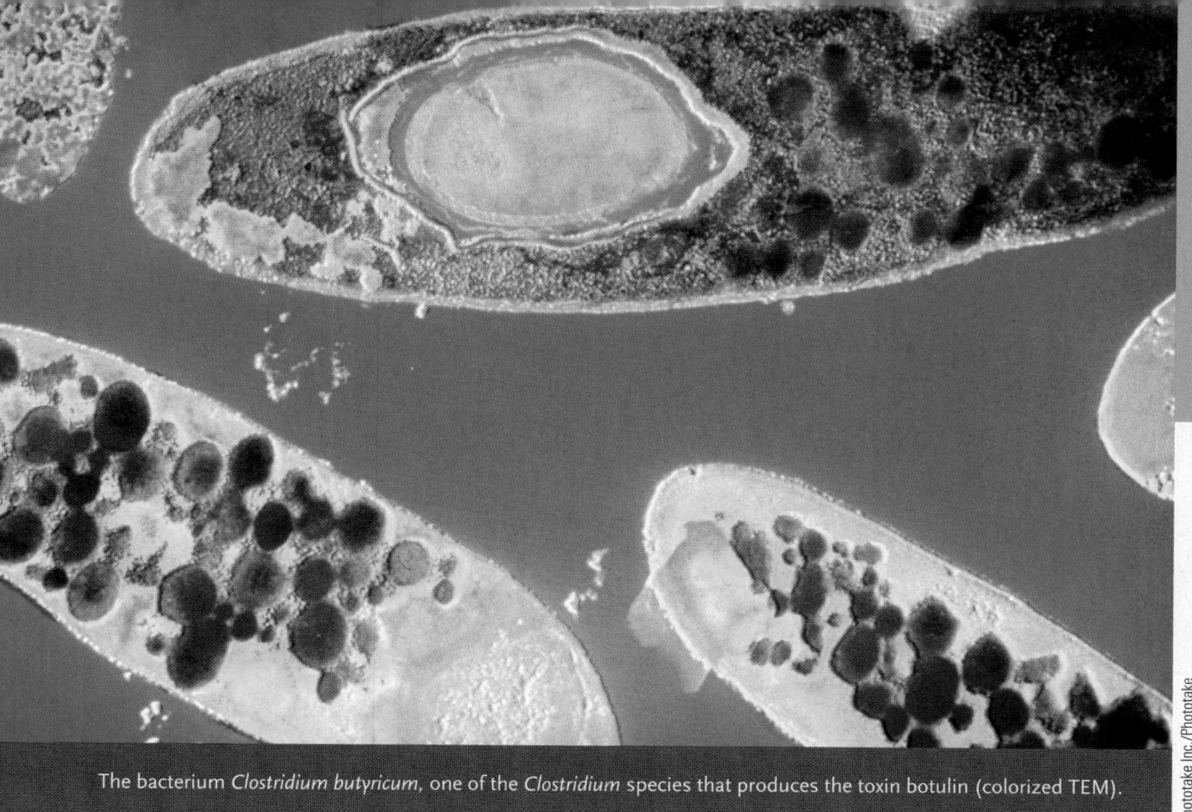

The bacterium *Clostridium butyricum*, one of the *Clostridium* species that produces the toxin botulin (colorized TEM).

Phototake Inc./Phototake

25

STUDY OUTLINE

Prokaryotes: Bacteria and Archaea

Why It Matters. . . You wait in line with anticipation at a fast-food restaurant, biding your time until you reach the counter and get your hamburger. Somewhere in the back of your mind may be the worry that the hamburger will contain bacteria that could make you sick or even kill you. The hamburger you receive will be well done, almost to the crispy stage, because of that fear. Not too many years ago, people were sickened, and a few even died, because their fast-food hamburgers were contaminated by a pathogenic strain of the bacterium *Escherichia coli,* the normally harmless bacteria that inhabit our intestinal tract. Since then, fast-food restaurants have cooked their hamburgers well beyond the point required to kill any lurking *E. coli* or other pathogenic bacteria.

The bacterium *E. coli* is a prokaryote, an organism lacking a true nucleus. Prokaryotes, the topic of this chapter, are the smallest organisms in the world **(Figure 25.1).** Few species are more than 1 to 2 μm long; from 500 to 1,000 of them would fit side by side across the dot above this letter "i."

Prokaryotes are small, but their total collective mass (their *biomass*) on Earth may be greater than that of all plant life. They colonize every niche on Earth that supports life, meaning that they are found essentially everywhere. Huge numbers of bacteria inhabit surfaces and cavities of the human body, including the skin, the mouth and nasal passages, the large intestine, and the vagina. Collectively bacteria in and on the human body outnumber the body's own cells by about tenfold.

Biologists classify prokaryotes into two of the three domains of life, the **Archaea** and the **Bacteria** (the third domain, the **Eukarya,** includes all eukaryotes). Bacteria are the prokaryotic organisms most familiar to us, including many types responsible for diseases of humans and other animals and many other types found in a wide variety of ecosystems. Many members of the domain

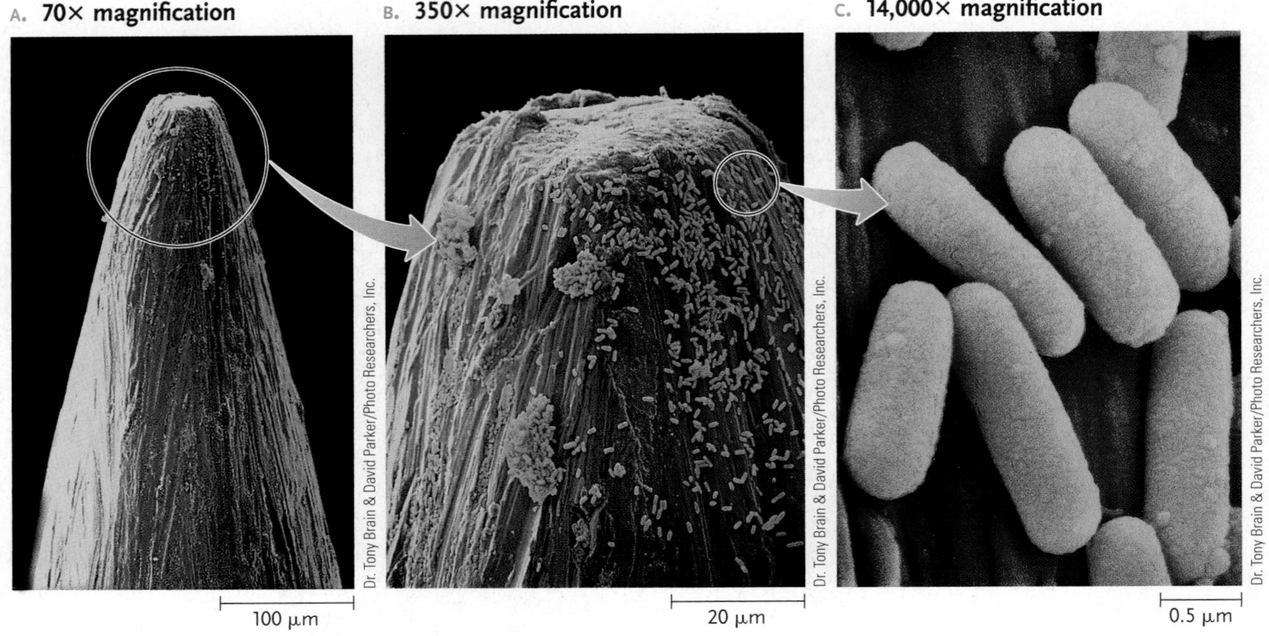
A. 70× magnification
B. 350× magnification
C. 14,000× magnification
100 μm
20 μm
0.5 μm
Dr. Tony Brain & David Parker/Photo Researchers, Inc.
Dr. Tony Brain & David Parker/Photo Researchers, Inc.
Dr. Tony Brain & David Parker/Photo Researchers, Inc.

FIGURE 25.1
Bacillus bacteria on the point of a pin.

Archaea (*arkhaios* = ancient) live under conditions so extreme, including high salinity, acidity, or temperature, that their environments cannot be tolerated by other organisms, including bacteria. In general, archaeans are less well studied, and therefore less well understood compared to bacteria.

As a group, prokaryotes have a wide range of metabolic capabilities. Their metabolic activities are crucial for maintenance of the biosphere. In particular, prokaryotes are the key players in the life-sustaining recycling of the elements carbon, nitrogen, and oxygen. For example, prokaryotes (along with other organisms) break down organic material in dead plants and animals, releasing carbon dioxide that is used for plant growth. In addition, prokaryotes are necessary to make nitrogen available to most forms of life. And a significant amount of the oxygen in the atmosphere originates from bacterial photosynthesis. An illustration of prokaryotes' importance is Biosphere 2, an attempt by scientists to build a completely closed ecosystem in Arizona. The attempt failed, in part because the researchers did not have a complete enough understanding of the activities of the microorganisms in the soil. Through respiration by soil microorganisms, the oxygen level in the Biosphere structure decreased to lower-than-expected levels and the ecosystem could no longer sustain itself. This small-scale example illustrates the essential role of prokaryotes in enabling life of all forms to exist.

Prokaryotes also have specific impacts on the lives of humans. Among other things, they are important for the production of certain foods; they carry out chemical reactions that are of importance in industry; they are used for the production of pharmaceutical products; they cause diseases; and they are used for bioremediation of polluted sites.

In Chapter 24 you learned how prokaryotic cells likely were produced from protocells. Evolution has generated the diversity of prokaryotes on Earth today. In this chapter you will learn about the structure and function of prokaryotes, and about the diverse organisms in the evolutionary branches of Bacteria and Archaea. <

25.1 Prokaryotic Structure and Function

Prokaryotes show great diversity in their ability to colonize areas that can sustain life. Their cells are small and, while prokaryotes show a greater structural diversity than eukaryotes, prokaryotic cells are not as complex in organization as eukaryotic cells. For instance, although prokaryotes do not have a membrane-bound nucleus or organelles, their DNA and some proteins are localized in particular places. They vary in how their cell membrane is protected, and some species have specialized surface structures that protect them from their environment or that enable them to move actively. Prokaryotes also show great diversity in the ways they obtain energy and in their metabolic activities.

The diversity of prokaryotes has arisen through rapid adaptation to their environments as a result of evolution by natural selection. Genetic variability in prokaryotic populations, the basis for this rapid adaptation, derives largely from mutation and to a lesser degree from the transfer of genes between organisms by transformation, transduction, and conjugation (see Chapter 17). Since prokaryotes have much shorter generation times than eukaryotes and smaller genomes (roughly 1,000 times smaller than an average eukaryote), prokaryotes have roughly 1,000 times more mutations per gene, per unit time, per individual, than do eukaryotes. Furthermore, prokaryotes typically have much larger population sizes than eukaryotes, which contrib-

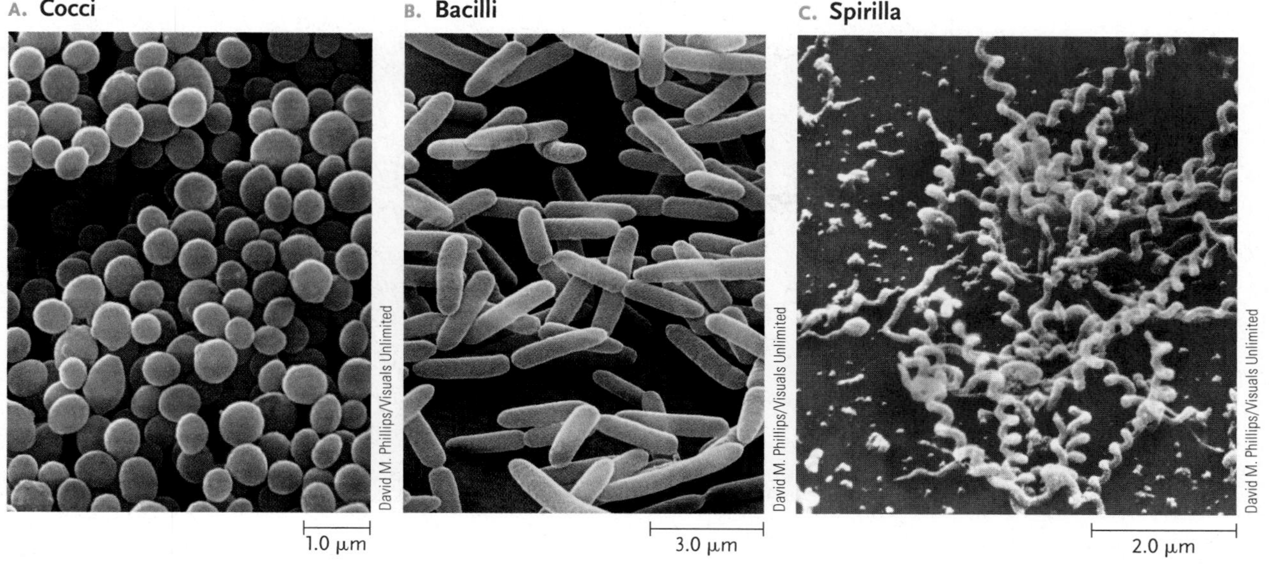
A. Cocci
B. Bacilli
C. Spirilla
1.0 µm
3.0 µm
2.0 µm
David M. Phillips/Visuals Unlimited
David M. Phillips/Visuals Unlimited
David M. Phillips/Visuals Unlimited

FIGURE 25.2
Common shapes among prokaryotes. **(A)** Scanning electron microscope (SEM) image of cocci of *Micrococcus* bacteria; **(B)** SEM of bacilli of *Salmonella* bacteria; **(C)** SEM of spirilla of *Spiroplasma* bacteria.

utes to their greater genetic variability. In short, prokaryotes have an enormous capacity to adapt, and this has been the key to their evolutionary success.

Prokaryotes Are Different in Structure and Organization from Eukaryotic Cells

Like eukaryotic cells, prokaryotic cells are bounded by a plasma membrane. Prokaryotic cells also have a cell wall which provides reinforcement and protection. Within the cell, the genetic material of prokaryotes is localized to a non-membrane enclosed area of the cell, rather than being in a membrane-bound nucleus as in eukaryotes. Many prokaryotic species contain few if any internal membranes; they lack, therefore, the membrane-bound organelles characteristic of eukaryotic cells, including mitochondria, endoplasmic reticulum, Golgi complex, and chloroplasts. With few exceptions, the reactions carried out by these organelles in eukaryotes are distributed between the cytoplasmic solution and the plasma membrane in prokaryotes.

As you learned in Section 5.3, eukaryotic cells have well-organized cytoskeletons. Cytoskeletons were once thought to be absent in prokaryotes, but recent research has revealed that prokaryotic cells have filamentous cytoskeleton structures with functions similar to those in eukaryotes. Prokaryotic cytoskeletons play important roles in creating and maintaining the proper shape of cells, in cell division and, in certain prokaryotes, in determining polarity of the cells.

Although prokaryotic cells appear relatively simple according to this description, their simplicity is deceptive. Most can use a variety of substances as energy and carbon sources, and they can synthesize almost all of their required organic molecules from simple inorganic raw materials. In many respects, prokaryotes are more biochemically versatile than eukaryotes.

The following specifics of prokaryotic structure illustrate some of the diversity seen in this group of organisms.

SHAPE Three shapes are common among prokaryotes: spheres, rods, and spirals **(Figure 25.2)**. The spherical prokaryotes are **cocci** (singular, *coccus*) (*kokkos* = berry). Cylindrical or rod-shaped prokaryotes are **bacilli** (singular, *bacillus*) (*bacillus* = small staff or rod). The spiral prokaryotes are the **spirilla** (singular, *spirillum*) (*speira* = coil, twist, wreath), which are twisted helically like a corkscrew, and **vibrios** (*vibrare* = to move quickly to and fro), which are curved and commalike. Among the prokaryotes of all structural types are some that live singly and others that link into chains or aggregates of cells.

INTERNAL STRUCTURES The genome of most prokaryotes consists of a single, circular DNA molecule called the *prokaryotic chromosome.* There are exceptions: a few bacterial species, for example the causative agent of Lyme disease *(Borrelia burgdorferi),* have a linear chromosome. Genome sequencing projects have shown that the range of genome sizes among bacteria and archaeans is about 20-fold, with the smallest genome of any organism that can be cultured in the laboratory, that of *Mycoplasma genitalium,* being about 580,000 bp. In all prokaryotes, the chromosome is packed into a central area of the cell called the **nucleoid.** There is no nucleolus in the nucleoid, and it has no boundary membranes equivalent to the nuclear envelope of eukaryotes (**Figure 25.3;** and see Sections 5.2 and 5.3). The arrangement of the genetic material in prokaryotes permits the translation of mRNA transcripts of protein-coding genes to begin while transcription is in progress. This is possible because the mRNA is immediately accessible to ribosomes, the cellular structures upon which translation takes place. Simultaneous transcription and translation is not possible in eukaryotes because the ribosomes are in the cyto-

FIGURE 25.3
The structures of a bacterial cell.

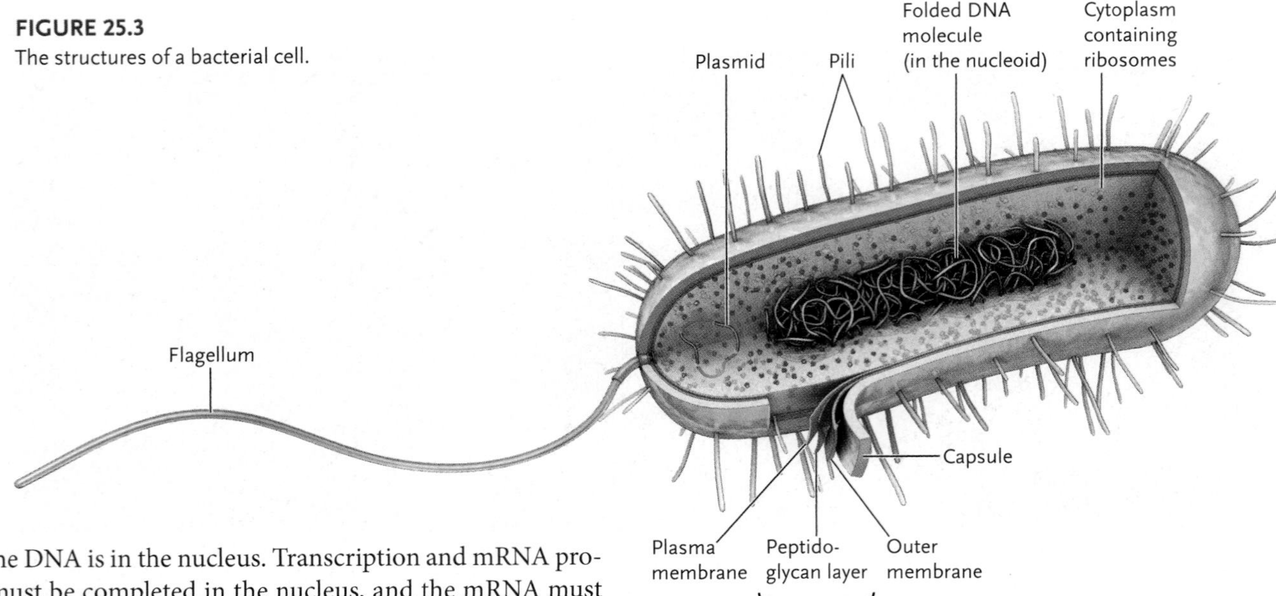

Folded DNA molecule (in the nucleoid)
Cytoplasm containing ribosomes
Plasmid
Pili
Flagellum
Capsule
Plasma membrane
Peptido-glycan layer
Outer membrane
Cell wall

sol and the DNA is in the nucleus. Transcription and mRNA processing must be completed in the nucleus, and the mRNA must enter the cytosol before translation can begin. (see Section 15.3).

Besides the DNA of the nucleoid, many prokaryotes also contain small circles of DNA called **plasmids,** which are distributed in the cytoplasm. The plasmids, which often contain genes with functions that supplement those in the nucleoid, contain a replication origin that allows them to replicate along with the nucleoid DNA and be passed on during cell division (see Section 14.5).

Prokaryotic ribosomes are smaller than eukaryotic ribosomes and contain fewer proteins and RNA molecules. Archaeal ribosomes resemble those of bacteria in size, but differ in structure. Scientists have demonstrated that, with some differences in detail, bacterial ribosomes carry out protein synthesis by the same mechanisms as those of eukaryotes (see Section 15.4). Interestingly, protein synthesis in archaeans is a combination of bacterial and eukaryotic processes, with some unique archaeal features. As a result, antibiotics that stop bacterial infections by targeting ribosome activity do not stop protein synthesis of archaeans.

Some prokaryotes are capable of photosynthesis. Photosynthetic bacteria and archaeans have complex layers of intercellular membranes formed by invaginations of the plasma membrane on which photosynthesis takes place. Those membranous structures correspond to those that carry out photosynthesis in plants, but they are organized differently.

The cytoplasm of many prokaryotes also contains storage granules holding glycogen, lipids, phosphates, or other materials. The stored material is used as an energy reserve or a source of building blocks for synthetic reactions.

PROKARYOTIC CELL WALLS The plasma membrane of prokaryotes must withstand both high intracellular osmotic pressures and the action of natural chemicals in the environment that have detergent properties. Most prokaryotes have one or more layers of materials coating the plasma membrane to form a cell wall that provides the necessary protection, although the structure of the wall is quite different in bacteria and archaeans.

Bacteria typically are surrounded by a cell wall. The primary structural molecule of bacterial cell walls is **peptidoglycan,** a polymer formed from a polysaccharide backbone tied together by short polypeptides. (Polysaccharides are described in Section 3.2.) Peptidoglycan varies in chemical structure among different bacterial species.

Differences in bacterial cell wall composition are of clinical importance. In 1882, Hans Christian Gram, a Danish physician, developed a staining method to distinguish two types of bacteria in body fluids, each of which could cause pneumonia. In this **Gram stain technique,** an investigator treats bacteria with the dye crystal violet and then with iodine, which fixes the dye to the cell wall. Next the bacteria are washed with alcohol, and then treated with a second stain, either fuchsin or safranin. Bacteria that appear purple after these steps have retained the crystal violet stain; they are **Gram-positive.** Bacteria that appear pink after these steps have lost the crystal violet stain in the alcohol wash and are stained pink with the second dye; they are **Gram-negative.** (Gram-positive cells also react with the second dye, but the stain does not affect the color imparted by the crystal violet.)

The staining difference reflects differences in the cell walls of the bacteria **(Figure 25.4).** The cell wall of typical Gram-positive bacteria consists of a thick peptidoglycan layer (see Figure 25.4A). In contrast, the cell wall of typical Gram-negative bacteria consists of a thin layer of peptidoglycan (see Figure 25.4B). Outside of the thin cell wall is an additional boundary membrane, called the **outer membrane,** which covers the peptidoglycan layer. The outer membrane contains **lipopolysaccharides,** assembled from lipid and polysaccharide subunits found nowhere else in nature. The outer membrane protects Gram-negative bacteria from potentially damaging substances in their environment. For example, the outer membrane of *E. coli* protects it from the detergent effects of bile released into the intestinal tract, which otherwise would lyse (break open) the bacterium and kill it.

Rapidly distinguishing between Gram-positive and Gram-negative bacteria is important for determining the first line of treatment for bacterial-caused human diseases. Most pathogenic

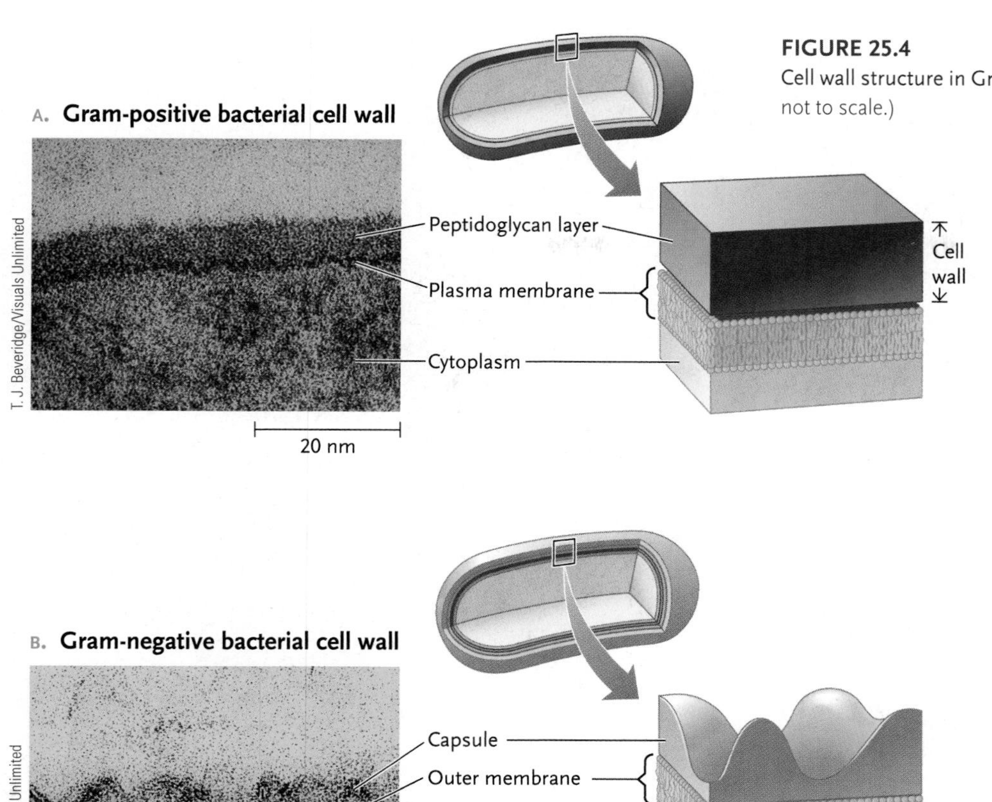
A. Gram-positive bacterial cell wall
Peptidoglycan layer
Plasma membrane
Cytoplasm
Cell wall
20 nm
T. J. Beveridge/Visuals Unlimited
B. Gram-negative bacterial cell wall
Capsule
Outer membrane
Peptidoglycan layer
Plasma membrane
Cytoplasm
Cell wall
20 nm
T. J. Beveridge/Visuals Unlimited

FIGURE 25.4
Cell wall structure in Gram-positive and Gram-negative bacteria. (Drawings not to scale.)

bacteria are Gram-negative species; their outer membrane protects them against the body's defense systems and blocks the entry of drugs such as antibiotics. For example, the antibiotic penicillin blocks new bacterial cell wall formation by inhibiting peptidoglycan crosslinking. The weakened cell wall soon leads to the death of the bacterium. Penicillin is effective against Gram-positive pathogens, but it is less effective against Gram-negative pathogens because their outer membrane inhibits entry of the antibiotic.

The walls of many Gram-positive and Gram-negative bacteria are coated with an external layer of polysaccharides called the **glycocalyx,** a slime coat typically composed of polysaccharides. When the glycocalyx is diffuse and loosely associated with the cells, it is a **slime layer;** when it is gelatinous and more firmly attached to the cells, it is a **capsule (Figure 25.5).** Depending on the species, the capsule ranges from a layer that is thinner than the cell wall to many times thicker than the entire cell. The glycocalyx helps protect cells from physical damage and desiccation, and to the effects of some antibiotics, and may enable a cell to attach to a surface.

In many bacteria, the capsule prevents bacterial viruses and molecules such as enzymes, antibiotics, and antibodies from reaching the cell surface. In many pathogenic bacteria, the presence or absence of the protective capsule differentiates infective from noninfective forms. For example, normal *Streptococcus pneumoniae* bacteria are capsulated and are virulent, causing severe pneumonia in humans and other mammals. Mutant *S. pneumoniae* without capsules are nonvirulent and can easily be eliminated by the body's immune system if they are injected into mice or other animals (see Section 14.1).

Archaeans differ from bacteria by having some unique plasma membrane and cell wall characteristics. The lipid molecules in archaean plasma membranes have a chemical bond between the hydrocarbon chains and glycerol that is unlike that found in the plasma membranes of all other organisms. The difference is significant because the exceptional linkage is more resistant to disruption, making the plasma membranes of archaeans more tolerant of the extreme environmental conditions under which many of these organisms live.

The cell walls of some archaeans are assembled from molecules related to the peptidoglycans, but that have different molecular components and bonding structure. Others have walls assembled from proteins or polysaccharides instead of peptidoglycans. The cell walls of archaeans are as resistant to physical disruption as the plasma membranes are; some archaeans can be boiled in strong detergents without disruption. Different archaeans stain as either Gram-positive or Gram-negative.

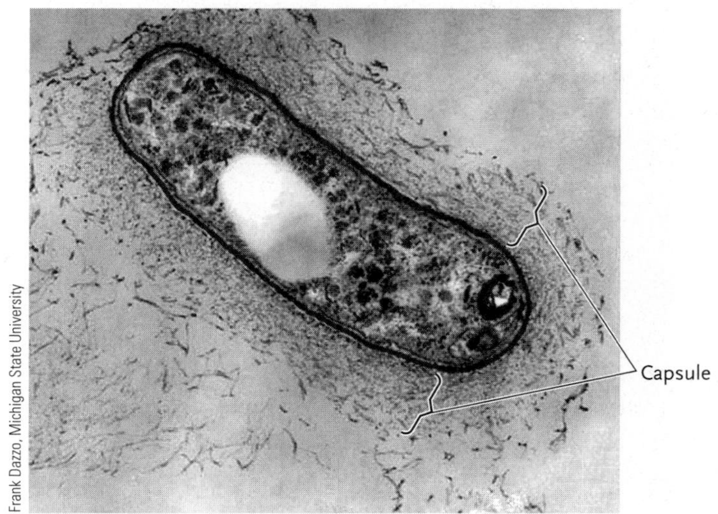
Capsule
Frank Dazzo, Michigan State University

FIGURE 25.5
The capsule surrounding the cell wall of *Rhizobium,* a Gram-negative, soil-dwelling bacterium.

FIGURE 25.6
A flagellum of a Gram-negative bacterium. A proton (H^+) gradient drives the motor, which rotates the flagellum in a counterclockwise direction.

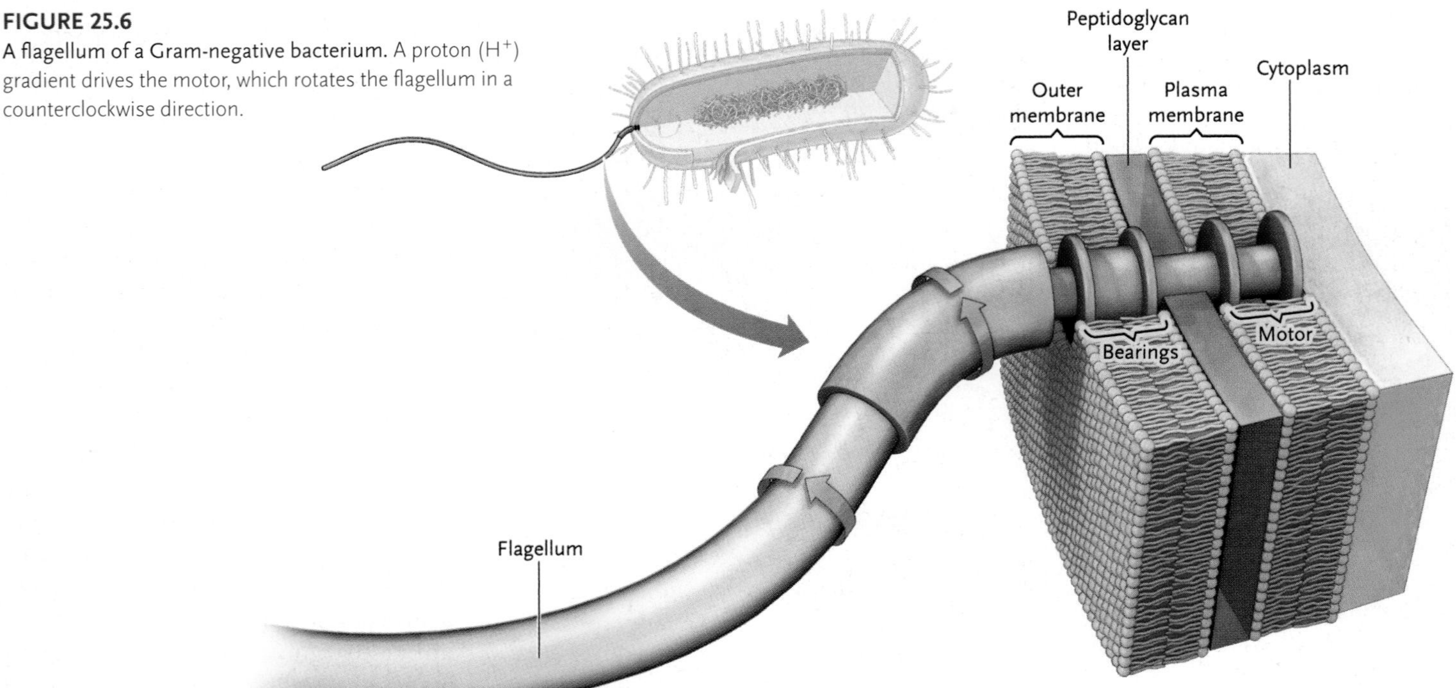
Peptidoglycan layer
Outer membrane
Plasma membrane
Cytoplasm
Bearings
Motor
Flagellum

FLAGELLA AND PILI Many bacteria and archaeans can move actively through liquids and across wet surfaces. The most common mechanism for movement involves the action of **flagella** (singular, *flagellum*), (*flagellum* = whip) extending from the cell wall **(Figure 25.6).** These flagella are much smaller and simpler than the flagella of eukaryotic cells and contain no microtubules (eukaryotic flagella are discussed in Section 5.3).

A bacterial flagellum consists of a helical fiber of protein that rotates in a socket in the cell wall and plasma membrane, much like the propeller of a boat. The rotation, produced by what is essentially a tiny electric motor, pushes the cell through liquid. The motor is powered by a gradient of hydrogen or sodium ions, which flow through it as positive charges, creating an electrical repulsion that makes the flagellum rotate. In some bacteria, flagella group together in bundles that rotate to propel the cell.

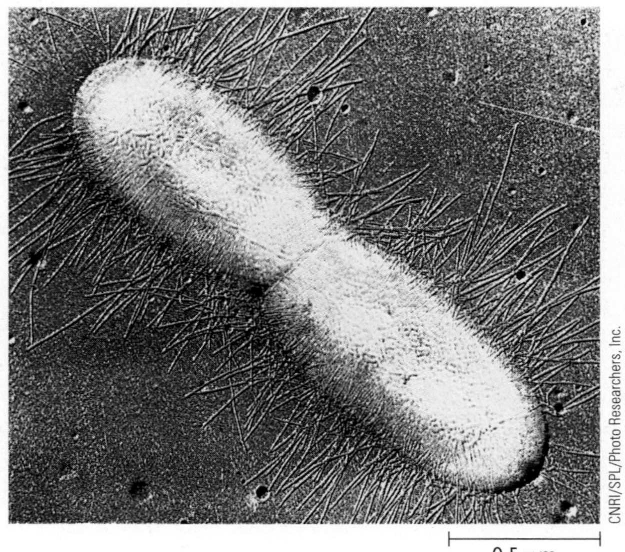
0.5 µm
CNRI/SPL/Photo Researchers, Inc.

FIGURE 25.7
Pili extending from the surface of a dividing *E. coli* bacterium.

Archaeal flagella are analogous, not homologous, to bacterial flagella. That is, they carry out the same function, but they differ in their structure and mechanisms of action. The genes for the two types of flagellar systems are also different.

Some bacteria and archaeans have rigid, hairlike shafts of protein called **pili** (singular, *pilus*) extending from their cell walls **(Figure 25.7).** Among bacteria, pili are characteristic primarily of Gram-negative bacteria; relatively few Gram-positive bacteria produce these structures. Recognition proteins at the tips of pili allow bacterial cells to adhere to other cells. One type, called *sex pili,* allows bacterial cells to adhere to each other as a prelude to conjugation, a joining of two cells that leads to transfer of genetic material from one to the other (see Section 17.1). Other types help bacteria to bind to animal cells. For example, *Neisseria gonorrhoeae,* the Gram-negative bacterium that causes gonorrhea, has pili that allow it to attach to cells of the throat, eye, urogenital tract, or rectum in humans.

Prokaryotes Have the Greatest Metabolic Diversity of All Living Organisms

All organisms take in carbon and energy in some form, but prokaryotes show the greatest diversity in their modes of securing these resources **(Figure 25.8).** Prokaryotes are classified on the basis of the source of carbon atoms they use. Some prokaryotes are **autotrophs** (*auto* = self; *trophe* = nourishment), meaning that they, like plants, obtain carbon from CO_2 (which is considered to be an inorganic molecule). Others are **heterotrophs,** meaning that they, like humans and other animals, obtain carbon from organic molecules. Prokaryotic heterotrophs obtain carbon

Carbon source		Energy source: Oxidation of molecules*	Energy source: Light
	CO_2	**CHEMOAUTOTROPH** Found in some bacteria and archaeans; not found in eukaryotes	**PHOTOAUTOTROPH** Found in some photosynthetic bacteria, in some protists, and in plants
	Organic molecules	**CHEMOHETEROTROPH** Include some bacteria and archaeans, and also in protists, fungi, animals, and plants	**PHOTOHETEROTROPH** Found in some photosynthetic bacteria

*Inorganic molecules for chemoautotrophs and organic molecules for chemoheterotrophs.

FIGURE 25.8
Modes of nutrition among Bacteria and Archaea. All four modes of nutrition occur in the Bacteria, with chemoheterotrophs as the most common type. Among the Archaea, chemoautotrophs are most common, while other archaeans are chemoheterotrophs.

from the organic molecules of living hosts, or from organic molecules in the products, wastes, or remains of dead organisms.

Prokaryotes are also classified according to the source of the energy they use to drive biological activities. **Chemotrophs** (*chemo* = chemical; *trophe* = nourishment) obtain energy by oxidizing inorganic or organic substances, while **phototrophs** obtain energy from light. Combining carbon source and energy gives us the following four types (see Figure 25.8):

1. **Chemoautotrophs:** Prokaryotic chemoautotrophs obtain energy by oxidizing inorganic substances such as hydrogen, iron, sulfur, ammonia, nitrites, and nitrates and use CO_2 as their carbon source. They use the electrons they remove in the oxidations to make organic molecules by reducing CO_2 or to provide the energy for ATP synthesis (using an electron transfer system embedded in the plasma membrane). Chemoautotrophy occurs widely among prokaryotes, including many bacteria and most archaeans, but is not found among eukaryotes.
2. **Chemoheterotrophs:** Prokaryotic chemoheterotrophs oxidize organic molecules as their energy source and obtain carbon in organic form. They include most of the bacteria that cause disease in humans, domestic animals, and plants and many bacteria that are responsible for decomposing matter. They are the largest prokaryotic group in terms of numbers of known species.
3. **Photoautotrophs:** Photoautotrophs are photosynthetic organisms that use light as their energy source and CO_2 as their carbon source. They include several groups of bacteria, for example, the *cyanobacteria,* the *green sulfur bacteria,* and the *purple sulfur bacteria,* as well as plants and many protists. The cyanobacteria use water (H_2O) as their source of electrons for reducing CO_2, while the two types of sulfur bacteria use sulfur or sulfur compounds.
4. **Photoheterotrophs:** Photoheterotrophs use light as their ultimate energy source but obtain carbon in organic form rather than as CO_2. Photoheterotrophs are limited to two groups of bacteria, the *green* and *purple nonsulfur bacteria.* "Nonsulfur" indicates they are unable to oxidize sulfur or other inorganic substances as an ultimate source of electrons for reductions; instead, they use a variety of substrates, including H_2, alcohols, or organic acids.

Prokaryotes Differ in Whether Oxygen Can Be Used in Their Metabolism

Prokaryotes also differ in how their metabolic systems function with respect to oxygen (see Chapter 8). **Aerobes** require oxygen for cellular respiration (in other words, oxygen is the final electron acceptor for that process); **obligate aerobes** cannot grow without oxygen. **Anaerobes** do not require oxygen to live. **Obligate anaerobes** are poisoned by oxygen, and survive either by fermentation, in which organic molecules are the final electron acceptors (see Section 8.5), or by a form of respiration in which inorganic molecules such as nitrate ions (NO_3^-) or sulfate ions (SO_4^{2-}) are used as final electron acceptors. **Facultative anaerobes** use O_2 when it is present, but under anaerobic conditions, they live by fermentation.

Prokaryotes Fix and Metabolize Nitrogen

Nitrogen is a component of amino acids and nucleotides and, hence, is of vital importance for the cell. Prokaryotes are able to metabolize nitrogen in many forms. For example, a number of bacteria and archaeans are able to reduce atmospheric nitrogen (N_2, the major component of Earth's atmosphere) to ammonia (NH_3), a process called **nitrogen fixation.** The ammonia is quickly ionized to ammonium (NH_4^+), which the cell then uses in biosynthetic pathways to produce nitrogen-containing molecules such as amino acids and nucleic acids. Nitrogen fixation is an exclusively prokaryotic process and is the only means of replenishing the nitrogen sources used by most microorganisms and by all plants and animals. In other words, all organisms use nitrogen that is fixed by bacteria. Examples of nitrogen-fixing bacteria are some of the cyanobacteria and *Azotobacter* species which are free-living, and *Rhizobium* species which are in symbiotic relationships with plants (see Chapter 33).

Not all bacteria convert fixed nitrogen directly into organic molecules. Some bacteria carry out **nitrification,** the conversion of ammonium (NH_4^+) to nitrate (NO_3^-). This is carried out in two steps by two types of *nitrifying bacteria.* One type of nitrifying bacterium converts ammonium to nitrite (NO_2^-) (for example, *Nitrosomonas* species), while the other converts nitrite to nitrate (for example, *Nitrobacter* species). Because of this specialization, both types of nitrifying bacteria are usually present in soils and water, with some converting ammonium to nitrite and others using that nitrite to produce nitrate. The nitrate can be used by plants and fungi to incorporate nitrogen into organic molecules. Animals obtain nitrogen in organic form by eating other organisms.

In sum, nitrification makes nitrogen available to many other organisms, including plants and animals and bacteria that cannot metabolize ammonia. You will learn more about nitrogen metabolism in connection with the nitrogen cycle (see Chapter 51). The metabolic versatility of the prokaryotes is one factor that accounts for their abundance and persistence on the planet; another factor is their impressive reproductive capacity.

Prokaryotes Pass Genes to Progeny by Asexual Reproduction or, Rarely, from Cell to Cell by Horizontal Gene Transfer

In prokaryotes, asexual reproduction is the normal mode of reproduction. In this process, a parent cell divides by binary fission into two daughter cells that are exact genetic copies of the parent (see Figure 10.18). In this process, genetic material moves from generation to generation, that is, by descent.

Conjugation, in which two parent cells join or "mate," occurs in some bacterial and archaeal species. Conjugation depends upon genes carried by a plasmid that replicates separately from the prokaryotic chromosome. Usually only the plasmid is passed on during conjugation, but in some bacteria, the plasmid integrates into the chromosome of the host so that host genes transfer from one parent (donor) to the other (recipient). Genetic recombination then occurs, thereby achieving a prokaryotic form of sexual reproduction. The recombinant cell divides to produce daughter cells that differ in genetic information from either parent. The movement of genetic material between organisms by means other than descent, as is the case with conjugation, is **horizontal gene transfer.** Horizontal gene transfer by conjugation between bacterial cells is described in Section 17.1).

A small number of bacteria can produce an **endospore,** so-called because it develops *within* the cell **(Figure 25.9).** The endospore, which typically develops when environmental conditions become unfavorable, is essentially a dormant cell containing the bacterial genome and some of the cytoplasm. It is metabolically inactive and highly resistant to heat, desiccation, and attack by enzymes or other chemical agents. In the first step of endospore formation, binary fission cuts the parent cell into two parts of unequal size. The larger cell then envelops the smaller one (which contains the genome) and surrounds it with a tough, chemically resistant protein coat; the smaller cell develops into the endospore. Rupture of the larger cell releases the endospore to the environment. If environmental conditions become favorable for growth, the spore germinates: it becomes permeable, water enters the cell, its surface coat breaks, and the cell is released in a metabolically active form.

No one is certain how long endospores can survive. There are claims that endospores survive for thousands or millions of years, but the data are controversial. They certainly can survive a very long time.

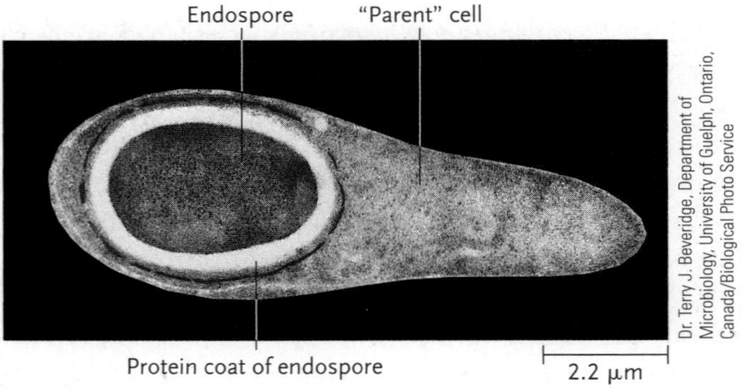
Endospore
"Parent" cell
Protein coat of endospore
2.2 μm
Dr. Terry J. Beveridge, Department of Microbiology, University of Guelph, Ontario, Canada/Biological Photo Service

FIGURE 25.9
A developing endospore within a cell of the bacterium *Clostridium tetani*, a dangerous pathogen that causes tetanus.

In Nature, Bacteria May Live in Communities Attached to a Surface

Researchers grow prokaryotes as individuals in liquid cultures or as isolated colonies on solid media. The results from studies using pure cultures have been crucial in developing an understanding of, among many other things, the nature of the genetic material, DNA replication, gene expression, and gene regulation. But, in nature, bacteria typically are not in pure cultures. Instead, there are mixed populations of bacterial species, perhaps also with archaeal and eukaryotic microbial species. Thus, information from pure cultures in the laboratory may not apply to populations of the same species in nature.

Researchers have discovered that, in nature, prokaryotes often live in communities where they interact in a variety of ways. The communities may consist of one or more species of bacteria, or archaeans, or both bacteria and archaeans. Eukaryotic microorganisms may also be in the communities. One important type of prokaryotic community is known as a **biofilm,** which consists of a complex aggregation of microorganisms attached to a surface. Benefits of biofilm formation to prokaryotes include adherence of the organisms to hospitable surfaces, the transfer of genes between species, and living off the products of other organisms in the biofilm. Biofilms form on any surface that has sufficient water and nutrients for prokaryotes to grow. For instance, they may be found on lake surfaces, on rocks in freshwater or marine environments (making the rocks slippery), surrounding plant roots and root hairs, and on animal tissues such as the intestinal mucosa and teeth (human dental plaque is a biofilm).

Biofilms have practical consequences for humans, which are both beneficial and detrimental. On the beneficial side, for example, biofilms on solid supports are used in sewage treatment plants for processing organic matter before the water is discharged. They can also be effective in the bioremediation (biological clean-up) of toxic organic molecules that contaminate groundwater. On the detrimental side, however, biofilms can be harmful to human health. For example, biofilms adhere to many kinds of surgical equipment and supplies, including catheters and synthetic implants such as pacemakers and artificial joints.

When pathogenic bacteria are involved, biofilms can cause infections which are difficult to treat, because pathogenic bacteria in a biofilm are up to 1,000 times more resistant to antibiotics than are the same bacteria in pure liquid cultures. Other examples of medical conditions resulting from the activities of biofilms include middle-ear infections, bacterial endocarditis (an infection of the heart's inner lining or the heart valves), and Legionnaire's disease (an acute respiratory infection caused by breathing in pieces of biofilms containing the pathogenic bacterium *Legionella*).

Imagine a surface, living or environmental, over which water that contains nutrients is flowing. The surface rapidly becomes coated with polymeric organic molecules from the liquid, such as polysaccharides or glycoproteins. **Figure 25.10** shows how a biofilm can form on such a surface.

Genomic and proteomic studies have shown that the changes in prokaryote physiology that accompany the formation of a biofilm result from marked changes in the prokaryote's gene expression pattern. In effect, the prokaryote becomes a significantly different organism. This discovery is especially important in the case of pathogenic bacteria, since most research on the control of those bacteria is done with liquid cultures. The challenge now is to devise new treatment strategies for biofilm-caused diseases. If we can gain a better understanding of the genetic changes involved in the transition from free-floating to biofilm state, then we may be able to devise treatments that will switch the bacteria back to the free-living state, where they are more susceptible to antibiotics.

In sum, we must recognize that rather than living as individuals as once was thought, prokaryotes typically live in communities in nature. Much remains to be learned about how bacteria form a biofilm, how the change in gene expression during the transition is regulated, and how they interact within the biofilm.

STUDY BREAK 25.1

1. **What distinguishes a prokaryotic cell from a eukaryotic cell?**
2. **What is the difference between a chemoheterotroph and a photoautotroph?**
3. **What is the difference between an obligate anaerobe and a facultative anaerobe?**
4. **What is the difference between nitrogen fixation and nitrification? Why are nitrogen-fixing prokaryotes important?**
5. **What is a biofilm? Give an example of a biofilm that is beneficial to humans and one that is harmful.**

THINK OUTSIDE THE BOOK

On your own or collaboratively, research a study in which genomic experiments have shown marked changes in gene expression pattern when a bacterium forms a biofilm. Outline how the experiment was done, and summarize the results of the experiments.

25.2 The Domain Bacteria

Figure 25.11 presents a simplified phylogenetic tree of the domains Bacteria and Archaea (the Eukarya is the third domain of life). This tree is a model based on genome sequence comparisons of a number of prokaryotes. It shows five major evolutionary groups; complete trees show several more groups. Undoubtedly the details of the tree in terms of the species represented will be refined and altered as more data are collected.

As you learned in Section 23.6, most systematists construct phylogenetic trees based on the analysis of molecular sequences,

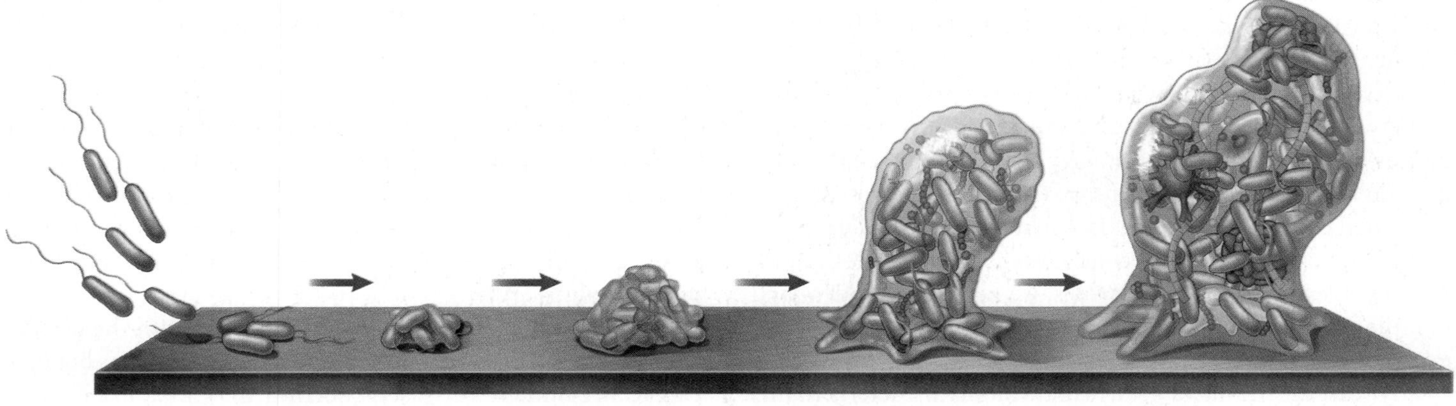

1 Reversible attachment of bacteria (seconds). Free bacteria adhere to a surface coated with organic molecules using van der Waals forces.
2 Irreversible attachment of bacteria (seconds–minutes). Bacteria not immediately leaving the surface become more strongly anchored using structures such as pili.
3 Growth and division of bacteria (hours–days). A bacterial colony forms.
4 Physiology of the bacteria changes and the cells secrete extracellular polymer substances (EPS), a slimy, gluelike substance similar to molecules found in bacterial slime layers. The EPS forms a matrix that binds cells to each other and anchors the complex to the surface, establishing the biofilm (hours–days).
5 EPS matrix entraps a variety of materials, including dead cells and insoluble minerals. Over time, other organisms are attracted to and join the biofilm. Depending on the environment, those organisms may include other bacterial species, algae, fungi, and/or protists, producing diverse microbial communities (days–months).

FIGURE 25.10
Steps in the formation of a biofilm.

such as DNA sequences and amino acid sequences of genes. Using molecular data on gene sequences in different organisms, researchers can hypothesize how the observed variations in genes may be explained in terms of evolutionary lineages—descent with modification, if you will. The key assumption in such analysis is vertical inheritance, namely that genes are passed from parents to offspring.

A drawback of gene-based analysis is that data from different genes sometimes yield different phylogenetic trees. A way around this problem is not to use one or a few genes for molecular phylogenetics, but to use the complete sequences of genomes. This is part of the research done by investigators in the field of comparative genomics (see Section 18.3).

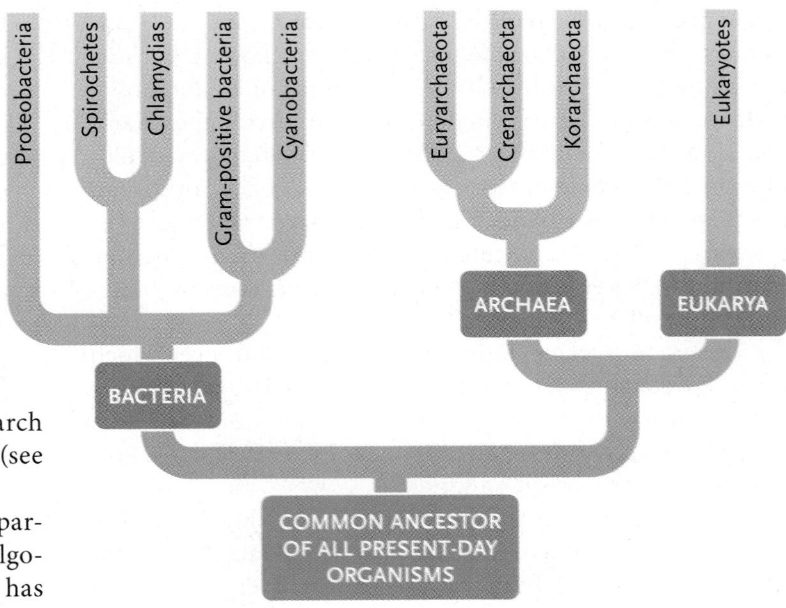
Proteobacteria
Spirochetes
Chlamydias
Gram-positive bacteria
Cyanobacteria
Euryarchaeota
Crenarchaeota
Korarchaeota
Eukaryotes
ARCHAEA
EUKARYA
BACTERIA
COMMON ANCESTOR OF ALL PRESENT-DAY ORGANISMS

FIGURE 25.11
An abbreviated and simplified phylogenetic tree of prokaryotes showing the five important evolutionary branches.

Many more prokaryotic genomes are available for comparison than is true for eukaryotes. Sophisticated computer algorithms are used to make such comparisons. This analysis has revealed that, at least in prokaryotes, there is much more to evolution than vertical inheritance. In particular, and most importantly, there is horizontal gene transfer (see Section 25.1). For scientists trying to reconstruct the phylogenetic history of a group of organisms on the basis of gene sequences, horizontal gene transfer muddies the waters significantly. That is, suppose a gene moved from ancient bacterium A to unrelated ancient bacterium B on a plasmid, and then subsequently moved to bacterium B's chromosome. Based on studying the sequences of the gene in present-day descendants, an investigator might conclude that the two bacterial species are closely related whereas genome level studies could show them to be distantly related.

Indeed the available data show that horizontal gene transfer was and is relatively common in prokaryotes. Even the gene for the small subunit of ribosomal RNA is known to have moved between prokaryotic organisms via horizontal gene transfer. As Section 23.6 pointed out, that gene played a key role in establishing the three-domain model for life on Earth.

Researchers are also aware of some horizontal gene transfer between prokaryotes and eukaryotes, adding yet other difficulties to establishing evolutionary lineages with confidence, particularly for the earliest forms of life. Some researchers argue that creating one Tree of Life that depicts the evolution of all life forms accurately and completely is not possible and, especially relevant to the prokaryotes, we may never understand completely how the earliest prokaryotes and eukaryotes evolved.

We discuss the major groups of the domain Bacteria in this section and of the domain Archaea in the next section.

The Five Major Evolutionary Groups of Bacteria have Diverse Properties

In this section we discuss the five major evolutionary groups of bacteria shown in Figure 25.11—the proteobacteria, spirochetes, the chlamydias, the Gram-positive bacteria, and the cyanobacteria.

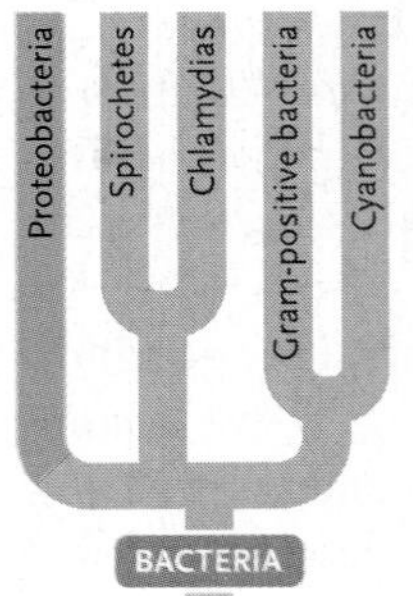
Proteobacteria
Spirochetes
Chlamydias
Gram-positive bacteria
Cyanobacteria
BACTERIA

PROTEOBACTERIA: PURPLE BACTERIA AND THEIR RELATIVES The proteobacteria are a highly diverse group of Gram-negative bacteria that scientists hypothesize derive from a purple, photosynthesizing evolutionary ancestor. Many present-day species retain those characteristics, carrying out photosynthesis as either photoautotrophs (the purple sulfur bacteria) or photoheterotrophs (the purple nonsulfur bacteria). "Purple" refers to the color given to the cells by their photosynthetic pigment, a type of chlorophyll distinct from that of plants. Proteobacteria carry out a type of photosynthesis that does not use water as an electron donor and does not release oxygen as a by-product of photosynthesis.

Other present-day proteobacteria are chemoheterotrophs that are thought to have evolved as an evolutionary branch following the loss of photosynthetic capabilities in an early proteobacterium. The evolutionary ancestors of mitochondria are considered likely to have been ancient nonphotosynthetic proteobacteria.

Among the chemoheterotrophs classified with the proteobacteria are bacteria that cause human diseases such as bubonic plague, Legionnaire's disease, gonorrhea, gastric (stomach) ulcer, and various forms of gastroenteritis and dysentery; bacterial plant pathogens that cause rots, scabs, and wilts; and the colon-inhabiting *E. coli* (shown dividing in Figure 25.7). (**Figure 25.12** describes the experiments demonstrating that a bacterium—*Helicobacter pylori*—causes peptic ulcers.) The proteobacteria also include both free-living and symbiotic nitrogen-fixing bacteria.

Among the more unusual nonphotosynthetic proteobacteria are the slime-secreting myxobacteria ("slime bacteria"). Myxobacteria glide actively over the slime, typically traveling

FIGURE 25.12 **Experimental Research**

Demonstration of the Bacterial Cause of Peptic Ulcers

Question: Do bacteria cause gastric and duodenal ulcers?

Experiment: A gastric ulcer (also called a peptic ulcer) is an erosion of the mucosal layer of the stomach that can potentially penetrate into the deeper layers of the stomach wall. Pepsin and HCl in the stomach irritate the exposed stomach wall, causing extreme pain. Similar ulcers, called duodenal ulcers, can form in the duodenum. In 1982, physicians "knew" that gastric ulcers were caused by excessive amounts of stomach acid that was caused by stress, smoking, personality issues, or susceptibility genes. They treated an ulcer with various kinds of antacids or drugs to inhibit acid production in the stomach. Typically the ulcer symptoms returned after treatment was stopped. Prior to that time, various researchers had observed the presence of spiral or curved bacteria in biopsies taken from the majority of patients with gastric and duodenal ulcers. The bacteria were considered either to be contaminants or not related to ulcer formation.

In 1982, Barry Marshall and J. Robin Warren of Royal Perth Hospital, Perth, Western Australia, tested the hypothesis that bacteria caused peptic ulcers. They took biopsies of peptic ulcers and duodenal ulcers and assayed for the presence of bacteria by Gram staining and culturing on Petri plates.

Results: The results of the biopsies, published in 1984, are shown in the table.

Association of bacteria with biopsy samples

Biopsy Appearance	Total Samples	Number (%) Associated with Bacteria
Gastric ulcer	22	18 (77%)
Duodenal ulcer	13	13 (100%)
Total	35	31 (89%)

The researchers' results showed that almost all cases of gastric or duodenal ulcers were associated with bacteria. The bacteria had a curved appearance, but the species was unknown at the time. It is now known to be the Proteobacterium *Helicobacter pylori*.

Helicobacter pylori

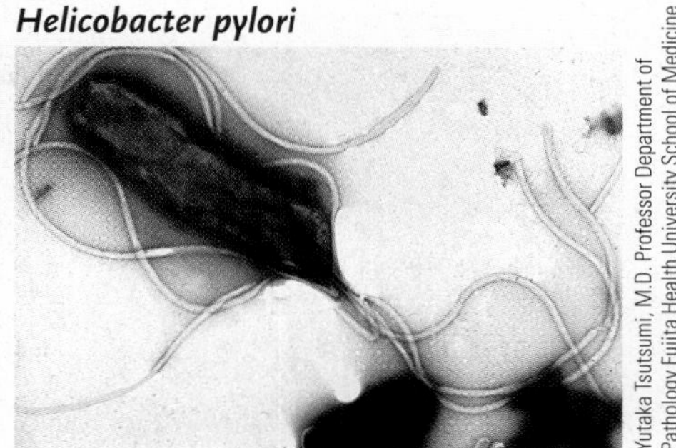

Yutaka Tsutsumi, M.D. Professor Department of Pathology Fujita Health University School of Medicine

Marshall and Warren's results were met with a great deal of skepticism, and logical reasoning did not persuade most physicians that gastric and duodenal ulcers were bacterial diseases. How could it be shown that *H. pylori* can cause gastric and duodenal ulcers? The accepted way to prove that a bacterium is a pathogen is to fulfill Koch's postulates, a set of four conditions proposed in the 19th century by Robert Koch. The postulates are as follows:

1. The bacteria must be present in every case of the disease.
2. The bacteria must be isolated from the infected host and grown in pure culture.
3. The specific disease must be induced when a pure culture of the bacteria is introduced into a healthy host.
4. The bacteria must be isolated from the now experimentally infected host.

Postulates 3 and 4 had not been fulfilled in the study. Attempts to use piglets to fulfill them failed. Therefore, in a bold and highly unusual move, Marshall decided to try to fulfill postulates 3 and 4 himself. He drank 200 mL of a *H. pylori* suspension, then fasted the rest of the day. On the fifth to eighth day after drinking the bacteria, Marshall began feeling very nauseated to the point of vomiting. On the tenth day, a colleague performed a biopsy, the analysis of which showed the presence of *H. pylori* cells. *Helicobacter pylori* was proven to be a pathogen.

Conclusion: Marshall and Warren's experiments had shown that *Helicobacter pylori* can cause gastric and duodenal ulcers. This bacterium probably accounts for more than 80% of cases of such ulcers in humans in the United States. In *H. pylori*-caused ulcers, the effective treatment is an antimicrobial agent such as an antibiotic. Marshall and Warren received a Nobel Prize in 2005 "for their discovery of the bacterium *Helicobacter pylori* and its role in gastritis and peptic ulcer disease."

Sources: B. J. Marshall, J. A. Armstrong, D. B. McGechie, and R. J. Glancy. 1985. Attempt to fulfil Koch's postulates for pyloric Campylobacter. *Medical Journal of Australia* 142:436–439; B. J. Marshall and J. R. Warren. 1984. Unidentified curved bacilli in the stomach of patients with gastritis and peptic ulceration. *The Lancet* 1(8390):1311–1315.

in swarms of cells that stay together using an intercellular signaling system. Enzymes secreted by the swarms digest "prey"—other bacteria, primarily—that become stuck in the slime. When environmental conditions become unfavorable, as when soil nutrients or water are depleted, myxobacteria form a *fruiting body* (**Figure 25.13**), which contains clusters of spores. When the fruiting body bursts, the spores disperse and form new colonies.

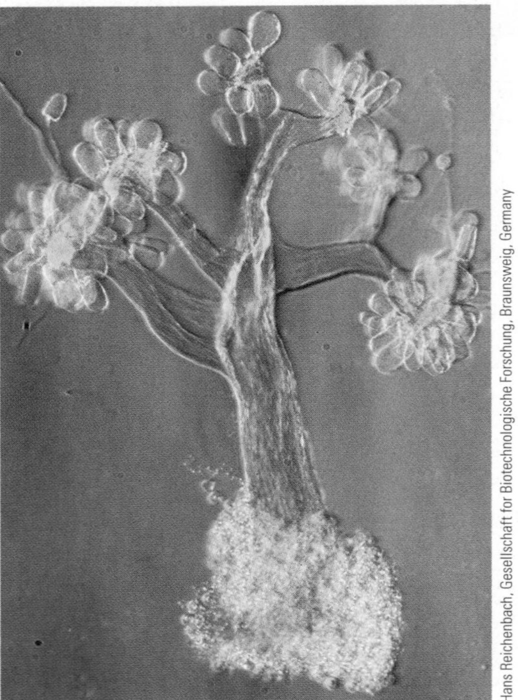

Hans Reichenbach, Gesellschaft für Biotechnologische Forschung, Braunsweig, Germany

FIGURE 25.13
The fruiting body of *Chondromyces crocatus*, a myxobacterium, a member of the Proteobacteria. Cells of this species collect together to form the fruiting body.

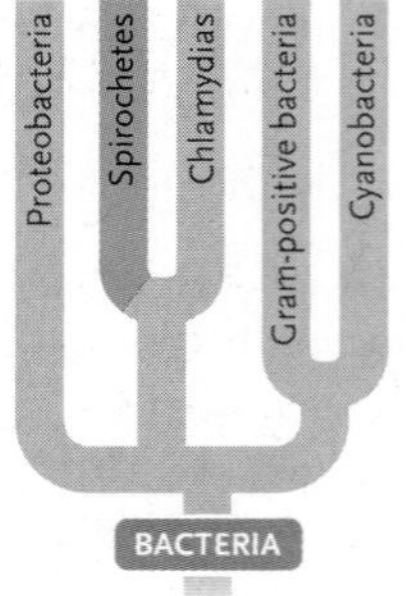
Proteobacteria
Spirochetes
Chlamydias
Gram-positive bacteria
Cyanobacteria
BACTERIA

SPIROCHETES The spirochetes are Gram-negative bacteria with helically spiraled bodies and an unusual form of movement in which their flagella, which are embedded in the region between the outer membrane and the cell membrane, cause the entire cell to twist in a corkscrew pattern. Their corkscrew movements enable them to move in viscous environments such as mud and sewage, where they are common. Two spirochetes, *Treponema denticola* and *Treponema vincentii,* are more or less harmless inhabitants of the human mouth; *Treponema pallidum* is the cause of syphilis. Other pathogenic spirochetes cause relapsing fever and Lyme disease. Beneficial spirochetes in termite intestines aid in the digestion of plant fiber.

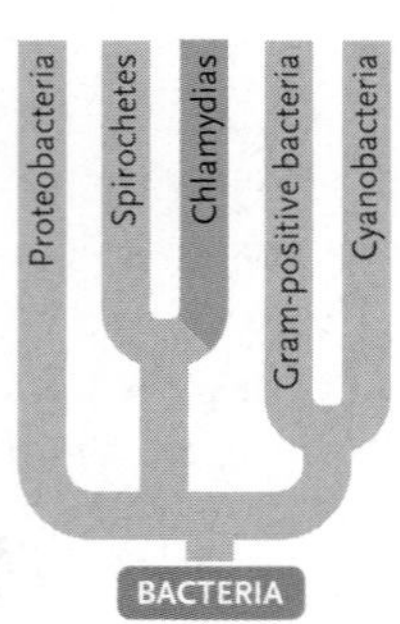
Proteobacteria
Spirochetes
Chlamydias
Gram-positive bacteria
Cyanobacteria
BACTERIA

CHLAMYDIAS The chlamydias are structurally unusual among bacteria because, although they are Gram-negative and have cell walls with a membrane outside of them, they lack peptidoglycan. All the known chlamydias are obligate intracellular parasites, meaning that they can only reproduce within other cells. Chlamydias cause various diseases in animals. One species in this group, *Chlamydia trachomatis* **(Figure 25.14),** is responsible for chlamydia, one of the most common sexually transmitted infections of the urinary and reproductive tracts of humans. The same bacterium causes trachoma, an infection of the cornea that is the leading cause of preventable blindness in humans.

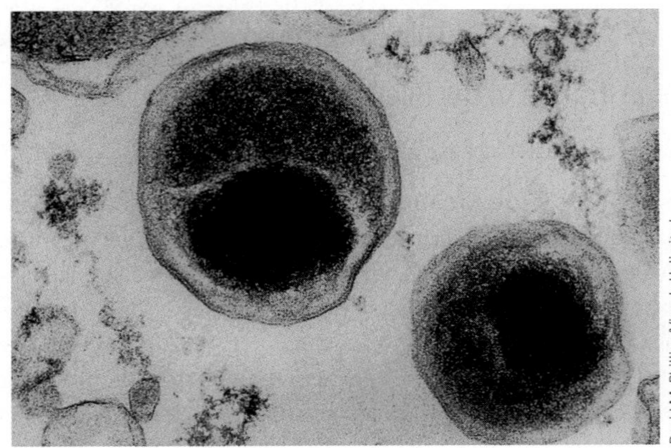

David M. Phillips/Visuals Unlimited

FIGURE 25.14
Cells of *Chlamydia trachomatis* inside a human cell. This bacterium is a major cause of infectious eye and genital diseases in humans.

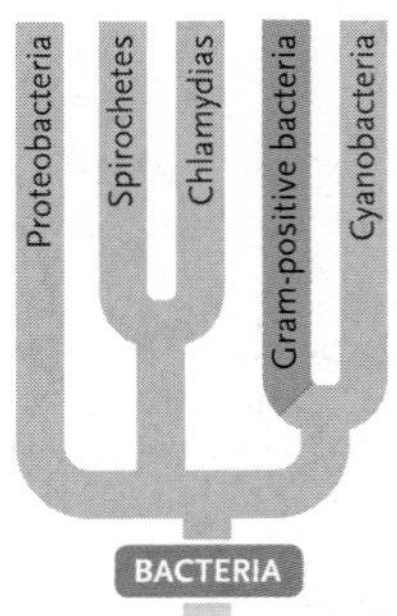
Proteobacteria
Spirochetes
Chlamydias
Gram-positive bacteria
Cyanobacteria
BACTERIA

GRAM-POSITIVE BACTERIA The large group of Gram-positive bacteria contains many species that live primarily as chemoheterotrophs. One species, *Bacillus subtilis,* is studied by biochemists and geneticists almost as extensively as is *E. coli.* A number of Gram-positive bacteria cause human diseases, including *Bacillus anthracis,* which causes anthrax; *Staphylococcus aureus,* which causes a range of illnesses from mild infections like pimples, boils, and abscesses, to life-threatening conditions such as toxic shock syndrome, pneumonia, meningitis and sepsis (an infection state involving a whole-body inflammation; commonly referred to as blood poisoning); and *Streptococcus pyogenes* **(Figure 25.15),** which is a cause of streptococcal pharyngitis ("strep throat"), scarlet fever, and rheumatic fever. Nevertheless, some Gram-positive bacteria are beneficial; *Lactobacillus* species, for example, carry out the lactic acid fermentation used in the production of pickles, sauerkraut, yogurt, and some other fermented foods. One unusual group of bacteria, the mycoplasmas, is classified in the group Gram-positive bacteria by molecular studies

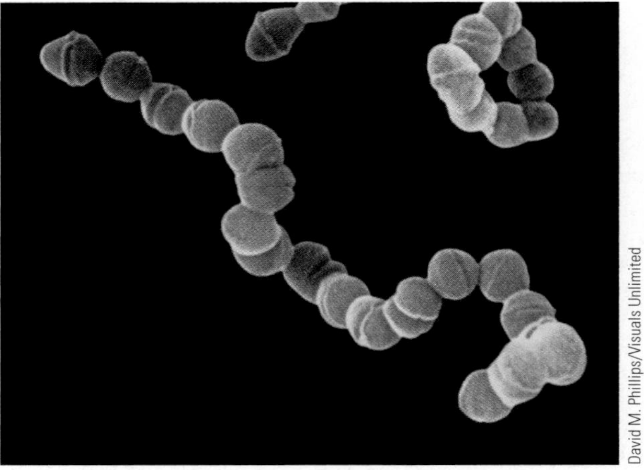

David M. Phillips/Visuals Unlimited

FIGURE 25.15
Streptococcus bacteria, a genus of Gram-positive bacteria, forming a long chain of cells which is typical of the many species in this genus.

even though they react negatively to the Gram stain. Their staining reaction indicates that they lack a cell wall. Because of this, they are insensitive to antibiotics such as penicillin, which target cell wall synthesis. Some species are pathogenic to humans; *Mycoplasma pneumonia,* for instance, causes a form of pneumonia.

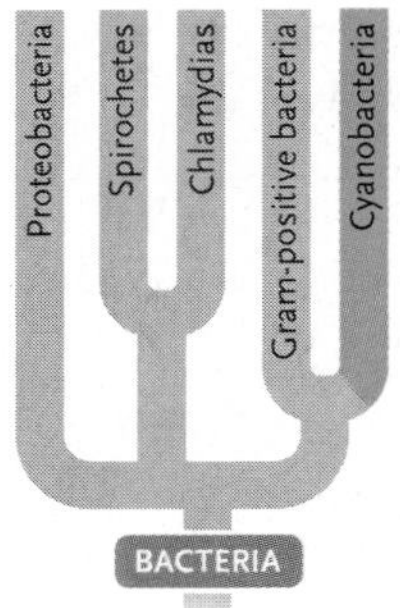
Proteobacteria
Spirochetes
Chlamydias
Gram-positive bacteria
Cyanobacteria
BACTERIA

CYANOBACTERIA The cyanobacteria **(Figure 25.16)** are Gram-negative photoautotrophs that have a blue-green color and carry out photosynthesis using the same pathways as eukaryotic algae and plants, and use the same chlorophyll that plants have as their primary photosynthetic pigment. They release oxygen as a by-product of photosynthesis.

The direct ancestors of present-day cyanobacteria were the first organisms to use the water-splitting reactions of photosynthesis. As such, they were critical to the appearance of abundant oxygen in the atmosphere, which allowed the evolutionary development of aerobic organisms (see Section 24.2). Chloroplasts probably evolved from early cyanobacteria that were incorporated into the cytoplasm of primitive eukaryotes, which eventually gave rise to the algae and higher plants (see Section 24.3). Besides releasing oxygen, present-day cyanobacteria help fix nitrogen into organic compounds in aquatic habitats and also in lichens that involve a cyanobacterium and a filamentous fungus living together symbiotically (see Chapter 28). (Most lichens involve a green alga and a filamentous fungus.)

A. A population of cyanobacteria covering the surface of a pond

Dr. Jeremy Burgess/SPL/Photo Researchers, Inc.

B. Chains of cyanobacterial cells. Some cells in the chains form spores. The heterocyst is a specialized cell that fixes nitrogen.

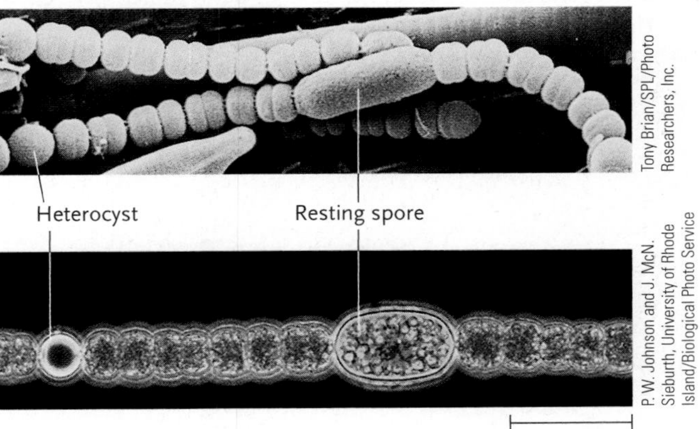
Heterocyst
Resting spore
6 μm

Tony Brian/SPL/Photo Researchers, Inc.

P. W. Johnson and J. McN. Sieburth, University of Rhode Island/Biological Photo Service

FIGURE 25.16
Cyanobacteria.

Bacteria Cause Diseases by Several Mechanisms

As you have just learned, some bacteria cause diseases while others are beneficial. Here we focus on pathogenic bacteria.

Bacteria vary in the pathways by which they cause diseases. A number of bacterial lineages produce **exotoxins,** toxic proteins that are secreted from the bacterium or are released from the cell when it lyses, and interfere with the biochemical processes of body cells in various ways. For example, the exotoxin of the Gram-positive bacterium *Clostridium botulinum* is found as a contaminant in poorly preserved foods, and causes botulism. The botulism exotoxin is one of the most poisonous substances known: a few nanograms of the toxin can cause illness, and a few hundred grams could kill every human on Earth. It acts by interfering with the transmission of nerve impulses. The muscle paralysis produced by the exotoxin can be fatal if the muscles that control breathing are affected. Interestingly, the botulinum exotoxin, under the brand name Botox, is used in low doses for the cosmetic removal of wrinkles, in the treatment of migraine headaches, of involuntary contraction of the eye muscles, and of some other medical conditions.

Some other bacteria cause disease through **endotoxins.** Endotoxins are not released by living cells as exotoxins are; instead, they are lipopolysaccharides released from the outer membrane surrounding the cell walls when the bacterium dies and lyses. Endotoxins are natural components of the outer membrane of all Gram-negative bacteria, which include *E. coli, Salmonella* species, and *Shigella* species. These lipopolysaccharides cause disease by overstimulating the host's immune system, often triggering inflammation. Endotoxin release has different effects, depending on the bacterial species and the site of infection, and include typhoid or other fevers, diarrhea, and, in severe cases, organ failure and death. For example, *Salmonella enterica typhi,* the cause of typhoid, enter the human intestines and penetrate the intestinal wall, eventually ending up in the lymph nodes. There they multiply, and some of the cells die and lyse, releasing endotoxins into the bloodstream. This both triggers the host's immune response and causes sepsis. If the infection is not successfully treated, the condition can progress to multiple organ system failure and, eventually, death.

Some bacteria release **exoenzymes,** enzymatic proteins that digest plasma membranes and cause cells of the infected host to rupture and die. Exoenzymes may also digest extracellular materials such as collagen, causing connective tissue diseases. Some exoenzymes attack red or white blood cells, leading to anemias, impairment of the immune response, or interference with blood clotting. Among the bacteria that release exoenzymes are particular species of *Streptococcus, Staphylococcus,* and *Clostridium.*

Necrotizing fasciitis (flesh-eating disease), the spectacularly destructive and rapid degeneration of subcutaneous tissues in the skin, is caused by an exoenzyme released by *Streptococcus pyogenes, Staphylococcus aureus,* and some other bacteria.

Some of the ill effects of bacteria have little to do with exotoxins, endotoxins, or exoenzymes, but are caused purely by the body's responses to infection. The severe pneumonia caused by *Streptococcus pneumoniae,* for example, results from massive accumulation of fluid and white blood cells in the lungs in response to the infection. The white blood cells have little effect on the bacteria, however, because of the bacterial cell's protective capsule. As the fluid, white blood cells, and bacteria continue to accumulate, they block air passages in the lungs and severely impair breathing.

Pathogenic Bacteria Commonly Develop Resistance to Antibiotics

Antibiotics are routinely used to treat bacterial infections. These substances, produced as defensive molecules by some bacteria and fungi, or by chemical synthesis, kill or inhibit the growth of other microbial species. For example, streptomycins, produced by soil bacteria, block protein synthesis in their target cells. Penicillins, which are produced by fungi, prevent formation of covalent bonds that hold bacterial cell walls together, weakening the wall and causing the cells to rupture.

Many pathogenic bacteria develop resistance to antibiotics through mutations that allow them to break down the drugs or otherwise counteract their effects (see *Why It Matters,* Chapter 20). Resistance is also acquired by horizontal gene transfer when resistance genes on plasmids transfer between bacteria by conjugation, or when DNA is brought into pathogens by other pathways such as transformation and transduction (see Section 17.1). Taking antibiotics routinely in mild doses, or failing to complete a prescribed dosage, contribute to the development of resistance by selecting strains that can survive in the presence of the drug. Overprescription of antibiotics for colds and other virus-caused diseases does not affect the virus, but can result in resistance among bacteria present in the body. Antibacterial agents that may result in resistance are also commonly included in such commercial products as soaps, detergents, and deodorants. Resistance is a form of evolutionary adaptation; antibiotics alter the bacterium's environment, conferring a reproductive advantage on those strains best adapted to the altered conditions.

The development of resistant strains has made tuberculosis, cholera, typhoid fever, gonorrhea, "staph," and other diseases caused by bacteria difficult to treat with antibiotics. For example, as recently as 1988, drug-resistant strains of *Streptococcus pneumoniae,* which causes pneumonia, meningitis, and middle-ear infections, were practically unheard of in the United States. Now, resistant strains of *S. pneumoniae* are common and increasingly difficult to treat with antibiotics.

In this section, you have seen that bacteria thrive in nearly every habitat on Earth, including the human body. However, some members of the second prokaryotic domain, the Archaea, the subject of the next section, live in habitats that are too forbidding even for the Bacteria.

STUDY BREAK 25.2

1. **What were the likely characteristics of the evolutionary ancestor of present-day proteobacteria?**
2. **What are the differences between the way photosynthesis is carried out by photosynthetic Proteobacteria and by cyanobacteria?**
3. **What is an exotoxin, an endotoxin, and an exoenzyme, and how do they differ with respect to how they cause disease?**

THINK OUTSIDE THE BOOK

Individually or collaboratively, research the Internet to identify an example of a phylogenetic tree for bacteria that does not have a traditional tree-like form. What does the alternate tree form say about the evolution of the group?

25.3 The Domain Archaea

The first-studied archaeans were found in extreme environments, such as hot springs, hydrothermal vents on the ocean floor, and salt lakes **(Figure 25.17)**. For that reason, these prokaryotes were

A. **Red-purple color produced by archaeans in highly saline water**

iStockphoto.com/David Crane

B. **Bright colors produced in hot, sulfur-rich water by the oxidative activity of archaeans that convert H_2S to elemental sulfur.**

Image copyright Sascha Burkard, 2010. Used under license from Shutterstock.com

FIGURE 25.17 Typically extreme archaean habitats.

INSIGHTS FROM THE Molecular Revolution

Extreme but Still in Between: Do archaeans belong in a separate classification of living organisms?

Many archaeans have a lifestyle clearly different from those of the bacteria and eukaryotes, and *Methanococcus jannaschii* is no exception. It was first discovered by the deep-sea submarine *Alvin* in a hot-water vent in the ocean at a depth of more than 2,600 m. It can grow in a temperature range from 48–94°C, and grows optimally at 85°C. It can tolerate pressure as high as 200 times the pressure of air at sea level. Based on biochemical and molecular comparisons with bacteria and eukaryotes, some researchers hypothesized that archaeans represented a separate domain of living organism.

Research Question

Are archaeans a separate domain of living organisms, distinct from Bacteria and Eukarya?

Experiment

In 1996 Carol J. Bult, Carl R. Woese, J. Craig Venter, and 37 other scientists at the Institute for Genomic Research obtained the complete DNA sequence of the genome of an archaean, *Methanococcus jannaschii*. It was the first archaean genome to be sequenced, and the fourth genome of any type of organism. The genome sequence was obtained by sequencing randomly chosen overlapping DNA fragments from a DNA library until the entire genome was completed (the whole-genome shotgun approach, described in Section 18.3).

The *M. jannaschii* genome sequence was compared with two bacterial genome sequences and the genome sequence of the brewer's yeast *Saccharomyces cerevisiae*, a eukaryote.

Results

The results of the genome sequencing analysis are summarized in the figure. The genome consists of one main chromosome and two small plasmids. Together they contain 1,738 protein-coding genes. At the time, only 38% of those genes could be confidently assigned functions based on sequence similarities with those of genes coding for known proteins in other organisms. Genes related to energy production, cell division, and metabolism are most similar to genes for those processes in bacteria, while most genes involved in replication, transcription, and translation are most similar to genes for those processes in eukaryotes. Other identified genes encode proteins unique to the Archaea, such as those encoding some enzymes and other proteins of the pathway reducing CO_2 to methane. Many other proteins with no known counterparts in the Bacteria or Eukarya were among the unidentified 62%, demonstrating the unique character of the Archaea.

Conclusion

Genome sequence analysis supported the hypothesis that the Archaea represent a domain of living organisms distinct from Bacteria and Eukarya. Analyses of many more bacterial, archaeal and eukaryote genomes since the time of this study have further supported this hypothesis.

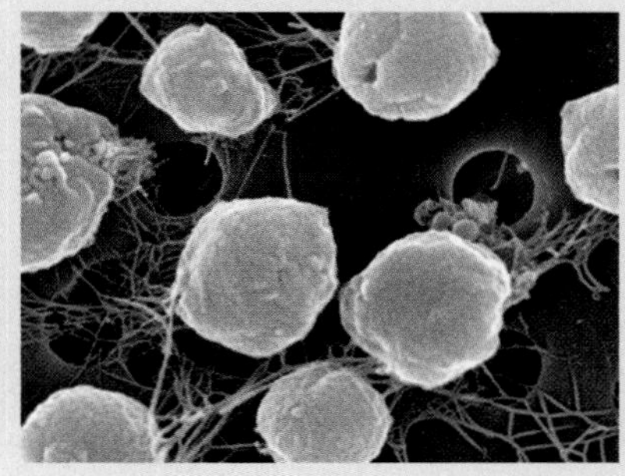
Chan B. Park, Dept. Chem. Eng., UC Berkeley

Methanococcus jannaschii

Genome sequencing

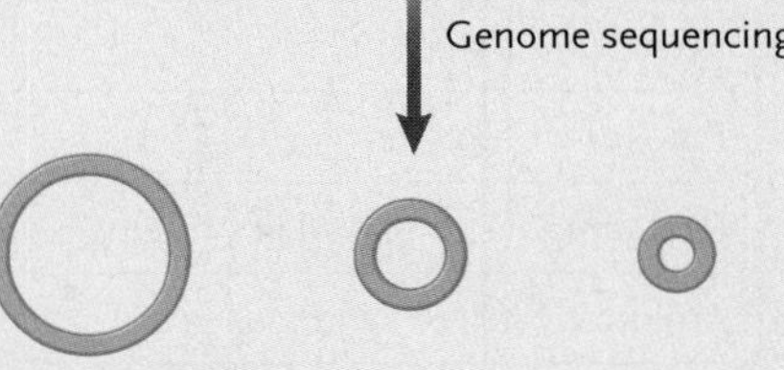

Main chromosome:
- 1.66 Mb
- 1,682 protein-coding genes

Plasmid 1:
- 58 kb
- 44 protein-coding genes

Plasmid 2:
- 16.5 kb
- 12 protein-coding genes

KEY

Mb = megabase = 10^6 base pairs
kb = kilobase = 10^3 base pairs

Summary of analysis of *Methanococcus jannaschii* genome sequence.

Source: C. J. Bult et al. 1996. Complete genome sequence of the methanogenic archaeon, *Methanococcus jannaschii*. *Science* 273:1058–1073.

called *extremophiles* ("extreme lovers"). Subsequently archaeans have also been found living in normal environments; like bacteria, these are *mesophiles.*

Many archaeans are chemoautotrophs that obtain energy by oxidizing inorganic substances, while others are chemoheterotrophs that oxidize organic molecules. No known member of the Archaea has been shown to be pathogenic.

Research has shown that they have some eukaryotic features, some bacterial features, and some features that are unique to the group (also discussed in Section 24.3; **Table 25.1** compares the characteristics of Bacteria, Archaea, and Eukarya). Based on phylogenetics analysis of small ribosomal subunit RNA sequences by Carl Woese and his colleagues that compared their DNA and rRNA sequences with those of other organisms, Archaea were subsequently classified as a separate domain of life. (*Insights from the Molecular Revolution* describes the research that first revealed the complete DNA sequence of an archaean.) Scientists use sequencing studies and the archeans' unique characteristics to identify the organisms in this group.

Molecular Studies Reveal Three Evolutionary Branches in the Archaea

Based on differences between the rRNA coding sequences in their genomes, the domain Archaea is divided into three groups (see Figure 25.11). Two major groups, the **Euryarchaeota** and the **Crenarchaeota,** contain archaeans that have been cultured and examined in the laboratory. The third group, the **Korarchaeota,** has been recognized solely on the basis of rRNA coding sequences in DNA taken from environmental samples. A fourth group, the **Nanoarchaeota,** was proposed based on rRNA sequence analysis of a thermophilic archaean found in a symbiotic relationship with another thermophilic archaean. Genome sequence comparisons

TABLE 25.1 Characteristics of the Bacteria, Archaea, and Eukarya

Characteristic	Bacteria	Archaea	Eukarya
DNA arrangement	Single, circular in most, but some linear and/or multiple	Single, circular	Multiple linear molecules
Chromosomal proteins	Prokaryotic histone-like proteins	Five eukaryotic histones	Five eukaryotic histones
Genes arranged in operons	Yes	Yes	No
Nuclear envelope	No	No	Yes
Mitochondria	No	No	Yes
Chloroplasts	No	No	Yes
Peptidoglycans in cell wall	Present	Present but modified, or absent	Absent
Membrane lipids	Unbranched; linked by ester linkages	Branched; linked by ether linkages	Unbranched; linked by ester linkages
RNA polymerase	One type	Multiple types	Multiple types
Ribosomal proteins	Prokaryotic	Some prokaryotic, some eukaryotic	Eukaryotic
First amino acid placed in proteins	Formylmethionine	Methionine	Methionine
Aminoacyl–tRNA synthetases	Prokaryotic	Eukaryotic	Eukaryotic
Cell division proteins	Prokaryotic	Prokaryotic	Eukaryotic
Proteins of energy metabolism	Prokaryotic	Prokaryotic	Eukaryotic
Sensitivity to chloramphenicol and streptomycin	Yes	No	No

have now shown that the Nanoarchaeota are most probably a subgroup of the Euryarchaeota.

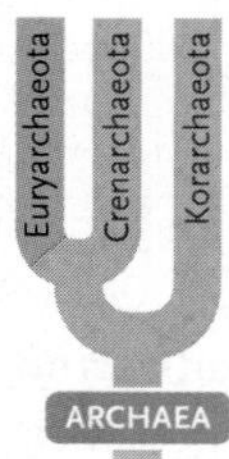

Euryarchaeota
Crenarchaeota
Korarchaeota
ARCHAEA

EURYARCHAEOTA The Euryarchaeota are found in different extreme environments. They include methanogens, which produce methane; extreme halophiles, which live in high concentrations of salt; and some extreme thermophiles, which live under high-temperature conditions.

Methanogens (methane generators) live in reducing environments **(Figure 25.18)**, and represent about one half of all known species of archaeans. All known methanogens belong to the Euryarchaeota. Examples are members of the genera *Methanococcus* and *Methanobacterium*. Methanogens are obligate anaerobes, meaning they are killed by oxygen. They are found in the anoxic (oxygen-lacking) sediments of swamps, lakes, marshes, and sewage works, as well as in more moderate environments, such as the rumen of cattle, sheep, and camels; the large intestine of dogs and humans; and the hindguts of insects such as termites and cockroaches. Methanogens generate energy by converting at least ten different substrates such as carbon dioxide and hydrogen gas, methanol, or acetate into methane gas (CH_4), which is released into the atmosphere. A single species may use two or three substrates, for example converting carbon dioxide and hydrogen into methane and water.

R. Robinson/Visuals Unlimited
5 μm

FIGURE 25.18
A colony of cells of the methanogenic archaean *Methanosarcina,* which live in the sulfurous, waterlogged soils of marshes and swamps.

Halophiles are salt-loving organisms. Extreme halophilic Archaea live in highly saline (salty) environments such as the Great Salt Lake or the Dead Sea, and on foods preserved by salting. Moreover, they require a high concentration of salt to live: they need a minimum NaCl concentration of about 1.5 M (about 9% solution), and can live in a fully saturated solution (5.5 M, or 32%). All known extreme halophilic Archaea belong to the Euryarchaeota. Most are aerobic chemoheterotrophs; they obtain energy from sugars, alcohols, and amino acids using pathways similar to those of bacteria. Examples are *Halo-*

bacterium and *Natrosobacterium* species. *Halobacterium* species, like a number of extreme halophiles, use light as a secondary energy source supplementing the oxidations that are their primary source of energy.

Extreme thermophiles live in extremely hot environments. Extreme thermophilic archaeans live in thermal areas such as ocean floor hydrothermal vents and hot springs such as those in Yellowstone National Park. Their optimal temperature range for growth is 70° to 95°C, approaching the boiling point of water. By comparison, no eukaryotic organism is known to live at a temperature higher than 60°C. Some extreme thermophiles are members of the Euryarchaeota. Some of them, such as *Pyrophilus* species, are obligate anaerobes, while others, such as *Thermoplasma* species, are facultative anaerobes that grow on a variety of organic compounds.

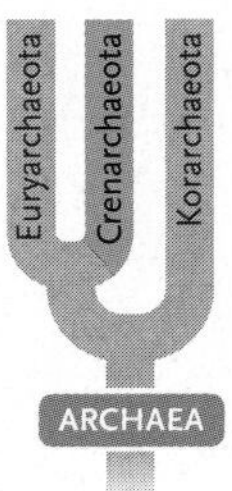
Euryarchaeota
Crenarchaeota
Korarchaeota
ARCHAEA

CRENARCHAEOTA The group Crenarchaeota contains most of the extreme thermophiles. Their optimal temperature range of 75° to 105°C is higher than that of the Euryarchaeota. Most are unable to grow at temperatures below 70°C. The most thermophilic member of the group, *Pyrolobus,* grows optimally at 106°C, but dies below 90°C. It can also grow at 113°C and survive an hour of autoclaving at 121°C! *Pyrolobus* species lives in ocean floor hydrothermal vents where the pressure makes it possible to have temperatures above 100°C, the boiling point of water on Earth's surface.

Also within this group are **psychrophiles** ("cold loving"), organisms that grow optimally in cold temperatures in the range of −10 to −20°C. These organisms are found mostly in the Antarctic and Arctic oceans, which are frozen most of the year, and in the intense cold of ocean depths.

Mesophilic members of the Crenarchaeota comprise a large part of plankton in cool, marine waters where they are food sources for other marine organisms. As yet, no individual species of these archaeans has been isolated and characterized.

Crenarchaeota archaeans exhibit a wide range of metabolism with regard to oxygen, and include obligate anaerobes, facultative anaerobes, and aerobes.

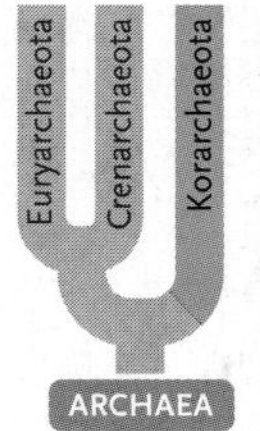
Euryarchaeota
Crenarchaeota
Korarchaeota
ARCHAEA

KORARCHAEOTA The group Korarchaeota has been recognized solely on the basis of analyzing rRNA sequences in DNA obtained from marine and terrestrial hydrothermal environments, such as the Emerald Pool at Yellowstone National Park (see Figure 25.17B). To date, no members of this group have been isolated and cultivated in the lab, and nothing is known about their physiology. They are the oldest lineage in the domain Archaea according to molecular data.

Thermophilic archaeans are important commercially. For example, enzymes from some species are used in basic and applied research, such as the thermostable DNA polymerase used in the polymerase chain reaction (PCR; see Section 18.1). Other enzymes from thermophilic archaeans are being tested for suitability as addition to detergents, where it is hoped that they will be active under high temperatures and an acidic pH.

From the highly varied prokaryotes, we now turn to eukaryotes, a group which does not equal the metabolic diversity or the conquest of extreme environments we have seen in the Bacteria and Archaea, but which does display new levels of physiological and behavioral complexity. The next six chapters discuss the eukaryotic kingdoms of protists, plants, fungi, and animals.

STUDY BREAK 25.3 <

1. **What distinguishes members of the Archaea from members of the Bacteria and Eukarya?**
2. **How does a methanogen obtain its energy? In which group or groups of Archaea are methanogens found?**
3. **Where do extreme halophilic archaeans live? How do they obtain energy? In which group or groups of Archaea are the extreme halophiles found?**
4. **What are extreme thermophiles and psychrophiles?**

UNANSWERED QUESTIONS

As you learned in this chapter, archaeans are morphologically similar to bacteria, yet they have a number of features that are more related to eukaryotes than to bacteria. More specifically, the DNA replication and transcription machineries of archaeans are related to those of eukaryotes and distinct from those found in bacteria. Also, recent work has revealed that, in some archaeans, aspects of cell division have close parallels in human cells but not in bacteria.

How is chromosome replication accomplished in species of the archaean genus *Sulfolobus*?

Analysis of the available genome sequences of archaeans show that the machinery that archaea use to replicate their DNA is, like the transcription machinery, related to that of eukaryotes and distinct from the bacterial apparatus. Recall from Chapter 14 that bacteria have a single replication origin per chromosome, whereas eukaryotes have many replication origins per chromosome. Some archaeans replicate their chromosomes from a single origin, but a number of species have multiple replication origins per chromosome. For example, species in the *Sulfolobus* genus (hyperthermophiles of the Crenarchaeota group of the domain Archaea) have three replication origins in their single chromosome. All three of these origins are used in every cell in every cell cycle, indicating that there is a sophisticated control system that ensures that each origin is used once, and only once, per cell division cycle. The nature of this control system is currently unknown.

The function of a replication origin is to recruit proteins to unwind double-stranded DNA which exposes single-stranded DNA for use as templates. The single-stranded regions produced at an origin are extended by a DNA helicase (see Section 14.3) which, in archaeans, is called the MCM helicase. This archaean helicase has been the subject of intense structural and biochemical study, and much is now known about its form and function. However, the way in which it is recruited to replication origins remains poorly understood. Finally, with three replication origins being used in every cell for every cell division cycle, a total of six replication forks are simultaneously used for DNA replication (two from each origin). How these replication forks are resolved when they collide on the circular chromosome to produce two copies of the archaeal chromosome is currently not known.

How does cell division take place in the archaean *Sulfobolus*?

Once replication is complete, *Sulfolobus* keeps two replicated chromosomes paired for a time prior to their segregation at cell division. Until recently, the cell division machinery used by *Sulfolobus* and relatives in the Crenarchaeota was unknown. Nearly all bacteria, members of other archaeal groups, and eukaryotes have proteins belonging to the tubulin and actin families of cytoskeletal proteins which play integral roles in cell division. However, *Sulfolobus* and its relatives lack these proteins yet they divide by binary fission. Recent work has established that cell division in *Sulfolobus* uses the so-called ESCRT machinery. Remarkably the ESCRT proteins are conserved throughout the Eukarya, where they play roles in a wide variety of membrane processes. Cell biological and genetic studies have revealed that the ESCRT system plays a key role in *Sulfolobus* cell division, hinting at the ancestral role of this conserved machinery. Many questions remain unanswered, including: How are the *Sulfolobus* chromosomes segregated prior to cell division? and How is the site of cell division defined? With recent advances in the genetic manipulation of archaeans, researchers anticipate that these questions and others will be readily addressed.

Think Critically

The ability to replicate DNA is fundamental to the propagation of all life on the planet. Yet it is apparent that bacteria use a different machinery to replicate their DNA than that employed by archaeans and eukaryotes. How can two distinct machineries for such a fundamentally important process have evolved?

Stephen D. Bell

Rachel Y. Samson

Stephen D. Bell and **Rachel Y. Samson** work at the Sir William Dunn School of Pathology, Oxford University. Steve is the Professor of Microbiology and his research group is focused on investigating transcription, DNA replication and cell division in the archaeal domain of life. Rachel joined Steve's group in 2007 and is researching the molecular and genetic basis of archaeal cell division. To learn more about their research, go to: http://users.path.ox.ac.uk/~sbell/Welcome.html.

REVIEW KEY CONCEPTS

Go to **CENGAGENOW** at www.cengage.com/login to access quizzing, animations, exercises, articles, and personalized homework help.

25.1 Prokaryotic Structure and Function

- Three shapes are common in prokaryotes: spherical, rodlike, and spiral (Figure 25.2).
- Prokaryotic genomes typically consist of a single, circular DNA molecule packaged into the nucleoid. Many prokaryotic species also contain plasmids, which replicate independently of the main DNA (Figure 25.3).
- Gram-positive bacteria have a cell wall consisting of a thick peptidoglycan layer. Gram-negative bacteria have a thin peptidoglycan layer. The thin cell wall is surrounded by an outer membrane (Figure 25.4).
- A polysaccharide capsule or slime layer surrounds many bacteria. This sticky, slimy layer both protects the bacteria and helps them adhere to surfaces (Figure 25.5).
- Some prokaryotes have flagella, corkscrew-shaped protein fibers that rotate like propellers, and pili, protein shafts that help bacterial cells adhere to each other or to eukaryotic cells (Figure 25.7).
- Prokaryotes show great diversity in their modes of obtaining energy and carbon. Chemoautotrophs obtain energy by oxidizing inorganic substrates and use carbon dioxide as their carbon source. Chemoheterotrophs obtain both energy and carbon from organic molecules. Photoautotrophs are photosynthetic organisms that use light as a source of energy and carbon dioxide as their carbon source. Photoheterotrophs use light as a source of energy and obtain their carbon from organic molecules (Figure 25.8).
- Some prokaryotes are capable of nitrogen fixation, the conversion of atmospheric nitrogen to ammonia; others are responsible for nitrification, the two-step conversion of ammonium to nitrate.
- Prokaryotes normally reproduce asexually by binary fission. Some prokaryotes are capable of conjugation, in which part of the DNA of one cell is transferred to another cell.
- In nature, prokaryotes may live in an interacting community, such as a biofilm (Figure 25.10). Biofilms have both detrimental and beneficial consequences; they can harm human health, but they also can be effective in, for example, bioremediation.

Animation: Prokaryotic body plan

Animation: Gram staining

Animation: Prokaryotic fission

Animation: Prokaryotic conjunction

25.2 The Domain Bacteria

- Bacteria are divided into several evolutionary branches (Figure 25.11).
- The proteobacteria are Gram-negative bacteria that include purple sulfur (photoautotrophic) and nonsulfur (photoheterotrophic) photosynthetic species, and nonphotosynthetic species. Free-living proteobacteria include the spore-forming myxobacteria and species that fix nitrogen (Figure 25.13).
- The chlamydias are Gram-negative intracellular parasites that cause various diseases in animals. They have cell walls with an outer membrane, but they lack peptidoglycans (Figure 25.14).

- The spirochetes are spiral-shaped bacteria that are propelled by twisting movements produced by the rotation of flagella.
- The Gram-positive bacteria are primarily chemoheterotrophs, and includes many pathogenic species (Figure 25.15).
- The cyanobacteria are Gram-negative photoautotrophs that carry out photosynthesis and release oxygen as a by-product (Figure 25.16).
- Bacteria cause disease through exotoxins, endotoxins, and exoenzymes.
- Pathogenic bacteria may develop resistance to antibiotics through mutation of their own genes, or by acquiring resistance genes or plasmids from other bacteria.

Animation: Examples of Bacteria

25.3 The Domain Archaea

- Members of the domain Archaea have some features that are like those of bacteria, other features that are like those of eukaryotes, and some other features that are uniquely archaean (Table 25.1).
- The archaean plasma membrane contains unusual lipid molecules. The cell walls of archaeans consist of distinct molecules similar to peptidoglycans, or of protein or polysaccharide molecules.
- The Archaea are classified into three groups. The Euryarchaeota include the methanogens, the extreme halophiles, and some extreme thermophiles. The Crenarchaeota contain most of the archaean extreme thermophiles, as well as psychrophiles and mesophiles. Obligate anaerobes, facultative anaerobes, and aerobes are found among the Crenarchaeota. The Korarchaeota are recognized only on the basis of sequences in DNA samples.

UNDERSTAND AND APPLY

Test Your Knowledge

1. A urologist identifies cells in a man's urethra as bacterial. Which of the following descriptions applies to the cells?
   a. They have sex pili, which give them motility.
   b. They have flagella, which allow them to remain in one position in the urethral tube.
   c. They are covered by a capsule, which enables them to multiply quickly.
   d. They are covered by pili, which keep them attached to the urethral walls.
   e. They contain a peptidoglycan cell wall, which gives them buoyancy to float in the fluids of the urethra.
2. A bacterium that uses nitrites as its only energy source was found in a deep salt mine. It is a:
   a. chemoautotroph.
   b. parasite.
   c. photoautotroph.
   d. heterotroph.
   e. photoheterotroph.
3. The ______ are all oxygen-producing photoautotrophs.
   a. spirochetes
   b. chlamydias
   c. cyanobacteria
   d. Gram-positive bacteria
   e. proteobacteria
4. At the health center, a fecal sample was taken from a feverish student. Organisms with corkscrew-like flagella and no endomembranes but with cell walls that lack peptidoglycan were isolated as the cause for the illness. These organisms probably belong to the group:
   a. chlamydias.
   b. spirochetes.
   c. Euryarchaeota.
   d. Cyanobacteria.
   e. Archaea.
5. Which of the following is *not* a property of an endospore?
   a. resistant to boiling—must be autoclaved to be killed
   b. metabolically inactive
   c. can potentially survive millions of years
   d. provides a method to preserve bacterial DNA under harsh conditions
   e. is the portion of the bacterial cell that undergoes conjugation
6. Bacterial cells are generally thought to act independently. An exception to this is bacterial cells involved in:
   a. biofilms.
   b. photosynthesis.
   c. peptidoglycan layering.
   d. toxin release.
   e. facultative anaerobic metabolism.
7. Penicillin, an antibiotic, inhibits the formation of cross-links between sugar groups in peptidoglycan. Bacteria treated with penicillin should be:
   a. aerobic.
   b. anaerobic.
   c. Gram-negative.
   d. Gram-positive.
   e. flagellated.
8. The best choice when using/prescribing antibiotics is to:
   a. increase the dosage when the original amount does not work.
   b. determine the kind of bacterium causing the problem.
   c. stop taking the antibiotic when you feel better but the prescription has not run out.
   d. ask the doctor to prescribe a drug as a precaution for an infection you do not have.
   e. choose soaps that are labeled "antibacterial."
9. Archaeans have:
   a. proteins of energy metabolism like bacteria, cell division proteins like bacteria, and mitochondria like eukaryotes.
   b. multiple types of RNA polymerases like eukaryotes, chromosomes like bacteria, and formylmethionine as the first amino acid placed in proteins.
   c. operons like bacteria, multiple types of RNA polymerases like eukaryotes, and cell division proteins like bacteria.
   d. no sensitivity to streptomycin, histones like eukaryotes, and no operons.
   e. histones like eukaryotes, a single type of RNA polymerase like bacteria, and operons like bacteria.
10. Methanogens, obligate anaerobes that generate methane, are:
    a. members of the Proteobacteria.
    b. members of the Cyanobacteria.
    c. members of the Spirochetes.
    d. members of the Euryarchaeota.
    e. members of the Korarchaeota.

Discuss the Concepts

1. The digestive tract of newborn chicks is free of bacteria until, for example, they eat food that has been exposed to the feces of adult chickens. The ingested bacteria establishes a population in the digestive tract that is beneficial for the digestion of food. However, if *Salmonella* are present in the adult feces, this bacterium, which can be pathogenic for humans who ingest it, may become established in the digestive tracts of the chicks. To eliminate the possibility that *Salmonella* might become established, should farmers feed newborn chicks a mixture of harmless known bacteria from a lab culture, or a mixture of unknown fecal bacteria from healthy adult chickens? Design an experiment to answer this question.
2. Investigators in Australia found that mats of pond scum formed by the bacterium *Botyrococcus braunii* decayed into a substance resembling crude oil when the ponds dried up. Formulate a hypothesis explaining how this process may have contributed to Earth's oil deposits.

Design an Experiment

You receive a sample of an unidentified prokaryote. Design an experiment to see if it is a bacterium or an archaean.

Interpret the Data

The proteobacteria in the genus *Pseudomonas* are important spoilers of food. The rate at which food spoils is dependent upon temperature. The data in the figure show the relationship between the square root of the growth rate (how fast the population is growing per day; unitless) and the temperature of the culture.

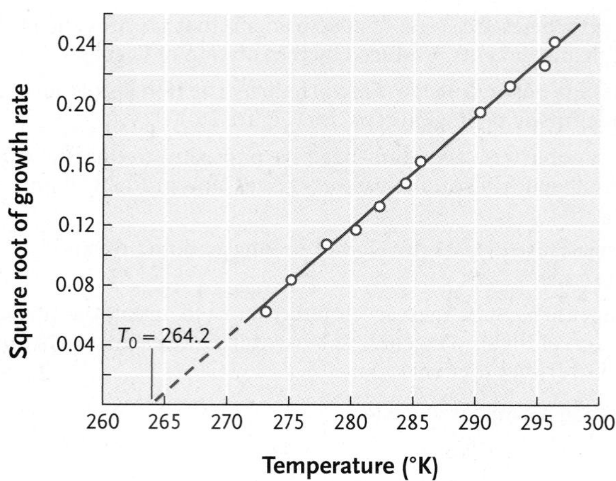
Square root of growth rate
0.24
0.20
0.16
0.12
0.08
0.04
T_0 = 264.2
260
265
270
275
280
285
290
295
300
Temperature (°K)

1. Convert the temperature of the data point at the lowest temperature from °K to °C. Why are there no data below this temperature.
2. Does typical refrigeration (temperature lowered to 4°C) appear sufficient to limit growth rate compared to room temperature (20°C)?

Source: Data from D. A. Ratkowsky et al. 1982. Relationship between temperature and growth rate of bacterial cultures. *Journal of Bacteriology* 149:1–5.

Apply Evolutionary Thinking

Does our understanding of horizontal gene transfer affect our application of the biological species concept to bacteria?

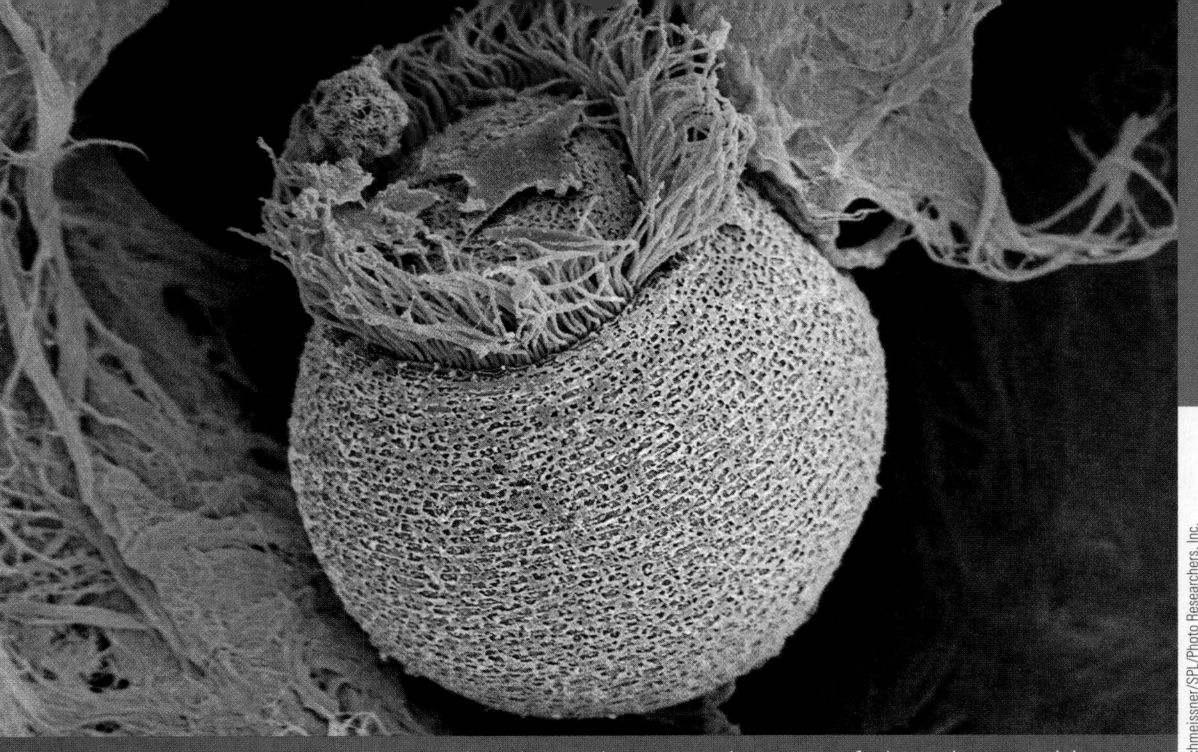

A ciliated protozoan, a type of protist (colorized SEM). This protozoan lives in water, feeding on bacteria and decaying organic matter.

Steve Gschmeissner/SPL/Photo Researchers, Inc.

26

STUDY OUTLINE

Protists

Why It Matters. . . Go for a swim just about anywhere in the natural world and you will share the water with multitudes of diverse organisms called protists. Like their most ancient ancestors, almost all of these eukaryotic species are aquatic. Structurally, single-celled protists are the simplest of all eukaryotes. Although most are microscopic in size, many have had or have a significant impact on the world. For example, the protist *Phytophthora infestans,* a water mold also referred to as a downy mildew, infects valuable crop plants such as potatoes. Potatoes were one of the main food staples in Europe in the nineteenth century, and *P. infestans* caused a lot of damage to potato crops in many countries. In Ireland, for instance, the growing seasons between 1845 and 1860 were cool and damp. Year after year, *P. infestans* spores spread along thin films of water on the plants. Late blight, a rotting of plant parts, became epidemic. In this so-called Irish potato famine, an estimated one million people starved to death or died of typhoid fever (a secondary effect), and another million people fled to other countries.

Today, related species threaten forests in the United States, Europe, and Australia. For example, when conditions favored its growth, *Phytophthora ramorum* started an epidemic of sudden oak death in California, during which tens of thousands of oak trees have died. As the name suggests, infected trees die rapidly. The first sign of infection is a dark red-to-black sap oozing from the bark surface. The pathogen has now jumped to madrones, redwoods, and certain other trees and shrubs. Cascading ecological changes resulting from tree death caused by this pathogen will reduce sources of food and shelter for forest species.

In this chapter, we describe several lineages of protists, a diverse array of organisms that are clearly not a monophyletic group. Also known as *protoctists,* protists are the products of the varied early branching of eukaryotic evolution. They are abundant on Earth and play key ecological, economic, and medical roles in the world's biological communities.

We begin this chapter with a discussion of the identity of the protists. As our discussion will show, the protists are so diverse that they are best defined as what they are not—that is, by contrasting them with other organisms. <

26.1 What Is a Protist?

Protists are easily the most varied of all Earth's creatures. **Figure 26.1** shows a number of protists, illustrating their great diversity. Protists include both microscopic single-celled and large multicellular organisms. They may inhabit aquatic environments, moist soils, or the bodies of animals and other organisms, and they may live as predators, photosynthesizers, parasites, or decomposers. The extreme diversity of the group has made the protists difficult to classify.

Protists Are Most Easily Classified by What They Are Not

The one reasonable certainty about protist classification is that they are not prokaryotes, fungi, plants, or animals. Because protists are eukaryotes, the boundary between them and prokaryotes is clear and obvious. Unlike prokaryotes, protists have a true nucleus, with multiple, linear chromosomes and cytoplasmic organelles, including mitochondria (in most but not all species), endoplasmic reticulum, Golgi complex, and chloroplasts (in some species). They reproduce asexually by mitosis or sexually by meiosis and union of sperm and egg cells, rather than by binary fission as do prokaryotes.

Figure 26.2 shows the phylogenetic relationships between protists and other eukaryotes. This tree represents a consensus based on molecular data, and particularly on phylogenomics analysis, which compares sequences of whole genomes, or large segments of genomes. The tree shows the most ancient bifurcation of eukaryotes into the **Unikonta (unikonts),** eukaryotes

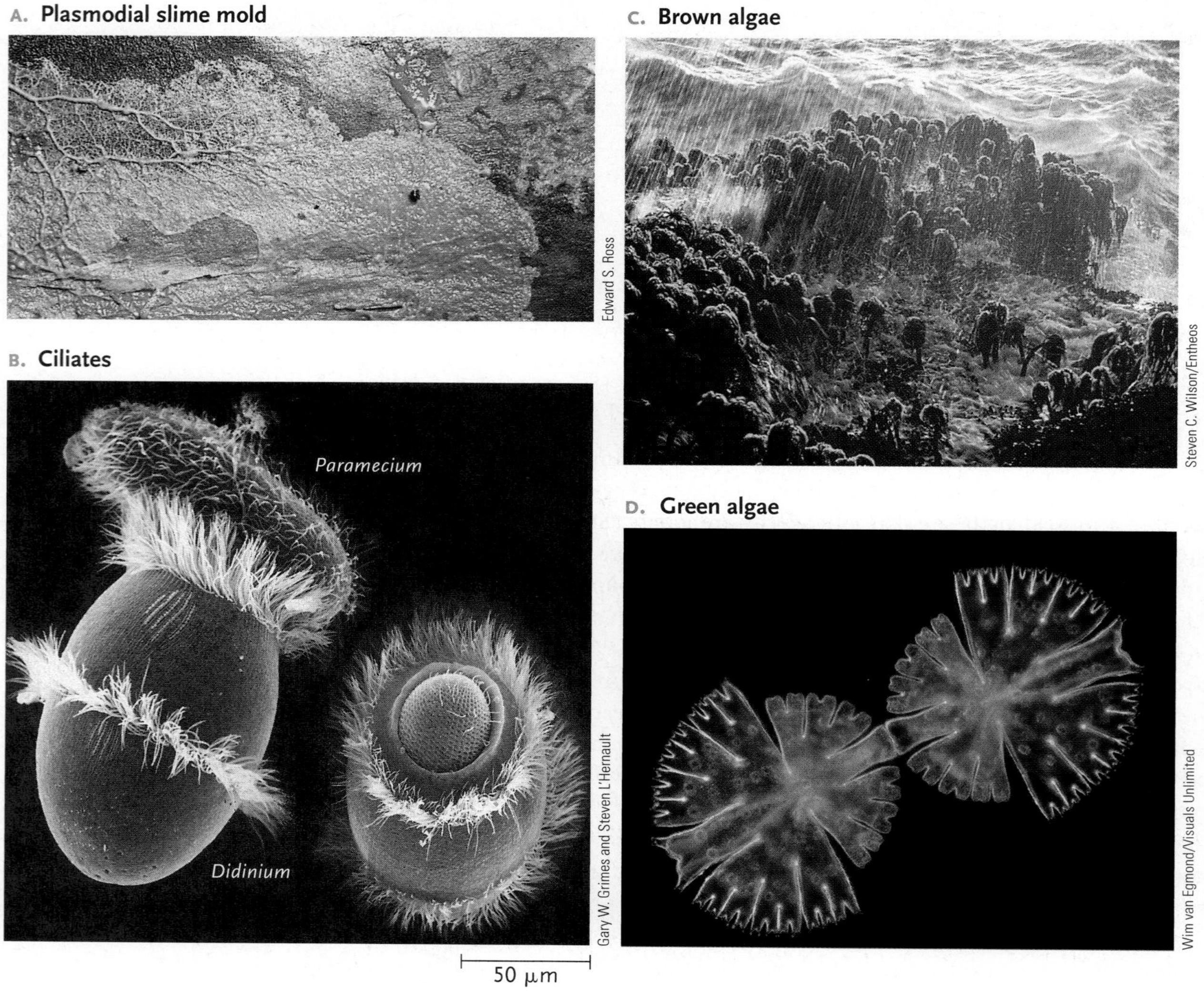

A. Plasmodial slime mold
Edward S. Ross
B. Ciliates
Paramecium
Didinium
Gary W. Grimes and Steven L'Hernault
50 μm
C. Brown algae
Steven C. Wilson/Entheos
D. Green algae
Wim van Egmond/Visuals Unlimited

FIGURE 26.1

A sampling of protist diversity. **(A)** *Physarum,* a plasmodial slime mold (yellow shape, lower part of figure) migrating over a rotting log. **(B)** *Didinium,* a ciliate, consuming another ciliate, *Paramecium.* **(C)** *Postelsia palmaeformis* (the sea palm), a brown alga, thriving in the surf pounding a California coast. **(D)** *Micrasterias,* a single-celled green alga, here shown dividing in two.

with a single flagellum, and **Bikonta (bikonts),** eukaryotes with two flagella. Further evolution produced several categories of eukaryotes. As you can see in Figure 26.2, all eukaryotic organisms except animals, land plants, and fungi—three groups that arose from protist ancestors—are protists. Although some protists have features that resemble those of the fungi, plants, or animals, several characteristics are distinctive. For instance, cell wall components in protists differ from those of the fungi (molds and yeasts). In contrast to land plants, protists lack highly differentiated structures equivalent to true roots, stems, and leaves; they also lack the protective structures that encase developing embryos in plants. Protists are distinguished from animals by their lack of highly differentiated structures such as limbs and a heart, and by the absence of features such as nerve cells, complex developmental stages, and an internal digestive tract. Protists also lack collagen, the characteristic extracellular support protein of animals. Thus, "protist" is a catchall group that includes all the eukaryotes that are not fungi, plants, or animals.

Until relatively recently, an accepted classification scheme was to group protists into a kingdom, the Protoctista (also called the Protista). Further, the protists were subclassified into phyla according to criteria such as body form, modes of nutrition and movement, and forms of meiosis and mitosis. However, comparisons of nucleic acid and amino acid sequences, now considered the most informative method for determining the evolutionary relationships of protists, show that most of the organisms previously grouped together do not share a common ancestry; that is, they are actually a polyphyletic assemblage of organisms. Further, many protists are no more closely related to each other than they are to the fungi, plants, or animals.

For simplicity, we will refer to these organisms as protists in this book, with the understanding that it is a collection of

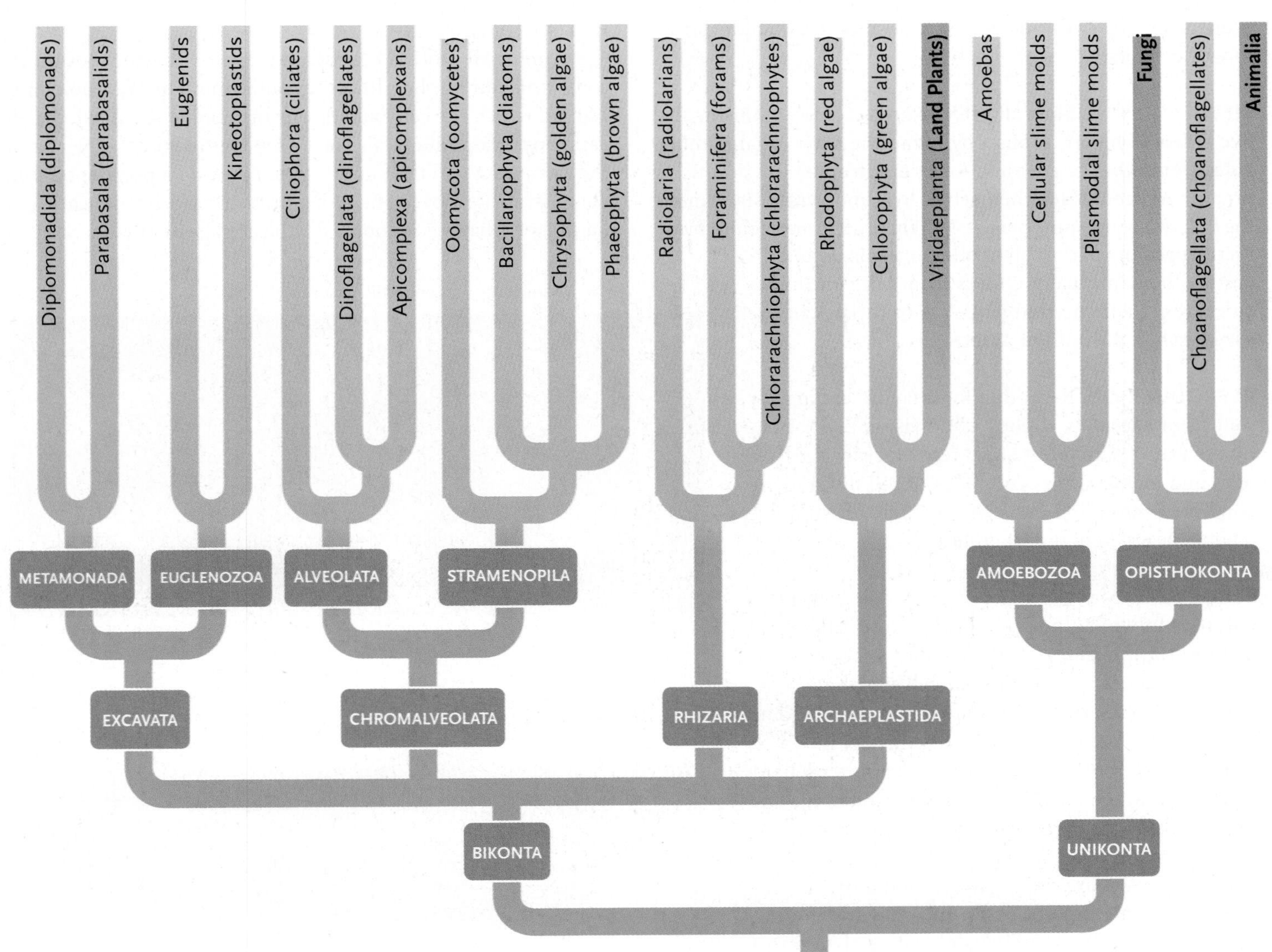

Diplomonadida (diplomonads)
Parabasala (parabasalids)
Euglenids
Kinetoplastids
Ciliophora (ciliates)
Dinoflagellata (dinoflagellates)
Apicomplexa (apicomplexans)
Oomycota (oomycetes)
Bacillariophyta (diatoms)
Chrysophyta (golden algae)
Phaeophyta (brown algae)
Radiolaria (radiolarians)
Foraminifera (forams)
Chlorarachniophyta (chlorarachniophytes)
Rhodophyta (red algae)
Chlorophyta (green algae)
Viridaeplanta (Land Plants)
Amoebas
Cellular slime molds
Plasmodial slime molds
Fungi
Choanoflagellata (choanoflagellates)
Animalia
METAMONADA
EUGLENOZOA
ALVEOLATA
STRAMENOPILA
AMOEBOZOA
OPISTHOKONTA
EXCAVATA
CHROMALVEOLATA
RHIZARIA
ARCHAEPLASTIDA
BIKONTA
UNIKONTA

FIGURE 26.2
The phylogenetic relationship between the evolutionary groups of the protists (green branches) and the relationship of the protists with other eukaryotes. The Archaeplastida includes the land plants of the kingdom Plantae (rust colored branch), and the Opisthokonta includes the animals of the kingdom Animalia and the fungi of the kingdom Fungi (rust colored branches).

largely unrelated organisms placed together for convenience. We will refer to the major evolutionary clusterings as groups (see Figure 26.2).

You may have heard two broad terms historically associated with large assemblages of protists, *protozoa* ("first animal") (also called *protozoans*) and *algae.* Neither of these assemblages corresponds to a taxonomic group of protists, however. That is, protozoa is a name for a diverse assemblage of heterotrophic protists, formerly considered to be animals, that are mobile and feed either by ingesting prey or absorbing organic molecules, whereas algae is a name for several sets of photosynthetic protists. The term "algae" will be used in this chapter in association with specific taxonomic groups of protists.

Protist Diversity Is Reflected in Their Metabolism, Reproduction, Structure, and Habitat

As you might expect from the broad range of organisms included, protists are highly diverse in metabolism, reproduction, structure, and habitat.

METABOLISM Almost all protists are aerobic organisms that live either as heterotrophs—by obtaining their organic molecules from other organisms—or as autotrophs—by producing organic molecules for themselves by photosynthesis. Among the heterotrophs, some protists obtain organic molecules by directly ingesting part or all of other organisms and digesting them internally. Others absorb organic molecules from their environment. A few protists can live as either heterotrophs or autotrophs.

REPRODUCTION Reproduction may be asexual by mitosis or sexual by meiotic cell division and formation of gametes. In protists that reproduce by both mitosis and meiosis, the two modes of cell division are combined into a life cycle that is highly distinctive among the different protist groups.

STRUCTURE Many protists live as single cells or as **colonies** in which individual cells show little or no differentiation and are potentially independent. Within colonies, individuals use cell signaling to cooperate on tasks such as feeding or movement. Some protists are large multicellular organisms, in which cells are differentiated and completely interdependent. For example, seaweeds are multicellular marine protists that include the largest and most differentiated organisms of the group; their structures include a holdfast to secure the organism to the rocks, leaflike fronds, and, in some cases, an air bladder for flotation. The giant kelp of coastal waters rival forest trees in size.

Some single-celled and colonial protists have complex intracellular structures, some found nowhere else among living organisms **(Figure 26.3).** For example, many freshwater protists have a mechanism that maintains water balance in and out of the cell to prevent lysis. Excess water entering cells by osmosis (see Section 6.3) is handled using a specialized cytoplasmic organelle, the **contractile vacuole.** The contractile vacuole gradually fills with water; when it reaches maximum size it moves to the plasma membrane and forcibly contracts, expelling the water to the outside through a pore in the membrane. Many protists also have **food vacuoles** that digest prey or other organic material engulfed by the cells. Enzymes secreted into the food vacuoles digest the organic molecules; any remaining undigested matter is expelled to the outside by a mechanism similar to the expulsion of water by contractile vacuoles.

The cells of some protists are supported by an external cell wall, or by an internal or external shell built up from organic or mineral matter; in some, the shell takes on highly elaborate forms. Other protists have a **pellicle,** a layer of supportive protein fibers located inside the cell, just under the plasma membrane, which provides strength and flexibility instead of a cell wall (see Figure 26.5).

Almost all protists have structures that provide motility at some time during their life cycle. Some move by amoeboid motion, in which the cell extends one or more lobes of cytoplasm called **pseudopodia** ("false feet"; see Figure 26.20). The rest of the cytoplasm and the nucleus then flow into a pseudopodium, completing the movement. Other protists move by the beating of flagella or cilia (see Section 5.3; cilia are essentially the same as

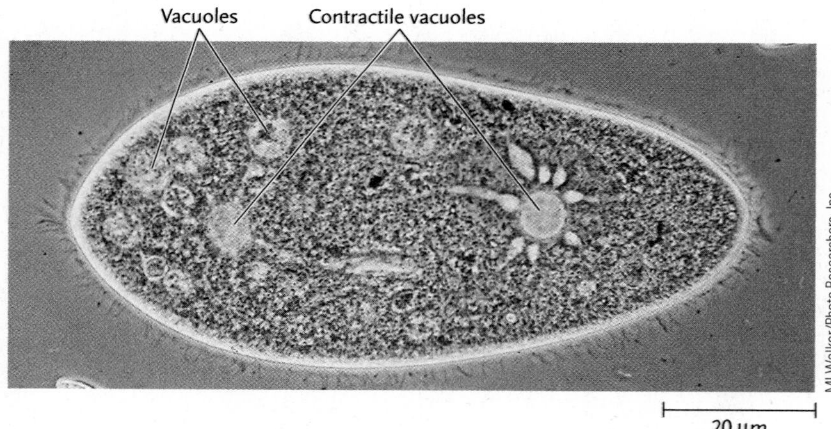
Vacuoles
Contractile vacuoles
MI Walker/Photo Researchers, Inc.
20 µm

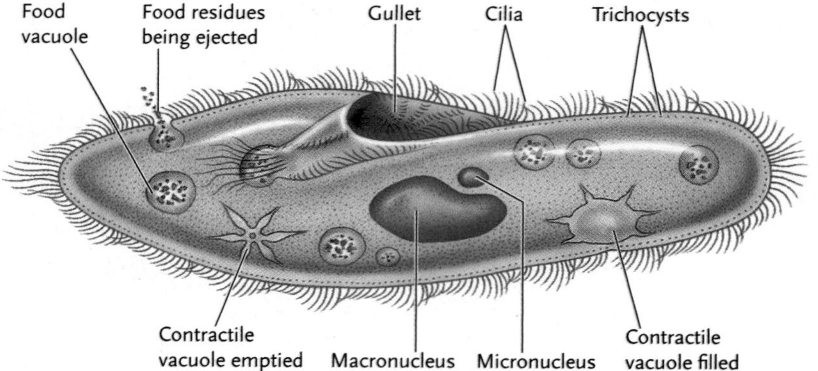
Food vacuole
Food residues being ejected
Gullet
Cilia
Trichocysts
Contractile vacuole emptied
Macronucleus
Micronucleus
Contractile vacuole filled

FIGURE 26.3
A ciliate, *Paramecium,* showing the cytoplasmic structures typical of many protists.

flagella, except that cilia are often shorter and occur in greater numbers on a cell). In some protists, cilia are arranged in complex patterns, with an equally complex network of microtubules and other cytoskeletal fibers supporting the cilia under the plasma membrane. Among the protists are the most complex single cells known because of the wide variety of cytoplasmic structures they have.

HABITAT Protists live in aqueous habitats, including aquatic or moist terrestrial locations such as oceans, freshwater lakes, ponds, streams, and moist soils, and within host organisms. In bodies of water, small photosynthetic protists collectively make up the **phytoplankton** (*phytos* = plant; *planktos* = drifting), the abundant organisms that capture the energy of sunlight in nearly all aquatic habitats. These photosynthetic protists provide organic substances and oxygen for heterotrophic bacteria and protists and for the small crustaceans and animal larvae that are the primary constituents of **zooplankton** (*zoe* = life, usually meaning animal life); although protists are not animals, biologists often include them among the zooplankton. The phytoplankton and the larger multicellular protists forming seaweeds collectively account for about half of the total organic matter produced by photosynthesis.

In the moist soils of terrestrial environments, protists play important roles among the detritus feeders that recycle matter from organic back to inorganic form. In their roles in phytoplankton, zooplankton, and as detritus feeders, protists are enormously important in the world ecosystem.

Protists that live in host organisms are parasites, obtaining nutrients from the host. Indeed, many of the parasites that have significant effects on human health are protists, causing diseases such as malaria, sleeping sickness, and giardiasis.

STUDY BREAK 26.1 <

What distinguishes protists from prokaryotes? What distinguishes them from fungi, plants, and animals?

>

THINK OUTSIDE THE BOOK

Phylogenomics employs comparisons of genome sequences or the sequences of many genes simultaneously to construct phylogenetic trees. Use the Internet or research literature to outline how recent phylogenomic analysis has provided new insights into the evolutionary relationships among the protists.

26.2 The Protist Groups

This section considers the biological features of each of the groups of protists in Figure 26.2. The diversity of the protists has made their classification challenging to say the least. Changes and refinements are likely as more sequence data are accumulated.

The Excavates Include Protists with Unique Flagella and Protists with Modified Mitochondria

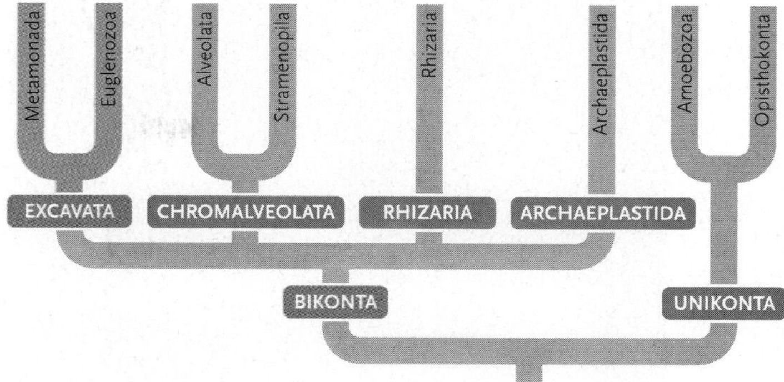

Metamonada
Euglenozoa
Alveolata
Stramenopila
Rhizaria
Archaeplastida
Amoebozoa
Opisthokonta
EXCAVATA
CHROMALVEOLATA
RHIZARIA
ARCHAEPLASTIDA
BIKONTA
UNIKONTA

All members of the **Excavata** are single-celled animal parasites that have greatly reduced mitochondria, or organelles derived from mitochondria, and move by means of flagella; most have a scooped out (excavated) feeding apparatus on the ventral surface of the cell. Because they do not possess functional mitochondria, excavates are incapable of aerobic respiration and are limited to glycolysis as an ATP source. However, the nuclei of excavates contain genes derived from mitochondria, meaning that the ancestors of these protists probably had mitochondria. The changes in their mitochondria may have occurred as an adaptation to the parasitic way of life, in which oxygen is in short supply. Two major subgroups of the Excavata are the **Metamonada** and **Euglenozoa.**

METAMONADA **Metamonads** consist of the Diplomonadida and Parabasala. *Diplomonad* cells have two nuclei and move by means of multiple freely beating flagella. In addition to lacking mitochondria, they also lack a clearly defined endoplasmic reticulum and Golgi complex. The best-known representative of the group, *Giardia lamblia* **(Figure 26.4A),** infects the mammalian intestinal tract, inducing severe diarrhea and abdominal cramps. *Giardia* is spread by water that is contaminated with feces, in which resistant cysts of the protist can be present in large numbers. So many streams and lakes in wilderness areas of the United States have become contaminated with *Giardia* cysts that hikers must boil water from these sources before drinking it, or pass it through filters able to remove particles as small as 1 μm. Treating water with chemicals such as chlorine or iodine does not kill the cysts.

Parabasalids also have freely beating flagella and, in addition, they have a sort of fin called an **undulating membrane,** formed by a flagellum buried in a fold of the cytoplasm. The buried flagellum allows parabasalids to move through thick and viscous fluids. Among the Parabasala are the trichomonads, including *Trichomonas vaginalis* **(Figure 26.4B),** a worldwide nuisance responsible for infections of the urinary and reproductive tracts in both men and women. The infective trichomonad is passed from person to person primarily, but not exclusively, by sexual intercourse. It lives in the vagina in women and in the urethra of both sexes. The infection is usually symptomless in men, but in women *T. vaginalis* can cause severe inflammation and irritation of the vagina and vulva. It is easily cured by drugs.

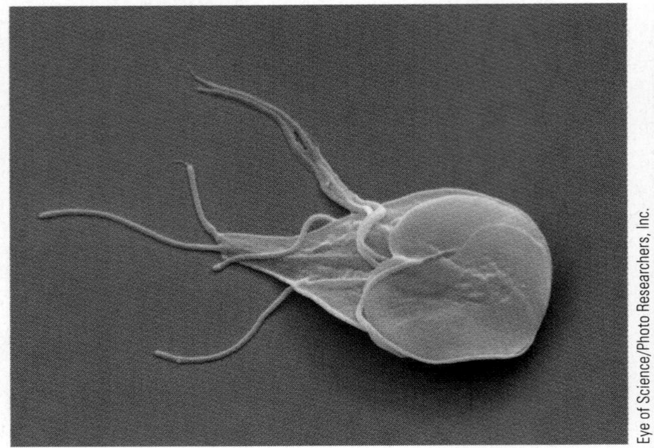
A. Giardia lamblia
Eye of Science/Photo Researchers, Inc.

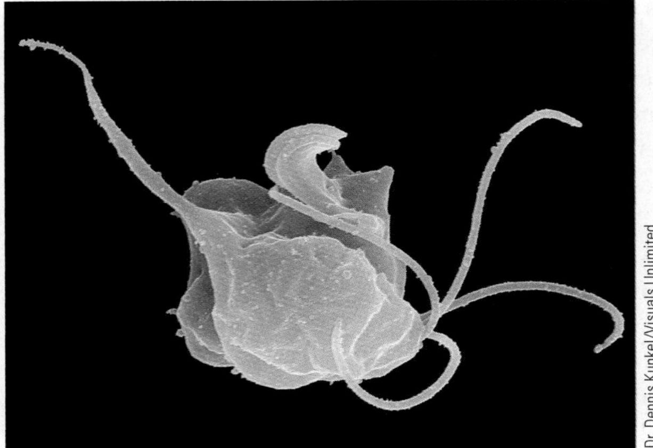
B. Trichomonas vaginalis
Dr. Dennis Kunkel/Visuals Unlimited

FIGURE 26.4
Metamonads of the excavates. **(A)** A diplomonad, *Giardia lamblia,* that causes intestinal disturbances. **(B)** A parabasalid, *Trichomonas vaginalis,* that causes a sexually transmitted disease, trichomoniasis.

EUGLENOZOA **Euglenozoans** include about 1,800 species, almost all single-celled, highly motile cells that swim by means of flagella. Their mitochondria are characterized by disc-shaped cristae (inner mitochondrial membranes). While most are photosynthetic, some are facultative heterotrophs; some can even alternate between photosynthesis and life as a heterotroph. The best-known members of this group are the *euglenids,* with *Euglena gracilis* **(Figure 26.5)** as the best-known species. Other members constitute the parasitic *kinetoplastids,* which includes some organisms responsible for human diseases.

Euglenids. **Euglenids** are free-living protists with anterior flagella. With the exception of a few marine species, the euglenids inhabit freshwater ponds, streams, and lakes. Most are autotrophs that carry out photosynthesis by the same mechanisms as plants, using the same photosynthetic pigments, including chlorophylls *a* and *b* and β-carotene. Many of the photosynthetic euglenids, including *E. gracilis,* can also live as heterotrophs by absorbing organic molecules through the plasma membrane. Some euglenids lack chloroplasts and live entirely as heterotrophs.

Euglena gracilis and other euglenids have a profusion of cytoplasmic organelles, including a contractile vacuole and, in photosynthetic species, chloroplasts (see Figure 26.5). Rather than an external cell wall, the euglenids have a spirally grooved pellicle formed from transparent, protein-rich material. Most of the photosynthetic euglenids, including *E. gracilis,* have an *eyespot* containing carotenoid pigment granules in association with a light-sensitive structure. The eyespot is part of a sensory

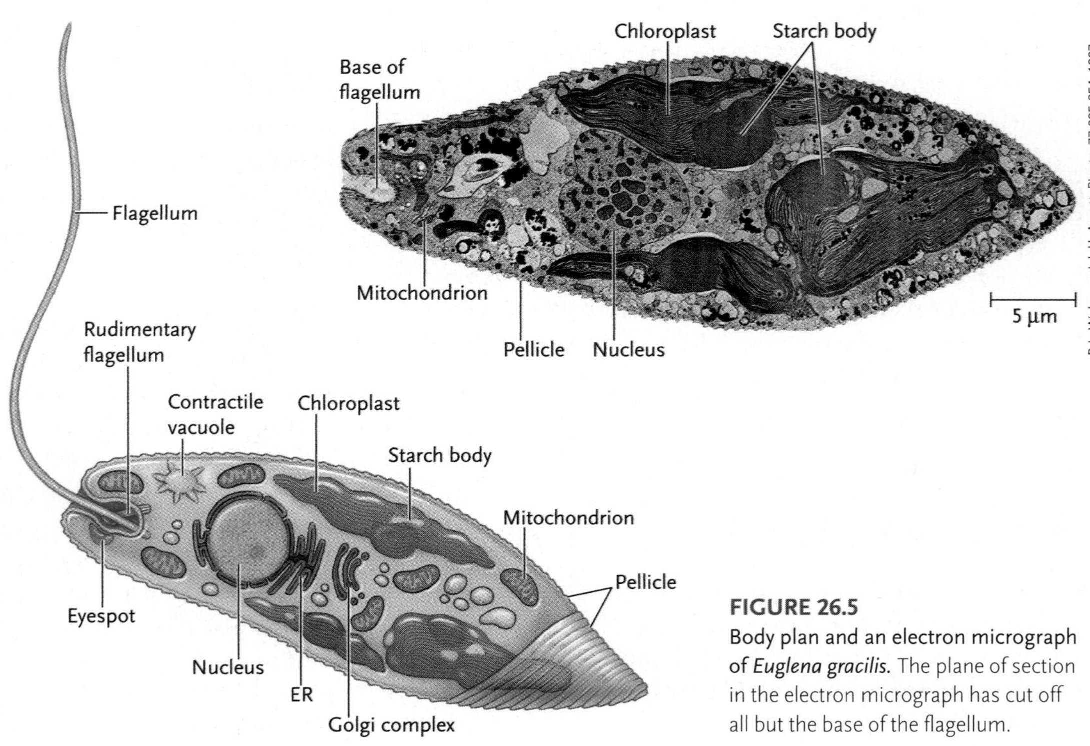
Chloroplast
Starch body
Base of flagellum
Flagellum
Mitochondrion
Pellicle
Nucleus
5 μm
P. L. Walne and J. H. Arnott, Planta, 77:325-354, 1967
Rudimentary flagellum
Contractile vacuole
Chloroplast
Starch body
Mitochondrion
Pellicle
Eyespot
Nucleus
ER
Golgi complex

FIGURE 26.5
Body plan and an electron micrograph of *Euglena gracilis.* The plane of section in the electron micrograph has cut off all but the base of the flagellum.

mechanism that stimulates cells to swim toward moderately bright light or away from intensely bright light so that the organism is in light conditions for optimal photosynthetic activity. The cells swim by whiplike movements of flagella that extend from a pocketlike depression at one end of the cell. Most have two flagella, one rudimentary and short, the other long.

Kinetoplastids. Kinetoplastids are a group of nonphotosynthetic, heterotrophic cells that live as animal parasites **(Figure 26.6).** Their name reflects the structure of the single mitochondrion in a cell of this group, which contains a large DNA–protein deposit called a *kinetoplast.* Most kinetoplastids have two flagella, which are used for movement. In some cases, as in Figure 26.6, one of the flagella is attached to the side of the cell, forming an undulating membrane that is often used to enable the organism to glide along or attach to surfaces.

The kinetoplastids include the trypanosomes, responsible for several diseases afflicting millions of humans in tropical regions. *Trypanosoma brucei* (see Figure 26.6) causes African sleeping sickness, transmitted from one host to another by bites of the tsetse fly. Early symptoms include fever, headaches, rashes, and anemia. Untreated, the disease damages the central nervous system, leading to a sleeplike coma and eventual death. The disease has proved difficult to control because the same trypanosome infects wild mammals, providing an inexhaustible reservoir for the parasite. Other trypanosomes, also transmitted by insects, cause Chagas disease in the southwestern United States and Central and South America, and leishmaniasis in the tropics. Humans with Chagas disease have an enlarged liver and spleen and may experience severe brain and heart damage; people with leishmaniasis have skin sores and ulcers that may become very deep and disfiguring, particularly to the face.

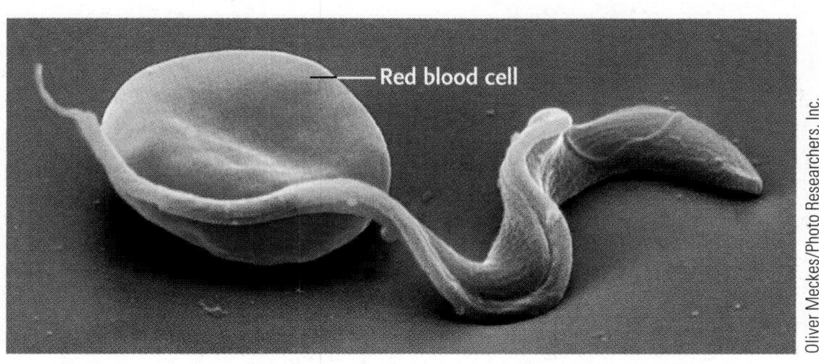
Red blood cell
Oliver Meckes/Photo Researchers, Inc.

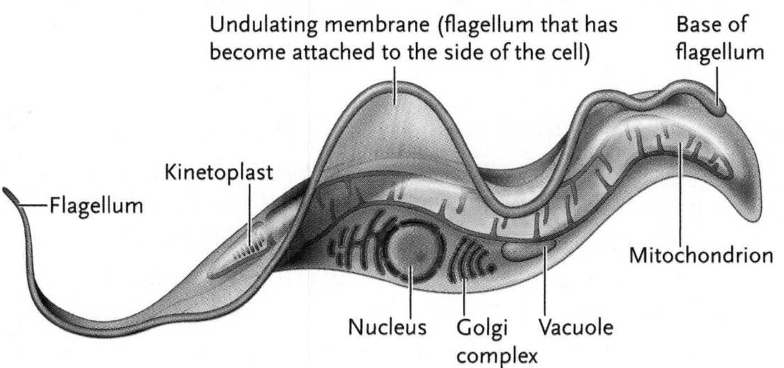
Undulating membrane (flagellum that has become attached to the side of the cell)
Base of flagellum
Kinetoplast
Flagellum
Mitochondrion
Nucleus
Golgi complex
Vacuole

FIGURE 26.6
Trypanosoma brucei, the parasitic kinetoplastid that causes African sleeping sickness.

The Chromalveolates Are a Heterogeneous Group of Protists with a Range of Forms and Life Styles

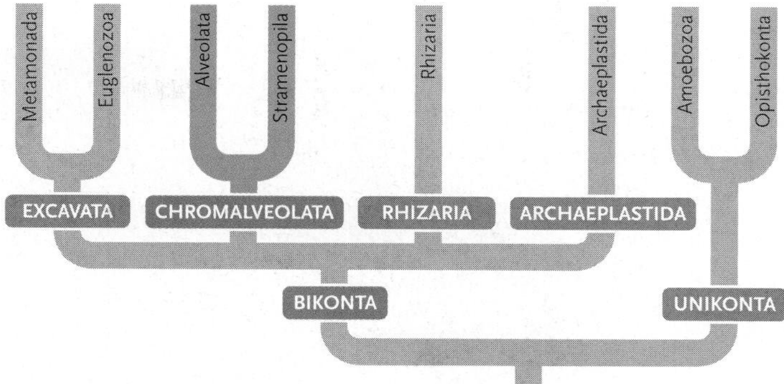
Metamonada
Euglenozoa
Alveolata
Stramenopila
Rhizaria
Archaeplastida
Amoebozoa
Opisthokonta
EXCAVATA
CHROMALVEOLATA
RHIZARIA
ARCHAEPLASTIDA
BIKONTA
UNIKONTA

The **Chromalveolata** consists of the **Alveolata** and the **Stramenopila.**

ALVEOLATA The **alveolates** are so called because they have small, flattened, membrane-bound vesicles called *alveoli* (*alvus* = belly) in a layer just under the plasma membrane. The alveolates include two motile, primarily free-living groups, the **Ciliophora** and **Dinoflagellata,** and a nonmotile, parasitic group, the **Apicomplexa.**

Ciliophora. Ciliophora—the **ciliates**—include nearly 10,000 known species of primarily single-celled but highly complex heterotrophic organisms that swim by means of cilia (see Figures 26.1B and 26.3). Ciliates were among the first organisms observed in the seventeenth century by the pioneering microscopist Anton van Leeuwenhoek (see Figure 5.1). Essentially any sample of pond water or bottom mud contains a wealth of these creatures.

The organisms in the Ciliophora have many highly developed organelles, including a mouthlike gullet lined with cilia; structures that exude mucins, toxins, or other defensive and offensive materials from the cell surface; contractile vacuoles; and complex systems of food vacuoles. A pellicle reinforces cell shape. A complex cytoskeletal network of microtubules and other fibers anchors the cilia just below the pellicle and coordinates the ciliary beating. The cilia can stop and reverse their beating in synchrony, allowing ciliates to stop, back up, and turn if they encounter negative stimuli. Evidence that the cytoskeletal network organizes ciliary beating comes from microsurgical experiments in which a segment of the body surface was cut out and reinserted in the opposite direction. The cilia in the reversed segment beat in the opposite direction to those on the rest of the organism.

The ciliates are the only eukaryotes that have two types of nuclei in each cell: one or more small nuclei called *micronuclei,* and a single larger *macronucleus* (see Figure 26.3). A **micronucleus** is a diploid nucleus that contains a complete complement of genes. It functions primarily in cellular reproduction, which may be asexual or sexual. The number of micronuclei present depends on the species. The **macronucleus** develops from a micronucleus, but loses all genes except those required for basic "housekeeping" functions of the cell and for ribosomal RNAs. These DNA

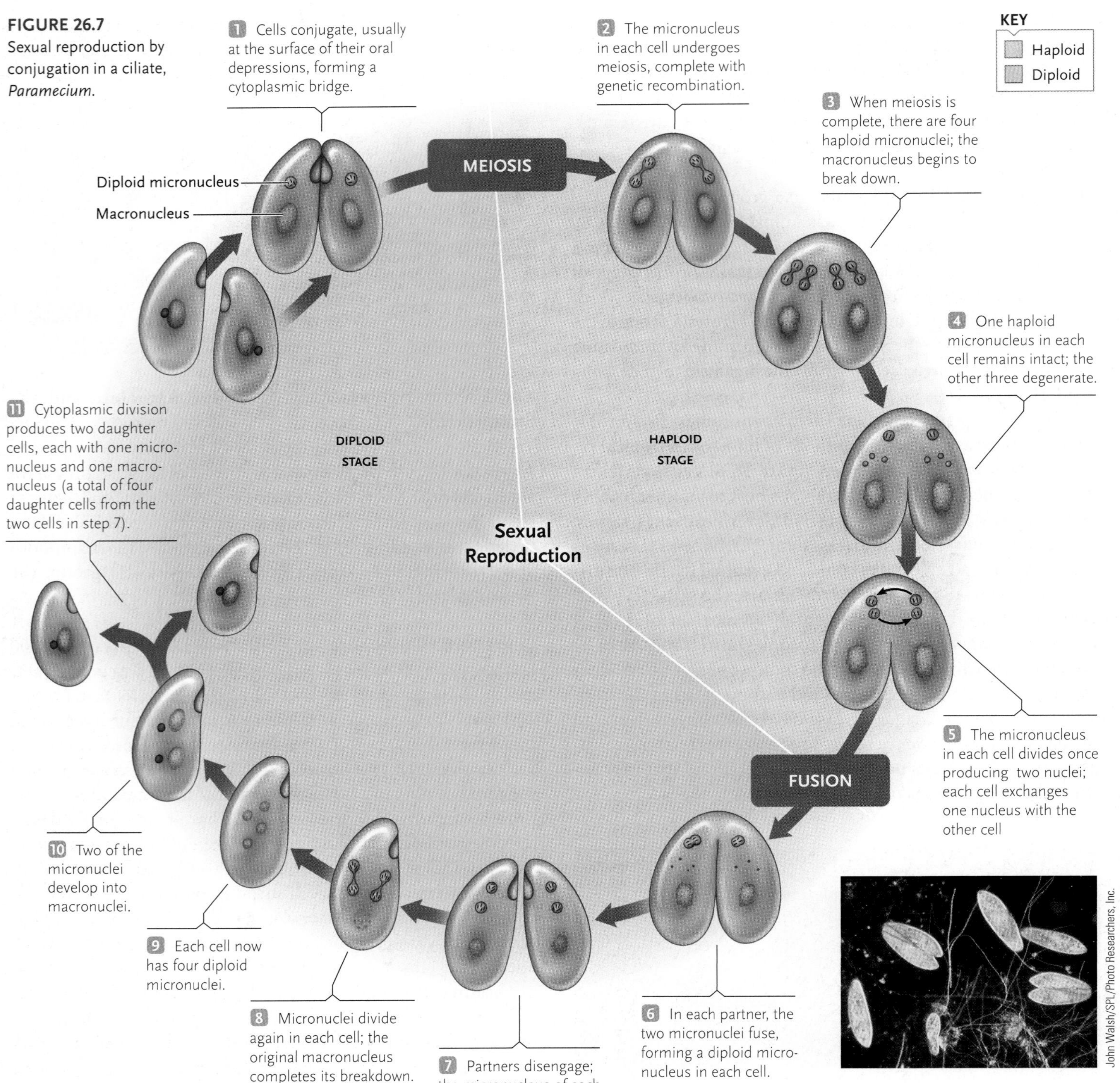
KEY
Haploid
Diploid
1 Cells conjugate, usually at the surface of their oral depressions, forming a cytoplasmic bridge.
Diploid micronucleus
Macronucleus
MEIOSIS
2 The micronucleus in each cell undergoes meiosis, complete with genetic recombination.
3 When meiosis is complete, there are four haploid micronuclei; the macronucleus begins to break down.
4 One haploid micronucleus in each cell remains intact; the other three degenerate.
5 The micronucleus in each cell divides once producing two nuclei; each cell exchanges one nucleus with the other cell
FUSION
6 In each partner, the two micronuclei fuse, forming a diploid micro-nucleus in each cell.
7 Partners disengage; the micronucleus of each divides mitotically.
8 Micronuclei divide again in each cell; the original macronucleus completes its breakdown.
9 Each cell now has four diploid micronuclei.
10 Two of the micronuclei develop into macronuclei.
11 Cytoplasmic division produces two daughter cells, each with one micro-nucleus and one macro-nucleus (a total of four daughter cells from the two cells in step 7).
DIPLOID STAGE
HAPLOID STAGE
Sexual Reproduction
John Walsh/SPL/Photo Researchers, Inc.

FIGURE 26.7 Sexual reproduction by conjugation in a ciliate, *Paramecium.*

sequences are duplicated many times, greatly increasing its capacity to transcribe the mRNAs needed for these functions, and the rRNAs needed to make ribosomes.

In asexual reproduction by mitosis, both types of nuclei replicate their DNA, divide, and are passed on to daughter cells. **Figure 26.7** illustrates sexual reproduction of *Paramecium* which is initiated when two cells **conjugate** and a cytoplasmic bridge forms between the cells.

Ciliates abound in freshwater and marine habitats, where they feed voraciously on bacteria, other protists, and each other. *Paramecium* is a typical member of the group (see Figure 26.3). Its rows of cilia drive it through its watery habitat, rotating the cell on its long axis while it moves forward, or backs and turns. The cilia also sweep water laden with prey and food particles into the gullet, where food vacuoles form. The ciliate digests food in the vacuoles and eliminates indigestible material through an anal pore. Contractile vacuoles with elaborate, raylike extensions remove excess water from the cytoplasm and expel it to the outside. When ciliates are under attack or otherwise stressed, surface organelles called **trichocysts** discharge many dartlike protein threads.

Some ciliates live individually while others are colonial. Certain ciliates are animal parasites; others live and reproduce in their hosts as mutually beneficial symbionts. (*Symbiosis* is the interaction between two organisms living together in close association, sometimes one inside another.) A compartment of the stomach of cattle and other grazing animals contains large numbers of symbiotic ciliates that digest the cellulose in their host's plant diet. The animals then digest the excess ciliates.

One ciliate, *Balantidium coli,* is a human intestinal parasite that causes diarrhea, with stools typically containing blood and pus. It is passed on when humans eat food contaminated by the feces of animals infected by *Balantidium,* particularly pigs. Less than 1% of the human population is infected worldwide.

Dinoflagellata. Dinoflagellata—dinoflagellates—consists of over 4,000 known species, most of which are single-celled organisms in marine phytoplankton. They live as heterotrophs or autotrophs; many can carry out both modes of nutrition. Some contain algae as symbionts. Typically, they have a shell formed from cellulose plates **(Figure 26.8).** The beating of flagella, which fit into grooves in the plates, makes dinoflagellates spin like a top (*dinos* = spinning).

The cytoplasmic structures of dinoflagellates include mitochondria, chloroplasts in photosynthetic species, and other internal membrane systems characteristic of eukaryotes. The photosynthetic dinoflagellates contain chlorophylls *a* and *c* along with accessory pigments that make them golden-brown or brown; algal symbionts give some a green, blue, or red color.

Their abundance in phytoplankton makes dinoflagellates a major primary producer of ocean ecosystems. Some species live as symbionts in the tissues of other marine organisms such as jellyfish, sea anemones, corals, and mollusks. For example, dinoflagellates in coral use the coral's carbon dioxide and nitrogenous waste, while supplying 90% of the coral's nutrition. The vast numbers of dinoflagellates living as photosynthetic symbionts in tropical coral reefs allow the reefs to reach massive size; without the dinoflagellates many coral species would die.

Some dinoflagellates are **bioluminescent**—they glow or release a flash of light, particularly when disturbed. The production of light depends on the enzyme *luciferase* and its substrate *luciferin,* which is similar to the system that produces light in fireflies. Dinoflagellate fluorescence can make the sea glow in the wake of a boat at night, and coat nocturnal surfers and swimmers with a ghostly light.

At times dinoflagellate populations grow to such large numbers that they color the seas red, orange, or brown. The resulting **red tides** are common in spring and summer months along the warmer coasts of the world, including all the U.S. coasts. Some red-tide dinoflagellates produce a toxin that interferes with nerve function in the animals that ingest these protists. Fish that feed on plankton, and birds that feed on the fish, may be killed in huge numbers by the toxin. Dinoflagellate toxin does not noticeably affect clams, oysters, and other mollusks, but it becomes concentrated in their tissues. Eating the tainted mollusks can cause respiratory failure and death in humans and other animals. The toxin is especially deadly for mammals because it paralyzes the diaphragm and other muscles required for breathing.

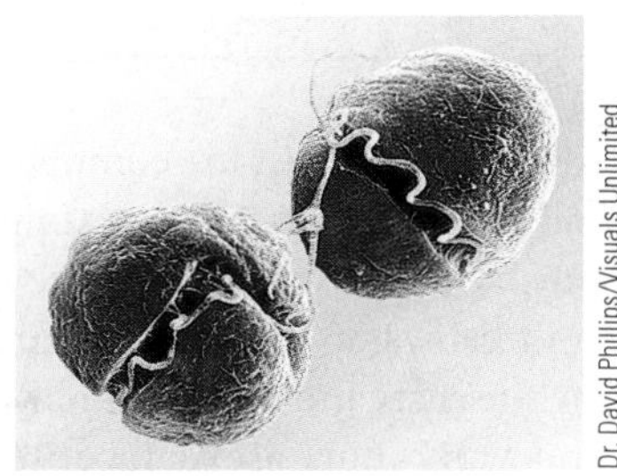

FIGURE 26.8
Karenia brevis, a toxin-producing dinoflagellate.

Dr. David Phillips/Visuals Unlimited

Apicomplexa. Apicomplexa—apicomplexans—are all nonmotile parasites of animals. They absorb nutrients through their plasma membranes rather than by engulfing food particles, and they lack food vacuoles. They get their name from the *apical complex,* a special grouping of fibrils, microtubules and organelles at one end of the cell that functions in attachment and invasion of host cells.

Typically, apicomplexan life cycles involve both asexual and sexual reproduction. All the apicomplexans, which includes almost 4,000 known species, produce infective sporelike stages called *sporozoites.* The sporozoites reproduce asexually in cells they infect, eventually bursting them, which releases the progeny to infect new cells. At some point they generate specialized cells that form gametes; fusion of gametes produces resistant cells known as *cysts.* Usually, a host is infected by ingesting cysts, which divide to produce sporozoites. This basic life cycle pattern varies considerably among the apicomplexans, and many of these organisms use more than one host species for different stages of their life cycle.

One apicomplexan genus, *Plasmodium,* is responsible for malaria, one of the most widespread and debilitating diseases of humans. The disease is transmitted by the bite of 60 different species of mosquitoes, all members of the genus *Anopheles.* Although the disease is now rare in the United States, *Anopheles* mosquitoes are common enough to spread malaria if *Plasmodium* is introduced by travelers from other countries. The infective cycle of *Plasmodium,* described in *Focus on Applied Research,* is representative of the complex life cycles of apicomplexans.

Another organism in this group, *Toxoplasma,* has a sexual phase of its life cycle in cats and asexual phases in humans, cattle, pigs, and other animals. Cysts of the parasite in the feces of infected cats are spread in household and garden dust. Humans ingesting or inhaling the cysts develop toxoplasmosis, a disease that is usually mild in adults but can cause severe brain damage or even death to a fetus. Because of the danger of toxoplasmosis, pregnant women should avoid emptying litter boxes or otherwise cleaning up after a cat.

STRAMENOPILA Stramenopila—stramenopiles—have two different flagella: one with hollow tripartite projections that give the flagellum a "hairy" appearance and a second one that is plain. The flagella occur only on reproductive cells such as eggs and sperm, except in the golden algae, in which cells are flagellated in all stages. The Stramenopila includes the **Oomycota** (water molds, white rusts, and mildews—formerly classified as fungi), **Bacillariophyta**

FOCUS ON Applied Research

Malaria and the *Plasmodium* Life Cycle

Although malaria is uncommon in the United States, it is a major epidemic in many other parts of the world. From 300 million to 500 million people get malaria each year in tropical regions, including Africa, India, Southeast Asia, the Middle East, Oceania, and Central and South America. Of these, about 2 million die each year, which is twice as many deaths as from AIDS, worldwide. It is particularly deadly for children younger than six years old. In many countries where malaria is common, people often get malaria repeatedly.

Four different species of the apicomplexan genus *Plasmodium* cause malaria. In the life cycle of the parasites (see **figure**), sporozoites develop in the female *Anopheles* mosquito, which transmits them by its bite to human or bird hosts. The infecting parasites divide repeatedly in their hosts, initially in liver cells and then in red blood cells. Their growth causes red blood cells to rupture in regular cycles every 48 or 72 hours, depending on the *Plasmodium* species. The ruptured red blood cells clog vessels and release the parasite's metabolic wastes, causing cycles of chills and fever.

The victim's immune system is ineffective because, during most of the infective cycle, the parasite is inside body cells and thus "hidden" from antibodies. Further, *Plasmodium* regularly changes its surface molecules, continually producing new forms that are not recognized by antibodies developed against a previous form. In this way, the parasite keeps one step ahead of the immune system, often making malarial infections essentially permanent.

Travelers in countries with high rates of malaria are advised to use antimalarial drugs such as chloroquine, quinine, or quinidine as a preventative. However, many *Plasmodium* strains in Africa, India, and Southeast Asia have developed resistance to the drugs. Vaccines have proved difficult to develop; because vaccines work by inducing the production of antibodies that recognize surface groups on the parasites, they are defeated by the same mechanisms the parasite uses inside the body to keep one step ahead of the immune reaction.

While in a malarial region, travelers should avoid exposure to mosquitoes by remaining indoors from dusk until dawn and sleeping inside mosquito nets treated with insect repellent. When out of doors, travelers should wear clothes that expose as little skin as possible and are thick enough to prevent mosquitoes from biting through the cloth. An insect repellent containing DEET should be spread on any skin that is exposed.

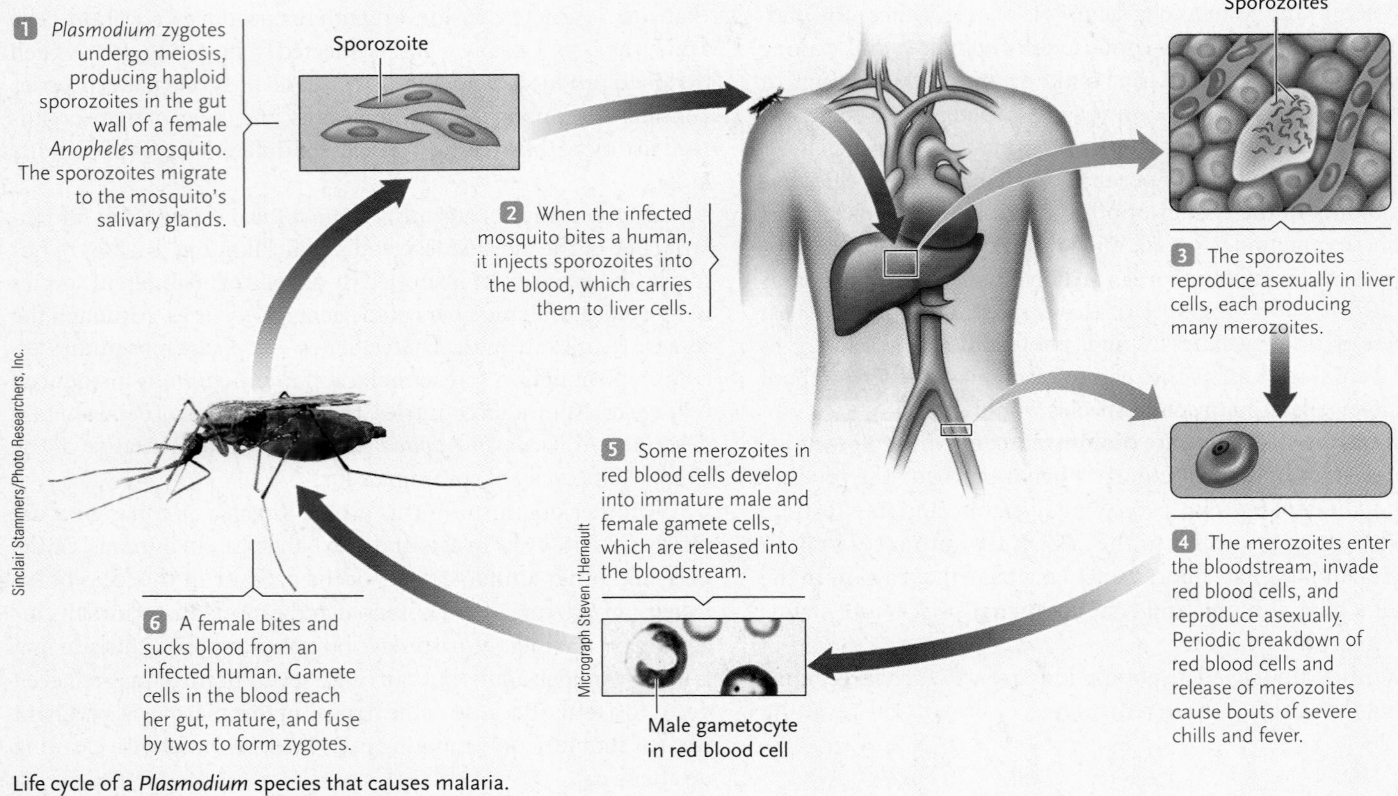
1 *Plasmodium* zygotes undergo meiosis, producing haploid sporozoites in the gut wall of a female *Anopheles* mosquito. The sporozoites migrate to the mosquito's salivary glands.
Sporozoite
2 When the infected mosquito bites a human, it injects sporozoites into the blood, which carries them to liver cells.
Sporozoites
3 The sporozoites reproduce asexually in liver cells, each producing many merozoites.
4 The merozoites enter the bloodstream, invade red blood cells, and reproduce asexually. Periodic breakdown of red blood cells and release of merozoites cause bouts of severe chills and fever.
5 Some merozoites in red blood cells develop into immature male and female gamete cells, which are released into the bloodstream.
Male gametocyte in red blood cell
6 A female bites and sucks blood from an infected human. Gamete cells in the blood reach her gut, mature, and fuse by twos to form zygotes.
Sinclair Stammers/Photo Researchers, Inc.
Micrograph Steven L'Hernault

Life cycle of a *Plasmodium* species that causes malaria.

(diatoms), **Chrysophyta** (golden algae), and **Phaeophyta** (brown algae).

Oomycota. **Oomycota—water molds, white rusts,** and **downy mildews**—are funguslike stramenopiles that lack chloroplasts and live as heterotrophs **(Figure 26.9).** Like fungi, they secrete enzymes that digest the complex molecules of surrounding dead or alive organic matter into simpler substances small enough to be absorbed into their cells. The water molds live almost exclusively in freshwater lakes and streams or moist terrestrial habitats; the white rusts and downy mildews are parasites of plants. Oomycota may reproduce asexually or sexually.

A. **Water mold**

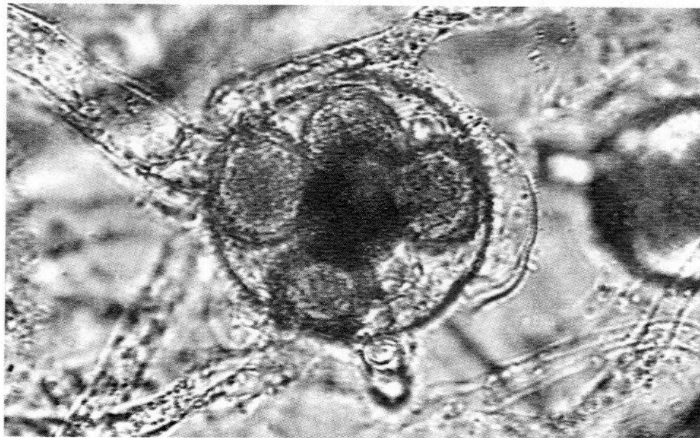

B. **Water mold infecting fish**

Heather Angel

C. **Downy mildew**

W. Merrill

FIGURE 26.9
Oomycota. **(A)** The water mold *Saprolegnia parasitica.* **(B)** *S. parasitica* growing as cottony white fibers on the tail of an aquarium fish. **(C)** A downy mildew, *Plasmopara viticola,* growing on grapes. At times it has nearly destroyed vineyards in Europe and North America.

Like fungi, many Oomycota grow as microscopic, nonmotile filaments called **hyphae** (singular, hypha), which form a network called a **mycelium (Figure 26.10).** Other features, however, set the Oomycota apart from the fungi; chief among them are differences in nucleotide sequence, which clearly indicate close evolutionary relationships to the stramenopiles rather than to the fungi. Further, nuclei in hyphae are diploid in the Oomycota, rather than haploid as in the fungi, and reproductive cells are flagellated and motile; fungi have no motile stages. Finally, the cell walls of most Oomycota contain cellulose (see Figure 3.9C); fungal cell walls instead contain a different polysaccharide, chitin (see Figure 3.9D)

Most water molds are key decomposers of both aquatic and moist terrestrial habitats. Dead animal or plant material immersed in water commonly becomes coated with cottony water molds. Other water molds parasitize living aquatic animals, such as the mold growing on the fish shown in Figure 26.9B. The white rusts and downy mildews are parasites of land plants (see Figure 26.9C).

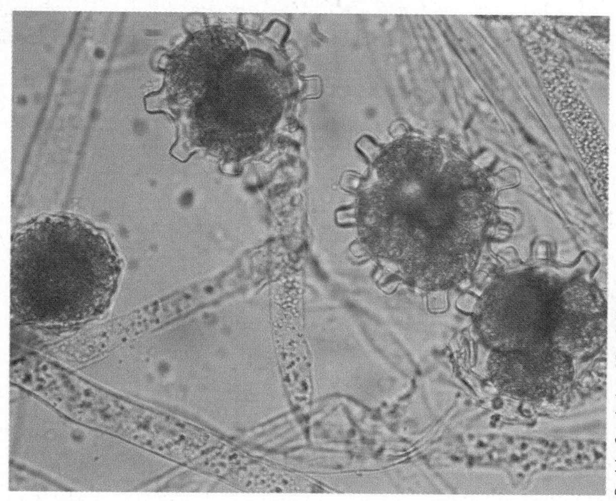

Dr. John Cunningham/Visuals Unlimited

FIGURE 26.10
The funguslike body form of the Oomycota, consisting of filaments called hyphae, which grow into a network called a mycelium.

Some water molds have had drastic effects on human history. *P. infestans,* a water mold that causes rotting of potato and tomato plants, was responsible for the Irish potato famine of 1845 to 1860 (see *Why It Matters*).

Bacillariophyta. Bacillariophyta—diatoms—are single-celled organisms that are covered by a glassy silica shell, which is intricately formed and beautiful in many species. The two halves of the shell fit together like the top and bottom of a candy box **(Figure 26.11).** Substances move to and from the plasma membrane through elaborately patterned perforations in the shell. Although flagella are present only in gametes, many diatoms move by an unusual mechanism in which a secretion released through grooves in the shell propels them in a gliding motion.

Diatoms are autotrophs that carry out photosynthesis by pathways similar to those of plants. The primary photosynthetic organisms of marine plankton, diatoms fix more carbon into organic material than any other planktonic organism. They are also abundant in freshwater habitats as both phytoplankton and bottom-dwelling species. Although most diatoms are free living, some are symbionts inside other marine protists. One diatom,

Jan Hinsch/SPL/Photo Researchers, Inc.

FIGURE 26.11
Diatom shells. Depending on the species, the shells are either radially or bilaterally symmetrical, as seen in this sample.

FIGURE 26.12 **Experimental Research**

The Evolutionary History of Diatoms as Revealed by Genome Analysis

Question: What do diatom genomes reveal about their evolutionary history?

Experiment: Diatoms are members of the chromalveolates. There are two major classes of diatoms, the radially symmetrical centrics and the bilaterally symmetric pennates. The earliest fossil centrics date to 180 million years ago, whereas the earliest fossil pennates date to 90 million years ago. Researchers have long been interested in how the two types of diatoms have evolved and diverged.

The genome sequence of a pennate diatom, *Phaeodactylum tricornutum*, was reported in 2008, and the genome sequence of a centric diatom, *Thalassiosira pseudonana*, was reported in 2004. Researchers involved in the 2008 report compared the genome sequences of the two diatoms, and compared the diatom genome sequences with the genome sequences of three metazoans, two plants, three green algae, one red alga, two fungi, and ten other chromalveolates.

Results: The table below presents the results of the genome sequence comparisons. In the table, three types of genes are distinguished:

- Core genes: Genes found in all of the eukaryotic groups with which the diatom sequences were compared.
- Diatom-specific genes: Genes found in both diatoms, but not in the other organisms.
- Unique genes: Genes found only in one or other of the two diatoms.

	Diatom species	
	P. tricornutum	*T. pseudonana*
Genome size	27.4 Mb[1]	32.4 Mb
Number of genes	10,402	11,776
Core genes	3,523	4,332
Diatom-specific genes	1,328	1,407
Unique genes	4,366	3,912

[1] Mb = 1 megabase = 10^6 base pairs

Notably, the results show that the two diatoms have only about 57% of their genes in common, illustrating a high degree of divergence since the split to the two diatom lineages.

The researchers made some other observations from the genome sequence comparisons:

- 171 of their chloroplast genes are of red algal origin; 108 of these genes are shared by the two species.
- 587 genes are of prokaryotic origin; 56% of these genes are found in both diatoms. These genes are presumed to have been obtained by horizontal gene transfer, the movement of genetic material between organisms other than by descent (see Sections 17.1 and 25.1). These genes represent more than 5% of the total genes of these diatoms, an unusually high percentage for a eukaryote. Most of the genes can be traced to proteobacteria, cyanobacteria, and various archaea. This observation is interpreted to mean that there has been a long-term association between diatoms and bacteria that has led to the transfer of useful genes.
- The diatom-specific genes are evolving faster than other genes in these diatoms or in eukaryotes in general. Although the exact reason for this fact is unknown, it is thought that it may correlate with the extensive diversification of diatoms.

Conclusion: Analysis of the two diatom genome sequences has provided insights into the evolutionary history of diatoms. The analysis provides interesting comparisons with other eukaryotic organisms, and supports the hypothesis that chromalveolates evolved from symbiosis between a photosynthetic red alga and a heterotrophic host (a concept discussed further later in the chapter). The evidence for the extent of horizontal gene transfer, and for the increased rate of evolution of diatom-specific genes, may help explain the high degree of diversity of diatoms and their adaptations for their life in the oceans.

Sources: C. Bowler et al. 2008. The *Phaeodactylum* genome reveals the evolutionary history of diatom genomes. *Nature* 456:239–244; E. V. Armbrust et al. 2004. The genome of the diatom *Thalassiosira pseudonana:* ecology, evolution, and metabolism. *Science* 306:79–86.

Pseudo-nitzschia, produces a toxic amino acid that can accumulate in shellfish. The amino acid, which acts as a nerve poison, causes amnesic shellfish poisoning when ingested by humans; the poisoning can be fatal.

Asexual reproduction in diatoms occurs by mitosis followed by a form of cytoplasmic division in which each daughter cell receives either the top or bottom half of the parent shell. The daughter cell then secretes the missing half, which becomes the smaller, inside shell of the box. The daughter cell receiving the larger top half grows to the same size as the parent shell, but the cell receiving the smaller bottom half is limited to the size of this shell. As asexual divisions continue, the cells receiving bottom halves become progressively smaller. Very small diatoms may switch to a sexual mode of reproduction; they enter meiosis and produce flagellated gametes, which lose their shells and fuse in pairs to form a zygote. The zygote grows to normal size before secreting a completely new shell with full-size top and bottom halves.

The shells of diatoms are common in fossil deposits. In fact, more diatoms are known as fossils than as living species—some 35,000 extinct species have been described as compared with 7,000 living species. For about 180 million years the shells of diatoms have been accumulating into thick layers of sediment at the bottom of lakes and seas. Since diatoms store food as oil, fossil diatoms may be a source of oil in many oil deposits. **Figure 26.12** presents the results of genome analysis experiments that provide some insights into the evolution of diatoms.

FIGURE 26.13
Golden and brown algae. **(A)** A microscopic, swimming colony of *Synura*, a golden alga. Each cell bears two flagella, which are not visible in this light micrograph. **(B)** A brown alga, *Macrocystis*. Gas bladders keep the blades floating. **(C)** The holdfast, stemlike stipes, and leaflike blades, as seen in another brown alga, the sea palm *Polstelsia palmaeformis*.

A. **Golden alga, *Synura***

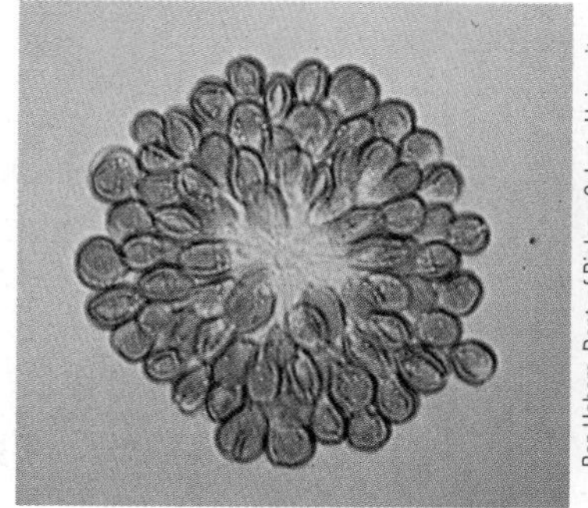

Ron Hoham, Dept. of Biology, Colgate University

B. **Brown alga, *Macrocystis***

C. **Brown alga, *Polstelsia palmaeformis***

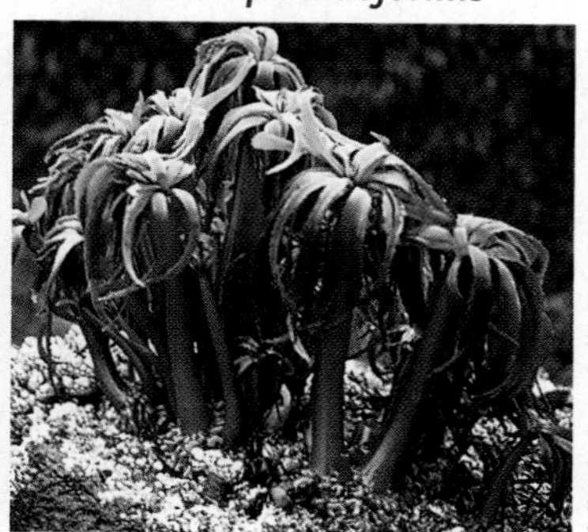

Jeffrey Levinton, State University of New York, Stony Brook

Grinding the fossilized shells into a fine powder produces *diatomaceous earth,* which is used in abrasives and filters, as an insulating material, and as a pesticide. Diatomaceous earth kills crawling insects by abrading their exoskeleton, causing them to dehydrate and die. Insect larvae are killed in the same way. Insects also die when they eat the powder but larger animals, including humans, are unaffected by it.

Chrysophyta. Chrysophyta—golden algae—mostly are colonial forms in which each cell of the colony bears a pair of flagella **(Figure 26.13A).** The golden algae have glassy shells, but in the form of plates or scales rather than in the candy-box form of the diatoms.

Nearly all chrysophytes are autotrophs and carry out photosynthesis using pathways similar to those of plants. Their color is due to a brownish carotenoid pigment, fucoxanthin, which masks the green color of the chlorophylls. Golden algae are important in freshwater habitats and in "nanoplankton," a community of marine phytoplankton composed of huge numbers of extremely small cells. During the spring and fall, "blooms" of golden algae can give a fishy taste and brownish color to the water.

Phaeophyta. Phaeophyta—brown algae (*phaios* = brown)—are photosynthetic autotrophs that range from microscopic forms to giant kelps that reach lengths of 50 m or more (Figure 26.1C and **Figure 26.13B and C**). Their color is also due to fucoxanthin. Their cell walls contain cellulose and a mucilaginous polysaccharide, alginic acid.

Nearly all of the 1,500 known phaeophyte species inhabit temperate or cool coastal marine waters. The kelps form vast underwater forests; fragments of these algae litter the beaches in coastal regions where they grow. Great masses of another brown alga, *Sargassum,* float in an area of the mid-Atlantic Ocean called the Sargasso Sea, which covers millions of square kilometers between the Azores and the Bahamas.

Kelps are the largest and most complex of all protists. Their tissues are differentiated into leaflike *blades,* stalklike *stipes,* and rootlike *holdfasts* that anchor them to the bottom. Hollow, gas-filled bladders give buoyancy to the stipes and blades and help keep them upright. The stalks of some kelps contain tubelike vessels, similar to the vascular elements of plants, which rapidly distribute dissolved sugars and other products of photosynthesis throughout the body of the alga.

Life cycles among the brown algae are typically complex and in many species consist of alternating haploid and diploid generations **(Figure 26.14).** The large structures that we recognize as kelps and other brown seaweeds are diploid **sporophytes,** so called because they give rise to haploid spores by meiosis. The spores germinate and divide by mitosis to form a small, independent, haploid **gametophyte** generation, which gives rise to haploid gametes. Variations occur in smaller brown algae, including some life cycles in which the sporophytes and gametophytes are the same size and some in which the gametophyte is larger than the sporophyte.

The alginic acid in brown algal cell walls, called **algin** when extracted, is an essentially tasteless and nontoxic substance used to thicken such diverse products as ice cream, pudding, salad dressing, jellybeans, cosmetics, paper, and floor polish. Brown algae are also harvested as food crops and as fertilizers.

The Rhizarians Consist of Amoebas with Filamentous Pseudopods

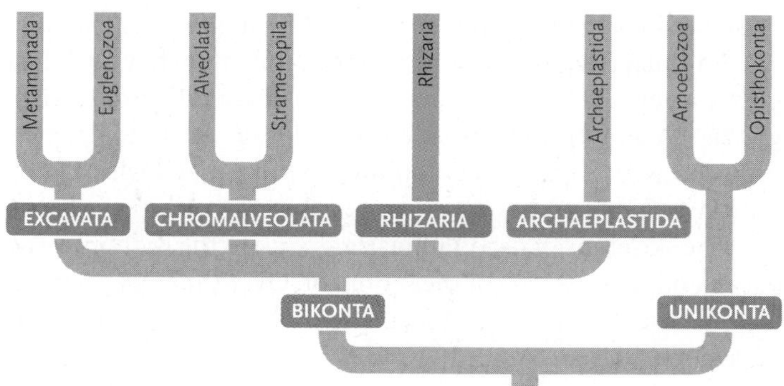

Metamonada
Euglenozoa
Alveolata
Stramenopila
Rhizaria
Archaeplastida
Amoebozoa
Opisthokonta
EXCAVATA
CHROMALVEOLATA
RHIZARIA
ARCHAEPLASTIDA
BIKONTA
UNIKONTA

Amoeba is a descriptive term for a single-celled protist that moves by means of temporary cellular projections called pseudopods. Several major groups of protists contain amoebas, which are similar in form but are not all closely related. The amoebas of the **Rhizaria** produce stiff, filamentous pseudopodia, and many produce hard outer shells, also called *tests.* We consider here two heterotrophic groups of rhizarian amoebas, the **Radiolaria** and the **Foraminifera,** and a third, photosynthesizing group, the **Chlorarachniophyta.**

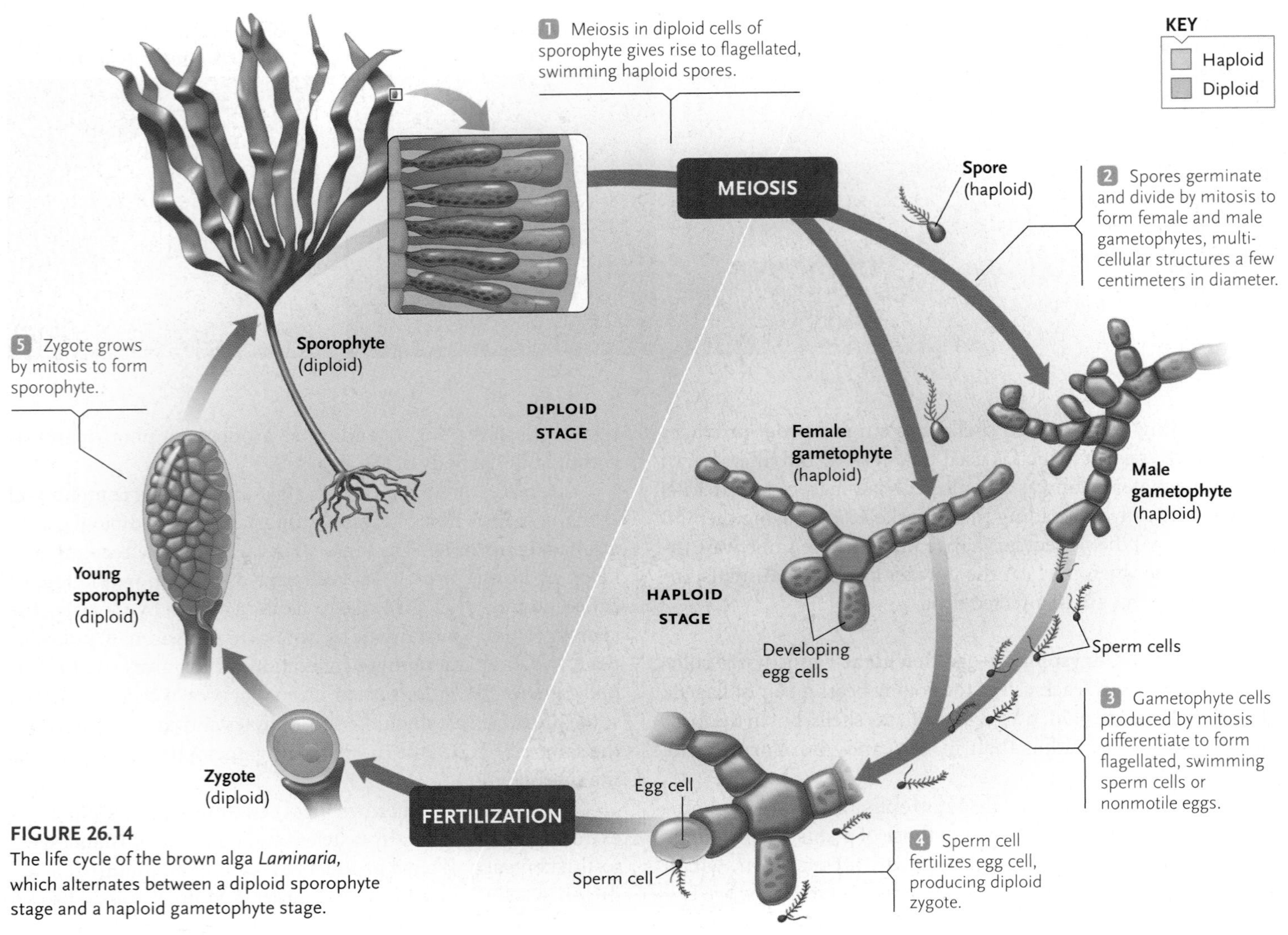
1 Meiosis in diploid cells of sporophyte gives rise to flagellated, swimming haploid spores.
KEY
Haploid
Diploid
MEIOSIS
Spore (haploid)
2 Spores germinate and divide by mitosis to form female and male gametophytes, multicellular structures a few centimeters in diameter.
5 Zygote grows by mitosis to form sporophyte.
Sporophyte (diploid)
DIPLOID STAGE
Female gametophyte (haploid)
Male gametophyte (haploid)
Young sporophyte (diploid)
HAPLOID STAGE
Developing egg cells
Sperm cells
3 Gametophyte cells produced by mitosis differentiate to form flagellated, swimming sperm cells or nonmotile eggs.
Zygote (diploid)
Egg cell
FERTILIZATION
Sperm cell
4 Sperm cell fertilizes egg cell, producing diploid zygote.

FIGURE 26.14
The life cycle of the brown alga *Laminaria,* which alternates between a diploid sporophyte stage and a haploid gametophyte stage.

RADIOLARIA Radiolaria—radiolarians—are distinguished by axopods, slender, raylike strands of cytoplasm supported internally by long bundles of microtubules. They engulf prey organisms that stick to the axopods and digest them in food vacuoles.

Radiolarians live in marine environments. They secrete a glassy internal skeleton from which the axopods project **(Figure 26.15A and B).** Just outside the skeleton, the cytoplasm is crowded with frothy vacuoles and lipid droplets, which provide buoyancy.

The skeletons of dead radiolarians sink to the bottom of the ocean and become part of the sediment. Over time, they harden into sedimentary rocks that form an important part of the geological record.

FORAMINIFERA: FORAMS Foraminifera—forams—live in marine environments. Their shells consist of organic matter which is reinforced by calcium carbonate **(Figure 26.15C and D).** Most foram shells are chambered, spiral structures that, although microscopic, resemble those of mollusks. Forams are identified and classified primarily by the form of the shell; about 250,000 species are known. Some species are planktonic, but they are most abundant on sandy bottoms and attached to rocks along the coasts. Their name comes from the perforations in their shells (*foramen* = little hole), through which extend long, slender strands of cytoplasm supported internally by a network of needlelike spines. The forams engulf prey that adhere to the strands and conduct them through the holes in the shell into the central cytoplasm, where they are digested in food vacuoles. Some forams have algal symbionts that carry out photosynthesis, allowing them to live as both heterotrophs and autotrophs.

Marine sediments are typically packed with the shells of dead forams. The sediments may be hundreds of feet thick; the White Cliffs of Dover in England are composed primarily of the shells of ancient forams. Most of the world's deposits of limestone and marble contain foram shells; the great pyramids and many other monuments of ancient Egypt are built from blocks cut from fossil foram deposits. Because distinct species lived during different geological periods, they are widely used to establish the age of sedimentary rocks containing their shells. They, along with radiolarian species, are also used as indicators by oil prospectors because layers of forams often overlie oil deposits.

CHLORARACHNIOPHYTA Chlorarachniophyta—chlorarachniophytes—are green, photosynthetic amoebas that also engulf food. They contain chlorophylls *a* and *b,* but they are phylogeneti-

A. **Radiolarian skeletons**

Wim van Egmond/Visuals Unlimited

B. **Living foram**

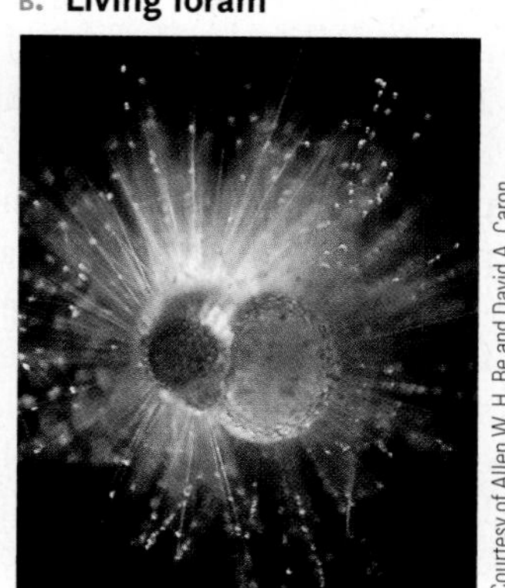
Courtesy of Allen W. H. Be and David A. Caron

C. **Foram shells**

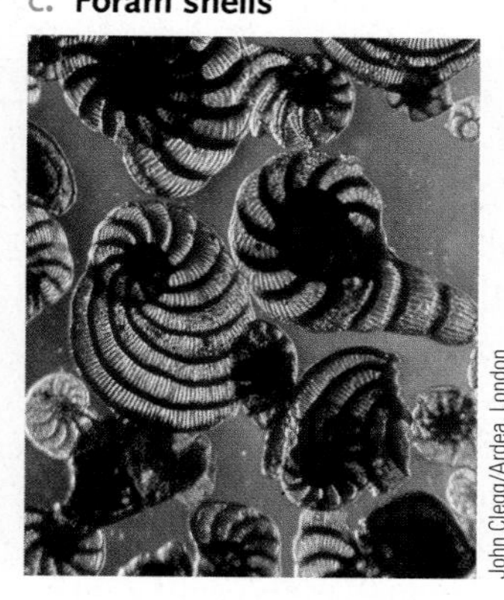
John Clegg/Ardea, London

D. **Foram body plan**

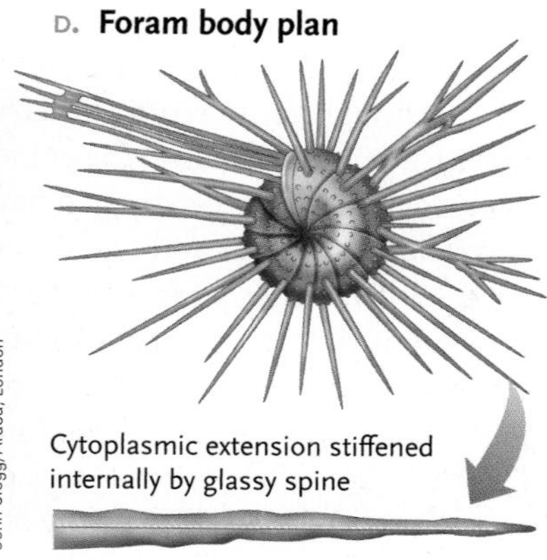
Cytoplasmic extension stiffened internally by glassy spine

FIGURE 26.15

Radiolarians and forams. **(A)** The internal skeletons of two radiolarian species, possibly *Pterocorys* and *Stylosphaera*. Bundles of microtubules support the cytoplasmic extensions of the radiolarians. **(B)** A living foram, showing the cytoplasmic strands extending from its shell. **(C)** Empty foram shells. **(D)** The body plan of a foram. Needlelike, glassy spines support the cytoplasmic extensions of the forams.

cally distinct from other chlorophyll *b*–containing eukaryotes. Many filamentous pseudopodia extend from the cell surface.

The Archaeplastids Include the Red and Green Algae, and Land Plants

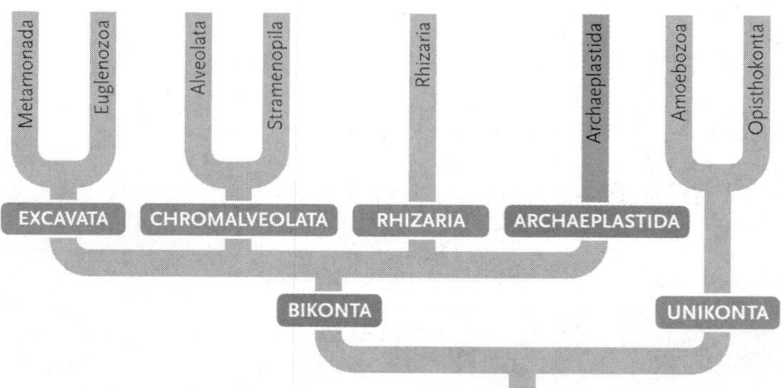
Metamonada
Euglenozoa
Alveolata
Stramenopila
Rhizaria
Archaeplastida
Amoebozoa
Opisthokonta
EXCAVATA
CHROMALVEOLATA
RHIZARIA
ARCHAEPLASTIDA
BIKONTA
UNIKONTA

The **Archaeplastida** consist of the **Rhodophyta** (red algae) and **Chlorophyta** (green algae), which are protists, and the land plants (the ***viridaeplantae***, or "true plants"), which comprise the kingdom Plantae. These three groups are descended from the first eukaryote that acquired a chloroplast from an endosymbiotic cyanobacterium (see Section 24.3 and later in this chapter), and they are all photosynthesizers. Here we describe the red and green algae; we discuss land plants in Chapter 27.

RHODOPHYTA: THE RED ALGAE Nearly all the 4,000 known species of red algae, which are also known as the Rhodophyta (*rhodon* = rose), are small marine seaweeds **(Figure 26.16).** Fewer than 200 species are found in freshwater lakes and streams or in soils. Most red algae grow attached to sandy or rocky substrates, but a few occur as plankton. Although most are free-living autotrophs, some are parasites that attach to other algae or plants.

Red algae are typically multicellular organisms, with plantlike bodies that are composed of interwoven filaments. The base of the body is differentiated into a holdfast, which anchors it to the bottom or other solid substrate, and into stalks with leaflike plates. Their cell walls contain cellulose and mucilaginous pectins that give them a slippery texture. In some species, the walls are hardened with stonelike deposits of calcium carbonate. Many

A. **Filamentous red alga**

Wim van Egmond/Visuals Unlimited

B. **Sheetlike red alga**

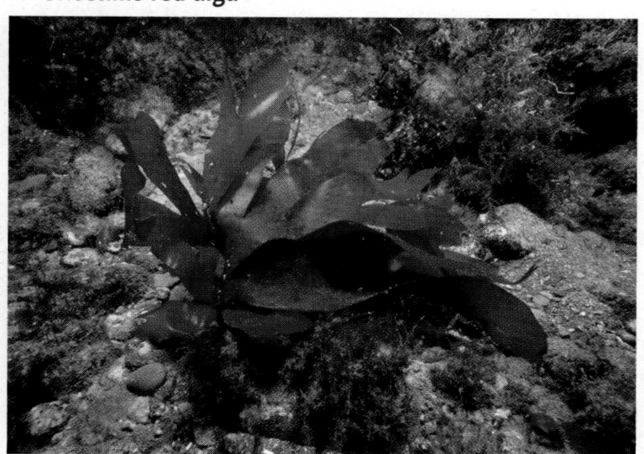
Andrew J. Martinez/Photo Researchers, Inc.

FIGURE 26.16

Red algae. **(A)** *Antithamnion plumula*, showing the filamentous and branched body form most common among red algae. **(B)** A sheetlike red alga growing on a tropical reef.

of the red algae with stony cell walls resemble corals and occur with corals in reefs and banks.

Although most red algae are reddish in color, some are greenish purple or black. The color differences are produced by accessory pigments, mainly *phycobilins,* which absorb green wavelengths of light and reflect red ones, thereby masking the green color of their chlorophylls. The phycobilins are unusual photosynthetic pigments with structures similar to the ring structure of hemoglobin. The accessory pigments of some red algae make them highly efficient in absorbing the shorter wavelengths of light that penetrate to the ocean depths, allowing them to grow at deeper levels than any other algae. Some red algae live at depths to 260 m if the water is clear enough to transmit light to these levels.

Red algae have complex reproductive cycles involving alternation between diploid sporophytes and haploid gametophytes. No flagellated cells occur in the red algae; instead, gametes are released into the water to be brought together by random collisions in currents.

Extracts containing the mucilaginous pectins of red algal cell walls are widely used in industry and science. Extracted **agar** is used as a moisture-preserving, inert agent in cosmetics and baked goods, as a setting agent for jellies and desserts, and as a culture medium in the laboratory. **Carrageenan,** extracted from the red alga *Eucheuma,* is used to thicken and stabilize paints, dairy products such as pudding and ice cream, and many other creams and emulsions.

Some red algae are harvested as food in Japan and China. *Porphyra,* one of these harvested algae, is used in sushi bars as the *nori* wrapped around fish and rice.

CHLOROPHYTA: THE GREEN ALGAE Chlorophyta (*chloros* = green)—green algae—are autotrophs that carry out photosynthesis using the same pigments as plants. They include single-celled, colonial, and multicellular species (**Figure 26.17**; see also Figure 26.1D). Most green algae are microscopic, but some range upward to the size of small seaweeds. Although the multicellular green algae have bodies that are filamentous, tubular, or leaflike, there is relatively little cellular differentiation as compared with the brown algae. However, the most complex green algae, such as the sea lettuce *Ulva* (see Figure 26.17C), have tissues differentiated into a leaflike body and a holdfast.

With at least 16,000 species, green algae show more diversity than any other algal group. Most live in freshwater aquatic habitats, but some are marine, or live on rocks and soil surfaces, on tree bark, or even on snow. The green, slimy mat that grows profusely in stagnant pools and ponds, for example, consists of filaments of a green alga. A few species live as symbionts in other protists or in fungi and animals. Lichens (see Figure 28.15) are the primary example of a symbiotic relationship between green algae and fungi. Many animal phyla, including some marine snails and sea anemones, contain green algal chloroplasts, or entire green algae, as symbionts in their cells.

Life cycles among the green algae are as diverse as their body forms. Many can reproduce either sexually or asexually, and some alternate between haploid and diploid generations. Gametes in different species may be morphologically identical flagellated cells, or cells that have differentiated into a flagellated sperm cell and a nonmotile egg cell. Most common is a life cycle with a multicellular haploid phase and a single-celled diploid phase (**Figure 26.18**).

Among all the algae, the nucleic acid sequences of green algae are most closely related to those of land plants. In addition, as we have noted, green algae use the same photosynthetic pigments as plants, including chlorophylls *a* and *b,* and have the same complement of carotenoid accessory pigments. In some green algae, the thylakoid membranes within chloroplasts are arranged into stacks resembling the grana of plant chloroplasts (see Section 5.4). As storage reserves, green algae contain starches of the same types as plants, and the cell walls of some green algal

FIGURE 26.17

Green algae. **(A)** A single-celled green alga, *Acetabularia,* which grows in marine environments. Each individual in the cluster is a large, large, multinucleate cell with a rootlike base, stalk, and cap. **(B)** A colonial green alga, *Volvox.* Each green dot in the spherical wall of the colony is a potentially independent, flagellated cell. Daughter colonies can be seen within the parent colony. **(C)** A multicellular green alga, *Ulva,* common to shallow seas around the world.

A. Single-celled green alga

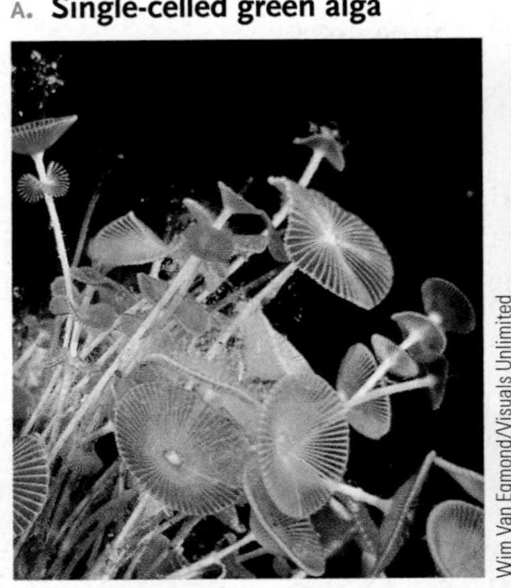
Wim Van Egmond/Visuals Unlimited

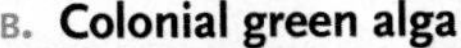
B. Colonial green alga

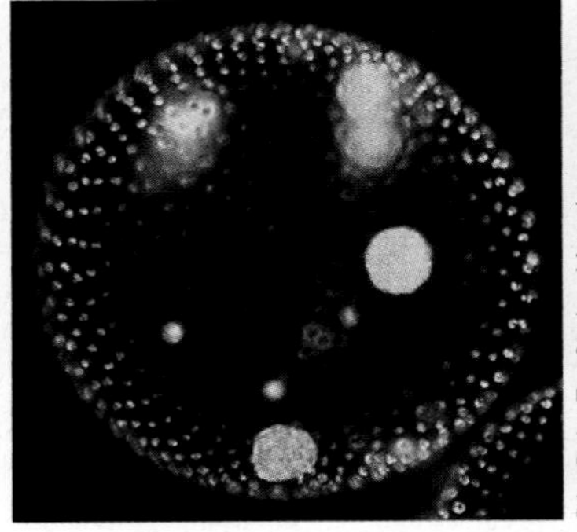
Brian Parker/Tom Stack and Associates

C. Multicellular green alga

Manfrage Kage/Peter Arnold, Inc.

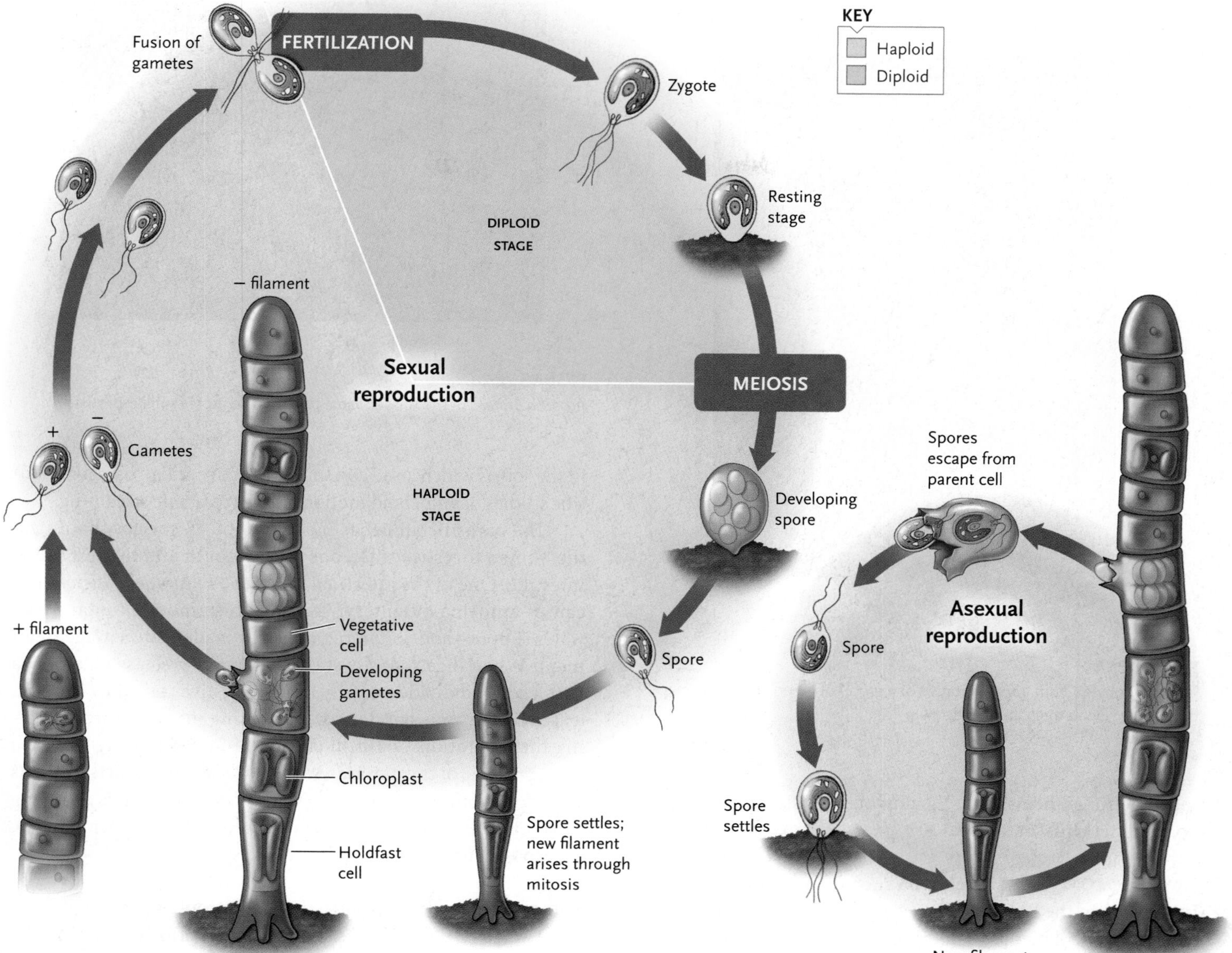

FIGURE 26.18
The life cycle of the green alga *Ulothrix.* In sexual reproduction, the haploid stage is multicellular and the diploid stage is a single cell, the zygote. "+" and "−" are morphologically identical mating types ("sexes") of the alga. In asexual reproduction, all stages are haploid.

species contain cellulose, pectins, and other polysaccharides like those of plants. On the basis of these similarities, many biologists propose that some ancient green alga gave rise to the evolutionary ancestors of modern-day plants.

What green alga might have been the ancestor of modern land plants? Many biologists consider a group known as the **charophytes** to be most similar to the algal ancestors of land plants. These organisms, including *Chara* **(Figure 26.19),** *Spirogyra, Nitella,* and *Coleochaete,* live in freshwater ponds and lakes. Their ribosomal RNA and chloroplast DNA sequences are more closely related to plant sequences than those of any other green alga. Further, the new cell wall separating daughter cells in charophytes is formed through development of a cell plate, by a mechanism closely similar to that of plants (see Section 10.2). The body form is distinctly plantlike, with a stemlike axis upon which whorls of leaflike blades occur at intervals.

The Unikonts Include Protists That Are Closely Related to Animals and Fungi

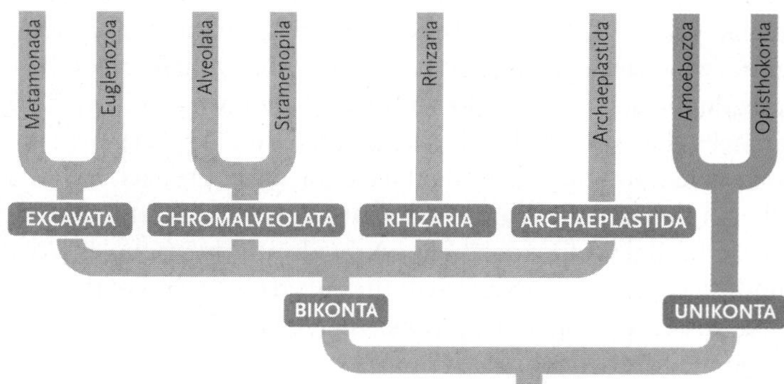

Metamonada
Euglenozoa
Alveolata
Stramenopila
Rhizaria
Archaeplastida
Amoebozoa
Opisthokonta
EXCAVATA
CHROMALVEOLATA
RHIZARIA
ARCHAEPLASTIDA
BIKONTA
UNIKONTA

The **Unikonta—unikonts**—include amoebas with unsupported pseudopods, protists closely related to animals or fungi, and ani-

Reproductive structures
Dr. John Clayton, National Institute of Water and Atmospheric Research, New Zealand

FIGURE 26.19
The charophyte *Chara,* representative of a group of green algae that may have given rise to the plant kingdom.

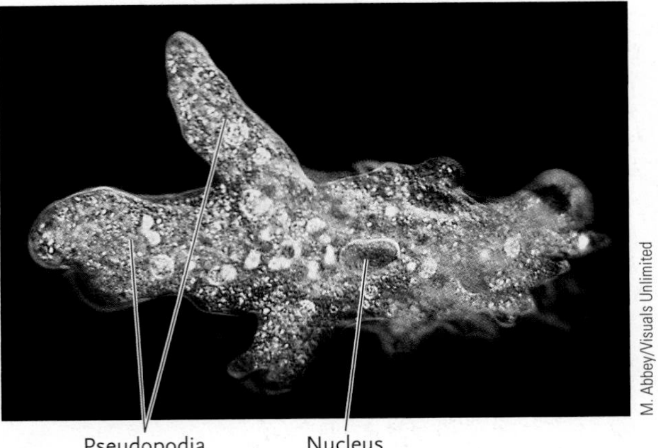
Pseudopodia
Nucleus
M. Abbey/Visuals Unlimited

FIGURE 26.20
Amoeba proteus of the Amoebozoa, perhaps the most familiar protist.

mals and fungi themselves. The unikonts consist of the **Amoebozoa** and the **Opisthokonta.**

AMOEBOZOA The **Amoebozoa** include most of the amoebas (others are in the Rhizaria) as well as the cellular and plasmodial slime molds. All members of this group use pseudopods for locomotion and feeding for all or part of their life cycles.

Amoebas. Amoebas of the Amoebozoa are single-celled organisms that are abundant in marine and freshwater environments and in the soil. They use pseudopods for locomotion and feeding. The pseudopods extend and retract at any point on their body surface and are unsupported by any internal cellular organization. This type of pseudopod—called a *lobose* ("lobelike") *pseudopod*—distinguishes these amoebas from those in the Rhizaria, which have stiff, supported pseudopods. As a result of their pseudopod activity, and the ability to flatten or round up, these amoebas have no fixed body shape. A number of species are parasites, but most species feed on bacteria, other protists, and bits of organic matter. The ingested matter is enclosed in food vacuoles and digested by enzymes secreted into the vacuoles. Any undigested matter is expelled to the outside by fusion of the vacuole with the plasma membrane. Their reproduction is entirely asexual, through mitotic divisions.

The most-studied amoeba of the amoebozoans is *Amoeba proteus* **(Figure 26.20).** Its natural habitat is in freshwater ponds and streams. Another member, *Acanthamoeba,* which lives in the soil, is widely used as a source of actin and myosin for scientific studies of amoeboid motion and cytoplasmic streaming.

The parasitic amoebas include some 45 species that infect the human digestive tract, one in the mouth and the rest in the intestine. One of the intestinal parasites, *Entamoeba histolytica,* causes amoebic dysentery. Cysts of this amoeba contaminate water supplies and soil in regions with inadequate sewage treatment. When ingested, a cyst breaks open to release an amoeba that feeds and divides rapidly in the digestive tract. Enzymes released by the amoebas destroy cells lining the intestine, producing the ulcerations, painful cramps, and debilitating diarrhea characteristic of the disease. Amoebic dysentery afflicts millions of people worldwide; in less-developed countries, it is a leading cause of death among infants and small children. Other parasitic amoebas cause less severe digestive upsets.

Slime Molds. **Slime molds** are heterotrophic protists that, at some stage of their life cycle, exist as individuals that move by amoeboid motion but the remainder of the time exist in more complex forms. They live on moist, rotting plant material such as decaying leaves and bark. The cells engulf particles of dead organic matter, and also bacteria, yeasts, and other microorganisms, and digest them internally. At one stage of their life cycles, they differentiate into a funguslike, stalked structure called a **fruiting body,** which forms spores by either asexual or sexual reproduction. Some species are brightly colored in hues of yellow, green, red, orange, brown, violet, or blue. The two major evolutionary lineages of slime molds, the *cellular slime molds* and the *plasmodial slime molds,* differ in cellular organization.

Cellular slime molds exist primarily as individual cells, either separately or as a coordinated mass. Among the 70 or so species of cellular slime molds, *Dictyostelium discoideum* is best known; its genome sequence was reported in 2005. **Figure 26.21** shows asexual reproduction (steps 1–5) and sexual reproduction (steps 6–7) in *D. discoideum.*

Plasmodial slime molds exist primarily as a large composite mass, the **plasmodium,** in which individual nuclei are suspended in a common cytoplasm that is surrounded by a single plasma membrane. (This is not to be confused with *Plasmodium,* the genus

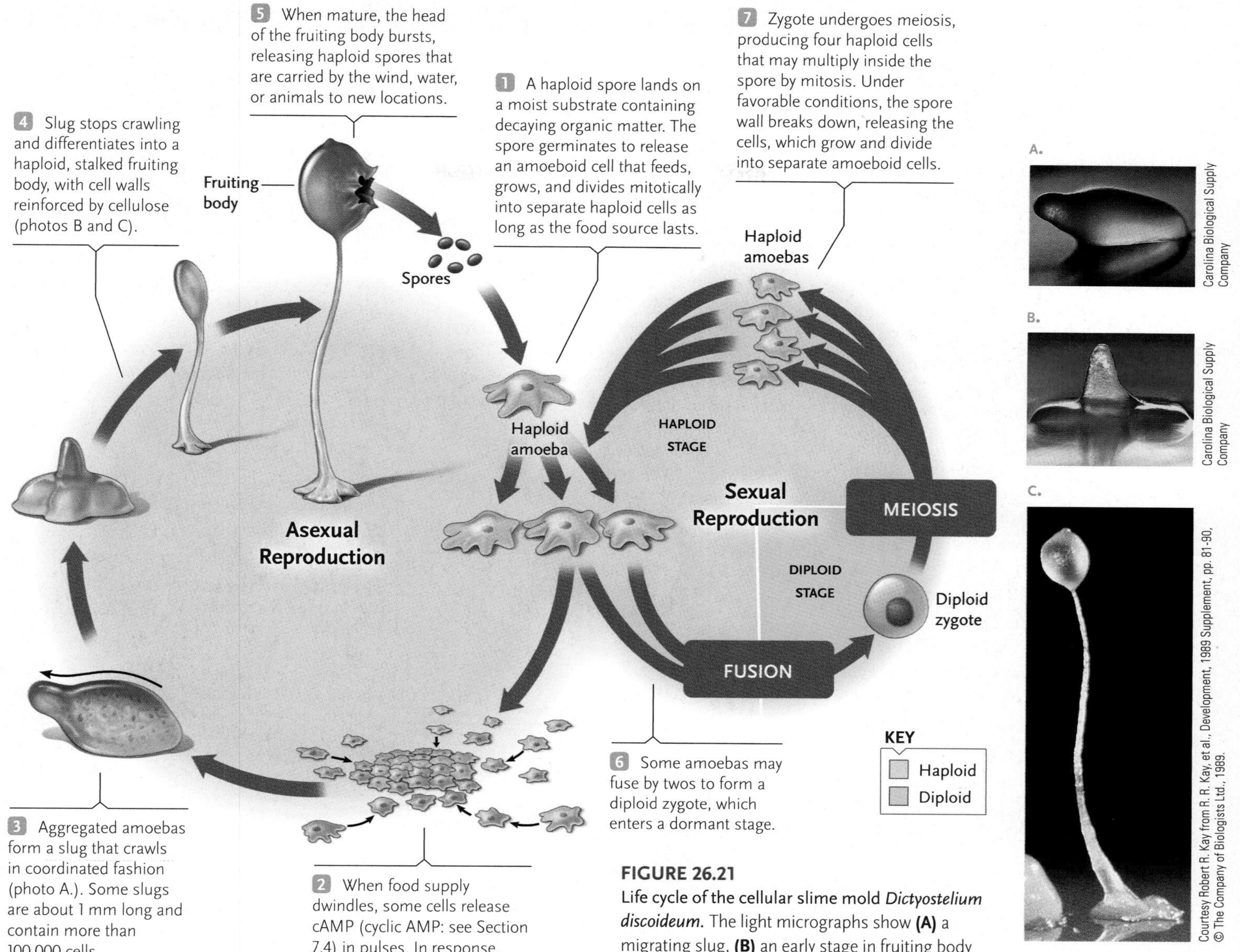
5 When mature, the head of the fruiting body bursts, releasing haploid spores that are carried by the wind, water, or animals to new locations.
7 Zygote undergoes meiosis, producing four haploid cells that may multiply inside the spore by mitosis. Under favorable conditions, the spore wall breaks down, releasing the cells, which grow and divide into separate amoeboid cells.
1 A haploid spore lands on a moist substrate containing decaying organic matter. The spore germinates to release an amoeboid cell that feeds, grows, and divides mitotically into separate haploid cells as long as the food source lasts.
4 Slug stops crawling and differentiates into a haploid, stalked fruiting body, with cell walls reinforced by cellulose (photos B and C).
Fruiting body
Spores
Haploid amoebas
Haploid amoeba
HAPLOID STAGE
Sexual Reproduction
MEIOSIS
Asexual Reproduction
DIPLOID STAGE
Diploid zygote
FUSION
KEY
Haploid
Diploid
6 Some amoebas may fuse by twos to form a diploid zygote, which enters a dormant stage.
3 Aggregated amoebas form a slug that crawls in coordinated fashion (photo A.). Some slugs are about 1 mm long and contain more than 100,000 cells.
2 When food supply dwindles, some cells release cAMP (cyclic AMP: see Section 7.4) in pulses. In response, amoebas aggregate together.
A.
B.
C.
Carolina Biological Supply Company
Carolina Biological Supply Company
Courtesy Robert R. Kay from R. R. Kay, et al., Development, 1989 Supplement, pp. 81-90, © The Company of Biologists Ltd., 1989.

FIGURE 26.21
Life cycle of the cellular slime mold *Dictyostelium discoideum*. The light micrographs show **(A)** a migrating slug, **(B)** an early stage in fruiting body formation, and **(C)** a mature fruiting body.

of apicomplexans that causes malaria.) There are about 500 known species of plasmodial slime molds. The main phase of the life cycle, the plasmodium (see Figure 26.1A), flows and feeds as a single huge amoeba—a single cell that contains thousands to millions or even billions of diploid nuclei surrounded by a single plasma membrane. Typically, a plasmodium, which may range in size from a few centimeters to more than a meter in diameter, moves in thick, branching strands connected by thin sheets. The movements occur by cytoplasmic streaming, driven by actin microfilaments and myosin (see Section 5.3). You may have seen one of these slimy masses crossing a lawn, moving over a mat of dead leaves, climbing a tree, or even in the movies—a slime mold in effect stars as a monster in the classic science fiction movie *The Blob.*

At some point, often in response to unfavorable environmental conditions, fruiting bodies form at sites on the plasmodium. At the tips of the fruiting bodies, nuclei become enclosed in separate cells, each surrounded by its own plasma membrane and cell wall. Depending on the species, either chitin or cellulose may reinforce the walls. These cells undergo meiosis, forming haploid, resistant spores that are released from the fruiting bodies and carried about by water or wind. If they reach a favorable environment, the spores germinate to form flagellated or unflagellated gametes, depending on the species, that fuse to form a diploid zygote. The zygote nucleus then divides repeatedly without an accompanying division of the cytoplasm, forming many diploid nuclei suspended in the common cytoplasm of a new plasmodium.

Both the cellular and plasmodial slime molds, particularly *Dictyostelium* (cellular) and *Physarum* (plasmodial; see Figure 26.1A), have been of great interest to scientists because of their ability to differentiate into fruiting bodies with stalks and spore-bearing structures. This differentiation is much simpler than the complex developmental pathways of other eukaryotes, providing a unique opportunity to study cell differentiation at its most fundamental level. One such study, examining the role of cAMP (cyclic AMP) in differentiation, is described in *Insights from the Molecular Revolution*.

Getting the Slime Mold Act Together: What molecules regulate development of *Dictyostelium*?

Development of differentiated structures can be followed at its simplest level in slime molds. In the cellular slime mold *Dictyostelium discoideum,* the aggregation of individual cells leading to differentiation begins when unfavorable living conditions induce some cells to secrete cyclic AMP (cAMP). Other *Dictyostelium* cells move toward the regions of highest cAMP concentration and aggregate into the slug stage. Further pulses of cAMP trigger differentiation into a stalk and spores.

Within the aggregating cells, the cAMP activates a cAMP-dependent protein kinase (PKA; see Section 7.4). The PKA, which is active only when cAMP is present, adds phosphate groups to target proteins in the cells. The target proteins, activated or deactivated by addition of the phosphate groups, trigger cellular developmental processes that lead to slug formation and differentiation of the stalk and spores.

Research Question

Which is more important to the process of development in *Dictyostelium,* cAMP or the PKA?

Experiments

Adam Kuspa and his graduate student Bin Wang at Baylor College of Medicine, Houston, Texas, set out to answer these questions. Their experimental approach is shown in the **figure** on the right.

Results

The researchers found that the cAMP-deficient cells with the modified PKA gene aggregated into slugs when the cultures were deprived of food (in this case, bacteria). The slugs then differentiated normally into fruiting bodies. No cAMP could be detected in the cells.

Conclusions

The results indicated that activated PKA by itself can trigger all the steps in the developmental pathway from individual cells to slugs. Thus the requirement for cAMP in normal slime mold development is primarily or exclusively to stimulate the PKA. And, because development can proceed with active PKA alone, this protein kinase is probably more central to the growth and differentiation processes of the slime mold than is cAMP.

These results are of more than passing interest because both cAMP and cAMP-dependent protein kinases are also active in animal development and intercellular signaling, including that of humans and other mammals. They also show that in *Dictyostelium,* a single molecule, the PKA normally activated by cAMP, can trigger all stages of development and differentiation.

Source: B. Wang and A. Kuspa. 1997. *Dictyostelium* development in the absence of cAMP. *Science* 277:251–254.

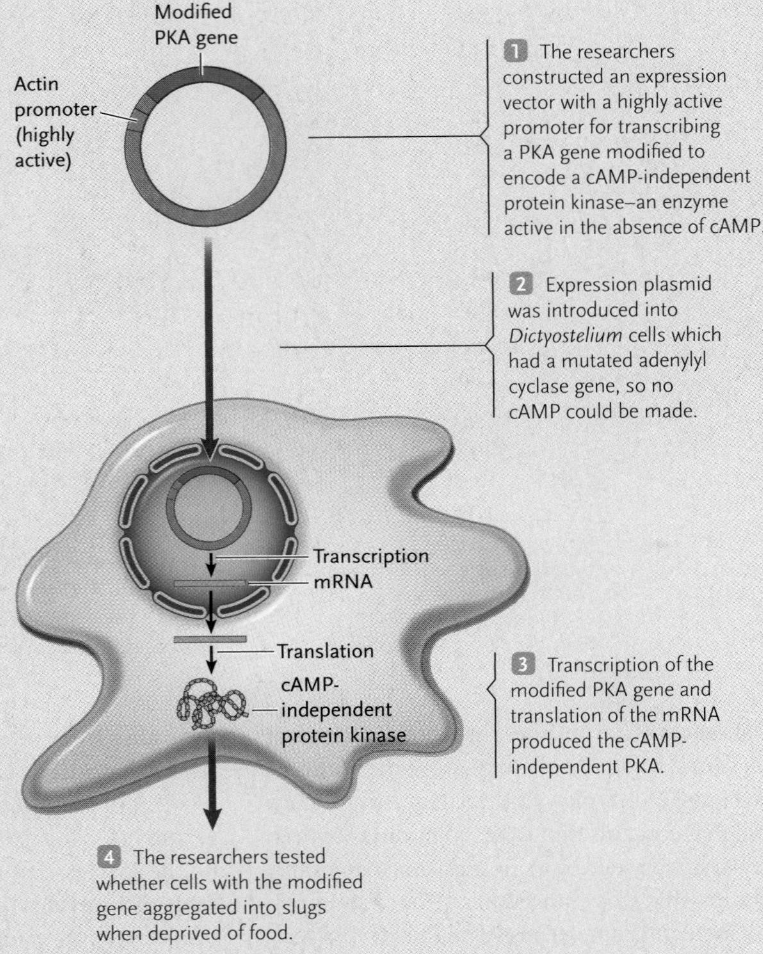
Modified PKA gene
Actin promoter (highly active)
1 The researchers constructed an expression vector with a highly active promoter for transcribing a PKA gene modified to encode a cAMP-independent protein kinase–an enzyme active in the absence of cAMP.
2 Expression plasmid was introduced into Dictyostelium cells which had a mutated adenylyl cyclase gene, so no cAMP could be made.
Transcription
mRNA
Translation
cAMP-independent protein kinase
3 Transcription of the modified PKA gene and translation of the mRNA produced the cAMP-independent PKA.
4 The researchers tested whether cells with the modified gene aggregated into slugs when deprived of food.

The plasmodial slime molds are particularly useful in this kind of research because they become large enough to provide ample material for biochemical and molecular analyses. Actin and myosin extracted from *Physarum polycephalum,* for example, have been much used in studies of actin-based motility. A further advantage of plasmodial slime molds is that the many nuclei of a plasmodium usually replicate and pass through mitosis in synchrony, making them useful in research tracking the changes that take place in the cell cycle.

OPISTHOKONTA The opisthokonts (*opisthen* = behind or posterior) are a broad group of eukaryotes that includes the choanoflagellates, protists thought to be the ancestors of fungi and animals. A single posterior flagellum is found at some stage in the life cycle of these organisms.

Choanoflagellata (*choanos* = collar)—**choanoflagellates**—are named for a collar of closely packed microvilli that surrounds the single flagellum by which these protists move and take in food **(Figure 26.22).** The collar resembles an upside-down

FIGURE 26.22
A choanoflagellate.

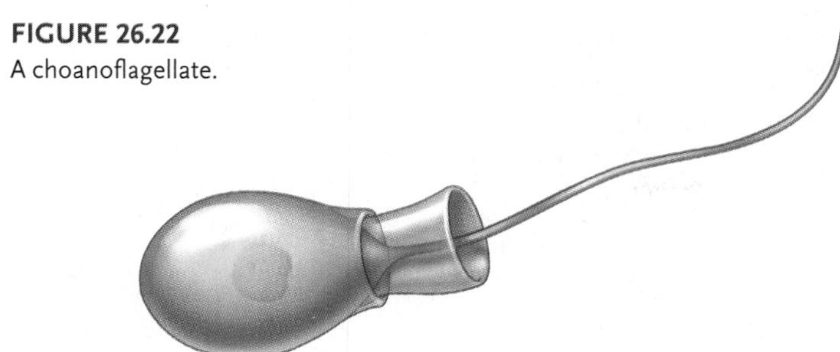

lampshade. There are about 150 species of choanoflagellates. They live in fresh and marine waters. Some species are mobile, with the flagellum pushing the cells along, as is the case with animal sperm, in contrast to most flagellates, which are pulled by their flagella. Most choanoflagellates, though, are *sessile;* that is, attached via a stalk to a surface. A number of species are colonial with a cluster of cells on a single stalk.

Choanoflagellates have the same basic structure as choanocytes (collar cells) of sponges, and they are similar to collared cells that act as excretory organs in organisms such a the flatworms and rotifers. These morphological similarities, as well as molecular sequence comparison data, indicate that a choanoflagellate type of protist is likely to have been the ancestor of animals and, of course, of present-day choanoflagellates.

In Several Protist Groups, Plastids Evolved from Endosymbionts

We have encountered chloroplasts in a number of eukaryotic organisms in this chapter: red algae, green algae, land plants, euglenids, dinoflagellates, stramenopiles, and chlorarachniophytes. How did these chloroplasts evolve?

In Section 24.3 we discussed the endosymbiotic theory for the origin of eukaryotes. In brief, an anaerobic prokaryote ingested an aerobic prokaryote, which survived as an endosymbiont (see Figure 24.5). Over many generations, the endosymbiont became an organelle, the mitochondrion, which was incapable of free living, and the result was a true eukaryotic cell. Cells of animals, fungi, and some protists derive from this ancestral eukaryote. The addition of plastids (the general term for chloroplasts and related organelles) through further endosymbiotic events produced the cells of all photosynthetic eukaryotes, including land plants, algae, and some other protists.

Figure 26.23 presents a model for the origin of plastids in eukaryotes through two major endosymbiosis events. First, in a single **primary endosymbiosis** event perhaps 600 million years ago, a eukaryotic cell engulfed a photosynthetic cyanobacterium (a photosynthetic prokaryote, remember). In some such cells, the cyanobacterium was not digested, but instead formed a symbiotic relationship with the engulfing host cell—it became an endosymbiont. Over time the symbiont lost genes no longer required for independent existence, and most of the remaining genes migrated from the prokaryotic genome to the host's nuclear genome. The symbiont had become an organelle—a chloroplast. All plastids subsequently evolved from this original chloroplast. Evidence for the single origin of plastids comes from a variety of sequence comparisons, including recent sequencing of the genomes of key protists: a red alga and a diatom.

The first photosynthesizing eukaryote was essentially an ancestral single-celled alga. The chloroplasts of the Archaeplastida—the red algae, green algae, and land plants—result from evolutionary divergence of this organism. Their chloroplasts, which originate from primary endosymbiosis, have two membranes, one from the plasma membrane of the engulfing eukaryote and the other from the plasma membrane of the cyanobacterium.

At least three **secondary endosymbiosis** events led to the plastids in other protists (see Figure 26.23). In each case, a nonphotosynthetic eukaryote engulfed a photosynthetic eukaryote, and new evolutionary lineages were produced. In one of these events, a red alga ancestor was engulfed and became an endosymbiont. In models accepted by a number of scientists, the transfer of functions that occurred over evolutionary time led to the chloroplasts of the stramenopiles, ciliates, and dinoflagellates. And, from the same photosynthetic ancestor, loss of chloroplast functions occurred in the lineage of the Apicomplexa, which have a remnant plastid. In an independent event, a nonphotosynthetic eukaryote engulfed a green alga ancestor. Subsequent evolution in this case produced the euglenids. In a different event, a similar endosymbiosis involving a green alga led to the chlorarachniophytes. In these protists, the chloroplast is contained still within the remnants of the original symbiont cell, with a vestige of the original nucleus (the nucleomorph) also present.

Note that secondary endosymbiosis has produced plastids with additional membranes acquired from the new host, or series of hosts. For example, euglenids have plastids with three membranes, while chlorarachniophytes have plastids with four membranes (see Figure 26.23). Sequencing the genomes of the chlorarachniophyte's nucleus, chloroplast, and vestigial nucleus is providing interesting information about the early endosymbiosis event that generated these organisms.

In sum, the protists are a highly diverse and ecologically important group of organisms. Their complex evolutionary relationships, which have long been the subject of contention, are now being revised as new information is discovered, including more complete genome sequences. A deeper understanding of protists is also contributing to a better understanding of their recent descendents, the fungi, plants, and animals. We turn to these descendents in the next four chapters, beginning with the fungi.

STUDY BREAK 26.2 <

1. **What is the evidence that the Excavates, which lack mitochondria, derive from ancestors that had mitochondria rather than from ancestors that were in lineages that never contained mitochondria?**
2. **In primary endosymbiosis, a nonphotosynthetic eukaryotic cell engulfed a photosynthetic cyanobacterium. How many membranes surround the chloroplast that evolved?**

FIGURE 26.23
The origin and distribution of plastids among the eukaryotes by primary and secondary endosymbiosis.

Ancestral eukaryote
Mitochondrion
Animals
Fungi
Select protists
Primary endosymbiosis
Photosynthetic bacterial endosymbiont
Red algae
Plastid
Green algae
Land plants
Plastid
Secondary endosymbiosis
Non-photosynthetic eukaryote
Red alga
Green alga
Plastid with multiple membranes
Nucleo-morph
Plastid with multiple membranes
Plastid
Plastid
Stramenopiles
Ciliates
Apicomplexans
Dinoflagellates
Chromalveolates
Chlorarachniophytes
Euglenids

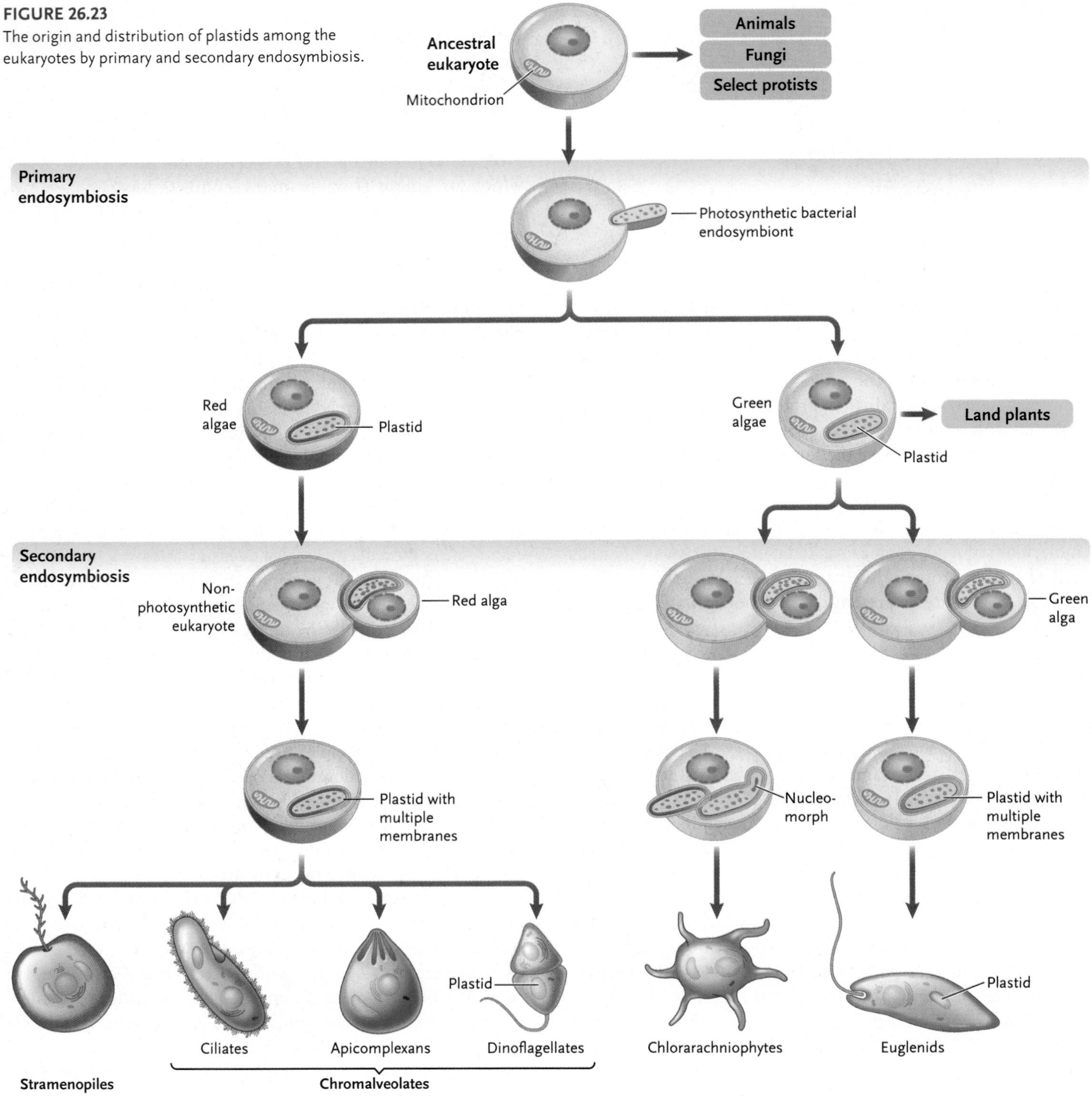

 UNANSWERED QUESTIONS

What was the first eukaryote?

Since prokaryotes precede eukaryotes in the fossil record, we assume that eukaryotes arose after prokaryotes. The first eukaryote would have been some sort of protist—a single-celled organism with a nucleus and some rudimentary organelles, perhaps even a half-tamed mitochondrion. One approach to identifying which of the surviving protists is the most ancient has been to infer evolutionary trees from gene sequence data. To determine the earliest branching eukaryote, these trees need to include the prokaryotes. But herein lies the problem—prokaryotes are very distant, evolutionarily speaking, from even the simplest eukaryotes, and the mathematical models used to construct evolutionary trees are not yet up to the job. Initially, these models suggested that some protist parasites, like the excavates *Giardia* and *Trichomonas*, might be the most ancient eukaryotes, and this idea fit nicely with the fact that these protists lacked mitochondria. Indeed, for a time it was thought that the excavates might actually have diverged from the eukaryotic branch of life before the establishment of mitochondria. Nowadays, we know that *Giardia* and *Trichomonas* did initially have mitochondria. The latest research shows that they even have a tiny relic of the mitochondrion, though exactly what it does in these oxygen-shunning parasites remains to be figured out. Thus, trees depicting *Giardia* and *Trichomonas* at the base of the great expansion of eukaryotic life must be viewed with some caution—these protists might be the surviving representatives of the earliest cells with a nucleus, but they might not be. We simply need better methods for identifying just what the first eukaryotes were like.

How many times did plastids arise by endosymbioses?

For many years researchers thought that the green algae, plants, and red algae were the only organisms to have primary endosymbiosis-derived plastids. However, a second, independent primary endosymbiosis has been recently discovered in which a shelled amoeba has captured and partially domesticated a cyanobacterium. This organism, known as *Paulinella,* is a vital window into the process by which autotrophic eukaryotes first arose some 600 million years ago. *Paulinella* has tamed the cyanobacterium sufficiently to have it divide and segregate in coordination with host cell division, but the endosymbiont is still very much a cyanobacterium and has undergone little of the modification and streamlining we see in the red or green algal plastids.

After a primary endosymbiosis was established, the second chapter in plastid acquisition could take place. Secondary endosymbiosis involves a eukaryotic host engulfing and retaining a eukaryotic alga. Essentially, secondary endosymbiosis can convert a heterotrophic organism into an autotroph by hijacking a photosynthetic cell and putting it to work as a solar-powered food factory. Secondary endosymbiosis results in plastids with three or four membranes, and we know that it occurred at least three times—once for the euglenids, once for the chlorarachniophytes, and once for the chromalveolates (a grouping of stramenopiles and alveolates). We can even tell what kind of endosymbiont was involved by the biochemistry and genetic makeup of the plastid: a green alga for euglenids and chlorarachniophytes, and a red alga for chromalveolates. The number of secondary endosymbioses is hotly debated, largely because not all protistologists support the existence of chromalveolates. Some contend that there were multiple, independent enslavements of different red algae to produce the dinoflagellates, stramenopiles, and apicomplexans. Understanding these events is crucial to confirming or refuting the proposed chromalveolate "supergroup."

A nice example of secondary endosymbiosis-in-action was recently discovered by Japanese scientists who found a flagellate, *Hatena,* with a green algal endosymbiont. *Hatena* has not yet assumed control of endosymbiont division and has to get new symbionts each time it divides, so it appears to be at a very early stage in establishing a relationship. We also want to know how secondary endosymbioses proceed because they have been a major driver in eukaryotic evolution. The stramenopiles, for instance, are the most important ocean phytoplankton and are key to ocean productivity and global carbon cycling. Knowing exactly how they got to be autotrophs in the first place is fundamental to understanding the world we live in.

Think Critically

1. What does the research examining the history of the earliest eukaryotes and endosymbioses reveal about evolution and the process of science?
2. Given that parasitic species are often considered to have free-living ancestors, and therefore are more evolutionarily recent than a closely related free-living form, is it possible that prokaryotes evolved from a single-celled eukaryote?

Geoff McFadden

Geoff McFadden is a professor of botany at the University of Melbourne. He studies the early evolution of eukaryotes, especially the origin and evolution of plastids and mitochondria. You can learn more about McFadden's research by visiting http://homepage.mac.com/fad1/McFaddenLab.html.

REVIEW KEY CONCEPTS

Go to **CENGAGENOW** at www.cengage.com/login to access quizzing, animations, exercises, articles, and personalized homework help.

26.1 What Is a Protist?

- Protists are eukaryotes that differ from fungi in having motile stages in their life cycles and distinct cell wall molecules. Unlike plants, they lack true roots, stems, and leaves. Unlike animals, protists lack collagen, nerve cells, and an internal digestive tract, and they lack complex developmental stages (Figures 26.1 and 26.2).
- Protists are aerobic organisms that live as autotrophs or heterotrophs, or by a combination of both nutritional modes. Some are parasites or symbionts living in or among the cells of other organisms.
- Protists live in aquatic or moist terrestrial habitats, or as parasites within animals as single-celled, colonial, or multicellular organisms, and range in size from microscopic to some of Earth's largest organisms.
- Reproduction may be asexual by mitotic cell divisions, or sexual by meiosis and union of gametes in fertilization.
- Many protists have specialized cell structures including contractile vacuoles, food vacuoles, eyespots, and a pellicle, cell wall, or shell. Most are able to move by means of flagella, cilia, or pseudopodia (Figure 26.3).

Animation: *Paramecium* body plan

26.2 The Protist Groups

- The most ancient bifurcation separates eukaryotes into the Unikonta and Bikonta. Protist groups are found in both the unikonts and bikonts (Figure 26.2).
- The Excavata includes the Metamonada, protists with unique flagella, and the Euglenozoa, protists with modified mitochondria.
- Metamonads consist of diplomonads and parabasalids, which have flagellated, single cells that lack mitochondria (Figure 26.4).

- Euglenozoans are almost all single-celled, autotrophic or heterotrophic (some are both), motile protists that swim using flagella. The free-living, photosynthetic forms—the euglenids—typically have complex cytoplasmic structures, including eyespots. The parasitic, nonphotosynthetic forms—the kinetoplastids—have characteristic mitochondrial structures (Figures 26.5 and 26.6).
- The Chromalveolata, protists with heterogeneous forms and life styles, includes the Alveolata and the Stramenopila.
- The alveolates include the ciliates, dinoflagellates, and apicomplexans. The ciliates swim using cilia and have complex cytoplasmic structures and two types of nuclei, the micronucleus and macronucleus. The dinoflagellates swim using flagella and are primarily marine organisms; some are photosynthetic. The apicomplexans are nonmotile parasites of animals (Figures 26.7 and 26.8).
- The stramenopiles include the funguslike groups, the Bacillariophyta (diatoms), Chrysophyta (golden algae), and Phaeophyta (brown algae). For most stramenopiles, flagella occur only on reproductive cells. Many Oomycota grow as masses of microscopic hyphal filaments and secrete enzymes that digest organic matter in their surroundings. Diatoms are single-celled organisms covered by a glassy silica shell; golden algae are colonial forms; brown algae are primarily multicellular marine forms that include large seaweeds with extensive cell differentiation (Figures 26.9–26.14).
- Rhizaria, amoebas with filamentous pseudopods supported by internal cellular structures, include the Radiolaria, the Foraminifera, and the Chlorarachniophyta. Many rhizarians produce hard outer shells.
- Radiolaria (radiolarians) are primarily marine organisms that secrete a glassy internal skeleton. They feed by engulfing prey that adhere to their axopods. Foraminifera (forams) are marine, single-celled organisms that form chambered, spiral shells containing calcium. They engulf prey that adhere to the strands of cytoplasm extending from their shells. Chlorarachniophytes engulf food using their pseudopodia (Figure 26.15).
- The Archaeplastida include the Rhodophyta (red algae) and Chlorophyta (green algae), as well as the land plants.
- Red algae are typically multicellular, primarily photosynthetic organisms of marine environments, with plantlike bodies composed of interwoven filaments. They have complex life cycles including alternation of generations, with no flagellated cells at any stage (Figure 26.16).
- Green algae are single-celled, colonial, and multicellular species that live primarily in freshwater habitats and carry out photosynthesis by mechanisms similar to those of plants; all produce flagellated gametes (Figures 26.17–26.19).
- The Unikonta includes the Amoebozoa, protists with unsupported pseudopods, and Opisthokonta, protists closely related to animals and fungi, and animals and fungi themselves.
- Amoebozoans include most amoebas and two heterotrophic slime molds, cellular (which move as individual cells) and plasmodial (which move as large masses of nuclei sharing a common cytoplasm). The amoebas in this group are heterotrophs abundant in marine and freshwater environments and in the soil. They move by extending pseudopodia (Figures 26.20 and 26.21).
- The opisthokonts are a broad group of eukaryotes that includes the choanoflagellates. These protists are characterized by a collar of microvilli surrounding a single flagellum. A choanoflagellate type of protist is considered likely to have been the ancestor of animals (Figure 26.22).
- Several groups of protists, as well as land plants, contain chloroplasts. Present-day chloroplasts and other plastids result from endosymbiosis events that took place millions of years ago: In a primary endosymbiosis event, a eukaryotic cell engulfed a cyanobacterium, which became an endosymbiont. Over time, the symbiont became an organelle, the chloroplast. This first photosynthesizing organism was a green alga. Evolutionary divergence produced the red algae, green algae, and land plants. By secondary endosymbiosis, in which a nonphotosynthetic eukaryote engulfed a photosynthetic eukaryote, the various photosynthetic protists were produced (Figure 26.23).

Animation: Body plan of *Euglena*

Animation: Ciliate conjugation

Animation: Apicomplexan life cycle

Animation: Red alga life cycle

Animation: Green alga life cycle

Animation: Amoeboid motion

Animation: Cellular slime mold life cycle

UNDERSTAND AND APPLY

Test Your Knowledge

1. Which of the following is a characteristic of protists that is also found in at least one other group?
   a. division by binary fission
   b. multicellular structures
   c. complex developmental stages
   d. peptidoglycan cell walls
   e. organelles and reproduction by meiosis/mitosis
2. Which of the following is *not* found among the protist groups?
   a. life cycles
   b. contractile vacuoles
   c. pellicles
   d. collagen
   e. pseudopodia
3. A member of this group can cause urinary infections and be spread by sexual intercourse. The group is characterized by a flagellum buried in a fold of the cytoplasm that allows the organisms to move through thick and viscous fluids of humans. The group is:
   a. Ciliophora.
   b. Apicomplexa.
   c. Amoebozoa.
   d. Parabasala.
   e. Diplomonadida.
4. When *Paramecium* conjugate:
   a. cytoplasmic division produces four daughter cells, each having two micronuclei and two macronuclei.
   b. one haploid micronucleus in each cell remains intact; the other three degenerate. The micronucleus of each cell divides once, producing two nuclei, and each cell exchanges one nucleus with the other cell. In each partner the two micronuclei fuse, forming a diploid zygote micronucleus in each cell.

c. the micronucleus of each of the disengaged partners divides meiotically. Macronuclei divide again in each cell and the original micronucleus breaks down. Each cell has two haploid micronuclei; one of the macronuclei develops into a micronucleus.
d. the mating cells join together at opposite sites of their oral depression.
e. the micronucleus in each cell undergoes mitosis. When mitosis is complete there are four diploid macronuclei; the micronucleus then breaks down.

5. The group Diplomonadida is characterized by:
   a. a mouthlike gullet and hairlike surface. *Paramecium* is an example.
   b. flagella and a lack of mitochondria. *Giardia* is an example.
   c. nonmotility, parasitism, and sporelike infective stages. *Toxoplasma* is an example.
   d. switching between autotrophic and heterotrophic life styles. *Euglena* is an example.
   e. large protein deposits and movement by two flagella, which are part of an undulating membrane. *Trypanosoma* is an example.
6. The greatest contributors to protist fossil deposits, and probable source of oil in many oil deposits, are:
   a. Oomycota.
   b. Chrysophyta.
   c. Bacillariophyta.
   d. Sporophyta.
   e. Alveolata.
7. The group with the distinguishing characteristic of gas-filled bladders and a cell wall composed of alginic acid, which is used by humans to thicken such diverse products as ice cream and floor polish, is:
   a. Chrysophyta.
   b. Phaeophyta.
   c. Oomycota.
   d. Bacillariophyta.
   e. none of the preceding.
8. *Plasmodium* is transmitted to humans by the bite of a mosquito *(Anopheles)*. Its life cycle is characterized by spores, gametes, and cysts that can "hide" in human cells. This infective protist belongs to the group:
   a. Apicomplexa.
   b. Archaeplastida.
   c. Dinoflagellata.
   d. Oomycota.
   e. Ciliophora.
9. Tripping on a rotten log, a hunter notices a mass resembling mucus that appears to be moving slowly toward what appear to be brightly colored fruiting bodies. The organisms in the mass are:
   a. amoebas in the group Rhizaria.
   b. slime molds.
   c. red algae.
   d. green algae.
   e. charophytes.
10. Endosymbioses that lead to the evolution of euglenoids, and separately, the evolution of chlorarachniophytes were the result of the combining of:
    a. two ancestral nonphotosynthetic prokaryotes.
    b. two ancestral photosynthetic prokaryotes.
    c. a nonphotosynthetic eukaryote with a photosynthetic eukaryote.
    d. a photosynthetic prokaryote with a nonphotosynthetic eukaryote.
    e. mitochondria with an already established plastid.

Discuss the Concepts

1. You decide to vacation in a developing country where sanitation practices and standards of personal hygiene are inadequate. Considering the information about protists covered in this chapter, would you consider it safe to drink water in that country? What treatments could make the water safe to drink? What kinds of foods might be best avoided? What kinds of preparation might make foods safe to eat?
2. The overreproduction of dinoflagellates, which produces red tides, is sometimes caused by fertilizer runoff into coastal waters. The red tides kill countless aquatic species, birds, and other wildlife. Would you consider drastic cutbacks in the use of fertilizers as a means to lessen the red tides? Why?

Design an Experiment

Design an experiment to demonstrate whether the flagellated protist *Euglena* is phototropic, that is, is attracted to and moves toward light. Also propose a follow-up experiment (on the assumption of a positive result) to determine the wavelength range and light intensity range sufficient to cause phototropic movement.

Interpret the Data

Marine diatoms are a crucial part of marine productivity because they are food for many organisms. Their productivity is dependent on how well they photosynthesize. The graph shows the response of photosynthesis in the marine diatom *Skeletonema costatum* to light.

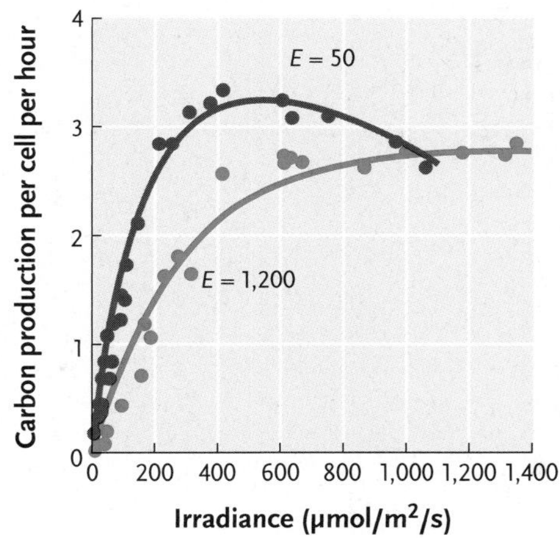
Carbon production per cell per hour
E = 50
E = 1,200
0
1
2
3
4
0
200
400
600
800
1,000
1,200
1,400
Irradiance (μmol/m²/s)

Photosynthesis is expressed on the *y* axis as carbon production per cell (P^{cell} in units of picograms (pg) of C per hour, with C as elemental carbon), and light is expressed on the *x* axis as the number of photons of irradiance that can activate photosynthesis (0 μmol m^{-2} s^{1} would be complete darkness, and 2,000 μmol m^{-2} s^{1} would be full sunlight at mid-day). The black symbols show a cell culture grown for five days at 50 μmol m^{-2} s^{1} of irradiance (the treatment irradiance), with an arrow showing the level of photosynthetic production at the end of this period. The white symbols show a different population of cells from the same culture grown for five days at 1,200 μmol m^{-2} s^{1} of irradiance (the treatment irradiance), with an arrow showing the level of photosynthetic production. Both cultures were then exposed to all of the possible irradiance conditions for 30-minute periods.

1. In each light treatment, at what irradiance was the photosynthetic rate maximized? Was this the same rate as the treatment irradiance; why or why not?
2. The response of photosynthesis to irradiance at low irradiance is linear. The rate of this linear increase is the same as the organism's photosynthesis efficiency with respect to light. Which treatment has higher photosynthesis efficiency?

Source: Data from T. Anning et al. 2000. Photoacclimation in the marine diatom *Skeletonema costatum. Limnology and Oceanography.* 45(8):1807–1817.

Apply Evolutionary Thinking

Use the Internet to research why studies of a molecular sensor, receptor tyrosine kinase (see Section 7.3), supports the hypothesis that a choanoflagellate type of protist is the ancestor of animals. Summarize your findings.

Express Your Opinion

The pathogen that causes sudden oak death has already infected 26 kinds of plants in California and Oregon. Some infected species are commonly sold as nursery stock. Should the states that are free of this pathogen be allowed to prohibit shipping of all plants from the states that are affected? Go to CengageNOW to investigate both sides of the issue and then vote.

A temperate forest with representatives of three major groups of land plants—mosses (bryophytes), conifers (gymnosperms), and flowering plants (angiosperms).

Tim Street-Porter/Getty Images

27

STUDY OUTLINE

Plants

Why It Matters. . . Ages ago, along the edges of the ancient supercontinent Laurentia, the only sound was the rhythmic, muffled crash of waves breaking in the distance. There were no birds or other animals, no plants with leaves rustling in the breeze. In the preceding eons, oxygen-producing photosynthetic cells had come into being and had gradually changed Earth's atmosphere. Solar radiation had converted much of the oxygen into a dense ozone layer—a shield against lethal doses of ultraviolet radiation, which had kept early organisms below the water's surface. Now, they could populate the land.

Cyanobacteria may have been the first to adapt to intertidal zones and then to spread into the freshwater environment of shallow, coastal streams. Later, green algae and fungi made the same journey. Seven to eight hundred million years ago, green algae living near the edge of a pond or stream became the ancestors of modern plants. Several lines of molecular and morphological evidence indicate that a series of adaptive shifts in charophyte algae introduced in Chapter 26 gave rise to the clade we recognize as the **Kingdom Plantae.** Today the plant kingdom—sometimes called the Viridiplantae, or "green plants"—encompasses at least 260,000 living species, organized in this textbook into 10 phyla. The vast majority of these extant species are flowering plants, a lineage that has become spectacularly diverse and profoundly important in human affairs. For example, flowering plants include crop species that supply roughly four-fifths of the calories humans consume, as well as numerous medicines, fibers used for clothing, and hardwood lumber. Far fewer species populate another lineage, the conifers, but they include the tallest and most massive individual land plants that have ever lived and hugely important sources of wood, resins, and fibers for paper. Mosses, horsetails, and ferns represent still other lines in which crucial adaptations to life on land arose **(Figure 27.1).** Remains of their ancestors formed most of the vast coal deposits that literally fu-

A. **Mosses growing on rocks**

B. **A ponderosa pine**

C. **An orchid**

FIGURE 27.1
Representatives of the Kingdom Plantae. **(A)** *Sphagnum* moss growing at the edge of a stream. Mosses evolved relatively soon after plants made the transition to land. **(B)** A ponderosa pine, *Pinus ponderosa*. This species and other conifers belonging to the phylum Coniferophyta represent the gymnosperms. **(C)** An orchid, *Cattleya rojo*, a showy example of a flowering plant.

eled the Industrial Revolution and continue to be major global energy sources.

All plants are multicellular. Most also are autotrophs, the primary producers that carry out photosynthesis, using solar energy, carbon dioxide, water, and dissolved minerals to manufacture their own food. Together with photosynthetic bacteria and "protists," plant tissues provide the nutritional foundation for nearly all communities of life.

As you might guess, while the ancestors of land plants were making the transition to a fully terrestrial life, some crucial adaptive changes unfolded. Eons of natural selection sorted out solutions to fundamental problems, among them avoiding desiccation, physically supporting the plant body in air, obtaining nutrients from soil, and reproducing sexually in environments where water would not be available for dispersal of eggs and sperm. With time, plants evolved features that not only addressed these problems but also provided access to nearly every land environment. Those ecological opportunities opened the way for a dramatic radiation of plant species—and for the radiation of animal species that eventually came to include our own. <

FIGURE 27.2
Chara, a stonewort. This representative of the charophyte lineage is known commonly as muskweed because of its skunky odor.

27.1 The Transition to Life on Land

Land plants and green algae share several fundamental traits: they have cellulose in their cell walls, they store energy captured during photosynthesis as starch, and their light-absorbing pigments include both chlorophyll *a* and chlorophyll *b*. Like other green algae, the charophyte lineage from which land plants evolved arose in water and has aquatic descendants today **(Figure 27.2)**. Yet because land environments pose very different challenges than aquatic environments, evolution in land plants produced a range of adaptations essential to survival on dry land.

The algal ancestors of plants probably had invaded land by about 450 million years ago (mya). We say "probably" because the fossil record is sketchy in pinpointing when the first truly terrestrial plants appeared. One reason for the dearth of information is that the earliest land plants lacked hard parts (such as wood) that fossilize well. A British and Arab research team working in the Middle East found fossilized tissue and spores from what appears to be a land plant in rocks dated to 475 mya. If the remains indeed are from a plant, they represent the earliest known plant fossils. Even in more recent deposits the most common finds of possible plant parts are microscopic bits and pieces. Obvious leaves, stems, roots, and reproductive parts seldom occur together, or if they do, it can be difficult to determine if the fossilized bits all belong to the same individual. Whole plants are extremely rare. Adding to the challenge, more than a few chemical and structural adaptations to life on land arose independently in several plant lineages. Consequently, a fossil may have some but not all the features of modern land plants, making it difficult to determine whether a given specimen was actually terrestrial, aquatic, or a transitional form. Despite these problems, plant scientists have been able to gain insight into a variety of innovations and overall trends in plant evolution.

Early Biochemical and Structural Adaptations Enhanced Plant Survival on Land

To survive on land, plants had to have protection against drying out, a demand that had not been imposed on algae in their aquatic habitats. The earliest land plants may have benefited from an inherited ability to make **sporopollenin,** a resistant polymer that surrounds the zygotes of modern charophytes. In land plants, capsules called **sporangia** ("spore chamber"; singular, sporangium) produce reproductive spores, and sporopollenin is a major

A. **Cuticle on the surface of a leaf**

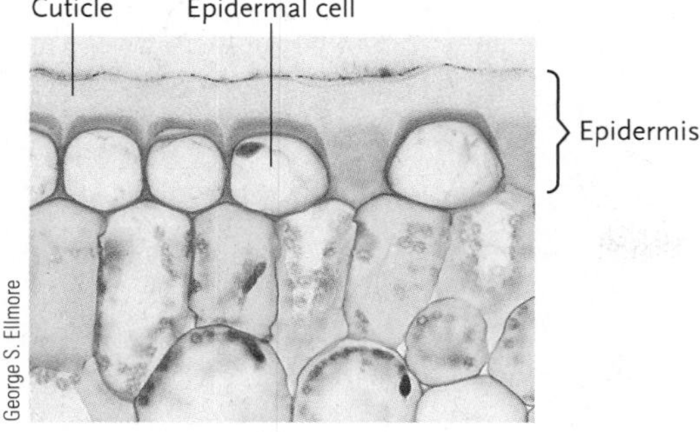
Cuticle
Epidermal cell
Epidermis

B. **Stomata**

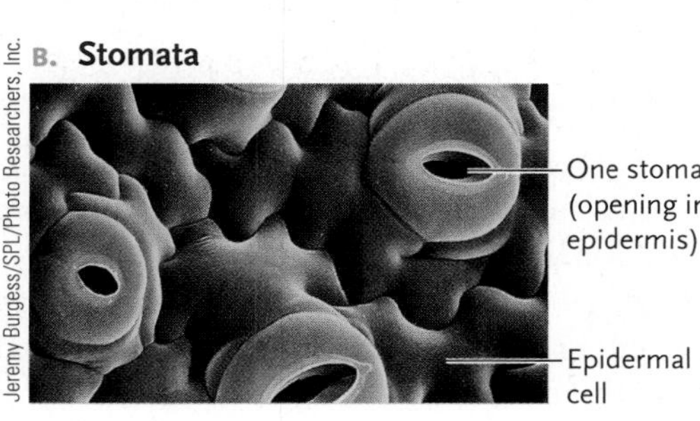
One stoma (opening in epidermis)
Epidermal cell

George S. Ellmore

Jeremy Burgess/SPL/Photo Researchers, Inc.

FIGURE 27.3
Land plant adaptations for limiting water loss. (A) A waxy cuticle, which covers the epidermis of land plants and helps reduce water loss. **(B)** Surface view of stomata in the epidermis (surface layer of cells) of a leaf. Stomata allow carbon dioxide to enter plant tissues and oxygen to leave while controlling loss of water.

component of the thick sporangium wall that protects spores from drying and other damage. Other reproductive adaptations in plants included multicellular chambers that protect developing gametes, and a dependent, multicellular embryo that is sheltered inside tissues of a parent plant. Botanists use the term **embryophyte** (*phyton* = plant) as a synonym for land plants because all land plants produce a protected embryo during their reproductive cycle.

Some of the first land plants also evolved an outer waxy layer called a **cuticle,** which slows water loss, helping to prevent desiccation **(Figure 27.3A)**. Another adaptation in most land plant lineages was the presence of **stomata** (singular, stoma; *stoma* = mouth), tiny openings in the cuticle-covered surfaces **(Figure 27.3B)**. A stoma is formed by a pair of cells that can change shape and so open up or close a space between them, and stomata became the main route for plants to take up carbon dioxide and control water loss by evaporation. The next unit describes stomata more fully.

If you had been a botanist about 470 million years ago, you would have noted two general categories of land plants. One included **nonvascular plants**—for historical reasons also called **bryophytes**—which lacked well-developed internal tubes for transporting water and organic compounds such as sugars. Mosses are in this category. As far as we know, nonvascular plants are not monophyletic; that is, the various phyla described shortly do not share a recent ancestor. In fact, they are extremely diverse. The other general category encompassed **vascular plants,** more closely related lineages in which specialized transport tissues had arisen. One of these tissues, called **xylem,** contained stacked water-conducting cells called *tracheids,* which is why vascular plants collectively are also known as **tracheophytes.** Another vascular tissue, called **phloem,** is an adaptation for transporting sugars and other metabolic products. Early on, no vascular plants produced seeds, but more recent evolutionary events resulted in some vascular plants that produce seeds as part of their reproductive cycle.

Internal transport tubes were merely one of the adaptations that roughly correlated with the divergence of nonvascular and vascular plants. Another was the capacity to synthesize **lignin,** a tough, rather inert polymer that strengthens the secondary walls of certain types of plant cells and thus helps vascular plants to grow taller and stay erect on land, giving photosynthetic tissues better access to sunlight. All vascular plants also have **apical meristems,** regions of unspecialized dividing cells near the tips of shoots and roots. Meristem tissue is the foundation for a vascular plant's extensively branching stem parts. Chapter 31 describes in detail how descendants of such unspecialized cells differentiate and form all mature plant tissues. As you'll read shortly, although mosses are classified as nonvascular, they have apical meristems and some other adaptive similarities to vascular plants.

Vascular Tissue Was an Adaptation for Transporting Substances within a Large Plant Body

The Latin *vas* means duct, and vascular plants have specialized tissues made up of cells arranged in lignified, tubelike structures that branch throughout the plant body, conducting water and solutes. As you've just read, the vascular tissue called xylem distributes water—including dissolved mineral ions—through plant parts. The other vascular tissue, phloem, distributes sugars manufactured during photosynthesis. Chapter 32 explains how tracheids and other specialized cells of xylem and phloem perform these key internal transport functions.

Most of the plants you are familiar with—flowering plants, conifers, ferns—are vascular plants. Supported by cells with lignin and serviced by a well-developed vascular system, the body of a plant can grow large. Extreme examples are the giant redwood trees of the northern California coast, some of which are more than 300 feet tall. By contrast, nonvascular plants lack lignified cells, and have very simple internal transport systems, or none at all. This is why modern nonvascular plants generally are small.

Shoot and Root Systems Were Adaptations for Support and Nutrition

A nonvascular plant has no true stems, structures that are fundamental adaptations for support of an erect plant body. In vascular plants, the evolution of sturdy stems became the basis of a large, aerial *shoot system.* This adaptive shift was correlated with the capacity to synthesize lignin.

The first unquestioned fossils of a vascular plant are specimens in the extinct genus *Cooksonia.* As **Figure 27.4A** shows, this plant had leafless "naked" stems with simple, forked branches of

FIGURE 27.4
Fossils from early vascular plants. **(A)** *Cooksonia,* which dates to about 420 mya. *Cooksonia* stems lacked leaves and were probably less than 3 cm long. The cup-shaped structures at the top of the stems produced reproductive spores. **(B)** Cross section of the stem of a rhyniophyte, *Rhynia gwynne-vaughnii,* about 3 mm in diameter. This fossil was embedded in chert approximately 400 mya. Still visible in it are traces of the transport tissues xylem and phloem, along with other specialized tissues.

A. **Fossil of *Cooksonia***

© Reprinted with permission from Elsevier

B. **Cross section of *Rhynia gwynne-vaughnii***

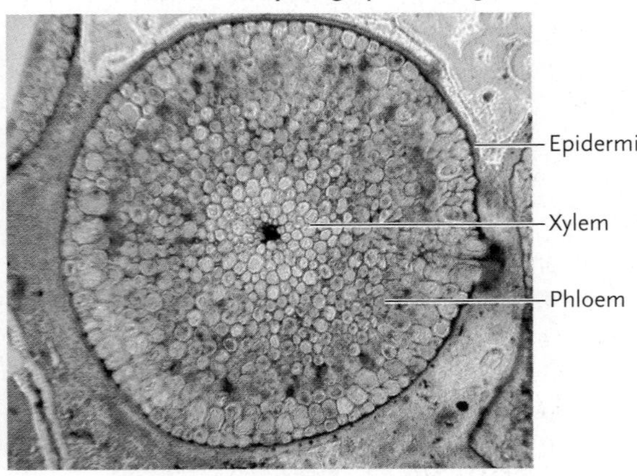
Epidermis
Xylem
Phloem
Hans Steur

equal size. Sporangia formed at the branch tips and simple xylem wove through the stems. Other species formerly lumped into the genus *Cooksonia* were early vascular plants called *rhyniophytes* (because numerous specimens are from rock outcrops near Rhynie, Scotland). *Cooksonia* and rhyniophyte fossils all date to about 420 mya and show similar features, including vascular tissue **(Figure 27.4B).** Most likely none of these species had roots, which are organs specialized to anchor a vascular plant and take up water and minerals. Instead these ancestral forms probably were supported physically by a **rhizome**—a horizontal, modified stem that can penetrate a substrate and anchor the plant. Cells called *rhizoids,* which can help anchor a plant and take up water, may have extended into the soil.

UPRIGHT GROWTH: LIGNIFIED SHOOT SYSTEMS Above ground, the simple stems of early land plants gradually became more specialized, evolving into the **shoot systems** we see in all living vascular plants. Shoot systems have stems and leaves that arise from apical meristems and that absorb light energy from the sun and carbon dioxide from the air. Lignin became incorporated in the water-conducting tracheids of xylem. Its added mechanical strength in these and other types of cells provided vascular plants with adaptive advantages. For instance, a strong, internal scaffold composed of tracheids could support upright, branching stems bearing leaves and other photosynthetic structures—and so help increase the surface area for intercepting sunlight. Also, reproductive structures such as sporangia borne on aerial stems might serve as platforms for more efficient launching of spores from the parent plant.

ENHANCED PHOTOSYNTHESIS: LEAVES WITH STOMATA AND VEINS Structures we think of as "leaves" arose several times during plant evolution. Leaves are modifications of stems and fall into two general categories. *Microphylls* are narrow and have a single strand of vascular tissue, or **vein.** *Megaphylls* are broad leaves with multiple veins. **Figure 27.5** illustrates the basic

A. **Development of microphylls as an offshoot of the main vertical axis**

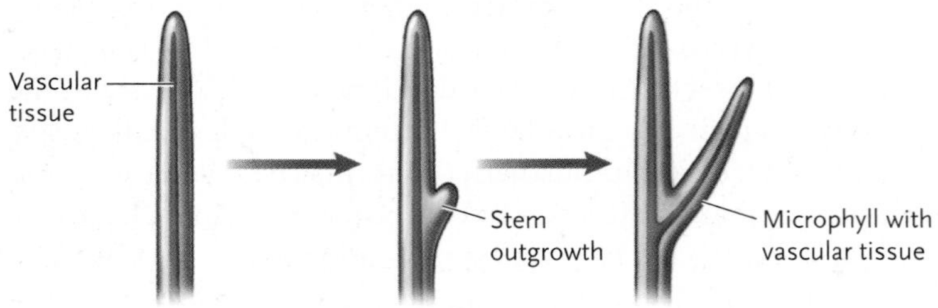
Vascular tissue
Stem outgrowth
Microphyll with vascular tissue

B. **Development of megaphylls in a branching pattern**

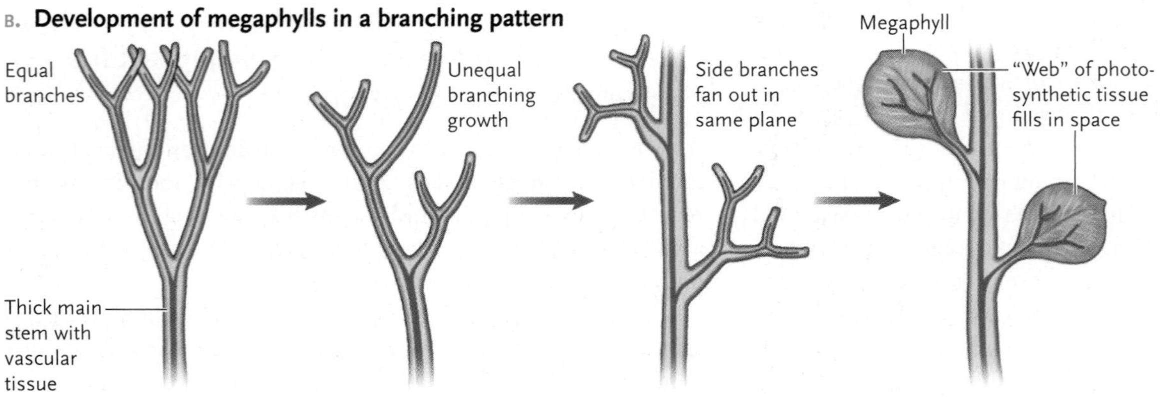
Equal branches
Unequal branching growth
Side branches fan out in same plane
Megaphyll
"Web" of photo-synthetic tissue fills in space
Thick main stem with vascular tissue

FIGURE 27.5
Evolution of leaves. **(A)** One type of early leaflike structure, the microphyll, may have evolved as an offshoot of the plant's main vertical axis. Each leaf contained a single vein (transport vessel). The seedless vascular plants known as lycophytes (club mosses) still have this type of leaf. **(B)** In other seedless vascular plants, the form of leaf called a megaphyll could have arisen in a series of steps that began when the main stem evolved a branching growth pattern. Small side branches then fanned out and photosynthetic tissue filled the space between them, becoming the leaf blade. With time the small branches became modified into veins. Ferns, horsetails, gymnosperms, and angiosperms all have this type of leaf.

steps of possible evolutionary pathways for these leaf forms. Among the lycophytes (Section 27.3), microphyll-like parts may have evolved as flaplike outgrowths of the plant's main vertical stem (see Figure 27.5A). Club mosses, the most common living lycophytes, still have this type of leaf. In contrast, megaphylls are thought to have arisen from modified branches, and by the late Paleozoic era (about 350 mya), they were appearing in several plant groups. Figure 27.5B shows a simplified scenario for this adaptive shift: a slender stem branches into two roughly equal sections in a recurring pattern (called dichotomous branching). As slender branched sections are overtopped by more robust ones, the sections flatten out and a web of tissue having cells with chloroplasts then fills in the gap between branches. Over evolutionary time the tissue gains stomata, which provide access to carbon dioxide for photosynthesis, and a simple network of veins that transport water.

There is a strong correlation between the density of venation in a plant's leaves and the plant's capacity for photosynthesis—and, ultimately, for producing energy for growth. Timothy Brodribb and Taylor Feild, plant physiologists who compared the leaves of more than 500 living and extinct flowering plants, documented a dramatic increase in the number of leaf veins over the course of angiosperm evolution. They discovered that by about 100 mya, leaves of later evolving flowering plant species had up to 10 times more veins than leaves of more ancient species! Brodribb and Feild calculated that even a tripling in the number of veins would have increased the capacity of photosynthesis by 100%. Their findings suggest strongly that the evolution of large leaves with numerous stomata and a dense system of veins was a major factor in the remarkable proliferation of flowering plants that has occurred over the past 150 million years.

SUPPORT AND NUTRITION: ROOTS As upright growth emerged as a basic feature of vascular plants, it required a means of anchoring aerial parts in the soil, as well as effective strategies for obtaining water and soil nutrients. The evolution of **roots**—anchoring organs that also absorb water and nutrients—fulfilled these requirements. Frustratingly, fossils of early vascular plants are not very helpful in this regard and we are left with more questions than answers. Exactly when roots appeared is uncertain, and it is quite possible that the feature arose independently in various groups. At some point, ancestral vascular plants did come to have roots, and ultimately vascular plants developed whole specialized **root systems,** which generally consist of underground, cylindrical absorptive structures with a large surface area that favors the rapid uptake of soil water and dissolved minerals.

In the Plant Life Cycle, the Diploid Phase Became Dominant

As early plants moved into drier habitats, their life cycles also were modified considerably. You may recall that in sexually reproducing organisms, meiosis in diploid cells produces haploid (n) reproductive cells (see Chapter 11). These cells may be gametes—sperm or eggs—or they may be **spores,** which can give rise to a new haploid individual asexually, without mating.

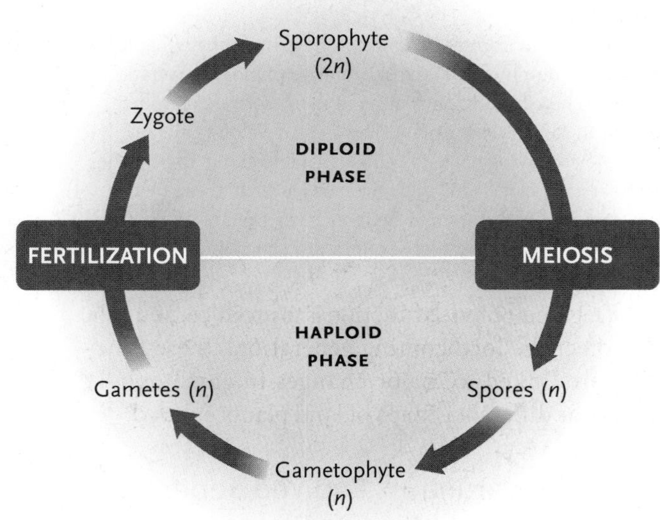

Sporophyte
(2n)
Zygote
DIPLOID
PHASE
FERTILIZATION
MEIOSIS
HAPLOID
PHASE
Gametes (n)
Spores (n)
Gametophyte
(n)

FIGURE 27.6
Overview of the alternation of generations, the basic pattern of the plant life cycle. The relative dominance of haploid and diploid phases is different for different plant groups (compare with Figure 27.7).

All sexually reproducing organisms form both diploid and haploid cells, but what happens after each cell type forms varies greatly. In sexually reproducing animals, for example, a multicellular body develops only when diploid cells divide by mitosis. Haploid cells do not divide but become gametes. In plants, by contrast, mitosis occurs in both diploid and haploid cells, so plants essentially have two multicellular phases, one diploid and one haploid. This phenomenon, found in all plants, is called **alternation of generations (Figure 27.6).** The diploid generation produces haploid spores and is called a **sporophyte** ("spore producer"). The haploid generation produces gametes and is called a **gametophyte** ("gamete producer").

Some green algae show alternation of generations, but most, like the charophytes, are entirely haploid organisms. They produce many swimming gametes, and their zygotes do not develop a multicellular diploid phase. The first land plants had long-lived haploid gametophytes and short-lived diploid sporophytes. Modern bryophtes still show this life cycle pattern, but as vascular plants evolved, the gametophyte phase became briefer and smaller and the sporophytes became the larger, dominant life phase. As plants evolved on land, the haploid gametophyte phase became physically smaller and less complex and had a shorter life span whereas just the opposite occurred with the diploid sporophyte phase. In mosses and other nonvascular plants the sporophyte is a little larger and longer-lived than in green algae, and in vascular plants the sporophyte clearly is larger and more complex and lives much longer than the gametophyte **(Figure 27.7).** When you look at a pine tree, for example, you see a large, long-lived sporophyte.

The sporophyte generation begins after fertilization, when the resulting zygote grows mitotically into a multicellular, diploid organism. Its body will eventually develop sporangia, which produce spores.

The haploid phase of the plant life cycle begins in the sporangia. There, meiosis produces haploid spores. The spores divide by mitosis and give rise to multicellular haploid gameto-

FIGURE 27.7
Evolutionary trend from dominance of the gametophyte (haploid) generation to dominance of the sporophyte (diploid) generation, represented here by existing species ranging from a charophyte alga *(Chara)* to a flowering plant. This trend developed as early plants were colonizing habitats on land. In general, the sporophytes of vascular plants are larger and more complex than those of bryophytes, and their gametophytes are reduced in size and complexity. In this diagram ferns represent seedless vascular plants.

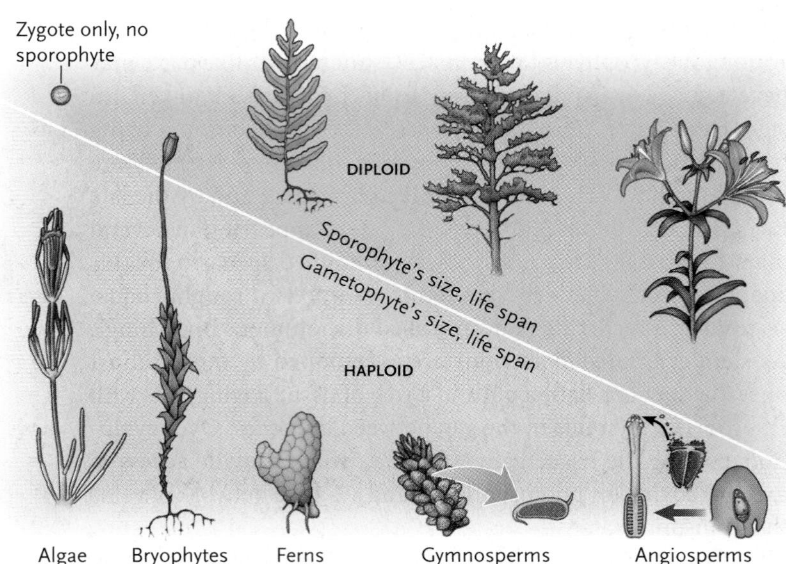
Zygote only, no sporophyte
DIPLOID
Sporophyte's size, life span
Gametophyte's size, life span
HAPLOID
Algae
Bryophytes
Ferns
Gymnosperms
Angiosperms

phytes. A gametophyte's function is to produce, nourish, and protect the forthcoming generation. These functions were linked to major changes in gametophyte structure as different groups of land plants evolved.

Some Vascular Plants Evolved Separate Male and Female Gametophytes

During sexual reproduction in plants, meiosis produces spores. When a plant makes only one type of spore it is said to be **homosporous** ("same spore"). A gametophyte that develops from such a spore is bisexual—it can produce both sperm and eggs.

The sperm have flagella and are motile, for they have to swim through liquid water in order to encounter female gametes. Other vascular plants, including gymnosperms and angiosperms, are **heterosporous.** They produce two types of spores in two different types of sporangia, and those spores develop into small, sexually different gametophytes. The smaller spore type develops into a male gametophyte—a *pollen grain.* The larger one develops into a female gametophyte, in which eggs form and fertilization occurs.

The evolution of pollen grains, pollination, and seeds helped spark the diversification of plants in the Devonian period, 408–360 mya. Technically seed plants are spermatophytes ("seed plant"), and as shown in Figure 27.8, seed plants as a whole belong to a grouping called the Spermatophyta. So many new fossils appear in Devonian rocks that paleobotanists (who study fossil plants) have thus far been unable to determine which fossil lineages gave rise to modern plant phyla. As each major lineage came into being, its characteristic adaptations included modifications of existing structures and functions (**Figure 27.8** and **Table 27.1**).

TABLE 27.1 Trends in Plant Evolution

Traits derived from algal ancestor: cell walls with cellulose, energy stored in starch, two forms of chlorophyll (*a* and *b*); possibly, sporopollenin in spore wall.

Bryophytes	Ferns and Their Relatives	Gymnosperms	Angiosperms	Functions in Land Plants
Cuticle ⟶				Protection against water loss, pathogens
Stomata ⟶				Regulation of water loss and gas exchange (CO_2 in, O_2 out)
Nonvascular ⟶	Vascular ⟶			Internal tubes that transport water, nutrients
	Lignin ⟶			Mechanical support for vertical growth
Apical meristem ⟶				Branching shoot system
	Roots, stems, leaves ⟶			Enhanced uptake, transport of nutrients and enhanced photosynthesis
Haploid phase dominant ⟶	Diploid phase dominant ⟶			Genetic diversity
One spore type (homospory) ⟶	Two spore types (heterospory) ⟶			Promotion of genetic diversity
Motile gametes ⟶			Nonmotile gametes ⟶	Protection of gametes within parent body
Seedless ⟶		Seeds ⟶		Protection of embryo

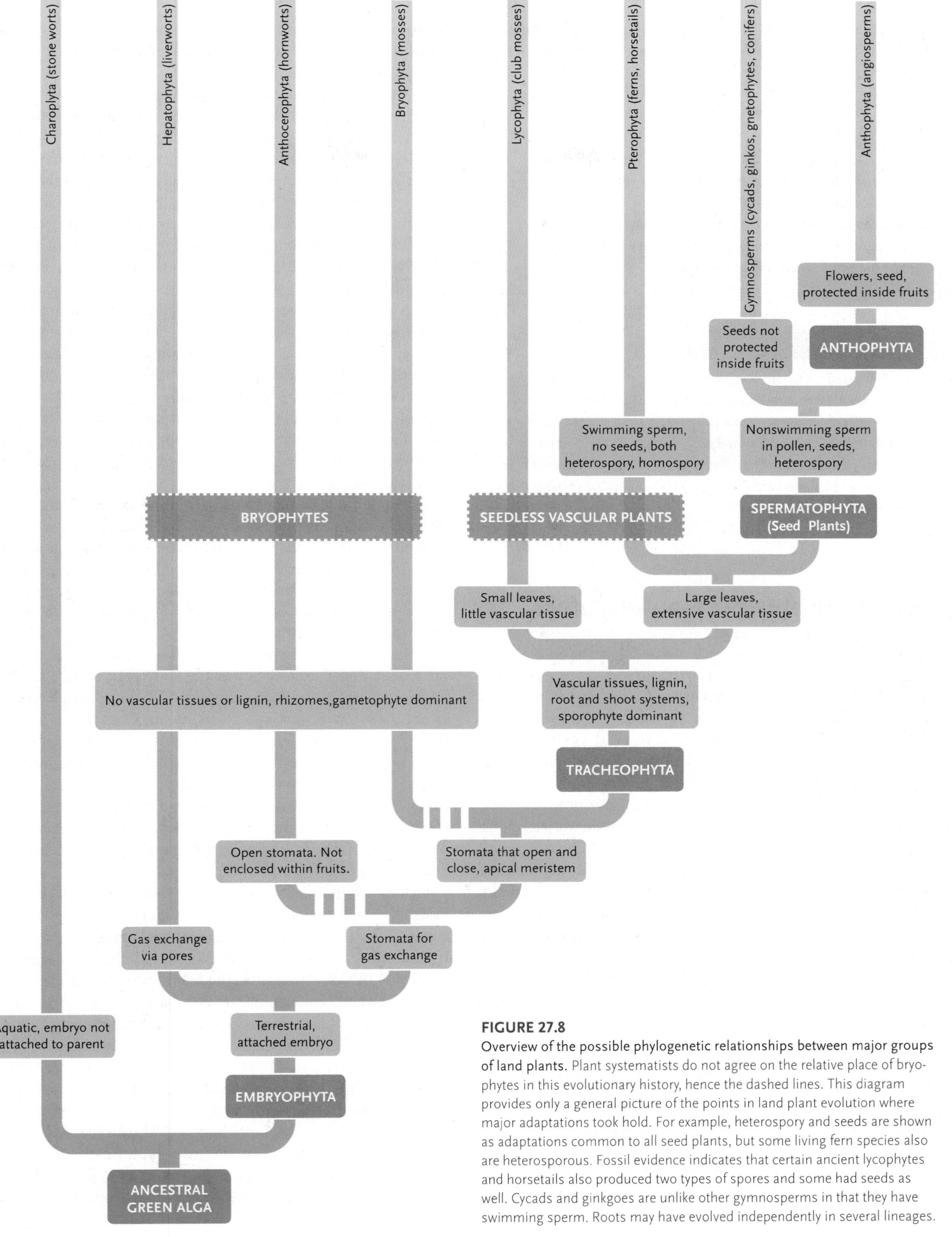

Charoplyta (stone worts)
Hepatophyta (liverworts)
Anthocerophyta (hornworts)
Bryophyta (mosses)
Lycophyta (club mosses)
Pterophyta (ferns, horsetails)
Gymnosperms (cycads, ginkos, gnetophytes, conifers)
Anthophyta (angiosperms)
Flowers, seed, protected inside fruits
Seeds not protected inside fruits
ANTHOPHYTA
Swimming sperm, no seeds, both heterospory, homospory
Nonswimming sperm in pollen, seeds, heterospory
BRYOPHYTES
SEEDLESS VASCULAR PLANTS
SPERMATOPHYTA (Seed Plants)
Small leaves, little vascular tissue
Large leaves, extensive vascular tissue
No vascular tissues or lignin, rhizomes,gametophyte dominant
Vascular tissues, lignin, root and shoot systems, sporophyte dominant
TRACHEOPHYTA
Open stomata. Not enclosed within fruits.
Stomata that open and close, apical meristem
Gas exchange via pores
Stomata for gas exchange
Aquatic, embryo not attached to parent
Terrestrial, attached embryo
EMBRYOPHYTA
ANCESTRAL GREEN ALGA

FIGURE 27.8
Overview of the possible phylogenetic relationships between major groups of land plants. Plant systematists do not agree on the relative place of bryophytes in this evolutionary history, hence the dashed lines. This diagram provides only a general picture of the points in land plant evolution where major adaptations took hold. For example, heterospory and seeds are shown as adaptations common to all seed plants, but some living fern species also are heterosporous. Fossil evidence indicates that certain ancient lycophytes and horsetails also produced two types of spores and some had seeds as well. Cycads and ginkgoes are unlike other gymnosperms in that they have swimming sperm. Roots may have evolved independently in several lineages.

The next sections fill out this general picture, beginning with the plants that most clearly resemble the plant kingdom's algal ancestors.

STUDY BREAK 27.1

1. **How did plant adaptations such as a root system, a shoot system, and a vascular system collectively influence the transition to terrestrial life?**
2. **Describe the difference between homospory and heterospory, and explain how heterospory paved the way for other reproductive adaptations in land plants.**

THINK OUTSIDE THE BOOK

In 2009 researchers studying seaweeds off the coast of California reported discovering a species of red alga with lignin in its tissues. The project's lead scientist was Patrick Martone, now at the University of British Columbia. Use the Web to learn more about Martone's work. Why was the discovery of lignin in a red alga so surprising, and what are its potential evolutionary implications?

27.2 Bryophytes: Nonvascular Land Plants

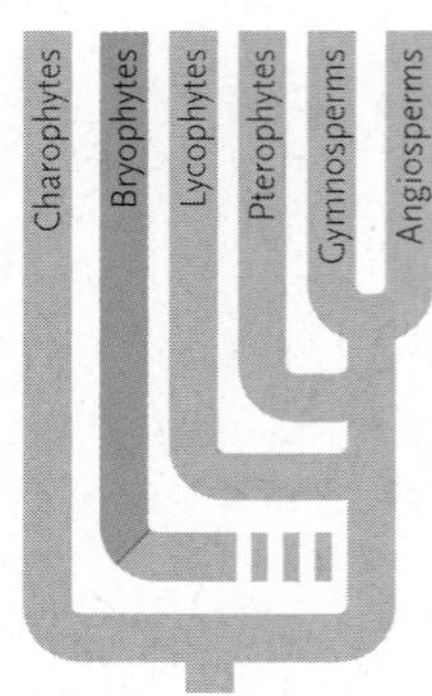
Charophytes
Bryophytes
Lycophytes
Pterophytes
Gymnosperms
Angiosperms

The bryophytes (*bryon* = moss)—liverworts, hornworts, and mosses—have a combination of traits that allow them to bridge aquatic and land environments. Bryophytes lack a well-developed system for conducting water and while some can withstand dessication, it is not surprising that they commonly grow on wet sites along creek banks or on rocks just above running water; in bogs, swamps, or the dense shade of damp forests; and on moist tree trunks or rooftops. Some species are **epiphytes** (*ep* = upon)—they grow independently (that is, not as a parasite) on another organism or some other moist place, ranging from the splash zone just above high tide on rocky shores to edges of snowbanks.

In general, bryophytes are strikingly algalike. They produce flagellated sperm that must swim through water to reach eggs, and they do not have a complex vascular system (although some have a primitive type of conducting tissue). Bryophytes do have parts that are rootlike, stemlike, and leaflike. However, the "roots" are rhizoids, and bryophyte "stems" and "leaves" did not evolve from the same structures as vascular plant stems and leaves did. (That is, stems and leaves are not homologous in the two groups.) Also, as already mentioned, bryophyte tissues do not contain lignified cells. The absence of this strengthening material and the lack of internal pipelines for efficient nutrient transport partly account for bryophytes' small size—typically less than 20 cm long—and for their tendency to grow sprawled along surfaces instead of upright.

Despite these limitations, bryophytes are clearly adapted to land. Along with their leaflike, stemlike, and fibrous, rootlike organs, sporophytes of some species have a water-conserving cuticle and stomata. Like most plants, bryophytes also have both sexual and asexual reproductive modes. And as is true of all plants, the life cycle has both gametophyte (n) and sporophyte ($2n$) phases, though the sporophyte is tiny and lives only a short time. **Figure 27.9** shows the green, leafy gametophyte of a moss plant, with miniscule diploid sporophytes attached to it by slender stalks. Bryophyte gametophytes produce gametes sheltered within a layer of protective cells called a **gametangium** (plural, *gametangia*). The gametangia in which bryophyte eggs form are flask-shaped structures called **archegonia** (*archi* = first, *gonos* = seed). Flagellated sperm form in rounded gametangia called **antheridia** (*antheros* = flowerlike). The sperm swim through a film of water on the plant to the archegonia and fertilize eggs. Each fertilized egg gives rise to a diploid embryo sporophyte, which stays attached to the gametophyte, produces spores—and the cycle repeats.

Despite these similarities with more complex plants, bryophytes are unique in several ways. Unlike in vascular species, the gametophyte is much larger than the sporophyte and obtains its nutrition independently of the sporophyte body. In fact, the comparatively tiny sporophyte remains attached to the gametophyte and depends on the gametophyte for much of its nutrition.

The position of bryophytes in plant evolution is still an open question. The basic bryophyte body plan may be similar to the ancestral condition from which higher plants evolved. In another view, bryophytes are a side shoot of evolution, completely separate from the path that led to vascular plants. The fossil record provides little help in resolving the issue, because the first undisputed bryophyte fossils appear in late-Devonian rocks 350 mya, after vascular plants were already on the scene. (Fossil remains that may resemble liverworts, however, have recently been discovered in rocks that are 50 to 100 million years older.)

Despite questions raised by recent fossil finds, most current molecular, biochemical, cellular, and morphological evidence supports the view that bryophytes are not a monophyletic group. Instead, the various bryophytes evolved as separate lineages, in parallel with vascular plants. The relationships are far from resolved, however. For example, molecular evidence can be interpreted to mean that liverworts diverged early on from the lineage that led to all other land plants, with hornworts diverging later and mosses later still. Until new discoveries and interpretive work clarify this picture, the classification scheme in this chapter places liverworts, hornworts, and mosses in separate phyla. Our survey of nonvascular plants begins with the liverworts and hornworts, the simplest of the group, and concludes with mosses—plants that not only are more familiar to most of us, but whose structure and physiology more closely resemble that of vascular plants.

FIGURE 27.9
Multicellular structures enclosing plant gametes, a bryophyte innovation. **(A)** The gametophyte and sporophyte phases of the moss *Mnium*. In this species the gametangia are embedded in tissue of the gametophyte. In some other bryophytes the gametangia are attached on the gametophyte's surface. The two types of moss gametangia are the **(B)** antheridium, containing cells from which sperm arise, and the **(C)** archegonium, containing an egg cell. When fertilized, the egg cell gives rise to sporophytes.

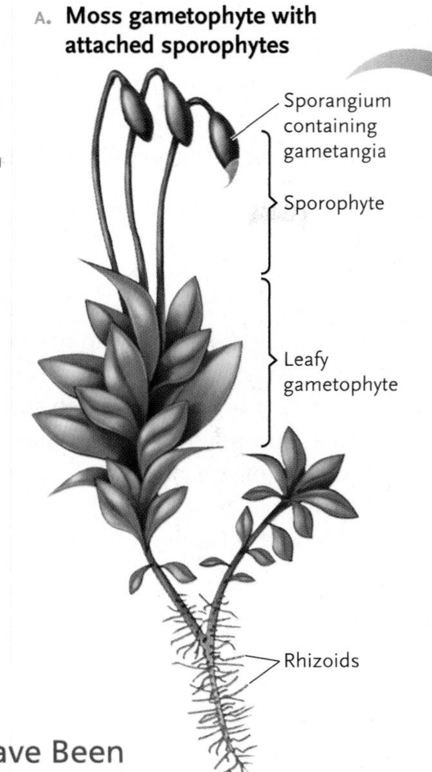
A. Moss gametophyte with attached sporophytes
Sporangium containing gametangia
Sporophyte
Leafy gametophyte
Rhizoids
B. Antheridium
C. Archegonium
Protective cell layer
Cells that produce sperm
Egg cell
Gametangia

Liverworts May Have Been the First Land Plants

Liverworts make up the phylum Hepatophyta, and early herbalists thought that these small plants were shaped like the lobes of the human liver (*hepat* = liver; *wort* = herb). The resemblance might be vague to modern eyes: many of the 6,000 species of liverworts consist of a flat, branching, ribbonlike plate of tissue closely pressed against damp soil. This simple body, called a *thallus* (plural, *thalli*) is the gametophyte generation. Threadlike rhizoids anchor the gametophytes to their substrate. About two-thirds of liverwort species have leaflike structures and some have stemlike parts. None have stomata, the openings that regulate gas exchange in most other land plants, although some species do have pores that open and close. Mitochondrial gene sequence data show that liverworts lack a few features (three introns) that are present in other bryophytes and in vascular plants. This finding has bolstered the hypothesis that liverworts were probably the first land plants.

In species of the liverwort genus *Marchantia* **(Figure 27.10)**, male and female gametophytes are separate plants. Male plants produce antheridia and female plants produce archegonia on specialized stalked organs (see Figure 27.10A, B). The sperm released from an antheridium must swim through surface water to encounter an egg inside an archegonium of a female gametophyte. After fertilization, a small, diploid sporophyte develops inside the archegonium, matures there, and produces haploid spores by meiosis. During meiosis, *Marchantia* sex chromosomes segregate, so some spores have the male genotype and others the female genotype. As in other liverworts, the spores develop inside jacketed sporangia that split open to release the spores. A spore that is carried by air currents to a suitable location germinates and gives rise to a haploid gametophyte, which is either male or female. *Marchantia* also can reproduce asexually by way of **gemmae** (*gem* = bud), small cell masses that form in cuplike growths on a thallus (see Figure 27.10C). Gemmae can grow into new thalli when rainwater splashes them out of the cups and onto an appropriately moist substrate.

Hornworts Have Both Plantlike and Algalike Features

About 100 species of hornworts make up the phylum **Anthocerophyta.** Simple stomata, which are permanently open, appear in this group. Like some liverworts, a hornwort gametophyte has a flat thallus, but the sporangium of the sporophyte phase is long and pointed, like a horn **(Figure 27.11).** The sporangia split into two or three ribbonlike sections when they release spores. Sexual reproduction occurs in basically the same way as in liverworts: free-swimming sperm fertilize eggs, which give rise to the sporophytes. Hornworts sometimes reproduce asexually by fragmentation as pieces of a thallus break off, form rhizoids, and develop into new individuals. Many hornwort species have cell features in common with green algae, including the presence in each cell of a single large chloroplast that contains protein bodies called pyrenoids.

FIGURE 27.10
The bryophyte *Marchantia*, the only liverwort to produce **(A)** male and **(B)** female gametophytes on separate plants. *Marchantia* also reproduces asexually by way of **(C)** gemmae, multicellular vegetative bodies that develop in tiny cups on the plant body. Gemmae can grow into new plants when splashing raindrops transport them to suitable sites.

A. Male plant

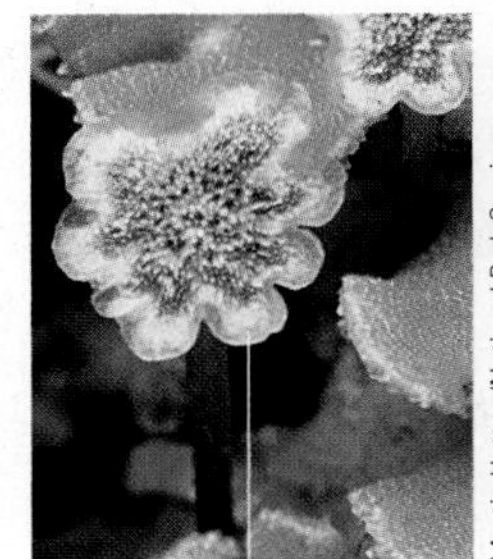
Male gametophyte

Martin Hutten/National Park Service

B. Female plant

Female gametophyte

Paul Stehr-green/National Park Service

C. Asexual reproduction

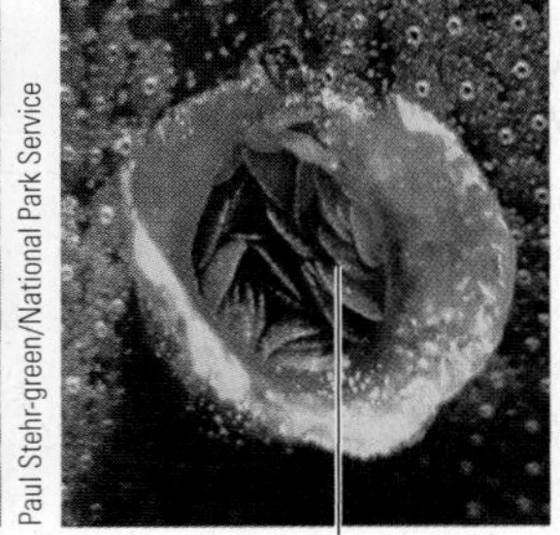
Gemmae

© Wayne P. Armstrong, Professor of Biology and Botany, Palomar College, San Marcos, California

FIGURE 27.11
The hornwort *Anthoceros.* The base of each long, slender sporophyte is embedded in the flattened, leafy gametophyte.

No other group of land plants has this feature, and some biologists have speculated that the distinction of "first land plant" should be assigned not to liverworts but to hornworts instead.

Mosses Most Closely Resemble Vascular Plants

Chances are that you have seen, touched, or sat upon at least several of the approximately 10,000 species of mosses, and the use of the name **Bryophyta** for this phylum underscores the fact that mosses are the best-known bryophytes. They also are structurally and functionally most similar to the vascular plants we will consider in following sections. Their spores, produced by the tens of millions in sporangia, give rise to threadlike, haploid gametophytes that grow into the familiar moss plants, which often form tufts or carpets of vegetation on the surface of rocks, soil, or bark.

The moss life cycle, diagrammed in **Figure 27.12,** begins when a haploid (n) spore lands on a wet soil surface. After the

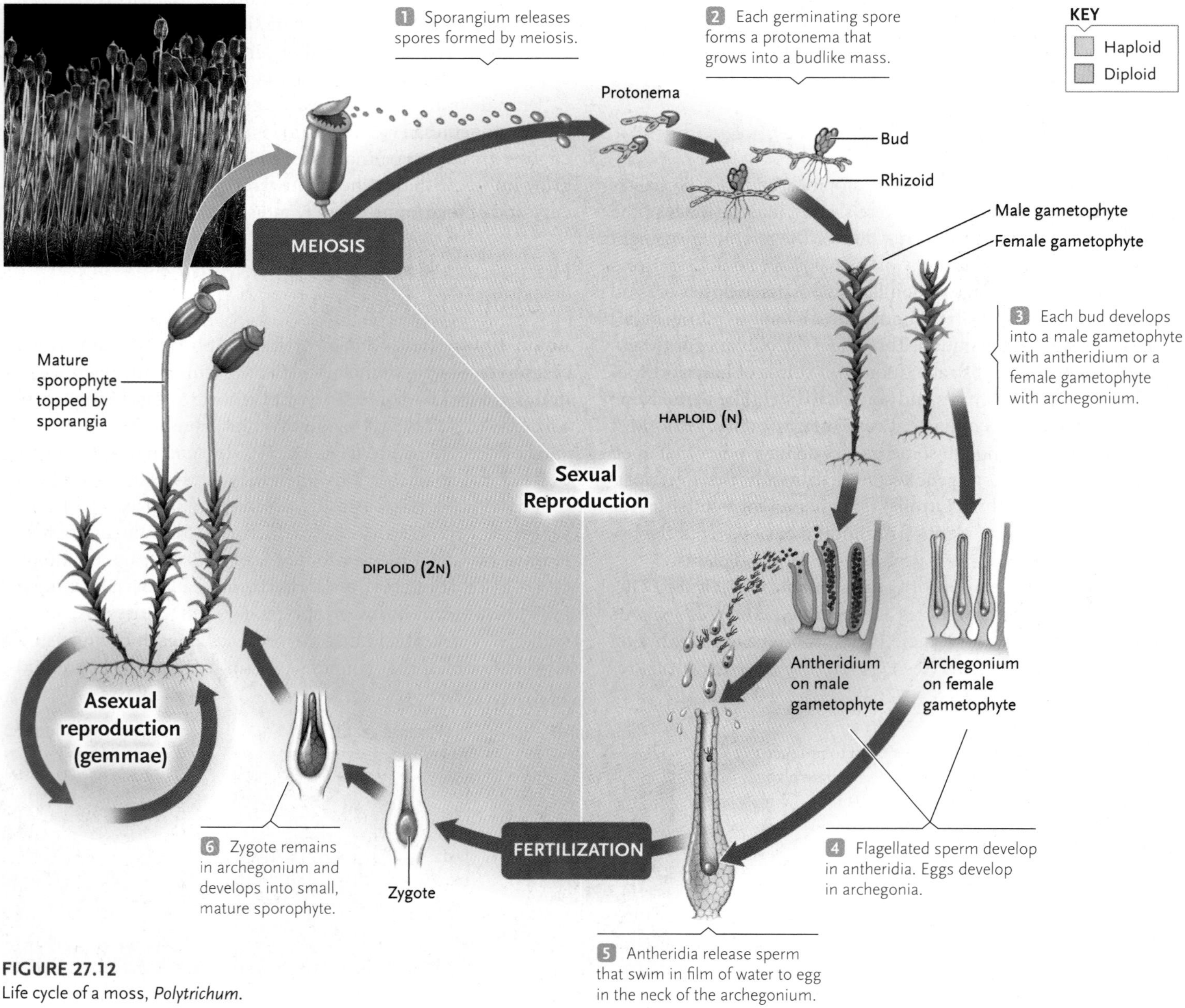

1 Sporangium releases spores formed by meiosis.

2 Each germinating spore forms a protonema that grows into a budlike mass.

KEY
Haploid
Diploid

Protonema
Bud
Rhizoid
Male gametophyte
Female gametophyte

3 Each bud develops into a male gametophyte with antheridium or a female gametophyte with archegonium.

MEIOSIS
HAPLOID (N)
Sexual Reproduction
DIPLOID (2N)

Mature sporophyte topped by sporangia

Asexual reproduction (gemmae)

Antheridium on male gametophyte
Archegonium on female gametophyte

FERTILIZATION
Zygote

6 Zygote remains in archegonium and develops into small, mature sporophyte.

4 Flagellated sperm develop in antheridia. Eggs develop in archegonia.

5 Antheridia release sperm that swim in film of water to egg in the neck of the archegonium.

FIGURE 27.12
Life cycle of a moss, *Polytrichum.*

spore germinates it elongates and branches into a filamentous web of tissue called a **protonema** ("first thread"), which can become dense enough to color the surface of soil, rocks, or bark visibly green. After several weeks of growth, the budlike cell masses on a protonema develop into leafy, green gametophytes anchored by rhizoids. A single protonema can be extremely prolific, producing bud after bud—and in this way giving rise to a dense clone of genetically identical gametophytes. Leafy mosses also may reproduce asexually by gemmae produced at the surface of rhizoids as well as on aboveground parts.

When a leafy moss is sexually mature, gametangia develop on its gametophytes and gametes form in them. In some moss genera, plants are unisexual and produce male *or* female gametangia—antheridia at the tips of male gametophytes and archegonia at the tips of female gametophytes. In other genera, plants are bisexual and produce both antheridia and archegonia. Propelled by a pair of flagella, sperm released from antheridia swim through a film of dew or rainwater and down a channel in the neck of the archegonium, attracted by a gradient of a chemical secreted by each egg. Fertilization produces the new sporophyte generation inside the archegonium, in the form of diploid zygotes that develop into small, mature sporophytes, each consisting of a sporangium on a stalk. Moss sporophytes may eventually develop chloroplasts and nourish themselves by photosynthesis, but initially they depend on the gametophytes for food. And even after a moss sporophyte begins photosynthesis, it still must obtain water, carbohydrates, and some other nutrients from the gametophyte.

Certain moss gametophytes are structurally complex, with features somewhat similar to those of higher plants. For example, mosses have stomata that open and close, and some species have a central strand of primitive conducting tissue. One kind of this tissue is made up of elongated, thin-walled, dead and empty cells that conduct water. These specialized cells, called hydroids, have oblique end walls that sometimes are partly dissolved or perforated with pores. Experiments with dyes show that water moves through them, as it does in similar xylemlike arrangements in vascular plants (see Chapters 31 and 32). In a few mosses, the water-conducting cells are surrounded by sugar-conducting tissue resembling the phloem of vascular plants.

Mosses and other bryophytes are important both ecologically and economically. As colonizers of bare land, their small bodies trap particles of organic and inorganic matter, helping to build soil on bare rock and stabilizing soil surfaces with a biological crust in harsh places like coastal dunes, inland deserts, and embankments created by road construction. Some hornworts harbor mutualistic nitrogen-fixing cyanobacteria, and so increase the amount of nitrogen available to other plants. In arctic tundras, bryophytes constitute as much as half the biomass, and they are crucial components of the food web that supports animals in that ecosystem. People have long used *Sphagnum* and other absorbent "peat" mosses (which typically grow in bogs) for everything from diapering babies and filtering whiskey to increasing the water-holding capacity of garden soil. Peat moss also is used as a fuel; each day the Rhode generating station in Ireland, one of several in that nation, burns 2,000 metric tons of peat to produce electricity.

In the next section we turn to the vascular plants, which have specialized tissues that can transport water, minerals, and sugars. Without the capacity to move these substances efficiently throughout the plant body, large sporophytes could not have evolved on land. Unlike bryophytes, modern vascular plants are monophyletic—all groups are probably descended from a common ancestor.

STUDY BREAK 27.2

1. **Give some examples of bryophyte features that bridge aquatic and terrestrial environments.**
2. **How do specific aspects of a moss plant's anatomy resemble those of vascular plants?**

27.3 Seedless Vascular Plants

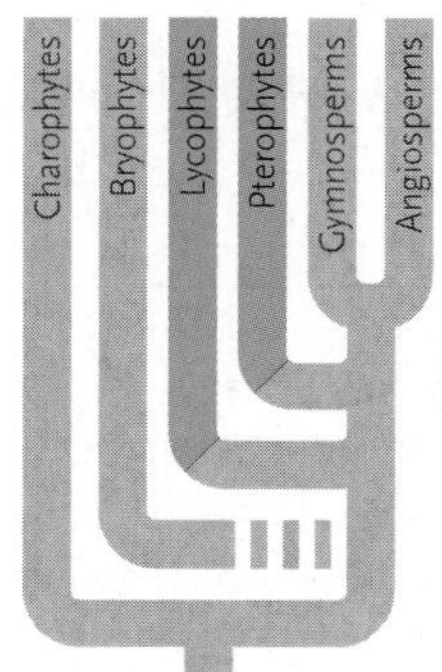
Charophytes
Bryophytes
Lycophytes
Pterophytes
Gymnosperms
Angiosperms

The first vascular plants, which did not "package" their embryos inside protective seeds, were the dominant plants on Earth for almost 200 million years, until seed plants became abundant. The fossil record shows that seedless vascular plants were well established by the late Silurian, some 428 mya, and they flourished until the end of the Carboniferous, about 250 mya. Some living seedless vascular plants have certain bryophyte-like traits, whereas others have some characteristics of seed plants. On one hand, like bryophytes, seedless vascular plants reproduce sexually by releasing spores, and they have swimming sperm that require free water to reach eggs. On the other hand, as in seed plants, the sporophyte of a seedless vascular plant separates from the gametophyte at a certain point in its development and has well-developed vascular tissues (xylem and phloem). Also, the sporophyte is the larger, longer-lived stage of the life cycle and the gametophytes are very small. Some gametophytes even lack chlorophyll. **Table 27.2** summarizes these characteristics and gives an overview of seedless vascular plant features within the larger context of modern plant phyla.

Seedless vascular plants once encompassed a huge number of diverse species of trees, shrubs, and herbs. In the late Paleozoic era, they were Earth's dominant vegetation. Some lineages have endured to the present, but collectively these survivors total fewer than 14,000 species. The taxonomic relationships between various lines are still under active investigation, and comparisons of gene sequences from the genomes in plastids, cell nuclei, and sometimes mitochondria are revealing previously unsuspected links between some of them. In this book we assign seedless vascular plants to two phyla, the Lycophyta (club mosses and their close relatives) and the Pterophyta (ferns, whisk ferns, and horsetails).

TABLE 27.2	Plant Phyla and Major Characteristics		
Phylum	**Common Name**	**Number of Species***	**Common General Characteristics**
Bryophytes: Nonvascular plants. Gametophyte dominant, free water required for fertilization, cuticle and stomata present in some.			
Hepatophyta	Liverworts	6,000	Leafy or simple flattened thallus with pores, rhizoids; spores in capsules; cuticle. Moist, humid habitats.
Anthocerophyta	Hornworts	100	Simple flattened thallus, rhizoids; stomata; hornlike sporangia. Moist, humid habitats.
Bryophyta	Mosses	10,000	Feathery or cushiony thallus, some have hydroids; spores in capsules. Moist, humid habitats; colonizes bare rock, soil, or bark.
Seedless vascular plants: Sporophyte dominant, swimming sperm, free water required for fertilization, cuticle and stomata present in all.			
Lycophyta	Club mosses	1,000	Simple leaves, true roots; most species have sporangia on sporophylls. Mostly wet or shady habitats.
Pterophyta	Ferns, whisk ferns, horsetails	13,000	*Ferns:* Finely divided leaves, woody stems in tree ferns; sporangia in sori. Habitats from wet to arid. *Whisk ferns:* Branching stem from rhizomes; sporangia on stem scales. Tropical to subtropical habitats. *Horsetails:* Hollow photosynthetic stem, scalelike leaves with silica in cell walls, sporangia in strobili. Swamps, disturbed habitats.
Gymnosperms: Vascular plants with "naked" seeds. Nonmotile sperm arise from male gametophytes in pollen grains.			
Cycadophyta	Cycads	185	Shrubby or treelike with palmlike leaves, pithy stems; male and female strobili on separate plants. Widely distributed in warm climates.
Ginkgophyta	Ginkgo	1	Woody-stemmed tree, deciduous fan-shaped leaves. Male, female structures on separate plants. Temperate areas of China.
Gnetophyta	Gnetophytes	70	Shrubs or woody vines; one has strappy leaves. Male and female strobili on separate plants. Limited to deserts, tropics.
Coniferophyta	Conifers	550	Mostly evergreen, woody trees and shrubs with needlelike or scalelike leaves; male and female cones usually on same plant.
Angiosperms (Anthophyta): Woody or herbaceous plants with flowers, and seeds protected inside fruits; nearly all land habitats, some aquatic.			
Major groups: Magnoliids, Monocots, Eudicots			
Magnoliids	Magnolias, laurels, avocado, black peppers, and others	8,000+	Pollen grains have a single groove; some species with three of more cotyledons.
Monocots	Grasses, palms, lilies, orchids, and others	60,000+	Pollen grains have a single groove; one cotyledon. Parallel-veined leaves common.
Eudicots	Most fruit trees, roses, cabbages, melons, beans, potatoes, and others	200,000+	Pollen grains have three grooves. Most species have two cotyledons; net-veined leaves common.

*Numbers of species are approximate.

Early Seedless Vascular Plants Flourished in Moist Environments

The extinct plant genus *Cooksonia* (see Figure 27.4) probably was one of the earliest ancestors of modern seedless vascular plants. It and similar plants dominated the mudflats and swamps of the Devonian period (408–360 mya). Like other members of its extinct phylum (Rhyniophyta) *Cooksonia* species were small, rootless, and leafless, but their simple stems had a central core of xylem, an arrangement seen in many existing vascular plants. Although this and other now-extinct phyla came and went, ancestral forms of both modern phyla of seedless vascular plants appeared. In botanical terms, the earliest seedless vascular plants were "herbs"—that is, they did not have woody, lignified tissue.

A. **Lycophyte tree *(Lepidodendron)***

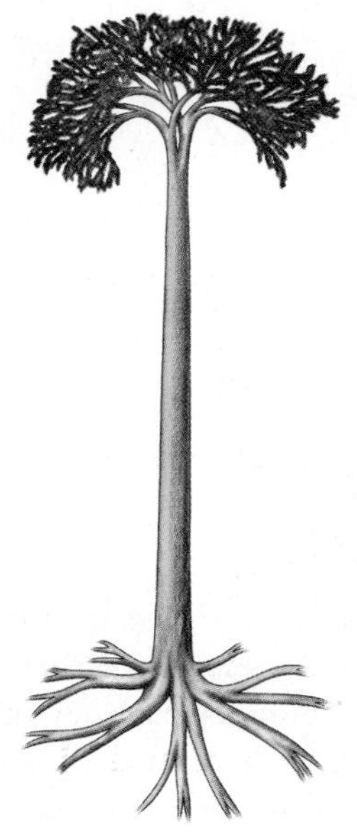

B. **Artist's depiction of a Coal Age forest**

© The Field Museum, #CSGEO75400c
Stem of a giant lycophyte (Lepidodendron)
Seed fern (Medullosa); probably related to the progymnosperms, which may have been among the earliest seed-bearing plants
Stem of a giant horsetail (Calamites)

FIGURE 27.13
Reconstruction of the lycophyte tree *Lepidodendron* and its environment. **(A)** Fossil evidence suggests that *Lepidodendron* grew to be about 35 m tall with a trunk 1 m in diameter. **(B)** Artist's depiction of a Coal Age forest.

By the start of the ensuing Carboniferous period, however, the small herbaceous Devonian plants had given rise to larger shrubby species and to trees—long-lived plants with a single main stem containing woody tissue. Other features of some of these early trees may have included bark, roots, leaves, and even seeds.

Carboniferous forests were lush, swampy places dominated by members of the phylum **Lycophyta,** a group in which roots evolved—an innovation that provided both firm anchoring in the soil and enhanced access to soil water and nutrients. One example is *Lepidodendron,* which had broad, straplike leaves and sporangia near the ends of the branches **(Figure 27.13A).** It also had xylem and several other types of tissues that are typical of all modern vascular plants, although probably not in the same proportions as seen today.

Carboniferous forests also were home to representatives of the phylum **Pterophyta,** including ferns such as *Medullosa* and giant horsetails such as *Calamites*—huge treelike plants that could have a trunk 30 cm in diameter. The strong, upright stems were attached to a system of rhizomes. Ferns populated the forest understory. Some early seed plants were present as well, including now-extinct fernlike plants, called seed ferns, which bore seeds at the tips of leaves **(Figure 27.13B).**

When leaves, branches, and old trees in swampy Carboniferous landscapes fell to the ground, they became buried in anaerobic sediments. Over geologic time, these buried remains were compressed and fossilized, and today they form much of the world's coal reserves. This is why coal is called a "fossil fuel," and the Carboniferous period is called the Coal Age. Characterized by a humid climate over much of the planet, and by the dominance of seedless vascular plants, the Carboniferous period continued for 150 million years, ending when climate patterns changed during the Paleozoic era (550–250 mya).

Most modern seedless vascular plants are ferns, and like their ancestors they are confined largely to wet or humid environments because they require external water for reproduction. Except for whisk ferns, their gametophytes do not rely on vascular tissues for water transport, and male gametes swim through water to reach eggs. In the wild, the few vascular seedless plants that are adapted to dry environments such as deserts can reproduce sexually only when seasonal rains make surface water available.

Modern Lycophytes Are Small and Have Simple Vascular Tissues

Unlike their ancient relatives, the most familiar of the 1,000 or so living species of lycophytes are small, ground-hugging species. They include spike mosses, quillworts, and the club mosses shown in **Figure 27.14A**—members of genera such as *Lycopodium* that grow on forest floors. A club moss sporophyte has stems and roots that contain a small amount of the vascular tissue xylem. Near the tips of stems are small green microphyll-type leaves called **sporophylls,** with a single sporangium tucked at the base of each one. A cluster of sporophylls forms a conical **strobilus** (plural, *strobili*). In some species the sporangia release haploid spores produced by meiosis **(Figure 27.14B).** If a spore germinates (sometimes several years after it is released), it forms a free-living gametophyte that

A. ***Lycopodium* sporophyte**

Strobilus

B. **Fossilized lycophyte spore**

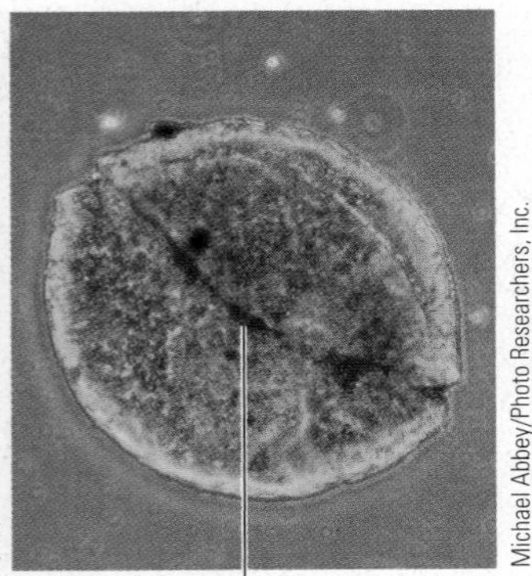
Trilete scar

FIGURE 27.14
Lycophytes. **(A)** *Lycopodium* sporophyte, showing the conelike strobili in which spores are produced. **(B)** A fossilized lycophyte spore bearing a characteristic Y-shaped mark called a trilete scar.

differs markedly from the sporophyte. Ranging in size from nearly invisible to several centimeters, the gametophyte easily becomes buried under decomposing plant litter. There rhizoids attach it to a substrate. It cannot photosynthesize, and instead obtains nutrients by way of mycorrhizae, a symbiotic interaction in which a plant's rhizoids or roots associate with a particular fungus. Although many lycophyte species are homosporous—gametophytes are bisexual and produce both eggs and sperm—some are heterosporous. Regardless, as with ancestral lycophytes, the sperm require water in which they can swim to the eggs. After fertilization, the life cycle comes full circle as the zygote develops into a diploid embryo that grows into a sporophyte.

Ferns, Whisk Ferns, Horsetails, and Their Relatives Make Up the Diverse Phylum Pterophyta

Second only to the flowering plants, the phylum Pterophyta (*pteron* = wing) contains a large and diverse group of vascular plants—the 13,000 or so species of ferns, whisk ferns, and horsetails. Most ferns, including some that are poplar houseplants, are native to tropical and temperate regions. Some floating species are less than 1 cm across, while some tropical "tree ferns" grow to 25 m tall. Other species are adapted to life in arctic and alpine tundras, salty mangrove swamps, and semi-arid deserts.

COMPLEX ANATOMICAL FEATURES IN FERNS The familiar plant body of a fern is the sporophyte phase **(Figure 27.15)**. It produces an aboveground clump of fern leaves, called fronds. Often finely divided and featherlike, and containing multiple strands of vascular tissue, fronds are an example of large, macrophyll-type leaves, a trait ferns share with seed plants. A typical frond has a well-developed epidermis with chloroplasts in the epidermal cells and stomata on the lower surface. Young fronds are tightly coiled, and as they emerge above the soil these "fiddleheads" (so named because they resemble the scroll of a violin) unroll and expand. Some people enjoy collecting and eating the unfurled fiddleheads of edible (noncarcinogenic) species.

Except for tropical tree ferns, the stems of most ferns are underground rhizomes. The stem's vascular system is organized into a complex, interconnecting network of bundles, each having a core of xylem surrounded by phloem. Roots descend along the length of the rhizomes. Remarkably, a rhizome can live for centuries, growing at its tip and extending outward horizontally through the soil, sometimes over a considerable area. In most ferns, the fronds arise from nodes positioned along the rhizome. A **node** is the point on a stem where one or more leaves are attached.

In most fern species, the sporophyte produces sporangia on the lower surface or margin of some leaves. Often, several sporangia are clustered into a rust-colored **sorus** (plural, *sori*). Sori may be exposed or they may be protected with a flap of tissue. Each sporangium is a delicate case, shaped rather like an old-fashioned pocket watch and covered by a layer of epidermal cells. In the layer, a row of thick-walled cells called the **annulus** ("ring") nearly encircles the sporangium.

Inside the sporangium, haploid spores arise by meiosis. Meanwhile, the sporangium slowly dries, and in the process the annulus contracts. Eventually the force of the contracting annulus rips open the sporangium, which snaps back on itself, flinging out the mature spores. In this way fern spores can be dispersed up to 2 m away from the parent plant. Wind may carry them much farther: on board the *Beagle,* Charles Darwin collected fern spores hundreds of miles from shore.

A germinating spore develops into a gametophyte, which is typically a small, heart-shaped plant anchored to the soil by rhizoids. Both antheridia and archegonia are present on the lower surface of each gametophyte, where moisture is trapped. Inside an antheridium is a globular packet of haploid cells, each of which develops into a helical sperm cell with many flagella. When water is present, the antheridium bursts, releasing the sperm. If mature archegonia are nearby, the sperm swim toward them, drawn by a chemical attractant that diffuses from the neck of the archegonium, which is open when free water is present.

After a sperm fertilizes an egg, the diploid zygote begins dividing and developing into an embryo, which at this stage obtains nutrients from the gametophyte. In a short time, however, the embryo develops into a young sporophyte that is larger than the gametophyte and has its own green leaf and a root system.

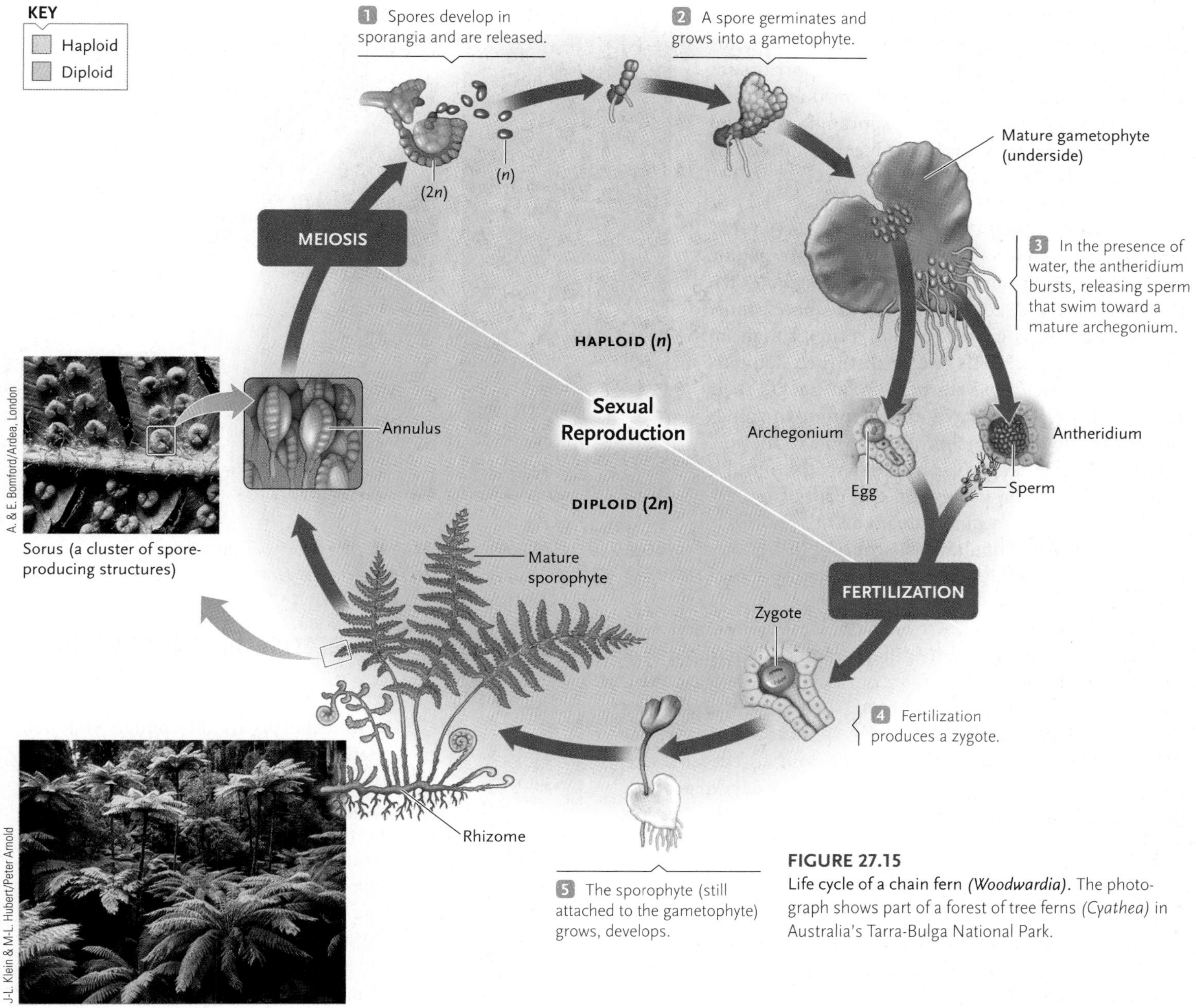

KEY

Haploid

Diploid

1 Spores develop in sporangia and are released.

2 A spore germinates and grows into a gametophyte.

Mature gametophyte (underside)

3 In the presence of water, the antheridium bursts, releasing sperm that swim toward a mature archegonium.

(2n)

(n)

MEIOSIS

HAPLOID (n)

Sexual Reproduction

DIPLOID (2n)

Annulus

Archegonium

Egg

Antheridium

Sperm

FERTILIZATION

Zygote

4 Fertilization produces a zygote.

Mature sporophyte

Rhizome

Sorus (a cluster of spore-producing structures)

A. & E. Bomford/Ardea, London

J-L. Klein & M-L. Hubert/Peter Arnold

5 The sporophyte (still attached to the gametophyte) grows, develops.

FIGURE 27.15
Life cycle of a chain fern *(Woodwardia)*. The photograph shows part of a forest of tree ferns *(Cyathea)* in Australia's Tarra-Bulga National Park.

The sporophyte now is nutritionally independent and the parent gametophyte degenerates and dies.

FEATURES OF EARLY VASCULAR PLANTS IN WHISK FERNS The whisk ferns and their relatives are represented by two genera, *Psilotum* (pronounced si-LOW-tum) and *Tmesipteris* (may-SIP-ter-is), with only about 10 species in all. These rather uncommon plants grow in tropical and subtropical regions, often as epiphytes. In the United States the range for *Psilotum* species **(Figure 27.16)** includes Hawaii, Gulf Coast states such as Florida and Louisiana, and parts of the West.

Whisk fern sporophytes are up to 60 cm tall and resemble the extinct *Cooksonia*. Like those early vascular plants, they lack roots and leaves. Instead, small scales adorn an upright branching stem that ascends from a horizontal rhizome system anchored by rhizoids. The rhizoids have mycorrhizal fungi associated with them, which provide enhanced access to some nutrients.

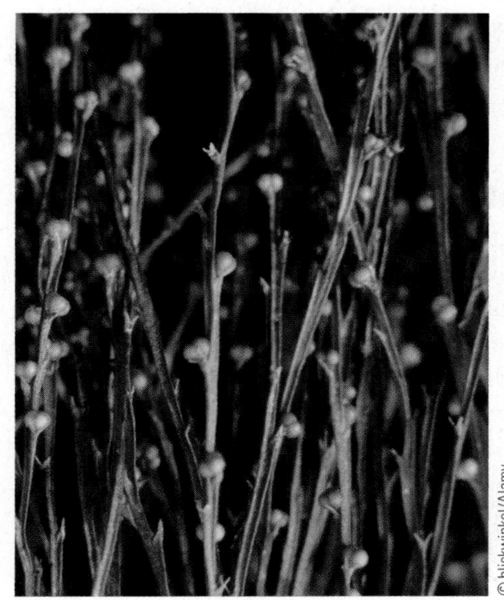

© blickwinkel/Alamy

FIGURE 27.16
Sporophytes of a whisk fern *(Psilotum)*, a seedless vascular plant. Three-lobed sporangia occur at the ends of stubby branchlets; inside the sporangia, meiosis gives rise to haploid spores.

A whisk fern's stem is structurally and functionally multifaceted. Its epidermal cells carry out photosynthesis, whereas its core has the transport tissues xylem and phloem and other anatomical features of more complex vascular plants. Sporangia rest atop some of the stem scales. Inside them, meiotic divisions of specialized cells produce haploid spores.

HORSETAILS, REMNANTS OF AN ANCIENT PAST

Only fifteen horsetail species in a single genus, *Equisetum,* have survived to the present **(Figure 27.17).** Unlike towering ancestors such as *Calamites,* most modern horsetails grow less than a meter high, in moist soil along streams and in disturbed habitats, such as roadsides and beds of railroad tracks. Their sporophytes typically have underground rhizomes along with roots that anchor the rhizome to the soil. Whorls of scalelike leaves are arranged around a photosynthetic stem that is stiff and gritty from the silica that horsetails accumulate in their tissues. They came to be known popularly as "scouring rushes" because American pioneers used them to scrub out pots and pans.

A. **Sporophyte stem**

William Ferguson

B. **Sporangia**

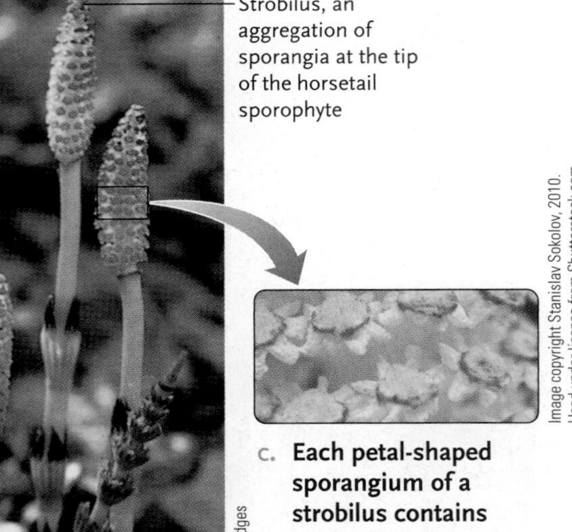
Strobilus, an aggregation of sporangia at the tip of the horsetail sporophyte
C. Each petal-shaped sporangium of a strobilus contains spores that formed by meiosis.
W. H. Hodges
Image copyright Stanislav Sokolov, 2010.
Used under license from Shutterstock.com

FIGURE 27.17
A species of *Equisetum*, the horsetails. **(A)** Vegetative stem. **(B)** Strobili, which bear sporangia. **(C)** Close-up of sporangia and associated structures on a strobilus.

Equisetum sporangia are borne in strobili on highly specialized stem structures quite different from the sporophylls of club mosses. Each stalked spore-bearing structure in a strobilus resembles an umbrella and is attached at right angles to a main axis. Haploid spores develop in sporangia attached near the edge of the "umbrella's" underside, and air currents disperse them. They must germinate within a few days to produce gametophytes, which are free-living plants about the size of a small pea.

STUDY BREAK 27.3

1. **Compare and contrast the lycophyte and bryophyte life cycles with respect to the sizes and longevity of gametophyte and sporophyte phases.**
2. **In ferns, whisk ferns, and horsetails, what kinds of structures fulfill the roles of roots and leaves?**
3. **How does the life cycle of a horsetail differ from that of a fern?**

THINK OUTSIDE THE BOOK

Seed ferns, an extinct group mentioned briefly in this section, are assigned to the extinct phylum Pteridospermophyta. They were nearly wiped out by the Permian–Triassic mass extinction event that occurred about 251 mya and became extinct soon thereafter. Seed-bearing conifers living at the time fared better, and conifers were the dominant land plants for at least another 100 million years. Using Web or library resources, develop a plausible explanation for why the seed fern and conifer lineages had such differing evolutionary fates.

27.4 Gymnosperms: The First Seed Plants

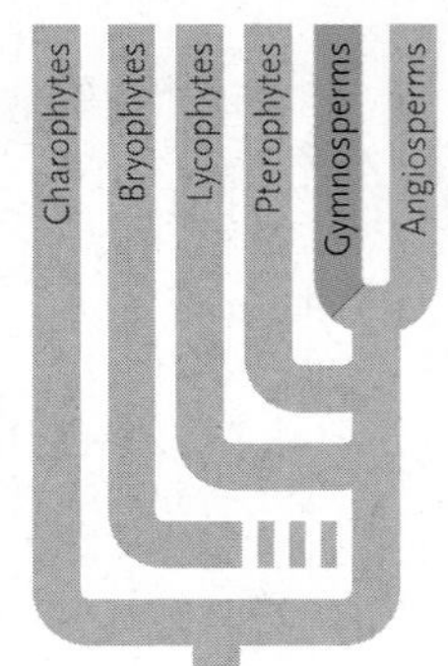
Charophytes
Bryophytes
Lycophytes
Pterophytes
Gymnosperms
Angiosperms

Gymnosperms are the conifers and their relatives. The earliest fossils identifiable as gymnosperms are found in Devonian rocks dating to about 390 mya. By the Carboniferous (360–300 mya) when seedless plants were dominant, many lines of gymnosperms had also arisen, and the first true conifers appeared. These radiated during the Permian period (300–250 mya), and the Mesozoic era that followed, 248 to 65 mya, was the age not only of the dinosaurs but of the gymnosperms as well.

The evolution of gymnosperms involved sweeping changes in plant structures related to reproduction. As a prelude to our survey of modern gymnosperms, we begin by considering some of these innovations, which opened new adaptive options for land plants.

Major Adaptations in Growth and Reproduction Occurred as Gymnosperms Evolved

Early gymnosperm evolution was marked by several innovations, all of which facilitated the radiation of gymnosperms into various land environments. They included increasing size and important reproductive adaptations—the ovule, pollen and pollination, and the seed. Although the fossil record is sketchy, it provides glimpses of the order and types of changes that accompanied this shift.

FIGURE 27.18
Fossil-based reconstruction of *Archaeopteris*, a large Devonian progymnosperm. It could grow 25 m tall and may have been a seed-forming ancestor of modern gymnosperms.

A NEW MERISTEM FOR SECONDARY GROWTH By about 390 mya, in the mid Devonian, a group called *progymnosperms* is thought to have given rise to the first true trees, some of which had woody trunks more than a meter in diameter. Such girth was the result of a new adaptation, a lateral meristem that produced *secondary growth*—the development of woody layers of xylem and phloem that takes place once the initial, primary plant body forms and makes roots and stems sturdier by increasing their diameter. **Figure 27.18** shows an artist's rendering of the progymnosperm *Archaeopteris,* a small seedless tree with a tapering trunk that was common over parts of Europe and North America and is thought to have reached a height of about 6 m. Chapter 31 discusses secondary growth in detail.

OVULES, POLLEN, AND SEEDS The word *gymnosperm* is derived from the Greek *gymnos,* meaning naked, and *sperma,* meaning seed. Progymnosperms did not have seeds, but from fossils we know that species in the plant group that included *Archaeopteris* were heterosporous: they produced two types of spores (megaspores and microspores) that were released from structurally different sporangia. In some progymnosperm fossils, megaspores evidently were retained instead of released, and were surrounded by a layer of tissue called an *integument*—an arrangement that has been interpreted as the foundation for an ovule that would develop into a seed. Other changes, perhaps happening simultaneously, resulted in the evolution of pollen grains from microspores. As described next, pollen grains have hard walls and fossilize relatively well. Some amazingly well-preserved specimens reveal that as the Devonian period came to a close, several different types of gymnosperm pollen grains existed.

POLLEN AND OVULES: SHELTER FOR SPORES Unlike bryophytes and seedless vascular plants, gymnosperm sporophytes do not disperse their spores. The sporophyte produces haploid spores by meiosis, but it retains these spores inside reproductive structures where they give rise to multicellular haploid gametophytes. As you've read, sperm develop inside a **pollen grain,** a male gametophyte produced from a microspore that typically has walls reinforced with the polymer sporopollenin. All but a few gymnosperms have nonmotile sperm. Usually, two of these nonswimming cells develop inside each pollen grain—very different from the flagellated, swimming sperm of algae and plants that do not produce seeds.

An **ovule** is a structure in a sporophyte in which a female gametophyte develops, complete with an egg. Physically connected to the sporophyte and surrounded by the ovule's protective layers, a female gametophyte no longer faces the same risks of predation or environmental assault that can threaten a free-living gametophyte.

Pollination is the transfer of pollen to female reproductive parts via air currents or on the bodies of animal pollinators. Pollen and pollination were enormously important adaptations for gymnosperms, because the shift to nonswimming sperm along with a means for delivering them to female gametes meant that reproduction no longer required liquid water. The only gymnosperms that have retained swimming sperm are the cycads and ginkgoes, described shortly, which have relatively few living species and are restricted to a smattering of native habitats.

SEEDS: PROTECTING AND NOURISHING PLANT EMBRYOS A **seed** is the structure that forms when an ovule matures after a pollen grain reaches it and a sperm fertilizes the egg. It consists of three basic parts: (1) the embryo sporophyte, (2) tissues around it containing carbohydrates, proteins, and lipids that nourish the embryo until it becomes established as a plantlet with leaves and roots, and (3) a tough, protective outer seed coat **(Figure 27.19).** This complex structure makes seeds ideal packages for sheltering an embryo from drought, cold, or other adverse conditions. As a result, seed-making plants enjoy a tremendous survival advantage over species that simply release spores to the environment. Encased in a seed, the embryo also can be transported far from its parent, as when ocean currents carry coconut seeds ("coconuts" protected in large, buoyant fruits) hundreds of kilometers across the sea. As discussed in Chapter 34, some plant embryos housed in seeds can remain dormant for months or years before environmental conditions finally stimulate them to germinate and grow.

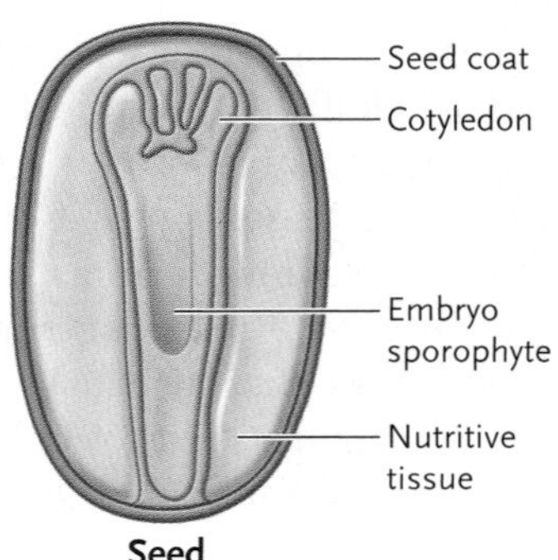
Seed coat
Cotyledon
Embryo sporophyte
Nutritive tissue
Seed

FIGURE 27.19
Generalized view of a seed—in this case, the seed of a pine, a gymnosperm.

Modern Gymnosperms Include Conifers and a Few Other Groups

Today there are about 800 gymnosperm species, including the largest and oldest living land plants. Giant sequoias *(Sequoiadendron giganteum)* may grow to more than 80 m, and coast redwoods *(Sequoia sempervirens)* more than 112 m. Both are native to California. Tree-ring analysis revealed that one specimen of another California native, the bristlecone pine *(Pinus aristata)* has survived almost 5,000 years, making it the world's oldest known continuously standing tree. The sporophytes of nearly all gymnosperms are sizable trees or shrubs, although a few are woody vines. The most widespread and familiar gymnosperms are the conifers (Coniferophyta). Others are the cycads (Cycadophyta), ginkgoes (Ginkgophyta), and gnetophytes (Gnetophyta).

Economically, gymnosperms, particularly conifers, are vital to human societies. They are sources of lumber, paper pulp, turpentine, and resins, among other products. They also have huge ecological importance. Their habitats range from tropical forests to deserts, but gymnosperms are most dominant in the cool-temperate zones of the northern and southern hemispheres. They flourish in poor soils where flowering plants do not compete as well. In North America, for example, conifer forests cover more than one-third of the continent's landmass—although in some areas, logging has significantly reduced the once-lush forest cover. Our survey of living gymnosperms begins, however, with the cycads, ginkgoes, and gnetophytes—the latter two groups remnants of lineages that have all but vanished from the modern scene.

Carlton Ray/Science Photo Library/Photo Researchers, Inc.

FIGURE 27.20
The cycad *Encephalartos.* Note the large, terminal female cone and fernlike leaves.

Cycads Are Restricted to Warmer Climates

During the Mesozoic era (250–65 mya), the **Cycadophyta** (*kykas* = palm), or cycads, flourished along with the dinosaurs. About 185 species have survived to the present, confined to the tropics and subtropics.

At first glance, you might mistake a cycad for a small palm tree **(Figure 27.20).** Some cycads have massive, cone-shaped strobili (clusters of sporophylls) that bear either pollen or ovules. Air currents or beetles transfer pollen from male plants to the developing gametophyte on female plants. Poisonous alkaloids that may help deter insect predators occur in various cycad tissues. In tropical Asia, some people consume cycad seeds and flour made from cycad trunks, but only after the toxic compounds have been rinsed away. Much in demand from fanciers of unusual plants, cycads in some countries are uprooted and sold in what amounts to a black-market trade—greatly diminishing their numbers in the wild.

A. **Ginkgo tree**

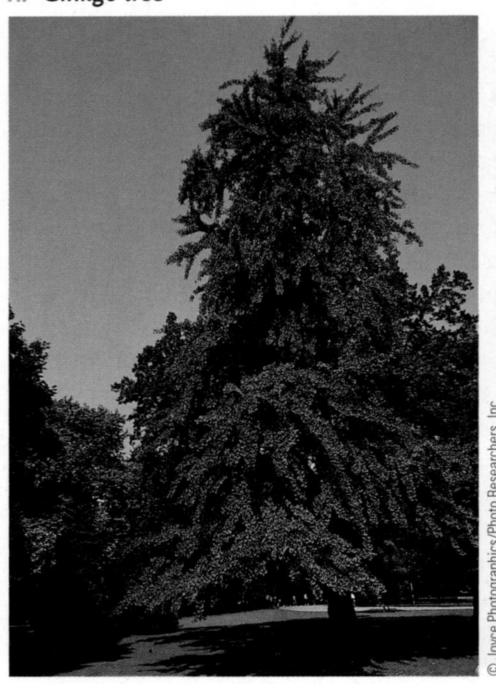

© Joyce Photographics/Photo Researchers, Inc.

B. **Fossil and modern gingko leaves**

© B. RUNK/S. SCHOENBERGER/Grant Heilman Photography

C. **Male cone**

William Ferguson

D. **Ginkgo seeds**

Kingsley R. Stern

FIGURE 27.21
Ginkgo biloba. **(A)** A ginkgo tree. **(B)** Fossilized ginkgo leaf compared with a leaf from a living tree. The fossil formed at the Cretaceous–Tertiary boundary. Even though 65 million years have passed, the leaf structure has not changed much. **(C)** Pollen-bearing cones and **(D)** fleshy-coated seeds of the *Ginkgo.*

Ginkgoes Are Limited to a Single Living Species

The phylum **Ginkgophyta** has one living species, the ginkgo (or maiden-hair) tree *(Ginkgo biloba),* which grows wild today only in warm-temperate forests of central China. Ginkgo trees are large, diffusely branching trees with characteristic fan-shaped leaves **(Figure 27.21)** that turn a brilliant yellow in autumn. Nursery-propagated male trees often are planted in cities because they are resistant to insects, disease, and air pollutants. The female trees are equally pollution-resistant, but gardeners shy away from them—their fleshy seeds synthesize butyric acid, which has a notoriously foul odor.

Gnetophytes Include Simple Seed Plants with Intriguing Features

The phylum Gnetophyta contains three genera—*Gnetum, Ephedra,* and *Welwitschia*—that together include about 70 species. Moist, tropical regions are home to about 30 species of *Gnetum,* which includes both trees and leathery leafed vines (lianas). About 35 species of *Ephedra* grow in desert regions of the world **(Figure 27.22A–C).**

Of all the gymnosperms, *Welwitschia* strikes many people as the most bizarre. It grows in the hot deserts of south and west Africa. The bulk of the plant is a deep-reaching taproot, with the only exposed part a woody disk-shaped stem that bears cone-shaped strobili and leaves. The plant never produces more than two strap-shaped leaves, which split lengthwise repeatedly as the plant grows older, producing a rather scraggly pile **(Figure 27.22D).**

Although gnetophytes are structurally and functionally simpler than most other seed plants, recent studies of sexual reproduction mechanisms in *Gnetum* and *Ephedra* species uncovered a two-step process of fertilization—which is a hallmark of angiosperms, the most advanced seed plants. This discovery raised some provocative evolutionary questions, even leading some investigators to propose that ancient gnetophytes may have given rise to flowering plants. This hypothesis has since been contradicted by molecular findings, such as those arrived at by a research team at the Academia Sinica in Taiwan (China). When the team compared 65 nuclear rRNA sequences from ferns, gymnosperms, and angiosperms, their analysis supported the hypothesis that cycads and ginkgoes represent the earliest gymnosperm lineage, with a divergent lineage of gnetophytes and conifers arising later. The team found no molecular evidence for a link between the Gnetophyta and angiosperms.

Conifers Are the Most Common Gymnosperms

About 80% of all living gymnosperm species are members of the phylum **Coniferophyta,** or conifers ("cone-bearers"). Conifer sporophytes are longer-lived, and anatomically and morphologically more complex, than any sporophyte phase we have discussed so far. For example, each year most of the 550 conifer species produce a new layer of woody secondary tissue (which you can see illustrated in Figures 31.25 and 31.26); in some species, such as the sequoias and Douglas fir *(Pseudotsuga menziesii)* the girth of mature trees can easily exceed 6 m, and some individuals measure more than 15 m in diameter. Conifers also form woody reproductive cones, and most species are woody trees or shrubs with needlelike or scalelike leaves, which are anatomically adapted to arid habitats. For instance, needles have a thick cuticle, sunken stomata, and a fibrous epidermis, all traits that reduce the loss of water vapor.

Most conifers are evergreens. That is, although they shed old leaves, often in autumn, they retain enough leaves so that they still look "green," unlike deciduous species like maples, which shed *all* their leaves as winter approaches. Familiar conifer examples are the pines, spruces, firs, hemlocks, junipers, cypresses, and redwoods. Conifers produce pollen in clusters of small strobili and ovules in larger, woody ones. Both types of strobili are often referred to as cones. Seeds develop on the shelf-like scales of the female cones.

Pines and many other gymnosperms produce resins, a mix of organic compounds that are by-products of metabolism. Resin accumulates and flows in long resin ducts through the wood, inhibiting the activity of wood-boring insects and certain microbes. Pine resin extracts are the raw material of turpentine and (minus the volatile terpenes) the sticky rosin used to treat violin bows.

We know a great deal about the pine life cycle **(Figure 27.23),** so it is a convenient model for gymnosperms. All but 1 of the 93 pine species are trees (*Pinus mugo,* native to high elevations in Europe, is a shrub). The male cones (strobili) are relatively small and delicate, only about 1 cm long, and are borne on the lower branches. Each one consists of many small scales, which are specialized leaves (called sporophylls) attached to the cone's axis in a

A. ***Ephedra* plant**

Edward S. Ross

B. ***Ephedra* male cone**

William Ferguson

C. ***Ephedra* female cone**

Robert & Linda Mitchell Photography

D. ***Welwitschia* plant with female cones**

Fletcher & Baylis/Photo Researchers, Inc.

FIGURE 27.22
Gnetophytes. **(A)** Sporophyte of *Ephedra,* with close-ups of **(B)** its pollen-bearing cones and **(C)** a seed-bearing cone, which develop on separate plants. **(D)** Sporophyte of *Welwitschia mirabilis,* with seed-bearing cones.

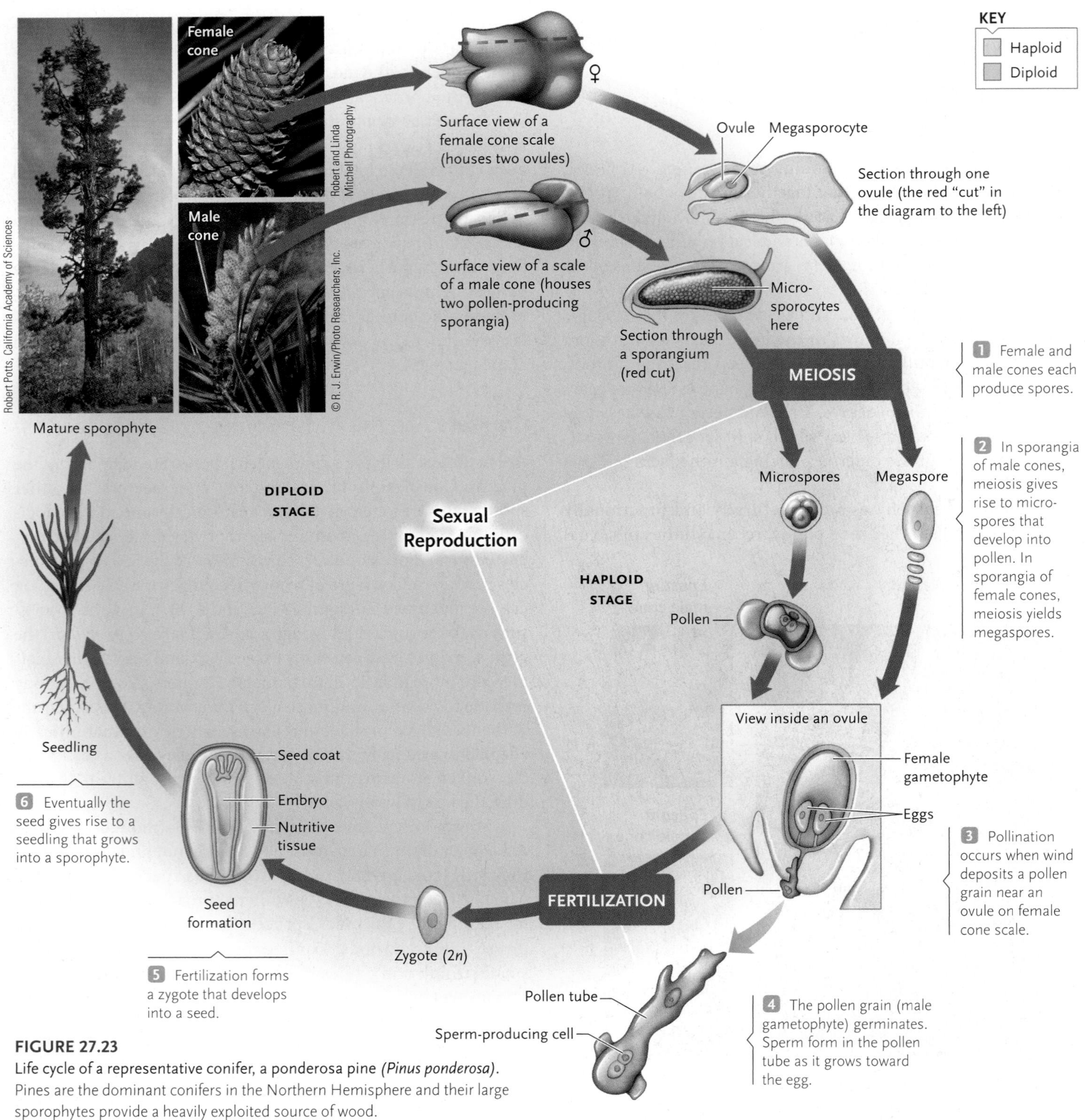
Female cone
Male cone
Robert Potts, California Academy of Sciences
Robert and Linda Mitchell Photography
© R. J. Erwin/Photo Researchers, Inc.
Mature sporophyte
Surface view of a female cone scale (houses two ovules)
Surface view of a scale of a male cone (houses two pollen-producing sporangia)
KEY
Haploid
Diploid
Ovule
Megasporocyte
Section through one ovule (the red "cut" in the diagram to the left)
Section through a sporangium (red cut)
Micro-sporocytes here
MEIOSIS
1 Female and male cones each produce spores.
2 In sporangia of male cones, meiosis gives rise to microspores that develop into pollen. In sporangia of female cones, meiosis yields megaspores.
Microspores
Megaspore
DIPLOID STAGE
Sexual Reproduction
HAPLOID STAGE
Pollen
View inside an ovule
Female gametophyte
Eggs
3 Pollination occurs when wind deposits a pollen grain near an ovule on female cone scale.
Pollen
FERTILIZATION
Seedling
6 Eventually the seed gives rise to a seedling that grows into a sporophyte.
Seed coat
Embryo
Nutritive tissue
Seed formation
Zygote (2n)
5 Fertilization forms a zygote that develops into a seed.
Pollen tube
Sperm-producing cell
4 The pollen grain (male gametophyte) germinates. Sperm form in the pollen tube as it grows toward the egg.

FIGURE 27.23
Life cycle of a representative conifer, a ponderosa pine *(Pinus ponderosa)*.
Pines are the dominant conifers in the Northern Hemisphere and their large sporophytes provide a heavily exploited source of wood.

spiral. Two sporangia, called *microsporangia,* develop on the underside of each scale. Inside the microsporangia, spore "mother cells" called microsporocytes undergo meiosis and give rise to haploid **microspores.** Each microspore then undergoes mitosis to develop into a winged pollen grain—an immature male gametophyte. At this stage the pollen grain consists of four cells, two that will degenerate and two that will function later in reproduction.

Young female cones develop higher in the tree, at the tips of upper branches. The cone scales bear ovules. Inside each ovule is a spore mother cell called a megasporocyte. Unlike microsporocytes, the megasporocyte in an ovule undergoes meiosis only when conditions are right and produces four haploid spores called **megaspores.** Only one megaspore survives, however, and it develops slowly, becoming a mature female gametophyte only when pollination is underway. In a pine, the process takes well over a year. The mature female gametophyte is a small oval mass of cells with several archegonia at one end, each containing an egg.

A. **Flowering plants in a desert**

B. **Alpine angiosperms**

C. **Triticale, a grass**

D. **A parasitic angiosperm**

FIGURE 27.24

Flowering plants. Diverse photosynthetic species are adapted to nearly all environments, ranging from **(A)** deserts to **(B)** snowlines of high mountains. **(C)** Triticale, a hybrid grain derived from parental stocks of wheat *(Triticum)* and rye *(Secale)*, is one example of the various grasses utilized by humans. **(D)** The parasitic flowering plant Indian pipe *(Monotropa uniflora)* having no chlorophyll of its own, obtains food by associating with mycorrhizae, which are in turn associated with the roots of photosynthetic plants.

STUDY BREAK 27.4

1. **What are the four major reproductive adaptations that evolved in gymnosperms?**
2. **What are the basic parts of a seed, and how is each one adaptive?**
3. **Describe some features that make conifers structurally more complex than other gymnosperms.**

Each spring, air currents lift vast numbers of pollen grains off male cones—by some estimates, a single pine may spew billions. The extravagant numbers assure that at least some pollen grains will land on female cones. The process is not as random as it might seem: studies have shown that the contours of female cones create air currents that can favor the "delivery" of pollen grains near the cone scales. After pollination, the pollen grain develops into a *pollen tube* that grows toward the female spore mother cell. As it does, sperm form in the tube and stimulate maturation of the female gametophyte and the production of eggs. When a pollen tube reaches an egg, the stage is set for fertilization, the formation of a zygote, and early development of the plant embryo. Often, fertilization occurs months to a year after pollination. Once an embryo forms, a pine seed—which, recall, includes the embryo, female gametophyte tissue, and seed coat—eventually is shed from the cone. The seed coat protects the embryo from drying out, and the female gametophyte tissue serves as its food reserve. This tissue makes up the bulk of a "pine nut."

27.5 Angiosperms: Flowering Plants

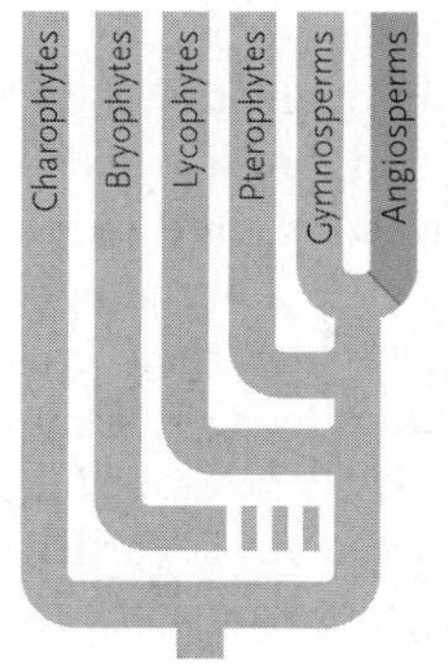

Charophytes
Bryophytes
Lycophytes
Pterophytes
Gymnosperms
Angiosperms

Of all plant phyla, the flowering plants, or angiosperms, are the most widespread today. At least 260,000 species are known (**Figure 27.24** shows a few examples), and researchers regularly discover new ones in previously unexplored regions of the tropics. The word angiosperm means "enclosed seed." It refers to the fact that a modified leaf called a **carpel** surrounds and protects the ovules and later, the seeds of angiosperms. Carpels are parts of flowers, reproductive structures that are a key defining feature of angiosperms. Another defining angiosperm feature is the fruit, which surrounds the angiosperm embryo and aids seed dispersal.

FIGURE 27.25
Fossil of *Archaefructus sinensis,* thought to have been an early flowering plant. The sketch shows what this small, possibly aquatic plant may have looked like.

***Archaefructus sinensis* fossil**

© David Dilcher, Florida Museum of Natural History/Paleobotany Laboratory

Sketch of *Archaefructus sinensis*

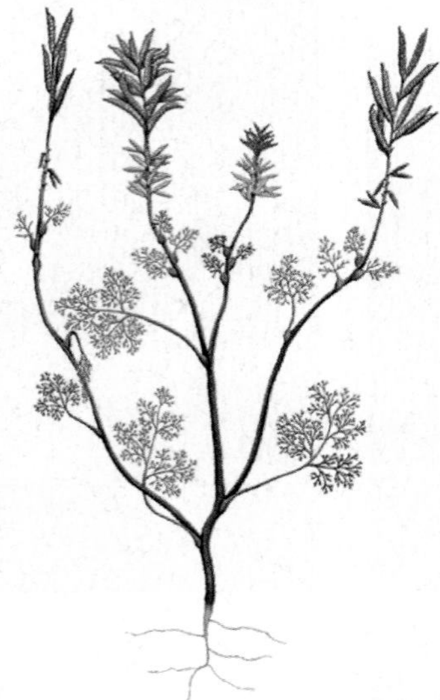

FIGURE 27.26
One proposed phylogenetic tree for flowering plants.

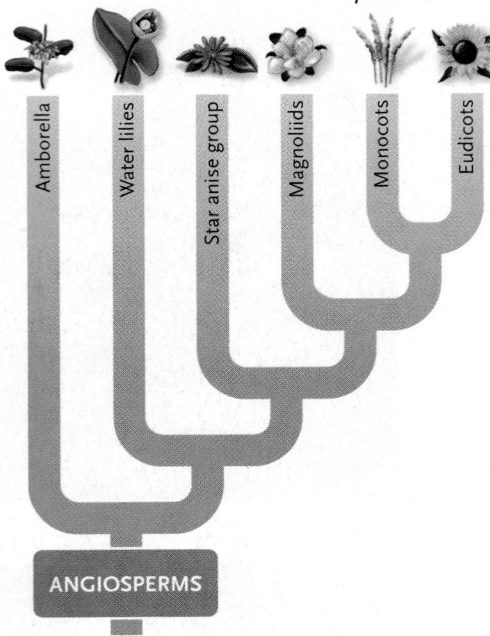
Majority of angiosperms
Amborella
Water lilies
Star anise group
Magnoliids
Monocots
Eudicots
ANGIOSPERMS

In addition to having flowers and fruits, angiosperms are the most ecologically diverse plants on Earth, growing on dry land and in wetlands, fresh water, and the seas. They range in size from tiny duckweeds about 1 mm long to towering *Eucalyptus* trees more than 100 m tall. Most are free-living photosynthesizers. Others lack chloroplasts and obtain nutrients through associations with fungi.

The Fossil Record Reveals Little about the Origin of Flowering Plants

The evolutionary origin of angiosperms has confounded plant biologists for well over a hundred years. Charles Darwin famously called it the "abominable mystery," because flowering plants seem to appear suddenly in the fossil record, without a fossil sequence that clearly links them to any other plant groups. The oldest well-documented fossil specimens date back about 125 million years to the early Cretaceous. Discovered in China, these remarkable fossils show complex and strikingly modern-looking plants that have leaves, stems, fruits, and seeds **(Figure 27.25).** Two species have been unearthed and have been assigned to the genus *Archaefructus,* representing a newly discovered, extinct angiosperm group.

The fossil record has yet to reveal obvious transitional organisms between flowering plants and either gymnosperms or seedless vascular plants. As with gymnosperms, attempts to reconstruct the earliest flowering plant lineages from morphological, developmental, and biochemical characteristics have produced several conflicting classifications and family trees. Some paleobotanists hypothesize that flowering plants arose in the Jurassic period (200–145 mya); others propose they arose earlier, in the Triassic (250–200 mya) from now-extinct gymnosperms or from seed ferns.

As the Mesozoic era ended and the modern Cenozoic era began, great extinctions occurred within both plant and animal kingdoms. Gymnosperms declined, and dinosaurs disappeared. Flowering plants, mammals, and social insects flourished, radiating into new environments. Today we live in what has been called "the age of flowering plants."

Angiosperms Are Subdivided into Several Clades, Including Monocots and Eudicots

Angiosperms are assigned to the phylum **Anthophyta,** a name that derives from the Greek *anthos,* meaning flower. **Figure 27.26** shows one current model of major clades within the phylum. The great majority of angiosperms are classified either as monocots or eudicots. **Monocots** are distinguished by the morphology of their embryos, which have a single seed leaf called a cotyledon ("cup-like hollow"). **Eudicots** ("true dicots"), which generally have two cotyledons, are set apart from other angiosperms by the structure of their pollen grains, which have three grooves **(Figure 27.27).** By contrast, the pollen of monocots and all other seed plants (including more than 8,500 species once lumped with eudicots under the term "dicots") have only a single groove. Paleobotanists use this clear structural difference not only to help establish the general type of plant that produced fossil pollen, but also what types of plants were present in fossil deposits of a particular age or geographic location.

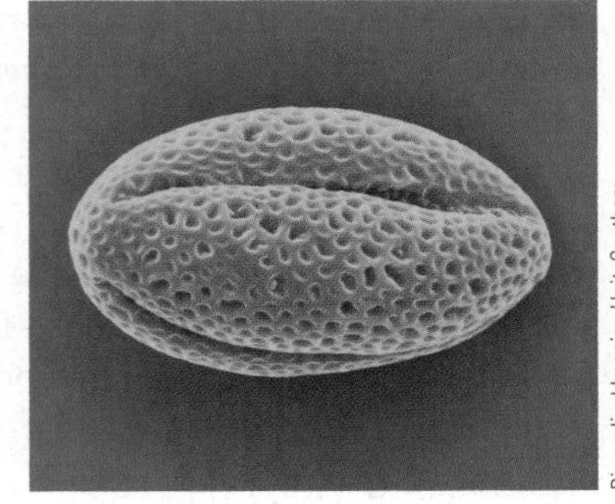

Biomedical Imaging Unit, Southam

FIGURE 27.27
Eudicot pollen grain. Eudicot pollen grains have three slitlike grooves, only two of which are visible here. Pollen made by all other seed plants, including monocots, have just one groove.

INSIGHTS FROM THE Molecular Revolution

The Powerful Genetic Toolkit for Studying Plant Evolution

Unlike most other eukaryotes, plants have three distinct sets of genes—in the cell nucleus, in mitochondria, and in chloroplasts. Chloroplast DNA, or cpDNA, has been especially useful for evolutionary studies, particularly the chloroplast *rbcL* gene. Mutations of the gene have occurred slowly, at about one-fourth to one-fifth the rate of genes in the nucleus. As a result, the DNA sequences of *rbcL* genes of different species diverge less than those of most other plant genes. Further, there are no introns—noncoding sequences—interrupting the coding sequence of the *rbcL* gene. Researchers can compare *rbcL* DNA from different species base by base, with no need to subtract introns. At the same time, the *rbcL* genes of different species are different enough to allow researchers to assemble evolutionary trees based on the degree of sequence variation.

Research Question

What were the major branch points in plant evolution?

Experiments

Studies using cpDNA have helped fuel several fundamental shifts in our understanding of branch points in plant evolution. For example, together with gene sequence data from nuclear DNA, analysis of *rbcL* genes provided evidence that charophyte green algae were the evolutionary forerunners of land plants. Similarly, in the late 1990s an international research team led by Yin-Long Qiu at the University of Massachusetts at Amherst correlated the loss of introns from two mitochondrial genes with the hypothesis that the first land plants were liverworts. Qiu and his colleagues carried out a genetic survey of more than 350 land plants representing all major lineages. They discovered that the noncoding sequences were present in all other bryophytes and all major lines of vascular plants, but were absent in liverworts, green algae, and all other eukaryotes. Analysis of *rbcL* sequences in various plant groups support the findings. Data from cpDNA and mtDNA analyses also underlie the hypothesis that, as land plants evolved, the ancient relatives of club mosses (lycophytes) were the forerunners of other vascular plants. Clearly, these varied molecular tools, and cpDNA in particular, are helping plant scientists explore evolutionary relationships across the whole spectrum of the Kingdom Plantae.

Conclusions

The evidence from cpDNA and mtDNA analysis points to liverworts as the first land plants, and ancient relatives of club mosses as the forerunners of other vascular plants.

Source: Y.-L. Qiu et al. 1998. The gain of three mitochondrial introns identifies liverworts as the earliest land plants. *Nature* 394:671–674.

Although most angiosperms can fairly easily be categorized as either monocots or eudicots, figuring out the appropriate classification for other angiosperms is an ongoing challenge and an extremely active area of plant research. The diagram in Figure 27.26 reflects a synthesis of evidence from both morphological and molecular studies, an approach examined in this chapter's *Insights from the Molecular Revolution*. Along with eudicots and monocots, botanists currently recognize four other clades **(Figure 27.28).** The **magnoliids,** a group that includes magnolias (see Figure 27.28A), laurels, and avocados are more closely related to monocots than to eudicots. Plants that are the sources of spices such as peppercorns, nutmeg, and cinnamon are also in the magnoliid clade. The other three clades are considered to be **basal angiosperms** representing the earliest branches of the flowering plant lineage. They include the star anise group (see Figure 27.28B), water lilies (see Figure 27.28C), and an intriguing ancient line represented by a single shrub, *Amborella trichopoda* (see Figure 27.28D). Found only in cloud forests of the South Pacific island of New Caledonia, *Amborella's* small white flowers and vascular system are structurally simpler than those of other angiosperms, and its female gametophyte differs as well. These morphological differences and a comparison of the nucleotide sequences of genes encoding the two angiosperm phytochromes (photoreceptors discussed in Chapter 35) make *Amborella* a candidate for the closest living relative of the first flowering plants **(Figure 27.29).**

Figure 27.30A gives some idea of the variety of living monocots, which include grasses, palms, lilies, and orchids. The world's major crop plants (wheat, corn, rice, rye, sugarcane, and barley) are domesticated grasses, and all are monocots. There are at least 60,000 species of monocots, including 10,000 grasses

FIGURE 27.28
Representatives of basal angiosperm clades. **(A)** Southern magnolia (*Magnolia grandiflora*), a magnoliid. **(B)** Star anise (*Illicium floridanum*). **(C)** Yellow pond lily (*Nuphar polysepala*). **(D)** *Amborella trichopoda.*

A. **Southern magnolia (*Magnolia grandiflora*), a magnoliid**

D. Harms/Peter Arnold, Inc.

B. **Star anise (*Illicium floridanum*)**

Rob and Ann Simpson/Visuals Unlimited

C. **Yellow pond lily (*Nuphar polysepala*), a water lily**

© Accent Alaska.com/Alamy

D. ***Amborella***

© Sangtae Kim, University of Florida

FIGURE 27.29 **Experimental Research**

Analyzing Phytochromes to Explore Angiosperm Evolution

Question: Did adaptive changes in light-sensitive pigments play a role in the early evolution of flowering plants?

Experiment: Many paleobotanists consider that the first angiosperms arose in the understory of moist land habitats dominated by large Mesozoic gymnosperms and ferns—a habitat not too unlike that where *Amborella* lives today. In this shaded environment, young plants that could gain a roothold despite the lack of bright light might have a distinct survival advantage. To investigate the possibility that strong selection pressure favoring adaptation to dim light played a part in angiosperm evolution, researchers looked at the genes coding for pigment proteins called *phytochromes* that allow plants to detect light of red and far-red wavelengths—wavelengths which predominate in dim light.

An activated phytochrome launches a signaling pathway that may stimulate or inhibit some aspect of plant growth. In seed plants, a small family of *PHY* genes encodes five main phytochromes, A through E, each with a distinct function. The nucleotide sequences of two of the genes, *PHYA* and *PHYC*, are about 50 percent identical. Plant scientists hypothesize that duplication of an ancestral *PHY* gene occurred before angiosperms arose. *PHYA* and *PHYC* evidently are the descendants of this duplicated ancestral gene. *PHYC* is sensitive to relatively bright light but apparently doesn't respond to dim light and *PHYA* is highly sensitive to dim light and is inactivated by bright light.

To investigate the hypothesis that strong selection pressure drove the divergence of these two genes in ancestral angiosperms, researchers obtained amino acid sequence data for key functional domains of the phytochrome proteins encoded by *PHYA* and *PHYC* in 45 plant species. Most species in the sample were angiosperms; several conifers represented the presumed ancestral gene. Analysis of the data focused both on the number of substitutions in the targeted phytochrome amino acid sequences, as well as on the biochemical effects of the substitutions. Among other information, the analysis produced a phylogenetic tree that displayed genetic divergences as branch points.

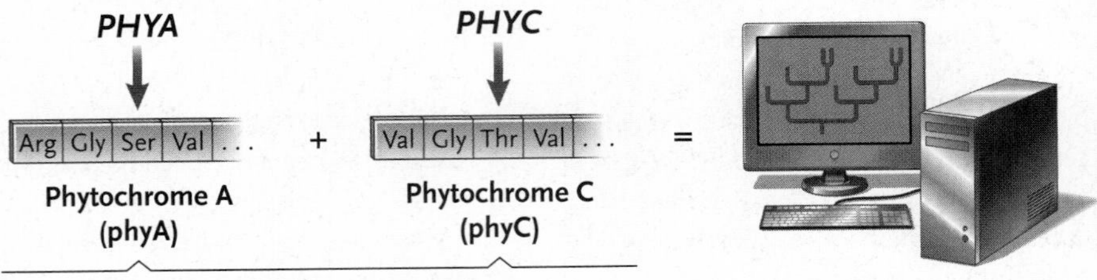
PHYA
Arg Gly Ser Val ...
Phytochrome A
(phyA)
+
PHYC
Val Gly Thr Val ...
Phytochrome C
(phyC)
=
Amino acid sequences of key functional domains of phyA and phyC proteins from conifer, multiple angiosperms.

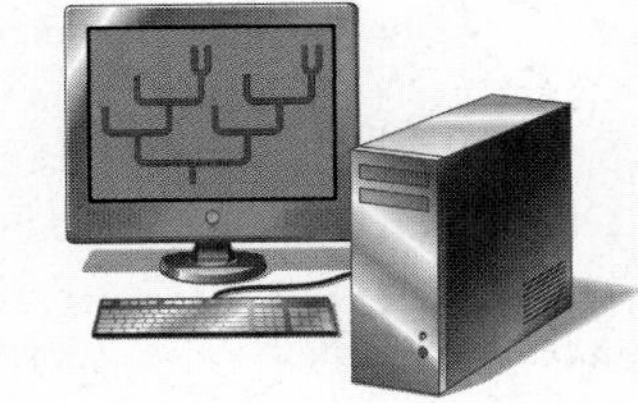
Computer analysis

Results: In the tree branch leading from the presumed ancestral *PHY* gene to *PHYA*, 32 amino acid substitutions occurred. Of these, 11 could be best interpreted as shifts associated with selection pressure. In the branch leading to *PHYC*, only seven substitutions occurred; four were best interpreted as associated with selection pressure.

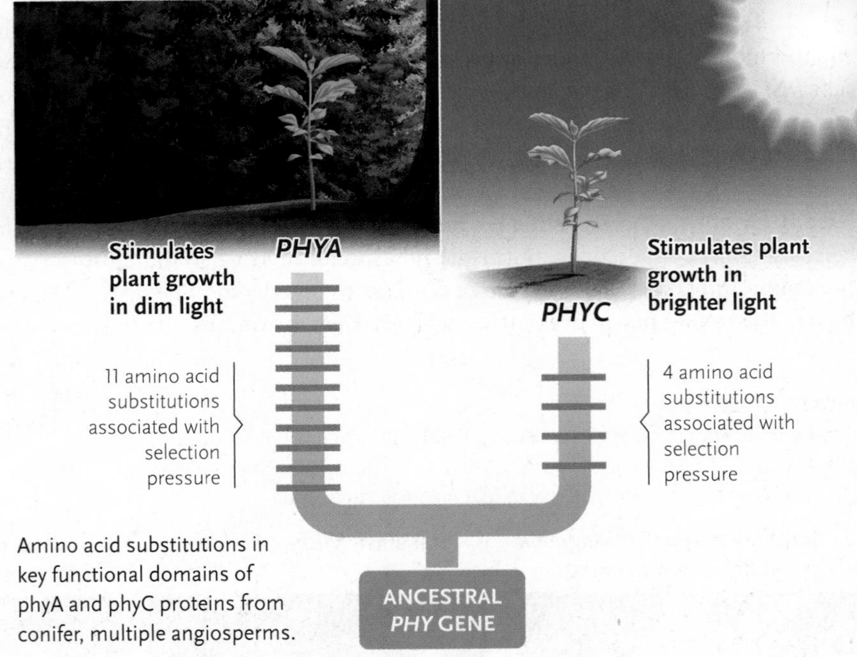
Stimulates plant growth in dim light
PHYA
PHYC
Stimulates plant growth in brighter light
11 amino acid substitutions associated with selection pressure
4 amino acid substitutions associated with selection pressure
Amino acid substitutions in key functional domains of phyA and phyC proteins from conifer, multiple angiosperms.
ANCESTRAL PHY GENE

Conclusion: These findings support the hypothesis that diverging molecular characteristics of *PHYA* and *PHYC* correlated with diverging functional features of the phytochromes the genes encode. Evidently, *PHYA* evolved under strong selection pressure. The availability of phytochrome phyA may have allowed early angiosperm seedlings to grow in mostly dim light conditions—an adaptive change that may have fostered the establishment of angiosperms in the shade of Mesozoic forests.

Source: Sarah Mathews, J. Gordon Burleigh, and Michael J. Donoghue. 2003. Adaptive evolution in the photosensory domain of phytochrome A in early angiosperms. *Molecular Biology and Evolution* 20:1087–1097.

A. **Representative monocots**

Wheat *(Triticum)*

Image copyright Florin C, 2010. Used under license from Shutterstock.com

Tulips *(Tulipa)*

Darrell Gulin/Getty Images

Eastern prairie fringed orchid *(Platanthera leucophaea)*

© Clint Farlinger/Alamy

FIGURE 27.30
Examples of monocots and eudicots.

B. **Representative eudicots**

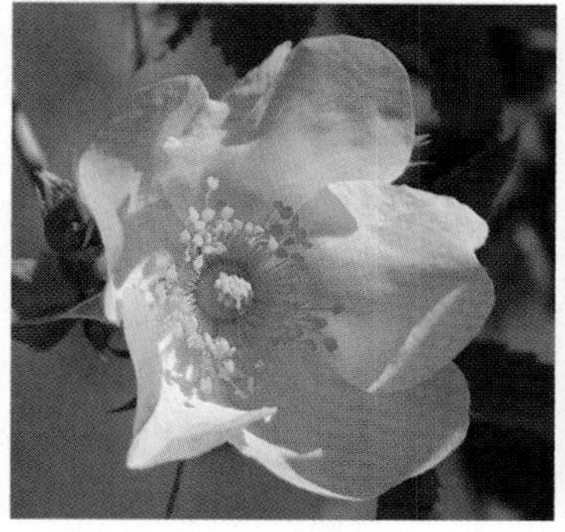

Rose *(Rosa)*

© Avico Ltd/Alamy

Yellow bush lupine *(Lupinus arboreus)*

Kennan Ward/Corbis

Cherry *(Prunus)*

Robert Essel NYC/Corbis

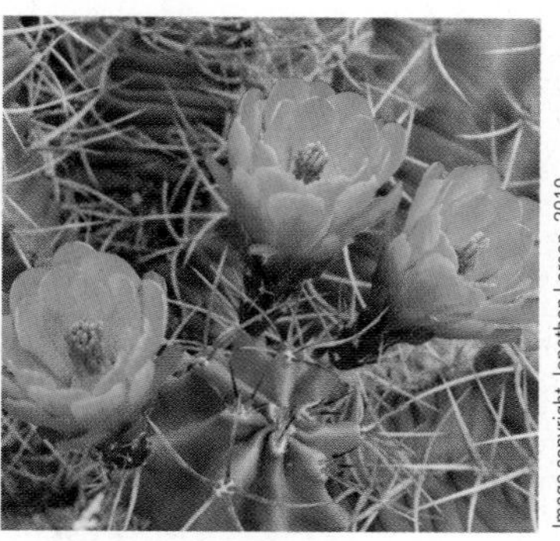

Claret cup cactus *(Echinocereus triglochidratus)*

Image copyright Jonathan Larsen, 2010. Used under license from Shutterstock.com

and 20,000 orchids. Eudicots are even more diverse, with nearly 200,000 species **(Figure 27.30B)**. They include flowering shrubs and trees, most nonwoody (herbaceous) plants, and cacti. **Figure 27.31** shows the life cycle of a monocot, a lily. The life cycle of a typical eudicot is described in detail in the next unit, which focuses on the structure and function of flowering plants.

Many Factors, Including Developmental Plasticity, Contributed to the Adaptive Success of Angiosperms

Throughout the whole panorama of land plant evolution, no group can match angiosperms in terms of one fundamental attribute: their *developmental plasticity.* No other plant group even comes close to equaling the sheer numbers of angiosperm species and the astonishing diversity of the sizes, shapes, habitats, growth habits, and nutrition modes of flowering plants. Whatever changes resulted in the evolution of the angiosperm lineage from one rooted in a gymnosperm heritage, they must have included a capacity to undergo (and survive) genetic changes that created new options for growth, development, and functioning. Here we consider just a small subset of basic shifts that helped distinguish angiosperms from gymnosperms.

MORE EFFICIENT TRANSPORT OF WATER AND NUTRIENTS Where gymnosperms have only one type of water-conducting cell (tracheids), angiosperms have an additional, more specialized type (called vessel elements). As a result, an angiosperm's xylem vessels move water more rapidly from roots to shoot parts. Also, modifications in angiosperm phloem tissue allow it to more efficiently transport sugars produced in photosynthesis through the plant body.

ENHANCED NUTRITION AND PHYSICAL PROTECTION FOR EMBRYOS Other changes in angiosperms made it more likely that reproduction would succeed. For example, a two-step *double fertilization* process in the seeds of flowering plants gives rise to both an embryo and a unique nutritive tissue (called endosperm) that nourishes the embryonic sporophyte. The ovule containing a female gametophyte is enclosed within an **ovary,** which develops from a carpel and shelters the ovule against desiccation and against attack by herbivores or pathogens. In turn, ovaries develop into the fruits that house angiosperm seeds. As noted earlier, a fruit not only protects seeds, but helps disperse them—for instance, when an animal eats a fruit, seeds may pass through the animal's gut none the worse for the journey and be released in a new location in the animal's feces. Above all, angiosperms have

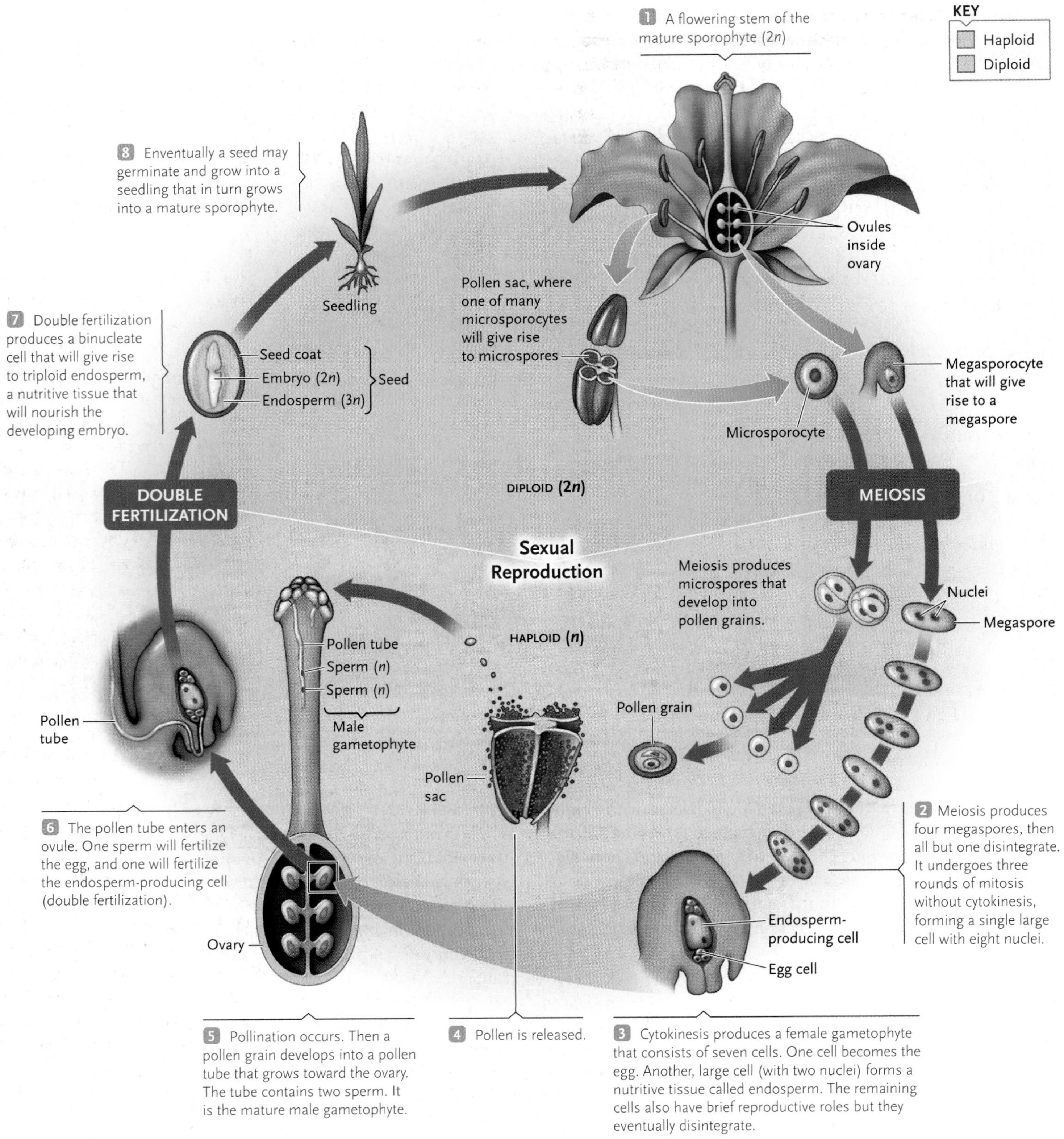

KEY
Haploid
Diploid
1 A flowering stem of the mature sporophyte (2n)
Ovules inside ovary
Pollen sac, where one of many microsporocytes will give rise to microspores
Megasporocyte that will give rise to a megaspore
Microsporocyte
MEIOSIS
DIPLOID (2n)
Sexual Reproduction
HAPLOID (n)
Meiosis produces microspores that develop into pollen grains.
Nuclei
Megaspore
Pollen grain
2 Meiosis produces four megaspores, then all but one disintegrate. It undergoes three rounds of mitosis without cytokinesis, forming a single large cell with eight nuclei.
Endosperm-producing cell
Egg cell
3 Cytokinesis produces a female gametophyte that consists of seven cells. One cell becomes the egg. Another, large cell (with two nuclei) forms a nutritive tissue called endosperm. The remaining cells also have brief reproductive roles but they eventually disintegrate.
Pollen sac
4 Pollen is released.
Pollen tube
Sperm (n)
Sperm (n)
Male gametophyte
Ovary
5 Pollination occurs. Then a pollen grain develops into a pollen tube that grows toward the ovary. The tube contains two sperm. It is the mature male gametophyte.
Pollen tube
6 The pollen tube enters an ovule. One sperm will fertilize the egg, and one will fertilize the endosperm-producing cell (double fertilization).
DOUBLE FERTILIZATION
Seed coat
Embryo (2n)
Endosperm (3n)
Seed
7 Double fertilization produces a binucleate cell that will give rise to triploid endosperm, a nutritive tissue that will nourish the developing embryo.
Seedling
8 Enventually a seed may germinate and grow into a seedling that in turn grows into a mature sporophyte.

FIGURE 27.31
Life cycle of a flowering plant. This diagram shows a generalized life cycle. Notice that it includes double fertilization: The male gametophyte delivers two sperm to an ovule. One sperm fertilizes the egg, forming the embryo, and the other fertilizes the endosperm-producing cell, which nourishes the embryo. Figure 34.5 shows the life cycle of a eudicot.

A. Bat pollinating a giant saguaro

Merlin d. Tuttle, Bat Conservation International

B. Hawk moth pollinating an orchid

© Photo by Marcel Lecoufle

C. Hummingbird visiting a hibiscus flower

Robert A. Tyrrell

D. Bee-attracting pattern of a marsh marigold

Visible light
UV light

Thomas Eisner/Cornell University

FIGURE 27.32

Coevolution of flowering plants and animal pollinators. The colors and configurations of some flowers, and the production of nectar or odors, have coevolved with specific animal pollinators. **(A)** At night, nectar-feeding bats sip nectar from flowers of the giant saguaro *(Carnegia gigantea)*, transferring pollen from flower to flower in the process. **(B)** The hawkmoth *Xanthopan morgani praedicta* has a proboscis long enough to reach nectar at the base of the equally long floral spur of the orchid *Angraecum sesquipedale.* **(C)** A Bahama woodstar hummingbird *(Calliphlox evelynae)* sipping nectar from a hibiscus blossom *(Hibiscus)*. The long narrow bill of hummingbirds coevolved with long, narrow floral tubes. **(D)** Under ultraviolet light, the bee-attracting pattern of a gold-petaled marsh marigold becomes visible to human eyes.

flowers, the unique reproductive organs that you will read much more about in the next unit.

COEVOLUTION WITH ANIMAL POLLINATORS The evolutionary success of angiosperms correlates not only with the adaptations just described, but also with efficient mechanisms of transferring pollen to female reproductive parts. While a conifer depends on air currents to disperse its pollen, angiosperms coevolved with pollinators—insects, bats, birds, and other animals that withdraw pollen from male floral structures (often while obtaining nectar) and inadvertently transfer it to female reproductive parts. **Coevolution** occurs when two or more species interact closely in the same ecological setting. A heritable change in one species affects selection pressure operating between them, and the other species evolves as well. Over time, plants that came to have distinctive flowers, scents, and sugary nectar coevolved with animals that could take advantage of the rich food source.

In general, a flower's reproductive parts are positioned so that visiting pollinators will brush against them. In addition, many floral features correlate with specific pollinators. For example, reproductive parts may be located above nectar-filled floral tubes the same length as the feeding structure of a preferred pollinator. Nectar-sipping bats **(Figure 27.32A)** and moths forage by night. They pollinate intensely sweet-smelling flowers with white or pale petals that are more visible than colored petals in the dark. Long, thin mouthparts of moths and butterflies reach nectar in narrow floral tubes or flora spurs. The Madagascar hawkmoth uncoils a mouthpart the same length—an astonishing 22 cm—as a narrow floral spur of an orchid it pollinates, *Angraecum sesquipedale* **(Figure 27.32B).** Red and yellow flowers attract birds **(Figure 27.32C),** which have good daytime vision but a poor sense of smell. Hence bird-pollinated plants do not squander metabolic resources to make fragrances. By contrast, flowers of species that are pollinated by beetles or flies may smell like rotten meat, dung, or decaying matter. Daisies and other fragrant flowers with distinctive patterns, shapes, and red or orange components attract butterflies, which forage by day.

Bees see ultraviolet light and visit flowers with sweet odors and parts that appear to humans as yellow, blue, or purple **(Figure 27.32D).** Produced by pigments that absorb ultraviolet light, the colors form patterns called "nectar guides" that attract bees—which may pick up or "drop off" pollen during the visit. Here, as in our other examples, flowers contribute to the reproductive success of plants that bear them.

Current Research Focuses on Genes Underlying Transitions in Plant Traits

Improvements in the ability of plant scientists to manipulate, analyze, and compare modern plant genomes, coupled with advances in the analysis of fossil plants, are having a profound impact on our understanding of the evolution of flowering plants. A case in point is research on the gene *LFY*, which encodes the regulatory protein LEAFY (Chapter 16 discusses regulatory proteins in detail). The LEAFY protein typically controls expression of several genes by binding to the genes' control sequences. All land plants carry the *LFY* gene, but its effects on phenotype vary markedly in different plant groups. In mosses, which arose almost 400 million years ago, the LEAFY protein regulates growth throughout the plant. In ferns and gymnosperms, which arose later, LEAFY controls growth in a subset of tissues. In angiosperms, LEAFY regulates gene expression only in the particular type of meristem tissue that gives rise to flowers (a topic of Chapter 34). Curious about the evolutionary shift from a general to a specific effect, Alexis Maizel and his team at the Max Planck Institute for Developmental Biology in Germany compared *LFY* sequences and their corresponding proteins in 14 species, including a moss, ferns, gymnosperms, and the angiosperms *Arabidopsis* (thale cress) and snapdragon. Remarkably, they discovered that the evolutionary honing of the effects of the LEAFY protein correlated with only a handful of changes in the base sequence of the *LFY* gene. Each change affected how—or if—the LEAFY protein regulated the expression of a given gene. Over time, LEAFY took on its highly specific, crucial role in angiosperms, helping to direct the developmental events that produce flowers.

Today some of the most exciting research in all of biology involves studies exploring the connections between genetic changes and key evolutionary transitions in plant form and functioning. As the genes of many more plant species are sequenced and correlated with evidence from comparative morphology and the fossil record, we can expect a steady stream of new insights about the evolutionary journey of all major plant lineages. In Chapter 28 a very different group of organisms, the fungi, takes center stage. Although many fungal species seem superficially plantlike, biologists today are avidly exploring evolutionary links between fungi and animals.

STUDY BREAK 27.5 <

1. **How has the relative lack of fossils of early angiosperms affected our understanding of this group?**
2. **Describe two basic features that distinguish monocots from eudicots, and give some examples of species in each clade.**
3. **List at least three adaptations that have contributed to the evolutionary success of angiosperms as a group.**

UNANSWERED QUESTIONS

Where did flowering plants come from?

Flowers are a unique feature of angiosperms, yet botanists still understand little of their evolutionary origin. When flowering plants appear in the Cretaceous fossil record, they appear suddenly and diversify immediately, a situation Darwin famously referred to as an "abominable mystery." What did the first flower look like, and what did it come from?

As described in this chapter, recent molecular analyses have converged on *Amborella trichopoda* as the living representative of the most ancient lineage in the angiosperm family tree. *Amborella* flowers have some features considered evolutionarily primitive, such as petals and sepals that are not distinctly different in form. But *Amborella* also has some features thought to have evolved much more recently, such as single-sex flowers that have only male or only female reproductive parts. Should we be surprised to find both primitive and advanced traits in this ancient lineage? Not at all. *Amborella* has existed on Earth for millions of years, and its flowers may have evolved new features over that time.

The puzzle of where angiosperms came from and what the first flowering plants looked like has not been solved by fossil studies, either. This chapter discusses the fossil species *Archaefructus*, which dates from the Cretaceous (145 mya) and is thought by some to be the oldest known fossil flower. It consists of an elongated axis with stamens (male reproductive structures) toward the base and carpels (female reproductive structures) toward the apex, and no sepals and petals. This elongated flower is unlike the flowers of any modern angiosperm, and its structure suggests that the earliest flowers may have been very different from what we see today. However, some paleobotanists have reinterpreted the *Archaefructus* "flower" as an inflorescence (a flower cluster), with male flowers at the base and female flowers toward the apex. A recently described third species, *A. eoflora*, has a bisexual flower between the male and female structures, prompting some workers to suggest that *Archaefructus* reproductive structures are a mix between flowers and inflorescences. Recent analyses suggest that this enigmatic plant is actually close to the water lilies (Nymphaeales) or buttercups (Ranunculaceae), and not representative of the oldest flowers. This debate continues.

Plant scientists also disagree about the ancestors of angiosperms. Some gnetophytes—gymnosperms that include *Welwitschia* and *Ephedra* species (refer to Figure 27.23)—have features similar to angiosperms. Botanists long speculated that the two groups might be closely related, with a common ancestor that had flowerlike features. However, recent analyses based on DNA sequence data suggest that gnetophytes are not closely related to angiosperms after all. There are also fossil gymnosperm taxa with features that might be forerunners of carpels or other flower parts, but paleobotanists disagree on these interpretations as well. Thus examinations of fossils and extant species have yet to resolve key questions about the evolution of angiosperms.

As much insight as these molecular studies give us, they have not brought us any closer to understanding the fundamental question of how angiosperms arose. Additional fossil data may help provide the answer, but it is also possible that the earliest angiosperms, or their direct ancestors, lived in habitats where fossils do not readily form. Additional molecular data may deepen our understanding of how changes in genes produced the first flower, but molecular data based on contemporary species will not help decipher what the first angiosperm and the first flower looked like. Thus, it is possible that the abominable mystery will live on.

Think Critically

1. Why might early flowers have looked different from contemporary flowers?
2. Many groups of early angiosperms went extinct. What are some reasons that modern plants might be more successful?

Amy Litt

Amy Litt is Director of Plant Genomics and Cullman Curator at the New York Botanical Garden, where she also earned her Ph.D. Her main interests lie in the evolution of plant form and how changes in gene function during the course of plant evolution have produced novel plant forms and functions—particularly new flower and fruit morphologies. Learn more about her work at http://sciweb.nybg.org/science2/Profile_106.asp.

REVIEW KEY CONCEPTS

Go to **CENGAGENOW** at www.cengage.com/login to access quizzing, animations, exercises, articles, and personalized homework help.

27.1 The Transition to Life on Land

- Plants are thought to have evolved from green algae between 425 and 490 million years ago (Figure 27.2).
- Key adaptations in early land plants include an outer cuticle that helps prevent desiccation, tissues containing lignified cells, spores protected by a wall containing sporopollenin, multicellular chambers that protect developing gametes, and an embryo sheltered inside a parent plant (Figure 27.3).
- Land plant evolution also included the development of vascular tissues, root and shoot systems, including stems containing lignified cells and leaves having stomata; a shift from dominance by a long-lived, larger haploid gametophyte to dominance of a long-lived, larger diploid sporophyte, and a shift from homospory to heterospory with separate male and female gametophytes (Figures 27.5–27.8).
- Male gametophytes (pollen) became specialized for dispersal without liquid water. Female gametophytes became specialized for enclosing embryo sporophytes in seeds.

Animation: Milestones in plant evolution

Animation: Haploid to diploid dominance

Animation: Evolutionary tree for plants

Animation: The importance of alternation of generations

27.2 Bryophytes: Nonvascular Land Plants

- Existing nonvascular land plants, or bryophytes, include the liverworts (Hepatophyta), hornworts (Anthocerophyta), and mosses (Bryophyta). Liverworts may have been the first land plants.
- Bryophytes release spores and produce flagellated sperm that swim through free water to reach eggs. They lack a vascular system; true roots, stems, and leaves; and lignified tissue. A larger, dominant gametophyte (n) phase alternates with a small, fleeting sporophyte ($2n$) phase (Figures 27.9–27.12).

Animation: *Marchantia*, a liverwort

Animation: Moss life cycle

27.3 Seedless Vascular Plants

- Existing seedless vascular land plants include the lycophytes (club mosses), whisk ferns, horsetails, and ferns. They release spores and have swimming sperm, but have well-developed vascular tissues. The sporophyte is the larger, longer-lived stage of the life cycle and develops independently of the small gametophyte.
- Club mosses (Lycophyta) have sporangia clustered at the bases of specialized leaves called sporophylls. Clusters of sporophylls form a strobilus (cone). Haploid spores germinate to form small gametophytes. The life cycle is similar in ferns, whisk ferns, and horsetails (Pterophyta) (Figures 27.13–27.17).
- Ferns are the most diverse group of seedless vascular plants. Most species do not have aboveground stems, only leaves that arise from nodes along an underground rhizome. Fern leaves typically have well-developed stomata, and the vascular system consists of bundles, each with xylem surrounded by phloem. Sporangia on sporophylls (fronds) release spores that develop into gametophytes. Sexual reproduction produces a much larger, long-lived sporophyte.

Animation: Seedless vascular plants

Animation: Fern life cycle

27.4 Gymnosperms: The First Seed Plants

- Gymnosperms (conifers and their relatives), together with angiosperms (flowering plants), are the seed-bearing vascular plants. During the Mesozoic era (248–65 mya), gymnosperms were the dominant land plants. Major reproductive adaptations include pollination, the ovule, and the seed. An ovule is a sporangium in which a large spore gives rise to a female gametophyte, which is attached to and protected by the sporophyte. Male gametophytes arise from a smaller spore type. Pollination and fertilization do not require water. Pollination takes place via air currents or animal pollinators. Fertilization occurs with the ovule and ultimately results in a seed. Gymnosperm seeds mainly protect, nourish, and help disperse embryonic sporophytes (Figures 27.18–27.23).

Animation: *Pinus* cones

Animation: Pine life cycle

27.5 Angiosperms: Flowering Plants

- Angiosperms (Anthophyta) have dominated the land for more than 100 million years and currently are the most diverse plant group. The two main angiosperm clades are monocots and eudicots. Other clades are represented by magnolias and their relatives (magnoliids), water lilies, the star anise group, and *Amborella,* a single species thought to be the most basal living angiosperm (Figures 27.30 and 27.31).
- Angiosperms have the most efficient plant vascular system (xylem and phloem). Reproductive adaptations include a protective ovary around the ovule, endosperm, fruits that aid seed dispersal, the complex organs called flowers, and the coevolution of flower characteristics with the structural and/or physiological characteristics of animal pollinators (Figures 27.24–27.32).

Animation: Flower parts

Animation: Monocot life cycle

UNDERSTAND AND APPLY

Test Your Knowledge

1. Which of the following is *not* an evolutionary trend among plants?
   a. developing vascular tissues
   b. becoming seedless
   c. having a dominant diploid generation
   d. producing nonmotile gametes
   e. producing two types of spores
2. As plants made the evolutionary transition to a terrestrial existence, they benefited from adaptations that:
   a. increased the motility of their gametes on dry land.
   b. flattened the plant body to expose it to the sun.
   c. reduced the number and distribution of roots to prevent drying.
   d. provided mechanisms for gaining access to nutrients in soil.
   e. allowed stems and leaves to absorb water from the atmosphere.
3. Land plants no longer required water as a medium for reproduction with the evolution of:
   a. fruits and roots.
   b. flowers and leaves.
   c. cell walls and rhizoids.
   d. lignified stems.
   e. seeds and pollen.
4. Which is the correct matching of phylum and plant group?
   a. Anthophyta: pines
   b. Bryophyta: gnetophytes
   c. Coniferophyta: angiosperms
   d. Hepatophyta: cycads
   e. Pterophyta: horsetails
5. A homeowner noticed moss growing between bricks on his patio. Closer examination revealed tiny brown stalks with cuplike tops emerging from green leaflets. These brown structures were:
   a. the sporophyte generation.
   b. the gametophyte generation.
   c. elongated haploid reproductive cells.
   d. archegonia.
   e. antheridia.
6. Horsetails are most closely related to:
   a. mosses and whisk ferns.
   b. liverworts and hornworts.
   c. cycads and ginkgos.
   d. club mosses and ferns.
   e. gnetophytes and gymnosperms.
7. Which feature(s) do ferns share with all other land plants?
   a. sporophyte and gametophyte life cycle stages
   b. gametophytes supported by a thallus
   c. dispersal of spores from a sorus
   d. asexual reproduction by way of gemmae
   e. water uptake by means of rhizoids
8. The evolution of true roots is first seen in:
   a. liverworts.
   b. seedless vascular plants.
   c. mosses.
   d. flowering plants.
   e. conifers.
9. Based solely on numbers of species, the most successful plants today are:
   a. angiosperms.
   b. ferns.
   c. gymnosperms.
   d. mosses.
   e. the bryophytes as a group.
10. Angiosperms and gymnosperms share the following characteristic(s):
    a. pollination by means of water.
    b. seeds protected within an ovary.
    c. embryonic cotyledons.
    d. a dominant sporophyte generation.
    e. a seasonal loss of all leaves.

Discuss the Concepts

1. Suggest adjustments in the angiosperm life cycle that would better suit plants to some future world where environments were generally hotter and more arid. Do the same for a colder and wetter environment.
2. Working in the field, you discover a fossil of a previously undescribed plant species. The specimen is small and may not be complete; the parts you have do not include any floral organs. What sorts of observations would you need in order to classify the fossil as a seedless vascular plant with reasonable accuracy? What evidence would you need in order to distinguish between a fossil lycopod and a fern?
3. Modern humans emerged about 100,000 years ago. How accurate is it to state that our species has lived in the Age of Wood? Explain.

4. Compare the size, anatomical complexity, and degree of independence of a moss gametophyte, a fern gametophyte, a Douglas fir female gametophyte, and a dogwood female gametophyte. Which one is the most protected from the external environment? Which trends in plant evolution does your work on this question bring to mind?

Design an Experiment

You are studying mechanisms that control the development of flowers, and your research to date has focused on eudicots. A colleague has suggested that you broaden your analysis to include representative basal angiosperms. Outline the rationale for this expanded approach and indicate which additional species or group(s) you plan to include. Discuss the type(s) of data you plan to gather and why you feel the information will make your study more complete.

Interpret the Data

Green plants, green algae, red algae, and microscopic freshwater algae called glaucophytes are all primary photosynthetic eukaryotes—they have plastids, including chloroplasts that carry out photosynthesis. Biologists have long accepted the hypothesis that plastids arose by way of endosymbiosis of a cyanobacterium during the evolution of lineages leading to primary photosynthetic eukaryotes (see Sections 24.3 and 26.2). Until recently, however, there was no unequivocal molecular evidence to support this assumption. If the assumption is correct, then a phylogenetic tree based on molecular data should support the prediction that all primary photosynthetic eukaryotes are descended from a common ancestor. Naiara Rodríguez-Ezpeleta at the University of Montreal led a team that tested this prediction. They already knew that proteins associated with plastids in cyanobacteria and photosynthetic eukaryotes share some amino acid sequences. In the experiment of interest here, they sequenced the amino acids in 143 proteins encoded by nuclear genes of species representing all the primary photosynthetic groups. They used a computerized statistical method called bootstrapping to analyze the large number of different data sets from the selected species. Basically, bootstrapping produced a series of phylogenetic trees based on comparisons of the amino acid sequences in the target proteins from each species. It then compared the trees to see where overlaps occurred, and generated a summary tree like the one in **Figure A.**

1. Does Figure A support or contradict the hypothesis that a single endosymbiosis event correlated with the evolution of organisms containing plastids?
2. Do the data support the prediction that glaucophytes and green plants are sister groups—that is, that they are more closely related to each other than to other groups in the phylogeny?

Source: N. Rodriguez-Ezpeleta et al. 2005. Monophyly of primary photosynthetic eukaryotes: green plants, red algae, and glaucophytes. *Current Biology* 15:1325–1330.

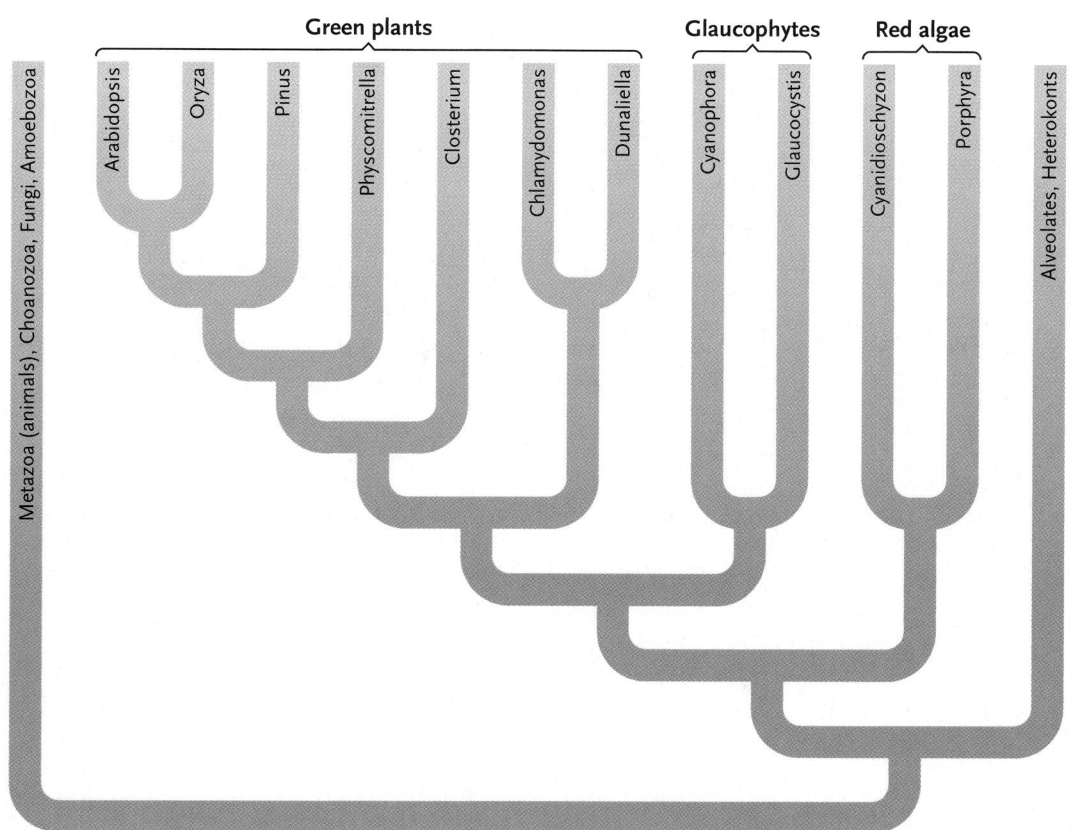

Green plants
Glaucophytes
Red algae
Metazoa (animals), Choanozoa, Fungi, Amoebozoa
Arabidopsis
Oryza
Pinus
Physcomitrella
Closterium
Chlamydomonas
Dunaliella
Cyanophora
Glaucocystis
Cyanidioschyzon
Porphyra
Alveolates, Heterokonts

FIGURE A
Phylogeny based on a dataset of 143 nuclear encoded proteins.

Apply Evolutionary Thinking

Evolutionary biologist Spencer C. H. Barrett has written that the reproductive organs of angiosperms are more varied than the equivalent structures of any other group of organisms. Explain why you agree or disagree with his view.

Express Your Opinion

Demand for paper is a big factor in deforestation. However, using recycled paper can add to the cost of a product. Are you willing to pay more for papers, books, and magazines that are printed on recycled paper? Go to www.cengage.com/login to investigate both sides of the issue and then vote.

Mushroom-forming fungi of the genus *Mycena*, which are commonly found on decaying wood.

Fritz Polking/Peter Arnold, Inc

28

STUDY OUTLINE

Fungi

Why It Matters. . . In most natural environments, decay is everywhere—rotting leaves, moldering branches, perhaps the disintegrating carcass of an insect or a small mammal. Gradually, this organic matter is broken down and its elements are recycled in a long-term process that has a huge impact on terrestrial ecosystems. For example, each year the cycling of nutrients returns at least 85 billion tons of carbon, in the form of carbon dioxide, to the atmosphere. Chief among the recyclers are the curious and, to many eyes, beautiful organisms of the **Kingdom Fungi**—about 70,000 described species of molds, mushroom-forming fungi, yeasts, and their relatives **(Figure 28.1)**, and an estimated 1.5 million more that are yet to be described.

Fungi are eukaryotes, most are multicellular, and all are heterotrophs, obtaining their nutrients by breaking down organic molecules that other organisms have synthesized. Molecular evidence suggests that fungi were present on land at least 500 million years ago, and possibly much earlier. In the course of the intervening millennia, evolution equipped fungi with a remarkable ability to break down organic matter, ranging from living and dead organisms and organic wastes to your groceries, clothing, paper and wood, even photographic film. Along with heterotrophic bacteria, they have become Earth's premier decomposers.

Fungi collectively also are a major cause of plant diseases, and a host of species causes disease in humans and other animals. Some even produce carcinogenic toxins. On the other hand, 90% of plants obtain needed minerals by way of a symbiotic relationship with a fungus. Humans have harnessed the metabolic activities of certain fungi to obtain substances ranging from cheeses and wine to therapeutic drugs such as penicillin and the immunosuppressant cyclosporin. And, as you know from previous chapters, species such as the yeast *Saccharomyces cerevisiae* and the mold *Neurospora crassa* are pivotal model organisms in studies of DNA structure and function, and the

FIGURE 28.1 Examples of fungi that hint at the rich diversity within the Kingdom Fungi.

Sulfur shelf fungus *Laetiporus*

Robert C. Simpson/Nature Stock

Big laughing mushroom *Gymnopilus*

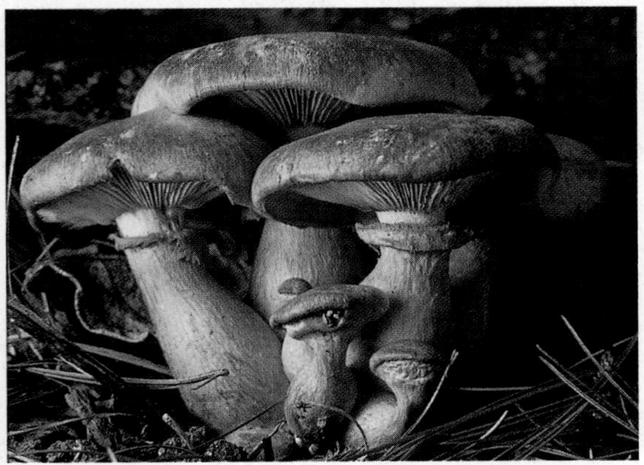

Robert C. Simpson/Nature Stock

Baker's yeast cells *Saccharomyces cerevisiae*

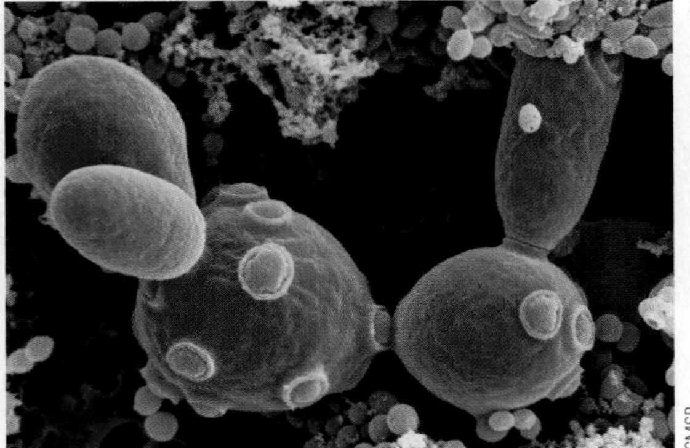

CMSP

yeast has also been important in the development of genetic engineering methods.

Despite their profound impact on ecosystems and other life forms, most of us have only a passing acquaintance with the fungi—perhaps limited to the mushrooms on our pizza or the invisible but annoying types that cause skin infections like athlete's foot. This chapter provides you with an introduction to mycology, the study of fungi (*mykes* = mushroom, *logia* = study). We begin with general characteristics of this kingdom and then discuss its major divisions. <

28.1 General Characteristics of Fungi

We begin our survey of fungi by examining how fungi differ from other forms of life, how fungi obtain nutrients, and the adaptations for reproduction and growth that enable fungi to spread far and wide through the environment.

Fungi May Be Single-Celled or Multicellular

Two basic body forms, single-celled and multicellular, emerged as the lineages of fungi evolved. Some fungi are single cells, a form called **yeast,** while others are multicellular organisms made up of threadlike filaments. Still others alternate between yeast and multicellular forms at different stages of the life cycle. Regardless, a rigid wall usually surrounds the plasma membrane of fungal cells. Generally the polysaccharide **chitin** provides this rigidity, the same function it serves in the external skeletons of insects and other arthropods.

A multicellular fungus exploits food sources by way of a cottony mesh of slender filaments that branch repeatedly as they grow over or into organic matter. Each filament is a **hypha** (*hyphe* = web) (plural, *hyphae*), and the combined mass of hyphae is called a **mycelium** (plural, *mycelia*). Hyphae generally are tubular **(Figure 28.2A),** and in most multicellular fungi they are partitioned by cross walls called **septa** (*saeptum* = partition, or fence) (singular, *septum*). The septa create cell-like compartments that contain organelles.

The unusual features of hyphae have led many mycologists to question whether "multicellular" is really an accurate description for most fungal architecture. For instance, depending on the species, hyphal cells may have more than one nucleus, and septa have pores that permit nuclei and other organelles to move between hyphal cells. These passages also allow cytoplasm to extend from one hyphal cell to the next, throughout the whole mycelium. By a mechanism called **cytoplasmic streaming,** cytoplasm containing nutrients can flow unimpeded through the hyphae, from food-absorbing parts of the fungal body to reproductive structures or other nonabsorptive parts. Fungi with cell-like compartments are still commonly referred to as "multicellular" but with the understanding that the term has a special meaning for these organisms.

A. **Fungal hyphae**

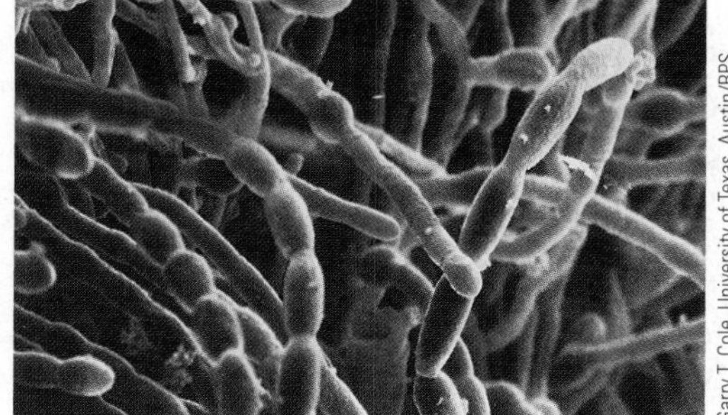

Garry T. Cole, University of Texas, Austin/BPS

B. **Multicellular fungus**

Sporocarp
Mycelium

C.

Gregory M. Filip/USDA

FIGURE 28.2
Fungal mycelia and rhizomorphs. (A) Micrograph of fungal hyphae. **(B)** Sketch of the mycelium of a mushroom forming fungus, which consists of branching septate hyphae. **(C)** Tangled rhizomorphs of *Armillaria ostoyae*, the honey mushroom that was uncovered by biologists working in an Oregon forest. Rhizomorphs are modified hyphae which typically extend over roots or through crevices between a tree's bark and the underlying wood. They are features of some species in the phylum Basidiomycota described in Section 28.2.

A multicellular fungus grows larger as its hyphae elongate and branch, and, in some cases, aggregate to form reproductive and other specialized structures such as mushrooms **(Figure 28.2B)**. Each hypha elongates at its tip as new wall polymers (delivered by vesicles) are incorporated and additional cytoplasm, including organelles, is synthesized. As the new hyphae elongate, then branch themselves, an extensive mycelium can form quickly. Each forming branch fills with cytoplasm that includes new nuclei produced by mitosis. Although the rapid branching of hyphae is what allows multicellular fungi to grow aggressively—sometimes increasing in mass many times over within a few days—researchers have only recently gained the tools to explore the mechanisms that underlie this phenomenon. Studies spurred by the sequencing of the genome of *Neurospora crassa* suggest that multiple steps involving a variety of genes and their interacting protein products determine where and when a new branch arises. Given that the rapid growth of fungal mycelia has such a tremendous impact in ecosystems and in the progression of fungal disease in plants and animals, this topic is a significant area of mycological research.

Beyond their role in nutrient transport, aggregations of hyphae are the structural foundation for all other parts that arise as a multicellular fungus develops. For example, in many fungi a subset of hyphae interweave tightly, becoming prominent reproductive structures sometimes called *sporocarps*. Grocery store mushrooms are examples. But while a mushroom or some analogous structure may be the most conspicuous part of a given fungus, it usually represents only a small fraction of the organism's total mass. The rest, often invisible to the unaided eye, penetrates the food source the fungus is slowly digesting. In some fungi, modified hyphae form rootlike variations on mycelial threads called *rhizomorphs*. These cordlike structures both anchor the fungus to its existing substrate and provide a means for rapidly colonizing new food sources. Dense tangles of the strands can so damage susceptible tree roots that large patches of forest trees die. In a famous example, U.S. Forest Service scientists found that rhizomorphs of a single individual of the parasitic honey mushroom *Armillaria ostoyae* **(Figure 28.2C)** cover an area equivalent to 1,665 football fields in an eastern Oregon forest. By one estimate, the mass of tangled strands extends an average of 1 m deep and nearly 6,000 m across, making the fungus perhaps one of the largest organisms on Earth.

Fungi Obtain Nutrients by Extracellular Digestion and Absorption

Some major selection pressures have shaped the adaptations by which fungi obtain nutrients. As heterotrophs, fungi must secure nutrients by breaking down organic substances formed by other organisms. Nearly all fungi are terrestrial, but unlike other land-dwelling heterotrophs (such as animals), fungi do not move about in their environment. They also lack mouths or appendages for seizing, handling, and dismantling food items. Instead, fungi have a very different suite of adaptations for obtaining nutrients.

Experiments have shown that most species of fungi can synthesize nearly all their required nutrients from a few raw materials, including water, some minerals and vitamins (especially B vitamins), and a sugar or some other organic carbon source. For many species, carbohydrates in dead organic matter are the carbon sources, and fungi with this mode of nutrition are called **saprobes** (*sapros* = rotten). Other fungi are parasites, which extract carbohydrates from tissues of a living host, harming it in the process. Most fungi that parasitize living plants produce hyphal branches called **haustoria** (*haustor* = drinker) that penetrate the walls of a host plant's cells and channel nutrients back to the fungal body. Parasitic fungi include those responsible for many devastating plant diseases, such as wheat rust and Dutch elm disease. Still other fungi are nourished by plants with which they have a mutually beneficial symbiotic association, a topic we return to later in this chapter.

Regardless of their nutritional mode, all fungi gain the raw materials required to build and maintain their cells by absorption from the environment. Fungi can absorb many small molecules directly from their surroundings and gain access to other nutrients through extracellular digestion. In this process, a fungus synthesizes digestive enzymes and packages them in secretory vesicles via the pathway described in Section 15.4. When released to the outside the enzymes digest nearby organic matter, breaking down larger molecules into absorbable fragments. Fungal species differ in the particular digestive enzymes they synthesize, so a substrate that is a suitable food source for one species may be unavailable to another. Although there are exceptions, in nature fungi typically thrive only in moist environments where they can directly absorb water, dissolved ions, simple sugars, amino acids, and other small molecules. When some of a mycelium's hyphal filaments contact a source of food, growth is channeled in the direction of the food source. Nutrients are absorbed by various parts of a hypha.

Small atoms and molecules pass readily through the tips, and then transport mechanisms move them through the underlying plasma membrane. Large organic molecules, such as the carbohydrate cellulose (see Section 3.2), *cannot* directly enter any part of a fungus. To use such substances as a food source, a fungus must secrete hydrolytic enzymes that break down the large molecules into smaller, absorbable subunits. Depending on the size of the subunit, further digestion may occur inside cells.

As described earlier, fungal adaptations for efficient extracellular digestion make them masters of the decay so vital to terrestrial ecosystems. For instance, in a single autumn one elm tree can shed 400 pounds of withered leaves; and in a tropical forest, a year's worth of debris may total 60 tons per acre. Without the metabolic activities of saprobic fungi and other decomposers such as bacteria, natural communities would rapidly become buried in their own detritus. As fungi digest dead tissues of other organisms, they also make a major contribution to the recycling of chemical elements those tissues contain. For instance, over time the degradation of organic compounds by saprobic fungi helps return key nutrients such as nitrogen and phosphorus to ecosystems. But the prime example of this recycling virtuosity involves carbon. The respiring cells of fungi and other decomposers give off carbon dioxide, liberating carbon that would otherwise remain locked in the tissues of dead organisms. Each year this activity recycles a vast amount of carbon to plants, the primary producers of nearly all ecosystems on Earth.

All Fungi Reproduce by Way of Spores, but Other Aspects of Reproduction Vary

Biologists have observed a striking number of reproductive variations in fungi, differences that are part of what makes them fascinating to study. As you will learn in the next section, fungi have traditionally been classified on the basis of their reproductive characteristics, although today evidence from molecular analysis also plays a prominent role.

Overall, most fungi have the capacity to reproduce both sexually and asexually. Although no single diagram can depict all the variations, **Figure 28.3** gives an overview of the life cycle stages that mycologists have observed in several groups of fungi. The figure illustrates two general points. First, the life cycle of multicellular fungi typically involves a haploid (n) stage (step 1 is the end point of this stage), a diploid ($2n$) stage (step 5), and between them a dikaryotic ("two nuclei") stage in which the fungus forms hyphae (and a mycelium) that are $n + n$—neither strictly haploid *nor* diploid (step 3). Depending on the type of fungus, this stage may be long lasting or extremely brief, and it is described more fully later in this section. Second, all fungi, whether they are multicellular or in a single-celled, yeast form, can reproduce via **fungal spores.** The spores are microscopic, and in all but one group they are nonmotile—that is, they are not propelled by flagella. Each spore is a walled single cell or multicellular structure that is dispersed from the parent body, often via wind or water. The spores of single-celled fungi form inside the parent cell, then escape when the cell wall breaks open. In multicellular fungi, spores arise in or on specialized hyphal structures and may develop thick walls that help them withstand cold temperatures or drying out after they are released.

Reproduction by way of spores is one of the crucial fungal adaptations. Most fungi are opportunists, obtaining energy by exploiting food sources that occur unpredictably in the environment. Spores can go dormant for long periods, germinating only when environmental conditions favor germination and growth. Having lightweight spores that are easily disseminated by air or by water also increases opportunities for finding suitable food sources that serve as substrates. And releasing vast numbers of spores, as some fungi do, improves the odds that at least a few spores will germinate and produce a new individual.

In nature generally, opportunistic organisms are adapted to reach new food sources quickly and utilize them rapidly. Fungi that are adept at degrading simple sugars and starches often are among the first decomposers to exploit a new source of food. They meet with keen competition from each other and from other decomposers. However, once fungal spores encounter potential food and favorable conditions, they can quickly develop into new individuals that simultaneously feed and rapidly make more spores.

Many opportunistic fungi develop rapidly, growing and reproducing before the food source is depleted. A common tradeoff for speed, however, is small, even microscopic, body size. Larger species of fungi are often adapted to move in later, exploiting food sources such as cellulose and lignin (a complex polymer in the walls of many plant cells), which their predecessors may have lacked the enzymatic machinery to digest efficiently. Some of these fungi may produce huge mycelia (and reproductive sporocarps such as mushrooms) by extracting nutrients from dead trees that contain enough organic material to sustain an extended period of growth.

FEATURES OF ASEXUAL REPRODUCTION IN FUNGI When a fungus produces spores asexually (see Figure 28.3), it may disperse billions of them into the environment. Some fungi (includ-

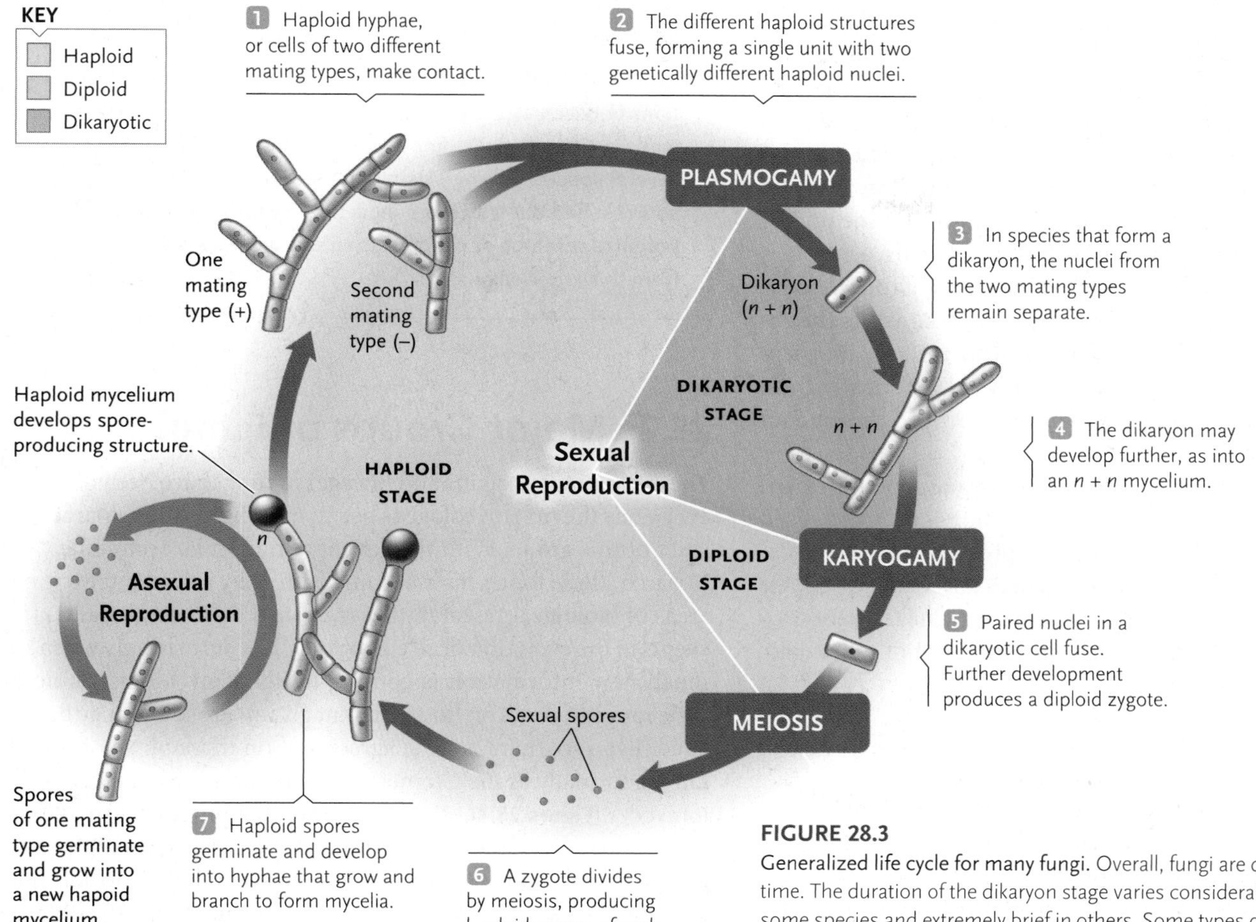
KEY
Haploid
Diploid
Dikaryotic
1 Haploid hyphae, or cells of two different mating types, make contact.
2 The different haploid structures fuse, forming a single unit with two genetically different haploid nuclei.
PLASMOGAMY
3 In species that form a dikaryon, the nuclei from the two mating types remain separate.
Dikaryon ($n + n$)
DIKARYOTIC STAGE
$n + n$
4 The dikaryon may develop further, as into an $n + n$ mycelium.
One mating type (+)
Second mating type (–)
Haploid mycelium develops spore-producing structure.
HAPLOID STAGE
Sexual Reproduction
n
Asexual Reproduction
DIPLOID STAGE
KARYOGAMY
5 Paired nuclei in a dikaryotic cell fuse. Further development produces a diploid zygote.
Sexual spores
MEIOSIS
Spores of one mating type germinate and grow into a new hapoid mycelium.
7 Haploid spores germinate and develop into hyphae that grow and branch to form mycelia.
6 A zygote divides by meiosis, producing haploid spores of each mating type.

FIGURE 28.3
Generalized life cycle for many fungi. Overall, fungi are diploid for only a short time. The duration of the dikaryon stage varies considerably, being lengthy for some species and extremely brief in others. Some types of fungi reproduce only asexually while in others shifts in environmental factors, such as the availability of key nutrients, can trigger a shift from asexual to sexual reproduction or vice versa. For still others sexual reproduction is the norm.

ing many yeasts) also can reproduce asexually by budding or fission, or, in multicellular types, when fragments of hyphae break away from the mycelium and grow into separate individuals. In still others, environmental factors may determine whether the fungus produces hyphal fragments *or* asexual spores. These asexual reproductive strategies all result in new individuals that are essentially clones of the parent fungus. They can be viewed as another adaptation for reproductive efficiency, because the alternative—sexual reproduction—requires the presence of a suitable partner and generally involves several more steps.

The asexual stage of many multicellular fungi—including the pale gray fuzz you might see on berries or bread—is often called a **mold.** The term can be confusing if you are attempting to keep track of taxonomic groupings. For example, the water molds and slime molds described in Chapter 26 are protists, although they were grouped with fungi until additional research revealed their true evolutionary standing. The mold visible on an overripe raspberry is actually a fungal mycelium with aerial structures bearing sacs of haploid spores at their tips.

FEATURES OF SEXUAL REPRODUCTION IN FUNGI Although asexual reproduction is the norm, quite a few fungi shift to sexual reproduction when environmental conditions (such as a lack of nitrogen) or other influences dictate. As you may remember from Chapter 11, in sexual reproduction two haploid cells unite, and in most species fertilization—the fusion of two gamete nuclei to form a diploid zygote nucleus—soon follows. In fungi, however, the partners in sexual union can be two hyphae, two gametes, or other types of cells; the particular combination depends on the species involved. And in sharp contrast to other life forms, many days, months, or even years may pass between the time fertilization gets underway and when it is completed.

During the initial sexual stage, called **plasmogamy** (*plasma* = a formed thing, *gamos* = marriage), the cytoplasms of two genetically different partners fuse. The resulting new cell, a **dikaryon** (two nuclei), contains two haploid nuclei, one from each parent. A dikaryon itself is not haploid (the condition of having one set of chromosomes) because it contains two nuclei. But neither is it diploid, because the nuclei are not fused. So, to be precise, we say that a dikaryon has an $n + n$ nuclear condition.

For example, one form of plasmogamy occurs when hyphal cells of two different **mating types** fuse. The uniting hyphae belong to mycelia of different individuals of the same species that happen to grow near one another. The fusion of different mating types ensures genetic diversity in new individuals.

Once a dikaryon forms, the amount of time that elapses before the next stage begins depends on the type of fungus, as described in the next section. Sooner or later, however, a second phase of fertilization unfolds: The nuclei in the dikaryotic cell fuse to make a $2n$ zygote nucleus. This process is called **karyogamy** ("nuclear union"). In fungi that form mushrooms, it occurs in the tips of hyphae that end in the delicate gills on the underside of a mushroom cap (see Figure 28.1). In animals, a zygote is the first cell of a new individual, but in the world of fungi the zygote has a different fate. After it forms, meiosis converts the zygote nucleus into four haploid (n) nuclei. Those nuclei are packaged into haploid "sexual spores," which vary genetically from each parent. Then the spores are released to spread throughout the environment.

To sum up, in fungi both asexual and sexual spores are haploid, and both can germinate into haploid individuals. However, asexual spores are genetically identical products of asexual reproduction, while sexual spores are genetically varied products of sexual reproduction. We turn now to current ideas on the evolutionary history of fungi, and a survey of the major taxonomic groups in this kingdom.

STUDY BREAK 28.1

1. **What features distinguish the two basic fungal body forms?**
2. **What is a fungal spore, and how does it function in reproduction?**
3. **Fungi reproduce sexually or asexually, but for many species the life cycle includes an unusual stage not seen in other organisms. What is this genetic condition, and what is its role in the life cycle?**

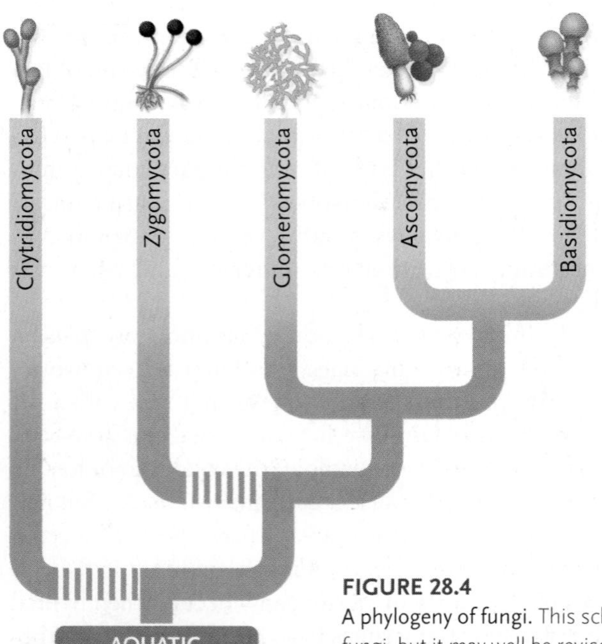
Chytridiomycota
Zygomycota
Glomeromycota
Ascomycota
Basidiomycota
AQUATIC PROTIST ANCESTOR

FIGURE 28.4
A phylogeny of fungi. This scheme represents a widely accepted view of the general relationships between major groups of fungi, but it may well be revised as new molecular findings provide more information. The dashed lines indicate that within two traditional groups, the chytrids and zygomycetes, lines of descent of some subgroups are not well understood and the phyla are probably not monophyletic.

THINK OUTSIDE THE BOOK

Section 18.3 introduced systems biology, the study of interactions between all the molecular components of a cell or organism. Complete genome sequences are now available for several species of fungi. Using the Internet or other resources, find out which sequences are available and describe potential research or benefits scientists are hoping to realize from this knowledge.

28.2 Major Groups of Fungi

The evolutionary origins and lineages of fungi have been obscure ever since the first mycologists began mulling over the characteristics of this group. With the advent of molecular techniques for research, these topics have become extremely active and exciting areas of biological research that may shed light on fundamental events in the evolution of all eukaryotes. Not surprisingly when so much new information is coming to light, mycologists hold a wide range of views on how different groups arose and may be related. Even so, there is wide agreement on three phyla of fungi, known formally as the Glomeromycota, Ascomycota, and Basidiomycota **(Figure 28.4)**. A fourth traditional phylum, a small but diverse array of fungi called the Zygomycota, probably is not monophyletic, and considerable research is underway to sort out the evolutionary relationships among the lineages typically included in the group. Molecular studies suggest that some members of a fifth traditional phylum, the Chytridiomycota, also should more properly be reclassified into two other, separate phyla. All chytrids share certain features, however, and here we consider them as a unit.

In a sixth group, termed conidial fungi, asexual reproduction produces spores called conidia. "Conidial" is not a true taxonomic classification, however. Rather, it serves as a holding station for fungal species that have not yet been assigned to one of the five phyla because no sexual reproductive phase has been observed. This is another instance in which the name for a fungal group can be confusing, because numerous species belonging to the Zygomycota, Ascomycota, and Basidiomycota also form conidia as part of their asexual reproductive cycle.

Finally, we will briefly consider a particularly odd group of single-celled parasites called **microsporidia.** Based on genetic studies, some researchers suspect they may have arisen from a zygomycete ancestor. Other mycologists have proposed that microsporidia make up yet another phylum within the Kingdom Fungi.

INSIGHTS FROM THE Molecular Revolution

There Was Probably a Fungus among Us

What is the taxonomic relationship between the fungi and other eukaryotes? The relationships of the fungi to protists, plants, and animals are buried so far back in evolutionary history that they have proved difficult to reconstruct.

Research Question

Are fungi most closely related to plants, protists, or animals?

Experiment

On the basis of morphological comparisons, for a long time taxonomists classified fungi as more closely related to the protists or plants than to animals. However, an investigation of ribosomal RNA (rRNA) sequences forced a significant revision of this hypothesis.

Patricia O. Wainwright, Gregory Hinkle, Mitchell L. Sogin, and Shawn K. Stickel of Rutgers University and the Marine Biology Laboratory, Woods Hole, carried out the analysis by comparing sequences of 18S rRNA, an rRNA molecule that forms part of the small ribosomal subunit in eukaryotes (see Section 15.4). The investigators began their work by sequencing the 18S rRNA molecules of species among the sponges, ctenophores, and cnidarians (see Chapter 29), which had never been sequenced before. These sequences were then compared with the 18S rRNA sequences of fungi, plants, and several protists, including protozoans and algae, which had been previously obtained by others. For the comparisons, the investigators used a computer program that sorts the rRNA sequences into related groups under the assumption that species with the greatest similarities in 18S rRNA sequence are most closely related. The sequence information was entered into the program in several different combinations; each time the analysis came up with the same family tree (see **figure**).

Conclusion

The family tree placed animals as the branch most closely related to fungi and indicates that the two groups probably share a common choanoflagellate ancestor not shared with any of the other groups. Other investigators have cited similarities in biochemical pathways in fungi and animals, such as pathways that make the amino acid hydroxyproline, the protein ferritin (which combines with iron atoms), and the polysaccharide chitin, which is a primary constituent of both fungal cell walls and arthropod exoskeletons. Studies of fungi called chytrids (p. 624) also are providing provocative insights on this topic.

Source: Patricia O. Wainwright, Gregory Hinkle, Mitchell L. Sogin, and Shawn K. Stickel, 1993. Monophyletic origins of the Metazoa: An evolutionary link with Fungi. *Science* 260: 340–342.

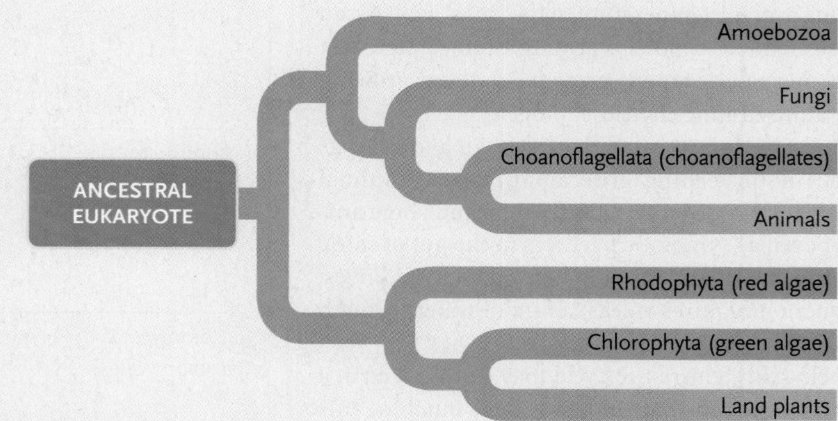
ANCESTRAL EUKARYOTE
Amoebozoa
Fungi
Choanoflagellata (choanoflagellates)
Animals
Rhodophyta (red algae)
Chlorophyta (green algae)
Land plants

Fungi Were Present on Earth by at Least 500 Million Years Ago

Many fungi look plantlike, and for many years fungi were classified as plants. As biologists learned more about the distinctive characteristics of fungi, however, it became clear that fungi merited a separate kingdom. The discovery of chitin in fungal cells, and recent comparisons of DNA and RNA sequences, all indicate that fungi and animals are more closely related to each other than they are to other eukaryotes (see *Insights from the Molecular Revolution*). Analysis of the sequences of several genes suggests that the lineages leading to animals and fungi may have diverged around 965 million years ago. Whenever the split developed, phylogenetic studies indicate that fungi first arose from a single-celled, flagellated protist—the sort of organism that does not fossilize well. Although traces of what may be fossil fungi exist in rock formations nearly 1 billion years old, the oldest fossils that we can confidently assign to the modern Kingdom Fungi appear in rock strata laid down about 500 million years ago.

Once They Appeared, Fungi Radiated into Several Major Lineages

Most likely, the first fungi were aquatic. When other kinds of organisms began to colonize land, they may well have brought fungi along with them. For example, researchers have discovered what appear to be mycorrhizae—symbiotic associations of a fungus and a plant—in fossils of some of the earliest known land plants (see Chapter 27). The final section of this chapter examines the nature of mycorrhizae more fully.

Over time, fungi diverged into the strikingly diverse groups that we consider in the rest of this section **(Table 28.1)**. As the lineages diversified, different adaptations associated with reproduction arose. For this reason, mycologists traditionally assigned fungi to phyla according to the type of structure that houses the final stages of sexual reproduction and releases sexual spores. These features can still be useful indicators of the phylogenetic standing of a fungus, although now the powerful tools of molecular analysis are bringing many revisions to our understanding of the evolutionary journey of fungi. Our survey begins with chytrids, which probably most closely resemble the fungal kingdom's most ancient ancestors.

Chytrids Produce Motile Spores That Have Flagella

Chytrids, including those in the phylum Chytridiomycota, encompass about a thousand species. Chytrids are the only fungi that produce motile spores, which swim by way of flagella. Nearly all chytrids are microscopic **(Figure 28.5A)**, and mycologists have recently begun paying significant attention to them, in part because their characteristics strongly suggest that the lineages that make up this group arose near the beginning of fungal evolution. Another reason for research interest is the discovery that the chytrid *Batrachochytrium dendrobatidis* is responsible for a disease epidemic that has wiped out an estimated two-thirds of the species of harlequin frogs *(Atelopus)* of the American tropics **(Figure 28.5B and C)**. The epidemic has correlated with the rising average temperature in the frogs' habitats, an increase credited to global warming. Studies show that the warmer environment provides optimal growing temperatures for the chytrid pathogen.

Most chytrids are aquatic, although a few live as saprobes in soil, feeding on decaying plant and animal matter; as parasites on insects, plants, and some animals or even as symbiotic partners in the gut of cattle and some other herbivores. Wherever a chytrid lives, reproduction requires at least a film of water through which the swimming spores can move.

A chytrid's entire life cycle may unfold within a matter of days, and, for most species, much of this brief lifetime is spent in asexual reproduction. Although individuals initially exist as a vegetative (nonreproductive) phase, the fungus soon shifts into a reproductive mode. Flagella-bearing spores are released

TABLE 28.1 Summary of Fungal Phyla

Phylum	Body Type	Key Feature
Chytridiomycota and other chytrids	One to several cells	Motile spores propelled by flagella; usually asexual
Zygomycota (zygomycetes)	Hyphal	Sexual stage in which a resistant zygospore forms for later germination
Glomeromycota (glomeromycetes)	Hyphal	Hyphae associated with plant roots, forming arbuscular mycorrhizae
Ascomycota (ascomycetes)	Hyphal	Sexual spores produced in sacs called asci
Basidiomycota (basidiomycetes)	Hyphal	Sexual spores (basidiospores) form in basidia of a prominent fruiting body (basidiocarp)

A. ***Chytriomyces hyalinus***

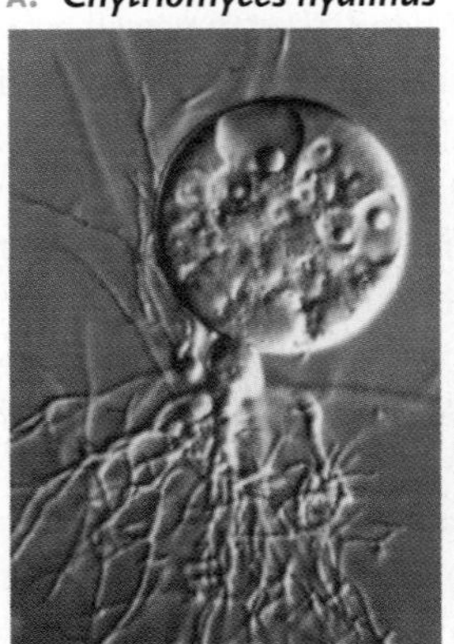

John Taylor/Visuals Unlimited

B. **Chytridiomycosis in a frog**

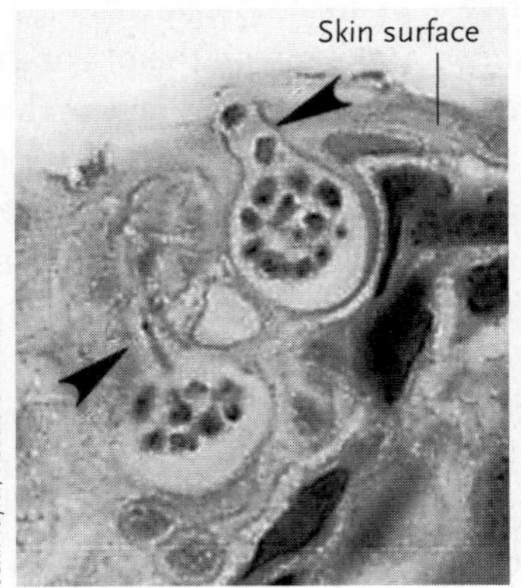
Skin surface

Center for Disease Control

C. **Harlequin frog**

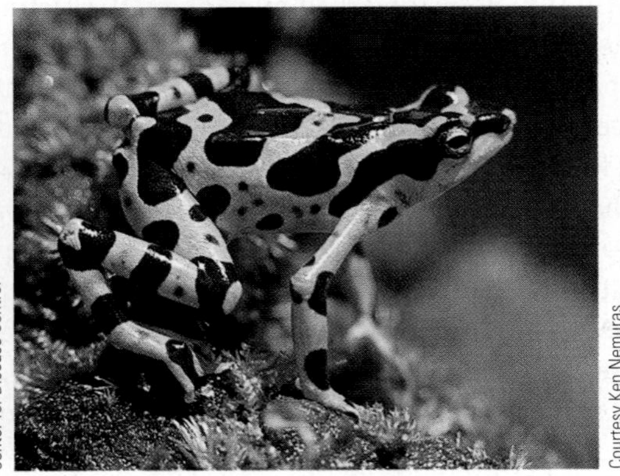

Courtesy Ken Nemuras

FIGURE 28.5
Chytrids. **(A)** *Chytriomyces hyalinus,* one of the few chytrids that reproduces sexually. **(B)** Chytridiomycosis, a fungal infection, shown here in the skin of a frog. The two arrows point to flask-shaped cells of the parasitic chytrid *Batrachochytrium dendrobatidis,* which has devastated populations of harlequin frogs **(C)**.

to the environment and each swims briefly until it comes to rest on a substrate and a tough cyst forms around it. Under proper conditions, it will soon germinate and launch the life cycle anew.

A few chytrids reproduce sexually. Mycologists have observed a remarkable variety of sexual modes, but in all of them spores of different mating types unite. Karyogamy directly follows plasmogamy to produce a $2n$ zygote. This cell may form a mycelium that gives rise to sporangia, or it may directly give rise to either asexual or sexual spores.

Zygomycetes Form Zygospores for Sexual Reproduction

Fewer than a thousand species are assigned to the phylum Zygomycota—fungi that reproduce sexually by way of structures called *zygospores* **(Figure 28.6)**. What zygomycetes lack in numbers, however, they make up for in impact on other organisms. Many zygomycetes are saprobes that live in soil, feeding on plant detritus. There, their metabolic activities release mineral nutrients in forms that plant roots can take up. Some zygomycetes are parasites of insects (and even other fungi), and some wreak havoc on human food supplies, spoiling stored grains, bread, fruits, and vegetables such as sweet potatoes. Others, however, have become major partners in commercial enterprises, where they are used in manufacturing products that range from industrial pigments to pharmaceuticals.

Most zygomycetes have aseptate hyphae, a feature that distinguishes them from the other multicellular fungi. Mycelia of many zygomycetes may occur in either a + or − mating type (Figure 28.6, step 1), and the nuclei of the two mating types are equivalent

FIGURE 28.6
Life cycle of the bread mold *Rhizopus stolonifer*, a zygomycete. Asexual reproduction is common, but different mating types (+ and −) also reproduce sexually. In both cases, haploid spores form and give rise to new mycelia.

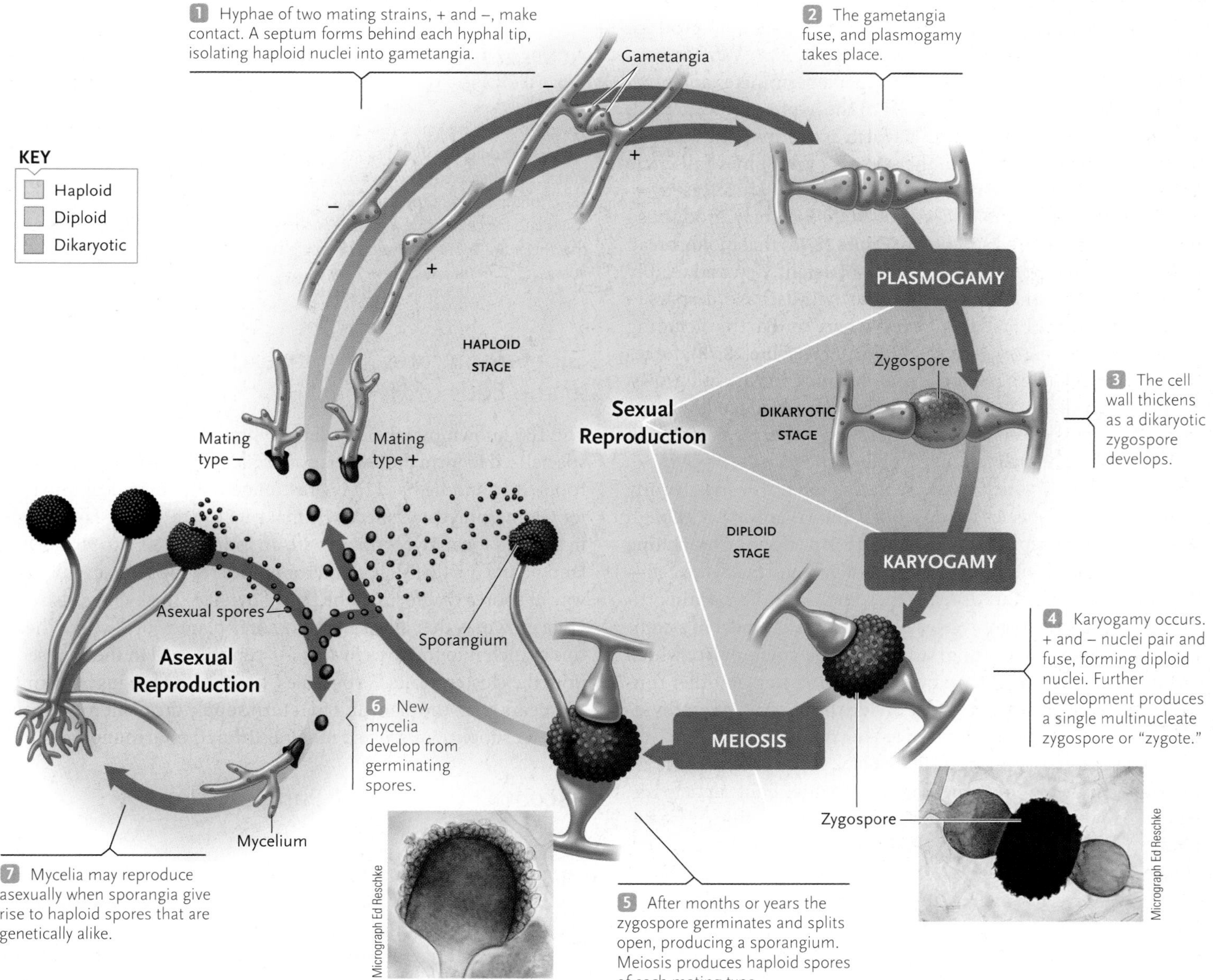

1 Hyphae of two mating strains, + and –, make contact. A septum forms behind each hyphal tip, isolating haploid nuclei into gametangia.
Gametangia
–
+
2 The gametangia fuse, and plasmogamy takes place.
KEY
Haploid
Diploid
Dikaryotic
–
+
PLASMOGAMY
HAPLOID STAGE
Zygospore
3 The cell wall thickens as a dikaryotic zygospore develops.
Sexual Reproduction
DIKARYOTIC STAGE
Mating type –
Mating type +
DIPLOID STAGE
KARYOGAMY
Asexual spores
Sporangium
4 Karyogamy occurs. + and – nuclei pair and fuse, forming diploid nuclei. Further development produces a single multinucleate zygospore or "zygote."
Asexual Reproduction
6 New mycelia develop from germinating spores.
MEIOSIS
Zygospore
Mycelium
7 Mycelia may reproduce asexually when sporangia give rise to haploid spores that are genetically alike.
Micrograph Ed Reschke
5 After months or years the zygospore germinates and splits open, producing a sporangium. Meiosis produces haploid spores of each mating type.
Micrograph Ed Reschke

to gametes. Each strain secretes steroidlike hormones that can stimulate the development of sexual structures in the complementary strain and cause sexual hyphae to grow toward each other. When + and − hyphae come into close proximity, a septum forms behind the tip of each hypha, producing a terminal **gametangium** that contains several haploid nuclei. When the gametangia of the two strains make contact, cellular enzymes digest the wall between them, yielding a single large, thin-walled cell that contains many nuclei from both parents (steps 2 and 3). In other words, plasmogamy has occurred, and this new cell is a dikaryon. Gradually a second, inner wall forms, thickens, and hardens. This structure, with the multinucleate cell inside it, is a **zygospore,** the structure that gives this fungal group its scientific name. It becomes dormant and can stay dormant for months or years.

Karyogamy follows plasmogamy and eventually the diploid zygospore ends its dormancy. The cell undergoes meiosis and produces a stalked sporangium (step 5). The sporangium contains haploid spores of each mating type, which are released to the outside world. When a spore later germinates, it produces either a + or a − mycelium, and the sexual cycle can continue.

Like other fungi, however, zygomycetes usually reproduce asexually, as shown at the lower left in Figure 28.6. When a haploid spore lands on a favorable substrate, it germinates and gives rise to a branching mycelium. Some of the hyphae grow upward, and saclike, thin-walled sporangia form at the tips of these aerial hyphae. Inside the sporangia the asexual cycle comes full circle as new haploid spores arise through mitosis and are released.

The black bread mold, *Rhizopus stolonifer,* may produce so many charcoal-colored sporangia **(Figure 28.7A)** that moldy bread looks black. The spores released are lightweight, dry, and readily wafted away by air currents. In fact, winds have dispersed *R. stolonifer* spores just about everywhere on Earth, including the Arctic. Another zygomycete, *Pilobolus* **(Figure 28.7B),** forcefully spews its sporangia away from the dung in which it grows. A grazing animal may eat a sporangium on a blade of grass; the spores then pass through the animal's gut unharmed and begin the life cycle again in a new dung pile.

Zygomycetes which have aseptate hyphae are structurally simpler than the species in most other fungal groups. Although septa wall off the reproductive structures, in effect the branching mycelium of each fungus is a single, huge, multinucleate cell—the same body structure as found in some algae and certain protists. Because such zygomycetes have numerous nuclei in a common cytoplasm, these fungi are said to be **coenocytic,** which means "contained in a shared vessel." By contrast, in other fungal groups septa at least partially divide the hyphae into individual cells, which typically contain two or more nuclei.

Presumably, having hyphae that lack septa confers some selective advantages. One benefit may be that without septa to impede the flow, nutrients can move freely from the absorptive hyphal tips to other hyphae where reproductive parts develop. Hence the fungus may be able to reproduce faster.

In zygomycetes, aggregations of "cooperating" hyphae may form body structures specialized for certain functions. However, such structures are more common in the three groups of more complex fungi that we consider next.

A. Sporangia of *Rhizopus stolonifer*

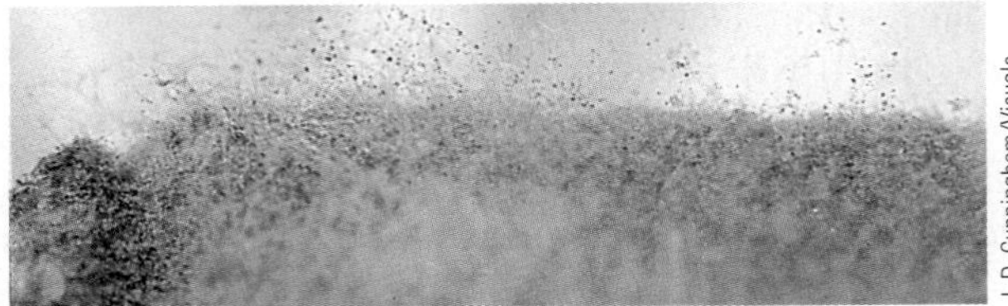

J. D. Cunningham/Visuals Unlimited

B. Sporangia (dark sacs) of *Pilobolus*

John Hodgin

500 μm

FIGURE 28.7

Two of the numerous strategies for spore dispersal by zygomycetes. **(A)** The sporangia of *Rhizopus stolonifer,* shown here on a slice of bread, release powdery spores that are easily dispersed by air currents. **(B)** In *Pilobolus,* the spores are contained in a sporangium (the dark sac) at the end of a stalked structure. When incoming rays of sunlight strike a light-sensitive portion of the stalk, turgor pressure (pressure against a cell wall due to the movement of water into the cell) inside a vacuole in the swollen portion becomes so great that the entire sporangium may be ejected outward as far as 2 m—a remarkable feat, given that the stalk is only 5 to 10 mm tall.

Glomeromycetes Form Asexual Spores, at the Ends of Hyphae

The 160 known members of the phylum Glomeromycota are all specialized to form the associations called mycorrhizae with plant roots. It would be hard to overestimate their ecological impact, for Glomeromycota collectively make up roughly half of the fungi in soil and form mycorrhizae with an estimated 80% to 90% of all land plants. Virtually all glomeromycetes reproduce asexually, by way of spores that form at the tips of hyphae. The hyphae also secrete enzymes that allow them to enter plant roots, where their tips branch into treelike clusters. As you will read in the next section, the clusters, called arbuscules, nourish the fungus by taking up sugars from the plant and in return supply the plant roots with a steady supply of dissolved minerals from the surrounding soil.

Ascomycetes, the Sac Fungi, Produce Sexual Spores in Saclike Asci

The phylum Ascomycota includes more than 30,000 species that produce reproductive structures called *asci* **(Figure 28.8).** A few ascomycetes prey upon various agricultural insect pests and thus have potential for use as "biological pesticides." Many more are destructive plant pathogens, including *Venturia inaequalis,*

A. **Ascocarp**

Ascospore (sexual spore)

Ascus

Spore-bearing hypha of this ascocarp

B. **Asci**

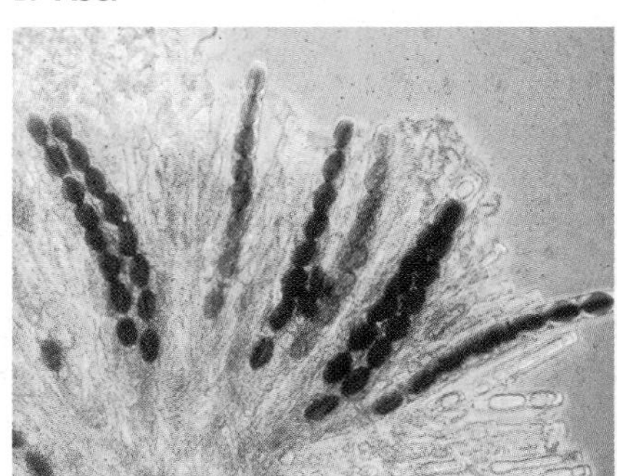

© North Carolina State University, Department of Plant Pathology

C. **Ascocarp**

© Michael Wood/mykoweb.com

D. **Morel**

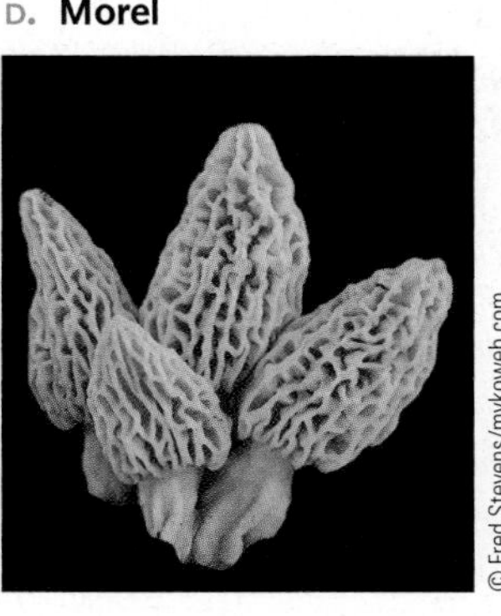

© Fred Stevens/mykoweb.com

FIGURE 28.8

A few of the ascomycetes, or sac fungi. The examples shown are multicellular species that form ascocarps as reproductive structures. **(A)** A cup-shaped ascocarp, composed of tightly interwoven hyphae. The spore-producing asci occur inside the cup. **(B)** Asci on the inner surface of an ascocarp. **(C)** The velvety appearance the Scarlet cup fungus *(Sarcoscypha)* is due to the thousands of asci in its ascocarp. **(D)** True morels *(Morchella esculenta)*, a prized edible fungus.

the fungus responsible for apple scab, and *Ophiostoma ulmi,* which causes Dutch elm disease. Several ascomycetes can be serious pathogens of humans. For example, *Claviceps purpurea,* a parasite on rye and other grains, causes ergotism, a disease marked by vomiting, hallucinations, convulsions, and, in severe cases, gangrene and even death. Other ascomycetes cause nuisance infections such as athlete's foot and ringworm. Species of *Aspergillus* grow in damp grain or peanuts; their metabolic wastes, known as aflatoxins, can cause cancer in people who eat the poisoned food over an extended period of time. A few ascomycetes even show trapping behavior, ensnaring small worms that they then digest **(Figure 28.9B)**. Yet some ascomycetes are valuable to humans: one species, the orange bread mold *Neurospora crassa,* has been important in genetic research, including the elucidation of the one gene–one enzyme hypothesis (see Section 15.1). And certain species of *Penicillium* **(Figure 28.9A)** are the source of the penicillin family of antibiotics, while others produce the aroma and distinctive flavors of Camembert and Roquefort cheeses. This multifaceted division also includes gourmet delicacies such as truffles *(Tuber melanosporum)* and the succulent true morel *Morchella esculenta.*

Although yeasts and filamentous fungi with a yeast stage in the life cycle occur in all fungal groups except chytrids and glomeromycetes, many of the best-known yeasts are ascomycetes. *Candida albicans* **(Figure 28.10)** commonly infects mucous membranes, especially of the mouth (where it causes a disorder called thrush) and the vagina. *Saccharomyces cerevisiae,* which produces the ethanol in alcoholic beverages and the carbon di-

A. **A *Penicillium* species**

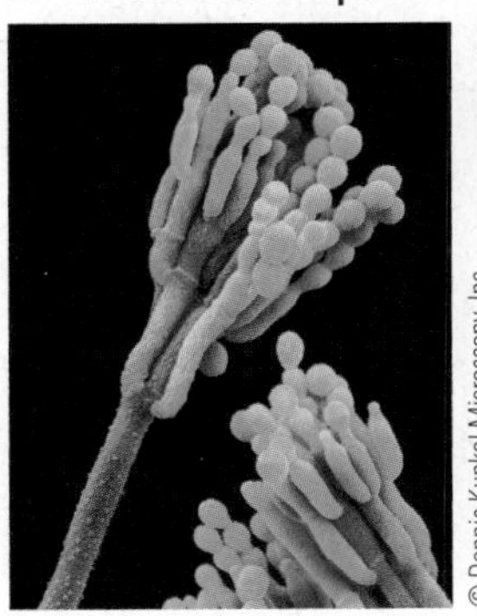

© Dennis Kunkel Microscopy, Inc.

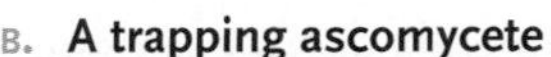

B. **A trapping ascomycete**

N. Allin and G. L. Barron

FIGURE 28.9

Other ascomycete representatives. **(A)** *Eupenicillium.* Notice the rows of conidia (asexual spores) atop the structures that produce them. **(B)** Hyphae of *Arthrobotrys dactyloides,* a trapping ascomycete, form nooselike rings. When the fungus is stimulated by the presence of a prey organism, rapid changes in ion concentrations draw water into the hypha by osmosis. The increased turgor pressure shrinks the "hole" in the noose and captures this nematode. The hypha then releases digestive enzymes that break down the worm's tissues.

Yeast cells

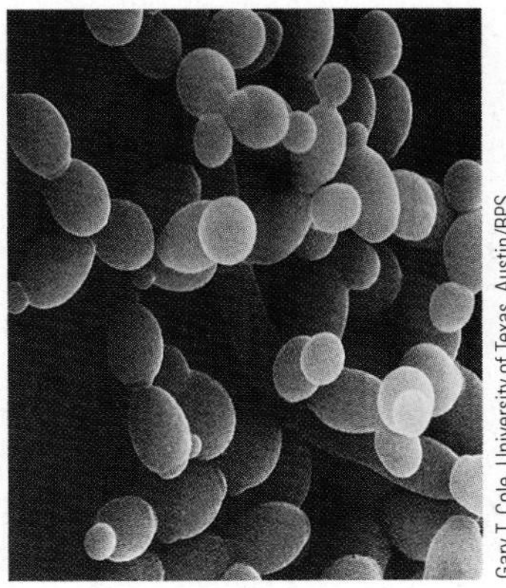

Gary T. Cole, University of Texas, Austin/BPS

FIGURE 28.10

Candida albicans, cause of yeast infections of the mouth and vagina.

oxide that leavens bread, has also been a model organism for genetic research. By one estimate it has been the subject of more genetic experiments than any other eukaryotic microorganism. Yeasts commonly reproduce asexually by fission or budding from the parent cell, but many also can reproduce sexually after the fusion of two cells of different mating types (analogous to the mating types described earlier). Many ascomycete yeasts are found naturally in the nectar of flowers and on fruits and leaves. At least 1,500 species have been described, and mycologists suspect that thousands more are yet to be identified.

Tens of thousands of ascomycetes, however, are not yeasts. They are multicellular, with tissues built up from septate hyphae. Although septa do slow the flow of nutrients (which, recall, can cross septa through pores), they also confer advantages. For example, septa present barriers to the loss of cytoplasm if a hypha is torn or punctured, whereas in an aseptate zygomycete, fluid pressure may force out a significant amount of cytoplasm before a breach can be sealed by congealing cytoplasm. In ways that are not well understood, septa can also limit the damage from toxins that are secreted by competing fungi.

As with zygomycetes, certain hyphae in ascomycetes are specialized for asexual reproduction. Instead of making spores inside sporangia, however, many ascomycetes produce asexual spores called **conidia** (*konis* = dust) (singular, *conidium*). In some of the species, the conidia form in chains that elongate from modified hyphal branches called **conidiophores.** In other ascomycetes, the conidia may pinch off from the hyphae in a series of "bubbles," a bit like a string of detachable beads. Either way, an ascomycete can form and release spores much more quickly than a zygomycete can. Each newly formed conidium contains a haploid nucleus and some of the parent hypha's cytoplasm. Conidia and conidiophores of some ascomycete species are visible as the white powdery mildew that attacks grapes, roses, grasses, and the leaves of squash plants.

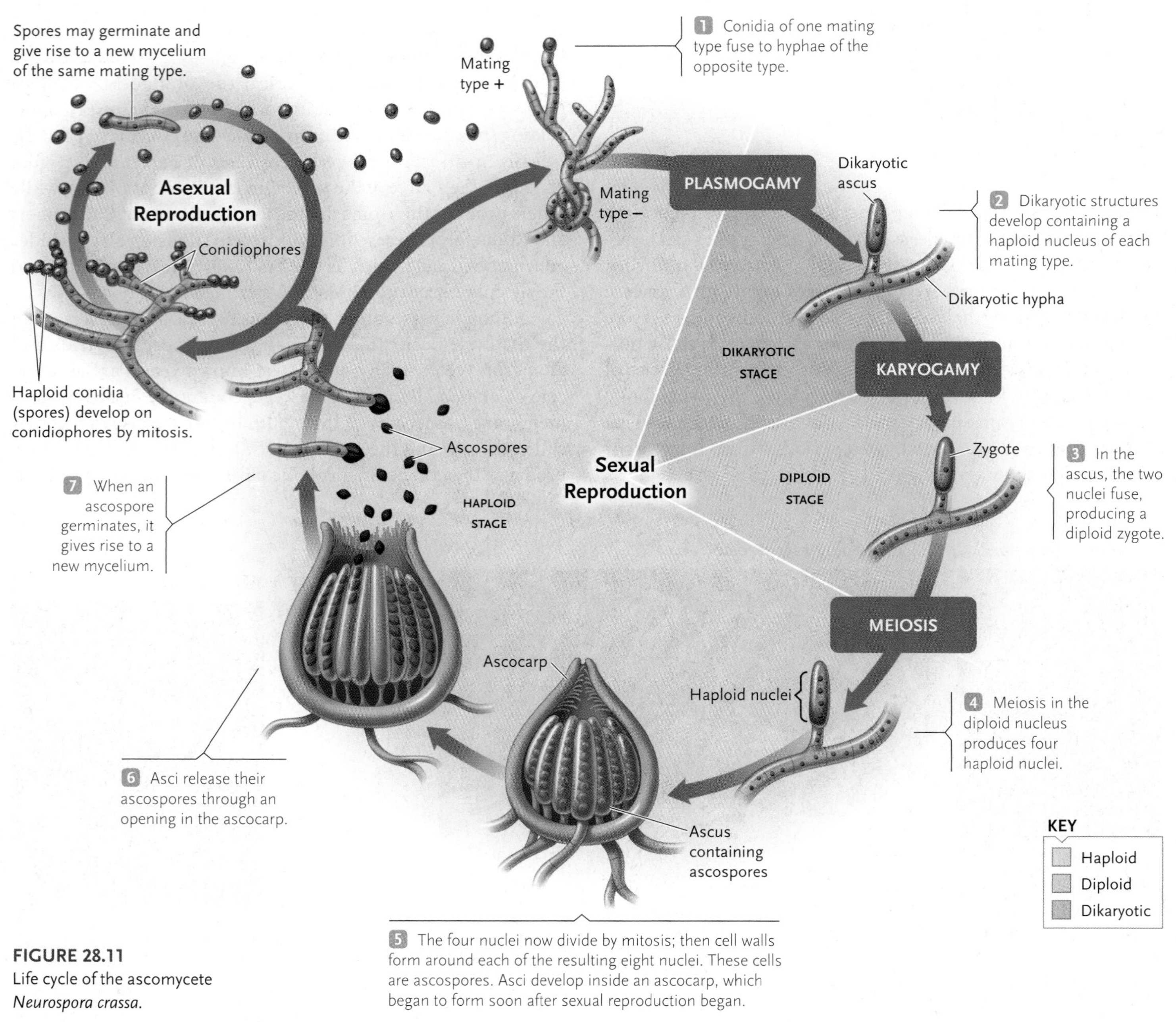

Spores may germinate and give rise to a new mycelium of the same mating type.
Mating type +
1 Conidia of one mating type fuse to hyphae of the opposite type.
Asexual Reproduction
Conidiophores
Mating type −
PLASMOGAMY
Dikaryotic ascus
2 Dikaryotic structures develop containing a haploid nucleus of each mating type.
Dikaryotic hypha
Haploid conidia (spores) develop on conidiophores by mitosis.
DIKARYOTIC STAGE
KARYOGAMY
Ascospores
Zygote
3 In the ascus, the two nuclei fuse, producing a diploid zygote.
Sexual Reproduction
DIPLOID STAGE
HAPLOID STAGE
7 When an ascospore germinates, it gives rise to a new mycelium.
MEIOSIS
Ascocarp
Haploid nuclei
4 Meiosis in the diploid nucleus produces four haploid nuclei.
6 Asci release their ascospores through an opening in the ascocarp.
Ascus containing ascospores
KEY
Haploid
Diploid
Dikaryotic
5 The four nuclei now divide by mitosis; then cell walls form around each of the resulting eight nuclei. These cells are ascospores. Asci develop inside an ascocarp, which began to form soon after sexual reproduction began.

FIGURE 28.11
Life cycle of the ascomycete *Neurospora crassa.*

Ascomycetes can also reproduce sexually and are commonly termed sac fungi because most of the events that generate haploid sexual spores occur in saclike cells called **asci** (*askos* = leather bag) (singular, *ascus*). In *Neurospora crassa* and other complex ascomycetes, reproductive bodies called **ascocarps** bear or contain the asci. Some ascocarps resemble globes, others flasks or open dishes. In *N. crassa,* an ascocarp begins to develop when two haploid mycelia of mating types + and − fuse (**Figure 28.11,** step 1). Plasmogamy then takes place, with the details differing from species to species. (In some species, hormonal signals cause the tip of one hypha to enlarge and form a "female" reproductive organ called an ascogonium, while the other hyphal tip develops into a "male" antheridium.) Paired nuclei, one from each mating type, migrate into the hyphae. During plasmogamy, the fused sexual structures give rise to dikaryotic hyphae. Asci form at the hyphal tips. Inside them, karyogamy takes place, producing a diploid zygote nucleus. It divides by meiosis, producing four haploid nuclei. In yeasts and some other ascomycetes cell division stops at this point, but in *N. crassa* and in many other species a round of mitosis ensues and results in eight nuclei. Regardless, the nuclei, other organelles, and a portion of cytoplasm then are incorporated into ascospores that may germinate on a suitable substrate and continue the life cycle.

Basidiomycetes, the Club Fungi, Form Sexual Spores in Club-Shaped Basidia

The 30,000 or so species of fungi in the phylum Basidiomycota include the mushroom-forming species, shelf fungi, coral fungi, bird's nest fungi, stinkhorns, smuts, rusts, and puffballs **(Figure 28.12).** The common name for this group is club fungi, so named because the spore-producing cells, called **basidia** (*basis* = base or foundation), usually are club shaped. Some species have enzymes for digesting cellulose and lignin and are important decomposers of woody plant debris. A surprising number of basidiomycetes, including the prized edible oyster mushrooms *(Pleurotus ostreatus),* also can trap and consume bacteria and small animals such as rotifers and nematodes by secreting paralyzing toxins or gluey substances that immobilize the prey. This adaptation gives the fungus access to a rich source of molecular nitrogen, an essential nutrient that is often scarce in terrestrial habitats.

Many basidiomycetes take part in vital mutualistic associations with the roots of forest trees, as discussed later in this chapter. Others, the rusts and smuts, are parasites that cause serious diseases in wheat, rice, and other plants. Still others produce millions of dollars' worth of the reproductive structures commonly called mushrooms.

A. **Coral fungus**

Robert C. Simpson/Nature Stock

B. **Shelf fungus**

Robert C. Simpson/Nature Stock

C. **White-egg bird's nest fungus**

Jeffrey Lepore/Photo Researchers, Inc.

D. **Fly agaric mushroom**

Jane Burton/Bruce Coleman Ltd.

E. **Scarlet hood**

Robert C. Simpson/Nature Stock

FIGURE 28.12
Representative basidiomycetes, or club fungi. **(A)** The light red coral fungus *Ramaria.* **(B)** The shelf fungus *Laetiporus.* **(C)** The white-egg bird's nest fungus *Crucibulum laeve.* Each tiny "egg" contains spores. Raindrops splashing into the "nest" can cause "eggs" to be ejected, thereby spreading spores into the surrounding environment. **(D)** The fly agaric mushroom *Amanita muscaria,* which is toxic to humans. **(E)** The scarlet hood *Hygrophorus.*

Amanita muscaria, the fly agaric mushroom (see Figure 28.12D), has been used as a fly poison, from which it gets its common name. Due to its euphoria-producing effects, *A. muscaria* also is used in the religious rituals of ancient societies in Central America, Russia, and India. (Less pleasant side effects can include significant nausea.) Other species of this genus, including the death cap mushroom *Amanita phalloides,* produce deadly toxins. In eukaryotic cells, the *A. phalloides* toxin, a short peptide called α-amanitin, halts the transcription of protein-coding genes by RNA polymerase, thereby inhibiting protein synthesis. Within 8 to 24 hours of ingesting as little as 5 mg of the toxin, vomiting and diarrhea begin. Later, kidney and liver cells start to degenerate; without intensive medical care, death can follow within a few days.

A few basidiomycetes generally reproduce only by asexual means, by budding or shedding a fragment of a hypha. One is *Cryptococcus neoformans,* which causes a form of meningitis in humans. In general, however, basidiomycetes do reproduce sexually, producing large numbers of haploid sexual spores. **Figure 28.13** shows the life cycle of a typical basidiomycete.

Basidia typically develop on a **basidiocarp,** which is the reproductive body of the fungus. A basidiocarp consists of tight clusters of hyphae; the feeding mycelium is buried in the soil or decaying wood. The shelflike bracket fungi visible on trees are basidiocarps, and about 10,000 species of club fungi produce the basidiocarps we call mushrooms. Each is a short-lived reproductive body consisting of a stalk and a cap. Basidia develop on "gills," which are the sheets of tissue on the underside of the cap.

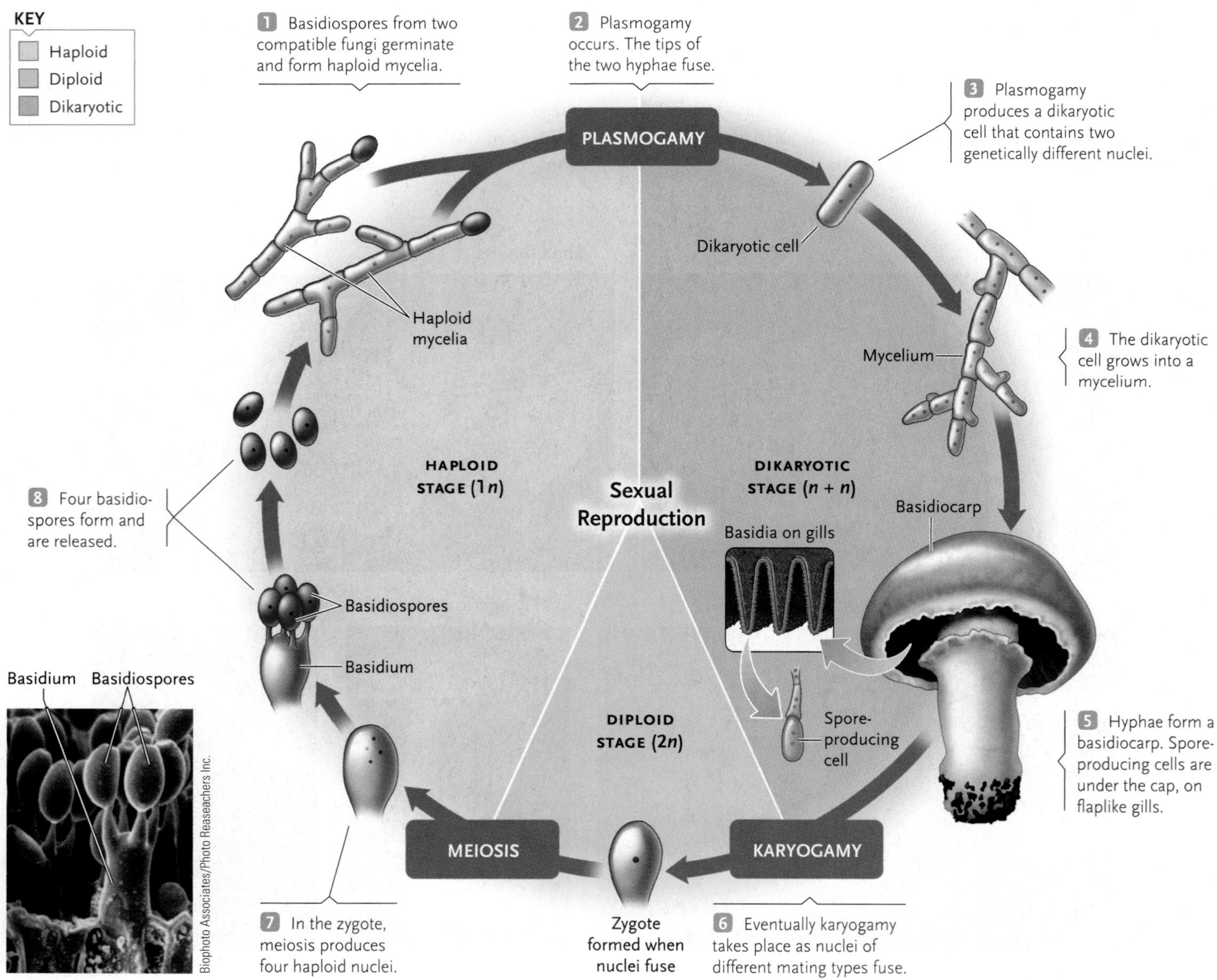

KEY
Haploid
Diploid
Dikaryotic
1 Basidiospores from two compatible fungi germinate and form haploid mycelia.
2 Plasmogamy occurs. The tips of the two hyphae fuse.
PLASMOGAMY
3 Plasmogamy produces a dikaryotic cell that contains two genetically different nuclei.
Dikaryotic cell
Haploid mycelia
Mycelium
4 The dikaryotic cell grows into a mycelium.
HAPLOID STAGE (1n)
Sexual Reproduction
DIKARYOTIC STAGE (n + n)
Basidiocarp
8 Four basidiospores form and are released.
Basidia on gills
Basidiospores
Basidium
Basidium Basidiospores
DIPLOID STAGE (2n)
Spore-producing cell
5 Hyphae form a basidiocarp. Spore-producing cells are under the cap, on flaplike gills.
MEIOSIS
KARYOGAMY
7 In the zygote, meiosis produces four haploid nuclei.
Zygote formed when nuclei fuse
6 Eventually karyogamy takes place as nuclei of different mating types fuse.
Biophoto Associates/Photo Reaseachers Inc.

FIGURE 28.13

Generalized life cycle of the basidiomycete *Agaricus arvensis,* a close relative of the common button mushroom sold in grocery stores. During the dikaryotic stage, *A. arvensis* cells contain two genetically different nuclei, shown here in different colors. Inset: Micrograph showing basidia and basidiospores.

The basidia undergo meiosis to produce microscopic, haploid **basidiospores** (Figure 28.13, inset) that disperse throughout the environment.

When a basidiospore lands on a suitable food source, it germinates and gives rise to a haploid mycelium. Two compatible mating types growing near each other may undergo plasmogamy. The resulting mycelium is dikaryotic, its cells containing one nucleus from each mating type. The dikaryotic stage of a basidiomycete is the feeding mycelium that can grow for years—a major departure from an ascomycete's short-lived dikaryotic stage. Accordingly, a basidiomycete has many more opportunities for producing sexual spores, and the mycelium can give rise to reproductive bodies many times.

After an extensive mycelium develops, and when environmental conditions such as moisture are favorable, basidiocarps grow from the mycelium and develop basidia. At first, each basidium in the mushroom or other reproductive body is dikaryotic, but then the two nuclei undergo karyogamy, fusing to form a diploid zygote nucleus. The zygote exists only briefly; meiosis soon produces haploid basidiospores, which are wafted away from the basidium by air currents. Basidia can produce huge numbers of spores—for many species, estimates run as high as 100 million spores *per hour* during reproductive periods, day after day.

Squirrels and many other small animals may eat mushrooms almost as soon as they appear, but in some species the underlying mycelium can live for many years. As such a mycelium grows, specialized mechanisms during cell division maintain the dikaryotic condition and the paired nuclei in each hyphal cell.

Conidial Fungi Are Species for Which No Sexual Phase Is Known

As noted earlier, fungi generally are classified on the basis of their structures for sexual reproduction. When a sexual phase is absent or has not yet been detected, the fungal species is lumped into a convenience grouping, the conidial fungi (recall that conidia are asexual spores). This classification is the equivalent of "unidentified." Other historical names for this grouping are "imperfect fungi" and deuteromycetes.

When researchers discover a sexual phase for a conidial fungus, or when molecular studies establish a clear relationship to a sexual species, the conidial fungus is reassigned to the appropriate phylum. Thus far, some have been classified as basidiomycetes, but most conidial fungi have turned out to be ascomycetes.

Microsporidia Are Single-Celled Sporelike Parasites

There are more than 1,200 species of the single-celled parasites called **microsporidia.** They are known to infect insects including honeybees and grasshoppers, and vertebrates including fish and humans—especially individuals with compromised immune systems such as people with AIDS. Microsporidia are rather mysterious organisms. Physically they resemble spores **(Figure 28.14),** but they lack mitochondria and have other puzzling characteristics. Molecular studies suggest that they are related to zygomycetes, and some researchers have proposed that the group may have lost many typical fungal features as it evolved a highly specialized parasitic lifestyle.

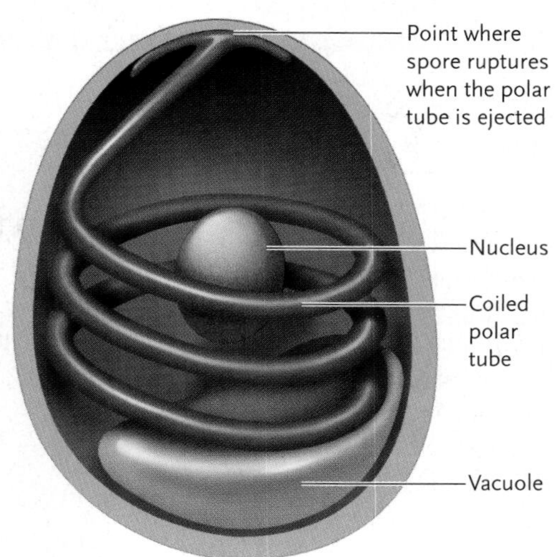
Point where spore ruptures when the polar tube is ejected
Nucleus
Coiled polar tube
Vacuole

FIGURE 28.14
Structure of microsporidia. When a spore germinates, its vacuole expands and forces the coiled "polar tube" outward and into a nearby, soon-to-be host cell. The nucleus and cytoplasm of the parasite enter the host through the tube, launching developmental steps that lead to the development of more microsporidia inside the host.

STUDY BREAK 28.2

1. **Name the major phyla and phylum-level categories of the Kingdom Fungi and describe the reproductive adaptations that distinguish each one.**
2. **In terms of structure, which are the simplest fungal groups? The most complex?**
3. **Describe some ways, positive or negative, that members of each fungal phylum interact with other life forms.**

THINK OUTSIDE THE BOOK

Humans are biologically attractive hosts for a wide range of fungi that colonize epidermal tissues—skin, hair, and nails. These fungal species, often described medically as dermatophytes, are categorized as anthrophilic, zoophilic, and geophilic species. Investigate this topic and make a list of the skin disorders associated with each type, as well as common sources of infection. What basic nutrition source do all dermatophytes share?

28.3 Fungal Associations

Many fungi are partners in mutually beneficial interactions with plants and other photosynthesizers, with insects, and even with mammals—associations that play major roles in the functioning

FOCUS ON Applied Research

Lichens as Monitors of Air Pollution's Biological Damage

Lichens have become reliable pollution-monitoring devices all over the world—in some cases, replacing costly electronic monitoring stations. Different species are vulnerable to specific pollutants. For example, *Ramalina* lichens are damaged by nitrate and fluoride salts. Elevated levels of sulfur dioxide (a major component of acid rain) cause old man's beard *(Usnea trichodea)* to shrivel and die, but strongly promote the growth of a crusty European lichen, *Lecanora conizaeoides.* The sensitivity of yellow *Evernia* lichens to SO_2 enabled the scientist who discovered its damage at remote Isle Royale in Michigan to point the finger northward to coal-burning furnaces at Thunder Bay, Canada. Conversely, healthy lichens on damaged trees of Germany's Black Forest lifted suspicion from French coal-burning power plants and allowed investigators to identify the true source of the tree damage: nitrogen oxides from automobile exhausts. The result was Germany's first auto emission standards, which went into effect in the 1990s.

Robert W. Domm/Getty Images

Usnea (old man's beard), a pendent (hanging) lichen.

of ecosystems. A **symbiosis** is a state such as parasitism or mutualism in which two or more species live together in close association. Chapter 52 discusses general features of symbiotic associations more fully; here we are interested in some examples of the symbioses fungi form with photosynthetic partners—cyanobacteria, green algae, and plants.

A Lichen Is an Association between a Fungus and a Photosynthetic Partner

You may be familiar with one type of lichen, the leathery patches of various colors growing on certain rocks. Technically, a **lichen** is a single vegetative body that is the result of an association between a fungus and a photosynthetic partner. The fungal partner in a lichen, called the **mycobiont,** usually makes up about 90% of the whole. The other 10% is the photosynthetic partner, called the **photobiont.** Most frequently, these are green algae of the genus *Trebouxia* or cyanobacteria of the genus *Nostoc.* Thousands of ascomycetes and a few basidiomycetes form this kind of symbiosis, but only about 100 photosynthetic species serve as photobionts.

Lichens often live in harsh, dry microenvironments, including on bare rock and wind-whipped tree trunks. Yet lichens have vital ecological roles and important human uses. Lichens secrete acid that eats away at rock, breaking it down and converting it to soil that can support larger plants. Some paleobiologists (who study ancient life forms) have suggested that lichens may have been some of the earliest land organisms, covering bare rocks during the Ordovician period (roughly 500 million to 425 million years ago). In this scenario, millennia of decaying lichens would have created the first soils in which the earliest land plants could grow. Today, lichens continue to enhance the survival of other life forms. For instance, in arctic tundra, where plants are scarce, reindeer and musk oxen can survive by eating lichens. Insects, slugs, and some other invertebrates also consume lichens, and they are nest-building materials for many birds and small mammals. People have derived dyes from lichens; they are even a component of garam masala, an ingredient in Indian cuisine. Some environmental chemists monitor air pollution by observing lichens, most of which cannot grow in heavily polluted air (see *Focus on Applied Research*).

Because lichens are composite organisms, it may seem odd to talk of lichen "species." Biologists do give lichens binomial names, however, based on the characteristics of the mycobiont. More than 13,500 lichens are recognized, each one a unique combination of a particular species of fungus and one or more species of photobiont. The relationship often begins when a fungal mycelium contacts a free-living cyanobacterium, algal cell, or both. The fungus parasitizes the photosynthetic host cell, sometimes killing it. If the host cell can survive, however, it multiplies in association with the fungal hyphae. The result is a tough, pliable body called a **thallus,** which can take a variety of forms **(Figure 28.15A).** Short, specialized hyphae penetrate algal cells of the thallus, which become the fungus's sole source of nutrients. Often, the mycobiont of a lichen absorbs up to 80% of the carbohydrates the photobiont produces.

Benefits for the photobiont are less clear-cut, in part because the drain on nutrients hampers its growth and reproduction. In one view, many and possibly most lichens are parasitic symbioses in which the photobiont does not receive equal benefit. On the other hand, it is relatively rare to find a lichen's photobiont species living independently in the same conditions under which the lichen survives, whereas as part of a lichen it may eke out an enduring existence; some lichens have been dated as being more than 4,000 years old! Studies have also revealed that at least some green algae do clearly benefit from the relationship. Such algae are sensitive to desiccation and intense ultraviolet radiation. Sheltered by a lichen's fungal tissues, a green alga can thrive in locales where alone it would perish. Clearly, we still have quite a bit to learn about the physiological interactions between lichen partners.

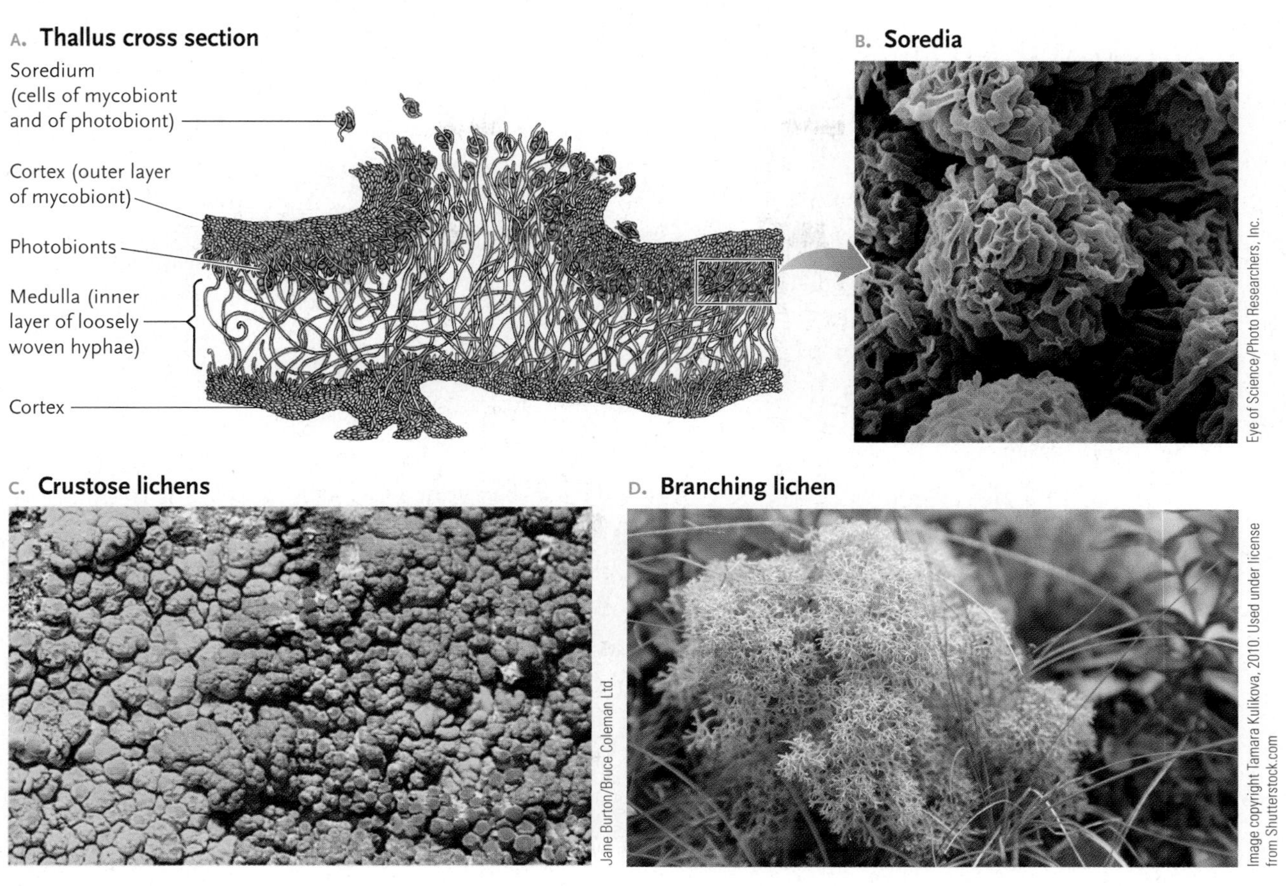
A. Thallus cross section
Soredium (cells of mycobiont and of photobiont)
Cortex (outer layer of mycobiont)
Photobionts
Medulla (inner layer of loosely woven hyphae)
Cortex
B. Soredia
Eye of Science/Photo Researchers, Inc.
C. Crustose lichens
Jane Burton/Bruce Coleman Ltd.
D. Branching lichen
Image copyright Tamara Kulikova, 2010. Used under license from Shutterstock.com

FIGURE 28.15
Lichens. **(A)** Sketch of a cross section through the thallus of the lichen *Lobaria verrucosa.* **(B)** The soredia, which contain both hyphae and algal cells, are a type of dispersal fragment by which lichens reproduce asexually. **(C)** Crustose lichens. **(D)** Erect, branching lichen, *Cladonia rangiferina.*

As you might expect with such a communal life form, reproduction has its quirky aspects. In lichens that involve an ascomycete, the fungus produces ascospores that are dispersed by the wind. The spores germinate to form hyphae that may colonize new photosynthetic cells and so establish new symbioses. A lichen itself can also reproduce in at least two ways. In some types, a section of the thallus detaches and grows into a new lichen. In about one-third of lichens, specialized regions of the thallus give rise asexually to reproductive cell clusters called **soredia** (*soros* = heap) (singular, *soredium*). Each cluster includes both algal and hyphal cells **(Figure 28.15B).** As the lichen grows, the soredia detach and are dispersed by water, wind, or passing animals.

Mycorrhizae Are Symbiotic Associations of Fungi and Plant Roots

A **mycorrhiza** ("fungus root") is a mutualistic symbiosis in which fungal hyphae associate intimately with plant roots. Mycorrhizae greatly enhance the plant's ability to extract various nutrients, especially phosphorus and nitrogen, from soil (see Chapter 33).

In **endomycorrhizae,** the fungal hyphae penetrate the walls of root cells. This kind of association occurs on the roots of nearly all flowering plants, and in most cases a glomeromycete is the fungal partner. The branched hyphae of endomycorrhizae are called arbuscules **(Figure 28.16),** and glomeromycetes are sometimes referred to as arbuscular mycorrhizal fungi.

Basidiomycetes are the usual fungal partners in **ectomycorrhizae (Figure 28.17),** in which hyphal tips grow between and around the young roots of trees and shrubs but never enter the root cells. Ectomycorrhizal associations with trees—often several of them—are very common. For instance, the extensive root system of a mature pine may be studded with ectomycorrhizae involving dozens of fungal species. The musky-flavored truffles *(Tuber melanosporum)* prized by gourmets are ascomycetes that form ectomycorrhizal associations with oak trees (genus *Quercus*).

Orchids are partners in a unique mycorrhizal relationship. The fungal partner, usually a basidiomycete, lives inside the orchid's tissues and provides the plant with a variety of nutrients. In fact, seeds of wild orchids germinate, and seedlings survive, only when such mycorrhizae are present.

In general, mycorrhizae represent a "win-win" situation for the partners. The fungal hyphae absorb carbohydrates made by the plant, along with some amino acids and perhaps growth factors as well. The growing plant in turn absorbs mineral ions made accessible to it by the fungus. Collectively, the fungal hyphae have a huge surface area for absorbing mineral ions from a large volume of the surrounding soil. Dissolved mineral ions accumu-

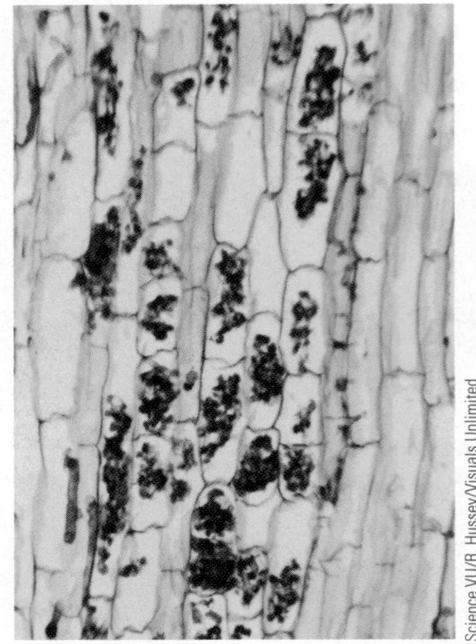
A. Leek root with endomycorrhizae (black)
Science VU/R. Hussey/Visuals Unlimited

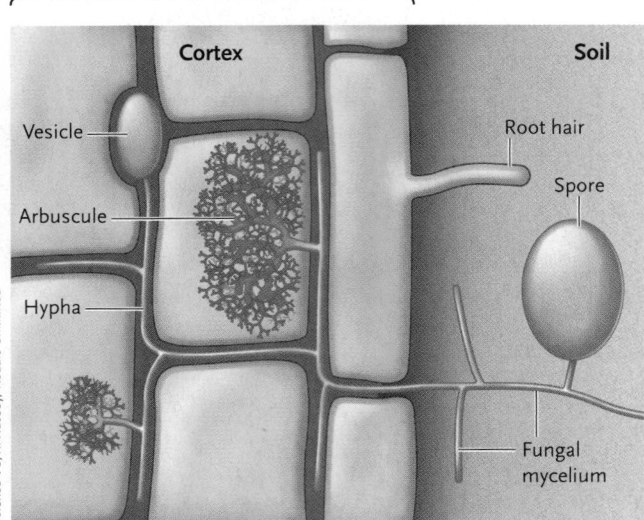
B. Arbuscule
Root
Cortex
Soil
Vesicle
Root hair
Spore
Arbuscule
Hypha
Fungal mycelium

FIGURE 28.16
Endomycorrhizae. **(A)** In this instance, the roots of leeks are growing in association with the glomeromycete *Glomus versiforme* (longitudinal section). Notice the arbuscules that have formed as the fungal hyphae branched after entering the leek root **(B).**

late in the hyphae when they are plentiful in the soil and are released to the plant when they are scarce. This service is a survival boon to a great many plants, especially species that cannot readily absorb mineral ions, particularly phosphorus **(Figure 28.18).** For plants that inhabit soils poor in mineral ions, such as in tropical rain forests, mycorrhizal associations are crucial for survival. Likewise, in temperate forests, species of spruce, oak, pine, and some other trees die unless mycorrhizal fungi are present. Plants that live in dry habitats often rely on specialized mycorrhizal hyphae that serve as conduits for water into the root. Like lichens, mycorrhizae are highly vulnerable to damage from pollutants, especially acid rain.

A. Lodgepole pine
Prof. D. J. Read, University of Sheffield

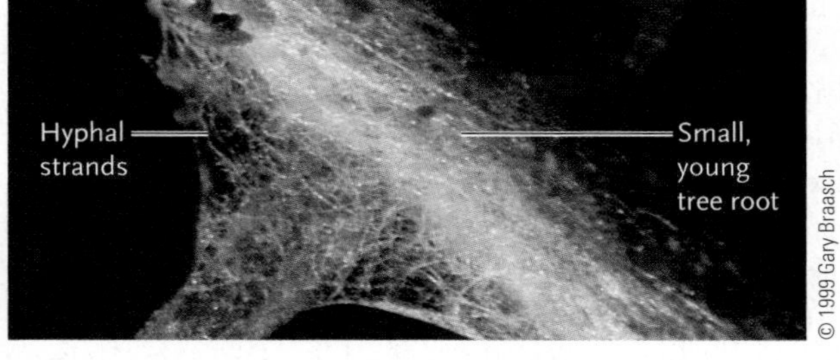
B. Mycorrhiza
Hyphal strands
Small, young tree root
© 1999 Gary Braasch

FIGURE 28.17
Ectomycorrhizae. **(A)** Seedling of pine, *Pinus contorta,* longitudinal section. Notice the extent of the mycorrhiza compared with the above-ground portion of the seedling, which is only about 4 cm tall. **(B)** Mycorrhiza of a hemlock tree.

Mycorrhizae have a long evolutionary history. Fossils show that endomycorrhizae were common among ancient land plants, and some biologists have speculated they might have been key for enhancing the transport of water and minerals to the plants. In that scenario, endomycorrhizae may have played a crucial role in allowing plants to make the transition to life on land.

STUDY BREAK 28.3 <

1. **Explain what a lichen is, and how each partner contributes to the whole.**
2. **Describe the biological and ecological roles of mycorrhizae.**
3. **How do endomycorrhizae and ectomycorrhizae differ?**

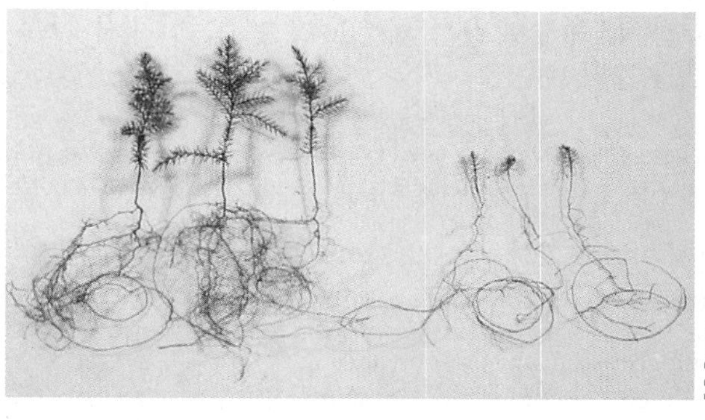
F. B. Reeves

FIGURE 28.18
Effect of mycorrhizal fungi on plant growth. The six-month-old juniper seedlings on the left were grown in sterilized low-phosphorus soil inoculated with a mycorrhizal fungus. The seedlings on the right were grown under the same conditions but without the fungus.

UNANSWERED QUESTIONS

How many species of fungi are there?

The majority of fungi exist as microscopic hyphae or yeasts, invisible to the unaided eye. Even those fungi that produce visible mushrooms generally have a smaller number of distinguishing morphological characters compared to many plants and animals. Because most fungal diversity is "cryptic," estimates of worldwide fungal species richness have been based not on direct observations, but rather on extrapolations of plant-to-fungus ratios. The most widely cited is David Hawksworth's estimate of over 1.5 million fungal species. Currently, DNA sequencing and environmental DNA analyses are providing a window into the previously hidden diversity of fungi. DNA analyses have revealed significant genetic divergence in geographically separated populations of several well-known mushroom-forming fungi, suggesting that many of these "species" are actually multispecies complexes. Environmental DNA sequencing has revealed previously undescribed diversity, even at the level of entire phylogenetic lineages. Is it possible, then, that there are actually more than 1.5 million species of fungi? Though it is too early to answer that question with certainty, even the figures cited at the beginning of this chapter (70,000 described species and over 1.5 million awaiting formal description) powerfully illustrate both the vast diversity of fungi and the limitations of our knowledge about this diversity.

Do different fungal species respond differently to environmental changes?

Ecologists Shigeo Yachi and Michel Loreau formulated the "biological insurance" hypothesis, which predicts that ecosystem functions will be maintained more reliably in natural communities having species that perform the same general ecological function (such as serving as decomposers) but that differ in their responses to environmental changes. When environmental conditions shift in such communities, a species that fills the same ecological role as another but is better suited to the new conditions can then "take over" for a species better adapted to the former conditions. Changes in the makeup of fungal communities resulting from climate change or other ecological perturbations could have important consequences for ecosystem functions such as the availability of soil nutrients, storage of carbon in organic material, and plant growth. For instance, experiments in which fungi were exposed to above-normal concentrations of nitrogen have demonstrated pronounced shifts in the composition of communities of mycorrhizal and saprobic fungi. Studies of CO_2 effects have shown changes in fungal community composition when the fungi were exposed to elevated CO_2 treatments, although several longer-term studies indicate that these differences wane over time. Gaining a better understanding of the mechanisms underlying these patterns and linking changes in community composition to changes in how ecosystems function are important directions for further research.

Think Critically

Ectomycorrhizal fungal communities often exhibit high local diversity, where the root systems of single trees may contain dozens—and a single-host forest stand may contain hundreds—of different fungal symbionts.

1. What factors might allow multiple species of fungi to coexist on a single tree?
2. What factors might favor the dominance of one symbiont over another?
3. How might a better understanding of these phenomena be beneficial to conservation planning?

Todd Osmundson

Todd Osmundson is a postdoctoral researcher at the University of California, Berkeley. His research areas include systematics, biodiversity, ecology, and conservation of fungi and microbes, with an emphasis on ectomycorrhizal basidiomycete fungi. Learn more about Osmundson's work at: http://nature.berkeley.edu/~tosmunds.

REVIEW KEY CONCEPTS

Go to **CENGAGENOW** at www.cengage.com/login to access quizzing, animations, exercises, articles, and personalized homework help.

28.1 General Characteristics of Fungi

- Fungi are key decomposers contributing to the recycling of carbon and some other nutrients. They occur as single-celled yeasts or multicellular filamentous organisms.
- The fungal mycelium consists of filamentous hyphae that grow throughout the substrate the fungus feeds upon (Figure 28.2). A wall containing chitin surrounds the plasma membrane, and in most species septa partition the hyphae into cell-like compartments.
- Pores in septa permit cytoplasm and organelles to move between hyphal cells. Aggregations of hyphae form all other tissues and organs of a multicellular fungus.
- Fungi gain nutrients by extracellular digestion and absorption. Saprobic species feed on nonliving organic matter. Parasitic types obtain nutrients from tissues of living organisms. Many fungi are partners in symbiotic relationships with plants.
- All fungi may reproduce via spores generated either asexually or sexually (Figure 28.3). Some types also may reproduce asexually by budding or fragmentation of the parent body.
- Sexual reproduction usually has two stages. First, in plasmogamy, the cytoplasms of two haploid cells fuse to become a dikaryon containing a haploid nucleus from each parent. Later, in karyogamy, the nuclei fuse and form a diploid zygote. Meiosis then generates haploid spores.

Animation: Mycelium

28.2 Major Groups of Fungi

- The main phyla of fungi are the Chytridiomycota (which have motile spores), Zygomycota (zygospore-forming fungi), Glomeromycota, Ascomycota (sac fungi), and Basidiomycota (club fungi) (Figure 28.4). The phyla traditionally have been distinguished mainly by the structures of sexual reproduction. When a sexual phase cannot be detected or is absent from the life cycle, the specimen is assigned to an informal grouping, the conidial fungi.
- Chytrids usually are microscopic. They are the only fungi that produce motile, flagellated spores. Many are parasites (Figure 28.5).
- Zygomycetes have aseptate hyphae and are coenocytic, with many nuclei in a common cytoplasm. They sometimes reproduce sexually by way of hyphae that occur in + and − mating types; haploid nuclei in the hyphae function as gametes. Further development produces the zygospore, which may become dormant.
- When the zygospore of a zygomycete breaks dormancy, it produces a stalked sporangium containing haploid spores of each mating type, which are released (Figures 28.6 and 28.7).
- Glomeromycetes form a distinct type of endomycorrhizae in association with plant roots. They reproduce asexually, by way of spores that form at the tips of hyphae.
- Most ascomycetes are multicellular (Figure 28.9). In asexual reproduction, chains of haploid asexual spores called conidia elongate or pinch off from the tips of conidiophores (modified aerial hyphae; Figure 28.10).
- In sexual reproduction of ascomycetes, haploid sexual spores called ascospores arise in saclike cells called asci. In the most complex species, reproductive bodies called ascocarps bear or contain the asci. Ascospores can give rise to a new haploid mycelium (Figures 28.8 and 28.11).
- Most basidiomycete species reproduce only sexually. Club-shaped basidia develop on a basidiocarp and bear sexual spores on their surface. When dispersed, these basidiospores may germinate and give rise to a haploid mycelium (Figure 28.13).
- Microsporidia are single-celled sporelike parasites of arthropods, fish, and humans (Figure 28.14).

Animation: Zygomycete life cycle

Animation: Sac fungi

Animation: Club fungus life cycle

28.3 Fungal Associations

- Many ascomycetes and a few basidiomycetes enter into symbioses with cyanobacteria or green algae to produce the communal life form called a lichen, which has a tough, pliable body called a thallus. The algal cells supply the lichen's carbohydrates, most of which are absorbed by the fungus.
- In some lichens a section of the thallus may detach and grow into a new individual. In others, specialized regions of the thallus give rise asexually to reproductive soredia that include both algal and hyphal cells (Figure 28.15).
- In the symbiosis called a mycorrhiza, fungal hyphae make mineral ions and sometimes water available to the roots of a plant partner. The fungus in turn absorbs carbohydrates, amino acids, and possibly other growth-enhancing substances from the plant (Figures 28.16–28.18).
- In endomycorrhizae, the fungal hyphae (usually of a glomeromycete) penetrate the cells of the root. With ectomycorrhizae, hyphae form a netlike structure that surrounds the cortical cells of young roots but hyphae do not enter root cells; the usual fungal partner is a basidiomycete.

Animation: Lichens

Animation: Mycorrhiza

UNDERSTAND AND APPLY

Test Your Knowledge

1. Which of the following statements does *not* reflect current understanding of phylogenetic relationships and features among fungi?
   a. Lineages leading to fungi diverged from those leading to plants about 965 mya.
   b. Chytrids are probably a paraphyletic grouping and may most closely resemble the earliest fungi.
   c. There are four broadly accepted fungal phyla, the Zygomycota, Glomeromycota, Ascomycota, and Basidiomycota.
   d. As fungal lineages diversified, each evolved distinctive reproductive adaptations.
   e. Conidial fungi do not belong to a recognized fungal phylum.

2. Which of the following events is/are a necessary part of a typical asexual cycle in fungal reproduction?
   a. formation of a dikaryon
   b. hyphae developing into a mycelium
   c. formation of a diploid zygote
   d. plasmogamy, which occurs when hyphae fuse at their tips
   e. production and release of large numbers of spores
3. A trait common to all fungi is:
   a. reproduction via spores.
   b. parasitism.
   c. septate hyphae.
   d. a dikaryotic phase inside a zygospore.
   e. plasmogamy after an antheridium and ascogonium come into contact.
4. The chief characteristic used to classify fungi into the major fungal phyla is:
   a. nutritional dependence on nonliving organic matter.
   b. recycling of nutrients in terrestrial ecosystems.
   c. adaptations for obtaining water.
   d. features of reproduction.
   e. cell wall metabolism.
5. At lunch George ate a mushroom, some truffles, a little Camembert cheese, and a bit of moldy bread. Which of the following groups was *not* represented in the meal?
   a. Basidiomycota
   b. Ascomycota
   c. conidial fungi
   d. chytrids
   e. Zygomycota
6. Which of the following fungal reproductive structures are diploid?
   a. basidiocarps
   b. ascospores
   c. conidia
   d. gametangia
   e. zygospores
7. A mushroom is:
   a. the food-absorbing region of an ascomycete.
   b. the food-absorbing region of a basidiomycete.
   c. a reproductive structure formed only by basidiomycetes.
   d. a specialized form of mycelium not constructed of hyphae.
   e. a collection of saclike cells called asci.
8. A zygomycete is characterized by:
   a. usually, aseptate hyphae.
   b. mostly sexual reproduction.
   c. absence of + and − mating types.
   d. the tendency to form mycorrhizal associations with plant roots.
   e. a life cycle in which karyogamy does not occur.
9. Which of the following statements apply/applies to lichens?
   a. It is a fungus that breaks down rock to provide nutrients for an alga.
   b. It colonizes bare rocks and slowly degrades them to small particles.
   c. It spends part of the life cycle as a mycobiont and part as a fungus.
   d. It is an association between a basidiomycete and an ascomycete.
   e. It is an association between a photobiont and a fungus.
10. In a college greenhouse a new employee observes fuzzy mycorrhizae in the roots of all the plants. Destroying no part of the plants, she carefully removes the mycorrhizae. The most immediate result of this "cleaning" is that the plants cannot:
   a. carry out photosynthesis.
   b. absorb water through their roots.
   c. transport water up their stems.
   d. extract as much nitrogen from soil.
   e. store carbohydrates in their roots.

Discuss the Concepts

1. A mycologist wants to classify a specimen that appears to be a new species of fungus. To begin the classification process, what kinds of information on body structures and/or functions must the researcher obtain in order to assign the fungus to one of the major fungal groups?
2. In a natural setting—a pile of horse manure in a field, for example—the sequence in which various fungi appear illustrates ecological succession, which is the replacement of one species by another in a community (see Chapter 50). The first fungi to appear are the most efficient opportunists, for they can form and disperse spores most rapidly. In what order would you expect representatives from each division of fungi to appear on the manure pile? Why?
3. As the text noted, conifers, orchids, and some other types of plants cannot grow properly if their roots do not form associations with fungi, which provide the plant with minerals such as phosphate and in return receive carbohydrates and other nutrients synthesized by the plant. In some instances, however, the plant receives proportionately more nutrients than the fungus does. Even so, biologists still consider this to be a mycorrhizal association. Explain why you agree or disagree.
4. Humans are fundamentally diploid organisms. Explain how this state of affairs compares with the fungal life cycle, then compare the two general life cycles in light of the two groups' overall reproductive strategies.

Design an Experiment

Experiments on the orange bread mold *Neurospora crassa,* an ascomycete, were pivotal in elucidating the concept that each gene encodes a single enzyme. As *N. crassa* ascospores arise through meiosis and then mitosis in an ascus, each ascospore occupies a particular position in the final string of eight spores the ascus contains:

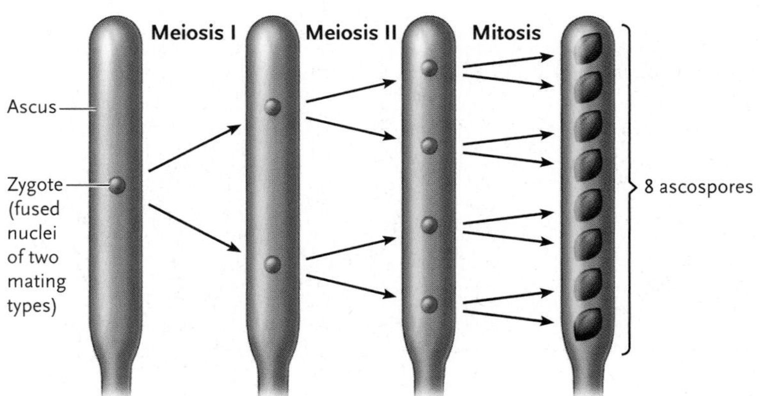
Meiosis I
Meiosis II
Mitosis
Ascus
Zygote (fused nuclei of two mating types)
8 ascospores

This quirk of ascospore development was extremely useful to early geneticists, because it vastly simplified the task of figuring out which alleles ended up in particular ascospores following meiosis. Recalling genetics topics discussed in Chapter 11, why was the analysis easier?

Interpret the Data

The fungus *Cryptococcus neoformans,* a basidiomycete yeast, is a dangerous pathogen in humans who have a weakened immune system, such as HIV/AIDS patients. The disease it causes, cryptococcosis, produces severe symptoms ranging from a pneumonia-like illness to encephalitis. *C. neoformans* thrives in dark places such as soil, dung, and caves, and light inhibits its sexual reproduction by preventing the development of a dikaryon stage. The recent sequencing of its genome has spurred efforts to identify genes that control this inhibition. Alexander Idnurm and Joseph Heitman of Duke University Medical Center performed a series of experiments using wild-type and mutant *C. neoformans* specimens. As described in Section 28. 2, a dikaryon develops when haploid hyphae of two mating types fuse, forming the $n + n$ structure. In the wild *C. neoformans* has two mating types, denoted α and **a.** In both, the normal gene BWC encodes the light-based inhibition response; there is no inhibition in strains having the mutant *bwc1*. In one series of experiments Idnurm and Heitman tested how efficiently fusion events occurred in crosses between matings of wild-type parental hyphae (denoted +) and mutant hyphae (denoted by Δ) that were exposed to different light wavelengths. Colored bars in the figure indicate the color (wavelength) of the light stimulus; black denotes darkness, and red denotes dim light at the red/far-red end of the visible spectrum. Green and blue represent wavelengths toward the other end of the visible spectrum. To interpret this graph you must read the mating combinations vertically.

1. In the dark, did matings in which both partners were either wild-type or mutant forms produce a different response to the varying light stimuli than crosses of a wild-type and a mutant?
2. Within the other categories of light stimuli, how did the makeup of the different mating pairs correlate with the response?

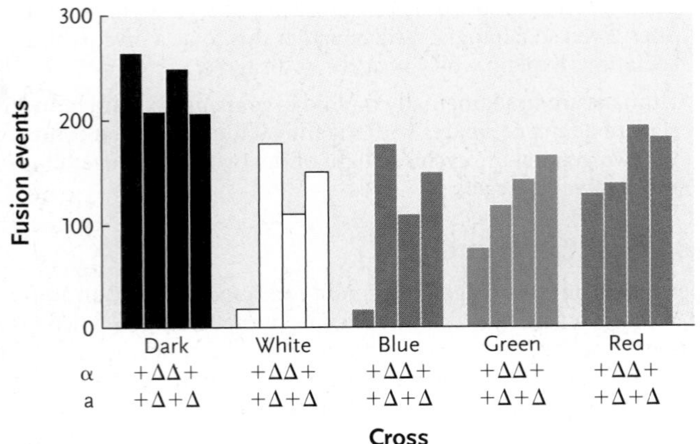
300
200
100
0
Fusion events
Dark
White
Blue
Green
Red
α
a
+ΔΔ+
+Δ+Δ
Cross

Source: A. Idnurm and J. Heitman. 2005. Light controls growth and development via a conserved pathway in the fungal kingdom. *PLoS Biology* 3:e95.

Apply Evolutionary Thinking

The hypothesis that fungi are more closely related to animals than to plants has received support from studies of fungus genomes. For instance, scientists have documented striking similarities in the structure of many fungal and human genes—similarities that may be especially important in medicine. One mycologist, John Taylor of the University of California, Berkeley, suggests that a close biochemical relationship between fungi and animals may explain why fungal infections are typically so resistant to treatment, and why it has proven rather difficult to develop drugs that kill fungi without damaging their human or other animal hosts. About 100 fungal genomes have been or soon will be sequenced, including genomes of several medically important species. If you are a researcher working to develop new antifungal drugs, how could you make use of this growing genetic understanding? Using Internet resources, can you find examples of antifungal drugs that exploit biochemical differences between animals and fungi?

Express Your Opinion

The disappearance of lichens and soil fungi may be an early indication that coal-fired power plants are emitting pollutants that also can endanger human health. Controlling emissions raises the cost of energy for consumers. Should pollution standards for these power plants be tightened? Go to www.cengage.com/login to investigate both sides of the issue, and then vote.

Weaver ants *(Oecophylla longinoda)* carry a leaf to repair their nest in Papua New Guinea.

Mark Moffett/Minden Pictures

29

STUDY OUTLINE

Animal Phylogeny, Acoelomates, and Protostomes

Why It Matters. . . In 1909, a lucky fossil hunter named Charles Walcott tripped over a rock on a mountain path in British Columbia, Canada. Under the force of his hammer, the rock split apart, revealing the discovery of a lifetime. As Walcott and other workers examined rocks at this locality, they soon found fossils of more than 120 species of previously unknown animals from the Cambrian period. These creatures had lived on the muddy sediments of a shallow ocean basin; about 530 million years ago, an underwater avalanche buried them in a rain of silt that was eventually compacted into finely stratified shale. Over millions of years, the shale was uplifted by tectonic activity and incorporated into the mountains of western Canada. It is now known as the Burgess Shale formation.

Some animals in the Burgess Shale were truly bizarre **(Figure 29.1).** For example, *Opabinia* was about as long as a tube of lipstick; it had five eyes on its head and a grasping organ that it may have used to capture prey. No living animals even remotely resemble *Opabinia.* The smaller *Hallucigenia* sported seven pairs of large spines on one side and seven pairs of soft organs on the other. Recent research suggests that *Hallucigenia* may belong in the phylum Onychophora, described in Section 29.7. Nevertheless, most species of the Burgess Shale left no descendants that are still alive today. Thus, this remarkable assemblage of fossils provides a glimpse of some evolutionary novelties that—whether through the action of natural selection or just plain bad luck—were ultimately unsuccessful.

Other animal lineages have shown much greater longevity. Zoologists have described nearly 2 million living species in the kingdom **Animalia.** The familiar **vertebrates,** animals with a backbone,

FIGURE 29.1 Animals of the Burgess Shale. *Opabinia* had five eyes and a grasping organ on its head. *Hallucigenia* had seven pairs of spines and soft protuberances.

Opabinia

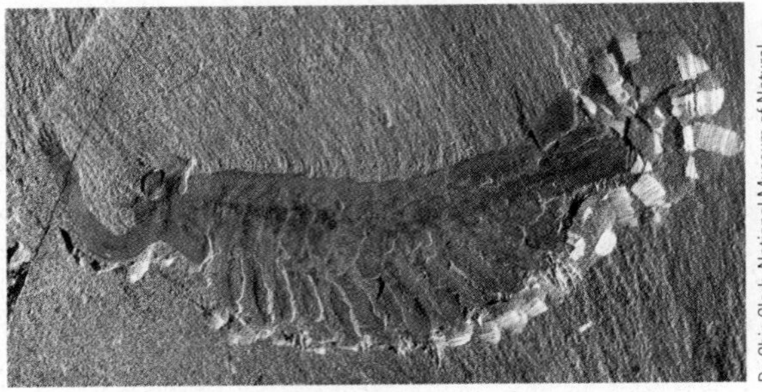

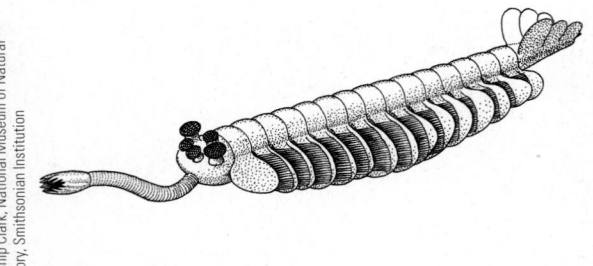

Dr. Chip Clark, National Museum of Natural History, Smithsonian Institution

Hallucigenia

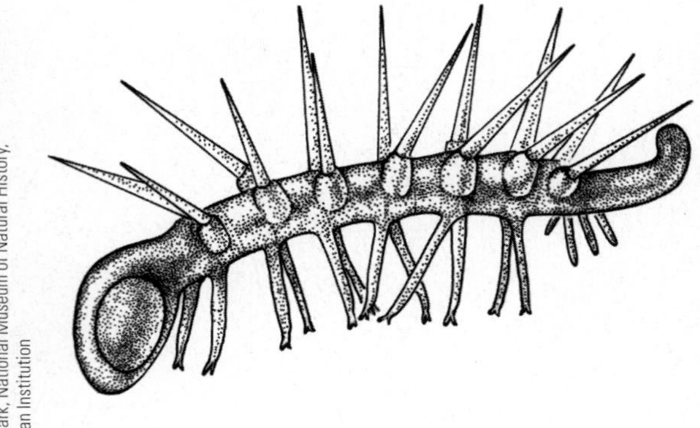

Dr. Chip Clark, National Museum of Natural History, Smithsonian Institution

encompass only a small fraction (about 47,000) of that total. The overwhelming majority of animals fall within the descriptive grouping of **invertebrates,** animals without a backbone.

The remarkable evolutionary diversification of animals resulted from their ability to consume other organisms as food and, for most groups, their ability to move from one place to another. Today, animals are important consumers in nearly every environment on Earth. Their diversification has been accompanied by the evolution of specialized tissues and organ systems as well as complex behaviors.

In this chapter, we introduce the general characteristics of animals and a phylogenetic hypothesis about their evolutionary history and classification. We also survey some of the major invertebrate phyla; a *phylum* is an ancient monophyletic lineage with a distinctive body plan. In Chapter 30 we examine the deuterostome lineage, which includes the vertebrates and their nearest invertebrate relatives. <

29.1 What Is an Animal?

Biologists recognize the Kingdom Animalia as a monophyletic group that is easily distinguished from the other kingdoms.

All Animals Share Certain Structural and Behavioral Characteristics

Animals are eukaryotic, multicellular organisms. Their cells lack cell walls, a trait that differentiates them from plants and fungi. Because the individual cells of most animals are similar in size, very large animals like elephants have many more cells than small ones like fleas. In large animals, most cells are far from the body surface, but specialized tissues and organ systems deliver nutrients and oxygen to them and carry wastes away.

All animals are **heterotrophs:** they eat other organisms to acquire energy and nutrients. Food is ingested (eaten), then digested (broken down), and its breakdown products are absorbed by specialized tissues. Animals use oxygen to metabolize the food they eat through the biochemical pathways of aerobic respiration, and most store excess energy as glycogen, oil, or fat.

All animals are **motile**—able to move from place to place—at some time in their lives. They travel through the environment to find food or shelter and to interact with other animals. Most familiar animals are motile as adults. However, in some species, such as mussels and barnacles, only the young are motile; they eventually settle down as **sessile** (unable to move from one place to another) adults. The advantages of motility have fostered the evolution of locomotor structures, including fins, legs, and wings that are powered by muscles, which are specialized contractile tissues that move individual body parts. Most animals also have sensory and nervous systems that allow them to receive, process, and respond to information about the environment through which they are traveling.

Animals reproduce either asexually or sexually; in some groups they switch from one mode to the other. Sexually reproducing species produce short-lived, haploid **gametes** (eggs and sperm), which fuse to form diploid **zygotes** (fertilized eggs). Animal life cycles generally include a period of development during which mitosis transforms the zygote into a multicelled **embryo;** an embryo develops into a reproductively immature juvenile or a free-living **larva,** which becomes a reproductively mature adult. Larvae often differ markedly from adults and may occupy different habitats and consume different foods.

The Animal Lineage Probably Arose from a Colonial Choanoflagellate Ancestor

An overwhelming body of morphological and molecular evidence indicates that all animal phyla had a common ancestor. For example, all animals share similarities in their cell-to-cell junctions and the molecules in their extracellular matrices (see Section 5.5) as well as similarities in the structure of their ribosomal RNAs.

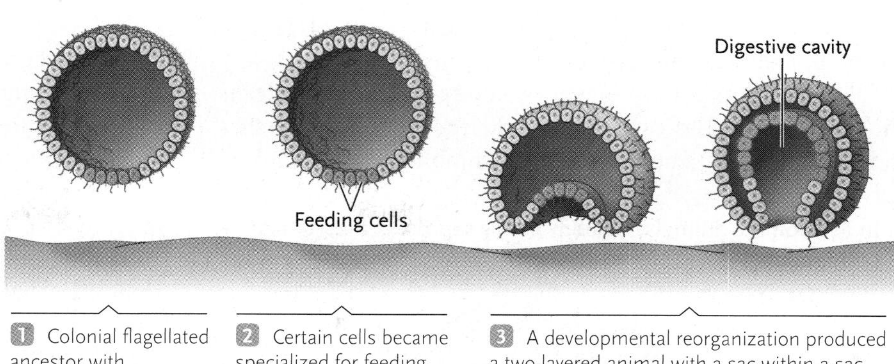
Digestive cavity
Feeding cells
1 Colonial flagellated ancestor with unspecialized cells
2 Certain cells became specialized for feeding and other functions.
3 A developmental reorganization produced a two-layered animal with a sac-within-a-sac body plan.

FIGURE 29.2
Animal origins. Many biologists believe that animals arose from a colonial, flagellated protist in which cells became specialized for specific functions and a developmental reorganization produced two cell layers. The cell movements illustrated here are similar to those that occur during the early development of many animals, as described in Chapter 48.

Most biologists agree that the common ancestor of all animals was probably a colonial, flagellated protist that lived at least 700 million years ago, during the Precambrian era. It may have resembled the minute, sessile choanoflagellates that live in both freshwater and marine habitats today (see Figure 26.22). In 1874 the German embryologist Ernst Haeckel proposed a colonial, flagellated ancestor, suggesting that it was a hollow, ball-shaped organism with unspecialized cells. According to his hypothesis, its cells became specialized for particular functions, and a developmental reorganization produced a double-layered, sac-within-a-sac body plan **(Figure 29.2).** As you will see in Chapter 48, the embryonic development of many living animals roughly parallels this hypothetical evolutionary transformation.

STUDY BREAK 29.1

1. **What characteristics distinguish animals from plants and fungi?**
2. **How does the ability of animals to move through the environment relate to their acquisition of nutrients and energy?**

29.2 Key Innovations in Animal Evolution

Once established, the animal lineage diversified quickly into an amazing array of body plans. Before the development of molecular sequencing techniques, biologists identified several key morphological innovations that they used to develop hypotheses about the evolutionary relationships of the major animal groups.

Tissues and Tissue Layers Appeared Early in Animal Evolution

The presence or absence of **tissues**—groups of cells that share a common structure and function—divides the animal kingdom into two distinct branches. One branch, the sponges, or Parazoa (*para* = alongside, *zoon* = animal), lacks tissues. All other animals, collectively grouped in the Eumetazoa (*eu* = true, *meta* = in common with), have tissues.

During the development of eumetazoans, embryonic tissues form as either two or three concentric **primary cell layers** (described further in Chapter 48). The innermost layer, the **endoderm,** eventually develops into the lining of the gut (digestive system) and, in some animals, respiratory organs. The outermost layer, the **ectoderm,** forms the external covering and nervous system. Between the two, the **mesoderm** forms the muscles of the body wall and most other structures between the gut and the external covering. Some animals have a **diploblastic** body plan that includes only two layers, endoderm and ectoderm. However, most animals are **triploblastic,** having all three primary cell layers.

Most Animals Exhibit either Radial or Bilateral Symmetry

The most obvious feature of an animal's body plan is its shape **(Figure 29.3).** Most animals are **symmetrical;** in other words, their bodies can be divided by a plane into mirror-image halves. The

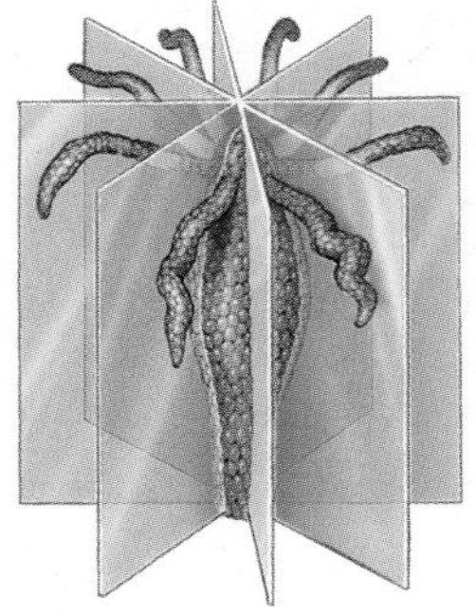

Radial symmetry
Dorsal
Posterior
Anterior
Ventral
Bilateral symmetry

FIGURE 29.3
Patterns of body symmetry. Most animals have either radial or bilateral symmetry.

rare exceptions include most sponges, which have irregular shapes and are therefore **asymmetrical.**

Eumetazoans exhibit one of two body symmetry patterns. The Radiata includes two phyla, Cnidaria (hydras, jellyfishes, and sea anemones) and Ctenophora (comb jellies), which have **radial symmetry.** Their body parts are arranged regularly around a central axis, like the spokes on a wheel. Thus, any cut down the long axis of a hydra divides it into matching halves. Radially symmetrical animals are usually sessile or slow moving and receive sensory input from all directions.

All other eumetazoan phyla fall within the Bilateria, animals that have **bilateral symmetry.** In other words, only a cut along the midline from head to tail divides them into left and right sides that are essentially mirror images of each other. Bilaterally symmetrical animals also have **anterior** (front) and **posterior** (back) ends as well as **dorsal** (upper) and **ventral** (lower) surfaces. As these animals move through the environment, the anterior end encounters food, shelter, or enemies first. Thus, in bilaterally symmetrical animals, natural selection also favored **cephalization,** the development of an anterior head where sensory organs and nervous system tissue are concentrated.

A. In acoelomate animals, no body cavity separates the gut and body wall.

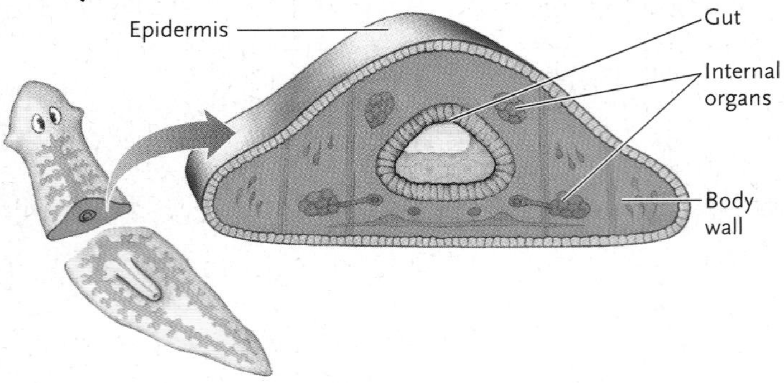
Epidermis
Gut
Internal organs
Body wall

B. In pseudocoelomate animals, the pseudocoelom forms between the gut (a derivative of endoderm) and the body wall (a derivative of mesoderm).

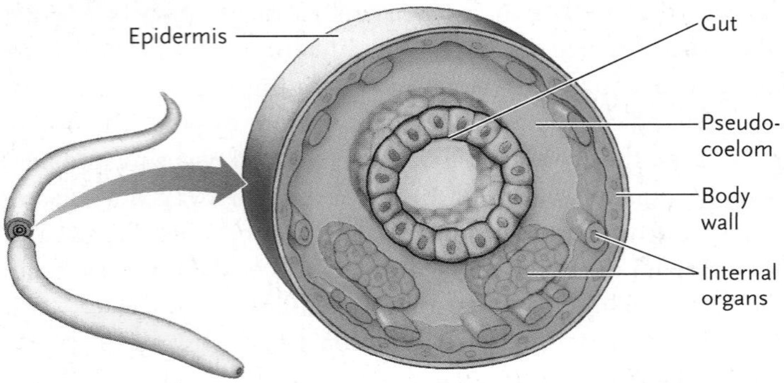
Epidermis
Gut
Pseudo-coelom
Body wall
Internal organs

C. In coelomate animals, the coelom is completely lined by peritoneum (a derivative of mesoderm).

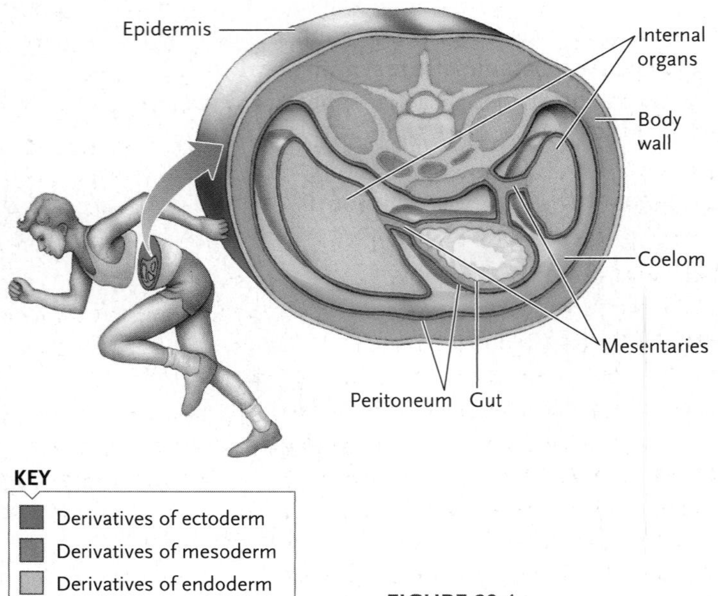
Epidermis
Internal organs
Body wall
Coelom
Mesentaries
Peritoneum
Gut

KEY
- Derivatives of ectoderm
- Derivatives of mesoderm
- Derivatives of endoderm
- Body cavity

FIGURE 29.4
Body plans of triploblastic animals.

Many Animals Have Body Cavities That Surround Their Internal Organs

The body plans of many bilaterally symmetrical animals include a body cavity that separates the gut from the muscles of the body wall **(Figure 29.4). Acoelomate** animals (*a* = not, *koilomat* = hollow), such as flatworms (phylum Platyhelminthes), lack such a cavity; a continuous mass of tissue, derived largely from mesoderm, packs the region between the gut and the body wall (see Figure 29.4A). **Pseudocoelomate** animals (*pseudo* = false), including the roundworms (phylum Nematoda) and wheel animals (phylum Rotifera), have a **pseudocoelom,** a fluid- or organ-filled space between the gut and the muscles of the body wall (see Figure 29.4B). Internal organs lie within the pseudocoelom and are bathed by its fluid. **Coelomate** animals have a **coelom,** a fluid-filled body cavity, also between the gut and body wall, that is completely lined by the **peritoneum,** a thin tissue derived from mesoderm (see Figure 29.4C). Membranous extensions of the inner and outer layers of the peritoneum, the **mesenteries,** surround the internal organs and suspend them within the coelom.

Biologists describe the body plan of pseudocoelomate and coelomate animals as a "tube within a tube"; the digestive system forms the inner tube, the body wall forms the outer tube, and the body cavity lies between them. The body cavity separates internal organs from the body wall, allowing the organs to function independently of whole-body movements. The fluid within the cavity also protects delicate organs from mechanical damage. And, because the volume of the body cavity is fixed, the incompressible fluid within it serves as a **hydrostatic skeleton,** which provides support. In some animals muscle contractions can shift the fluid, changing the animals' shape and allowing them to move from place to place (see Section 41.2).

Developmental Patterns Mark a Major Divergence in Animal Ancestry

Embryological and molecular evidence suggests that bilaterally symmetrical animals are divided into two lineages: the protostomes, which includes most phyla of invertebrates, and the deuterostomes, which includes the vertebrates and their nearest invertebrate relatives. Protostomes and deuterostomes differ in several developmental characteristics **(Figure 29.5)**.

Shortly after fertilization, an egg undergoes a series of mitotic divisions called **cleavage** (see Section 48.2). The first two cell divisions divide a zygote as you might slice an apple, cutting it into four wedges from top to bottom. In many protostomes, subsequent cell divisions produce daughter cells that lie *between* the pairs of cells below them; this pattern is called **spiral cleavage** (left side of Figure 29.5A). In deuterostomes, by contrast, subsequent cell divisions produce a mass of cells that are stacked directly above and below one another; this pattern is called **radial cleavage** (right side of Figure 29.5A).

Protostomes and deuterostomes often differ in the timing of important developmental events. During cleavage, certain genes are activated at specific times, determining a cell's developmental path and ultimate fate. Many protostomes undergo **determinate cleavage:** each cell's developmental path is determined as the cell is produced. Thus, one cell isolated from a two- or four-cell protostome embryo cannot develop into a functional embryo or larva. By contrast, many deuterostomes have **indeterminate cleavage:** the developmental fates of cells are determined later. A cell isolated from a four-cell deuterostome embryo will develop into a functional, although smaller than usual, embryo or larva. Like other deuterostomes, humans have indeterminate cleavage; thus, the two cells produced by the first cleavage division sometime separate and develop into identical twins.

As development proceeds after cleavage, some cells migrate to the inside of the embryo through the **blastopore,** an opening on its surface. These cells become the endoderm, which forms the **archenteron,** the developing gut (see Figure 29.5B). Later in development, a second opening at the opposite end of the embryo transforms the pouchlike archenteron into a digestive tube (see Figure 29.5C). In protostomes (*protos* = first, *stoma* = mouth), the blastopore develops into the mouth, and the second opening forms the anus. In some deuterostomes (*deuteros* = second), the blastopore develops into the anus, and the second opening becomes the mouth.

Protostomes and deuterostomes also differ in the origin of mesoderm and the coelom (see Figure 29.5B). In most protostomes, mesoderm originates from a few specific cells near the blastopore. As the mesoderm grows and develops, it splits into inner and outer layers. The space between the layers is called a **schizocoelom** (*skhizein* = split). In deuterostomes, mesoderm often forms from outpocketings of the archenteron. The space pinched off from the archenteron by the outpocketing mesoderm is called an **enterocoelom** (*enteron* = intestine).

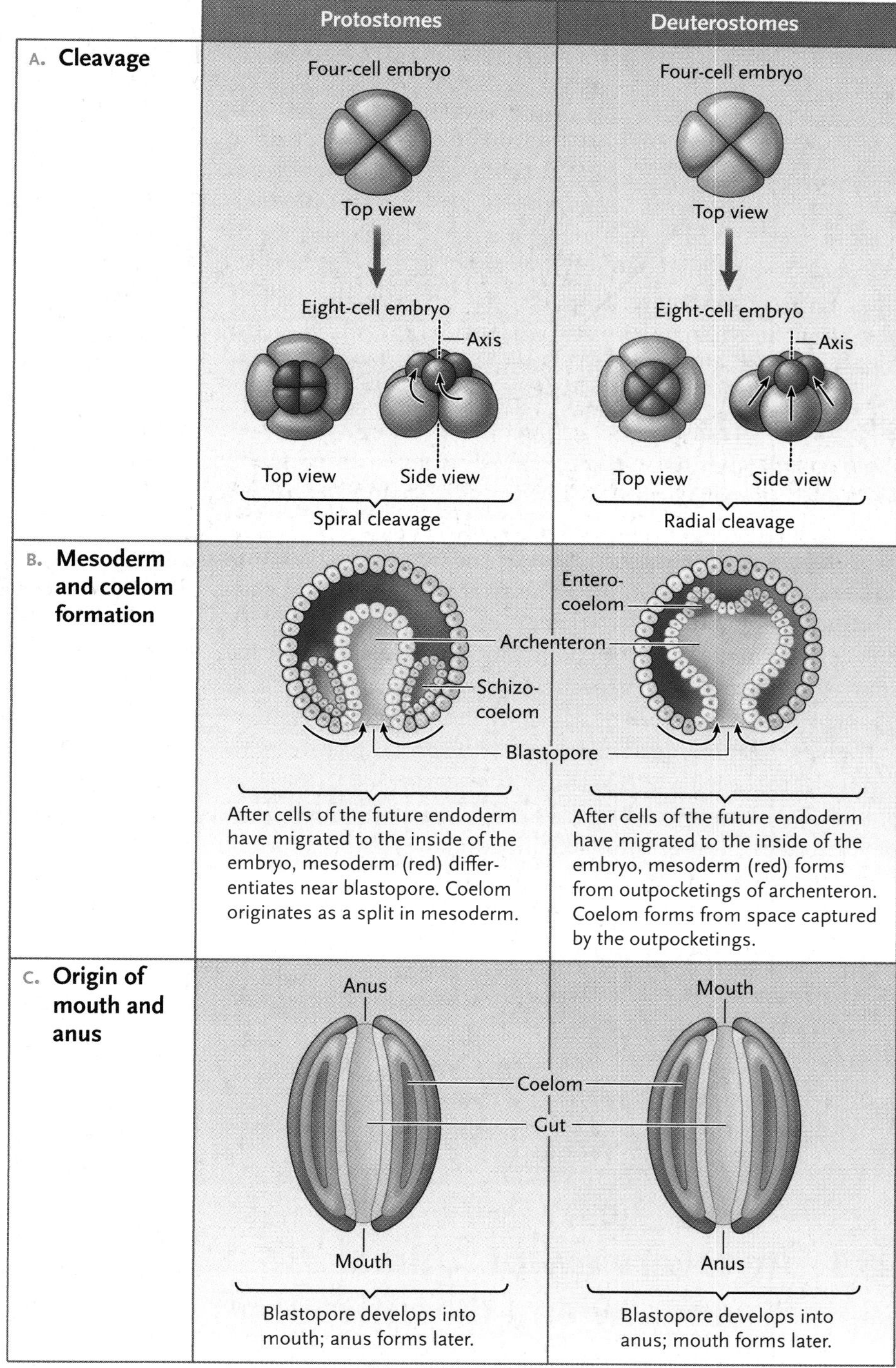

Protostomes
Deuterostomes
A. Cleavage
Four-cell embryo
Four-cell embryo
Top view
Top view
Eight-cell embryo
Eight-cell embryo
Axis
Axis
Top view
Side view
Top view
Side view
Spiral cleavage
Radial cleavage
B. Mesoderm and coelom formation
Entero-coelom
Archenteron
Schizo-coelom
Blastopore
After cells of the future endoderm have migrated to the inside of the embryo, mesoderm (red) differentiates near blastopore. Coelom originates as a split in mesoderm.
After cells of the future endoderm have migrated to the inside of the embryo, mesoderm (red) forms from outpocketings of archenteron. Coelom forms from space captured by the outpocketings.
C. Origin of mouth and anus
Anus
Mouth
Coelom
Gut
Mouth
Anus
Blastopore develops into mouth; anus forms later.
Blastopore develops into anus; mouth forms later.
KEY
Derivatives of ectoderm
Derivatives of mesoderm
Derivatives of endoderm
Coelom (body cavity)

FIGURE 29.5
Protostomes and deuterostomes. The two lineages of coelomate animals differ in **(A)** cleavage patterns, **(B)** the origin of mesoderm and the coelom, and **(C)** the origin of the mouth and the anus.

Several other characteristics also differ in adult protostomes and deuterostomes. For example, the central nervous system of protostomes is generally positioned on the ventral side of the body, and their brain surrounds the anterior section of the digestive tract. By contrast, the nervous system and brain of deuterostomes lie on the dorsal side of the body.

Segmentation Divides the Bodies of Some Animals into Repeating Units

Some phyla in both the protostome and deuterostome lineages exhibit varying degrees of **segmentation,** the production of body parts as repeating units. During development, segmentation first arises in the mesoderm, the middle tissue layer that produces most of the body's bulk. In humans and other vertebrates, we see evidence of segmentation in the vertebral column (backbone), ribs, and associated muscles, such as the "six-pack abs" that sit-ups accentuate. In some animals, segmentation is also reflected in structures derived from the endoderm and ectoderm. For example, the ring-like pattern on an earthworm or a caterpillar matches the underlying segments.

Segmentation provides several advantages. In markedly segmented animals, such as earthworms and their relatives, each segment may include a complete set of important organs, including respiratory surfaces and parts of the nervous, circulatory, and excretory systems. Thus, a segmented animal may survive damage to the organs in one segment, because those in other segments perform the same functions. Segmentation also improves control over movement, especially in wormlike animals. Each segment has its own set of muscles, which can act independently of those in other segments. Thus, an earthworm can move its anterior end to the left while it swings its posterior end to the right. Similarily, the segmented backbone and body wall musculature of vertebrates allow greater flexibility of movement than would unsegmented structures.

STUDY BREAK 29.2 <

1. **What is a tissue, and what three primary tissue layers are present in the embryos of most animals?**
2. **What type of body symmetry do humans have?**
3. **What is the functional significance of the coelom?**
4. **What does having a segmented body allow some animals to do?**

29.3 An Overview of Animal Phylogeny and Classification

For many years, biologists used the morphological innovations and embryological patterns described earlier to trace the phylogenetic history of animals. These efforts were sometimes hampered by the difficulty of identifying homologous structures in different phyla and by morphological data that led to contradictory interpretations. More recently, researchers have used molecular sequence data to reanalyze animal relationships. Although biologists now recognize nearly 40 animal phyla, in this chapter we focus primarily on the phyla that include substantial numbers of species.

Molecular Analyses Have Refined Our Understanding of Animal Phylogeny

Molecular analyses of animal relationships are often based on nucleotide sequences in small subunit ribosomal RNA and mitochondrial DNA (see Chapter 15). Recent analyses of *Hox* gene sequences provide similar results. (*Hox* genes are described in Sections 16.4 and 22.6.) These molecular analyses are still based on studies of relatively few genes, rather than entire genomes. Thus, the phylogenetic tree based on molecular sequences **(Figure 29.6)** represents a working hypothesis; its details will likely change as researchers accumulate more data.

The phylogenetic tree based upon molecular characters includes all of the major lineages that biologists had defined using morphological innovations and embryological characters. For example, molecular data confirm the distinctions between the Parazoa and the Eumetazoa and between the Radiata and the Bilateria. They also confirm the separation of the deuterostome phyla from all others within the Bilateria.

However, the molecular phylogeny groups many other phyla—including the acoelomate animals, pseudocoelomate animals, protostomes, and a few others—into one taxon, the Protostomia. This group is, in turn, subdivided into two major lineages, the Lophotrochozoa and the Ecdysozoa, groups that were not previously recognized. The name Lophotrochozoa (*lophos* = crest, *trochos* = wheel) refers to both the lophophore, a feeding structure found in three phyla (illustrated in Figure 29.15), and the trochophore, a type of larva found in annelids and mollusks (illustrated in Figure 29.23). The name Ecdysozoa (*ekdysis* = a stripping or casting off) refers to the cuticle or external skeleton that species within this group secrete and periodically molt (or "cast off") when they experience a growth spurt or begin a different stage of the life cycle (illustrated in Figure 29.34); the molting process is called **ecdysis.**

The Molecular Phylogeny Reveals Some Surprising Patterns in the Evolution of Key Morphological Innovations

Phylogenetic trees contain explicit hypotheses about evolutionary change, and the molecular phylogeny has forced biologists to reevaluate the evolution of several important morphological innovations. For example, traditional phylogenies based upon morphology and embryology usually inferred that the absence of a body cavity, the acoelomate condition, was an ancestral state and that the presence of a body cavity, the pseudocoelomate or coelomate condition, was derived. But the molecular tree provides a very different view. Because most protostome phyla have a schizocoelom, the molecular tree suggests that this trait is ancestral within the lineage, having evolved in the common ancestor of the lineage. If that hypothesis is correct, then the acoelomate condition of flatworms (phylum Platyhelminthes) represents the evolutionary *loss* of the schizocoelom, *not* an ancestral

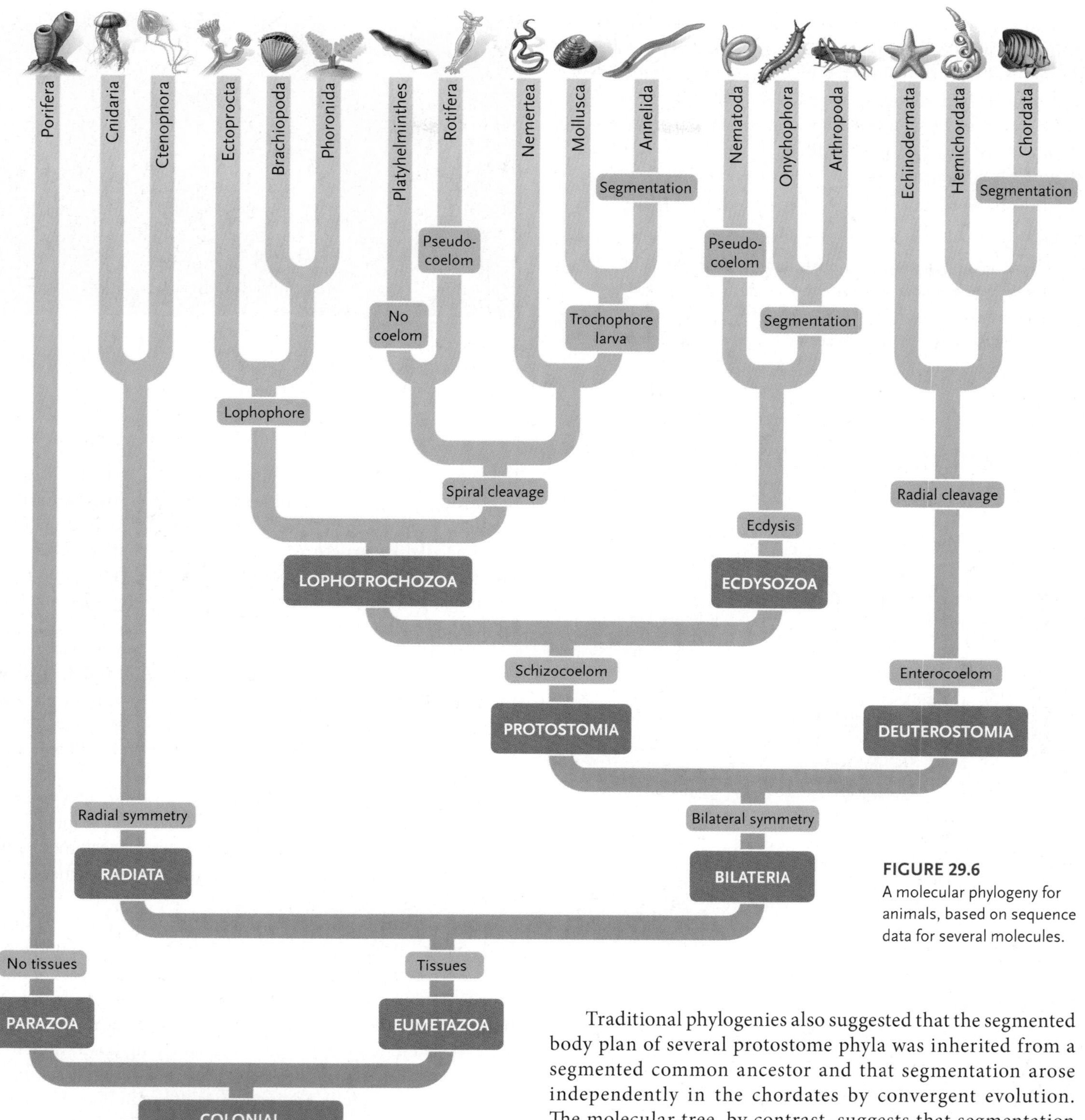
Porifera
Cnidaria
Ctenophora
Ectoprocta
Brachiopoda
Phoronida
Platyhelminthes
Rotifera
Nemertea
Mollusca
Annelida
Nematoda
Onychophora
Arthropoda
Echinodermata
Hemichordata
Chordata
Segmentation
Segmentation
Pseudo-coelom
Pseudo-coelom
No coelom
Trochophore larva
Segmentation
Lophophore
Spiral cleavage
Radial cleavage
Ecdysis
LOPHOTROCHOZOA
ECDYSOZOA
Schizocoelom
Enterocoelom
PROTOSTOMIA
DEUTEROSTOMIA
Radial symmetry
Bilateral symmetry
RADIATA
BILATERIA
No tissues
Tissues
PARAZOA
EUMETAZOA
COLONIAL CHOANOFLAGELLATE ANCESTOR

FIGURE 29.6
A molecular phylogeny for animals, based on sequence data for several molecules.

condition. Similarly, the molecular tree hypothesizes that the pseudocoelom evolved independently in rotifers (Lophotrochozoa, phylum Rotifera) and in roundworms (Ecdysozoa, phylum Nematoda) as modifications of the ancestral schizocoelom. Thus, according to the molecular tree, the pseudocoelomate condition of these organisms is the product of convergent evolution.

Traditional phylogenies also suggested that the segmented body plan of several protostome phyla was inherited from a segmented common ancestor and that segmentation arose independently in the chordates by convergent evolution. The molecular tree, by contrast, suggests that segmentation evolved independently in *three* lineages—segmented worms (Lophotrochozoa, phylum Annelida), arthropods and velvet worms (Ecdysozoa, phyla Arthropoda and Onychophora), and chordates (Deuterostomia, phylum Chordata)—rather than in just two lineages.

Despite these surprising findings about morphological evolution, most biologists now embrace the hypothesis provided by molecular sequence studies. In the future, new data will undoubtedly foster active discussions, heated disputes, and revisions of the phylogeny. You, the student may be understandably frustrated by the lack of consensus among experts and an ever-changing phylogeny for animals. But these disputes highlight the uniqueness of science as a way of knowing the natural world through the process of collecting evidence and rigorously challenging accepted hypotheses. The phylogenetic tree based on molecular sequence data is truly revolutionary, and the dust has yet to settle on the disagreements that new analyses provoke.

STUDY BREAK 29.3

1. **Which major groupings of animals defined on the basis of morphological characters have been confirmed by molecular sequence studies?**
2. **What type of body cavity is ancestral within the Protostomes?**

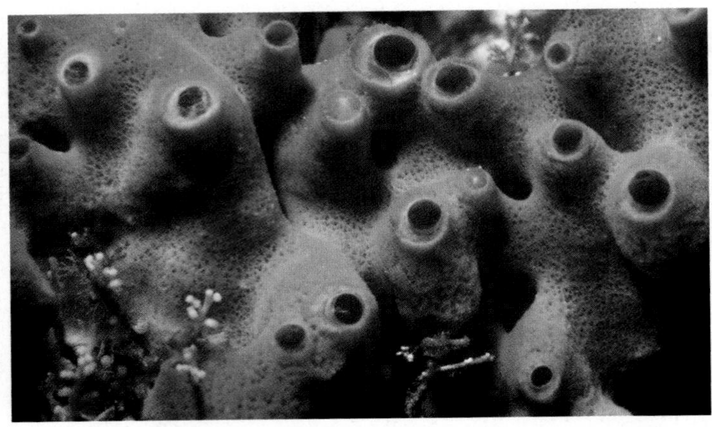
© WATKINS, BRUCE/Animals Animals

FIGURE 29.7
Asymmetry in sponges. Sponges, such as this *Haliclona* species from the Philippines, often have an asymmetrical shape.

29.4 Animals without Tissues: Parazoa

The Parazoa is a lineage that includes just one group of animals, the sponges.

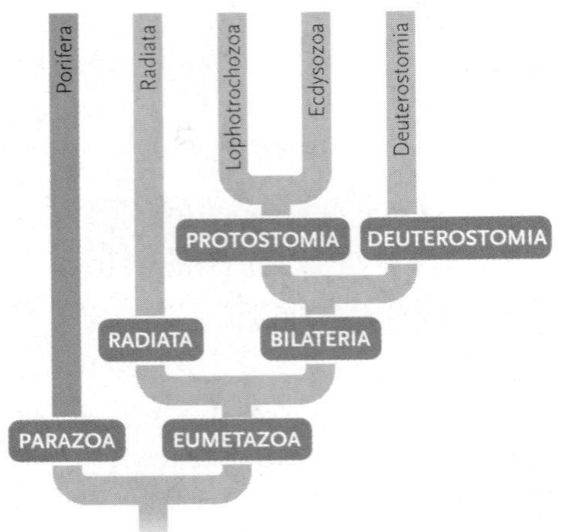
Porifera
Radiata
Lophotrochozoa
Ecdysozoa
Deuterostomia
PROTOSTOMIA
DEUTEROSTOMIA
RADIATA
BILATERIA
PARAZOA
EUMETAZOA

Sponges Have Simple Body Plans and Lack Tissues

Sponges, phylum Porifera (meaning "pore bearers"), lack true tissues: during development, their cells do not form the layers typical of other phyla. Mature sponges are sessile, and their shapes are less fixed than those of other animals, because mobile cells allow them to change shape in response to local conditions **(Figure 29.7).** Sponges have been abundant since the Cambrian, especially in shallow coastal areas. Most of the 8,000 living species of sponges are marine and range in size from 1 cm to 2 m at maturity.

Sponges have simple body plans **(Figure 29.8).** Flattened cells form an outer layer, the **pinacoderm.** The inner surface of saclike sponges is lined by collar cells, called **choanocytes,** each equipped with a beating flagellum and a surrounding "collar" of modified microvilli. (Choanocytes resemble the cells of choanoflagellates, the hypothesized ancestor of all animals.) Amoeboid cells wander through the gelatinous **mesohyl** between the two layers; they secrete a supporting skeleton of a fibrous protein and *spicules,* small needlelike structures of calcium carbonate or silica (see Figure 29.8D). The natural sponges that are used for bathing and washing are the fibrous remains of the bath sponge (*Spongia* species), which lacks mineralized parts.

A sponge's body is an elaborate system for filtering food particles from the surrounding water. Water flows through pores in the pinacoderm into a central chamber, the **spongocoel,** and then out of the sponge through one or more openings called **oscula** (singular, *osculum*). The beating flagellae of the choanocytes maintain a constant flow of water, and contractile pore cells *(porocytes)* adjust the flow rate. Even a small sponge may filter as much as 20 liters of water per day. The choanocytes capture suspended particles and microorganisms from the water and pass this food to mobile amoeboid cells, which carry nutrients to cells of the pinacoderm.

Most sponges are **hermaphroditic:** individuals produce both sperm and eggs. Sperm are released into the spongocoel and then out into the environment through oscula; eggs remain in the mesohyl, where sperm from other sponges, drawn in with water, fertilize them. Zygotes develop into flagellated larvae that are expelled to fend for themselves. Surviving larvae attach to substrates and undergo **metamorphosis** (a reorganization of form) into sessile adults. Some sponges also reproduce asexually; small fragments break off an adult and grow into new sponges. Many species also produce *gemmules,* clusters of cells with a resistant covering that allows them to survive unfavorable conditions; gemmules germinate into new sponges when conditions improve.

STUDY BREAK 29.4

1. **What type of body symmetry do sponges exhibit?**
2. **How does a sponge gather food from its environment?**

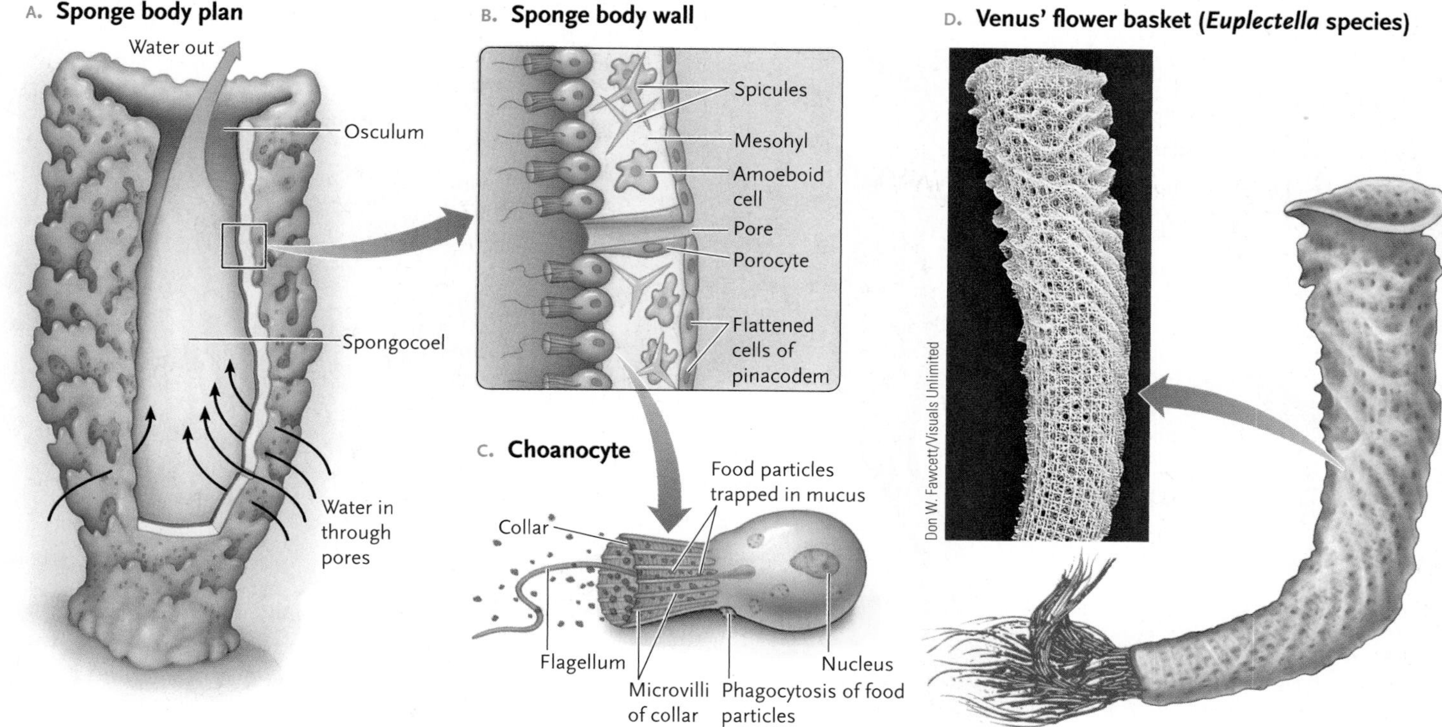
A. Sponge body plan
Water out
Osculum
Spongocoel
Water in through pores
B. Sponge body wall
Spicules
Mesohyl
Amoeboid cell
Pore
Porocyte
Flattened cells of pinacodem
C. Choanocyte
Food particles trapped in mucus
Collar
Flagellum
Microvilli of collar
Phagocytosis of food particles
Nucleus
D. Venus' flower basket (Euplectella species)
Don W. Fawcett/Visuals Unlimited

FIGURE 29.8

The body plan of sponges. Most sponges have **(A)** simple body plans and **(B)** relatively few cell types. **(C)** Beating flagella on the choanocytes create a flow of water through incurrent pores, into the spongocoel, and out through the osculum. **(D)** Venus' flower basket, a marine sponge, has spicules of silica fused into a rigid framework.

29.5 Eumetazoans with Radial Symmetry

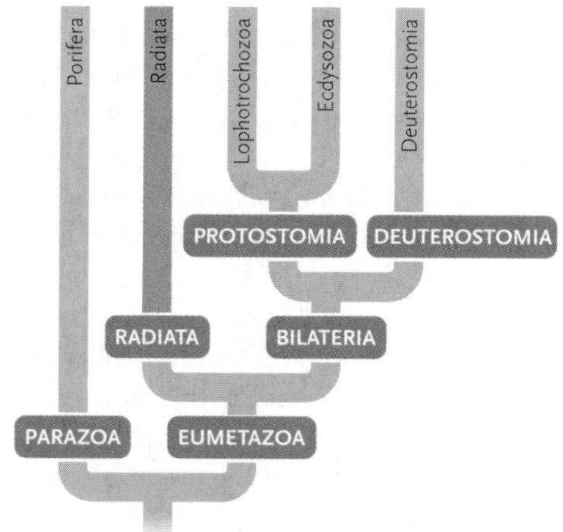
Porifera
Radiata
Lophotrochozoa
Ecdysozoa
Deuterostomia
PROTOSTOMIA
DEUTEROSTOMIA
RADIATA
BILATERIA
PARAZOA
EUMETAZOA

Unlike sponges, eumetazoans have true tissues, which develop from distinct layers in the embryo. Working together, the cells of a tissue perform complex functions beyond the capacity of individual cells. For example, nerve tissue transmits information rapidly through an animal's body, and epithelial tissue forms barriers that surround the body and line body cavities. In this section, we describe eumetazoans with radial symmetry, which enables them to sense stimuli from all directions, an effective adaptation for life in open water.

Two phyla of soft-bodied organisms, Cnidaria and Ctenophora, have radial symmetry and tissues, but they lack organ systems and a coelom. Species in both phyla possess a **gastrovascular cavity** that serves both digestive and circulatory functions. It has a single opening, the mouth. Gas exchange and excretion occur by diffusion because no cell is far from a body surface.

The radiate phyla have a diploblastic body plan with only two tissue layers, the inner *gastrodermis* (an endoderm derivative) and the outer *epidermis* (an ectoderm derivative). Most species also possess a gelatinous *mesoglea* (*mesos* = middle, *glia* = glue) between the two layers. The mesoglea contains widely dispersed fibrous and amoeboid cells.

Cnidarians Use Nematocysts to Stun or Kill Prey

Nearly all of the 8,900 species in the phylum Cnidaria (*knide* = stinging nettle, a plant with irritating surface hairs) live in the sea **(Figure 29.9).** Their body plan is organized around a saclike gastrovascular cavity; the mouth is ringed with tentacles, which push food into it. Cnidarians may be vase-shaped, upward-pointing **polyps** or bell-shaped, downward-pointing **medusae** (see Figure 29.9A). Most polyps attach to a substrate at the *aboral* (opposite the mouth) end; medusae are unattached and float.

Cnidarians are the simplest animals that exhibit a division of labor among specialized tissues (see Figure 29.9B and C). (Sponges have specialized cells, but no tissues.) The gastrodermis includes gland cells and phagocytic nutritive cells. Gland cells secrete enzymes into the gastrovascular cavity for the extracellular digestion of food, which is then engulfed by nutritive cells and exposed to intracellular digestion. The epidermis includes nerve cells, sensory cells, contractile cells, and cells specialized

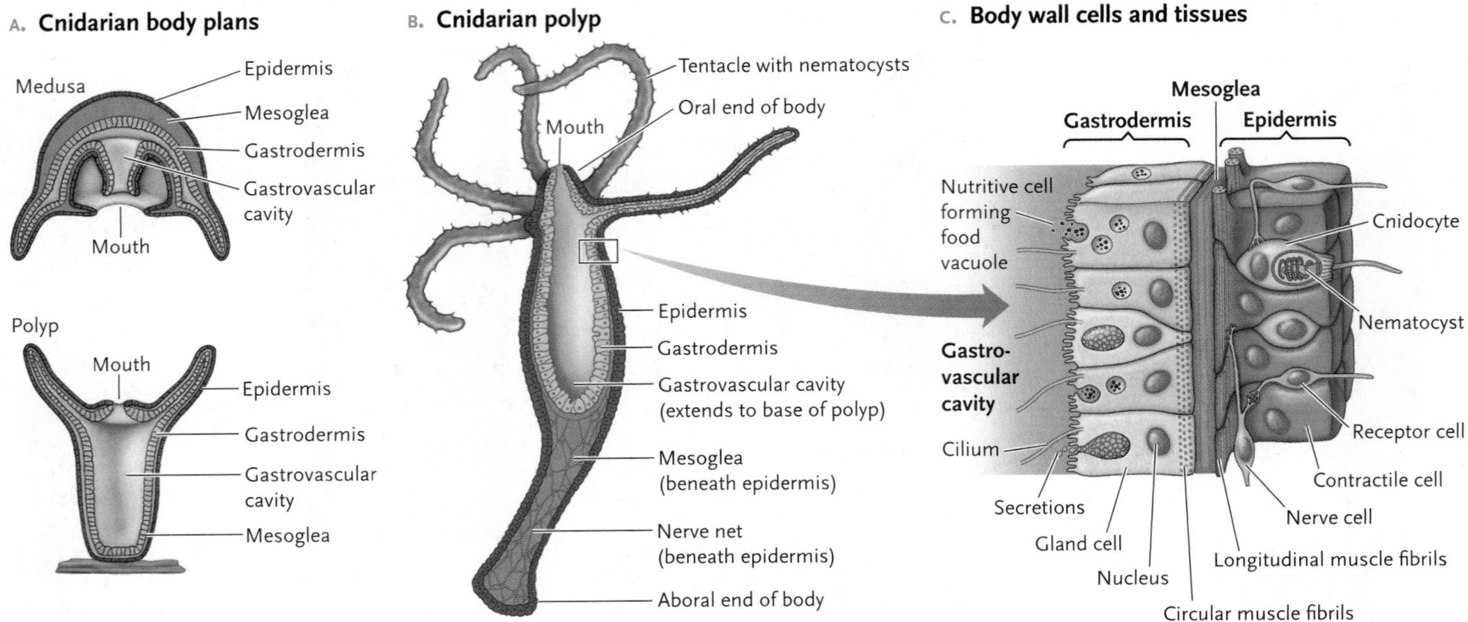
A. Cnidarian body plans
Medusa
Epidermis
Mesoglea
Gastrodermis
Gastrovascular cavity
Mouth
Polyp
Mouth
Epidermis
Gastrodermis
Gastrovascular cavity
Mesoglea
B. Cnidarian polyp
Tentacle with nematocysts
Oral end of body
Mouth
Epidermis
Gastrodermis
Gastrovascular cavity (extends to base of polyp)
Mesoglea (beneath epidermis)
Nerve net (beneath epidermis)
Aboral end of body
C. Body wall cells and tissues
Mesoglea
Gastrodermis
Epidermis
Nutritive cell forming food vacuole
Cnidocyte
Nematocyst
Gastro-vascular cavity
Receptor cell
Cilium
Contractile cell
Secretions
Nerve cell
Gland cell
Longitudinal muscle fibrils
Nucleus
Circular muscle fibrils

FIGURE 29.9
The cnidarian body plan. **(A)** Cnidarians exist as either polyps or medusae. **(B)** The body of both forms is organized around a gastrovascular cavity, which extends all the way to the aboral end of the animal. **(C)** The two tissue layers in the body wall, the gastrodermis and the epidermis, include a variety of cell types.

for prey capture. A layer of acellular mesoglea separates the gastrodermis from the epidermis.

Cnidarians prey on crustaceans, fishes, and other animals. The epidermis includes unique cells, **cnidocytes,** each armed with a stinging **nematocyst (Figure 29.10).** The nematocyst is an encapsulated, coiled thread that is fired at prey or predators, sometimes releasing a toxin through its tip. Discharge of nematocysts may be triggered by touch, vibrations, or chemical stimuli. The toxin can paralyze small prey by disrupting nerve cell membranes. The painful stings of some jellyfishes and corals result from the discharge of nematocysts.

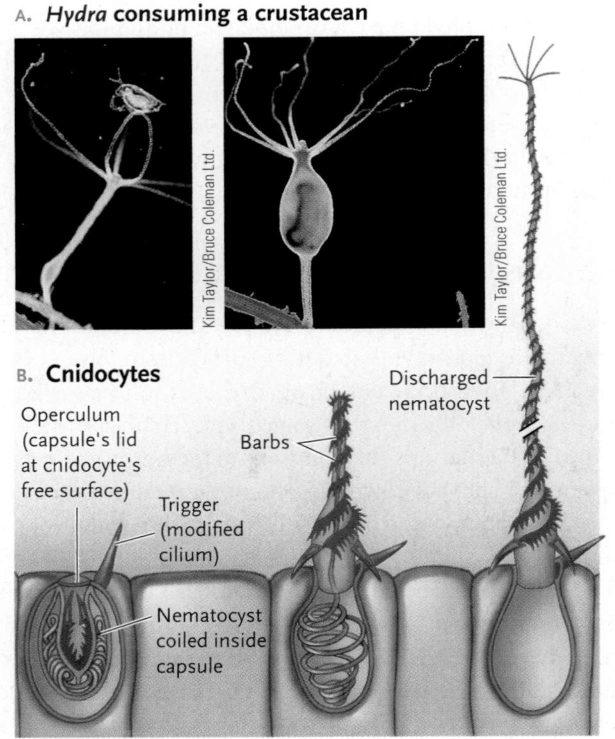
A. *Hydra* consuming a crustacean
Kim Taylor/Bruce Coleman Ltd.
Kim Taylor/Bruce Coleman Ltd.
B. Cnidocytes
Operculum (capsule's lid at cnidocyte's free surface)
Trigger (modified cilium)
Barbs
Discharged nematocyst
Nematocyst coiled inside capsule

FIGURE 29.10
Predation by cnidarians. **(A)** A polyp of a freshwater *Hydra* captures a small crustacean with its tentacles and swallows it whole. **(B)** Cnidocytes, special cells on the tentacles, encapsulate nematocysts, which are discharged at prey.

Cnidarians engage in directed movements by contracting specialized cells in the epidermis. In medusae, the mesogleal jelly serves as a deformable skeleton against which contractile cells act. Rapid contractions narrow the bell, forcing out jets of water that propel the animal. Polyps use their water-filled gastrovascular cavity as a hydrostatic skeleton. When some cells contract, fluid within the chamber is shunted about, changing the body's shape and moving it in a particular direction.

The **nerve net,** which threads through both tissue layers, coordinates responses to stimuli, but it has no central control organ or brain. Impulses initiated by sensory cells are transmitted in all directions from the site of stimulation.

Many cnidarians exist in only the polyp or the medusa form, but some have a life cycle that alternates between them **(Figure 29.11).** In the latter type, the polyp often produces new individuals asexually from buds that break free of the parent (see Figure 47.2). The medusa is often the sexual stage, producing sperm and eggs, which are released into the water. The four lineages of Cnidaria differ in the form that predominates in the life cycle.

HYDROZOA Most of the 2,700 species in the Hydrozoa have both polyp and medusa stages in their life cycles (see Figure 29.11). The polyps form sessile colonies that develop asexually from one individual. A colony can include thousands of polyps, which may be specialized for feeding, defense, or reproduction. They share food through their connected gastrovascular cavities.

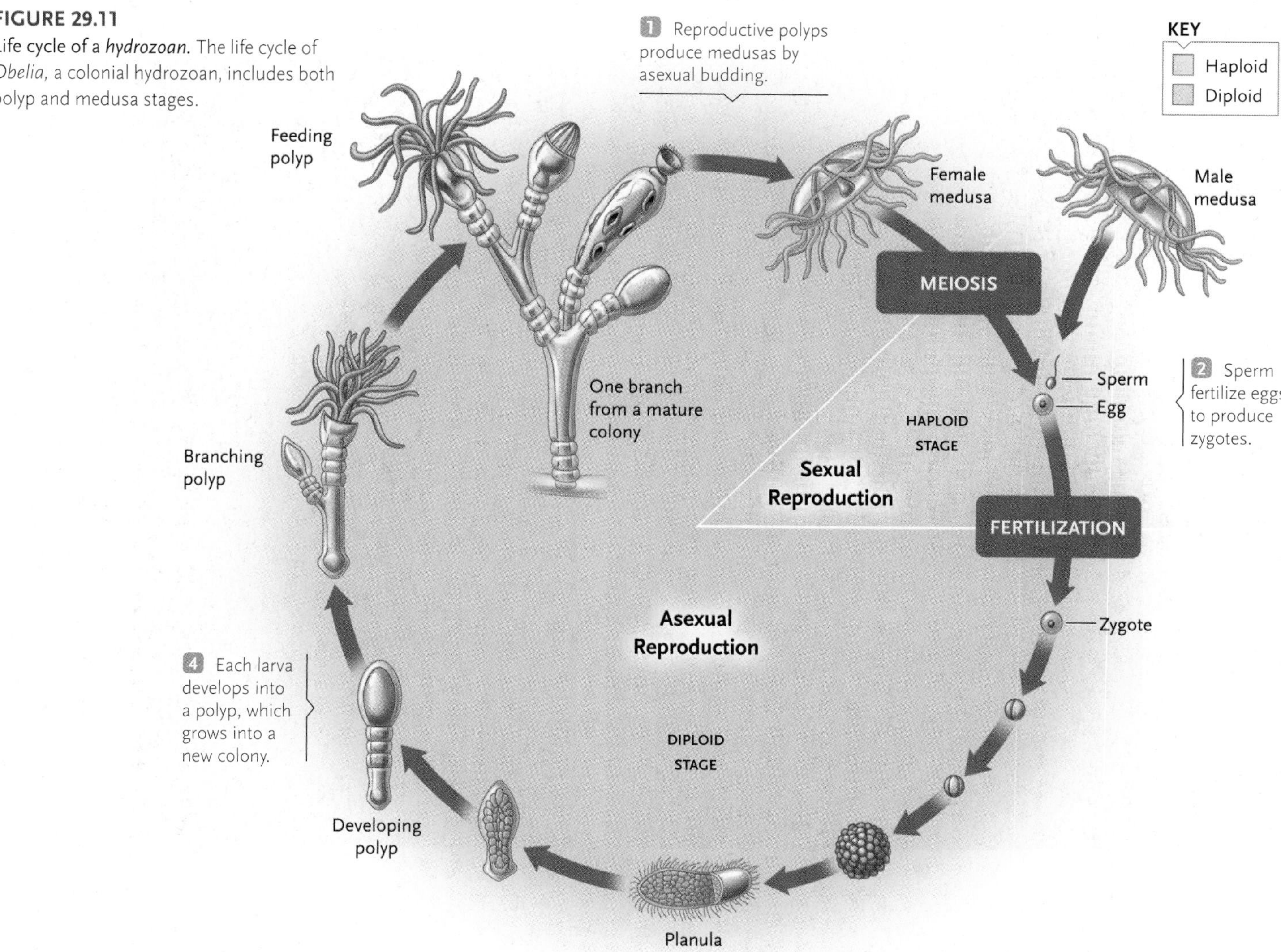
1 Reproductive polyps produce medusas by asexual budding.
KEY
Haploid
Diploid
Feeding polyp
Female medusa
Male medusa
MEIOSIS
One branch from a mature colony
Sperm
Egg
2 Sperm fertilize eggs to produce zygotes.
HAPLOID STAGE
Sexual Reproduction
Branching polyp
FERTILIZATION
Zygote
Asexual Reproduction
4 Each larva develops into a polyp, which grows into a new colony.
DIPLOID STAGE
Developing polyp
Planula
3 The zygote develops into a crawling or swimming planula larva.

FIGURE 29.11
Life cycle of a *hydrozoan.* The life cycle of *Obelia,* a colonial hydrozoan, includes both polyp and medusa stages.

Unlike most hydrozoans, freshwater species of *Hydra* (see Figure 29.10A) live as solitary polyps that attach temporarily to rocks, twigs, and leaves. Under favorable conditions, hydras reproduce by budding. Under adverse conditions they produce eggs and sperm; the zygotes, which are encapsulated in a protective coating, develop and grow when conditions improve.

SCYPHOZOA The medusa stage predominates in the 200 species within the Scyphozoa **(Figure 29.12A),** which are commonly known as jellies or jellyfishes. Scyphozoans range from 2 cm to more than 2 m in diameter. Nerve cells near the margin of the bell control their tentacles and coordinate the rhythmic activity of contractile cells, which move the animal. Specialized sensory cells are clustered at the edge of the bell: **statocysts** sense gravity and **ocelli** are sensitive to light. Scyphozoan medusae are either male or female; they release gametes into the water where fertilization takes place.

CUBOZOA The 20 known species of box jellyfish, the Cubozoa **(Figure 29.12B),** exist primarily as cube-shaped medusas only a few centimeters tall; the largest species grows to 25 cm in height. Nematocyst-rich tentacles grow in clusters from the four corners of the boxlike medusa, and groups of light receptors and image-forming eyes occur on the four sides of the bell. Unlike the true jellyfish, cubozoans are active swimmers. They feed on small fishes and invertebrates, immobilizing their prey with one of the deadliest toxins produced by animals. Cubozoans live in tropical and subtropical coastal waters, where they sometimes pose a serious threat to swimmers.

ANTHOZOA The Anthozoa includes 6,000 species of corals and sea anemones **(Figure 29.13).** Anthozoans exist only as polyps, and often reproduce by budding or fission; most also reproduce sexually. Corals are always sessile and usually colonial. Most species build calcium carbonate skeletons, which sometimes accumulate into gigantic underwater reefs. The energy needs of many corals are partly fulfilled by the photosynthetic activity of symbiotic protists that live within the corals' cells. For this reason, corals are restricted to shallow water where sunlight can penetrate. Sea anemones, by contrast, are soft-bodied, solitary polyps, ranging from 1 cm to 10 cm in diameter. They occupy shallow coastal waters. Most species are essentially sessile, but some move by crawling slowly or by using their gastrovascular cavity as a hydrostatic skeleton.

A. **Scyphozoan *(Chrysaora quinquecirrha)***

Michael Durham/Minden Pictures

B. **Cubozoan *(Chironex fleckeri)***

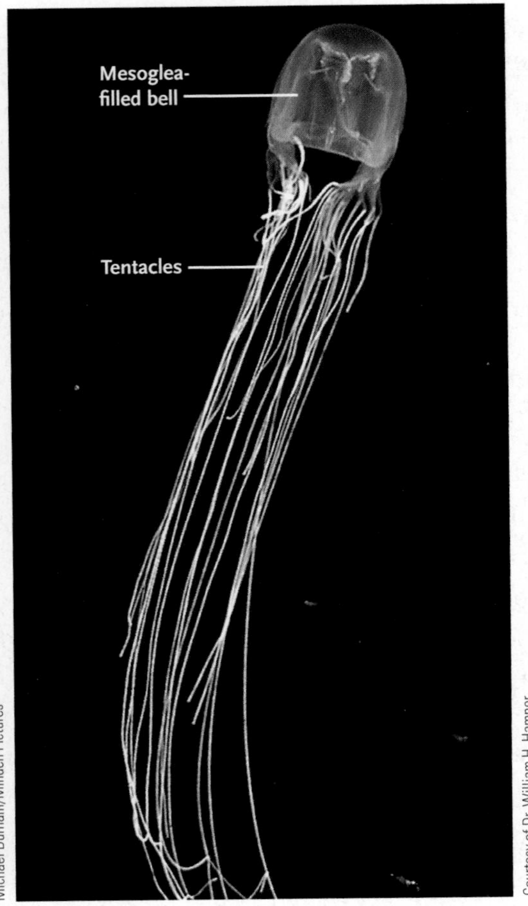
Mesoglea-filled bell
Tentacles
Courtesy of Dr. William H. Hamner

FIGURE 29.12
Scyphozoans and cubozoans. **(A)** Most scyphozoans, like the sea nettle, live as floating medusae. Their tentacles trap prey, and the long oral arms transfer it to the mouth on the underside of the bell. **(B)** Cubozoans, like the sea wasp, are strong swimmers that actively pursue small fishes and invertebrates.

A. **Staghorn coral *(Acropora cervicornis)***

Tentacle of one polyp
Interconnected skeletons of polyps of a colonial coral
Christian DellaCorte

B. **Sea anemone *(Urticina lofotensis)* escape behavior**

F. S. Westmorland

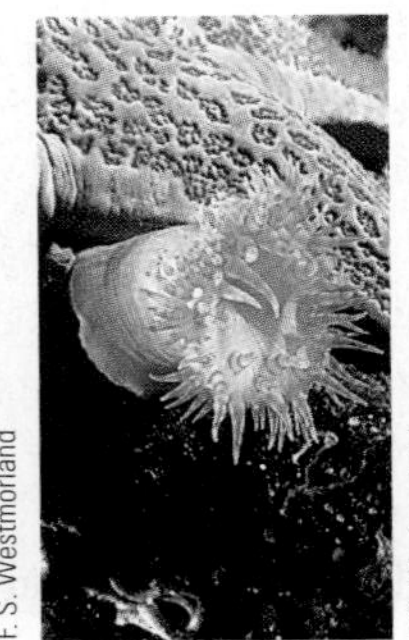
F. S. Westmorland

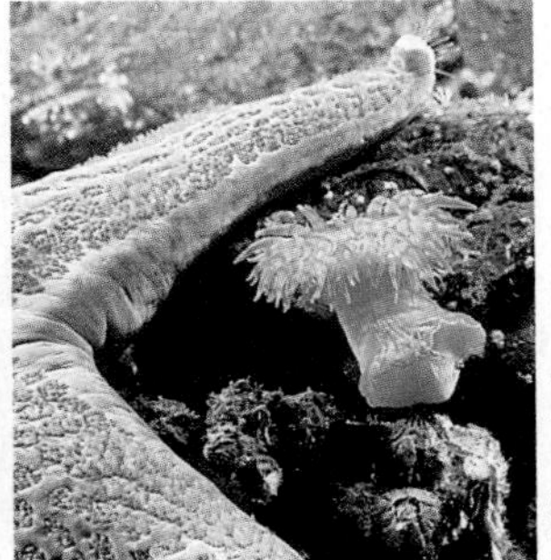
F. S. Westmorland

FIGURE 29.13
Anthozoans. **(A)** Many corals are colonial, and their polyps build a hard skeleton of calcium carbonate. The skeletons accumulate to form coral reefs in shallow tropical waters (see Figure 49.29B). **(B)** A white-spotted sea anemone detaches from its substrate to escape from a predatory sea star.

Ctenophores Use Tentacles to Feed on Microscopic Plankton

Like the cnidarians, the 100 species of comb jellies in the phylum Ctenophora (*ktenos* = comb, *-phoros* = bearing) have radial symmetry, mesoglea, and feeding tentacles. However, they differ from cnidarians in significant ways: they lack nematocysts; they expel some waste through anal pores opposite the mouth; and certain tissues appear to be of mesodermal origin. These transparent, and often luminescent, animals range in size from a few millimeters to a few meters **(Figure 29.14).** They live primarily in coastal regions of the oceans.

Ctenophores move by beating cilia arranged on eight longitudinal plates that resemble combs. They are the largest animals to use cilia for locomotion, but they are feeble swimmers. Nerve cells connected to the cilia coordinate the animals' movements, and a gravity-sensing statocyst helps them maintain an upright position. They capture microscopic plankton with their two tentacles, which have specialized cells that discharge sticky filaments; food particles are ingested as the food-laden tentacles are drawn across the mouth. Ctenophores are hermaphroditic, producing gametes in cells that line the gastrovascular cavity. Eggs and sperm are expelled through the mouth or from special pores, and fertilization occurs in the open water.

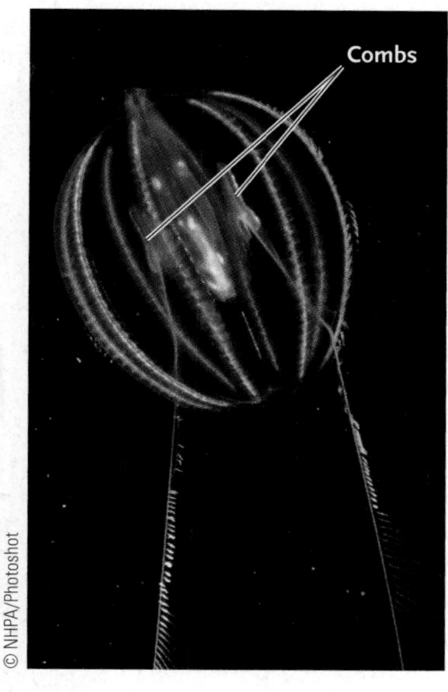
Combs
© NHPA/Photoshot

FIGURE 29.14
Ctenophores. Comb jellies, like this sea gooseberry (*Pleurobrachia pileus*) from the Gulf of Maine, collect microscopic prey on their long, sticky tentacles and then wipe the food-laden tentacles across their mouths.

STUDY BREAK 29.5 <

1. How do cnidarians capture, consume, and digest their prey?
2. Which group of cnidarians has only a polyp stage in its life cycle?
3. What do ctenophores eat, and how do they collect their food?

29.6 Lophotrochozoan Protostomes

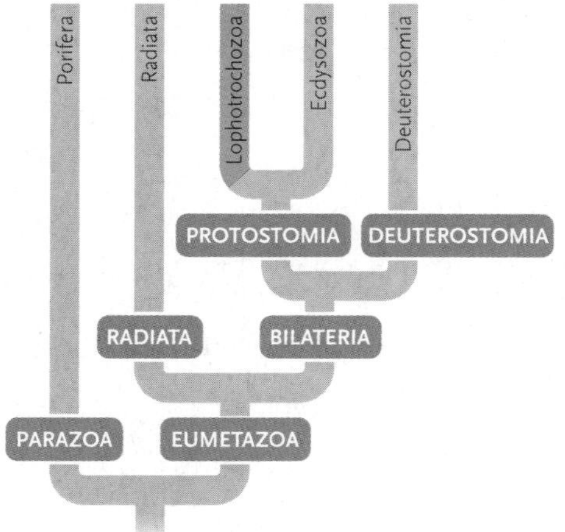
Porifera
Radiata
Lophotrochozoa
Ecdysozoa
Deuterostomia
PROTOSTOMIA
DEUTEROSTOMIA
RADIATA
BILATERIA
PARAZOA
EUMETAZOA

The remaining organisms described in this chapter fall within the group Bilateria: they have bilateral symmetry and a greater variety of tissues than do the Radiata. Bilaterians also have organ systems, structures that include two or more tissue types that perform specific functions. Most of them also possess a coelom or pseudocoelom. With bilateral symmetry and sensory organs concentrated at the anterior end of the body, most bilaterians engage in highly directed, often rapid movements in pursuit of food, mates, or to escape danger. And their complex organ systems accomplish tasks more efficiently than simple tissues. For example, animals that have a tubular digestive system surrounded by a space (the coelom or pseudocoelom) use muscular contractions of the digestive system to move ingested food past specialized epithelia that break it down and absorb the breakdown products.

Molecular analyses group eight of the phyla that we consider into the Lophotrochozoa, one of the two main protostome lineages (see Figure 29.6).

The Lophophorate Phyla Share a Distinctive Feeding Structure

Three small groups of aquatic, coelomate animals—the phyla Ectoprocta, Brachiopoda, and Phoronida—possess a **lophophore,** a circular or U-shaped fold with one or two rows of hollow, ciliated tentacles surrounding the mouth **(Figure 29.15).** Molecular sequence data as well as the lophophore suggest that these phyla share a common ancestry.

The lophophore, which looks like a crown of tentacles at the anterior end of the animal, serves as a site for gas exchange and waste elimination as well as for food capture. Most lophophorates are sessile suspension-feeders as adults: movement of cilia on the tentacles brings food-laden water toward the lophophore, the tentacles capture small organisms and debris, and the cilia transport them to the mouth. The lophophorates have a complete digestive system, which is U-shaped in most species, with the anus lying outside the ring of tentacles.

PHYLUM ECTOPROCTA The Ectoprocta (sometimes called Bryozoa) are tiny colonial animals that mainly occupy marine habitats (see Figure 29.15A). They secrete a hard covering over their soft bodies and feed by extending the lophophore through a hole. Each colony, which may include more than a million individuals, is produced asexually by a single animal. Ectoprocta colonies are permanently attached to solid substrates, where they form encrusting mats, bushy upright growths, or jellylike blobs. Nearly 4,000 living species are known.

FIGURE 29.15
Lophophorate animals. Although the lophophorate animals differ markedly in appearance, they all use a lophophore—the feathery structures in the photos—to acquire food.

A. **Ectoprocta (*Plumatella repens*)**

© blickwinkel/Hecker/Alamy

B. **Brachiopoda (*Terebraulina septentrionalis*)**

© Andrew J. Martinez/Photo Researchers, Inc.

C. **Phoronida (*Phoronopsis californica*)**

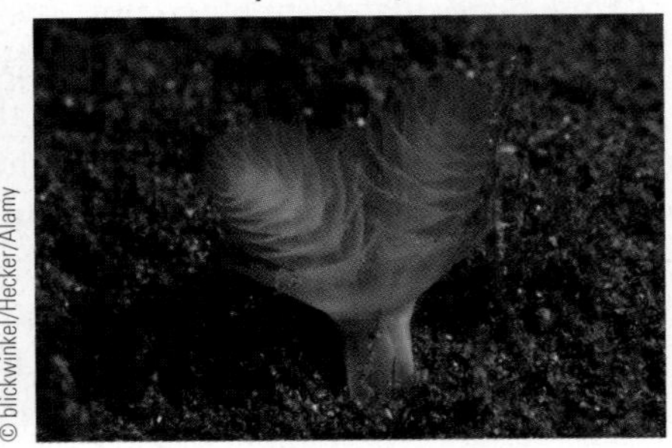

© Lawrence Naylor/Photo Researchers, Inc.

PHYLUM BRACHIOPODA The Brachiopoda, or lampshells, have two calcified shells that develop on the animal's dorsal and ventral sides (see Figure 29.15B). Most species attach to substrates with a stalk that protrudes through one of the shells. The lophophore is held within the two shells, and the animal feeds by opening its shell and drawing water over its tentacles. Although only 250 species of brachiopods live today, more than 30,000 extinct species are known from the fossil record, mostly from Paleozoic seas.

PHYLUM PHORONIDA The 18 or so species of phoronid worms vary in length from a few millimeters to 25 cm (see Figure 29.15C). They usually build tubes in soft ocean sediments or on hard substrates, and feed by protruding the lophophore from the top of the tube. The animal can withdraw into the tube when disturbed. Phoronida reproduce both sexually and by budding.

Flatworms Have Digestive, Excretory, Nervous, and Reproductive Systems, but Lack a Coelom

The 13,000 flatworm species in the phylum Platyhelminthes (*platys* = flat; *helminth-* = worm) live in aquatic and moist terrestrial habitats. Like cnidarians, flatworms can swim or float in water, but they are also able to crawl over surfaces. They range from less than 1 mm to more than 60 cm in length, but most are just a few millimeters thick. Free-living species eat live prey or decomposing carcasses, whereas parasitic species consume the tissues of living hosts.

Like the radiate phyla, flatworms are acoelomate, but they have a complex structural organization that reflects their triploblastic construction **(Figure 29.16).** Endoderm lines the digestive cavity with cells specialized for the chemical breakdown and absorption of ingested food. Mesoderm, the middle tissue layer, produces muscles and reproductive organs. Ectoderm produces a ciliated epidermis, the nervous system, and the *flame cell system,* a simple excretory system; flame cells regulate the concentrations of salts and water within body fluids, allowing free-living flatworms to live in freshwater habitats. Flatworms do not have circulatory or respiratory systems, but, because all cells of their dorsoventrally (top-to-bottom) flattened bodies are near an interior or exterior surface, diffusion supplies them with nutrients and oxygen.

The flatworm nervous system includes two or more longitudinal ventral nerve cords interconnected by numerous smaller nerve fibers, like rungs on a ladder. An anterior **ganglion,** a concentration of nervous system tissue that serves as a primitive brain, integrates their behavior. Most free-living species have *ocelli,* or eye spots, that distinguish light from dark and tiny chemoreceptor organs that sense chemical cues.

FIGURE 29.16
Flatworms. The phylum Platyhelminthes, exemplified by a freshwater planarian, have well-developed digestive, excretory, nervous, and reproductive systems. Because flatworms are acoelomate, their organ systems are embedded in a solid mass of tissue between the gut and the epidermis.

Digestive system

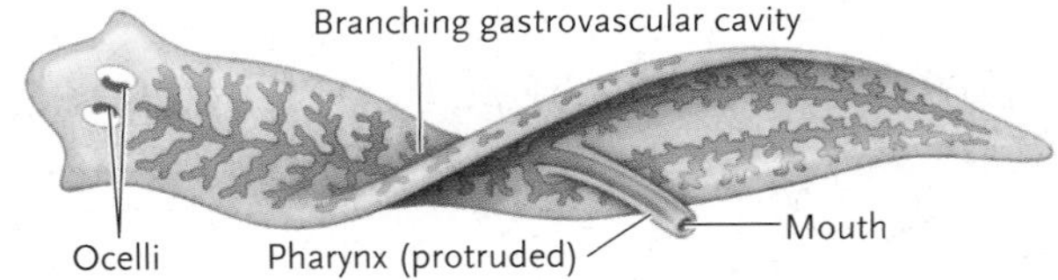
Branching gastrovascular cavity
Ocelli
Pharynx (protruded)
Mouth

Nervous system

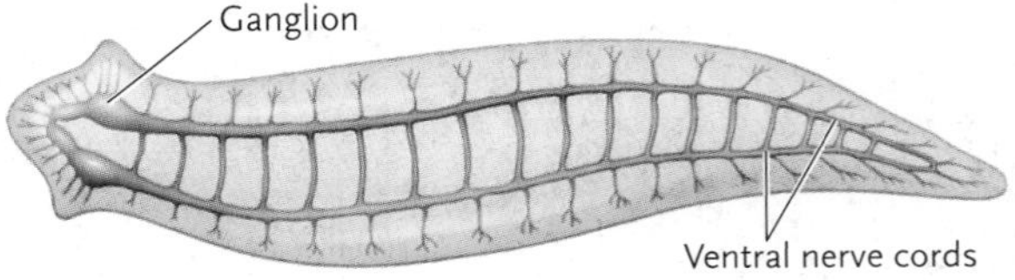
Ganglion
Ventral nerve cords

Reproductive system

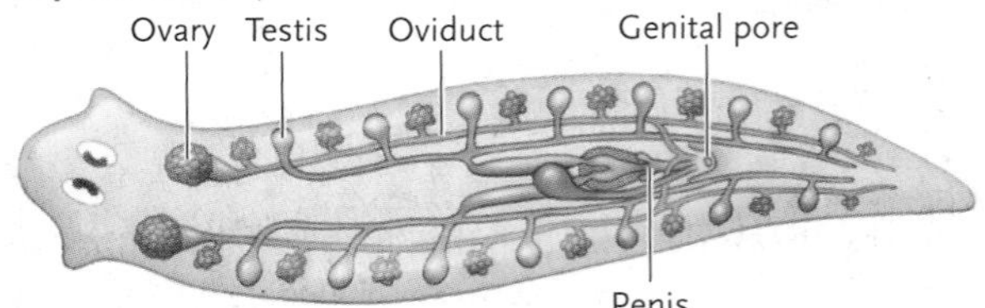
Ovary
Testis
Oviduct
Genital pore
Penis

Excretory system

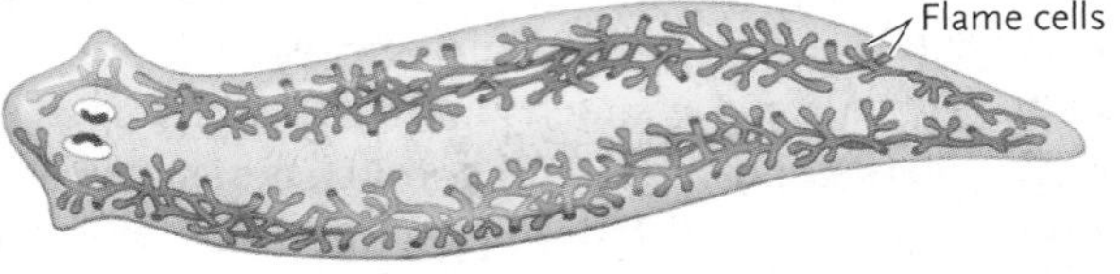
Flame cells

© Cory Gray

FIGURE 29.17
Turbellaria. A few turbellarians, such as *Pseudoceros dimidiatus*, are colorful marine worms.

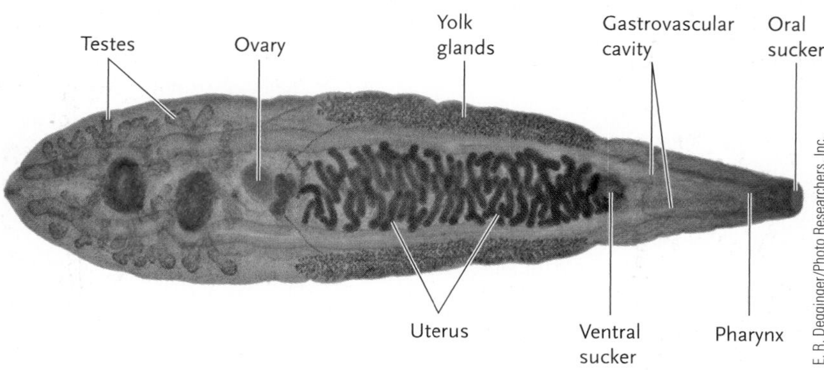
Testes
Ovary
Yolk glands
Gastrovascular cavity
Oral sucker
Uterus
Ventral sucker
Pharynx
E. R. Degginger/Photo Researchers, Inc.

FIGURE 29.18
Trematoda. The hermaphroditic Chinese liver fluke *(Opisthorchis sinensis)* uses a well-developed reproductive system to produce thousands of eggs.

The phylum Platyhelminthes includes four lineages, defined largely by their anatomical adaptations to free-living or parasitic habits.

TURBELLARIA Most free-living flatworms (Turbellaria) live in the sea **(Figure 29.17)**, but the familiar planarians and a few others live in fresh water or on land. Turbellarians swim by undulating the body wall musculature, or they crawl across surfaces by using muscles and cilia to glide on mucous trails produced by the ventral epidermis.

The gastrovascular cavity in free-living flatworms is similar to that in cnidarians. Food is ingested and wastes eliminated through a single opening, the mouth, located on the ventral surface. Food passes from the mouth to a muscular **pharynx** that leads to the digestive cavity (see Figure 29.16). Chemicals secreted into the saclike cavity digest the food into particles; then cells throughout the gastrovascular surface engulf the particles and subject them to intracellular digestion. In some species, the digestive cavity is highly branched, increasing the surface area for digestion and absorption.

Nearly all turbellarians are hermaphroditic, with complex reproductive systems (see Figure 29.16). When they mate, each partner functions simultaneously as a male and a female. Many free-living species also reproduce asexually by separating their anterior half from their posterior half. Both halves subsequently regenerate the missing parts.

TREMATODA AND MONOGENOIDEA Flukes (Trematoda and Monogenoidea) are parasites that obtain nutrients from tissues of a living host **(Figure 29.18)**. Most adult trematodes are **endoparasites,** living in the gut, liver, lungs, bladder, or blood vessels of vertebrates. Monogenes are **ectoparasites,** attaching to the gills or skin of aquatic vertebrates. Flukes are structurally specialized for a parasitic existence. They use suckers or hooks to attach to hosts, and a tough outer covering protects them from chemical attack. They produce numerous eggs that can readily infect new hosts. Monogene flukes usually have simple life cycles with a single host species. Trematodes, by contrast, have complex life cycles and multiple hosts. Humans suffer potentially fatal infections by many flukes, as discussed in *Focus on Applied Research.*

CESTODA Tapeworms (Cestoda) develop, grow, and reproduce within the intestines of vertebrates **(Figure 29.19)**. Through evolution, they have lost their mouths and digestive systems and absorb nutrients directly through their body wall. Tapeworms have a specialized structure, the *scolex,* with hooks and suckers that attach to the host's intestine; like the flukes, they also have a protective covering resistant to digestive enzymes.

Most of a tapeworm's body, which may be up to 20 m long, consists of a series of identical structures, *proglottids,* that con-

A. **Tapeworm**

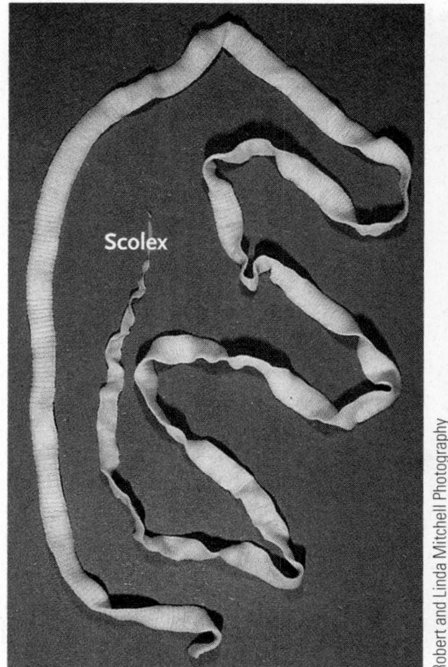
Scolex
Robert and Linda Mitchell Photography

B. **Scolex**

Cath Ellis, University of Hull/SPL/Photo Researchers, Inc.

FIGURE 29.19
Cestoda. (A) Tapeworms have long bodies comprised of a series of proglottids that each produce thousands of fertilized eggs. **(B)** The anterior end is a scolex with hooks and suckers that attach to the host's intestinal wall.

FOCUS ON Applied Research

Breaking the Life Cycles of Parasitic Worms

Many parasitic flatworms (phylum Platyhelminthes) and roundworms (phylum Nematoda) call the human body home, frequently causing disfiguring or life-threatening infections. The effort to control or eliminate these infections often begins when parasitologists or public health researchers study the worms' life cycles and ecology. This approach is successful because the worms often have more than one host: humans may be the *primary host,* harboring the sexually mature stage of the parasites' life cycles, but other animals serve as *intermediate hosts* to the larval stages. If researchers can learn the details of a parasite's life cycle, they can identify ways to cut it short before the parasite infects a human host.

Roughly 200 million people, most of them in sub-Saharan Africa, suffer from *schistosomiasis,* a disease caused by several species of flatworms called blood flukes (Trematoda). Japanese blood flukes *(Schistosoma japonicum)* mature and mate in blood vessels of the human intestine **(Figure 1).** Sharp spines on their eggs rupture the blood vessels, releasing the eggs into the lumen of the gut, from which they are passed with feces. When the infected human waste enters standing fresh water, the eggs hatch into ciliated larvae, which burrow into certain aquatic snails. The larvae feed on the snail's tissues and reproduce asexually. Their offspring leave the snail and, when they contact human skin, bore inward to a blood vessel. They eventually reach the intestine, where they complete their complex life cycle and produce fertilized eggs. Infected humans mount an immune response against flukes, but it is always a losing battle. Severe infections cause coughs, rashes, pain, and eventually diarrhea, anemia, and permanent damage to the intestines, liver, spleen, bladder, and kidneys. Death often results. Drug therapy can reduce the symptoms and limit the spread of these parasites, but schistosomiasis is most common in countries where people have limited access to medical care. Research has demonstrated that the disease can be controlled by proper sanitation and the elimination of the snails that serve as intermediate hosts.

Tapeworms (Cestoda), another group of flatworms, rely primarily on vertebrate hosts. Fishes, hogs, or cattle become intermediate hosts by inadvertently eating tapeworm eggs **(Figure 2).** When the host's digestive enzymes dissolve the protective covering on the eggs, the newly freed larvae bore through the intestinal wall and travel through the bloodstream to muscles or other tissues, where they *encyst* (produce a protective covering and enter a resting stage). Humans and other carnivores ingest living cysts when they eat the undercooked flesh of these animals. The scolex of the tapeworm larva then attaches to the primary host's intestinal wall, and the worm begins to grow. When mature, its proglottids produce huge numbers of eggs, which are released with feces to begin the next generation. Tapeworm infection can result in malnutrition of the host, because tapeworms consume much of the nutrients that the host ingests. They can also grow large enough to cause intestinal blockage. Research on the tapeworm life cycle suggests that careful inspection and adequate cooking of meat can prevent infection.

The roundworm *Trichinella spiralis* causes the painful and sometimes fatal symptoms of *trichinosis.* Adult trichinas breed in the small intestine of their hosts, including hogs and some game animals. Female worms release juveniles, which burrow into blood vessels and travel to various organs, where they become encysted **(Figure 3).** Humans become infected when they consume insufficiently cooked pork or other meat that contains encysted larvae. Once in the human digestive tract, the encysted worms complete their development, mate, and produce larvae. Most of the awful symptoms of trichinosis, including severe pain, high fever, and debilitating anemia, are produced by the migration of millions of larvae throughout the primary host's tissues. Once begun, the infection is difficult or impossible to control, but, as with tapeworms, thorough cooking of meat can prevent infection.

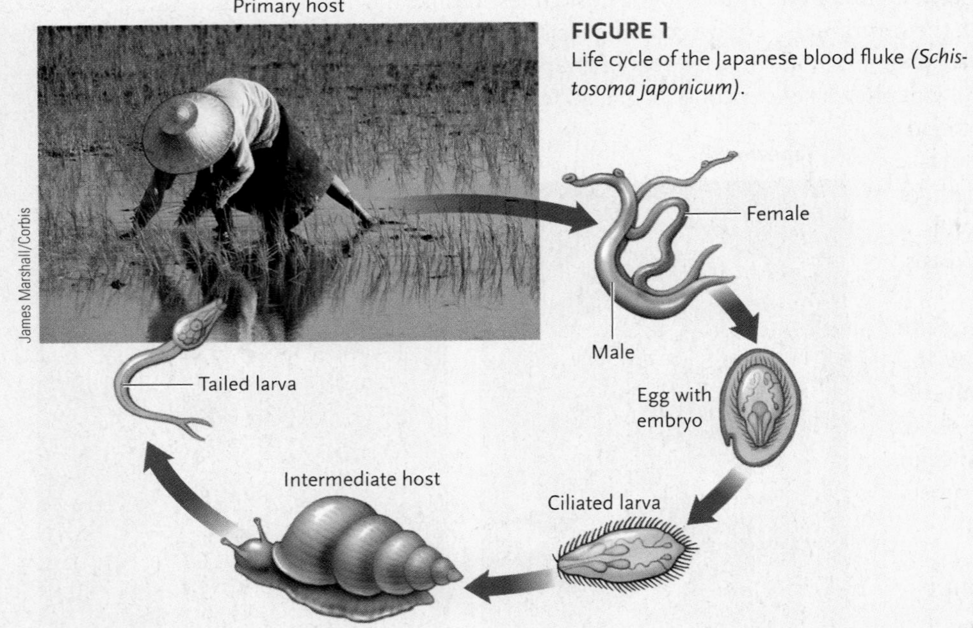

Primary host
Female
Male
Egg with embryo
Ciliated larva
Intermediate host
Tailed larva
James Marshall/Corbis

FIGURE 1
Life cycle of the Japanese blood fluke *(Schistosoma japonicum).*

tain little more than male and female reproductive systems. New proglottids are generated near the scolex; older proglottids, each carrying as many as 80,000 fertilized eggs, break off from the tapeworm's posterior end and leave the host's body in feces. As described in *Focus on Applied Research,* tapeworms are important parasites on humans, pets, and livestock.

Rotifers Are Tiny Pseudocoelomates with a Jawlike Feeding Apparatus

Most of the 1,800 species in the phylum Rotifera (*rota* = wheel; *ferre* = to carry) live in fresh water **(Figure 29.20).** All are microscopic—about the size of a ciliate protist—but they have well-developed

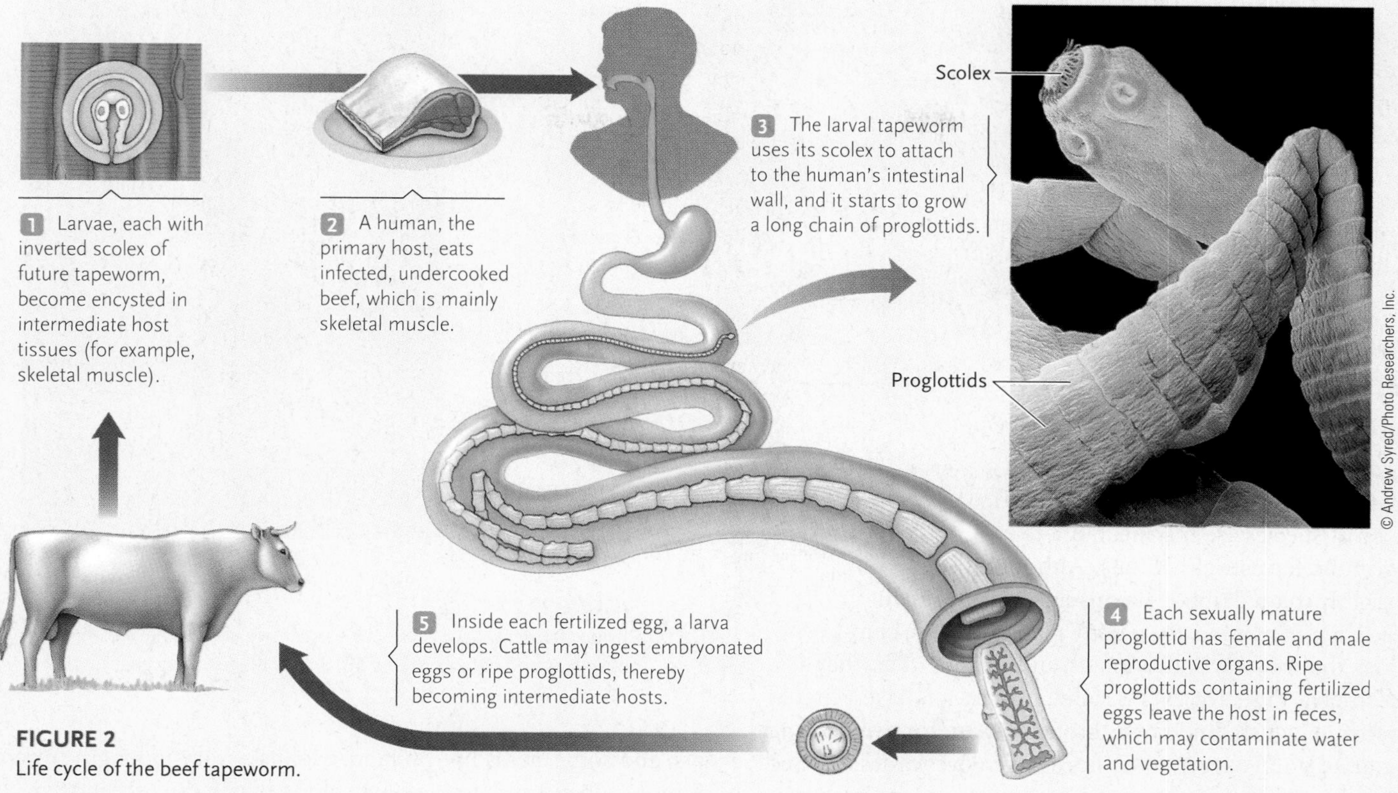
Scolex
3 The larval tapeworm uses its scolex to attach to the human's intestinal wall, and it starts to grow a long chain of proglottids.
Proglottids
© Andrew Syred/Photo Researchers, Inc.
1 Larvae, each with inverted scolex of future tapeworm, become encysted in intermediate host tissues (for example, skeletal muscle).
2 A human, the primary host, eats infected, undercooked beef, which is mainly skeletal muscle.
4 Each sexually mature proglottid has female and male reproductive organs. Ripe proglottids containing fertilized eggs leave the host in feces, which may contaminate water and vegetation.
5 Inside each fertilized egg, a larva develops. Cattle may ingest embryonated eggs or ripe proglottids, thereby becoming intermediate hosts.

FIGURE 2
Life cycle of the beef tapeworm.

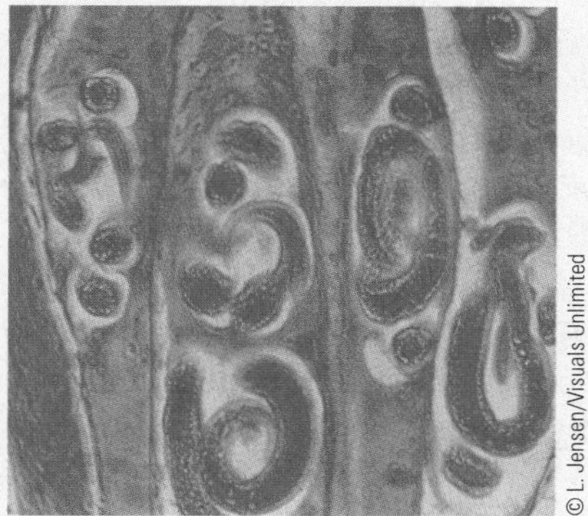
© L. Jensen/Visuals Unlimited

FIGURE 3
Trichinella spiralis juveniles in muscle tissue.

The nematodes called filarial worms (*Wuchereria* and *Brugia*) cause another debilitating infection. These large roundworms (up to 10 cm long) live in the lymphatic system, where they obstruct the normal flow of lymphatic fluid to the bloodstream. Female worms release first-stage larvae, which are acquired by mosquitoes, the intermediate host, when they feed on human blood. The larvae develop into second-stage larvae in mosquitoes, and when a mosquito bites another human, it may transmit those larvae to a new host. If victims experience severe filarial worm infection, their lymphatic vessels can be so obstructed that surrounding tissues swell grotesquely, a condition known as *elephantiasis* **(Figure 4).** Public health programs that reduce or eliminate mosquito populations lower the incidence of this disease.

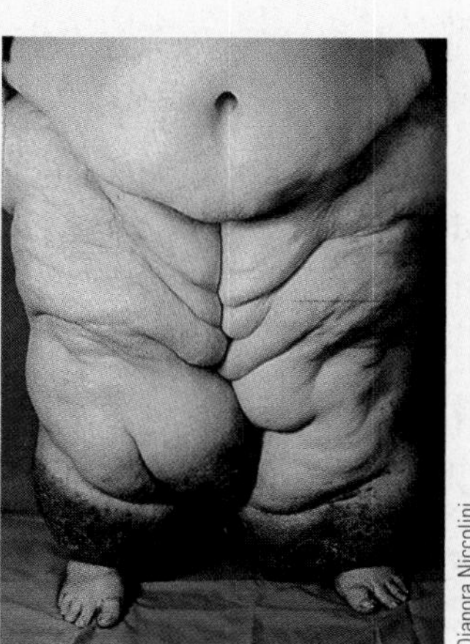
Dianora Niccolini

FIGURE 4
Elephantiasis caused by *Wuchereria bancrofti*.

digestive, reproductive, excretory, and nervous systems as well as a pseudocoelom. In some habitats, rotifers make up a large part of the zooplankton, tiny animals that float in open water.

Rotifers use coordinated movements of cilia, arranged in a wheel-like *corona* around the head, to propel themselves in the environment. Cilia also bring food-laden water to their mouths. Ingested microorganisms are conveyed to the *mastax,* a toothed grinding organ, and then passed to the stomach and intestine. Rotifers have a **complete digestive tract:** food enters through the mouth, and undigested waste is voided through the anus.

The life history patterns of some rotifers are adapted to the ever-changing environments in small bodies of water. During most months, rotifer populations include only females that reproduce by **parthenogenesis,** a form of asexual reproduction in which unfertilized eggs develop into diploid females (see Section 47.1). When environmental conditions deteriorate, females produce eggs that develop into haploid males. The males fertilize haploid eggs to produce diploid female zygotes. The fertilized eggs have durable shells and food reserves to survive drying or freezing.

A. **Rotifer (*Philodina roseola*) body plan**

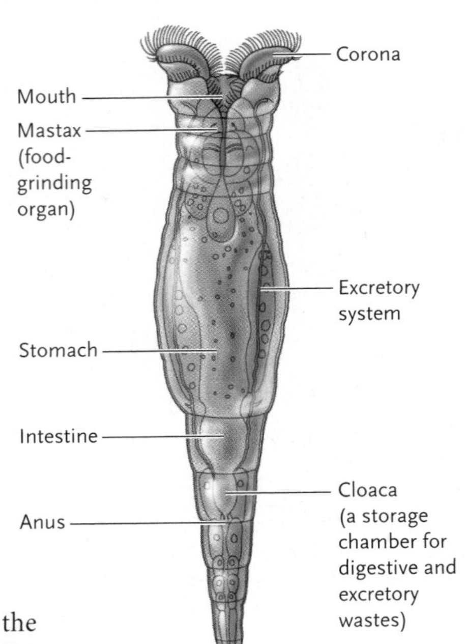
Corona
Mouth
Mastax (food-grinding organ)
Excretory system
Stomach
Intestine
Cloaca (a storage chamber for digestive and excretory wastes)
Anus

B. **Rotifer (*Philodina* species) laying eggs**

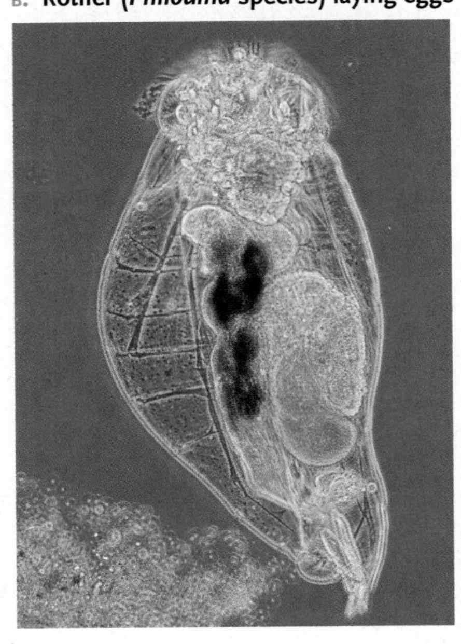

Herve Chaumeton/Agence Nature

FIGURE 29.20
Phylum Rotifera. **(A)** Despite their small size, rotifers have complex body plans and organ systems. **(B)** This rotifer is laying eggs.

Ribbon Worms Use a Proboscis to Capture Food

The 650 species of ribbon worms or proboscis worms (phylum Nemertea) vary from less than 1 cm to 30 m in length **(Figure 29.21).** Most species are marine, but a few occupy moist terrestrial habitats. Although the often brightly colored ribbon worms superficially resemble free-living flatworms, their body plans are more complex. First, they possess both a mouth and an anus; thus, they have a complete digestive tract. Second, nemerteans have a circulatory system in which fluid flows through **circulatory vessels** that carry nutrients and oxygen to tissues and remove wastes (see Section 42.1). Finally, they have a muscular, mucus-covered proboscis, a tube that can be everted (turned inside out) through the mouth or a separate pore to capture prey. The proboscis is housed within a chamber, the *rhynchocoel,* which is unique to this phylum.

Mollusks Have a Muscular Foot and a Mantle That Secretes a Shell or Aids in Locomotion

Most of the 100,000 species in the phylum Mollusca (*mollis* = soft)—including clams, snails, octopuses, and their relatives—are marine. However, many clams and snails occupy freshwater habitats, and some snails live on land. Mollusks vary in length from clams less than 1 mm to the giant squids, which can exceed 18 m.

In mollusks, the body is divided into three regions: the visceral mass, the head–foot, and the mantle **(Figure 29.22).** The **visceral mass** contains digestive, excretory, and reproductive systems and the heart. The muscular **head–foot** often provides the major means of locomotion. In more active groups, the head region is well-defined and carries sensory organs and a brain. The mouth often includes a toothed **radula,** which scrapes food into small particles or drills through the shells of prey. Many mollusks are covered by a protective shell of calcium carbonate secreted by the **mantle,** which is one or two folds of the body wall that often enclose the visceral mass. The mantle also defines a space, the *mantle cavity,* which houses the *gills,* delicate respiratory structures with enormous surface area. In most mollusks, cilia on the mantle and gills generate a steady flow of water into the mantle cavity.

FIGURE 29.21
Phylum Nemertea. **(A)** The flattened, elongated bodies of ribbon worms are often brightly colored. **(B)** Ribbon worms have a complete digestive system and a specialized cavity, the rhynchocoel, that houses a protrusible proboscis.

A. **Ribbon worm (*Lineus* species)**

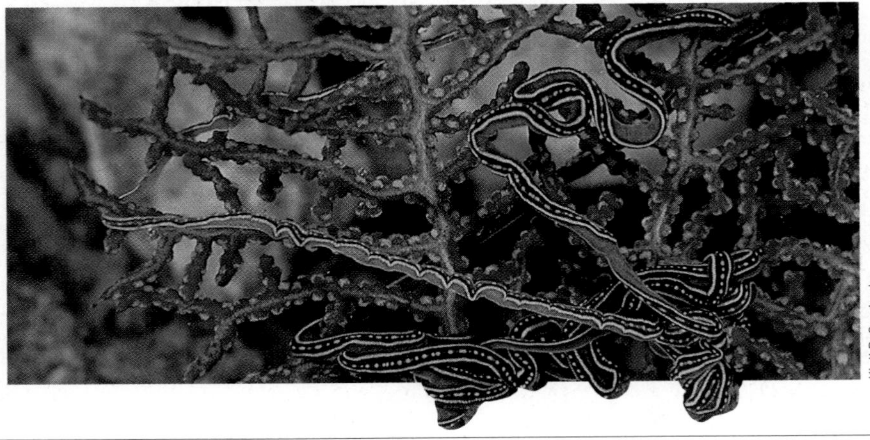

Kjell B. Sandved

B. **Ribbon worm anatomy**

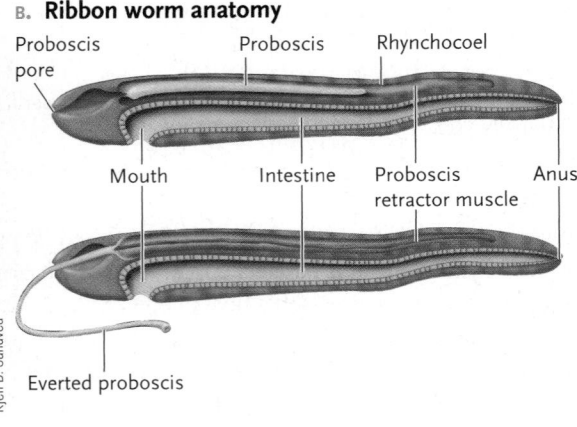

Proboscis pore
Proboscis
Rhynchocoel
Mouth
Intestine
Proboscis retractor muscle
Anus
Everted proboscis

The large size of mollusks—as well as their possession of a coelom—requires a circulatory system to maintain cells that are far from the body surface. Most mollusks have an **open circulatory system** in which **hemolymph,** a bloodlike fluid, leaves the circulatory vessels and bathes tissues directly (see Figure 42.3A). Hemolymph pools in spaces called *sinuses,* and then drains into vessels that carry it back to the heart.

The sexes are usually separate, although many snails are hermaphroditic. Fertilization may be internal or external. The zygotes of marine species often develop into free-swimming, ciliated **trochophore** larvae **(Figure 29.23),** typical of both this phylum and the phylum Annelida, which we describe next. In some mollusks, the trochophore develops into a second larval stage, called a *veliger,* before metamorphosing into an adult. Some snails as well as octopuses and squids have direct development: embryos develop into miniature replicas of the adults.

Mollusca includes eight lineages. In the following sections, we examine the four that are most commonly encountered.

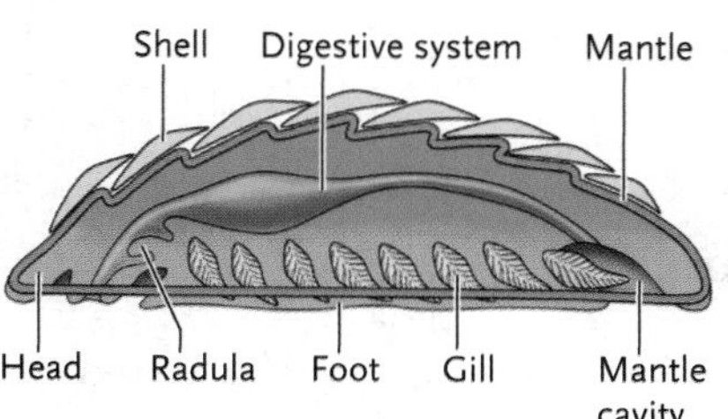

Chitons
Shell
Digestive system
Mantle
Head
Radula
Foot
Gill
Mantle cavity

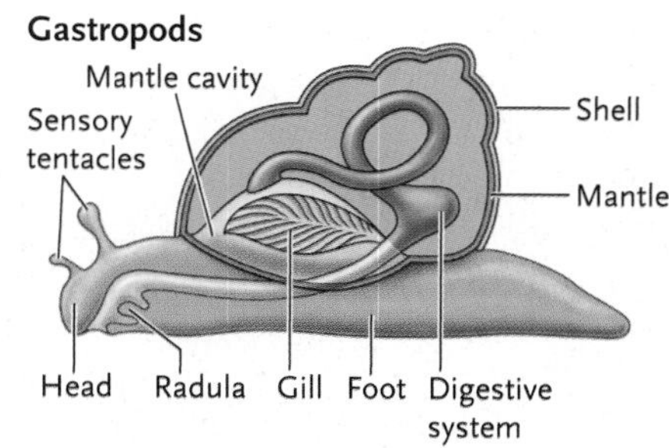

Gastropods
Mantle cavity
Sensory tentacles
Shell
Mantle
Head
Radula
Gill
Foot
Digestive system

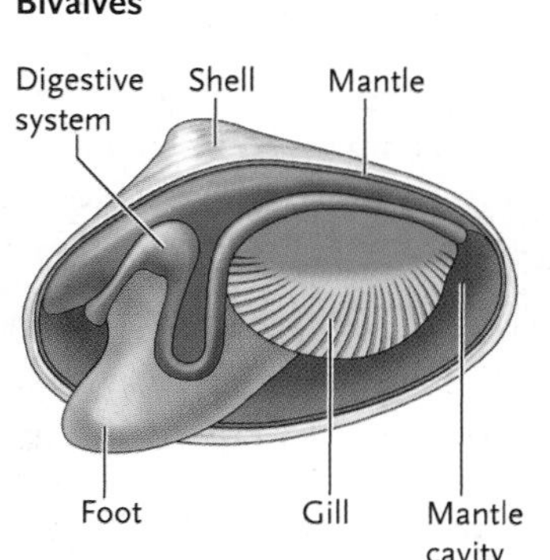

Bivalves
Digestive system
Shell
Mantle
Foot
Gill
Mantle cavity

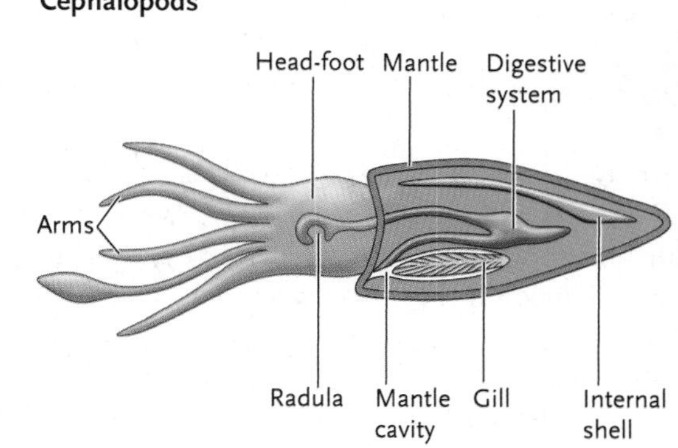

Cephalopods
Head-foot
Mantle
Digestive system
Arms
Radula
Mantle cavity
Gill
Internal shell

KEY
- Head-foot
- Visceral mass
- Mantle

FIGURE 29.22
Molluscan body plans. The bilaterally symmetrical body plans of most mollusks include a muscular head-foot, a visceral mass, and a mantle. Bivalves have no head, and cephalopods have arms on the head-foot.

POLYPLACOPHORA The 600 species of chitons (Polyplacophora: *poly* = many; *plax* = flat surface) are sedentary mollusks that graze on algae along rocky marine coasts. The oval, bilaterally symmetrical body has a dorsal shell divided into eight plates that allow it to conform to irregularly shaped surfaces **(Figure 29.24).** When a chiton is disturbed or exposed to strong wave action, the muscles of its broad foot maintain a tenacious grip, and the mantle's edge functions like a suction cup to hold fast to the substrate.

GASTROPODA Snails and slugs (Gastropoda: *gaster* = belly; *podos* = foot) are the largest molluscan group, numbering 40,000 species **(Figure 29.25).** Aquatic species use gills to acquire oxygen, but in terrestrial species a modified mantle cavity functions as an

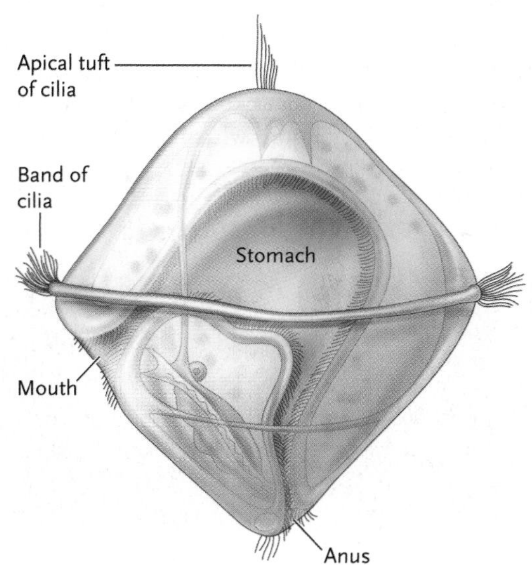

Apical tuft of cilia
Band of cilia
Stomach
Mouth
Anus

FIGURE 29.23
Trochophore larva. At the conclusion of their embryological development, both mollusks and annelids typically pass through a trochophore stage. The top-shaped trochophore larva has a band of cilia just anterior to its mouth.

Jeff Foott/Tom Stack & Associates

FIGURE 29.24
Polyplacophora. Chitons live on rocky shores, where they use their foot and mantle to grip rocks and other hard substrates. This bristled chiton *(Mopalia ciliata)* lives in Monterey Bay, California.

A. Gastropod body plan

Gill
Excretory organ
Anus
Mantle cavity
Head
Heart
Digestive gland
Stomach
Shell
Mantle
Foot
Radula
Mouth
Anus

Danielle C. Zacheri with John McNulty

B. Edible snail (*Helix pomatia*)

J. Kottmann/Peter Arnold, Inc.

C. Marine nudibranchs (*Flabellina iodinea*)

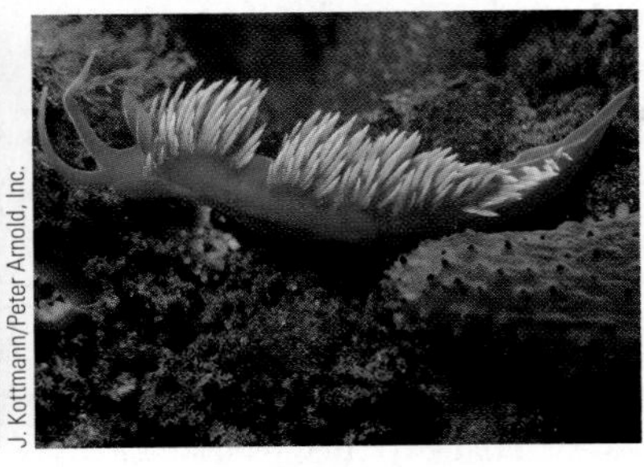

iStockphoto.com/Thornberry

FIGURE 29.25
Gastropoda. **(A)** Most gastropods have a coiled shell that houses the visceral mass. A developmental process called torsion causes the digestive and excretory systems to eliminate wastes into the mantle cavity, near the animal's head. **(B)** The edible snail is a terrestrial gastropod. **(C)** Nudibranchs, like these Spanish shawl nudibranchs, are shell-less marine snails.

air-breathing lung. Gastropods feed on algae, vascular plants, or animal prey. Some are scavengers, and a few are parasites.

The visceral mass of most snails is housed in a coiled or cone-shaped shell that is balanced above the rest of the body, much as you balance a backpack full of books (see Figure 29.25A and B). Most shelled snails undergo **torsion,** a curious realignment of body parts that is independent of shell coiling. Muscle contractions and differential growth twist the visceral mass and mantle 180° relative to the head–foot. This rearrangement moves the mantle cavity forward so that the head can be withdrawn into the shell in times of danger. It also brings the gills, anus, and excretory openings above the head—a potentially messy configuration, were it not for cilia that sweep away wastes.

Some gastropods, including terrestrial slugs and colorful nudibranchs (sea slugs), are shell-less, a condition that leaves them vulnerable to predators (see Figure 29.25C). Some nudibranchs consume cnidarians and then transfer undischarged nematocysts to projections on their dorsal surface, where these "borrowed" stinging capsules provide protection.

The nervous and sensory systems of gastropods are well developed. Tentacles on the head include chemical and touch receptors; the eyes detect changes in light intensity but don't form images.

BIVALVIA The 8,000 species of clams, scallops, oysters, and mussels (Bivalvia: *bi* = two; *valva* = folding door) are restricted to aquatic habitats. They are enclosed within a pair of shells, hinged together dorsally by a ligament **(Figure 29.26).** Contraction of the ligament opens the shell by pulling the two sides apart, and contraction of one or two **adductor muscles** closes it by pulling them together (see Figure 29.26A). Although some bivalves are tiny, giant clams of the South Pacific can be more than 1 m across and weigh 225 kg.

Adult mussels and oysters are sessile and permanently attached to hard substrates, but many clams are mobile and use their muscular foot to burrow in sand or mud. Some bivalves, such as young scallops, swim by rhythmically clapping their valves (shells) together, forcing a current of water out of the mantle cavity (see Figure 29.26B). The "scallops" that we eat are the scallops' well-developed adductor muscles.

Bivalves have a reduced head, and they lack a radula. Part of the mantle forms two tubes called *siphons* (see Figure 29.26C). Beating of cilia on the gills and mantle carry water into the mantle cavity through the incurrent siphon and out through the excurrent siphon. Incurrent water carries dissolved oxygen and food particles to the gills, where oxygen is absorbed. Mucus strands on the gills trap the food, which is then transported by

A. Bivalve body plan

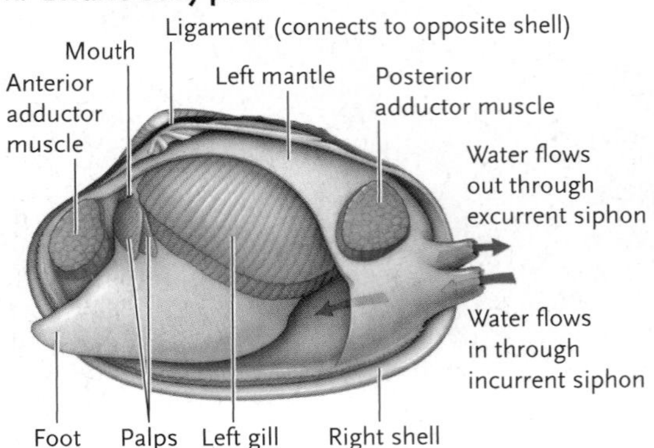
Ligament (connects to opposite shell)
Mouth
Anterior adductor muscle
Left mantle
Posterior adductor muscle
Water flows out through excurrent siphon
Water flows in through incurrent siphon
Foot
Palps
Left gill
Right shell

B. Bivalve locomotion

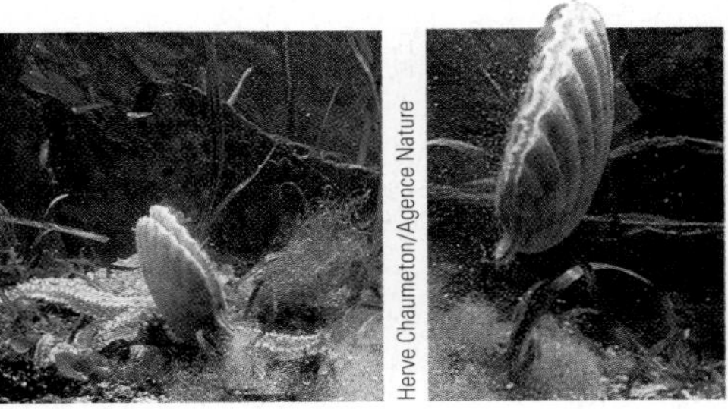

Herve Chaumeton/Agence Nature

Herve Chaumeton/Agence Nature

C. Geoduck (*Panope generosa*)

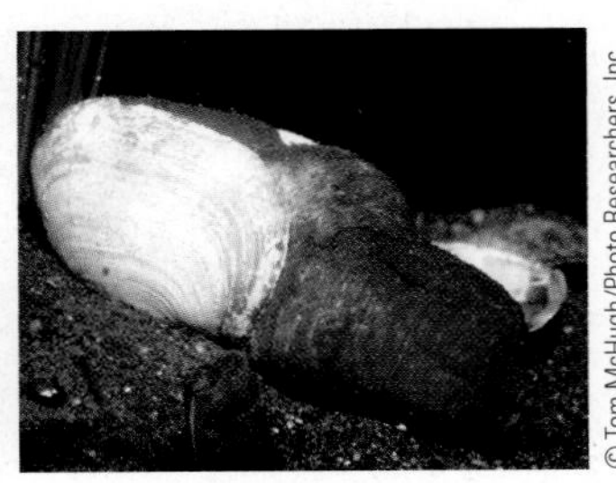

© Tom McHugh/Photo Researchers, Inc.

FIGURE 29.26
Bivalvia. **(A)** Bivalves are enclosed in a hinged two-part shell. Part of the mantle forms a pair of water-transporting siphons. **(B)** When threatened by a predator (in this case a sea star), some scallops clap their shells together rapidly, propelling the animal away from danger. **(C)** The geoduck is a clam with enormous muscular siphons.

cilia to *palps,* where final sorting takes place; edible tidbits are carried to the mouth. The excurrent water carries away metabolic wastes and feces.

Despite their sedentary existence, bivalves have moderately well-developed nervous systems; sensory organs that detect chemicals, touch, and light; and statocysts to sense their orientation. When they encounter pollutants, many bivalves stop pumping water and close their shells. When confronted by a predator, some burrow into sediments or swim away.

CEPHALOPODA The 600 species of octopuses, squids, and nautiluses (Cephalopoda: *kephale* = head) are active marine predators, including the fastest and most intelligent invertebrates **(Figure 29.27).** They vary in length from a few centimeters to 18 m. Giant squids, the largest invertebrates known, may be the source of "sea monster" stories.

The cephalopod body has a fused head and foot (see Figure 29.27D). The head comprises the mouth and eyes. The ancestral "foot" forms a set of arms and tentacles, equipped with suction pads, adhesive structures, or hooks. Cephalopods use these structures to capture prey and a pair of beaklike jaws to bite or crush it. Venomous secretions often speed the captive's death. Some species use their radula to drill through the shells of other mollusks.

Cephalopods have a highly modified shell. Octopuses have no remnant of a shell at all. In squids and cuttlefishes, it is reduced to a stiff internal support. Only the chambered nautilus and its relatives retain an external shell.

Squids are the most mobile cephalopods, moving rapidly by a kind of jet propulsion. When muscles in the mantle relax, water is drawn into the mantle cavity. When they contract, a jet of water is squeezed out through a funnel-shaped excurrent siphon. By manipulating the position of the mantle and siphon, the animal can control the rate and direction of its locomotion. When threatened, a squid can make a speedy escape. Many species simultaneously release a dark fluid, commonly called "ink," that obscures their direction of movement. Being highly active, cephalopods need lots of oxygen, and they alone among the mollusks have a **closed circulatory system:** hemolymph is confined within the walls of hearts and blood vessels, providing increased pressure to the vascular fluid. Moreover, accessory hearts speed the flow of hemolymph through the gills, enhancing the uptake of oxygen and the release of carbon dioxide.

FIGURE 29.27
Cephalopoda. **(A)** Squids and **(B)** octopuses are the most familiar cephalopods. **(C)** The chambered nautilus and its relatives retain an external shell. **(D)** Like other cephalopods, the squid body includes a fused head and foot; most organ systems are enclosed by the mantle.

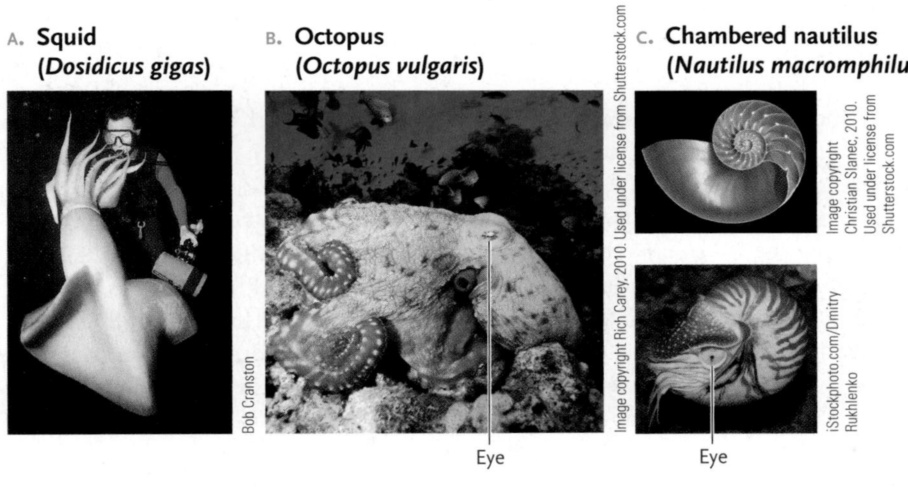
A. **Squid (*Dosidicus gigas*)**
B. **Octopus (*Octopus vulgaris*)**
C. **Chambered nautilus (*Nautilus macromphilus*)**
Bob Cranston
Image copyright Rich Carey, 2010. Used under license from Shutterstock.com
Image copyright Christian Slanec, 2010. Used under license from Shutterstock.com
iStockphoto.com/Dmitry Rukhlenko
Eye
Eye

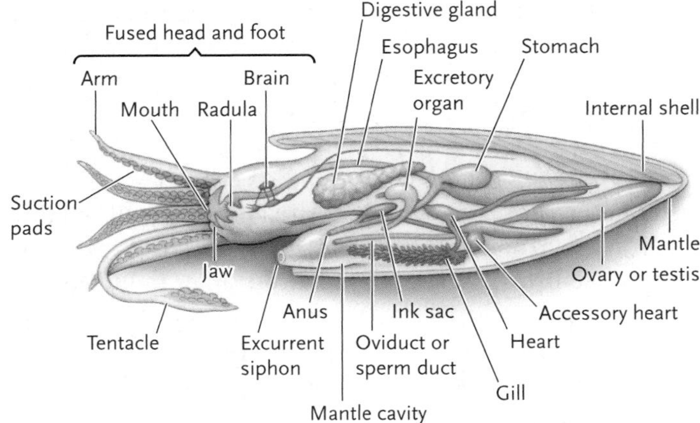
D. **Internal anatomy of squid**
Fused head and foot
Arm
Brain
Mouth
Radula
Digestive gland
Esophagus
Stomach
Excretory organ
Internal shell
Suction pads
Mantle
Jaw
Ovary or testis
Anus
Ink sac
Accessory heart
Tentacle
Excurrent siphon
Oviduct or sperm duct
Heart
Gill
Mantle cavity

Compared with other mollusks, cephalopods have large and complex brains. Giant nerve fibers connect the brain with the muscles of the mantle, enabling quick responses to food or danger. Their image-forming eyes are similar to those of vertebrates. Cephalopods are also highly intelligent. Octopuses, for example, learn to recognize objects with distinctive shapes or colors, and they can be trained to approach or avoid them.

Cephalopods have separate sexes and elaborate courtship rituals. Males store sperm within the mantle cavity and use a specialized tentacle to transfer packets of sperm into the female's mantle cavity, where fertilization occurs. Fertilized eggs, wrapped in a protective jelly, are attached to objects in the environment. The young hatch with an adult body form.

Annelids Exhibit a Serial Division of the Body Wall and Some Organ Systems

The 15,000 species of segmented worms (phylum Annelida: *anellus* = little ring) occupy marine, freshwater, and moist terrestrial habitats. Terrestrial annelids eat organic debris; aquatic species consume algae, microscopic organisms, detritus, or other animals. They range from a few millimeters to several meters in length.

The annelid body is highly segmented: the body wall muscles and some organs—including respiratory surfaces, parts of the nervous, circulatory, and excretory systems, and the coelom itself—are divided into similar repeating units **(Figure 29.28).** Body segments are separated by transverse partitions called

A. **Digestive system**

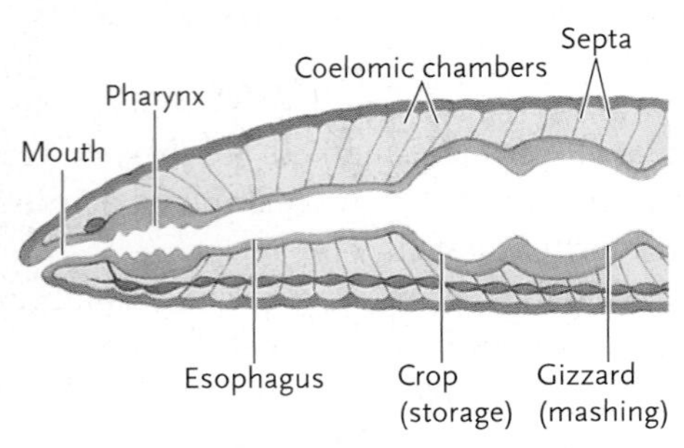

Septa
Coelomic chambers
Pharynx
Mouth
Esophagus
Crop
(storage)
Gizzard
(mashing)

B. **Circulatory system**

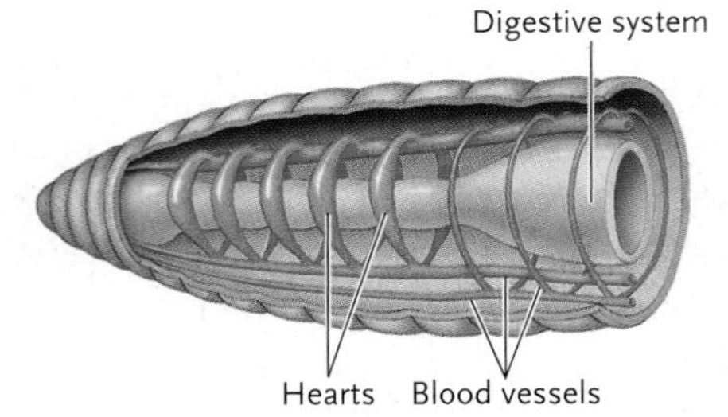

Digestive system
Hearts
Blood vessels

C. **Nervous system**

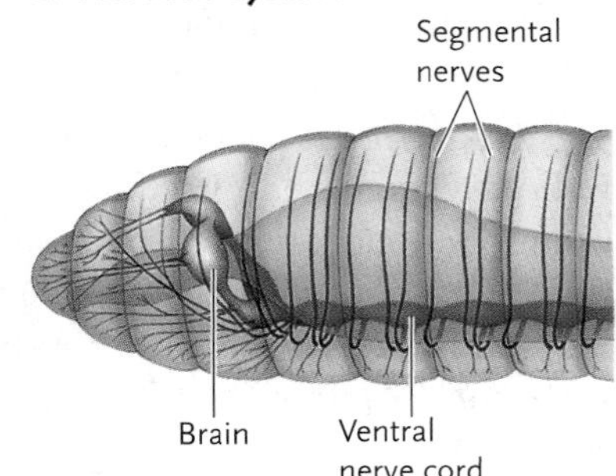

Segmental
nerves
Brain
Ventral
nerve cord

D. **Excretory system (metanephridium)**

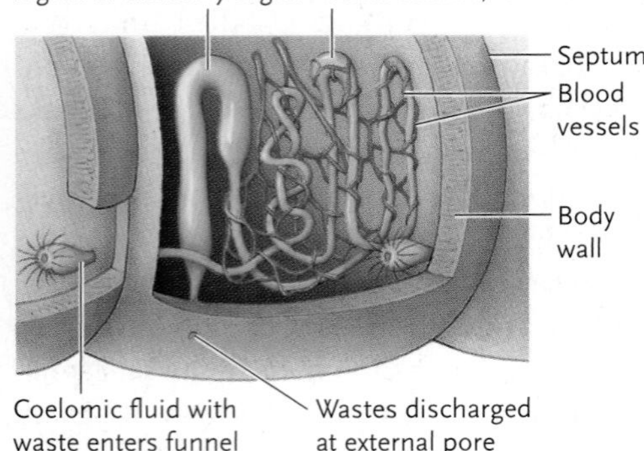

Bladderlike storage
region of excretory organ
Excretory organ's thin loop reabsorbs
some solutes, returns them to blood
Septum
Blood
vessels
Body
wall
Coelomic fluid with
waste enters funnel
Wastes discharged
at external pore

E. **Cross section**

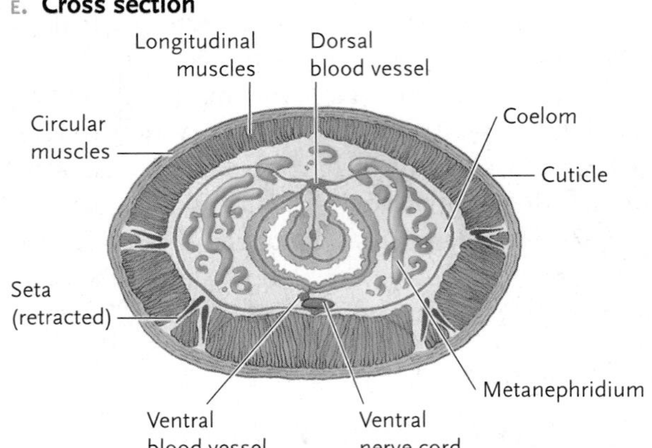

Longitudinal
muscles
Dorsal
blood vessel
Circular
muscles
Coelom
Cuticle
Seta
(retracted)
Metanephridium
Ventral
blood vessel
Ventral
nerve cord

FIGURE 29.28
Segmentation in the phylum Annelida. Although the digestive system **(A),** the longitudinal blood vessels **(B),** and the ventral nerve cord **(C)** are not segmented, the coelom **(A),** blood vessels **(B),** nerves **(C),** and excretory organs **(D)** appear as repeating structures in most segments. The body musculature **(E)** includes both circular and longitudinal layers that allow these animals to use the coelomic chambers as a hydrostatic skeleton.

septa. The digestive system and major blood vessels are not segmented and run the length of the animal.

The body wall muscles of annelids have both circular and longitudinal layers. Alternate contractions of these muscle groups allow annelids to make directed movements, using the pressure of the fluid in the coelom as a hydrostatic skeleton. All annelids except leeches also have chitin-reinforced bristles, called **setae** (sometimes written *chaetae*), which protrude outward from the body wall. Setae anchor the worm against the substrate, providing traction.

Annelids have a complete digestive system and a closed circulatory system. However, they lack a discrete respiratory system; oxygen and carbon dioxide diffuse through the skin. The excretory system is composed of paired **metanephridia,** which usually occur in all body segments posterior to the head. The nervous system is highly developed, with ganglia (local control centers) in every segment, a simple brain in the head, and sensory organs that detect chemicals, moisture, light, and touch.

Most freshwater and terrestrial annelids are hermaphroditic, and worms exchange sperm when they mate. Newly hatched worms have an adult morphology. Some terrestrial annelids also reproduce asexually by fragmenting and regenerating missing parts. Marine annelids usually have separate sexes, and release gametes into the sea for fertilization. The zygotes develop into trochophore larvae that add segments, gradually assuming an adult form.

Annelida includes three lineages, each of which is largely restricted to one environment.

POLYCHAETA The 10,000 species of bristle worms (Polychaeta: *chaite* = bristles or hair) are primarily marine **(Figure 29.29).** Many live under rocks or in tubes constructed from mucus, calcium carbonate secretions, grains of sand, and small shell fragments. Their setae project from well-developed **parapodia** (singular, *parapodium* = closely resembling a foot), fleshy lateral extensions of the body wall used for locomotion and gas exchange. Sense organs are concentrated on a well-developed head.

Crawling or swimming polychaetes are often predatory; they use sharp jaws in a protrusible muscular pharynx to grab small invertebrate prey. Other species graze on algae or scavenge organic matter. A few tube dwellers draw food-laden water into the tube by beating their parapodia; most others collect food by extending feathery, ciliated, mucus-coated tentacles.

OLIGOCHAETA Most of the 3,500 species of oligochaete worms (Oligochaeta: *oligos* = few) are terrestrial **(Figure 29.30),** but they are restricted to moist habitats because they quickly dehydrate in dry air or soil. They range in length from a few millimeters to more than 3 m. Terrestrial oligochaetes, the earthworms, are nocturnal, spending their days in burrows that they excavate. Aquatic species live in mud or detritus at the bottom of lakes, rivers, and estuaries.

Earthworms are scavengers on decomposing organic matter. In his book *The Formation of Vegetable Mould through the Action of Worms,* Darwin noted that earthworms can ingest their own weight in soil every day. He calculated that a typical population of 16,000 worms per hectare (10,000 sq m) consumes more than 20 tons of soil in a year. This impressive activity aerates soil and makes nutrients available to plants by mixing the subsoil with the topsoil. Earthworms have complex organ systems (see Figure 29.28), and they sense light and touch at both ends of the body. In addition, they have moisture receptors, an important adaptation in organisms that must stay wet to allow gas exchange across the skin.

HIRUDINEA The 500 species of leeches (Hirudinea: *hirudo* = leech) are mostly freshwater parasites. They have dorsoventrally flattened, tapered bodies with a sucker at each end. Although the body wall is segmented, the coelom is reduced and not partitioned. Many leeches are ectoparasites of vertebrates, but some attack small invertebrate prey.

Parasitic leeches feed on the blood of their hosts. Most attach to the host with the posterior sucker, and then use their sharp jaws to make a small, often painless, triangular incision. A sucking apparatus draws blood from the prey, while a special secretion, *hirudin,* maintains the flow by preventing the host's blood from coagulating. Leeches have a highly branched gut that allows them to consume huge blood meals **(Figure 29.31).** For centuries, doctors used medicinal leeches *(Hirudo medicinalis)* to "bleed" patients; today, surgeons use them to drain excess fluid from tissues after reconstructive surgery, reducing swelling until the patient's blood vessels regenerate and resume this function.

A. **Fan worm (*Sabella melanostigma*)**

Kjell B. Sandved

B. ***Nereis* feeding structures**

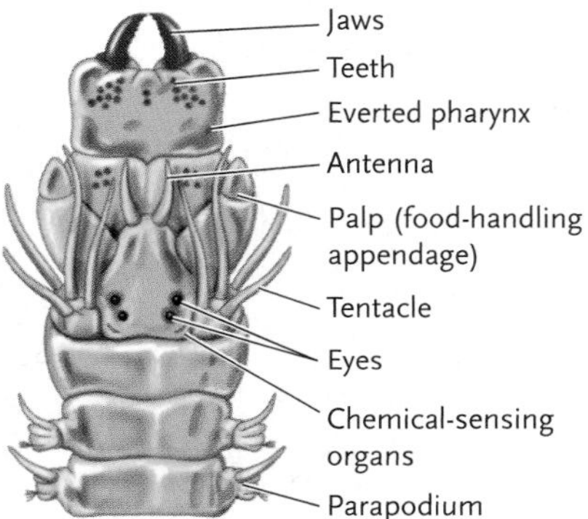
Jaws
Teeth
Everted pharynx
Antenna
Palp (food-handling appendage)
Tentacle
Eyes
Chemical-sensing organs
Parapodium

C. ***Proceraea cornuta* setae**

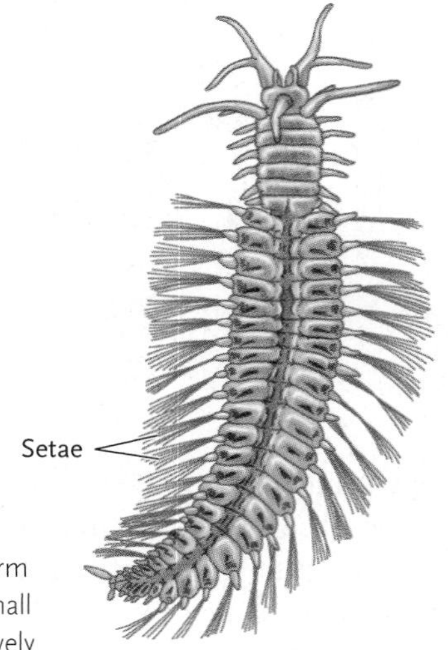
Setae

FIGURE 29.29
Polychaeta. (A) The tube-dwelling fan worm has mucus-covered tentacles that trap small food particles. **(B)** Some polychaetes actively seek food; when they encounter a suitable tidbit, they evert their pharynx, exposing sharp jaws that grab the prey and pull it into the digestive system. **(C)** Many marine polychaetes have numerous setae, which they use for locomotion.

STUDY BREAK 29.6

1. **What organs systems are present in free-living flatworms (Turbellaria)? Which of these organ systems is absent in tapeworms (Cestoda)?**
2. **What characteristic reveals the close evolutionary relationship of ectoprocts, brachiopods, and phoronid worms?**
3. **What anatomical structures and physiological systems allow squids and other cephalopods to be much more active than other types of mollusks?**
4. **Which organs systems exhibit segmentation in most annelid worms?**

THINK OUTSIDE THE BOOK

Search the records of your local health department, either online or in person, to discover whether people in your town or city routinely suffered from parasitic worm infections in the past. What public health measures stopped the outbreak?

FIGURE 29.30
Oligochaeta. Earthworms *(Lumbricus terrestris)* generally move across the ground surface at night.

Image copyright mashe, 2010. Used under license from Shutterstock.com

Leech before feeding

Leech after feeding

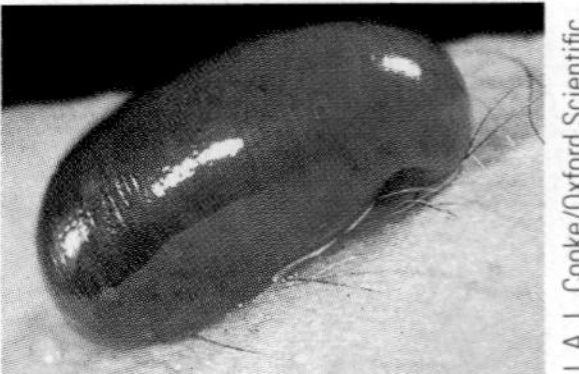

J. A. L. Cooke/Oxford Scientific Films/Photolibrary.com

J. A. L. Cooke/Oxford Scientific Films/Photolibrary.com

FIGURE 29.31
Hirudinea. Parasitic leeches consume huge blood meals, as shown by these before and after photos of a medicinal leech *(Hirudo medicinalis)*. Because suitable hosts are often hard to locate, gorging allows a leech to take advantage of any host it finds.

29.7 Ecdysozoan Protostomes

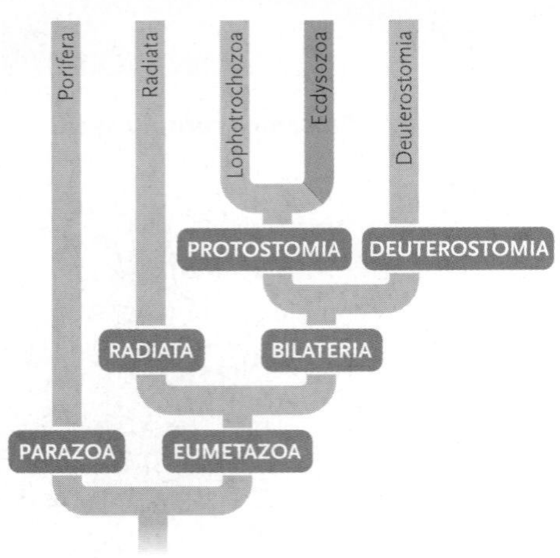
Porifera
Radiata
Lophotrochozoa
Ecdysozoa
Deuterostomia
PROTOSTOMIA
DEUTEROSTOMIA
RADIATA
BILATERIA
PARAZOA
EUMETAZOA

The three phyla in the protostome group Ecdysozoa all have an external covering that they shed periodically. The outer covering protects these animals from harsh environmental conditions, and it helps parasitic species resist host defenses. Although many of these animals live in either aquatic or moist terrestrial habitats, a tough exoskeleton allows one group, the insects, to thrive on dry land and in the air.

Nematodes Are Unsegmented Worms Covered by a Flexible Cuticle

Roundworms (phylum Nematoda: *nema* = thread) are perhaps the most abundant animals on Earth. A cupful of rich soil, a dead earthworm, or a rotting fruit may contain thousands of them. Although 80,000 species have been described, experts estimate that more than half a million exist. Many nematodes are almost microscopically small, but some species are a meter or more long. They occupy nearly every freshwater, marine, and terrestrial habitat on Earth, consuming detritus, microorganisms, plants, or animals.

The roundworm body is cylindrical and usually tapered at both ends **(Figure 29.32)**. None of the cells have cilia or flagella. Roundworms are covered in a tough but flexible, water-resistant **cuticle,** which is replaced by the underlying epidermis as the worm grows. The cuticle prevents the animal from dehydrating in dry environments, and, in parasitic species, it resists attack by acids and enzymes in a host's digestive system. Beneath the cuticle and epidermis, a layer of longitudinal muscles extends the length of the body. Nematodes use alternating muscle contractions to push against the substrate and propel themselves forward, usually with a thrashing motion.

© 2006 Alistair Dove/Image Quest Marine

FIGURE 29.32
Phylum Nematoda. Some roundworms, like these *Anguillicola* crassus in the swim bladder of an eel, are parasites of plants or animals. Others are important consumers of dead organisms in most ecosystems.

The adults of one soil-dwelling species, *Caenorhabditis elegans,* are transparent and contain fewer than 1,000 cells. As *Focus on Model Organisms* explains, biologists have studied this worm inside and out.

Nematodes reproduce sexually, and the sexes are separate in most species. In some, internal fertilization produces many thousands of fertile eggs per day. The eggs of many species can remain dormant if environmental conditions are unsuitable.

Because of their great numbers, nematodes have enormous ecological, agricultural, and medical significance. Free-living species are responsible for decomposition and nutrient recycling in many habitats. Parasitic nematodes attack the roots of plants, causing tremendous crop damage. Roundworms also parasitize animals, including humans. Although some, like pinworms *(Enterobius),* are more of a nuisance than a danger, others, like trichinas or filarial worms, can cause serious disease, disfigurement, or even death (see *Focus on Applied Research* on pages 654–655). More than 1 billion people worldwide suffer from debilitating and life-threatening nematode infections.

Velvet Worms Have Segmented Bodies and Numerous Unjointed Legs

The 65 living species of velvet worms (phylum Onychophora: *onyx* = claw) live under stones, logs, and forest litter in the tropics and in moist temperate habitats in the southern hemisphere. They range in size from 15 mm to 15 cm and feed on small invertebrates.

Onychophorans have a flexible cuticle, segmented bodies, and numerous pairs of unjointed legs **(Figure 29.33)**. Like the annelids, they have pairs of excretory organs in most segments. But unlike annelids, they have an open circulatory system, a special-

FOCUS ON Model Organisms

Caenorhabditis elegans

Researchers studying the tiny, free-living nematode *Caenorhabditis elegans* have made many recent advances in molecular genetics, animal development, and neurobiology. It is so popular as a model research organism that most workers simply refer to it as "the worm."

Several attributes make *C. elegans* a model research organism. It has an adult size of about 1 mm and thrives on cultures of *E. coli* or other bacteria; thus, thousands can be raised in a culture dish. It is hermaphroditic and often self-fertilizing, which allows researchers to maintain pure genetic strains. It completes its life cycle from egg to reproductive adult within 3 days at room temperature. Furthermore, stock cultures can be kept alive indefinitely by freezing them in liquid nitrogen or in an ultra-cold freezer set to minus 80°C. Researchers can therefore store new mutants for later research without having to clean, feed, and maintain active cultures.

Best of all, the worm is anatomically simple (see figure); an adult contains just 959 cells (excluding the gonads). Having a fixed cell number is relatively uncommon among animals, and developmental biologists have made good use of this trait. The eggs, juveniles, and adults of the worm are completely transparent, and researchers can observe cell divisions and cell movements in living animals with straightforward microscopy techniques. There is no need to kill, fix, and stain specimens for study. And virtually every cell in the worm's body is accessible for manipulation by laser microsurgery, microinjection, and similar approaches.

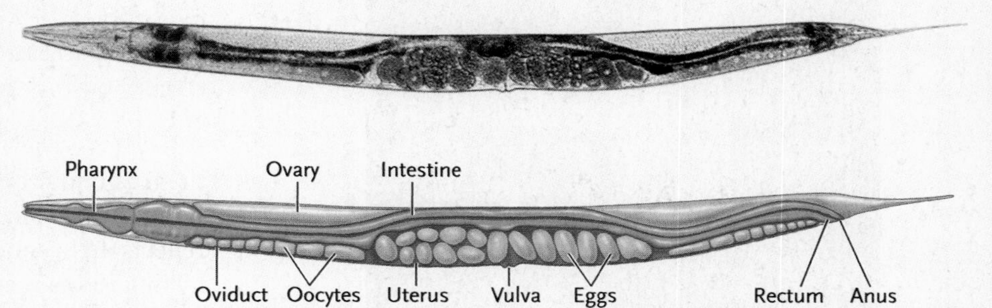

Pharynx
Ovary
Intestine
Oviduct
Oocytes
Uterus
Vulva
Eggs
Rectum
Anus

J. Sulston, MRC Laboratory of Molecular Biology

James Christensen/Minden Pictures

FIGURE 29.33
Phylum Onychophora. Species in the small phylum Onychophora have segmented bodies and unjointed appendages.

ized respiratory system, relatively large brains, jaws, and tiny claws on their feet. Many produce live young, which, in some species, are nourished within a uterus. Fossil evidence indicates that the onychophoran body plan has not changed much over the last 500 million years.

Arthropods Are Segmented Animals with a Hard Exoskeleton and Jointed Appendages

The more than 1 million known species of arthropods (phylum Arthropoda: *arthron* = joint) include more than half the animal species on Earth—and only a fraction of the living arthropods have been described. This huge lineage includes insects, spiders, crustaceans, millipedes, centipedes, the extinct trilobites, and their relatives.

Arthropods have a segmented body encased in a rigid **exoskeleton.** This external covering is made of chitin, a mix of polysaccharide fibers glued together with glycoproteins, as well as waxes and lipids that block the passage of water. In some marine groups, such as crabs and lobsters, it is hardened with calcium carbonate. The exoskeleton probably first evolved in marine species, providing protection against predators. In terrestrial habitats, it provides support against gravity and protection from dehydration, which contributes to the success of insects in even the driest places on Earth. The exoskeleton is especially thin and flexible at the joints between body segments and in the appendages. Contractions of muscles attached to the inside of the exoskeleton move individual body parts like levers, allowing highly coordinated movements and patterns of locomotion that are more precise than those in soft-bodied animals with hydrostatic skeletons.

Although the exoskeleton has obvious advantages, it is nonexpandable and therefore could limit growth of the animal. But, like other Ecdysozoa, arthropods grow and periodically develop a soft, new exoskeleton beneath the old one, which they shed in the complex process of ecdysis **(Figure 29.34).** After shedding the old exoskeleton, aquatic species swell with water and terrestrial species swell with air before the new one hardens. They are especially vulnerable to predators at these times. "Soft-shelled" crabs, prized as food in many countries, are crabs that have recently molted.

As arthropods evolved, body segments became fused in various ways, reducing the overall number of segments. Each region of the body, along with its highly modified paired appendages, is spe-

Mitsuhiko Imamori/Minden Pictures

FIGURE 29.34
Ecdysis in insects. Like all other arthropods, this cicada *(Graptopsaltria nigrofuscata)* sheds its old exoskeleton as it grows and when it undergoes metamorphosis into a winged adult.

cialized, but the structure and function vary greatly among groups. In insects (see Figure 29.43), which have three body regions, the **head** includes a brain, sensory structures, and some sort of feeding apparatus. The segments of the **thorax** bear walking legs and, in some insects, wings. The **abdomen** includes much of the digestive system and sometimes part of the reproductive system.

The coelom of arthropods is greatly reduced, but another cavity, the *hemocoel,* is filled with bloodlike hemolymph. The heart pumps the hemolymph through an open circulatory system, bathing tissues directly.

Arthropods are active animals and require substantial quantities of oxygen. Different groups have distinctive mechanisms for gas exchange, because oxygen cannot cross the impermeable exoskeleton. Marine and freshwater species, like crabs and lobsters, rely on diffusion across gills. The terrestrial groups—insects and spiders—have developed unique and specialized respiratory systems.

High levels of activity also require intricate sensory structures. Many arthropods are equipped with a highly organized central nervous system, touch receptors, chemical sensors, **compound eyes** that include multiple image-forming units, and in some, hearing organs.

Arthropod systematics is an active area of research, and scientists are currently using molecular, morphological, and developmental data to reexamine relationships within this immense group. As *Insights from the Molecular Revolution* explains, hypotheses about the phylogeny and classification of arthropods are in a state of flux. Some researchers even argue for splitting

Dr. Chip Clark

FIGURE 29.35
Subphylum Trilobita. Trilobites, like *Olenellus gilberti,* bore many pairs of relatively undifferentiated appendages.

them into four or more phyla. We follow the traditional definition of five *subphyla,* partly because this classification adequately reflects arthropod diversity and partly because no alternative hypothesis has been widely adopted by experts.

SUBPHYLUM TRILOBITA The trilobites (subphylum Trilobita: *tri* = three, *lobos* = lobe), now extinct, were among the most numerous animals in shallow Paleozoic seas. They disappeared in the Permian mass extinction, but the cause of their demise is unknown. Most trilobites were ovoid, dorsoventrally flattened, and heavily armored, with two deep longitudinal grooves that divided the body into the one median and two lateral lobes for which the group is named **(Figure 29.35).** Their segmented bodies were organized into three body regions: the head, which included a pair of sensory **antennae** (chemosensory organs) and two compound eyes, and a thorax and an abdomen, both of which bore pairs of walking legs.

The position of trilobites in the fossil record indicates that they were among the earliest arthropods. Thus, biologists are confident that their three body regions and numerous unspecialized appendages—one pair per segment—represent ancestral traits within the phylum. As you will learn as you read about the other four subphyla, the subsequent evolution of the different arthropod groups included dramatic remodeling of the major body regions as well as modifications of the ancestral, unspecialized paired appendages into highly modified structures that perform different functions.

SUBPHYLUM CHELICERATA In spiders, ticks, mites, scorpions, and horseshoe crabs (subphylum Chelicerata: *chela* = claw, *keras* = horn), the first pair of appendages, the **chelicerae,** are fanglike structures used for biting prey. The second pair of appendages, the *pedipalps,* serve as grasping organs, sensory organs, or walking legs. All chelicerates have two major body regions, the **cephalothorax** (a fused head and thorax) and the abdomen. The group originated in shallow Paleozoic seas, but most living species are terrestrial. They vary in length from less than a millimeter to 20 cm; all are predators or parasites.

The 60,000 species of spiders, scorpions, mites, and ticks (Arachnida) represent the vast majority of chelicerates **(Figure 29.36).** Arachnids have four pairs of walking legs on the

INSIGHTS FROM THE Molecular Revolution

Unscrambling the Arthropods: Morphological versus molecular characteristics?

For years, biologists disagreed about the evolutionary relationships among the major lineages of arthropods. Some researchers grouped hexapods and myriapods together **(Figure 1, left)** because they share certain morphological characteristics, which may indicate a common ancestry: unbranched appendages, a tracheal system for gas exchange, Malpighian tubules for excretion, and one pair of antennae. However, other researchers noted that hexapods and crustaceans shared a different set of important morphological characters: jawlike mandibles on the fourth head segment, similar compound eyes, comparable development of the nervous system, and similarities in the structure of thoracic appendages.

Research Question

Given the variety of morphological characteristics that hexapods share with myriapods and those that they share with crustaceans, what are the evolutionary relationships among these major lineages of arthropods?

Experiment

A molecular study by Jeffry L. Boore and his colleagues at the University of Michigan lends support to the hypothesis that hexapods may indeed be more closely related to crustaceans than to myriapods. Boore and his coworkers compared the arrangement of genes in the mitochondrial DNA of representative arthropod species. Mitochondrial DNA (mtDNA) was chosen for study because it has been subject to many random rearrangements during the evolutionary history of the arthropods, producing wide variations in the order and placement of the 36 or 37 genes typically present. As long as the rearrangements do not disrupt gene function, they appear to have little or no effect on fitness; therefore, they may not be affected by natural selection. As Boore and his colleagues noted, the large number of possible rearrangements of the mtDNA genes makes it unlikely that the same gene order would appear by chance in any two groups; thus, groups with the same arrangement are likely to share a common ancestor in which the rearrangement first appeared.

The investigators sequenced the mtDNA of a chelicerate, two crustaceans, and a myriapod (listed in the figure caption). They also sequenced the mtDNAs of a mollusk and an annelid, animals that are only distantly related to arthropods, to provide outgroup comparisons. They compared these sequences with those in 14 other invertebrates, including four hexapods and one crustacean, already obtained by other workers.

Results

The locations of two genes encoding leucine-carrying tRNAs provide the strongest clue to the evolutionary relationships of these organisms. These genes are positioned next to each other in the mtDNA of the chelicerate, the mollusk, and the annelid. This arrangement appears to reflect the ancestral gene order present in the common ancestor of annelids, mollusks, and arthropods. However, in the hexapods and the crustaceans studied, one of the leucine tRNA genes is located at a different position, between two genes coding for electron transfer proteins. Moreover, this rearrangement does not appear in the myriapod examined. It is extremely unlikely that the same translocation of the leucine tRNA gene occurred independently in hexapods and crustaceans. Instead, this gene rearrangement probably occurred in an ancestor shared by hexapods and crustaceans after this ancestor diverged from the lineage that includes the chelicerates and myriapods.

Conclusion

The shared gene arrangement in insects and crustaceans and the absence of a matching arrangement in myriapods suggests that hexapods and crustaceans are closely related, and that hexapods and myriapods are not, in spite of their morphological similarities (see figure, New tree). Other investigators, who compared the sequences of ribosomal RNA genes in a variety of arthropod species, reached the same conclusions independently.

Source: J. L. Boore et al. 1995. Deducing the pattern of arthropod phylogeny from mitochondrial DNA rearrangements. *Nature* 376:163–165.

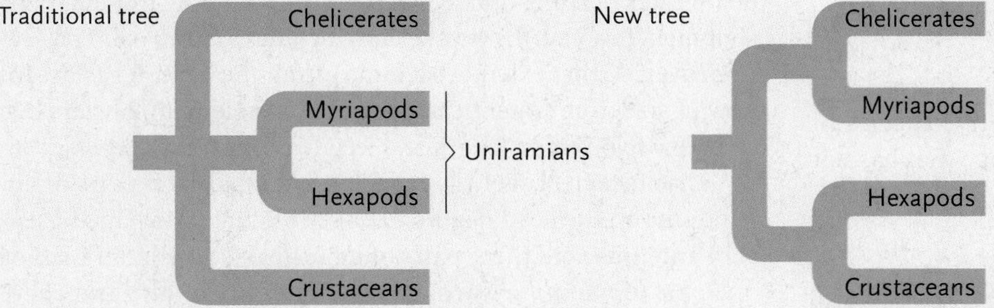
Traditional tree
Chelicerates
Myriapods
Uniramians
Hexapods
Crustaceans
New tree
Chelicerates
Myriapods
Hexapods
Crustaceans

The traditional phylogenetic tree for arthropods (left) differs from that suggested by the research described in this box. The DNA sequences used to construct the new phylogenetic tree (right) were obtained from the chelicerate *Limulus*, the myriapod *Thyrophygus*, the hexapod *Drosophila*, and the crustaceans *Daphnia* and *Homarus*.

cephalothorax and highly modified chelicerae and pedipalps. In some spiders, males use their pedipalps to transfer packets of sperm to females. Scorpions use them to shred food and to grasp one another during courtship. Many predatory arachnids have excellent vision, provided by simple eyes on the cephalothorax. Scorpions and some spiders also have unique pocketlike respiratory organs, called **book lungs,** which are derived from abdominal appendages.

Spiders, like most other arachnids, subsist on a liquid diet. They use their chelicerae to inject paralyzing poisons and digestive enzymes into prey and then suck up the partly digested tissues. Many spiders are economically important predators, helping to control insect pests. Only a few are a threat to humans. The toxin of a black widow *(Latrodectus mactans)* causes paralysis, and the toxin of the brown recluse *(Loxosceles reclusa)* destroys tissues around the site of the bite.

Although many spiders hunt actively, others capture prey on silken threads secreted by **spinnerets,** which are modified abdominal appendages. Some species weave the threads into complex, netlike webs. The silk is secreted as a liquid, but quickly

A. **Wolf spider (*Lycosa* species)**

P. J. Bryant, University of California-Irvine

B. **Spider anatomy**

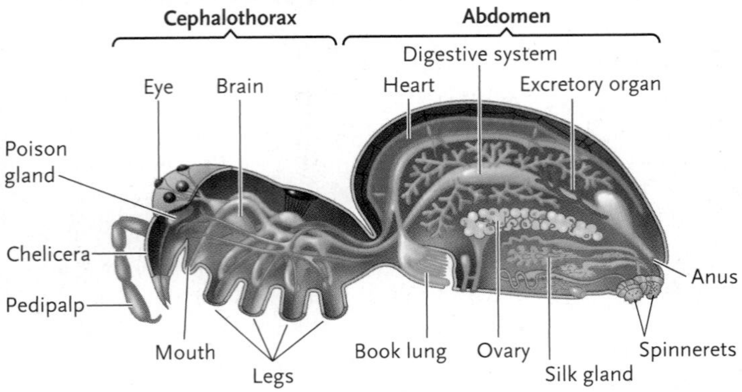
Cephalothorax
Abdomen
Digestive system
Eye
Brain
Heart
Excretory organ
Poison gland
Chelicera
Pedipalp
Mouth
Legs
Book lung
Ovary
Silk gland
Spinnerets
Anus

C. **Scorpion (*Centruroides sculpuratus*)**

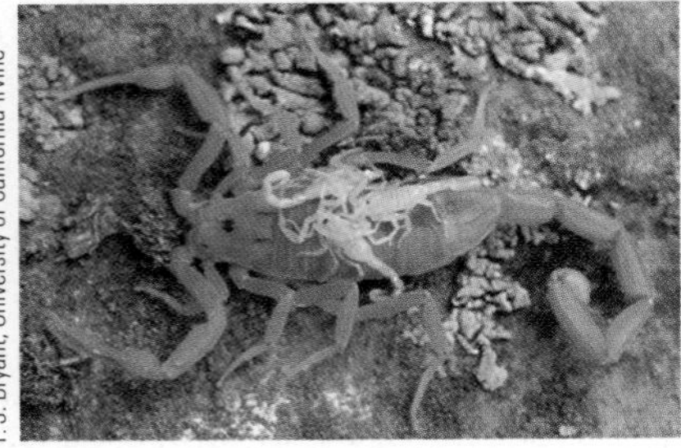
P. J. Bryant, University of California-Irvine

D. **House dust mite (*Dermatophagoides pteronyssinus*)**

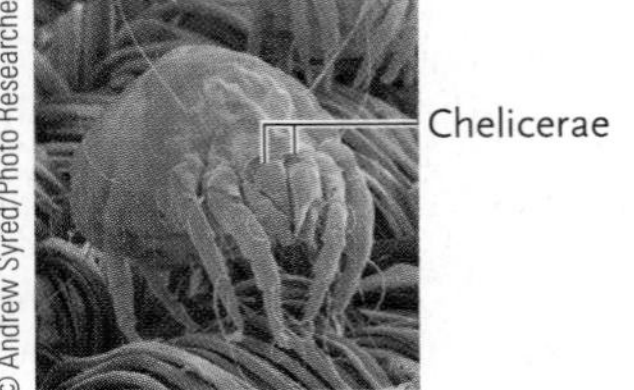
Chelicerae
© Andrew Syred/Photo Researchers, Inc.

FIGURE 29.36
Subphylum Chelicerata, subgroup Arachnida. **(A)** The wolf spider is harmless to humans. **(B)** The arachnid body plan includes a cephalothorax and abdomen. **(C)** Scorpions have a stinger at the tip of the segmented abdomen. Many protect their eggs and young. **(D)** House dust mites, shown in a scanning electron micrograph, feed on microscopic debris.

hardens on contact with air. Spiders also use silk to make nests, to protect their egg masses, as a safety line when moving through the environment, and to wrap prey for later consumption.

Most mites are tiny, but they have a big impact. Some are serious agricultural pests that feed on the sap of plants. Others cause mange (patchy hair loss) or painful and itchy welts on animals. House dust mites cause allergic reactions in many people. Ticks, which are generally larger than mites, are blood-feeding ectoparasites that often transmit pathogens, such as the bacteria that cause Rocky Mountain spotted fever and Lyme disease.

The subphylum Chelicerata also includes five species of horseshoe crabs (Merostomata), an ancient lineage that has not changed much over its 350 million year history **(Figure 29.37).**

Darlyne A. Murawski

FIGURE 29.37
Marine chelicerates. Horseshoe crabs, like *Limulus polyphemus,* are included in the Merostomata.

Horseshoe crabs are carnivorous bottom feeders in shallow coastal waters. Beneath their characteristic shell, they have one pair of chelicerae, a pair of pedipalps, four pairs of walking legs, and a set of paperlike gills that are derived from ancestral walking legs.

SUBPHYLUM CRUSTACEA The 35,000 species of shrimps, lobsters, crabs, and their relatives (subphylum Crustacea: *crusta* = crust or hard shell) represent a lineage that emerged more than 500 million years ago. They are abundant in marine and freshwater habitats. A few species, such as sowbugs and pillbugs, live in moist, sheltered terrestrial environments. In many crustaceans two, or even all three, of the arthropod body regions—head, thorax, and abdomen—are fused; a fused cephalothorax and a separate abdomen is an especially common pattern. The edible "tail" of a lobster, shrimp, or crayfish is actually a highly muscularized abdomen. In some, the exoskeleton includes a **carapace,** a protective covering that extends backward from the head. Crustaceans vary in size from water fleas less than 1 mm long to lobsters that can grow to 60 cm in length and weigh as much as 20 kg.

Crustaceans generally have five characteristic pairs of appendages on the head **(Figure 29.38).** Most have two pairs of sensory antennae and three pairs of mouthparts. The latter include one pair of *mandibles,* which move laterally to bite and chew, and two pairs of *maxillae,* which hold and manipulate food. Numerous paired appendages posterior to the mouthparts vary among groups.

Most crustaceans are active animals that exhibit complex patterns of movement. Their activities are coordinated by elaborate sensory and nervous systems, including chemical and touch receptors in the antennae, compound eyes, statocysts on the head, and sensory hairs embedded in the exoskeleton throughout the body. The nervous system is similar to that in annelids, but the ganglia are larger and more complex, allowing a finer level of motor control. High levels of activity require substantial amounts of oxygen, and larger species have complex, feathery gills tucked beneath the carapace. Activity also produces abundant metabolic wastes that are excreted by diffusion across the gills or, in larger species, by glands located in the head.

A. Crab (Ocypode species)
Please Image copyright EcoPrint, 2010. Used under license from Shutterstock.com

B. Lobster (Homarus americanus)
Herve Chaumeton/Agence Nautr

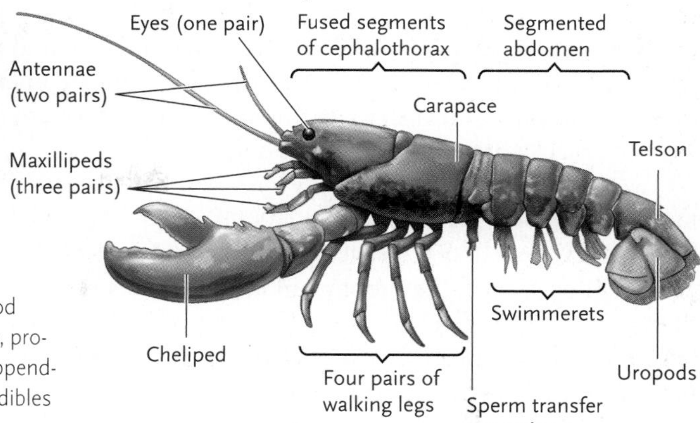
C. Lobster external anatomy
Eyes (one pair)
Fused segments of cephalothorax
Segmented abdomen
Antennae (two pairs)
Carapace
Telson
Maxillipeds (three pairs)
Swimmerets
Cheliped
Four pairs of walking legs
Uropods
Sperm transfer appendage

FIGURE 29.38
Decapod crustaceans. **(A)** Crabs, like this ghost crab, and **(B)** lobsters are typical decapod crustaceans. The abdomen of a crab is shortened and wrapped under the cephalothorax, producing a compressed body. **(C)** Lobsters bear 19 pairs of distinctive appendages; only appendages on the left side of the animal are illustrated in this lateral view, and one pair of mandibles and two pairs of maxillae are not shown.

FIGURE 29.39
Copepods. Tiny crustaceans, like these copepods (*Cyclops* species), occur by the billions in freshwater and marine plankton.

Oxford Scientific/Getty Images

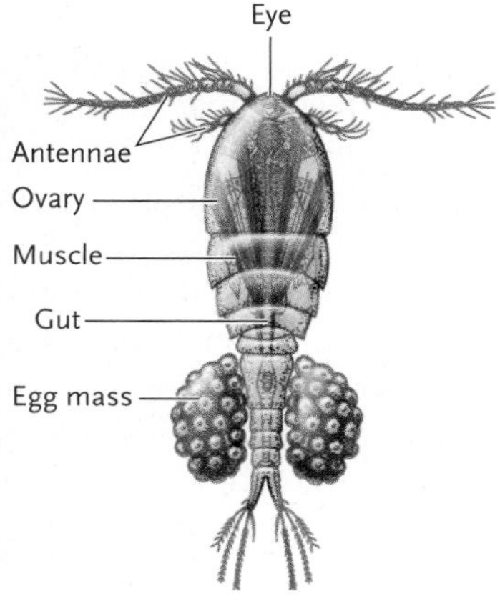
Eye
Antennae
Ovary
Muscle
Gut
Egg mass

FIGURE 29.40
Barnacles. Gooseneck barnacles *(Lepas anatifera)* attach to the underside of floating debris. Like other barnacles, they open their shells and extend their feathery legs to collect particulate food from seawater.

Runk/Schoenberger/Grant Heilman

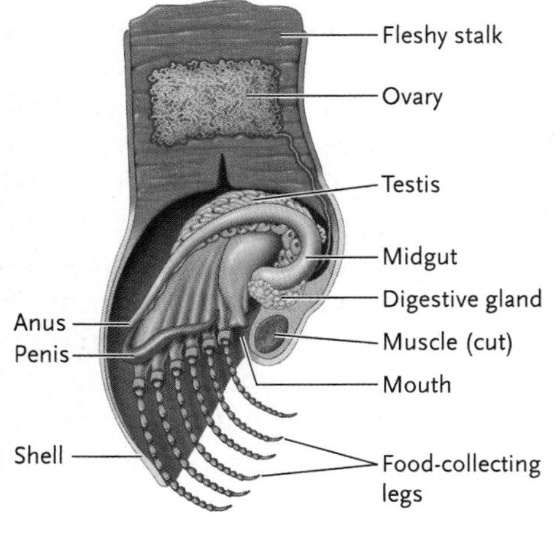
Fleshy stalk
Ovary
Testis
Midgut
Digestive gland
Anus
Muscle (cut)
Penis
Mouth
Shell
Food-collecting legs

The sexes are typically separate, and courtship rituals are often complex. Eggs are usually brooded on the surface of the female's body or beneath the carapace. Many have free-swimming larvae that, after undergoing a series of molts, gradually assume an adult form.

The subphylum includes so many different body plans that it is usually divided into six major groups with numerous subgroups. The crabs, lobsters, and shrimps (Decapoda, meaning "10 feet," which is a subgroup of the Malacostraca) number more than 10,000 species. The vast majority of decapods are marine, but a few shrimps, crabs, and crayfishes occupy freshwater habitats. Some crabs also live in moist terrestrial habitats, where they scavenge dead vegetation, clearing the forest floor of debris.

All decapods exhibit extreme specialization of their appendages. In the American lobster, for example, each of the 19 pairs of appendages is different (see Figure 29.38C). Behind the antennae, mandibles, and maxillae, the thoracic segments have three pairs of *maxillipeds,* which shred food and pass it up to the mouth, a pair of large *chelipeds* (pinching claws), and four pairs of walking legs. The abdominal appendages include a pair specialized for sperm transfer (in males only), *swimmerets* for locomotion and for brooding eggs, and *uropods,* a pair of appendages that, combined with the *telson,* the tip of the abdomen, form a fan-shaped tail. If any appendage is damaged, the animal can autotomize (drop) it and begin growing a new one before its next molt.

Representatives of several crustacean groups—fairy shrimps, amphipods, water fleas, krill, ostracods, and copepods **(Figure 29.39)**—live as plankton in the upper waters of oceans and lakes. Most are only a few millimeters long, but are present in huge numbers. They feed on microscopic algae or detritus and are themselves food for larger invertebrates, fishes, and some suspension-feeding marine mammals like the baleen whales. Planktonic crustaceans are among the most abundant animals on Earth.

Adult barnacles (Cirripedia, meaning "hairy footed," a subgroup of the Maxillopoda) are sessile, marine crustaceans that live within a strong, calcified cup-shaped shell **(Figure 29.40).** Their free-swimming larvae attach permanently to substrates—rocks, wooden pilings, the hulls of ships, the shells of mollusks, even the skin of whales—and secrete the shell, which is actually a modified exoskeleton. To feed, barnacles open the shell and extend six pairs of feathery legs. The beating legs capture microscopic plankton and transfer it the mouth. Unlike most crustaceans, barnacles are hermaphroditic.

SUBPHYLUM MYRIAPODA The 3,000 species of centipedes (Chilopoda) and 10,000 species of millipedes (Diplopoda) are classified together in the subphylum Myriapoda (*myrias* = ten thousand, *pod-* = feet). Myriapods have two body regions, a head and a segmented trunk **(Figure 29.41)**. The head bears one pair of antennae, and the trunk bears one (centipedes) or two (millipedes) pairs of walking legs on most of its many segments. Myriapods are terrestrial, and many species live under rocks or dead leaves. Centipedes are fast and voracious predators, using powerful toxins to kill their prey; they generally feed on invertebrates, but some eat small vertebrates. The bite of some species is harmful to humans. Although most species are less than 10 cm long, some grow to 25 cm. The millipedes are slow but powerful herbivores or scavengers. The largest species attain a length of nearly 30 cm. Although they lack a poisonous bite, they curl into a ball and exude noxious liquids when disturbed.

SUBPHYLUM HEXAPODA In terms of sheer numbers, diversity, and the range of habitats they occupy, the 1,000,000 or more species of insects and their closest relatives (subphylum Hexapoda, *hexapoda* = six feet) are the most successful animals on Earth. They were among the first animals to colonize terrestrial habitats, where most species still live. The oldest insect fossils date from the Devonian, 380 million years ago. Insects have one pair of antennae on the head, a pair of mandibles for feeding, and unbranched appendages. Insects are generally small, ranging from 0.1 mm to 30 cm in length. The lineage is divided into about 30 subgroups **(Figure 29.42)**. The insect body plan always includes a head, thorax, and abdomen **(Figure 29.43)**. The head is equipped with multiple mouthparts, a pair of compound eyes, and one pair of sensory antennae. The thorax has three pairs of walking legs and often one or two pairs of wings. Insects are the only invertebrates capable of flight. Their wings, which are made of lightweight but durable sheets of chitin and sclerotin, arise embryonically from the body wall; unlike the wings of birds and bats, insect wings are not derived from ancestral appendages.

Studies in evolutionary developmental biology have begun to unravel the genetic changes that fostered certain aspects of the insect body plan. For example, the *Distal-less* gene (*Dll* for short) is a highly conserved tool-kit gene (see Section 22.6) that triggers the development of appendages in all sorts of animals—the legs of chickens, the fins of fishes, the parapodia of polychaete worms, and the diverse appendages of arthropods. All arthropods also have a gene called *Ultrabithorax* (*Ubx* for short). It is one of the *Hox* genes that control the overall body plans of animals. In insects, the *Ubx* gene contains a unique mutation, not found in other arthropods, that causes the protein for which it codes to repress *Dll,* thereby preventing the formation of appendages wherever *Ubx* is expressed. And because insects express *Ubx* in their abdomen, they do not grow abdominal appendages. All other arthropods, which have the ancestral, nonrepressing form of the *Ubx* gene, have appendages in the posterior region of their body. Thus, one mutation in a *Hox*-family gene has fostered the evolution of a highly distinctive morphological trait in insects—having legs on the thorax, but not on the abdomen.

Insects exchange gases through a specialized **tracheal system,** a branching network of tubes that carry oxygen from small openings in the exoskeleton to tissues throughout the body (see Figure 44.5). Insects excrete nitrogenous wastes through specialized **Malpighian tubules** that transport wastes to the digestive system for disposal with feces (see Figure 46.6). Both of these organ systems are unique among animals. Insect sensory systems are diverse and complex. Besides image-forming compound eyes (see Figure 39.10), many insects have light-sensing ocelli on their heads. Many also have hairs, sensitive to touch, on their antennae, legs, and other regions of the body. Chemical receptors are particularly common on the legs, allowing the identification of food. And many groups of insects have hearing organs to detect predators and potential mates. The familiar chirping of crickets, for example, is a mating call emitted by males that may repel other males and attract females.

As a group, insects feed in every conceivable way and on most other organisms. Species that eat plants, such as grasshoppers, have a pair of rigid mandibles, which chew food before it is ingested. Behind the mandibles is a pair of maxillae, which may

A. **Millipede (*Spirobolus* species)**

Steve Martin/Tom Stack & Associates

B. **Centipede (*Scolopendra subspinipes*)**

Z. Leszczynski/Animals Animals-Earth Scenes

FIGURE 29.41

Millipedes and centipedes. (A) Millipedes feed on living and decaying vegetation. They have two pairs of walking legs on most segments. **(B)** Like all centipedes, this Southeast Asian species, shown feeding on a small frog, is a voracious predator. Centipedes have one pair of walking legs per segment.

A. Silverfish (Thysanura, *Lepisma saccharina*) are primitive wingless insects.

Barry Runk/Grant Heilman Photography

B. Dragonflies, like the flame skimmer (Odonata, *Epitheca cynosura*), have aquatic larvae that are active predators; adults capture other insects in mid-air.

Image copyright Steven Russell Smith Photos, 2010. Used under license from Shutterstock.com

C. Male praying mantids (Mantodea, *Mantis religiosa*) are often eaten by the larger females during or immediately after mating.

© BIOS Borrell Bartomeu/Peter Arnold, Inc.

D. This rhinoceros beetle (Coleoptera, *Lucanus cervus*) is one of more than 250,000 beetle species that have been described.

Image copyright Henrik Larsson, 2010. Used under license from Shutterstock.com

E. Fleas (Siphonoptera, *Hystrichopsylla dippiei*) have strong legs with an elastic ligament that allows these parasites to jump on and off their animal hosts.

Nigel Cattlin/Photo Researchers, Inc.

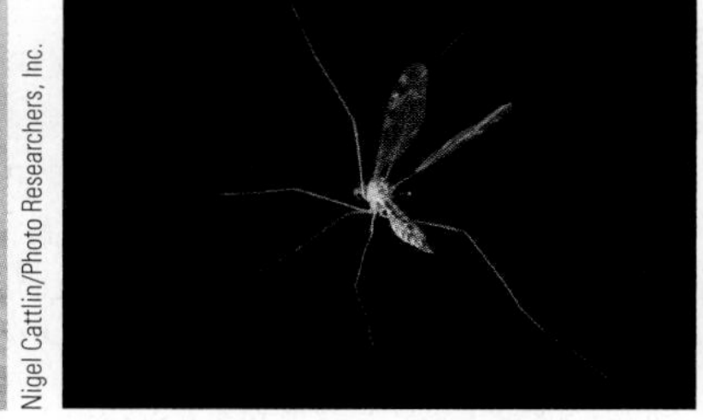

F. Crane flies (Diptera, *Tipula* species) look like giant mosquitoes, but their mouthparts are not useful for biting other animals; the adults of most species live only a few days and do not feed at all.

©Michael Durham/Minden Pictures

G. The luna moth (Lepidoptera, *Actias luna*), like other butterflies and moths, has wings that are covered with colorful microscopic scales.

iStockphoto.com/William Perry

H. Like many other ant species, fire ants (Hymenoptera, *Solenopsis invicta*) live in large cooperative colonies. Fire ants—named for their painful sting—were introduced into southeastern North America, where they are now serious pests.

iStockphoto.com/Scott Harms

FIGURE 29.42
Insect diversity. Insects are grouped into about 30 orders, 8 of which are illustrated here.

FIGURE 29.43
The insect body plan. Insects have distinct head, thorax, and abdomen. Of all the internal organ systems, only the dorsal blood vessel, ventral nerve cord, and some muscles are strongly segmented.

Internal anatomy of a female grasshopper

Brain
Dorsal blood vessel
Heart
Ovary
Mouth
Ventral nerve cord with ganglia
Digestive system
Malpighian tubules

External anatomy of a grasshopper

Antenna
Head
Thorax
Abdomen
Tympanum (hearing organ)
Wing
Compound eye
Mouth parts
Legs

FIGURE 29.44
Specialized insect mouthparts. The **(A)** ancestral chewing mouthparts have been modified by evolution, allowing different insects to **(B)** sponge up food, **(C)** drink nectar, and **(D)** pierce skin to drink blood.

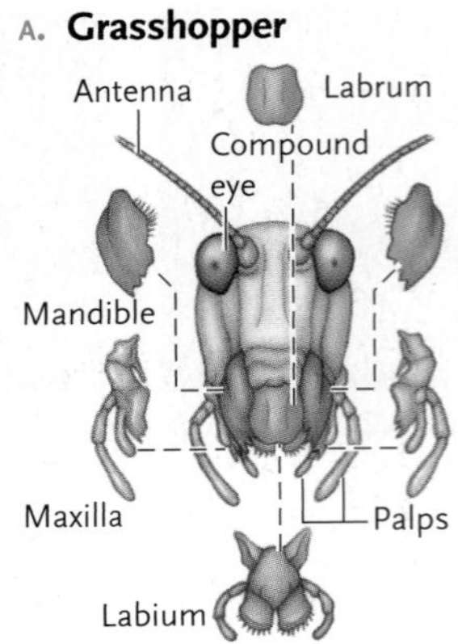
A. Grasshopper
Antenna
Labrum
Compound eye
Mandible
Maxilla
Palps
Labium

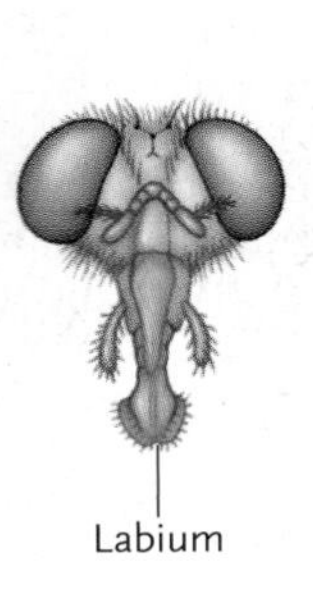
B. Housefly
Labium

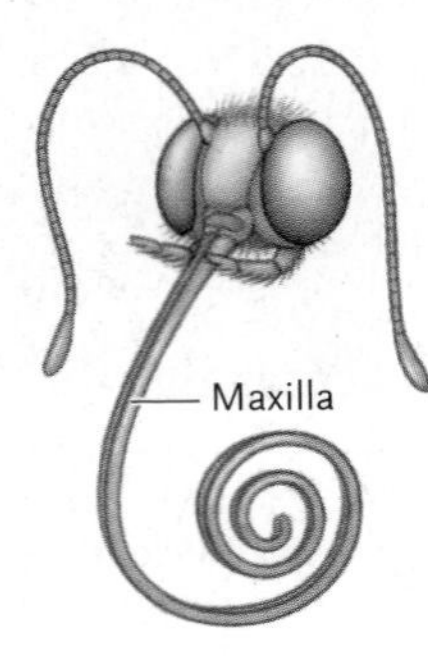
C. Butterfly
Maxilla

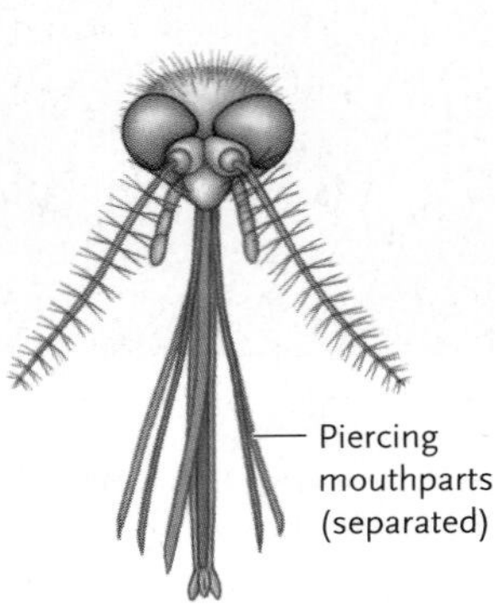
D. Mosquito
Piercing mouthparts (separated)

also aid in food acquisition. Insects also have inflexible upper and lower lips, the *labrum* and *labium*, respectively. A tongue-like structure, just dorsal to the labium in chewing insects, houses the openings of the salivary glands. But evolution has modified this ancestral mandibulate pattern in numerous ways **(Figure 29.44)**. In houseflies, the mouthparts are adapted for sopping up liquid food. In butterflies and moths the mouthparts include a long proboscis to drink nectar. And in some biting flies, like mosquitoes and blackflies, the mouthparts have evolved into piercing structures.

After it hatches from an egg, an insect passes through a series of developmental stages called *instars*. Several hormones control development and ecdysis, which marks the passage from one instar to the next. Insects exhibit one of three basic patterns of postembryonic development **(Figure 29.45)**. Primitive, wingless species simply grow and shed their exoskeleton without undergoing major changes in morphology. Other species undergo **incomplete metamorphosis;** they hatch from the egg as a *nymph*, which lacks functional wings. In many species, such as grasshoppers (order Orthoptera), the nymphs resemble the adults. In other insects, such as dragonflies (order Odonata), the aquatic nymphs are morphologically very different from the adults.

Most insects undergo **complete metamorphosis:** the larva that hatches from the egg differs greatly from the adult. Larvae and adults often occupy different habitats and consume different food. The larvae (caterpillars, grubs, or maggots) are often worm-shaped, with chewing mouthparts. They grow and molt several times, retaining their larval morphology. Before they transform into sexually mature adults, they spend a period of time as a sessile **pupa.** During this stage, the larval tissues are drastically reorganized. The adult that emerges is so different from the larva that it is often hard to believe that they are of the same species. Moths, butterflies, beetles, and flies are examples of insects with complete metamorphosis. Their larval stages specialize in feeding and growth, whereas the adults are adapted for dispersal and reproduction. In some species, the adults never feed, relying on the energy stores accumulated during the larval stage.

The 240-million-year history of insects has been characterized by innovations in morphology, life cycle patterns, locomotion, feeding, and habitat use. Their well-developed nervous systems govern exceptionally complex patterns of behavior, including parental care, a habit that reaches its zenith in the colonial social insects, the ants, bees, and wasps (see Chapter 55). The factors that contribute to the insects' success also make them our most aggressive competitors. They destroy vegetable crops, stored food, wool, paper, and timber. They feed on blood from

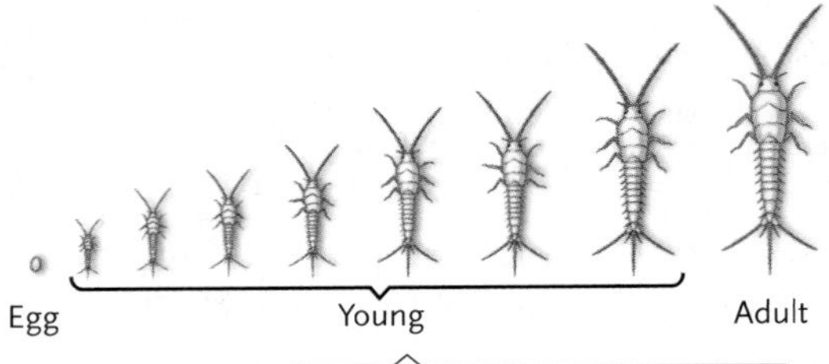
A. No metamorphosis
Egg
Young
Adult

Some wingless insects, like silverfish (order Thysanura), do not undergo a dramatic change in form as they grow.

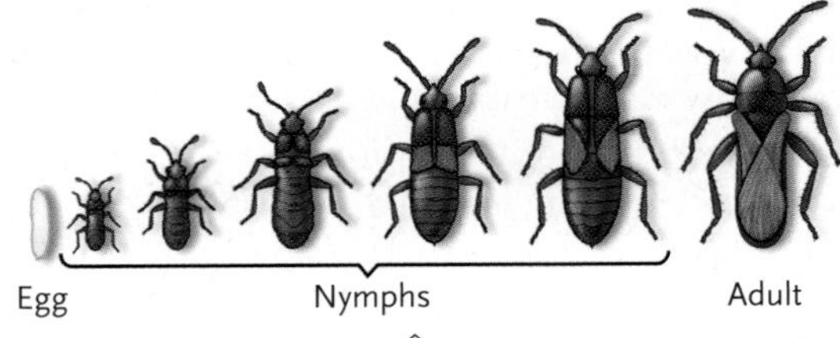
B. Incomplete metamorphosis
Egg
Nymphs
Adult

Other insects, such as true bugs (order Hemiptera), have incomplete metamorphosis; they develop from nymphs into adults with relatively minor changes in form.

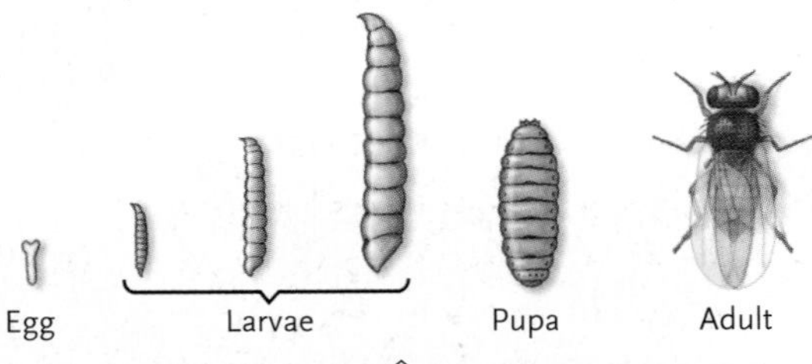
C. Complete metamorphosis
Egg
Larvae
Pupa
Adult

Fruit flies (order Diptera) and many other insects have complete metamorphosis; they undergo a total reorganization of their internal and external anatomy when they pass through the pupal stage of the life cycle.

FIGURE 29.45
Patterns of postembryonic development in insects.

humans and domesticated animals, sometimes transmitting disease-causing pathogens as they do so. Nevertheless, insects are essential members of terrestrial ecological communities. Many species pollinate flowering plants, including important crops. Many others attack or parasitize species that are harmful to human activities. And most insects are a primary source of food for other animals. Some even make useful products like honey, beeswax, and silk.

In the next chapter we consider the lineage of deuterostomes, which includes the vertebrates and their closest invertebrate relatives.

STUDY BREAK 29.7

1. **What part of a parasitic nematode's anatomy protects it from the digestive enzymes of its host?**
2. **If an arthropod's rigid exoskeleton cannot be expanded, how does the animal grow?**
3. **How do the number of body regions and the appendages on the head differ among the four subphyla of living arthropods?**
4. **How do the life stages differ between insects that have incomplete metamorphosis and those that have complete metamorphosis?**

UNANSWERED QUESTIONS

What are the evolutionary relationships among the invertebrate lineages?

If we step back from our vertebrate and terrestrial biases, invertebrates are the dominant form of life on Earth. Many students of natural history are overwhelmed by the sheer numbers of organisms that one encounters, particularly in the aquatic environment, and the challenge of categorizing them taxonomically. Sorting them out in terms of their ecological roles is equally daunting. The problem we face is that invertebrates do it all—they are predators, herbivores, parasites, detritivores, and the primary symbiotic organisms on Earth. This adaptability of form and function is a fascinating hallmark of the animal way of life, but it poses a legion of questions, many still unanswered. Recent advances in genetic technology have advanced our understanding, but there is much left to do.

In the past two decades, our ability to compare genetic information from various invertebrate groups has led to a remarkable reshuffling of the long-established categories used to classify these organisms. The first categories to be eliminated were groups based on superficial phenotypic resemblances, such as the "pseudocoelomates," which had plagued student understanding of diversity. Today, we have a much deeper knowledge of the evolutionary relationships of these organisms. Perhaps the most exciting discovery is that much of the diversity we see is not the product of slow changes in protein-coding gene sequences, but rather the result of variations in the timing and location of the expression of genes that affect development. Evo-devo, the melding of evolutionary and developmental biology—mostly made possible by intensive studies of two model invertebrates, *Caenorhabditis elegans* and *Drosophila melanogaster*—has revealed that changes in the expression of relatively simple sets of genes have brought about the myriad forms of life we see among the invertebrates. As systematists incorporate these new discoveries in their analyses, our understanding of the evolutionary relationships among the invertebrates will surely change.

What is the genetic basis of the diversity of form and function observed among invertebrates?

Because of advances in genetics research, we are on the cusp of being able to answer some fundamental questions. How does an animal's body develop either radial or bilateral symmetry? How does an organism develop a head with a concentration of nervous system tissue, and how do all the exquisite sensory systems associated with a big brain develop? From where do the respiratory pigments, which increase the capacity of the hemolymph or blood to carry oxygen, increasing an organism's capacity for activity, arise? How does the immune system develop, and what genetic change fosters the quantum leap from a nonspecific defense system to one that responds specifically to foreign invaders? More practically, are there genetic switches that we can manipulate? Can we make blood-feeding invertebrates, such as mosquitoes, or parasitic species, like tapeworms or filarial worms, innocuous? Is it possible to use genetic engineering to reduce the ability of the mosquito *Anopheles gambiae* (the "deadliest organism on Earth") to transmit malaria? Will it be possible to forestall or reverse the global decline of coral reefs, one of the richest habitats on Earth?

The answers to these and many more basic and applied questions lie within a deep knowledge and understanding of the invertebrates and the roles they play on our planet. If one looks at invertebrates as dynamic systems, as rich sources of clues to life on Earth, the questions they pose easily provide a lifetime of investigation and reward.

Think Critically

1. Based on your reading of this chapter, provide two examples of how recent genetic comparisons of invertebrates have confirmed groupings based on morphology and embryology.
2. Based on the content of this chapter, which genes would you study if you wanted to learn about the genetic basis of diversity among invertebrates?

David Kramer

William S. Irby is an associate professor in the Department of Biology at Georgia Southern University in Statesboro. His research focuses on the ecology and evolution of blood-feeding behavior in mosquitoes. To learn more about Irby's work, go to http://www.bio.georgiasouthern.edu/amain/faclist.html.

REVIEW KEY CONCEPTS

Go to **CENGAGENOW** at www.cengage.com/login to access quizzing, animations, exercises, articles, and personalized homework help.

29.1 What Is an Animal?

- Animals are eukaryotic, multicellular, heterotrophs that are motile at some time in their lives.
- Animals probably arose from a colonial flagellated ancestor during the Precambrian era (Figure 29.2).

29.2 Key Innovations in Animal Evolution

- All animals except sponges have tissues, which are organized into either two or three tissue layers.
- Although some animals exhibit radial symmetry, most exhibit bilateral symmetry (Figure 29.3).
- Acoelomate animals have no body cavity. Pseudocoelomate animals have a body cavity between the derivatives of endoderm and mesoderm. Coelomate animals have a body cavity that is entirely lined by derivatives of mesoderm (Figure 29.4).
- Two lineages of bilaterally symmetrical animals differ in developmental patterns (Figure 29.5). Most protostomes exhibit spiral, determinate cleavage, and their coelom forms from a split in a solid mass of mesoderm. Most deuterostomes have radial, indeterminate cleavage, and the coelom usually forms within outpocketings of the primitive gut.
- Four animal phyla exhibit segmentation.

Animation: Types of body symmetry

Animation: Types of body cavities

Animation: Developmental differences between protostomes and deuterostomes

29.3 An Overview of Animal Phylogeny and Classification

- Analyses of molecular sequence data have refined our view of animal evolutionary history (Figure 29.6). The molecular phylogeny recognizes some major lineages that had been identified on the basis of morphological and embryological characters. Sponges are grouped in the Parazoa. All other lineages are grouped in the Eumetazoa. Among the Eumetazoa, the Radiata includes animals with two tissue layers and radial symmetry, and the Bilateria includes animals with three tissue layers and bilateral symmetry.
- Bilateria is further subdivided into Protostomia and Deuterostomia. The new phylogeny divides the Protostomia into the Lophotrochozoa and the Ecdysozoa.
- The molecular phylogeny suggests that ancestral protostomes had a coelom and that segmentation arose independently in three lineages.

29.4 Animals without Tissues: Parazoa

- Sponges (phylum Porifera) are asymmetrical animals with limited integration of cells in their bodies (Figure 29.7).
- The body of many sponges is a water-filtering system (Figure 29.8). Flagellated choanocytes draw water into the body and capture particulate food.

Animation: Body plan of a sponge

29.5 Eumetazoans with Radial Symmetry

- The two major radiate phyla, Cnidaria and Ctenophora, have two well-developed tissue layers with a gelatinous mesoglea between them (Figure 29.9). Members of both phyla lack organ systems. All are aquatic.
- Cnidarians capture prey with tentacles and stinging nematocysts (Figures 29.10, 29.12, and 29.13). Their life cycles may include polyps, medusae, or both (Figure 29.11).
- Ctenophores use long tentacles to capture particulate food and use rows of cilia for locomotion (Figure 29.14).

Animation: Cnidarian body plans

Animation: Nematocyst action

Animation: Cnidarian life cycle

29.6 Lophotrochozoan Protostomes

- The taxon Lophotrochozoa includes eight phyla.
- Three small phyla (Ectoprocta, Brachiopoda, and Phoronida) use a lophophore to feed on particulate matter (Figure 29.15).
- Free-living flatworm species (phylum Platyhelminthes) have well-developed digestive, excretory, reproductive, and nervous systems (Figures 29.16 and 29.17). Parasitic species attach to their animal hosts with suckers or hooks (Figures 29.18 and 29.19).
- The rotifers (phylum Rotifera) are tiny and abundant inhabitants of freshwater and marine ecosystems (Figure 29.20). Movements of cilia in the corona control their locomotion and bring food to their mouths.
- The ribbon worms (phylum Nemertea) are elongate and often colorful animals with a proboscis housed in a rhynchocoel (Figure 29.21).
- Mollusks (phylum Mollusca) have fleshy bodies that are often enclosed in a hard shell. The molluscan body plan includes a head–foot, visceral mass, and mantle (Figures 29.22 and 29.24–29.27).
- Segmented worms (phylum Annelida) generally exhibit segmentation of the coelom and of the muscular, circulatory, excretory, respiratory, and nervous systems. They use the coelom as a hydrostatic skeleton for locomotion (Figures 29.28–29.31).

Animation: Planarian organ systems

Animation: Blood fluke life cycle

Animation: Tapeworm life cycle

Animation: Earthworm body plan

Animation: Molluscan groups

Animation: Snail body plan

Animation: Torsion in gastropods

Animation: Clam body plan

Animation: Cuttlefish body plan

29.7 Ecdysozoan Protostomes

- The taxon Ecdysozoa includes three phyla that periodically shed their cuticle or exoskeleton.
- Roundworms (phylum Nematoda) feed on decaying organic matter or parasitize plants or animals (Figure 29.32). They move by contracting longitudinal muscles of the body wall.

- The velvet worms (phylum Onychophora) have segmented bodies and unjointed legs (Figure 29.33). Some species bear live young, which develop in a uterus.
- The segmented bodies of the arthropods (phylum Arthropoda) have specialized appendages for feeding, locomotion, or reproduction. Arthropods shed their exoskeleton as they grow or enter a new stage of the life cycle (Figure 29.34). They have an open circulatory system, a complex nervous system, and, in some groups, highly specialized respiratory and excretory systems.
- Arthropods are divided into five subphyla. The extinct trilobites (subphylum Trilobita), with three-lobed bodies and relatively undifferentiated appendages, were abundant in Paleozoic seas (Figure 29.35). Chelicerates have a cephalothorax and abdomen; appendages on the head serve as pincers or fangs and pedipalps (Figures 29.36 and 29.37). Crustaceans have a carapace that covers the cephalothorax as well as highly modified appendages, including antennae and mandibles (Figures 29.38–29.40). Myriapods have a head and an elongate, segmented trunk (Figure 29.41). Hexapods have three body regions, three pairs of walking legs on the thorax, and three pairs of feeding appendages on the head (Figures 29.42–29.44). Insects exhibit three patterns of postembryonic development (Figure 29.45).

Animation: Roundworm body plan

Animation: Crab life cycle

Animation: Chelicerates

Animation: Insect head parts

Animation: Insect development

UNDERSTAND AND APPLY

Test Your Knowledge

1. Which of the following characteristics is *not* typical of most animals?
   a. heterotrophic
   b. sessile
   c. bilaterally symmetrical
   d. multicellular
   e. motile at some stage of life cycle
2. A body cavity that separates the digestive system from the body wall but is *not* completely lined with mesoderm is called a:
   a. schizocoelom.
   b. mesentery.
   c. peritoneum.
   d. pseudocoelom.
   e. hydrostatic skeleton.
3. Protostomes and deuterostomes typically differ in:
   a. their patterns of body symmetry.
   b. the number of germ layers during development.
   c. their cleavage patterns.
   d. the size of their sperm.
   e. the size of their digestive systems.
4. The nematocysts of cnidarians are used primarily for:
   a. capturing prey.
   b. detecting light and dark.
   c. courtship.
   d. sensing chemicals.
   e. gas exchange.
5. Which organ system is absent in flatworms (phylum Platyhelminthes)?
   a. nervous system
   b. reproductive system
   c. circulatory system
   d. digestive system
   e. excretory system
6. Which part of a mollusk secretes the shell?
   a. visceral mass
   b. radula
   c. trochophore
   d. head–foot
   e. mantle
7. What is the major morphological innovation seen in annelid worms?
   a. a complete digestive system
   b. image-forming eyes
   c. a respiratory system
   d. an open circulatory system
   e. body segmentation
8. Which phylum includes the most abundant animals in soil?
   a. Nematoda
   b. Rotifera
   c. Mollusca
   d. Annelida
   e. Brachiopoda
9. Which body region of an insect bears the walking legs?
   a. head
   b. carapace
   c. abdomen
   d. thorax
   e. trunk
10. Ecdysis refers to a process in which:
   a. bivalves use siphons to pass water across their gills.
   b. arthropods shed their old exoskeletons.
   c. cnidarians build skeletons of calcium carbonate.
   d. rotifers produce unfertilized eggs.
   e. squids escape from predators in a cloud of ink.

Discuss the Concepts

1. Many invertebrate species are hermaphroditic. What selective advantages might this characteristic offer? In what kinds of environments might it be most useful?
2. People who eat raw clams and oysters harvested from sewage-polluted waters often develop mild to severe gastrointestinal infections. These mollusks are suspension feeders. Develop a hypothesis about why people who eat them raw may be at risk.
3. On a voyage to the ocean bottom, a biologist discovers a worm that appears to be new to science. What characteristics of this animal should the biologist examine to determine whether or not she has discovered a previously undescribed phylum?

4. The phylogenetic tree and classification based on molecular sequence data suggest that segmentation evolved independently in Lophotrochozoa (phylum Annelida), Ecdysozoa (phyla Onychophora and Arthropoda), and Deuterostomia (phylum Chordata). What morphological evidence would you try to collect to confirm that segmentation is not homologous in these three groups?
5. What are the relative advantages and disadvantages of radially symmetrical and bilaterally symmetrical body plans?

Design an Experiment

Design an experiment to test the hypothesis that the cuticle of parasitic nematodes protects them from the acids and enzymes present in the digestive systems of their hosts. Your design must include both experimental and control treatments.

Interpret the Data

A research team led by Andreas Michalson of the Kliniken Essen, a hospital in Germany, compared the relative effectiveness of "leech therapy" and the topical application of diclofenac, an anti-inflammatory drug, for the treatment of pain caused by osteoarthritis in the thumb. Leech saliva contains more than 30 biologically active compounds, including some with anti-inflammatory effects. For the group receiving drug therapy, the patients applied the drug to their thumbs at least twice a day for up to 30 days. For the group receiving leech therapy, researchers applied two or three leeches to the affected joint and allowed them to feed until they were satiated; the average feeding time was 50 minutes. Patients in the leech therapy group received just one treatment. The researchers asked patients to score their pain (0 = no pain; 100 = severe pain) before treatment (day 0 in the figure) and again one week, one month, and two months after treatment. The results are presented in the accompanying bar chart. Which treatment had the greater effect in reducing the pain of arthritis patients in this study? Did leech therapy have a prolonged effect?

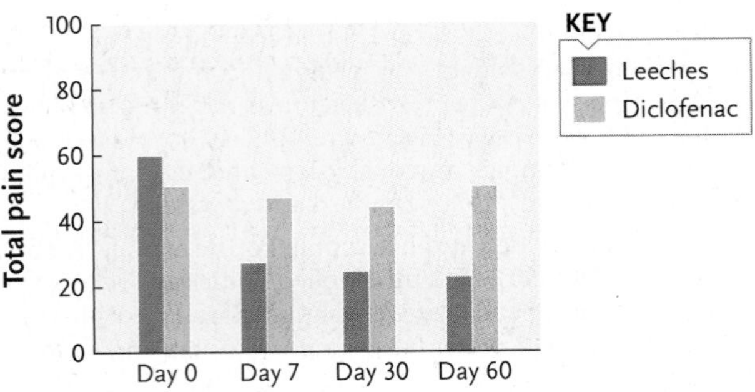

Total pain score
100
80
60
40
20
0
Day 0
Day 7
Day 30
Day 60
KEY
Leeches
Diclofenac

Apply Evolutionary Thinking

Many insects have a larval stage that is morphologically different from the adult and that feeds on different foods. What selection pressures may have fostered the evolution of a life cycle with such distinctive life stages? Your answer should address the different biological activities that characterize each life cycle stage.

Express Your Opinion

Cone snails are diverse but most species have limited geographical ranges in tropical coral reefs, which makes them highly vulnerable to extinction. We do not know how many are harvested, because no one monitors the trade. Should the United States push to extend regulations on trade in endangered species to cover any species captured from the wild? Go to www.cengage.com/login to investigate both sides of the issue and then vote.

30

Snow monkeys *(Macaca fuscata)*. These snow monkeys, which have the northernmost distribution of any nonhuman primate, are soaking in a hot spring in Japan.

Deuterostomes: Vertebrates and Their Closest Relatives

STUDY OUTLINE

Why It Matters. . . In 1798, naturalists at the British Museum skeptically probed a curious specimen that had been sent from Australia. The furry creature—about the size of a housecat—had webbed front feet, a ducklike bill, and a flat, paddlelike tail **(Figure 30.1)**. The scientists eagerly searched for evidence that a prankster had stitched together parts from wildly different animals, but they found no signs of trickery and soon accepted the duck-billed platypus *(Ornithorhynchus anatinus)* as a genuine zoological novelty.

Further study has revealed that the platypus is even stranger than those scientists could have imagined. Like other mammals, the platypus is covered with hair, and females produce milk that the offspring lick off the fur on their mother's belly. But like turtles and birds, a platypus has no teeth, and it reproduces by laying eggs instead of giving birth to its offspring. And like turtles, birds, lizards, snakes, and crocodilians, it has a cloaca, a multipurpose chamber through which it releases feces, urine, and eggs. Scientists had never before seen such a weird combination of traits, and they didn't quite know what to make of them.

Studies of the platypus under natural conditions have helped biologists make sense of its characteristics. The platypus inhabits streams and lagoons in Australia and Tasmania. It rests in streamside burrows during the day, but at night it slips into the water to hunt for invertebrates. Its dense fur keeps its body warm and dry under water, and its tail serves both as a rudder and as a storehouse for energy-rich fat. It uses its bill to scoop up food and the horny pads that line its jaws to grind up prey. While underwater, the platypus clamps shut its eyes, ears, and nostrils, relying on roughly 800,000 sensory receptors in its bill to detect the movements and weak electrical discharges of nearby prey.

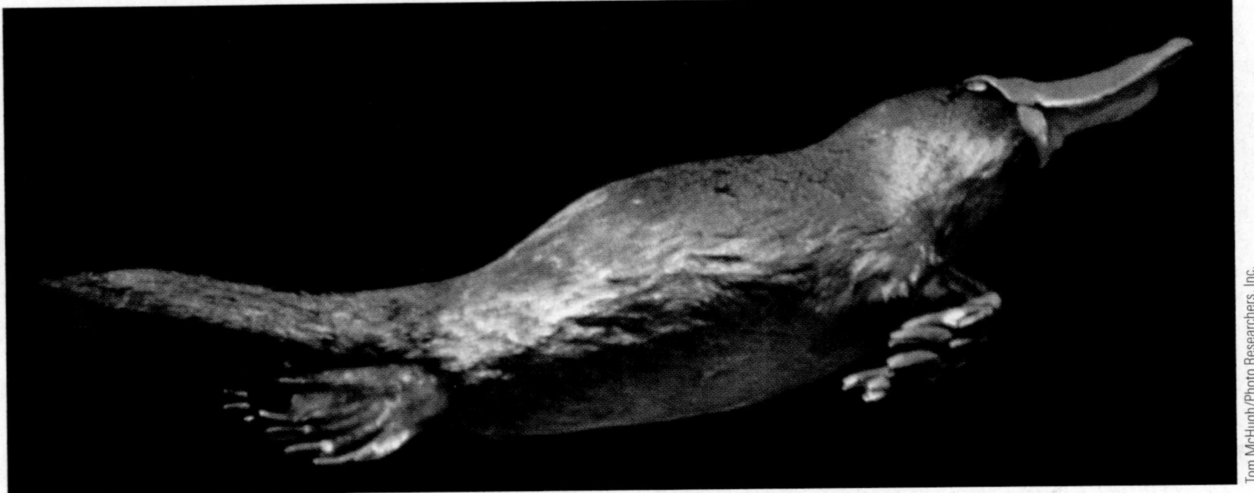

Tom McHugh/Photo Researchers, Inc.

FIGURE 30.1
A puzzling animal. Because of its strange mixture of traits, the platypus *(Ornithorhynchus anatinus)* amazed the first European zoologists who saw it.

The platypus, with its strange combination of characteristics, illustrates the remarkable diversity of adaptations that enable vertebrates—animals with backbones—to occupy nearly every habitat on Earth. Despite the platypus's mixed characteristics, biologists eventually classified it as a member of the mammal lineage because, like all other mammals, it has hair on its body and produces milk to nourish its offspring. Today biologists know that it is one of just a few remaining survivors of an early lineage of egg-laying mammals.

In this chapter, we survey the Deuterostomia, a monophyletic lineage of animals that dates to the Paleozoic. The deuterostomes are defined by features of early embryological development and molecular sequence data (see Chapter 29). There are three living phyla of deuterostomes; we briefly consider two phyla of invertebrate deuterostomes before focusing on the Phylum Chordata, which includes a few thousand species of invertebrates as well as many living species of vertebrates. <

30.1 Invertebrate Deuterostomes

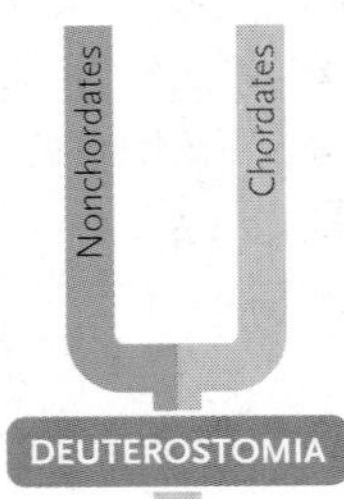

Nonchordates
Chordates
DEUTEROSTOMIA

Deuterostome body plans have been so modified by evolution that a casual observer would not readily group the two phyla of invertebrate deuterostomes—Echinodermata and Hemichordata—together with the Phylum Chordata. However, embryological and molecular analyses agree that all three are indeed closely related.

Echinoderms Have Secondary Radial Symmetry and an Internal Skeleton

The phylum Echinodermata (*echinos* = spiny, *derma* = skin) includes 6,600 species of sea stars, sea urchins, sea cucumbers, brittle stars, and sea lilies. These slow moving or sessile, bottom-dwelling animals are important herbivores and predators in shallow coastal waters and the ocean depths. Although the phylum was diverse in the Paleozoic, only a remnant of that fauna remains. Living species vary in size from less than 1 cm to more than 50 cm in diameter.

Echinoderms develop from a bilaterally symmetrical, free-swimming larva. But as a larva develops, it assumes a secondary radial symmetry, often organized around five rays or "arms" **(Figure 30.2)**. Many echinoderms have an *oral surface,* with the mouth facing the substrate, and an *aboral surface* facing in the opposite direction. Virtually all echinoderms have an internal skeleton made of calcium-stiffened *ossicles* that develop from mesoderm. In some groups, fused ossicles form a rigid container called a *test.* In most, spines or bumps project from the ossicles.

The internal anatomy of echinoderms is unique among animals **(Figure 30.3)**. They have a well-defined coelom and a complete digestive system (see Figure 30.3A), but no excretory or respiratory systems, and most have only a minimal circulatory system. In many, gases are exchanged and metabolic wastes eliminated through projections of the epidermis and peritoneum near the base of the spines. Given their radial symmetry, there is no head or central brain; the nervous system is organized around nerve cords that encircle the mouth and branch into the rays. Sensory cells are abundant in the skin.

Echinoderms have a unique locomotor system, the *water vascular system,* which consists of fluid-filled canals (see Figure 30.3B). In a sea star, for example, water enters the system through the *madreporite,* a sievelike plate on the aboral surface. A short tube connects it to the *ring canal,* which surrounds the esophagus. The ring canal branches into five *radial canals* that extend into the arms. Each radial canal is connected to numerous *tube feet* that protrude through holes in the ossicles (see Figure 30.3C). Each tube foot has a mucus-covered, suckerlike tip and a small muscular bulb, the *ampulla,*

FIGURE 30.2
Echinoderm diversity. Echinoderms exhibit secondary radial symmetry, usually organized as five rays around an oral–aboral axis.

Herve Chaumeton/Agence Nature

A. Asteroidea: This sea star *(Fromia milleporella)* lives in the intertidal zone.

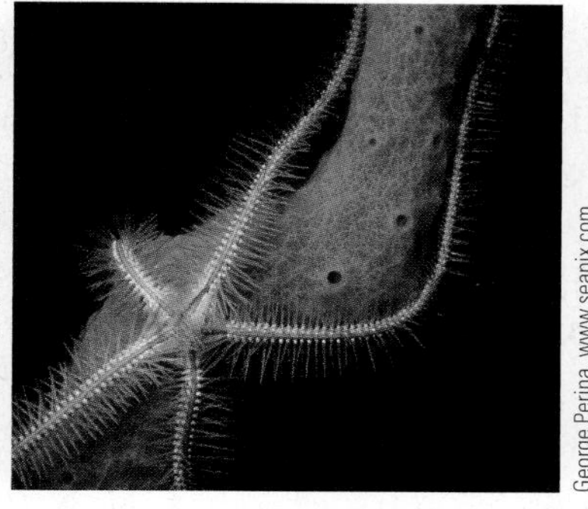
George Perina, www.seapix.com

B. Ophiuroidea: A brittle star *(Ophiothrix swensonii)* perches on a coral branch.

Edward Snow/Bruce Coleman USA

C. Echinoidea: A sea urchin *(Strongylocentrotus purpuratus)* grazes on algae.

Jan Haaga, Kodiak Lab, AFSC/NMFS

D. Holothuroidea: A sea cucumber *(Cucumaria miniata)* extends its tentacles, which are modified tube feet, to trap particulate food.

Fred Bavendam/Minden Pictures

E. Crinoidea: A feather star *(Oxycomanthus bennetti)* feeds by catching small particles with its numerous arms.

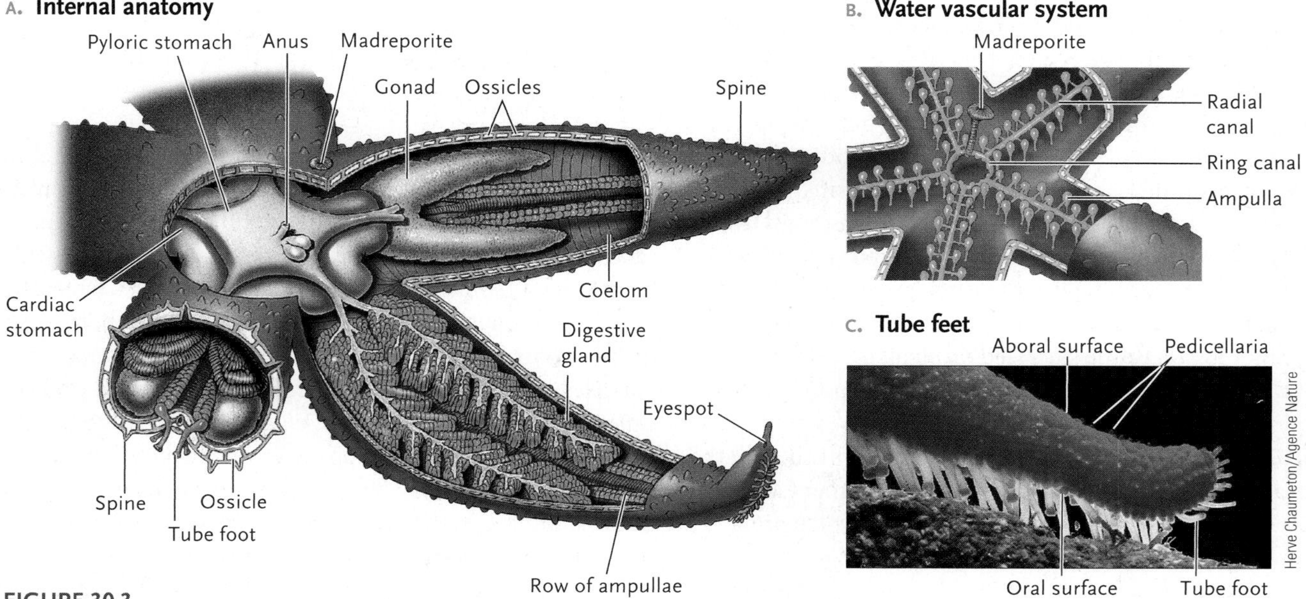
A. Internal anatomy
Pyloric stomach
Anus
Madreporite
Gonad
Ossicles
Spine
Cardiac stomach
Coelom
Digestive gland
Eyespot
Spine
Ossicle
Tube foot
Row of ampullae
B. Water vascular system
Madreporite
Radial canal
Ring canal
Ampulla
C. Tube feet
Aboral surface
Pedicellaria
Oral surface
Tube foot
Herve Chaumeton/Agence Nature

FIGURE 30.3
Internal anatomy of a sea star. (A) The coelom is well developed in echinoderms, as illustrated by this cutaway diagram of a sea star. **(B)** The water vascular system, unique in the animal kingdom, operates the tube feet **(C),** which are responsible for locomotion. Note the pedicellariae on the upper surface of the sea star's arm **(C).**

which lies inside the body. When an ampulla contracts, fluid is forced into the tube foot, causing it to lengthen and attach to the substrate. The tube foot then contracts, pulling the animal along. As the tube foot shortens, water flows back into the ampulla, and the tube foot releases its grip on the substrate. The tube foot can then take another step forward, reattaching to the substrate. Although each tube foot has limited strength, the coordinated action of hundreds or thousands of them is so strong that they can hold an echinoderm to a substrate even against strong wave action.

Echinoderms have separate sexes, and most reproduce by releasing gametes into the water. Radial cleavage is so clearly apparent in the transparent eggs of some sea urchins that they are commonly used for demonstrations of cleavage in introductory biology laboratories. A few echinoderms also reproduce asexually by splitting in half and regenerating the missing parts; some can regenerate body parts lost to predators.

Echinoderms are divided into six groups, one of which, the sea daisies (Concentricycloidea) was discovered only in 1986. These small, medusa-shaped animals occupy sunken, waterlogged wood in the deep sea. Here we describe the five other groups, which are more diverse and better known.

ASTEROIDEA The 1,500 species of sea stars (Asteroidea, from *asteroeides* = starlike) live from rocky shorelines to depths of 10,000 m. The asteroid body plan consists of a central disk surrounded by five to 20 radiating "arms" (see Figure 30.2A), with the mouth centered on the oral surface. The ossicles of the endoskeleton are not fused, permitting flexibility of the arms and disk. Small pincers, **pedicellariae,** at the base of short spines remove debris that falls onto the animal's surface (see Figure 30.3C). Many sea stars feed on soft-bodied invertebrates and small fishes. Other species consume bivalve mollusks: after using their tube feet to pry apart the two shells, they evert their stomachs and secrete digestive enzymes onto the mollusk's flesh. Some sea stars are destructive predators of corals, endangering many reefs.

OPHIUROIDEA The 2,000 species of brittle stars and basket stars (Ophiuroidea, from *ophioneos* = snakelike) occupy roughly the same range of habitats as sea stars. Their bodies have a well-defined central disk and slender, elongate arms that are sometimes branched (see Figure 30.2B). Ophiuroids can crawl fairly swiftly across substrates by moving their arms in coordinated fashion. As their common name implies, the arms are delicate and easily broken, an adaptation that allows them to escape from predators with only minor losses. Brittle stars feed on small prey, suspended plankton, or detritus that they extract from muddy deposits.

ECHINOIDEA The 950 species of sea urchins and sand dollars (Echinoidea, *ekhinos* = porcupine) lack arms altogether (see Figure 30.2C). Their ossicles are fused into solid tests, which provide excellent protection but restrict flexibility. The test is spherical in sea urchins and flattened in sand dollars. Five rows of tube feet, used primarily for locomotion, emerge through pores in the test. Most echinoids have movable spines, some with poison glands that protect them from predators; a jab from certain tropical species can cause severe pain and inflammation to a careless swimmer. Echinoids graze on algae and other organisms that cling to marine surfaces. In the center of an urchin's oral surface is a five-part nipping jaw that is controlled by powerful muscles. Some species damage kelp beds, disrupting the habitat of young lobsters and other crustaceans.

HOLOTHUROIDEA The 1,500 species of sea cucumbers (Holothuroidea, from *holo-thourion* = water polyp) are elongate animals that lie on their sides on the ocean bottom (see Figure 30.2D). Although they have five rows of tube feet, their endoskeleton is reduced to widely separated microscopic plates. The body, which is elongated along the oral-aboral axis, is soft and fleshy, with a tough, leathery covering. Modified tube feet form a ring of tentacles around the mouth, which points to the side or upward. Some species secrete a mucous net that traps plankton or other food particles. Other species extract food from bottom sediments. Many sea cucumbers exchange gases through an extensively branched *respiratory tree* that arises from the rectum, the part of the digestive system just inside the anus at the aboral end of the animal. A well-developed circulatory system distributes oxygen and nutrients to tissues throughout the body.

CRINOIDEA The 600 living species of sea lilies and feather stars (Crinoidea, from *krinon* = lily) are the surviving remnants of a fauna that was diverse and abundant 500 million years ago (see Figure 30.2E). Most species occupy marine waters of medium depth. The central disk and mouth point upward rather than toward the substrate. Between five and several hundred branched arms surround the disk; new arms are added as a crinoid grows larger. The branches of the arms are covered with tiny mucus-coated tube feet, which trap suspended microscopic organisms. The sessile sea lilies have the central disk attached to a flexible stalk that can reach a meter in length. Adult feather stars can swim or crawl weakly, attaching temporarily to substrates.

Acorn Worms Use A Pharynx With Branchial Slits To Acquire Food And Oxygen

The 80 species of acorn worms (phylum Hemichordata, *hemi* = half, *chorda,* referring to the phylum Chordata) are sedentary marine animals that live in U-shaped tubes or burrows in coastal sand or mud. Their soft bodies, which range from 2 cm to 2 m in length, are organized into an anterior proboscis, a tentacled collar, and an elongate trunk **(Figure 30.4).** They use their muscular, mucus-coated proboscis to construct burrows and trap food particles. Acorn worms also have pairs of **branchial slits** (sometimes called *pharyngeal slits* or *gill slits*) in the pharynx, the part of the digestive system just posterior to the mouth. Beating cilia create a flow of water, which enters the pharynx through the mouth and exits through these openings. As water passes through, suspended food is trapped and shunted into the digestive system, and gases are exchanged across the partitions between the slits. This coupling of feeding and respiration reflects the evolutionary relationship between hemichordates and chordates, the phylum that we consider next.

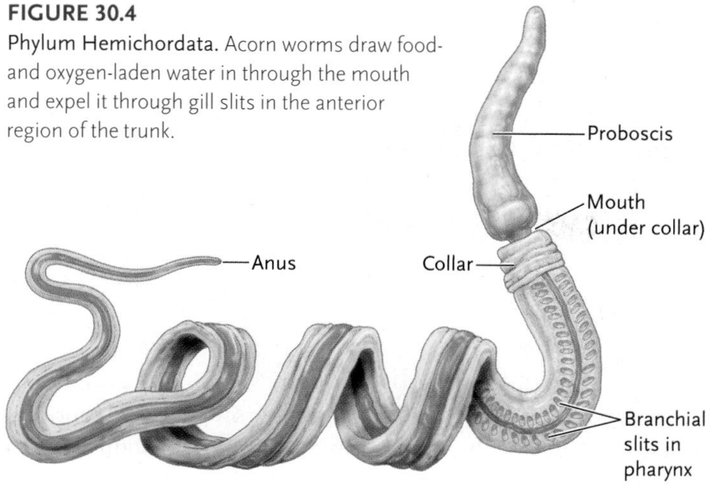

FIGURE 30.4
Phylum Hemichordata. Acorn worms draw food- and oxygen-laden water in through the mouth and expel it through gill slits in the anterior region of the trunk.

STUDY BREAK 30.1

1. **What organ system is unique to echinoderms, and what is its function?**
2. **How does a perforated pharynx enable hemichordates to acquire food and oxygen from seawater?**

30.2 Overview of the Phylum Chordata

The phylum Chordata contains three subphyla: two lineages of invertebrates, Cephalochordata and Urochordata, and the diverse lineage of vertebrates, Vertebrata.

Key Morphological Innovations Distinguish Chordates from Other Deuterostome Phyla

Chordates are distinguished from other deuterostomes by a set of key morphological innovations: a *notochord, segmental muscles in the body wall and tail,* a *dorsal hollow nerve chord,* and a *perforated pharynx* **(Figure 30.5).** These structures foster higher levels of activity, unique modes of aquatic locomotion, and more efficient feeding and oxygen acquisition.

NOTOCHORD Early in chordate development, mesoderm that is dorsal to the developing digestive system forms a **notochord** (*noton* = the back; *chorda* = string). This flexible rod, constructed of fluid-filled cells surrounded by tough connective tissue, supports the embryo from head to tail. The notochord later forms the skeleton of invertebrate chordates. Body wall muscles are anchored to the notochord, and when these muscles contract, the notochord bends, but it does not shorten. As a result, the chordate body swings left and right during locomotion, propelling the animal forward; unlike annelids and other nonchordate invertebrates, the chordate body does not shorten when the animal is moving. Remnants of the notochord persist as gelatinous disks in the backbones of adult vertebrates.

SEGMENTAL BODY WALL AND TAIL MUSCLES Chordates evolved in water, and they swim by contracting segmentally arranged blocks of muscles in the body wall and tail. The chordate tail, which is posterior to the anus, provides much of the propulsion in aquatic species. Segmentation allows each muscle block to contract independently; waves of contractions pass down one side of the animal and then down the other, sweeping the body and tail back and forth in a smooth and continuous movement.

DORSAL HOLLOW NERVE CORD The central nervous system of chordates is a hollow nerve cord on the dorsal side of the animal (see Chapter 38). By contrast, most nonchordate invertebrates have solid nerve cords on the ventral side. In vertebrates, an anterior enlargement of the nerve cord forms the brain; in invertebrates, the anterior concentration of nervous system tissue is described as a *ganglion.*

PERFORATED PHARYNX Like the hemichordates described above, most chordates have outpocketings, perforations, or slits in the pharynx during some stage of the animal's life cycle. These paired openings originated as exit holes for water that carried particulate food into the mouth. Invertebrate chordates exchange gases with water as it passes by the walls of the pharynx, acquiring oxygen and eliminating carbon dioxide. Invertebrate chordates and fishes retain a perforated pharynx throughout their lives. In most air-breathing terrestrial vertebrates, the outpocketings or slits are present only during embryonic development and in larvae.

Invertebrate Chordates Are Small, Marine Suspension Feeders

Two subphyla of invertebrate chordates exhibit the basic chordate body plan in its simplest form.

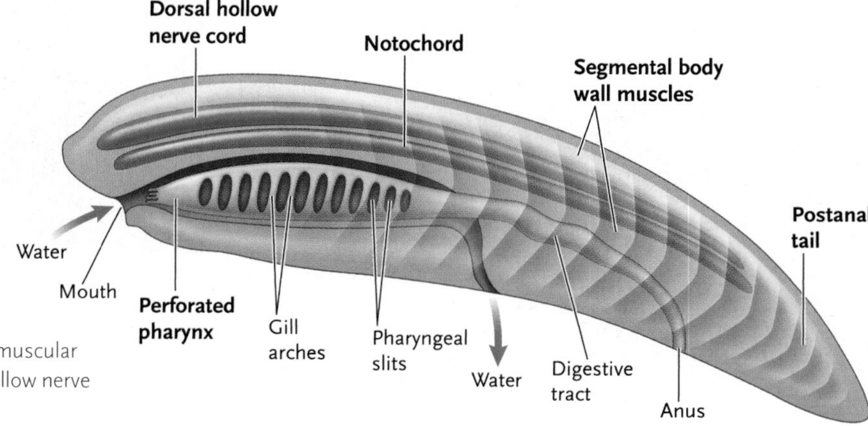
Dorsal hollow nerve cord
Notochord
Segmental body wall muscles
Postanal tail
Water
Mouth
Perforated pharynx
Gill arches
Pharyngeal slits
Water
Digestive tract
Anus

FIGURE 30.5
Diagnostic chordate characters. Chordates have a notochord, a muscular postanal tail, a segmental body wall and tail muscles, a dorsal, hollow nerve cord, and a perforated pharynx.

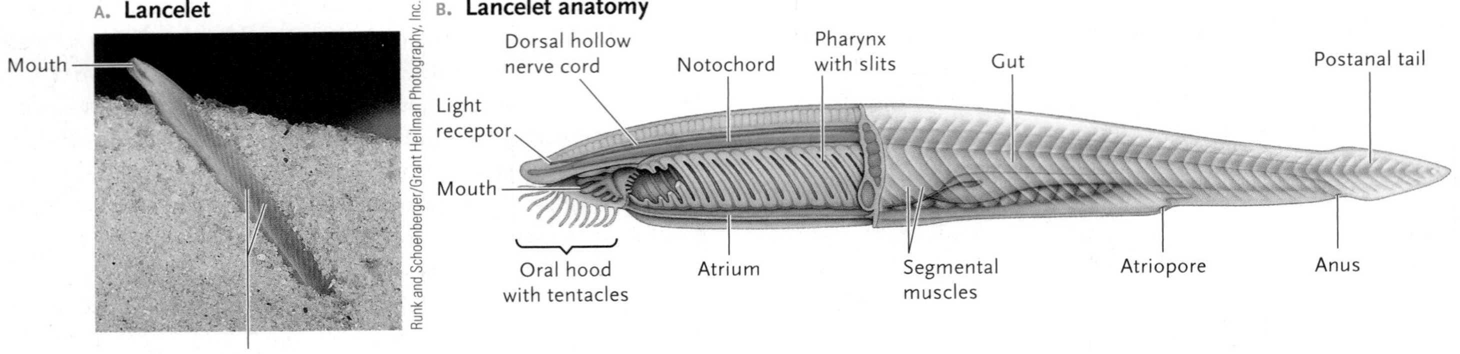
A. Lancelet
Mouth
Segmental muscles
Runk and Schoenberger/Grant Heilman Photography, Inc.
B. Lancelet anatomy
Dorsal hollow nerve cord
Notochord
Pharynx with slits
Gut
Postanal tail
Light receptor
Mouth
Oral hood with tentacles
Atrium
Segmental muscles
Atriopore
Anus

FIGURE 30.6
Cephalochordates. **(A)** The unpigmented skin of adult lancelets reveals their segmental body wall muscles. A cutaway view **(B)** illustrates their internal anatomy.

SUBPHYLUM CEPHALOCHORDATA The 28 lancelet species (subphylum Cephalochordata, *kephale* = head) occupy warm, shallow marine habitats where they lie mostly buried in sand **(Figure 30.6)**. Although generally sedentary, they have well-developed body wall muscles and a prominent notochord. Most species are included in the genus *Branchiostoma* (formerly *Amphioxus*). Lancelet bodies, which are 5 to 10 cm long, are pointed at both ends like the double-edged surgical tools for which they are named. Adults have light receptors on the head as well as chemical sense organs on tentacles that grow from the **oral hood.** Lancelets use cilia to draw food-laden water through hundreds of pharyngeal slits; water flows into the atrium and is expelled through the **atriopore.** Most gas exchange occurs across the skin.

SUBPHYLUM UROCHORDATA The 2,500 species of tunicates, sometimes called sea squirts (subphylum Urochordata, from *oura* = tail), float in surface waters or attach to substrates in shallow marine habitats. The sessile adults of many species secrete a gelatinous or leathery "tunic" around their bodies and squirt water through a siphon when disturbed; adults grow to several centimeters **(Figure 30.7)**. In the most common group of sea squirts (Ascidiacea), the swimming larvae possess the defining chordate features. Larvae eventually attach to substrates and transform into sessile adults. During metamorphosis, they lose most traces of the notochord, dorsal nerve cord, and tail, and their basketlike pharynx enlarges. In adults, beating cilia pull water into the pharynx through an **incurrent siphon.** A mucous net traps particulate food, which is carried, with the mucus, to the gut. Water passes through the pharyngeal slits, enters a chamber called the **atrium,** and is expelled—along with digestive wastes and carbon dioxide—through the **atrial siphon.** Oxygen is absorbed across the walls of the pharynx.

Vertebrates Possess Several Unique Tissues, Including Bone and Neural Crest

The most distinctive anatomical characteristic of the subphylum Vertebrata is an internal skeleton that provides structural support for muscles and protection for the nervous system and other organs. The skeleton and the muscles attached to it enable most vertebrates to move rapidly through the environment. A vertebrate's skeleton is composed of many separate, bony elements. Indeed, vertebrates are the only animals that have **bone,** a con-

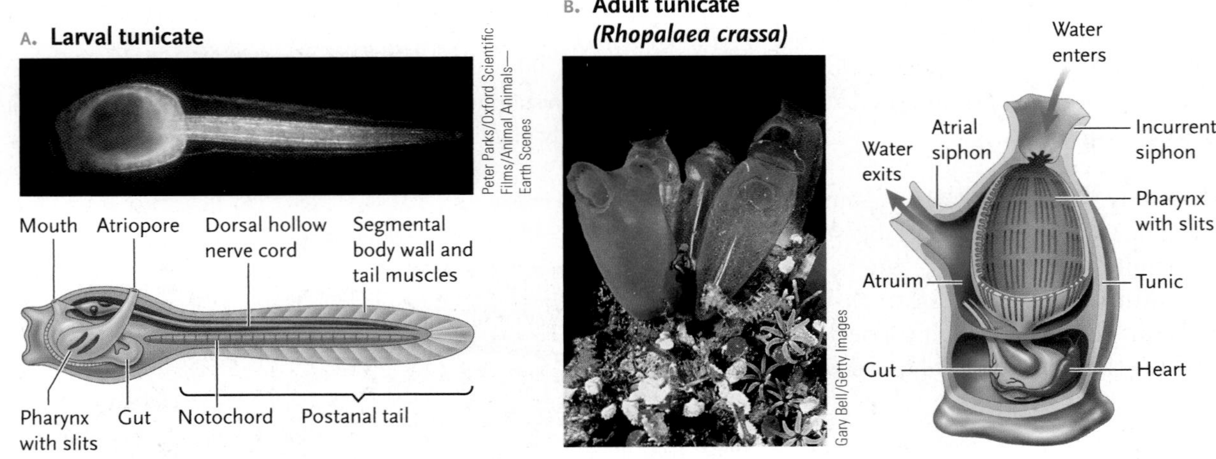
A. Larval tunicate
Peter Parks/Oxford Scientific Films/Animal Animals—Earth Scenes
Mouth
Atriopore
Dorsal hollow nerve cord
Segmental body wall and tail muscles
Pharynx with slits
Gut
Notochord
Postanal tail
B. Adult tunicate (Rhopalaea crassa)
Gary Bell/Getty Images
Water enters
Atrial siphon
Water exits
Incurrent siphon
Pharynx with slits
Atruim
Tunic
Gut
Heart

FIGURE 30.7
Urochordates. **(A)** A tadpolelike tunicate larva metamorphoses into **(B)** a sessile adult. After a larva attaches to a substrate at its anterior end, the tail, notochord, and most of the nervous system are recycled to form new tissues. Slits in the pharynx multiply, the mouth becomes the incurrent siphon, and the atriopore becomes the atrial siphon.

nective tissue in which living cells secrete the mineralized matrix that surrounds them (see Figure 36.4D). The **vertebral column,** made up of individual **vertebrae,** surrounds and protects the dorsal nerve cord, and a bony **cranium** surrounds the brain. The cranium, vertebral column, ribs, and sternum (breastbone) make up the **axial skeleton.** Most vertebrates also have a **pectoral girdle** anteriorly and a **pelvic girdle** posteriorly that attach bones in the fins or limbs to the axial skeleton. Bones of the two girdles and the appendages constitute the **appendicular skeleton.** One vertebrate lineage, Chondrichthyes, has lost its bone over evolutionary time; its skeleton is made of cartilage, a dense but flexible connective tissue that is often a developmental precursor of bone (see Section 36.2).

Vertebrates also possess a unique cell type, **neural crest,** which is distinct from endoderm, mesoderm, and ectoderm. Neural crest cells arise next to the developing nervous system, but later migrate throughout a vertebrate's body. They ultimately contribute to many uniquely vertebrate structures, including parts of the cranium, teeth, sensory organs, cranial nerves, and the medulla (that is, the interior part) of the adrenal glands.

Finally, the brains of vertebrates are much larger and more complex than those of invertebrate chordates. Moreover, the vertebrate brain is divided into three regions—the forebrain, midbrain, and hindbrain—each of which governs distinct nervous system functions (see Section 38.1).

STUDY BREAK 30.2 <

1. **On a field trip to a lake, a college student captures a worm-shaped animal with segmented body-wall muscles. While examining the specimen in the laboratory the following day, she determines that the main nerve cord runs along the ventral side of the animal. Is this animal a chordate?**
2. **What structures distinguish vertebrates from invertebrate chordates?**

30.3 The Origin and Diversification of Vertebrates

Biologists use embryological, molecular, and fossil evidence to trace the origin of vertebrates and to chronicle their evolutionary diversification.

Vertebrates Probably Arose from an Invertebrate Chordate Ancestor through the Duplication of Genes That Regulate Development

Recent genetic sequence studies suggest that vertebrates are more closely related to urochordates than to cephalochordates. Nevertheless, the fishlike form of adult lancelets probably better represents the anatomy of the earliest vertebrates than does the highly specialized morphology of adult tunicates. The evolution of vertebrates from an invertebrate chordate ancestor was marked by the emergence of neural crest, bone, and other typically vertebrate traits. What genetic changes were responsible for these remarkable developments? Biologists now hypothesize that an increase in the number of homeotic—structure determining—genes may have made the development of more complex anatomy possible. (Homeotic genes are described further in Sections 16.4 and 22.6.)

In animals, one group of homeotic genes, the *Hox* genes, influences the three-dimensional shape of the animal and the locations of important structures—such as eyes, wings, and legs—particularly along the head to tail axis of the body. *Hox* genes are arranged on the chromosomes in a particular order, forming what biologists call the *Hox* gene complex (see Section 22.6 and Chapter 34). Each gene in the complex governs the development of particular structures. Animal groups with the simplest structure, such as cnidarians, have two *Hox* genes. Those with more complex anatomy, such as insects, have 10. Chordates have as many as 13 or 14. Thus, lineages with many *Hox* genes generally have more complex anatomy than do those with fewer *Hox* genes.

Molecular analyses also reveal that the entire *Hox* gene complex was duplicated several times in the evolutionary history of vertebrates, producing multiple copies of all its genes **(Figure 30.8).** The cephalochordate *Branchiostoma* has just one *Hox* gene complex, but the most primitive living vertebrates, the jawless hagfishes described later, have two. All vertebrates that possess jaws, a derived characteristic, have at least four sets, and some fishes have as many as seven. Evolutionary developmental biologists hypothesize that the duplication of *Hox* genes and other tool-kit genes allowed the evolution of new structures: while the original copies of these genes maintained their ancestral functions, the duplicate copies assumed *new* functions, directing the development of novel structures, such as the vertebral column and jaws.

Early Vertebrates Diversified into Numerous Lineages with Distinctive Adaptations

The oldest known vertebrate fossils were discovered in the late 1990s, when scientists in China described several species from the early Cambrian period, about 550 million years ago. Both *Myllokunmingia* and *Haikouichthys* were fish-shaped animals about 3 cm long **(Figure 30.9).** In both species the brain was surrounded by a cranium, which, in these cases, was formed of fibrous connective tissue or cartilage. They also had segmented body-wall muscles and fairly well-developed fins, but neither shows any evidence of bone.

The early vertebrates gave rise to numerous descendants, which varied greatly in anatomy, physiology, and ecology. New feeding mechanisms and locomotor structures were often crucial to their success. Today, vertebrates occupy nearly every habitat and feed on virtually all other organisms. Here we briefly introduce the major vertebrate lineages **(Figure 30.10).**

Although biologists use four key morphological innovations—a cranium, vertebrae, bone, and neural crest cells—to

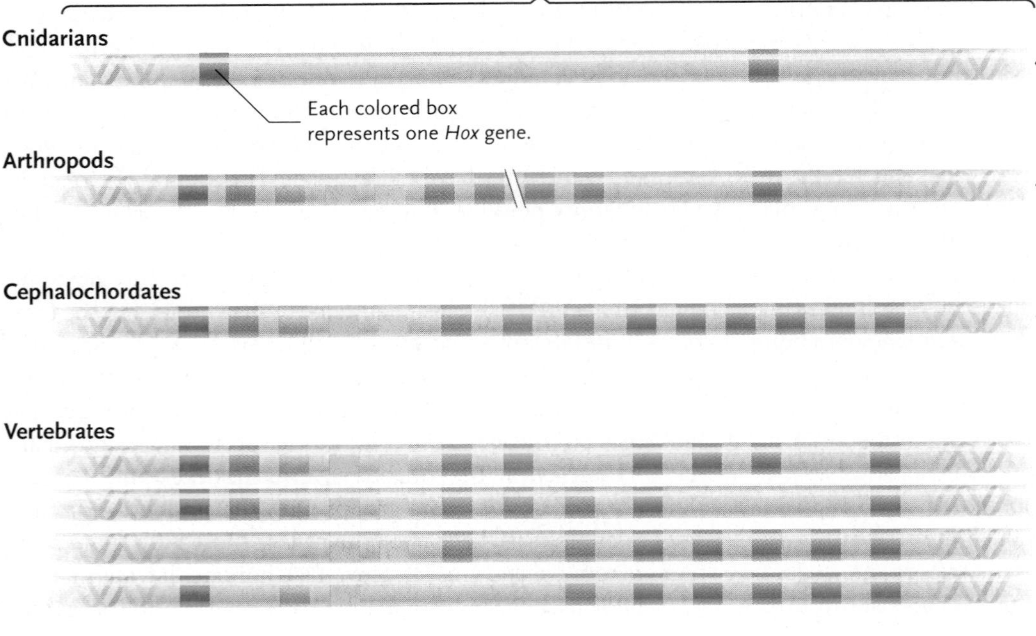
Each row of colored boxes represents one Hox gene complex.
Cnidarians
Each colored box represents one Hox gene.
Arthropods
Cephalochordates
Vertebrates
A. Invertebrates with simple anatomy, such as cnidarians, have a single Hox gene complex that includes just a few Hox genes.
B. Invertebrates with more complicated anatomy, such as arthropods, have a single Hox gene complex, but with a larger number of Hox genes.
C. Invertebrate chordates, such as cephalochordates, also have a single Hox gene complex, but with even more Hox genes than are found in nonchordate invertebrates.
D. Vertebrates, such as the laboratory mouse, have numerous Hox genes, arranged in two to seven Hox gene complexes. The additional Hox gene complexes are products of wholesale duplications of the ancestral Hox gene complex. The additional copies of Hox genes probably specify the development of uniquely vertebrate characteristics, such as the cranium, vertebral column, and neural crest cells.

FIGURE 30.8

Hox genes and the evolution of vertebrates. The *Hox* genes in different animals appear to be homologous, indicated here by their color and position in the complex. Vertebrates have many more individual *Hox* genes than most invertebrates do, and the entire *Hox* gene complex was duplicated in the vertebrate lineage.

identify vertebrates, these structures did not arise all at once. Instead, they appeared somewhat independently of one another as new groups arose. Some researchers and textbooks present a phylogeny and classification that places the "vertebrates" (animals that have vertebrae) within a larger lineage called the "craniates" (animals that have a cranium). But only one small group, the hagfishes (Myxinoidea, described later), has a cranium but no vertebrae, and some recent molecular analyses do not support its separation from the other vertebrates. Thus, for the sake of simplicity, we describe organisms that possessed any of the four key innovations as "vertebrates."

Several groups of early jawless vertebrates are *described* as "agnathans" (*a* = not, *gnathos* = jaw), but they do not form a monophyletic group (that is, one that includes an ancestor and all of its descendants). Although most became extinct in the Paleozoic era, two ancient lineages, Myxinoidea and Petromyzontoidea, still live today. All other vertebrates possess moveable jaws; they are members of the monophyletic lineage **Gnathostomata** ("jawed mouth"). The first jawed fishes, the Acanthodii and Placodermi, are now extinct. But three lineages of jawed fishes are still abundant in aquatic habitats: the Chondrichthyes, such as sharks and skates, have cartilaginous skeletons; the Actinopterygii and Sarcopterygii have bony skeletons. All jawless vertebrates and jawed fishes are restricted to aquatic habitats, and they use gills to extract oxygen from the water that surrounds them.

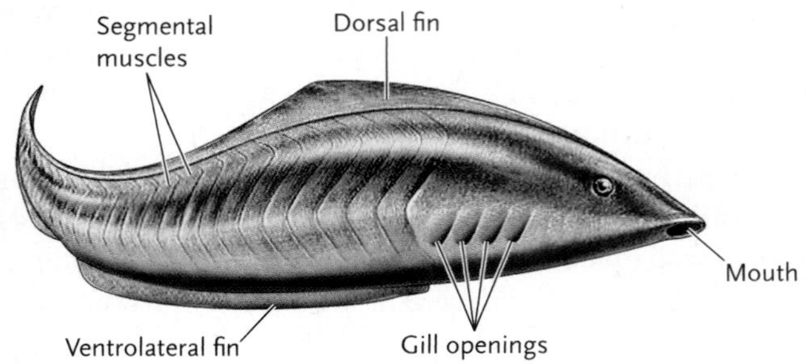
Segmental muscles
Dorsal fin
Mouth
Ventrolateral fin
Gill openings

FIGURE 30.9

An early vertebrate. *Myllokunmingia*, one of the earliest vertebrates yet discovered, had no bones; it was about 3 cm long.

The Gnathostomata also includes the monophyletic lineage **Tetrapoda** (*tetra* = four, *pod* = foot); most tetrapods use four limbs for locomotion. Many tetrapods are semiterrestrial or terrestrial, although some, like sea turtles and porpoises, have secondarily returned to aquatic habitats. Adult tetrapods generally use air-breathing lungs for gas exchange. Within the Tetrapoda, one lineage, the amphibians, includes animals, such as frogs and salamanders, that typically need standing water to complete their life cycles. Another lineage, the **Amniota,** comprises animals with specialized eggs that can develop on land. Since their appearance, the amniotes diversified into lineages that include mammals, turtles, lizards, snakes, alligators, and birds. We consider the detailed evolutionary history of the amniotes in Section 30.7.

STUDY BREAK 30.3

1. **How do the *Hox* genes of vertebrates differ from those of cephalochordates?**
2. **Which of the taxonomic groups Amniota, Gnathostomata, and Tetrapoda includes the largest number of species? Which includes the fewest?**

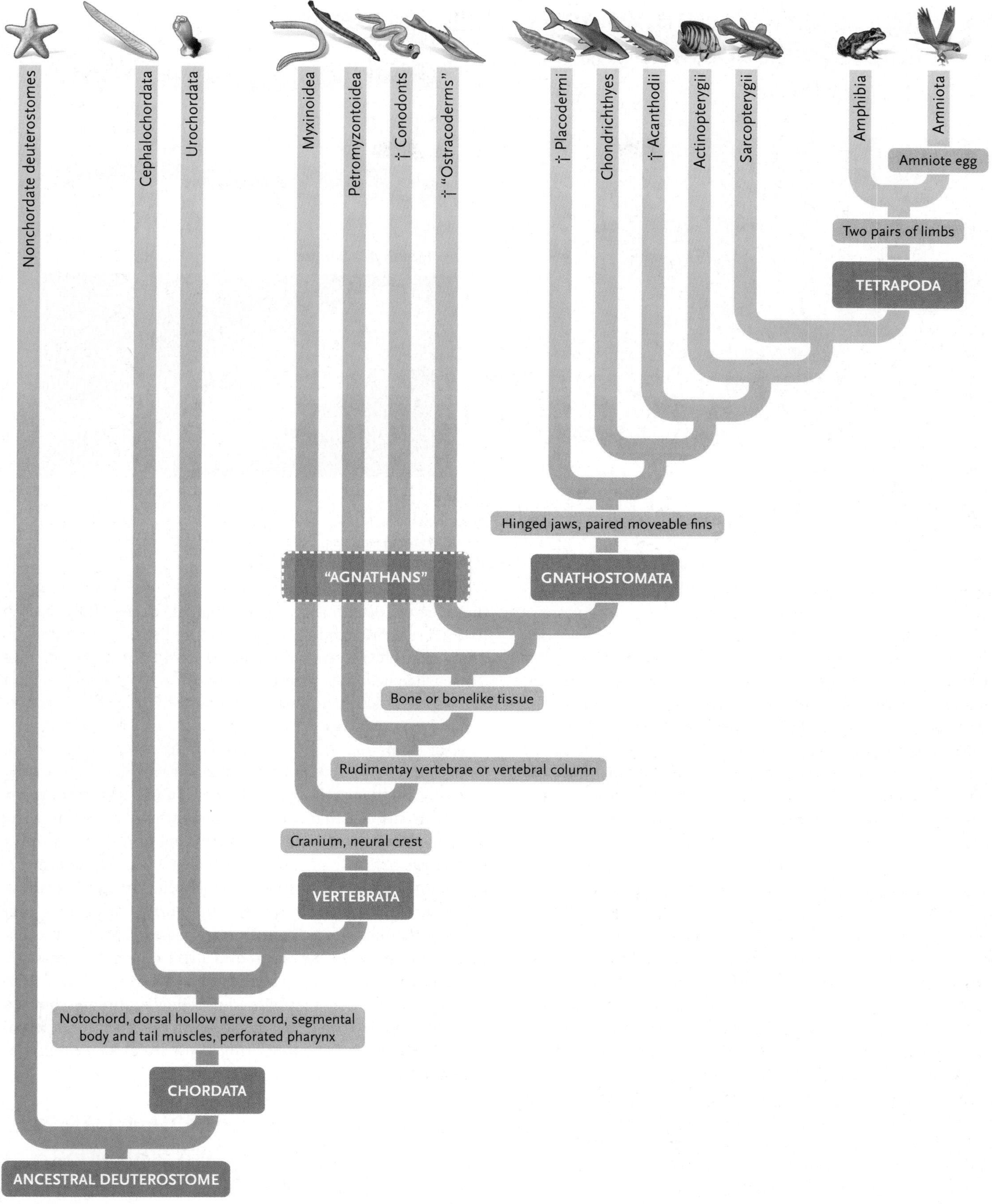
Nonchordate deuterostomes
Cephalochordata
Urochordata
Myxinoidea
Petromyzontoidea
† Conodonts
† "Ostracoderms"
† Placodermi
Chondrichthyes
† Acanthodii
Actinopterygii
Sarcopterygii
Amphibia
Amniota
Amniote egg
Two pairs of limbs
TETRAPODA
Hinged jaws, paired moveable fins
"AGNATHANS"
GNATHOSTOMATA
Bone or bonelike tissue
Rudimentay vertebrae or vertebral column
Cranium, neural crest
VERTEBRATA
Notochord, dorsal hollow nerve cord, segmental body and tail muscles, perforated pharynx
CHORDATA
ANCESTRAL DEUTEROSTOME

FIGURE 30.10

Chordate phylogeny. The cladogram illustrates the evolutionary relationships and defining characteristics of the major lineages of vertebrates. The figure does not illustrate the times at which groups first appeared. The terms "agnathan" and "ostracoderm" are descriptive and do not identify monophyletic groups. Extinct groups are marked with a dagger.

30.4 "Agnathans": Hagfishes and Lampreys, Conodonts and Ostracoderms

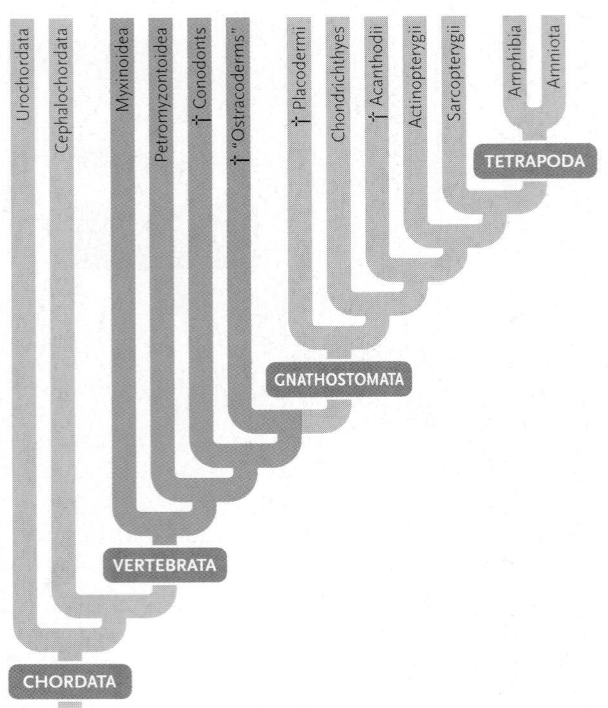

Urochordata
Cephalochordata
Myxinoidea
Petromyzontoidea
† Conodonts
† "Ostracoderms"
† Placodermi
Chondrichthyes
† Acanthodii
Actinopterygii
Sarcopterygii
Amphibia
Amniota
TETRAPODA
GNATHOSTOMATA
VERTEBRATA
CHORDATA

A. **Living jawless fishes**

Hagfish

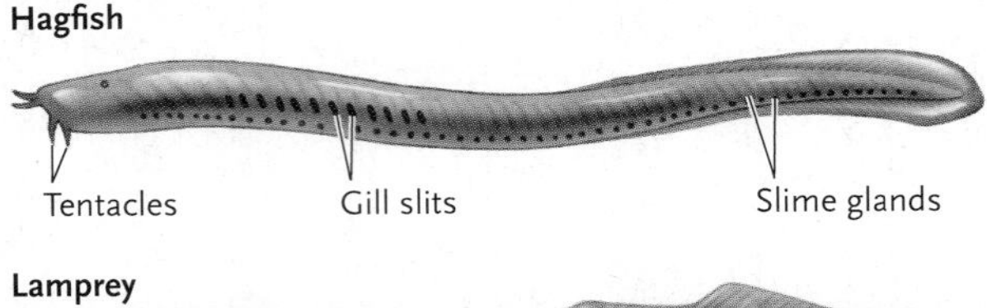

Tentacles
Gill slits
Slime glands

Lamprey

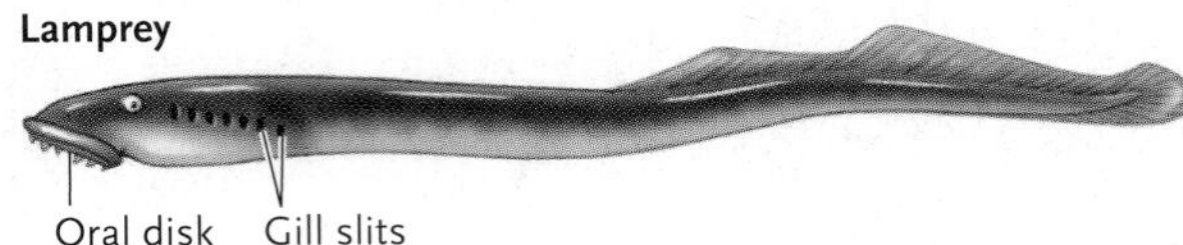

Oral disk
Gill slits

B. **Mouth of a lamprey**

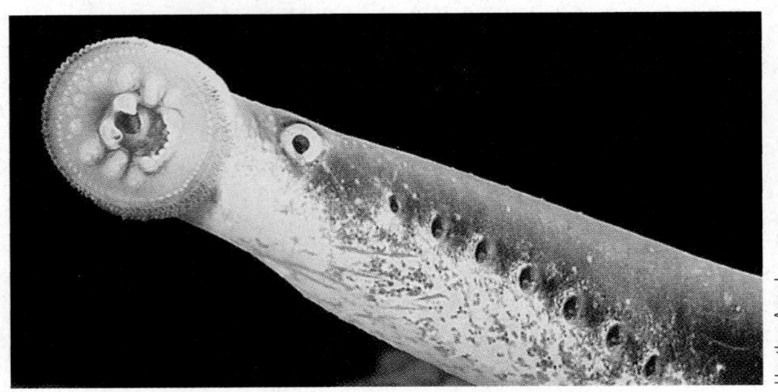

Heather Angel

FIGURE 30.11
Living agnathans. **(A)** Two groups of jawless fishes, hagfishes and lampreys, survive today. **(B)** Lampreys use a toothed oral disk to attach to a host and feed on its blood and soft tissues.

The earliest vertebrates lacked jaws, but they had a muscular pharynx, which they used to suck edible tidbits into their mouths. Although they do not form a monophyletic group, we describe four groups of jawless vertebrates together because their ancestral jawless condition distinguishes them from the monophyletic Gnathostomata. The two living groups of agnathans as well as species that flourished in the Paleozoic vary greatly in size and shape as well as in the number of vertebrate characters they possess.

Hagfishes and Lampreys Are the Living Descendants of Ancient Agnathan Lineages

Two apparently separate lineages of jawless vertebrates, hagfishes (Myxinoidea) and lampreys (Petromyzontoidea), still live today. Both have skeletons composed entirely of cartilage. Although scientists have found no fossilized hagfishes or lampreys from the early Paleozoic era, the absence of both jaws and bone in these groups suggests that their lineages arose early in vertebrate history, before the evolution of bone. Hagfishes and lampreys have a well-developed notochord, but no true vertebrae or paired fins, and their skin has no scales. Individuals grow to a maximum length of about 1 m **(Figure 30.11)**.

The axial skeleton of the 60 living species of hagfishes includes only a cranium and a notochord; it has no specialized structures surrounding the dorsal nerve cord. Some biologists do not even include hagfishes among the Vertebrata, because they lack any sign of vertebrae. Hagfishes are marine scavengers that burrow in sediments on continental shelves. They feed on invertebrate prey and on dead or dying fishes. In response to predators, they secrete an immense quantity of sticky, noxious slime; when no longer threatened, a hagfish ties itself into a knot and wipes the slime from its body. Hagfish life cycles are simple and lack a larval stage.

The 40 or so living species of lampreys have traces of an axial skeleton. Their notochord is surrounded by dorsally pointing cartilages that partially cover the nerve cord; many biologists suspect that this arrangement may reflect an early stage in the evolution of the vertebral column. Most lamprey species are parasitic as adults. They have a circular mouth surrounded by a sucking disk with which they attach to a fish or other vertebrate host; they feed on a host's body fluids after rasping through its skin. In most species, sexually mature adults migrate from the ocean or a lake to the headwaters of a stream, where they lay eggs and then die. Their suspension-feeding larvae, which resemble adult cephalochordates, burrow into mud and develop for as long as seven years before undergoing metamorphosis and migrating to the sea or a lake to live as parasitic adults.

Conodonts and Ostracoderms Were Early Jawless Vertebrates with Bony Structures

Mysterious bonelike fossils, most less than 1 mm long, have long been known in oceanic rocks dating from the early Paleozoic era through the early Mesozoic era. Called **conodont** ("cone tooth") elements, these abundant fossils were once described as the support structures of marine algae or the feeding structures of an-

FIGURE 30.12
Extinct agnathans. **(A)** Conodonts were elongate, soft-bodied animals with bonelike feeding structures in the mouth and pharynx. **(B)** Some ostracoderms had large bony plates on the head and small bony scales on the rest of the body; this species was about 6 cm long.

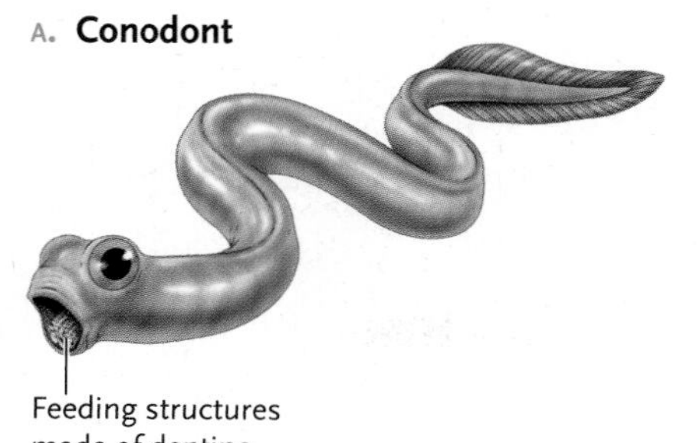
A. Conodont
Feeding structures made of dentine

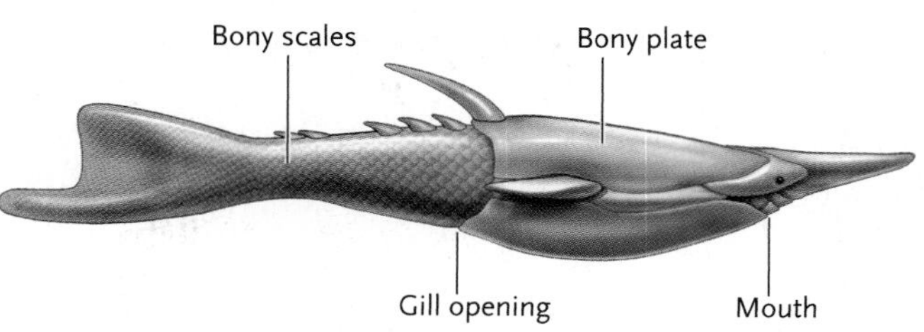
B. Ostracoderm (Pteraspis)
Bony scales
Bony plate
Gill opening
Mouth

cient invertebrates. However, recent analyses of their mineral composition reveal that they were made of dentine, a bonelike component of vertebrate teeth. In the 1980s and 1990s, several research teams discovered fossils of intact conodont animals with these elements in place.

Conodonts were elongate, soft-bodied animals; most were 3 to 10 cm long. They had a notochord, cranium, segmental body wall muscles, and large, moveable eyes **(Figure 30.12A)**. The conodont elements at the front of the mouth were forward pointing, hook-shaped structures that apparently functioned to collect food; those in the pharynx were stouter, suitable for crushing small items that had been consumed. Paleontologists now classify conodonts as vertebrates—the earliest vertebrates with bonelike structures.

An assortment of jawless fishes, representing several evolutionary lineages and collectively described as **ostracoderms** (*ostrakon* = shell), was abundant from the Ordovician through the Devonian periods **(Figure 30.12B)**. Like their invertebrate chordate ancestors, ostracoderms used their pharynx to extract small food particles from mud and water. However, the ostracoderms' muscular pharynx enabled them to *suck* mud and water into their mouths, providing a much stronger flow than the cilia-driven currents of invertebrate chordates. The greater flow rate allowed ostracoderms to collect food more rapidly. It also supported a larger body size: although most species were small, some ostracoderms reached a length of 2 m.

The skin of ostracoderms was heavily armored with bony plates and scales. Although some ostracoderms had paired lateral extensions of their armor, they could not move them the way living fishes move their paired fins. Ostracoderms lacked a true vertebral column, but some had rudimentary support structures surrounding the nerve cord. They also had other distinctly vertebrate-like characteristics. For example, imprints in the head shields of some species indicate that their brains had the three regions—forebrain, midbrain, and hindbrain—typical of all later vertebrates (see Section 38.1).

STUDY BREAK 30.4 <

1. **What characteristics of the living hagfishes and lampreys suggest that their lineages arose very early in vertebrate evolution?**
2. **What traits in conodonts and ostracoderms are derived relative to those in hagfishes and lampreys?**

30.5 Gnathostomata: The Evolution of Jaws

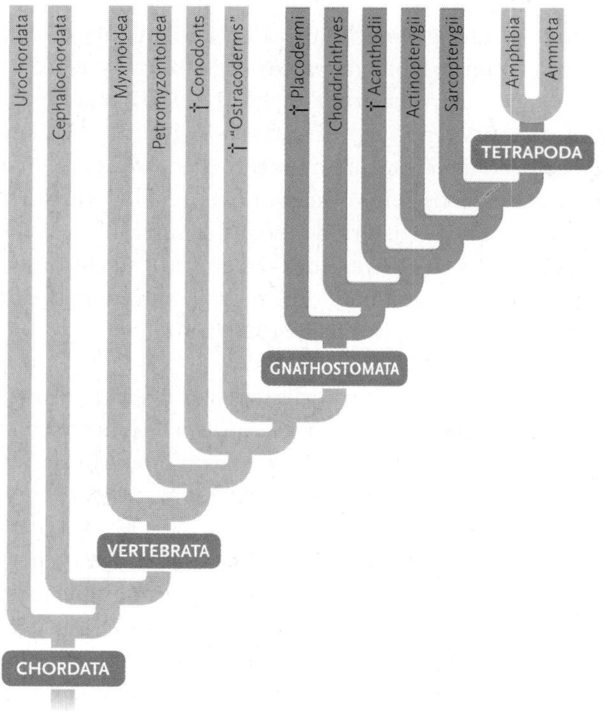
Urochordata
Cephalochordata
Myxinoidea
Petromyzontoidea
† Conodonts
† "Ostracoderms"
† Placodermi
Chondrichthyes
† Acanthodii
Actinopterygii
Sarcopterygii
Amphibia
Amniota
TETRAPODA
GNATHOSTOMATA
VERTEBRATA
CHORDATA

The first gnathostomes were jawed fishes. Key derived traits made their feeding and locomotion more efficient than those of their ancestors.

Jawed Fishes First Appeared in the Paleozoic Era

The renowned anatomist and paleontologist Alfred Sherwood Romer of Harvard University described the evolution of jaws as "perhaps the greatest of all advances in vertebrate history." Hinged jaws allow vertebrates to grasp, kill, shred, and crush large food items. Some living species also use their jaws for defense, for grooming, to construct nests, and to transport their young.

THE ORIGIN OF JAWS AND FINS Embryological evidence suggests that jaws evolved from paired **gill arches**—cartilaginous or bony supporting structures between the gill slits—in the pharynx of a jawless ancestor **(Figure 30.13)**. One pair of ancestral gill arches

FIGURE 30.13
The evolution of jaws.

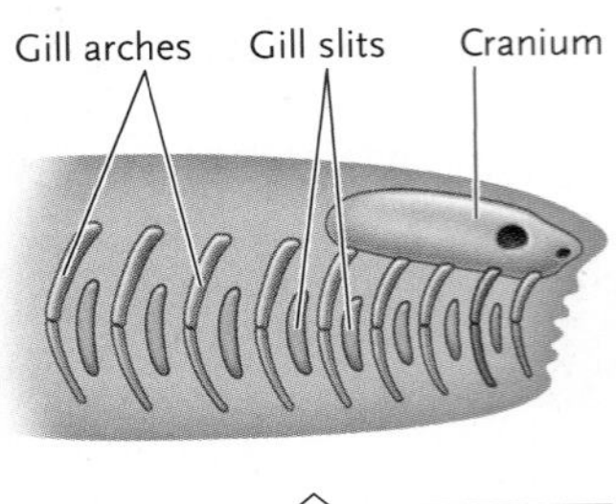
Gill arches
Gill slits
Cranium

A. Jaws evolved from gill arches in the pharynx of jawless fishes.

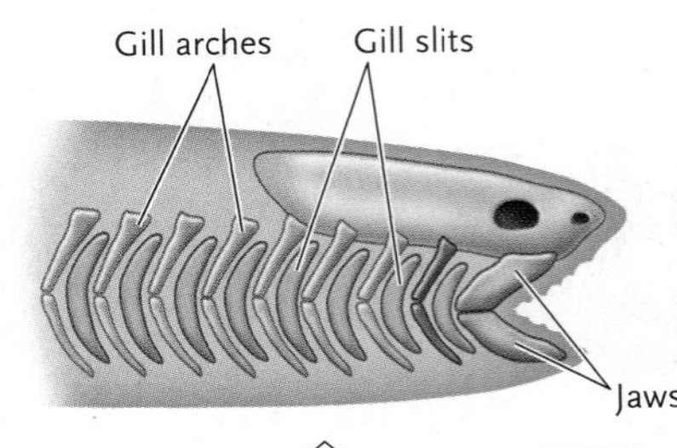
Gill arches
Gill slits
Jaws

B. In early jawed fishes, the upper jaw was firmly attached to the cranium.

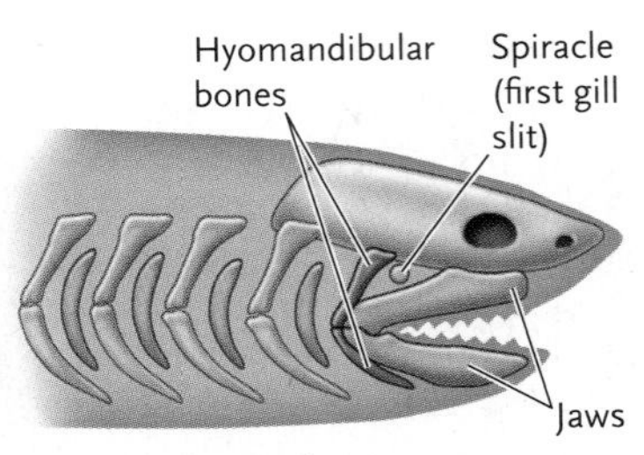
Hyomandibular bones
Spiracle (first gill slit)
Jaws

C. In later jawed fishes, the jaws were supported by the hyomandibular bones, which were derived from a second pair of gill arches.

formed bones in the upper and lower jaws, while a second pair was transformed into the hyomandibular bones that braced the jaws against the cranium. Nerves and muscles of the ancestral suspension-feeding pharynx control and move the jaws.

Innovative locomotor mechanisms have often appeared at roughly the same time as innovative feeding mechanisms in the vertebrate lineage; selection probably favored individuals that had the ability to pursue and capture a large range of animal prey as well as feed on them. Thus, many early jawed fishes also had fins. The earliest fins were folds of skin and moveable spines that stabilized locomotion and deterred predators. Moveable fins appeared independently in several lineages, and by the Devonian period, most fishes had unpaired (dorsal, anal, and caudal) and paired (pectoral and pelvic) fins **(Figure 30.14).**

EARLY JAWED FISHES In two early lineages of jawed fishes, spiny sharks and placoderms, the upper jaw was firmly attached to the cranium (see Figure 30.13B); their inflexible mouths simply snapped open and shut **(Figure 30.15).** Both groups also show evidence of an internal skeleton.

Spiny sharks (Acanthodii, *akantha* = thorn), which persisted from the late Ordovician through the Permian periods, were less than 20 cm long. Their small, light scales, streamlined bodies, well-developed eyes, large jaws, and numerous teeth suggest that they were fast swimmers and efficient predators. Most had a row of ventral spines and fins with internal skeletal support on each side of the body. Acanthodian anatomy suggests that they are closely related to the bony fishes living today.

Placoderms (Placodermi, *plax* = flat surface) appeared in the Silurian and diversified in the Devonian and Carboniferous periods, but they left no direct descendants. Some, like *Dunkleosteus,* reached a length of 10 m. The anterior part of the placoderm body was covered by heavy bony plates and the posterior part by smaller, bony scales. Their jaws had sharp cutting edges, but not separate teeth, and their paired fins had internal skeletons and powerful muscles.

Chondrichthyes Includes Fishes with Cartilaginous Endoskeletons

The 850 living species in the Chondrichthyes (*khondros* = cartilage, *ikhthys* = fish) have skeletons composed entirely of cartilage, which is much lighter than bone. In most vertebrates, the internal skeleton is first formed of cartilage, but the cartilage is gradually replaced by bone during embryonic development (see Section 36.2). Apparently, some mutation or mutations eliminated the second, bone-forming step in this process when this lineage arose. Because all fishes that lived before the appearance of Chondrichthyes had either bony armor or bony endoskeletons, biologists conclude that the absence of bone in Chondrichthyes is a derived, not an ancestral, trait.

Most living chondrichthyans are grouped in the Elasmobranchii, which includes the skates, rays, and sharks; nearly all

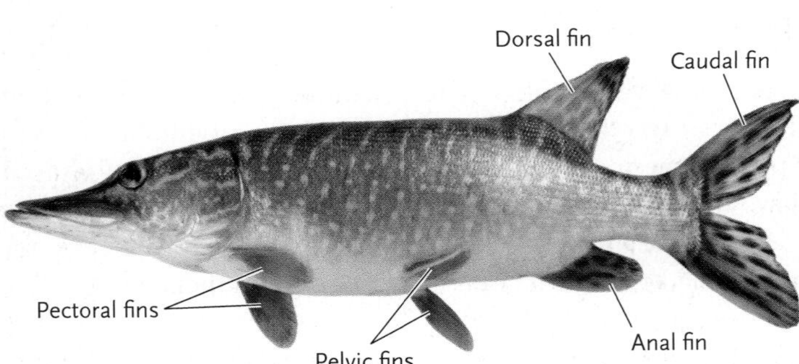
Dorsal fin
Caudal fin
Pectoral fins
Pelvic fins
Anal fin

FIGURE 30.14
Fish fins. Most fishes have both paired and unpaired fins. Image copyright Andrey Burmakin, 2010. Used under license from Shutterstock.com

A. **Spiny shark *(Climatius)***

B. **Placoderm *(Dunkleosteus)***

FIGURE 30.15
Early gnathostomes. **(A)** Most acanthodians were small, reaching a total length of no more than 20 cm. **(B)** Some placoderms were gigantic, up to 10 m in length. Some acanthodians had teeth on their jaws, but placoderms had only sharp, cutting edges.

A. **Manta ray *(Manta birostris)***

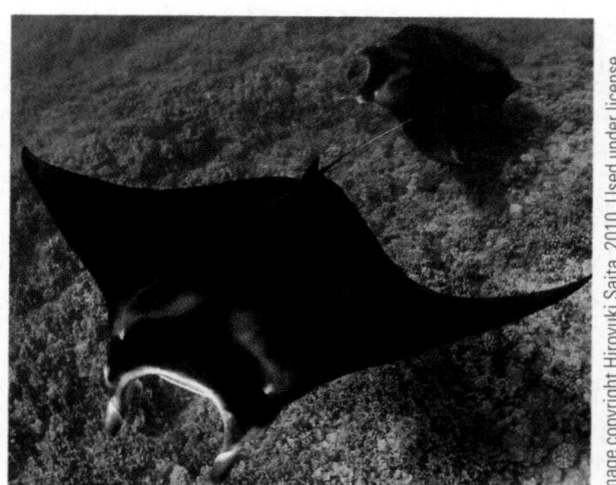
Image copyright Hiroyuki Saita, 2010. Used under license from Shutterstock.com

B. **Grey reef shark *(Carcharhinus amblyrhynchos)***

iStockphoto.com/Josh Friedman

C. **Swell shark *(Cephaloscylium ventricosum)* egg case**

Alex Kerstitch/Visuals Unlimited

D. **Sand Tiger Shark tooth whorls *(Eugomphodus taurus)***

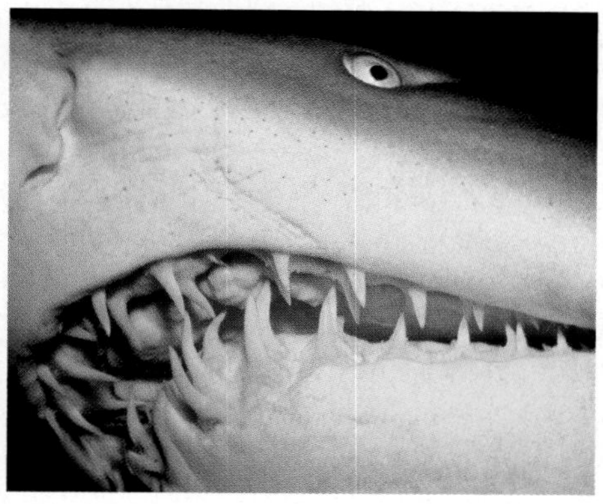
Jeff Rotman

FIGURE 30.16
Chondrichthyes. **(A)** Skates and rays as well as **(B)** sharks are grouped in the Elasmobranchii. **(C)** Many shark egg cases include a large yolk that nourishes the developing embryo. **(D)** The teeth of elasmobranchs develop in whorls, with new teeth ready to migrate forward to replace older, worn teeth.

are marine predators **(Figure 30.16).** Skates and rays are dorsoventrally flattened (see Figure 30.16A). They swim by undulating their enlarged pectoral fins. Most are bottom dwellers that often lie partly buried in sand. They feed on hard-shelled invertebrates, which they crush with massive, flattened teeth. The largest species, the manta ray *(Manta birostris),* which measures 6 m across, feeds on plankton in the open ocean. Some rays have electric organs that stun prey with as much as 200 volts.

Sharks (see Figure 30.16B) are among the ocean's dominant predators. Flexible fins, lightweight skeletons, streamlined bodies, and the absence of heavy body armor allow most sharks to pursue prey rapidly. Their large livers contain copious amounts of oil, which is lighter than water, increasing their buoyancy. The great white shark *(Carcharodon carcharias)* is the largest predatory species, attaining a length of 10 m. The whale shark *(Rhincodon typus),* which grows to 18 m, is the largest fish; it feeds on plankton.

Elasmobranchs—including sharks, skates, and rays—exhibit remarkable adaptations for acquiring and processing food. Their teeth develop in whorls under the fleshy parts of the mouth (Figure 30.16D). New teeth migrate forward as old, worn teeth break free. In many sharks, the upper jaw is loosely attached to the cranium, and it swings down during feeding. As the jaws open, the mouth spreads widely, sucking in large, hard-to-digest chunks of prey, which are swallowed intact. Although the elasmobranch digestive system is short, it includes a corkscrew-shaped **spiral valve,** which slows the passage of material and increases the surface area available for digestion and absorption.

Elasmobranchs also have well-developed sensory systems. In addition to vision and olfaction, they use **electroreceptors** to detect weak electric currents produced by other animals. And their **lateral-line system,** a row of tiny sensors in canals along both sides of the body, detects vibrations in water (see Figure 39.4).

Chondrichthyans exhibit numerous reproductive specializations. Males have a pair of organs, the **claspers,** on the pelvic fins, which help transfer sperm into the female's reproductive tract. Fertilization occurs internally. In many species, females produce large yolky eggs with tough leathery shells (see Figure 30.16C). Others retain the eggs within the oviduct until the young hatch. A few species nourish young within a uterus.

The Actinopterygii and Sarcopterygii Are Fishes with Bony Endoskeletons

In terms of diversity and sheer numbers, the fishes with bony endoskeletons—a cranium, vertebral column with ribs, and bones supporting their moveable fins—are the most successful of all vertebrates. The endoskeleton provides lightweight support, particularly compared with the heavy bony armor of ostracoderms and placoderms, and enhances their locomotor efficiency. Fishes with bony endoskeletons first appeared in the Silurian period and rapidly diversified into two lineages. The ray-finned fishes (Actinopterygii, *aktis* = ray, *pteron* = wing) have fins that are supported by thin and flexible bony rays. The fleshy-finned fishes (Sarcopterygii, from *sarco* = flesh) have fins that are supported by muscles and an internal bony skeleton. Ray-finned fishes have always been more diverse, and they vastly outnumber the fleshy-finned fishes today. The 21,000 living species of bony fishes occupy nearly every aquatic habitat and represent more than 95% of living fish species. Adults range from 1 cm to more than 6 m in length.

Bony fishes have numerous adaptations that increase their swimming efficiency. In many modern ray-finned fishes, a gas-filled **swim bladder** serves as a hydrostatic organ that increases buoyancy (see Figure 30.18A). The swim bladder is derived from an ancestral air-breathing lung that allowed early actinopterygians to gulp air, supplementing their gill respiration in aquatic habitats where dissolved oxygen concentration was low. The scales of most bony fishes are small, smooth, and lightweight. And their bodies are covered with a protective coat of mucus, which retards bacterial growth and smoothes the flow of water.

ACTINOPTERYGII The most primitive living actinopterygians, sturgeons and paddlefishes, have mostly cartilaginous skeletons **(Figure 30.17A).** These large fishes live in rivers and lakes of the northern hemisphere. Sturgeons feed on detritus and invertebrates; paddlefish consume plankton. Gars and bowfins are remnants of a more recent radiation **(Figure 30.17B).** They occur only in the eastern half of North America, where they feed on fishes and other prey. Gars are protected from predators by a heavy coat of bony scales.

Teleosts, the latest radiation of Actinopterygii, are the most diverse, successful, and familiar bony fishes. Evolution has produced a wide range of body forms **(Figure 30.18).** Teleosts have an internal skeleton made almost entirely of bone. On either side of the head, a flap of the body wall, the **operculum,** covers a chamber that houses the gills (that is, delicate respiratory organs). Sensory systems generally include large eyes, a lateral-line system, sound receptors, chemoreceptive nostrils, and taste buds. Variations in jaw structure allow different teleosts to consume plankton, seaweed, invertebrates, or other vertebrates.

Teleosts exhibit remarkable feeding and locomotor adaptations. When some teleosts open their mouths, bones at the front of the jaws swing forward to create a circular opening. Folds of skin extend backward, forming a tube through which they suck food (see Figure 30.18F). Many also have symmetrical caudal fins, posterior to the vertebral column, which provide power for locomotion. And their pectoral fins lie high on the sides of the body, providing fine control over swimming. Some species use their pectoral fins for acquiring food, for courtship, and for care of eggs and young. Some teleosts even use them for crawling on land or gliding in air.

Most marine species produce small eggs that hatch into planktonic larvae. Eggs of freshwater teleosts are generally larger and hatch into tiny versions of the adults. Parents often care for their eggs and young, fanning oxygen-rich water over them, removing fungal growths, and protecting them from predators. Some freshwater species, such as guppies, give birth to live young.

SARCOPTERYGII Two groups of fleshy-finned fishes (Sarcopterygii), the lobe-finned fishes and lungfishes, are now represented by only eight living species **(Figure 30.19).**

Although lobe-finned fishes were once thought to have been extinct for 65 million years, a living coelacanth *(Latimeria chalumnae)* was discovered in 1938 near the Comoros Islands, off the southeastern coast of Africa. We now know that a population of this meter-long fish lives at depths of 70 to 600 m, feeding on fishes and squid. Remarkably, a second population of coelacanths was discovered in 1998, when a specimen was found in an Indonesian fish market, 10,000 km east of the Comoros population. Based on analyses of its DNA, it is a distinct species *(Latimeria menadoensis).*

Lungfishes have changed relatively little over the last 200 million years. Six living species are distributed on southern continents. The Australian lungfishes, which live in rivers and pools, use their lungs to supplement gill respiration when dissolved oxygen concentration is low. The South American and African species, which live in swamps, use their lungs to collect oxygen during the annual dry season, which they spend encased in a mucus-lined burrow in the dry mud. When the rains begin, water fills the burrow and the fishes awaken from dormancy.

STUDY BREAK 30.5 <

1. **What characteristics of sharks and rays make them more efficient predators than the acanthodians or placoderms?**
2. **How do the air bladder and fins of ray-finned bony fishes increase their locomotor abilities?**
3. **How do the lungs of lungfishes allow them to survive in stressful environments?**

A. Lake sturgeon *(Accipenser fulvescens)*

Ken Lucas/Visuals Unlimited

B. Long-nosed gar *(Lepisosteus osseus)*

© Shedd Aquarium/Patrice Ceisel

FIGURE 30.17
Primitive actinopterygians. **(A)** Sturgeons and **(B)** gars are living representatives of early actinopterygian radiations.

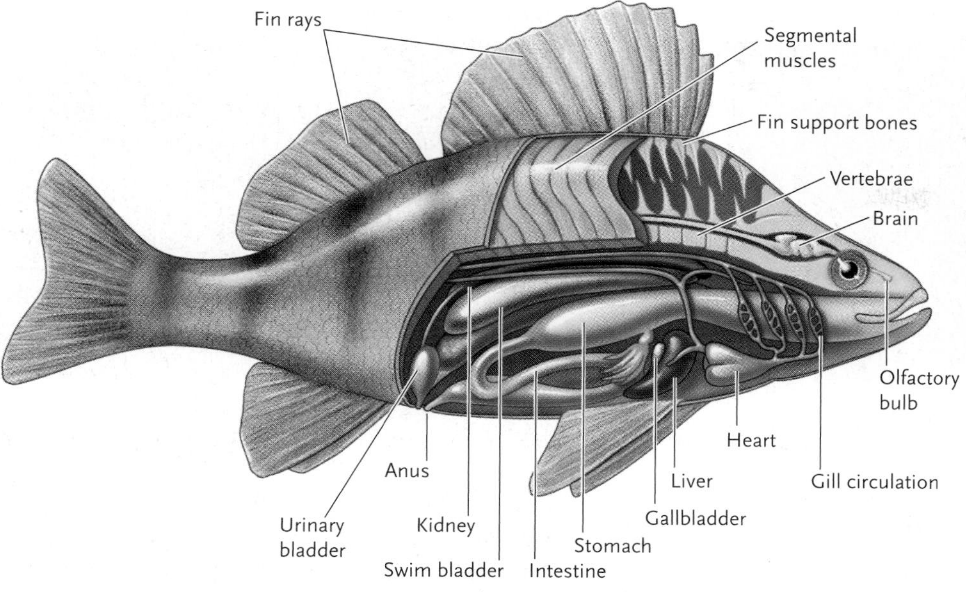
Fin rays
Segmental muscles
Fin support bones
Vertebrae
Brain
Olfactory bulb
Heart
Gill circulation
Liver
Gallbladder
Anus
Urinary bladder
Kidney
Stomach
Swim bladder
Intestine

A. Teleost internal anatomy

B. Sea horses, like the northern sea horse *(Hippocampus hudsonius)*, use a prehensile tail to hold on to substrates; they are weak swimmers.

Digital Vision/Getty Images Inc.

Kit Kittle/Corbis

C. The long, flexible body of a spotted moray eel *(Gymnothorax moringa)* can wiggle through the nooks and crannies of a reef.

Image copyright Lynsey Allan, 2010. Used under license from Shutterstock.com

D. Flatfishes, like this European flounder *(Platichthys flesus)*, lie on one side and leap at passing prey.

Operculum

Brandon Cole/Visuals Unlimited

E. Open ocean predators, like the yellowfin tuna *(Thunnus albacares)*, have strong, torpedo-shaped bodies and powerful caudal fins.

Arthur W. Ambler/Photo Researchers, Inc.

F. Kissing Gouramis *(Helostoma temmincki)* extend their jaws into a tube that sucks food into the mouth.

FIGURE 30.18

Teleost diversity. Although all teleosts share similar internal features, their diverse shapes adapt them to different diets and types of swimming.

A. Coelacanth *(Latimeria chalumnae)*

Peter Scoones/Getty Images

B. African lungfish *(Protopterus annectens)*

Tom McHugh

FIGURE 30.19
Sarcopterygians. **(A)** Two species of lobe-finned fishes, coelocanths, still live in the oceans today. **(B)** The African lungfish is one of only six living lungfish species.

30.6 Tetrapoda: The Evolution of Limbs

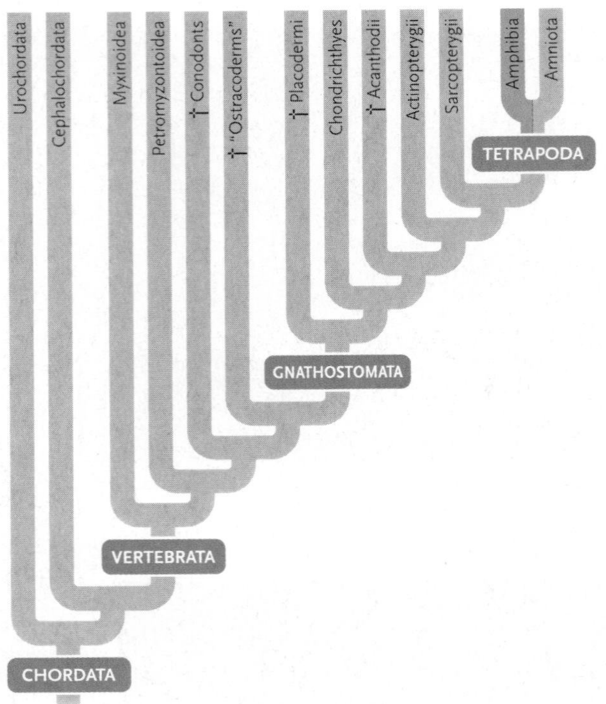

Urochordata
Cephalochordata
Myxinoidea
Petromyzontoidea
† Conodonts
† "Ostracoderms"
† Placodermi
Chondrichthyes
† Acanthodii
Actinopterygii
Sarcopterygii
Amphibia
Amniota
TETRAPODA
GNATHOSTOMATA
VERTEBRATA
CHORDATA

The fossil record suggests that tetrapods arose from a group of fleshy-finned fishes, the *osteolepiforms,* in the late Devonian period. Osteolepiforms and early tetrapods shared several derived characteristics: both had curious infoldings of their tooth surfaces, a trait with unknown function, and the shapes and positions of bones of the skulls and appendages were similar **(Figure 30.20).**

Key Adaptations Facilitated the Transition to Land

Fishes are not adapted to live on land, and the first tetrapods faced serious environmental challenges. First, because air is less dense than water, it provides less support for an animal's body. Second, animals exposed to air inevitably lose body water by evaporation. Third, the sensory systems of fishes, which work well under water, do not function well in air. However, swampy late-Devonian habitats also offered distinct advantages. Land plants, soft-bodied invertebrates, and arthropods provided abundant food, oxygen was more readily available in air than in water, and predators did not yet live in these new habitats.

In some ways, osteolepiforms had characteristics that may have facilitated the transition to land (see Figure 30.20A). Most had strong, stout fins that enabled them to crawl on the muddy bottom of shallow pools, and their vertebral column included crescent-shaped bones that provided good support. They had nostrils leading to sensory pits that housed olfactory (smell) receptors. And they almost certainly had lungs to augment gill respiration in the swampy, oxygen-poor waters where they lived.

The earliest tetrapod for which we have nearly complete skeletal data is the semiterrestrial, meter-long *Ichthyostega* (see Figure 30.20B). Compared with its fleshy-finned ancestors, *Ichthyostega* had a stronger vertebral column, sturdier girdles and appendages, a rib cage that protected its internal organs (including lungs), and a neck. Fishes have no neck; the pectoral girdle is fused to the cranium. But several vertebrae separated these structures in *Ichthyostega,* allowing it to move its head to scan the environment and to capture food. However, *Ichthyostega* retained a fishlike lateral-line system, caudal fin, and scaly covering on its body.

Life on land also required changes in sensory systems. In fishes, for example, the body wall picks up sound vibrations and transfers them to sensory receptors directly. But sound waves are harder to detect in air. Early tetrapods developed a **tympanum,** a specialized membrane on either side of the head that is vibrated by airborne sounds. The tympanum connects to the **stapes,** a bone that is homologous to the hyomandibula, which had supported the jaws of fishes (see Figure 23.6). The stapes, in turn, transfers vibrations to the sensory cells of an inner ear.

Modern Amphibians Are Very Different from Their Paleozoic Ancestors

Most of the more than 6,000 species of living amphibians—including frogs, salamanders, and caecelians—are small, and their skeletons contain fewer bones than those of Paleozoic tetrapods

A. ***Eusthenopteron*, an osteolepiform fish**

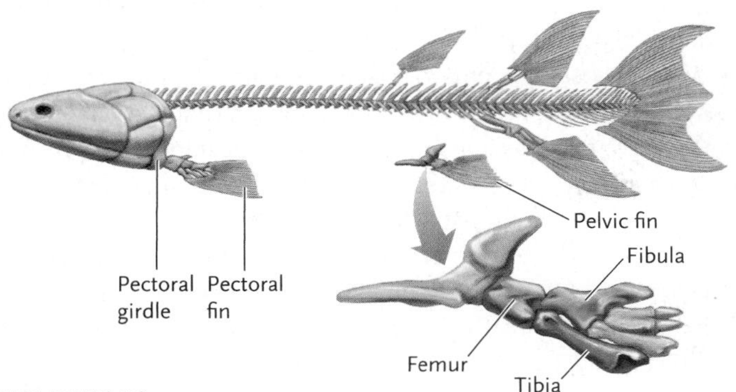
Pelvic fin
Fibula
Pectoral girdle
Pectoral fin
Femur
Tibia

B. ***Ichthyostega*, an early tetrapod**

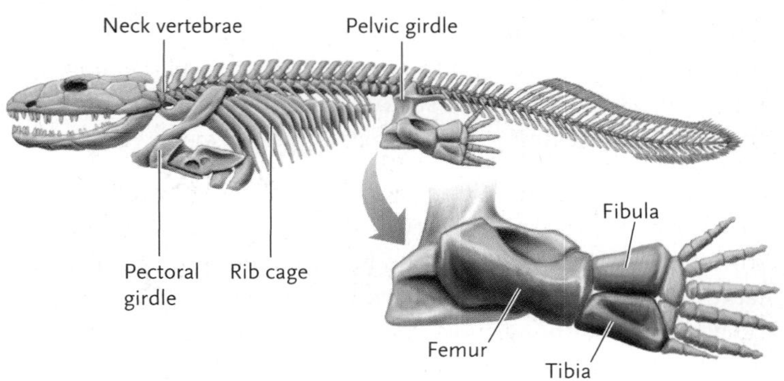
Neck vertebrae
Pelvic girdle
Fibula
Pectoral girdle
Rib cage
Femur
Tibia

FIGURE 30.20
Evolution of tetrapod limbs. **(A)** The limb skeleton of osteolepiform fishes is homologous to that of **(B)** early tetrapods. Although *Ichthyostega* retained many fishlike characteristics, its pectoral girdle was completely freed from the cranium and it had a heavy rib cage. Fossils of its forefoot have not yet been discovered.

like *Ichthyostega.* All living amphibians are carnivorous as adults, but the aquatic larvae of some species are herbivores.

Most living amphibians have a thin, scaleless skin, well supplied with blood vessels, that is a major site of gas exchange. Because gases must enter the body across a thin layer of water, the skin of most amphibians must remain moist, restricting them to aquatic or wet terrestrial habitats. Adults of some species also acquire oxygen through saclike lungs. The evolution of lungs was accompanied by modifications of the heart and circulatory system that increase the efficiency with which oxygen is delivered to body tissues (see Section 42.1).

The life cycles of many amphibians (*amphi* = of both kinds, *bios* = life) include both larval and adult stages. Eggs are laid and fertilized in water, where they hatch into larvae, such as the tadpoles of frogs, which eventually metamorphose into adults (see Figure 40.9). Although the larvae of many species are aquatic, adults may be aquatic, amphibious, or terrestrial. Some salamanders are paedomorphic; the larval stage attains sexual maturity without changing its form or moving to land. By contrast, some frogs and salamanders reproduce on land and skip the larval stage altogether. But even though they are terrestrial breeders, their eggs dry out quickly unless they are laid in moist places.

Modern amphibians are represented by three lineages **(Figure 30.21).** Populations of practically all amphibian species have declined rapidly in recent years, probably because of exposure to environmental pollutants and fungal infections (see Chapter 53).

ANURA The 6,400 species of frogs and toads have short, compact bodies, and adults lack tails. Their elongate hind legs and webbed feet allow them to hop on land or swim. A few species are adapted to dry habitats, withstanding periods of drought by encasing themselves in mucous cocoons.

A. **Northern leopard frog *(Rana pipiens)***

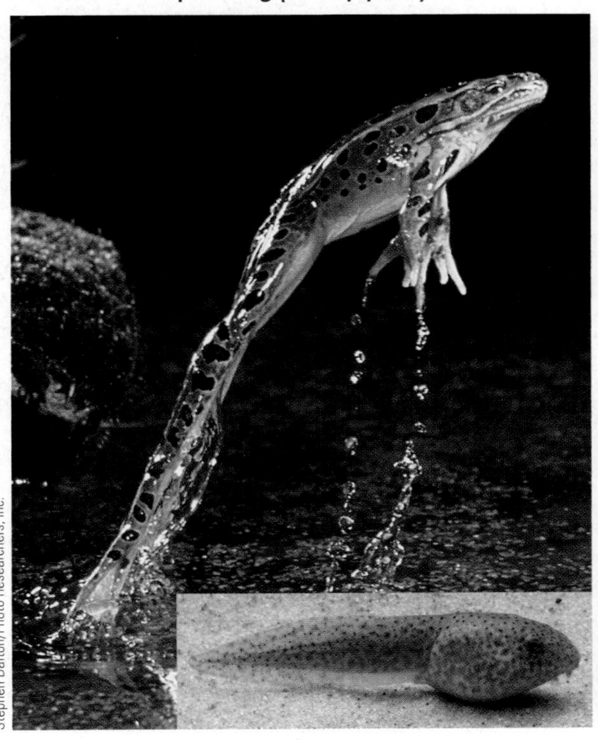
Stephen Dalton/Photo Researchers, Inc.
Harry Rogers/Photo Researchers, Inc.

B. **Red-spotted newt *(Notophthalmus viridescens)***

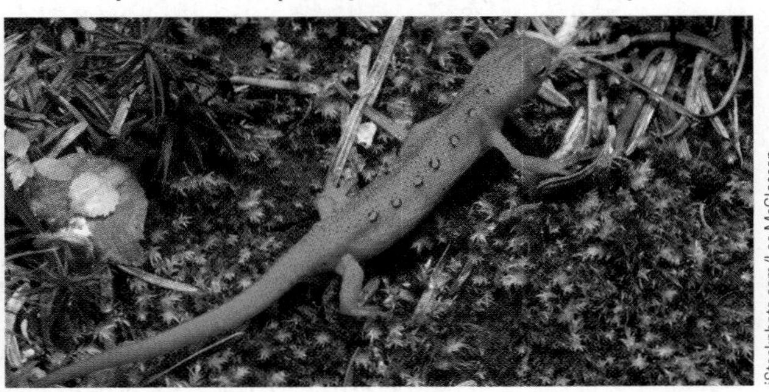
iStockphoto.com/Les McGlasson

C. **A caecelian *(Caecelia nigricans)***

Juan M. Renjifo/Animals Animals—Earth Scenes

FIGURE 30.21
Living amphibians. **(A)** Frogs and toads have compact bodies and long hind legs; their larvae (inset), called tadpoles, are legless and rely on their tails for locomotion. **(B)** Salamanders and newts have an elongate body and four legs. **(C)** Caecelians are legless burrowing amphibians.

CAUDATA Most of the 580 species of salamanders and newts have an elongate, tailed body and four legs. They walk by alternately contracting muscles on either side of the body much the way fishes swim. Species in the most diverse group, the lungless salamanders, are fully terrestrial throughout their lives, using their skin and the lining of the throat for gas exchange.

GYMNOPHIONA The 170 species of caecelians (Gymnophiona, from *gymnos* = naked, *ophioneos* = snakelike) are legless burrowing animals with wormlike bodies. They occupy tropical habitats throughout the world. Unlike other modern amphibians, caecelians have small bony scales embedded in their skin. Fertilization is internal, and females give birth to live young.

STUDY BREAK 30.6 <

1. **For the first tetrapods, what were the advantages and disadvantages of moving onto the land?**
2. **What parts of the life cycle in most modern amphibians are dependent on water or very moist habitats?**

30.7 Amniota: The Evolution of Fully Terrestrial Vertebrates

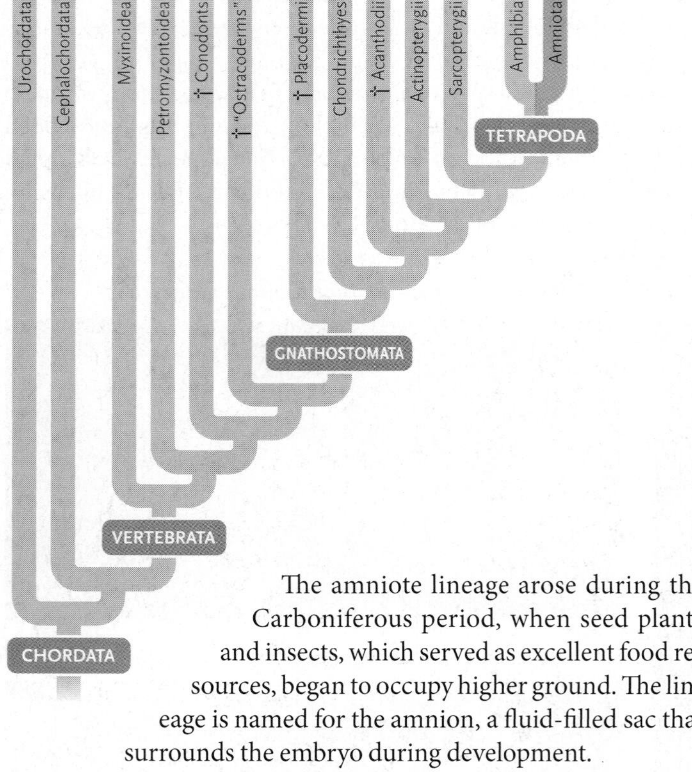
Urochordata
Cephalochordata
Myxinoidea
Petromyzontoidea
† Conodonts
† "Ostracoderms"
† Placodermi
Chondrichthyes
† Acanthodii
Actinopterygii
Sarcopterygii
Amphibia
Amniota
TETRAPODA
GNATHOSTOMATA
VERTEBRATA
CHORDATA

The amniote lineage arose during the Carboniferous period, when seed plants and insects, which served as excellent food resources, began to occupy higher ground. The lineage is named for the amnion, a fluid-filled sac that surrounds the embryo during development.

Key Adaptations Allow Amniotes to Live a Fully Terrestrial Life

Based on the abundance and diversity of their fossils, amniotes were extremely successful; they quickly replaced many nonamniote species in terrestrial habitats. Although the fossil record includes abundant skeletons of early amniotes, it provides little direct information about their soft body parts and physiology. For amniotes living today, three key adaptations allow them to live in dry habitats, freeing them from a dependency on moist surroundings and standing water. First, they have a tough, dry skin. The skin cells are filled with keratin and lipids, substances that are relatively impermeable to water. Thus, amniotes do not dehydrate in air as quickly as amphibians do.

Second, many amniotes produce an **amniote egg,** which can survive and develop on dry land. The eggs of modern reptiles and birds have four specialized membranes and a hard or leathery shell **(Figure 30.22).** The membranes protect the developing embryo and facilitate gas exchange and excretion; the shell, which is perforated by microscopic pores, mediates the exchange of air and water between the egg and its environment. The egg also includes generous supplies of **yolk,** the embryo's main energy source, and **albumin,** a source of nutrients and water. Compared with those of amphibians, amniote eggs are large; and, as amniotes lack a larval stage, the young hatch as miniature versions of the adult. In contrast to reptiles and birds, the eggs of virtually all mammals lack a shell. Instead, mammalian embryos which have the same four membranes associated with them, implant in the wall of the mother's uterus and receive nutrients and oxygen directly from her.

Third, many amniotes produce uric acid as a waste product of nitrogen metabolism (see Chapter 46). By contrast, fishes and amphibians produce ammonium ions or urea, toxic materials that require lots of water to flush them from body tissues. Because uric acid is less toxic than these other compounds, it can be excreted as a semisolid paste, which conserves body water.

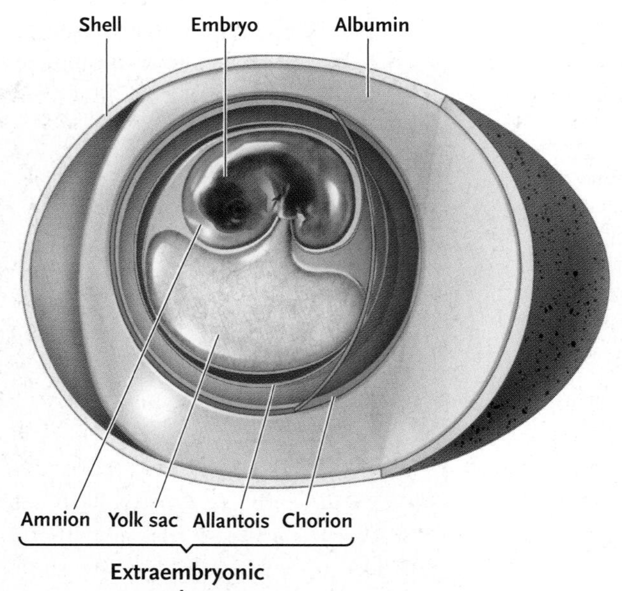
Shell
Embryo
Albumin
Amnion
Yolk sac
Allantois
Chorion
Extraembryonic membranes

FIGURE 30.22
The amniote egg. A water-retaining egg with four specialized membranes (the amnion, allantois, chorion, and yolk sac) and a hard or leathery shell allowed amniotes and their descendants to reproduce in dry terrestrial environments.

The Fossil Records Suggests That Amniotes Diversified into Three Major Lineages

A traditional view, based largely on the morphology of extinct species, suggests that amniotes differentiated into three lineages during the Carboniferous and Permian periods: synapsids, anapsids, and diapsids **(Figure 30.23).** These three groups are distinguished by differences in the number of openings, called **temporal fenestrae** (*fenestra* = window), and the **bony temporal arches** that border them on the sides of the skull. In the animals that have them, temporal fenestrae provide space for large and powerful jaw muscles. We briefly describe the three traditionally recognized lineages and examine diversification within the diapsids before considering alternative views of amniote relationships.

SYNAPSIDA Nearly all researchers agree that **Synapsida** (*syn* = with; *apsis* = arch), a group of small terrestrial predators, was the earliest lineage to emerge from the ancestral amniotes. They had one temporal fenestra and one temporal arch on either side of the skull. Synapsids appeared late in the Permian period; mammals are their living descendants.

ANAPSIDA According to the traditional view, **Anapsida** ("no arch") was the second amniote lineage to arise; they lacked temporal fenestrae and temporal arches, a condition that scientists recognize as the ancestral amniote skull morphology. Traditionally, biologists hypothesized that turtles are the only living representatives of this lineage, because all extinct and living turtles lack temporal fenestrae and temporal arches.

DIAPSIDA Most Mesozoic amniotes belong to the third lineage, **Diapsida** ("two arches"); these animals had two temporal fenestrae and two temporal arches. Their living descendants include lizards and snakes, crocodilians, and birds. Morphological and genetic sequence data from living species confirm that the diapsids form a monophyletic lineage.

Diapsids Diversified Wildly during the Mesozoic Era

Fossil evidence and molecular sequence data from living species reveal that the early diapsids differentiated into two major lineages, the **Archosauromorpha** (*arkhon* = ruler, *sauros* = lizard, *morphe* = form) and the **Lepidosauromorpha** (*lepis* = scale), which differed in many skeletal characteristics.

The archosauromorphs (commonly called archosaurs), or "ruling reptiles," include crocodilians, pterosaurs, dinosaurs, and birds. Crocodilians, which first appeared during the Triassic period, have bony armor and a laterally flattened tail that propels them through water. Pterosaurs, now extinct, were flying predators of the Jurassic and Cretaceous periods. Their wings were composed of thin sheets of skin attached to the sides of the body and supported by an elongate finger. Small pterosaurs may have been active fliers, but large ones, with wingspans as wide as 13 m, probably soared on air currents as vultures do today.

Two lineages of dinosaurs, "lizard-hipped" saurischians and "bird-hipped" ornithischians proliferated in the Triassic and Jurassic periods. As their names imply, they differed in the anatomy of their pelvic girdles. The saurischian lineage included bipedal carnivores and quadrupedal herbivores. Most carnivorous saurischians were swift runners. Their forelimbs, however, were often ridiculously short. *Tyrannosaurus,* which was 15 m long and stood 6 m high, is the most familiar carnivorous saurischian, but most species were much smaller. One group of small carnivorous saurischians was ancestral to birds. By the Cretaceous period, some herbivorous saurischians had also attained gigantic size, and many had long, flexible necks. For example, *Apatosaurus* (previously known as *Brontosaurus*) was 25 m long and may have weighed 50,000 kg.

The largely herbivorous ornithischian dinosaurs had enormous, stocky bodies. This lineage included the armored or plated dinosaurs (*Ankylosaurus* and *Stegosaurus*), the duck-billed dinosaurs *(Hadrosaurus),* horned dinosaurs *(Styracosaurus),* and some with remarkably thick skulls *(Pachycephalosaurus).* The ornithischians were most abundant in the Jurassic and Cretaceous periods.

The second major lineage of diapsids was the lepidosauromorphs (commonly called lepidosaurs), a diverse group that included both marine and terrestrial animals. Plesiosaurs were marine, fish-eating creatures that used long, paddle-shaped limbs to row through the water. Ichthyosaurs were porpoise-like animals with laterally flattened tails. They were so highly specialized for marine life that they could not venture onto land, even to reproduce. Instead, they gave birth to live young under water as porpoises and whales do today. A third important group within this lineage is the squamates, which includes the living lizards and snakes.

The Phylogenetic Position of Turtles is a Subject of Ongoing Debate

Scientists have debated the evolutionary relationships among the amniotes for more than a century, and they still have not reached a consensus. The most persistent problem is how turtles fit into the amniote phylogenetic tree. The traditional hypothesis (described previously and pictured in Figure 30.23) regards turtles as the living descendants of an ancient anapsid lineage. According to this hypothesis, turtle skulls lack temporal fenestrae and arches because this lineage retained the ancestral anapsid skull condition.

In the 1990s, however, researchers who conducted a detailed morphological analysis of some amniote groups suggested that turtles are closely related to lepidosaurs and therefore part of the diapsid lineage. Hoping to resolve the ongoing dispute about amniote relationships, biologists turned to molecular analyses. But molecular analyses have not produced a clear conclusion: although much molecular research suggests that turtles are diapsids, some studies indicate that they are

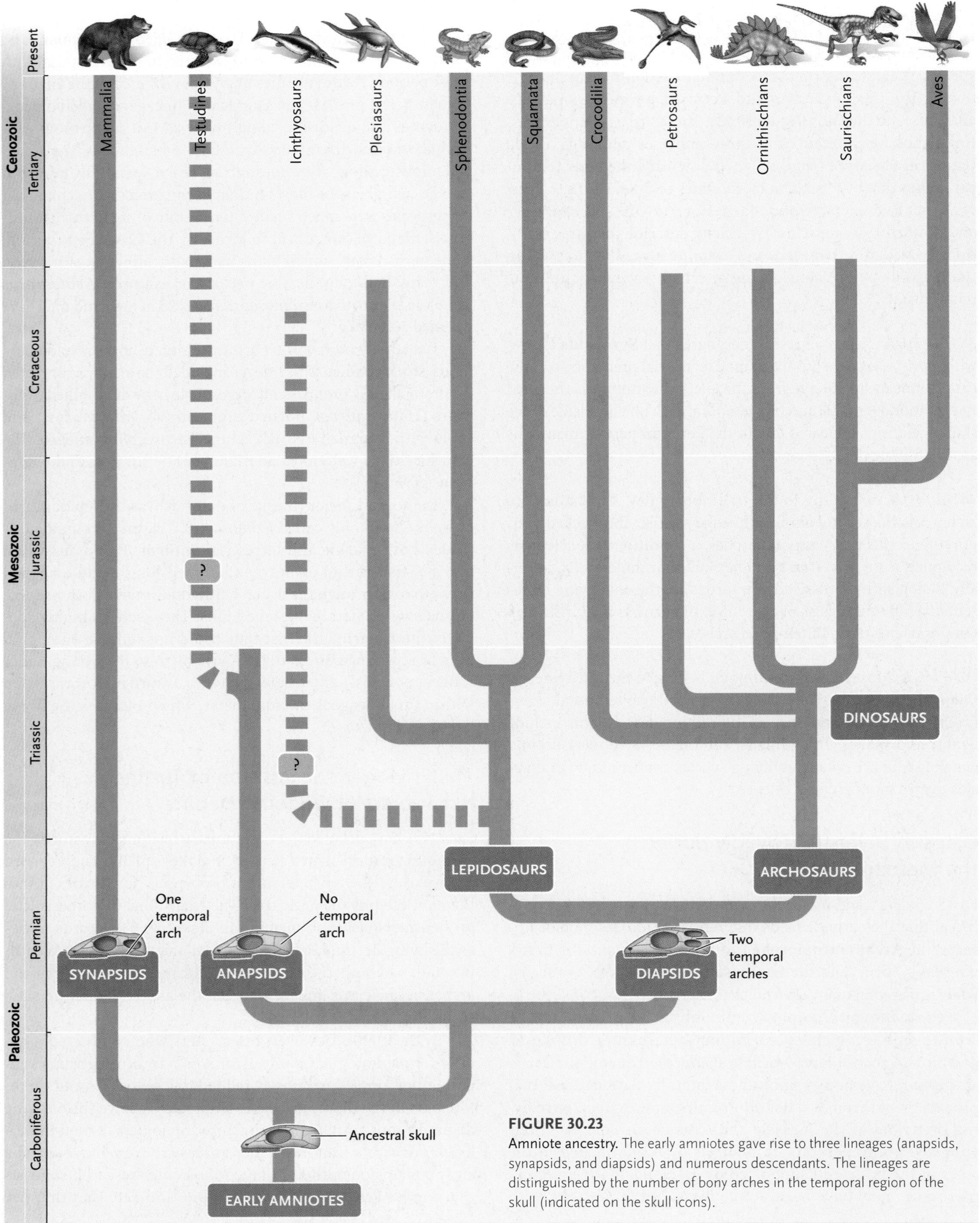

Present
Cenozoic
Tertiary
Cretaceous
Mesozoic
Jurassic
Triassic
Permian
Paleozoic
Carboniferous
Mammalia
Testudines
Ichthyosaurs
Plesiasaurs
Sphenodontia
Squamata
Crocodilia
Petrosaurs
Ornithischians
Saurischians
Aves
?
?
DINOSAURS
LEPIDOSAURS
ARCHOSAURS
One temporal arch
No temporal arch
Two temporal arches
SYNAPSIDS
ANAPSIDS
DIAPSIDS
Ancestral skull
EARLY AMNIOTES

FIGURE 30.23

Amniote ancestry. The early amniotes gave rise to three lineages (anapsids, synapsids, and diapsids) and numerous descendants. The lineages are distinguished by the number of bony arches in the temporal region of the skull (indicated on the skull icons).

closely related to archosaurs, whereas other studies indicate a close relationship to lepidosaurs. According to hypotheses that include turtles within the diapsids (that is, that they are closely related either to archosaurs or to lepidosaurs), a turtle ancestor must have had the two temporal fenestrae and two temporal arches that are characteristic of the diapsid lineage. If that was the case, the turtle lineage must have subsequently lost those characteristics and returned to what appears to be the ancestral anapsid condition. To confuse matters even further, other analyses that include data from the fossil record as well as morphological and genetic sequence data on living species provide as much support for the traditional hypothesis that turtles are part of the anapsid lineage as for the newer hypotheses that place turtles among the diapsids. Thus, although it is easy to recognize turtles when we see them, the details of their evolutionary history are not yet understood.

STUDY BREAK 30.7

1. **How did the evolution of the amniote egg free amniotes from a dependence on standing water?**
2. **What groups of animals are included in each of the three traditionally defined amniote lineages?**
3. **Based upon the evolutionary history of the diapsid amniotes, are crocodilians more closely related to lizards or to birds?**

THINK OUTSIDE THE BOOK

Visit a local natural history museum or search the Internet to learn more about extinct diapsids, such as plesiosaurs, ichthyosaurs, pterosaurs, and non-avian dinosaurs. What types of habitats did they occupy, and what foods did they eat?

30.8 Testudines: Turtles

Turtles Have Bodies Encased in a Bony Shell

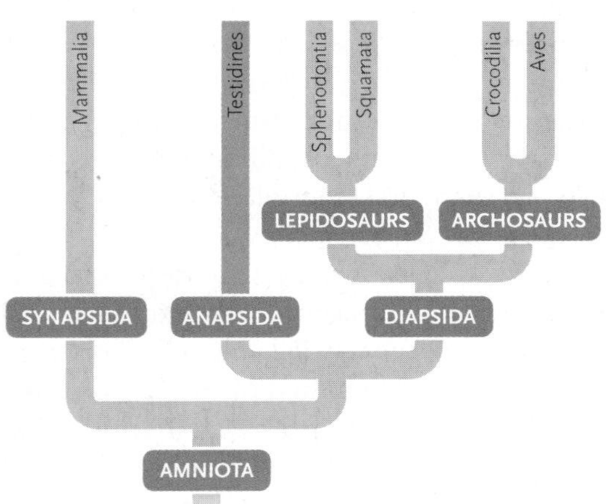
Mammalia
Testudines
Sphenodontia
Squamata
Crocodilia
Aves
LEPIDOSAURS
ARCHOSAURS
SYNAPSIDA
ANAPSIDA
DIAPSIDA
AMNIOTA

A. The turtle skeleton

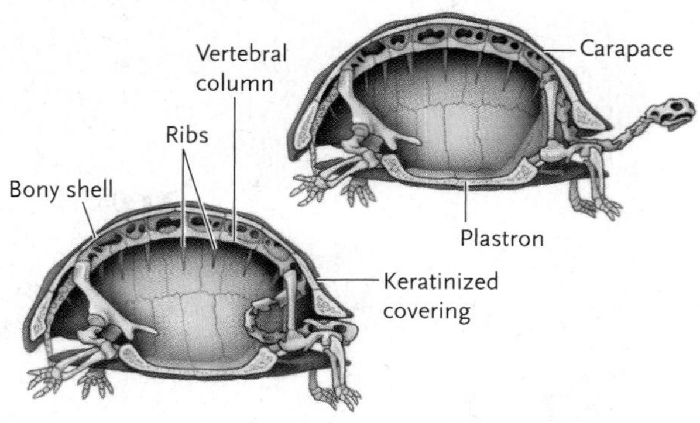
Vertebral column
Carapace
Ribs
Bony shell
Plastron
Keratinized covering

B. Eastern painted turtle (*Chrysemys picta*)

iStockphoto.com/Bruce MacQueen

FIGURE 30.24
Testudines. **(A)** Most turtles can withdraw their heads and legs into a bony shell. **(B)** Aquatic turtles often bask in the sun to warm up. The sunlight may also help eliminate parasites that cling to the turtle's skin.

The turtle body plan, largely defined by a bony, boxlike shell, has changed little since the group first appeared during the Triassic period **(Figure 30.24).** The shell includes a dorsal **carapace** and a ventral **plastron.** A turtle's ribs are fused to the inside of the carapace, and, in contrast to other tetrapods, the pectoral and pelvic girdles lie within the ribcage. Large keratinized scales cover the bony plates that form the shell.

The 250 living species of turtles occupy terrestrial, freshwater, and marine habitats. They range from 8 cm to 2 m in length. All species lack teeth, but they use a keratinized beak and powerful jaw muscles to feed on plants or animal prey. When threatened, most species retract into their shells. Many species are now highly endangered because adults are hunted for meat, their eggs are consumed by humans and other predators, and their young are collected for the pet trade.

STUDY BREAK 30.8

1. **How does the overall structure of turtles distinguish them from other amniotes?**

Pete & Judy Morrin/Ardea London

A. Sphenodontia includes the tuatara *(Sphenodon punctatus)* and one other species.

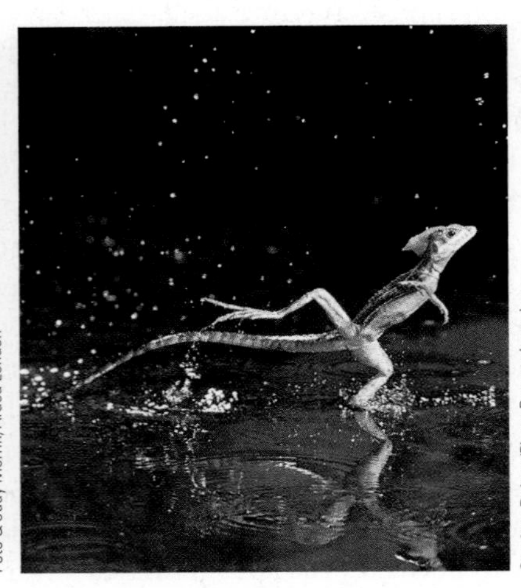
Stephen Dalton/Photo Researchers, Inc.

B. Basilisk lizards *(Basiliscus basiliscus)* escape from predators by running across the surface of streams.

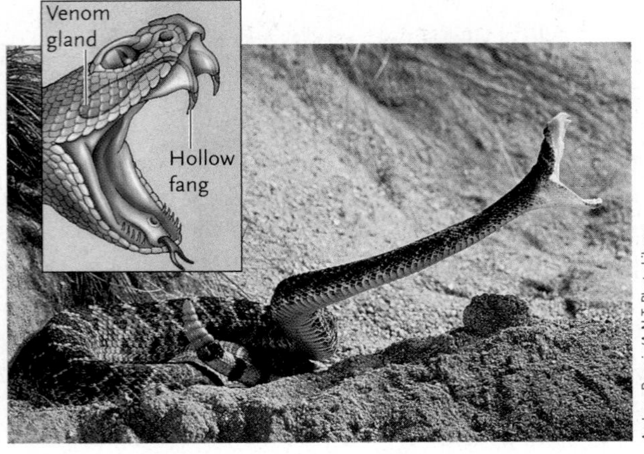
Venom gland
Hollow fang
Andrew Dennis/A.N.T. Photo Library

C. A western diamondback rattlesnake *(Crotalus atrox)* of the American southwest bares its fangs with which it injects a powerful toxin into prey.

FIGURE 30.25
Living lepidosaurs.

30.9 Living Lepidosaurs: Sphenodontids and Squamates

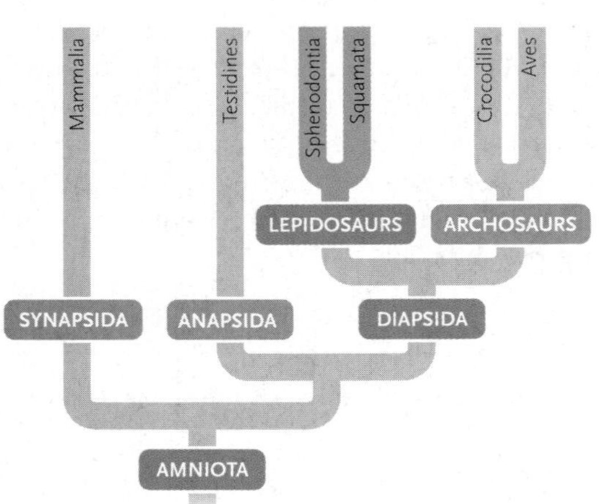
Mammalia
Testidines
Sphenodontia
Squamata
Crocodilia
Aves
LEPIDOSAURS
ARCHOSAURS
SYNAPSIDA
ANAPSIDA
DIAPSIDA
AMNIOTA

Two groups of lepidosaurs—Sphenodontida and Squamata—still live today. In addition to sharing certain skeletal characteristics, their bodies are encased in a dry, scaly skin.

Living Sphenodontids Are Remnants of a Diverse Mesozoic Lineage

The tuatara *(Sphenodon punctatus)* is one of two living representatives of the sphenodontids (*sphen* = wedge, *odont* = tooth), a diverse Mesozoic lineage **(Figure 30.25A).** These lizard-shaped animals survive on a few islands off the coast of New Zealand. Adults are about 60 cm long. They live in dense colonies, where males and females defend small territories using vocal and visual displays. They often share underground burrows with seabirds, feeding mainly on invertebrates and small vertebrates. They are primarily nocturnal and maintain low body temperatures during periods of activity.

Squamates—Lizards and Snakes—Are Covered by Overlapping, Keratinized Scales

The skin of lizards and snakes (Squamata, *squama* = scale) is composed of overlapping, keratinized scales that protect against dehydration. Squamates periodically shed their skin as they grow, much the way arthropods shed their exoskeletons (see Section 29.7). Most squamates regulate their body temperature behaviorally; they are active only when weather conditions are favorable, and they shuttle between sunny and shady places when they need to warm up or cool down (see Section 46.7).

Most of the 4,800 lizard species are less than 15 cm long **(Figure 30.25B).** However, the Komodo dragon *(Varanus komodoensis)* grows to nearly 3 m. Lizards occupy a wide range of habitats, but they are especially common in deserts and the tropics; one species *(Zootoca vivipara)* occurs within the Arctic Circle. Most lizards feed on insects, although some eat leaves or meat. The diverse tropical genus *Anolis* has become a frequent subject of research, as described in *Focus on Model Research Organisms.*

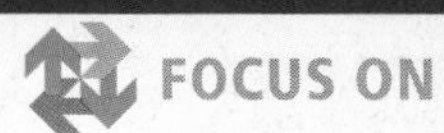

FOCUS ON **Model Research Organisms**

Anolis Lizards of the Caribbean

The lizard genus *Anolis* has been a model system for studies in ecology and evolutionary biology since the 1960s, when Ernest E. Williams of Harvard University's Museum of Comparative Zoology first began studying it. With roughly 400 known species—and new ones being described all the time—*Anolis* is one of the most diverse vertebrate genera. Most anoles are less than 10 cm long, not including the tail, and many occur at high densities, making it easy to collect lots of data in a relatively short time. Male anoles defend territories, and their displays make them conspicuous even in dense forests.

Anolis species are widely distributed in South America and Central America, but nearly 40% occupy Caribbean islands. The number of species on an island is generally proportional to the island's size. Cuba, the largest island, has more than 50 species, whereas small islands have just one or two.

Studies by Williams and others suggest that the anoles on some large islands are the products of independent adaptive radiations. Eight of the 10 *Anolis* species now found on Puerto Rico probably evolved on that island from a common ancestor. Similarly, the seven *Anolis* species on Jamaica shared a common ancestor, which was different from the ancestor of the Puerto Rican species. The anole faunas on Cuba and Hispaniola are the products of several independent radiations on each island.

Williams discovered that these independent radiations had produced similar-looking species on different islands. He developed the concept of the *ecomorph,* a group of species that have similar morphological, behavioral, and ecological characteristics even though they are not closely related within the genus. Williams named the ecomorphs after the vegetation that they commonly used (see **figure**). For example, grass anoles are small, slender species that usually perch on low, thin vegetation. Trunk-ground anoles have chunky bodies and large heads, and they perch low on tree trunks, frequently jumping to the ground to feed. Although the grass anoles or the trunk-ground anoles on different islands are similar in many ways, they are not closely related to each other. Their resemblances are the products of convergent evolution.

Ecomorphs exist because evolutionary processes have accentuated the morphological differences among species that occupy different types of vegetation. Jon Losos of Harvard University has demonstrated that trunk-ground anoles, which have relatively long legs and tails, can run faster on wide surfaces and jump farther than species with relatively short legs. And in nature the trunk-ground anoles run and jump more frequently than the other ecomorphs do.

Different ecomorphs on an island use different parts of their habitats by choosing different perch sites (grass, tree trunks, rocks). When two or more species of the same ecomorph inhabit the same island, they occupy habitats with different temperature and shade conditions (see figure). For example, in Puerto Rico, one species of trunk-ground anole *(Anolis gundlachi)* occupies cool, shady uplands; another *(Anolis cristatellus)* lives in warm, fairly open lowland habitats; and a third species *(Anolis cooki)* lives in desert habitats. Other species in Puerto Rico exhibit similar differences in their distributions. These differences in geographical distribution and habitat use presumably allow the different species to avoid competition with each other and gain access to the resources they need to survive and reproduce.

Evolutionary processes have also fostered physiological differences that reinforce the ecological separation established by the lizards' use of different habitats. For example, *A. cristatellus* maintains higher body temperatures than *A. gundlachi,* and neither is physiologically adapted to the environment of the other: *A. cristatellus* dies in the high altitude forests where *A. gundlachi* thrives, while *A. gundlachi* suffers heat stress at body temperatures that are typical for *A. cristatellus.*

Researchers throughout the Americas continue to explore the ecology and evolution of anoles. Some unravel their biogeography and systematic relationships; others focus on the ecology of populations and communities; still others study their social behavior or sensory physiology. With so many species distributed across hundreds of Caribbean islands, the lizard genus *Anolis* provides fertile ground for testing hypotheses about nearly every aspect of vertebrate biology.

A. cooki

Manuel Leal, Duke University

A. poncensis

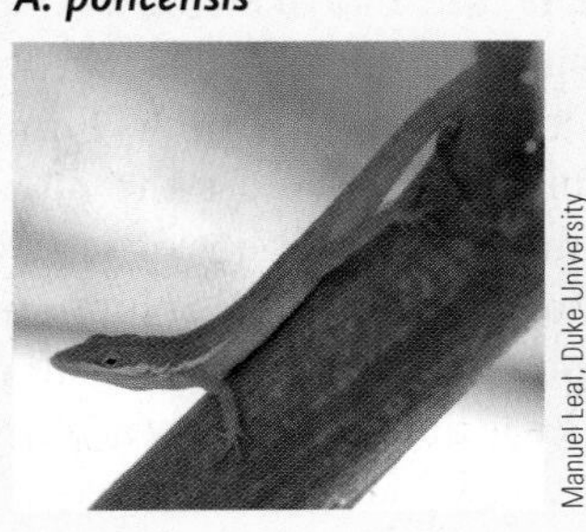

Manuel Leal, Duke University

Sunny lowlands | Shaded uplands

A. cuvieri

A. occultus

Desert habitats

A. stratulus

A. evermanni

A. gundlachi

A. cristatellus

A. cooki

A. pulchellus

A. krugi

A. poncensis

A. gundlachi

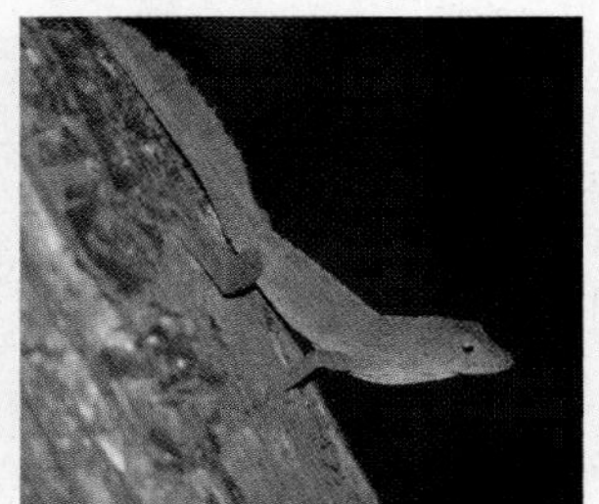

Manuel Leal, Duke University

A. krugi

Manuel Leal, Duke University

The 2,900 species of snakes evolved from a lineage of lizards that lost their legs over evolutionary time (see Section 19.3). Streamlined bodies make snakes efficient burrowers or climbers **(Figure 30.25C)**. Many subterranean species are only 10 or 15 cm long, but the giant constrictors may grow to 10 m. Unlike lizards, all snakes are predators that swallow prey whole. Snakes have thinner skull bones than their lizard ancestors did, and the bones are connected to each other by elastic ligaments that stretch remarkably, allowing some snakes to swallow food that is larger than their heads. Snakes also have well-developed sensory systems for detecting prey. The flicking tongue carries airborne molecules to sensory receptors in the roof of the mouth. Most snakes can detect vibrations on the ground, and some, like rattlesnakes, have heat-sensing organs (see Figure 39.22). Many snakes kill by constriction, which suffocates prey, and several groups produce toxins that immobilize, kill, and partially digest the prey before it is swallowed.

STUDY BREAK 30.9 <

1. **In addition to losing their legs over evolutionary time, how do snakes differ from their lizard ancestors?**
2. **How do the diets of lizards and snakes differ?**

30.10 Living Archosaurs: Crocodilians and Birds

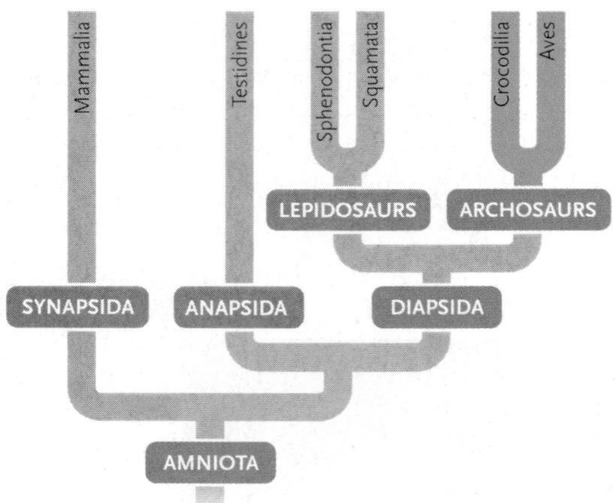
Mammalia
Testidines
Sphenodontia
Squamata
Crocodilia
Aves
LEPIDOSAURS
ARCHOSAURS
SYNAPSIDA
ANAPSIDA
DIAPSIDA
AMNIOTA

Two groups of archosaurs—Crocodilia and Aves—have survived since the Mesozoic era. Although they bear little resemblance to each other in their external morphology, similarities in their heart and respiratory structures as well as molecular sequence data confirm their close relationship.

Crocodilians Are Semiaquatic, Predatory Archosaurs

In addition to birds, the 21 living species of alligators and crocodiles (Crocodilia, from *crocodilus* = crocodile) are the remnants of a once-diverse archosaur lineage **(Figure 30.26)**. The largest species, the Australian saltwater crocodile *(Crocodylus porosus)*, grows to 7 m. Crocodilians are aquatic predators that consume other vertebrates. Striking anatomical adaptations distinguish them from living lepidosaurs, including a four-chambered heart and a one-way flow of air through their lungs, characters that are homologous to those of birds.

Mary Ann McDonald/Corbis

FIGURE 30.26
Living non-feathered archosaurs. Crocodilia includes semiaquatic predators, like this resting African Nile crocodile *(Crocodylus niloticus)*, that frequently bask in the sun.

American alligators *(Alligator mississippiensis)* exhibit strong maternal behavior, which also reflects their relationship to birds. Females guard their nests ferociously and help their young break out of their eggshells. The young stay close to the mother for about a year, feeding on scraps that fall from her mouth and living under her watchful protection. Most alligator and crocodile species are highly endangered. Their habitats have been disrupted by human activities, and they have been hunted for meat and leather. Protection efforts have been extremely successful, however. American alligators, for example, have recovered from the brink of extinction.

Birds Have Key Adaptations That Reduce Body Weight and Power Flight

Birds (Aves) appeared in the Jurassic period as descendants of carnivorous, bipedal dinosaurs. Thus, they are full-fledged members of the archosaur lineage. Their evolutionary relationship to dinosaurs is evident in their skeletal anatomy, the scales on their legs and feet, and their posture when walking. However, powered flight gave birds access to new adaptive zones, setting the stage for their astounding evolutionary success **(Figure 30.27A)**.

The skeletons of birds are both lightweight and strong **(Figure 30.27B)**. For example, the endoskeleton of the frigate bird, which has a 1.5 m wingspan, weighs little more than 100 g, far less than its feathers. Most birds have hollow limb bones with small supporting struts that crisscross the internal cavities. Evolution has also reduced the number of separate bony elements in the wings, skull, and vertebral column (especially the tail), making the skeleton light and rigid. And all modern birds lack teeth, which are dense and heavy; they acquire food with a lightweight, keratinized bill. Many species have a long, flexible

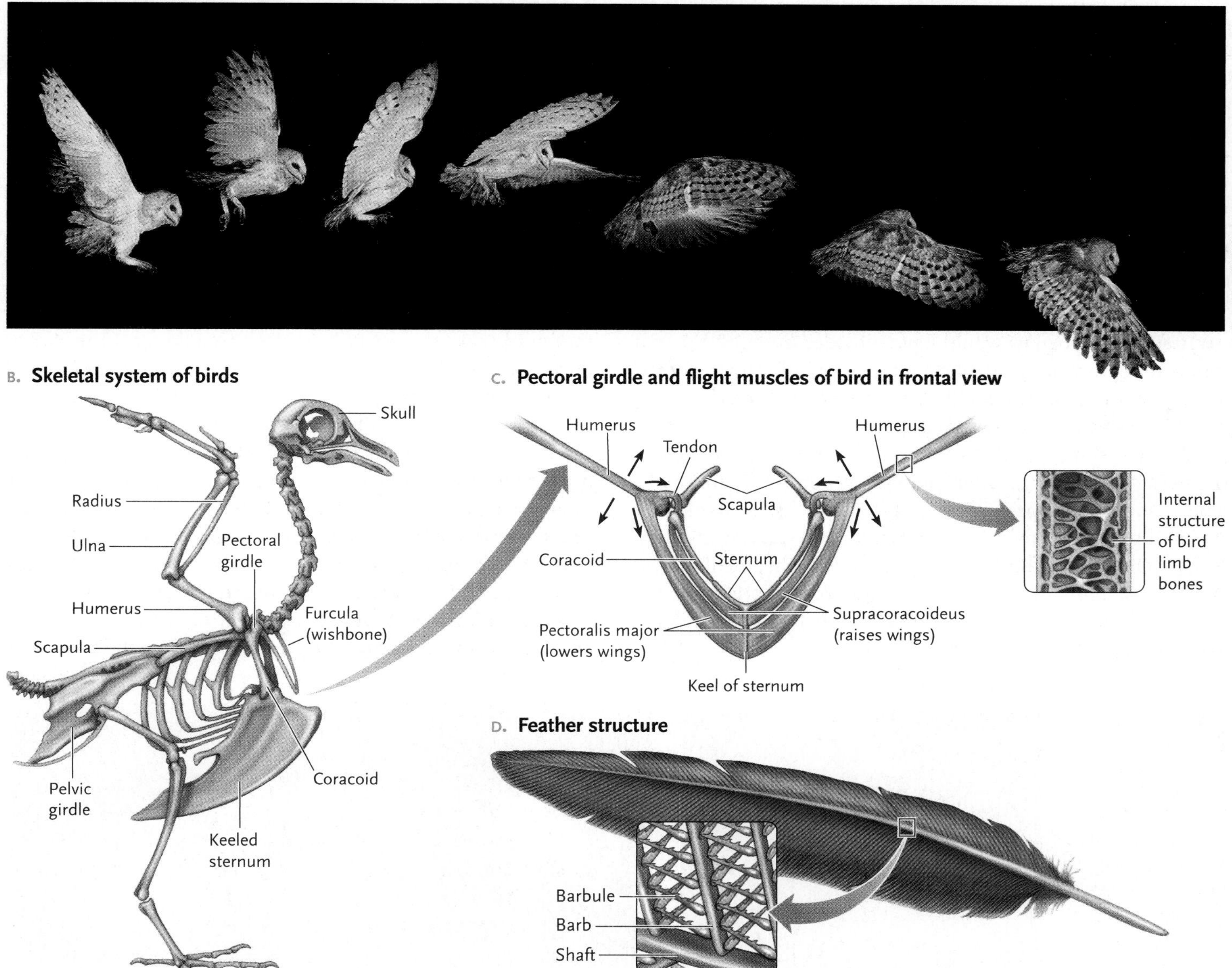
A. Wing movements of an owl during flight
Gerard Lacz/ANT Photolibrary
B. Skeletal system of birds
Skull
Radius
Ulna
Pectoral girdle
Humerus
Furcula (wishbone)
Scapula
Pelvic girdle
Coracoid
Keeled sternum
C. Pectoral girdle and flight muscles of bird in frontal view
Humerus
Tendon
Humerus
Scapula
Coracoid
Sternum
Pectoralis major (lowers wings)
Supracoracoideus (raises wings)
Keel of sternum
Internal structure of bird limb bones
D. Feather structure
Barbule
Barb
Shaft

FIGURE 30.27
Adaptations for flight in birds. **(A)** The flapping movements of a bird's wing provide *thrust* for forward movement and *lift* to counteract gravity. **(B)** The bird skeleton includes a boxlike trunk, short tail, long neck, lightweight skull and beak, and well-developed limbs. In large birds, limb bones are hollow. **(C)** Two sets of flight muscles attach to a keeled sternum; one set raises the wings, and the other lowers it. **(D)** Flexible feathers form an airfoil on the wing surface.

neck, which allows them to use their bills for feeding, grooming, nest-building, and social interactions.

The bones associated with flight are generally large. The forelimb and forefoot are elongate, forming the structural support for the wing. And most modern birds possess a **keeled sternum** (breastbone) to which massive flight muscles are attached **(Figure 30.27C).** Not all birds are strong fliers, however; ostriches and other bipedal runners have strong, muscular legs but small wings and flight muscles (see Figure 19.2).

Like the skeleton, soft internal organs are modified in ways that reduce weight. Most birds lack a urinary bladder; uric acid paste is eliminated with digestive wastes. Females have only one ovary and never carry more than one mature egg; eggs are laid as soon as they are shelled.

All birds also possess **feathers (Figure 30.27D),** sturdy, lightweight structures derived from scales in the skin of their ancestors. Each feather has numerous barbs and barbules with tiny hooks and grooves that maintain the feathers' structure, even during vigorous activity. Flight feathers on the wings provide lift, contour feathers streamline the surface of the body, and down feathers form an insulating cover close to the skin. Worn feathers are molted and replaced once or twice each year.

Other adaptations for flight allow birds to harness the energy needed to power their flight muscles. Their metabolic rates are eight to ten times higher than those of other comparably sized diapsids, and they process energy-rich food rapidly. A complex and efficient respiratory system (see Figure 44.7) and four-chambered heart (see Figure 42.5D) enable them to consume and distribute oxygen efficiently. As a consequence of high rates of metabolic heat production, birds maintain a high and constant body temperature (see Section 46.8).

Flying Birds Were Abundant by the Cretaceous Period

Although the earliest known bird, the pigeon-sized *Archaeopteryx,* had feathers, its skeleton was essentially that of a small dinosaur (see Figure 19.13). It had digits and claws on the forelimbs, teeth on its jaws, many bones in its wings and vertebral column, and only a poorly developed sternum. How could flight evolve in so unbirdlike an animal? Some biologists hypothesize that *Archaeopteryx* ran after prey, using its feathered wings like fly swatters. Larger wings would have provided extra lift when it jumped at prey, and gradual evolutionary modifications of the wing bones and muscles could have led to powered flight.

Crow-sized birds with full flight capability appeared by the early Cretaceous period. They had a keeled sternum and other modern skeletal features. The modern groups of wading birds and seabirds first appear in late Cretaceous rocks; fossils of other modern groups are found in slightly later deposits. Woodpeckers, perching birds, birds of prey, pigeons, swifts, the flightless ratites, penguins, and some other groups were all present by the end of the Oligocene; birds continued to diversify through the Miocene (see Table 22.1).

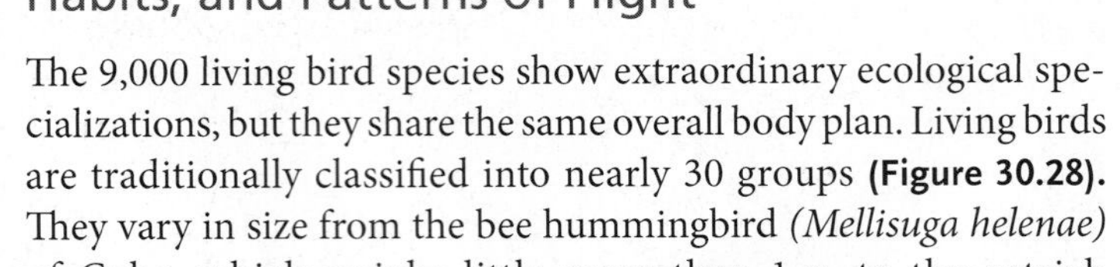

Modern Birds Vary in Morphology, Diet, Habits, and Patterns of Flight

The 9,000 living bird species show extraordinary ecological specializations, but they share the same overall body plan. Living birds are traditionally classified into nearly 30 groups **(Figure 30.28).** They vary in size from the bee hummingbird *(Mellisuga helenae)* of Cuba, which weighs little more than 1 g, to the ostrich *(Struthio camelus),* which can weigh as much as 150 kg.

The structure of the bill usually reflects a bird's diet. Seed and nut eaters, such as finches and parrots, have deep, stout bills that crack hard shells. Carnivorous hawks and carrion-eating vultures have sharp bills to rip flesh, and nectar-feeding hummingbirds have slender bills to reach into flowers. The bills of ducks are modified to extract particulate matter from water, and many perching birds have slender bills to feed on insects.

Birds also differ in the structure of their feet and wings. Predators have large, strong talons (claws), whereas ducks and other swimming birds have webbed feet that serve as paddles. Long-distance fliers like albatrosses have narrow wings; those that hover at flowers, such as hummingbirds, have wide ones. The wings of some species, like penguins, are so specialized for swimming that they are incapable of aerial flight.

All birds have well-developed sensory and nervous systems, and their brains are proportionately larger than those of other

FIGURE 30.28
Bird diversity. Birds, the living feathered archosaurs, have diversified into lineages that are adapted to live in different habitats and feed on different foods.

Kevin Schafer/Corbis

A. The Laysan albatross (Procellariiformes, *Phoebastria immutabilis*) has the long thin wings typical of birds that fly great distances.

Arthur Morris/Visuals Unlimited

B. The roseate spoonbill (Ciconiformes, *Ajaia ajaja*) uses its bill to strain food particles from water.

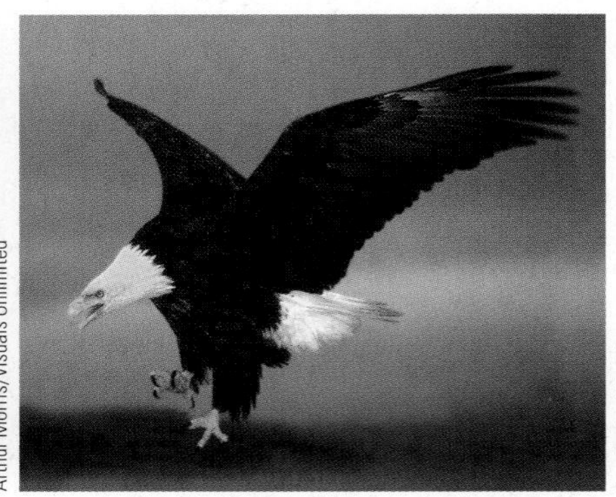

Ron Sanford/Corbis

C. The bald eagle (Falconiformes, *Haliaeetus leucocephalus*) uses its sharp bill and talons to capture and tear apart prey.

WIM KLOMP/FOTO NATURA/Minden Pictures

D. A European nightjar (Caprimulgiformes, *Caprimulgus europaeus*) uses its wide mouth to capture flying insects.

iStockphoto.com/Steve Byland

E. A ruby-throated hummingbird *(Archilochus colubris)* hovers before a flower to drink its nectar.

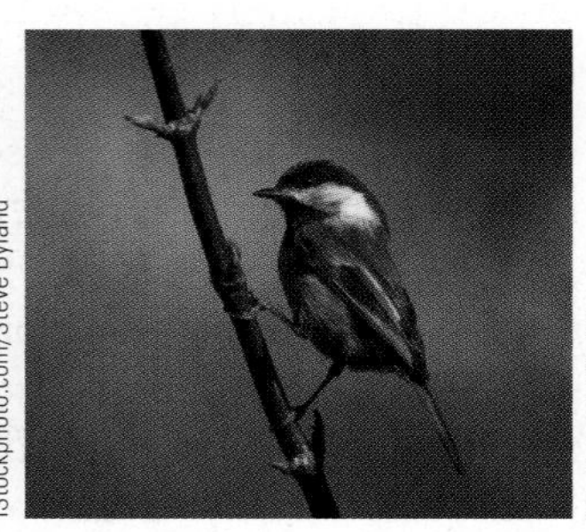

Tim Zurowski/Corbis

F. The chestnut-backed chickadee (Passeriformes, *Parus rufescens*) uses its thin bill to probe for insects in dense vegetation.

diapsids of comparable size. Large eyes provide sharp vision, and most species also have good hearing, which nocturnal hunters like owls use to locate prey. Some vultures and other species have a good sense of smell, which they use to find food. Migrating birds use polarized light, changes in air pressure, and Earth's magnetic field for orientation.

Most birds exhibit complex social behavior, including courtship, territoriality, and parental care. Many species communicate with vocalizations and visual displays to challenge other individuals or attract mates. Most raise their young in a nest, using body heat to incubate eggs. The nest may be a simple depression on a gravelly beach, a cup woven from twigs and grasses, or a feather-lined hole in a tree.

Many bird species embark on a semiannual long-distance migration (see Section 55.1). The golden plover *(Pluvialis dominica),* for example, migrates 20,000 km twice each year. Migrations are a response to seasonal changes in climate. Birds travel toward the tropics as winter approaches; in spring, they return to high latitudes using seasonally available food sources.

STUDY BREAK 30.10

1. **What anatomical and behavioral characteristics of crocodilians demonstrate their relatively close relationship to birds?**
2. **What specific adaptations allow birds to fly?**
3. **How do the structures of a bird's bill, wings, and feet reflect its dietary and habitat specializations?**

30.11 Mammalia: Monotremes, Marsupials, and Placentals

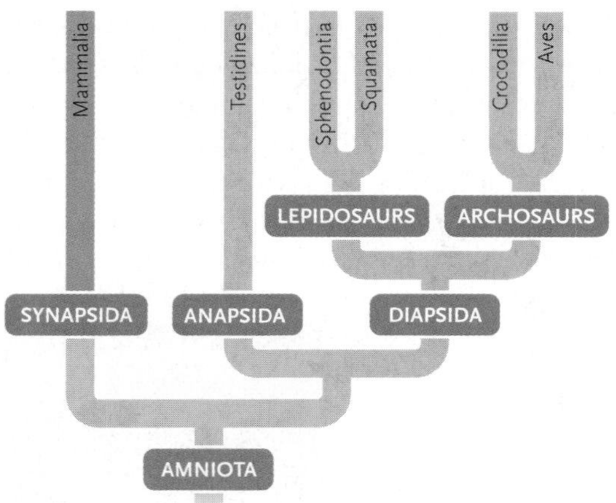
Mammalia
Testidines
Sphenodontia
Squamata
Crocodilia
Aves
LEPIDOSAURS
ARCHOSAURS
SYNAPSIDA
ANAPSIDA
DIAPSIDA
AMNIOTA

The synapsid lineage, which includes the living mammals, was the first group of amniotes to diversify broadly on land. Indeed, during the late Paleozoic era, medium- to large-sized synapsids were the most abundant predators in terrestrial habitats. One particularly successful and persistent branch of the synapsid lineage, the *therapsids,* exhibited many mammal-like characteristics in their legs, skulls, jaws, and teeth. And by the end of the Triassic period, the earliest mammals—most of them no bigger than a rat—had emerged from therapsid ancestors. Several mammalian lineages coexisted with dinosaurs and other diapsids throughout the Mesozoic, but paleontologists hypothesize that most Mesozoic mammals were active only at night to avoid predatory dinosaurs, which were active during the day. Two mammalian lineages, the egg-laying Prototheria (or monotremes) and the live-bearing Theria (marsupials and placentals), survived the mass extinction that eliminated most dinosaurs at the end of the Mesozoic. The Theria diversified into the mammalian groups that are most familiar today.

Mammals Exhibit Key Adaptations in Anatomy, Physiology, and Behavior

Four sets of key adaptations fostered the success of mammals.

HIGH METABOLIC RATE AND BODY TEMPERATURE Like birds, mammals have high metabolic rates that release enough energy from food to maintain high activity levels and enough heat to maintain high body temperatures (see Section 46.8). An outer covering of fur and a layer of subcutaneous fat help retain body heat. Using metabolic heat to stay warm requires lots of oxygen, and mammals have a muscular organ, the diaphragm, that fills their lungs with air (see Figure 44.8). Four-chambered hearts and complex circulatory systems deliver oxygen to active tissues (see Figure 42.5D).

SPECIALIZATIONS OF THE TEETH AND JAWS Mammals also have anatomical features that allow them to feed efficiently. Ancestrally, mammals have four types of teeth (see Figure 45.17): flattened **incisors** nip and cut food; pointed **canines** pierce and kill prey; and two sets of cheek teeth, **premolars** and **molars,** grind and crush food. Moreover, teeth in the upper and lower jaws occlude (that is, fit together) tightly as the mouth is closed; thus, mammals can use their large jaw muscles to chew food thoroughly. The occlusion of premolars and molars allows some mammals to feed on exceptionally tough plant material, such as grasses and twigs.

PARENTAL CARE Mammals provide substantial parental care to their young. In most species, young complete development within a female's uterus, deriving nourishment through the **placenta,** a specialized organ that mediates the delivery of oxygen and nutrients (see Section 47.2). Females also have **mammary glands,** specialized structures that produce energy-rich milk, a watery mixture of fats, sugars, proteins, vitamins, and minerals. This perfectly balanced diet is the sole source of nutrients for newborn offspring.

COMPLEX BRAINS Finally, mammals have larger brains than other tetrapods of equivalent body size; the difference lies primarily in the **cerebral cortex,** the part of the forebrain responsible for information processing and learning (see Figure 38.6). Extensive postnatal care provides opportunities for offspring to learn from older individuals. Thus, mammalian behavior is strongly influenced by past experience as well as by genetically programmed instincts.

The Major Groups of Modern Mammals Differ in Their Reproductive Adaptations

Biologists recognize a primary distinction between two lineages of modern mammals: the egg-laying Prototheria (*protos* = first, *therion* = wild beast), or monotremes, and the live-bearing Theria. The Theria, in turn, diversified into two sublineages: the Metatheria (*meta* = between), or marsupials, and the Eutheria (*eu* = true), or placentals, which also differ in their reproductive adaptations.

MONOTREMES The three living species of monotremes (Prototheria), which are limited to the Australian region, reproduce with a leathery-shelled egg **(Figure 30.29)**. Newly hatched young lap up milk secreted by modified sweat glands on the mother's belly. The two species of echidnas, or spiny anteaters (*Tachyglossus aculeatus* and *Zaglossus bruijni*), feed on ants or termites. The duck-billed platypus *(Ornithorhynchus anatinus)* lives in burrows along riverbanks and feeds on aquatic invertebrates.

A. **Short-nosed echidna *(Tachyglossus aculeatus)***

D. & V. Blagden/ANT Photo Library

B. **Duck-billed platypus *(Ornithorhynchus anatinus)***

Jean Phillippe Varin/Jacana/Photo Researchers, Inc.

FIGURE 30.29
Monotremes. **(A)** The short-nosed echidna is terrestrial. **(B)** The duck-billed platypus raises its young in a streamside burrow.

MARSUPIALS The 240 species of marsupials (Metatheria) have short gestation; the young are nourished through a placenta very briefly—sometimes only for 8 to 10 days—before birth. Newborns use their forelimbs to drag themselves across the mother's belly fur and enter her abdominal pouch, the **marsupium,** where they complete development attached to a teat. Marsupials are the dominant native mammals of Australia and a minor component of the South American fauna **(Figure 30.30)**; only one marsupial species, the opossum *(Didelphis virginiana)*, occurs in North America.

PLACENTALS The 4,000 species of placental mammals (Eutheria) are the dominant mammals today. They complete embryonic development in the mother's uterus, nourished through a placenta until they reach a fairly advanced stage of development. Some species, like humans, are helpless at birth, but others, such as horses, are quickly mobile.

Biologists divide eutherians into about 18 groups, traditionally identified as orders, only eight of which include more than 50 living species **(Figure 30.31)**. Rodents (Rodentia) make up about 45% of eutherian species, and bats (Chiroptera) comprise another 22%. Our own group, Primates, is represented by fewer than 250 living species (less than 5% of all mammalian species), many of which are highly endangered. Analyses of molecular sequence data have allowed researchers to develop new insights into the details of mammalian relationships, as illustrated by one example in *Insights from the Molecular Revolution.*

Some eutherians have highly specialized locomotor structures. For example, whales and dolphins (Cetacea) are descended from terrestrial ancestors, but their appendages do not function on land; they are now restricted to aquatic habitats. By contrast, seals and walruses (Carnivora) feed underwater but rest and breed on land. Bats (Chiroptera) use wings for powered flight.

Although early mammals appear to have been insectivorous, the diets of modern eutherians are diverse. Odd-toed un-

Milse, T./Arco Images/Peter Arnold

FIGURE 30.30
Marsupials. An Eastern gray kangaroo *(Macropus giganteus)* carries her "joey" in her pouch.

A. Capybaras (Rodentia, *Hydrochoerus hydrochaeris*), the largest rodents, feed on vegetation in South American wetlands.

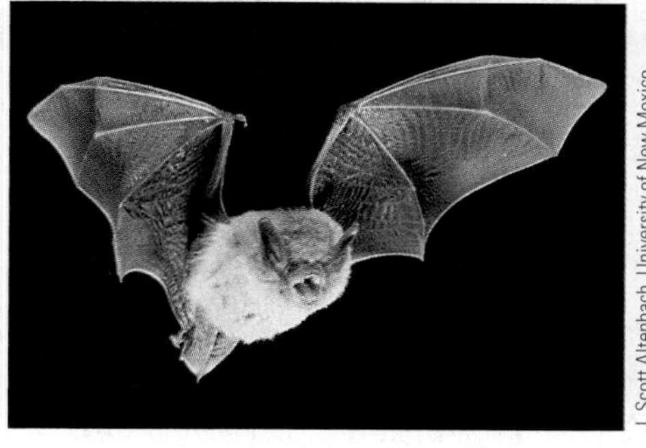

Theo Allofs/Photonica/Getty Images Inc.

B. Most bats, like the Yuma Myotis (Chiroptera, *Myotis yumanensis*), are nocturnal predators on insects.

J. Scott Altenbach, University of New Mexico

Leonard Lee Rue III

C. Walruses (Carnivora, *Obodenus rosmarus*) feed primarily on marine invertebrates in frigid arctic waters.

D. Narwhals (Cetacea, *Monodon monoceros*), which have one incisor that grows into a spiral tusk, may be a source of unicorn legends.

FIGURE 30.31

Eutherian diversity. Among the traditionally defined orders of mammals, only eight include more than 50 living species. Seven of these orders are pictured here. The eighth order, Primates, is described later in this chapter.

Paul Nicklen/Getty Images

E. Arabian camels (Artiodactyla, *Camelus dromedarius*) use enlarged foot pads to cross hot desert sands.

David Parker/SPL/Photo Researchers, Inc.

iStockphoto.com/DAMIAN KUZDAK

F. European hares (Lagomorpha, *Lepus europaeus*) are herbivores that feed in open fields and grasslands.

© David Cole/Alamy

G. Common moles (Lipotyphla, *Talpa europaea*) feed largely on insects that they find underground.

gulates (*ungula* = hoof) like horses and rhinoceroses (Perissodactyla), even-toed ungulates like cows and camels (Artiodactyla), and rabbits and hares (Lagomorpha) feed on vegetation. Lions and wolves (Carnivora) consume other animals. Many hedgehogs, moles, and shrews (Lipotyphla) and bats eat insects, but some feed on flowers, fruit, and nectar. Many whales and dolphins prey on fishes and other animals, but some eat plankton. And some groups, including rodents and primates, feed opportunistically on both plant and animal matter.

STUDY BREAK 30.11 <

1. **During the Mesozoic era, why were most mammals active only at night?**
2. **Which key adaptations in mammals allow them to be active under many types of environmental conditions?**
3. **On what basis are the three major groups of living mammals distinguished?**

30.12 Nonhuman Primates

We now focus our attention on Primates, the mammalian lineage that includes humans, apes, monkeys, and their close relatives. The first Primates appeared early in the Eocene epoch, about 55 million years ago, in forested habitats in North America, Europe, Asia, and North Africa.

The Guinea Pig Is Not a Rat

In the Linnaean classification of mammals, based on morphological characters, Rodentia was the largest mammalian order. It included more than 1,800 species of squirrels, rats, mice, porcupines, capybaras—and guinea pigs. The placement of these animals in one order implies that Rodentia is a monophyletic lineage descended from a common ancestor. Biologists accepted this classification until 1991, when researchers compared the amino acid sequences of 15 proteins encoded in the nuclear DNA of various "rodent" species. The comparisons revealed differences suggesting that guinea pigs are not rodents. Subsequent studies of nuclear genes produced contradictory results.

Research Question

Are guinea pigs part of the rodent lineage?

Experiment

A cooperative study by Anna Maria D'Erchia and colleagues at universities and institutes in Italy and Sweden used mitochondrial DNA (mtDNA) sequences to answer this question. They used mtDNA sequences because they are easy to isolate and purify, and typically undergo many random mutations and rearrangements that have no apparent effect on gene function. Thus the changes observed in mtDNA are expected to reflect more faithfully the ticking of the molecular clock that tracks the time course of evolutionary events.

The researchers sequenced mtDNA of guinea pigs and rabbits (order Lagomorpha), mammals that are thought to be closely related to rodents. Their analysis also included data on the mtDNA, previously sequenced by other workers, of 14 other mammals in eight monophyletic lineages: primates (Primates), seals (Carnivora), cows (Artiodactyla), whales (Cetacea), horses (Perissodactyla), mice and rats (Rodentia), hedgehogs (Insectivora, now called Lipotyphla), and opossums (Marsupialia).

Results

D'Erchia and her colleagues evaluated these sequences with three different statistical programs that use similarities and differences in mtDNA sequences to construct phylogenetic trees. They also conducted analyses using nuclear DNA, including separate evaluations of the entire nuclear genome, the protein-encoding sequences, and the DNA encoding ribosomal RNA. Significantly, all the methods produced essentially the same family tree (shown in the **figure**).

The phylogenetic tree indicates that guinea pigs are not rodents, and it places them in a group of their own. Furthermore, the guinea pig group shared a more recent common ancestor with all of the other mammalian orders examined than it did with the lineages that include rodents, hedgehogs, and opossums.

Conclusion

Thus, guinea pigs merit placement in a separate group from rodents. The results for rabbits also indicate that they are more closely related to other mammals than they are to mice and rats. Incidentally, the tree also supports the conclusion from other molecular studies that cows and whales are more closely related to each other than to other mammals (see *Insights from the Molecular Revolution* in Chapter 23).

Source: A. M. D'Erchia et al. 1996. The guinea-pig is not a rodent. *Nature* 381:597–600.

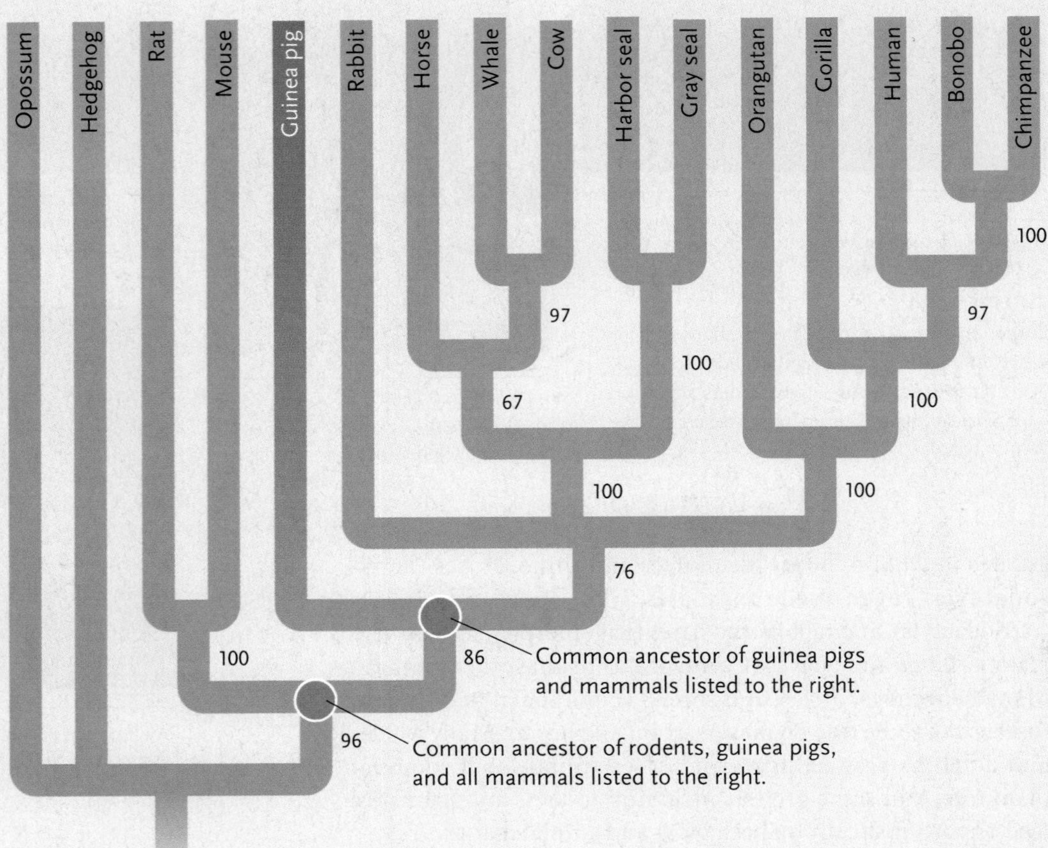
Opossum
Hedgehog
Rat
Mouse
Guinea pig
Rabbit
Horse
Whale
Cow
Harbor seal
Gray seal
Orangutan
Gorilla
Human
Bonobo
Chimpanzee
100
97
97
100
100
67
100
100
100
76
100
86
96
Common ancestor of guinea pigs and mammals listed to the right.
Common ancestor of rodents, guinea pigs, and all mammals listed to the right.

FIGURE 1
Molecular phylogeny for guinea pigs and other mammals. The numbers next to the branches in the tree indicate the probability that the branch is correct, on a scale of 0 to 100.

Key Derived Traits Enabled Primates to Become Arboreal, Diurnal, and Highly Social

Several derived traits allow primates to be arboreal (to live in trees rather than on the ground). For example, most primates have a more erect posture than other mammals, and they have flexible hip and shoulder joints, which allow a variety of locomotor activities. They can grasp objects with their hands and feet, because they have nails, not claws, on their fingers and toes; their fingertips are well endowed with sensory nerves that enhance the sense of touch. Unlike other mammals, most primates have an opposable big toe, which can touch the tips of other digits and the sole of the foot; many species also have an opposable thumb.

Most primates are diurnal (active during daylight hours), and, unlike most mammals, they rely more on vision than on their sense of smell. Thus, they generally have short snouts and small olfactory lobes of the brain. Most species have forward-facing eyes with overlapping fields of vision, providing excellent depth perception, which comes in handy when moving through trees. Many species have color vision.

Primate brains—especially the regions that integrate information—are large and complex. As a result, they have an exceptional capacity to learn. Most species live in social groups; thus, young primates, which mature slowly, can interact with and learn from their elders and peers during an extended period of parental care. Females give birth to only one or two young at a time, allowing them to devote substantial attention to each offspring.

Living Primates Include Two Major Lineages

Primatologists recognize two lineages within the Primates **(Figure 30.32)**, the Strepsirhini (*streptos* = twisted or turned, *rhinos* = nose) and the Haplorhini (*haplos* = single or simple).

STREPSIRHINI The 36 living species of Strepsirhini—lemurs, lorises, and galagos—possess many ancestral morphological traits, including moist, fleshy noses and eyes that are positioned somewhat laterally on their heads **(Figure 30.33).** Strepsirhines generally have short gestation periods and rapid maturation. Today, 22 lemur species survive on Madagascar, a large island off the east coast of Africa; they are ecologically diverse and range in size from 40 g to 7 kg. Some lemurs are arboreal, whereas others spend substantial time on the ground. The 12 species of lorises and galagos occupy tropical forest and subtropical woodlands in Africa, India, and Southeast Asia; they are all arboreal and nocturnal.

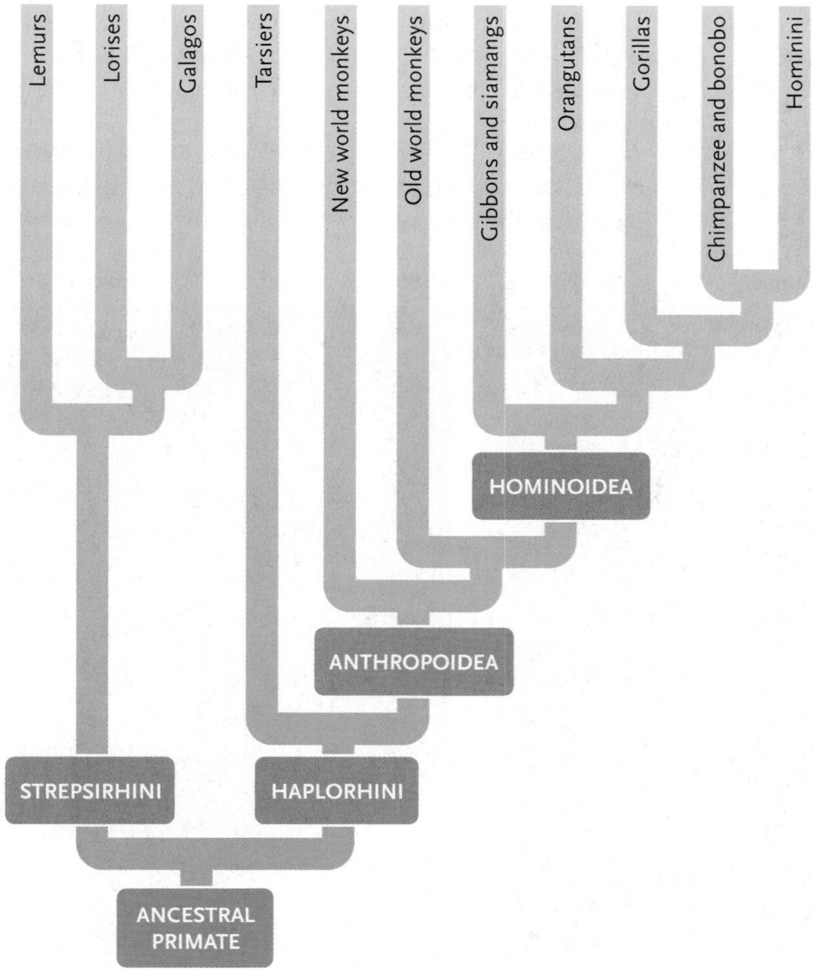

Lemurs
Lorises
Galagos
Tarsiers
New world monkeys
Old world monkeys
Gibbons and siamangs
Orangutans
Gorillas
Chimpanzee and bonobo
Hominini
HOMINOIDEA
ANTHROPOIDEA
STREPSIRHINI
HAPLORHINI
ANCESTRAL PRIMATE

FIGURE 30.32
Primate phylogeny. A phylogenetic tree for the Primates illustrates the two main lineages: Strepsirhini and Haplorhini. Note that chimpanzees and bonobos are the closest living relatives of humans.

Cagan Sekercioglu/Visuals Unlimited

FIGURE 30.33
Strepsirhines. The ring-tailed lemur *(Lemur catta)* of Madagascar has ancestral primate characteristics, such as a long snout and wet nose.

FIGURE 30.34
Tarsiers. Tarsiers *(Tarsius syrichta)* are classified as haplorhines, but they retain many ancestral characteristics.

iStockphoto.com/Holger Mette

HAPLORHINI Most species in the Haplorhini—the familiar monkeys and apes—have many derived primate characteristics, including compact, dry noses, and forward-facing eyes.

However, five species of tarsiers, which are restricted to tropical forests on the islands of Southeast Asia, exhibit several ancestral traits: small body size (about 100 g), large eyes and ears, and two grooming claws on each foot **(Figure 30.34).** But they share the derived characteristics of dry noses and forward-facing eyes with the other haplorhines; DNA sequence data link them to the monkeys and apes and not to the strepsirhines described earlier.

The 130 or so species of monkeys, 13 species of apes, and humans constitute the monophyletic haplorhine lineage **Anthropoidea,** which probably arose in Africa. Fossils of a diverse and abundant radiation of forest-dwelling anthropoids, dating from the late Eocene epoch, have been discovered in northern Egypt. Continental drift then established long-term geographical and evolutionary separation of anthropoids in the New World and Old World **(Figure 30.35).**

By the middle of the Oligocene epoch, about 30 million years ago, the ancestors of the New World monkeys had arrived in South America and begun to diversify there. They probably rafted across the Atlantic, which was narrower at that time, on trees or other storm debris. New World monkeys now live in Central and South America (see Figure 30.35A). They range in size from tiny marmosets and tamarins (350 g), to hefty howler monkeys (10 kg). Most are exclusively arboreal and diurnal. The larger species may hang below branches by their arms, and some use a prehensile (grasping) tail as a fifth limb.

Anthropoids diversified most spectacularly in the Old World, however, eventually giving rise to two lineages—one ancestral to Old World monkeys and the other to apes and humans. Although many people assume that the apes are descended from Old World monkeys, the fossil record contradicts that impression. The earliest hominoid (ape) fossils date to the early Miocene, roughly 23 million years ago, but the oldest known Old World monkeys appeared several million years later.

Old World monkeys, which occupy habitats ranging from tropical rain forests to deserts in Africa and Asia, may grow as large as 35 kg (see Figure 30.35B). Many species are sexually dimorphic; in other words, males and females attain different adult sizes (see Section 20.3). Arboreal species use all four limbs for locomotion, but none has a prehensile tail. Some species, such as baboons, often walk or run on the ground.

Within the anthropoid lineage, the **Hominoidea** ("human-like") is a monophyletic group that includes apes and humans. The climate of the early Miocene was wetter than it is today, and eastern Africa, where many early hominoid fossils are found, was covered with extensive forests. A shift to a cooler and drier climate in the middle Miocene, between 17 and 14 million years ago, converted dense forests into more open woodlands. Hominoids probably adopted a more terrestrial existence and shifted their diets to include leaves and hard foods. Miocene hominoids ranged in size from 4 kg to 80 kg. They occupied both forest and open woodland habitats; some may have been at least partially ground dwelling.

Although hominoids are closely related to Old World monkeys, several characteristics distinguish them. Apes lack a tail, and great apes (orangutans, gorillas, chimpanzees, and bono-

A. **Spider monkey *(Ateles geoffroyi)***

Frans Lanting/Corbis

B. **Hamadryas baboon *(Papio hamadryas)***

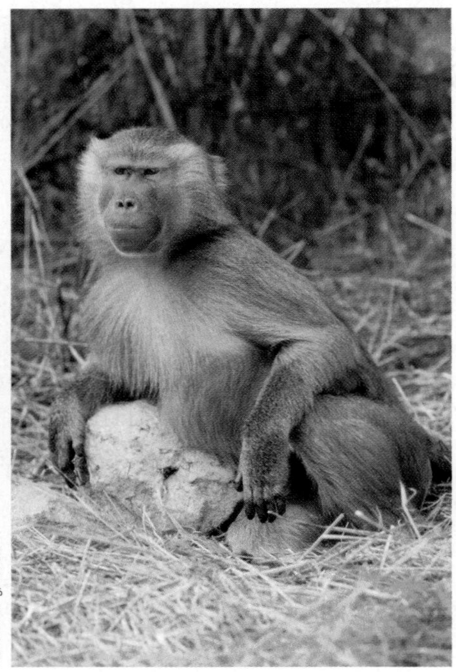

Joseph Van Os/Getty Images Inc.

FIGURE 30.35
New World and Old World monkeys. (A) Many New World monkeys, such as the spider monkey, have prehensile tails, which they use as a fifth limb. Old World monkeys lack prehensile tails, and many, such as **(B)** the Hamadryas baboon, are largely terrestrial.

bos) are much larger than monkeys. Moreover, the posterior region of the vertebral column is shorter and more stable in apes. Apes also show more complex behavior.

The gibbons and siamangs, which live in tropical forests in Southeast Asia, are the smallest of the apes, ranging in weight from 6 to 11 kg. With extremely long arms and strong shoulders, they hang below branches by their arms and swing themselves forward, a pattern of locomotion called **brachiation (Figure 30.36A).** The much larger orangutan *(Pongo pygmaeus),* now restricted to forested areas on the islands of Borneo and Sumatra, can grow to 90 kg. Orangutans use both hands and feet to climb trees; they sometimes venture onto the ground on all fours.

Gorillas *(Gorilla gorilla),* which are currently restricted to two large central African forests, are the largest of the living primates. Males can weigh 180 kg; females are about half that size. Because of their size, gorillas spend most of their time on the ground. They often use "knuckle-walking" locomotion, leaning forward and supporting part of their weight on the backs of their hands. Gorillas are almost exclusively vegetarian.

Chimpanzees *(Pan troglodytes)* are also forest dwellers, weighing up to 45 kg **(Figure 30.36B).** Like gorillas, they spend most of their waking hours on the ground; they often knuckle-walk, but sometimes adopt a **bipedal** (two-legged) stance and swagger short distances. Groups of related males form loosely defined communities of up to 50 individuals, which may cooperate in hunts and foraging. Bonobos *(Pan paniscus),* sometimes called pygmy chimpanzees, are restricted to a small area in central Africa. Somewhat smaller than chimps, they have longer legs and smaller heads.

The Primates also includes humans *(Homo sapiens),* which occupy virtually all terrestrial habitats. Humans have adaptations that allow an upright posture and bipedal locomotion. They are ground-dwelling animals with extremely broad diets and complex social behavior.

A. **Black-handed gibbon *(Hylobates agilis)***

Art Wolfe/Photo Researchers, Inc.

B. **Chimpanzee *(Pan troglodytes)***

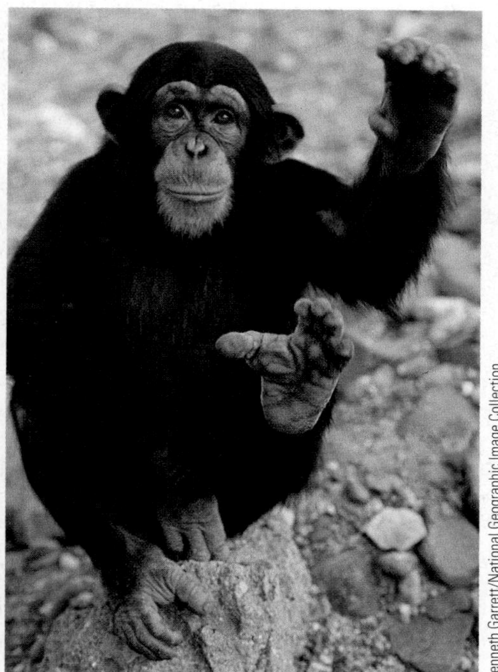

Kenneth Garrett/National Geographic Image Collection

FIGURE 30.36

Apes. **(A)** Small-bodied apes, such as the black-handed gibbon are agile brachiators that swing through the trees with ease. **(B)** Among the large-bodied apes, chimpanzees have opposable thumbs and big toes.

STUDY BREAK 30.12

1. **What characteristics of primates allow them to spend a great deal of time in trees?**
2. **What is the lowest taxonomic group that includes monkeys, apes, and humans? What is the lowest taxonomic group that includes only apes and humans?**
3. **Which ape species spend the most time on the ground?**

30.13 The Evolution of Humans

Genetic analyses of living hominoid species indicate that African hominoids diverged into several lineages between 10 million and 5 million years ago; one lineage, the **hominins,** includes modern humans and our bipedal ancestors. (Until recently, biologists used the term "hominid" to describe humans and their bipedal ancestors, but that term now defines the group that includes humans, bipedal human ancestors, and the great apes.)

Hominins First Walked Upright in East Africa about 6 Million Years Ago

Upright posture and bipedal locomotion are key adaptations that distinguish hominins from apes. Researchers infer these capabilities in early hominin fossils from the anatomy of the skull, spine, pelvis, knees, ankles, and feet **(Figure 30.37).** As a consequence of bipedal locomotion, the hands were no longer used for locomotor functions, allowing them to become specialized for other activities, such as tool use. Hominins also eventually developed larger brains.

Since the mid-1990s, paleontologists have uncovered an astounding array of fossils of hominin species that lived in East Africa and South Africa between 7 million and 1 million years ago **(Figure 30.38).** In 2000, researchers found 13 fossils of *Orrorin tugenensis* ("first man" in a local African language), a species that lived about 6 million years ago in East African forests. Shortly thereafter, researchers described the skull of an even older fossil, *Sahelanthropus tchadensis,* but because the pelvis and legs of this species are still unknown, scientists are unsure about its ability to walk bipedally. Indeed, the fossilized remains of most hominins are fragmentary.

FIGURE 30.37
Adaptations for bipedal locomotion. Differences in the posture, skeleton, and muscles of monkeys, great apes, and humans illustrate the anatomical changes that accompanied upright, bipedal locomotion. Evolutionary changes in the spine, pelvis, hip, knee, ankle, and foot were accompanied by changes in the sizes of leg muscles and their points of attachment to the bones they move.

Monkey
Ape
Human

FIGURE 30.38
Hominin time line. Several species of hominins lived simultaneously at sites in eastern and southern Africa. The time line for each species and genus reflects the ages of known fossils. The question marks indicate uncertainty about the precise timing of the origin and extinction of various hominin species as well as evolutionary relationship between *Australopithecus* and *Homo*. Some of the skulls pictured are reconstructed from fragmentary fossils.

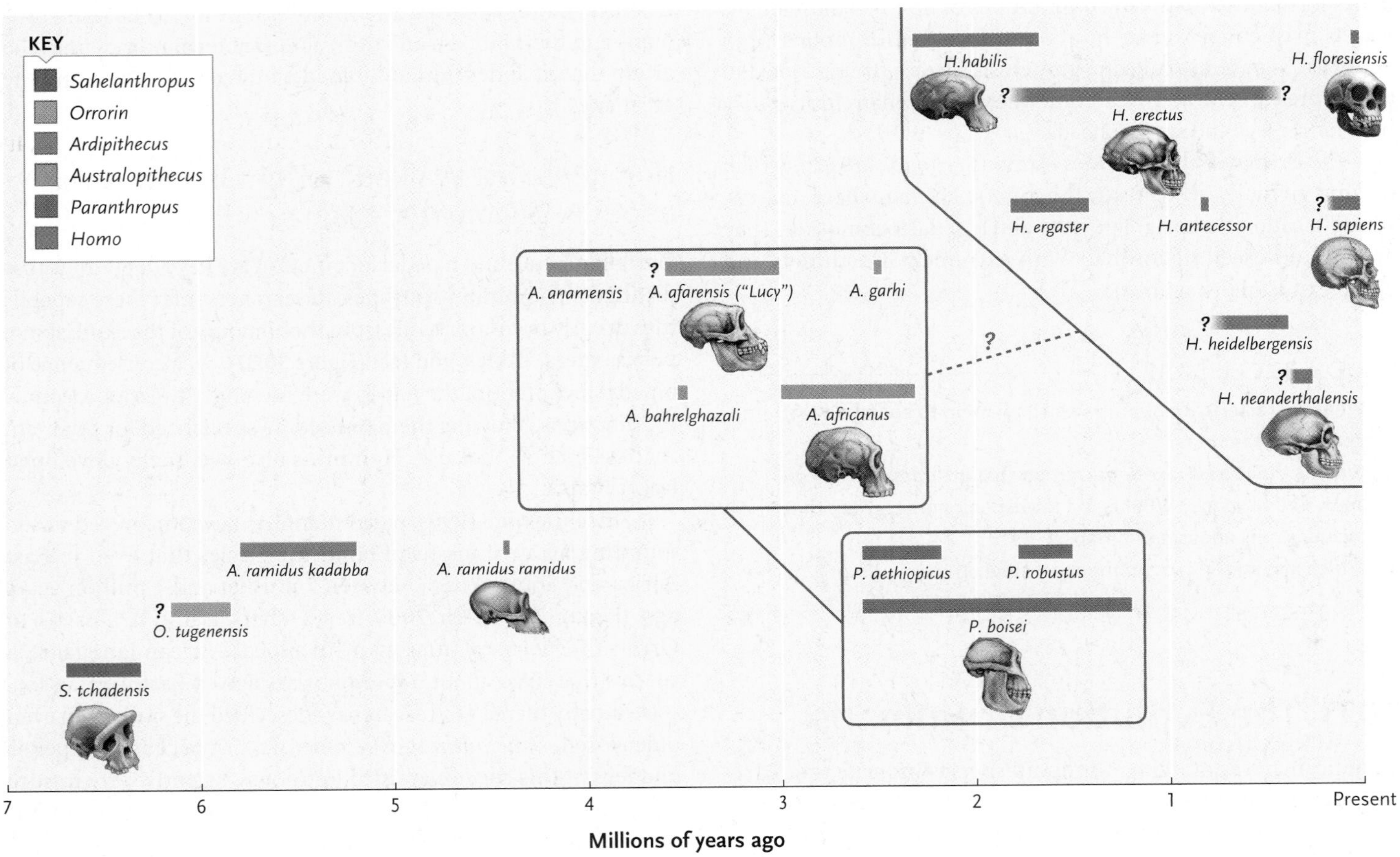

KEY
Sahelanthropus
Orrorin
Ardipithecus
Australopithecus
Paranthropus
Homo

H. habilis
? *H. erectus* ?
H. floresiensis
H. ergaster
H. antecessor
? *H. sapiens*
? *H. heidelbergensis*
? *H. neanderthalensis*
A. anamensis
? *A. afarensis* ("Lucy")
A. garhi
A. bahrelghazali
A. africanus
?
A. ramidus kadabba
A. ramidus ramidus
? *O. tugenensis*
S. tchadensis
P. aethiopicus
P. robustus
P. boisei

7 6 5 4 3 2 1 Present
Millions of years ago

The most spectacular discovery of early hominin fossils dates to 1994, when Tim White of the University of California, Berkeley and two Ethiopian colleagues described the remains of 36 individuals of *Ardipithecus ramidus* discovered in the Awash Valley of central Ethiopia. The animals had lived in a closed canopy woodland about 4.4 million years ago. The fossils were badly deformed and so fragile that they turned to dust in the researchers' hands. After the fossils were carefully excavated and transported to a laboratory at the National Museum of Ethiopia, an international and interdisciplinary team of 47 researchers labored for 15 years to analyze them. They published their findings in a series of 11 papers in *Science* magazine. The work was later hailed by the editors of *Science* as the "breakthrough of the year" in 2009.

Among the fossils that White's team uncovered, the researchers found 125 pieces of one female's skeleton, including bones from her arm and hand, pelvis, leg, ankle, and foot, lower jaw with teeth, and cranium **(Figure 30.39)**. Nicknamed "Ardi," she had the body size and brain size of a modern chimpanzee; stood about 120 cm tall and weighed about 50 kg. Studies of Ardi's anatomy revealed that her face protruded less and her upper canine teeth were shorter and duller than those of modern chimpanzees. The upper parts of her pelvis are shorter and broader than those of living apes, and the anatomy of her hand suggests that she did not knuckle-walk as modern chimps and gorillas do. These observations suggest that she was able to walk bipedally on the ground, although perhaps not as well as later hominins. Nevertheless, Ardi's foot had an opposable big toe, and the bones of her fingers were long and curved; these characters suggest that she spent considerable time climbing trees, perhaps to escape predators, seek food, or build nests for sleeping. White and his team described Ardi as having an anatomy that was intermediate between that of chimps and later groups of hominins. Some researchers have taken issue with some of the interpretations that White's team published, partly because the fossils were in terrible condition when they were found and partly because, at the time of this writing, few other researchers have yet had the opportunity to study the fossils in detail.

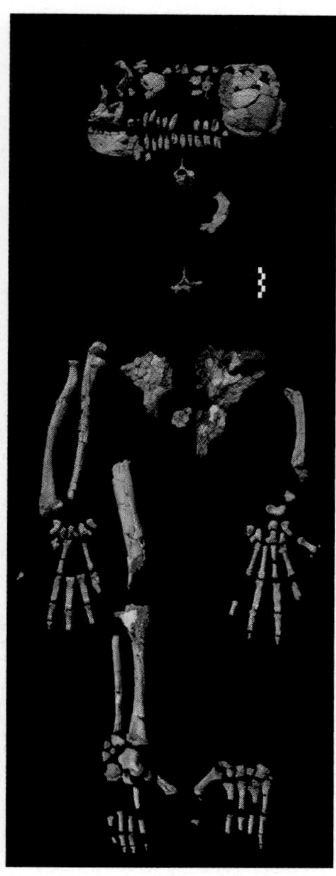

FIGURE 30.39
Ardipithecus. Scientists spent 15 years analyzing and reassembling this fairly complete skeleton of *Ardipithecus ramidus.*

Hominin fossils from 4.2 million to 1.2 million years ago are well known from many sites in East, Central, and South Africa. They are currently assigned to the genera *Australopithecus* (*australis* = southern, *pithekos* = ape) and *Paranthropus* (*para* = beside, and *anthropos* = human being). With their large faces, protruding jaws, and small skulls and brains, most of these hominins had an apelike appearance (see Figure 30.38). *Australopithecus anamensis,* which lived in East Africa around 4 million years ago, is the oldest known species. It had thick enamel covering on its teeth, a derived hominin characteristic; the structure of a fossilized leg bone suggests that it was bipedal.

Specimens of more than 60 individuals of *Australopithecus afarensis* have been found in northern Ethiopia, including about 40% of a female skeleton, named "Lucy" by its discoverers **(Figure 30.40)**. *A. afarensis* lived 3.5 million to 3 million years ago, but it retained several ancestral characteristics. For example, it still had moderately large and pointed canine teeth, and a relatively small brain. Males and females were 150 cm and 120 cm tall, respectively. Skeletal analyses suggest that *A. afarensis* was fully bipedal, a conclusion supported by fossilized footprints preserved in a layer of volcanic ash.

Other species of *Australopithecus* and *Paranthropus* lived in East Africa or South Africa between 3.7 million and 1 million years ago. Adult males ranged from 40 to 50 kg in weight and from 130 to 150 cm in height; females were smaller. Most species had deep jaws and large molars. Several species had a crest of bone along the midline of the skull, providing a large surface for the attachment of jaw muscles. These anatomical features suggest that they fed on hard food, such as nuts, seeds, and other vegetable products. One species, *Australopithecus africanus,* known only from South Africa, had small jaws and teeth, indicating that it probably consumed a softer diet. The phylogenetic relationships of the species classified as *Australopithecus* and *Paranthropus*—and their exact relationships to later hominins—are not yet fully understood. But most scientists agree that *Australopithecus* was ancestral to humans, which are classified in the genus *Homo.*

A. "Lucy" *(Australopithecus afarensis)*

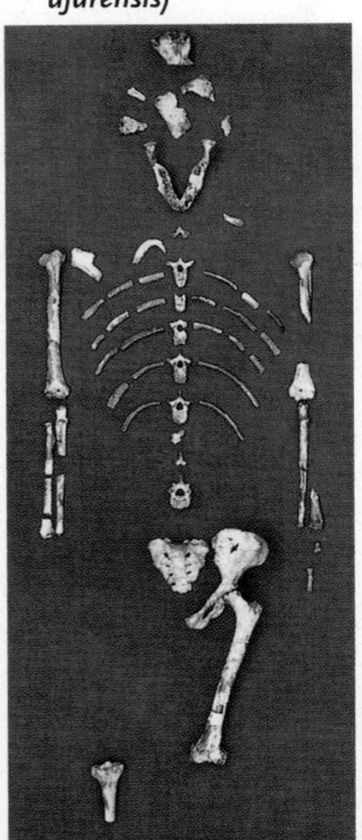

Dr. Donald Johanson, Institute of Human Origins

B. Australopithecine footprints

J. Reader/Photo Researchers, Inc.

FIGURE 30.40
Australopithecines. **(A)** Researchers named the most complete fossil of *Australopithecus afarensis* "Lucy." **(B)** Mary Leakey discovered australopithecine footprints, made in soft, damp volcanic ash about 3.7 million years ago. The footprints indicate that australopithecines were fully bipedal.

Homo habilis Was Probably the First Hominin to Manufacture Stone Tools

Pliocene fossils of the earliest humans, which may have included several species, are fragmentary. They are also widely distributed in space and time, complicating analyses of their relationships. For the sake of simplicity, we describe them as belonging to one species, *Homo habilis* ("handy man").

From 2.3 million to 1.7 million years ago, *H. habilis* occupied the woodlands and savannas of eastern and southern Africa, sharing these habitats with various species of *Paranthropus*. The two genera are easy to tell apart because the brains of *H. habilis* were at least 20% larger, and they had larger incisors and smaller molars than their hominin cousins. Their diet included hard-shelled nuts and seeds as well as soft fruits, tubers, leaves, and insects. They may also have hunted small prey or scavenged carcasses left by large predators.

Researchers have found numerous tools dating to the time of *H. habilis,* but they are not sure which species used them. Many of the hominin species alive at the time probably cracked marrowbones with rocks or scraped flesh from bones with sharp stones. Paleoanthropologist Louis Leakey was the first to discover evidence of tool *making* at East Africa's Olduvai Gorge, which cuts through a great sequence of sedimentary rock layers. The oldest tools at this site are crudely chipped pebbles, which were probably manufactured by *H. habilis*.

Homo erectus Dispersed from Africa to Other Continents

Early in the Pleistocene epoch, about 1.8 million years ago, new species of humans appeared in East Africa. Most fossils are fragmentary. For convenience, we describe them all as *Homo erectus* ("upright man"), recognizing that they probably represent several species. One nearly complete skeleton suggests that *H. erectus* was taller than its ancestors, had a much larger brain, a thicker skull, and protruding brow ridges **(Figure 30.41)**.

H. erectus made fairly sophisticated tools, including the hand axe (see Figure 30.41B), which they apparently used to cut food and other materials, to scrape meat from bones, and to dig for roots. *H. erectus* probably fed on both plants and animals; they may have hunted and scavenged animal prey. Archaeological data point to their use of fire to process food and to keep themselves warm.

The pressure of growing populations apparently forced groups of *H. erectus* to move out of Africa about 1.5 million years ago. They dispersed northward from East Africa into both northwestern Africa and Eurasia. Some moved eastward through Asia as far as the island of Java. Recent discoveries in Spain indicate that *H. erectus* also occupied parts of Western Europe.

Modern Humans Are the Only Surviving Descendants of *Homo erectus*

Judging from its geographical distribution, *Homo erectus* was successful in many environments. Fossils from Africa, Asia, and Europe indicate that it gave rise to several species that first appeared more than 400,000 years ago. These early descendants, described as archaic humans, generally had a larger brain, rounder skull, and smaller molars than *H. erectus.* Like *H. erectus* before them, archaic humans left Africa and established populations in the Middle East, Asia, and Europe.

A. *Homo erectus*

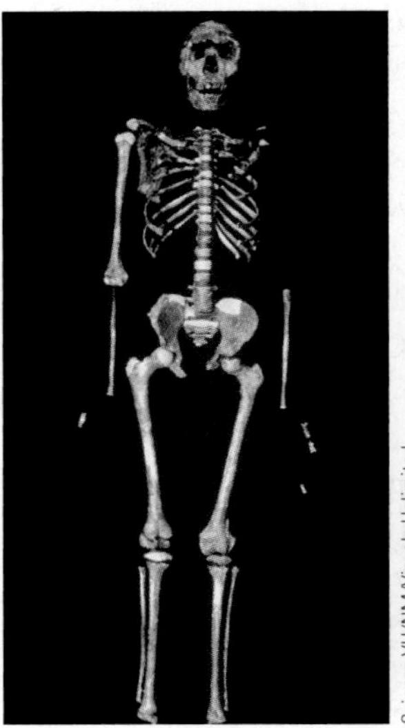

Science VU/NM/Visuals Unlimited

B. Hand axe

AAAC/Topham/The Image Works

FIGURE 30.41
Homo erectus. **(A)** A nearly complete skeleton of *Homo erectus* was discovered in Kenya. **(B)** Hand axes are frequently found at *H. erectus* sites.

Some time later, 200,000 to 100,000 years ago, modern humans, *Homo sapiens* ("wise man"), arose in Africa. These modern humans also migrated into Europe and Asia and may have eventually driven the archaic humans to extinction, perhaps through competition for resources rather than hostile interactions.

ARCHAIC HUMANS The Neanderthals *(Homo neanderthalensis),* who occupied Europe and western Asia from 150,000 to 28,000 years ago, are the best-known archaic humans. Compared with modern humans, they had a heavier build, more-pronounced brow ridges, and slightly larger brains (see Figure 30.38). Neanderthals were culturally and technologically sophisticated. They made complex tools, including wooden spears, stone axes, and flint scrapers and knives. At some sites they built shelters of stones, branches, and animal hides, and they routinely used fire. They were successful hunters and probably consumed nuts, berries, fishes, and bird eggs. Some groups buried their dead, and they may have used rudimentary speech.

Researchers once classified Neanderthals as a subspecies of *H. sapiens,* but most scientists now recognize them as a separate species. In 1997 two teams of researchers, Matthias Kring and Svante Pääbo of the University of Munich and Anne Stone and Mark Stoneking of Pennsylvania State University, independently analyzed short segments of mitochondrial DNA (mtDNA) extracted from the fossilized arm bone of a Neanderthal. Unlike nuclear DNA, which individuals inherit from both parents, only mothers pass mtDNA to offspring. It does not undergo genetic recombination (see Section 13.5), and it has a high mutation rate, making it useful for phylogenetic analyses. Many scientists believe that mutation rates in mtDNA are fairly constant, allowing this molecule to serve as a molecular clock (see Section 23.6). Comparing the Neanderthal sequence with mtDNA from 986 living humans, the researchers discovered three times more differences between the Neanderthals and modern humans than between pairs of modern humans in their sample. These results suggest that Neanderthals and modern humans are different species that diverged from a common ancestor 690,000 to 550,000 years ago—hundreds of thousands of years before modern humans appeared.

In 2003, researchers discovered eight skeletons of a tiny hominin, tentatively described as *Homo floresiensis,* in a cave on the Indonesian island Flores. Remarkably, these individuals stood just 1 m tall and are estimated to have weighed about 25 kg. The researchers who discovered these remains have hypothesized that they represent a new archaic hominin species, a descendant of *H. erectus,* that lived in Southeast Asia until roughly 12,000 years ago! Some skeptics suggested that these individuals represent a population of modern humans who suffered from some genetic disease or condition that caused their small size. But subsequent analyses of their skeletal remains indicate that *H. floresiensis* branched off from *H. erectus* even earlier than *H. neanderthalensis.* Their diminutive stature may have resulted from selection for a reduction in size, a phenomenon observed in other island-dwelling mammals. Indeed, fossils of miniature elephants have been discovered nearby. The unresolved questions about this new archaic human include how and where it first evolved and how it may have interacted with modern humans who lived nearby.

MODERN HUMANS Modern humans *(Homo sapiens)* differ from Neanderthals and other archaic humans in having a slighter build, less-protruding brow ridges, and a more prominent chin. The earliest fossils of modern humans found in Africa and Asia are roughly 150,000 years old; those from the Middle East are 100,000 years old. Fossils from about 20,000 years ago are known from Western Europe, the most famous being those of the Cro-Magnon deposits in southern France. The widespread appearance of modern humans roughly coincided with the demise of Neanderthals in Western Europe and the Middle East 40,000 to 28,000 years ago. Although the two species apparently coexisted in some regions for thousands of years, until recently, we have had little concrete evidence about their interactions.

In May 2010, an international team of more than 50 scientists, led by Svante Pääbo, reported a draft sequence of the Neanderthal genome in the journal *Science.* The researchers collected genetic data from fossils of three Neanderthal females who had lived more than 38,000 years ago in what is now Croatia. They compared the Neanderthal DNA sequences to those of five living humans—one each from southern Africa, West Africa, China, Papua New Guinea, and France. Remarkably, the European and the Asians shared between 1% and 4% of their recently derived (that is, newly evolved) nuclear DNA sequences with Neanderthals, but Africans and Neanderthals shared no derived nuclear DNA sequences at all. These data strongly suggest that Neanderthals interbred with modern humans outside of Africa before modern humans expanded their geographical range into distant regions of Europe and Asia. Although the analysis is far from complete at the time of this writing, the genes that modern Europeans and Asians share with Neanderthals appear to be involved in cognitive development, skeletal development, and metabolism. Thus, there appears to be a little Neanderthal in all of us who trace our ancestry to Europe or Asia.

Many researchers believe that all modern human populations are descended from an ancestral population that originated in Africa. DNA sequence data from modern humans confirm this hypothesis. In 1987, Rebecca Cann, Mark Stoneking, and Allan Wilson of the University of California, Berkeley and their colleagues published an analysis of mtDNA sequences from more than 100 ethnically diverse humans on four continents. They found that contemporary African populations contain the greatest variation in mtDNA. One explanation for this observation is that neutral mutations have been accumulating in African populations longer than in others, marking the African populations as the oldest on Earth. They also found that all human populations contain at least one mtDNA sequence of African origin, suggesting an African ancestry for all modern humans. Cann and her colleagues named the ancestral population, which lived approximately 200,000 years ago, the "mitochondrial Eve."

Other researchers have examined genetic material that males inherit only from their fathers. In 1995, L. Simon Whitfield and his colleagues at Cambridge University published a study on an 18,000-base-pair sequence from the Y chromosome, which does not undergo recombination with the X chromosome. Because the sequence contains no genes, it should not be subject to natural selection. Thus, sequence variations should result only from random mutations, which can serve as a molecular clock. The researchers discovered only three sequence mutations among the five subjects examined—a surprising result, given that the sample included a European, a Melanesian, a South American, and two Africans. By contrast, a chimpanzee exhibited 207 differences from the human version of the sequence. Using a sophisticated statistical analysis, the Whitfield team calculated that the common ancestor of these five diverse humans, dubbed the "African Adam," lived between 37,000 and 49,000 years ago. The limited genetic diversity and relatively recent origin of a common ancestor clearly support the African origin of *H. sapiens*. Follow-up studies on the Y chromosomes of thousands of men from Africa, Europe, Asia, Australia, and the Americas have confirmed that all modern humans are the descendants of a single migration out of Africa.

Ever since Darwin first included humans in his analyses about the diversity of life, the study of human evolution has been a controversial subject. Some disputes arise because the fossil record for humans gives up its secrets slowly and in piecemeal fashion. Other disagreements result from the assumptions that researchers must make about the sizes and geographical ranges of ancient populations, the amount of gene flow they experienced, and how natural selection may have affected them. But intellectual disputes are routine in science, and they challenge researchers to refine their hypotheses and to test them in new ways. Questions about the details of human origins are at the center of one of the liveliest debates in evolutionary biology today; additional research will surely clarify the evolutionary history of our species.

STUDY BREAK 30.13 <

1. **What trait allows researchers to distinguish between apes and humans?**
2. **What evidence suggests that Neanderthals and modern humans represent two distinct species?**
3. **What evidence suggests that modern humans originated in Africa?**

UNANSWERED QUESTIONS

What causes the evolution of diversity?

In this chapter, you have read about the extensive diversity—in size, shape, color, structure, lifestyle, and habitats—of the vertebrate animals. Their mechanisms for maintaining themselves—including behaviors such as moving, feeding, reproducing, hiding, fighting, and sleeping—are equally varied. But nearly all vertebrates share some fundamental features. Recent research shows us that this generality is especially true of early development, but it is also apparent in many of the basic homeostatic mechanisms that vertebrates share—features of digestion and metabolism, respiration, and other characteristics regulated by the products of genetic networks and cascades. Biologists had long thought that once we understood the genetics of a variety of species, we would also understand the basis for their evolution and the maintenance of diversity. But now that biologists have sequenced a number of animal genomes, we have actually learned more about the genetic information that is shared by all animal species than we have learned about the genes that promote diversification. One of the great unanswered questions in biology, therefore, is how diversity arises and how it is maintained, given that so much genetic information is shared among even distantly related species. Equally important is why the same kinds of features evolved almost identically time after time in many unrelated lineages.

The ever-increasing body of genetic information is opening the "black box" of why there are so many species, and how they came to be. For example, we've long known that limbs evolved from fins, based on fossil evidence (which continues to accumulate, providing additional supportive evidence). But we haven't known what the *mechanisms* for forming a fin or a limb are, and how selection works to modify such structures. Now, with our knowledge of *Hox* and other tool-kit genes that control embryonic development, we understand how fins and limbs are formed. We can also experimentally manipulate development and perform selection experiments to test possible pathways of evolution—the "how they came to be."

Why do we see the recurrent, independent evolution of common themes?

A phenomenon that we often see, but don't yet fully understand, is why certain themes—such as body elongation, limblessness, and tooth modification—recur among distantly related vertebrates. The evolution of viviparity, live-bearing reproduction, is one such theme. In the vast majority of animals, reproduction occurs when a female lays her eggs in water and a male sprays sperm over them; typically both parents then abandon the fertilized eggs. But some species in many separate vertebrate lineages have evolved forms of viviparity. Cartilaginous and bony fishes exhibit diverse modes of embryonic nutrition, and in one group, the sea horses, it is the males that become pregnant. Amphibians also exhibit diverse patterns of viviparity: some have pregnant fathers and others have mothers that brood embryos in the skin of their backs, in their stomachs, or in their oviducts. And many squamate reptiles grow placentas similar to those of mammals. All mammals except monotremes are live-bearers with maternal nutrition. In fact, among the living vertebrate groups, birds are the only group that has not evolved the live-bearing habit. Can you think of some reasons why?

In some fishes, amphibians, and squamates, the mother supplies all the nutrition for the developing young through her investment in the egg; but in other species, females resorb their yolks and provide nutrients directly to offspring that are born as fully metamorphosed juveniles. Given that viviparity has evolved independently in approximately 200 vertebrate groups, it is not surprising that we see such a variety of reproductive patterns. Biologists are just beginning to understand the evolution of viviparity and to determine which patterns are shared among different evolutionary lineages and which are not. It appears that the hormonal basis for viviparity is similar in many groups, although the timing, the receptors involved, and the physiological responses vary. Researchers are now identifying candidate genes in the hope of unraveling the genetic networks that have fostered the different modes of viviparity. Ecological studies are revealing the interactions of potential selection regimes that influence reproductive modes. However, we still have much to learn about how genetics, development, physiology, and ecology interact in the evolution of diversity and common recurring themes.

Think Critically

If viviparity has evolved multiple times in different vertebrate lineages, it must have been favored by natural selection. What might some of the benefits of viviparity be?

Marvalee H. Wake

Marvalee H. Wake is a professor in the Department of Integrative Biology and the graduate school at the University of California, Berkeley. She studies vertebrate evolutionary morphology, development, and reproductive biology, with the goal of understanding evolutionary patterns and processes. To learn more about Dr. Wake's research, go to http://ibB.edu/research/interests/research_profile.php?person=236.

REVIEW KEY CONCEPTS

Go to **CENGAGENOW** at www.cengage.com/login to access quizzing, animations, exercises, articles, and personalized homework help.

30.1 Invertebrate Deuterostomes

- Echinoderms have secondary radial symmetry, a five-part body plan, and a unique water vascular system that is used in locomotion and feeding (Figures 30.2 and 30.3).
- Hemichordates collect oxygen and particulate food from seawater that enters the mouth and exits the pharynx through the branchial slits (Figure 30.4).

30.2 Overview of the Phylum Chordata

- Chordates share several derived characteristics: a notochord; postanal tail; segmentation of body wall and tail muscles; a dorsal, hollow nerve cord; and a perforated pharynx at some stage of the life cycle (Figure 30.5).

- Two subphyla of invertebrate chordates use their perforated pharynx to collect particulate food (Figures 30.6 and 30.7).
- The subphylum Vertebrata includes animals with a bony endoskeleton, structures derived from neural crest cells, and a brain divided into three regions.

Animation: Tunicate body plan

Animation: Lancelet body plan

30.3 The Origin and Diversification of Vertebrates

- Vertebrates evolved from an invertebrate chordate ancestor, probably through duplication of the *Hox* gene complex (Figure 30.8).
- The earliest vertebrates were fish-shaped animals with only some vertebrate characteristics (Figure 30.9).
- Vertebrates diversified into numerous lineages (Figure 30.10). Gnathostomata includes all jawed vertebrates. Tetrapoda includes all lineages that ancestrally had four legs. Amniota includes groups descended from animals that produced an amniote egg.

Animation: Vertebrate evolution

30.4 "Agnathans": Hagfishes and Lampreys, Conodonts and Ostracoderms

- Living agnathans—hagfishes and lampreys—are jawless fishes that lack vertebrae and paired fins (Figure 30.11).
- Conodonts had bonelike elements in the pharynx. Ostracoderms were heavily armored, jawless fishes that sucked particulate food into their mouths (Figure 30.12).

Animation: Jawless fishes

30.5 Gnathostomata: The Evolution of Jaws

- Jaws arose through the evolutionary modification of gill arches, which supported the pharynx of ostracoderms (Figure 30.13). Fins arose at the same time (Figure 30.14).
- The first jawed fishes, Acanthodii and Placodermi, are now extinct (Figure 30.15). Chondrichthyans have a skeleton composed of cartilage (Figure 30.16). Actinopterygians and Sarcopterygians have a skeleton composed of bone. Actinopterygians have fins supported by bony rays (Figures 30.17 and 30.18). Sarcopterygians have fins supported by muscles and a bony endoskeleton (Figure 30.19).

Animation: Evolution of jaws

Animation: Cartilaginous fishes

Animation: Bony fish body plan

30.6 Tetrapoda: The Evolution of Limbs

- Tetrapods arose in the late Devonian (Figure 30.20). Key tetrapod adaptations include a strong vertebral column, girdles, limbs, and modified sensory systems.
- Modern amphibians are generally restricted to moist habitats. Their life cycles often include larval and adult stages. Caudata (salamanders) are elongate, tailed amphibians. Anura (frogs) have compact bodies, long legs, and no tails. Gymnophiona (caecelians) are legless burrowers (Figure 30.21).

Animation: Evolution of limb bones

Animation: Salamander locomotion

30.7 Amniota: The Evolution of Fully Terrestrial Vertebrates

- Key adaptations in amniotes, the first fully terrestrial vertebrates, included a water-resistant skin, amniote eggs (Figure 30.22), and the excretion of nitrogen wastes as uric acid.
- A traditional view suggests that early amniotes diversified into three lineages (anapsids, synapsids, and diapsids) distinguished by the number of temporal fenestrae and temporal arches on the skull (Figure 30.23).
- Diapsids split into two lineages, archosaurs and lepidosaurs.
- The phylogenetic position of turtles is the subject of disagreement.

Animation: Amniote egg

30.8 Testudines: Turtles

- The turtle body plan includes a bony shell, with a dorsal carapace and ventral plastron (Figure 30.24).

Animation: Tortoise shell and skeleton

30.9 Living Lepidosaurs: Sphenodontids and Squamates

- Sphenodontids are remnants of a once-diverse lineage (Figure 30.25A).
- Squamates have skin composed of overlapping, keratinized scales (Figure 30.25B and C). Lizards consume a variety of foods, but all snakes are predators on other animals.

30.10 Living Archosaurs: Crocodilians and Birds

- Crocodilians are semiaquatic predators (Figure 30.26).
- Birds have adaptations that reduce their weight and generate power for flight (Figure 30.27).
- Modern birds exhibit adaptations of their bills, feet, wings, and behavior (Figure 30.28).

Animation: Crocodile body plan

Animation: Feather development

Animation: Avian bone and muscle structure

30.11 Mammalia: Monotremes, Marsupials, and Placentals

- Key adaptations of mammals include endothermy, which allows high levels of activity; modification of the teeth and jaws; extensive parental care of offspring; and large and complex brains.
- There are three major groups of mammals: the Prototheria, to which the monotremes belong, and two lineages of Theria, to which the marsupials and the placentals belong. The three groups differ in their reproductive patterns.
- Monotremes are restricted to Australia and New Guinea (Figure 30.29). Marsupials are abundant in Australia and occur in South America and North America (Figure 30.30). Most living mammals are placentals, occupying nearly all terrestrial and aquatic habitats (Figure 30.31).

Animation: Mammalian dentition

Animation: Mammalian radiations

Animation: Structure of the placenta

30.12 Nonhuman Primates

- Key adaptations allow Primates to be arboreal and diurnal: upright posture and flexible limbs; good depth perception; and a large and complex brain.
- The Primates includes two major lineages (Figure 30.32). The Strepsirhini have many ancestral primate characteristics (Figure 30.33). The Haplorhini have many derived primate characteristics (Figures 30.34 and 30.35).
- Primates arose in forests about 55 mya. The hominoid lineage, which includes apes and humans, arose in Africa about 23 mya (Figure 30.36).

Animation: Primate skeletons

Animation: Skulls of extinct primates

30.13 The Evolution of Humans

- Hominins, the lineage that includes humans, arose in Africa between 10 mya and 5 mya. Hominin anatomy permits bipedal locomotion (Figure 30.37). Over time, hominins developed larger brains and tool-making behavior. Several genera of hominins occupied sub-Saharan Africa for several million years (Figures 30.38–30.40).
- *Homo habilis* was the first hominin species to make stone tools. *Homo erectus,* which arose in East Africa about 1.8 mya, made sophisticated stone tools (Figure 30.41).
- The early descendants of *H. erectus* left Africa in waves, populating Asia and Europe. Neanderthals, the best known of these groups, became extinct about 30,000 years ago.
- Modern humans, *Homo sapiens,* arose approximately 150,000 years ago and migrated out of Africa, eventually replacing archaic humans in Europe and Asia.

Animation: Fossils of australopiths

Animation: *Homo* skulls

Animation: Primate phylogenetic tree

Animation: Genetic distance between human groups

UNDERSTAND AND APPLY

Test Your Knowledge

1. Which phylum includes animals that have a water vascular system?
   a. Echinodermata
   b. Hemichordata
   c. Chordata
   d. Tetrapoda
   e. Amniota
2. Which of the following is *not* a characteristic of all chordates?
   a. notochord and postanal tail
   b. segmental body wall and tail muscles
   c. segmented nervous system
   d. dorsal hollow nerve cord
   e. perforated pharynx
3. Which group of vertebrates has adaptations that allow it to reproduce on land?
   a. agnathans
   b. tetrapods
   c. gnathostomes
   d. amniotes
   e. ichthyosaurs
4. Which group of fishes has the most living species today?
   a. sarcopterygians
   b. actinopterygians
   c. chondrichthyans
   d. acanthodians
   e. ostracoderms
5. Modern amphibians:
   a. closely resemble their Paleozoic ancestors.
   b. always occupy terrestrial habitats as adults.
   c. never occupy terrestrial habitats as adults.
   d. are generally larger than their Paleozoic ancestors.
   e. are generally smaller than their Paleozoic ancestors.
6. Which of the following key adaptations allows amniotes to occupy terrestrial habitats?
   a. the production of carbon dioxide as a metabolic waste product
   b. an unshelled egg that is protected by jellylike material
   c. a dry skin that is largely impermeable to water
   d. a lightweight skeleton with hollow bones
   e. feathers or fur that provide insulation against cold weather
7. Which of the following characteristics does *not* contribute to powered flight in birds?
   a. a lightweight skeleton
   b. efficient respiratory and circulatory systems
   c. enlarged forelimbs and a keeled sternum
   d. a high metabolic rate that releases energy from food rapidly
   e. scaly skin on the legs and feet
8. Which of the following characteristics did *not* contribute to the evolutionary success of mammals?
   a. extended parental care of young
   b. an erect posture and flexible hip and shoulder joints
   c. specializations of the teeth and jaws
   d. enlargement of the brain
   e. high metabolic rate and high body temperature
9. The Hominoidea is a monophyletic group that includes:
   a. apes and monkeys
   b. apes only
   c. humans and human ancestors
   d. apes and humans
   e. monkeys, apes, and humans
10. Which of the following hominin species was the earliest?
   a. *Ardipithecus ramidus*
   b. *Australopithecus afarensis*
   c. *Homo habilis*
   d. *Homo erectus*
   e. *Homo neanderthalensis*

Discuss the Concepts

1. Most sharks and rays are predatory, but the largest species feed on plankton. Construct a hypothesis to explain this observation. How would you test your hypothesis?
2. When tetrapods first ventured onto the land, what new selection pressures did they face? What characteristics might have fostered the success of these animals as they made the transition from aquatic to terrestrial habitats?
3. Use a pair of binoculars to observe several species of birds that live in different types of environments, such as lakes and forests. How are their beaks and feet adapted to their habitats and food habits?

4. Imagine that you unearthed the complete fossilized remains of a mammal. How would you determine the food habits of this now extinct animal?
5. Many myths about human evolution are embraced by popular culture. Using the information you have learned about human evolution, argue against each of the following myths.
   a. Humans evolved from chimpanzees.
   b. Evolution occurred in a steady linear progression from primitive primate to anatomically modern humans.
   c. All human characteristics, such as bipedal locomotion and an enlarged brain, evolved simultaneously and at the same rate.

Design an Experiment

Walking along a rocky coast one day, you discover two small creatures—one lumpy and the other worm-shaped. What anatomical studies would you conduct to determine whether or not they are chordates? What genetic studies might provide supplementary evidence?

Interpret the Data

The phylogenetic tree for vertebrates depicted below was constructed from sequence data for two rRNA mitochondrial genes (12S and 16S). How do the results of this analysis compare with the traditional view of turtles as living representatives of an anapsid lineage?

Apply Evolutionary Thinking

Birds and crocodiles are both descended from an ancestral archosaur. What shared anatomical and behavioral characteristics reflect this common ancestry? Explain why dinosaurs, which were also members of the archosaur lineage, may have shared these traits as well. Review Figure 30.23 before formulating your answer.

Express Your Opinion

Private collectors find and protect fossils, but a private market for rare vertebrate fossils raises the cost to museums and encourages theft from protected fossil beds. Should private collecting of vertebrate fossils be banned? Go to www.cengage.com/login to investigate both sides of the issue and then vote.

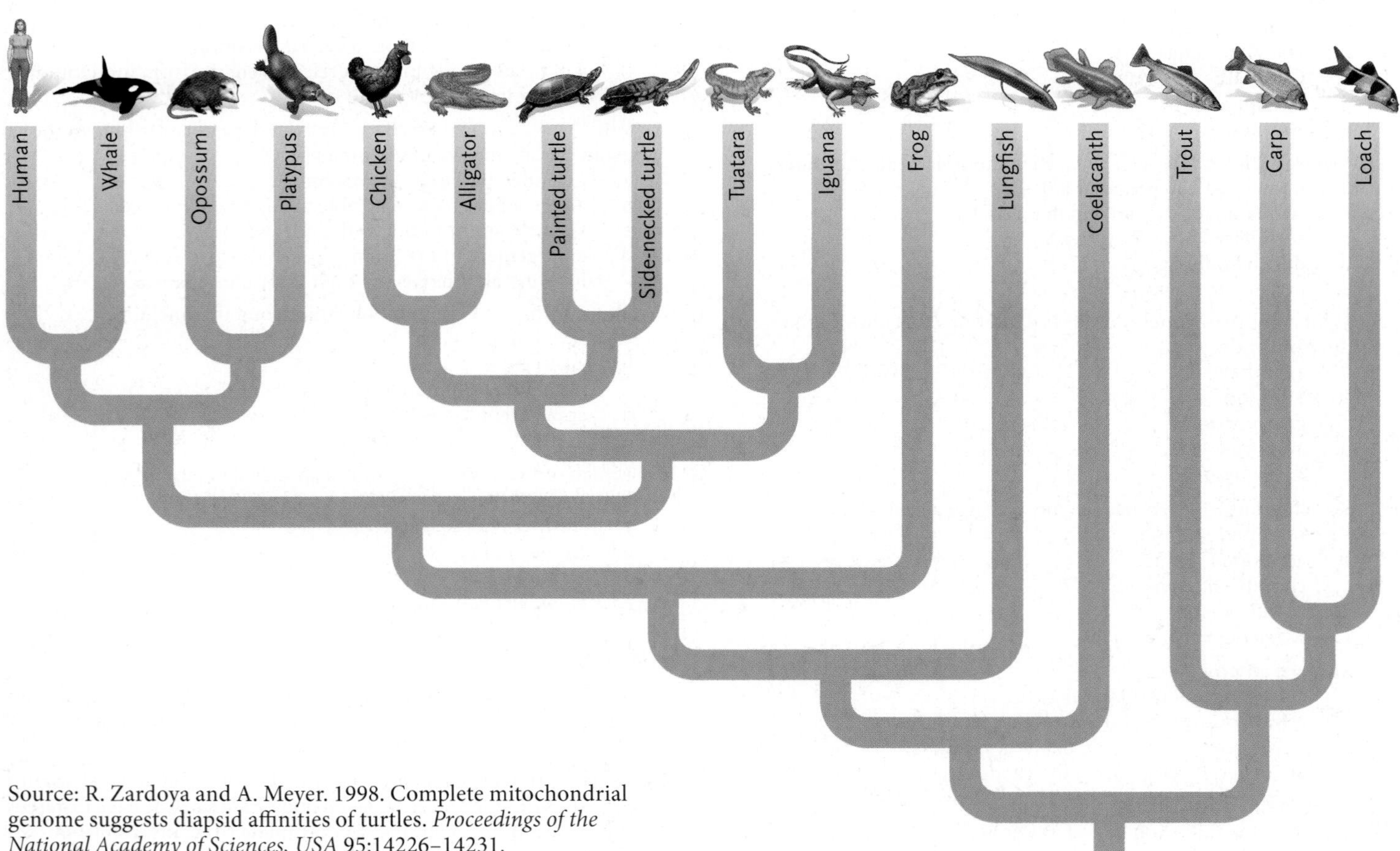

Human
Whale
Opossum
Platypus
Chicken
Alligator
Painted turtle
Side-necked turtle
Tuatara
Iguana
Frog
Lungfish
Coelacanth
Trout
Carp
Loach

Source: R. Zardoya and A. Meyer. 1998. Complete mitochondrial genome suggests diapsid affinities of turtles. *Proceedings of the National Academy of Sciences, USA* 95:14226–14231.

Basic parts of the shoot system of an apple tree *(Malus domestica)*, including leaves, stems, and vividly colored fruits.

31

STUDY OUTLINE

The Plant Body

Why It Matters. . . Fossil teeth discovered in the East African Rift Valley indicate that the early ancestors of humans likely consumed a variety of plant parts—hard-shelled nuts, dry seeds, soft fruits, and leaves. By about 11,000 years ago agriculture got its start when humans began domesticating seed plants. In fact, directly or indirectly, plant leaves, stems, roots, flowers, seeds, and fruits are the basic sources of energy for nearly all of Earth's animal life **(Figure 31.1A)**. Plants also provide the materials for building human dwellings, beaver lodges, and bird nests, not to mention raw materials for making paper, dozens of pharmaceuticals, and a host of essentials of modern life **(Figure 31.1B–D)**.

FIGURE 31.1
Examples of plant parts that provide food or other resources for animals. **(A)** A pronghorn *(Antilocapra americana)*, which consumes leaves and grasses. **(B)** Leafcutter ants, which harvest bits of leaves to nourish a fungus that the ant colony uses as food. **(C)** A hummingbird (Trochilidae). **(D)** Lumber being used in home construction.

A. **Pronghorn feeding on leaves**

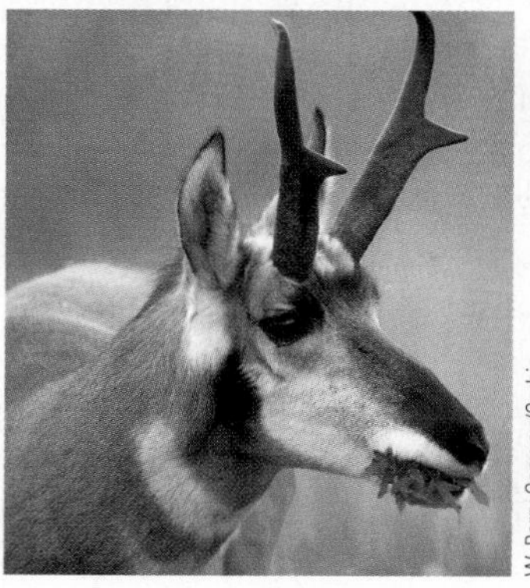

B. **Ants harvesting leaves**

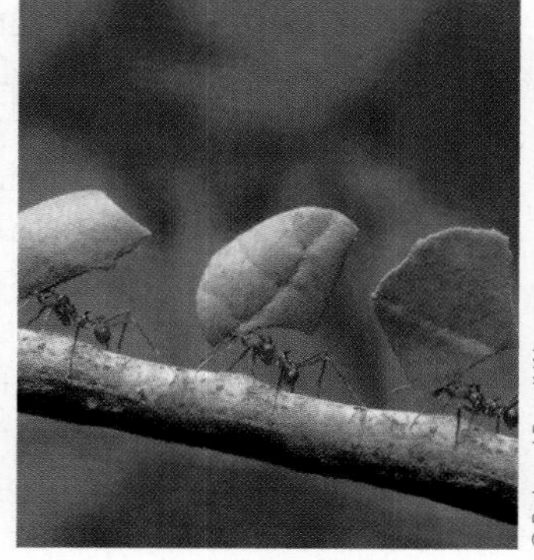

C. **Hummingbird in its nest**

D. **Humans building with lumber**

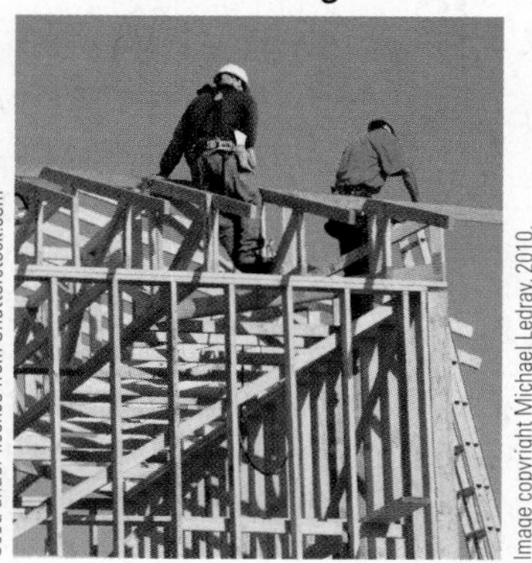

As described in Chapter 27, plants that made the transition from aquatic to terrestrial life did so only as evolutionary adaptations in form and function helped solve problems posed by the terrestrial environment. These adaptations included a shoot system that helps support leaves and other body parts in air, a root system that anchors the plant in soil and provides access to soil nutrients and water, tissues for internal transport of nutrients and water, and specializations for preventing water loss.

This chapter launches this textbook's survey of the structure and functioning of plants—their morphology, anatomy, and physiology. *Morphology* refers to the shape and overall appearance of plant parts, such as leaves. *Anatomy* is the physical structure of plant parts, and their function is the concern of plant *physiology*. Our focus throughout this unit is the plant phylum Anthophyta—angiosperms, or flowering plants. In terms of distribution and sheer numbers of species, angiosperms are the most successful plants on Earth. <

31.1 Plant Structure and Growth: An Overview

Because plants are photosynthetic autotrophs, they require access to sunlight and the carbon dioxide in air, as well as the water available in soil. In addition, many plants require nutrients that are usually available only in soil, and their aboveground parts may need the physical support of structures anchored in the ground. The evolutionary responses to these functional challenges produced a plant body consisting of two connected but quite different structural components—a photosynthetic *shoot system* extending upward into the air and a nonphotosynthetic *root system* extending down into the soil. Each system consists of various **organs**—body structures that contain two or more types of tissues and that have a definite form and function. Plant organs include leaves, stems, and roots, among others. A **tissue** is a group of cells and intercellular substances that function together in one or more specialized tasks. For example, a soft leaf consists of tissue called parenchyma that includes cells specialized for functions such as photosynthesis and gas exchange.

In Plant Tissues the Cells Share Some General Features

All plant cells share certain features that were introduced in Chapter 5. New plant cells develop a primary cell wall around the **protoplast,** the botanical term for the cell's cytoplasm, organelles, and plasma membrane. The primary wall is made up of cellulose fibers embedded in a matrix of hemicellulose and other branched carbohydrates (see Section 5.4). This combination helps make the wall rigid but flexible. The polysaccharide pectin also is abundant in the primary wall and in the middle lamella, the layer between the primary walls of neighboring cells that helps bind cells together in tissues. (Plant pectin is often used to congeal jams and jellies.) As a plant grows, the protoplast of many types of plant cells deposits additional cellulose and other materials inside the primary wall, forming a strong secondary cell wall (see Figure 5.25).

As in animals, as each plant cell matures and *differentiates* (becomes specialized for a particular function), specific genes become active. But while nearly all fully differentiated animal cells must be alive to perform their functions, some plant cells, especially in vascular tissues, must die and their protoplasts must disintegrate before the tissue of which they are a part can carry out its key functions.

In some plants, lignin, a water-insoluble, inert polymer, becomes incorporated into cell walls. This **lignification** anchors and stiffens the cellulose fibers in the walls and protects the other wall components from physical or chemical damage. Water cannot penetrate lignified cell walls. Many biologists believe that the evolution of large vascular plants became possible when certain cells developed biochemical pathways leading to lignification and could therefore become organized into watertight conducting channels.

Water and minerals pass from one lignified cell to another across *pits,* narrow regions where the secondary wall is absent and the primary wall is thinner and more porous than elsewhere. Solutes such as amino acids and sugars, as well as other substances, move in the plasmodesmata linking living plant cells (see Figure 5.25).

Shoot and Root Systems Perform Different but Integrated Functions

A plant's **shoot system** is basically a series of modules—repeating components such as stems, leaves, and in angiosperms, flowers **(Figure 31.2).** This modular organization correlates with physiological flexibility that is one of the most striking characteristics of plants. Unlike animals, plants have the genetic capacity to modify the development of organs as environmental conditions shift. In general, during the course of a plant's life cycle embryonic shoots called *buds* give rise to leaves, flowers, or both, as different gene-guided developmental pathways unfold. Section 31.3 looks more closely at the growth of stems and leaves, and Chapter 34 describes the development of flowers and other reproductive structures.

The shoot system grows aboveground, and is highly adapted for photosynthesis. Leaves greatly increase a plant's surface area and thus its exposure to light. Stems position leaves for maximum light exposure—an adaptation for maximizing photosynthesis—and in angiosperms they position flowers for pollination. As you will read shortly, some parts of the shoot system also store carbohydrates manufactured during photosynthesis.

The **root system** usually grows below ground. It anchors the plant, and in most land plants provides structural support for a plant's upright parts. The root system also absorbs water and dissolved minerals from soil and stores carbohydrates. Adaptations in the structure and function of plant cells and tissues were an integral part of the evolution of both shoots and roots. For example, vascular tissues became specialized as pipelines that conduct water, minerals, and organic substances

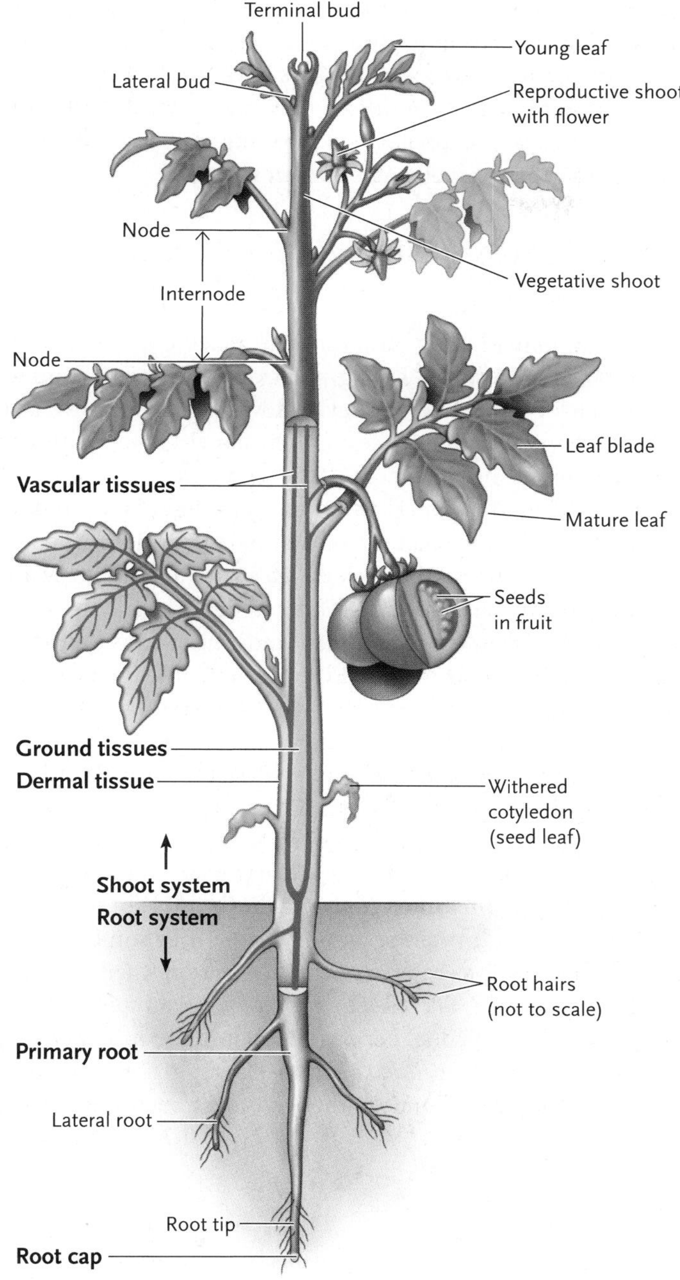
Terminal bud
Young leaf
Lateral bud
Reproductive shoot
with flower
Node
Internode
Vegetative shoot
Node
Leaf blade
Vascular tissues
Mature leaf
Seeds
in fruit
Ground tissues
Dermal tissue
Withered
cotyledon
(seed leaf)
Shoot system
Root system
Root hairs
(not to scale)
Primary root
Lateral root
Root tip
Root cap

FIGURE 31.2
Basic body plan for the tomato plant Solanum lycopersicum, a typical angiosperm. Vascular tissues (purple) conduct water, dissolved minerals, and organic substances. They thread through ground tissues, which make up most of the plant body. Dermal tissues (epidermis, in this case) cover the surfaces of the root and shoot systems.

throughout the plant. The root hairs sketched in Figure 31.2 are surface cells specialized for absorbing water from soil.

Meristems Produce New Tissues Throughout a Plant's Life

As you know from experience, animals generally grow to a certain size, and then their growth slows dramatically, or stops. This pattern is called **determinate growth.** In contrast, most plant species have parts that can grow throughout the plant's life, a pattern called **indeterminate growth.** Certain plant parts such as leaves, flowers, and fruits exhibit determinate growth, but every plant's shoot and root system also includes clumps of self-perpetuating embryonic tissue called **meristems** (*merizein* = to divide). Plant hormones stimulate meristems to produce new tissues more or less continuously while a plant is alive. A capacity for indeterminate growth gives plants a great deal of flexibility—or what biologists often call *plasticity*—in their repertoire of responses to changes in environmental factors such as light, temperature, water, and the supply of nutrients. This plasticity has major adaptive benefits for an organism that cannot move about, as most animals can. For example, if external factors change the direction of incoming light for photosynthesis, stems can "shift gears" and grow in that direction. Likewise, roots can grow outward toward water.

Animals grow mainly by mitosis, which increases the number of body cells. Plants, however, grow by two mechanisms—an increase in the number of cells by mitotic cell divisions in meristems, *and* an increase in the size of individual cells. The daughter cells resulting from mitosis in meristems rapidly increase in size—especially in length—usually becoming much larger than the parent cell. In contrast, when animal cells divide mitotically the daughter cells increase only to the size of the parent cell.

Meristems Are Responsible for Growth in Both Height and Girth

All vascular plants have a type of meristem that adds length, and some also have a different type of meristem for increasing girth **(Figure 31.3).** An increase in length arises from **apical meristems,** clumps of self-perpetuating tissue at the tips of their buds, stems, and roots (see Figure 31.3A). Tissues that develop from apical meristems are called **primary tissues** and make up the **primary plant body.** Growth of the primary plant body is called **primary growth.**

Some species of plants—grasses and marigolds, for example—show only primary growth, which occurs at the tips of roots and shoots. Others, particularly plants such as trees that have a woody body, show **secondary growth,** which originates at self-perpetuating cylinders of tissue called **lateral meristems.** Secondary growth increases the diameter of older roots and stems (see Figure 31.3B). The tissues that develop from lateral meristems, called **secondary tissues,** make up the woody **secondary plant body** we see in trees and shrubs.

Primary and secondary growth can go on simultaneously in a single plant, with primary growth increasing the length of shoot parts and secondary growth adding girth. Each spring, for example, a maple tree undergoes primary growth at each of its root and shoot tips, while secondary growth increases the diameter of its older woody parts. Plant hormones govern these growth processes and other key events that are described in Chapter 35.

A. Plants increase in length by cell divisions in apical meristems and by elongation of the daughter cells.

Shoot apical meristem
Dividing cells near all shoot tips are responsible for a shoot's primary tissues and growth.

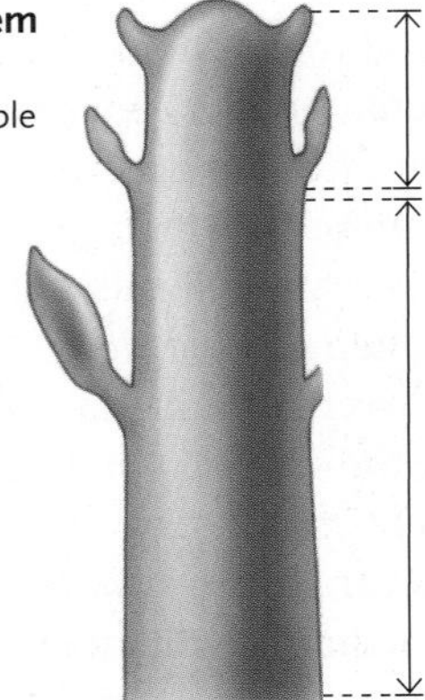
Cells divide in shoot apical meristem
New cells elongate and start to differentiate into primary tissues.

Root apical meristem
Dividing cells near root tips are responsible for a root's primary tissues and growth.

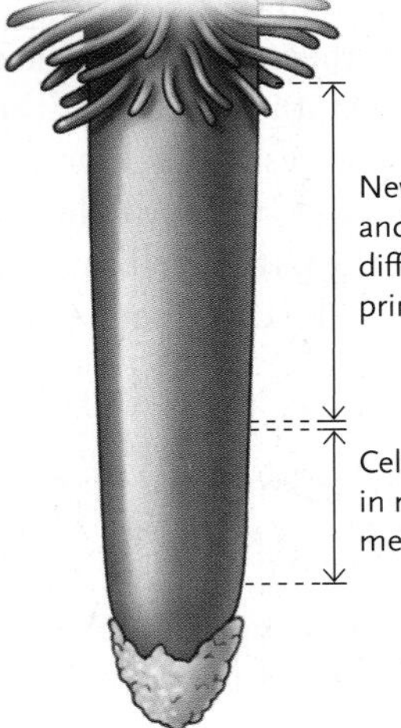
New cells elongate and start to differentiate into primary tissues.
Cells divide in root apical meristem

B. Some plants increase in girth by way of cell divisions in lateral meristems.

Lateral meristems
Dividing cells are responsible for the increase in diameter of shoots and roots.

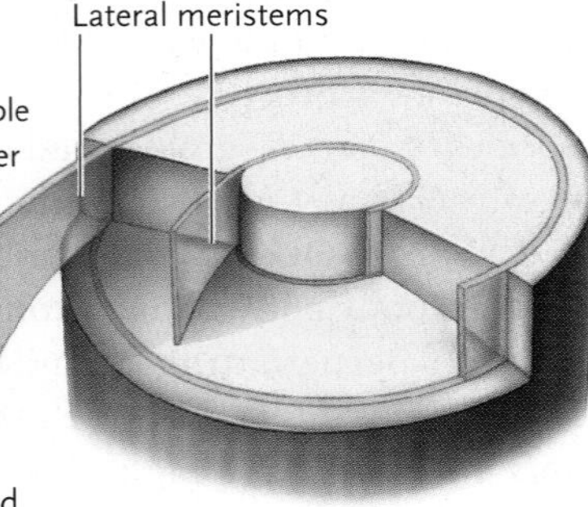
Lateral meristems

FIGURE 31.3 Approximate locations of types of meristems that are responsible for increases in the length and diameter of the shoots and roots of a vascular plant.

Monocots and Eudicots Are the Two General Structural Forms of Flowering Plants

Several broad categories of body architecture arose as flowering plants evolved. The two major ones are the monocot and eudicot lineages discussed in Chapter 27. Grasses, irises, cattails, and palms are examples of monocots. Eudicots include nearly all familiar angiosperm trees and shrubs, as well as many other types of plants. Examples are maples, willows, oaks, cacti, roses, poppies, sunflowers, and garden beans and peas.

Although monocots and eudicots have similar types of tissues, their body structures differ in distinctive ways **(Table 31.1).** As we discuss the morphology of flowering plants, we will refer frequently to these structural differences.

Flowering Plants Can Be Grouped According to Type of Growth and Lifespan

The distinctions in body structure and development between monocot and eudicot flowering plants are important from an evolutionary perspective. However, flowering plants have also evolved many other differences in their patterns of growth and reproduction.

A major distinction is whether a species has secondary growth—a trait that demands a substantial investment of energy and other resources, and that correlates with a long life. Many eudicots (and all gymnosperms) show secondary growth, which produces woody tissue. By contrast, most monocots and some eudicots show little or no secondary growth. They mature and reproduce quickly. The term *herbaceous* refers to such plants, which do not produce woody tissue.

We can also distinguish plants by lifespan. **Annuals** are herbaceous plants that complete their life cycle in one growing season. With little or no secondary growth, annuals usually have only apical meristems. Examples are marigolds (a eudicot) and corn (a monocot). **Biennials** complete the life cycle in two growing seasons, with limited secondary growth in some species. Roots, stems, and leaves form in the first season, then the plant flowers, forms seeds, and dies in the second. Examples are carrots and celery (eudicots). **Perennials** grow and reproduce year after year. Many perennials, such as trees, shrubs, and some vines, have secondary tissues, although others, such as irises and daffodils, do not.

STUDY BREAK 31.1 <

1. **Compare and contrast the components and functions of a land plant's shoot and root systems.**
2. **Explain what meristem tissue is, and name and describe the functions of the basic types of meristems.**

31.2 The Three Plant Tissue Systems

Meristems produce three tissue systems that provide the foundation for the various plant organs. The **ground tissue system,** which makes up most of the primary plant body, functions in metabolism, storage, and support. The **vascular tissue system** consists of interconnecting cells that form channels that transport water and nutrients throughout the plant. The tissues are organized in bundles that are dispersed through the ground tissues. The **dermal tissue system** serves as a skinlike protective covering

TABLE 31.1 Eudicots and Monocots Compared

Charactero	Eudicots	Monocots
Cotyledons	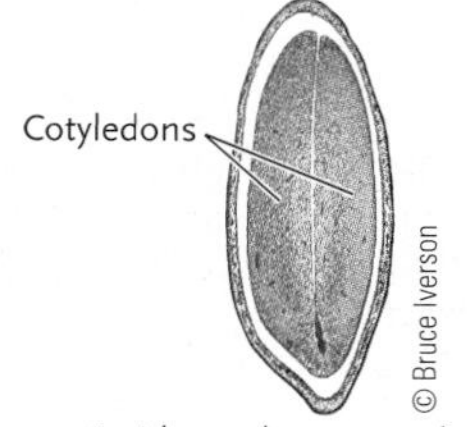 Cotyledons © Bruce Iverson Inside seeds, two cotyledons (seed leaves of embryo)	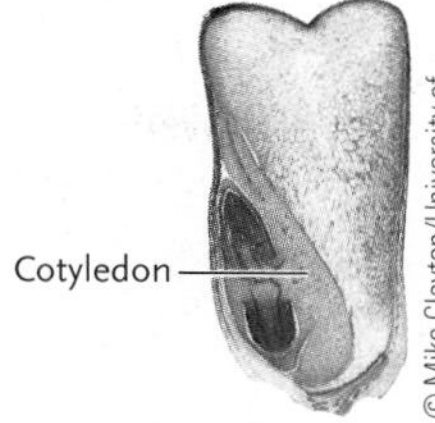 Cotyledon © Mike Clayton/University of Wisconsin Department of Botany Inside seeds, one cotyledon (seed leaf of embryo)
Floral parts	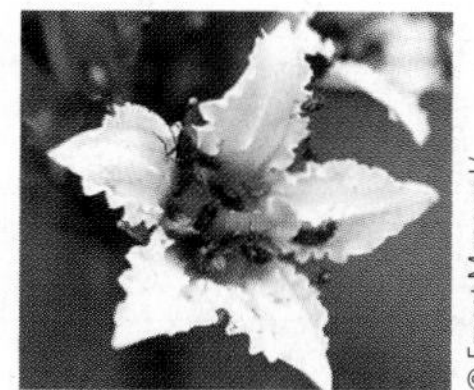 © Ernest Manewal/ Index Stock Imagery Usually four or five floral parts (or multiples of four or five)	 Image copyright Sarah Spencer, 2010. Used under license from Shutterstock.com Usually three floral parts (or multiples of three)
Leaf veins	 © Simon Fraser/ Photo Researchers, Inc. Leaf veins usually in a netlike array	 Gary Head Leaf veins usually running parallel with one another
Pollen pores and furrows	 © Andrew Syred/ Photo Researchers, Inc. Three pores or furrows (or furrows with pores) in pollen grains	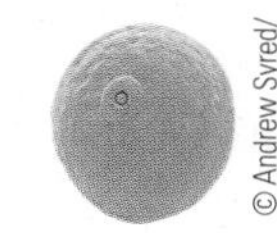 © Andrew Syred/ Photo Researchers, Inc. One pore or furrow in the pollen grain surface
Location of vascular bundles	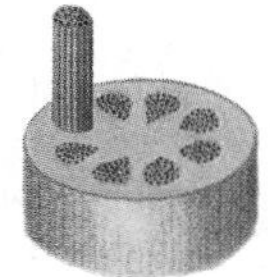 Vascular bundles organized as a ring in ground tissue	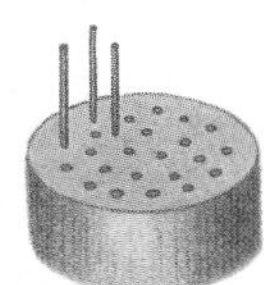 Vascular bundles distributed throughout ground tissue
Root system	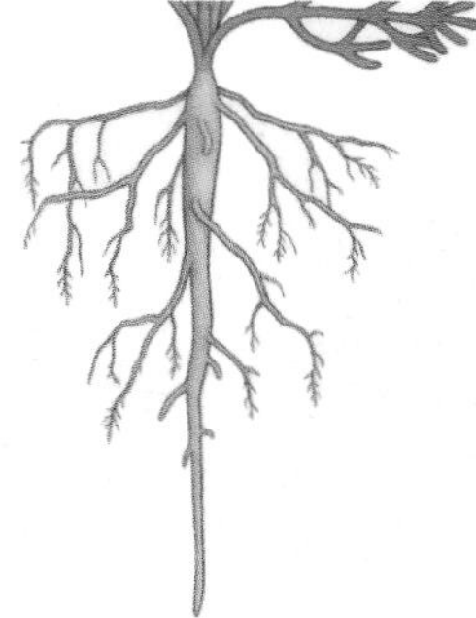 Usually a main taproot with smaller lateral roots	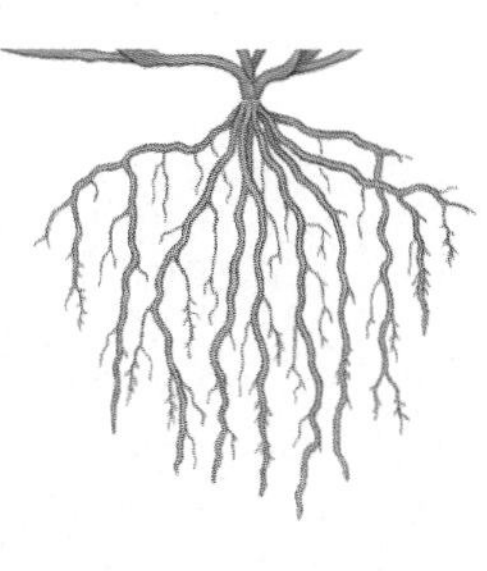 Usually a branching fibrous root system

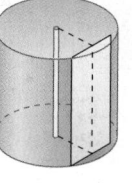

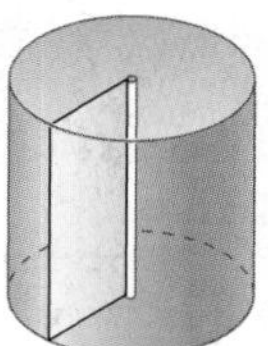
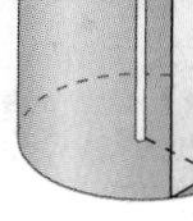
Radial
Tangential
Transverse

FIGURE 31.4
Terms that identify how tissue specimens are cut from a plant. Along the radius of a stem or root, longitudinal cuts give radial sections. Cuts at right angles to a root or stem radius give tangential sections. Cuts perpendicular to the long axis of a stem or root give transverse sections (cross sections).

for the plant body. Figure 31.2 shows the general location of each system in the shoot and root.

Each tissue system includes several types of tissue, and each tissue is made up of cells with specializations for different functions **(Table 31.2).** *Simple* tissues have only one type of cell. Other tissues are *complex,* with organized arrays of two or more types of cells. **Figure 31.4** will help you interpret micrographs of plant tissues, beginning with the tissues in a transverse section of a stem shown in **Figure 31.5A.**

Ground Tissues Are All Structurally Simple, but Differ in Important Ways

Plants have three types of ground tissue, each with a distinct structure and function—*parenchyma, collenchyma,* and *sclerenchyma* **(Figure 31.5B–D).** Each type is structurally simple, being composed mainly of one kind of cell. In a very real sense the cells in ground tissues are the "worker bees" of plants, carrying out photosynthesis, storing carbohydrates or water, providing mechanical support for the plant body, and performing other basic functions. Each kind of cell has a distinctive wall structure, and some also have variations in the protoplast.

PARENCHYMA: SOFT PRIMARY TISSUES Most of the soft, moist primary growth of roots, stems, leaves, flowers, and fruits is parenchyma (*para* = around, *khein* = fill in, or pour). Parenchyma cells can occur both as part of parenchyma tissue and as individual cells in other tissues.

Most parenchyma cells have only a thin primary wall (and no lignin), so they are pliable and permeable to water. Often the cells are round or many-sided, although they also can be elongated like a sausage, as in Figure 31.5B. Sometimes there are air spaces between parenchyma cells, especially in leaves (see Figure 31.16).

On the whole, parenchyma cells are considered to be relatively unspecialized. Yet subsets of them do become differentiated for specialized roles. For example, photosynthesis occurs in parenchyma cells in which a large number of chloroplasts develop. The flesh of a potato tuber or an apple is mainly parenchyma specialized for carbohydrate storage. And in many plant species, some parenchyma cells are specialized for short-distance transport of solutes. These cells are common in tissues in which water and solutes must be rapidly

TABLE 31.2	Summary of Flowering Plant Tissues and Their Components		
System	**Tissue**	**Cell Types in Tissue**	**Function**
Ground tissue	Parenchyma	Parenchyma cells	Photosynthesis, respiration, storage, secretion
	Collenchyma	Collenchyma cells	Flexible strength for growing plant parts
	Sclerenchyma	Fibers or sclereids	Rigid support, physical/mechanical protection
Vascular tissue	Xylem	Conducting cells (tracheids, vessel members); parenchyma cells; sclerenchyma cells	Transport of water and dissolved minerals
	Phloem	Conducting cells (sieve tube members); parenchyma cells; sclerenchyma cells	Sugar transport
Dermal tissue	Epidermis	Fairly unspecialized epidermal cells; specialized cells such as guard cells	Control of gas exchange; water loss; protection
	Periderm	Cork; cork cambium; secondary cortex	Protection

moved from cell to cell—for example, in vascular tissues and in tissues that secrete nectar. Parenchyma cells usually remain alive and capable of dividing when they are mature. In fact, their mitotic divisions produce the new cells that often heal wounds in plant parts.

COLLENCHYMA: FLEXIBLE SUPPORT The ground tissue called **collenchyma** (see Figure 31.5C) helps support plant parts (*kolla* = glue). Collenchyma cells typically are elongated. Collectively they often form strands or a sheath-like cylinder under the dermal tissue of growing shoot regions and the stem-like *petiole* at the base of a leaf. The "strings" of celery include collenchyma. Like parenchyma cells, collenchyma cells remain alive and metabolically active as they mature.

Collenchyma is an especially important adaptation in parts such as elongating stems that require structural support as they grow. Each collenchyma cell develops only a primary wall built of layers of cellulose and pectin, but over time it thickens as the cell synthesizes these components. The result is a strong primary wall that can stretch as the cell elongates. This sturdy but elastic

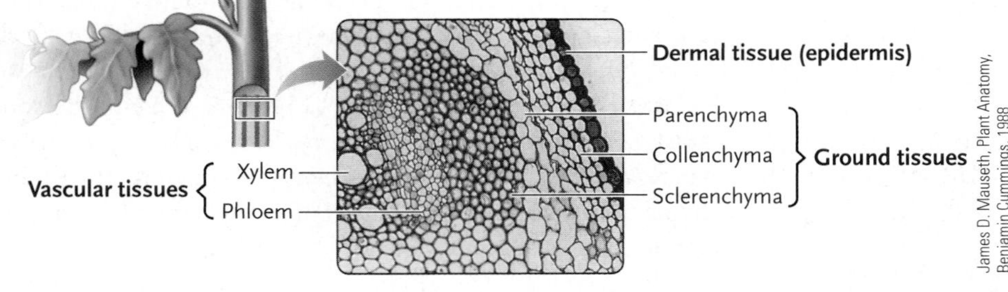
A. Location of tissues in stem
Dermal tissue (epidermis)
Parenchyma
Collenchyma
Sclerenchyma
Ground tissues
Vascular tissues
Xylem
Phloem
James D. Mauseth, Plant Anatomy, Benjamin Cummings, 1988

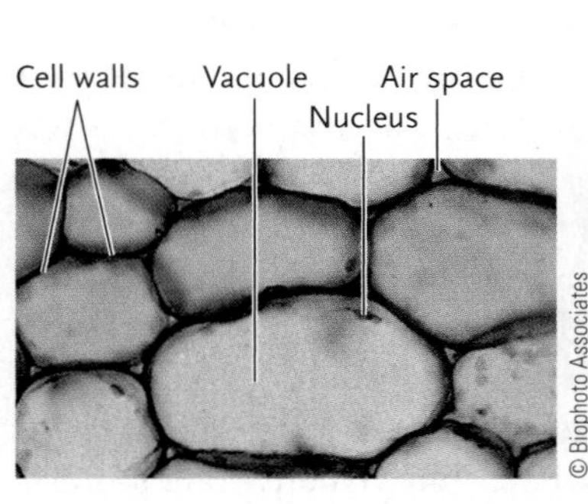
Cell walls
Vacuole
Air space
Nucleus
© Biophoto Associates

B. Parenchyma tissues consist of soft, living cells specialized for storage, other functions.

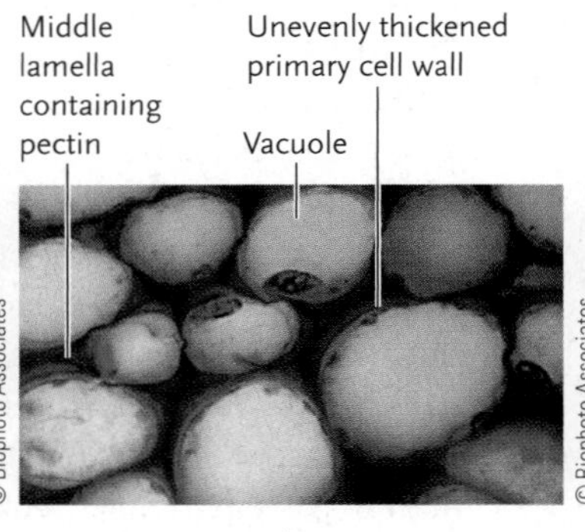
Middle lamella containing pectin
Unevenly thickened primary cell wall
Vacuole
© Biophoto Associates

C. Collenchyma tissues provide flexible support.

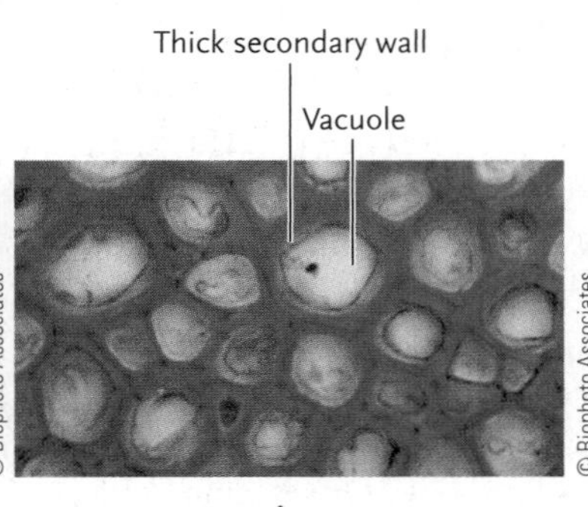
Thick secondary wall
Vacuole
© Biophoto Associates

D. Sclerenchyma tissues provide rigid support and protection.

FIGURE 31.5 Locations and examples of ground, vascular, and dermal tissues in flowering plants. **(A)** Ground, vascular, and dermal tissues in a buttercup stem *(Ranunculus)*, transverse section. Ground tissues are simple tissues, while vascular and dermal tissues are complex, containing various types of specialized cells. **(B–D)** Examples of ground tissues from the stem of a sunflower plant *(Helianthus annuus)*.

structure is an adaptation that ensures that collenchyma is not so rigid as to hamper the growth of plant parts it supports.

SCLERENCHYMA: RIGID SUPPORT AND PROTECTION Mature plant parts gain additional mechanical support and protection from **sclerenchyma** (*skleros* = hard). Most cells of this ground tissue develop thick secondary walls (see Figure 31.5D) that become heavily lignified. Once a sclerenchyma cell is encased in lignin, it dies because its protoplast can no longer take up nutrients or exchange gases with the environment. The walls remain, however, providing protection and support for the life of the plant.

There are two types of sclerenchyma cells, *sclereids* and *fibers*. **Sclereids** take a variety of shapes. Clumped, roughly cube-shaped sclereids give pears their slightly gritty texture **(Figure 31.6A).** Other sclereids form a thick protective coat around some kinds of seeds. Peach and cherry pits and the hard shells of coconuts, walnuts, and brazil nuts are examples. **Fibers (Figure 31.6B)** are elongated sclerenchyma cells that are sometimes compared to heavy rubber because they are somewhat pliable but resist stretching. Cotton fabric is woven from fibers that develop around the seeds of cotton plants *(Gossypium),* while linen cloth is made from fibers in the stems of flax plants. Tree bark gains a great deal of tensile strength from fibers that are incorporated in it. Fibers also occur in a plant's vascular tissues, the topic we turn to next.

A. **Sclereids**

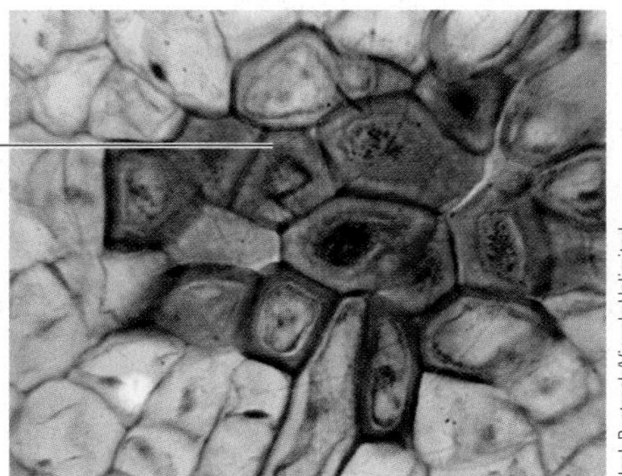
Thick secondary wall
Jack Bostrack/Visuals Unlimited

B. **Fibers**

D. E. Akin and I. L. Rigsby, Richard B. Russel Agricultural Research Service, U. S. Department of Agriculture, Athens, Georgia

FIGURE 31.6
Examples of sclerenchyma cells. **(A)** From the flesh of a pear *(Pyrus),* sclereids called stone cells. Each has a thick, lignified wall. **(B)** Strong fibers from stems of a flax plant *(Linum).*

Vascular Tissues Are Specialized for Conducting Fluids

Plant vascular tissues are complex tissues composed of specialized conducting cells, parenchyma cells, and fibers. *Xylem* and *phloem,* the two kinds of vascular tissues in flowering plants, are organized into bundles of interconnected cells that extend throughout the plant.

XYLEM: TRANSPORTING WATER AND MINERALS Xylem (*xylon* = wood) conducts water and minerals absorbed from the soil upward from a plant's roots to the shoot. As you read in Chapter 27, xylem was a key adaptation in early land plants. Xylem contains two types of conducting cells: *tracheids,* which evolved first, and *vessel members,* which arose in most lineages of angiosperms. When both types of conducting cells reach maturity, they lay down secondary cell walls that are thickened and strengthened by lignin and cellulose. The cell's protoplast then disintegrates and the cell dies—leaving arrays of abutting, empty cells that can serve as pipelines for water and minerals.

Tracheids are elongated cells, with a rather narrow diameter and tapered, overlapping ends **(Figure 31.7A).** Water can move from cell to cell through openings called pits. Usually, a pit in one cell is opposite a pit of an adjacent cell, so water seeps laterally from tracheid to tracheid.

Vessel members are joined end to end in tubelike multicellular columns called **vessels (Figure 31.7B).** Like tracheids, vessel members have pits through which water can move from cell to cell. However, they also have other structural adaptations that enhance water flow. They have a wider inside diameter than tracheids, and as vessel members mature, enzymes break down portions of their end walls, producing perforations. Some vessel members have a single, large perforation, so that the end is completely open. Others have a cluster of small, round perforations or ladderlike bars that extend across the open end **(Figure 31.7C).** The predictability of the perforation patterns suggests that this process is under precise genetic control.

Fossils show that the forerunners of modern plant species had only tracheids for water transport. Today, tracheids are still the only conducting cells in most ferns, all gymnosperms, and basal angiosperms such as *Amborella* and some water lilies. Most other angiosperms have both tracheids and vessel members, however. Presumably, the evolution of vessel members conferred an adaptive advantage by increasing the efficiency with which water and minerals move throughout the plant.

Xylem also contains parenchyma cells and sclerenchyma fibers. The parenchyma cells help transport minerals through vessel members and tracheids. Sclerenchyma fibers function like

A. **Tracheids, tangential section**

Pits in tracheid

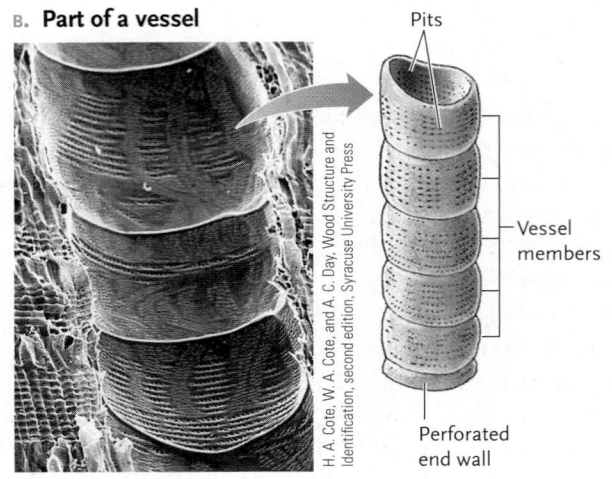

Alison W. Roberts, University of Rhode Island

B. **Part of a vessel**

Pits

Vessel members

Perforated end wall

H. A. Cote, W. A. Cote, and A. C. Day, Wood Structure and Identification, second edition, Syracuse University Press

C. **Scalariform perforation plate**

H. A. Core, W. A. Cote and A. C. Day, Wood Structure and Identification, 2nd Ed., Syracuse University Press, 1979

FIGURE 31.7
Representative tracheids and vessel members from woody stems, elements in xylem that conduct water and dissolved mineral salts through the body of a vascular plant. The electron micrographs show **(A)** tracheids from a pine *(Pinus)* and **(B)** a vessel from a red oak *(Quercus rubra)*. In the vessels of oaks and many other plant species the perforation plates between vessel members are simple, roughly oval openings. Another common form is the ladderlike scalariform perforation plate shown in **(C).**

steel cables in concrete, helping keep the xylem tissue fairly rigid and lending overall structural support to the plant.

PHLOEM: A TISSUE THAT TRANSPORTS SUGARS AND OTHER SOLUTES The vascular tissue **phloem** (*phloios* = tree bark) transports solutes, notably the sugars made in photosynthesis, throughout the plant. The main conducting cells of phloem are **sieve tube members (Figure 31.8),** which connect end to end, forming a **sieve tube.** As the name implies, their end walls, called sieve plates, are laced with openings. In flowering plants the phloem is strengthened by fibers and sclereids.

Immature sieve tube members contain the usual plant organelles. Over time, however, the cell nucleus and internal membranes in plastids break down, mitochondria shrink, and the cytoplasm is reduced to a thin layer lining the interior surface of the cell wall. Even without a nucleus, sieve tube members live up to several years in most plants, and much longer in some trees.

In many flowering plants, specialized parenchyma cells called **companion cells** are connected to mature sieve tube members by plasmodesmata. Unlike sieve tube members, companion cells retain their nucleus when mature. They assist sieve tube members both with the uptake of sugars and with the unloading of sugars in tissues that are growing or that store food. They may also help regulate the metabolism of mature sieve tube members. Chapter 32 returns to the functions of phloem cells.

FIGURE 31.8
Structure of sieve tube members. (A) Micrograph showing sieve tube members in longitudinal section. The arrows point to cells that may be companion cells. Long tubes of sieve tube members conduct sugars and other organic compounds. **(B)** SEM of sieve plates of sieve tube members in phloem.

A. **Sieve-tube members**

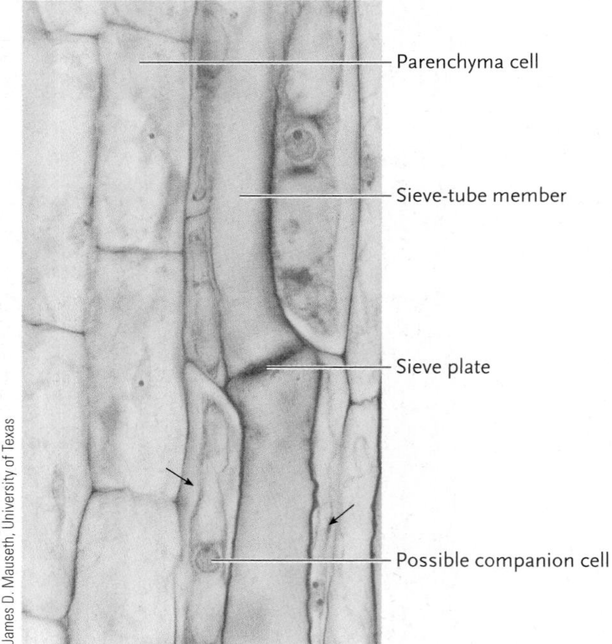
Parenchyma cell
Sieve-tube member
Sieve plate
Possible companion cell

James D. Mauseth, University of Texas

B. **Sieve plate**

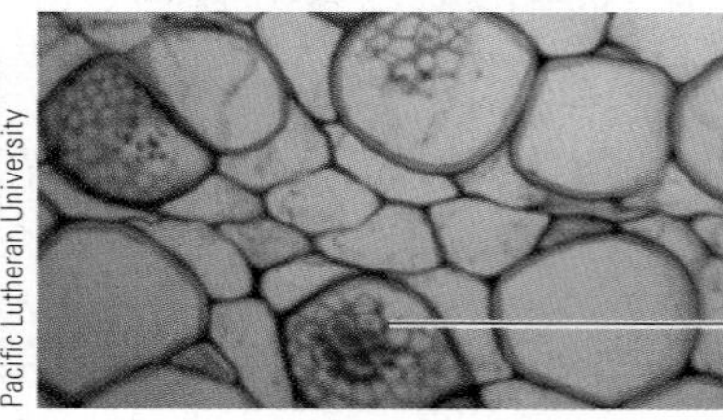
Sieve plate with openings

Courtesy of Professor John Main, Pacific Lutheran University

Dermal Tissues Protect Plant Surfaces

A complex tissue called **epidermis** covers the primary plant body in a single continuous layer **(Figure 31.9A)** or sometimes in multiple layers of tightly packed cells. The outer surface of epidermal cell walls is coated with waxes that are embedded in cutin, a network of chemically linked fats. Epidermal cells secrete this coat-

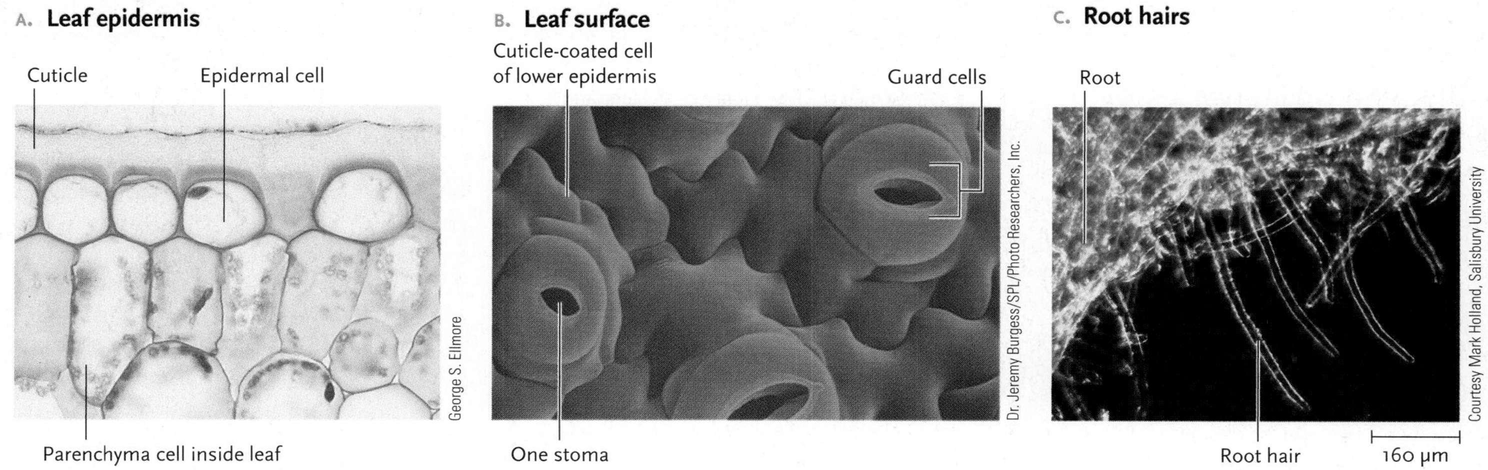
A. Leaf epidermis
Cuticle
Epidermal cell
Parenchyma cell inside leaf
George S. Ellmore
B. Leaf surface
Cuticle-coated cell of lower epidermis
Guard cells
One stoma
Dr. Jeremy Burgess/SPL/Photo Researchers, Inc.
C. Root hairs
Root
Root hair
160 μm
Courtesy Mark Holland, Salisbury University

FIGURE 31.9

Structure and examples of epidermal tissue. (A) Cross section of leaf epidermis from a bush lily *(Clivia miniata)*. **(B)** Scanning electron micrograph of a leaf surface, showing cuticle-covered epidermal cells and stomata. **(C)** Root hairs, trichomes which develop from root epidermis.

ing, or **cuticle,** which resists water loss and helps fend off attacks by microbes. A cuticle coats all external plant parts except the very tips of the shoot and most absorptive parts of roots. Other root regions have an extremely thin cuticle.

Some epidermal cells are modified in ways that represent major adaptations for plants. Young stems, leaves, flower parts, and even some roots have pairs of crescent-shaped **guard cells (Figure 31.9B).** Unlike other epidermal cells, guard cells contain chloroplasts and so can carry out photosynthesis. In addition, the porelike openings between guard cells, termed **stomata** (singular, *stoma*), play another crucial role. Carbon dioxide for photosynthesis enters plants through open stomata, while water vapor and oxygen exit the plant by the same route. As described more fully in Chapter 32, mechanisms that regulate the opening and closing of stomata are essential to maintaining an appropriate balance between the intake of CO_2 and the loss of water vapor from plant tissues.

DISCOVERING HOW GENES CONTROL STOMATA With their exact spacing and vital role in regulating gas exchange between a plant and its environment, stomata have captured the interest of geneticists probing the molecular underpinnings of plant development. Working with *Arabidopsis thaliana* (thale cress), researchers have identified an enzyme—encoded by the gene *YDA*—that appears to exert overall control over where and how many stomata form. In mutant plants with a defective enzyme, the epidermis is blanketed with stomata packed side by side. The plants often die early in development or are stunted and appear fuzzy—hence the enzyme's name, YODA, recalling the short, hairy Star Wars character. In nonmutated wild-type plants, unequal divisions of precursor cells produce one smaller and one larger daughter cell, and the smaller one gives rise to a stoma's two guard cells. (The larger cell either divides again or becomes an underlying epidermal cell.) When the YODA enzyme is activated (by phosphorylation), it triggers a cascade of reactions that, by some as-yet-unknown mechanism, either promote or restrict these asymmetric divisions.

Many plants also can apparently modify their stomata in response to changing levels of carbon dioxide in the environment. A flurry of research on this topic began when plant physiologist F. Ian Woodward of Sheffield University hypothesized that a rising concentration of CO_2 in the atmosphere—a hallmark of global climate change—acts as a trigger for the development of fewer stomata in leaves, because under such conditions plants would not require as many stomata to obtain adequate CO_2. When Woodward examined herbarium specimens collected over a recent 200-year period during which atmospheric CO_2 has steadily increased, the number of stomata in leaves of the test specimens had indeed declined. Subsequent research has confirmed that plants can reduce or increase the number of stomata in response to external CO_2 levels. Apparently, a biochemical signal from a plant's mature leaves influences the development of new leaves, adjusting the number of stomata to meet demand. Experiments have identified at least one gene that has a central role in these events, and more studies are underway to elucidate the control pathways.

OTHER EPIDERMAL SPECIALIZATIONS Single-celled or multicellular outgrowths, collectively called **trichomes,** give the stems or leaves of some plants a hairy appearance. Some trichomes exude sugars that attract insect pollinators. Leaf trichomes of members of the genus *Urtica,* the stinging nettles, provide protection by injecting an irritating toxin into the skin of animals that brush against the plant, or try to eat it. **Root hairs,** which develop as extensions in the outer wall of root epidermal cells **(Figure 31.9C),** are also trichomes. As we noted earlier, root hairs absorb much of a plant's water and minerals from the soil.

The epidermal cells of flower petals (which are modified leaves) synthesize pigments that are partly responsible for a blossom's colors. However, molecular studies have revealed that flower colors and their intensity or brightness also depend on the shape of the epidermal cells, as described in *Insights from the Molecular Revolution.*

INSIGHTS FROM THE **Molecular Revolution**

Shaping Up Flower Color

Different pigments in flowers produce different colors, but are pigments the whole color story? A molecular study of flower color in snapdragons *(Antirrhinum majus)* provided a surprising answer. Ken-ichi Noda of the Nippon Oil Company in Japan and Beverley J. Glover and her colleagues at the John Innes Centre in England were interested in a mutant snapdragon called *mixta*, which produces pale red flower petals with a dull, flat surface **(Figure 1)** rather than the deep red, velvety petals of wild-type plants **(Figure 2).**

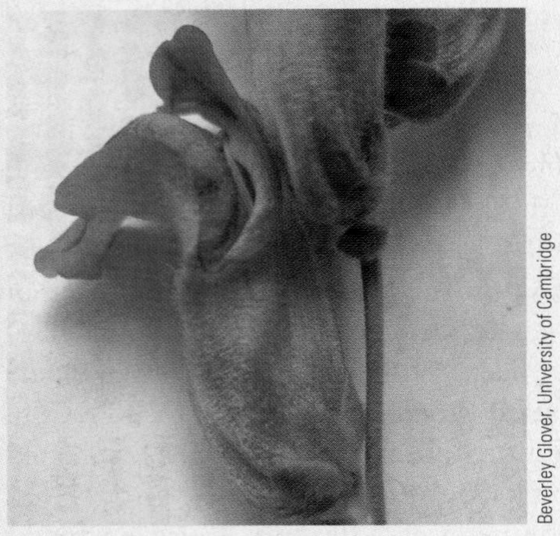

FIGURE 1
Mutant snapdragon, which has flowers with pale, flat-looking petals

FIGURE 2
Wild-type snapdragon, which has flowers with deep red petals.

The investigators isolated, cloned, and sequenced the *mixta* gene. Its sequence closely resembles a regulatory gene that activates genes in some other plants. The similarities suggested that the normal snapdragon gene also codes for a regulatory protein that produces normal flower color. When the transposable element inserts in the gene, the regulatory protein is lost (see Section 15.5).

Research Question

How does the regulatory protein govern flower color?

Experiments

At first Noda and his colleagues thought the protein regulated production of anthocyanin, a pigment that gives flowers a red color, and that loss of the protein in mutant plants hampered anthocyanin production. They discarded this hypothesis when both the wild-type and *mixta* plants were found to have normal levels of anthocyanin. However, microscopic examination of flower petals revealed that wild-type and *mixta* epidermal cells are shaped differently. Normal plants have conical epidermal cells, with the tip of the cone pointing outward and giving the petals a velvety appearance. Epidermal cells of *mixta* mutants have a flat, irregular surface that produces a dull appearance. **Figure 3** shows the surface of a variegated flower petal, which has both conical and flat cells. This structural difference suggested that in wild-type petals, the cone tips act as prisms that make the red pigment clearly visible, while the irregular surface of the mutant cells scatters light and masks the pigment color.

As a test, the research team removed the cell walls from the epidermal petal cells, a step that eliminated differences in cell shape. Both the normal and mutant cells had the same intense, red color.

Conclusion

Based on their investigations, the researchers proposed that the regulatory protein encoded in the normal *MIXTA* gene activates other genes whose protein products in some way produce the conical cell shape.

Source: Ken-ichi Noda et al. 1994. Flower colour intensity depends on specialized cell shape controlled by a Myb-related transcription factor. *Nature* 369:661–664.

Flat cells
Conical cells

FIGURE 3
Scanning electron micrograph of the surface of a variegated snapdragon petal, showing conical cells in the bright red colored areas and flat cells in the pale colored areas. The genetic events that produce variegated petals were clues that helped the research team identify the *mixta* gene.

STUDY BREAK 31.2

1. **Describe the defining features, cellular components, and functions of the ground tissue system.**
2. **What are the functions of xylem and phloem?**
3. **What are the cellular components and functions of the dermal tissue system?**

31.3 Primary Shoot Systems

A flowering plant's primary shoot system consists of the main stem and leaves. In this section we take a closer look at how the constituent parts of the primary shoot system are organized and how they grow and function.

Stems Are Adapted to Provide Support, Routes for Vascular Tissues, Storage, and New Growth

Stems incorporate evolutionary adaptations that serve four main functions. First, stems provide mechanical support, generally along a vertical (upright) axis, for body parts such as leaves and flowers. Second, stems house the vascular tissues (xylem and phloem), which transport water and dissolved minerals, hormones, products of photosynthesis, and other substances throughout the plant. Third, stems often have modifications for storing water and food. And finally, buds and specific stem regions contain meristematic tissue that gives rise to new cells of the shoot.

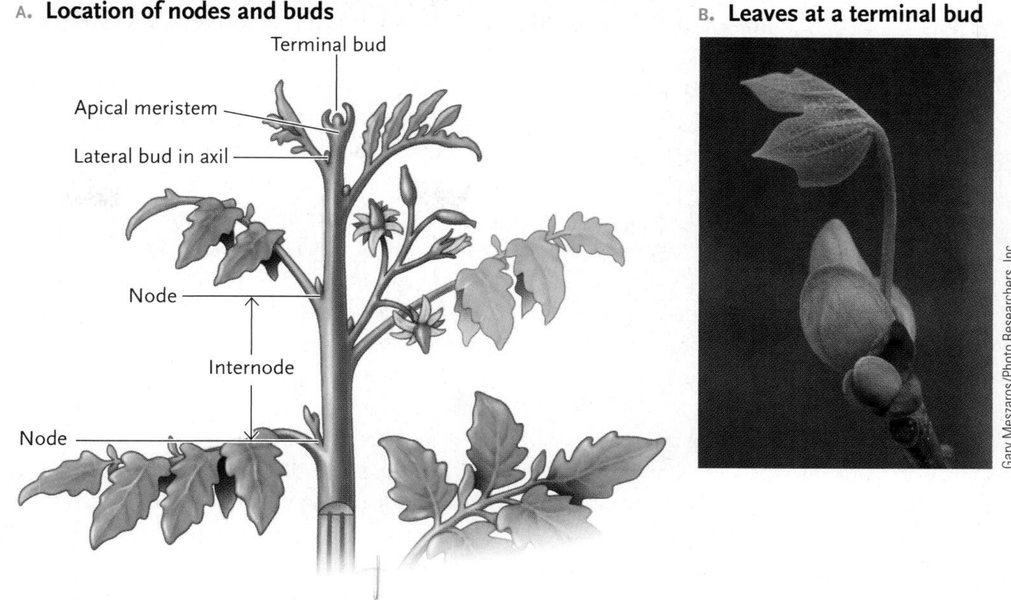

A. Location of nodes and buds
Terminal bud
Apical meristem
Lateral bud in axil
Node
Internode
Node
B. Leaves at a terminal bud
Gary Meszaros/Photo Researchers, Inc.

FIGURE 31.10
Modular structure of a stem. (A) The arrangement of nodes and buds on a plant stem. **(B)** Formation of leaves at a terminal bud of a dogwood *(Cornus)*.

THE MODULAR ORGANIZATION OF A STEM As Section 31.1 noted, we can think of a stem as a set of modules. **Figure 31.10A** shows the two basic elements of each stem module, a *node* and an *internode*. A **node** is where one or more leaves attach to a stem. An **internode** is the area between two nodes. The upper angle between the stem and an attached leaf is an **axil.** New primary growth occurs in the embryonic shoots called buds. A **terminal bud** occurs at the apex of each stem **(Figure 31.10B)**. **Lateral buds** in the leaf axils produce new branches or shoots that give rise to flowers.

OVERVIEW OF STEM GROWTH Several general patterns of growth evolved in flowering plants. In eudicots, most growth in a stem's length occurs directly below the apical meristem, as internode cells divide and elongate. Internode cells nearest the apex are most active, so the most visible new growth occurs at the ends of stems. By contrast, in grasses and some other monocots the upper cells of an internode stop dividing as the newly formed internode elongates, and cell division is limited to a meristematic region at the base of the internode. The stems of bamboo and other grasses elongate as the growth of such meristems "pushes up" the internodes. This adaptation allows grasses to rapidly grow back after grazing by herbivores (or being chopped off by a lawnmower), because the meristem is not removed.

Terminal buds release a hormone that inhibits the growth of nearby lateral buds, a phenomenon called **apical dominance.** Gardeners who want a bushier plant can stimulate lateral bud growth by periodically cutting off the terminal bud. The flow of hormone signals from the apical meristem then dwindles to a level low enough that lateral buds begin to grow. In nature, apical dominance is an adaptation that directs the plant's resources into growing up toward the light.

PRIMARY GROWTH AND STRUCTURE OF A STEM Primary growth of a shoot and its parts begins in the shoot apical meristem. In addition to giving rise to stem tissue, the shoot apical meristem also periodically generates shoot primordia, bulges that are the first developmental stages of leaves, additional shoots, and reproductive structures such as flowers.

A shoot apical meristem is a dome-shaped mass of cells. When one cell divides, one of its daughter cells becomes an **initial** that remains as part of the meristem. The other daughter cell, called a **derivative,** will be the source of specialized cells. In this way initials function much like stem cells in animals. They replenish the supply of initials in the meristem and also provide derivatives that result in a plant's new growth. **Figure 31.11** shows the sequence for a eudicot shoot.

As derivatives differentiate, they give rise to three **primary meristems:** *protoderm, procambium,* and *ground meristem* (see Figure 31.11). These meristems are relatively unspecialized tissues producing cells that differentiate in turn into specialized cells and tissues. In eudicots, the primary meristems are also responsible for elongation of the plant body. Some monocots, such as palms, have a *primary thickening meristem* (just under the protoderm) that contributes to both stem elongation and its lateral growth.

How do the genetically identical cells of an apical meristem give rise to three types of primary meristem cells, and ultimately to all the specialized cells of the plant? *Focus on Basic Research* describes some experiments that are probing the genetic mechanisms underlying meristem activity.

Each primary meristem occupies a different position in the shoot tip, as shown in Figure 31.11A. Outermost is **protoderm,** a meristem that will produce the stem's epidermis. While protoderm cells divide and the resulting derivatives are maturing, the shoot tip continues to grow. Eventually, the protoderm cells dif-

FOCUS ON Basic Research

Homeobox Genes: How the Meristem Gives Its Marching Orders

How do descendents of some dividing cells in a shoot apical meristem (SAM) "know" to become stem tissues, while others embark on the developmental path that produces leaves or other shoot parts? The full answer to this question is not yet known, but research teams around the world are studying a genetic mechanism that appears to guide the process.

Working with SAM tissue from maize (*Zea mays*, generally known in North America as corn), investigators have identified more than a dozen regulatory genes whose protein products activate groups of other genes in differentiating cells. Some genes that act in this way to guide development along a particular path are called *homeotic genes*, because they contain a nucleotide sequence called the homeobox. As described in Chapter 16, the homeobox is the part of a homeotic gene that codes for a homeodomain—a sequence of amino acids in a transcription factor. The homeodomain binds to regulatory regions of certain genes; this binding regulates transcription of such genes, turning them on or off. Homeobox genes were fi rst discovered in studies of how legs, antennae, and other structures develop in the fruit fly *Drosophila melanogaster*.

Sarah Hake of the Plant Gene Expression Center (United States Department of Agriculture) was curious about the action of a homeotic gene in maize that is known as *knotted-1* (*KN-1*). Normally, the *KN-1* gene is expressed in apical meristems, where it keeps the meristem in an undifferentiated state. When a mutated form of the gene, *kn-1*, is expressed, however, the mutation causes abnormal knobby growths on leaves—hence the gene's name. Hake's research helped establish that *KN-1* defines developmental pathways that unfold in meristems. For example, when Hake cloned the *KN-1* gene and inserted it into tobacco leaf cells, the cells *de*differentiated and began acting like meristem cells. As they divided, they produced lines that could differentiate into leaves and stems.

Subsequent studies of *KN-1* in species as diverse as sunflowers and garden peas have led to the identification of the family of what are now called knotted-1-like genes, all of which encode regulatory proteins that influence developmental pathways. As in maize, some are typically expressed in SAM tissue. In sunflower, tomato, and perhaps other species, knotted-1-like genes also appear to be expressed in differentiated plant parts including leaves, flowers, stems, and even roots. The early work on SAM tissue and homeobox genes in maize has blossomed into a wide-ranging investigation of the molecular signals that shape plant architecture.

ferentiate into specific types of epidermal cells, including guard cells and trichomes.

Inward from the protoderm is the **ground meristem,** which will give rise to ground tissue, most of which is parenchyma. **Procambium,** which produces the primary vascular tissues, exists as threadlike strands of cells within the ground meristem. Procambial cells are long and thin, and their spatial orientation foreshadows the future function of the tissues they produce. In

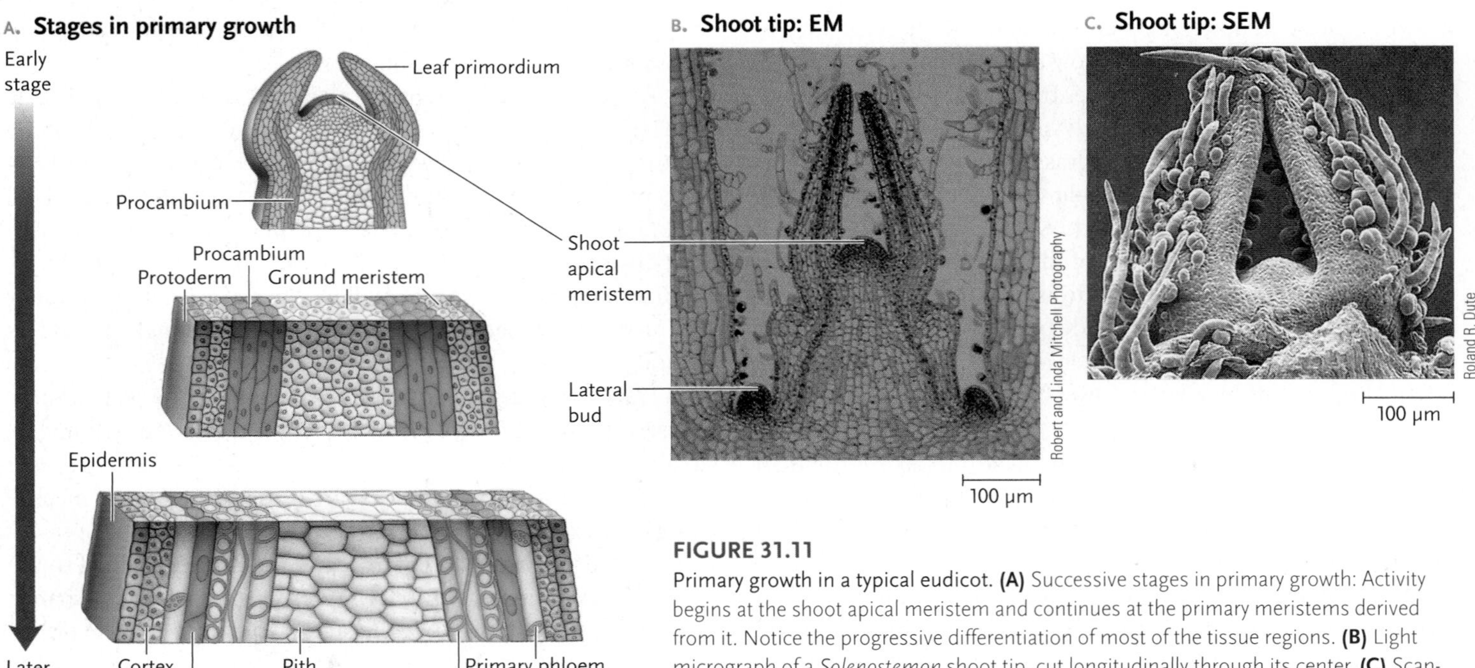
A. Stages in primary growth
Early stage
Leaf primordium
Procambium
Procambium
Protoderm
Ground meristem
Epidermis
Later stage
Cortex
Procambium
Pith
Primary phloem
Primary xylem
B. Shoot tip: EM
Shoot apical meristem
Lateral bud
100 μm
Robert and Linda Mitchell Photography
C. Shoot tip: SEM
100 μm
Roland R. Dute

FIGURE 31.11
Primary growth in a typical eudicot. **(A)** Successive stages in primary growth: Activity begins at the shoot apical meristem and continues at the primary meristems derived from it. Notice the progressive differentiation of most of the tissue regions. **(B)** Light micrograph of a *Solenostemon* shoot tip, cut longitudinally through its center. **(C)** Scanning electron micrograph of its surface.

most plants, inner procambial cells give rise to xylem and outer procambial cells to phloem. In plants with secondary growth, a thin region of procambium between the primary xylem and phloem remains undifferentiated. As described shortly, later on it will give rise to a lateral meristem.

The developing vascular tissues become organized into **vascular bundles,** multistranded cords of primary xylem and phloem. Wrapped or capped by sclerenchyma, the bundles thread lengthwise through the parenchyma, forming a **stele** (Greek *stele* = pillar; some botanists prefer to call vascular bundles in the shoot the *vascular cylinder*). The stele runs vertically. The ground tissue outside it forms a **cortex,** while tissue inside it is called **pith (Figure 31.12A).** As leaves and buds appear along a stem, some vascular bundles in the stem branch off into these developing tissues.

Both cortex and pith are mainly parenchyma, and in some plants the pith parenchyma stores starch. In the stems of most monocots, vascular bundles are dispersed through the ground tissue **(Figure 31.12B),** so separate cortical and pith regions do not form. In some monocots, including bamboo, the pith breaks down, leaving the stem with a hollow core. The hollow stems of some hard-walled bamboo species are used to make bamboo flutes.

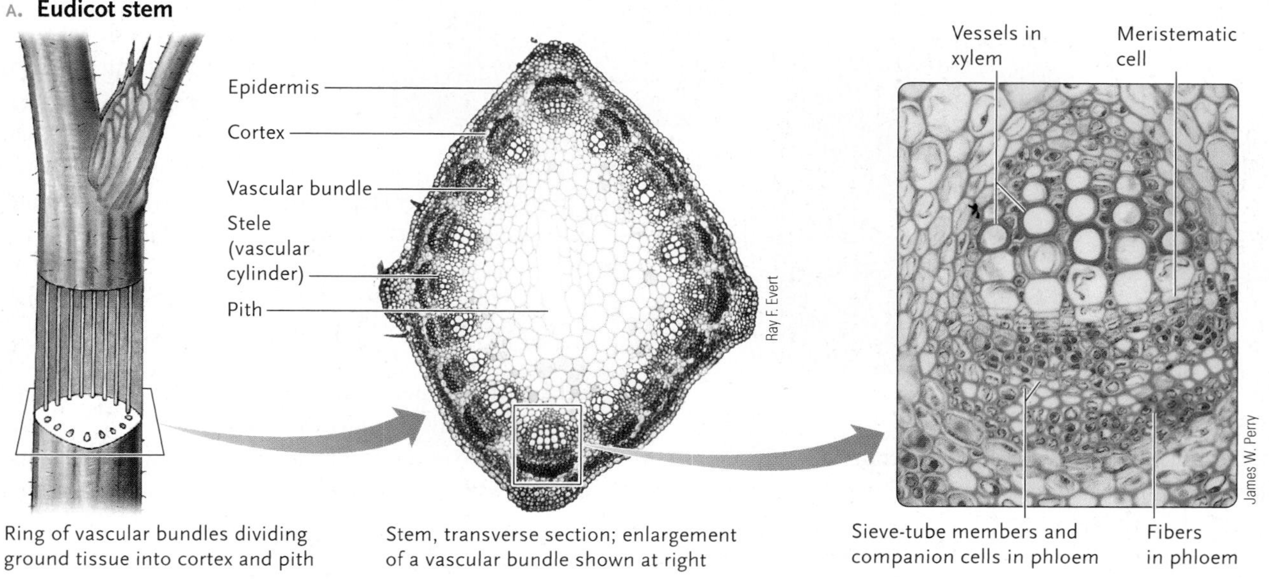
A. Eudicot stem
Epidermis
Cortex
Vascular bundle
Stele (vascular cylinder)
Pith
Ray F. Evert
Vessels in xylem
Meristematic cell
James W. Perry
Ring of vascular bundles dividing ground tissue into cortex and pith
Stem, transverse section; enlargement of a vascular bundle shown at right
Sieve-tube members and companion cells in phloem
Fibers in phloem

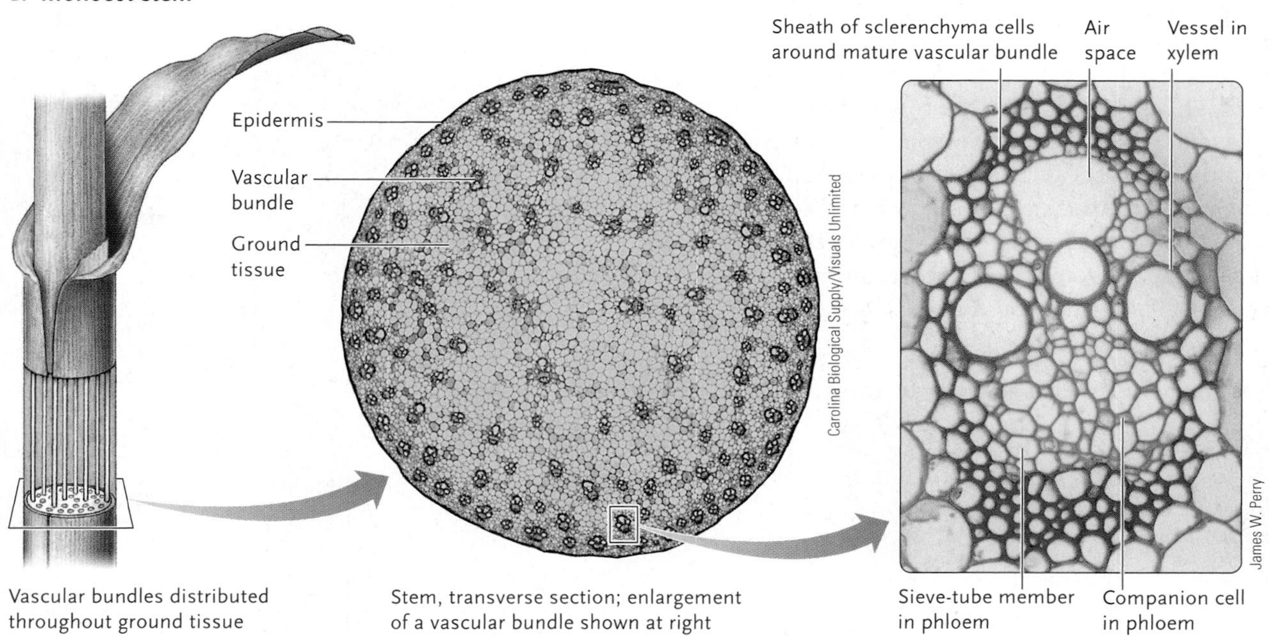
B. Monocot stem
Epidermis
Vascular bundle
Ground tissue
Carolina Biological Supply/Visuals Unlimited
Sheath of sclerenchyma cells around mature vascular bundle
Air space
Vessel in xylem
James W. Perry
Vascular bundles distributed throughout ground tissue
Stem, transverse section; enlargement of a vascular bundle shown at right
Sieve-tube member in phloem
Companion cell in phloem

FIGURE 31.12
Organization of cells and tissues inside the stem of a eudicot and a monocot. **(A)** Part of a stem from alfalfa *(Medicago)*, a eudicot. In many species of eudicots and conifers, the vascular bundles develop in a more or less ringlike array in the ground tissue system, as shown here. **(B)** Part of a stem from corn *(Zea mays)*, a monocot. In most monocots and some herbaceous eudicots, vascular bundles are scattered through the ground tissue, as shown here.

A. **Onion bulb**

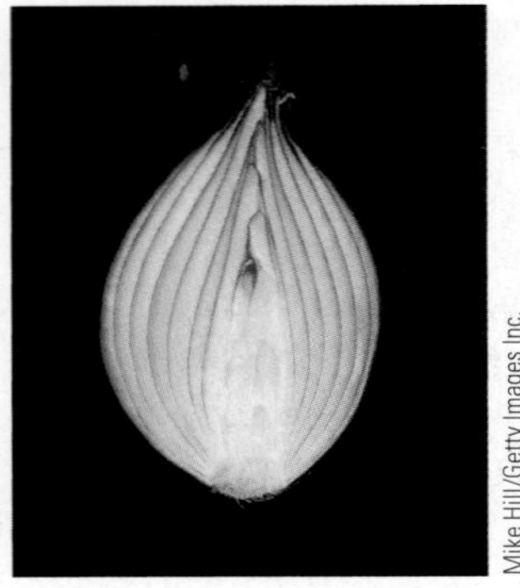
Mike Hill/Getty Images Inc.

B. **Potato tuber**

Wally Eberhart/Visuals Unlimited

C. **Ginger rhizome**

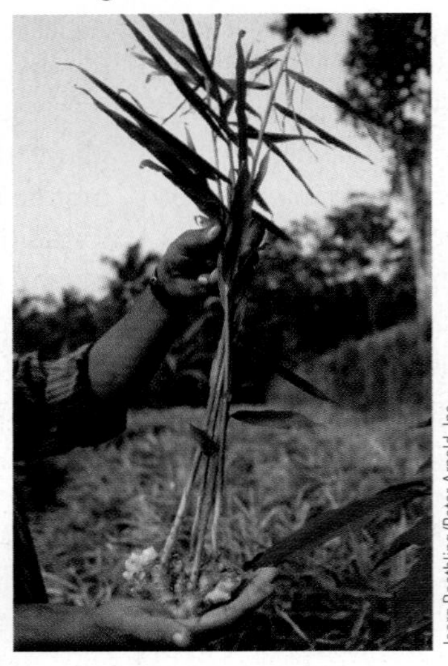
Joerg Boethling/Peter Arnold, Inc.

D. **Crocus corm**

Alan & Linda Detrick/
Photo Researchers, Inc.

E. **Strawberry stolons**

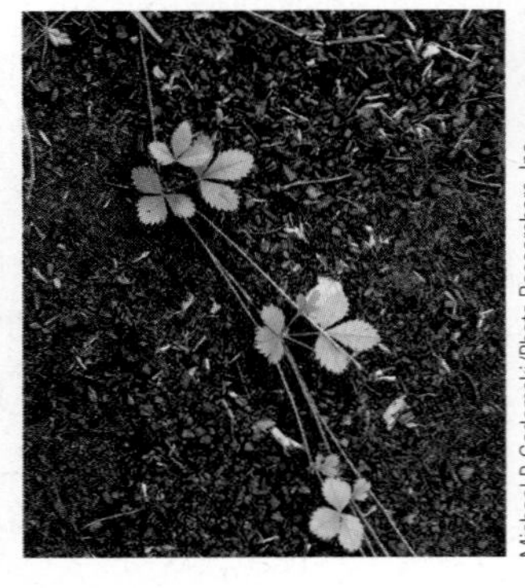
Michael P. Gadomski/Photo Researchers, Inc.

FIGURE 31.13
A selection of modified stems. **(A)** The fleshy bulbs of onions *(Allium cepa)* are modified leaves in which the plant stores starch. **(B)** A potato *(Solanum tuberosum)*, a tuber. **(C)** Ginger "root," the pungent, starchy rhizome of the ginger plant *(Zingiber officinale)*. **(D)** Crocus plants (genus *Crocus*) typically grow from a corm. **(E)** A strawberry plant *(Fragaria ananassa)* and stolon.

STEM MODIFICATIONS Evolution has produced a range of stem specializations, including structures modified for reproduction, food storage, or both **(Figure 31.13)**. An onion or a garlic head is a *bulb,* a modified shoot that consists of a bud with fleshy leaves. *Tubers* are stem regions enlarged by the presence of starch-storing parenchyma cells; examples of plants that form tubers are the potato and the cassava (the source of tapioca). The "eyes" of a potato are buds at nodes of the modified stem, and the regions between eyes are internodes. Many grasses, ground ivies, and some weeds are difficult to eradicate because they have *rhizomes*—long underground stems that can extend as much as half a meter deep into the soil and rapidly produce new shoots when existing ones are pulled out. The pungent, starchy "root" of ginger is a rhizome also. Crocuses and some other ornamental plants develop elongated, fleshy underground stems called *corms,* another starch-storage adaptation. Tubers, rhizomes, and corms all have meristematic tissue at nodes from which new plants can be propagated—a vegetative (asexual) reproductive mode. Other plants, including the strawberry, reproduce vegetatively via slender stems called *stolons,* which grow along the soil surface. New plants arise at nodes along the stolon.

Leaves Are Adapted for Photosynthesis, Gas Exchange, and Conserving Water

Each spring a mature maple tree heralds the new season by unfurling roughly 100,000 leaves. Some other tree species produce leaves by the millions. For these and most other plants, leaves are multifaceted organs with evolutionary adaptations that make them the main sites of photosynthesis and permit them to exchange gases and limit the loss of water by evaporation.

A. **Common forms of eudicot and monocot leaves**

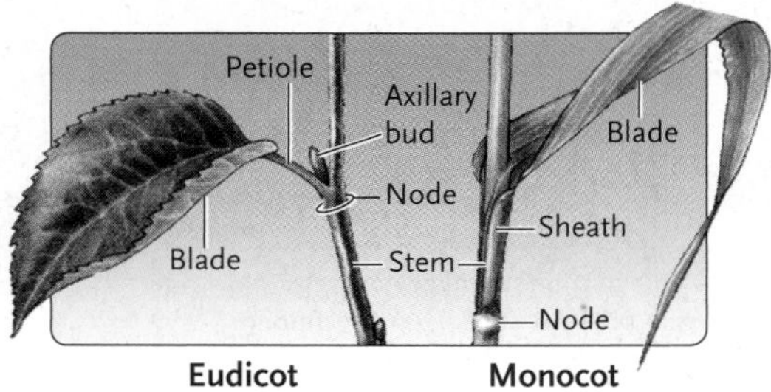
Petiole
Axillary bud
Node
Blade
Stem
Blade
Sheath
Node
Eudicot
Monocot

B. **Simple leaves (eudicot)**

Poplar
(Populus)
Oak
(Quercus)
Maple
(Acer)

C. **Compound leaves (eudicot)**

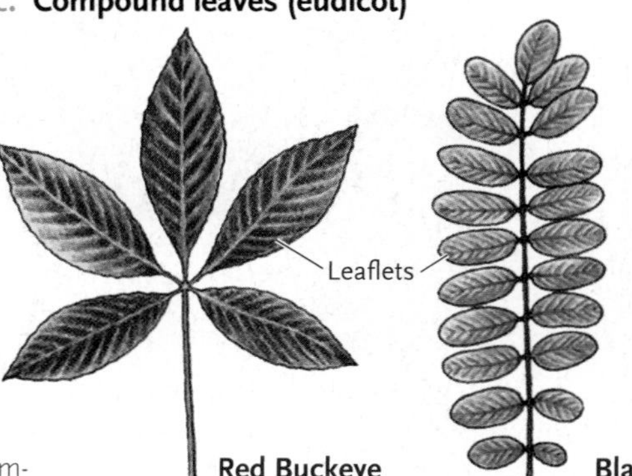
Leaflets
Red Buckeye
(Aesculus)

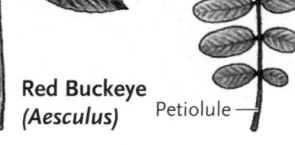

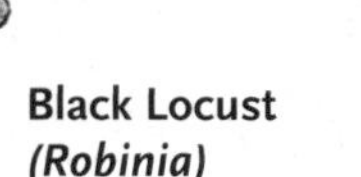
Petiolule
Black Locust
(Robinia)

Honey Locust
(Gleditsia)

FIGURE 31.14
Leaf forms. **(A)** Common forms of eudicot and monocot leaves. **(B)** Examples of simple eudicot leaves. **(C)** Examples of compound eudicot leaves.

LEAF MORPHOLOGY AND ANATOMY Externally, the most conspicuous part of a leaf is its thin, flattened **blade,** which provides a large surface area for absorbing sunlight and carbon dioxide **(Figure 31.14A).** In many eudicot leaves, the blade narrows to a stalklike **petiole,** which attaches the leaf to a stem. Petioles can be long, short, or in between, depending on the species. The stalks of celery and rhubarb are especially fleshy petioles. Unless a leaf's petiole is very short, it holds the leaf away from the stem and helps prevent individual leaves from shading one another. Most monocot leaves lack a petiole. Instead, in species such as rye grass or corn, the blade is long and narrow and its base simply forms a sheath around the stem.

Studies show that in general, leaves of flowering plants are oriented on the stem axis so that they can capture the maximum amount of sunlight. Chapter 35 describes how the stems and leaves of some plants change position to follow the sun's movement during the course of a day.

LEAF MODIFICATIONS Leaf forms are based on two basic patterns: simple leaves, which have a single blade **(Figure 31.14B),** and compound leaves, in which the blade is divided into smaller leaflets **(Figure 31.14C).** Leaf edges or margins may be smooth, toothed, or lobed. As with other plant parts, however, the adaptation of land plants to different environments has produced tremendous variety in leaf morphology. For instance, spiny margins on the leaves of the carnivorous Venus flytrap *(Dionaea muscipula)* **(Figure 31.15A)** prevent the escape of insects that become trapped when the seemingly hinged leaves snap shut around them—a movement that takes only about a tenth of a second.

Cactus leaves are modified as spines, while leaves of other plants have trichomes that take the form of hairs or hooks **(Figure 31.15B)**—all possible adaptations for defense against grazing by herbivores. Leaves or parts of leaves also may be modified into tendrils, like those of the sweet pea **(Figure 31.15C),** or other structures. Epidermal cells on the leaves of the saltbush *Atriplex spongiosa*

FIGURE 31.15

A few adaptations of leaves. **(A)** Margins on the leaves of the Venus flytrap *(Dionaea muscipula)* are modified into long, interlocking spines. **(B)** Specializations on a tomato leaf include hooklike hairs and lobed glandular structures that release an insect-deterring chemical. **(C)** The tendrils of a sweet pea *(Lathyrus odoratus)* help to support the climbing plant's stem. **(D)** Light micrograph of salt bladders that appear on the leaves of a saltbush plant *(Atriplex spongiosa)*. The "bladders" are trichomes, specialized outgrowths of the leaf epidermis in which excess salt from the plant's tissue fluid accumulates. The salt-laden trichomes eventually burst or slough off.

A. **Interlocking spines of Venus flytrap leaves**

iStockphoto.com/Ryan Poling

B. **Hairs and glandular structures on a tomato leaf**

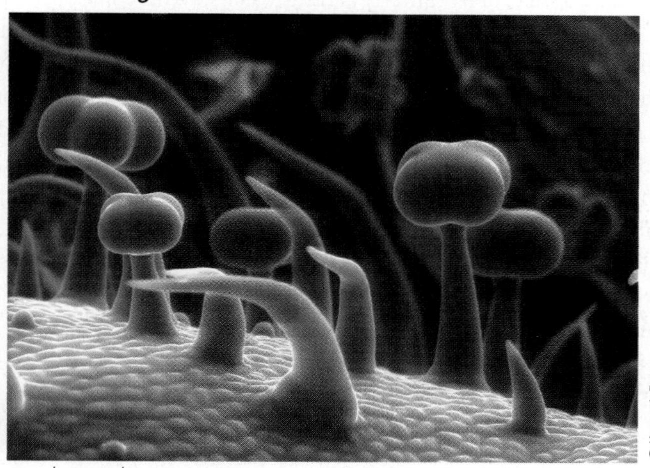
50 μm

© Kenneth Bart

C. **Tendrils of a sweet pea**

Maxine Adcock/Photo Researchers, Inc.

D. **Salt bladder of a saltbush plant**

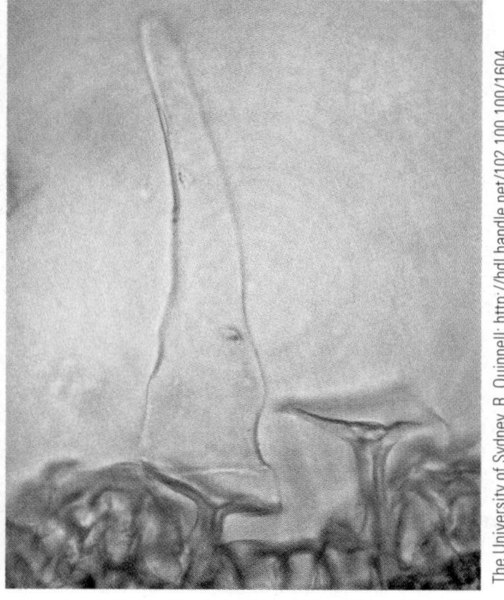

The University of Sydney, R. Quinnell: http://hdl.handle.net/102.100.100/1604

form balloon-like structures **(Figure 31.15D)** that contain concentrated Na^+ and Cl^- taken up from the salty soil. Eventually, the salt-filled epidermal cells burst or fall off the leaf, releasing the salt to the outside. This adaptation helps control the salt concentration in the plant's tissues—another example of the link between structure, function, and the environment in which a plant lives.

LEAF PRIMARY GROWTH AND INTERNAL STRUCTURE In both flowering plants and gymnosperms, leaves develop on sides of the shoot apical meristem. Initially, meristematic cells near the apex divide and their derivatives elongate. The resulting bulge enlarges into a thin, rudimentary leaf, or **leaf primordium** (see Figure 31.11). As the plant grows and internodes elongate, the leaves that form from leaf primordia become spaced at intervals along the length of a stem.

Leaf tissues typically have several layers **(Figure 31.16)**. Uppermost is epidermis, with cuticle covering its outer surface. Just beneath the epidermis is **mesophyll** (*mesos* = middle; *phyllon* = leaf), ground tissue composed of loosely packed parenchyma cells that contain chloroplasts. The leaves of many plants, especially eudicots, contain two layers of mesophyll. *Palisade mesophyll* cells contain more chloroplasts and are arranged in compact columns with smaller air spaces between them, typically toward the upper leaf surface. *Spongy mesophyll,* which tends to be located toward the underside of a leaf, consists of irregularly arranged cells with a conspicuous network of air spaces that gives it a spongy appearance. Air spaces between mesophyll cells enhance the uptake of carbon dioxide and release of oxygen during photosynthesis and account for 15% to 50% of a leaf's volume. Mesophyll also contains collenchyma and sclerenchyma cells, which support the leaf blade.

Below the mesophyll is another cuticle-covered epidermal layer. Except in grasses and a few other plants, this layer contains most of the stomata through which water vapor exits the leaf and carbon dioxide enters. For example, the upper surface of an apple leaf has no stomata, while a square centimeter of the lower surface has more than 20,000. A square centimeter of the upper epidermis of a tomato leaf has about 1,200 stomata, whereas the same area of the lower epidermis has 13,000. The positioning of stomata on the side of the leaf that faces away from the sun may be an adaptation limiting water loss by evaporation through stomatal openings.

Vascular bundles form a lacy network of **veins** throughout the leaf. Eudicot leaves typically have a branching vein pattern; in monocot leaves, veins tend to run in parallel arrays.

In temperate regions, most leaves are temporary structures. In deciduous (*deciduus* = that which falls off) species such as birches and maples, hormonal signals cause the leaves to drop from the stem as days shorten in autumn. Other temperate species, such as camellias or hollies, as well as conifers, also drop leaves, but they appear "evergreen" because the leaves may persist for several years and do not all drop at the same time.

Plant Shoots May Have Juvenile and Adult Forms

Leaf shape and other shoot characteristics can mirror the progress of a long-lived plant through its life cycle. Plants that live many years may spend part of their lives in a juvenile phase, then shift to a mature, or adult phase. The differences between juveniles and adults often are reflected in leaf size and shape, in the arrangement of leaves on the stem, or in a change from vegetative growth to a reproductive stage—or sometimes all three. Plants in the genus *Eucalyptus* display a wide range of such variations. For example, in the yellow gum (*E. doratoxylon* or *E. erythrocorys*) the leaves of a mature tree are bluish-green, long and slender with long petioles, and occur in alternating positions on a stem **(Figure 31.17)**. By contrast, the leaves of a growing seedling are oval with a pale grayish

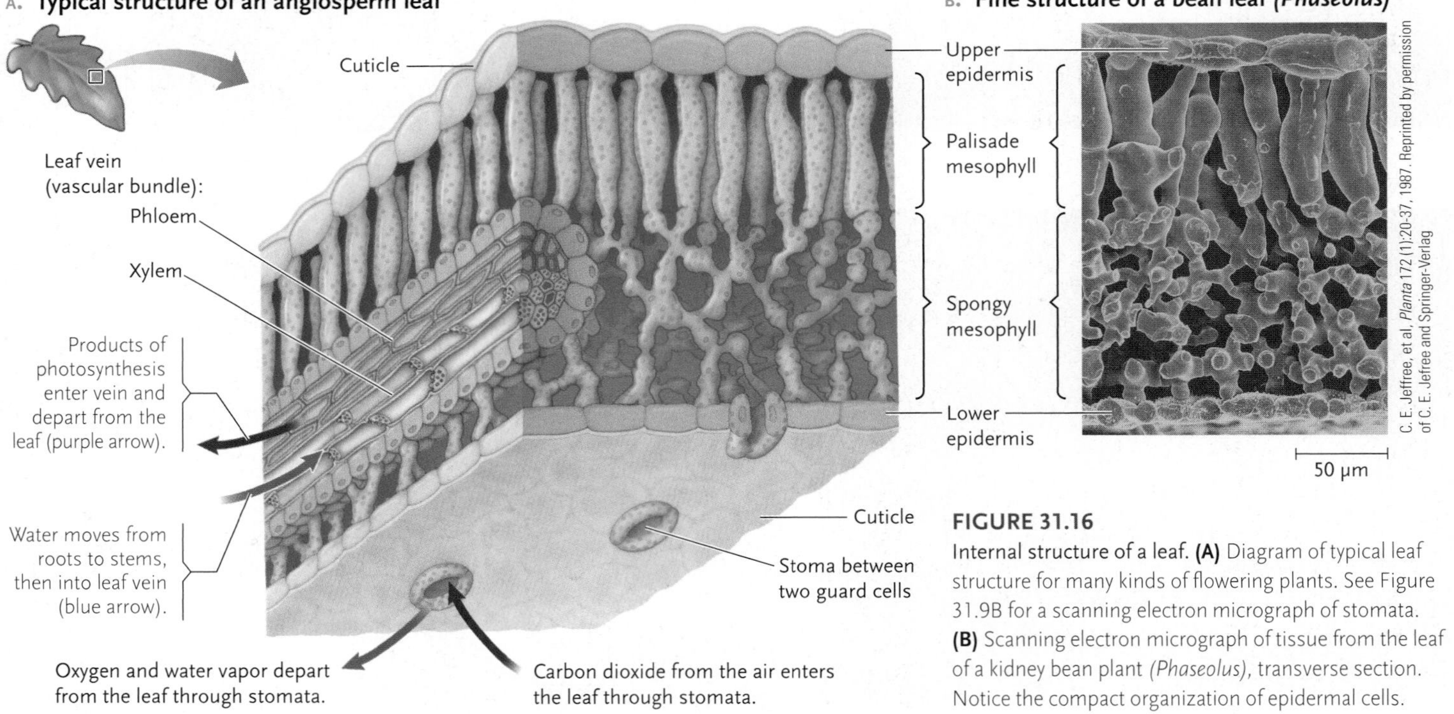

A. Typical structure of an angiosperm leaf
Cuticle
Leaf vein
(vascular bundle):
Phloem
Xylem
Products of photosynthesis enter vein and depart from the leaf (purple arrow).
Water moves from roots to stems, then into leaf vein (blue arrow).
Cuticle
Stoma between two guard cells
Oxygen and water vapor depart from the leaf through stomata.
Carbon dioxide from the air enters the leaf through stomata.
B. Fine structure of a bean leaf (Phaseolus)
Upper epidermis
Palisade mesophyll
Spongy mesophyll
Lower epidermis
50 μm
C. E. Jeffree, et al, Planta 172 (1):20-37, 1987. Reprinted by permission of C. E. Jeffree and Springer-Verlag

FIGURE 31.16
Internal structure of a leaf. **(A)** Diagram of typical leaf structure for many kinds of flowering plants. See Figure 31.9B for a scanning electron micrograph of stomata. **(B)** Scanning electron micrograph of tissue from the leaf of a kidney bean plant *(Phaseolus)*, transverse section. Notice the compact organization of epidermal cells.

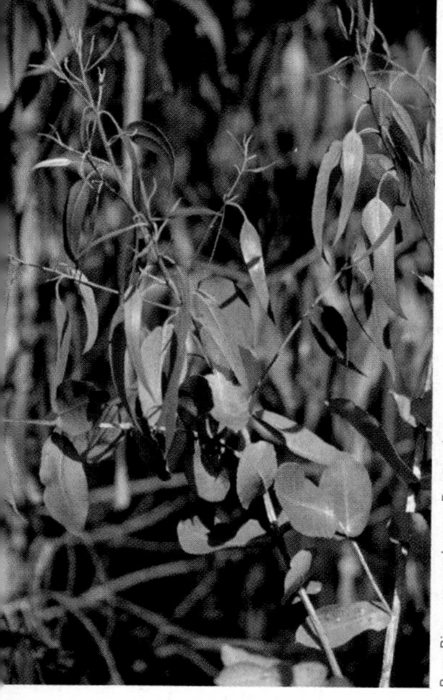
Barry Rice, sarracenia.com, Bugwood.org

FIGURE 31.17
Age-related phase changes in *Eucalyptus*. Juvenile leaves are oval; mature leaves are elongated.

surface, have little or no petiole, and occur opposite one another on the stem. A Southern magnolia tree (*Magnolia grandiflora*) doesn't flower until its juvenile phase ends, which can be 20 years or more from the time the *M. grandiflora* seed sprouts. In fact, most woody plants must attain a certain size before their meristem tissue can respond to the hormonal signals that govern flower development, a topic we consider in Chapter 35.

Phase changes provide more examples of the plasticity that characterizes plant development. They almost certainly are associated with changes in the expression of genes that control the development of stem nodes, leaf and flower buds, and other basic aspects of plant growth.

STUDY BREAK 31.3

1. **Describe the functions of stems and stem structure, and list the basic steps in primary growth of stems.**
2. **Explain the general function of leaves and how leaf anatomy supports this role in eudicots and monocots.**
3. **Describe the steps in primary growth of a leaf and the structures that result from the process.**
4. **Describe two examples of the life phases of long-lived plant species.**

THINK OUTSIDE THE BOOK

With a group of your classmates, make a list of at least five vegetables and five fruits that have not been discussed in this chapter and that you have eaten or seen in the market. Using library resources or the Web, identify the source plant's genus and species, then describe which anatomical part of the plant each fruit or vegetable is, and identify which tissue(s) make up the fruit or vegetable. Then describe each item's function in the plant.

31.4 Root Systems

Plants have to absorb enough water and dissolved minerals to sustain growth and routine cellular maintenance, a task that requires a tremendous root surface. In one study, measurements of the root system of a rye plant (*Secale cereale*) that had been growing for only four months had a surface area of more than 700 m^2—about 130 times greater than the surface area of its shoot system. The roots of carrots, sugar beets, and most other plants also store nutrients produced in photosynthesis, some to be used by root cells and some to be transported later to cells of the shoot. As a root system penetrates downward and spreads out, it also anchors the aboveground parts.

Taproot and Fibrous Root Systems Are Specialized for Particular Functions

Most eudicots begin life with a **taproot system**—a single main root, or taproot, that is adapted for storage and smaller branching roots called **lateral roots (Figure 31.18A).** As the main root grows downward, its diameter increases, and the lateral roots emerge along the length of its older, differentiated regions. The youngest lateral roots are near the root tip. Carrots and dandelions have a taproot system, as do many pines and other conifers. The taproot

FIGURE 31.18
Types of roots. **(A)** Taproot system of a California poppy (*Eschscholzia californica*). **(B)** Fibrous root system of a grass plant. **(C)** The prop roots of red mangrove trees (*Rhizophora*), examples of adventitious roots.

A. **Taproot system**

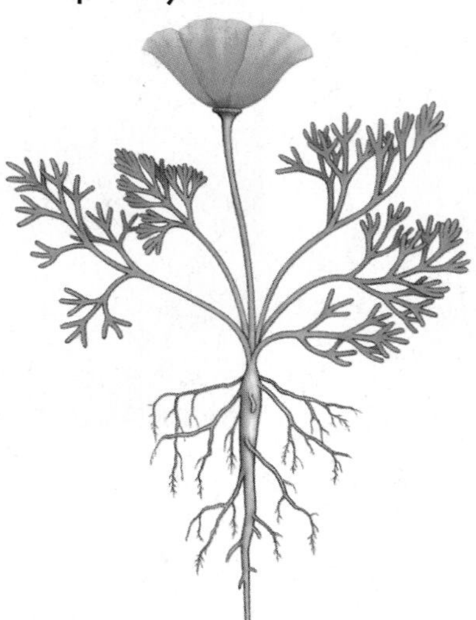

B. **Fibrous root system**

C. **Adventitious roots**

Beth Davidow/Visuals Unlimited

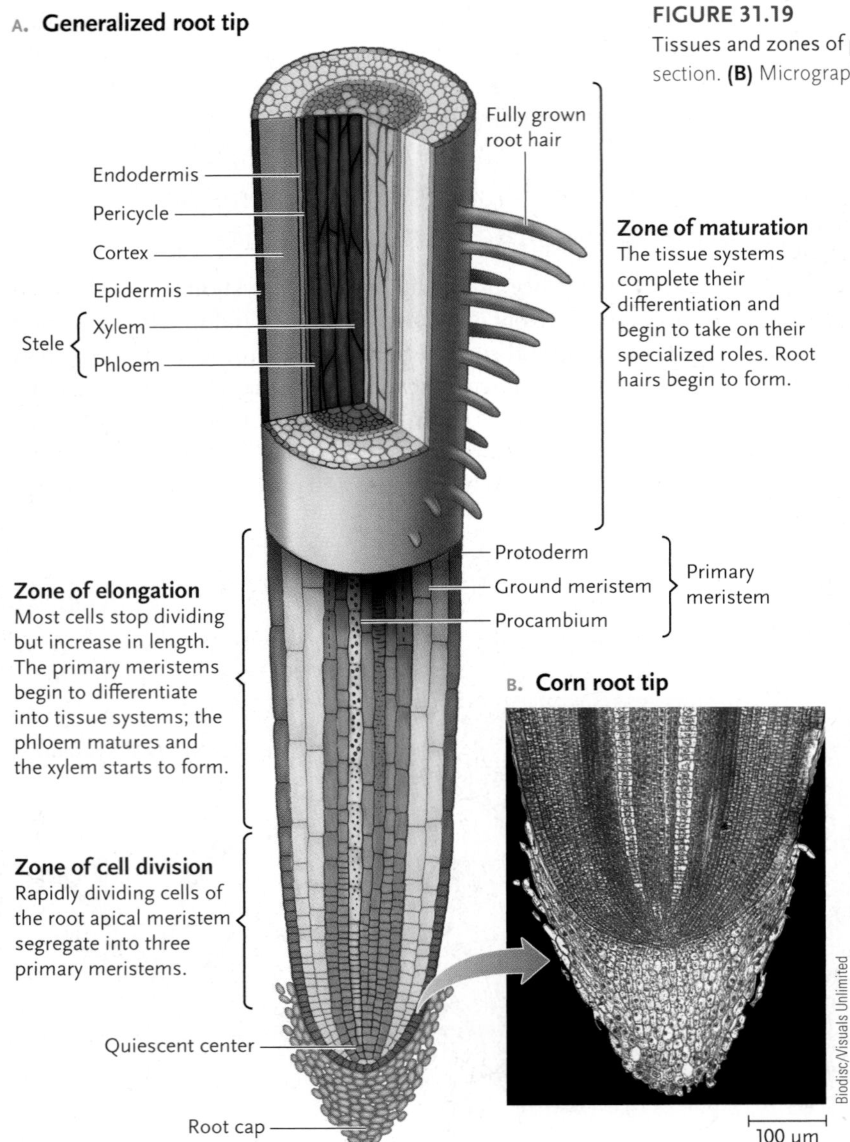
A. Generalized root tip
Fully grown root hair
Endodermis
Pericycle
Cortex
Epidermis
Stele
Xylem
Phloem
Zone of maturation
The tissue systems complete their differentiation and begin to take on their specialized roles. Root hairs begin to form.
Protoderm
Ground meristem
Procambium
Primary meristem
Zone of elongation
Most cells stop dividing but increase in length. The primary meristems begin to differentiate into tissue systems; the phloem matures and the xylem starts to form.
Zone of cell division
Rapidly dividing cells of the root apical meristem segregate into three primary meristems.
Quiescent center
Root cap
B. Corn root tip
100 µm
Biodisc/Visuals Unlimited

FIGURE 31.19
Tissues and zones of primary growth in a root tip. **(A)** Generalized root tip, longitudinal section. **(B)** Micrograph of a corn root tip, longitudinal section.

system of a longleaf pine *(Pinus palustris)* can penetrate 4 meters or more into the soil.

Grasses and many other monocots develop a **fibrous root system** in which several main roots branch to form a dense mass of smaller roots **(Figure 31.18B)**. Fibrous root systems are adapted to absorb water and nutrients from the upper layers of soil, and tend to spread out laterally from the base of the stem. Fibrous roots are important ecologically because dense root networks help hold topsoil in place and prevent erosion. During the 1930s, drought, overgrazing by livestock, and intensive farming in the American Midwest destroyed hundreds of thousands of acres of native prairie grasses, contributing to soil erosion on a massive scale. Swirling clouds of soil particles prompted journalists to name the area the Dust Bowl.

In some plants, **adventitious roots** arise from the stem of the young plant. "Adventitious" refers to any structure arising at an unusual location, such as roots that grow from stems or leaves. Adventitious roots and their branchings all are about the same length and diameter. Those of some climbing plants produce a gluey substance (from trichomes) that allows them to cling to vertical surfaces. The *prop roots* of a corn plant are adventitious roots that develop from the shoot node nearest the soil surface; they both support the plant and absorb water and nutrients. Mangroves and other trees that grow in marshy habitats often have huge prop roots, which develop from branches as well as from the main stem **(Figure 31.18C)**.

Root Structure Is Specialized for Underground Growth

Like shoots, roots have distinct anatomical parts, each with a specific function. In most plants, primary growth of roots begins when an embryonic root (called a *radicle*) emerges from a germinating seed and its meristems become active. **Figure 31.19** shows the structure of a root tip. Notice that the root apical meristem terminates in a dome-shaped cell mass, the **root cap.** The meristem produces the cap, which in turn surrounds and protects the meristem as the root elongates through the soil. Certain cells in the cap respond to gravity, which guides the root tip downward. Cap cells also secrete a polysaccharide-rich substance that lubricates the tip and eases the growing root's passage through the soil. Outer root cap cells are continually abraded off and replaced by new cells at the cap's base.

ZONES OF PRIMARY GROWTH IN THE ROOT Primary growth in roots takes place in successive stages, beginning at the root tip and progressing upward. Just inside the root cap some roots have a small clump of apical meristem cells called the **quiescent center.** Unlike other meristematic cells, cells of the quiescent center divide very slowly unless the root cap or the apical meristem is injured, then they become active and can regenerate the damaged part. The quiescent center also may include cells that synthesize plant hormones that control root development.

The root apical meristem and the actively dividing cells behind it form the **zone of cell division.** As in the stem, cells of the apical meristem segregate into three primary meristems. Cells in the center of the root tip become the procambium; those just outside the procambium become ground meristem; and those on the periphery of the apical meristem become protoderm.

The zone of cell division merges into the **zone of elongation.** Most of the increase in a root's length comes about here as cells become longer as their vacuoles fill with water. This "hydraulic" elongation pushes the root cap and apical meristem through the soil as much as several centimeters a day.

Above the zone of elongation, cells do not increase in length but they may differentiate further and take on specialized roles in the **zone of maturation.** For instance, epidermal cells in this

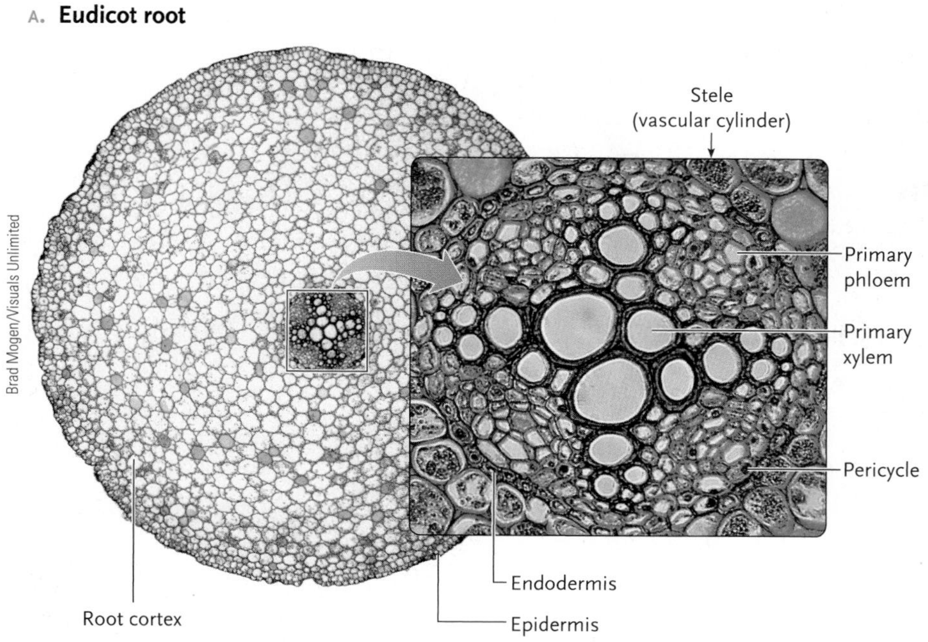

A. **Eudicot root**

Stele (vascular cylinder)

Primary phloem

Primary xylem

Pericycle

Endodermis

Epidermis

Root cortex

Brad Mogen/Visuals Unlimited

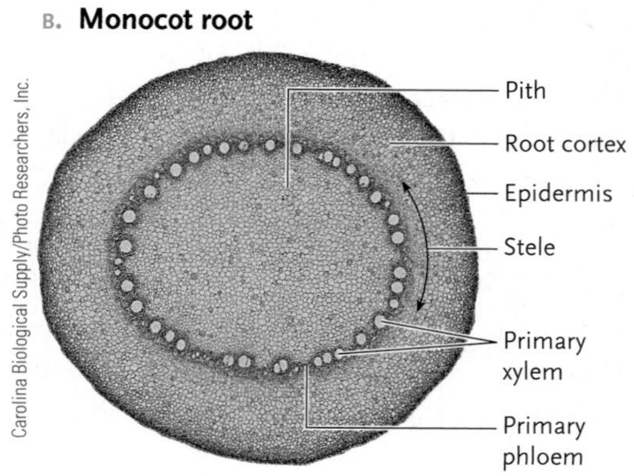

B. **Monocot root**

Pith

Root cortex

Epidermis

Stele

Primary xylem

Primary phloem

Carolina Biological Supply/Photo Researchers, Inc.

FIGURE 31.20
Stele structure in eudicot and monocot roots.
(A) A young root of the buttercup *Ranunculus*, a eudicot. The close-up shows details of the stele. **(B)** Root of a corn plant *(Zea mays)*, a monocot. Notice how the stele divides the ground tissue into cortex and pith. Both roots are shown in transverse section.

zone give rise to root hairs, and the procambium, ground meristem, and protoderm complete their differentiation here.

TISSUES OF THE ROOT SYSTEM Coupled with primary growth of the shoot, primary root growth produces a unified system of vascular pipelines extending from root tip to shoot tip. The root procambium produces cells that mature into the root's xylem and phloem **(Figure 31.20)**. Ground meristem gives rise to the root's cortex, its ground tissue of starch-storing parenchyma cells that surround the stele. In eudicots, the stele runs through the center of the root (see Figure 31.20A). In corn and some other monocots, the stele forms a ring that divides the ground tissue into cortex and pith (see Figure 31.20B).

The cortex contains air spaces that allow oxygen to reach all of the living root cells. Numerous plasmodesmata connect the cytoplasm of adjacent cells of the cortex. In many flowering plants, the outer root cortex cells give rise to an **exodermis,** a narrow band of cells beneath the epidermis. Among other functions, the exodermis may limit water losses from roots and help regulate the absorption of ions. The innermost layer of the root cortex is the **endodermis,** a thin, selectively permeable barrier that helps control the movement of water and dissolved minerals into the stele. Chapter 32 looks in more detail at the roles of exodermis and endodermis.

Between the stele and the endodermis is the **pericycle,** consisting of one or more layers of parenchyma cells that can still function as meristem. The pericycle gives rise to lateral roots **(Figure 31.21)**. In response to chemical growth regulators, rudimentary roots, or **root primordia,** arise at specific sites in the pericycle. Gradually, the lateral roots emerge and grow out through the cortex and epidermis, aided by enzymes released by the root primordium that help break down the intervening cells. The distribution and frequency of lateral root formation partly control the overall shape of the root system—and the extent of the soil area it can penetrate.

In some cells in the developing root epidermis the outer surface becomes extended into root hairs (see Figure 31.19). Root hairs can be more than a centimeter long and can form in less than a day. Collectively, the thousands or millions of them on a plant's roots greatly increase the plant's absorptive surface. Root hair structure supports this essential function. Each hair is a slender tube with thin walls made sticky on their surface by a coating of pectin. Soil particles tend to adhere to the wall, providing an intimate association between the hair and the sur-

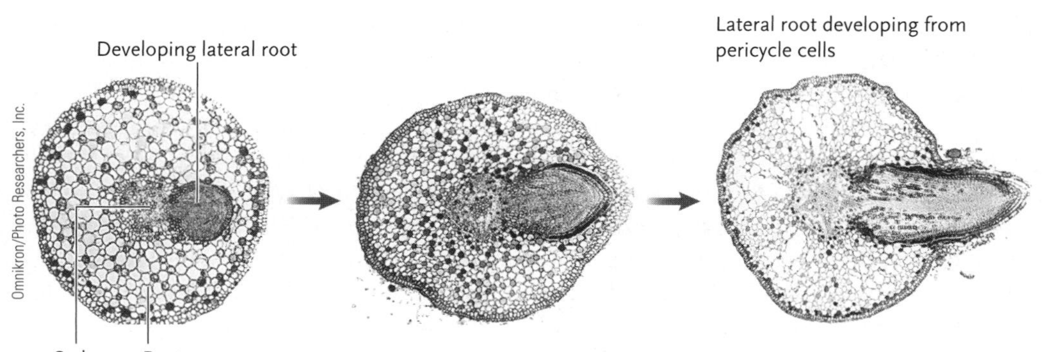

Developing lateral root
Lateral root developing from pericycle cells
Stele
Root cortex
Omnikron/Photo Researchers, Inc.

FIGURE 31.21
Micrographs showing the formation of a lateral root from the pericycle of a willow tree *(Salix)*. These micrographs show transverse sections.

rounding earth, thus facilitating the uptake of water molecules and mineral ions from soil. When plants are transplanted, rough handling can tear off much of the fragile absorptive surface. Unable to take up enough water and minerals, the transplant may die before new root hairs can form.

STUDY BREAK 31.4

1. **Compare the two general types of root systems.**
2. **Describe the zones of primary growth in roots.**
3. **Describe the various tissues that arise in a root system and their functions.**

31.5 Secondary Growth

Primary growth produces the primary body parts of a seed plant—its root and stem systems. In plants that have secondary growth, the tissues we know as wood and bark add girth to roots and stems over two or more growing seasons. In such plant species, including trees and shrubs, every so often mitosis is reactivated in two types of lateral meristems, or *cambia* (singular, cambium). Over time, their activity makes older stems and roots more massive. One of the lateral meristems, **vascular cambium,** produces secondary xylem and phloem. The other lateral meristem, called **cork cambium,** produces **cork,** a secondary epidermis that is a major ingredient in bark.

Primary and secondary growth in a stem

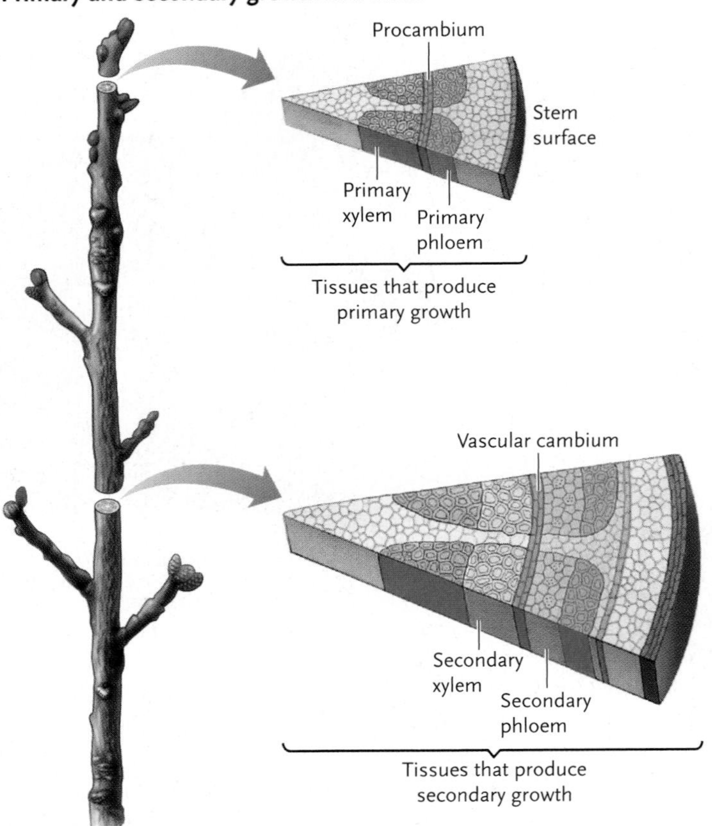
Procambium
Stem surface
Primary xylem
Primary phloem
Tissues that produce primary growth
Vascular cambium
Secondary xylem
Secondary phloem
Tissues that produce secondary growth

FIGURE 31.22
Secondary and primary growth compared. In a woody plant, primary growth resumes each spring at the terminal and lateral buds. Secondary growth resumes at the vascular cambium inside the stem.

Vascular Cambium Gives Rise to Secondary Growth in Stems

After a woody plant's stem completes its primary growth, each vascular bundle contains a layer of undifferentiated cells between the primary xylem and phloem. These cells, along with parenchyma cells between the bundles, eventually give rise to a cylinder of vascular cambium that encircles the xylem and pith of the stem **(Figure 31.22).** Vascular cambium consists of two types of cells, called *fusiform initials* and *ray initials.* Secondary growth takes place mainly as these cells divide in a plane parallel to the vascular cambium, adding girth to the stem. Fusiform initials give rise to secondary xylem and phloem cells that mainly conduct fluid vertically through a stem. Secondary xylem forms on the inner face of the vascular cambium and secondary phloem forms on the outer face. Dividing ray initials produce rays—horizontal columns of living parenchyma cells arranged like spokes of a wheel. Ray parenchyma stores carbohydrates or helps move solutes horizontally across the stem. In conifers, rays help transport the sticky resins commonly called *pitch.*

With time, a great deal of secondary xylem forms inside the ring of vascular cambium. This hard secondary xylem is **wood.** Outside the vascular cambium, some secondary phloem also is added each year **(Figure 31.23).** (The primary phloem cells, which have thin walls, are destroyed as they are pushed outward by secondary growth.) As a stem's diameter increases, the enlarging mass of new tissue eventually ruptures the cortex. Parts of it split away and carry epidermis with them. Cork cambium replaces the lost epidermis with cork.

Bark consists of the tissues sandwiched between the vascular cambium and the stem surface. It includes the secondary phloem and the **periderm** (*peri* = surrounding; *derma* = skin), which consists of cork, cork cambium, and secondary cortex **(Figure 31.24).** Tubular openings (called *lenticels*) develop in the periderm. They function a bit like snorkels, permitting exchanges of oxygen and carbon dioxide between the living tissues and the outside air.

Girdling a tree by removing a belt of bark around the trunk is lethal if it destroys the vascular cambium and secondary phloem layer, preventing nutrients from photosynthesis in leaves from reaching the tree's roots. Natural corks used to seal bottles are manufactured from the especially thick outer bark of the cork oak, *Quercus suber.*

As a tree ages, changes unfold in the appearance and function of the wood itself. In the center of its older stems and roots is **heartwood,** dry tissue that no longer transports water and solutes and is a storage depot for some defensive compounds. In time, these compounds—including resins, oils, gums, and tannins—clog and fill in the oldest xylem pipelines. Typically they darken heartwood, strengthen it, and make it more aromatic and resistant to decay or wood-

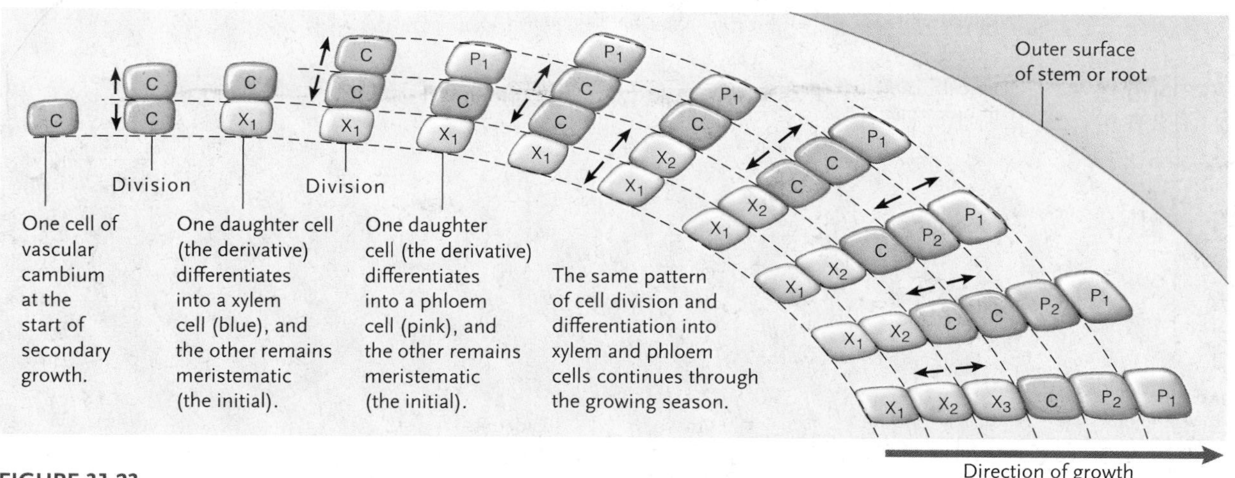
C
C
C
X1
Division
Division
One cell of vascular cambium at the start of secondary growth.
One daughter cell (the derivative) differentiates into a xylem cell (blue), and the other remains meristematic (the initial).
One daughter cell (the derivative) differentiates into a phloem cell (pink), and the other remains meristematic (the initial).
The same pattern of cell division and differentiation into xylem and phloem cells continues through the growing season.
Outer surface of stem or root
X1 X2 X3 C P2 P1
Direction of growth

FIGURE 31.23

Relationship between the vascular cambium and its derivative cells (secondary xylem and phloem). The drawing shows stem growth through successive seasons. Notice how the ongoing divisions displace the cambial cells, moving them steadily outward even as the core of xylem increases the stem or root thickness.

FIGURE 31.24

Structure of a woody stem showing extensive secondary growth. Heartwood, the mature tree's core, has no living cells. Sapwood, the cylindrical zone of xylem between the heartwood and vascular cambium, contains some living parenchyma cells among the nonliving vessels and tracheids. Everything outside the vascular cambium is bark. Everything inside it is wood.

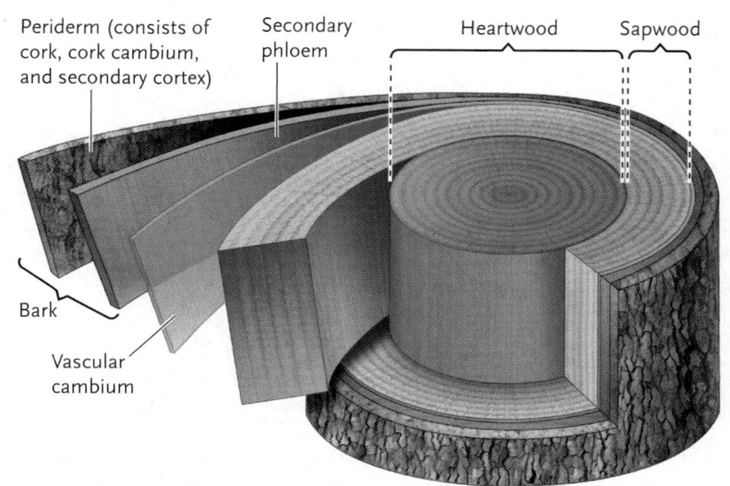
Periderm (consists of cork, cork cambium, and secondary cortex)
Secondary phloem
Heartwood
Sapwood
Bark
Vascular cambium

boring insects. **Sapwood** is located between heartwood and the vascular cambium. Compared with heartwood, it is wet and not as strong.

In regions that have large seasonal differences in temperature, trees produce secondary xylem seasonally, with larger-diameter cells produced in spring and smaller-diameter cells in summer. This "spring wood" and "summer wood" reflect light differently, and it is possible to identify them as alternating light and dark bands that represent annual growth layers—the "tree rings" pictured in **Figure 31.25.**

FIGURE 31.25

Secondary growth and tree ring formation. (A) Radial cut through a woody stem that has three annual rings, corresponding to secondary growth in years 2–4. **(B)** Growth rings of an elm *(Ulmus)*. Each ring corresponds to one growing season. **(C)** Growth rings of a pine, including dark heartwood and lighter sapwood. Pine trees are fast-growing conifers and generally have wider rings than species that grow more slowly.

A. **A woody stem**

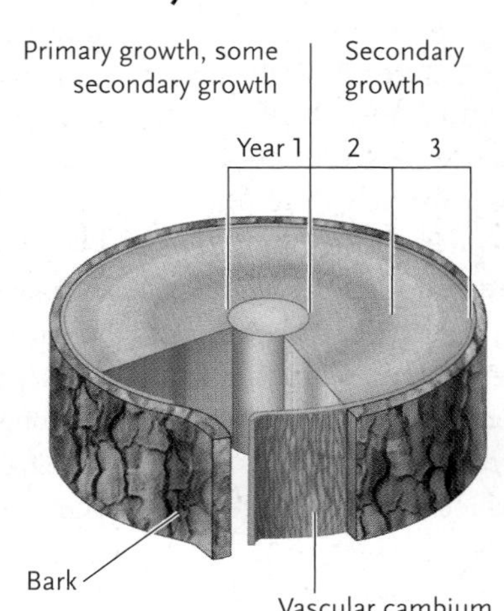
Primary growth, some secondary growth
Secondary growth
Year 1
2
3
Bark
Vascular cambium

B. **Growth rings of American elm (*Ulmus americana*)**

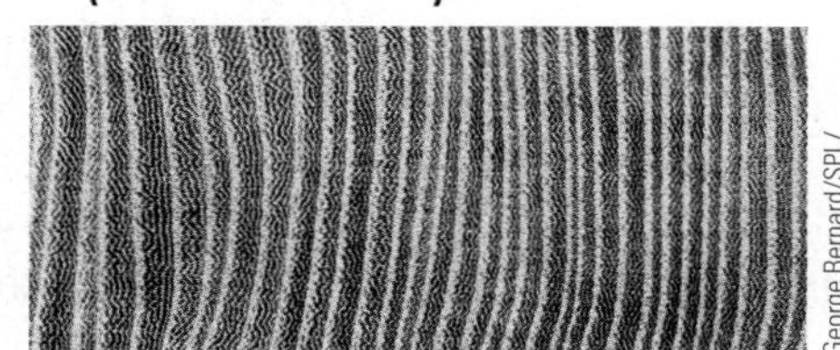

George Bernard/SPL/ Photo Researchers, Inc.

C. ***Pinus* growth rings**

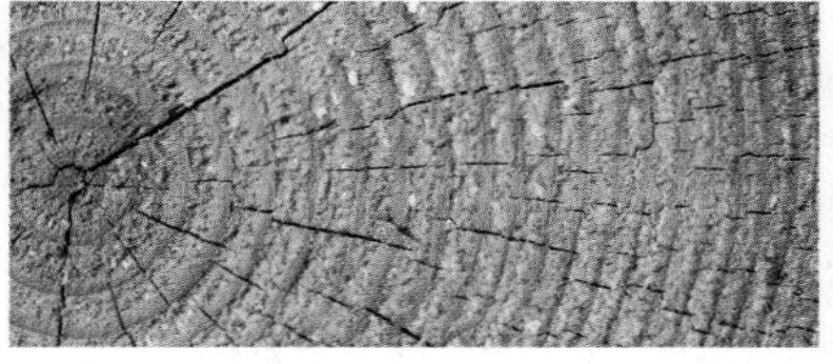

Image copyright Zibedik, 2010. Used under license from Shutterstock.com

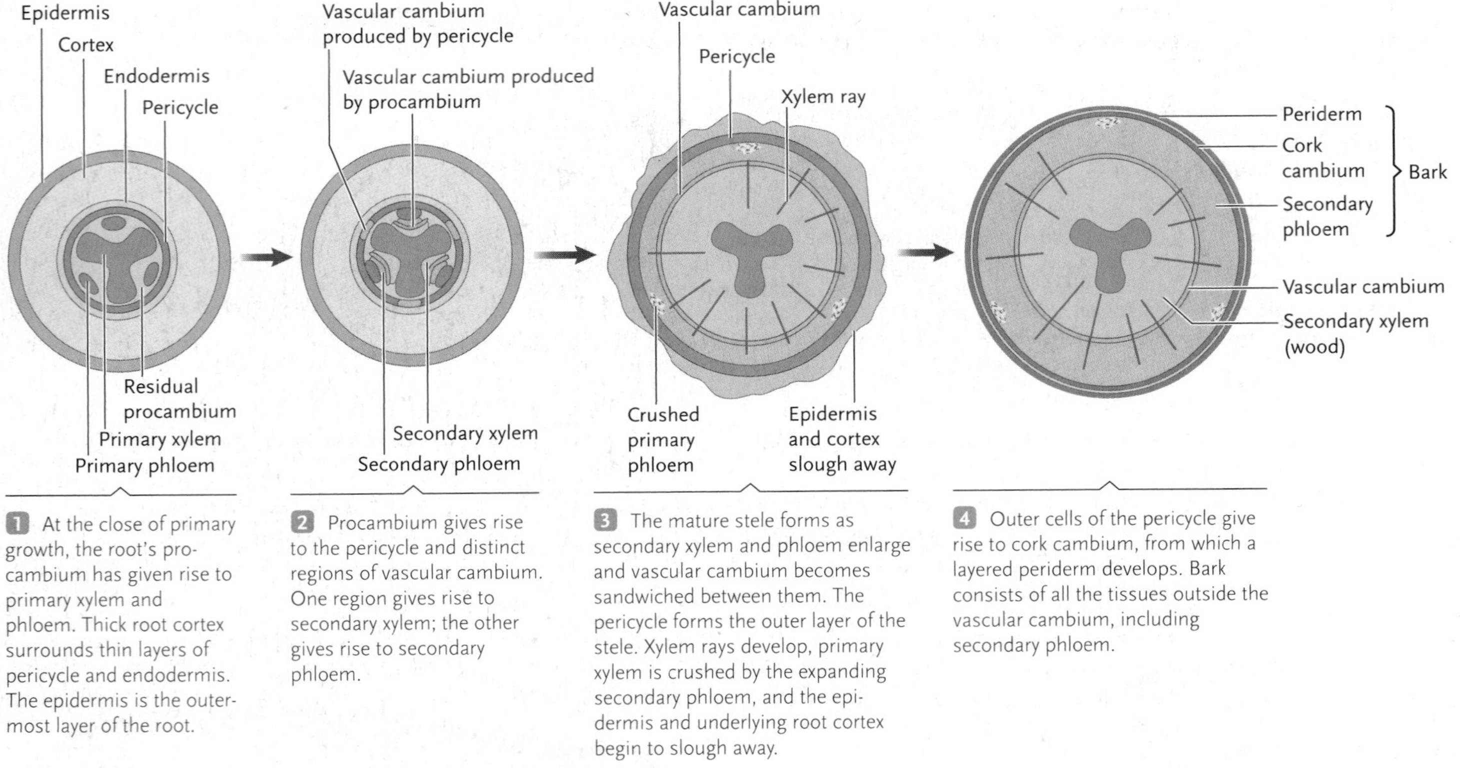
Epidermis
Cortex
Endodermis
Pericycle
Residual procambium
Primary xylem
Primary phloem
Vascular cambium produced by pericycle
Vascular cambium produced by procambium
Secondary xylem
Secondary phloem
Vascular cambium
Pericycle
Xylem ray
Crushed primary phloem
Epidermis and cortex slough away
Periderm
Cork cambium
Secondary phloem
Bark
Vascular cambium
Secondary xylem (wood)
1 At the close of primary growth, the root's procambium has given rise to primary xylem and phloem. Thick root cortex surrounds thin layers of pericycle and endodermis. The epidermis is the outermost layer of the root.
2 Procambium gives rise to the pericycle and distinct regions of vascular cambium. One region gives rise to secondary xylem; the other gives rise to secondary phloem.
3 The mature stele forms as secondary xylem and phloem enlarge and vascular cambium becomes sandwiched between them. The pericycle forms the outer layer of the stele. Xylem rays develop, primary xylem is crushed by the expanding secondary phloem, and the epidermis and underlying root cortex begin to slough away.
4 Outer cells of the pericycle give rise to cork cambium, from which a layered periderm develops. Bark consists of all the tissues outside the vascular cambium, including secondary phloem.

FIGURE 31.26
Secondary growth in the root of one type of woody plant.

Secondary Growth Can Also Occur in Roots

The roots of grasses, palms, and other monocots are almost always the product of primary growth alone. In plants with roots that have secondary growth, the continuous ring of vascular cambium develops differently than it does in stems. When their primary growth is done, these roots have a layer of residual procambium between the xylem and phloem of the stele (**Figure 31.26**, step 1). The vascular cambium arises in part from this residual cambium, and in part from the pericycle (step 2). Eventually, the cambial tissues arising from the procambium and those arising from the pericycle merge into a complete cylinder of vascular cambium (step 3). The vascular cambium functions in roots as it does in stems, giving rise to secondary xylem to the inside and secondary phloem to the outside. As secondary xylem accumulates, older roots can become extremely thick and woody. Their ongoing secondary growth is powerful enough to break through concrete sidewalks and even dislodge the foundations of homes.

Cork cambium also forms in roots, where it is produced first by the pericycle and later by initials in secondary phloem. In many woody eudicots and all gymnosperms, most of the root epidermis and cortex fall away, and the surface consists entirely of periderm (step 4).

Secondary Growth Is an Adaptive Response

Plants, like all living organisms, compete for resources and woody stems and roots confer some advantages. Plants with masses of leaves supported by taller, stronger stems that defy the pull of gravity can intercept more of the light energy from the sun. With a greater energy supply for photosynthesis, they have the metabolic means to increase their root and shoot systems, and thus are better able to acquire resources—and ultimately to reproduce successfully.

During both primary and secondary growth (**Figure 31.27**), the development of new tissues and organs maintains a balance between the shoot system and root system. Leaves and other photosynthetic parts of the shoot must supply root cells with enough sugars to support their metabolism, and roots must provide shoot structures with water and minerals. As long as a plant is growing, this balance is maintained, even as the complexity of the root and shoot systems increases, whether the plant lives for only a few months or—like some bristlecone pines—for 6,000 years.

STUDY BREAK 31.5

1. **Explain the nature of secondary growth and where it typically occurs in plants.**
2. **Describe the components of vascular cambium and their roles in secondary growth in stems, including the development of tissues such as bark, cork, and wood.**
3. **Compare secondary growth in stems and in roots.**

FIGURE 31.27
Summary of the primary and secondary growth stages of plants.

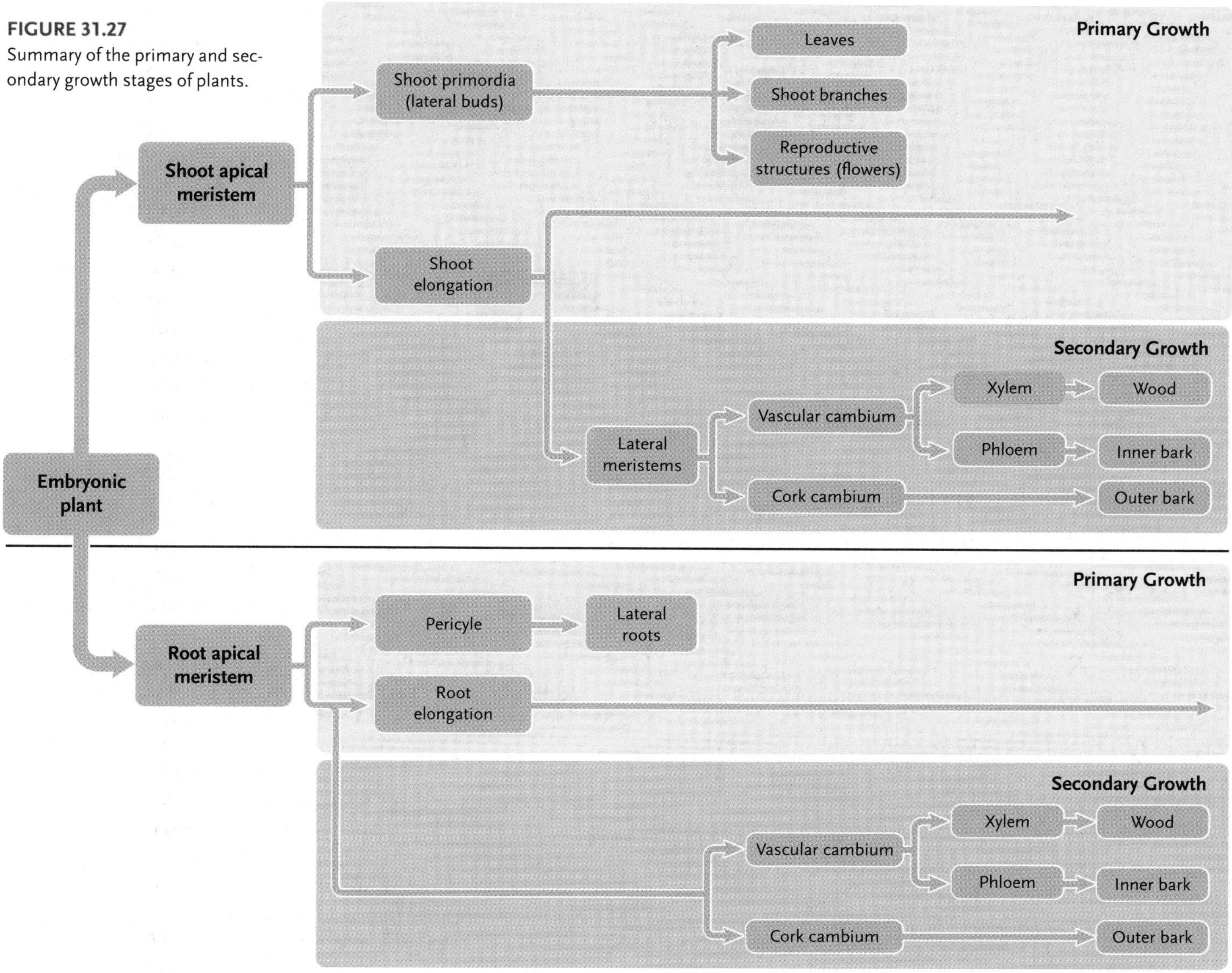

Primary Growth
Embryonic plant
Shoot apical meristem
Shoot primordia (lateral buds)
Leaves
Shoot branches
Reproductive structures (flowers)
Shoot elongation
Secondary Growth
Lateral meristems
Vascular cambium
Xylem
Wood
Phloem
Inner bark
Cork cambium
Outer bark

Primary Growth
Root apical meristem
Pericyle
Lateral roots
Root elongation
Secondary Growth
Vascular cambium
Xylem
Wood
Phloem
Inner bark
Cork cambium
Outer bark

 UNANSWERED QUESTIONS

How are stem cell reservoirs sustained?

As you have learned in this chapter, a special feature of plants is their ability to continuously grow and generate new organs such as leaves and flowers throughout their life cycle. This unique developmental strategy, which enables plants to flexibly adapt to changing environmental conditions, requires that they continuously produce a supply of new cells from meristems at their growing shoot and root tips. The source of new cells within each meristem is the stem cell reservoir, a small pool of cells that are not committed to a particular organ or tissue fate. During the normal course of plant development, the stem cells slowly divide to replenish the reservoir while simultaneously producing daughter cells that form the different types of tissues and organs.

Because plants such as sequoias can grow and develop for hundreds or even thousands of years, stem cell reservoirs are long lasting and must be highly stable over time. Although it has been demonstrated that mutations that cause excess loss or accumulation of stem cells profoundly affect plant development and architecture, it is not well understood how plant stem cell reservoirs are sustained. Research has shown that the unique identity of shoot stem cell is conferred by a signal from the cells beneath them, which are known as niche cells, because they provide a supportive microenvironment for their stem cell neighbors. In turn the stem cells send a negative signal to the niche cells, via an extracellular signal transduction pathway, to limit their activity. This feedback loop maintains the stem cell reservoir at a relatively constant size despite the continuous departure of daughter cells to form tissues and organs. Yet despite its critical importance for plant development, many gaps in our understanding of this regulatory process remain. Key objectives for the future will be to identify all of the components of the stem cell signaling pathways and their outputs at the level of meristem cell division and proliferation, and to determine how additional regulatory elements such as hormones and epigenetic factors modulate the activity of the feedback loop.

What is the nature of shoot stem cell identity?

Plant stem cells have unique features and functions. Shoot stem cells are characterized by their slow but continuous cell divisions, their ability to perpetuate themselves throughout the life of the plant, and their ability to produce daughter cells that differentiate into different tissues. Evidence indicates that stem cells express particular combinations of genes that give them their identity, and identifying their gene expression profiles is fundamental to defining the nature of this essential cell type. Although a specific suite of genes has been associated with stem cell identity in animals, the array of gene products, or molecular signature, that defines a plant stem cell is not known. There are several reasons for this knowledge gap. One is that the shoot stem cell reservoir is made up of only a few dozen cells in an inaccessible location at the shoot tip. Another is that plant cells are physically connected to one another via their cell walls, so that plant stem cells cannot be isolated and cultured as animal stem cells are. Such obstacles are falling, however, as plant scientists are able to apply sophisticated new methods (such as laser microdissection and cell sorting with high-throughput gene expression profiling) that provide an unprecedented opportunity to precisely determine the molecular characteristics of plant stem cells.

Think Critically

Root meristems generate the underground root system whereas shoot meristems produce the aboveground stems, leaves, and flowers. Considering that they give rise to different tissues, how much overlap might exist between the molecular signatures of stem cells in the root and shoot?

David Kramer

Jennifer Fletcher is a research geneticist at the USDA-ARS Plant Gene Expression Center and an adjunct associate professor of plant biology at the University of California, Berkeley. Her research focuses on the molecular genetics of shoot and flower development in *Arabidopsis*. To learn more about Fletcher's work, visit http://www.pgec.usda.gov/Fletcher/fletchlab.html.

REVIEW KEY CONCEPTS

Go to **CENGAGENOW** at www.cengage.com/login to access quizzing, animations, exercises, articles, and personalized homework help.

31.1 Plant Structure and Growth: An Overview

- The vascular plant body consists of an aboveground shoot system and an underground root system (Figure 31.2).
- Meristems (Figure 31.3) give rise to the plant body and are responsible for a plant's lifelong growth. When a meristem cell divides, one daughter remains an initial that continues to function as meristem. The other daughter is a derivative that differentiates.
- Primary growth of roots and shoots originates at apical meristems. In plants that show secondary growth lateral meristems increase the diameter of stems and roots.
- The two major classes of flowering plants (angiosperms) are monocots and eudicots (Table 31.1).

Animation: Tissue systems of a tomato plant

Animation: Apical meristems

31.2 The Three Plant Tissue Systems

- Growing plant cells form secondary walls inside the primary walls. Maturing cells become specialized for specific functions, some of which are carried out by walls of dead cells.
- Ground tissues make up most of the plant body, vascular tissues serve in transport, and dermal tissue forms a protective cover. Of the three types of ground tissues, parenchyma is active in photosynthesis, storage, and other tasks. Collenchyma and sclerenchyma provide mechanical support (Figure 31.5).
- Xylem and phloem are vascular tissues. Xylem conducts water and solutes and consists of conducting cells called tracheids and vessel members (Figure 31.7). In phloem, the food-conducting tissue, living cells called sieve tube members join end to end in sieve tubes (Figure 31.8).
- The dermal tissue, epidermis (Figure 31.9), is coated with a waxy cuticle that restricts water loss. Water vapor and other gases enter and leave the plant through pores called stomata, which are flanked by specialized epidermal cells called guard cells. Epidermal specializations also include trichomes such as root hairs.

31.3 Primary Shoot Systems

- The primary shoot system consists of the main stem, leaves, and buds, plus any attached flowers and fruits. Stems provide mechanical support, house vascular tissues, and may store food and fluid.
- Stems are organized into modular segments. Nodes are points where leaves and buds are attached; internodes fall between nodes (Figure 31.10A). The terminal bud at a shoot tip consists of shoot apical meristem. Meristem tissue in buds gives rise to leaves, flowers, or both (Figure 31.10B).
- Derivatives of the apical meristem produce three primary meristems (Figure 31.11). Protoderm makes the stem's epidermis, procambium gives rise to primary xylem and phloem, and ground meristem gives rise to ground tissue.
- Vascular tissues are organized into vascular bundles, with phloem surrounding xylem in each bundle (Figure 31.12). The bundles form a stele (vascular cylinder). Monocot and eudicot leaves have blades of different forms, all providing a large surface area for absorbing sunlight and carbon dioxide (Figure 31.14 and 31.16). Leaf modifications are adaptive responses to environmental selection pressures (Figure 31.15). Leaf characteristics such as shape or arrangement may change over the life cycle of a long-lived plant (Figure 31.17).

Animation: Ground tissues

Animation: Vascular tissues

Animation: Monocot and dicot leaves

Animation: Simple and compound leaves

Animation: Leaf organization

31.4 Root Systems

- Roots absorb water and minerals and conduct them to aerial plant parts; they anchor and sometimes support the plant and often store food. Root morphologies include taproot systems, fibrous root systems, and adventitious roots (Figure 31.18).
- During primary growth of a root, the primary meristem and actively dividing cells make up the zone of cell division, which merges into the zone of elongation. Past the zone of elongation, cells may differentiate in the zone of cell maturation (Figure 31.19).
- A root's xylem and phloem usually are arranged as a central stele (Figure 31.20). Parenchyma outside the stele forms the root cortex. The root endodermis also wraps around the stele. Inside it the pericycle contains parenchyma that can function as meristem. The pericycle gives rise to root primordia from which lateral roots emerge (Figure 31.21). Root hairs greatly increase the absorptive surface of roots.

Animation: Root organization

Animation: Root cross section

Animation: Root systems

31.5 Secondary Growth

- Secondary growth makes older stems and roots woody and more massive via the activity of vascular cambium and cork cambium.
- Vascular cambium consists of fusiform initials, which generate secondary xylem and phloem, and ray initials, which produce parenchyma cells aligned in horizontal xylem rays (Figures 31.22 and 31.23). Secondary growth takes place as these cells divide.
- Cork cambium produces cork, which replaces epidermis lost when stems increase in diameter. Together, cork cambium and cork make up the periderm (Figure 31.24), the outer portion of bark.
- In root secondary growth, a thin layer of procambium cells between the xylem and phloem differentiates into vascular cambium (Figure 31.26), which gives rise to secondary xylem and phloem. The pericycle produces root cork cambium.

Animation: Secondary growth

Animation: Secondary growth in a root

Animation: Growth in a walnut twig

Animation: Layers in a woody stem

Animation: Annual rings

UNDERSTAND AND APPLY

Test Your Knowledge

1. With respect to growth, plants differ from animals in that:
   a. plant growth involves only an increase in the total number of the organism's cells.
   b. plant cells remain roughly the same size after cell division, whereas animal cells increase in size after they form.
   c. all plants form woody tissues during growth.
   d. plants have indeterminate growth; animals have determinate growth.
   e. plants can grow only when young; animals grow throughout their life.
2. Identify the correct pairing of a plant tissue and its function.
   a. epidermis: rigid support
   b. xylem: sugar transport
   c. parenchyma: photosynthesis, respiration
   d. phloem: water and mineral transport
   e. periderm: control of gas exchange
3. Identify the correct pairing of a structure and its component(s).
   a. epidermis: companion cells
   b. phloem: sieve tube members
   c. sclerenchyma: nonlignified cell walls
   d. secondary cell wall: cuticle
   e. parenchyma: sclereids
4. Which of the following is *not* part of a stem?
   a. petiole
   b. pith
   c. xylem
   d. procambium
   e. ground meristem
5. Which of the following would be absent in a eudicot leaf?
   a. spongy mesophyll
   b. palisade mesophyll
   c. pericycle
   d. vascular bundles
   e. stoma
6. A student left a carrot in her refrigerator. Three weeks later she noticed slender white fibers growing from its surface. They were not a fungus. Instead they represented:
   a. lateral roots on a taproot.
   b. adventitious roots.
   c. root hairs on a fibrous root.
   d. root hairs on a lateral root.
   e. young prop roots.
7. Which of the following is *not* a structure that results from secondary plant growth?
   a. periderm
   b. sapwood
   c. cork
   d. pith
   e. heartwood
8. Which characteristic do monocots and eudicots share?
   a. the position of the vascular bundles
   b. the pattern of leaf veins
   c. the number of grooves in the pollen grains
   d. the number of cotyledons
   e. the formation of flowers
9. A student forgets to water his plant and the leaves start to droop. The structures first affected by water loss and now not functioning are the:
   a. sieve tubes.
   b. sclereids and fibers.
   c. vessel members and tracheids.
   d. companion cells.
   e. guard cells and stoma.
10. The greatest mitotic activity in a root takes place in the:
    a. zone of maturation.
    b. zone of cell division.
    c. zone of elongation.
    d. root cap.
    e. endodermis.

Discuss the Concepts

1. The spines of cacti are modified leaves. Cacti are adapted to arid habitats in which relatively few other plant species grow. What kinds of selection pressures may have operated to favor the evolution of spinelike cactus leaves?
2. While camping you notice a "Do Not Litter" sign nailed onto the trunk of a mature oak tree about 7 feet off the ground. When you return five years later, will the sign be at the same height, or will the tree's growth have raised it higher?
3. Peaches, cherries, and other fruits with pits are produced only on secondary branches that are one year old. To renew the fruiting wood on a peach tree, how often would you prune it? Where on a branch would you make the cut, and why?
4. African violets and some other flowering plants are propagated commercially using leaf cuttings. Initially, a leaf detached from a parent plant is placed in a growth medium. In time, adventitious shoots and roots develop from the leaf blade, producing a new plant. Which cells in the original leaf tissue are the most likely to give rise to the new structures? What property of the cells makes this propagation method possible?

Design an Experiment

The sticky cinquefoil *(Potentilla glandulosa)* is a small, deciduous plant with bright yellow flowers that lives throughout the American West, and its leaf phenotype can vary depending on environmental conditions. Inland, where there are dramatic seasonal temperature swings and unpredictable droughts, plants shed their large "summer leaves" in autumn when the temperature begins to drop. In the spring new leaves are smaller and develop in a compact rosette. This phenotype persists for several months and is thought to be an adaptation that makes the plants drought-resistant (because less water evaporates from reduced leaf surfaces). By contrast, the leaves of *P. glandulosa* plants growing in a coastal climate are always large. In their habitat, seasonal temperature swings are not as great and the annual cycle of winter rain and summer drought is highly predictable. Suppose you decide to explore the hypothesis that the coastal population is genetically capable of exhibiting the same seasonal shift in leaf morphology as the inland plants. Would you need access to a greenhouse where you can control variables, or would it be just as easy to do experiments in the wild? Explain your reasoning and outline your experimental design—including the variable or variables you will test in the first experiment.

Interpret the Data

Rings from long-lived trees may track centuries of annual changes in climatic conditions such as temperature, rainfall, or both. Dendrochronology studies show that Siberian pines *(Pinus sibirica)* growing at high elevations (2,450 m) in the Tarbagatay Pass region of Mongolia may live 500 years, or longer. Growth is faster and rings are wider in warmer years. The reverse is true in cooler years. A research team led by Gordon C. Jacoby of the Tree-Ring Laboratory at the Lamont-Doherty Earth Observatory examined 450 years of *P. sibirica* tree ring data to determine whether the rings provided evidence of a recent climate warming trend in the area. **Figure 1** shows the growth fluctuations in a single tree between the years of 1550 and 1995. The upper line in **Figure 2** shows the aggregated growth fluctuations of a larger sample of the trees during the same period. **Figure 3** correlates the aggregated growth information (solid line) with a reconstruction of annual temperature fluctuations in the region (dotted line).

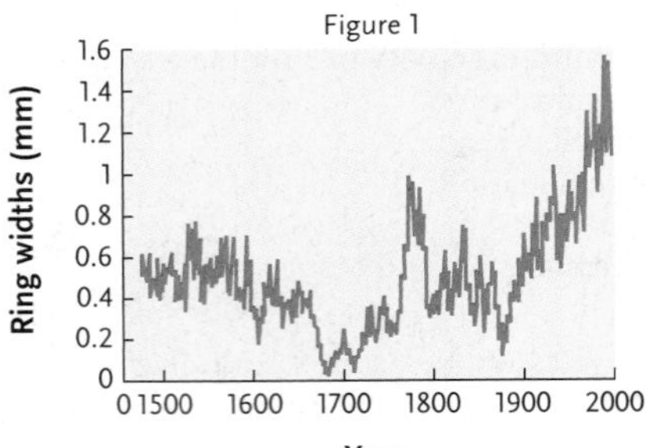
Figure 1
Ring widths (mm)
1.6
1.4
1.2
1
0.8
0.6
0.4
0.2
0
0 1500
1600
1700
1800
1900
2000
Year

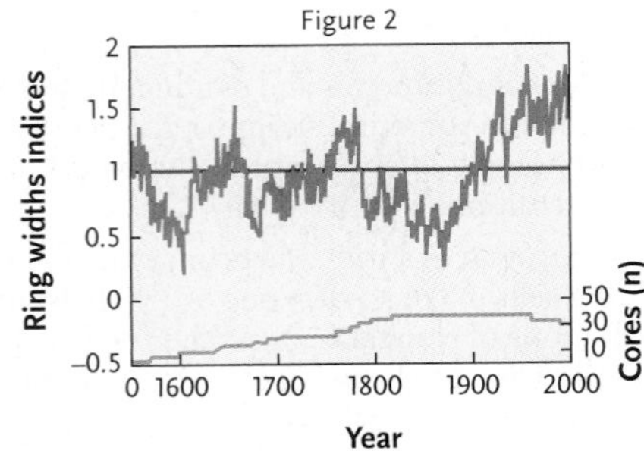
Figure 2
Ring widths indices
2
1.5
1
0.5
0
−0.5
0 1600
1700
1800
1900
2000
Year
50
30
10
Cores (n)

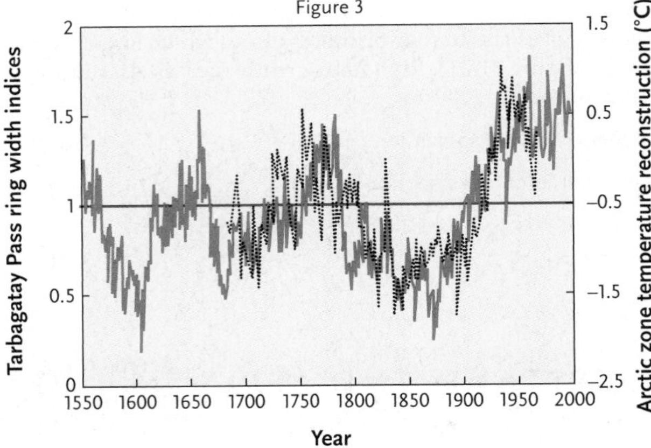
Figure 3
Tarbagatay Pass ring width indices
2
1.5
1
0.5
0
1550
1600
1650
1700
1750
1800
1850
1900
1950
2000
Year
1.5
0.5
−0.5
−1.5
−2.5
Arctic zone temperature reconstruction (°C)

1. Based on the graphed data, what do you think the research team concluded from this experiment?
2. The researchers included a graph of growth data for a single tree in their research report. Why would such information be relevant?
3. In Figure 2, what does the rising trend of the lower line signify? Speculate about why the researchers included that information.
4. In their published report, the researchers included a note that several major volcanic eruptions occurred in the middle of the nineteenth century, spewing massive amounts of smoke and ash into Earth's atmosphere. Why is that fact important for interpreting the graph data?

Source: G. Jacoby *et al.* 1996. Mongolian tree rings and 20th-century warming. *Science* 273:771–777.

Apply Evolutionary Thinking

The first angiosperms may originally have been small, treelike plants in tropical regions, but eventually they began diversifying rapidly into other habitats where early gymnosperms flourished. South African botanist William Bond proposed the "slow seedling" hypothesis to help explain this evolutionary change, and botanists continue to refine it. The hypothesis proposes that angiosperms were able to encroach on many habitats where ancient gymnosperms lived, in part because flowering species increasingly evolved adaptations that made them fast-growing herbaceous plants. Gymnosperms grow more slowly. Based on your reading of this chapter and Chapter 27, what are some structural and biochemical features of gymnosperms (such as conifers) that might result in slower growth, putting them at a competitive disadvantage in this scenario?

Express Your Opinion

Large-scale farms and large cities compete for clean, fresh water, which is becoming scarcer as human population growth skyrockets. Should cities restrict urban growth to reduce conflicts over water supplies? Go to www.cengage.com/login to investigate both sides of the issue and then vote.

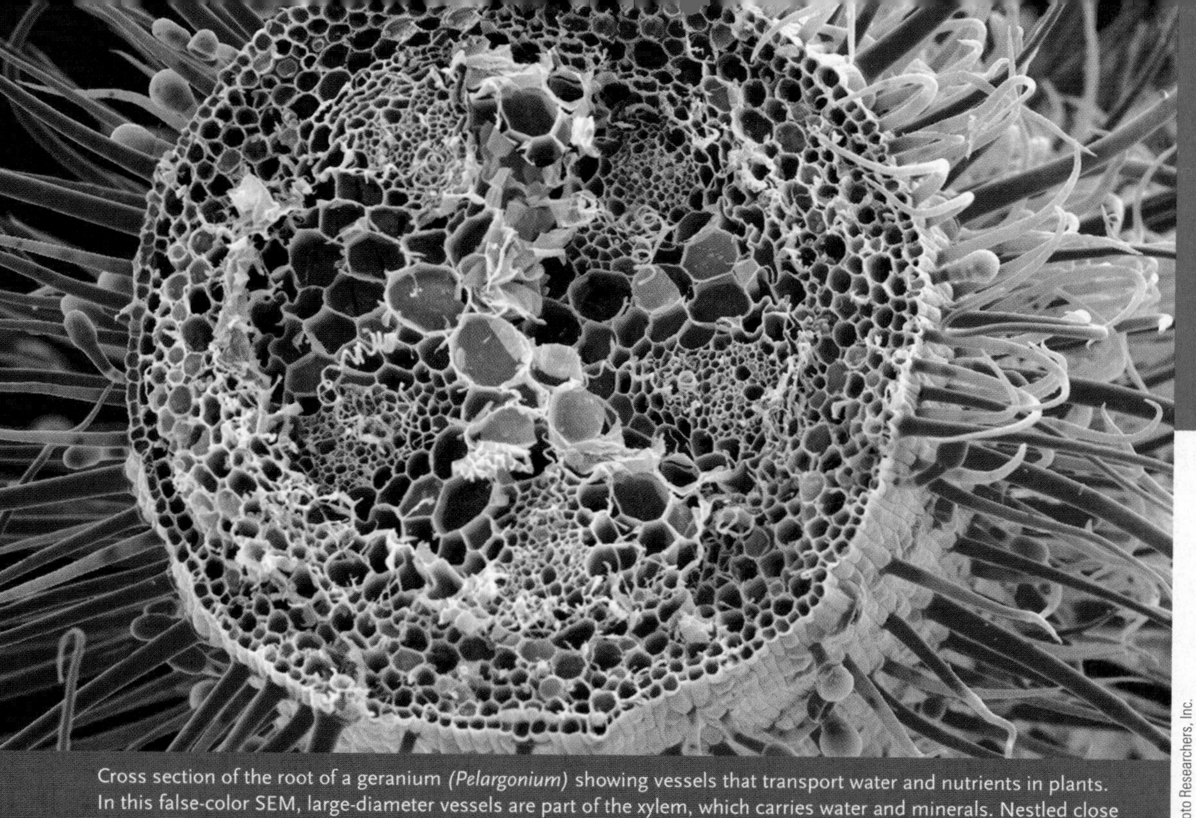

Cross section of the root of a geranium *(Pelargonium)* showing vessels that transport water and nutrients in plants. In this false-color SEM, large-diameter vessels are part of the xylem, which carries water and minerals. Nestled close to xylem vessels are bundles of the smaller transport tubes of phloem, called sieve tubes, which transport sugars.

SPL/Photo Researchers, Inc.

32

STUDY OUTLINE

Transport in Plants

FIGURE 32.1
Redwoods *(Sequoia sempervirens)* such as this tree growing in coastal California have reached recorded heights of over 100 m during life spans of more than 2,000 years. Such extremely tall trees exemplify the ability of plants to move water and solutes from roots to shoots over amazingly long distances.

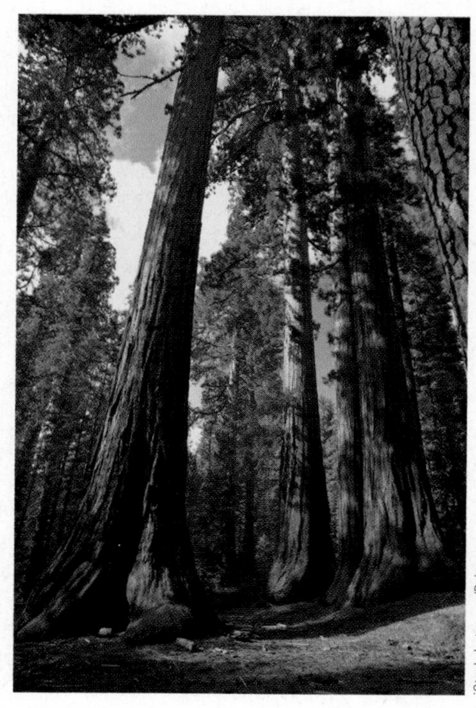

iStockphoto.com/Pgiam

Why It Matters. . . The coast redwood, *Sequoia sempervirens,* takes life to extremes. Redwood trees can live for more than 2,000 years, and they can grow taller than any other organism on Earth—110 meters (361 feet) or more **(Figure 32.1)**. Scientists estimate that each day these giants move thousands of liters of water—with its cargo of dissolved nutrients—between roots and leaves against the pull of gravity and without any sort of pump. Even in much smaller plants, such a physiological feat seems to challenge the laws of physics.

In fact, however, just the opposite is true. The evolutionary adaptations by which redwoods and other plants move water, sugar, and other solutes through their bodies rely on mechanisms common in living systems, from diffusion and active transport to the cumulative effects of physical interactions such as cohesion and evaporation. In this chapter you will learn how such mechanisms and the two specialized transport tissues in vascular plants, xylem and phloem, jointly solve a fundamental biological problem—the need to acquire materials from the environment and distribute them throughout the plant body.

Our discussion begins with a brief review of the principles of water and solute movement in plants, a topic introduced in Chapter 6. Next we examine how those principles operate as water and minerals move in the xylem from roots to shoot parts. Last we consider how the phloem transports sugar and other organic substances from their sources to the locations where they are used in a plant's diverse metabolic activities. <

32.1 Principles of Water and Solute Movement in Plants

In plants, as in all organisms, the movement of water and solutes begins at the level of individual cells and relies on mechanisms such as osmosis and the operation of transport proteins in the plasma membrane. Once water and nutrients enter a plant's vascular tissues, other mechanisms carry them between various regions of the plant body in response to changing demands for those substances. Ultimately, these movements of materials result from the integrated activities of the individual cells, tissues, and organs of a single, smoothly functioning organism—the whole plant.

Plant transport mechanisms fall into two general categories—those for long-distance transport throughout the plant, and those for moving substances short distances, as between cells (**Figure 32.2**). Long-distance transport mechanisms move substances in vascular pipelines between roots and shoot parts. Thus xylem carries water and dissolved minerals from roots to other plant parts, including leaves. Likewise, carbohydrates move in the phloem from photosynthesizing leaves and stems into roots and other structures. Phloem also carries a variety of other substances including hormones, amino acids, fatty acids, and the nucleic acid RNA, which may be an important long-distance signaling molecule in plants (see Chapter 35).

Short-distance transport mechanisms move substances to and from vascular tissues and into and between cells across membranes. For example, water and dissolved minerals enter roots by crossing the cell membranes of root hairs. Similarly, most of the mineral ions and water taken up by roots enter the xylem, whereas a smaller portion moves laterally into the cells that make up root and stem tissues. Some absorbed ions that reach leaves also cross laterally into the phloem in leaf veins. Other examples of short distance transport include sugar molecules in phloem moving into metabolically active cells across plasma membranes and the uptake by leaf cells of minerals in applied fertilizers.

Before looking in more detail at long-distance transport in xylem and phloem, we consider the short-distance mechanisms that move water and solutes into and out of specific cells in roots, leaves, and stems.

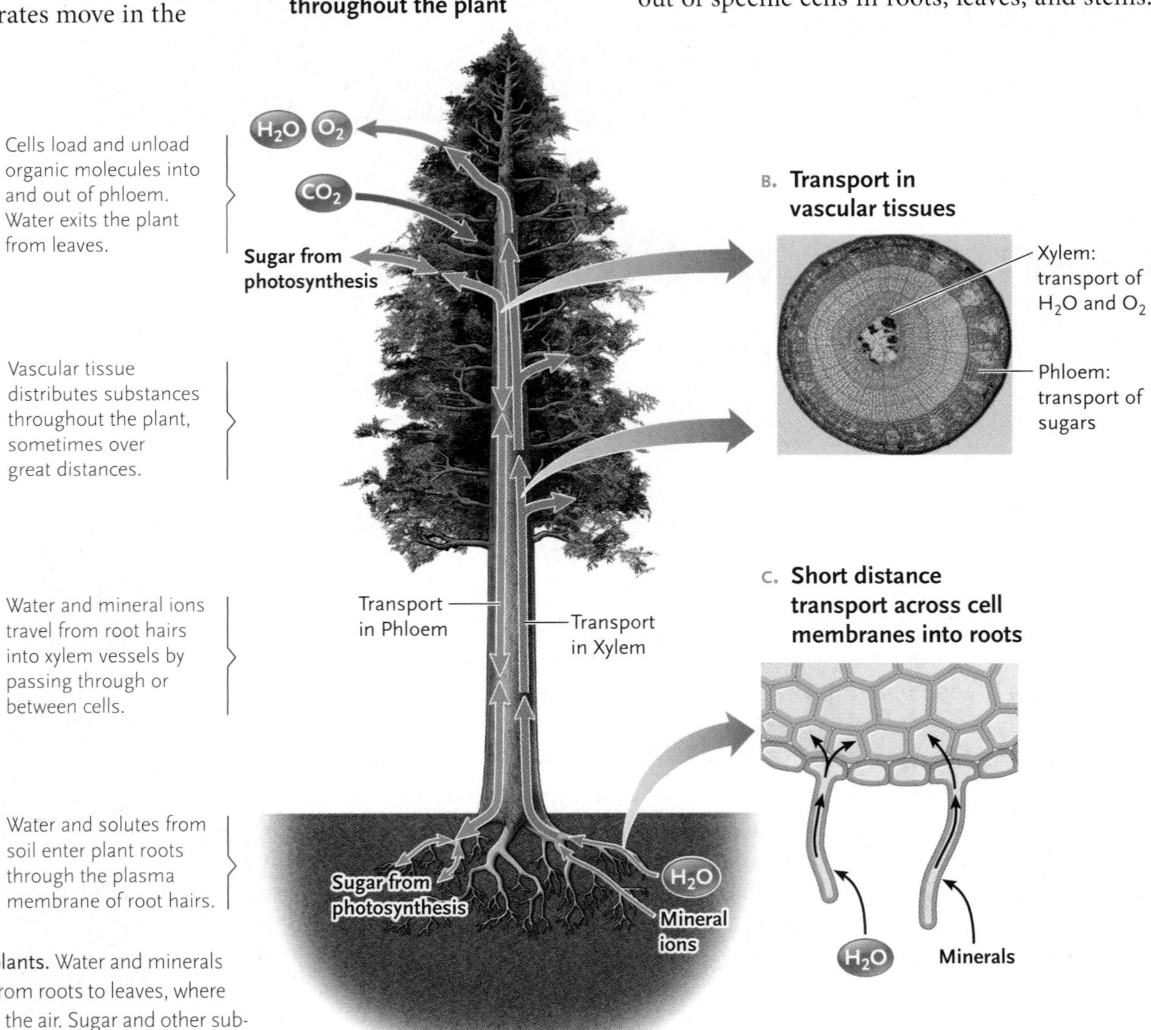
A. Long distance transport throughout the plant
Cells load and unload organic molecules into and out of phloem. Water exits the plant from leaves.
H_2O
O_2
CO_2
Sugar from photosynthesis
Vascular tissue distributes substances throughout the plant, sometimes over great distances.
Water and mineral ions travel from root hairs into xylem vessels by passing through or between cells.
Transport in Phloem
Transport in Xylem
Water and solutes from soil enter plant roots through the plasma membrane of root hairs.
Sugar from photosynthesis
H_2O
Mineral ions
B. Transport in vascular tissues
Xylem: transport of H_2O and O_2
Phloem: transport of sugars
Carolina Biological Supply Company/Phototake
C. Short distance transport across cell membranes into roots
H_2O
Minerals

FIGURE 32.2
Overview of transport routes in plants. Water and minerals are transported upward in xylem from roots to leaves, where some of the water evaporates into the air. Sugar and other substances that nourish the plant are transported in phloem in many directions from sites where they are made or stored to sites where they are needed. For example, sugar produced in leaves by photosynthesis flows to any active or growing tissues where it is needed. **(A)** Xylem: Long distance transport of water and minerals throughout the plant, mainly from roots to shoots. **(B)** Phloem: Long distance transport of carbohydrates from leaves and stems into other tissues, and transport of other substances including hormones. **(C)** Short distance transport of water and minerals across cell membranes into the xylem of the roots. Short distance transport includes movement of water, minerals, sugars, and other materials through and between cells and into and out of xylem and phloem.

Keep in mind that the primary plant cell wall does not prevent most solutes from moving into plant cells. The majority can diffuse through the wall, and some—including small molecules and some large ones, including proteins, RNAs, and virus particles—cross it by way of the plasmodesmata that connect adjacent cells (see Section 5.4).

Both Passive and Active Mechanisms Move Substances into and out of Plant Cells

Recall from Chapter 6 that two general mechanisms transport water and solutes across the cell plasma membrane. To recap that discussion, in *passive transport,* substances move down a concentration gradient or, if the substance is an ion, down an electrochemical gradient (see Section 6.3). *Active transport* requires the cell to expend energy in moving substances *against* a gradient, usually by hydrolysis of ATP. Ions and some larger molecules cross cell membranes assisted by transport proteins embedded in the membrane.

An electrochemical gradient exists across a cell membrane when there is both a concentration difference and an electrical charge difference on the two sides of the membrane. This situation occurs when an ion is more concentrated on one side of a cell's plasma membrane (see Section 6.4). The ionic makeup of a plant cell's cytoplasm is such that the cytoplasm is slightly more negative than the fluid outside the cell. This charge difference, the *membrane potential,* is a potential source of energy—that is, ion movements down an electrochemical gradient can perform cellular work.

ATP drives the active transport of substances into and out of cells, including plant cells. To review this mechanism briefly, hydrogen ions (protons), which tend to be more concentrated outside the cell than in the negatively charged cytoplasm, play a central role in the process. First, a proton pump pushes H^+ across the plasma membrane against its electrochemical gradient, from the inside to the outside of the cell **(Figure 32.3A).** As protons accumulate outside the cell, the electrochemical gradient becomes steeper and significant potential energy is available. Crucial solutes such as cations (positively charged ions) often are more concentrated in the extracellular fluid. One result of the increased charge difference created by proton pumping is that cations move into the cell through channel proteins following the electrochemical gradient **(Figure 32.3B).** These cations include mineral ions that have essential roles in plant-cell metabolism.

The H^+ gradient also powers *secondary active transport,* in which an ion's concentration gradient is the energy source for active transport of another substance. The two secondary mechanisms—*symport* and *antiport*—actively transport ions, sugars, and amino acids into and out of plant cells against their concentration gradients. In **symport,** the potential energy released as H^+ follows its gradient into the cell is coupled to the simultaneous uptake of another ion or molecule **(Figure 32.3C).** In this way, plant cells can take up metabolically important anions such as nitrate (NO_3^-) and potassium (K^+). Most organic substances that enter plant cells move in by symport as well.

In **antiport,** the energy released as H^+ diffuses into the cell powers the active transport of a second molecule, such as Ca^{2+}, in the opposite direction, *out of* the cell **(Figure 32.3D).** One of antiport's key functions is to remove excess Na^+, which readily moves into plant cells by facilitated diffusion through channel proteins. If the Na^+ were not eliminated, it would quickly build up to toxic levels.

FIGURE 32.3
Review of active transport across the plasma membrane of a plant cell.

A. **H^+ pumped against its electrochemical gradient**

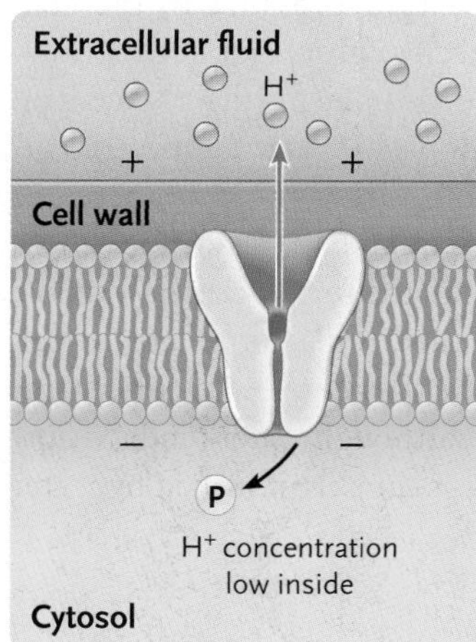
Extracellular fluid
H+
Cell wall
P
H+ concentration low inside
Cytosol

A proton pump moves hydrogen ions (H^+) out of the cytoplasm, creating an H^+ gradient that becomes a source of energy for transporting other ions and molecules such as sugar into the plant cell.

B. **Uptake of cations**

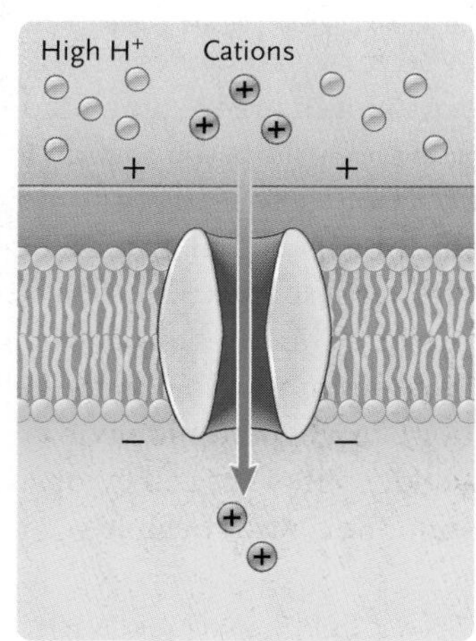
High H+
Cations

Some cations, such as NH_4^+, enter the cell through selective channel proteins, following the electrochemical gradient created by H^+ pumping.

C. **Symport**

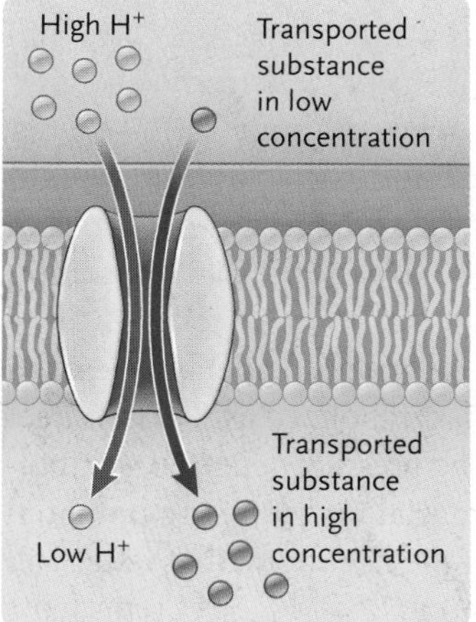
High H+
Transported substance in low concentration
Transported substance in high concentration
Low H+

In symport, the inward diffusion of H^+ is coupled with the simultaneous active transport of another substance into the cell.

D. **Antiport**

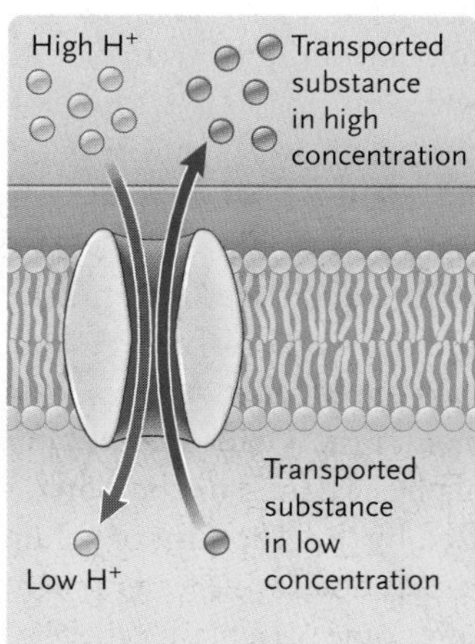
High H+
Transported substance in high concentration
Transported substance in low concentration
Low H+

In antiport, H^+ moving into the cell powers the movement of another solute in the opposite direction.

Both passive and active transport are selective—they transport specific substances. Two factors govern this specificity. One is the size of the interior channel, which allows only molecules in a particular size range to pass through. The other factor is the distribution of charges along the inside of the channel. A channel that permits cations such as Na^+ to pass through easily may completely bar anions such as Cl^- and vice versa.

Relatively speaking, only small amounts of mineral ions and other solutes move into and out of plant cells. As we see next, H_2O is another matter. Throughout a plant's life large volumes of water enter and exit its cells and tissues by way of osmosis.

In Plants Water Moves by Bulk Flow and by Osmosis

The movement of water into and through plant cells and tissues is one of the most important aspects of plant physiology. Inside a plant's tubelike vascular tissues, large amounts of water or any other fluid travel by **bulk flow**—the group movement of molecules in response to a difference in pressure between two locations, like water in a closed plumbing system gushing from an open faucet. For example, the dilute solution of water and ions that flows in the xylem, called **xylem sap,** moves by bulk flow from roots to shoot parts. The solution is pulled upward through the plant body in a process that relies on the cohesion of water molecules. We will consider that process more fully later in this chapter. In the meantime, we turn to osmosis, the short-distance mechanism for moving water.

You may recall from Section 6.3 that **osmosis** moves water into and out of cells across a selectively permeable membrane. The potential energy in water underlies the phenomenon of osmosis. "Potential energy," remember, is simply stored energy. In water, potential energy is associated with the inherent capacity of water molecules to move from one site to another when conditions dictate. This potential energy, called **water potential,** is symbolized by the Greek letter Ψ (*psi,* pronounced "sye"). By convention, pure water at normal atmospheric pressure and temperature has a Ψ value of zero. Shifts away from this baseline propel water into or out of cells. For example, if Ψ is higher outside the cell than inside, water tends to flow into the cell. In plants, two factors that strongly influence water potential shifts—and thus water movements—are the presence of solutes and physical pressure.

The effect of dissolved solutes on water's tendency to cross a membrane is called *solute potential,* symbolized by Ψ_S. The effect of physical pressure on water is called *pressure potential,* symbolized by Ψ_P. The sum of solute potential and pressure potential largely determine water potential—the likelihood that water will enter or leave a plant cell. The equation $\Psi = \Psi_P + \Psi_S$ represents these combined effects.

The relationship between water potential and solute potential is important to understanding transport phenomena in plants because water tends to move from regions where there are relatively few solutes and water potential is higher to regions where there are more solutes and water potential is lower. Solutes are usually more concentrated inside plant cells than in the fluid surrounding them. This means the water potential is higher outside plant cells than inside them, so water tends to enter the cells. This in fact is the mechanism that draws soil water into a plant's roots.

The osmotic flow of water into a plant cell cannot continue indefinitely, however, because eventually physical pressure counterbalances it. The wall around a plant cell strictly limits how much the cell can expand. Accordingly, as water moves into plant cells, the pressure inside them increases. This pressure, called **turgor pressure,** represents the pressure potential inside a plant cell. It rises until it is high enough to prevent more water from entering a cell by osmosis. In effect, turgor pressure has increased the water potential inside the cell until it equals the potential of the water outside the cell.

Plant physiologists measure water potential in units of pressure called **megapascals** (MPa). The baseline is standard atmospheric pressure, so the water potential of pure water at standard atmospheric pressure is written as 0 MPa. Measurement of MPa can be used to describe how different conditions of solute potential and pressure potential affect water flow **(Figure 32.4).** As the figure shows, adding solutes (such as sucrose) to a quantity of pure water reduces the water's MPa (because its relative concentration is lower). On the other hand, increasing pressure increases MPa, and the balance of these two factors determines the MPa of any particular water sample. When a barrier that admits water but not solutes separates two water samples of unequal MPa, water will flow from a solution of higher MPa to a solution of lower MPa. With these principles in mind, consider now how they operate in living plant cells.

Recall from Section 5.4 that a plant cell has relatively little cytoplasm, in a thin layer next to the plasma membrane. Filling most of the cell's volume is at least one large **central vacuole** that is surrounded by a vacuolar membrane, the **tonoplast.** The central vacuole contains a dilute solution of sugars, salts, proteins, and other organic molecules. The water in this solution is essential for maintaining appropriate turgor pressure in the cell. When external solutes enter a plant cell's cytoplasm, active transport moves many solutes from the cytoplasm into the central vacuole through channels in the tonoplast. Water follows the solutes into the vacuole by osmosis until turgor pressure increases the water potential inside the cell to equal that outside the cell, as already described. On the other hand, the turgid central vacuole can be a water reservoir when the medium around a plant cell becomes hypertonic to the cell (has a higher solute concentration). In a hypertonic environment, although water tends to flow rapidly out of the cell cytoplasm, it is quickly replaced by water from the central vacuole.

The Discovery of Aquaporins Explained Rapid Water Movements across Cell Membranes

A plant cell must compensate fairly quickly for water gains or losses caused by changes in osmotic flow. This means that when osmotic conditions change, a large amount of water must move rapidly across plant cell membranes. How does water move so

A. **Start with pure water**

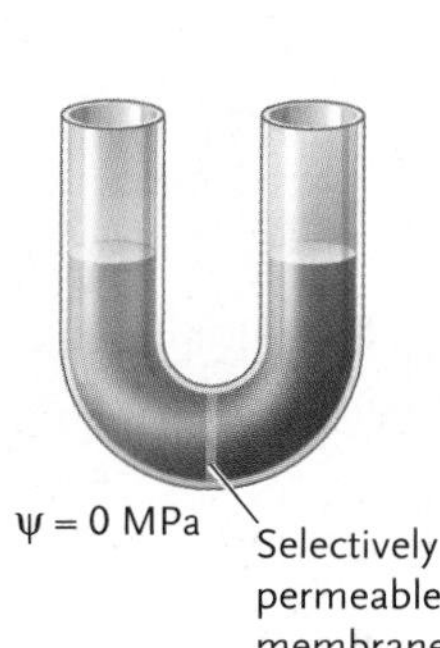
ψ = 0 MPa
Selectively permeable membrane

Pure water in a curved tube with compartments separated by a selectively permeable membrane

B. **Add sucrose to one side**

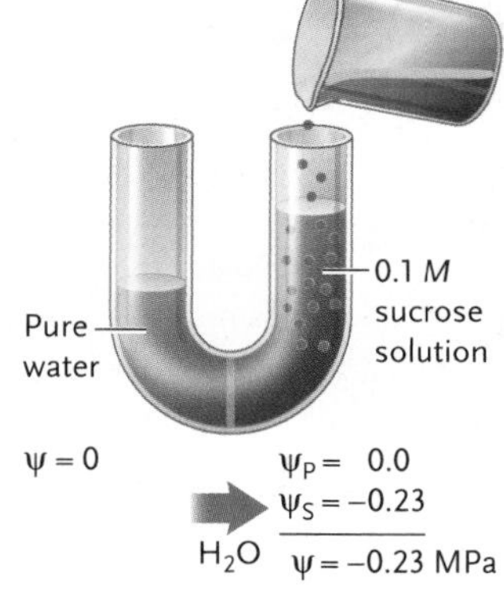
Pure water
0.1 M sucrose solution
ψ = 0
ψP = 0.0
ψS = −0.23
H2O
ψ = −0.23 MPa

When sucrose is added to the water on one side to form a 0.1 *M* sucrose solution, the water potential on that side falls. Water moves into the solution by osmosis.

C. **Apply pressure**

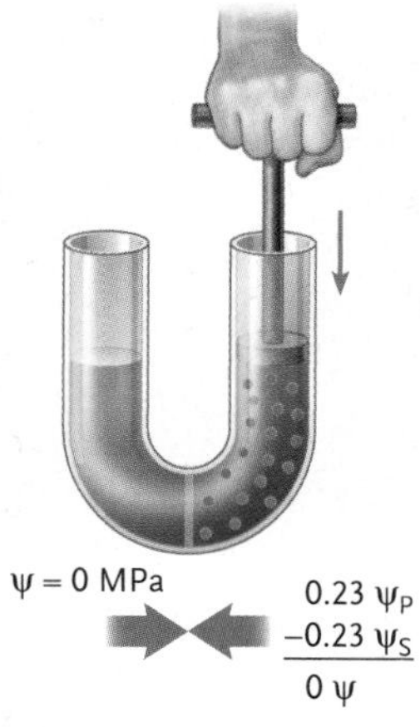
ψ = 0 MPa
0.23 ψP
−0.23 ψS
0 ψ

By applying enough pressure (ψ_P) to the solution to balance the osmotic pressure, water potential can be increased to zero, equaling that on the pure-water side of the membrane. Now there is no net movement of water across the membrane.

D. **Increase pressure**

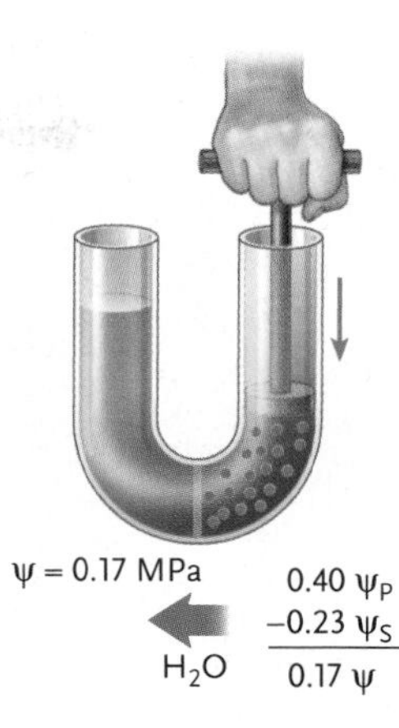
ψ = 0.17 MPa
0.40 ψP
−0.23 ψS
0.17 ψ
H2O

Increasing pressure further increases the water potential of the sucrose solution, so water moves back across the membrane into the compartment containing pure water.

E. **Reduce pressure on opposite side**

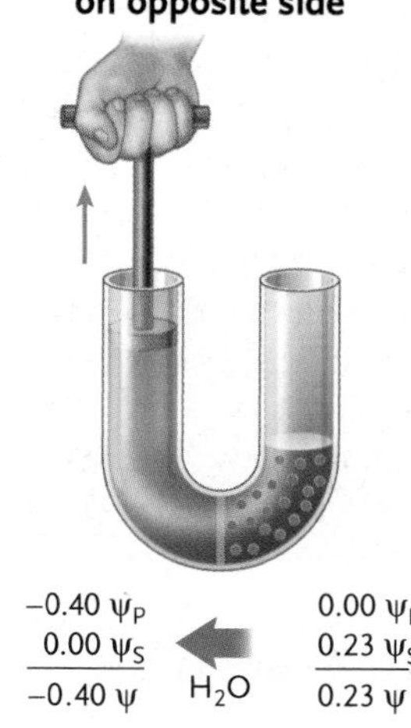
−0.40 ψP
0.00 ψS
−0.40 ψ
H2O
0.00 ψP
0.23 ψS
0.23 ψ

Water potential in a system decreases under tension (negative pressure)–suggested here by pulling up on the plunger. As the ψ of the pure water falls, even more water leaves the sucrose solution.

Plant physiologists assign a value of 0 MPa to the water potential (ψ) of pure water in an open container under normal atmospheric pressure and temperature.

FIGURE 32.4
The relationship between osmosis and water potential. If the water potential is higher on one side of a selectively permeable membrane, water will cross the membrane to the area of lower water potential. This diagram shows pure water on one side of a selectively permeable membrane and a simple sucrose solution on the other side. In an organism, however, the selectively permeable membranes of cells are rarely if ever in contact with pure water.

rapidly across cell membranes? Researchers working with red blood cells found channel proteins that allow the bulk flow of water across cell membranes—a mechanism much speedier than normal osmosis. Chapter 6 introduced these channels, which were named **aquaporins.** As described in *Insights from the Molecular Revolution,* the discovery of aquaporins in animal cells led plant scientists to investigate similar proteins in plant cells, and we now know that various aquaporins occur in the tonoplast and the plasma membrane. Most have major roles in maintaining water balance in cells. In roots they facilitate the uptake of water and its movement into the xylem. Accumulating research evidence suggests that plant aquaporins are important in a wide variety of physiological functions.

Water Stress Can Lead to Wilting

The water mechanics we have been discussing have major implications for land plants. For instance, the drooping of leaves and stems called **wilting** occurs when environmental conditions cause a plant to lose more water than it gains. Conditions that lead to wilting include dry soil, in which case the water potential in the soil falls below that in the plant. Then the turgor pressure inside the cells falls, and the protoplast shrinks away from the cell wall. Similarly, placing a cell in a sucrose solution can cause the protoplast to shrink **(Figure 32.5A)**. By contrast, as long as the Ψ of soil is higher than that in root epidermal cells, water will follow the Ψ gradient and enter root cells, making them turgid, or firm **(Figure 32.5B)**. As we see in the next section, water and solutes entering roots may move through the plant body by several routes.

STUDY BREAK 32.1 <

1. **Explain the role(s) of a gradient of protons in moving substances across a plant cell's plasma membrane.**
2. **What is "water potential," and why is it important with respect to plant cells?**

32.2 Transport in Roots

Soil around roots provides a plant's water and minerals, but roots do not simply "soak up" these essential substances. Instead, water and minerals that enter roots first travel laterally through the root cortex to the root xylem. Only then do they begin their journey upward to stems, leaves, and other tissues.

Water Travels to the Root Xylem by Three Pathways

Soil water always enters a root through the root epidermis. Once inside a root, however, water may take one of three routes into the root xylem, traveling either through living cells or in extracellular

TIPs and PIPs—the Osmosis Express

Researchers working with red blood cells discovered that the plasma membranes of these animal cells contained channel proteins that allowed water to move across the membrane. The flow occurred osmotically, from higher to lower concentration, but much more rapidly than in normal osmosis. The channels were named aquaporins. Plant physiologists became intrigued by the presence of a structurally similar protein called TIP (tonoplast intrinsic protein) in the tonoplast, the cell membrane enclosing the vacuole. Such channels would explain how water can flow rapidly into and out of the vacuole.

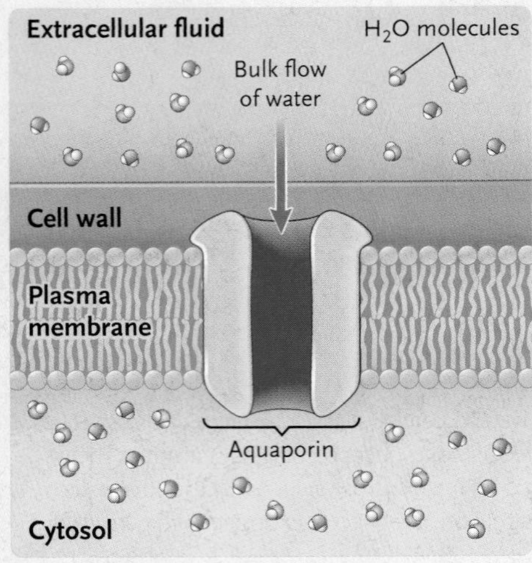
Extracellular fluid
H_2O molecules
Bulk flow of water
Cell wall
Plasma membrane
Aquaporin
Cytosol

Research Question:

Are TIPs plant aquaporins?

Experiments:

Christophe Maurel and his colleagues at the University of California, San Diego designed an experiment to test the hypothesis that TIPs function as aquaporins in plants. Their experimental design included several steps.

1. The team began by isolating a gene that encodes a TIP in *Arabidopsis thaliana* plants and cloned the gene's coding sequence. They then used genetic engineering techniques to generate mRNAs from the sequence.
2. To ensure that no other plant protein could affect the outcome of their experiments, they next inserted the TIP mRNA into oocytes of *Xenopus laevis* (the African clawed frog), which normally are only slightly permeable to water.
3. When the test oocytes were placed in a hypotonic medium they quickly swelled and ruptured, a response that did not occur in controls lacking the TIP mRNA.

Conclusion:

These results supported the conclusion that the TIP protein forms an aquaporin. In its normal location in the plant cell tonoplast, TIP apparently allows water to move readily in either direction as osmotic conditions shift.

At this writing dozens of plant aquaporins are known in a variety of species, including an assortment of them in the plasma membrane (PIPs, or plasma membrane intrinsic proteins). In *Arabidopsis* alone, 10 TIPs and 13 PIPs have been identified. Different PIP genes are expressed in cells of different plant parts, and different environmental influences—such as drought and high salinity—have different effects on how the channels function.

Source: C. Maurel et al. 1993. The vacuolar membrane protein gamma-TIP creates water specific channels in *Xenopus* oocytes. *EMBO Journal* 12:2241–2247.

The experimenter begins with flaccid plant cells at atmospheric pressure and temperature. The cells contain enough water to prevent the plasma membrane from shrinking away from the cell wall, but lack turgor.

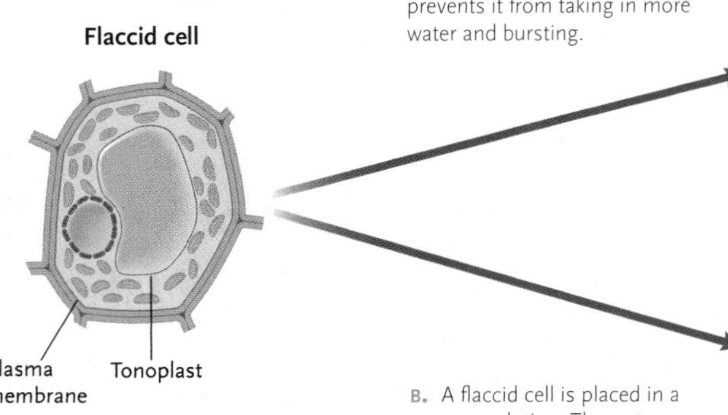
Flaccid cell
Plasma membrane
Tonoplast

A. A flaccid cell is placed in distilled water, which has a water potential of zero—much greater than the negative water potential inside the cell. The cell gains water by osmosis and swells until it is turgid. The cell wall prevents it from taking in more water and bursting.

Pure water

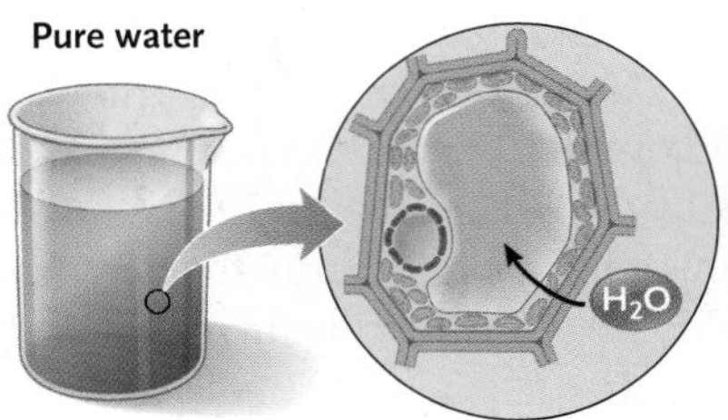
H_2O

Distilled water

$\psi_P = 0$ MPa
$\psi_S = 0$ MPa

Turgid cell at equilibrium with its environment

$\psi_P = 0.7$ MPa
$\psi_S = -0.7$ MPa
$\psi = 0.0$ MPa

Turgid cells from an iris petal *(Iris)*

Perennou Nuridsany/Photo Researchers, Inc.

B. A flaccid cell is placed in a sucrose solution. The water potential inside the cell is much greater than that in the solute-rich solution, and the cell loses water until the vacuole shrinks and the protoplast shrinks away from the cell wall. This outcome of the experiment is called plasmolysis.

Sucrose solution

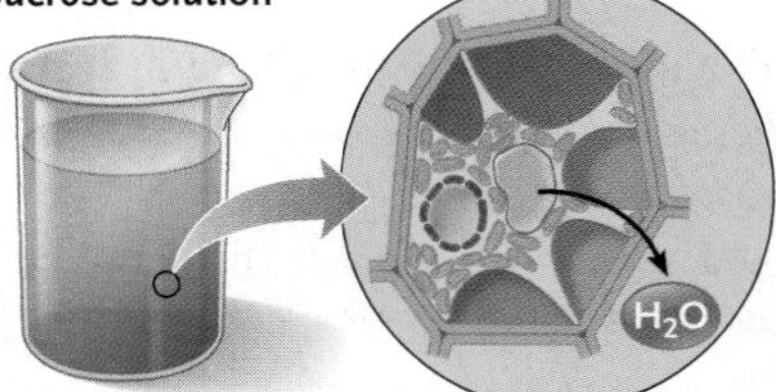
H_2O

0.4 *M* sucrose

$\psi_P = 0.0$ MPa
$\psi_S = -0.9$ MPa
$\psi = -0.9$ MPa

Plasmolyzed cell at equilibrium with its environment

$\psi_P = 0.0$ MPa
$\psi_S = -0.9$ MPa
$\psi = -0.9$ MPa

Plasmolyzed cells from a wilted iris petal

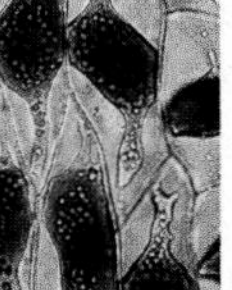

Perennou Nuridsany/Photo Researchers, Inc.

FIGURE 32.5
An experiment to test the effects of different osmotic environments on plant cells. Notice that in both **(A)** and **(B)** the final condition is the same: the water potential of the plant cell and its environment become equal.

A. Apoplastic, symplastic, and transmembrane pathways

In the **apoplastic pathway** (red), water moves through nonliving regions–the continuous network of adjoining cell walls and tissue air spaces. However, when it reaches the endodermis, it passes through one layer of living cells.

In the **symplastic pathway** (green), water passes into and through living cells. After being taken up into root hairs water diffuses through the cytoplasm and passes from one living cell to the next through plasmodesmata.

In the **transmembrane pathway** (black), water that enters the cytoplasm moves between living cells by diffusing across cell membranes, including the plasma membrane and perhaps the tonoplast.

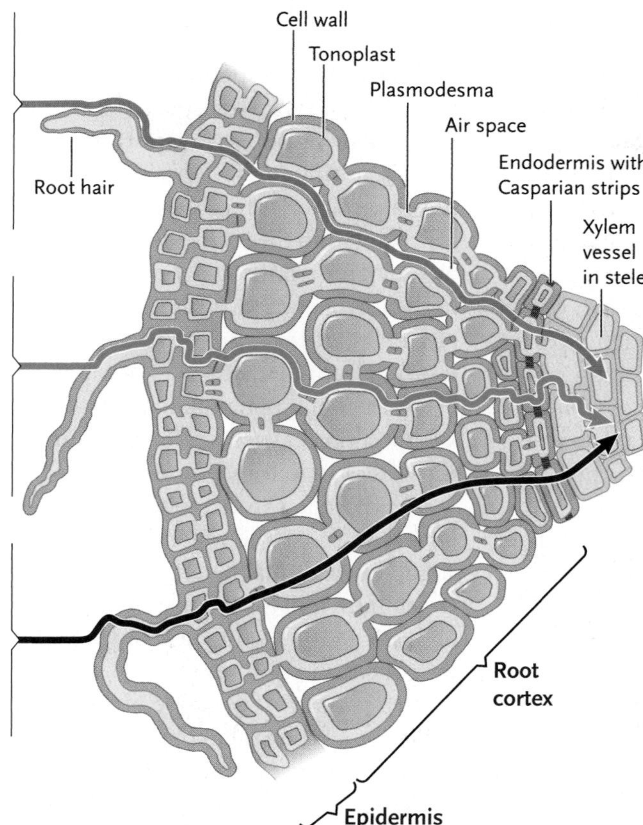

Cell wall
Tonoplast
Plasmodesma
Air space
Endodermis with Casparian strips
Root hair
Xylem vessel in stele
Root cortex
Epidermis

B. Casparian strip

Waxy, water-impervious Casparian strip (brown) in abutting walls of endodermal cells that control water and nutrient uptake

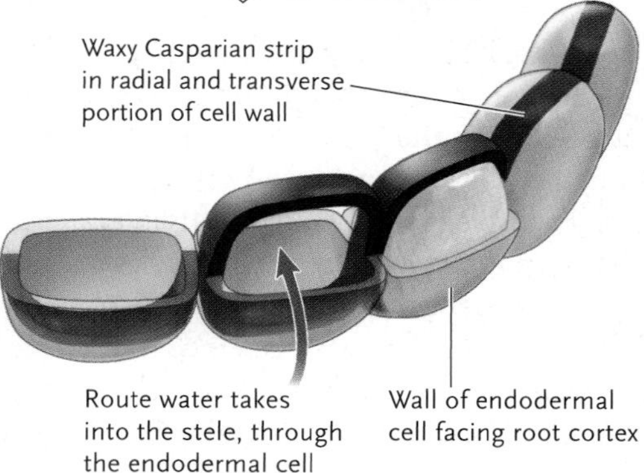

Waxy Casparian strip in radial and transverse portion of cell wall
Route water takes into the stele, through the endodermal cell plasma membrane (symplastic pathway)
Wall of endodermal cell facing root cortex

FIGURE 32.6
Pathways for the movement of water into roots. **(A)** Water can enter roots by way of the apoplast or symplast, or by a transmembrane pathway. Ions also enter roots via these three pathways, but must be actively transported into a cell—and the symplastic pathway—when they reach the impervious Casparian strips of the endodermis **(B).** The roots of most flowering plants have both an endodermis surrounding the stele and an exodermis just beneath the epidermis. Both cell layers have a Casparian strip, which helps control the uptake of water and dissolved nutrients.

areas of the root **(Figure 32.6A).** The extracellular route passes through a continuous network of adjoining cell walls and air spaces in root tissue but does not cross cell membranes. This route is called the *apoplast* (roughly, "not including cells"), and water in them follows an **apoplastic pathway.** Plant physiologists refer to a plant's living parts as the *symplast* ("within cells") and water moving through roots in the **symplastic pathway** moves from cell to cell through the open channels of plasmodesmata. Water also can enter root cells through channels in the cell plasma membranes, a **transmembrane pathway.** Water may cross the tonoplast of the central vacuole in this way as well.

When water enters a root, some diffuses into epidermal cells, entering the symplast. But a great deal of the water taken up by plant roots moves into the apoplast, moving along through cell walls and intercellular spaces. This apoplastic water (and any solutes dissolved in it) travels rapidly inward until it encounters the endodermis, the sheetlike single layer of cells that separates the root cortex from the stele. Cells in the root cortex generally have air spaces between them (which helps aerate the tissue), but endodermal cells are tightly packed. Each endodermal cell also has a **Casparian strip** in its radial and transverse walls, positioned somewhat like a band of packing material around the contents of a package **(Figure 32.6B).** The strip is impregnated with suberin, a waxy substance impermeable to water. Thus the Casparian strip blocks the apoplastic pathway at the endodermis, preventing water and solutes in the apoplast from automatically passing on into the stele. Instead, if molecules are to move into the stele, they must detour across the plasma membranes of endodermal cells, entering the cells (and the symplast) where the wall is not blanketed by a Casparian strip. From there water and solutes can pass through plasmodesmata to cells in the outer layer of the stele (the pericycle), then on into the xylem.

Although water molecules can easily cross an endodermal cell's plasma membrane, the semipermeable membrane allows only a subset of the solutes in soil water to cross. Undesirable solutes may be barred, whereas desirable ones may move into the cell by facilitated diffusion or active transport. Conversely, the endodermis prevents needed substances in the xylem from leaking out, back into the root cortex. In this way the endodermis provides important control over which substances enter and leave a plant's vascular tissue. The roots of most flowering plants also have a second layer of cells with Casparian strips just inside the root epidermis. This layer, the exodermis, functions like the endodermis.

Roots Take Up Ions by Active Transport

Mineral ions in soil water also enter roots through the epidermis. Some enter the apoplast along with water, but most ions important for plant nutrition tend to be much more concentrated in roots than in the surrounding soil, so they cannot follow a concentration gradient into root epidermal cells. Instead the epidermal cells actively transport ions inward—that is, ions enter the symplast immediately. They travel to the xylem via the symplastic or transmembrane pathways. Other ions can still move inward following the apoplastic pathway until they reach the Casparian strip of the endodermis. To contribute to the plant's nutrition, however, they must be actively transported from the exodermis into cells of the root cortex and, as just described, from the endodermis into the stele. In short, mechanisms that control which solutes will be absorbed by root cells ultimately determine which solutes will be distributed through the plant.

Once an ion reaches the stele, it diffuses from cell to cell until it is "loaded" into the xylem. Experiments to determine whether the loading is passive (by diffusion) or active have been inconclusive, so the details of this final step are not entirely clear. Because the xylem's conducting elements are not living, water and ions in effect reenter the apoplastic pathway when they reach either tracheids or vessels. Once in the xylem, water can move laterally to and from tissues or travel upward in the conducting elements. Minerals are distributed to living cells and taken up by active transport. The following section examines how this "distribution of the mineral wealth" takes place.

STUDY BREAK 32.2

1. **Explain two key differences in how the apoplastic and symplastic pathways route substances laterally in roots.**
2. **How does an ion enter a root hair and then move to the xylem?**

32.3 Transport of Water and Minerals in the Xylem

We return now to the question that opened this chapter: How does the solution of water and minerals called xylem sap move—100 m or more in the tallest trees—from roots to stems, then into leaves? Xylem sap is mostly water, and we know that it moves upward by bulk flow through the tracheids and vessels in xylem. Yet because mature xylem cells are dead, they cannot expend energy to move water into and through the plant shoot. Instead, the driving force for the upward movement of xylem sap from root to shoot is the water potential gradient between leaves and the air. Sunlight and other factors affect the gradient, which causes water to evaporate from leaves and other aerial plant parts. Experiments show that only a small fraction of the water in xylem sap is used in a plant's growth and metabolism. The rest evaporates from the epidermis of aboveground plant parts. This loss of water vapor from aerial plant parts is called **transpiration.** As described next, transpiration drives the ascent of sap.

The Mechanical Properties of Water Have Key Roles in Its Transport

Chapter 2 introduced several biologically important mechanical properties of water. Two of them interest us here. First, water molecules are strongly *cohesive:* they tend to form hydrogen bonds with one another. Second, water molecules are *adhesive:* they form hydrogen bonds with molecules of other substances, including the carbohydrates in plant cell walls. Water's cohesive and adhesive forces jointly pull water molecules into exceedingly small spaces, such as crevices in cell walls or narrow tubes such as xylem vessels in roots, stems, and leaves. In 1914, plant physiologist Henry Dixon explained the ascent of sap in terms of the relationship between transpiration and water's mechanical properties. His model of xylem transport is now called the **cohesion–tension mechanism of water transport (Figure 32.7).**

According to the cohesion–tension model, water transport begins as water evaporates from the walls of mesophyll cells of leaves and into the intercellular spaces. This water vapor escapes by transpiration through open stomata, the minute passageways in the leaf surface. As water molecules exit the leaf, they are replaced by others from the mesophyll cell cytoplasm. The water loss gradually reduces the water potential in a transpiring cell below the water potential in the leaf xylem. Now, water from the xylem in the leaf veins follows the gradient into cells, replacing the water lost in transpiration.

In the xylem, water molecules are confined in narrow, tubular xylem cells. The water molecules form a long chain, like a string of weak magnets, held together by hydrogen bonds between individual molecules. When a water molecule moves out of a leaf vein into the mesophyll, its hydrogen bonds with the next molecule in line stretch but do not break. The stretching creates *tension*—a negative pressure gradient—in the column. Adhesion of the water column to xylem vessel walls adds to the tension. Under continuous tension from above, the entire column of water molecules in xylem is drawn upward, in a fashion somewhat analogous to the way water moves up through a drinking straw. Botanists refer to this root-to-shoot flow as the *transpiration stream.*

Transpiration continues regardless of whether evaporating water is replenished by water rapidly taken up from the soil. Wilting is visible evidence that the water-potential gradient between soil and a plant's shoot parts has shifted. Remember that as soil dries out, the remaining water molecules are held ever more tightly by the soil particles. In effect, the action of soil particles reduces the water potential in the soil surrounding plant roots, and as this happens the roots take up water more slowly. However, because the water that evaporates from the plant's leaves is no longer being fully replaced, the leaves wilt as turgor pressure drops. Reducing the water potential in soil by adding solutes such as NaCl and other salts can have the same wilting effect. When the water potential in the soil finally equals that in leaf cells, a gradient no longer exists. Then movement of water from the soil into roots and up to the leaves comes to a halt.

Leaf Anatomy Contributes to Cohesion–Tension Forces

Leaf anatomy is key to the processes that move water upward into plants. To begin with, as much as two-thirds of a leaf's volume consists of air spaces. Thus there is a large internal surface area for evaporation. Leaves also may have thousands to millions of stomata, through which water vapor can escape. Both these factors increase transpiration. Also, every square centimeter of a leaf contains thousands of tiny xylem veins, so that most leaf cells lie within half a millimeter of a vein. This close proximity ensures a

FIGURE 32.7

Cohesion–tension mechanism of water transport. Transpiration, the evaporation of water from shoot parts, creates tension on the water in xylem sap. This tension, which extends from leaf to root, pulls upward columns of water molecules that are hydrogen-bonded to one another. Water molecules are not shown to scale.

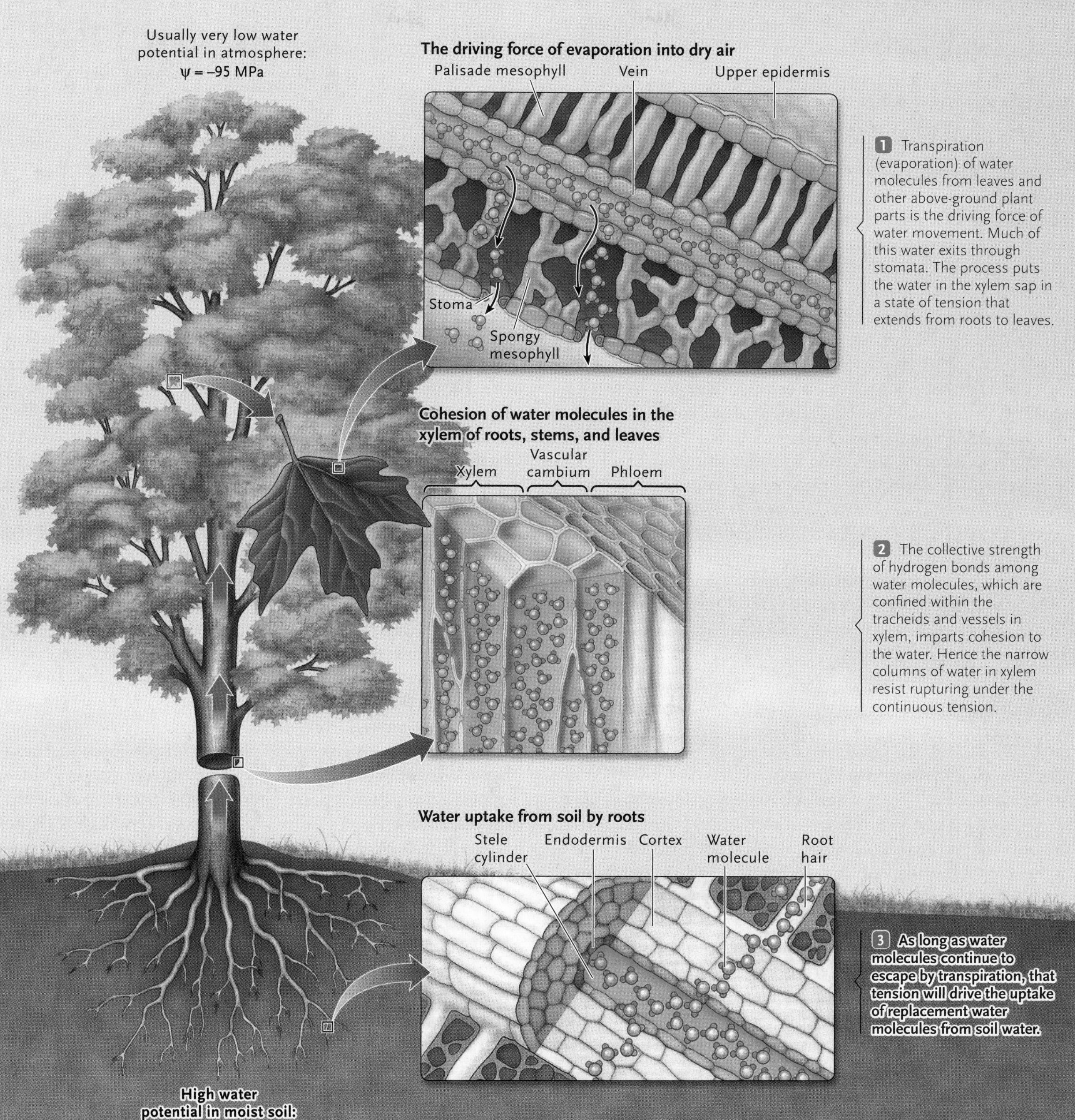

Usually very low water potential in atmosphere: ψ = −95 MPa
The driving force of evaporation into dry air
Palisade mesophyll
Vein
Upper epidermis
Stoma
Spongy mesophyll
1 Transpiration (evaporation) of water molecules from leaves and other above-ground plant parts is the driving force of water movement. Much of this water exits through stomata. The process puts the water in the xylem sap in a state of tension that extends from roots to leaves.
Cohesion of water molecules in the xylem of roots, stems, and leaves
Xylem
Vascular cambium
Phloem
2 The collective strength of hydrogen bonds among water molecules, which are confined within the tracheids and vessels in xylem, imparts cohesion to the water. Hence the narrow columns of water in xylem resist rupturing under the continuous tension.
Water uptake from soil by roots
Stele cylinder
Endodermis
Cortex
Water molecule
Root hair
3 As long as water molecules continue to escape by transpiration, that tension will drive the uptake of replacement water molecules from soil water.
High water potential in moist soil: ψ = >0 MPa

ready supply of water to cells and the spaces between them, sites from which the water can readily evaporate.

As water evaporates from a leaf, surface tension at the interface between the water film and the air in the leaf space translates into negative pressure that draws water from the leaf veins. This tension is multiplied many times over in all of the leaves and xylem veins of a plant. It increases further as the plant's metabolically active cells take up xylem sap.

In the Tallest Trees, the Cohesion–Tension Mechanism May Reach Its Physical Limit

A variety of experiments have tested the premises of the cohesion–tension model, and thus far the data strongly support it. For example, the model predicts that xylem sap will begin to move upward at the top of a tree early in the day when water begins to evaporate from leaves. Experiments with several different tree species have confirmed that this is the case. The experiments also showed that sap transport peaks at midday when evaporation is greatest, then tapers off in the evening as evaporative water loss slows.

Other experiments have probed the relationship between xylem transport and tree height. George W. Koch at Northern Arizona University and his colleagues studied several of the tallest living redwoods, including one that towers nearly 113 m above the forest floor. When the scientists measured the maximum tension exerted in the xylem sap in twigs at the tops of the trees, they discovered that it approached the known physical limit at which the bonds between water molecules in a column of water in a conifer's xylem will rupture. Based on this finding and other evidence, the team has predicted that the maximum height for a healthy redwood tree is 122 to 130 m—so the tallest living redwoods may grow taller still.

Root Pressure Contributes to Upward Water Movement in Some Plants

The cohesion–tension mechanism accounts for upward water movement in tall trees. In some nonwoody plant species, however—lawn grasses, for instance—a positive pressure can develop in roots and force xylem sap upward. This **root pressure** operates under conditions that reduce transpiration, such as high humidity or low light. In fact, the mechanism that produces root pressure often operates at night, when transpiration slows or stops. Then, active transport of ions into the stele sets up a water potential gradient across the endodermis. Because the Casparian strip of the endodermis tends to prevent ions from moving back into the root cortex, the water potential difference becomes quite large. It can move enough water and dissolved solutes into the xylem to produce a relatively high positive pressure. Although not sufficient to force water to the top of a very tall plant, in some smaller plant species root pressure is strong enough to force water out of leaf openings, in a process called **guttation (Figure 32.8).** Pushed up and out of vein endings by root pressure, tiny droplets of water that look like dew in the early morning emerge from modified stomata at the margins of leaves.

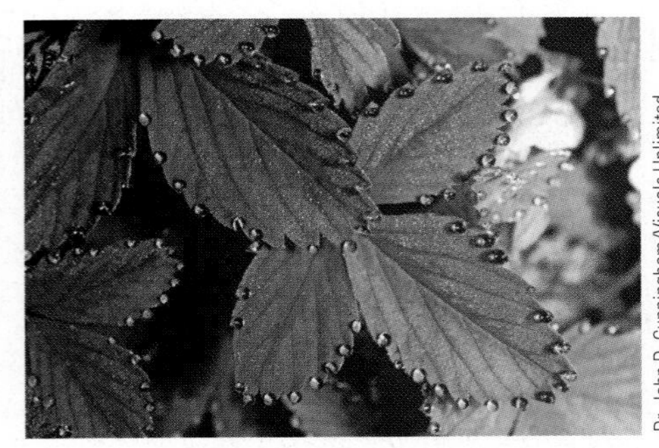
Dr. John D. Cunningham/Visuals Unlimited

FIGURE 32.8 Guttation, caused by root pressure. The drops of water appear at the endings of xylem veins along the leaf edges of a strawberry plant *(Fragaria)*.

Stomata Regulate the Loss of Water by Transpiration

Three environmental conditions have major effects on the rate of transpiration: relative humidity, air temperature, and air movement. Relative humidity is a measure of the amount of water vapor in air. The less water vapor in the air, the more evaporates from leaves (because the water potential is higher in the leaves than in the dry air). Rising air temperature at the leaf surface also speeds evaporation. Although evaporation cools the leaf somewhat, the amount of water lost can double for each 10°C-rise in air temperature. Air moving across the leaf surface carries water vapor away from the surface and so makes a steeper gradient. Together these factors explain why on extremely hot, dry, breezy days, the leaves of certain plants must completely replace their water each hour.

Even when conditions are not so drastic, more than 90% of the water moving into a leaf can be lost through transpiration. About 2% of the water remaining in the leaf is used in photosynthesis and other activities. These measurements emphasize the need for controls over transpiration, for if water loss exceeds water uptake by roots, the resulting dehydration of plant tissues interferes with normal functioning, and the plant may wilt and die.

The cuticle-covered epidermis of leaves and stems reduces the rate of water loss from aboveground plant parts, but it also limits the rate at which CO_2 for photosynthesis can diffuse into the leaf. The functioning of stomata also affects a plant's water balance. When stomata are open, carbon dioxide can be absorbed, but unless the relative humidity of external air is 100%, water always moves out. However, plants have evolved adaptations that balance water loss with CO_2 uptake. This "transpiration–photosynthesis compromise" involves the regulation of transpiration and gas exchange by opening and closing stomata as environmental conditions change.

OPENING AND CLOSING OF STOMATA Two guard cells flank each stomatal opening **(Figure 32.9).** Their elastic walls are reinforced by parallel strands of cellulose microfibrils that wrap around the walls in a radial pattern, an arrangement that often is likened to the steel belts in a radial tire. The inner walls are thicker and less elastic than the outer walls.

The opening and closing of stomata provide good examples of a symport mechanism (see Figure 32.3C). Stomata open when potassium ions (K^+) flow into the guard cells through ion channels **(Figure 32.10).** As a first step, an active transport pump in the

FIGURE 32.9

Guard cells and stomatal action. (A) An open stoma. Water entered collapsed guard cells, which swelled under turgor pressure and moved apart, thus forming the stoma in the needlelike leaf of the rock needlebush *(Hakea gibbosa)*. **(B)** A closed stoma. Water exited the swollen guard cells, which collapsed against each other and closed the stoma.

A. **Open stoma**

Guard cell
Guard cell
Stoma
Courtesy of E. Raveh

B. **Closed stoma**

Chloroplast (guard cells are the only epidermal cells that have these organelles)
Courtesy of E. Raveh
20 μm

plasma membrane begins pumping H^+ ions out of the guard cells. Recall from Section 32.1 that H^+ pumped out of the cell can then follow its concentration gradient back into the cell. This inward flow of H^+ powers the active transport of K^+ into the guard cell. As a result, the K^+ concentration in turgid guard cells may be four to eight times higher than that in flaccid (limp) guard cells. Water follows inward by osmosis. As turgor pressure builds in the guard cells, the supportive "belts" of cellulose microfibrils in the outer cell walls prevent the cells from expanding to the side. Instead each cell elongates as the microfibrils move farther apart. As a result, the two elongating cells bow away from each other and create a stoma ("mouth") between them. Stomata close when the H^+ active transport protein stops pumping. K^+ flows passively out of the guard cells, and water follows by osmosis. When the water content of the guard cells dwindles, turgor pressure drops. The guard cells collapse against each other, closing the stomata.

In most plants, stomata open at first light, stay open during daylight, and close at night. Experiments have shown that guard cells respond to a number of environmental and chemical signals, any of which can induce the ion flows that open and close stomata. These signals include light, CO_2 concentration in the air spaces inside leaves, and the amount of water available to the plant.

LIGHT AND CO_2 CONCENTRATION Light induces stomata to open through stimulation of blue-light receptors, probably located in the plasma membrane of guard cells. When stimulated, the receptors start the chain of events leading to stomatal opening by triggering activity of the H^+ pumps. Also, as photosynthesis begins in response to light, CO_2 concentration drops in the leaf air spaces as chloroplasts use the gas in carbohydrate production. In some way, this drop in CO_2 concentration sets off the series of events increasing the flow of K^+ into guard cells and furthers stomatal opening. The effects of reduced CO_2 concentration have been tested by placing plants in the dark in air containing no CO_2. Even in the absence of light, as the CO_2 concentration falls in leaves, guard cells swell and the stomata open.

Normally, when the sun goes down, a plant's demand for CO_2 drops as photosynthesis comes to a halt. Yet aerobic respiration continues to produce CO_2, which accumulates in leaves. As CO_2 concentration rises, and the blue-light wavelengths that activated the H^+ pumps wane, K^+ is lost from the guard cells and they collapse, closing the stomata. Thus, at night transpiration is reduced and water is conserved.

A. **Open stomata, with potassium mostly in guard cells**

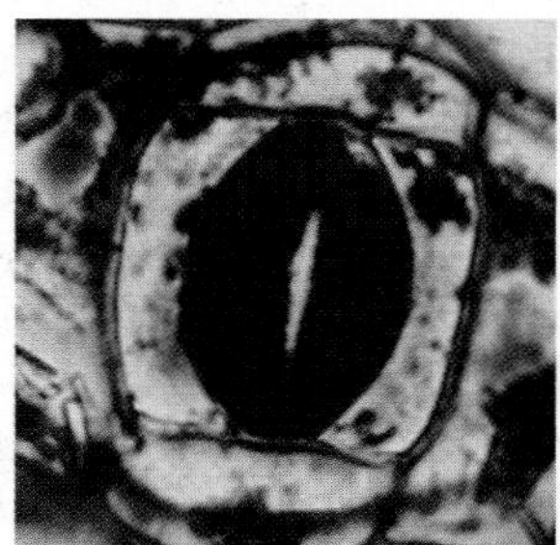
T. A. Masefield

B. **Closed stomata, with potassium mostly in epidermal cells**

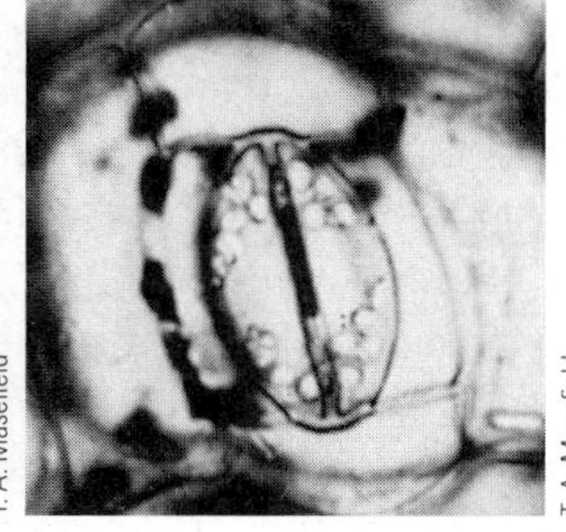
T. A. Masefield

FIGURE 32.10

Evidence for potassium accumulation in stomatal guard cells undergoing expansion. Strips from the leaf epidermis of a dayflower *(Commelina communis)* were immersed in a solution containing a stain that binds preferentially with potassium ions. **(A)** In leaf samples with open stomata, most of the potassium was concentrated in the guard cells. **(B)** In leaf samples with closed stomata, little potassium was in guard cells; most was present in adjacent epidermal cells.

WATER STRESS As long as water is readily available to a plant's roots, the stomata remain open during daylight. However, if water loss stresses a plant, the stomata close or open only slightly, regardless of light intensity or CO_2 concentration. Experiments have shown that the stress-related closing of stomata depends on a hormone, abscisic acid (ABA) that is released by roots when water is unavailable. This mechanism seems especially important in nonwoody plants. In some simple but elegant studies, test plants were suspended in containers so that only one-half the root system received water. Even though the roots with access to water could absorb enough water to satisfy the needs of all the plants' leaves, the stomata still closed. Tissue analysis revealed that water-stressed roots rapidly synthesize ABA. Transported through the xylem, this hormone stimulates K^+ loss by guard cells, and water moves out of the cell by osmosis (enhanced by aquaporins) so the stomata close **(Figure 32.11)**. Mesophyll cells also take up ABA from the xylem and release it, with the same effects on stomata, when their turgor pressure falls because of excessive water loss. ABA can also cause stomata to close when the hormone is added experimentally to leaves.

THE BIOLOGICAL CLOCK Besides responding to light, CO_2 concentration, and water stress, stomata apparently open and close on a regular daily schedule imposed by a biological clock. Even when plants are placed in continuous darkness, their stomata

A. **Stoma is open; water and solutes have moved in.**

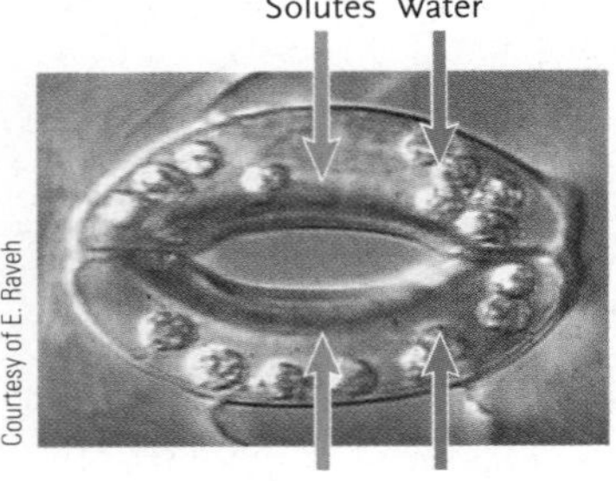
Solutes Water
Courtesy of E. Raveh

B. **Stoma is closed; water and solutes have moved out.**

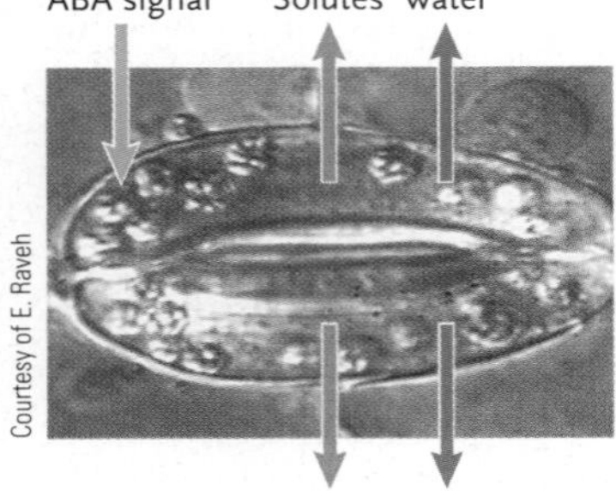
ABA signal Solutes Water
Courtesy of E. Raveh

FIGURE 32.11
Hormonal control of stomatal closing. **(A)** When a stoma is open, high solute concentrations in the cytoplasm of both guard cells have raised the turgor pressure, keeping the cells swollen open. **(B)** In a water-stressed plant, the hormone abscisic acid (ABA) binds to receptors on the guard cell plasma membrane. Binding activates a signal transduction pathway that lowers solute concentrations inside the cells, which lowers the turgor pressure—so the stoma closes.

A. **Oleanders** B. **Oleander leaf**

Cuticle
Multilayer epidermis
Recessed stoma

iStockphoto.com/lubilub
Thomas L. Rost

C. **Spines (modified leaves) on a cactus stem** D. **CAM plant**

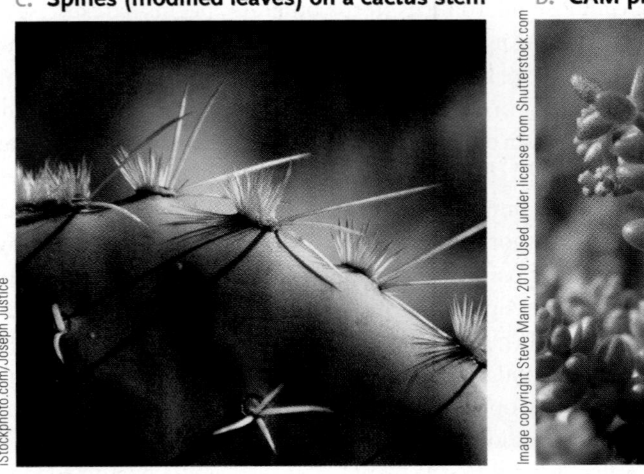
iStockphoto.com/Joseph Justice
Image copyright Steve Mann, 2010. Used under license from Shutterstock.com

FIGURE 32.12
Some adaptations that enable plants to survive water stress. **(A)** Oleanders *(Nerium oleander)* are adapted to arid conditions. **(B)** As shown in the micrograph, oleander leaves have recessed stomata on their lower surface and a multilayer epidermis covered by a thick cuticle on the upper surface. **(C)** Like many other cacti, the leaves of the Graham dog cactus *(Opuntia grahamii)* are modified into spines that protrude from the underlying stem. Transpiration and photosynthesis occur in the green stems. **(D)** *Sedum,* a CAM plant, in which the stomata open only at night.

open and close (for a time) in a cycle that roughly matches the day/night cycle of Earth. Such *circadian rhythms* (*circa* = around; *dies* = day) are also common in animals, and several, including wake/sleep cycles in mammals, are known to be controlled by hormones—a topic pursued in Chapter 40.

In Dry Climates, Plants Exhibit Various Adaptations for Conserving Water

Many plants have other evolutionary adaptations that conserve water, including modifications in structure or physiology **(Figure 32.12).** Oleanders, for example, have stomata on the underside of the leaf at the bottom of pitlike invaginations of the leaf epidermis lined by hairlike trichomes (Figure 32.12B). Sunken stomata are less exposed to drying breezes, and trichomes help retain water vapor at the pore opening, so that water evaporates from the leaf much more slowly.

The leaves of *xerophytes*—plants adapted to hot, dry environments in which water stress can be severe—have a thickened cuticle that gives them a leathery feel and provides enhanced protection against evaporative water loss. An example is mesquite *(Prosopis).* In still other plants that inhabit arid landscapes, such as cacti, stems are thick, leaflike pads covered by sharp spines that actually are modified leaves (Figure 32.12C). These structural alterations reduce the surface area for transpiration. In some plants coping with water stress includes a capacity to resist *cavitation,* the formation of physiologically dangerous air bubbles in xylem mentioned in Section 31.2.

One intriguing variation on water-conservation mechanisms occurs in CAM plants, including cacti, orchids, and most succulents. As discussed in Section 9.4, **crassulacean acid metabolism** (CAM) is a biochemical variation of photosynthesis that was discovered in a member of the family Crassulaceae. CAM plants generally have fewer stomata than other types of plants, and their stomata typically follow a reversed schedule. They are closed during the day when temperatures are higher and the relative humidity is lower, and open at night. At night, the plant temporarily fixes carbon dioxide by converting it to malate, an organic acid. In the daytime, the CO_2 is released from malate and diffuses into chloroplasts, so photosynthesis takes place even though a CAM plant's stomata are closed. This adaptation prevents heavy evaporative water losses during the heat of the day.

STUDY BREAK 32.3

1. **Explain the key steps in the cohesion–tension mechanism of water transport in a plant.**
2. **How and when do stomata open and close? In what ways is their functioning important to a plant's ability to manage water loss?**

THINK OUTSIDE THE BOOK

Cavitation, the formation of a bubble-like space in a tracheid or xylem vessel, is a serious risk for plants under water stress. Collaboratively or individually, research the causes and effects of cavitation. Can it be caused by environmental factors other than water stress? Is it possible for such a break in the column of water molecules to "heal"?

32.4 Transport of Organic Substances in the Phloem

A plant's phloem is another major long-distance transport system, and a superhighway at that: it carries huge amounts of carbohydrates, lesser but vital amounts of amino acids, fatty acids, and other organic compounds, and still other essential substances such as hormones. And unlike the xylem's unidirectional upward flow, the phloem transports substances throughout the plant to wherever they are used or stored. Organic compounds and water in the sieve tubes of phloem are under pressure and driven by concentration gradients.

Organic Compounds Are Stored and Transported in Different Forms

Plants synthesize various kinds of organic compounds, including large amounts of carbohydrates that are stored mainly as starch. Plant cells must often export these compounds for use in distant cells. Yet, with a few exceptions, macromolecules such as starches, proteins, and fat molecules are too large to cross cell membranes and leave the cells where they are made. They also may be too insoluble in water to be transported to other regions of the plant body. Consequently, in leaves and other plant parts, specific reactions convert organic compounds to transportable forms. For example, hydrolysis of starch liberates glucose units, which combine with fructose to form sucrose—the main form in which sugars are transported through the phloem of most plants. Proteins are broken down into amino acids, and lipids converted into fatty acids. These forms are also better able to cross cell membranes by passive or active mechanisms.

Organic Solutes Move by Translocation

In plants, the long-distance transport of substances is called **translocation.** Botanists most often use this term to refer to the distribution of sucrose and other organic compounds by phloem, and they understand the mechanism best in flowering plants. The phloem of flowering plants contains interconnecting sieve tubes formed by living sieve tube member cells (see Figure 31.8). Sieve tubes lie end to end within vascular bundles, and they extend through all parts of the plant. Water and organic compounds, collectively called **phloem sap,** flow rapidly through large pores on the sieve tubes' end walls—another example of a structural adaptation that suits a particular function.

Phloem Sap Moves from Source to Sink under Pressure

Over the decades, plant physiologists have proposed several mechanisms of translocation, but it was the tiny aphid, an annoying garden insect, that helped demonstrate that organic compounds flow under pressure in the phloem. An aphid attacks plant leaves and stems, forcing its needlelike stylet (a mouthpart) into sieve tubes to obtain the dissolved sugars and other nutrients inside. Numerous experiments with aphids have shown that in most plant species, sucrose is the main carbohydrate being translocated through the phloem. Studies also verify that the contents of sieve tubes are under high pressure, often five times as much as in an automobile tire. **Figure 32.13** explains a simple and innovative experiment that provided direct confirmation that phloem sap flows under pressure. When a live aphid feeds on phloem sap, this pressure forces the fluid through the aphid's gut and (minus nutrients absorbed) out its anus as "honeydew." A car parked under an aphid-infested tree might get spattered with sticky honeydew droplets, thanks to the high fluid pressure in the tree's phloem.

Much of what plant physiologists know about the transport of phloem sap has come from studies of sucrose transport in flowering plants. A fundamental discovery is that in flowering plants sucrose-laden phloem sap flows from a starting location, called the *source,* to another site, called the *sink,* along gradients of decreasing solute concentration and pressure. A **source** is any region of the plant where organic substances are produced and loaded into the phloem's sieve tube system. A **sink** is any region where organic substances are used or stored and so are unloaded from the sieve tube system. What causes the sugar sucrose and other solutes produced in leaf mesophyll to flow from a source to a sink? In flowering plants, the **pressure flow mechanism** builds up at the source end of a sieve tube system and pushes those solutes by bulk flow toward a sink, where they are removed. **Figure 32.14** summarizes this mechanism, using sucrose as an example.

FIGURE 32.13 **Experimental Research**

Translocation Pressure

Hypothesis: High pressure forces phloem sap to flow through sieve tubes from a source to a sink.

Experiment: In the late 1970s, John Wright and Donald Fisher at the University of Georgia devised an experiment to directly measure the turgor pressure in sieve tubes of weeping willow saplings *(Salix babylonica)* under nondestructive conditions, using aphids that feed on *S. babylonica* in the wild. Weeping willow saplings were grown in a greenhouse under natural conditions of light and moisture. Aphids were placed on the trees and allowed to begin feeding by inserting their stylets into sieve tubes in the normal fashion. After being anesthetized by exposure to high concentrations of carbon dioxide, the aphids' bodies were cut away and only their stylets were left embedded in the sieve tubes. A tiny pressure-measuring device called a micromanometer then was glued over the end of each stylet. The micromanometer registered the volume and pressure of phloem sap as it was exuded from the stylet over time periods ranging from 30 to 90 minutes.

A. **Aphid releasing honeydew**

Martin H. Zimmermann

B. **Micrograph of aphid stylet in sieve tube**

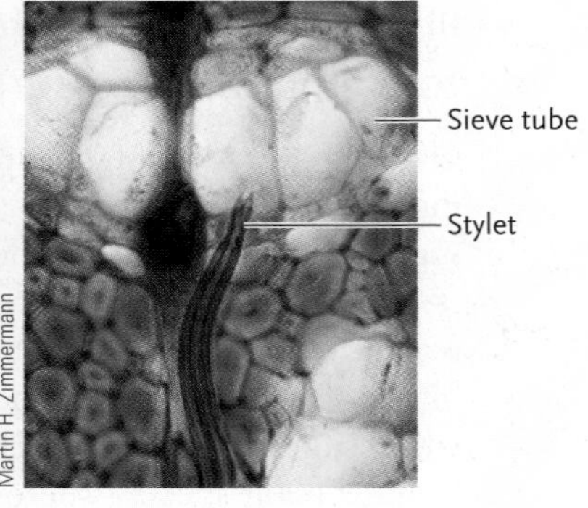
Sieve tube
Stylet
Martin H. Zimmermann

Results: In nearly all cases, a high volume of pressurized sap flowed through the severed stylets into the micromanometer during the test periods.

Conclusion: The evidence supports pressure flow as the mechanism that moves phloem sap through sieve tubes.

Other experiments have confirmed that both turgor pressure and the concentration of sucrose are highest in sieve tubes closest to the sap source. Phloem sap also moves most rapidly closest to the source, where pressure is highest.

Source: J. P. Wright and D. B. Fisher. 1980. The direct measurement of sieve tube turgor pressure using severed aphid stylets. *Plant Physiology* 65:1133–1135.

Mature photosynthesizing leaves are sources. Another example is a tulip bulb. In spring, stored food is mobilized for transport upward to growing plant parts, but after the plants bloom, the bulb becomes a sink as sugars manufactured in the tulip plant's leaves are translocated into it for storage. Young leaves, roots, and developing fruits generally start out as sinks, becoming sources later when the season changes or the plant enters a new developmental phase. In general, sinks receive organic compounds from sources closest to them. Hence, the lower leaves on a rose bush may supply sucrose to roots, whereas leaves farther up the shoot supply the shoot tip.

Substances carried in phloem must be loaded into sieve tube members at sources and unloaded from them at sinks, and both processes have prompted a great deal of research. We know, for example, that in some plants the solutes produced in leaf mesophyll move between the symplast (from cell to cell via plasmodesmata) and the apoplast (adjoining cell walls and air spaces). In addition, active transport mechanisms have a major role in the movement of solutes into and out of cells. Consider sucrose. After this sugar forms inside leaf mesophyll cells it is exported and travels to sieve tubes and their adjacent companion cells, which interact so closely they often are considered a functional unit. Experiments with tobacco and some other species show that exported sucrose soon leaves the symplast and enters the apoplast next to a small phloem vein. Next, it is actively pumped into companion cells by symport, in which H^+ ions move into the cell through the same carrier that takes up the sugar molecules. Companion cells shuttle the sugar into sieve tube members.

In some plants, companion cells become modified into **transfer cells** that facilitate the short-distance transport of various organic solutes from the apoplast into the symplast. Transfer cells generally form when large amounts of solutes must be loaded or unloaded into the phloem, and they move substances through plasmodesmata to sieve tube members. As a transfer cell is forming, parts of the cell wall grow inward like pleats. This structural feature increases the surface area across which solutes can be taken up. The underlying plasma membrane, packed with transport proteins covers the ingrowths. Transfer cells also enhance solute transport between living cells in the xylem, and they occur in glandlike tissues that secrete nectar. Plant physiologists have discovered transfer cells in species from most taxonomic groups in the plant kingdom, as well as in fungi and algae. In part because they arise from differentiated cells (instead of from meristem cells like other plant types of plant cells), researchers are working to define the molecular mechanisms that trigger their development.

When sucrose is loaded into sieve tubes its concentration rises inside the tubes. Thus the water potential falls, and water flows into the sieve tubes by osmosis. In fact, the phloem typically carries a great deal of water. As water enters sieve tubes, turgor pressure in the tubes increases, and the sucrose-rich fluid moves by bulk flow into the increasingly larger sieve tubes of larger veins. Eventually, the fluid is pushed out of the leaf into the stem and toward a sink. When sucrose is unloaded at the sink, the water potential in sieve tubes rises, so water tends to flow out of the tubes by osmosis into the surrounding cells, where the water potential is lower. Some of this water "follows the solutes" into sink cells, whereas much more enters the xylem and is recirculated.

A. **Loading at a source**

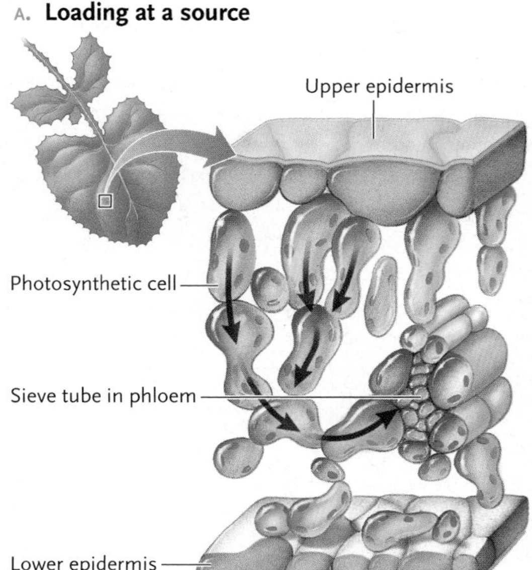
Upper epidermis
Photosynthetic cell
Sieve tube in phloem
Lower epidermis

Photosynthetic cells in leaves are a common source of carbohydrates that must be distributed through a plant. Small, soluble forms of these compounds move from the cells into phloem (in a leaf vein).

B. **Bulk flow from source to sink**

Xylem vessel
Sieve tube
Source
Photosynthesizing leaf cell
High turgor pressure
Companion cell
Sugar (sucrose)
Water
Phloem sap
Companion cell
Low turgor pressure
Sink
Root cell
Water in the transportation stream moves upward in xylem
5 As sucrose enters cells of the sink, the water potential in the sieve tube rises. Osmosis moves water out of the sieve tube—some into sink cells, more into the xylem.
1 Phloem sap forms as active transport loads sucrose into companion cells and then into sieve members, against concentration gradients.
2 As sucrose becomes more concentrated in the sieve tube, the water potential in the sieve tube falls, so water from xylem enters the tube by osmosis, increasing turgor pressure.
3 Under high pressure, phloem sap moves by bulk flow between a source and a sink. Water moves into and out of the system all along the way.
4 Pressure and sucrose concentration gradually decreases as the sink takes up sucrose from phloem, by active transport from sieve tube members into companion cells and then into sink cells.

FIGURE 32.14
Summary of the pressure flow mechanism in the phloem of flowering plants. Organic solutes are loaded into sieve tubes at a source, such as a leaf, and move by bulk flow toward a sink, such as roots or rapidly growing stem parts.

Sieve tubes are mostly passive conduits for translocation. The system works because companion cells supply much of the energy that loads sucrose and other solutes at the source, and because solutes are removed at their sinks. As sucrose (and water) enters a sink, for example, its concentration in sieve tubes falls, as does the pressure potential. Thus for sucrose and other solutes transported in the phloem, there is always a gradient of concentration from source to sink—and a pressure gradient that keeps the solute moving along.

As noted previously, phloem sap moving through a plant carries a wide variety of substances, including hormones, amino acids, organic acids, and agricultural chemicals. The phloem also transports organic nitrogen compounds and mineral ions that are withdrawn from dying leaves and stored for reuse in root tissue.

The transport functions of xylem and phloem are closely integrated with phenomena discussed later in this unit—reproduction and embryonic development, and the hormonal regulation of plant growth.

STUDY BREAK 32.4 <

1. **Compare and contrast translocation and transpiration.**
2. **Using sucrose as your example, summarize how a substance moves from a source into sieve tubes and then is unloaded at a sink. What is this mechanism called, and why?**

>

THINK OUTSIDE THE BOOK

The direction of phloem transport to or from a leaf may shift several times over the course of its lifespan. Collectively or on your own, research these types of changes. When are leaves sinks, and what triggers a leaf's shift to become a source? How does the development of flowers and fruit beyond a leaf affect the direction of phloem transport?

UNANSWERED QUESTIONS

What are plasmodesmata made of, and exactly how do they function? Plasmodesmata, the cytoplasmic channels through plant cell walls, connect plant cells to each other. Yet two fundamental questions about plasmodesmata remain unanswered: Exactly how do plasmodesmata function, and what are their structural components?

As described in this chapter and in Chapter 5, botanists have long assumed that nutrients, water, and small molecules that serve as growth regulators move through plasmodesmata, which form part of the symplastic pathway in plant tissues. Recent studies have demonstrated that larger molecules, including viruses and important proteins involved in plant growth and development, also move from cell to cell through plasmodesmata. For example, Patricia Zambryski and K. M. Crawford at the University of California at Berkeley reported that proteins, including transcription factors, travel via plasmodesmata from the cell that produces the proteins to adjacent cells where the factors promote or inhibit the expression of particular genes.

Although the normal functions of plasmodesmata in plant growth and development still are not well understood, ongoing research by Zambryski and other plant scientists has begun to shed light on the workings of these vital channels. For instance, a variety of studies of the processes by which viruses spread through plant tissues have revealed that plasmodesmata are not simply static, open channels. Instead they are dynamic structures with the capacity to close, reopen, widen, and narrow. This capacity for structural change is not triggered by viral infection: rather, it seems that viruses simply take over the plant's natural mechanism for moving molecules from one cell to another.

Plasmodesmata were first observed using electron microscopy several decades ago, and they appear to be lined with proteins as well as membranes. Multiple biochemical approaches have failed to identify the proteins, probably because of the difficulty of purifying proteins that are associated with both a membrane and the cell wall. Genetic screens to identify plasmodesmata proteins, as well as the genes that regulate the functioning of plasmodesmata, are currently under way and may finally reveal details of plasmodesmata structure. As our understanding of the architecture of plasmodesmata and how they function grows, so will insights into the mechanisms of plant development, how plants interact with viral pathogens, and other questions as well.

Think Critically

When a recent study tracked the movement of fluorescent tracers through the plasmodesmata of tobacco leaf trichomes, the results suggested that the tracers moved only in one direction and that the movement was not caused by bulk flow. The research team also found that sodium azide, a potent inhibitor of cell metabolism, halted tracer movement. Do these findings support or contradict a hypothesis that active transport may sometimes move solutes through plasmodesmata?

Beverly McMillan

Beverly McMillan

REVIEW KEY CONCEPTS

Go to **CENGAGENOW** at www.cengage.com/login to access quizzing, animations, exercises, articles, and personalized homework help.

32.1 Principles of Water and Solute Movement in Plants

- Plants have mechanisms for moving water and solutes long-distance from the root to shoot or vice versa, and over short distances from cell to cell (Figure 32.2).
- Both passive and active mechanisms move substances into and out of plant cells. Solutes generally are transported by carriers, either passively down a concentration or electrochemical gradient (in the case of ions), or actively against a gradient, which requires cellular energy. An H^+ gradient creates the membrane potential that drives the cross-membrane transport of many ions or molecules (Figure 32.3).
- Most organic substances enter plant cells by symport, in which the energy of the H^+ gradient is coupled with uptake of a different solute. Some substances cross the plant cell membrane by antiport, in which energy of the H^+ gradient powers movement of a second solute out of cells.
- Driven by water potential (Ψ), water crosses plant cell membranes by osmosis, moving from regions where water potential is higher to regions where it is lower.
- Water potential is the sum of turgor pressure and solute potential. Water potential is measured in megapascals (MPa) (Figures 32.4 and 32.5).
- Water and solutes also move by osmosis into and out of the cell's central vacuole, transported from the cytoplasm across the tonoplast. Aquaporins enhance osmosis across the tonoplast and plasma membrane. Water in the central vacuole is vital for maintaining turgor pressure inside a plant cell.
- Bulk flow of fluid occurs when pressure at one point in a system changes with respect to another point in the system.

32.2 Transport in Roots

- Water and mineral ions entering roots travel laterally through the root cortex to the root xylem, following one or more of three major routes: the apoplastic pathway, the symplastic pathway, and the transmembrane pathway (Figure 32.6A).
- In the apoplastic pathway, water entering roots diffuses between root epidermal cells. Absorbed water and solutes then enter either the symplastic or transmembrane pathway, both of which pass through cells.
- Casparian strips form a barrier that forces water and ions in the apoplast to pass through cells to enter the stele. Ions diffuse from cell to cell to reach the xylem. Roots of many flowering plants have an exodermis, a second cell layer with Casparian strips, just inside the root epidermis (Figure 32.6B).

Animation: Water absorption

Animation: Root functioning

32.3 Transport of Water and Minerals in the Xylem

- In xylem, tension generated by transpiration extends down from leaves to roots. By the cohesion–tension mechanism of water transport, water molecules are pulled upward by tension created as water exits a plant's leaves (Figure 32.7).
- In tall trees, negative pressure generated in the shoot drives bulk flow of xylem sap. In some herbaceous species, positive pressure may develop in roots and force xylem sap upward (Figure 32.8).
- Transpiration and CO_2 uptake occur mostly through stomata. Relative humidity, air temperature, and air movement at the leaf surface affect the transpiration rate (Figure 32.9).
- Most plants lose water and take up CO_2 during the day, when stomata are open. At night, when stomata close, plants conserve water and the inward movement of CO_2 falls.
- Stomata open in response to falling levels of CO_2 in leaves and also to incoming light wavelengths that activate photoreceptors in guard cells.
- Activation of photoreceptors triggers active transport of K^+ into guard cells. Simultaneous entry of anions such as Cl^- and synthesis of negatively charged organic acids increase the solute concentration, lowering the water potential so that water enters by osmosis. As turgor pressure builds, guard cells swell and draw apart, producing the stomatal opening (Figure 32.10).
- Guard cells close when light wavelengths used for photosynthesis wane. The stomata of water-stressed plants close regardless of light or CO_2 needs, possibly under the influence of the plant hormone ABA. The leaves of species native to arid environments typically have water-conserving adaptations (Figures 32.11 and 32.12).

Animation: Stomata

Animation: Transpiration

Animation: Interdependent processes

32.4 Transport of Organic Substances in the Phloem

- In flowering plants, phloem sap is translocated in sieve tube members. Differences in pressure between sources and sink regions drive the flow (Figure 32.13).
- In leaves, the sugar sucrose is actively transported into companion cells next to sieve tube members, then loaded into the sieve tubes through plasmodesmata. In some plants, transfer cells take up materials and pass them to sieve tube members. Transfer cells in xylem enhance the transport of solutes between tissues.
- As the sucrose concentration increases in the sieve tubes, water potential decreases. The resulting influx of water increases pressure inside the sieve tubes, so phloem sap flows in bulk toward the sink, where sucrose and water are unloaded (Figure 32.14).

UNDERSTAND AND APPLY

Test Your Knowledge

1. Antiport transport mechanisms:
   a. move dissolved materials by osmosis.
   b. transport molecules in the opposite direction of H^+ transported by proton pumps.
   c. transport molecules in the same direction as H^+ is pumped.
   d. are not affected by the size of molecules to be transported.
   e. are not affected by the charge of molecules to be transported.
2. All the following have roles in transporting materials between plant cells except:
   a. the stele.
   b. symport.
   c. the cell membrane.
   d. stomata.
   e. transport proteins.
3. Turgor pressure is best expressed as the:
   a. movement of water into a cell by osmosis.
   b. driving force for osmotic movement of water (Ψ).
   c. group movement of large numbers of molecules because of a difference in pressure between two locations.
   d. equivalent of water potential.
   e. pressure exerted by fluid inside a plant cell against the cell wall.
4. Water potential is:
   a. the driving force for the osmotic movement of water into plant cells.
   b. higher in a solution that has more solute molecules relative to water molecules.
   c. a measure of the physical pressure required to halt osmotic water movement across a membrane.
   d. a measure of the combined effects of a solution's pressure potential and its solute potential.
   e. the functional equivalent of turgor pressure.
5. To regulate the flow of water and minerals in the root, the:
   a. Casparian strip of endodermal cells blocks the apoplastic pathway, forcing water and solutes to cross cell plasma membranes in order to pass into the stele.
   b. apoplastic pathway is expanded, allowing a greater variety of substances to move into the stele.
   c. symplastic pathway is modified in ways that make plasma membranes of root cortex cells more permeable to water and solutes.
   d. symplastic pathway shuts down entirely so that substances can move only through the apoplast.
   e. transmembrane pathway augments transport via the apoplast, shunting substances around cells.
6. An indoor gardener leaving for vacation completely wraps a potted plant with clear plastic. Temperature and light are left at low intensities. The effect of this strategy is to:
   a. halt photosynthesis.
   b. reduce transpiration.
   c. cause guard cells to shrink and stomata to open.
   d. destroy cohesion of water molecules in the xylem.
   e. increase evaporation from leaf mesophyll cells.
7. Stomata open when:
   a. water has moved out of the leaf by osmosis.
   b. K^+ flows out of guard cells.
   c. turgor pressure in the guard cells lessens.
   d. the H^+ active transport protein stops pumping.
   e. outward flow of H^+ sets up a concentration gradient that moves K^+ in via symport.
8. A factor that contributes to the movement of water up a plant stem is:
   a. active transport of water into the root hairs.
   b. an increase in the water potential in the leaf's mesophyll layer.
   c. cohesion of water molecules in stem and leaf xylem.
   d. evaporation of water molecules from the walls of cells in root epidermis and cortex and in the stele.
   e. absorption of raindrops on a leaf's epidermis.

9. In translocation of sucrose-rich phloem sap:
   a. the sap flows toward a source as pressure builds up at a sink.
   b. crassulacean acid metabolism reduces the rate of photosynthesis.
   c. companion cells use energy to load solutes at a source and the solutes then follow their concentration gradients to sinks.
   d. sucrose diffuses into companion cells whereas H^+ simultaneously leaves the cells by a different route.
   e. companion cells pump sucrose into sieve tube members.
10. In Vermont in early spring, miles of leafless sugar maple trees have buckets hanging from "spigots" tapped into their xylem to capture the sugary xylem sap, which is the fluid raw material for making maple syrup. Select the statement(s) below that best applies to this phenomenon.
   a. Transpiration is pulling the sap upward from the roots.
   b. Just as phloem can contain a great deal of water, xylem typically contains a great deal of sucrose.
   c. Root pressure probably is forcing the xylem sap upward in the tree.
   d. Transfer cells move excess sucrose from phloem into the xylem.

Discuss the Concepts

1. Many popular houseplants are native to tropical rain forests. Among other characteristics, many nonwoody species have extraordinarily broad-bladed leaves, some so ample that indigenous people use them as umbrellas. What environmental conditions might make a broad leaf adaptive in tropical regions, and why?
2. Insects such as aphids that prey on plants by feeding on phloem sap generally attack only young shoot parts. Other than the relative ease of piercing less mature tissues, suggest a reason why it may be more adaptive for these animals to focus their feeding effort on younger leaves and stems.
3. So-called systemic insecticides often are mixed with water and applied to the soil in which a plant grows. The chemicals are effective against sucking insects no matter which plant tissue the insects attack, but often do not work as well against chewing insects. Propose a reason for this difference.
4. Concerns about global climate change and the greenhouse effect center on rising levels of greenhouse gases, including atmospheric carbon dioxide. Plants use CO_2 for photosynthesis, and laboratory studies suggest that increased CO_2 levels could cause a rise in photosynthetic activity. However, as one environmentalist noted, "What plants do in environmental chambers may not happen in nature, where there are many other interacting variables." Strictly from the standpoint of physiological effects, what are some possible ramifications of a rapid doubling of atmospheric CO_2 on plants in temperate environments? In arid environments?

Design an Experiment

CAM plants typically open their stomata at night instead of during the daytime, in part to limit transpiration on hot days. They include some, but not all, species of succulents (such as cacti and members of the genus *Sedum*). One mild spring afternoon while working in a mountain desert of eastern Oregon you discover a new species of succulent, and although you immediately assume that it is a CAM plant, a quick look with your field microscope reveals that its stomata are wide open. Then you remember that when CAM plants have access to plenty of moisture and are exposed to mild nighttime temperatures, they may shift temporarily to a more common mode of photosynthesis (see Section 9.3). During this period their stomata open during the day and close at night. Your collecting permit allows you to gather a few specimens, which you take back to your lab for testing. Design a simple experiment to determine the basic photosynthetic strategy of the new species, and explain how it will provide the information you seek.

Interpret the Data

As you already know, photosynthesizing plant cells require water delivered in the xylem. Recent experiments have revealed that the concentration of ions, especially potassium (K^+) in xylem water influences the velocity of water flow through the xylem, possibly by affecting the pit membranes between xylem vessels.

Roots take up K^+ in soil water, but phloem sap also contains K^+ that "recycles" back to the xylem. M.A. Zwieniecki and his colleagues hypothesized that changes in the flow of ions from phloem to xylem can alter the velocity (flow rate) at which water moves through the xylem. Working with red maples *(Acer rubrum)* and sugar maples *(Acer saccharum)*, they devised a girdling experiment that would prevent the recycling of phloem K^+ to xylem without disrupting the movement of xylem water. (Recall that girdling a tree or branch stops the movement of phloem sap beyond the cut.) The experimental design included an apparatus for maintaining normal pressure in the xylem and for adding either deionized water or water containing potassium chloride (KCl), a source of ions, to it. After experimental testing on 43 tree branches, they obtained results for xylem sap flow through the branches as shown in the graphs below.

1. What do the different colored blocks represent?
2. Do the results support or not support a hypothesis that the flow rate of xylem sap slows in response to a decline in ion concentration (KCl)?
3. Were results substantially similar or substantially different for the two species used in the experiment?

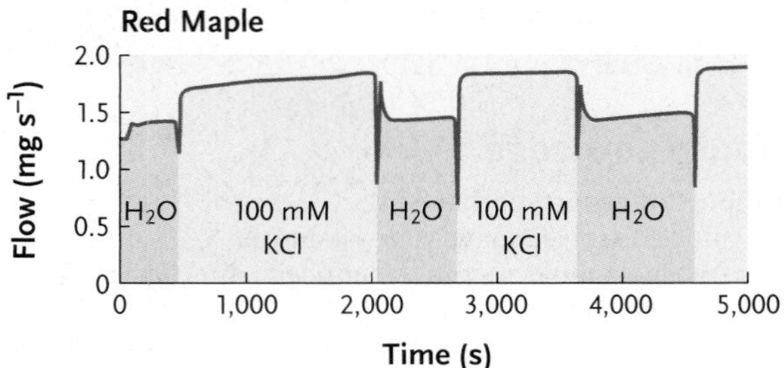

Red Maple
Flow (mg s⁻¹)
2.0
1.5
1.0
0.5
0
H2O
100 mM KCl
H2O
100 mM KCl
H2O
0
1,000
2,000
3,000
4,000
5,000
Time (s)

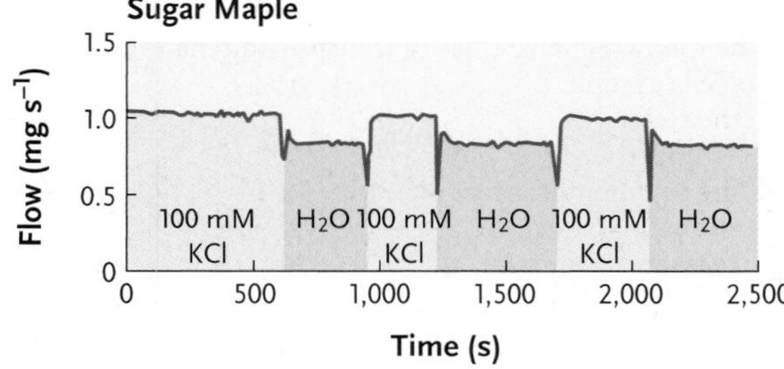

Sugar Maple
Flow (mg s⁻¹)
1.5
1.0
0.5
0
100 mM KCl
H2O
100 mM KCl
H2O
100 mM KCl
H2O
0
500
1,000
1,500
2,000
2,500
Time (s)

Apply Evolutionary Concepts

A variety of structural features of land plants reflect the conflicting demands for conserving water and taking in carbon dioxide for photosynthesis. Identify at least four fundamental structural adaptations that help resolve this dilemma and explain how each one contributes to a land plant's survival.

Express Your Opinion

Phytoremediation using genetically engineered plants can increase the efficiency with which a contaminated site is cleaned up. Do you support planting genetically engineered plants for such projects? Go to www.cengage.com/login to investigate both sides of the issue and then vote.

Lush azaleas *(Rhododendron)* and a stately Southern live oak *(Quercus virginiana)* draped with the unusual flowering plant called Spanish moss *(Tillandsia usneoides)*. The roots of shrubs, trees, and most other plants take up water and minerals from soil, but Spanish moss is an epiphyte—it lives independently on other plants and obtains nutrients by way of absorptive hairs on its leaves and stems.

© Ellen McKnight/Alamy

33

STUDY OUTLINE

Plant Nutrition

Why It Matters. . . Tropical rainforests are remarkable for many reasons, but for biologists the key one may be that they are among the most biologically diverse ecosystems on Earth. In addition to containing countless thousands of species of animals, fungi, protists, and prokaryotes, these lush domains are dense with broadleaved, evergreen trees, sinuous vines, and other vegetation. With rain a near-daily event, it may not seem surprising that the trees' foliage is a deep, luxuriant green **(Figure 33.1).**

Yet tropical rainforests are demanding places for plants to survive, in large part because the soil is chronically deficient in nutrients necessary for plant metabolism. This nutrient scarcity is a direct outcome of the incessant rain and the high acidity of tropical rainforest soil. There is ample moisture in the upper layer of soil, but in acidic soil minerals vital to plant metabolism, such as potassium, calcium, magnesium, and phosphorus, are subject to **leaching**—being washed into deeper soil levels that are not as accessible to plant roots. In addition, in the warm, moist environment of a tropical rainforest, bacteria and fungi speedily decompose fallen leaves and other organic remains. Just as rapidly, established trees and vines take up any nutrients these decomposers have released, leaving few or none to enrich the soil. As falling rain dissolves some atmospheric CO_2, it creates carbonic acid—a type of "acid rain"—that exacerbates the leaching problem even more.

Such poor soil and the near perpetual twilight at the forest floor make it extremely difficult for small shrubs and herbaceous plants to survive there. Nearly all such plants climb upward as vines using the tree trunks for mechanical support, or they live attached to the upper branches of taller species, where they can absorb needed minerals from falling dust or from the surfaces of other

FIGURE 33.1
A lush tropical rain forest growing in Southeast Asia.

plants. These intricate adaptations to a particular environment allow the plants to secure energy and raw materials and to use both for growth and development.

Tropical rainforests are not unique in posing nutritional challenges for plants. In fact, plants rarely have ready access to a full complement of necessary resources. In a rainforest, the carbon, hydrogen, and oxygen plants need for photosynthesis are relatively easy to come by: plants there usually get enough carbon from the CO_2 in air, and their roots can take up enough water to gain the necessary hydrogen and oxygen. But soils in other environments are frequently dry, making water a limited resource, and almost nowhere in nature do soils hold lavish amounts of dissolved minerals such as nitrogen, calcium, and others that are vital for a plant's survival. In response to the challenge of obtaining nutrients, plants have evolved the range of structural and physiological adaptations that we consider in this chapter. <

33.1 Plant Nutritional Requirements

No organism grows normally when deprived of a chemical element essential for its metabolism. In the latter half of the nineteenth century, plant physiologists exploited rapid advances in chemistry to probe both the chemical composition of plants and the essential nutrients plants need to survive. In recent times researchers have brought to bear sophisticated methods to expand our understanding of the range of plant nutrients, including those required only in trace amounts.

Plants Require Macronutrients and Micronutrients for Their Metabolism

By weight, the tissues of most plants are more than 90% water. Early researchers could obtain a rough idea of the composition of a plant's dry weight by burning the plant and then analyzing the ash. This method typically yielded a long list of elements, but the results were flawed. Chemical reactions during burning can dissipate quantities of some important elements, such as nitrogen. Also, plants take up a variety of ions that they do not use; depending on the minerals present in the soil where a plant grows, a plant's tissues can contain nonnutritive elements such as gold, lead, arsenic, and uranium.

STUDYING PLANT NUTRITION USING HYDROPONICS In 1860, German plant physiologist Julius von Sachs pioneered an experimental method for identifying the minerals absorbed into plant tissues that are essential for plant growth. Sachs carefully measured amounts of compounds containing specific minerals and mixed them in different combinations with pure water. He then grew plants in the solutions, a method now called **hydroponic culture** (*hydro* = water; *ponos* = work). By eliminating one element at a time and observing the results, Sachs deduced a list of six essential plant nutrients, in descending order of the amount required: nitrogen, potassium, calcium, magnesium, phosphorus, and sulfur.

Sachs's innovative research paved the way for decades of increasingly sophisticated studies of plant nutrition, and the eventual identification of many more essential plant nutrients. In the spirit of his work, one basic experimental method involves growing a plant in a solution containing a complete spectrum of known and possible essential nutrients **(Figure 33.2A)**. The healthy plant is then transferred to a solution that is identical, except that it lacks one element having an unknown nutritional role **(Figure 33.2B)**. Abnormal growth of the plant in this solution is evidence that the missing element is essential. If the plant grows normally, the missing element may not be essential; however, only further experimentation can confirm this hypothesis.

In a typical modern hydroponic apparatus, the nutrient solution is refreshed regularly, and air is bubbled into it to supply oxygen to the roots. Without sufficient oxygen for respiration, the plants' roots do not absorb nutrients efficiently. (The same effect occurs in poorly aerated soil.) Variations of this technique are used on a commercial scale to grow some vegetables, such as lettuce and tomatoes.

ESSENTIAL MACRONUTRIENTS AND MICRONUTRIENTS Hydroponics research has revealed that plants generally require 17 essential elements **(Table 33.1)**. By definition, an **essential element** is necessary for normal growth and reproduction, cannot be functionally replaced by a different element, and has one or more roles in plant metabolism. With enough sunlight and the 17 essential elements, plants can synthesize all the compounds they need.

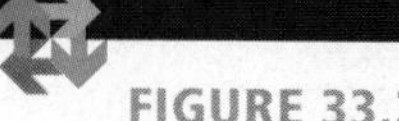

FIGURE 33.2 **Research Method**

Hydroponic Culture

Purpose: In studies of plant nutritional requirements, using hydroponic culture allows a researcher to manipulate and precisely define the types and amounts of specific nutrients that are available to test plants.

Protocol: In a typical hydroponic apparatus, many plants are grown in a single solution containing pure water and a defined mix of mineral nutrients. The solution is replaced or refreshed as needed and is aerated with a bubbling system.

A. **Basic components of a hydroponic apparatus**

Plant support
Nutrient solution
Air pumped into bubbling system

B. **Procedure for identifying elements essential for proper plant nutrition**

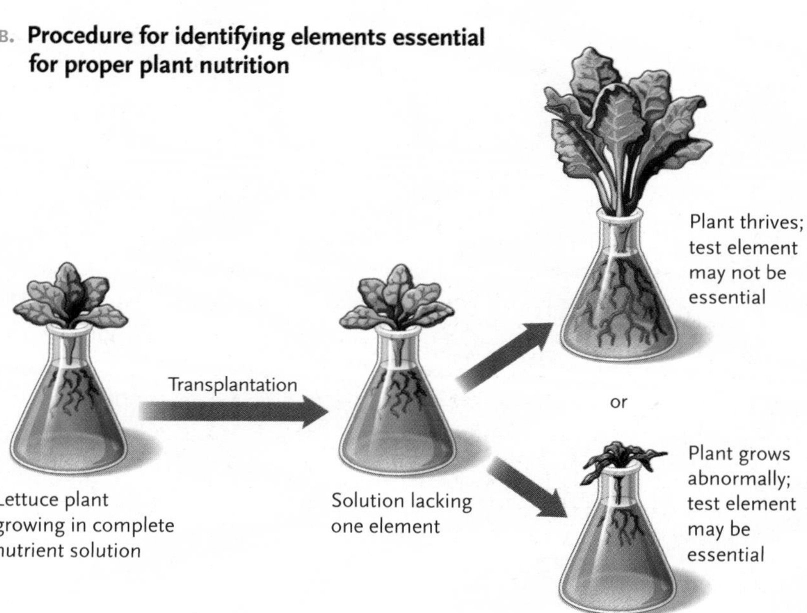

Plant thrives; test element may not be essential
Transplantation
or
Plant grows abnormally; test element may be essential
Lettuce plant growing in complete nutrient solution
Solution lacking one element

A "complete" solution contains all the known and suspected essential plant nutrients. An "incomplete" solution contains all but one of the same nutrients, in the same amounts. For experiments, researchers first grow plants in a complete solution, then transplant some of the plants to an incomplete solution.

Interpreting the Results: Normal growth of test plants suggests that the missing nutrient is not essential, whereas abnormal growth is evidence that the missing nutrient may be essential.

Nine of the essential elements are **macronutrients,** meaning that plants incorporate relatively large amounts of them into their tissues. Three of these elements—carbon, hydrogen, and oxygen—account for about 96% of a plant's dry mass. Together, these three elements are the key components of lipids and of carbohydrates such as cellulose; with the addition of nitrogen, they form the basic building blocks of proteins and nucleic acids. Plants also use phosphorus in constructing nucleic acids, ATP, and phospholipids, and they use potassium for functions ranging from enzyme activation to mechanisms that control the opening and closing of stomata. Rounding out the list of macronutrients are calcium, sulfur, and magnesium. All macronutrients except carbon, hydrogen, and oxygen are classified as minerals, which chemists usually define as elements or compounds formed by geological processes and that have a crystalline structure. Minerals are available to plants through the soil as ions dissolved in water, and most minerals that serve as nutrients in plants are derived from the weathering of rocks and inorganic particles in the Earth's crust.

The other elements essential to plants are also minerals, and are classed as **micronutrients** because plants require them only in trace amounts. Nevertheless, they are just as vital as macronutrients to a plant's health and survival. For example, 5 metric tons of potatoes contain roughly the amount of copper in a single (copper-plated) penny—yet without it, potato plants are sickly and do not produce normal tubers.

Chlorine, generally present in soil in its anionic form Cl^- (chloride), was identified as a micronutrient nearly a century after Sachs's experiments. Chloride functions in some reactions of photosynthesis and (along with K^+) in the opening and closing of stomata, among other roles. The researchers who discovered its importance in plant nutrition performed hydroponic culture experiments in a California laboratory near the Pacific Ocean, where the air, like coastal air everywhere, contains sodium chloride. The investigators found that their test plants could obtain tiny but sufficient quantities of chloride from the air, as well as from sweat (which also contains NaCl) on the researchers' own hands. Great care had to be taken to exclude chlorine from the test plants' growing environment to prove that it was essential.

In some cases, plant seeds contain enough of certain trace minerals to sustain the adult plant. For example, nickel (Ni^{2+}) is a component of urease, the enzyme required to hydrolyze urea. Urea is a toxic by-product of the breakdown of nitrogenous compounds, and it will kill cells if it accumulates. In the late 1980s investigators found that barley seeds contain enough nickel to sustain two complete generations of barley plants. Plants grown in the absence of nickel did not begin to show signs of nickel deficiency until the third generation.

TABLE 33.1 Essential Plant Nutrients and Their Functions

Element	Commonly Absorbed Forms	Some Known Functions	Some Deficiency Symptoms
Macronutrients			
Carbon*	CO_2	Raw materials for photosynthesis	Rarely deficient
Hydrogen*	H_2O		No symptoms; available from water
Oxygen*	O_2, H_2O, CO_2		No symptoms; available from water and CO_2
Nitrogen	NO_3^-, NH_4^+	Component of proteins, nucleic acids, coenzymes, chlorophylls	Stunted growth; light-green newer leaves; older leaves yellow and die (chlorosis)
Phosphorus	$H_2PO_4^-$, HPO_4^{2-}	Component of nucleic acids, phospholipids, ATP, several coenzymes	Purplish veins; stunted growth; fewer seeds, fruits
Potassium	K^+	Activation of enzymes; key role in maintaining water-solute balance and so influences osmosis	Reduced growth; curled, mottled, or spotted older leaves; burned leaf edges; weakened plant
Calcium	Ca^{2+}	Roles in formation and maintenance of cell walls and in membrane permeability; enzyme cofactor	Leaves deformed; terminal buds die; poor root growth
Sulfur	SO_4^{2-}	Component of most proteins, coenzyme A	Light-green or yellowed leaves; reduced growth
Magnesium	Mg^{2+}	Component of chlorophyll; activation of enzymes	Chlorosis; drooping leaves
Micronutrients			
Chlorine	Cl^-	Role in root and shoot growth, and in photosynthesis	Wilting; chlorosis; some leaves die (deficiency not seen in nature)
Iron	Fe^{2+}, Fe^{3+}	Roles in chlorophyll synthesis, electron transport; component of cytochrome	Chlorosis; yellow and green striping in grasses
Boron	H_3BO_3	Roles in germination, flowering, fruiting, cell division, nitrogen metabolism	Terminal buds, lateral branches die; leaves thicken, curl, and become brittle
Manganese	Mn^{2+}	Role in chlorophyll synthesis; coenzyme action	Dark veins, but leaves whiten and fall off
Zinc	Zn^{2+}	Role in formation of auxin, chloroplasts, and starch; enzyme component	Chlorosis; mottled or bronzed leaves; abnormal roots
Copper	Cu^+, Cu^{2+}	Component of several enzymes	Chlorosis; dead spots in leaves; stunted growth
Molybdenum	MoO_4^{2-}	Component of enzyme used in nitrogen metabolism	Pale green, rolled or cupped leaves
Nickel	Ni^{2+}	Component of enzyme required to break down urea generated during nitrogen metabolism	Dead spots on leaf tips (deficiency not seen in nature)

*Carbon, hydrogen, and oxygen are the nonmineral plant nutrients. All others are minerals.

Besides the 17 essential elements, some species of plants may require additional micronutrients. Experiments suggest that many, perhaps most, plants adapted to hot, dry conditions require sodium; many plants that photosynthesize by the C_4 pathway (see Section 9.4) appear to be in this group. A few plant species require selenium, which is also an essential micronutrient for animals. Horsetails *(Equisetum)* require silicon, and some grasses (such as wheat) may also need it. Scientists continue to discover additional micronutrients for specific plant groups.

Both micronutrients and macronutrients play vital roles in plant metabolism. Many function as cofactors or coenzymes in protein synthesis, starch synthesis, photosynthesis, and aerobic respiration. As you read in Section 32.1, some also have a role in creating solute concentration gradients across plasma membranes, which are responsible for the osmotic movement of water.

Nutrient Deficiencies Cause Abnormalities in Plant Structure and Function

Plants differ in the quantity of each nutrient they require—the amount of an essential element that is adequate for one plant species may be insufficient for another. Lettuce and other leafy plants require more nitrogen and magnesium than do other plant types, for example, and alfalfa requires significantly more potassium than do lawn grasses. An adequate amount of an essential element for one plant may even be harmful to another. For example, the amount of boron required for normal growth of sugar beets is toxic for soybeans. For these reasons, the nutrient content of soils is an important factor in determining which plants will grow well in a given location.

Plants that are deficient in one or more of the essential elements develop characteristic symptoms (Table 33.1 lists some observable symptoms of nutrient deficiencies). The symptoms give some indication of the metabolic roles the missing elements play. Deficiency symptoms typically include stunted growth, abnormal leaf color, dead spots on leaves, or abnormally formed stems **(Figure 33.3)**. For instance, iron is a component of the cytochromes on which the cellular electron transfer system depends, and it plays a role in reactions that synthesize chlorophyll. Iron deficiency causes **chlorosis,** a yellowing of plant tissues that results from a lack of chlorophyll (see Figure 33.3B). Because ionic iron (Fe^{3+}) is relatively insoluble in water, gardeners often fertilize plants with a soluble iron compound called chelated iron to stave off or cure chlorosis. Similarly, because magnesium is a necessary component of chlorophyll, a plant deficient in this element has fewer chloroplasts than normal in its leaves and other photosynthetic parts. It appears paler green than normal, and its growth is stunted because of reduced photosynthesis (see Figure 33.3C).

Plants that lack adequate nitrogen may also become chlorotic (see Figure 33.3D), with older leaves yellowing first because the nitrogen is preferentially shunted to younger, actively growing plant parts. This adaptation is not surprising, given nitrogen's central role in the synthesis of amino acids, chlorophylls, and other compounds vital to plant metabolism. With some other mineral deficiencies, young leaves are the first to show symptoms. These kinds of observations underscore the point that plants use different nutrients in specific, often metabolically complex ways.

Soils are more likely to be deficient in nitrogen, phosphorus, potassium, or some other essential mineral than to contain too much, and farmers and gardeners typically add nutrients to suit the types of plants they wish to cultivate. They may observe the deficiency symptoms of plants grown in their locale or have soil tested in a laboratory, then choose a fertilizer with the appropriate balance of nutrients to compensate for the deficiencies. Packages of commercial fertilizers use a numerical shorthand (for example, 15-30-15) to indicate the percentages of nitrogen, phosphorus, and potassium they contain.

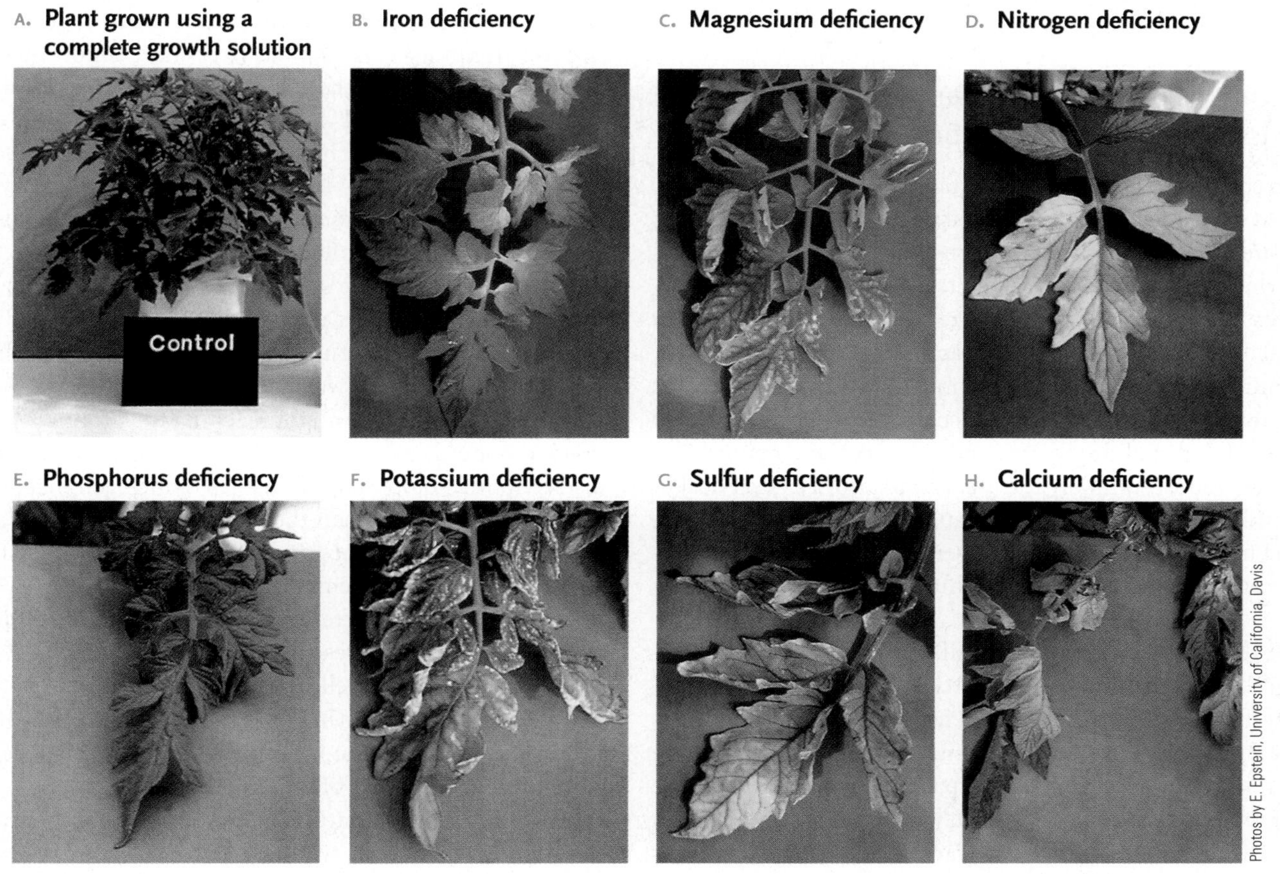
A. Plant grown using a complete growth solution
B. Iron deficiency
C. Magnesium deficiency
D. Nitrogen deficiency
Control
E. Phosphorus deficiency
F. Potassium deficiency
G. Sulfur deficiency
H. Calcium deficiency
Photos by E. Epstein, University of California, Davis

FIGURE 33.3
Leaves and stems of tomato plants showing visual symptoms of seven different mineral deficiencies. The plants were grown in the laboratory, where the experimenter could control which nutrients were available.

STUDY BREAK 33.1

1. What are the two main categories of the essential elements plants need? Give several examples of each.
2. Do all plants require the same basic nutrients in the same amounts? Explain.

THINK OUTSIDE THE BOOK

Symptoms of a plant nutrient deficiency rarely if ever show up in wild plants. Instead they generally are limited to species cultivated as food crops or common house and garden plants. On your own or with fellow students, develop a hypothesis to explain why this difference exists and describe an experiment that would test your hypothesis.

33.2 Soil

Soil anchors plant roots and is the main source of the inorganic nutrients plants require. It also is the source of water for most plants, and of oxygen for respiration in root cells. The physical texture of soil is a factor in whether root systems have access to sufficient water and dissolved oxygen. Together, physical and chemical properties of soils have a major impact on the ability of plants to grow, survive, and reproduce in particular habitats.

The Components of a Soil and the Size of the Particles Determine Its Properties

Soil is a complex mix of mineral particles, chemical compounds, ions, decomposing organic matter, air, water, and assorted living organisms. Most soils develop from the physical or chemical weathering of rock (which also liberates mineral ions). The different kinds of soil particles range in size from sand (2.0–0.02 mm) to silt (0.02–0.002 mm) and clay (diameter less than 0.002 mm). These mineral particles usually are mixed with various organic components, including **humus**—decomposing parts of plants and animals, animal droppings, and other organic matter. Dry humus has a loose, crumbly texture. It can absorb a great deal of water and thus contributes to the capacity of soil to hold water. Organic molecules in humus are reservoirs of nutrients, including nitrogen, phosphorus, and sulfur, that are vital to living plants.

The relative proportions of the different sizes of mineral particles give soil its basic texture—gritty if the soil is largely sand, smooth if silt predominates, and dense and heavy if clay is the major component. A soil's texture in turn helps determine the number and volume of pores—air spaces—that it contains. The relative amounts of sand, silt, and clay determine whether a soil is sticky when wet, with few air spaces (mostly clay), or dries quickly and may wash or blow away (mostly sand). Clay soils are more than 30% clay, whereas sandy soils contain less than 20% clay or silt.

The piles of bagged humus for sale at garden centers each spring reflect the fact that the amount of humus in a soil also affects plant growth. Its plentiful organic material feeds decomposers whose metabolic activities in turn release minerals that plant roots can take up, but that is not its only value in soil. Humus helps retain soil water and, with its loose texture, helps aerate soil as well. Well-aerated soils containing roughly equal proportions of humus, sand, silt, and clay are **loams,** and they are the soils in which most plants do best.

In Turn, Plants and Other Organisms Influence Soil Features

A square meter of fertile soil contains trillions of bacteria, hundreds of millions of fungi, and several million nematodes, plus an array of other worms and insects. It also contains dead plant roots, leaves, and other parts. Bacteria and fungi decompose this and other organic matter on and in the soil and burrowing creatures such as earthworms aerate the soil. On the other hand, the roots and other tissues of plants may play a key role in shaping the characteristics and composition of soil, including the abundance of soil-dwelling organisms.

Experiments document these soil-shaping activities. For example, Edward Ayres and his colleagues at Colorado State University's Natural Resource Ecology Laboratory studied soil properties in Colorado's San Juan Mountains, where stands of trembling aspen *(Populus tremuloides),* lodgepole pine *(Pinus contorta),* or Engelmann spruce *(Picea engelmannii)* live in close proximity. *P. tremuloides* trees have a more open growth form than pines and spruce trees do, all their leaves drop each year in autumn, and previous research had shown that *P. tremuloides* leaf litter has about twice the nitrogen content of the other two species. With these facts in mind, the Ayres team hypothesized that in their four study areas, such species-specific characteristics would influence the physical, chemical, and biological properties of the soil. The data they gathered supported parts of their hypothesis and also raised questions. For example, they found that in all study areas, the soil in which the aspens grew was significantly warmer—a difference that the team attributed to increased sunlight reaching the ground through the relatively open aspen canopies. The soil littered with aspen leaves also contained more nitrate—a form of nitrogen that plant roots can readily take up—than the nearby soil where the lignin-rich needlelike leaves of pines and spruces accumulated. The study was not designed to attempt a comprehensive analysis of the diversity of the soil's bacterial, fungal, and microscopic animal communities, but the researchers did document markedly different arrays of soil-dwelling organisms associated with the aspens, pines, and spruces in each study area. Clearly, we have a lot more to learn about the intricate interactions between plants, soils, and communities of soil organisms.

As soils develop naturally, they tend to take on a characteristic vertical profile, with a series of layers or **horizons (Figure 33.4).** Each horizon has a distinct texture and composition that varies with soil type. The top layer of surface litter—organic matter like twigs and leaves, animal dung, and fungi—

is accordingly called the *O horizon.* **Topsoil,** the most fertile layer, occurs just below and forms the *A horizon*. This fairly loose layer may be less than a centimeter deep on steep slopes to more than a meter deep in grasslands. It consists of humus mixed with mineral particles and is where the roots of most herbaceous plants are located. Below the topsoil is the **subsoil** or *B horizon,* a layer of larger soil particles containing relatively little organic matter. Mineral ions, including those that serve as nutrients in plants, tend to accumulate in the B horizon, and mature tree roots generally extend down into this layer. Under it is the *C horizon,* a layer of mineral particles and rock fragments that extends down to bedrock.

Regions where the topsoil is naturally deep and rich in humus are ideal for agriculture. Prime examples are the vast former grasslands of the North American Midwest and Ukraine now converted to fields of corn, wheat, soybeans, and other crops. Modern cultivation practices that sharply reduce erosion have helped maintain this soil resource. In arid regions, chronic, sparse rainfall generally correlates with low natural soil humus, and agriculture is only possible with intensive irrigation and soil management. Nor can agriculture flourish for long on land cleared of a tropical rainforest, because of the soil leaching and lack of nutrients described in the chapter introduction.

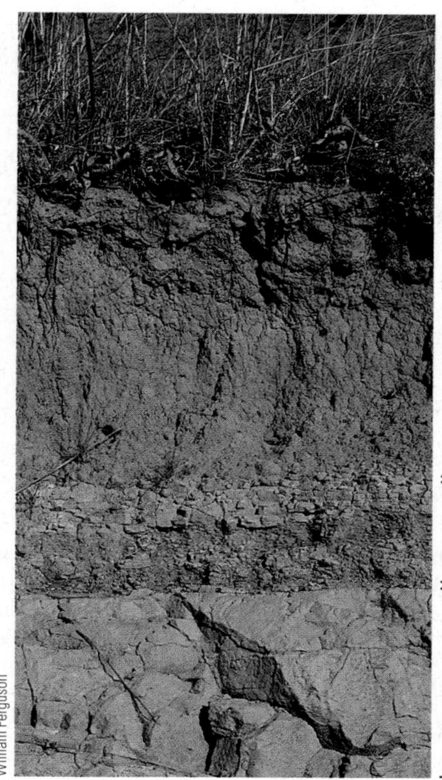

William Ferguson
O horizon
Fallen leaves and other organic material littering the surface of mineral soil
A horizon
Topsoil, which contains some percentage of decomposed organic material and which is of variable depth; here it extends about 30 cm below the soil surface
B horizon
Subsoil; larger soil particles than the A horizon, not much organic material, but greater accumulation of minerals; here it extends about 60 cm below the A horizon
C horizon
No organic material, but partially weathered fragments and grains of rock from which soil forms; extends to underlying bedrock
Bedrock

FIGURE 33.4
Soil horizons in a grassland.

The Characteristics of Soil Affect Root–Soil Interactions

Roots are superbly adapted to penetrate soil and extract needed nutrients from it, but they also are quite sensitive to variations in the properties of soil. In the following section we consider some adaptations plants have evolved in many otherwise inhospitable soil environments. First, however, we consider the general ways in which soil composition influences the ability of plant roots to obtain water and minerals.

WATER AVAILABILITY As water flows into and through soil, gravity pulls much of the water down through the spaces between soil particles into deeper soil layers. This available water is part of the **soil solution (Figure 33.5),** a combination of water and dissolved substances that coats soil particles and partially fills pore spaces. The solution develops through ionic interactions between water molecules and soil particles. Clay particles and the organic components in soil (especially proteins) often bear negatively charged ions on their surfaces. The negative charges attract the polar water molecules, which form hydrogen bonds with the soil particles (see Section 2.4).

Unless a soil is irrigated, the amount of water in the soil solution depends largely on the amount and pattern of precipitation (rain or snow) in a region. How much of this water is actually available to plants depends on the soil's composition—the size of the air spaces in which water can accumulate and the proportions of water-attracting particles of clay and organic matter. By volume, soil is about one-half solid particles and one-half air space.

The type and size of the particles in a given soil has a major effect on how well plants will grow there. Sand particles are small and sandy soil has relatively large air spaces, so water drains rapidly below the top two soil horizons where most plant roots are located. Soils rich in clay or humus often hold quite a bit of water,

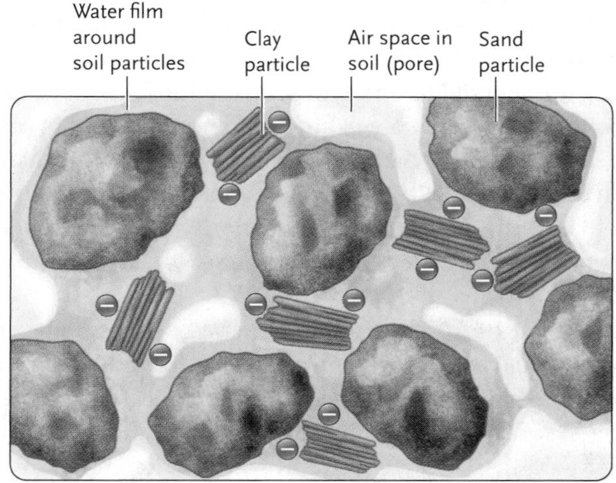

Water film around soil particles
Clay particle
Air space in soil (pore)
Sand particle

FIGURE 33.5
Location of the soil solution. Negatively charged ions on the surfaces of soil particles attract water molecules, which coat the particles and fill spaces between them (blue). Hydrogen bonds between water and soil components counteract the pull of gravity and help hold some water in the soil spaces.

TABLE 33.2	Water Potentials of Soils	
	Water Potential (MPa)	
Soil Type	10% Moisture	30% Moisture
Sand	−0.05	−0.001
Loam	−0.5	−0.005
Clay	−10.0	−0.1

Source: Mauseth *Botany* 4e, Jones and Bartlett.

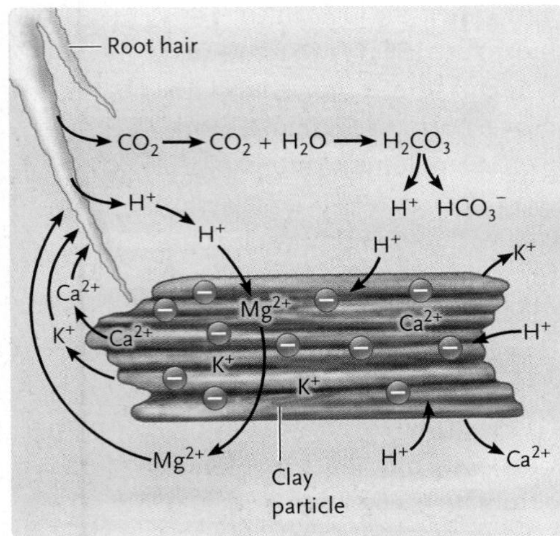

Root hair
$CO_2 \rightarrow CO_2 + H_2O \rightarrow H_2CO_3$
H^+ HCO_3^-
H^+
Ca^{2+}
K^+
Mg^{2+}
Clay particle

FIGURE 33.6
Cation exchange on the surface of a clay particle. When cations come into contact with the negatively charged surface of the particle, they become adsorbed. As one type of cation, such as H^+, becomes adsorbed, other ions are liberated and can be taken up by plant roots.

but in the case of clay, ample water is not necessarily an advantage for plants. Whereas a humus-rich soil contains lots of air spaces, the closely layered particles in clay allow few air spaces—and what spaces there are tend to hold tightly the water that enters them. The lack of air spaces in clay soils also severely limits supplies of oxygen available to roots for cellular respiration, and the plant's metabolic activity suffers. Thus, few plants can flourish in clay soils, even when water content is high. (Overwatered houseplants die because their roots are similarly "smothered" by water.) Plants do not fare much better in drier clay-rich soils, because roots cannot extract the existing water and cannot easily penetrate the densely packed clay. These characteristics explain why good agricultural soils tend to be sandy or silty loams, which contain a mix of humus and coarse and fine particles.

As you learned in Chapter 31, root hairs are specialized extensions of root epidermal cells; they directly contact the soil solution and allow roots to absorb water (and dissolved ions). And as you saw in Section 32.2, differences in water potential govern the osmotic movement of water into plant roots. The soil solution usually contains fewer dissolved solutes than does the water in the cells of plant roots. Accordingly, water tends to move from wet soil, where water potential is higher, into roots, where the water potential is lower. Notice in **Table 33.2** that the water potential in clay soils is significantly lower than in other soil types, even when clay is relatively wet. As roots extract water from clay soil, the water potential may fall below that in a plant's roots, making it impossible for water to diffuse into the roots. Water still continues to evaporate from leaves and to be used in photosynthesis, however, so the plant eventually wilts. Plants that survive in deserts or in salty soils have adaptations that permit their roots to absorb water even when osmotic conditions in soil do not favor water movement into the plant.

MINERAL AVAILABILITY Some mineral nutrients enter plant roots as cations (positively charged ions) and some as anions (negatively charged ions). Although both cations and anions may be present in soil solutions, they are not equally available to plants.

Cations such as magnesium (Mg^{2+}), calcium (Ca^{2+}), and potassium (K^+) cannot easily enter roots because they are attracted by the net negative charges on the surfaces of soil particles. To varying degrees, they become reversibly bound to negative ions on the surfaces. Attraction in this form is called *adsorption*. Roots do acquire cations, however, through **cation exchange.** In this mechanism one cation, usually H^+, replaces a soil cation **(Figure 33.6).** The protons (H^+) come from two main sources. Respiring root cells release carbon dioxide, which dissolves in the soil solution, yielding carbonic acid (H_2CO_3). Subsequent reactions ionize H_2CO_3 to produce bicarbonate (HCO_3^-) and H^+. Reactions involving organic acids inside roots also produce H^+, which is excreted. As H^+ enters the soil solution, it displaces adsorbed mineral cations attached to clay and humus, freeing them to move into roots. Other types of cations may also participate in this type of exchange, as shown in Figure 33.6.

By contrast, anions in the soil solution, such as nitrate (NO_3^-), sulfate (SO_4^{2-}), and phosphate (PO_4^{3-}), are only weakly bound to soil particles, and so they generally move fairly freely into root hairs. However, because they are so weakly bound compared with cations, anions are more subject to loss from soil by leaching.

The pH of soil affects the availability of some mineral ions. Soil pH is a function of the balance between cation exchange and other processes that raise or lower the concentration of H^+ in soil. As noted earlier, in areas that receive heavy rainfall, soils tend to become acidic (that is, they have a pH of less than 7). This acidification occurs in part because moisture promotes the rapid decay of organic material in humus; as the material decomposes it releases its organic acids. Acid precipitation, which results from the release of sulfur and nitrogen oxides into the air (in large measure from the burning of fossil fuels and industrial emissions), also contributes to soil acidification. By contrast, the soil in arid regions, where precipitation is low, often is alkaline (the pH is greater than 7).

Although most plants are not directly sensitive to soil pH, chemical reactions in very acid (pH <5.5) or very alkaline (pH >9.5) soils can have a major impact on whether plant roots take up various mineral cations. For example, experiments have showed that in the presence of OH^- in alkaline

FOCUS ON Applied Research

Plants Poised for Environmental Cleanup

Scientists have long been seeking efficient modes of *phytoremediation,* the use of plants to remove pollutants from the environment. One target is the highly toxic compound methylmercury (MeHg), which is present in coastal soils and wetlands contaminated by industrial wastes containing an ionic form of the element mercury called Hg(II). Bacteria in contaminated sediments metabolize Hg(II) and generate MeHg as a metabolic by-product. Once MeHg forms, it enters the food web and eventually becomes concentrated in tissues of fishes and other animals. In humans, MeHg can lead to degeneration of the nervous system. It causes most cases of mercury poisoning from consuming contaminated fish.

In the 1990s a team of scientists including Scott Bizily and Richard Meagher at the University of Georgia decided to try to create genetically modified plants capable of detoxifying mercury-contaminated soil and wetlands. It was already known that bacteria in contaminated sediments possess two genes, *merA* and *merB,* which encode enzymes that convert MeHg into elemental mercury (Hg)—a relatively inert, much less dangerous substance. Clones of both these bacterial mercury-resistance genes were available. After modifying the cloned genes so that plants could express them, the team used a vector (the bacterium *Rhizobium radiobacter*) to introduce each gene into several different sets of *Arabidopsis thaliana* plants (thale cress). They obtained three groups of transgenic plants: some that were *merA* only, some that were *merB* only, and some that were *merA* and *merB.* Seeds from each group were grown (along with wild-type controls) in five different growth media—one containing no mercury and the other four containing increasing concentrations of methylmercury. Wild-type and *merA* seeds germinated and grew only in the mercury-free growth medium. The *merB* seedlings fared somewhat better: they germinated and grew briefly even at the highest concentrations of MeHg, but then died. By contrast, seeds with the *merA/merB* genotype germinated at high MeHg concentrations, producing seedlings that grew into robust plants **(Figures 1 and 2).** In later tests *merA/merB* plants were grown in chambers in which the chemical composition of the air was monitored. This study revealed that the doubly transgenic plants also were transpiring large amounts of Hg. These findings implied that *A. thaliana* plants having both *merA* and *merB* genes were able to take up the toxic methylmercury and convert it to a harmless form. Meagher and his colleagues now are experimenting with ways of increasing the efficiency of phytoremediating enzymes when plant cells express *merA* and *merB.* They also are studying the mechanisms by which ionic mercury taken up by roots may be transported via the xylem to leaves and other shoot parts. The goal is to engineer plants that accumulate large quantities of mercury in aboveground tissues that can be harvested, leaving the living plant to continue its "work" of detoxifying a contaminated landscape.

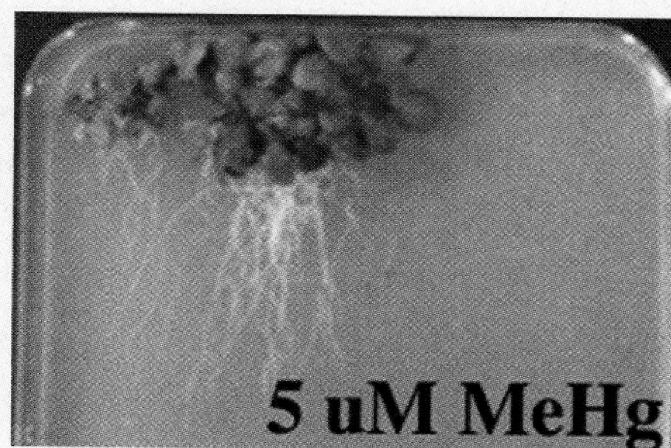
5 uM MeHg

FIGURE 1
Transgenic *A. thaliana* plants and wild-type controls growing in a medium containing 5 micromoles of methylmercury.

10 uM MeHg

FIGURE 2
When the MeHg concentration is increased to 10 micromoles, only the MerA/MerB plants grow.

soil, calcium and phosphate ions react to form insoluble calcium phosphates. The phosphate captured in these compounds is as unavailable to roots as if it were completely absent from the soil.

For a soil to sustain plant life over long periods, the mineral ions that plants take up must be replenished naturally or artificially. Over the long run, some mineral nutrients enter the soil from the ongoing weathering of rocks and smaller bits of minerals. In the shorter run, minerals, carbon, and some other nutrients are returned to the soil by the decomposition of organisms and their parts or wastes. Airborne compounds, such as sulfur in volcanic and industrial emissions, may enter soil when they dissolve in rain and fall to earth. Minerals, including compounds of nitrogen and phosphorus, may also enter soil in fertilizers.

Although the use of commercial fertilizers maintains high crop yields, agricultural chemicals do not add humus to the soil. Their use can also cause serious pollution problems, as when nitrogen-rich runoff from agricultural fields promotes the serious overgrowth of algae in lakes and bays. In many parts of the world, industrial pollutants such as cadmium, lead, and mercury are increasingly serious soil contaminants. The use of plants to remove such materials from soil, called *phytoremediation,* is the topic of this chapter's *Focus on Applied Research.*

STUDY BREAK 33.2 <

1. **Why is humus an important component of fertile soil?**
2. **How does the composition of a soil affect a plant's ability to take up water?**
3. **What factors affect a plant's ability to absorb minerals from the soil?**

33.3 Obtaining and Absorbing Nutrients

Soil managed for agriculture can be plowed, precisely irrigated, and chemically adjusted to provide air, water, and nutrients in optimal quantities for a particular crop. By contrast, in natural habitats, wide variations in soil minerals, humus, pH, the presence of other organisms, and other factors influence the availability of essential elements. Although adequate carbon, hydrogen, and oxygen are typically available, other essential elements may not be as abundant. In particular, nitrogen, phosphorus, and potassium are often relatively scarce. The evolutionary solutions to these challenges include an array of adaptations in the structure and functioning of plant roots.

Root Systems Allow Plants to Locate and Absorb Essential Nutrients

Immobile organisms such as plants must locate nutrients in their immediate environment, and for plants the adaptive solution to this problem is an extensive root system. Roots make up 20% to 50% of the dry weight of many plants, and even more in species growing where water or nutrients are especially scarce, such as arctic tundra. As long as a plant lives, its root system continues to grow, branching out through the surrounding soil. Roots do not necessarily grow *deeper* as a root system branches out, however. In arid regions, a shallow-but-broad root system may be better positioned to take up water from occasional rains that may never penetrate below the first few inches of soil.

You may recall from Section 31.4 that roots take up ions in the regions just behind the root tips, where root hairs are present. These diminutive absorptive structures, shown in Figure 31.10C, are a major adaptation for the uptake of mineral ions and water. Over successive growing seasons, long-lived plants such as trees can develop millions, even billions, of root tips, each one a potential absorption site. In a plant such as a mature red oak *(Quercus rubra)*, which has a vast root system, the total number of root hairs is astronomical. Even in young plants, root hairs greatly increase the root surface area available for absorbing water and ions.

Chapter 32 also mentioned another plant adaptation for gaining access to mineral ions—ion-specific transport proteins in plant cell membranes by which the cells selectively absorb ions from soil. For example, from studies of plants such as *Arabidopsis thaliana*, a weed that has become a key model organism for plant research, we know that transport channels for potassium ions (K^+) are embedded in the cell membranes of root cortical cells. Such ion transporters absorb more or less of a particular ion depending on chemical conditions in the surrounding soil.

The Discovery That Roots Also Secrete Substances into Soil Expands Our Understanding of Plant Adaptations for Obtaining Nutrients

In addition to acquiring substances from the surrounding soil, roots of various plant species also release into soil a long list of organic compounds, including carbohydrates, amino acids, various organic and fatty acids, as well as enzymes and other proteins. Today experiments are revealing the details of how such "root exudates" may improve a plant's access to particular nutrients. For example, a study headed by Corey D. Broeckling of Colorado State University showed that the roots of *A. thaliana* and *Medicago trunculata* (a legume commonly called barrel medic) secrete organic compounds that help determine which species of soil fungi can thrive near the roots. As you will read shortly, such fungi are partners in symbiotic relationships that help nourish many plant species. Similarly, Eric Paterson and his coworkers at the University of Aberdeen in Scotland found that roots of barley plants *(Hordeum vulgare)* exposed to above-normal nitrogen increased their release of organic substances that promote the growth of soil organisms that convert nitrogen into a chemical form plant roots can absorb. Other compounds released by roots enhance the uptake of phosphate.

Nutrients Move into and through the Plant Body by Several Routes

Most mineral ions enter plant roots passively along with the water in which they are dissolved. Some enter root cells immediately. Others travel in solution *between* cells—in the apoplast—until they meet the endodermis sheathing the root's stele (see Figure 32.6). At the endodermis, the ions are actively transported into the endodermal cells and then into the xylem for transport to cells throughout the plant. Inside cells, most mineral ions enter vacuoles or remain in the cytoplasm, where they are immediately available for metabolic reactions.

Some nutrients, such as nitrogen-containing ions, move in phloem from site to site in the plant, as dictated by growth and seasonal needs. In plants that shed their leaves in autumn, before the leaves age and fall significant amounts of nitrogen, phosphorus, potassium, and magnesium move out of them and into twigs and branches. This evolutionary adaptation conserves the nutrients, which will be used in new growth the next season. Likewise, in late summer, mineral ions move to the roots and lower stem tissues of perennial range grasses that typically die back during the winter. These activities are regulated by hormonal signals, which are the topic of Chapter 35.

Given the essential role of roots in a plant's nutritional survival, it is not surprising that research is uncovering an ever-growing list of root adaptations for exploiting and enhancing nutrient resources in soil. We now take a closer look at two of these adaptations—the associations called *mycorrhizae* and interactions with nitrogen-fixing microorganisms.

Mycorrhizae and Nitrogen-Fixing Bacteria: Partners That Increase Plant Access to Scarce Nutrients

Mycorrhizae are crucial symbiotic associations between a fungus and the roots of a plant (see Section 28.3). They promote the uptake of water and nutritionally vital ions—especially phosphate—in most species of plants. As shown in Figure 28.17, the fungal partner in the association often grows as a network of hyphal filaments around and beyond the plant's roots. Collectively, the hyphae provide a tremendous surface area for absorbing ions from a

INSIGHTS FROM THE Molecular Revolution

Getting to the Roots of Plant Nutrition

One way that mycorrhizal fungi benefit their host plants is by increasing the amount of phosphate plants take up from the soil.

Research Question

At the molecular level, how do the fungi accomplish this beneficial process?

Experiments

Maria J. Harrison and Marianne L. van Buuren at the Samuel Roberts Noble Foundation in Ardmore, Oklahoma, were able to identify a gene in one of these fungi that encodes a phosphate transport protein.

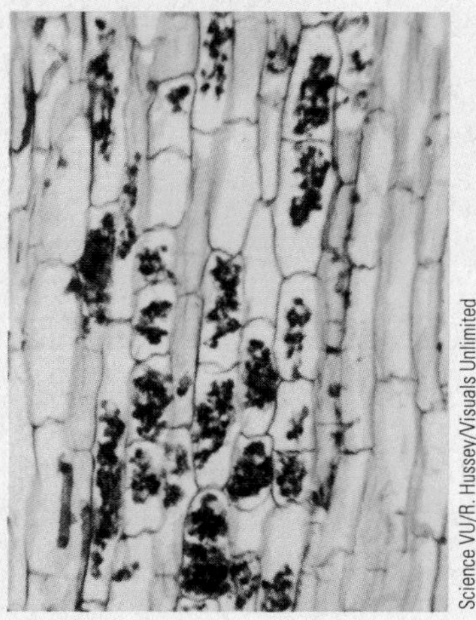

Science VU/R. Hussey/Visuals Unlimited

FIGURE 1
Photomicrograph of *Glomus versiforme* hyphae penetrating a leek root.

Starting with a cDNA library prepared from a plant that had been colonized by the mycorrhizal fungus *Glomus versiforme* **(Figure 1),** Harrison and van Buuren probed the plant-derived cDNA with a gene that encodes a phosphate transporter in yeast. They hoped to determine whether any cDNA sequences were similar to the yeast gene. (As Section 18.1 describes, a cDNA library is a cloned collection of DNA sequences derived from mRNAs, so it represents sequences that encode proteins.) The transport protein encoded by the yeast gene is embedded in the plasma membrane, where it moves phosphate ions into yeast cells by active transport.

When mixed with the plant-derived cDNA sequences, the probe did indeed pair with one of the cDNA sequences. Subsequent sequencing of the segment revealed that the cDNA coded for a protein with a structure typical of many eukaryotic and prokaryotic membrane transport proteins.

To eliminate the possibility that the probe was identifying a plant cDNA in the library rather than one from the mycorrhizal fungus, the investigators next used the identified cDNA to probe a preparation containing all the DNA of a plant that had not been colonized by *Glomus.* No pairing occurred with any of the plant DNA fragments, confirming that the cDNA represented a gene from the fungus. Additional experiments supported this finding.

Harrison and van Buuren carried their investigation further to see whether the fungal gene actually encoded a phosphate transport protein. For this set of experiments, the investigators used a yeast mutant with a nonfunctional phosphate transporter. Because these mutant yeast cells can't readily take in phosphate, they grow very slowly, even in a phosphate-rich culture medium.

The researchers added the *Glomus* gene to the mutants under conditions that increased the chances that the yeast cells would take up and incorporate the DNA. The yeast cells then began to grow normally, indicating that they could now synthesize a functional phosphate transporter. When radioactive phosphate ions were added to the culture, the cells rapidly became labeled, confirming that they were taking up phosphate ions at a much greater rate than untreated mutants.

Conclusion

This study was the first to reveal the molecular basis of phosphate transport by the mycorrhizal fungi. More recent work with potato plants *(Solanum tuberosum)* has identified a gene encoding a phosphate transporter protein that is expressed in parts of potato roots where mycorrhizae form. These lines of research may lead to methods for reducing the amount of phosphate fertilizers added to crop plants by identifying mycorrhizal fungi providing the most efficient phosphate uptake—or by engineering crop plants with an improved capacity to take in this essential nutrient.

Source: M. J. Harrison and M. L. van Buuren. 1995. A phosphate transporter from the mycorrhizal fungus *Glomus versiforme. Nature* 378:626–629.

large volume of soil. As with plant roots, transport proteins shepherd ions into hyphae. Experiments have verified that hyphal transport proteins are encoded by the DNA of the fungus, not that of the plant (as described in *Insights from the Molecular Revolution*). Some of the plant's sugars and nitrogenous compounds nourish the fungus, and as the root grows, it uses some of the minerals that the fungus has secured. In other types of mycorrhizae, the fungus actually lives inside cells of the root cortex. Orchids, for example, depend on this type of mutualistic association. And, as you will see shortly, some other plants gain access to adequate nitrogen by way of mutually beneficial associations with bacteria.

Plants Depend on Bacteria for an Adequate Supply of Usable Nitrogen

It might seem that plants live surrounded by nitrogen. For example, nitrogen steadily enters the soil in organic compounds released when dead organisms and animal wastes decompose. Dried blood is about 12% nitrogen by weight—although the nitrogen is bound up in complex organic molecules such as amino acids and proteins. Air contains plenty of gaseous nitrogen—this N_2 is almost 80% by volume—but plants cannot extract it because they lack the enzyme necessary to break apart the three covalent bonds in each N_2 molecule ($N{\equiv}N$). Plants can absorb nitrogen from the atmosphere that reaches the soil in the form of nitrate, NO_3^-, and ammonium ion, NH_4^+, and experiments show that roots of at least some plant species directly take up amino acids. Even so, lack of nitrogen is the single most common limit to plant growth because there usually is not nearly enough nitrogen available in these forms to meet plants' ongoing needs.

Instead, the main natural processes that replenish soil nitrogen and convert it to an absorbable form are carried out by bacteria. These processes, which we'll now consider, are part of the *nitrogen cycle,* the global movement of nitrogen in its various chemical forms from the environment to organisms and back to the environment, which is described in Chapter 52.

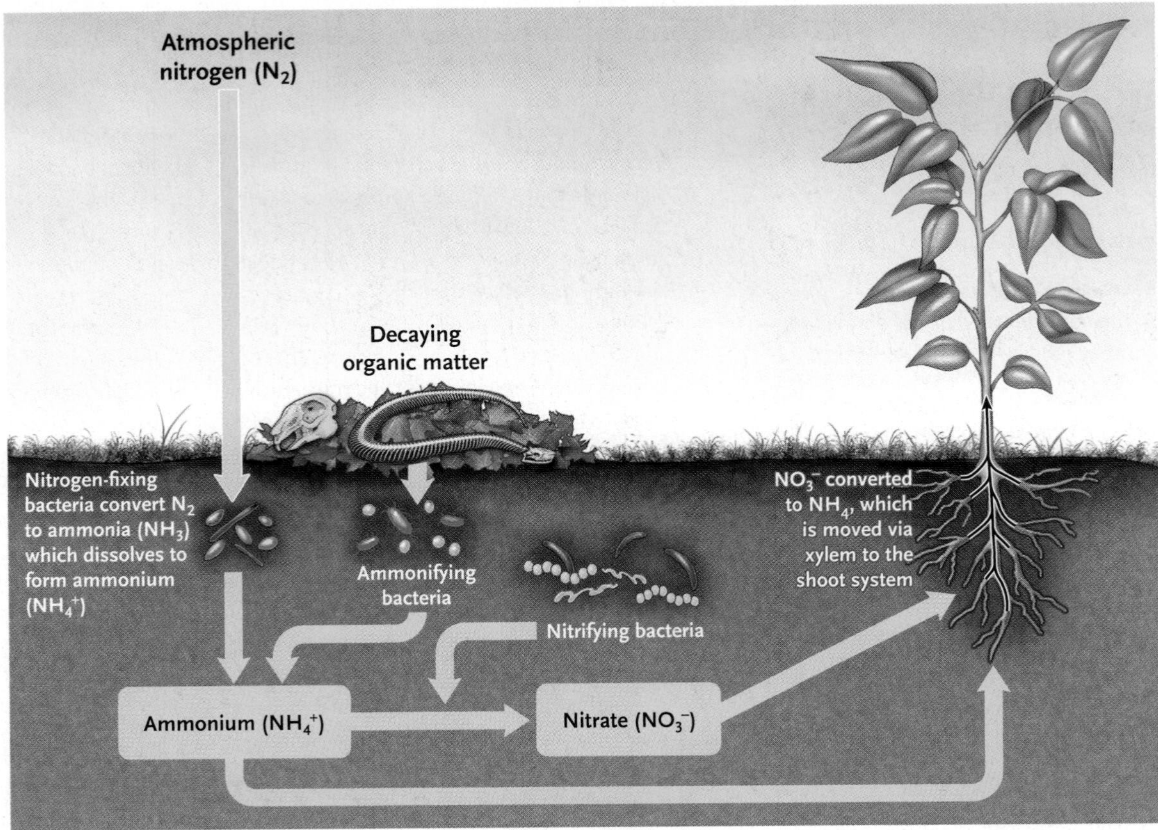
Atmospheric nitrogen (N_2)
Decaying organic matter
Nitrogen-fixing bacteria convert N_2 to ammonia (NH_3) which dissolves to form ammonium (NH_4^+)
Ammonifying bacteria
Nitrifying bacteria
NO_3^- converted to NH_4, which is moved via xylem to the shoot system
Ammonium (NH_4^+)
Nitrate (NO_3^-)

FIGURE 33.7 How plants obtain nitrogen from soil. Many commercial nitrogen fertilizers are in the chemical form of nitrate, which plant roots readily take up, or in the form of ammonium, which nitrifying bacteria convert to nitrate.

PRODUCTION AND ASSIMILATION OF AMMONIUM AND NITRATE The incorporation of atmospheric nitrogen into compounds that plants can take up is called **nitrogen fixation. Figure 33.7** summarizes the basic steps. Metabolic pathways of *nitrogen-fixing bacteria* living in the soil or in mutualistic association with plant roots add hydrogen to atmospheric N_2, producing two molecules of NH_3 (ammonia) and one H_2 for each N_2 molecule. The process requires a substantial input of ATP and is catalyzed by the enzyme nitrogenase. In a final step, H_2O and NH_3 react, forming NH_4^+ (ammonium) and OH^-.

Another bacterial process, called **ammonification,** also produces NH_4^+ when soil bacteria known as *ammonifying bacteria* break down decaying organic matter. In this way, nitrogen already incorporated into plants and other organisms is recycled.

Although plants use NH_4^+ to synthesize organic compounds, most plants absorb nitrogen in the form of nitrate, NO_3^-. Nitrate is produced in soil by **nitrification,** in which NH_4^+ is oxidized to NO_3^-. Soils generally teem with *nitrifying bacteria,* which carry out this process. Because of ongoing nitrification, nitrate is far more abundant than ammonium in most soils. Usually, the only soils from which plant roots take up ammonium directly are highly acidic, such as in bogs, where the low pH is toxic to nitrifying bacteria.

NITROGEN ASSIMILATION Once inside root cells, absorbed NO_3^- is converted by a multistep process back to NH_4^+. In this form, nitrogen is rapidly used to synthesize organic molecules, mainly amino acids. These molecules pass into the xylem, which transports them throughout the plant. In some plants, the nitrogen-rich precursors for needed substances travel in xylem to leaves, where different organic molecules are synthesized. Those molecules travel to other plant cells in the phloem.

NITROGEN FIXATION IN PLANT–BACTERIA ASSOCIATIONS Although some nitrogen-fixing bacteria live free in the soil, by far the largest percentage of nitrogen is fixed by species of *Rhizobium* and *Bradyrhizobium,* which form mutualistic associations with the roots of plants in the legume family. The host plant supplies organic molecules that the bacteria use for cellular respiration, and the bacteria supply NH_4^+ that the plant uses to produce proteins and other nitrogenous molecules. In legumes—a large family that includes peas, beans, clover, alfalfa, and alders, among others—the nitrogen-fixing bacteria reside in **root nodules,** localized swellings on roots **(Figure 33.8).** Farmers may exploit root nodules to increase soil nitrogen by rotating crops (for example, planting soybeans and corn in alternating years). When the legume crop is harvested, the root nodules and other tissues remaining in the soil enrich its nitrogen content.

For a plant, an association with nitrogen-fixing bacteria offers the selective advantage of a steady source of absorbable nitrogen. Decades of research have revealed the details of how this remarkable relationship unfolds. Usually, a single species of nitrogen-fixing bacteria colonizes a single legume species, drawn to the plant's roots by chemical attractants—primarily compounds called flavonoids—that the roots secrete. Through a sequence of exchanged molecular signals, bacteria are able to penetrate a root hair and form a colony inside the root cortex.

An association between a soybean plant *(Glycine max)* and a bacterium *(Bradyrhizobium japonicum)* illustrates the process. In response to a specific flavonoid released by soybean

A. **Root nodules**

B. **Field experiment with soybeans *(Glycine max)* and *Rhizobium***

C. **Bacteroids**

Root nodule

Adrian P. Davies/Bruce Coleman

NifTAL Project, University of Hawaii, Maui

Dr. Jeremy Burgess/SPL/Photo Researchers, Inc.

FIGURE 33.8
The beneficial effect of root nodules. **(A)** Root nodules on a soybean plant *(Glycine max)*. **(B)** Soybean plants growing in nitrogen-poor soil. The plants on the right were inoculated with *Rhizobium* cells and developed root nodules. **(C)** False-color transmission electron micrograph showing membrane-bound bacteroids (red) in a root nodule cell. Membranes that enclose the bacteroids appear blue. The large yellow-green structure is the cell's nucleus.

roots, bacterial genes called *nod* genes (for *nodule*) begin to be expressed **(Figure 33.9A)**. Products of the *nod* gene cause the tip of the root hair to curl toward the bacteria and trigger the release of bacterial enzymes that break down the root hair cell wall **(Figure 33.9B)**. As bacteria enter the cell and multiply, the plasma membrane forms a tube called an **infection thread** that extends into the root cortex, allowing the bacteria to invade cortex cells **(Figure 33.9C)**. The enclosed bacteria, now called **bacteroids,** enlarge and become immobile. Stimulated by still other *nod* gene products, cells of the root cortex begin to divide. This region of proliferating cortex cells forms the root nodule **(Figure 33.9D)**. Typically, each cell in a root nodule contains several thousand bacteroids; the plant takes up some of the nitrogen fixed by the bacteroids, and the bacteroids use some compounds produced by the plant.

Inside bacteroids, N_2 is reduced to NH_4^+ (ammonium) using ATP produced by cellular respiration. The process is catalyzed by nitrogenase. Ammonium is highly toxic to cells if it accumulates, however. Thus, NH_4^+ is moved out of bacteroids into the surrounding nodule cells immediately and converted to other compounds, such as the amino acids glutamine and asparagine.

One factor encoded by the bacterial *nod* genes stimulates plant nodule cells to produce a protein called **leghemoglobin** ("legume hemoglobin"). Like the hemoglobin of animal red blood cells, leghemoglobin contains a reddish, iron-containing heme group that binds oxygen. Its color gives root nodules a pinkish cast (see Figure 33.8). Leghemoglobin picks up oxygen at the cell surface and shuttles it inward to the bacteroids. This method of oxygen delivery is vital, because nitrogenase, the enzyme responsible for nitrogen fixation, is irreversibly inhibited by excess O_2. Leghemo-

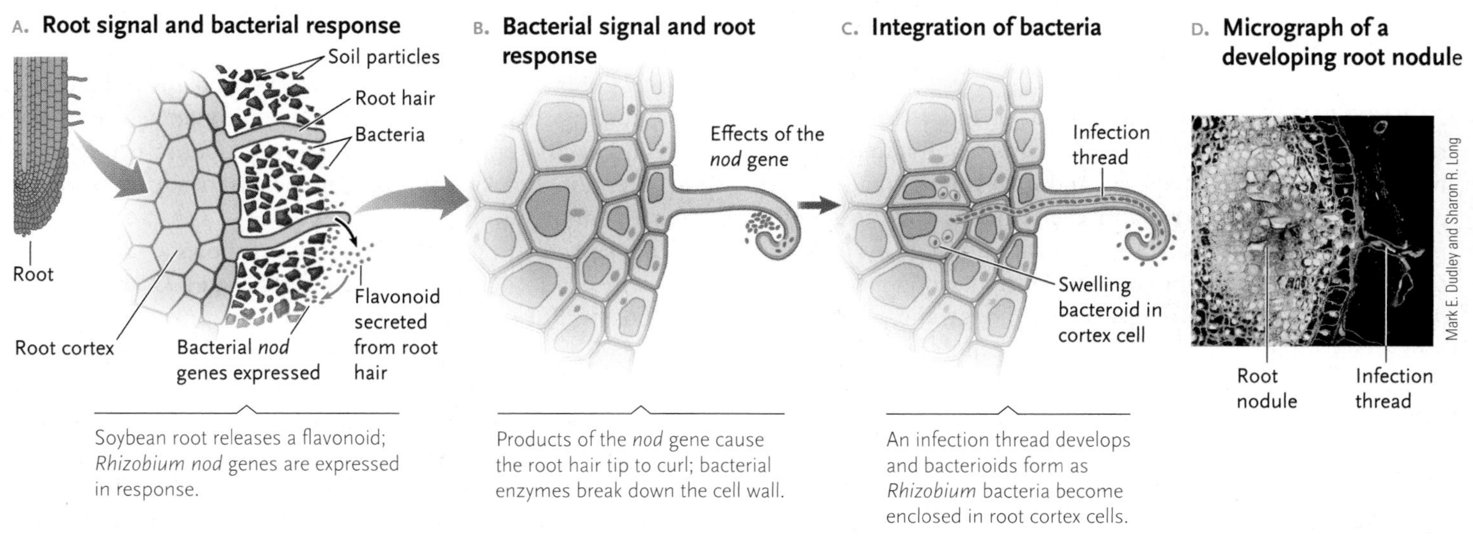

A. Root signal and bacterial response
Soil particles
Root hair
Bacteria
Root
Root cortex
Bacterial nod genes expressed
Flavonoid secreted from root hair
Soybean root releases a flavonoid; Rhizobium nod genes are expressed in response.
B. Bacterial signal and root response
Effects of the nod gene
Products of the nod gene cause the root hair tip to curl; bacterial enzymes break down the cell wall.
C. Integration of bacteria
Infection thread
Swelling bacteroid in cortex cell
An infection thread develops and bacterioids form as Rhizobium bacteria become enclosed in root cortex cells.
D. Micrograph of a developing root nodule
Root nodule
Infection thread
Mark E. Dudley and Sharon R. Long

FIGURE 33.9
Root nodule formation in legumes, which form a mutualistic association with the nitrogen-fixing bacteria *Rhizobium* and *Bradyrhizobium*.

A. **Cobra lily *(Darlingtonia californica)***

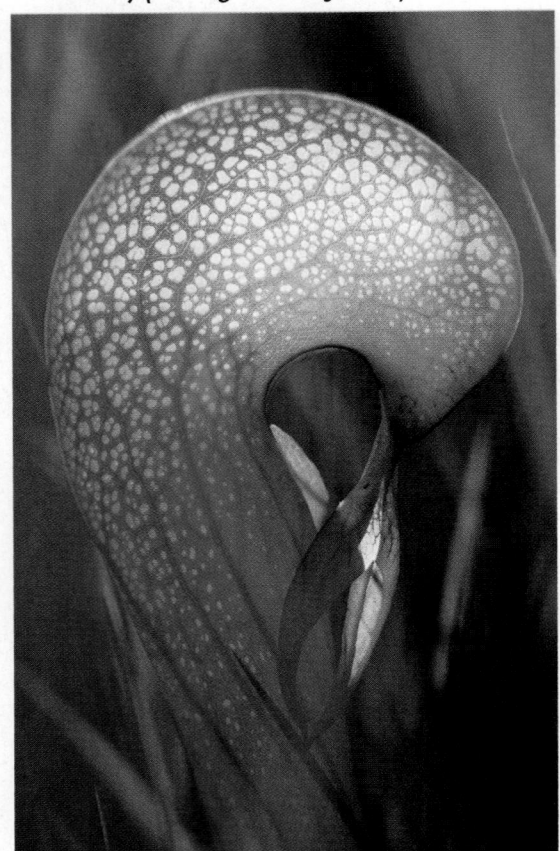

B. **Dodder *(Cuscuta)***

C. **Snow plant *(Sarcodes sanguinea)***

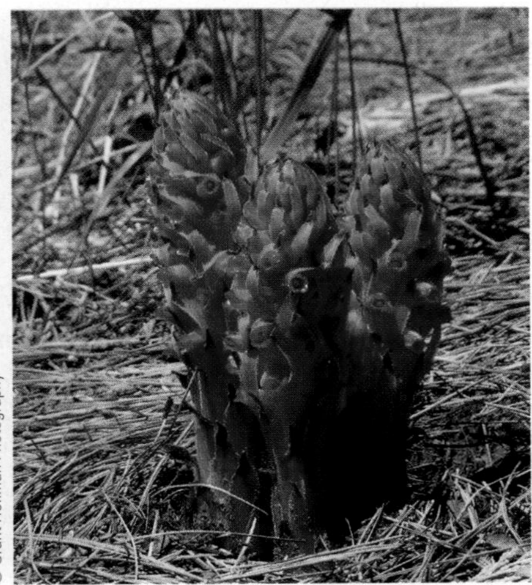

D. **Lady-of-the-night orchid *(Brassavola nodosa)***

FIGURE 33.10

Some plants with unusual adaptations for obtaining nutrients. **(A)** Cobra lily *(Darlingtonia californica)*, a carnivorous plant. The patterns formed by light shining through the plant's pitcherlike leaves are thought to confuse insects that have entered the pitcher, making an exit more difficult. **(B)** A parasitic dodder, one of the more than 150 *Cuscuta* species. Dodders have slender yellow-orange stems that twine around the host plant before producing haustorial roots that absorb nutrients and water from the host's xylem and phloem. **(C)** Snow plant *(Sarcodes sanguinea)*, which pops up in the deep humus of shady conifer forests after snow has melted in spring. This species lacks chlorophyll and does not photosynthesize. Instead its roots intertwine with hyphae of soil fungi that also form associations with the roots of nearby conifers. Radiocarbon studies have shown that the fungi take up sugars and other nutrients from the trees and pass a portion of this food to the snow plant. **(D)** The lady-of-the-night orchid *(Brassavola nodosa)*, a tropical epiphyte.

globin delivers just enough oxygen to maintain bacteroid respiration without shutting down the action of nitrogenase.

Some Plants Obtain Nutrients in Unusual Ways

The Venus flytrap, the cobra lily and various species of sundews are members of a curious group of plants that obtain nitrogen and other nutrients by trapping and digesting animals. A few species of tropical pitcher plants even capture and digest mice and small rats. Such "carnivorous" (meat eating) plants have become adapted to survive in nutrient-deficient, and especially nitrogen-deficient, environments such as boggy and sandy areas through elaborate mechanisms for extracellular digestion and absorption. The cobra lily (*Darlingtonia californica;* **Figure 33.10A**) is a good example. Its leaves form a "pitcher" that is partly filled with digestive enzymes. Insects lured in by attractive odors often wander deeper into the pitcher, encountering downward-pointing leaf hairs that have a slick, waxy coating and speed the insect's descent into the pool of enzymes. The plant then absorbs monomers released as the animal tissues are digested.

Dodders (**Figure 33.10B**) and thousands of other species of flowering plants are parasites that obtain some or all of their nutrients from the tissues of other plants. Parasitic species develop *haustorial roots* (similar to the haustoria of fungi described in Chapter 28) that penetrate deep into the host plant and tap into its vascular tissues. Although some parasitic plants, like mistletoe, contain chlorophyll and thus can photosynthesize, dodders and other nonphotosynthesizers rob the host of sugars as well as water and minerals.

The snow plant *(Sarcodes sanguinea)* shows a variation on this theme. As its deep red color suggests (**Figure 33.10C**), it lacks chlorophyll, but it doesn't have haustorial roots. Instead, the snow plant's roots take up nutrients from mycorrhizae they "share" with the roots of nearby conifers.

Epiphytes, such as the tropical orchid pictured in **Figure 33.10D,** are not parasitic even though they grow on other plants. Some trap falling debris and rainwater among their leaves, whereas their roots (including mycorrhizae, in the case of the orchid) invade the moist leaf litter and absorb nutrients from it as the litter decomposes. In temperate forests, many mosses and lichens are epiphytes.

These and other strategies plants have evolved for obtaining nutrients and water are only part of the survival equation, however. Plants use nutrients not only for growth and maintenance, but also, of course, for building structures such as pollen, flowers, and seeds used in reproduction—our topic in Chapter 34.

STUDY BREAK 33.3

1. **What is a mycorrhiza, and why are mycorrhizal associations so vital to many plants?**
2. **Distinguish between nitrogen fixation, ammonification, and nitrification.**
3. **Summarize the mechanism by which associations with bacteria supply nitrogen to plants such as legumes.**

THINK OUTSIDE THE BOOK

Botanists have identified a large number of plants that trap and digest small animals. Working with fellow students, do some research online or in the library to develop a list of at least five different variations in the trapping mechanism. Which plant structure or structures have been modified to form each type of trap?

UNANSWERED QUESTIONS

What is the role of organic nitrogen in plant nutrition and nitrogen cycling?

The classical model of nitrogen (N) cycling in plant ecology has been that plants take up nitrogen in inorganic forms—nitrate and ammonium—made available by microbial conversion of organic N to mineral N. This process of nitrogen mineralization involves the decomposition of organic nitrogen-containing substances released into the soil from dead organisms. Such substances include proteins, chitin and peptidoglycan (components of fungal and bacterial cell walls), and DNA and RNA. A fundamental assumption of this classical model is that plants cannot compete with soil microorganisms for mineral nitrogen, but rather only have access to what remains after microbial decomposers satisfy their own demand for nitrogen and release excess nitrogen extracted from the organic matter they consume. In recent years, however, studies on nitrogen mineralization in low N soils, such as those in many arctic tundra ecosystems, have shown that mineralization rates can be too low to account for the amount of nitrogen taken up by plants in nitrogen-poor environments. This finding raised a fundamental question about nitrogen cycling in arctic tundra and other low-nitrogen environments: How do plants acquire nitrogen from soils where nitrogen mineralization rates are too low to account for plant demand?

Since the mid 1990s evidence has mounted that plants can bypass the microbial nitrogen mineralization step of the classical nitrogen cycling model. Instead of relying solely on inorganic forms of nitrogen such as nitrate, plants may also derive nitrogen from organic molecules such as amino acids released into the soil solution as decomposer microorganisms degrade proteinaceous organic matter. The finding that many plants indeed can take up amino acids has raised the possibility that the uptake of organic nitrogen may be an important component of plant nitrogen budgets in low N environments, such as arctic tundra. First discovered in arctic tundra plants, amino acid uptake has since been reported for plants in a variety of habitats. Further, we now know that roots possess specific amino acid transporters that can take up a variety of amino acids, and recent studies suggest that root amino acid transport is a trait that can be induced by low nitrogen availability.

Although these discoveries have forced researchers to develop more complete models of nitrogen cycling and uptake by plants, much remains to be learned. For example, what controls amino acid availability in soil? Can plants take up organic nitrogen forms other than amino acids? What controls the availability of these other N forms? In my laboratory we are studying a range of phenomena that bear on all these questions, including the extent to which major organic nitrogen sources such as proteins and chitin contribute to nitrogen availability in soils in different places at different times. For this research, we are measuring the activity of microbial enzymes responsible for breaking down different organic nitrogen sources in soils from a variety of environments, and comparing the enzyme activities to overall nitrogen mineralization rates. We are also examining basic aspects of soil chemistry, such as acidity, to try to determine the specific organic nitrogen forms microbes are consuming and mineralizing under different conditions. Our initial findings indicate that proteins, once thought to be the predominant form of organic nitrogen that microorganisms degrade, may be a less important source of nitrogen in many environments, particularly in acidic soils. Instead, in such low pH soil environments, the breakdown of chitin may be much more important in nitrogen cycling. Acidic soils are fairly common in nature, and this finding raises the possibility that plants growing in such soils may be able to acquire nitrogen by taking up chitin breakdown products. This hypothesis remains untested. As researchers in my laboratory and elsewhere probe it and other questions, the answers will greatly expand our understanding of nitrogen cycling in general, and of specific plant adaptations for gaining access to this essential nutrient.

Think Critically

1. Soil microorganisms consume nitrogen-containing organic compounds, and mineralize nitrogen in the process, even after their own nitrogen demands are satisfied. Why don't these microorganisms stop breaking down nitrogenous organic compounds once their metabolic needs are met?
2. Can you think of any reasons why microorganisms might prefer to degrade chitin rather than proteins under acidic conditions?

Michael Weintraub

Michael Weintraub is an assistant professor of soil ecology in the University of Toledo's Department of Environmental Sciences. His research focuses on developing a mechanistic understanding of key soil processes, to gain insight into how terrestrial ecosystems function, and to predict how they will respond to disturbances such as climate change and nitrogen deposition. To learn more about his work, go to: http://www.eeescience.utoledo.edu/Faculty/weintraub/ESELab.htm.

REVIEW KEY CONCEPTS

Go to **CENGAGENOW** at www.cengage.com/login to access quizzing, animations, exercises, articles, and personalized homework help.

33.1 Plant Nutritional Requirements

- Plants require 17 essential nutrients (Table 33.1). With enough sunlight and these nutrients, plants can synthesize all the compounds they require to survive.
- Nine essential elements are macronutrients, required in relatively large amounts. Of these, carbon, hydrogen, oxygen, and nitrogen are the main building blocks in the synthesis of carbohydrates, lipids, proteins, and nucleic acids. Macronutrients dissolved in the soil solution are nitrogen, potassium, calcium, magnesium, phosphorus, and sulfur.
- Plants require essential micronutrients in much smaller amounts. Known micronutrients are chlorine, iron, boron, manganese, zinc, copper, molybdenum, and nickel.
- Each plant species requires specific amounts of specific nutrients. Typical deficiency symptoms are stunted growth, yellowing or other abnormal changes in leaf color, dead spots on leaves, or abnormally formed stems (Figure 33.3).
- Most mineral ions enter plant roots dissolved in water. Inside cells, most mineral ions enter vacuoles or the cell cytoplasm, where they are available for metabolism. Some elements, such as nitrogen and potassium, are mobile—they can move from site to site in phloem as the plant grows.

33.2 Soil

- Soil consists of sand, silt, and clay particles, usually held together by humus and other organic components. Humus absorbs considerable water and contributes to the water-holding capacity of soil.
- The relative proportions of various soil mineral particles and humus give soil its basic texture and structure. The best agricultural soils are loams that contain clay, sand, silt, and humus in roughly equal proportions. Topsoil is the most fertile soil layer (Figure 33.4).
- Soil particles are thinly coated by the soil solution, a mixture of water and solutes (Figure 33.5). Root hairs and other root epidermal cells absorb water and solutes from this solution.
- The amount of water available to plant roots depends mainly on the relative proportions of different soil components. Water moves quickly through sandy soils, whereas soils rich in clay and humus tend to hold water.
- Cations are adsorbed on the negatively charged surfaces of soil particles, potentially limiting their uptake by roots. Cation exchange, in which mineral cations are replaced by H^+, helps make these nutrients available to plants (Figure 33.6). Anions are more weakly bound to soil particles; they move more readily into root hairs but also are more apt to leach out of topsoil. In nature, the soil solution surrounding plant roots generally contains only tiny amounts of essential mineral ions.
- Plants influence the physical, chemical, and biological features of soil, as when dead plant parts decompose and substances they contain enter the soil (Figure 33.7).

Animation: Soil profile

33.3 Obtaining and Absorbing Nutrients

- Numerous adaptations help plants solve the problems of obtaining and absorbing nutrients. Roots penetrate the soil towards nutrients and water; huge numbers of root hairs increase the root's absorptive surface. Ion-specific transporters in root cortical cells adjust the plant's uptake of particular ions. Mycorrhizal associations between fungi and plant roots enhance the absorption of nutrients, notably phosphorus.
- Nitrogen usually is the scarcest nutrient in soil, and nitrogen-fixing bacteria produce much of the usable soil nitrogen. Nitrogen fixation reduces atmospheric N_2 to NH_4^+ (ammonium) in a reaction that requires nitrogenase as a catalyst. Nitrifying bacteria rapidly convert NH_4^+ to nitrate, the form in which the roots of most plants absorb nitrogen (Figure 33.7).
- In legumes and a few other species, nitrogen-fixing bacteria reside in root nodules in a mutualistic association (Figure 33.8).
- Bacteria (bacteroids) enclosed in a root nodule reduce N_2 to NH_4^+ (Figure 33.9). The toxic NH_4^+ is moved out of the bacteroids and converted to nitrogen-rich, nontoxic compounds such as amino acids. In plants that do not form root nodules, nitrate absorbed by roots is reduced to ammonium, which then is converted to nontoxic forms.
- In many plant species, root cells synthesize amino acids and other organic nitrogenous compounds, and these molecules are transported in phloem throughout the plant. In some plants, the nitrogen-rich precursors travel in xylem to leaves, where different organic molecules are synthesized. Those molecules move to other cells in phloem.
- A few plant species have evolved alternative mechanisms for obtaining some or all of their nutrients (Figure 33.10). Carnivorous plants have structures that physically trap insects or other small animals, and produce solutions of enzymes that digest the animal tissues, releasing absorbable nutrients.
- Some plant species parasitize other plants. The parasite may or may not contain chlorophyll and carry out photosynthesis; species that do not photosynthesize obtain all of their nutrition from the host. Epiphytes grow on other plants but obtain nutrients independently.

Animation: Uptake of nutrients by plants

UNDERSTAND AND APPLY

Test Your Knowledge

1. Which best applies to a micronutrient?
   a. It makes up 96% of the plant's dry mass.
   b. It cannot be replaced artificially.
   c. It is early on the periodic chart compared with macronutrients.
   d. It is required in large amounts during sunlight hours.
   e. It is an essential element.
2. Nutrient runoff from fertilizing lush lawns often causes "algal blooms" in nearby lakes, making swimming impossible. The fertilizer components most likely to have caused the blooms are:
   a. iron, magnesium, and nitrogen.
   b. nitrogen, phosphorus, and sulfur.
   c. nitrogen, potassium, and phosphorus.
   d. selenium, magnesium, and potassium.
   e. nitrogen, magnesium, and nickel.
3. Which of the following are *not* among the ideal soil conditions for growing crops?
   a. extremely large air spaces
   b. sandy or silty loam
   c. blend of sand and clay
   d. less than 5% humus
   e. thick top soil
4. Which of the following processes contributes to the uptake of mineral ions by plant roots?
   a. chlorosis
   b. osmosis
   c. cation exchange
   d. anion leaching
   e. growth of root hairs
5. Which of the following does *not* influence soil pH?
   a. rainfall
   b. hydroponic growth
   c. release of sulfur and nitrogen oxides into the air
   d. decomposition of organisms
   e. weathering of rock
6. Which of the following is a common process that makes usable nitrogen available to plants?
   a. nitrogen-fixing bacteria synthesizing nitrate
   b. ammonifying bacteria using ammonium to produce nitrate
   c. nitrifying bacteria converting NH_4^+ to NO_3^-
   d. the direct absorption of NH_4^+ by root hairs
   e. the absorption of atmospheric N_2 into the xylem
7. The *nod* genes in the bacteria in soybean nodules allow the bacteria to fix nitrogen. Which of the following, if any, is *not* a step in this process?
   a. The products of *nod* genes cause cells of the root cortex to divide and become the root nodule in which bacteroids fix nitrogen for the plant.
   b. In the cortex cells bacteria enlarge and become immobile, forming bacteroids.
   c. Bacteria enter the root hair cell and multiply, causing the cell plasma membrane to form an infection thread that extends into the root cortex.
   d. Roots release flavonoid, which turns on the expression of bacterial *nod* genes. Products of *nod* genes cause the tip of the root hair to curl toward the bacteria.
   e. Root hairs trigger release of bacterial enzymes that break down root hair cell walls.
8. Being "carnivorous" is a plant adaptation to obtain:
   a. oxygen.
   b. phosphorus.
   c. potassium.
   d. nitrogen.
   e. carbon.
9. Haustorial roots are characteristic of plants that are:
   a. parasites.
   b. epiphytes.
   c. nitrate fixers.
   d. leghemoglobin users.
   e. carnivorous.
10. Identify the correct match of a nutrient with its function.
   a. chlorine: component of several enzymes
   b. potassium: component of nucleic acids
   c. phosphorus: component of most proteins
   d. manganese: role in shoot and root growth
   e. calcium: maintenance of cell walls and membrane permeability

Discuss the Concepts

1. If you want to study factors that affect plant nutrition in nature, what would be the advantages and disadvantages of using a hydroponic culture method?
2. Gardeners often add a humus-rich "soil conditioner" to garden plots before they plant. Adding the conditioner helps aerate the soil, and the decomposing organic materials in humus provide nutrients. If the plot is for annual plants, it often must be reconditioned year after year, even though the gardener faithfully pulls weeds, fertilizes seedlings, applies chemicals to curtail disease-causing soil microbes, and immediately tosses out the mature plants (along with any plant debris) when they have finished bearing. Suggest some reasons why reconditioning is necessary in this scenario, and some strategies that could help limit the need for it.
3. One effect of acid rain is to dissolve rock, liberating minerals into soil. Accordingly, can a case be made that acid rain confers environmental benefits as well as doing harm? What are some other factors, especially with regard to plant adaptations for gaining nutrients, that bear on this question?
4. Using Table 33.1 as a guide, describe some of the known roles of nitrogen, phosphorus, and potassium in plant function. What are some of the signs that a plant suffers a deficiency in those elements?

Design an Experiment

A plant in your garden is undersized and develops chlorotic leaves even though you fertilize it with a mixture that contains nitrogen, potassium, and phosphorus. After determining that the plant receives enough sunlight for photosynthesis, you next decide to test whether its mineral nutrition is adequate. What specific hypothesis will your experiment test? How will your experimental design test the hypothesis?

Interpret the Data

Numerous studies have shown that the particular plant species growing in a soil influence the soil's chemical properties. With this in mind, in the study of Colorado mountain soil mentioned in Section 33.2, the researchers expected to find differences in the study area's soil chemistry depending on which of the three main tree species—trembling aspen *(Populus tremuloides)*, lodgepole pine *(Pinus contorta)*, and Engelmann spruce *(Picea engelmannii)*—was dominant at a particular site. One factor that affects soil chemistry is the chemical composition of the leaf litter that falls to the ground and decomposes. The table that follows shows values for leaf litter quality as a function of the litter's nitrogen, carbon, cellulose, and lignin content. For each species it also shows the biomass of fine roots (less than 1 mm diameter) and coarse roots (more than 1 mm diameter) in the upper 10 cm of soil.

1. Based on these values, summarize the findings for aspen trees, compared with the pines and spruce trees in the sample.
2. Overall, how does the ratio of lignin to nitrogen in leaf litter seem to correlate with total litter nitrogen content?
3. Did the root biomass among the three tree species differ significantly?

Source: Adapted from E. Ayres et al. 2009. Tree species traits influence soil physical, chemical, and biological properties in high elevation forests. PLoS ONE 4(6): e5964.

Apply Evolutionary Thinking

This chapter's *Focus on Applied Research* discusses phytoremediation, the use of plants to remove environmental pollutants such as heavy metals. Some plant species are "hyperaccumulaters" that take up arsenic and other metallic contaminants and sequester such toxins in shoot parts. How might this activity confer a selective advantage?

Leaf Litter Quality and Root Biomass of Tree Species in a High Elevation Forest in Colorado

	Litter N (%)	Litter C (%)	Litter C:N	Litter cellulose (%)	Litter lignin (%)	Litter lignin:N	Fine root biomass (g/m²)	Coarse root biomass (g/m²)
Aspen	0.84±0.08	49.4±0.4	60.4±5.2	42.1±3.0	20.2±2.2	24.1±1.2	321±47	422±134
Pine	0.47±0.03	52.0±0.3	112.1±8.3	59.0±1.0	33.8±0.5	73.2±5.7	215±25	232±87
Spruce	0.41±0.01	48.7±0.1	118.3±4.2	41.1±1.1	19.2±0.7	46.6±1.7	249±28	319±62

The reproductive structures of an ornamental poppy *(Papaver rhoeas)*. Bright yellow male parts, where pollen grains develop, surround female parts that produce eggs. Within them fertilization occurs and seeds develop (photographer's close-up).

34

STUDY OUTLINE

Reproduction and Development in Flowering Plants

Why It Matters. . . Chocolate, a confection beloved by humans with a sweet tooth, gets its start within the fruits of *Theobroma cacao,* a small flowering tree *T. cacao* evolved in the undergrowth of tropical rain forests in Central America, where the Maya and Aztec peoples domesticated it and incorporated it into their religious and cultural life. Today cacao trees flourish on vast plantations in the tropical lowlands of Central and South America, the West Indies, West Africa, and New Guinea. Unlike most angiosperms, which produce flowers at the tips of floral shoots, *T. cacao* flowers grow directly from buds on the tree trunk. Small flying insects called midges pollinate the flowers. Pollination, in turn, is the first step toward fertilization of the eggs, and within about six months, large, podlike fruits develop from them **(Figure 34.1).** Each fruit contains from 20 to 60 seeds—the cacao "beans" that manufacturers process into cocoa, eating chocolate, and other products.

Angiosperms have elaborate reproductive systems housed in their flowers, systems that produce, protect, and nourish sperm, eggs, and developing embryos. As with *T. cacao,* the flowers of many species also attract animal pollinators, which function in bringing sperm and egg together. Once a new individual forms and begins to grow, finely regulated gene interactions guide the development of all a plant's parts.

Like the seeds of other flowering plants, cacao seeds result from sexual reproduction—the union of sperm and eggs that are produced by meiosis (see Section 11.1). As Chapter 16 describes in greater detail, this reproductive mode yields genetically diverse offspring. This diversity, in turn, pro-

Wildlife/Peter Arnold, Inc.

Image copyright Norman Chan, 2010.
Used under license from Shutterstock.com

FIGURE 34.1
Flowers and fruits growing from the trunk of a cacao tree *(Theobroma cacao)* in Central America. Each fruit is the mature ovary of a *T. cacao* flower.

vides an adaptive advantage in a changing environment, because it increases the chance that as change occurs—perhaps a shift in climate, or the appearance of a new predator or pathogen—at least some offspring will grow to maturity and reproduce successfully. *T. cacao* and many other plants are reproductively flexible, however. Under certain circumstances, they can also reproduce asexually, so that individuals of the new generation are genetically identical (clones) of their parents.

In the first three sections of this chapter we will focus on sexual reproduction, which dominates the life cycle of flowering plants. We then consider asexual reproduction and conclude with a discussion of early plant development. <

34.1 Overview of Flowering Plant Reproduction

In plants as in animals, sexual reproduction occurs when male and female haploid gametes unite to create a diploid fertilized egg. This fertilized egg—the zygote—then embarks on its gene-guided developmental course. A new plant grows and develops as the zygote and subsequent cells divide by mitosis, enlarge, and differentiate, becoming specialized for particular functions.

Once an angiosperm zygote has formed, early development generates an embryo enclosed within a seed. This embryo already contains early versions of the basic plant tissue systems, so the embryo technically is already a **sporophyte**—the diploid, spore-producing body of a plant described in Section 27.1. When most people look at a plant—say a cherry tree or a rose bush—what they think of as "the plant" is the sporophyte **(Figure 34.2).**

At some point during an angiosperm sporophyte's growth and development, one or more of its vegetative shoots undergo changes in structure and function and become reproductive shoots. These *floral shoots* will give rise to a flower or inflorescence (a group of flowers on the same floral shoot), and certain cells in the flowers will divide by meiosis. Unlike in animal meiosis, however, meiosis in plants does not yield haploid gametes. Instead, meiosis gives rise to haploid **spores,** walled cells that divide by mitosis. Each grows into a multicellular—but still haploid—**gametophyte**

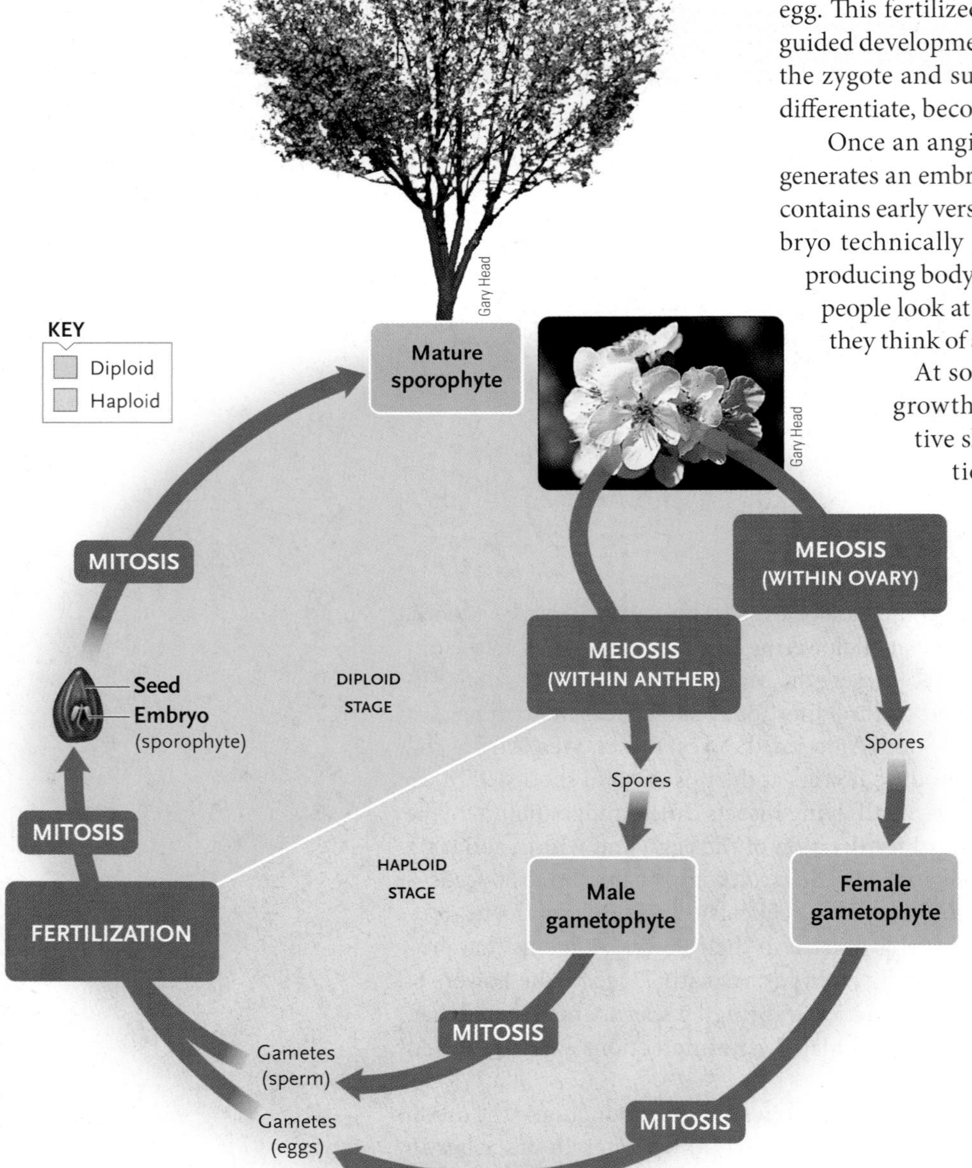
Gary Head
Gary Head
KEY
Diploid
Haploid
Mature sporophyte
MITOSIS
MEIOSIS (WITHIN OVARY)
MEIOSIS (WITHIN ANTHER)
Seed
Embryo (sporophyte)
DIPLOID STAGE
Spores
Spores
MITOSIS
HAPLOID STAGE
FERTILIZATION
Male gametophyte
Female gametophyte
MITOSIS
Gametes (sperm)
Gametes (eggs)
MITOSIS

FIGURE 34.2
Overview of the flowering plant life cycle, using the cherry *(Prunus)* as an example. This type of reproductive cycle, alternation of generations, has a haploid phase in which multicellular gametophytes produce gametes and a diploid phase that begins when two gametes fuse to form a zygote. This zygote develops into a multicellular embryo within a seed and then into a mature sporophyte. Meiotic divisions in the flower of the sporophyte produce spores, which give rise to new gametophytes.

("gamete-producing plant"). The gametophyte of a vascular plant is a tiny structure that produces haploid sex cells, the gametes, also by mitosis. Male gametophytes produce sperm cells, the male gametes of flowering plants; female gametophytes produce eggs. This division of a life cycle into a diploid, spore-producing generation and a haploid, gamete-producing one is called **alternation of generations.**

In virtually all plants, the gametophyte and sporophyte are strikingly different from one another in both function and structure. As Chapter 27 describes in detail, the course of plant evolution has been marked by a trend toward a larger and more nutritionally independent sporophyte. For instance, in bryophytes (mosses and liverworts) the sporophyte is usually smaller than the gametophyte, growing out of the gametophyte and obtaining nutrients from it (see Section 27.2). This relationship changed as vascular plants evolved. In ferns, which are seedless vascular plants, the gametophyte is much smaller than the sporophyte and is free-living for much of its life, and in most fern species the gametophyte nourishes itself by photosynthesis. In angiosperms and other seed plants, gametophytes are small structures that are retained *inside* the sporophyte for all or part of their lives. The female gametophyte of a flowering plant usually consists of only seven cells that are embedded in floral tissues, as you will read shortly. Male gametophytes consist of only three cells. They are released into the environment as pollen grains so small that they are measured in micrometers. The pollen grain matures when it reaches a compatible ovule. When fertilization ensues, an embryo develops within a seed.

As you will read later on in this chapter, sporophytes may also reproduce asexually by several means, all of which produce genetically identical offspring. We turn now to our consideration of sexual reproduction in angiosperms, beginning with the crucial step in which flowers develop.

STUDY BREAK 34.1

1. **What are the two "alternating generations" of plants?**
2. **How do these two life phases differ in structure and function?**

34.2 The Formation of Flowers and Gametes

Flowering marks a developmental shift for an angiosperm. Chemical signals—triggered in part by environmental cues such as day length and temperature—travel to the apical meristem of a shoot and trigger changes in cell activity there. The shoot stops its vegetative growth and undergoes further changes that transform it into a floral shoot capable of giving rise to floral organs.

In Angiosperms, Flowers Contain the Organs for Sexual Reproduction

A flower develops from the end of the floral shoot, called the **receptacle.** In a "typical" flower, cells in the receptacle differentiate to produce four types of concentric tissue regions called *whorls*

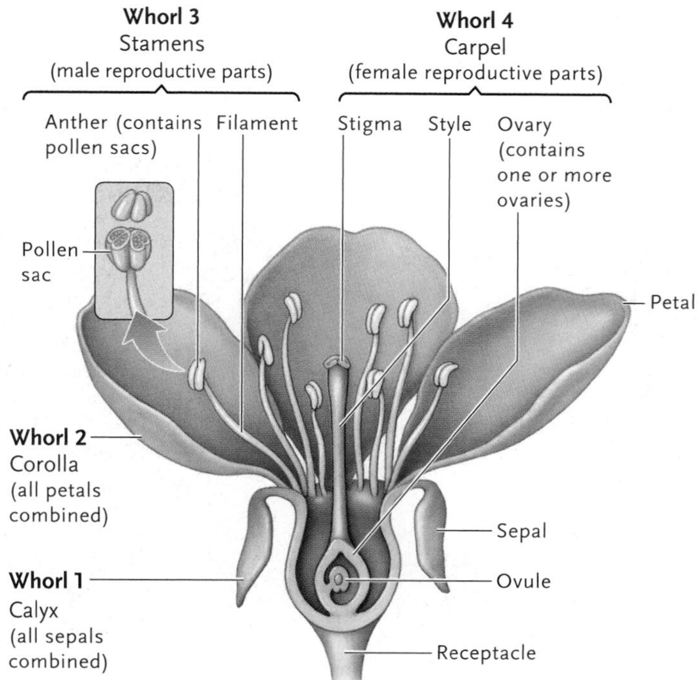
Whorl 3
Stamens
(male reproductive parts)
Whorl 4
Carpel
(female reproductive parts)
Anther (contains pollen sacs)
Filament
Stigma
Style
Ovary (contains one or more ovaries)
Pollen sac
Petal
Whorl 2
Corolla
(all petals combined)
Sepal
Ovule
Whorl 1
Calyx
(all sepals combined)
Receptacle

FIGURE 34.3
Structure of a cherry *(Prunus)* flower, with the four whorls indicated. Like the flowers of many angiosperms, it has a single carpel and several stamens. The anthers of each stamen produce haploid pollen. The stigma of the carpel receives the pollen, and the ovule inside the ovary contains the haploid egg. In this diagram, two of the cherry flower's five petals have been removed.

(Figure 34.3). Flowers with all four whorls are called **complete flowers (Figure 34.4A).** The two outer whorls consist of nonfertile, vegetative structures. The outermost whorl (whorl 1), the **calyx,** is made up of leaflike **sepals.** The calyx is usually green, and, early in the flower's development, it encloses all the other parts, as in an unopened rose bud. The next whorl, the **corolla,** includes the **petals.** Corollas usually are the "showy" parts of flowers with distinctive colors, patterning, and shapes. These features often function in attracting bees and other animal pollinators.

FIGURE 34.4
Examples of complete and incomplete flowers. (A) Complete flower of a dog rose *(Rosa canina)*, which has all four floral whorls. **(B)** Incomplete flower of a squash plant *(Cucurbita)*—in this case, a "female" plant in which the flowers contain only carpels.

A. **Dog rose**

B. **Squash flower**

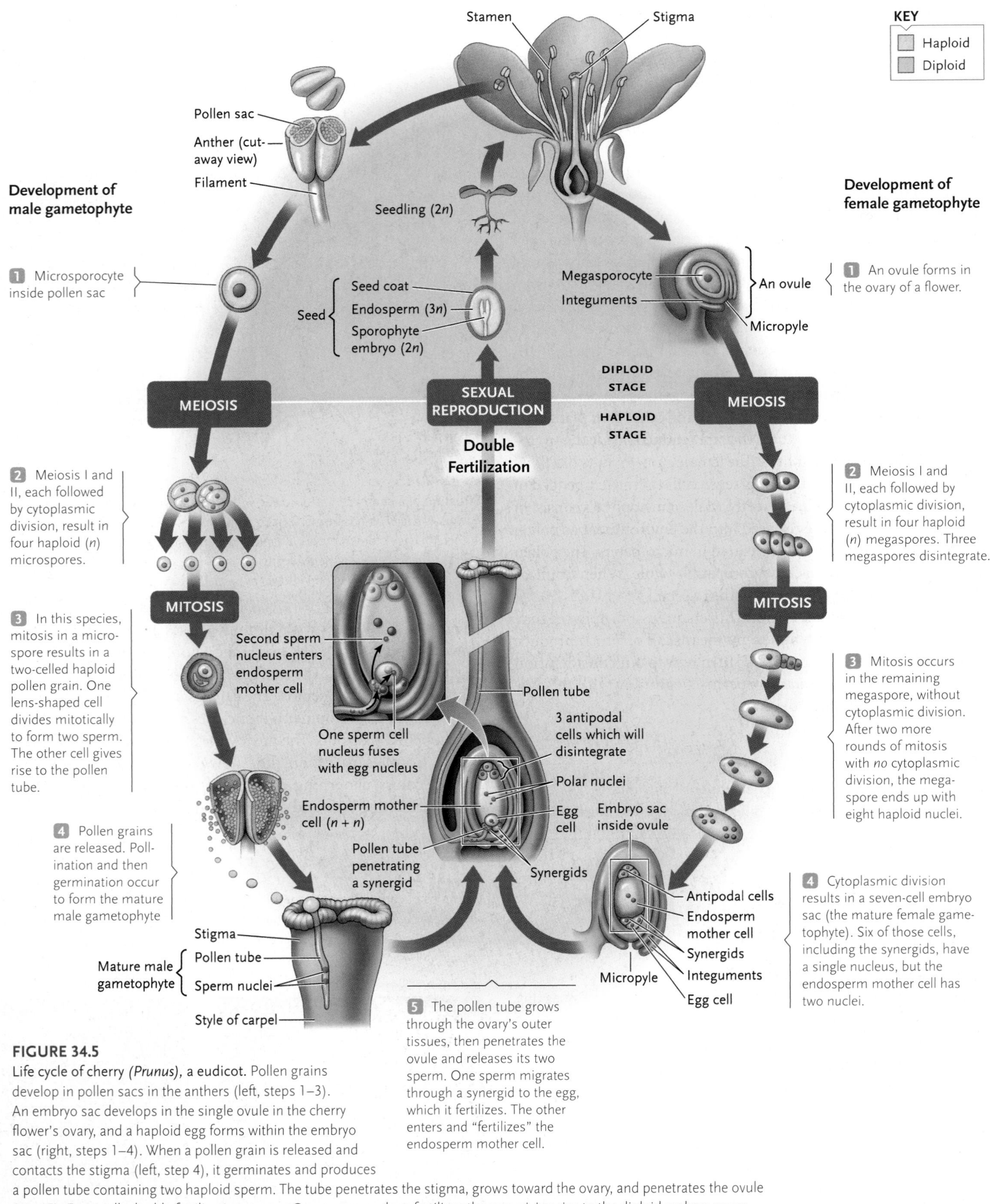
KEY
Haploid
Diploid
Stamen
Stigma
Pollen sac
Anther (cut-away view)
Filament
Development of male gametophyte
Seedling (2n)
Development of female gametophyte
1 Microsporocyte inside pollen sac
Seed coat
Endosperm (3n)
Seed
Sporophyte embryo (2n)
Megasporocyte
Integuments
An ovule
Micropyle
1 An ovule forms in the ovary of a flower.
DIPLOID STAGE
MEIOSIS
SEXUAL REPRODUCTION
HAPLOID STAGE
MEIOSIS
Double Fertilization
2 Meiosis I and II, each followed by cytoplasmic division, result in four haploid (n) microspores.
2 Meiosis I and II, each followed by cytoplasmic division, result in four haploid (n) megaspores. Three megaspores disintegrate.
MITOSIS
MITOSIS
3 In this species, mitosis in a microspore results in a two-celled haploid pollen grain. One lens-shaped cell divides mitotically to form two sperm. The other cell gives rise to the pollen tube.
Second sperm nucleus enters endosperm mother cell
Pollen tube
3 Mitosis occurs in the remaining megaspore, without cytoplasmic division. After two more rounds of mitosis with no cytoplasmic division, the megaspore ends up with eight haploid nuclei.
One sperm cell nucleus fuses with egg nucleus
3 antipodal cells which will disintegrate
Polar nuclei
Endosperm mother cell (n + n)
Egg cell
Embryo sac inside ovule
Pollen tube penetrating a synergid
Synergids
4 Pollen grains are released. Pollination and then germination occur to form the mature male gametophyte
Antipodal cells
Endosperm mother cell
Synergids
Integuments
Egg cell
Micropyle
4 Cytoplasmic division results in a seven-cell embryo sac (the mature female gametophyte). Six of those cells, including the synergids, have a single nucleus, but the endosperm mother cell has two nuclei.
Stigma
Mature male gametophyte
Pollen tube
Sperm nuclei
Style of carpel
5 The pollen tube grows through the ovary's outer tissues, then penetrates the ovule and releases its two sperm. One sperm migrates through a synergid to the egg, which it fertilizes. The other enters and "fertilizes" the endosperm mother cell.

FIGURE 34.5
Life cycle of cherry *(Prunus)*, a eudicot. Pollen grains develop in pollen sacs in the anthers (left, steps 1–3). An embryo sac develops in the single ovule in the cherry flower's ovary, and a haploid egg forms within the embryo sac (right, steps 1–4). When a pollen grain is released and contacts the stigma (left, step 4), it germinates and produces a pollen tube containing two haploid sperm. The tube penetrates the stigma, grows toward the ovary, and penetrates the ovule (step 5). Eventually double fertilization occurs. One sperm nucleus fertilizes the egg, giving rise to the diploid embryo sporophyte, and the other sperm nucleus fuses with the endosperm mother cell, forming a triploid (3*n*) cell from which nutritive endosperm develops. A seed forms as both the embryonic sporophyte and the endosperm become encased in a seed coat.

A flower's two inner whorls are specialized for making gametes. Inside the corolla is the whorl of **stamens** (whorl 3), in which male gametophytes form. In almost all living flowering plant species, a stamen consists of a slender **filament** (stalk) capped by a bilobed **anther.** Each anther contains four **pollen sacs,** in which pollen develops.

The innermost whorl (whorl 4) consists of parts associated with the formation and fertilization of eggs. These parts, sometimes collectively called the *pistil,* include one or more **carpels,** in which female gametophytes form. The lower part of a carpel is the **ovary.** Inside it is one or more **ovules,** in which an egg develops and fertilization takes place. In many flowers that have more than one carpel, the carpels fuse into a single, common ovary containing multiple ovules. Typically, the carpel's slender **style** widens at its upper end, terminating in the **stigma,** which serves as a landing platform for pollen. Fused carpels may share a single stigma and style, or each may retain separate ones. The name angiosperm ("seed vessel") refers to the carpel.

The arrangement and number of whorl types varies in different species. In some species flowers lack one or more of the whorls, and thus botanists describe them as **incomplete flowers** **(Figure 34.4B).** Botanists also distinguish flowers on the basis of the sexual parts they contain. Most angiosperms produce **perfect flowers,** which have both kinds of sexual parts—that is, both stamens and carpels. **Imperfect flowers** have stamens or carpels, but not both. (Notice that all imperfect flowers are also incomplete because they lack one of the whorls.) Species with imperfect flowers are further divided according to whether individual plants produce both sexual types of flowers, or only one. In **monoecious** (mono- = one; oikia = house) species, such as oaks, each plant has some "male" flowers with only stamens and some "female" flowers with only carpels. In **dioecious** ("two houses") species, such as willows, a given plant produces flowers having only stamens or only carpels. With this basic angiosperm reproductive anatomy in mind, we can consider the processes by which male and female gametes come into being.

Pollen Grains Arise from Microspores in Anthers

Most of a flowering plant's reproductive life cycle, from production of sperm and eggs to production of a mature seed, takes place within its flowers. **Figure 34.5** shows this cycle as it unfolds in a perfect flower. The microspores that give rise to male gametophytes are produced in a flower bud's anthers (see Figure 34.5, left). The pollen sacs inside each anther hold diploid microsporocytes (or *microspore mother cells*); each microsporocyte undergoes meiosis and eventually produces four small haploid **microspores.** Like most plant cells, the microspores are walled, and inside its wall each microspore divides again, this time by mitosis. The result is an immature, haploid male gametophyte—a **pollen grain.**

Of the two nuclei produced by the mitotic division of a microspore, one again divides. After this second round of mitosis the male gametophyte consists of three cells—two sperm cells plus a third cell that controls the development of a **pollen tube** **(Figure 34.6A).** When pollen lands on a stigma, this tube grows through the tissues of a carpel and carries the sperm cells to the ovary. A mature male gametophyte consists of the pollen tube and sperm cells—the male gametes.

The walls of pollen grains are hardened by the decay-resistant polymer *sporopollenin,* and are tough enough to protect the male gametophyte during the somewhat precarious journey from anther to stigma. These walls are so distinctive that the family to which a plant belongs often can be identified from pollen alone—based on the size and wall sculpturing of the grains, as well as the number of pores in the wall **(Figure 34.6B–D).** Because they withstand decay, pollen grains fossilize well and can provide revealing clues about the evolution of seed plants and the ecological communities that lived in the past.

Eggs and Other Cells of Female Gametophytes Arise from Megaspores

Meanwhile, in the ovary of a flower, one or more dome-shaped masses form on the inner wall. Each mass becomes an ovule (see Figure 34.5, right), which, if all goes well, develops into a seed. Only one ovule forms in the carpel of some flowers, such as the cherry. Dozens, hundreds, or thousands may form in the carpels of other flowers, such as those of a bell pepper plant *(Capsicum annuum).* At one end, the ovule has a small opening, called the **micropyle.**

Inside the cell mass, a diploid megasporocyte (or *megaspore mother cell*) divides by meiosis, forming four haploid **megaspores.** In most species, three of these megaspores disintegrate. The remaining megaspore enlarges and develops into the female gametophyte in a sequence of steps tracked in Figure 34.5.

First, three rounds of mitosis occur *without* cytoplasmic division; the result is a single cell with eight nuclei arranged in two

FIGURE 34.6
Examples of a pollen tube and pollen grain diversity. **(A)** Scanning electron micrograph of a pollen tube (orange) extending from a germinating pollen grain of prairie gentian *(Gentiana)* Three other SEMs show pollen grains from **(B)** a grass, **(C)** chickweed *(Stellaria),* and **(D)** ragweed *(Ambrosia)* plants.

A. **Pollen tube**

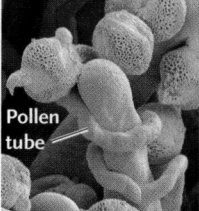

Pollen tube

Susumu Nishinaga

B. **Grass pollen grains**

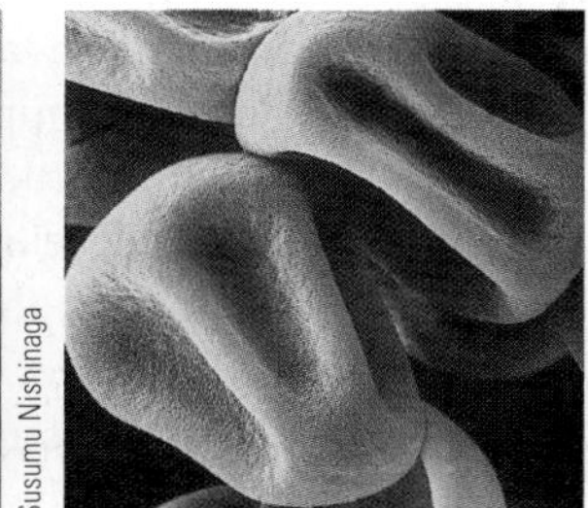

David M. Phillips/Visuals Unlimited

C. **Chickweed pollen grain**

Dr. Jeremy Burgess/SPL/Photo Researchers, Inc.

D. **Ragweed pollen grains**

David Scharf/Peter Arnold, Inc.

groups of four. Next, one nucleus in each group migrates to the center of the cell; these two **polar nuclei** ("polar" because they migrate from opposite ends of the cell) may fuse or remain separate. The cytoplasm then divides, and a cell wall forms around the two polar nuclei, forming a single large endosperm mother cell (also called the *central cell*). A wall also forms around each of the other nuclei. Three of these walled nuclei become *antipodal cells,* which eventually disintegrate. Three others form a cluster (called the "egg apparatus") near the micropyle; one of them is an **egg cell** that may eventually be fertilized. The other two, called *synergids,* will have a role in fertilization. The eventual result of all these events is an **embryo sac** containing seven cells and eight nuclei. This embryo sac is the female gametophyte.

DISCOVERING HOW CHEMICAL CUES GUIDE EMBRYO SAC DEVELOPMENT How do the four different types of embryo sac cells become differentiated, allowing them to fulfill their respective reproductive functions? A study led by Gabriela C. Pagnussat at the University of California, Davis has provided some insights into this mechanism. Working with *Arabidopsis thaliana* plants, Pagnussat and her colleagues demonstrated that the plant hormone auxin, already known for other growth effects in plants (described in Chapter 35), is a major factor in determining the identity and roles of cells and nuclei within the embryo sac. The team determined that when mitosis begins in a megaspore, an auxin gradient is established along the future axis of the developing embryo sac. The highest auxin concentration occurs near the micropyle, where the pollen tube will enter the ovule. As mitotic cell divisions continue, each of the eight products develops as an egg, synergid, polar nucleus, or antipodal cell depending on its position within the auxin gradient. Thus the egg cell and synergids develop where auxin is most concentrated, the antipodal cells develop where auxin is lowest, and the polar nuclei develop where there is an intermediate auxin concentration (Figure 34.7). Auxin can be said to provide *positional information* for embryo sac development, a topic we return to in Section 34.5.

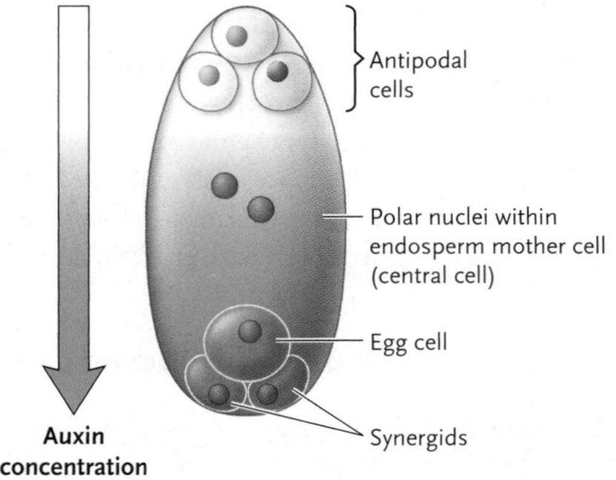
Antipodal cells
Polar nuclei within endosperm mother cell (central cell)
Egg cell
Synergids
Auxin concentration

FIGURE 34.7
Model for patterning of the female gametophyte by an auxin gradient. A study by Gabriela Pagnussat and coworkers showed that the concentration of auxin (indicated in red) determined the fate of cells in the developing gametophyte. Where auxin was most concentrated, cells became specialized as synergids. At the lowest concentration, they became antipodal cells. Egg cells developed in between these two extremes.

OTHER OPTIONS FOR EMBRYO SAC DEVELOPMENT In about a third of flowering plants, biologists have observed variations in the events that produce a female gametophyte. In lilies, for example, changes in the sequence of cell divisions produce several cells with triploid nuclei (see Figure 27.31). The egg cell is not involved, however, so such differences do not affect reproduction. They may be important in the development and functioning of other embryonic tissues.

The formation of gametophytes sets the stage for events by which a seed plant may complete its life cycle. As we see next, pollination is the next crucial step.

STUDY BREAK 34.2

1. **What is the biological role of flowers, and what fundamental developmental shift must occur before an angiosperm can produce a flower?**
2. **Explain the steps leading to the formation of a mature male gametophyte, beginning with microsporocytes in a flower's anthers. Which structures are diploid and which haploid?**
3. **Trace the development of a female gametophyte, beginning with the megasporocyte in an ovule of a flower's ovary. Which structures are diploid and which haploid?**

34.3 Pollination, Fertilization, and Germination

The process by which plants produce seeds—which have the potential to give rise to new individuals—begins with **pollination,** when pollen grains make contact with the stigma of a flower. Air or water currents, birds, bats, insects, or other agents make the transfer. (Section 27.5 discussed the complex relationship between some flowering plants and their animal pollinators.)

Pollination is the first in a series of events leading to *fertilization,* the fusion of the haploid nuclei of an egg and sperm inside the flower's ovary. The resulting embryo and its ovule mature into a seed housing a young diploid sporophyte, and when the seed sprouts, or *germinates,* the sporophyte begins to grow.

Pollination Requires Compatible Pollen and Female Tissues

Even after pollen reaches a stigma, in most cases pollination and fertilization can take place only if the pollen and stigma are compatible. For example, if pollen from one species lands on a stigma from another, chemical incompatibilities usually prevent pollen tubes from developing.

Even when the sperm-bearing pollen and a stigma are from the same species, pollination may not lead to fertilization unless the pollen and stigma belong to genetically distinct individuals. For instance, when pollen from a given plant lands on that plant's

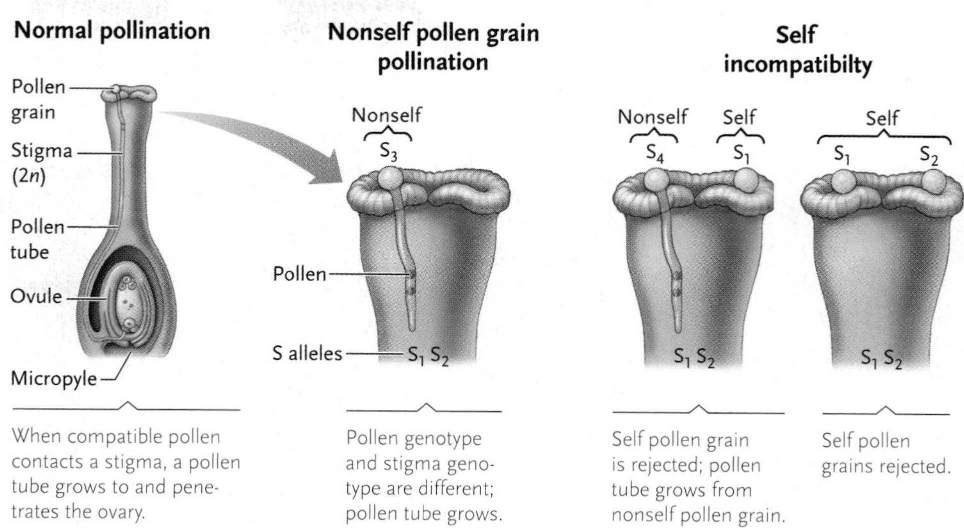
Normal pollination
Pollen grain
Stigma (2n)
Pollen tube
Ovule
Micropyle
When compatible pollen contacts a stigma, a pollen tube grows to and penetrates the ovary.
Nonself pollen grain pollination
Nonself
S3
Pollen
S alleles
S1 S2
Pollen genotype and stigma genotype are different; pollen tube grows.
Self incompatibilty
Nonself
S4
Self
S1
S1 S2
Self pollen grain is rejected; pollen tube grows from nonself pollen grain.
Self
S1
S2
S1 S2
Self pollen grains rejected.

FIGURE 34.8
Self-incompatibility. When a pollen grain has an *S* allele that matches one in the stigma (which is diploid), the result is a biochemical response that prevents fertilization—in this illustration, by preventing the growth of a pollen tube.

own stigma, a pollen tube may begin to develop, but stop before reaching the embryo sac. This **self-incompatibility** is a biochemical recognition and rejection process that prevents self-fertilization, and it apparently results from interactions between proteins encoded by *S* (self) genes.

Research has shown that *S* genes usually have multiple alleles—in some species there may be hundreds—and a common type of incompatibility occurs when pollen and stigma carry an identical *S* allele. The result is a biochemical signal that prevents proper formation of the pollen tube **(Figure 34.8).** For example, studies on plants of the mustard family have revealed that pollen contacting an incompatible stigma produces a protein that prevents the stigma from hydrating the relatively dry pollen grain, an essential step if the pollen tube is to grow. A wide range of self-incompatibility responses has been discovered, however. In cacao, for instance, when incompatible pollen contacts a stigma, a pollen tube grows normally but a hormonal response soon causes the flower to drop off the plant, preventing fertilization.

Self-incompatibility prevents inbreeding and promotes genetic variation, which is the raw material for natural selection and adaptation. Even so, many flowering plants do self-pollinate, either partly or exclusively, because that mode, too, has benefits in some circumstances. (Mendel's peas are a classic example.) For instance, "selfing" may help preserve adaptive traits in a population. It also reduces or eliminates a plant's reliance on wind, water, or animals for pollination, and thus ensures that seeds will form when conditions for cross-pollination are unfavorable, such as when pollinators or potential mates are scarce.

Double Fertilization Occurs in Flowering Plants

If a pollen grain lands on a compatible stigma, it absorbs moisture and germinates a pollen tube, which burrows through the stigma and style toward an ovule (see Figure 34.5). Chemical cues from the two synergid cells lying close to the egg cell help guide the pollen tube toward its destination. Before or during these events, the pollen grain's haploid sperm-producing cell divides by mitosis, forming two haploid sperm. When the pollen tube reaches the ovule, it enters through the micropyle and an opening forms in its tip. By this time one synergid has begun to die (an example of programmed cell death), and the two sperm are released into the disintegrating cell's cytoplasm. Experiments suggest that elements of the synergid's cytoskeleton guide the sperm cells onward, one to the egg cell and the other to the central cell.

Next there occurs a remarkable sequence of events called **double fertilization,** which has been observed only in flowering plants and (in a somewhat different version) in the gnetophyte *Ephedra* (see Section 27.4). Typically, one sperm nucleus fuses with the egg to form a diploid ($2n$) zygote. The other sperm nucleus fuses with the central cell, forming a cell with a triploid ($3n$) nucleus. Tissues derived from that $3n$ cell are called **endosperm** (*endon* = within; *sperma* = seed). The endosperm nourishes the embryo and, in monocots, the seedling, until its leaves form and photosynthesis has begun.

Embryo-nourishing endosperm forms only in flowering plants, and its evolution coincided with a reduction in the size of the female gametophyte. In other land plants, such as gymnosperms and ferns, the gametophyte itself contains enough stored food to nourish the embryonic sporophytes.

The Embryonic Sporophyte Develops inside a Seed

When the zygote first forms, it starts to develop and elongate even before mitosis begins. For example, in shepherd's purse *(Capsella),* shown in **Figure 34.9,** most of the organelles in the zygote, including the nucleus, become housed in the top half of the cell, while a vacuole takes up most of the lower half (see Figure 34.9, step 1). The first round of mitosis divides the zygote into an upper *apical cell* and a lower *basal cell.* The apical cell then gives rise to the multicellular embryo, while most descendants of the basal cell form a simple row of cells, the **suspensor,** which transfers nutrients from the parent plant to the embryo (step 2).

The first apical cell divisions produce a globe-shaped structure attached to the suspensor. As they continue to grow, embryos of *Capsella* and other eudicots become heart-shaped (step 3). Each lobe of the "heart" is a developing cotyledon (seed leaf), which will provide nutrients for growing tissues in a germinat-

FIGURE 34.9
Stages in the embryonic development of shepherd's purse *(Capsella bursa-pastoris)*, a eudicot. The micrographs are not to the same scale. Figure 34.17 shows more detail of the development of early plant embryos.

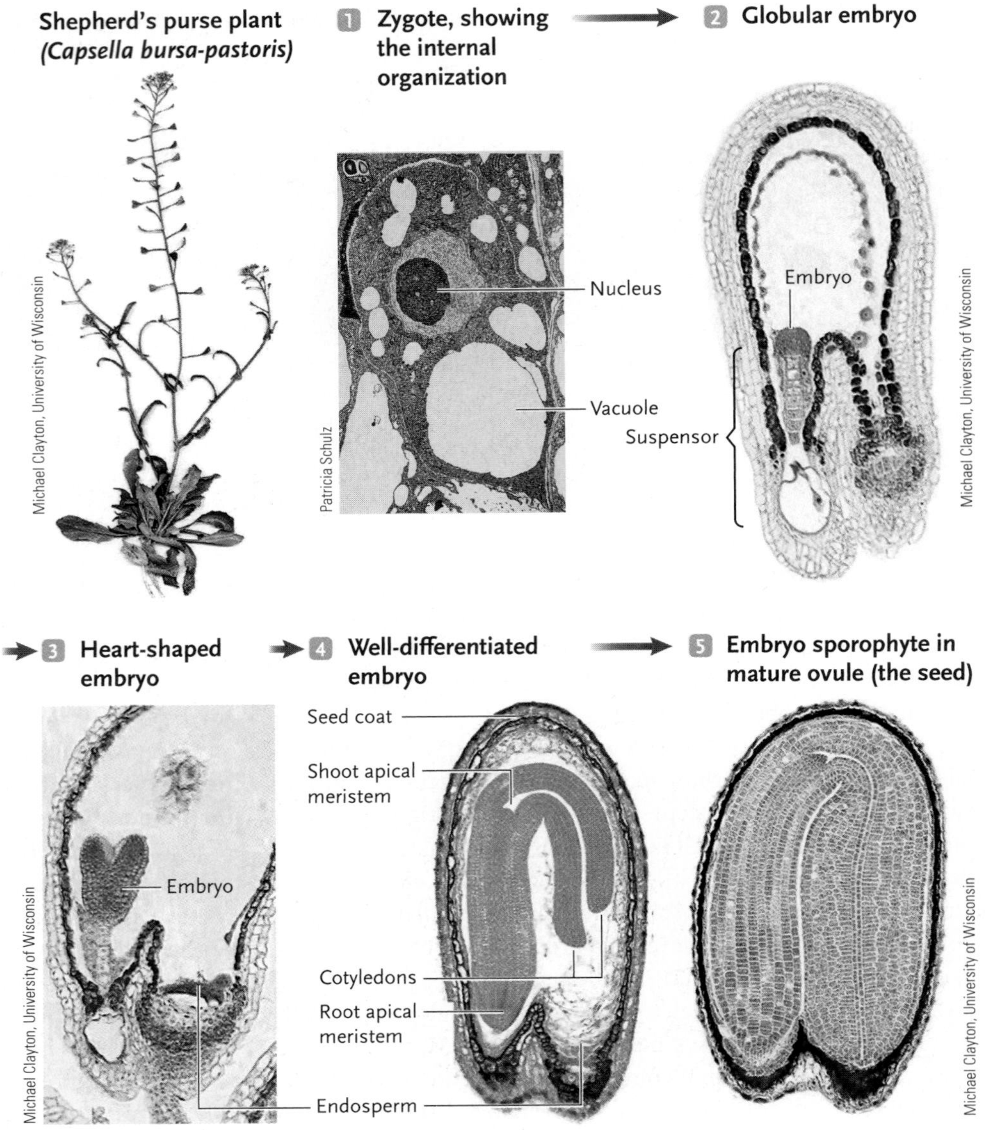

Shepherd's purse plant (Capsella bursa-pastoris)
Michael Clayton, University of Wisconsin
1 Zygote, showing the internal organization
Nucleus
Vacuole
Patricia Schulz
2 Globular embryo
Embryo
Suspensor
Michael Clayton, University of Wisconsin
3 Heart-shaped embryo
Embryo
Michael Clayton, University of Wisconsin
4 Well-differentiated embryo
Seed coat
Shoot apical meristem
Cotyledons
Root apical meristem
Endosperm
5 Embryo sporophyte in mature ovule (the seed)
Michael Clayton, University of Wisconsin

ing seedling. Typically, the two cotyledons absorb much of the nutrient-storing endosperm and become plump and fleshy. For instance, mature seeds of a sunflower *(Helianthus annuus)* have no endosperm at all. In some eudicots, however, the cotyledons remain as slender structures; they produce enzymes that digest the seed's ample endosperm and transfer the liberated nutrients to the seedling. Monocots have one, large cotyledon; in many monocot species, especially grasses such as corn and rice, the cotyledon absorbs the endosperm after germination, when the embryo inside the seed begins to grow.

A **seed** is a mature ovule. By the time the ovule is mature it has become encased by a protective **seed coat.** The embryo inside the seed has a lengthwise axis with a root apical meristem at one end and a shoot apical meristem at the other (steps 4 and 5).

Figure 34.10A and **Figure 34.10B** show the structural organization of seeds of two eudicots, the kidney bean *(Phaseolus vulgaris)* and the castor bean plant *(Ricinus communis).* The embryo makes up nearly all of both types of seeds. A mature kidney bean embryo has broad, fleshy cotyledons, and none of the seed's endosperm remains to nourish it. A castor bean embryo has much thinner cotyledons that are surrounded by endosperm, but in other ways the embryos are quite similar. The **radicle,** or embryonic root, is located near the micropyle, where the pollen tube entered the ovule prior to fertilization. The radicle attaches to the cotyledon at a region of cells called the **hypocotyl.** Beyond the hypocotyl is the **epicotyl,** which has the shoot apical meristem at its tip and which often bears a cluster of tiny foliage leaves, the **plumule.** At germination, when the root and shoot first elongate and emerge from the seed, the cotyledons are positioned at the first stem node with the epicotyl above them and the hypocotyl below them.

Deconstructing the seed of a monocot such as corn can be a bit more complex. To begin with, a corn kernel is not a seed but a type of fruit called a *grain.* It contains a seed, much of which is endosperm **(Figure 34.10C).** The embryo has only one very large cotyledon, a shield-shaped mass called a **scutellum,** which absorbs nutrients from the endosperm. In addition, protective tissues blanket the root and shoot apical meristems of monocot embryos. The shoot apical meristem and plumule are covered by a **coleoptile,** a sheath of cells that protects them during upward growth through the soil. A similar covering, the **coleorhiza,** sheathes the radicle until it breaks out of the seed coat and enters the soil as the young plant's primary root.

Fruits Protect Seeds and Aid Seed Dispersal

Most angiosperm seeds are housed inside fruits, which provide protection and often aid seed dispersal. A **fruit** is a matured or ripened ovary. Usually, a fruit begins to develop after pollination and the subsequent fertilization of an egg in a flower's ovule. The start of ovule growth after fertilization is called "fruit set." The fruit wall, called the **pericarp,** develops from the ovary wall and can have several layers. Hormones in pollen grains provide the initial stimulus that turns on the genetic machinery leading to fruit development; additional signals come from hormones produced by the developing seeds.

A. **Kidney bean *(Phaseolus vulgaris)***

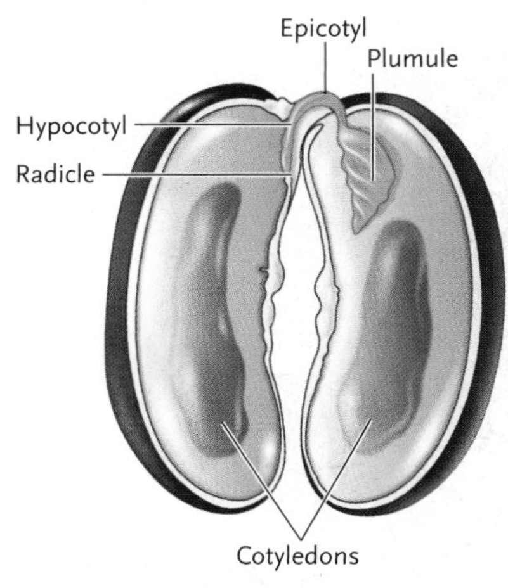
Epicotyl
Plumule
Hypocotyl
Radicle
Cotyledons

B. **Castor bean *(Ricinus communis)***

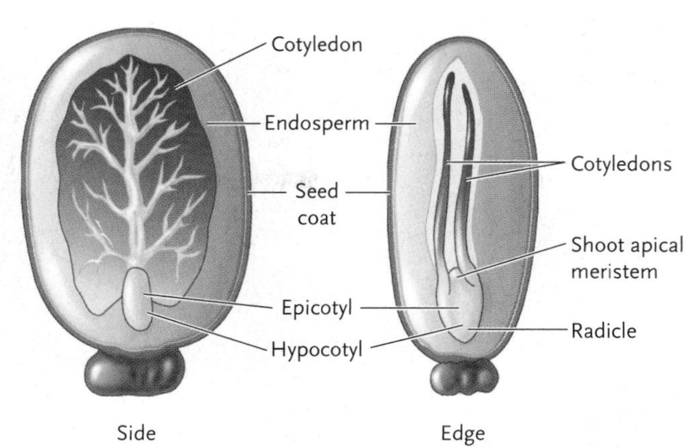
Cotyledon
Endosperm
Seed coat
Epicotyl
Hypocotyl
Cotyledons
Shoot apical meristem
Radicle
Side
Edge

C. **Corn *(Zea mays)***

Seed coat (fused with pericarp)
Endosperm in seed
Scutellum (cotyledon)
Coleoptile
Plumule
Shoot apical meristem
Radicle (root apical meristem)
Coleorhiza
Embryo

FIGURE 34.10
The structure of eudicot and monocot seeds. Eudicot seeds have two cotyledons, which store food absorbed from the endosperm, but the timing of this function varies in different species. **(A)** The cotyledons of a kidney bean *(Phaseolus vulgaris)* take up nutrients from endosperm while the seed develops, becoming fleshy. **(B)** In the castor bean *(Ricinus communis)*, the endosperm is thick and the cotyledons are thin until the seed germinates, when the cotyledons begin to take up endosperm nutrients. **(C)** A kernel of corn *(Zea mays)* contains a representative monocot seed, which has been isolated here and is shown in longitudinal section. Monocot seeds have a single cotyledon, which develops into a shield-shaped scutellum that absorbs nutrients from endosperm.

Fruits are extremely diverse, and biologists classify them into types based on combinations of structural features. A major defining feature is the nature of the pericarp, which may be fleshy (as in peaches) or dry (as in a hazelnut). A fruit also is classified according to the number of ovaries or flowers from which it develops. **Simple fruits** develop from a single ovary. In many of them, such as peaches, tomatoes, and the cacao fruits pictured in Figure 34.1, at least one layer of the pericarp is fleshy and juicy. Other simple fruits, including grains and nuts, have a thin, dry pericarp, which may be fused to the seed coat (like it is in corn). The garden pea *(Pisum sativa)* is a simple fruit, the peas being the seeds and the surrounding shell the pericarp. **Aggregate fruits** are formed from several ovaries in a single flower. Examples are raspberries and strawberries, which develop from clusters of individual ovaries. Strawberries also qualify as *accessory* fruits, in which floral parts in addition to each ovary become incorporated as the fruit develops. For instance, anatomically, the fleshy part of a strawberry is an expanded receptacle (the end of the floral shoot) and the strawberry fruits are the tiny, dry nubbins (called *achenes*) you see embedded in the fleshy tissue of each berry. *Multiple fruits* develop from several ovaries in multiple flowers. For example, a pineapple is a multiple fruit that develops from the enlarged ovaries of several flowers clustered together in an inflorescence. **Figure 34.11** shows examples of some different types of fruits.

Fruits not only protect seeds—they also aid seed dispersal in specific environments. For example, maple fruits have winglike extensions for dispersal (see Figure 34.11E). When the fruit drops, the wings cause it to spin sideways and downward and also can carry it away on a breeze. This aerodynamic property propels maple seeds to new locations, where they will not have to compete with the parent tree for water and minerals. Fruits also may have hooks, spines, hairs, or sticky surfaces, and they are ferried to new locations when they adhere to feathers, fur, or blue jeans of animals that brush against them. Fleshy fruits such as blueberries are nutritious food for many animals, and their seeds are adapted for surviving digestive enzymes in the animal gut. The enzymes remove just enough of the hard seed coats to increase the chance of successful germination when the seeds are expelled from the animal's body in feces.

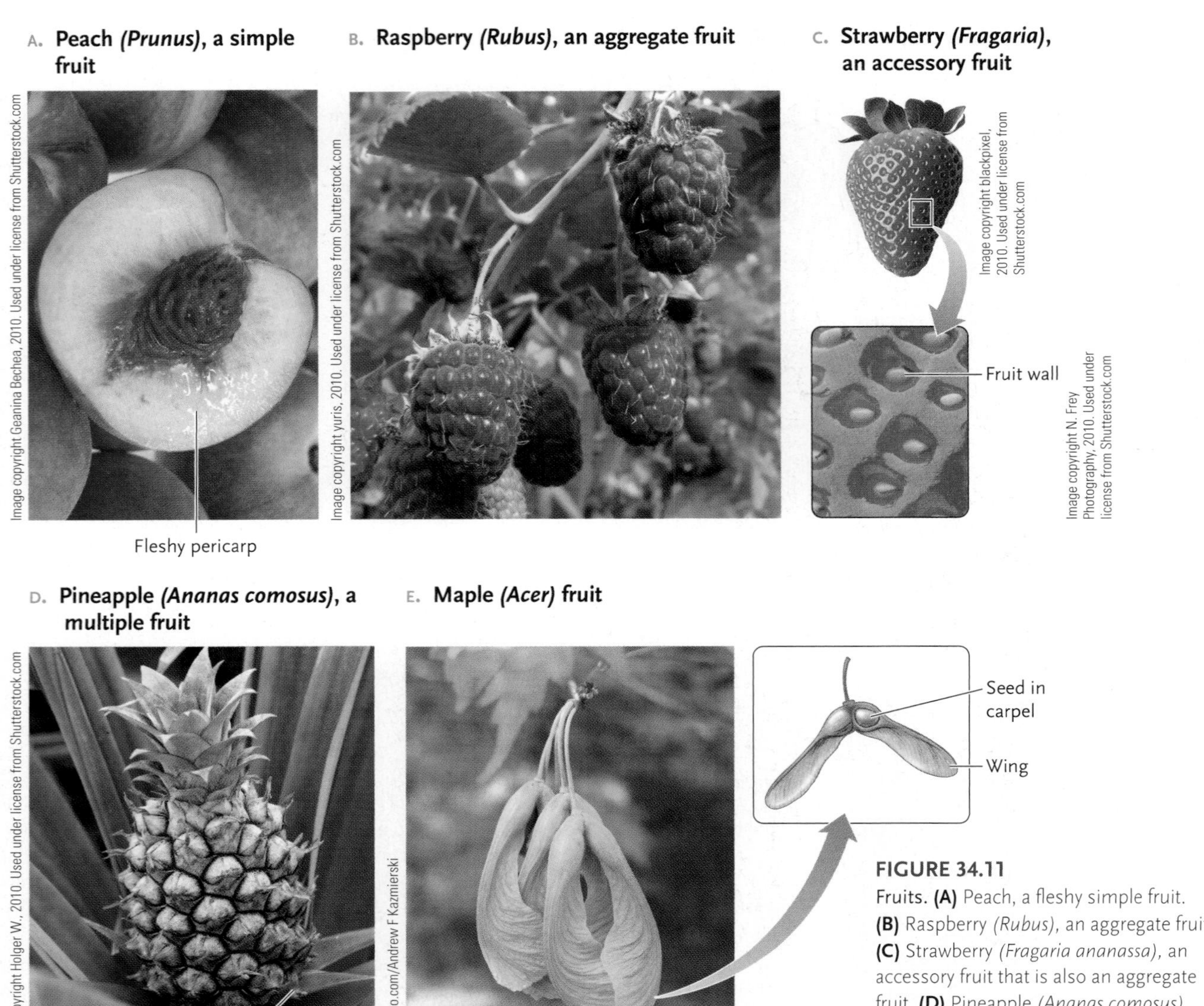
A. Peach (Prunus), a simple fruit
B. Raspberry (Rubus), an aggregate fruit
C. Strawberry (Fragaria), an accessory fruit
Fruit wall
Fleshy pericarp
D. Pineapple (Ananas comosus), a multiple fruit
E. Maple (Acer) fruit
Seed in carpel
Wing
Flower
Image copyright Geanina Bechea, 2010. Used under license from Shutterstock.com
Image copyright yuris, 2010. Used under license from Shutterstock.com
Image copyright blackpixel, 2010. Used under license from Shutterstock.com
Image copyright N. Frey Photography, 2010. Used under license from Shutterstock.com
Image copyright Holger W., 2010. Used under license from Shutterstock.com
iStockphoto.com/Andrew F Kazmierski

FIGURE 34.11
Fruits. **(A)** Peach, a fleshy simple fruit. **(B)** Raspberry *(Rubus)*, an aggregate fruit. **(C)** Strawberry *(Fragaria ananassa)*, an accessory fruit that is also an aggregate fruit. **(D)** Pineapple *(Ananas comosus)*, a multiple fruit. **(E)** Winged fruits of maple *(Acer)*.

Seed Germination Continues the Life Cycle

A mature seed is essentially dehydrated. On average, water accounts for only about 10% of its weight—too little for cell expansion or metabolism. Germination gets underway when the seed begins to soak up water. The hydrated embryo then resumes growth. Ideally, a seed germinates when external conditions favor the survival of the embryo and growth of the new sporophyte. This timing is important, because once germination is underway the embryo loses the protection of the seed coat and other structures that surround it. Overall, the amount of soil moisture and oxygen, the temperature, day length, and other environmental factors influence when germination takes place.

In some species, the life cycle may include a period of seed **dormancy** (*dormire* = to sleep), in which the seed's metabolic activity is suspended. The conditions required for dormant seeds to germinate vary widely. Depending on the species, seeds may require minimum periods of daylight or darkness, repeated soaking, mechanical abrasion, or exposure to certain enzymes, the high heat of a fire, or a freeze–thaw cycle before they finally break dormancy. A control system based on the hormone abscisic acid, or ABA, underlies the shift from active metabolism to dormancy, and then reactivation and germination. When environmental conditions—perhaps too-cold temperature or drought—do not favor the survival of a potential seedling, seeds produce high levels of ABA. The hormone in turn stimulates synthesis of a transcription factor that turns on germination-inhibiting genes. When conditions eventually favor a seedling's survival, the genes switch off and dormancy ends. Kenji Miura of the University of Tsukuba in Japan and Mike Hasegawa at Purdue University recently discovered that this reversal occurs when a regulatory protein binds the transcription factor. Then, the germination-inhibiting genes are no longer expressed and germination can begin.

In some desert plants, hormones in the seed coat inhibit growth of a seedling until heavy rains flush them away. This adaptation ensures that seeds germinate only when there is enough water in the soil to support growth of the plant through the flow-

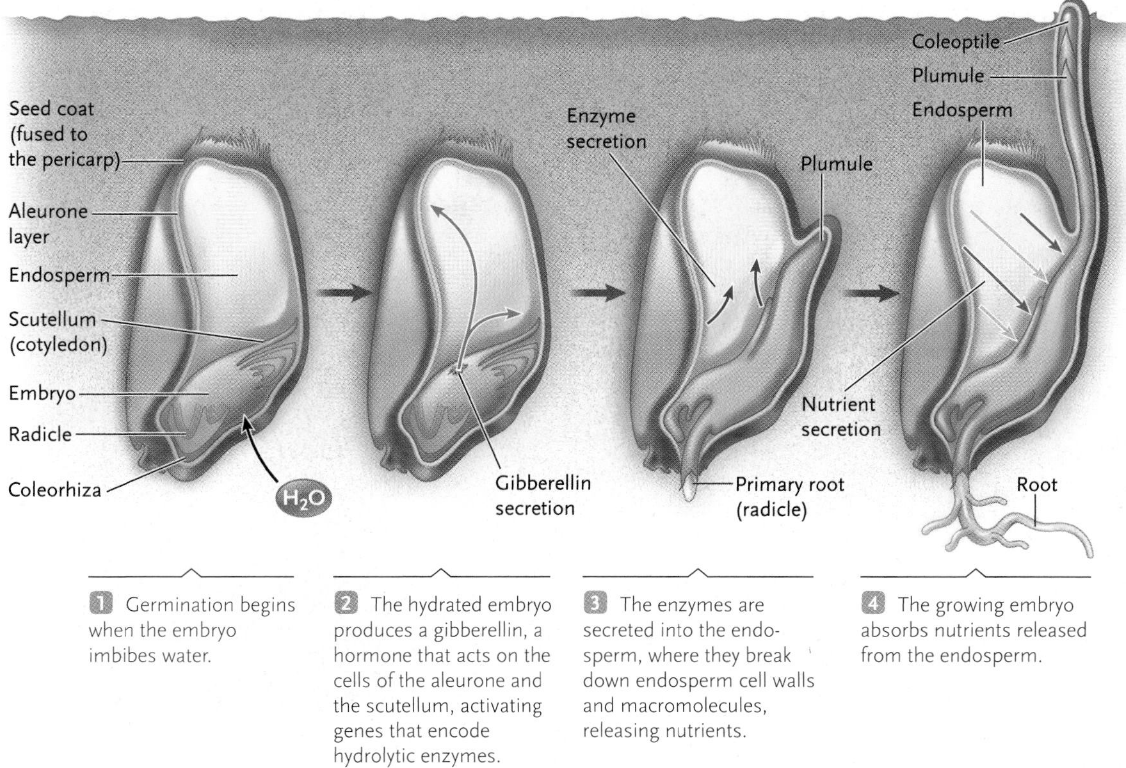
Seed coat (fused to the pericarp)
Aleurone layer
Endosperm
Scutellum (cotyledon)
Embryo
Radicle
Coleorhiza
H_2O
Enzyme secretion
Gibberellin secretion
Plumule
Nutrient secretion
Primary root (radicle)
Coleoptile
Plumule
Endosperm
Root

1 Germination begins when the embryo imbibes water.

2 The hydrated embryo produces a gibberellin, a hormone that acts on the cells of the aleurone and the scutellum, activating genes that encode hydrolytic enzymes.

3 The enzymes are secreted into the endosperm, where they break down endosperm cell walls and macromolecules, releasing nutrients.

4 The growing embryo absorbs nutrients released from the endosperm.

FIGURE 34.12
How food reserves are mobilized in a germinated seed within a grain of barley *(Hordeum vulgare)*, a monocot. Each grain is a fruit containing a single large seed.

ering and seed production stages before the soil dries again. Many plants in harsh environments, such as deserts, cycle from germination to growth, flowering, and seed development in the space of a few weeks and their offspring remain dormant as seeds until conditions once again favor germination and growth.

Seeds of some species appear to remain viable for amazing lengths of time. Thousand-year-old lotus seeds *(Nelumbo lutea)* discovered in a dry lakebed have germinated trouble-free. And in one startling case, seeds of arctic lupine *(Lupinus arcticus)* were discovered along with fecal material in 10,000-year-old lemming burrows. When the seeds were thawed, they germinated readily as well.

Germination begins with **imbibition,** in which water molecules move into the seed, attracted to hydrophilic groups of stored proteins. As water enters, the seed swells, the coat ruptures, and the radicle begins its downward growth into the soil. Within this general framework, however, there are many variations among plants.

Once the seed coat splits, water and oxygen move more easily into the seed. Metabolism switches into high gear as cells divide and elongate to produce the seedling. Stable enzymes that were synthesized before dormancy become active; other enzymes are produced as the genes encoding them begin to be expressed. Among other roles, the increased gene activity and enzyme production mobilize the seed's food reserves in cotyledons or endosperm. Nutrients released by the enzymes sustain the rapidly developing seedling until its root and shoot systems are established.

The events of seed germination have been studied extensively in grains, and **Figure 34.12** illustrates them in barley. The seed's endosperm is separated from the pericarp by a thin layer of cells called the **aleurone.** As a hydrating seed imbibes water, the embryo produces a *gibberellin,* a hormone that stimulates aleurone cells to manufacture and secrete hydrolytic enzymes. Some of these enzymes digest components of endosperm cell walls; others digest proteins, nucleic acids, and starch of the endosperm, releasing nutrient molecules for use by cells of the young root and shoot. Although it is clear that nutrient reserves are also mobilized by metabolic activity in eudicots and in gymnosperms, the details of the process are not well understood.

Inside a germinating seed, embryonic root cells are generally the first to divide and elongate, giving rise to the radicle. When the radicle emerges from the seed coat as the primary root, germination is complete. **Figure 34.13** and **Figure 34.14** depict the stages of early development in a kidney bean, a eudicot, and in corn, a monocot. As the young plant grows, its development continues to be influenced by interactions of hormones and environmental factors, as you will read in Chapter 35.

Most plants give rise to large numbers of seeds because, in the wild, only a tiny fraction of seeds survive, germinate, and eventually grow into another mature plant. Also, flowers, seeds, and fruits represent major investments of plant resources. Asexual reproduction, discussed next, is a more "economical" means by which many plants can propagate themselves.

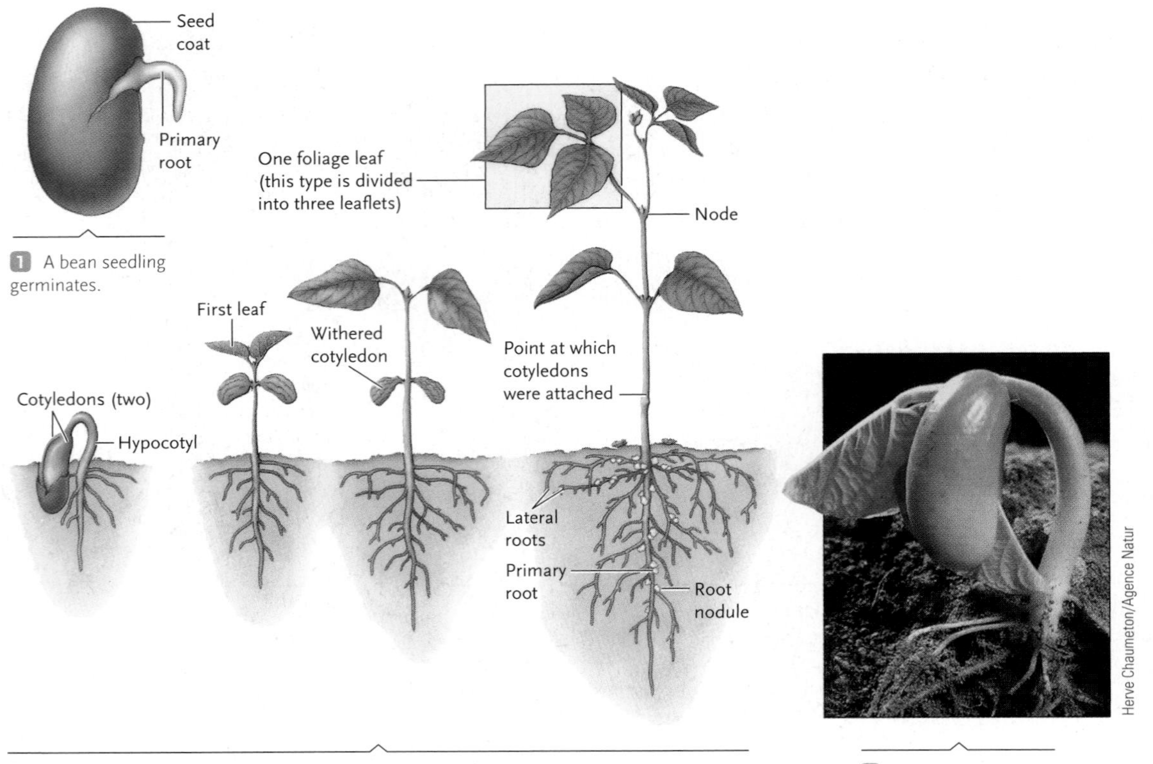
Seed coat
Primary root
One foliage leaf (this type is divided into three leaflets)
Node
First leaf
Withered cotyledon
Point at which cotyledons were attached
Cotyledons (two)
Hypocotyl
Lateral roots
Primary root
Root nodule
Herve Chaumeton/Agence Natur

1 A bean seedling germinates.

2 Food-storing cotyledons are lifted above the soil surface when cells of the hypocotyl elongate. The hypocotyl becomes hook-shaped and forces a channel through the soil as it grows. At the soil surface, the hook straightens in response to light. For several days, cells of the cotyledons carry out photosynthesis; then the cotyledons wither and drop off. Photosynthesis is taken over by the first leaves that develop along the stem and later by foliage leaves.

3 Leaves break through the seed coat.

FIGURE 34.13
Stages in the development of a representative eudicot, the kidney bean *(Phaseolus vulgaris)*.

FIGURE 34.14
Stages in the development of a representative monocot, the corn plant *(Zea mays)*.

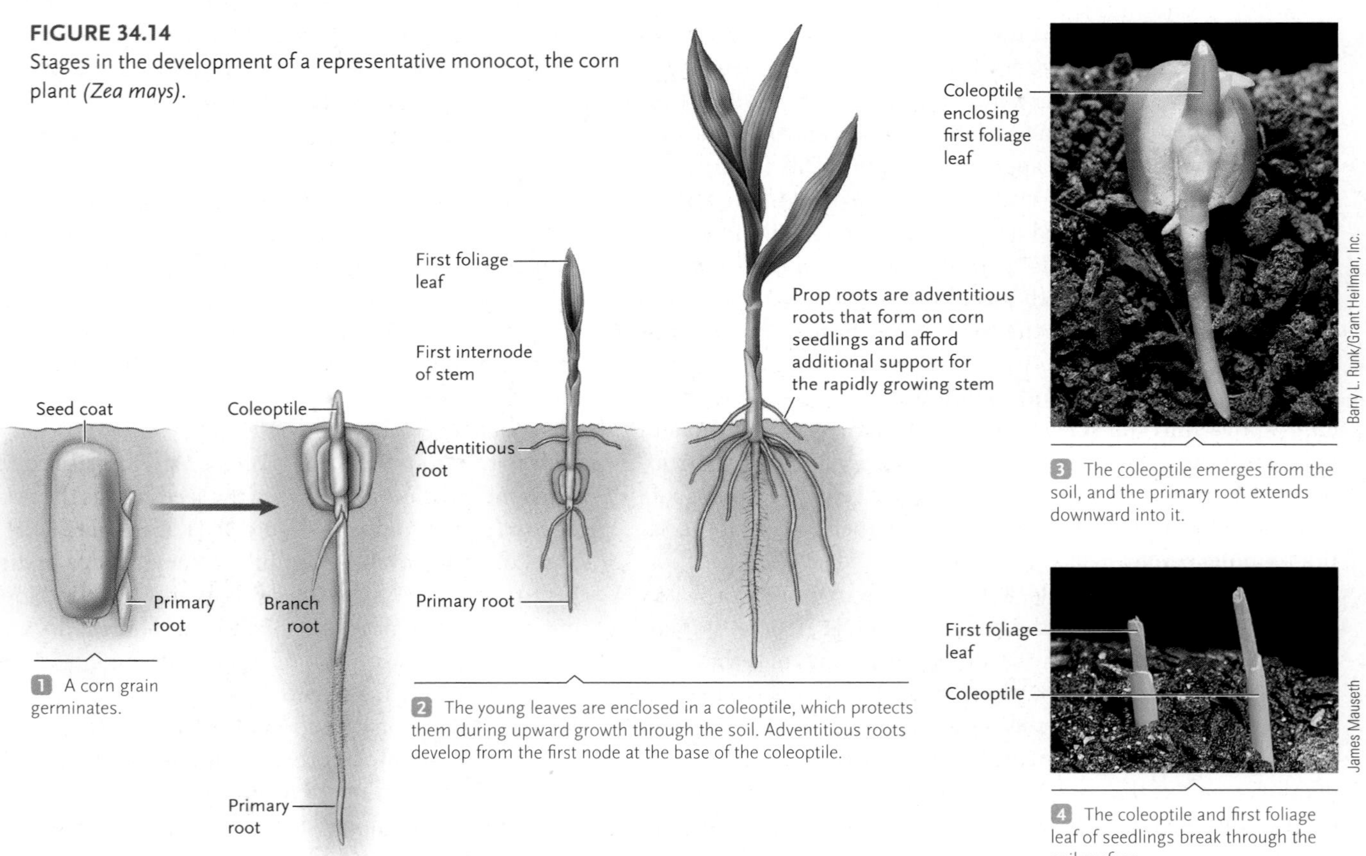
Coleoptile enclosing first foliage leaf
First foliage leaf
First internode of stem
Prop roots are adventitious roots that form on corn seedlings and afford additional support for the rapidly growing stem
Seed coat
Coleoptile
Adventitious root
Primary root
Branch root
Primary root
Primary root
First foliage leaf
Coleoptile
Barry L. Runk/Grant Heilman, Inc.
James Mauseth

1 A corn grain germinates.

2 The young leaves are enclosed in a coleoptile, which protects them during upward growth through the soil. Adventitious roots develop from the first node at the base of the coleoptile.

3 The coleoptile emerges from the soil, and the primary root extends downward into it.

4 The coleoptile and first foliage leaf of seedlings break through the soil surface.

STUDY BREAK 34.3 <

1. Explain the sequence of events in a flowering plant that begins with formation of a pollen tube and culminates with the formation of a diploid zygote and the $3n$ cell that will give rise to endosperm in a seed.
2. Early angiosperm embryos undergo a series of general changes as a seed matures. Summarize this sequence, then describe the structural differences that develop in the seeds of monocots and eudicots.
3. Germination begins when a seed imbibes water. What are the next key biochemical and developmental events that bring an angiosperm's life cycle full circle?

>

THINK OUTSIDE THE BOOK

For many perennial plant species in temperate and high-latitude regions, dormancy is a major adaptation for surviving low winter temperatures. However, cold winters do not occur at tropical latitudes. Using library or Internet research sources, find out if there are environmental conditions that trigger automatic dormancy in wild-growing tropical plants. What implications do your findings have for gardeners in your area who want to grow tropical species year round?

34.4 Asexual Reproduction of Flowering Plants

As we mentioned in Section 34.1, many flowering plants have the capacity for both sexual and asexual reproduction. Some flowering plants, including some citrus species and the grass variety known as Kentucky blue grass *(Poa pratensis),* can reproduce asexually through a mechanism called **apomixis,** in which there is no union of haploid maternal and paternal gametes formed by meiosis. Typically, an embryo develops from an unfertilized egg or from diploid cells in the ovule tissue around the embryo sac. The resulting seed is said to contain a **somatic embryo,** which is genetically identical to the parent.

Another asexual alternative is **vegetative reproduction,** in which a new, genetically identical individual develops from a parent plant's non-reproductive tissue, usually a bit of meristematic tissue in a bud on the root or stem. Some of the stem modifications described in Section 31.3 function in this way. For instance, although strawberries can reproduce sexually by way of seeds, new roots and shoots also develop at each node along the stolons ("runners") a strawberry plant sends out. Likewise, new plantlets develop from "suckers" that sprout from the roots of blackberry bushes and "eyes" in the tubers of potatoes, or from buds on the short underground stems of onions and lilies.

Vegetative reproduction relies on an intriguing property of plants—namely, that many fully differentiated plant cells are **totipotent** (*totus* = all; *potens* = powerful). That is, they have the genetic potential to develop into a whole, fully functional plant. Under appropriate conditions, a totipotent cell can *dedifferentiate;* it returns to an unspecialized embryonic state, and the genetic program that guides the development of a new individual is turned on.

Vegetative Reproduction Is Common in Nature

Various plant species have developed different mechanisms for vegetative reproduction. In **fragmentation,** cells in a piece of the parent plant dedifferentiate and then can regenerate missing plant parts. Many gardeners have discovered to their frustration that a chunk of dandelion root left in the soil can rapidly grow into a new dandelion plant in this way. When a leaf falls or is torn away from a jade plant (*Crassula* species), a new plant can develop from meristematic tissue that is adjacent to the wound surface in the detached leaf. The "mother of thousands" plant, *Kalanchoe daigremontiana,* produces a wealth of seeds, while more or less simultaneously meristematic tissue in notches along the leaf margin gives rise to tiny plantlets **(Figure 34.15)** that eventually fall to the ground, where they can sprout roots and grow to maturity.

In wild plant species, most types of asexual reproduction result in offspring located near the parent. As already noted, these clonal populations lack the variability provided by sexual reproduction, variation that enhances the odds for survival when environmental conditions change. Yet asexual reproduction offers an advantage in some situations. It usually requires less energy than producing complex reproductive structures such as seeds and showy flowers to attract pollinators. Moreover, clones are likely to be well suited to the environment in which the parent grows.

Many Commercial Growers and Gardeners Use Artificial Vegetative Reproduction

For centuries, gardeners and farmers have used asexual plant propagation to grow particular crops and trees and some ornamental plants. They routinely use *cuttings,* pieces of stems or leaves, to generate new plants. Placed in water or moist soil, a cut-

Ed Reschke/Peter Arnold

FIGURE 34.15
Kalanchoe daigremontiana, the mother of thousands plant. Each tiny plant growing from the leaf margin can become a new, independent adult plant.

ting may sprout roots within days or a few weeks. Virtually the entire world supply of commercially sold bananas is harvested from plants grown from cuttings of a single variety called the "Cavendish banana." Because the tens of millions of plants that produce this banana are genetically uniform, the plants lack genetic diversity that would provide some measure of adaptive insurance against harmful fungi and other pests. As a result, by some estimates bananas are the most heavily pesticide-treated fruit in the global marketplace.

Trees and wine grapes often are propagated by grafting a bud or branch from a plant with desirable fruit traits—the *scion*—and joining it to a root or stem from a plant with useful root traits—the *stock*. A grafted plant usually produces flowers and fruit identical to those of the scion's parent plant. The scion of a grafted wine grape variety may be chosen for the quality of its fruit, and the stock for its hardy, disease-resistant root system. Vegetative propagation can also be used to grow plants from single cells. Rose bushes and fruit trees from nurseries and commercially important fruits and vegetables such as Bartlett pears, McIntosh apples, Thompson seedless grapes, and asparagus come from plants produced vegetatively in tissue culture conditions that cause their cells to dedifferentiate to an embryonic stage.

VEGETATIVE PROPAGATION IN TISSUE CULTURE In groundbreaking experiments in the 1950s, Frederick C. Steward explored the totipotency of plant cells. Together with his coworkers at Cornell University, Steward propagated whole carrot plants in the laboratory by culturing carrot root phloem. Later researchers confirmed that almost any plant cell that has a nucleus and lacks a secondary cell wall may be totipotent.

The method of plant tissue culture Steward pioneered is simple in its general outlines **(Figure 34.16).** Bits of tissue are excised from a plant and grown in a nutrient medium. The procedure disrupts normal interactions between cells in the tissue sample, and the cells dedifferentiate and form an unorganized, white cell mass called a **callus.** When cultured with nutrients and growth hormones, some cells of the callus regain totipotency and develop into plantlets with roots and shoots.

Steward's work laid the foundation for large-scale commercial applications of plant tissue culture, as well as for a whole new field of research on *somatic embryogenesis* in plants. Single cells derived from a callus generated from shoot meristem are placed in a medium containing nutrients and hormones that promote cell differentiation. With some species, totipotent cells in the sample eventually give rise to diploid somatic embryos that can be packaged with nutrients and hormones in artificial "seeds" (see Figure 34.16). Endowed with the same traits as their parent, crop plants grown from somatic embryos are genetically uniform.

However, mutations often occur in the DNA of somatic embryos derived from callus culture. Screening techniques can identify such *somaclonal* mutants with desirable traits—for example, resistance to a disease that attacks wild-type plants of the same species. In plants that are infected with viruses, callus cultures can be restricted to virus-free cells and thus generate virus-free clones. Tissue culture propagation can then produce hundreds or thousands of identical plants from a single specimen. This technique, called **somaclonal selection,** is now a staple tool in efforts to improve major food crops, such as corn, wheat, rice, and soybeans. It is also being used to rapidly increase stocks of hybrid orchids, lilies, and other valued ornamental plants. The yellow and orange tomatoes that have become common in produce markets are the fruits of plants developed by somaclonal selection.

Research on tobacco and some other species has shown that plants can also be regenerated by **protoplast fusion.** In this method, the walls of living cells in solution are first digested away by enzymes, leaving the protoplasts. Then the protoplasts are induced to fuse, either by applying an electric current, a laser beam, or chemical additives to the solution, which briefly "loosen" the plasma membranes. The resulting cell (now $4n$, or tetraploid) is transferred to a solid nutrient medium and allowed to develop into a callus, then individual callus cells are stimulated to develop into embryos. If the fused protoplasts come from somatic cells of a single species, the embryos often grow into fertile plants. It has proven more difficult to grow fertile hybrids from fused protoplasts of different species and genera, probably because there are species-specific signals that govern key physiological functions. Even so, this method has produced the pomato, a cross between a potato and a tomato.

Regardless of how it comes into being, an embryonic sporophyte changes significantly as it begins the developmental journey toward maturity, when it will be capable of reproducing. Next we explore what researchers are learning about these developmental changes.

STUDY BREAK 34.4 <

1. **Describe three modes of asexual reproduction that occur in flowering plants.**
2. **What is totipotency, and how do methods of tissue culture exploit this property of plant cells?**

34.5 Early Development of Plant Form and Function

Unlike animals, plants have specialized body parts such as leaves and flowers that may arise from meristems throughout an individual's life—sometimes for thousands of years. Accordingly, in plants the biological role of embryonic development is not to generate the tissues and organs of the adult, but to establish a basic body plan—the root–shoot axis and the radial, "outside-to-inside" organization of epidermal, ground, and vascular tissues (see Section 31.1)—and the precursors of the primary meristems. Though they may sound simple, these fundamentals and the stages beyond them all depend on an intricately orchestrated sequence of molecular events that plant scientists are exploring through sophisticated experimentation.

One of the most fruitful approaches has been the study of plants with natural or induced gene mutations that block or

FIGURE 34.16 **Research Method**

Plant Cell Culture

Purpose: To determine if differentiated plant cells can be returned to a totipotent state.

Protocol: Fragments of tissue are excised from a plant and grown in a medium that contains all required nutrients. The procedure disrupts normal interactions among the excised cells. They dedifferentiate, forming an unorganized callus. When cells removed from the callus are grown in a medium containing nutrients and hormones, some develop into plantlets.

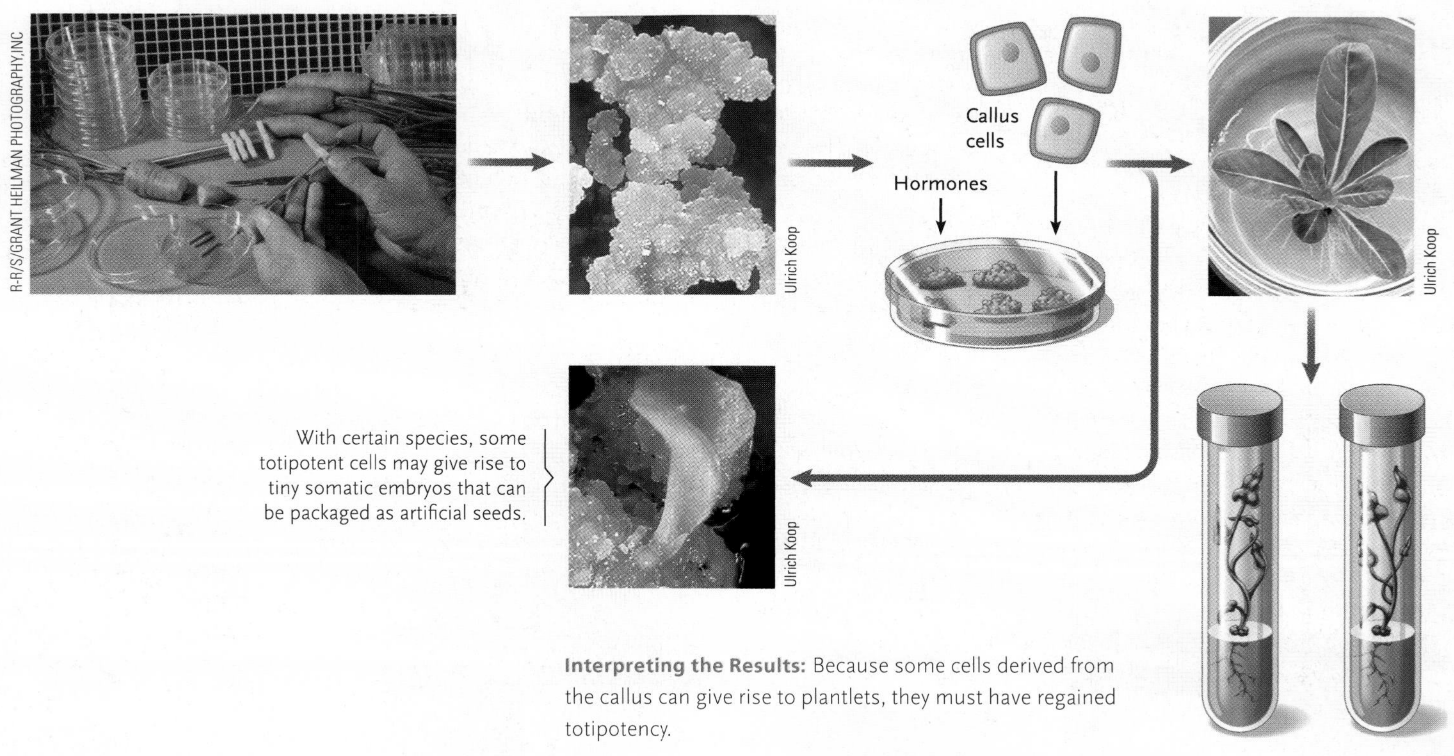
R-R/S/GRANT HEILMAN PHOTOGRAPHY,INC
Ulrich Koop
Callus cells
Hormones
Ulrich Koop
With certain species, some totipotent cells may give rise to tiny somatic embryos that can be packaged as artificial seeds.
Ulrich Koop

Interpreting the Results: Because some cells derived from the callus can give rise to plantlets, they must have regained totipotency.

otherwise affect steps in development—and accordingly give researchers insight into the developmental roles of the normal, wild-type versions of those abnormal genes. Some of these genes are **homeotic genes,** regulatory genes in the genome of an organism that encode transcription factors. The transcription factors are proteins that control the expression of other genes, which in turn direct events in development (see *Focus on Research* in Chapter 31). While researchers work with various species to probe the genetic underpinnings of early plant development, the thale cress *(Arabidopsis thaliana)* has become a favorite model organism for plant genetic research (see *Focus on Basic Research*).

Within Hours, an Early Plant Embryo's Basic Body Plan Is Established

The entire *A. thaliana* genome has been sequenced, providing a powerful molecular "database" for determining how various genes contribute to shaping the plant body. Experimenters' ability to trace the expression of specific genes has shed considerable light on how the root–shoot axis is set and how the three basic plant tissue systems arise.

THE ROOT–SHOOT AXIS Shortly after fertilization gives rise to an *A. thaliana* zygote, the single cell divides. Electron microscopy shows that, as with the *Capsella* zygote described earlier, this first round of mitosis produces a small apical cell and a larger basal cell **(Figure 34.17A).** The apical cell receives the lion's share of the cytoplasm, while the basal cell receives the zygote's large vacuole and less cytoplasm. Researchers have confirmed that in this asymmetrical division of the zygote, the daughter cells receive different mixes of mRNAs—the gene transcripts that will be translated into proteins.

Translation of differing mRNAs produces proteins that include several transcription factors. As these factors trigger the expression of differing genes, distinct biochemical pathways unfold in the two cells in ways that establish the organization of the plant body. For example, a basal cell initially exports a signaling molecule (a hormone of the auxin family, discussed in Chapter 35) to the apical cell, and this sets in motion steps leading to the development of the various embryonic shoot features (Figure 34.17B).

FOCUS ON **Basic Research**

Model Research Organisms: *Arabidopsis thaliana*

For plant geneticists, the little white-flowered thale cress, *Arabidopsis thaliana,* has attributes that make it a prime subject for genetic research. A tiny member of the mustard family Brassicaceae, *Arabidopsis thaliana,* is revealing answers to some of the biggest questions in plant development and physiology.

Each plant grows only a few centimeters tall, so little laboratory space is required to house a large population. As long as *A. thaliana* is provided with damp soil containing basic nutrients, it grows easily and rapidly in artificial light. Like Mendel's peas, *A. thaliana* is self-compatible and self-fertilizing, and the flowers of a single plant can yield thousands of seeds per mating. Seeds grow to mature plants in just over a month and then flower and reproduce themselves in another 3 to 4 weeks. This permits investigators to perform desired genetic crosses and obtain large numbers of offspring having known, desired genotypes with relative ease. Individual *A. thaliana* cells also grow well in culture.

Arabidopsis thaliana

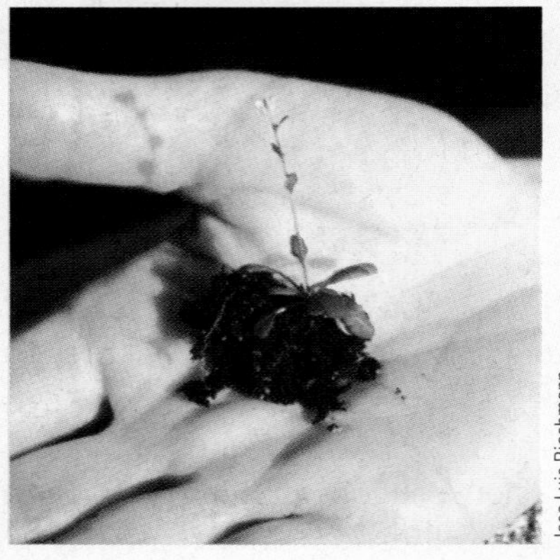

Jose Luis Riechmann

The *A. thaliana* genome was the first complete plant genome to be sequenced; at this writing researchers have identified approximately 28,000 genes arranged on five pairs of chromosomes. The genome contains relatively little repetitive DNA, so it is fairly easy to isolate *A. thaliana* genes, which can then be cloned using genetic engineering techniques. Cloned genes are inserted into bacterial plasmids and the recombinant plasmids transferred to the bacterial species *Agrobacterium tumefaciens,* which readily infects *A. thaliana* cells. Amplified by the bacteria, the genes and their protein products can be sequenced or studied in other ways.

Typically, researchers use chemical mutagens or recombinant bacteria to introduce changes in the *A. thaliana* genome. These mutants have become powerful tools for exploring molecular and cellular mechanisms that operate in plant development—for example, elucidation of the homeotic genes responsible for flower development described in this chapter. *A. thaliana* mutants are also being used to probe fundamental questions such as how plant cells respond to gravity, and what roles the pigments called phytochromes play in plant responses to light.

Several years ago an ambitious, multinational research effort was launched to determine the functions of all *A. thaliana* genes. This work is providing a comprehensive genetic portrait of a flowering plant—how each gene affects the functioning of not only individual cells but of the plant as a whole. The Arabidopsis Information Resource (TAIR) tallies the percentages of *A. thaliana* genes in different functional categories **(Figure 1).**

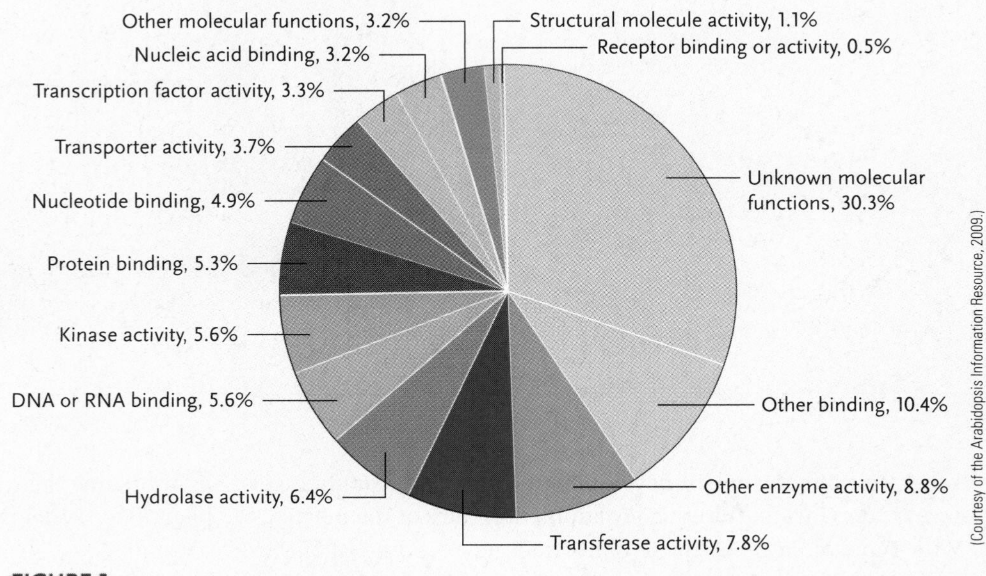
Other molecular functions, 3.2%
Nucleic acid binding, 3.2%
Transcription factor activity, 3.3%
Transporter activity, 3.7%
Nucleotide binding, 4.9%
Protein binding, 5.3%
Kinase activity, 5.6%
DNA or RNA binding, 5.6%
Hydrolase activity, 6.4%
Structural molecule activity, 1.1%
Receptor binding or activity, 0.5%
Unknown molecular functions, 30.3%
Other binding, 10.4%
Other enzyme activity, 8.8%
Transferase activity, 7.8%
(Courtesy of the Arabidopsis Information Resource, 2009.)

FIGURE 1
The percentages of *A. thaliana* genes that influence different functional categories.

Later, gene expression and the flow of chemical signals shift in ways that promote the development of specific structures in the basal cell, including portions of the root apical meristem.

Several of the genes that influence root–shoot polarity have been identified, and when any of them is disrupted, the result can be a serious defect. For example, when an embryo receives two copies of a mutant gene called *gnom,* the embryo doesn't develop distinct root and shoot regions. Instead it remains a lumpy blob.

RADIAL ORGANIZATION OF TISSUE LAYERS A day or so after an *A. thaliana* egg cell is fertilized, the embryo consists of eight cells. Even at this early stage both the root–shoot axis and the beginnings of tissue systems are present. When an embryo reaches the so-called torpedo stage, cells representing all three basic tissue systems are in place **(Figure 34.17C).** Again working with mutant plants, investigators have identified several *A. thaliana* genes that help govern early development of tissue systems. For example, a gene called *SCR* encodes a protein that apparently regulates mitotic divisions that produce the first cells of a developing root's cortex and endoderm tissue layers (Figure 31.20 shows the locations of these tissues in a mature root). The roots of a mutant *scr* seedling contain cells with jumbled characteristics of both tissue layers.

A. **Developing embryos**

Apical cell

Basal cell

25 μm

Kelly Yee and John J. Harada

B. **Seedling polarity**

Seedlings with normal polarity

Damien Lovegrove/SPL/Photo Researchers, Inc.

C. **Beginnings of plant tissue systems**

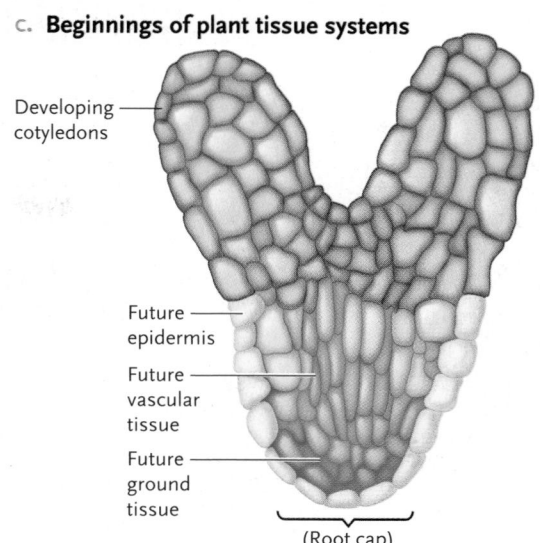

Developing cotyledons

Future epidermis

Future vascular tissue

Future ground tissue

(Root cap)

FIGURE 34.17
Stages in the development of the basic body plan of a plant embryo. (A) After a zygote forms, the first round of cell division produces an embryo with an apical cell that contains much of the zygote's cytoplasm, and a larger basal cell that receives the vacuole and less cytoplasm. The division allots different transcription factors to each cell and establishes the plant's root–shoot axis. **(B)** Normal *A. thaliana* seedlings, in which the root–shoot polarity has become established. **(C)** The approximate locations of early embryonic cells that are the forerunners of epidermal, ground, and vascular tissue systems, respectively.

No matter what the species, nearly all new plant embryos have the general body plan we have been discussing. As development proceeds, cells at different sites become specialized in prescribed ways as a particular set of genes is expressed in each type of cell—a process known as *differentiation*. Differentiated cells in turn are the foundation of specialized tissues and organs, which come about through processes we consider next.

Key Developmental Cues Are Based on a Cell's Position

Although many of the specifics of development differ in animals and plants, one fundamental holds true for both: Normal development produces ordered spatial arrangements of differentiated tissues. Examples in plants include root and shoot apical meristems at opposite ends of the root–shoot axis, the cotyledons that divide the shoot into an upper epicotyl and a lower hypocotyl, and the nested layers of vascular, ground, and epidermal tissue systems. Developmental biologists call this progressive ordering of parts **pattern formation,** and a wealth of research has shown that it is guided by the position of cells relative to one another. Like the information provided by a hormone gradient mentioned in Section 34. 2, this positional information provides cues that "tell" cells where they are in the developing embryo and thus lay the groundwork for an appropriate genetic response.

Numerous researchers have explored how cells in a developing plant or plant part receive and respond to positional information. Experiments have demonstrated, for example, that only certain cells in the epidermis of an embryonic root will give rise to root hairs, the type of trichomes that take up water and minerals from soil (see Section 31.2). These specialized root epidermal cells all share the same position with respect to the underlying root cortex—each abuts two cortical cells. By contrast, no root hair extension will develop from an epidermal cell that lines up against only one cortical cell. **Figure 34.18** diagrams one model of what happens next. In this scenario, one or more chemical signals may cross from cortical to epidermal cells by way of plasmodesmata. When an epidermal cell receives signals from a single cortex cell, a series of genes are expressed in a cascade of effects that culminate in the expression of a gene called GL2 (or GLABRA2). The product of GL2 blocks the formation of root hairs. If, on the other hand, an epidermal cell aligns with two cortex cells, it receives signals from both and the cascade of gene effects blocks expression of GL2—

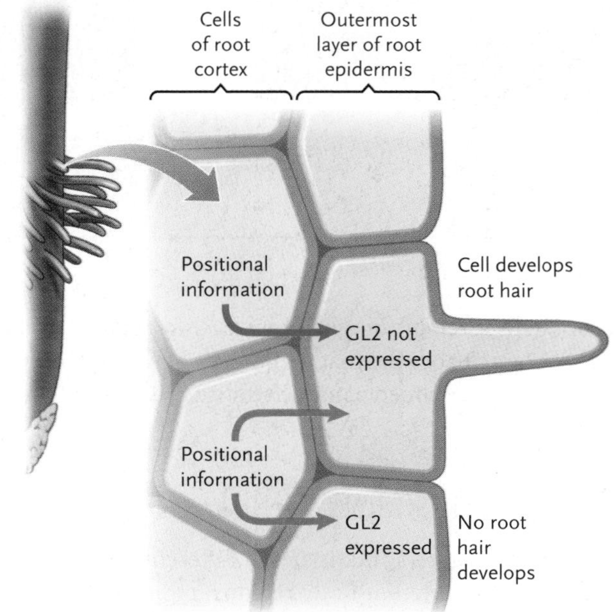

Cells of root cortex
Outermost layer of root epidermis
Positional information
Cell develops root hair
GL2 not expressed
Positional information
GL2 expressed
No root hair develops

FIGURE 34.18
One model of how positional information influences the development of root hairs. In this model, the only epidermal cells that develop root hairs are those whose inner wall is in contact with two root cortex cells. Such positioning gives rise to signals that block the expression of the GL2 (GLABRA2) gene. When GL2 is expressed, a root epidermal cell will not develop a root hair.

INSIGHTS FROM THE Molecular Revolution

Trichomes: Window on development in a single plant cell

The delicate plant cell extensions called trichomes are helping to illuminate developmental processes that go on in a single plant cell as it differentiates—that is, as it acquires its ultimate specialized structure and function. In *A. thaliana,* each of these minute protuberances consists of a single cell with a branching tripartite pattern **(Figure 1).**

As a trichome differentiates, increases in size, and extends branches in different directions, its chromosomes—and the cell's DNA—are duplicated several times over without mitosis. As a result of the process, called *endoduplication,* the cell has multiple copies of its chromosomes. Experiments that isolate the effects of different mutants have helped confirm that the overall amount of DNA in the cell strongly influences the cell's structure, and also that several genes interact to determine the structure. One of these genes is called *TRY* (for *TRIPTYCHON*). When it is mutated the affected plant's trichomes have a double complement of DNA and develop five branches **(Figure 2).** But genes that influence endoduplication are only part of the story. Experiments with other mutants show that several other genes also help produce the characteristic three-pronged trichome branching. For example, when a gene called *STICHEL* is mutated, *sti* mutants don't develop any branches **(Figure 3).** The underlying reason for this phenotype is not yet understood.

These examples underscore how complex molecular interactions that affect multiple aspects of a cell's functioning ultimately determine a cell's form and function. Because the genes that operate in trichomes are also involved in the development of other types of plant cells, studying them promises to shed light on processes that generate differentiated cells throughout the plant body.

Source: Sven Schellman and Martin Hülskamp, 2005. Epidermal differentiation: Trichomes in *Arabidopsis* as a Model System. *Int. J. Dev. Biol.* 49: 579–584.

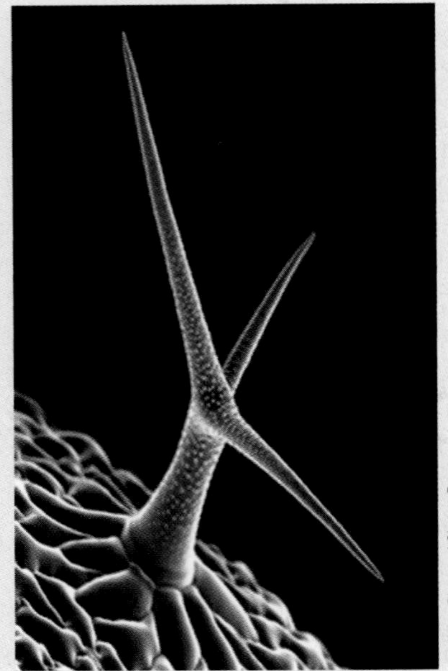

Jonathan Plett and Sharon Regan

FIGURE 1
Normal trichome from the epidermis of a leaf of *Arabidopsis thaliana.*

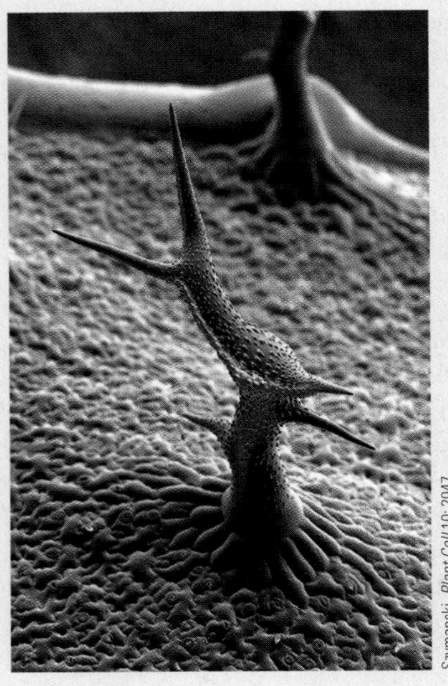

Szymanski, *Plant Cell* 10: 2047

FIGURE 2
Five-pronged trichome from a *try* mutant.

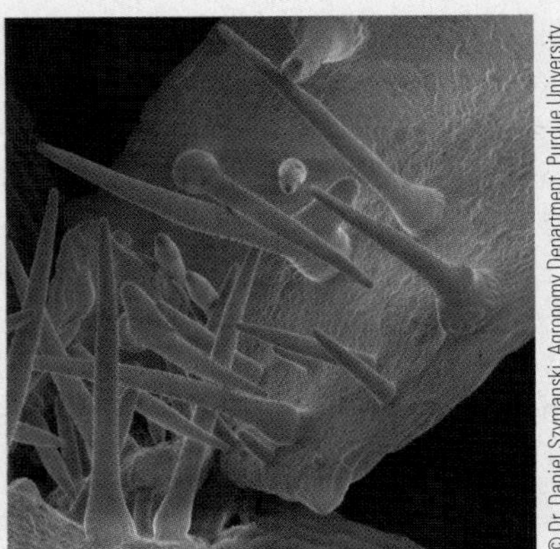

© Dr. Daniel Szymanski, Agronomy Department, Purdue University

FIGURE 3
The unbranched trichomes of an *sti* mutant.

and a root hair develops. *Insights from the Molecular Revolution* gives more examples of ways that trichomes such as root hairs have become popular experimental models for studying the differentiation of plant cells.

Morphogenesis Shapes the Plant Body

As a plant embryo grows and tissues of differentiated cells form, the stage is set for different body regions to develop characteristic shapes and structures that correlate with their function. This process, called **morphogenesis,** shapes the new shoot and root parts produced by dividing cells in meristems. In animals, morphogenesis involves localized cell division and growth, as well as migration of cells and entire tissues from one site to another (see Chapter 48). Plant cells, however, are enclosed within thick walls and usually cannot move. Thus morphogenesis in plants relies on mechanisms that don't require mobility. One of these mechanisms is **oriented cell division,** which establishes the overall shape of a plant organ, and another is **cell expansion,** which enlarges the cells in specific directions in a developing organ.

ORIENTED CELL DIVISION As described in Chapter 31, roots and stems grow lengthwise as the division and expansion of cells produce columns of cells parallel to the root–shoot axis. The cell divisions occur in a transverse plane—that is, a new cell plate, and then new cell walls, form so that the cells become stacked one atop the other like wooden blocks **(Figure 34.19A).** A plant adds girth—increases in circumference—by way of cell divisions in other planes. For instance, new cell walls may form parallel to the

External surface (for example, of a stem)

Division

New cells formed by division in a transverse plane

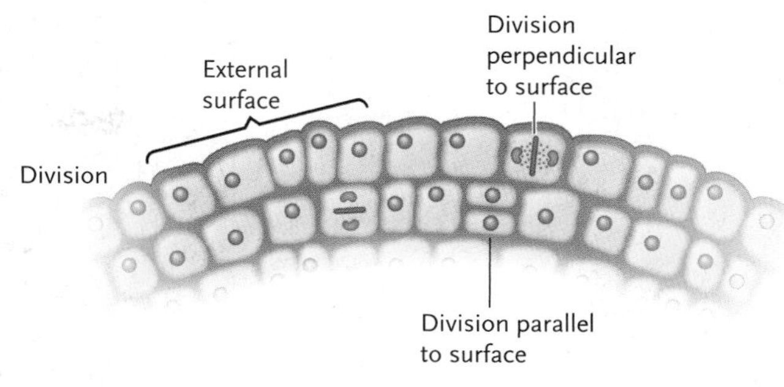

Division perpendicular to surface
External surface
Division
Division parallel to surface

FIGURE 34.19
Plant cell division in different planes. The external surface nearest the dividing cell is the reference point for establishing division planes.

A. Adding length: Division in a transverse plane parallel to the root-shoot axis produces daughter cells stacked like bricks.

B. Adding girth: Cells may divide so that the daughter cell forms parallel to the plant's nearest surface or so that the new cell wall forms perpendicular to the nearest surface.

FIGURE 34.20
How the plane of cell division is determined in a plant cell. This series of micrographs shows events in onion *(Allium cepa)* root tip cells, starting with preprophase of mitosis.

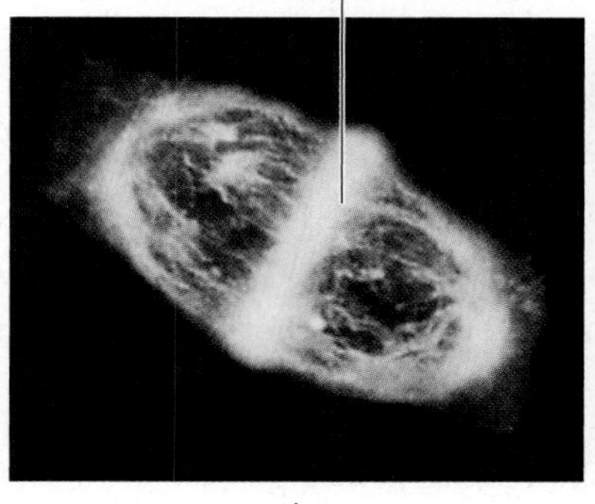

Preprophase band

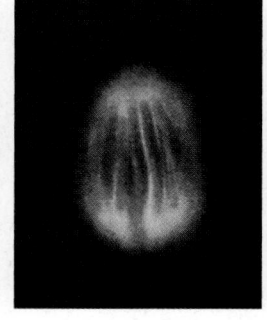

A. In preprophase of mitosis, the microtubules break their links to the plasma membrane and assemble into a thick preprophase band.

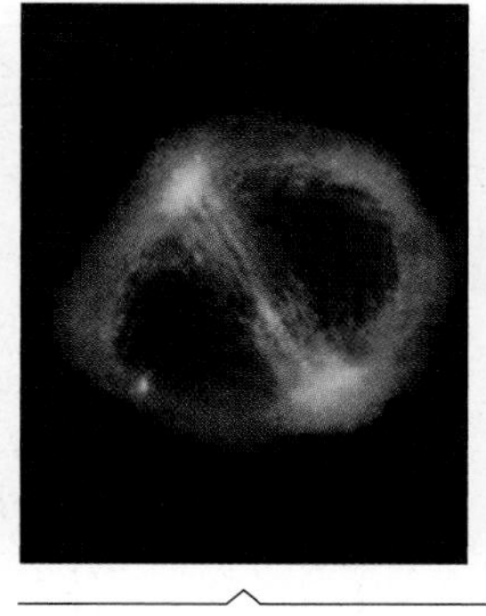

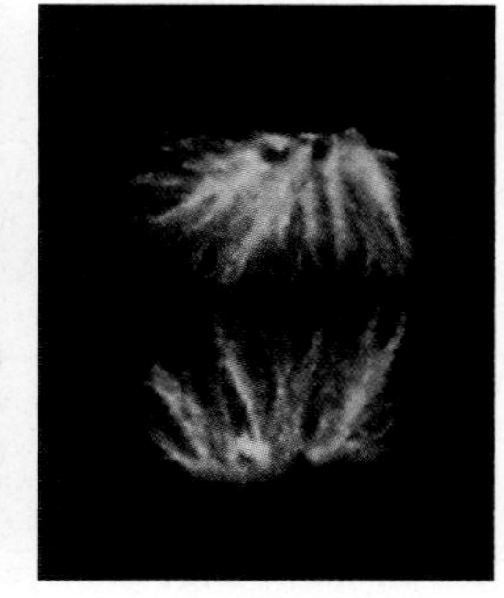

B. As cell division begins, the preprophase band disassembles and microtubules organize into a spindle.

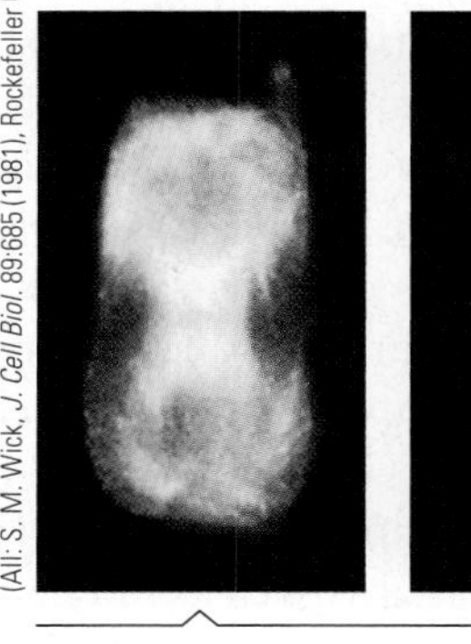

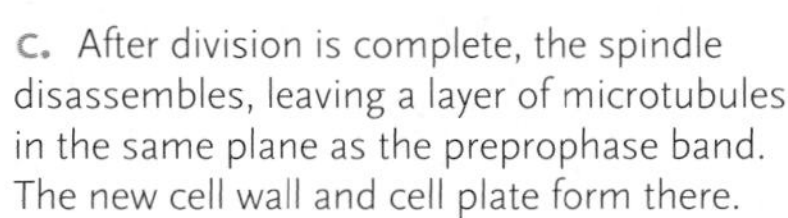

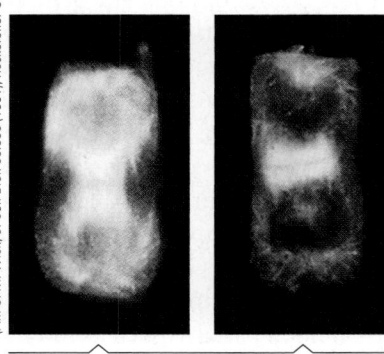

C. After division is complete, the spindle disassembles, leaving a layer of microtubules in the same plane as the preprophase band. The new cell wall and cell plate form there.

(All: S. M. Wick, *J. Cell Biol.* 89:685 (1981), Rockefeller University Press.)

nearest plant surface, or perpendicular both to the nearest surface and to the transverse plane **(Figure 34.19B)**.

You may remember from Chapter 10 that the cell plate forms during the cytokinesis phase of mitosis. This step establishes the plane of the middle lamella that will eventually separate the parent and daughter cells. The capacity of dividing plant cells to synthesize a new cell plate in a different plane from the old one underlies morphogenesis in nearly all plant groups. In meristematic tissue, changes in the plane of cell division establish the direction in which structures such as lateral roots, branches, and leaf and flower buds will grow, and so gives the plant body its overall form.

Figure 34.20 shows how the plane of a cell plate is established. While a plant cell destined to divide is still in interphase, hours before mitosis begins, the cell nucleus migrates to a particular location in the cell. The nucleus becomes surrounded by a layer of microtubules and microfilaments that radiate outward from it. Where the layer contacts the cell wall, a belt of microtubules and microfilaments called the *preprophase band* forms briefly, encircling the cell cytoplasm. The band usually disappears as mitosis gets underway in the cell, but its position marks the site where the cell plate forms during cytokinesis. Remnants of microfilaments may guide the edges of the developing cell plate into the proper position against the cell wall.

CELL EXPANSION Once a cell has divided, the daughter cells expand to mature size. Yet plant cells are encased in a primary wall of nonliving material. Botanists are beginning to learn how the cell wall expands to accommodate the enlarging cell within.

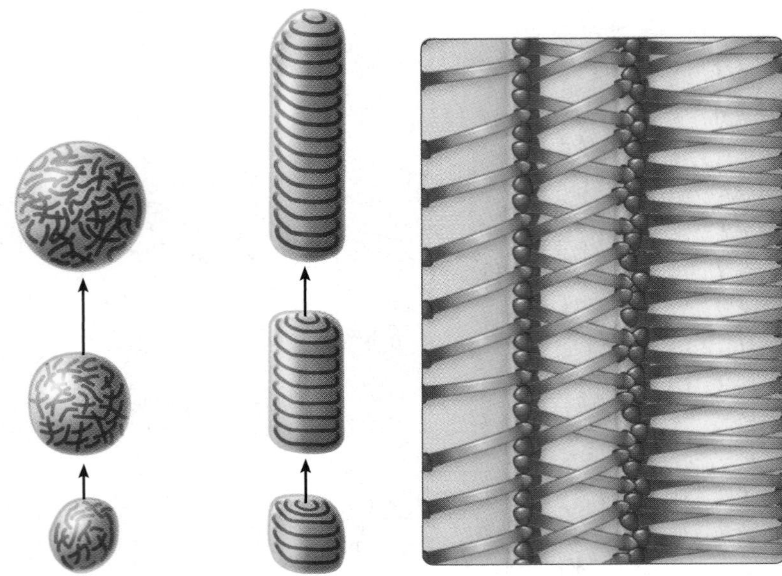

A. When microfibrils are oriented at random, the primary wall is elastic all over, so the cell can grow in all directions.

B. When microfibrils are oriented transversely, the cell can grow only longitudinally.

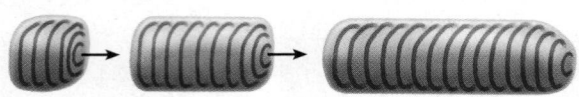

C. When microfibrils are oriented longitudinally, the cell can grow only laterally.

FIGURE 34.21
Cell expansion and the orientation of cellulose microfibrils. In each cell, microtubules inherited from the parent cell are already oriented in prescribed patterns that govern how cellulose microfibrils will be oriented in the cell wall. Their orientation in turn governs the direction in which a cell can expand.

Primary cell walls consist of a loose mesh of cellulose microfibrils embedded in a gel-like matrix. As plant cells mature, they may elongate to as much as 100 times their embryonic lengths. During this elongation, the cellulose meshwork is first loosened and then stretched. Turgor pressure supplies the force for stretching. The exact mechanism that loosens the wall structure is not known, although experiments indicate that it depends on a dramatic drop in pH. Some researchers suggest that an auxin in the cell cytoplasm may stimulate a plasma membrane proton pump that moves H^+ into the cell wall (see Section 6.4). The acidic wall conditions may activate hydrolytic enzymes that break bonds between wall components, or they may promote loosening in some other way.

During expansion, enzyme complexes in the cell's plasma membrane synthesize new cellulose microfibrils from glucose in the cytoplasm. When each microfibril is fully formed, it is bound in place in the growing wall by pectins and other wall components.

The direction of cell expansion depends on the orientation of the newly formed cellulose microfibrils **(Figure 34.21)**. If the microfibrils are randomly oriented, the cell expands equally in all directions. If they are oriented at right angles to the cell's long axis, the cell expands lengthwise. And if new fibrils are deposited parallel to the long axis of the cell, the cell expands laterally.

PATTERNS OF CELL DIVISION DURING EARLY GROWTH Like the first mitotic division in an *A. thaliana* zygote, it's not uncommon for cell divisions in a growing plant to be asymmetrical, so that one daughter cell ends up with more cytoplasm than the other. The unequal distribution of cytoplasm means that the daughter cells differ in their composition and structure, and the differences affect how they interact with their neighbors during growth, even though all cells carry the same genes. Their cytoplasmic differences and interactions with one another trigger selective gene expression. Such events seal the developmental fate of particular cell lineages. Their descendant cells divide in prescribed planes and expand in set directions, producing plant parts with diverse shapes and functions.

Regulatory Genes Guide the Development of Floral Organs

Research with several plant species has shed light on the genetic mechanisms that govern the formation of the parts of a flower. For example, experiments with *A. thaliana* carried out by Elliot Meyerowitz and his colleagues at the California Institute of Technology showed that *floral organ homeotic genes* regulate the development of the sepals, petals, stamens, and carpels in flowers.

The Meyerowitz team studied plants with various mutations in floral organs. By observing the effects of specific mutations on the structure of *A. thaliana* flowers, the investigators eventually identified three classes of homeotic gene activity—which they named A, B, and C—that regulate different aspects of normal flower development. Subsequent studies by other scientists identified an E class of gene activity that appears to be an essential partner in the functioning of A, B, and C class genes. The effects of the genes overlap, so that A, B, and C class genes are expressed in two whorls, and E class genes in three: Class A genes are expressed in whorls 1 and 2 (sepals and petals), class B genes in whorls 2 and 3 (petals and stamens), class C genes in whorls 3 and 4 (stamens and carpels), and class E genes in whorls 2, 3, and 4 **(Figure 34.22)**.

Abnormal floral patterns such as those in Figure 34.22B–D show how mutations in the floral organ homeotic genes can affect the identity of flower parts. For example, a mutation that deactivates the A-class gene *APETALA2* produces a flower with carpels and stamens in whorl 1. Another intriguing finding is that the A and C activity classes normally oppose each other. When no A gene is expressed, C activity spreads into whorls where the A usually occurs, and vice versa. Subsequent studies have examined many other floral homeotic genes, as well as the genes that control the various gene classes.

As the genes governing flower development are isolated, they can be cloned and their nucleotide sequences defined and

FIGURE 34.22 **Experimental Research**

Probing the Roles of Floral Organ Identity Genes

Question: What are the genetic mechanisms that govern the formation of the parts of a flower?

Experiments: Meyerowitz and his colleagues grew *A. thaliana* plants having mutated, inactivated versions of the genes that were suspected of controlling the proper development of floral organs. They compared the types and arrangements of floral organs in the test plants with the organs present in normal, wild-type *A. thaliana* flowers.

A. **Normal *A. thaliana***

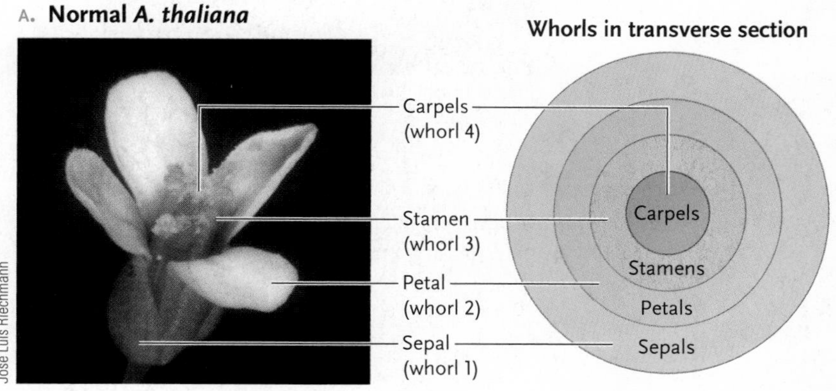

Whorls in transverse section
Carpels (whorl 4)
Stamen (whorl 3)
Petal (whorl 2)
Sepal (whorl 1)
Carpels
Stamens
Petals
Sepals
Jose Luis Riechmann

Normal arrangement of organs: carpels in whorl 4, stamens in whorl 3, petals in whorl 2, and sepals in whorl 1

Results: At least three classes of homeotic genes (A, B, and C) regulate different aspects of normal floral organ development.

B. **When mutation inactivates the *APETALA2* gene, class A genes are not expressed.**

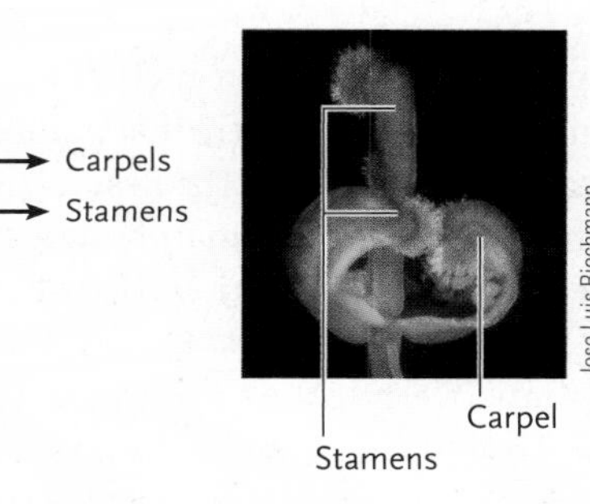

Sepals → Carpels
Petals → Stamens
Stamens
Carpel
Jose Luis Riechmann

Stamens replace petals and carpels replace sepals. Organ identity in the other whorls does not change.

C. **When mutation inactivates the *APETALA3* or *PISTILLATA* gene, class B genes are not expressed.**

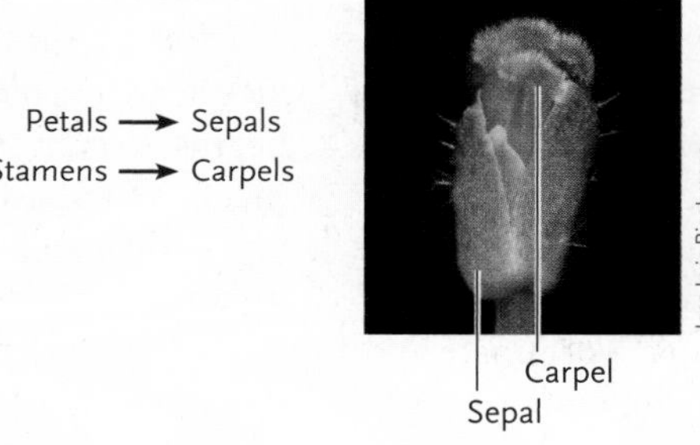

Petals → Sepals
Stamens → Carpels
Sepal
Carpel
Jose Luis Riechmann

Carpels replace stamens and sepals replace petals. Organ identity in the other whorls does not change.

D. **When mutation inactivates the *AGAMOUS* gene, class C genes are not expressed.**

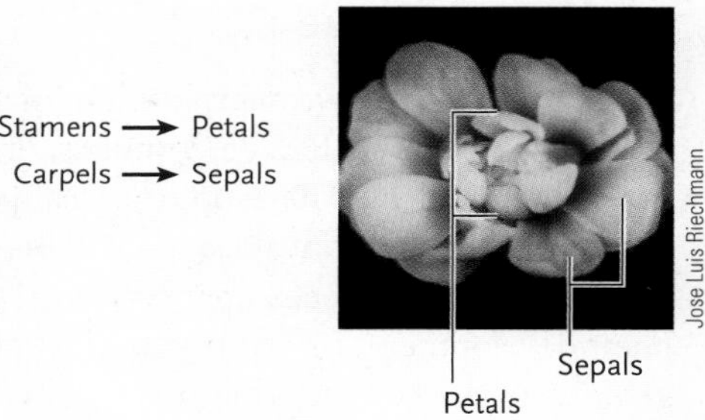

Stamens → Petals
Carpels → Sepals
Petals
Sepals
Jose Luis Riechmann

No carpels; instead petals develop in whorl 3 and a version of a floral meristem develops where whorl 4 would normally be. It gives rise to extra petals and sepals. Organ identity in other whorls does not change.

Further experiments with double mutants (inactivation of both A and B, A and C, and B and C) all produce abnormal flowers. For example, only carpels develop in mutants having only an active C class gene, and only sepals develop in plants having only an active B class gene.

E. **Overlapping activity fields of floral organ identity genes**

Carpels			C	E	**Whorl 4**
Stamens		B	C	E	**Whorl 3**
Petals	A	B		E	**Whorl 2**
Sepals	A				**Whorl 1**

Conclusions: In *A. thaliana,* A, B, and C activity genes, expressed alone or in pairs, underlie the development of a normal pattern of floral organs. The fields of activity overlap. In addition, A and C activity apparently counteract each other, and if one is absent the other can spread beyond the whorls where it normally appears. Subsequent research revealed that a fourth E class of gene activity is required for proper expression of other organ identity genes.

Source: This box summarizes several experiments: D. Weigel and E. M. Meyerowitz, 1993. Activation of floral homeotic genes in *Arabidopsis. Science* 261, 1723–1726.
J. L. Bowman, J. Alvarez, D. Weigel, E. M. Meyerowitz, and D. R. Smyth, 1993. Control of flower development in *Arabidopsis thaliana* by *APETALA1* and interacting genes. *Development* 119, 721–743.

A. **Leaf primordia of *Coleus***

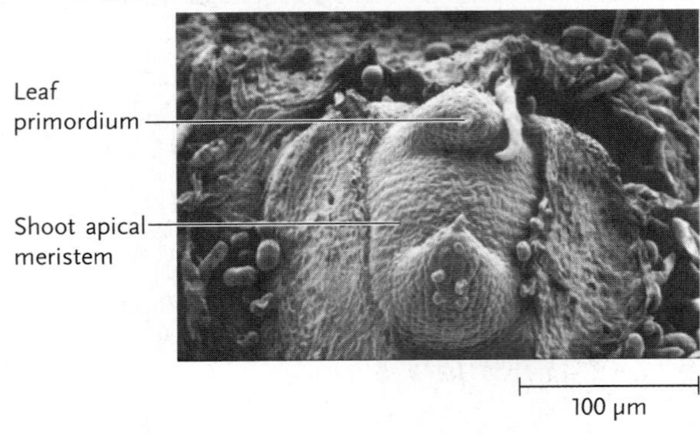
Leaf primordium
Shoot apical meristem
100 μm

B. **Early stages of leaf development**

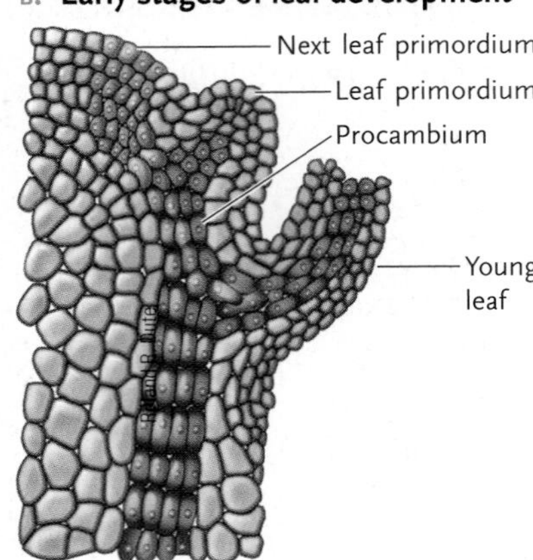
Next leaf primordium
Leaf primordium
Procambium
Young leaf

FIGURE 34.23
Early stages in leaf development. (A) Leaf primordia at the shoot tip of a member of *Coleus,* a genus that includes several popular house plants. **(B)** Diagram showing leaf primordia in different stages of development. See also Figure 34.13, which shows the progression of leaves that form during the early growth and development of a eudicot.

manipulated. Such cloned genes already are of keen interest in plant genetic engineering, because food grains such as wheat and many other vital agricultural commodities come directly or indirectly from flowers.

Leaves Arise from Leaf Primordia in a Closely Regulated Sequence

A mature leaf may have many millions of differentiated cells organized into tissues such as epidermis and mesophyll. As described in Chapter 31, leaves develop from leaf primordia that arise just behind the tips of shoot apical meristems **(Figure 34.23)**.

Clonal analysis has opened a window on many aspects of plant development, including how leaf primordia originate and give rise to leaves. In this method, the investigator cultures meristematic tissue that contains a mutated embryonic cell having a readily observable trait, such as the absence of normal pigment. (In the laboratory, this kind of mutation can be induced by chemicals or radiation.) The unusual trait then serves as a marker that identifies the mutant cell's clonal descendants, making it possible to map the growing structure. Researchers have used clonal analysis to study leaf development in garden peas, tomatoes, grasses, and tobacco, among others.

Like flowers, leaves arise through a developmental program that begins with gene-regulated activity in meristematic tissue. Hormones or other signals may arrive at target cells via the stem's vascular tissue, activating genes that regulate development. Studies show that small phloem sieve tubes penetrate a young leaf primordium almost immediately after it begins to bulge out from the underlying meristematic tissue, and xylem soon follows. The early phloem connections are especially vital to the leaf's survival, because the leaf does not begin photosynthesis until it attains one-third of its mature size.

A growing primordium becomes cone-shaped, wider at its base than at its tip. At a certain point, mitosis speeds up in cells along the flanks of the lengthening cone. In eudicots, the rapid cell divisions occur perpendicular to the surface and produce the leaf blade. In monocot grasses, which have long, narrow leaves, vertical "files" of cells develop as cells in the meristem at the base of the cone divide in a plane parallel to the surface. The ultimate shape of a leaf depends in part on variations in the plane and rate of these cell divisions.

Cells at the leaf tip, and those of xylem and phloem that service it, are the oldest in the leaf, and it is here that photosynthesis begins. Commonly, leaf tip cells also are the first to stop dividing. By the time a leaf has expanded to its mature size, all mitosis has ended and the leaf is a fully functional photosynthetic organ.

In nature, genes that govern leaf and flower development switch on or off in response to changing environmental conditions. In many perennials, new leaves begin to develop inside buds in autumn, then become dormant until the following spring, when external conditions favor further growth. Environmental cues stimulate the gene-guided production of hormones that travel through the plant in xylem and phloem, triggering renewed leaf growth and expansion. Leaves and other shoot parts also age, wither, and fall away from the plant as hormonal signals change. The far-reaching effects of plant hormones on growth and development are the subject of Chapter 35.

STUDY BREAK 34.5 <

1. **What is a homeotic gene? Give at least two examples of plant tissues such genes might govern in a species such as *A. thaliana*.**
2. **What are the two basic mechanisms of morphogenesis in plants? Describe the patterns of cell division by which a plant part (a) grows longer and (b) adds girth.**
3. **Summarize the gene-guided developmental program that gives rise to a leaf.**

UNANSWERED QUESTIONS

What are the signaling events that mediate pollen-tube guidance during compatible pollination?

As you learned in this chapter, after pollen lands on the surface of the carpel, it forms a pollen tube, which then invades the carpel, migrates past several different cell types, and enters the micropyle to fertilize an egg and central cell. Recent work has shown that pollen-tube migration is mediated by a series of cell–cell interactions such as attraction, repulsion, and adhesion. However, most of these conclusions have been based on analyzing static images and fixed tissues, and it has been hypothesized that many more subtle, dynamic interactions between the pollen tube and female cells exist to ensure compatible pollination. It has been shown, for example, that pollen tubes gain the competence to successfully find the ovules only when they grow on the pistil tissues, a process that is functionally analogous to the transformation that mammalian spermatozoa undergo after residence for a finite amount of time in the female reproductive tract. It was also recently demonstrated that once a pollen tube penetrates the ovule, additional pollen tubes are prevented from gaining access into the targeted ovule, a process that is functionally analogous to the prevention of polyspermy by a fertilized mammalian egg. These novel signaling events are spurring researchers to take a closer look at pollen-tube guidance to ovules. Only when such signaling events are described can appropriate efforts be taken to identify the cues that mediate these events (see next question).

What kinds of approaches are being taken to identify novel signaling events? Researchers usually first identify a mutant plant that is defective in any of the pollen-tube guidance steps that are essential for successful fertilization. Subsequently, they analyze the defects to learn more about how this process normally happens. Other researchers develop microscopy-based real-time assays to directly observe pollen-tube behavior with a variety of female tissues.

What are the chemical cues from female tissues that facilitate compatible pollination?

As you learned in this chapter, if a pollen tube lands on a compatible stigma, chemical cues produced by the female tissue then guide the pollen tube from the stigma to the embryo sac of an ovule. Several lines of evidence point to a role for sporophytic and gametophytic tissues in proper guidance of pollen tubes to ovules. Recently, it was shown that cysteine-rich polypeptides (LUREs) secreted from synergid cells facilitate pollen tube entry into the ovules of *Torenia fournieri* (wishbone flower). Despite these advances, the identities of the cues released by other pistil tissues remain unknown. What hurdles have hampered the efforts to uncover pollen-tube navigation cues of even the known signaling events described above? First, the pistil tissue within which the pollen tube elongates to the ovule is comprised of several types of tissue, including stigma, style, and transmitting tract, and these tissues are not readily accessible. Second, analyzing the dynamic responses of pollen tubes is difficult given that pollen-tube navigation occurs well within opaque pistils. Third, it appears that multiple, stage-specific, short-range, and readily labile signals produced in minute quantities mediate pollen-tube guidance. However, recent development of highly sensitive global approaches and assays that directly monitor pollen-tube elongation offer hope that guidance cues will be uncovered sooner rather than later.

Think Critically

1. Do you think pollen-tube guidance signals from different plant species will be similar?
2. What are some of the practical applications of deciphering signals that mediate pollen-tube guidance?

Ravi Palanivelu

Ravi Palanivelu is an assistant professor in the Department of Plant Sciences at the University of Arizona. His current research focuses on the isolation and characterization of pollen-tube guidance signals during *Arabidopsis* reproduction, with the long-term goal of understanding the molecular basis of how cells communicate with each other. To learn more about Palanivelu's research, go to http://ag.arizona.edu/research/ravilab.

REVIEW KEY CONCEPTS

Go to **CENGAGENOW** at www.cengage.com/login to access quizzing, animations, exercises, articles, and personalized homework help.

34.1 Overview of Flowering Plant Reproduction

- In most flowering plant species, a multicellular diploid sporophyte (spore-producing plant) stage alternates with a multicellular haploid gametophyte (gamete-producing plant) stage. The sporophyte develops roots, stems, leaves, and, at some point, flowers. This life cycle pattern is called alternation of generations (Figure 34.2).

Animation: Flowering plant life cycle

34.2 The Formation of Flowers and Gametes

- A flower develops at the tip of a floral shoot and has up to four whorls which are supported by the receptacle. The calyx and corolla consist of the sepals and petals, respectively. The third whorl consists of stamens, and carpels make up the innermost whorl (Figure 34.3).
- The anther of a stamen contains sacs where pollen grains develop. Fertilization occurs if pollen lands on a compatible stigma, which contains an ovary in which eggs develop.

- A complete flower has both male and female reproductive parts. Monoecious species bear both types of flowers on each plant; in dioecious species the "male" and "female" flowers are on different plants (Figure 34.4).
- In pollen sacs, pollen grains (immature male gametophytes) develop inside microspores. In a pollen grain, one cell develops into two sperm cells (male gametes). Another cell produces the pollen tube (Figures 34.5 and 34.6).
- An ovule forms inside a carpel. In the ovule, meiosis produces megaspores, of which one survives and develops further, producing the female gametophyte: the seven-celled embryo sac, one cell of which is the haploid egg. A cell containing polar nuclei helps give rise to endosperm (Figure 34.7).

Animation: Floral structure and function

Animation: Flower parts

Animation: Microspores to pollen

34.3 Pollination, Fertilization, and Germination

- Upon pollination, the pollen grain resumes growth, a pollen tube develops, and mitosis of the male gametophyte's sperm-producing cell produces two sperm nuclei (Figure 34.8).
- In double fertilization, one sperm nucleus fuses with one egg nucleus to form a diploid ($2n$) zygote. The other sperm nucleus and the two polar nuclei formed in the embryo sac also fuse, forming a $3n$ cell that will give rise to triploid ($3n$) endosperm in the seed (Figure 34.9).
- After the endosperm forms, the ovule expands, and the embryonic sporophyte develops. This mature ovule is a seed. Inside it, the embryo has a lengthwise axis with a root apical meristem at one end and a shoot apical meristem at the other.
- Eudicot embryos have two cotyledons. The embryonic shoot consists of an upper epicotyl and a lower hypocotyl; also present is an embryonic root, the radicle. The single cotyledon of a monocot forms a scutellum that absorbs nutrients from endosperm. Apical meristems of a monocot embryo are protected by a coleoptile over the shoot tip and a coleorhiza over the radicle (Figure 34.10).
- A fruit is a matured or ripened ovary. Fruits protect seeds and aid in their dispersal by animals, wind, or water. They may be simple, aggregate, or multiple, depending on the number of flowers or ovaries from which they develop. Fruits also vary in the characteristics of the pericarp that surrounds the seed (Figure 34.11).
- The seeds of many plants remain dormant until external conditions such as moisture, temperature, and day length favor the survival of the embryo and the development of a new sporophyte (Figures 34.12–34.14).

Animation: Eudicot life cycle

Animation: Eudicot seed development

Animation: Apple fruit structure

Animation: Plant development

34.4 Asexual Reproduction of Flowering Plants

- Many flowering plants also reproduce asexually, with new plants arising by three general modes: by mitosis at nodes or buds along modified stems, by vegetative propagation (Figure 34.15), or by tissue culture from a parent plant's somatic (non-reproductive) cells (Figure 34.16).

34.5 Early Development of Plant Form and Function

- In plant sexual reproduction, development starts at fertilization. Early on, a new embryo acquires its root–shoot axis, and cells in different regions begin to become specialized for particular functions (Figure 34.17). In morphogenesis, body regions develop characteristic shapes and structures that correlate with their function (Figure 34.18).
- Dividing plant cells can synthesize a new cell plate in a different plane from the old one. Such changes establish the direction in which structures such as lateral roots, branches, and leaf and flower buds grow (Figures 34.19 and 34.20).
- Chemical signals that help guide morphogenesis appear to act on certain cells in meristematic tissue, activating homeotic genes that ultimately regulate cell division and differentiation (Figures 34.21 and 34.22).

Animation: ABC model for flowering

UNDERSTAND AND APPLY

Test Your Knowledge

1. An angiosperm life cycle includes:
   a. meiosis within the male gametophyte to produce sperm.
   b. meiosis within the female gametophyte to produce eggs.
   c. meiosis within the ovary to produce megaspores.
   d. fertilization leading to development of microspores.
   e. fertilization leading to development of megaspores.
2. In a flower:
   a. the ovary contains the ovule.
   b. the stamens support the petals.
   c. the anther contains the megaspores.
   d. the carpel includes the sepals.
   e. the corolla includes the receptacle.
3. Double fertilization in a flower means:
   a. six sperm fertilize two groups of three eggs each.
   b. one sperm fertilizes the egg; a second sperm fertilizes the mother cell.
   c. one microspore becomes a pollen grain; the other microspore becomes a sperm-producing cell.
   d. one pollen grain can make sperm nuclei and a pollen tube.
   e. one sperm can fertilize two endosperm mother cells.
4. A seed is best described as a (an):
   a. epicotyl.
   b. endosperm.
   c. ovary.
   d. mature spore.
   e. mature ovule.
5. The primary root develops from the embryonic:
   a. epicotyl.
   b. hypocotyl.
   c. coleoptile.
   d. radicle.
   e. plumule.
6. Which of the following is *not* a step in the germination of a monocot seed?
   a. Enzymes secreted into the endosperm digest the endosperm cell wall and macromolecules.
   b. The embryo imbibes water and then produces gibberellin.
   c. The embryo absorbs nutrients released from the endosperm.
   d. Endosperm develops as a food reserve.
   e. Gibberellin acts on the cells of the aleurone and scutellum to encode hydrolytic enzymes.

7. A student cuts off a leaflet from a plant and places it in a glass of water. Within a week roots appear on the base of the cutting. A month later she places the growing cutting into soil and it grows to the full size of the "parent" plant. This is an example of:
   a. parthenocarpy.
   b. fragmentation.
   c. grafting.
   d. vegetative reproduction.
   e. tissue culture propagation.
8. Which of the following is *not* an example of pattern formation in developing plants?
   a. an epidermal cell receiving developmental signals from a cortical cell
   b. the loosening of the cell wall to allow the elongation of selected cells to reach mature size
   c. regulation by homeotic genes of the position of different flower parts
   d. oriented cell division that establishes the shape of an organ
   e. cell expansion that directs specific cells to undergo mitosis at a given time and place
9. During the development of a leaf:
   a. mitotic cell divisions occur on planes specific to different plant groups.
   b. xylem vessels are the first to penetrate the leaf primordium.
   c. the growing leaf primordium becomes wider at its base than at its tip.
   d. the leaf primordium bulges from the region behind the shoot apical meristem.
   e. All of the above occur during leaf development.
10. In spring a lone walnut tree in your backyard develops attractive white flowers, and by the end of summer roughly half the flowers have given rise to the shelled fruits we know as walnuts. Walnut trees are self-pollinating. Assuming that pollination was 100% efficient in the case of your tree, which of the following statements best describes your tree's reproductive parts?
   a. Its flowers are in the botanical category of "perfect" flowers.
   b. The tree is monoecious.
   c. The tree is dioecious.
   d. The tree has imperfect, monoecious flowers.
   e. a and b together provide the best description of the tree's flowers.

Discuss the Concepts

1. A plant physiologist has succeeded in cloning a gene for pest resistance into petunia cells. How can she use tissue culture to propagate a large number of petunia plants that have the gene?
2. A large tree may have tens of thousands of shoot tips, and the cells in each tip can differ genetically from cells in other tips, sometimes substantially. Propose a hypothesis to explain this finding, and speculate about how it might be beneficial to the plant. How might this sort of natural variation be useful to human society?
3. Grocery stores separate displays of fruits and vegetables according to typical uses for these plant foods. For instance, bell peppers, cucumbers, tomatoes, and eggplants are in the vegetable section, while apples, pears, and peaches are displayed with other fruits. How does this practice relate to the biological definition of a fruit?

Design an Experiment

The developmental genetics of flowers are of keen interest in plant biotechnology, especially with regard to food plants such as wheat and rice. Outline a research program for a crop species that would exploit the genetics of flower development, including the effects of homeotic genes, to engineer a more productive variety.

Interpret the Data

As Chapter 35 discusses, the plant hormone ABA (abscisic acid) inhibits seed germination, while any of several GAs (gibberellins) stimulate it. For hundreds of plant species, exposure to a substance in wood smoke—one of a family of plant growth regulators called karrikins, or KARs—markedly improves the germination rate. When flames destroy the existing vegetation at a fire scene, more sunlight reaches the ground. In the presence of ample sunlight, KARs apparently promote germination by enhancing the expression of genes that encode GA.

The graph below summarizes experiments by Australian researchers studying KAR activity in mutant *A. thaliana* plants (called ga1-3), which do not synthesize GA. The graph tracks germination activity after two groups of identical test seeds began imbibing water in a medium that contained the gibberellin GA_3. All seeds were exposed to identical levels of a karrikin called KAR_1. The red symbols designate seeds of mutant ABA-deficient plants that were supplied with supplementary ABA. The white symbols are ABA-deficient seeds that did not receive the supplement. The horizontal bars indicate a statistical "confidence interval"—basically, how many of the test plants in each group were scored at each data point shown.

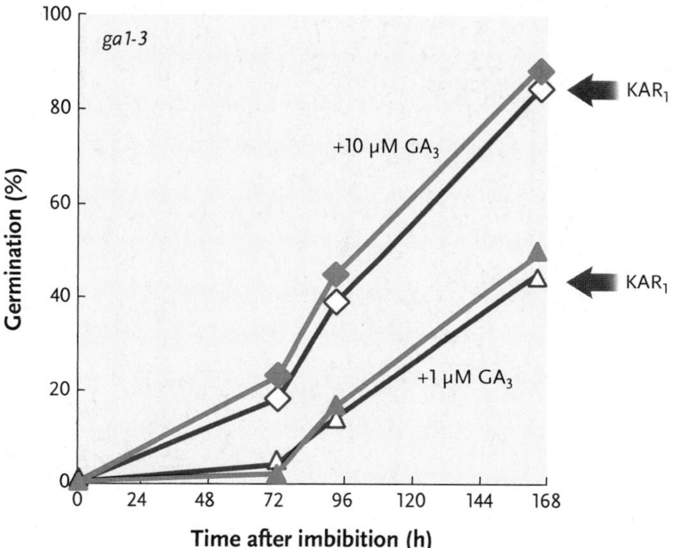

ga1-3
Germination (%)
100
80
60
40
20
0
+10 μM GA3
+1 μM GA3
KAR1
KAR1
0
24
48
72
96
120
144
168
Time after imbibition (h)

Response of phytohormone mutants to KAR_1: Germination of *A. thaliana* ga1-3 mutants in the presence of 1 or 10 μM GA_3 ± 1 μM KAR_1.

Based on this graph:

1. What were the experimental variables in the germination experiment?
2. Did the presence or absence of ABA have a significant effect on the experiment's results?
3. Comparing the upper and lower data sets, what are the *two* basic observations you can make about the effects of the seed exposure to differing GA concentrations?

Source: David C. Nelson, Julie-Anne Riseborough, Gavin R. Flematti, Jason Stevens, Emilio L. Ghisalberti, Kingsley W. Dixon and Steven M. Smith, 2009. Karrikins discovered in smoke trigger *Arabidopsis* seed germination by a mechanism requiring gibberellic acid synthesis and light. *Plant Physiology:* 149:863–873.

Apply Evolutionary Thinking

Botanists estimate that half or more of angiosperm species may be polyploids that arose initially through hybridization. *Polyploidy*—having more than a diploid set of the parental chromosomes—can result from nondisjunction of homologous chromosomes during meiosis, or when cytokinesis fails to occur in a dividing cell. *Hybridization* is the successful mating of individuals from two different species. Such an interspecific hybrid is likely to be sterile because it has uneven numbers of parental chromosomes, or because the chromosomes are too different to pair during meiosis. A sterile hybrid may reproduce asexually, however, and if by chance its offspring should become polyploid, that plant will be fertile because the original set of chromosomes will have homologs that can pair normally during meiosis. Explain why both the hybrid parent and fertile polyploid offspring may be considered a new species, and describe at least two ways in which this route to speciation differs from speciation in the animal kingdom.

Sunflower plants *(Helianthus)* with flower heads oriented toward the sun's rays—an example of a plant response to the environment.

35

STUDY OUTLINE

Plant Responses to the Environment

Why It Matters. . . For at least 4,000 years, humans have been using barley *(Hordeum vulgare)* to produce beer—a brew, it happens, made possible in part by a plant hormone. Like other seeds, those within barley grains synthesize one of the hormones called gibberellins (GA). Like all hormones, gibberellins are intercellular signaling molecules. Specifically, GA is a growth regulator synthesized in various plant parts. As a barley seed starts to germinate, it makes minute amounts of gibberellin, which stimulates the enzyme action that hydrolyzes starch in the grain, making nutrients available to the growing seedling. During the brewing step called malting, hydrolysis of the starch in barley is a prelude to the fermentation that ultimately yields beer. To make the brewing process more efficient, modern brewers add synthetic GA to germinating barley grains during malting. The GA supplement speeds up starch hydrolysis, allowing fermentation to get underway sooner.

Gibberellins also are widely used in the production of some major agricultural crops. Applied to developing seedless grapes (which lack GA from a seed), the hormone both increases fruit size and produces more open clusters **(Figure 35.1)**, which gives the grapes greater consumer appeal and helps prevent destructive mold. Commercial growers also use GA to enhance the size of lemons, limes, and Bing cherries, improve fruit set in apple and pear orchards, and to increase the amount of sucrose in sugar cane.

In nature GA and other plant hormones are part of a suite of adaptations essential to a plant's evolutionary fitness—its capacity to survive and reproduce. Collectively these adaptations are the

Normal
GA treated
Sylvan Wittwer/Visuals Unlimited

FIGURE 35.1
Effect of gibberellin on seedless grapes *(Vitis vinifera)*. Commercial grape growers use a gibberellin to lengthen the stems on which fruits develop, allowing space for individual grapes to grow larger. The grapes on the right are an example, having developed on vines that were treated with a gibberellin.

evolutionary solution to a key plant "problem"—the need to respond to changes in the environment despite being rooted in one place. And although many of the details remain to be worked out, plant scientists have found ample evidence that an elaborate system of hormonal signals regulates many, if not most aspects of plant growth and development.

At a fundamental level, hormones control the plant life cycle. They serve as triggers for seed germination and govern the development of a plant's body form, the shift from a phase of vegetative growth to a reproductive phase, and the timed death of flowers, leaves, and other parts. We also know that hormones change patterns of plant growth, cell metabolism, and morphogenesis in response to changing environmental rhythms, including seasonal changes in day length and temperature and the daily rhythms of light and dark. They also adjust those patterns in response to environmental conditions, such as the amount of sunlight or shade, moisture, soil nutrients, and other factors. Some hormones govern growth responses to directional stimuli, such as light, gravity, or the presence of nearby structures. Often, hormonal effects involve changes in gene expression, although sometimes other mechanisms are at work.

We begin by surveying the different groups of plant hormones and other signaling molecules, and then turn our attention to the remarkable diversity of plant responses to both internal and environmental signals. <

35.1 Plant Hormones

In plants, a **hormone** (*horman* = to stimulate) is a signaling molecule that regulates or helps coordinate some aspect of the plant's growth, metabolism, or development. Plant hormones act in response to two general types of cues: internal chemical conditions related to growth and development, and circumstances in the external environment that affect plant growth, such as light and the availability of water. Some plant hormones are transported from the tissue that produces them to another plant part, whereas others exert their effects in the tissue where they are synthesized.

All plant hormones are rather small organic molecules, and all are active in extremely low concentrations. Hormones that have effects outside the tissue where they are produced typically travel to their target sites in vascular tissues, or they diffuse to their target. Plant hormones vary greatly in their effects, although each hormone affects a given tissue in a particular way. For instance, some stimulate one or more aspects of the plant's growth or development, whereas others have an inhibiting influence. A given hormone also can have different effects in different tissues, and the effects can differ depending on a target tissue's stage of development. Adding to the complexity, many physiological responses result from the interaction of two or more hormones.

Biologists recognize at least seven major classes of plant hormones **(Table 35.1):** gibberellins, auxins, cytokinins, ethylene, brassinosteroids, abscisic acid (ABA), and jasmonates. Recent discoveries have added other hormonelike signaling agents to this list. Auxins were the first plant hormones to be identified, and we start with them as we consider each major class of plant hormones and discuss some of the newly discovered signaling molecules as well.

Auxins Promote Growth

Auxins are synthesized mainly in the shoot apical meristem and young stems and leaves. Among other effects, they stimulate plant growth by promoting cell elongation in stems and coleoptiles. Auxins also govern growth responses to light and gravity. Indoleacetic acid (IAA) is the most important natural auxin. Botanists often use the general term "auxin" to refer to IAA, a practice we follow here.

EXPERIMENTS LEADING TO THE DISCOVERY OF AUXINS The path to the discovery of auxins began in the late nineteenth century in the library of Charles Darwin's home in the English countryside (see *Focus on Basic Research* in Chapter 19). Among his many interests, Darwin was fascinated by plant **tropisms**—movements such as the bending of a houseplant toward light. This growth response, triggered by exposure to a directional light source, is an example of a **phototropism.**

Working with his son Francis, Darwin explored phototropisms by germinating seeds of two species of grasses, oats *(Avena sativa)* and canary grass *(Phalaris canariensis),* in pots on the sill of a sunny window. Recall from Chapter 34 that the shoot apical meristem and plumule of grass seedlings are sheathed by a coleoptile—a protective structure that is extremely sensitive to light. Darwin did not know this detail, but he observed that as the emerging shoots grew, within a few days they bent toward the light. He hypothesized that the tip of the shoot detected light and communicated that information to the coleoptile. Darwin

TABLE 35.1 Major Plant Hormones and Signaling Molecules

Hormone/Signaling Compound	Where Synthesized	Tissues Affected	Effects
Auxins	Apical meristems, developing leaves and embryos	Growing tissues, buds, roots, leaves, fruits, vascular tissues	Promote growth and elongation of stems; promote formation of lateral roots and dormancy in lateral buds; promote fruit development; inhibit leaf abscission; orient plants with respect to light, gravity
Gibberellins	Root and shoot tips, young leaves, developing embryos	Stems, developing seeds	Promote cell divisions and growth and elongation of stems; promote seed germination and bolting
Cytokinins	Mainly in root tips	Shoot apical meristems, leaves, buds	Promote cell division; inhibit senescence of leaves; coordinate growth of roots and shoots (with auxin)
Ethylene	Shoot tips, roots, leaf nodes, flowers, fruits	Seeds, buds, seedlings, mature leaves, flowers, fruits	Regulates elongation and division of cells in seedling stems, roots; in mature plants regulates senescence and abscission of leaves, flowers, and fruits
Brassinosteroids	Young seeds; shoots and leaves	Mainly shoot tips, developing embryos	Stimulate cell division and elongation, differentiation of vascular tissue
Abscisic acid	Leaves	Mainly shoot tips, developing embryos	Stimulate cell division and elongation, differentiation of vascular tissue
Jasmonates	Roots, seeds, probably other tissues	Various tissues, including damaged ones	In defense responses, promote transcription of genes encoding protease inhibitors; possible role in plant responses to nutrient deficiencies
Oligosaccharins	Cell walls	Damaged tissues; possibly active in most plant cells	Promote synthesis of phytoalexins in injured plants; may also have a role in regulating growth
Systemin	Damaged tissues	Damaged tissues	To date known only in tomato; roles in defense, including triggering jasmonate-induced chemical defenses
Salicylic acid	Damaged tissues	Many plant parts	Triggers synthesis of pathogenesis-related (PR) proteins, other general defenses

tested this idea in several ways **(Figure 35.2)** and concluded that when seedlings are illuminated from the side, "some influence is transmitted from the upper to the lower part, causing them to bend."

The Darwins' observations spawned decades of studies that illustrate how scientific understanding typically advances step-by-step, as one set of experimental findings stimulates new research. First, scientists in Denmark and Poland showed that the bending of a shoot toward a light source was caused by something that could move through agar (a jellylike culture material derived from certain red algae) but not through a sheet of the mineral mica. This finding prompted experiments establishing that indeed the stimulus was a chemical produced in the shoot tip. Soon afterward, in 1926, experiments by the Dutch plant physiologist Frits Went confirmed that the growth-promoting chemical diffuses downward from the shoot tip to the stem below **(Figure 35.3).** Using oat seeds, Went first sliced the tips from young shoots that had been grown under normal light conditions. He then placed the tips on agar blocks and left them there long enough for diffusible substances to move into the agar. Meanwhile, the decapitated stems stopped growing, but growth quickly resumed in seedlings that Went "capped" with the agar blocks (see Figure 35.3A). Clearly, a growth-promoting substance in the excised shoot tips had diffused into the agar, and from there into the seedling stems. Went also attached an agar block to one side of a decapitated shoot tip; when the shoot began growing again it bent away from the agar (see Figure 35.3B). Importantly, Went performed his experiments in total darkness, to avoid any "contamination" of his results by the possible effects of light.

Went did not determine the mechanism—differential elongation of cells on the shaded side of a shoot—by which the growth promoter controlled phototropism. However, he did develop a test that correlated specific amounts of the substance, later named auxin (*auxein* = to increase), with particular growth effects. This careful groundwork culminated several years later when other researchers identified auxin as indoleacetic acid (IAA).

EFFECTS OF AUXINS Auxin is one of the first chemical signals to help shape the plant body. When the zygote first divides, forming an embryo that consists of a basal cell and an apical cell (see

FIGURE 35.2 **Experimental Research**

The Darwins' Experiments on Phototropism

Question: Why does a plant stem bend toward the light?

Experiment 1: The Darwins observed that the first shoot of an emerging grass seedling, which is sheathed by a coleoptile, bends toward sunlight shining through a window. They removed the shoot tip from a seedling and illuminated one side of the seedling.

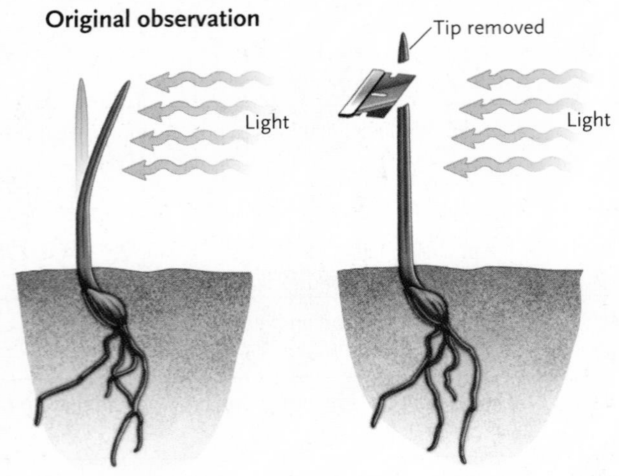

Original observation
Light
Tip removed
Light

Result: The seedling neither grew nor bent.

Experiment 2: The Darwins divided seedlings into two groups. They covered the shoot tips of one group with an opaque cap and the shoot tips of the other group with a translucent cap. All the seedlings were illuminated from the same side.

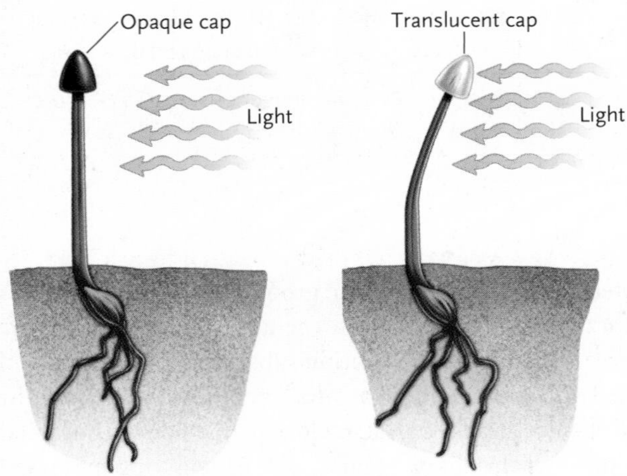

Opaque cap
Light
Translucent cap
Light

Result: The seedlings with opaque caps grew but did not bend. Those with translucent caps both grew *and* bent toward the light.

Conclusion: When seedlings are illuminated from one side, an unknown factor transmitted from a seedling's tip to the tissue below causes it to bend toward the light.

Source: C. R. Darwin. 1880. *The Power of Movement in Plants*. London: John Murray.

Section 34.3), auxin exported by the basal cell to the apical cell helps guide the development of the various features of the embryonic shoot. As the embryo develops, IAA is produced mainly by the leaf primordium of the young shoot (see Figure 34.22). While the developing shoot is underground, IAA is actively transported downward, stimulating the primary growth of the stem and root **(Figure 35.4).** Once an elongating shoot breaks through the soil surface, its tip is exposed to sunlight, and the first leaves unfurl and begin photosynthesis. Shortly thereafter the leaf tip stops producing IAA and that task is assumed first by cells at the leaf edges, then by cells at base of the young leaf. Even so, as described in Section 35.3, IAA continues to influence a plant's responses to light and plays a role in its growth responses to gravity as well. IAA also stimulates cell division in the vascular cambium and promotes the formation of secondary xylem, as well as the formation of new root apical meristems, including lateral meristems. Not all of auxin's effects promote growth, however. IAA also maintains apical dominance, which inhibits growth of lateral meristems on shoots and restricts the formation of branches (see Section 31.3). Hence, auxin is a signal that the shoot apical meristem is present and active.

Commercial orchardists spray synthetic IAA on fruit trees because it promotes uniform flowering, helps set the fruit, and also helps prevent fruit from dropping off the plant prematurely. Some synthetic auxins are used as **herbicides** (generally, weed killers), essentially stimulating a target plant to "grow itself to death." The most widely used herbicide in the world is the synthetic auxin 2,4-D (2,4-dichlorophenoxyacetic acid). This chemical kills broadleaf eudicot plants but spares monocots such as grasses. It is used extensively to prevent broadleaf weeds (which are eudicots) from growing in fields of cereal crops such as corn (which are monocots).

AUXIN TRANSPORT To exert their far-reaching effects on plant tissues, auxins must travel away from their main synthesis sites in shoot meristems and young leaves. Although IAA moves through plant tissues slowly—roughly 1 cm/hr—this rate is 10-times faster than could be explained by simple diffusion. How, then, is auxin transported?

Plant physiologists adapted the agar block method pioneered by Went to trace the direction and rate of auxin movements in different kinds of tissues. A research team led by Winslow Briggs at Stanford University determined that the shaded side of a shoot tip contains more IAA than the illuminated side. Hypothesizing that light causes IAA to move laterally from the illuminated to the shaded side of a shoot tip, the team then inserted a vertical barrier (a thin slice of mica) between the shaded and illuminated sides of a shoot tip. IAA could not cross the barrier, and when the shoot tip was illuminated it did not bend. In addition, the con-

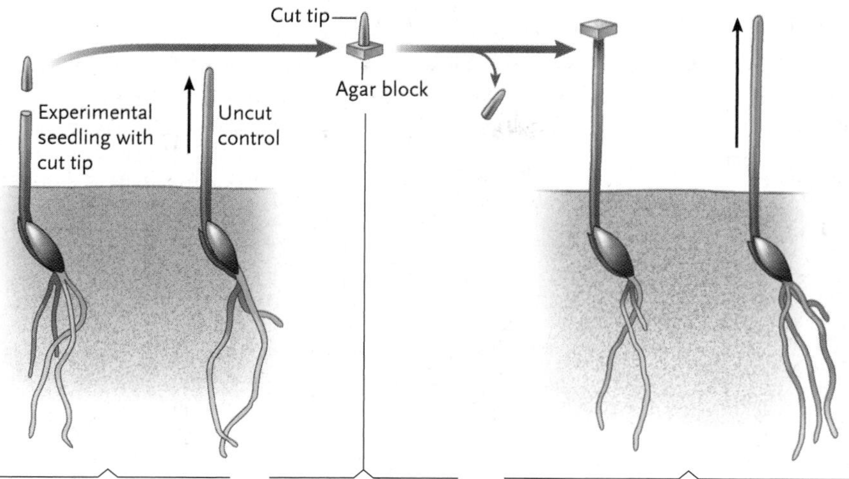

A. The procedure showing that IAA promotes elongation of cells below the shoot tip
Cut tip
Agar block
Experimental seedling with cut tip
Uncut control

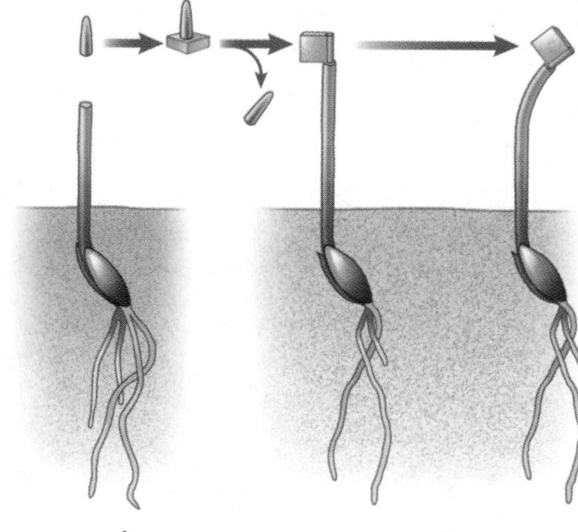

B. The procedure showing that cells in contact with IAA grow faster than those farther away

A.

1 After Went cut off the tip of an oat seedling, the shoot stopped elongating, while a control seedling with an intact tip continued to grow.

2 He placed the excised tip on an agar block for 1–4 hours. During that time, IAA diffused into the agar block from the cut tip.

3 Went then placed the agar block containing auxin on another detipped oat shoot, and the shoot resumed elongation, growing about as rapidly as in a control seedling with an intact shoot tip.

B.

1 Went removed the tip of a seedling and placed it on an agar block.

2 He placed the agar block containing auxin on one side of the shoot tip. Auxin moved into the shoot tip on that side, causing it to bend away from the hormone.

FIGURE 35.3
Two experiments by Frits Went demonstrating the effect of IAA on an oat coleoptile. Went carried out the experiments in darkness to prevent effects of light from skewing the results.

centrations of IAA in the two sides of the shoot tip remained about the same. When the barrier was shortened so that the separated sides of the tip again touched, the IAA concentration in the shaded area increased significantly, and the tip *did* bend. The study confirmed that IAA initially moves laterally in the shoot tip, from the illuminated side to the shaded side, where it triggers the elongation of cells and curving of the tip toward light. Subsequent research showed that IAA then moves downward in a shoot by way of a top-to-bottom mechanism called **polar transport.** That is, IAA in a coleoptile or shoot tip travels from the apex of the tissue to its base, such as from the tip of a developing leaf to the stem. **Figure 35.5** outlines the experimental method that demonstrated polar transport. When IAA reaches roots, it moves toward the root tip.

In a stem, parenchyma cells next to vascular bundles apparently transport IAA. The hormone moves through and between cells, apparently traveling by polar transport. Figure 35.5D diagrams one widely accepted model for this process. As you can see, the IAA enters at one end by diffusing passively through cell walls, driven by concentration and electrochemical gradients produced by H^+ pumps in the plasma membrane. The hormone exits at the opposite end by active transport across the plasma membrane.

Although auxin typically is not found in xylem, there is increasing evidence that it sometimes moves from parenchyma cells into phloem and travels rapidly through plants. As this work continues, researchers will undoubtedly gain a clearer understanding of how plants distribute this crucial hormone to their growing parts.

Kingsley R. Stern

Treated with auxin **Untreated**

FIGURE 35.4
The effect of IAA treatment on a gardenia *(Gardenia)* cutting. Four weeks after an auxin was applied to the base of the cutting on the left, its stem and roots have elongated, but the number of leaves is unchanged. The plant cutting on the right was not treated.

POSSIBLE MECHANISMS OF IAA ACTION Ever since auxin was discovered, researchers have sought to understand how IAA stimulates plant cells to elongate. As you may recall from Section 34.5, in an elongating plant cell the cellulose meshwork of the cell wall is first loosened and then stretched by turgor pressure. Several hormones, and auxin especially, apparently increase the plasticity (stretching) of the cell wall. Two major hypotheses seek to explain this effect, and both may be correct.

Plant cell walls grow much faster in an acidic environment—that is, when the pH is less than 7. The **acid-growth hypothesis** suggests that auxin causes cells to secrete acid (H^+) into the cell wall by stimulating the plasma membrane H^+ pumps to move protons from the cell interior into the cell wall; the increased acidity activates proteins called *expansins,* which penetrate the

FIGURE 35.5 **Experimental Research**

Tracking the Polar Transport of Auxin in Plant Shoots

Question: Can IAA move both upward and downward in a shoot?

Experiment: Many studies of the directional movement of IAA in plants have used excised sections of shoot tips (the coleoptile) from oat seedlings *(Avena sativa)*. The test sections were manipulated using a donor–receiver agar block method in which the blocks of agar contained different concentrations of IAA. Experiments using IAA labeled with radioactive carbon-14 made it easy to track IAA movements.

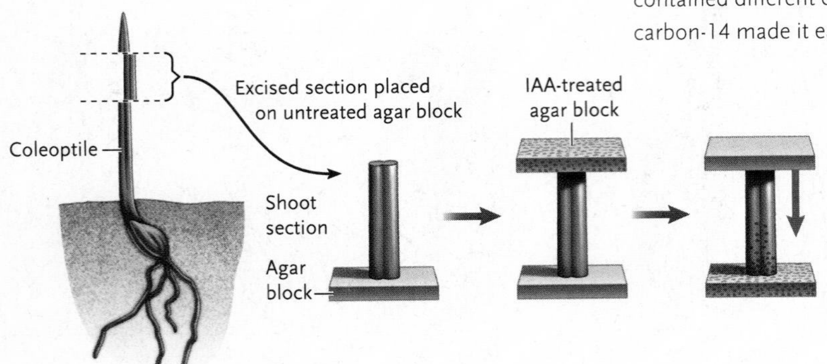

Excised section placed on untreated agar block
IAA-treated agar block
Coleoptile
Shoot section
Agar block

1. An upright coleoptile section was positioned on an untreated agar block, and then a second, "donor" agar block containing labeled IAA was placed atop the section.

Result: IAA traveled from the upper block to the lower one, indicating that the hormone moved downward through the vertical shoot tip.

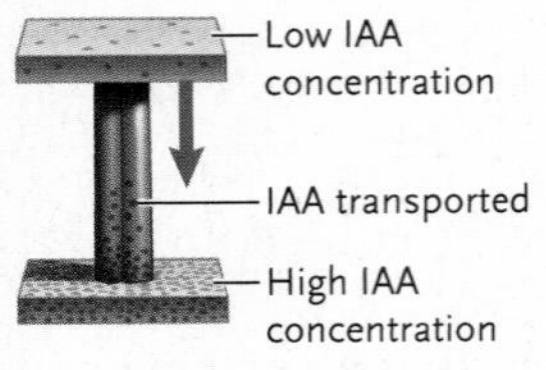

Low IAA concentration
IAA transported
High IAA concentration

2. An upright shoot tip section was positioned on an agar block containing a high concentration of labeled IAA. A donor block containing a lower concentration of IAA was placed atop the section.

Result: Even against its concentration gradient, IAA was transported from the upper block to the lower one.

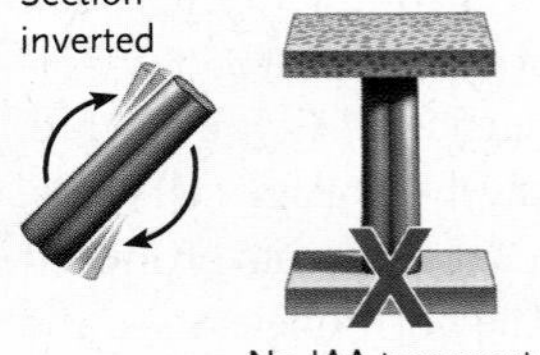

Section inverted
No IAA transport through inverted shoot section

3. The shoot tip from step 2 was inverted and the same procedure was performed.

Result: No IAA was transported downward.

Conclusion: In plant shoot tips, IAA is transported only in one direction, from the shoot tip downward to plant parts below.

How does the model for polar auxin transport shown at right support the result obtained in Step 2?

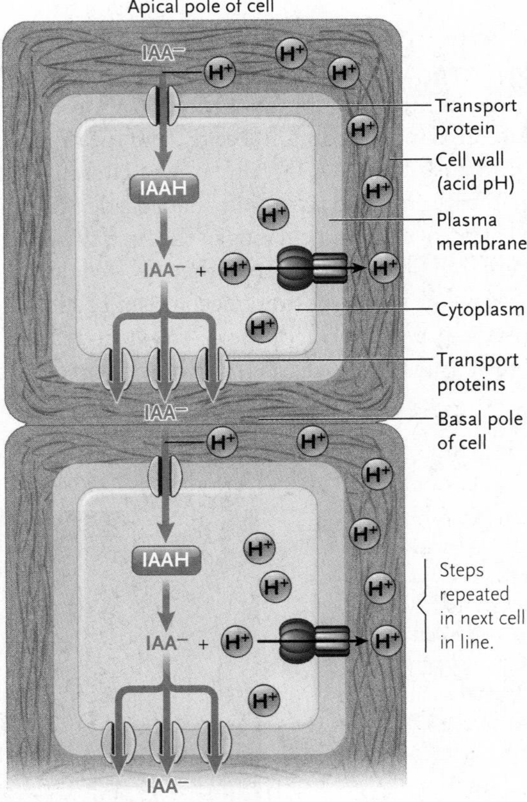

Apical pole of cell
IAA⁻
H⁺
IAAH
IAA⁻ + H⁺
Transport protein
Cell wall (acid pH)
Plasma membrane
Cytoplasm
Transport proteins
Basal pole of cell
Steps repeated in next cell in line.

A model for polar auxin transport. A plasma membrane H^+ pump maintains gradients of pH and electrical charge across the membrane, moving H^+ out of the cell using energy from ATP hydrolysis. Following the gradients, at the basal pole of a cell IAA diffuses through transport proteins into the cell wall, then (as IAAH) into the next cell in line.

Source: T. L. Lomax, G. K. Muday, and P. H. Rubery. 1995. Auxin transport. In *Plant Hormones: Physiology, Biochemistry and Molecular Biology*, P. J. Davies, ed. Dordrecht: Kluwer Academic Publishers, pp. 509–530.

FIGURE 35.6 How auxin may regulate expansion of plant cells. According to the acid-growth hypothesis, plant cells secrete acid (H^+) when auxin stimulates the plasma membrane H^+ pumps to move protons into the cell wall; the increased acidity activates enzymes called *expansins*, which disrupt bonds between cellulose microfibrils in the wall. As a result, the wall becomes extensible and the cell can expand.

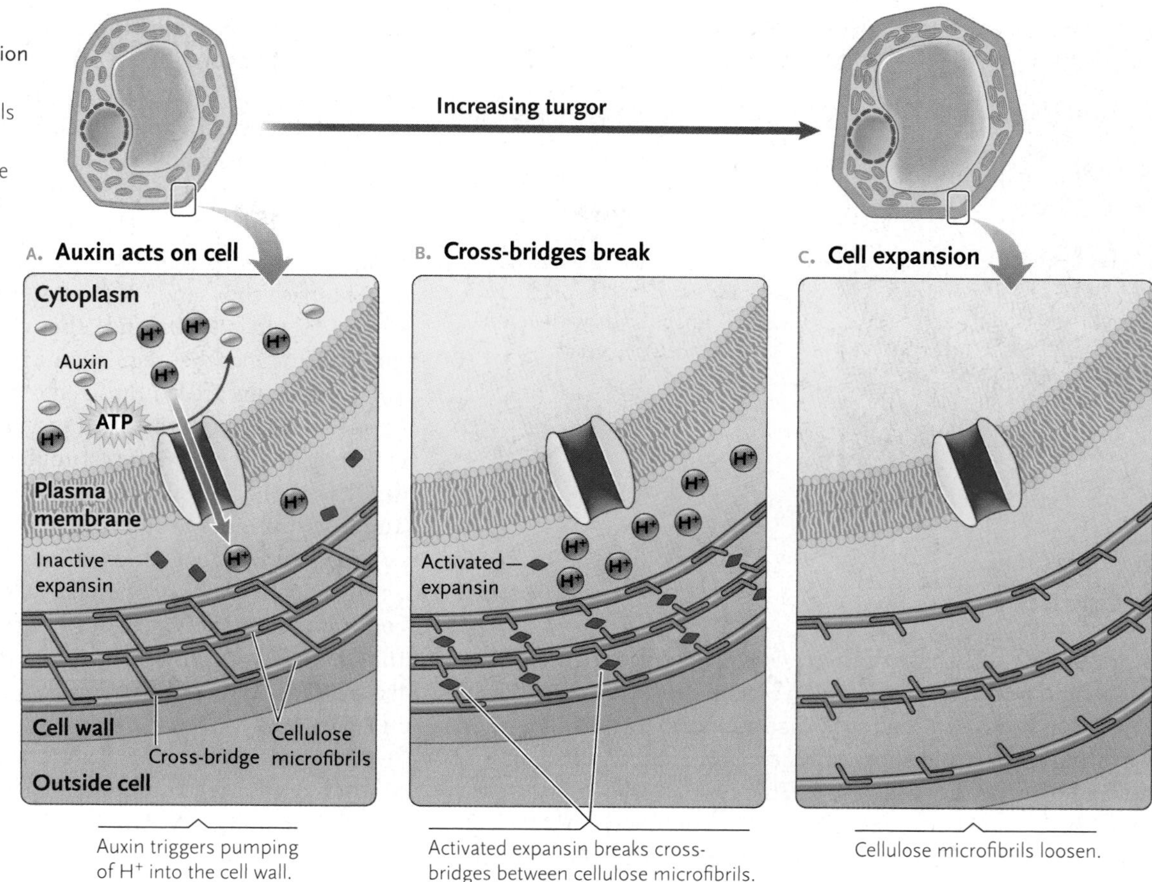

Increasing turgor
A. Auxin acts on cell
Cytoplasm
Auxin
ATP
Plasma membrane
Inactive expansin
Cell wall
Cross-bridge
Cellulose microfibrils
Outside cell
Auxin triggers pumping of H+ into the cell wall.
B. Cross-bridges break
Activated expansin
Activated expansin breaks cross-bridges between cellulose microfibrils.
C. Cell expansion
Cellulose microfibrils loosen.

cell wall and disrupt bonds between cellulose microfibrils in the wall **(Figure 35.6).** Activation of the plasma membrane H^+ pump also produces a membrane potential that pulls K^+ and other cations into the cell; the resulting osmotic gradient draws water into the cell, increasing turgor pressure and helping to stretch the "loosened" cell walls.

It is also possible that auxin triggers the expression of genes encoding enzymes that play roles in the synthesis of new wall components. Plant cells exposed to IAA do not show increased growth if they are treated with a chemical that inhibits protein synthesis. However, researchers have identified mRNAs that rapidly increase in concentration within 10–20 min after stem sections have been treated with auxin, although they still do not know exactly which proteins these mRNAs encode.

Gibberellins Also Stimulate Growth, Including the Elongation of Stems

Gibberellins stimulate various aspects of plant growth. Collectively they make up the largest class of plant hormones, with more than 130 recognized chemical variations. Gibberellins have been isolated from fungi and from flowering plants, including eudicots and some monocots. They may also exist in other plant groups as well. Among other effects, they trigger stem elongation and prompt seeds and buds to break dormancy. Research on barley embryos showed that a gibberellin provides signals during germination that lead to the enzymatic breakdown of endosperm, releasing nutrients that nourish the developing seedling (see Section 34.3).

Perhaps most apparent to humans is the ability of gibberellins to promote the lengthening of plant stems by stimulating both cell division and cell elongation. Synthesized in shoot and root tips and young leaves, gibberellins, like auxin, modify the properties of plant cell walls in ways that promote expansion (although the gibberellin mechanism does not involve acidification of the cell wall). It may be that the two hormones both affect expansins, or are functionally linked in some other way.

In most plant species that have been analyzed, the main controller of stem elongation is the gibberellin called GA_1. Normally, GA_1 is synthesized in small amounts in young leaves and transported throughout the plant in the phloem. When GA_1 synthesis goes awry, the outcome is a dramatic change in the plant's stature. For example, experiments with a dwarf variety of peas *(Pisum sativum)* and some other species show that these plants and their taller relatives differ at a single gene locus. Normal plants make an enzyme required for gibberellin synthesis; dwarf plants of the same species lack the enzyme, and their internodes elongate very little.

Another stark demonstration of the effect gibberellins can have on internode growth is **bolting,** growth of a floral stalk in plants that form vegetative rosettes, such as cabbages *(Brassica oleracea),* iceberg lettuce *(Lactuca sativa),* and tarweeds such as the silversword *(Argyroxiphium sandwicense).* In a rosette plant, stem internodes are so short that the leaves appear to arise from

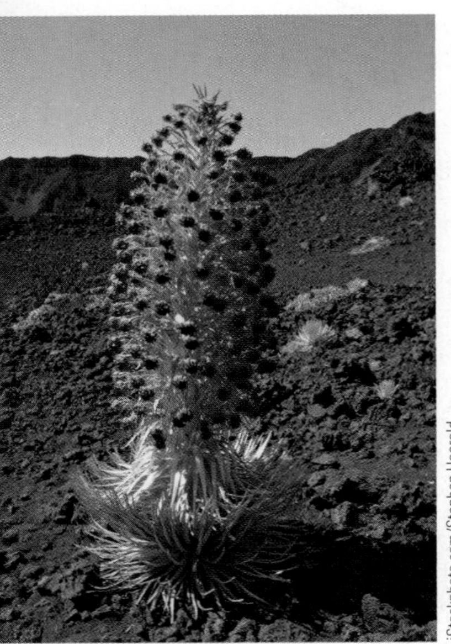
iStockphoto.com/Stephan Hoerold

FIGURE 35.7
Bolting. This Haleakala silversword, one of 28 endangered tarweed species that evolved in the isolation of the Hawaiian Islands, grows only along the upper fringe of Maui's volcanic Haleakala Crater. The plants may grow in a compact rosette form for more than half a century before several weeks of bolting produce a tall floral shoot, as shown here. After flowering, the plant dies.

a single node. When these plants flower, however, the stem elongates rapidly and flowers develop on the new stem parts **(Figure 35.7).** In nature, external cues such as increasing day length or warming after a cold snap stimulate gibberellin synthesis, and bolting occurs soon afterward. This observation supports the hypothesis that in rosette plants and possibly some others, gibberellins switch on internode lengthening when environmental conditions favor a shift from vegetative growth to reproductive growth.

Beyond the effects just mentioned, in monoecious species (having flowers of both sexual types on the same plant), applications of a gibberellin seems to encourage proportionately more "male" flowers to develop. As a result, there may be more pollen available to pollinate "female" flowers and, eventually, more fruit produced.

Cytokinins Enhance Growth and Retard Aging

Cytokinins play a major role in stimulating cell division (hence the name, which refers to cytokinesis) and many of their effects appear to come about as they interact with auxins. They were first discovered during experiments designed to define the nutrient media required for plant tissue culture. Researchers found that in addition to a carbon source such as sucrose or glucose, minerals, and certain vitamins, cells in culture also required two other substances. One was auxin, which promoted the elongation of plant cells but did not stimulate the cells to divide. The other substance could be coconut milk, which is actually liquid endosperm, or it could be DNA that had been degraded into smaller molecules by boiling. When either was added to a culture medium along with an auxin, the cultured cells would begin dividing and grow normally.

We now know that the active ingredients in both boiled DNA and endosperm are cytokinins, which have a chemical structure similar to that of the nucleic acid base adenine. The most abundant natural cytokinin is zeatin, so-called because it was first isolated from the endosperm of young corn seeds *(Zea mays)*. In endosperm, zeatin probably promotes the burst of cell division that takes place as a fruit matures. As you might expect, cytokinins also are abundant in the rapidly dividing meristem tissues of root and shoot tips. Cytokinins occur not only in flowering plants but also in many conifers, mosses, and ferns. They are also synthesized by many soil-dwelling bacteria and fungi and may be crucial to the growth of mycorrhizae, which help nourish thousands of plant species (see Section 33.3). Conversely, *Agrobacterium* and other microbes that cause plant tumors carry genes that regulate the production of cytokinins.

Cytokinins are synthesized mainly in root tips and apparently travel through the plant in xylem sap. Besides promoting cell division generally, they stimulate the growth of lateral buds and have other developmental and metabolic effects. For example, cytokinins promote expansion of young leaves (as leaf cells expand), cause chloroplasts to mature, and retard leaf aging. In concert with auxin they also coordinate the growth of roots and shoots. Investigators culturing tobacco tissues found that the relative amounts of auxin and a cytokinin strongly influenced not only growth, but also development, as illustrated in **Figure 35.8.**

Natural cytokinins can prolong the life of stored vegetables. Similar synthetic compounds are already widely used to prolong the shelf life of lettuces and mushrooms and to keep cut flowers fresh.

Ethylene Regulates Plant Aging and Some Other Responses

Most parts of a plant can produce **ethylene,** which is present in fruits, flowers, seeds, leaves, and roots and helps regulate a wide variety of plant physiological responses. Ethylene also is an unusual hormone, in part because it is structurally simple (see Table 35.1) and in part because it is a gas at normal temperature and pressure.

ETHYLENE AND PLANT SENESCENCE In plants, the aging of plant parts is termed **senescence.** Senescence is a closely controlled process of deterioration that leads to the death of plant cells. In autumn the leaves of deciduous trees senesce, often turning yellow or red as chlorophyll and proteins break down, allowing other pigments to become more noticeable. Ethylene triggers the expression of genes leading to the synthesis of chlorophyllases and proteases, enzymes that launch the breakdown process. In many plants, senescence is associated with **abscission,** the dropping of flowers, fruits, and leaves in response to environmental signals. In this process, ethylene apparently stimulates the activity of enzymes that digest cell walls in an abscission zone—a localized region at the base of the petiole. The petiole detaches from the stem at that point **(Figure 35.9).**

Senescence appears to require a range of cues. For some species, the funneling of nutrients into reproductive parts may help to trigger senescence of leaves, stems, and roots. When the drain of nutrients is halted by removing each newly emerging flower or seed pod, a plant's leaves and stems stay green and vigorous much longer **(Figure 35.10).** Gardeners routinely remove flower buds from many plants to maintain vegetative growth. Evidence suggests

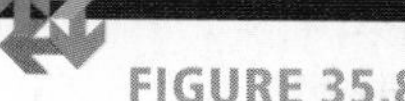

FIGURE 35.8 **Experimental Research**

Evidence That Auxin and Cytokinin Jointly Influence Plant Growth and Development

Question: Is an interaction between auxin and cytokinin important in the development of plant roots and shoots?

Experiment: Folke Skoog and Carlos Miller used callus tissue cultured from tobacco pith (spongy parenchyma from a stem) in a series of experiments to shed light on how shifts in the relative concentrations of auxin and cytokinin affect plant development.

A control culture was grown on a medium in which the auxin-to-cytokinin ratio was 10:1. Under these conditions, the growing tissue did not differentiate but remained as a callus.

1. Auxin is significantly reduced and cytokinin is increased slightly.

Results: The callus continues to grow but no differentiation occurs.

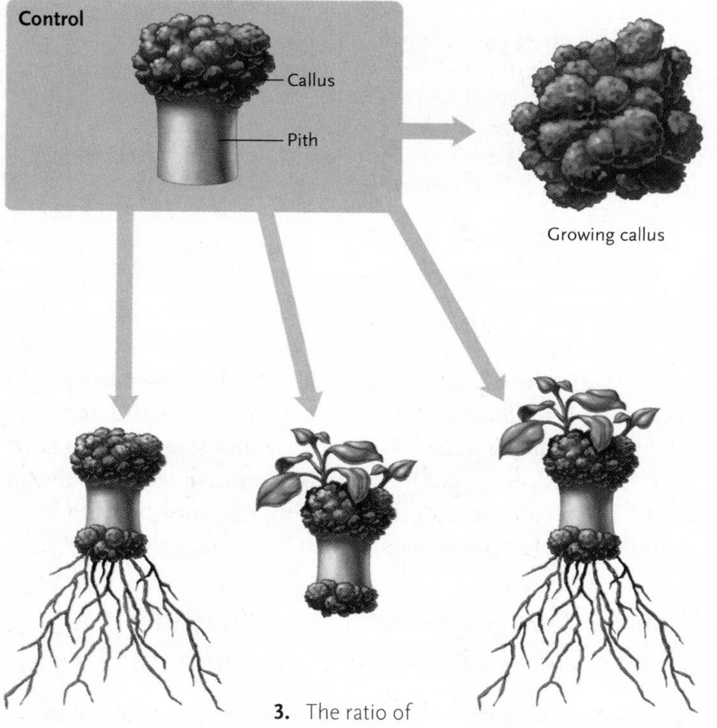

Control
Callus
Pith
Growing callus

2. The auxin-to-cytokinin ratio is increased to greater than 10:1.

Results: The cultured tissue produces roots but no differentiated shoot.

3. The ratio of cytokinin to auxin is increased.

Results: Only shoots develop.

4. The ratio of auxin to cytokinin is intermediate between the high and low values.

Results: Both shoots and roots develop.

Conclusion: In tobacco, normal development of root and shoot systems depends on the relative amounts of auxin and cytokinin present in a growing plant.

Source: F. Skoog and C.O. Miller. 1957. Chemical regulation of growth and organ formation in plant tissues cultured in vivo. *Symposia of the Society for Experimental Biology* 54(11): 118–130.

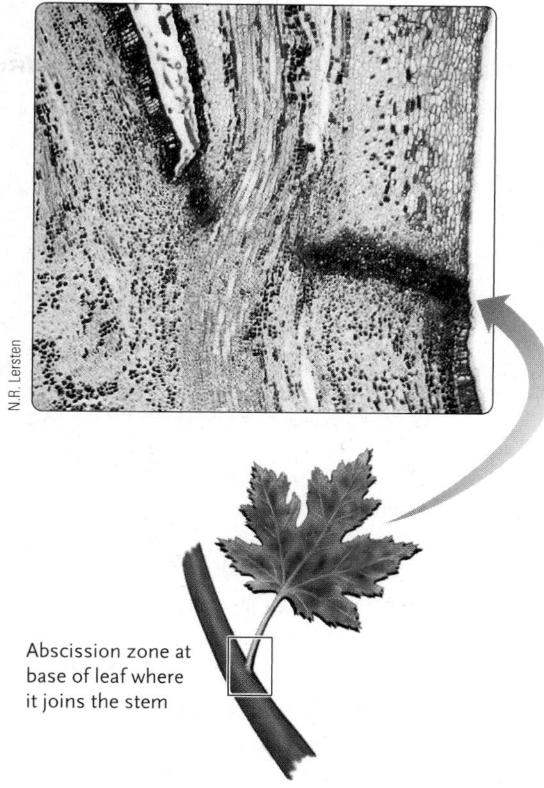

N.R. Lersten
Abscission zone at base of leaf where it joins the stem

FIGURE 35.9
Abscission zone in a maple *(Acer)*. **(A)** This longitudinal section at the left is through the base of the petiole of a leaf. **(B)** Senescing maple leaves that will soon fall from the tree.

that other cues are important as well, however. For instance, when a cocklebur is induced to flower under winterlike conditions, its leaves turn yellow regardless of whether the nutrient-demanding young flowers are left on or pinched off. It is as if a "death signal" forms that leads to flowering and senescence when there are fewer hours of daylight (typical of winter days). This observation underscores the general theme that many plant responses to the environment involve the interaction of multiple molecular signals.

FRUIT RIPENING: A FORM OF SENESCENCE Although the precise mechanisms are not well understood, ripening begins when a fruit starts to synthesize ethylene. The ripening process may involve the conversion of starch or organic acids to sugars, the softening of cell walls, or the rupturing of the cell membrane and loss of cell fluid. The same kinds of events occur in wounded plant tissues, which also synthesize ethylene.

Ethylene from an outside source can stimulate senescence responses, including ripening, when it binds to specific protein receptors on plant cells. The ancient Chinese observed that they could induce picked fruit to ripen faster by burning incense; later, it was found that the incense smoke contains

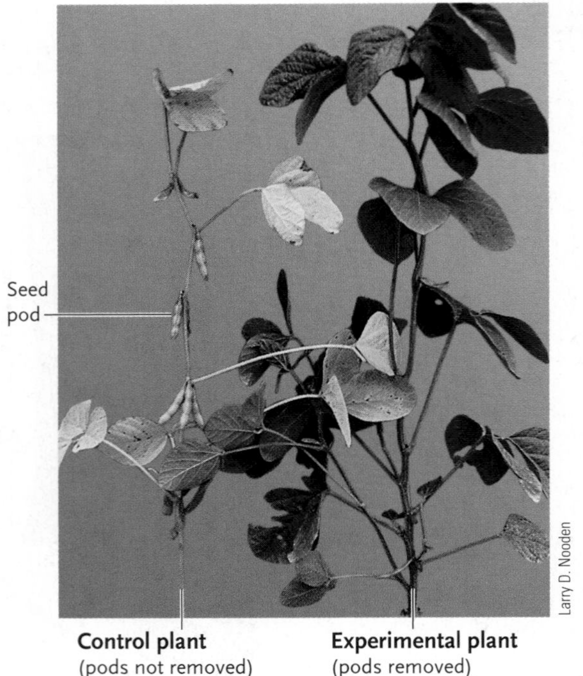
Seed pod
Control plant
(pods not removed)
Experimental plant
(pods removed)
Larry D. Nooden

FIGURE 35.10
Experimental results showing that the removal of seed pods from a soybean plant *(Glycine max)* delays its senescence.

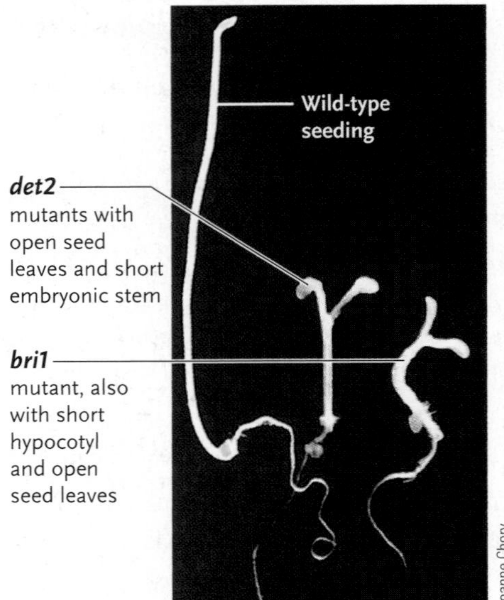
Wild-type seeding
det2
mutants with open seed leaves and short embryonic stem
bri1
mutant, also with short hypocotyl and open seed leaves
Joanne Chory

FIGURE 35.11
Experimental evidence that brassinosteroids can mediate a plant's responses to light by regulating gene expression. In *Arabidopsis,* wild-type seedlings synthesize a protein (encoded by the *DET2* gene) that prevents leaves from developing *(seedling at left)* until photosynthesis is possible, after the seedling breaks out of the dark environment of soil. When the gene is defective, a mutant *det2* plant *(center)* will develop a short hypocotyl (embryonic stem) and open seed leaves (cotyledons) even when there is no light for photosynthesis. Experiments with *bri1* mutants, which lack functioning receptors for a brassinosteroid, resulted in a similar phenotype *(right)*. These findings supported the hypothesis that a brassinosteroid is necessary for normal expression of the *DET2* gene.

ethylene. Today ethylene gas is widely used to ripen tomatoes, pineapples, bananas, honeydew melons, mangoes, papayas, and other fruit that has been picked and shipped while still green. Ripening fruit itself gives off ethylene, which is why placing a ripe banana in a closed sack of unripe peaches (or some other green fruit) often can cause the fruit to ripen. Oranges and other citrus fruits may be exposed to ethylene to brighten the color of their rind. Conversely, limiting fruit exposure to ethylene can delay ripening. Apples will keep for months without rotting if they are exposed to a chemical that inhibits ethylene production or if they are stored in an environment that inhibits the hormone's effects—including low atmospheric pressure and a high concentration of CO_2, which may bind ethylene receptors.

Brassinosteroids Regulate Plant Growth Responses

The dozens of steroid hormones classed as **brassinosteroids** all appear to be vital for normal growth in plants, for they stimulate cell division and elongation in a wide range of plant-cell types. Confirmed as plant hormones in the 1980s, brassinosteroids now are the subject of intense research on their sources and effects. Although brassinosteroids have been detected in a wide variety of plant tissues and organs, the highest concentrations are found in shoot tips and in developing seeds and embryos—all examples of young, actively developing parts. In laboratory studies, the hormones have different effects depending on the tissue where they are active. They have promoted cell elongation, differentiation of vascular tissue, and elongation of a pollen tube after a flower is pollinated. By contrast, they inhibit the elongation of roots. First isolated from pollen of a plant in the mustard family, *Brassica napus* (a type of canola), brassinosteroids seem to regulate the expression of genes associated with a plant's growth responses to light. This role was underscored by the outcomes of experiments using mutant *Arabidopsis* plants that were homozygous for a defective gene called *bri1* (for brassinosteroid-insensitive receptor) **(Figure 35.11).** The results provided convincing evidence that brassinosteroids mediate growth responses to light.

Abscisic Acid Suppresses Growth and Influences Responses to Environmental Stress

The hormone **abscisic acid** (ABA) has a variety of effects, many of which represent evolutionary adaptations to environmental challenges. Plants apparently synthesize ABA from carotenoid pigments inside plastids in leaves and possibly other plant parts. Several ABA receptors have been identified, and in general, we can group ABA effects into changes in gene expression that result in long-term inhibition of growth, and rapid, short-term physiological changes that are responses to immediate stresses, such as a lack of water, in a plant's surroundings. As its name suggests, at one time ABA was thought to play a major role in abscission. As already described, however, we now know that ethylene plays the major role in abscission.

SUPPRESSING GROWTH IN BUDS AND SEEDS Operating as a counterpoint to growth-stimulating hormones like gibberellins, ABA inhibits growth in response to environmental cues, such as seasonal changes in temperature and light. This growth suppression can last for many months or even years. For example, one of ABA's major growth-inhibiting effects is apparent in perennial plants, in which the hormone promotes dormancy in leaf buds—an important adaptive advantage in places where winter cold can damage young leaves. If ABA is applied to a growing leaf bud, the bud's normal development stops, and instead protective *bud scales*—modified, nonphotosynthetic leaves that are small, dry, and tough—form around the apical meristem and insulate it from the elements **(Figure 35.12)**. After the scales develop, most cell metabolic activity shuts down and the leaf bud becomes dormant.

In some plants that produce fleshy fruits, such as apples and cherries, abscisic acid is associated with the dormancy of seeds as well. As the seed develops, ABA accumulates in the seed coat, and the embryo does not germinate even if it becomes hydrated. The buildup of ABA in developing seeds does more than simply inhibit development, however. As early development draws to a close, ABA stimulates the transcription of certain genes, and large amounts of their protein products are synthesized. These proteins are thought to store nitrogen and other nutrients that the embryo will use when it eventually does germinate. Before such a seed can germinate, it usually will require a long period of cool, wet conditions, which stimulate the breakdown of ABA. Commercial growers often apply ABA and related growth inhibitors to plants slated to be shipped to plant nurseries. Dormant plants suffer less shipping damage, and the effects of the inhibitors can be reversed by applying a gibberellin.

RESPONSES TO ENVIRONMENTAL STRESS ABA also triggers plant responses to various environmental stresses, including cold snaps, high soil salinity, and drought. A great deal of research has focused on how ABA influences plant responses to a lack of water. When a plant is water-stressed, ABA helps prevent excessive water loss by stimulating stomata to close. As described in Section 32.3, flowering plants depend heavily on the proper functioning of stomata. When a lack of water leads to wilting, mesophyll cells in wilted leaves rapidly synthesize and secrete ABA. The hormone diffuses to guard cells, where an ABA receptor binds it. Binding stimulates the release of K^+ and water from the guard cells, and within minutes the stomata close.

Once bound to its receptor, ABA may exert its effects through a cascade of signals that includes phosphorylated proteins. Experiments have shown that an *Arabidopsis* mutant unable to respond to ABA lacks an enzyme that removes phosphate groups from certain proteins. This condition suggests that cleaving phosphates is one step in the ABA response. *Insights from the Molecular Revolution* highlights recent research filling in other steps in the ABA-induced response pathway.

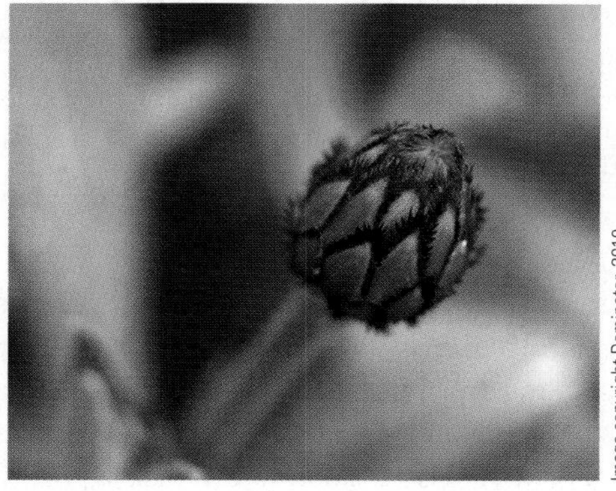

FIGURE 35.12
Bud scales, here on the bud of a perennial cornflower *(Centaurea montana)*.

Jasmonates and Oligosaccharins Regulate Growth and Function in Defense

In recent years, studies of plant growth and development have helped define the roles—or revealed the existence—of several other hormonelike compounds in plants. Like the well-established plant hormones just described, these substances are organic molecules, and only tiny amounts are required to alter some aspect of a plant's functioning. Some have long been known to exist in plants, but the extent of their signaling roles has only recently become better understood. This group includes **jasmonates** (JA), a family of about 20 compounds derived from fatty acids. Experiments with *Arabidopsis* and other plants have revealed numerous genes that respond to JA, including genes that help regulate root growth and seed germination. JA also appears to help plants "manage" stresses caused by deficiencies of certain nutrients (such as K^+). The JA family is best known, however, as part of the plant arsenal to limit damage by pathogens and predators, the topic of the following section.

Some other substances also are drawing keen interest from plant scientists, but because their signaling roles are still poorly understood they are not widely accepted as confirmed plant hormones. A case in point involves the complex carbohydrates that are structural elements in the cell walls of plants and some fungi. Several years ago, researchers observed that in some plants, some of these oligosaccharides could serve as signaling molecules. Such compounds were named **oligosaccharins,** and one of their known roles is to defend the plant against pathogens. In addition, oligosaccharins have been proposed as growth regulators that adjust the growth and differentiation of plant cells, possibly by modulating the influences of growth-promoting hormones such as auxin. At this writing, researchers in many laboratories are pursuing a deeper understanding of this curious subset of plant signaling molecules.

STUDY BREAK 35.1

1. **Which plant hormones promote growth and which inhibit it?**
2. **Give examples of how some hormones have both promoting and inhibiting effects on growth in different parts of the plant at different times of the life cycle.**

THINK OUTSIDE THE BOOK

This section discussed several target tissues for auxin, including cells just below the apical meristem of a shoot part, young lateral meristems, and vascular cambium. Collaboratively or on your own, write a short explanation of why each of these tissues responds differently to the hormone. You may find it helpful to reread the introduction to Chapter 16, which gives an overview of some early events in development.

INSIGHTS FROM THE Molecular Revolution

Stressing Out in Plants and People

Recent molecular work shows that plant responses to stress imposed by drought and cold involve some of the same chemical steps in plants and humans, indicating an ancient link to a common evolutionary ancestor. The research may also point the way to genetic engineering strategies to modify major crop plants for earlier and better responses to stress.

Many plant-stress responses are triggered by the hormone abscisic acid (ABA). Although individual steps in the response pathway are unclear, it is known that calcium ions increase in concentration in the cytoplasm when plant cells are exposed to ABA. Soon after the rise in Ca^{2+}, genes are activated that compensate for the stressful situation.

Nam-Hai Chua and his colleagues at the Rockefeller University and the University of Minnesota were interested in piecing together the molecular steps in the plant pathway. One substance they thought might be involved is *cyclic ADP-ribose (cADPR)*, a signaling molecule that was first implicated in calcium release pathways in animal cells.

Research Question: Is *cADPR* involved in the plant response to stress?

Experiments: In a series of experiments with tomato plants **(Figure 1)** the team demonstrated cADPR's role in the pathway.

Another experiment determined whether protein phosphorylation might be part of the pathway. To accomplish this, the investigators injected an inhibitor of protein kinases, the enzymes that phosphorylate proteins as a part of many cellular response pathways. After the inhibitor was added, injecting ABA, cADPR, or Ca^{2+} failed to activate *rd29A* and *kin2*, indicating that protein phosphorylation is a crucial step following these elements in the pathway.

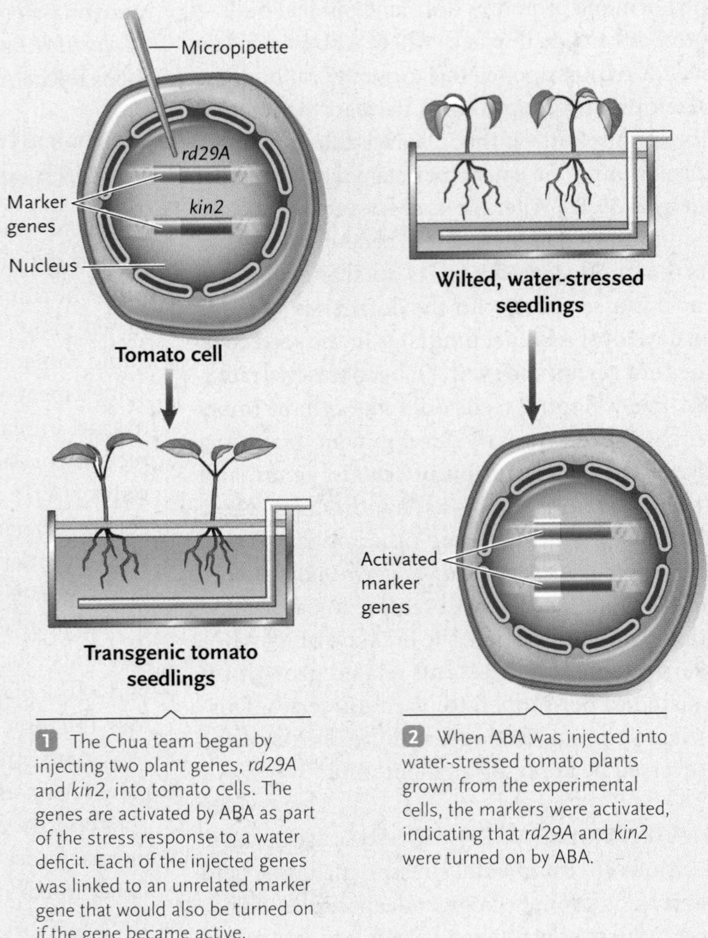
Micropipette
rd29A
kin2
Marker genes
Nucleus
Tomato cell
Transgenic tomato seedlings
Wilted, water-stressed seedlings
Activated marker genes
1 The Chua team began by injecting two plant genes, rd29A and kin2, into tomato cells. The genes are activated by ABA as part of the stress response to a water deficit. Each of the injected genes was linked to an unrelated marker gene that would also be turned on if the gene became active.
2 When ABA was injected into water-stressed tomato plants grown from the experimental cells, the markers were activated, indicating that rd29A and kin2 were turned on by ABA.

FIGURE 1
Piecing together the role of cADPR in the plant response to water stress.

Conclusion: From these results the Chua team was able to reconstruct a major part of the pathway:

ABA → ? →cADPR → Ca^{2+}↑ → protein kinases → ? → stress gene activation

The question marks indicate one or more unknown steps. Most significant is the ABA receptor that carries out the first step in the response pathway. A recently identified ABA receptor that is involved in the events that cause stomata to close (among other effects) may be this "missing link."

Because of the Chua team's research, experiments by others have begun to shed light on signal pathways in animals in which cADPR plays a part. For example, a study by researchers in Italy found that certain human white blood cells released ABA as part of an inflammation response. These and other discoveries are helping fill in steps in both

35.2 Plant Chemical Defenses

Plants do not have immune systems like those that have evolved in animals (the subject of Chapter 43). Even so, over the millennia, in higher plants virtually constant exposure to predation by herbivores and the onslaught of pathogens have resulted in a striking array of chemical defenses that ward off or reduce damage to plant tissues from infectious bacteria, fungi, worms, or plant-eating insects **(Table 35.2)**. You will discover in this section that as with the defensive strategies of animals, plant defenses include both general responses to any type of attack and specific responses to particular threats. Some get underway almost as soon as an attack begins, whereas others help promote the plant's long-term survival. And more often than not, multiple chemicals interact as the response unfolds.

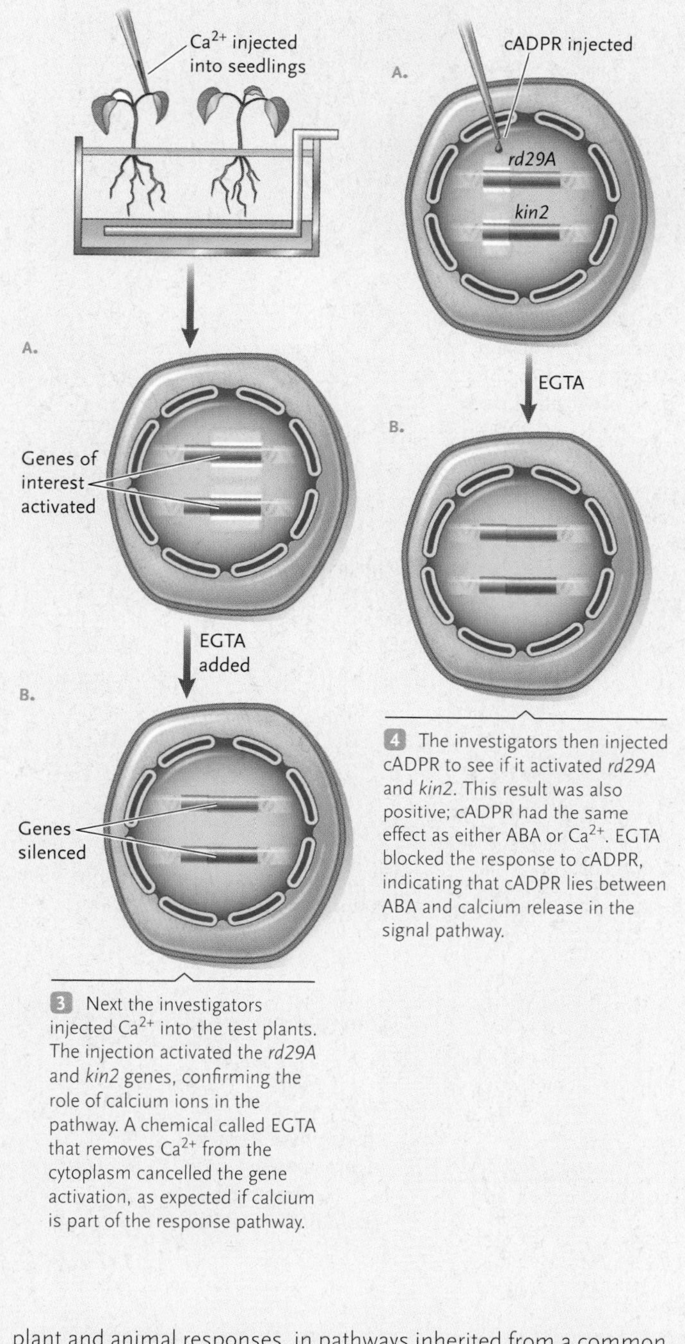
Ca^{2+} injected into seedlings
A.
Genes of interest activated
EGTA added
B.
Genes silenced
cADPR injected
A.
rd29A
kin2
EGTA
B.

3 Next the investigators injected Ca^{2+} into the test plants. The injection activated the *rd29A* and *kin2* genes, confirming the role of calcium ions in the pathway. A chemical called EGTA that removes Ca^{2+} from the cytoplasm cancelled the gene activation, as expected if calcium is part of the response pathway.

4 The investigators then injected cADPR to see if it activated *rd29A* and *kin2*. This result was also positive; cADPR had the same effect as either ABA or Ca^{2+}. EGTA blocked the response to cADPR, indicating that cADPR lies between ABA and calcium release in the signal pathway.

plant and animal responses, in pathways inherited from a common ancestor predating both plants and people.

Source: Y. Wu et al. 1997. Abscisic acid signaling through cyclic ADP-ribose in plants. *Science* 278:2126–2130.

TABLE 35.2 Summary of Plant Chemical Defenses

Type of Defense	Effects
General Defenses	
Jasmonate (JA) responses to wounds/injury by pathogens; pathways often include other hormones such as ethylene	Synthesis of defensive chemicals such as protease inhibitors
Hypersensitive response to infectious pathogens (e.g., fungi, bacteria)	Physically isolates infection site by surrounding it with dead cells
PR (pathogenesis-related) proteins	Enzymes, other proteins that degrade cell walls of pathogens
Salicylic acid (SA)	Mobilized during other responses and independently; induces the synthesis of PR proteins, operates in systemic acquired resistance
Systemin (in tomato)	Triggers JA response
Secondary metabolites	
Phytoalexins	Antibiotic
Oligosaccharins	Trigger synthesis of phytoalexins
Systemic acquired resistance (SAR)	Long-lasting protection against some pathogens; components include SA and PR proteins that accumulate in healthy tissues
Specific Defenses	
Gene-for-gene recognition of chemical features of specific pathogens (by binding with receptors coded by R genes)	Triggers defensive response (e.g., hypersensitive response, PR proteins) against pathogens
Other	
Heat-shock responses (encoded by heat-shock genes)	Synthesis of chaperone proteins that reversibly bind other plant proteins and prevent denaturing caused by heat stress
"Antifreeze" proteins	In some species, stabilize cell proteins under freezing conditions

Jasmonates and Other Compounds Interact in a General Response to Wounds

When an insect begins feeding on a leaf or some other plant part, the plant may respond to the resulting wound by launching what in effect is a cascade of chemical responses. These complex signaling pathways often rely on interactions among jasmonates, ethylene, or some other plant hormone. As the pathway unfolds it triggers expression of genes leading to chemical and physical defenses at the wound site. For example, in some plants jasmonate induces a response leading to the synthesis of protease inhibitors, which disrupt an insect's capacity to digest proteins in the plant tissue. The protein deficiency in turn hampers the insect's growth and functioning.

A plant's capacity to recognize and respond to the physical damage of a wound apparently has been subject to strong selection pressure during plant evolution. When a plant is wounded experimentally, numerous defensive chemicals can be detected in its tissues in relatively short order. One of these, **salicylic acid,** or **SA** (a compound similar to aspirin, which is acetylsalicylic acid), seems to have multiple roles in plant defenses, including interacting with jasmonates in signaling cascades.

Researchers are regularly discovering new variations of hormone-induced wound responses in plants. For example, experiments have elucidated some of the steps in an unusual pathway that thus far is known only in tomato *(Lycopersicum esculentum)* and a few other plant species. As diagrammed in **Figure 35.13,** the wounded plant rapidly synthesizes **systemin,** the first peptide hormone to be discovered in plants. (Various animal hormones are peptides, a topic covered in Chapter 40.) Systemin enters the phloem and is transported throughout the plant. Although various details of the signaling pathway have yet to be worked out, when receptive cells bind systemin, their plasma membranes release a lipid that is the chemical precursor of jasmonate. Next jasmonate is synthesized, and it in turn sets in motion the expression of genes that encode protease inhibitors, which protect the plant against attack, even in parts remote from the original wound.

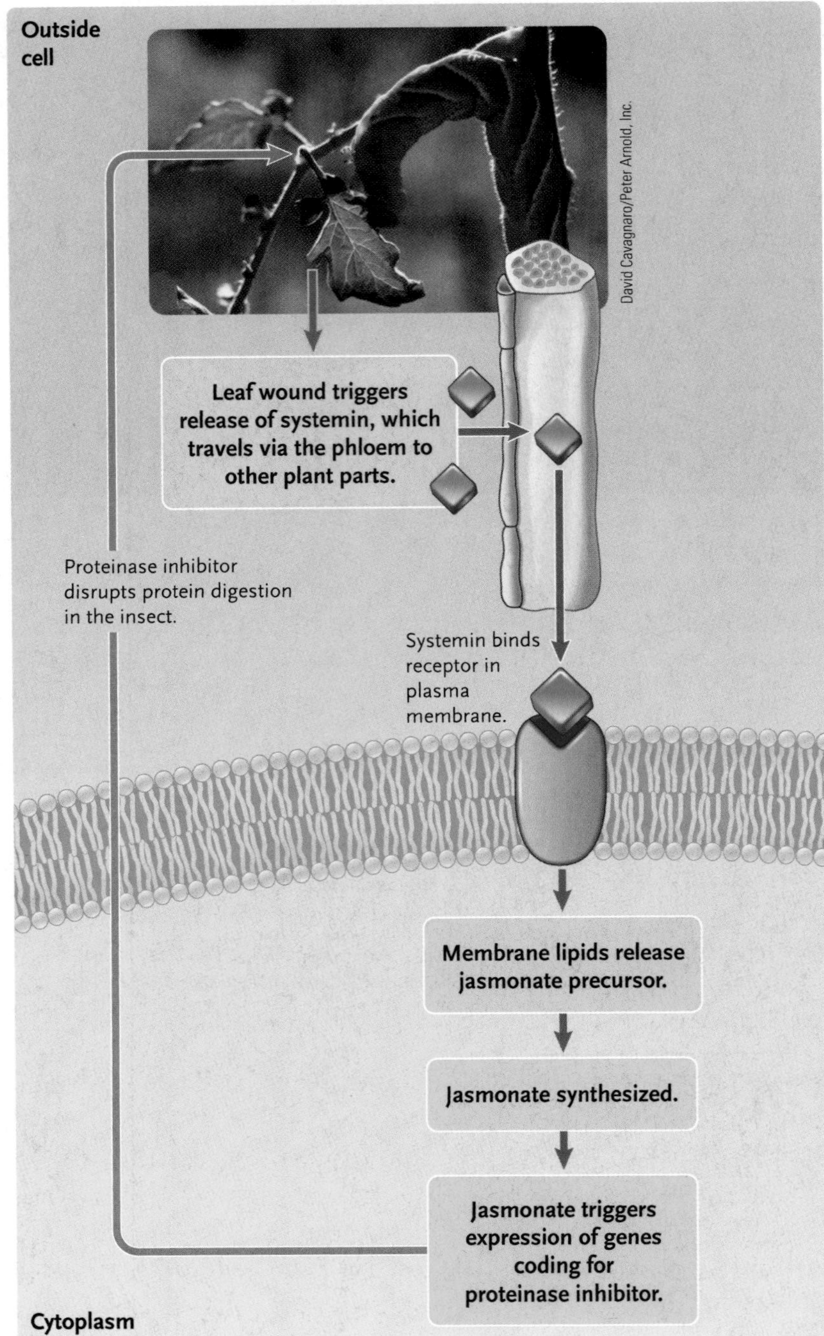
Outside cell
David Cavagnaro/Peter Arnold, Inc.
Leaf wound triggers release of systemin, which travels via the phloem to other plant parts.
Proteinase inhibitor disrupts protein digestion in the insect.
Systemin binds receptor in plasma membrane.
Membrane lipids release jasmonate precursor.
Jasmonate synthesized.
Jasmonate triggers expression of genes coding for proteinase inhibitor.
Cytoplasm

FIGURE 35.13
The systemin response to wounding. When a plant is wounded, it responds by releasing the protein hormone systemin. Transported through the phloem to other plant parts, in receptive cells systemin sets in motion a sequence of reactions that lead to the expression of genes encoding protease inhibitors—substances that can seriously disrupt an insect predator's capacity to digest protein.

The Hypersensitive Response and PR Proteins Are Other General Defenses

Often, a plant that becomes infected by pathogenic bacteria or fungi counters the attack by way of a **hypersensitive response**—a defense that physically cordons off an infection site by surrounding it with dead cells. Initially, cells near the site respond by producing a burst of highly reactive oxygen-containing compounds (such as hydrogen peroxide, H_2O_2) that can break down nucleic acids, inactivate enzymes, or have other toxic effects on cells. The burst is catalyzed by enzymes in the plant cell's plasma membrane. It may begin the process of killing cells close to the attack site and, as the response advances, programmed cell death may also come into play. In short order, the "sacrificed" dead cells wall off the infected area from the rest of the plant. Thus denied an ongoing supply of nutrients, the invading pathogen dies. A common sign of a successful hypersensitive response is a dead spot surrounded by healthy tissue **(Figure 35.14).**

While the hypersensitive response is underway, salicylic acid triggers other defensive responses by an infected plant. One of its effects is to induce the synthesis of **pathogenesis-related proteins,** or **PR proteins.** Some PR proteins are hydrolytic enzymes that break down components of a pathogen's cell wall. Examples are chitinases that dismantle the chitin in the cell walls of fungi and so kill the cells. In some cases, plant cell receptors also detect the presence of fragments of the disintegrating wall and set in motion additional defense responses.

Secondary Metabolites Defend against Pathogens and Herbivores

Many plants counter bacteria and fungi by making **phytoalexins,** biochemicals of various types that function as antibiotics. When an infectious agent breaches a plant part, genes encoding phytoalexins begin to be transcribed in the affected tissue. For instance, when a fungus invades plant tissues, the enzymes it secretes may

Nigel Cattlin/Photo Researchers Inc.

FIGURE 35.14
Evidence of the hypersensitive response. The dead brown spots on these leaves of a strawberry plant *(Fragaria)* are sites where a pathogen invaded, triggering the defensive destruction of the surrounding cells.

trigger the release of oligosaccharins. In addition to their roles as growth regulators (described in Section 35.1), these substances also can promote the production of phytoalexins, which have toxic effects on a variety of fungi. Plant tissues may also synthesize phytoalexins in response to attacks by viruses.

Phytoalexins are among many *secondary metabolites* produced by plants. Such substances are termed "secondary" because they are not routinely synthesized in all plant cells as part of basic metabolism. A wide range of plant species deploy secondary metabolites as defenses against feeding herbivores. Examples are alkaloids such as caffeine, cocaine, and the poison strychnine (in seeds of the *nux vomica* tree, *Strychnos nux-vomica*), tannins such as those in oak acorns, and various terpenes. The terpene family includes insect-repelling substances in conifer resins and cotton, and essential oils produced by sage and basil plants. Because these terpenes are volatile—they easily diffuse out of the plant into the surrounding air—they also can provide indirect defense to a plant. Released from the wounds created by a munching insect, they attract other insects that prey on the herbivore. Chapter 51 looks in detail at the interactions between plants and herbivores.

Gene-for-Gene Recognition Allows Rapid Responses to Specific Threats

One of the most interesting questions with respect to plant defenses is how plants first sense that an attack is underway. In some instances plants apparently can detect an attack by a specific predator through a mechanism called **gene-for-gene recognition.** This term refers to a matchup between the products of dominant alleles of two types of genes: a so-called ***R* gene** (for "resistance") in a plant, and an ***Avr* gene** (for "avirulence") in a particular pathogen. Thousands of R genes have been identified in a wide range of plant species. Dominant R alleles confer enhanced resistance to plant pathogens including bacteria, fungi, and nematode worms that attack roots.

The basic mechanism of gene-for-gene recognition is simple: The dominant R allele encodes a receptor in plasma membranes of a plant's cells, and the dominant pathogen *Avr* allele encodes a molecule that can bind the receptor. "Avirulence" implies that the pathogen becomes "not virulent," because binding of its Avr gene product triggers an immediate defense response in the plant. Trigger molecules run the gamut from proteins to lipids to carbohydrates that have been secreted by the pathogen or released from its surface **(Figure 35.15).** Experiments have demonstrated a rapid-fire sequence of early biochemical changes that follow binding of the Avr-encoded molecule; these include changes in ion concentrations inside and outside plant cells and

Required precondition
A plant has a dominant R gene encoding a receptor that can bind the product of a specfic pathogen dominant Avr gene.

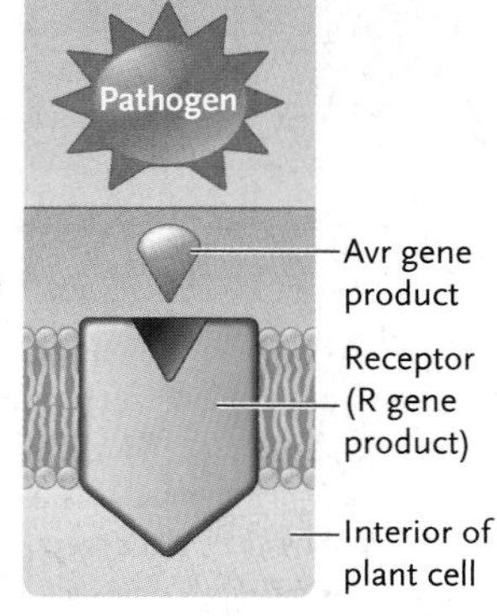
Pathogen
Avr gene product
Receptor (R gene product)
Interior of plant cell

1 When the R-encoded receptor binds its matching Avr product, the binding triggers signaling pathways, leading to various defense responses in the plant.

2 Fluxes of ions and enzyme activity at the plasma membrane contribute to the hypersensitive response. Soon PR proteins, phytoalexins, and salicyic acid (SA) are synthesized. The PR proteins and phytoalexins combat pathogens directly. SA promotes systemic acquired resistance.

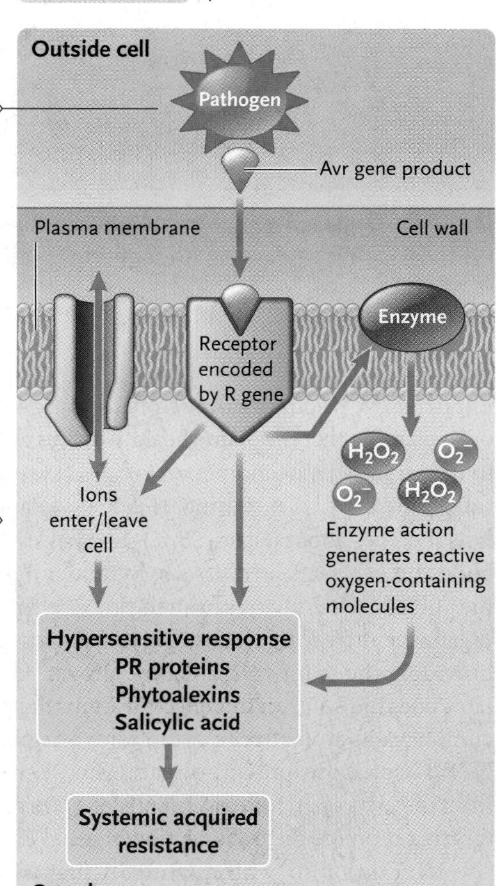
Outside cell
Pathogen
Avr gene product
Plasma membrane
Cell wall
Enzyme
Receptor encoded by R gene
Ions enter/leave cell
H_2O_2
O_2^-
Enzyme action generates reactive oxygen-containing molecules
Hypersensitive response
PR proteins
Phytoalexins
Salicylic acid
Systemic acquired resistance
Cytoplasm

FIGURE 35.15
Model of how gene-for-gene resistance may operate. For resistance to develop, the plant must have a dominant R gene and the pathogen must have a corresponding dominant *Avr* gene. Products of such "matching" genes can interact physically, rather like the lock-and-key mechanism of an enzyme and its substrate. Most R genes encode receptors at the plasma membranes of plant cells. As diagrammed in step 1, when one of these receptors binds an *Avr* gene's product, the initial result may be changes in the movements of specific ions into or out of the cell and the activation of membrane enzymes that catalyze the formation of highly reactive oxygen-containing molecules. Such events help launch other signaling pathways that lead to a variety of defensive responses, including the hypersensitive response (step 2).

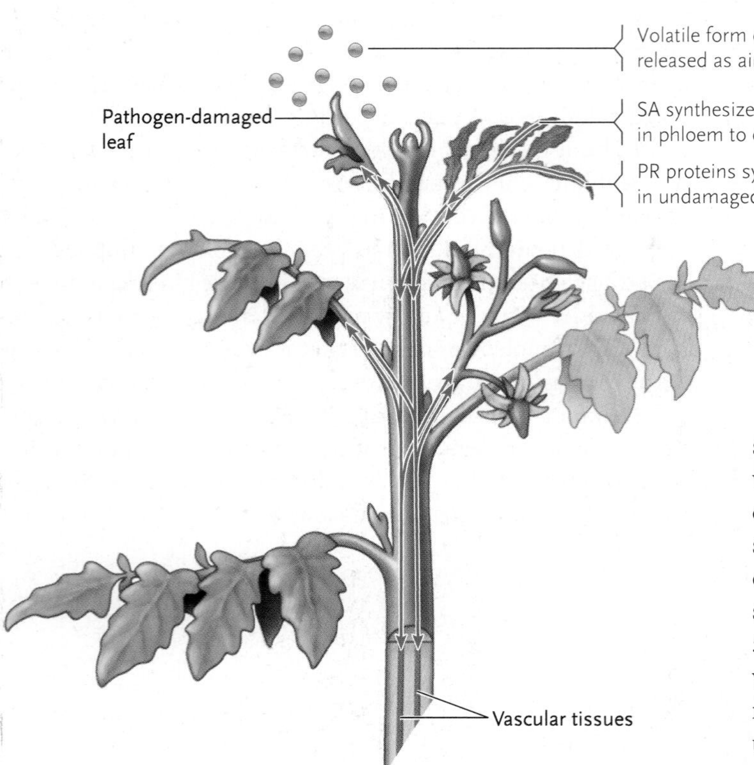
Volatile form of SA (methyl salicylate) released as airborne signal
Pathogen-damaged leaf
SA synthesized and transported in phloem to other organs
PR proteins synthesized in undamaged new growth
Vascular tissues

FIGURE 35.16
A proposed mechanism for systemic acquired resistance. When a plant successfully fends off a pathogen, the defensive chemical salicylic acid (SA) is transported in the phloem to other plant parts, where it may help protect against another attack by stimulating the synthesis of PR proteins. In addition, the plant synthesizes and releases a slightly different, more volatile form of SA called methyl salicylate. It may serve as an airborne signal to other parts of the plant as well as to neighboring plants.

the production of biologically active oxygen compounds that heralds the hypersensitive response. In fact, of the instances of gene-for-gene recognition plant scientists have observed thus far, most trigger the hypersensitive response and the ensuing synthesis of PR proteins, with their antibiotic effects.

Systemic Acquired Resistance Can Provide Long-Term Protection

The defensive response to a microbial invasion may spread throughout a plant, so that the plant's healthy tissues become less vulnerable to infection. This phenomenon is called **systemic acquired resistance,** and experiments using *Arabidopsis* plants have shed light on how it comes about **(Figure 35.16).** In a key early step, salicylic acid builds up in the affected tissues. By some route, probably through the phloem, the SA passes from the infected organ to newly forming organs such as leaves, which begin to synthesize PR proteins—again, providing the plant with a "home-grown" antimicrobial arsenal. How does the SA exert this effect? It seems that when enough SA accumulates in a plant cell's cytoplasm, a regulatory protein called NPR-1 (for *n*onexpressor of *p*athogenesis-*r*elated genes) moves from the cytoplasm into the cell nucleus. There it interacts with factors that promote the transcription of genes encoding PR proteins.

In addition to synthesizing SA that will be transported to other tissues by a plant's vascular system, the damaged leaf also synthesizes a chemically similar compound, methyl salicylate. This substance is volatile, and researchers speculate that it may serve as an airborne "harm" signal, promoting defense responses in the plant that synthesized it and possibly in nearby plants as well.

Extremes of Heat and Cold Also Elicit Protective Chemical Responses

Plant cells also contain **heat-shock proteins (HSPs),** a type of chaperone protein (see Section 3.4) found in cells of many species. HSPs bind and stabilize other proteins, including enzymes, which might otherwise stop functioning if they were to become denatured by rising temperature. Plant cells may rapidly synthesize HSPs in response to a sudden temperature rise. For example, experiments with cells and seedlings of soybean *(Glycine max)* showed that when the temperature rose 10°–15°C, in less than 5 min smRNA transcripts coding for as many as 50 different HSPs were present in cells. When the temperature returns to a normal range, HSPs release bound proteins, which can then resume their usual functions. Further studies have revealed that heat-shock proteins help protect plant cells subjected to other environmental stresses as well, including drought, salinity, and cold.

Like extreme heat, freezing can also be lethal to plants. If ice crystals form in cells they can literally tear the cell apart. In many cold-resistant species, dormancy (discussed in Section 35.4) is the long-term strategy for dealing with cold, but in the short term, such as an unseasonable cold snap, some species also undergo a rapid shift in gene expression that equips cold-stressed cells with so-called antifreeze proteins. Like heat-shock proteins, these molecules are thought to help maintain the structural integrity of other cell proteins.

STUDY BREAK 35.2 <

1. **Which plant chemical defenses are general responses to attack, and which are specific to a particular pathogen?**
2. **Why is salicylic acid considered to be a general systemic response to damage?**
3. **How is the hypersensitive response integrated with other chemical defenses?**

THINK OUTSIDE THE BOOK

Experiments by a team at Pennsylvania State University in State College have provided evidence that dodder (*Cuscuta* spp.; see Figure 33.10), a parasitic weed, uses airborne chemical cues to locate suitable host plants. The findings have added fuel to an ongoing debate about the extent to which airborne molecules may serve as a "language" that also operates in plant defenses. Research this topic and summarize the various arguments for and against the existence of "chemical crosstalk" among plants.

A. Seedlings bend toward light.

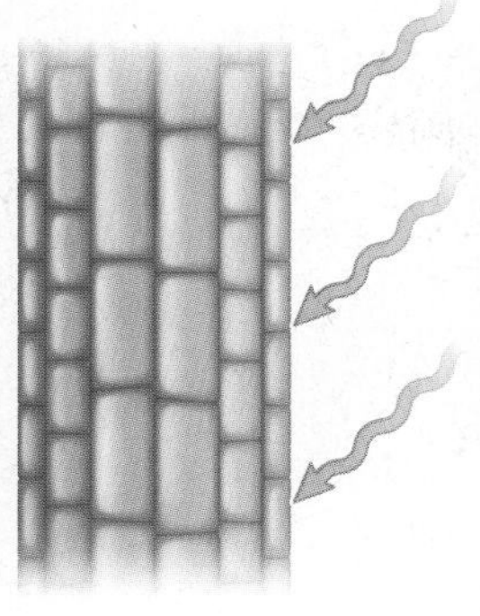

B. Rays from the sun strike one side of a shoot tip.

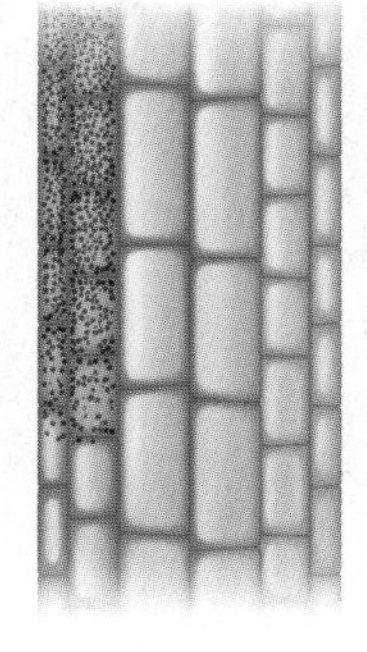

C. Auxin (red) diffuses down from the shoot tip to cells on its shaded side.

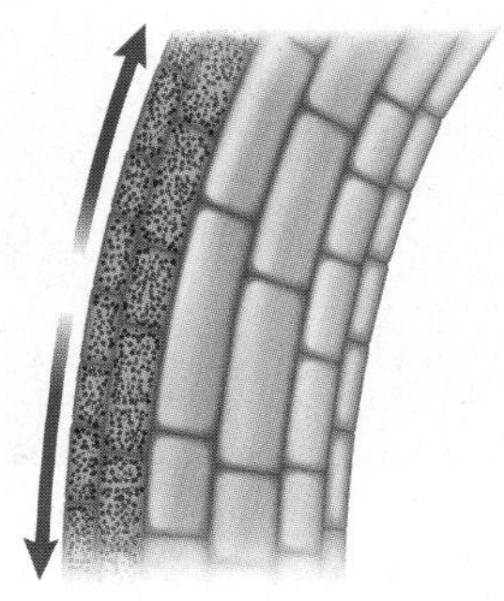

D. The auxin-stimulated cells elongate more quickly, causing the seedling to bend.

FIGURE 35.17
Phototropism in seedlings. (A) Tomato seedling grown in darkness; its right side was illuminated for a few hours before it was photographed.
(B–D) Hormone-mediated differences in the rates of cell elongation bring about the bending toward light. (Auxin is shown in red.)

35.3 Plant Movements

Although a plant cannot move from place to place as external conditions change, plants do alter the orientation of their body parts in response to environmental stimuli. As noted earlier in the chapter, growth toward or away from a unidirectional stimulus, such as light or gravity, is called a tropism. Tropic movement involves permanent changes in the plant body because cells in particular areas or organs grow differentially in response to the stimulus. Plant physiologists do not fully understand how tropisms occur, but they are fascinating examples of the complex abilities of plants to adjust to their environment. This section will also touch upon two other kinds of movements—developmental responses to physical contact, and changes in the position of plant parts that are not related to the location of the stimulus.

Phototropisms Are Responses to Light

Light is a key environmental stimulus for many kinds of organisms. Phototropisms, which we have already discussed in the section on auxins, are growth responses to a directional light source. As the Darwins discovered, if light is more intense on one side of a stem, the stem may curve toward the light **(Figure 35.17A).** Phototropic movements are extremely adaptive for photosynthesizing organisms because they help maximize the exposure of photosynthetic tissues to sunlight.

How do auxins influence phototropic movements? In a coleoptile that is illuminated from one side, IAA moves by polar transport into the cells on the shaded side **(Figure 35.17B–D).** Phototropic bending occurs because cells on the shaded side elongate more rapidly than do cells on the illuminated side.

The main stimulus for phototropism is light of blue wavelengths. Experiments on corn coleoptiles have shown that a large, yellow pigment molecule called phototropin can absorb blue wavelengths, and it may play a role in stimulating the initial lateral transport of IAA to the dark side of a shoot tip. Studies with *Arabidopsis* suggest there is more than one blue-light receptor, however. One is a light-absorbing protein called **cryptochrome,** which is sensitive to blue light and may also be an important early step in the various light-based growth responses. As you will read later, cryptochrome appears to have a role in other plant responses to light as well.

Gravitropism Orients Plant Parts to the Pull of Gravity

Plants show growth responses to Earth's gravitational pull, a phenomenon called **gravitropism.** After a seed germinates, the primary root curves down, toward the "pull" (positive gravitropism), and the shoot curves up (negative gravitropism).

Several hypotheses seek to explain how plants respond to gravity. The most widely accepted hypothesis proposes that plants detect gravity much as animals do—that is, particles called **statoliths** in certain cells move in the direction gravity pulls them. In the semicircular canals of human ears, tiny calcium carbonate crystals serve as statoliths; in most plants the statoliths are amyloplasts, modified plastids that contain starch grains (see Section 5.4). In eudicot angiosperm stems, amyloplasts often are present in one or two layers of cells just outside the vascular bundles. In monocots such as cereal grasses, amyloplasts are located in a region of tissue near the base of the leaf sheath. In roots, amyloplasts occur in the root cap. If the spatial orientation of a plant cell is shifted experimentally, its amyloplasts sink through the cytoplasm until they come to rest at the bottom of the cell **(Figure 35.18).**

How do amyloplast movements translate into an altered growth response? The full explanation appears to be complex, and there is evidence that somewhat different mechanisms operate in stems and in roots. In stems, the sinking of amyloplasts may provide a mechanical stimulus that triggers a gene-guided

A. **Root oriented vertically**

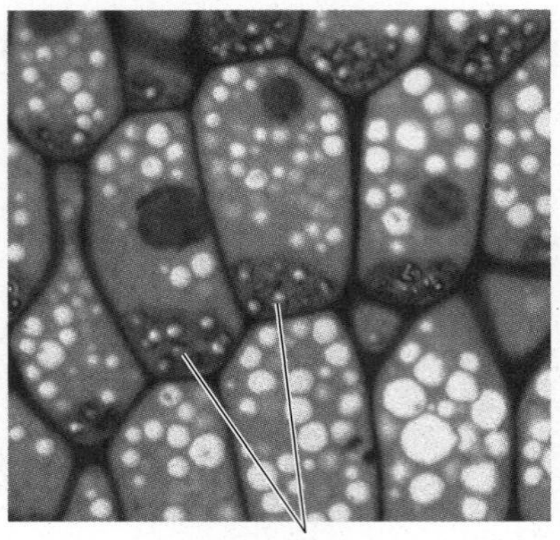
Statoliths

B. **Root oriented horizontally**

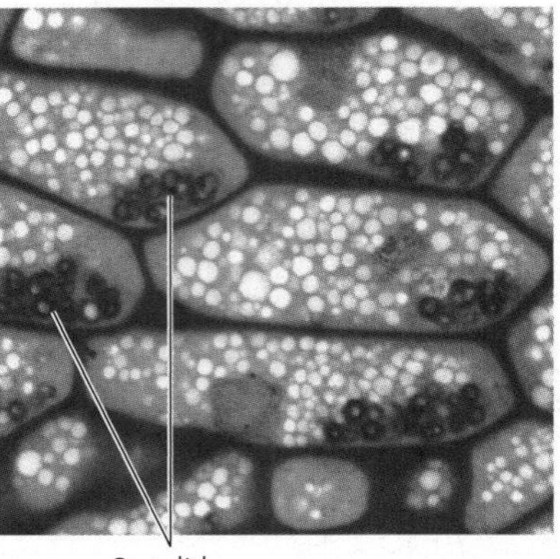
Statoliths
Micrographs courtesy of Randy Moore

FIGURE 35.18
Evidence that supports the statolith hypothesis. When a corn root was laid on its side, amyloplasts—statoliths—in cells from the root cap settled to the bottom of the cells within 5–10 min. Statoliths may be part of a gravity-sensing mechanism that redistributes auxin through a root tip.

redistribution of IAA. For example, when a potted sunflower seedling is turned on its side in a dark room, within 15–20 minutes cell elongation decreases markedly on the upper side of the growing horizontal stem, but increases on the lower side. With the adjusted growth pattern, the stem curves upward, even in the absence of light. Using different types of tests, researchers have been able to document the shifting of IAA from the top to the bottom side of the stem. The changing auxin gradient correlates with the altered pattern of cell elongation.

In roots, a high concentration of auxin has the opposite effect—it inhibits cell elongation. If a root is placed on its side, amyloplasts in the root cap accumulate near the side wall that now is the bottom side of the cap. In some way this stimulates cell elongation in the opposite wall, and within a few hours the root once again curves downward. In root tips of many plants, however, especially eudicots, researchers have not been able to detect a shift in IAA concentration that correlates with the changing position of amyloplasts. One hypothesis is that IAA is redistributed over extremely short distances in root cells, and therefore is difficult to measure. Root cells are much more sensitive to IAA than are cells in stem tissue, and even a tiny shift in IAA distribution could significantly affect their growth.

Along with IAA, calcium ions (Ca^{2+}) appear to play a major role in gravitropism. For example, if Ca^{2+} is added to an otherwise untreated agar block that is then placed on one side of a root cap, the root will bend toward the block. In this way, experimenters have been able to manipulate the direction of growth so that the elongating root forms a loop. Similarly, if an actively bending root is deprived of Ca^{2+}, the gravitropic response abruptly stops. By contrast, the negative gravitropic response of a shoot tip is inhibited when the tissue is exposed to excess calcium.

Just how Ca^{2+} interacts with IAA in gravitropic responses is unknown. One hypothesis posits that calcium functions as an activator. Calcium binds to a small protein called *calmodulin,* activating it in the process. Activated calmodulin in turn can activate a variety of key cell enzymes in many organisms, both plants and animals. One possibility is that calcium-activated calmodulin stimulates cell membrane pumps that enhance the flow of both IAA and calcium through a gravity-stimulated plant tissue.

Some of the most active research in plant biology focuses on the intricate mechanisms of gravitropism **(Figure 35.19)**. For example, there is increasing evidence that in many plants, cells in different regions of stem tissue differ in their sensitivity to IAA, and that gravitropism is linked in some fundamental way to these differences. In a few plants, including some cultivated varieties of corn and radish, the direction of the gravitropic response by a seedling's primary root is influenced by light. Clearly there is much more to be learned.

Michael Clayton, University of Wisconsin

FIGURE 35.19
Gravitropism in corn seedlings. The two corn grains shown here were oriented in opposite directions with respect to gravity. Regardless, the primary root of each one grew downward through the soil and each stem grew upward.

Thigmotropism and Thigmomorphogenesis Are Responses to Physical Contact

Varieties of peas, grapes, and some other plants demonstrate **thigmotropism** (*thigma* = touch), which is growth in response to contact with a solid object. Thigmotropic plants typically have long, slender stems and cannot grow upright without physical support. They often have *tendrils,* modified stems or leaves that can rapidly curl around a fencepost or the sturdier stem of a neighboring plant. If one side of a grape vine stem grows against a trellis, for example, specialized epidermal cells on that side of the stem tendril shorten whereas cells on the other side of the tendril rapidly elongate. Within minutes the tendril starts to curl around the trellis, forming tight coils that provide strong support for the vine stem. **Figure 35.20** shows thigmotropic twisting in the passionflower *(Passiflora).* Auxin and ethylene may be involved in thigmotropism, but most details of the mechanism remain elusive.

The rubbing and bending of stems caused by frequent strong winds, rainstorms, grazing animals, and even farm machinery can inhibit the overall growth of plants and can alter their growth patterns. In this phenomenon, called **thigmomorphogenesis,** a stem stops elongating and instead adds girth when it is regularly

Image copyright Andrew Ferguson, 2010. Used under license from Shutterstock.com

FIGURE 35.20
Thigmotropism in a passion flower *(Passiflora)* tendril, which is twisted around a support.

Wally Eberhart/Visuals Unlimited, Inc.

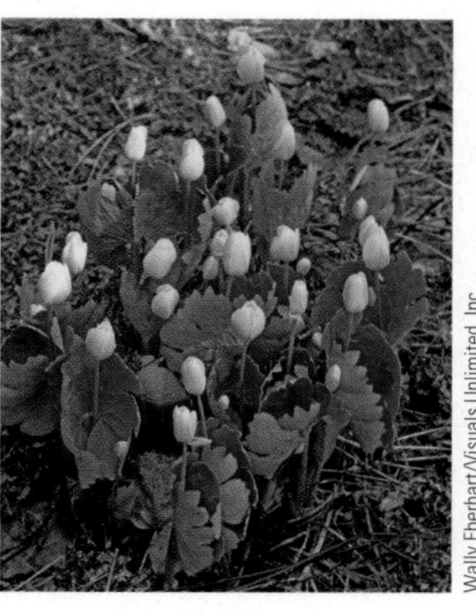
Wally Eberhart/Visuals Unlimited, Inc.

FIGURE 35.22
Nastic sleep movements in a bloodroot plant *(Sanguinaria canadensis)*.

subjected to mechanical stress. Merely shaking some plants daily for a brief period will inhibit their upward growth **(Figure 35.21)**. But although such plants may be shorter, their thickened stems will be stronger. Thigmomorphogenesis helps explain why plants growing outdoors are often shorter, have somewhat thicker stems, and are not as easily blown over as plants of the same species grown indoors. Trees growing near the snowline of windswept mountains show an altered growth pattern that reflects this response to wind stress.

Research on the cellular mechanisms of thigmomorphogenesis has begun to yield tantalizing clues. In one study, investigators repeatedly sprayed *Arabidopsis* plants with water and imposed other mechanical stresses, then sampled tissues from the stressed plants. The samples contained as much as double the usual amount of mRNA for at least four genes, which had been activated by the stress. The mRNAs encoded calmodulin and several other proteins that may have roles in altering *Arabidopsis* growth responses. The test plants were also short, generally reaching only half the height of unstressed controls.

Cary Mitchell

A. B. C.

FIGURE 35.21
Effect of mechanical stress on tomato plants *(Lycopersicon esculentum)*.
(A) This plant was the control; it was grown in a greenhouse, protected from wind and rain. **(B)** Each day for 28 days this plant was mechanically shaken for 30 sec at 280 rpm. **(C)** This plant received the same shaking treatment, but twice a day for 28 days.

Nastic Movements Are Nondirectional

Tropisms are responses to directional stimuli, such as light striking one side of a shoot tip, but many plants also exhibit **nastic movements** (*nastos* = pressed close together)—reversible responses to nondirectional stimuli, such as mechanical pressure or humidity. We see nastic movements in leaves, leaflets, and even flowers. For instance, certain plants exhibit nastic sleep movements, holding their leaves (or flower petals) in roughly horizontal positions during the day but folding them closer to the stem at night **(Figure 35.22)**. Tulip flowers "go to sleep" in this way.

Many nastic movements are temporary and result from changes in cell turgor. For example, the daily opening and closing of stomata in response to changing light levels are nastic movements, as is the traplike closing of the lobed leaves of the Venus flytrap when an insect brushes against hairlike sensory structures on the leaves. The leaves of *Mimosa pudica*, the sensitive plant, also close in a nastic response to mechanical pressure. Each *Mimosa* leaf is divided into pairs of leaflets **(Figure 35.23A)**. Touching even one leaflet at the leaf tip triggers a chain reaction in which each pair of leaflets closes up within seconds **(Figure 35.23B)**.

In many turgor-driven nastic movements, water moves into and out of the cells in **pulvini** (*pulvinus* = cushion), thickened pads of tissue at the base of a leaf or petiole. Stomatal movements depend on changing concentrations of ions within guard cells, and pulvinar cells drive nastic leaf movements in *Mimosa* and numerous other plants by the same mechanism **(Figure 35.23C)**.

How is the original stimulus transferred from cells in one part of a leaf to cells elsewhere? The answer lies in the polarity of charge across cell plasma membranes (see Section 6.4). Touching a *Mimosa* leaflet triggers an **action potential**—a brief reversal in the polarity of the membrane charge. When an action potential occurs at the plasma membrane of a pulvinar cell, the change in polarity causes potassium ion (K^+) channels to open, and ions flow out of the cell, setting up an osmotic gradient that draws water out as well. As water leaves by osmosis, turgor pressure falls, pulvinar cells become flaccid, and the leaflets move together. Later, when the process is reversed, the pulvinar cells regain turgor and the leaflets spread apart. Action potentials

FIGURE 35.23
Nastic movements in leaflets of *Mimosa pudica*, the sensitive plant. **(A)** In an undisturbed plant the leaflets are open. If a leaflet near the leaf tip is touched, changes in turgor pressure in pulvini at the base cause the leaf to fold closed. **(B, C).** The diagram sketches this folding movement in cross section. Other leaflets close in sequence as action potentials transmit the stimulus along the leaf.

A. **Undisturbed plant**

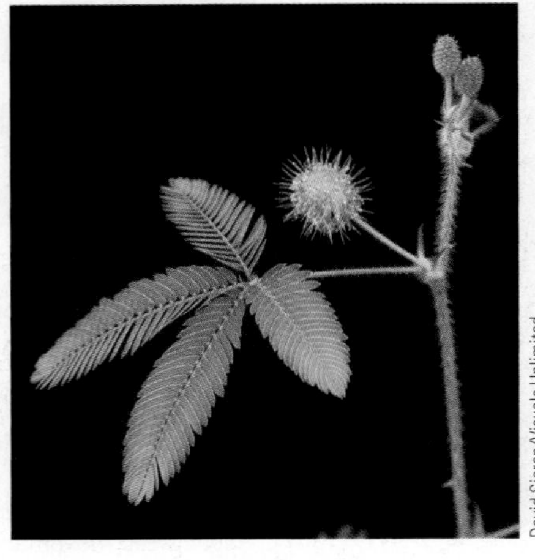
David Sieren/Visuals Unlimited

B. **Plant response to touch**

David Sieren/Visuals Unlimited

C. **Leaf folding mechanism**

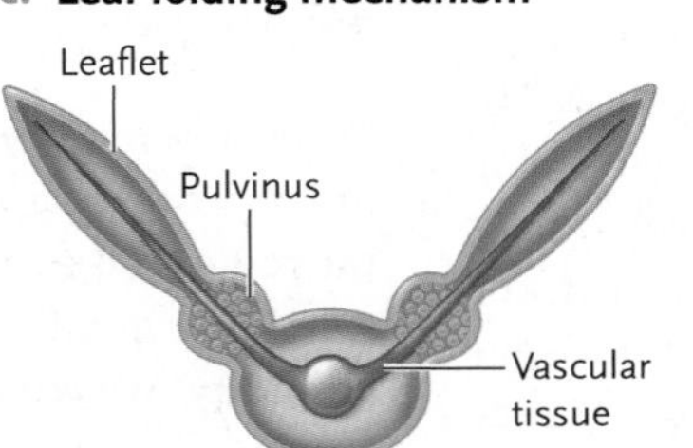
Leaflet
Pulvinus
Vascular tissue

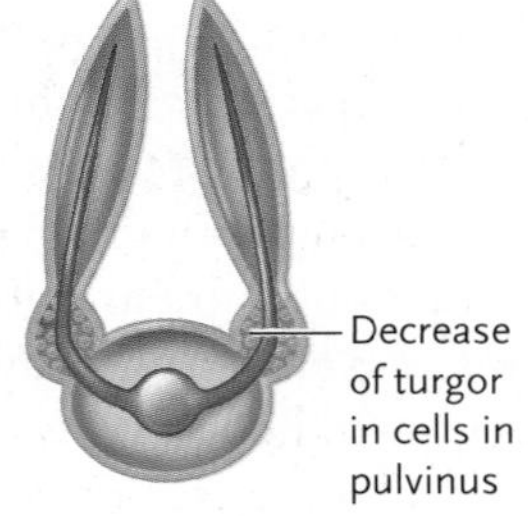
Decrease of turgor in cells in pulvinus

travel between parenchyma cells in the pulvini via plasmodesmata at the rate of about 2 cm/sec. Animal nerves conduct similar changes in membrane polarity along their plasma membranes (see Chapter 37). These changes in polarity, which are also called action potentials, occur much more rapidly—at velocities between 1 and 100 m/sec.

Stimuli other than touch also can trigger action potentials leading to nastic movements. Cotton, soybean, sunflower, and some other plants display *solar tracking,* nastic movements in which leaf blades are oriented toward the east in the morning, then steadily change their position during the day, following the sun across the sky. Such movements maximize the amount of time that leaf blades are perpendicular to the sun, which is the angle at which photosynthesis is most efficient.

STUDY BREAK 35.3

1. **What is the direct stimulus for phototropisms? For gravitropism?**
2. **Explain how nastic movements differ from tropic movements.**

35.4 Plant Biological Clocks

Like all eukaryotic organisms, plants have internal time-measuring mechanisms called **biological clocks** that adapt the organism to recurring environmental changes. In plants biological clocks help adjust both daily and seasonal activities.

Circadian Rhythms Are Based on 24-Hour Cycles

Some plant activities occur regularly in cycles of about 24 hours, even when environmental conditions remain constant. These are **circadian rhythms** (*circa* = around, *dies* = day). In Chapter 32, we noted that stomata open and close on a daily cycle, even when plants are kept in total darkness. Nastic sleep movements, described earlier, are another example of a circadian rhythm. Even when a plant that exhibits such movements is kept in constant light or darkness for a few days, it folds its leaves into the "sleep" position at roughly 24-hour intervals. In some way, the plant measures time without sunrise (light) and sunset (darkness). Such experiments demonstrate that internal controls, rather than external cues, largely govern circadian rhythms.

Circadian rhythms and other activities regulated by a biological clock help ensure that plants of a single species do the same thing, such as flowering, at the same time. For instance, flowers of the aptly named four-o'clock plant *(Mirabilis jalapa)* open predictably every 24 hours—in nature, in the late afternoon. Such coordination can be crucial for successful pollination. Although some circadian rhythms can proceed without direct stimulus from light, many biological clock mechanisms are influenced by the relative lengths of day and night.

Photoperiodism Involves Seasonal Changes in the Relative Length of Night and Day

Obviously, environmental conditions in a 24-hour period are not the same in summer as they are in winter. In North America, for instance, winter temperatures are cooler and winter day length is shorter. Experimenting with tobacco and soybean plants in the early 1900s, two American botanists, Wightman Garner and Henry Allard, elucidated a phenomenon they called **photoperiodism,** in which plants respond to changes in the relative lengths of light and dark periods in their environment during each 24-hour period. Through photoperiodism, the biological clocks of plants (and animals) make seasonal adjustments in their patterns of growth, development, and reproduction.

In plants, we now know that a family of blue-green pigments collectively called **phytochrome** often serves as a switching mechanism in the photoperiodic response, signaling the plant to make seasonal changes. Plants synthesize phytochrome in an in-

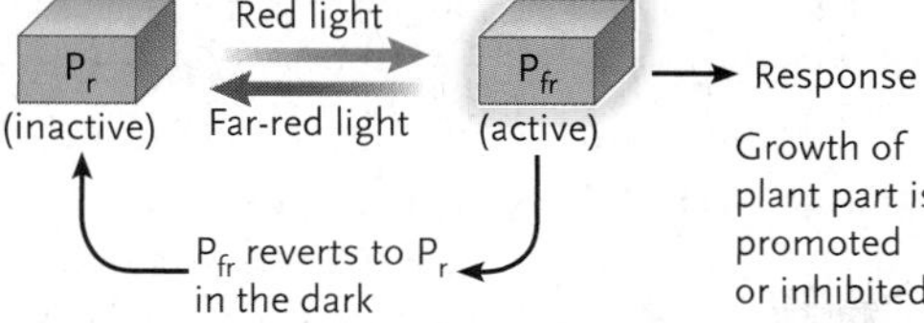
A. Interconversion of phytochrome
Red light
P_r
P_{fr}
(inactive)
Far-red light
(active)
Response
Growth of plant part is promoted or inhibited
P_{fr} reverts to P_r in the dark

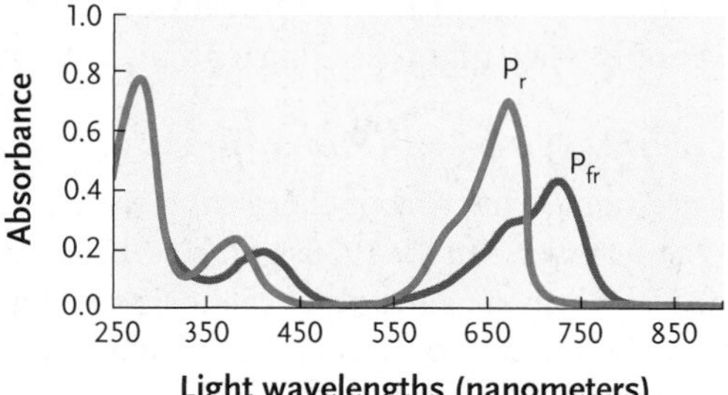
B. Changing absorption spectra
Absorbance
1.0
0.8
0.6
0.4
0.2
0.0
250
350
450
550
650
750
850
P_r
P_{fr}
Light wavelengths (nanometers)

FIGURE 35.24
The phytochrome switching mechanism, which can promote or inhibit growth of different plant parts. **(A)** Interconversion of phytochrome from the active form (P_{fr}) to the inactive form (P_r). **(B)** The absorption spectra associated with the interconversion of P_r and P_{fr}.

active form, P_r, which absorbs light of shorter wavelengths (about 660 nm) at the "red" end of the spectrum (see Figure 9.4). Sunlight contains relatively more red light than far-red light, which has a longer wavelength (about 730 nm). During daylight hours when red wavelengths dominate, P_r absorbs red light. Absorption of red light triggers the conversion of phytochrome to an active form designated P_{fr}, which absorbs light of far-red wavelengths. At sunset, at night, or even in shade, where far-red wavelengths predominate, P_{fr} reverts to P_r **(Figure 35.24).**

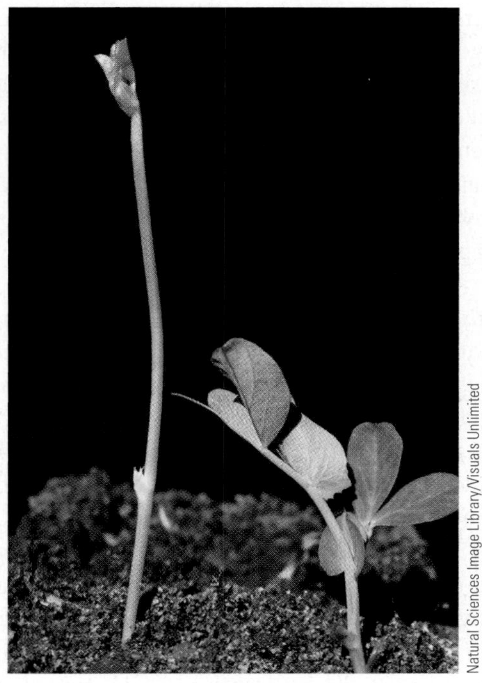
Natural Sciences Image Library/Visuals Unlimited

FIGURE 35.25
Effects of the absence of light on young bean plants *(Phaseolus)*. The seedling on the left was grown in darkness for several days. Its leaves are yellow because it could form carotenoids but not chlorophyll in darkness. It also has a larger stem, smaller leaves, and a smaller root system than the seedling on the right, which was grown in the light.

In nature a high concentration of P_{fr} "tells" a plant that it is exposed to sunlight. The plant's ability to sense sunlight is vital given that over time sunlight provides favorable conditions for leaf growth, photosynthesis, and flowering. The exact mechanism of this crucial transfer of environmental information still is not fully understood. Phytochrome activation may stimulate plant cells to take up Ca^{2+} ions, or it may induce certain plant organelles to release them. Either way, when free calcium ions combine with calcium-binding proteins (such as calmodulin) they may initiate at least some responses to light. Botanists suspect that P_{fr} controls the types of enzymes being produced in particular cells—and different enzymes are required for seed germination, stem elongation and branching, leaf expansion, and the formation of flowers, fruits, and seeds. When plants adapted to full sunlight are grown in darkness, they put more resources into stem elongation and less into leaf expansion or stem branching **(Figure 35.25).**

Cryptochrome—which, recall, is sensitive to blue light and appears to influence light-related growth responses—also interacts with phytochromes in producing circadian responses. Researchers have recently discovered that cryptochrome occurs not only in plants but also in animals such as fruit flies and mice. Does it act as a circadian photoreceptor in both kingdoms? Only further study will provide the answer.

Cycles of Light and Dark Often Influence Flowering

Photoperiodism is especially apparent in the flowering process. Like other plant responses, flowering is often keyed to changes in day length through the year and to the resulting changes in environmental conditions. Corn, soybeans, peas, and other annual plants begin flowering after only a few months of growth. Roses and other perennials typically flower every year or after several years of vegetative growth. Carrots, cabbages, and other biennials typically produce roots, stems, and leaves the first growing season, die back to soil level in autumn, then grow a new flower-forming stem the second season.

In the late 1930s Karl Hamner and James Bonner grew cocklebur plants *(Xanthium strumarium)* in chambers in which the researchers could carefully control environmental conditions, including photoperiod. And they made an unexpected discovery: Flowering occurred only when the test plants were exposed to a single night of 8.5 hours of uninterrupted darkness. The length of the "day" in the growth chamber did not matter, but if light interrupted the dark period for even a minute or two, the plant would not flower at all. Subsequent research confirmed that for most angiosperms, it is the length of darkness, not light, that controls flowering.

KINDS OF FLOWERING RESPONSES The photoperiodic responses of flowering plants are so predictable that botanists have long used them to categorize plants **(Figure 35.26).** The categories,

Flowers
Jan Zeevaart

FIGURE 35.26
Effect of day length on spinach *(Spinacia oleracea)*, a long-day plant. The plant on the left was not exposed to light long enough to trigger flowering. The plant on the right did flower after light exposure that mimicked the long days of spring.

which refer to day length, reflect the fact that scientists recognized the phenomenon of photoperiodic flowering responses long before they understood that darkness, not light, was the cue. **Long-day plants,** such as irises, daffodils, and corn, usually flower in spring when dark periods become shorter and day length becomes longer than some critical value—usually 9–16 hr. **Short-day plants,** including cockleburs, chrysanthemums, and potatoes, flower in late summer or early autumn when dark periods become longer and day length becomes shorter than some critical value. **Intermediate-day plants,** such as sugarcane, flower only when day length falls in between the values for long-day and short-day plants. **Day-neutral plants,** such as dandelions and roses, flower whenever they become mature enough to do so, without regard to photoperiod.

Experiments demonstrate what happens when plants are grown under the "wrong" photoperiod regimes. For instance, spinach, a long-day plant, flowers and produces seeds only if it is exposed to no more than 10 hr of darkness each day for 2 weeks

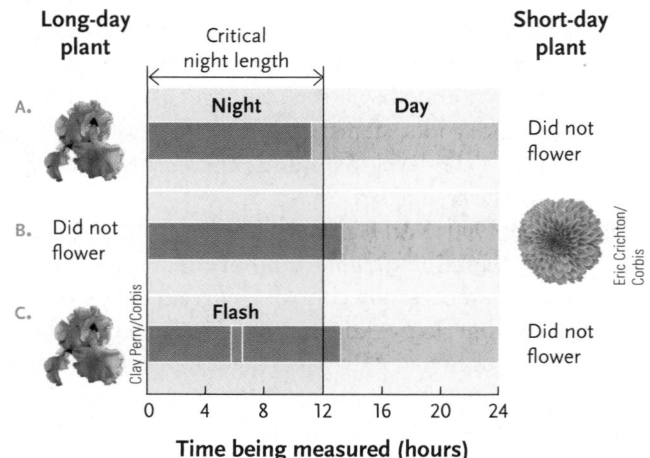
Long-day plant
Critical night length
Short-day plant
A.
Night
Day
Did not flower
B.
Did not flower
C.
Flash
Did not flower
Clay Perry/Corbis
Eric Crichton/Corbis
0
4
8
12
16
20
24
Time being measured (hours)

FIGURE 35.27
Experiments showing that short-day and long-day plants flower by measuring night length. Each horizontal bar signifies 24 hours. Blue bars represent night, and yellow bars day. **(A)** Long-day plants such as bearded irises flower when the night is shorter than a critical length, whereas **(B)** short-day plants such as chrysanthemums flower when the night is longer than a critical value. **(C)** When an intense red flash interrupts a long night, both kinds of plants respond as if it were a short night; the irises flowered but the chrysanthemums did not.

(see Figure 35.26). **Figure 35.27** illustrates the results of an experiment to test the responses of short-day and long-day plants to night length. In this experiment, bearded iris plants (*Iris* species), which are long-day plants, and chrysanthemums, which are short-day plants, were exposed to a range of light conditions. In each case, when the researchers interrupted a critical dark period with a pulse of red light, the light reset the plants' clocks. The experiment provided clear evidence that short-day plants flower only when nights are longer than a critical value—and long-day plants flower only when nights are shorter than a critical value.

CHEMICAL SIGNALS FOR FLOWERING When photoperiod conditions are right, what sort of chemical message stimulates a plant to develop flowers? In the 1930s botanists began postulating the existence of "florigen," a hypothetical hormone that served as the flowering signal. In a somewhat frustrating scientific quest, researchers spent the rest of the twentieth century seeking this substance in vain. Recently, however, molecular studies using *Arabidopsis* plants have defined a sequence of steps that may collectively provide the internal stimulus for flowering. Here again, we see one of the recurring themes in plant development—major developmental changes guided by several interacting genes.

Figure 35.28 traces the steps of the proposed flowering signal. To begin with, a gene called *CONSTANS* is expressed in a plant's leaves in tune with the daily light/dark cycle, with expression peaking at dusk (step 1). The gene encodes a regulatory protein called CO (not to be confused with carbon monoxide). As days lengthen in spring, the concentration of CO rises in leaves, and as a result a second gene is activated (step 2). The product of this gene, a regulatory protein called FT (for Flowering locus T), travels in the phloem to shoot tips (step 3). Once there, FT interacts with a second regulatory protein (step 4) that is synthesized only in shoot apical meristems (step 5). The encounter apparently sparks the development of a flower by promoting the expression of floral organ identity genes in the meristem tissue (see Section 34.5). Key experiments that uncovered this pathway all relied on analysis of DNA microarrays, a technique introduced in Section 18.3 and featured in this chapter's *Focus on Basic Research.*

VERNALIZATION AND FLOWERING Flowering is more than a response to changing night length. Temperatures also change with the seasons in most parts of the world, and they too influence flowering. For instance, unless buds of some biennials and perennials are exposed to low winter temperatures, flowers do not form on stems in spring. Low-temperature stimulation of flowering is called **vernalization** ("making springlike").

In 1915 the plant physiologist Gustav Gassner demonstrated that it was possible to influence the flowering of cereal plants by controlling the temperature of seeds while they were germinating. In one case, he maintained germinating seeds of winter rye *(Secale cereale)* at just above freezing (1°C) before planting them. In nature, winter rye seeds in soil germinate during the winter, giving rise to a plant that flowers months later, in summer. Plants grown from Gassner's test seeds, however, flowered the same summer even when the seeds were planted in the late spring. Home gardeners can induce flowering of daffodils and tulips by

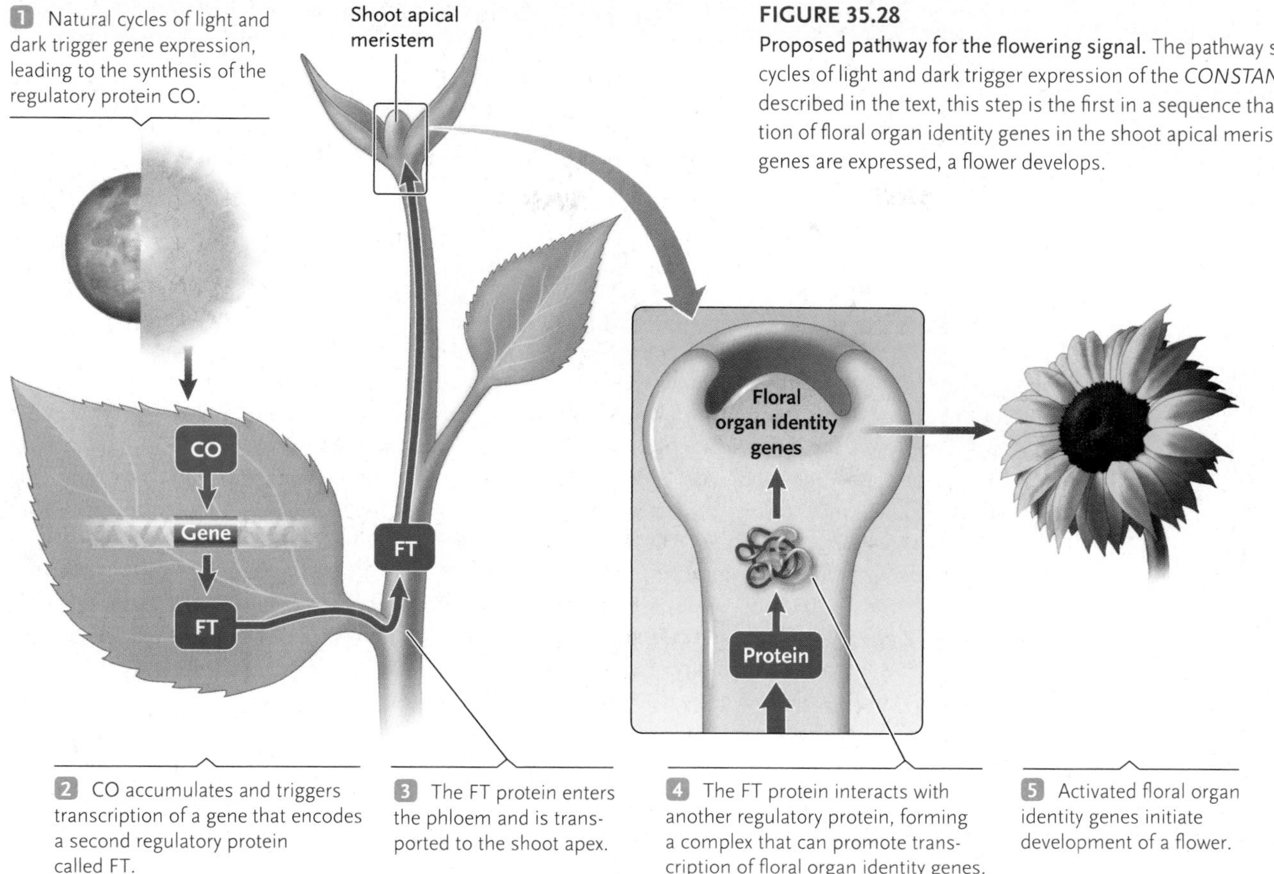
1 Natural cycles of light and dark trigger gene expression, leading to the synthesis of the regulatory protein CO.
Shoot apical meristem
CO
Gene
FT
FT
Floral organ identity genes
Protein
2 CO accumulates and triggers transcription of a gene that encodes a second regulatory protein called FT.
3 The FT protein enters the phloem and is transported to the shoot apex.
4 The FT protein interacts with another regulatory protein, forming a complex that can promote transcription of floral organ identity genes.
5 Activated floral organ identity genes initiate development of a flower.

FIGURE 35.28
Proposed pathway for the flowering signal. The pathway starts as shifting cycles of light and dark trigger expression of the *CONSTANS* gene. As described in the text, this step is the first in a sequence that leads to the activation of floral organ identity genes in the shoot apical meristem. When these genes are expressed, a flower develops.

putting the bulbs (technically, *corms*) in a freezer for several weeks before early spring planting. Commercial growers use vernalization to induce millions of plants, such as Easter lilies, to flower just in time for seasonal sales.

Dormancy Is an Adaptation to Seasonal Changes or Stress

As autumn approaches and days grow shorter, growth slows or stops in many plants even if temperatures are still moderate, the sky is bright, and water is plentiful. When a perennial or biennial plant stops growing under conditions that seem (to us) quite suitable for growth, it has entered a state of **dormancy.** Ordinarily, its buds will not resume growth until early spring.

Short days and long nights—conditions typical of winter—are strong cues for dormancy. In one experiment, in which a short period of red light interrupted the long dark period for Douglas firs, the plants responded as if nights were shorter and days were longer; they continued to grow taller **(Figure 35.29).** Conversion of P_r to P_{fr} by red light during the dark period prevented dormancy. In nature, buds may enter dormancy because less P_{fr} can form when day length shortens in late summer. Other environmental cues are at work also. Cold nights, dry soil, and a deficiency of nitrogen apparently also promote dormancy.

The requirement for multiple dormancy cues has adaptive value. For example, if temperature were the only cue, plants

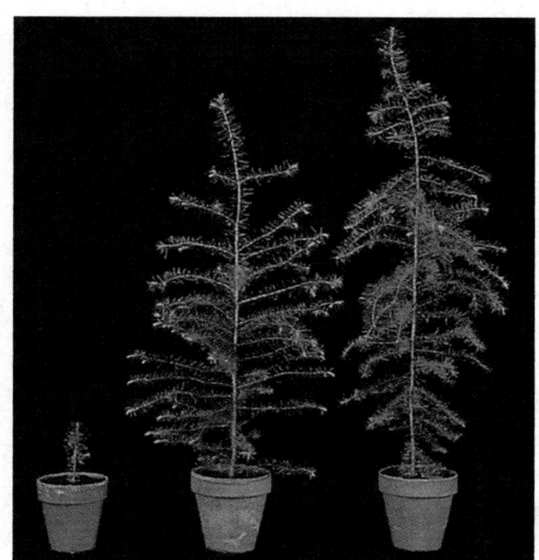
R. J. Downs

FIGURE 35.29
Effect of the relative length of day and night on the growth of Douglas firs *(Pseudotsuga menziesii)*. The young tree at the left was exposed to alternating periods of 12 hours of light followed by 12 hours of darkness for a year; its buds became dormant because day length was too short. The tree at the right was exposed to a cycle of 20 hours of light and 4 hours of darkness; its buds remained active and growth continued. The middle plant was exposed each day to 12 hours of light and 11 hours of darkness, with a 1-hour light in the middle of the dark period. This light interruption of an otherwise long dark period also prevented buds from going dormant.

FOCUS ON **Basic Research**

Using DNA Microarray Analysis to Track Down "Florigen"

The more plant scientists learn about plant genomes, the more they are relying on DNA microarray assays to elucidate the activity of plant genes.

Recall from Section 18.3 that a DNA microarray, also called a DNA chip, allows an investigator to explore questions such as how the expression of a particular gene differs in different types of cells. To quantify the expression of specific genes in particular types of cells, mRNA transcripts are isolated from the cells; then a cDNA library is created from each mRNA sample, using nucleotides labeled with fluorescent dyes. Probes (nucleotide sequences) representing every gene in the organism's genome are fixed onto a slide; when the labeled cDNAs are added to the slide, each will hybridize to the gene that expressed the mRNA from which it was made. Next, the DNA microarray is scanned with a laser that can detect fluorescence. When a gene is expressed in a cell, the dye fluoresces and gives a color that is in accord with the degree of its expression. The procedure can be manipulated to reveal the relative amounts of expression of more than one of a cell's genes.

Philip A. Wigge and his colleagues used this method to learn more about the signaling pathway that causes a plant's apical meristem to give rise to flowers. Previous research had established that in leaves, lengthening spring days coincided with rising concentrations of CO, a regulatory protein encoded by the *CONSTANS* gene. But what did CO regulate? Working with *Arabidopsis thaliana,* Wigge's group was able to narrow down the field to four genes, and using microarray analysis of DNA from leaf cells they pinpointed one called FT (for flowering locus T). The researchers found that in leaves, CO causes strong expression of FT: When enough CO is present, FT mRNA is rapidly transcribed, then enters the phloem. (The transport of mRNA in phloem is not unusual.) By contrast, when they tested CO's effects in shoot apex cells, they found that it triggers far less gene expression there. Clearly, CO was not directly triggering the development of flowers. However, FT mRNA moves in the phloem to the shoot apex, where it is translated into protein. Was that protein the direct flowering signal? Other studies had implicated a regulatory protein called FD, which microarray analysis had shown was expressed *only*—but very strongly—in the shoot apex.

To sort out this final piece of the puzzle, the Wigge team examined flowering responses in normal *A. thaliana* plants as well as in mutants having a normal FT protein but a defective *fd,* and vice versa. Flowering was abnormal in both types of mutants, possibly because the mutated "partner" suppressed some aspect of the functioning of the normal protein. On the other hand, in wild-type plants, which had a functioning FD protein, expression of FT triggered a marked increase in the expression of the floral organ gene *APETALA1* **(Figure 1).** These results have two major implications. First, they support the hypothesis that FT and FD interact in a normal flowering response. Second, the study suggests that FT, the CO-induced signal from leaves, conveys the environmental signal that it is time for a plant to flower. In that sense, FT may be the long sought "florigen." However, only by interacting with FD does FT "know" where to deliver its flowering signal—in the apical meristems of shoots.

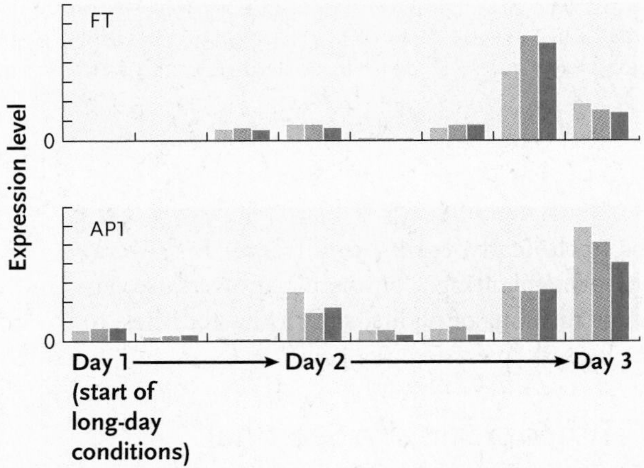
FT
AP1
Expression level
0
0
Day 1 (start of long-day conditions)
Day 2
Day 3

FIGURE 1

Effect of the FT protein on expression of the *APETALA1 (AP1)* floral organ identity gene. In nature, *Arabidopsis thaliana* is a long-day plant, and the experiment was carried out under long-day (that is, short-night) conditions. Three groups of replicates shown here in yellow, orange, and red respectively, were monitored for both AP1 and FT. After a brief delay, the expression of AP1 closely tracked the appearance of the FT regulatory protein, which had been activated by its interaction with the FD protein.

might flower and seeds might germinate in warm autumn weather—only to be killed by winter frost.

A dormancy-breaking process is at work between fall and spring. Depending on the species, breaking dormancy probably involves gibberellins and abscisic acid, and it requires exposure to low winter temperatures for specific periods **(Figure 35.30).** The temperature needed to break dormancy varies greatly among species. For example, the Delicious variety of apples grown in Utah requires 1,230 hours near 43°F (6°C); apricots grown there require only 720 hours at that temperature. Generally, trees growing in the southern United States or in Italy require less cold exposure than those growing in Canada or in Sweden.

STUDY BREAK 35.4 <

1. **Summarize the switching mechanism that operates in plant responses to changes in photoperiod.**
2. **Give some examples of how relative lengths of dark and light can influence flowering.**
3. **Explain why dormancy is an adaptive response to a plant's environment.**

Eric Welzel/Fox Hill Nursery, Freeport, Maine
Potted plant grown inside a greenhouse did not flower.
Branch exposed to cold outside air flowered.

FIGURE 35.30
Effect of cold temperature on dormant buds of a lilac *(Syringa vulgaris)*. In this experiment, a lilac plant was grown in winter inside a warm greenhouse with one branch extending through a hole to the outdoors. Only the buds on the branch exposed to low outside temperatures resumed growth in spring. This experiment suggests that low-temperature effects are localized in plants.

35.5 Signal Responses at the Cellular Level

Environmental stimuli such as changing light, temperature, or chemicals on the surface of an attacking pathogen are cues that signal a plant to alter its growth or physiology. For decades plant physiologists have looked avidly for clues about how those signals are converted into a chemical message that produces a change in a cell's growth, metabolism, or some other aspect of its functioning. Section 7.2 introduced basic mechanisms through which animal cells respond to external signals, and as we see next at least some of these mechanisms also apply to plant cells. We know the most about signaling pathways involving auxin, ethylene, salicylic acid, and blue light.

Several Signal Response Pathways Operate in Plants

Hormones and environmental stimuli alter the behavior of target cells, which have receptors to which specific signal molecules can bind and elicit a cellular response. In general, this mechanism unfolds in three steps introduced in Section 7.1. First, a target cell receptor receives the signal, which may be a molecule or an environmental cue such as sunlight. Next, the signal is transduced—that is, its "message" is changed into a form that can trigger the cellular response. In the third step, the transduced signal causes the cellular response **(Figure 35.31A).** Some transduced signals activate or turn off genes and so alter protein synthesis; others set in motion events that modify existing cell proteins.

Some plant hormones and growth factors bind to receptors at the target cell's plasma membrane. Others cross the plasma membrane and bind to receptors inside the cell, on its endoplasmic reticulum (ER), in the cytoplasm, or in the nucleus. Examples of the latter are ethylene receptors on the ER, and auxin receptors, which are a family of proteins in the cytoplasm often called simply TIR1 after the first one to be discovered. Research in several laboratories confirmed that auxin binds to TIR1 in the nucleus, setting in motion events that release inhibition of the transcription of particular genes. As a result, the previously repressed genes are turned on. In many cases (although not with TIR1), hormone binding causes the receptor to change shape. Regardless, binding of a hormone or growth factor triggers a complex pathway that leads to the cell response—the opening of ion channels, activation of transport proteins, or some other event **(Figure 35.31B).** Only cells with the appropriate receptor can respond to a particular signaling molecule. For example, certain cells in ripening fruits and developing seeds have ethylene receptors, but few if any cells in stems do.

We can think of plant hormones and other signaling molecules as external first messengers that deliver the initial physiological signal to a target cell. Often, binding of the signal molecule triggers the synthesis of internal second messengers (introduced in Section 7.4). These go-between molecules diffuse rapidly through the cytoplasm and provide the main chemical signal that alters cell functioning.

Second Messenger Systems Enhance the Plant Cell's Response to a Hormone's Signal

Second messengers usually are synthesized in the sequence of chemical reactions that converts an external signal into internal cell activity (Figure 35.31B, step 2). For many years the details of plant second-messenger systems were sketchy and hotly debated. Fairly early on, calcium ions were found to play a second messenger role in some hormonal responses. Recent experimental evidence indicates that auxin's hormonal signal is conveyed by cAMP (cyclic adenosine monophosphate), a major second messenger in cells of animals and other organisms. Inositol triphosphate (IP_3), a second messenger in plants, fungi, and animals, is involved in the reactions that close plant stomata in response to a signal from abscisic acid, ABA (Section 35.1).

It is probably common for signal response pathways in plant cells to include many steps in which different types of proteins and other molecules are mobilized. ABA's role in stomatal closure—triggered by water stress or some other environmental cue—begins when the hormone activates a receptor in the plant cell plasma membrane. Experiments have shown that this binding activates G proteins that in turn activate phospholipase C (see Figure 7.11). This enzyme stimulates the synthesis of second messengers such as inositol triphosphate (IP_3). The second messenger diffuses through the cytoplasm and binds with calcium channels in structures such as the endoplasmic reticulum and tonoplast. The bound channels then open, releasing calcium ions that activate protein kinases in the cytoplasm. In turn, these enzymes activate their target proteins by phosphorylating them. As described in Section 7.2, the original signal is greatly amplified by a cascade of activated protein kinases, each one of which can activate a large number of target proteins.

In addition to the basic signal transduction pathways described here, other routes may exist that are unique to plant cells. Light is the driving force for photosynthesis, and it may not be far-fetched to suppose that plants have evolved other unique light-related biochemical pathways as well. For instance, experi-

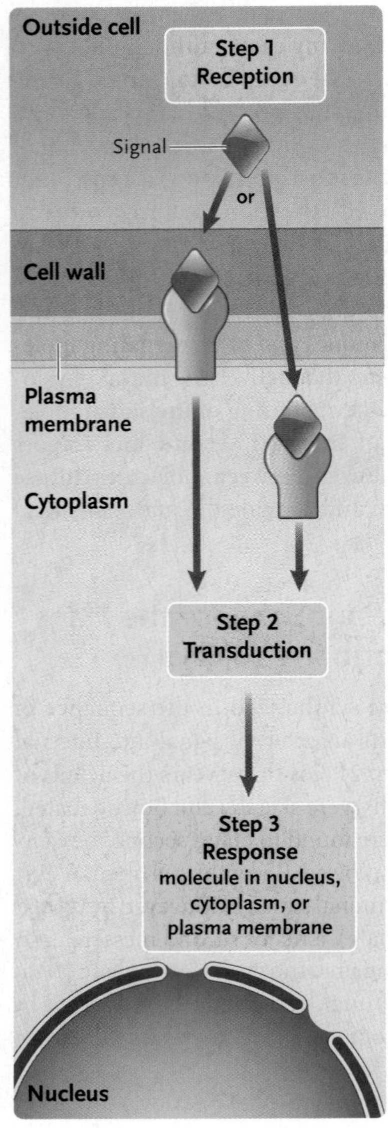
A. Stages of transduction
Outside cell
Step 1
Reception
Signal
or
Cell wall
Plasma membrane
Cytoplasm
Step 2
Transduction
Step 3
Response
molecule in nucleus, cytoplasm, or plasma membrane
Nucleus

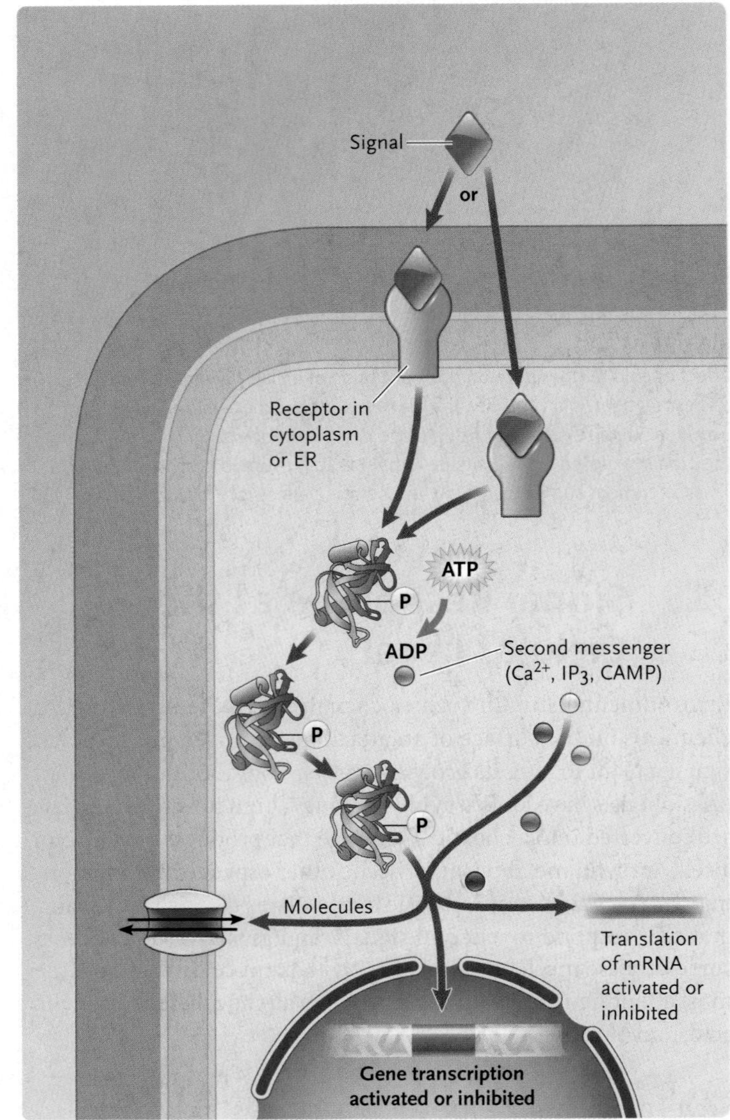
B. Second messengers in transduction
1 An arriving signal binds and activates a receptor in the plasma membrane, cytoplasm, ER, or nucleus. In most cases, the receptor changes shape, which triggers the transduction pathway inside the cell.
2 As the transduction pathway unfolds, receptor activation leads to activation of one or more proteins. This activation step may set in motion a cascade of protein phosphorylation, or it may mobilize second messenger molecules.
3 Phosphorylated proteins or second messenger molecules trigger the cell response, such as a change in ion flow into or out of the cell, a shift in translation of mRNA, or altering gene transcription in the nucleus.
Signal
or
Receptor in cytoplasm or ER
ATP
P
ADP
Second messenger (Ca^{2+}, IP_3, CAMP)
P
P
Molecules
Translation of mRNA activated or inhibited
Gene transcription activated or inhibited

FIGURE 35.31

Summary of signal response pathways in plant cells. **(A)** The three stages of a signal response pathway are reception of the signal, transduction into a form the cell can recognize, and the cell's response. **(B)** Proteins or second messenger molecules may be the intermediaries in the transduction stage of a signal response pathway.

ments are extending our knowledge of how plant cells respond to blue light, which, as we have discussed, triggers some photoperiod responses such as the opening and closing of stomata. Research is also revealing increasing complexity in the signal transduction pathways by which plant cells respond to auxin and defensive hormones such as brassinosteroids. Clearly, much remains to be discovered about this and many other aspects of plant functioning.

STUDY BREAK 35.5 <

1. **Summarize the various ways that chemical signals reaching plant cells are converted to changes in cell functioning.**
2. **What basic task does a second messenger accomplish?**
3. **Thinking back to Chapter 7, can you describe parallels between signal transduction mechanisms in the cells of plants and animals?**

UNANSWERED QUESTIONS

How does the plant genome respond to the external environment?

Genome size across the plant kingdom can vary more than 1,000-fold. For instance, the genome of *Arabidopsis thaliana* runs about 125 million base pairs, that of *Pinus taeda,* the loblolly pine, roughly 21 billion bp. Even more remarkably, genome size can vary markedly among individuals of the same species. Among inbred maize *(Zea mays)* plants, for example, the genome size of different individuals may vary by as much as 30%.

What molecular mechanisms underlie these genomic changes? In my laboratory at Case Western Reserve University, my coworkers and I have spent more than three decades investigating this question, examining environmentally induced heritable changes in flax *(Linum)*. Certain flax varieties respond to specific environmental stresses by physically restructuring specific portions of their genome, with the variant genomes being stably transmitted over many generations. Fortunately, the changes occur within a single generation and therefore can be readily tracked. We now know, for example, that the genomic changes occur in the vegetative apical meristem. This tissue is where all new growth occurs, and it also gives rise to the cells (in flowers) that ultimately produce flax gametes. This means that the environment can act as a selective agent to establish advantageous variants that may then be transmitted to the next generation.

We have identified one DNA fragment of interest—*Linum* insertion sequence-1 (LIS-1)—and followed it as responsive flax varieties grow in different environments. LIS-1 is assembled while the plant is growing under inducing conditions and is then (or simultaneously) inserted into a specific genome site with very high frequency. Under the inducing growth conditions, this event occurs in all the test plants, and all their progeny inherit the altered genomic site. Under different conditions, although the insertion occasionally occurs, none of the progeny inherit the altered site. Some other regions of the genome also frequently alter, but in our experiments only insertion of LIS-1 is strongly associated with a specific change in the plant's environment.

The high frequency with which LIS-1 arises and is transmitted to progeny in one environment but not others accords with a hypothesis that the insertion is directly adaptive or is closely linked to an adaptive change. Not all stresses result in the same genomic rearrangements, and the particular genome regions that are restructured under a particular stress are related to the physiological effects of that stress. We are currently investigating the complete genome sequences of the original and induced lines to identify the variable loci and will be using the genome reorganization to point us toward a molecular mechanism. We are also looking at intermediates that may be formed during the induction of the changes. Elucidating these and related mechanisms in flax can help clarify some basic questions—including whether genetic adaptations to stress are an integrated genome-wide reconfiguration rather than single gene mutations. We will also better understand the precise circumstances under which a stressful environment can trigger rapid, limited restructuring of a plant genome. And finally, such studies may also reveal how a stress can then act as the selective force determining which genome variations are transmitted into the next generation.

Think Critically

1. How does genome reconfiguration differ from mutation systems that normally operate in plants?
2. Given data on environmentally induced heritable changes in flax, how might plant scientists need to modify their thinking about the role of the environment as a mutagen, rather than simply as a selective agent?

Christopher A. Cullis

Christopher A. Cullis is the Frances Hobart Herrick Professor of Biology at Case Western Reserve University. His research focuses on the mechanisms by which DNA within the cell can change rapidly, particularly in response to external stimuli. He uses flax as a model organism and is interested in developing a flax genome project. Learn more about his research at http://www.case.edu/artsci/biol/people/cullis.html.

REVIEW KEY CONCEPTS

Go to **CENGAGENOW** at www.cengage.com/login to access quizzing, animations, exercises, articles, and personalized homework help.

35.1 Plant Hormones

- At least seven classes of hormones govern flowering plant development, including germination, growth, flowering, fruit set, and senescence (Table 35.1).
- Auxins, mainly IAA, promote elongation of cells in the coleoptile and stem among other effects (Figures 35.2–35.6).
- Gibberellins promote stem elongation and help seeds and buds break dormancy (Figures 35.1, 35.7).
- Cytokinins stimulate cell division, promote leaf expansion, and retard leaf aging (Figure 35.8).
- Ethylene promotes senescence, as well as fruit ripening and abscission (Figures 35.9 and 35.10).
- Brassinosteroids stimulate cell division and elongation (Figure 35.11).
- Abscisic acid (ABA) promotes stomatal closure and may trigger seed and bud dormancy (Figure 35.10).
- Jasmonates regulate growth and have roles in defense.

Animation: Auxin's effects

35.2 Plant Chemical Defenses

- Plants have diverse chemical defenses that limit damage from bacteria, fungi, worms, or plant-eating insects (Figure 35.13).
- The hypersensitive response isolates an infection site by surrounding it with dead cells (Figure 35.14).
- Oligosaccharins can trigger the synthesis of phytoalexins, secondary metabolites that function as antibiotics.

- Gene-for-gene recognition enables a plant to chemically recognize a specific pathogen and mount a defense (Figure 35.15).
- Systemic acquired resistance provides long-term protection against some pathogens. Salicylic acid has a central role in this response (Figure 35.16).
- Heat-shock proteins can reversibly bind enzymes and other proteins in plant cells and prevent them from denaturing when the plant is under heat stress.
- Some plants can synthesize "antifreeze" proteins that stabilize cell proteins when cells are threatened with freezing.

35.3 Plant Movements

- Plants adjust their growth patterns in response to environmental rhythms and unique environmental circumstances. These responses include tropisms.
- Phototropisms, mainly stimulated by blue light, are growth responses to a directional light source. (Figures 35.3 and 35.17).
- Gravitropism is a growth response to Earth's gravitational pull. Stems exhibit negative gravitropism (growing upward) whereas roots show positive gravitropism (Figures 35.18 and 35.19).
- Some plants or plant parts demonstrate thigmotropism, growth in response to contact with a solid object (Figure 35.20).
- Mechanical stress can cause thigmomorphogenesis, which causes the stem to add girth (Figure 35.21).
- Some plant species show nastic leaf movements in response to certain environmental cues. Changes in fluid pressure in cells of a pulvinus, a pad of tissue at the base of a leaf or petiole, cause the movements (Figures 35.22 and 35.23).

Animation: Phototropism

Animation: Gravity and statolith distribution

Animation: Gravitropism

35.4 Plant Biological Clocks

- Plants have biological clocks, internal time-measuring mechanisms with a biochemical basis. Environmental cues can "reset" the clocks, enabling plants to make seasonal adjustments in growth, development, and reproduction.
- In photoperiodism, plants respond to a change in the relative length of daylight and darkness in a 24-hour period. A switching mechanism involving the pigment phytochrome promotes or inhibits germination, growth, and flowering and fruiting. (Figure 35.24).
- Long-day plants flower when day length is long relative to night. Short-day plants flower when day length is relatively short, and intermediate-day plants flower when day length falls in between the values for long-day and short-day plants. Flowering of day-neutral plants is not regulated by light. In vernalization, a period of low temperature stimulates flowering (Figures 35.25–35.27).
- The direct trigger for flowering may begin in leaves, when the regulatory protein CO triggers the expression of the FT gene. The resulting mRNA transcripts move in phloem to apical meristems where translation of the mRNAs yields a second regulatory protein, which in turn interacts with a third. This final interaction activates genes that encode the development of flower parts (Figure 35.28).
- Senescence is the sum of processes leading to the death of a plant or plant structure.
- Dormancy is a state in which a perennial or biennial stops growing even though conditions appear to be suitable for continued growth (Figures 35.29 and 35.30).

Animation: Phytochrome conversions

Animation: Flowering response experiments

Animation: Vernalization

Animation: Day length and dormancy

35.5 Signal Responses at the Cellular Level

- Hormones and environmental stimuli alter the behavior of target cells, which have receptors to which signal molecules can bind. By means of a response pathway that ultimately alters gene expression, a signal can induce changes in the cell's shape or internal structure or influence its metabolism or the transport of substances across the plasma membrane (Figure 35.31).
- Some plant hormones and growth factors may bind to receptors at the target cell's plasma membrane, changing the receptor's shape. This binding often triggers the release of internal second messengers that diffuse through the cytoplasm and provide the main chemical signal that alters gene expression.
- Second messengers usually act by way of a reaction sequence that amplifies the cell's response to a signal. The sequence activates a series of proteins, including G proteins and enzymes that stimulate the synthesis of second messengers (such as IP_3) that bind ion channels on endoplasmic reticulum. Binding releases calcium ions, which enter the cytoplasm and activate protein kinases, enzymes that activate specific proteins that produce the cell response.

UNDERSTAND AND APPLY

Test Your Knowledge

1. Which of the following plant hormones does *not* stimulate cell division?
   a. auxins
   b. cytokinins
   c. ethylene
   d. gibberellins
   e. abscisic acid
2. Which is the correct pairing of a plant hormone and its function?
   a. salicylic acid: triggers synthesis of general defense proteins
   b. brassinosteroids: promote responses to environmental stress
   c. cytokinins: stimulate stomata to close in water-stressed plants
   d. gibberellins: slow seed germination
   e. ethylene: promotes formation of lateral roots
3. A characteristic of auxin (IAA) transport is:
   a. IAA moves by polar transport from the base of a tissue to its apex.
   b. IAA moves laterally from a shaded to an illuminated side of a plant.
   c. IAA enters a plant cell in the form of IAAH, an uncharged molecule that can diffuse across cell membranes.
   d. IAA exits one cell and enters the next by means of transporter proteins clustered at both the apical and basal ends of the cells.
   e. All of the above are characteristics of auxin transport in different types of cells.

4. Hanging wire fruit baskets have many holes or open spaces. The major advantage of these spaces is that they:
   a. prevent gibberellins from causing bolting or the formation of rosettes on the fruit.
   b. allow the evaporation of ethylene and thus slow ripening of the fruit.
   c. allow oxygen in the air to stimulate the production of ethylene, which hastens the abscission of fruits.
   d. allow oxygen to stimulate brassinosteroids, which hasten the maturation of seeds in/on the fruits.
   e. allow carbon dioxide in the air to stimulate the production of cytokinins, which promotes mitosis in the fruit tissue and hastens ripening.
5. Which of the following is *not* an example of a plant chemical defense?
   a. ABA inhibits leaves from budding if conditions favor attacks by sap-sucking insects.
   b. Jasmonate activates plant genes encoding protease inhibitors that prevent insects from digesting plant proteins.
   c. Acting against fungal infections, the hypersensitive response allows plants to produce highly reactive oxygen compounds that kill selected tissue, thus forming a dead tissue barrier that walls off the infected area from healthy tissues.
   d. Chitinase, a PR hydrolytic protein produced by plants, breaks down chitin in the cell walls of fungi and thus halts the fungal infection.
   e. Attack by fungi or viruses triggers the release of oligosaccharins, which in turn stimulate the production of phytoalexins having antibiotic properties.
6. Which of the following statements about plant responses to the environment is true?
   a. The heat-shock response induces a sudden halt to cellular metabolism when an insect begins feeding on plant tissue.
   b. In gravitropism, amyloplasts sink to the bottom of cells in a plant stem, causing the redistribution of IAA.
   c. The curling of tendrils around a twig is an example of thigmotropism.
   d. Phototropism results when IAA moves first laterally, then downward in a shoot tip when one side of the tip is exposed to light.
   e. Nastic movements, such as the sudden closing of the leaves of a Venus flytrap, are examples of a plant's ability to respond to specific directional stimuli.
7. In nature the poinsettia, a plant native to Mexico, blooms only in or around December. This pattern suggests:
   a. The long daily period of darkness (short day) in December stimulates the flowering.
   b. Vernalization stimulates the flowering.
   c. The plant is dormant for the rest of the year.
   d. Phytochrome is not affecting the poinsettia flowering cycle.
   e. A circadian rhythm is in effect.
8. Which of the following steps is *not* part of the sequence that is thought to trigger flowering?
   a. Cycles of light and dark stimulate the expression of the *CONSTANS* gene in a plant's leaves.
   b. CO proteins accumulate in the leaves and trigger expression of a second regulatory gene.
   c. mRNA transcribed during expression of a second regulatory gene moves via the phloem to the shoot apical meristem.
   d. Interactions among regulatory proteins promote the expression of floral organ identity genes in meristem tissue.
   e. CO proteins in the floral meristem interact with florigen, a so-called flowering hormone, which provides the final stimulus for expression of floral organ identity genes.
9. Damage from an infectious bacterium, fungus, or worm may trigger a plant defensive response when the pathogen or a substance it produces binds to:
   a. a receptor encoded by the plant's *avirulence (Avr)* gene.
   b. an *R* gene in the plant cell nucleus.
   c. a receptor encoded by a dominant R gene.
   d. PR proteins embedded in the plant cell plasma membrane.
   e. salicylic acid molecules released from the besieged plant cell.
10. In the sequence that unfolds after molecules of a hormone such as ABA bind to receptors at the surface of a target plant cell:
   a. first messenger molecules in the cytoplasm are mobilized, then G proteins carry the signal to second messengers such as protein kinases, which alter the activity of cell proteins such as IP_3.
   b. binding activates G proteins, which in turn activate second messengers such as IP_3; subsequent steps are thought to involve activation of genes that encode protein kinases.
   c. binding activates phospholipase C, which in turn activates G proteins, which then activate molecules of IP_3, a step that leads to the synthesis of protein kinases.
   d. binding stimulates G proteins to activate protein kinases, which then bind calcium channels in ER; the flux of calcium ions activates second messenger molecules that alter the activity of cell proteins or enter the cell nucleus and alter the expression of target genes.
   e. binding activates G proteins, which in turn activate phospholipase C; this substance then stimulates the synthesis of second messenger molecules, the second messengers bind calcium channels in the cell's ER, and finally protein kinases alter the activity of proteins by phosphorylating them.

Discuss the Concepts

1. You work for a plant nursery and are asked to design a special horticultural regimen for a particular flowering plant. The plant is native to northern Spain, and in the wild it grows a few long, slender stems that produce flowers each July. Your boss wants the nursery plants to be shorter, with thicker stems and more branches, and she wants them to bloom in early December in time for holiday sales. Outline your detailed plan for altering the plant's growth and reproductive characteristics to meet these specifications.
2. Synthetic auxins such as 2,4-D can be weed killers because they cause an abnormal growth burst that kills the plant within a few days. Suggest reasons why such rapid growth might be lethal to a plant.
3. In some plant species, an endodermis is present in both stems and roots. In experiments, the shoots of mutant plants lacking differentiated endodermis in their root and shoot tissue do not respond normally to gravity, but roots of such plants do respond normally. Explain this finding, based on your reading in this chapter.
4. In *A. thaliana* plants carrying a mutation called *pickle (pkl)*, the primary root meristem retains characteristics of embryonic tissue—it spontaneously regenerates new embryos that can grow into mature plants. However, when the mutant root tissue is exposed to a gibberellin (GA), this abnormal developmental condition is suppressed. Explain why this finding suggests that additional research is needed on the fundamental biological role of GA.

Design an Experiment

Tiny, thornlike trichomes on leaves are a common plant adaptation to ward off insects. Those trichomes develop very early on, as outgrowths of a seedling's epidermal cells. Biologists have observed, however, that many mature plants develop more leaf trichomes after the fact, as a *response* to insect damage. Researchers at the University of Chicago decided to study this phenomenon, and specifically wanted to determine the effects, if any, of jasmonate, salicylic acid, and gibberellin in stimulating trichome development. Keeping in mind that plant hormones often interact, how many separate experiments, at a minimum, would the research team have had to carry out to obtain useful initial data? Do you suppose they used mutant plants for some or all of the tests? Why or why not?

Interpret the Data

Jennifer Nemhauser, Todd Mockler, and Joanne Chory at the Plant Biology Laboratory of the Salk Institute for Biological Studies performed experiments to learn more about the interaction of auxin and brassinosteroids (BR) in regulating the growth of *Arabidopsis* seedlings. In a 2004 research report on this work, graphics for displaying experimental data included the Venn diagram shown here (Figure 1). Venn diagrams are useful for giving a snapshot of the interrelationships between sets of data, often using overlapping circles. In this diagram, "up" and "down" refer to hormone effects that stimulate or inhibit gene expression. The numerals in each circle represent numbers of seedlings.

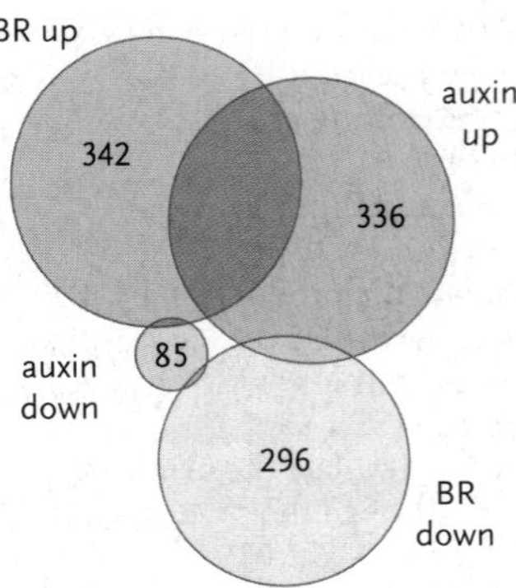
BR up
auxin up
342
336
85
auxin down
296
BR down

FIGURE 1
Venn diagram showing the overlap between *Yucca* genes responsive to auxin and BR.

1. Considering the two upper overlapping circles, what general finding do the data support?
2. Describe at least two other general findings summarized in the diagram.
3. Why isn't there any overlap between the "up" and "down" bubbles for each hormone?

Source: J. Nemhauser, T. Mockler, and J. Chory. 2004. Interdependency of brassinosteroids and auxin in *Arabidopsis* seedling growth. *PLoS Biology* 2:1460–1471.

Apply Evolutionary Thinking

Cryptochrome occurs in plants and animals. If it was inherited from their shared ancestor, what other major groups of organisms might also have it?

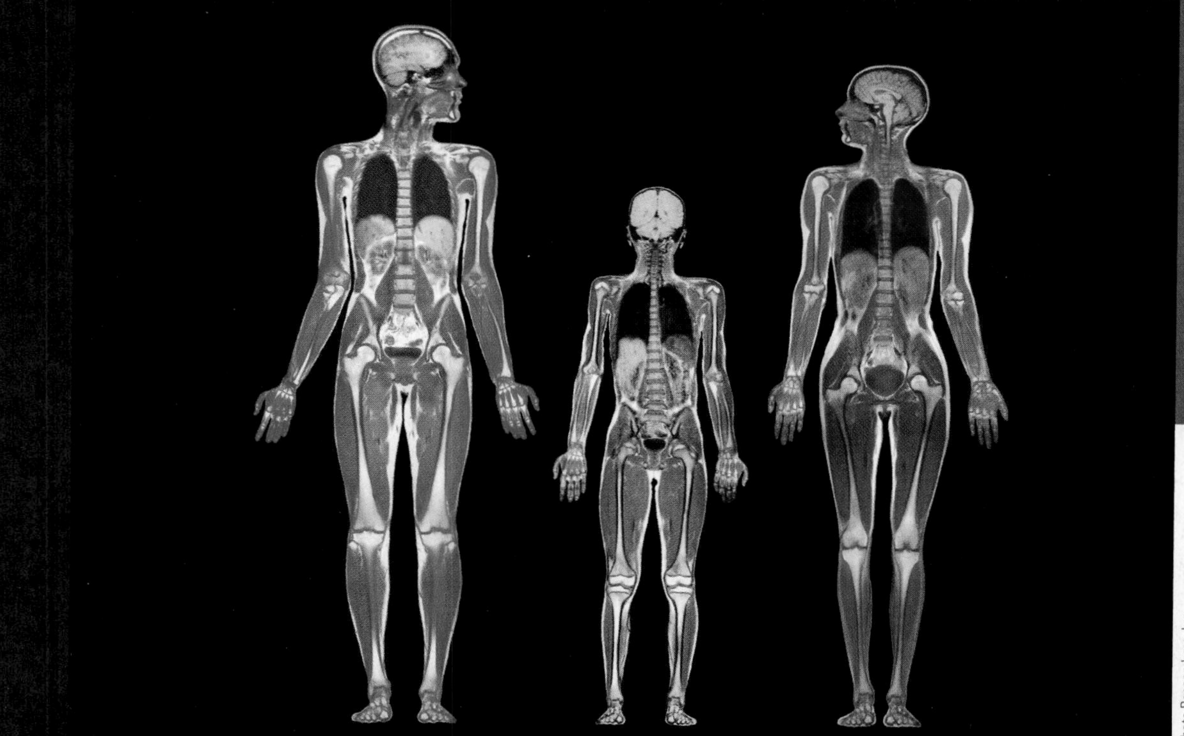

Magnetic resonance imaging (MRI) whole body scans of a man (left), a 9-year-old boy (middle), and a woman. Various organs can be seen in the scans: the whitish skeleton throughout the bodies, the brains within the skulls, lungs (dark) in the chests, lobes of the liver (green and brown ovals) in the abdomens, and bladders (dark ovals) in the lower abdomens.

Simon Fraser/SPL/Photo Researchers, Inc.

36

STUDY OUTLINE

Introduction to Animal Organization and Physiology

Why It Matters. . . On April 10, 1912, the RMS *Titanic* left Southampton, England, on her maiden voyage, bound for New York. A total of 2,223 people were on board, including 899 crew. Four days out, near midnight, lookouts spotted a large iceberg directly in the path of the ship. Emergency measures prevented a head-on collision, but the iceberg brushed the starboard side of the ship. Some parts of the hull buckled, and rivets popped out below the waterline. As a result, the *Titanic* began to take on water. The captain ordered a distress signal to be sent and, as the ship continued to take on water, he ordered passengers and crew to the lifeboats. The ship's lifeboats could hold a maximum of 1,178 people, far fewer than the number of people on board. Two hours and forty minutes after striking the iceberg, the *Titanic* sank. Of the people on board, 1,517 died, most of them succumbing to hypothermia in the −2°C water. At that temperature, death likely occurred within 15 minutes. *Hypothermia* is a condition in which the core temperature of the body falls significantly below normal for a prolonged period. In humans, a drop in core temperature of a few degrees affects brain function and leads to confusion; continued hypothermia, as in the case of the *Titanic,* can lead to death.

Human body temperature is normally regulated within a narrow range so as to provide optimal conditions for cellular life. Body temperature is just one aspect of an animal's internal environment that must be regulated. The maintenance of the internal environment in a relatively stable state is called **homeostasis** (*homeo* = similar; *stasis* = standing or stopping). The processes and activities responsible for homeostasis are called **homeostatic control systems.** These control systems compensate both for the external environmental changes that a human or other animal encounters, such as differences in temperature and humidity, and for changes in their own body systems, such as the availability of nutrients for cells and tissues. As we saw for the unfortunate *Titanic* passengers and crew, the homeostatic control systems for temperature regulation are limited in their ability to maintain the stable state, after which they break down. Not only can animals die from hypothermia after prolonged exposure to the cold, they can also die from hyperthermia after prolonged exposure to excessive heat. Similarly, an imbalance between energy input (in the form of nutrients) and energy output can be lethal.

All animals have body systems for acquiring and digesting nutrients to provide energy for life, growth, reproduction, and movement. Biologists are interested in the structures and functions of these systems. **Anatomy** is the study of the structures of organisms, and **physiology** is the study of their functions—the physico-chemical processes of organisms.

In this chapter we begin with the organization of individual cells into tissues, organs, and organ systems, the major body structures that carry out animal activities. Our discussion continues with a look at how the processes and activities of organ systems coordinate to accomplish homeostasis. The other chapters in this unit discuss the individual organ systems that carry out major body functions such as digestion, movement, and reproduction. Although we emphasize vertebrates throughout the unit, with particular reference to human physiology, we also make comparisons with invertebrates, to keep the structural and functional diversity of the animal kingdom in perspective and to understand the evolution of the structures and processes involved. <

36.1 Organization of the Animal Body

The individual cells of animals have the same requirements as cells of any kind. They must be surrounded by an aqueous solution that contains ions and molecules required by the cells, including complex organic molecules that can be used as an energy source. The concentrations of these molecules and ions must be balanced to keep cells from shrinking or swelling excessively because of osmotic water movement. Most animal cells also require oxygen to serve as the final acceptor for electrons removed in oxidative reactions. Animal cells must be able to release waste molecules and other by-products of their activities, such as carbon dioxide, to their environment. The physical conditions of the cellular environment, such as temperature, must also remain within tolerable limits.

The evolution of multicellularity (see Section 24.3) made it possible for organisms to create an **internal environment** of fluid that supplies all the needs of individual cells, including nutrient supply, waste removal, and osmotic balance. This internal environment allows multicellular organisms to occupy diverse **external environments,** including dry terrestrial habitats that would be lethal to single cells. Multicellular organisms can also become relatively large because their individual cells remain small enough to exchange ions and molecules with the internal fluid.

The evolution of multicellularity also allowed major life functions to be subdivided among specialized groups of cells, with each group concentrating on a single activity. In animals, some groups of cells became specialized for movement, others for food capture, digestion, internal circulation of nutrients, excretion of wastes, reproduction, and other functions. Specialization greatly increases the efficiency by which animals carry out these functions.

In most animals, specialized groups of cells are organized into tissues, the tissues into organs, and the organs into organ systems **(Figure 36.1)**. A **tissue** is a group of cells with the same structure and function, working together as a unit to carry out one or more specialized activities; four types of tissues are shown in the figure. An **organ** integrates two or more different tissues into a structure that carries out a specific function. The eye, liver, and stomach are examples of organs. An **organ system** (also called a **body system**) coordinates the activities of two or more organs to carry out a major body function such as movement, digestion, or reproduction. The nervous system, for example, is the main regulatory system of the body and consists of the brain, spinal cord, peripheral nerves, and sensory organs.

STUDY BREAK 36.1 <

1. **What are some advantages for an organism being multicellular?**
2. **What is the difference between a tissue, an organ, and an organ system?**

36.2 Animal Tissues

Although the most complex animals may contain hundreds of distinct cell types, all can be classified into one of four basic tissue groups: *epithelial, connective, muscle,* and *nervous* (see Figure 36.1). Each tissue type is assembled from individual cells. The properties of those cells determine the structure and, therefore, the function of the tissue. More specifically, the structure and integrity of a tissue depend on the structure and organization of the cytoskeleton within the cell, the type and organization of the extracellular matrix (ECM) surrounding the cell, and the junctions holding cells together (see Section 5.5).

Junctions of various kinds link cells into tissues (see Figure 5.29). *Anchoring junctions* form buttonlike spots or belts that weld cells together. They are most abundant in tissues subject to stretching, such as skin and heart muscle. *Tight junctions* seal the spaces between cells, keeping molecules and even ions from leaking between cells. For example, tight junctions in the tissue lining the urinary bladder prevent waste molecules and ions from leaking out of the bladder into other body tissues. *Gap junctions* open channels between cells in the same tissue, allowing ions and small molecules to flow freely from one to another. For example, gap junctions between muscle cells help muscle tissue function as a unit.

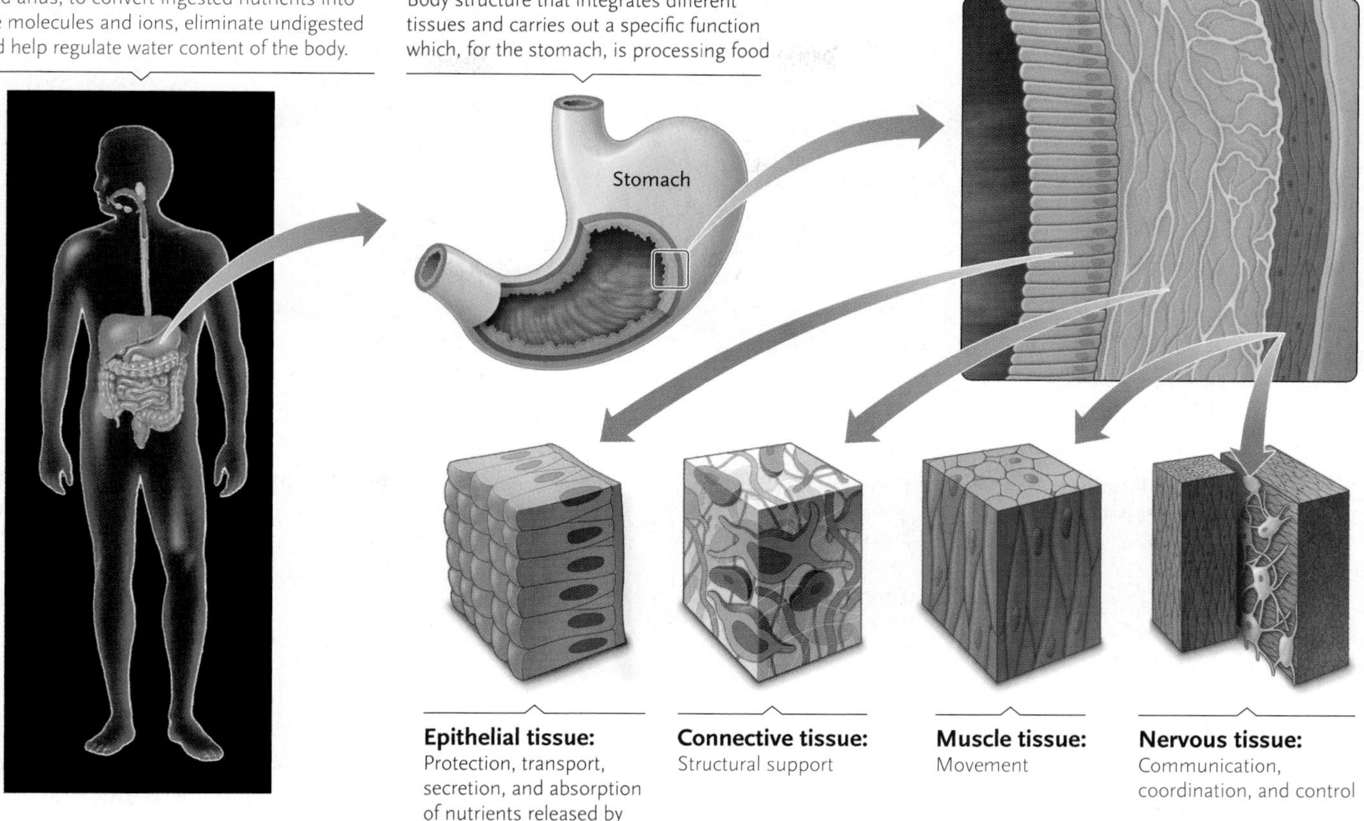
Organ system:
A set of organs that interacts to carry out a major body function. The digestive system coordinates the activities of organs, including the mouth, esophagus, stomach, small and large intestines, liver, pancreas, rectum, and anus, to convert ingested nutrients into absorbable molecules and ions, eliminate undigested matter, and help regulate water content of the body.
Organ:
Body structure that integrates different tissues and carries out a specific function which, for the stomach, is processing food
Stomach
Epithelial tissue:
Protection, transport, secretion, and absorption of nutrients released by digestion of food
Connective tissue:
Structural support
Muscle tissue:
Movement
Nervous tissue:
Communication, coordination, and control

FIGURE 36.1
Organization of animal cells into tissues, organs, and organ systems, exemplified here by the digestive system.

Epithelial Tissue Forms Protective, Secretory, and Absorptive Coverings and Linings of Body Structures

Epithelial tissue (*epi* = over; *thele* = covering) consists of sheetlike layers of cells that are usually connected by tight junctions, with little ECM material between them **(Figure 36.2).** Also called *epithelia* (singular, *epithelium*), these tissues cover body surfaces and the surfaces of internal organs, as well as line cavities and ducts within the body. They protect body surfaces from abrasion or from invasion by bacteria and viruses and secrete or absorb substances. For example, the epithelium covering a fish's gill structures serves as a barrier to bacteria and viruses and exchanges oxygen, carbon dioxide, and ions with the aqueous environment. In the epidermis of vertebrates, some epithelial cells contain a network of keratin, a family of fibrous proteins. Keratin forms protective structures: the fingernails, hair, claws, hooves, and horns of mammals, the feathers of birds, and the scales of reptiles and fish. In arthropods, the surface epithelium secretes a tough cuticle that forms a barrier to the environment and serves as the animal's skeleton.

Some epithelia, such as those lining the capillaries of the circulatory system, act as filters, allowing ions and small molecules to leak from the blood into surrounding tissues while barring the passage of blood cells and large molecules such as proteins.

Because epithelia form coverings and linings, they have a free (or outer) surface and an inner surface. The outer, **apical surface** may be exposed to water, air, or fluids within the body. The inner, **basal surface** adheres to a layer of ECM secreted by the epithelial cells called the **basal lamina,** which fixes the epithelium to underlying tissues, often connective tissues.

In internal cavities and ducts, the apical surface is often covered with *cilia,* which beat like oars to move fluids through the cavity or duct. The epithelium lining the oviducts in mammals, for example, is covered with cilia that generate fluid currents to move eggs from the ovaries to the uterus. In some epithelia, including the lining of the small intestine, the free surface is crowded with *microvilli,* fingerlike extensions of the plasma membrane that increase the area available for secretion or absorption.

TYPES OF EPITHELIA Epithelia are classified as *simple*—formed by a single layer of cells—or *stratified*—formed by multiple cell layers (see Figure 36.2). The shapes of cells within an epithelium may be *squamous* (mosaic, flattened, and spread out), *cuboidal* (shaped roughly like dice or cubes), or *columnar* (elongated, with the long axis perpendicular to the epithelial layer). Four principal types of epithelia are found in the body (see Figure 36.2).

The cells of some epithelia, such as those forming the skin and the lining of the intestine, divide constantly to replace worn and

A. Simple squamous epithelium

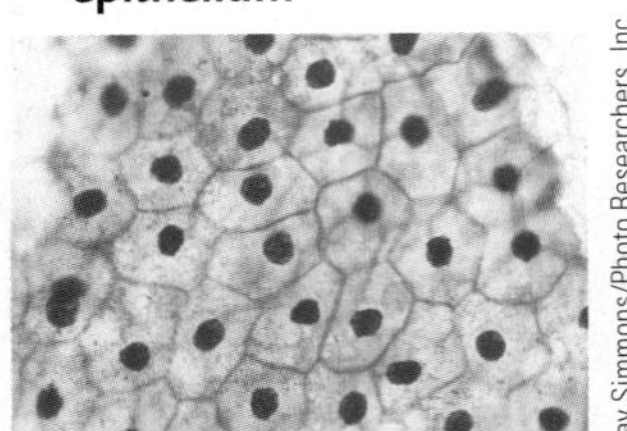
Ray Simmons/Photo Researchers, Inc.

Apical surface
Basal surface
Basal lamina

Description: Layer of flattened cells

Common locations: Blood vessel walls; air sacs of lungs

Function: Diffusion

B. Stratified squamous epithelium

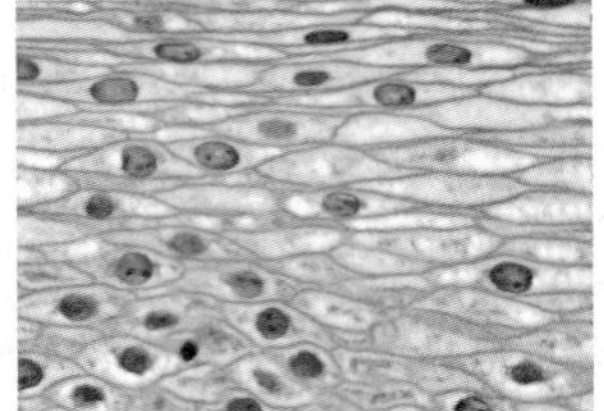
Dr. Gladden Willis/Getty Images

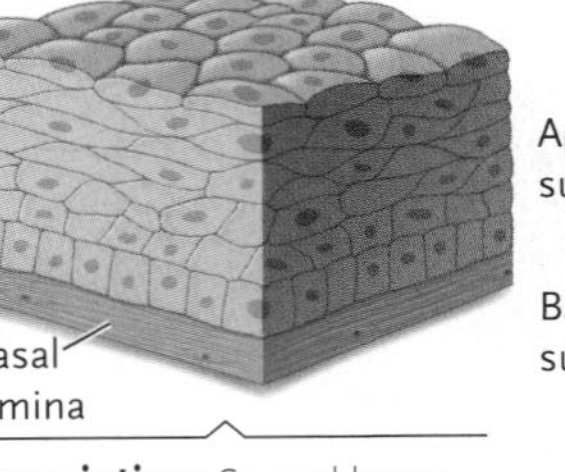
Apical surface
Basal surface
Basal lamina

Description: Several layers of flattened cells

Common locations: Skin and other surfaces subject to abrasion, such as the mouth, esophagus, and vagina

Function: Protection against abrasion; typically not involved in secretion or absorption

C. Cuboidal epithelium

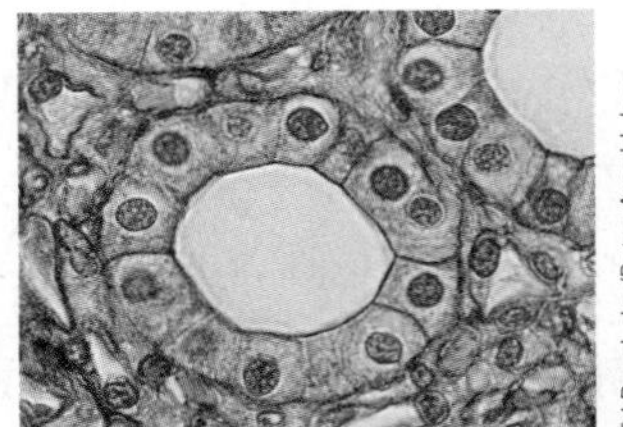
Ed Reschke/Peter Arnold, Inc.

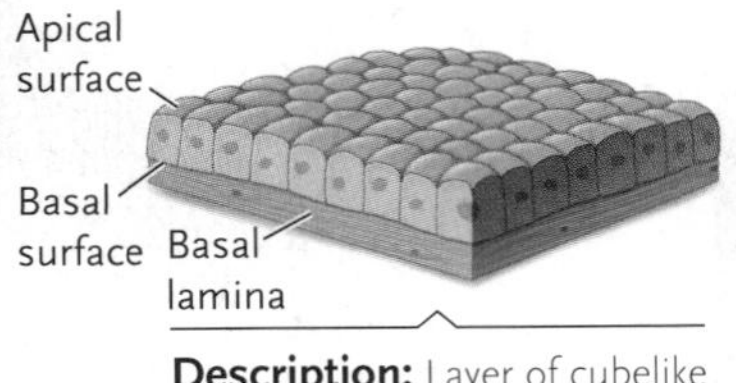
Apical surface
Basal surface
Basal lamina

Description: Layer of cubelike cells; free surface may have microvilli

Common locations: Glands and tubular parts of nephrons in kidneys

Function: Secretion, absorption

D. Columnar epithelium

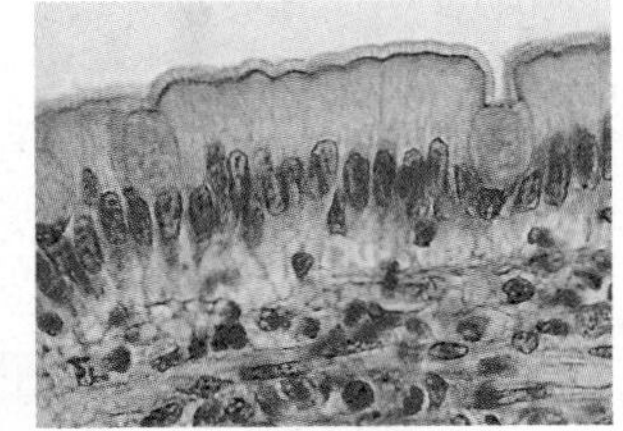
Don Fawcett/Visuals Unlimited

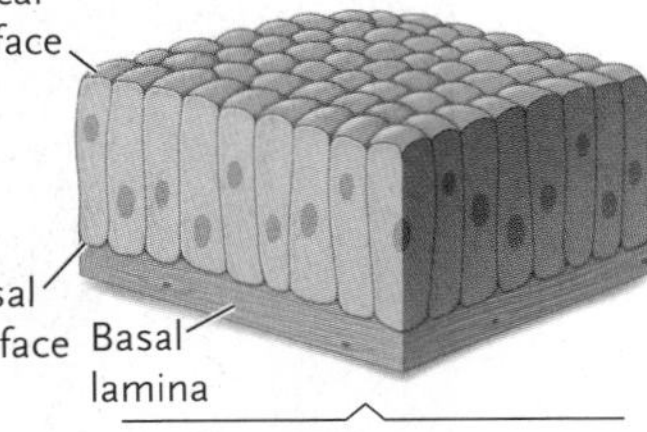
Apical surface
Basal surface
Basal lamina

Description: Layer of tall, slender cells; free surface may have microvilli

Common locations: Lining of gut and respiratory tract

Function: Secretion, absorption

FIGURE 36.2
Principal types of epithelia.

FIGURE 36.3
Exocrine and endocrine glands. The poison secreted by the blue poison dart frog *(Dendrobates azureus)* is one of the most lethal glandular secretions known.

Image copyright Michael Lynch, 2010. Used under license from Shutterstock.com

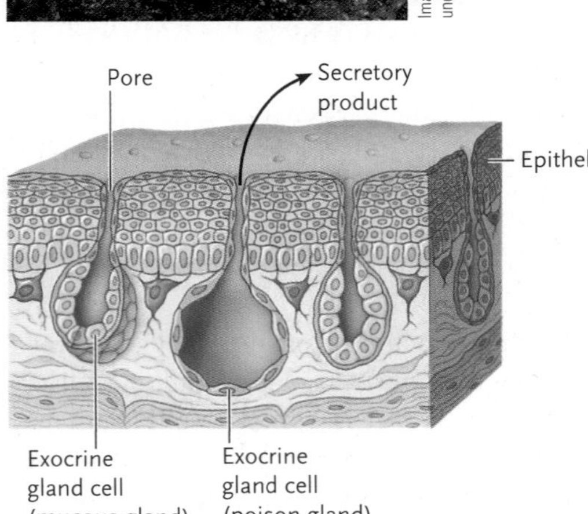
Pore
Secretory product
Epithelium
Exocrine gland cell (mucous gland)
Exocrine gland cell (poison gland)

A. Examples of exocrine glands: The mucus- and poison-secreting glands in the skin of a blue poison dart frog

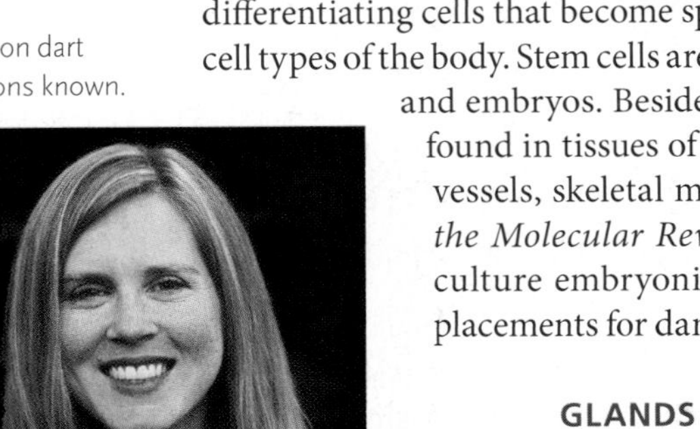
Thyroid

Epithelium
Endocrine gland cell
Blood vessel

B. Example of an endocrine gland: The thyroid gland, which secretes hormones that regulate the rate of metabolism and other body functions

dying cells. New cells are produced through division of stem cells in the basal layer of the skin. *Stem cells* are undifferentiated (unspecialized) cells that divide to produce either more stem cells or differentiating cells that become specialized into one of the many cell types of the body. Stem cells are found in both adult organisms and embryos. Besides the skin, adult stem cells are found in tissues of the brain, bone marrow, blood vessels, skeletal muscle, and liver. (*Insights from the Molecular Revolution* describes an effort to culture embryonic stem cells as a source of replacements for damaged tissues and organs.)

GLANDS FORMED BY EPITHELIA Epithelia typically contain or give rise to cells that are specialized for secretion. Some of these secretory cells are scattered among nonsecretory cells within the epithelium. Others form structures called **glands,** which are derived from pockets of epithelium during embryonic development.

Some glands, called **exocrine glands** (*exo* = external; *crine* = secretion), remain connected to the epithelium by a duct, which empties their secretions at the epithelial surface. Exocrine secretions include mucus, saliva, digestive enzymes, sweat, earwax, oils, milk, and venom (**Figure 36.3A**

INSIGHTS FROM THE Molecular Revolution

Culturing Human Embryonic Stem Cells

Embryonic stem (ES) cells—cells derived from embryos—are totipotent, meaning that they have the potential to develop into any tissue. Human ES cells are highly useful for research and hold great promise for biomedical applications. If stem cells can be directed to become any kind of differentiated cells in the test tube, those cells can potentially be used for replacing dead or damaged tissue in the body with healthy cells. Further, researchers forecast the possibility of growing customized tissues and replacement organs from stem cells, a process referred to as tissue engineering. The key to the value of human ES cells in biomedical science is the ability to maintain them indefinitely in an undifferentiated state, that is, without becoming specialized.

Research Question

How can human ES cells be cultured?

Experiment

Just a few years ago, James A. Thomson and his coworkers at the University of Wisconsin and collaborator at the Rambam Medical Center, Haifa, Israel, developed a successful method for culturing stem cells, which is outlined in the figure. Their use of mouse fibroblast cells derived from similar experiments in which nonhuman primate ES cells were successfully cultured. It was unknown at the outset if this experimental step would work with human cells.

Results

For 5 of 14 inner cell masses isolated, the human cells multiplied on the fibroblasts without differentiating. The division of the cells produced clumps. Cells in the clumps remained undifferentiated as long as the clumps did not contain more than 50 to 100 cells. To continue culturing the ES cells, some of the cells were taken and added to fresh culture dishes over mouse fibroblast cells. Using this technique, the five human ES cell cultures were maintained for as long as 8 months in the laboratory with no signs of differentiation or deterioration. They could be frozen, stored, and returned to active cultures at will.

Critically, the researchers demonstrated by several tests that the cultured cells retained full stem cell function:

1. Stem cells have characteristic surface molecules that change when the cells begin to differentiate into adult tissues. The cultured cells had appropriate surface molecules.
2. Stem cells contain active telomerase, an enzyme that helps maintain chromosomes at their normal length during rapid cycles of DNA replication and cell division (see Section 14.3). The enzyme becomes inactive in most cells as they differentiate into adult form. Telomerase in the cultured cells was fully active.
3. Stem cells can differentiate into a wide range of tissues. When mice were injected with samples of cultured cells, the cells grew into balls of tissue that included skin cells, gut epithelium, cartilage, bone, smooth and striated muscle, and nerve cells. The cultured cells seemed to differentiate into adult tissues when stimulated to do so, and thus to have all the characteristics of stem cells.

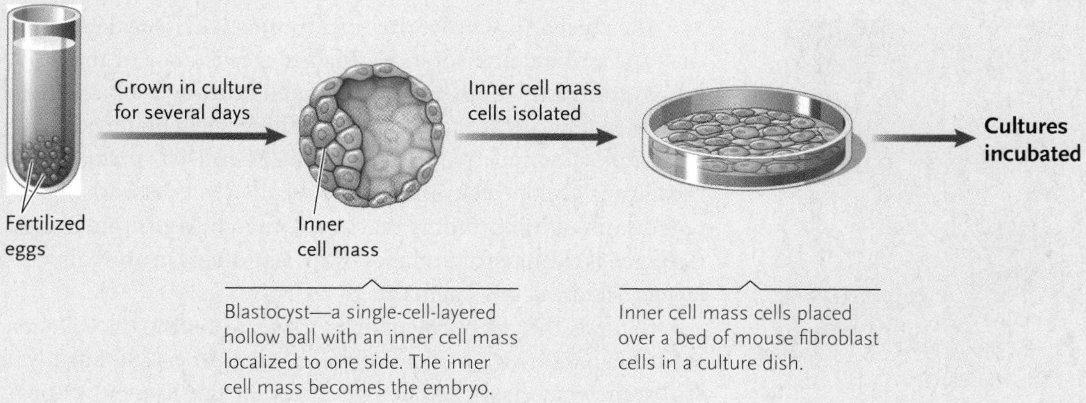

Fertilized eggs

Grown in culture for several days

Inner cell mass

Blastocyst—a single-cell-layered hollow ball with an inner cell mass localized to one side. The inner cell mass becomes the embryo.

Inner cell mass cells isolated

Inner cell mass cells placed over a bed of mouse fibroblast cells in a culture dish.

Cultures incubated

Conclusions

The researchers demonstrated that human ES cells could be cultured indefinitely while retaining all their stem cell properties. This was pioneering research, the first human ES cell line to be isolated.

Source: J. A. Thomson et al. 1998. Embryonic stem cell lines derived from human blastocysts. *Science* 282:1145–1147.

shows an exocrine gland in the skin of a poisonous tree frog). Other glands, called **endocrine glands** (*endo* = inside), become suspended in connective tissue underlying the epithelium, with no ducts leading to the epithelial surface. These ductless glands, such as the pituitary gland, adrenal gland, and thyroid gland **(Figure 36.3B),** release their products—hormones (see Chapters 7 and 40)—directly into the interstitial fluid, to be picked up and distributed by the circulatory system.

Some glands act as both exocrine glands and endocrine glands. The pancreas, for instance, has an exocrine function of secreting pancreatic juice through a duct into the small intestine where it plays an important role in food digestion, and an endocrine function of secreting the hormones insulin and glucagon into the bloodstream to help regulate glucose levels in the blood.

Connective Tissue Supports Other Body Tissues

Most animal body structures contain one or more types of **connective tissues.** Connective tissues support other body tissues, transmit mechanical and other forces, and in some cases act as filters. They consist of cells that form networks or layers in and around body structures and that are separated by nonliving material, specifically the ECM secreted by the cells of the tissue (see Section 5.5). Many forms of connective tissue have more ECM material (both by weight and by volume) than cellular material.

A. **Loose connective tissue**

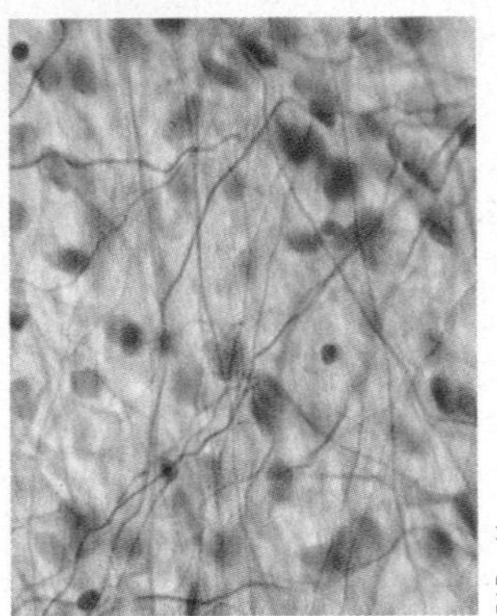

Ed Reschke

Collagen fiber
Fibroblast
Elastin fiber

Description: Fibroblasts and other cells surrounded by collagen and elastin fibers forming a glycoprotein matrix

Common locations: Under the skin and most epithelia; around blood vessels, nerves, and some internal organs

Function: Support, elasticity, diffusion

B. **Fibrous connective tissue**

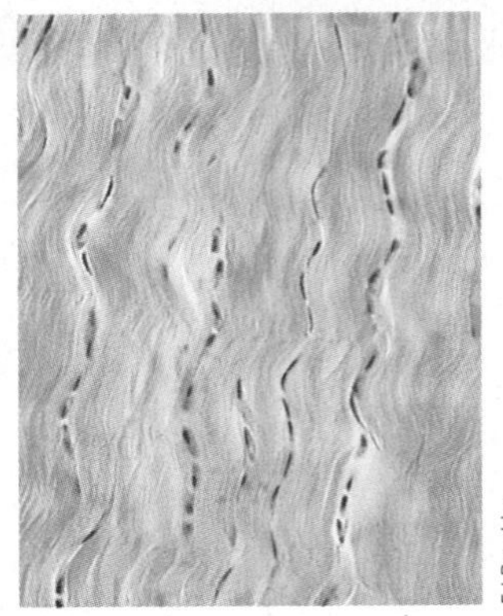

Ed Reschke

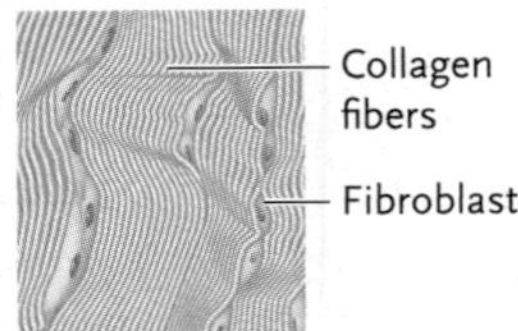

Collagen fibers
Fibroblast

Description: Long rows of fibroblasts surrounded by collagen and elastin fibers in parallel bundles with a dense ECM

Common locations: Tendons, ligaments

Function: Strength, elasticity

C. **Cartilage**

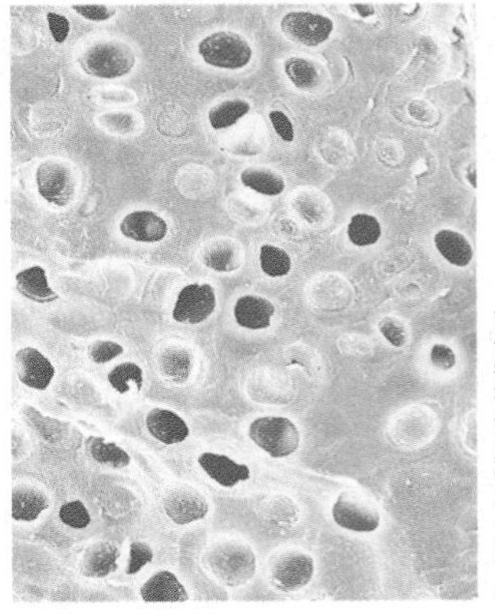

Fred Hossler/Visuals Unlimited

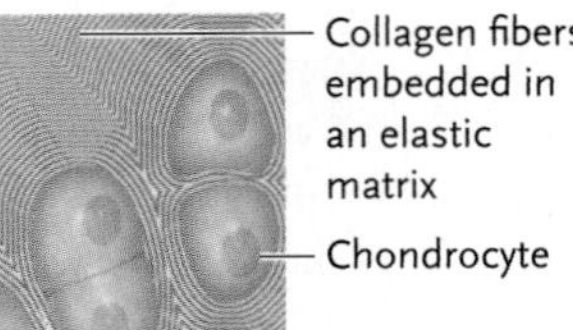

Collagen fibers embedded in an elastic matrix
Chondrocyte

Description: Chondrocytes embedded in a pliable, solid matrix of collagen and chondroitin sulfate

Common locations: Ends of long bones, ears, nose, parts of airways, skeleton of vertebrate embryos

Function: Support, flexibility, low-friction surface for joint movement

D. **Bone tissue**

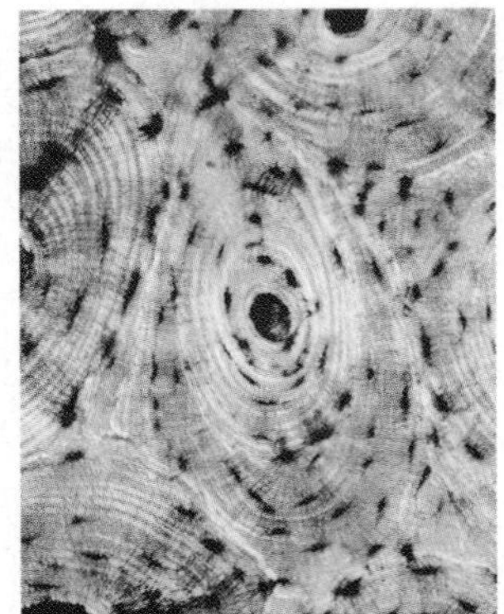

Ed Reschke

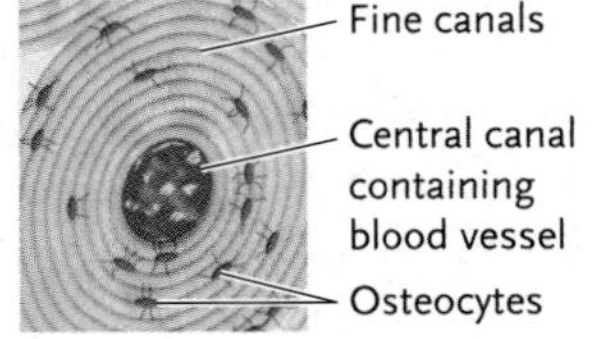

Fine canals
Central canal containing blood vessel
Osteocytes

Description: Osteocytes in a matrix of collagen and glycoproteins hardened with hydroxyapatite

Common locations: Bones of vertebrate skeleton

Function: Movement, support, protection

E. **Adipose tissue**

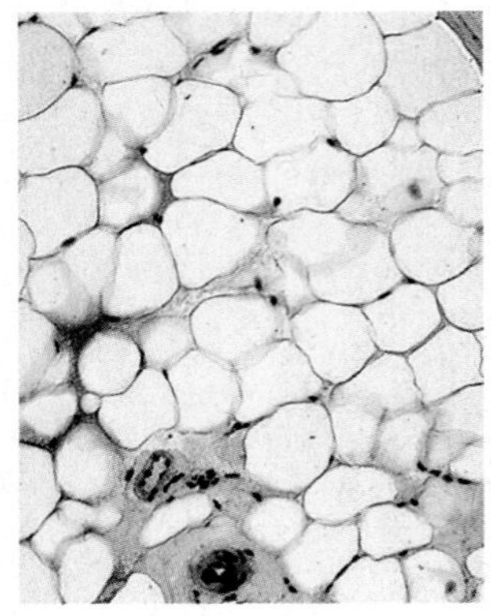

Ed Reschke

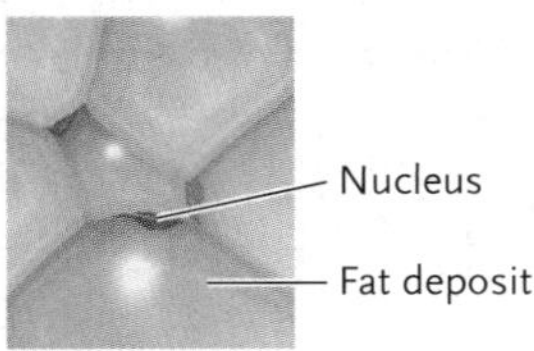

Nucleus
Fat deposit

Description: Large, tightly packed adipocytes with little ECM

Common locations: Under skin; around heart, kidneys

Function: Energy reserves, insulation, padding

F. **Blood**

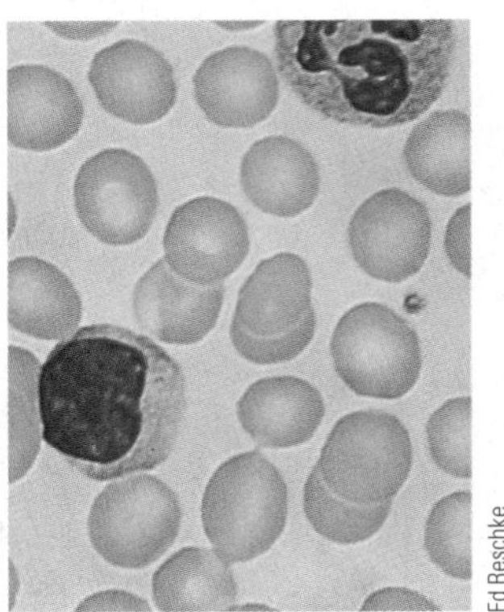

Ed Reschke

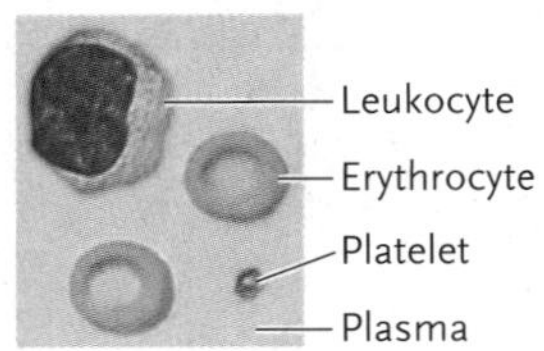

Leukocyte
Erythrocyte
Platelet
Plasma

Description: Leukocytes, erythrocytes, and platelets suspended in plasma, a fluid ECM

Common locations: Circulatory system

Function: Transport of substances

FIGURE 36.4
The six major types of connective tissues in vertebrates.

The mechanical properties of a connective tissue depend on the type and quantity of its ECM. The consistency of the ECM ranges from fluid (as in blood and lymph), through soft and firm gels (as in tendons), to hard and crystalline (as in bone).

In most connective tissues, the ECM consists primarily of the fibrous glycoprotein **collagen** embedded in a network of proteoglycans—glycoproteins that are very rich in carbohydrates. Collagen is the most abundant family of proteins in animals; collagens occur in all Eumetazoa.

In bone, the glycoprotein network surrounding the collagen is impregnated with mineral deposits that produce a hard, yet still somewhat elastic, structure. Another family of glycoproteins, **fibronectin,** aids in the attachment of cells to the ECM and helps hold the cells in position.

In some connective tissues another rubbery protein, **elastin,** adds elasticity to the ECM—it is able to return to its original shape after being stretched, bent, or compressed. Elastin fibers, for example, help the skin return to its original shape when pulled or stretched, and give the lungs the elasticity required for their alternating inflation and deflation. Related to elastin, the protein **resilin** is found only in insects and some crustaceans and is the most elastic material known. Resilin is the basis for the jumping of fleas, crickets, and locusts.

Vertebrates have six major types of connective tissues: **loose connective tissue, fibrous connective tissue, cartilage, bone, adipose tissue,** and **blood.** Each type has a characteristic function correlated with its structure **(Figure 36.4).**

LOOSE CONNECTIVE TISSUE **Loose connective tissue** consists of sparsely distributed **fibroblast** cells surrounded by a more or less open network of collagen and other glycoprotein fibers (Figure 36.4A); the fibroblasts secrete most of these proteins. Loose connective tissues support epithelia and form a corsetlike band around blood vessels, nerves, and some internal organs; they also reinforce deeper layers of the skin. Sheets of loose connective tissue, covered on both surfaces with epithelial cells, form the **mesenteries,** which hold the abdominal organs in place and provide lubricated, smooth surfaces that prevent chafing or abrasion between adjacent structures as the body moves.

FIBROUS CONNECTIVE TISSUE In **fibrous connective tissue,** fibroblasts are sparsely distributed among dense masses of collagen and elastin fibers that are lined up in highly ordered, parallel bundles (Figure 36.4B). The parallel arrangement produces maximum tensile strength and elasticity. Examples include **tendons,** which attach muscles to bones, and **ligaments,** which connect bones to each other at a joint. The cornea of the eye is a transparent fibrous connective tissue formed from highly ordered collagen molecules.

CARTILAGE **Cartilage** consists of sparsely distributed cartilage-producing cells called **chondrocytes,** surrounded by networks of collagen fibers embedded in a tough but elastic matrix of the glycoprotein *chondroitin sulfate* (Figure 36.4C). Elastin is also present in some forms of cartilage.

The elasticity of cartilage allows it to resist compression and stay resilient, like a piece of rubber. Bending your ear or pushing the tip of your nose, which are supported by a core of cartilage, gives a good idea of the flexible nature of this tissue. In humans, cartilage also supports the larynx, trachea, and smaller air passages in the lungs. It forms the disks cushioning the vertebrae in the spinal column and the smooth, slippery capsules around the ends of bones in joints such as the hip and knee. Cartilage also serves as a precursor to bone during embryonic development; in sharks and rays and their relatives, almost the entire skeleton remains as cartilage in adults.

BONE The densest form of connective tissue, **bone,** forms the skeleton, which supports the body, protects softer body structures such as the brain, and contributes to body movements.

Mature bone consists primarily of cells called **osteocytes** (*osteon* = bone) embedded in an ECM containing collagen fibers and glycoproteins impregnated with *hydroxyapatite,* a calcium-phosphate mineral (Figure 36.4D). The collagen gives bone tensile strength and elasticity; the hydroxyapatite resists compression and allows bones to support body weight. Cells called **osteoblasts** (*blast* = bud or sprout) produce the collagen and mineral of bone—as much as 85% of the weight of bone is mineral deposits. Osteocytes, in fact, are osteoblasts that have become trapped and surrounded by the bone materials they themselves produce. **Osteoclasts** (*clast* = break) are cells that remove the minerals and recycle them through the bloodstream. Bone is not a stable tissue; it is reshaped continuously by the bone-building osteoblasts and the bone-degrading osteoclasts.

Although bones appear superficially to be solid, they are actually porous structures, with microscopic spaces and canals. The structural unit of bone, the **osteon,** consists of a minute central canal surrounded by osteocytes embedded in concentric layers of mineral matter (see Figure 36.4D). A blood vessel and extensions of nerve cells run through the central canal, which is connected to the spaces containing cells by very fine, radiating canals filled with interstitial fluid. The blood vessels supply nutrients to the cells with which the bone is built, and the nerve cells hook up the bone cells to the body's nervous system.

ADIPOSE TISSUE The connective tissue called *adipose tissue* mostly contains large, densely clustered cells called **adipocytes** that are specialized for fat storage (Figure 36.4E). It has little ECM. Adipose tissue also cushions the body and, in mammals, forms an especially important insulating layer under the skin.

The animal body stores limited amounts of carbohydrates, primarily in muscle and liver cells. Excess carbohydrates are converted into the fats stored in adipocytes. The storage of chemical energy as fats offers animals a weight advantage. For example, the average human would weigh about 45 kg (100 pounds) more if the same amount of chemical energy was stored as carbohydrates instead of fats. Adipose tissue is richly supplied with blood vessels, which move fats or their components to and from adipocytes.

BLOOD **Blood** (see Figure 36.5F) is considered a connective tissue because its cells are suspended in a fluid ECM, plasma. The straw-colored **plasma** is a solution of proteins, nutrient molecules, ions, and gases.

Blood contains two primary cell types, **erythrocytes** (red blood cells; *erythros* = red) and **leukocytes** (white blood cells; *leukos* = white). Erythrocytes are packed with hemoglobin, a protein that can bind and transport oxygen. There are several types of leukocytes—all help to protect the body against invading viruses, bacteria, and other disease-causing agents. The blood plasma also contains **platelets,** membrane-bound fragments of specialized blood cells, which take part in the reactions that seal wounds with blood clots.

Blood is the major transport vehicle of the body. It carries oxygen and nutrients to body cells, removes wastes and by-products such as carbon dioxide, and maintains the internal fluid environment, including the osmotic balance between cells and the interstitial fluid. Blood also transports hormones and other signal molecules that coordinate body responses. (The components and roles of blood are described in Chapter 42.)

Muscle Tissue Produces the Force for Body Movements

Muscle tissue consists of cells that have the ability to contract (shorten). The contractions, which depend on the interaction of two proteins—**actin** and **myosin**—move body limbs and other structures, pump the blood, and produce a squeezing pressure in organs such as the intestine and uterus. Three types of muscle tissues, *skeletal, cardiac,* and *smooth,* produce body movements in vertebrates **(Figure 36.5).**

FIGURE 36.5
Structure of skeletal, cardiac, and smooth muscle.

A. **Skeletal muscle**

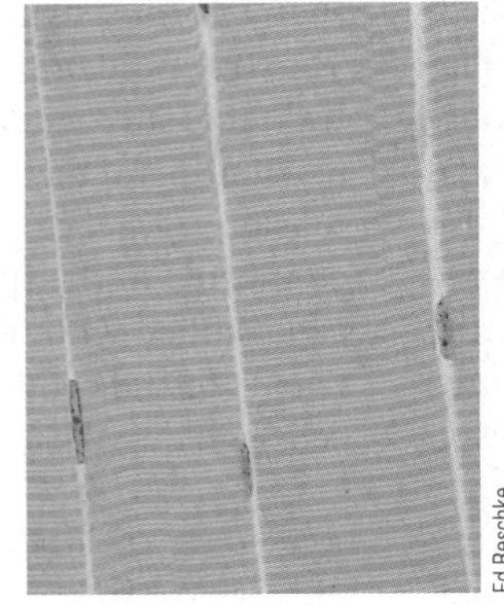
Ed Reschke

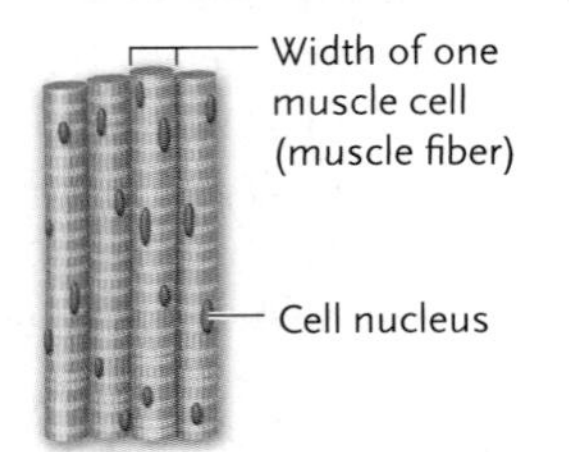
Width of one muscle cell (muscle fiber)
Cell nucleus

Description: Bundles of long, cylindrical, striated, contractile, multinucleate cells called muscle fibers

Typical location: Attached to bones of skeleton

Function: Locomotion, movement of body parts

B. **Cardiac muscle**

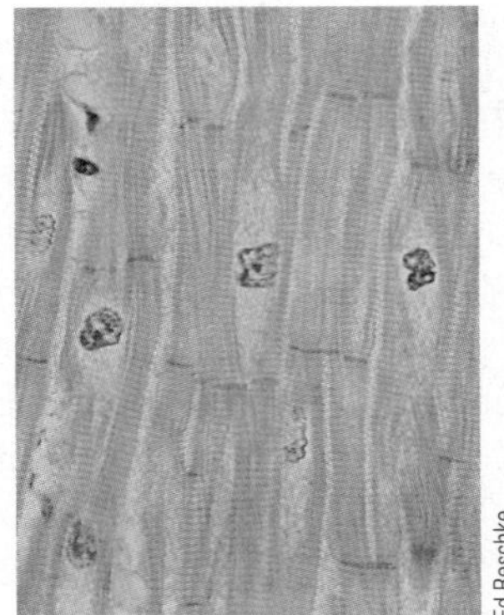
Ed Reschke

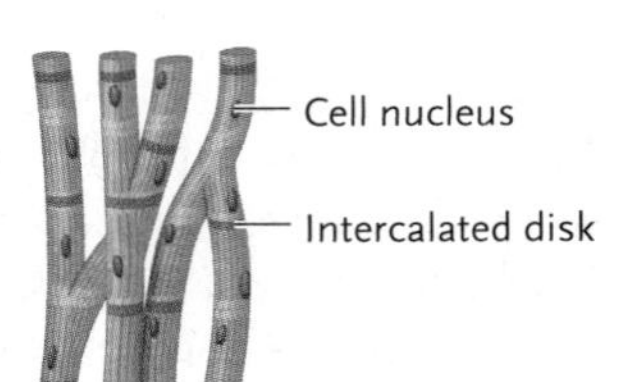
Cell nucleus
Intercalated disk

Description: Interlinked network of short and branched cylindrical, striated cells stabilized by anchoring junctions and gap junctions

Location: Wall of heart

Function: Pumping of blood within circulatory system

C. **Smooth muscle**

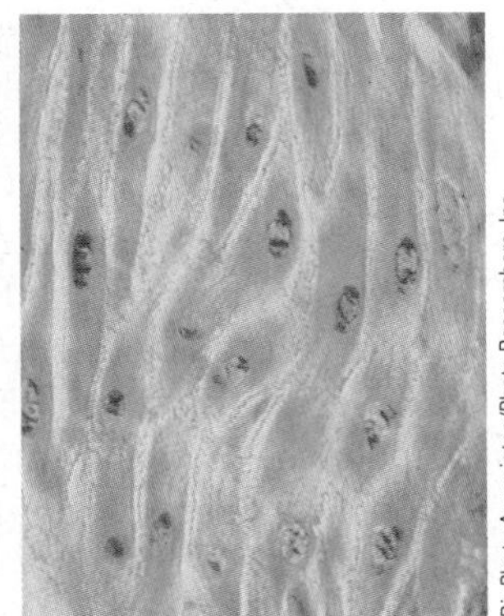
BioPhoto Associates/Photo Researchers, Inc.

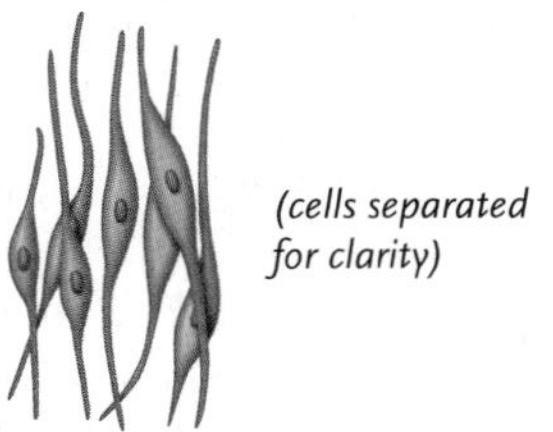
(cells separated for clarity)

Description: Loose network of contractile cells with tapered ends

Typical location: Wall of internal organs, such as stomach

Function: Movement of internal organs

SKELETAL MUSCLE **Skeletal muscle** is so called because most muscles of this type are attached by tendons to the skeleton. Skeletal muscle cells are also called **muscle fibers** because each is an elongated cylinder (Figure 36.5A). These cells are multinucleate (contain many nuclei in the same cytoplasm) and are packed with actin and myosin molecules arranged in highly ordered, parallel units that give the tissue a banded or striated appearance when viewed under a microscope. Muscle fibers packed side by side into parallel bundles surrounded by sheaths of connective tissue form many body muscles, such as the biceps.

Skeletal muscle contracts in response to signals carried by the nervous system. The contractions of skeletal muscles, which are characteristically rapid and powerful, move body parts and maintain posture. The contractions also release heat as a by-product of cellular metabolism. This heat helps mammals, birds, and some other vertebrates maintain their body temperatures when environmental temperatures fall. (Skeletal muscle is discussed further in Chapter 41.)

CARDIAC MUSCLE **Cardiac muscle** is the contractile tissue of the heart (Figure 36.5B). Cardiac muscle has a striated appearance because it contains actin and myosin molecules arranged like those in skeletal muscle. However, cardiac muscle cells are short and branched, with each cell connecting to several neighboring cells; the joining point between two such cells is called an **intercalated disk.** Cardiac muscle cells thus form an interlinked network, which is stabilized by anchoring junctions and gap junctions. This network enables heart muscle to contract in all directions, producing a squeezing or pumping action rather than the lengthwise, unidirectional contraction characteristic of skeletal muscle.

SMOOTH MUSCLE **Smooth muscle** is found in the walls of tubes and cavities in the body, including blood vessels, the stomach and intestine, the bladder, and the uterus. Smooth muscle cells are relatively small and spindle-shaped (pointed at both ends), and their actin and myosin molecules are arranged in a loose network rather than in bundles (Figure 36.5C). This loose network makes the cells appear smooth rather than striated when viewed under a microscope. Smooth muscle cells are enclosed by a mesh of connective tissue, and some of them are connected by gap junctions. The gap junctions transmit ions that make smooth muscles contract as a unit, typically producing a squeezing motion. Although smooth muscle contracts more slowly than skeletal and cardiac muscle do, its contractions can be maintained at steady levels for a much longer time. These contractions move and mix the stomach and intestinal contents, constrict blood vessels, and push the infant out of the uterus during childbirth.

Nervous Tissue Receives, Integrates, and Transmits Information

Nervous tissue contains cells called **neurons** (also called *nerve cells*) that serve as lines of communication and control between body parts. Billions of neurons are packed into the human brain; others extend throughout the body. Nervous tissue also contains **glial cells** (*glia* = glue), which physically support and provide nutrients to neurons, provide electrical insulation between them, and scavenge cellular debris and foreign matter.

A neuron consists of a *cell body*, which houses the nucleus and organelles, and two types of cell extensions, dendrites and axons **(Figure 36.6).** *Dendrites* are usually highly branched, whereas *axons* are usually unbranched except at their terminals. Depending on the type of neuron and its location in the body, its axon may extend from a few micrometers or millimeters to more than a meter.

The cell body or the dendrites of a neuron receive chemical or electrical signals from other neurons (the signal depends on the neuron type). The signals are converted into an electrical signal that is transmitted along the axons to the axon terminals, or endings. Depending on the type of neuron, axons ei-

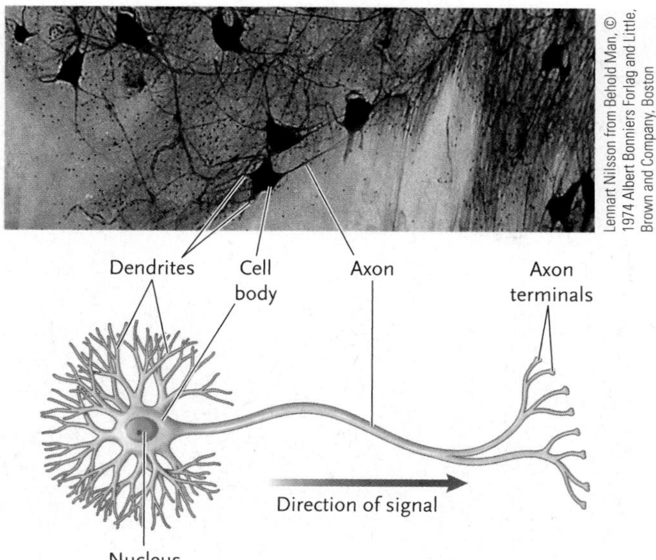
Lennart Nilsson from Behold Man, © 1974 Albert Bonniers Forlag and Little, Brown and Company, Boston
Dendrites
Cell body
Axon
Axon terminals
Direction of signal
Nucleus

FIGURE 36.6
Neurons and their structure. The micrograph shows a network of motor neurons, which relay signals from the brain or spinal cord to muscles and glands.

ther convert the electrical signal at their terminals to a chemical signal that stimulates a response in nearby muscle cells, gland cells, or other neurons or transfers the electrical signal directly to another neuron through gap junctions. (Neurons and their organization in body structures are discussed further in Chapters 37, 38, and 39.)

All four major tissue types—epithelial, connective, muscle, and nervous—combine to form the organs and organ systems of animals. The next section depicts the major organs and organ systems of vertebrates, and outlines their main tasks.

STUDY BREAK 36.2

1. **Distinguish between exocrine and endocrine glands. What is the tissue type of each of these glands?**
2. **What are the six major types of connective tissue in vertebrates?**
3. **What three types of muscle tissue produce body movements?**

THINK OUTSIDE THE BOOK

Individually or collaboratively, determine the type or types of muscle tissues found in invertebrates and give examples of each type.

36.3 Coordination of Tissues in Organs and Organ Systems

In the tissues, organs, and organ systems of an animal, each cell engages in the basic metabolic activities that ensure its own survival, and performs one or more functions of the system to which it belongs. All vertebrates (and most invertebrates) have eleven major organ systems, which are summarized in **Figure 36.7**, and discussed in the rest of this unit.

The functions of all these organ systems are coordinated and integrated to accomplish collectively a series of tasks that are vital to all animals, whether a flatworm, a salmon, a platypus, a moose, or a human. These functions include:

1. Acquiring nutrients and other required substances such as oxygen, coordinating their processing, distributing them throughout the body, and disposing of wastes.
2. Synthesizing the protein, carbohydrate, lipid, and nucleic acid molecules required for body structure and function.
3. Sensing and responding to changes in the environment, such as temperature, pH, and ion concentrations.
4. Protecting the body against injury or attack from other animals, and from viruses, bacteria, and other disease-causing agents.
5. Reproducing and, in many instances, nourishing and protecting offspring through their early growth and development.

Together these tasks maintain homeostasis, preserving the internal environment required for survival of the body. Homeostasis is the topic of the next section.

STUDY BREAK 36.3

What are the major functions of each of the eleven organ systems of the vertebrate body?

36.4 Homeostasis

To live, cells of all organisms must take in nutrients and O_2 from the external environment and eliminate wastes such as CO_2 and particular chemicals to that external environment. A single-celled organism such as an amoeba is in direct contact with the external environment. Although most cells of a multicellular animal are isolated from direct contact with the external environment, those cells have the same needs of nutrient and O_2 input and waste elimination. Those needs are met by the internal environment, namely the aqueous **extracellular fluid (ECF) (Figure 36.8).** The ECF has two components:

- **Plasma,** the fluid portion of blood.
- **Interstitial fluid** (*inter* = between; *stitial* = that which stands), the fluid that surrounds the cells.

The ECF connects all cells to the external environment. Thus, no matter where a cell is within the body, it can make the exchanges essential to its life with the interstitial fluid. Particular organ systems enable these exchanges between the external and internal environments. The digestive system processes incoming food and transfers the resulting nutrients into the plasma of the blood. The nutrients reach all parts of the body by the action of the circulatory system, along with O_2 which enters the blood by the action of the respiratory system. The nutrients and O_2 in the

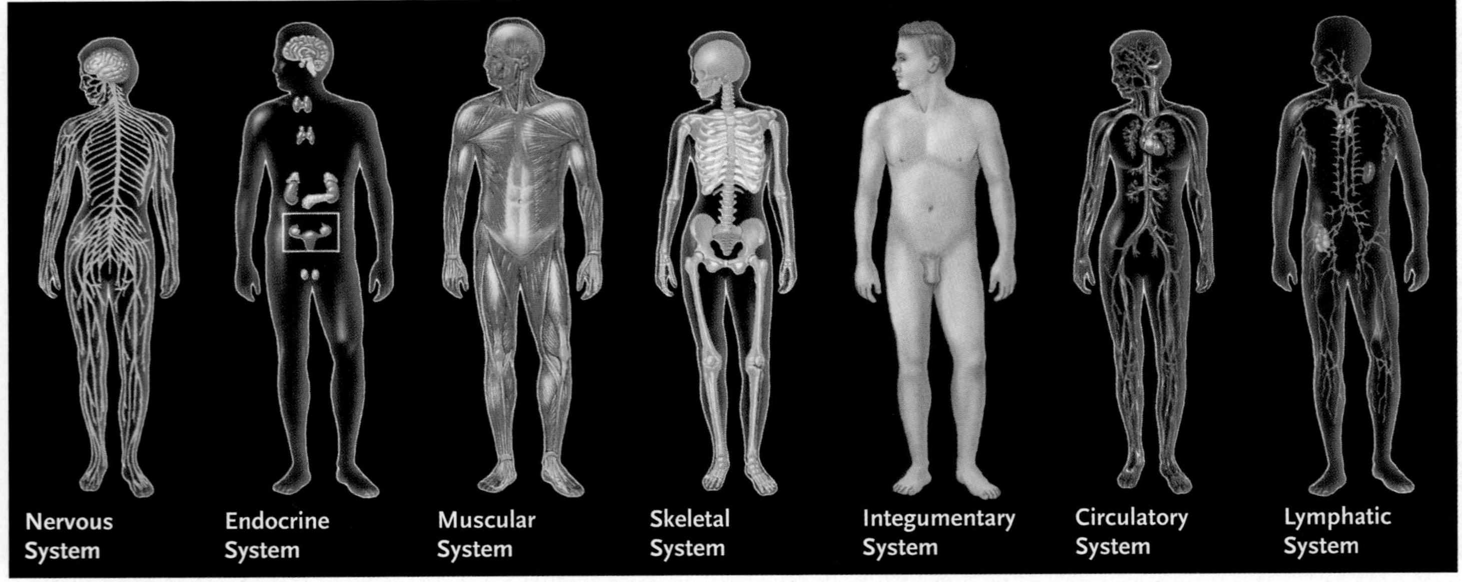
Nervous System
Endocrine System
Muscular System
Skeletal System
Integumentary System
Circulatory System
Lymphatic System

	Nervous System	Endocrine System	Muscular System	Skeletal System	Integumentary System	Circulatory System	Lymphatic System
Main structures:	Brain, spinal cord, peripheral nerves, sensory organs	Pituitary, hypothalamus, thyroid, adrenal, pancreas, and other hormone-secreting glands	Skeletal, cardiac, and smooth muscle	Bones, tendons, ligaments, cartilage	Skin, sweat glands, hair, nails	Heart, blood vessels, blood	Lymph nodes, lymph ducts, spleen, thymus
Main functions:	Principal regulatory system; monitors changes in internal and external environments and formulates compensatory responses; coordinates body activities	Regulates and coordinates body activities through secretion of hormones	Moves body parts; helps run bodily functions; generates heat; moves intestinal lumen contents	Supports and protects body parts; provides leverage for body movements; stores minerals	Covers external body surfaces and protects against injury and infection; helps regulate water content and body temperature	Distributes water, nutrients, oxygen, hormones, and other substances throughout body and carries away carbon dioxide and other metabolic wastes; helps stabilize internal temperature and pH	Returns excess fluid to the blood; defends body against invading viruses, bacteria, fungi, and other pathogens as part of immune system

FIGURE 36.7
Organ systems of the human body. The lymphatic system has functions that are part of the immune system. Other parts of the immune system, which are primarily cellular in nature, are not shown in this figure.

plasma reach the interstitial fluid through the capillaries and, from there, they enter the cells as needed. Waste moves in the opposite direction: from the cells into the interstitial fluid, and then to the plasma. The respiratory system handles removal of CO_2, and the excretory system handles the metabolic wastes.

FIGURE 36.8
Nature of the extracellular fluid (ECF).

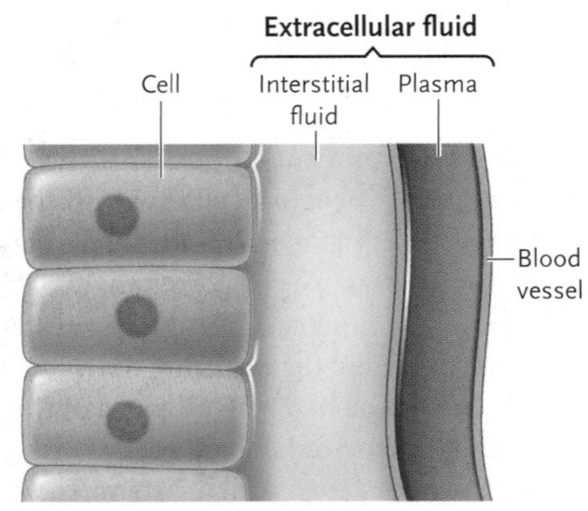
Extracellular fluid
Cell
Interstitial fluid
Plasma
Blood vessel

For optimal function of these systems, the composition and state of the ECF must be maintained within a narrow range so that cells have available the necessary nutrients and O_2 and wastes can be eliminated. Further, other aspects of the internal environment that are important for cellular (and therefore organism) life, such as temperature, must also be regulated within a narrow range (at least in warm-blooded animals).

As introduced in *Why It Matters,* the maintenance of the internal environment in a relatively stable state is **homeostasis.** Note that, although the *stasis* part of homeostasis might suggest a static, unchanging process, homeostasis is actually a *dynamic* process, in which internal adjustments are made continuously to compensate for external changes. For example, internal adjustments are needed for homeostasis during exercise or hibernation. The factors controlled by homeostatic mechanisms all require energy that must be constantly acquired from the external environment.

Homeostatic control systems are found in all organisms. Humans and other mammals maintain a consistent internal environment, but many kinds of multicellular animals show much more

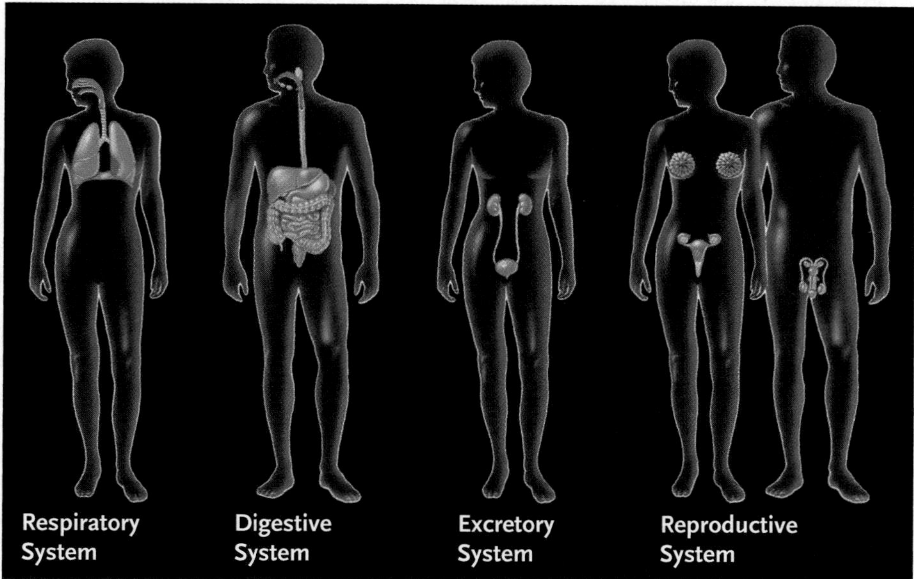

Respiratory System
Digestive System
Excretory System
Reproductive System

Respiratory System	Digestive System	Excretory System	Reproductive System
Main structures: Lungs, diaphragm, trachea, and other airways	**Main structures:** Oral cavity, pharynx, esophagus, stomach, intestines, liver, pancreas, rectum, anus	**Main structures:** Kidneys, bladder, ureter, urethra	**Main structures:** *Female*: ovaries, oviducts, uterus, vagina, mammary glands *Male*: testes, sperm ducts, accessory glands, penis
Main functions: Exchanges gases with the environment, including uptake of oxygen and release of carbon dioxide	**Main functions:** Converts ingested matter into molecules and ions that can be absorbed into body; eliminates undigested matter; helps regulate water content	**Main functions:** Removes and eliminates excess water, ions, and metabolic wastes from body; helps regulate internal osmotic balance and pH; helps regulate blood pressure	**Main functions:** Maintains the sexual characteristics and passes on genes to the next generation

variability in their internal environment than mammals do. Animals fall into two major categories in this regard: **regulators** maintain factors of the internal environment in a relatively constant state, and **conformers** have internal environments that match the external environment. For instance, mammals and birds are thermoregulators, meaning that they maintain their body temperature at a relatively constant value regardless of the temperature of the external environment. However, most animals—for example, fishes, reptiles, and insects—are thermoconformers, meaning that their body temperature matches that of the external environment. We will use mammals, and especially humans, as primary examples in the discussions throughout this unit.

Many Factors of the Internal Environment Are Homeostatically Regulated

Factors of the internal environment that are regulated homeostatically include:

1. *Nutrient concentration.* Energy production by cells requires a constant supply of nutrient molecules. The energy generated by catabolizing the nutrients is used for basic cellular processes and any specialized activities of the cell.
2. *Concentration of O_2.* Cellular respiration (see Chapter 8), the process that generates energy from catabolic reactions, requires a constant supply of O_2 for optimal productivity.
3. *Concentration of CO_2.* The CO_2 produced by the catabolic reactions of cellular respiration must be removed as waste or else the ECF would become increasingly acidic.
4. *Concentration of waste chemicals.* Particular biochemical reactions in the cell generate products that would be toxic to the cell if not removed as waste.
5. *Concentration of water and NaCl.* The relative concentrations of NaCl and water in the ECF affect how much water enters or leaves a cell and hence the cell's volume (see Chapter 6). These concentrations must be regulated to maintain a cell volume that is optimal for function; swollen or shrunken cells are typically functionally impaired.
6. *pH.* Changes in pH of the ECF can adversely affect enzymatic activities within cells, as well as nerve cell functions.
7. *Volume and pressure of plasma.* Both the volume and pressure in the vessels of the plasma component of the ECF must be maintained at levels adequate to distribute the fluid throughout the body. This circulation is vitally important for supplying cells with their needs and removing their wastes.
8. *Temperature.* Body cells (of warm-blooded animals) function optimally within a fairly narrow temperature range. Outside of that range, chemical reactions change their rates and may be completely inhibited. If cells become too cold, the rates of enzymatic reactions decrease too much and, if cells become too hot, structural and enzymatic proteins can be denatured and, therefore, become inactive.

Most Homeostatic Control Systems Operate Throughout the Body

Some homeostatic control systems operate locally whereas most operate systemically, that is, throughout the body. **Local homeostatic controls** operate only within an organ where a change in the internal environment needs to be addressed. For example, when you exercise, a skeletal muscle increases its use of O_2, which reduces the concentration of O_2 in the muscle cells and sets up a requirement for increased O_2 uptake from the interstitial fluid. The local homeostatic control in this instance responds to the decreased O_2 concentration by triggering the relaxation of the smooth muscle in the walls of the blood vessels supplying the exercising muscle. As a result, the blood vessels dilate (expand in diameter), increasing the blood flow to the exercising muscle, which brings more O_2. The end result is maintenance of an optimal O_2 concentration for the exercising skeletal muscle.

Systemic homeostatic controls are initiated outside of an organ or organ system to control that organ's or organ system's activity. The body's two major regulatory systems, the nervous system and the endocrine system, are responsible for most systemic homeostatic control of organs and organ systems. This type of control enables the regulation of several organs or organ systems to be coordinated toward a common goal. For example, if blood pressure drops too low, the nervous system acts on the heart to increase contraction strength and on blood vessels to constrict them so as to increase pressure in the system. Blood pH is controlled by both the nervous and endocrine systems, blood glucose by the endocrine system, internal temperature by the nervous and endocrine systems, and oxygen and carbon dioxide concentrations by the nervous system.

Homeostasis Is Accomplished by Negative Feedback Control Systems

The primary mechanism of homeostasis is **negative feedback** **(Figure 36.9)**. The components of a negative feedback control system maintaining homeostasis are:

- **Stimulus**—an environmental change that triggers a response.
- **Sensor**—the body component that detects the environmental change.
- **Integrator**—a control center that receives information from the sensor and compares it with the **set point,** the normal level for the condition being controlled. The integrator, typically part of the brain or endocrine system, sends out commands to correct any significant change from the set point detected.
- **Effector**—a system that is activated by the integrator to bring the condition under control back to the set point. Effectors may include parts of essentially any body tissue or organ.

FIGURE 36.9
Components of a negative feedback control system maintaining homeostasis. The sensor, integrator, and effector(s) are physical components in the body, such as a body structure or chemical.

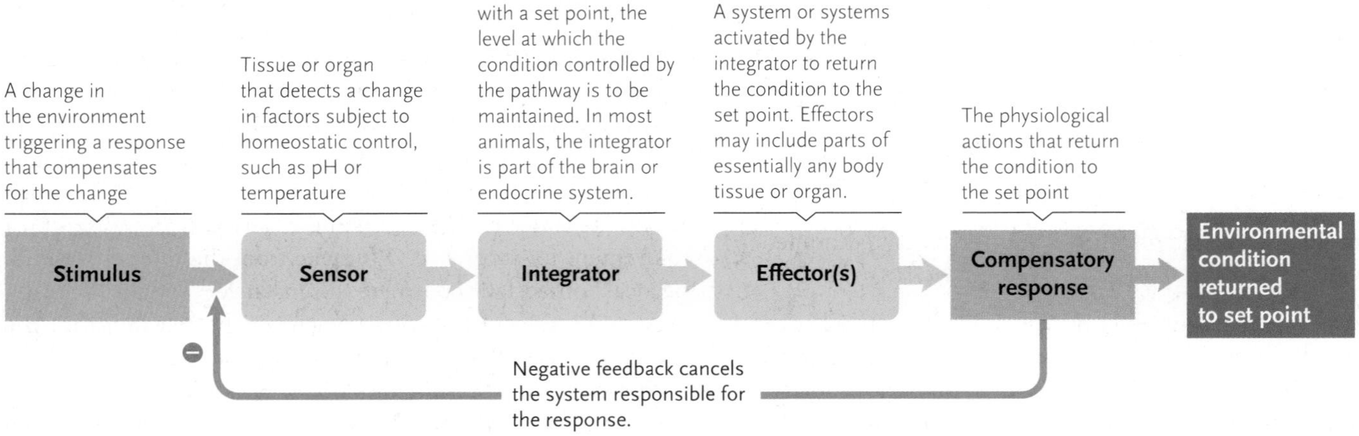

A change in the environment triggering a response that compensates for the change
Stimulus
Tissue or organ that detects a change in factors subject to homeostatic control, such as pH or temperature
Sensor
A control center that receives information from the sensor and initiates steps to correct the environmental change by comparing the detected change with a set point, the level at which the condition controlled by the pathway is to be maintained. In most animals, the integrator is part of the brain or endocrine system.
Integrator
A system or systems activated by the integrator to return the condition to the set point. Effectors may include parts of essentially any body tissue or organ.
Effector(s)
The physiological actions that return the condition to the set point
Compensatory response
Environmental condition returned to set point
Negative feedback cancels the system responsible for the response.

THE THERMOSTAT AS A NEGATIVE FEEDBACK CONTROL SYSTEM The concept of negative feedback may be most familiar in systems designed by human engineers. The thermostat maintaining temperature at a chosen level in a house provides an example:

- Stimulus—change in room temperature up or down more than a degree or so from the set point, the temperature you set in the thermostat.
- Sensor—thermometer circuitry within the thermostat that measures the temperature.
- Integrator—circuitry within the thermostat that activates the effector.
- Effector—component of the heating–cooling system that returns the room temperature to the set point. If the temperature has fallen below the set point, the effector is the furnace, which adds heat to the house until the temperature rises to the set point. If the temperature has risen above the set point, the effector is the air conditioner, which removes heat from the room until the temperature falls to the set point.

NEGATIVE FEEDBACK MECHANISMS IN ANIMALS Mammals and birds—warm-blooded vertebrates—also have a homeostatic control system that maintains body temperature within a relatively narrow range around a set point:

- Stimulus—change in body temperature beyond the normally controlled range around the set point.
- Sensor—temperature-monitoring nerve cells throughout the body.
- Integrator—temperature control center in a region of the brain called the *hypothalamus* (see Chapter 38) that detects changes in the temperature of the brain and the rest of the body, and compares it with a set point. For humans, the set point has a relatively narrow range centered at about 37°C.
- Effectors—physiological or behavioral responses that function to return body temperature to the set point.

If the temperature falls below the lower limit, as was the case for the unfortunate passengers and crew of the *Titanic* who ended up in the water, the hypothalamus activates effectors that constrict the blood vessels in the skin. The reduction in blood flow means that less heat is conducted from the blood through the skin to the environment, reducing heat loss from the skin. Other effectors may induce shivering, a physical mechanism to generate body heat. Also, integrating neurons in the brain, stimulated by signals from the hypothalamus, make us consciously sense a chill, which we may counteract behaviorally by, if possible, putting on more clothes or moving to a warmer area.

Conversely, if blood temperature rises above the set point by exposure to high external temperature or by vigorous exercise, the hypothalamus triggers effectors that dilate the blood vessels in the skin, increasing blood flow and heat loss from the skin. Other effectors induce sweating, which cools the skin and the blood flowing through it as the sweat evaporates. And again, through integrating neurons in the brain, we may consciously sense being overheated, which we may counteract behaviorally by shedding clothes, moving to a cooler location, stopping exercise, or taking a dip in a pool or a cool shower.

All mammals have similar homeostatic control systems that maintain or adjust body temperature. **Figure 36.10** illustrates how a dog responds to high environmental temperatures by the mechanisms we have discussed and, in addition, by panting. Birds also pant to reduce their body temperatures. Many terrestrial animals enter or splash water over their bodies to cool off.

Sometimes the temperature set point changes, and the negative feedback control systems then operate to maintain body temperature at the new set point. For example, if you become infected by certain viruses and bacteria, the temperature set point increases to a higher level, producing a fever to help overcome the infection. Once the infection is combated, the set point is readjusted down again to its normal level.

The homeostatic control system for temperature can also be overwhelmed by prolonged exposure to a particularly high or low temperature. In either case, the end result will be death unless the external temperature returns to a more moderate value. As the *Titanic* story describes, only 15 minutes in about-freezing water can kill a person. Likewise, homeostatic control systems for other regulated factors can also be overwhelmed or be disrupted. For example, the inability to homeostatically regulate blood glucose concentration appropriately leads to the disease state of diabetes.

Whereas mammals and birds regulate their internal body temperature within a narrow range around a set point, certain other vertebrates regulate over a broader range. These vertebrates use other, less precise negative feedback mechanisms for their temperature regulation. Snakes and lizards, for example, respond behaviorally to compensate for variations in environmental temperatures. They may absorb heat by basking on sunny rocks in the cool early morning and move to cooler, shaded spots in the heat of the afternoon.

Some invertebrates, such as dragonflies, moths, and butterflies, use muscular contractions equivalent to shivering when their body temperature falls below the level required for flight. The shivering contractions warm the muscles to flying temperature. All of these physiological and behavioral responses depend on negative feedback mechanisms involving sensors, integrators, and effectors.

Animals Also Have Positive Feedback Mechanisms That Do Not Result in Homeostasis

Under certain circumstances, animals respond to a change in internal or external environmental condition by a **positive feedback** mechanism that intensifies or adds to the change. Such mechanisms, with some exceptions, do not result in homeostasis. They operate when the animal is responding to life-threatening conditions (an attack, for instance), or as part of reproductive processes.

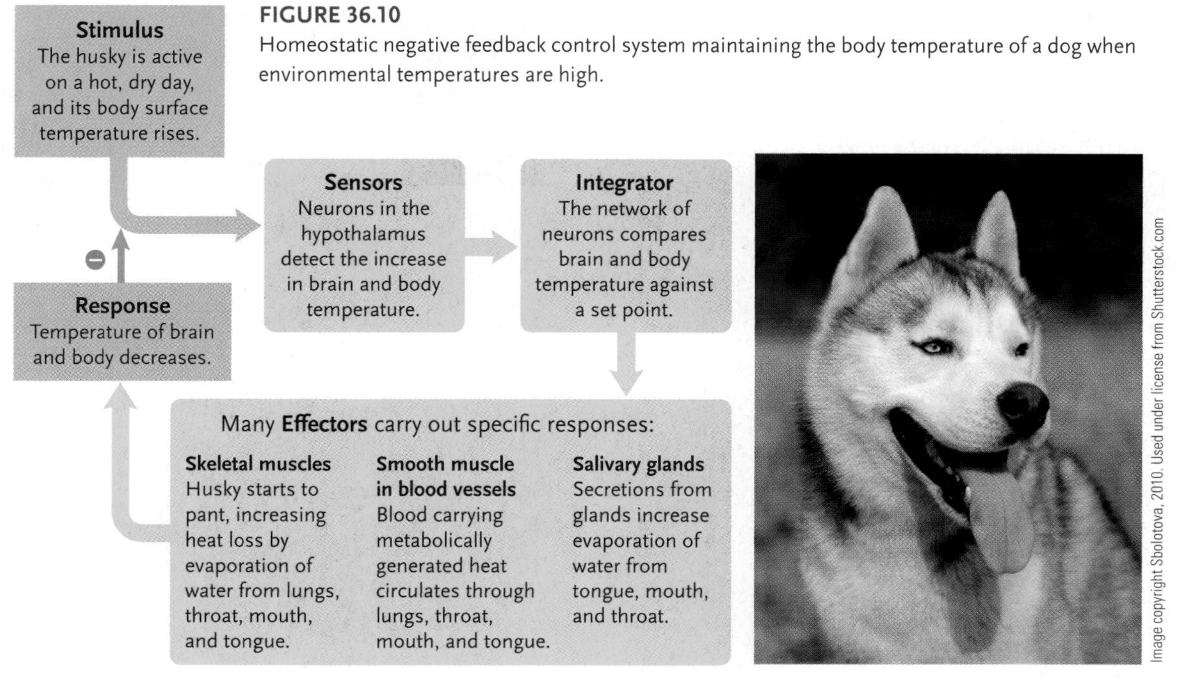
Stimulus
The husky is active on a hot, dry day, and its body surface temperature rises.
Sensors
Neurons in the hypothalamus detect the increase in brain and body temperature.
Integrator
The network of neurons compares brain and body temperature against a set point.
Response
Temperature of brain and body decreases.
Many Effectors carry out specific responses:
Skeletal muscles
Husky starts to pant, increasing heat loss by evaporation of water from lungs, throat, mouth, and tongue.
Smooth muscle in blood vessels
Blood carrying metabolically generated heat circulates through lungs, throat, mouth, and tongue.
Salivary glands
Secretions from glands increase evaporation of water from tongue, mouth, and throat.
Image copyright Sbolotova, 2010. Used under license from Shutterstock.com

FIGURE 36.10
Homeostatic negative feedback control system maintaining the body temperature of a dog when environmental temperatures are high.

The birth process in mammals is a prime example. During human childbirth, initial contractions of the uterus push the head of the fetus against the cervix, the opening of the uterus into the vagina. The pushing causes the cervix to stretch. Sensors that detect the stretching signal the hypothalamus to release a hormone, oxytocin, from the pituitary gland. Oxytocin increases the uterine contractions, intensifying the squeezing pressure on the fetus and further stretching the cervix. The stretching results in more oxytocin release and stronger uterine contraction, repeating the positive feedback circuit and increasing the squeezing pressure until the fetus is pushed entirely out of the uterus.

Because positive feedback mechanisms such as the one triggering childbirth do not result in homeostasis, they occur less commonly than negative feedback in animals. They also operate as part of larger, more inclusive negative feedback mechanisms that ultimately shut off the positive feedback pathway and return conditions to normal limits.

In conclusion, we learned in this chapter about the various tissues and organ systems of the body, and of the involvement of organ systems in homeostasis. Next, we begin a series of chapters describing the organ systems in detail, starting with the nervous system.

STUDY BREAK 36.4 <

What are the components of a negative feedback control system that results in homeostasis?

UNANSWERED QUESTIONS

As described in the chapter, all animals have homeostatic control systems through which they maintain a consistent internal environment. The physiological processes underlying many of these control systems are shaped by evolution to adapt to daily changes in the external environment. In mammals, many physiological processes display 24-hour rhythms that are generated by an internal gene-based clock called a circadian clock (*circa* = approximately, *dies* = day). Some malfunctions of systems that are normally under homeostatic control also seem to exhibit a circadian cycle.

Why do so many strokes and heart attacks occur in the morning?

Stroke and heart attack occur most frequently at a particular time of the day—in the morning—exhibiting a 24-hour, or circadian, rhythm. Circadian rhythms are generated through a molecular signaling mechanism called the circadian clock. A "central" circadian clock, in the brain, acts to control circadian rhythm in sleep-wake cycles. These same proteins are also found in blood vessels (vascular clock), where they influence vascular-specific functions by influencing molecules like nitric oxide. However, whether a broken clock contributes to the onset of heart attack and stroke remains an unanswered question. Current research in my laboratory and others is addressing these questions.

Location, location, location: Where in the body does the circadian clock malfunction to cause vascular disease?

Though heart attacks and strokes are the culmination of ongoing vascular disease, their onset is acute, often occurring in the morning around 8 AM. In contrast, the progression to vascular disease is chronic, occurring in a time-span of years. Our laboratory and others have now demonstrated that the genes comprising the circadian clock can contribute to vascular disease. This finding suggests that, in addition to daily cycles, the circadian clock controls aspects of vascular biology that are also important in the long-term. However, it remains unclear where in the body the circadian clock acts to control the vasculature. Is the central clock important or is the vascular clock the key? Indeed, there is active research ongoing to address these questions.

Research to address the role of the vascular clock may ultimately change the way we understand and treat arteriosclerosis, hypertension, and heart attack.

Think Critically

1. How might circadian rhythms have been important for survival in early organisms?
2. How is it advantageous for blood vessels to have their own circadian rhythm?

R. Daniel Rudic

R. Daniel Rudic is an Associate Professor in the Department of Pharmacology and Toxicology at the Medical College of Georgia. To learn more about his research on circadian rhythms and vascular biology, go to http://www.mcg.edu/som/phmtox/RudicLab/index.asp.

REVIEW KEY CONCEPTS

Go to **CENGAGENOW** at www.cengage.com/login to access quizzing, animations, exercises, articles, and personalized homework help.

36.1 Organization of the Animal Body

- In most animals, cells are specialized and organized into tissues, tissues into organs, and organs into organ systems. A tissue is a group of cells with the same structure and function, working as a unit to carry out one or more activities. An organ is an assembly of tissues integrated into a structure that carries out a specific function. An organ system is a group of organs that carry out related steps in a major physiological process (Figure 36.1).

36.2 Animal Tissues

- Animal tissues are classified as epithelial, connective, muscle, or nervous (Figure 36.1). The properties of the cells of these tissues determine the structures and functions of the tissues.
- Various kinds of junctions link cells in a tissue. Anchoring junctions "weld" cells together. Tight junctions seal the cells into a leakproof layer. Gap junctions form direct avenues of communication between the cytoplasm of adjacent cells in the same tissue.
- Epithelial tissue consists of sheetlike layers of cells that cover body surfaces and the surfaces of internal organs and line cavities and ducts within the body (Figure 36.2).
- Glands are secretory structures derived from epithelia. They may be exocrine (connected to an epithelium by a duct that empties on the epithelial surface) or endocrine (ductless, with no direct connection to an epithelium) (Figure 36.3).
- Connective tissue consists of cell networks or layers and an extracellular matrix (ECM). It supports other body tissues, transmits mechanical and other forces, and in some cases acts as a filter (Figure 36.4).
- Loose connective tissue consists of sparsely distributed fibroblasts surrounded by an open network of collagen and other glycoproteins. It supports epithelia and organs of the body and forms a covering around blood vessels, nerves, and some internal organs.
- Fibrous connective tissue contains sparsely distributed fibroblasts in a matrix of densely packed, parallel bundles of collagen and elastin fibers. It forms high tensile-strength structures such as tendons and ligaments.
- Cartilage consists of sparsely distributed chondrocytes surrounded by a network of collagen fibers embedded in a tough but highly elastic matrix of branched glycoproteins. Cartilage provides support, flexibility, and a low-friction surface for joint movement.
- In bone, osteocytes are embedded in a collagen matrix hardened by mineral deposits. Osteoblasts secrete collagen and minerals for the ECM; osteoclasts remove the minerals and recycle them into the bloodstream.
- Adipose tissue consists of cells specialized for fat storage. It also cushions and rounds out the body and provides an insulating layer under the skin.
- Blood consists of a fluid matrix, the plasma, in which erythrocytes and leukocytes are suspended. The erythrocytes carry oxygen to body cells; the leukocytes produce antibodies and initiate the immune response against disease-causing agents.
- Muscle tissue contains cells that have the ability to contract forcibly (Figure 36.5). Skeletal muscle, containing long cells called muscle fibers, moves body parts and maintains posture.
- Cardiac muscle, which contains short contractile cells with a branched structure, forms the heart.
- Smooth muscle consists of spindle-shaped contractile cells that form layers surrounding body cavities and ducts.
- Nervous tissue contains neurons and glial cells. Neurons communicate information between body parts in the form of electrical and chemical signals (Figure 36.6). Glial cells support the neurons or provide electrical insulation between them.

Animation: Cell junctions

Animation: Structure of an epithelium

Animation: Types of simple epithelium

Animation: Soft connective tissues

Animation: Specialized connective tissues

Animation: Muscle tissues

Animation: Functional zones of a motor neuron

Animation: Structure of human skin

Practice: Differences between cell and tissue types

36.3 Coordination of Tissues in Organs and Organ Systems

- Organs and organ systems are coordinated to carry out vital tasks, including maintenance of internal body conditions; nutrient acquisition, processing, and distribution; waste disposal; molecular synthesis; environmental sensing and response; protection against injury and disease; and reproduction.
- In vertebrates and most invertebrates, the major organ systems that accomplish these tasks are the nervous, endocrine, muscular, skeletal, integumentary, circulatory, lymphatic, immune, respiratory, digestive, excretory, and reproductive systems (Figure 36.7).

Animation: Human organ systems

36.4 Homeostasis

- Homeostasis is the process by which animals maintain their internal environment—the extracellular fluid (Figure 36.8)—under conditions their cells can tolerate. It is a dynamic state, in which internal adjustments are made continuously to compensate for environmental changes.
- Multicellular organisms fall into two categories with respect to homeostatic regulation of the internal environment: regulators maintain factors of the internal environment in a relatively constant state, whereas conformers have internal environments that match the external environment.
- Homeostatically regulated factors of the internal environment include nutrient concentration; O_2 and CO_2 concentrations; concentrations of waste chemicals, water and NaCl; pH; and volume and pressure of plasma.
- Most homeostatic control systems operate systemically, whereas only some operate locally.
- Homeostasis is accomplished by negative feedback mechanisms that include a sensor, which detects a change in an external or internal condition; an integrator, which compares the detected change with a set point; and an effector or effectors, which return the condition to the set point if it has varied (Figure 36.9).
- Animals also have positive feedback mechanisms, in which a change in an internal or external condition triggers a response that intensifies the change, and typically does not result in homeostasis.

UNDERSTAND AND APPLY

Test Your Knowledge

1. Which tissue type consisting of sheetlike layers of cells can both exchange oxygen and act as a barrier to bacteria?
   a. nervous
   b. epithelial
   c. connective
   d. muscle
   e. heart
2. Which of the following is a constant source of adult stem cells in a mammal?
   a. bone marrow
   b. pancreas
   c. basal lamina
   d. heart muscle
   e. kidney
3. A flexible, rubbery protein in connective tissue is called ______, whereas a more fibrous, less flexible glycoprotein is called ______.
   a. adipose; cartilage
   b. endocrine; exocrine
   c. sweat; hormones
   d. chondroitin sulfate; hydroxyapatite
   e. elastin; collagen
4. Adipose tissue:
   a. gives elasticity under epithelium.
   b. gives strength to tendons.
   c. insulates and is an energy reserve.
   d. provides movement, support, and protection.
   e. supports the nose and airways.
5. The bones of an elderly woman break more easily than those of a younger person. You would surmise that with aging, the cell type that diminishes in activity is the:
   a. osteocyte.
   b. osteoblast.
   c. osteoclast.
   d. chondrocyte.
   e. fibroblast.
6. Lifting weights will most increase the size of:
   a. skeletal muscle.
   b. smooth muscle.
   c. cardiac muscle.
   d. involuntary muscle.
   e. interlinked, branched muscle.
7. Which muscle types appear striated under a microscope?
   a. skeletal muscles only
   b. cardiac muscles only
   c. skeletal muscles and cardiac muscles
   d. smooth muscles only
   e. skeletal muscles and smooth muscles
8. Which of the following is *not* a homeostatic response?
   a. In a contest, a student eats an entire chocolate cake in 10 minutes, and his blood glucose level does not change dramatically.
   b. A jogger is sweating after a two-mile jog.
   c. The pupils in humans' eyes constrict when looking at a light.
   d. Physical activity increases carbon dioxide blood levels, which lowers blood pH, and increases breathing.
   e. Oxytocin is released by the hypothalamus during human childbirth.
9. When you exercise, a skeletal muscle increases its use of O_2, which reduces the concentration of O_2 in the muscle cells and sets up a requirement for increased O_2 uptake from the interstitial fluid. This is an example of:
   a. osmolarity.
   b. environmental sensing.
   c. integration.
   d. systemic homeostatic control.
   e. local homeostatic control.
10. The system that coordinates other organ systems is the:
    a. skeletal system.
    b. reproductive system.
    c. muscular system.
    d. nervous system.
    e. integumentary system.

Discuss the Concepts

1. Blood is often described as an atypical connective tissue. If you had to argue that blood is a connective tissue, what reasons would you include? What reasons would you include if you had to argue that blood is not a connective tissue?
2. What effect do you think a program of lifting weights would have on the bones of the skeleton? How would you design an experiment to test your prediction?
3. Positive feedback mechanisms are rare in animals compared with negative feedback mechanisms. Why do you think this is so?
4. Near the time of childbirth, collagen fibers in the connective tissue of the cervix break down, and gap junctions between the smooth muscle cells of the uterus increase in number. What do you think is the significance of these tissue changes?
5. Explain how, when driving, you control the car's speed by a typical negative feedback mechanism.

Design an Experiment

The regulation of temperature in mammals and birds is an example of homeostasis. Design an experiment to observe and measure processes involved in temperature homeostasis in sedentary versus athletic humans during exercise.

Interpret the Data

Vasopressin is an important regulator of blood pressure. Taurine has been shown to inhibit vasopressin release. The data below are from a study hypothesizing that taurine is part of a negative feedback system that regulates vasopressin. This figure shows the release of taurine from rat pituitary cells in the presence of vasopressin (10 nM). Are these data consistent with the hypothesis that taurine acts as part of a negative feedback system for vasopressin? Why?

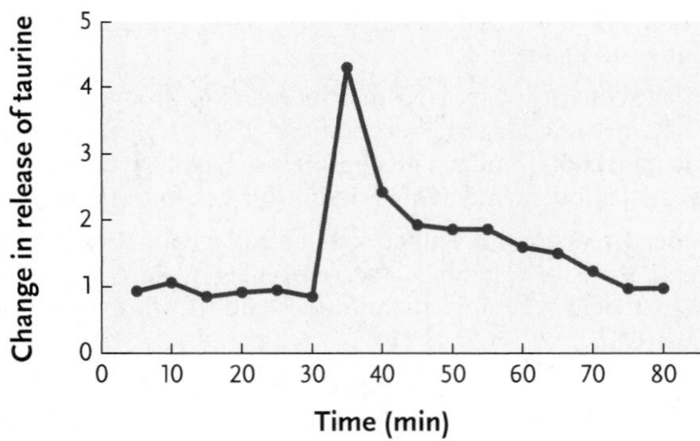

Change in release of taurine
0
1
2
3
4
5
0
10
20
30
40
50
60
70
80
Time (min)

Source: L. Rosso et al. 2004. Vasopressin-induced taurine efflux from rat pituicytes: a potential negative feedback for hormone secretion. *The Journal of Physiology* 554:731–742.

Apply Evolutionary Thinking

Large animals (like many of the vertebrates described in this chapter) often have more types of specialized tissues and organ systems than do small animals (such as flatworms). Why would natural selection favor the evolution of additional tissue types and organ systems in larger organisms?

Section through the cerebellum, a part of the brain that integrates signals coming from particular regions of the body (confocal light micrograph). Neurons, the cells that send and receive signals, are red; glial cells, which provide structural and functional support for neurons, are yellow; and nuclei are purple.

37

STUDY OUTLINE

Information Flow and the Neuron

Why It Matters. . . The dog stands alert, muscles tense, motionless except for a wagging tail. His eyes are turned toward his master, a boy, poised to throw a Frisbee for him to catch. Even before the Frisbee is released, the dog has anticipated the direction of its flight from the eyes and stance of the boy.

With a snap of the wrist, the boy throws the Frisbee, and the dog springs into action. Legs churning, eyes following the Frisbee, the dog runs beneath its track, closing the distance as the Frisbee reaches the peak of its climb and begins to descend. All this time, parts of the dog's brain have been processing information received through various sensory inputs. The eyes report his travel over the ground and the speed and arc of the Frisbee. Sensors in the inner ears, muscles, and joints detect the position of the dog's body, and his brain sends out signals that keep his movements on track and in balance. Other parts of the brain register inputs from sensors monitoring body temperature and carbon dioxide levels in the blood, and send signals that adjust heart and breathing rates accordingly.

At just the right instant, a burst of signals from the dog's brain causes trunk and leg muscles to contract in a coordinated pattern, and the dog leaps to intercept the Frisbee in midair with an assured snap of his jaws **(Figure 37.1)**. Now the animal twists, turning his head and eyes toward the ground as his brain calculates the motions required to land on his feet and in balance. The dog makes a perfect landing and trots happily back to his master, ready to repeat the entire performance.

The functions of the dog's nervous system in the chase and capture are astounding in the amount and variety of sensory inputs, the rate and complexity of the brain's analysis and integration of incom-

FIGURE 37.1
With perfect timing, a dog leaps to catch a Frisbee. The coordinated leap involves processing and integration of information by the dog's nervous system.

ing information, and the flurry of signals the brain sends to make compensating adjustments in body activities. Yet they are ordinary in the sense that the same activities take place countless times each day in the nervous system of all but the simplest animals.

All these activities, no matter how complex, depend on the functions of only two major cell types: *neurons* and *glial cells.* In most animals, these cells are organized into complex networks called *nervous systems.* This chapter describes neuron structure and tells how neurons send and receive signals with the aid and support of glial cells. The next chapter considers the neural networks of the brain and its associated structures. Chapter 39 discusses the sensory receptors that detect environmental changes and convert that information into signals for integration by the nervous system. <

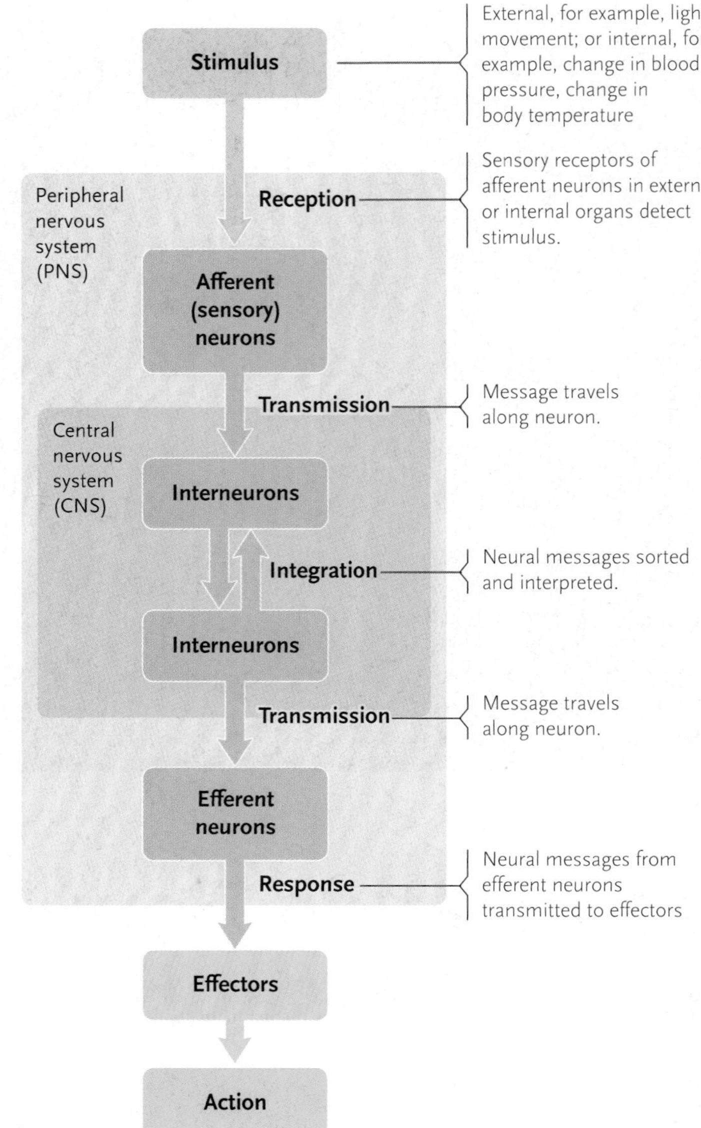
Stimulus
External, for example, light, movement; or internal, for example, change in blood pressure, change in body temperature
Peripheral nervous system (PNS)
Reception
Sensory receptors of afferent neurons in external or internal organs detect stimulus.
Afferent (sensory) neurons
Transmission
Message travels along neuron.
Central nervous system (CNS)
Interneurons
Integration
Neural messages sorted and interpreted.
Interneurons
Transmission
Message travels along neuron.
Efferent neurons
Response
Neural messages from efferent neurons transmitted to effectors
Effectors
Action

FIGURE 37.2
Neural signaling: the information-processing steps in the nervous system.

37.1 Neurons and Their Organization in Nervous Systems

An animal constantly receives stimuli from both internal and external sources. **Neural signaling,** communication by neurons, is the process by which an animal responds appropriately to a stimulus **(Figure 37.2).** In most animals, the four components of neural signaling are *reception, transmission, integration,* and *response.* **Reception,** the detection of a stimulus, is performed by **neurons,** the cellular components of nervous systems, and by specialized sensory receptors such as those in the eye and skin. **Transmission** is the sending of a message along a neuron, and then to another neuron or to a muscle or gland. **Integration** is the sorting and interpretation of neural messages and the determination of the appropriate response(s). **Response** is the "output" or action resulting from the integration of neural messages. For a dog catching a Frisbee, for instance, sensors in the eye receive light stimuli from the environment, and internal sensors receive stimuli from all the animal's organ systems. The neural messages generated are transmitted through the nervous system and integrated to determine the appropriate response, in this case stimulating the muscles so the dog jumps into the air and catches the Frisbee.

Neurons Are Cells Specialized for the Reception and Transmission of Informational Signals

Neural signaling involves three functional classes of neurons (the blue boxes in Figure 37.2). **Afferent** (*afferre* = to bring to) **neurons** (also called **sensory neurons**) transmit stimuli collected by their sensory receptors to **interneurons,** which integrate the information to formulate an appropriate response. In humans and some other primates, 99% of neurons are interneurons. **Efferent** (*efferre* = to carry away) **neurons** carry the signals indicating a response away from the interneuron networks to the **effectors,** the muscles and glands. Efferent neurons that carry signals to skeletal muscle are called **motor neurons.** The information-processing steps in the nervous system, therefore, are: (1) sensory receptors on afferent neurons receive a stimulus; (2) afferent neurons transmit the information to interneurons; (3) interneurons integrate the neural messages; and (4) efferent neurons transmit the neural messages to effectors, which act in a way appropriate to the stimulus.

Neurons vary widely in shape and size. All have an enlarged cell body and two types of extensions or processes, called dendrites and axons **(Figure 37.3).** The **cell body,** which contains the nucleus and the majority of cell organelles, is the site of synthesis of most of the proteins, carbohydrates, and lipids of the neuron. Dendrites and axons conduct electrical signals, which are produced by ions flowing down concentration gradients through channels in the plasma membrane of the neuron. **Dendrites** (*dendron* = tree) receive the signals and transmit them toward the cell body. Dendrites are generally highly branched, forming a treelike outgrowth at one end of the neuron. **Axons** conduct signals away from the cell body to another neuron or an effector. Neurons typically have a single axon, which arises from a junc-

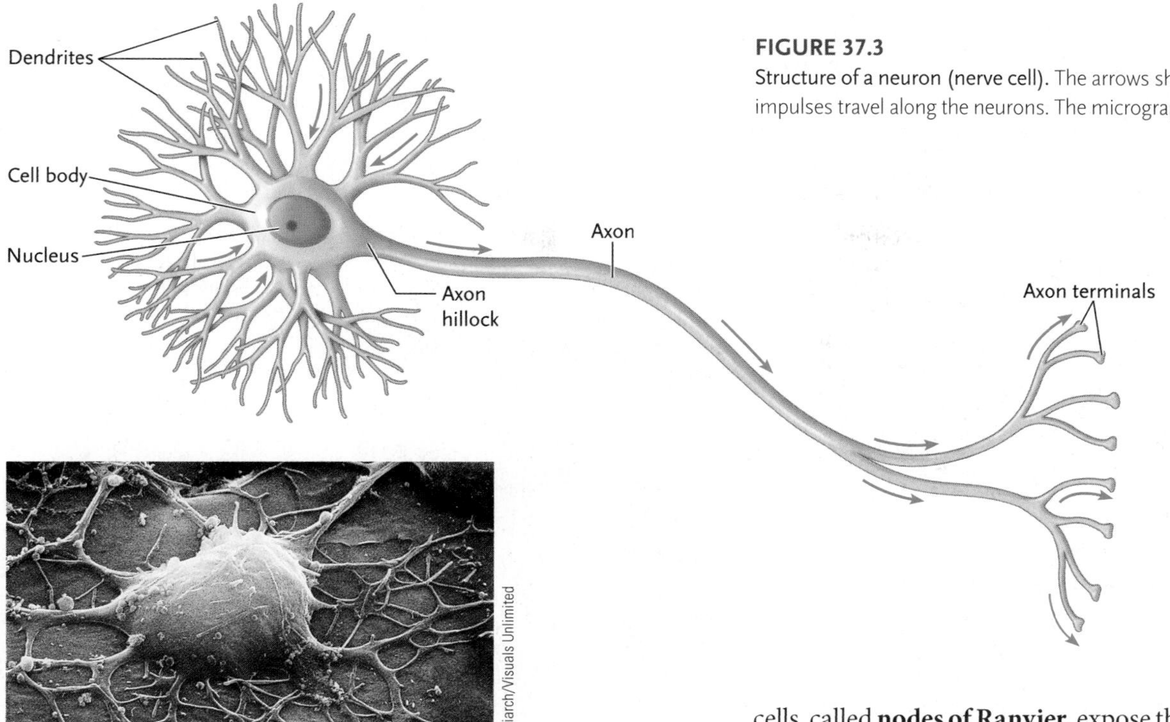
Dendrites
Cell body
Nucleus
Axon hillock
Axon
Axon terminals
10 μm
Triarch/Visuals Unlimited

FIGURE 37.3
Structure of a neuron (nerve cell). The arrows show the direction electrical impulses travel along the neurons. The micrograph is an SEM of a motor neuron.

tion with the cell body called an **axon hillock.** The axon has branches at its tip that end as small, buttonlike swellings called **axon terminals.** The more terminals from different presynaptic cells that contact a neuron, the more incoming information it can integrate.

Connections between axon terminals of one neuron and the dendrites or cell body of a second neuron contribute to the formation of **neural circuits.** A typical neural circuit contains an afferent (sensory) neuron, one or more interneurons, and an efferent neuron. The circuits combine into networks that interconnect the parts of the nervous system. In vertebrates, the afferent neurons and efferent neurons collectively form the **peripheral nervous system (PNS).** The interneurons form the brain and spinal cord, called the **central nervous system (CNS).** As depicted in Figure 37.2, afferent information is ultimately transmitted to the CNS where efferent information is initiated. The nervous systems of most invertebrates are also divided into central and peripheral divisions.

Neurons Are Supported Structurally and Functionally by Glial Cells

Glial cells are nonneuronal cells that provide nutrition and support to neurons. One type, called **astrocytes** because they are star-shaped **(Figure 37.4),** occurs only in the vertebrate CNS, where they closely cover the surfaces of blood vessels. Astrocytes provide physical support to neurons and help maintain the concentrations of ions in the interstitial fluid surrounding them. Two other types of glial cells—**oligodendrocytes** in the CNS and **Schwann cells** in the PNS—wrap around axons in a jelly roll fashion to form myelin sheaths **(Figure 37.5).** Myelin sheaths have a high lipid content because of the many layers of plasma membranes of the myelin-forming cells. Because of their high lipid content, the myelin sheaths act as electrical insulators. The gaps between Schwann cells, called **nodes of Ranvier,** expose the axon membrane directly to extracellular fluids. This structure speeds the rate at which electrical impulses move along the axons covered by glial cells.

Unlike most neurons, glial cells retain the capacity to divide throughout the life of the animal. This capacity allows glial tissues to replace damaged or dead cells, but also makes them the source of almost all brain tumors, produced when regulation of glial cell division is lost.

Neurons Communicate via Synapses

A **synapse** (*synapsis* = juncture) is a site where a neuron makes a communicating connection with another neuron or with an effector such as a muscle fiber or gland. On one side of the synapse is an axon terminal of the **presynaptic cell,** the neuron that transmits the signal. On the other side is a dendrite (in most cases) or a cell body of the **postsynaptic cell,** the neuron or the surface of an effector that receives the signal. Communication across a synapse may occur by the direct flow of an electrical signal or by means of a **neurotransmitter,** a chemical released by an axon terminal at a synapse. The vast majority of vertebrate neurons communicate by means of neurotransmitters.

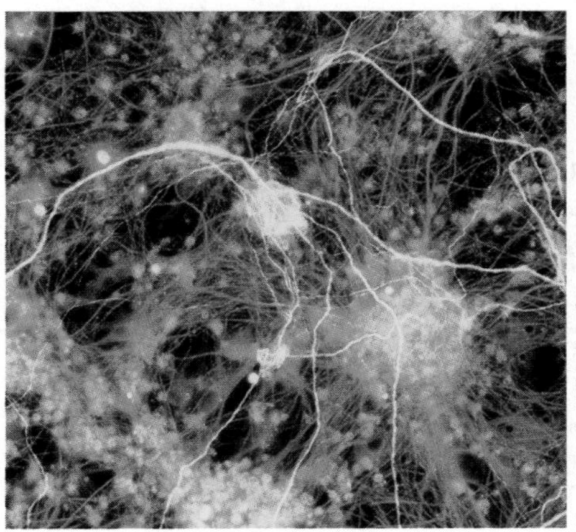
Nancy Kedersha/UCLA/Photo Researchers, Inc.

FIGURE 37.4
Astrocytes (orange), a type of glial cell, and a neuron (yellow) in brain tissue.

FIGURE 37.5
A Schwann cell, showing its myelin sheath, which acts as an electrical insulator. As many as 300 overlapping layers of the Schwann cell plasma membrane wind around an axon like a jelly roll.

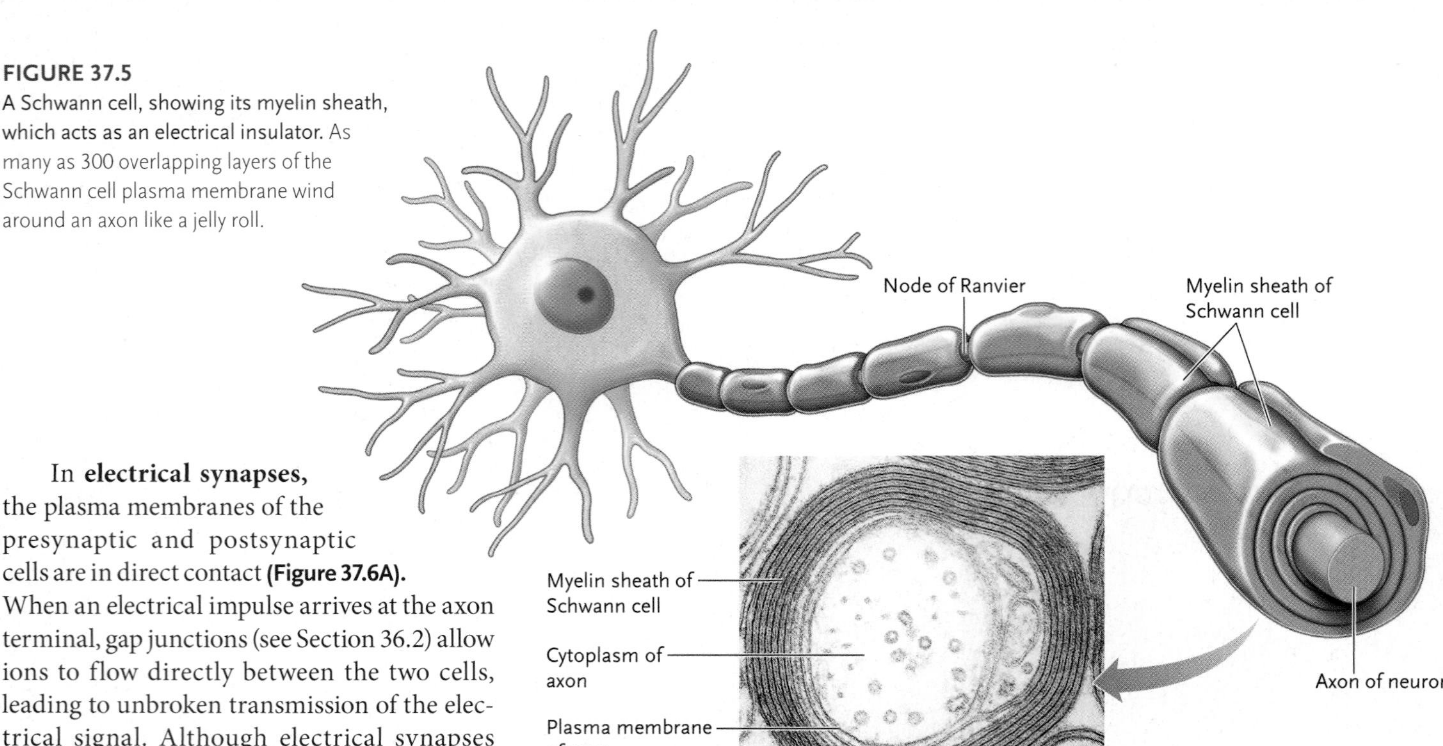

Node of Ranvier
Myelin sheath of Schwann cell
Myelin sheath of Schwann cell
Cytoplasm of axon
Plasma membrane of axon
Axon of neuron
C. Raines/Visuals Unlimited

In **electrical synapses,** the plasma membranes of the presynaptic and postsynaptic cells are in direct contact **(Figure 37.6A).** When an electrical impulse arrives at the axon terminal, gap junctions (see Section 36.2) allow ions to flow directly between the two cells, leading to unbroken transmission of the electrical signal. Although electrical synapses allow the most rapid conduction of signals, this type of connection is essentially "on" or "off" and unregulated. In humans and other animals, electrical synapses typically occur as neural circuit elements where synchrony of firing and speed of response are paramount. Electrical synapses are found, for instance, in cardiac muscle, in the retina of the eye, and in the pulp of a tooth.

In **chemical synapses,** the plasma membranes of the presynaptic and postsynaptic cells are separated by a narrow gap, about 25 nm wide, called the **synaptic cleft (Figure 37.6B).** When an electrical impulse arrives at an axon terminal, it causes the release of a neurotransmitter into the synaptic cleft. The neurotransmitter diffuses across the synaptic cleft and binds to a re-

FIGURE 37.6
The two types of synapses by which neurons communicate with other neurons or effectors.

A. **Electrical synapse**

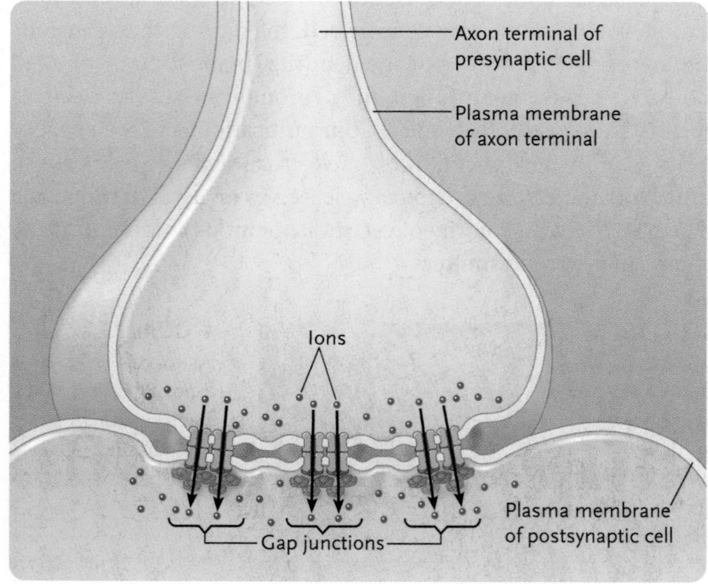

Axon terminal of presynaptic cell
Plasma membrane of axon terminal
Ions
Gap junctions
Plasma membrane of postsynaptic cell

In an electrical synapse, channel proteins in the plasma membranes of the presynaptic and postsynaptic cells form gap junctions. Ions flow through the gap junctions, allowing impulses to pass directly to the postsynaptic cell.

B. **Chemical synapse**

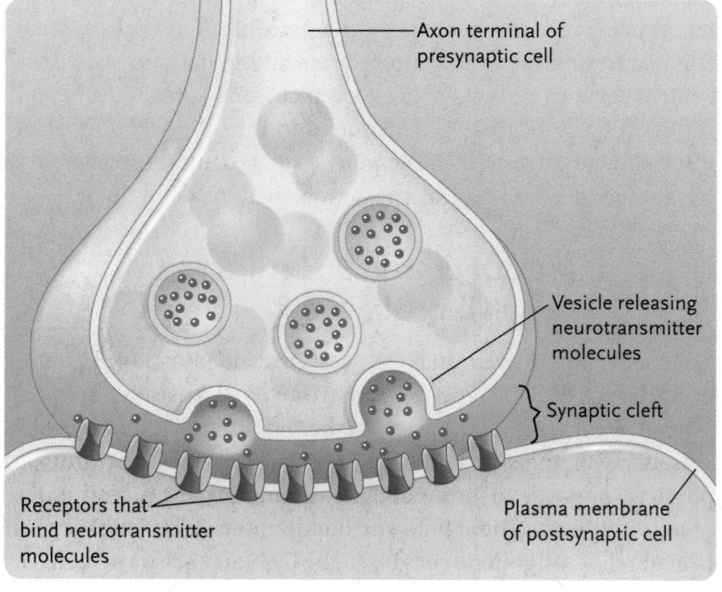

Axon terminal of presynaptic cell
Vesicle releasing neurotransmitter molecules
Synaptic cleft
Receptors that bind neurotransmitter molecules
Plasma membrane of postsynaptic cell

In a chemical synapse, the plasma membranes of the presynaptic and postsynaptic cells are separated by a narrow synaptic cleft. Neurotransmitter molecules diffuse across the cleft and bind to receptors in the plasma membrane of the postsynaptic cell. The binding opens channels to ion flow that may generate an impulse in the postsynaptic cell.

FIGURE 37.7 **Experimental Research**

Demonstration of Chemical Transmission of Nerve Impulses at Synapses

Question: How do neurons transmit signals across synapses?

Experiment: In the first part of the twentieth century, scientists were divided about how neurons transmit signals across synapses. Many scientists thought the signal was electrical, like the nerve signal itself; others thought that chemicals were involved. In 1921, Otto Loewi of Graz University, Austria, performed a simple, but elegant, experiment demonstrating that chemicals can transmit a signal across the synapse.

Results: Loewi isolated the hearts from two frogs, putting each into a separate container filled with a warm solution that kept the hearts alive. The two containers were linked to allow interchange of solutions. In this set up, the hearts could continue to beat for several hours. Loewi stimulated the vagus nerve of heart 1, which resulted in a rapid decrease in the strength and rate of its beating. Loewi observed that, after a short delay, the strength and rate of beating of heart 2 in the connected container also decreased.

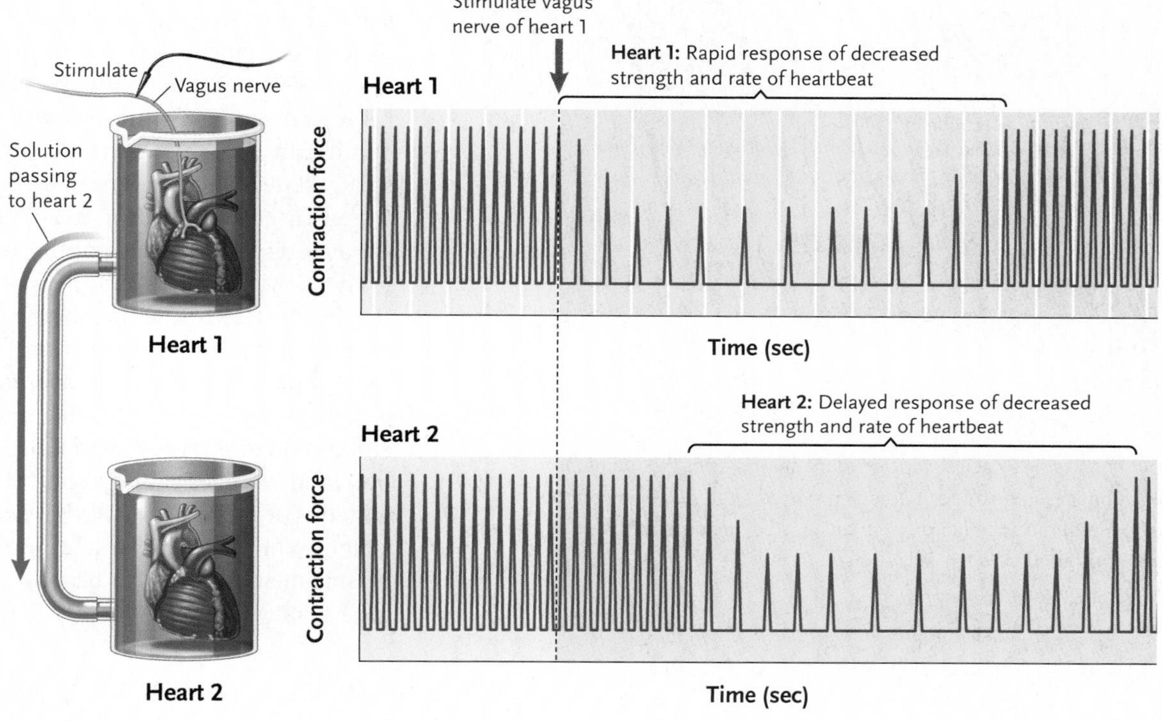
Stimulate
Vagus nerve
Solution passing to heart 2
Heart 1
Heart 2
Stimulate vagus nerve of heart 1
Heart 1
Heart 1: Rapid response of decreased strength and rate of heartbeat
Contraction force
Time (sec)
Heart 2
Heart 2: Delayed response of decreased strength and rate of heartbeat
Contraction force
Time (sec)

Conclusion: Loewi made the only possible conclusion from the data: that when the nerve was stimulated, some chemical substance was produced at the synapses of heart 1 that could diffuse through the solution and cause the same physiological effect in heart 2. The delayed response of heart 2 was the result of the time it took for the chemical to diffuse through the solution. Loewi called the chemical Vagusstoff (vagal substance), to relate it to the vagus nerve stimulation. Subsequently, Vagusstoff was shown to be acetylcholine.

Source: O. Loewi. 1921. Über humorale Übertragbarkeit der Herznervenwirkung. I. *Pflügers Archiv* 189:239–242.

ceptor in the plasma membrane of the postsynaptic cell. If enough neurotransmitter molecules bind to these receptors, the postsynaptic cell generates a new electrical impulse, which travels along its axon to reach a synapse with the next neuron or effector in the circuit. A chemical synapse is more than a simple on-off switch because many factors can influence the generation of a new electrical impulse in the postsynaptic cell, including neurotransmitters that inhibit that cell rather than stimulating it. The balance of stimulatory and inhibitory effects in chemical synapses contributes to the integration of incoming information in a receiving neuron.

Figure 37.7 shows the first experiment to demonstrate the chemical transmission of a nerve impulse across a synapse.

STUDY BREAK 37.1

1. **Distinguish between a dendrite and an axon.**
2. **Distinguish between the functions and locations of afferent neurons, efferent neurons, and interneurons.**
3. **What is the difference between an electrical synapse and a chemical synapse?**

FIGURE 37.8 **Research Method**

Measuring Membrane Potential

Purpose: To determine the membrane potentials of unstimulated and stimulated neurons and muscle cells.

Protocol: Prepare a microelectrode by drawing out a glass capillary tube to a tip with a diameter much smaller than that of a cell and filling it with a salt solution that can conduct an electric current. Under a microscope, use a micromanipulator (mechanical positioning device) to insert the tip of the microelectrode into an axon. Place a reference electrode in the solution outside the cell. Use an oscilloscope or voltmeter to measure the voltage between the microelectrode tip in the axon and the reference electrode outside the cell.

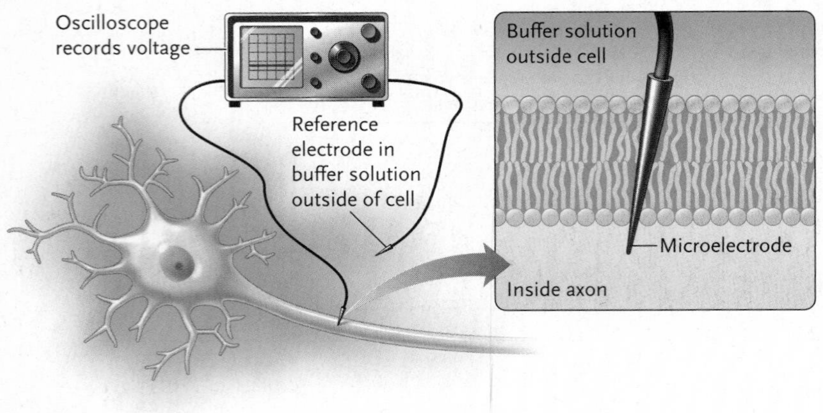

Oscilloscope records voltage
Reference electrode in buffer solution outside of cell
Buffer solution outside cell
Microelectrode
Inside axon

Interpreting the Results: The oscilloscope or voltmeter indicates the membrane potential in volts. Changes in membrane potential caused by stimuli or chemical treatments can be measured and recorded. For an isolated, unstimulated neuron (shown above) the membrane potential is typically about −70 mV.

37.2 Signaling by Neurons

All cells of an animal have a **membrane potential,** a separation of positive and negative charges across the plasma membrane. Outside the cell the net charge is positive, and inside the cell it is negative. This charge separation produces *voltage*—an electrical potential difference—across the plasma membrane.

Selective Permeability of the Plasma Membrane to Ions and Other Charged Molecules Produces the Membrane Potential

Membrane potential is the result of the selective permeability of the plasma membrane to charged atoms (ions) and molecules. Because of the action of the Na^+/K^+ pump in the membrane, Na^+ is removed from the cytoplasm and K^+ is brought into the cytoplasm (see Section 6.4; the ATPase of the pump hydrolyzes ATP to provide the energy for the pump's activity). This results in a concentration gradient of Na^+ strongly favoring its diffusion into the cell and the opposite for K^+.

Two other aspects of cellular membrane permeability are also crucial to the establishment of the resting membrane potential. First, the membrane is more permeable to passive diffusion of K^+ than of Na^+, about 50 times more permeable. Second, the cytoplasm is rich in large anionic molecules—amino acids, proteins, nucleic acids—that cannot cross the cell membrane. The distribution and mobility of these various ionic species produce the steady-state resting potential.

Because of its concentration gradient, K^+ attempts to diffuse from the cell; but, when it does, large anion molecules are left behind. These anions are not effectively neutralized by cations because the cell membrane does not permit significant Na^+ movement. At steady state, the cell has a negative internal potential produced by the unpaired organic anions that is just strong enough to prevent further K^+ exit. Thus, this steady state forms a balanced disequilibrium between the concentration gradient for K^+ and the electrical gradient counteracting further K^+ movement.

In most cells, the membrane potential does not change. However, neurons and muscle cells use the membrane potential in a specialized way. That is, in response to electrical, chemical, mechanical, and certain other types of stimuli, their membrane potential changes rapidly and transiently. Cells with this property are said to be excitable cells. Excitability, produced by a sudden ion flow across the plasma membrane, is the basis for nerve impulse generation.

Resting Potential Is the Unchanging Membrane Potential of an Unstimulated Neuron

The membrane of a neuron (or another excitable cell) that is not conducting an impulse exhibits a steady negative membrane potential, called the **resting potential.** Resting potentials are typically in the −50 to −70 millivolt (mV) range in isolated neurons **(Figure 37.8).** A neuron exhibiting the resting potential is said to be *polarized.*

The distribution of ions inside and outside an axon that produces the resting potential is shown in **Figure 37.9.** As described earlier in this section, the Na^+/K^+ pump is responsible for creating both the imbalance of Na^+ and K^+ inside and outside of the cell and the unpaired charge state between intracellular anions and extracellular positive charges. Voltage-gated ion channels for Na^+ and K^+ open and close when a neuron is stimulated, causing the membrane potential changes.

As noted previously, the tendency for K^+ to diffuse from the cell and leave behind unpaired negative charges is largely responsible for the negative resting potential. The electrical potential necessary to balance this diffusional potential of K^+ can be calculated and is called the **equilibrium potential** for K^+ (E_K).

Similarly, equilibrium potentials can be calculated for other ions distributed unequally across the cell membrane. If K^+ were the only ion moving across the membrane, the E_K would be equal to the resting potential. Typically, the resting potential is a little more positive than E_K, indicating that some positive ion may be leaking into the cell. The obvious candidate is Na^+ because of the strong concentration gradient for this ion between the outside and the inside of the cell. The reason that the positive charge of Na^+ does not swamp the negative charge within the cell is the extremely low permeability of the membrane to Na^+. Indeed, the equilibrium potential for Na^+ (E_{Na}) is around +50 mV; in other words, if Na^+ diffused freely across the membrane, the charge inside of the cell would have to be raised to +50 mV to maintain the disequilibrium of Na^+ concentration between the inside of the cell and the outside.

Equilibrium potentials provide a convenient way to predict ion movement across the cell membrane. If an equilibrium potential is close to the resting potential, there will be little ion movement even if the membrane is freely permeable to the ion. For example, if a K^+ channel opens in a membrane with a resting membrane potential in the −70 mV range, some small amount of K^+ will leave the cell to bring the membrane potential closer to the E_K, which is generally somewhat more negative than −70 mV. By contrast, if a Na^+ channel opens, both the concentration gradient for Na^+ and the attraction of the intracellular anions will act together to promote significant Na^+ entry into the cell. The effect will be to **depolarize** the cell—make it less negative.

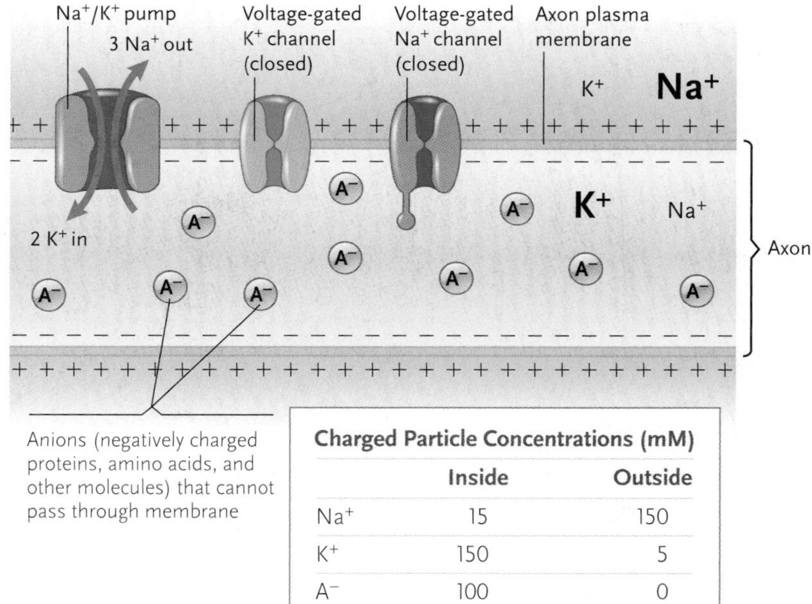
Na⁺/K⁺ pump
3 Na⁺ out
2 K⁺ in
Voltage-gated K⁺ channel (closed)
Voltage-gated Na⁺ channel (closed)
Axon plasma membrane
K⁺
Na⁺
A⁻
Axon
Anions (negatively charged proteins, amino acids, and other molecules) that cannot pass through membrane

Charged Particle Concentrations (mM)		
	Inside	Outside
Na^+	15	150
K^+	150	5
A^-	100	0

FIGURE 37.9
The distribution of ions inside and outside an axon that produces the resting potential, −70 mV. The distribution of ions that do not directly affect the resting potential, such as Cl^-, is not shown. The voltage-gated ion channels open and close when the membrane potential changes.

The Membrane Potential Changes from Negative to Positive during an Action Potential

When a neuron transmits an electrical impulse, an abrupt and transient change in membrane potential occurs; this is called the **action potential.** An action potential begins as a stimulus that causes positive charges from outside the neuron to flow inward, making the cytoplasmic side of the membrane less negative **(Figure 37.10).** As the membrane potential becomes less negative, the membrane (which was polarized at rest) becomes depolarized. Depolarization proceeds relatively slowly until it reaches a level known as the **threshold potential,** typically 10–20 mV more positive than the resting potential. Once the threshold is reached, the action potential fires—and the membrane potential suddenly increases. In less than 1 msec (millisecond, one-thousandth of a second), it rises so high that the inside of the plasma membrane becomes positive due to an influx of positive ions across the cell membrane, momentarily reaching a value of +30 mV or more. The potential then falls again, in many cases dropping to about −80 mV before rising again to the resting potential. When the potential is below the resting value, the membrane is said to be **hyperpolarized.** The entire change, from initiation of the action potential to the return to the resting potential, may take as little as 1 msec in the fastest neurons. Action potentials take the same basic form in neurons of all types, with differences in the values of the resting potential and the peak of the action potential, and in the time required to return to the resting potential.

An action potential is produced only if a stimulus that causes positive charges from outside the neurons to flow inward is strong enough to cause depolarization that reaches the threshold. This is referred to as the **all-or-nothing principle;** once triggered, the changes in membrane potential take place independently of the strength of the stimulus.

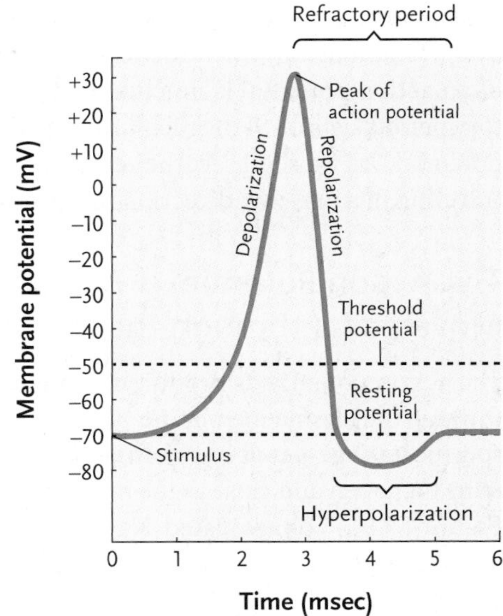
Refractory period
Peak of action potential
Depolarization
Repolarization
Threshold potential
Resting potential
Stimulus
Hyperpolarization
Membrane potential (mV)
+30
+20
+10
0
−10
−20
−30
−40
−50
−60
−70
−80
0 1 2 3 4 5 6
Time (msec)

FIGURE 37.10
Changes in membrane potential during an action potential.

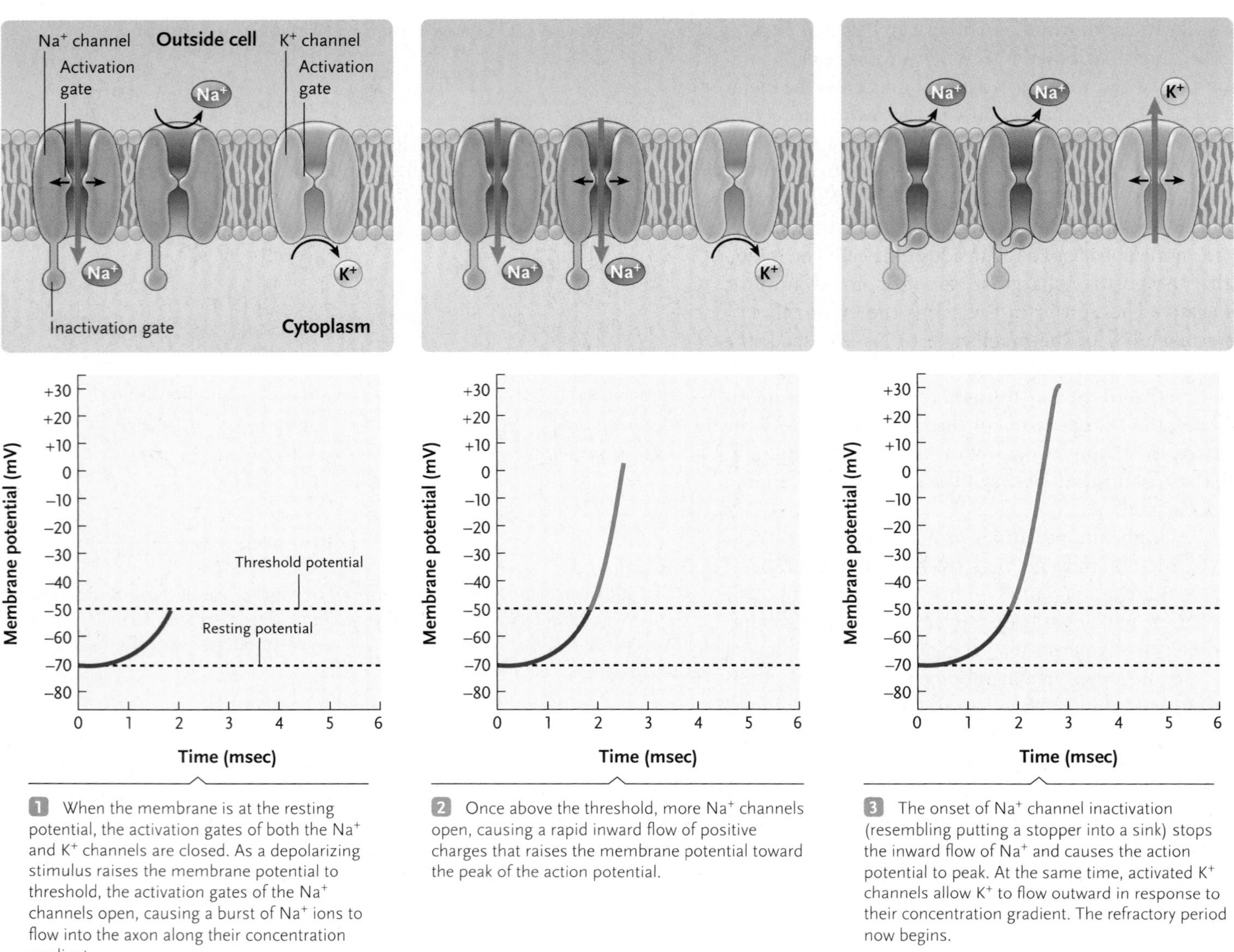
Na⁺ channel
Outside cell
K⁺ channel
Activation gate
Activation gate
Na⁺
K⁺
Inactivation gate
Cytoplasm
Membrane potential (mV)
+30
+20
+10
0
−10
−20
−30
−40
−50
−60
−70
−80
Threshold potential
Resting potential
Time (msec)
0
1
2
3
4
5
6

1 When the membrane is at the resting potential, the activation gates of both the Na^+ and K^+ channels are closed. As a depolarizing stimulus raises the membrane potential to threshold, the activation gates of the Na^+ channels open, causing a burst of Na^+ ions to flow into the axon along their concentration gradient.

2 Once above the threshold, more Na^+ channels open, causing a rapid inward flow of positive charges that raises the membrane potential toward the peak of the action potential.

3 The onset of Na^+ channel inactivation (resembling putting a stopper into a sink) stops the inward flow of Na^+ and causes the action potential to peak. At the same time, activated K^+ channels allow K^+ to flow outward in response to their concentration gradient. The refractory period now begins.

FIGURE 37.11
Changes in voltage-gated Na^+ and K^+ channels that produce the action potential.

Beginning at the peak of an action potential, the membrane enters a **refractory period** that consists of two phases. During the initial phase, the cell cannot be restimulated. This is followed by a period during which the threshold required for generation of an action potential is much higher than normal. The refractory period lasts until the membrane has stabilized at the resting potential. As we shall see, the refractory period keeps impulses traveling in a one-way direction in neurons.

The Action Potential Is Produced by Ion Movements through the Plasma Membrane

The action potential is produced by movements of Na^+ and K^+ through the plasma membrane. The movements are controlled by specific **voltage-gated ion channels,** membrane-embedded proteins that open and close as the membrane potential changes (see Figure 37.9). Voltage-gated Na^+ channels have two gates, an *activation gate* and an *inactivation gate,* whereas voltage-gated K^+ channels have one gate, an *activation gate.*

Figure 37.11 shows how the two voltage-gated ion channels operate to generate an action potential. At the end of an action potential, the membrane potential has returned to its resting state, but the ion distribution has changed slightly. That is, some Na^+ ions have entered the cell, and some K^+ ions have left the cell—but not many, relative to the total number of ions, and the distribution is not altered enough to prevent other action potentials from occurring. In the long term, the Na^+/K^+ active transport pumps restore the Na^+ and K^+ ions to their original locations.

Some of what is known about how ion flow through channels can change membrane potential has come from experiments using the *patch-clamp* technique. In the patch part of the technique, a micropipette with a tip 1 to 3 μm in diameter is touched to the plasma membrane of a neuron (or other cell type). The contact seals the membrane to the micropipette and, when the micropipette is pulled away, a patch of membrane with one or a few ion channels comes with it. The clamp part of the technique refers to a voltage clamp, in which an electronic device holds the

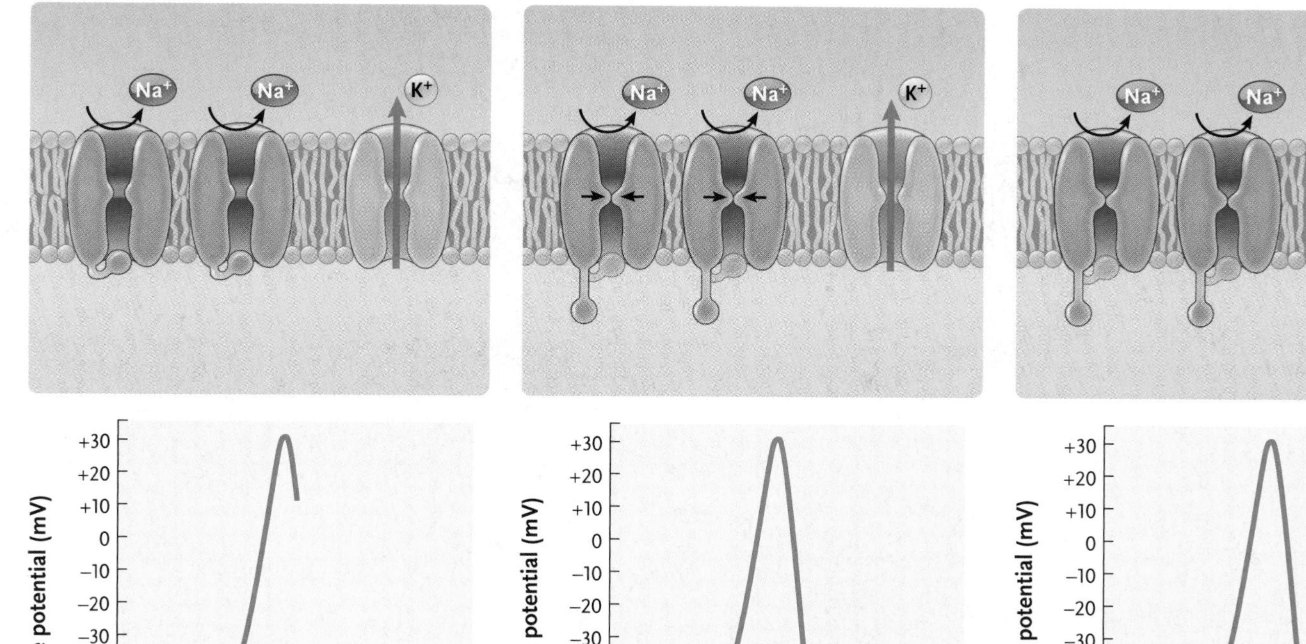

Na^+
Na^+
K^+
Membrane potential (mV)
+30
+20
+10
0
−10
−20
−30
−40
−50
−60
−70
−80
0 1 2 3 4 5 6
Time (msec)

4 The outward flow of K^+ along its concentration gradient compensates for the inward movement of Na^+ ions and causes the membrane potential to begin to fall.

5 As the membrane potential reaches the resting value, the activation gate of the Na^+ channel closes and the inactivation gate opens. In most neurons, the K^+ activation gate remains open and K^+ continues to flow outward for a brief time after the membrane reaches the resting potential. This excess outward flow causes hyperpolarization, in which the membrane potential dips briefly below the resting potential (the dip is seen in step 6).

6 Closure of the K^+ activation gate stabilizes the membrane potential at the resting value. The refractory period has now ended and the membrane is ready for another action potential.

membrane potential of the patch at a steady value chosen by the investigator. The investigator can add a stimulus that is expected to open or close ion channels. The amount of current the clamping device needs to keep the voltage constant is directly related to the number and charge of the ions moving through the channels and, hence, measures channel activity.

Neural Impulses Move by Propagation of Action Potentials

Once an action potential is initiated in a neuron, it passes along the surface of the axon as an automatic wave of depolarization traveling away from the stimulation point **(Figure 37.12).** The action potential does not need further trigger events to be propagated along the axon to the terminals. In a segment of an axon that is generating an action potential, the outside of the membrane becomes temporarily negative and the inside positive. Because opposites attract, as the region outside becomes negative, local current flow occurs between the area undergoing an action potential and the adjacent downstream inactive area both inside and outside the membrane (arrows, Figure 37.12). This current flow makes nearby regions of the axon membrane less positive on the outside and more positive on the inside; in other words, they depolarize the membrane.

The depolarization is large enough to push the membrane potential past the threshold, opening the voltage-gated Na^+ and K^+ channels and starting an action potential in the downstream adjacent region. In this way, each segment of the axon stimulates the next segment to fire, and the action potential moves rapidly along the axon as a nerve impulse.

The refractory period keeps an action potential from reversing direction at any point along an axon; only the region in front of the action potential can fire. The refractory period results from the properties of the voltage-gated ion channels. Once they have been opened to their activated state, the upstream voltage-gated ion channels need time to reset to their original positions before they can open again. Therefore, only downstream voltage-gated ion channels are able to open, ensuring the one-way movement

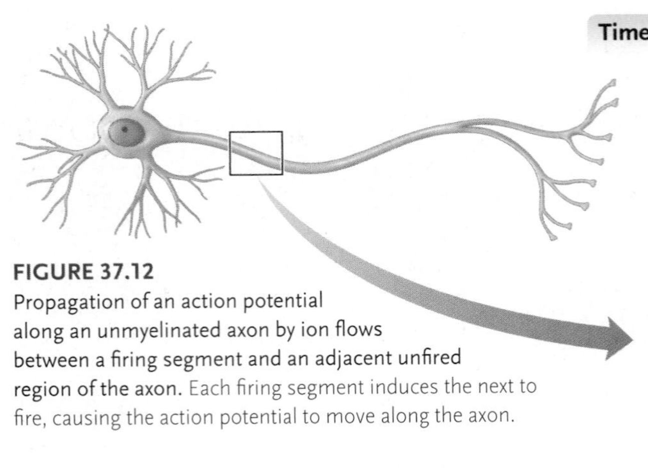
Direction of propagation of action potential
Time = 0
Active area at peak of action potential
Adjacent inactive area into which depolarization is spreading; will soon reach threshold
Remainder of axon still at resting potential
Na+
Local current flow that depolarizes adjacent inactive area from resting potential to threshold potential
Membrane potential (mV)
+30
0
–50
–70
Time = 1
Previous active area returning to resting potential; no longer active because of refractory period
Adjacent area that was brought to threshold by local current flow; now active at peak of action potential
New adjacent inactive area into which depolarization is spreading; will soon reach threshold
Remainder of axon still at resting potential
K+
Time = 2
At resting potential

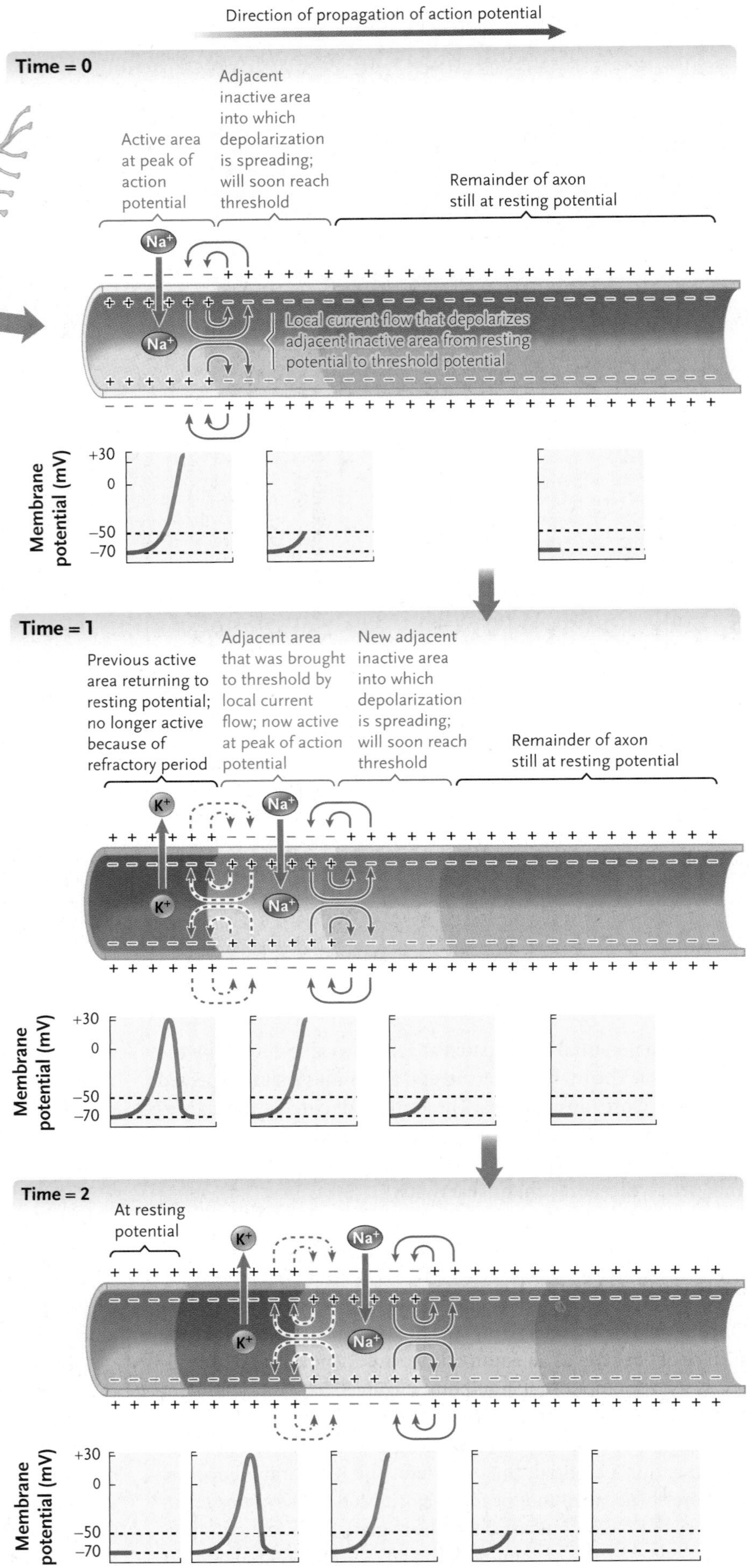

FIGURE 37.12
Propagation of an action potential along an unmyelinated axon by ion flows between a firing segment and an adjacent unfired region of the axon. Each firing segment induces the next to fire, causing the action potential to move along the axon.

of the action potential along the axon toward the axon tips. By the time the refractory period ends in a membrane segment that has just fired an action potential, the action potential has moved too far away to cause a second action potential to develop in the same segment.

The magnitude of an action potential stays the same as it travels along an axon, even where the axon branches at its tips. Thus the propagation of an action potential resembles a burning fuse, which burns with the same intensity along its length, and along any branches, once it is lit at one end. Unlike a fuse, however, an axon can fire another action potential of the same intensity within a few milliseconds after an action potential passes through.

Due to the all-or-nothing principle of action potential generation, the intensity of a stimulus is reflected in the *frequency* of action potentials—the greater the stimulus, the more action potentials per second, up to a limit depending on the axon type—rather than by the change in membrane potential. For most neuron types, the limit lies between 10 and 100 action potentials per second.

Both natural and synthetic substances target specific parts of the mechanism generating action potentials. Local anesthetics, such as procaine and lidocaine, bind to voltage-gated Na^+ channels and block their ability to transport ions; thus, sensory nerves in the anesthetized region cannot transmit pain signals. The potent poison of the pufferfish, tetrodotoxin, also blocks voltage-gated Na^+ channels in neurons, potentially causing muscle paralysis and death. The pufferfish is highly prized as a delicacy in Japan, and it is eaten only after careful preparation to remove organs carrying the

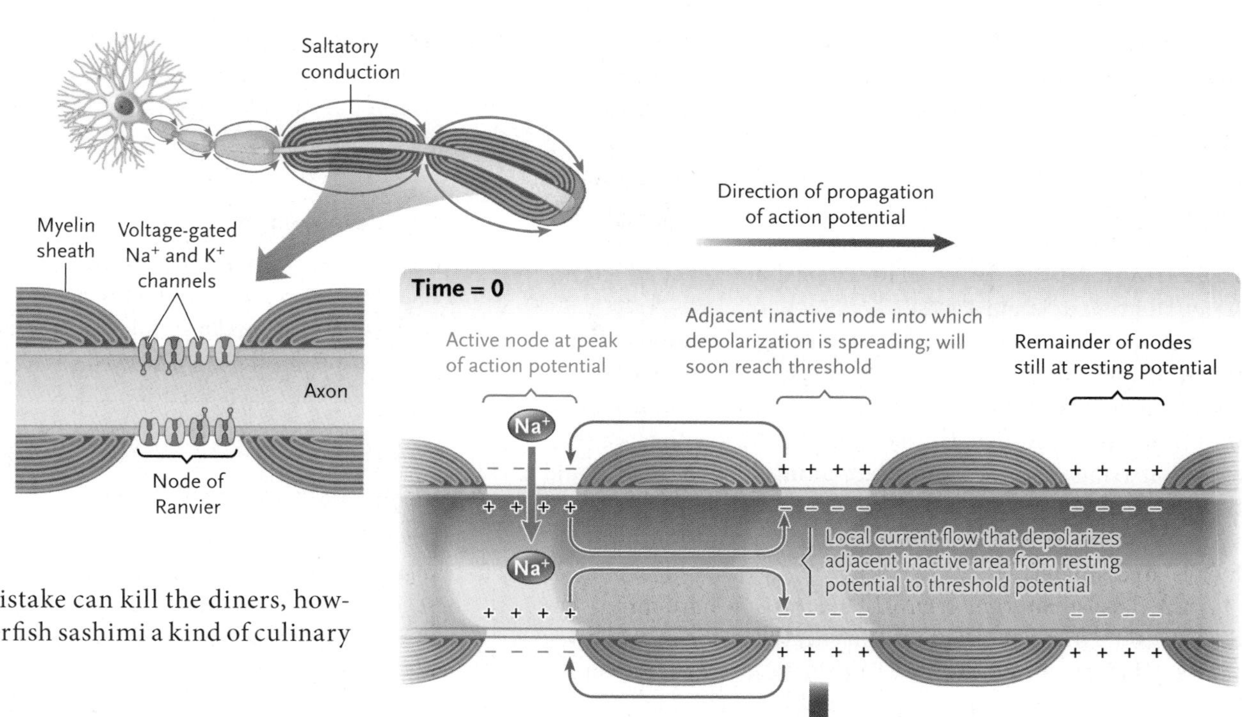
Saltatory
conduction
Myelin
sheath
Voltage-gated
Na+ and K+
channels
Axon
Node of
Ranvier
Direction of propagation
of action potential

tetrodotoxin. A mistake can kill the diners, however, making pufferfish sashimi a kind of culinary Russian roulette.

Saltatory Conduction Increases Propagation Rate in Small-Diameter Axons

In the propagation pattern shown in Figure 37.12, an action potential spreads along every segment of the membrane along the length of the axon. For this type of action potential propagation, the rate of conduction increases with the diameter of the axon. Axons with a very large diameter have evolved in invertebrates such as lobsters, earthworms, and squids as well as a few marine fishes. Giant axons typically carry signals that produce an escape or withdrawal response, such as the sudden flexing of the tail (abdomen) in lobsters that propels the animal backward. The largest known axons, 1.7 mm in diameter, occur in fanworms *(Myxicola)*. The signals they carry contract a muscle that retracts the fanworm's body into a protective tube when the animal is threatened.

Although large-diameter axons can conduct impulses as rapidly as 25 m/sec (over twice the speed of the world record 100-meter dash), they take up a great deal of space. In complex vertebrates, natural selection has led to a mechanism that allows small-diameter axons to conduct impulses rapidly. The mechanism, called **saltatory conduction** (*saltare* = to hop, leap), allows action potentials to "hop" rapidly along axons instead of burning smoothly like a fuse.

Saltatory conduction depends on the insulating myelin sheath that forms around some axons and in particular on the nodes of Ranvier, exposing the axon membrane to extracellular fluids. Voltage-gated Na^+ and K^+ channels crowded into the nodes allow action potentials to develop at these positions **(Figure 37.13)**. The inward move-

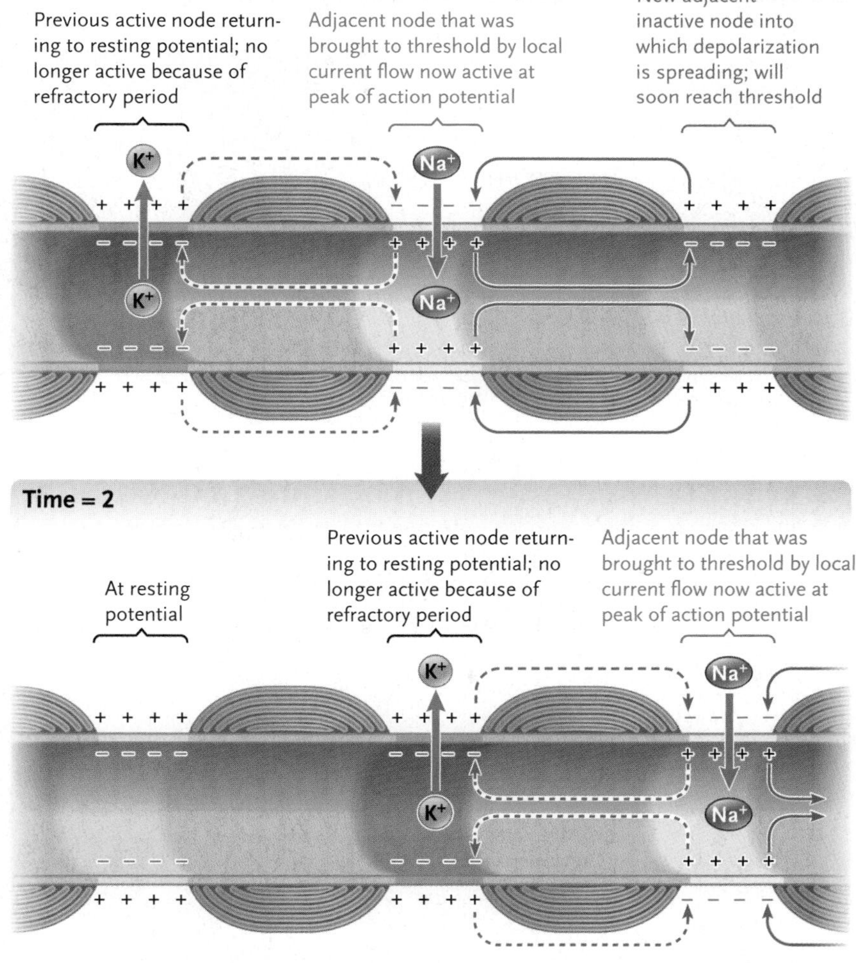
Time = 0
Active node at peak
of action potential
Adjacent inactive node into which
depolarization is spreading; will
soon reach threshold
Remainder of nodes
still at resting potential
Local current flow that depolarizes
adjacent inactive area from resting
potential to threshold potential
Time = 1
Previous active node returning to resting potential; no
longer active because of
refractory period
Adjacent node that was
brought to threshold by local
current flow now active at
peak of action potential
New adjacent
inactive node into
which depolarization
is spreading; will
soon reach threshold
Time = 2
At resting
potential
Previous active node returning to resting potential; no
longer active because of
refractory period
Adjacent node that was
brought to threshold by local
current flow now active at
peak of action potential

FIGURE 37.13
Saltatory conduction of the action potential by a myelinated axon. The action potential jumps from node to node, greatly increasing the speed at which it travels along the axon.

ment of Na^+ ions produces depolarization, but the excess positive ions are unable to leave the axon through the membrane regions covered by the myelin sheath. Instead, they diffuse rapidly to the next node where they cause depolarization, inducing an action potential at that node. As this mechanism repeats, the action potential jumps rapidly along the axon from node to node. Saltatory conduction proceeds at rates up to 130 m/sec while an unmyelinated axon of the same diameter conducts action potentials at about 1 m/sec.

Saltatory conduction allows thousands to millions of fast-transmitting axons to be packed into a relatively small diameter. For example, in humans the optic nerve leading from the eye to the brain is only 3 mm in diameter but is packed with more than a million axons. If those axons were unmyelinated, each would have to be about 100 times thicker to conduct impulses at the same velocity, producing an optic nerve about 300 mm (12 inches) in diameter.

The disease *multiple sclerosis* (*sklerosis* = hardening) underscores the importance of myelin sheaths to the operation of the vertebrate nervous system. In this disease, myelin is progressively lost from axons and replaced by hardened scar tissue. The changes block or slow the transmission of action potentials, producing numbness, muscular weakness, faulty coordination of movements, and paralysis that worsens as the disease progresses.

STUDY BREAK 37.2

1. **What mechanism ensures that an electrical impulse in a neuron is conducted in only one direction down the axon?**
2. **How does having a myelin sheath affect the conduction of impulses in neurons?**

THINK OUTSIDE THE BOOK

Individually or collaboratively, explore the research literature to find two examples of neurophysiological research that used tetrodotoxin (TTX) in the experiments. For the two projects, outline the hypothesis being tested or question being asked, how TTX was used, and the key results and conclusions.

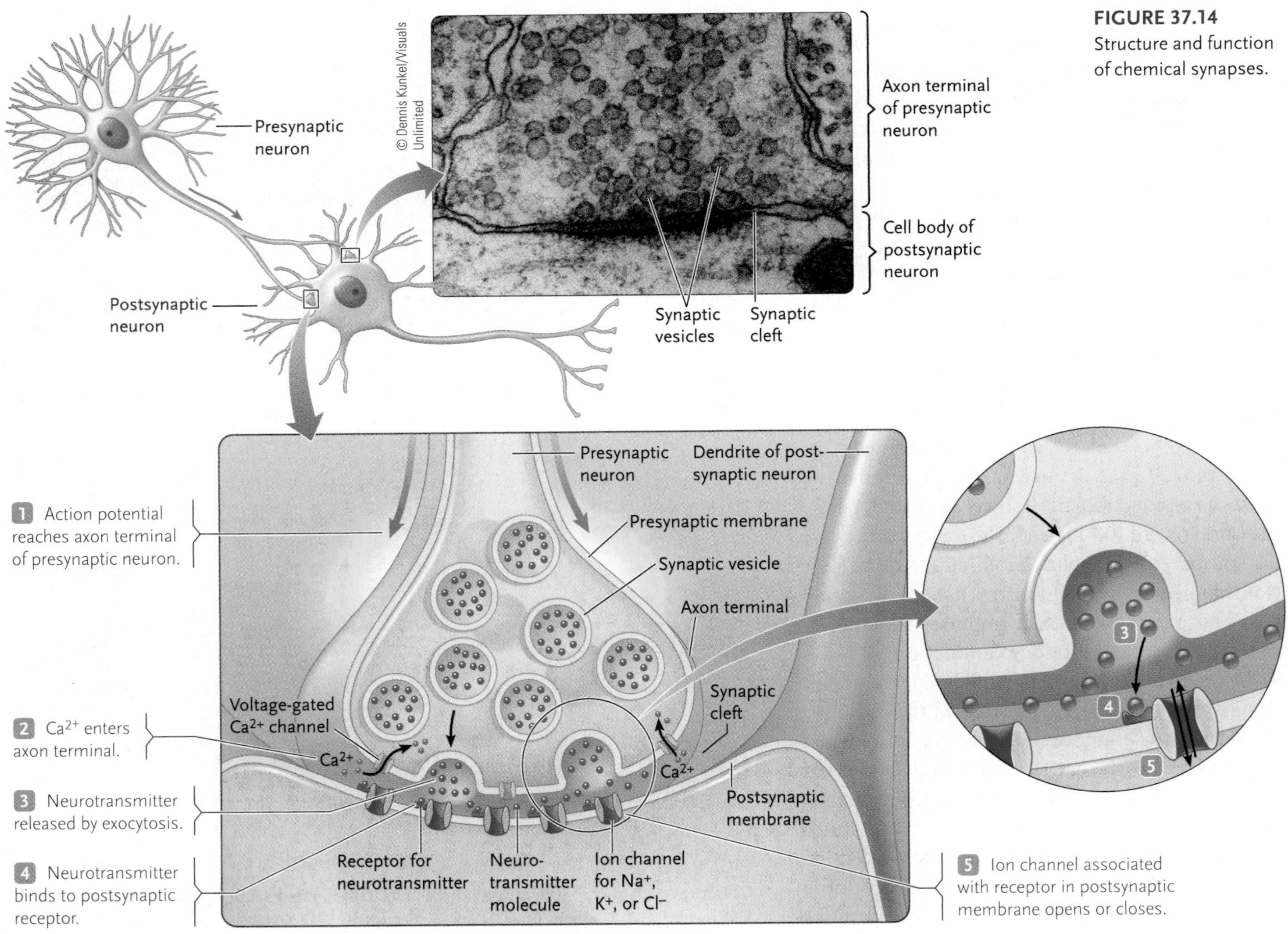

Presynaptic neuron
Postsynaptic neuron
© Dennis Kunkel/Visuals Unlimited
Axon terminal of presynaptic neuron
Cell body of postsynaptic neuron
Synaptic vesicles
Synaptic cleft
Presynaptic neuron
Dendrite of postsynaptic neuron
Presynaptic membrane
Synaptic vesicle
Axon terminal
Synaptic cleft
Voltage-gated Ca^{2+} channel
Ca^{2+}
Ca^{2+}
Postsynaptic membrane
Receptor for neurotransmitter
Neurotransmitter molecule
Ion channel for Na^+, K^+, or Cl^-
1 Action potential reaches axon terminal of presynaptic neuron.
2 Ca^{2+} enters axon terminal.
3 Neurotransmitter released by exocytosis.
4 Neurotransmitter binds to postsynaptic receptor.
5 Ion channel associated with receptor in postsynaptic membrane opens or closes.
3
4
5

FIGURE 37.14 Structure and function of chemical synapses.

37.3 Transmission across Chemical Synapses

Action potentials are transmitted directly across electrical synapses, but they cannot jump across the cleft in a chemical synapse. Instead, the arrival of an action potential causes neurotransmitter molecules to be released by the plasma membrane of the axon terminal, called the presynaptic membrane **(Figure 37.14)**. The neurotransmitter diffuses across the cleft and alters ion conduction by activating receptors in the postsynaptic membrane, the plasma membrane of the postsynaptic cell. Neurotransmitter communication from presynaptic to postsynaptic cells is a specialized case of cell-to-cell communication and signal transduction, which you learned about in Chapter 7. Neurobiologists study this phenomenon to understand the function of neurons, but some of the details are similar in many other types of cells.

Two types of receptors bind neurotransmitters, with different consequences:

- For **ionotropic neurotransmitter receptors**, a neurotransmitter binds directly to the receptor in the postsynaptic membrane, causing an associated ion channel gate to open or close, thereby altering the flow of a specific ion or ions in the postsynaptic cell. Figure 37.14 illustrates this type of receptor. (This is an example of a *ligand-gated ion channel,* a channel that opens or closes when a specific chemical, here a neurotransmitter, binds to the channel.) The time between arrival of an action potential at an axon terminal and alteration of the membrane potential in the postsynaptic cell may be as little as 0.2 msec.
- For **metabotropic neurotransmitter receptors**, a neurotransmitter works indirectly and, hence, more slowly (on the order of hundreds of milliseconds). For this type of receptor, which is typically of the G-protein–coupled receptor type, the neurotransmitters act as *first messengers.* That is, the neurotransmitters bind to the receptor, which activates it and triggers generation of a second messenger such as cyclic AMP or other processes (see Section 7.4). The cascade of *second-messenger* reactions opens or closes ion-conducting channels in the postsynaptic membrane. Neurotransmitters that bind to metabotropic neurotransmitter receptors typically have effects that may last for minutes or hours.

Some neurotransmitters bind only to one or other of the receptor types, whereas other neurotransmitters bind to both receptor types, bringing about different responses depending on which receptor is involved.

The time required for the release, diffusion, and binding of neurotransmitters across chemical synapses delays transmission as compared with the almost instantaneous transmission of impulses across electrical synapses. However, communication through chemical synapses allows neurons to receive inputs from hundreds to thousands of axon terminals at the same time. Some neurotransmitters have stimulatory effects, and others have inhibitory effects. All of the information received at a postsynaptic membrane is integrated to produce a response. Thus, by analogy, communication by electrical synapses resembles the effect of simply touching one wire to another; communication by neurotransmitters through both direct and indirect receptor responses resembles the integration of multiple inputs by a computer chip.

Neurotransmitters Are Released by Exocytosis

Neurotransmitters are stored in secretory vesicles called synaptic vesicles in the cytoplasm of an axon terminal. The arrival of an action potential at the terminal releases the neurotransmitters by *exocytosis:* the vesicles fuse with the presynaptic membrane and release the neurotransmitter molecules into the synaptic cleft.

The release of synaptic vesicles depends on voltage-gated Ca^{2+} channels in the plasma membrane of an axon terminal (see Figure 37.14). Ca^{2+} ions are constantly pumped out of all animal cells by an active transport protein in the plasma membrane, keeping their concentration higher outside than inside. (That is, the equilibrium potential for Ca^{2+}, E_{Ca}, is strongly positive.) As an action potential arrives, the change in membrane potential opens the Ca^{2+} channel gates in the axon terminal, allowing Ca^{2+} to flow back into the cytoplasm. The rise in Ca^{2+} concentration triggers a protein in the membrane of the synaptic vesicle that allows the vesicle to fuse with the plasma membrane, releasing neurotransmitter molecules into the synaptic cleft.

Each action potential arriving at a synapse typically causes approximately the same number of synaptic vesicles to release their neurotransmitter molecules. For example, arrival of an action potential at one type of synapse causes about 300 synaptic vesicles to release a neurotransmitter called acetylcholine. Each vesicle contains about 10,000 molecules of the neurotransmitter, giving a total of some 3 million acetylcholine molecules released into the synaptic cleft by each arriving action potential.

When a stimulus is no longer present, action potentials are no longer generated and a response is no longer needed. In this case, a series of events prevents continued transmission of the signal. When action potentials stop arriving at the axon terminal, the voltage-gated Ca^{2+} channels in the axon terminal close and the Ca^{2+} in the axon cytoplasm is quickly pumped to the outside. The drop in cytoplasmic Ca^{2+} stops vesicles from fusing with the presynaptic membrane, and no further neurotransmitter molecules are released. Any free neurotransmitter molecules remaining in the cleft quickly diffuse away, are broken down by enzymes in the cleft, or are pumped back into the axon terminals or into glial cells by active transport. Transmission of impulses across the synaptic cleft ceases within milliseconds after action potentials stop arriving at the axon terminal.

Most Neurotransmitters Alter Ion Flow through Na^+ or K^+ Channels

Neurotransmitters work by opening or closing ion channels; most of these channels conduct Na^+ or K^+ across the postsynaptic membrane, although some regulate chloride ions (Cl^-). The altered ion flow in the postsynaptic cell that results from the opening or closing of the gates may stimulate or inhibit the generation of action potentials by that cell. For example, if Na^+ channels are opened, the inward Na^+ flow brings the membrane po-

TABLE 37.1 Examples of Neurotransmitters and Drugs That Affect Neurotransmission

Neurotransmitter	Site(s) of Action	Type of Action	Drugs and Other Molecules That Affect Neurotransmission
Acetylcholine	Between some neurons of CNS and at neuromuscular junctions in PNS; acetylcholine-releasing neurons in the brain degenerate in people with Alzheimer disease, in which memory, speech, and perceptual abilities decline	Mostly excitatory; inhibitory at some sites	• Curare—blocks release; blocks muscle contraction and produces paralysis, potentially leading to death • Atropine—blocks receptors; used to relax iris muscles to dilate pupils for eye exam • Nicotine—activates receptors, causing increased attention and decreased stress and irritability
Monoamines (biogenic amines)			
Norepinephrine	CNS interneurons involved in diverse brain and body functions, such as memory, mood, sensory perception, muscle movements, maintenance of blood pressure and sleep; also released into body circulation as a hormone	Excitatory or inhibitory	• Amphetamines—stimulate release of norepinephrine and dopamine and block their reuptake; cause increased attention and focus and decreased fatigue • Methylphenidate (Ritalin)—increases release; used to treat attention deficit hyperactivity disorder (ADHD) • Certain antidepressants—prevent reuptake; used to treat depression and obsessive-compulsive disorder
Dopamine	CNS interneurons involved in many pathways similar to norepinephrine, but especially involved in reinforcements and addiction; not released into circulation	Mostly excitatory	• Amphetamines—stimulate release of norepinephrine and dopamine and block their reuptake; cause increased attention and focus and decreased fatigue • Cocaine—stimulates release of norepinephrine and dopamine and blocks dopamine reuptake; produces intense euphoria followed by depression
Serotonin	CNS interneurons in a number of pathways, including those regulating appetite, reproductive behavior, muscular movement, sleep, and emotional states such as anxiety	Inhibitory or modulatory	• Fluoxetine (Prozac), sertraline (Zoloft), paroxetine (Paxil)—block serotonin reuptake; used to treat depression and obsessive-compulsive disorder
Amino Acids			
Glutamate	Many CNS pathways, including those involved in vital brain functions such as memory and learning	Excitatory	• Phencyclidine (PCP or angel dust)—blocks receptor; causes feelings of power and confusion; can cause violent behavior
Gamma-aminobutryic acid (GABA)	Many CNS pathways; often acts in same circuits as glutamate	Inhibitory (main inhibitory neurotransmitter in mammalian CNS)	• Alcohol (ethanol)—stimulates GABA neurotransmission, increases dopamine neurotransmission, and inhibits glutamate neurotransmission; causes relaxation, a sense of euphoria, and sleepiness • Some antianxiety/sedative drugs such as diazepam (Valium), alprazolam (Xanax), and flunitrazepam (Rohypnol, the "date rape" drug)—bind to GABA receptors and increase GABA neurotransmission; result in sedation, sleepiness, and possibly amnesia • Tetanus toxin—released by bacterium *Clostridium tetani;* blocks GABA release in synapses that control muscle contraction, causing muscles to contract forcibly (once respiratory muscles are affected, victim dies quickly)

tential of the postsynaptic cell toward the threshold (the membrane becomes depolarized). If K^+ channels are opened, the outward flow of K^+ has the opposite effect (the membrane becomes hyperpolarized). The combined effects of the various stimulatory and inhibitory neurotransmitters at all the chemical synapses of a postsynaptic neuron or muscle cell determine whether the postsynaptic cell triggers an action potential. (*Insights from the Molecular Revolution* describes experiments that worked out the structure and function of an ion channel gated directly by a neurotransmitter.)

Neurotransmitter	Site(s) of Action	Type of Action	Drugs and Other Molecules That Affect Neurotransmission
Purines			
ATP	Released simultaneously with many neurotransmitters; may have independent actions, an area under active investigation	Modulates action of other neurotransmitters	
Neuropeptides			
Endorphins ("endogenous morphines")	Most act on CNS and PNS as well as on effectors such as muscle, reducing pain and, in some cases, also inducing euphoria; released during periods of pleasurable experience (such as eating or sexual intercourse), or physical stress (such as childbirth or extended physical exercise)	Inhibitory—modulate pain response	
Enkephalins (subclass of endorphins)	Active only in CNS	Inhibitory—modulate pain response	
Substance P	Released by special, unmyelinated sensory neurons in spinal cord, which produce rapid conduction for resulting signals; increases neural signals associated with intense, persistent, or severe pain. Endorphins are antagonistic to substance P, reducing the perception of pain.	Excitatory	
Gaseous neurotransmitters			
Nitrous oxide (NO)	Diffuses across cell membranes in PNS rather than being released at synapses; relaxes smooth muscles in walls of blood vessels, causing vessels to dilate, which increases blood flow; contributes to many nervous system functions such as learning, sensory responses, and muscle movements	Modulatory	• Sildenafil citrate (Viagra, an impotency drug)—aids erection by inhibiting enzyme that normally reduces effects of NO in vascular beds of penis

Many Different Molecules Act as Neurotransmitters

In all, nearly 100 different substances are now known or suspected to be neurotransmitters. Most of them are relatively small molecules that diffuse rapidly across the synaptic cleft. Some axon terminals release only one type of neurotransmitter; others release several types. Depending on the type of receptor to which it binds, the same neurotransmitter may stimulate or inhibit the generation of action potentials in the postsynaptic cell. **Table 37.1** lists a number of neurotransmitters, the types of molecules they repre-

INSIGHTS FROM THE Molecular Revolution

Dissecting Neurotransmitter Receptor Functions

Many ionotropic neurotransmitter receptors are ion channels that are opened or closed by the binding of a neurotransmitter molecule. Each of these receptors has two regions: a hydrophilic portion on the outside surface of the plasma membrane that binds the neurotransmitter, and a hydrophobic transmembrane portion that anchors the receptor in the plasma membrane and forms the ion-conducting channel.

Research Question

Are the two primary activities of these receptors—binding neurotransmitters and conducting ions—dependent on parts of the protein that work independently, or do they each require the entire protein structure?

Experiment

To answer the question, Jean-Luc Eiselé and his coworkers at Institut Pasteur, Paris, and colleagues at Centre Médical Universitaire, Geneva, Switzerland, constructed artificial receptors using regions of the ionotropic receptors for two different neurotransmitters. The two receptors are not closely related in amino acid sequence, bind different neurotransmitters, and react differently to calcium ions. One receptor, the nicotinic acetylcholine receptor (nAChR) is activated by acetylcholine and by nicotine, and ion conduction by the receptor is enhanced by Ca^{2+}. The other receptor, the serotonin receptor ($5HT_3$) is activated by serotonin; Ca^{2+} ions block the channel of the serotonin receptor and stop ion conduction. Both receptor proteins have two domains: a hydrophilic domain on the plasma membrane surface and a hydrophobic transmembrane channel embedded in the membrane.

To create the artificial receptors, the investigators used molecular techniques to break the genes encoding the nicotinic acetylcholine and serotonin receptors into two parts. They then reassembled the parts so that in the protein encoded by the chimeric gene, the part of the acetylcholine receptor located on the membrane surface was joined to the transmembrane channel of the serotonin receptor **(Figure 1).** Five versions of the artificial gene, encoding proteins in which the two parts were joined at different positions in the amino acid sequence, were then cloned to increase their quantity and injected into oocytes of the African clawed frog, *Xenopus laevis.* Once in the oocytes, the genes were expressed to produce the chimeric receptor proteins, which became inserted into the oocyte plasma membranes. The properties of the chimeric receptors were then examined.

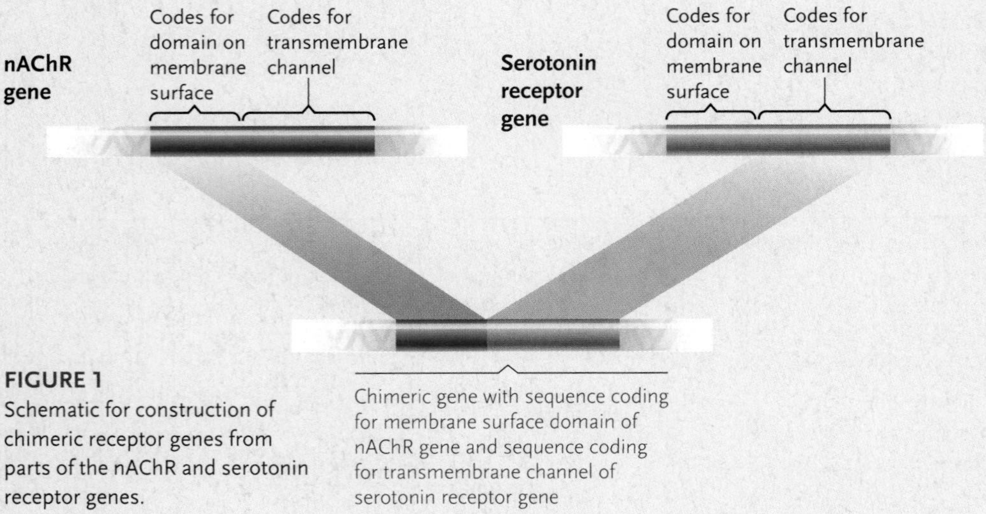

nAChR gene
Codes for domain on membrane surface
Codes for transmembrane channel
Serotonin receptor gene
Codes for domain on membrane surface
Codes for transmembrane channel
Chimeric gene with sequence coding for membrane surface domain of nAChR gene and sequence coding for transmembrane channel of serotonin receptor gene

FIGURE 1
Schematic for construction of chimeric receptor genes from parts of the nAChR and serotonin receptor genes.

Results

Of the five artificial receptors, all were able to bind acetylcholine, but only two were able to conduct ions in response to binding the neurotransmitter, as measured by an increase in the electrical current flowing across the plasma membrane of the oocytes. Agents that inhibit the normal acetylcholine receptor, such as curare, also inhibited the artificial receptors. Serotonin, in contrast, was not bound and did not open the receptor channels, and agents that inhibit the normal serotonin receptor had no effect on the artificial receptors. However, elevated Ca^{2+} concentrations blocked the channel, as in the normal serotonin receptor.

Conclusion

The remarkable research by Eiselé and his coworkers showed that the parts of a receptor binding a neurotransmitter and conducting ions function independently. Their work also demonstrates the feasibility of constructing chimeric receptors as a means for dissecting the functions of subregions of the receptors.

Source: J.-L. Eiselé et al. 1993. Chimaeric nicotinic-serotonergic receptor combines distinct ligand binding and channel specificities. *Nature* 366:479–483.

sent, their sites and type of action, and some drugs and other molecules that affect neurotransmission.

STUDY BREAK 37.3

1. **What features characterize a substance as a neurotransmitter?**
2. **Describe how a neurotransmitter that binds to an ionotropic neurotransmitter receptor in a presynaptic neuron controls action potentials in a postsynaptic neuron.**

37.4 Integration of Incoming Signals by Neurons

Most neurons receive a multitude of stimulatory and inhibitory signals carried by both direct and indirect neurotransmitters. These signals are integrated by the postsynaptic neuron into a response that reflects their combined effects. The integration depends primarily on the patterns, number, types, and activity of the synapses the postsynaptic neuron makes with presynaptic neurons. Inputs from other sources, such as indirect neurotransmitters and other signal molecules, can modify the integration. The response of the postsynaptic neuron is elucidated by the frequency of action potentials it generates.

Integration at Chemical Synapses Occurs by Summation

As mentioned earlier, depending on the type of receptor to which it binds, a neurotransmitter may stimulate or inhibit the generation of action potentials in the postsynaptic neuron. If a neurotransmitter opens an Na^+ channel, Na^+ enters the cell, causing a depolarization. This change in membrane potential pushes the neuron closer to threshold; that is, it is excitatory and is called an **excitatory postsynaptic potential,** or **EPSP.** However, if a neurotransmitter opens an ion channel that allows Cl^- to flow into the cell and K^+ to flow out, hyperpolarization occurs. This change in membrane potential pushes the neuron farther from threshold; that is, it is inhibitory and is called an **inhibitory postsynaptic potential,** or **IPSP.** In contrast to the all-or-nothing operation of an action potential, EPSPs and IPSPs are **graded potentials,** in which the membrane potential increases or decreases without necessarily triggering an action potential. And there are no refractory periods for EPSPs and IPSPs.

A neuron typically has hundreds to thousands of chemical synapses formed by axon terminals of presynaptic neurons contacting its dendrites and cell body **(Figure 37.15).** The events that occur at a single synapse produce either an EPSP or an IPSP in that postsynaptic neuron. But how is an action potential produced if a single EPSP is not sufficient to push the postsynaptic neuron to threshold? The answer involves summation of the inputs received through those many chemical synapses formed by presynaptic neurons. At any given time, some or many of the presynaptic neurons may be firing, producing EPSPs and/or IPSPs in the postsynaptic neuron. The sum of all the EPSPs and IPSPs at a given time determines the total potential in the postsynaptic neuron and, therefore, how that neuron responds. **Figure 37.16** shows, in a greatly simplified way, the effects of EPSPs and IPSPs on membrane potential, and how summation of inputs over space and time brings a postsynaptic neuron to threshold. The postsynaptic neuron in the figure has three presynaptic neurons forming synapses with its dendrites and cell body; Ex1 and Ex2 are excitatory neurons, and In1 is an inhibitory neuron. The summation point for EPSPs and IPSPs is the axon hillock of the postsynaptic neuron. The greatest density of voltage-gated Na^+ channels occurs in that region, resulting in the lowest threshold potential in the neuron.

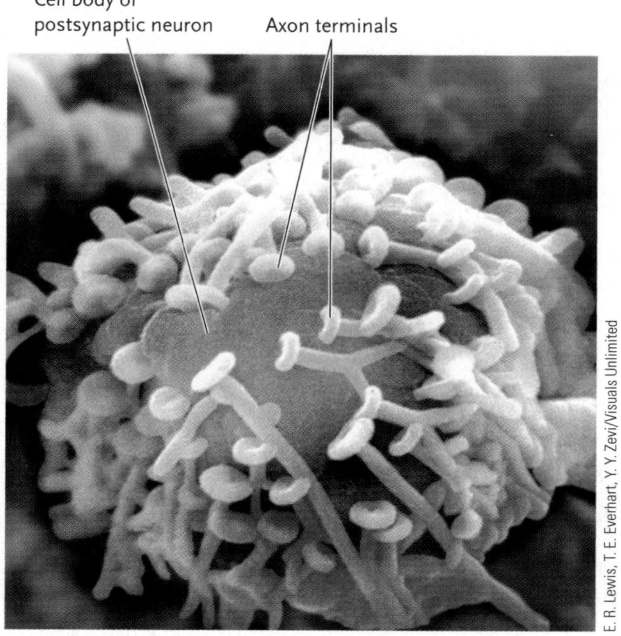
Cell body of postsynaptic neuron
Axon terminals
E. R. Lewis, T. E. Everhart, Y. Y. Zevi/Visuals Unlimited

FIGURE 37.15
The multiple chemical synapses relaying signals to a neuron. The drying process used to prepare the neuron for electron microscopy has toppled the axon terminals and pulled them away from the neuron's surface.

The Patterns of Synaptic Connections Contribute to Integration

The total number of connections made by a neuron may be very large—some single interneurons in the human brain, for example, form as many as 100,000 synapses with other neurons. The synapses are not absolutely fixed; they can change through modification, addition, or removal of synaptic connections—or even entire neurons—as animals mature and experience changes in their environments. The combined activities of all the neurons in the nervous system, constantly integrating information from sensory receptors and triggering responses by effectors, control the internal activities of animals and regulate their behavior. This behavior ranges from the simple reflexes of a flatworm to the complex behavior of mammals, including consciousness, emotions, reasoning, and creativity in humans.

Although researchers do not yet understand how processes such as ion flow, synaptic connections, and neural networks produce complex mental activities, they continue to find correspondences between them and the types of neuronal communication described in this chapter. In the next chapter we learn about how nervous systems of animals are organized, and how higher functions such as memory, learning, and consciousness are produced.

STUDY BREAK 37.4

How does a postsynaptic neuron integrate signals carried by direct and indirect neurotransmitters?

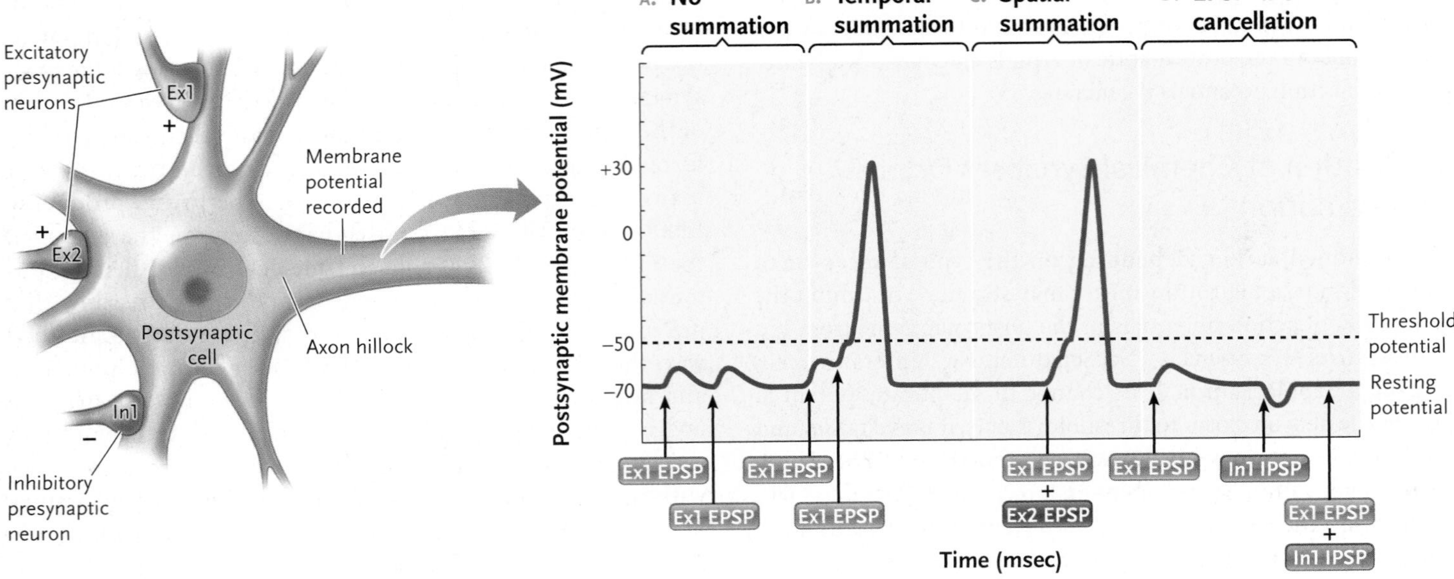
Excitatory
presynaptic
neurons
Ex1
+
+
Ex2
Membrane
potential
recorded
Postsynaptic
cell
Axon hillock
In1
–
Inhibitory
presynaptic
neuron
A. No summation
B. Temporal summation
C. Spatial summation
D. EPSP–IPSP cancellation
Postsynaptic membrane potential (mV)
+30
0
–50
–70
Threshold potential
Resting potential
Ex1 EPSP
Ex1 EPSP
Ex1 EPSP
Ex1 EPSP
Ex1 EPSP + Ex2 EPSP
Ex1 EPSP
In1 IPSP
Ex1 EPSP + In1 IPSP
Time (msec)

A. **No summation:** The axon of Ex1 releases a neurotransmitter, which produces an EPSP in the postsynaptic cell. The membrane depolarizes, but not enough to reach threshold. If Ex1 input causes a new EPSP after the first EPSP has died down, it will be of the same magnitude as the first EPSP and no progression toward threshold has taken place—no summation has occurred.

B. **Temporal summation:** If instead, Ex1 input causes a new EPSP before the first EPSP has died down, the second EPSP will sum with the first and a greater depolarization will have taken place. This summation of two (or more) EPSPs produced by successive firing of a single presynaptic neuron over a short period of time is temporal summation. If the total depolarization achieved in this way reaches threshold, an action potential is produced in the postsynaptic neuron.

C. **Spatial summation:** The postsynaptic neuron may be brought to threshold by **spatial summation**, the summation of EPSPs produced by the simultaneous firing of two different excitatory presynaptic neurons, such as Ex1 and Ex2.

D. **Summation resulting in cancellation:** EPSPs and IPSPs can sum to cancel each other out. In the example, firing of the excitatory presynaptic neuron Ex1 alone produces an EPSP, firing of presynaptic inhibitory neuron In3 alone produces an IPSP, while firing of Ex1 and In3 simultaneously produces no change in the membrane potential.

FIGURE 37.16
The effects of EPSPs and IPSPs on membrane potential, and how summation of inputs over space and time brings a postsynaptic neuron to threshold.

UNANSWERED QUESTIONS

What is the basic wiring diagram of the brain?

To understand how the brain processes information, it is important first to understand the wiring that handles this information. There are about 100,000,000,000 (10^{11}) neurons in the human brain. This is a daunting number, but it can be managed by grouping the neurons into classes or categories. It has been estimated that there are fewer than 10,000 different neuronal classes in the entire brain. Still, each neuron makes and receives synaptic contacts with about 1,000 other neurons on average, making the wiring diagram complex.

Many techniques have been used to unravel the wiring of the brain. The Spanish anatomist Santiago Ramon-y-Cajal used the Golgi silver staining technique to identify cell classes in the early 1900s. He intuited the direction of information flow and worked out circuits. More modern techniques have used dyes or retroviruses that transport along axons. Also, electrophysiology has been used to determine which neurons talk with each other. New molecular techniques are being developed that allow different fluorescent proteins to be expressed in different neuron classes. Because the neurons can glow in all colors of the rainbow, this technique has been called "brainbow".

Despite all of the advances for gathering information about the brain's wiring diagram, we do not have an adequate means to represent this complexity, nor do we have a means of understanding how information flows through such a complex network. Just as modern DNA sequencing technology allowed a revolution in the field of genomics (the categorization of all of the genes in an organism); similar breakthroughs will need to occur in the field of neuromics (the categorization of all of the neurons and their interactions). These breakthroughs will have to be in data management, computational simulations, and multisite recording techniques. I have been collaborating with computer scientists and database specialists to develop tools for cataloging neurons and their synaptic connections. The hope is that one day we will have a way to represent the complex wiring in the brain.

How did the brain evolve?

Given that the brain is so complex, how could it have evolved? If you look at human brains and the brains of other primates, such as monkeys, there is not much difference, except size. Is size all that matters, or are there differences in the way neural circuits function? In my lab, we have been investigating the evolution of neural circuits in simpler animals: sea slugs. These creatures have just 10,000 neurons in their brains, so it is easier to understand the circuitry. We are examining how the same sets of neurons produce different types of swimming behaviors in different species of sea slug.

Are there types of neurotransmitters that we haven't discovered yet?

Our understanding of how neurons communicate with each other is always advancing. Cajal was one of the first proponents of the neuron doctrine, the idea that neurons are separate cells that make contacts at specialized locations (synapses).There was a debate as to whether the information is passed from one neuron to another in the form of a chemical neurotransmitter or through electrical connections. Now we know that both occur. For chemical synapses, it was found that neurotransmitters are packaged into vesicles and released. But then it was discovered that some neurotransmitters are not released from vesicles. For example, the gas nitric oxide (NO) is synthesized in the cytoplasm and diffuses through cell membranes to activate enzymes in the cytoplasm of other neurons. Components of the membrane itself can be metabolized to act as neurotransmitters; anandamide is a fatty acid that is produced in the brain and acts on cannabinoid receptors, such as the CB1 receptor, which can modify synaptic release of neurotransmitters. Are there other types of neurotransmitters that the brain uses, but that haven't been discovered yet? What are they? How will we find them?

How does processing in the brain lead to consciousness?

Even if we figure out how to represent all of the neurons in the brain and all of the ways that they communicate, we still need to answer the ultimate question, "How does neural activity cause the sensation that we all feel of being conscious and alive?" In every age, it is hard to imagine the things that we do not know. For example, imagine trying to understand how the brain worked before electricity was understood; it would have seemed like a magical force.

Think Critically

What would it mean to understand the biological nature of conscious thought?

Paul S. Katz

Paul S. Katz is a professor in the Neuroscience Institute and the director of the Center for Neuromics at Georgia State University. His research interests include neuromics, neuromodulation, and the evolution of neuronal circuits underlying behavior. Learn more about Dr. Katz's work at http://neuroscience.gsu.edu/pkatz.html.

REVIEW KEY CONCEPTS

Go to **CENGAGENOW** at www.cengage.com/login to access quizzing, animations, exercises, articles, and personalized homework help.

37.1 Neurons and Their Organization in Nervous Systems

- The nervous system of an animal (1) receives information about conditions in the internal and external environment, (2) transmits the message along neurons, (3) integrates the information to formulate an appropriate response, and (4) sends out signals to muscles or glands that accomplish the response (Figure 37.2).
- Neurons have dendrites, which receive information and conduct signals toward the cell body, and axons, which conduct signals away from the cell body to another neuron or an effector (Figure 37.3).
- Afferent neurons conduct information from sensory receptors to interneurons, which integrate the information into a response. The response signals are passed to efferent neurons, which activate the effectors carrying out the response (Figure 37.2).
- The combination of an afferent neuron, an interneuron, and an efferent neuron makes up a basic neuronal circuit. The circuits combine into networks that interconnect the peripheral and central nervous systems.
- Glial cells help maintain the balance of ions surrounding neurons and form insulating layers around the axons (Figure 37.5).
- Neurons make connections by two types of synapses. In an electrical synapse, impulses pass directly from the sending to the receiving cell. In a chemical synapse, neurotransmitter molecules released by the presynaptic cell diffuse across a narrow synaptic cleft and bind to receptors in the plasma membrane of the postsynaptic cell (Figure 37.6).

Animation: Neuron structure and function

Animation: Impulse traveling through a nerve

37.2 Signaling by Neurons

- The membrane potential of a cell depends on the unequal distribution of positive and negative charges on either side of the membrane, which establishes a potential difference across the membrane.
- Three primary conditions contribute to the resting potential of neurons: (1) an Na^+/K^+ active transport pump that sets up concentration gradients of Na^+ ions (higher outside) and K^+ ions (higher inside); (2) an open channel that allows K^+ to flow out freely; and (3) negatively charged proteins and other molecules inside the cell that cannot pass through the membrane (Figure 37.9).
- An action potential is generated when a stimulus pushes the resting potential to the threshold value at which voltage-gated Na^+ and K^+ channels open in the plasma membrane. The inward flow of Na^+ changes membrane potential abruptly from negative to a positive peak. The potential falls to the resting value again as the gated K^+ channels allow this ion to flow out (Figure 37.11).

- Action potentials move along an axon as the ion flows generated in one segment depolarize the potential in the next segment (Figure 37.12).
- Action potentials are prevented from reversing direction by a brief refractory period, during which a segment of membrane that has just generated an action potential cannot be stimulated to produce another for a few milliseconds.
- In myelinated axons, ions can flow across the plasma membrane only at nodes where the myelin sheath is interrupted. As a result, action potentials skip rapidly from node to node by saltatory conduction (Figure 37.13).

Animation: Ion concentrations

Animation: Measuring membrane potential

Animation: Nerve structure

Animation: Action potential propagation

Animation: Ion flow in myelinated axons

Animation: Stretch reflex

37.3 Transmission across Chemical Synapses

- Neurotransmitters released into the synaptic cleft bind to receptors in the plasma membrane of the postsynaptic cell, altering the flow of ions across the plasma membrane of the postsynaptic cell and pushing its membrane potential toward or away from the threshold potential (Figure 37.14).
- Neurotransmitters that bind to ionotropic neurotransmitter receptors act directly. Binding to a receptor in the postsynaptic membrane activates it, leading to opening or closing of the channel in the associated ion channel.
- Neurotransmitters that bind to metabotropic neurotransmitter receptors act indirectly. Binding to a receptor in the postsynaptic membrane triggers generation of a second messenger, which causes an ion-conducting channel to open or close.
- Neurotransmitters are released from synaptic vesicles into the synaptic cleft by exocytosis, which is triggered by entry of Ca^{2+} ions into the cytoplasm of the axon terminal through voltage-gated Ca^{2+} channels opened by the arrival of an action potential.
- Neurotransmitter release stops when action potentials cease arriving at the axon terminal. Neurotransmitters remaining in the synaptic cleft are broken down by enzymes or taken up by the axon terminal or glial cells.
- Types of neurotransmitters include acetylcholine, amino acids, biogenic amines, neuropeptides, and gases such as NO (Table 37.1). Many of the biogenic amines and neuropeptides are also released into the general body circulation as hormones.

Animation: Chemical synapse

37.4 Integration of Incoming Signals by Neurons

- Neurons carry out integration by summing excitatory postsynaptic potentials (EPSPs) and inhibitory postsynaptic potentials (IPSPs); the summation may push the membrane potential of the postsynaptic cell toward or away from the threshold for an action potential (Figure 37.16).
- The combined effects of summation in all the neurons in the nervous system control behavior in animals and underlie complex mental processes in mammals.

Animation: Synaptic integration

UNDERSTAND AND APPLY

Test Your Knowledge

1. Nerve signals travel in the following manner:
   a. A dendrite of a sensory neuron receives the signal; its cell body transmits the signal to a motor neuron's axon, and the signal is sent to the target.
   b. An axon of a motor neuron receives the signal; its cell body transmits the signal to a sensory neuron's dendrite, and the signal is sent to the target.
   c. Efferent neurons conduct nerve impulses toward the cell body of sensory neurons, which send them on to interneurons and, ultimately, to afferent motor neurons.
   d. A dendrite of a sensory neuron receives the signal; the cell's axon transmits the signal to an interneuron, and the signal is transmitted by efferent neurons to effectors.
   e. The axons of oligodendrocytes transmit nerve impulses to the dendrites of astrocytes.
2. Glial cells:
   a. are unable to divide after an animal is born.
   b. called Schwann cells form the insulating myelin sheath around axons.
   c. called astrocytes form the nodes of Ranvier in the brain.
   d. called oligodendrocytes cover the surfaces of blood vessels in the PNS.
   e. are neuronal cells that provide nutrition and support to nonneuronal cells.
3. An example of a synapse could be the site where:
   a. neurotransmitters released by an axon travel across a gap and are picked up by receptors on a muscle cell.
   b. an electrical impulse arrives at the end of a dendrite causing ions to flow onto axons of presynaptic neurons.
   c. postsynaptic neurons transmit a signal across a cleft to a presynaptic neuron.
   d. oligodendrocytes contact the dendrites of an afferent neuron directly.
   e. an on-off switch stimulates an electrical impulse in a presynaptic cell to stimulate other presynaptic cells.
4. The resting potential in neurons requires:
   a. membrane transport channels to be constantly open for Na^+ and K^+ flow.
   b. the inside of neurons to be positive relative to the outside.
   c. the diffusion of K^+ out of the cell and a charge difference between the inside and outside of the axon set up by this movement of K^+.
   d. an active Na^+/K^+ pump, which pumps Na^+ and K^+ into the neuron.
   e. three Na^+ ions to be pumped through three Na^+ gates and two K^+ ions to be pumped through two K^+ gates.
5. The major role of the Na^+/K^+ pump is to:
   a. cause a rapid firing of the action potential so the inside of the membrane becomes momentarily positive.
   b. decrease the resting potential to zero.
   c. hyperpolarize the membrane above resting value.
   d. cause an action potential to enter a refractory period.
   e. maintain the resting potential at a constant negative value.

6. In the propagation of a nerve impulse:
    a. the refractory period begins as the K^+ channel opens, allowing K^+ ions to flow outward along their concentration gradient.
    b. Na^+ ions flow out of the axon with their concentration gradient.
    c. positive charges lower the membrane potential to its lowest action potential.
    d. gated K^+ channels open at the same time as the activation gate of Na^+ channels closes.
    e. the depolarizing stimulus lowers the membrane potential to open the Na^+ gates.
7. Which of the following does *not* contribute to propagation of action potentials?
    a. As the area outside the membrane becomes negative, it attracts ions from adjacent regions; as the inside of the membrane becomes positive, it attracts negative ions from nearby in the cytoplasm. These events depolarize nearby regions of the axon membrane.
    b. The refractory period allows the impulse to travel in only one direction.
    c. Each segment of the axon prevents the adjacent segments from firing.
    d. The magnitude of the action potential stays the same as it travels down the axon.
    e. Up to a limit, increasing the intensity of the stimulus increases the number of action potentials.
8. Which of the following statements best describes saltatory conduction?
    a. It inhibits direct neurotransmitter release.
    b. It transmits the action potential at the nodes of Ranvier and thus speeds up impulses on myelinated axons.
    c. It increases neurotransmitter release at the presynaptic membrane.
    d. It decreases neurotransmitter uptake at chemically gated postsynaptic channels.
    e. It removes neurotransmitters from the synaptic cleft.
9. Transmission of a nerve impulse to its target cell requires:
    a. endocytosis of neurotransmitters by excitatory presynaptic vesicles.
    b. thousands of molecules of neurotransmitter that had been stored in the postsynaptic cell to be released into the synaptic cleft.
    c. Ca^{2+} ions to diffuse through voltage-gated Ca^{2+} channels.
    d. a fall in Ca^{2+} in the cytoplasm to trigger a protein that causes the presynaptic vesicle to fuse with the plasma membrane.
    e. an action potential to open the Ca^{2+} gates so that Ca^{2+} ions, in higher concentration outside the axon, can flow back into the cytoplasm of the neuron.
10. Which of the following matches between neurotransmitter and site(s) of action is correct?
    a. Acetylcholine: Neuromuscular junctions in PNS
    b. Norepinephrine: Mostly acts on CNS and PNS as well as on effectors such as muscle
    c. Serotonin: Many CNS pathways, including those involved in vital brain functions such as memory and learning
    d. Endorphins: CNS interneurons in a number of pathways, including those regulating appetite, reproductive behavior, muscular movement, sleep, and emotional states such as anxiety
    e. Nitrous oxide: Released by special, unmyelinated sensory neurons in spinal cord

Discuss the Concepts

1. In some cases of ADHD (attention deficit hyperactivity disorder) the impulsive, erratic behavior typical of affected people can be calmed with drugs that stimulate certain brain neurons. Based on what you have learned about neurotransmitter activity in this chapter, can you suggest a neural basis for this effect?
2. Most sensory neurons form synapses either on interneurons in the spinal cord or on motor neurons. However, in many vertebrates, certain sensory neurons in the nasal epithelium synapse directly on brain neurons that activate behavioral responses to odors. Suggest at least one reason why natural selection might favor such an arrangement.
3. How did evolution of chemical synapses make higher brain functions possible?
4. Search for Internet resources using the term "Pediatric Neurotransmitter Disease" and, for one such disease, explain how the symptoms relate to neurotransmitter function.

Design an Experiment

Design an experiment to test whether neurons are connected via electrical or chemical synapses.

Interpret the Data

You learned in this chapter that Na^+/K^+ active transport pumps in the plasma membrane of the axons are responsible for creating the imbalance between Na^+ and K^+ inside and outside of the neuron that produces the resting potential.

In early research studying the role of ions, and the involvement of active transport of ions in signaling in neurons, investigators used the giant axon of a squid as a model. The diameter of a giant axon is far greater than that of a mammalian axon, which enabled researchers to isolate it easily and use it in in vitro experiments. In one early experiment, researchers investigated the active transport of Na^+ out of the axon in response to the presence of cyanide. Experimentally they hooked up a section of axon to a syringe, immersed the axon in artificial seawater, introduced radioactive ^{22}Na (as $^{22}NaCl$) into the axon, and then quantified the transport of ^{22}Na out through the axon's plasma membrane. The rate of ^{22}Na transport out of the axon was determined by measuring the radioactivity released into the fluid surrounding the axon over a period of time. The figure below shows the results of the experiment.

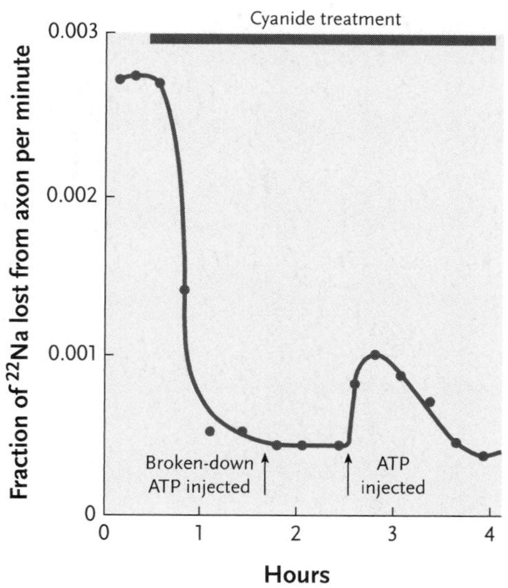
Cyanide treatment
Fraction of 22Na lost from axon per minute
0.003
0.002
0.001
0
Broken-down ATP injected
ATP injected
0
1
2
3
4
Hours

1. What is the effect of cyanide on Na^+ transport out of the squid axon? How do the data show the effect?
2. What is the effect of injecting broken-down ATP (ATP hydrolyzed to break it down into its component parts) on the transport of Na^+ out of the cyanide-treated axon?
3. What is the effect of injecting intact ATP on the transport of Na^+ out of the cyanide-treated axon? What does this result mean with respect to Na^+ active transport?

Source: P. C. Caldwell et al. 1960. The effects of injecting 'energy-rich' phosphate compounds on the active transport of ions in the giant axons of *Loligo*. *Journal of Physiology* 152:561–590.

Apply Evolutionary Thinking

A biologist hypothesized that the mechanism for the propagation of action potentials down a neuron evolved only once. What evidence would you collect from animals living today to support or refute that hypothesis?

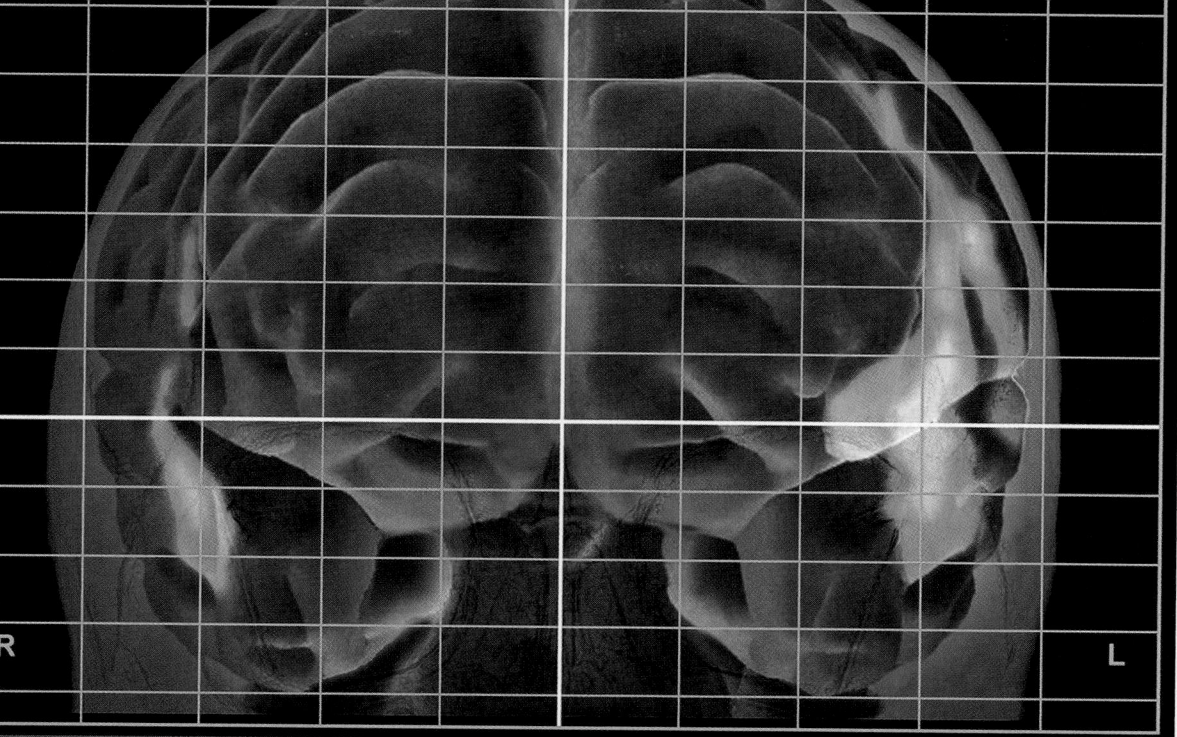
R
L

Activity in the human brain while reading aloud. The image combines an fMRI of a male brain with a PET scan, which shows that blood circulation increases in the language, hearing, and vision areas of the brain, especially in the left hemisphere.

38

STUDY OUTLINE

Nervous Systems

Why It Matters. . . The conductor's baton falls and the orchestra plays the first notes of a Mozart symphony. Unaware of the complex interactions of their nervous systems, the musicians translate printed musical notation into melodious sounds played on their instruments. Although their fingers and arms move to produce precise harmonies, the musicians are only vaguely conscious of these movements, learned through years of practice. Their only conscious endeavor is to interpret the music in line with the conductor's directions.

From the back of the hall, a common housefly, *Musca domestica,* moves in random twists and turns that bring it toward the stage. Although far less complex than that of a human, the fly's nervous system contains networks of neurons that work in the same way, in patterns adapted to its lifestyle.

The fly does not register the sounds reverberating through the hall as a significant sensory input. However, some of its receptors are exquisitely sensitive to the presence of potential food molecules, including those in the sweat on the conductor's face. The fly's swoops and turns bring it closer to the conductor; soon it alights on the tip of his nose. When sensory receptors in the fly's footpads detect organic matter on the surface of the nose, they trigger an automated feeding response: the fly's proboscis lowers and its gut begins contractions that suck up the nutrients.

The conductor's eyes notice the insect's approach, and sensory receptors in his skin pinpoint the spot where it lands. Without missing a beat, the conductor's hand flicks toward that exact spot. But his nervous system and effectors, although highly sophisticated, are no match for the escape reflexes of the fly. The fly's sensory receptors detect the motion of the fingers, sending impulses to the fly's leg and wing muscles that launch it into flight long before the fingers reach the nose.

The fly wanders into the orchestra, attracted to potential nutrients on various musicians, who respond with flicking movements that are no more successful than those of the conductor. At last, the fly lands on the left hand of the timpanist, who is listening with pleasure to the music while he awaits his entrance late in the first movement. His right hand holds a mallet. With a skill born of long practice in hitting drums, gongs, and bells with speed and precision, the timpanist deftly swings his mallet and dispatches the fly, ending the latest contest between mammalian and arthropod nervous systems.

The nervous systems underlying these behaviors are one of the features that set animals apart from other organisms. As animals evolved, the need to find food, living space, and mates, and to escape predators and other dangers, provided a powerful selection pressure for increasingly complex and capable nervous systems. Neurons, described in the previous chapter, provide the structural and functional basis for all these systems. We can trace some of the developments along this extended evolutionary pathway by examining the nervous systems of living animals, from invertebrates to mammals, and especially humans. <

38.1 Invertebrate and Vertebrate Nervous Systems Compared

The nervous systems of most invertebrates are relatively simple, typically containing fewer neurons, arranged in less complex networks, than vertebrate systems. As animal groups evolved, their nervous systems became more elaborate, providing the ability to integrate more sensory information and to formulate more complex responses. Our comparative survey of nervous systems begins with the simplest invertebrates.

Cnidarians and Echinoderms Have Nerve Nets

Cnidarians and echinoderms are radially symmetrical animals with body parts arranged regularly around a central axis, like the spokes of a wheel. Their nervous systems, called **nerve nets,** are loose meshes of neurons organized within that radial symmetry.

The nerve nets of cnidarians such as sea anemones extend into each "spoke" of the body **(Figure 38.1A).** Their neurons lack clearly differentiated dendrites and axons. When part of the animal is stimulated, impulses are conducted through the nerve net in all directions from the point of stimulation. Although there is no cluster of neurons that plays the coordinating role of a brain, nerve cells may be more concentrated in some regions. For example, in scyphozoan jellyfish, which swim by rhythmic contractions of their bells, neurons are denser in a ring around the margin of the bell, in the same area as the contractile cells that produce the swimming movements.

In echinoderms, including sea stars, the nervous system is a modified nerve net, with some neurons organized into **nerves,** bundles of axons enclosed in connective tissue and following the same pathway. A *nerve ring* surrounds the centrally located mouth, and a *radial nerve* that is connected to nerve nets branches throughout each arm **(Figure 38.1B).** If the radial nerve serving an arm is cut, the arm can still move, but not in coordination with the other arms.

More Complex Invertebrates Have Cephalized Nervous Systems

More complex invertebrates have neurons with clearly defined axons and dendrites, and more specialized functions. Some neurons are concentrated into functional clusters called **ganglia** (singular, *ganglion*). A key evolutionary development in invertebrates is a trend toward *cephalization,* the formation of a distinct head region containing both ganglia that constitute a **brain,** the control center of the nervous system, and major sensory structures. One or more solid **nerve cords**—bundles of nerves—extend from the central ganglia to the rest of the body; they are connected to smaller nerves. Another evolutionary trend is toward bilateral symmetry of the body and the nervous system, in which body parts are mirror images on left and right sides. These trends toward cephalization and bilateral symmetry are illustrated here in flatworms, arthropods, and mollusks.

In flatworms, a small brain consisting of a pair of ganglia at the anterior end is connected by two or more longitudinal nerve cords to nerve nets in the rest of the body **(Figure 38.1C).** The brain integrates inputs from sensory receptors, including a pair of anterior eyespots with receptors that respond to light. The brain and longitudinal nerve cords constitute the flatworm's **central nervous system (CNS),** the simplest one known, while the nerves from the CNS to the rest of the body constitute the **peripheral nervous system (PNS).**

Arthropods such as insects have a head region that contains a brain consisting of dorsal and ventral pairs of ganglia, and major sensory structures, usually eyes and antennae **(Figure 38.1D).** The brain exerts centralized control over the remainder of the animal. A ventral nerve cord enlarges into a pair of ganglia in each body segment. In arthropods with fused body segments, as in the thorax of insects, the ganglia are also fused into larger masses forming secondary control centers.

Although different in basic plan from the arthropod system, the nervous systems of mollusks (such as clams, snails, and octopuses) also rely on neurons clustered into paired ganglia and connected by major nerves. Different mollusks have varying degrees of cephalization, with cephalopods having the most pronounced cephalization of any invertebrate group. In the head of an octopus, for example, a cluster of ganglia fuses into a complex, lobed brain with clearly defined sensory and motor regions. Paired nerves link different lobes with muscles and sensory receptors, including prominent optic lobes linked by nerves to large, complex eyes **(Figure 38.1E).** Octopuses are capable of rapid movement to hunt prey and to escape from predators, behaviors that rely on rapid, sophisticated processing of sensory information.

Vertebrates Have the Most Specialized Nervous Systems

In vertebrates, the CNS consists of the brain and spinal cord, and the PNS consists of all the nerves and ganglia that connect the brain and spinal cord to the rest of the body **(Figure 38.1F).** All vertebrate nervous systems are highly cephalized, with major con-

A. **Cnidarian (sea anemone)**

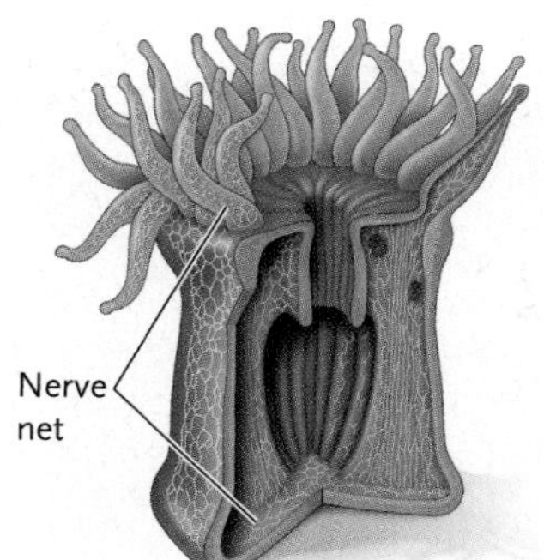
Nerve net

B. **Echinoderm (sea star)**

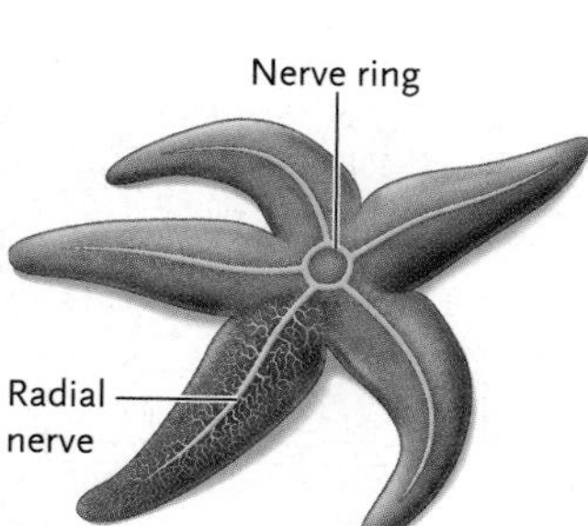
Nerve ring
Radial nerve

C. **Planarian (flatworm)**

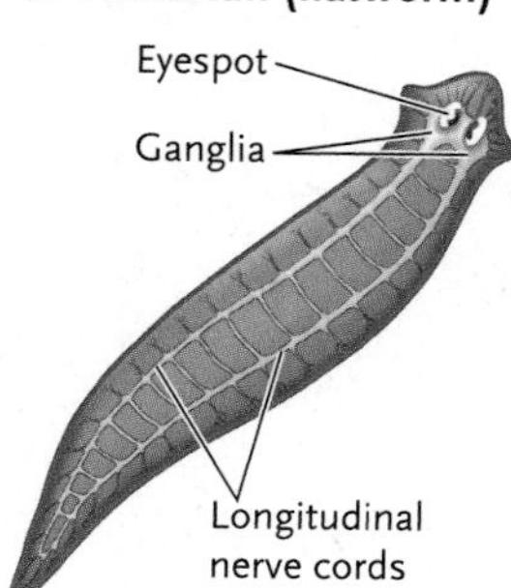
Eyespot
Ganglia
Longitudinal nerve cords

D. **Arthropod (grasshopper)**

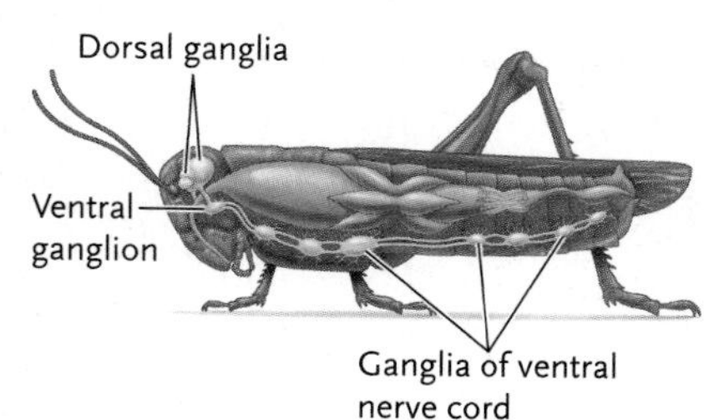
Dorsal ganglia
Ventral ganglion
Ganglia of ventral nerve cord

E. **Mollusk (octopus)**

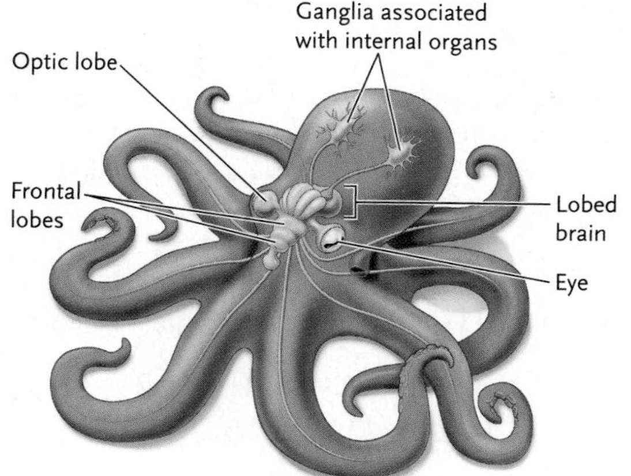
Ganglia associated with internal organs
Optic lobe
Frontal lobes
Lobed brain
Eye

F. **Chordate (salamander)**

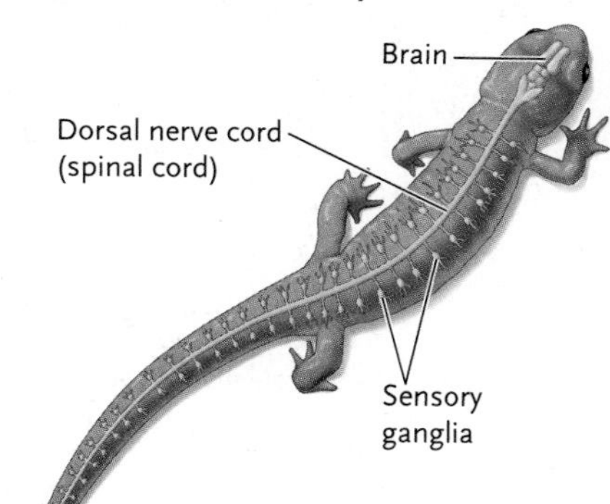
Brain
Dorsal nerve cord (spinal cord)
Sensory ganglia

FIGURE 38.1
Invertebrate and vertebrate nervous systems compared, showing increasing cephalization. The diagrams are not drawn to the same scale.

centrations of neurons in a brain located in the head. In contrast to invertebrate nervous systems, which have solid nerve cords located ventrally, the brain and nerve cord of vertebrates are hollow, fluid-filled structures located dorsally. The head contains specialized sensory organs, which are connected directly to the brain by nerves. Compared with invertebrates, the ganglia are greatly reduced in mass and functional activity except in the gut, which contains extensive interneuron networks.

The structure of the vertebrate nervous system reflects its pattern of development. The nervous system of a vertebrate embryo begins as the hollow **neural tube** (discussed more in Chapter 48), the anterior end of which develops into the brain and the rest into the **spinal cord.** The cavity of the neural tube becomes the fluid-filled **ventricles** of the brain and the **central canal** through the spinal cord. Adjacent tissues give rise to nerves that connect the brain and spinal cord with all body regions.

Early in embryonic development, the anterior part of the neural tube enlarges into three distinct regions: the **forebrain, midbrain,** and **hindbrain (Figure 38.2).** A little later, the embryonic hindbrain subdivides into the *metencephalon* and *myelencephalon;* the midbrain develops into the *mesencephalon;* and the forebrain subdivides into the *telencephalon* and *diencephalon.*

The metencephalon gives rise to the *cerebellum,* which integrates sensory signals from the eyes, ears, and muscle spindles with motor signals from the telencephalon, and the *pons,* a major traffic center for information passing between the cerebellum and the higher integrating centers of the adult telencephalon. The myelencephalon gives rise to the *medulla oblongata* (commonly shortened to medulla), which controls many vital involuntary tasks such as respiration and blood circulation. The mesencephalon gives rise to the (adult) midbrain, which with the pons and the medulla constitutes the brain stem. The midbrain has centers for coordinating reflex responses (involuntary reactions) to visual and auditory (hearing) input and relays signals to the telencephalon.

The embryonic telencephalon develops into the *cerebrum* (or adult telencephalon), the largest part of the brain. The cerebrum controls higher functions such as thought, memory, language, and emotions, as well as voluntary movements. The diencephalon gives rise to the *thalamus,* a center that receives sensory input and relays it to the regions of the cerebral cortex concerned with motor responses to such input, and to the *hypothalamus,* the primary center for homeostatic control over the internal environment. In fishes, the cerebrum is little more than a relay station for olfactory (sense of smell) information. In amphibians, reptiles, and birds, it becomes progressively larger and contains greater concentrations of integrative functions. In mammals, the cerebrum is the major integrative structure of the brain.

In the following sections, we examine vertebrate nervous systems, and the human nervous system in particular, beginning with the peripheral nervous system.

STUDY BREAK 38.1 <

1. **Distinguish between a nerve net, nerves, and nerve cords.**
2. **What is cephalization?**

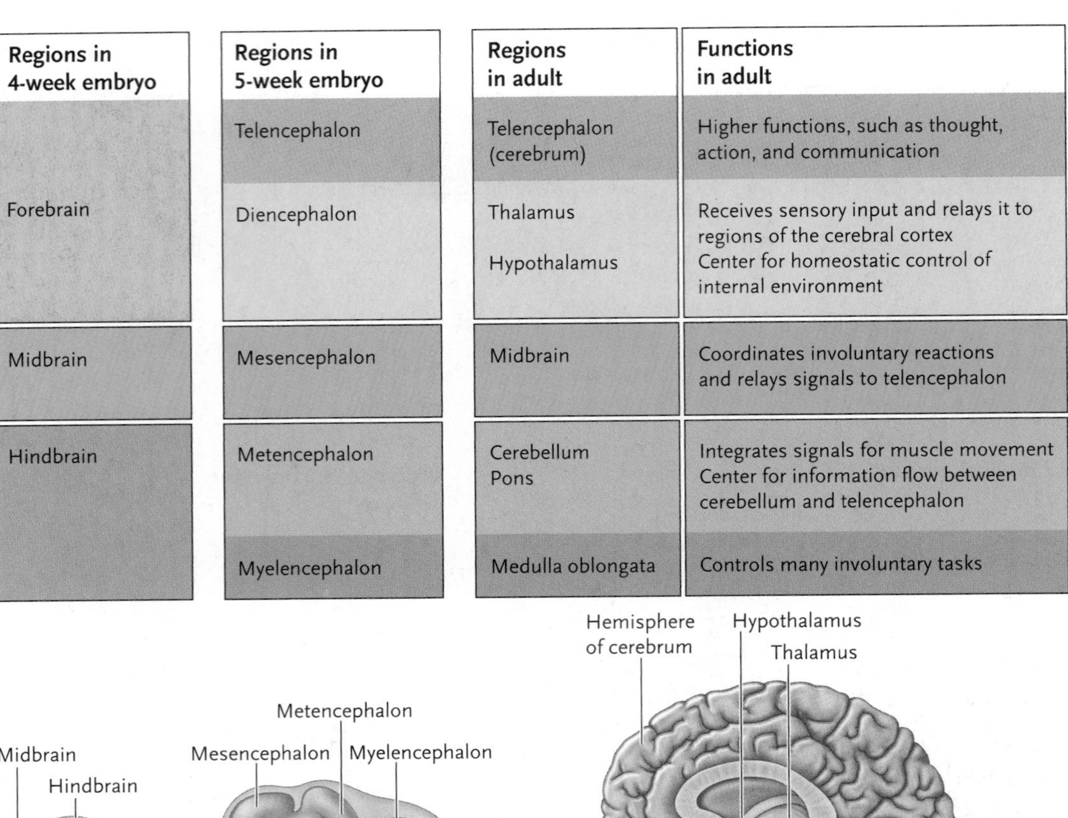

Regions in 4-week embryo	Regions in 5-week embryo	Regions in adult	Functions in adult
Forebrain	Telencephalon	Telencephalon (cerebrum)	Higher functions, such as thought, action, and communication
	Diencephalon	Thalamus	Receives sensory input and relays it to regions of the cerebral cortex
		Hypothalamus	Center for homeostatic control of internal environment
Midbrain	Mesencephalon	Midbrain	Coordinates involuntary reactions and relays signals to telencephalon
Hindbrain	Metencephalon	Cerebellum	Integrates signals for muscle movement
		Pons	Center for information flow between cerebellum and telencephalon
	Myelencephalon	Medulla oblongata	Controls many involuntary tasks

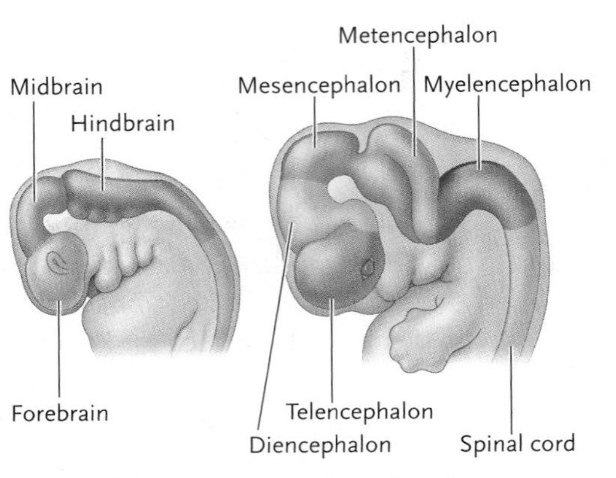
Midbrain
Hindbrain
Forebrain
4-week embryo
Metencephalon
Mesencephalon
Myelencephalon
Telencephalon
Diencephalon
Spinal cord
5-week embryo

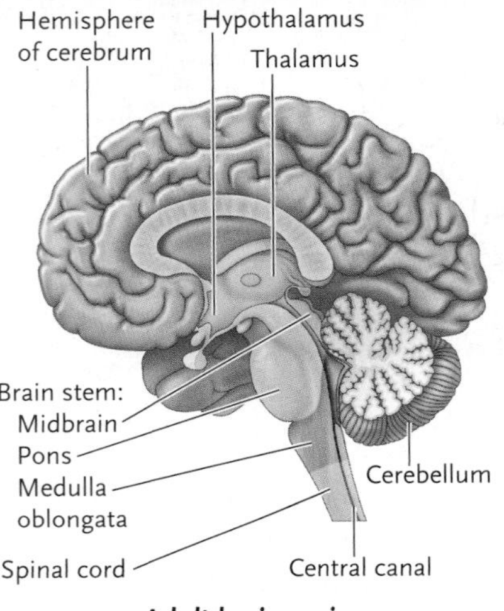
Hemisphere of cerebrum
Hypothalamus
Thalamus
Brain stem:
Midbrain
Pons
Medulla oblongata
Cerebellum
Spinal cord
Central canal
Adult brain regions

FIGURE 38.2
Development of the human brain from the anterior end of an embryo's neural tube.

38.2 The Peripheral Nervous System

Afferent neurons in the peripheral nervous system transmit signals to the CNS, and signals from the CNS are sent via efferent neurons in the PNS to the effectors that carry out responses **(Figure 38.3).** The afferent part of the system includes all the neurons that transmit sensory information from their receptors. The efferent part of the system consists of the axons of neurons that carry signals to the muscles and glands, acting as effectors. In mammals, 31 pairs of **spinal nerves** carry signals between the spinal cord and the body trunk and limbs, and 12 pairs of **cranial nerves** connect the brain directly to the head, neck, and body trunk. The efferent part of the PNS is further divided into somatic and autonomic systems (see Figure 38.3).

The Somatic System Controls the Contraction of Skeletal Muscles, Producing Body Movements

The **somatic nervous system** controls body movements that are primarily conscious and voluntary. Its neurons, called motor neurons, carry efferent signals from the CNS to the skeletal muscles. The dendrites and cell bodies of motor neurons are located in the spinal cord; their axons extend from the spinal cord to the skeletal muscle cells they control. As a result, the somatic portions of the cranial and spinal nerves consist only of axons.

Although the somatic system is primarily under conscious, voluntary control, some contractions of skeletal muscles are unconscious and involuntary. These include the reflexes, shivering, and the constant muscle contractions that maintain body posture and balance.

The Autonomic System Is Divided into Sympathetic and Parasympathetic Divisions

The **autonomic nervous system** controls largely involuntary processes including digestion, secretion by sweat glands, circulation of the blood, many functions of the reproductive and excretory systems, and contraction of smooth muscles in all parts of the body. It is organized into *sympathetic* and *parasympathetic* divisions, which are always active, and have opposing effects on the organs they affect, thereby enabling precise control **(Figure 38.4).** For example, in the circulatory system, sympathetic neurons stimulate the force and rate of the heartbeat, and parasympathetic

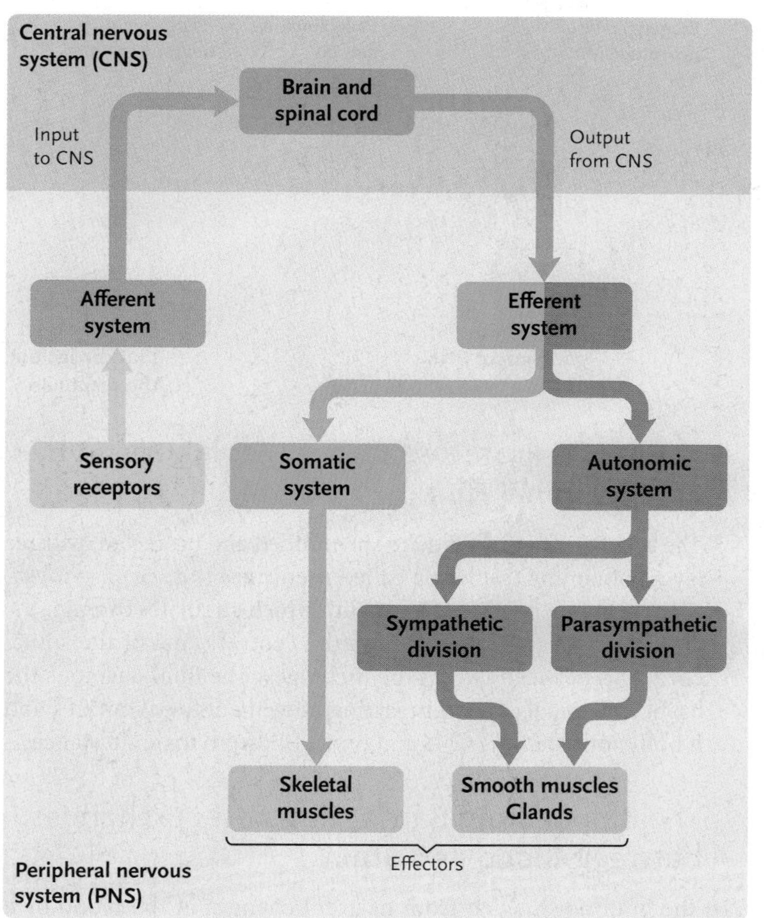
Central nervous system (CNS)
Brain and spinal cord
Input to CNS
Output from CNS
Afferent system
Efferent system
Sensory receptors
Somatic system
Autonomic system
Sympathetic division
Parasympathetic division
Skeletal muscles
Smooth muscles Glands
Effectors
Peripheral nervous system (PNS)

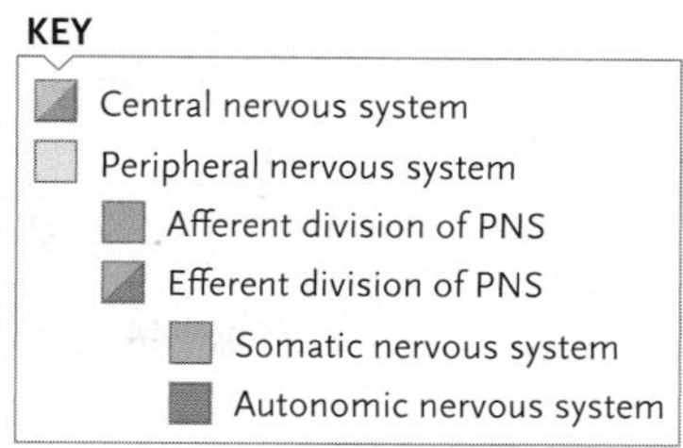
KEY
Central nervous system
Peripheral nervous system
Afferent division of PNS
Efferent division of PNS
Somatic nervous system
Autonomic nervous system

FIGURE 38.3
The central nervous system (CNS), the peripheral nervous system (PNS), and their subsystems.

FIGURE 38.4
Effects of the parasympathetic and sympathetic divisions of the central nervous system on organ and gland function.
Only one side of each division is shown; both are duplicated on the left and right sides of the body.

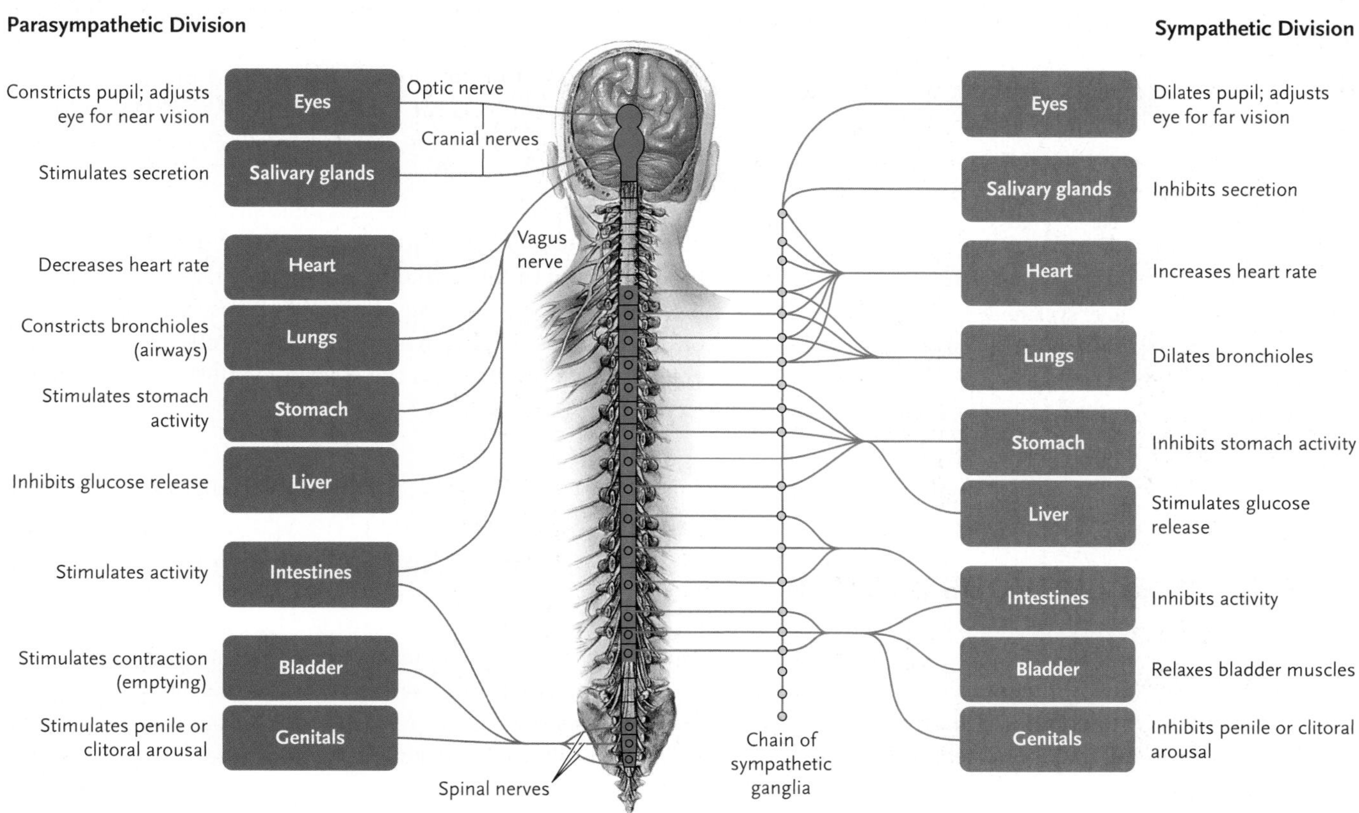
Parasympathetic Division
Sympathetic Division
Constricts pupil; adjusts eye for near vision
Eyes
Optic nerve
Cranial nerves
Stimulates secretion
Salivary glands
Vagus nerve
Decreases heart rate
Heart
Constricts bronchioles (airways)
Lungs
Stimulates stomach activity
Stomach
Inhibits glucose release
Liver
Stimulates activity
Intestines
Stimulates contraction (emptying)
Bladder
Stimulates penile or clitoral arousal
Genitals
Spinal nerves
Chain of sympathetic ganglia
Eyes
Dilates pupil; adjusts eye for far vision
Salivary glands
Inhibits secretion
Heart
Increases heart rate
Lungs
Dilates bronchioles
Stomach
Inhibits stomach activity
Liver
Stimulates glucose release
Intestines
Inhibits activity
Bladder
Relaxes bladder muscles
Genitals
Inhibits penile or clitoral arousal

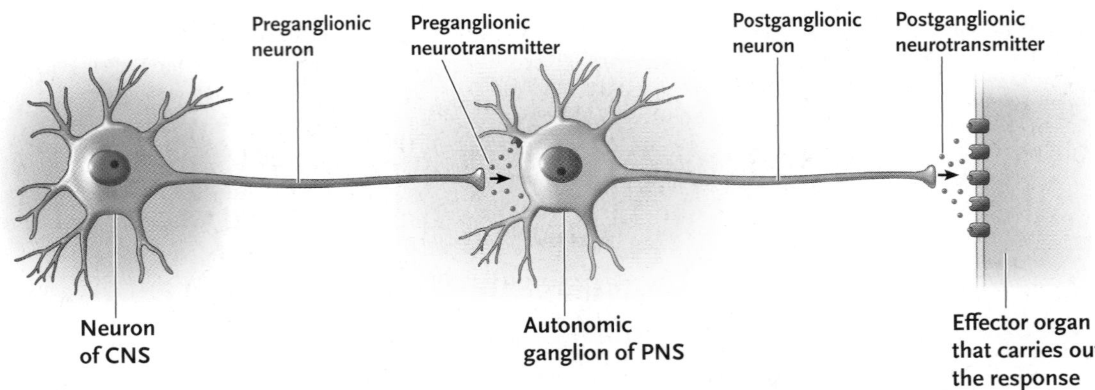
Preganglionic
neuron
Preganglionic
neurotransmitter
Postganglionic
neuron
Postganglionic
neurotransmitter
Neuron
of CNS
Autonomic
ganglion of PNS
Effector organ
that carries out
the response

FIGURE 38.5
An autonomic nervous system pathway.

neurons inhibit these activities. In the digestive system, sympathetic neurons inhibit the smooth muscle contractions that move materials through the small intestine, and parasympathetic neurons stimulate the same activities. These opposing effects control involuntary body functions precisely.

The pathways of the autonomic nervous system include two neurons **(Figure 38.5)**. The first neuron has its dendrites and cell body in the CNS, and its axon extends to a ganglion in the PNS. There, it synapses with the dendrites and cell body of the second neuron in the pathway. The axon of the second neuron extends from the ganglion to the effector carrying out the response.

The **sympathetic division** predominates in situations involving stress, danger, excitement, or strenuous physical activity. Signals from the sympathetic division increase the force and rate of the heartbeat, raise the blood pressure by constricting selected blood vessels, dilate air passages in the lungs, induce sweating, and open the pupils wide. Activities that are less important in an emergency, such as digestion, are suppressed by the sympathetic system. The **parasympathetic division,** in contrast, predominates during quiet, low-stress situations, such as while relaxing. Under the influence of the parasympathetic division, the effects of the sympathetic division, such as rapid heartbeat and elevated blood pressure, are reduced and "housekeeping" (maintenance) activities such as digestion predominate.

STUDY BREAK 38.2 <

Which of the two autonomic nervous system divisions predominates in the following scenarios? (a) You are hiking on a trail and suddenly a bear appears in your path. (b) It is a hot sunny day. You find a shady tree and sit down. Leaning against its trunk, you feel your eyes becoming heavy.

38.3 The Central Nervous System and Its Functions

The central nervous system (CNS) integrates incoming sensory information from the PNS into compensating responses, thus managing body activities. Recall that the vertebrate CNS consists of the brain and the spinal cord.

The CNS Is Protected by the Meninges and by Cerebrospinal Fluid

The brain and spinal cord are surrounded and protected by three layers of connective tissue called **meninges** (*meninga* = membrane), and by **cerebrospinal fluid,** which circulates through the ventricles of the brain, through the central canal of the spinal cord, and between two of the meninges. The fluid cushions the brain and spinal cord from jarring movements and impacts, and it both nourishes the CNS and protects it from toxic substances.

The Blood–Brain Barrier Regulates Exchanges between Blood and Brain

The brain is shielded from harmful changes in the blood by a highly selective **blood–brain barrier.** That is, unlike the epithelial cells forming capillary walls elsewhere in the body, which allow small molecules and ions to pass freely from the blood to surrounding fluids, those forming capillaries in the brain are sealed together by tight junctions (see Figure 5.29). The tight junctions set up a blood–brain barrier that prevents most substances dissolved in the blood from entering the cerebrospinal fluid and thus protects the brain and spinal cord from viruses, bacteria, and toxic substances that may be circulating in the blood.

A few types of molecules and ions, such as oxygen, carbon dioxide, alcohol, and anesthetics, can move directly across the lipid bilayer of the epithelial cell membranes by diffusion. A few other substances—most significantly glucose and ketones, the only molecules that brain and spinal cord cells can oxidize for energy—are moved across the plasma membrane by highly selective transport proteins.

The Brain Integrates Sensory Information and Formulates Responses; The Spinal Cord Relays Signals between the PNS and the Brain and Controls Reflexes

The *brain* is the major center that receives, integrates, stores, and retrieves information in vertebrates. Its interneuron networks generate responses that provide the basis for our voluntary movements, consciousness, behavior, emotions, learning, reasoning, language, and memory, among many other complex activities.

You learned earlier that the three major divisions of the embryonic neural tube—forebrain, midbrain, and hindbrain—give rise to the structures of the adult brain **(Figure 38.6)**. Each brain structure contains both **gray matter,** consisting of nerve cell bodies and dendrites, and **white matter,** consisting of axons, many of them surrounded by myelin sheaths. (The myelination of the axons gives white matter its color.)

The hindbrain develops into three major structures in the adult brain: the *cerebellum,* the *pons,* and the *medulla oblongata* (the *medulla*) (see Figure 38.2). The pons and medulla, along with the midbrain, form a stalklike structure known as the **brain stem,** which connects the forebrain with the spinal cord. All but two of the twelve pairs of cranial nerves also originate from the brain stem. The cerebellum, with its deeply folded surface, is an outgrowth of the pons.

The forebrain, which makes up most of the mass of the brain in humans, forms the *telencephalon (cerebrum).* The cerebrum, the largest part of the brain in humans, is organized into the left and right *cerebral hemispheres,* which have many fissures and folds (see Figure 38.6). Each hemisphere consists of **cerebral cortex,** a thin outer shell of gray matter covering a thick core of white matter. The *basal nuclei,* consisting of several regions of gray matter, are located deep within the white matter.

The *spinal cord,* which extends dorsally from the base of the brain, carries impulses between the brain and the PNS and contains the interneuron circuits that control motor reflexes. In cross section, the spinal cord has a butterfly-shaped core of gray matter surrounded by white matter. Pairs of spinal nerves connect with the spinal cord at spaces between the vertebrae.

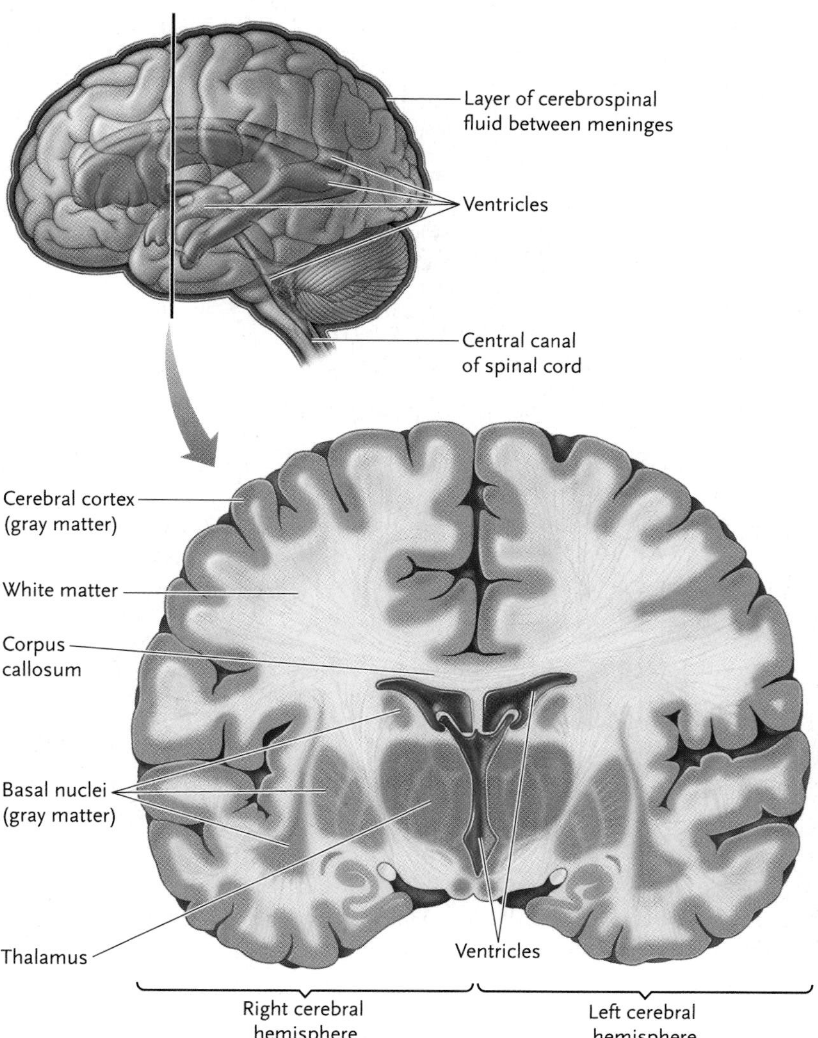
Layer of cerebrospinal fluid between meninges
Ventricles
Central canal of spinal cord
Cerebral cortex (gray matter)
White matter
Corpus callosum
Basal nuclei (gray matter)
Thalamus
Ventricles
Right cerebral hemisphere
Left cerebral hemisphere

FIGURE 38.6
The human brain, illustrating the distribution of gray matter and white matter, and the locations of the four ventricles (in blue) with their connection to the central canal of the spinal cord.

The afferent axons entering the spinal cord make synapses with interneurons in the gray matter, which send axons upward through the white matter of the spinal cord to the brain. Conversely, axons from interneurons of the brain pass downward through the white matter of the cord and make synapses with the dendrites and cell bodies of efferent neurons in the gray matter of the cord. The axons of these efferent neurons exit the spinal cord through the spinal nerves.

The gray matter of the spinal cord also contains pathways involved in **reflexes,** programmed movements that take place without conscious effort; an example is the knee-jerk response, in which a tap to the tendon just below the knee cap (patella) causes the leg to kick **(Figure 38.7).** The existence of only a single synapse between the efferent and afferent neurons in the reflex facilitates a rapid response—about 50 ms between tap and the start of the kick.

Similarly, a reflex pathway is involved for the rapid withdrawal of your hand if you touch a hot surface. You may recall from experience that when a reflex movement withdraws your hand from a hot surface or other damaging stimulus, you feel the pain shortly *after* the hand is withdrawn. This is the extra time required for impulses to travel from the neurons of the reflex via interneurons to the brain (see substance P in Table 37.1).

The Brain Stem Regulates Many Vital Housekeeping Functions of the Body

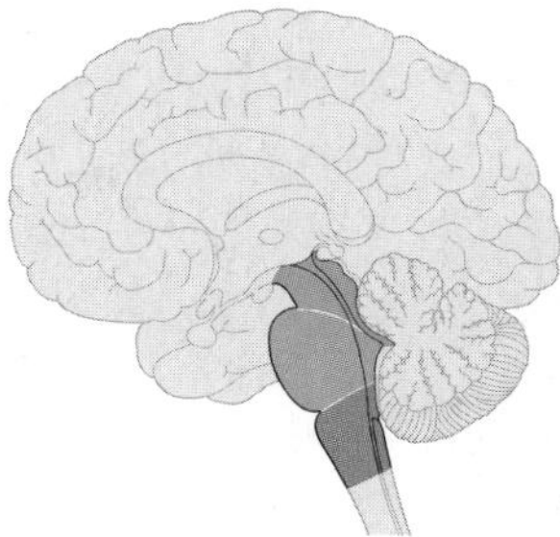

Gray-matter centers in the brain stem control many vital body functions without conscious involvement or control by the cerebrum. Among these functions are the heart and respiration rates, blood pressure, constriction and dilation of blood vessels, coughing, and reflex activities of the digestive system such as vomiting. Damage to the brain stem has serious and sometimes lethal consequences.

A complex network of interconnected neurons known as the **reticular formation** (*reticulum* = little net) runs through the length of the brain stem, connecting to the thalamus at the anterior end and to the spinal cord at the posterior end **(Figure 38.8).** All incoming sensory input goes to the reticular formation,

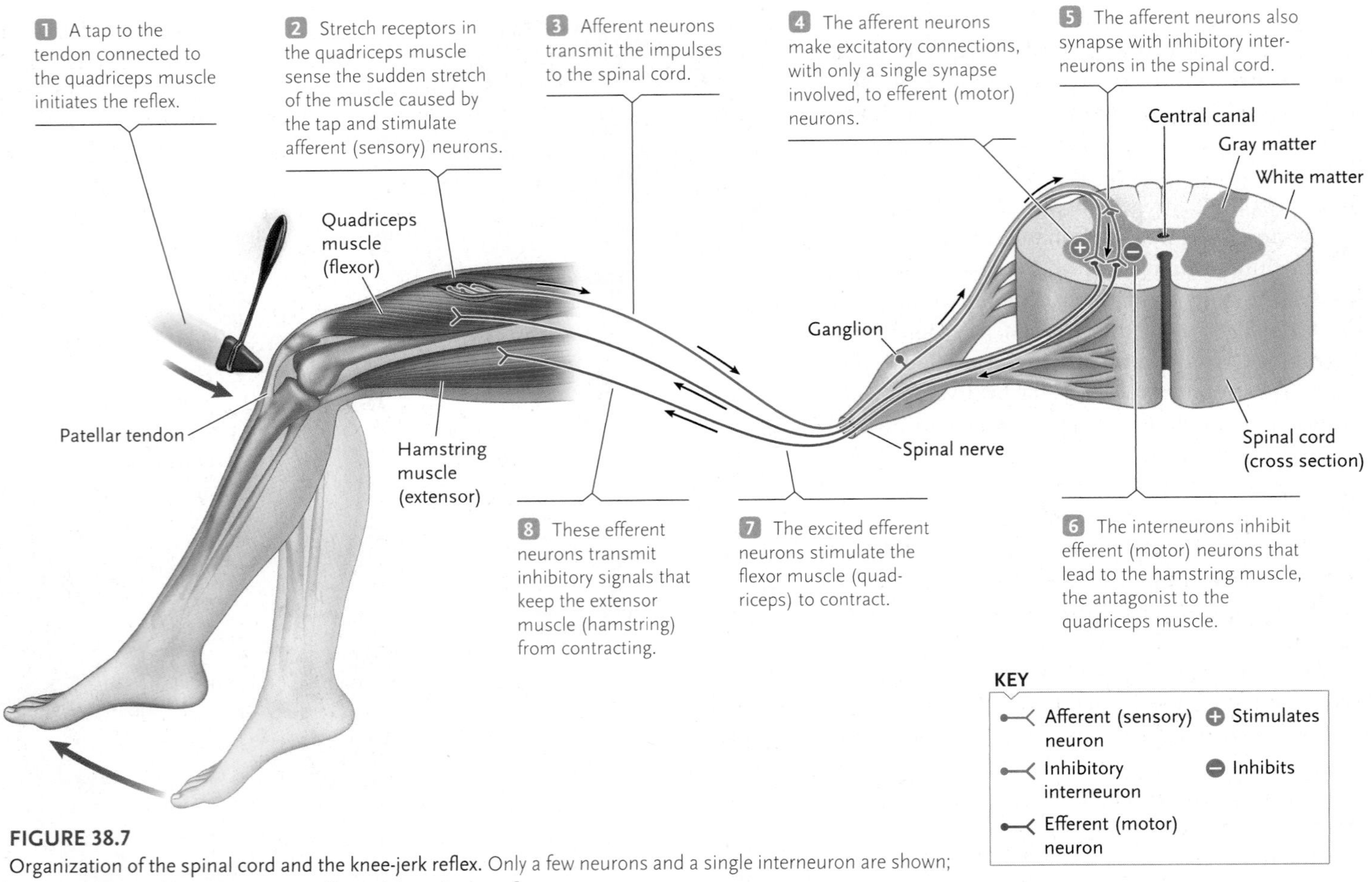

1 A tap to the tendon connected to the quadriceps muscle initiates the reflex.

2 Stretch receptors in the quadriceps muscle sense the sudden stretch of the muscle caused by the tap and stimulate afferent (sensory) neurons.

3 Afferent neurons transmit the impulses to the spinal cord.

4 The afferent neurons make excitatory connections, with only a single synapse involved, to efferent (motor) neurons.

5 The afferent neurons also synapse with inhibitory interneurons in the spinal cord.

Central canal
Gray matter
White matter
Quadriceps muscle (flexor)
Ganglion
Patellar tendon
Hamstring muscle (extensor)
Spinal nerve
Spinal cord (cross section)

8 These efferent neurons transmit inhibitory signals that keep the extensor muscle (hamstring) from contracting.

7 The excited efferent neurons stimulate the flexor muscle (quadriceps) to contract.

6 The interneurons inhibit efferent (motor) neurons that lead to the hamstring muscle, the antagonist to the quadriceps muscle.

KEY
Afferent (sensory) neuron
Inhibitory interneuron
Efferent (motor) neuron
Stimulates
Inhibits

FIGURE 38.7
Organization of the spinal cord and the knee-jerk reflex. Only a few neurons and a single interneuron are shown; many neurons and interneurons actually participate in the reflex.

which integrates the information and then sends signals to other parts of the CNS. The reticular formation has two parts. The **ascending reticular formation,** also called the **reticular activating system,** contains neurons that convey stimulatory signals via the thalamus to arouse and activate the cerebral cortex. It is responsible for the sleep–wake cycle; depending on the level of stimulation of the cortex, various levels of alertness and consciousness are produced. Lesions in this part of the brain stem result in coma. The other part, the **descending reticular formation,** receives information from the hypothalamus and connects with interneurons in the spinal cord that control skeletal muscle contraction, thereby controlling muscle movement and posture. The reticular formation filters incoming signals, helping to discriminate between important and unimportant ones. Such filtering is necessary because the brain is unable to process every one of the signals from millions of sensory receptors. For example, the action of the reticular formation enables you to sleep through many sounds but waken to specific ones, such as a cat meowing to be let out, or a baby crying.

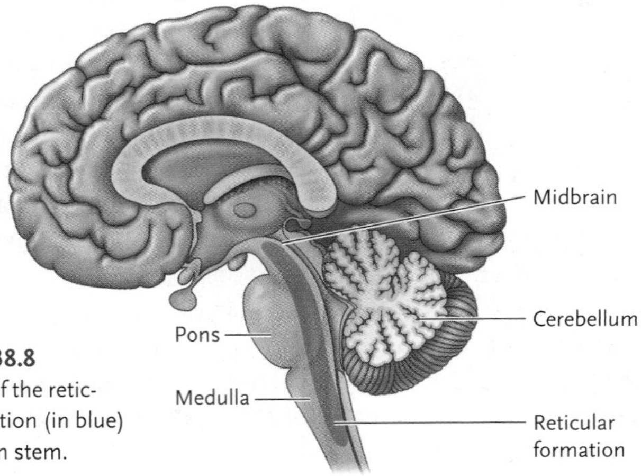

Midbrain
Cerebellum
Pons
Medulla
Reticular formation

FIGURE 38.8
Location of the reticular formation (in blue) in the brain stem.

The Cerebellum Integrates Sensory Inputs to Coordinate Body Movements

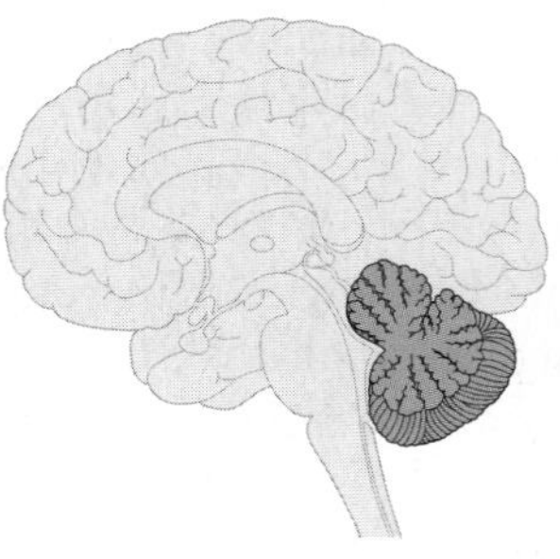

Although the cerebellum is an outgrowth of the pons (see Figure 38.8), it is separate in structure and function from the brain stem. Through its extensive connections with other parts of the brain, the cerebellum receives sensory input from receptors in muscles and joints, from balance receptors in the inner ear, and from the receptors of touch, vision, and hearing. These signals convey information about how the body trunk and limbs are positioned, the degree to which different muscles are contracted or relaxed, and the direction in which the body or limbs are moving. The cerebellum integrates these sensory signals and compares them with signals from the cerebrum that control voluntary body movements. Outputs from the cerebellum to the cerebrum, brain stem, and spinal cord modify and fine-tune the movements to keep the body in balance and directed toward targeted positions in space. The cerebellum of all mammals has essentially the same capabilities and works in the same way. The human cerebellum also contributes to the learning and memory of motor skills such as playing the guitar.

Gray-Matter Centers Control a Variety of Functions

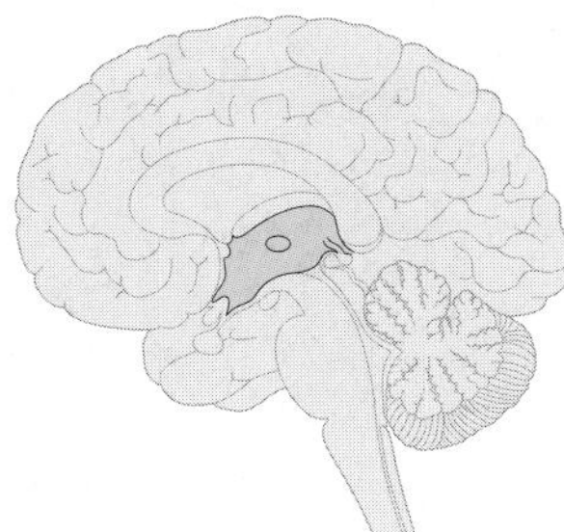

Gray-matter centers derived from the embryonic forebrain include the thalamus, hypothalamus, and basal nuclei **(Figure 38.9)**. These centers contribute to the control and integration of voluntary movements, body temperature and glandular secretions, osmotic balance of the blood and extracellular fluids, wakefulness, and the emotions, among other functions. Some of the gray-matter centers route information to and from the cerebral cortex, and between the forebrain, brain stem, and cerebellum.

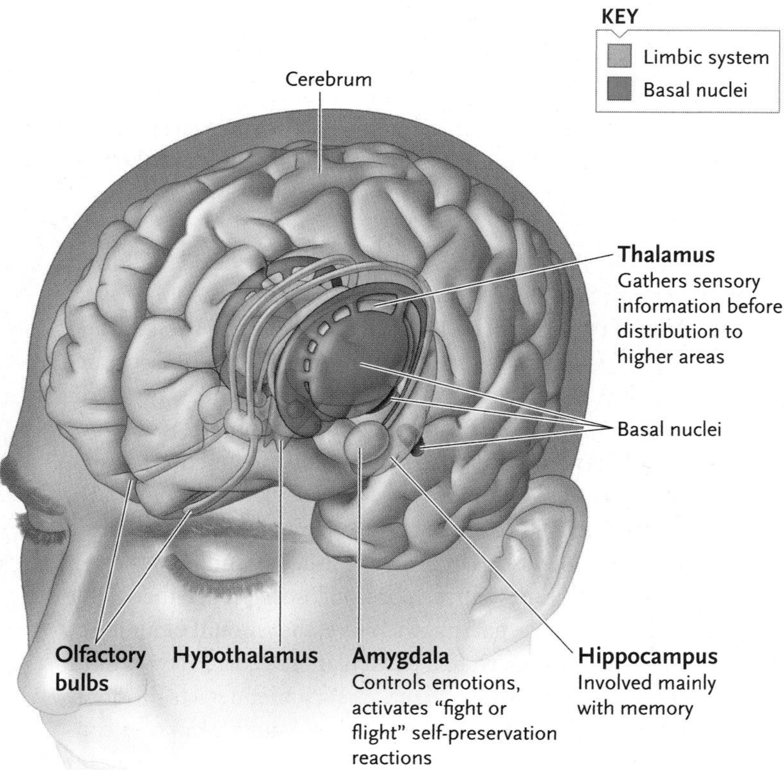
KEY
Limbic system
Basal nuclei
Cerebrum
Thalamus
Gathers sensory information before distribution to higher areas
Basal nuclei
Olfactory bulbs
Hypothalamus
Amygdala
Controls emotions, activates "fight or flight" self-preservation reactions
Hippocampus
Involved mainly with memory

FIGURE 38.9
Basal nuclei, thalamus, and hypothalamus gray-matter centers.

The **thalamus** (see Figure 38.9) forms a major switchboard that receives sensory information and relays it to the regions of the cerebral cortex concerned with motor responses to sensory information of that type. Part of the thalamus near the brain stem cooperates with the reticular formation in alerting the cerebral cortex to full wakefulness, or in inducing drowsiness or sleep.

The **hypothalamus** is a region of the brain located in the floor of the cerebrum (see Figure 38.9) that contains clusters of neurons known as *nuclei*. Suspended just below it and connected to it by a stalk of tissue is the *pituitary gland*, consisting of two fused lobes, the *anterior pituitary* and the *posterior pituitary*, which release hormones. The hypothalamus regulates basic homeostatic functions of the body both directly and through the release of hormones.

Some nuclei in the hypothalamus set and maintain body temperature, most notably by regulating basal metabolic rate. This is an interesting example of an endocrine circuit, in which a hypothalamic hormone causes the release of a hormone from the anterior pituitary that in turn causes the release of a thyroid hormone. Other nuclei monitor the osmotic balance of the blood plasma. If osmotic pressure rises above normal levels, these neurons secrete a hormone from their terminals in adjacent areas of the hypothalamus and from the posterior pituitary to increase thirst and recover more water from urine. (More detail on the structure and functions of the hypothalamus and pituitary and endocrine circuits is given in Chapter 40.)

These and other nuclei of the hypothalamus are able to detect changes in blood temperature, ionic and nutrient composition, and the presence of hormones released in other areas of the body because they are *not* protected by the blood–brain barrier. The wide-ranging influence of the hypothalamic nuclei can be seen in their coordination of autonomic system functions, such as the control of the heartbeat, contraction of smooth muscle cells in the digestive system, and glandular secretions. Indeed, a particular hypothalamic structure, the *suprachiasmatic nucleus*, is the main generator of the biological clock that times our myriad daily behavioral and metabolic rhythms.

The **basal nuclei** are gray-matter centers that surround the thalamus on both sides of the brain (see Figure 38.9). They moderate voluntary movements directed by motor centers in the cerebrum. Damage to the basal nuclei can affect the planning and fine-tuning of movements, leading to stiff, rigid motions of the limbs and unwanted or misdirected motor activity, such as tremors of the hands and inability to start or stop intended movements at the intended place and time. Parkinson disease, in which affected individuals exhibit all of these symptoms, results from degeneration of centers in and near the basal nuclei.

Parts of the thalamus, hypothalamus, and basal nuclei, along with other nearby gray-matter centers—the amygdala, hippocampus, and olfactory bulbs—form a functional network called the **limbic system** (*limbus* = edge, border), sometimes called our "emotional brain" (see Figure 38.9). The **amygdala** works as a switchboard, routing information about experiences that have an emotional component through the limbic system. The **hippocampus** is involved in sending information to the frontal lobes, and the **olfactory bulbs** relay inputs from odor receptors to both the

FIGURE 38.10
The lobes of the cerebrum, showing major regions and association areas of the cerebral cortex.

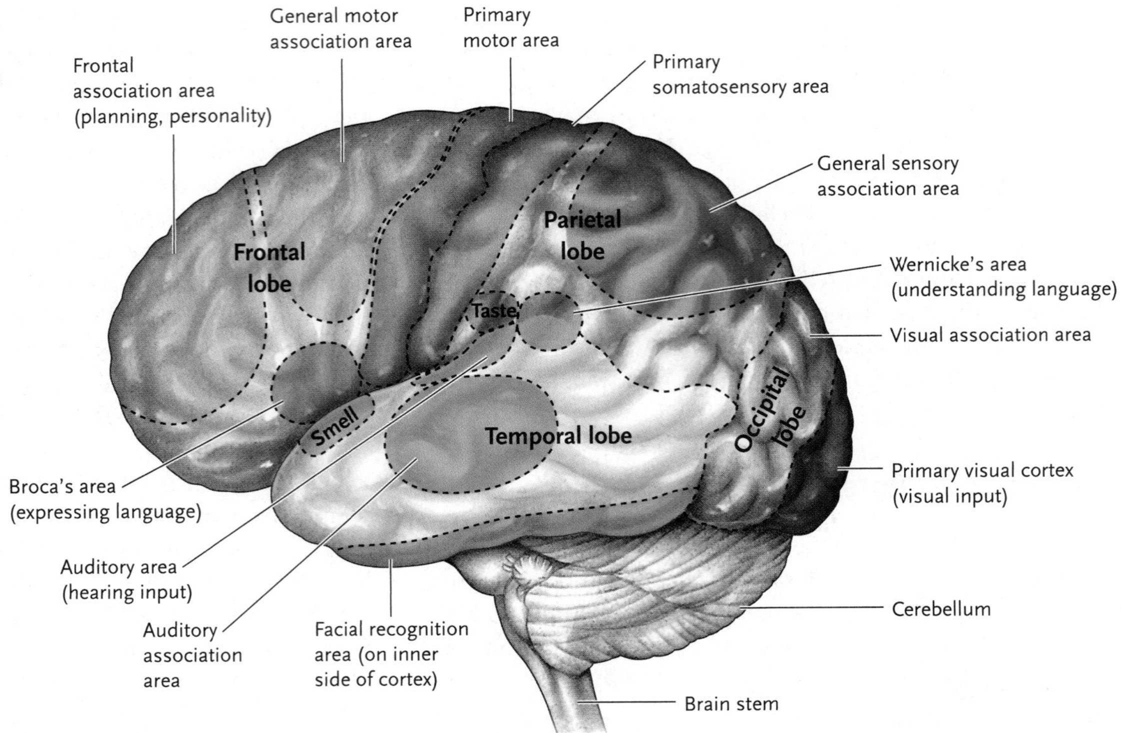

General motor association area
Primary motor area
Primary somatosensory area
Frontal association area (planning, personality)
General sensory association area
Frontal lobe
Parietal lobe
Wernicke's area (understanding language)
Taste
Visual association area
Occipital lobe
Smell
Temporal lobe
Broca's area (expressing language)
Primary visual cortex (visual input)
Auditory area (hearing input)
Cerebellum
Auditory association area
Facial recognition area (on inner side of cortex)
Brain stem

cerebral cortex and the limbic system. The olfactory connection to the limbic system may explain why certain odors can evoke particular, sometimes startlingly powerful emotional responses.

The limbic system controls emotional behavior and influences the basic body functions regulated by the hypothalamus and brain stem. Stimulation of different parts of the limbic system produces anger, anxiety, fear, satisfaction, pleasure, or sexual arousal. Connections between the limbic system and other brain regions bring about emotional responses such as smiling, blushing, or laughing.

The Cerebral Cortex Carries Out All Higher Brain Functions in Humans

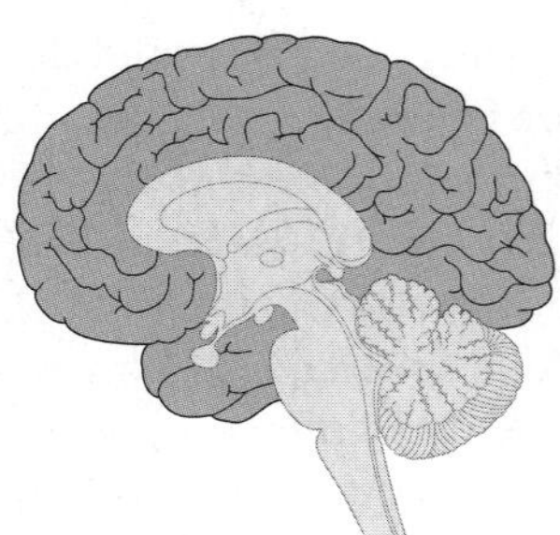

The gray matter of each hemisphere, the cerebral cortex, contains the processing centers for the integration of neural input and the initiation of neural output. The white matter of the cerebral hemispheres, by contrast, contains the neural routes for signal transmission between parts of the cerebral cortex, or from the cerebral cortex to other parts of the CNS. No information processing occurs in the white matter.

Over the course of evolution, the surface area of the cerebral cortex increased by continually folding in on itself, thereby expanding the structure into sophisticated information encoding and processing centers. Primates have cerebral cortexes with the largest number of convolutions. In humans, each cerebral hemisphere is divided by surface folds into *frontal, parietal, temporal,* and *occipital* lobes **(Figure 38.10).** Uniquely in mammals, the cerebral cortex of the cerebral hemispheres is organized into six layers of neurons; these layers are the newest part of the cerebral cortex in an evolutionary sense.

The two cerebral hemispheres can function separately, and each has its own communication lines internally and with the rest of the CNS and the body. The left cerebral hemisphere responds primarily to sensory signals from, and controls movements in, the right side of the body. The right hemisphere has the same relationships to the left side of the body. This opposite connection and control reflects the fact that the nerves carrying afferent and efferent signals cross from left to right within the spinal cord or brain stem. Thick axon bundles, forming a structure called the **corpus callosum** (see Figure 38.6), connect the two cerebral hemispheres and coordinate their functions.

STUDYING THE FUNCTIONS OF THE CEREBRAL CORTEX Physicians and researchers have learned much about the functions of various regions of the cerebral cortex by studying normal subjects, and patients with brain damage from stroke, infection, tumors, or mechanical disturbance. Scanning techniques such as **positron emission tomography (PET)** and **functional magnetic resonance imaging (fMRI)** allow researchers to identify the functions of specific brain regions in noninvasive ways. For a PET scan **(Figure 38.11),** an individual is given a dose of a radioactive compound, typically a modified form of glucose if brain activity is to be studied. When an area of the brain becomes active, more glucose is needed there to be broken down to provide energy. Therefore, the amount of radioisotope that becomes localized to a particular area correlates with the amount of brain activity in that area. The scanner detects the decay of the radioisotope of the modified glucose, and a computer converts the results into multicolored two- or three-dimensional images. The colors reflect the amount of radioactivity detected, and provide a quantitative view of the amount of brain activity.

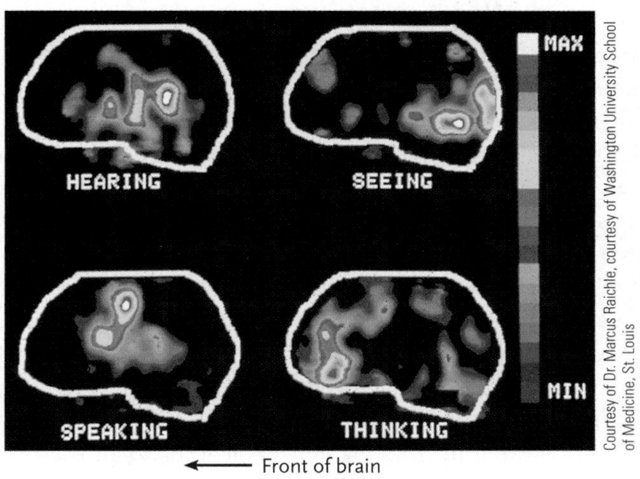
MAX
MIN
HEARING
SEEING
SPEAKING
THINKING
Front of brain
Courtesy of Dr. Marcus Raichle, courtesy of Washington University School of Medicine, St. Louis

FIGURE 38.11
PET scans showing regions of the brain active when a person performs specific mental tasks. The colors show the relative activity of the sections, with white being the most active.

In fMRI, a doughnut-shaped magnet surrounding the subject's head creates a magnetic field about 10,000 times greater than that of Earth. When an area of the brain becomes active, the blood flow to that area suddenly increases, bringing with it oxygen carried by hemoglobin in the red blood cells. This oxygen is released to supply oxygen to the tissues, producing deoxygenated hemoglobin. An fMRI scanner detects this deoxygenation of hemoglobin as oxygen is used by the neurons because oxygenated and deoxygenated hemoglobin have different properties—the oxygenated form slightly repels a magnetic field, whereas the deoxygenated form becomes slightly magnetic when a magnetic field is applied. The chapter opener image combines an fMRI result with a PET scan to show the brain activity associated with reading aloud.

SENSORY REGIONS OF THE CEREBRAL CORTEX Areas that receive and integrate sensory information are distributed over the cerebral cortex. In each hemisphere, the **primary somatosensory area,** which registers information on touch, pain, temperature, and pressure, runs in a band across the parietal lobes of the brain (see Figure 38.10). Experimental stimulation of this band in one hemisphere causes prickling or tingling sensations in specific parts on the opposite side of the body, beginning with the toes at the top of each hemisphere and running through the legs, trunk, arms, and hands, to the head **(Figure 38.12).**

Other sensory regions of the cerebral cortex have been identified with hearing, vision, smell, and taste (see Figure 38.10). Regions of the temporal lobes on both sides of the brain receive auditory inputs from the ears, while inputs from the eyes are processed in the primary visual cortex in both occipital lobes. Olfactory input from the nose is processed in the olfactory bulbs, located on the ventral side of the temporal lobes. Regions in the parietal lobes receive inputs from taste receptors on the tongue and other locations in the mouth.

MOTOR REGIONS OF THE CEREBRAL CORTEX The **primary motor area** of the cerebral cortex runs in a band just in front of the primary somatosensory area (see Figure 38.10). Experimental stimulation of points along this band in one hemisphere causes movement of specific body parts on the opposite side of the body,

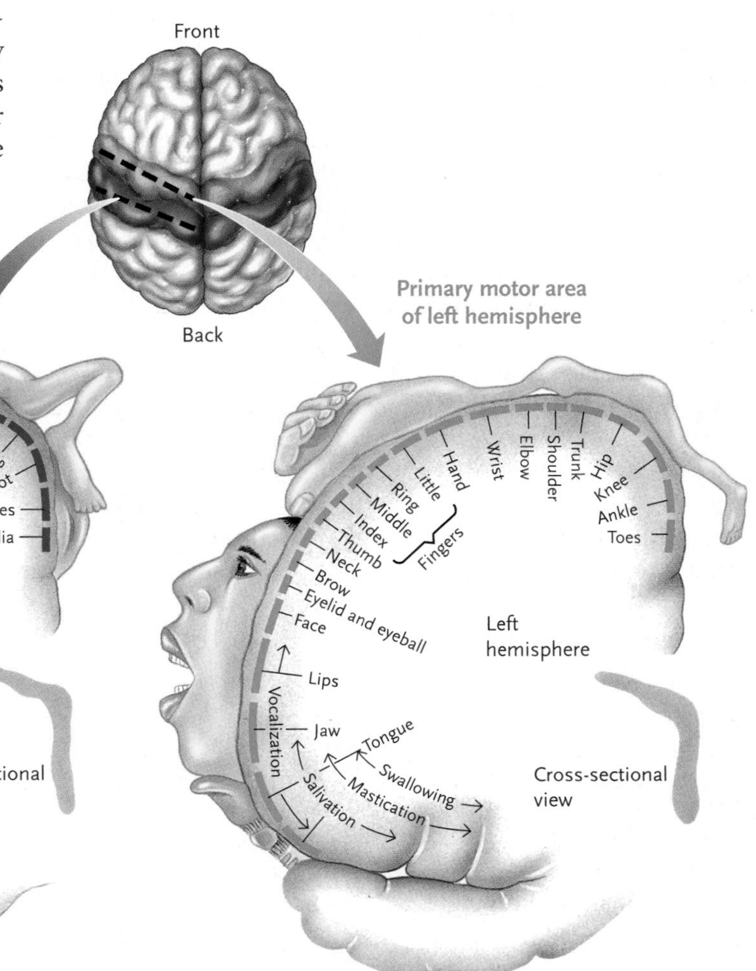
Front
Back
Primary somatosensory area of left hemisphere
Primary motor area of left hemisphere
Wrist
Hand
Forearm
Elbow
Arm
Shoulder
Head
Neck
Trunk
Hip
Leg
Foot
Toes
Genitalia
Little
Ring
Middle
Index
Thumb
Fingers
Eye
Nose
Face
Upper lip
Lips
Lower lip
Teeth, gums, and jaw
Tongue
Pharynx
Intraabdominal organs
Left hemisphere
Cross-sectional view
Hand
Wrist
Elbow
Shoulder
Trunk
Hip
Knee
Ankle
Toes
Little
Ring
Middle
Index
Thumb
Fingers
Neck
Brow
Eyelid and eyeball
Face
Lips
Vocalization
Jaw
Tongue
Swallowing
Mastication
Salivation
Left hemisphere
Cross-sectional view

FIGURE 38.12
The primary somatosensory and motor areas of the cerebrum. The distorted representations of the human body show the relative areas of the sensory and motor cortex devoted to different body regions.

corresponding generally to the parts registering in the primary somatosensory area at the same level (see Figure 38.11). Other areas that integrate and refine motor control are located nearby.

In both the primary somatosensory and primary motor areas, some body parts, such as the lips and fingers, are represented by large regions, and others, such as the arms and legs, are represented by relatively small regions. As shown in Figure 38.12, the relative sizes produce a distorted image of the human body that is quite different from the actual body proportions. The differences are reflected in the precision of touch and movement in structures such as the lips, tongue, and fingers. Such differences are also seen in many other animals, reflecting their adaptations to sensing the environment. For example, Figure 54.13B compares the anatomical and cerebral cortex proportions for a star-nosed mole; note how much of the cerebral cortex is devoted to the fleshy tentacles of the mole's nose, which contain tactile receptors, and its front, digging feet.

ASSOCIATION AREAS The sensory and motor areas of the cerebral cortex are surrounded by **association areas** (see Figure 38.10), which integrate information from the sensory areas, formulate responses, and pass them on to the primary motor area. Two of the most important association areas are *Wernicke's area* and *Broca's area* (see Figure 38.10), which function in spoken and written language. They are usually present on only one side of the brain—in the left hemisphere in 97% of the human population. Comprehension of spoken and written language depends on Wernicke's area, which coordinates inputs from the visual, auditory, and general sensory association areas. Interneuron connections lead from Wernicke's area to Broca's area, which puts together the motor program for coordination of the lips, tongue, jaws, and other structures producing the sounds of speech, and passes the program to the primary motor area. The brain-scan images in Figure 38.11 dramatically illustrate how these brain regions participate as a person performs different linguistic tasks.

People with damage to Wernicke's area have difficulty comprehending spoken and written words, even though their hearing and vision are unimpaired. Although they can speak, their words usually make no sense. People with damage to Broca's area have normal comprehension of written and spoken language, and know what they want to say, but are unable to speak except for a few slow and poorly pronounced words. Often, such people are also unable to write. Other areas of the brain are also involved in language functions.

Some Higher Functions Are Distributed in Both Cerebral Hemispheres; Others Are Concentrated in One Hemisphere

Most of the other higher functions of the human brain—such as abstract thought and reasoning; spatial recognition; mathematical, musical, and artistic ability; and the associations forming the basis of personality—involve the coordinated participation of many regions of the cerebral cortex. Some of these regions are equally distributed in both cerebral hemispheres, and some are more concentrated in one hemisphere.

Among the functions more or less equally distributed between the two hemispheres is the ability to recognize faces. This function is concentrated along the bottom margins of the occipital and temporal lobes (see Figure 38.10). People with damage to these lobes are often unable to recognize even close relatives by sight but can recognize voices immediately. Functions such as consciousness, the sense of time, and recognizing emotions also seem to be distributed in both hemispheres.

Typically some brain functions are more localized in one of the two hemispheres, a phenomenon called **lateralization.** The unequal distribution of these functions was originally worked out in the 1960s by Roger Sperry and Michael S. Gazzaniga of the California Institute of Technology (Sperry received a Nobel Prize for his research in 1981) in subjects who had had their corpus callosum (see Figure 38.6) cut surgically **(Figure 38.13).**

Studies of people with split hemispheres as well as surveys of brain activity by PET and fMRI have confirmed that, for the vast majority of people, the left hemisphere specializes in spoken and written language, abstract reasoning, and precise mathematical calculations. The right hemisphere specializes in nonverbal conceptualizing, intuitive thinking, musical and artistic abilities, and spatial recognition functions such as fitting pieces into a puzzle. The right hemisphere also handles mathematical estimates and approximations that can be made by visual or spatial representations of numbers. Thus the left hemisphere in most people is verbal and mathematical, and the right hemisphere is intuitive, spatial, artistic, and musical.

STUDY BREAK 38.4

1. **What is the blood–brain barrier, and what is its function?**
2. **What is the difference between gray matter and white matter?**
3. **What is the function of the brain stem?**
4. **Distinguish the structure and functions of the cerebellum from those of the cerebral cortex.**

THINK OUTSIDE THE BOOK

The most common cause of brain damage is a cerebrovascular accident (CVA), commonly called a stroke. Use the Internet or research literature to outline what the biological cause of a stroke is, and why a stroke produces different symptoms in different stroke victims.

38.4 Memory, Learning, and Consciousness

We set memory, learning, and consciousness apart from the other functions because they appear to involve coordination of structures from the brain stem to the cerebral cortex. **Memory** is the storage and retrieval of a sensory or motor experience, or a thought. **Learning** involves a change in the response to a stimulus based on information or experiences stored in memory.

FIGURE 38.13 **Experimental Research**

Investigating the Functions of the Cerebral Hemispheres

Question: Do the two cerebral hemispheres have different functions?

Experiment: Roger Sperry and Michael Gazzaniga studied split-brain individuals, in whom the corpus callosum connecting the two cerebral hemispheres had been surgically severed to relieve otherwise uncontrollable epileptic convulsions. In one experiment, they tested how subjects perceived words that were projected onto a screen in front of them.

The retinas of the eyes gather visual information and send signals via the optic nerves to the cerebral hemispheres (Figure A). Light from the *left* half of the visual field reaches light receptors on the *right* sides of the retinas, and parts of the two optic nerves carry signals to the *right* cerebral hemisphere. Light from the *right* half of the visual field reaches light receptors on the *left* sides of the retinas, and signals are sent to the *left* cerebral hemisphere.

The researchers projected words such as COWBOY in such a way that the subjects could see only the left half of the word (COW) with the left eye, and the right half of the word (BOY) with the right eye (Figure B). Sperry asked the subjects to say what word they saw, and he asked them to write the perceived word with the left hand—a hand that was deliberately blocked from the subject's view.

A. **Pathway of visual information from eyes to cerebral hemisphere**

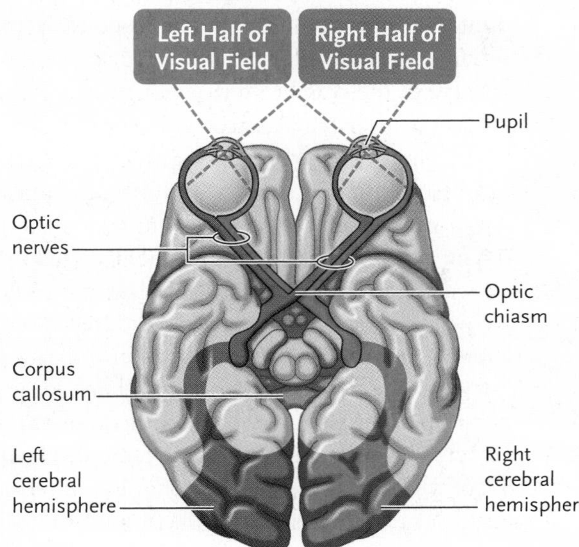
Left Half of Visual Field
Right Half of Visual Field
Pupil
Optic nerves
Optic chiasm
Corpus callosum
Left cerebral hemisphere
Right cerebral hemisphere

B. **Experimental set up—COW seen by right sides of retinas and BOY by left sides**

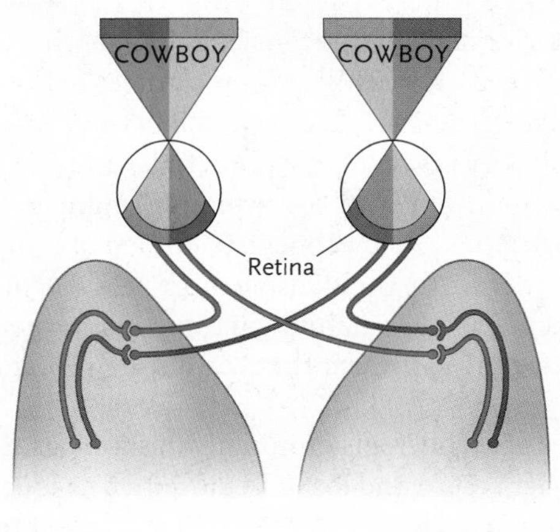
COWBOY
COWBOY
Retina

Results: The split-brain subjects said the word in the right half of the visual field (BOY), but wrote the word in the left half of the visual field (COW) (Figure C).

C. **Results given by split-brain subjects**

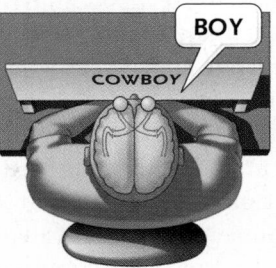

COW
COWBOY
BOY
COWBOY

Conclusions: The studies showed that the left and right hemispheres are specialized in different tasks. The left hemisphere processes language and was able to recognize BOY but received no information about COW. The right hemisphere directs motor activity on the left side of the body and was able to direct the left hand to write COW. However, the subjects could not say what word they wrote. That is, cutting the corpus callosum interrupted communication between the two halves of the cerebrum. In effect, one cerebral hemisphere did not know what the other was doing, and information stored in the memory on one side was not available to the other. In normal individuals, information is shared across the corpus callosum; they would see COWBOY and be able to speak and write the entire word.

Source: M. S. Gazzaniga and R. W. Sperry. 1967. Language after section of the central commissures. *Brain* 90:131–148.

Consciousness may be defined as awareness of ourselves, our identity, and our surroundings, and an understanding of the significance and likely consequences of events that we experience.

Memory Takes Two Forms, Short Term and Long Term

Psychology research and our everyday experience indicate that humans have at least two types of memory. **Short-term memory** stores information for seconds, minutes, or at most an hour or so. **Long-term memory** stores information from days to years or even for life. Short-term memory, but not long-term memory, is usually erased if a person experiences a disruption such as a sudden fright, a blow, a surprise, or an electrical shock. For example, a person knocked unconscious by an accident typically cannot recall the accident itself or the events just before it, but longstanding memories are not usually disturbed.

To explain these differences, investigators propose that short-term memories depend on transient changes in neurons that can be erased relatively easily, such as changes in the membrane potential of interneurons caused by EPSPs and IPSPs (excitatory and inhibitory postsynaptic potentials) and the action of neurotransmitters that lead to reversible changes in ion transport (see Section 37.3). By contrast, storage of long-term memory is considered to involve more or less permanent molecular, biochemical, or structural changes in interneurons, which establish signal pathways that cannot be switched off easily.

All memories probably register initially in short-term form. They are then either erased and lost, or committed to long-term form. The intensity or vividness of an experience, the attention focused on an event, emotional involvement, or the degree of repetition may all contribute to the conversion from short-term to long-term memory.

The storage pathway typically starts with an input at the somatosensory cortex that then flows to the amygdala, which relays information to the limbic system, and to the hippocampus, which sends information to the frontal lobes, a major site of long-term memory storage. People with injuries to the hippocampus cannot remember information for more than a few minutes; long-term memory is limited to information stored before the injury occurred.

How are neurons and neuron pathways permanently altered to create long-term memory? One change that has been much studied is **long-term potentiation:** a long-lasting increase in the strength of synaptic connections in activated neural pathways following brief periods of repeated stimulation. The synapses become increasingly sensitive over time, so that a constant level of presynaptic stimulation is converted into a larger postsynaptic output that can last hours, weeks, months, or years. (*Insights from the Molecular Revolution* describes experiments investigating the basis of long-term potentiation in neurons of the hippocampus.) Other changes consistently noted as part of long-term memory include more or less permanent alterations in the number and the area of synaptic connections between neurons, in the number and branches of dendrites, and in gene transcription and protein synthesis in interneurons.

Experiments have shown that protein synthesis is critical to long-term memory storage in animals as varied as *Drosophilia* and rats. For example, goldfish were trained to avoid an electrical shock by swimming to one end of an aquarium when a light was turned on. The fish could remember the training for about a month under normal conditions; if exposed to a protein synthesis inhibitor while being trained, they forgot the training within a day.

Learning Involves Combining Past and Present Experiences to Modify Responses

As with memory, all animals appear to be capable of learning to some degree. Learning involves three sequential mechanisms: (1) storing memories, (2) scanning memories when a stimulus is encountered, and (3) modifying the response to the stimulus in accordance with the information stored as memory.

One of the simplest forms of memory is **sensitization**—increased responsiveness to mild stimuli after experiencing a strong stimulus. The process was nicely illustrated by Eric Kandel of Columbia University and his associates in experiments with a shell-less marine snail known as the Pacific sea hare, *Aplysia californica,* which is frequently used in research involving reflex behavior, memory, and learning. Many of its neuron circuits have been completely worked out, allowing investigators to follow the reactions of each neuron active in pathways such as learning. The first time the researchers administered a single sharp tap to the siphon (which admits water to the gills), the slug retracted its gills by a reflex movement. However, at the next touch, whether hard or gentle, the siphon retracted much more quickly and vigorously. Sensitization in *Aplysia* has been shown to involve changes in synapses, which become more reactive when more serotonin is released by action potentials. Kandel received a Nobel Prize in 2000 for his research.

Learning skills or procedures, such as tying one's shoes, juggling, or playing a musical instrument, involve additional regions of the brain, particularly the cerebellum, where motor activity is coordinated. As we learn such skills, the process gradually becomes automated so that we do not think consciously about each step. (Learning and its relationship to animal behavior are considered further in Chapter 54.)

Consciousness Involves Different States of Awareness

The spectrum of human consciousness ranges from alert wakefulness to daydreaming, dozing, and sleep. Even during sleep there is some degree of awareness, because sleepers can respond to stimuli and waken, unlike someone who is unconscious. Moving between the states of consciousness has been found to involve changes in neural activity over the entire surface of the telencephalon. These changes can be seen using an *electroencephalogram* (EEG), which records voltage changes detected by electrodes placed on the scalp.

INSIGHTS FROM THE Molecular Revolution

Knockout Mice with a Bad Memory

Long-term potentiation (LTP) in neurons of the hippocampus is thought to be central to the conversion of short-term to long-term memory. The neurotransmitter glutamate is most often involved in the process. One of the receptors that binds glutamate in postsynaptic membranes of the hippocampus is the *NMDA (N-methyl-D-aspartate) receptor.* A prolonged series of stimuli causes glutamate to bind to the receptor, which opens an ion channel that forms part of the receptor's structure. Among other effects, Ca^{2+} ions flowing inward through the channel activate a protein kinase in the cytoplasm called *CaMKII.* (Recall that protein kinases are enzymes that, when activated, add phosphate groups to target proteins; addition of the phosphate groups increases or reduces activity of the proteins.)

One of the target proteins for the activated CaMKII is itself. After it adds phosphate groups to its own structure, CaMKII no longer needs to be activated by calcium—it remains turned on at elevated levels. Its more or less permanent activation makes the neuron more sensitive to incoming signals and increases the number of the signals it sends to other neurons. Researchers hypothesize that this change is a major contributor to LTP in the neuron. Among the supporting evidence is the observation that chemical blockage of the NMDA receptor impairs spatial learning and long-term memory in mice.

Research Question

Does activation of CaMKII contribute to LTP in neurons of the hippocampus?

Experiment

A research group at the Massachusetts Institute of Technology led by a Nobel Prize–winning scientist, Susumu Tonegawa, applied molecular techniques to answer this question. Tonegawa's group created a strain of "knockout" mice (see Section 18.2), from which the gene encoding the NMDA receptor has been deleted. The knockout mice lack NMDA receptors in the hippocampus. The researchers then compared normal and knockout mice in two ways.

1. To test for LTP in hippocampal neurons, the investigators dissected out brain slices and used microelectrodes to repeatedly stimulate the neurons.
2. To test for effects on spatial learning and long-term memory, the researchers placed the knockout mice in a pool in which they had to swim until they could find a submerged platform to rest on.

Results

The results are shown in the table below:

Conclusion

The data show that the lack of NMDA receptors in the hippocampus is sufficient to produce significant memory defects. The results support the hypothesis that the NMDA receptor is involved in the conversion from short-term to long-term memory, and that the hippocampus has a central role in this activity. The novel approach used by the Tonegawa group provided a new, molecular research method by which to analyze higher brain function.

Sources: J. Z. Tsien et al. 1996. Subregion- and cell type-restricted gene knockout in mouse brain. *Cell* 87:1317–1326; J. Z. Tsien et al. 1996. The essential role of hippocampal CA1 NMDA receptor-dependent synaptic plasticity on spatial memory. *Cell* 87:1327–1338.

	Effect of repeated stimulation of hippocampal neurons	Spatial learning and long-term memory
Normal mouse with NMDA receptors	Long-term potentiation	Normal—easily found platform in initial trials and readily able to remember its location in later trials
Knockout mouse—no NMDA receptors	No long-term potentiation	Slower than normal mice in finding platform in initial trials and unable to remember its location in later trials

When an individual is fully awake, the EEG records a pattern of rapid, irregular *beta waves* **(Figure 38.14).** With mind at rest and eyes closed, the person's EEG pattern changes to slower and more regular *alpha waves.* As drowsiness and light sleep come on, the wave trains gradually become larger, slower, but again less regular; these slower pulsations are called *theta waves.* During the transition from drowsiness to deep sleep, the EEG pattern shifts to even slower *delta waves.* The heart and breathing rates become slower and the skeletal muscles increasingly relaxed, although the sleeper may still change position and move the arms and legs.

Periodically during deep sleep, the delta wave pattern is replaced by the rapid, irregular beta waves characteristic of the waking state. The person's heartbeat and breathing rate increase, the limbs twitch, and the eyes move rapidly behind the closed eyelids, giving this phase its name of **rapid-eye-movement (REM) sleep.** The REM sleep phase occurs about every 1.5 hours while a healthy adult is sleeping, and lasts for 10 to 15 minutes. Sleepers do most of their dreaming during REM sleep, and most research subjects awakened from REM sleep report they were experiencing vivid dreams.

As mentioned earlier, the reticular activating system controls the sleep-wake cycle. It sends signals to the spinal cord, cerebellum, and cerebral cortex, and receives signals from the same locations. The flow of signals along these circuits determines whether we are awake or asleep.

Many other animals also alternate periods of wakefulness and sleep or inactivity. Although sleep obviously has restorative

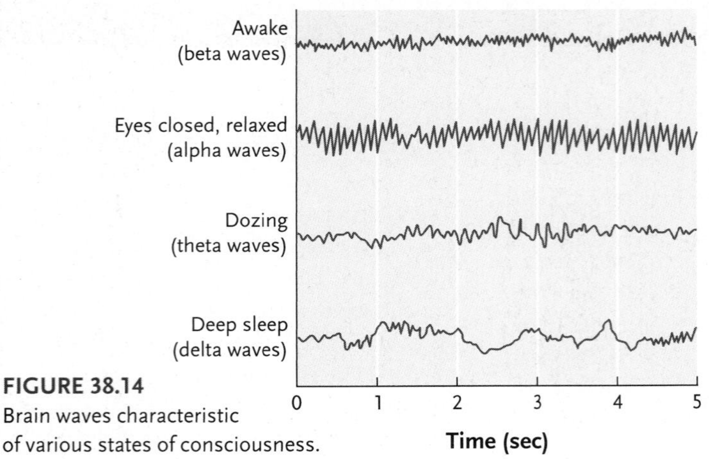
Awake
(beta waves)
Eyes closed, relaxed
(alpha waves)
Dozing
(theta waves)
Deep sleep
(delta waves)
0
1
2
3
4
5
Time (sec)

FIGURE 38.14
Brain waves characteristic of various states of consciousness.

effects on mental and physical functions, the physiological basis of these effects remains unknown.

In the previous chapter we learned about neurons, and in this chapter we have discussed the organization of neurons into nervous systems, as well as the structures of the brain and their functions. In the next chapter we consider the sensory systems that provide input for the brain to process.

STUDY BREAK 38.4 <

An aging person often experiences a progressive decline in cognitive function. This typically begins with short-term memory loss and the inability to learn new information. What parts of the brain might be changing?

UNANSWERED QUESTIONS

Did memory and learning evolve once or multiple times within the animal kingdom?

Memory is highly adaptive because it enables animals to avoid stimuli that previously caused pain or danger, and to seek out environments that lead to rewarding or pleasurable experiences. Of course, some of this ability is hard wired—we don't need to learn that bitter tastes are unpleasant and sweet tastes are pleasant. But the environmental stimuli that predict where to expect sweet versus bitter or hot versus cold stimuli can vary. So it makes sense that animals capable of movement can learn and remember the stimuli that tend to be together in the environment. But did this ability emerge independently in different phyla? If so, we might expect the underlying mechanisms to be quite different in the different phylogenetic groups, even if evolutionary adaptations have converged onto similar memory abilities (see Section 22.3 on convergent evolution). Or did memory evolve one time in a common ancestor before the various animal phyla diverged? In this case, we might expect to find shared underlying mechanisms no matter how distantly related the organisms. We do not know the answer to this question. But given what we do know about memory mechanisms, a common origin seems likely even for animals as distantly related as invertebrates and vertebrates.

One of the most remarkable findings from genetic studies is that many of the same genes and biochemical signaling pathways are utilized for memory formation and storage no matter what species one examines. In several cases, genetic pathways relevant to human cognition also appear to underlie memory in animals as distant from us as fruit flies. Cyclic AMP (cAMP) signaling is one of the best examples; this signaling pathway was first connected to memory in invertebrates of the genus *Aplysia* (sea hares) and *Drosophila* (fruit flies), then in rodents, and more recently in humans. In fact, the conversion from short-term to long-term memory in fruit flies, sea hares, bees, mice, rats, and probably humans involves activation of a gene called the cAMP responsive enhancer binding protein (CREB).

Similarities at the neural circuit level between humans and mice or rats are relatively easy to detect because the overall organization of rodent and human brains is similar. Thus, we can compare the requirements for neural circuits that are likely to be related through evolution. The similarities are harder to establish when we compare invertebrates and vertebrates because their brains are so different in size and structure. But even here, there are some hints that the way neural circuits store memories is at least analogous. A good example comes from studies of the neural circuits involved in conversion of short-term to long-term memory. An emerging theme is that different neural circuits may be required for short-term and long-term memory. Initially, this "dissection" of neural circuit function came from precise surgical lesion of brain regions in rodents and chickens. In certain cases, these lesions caused disruptions of either short-term or long-term memory, but not both. Similar effects are seen in people with localized brain damage.

More modern genetic approaches now permit disruptions that are specific not only to brain regions, but to individual neuronal cell types within a region. In our lab, we recently used this approach to determine which neurons are involved in short-term and long-term memory formation in fruit flies. We used a Pavlovian learning test in which the flies learn that a specific odor predicts that they will receive an electric shock. We then used precise genetic manipulations to test which neuron cell types require cAMP signaling for either short-term or long-term memory. We focused on a brain region called the mushroom bodies because previous work already had demonstrated that this is an olfactory learning center in insects. To our surprise, we found that two different neuronal cell types are required for short-term versus long-term memory.

We do not know why brains have evolved specialized circuits for short-term and long-term storage of the same memory. Perhaps it reflects an important feature for storing information within a neural network. The question is whether this feature emerged independently in different phyla or once in a common ancestor.

Think Critically

You come across two related species of beetles and you observe their behaviors. One beetle species seems always to feed on a single food source from a single plant species; we call it a "specialist." The second beetle species forages widely across many different food sources; it is a "generalist." You decide to study their learning abilities in the lab, and you design a fairly artificial learning task. You try to teach these animals that a pure chemical odor predicts an electric shock. If you had to guess (and you do), which beetle species will exhibit better learning in your experiment? Why?

Josh Dubnau

Josh Dubnau is an associate professor at Cold Spring Harbor Laboratory where his research focuses on memory in fruit flies. One theme in Dr. Dubnau's work is an attempt to use both reductionist methods and more holistic observations at the organismal level. You can learn more about Dubnau's research by visiting http://dubnaulab.cshl.edu/.

REVIEW KEY CONCEPTS

Go to **CENGAGENOW** at www.cengage.com/login to access quizzing, animations, exercises, articles, and personalized homework help.

38.1 Invertebrate and Vertebrate Nervous Systems Compared

- The simplest nervous systems are the nerve nets of cnidarians. Echinoderms have modified nerve nets, with some neurons grouped into nerves (Figure 38.1A and B).
- Flatworms, arthropods, and mollusks have a simple central nervous system (CNS), consisting of ganglia in the head region (a brain), and a peripheral nervous system (PNS), consisting of nerves from the CNS to the rest of the body (Figure 38.1C–E).
- In vertebrates, the CNS consists of a large brain located in the head and a hollow spinal cord, and the PNS consists of all the nerves and ganglia connecting the CNS to the rest of the body (Figure 38.1F).
- In the vertebrate embryo, the anterior end of the hollow neural tube develops into the brain, and the rest develops into the spinal cord. The embryonic brain enlarges into the forebrain, midbrain, and hindbrain, which develop into the adult structures (Figure 38.2).

Animation: Comparisons of animal nervous systems

Animation: Bilateral nervous systems

Animation: Regions of the vertebrate brain

Animation: Human brain development

38.2 The Peripheral Nervous System

- Afferent neurons in the PNS conduct signals to the CNS, and signals from the CNS travel via efferent neurons to the effectors—muscles and glands—that carry out responses (Figure 38.3).
- The somatic system of the PNS controls the skeletal muscles, producing voluntary body movements as well as involuntary muscle contractions that maintain balance, posture, and muscle tone.
- The autonomic system of the PNS, which controls involuntary functions, is organized into the sympathetic division and the parasympathetic division. The pathways of the autonomic system include two neurons (Figures 38.4 and 38.5).

Animation: Vertebrate nervous system divisions

Animation: Autonomic nerves

38.3 The Central Nervous System and Its Functions

- The brain and the spinal cord are surrounded and protected by the meninges. Cerebrospinal fluid provides nutrients and cushions the CNS. A blood–brain barrier allows only selected substances to enter the cerebrospinal fluid.
- Each adult brain structure contains gray matter and white matter. Functionally, the brain integrates sensory information and generates responses. The cerebrum is divided into right and left cerebral hemispheres, which are connected by a thick band of nerve fibers, the corpus callosum. Each hemisphere consists of the cerebral cortex, a thin layer of gray matter covering a thick core of white matter. Other collections of gray matter, the basal nuclei, are deep in the telencephalon (Figure 38.6).
- The spinal cord carries signals between the brain and the PNS. Its neuron circuits control reflex muscular movements and some autonomic reflexes (Figure 38.7).
- The medulla, pons, and midbrain form the brain stem, which connects the cerebrum, thalamus, and hypothalamus with the spinal cord.
- Gray-matter centers in the pons and medulla control involuntary functions. Centers in the midbrain coordinate responses to visual and auditory sensory inputs.
- The reticular formation receives sensory inputs from all parts of the body and sends outputs to the cerebral cortex that help maintain balance, posture, and muscle tone. It also regulates states of wakefulness and sleep (Figure 38.8).
- The cerebellum integrates sensory inputs on the positions of muscles and joints, along with visual and auditory information, to coordinate body movements.
- The telencephalon's subcortical gray-matter centers control many functions. The thalamus receives, filters, and relays sensory and motor information to and from regions of the cerebral cortex. The hypothalamus regulates basic homeostatic functions of the body and contributes to the endocrine control of body functions. The basal nuclei affect the planning and fine-tuning of body movements (Figure 38.9).
- The limbic system includes parts of the thalamus, hypothalamus, and basal nuclei, as well as the amygdala and hippocampus. It controls emotions and influences the basic body functions controlled by the hypothalamus and brain stem (Figure 38.9).
- The primary somatosensory areas of the cerebral cortex register incoming information on touch, pain, temperature, and pressure from all parts of the body. In general, the right cerebral hemisphere receives sensory information from the left side of the body and vice versa (Figures 38.10 and 38.12).
- The primary motor areas control voluntary movements of skeletal muscles (Figures 38.10 and 38.12).
- The association areas integrate sensory information and formulate responses that are passed on to the primary motor areas. Wernicke's area integrates visual, auditory, and other sensory information into the comprehension of language; Broca's area coordinates movements of the lips, tongue, jaws, and other structures to produce the sounds of speech (Figure 38.10).
- Long-term memory and consciousness are equally distributed between the two cerebral hemispheres. Spoken and written language, abstract reasoning, and precise mathematical calculations are left hemisphere functions; nonverbal conceptualizing, mathematical estimation, intuitive thinking, spatial recognition, and artistic and musical abilities are right hemisphere functions (Figure 38.13).

Animation: Organization of the spinal cord

Animation: Sagittal view of a human brain

Animation: Receiving and integrating areas

Animation: Primary motor cortex

38.4 Memory, Learning, and Consciousness

- Memory is the storage and retrieval of a sensory or motor experience or a thought. Short-term memory involves temporary storage of information, whereas long-term memory is essentially permanent.
- Learning involves modification of a response through comparisons made with information or experiences that are stored in memory.
- Consciousness is the awareness of ourselves, our identity, and our surroundings. It varies through states from full alertness to sleep and is controlled by the reticular activating system (Figure 38.14).

Animation: Structures involved in memory

UNDERSTAND AND APPLY

Test Your Knowledge

1. Ganglia first became enlarged and fused into a lobed brain in the evolution of:
   a. vertebrates.
   b. annelids.
   c. flatworms.
   d. cephalopods.
   e. mammals.
2. The metencephalon develops into the:
   a. spinal cord.
   b. cerebellum.
   c. mesencephalon.
   d. medulla oblongata.
   e. cerebrum.
3. The autonomic nervous system is subdivided into:
   a. afferent and efferent systems.
   b. sympathetic and parasympathetic divisions.
   c. skeletal and smooth muscle innervations.
   d. voluntary and involuntary controls.
   e. peripheral and central systems.
4. People with severe insect-sting allergies carry an Epipen, an auto-injector containing medication that they can use in an emergency. The medication causes smooth muscles in the lung passages to relax so they can breathe, but causes their hearts to pound rapidly. This is an example of stimulation of the:
   a. parasympathetic system.
   b. sympathetic system.
   c. somatic nervous system.
   d. limbic system.
   e. voluntary system.
5. Which of the following statements about the blood–brain barrier is *incorrect*?
   a. It is formed of capillary walls that are composed of tight junctions.
   b. It transports glucose to brain cells by means of transport proteins.
   c. It allows alcohol to pass through its lipid bilayer.
   d. Oxygen can move through the lipid bilayer.
   e. It reduces blood supply to brain cells compared with other body cells.
6. Which one of the following structures participates in a reflex?
   a. the gray matter of the brain
   b. the white matter of the brain
   c. the gray matter of the spinal cord
   d. an interneuron that stimulates an afferent neuron
   e. an interneuron that inhibits an afferent neuron
7. A segment of the brain stem that coordinates spinal reflexes with higher brain centers and regulates breathing and wakefulness is the:
   a. reticular formation.
   b. white matter of the pons.
   c. white matter of the medulla.
   d. hypothalamus.
   e. cerebellum.
8. Cushioning and nourishing the brain and spinal cord and filling the ventricles of the brain is (are):
   a. meninges.
   b. myelin.
   c. cerebrospinal fluid.
   d. ganglia.
   e. glucose.
9. Which structure and function are correctly paired below?
   a. thalamus: relays emotion signals through the limbic system
   b. basal nuclei: relay inputs from odor receptors to the cerebrum
   c. hypothalamus: releases hormones; sets up daily rhythms
   d. amygdala: relays sensory information to the cerebrum
   e. olfactory bulbs: moderate motor centers in the cerebrum
10. A patient had a tumor in Wernicke's area. It was initially diagnosed when the patient could not:
    a. understand the morning newspaper.
    b. hear a child crying.
    c. see the traffic light turn red.
    d. speak.
    e. feel if the car heater was on.

Discuss the Concepts

1. Meningitis is an inflammation of the meninges, the membranes that cover the brain and spinal cord. Diagnosis involves using a needle to obtain a sample of cerebrospinal fluid to analyze for signs of infection. Why analyze this fluid and not blood?
2. An accident victim arrives at the emergency room with severe damage to the reticular formation. Based on information in this chapter, describe some of the symptoms that the examining physician might discover.
3. In the 1930s and 1940s prefrontal lobotomy, in which neural connections in the frontal lobes of both cerebral hemispheres were severed, was used to treat behavioral conditions such as extreme anxiety and rebelliousness. Although the procedure calmed patients, it had side effects such as apathy and a seriously disrupted personality. In view of the information presented in this chapter, why do you think the operation had these effects?

Design an Experiment

How would you demonstrate, using mice, that gene activity in the brain is altered by aging?

Interpret the Data

Animal studies are often used to assess the effects of prenatal exposure to drugs. For example, Jack Lipton used rats to study the behavioral effect of prenatal exposure to MDMA (3,4-methylenedioxymethamphetamine), the active ingredient in the illicit drug ecstasy. He injected female rats with either MDMA or saline solution when they were 14 to 20 days pregnant, and their offspring's brains were forming. When those offspring were 21 days old, Lipton tested their response to a new environment. He placed each young rat in a new cage and used a photobeam system to record how much each rat moved around before settling down. The figure shows his results.

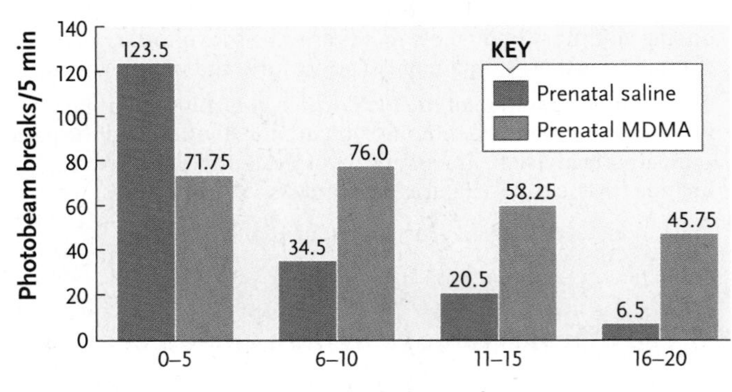
Photobeam breaks/5 min
140
120
100
80
60
40
20
0
123.5
71.75
34.5
76.0
20.5
58.25
6.5
45.75
0–5
6–10
11–15
16–20
5-min intervals
KEY
Prenatal saline
Prenatal MDMA

1. Which rats moved most (caused the most photobeam breaks) during the first 5 min in a new cage, those prenatally exposed to MDMA or the controls?
2. How many photobeam breaks did the MDMA-exposed rats make during their second 5 min in the new cage?
3. Which rats moved most during the last 5 min of the study?
4. Does this study support the hypothesis that exposure to MDMA affects a developing rat's brain?

Source: J. B. Koprich et al. 2003. Prenatal 3,4-methylenedioxymethamphetamine (ecstasy) alters exploratory behavior, reduces monoamine metabolism, and increases forebrain tyrosine hydroxylase fiber density of juvenile rats. *Neurotoxicology and Teratology* 25:509–517.

Apply Evolutionary Thinking

How do paleontologists contribute to our understanding of the evolution of the brain?

A greater horseshoe bat *(Rhinolophus ferrumequinum)* hunting a moth. The bat uses its sensory system to pursue prey, and the moth uses its sensory system in attempting to evade capture.

39

STUDY OUTLINE

Sensory Systems

Why It Matters. . . An insectivorous bat leaves its cave to look for food. As it flies, the bat emits a steady stream of ultrasonic clicking noises. Receptors in the bat's ears detect echoes of the clicks bouncing off objects in the environment and send signals to the brain, where they are integrated into a sound map that the animal uses to navigate through the complex environment. This ability, called **echolocation,** is so keenly developed that a bat can detect and avoid a thin wire in the dark.

Besides recognizing obstacles, the bat's auditory system is keenly tuned to the distinctive pattern of echoes from the fluttering wings of its favorite food, a moth. Although the slow-flying moth would seem doomed to become a meal for the foraging bat, natural selection has provided some species of moths with an astoundingly sensitive auditory sense as well. On each side of its abdomen is an "ear," a thin membrane that resonates at the frequencies of the clicks emitted by the bat. The moth's ears register the clicks while the bat is still about 30 m away and initiate a response that turns its flight path directly away from the source of the clicks.

In spite of the moth's evasive turns, if the bat approaches within about 6 m of it, echoes from the moth begin to register in the bat's auditory system, and the bat increases the frequency of its clicks, enabling it to pinpoint the moth's position.

The moth has not exhausted its evasive tactics, however. As the bat closes in, the increased frequency of the clicks sets off another programmed response that alters the moth's flight into sudden loops and turns, ending with a closed-wing, vertical fall toward the ground. After dropping a few feet, the moth resumes its fluttering flight and may again be detected by the bat, and so it goes.

Echolocation is not confined to bats. Porpoises and dolphins use echolocation to locate food fishes in murky waters, and whales use echolocation to keep track of the sea bottom and rocky obstacles.

Two bird species, the oilbird and the cave swiftlet, use echolocation to avoid obstacles and find their nests in dark caves.

Natural selection has produced highly adaptive sensory receptors in all animals. These systems, the subject of this chapter, provide animals with a steady stream of information about their internal and external environments. After integrating the information in the central nervous system (CNS), animals respond in ways that enable them to survive and reproduce.

We begin this chapter with a survey of animal sensory systems and the ways in which they work. Then we examine several individual receptor types and their characteristics. <

39.1 Overview of Sensory Receptors and Pathways

A **stimulus** is a change detected by the body. A variety of energy forms are stimuli, including heat, light, sound, and pressure. Stimuli are detected by **sensory receptors.** Sensory receptors associated with eyes, ears, skin, and other surface organs detect stimuli from the external environment. Sensory receptors associated with internal organs detect stimuli arising in the body interior. The receptors occur in three structural forms **(Figure 39.1).** Two of the forms involve peripheral endings of an afferent neuron, while the third form is a specialized cell that synapses with an afferent neuron.

Each sensory receptor has a defined **receptive field,** a region surrounding the receptor within which the receptor responds to a stimulus. Sensory receptors respond to stimuli in their receptive fields by undergoing a change in membrane potential. The change, called a **receptor potential,** varies in magnitude with the magnitude of the stimulus; thus, it is a graded potential (see Section 37.4). In most receptors, the change is caused by changes in the rate at which channels conduct positive ions such as Na^+, K^+, or Ca^{2+} across the plasma membrane. The conversion of a stimulus into a receptor potential is called **sensory transduction.** If the receptor potential is large enough, it will trigger an action potential in the afferent neuron that travels along the axon into the interneuron networks of the CNS. These interneurons integrate the sensory stimuli, and the brain formulates a compensating response (see Section 38.3).

Five Basic Types of Receptors Are Common to Almost All Animals

Many sensory receptors are positioned individually in body tissues. Others are part of complex sensory organs, such as the eyes or ears, which are specialized for reception of physical or chemi-

A. Sensory receptor consisting of free nerve endings—dendrites of an afferent neuron

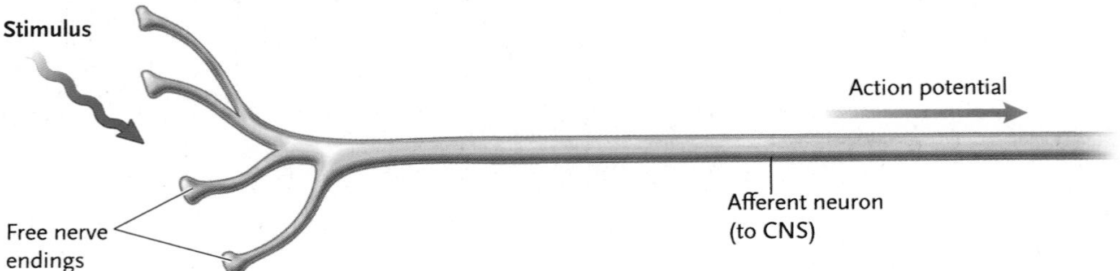
Stimulus
Action potential
Afferent neuron
(to CNS)
Free nerve
endings

In sensory receptors consisting of the dendrites of afferent neurons, a stimulus causes a change in membrane potential that generates action potentials in the axon of the neuron. Examples are pain receptors and some mechanoreceptors.

B. Sense organ—sensory receptor involving nerve endings of an afferent neuron enclosed in a specialized structure

Stimulus
Action potential
Nerve ending

In sensory receptors involving nerve endings enclosed in a specialized structure, a stimulus affecting the structure triggers an action potential in the afferent neuron. Some mechanoreceptors are of this type.

C. Sensory receptor formed by a cell that synapses with an afferent neuron

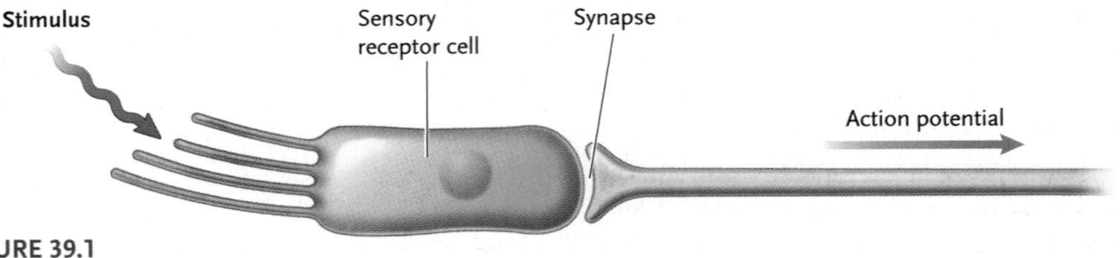
Stimulus
Sensory
receptor cell
Synapse
Action potential

In sensory receptors consisting of separate cells, a stimulus causes a change in membrane potential that releases a neurotransmitter from the cell. The neurotransmitter triggers an action potential in the axon of an afferent neuron to which the sensory receptor cell is synapsed. Examples are photoreceptors, chemoreceptors, and some mechanoreceptors.

FIGURE 39.1
Three forms of sensory receptors.

cal stimuli. Commonly, sensory receptors are classified into five major types, based on the type of stimulus that each detects:

1. **Mechanoreceptors** detect mechanical energy when it deforms membranes. Changes in pressure, body position, or acceleration are detected by mechanoreceptors, for instance. The auditory receptors in the ears are examples of mechanoreceptors.
2. **Photoreceptors** detect the energy of light. In vertebrates, photoreceptors are mostly located in the retina of the eye.
3. **Chemoreceptors** detect specific molecules, or chemical conditions such as acidity. The taste buds on the tongue are examples of chemoreceptors.
4. **Thermoreceptors** detect the flow of heat energy. Receptors of this type are located in the skin.
5. **Nociceptors** detect tissue damage or noxious chemicals; their activity registers as pain. Pain receptors are located in the skin, and also in some internal organs.

In addition to these major types, some animals have receptors that can detect electrical or magnetic fields.

Although humans are traditionally said to have five senses—vision, hearing, taste, smell, and touch—our sensory receptors actually detect more than four times as many kinds of environmental stimuli. Among these are external heat, internal temperature, gravity, acceleration, the positions of muscles and joints, body balance, internal pH, and the internal concentration of substances such as oxygen, carbon dioxide, salts, and glucose.

Afferent Neurons Link Receptors to the CNS

Sensory pathways begin at a sensory receptor and proceed by afferent neurons to a particular location in the CNS. For example, action potentials arising in the retina of the eye travel along the optic nerve to the visual cortex where they are interpreted by the brain as differences in the pattern, color, and intensity of light.

One way in which the intensity and extent of a stimulus is registered is by the frequency (number per unit time) of action potentials traveling along each axon of an afferent pathway. That is, the stronger the stimulus, the more frequently afferent neurons fire action potentials (see Section 37.2). A light touch to the hand, for example, causes action potentials to flow at low frequencies along the axons leading to the primary somatosensory area of the cerebral cortex. As the pressure increases, the frequency of action potentials rises in proportion; in the brain, the increase is interpreted as greater pressure on the hand.

The second way in which the intensity and extent of a stimulus are registered is by the number of afferent neurons that the stimulus activates to generate action potentials in the pathway. The more sensory receptors that are activated, the larger the number of axons that carry information to the brain. A light touch activates a relatively small number of receptors in a small area near the surface of the finger, for example. As the pressure increases, the resulting indentation of the finger's surface increases in area and depth, activating more and different types of receptors. In the appropriate somatosensory area of the brain, the larger number of axons carrying action potentials is interpreted as an increase in the area of the finger being touched.

Many Receptor Systems Reduce Their Response When Stimuli Remain Constant

In many systems, the effect of a stimulus is reduced if it continues at a constant level. The reduction, called **sensory adaptation,** reduces the frequency of action potentials generated in afferent neurons when the intensity of a stimulus remains constant.

For example, when you go to bed, you are initially aware of the touch and pressure of the covers on your skin. Within a few minutes, the sensations lessen or are lost even though your position remains the same. The loss reflects adaptation of mechanoreceptors in your skin. In contrast, nociceptors do not adapt to painful stimuli; the frequency of action potentials remains constant as long as the stimulus is detected.

Sensory adaptation also increases the sensitivity of receptor systems to *changes* in environmental stimuli, which may be more important to survival than keeping track of environmental factors that remain constant. You may have noticed a cat sitting motionless, focused on its prey, a mouse. As long as the environmental stimuli are constant, the cat's position remains fixed. However, if the mouse moves, the cat will respond rapidly and pounce.

Many prey animals take advantage of adaptation in predators as a means for concealment or defense. These animals instinctively become motionless when they sense a predator in their environment, which frequently allows them to remain undetected by the adapted senses of their predator.

Nonadapting receptors, such as those detecting pain, are also essential for survival. Pain signals a potential danger to some part of the body, and the signals are maintained until a response by the animal compensates for the stimulus causing the pain.

Perception of a Stimulus Results from Interpretation of Sensory Input

Perception is the conscious awareness of our external and internal environments derived from the processing of sensory input. That is, the action potentials from sensory receptors are the signals the brain uses to generate an interpretation—the perception—of the external and internal environments. Consider your perception of the world; it is different from reality for several reasons: (1) you lack receptors for particular types of energy, such as X-rays—those energy forms are invisible to you; (2) during processing of sensory input before it reaches the cerebral cortex, some features of the stimuli may be enhanced and others may be decreased or ignored; (3) input is further processed by the cerebral cortex, including comparison of the particular input with other incoming sensory input and with memories of similar situations. Each individual (perhaps with the exception of identical twins) has slightly different perceptions of the world. Certainly, human perception of the world is significantly different from the perceptions of other organisms. Examples include the detection of much higher sound frequencies by dogs, and the "seeing" of ultraviolet light by some species of insects.

STUDY BREAK 39.1

1. Define sensory transduction.
2. Sensory receptors send signals to the CNS in the same way. How are stimuli of different kinds perceived as being different?

39.2 Mechanoreceptors and the Tactile and Spatial Senses

Mechanical stimuli such as touch and movement are detected by mechanoreceptors. The mechanical forces of a stimulus create tension in the plasma membrane of a receptor, which causes ion channels to open, producing a receptor potential. If the receptor potential is large enough, an action potential is triggered in the afferent neuron leading to the CNS. Sensory information from the receptors informs the brain of the body's contact with objects in the environment, provides information on the movement, position, and balance of body parts, and underlies the sense of hearing.

Receptors for Touch and Pressure Occur throughout the Body

In vertebrates, mechanoreceptors detecting touch and pressure are embedded in the skin and other surface tissues, in skeletal muscles, in the walls of blood vessels, and in internal organs. In humans, touch receptors in the skin are concentrated in greatest numbers in the fingertips, lips, and tip of the tongue, giving these regions the greatest sensitivity to mechanical stimuli. In other areas, such as the skin of the back, arms, and legs, the receptors are more widely spaced.

You can compare the receptive fields of touch receptors by pressing two pins lightly against a fingertip and then against the skin of your arm or leg. On your fingertip, the pins can be quite close together—separated by only a millimeter or so—and still be discerned as two separate points. On your arm or leg, they must be nearly 5 cm (almost 2 inches) apart to be distinguished.

Human skin contains several types of touch and pressure receptors **(Figure 39.2)**. Some are free nerve endings, the dendrites of afferent neurons with no specialized structures surrounding them. Free nerve endings wrapped around hair follicles respond when the hair is bent, making you instantly aware, for example, of a breath of air, or a spider on your arm. Other mechanoreceptors, such as Pacinian corpuscles, have structures surrounding the nerve endings that contribute to reception of stimuli.

Proprioceptors Provide Information about Movements and Position of the Body

Mechanoreceptors called **proprioceptors** (*proprius* = one's own) detect stimuli that are used in the CNS to maintain body balance and equilibrium and to monitor the position of the head and limbs. The activity of these receptors allows you to touch the tip of your nose with your eyes closed, for example, or reach and brush away that spider on your arm.

TENSION RECEPTORS IN VERTEBRATES In the muscles and tendons of vertebrates, proprioceptors detect the position and movement of the limbs. The sensory receptors that detect the length of a muscle and its contraction are muscle spindles, bundles of small, specialized muscle cells wrapped with the dendrites of afferent neurons and enclosed in connective tissue **(Figure 39.3)**. When the muscle stretches, the spindle stretches also, stimulating the dendrites and triggering the production of action potentials. The strength of the response of stretch receptors to stimulation depends on how much and how fast the muscle is stretched. The proprioceptors of tendons, called **Golgi tendon organs**, are dendrites that branch within the fibrous connective tissue bundles that make up the tendon (see Figure 39.3). When a muscle contracts, the pull on the tendon tightens the connective tissue bundles, which in turn increases the tension on the bone to which the tendon is attached. The Golgi tendon organ responds to the ten-

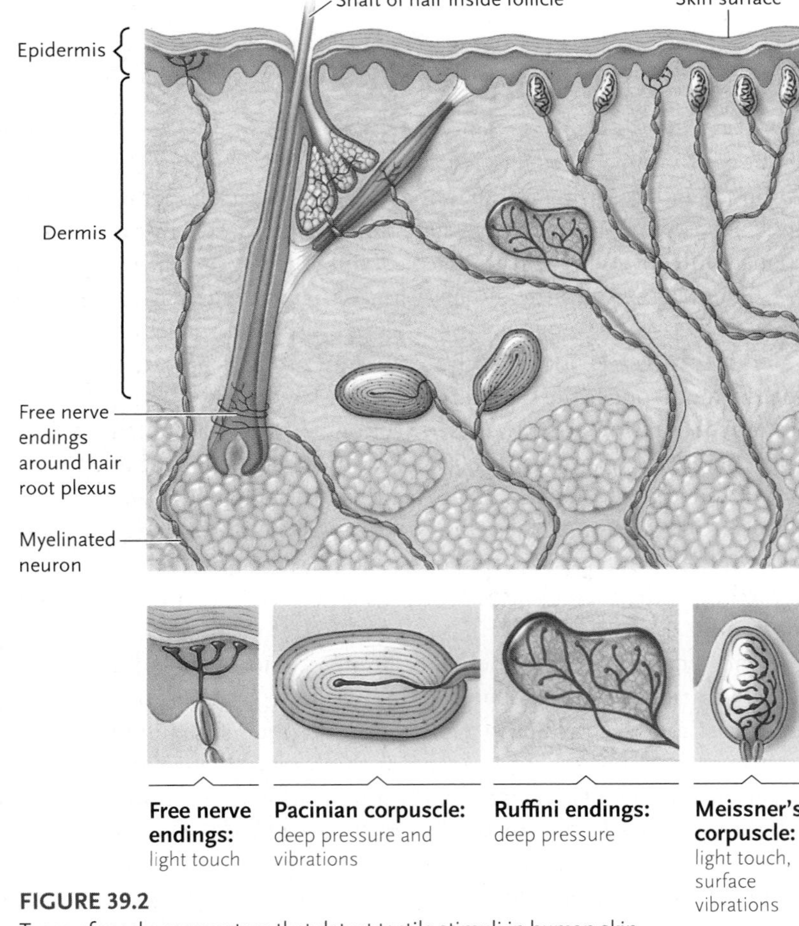
Shaft of hair inside follicle
Skin surface
Epidermis
Dermis
Free nerve endings around hair root plexus
Myelinated neuron
Free nerve endings: light touch
Pacinian corpuscle: deep pressure and vibrations
Ruffini endings: deep pressure
Meissner's corpuscle: light touch, surface vibrations

FIGURE 39.2
Types of mechanoreceptors that detect tactile stimuli in human skin.

sion by producing action potentials in the afferent neuron at a rate related to the amount of tension.

Proprioceptors allow the CNS to monitor the body's position and help keep the body in balance. They also allow muscles to apply constant force under a constant load, and to adjust almost instantly if the load changes. When you hold a cup while someone fills it with coffee, for example, the muscle spindles in your biceps muscle detect the additional stretch as the cup becomes heavier. Signals from the spindles allow you to compensate for the additional weight by increasing the contraction of the muscle, keeping the cup level with no conscious effort on your part. Proprioceptors are typically slow to adapt, so that the body's position and balance are constantly monitored.

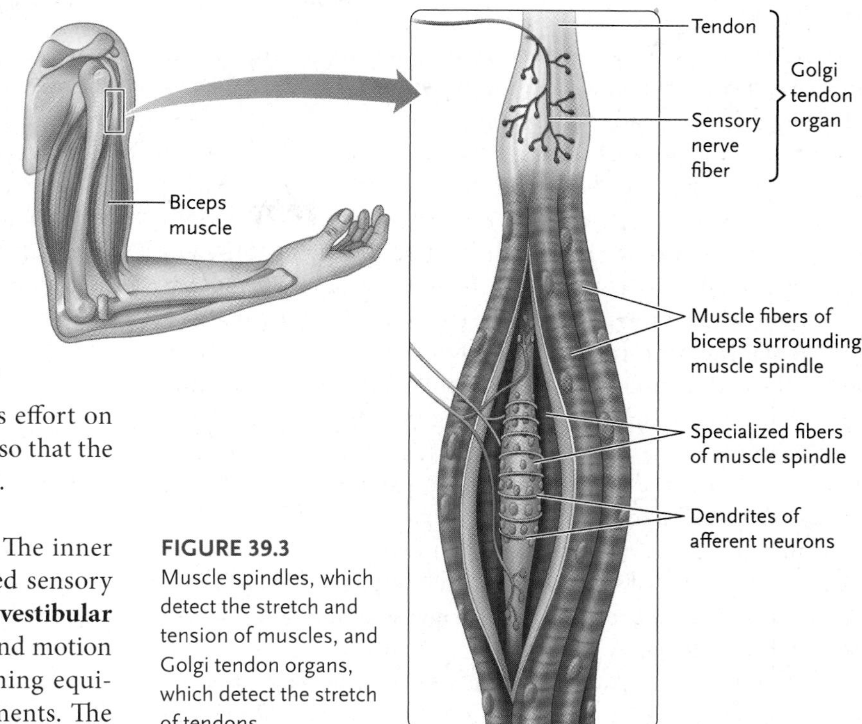

Biceps muscle
Tendon
Sensory nerve fiber
Golgi tendon organ
Muscle fibers of biceps surrounding muscle spindle
Specialized fibers of muscle spindle
Dendrites of afferent neurons

FIGURE 39.3
Muscle spindles, which detect the stretch and tension of muscles, and Golgi tendon organs, which detect the stretch of tendons.

THE VESTIBULAR APPARATUS IN VERTEBRATES The inner ear of most terrestrial vertebrates has two specialized sensory structures, the *vestibular apparatus* and the *cochlea*. The **vestibular apparatus** is responsible for perceiving the position and motion of the head and, therefore, is essential for maintaining equilibrium and for coordinating head and body movements. The cochlea is used in hearing, which we discuss later in this chapter.

The vestibular apparatus **(Figure 39.4)** consists of three **semicircular canals,** the **utricle,** and the **saccule,** which are all

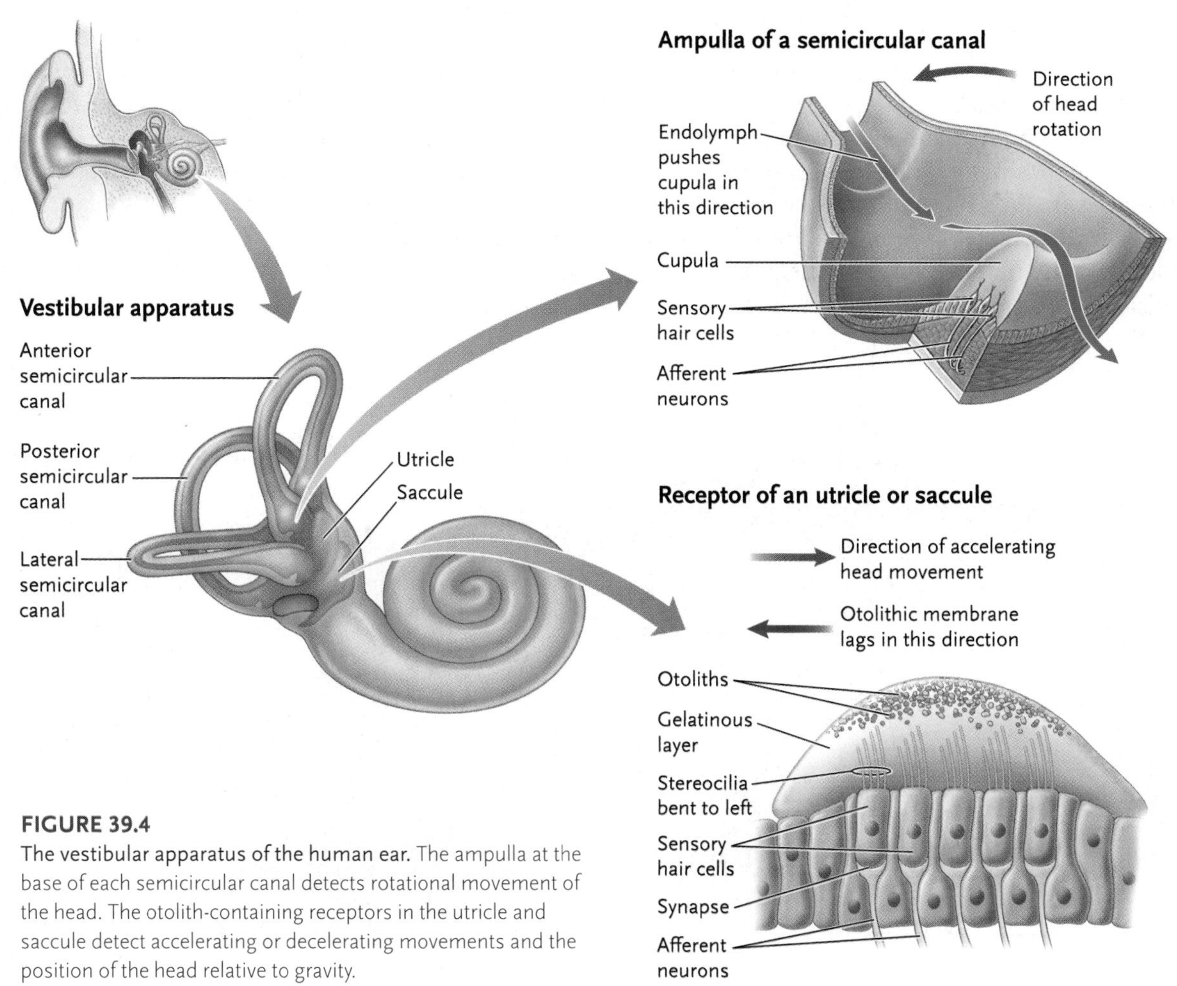

Vestibular apparatus
Anterior semicircular canal
Posterior semicircular canal
Lateral semicircular canal
Utricle
Saccule
Ampulla of a semicircular canal
Direction of head rotation
Endolymph pushes cupula in this direction
Cupula
Sensory hair cells
Afferent neurons
Receptor of an utricle or saccule
Direction of accelerating head movement
Otolithic membrane lags in this direction
Otoliths
Gelatinous layer
Stereocilia bent to left
Sensory hair cells
Synapse
Afferent neurons

FIGURE 39.4
The vestibular apparatus of the human ear. The ampulla at the base of each semicircular canal detects rotational movement of the head. The otolith-containing receptors in the utricle and saccule detect accelerating or decelerating movements and the position of the head relative to gravity.

filled with a fluid called *endolymph*. The semicircular canals, which are positioned at angles corresponding to the three planes of space, detect rotational (spinning) motions. Each canal has a swelling at its base called an *ampulla,* which is topped with sensory hair cells that synapse with afferent neurons. The surface of a hair cell at the other end from the synapse is covered with **stereocilia** (singular, *stereocilium*), which are actually microvilli (cell processes reinforced by bundles of microfilaments). The stereocilia extend into a gelatinous structure, the **cupula** (*cupula* = little cup), which protrudes into the endolymph of the canals.

When the head rotates horizontally, vertically, or diagonally, the endolymph in the semicircular canal corresponding to that direction lags behind, pulling the cupula with it. The displacement of the cupula bends the sensory hair cells. Channels at the tips of the stereocilia open and K^+ enters, causing the sensory hair cell to depolarize. The depolarization opens Ca^{2+} channels in the main part of the hair cell, and entry of Ca^{2+} causes the release of neurotransmitter from the hair cell into the synapse. The neurotransmitter triggers action potentials in the afferent neuron.

The utricle and saccule provide information about the position of the head with respect to gravity (up versus down), as well as changes in the rate of linear movement of the head. They are saclike structures in a bony chamber located between the semicircular canals and the cochlea. Oriented approximately 30° to each other, both utricle and saccule contain sensory hair cells with stereocilia. The hair cells are covered with a gelatinous *otolithic membrane* (which is similar to a cupula) in which **otoliths** (*oto* = ear; *lithos* = stone), small crystals of calcium carbonate, are embedded (see Figure 39.4).

When an animal is upright, the sensory hairs in the utricle are oriented vertically, and those in the saccule are oriented horizontally. When the head is tilted in any direction other than straight up and down, or when there is a change in linear motion of the body, the otolithic membrane of the utricle moves and bends the sensory hairs. Depending on the direction of movement, the hair cells release more or less neurotransmitter, which changes the rate of action potentials in the synapsed afferent neurons. The brain integrates the signals it receives and generates a perception of the movement. The utricle and saccule adapt quickly to the head's motion, decreasing their response when there is no change in the rate and direction of movement. In other words, the body adapts to the new position. For instance, when you move your head to the left, that new position becomes the "norm." Then, if you move your head again, signals from the utricle and saccule tell your brain that your head is moving to a new position.

THE LATERAL LINE SYSTEM IN AMPHIBIANS AND FISHES

Fishes and some aquatic amphibians detect vibrations and currents in the water through a series of mechanoreceptors along the length of the body called the **lateral line system (Figure 39.5).** In fish, the mechanoreceptors, known as *neuromasts,* also provide information about the fish's orientation with respect to gravity and its swimming velocity. Each dome-shaped neuromast has sensory hair cells clustered in its base. Stereocilia on the hair cells extend into a cupula similar to that of the vertebrate vestibular apparatus, which moves with pressure changes in the surrounding water. Movement of the cupula bends the stereocilia, which leads to the triggering of action potentials in associated afferent neurons (as described for the vertebrate vestibular apparatus).

Vibrations detected by the lateral line enable fishes to avoid obstacles, orient in a current, and monitor the presence of other moving objects in the water. The system is also responsible for the ability of schools of fish to move in unison, turning and diving in what appears to be a perfectly synchronized aquatic ballet. In actuality, the movement of each fish creates a pressure wave in the water that is detected by the lateral line systems of other fishes in the school. Schooling fish can still swim in unison even if blinded, but if the nerves leading from the lateral line system to the brain are severed, the ability to school is lost.

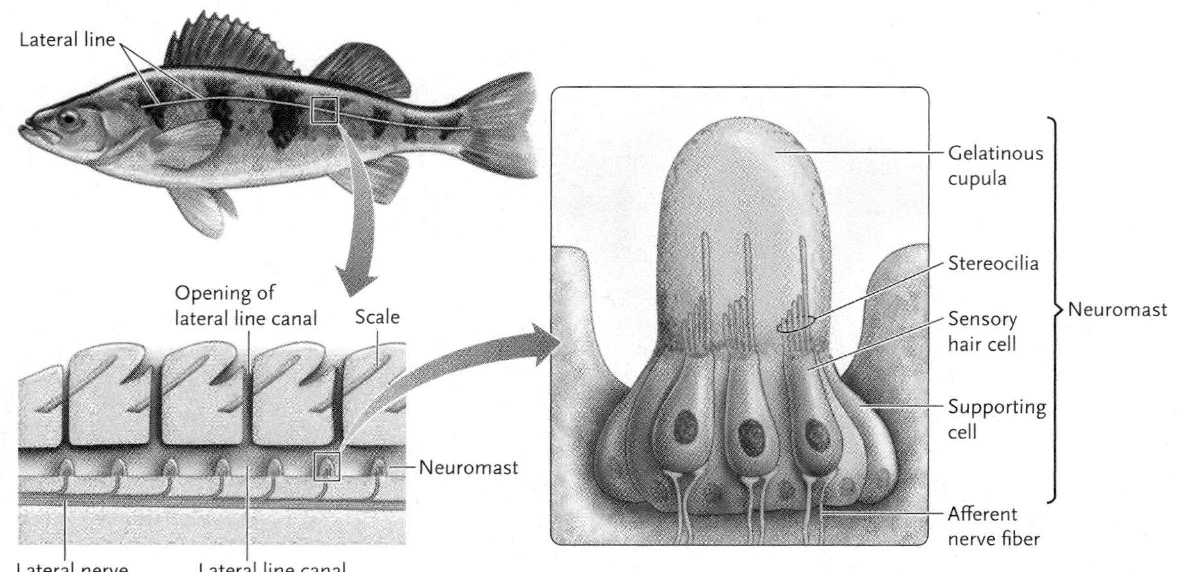

Lateral line
Opening of lateral line canal
Scale
Neuromast
Lateral nerve
Lateral line canal
Gelatinous cupula
Stereocilia
Sensory hair cell
Supporting cell
Afferent nerve fiber
Neuromast

FIGURE 39.5
The lateral line system of fishes. The sensory receptor of the lateral line, the neuromast, has a gelatinous cupula that is pushed and pulled by vibrations and currents transmitted through the lateral line canal. As the cupula moves, the stereocilia of the sensory hair cells are bent, generating action potentials in afferent neurons that lead to the brain.

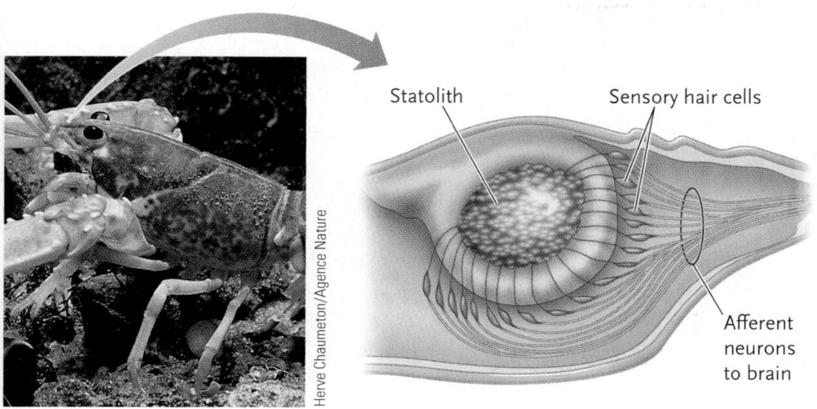
Statolith
Sensory hair cells
Afferent neurons to brain
Herve Chaumeton/Agence Nature

FIGURE 39.6
A statocyst, an invertebrate organ of equilibrium, and its location at the base of an antenna in a lobster. The statoliths inside are usually formed from fused grains of sand, as they are in the lobster, or from calcium carbonate.

STATOCYSTS IN INVERTEBRATES Many aquatic invertebrates, including jellyfishes, some gastropods, and some arthropods, have organs of equilibrium called **statocysts** (*statos* = standing; *kystis* = bladder, pouch). Most statocysts are fluid-filled chambers with walls that contain sensory hair cells enclosing one or more movable stonelike bodies called **statoliths (Figure 39.6)**; statoliths are similar to vertebrate otoliths. When the animal moves, the statoliths lag behind the movement, bending the sensory hairs and triggering action potentials in afferent neurons. In this way, the statocysts signal the brain about the body's position and orientation with respect to gravity.

STUDY BREAK 39.2

1. What is the function of proprioceptors?
2. What properties qualify proprioceptors as mechanoreceptors?

39.3 Mechanoreceptors and Hearing

Sounds are vibrations that travel as waves produced by the alternating compression and rarefaction of the air. The loudness, or *intensity,* of a sound depends on the amplitude (height) of the wave. The *pitch* of a sound—whether a musical tone, for example, is a high note or a low note—depends on the frequency of the waves, measured in hertz (cycles per second). The more cycles per second, the higher the pitch. Some animals, such as the bat in the introduction to this chapter, can hear sounds well above 100,000 hertz. Humans can hear sounds between about 20 and 20,000 hertz, which is why we cannot hear the bat's sonar clicks.

Invertebrates Have Varied Vibration-Detecting Systems

Most invertebrates detect sound and other vibrations through mechanoreceptors in their skin or other surface structures. An earthworm, for example, quickly retracts into its burrow at the smallest vibration of the surrounding earth, even though it has no specialized structures serving as ears. Cephalopods such as squids and octopuses have a system of mechanoreceptors on their head and tentacles, similar to the lateral line of fishes, which detects vibrations in the surrounding water. Many insects have sensory receptors in the form of hairs or bristles that vibrate in response to sound waves, often at particular frequencies.

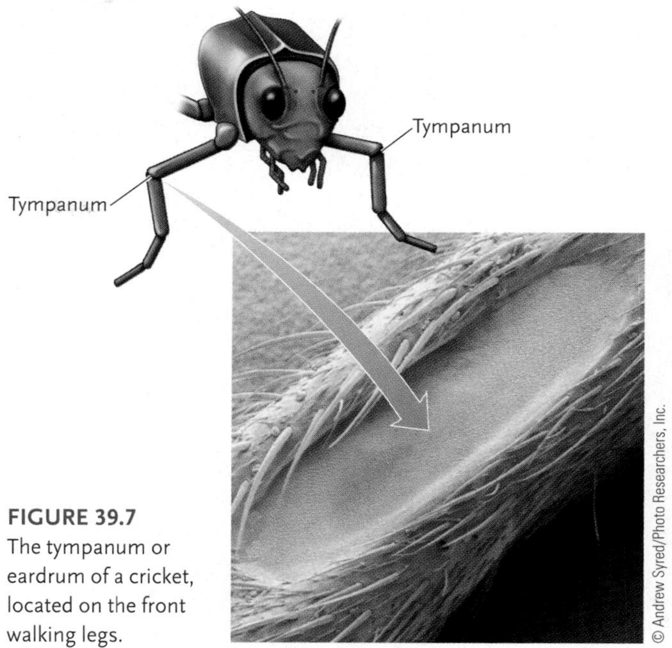
Tympanum
Tympanum
© Andrew Syred/Photo Researchers, Inc.

FIGURE 39.7
The tympanum or eardrum of a cricket, located on the front walking legs.

Some insects, such as moths, grasshoppers, and crickets, have complex auditory organs on either side of the abdomen or on the first pair of walking legs **(Figure 39.7)**. These "ears" consist of a thinned region of the insect's exoskeleton that forms a tympanum (*tympanum* = drum) over a hollow chamber. Sounds reaching the tympanum cause it to vibrate; mechanoreceptors connected to the tympanum translate the vibrations into nerve impulses.

Human Ears Are Representative of the Auditory Structures of Mammals

The auditory structures of terrestrial vertebrates transmit the vibrations of sound waves to sensory hair cells that respond by triggering action potentials. Here, we describe the human ear, which is representative of mammalian auditory structures **(Figure 39.8)**.

The **outer ear** has an external structure, the **pinna** (*pinna* = feather, wing), which concentrates and focuses sound waves. The sound waves enter the auditory canal, which leads from the exterior, and strike a thin sheet of tissue, the **tympanic membrane** (eardrum), which vibrates back and forth in response.

Behind the tympanic membrane is the **middle ear,** an air-filled cavity containing three small, interconnected bones: the **malleus** (hammer), the **incus** (anvil), and the **stapes** (stirrup). The stapes fits within a thin, elastic membrane, the **oval window,** and is held in place by a ligament. The vibrations of the tympanic membrane, transmitted by the malleus and incus, push the stapes back and forth within the oval window. The levering action

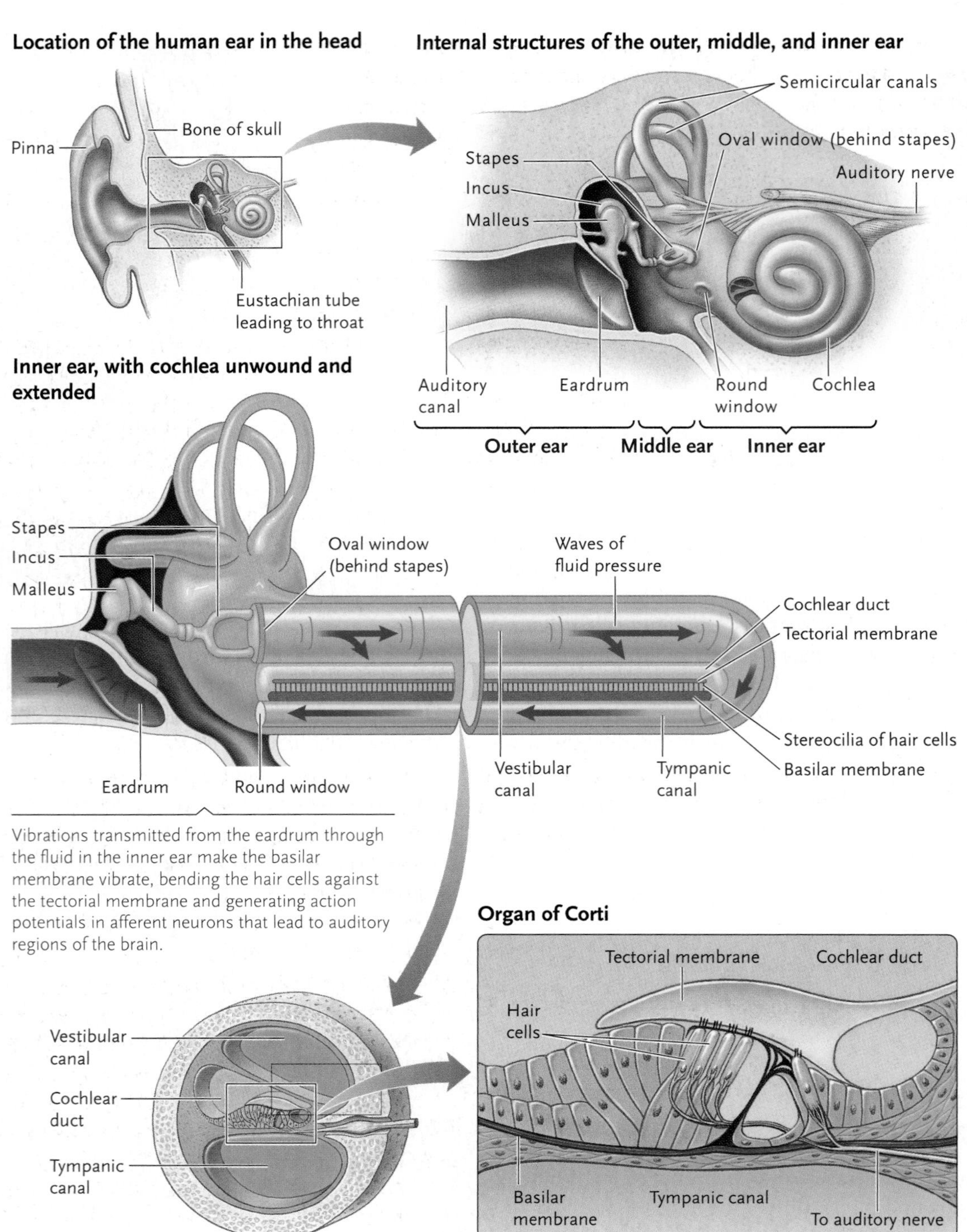
Location of the human ear in the head
Bone of skull
Pinna
Eustachian tube leading to throat
Internal structures of the outer, middle, and inner ear
Semicircular canals
Oval window (behind stapes)
Auditory nerve
Stapes
Incus
Malleus
Auditory canal
Eardrum
Round window
Cochlea
Outer ear
Middle ear
Inner ear
Inner ear, with cochlea unwound and extended
Stapes
Incus
Malleus
Oval window (behind stapes)
Waves of fluid pressure
Cochlear duct
Tectorial membrane
Stereocilia of hair cells
Basilar membrane
Eardrum
Round window
Vestibular canal
Tympanic canal
Vibrations transmitted from the eardrum through the fluid in the inner ear make the basilar membrane vibrate, bending the hair cells against the tectorial membrane and generating action potentials in afferent neurons that lead to auditory regions of the brain.
Organ of Corti
Tectorial membrane
Cochlear duct
Hair cells
Basilar membrane
Tympanic canal
To auditory nerve
Vestibular canal
Cochlear duct
Tympanic canal

FIGURE 39.8 Structures of the human ear.

of the bones, combined with the much larger size of the tympanic membrane compared to the oval window, amplifies the vibrations transmitted to the oval window by more than 20 times.

Inside the oval window is the **inner ear.** It contains several fluid-filled compartments, including the semicircular canals, utricle, saccule, and a spiraled tube, the **cochlea** (*cochlea* = snail shell). The cochlea twists through about two and a half turns; if stretched out flat, it would be about 3.5 cm long. Thin membranes divide the cochlea into three longitudinal chambers, the *vestibular canal* at the top, the *cochlear duct* in the middle, and the *tympanic canal* at the bottom (see Figure 39.8). The vestibular canal and the tympanic canal join at the outer tip of the cochlea, so that the fluid within them is continuous. Within the cochlear duct is the **organ of Corti** (also called **spiral organ**); it contains the sensory hair cells that detect sound vibrations transmitted to the inner ear (see Figure 39.8).

The vibrations of the oval window pass through the fluid in the vestibular canal, make the turn at the end, and travel back through the fluid in the tympanic canal. At the end of the tympanic canal, they are transmitted to the **round window,** a thin membrane that faces the middle ear.

The vibrations traveling through the inner ear cause the basilar membrane to vibrate in response. The **basilar membrane,** which forms part of the floor of the cochlear duct, anchors the sensory hair cells in the organ of Corti. The stereocilia of these cells are embedded in the **tectorial membrane,** which extends the length of the cochlear canal. When the basilar membrane vibrates, the stereocilia of the hair cells are bent back and forth in relation to the stationary tectorial membrane. The back and forth bending of the sensory hair cells alternately opens and closes ion channels in the stereocilia of the cells, causing alternating depolarization and hyperpolarization. Neurotransmitter release from the sensory hair cells triggers action potentials in the associated afferent neurons at the same frequency.

Pitch discrimination, the ability to distinguish between various frequencies of incoming sound stimuli, depends on the shape and properties of the basilar membrane. The basilar membrane is narrow at the oval window end and wide at its outer end. Due to its shape, different sound frequencies cause different regions of the basilar membrane to vibrate maximally: the narrow end vibrates most with high-frequency pitches, and the wide end vibrates most with low-frequency pitches. Wherever the basilar membrane vibrates, the sensory hair cells in that area move back and forth and trigger action potentials. In essence, the sensory hair cells in a particular location in the basilar membrane are tuned to a particular sound frequency.

More than 15,000 hair cells are distributed in small groups along the basilar membrane, enabling a high degree of pitch discrimination. Each group of hairs is connected by synapses to afferent neurons which in turn are bundled together in the *cochlear nerve,* a cranial nerve that carries information through intermediate regions and into the thalamus. From there, the signals are routed to specific regions in the temporal lobe, which integrates the information into the perception of sound at a corresponding pitch and loudness.

Another system protects the tympanic membrane from damage by changes in environmental atmospheric pressure. The system depends on the *Eustachian tube* (also called *auditory tube*), a duct that leads from the air-filled middle ear to the throat (see Figure 39.8). As we swallow or yawn, the tube opens, allowing air to flow into or out of the middle ear to equalize the pressure on both sides of the tympanic membrane. When swelling or congestion due to infections prevents the tube from admitting air, we complain of having stopped-up ears—we can sense that a pressure difference between the outer and middle ear is bulging the tympanic membrane, interfering with the transmission of sounds and causing pain.

STUDY BREAK 39.3

1. **What vibration-detecting systems are found in cephalopods and insects?**
2. **How are sounds of particular frequencies distinguished and "heard" by humans?**

39.4 Photoreceptors and Vision

The great majority of animals have receptors that can detect and respond to light. As animals evolved and became more complex, the complexity of their visual sensory receptors increased, leading to the highly developed eyes of cephalopods and vertebrates.

Vision Involves Detection and Perception of Radiant Energy

Photoreceptors detect light at particular wavelengths, and centers in a brain or central ganglion integrate signals arriving from the receptors into a perception of light. All animals use forms of a single lipidlike pigment, *retinal* (synthesized from vitamin A), in photoreceptors to absorb light energy. The simplest eyes are capable only of distinguishing light from dark; the most complex eyes distinguish shapes and colors and focus an accurate image of objects being viewed onto a layer of photoreceptors.

Invertebrate Eyes Take Many Forms

Some invertebrates, such as earthworms, do not have visual organs; instead, photoreceptors in their skin allow them to sense and respond to light. Earthworms respond negatively to light, as you can easily discover by shining a flashlight on an earthworm outside its burrow at night.

The eyes of other invertebrates are diverse, ranging from collections of photoreceptors with no lens and no image-forming capability to eyes remarkably like those of vertebrates. The photoreceptors of invertebrates are depolarized when they absorb light, and generate action potentials or increase their release of neurotransmitter molecules when they are stimulated. Vertebrate photoreceptors function differently, as we will see.

The simplest eye is the **ocellus** (plural, *ocelli;* also called an *eyespot* or *eyecup*). An ocellus, which detects light but does not form an image, consists of fewer than 100 photoreceptor cells lining a cup or pit. In planarians, for example, photoreceptor cells in a cuplike depression below the epidermis are connected to the dendrites of afferent neurons, which are bundled into nerves that travel from the ocelli to the cerebral ganglion **(Figure 39.9).** Each

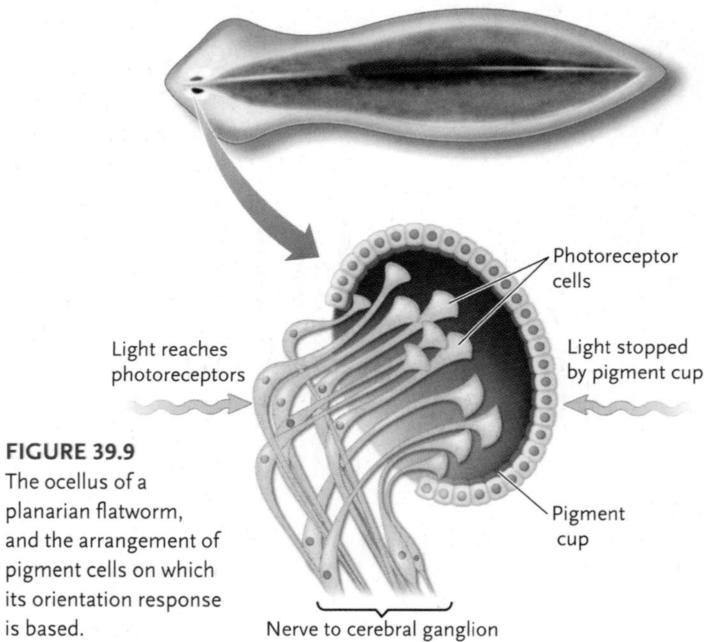
Photoreceptor cells
Light reaches photoreceptors
Light stopped by pigment cup
Pigment cup
Nerve to cerebral ganglion

FIGURE 39.9 The ocellus of a planarian flatworm, and the arrangement of pigment cells on which its orientation response is based.

ocellus is covered on one side by a layer of pigment cells that blocks most of the light rays arriving from the opposite side of the animal. As a result, most of the light received by the pigment cells enters the ocellus from the side that it faces. Through integration of information transmitted to the cerebral ganglion from the eyecups, planarians orient themselves so that the amount of light falling on the two ocelli is equal and diminishes as they swim. This reaction carries them directly away from the source of the light. Similar ocelli are found in a variety of animals, including insects, arthropods, and mollusks.

Two main types of image-forming eyes have evolved in invertebrates: compound eyes and single-lens eyes. The **compound eye** of insects, crustaceans, and a few annelids and mollusks contains hundreds to thousands of faceted visual units called **ommatidia** (*omma* = eye) fitted closely together **(Figure 39.10).** Each ommatidium samples a small part of the visual field. In insects, light entering an ommatidium is focused by a transparent **cornea** and a *crystalline cone* (just below the cornea) onto a bundle of photoreceptor cells. Microvilli of these cells interdigitate like the fingers of clasped hands, forming a central axis rich in rhodopsin, a retinal-containing **photopigment** (light-absorbing pigment). Absorption of light by rhodopsin causes action potentials to be generated in afferent neurons connected to the base of the ommatidium. From these signals, the brain receives a mosaic image of the world. Because even the slightest motion is detected simultaneously by many ommatidia, compound eyes are extraordinarily adept at detecting movement—a lesson soon learned by fly-swatting humans.

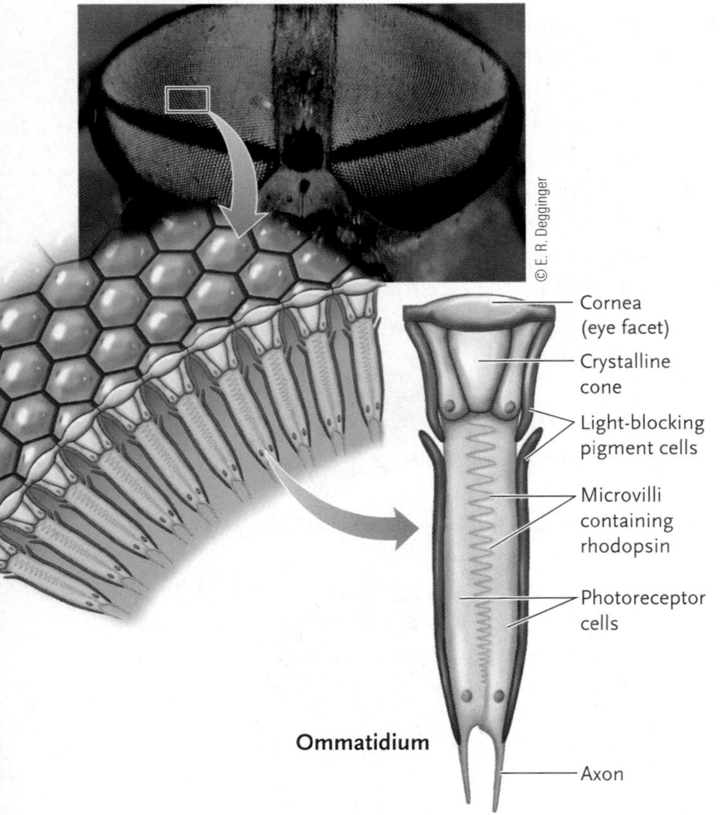
© E. R. Degginger
Cornea
(eye facet)
Crystalline
cone
Light-blocking
pigment cells
Microvilli
containing
rhodopsin
Photoreceptor
cells
Ommatidium
Axon

FIGURE 39.10
The compound eye of a deer fly (*Chrysops* species). Each ommatidium has a cornea that directs light into the crystalline cone; in turn, the cone focuses light on the photoreceptor cells. A light-blocking pigment layer at the sides of the ommatidium prevents light from scattering laterally in the compound eye.

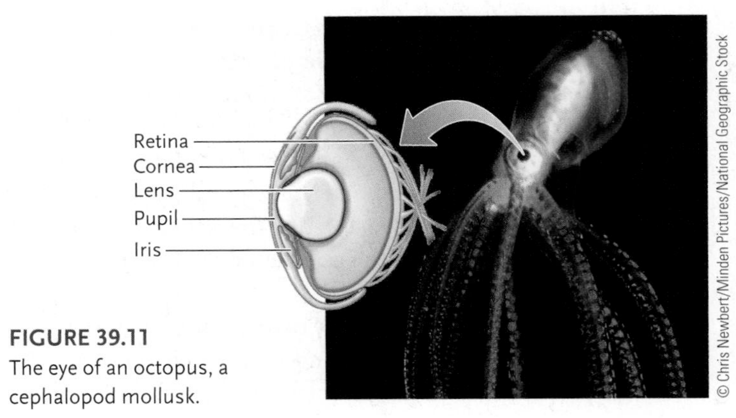
Retina
Cornea
Lens
Pupil
Iris
© Chris Newbert/Minden Pictures/National Geographic Stock

FIGURE 39.11
The eye of an octopus, a cephalopod mollusk.

The **single-lens eye** of cephalopods **(Figure 39.11)** resembles a vertebrate eye in that both types operate like a camera. In the cephalopod eye, light enters through the transparent cornea, a **lens** concentrates the light, and a layer of photoreceptors at the back of the eye, the **retina,** records the image. Behind the cornea is the **iris,** which surrounds the **pupil,** the opening through which light enters the eye. Muscles in the iris adjust the size of the pupil to vary the amount of light entering the eye. When the light is bright, circular muscles in the iris contract, shrinking the size of the pupil and reducing the amount of light that enters. In dim light, radial muscles contract and enlarge the pupil, increasing the amount of light that enters the eye. Muscles move the lens forward and back with respect to the retina to focus the image. This is an example of **accommodation,** a process by which the lens changes to enable the eye to focus on objects at different distances.

A neural network lies under the retina, meaning that light rays do not have to pass through the neurons to reach the photoreceptors. The vertebrate eye has the opposite arrangement. This and other differences in structure and function indicate that cephalopod and vertebrate eyes evolved independently.

Vertebrate Eyes Have a Complex Structure

The human eye **(Figure 39.12)** has similar structures—cornea, iris, pupil, lens, and retina—to those of the cephalopod eye just described. Light entering the eye through the cornea passes through the iris and then the lens. The lens focuses an image on the retina, and the axons of afferent neurons originating in the retina converge to form the optic nerve leading from the eye to the brain.

A clear fluid called the **aqueous humor** fills the space between the cornea and lens. This fluid carries nutrients to the lens and cornea, which do not contain any blood vessels. The main chamber of the eye, located between the lens and the retina, is filled with the jellylike **vitreous humor** (*vitrum* = glass). The outer wall of the eye contains a tough layer of connective tissue (the *sclera*). Inside it is a darkly pigmented layer (the *choroid*) that prevents light from entering except through the pupil. It also contains the blood vessels nourishing the retina.

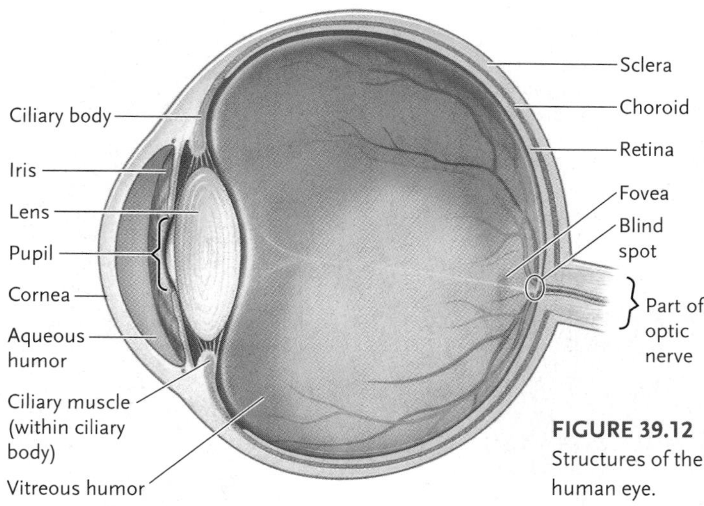
Sclera
Choroid
Ciliary body
Retina
Iris
Fovea
Lens
Blind spot
Pupil
Cornea
Part of optic nerve
Aqueous humor
Ciliary muscle (within ciliary body)
Vitreous humor

FIGURE 39.12 Structures of the human eye.

Two types of photoreceptors, rods and cones, occur in the retina along with layers of neurons that carry out an initial integration of visual information before it is sent to the brain. The **rods** are specialized for detection of light at low intensities; the **cones** are specialized for detection of different wavelengths (colors).

Accommodation does not occur by forward and back movement of the lens, as described for cephalopods. Rather, the lens of most terrestrial vertebrates is focused by changing its shape. The lens is held in place by fine ligaments that anchor it to a surrounding layer of connective tissue and muscle, the **ciliary body.** These ligaments keep the lens under tension when the ciliary muscle is relaxed. The tension flattens the lens, which is soft and flexible, and focuses light from distant objects on the retina **(Figure 39.13A).** When the ciliary muscles contract, they relieve the tension of the ligaments, allowing the lens to assume a more spherical shape and focusing light from nearby objects on the retina **(Figure 39.13B).**

FIGURE 39.13 Accommodation in terrestrial vertebrates: the lens changes shape rather than moving forward and back to focus on **(A)** distant and **(B)** near objects.

A. **Focusing on distant object**

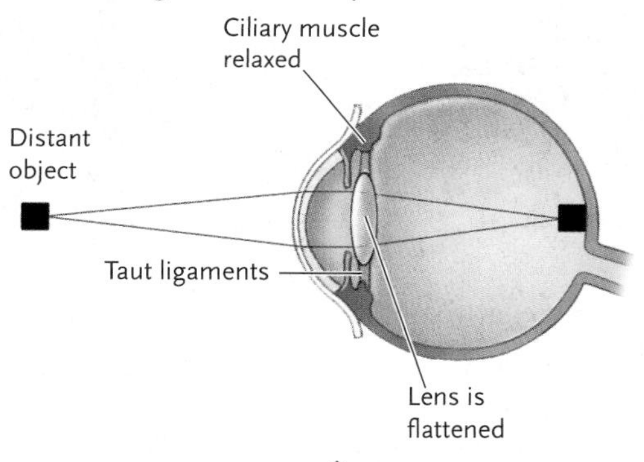
Ciliary muscle relaxed
Distant object
Taut ligaments
Lens is flattened

When the eye focuses on a distant object, the ciliary muscles relax, allowing the ligaments that support the lens to tighten. The tightened ligaments flatten the lens, bringing the distant object into focus on the retina.

B. **Focusing on near object**

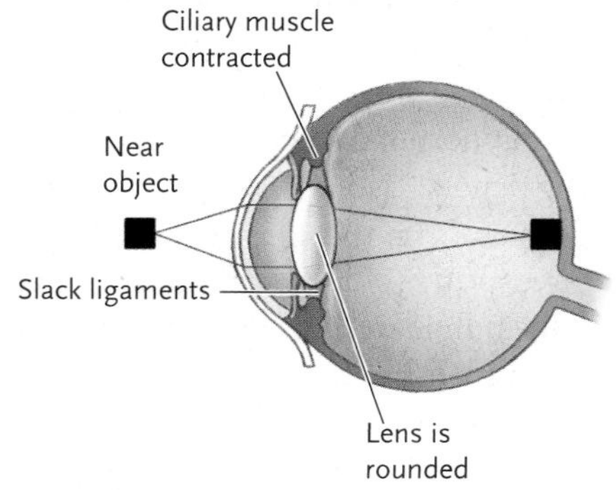
Ciliary muscle contracted
Near object
Slack ligaments
Lens is rounded

When the eye focuses on a near object, the ciliary muscles contract, loosening the ligaments and allowing the lens to become rounder. The rounded lens focuses a near object on the retina.

The Retina of Mammals and Birds Contains Rods and Cones and a Complex Network of Neurons

The retina of a human eye contains about 120 million rods and 6 million cones organized into a densely packed, single layer. Neural networks of the retina are layered on top of the photoreceptor cells, so that light rays focused by the lens on the retina must pass through the neurons before reaching the photoreceptors. The light must also pass through a layer of fine blood vessels that covers the surface of the retina.

In mammals and birds with eyes specialized for daytime vision, cones are concentrated in and around a small region of the retina, the **fovea** (see Figure 39.12). The image focused by the lens is centered on the fovea, which is circular and less than a millimeter in diameter in humans. The rods are spread over the remainder of the retina. We can see distinctly only the image focused on the fovea; the surrounding image is what we term *peripheral vision.* Mammals and birds with eyes specialized for night vision have retinas containing mostly rods, and lacking a defined fovea. Some fishes and many reptiles have cones generally distributed throughout their retina and very few rods.

The rods of mammals are much more sensitive than the cones to low-intensity light; in fact, they can respond to a single photon of light. This is why, in dim light, we can detect objects better by looking slightly to the side of the object. This action directs the image away from the cones in the fovea to the highly light-sensitive rods in surrounding regions of the retina.

SENSORY TRANSDUCTION BY RODS AND CONES Photoreceptors have three parts: (1) an outer segment consisting of stacked, flattened, membranous discs; (2) an inner segment where the cell's metabolic activities occur; and (3) the synaptic terminal, where neurotransmitter molecules are stored and released **(Figure 39.14A).** The light-absorbing pigment of rods and cones, retinal, is bonded covalently to **opsins** to produce **photopigments.** The photopigments are embedded in the membranous discs of the photoreceptors' outer segments **(Figure 39.14B).** The retinal-opsin photopigment in rods is **rhodopsin.**

In the dark, the retinal of rhodopsin is in an inactive form known as *cis*-retinal (see Figure 39.14B), and the rods steadily release the neurotransmitter glutamate. When rhodopsin absorbs a photon of light, retinal converts to its active form, *trans*-retinal (see Figure 39.14B), and the rods *decrease* the amount of glutamate they release. This will be discussed in the next section.

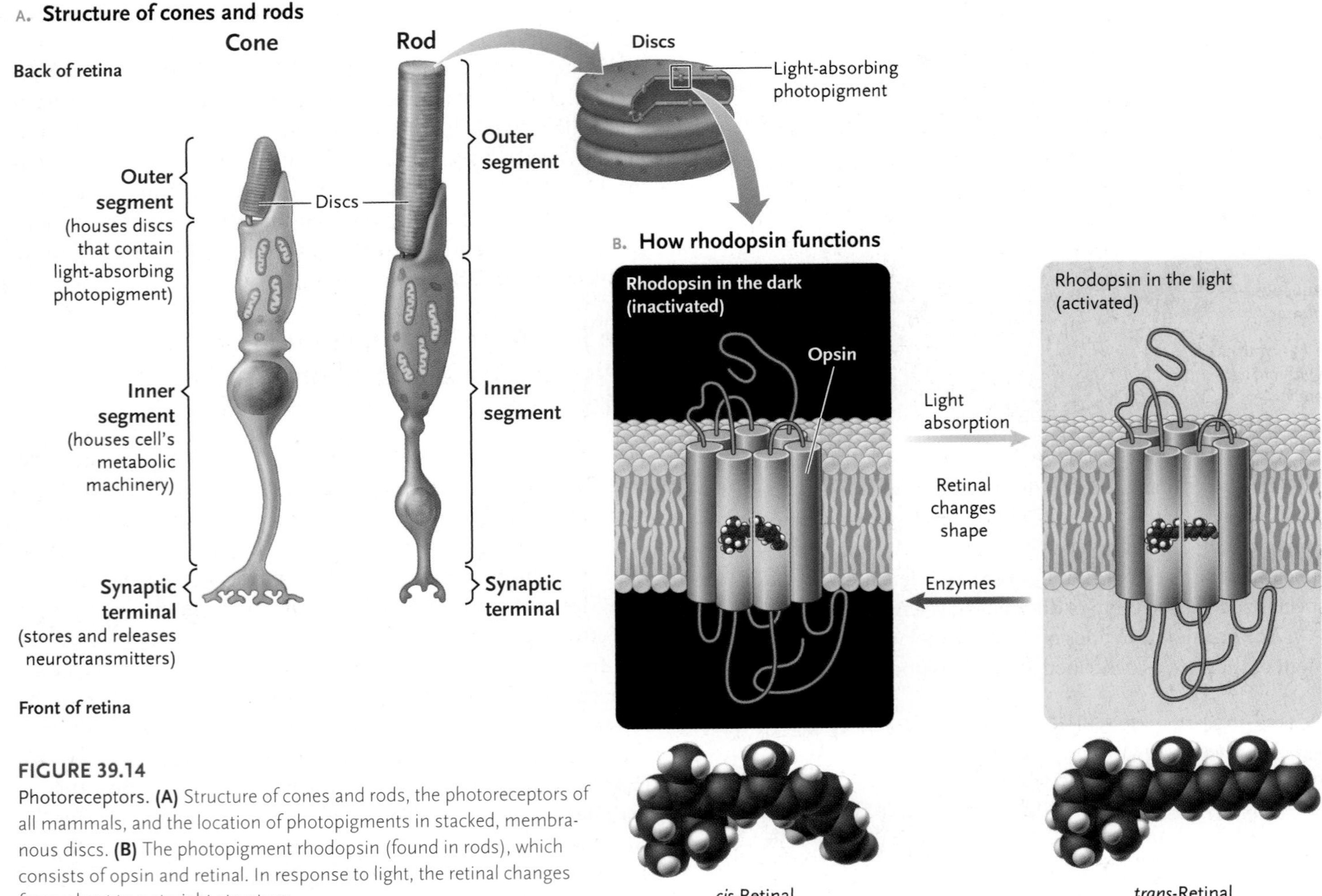
A. Structure of cones and rods
Cone
Rod
Back of retina
Outer segment (houses discs that contain light-absorbing photopigment)
Discs
Outer segment
Inner segment (houses cell's metabolic machinery)
Inner segment
Synaptic terminal (stores and releases neurotransmitters)
Synaptic terminal
Front of retina
Discs
Light-absorbing photopigment
B. How rhodopsin functions
Rhodopsin in the dark (inactivated)
Opsin
Light absorption
Retinal changes shape
Enzymes
Rhodopsin in the light (activated)
cis-Retinal
trans-Retinal

FIGURE 39.14
Photoreceptors. **(A)** Structure of cones and rods, the photoreceptors of all mammals, and the location of photopigments in stacked, membranous discs. **(B)** The photopigment rhodopsin (found in rods), which consists of opsin and retinal. In response to light, the retinal changes from a bent to a straight structure.

Rhodopsin is a membrane-embedded G-protein–coupled receptor (see Section 7.4). Recall that an extracellular signal received by a G-protein–coupled receptor activates the receptor, which triggers a signal transduction pathway within the cell, leading to a cellular response. Here, activated rhodopsin triggers a pathway that leads to the closure of Na^+ channels in the plasma membrane **(Figure 39.15).** Closure of the channels hyperpolarizes the photoreceptor's membrane, thereby decreasing neurotransmitter release. The response is graded: as light absorption increases, the amount of neurotransmitter released is reduced proportionately; if light absorption decreases, neurotransmitter release by the photoreceptor increases proportionately. Note that transduction in rods works in the opposite way from most sensory receptors, in which a stimulus increases neurotransmitter release.

VISUAL PROCESSING IN THE RETINA In the retina of all vertebrates, the two types of photoreceptors are linked to a network of neurons that carries out initial processing of visual information. The retina of mammals contains four types of neurons **(Figure 39.16).** The rods and cones form synaptic connections with bipolar cells, where the processing of visual information begins. **Ganglion cells** receive signals from the **bipolar cells,** with which they are synapsed. The axons of the ganglion cells extend over the retina and collect at the back of the eyeball to form the optic nerve, which carries action potentials to the brain. The point where the optic nerve exits the eye lacks photoreceptors, resulting in a *blind spot* several millimeters in diameter. Two other types of neurons form lateral connections in the retina: **horizontal cells** receive and integrate inputs from multiple photoreceptors located lateral to the region of light reception; and **amacrine cells** receive inputs from bipolar cells and excite ganglion cells.

In the dark, the steady release of glutamate from rods and cones depolarizes some of the bipolar cells and hyperpolarizes others. In the light, the decrease in neurotransmitter release from rods and cones results in the depolarized bipolar cells becoming hyperpolarized, and hyperpolarized bipolar cells becoming depolarized.

Signals from the rods and cones move vertically from the photoreceptors to bipolar cells and then to ganglion cells. Because the human retina has over 120 million photoreceptors, but only about 1 million ganglion cells, each ganglion cell receives signals from a clearly defined set of photoreceptors that constitute the receptive field for that cell. Therefore, stimulating numerous photoreceptors in a ganglion cell's receptive field results in only a single message to the brain from that cell. Receptive fields are typically circular and are of different sizes. Smaller receptive fields result in sharper images because they send more

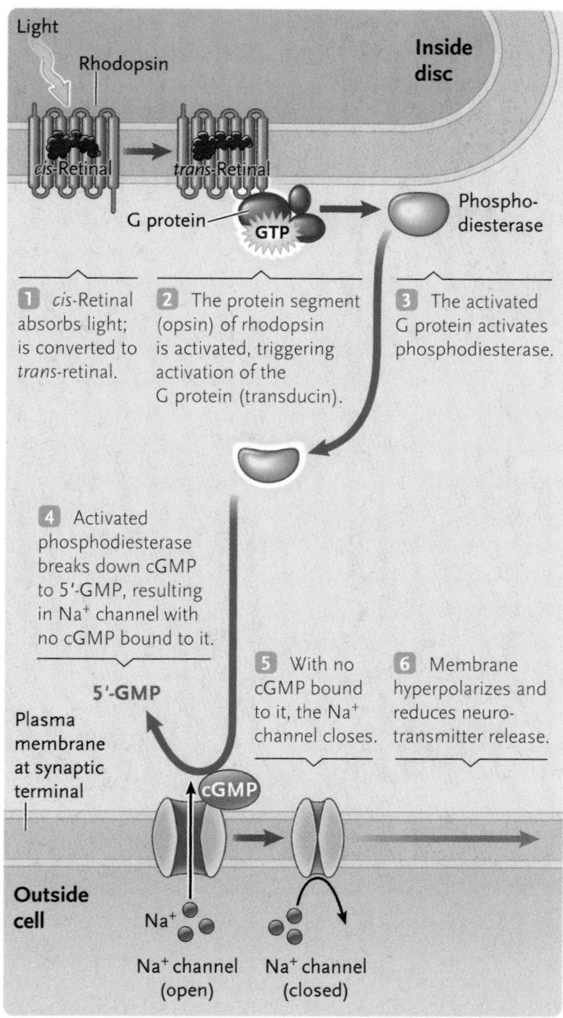
Light
Rhodopsin
Inside disc
cis-Retinal
trans-Retinal
G protein
GTP
Phospho-diesterase
1 cis-Retinal absorbs light; is converted to trans-retinal.
2 The protein segment (opsin) of rhodopsin is activated, triggering activation of the G protein (transducin).
3 The activated G protein activates phosphodiesterase.
4 Activated phosphodiesterase breaks down cGMP to 5'-GMP, resulting in Na+ channel with no cGMP bound to it.
5 With no cGMP bound to it, the Na+ channel closes.
6 Membrane hyperpolarizes and reduces neuro-transmitter release.
5'-GMP
Plasma membrane at synaptic terminal
cGMP
Outside cell
Na+
Na+ channel (open)
Na+ channel (closed)

FIGURE 39.15
The signal transduction pathway that closes Na^+ channels in photoreceptor plasma membranes when rhodopsin absorbs light.

precise information to the brain regarding the location in the retina where the light was received.

Signals may also move laterally from a rod or cone through a horizontal cell and continue to bipolar cells through the horizontal cell's inhibitory connections. To understand this, consider a spot of light falling on the retina. Photoreceptors detect the light and send a signal to bipolar cells and horizontal cells. The horizontal cells inhibit more distant bipolar cells that are outside the spot of light, causing the light spot to appear lighter and its surrounding dark area to appear darker. This type of visual processing is called **lateral inhibition** and serves to both sharpen the edges of objects and enhance contrast in an image.

Three Kinds of Opsin Pigments Underlie Color Vision

Many invertebrates and some species in each class of vertebrates have color vision. Color vision depends on the cone photoreceptors. Most mammals have only two types of cones, but humans and other primates have three types. Each human or primate cone cell contains one of three different photopigments, collectively called **photopsins,** in which retinal is combined with different opsins. The three photopsins absorb light over different, but overlapping, wavelength ranges, with peak absorptions at 445 nm (blue light), 535 nm (green light), and 570 nm (red light). The farther a wavelength is from the peak color absorbed, the less strongly the cone responds.

Having overlapping wavelength ranges for the three photoreceptors means that light at any visible wavelength will stimulate at least two of the three types of cones. However, because the maximal absorption of each type of cone is a different wavelength, it is stimulated to a different extent by light at a given wavelength. The differences, relayed to the visual centers of the brain, are integrated into the perception of a color correspond-

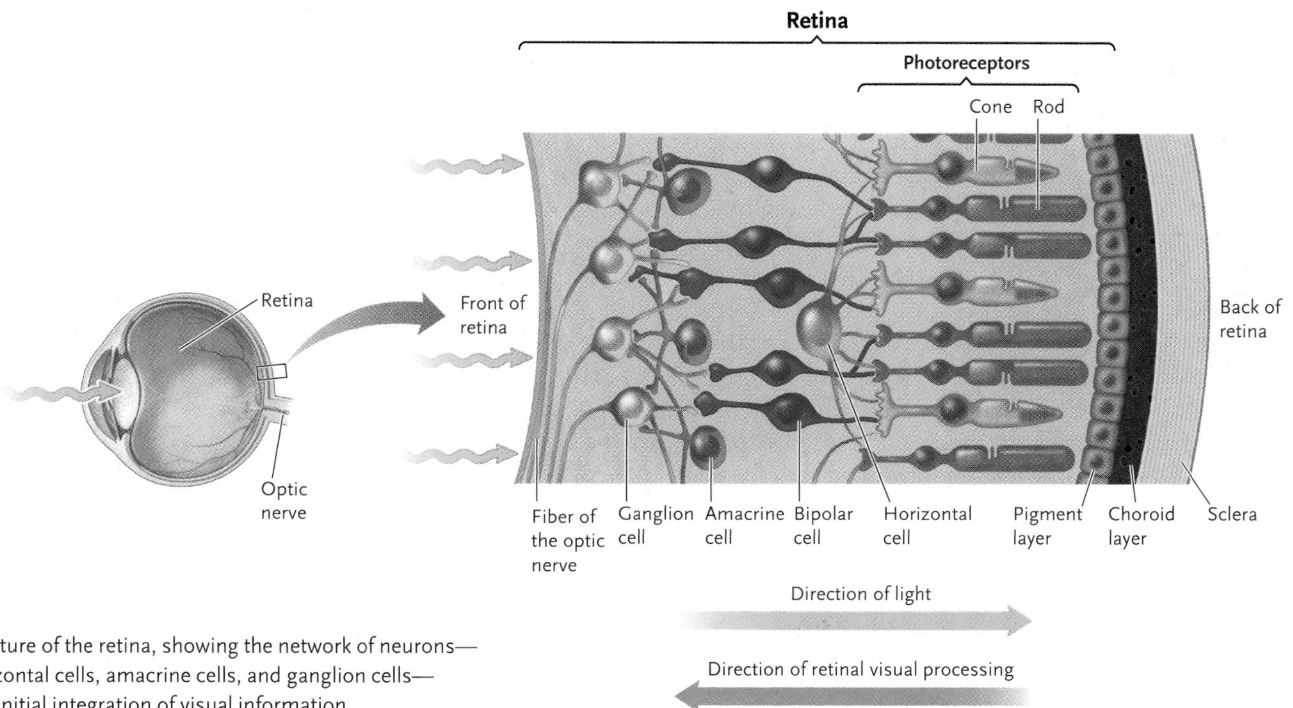
Retina
Photoreceptors
Cone
Rod
Retina
Optic nerve
Front of retina
Back of retina
Fiber of the optic nerve
Ganglion cell
Amacrine cell
Bipolar cell
Horizontal cell
Pigment layer
Choroid layer
Sclera
Direction of light
Direction of retinal visual processing

FIGURE 39.16
Microscopic structure of the retina, showing the network of neurons—bipolar cells, horizontal cells, amacrine cells, and ganglion cells—that carry out the initial integration of visual information.

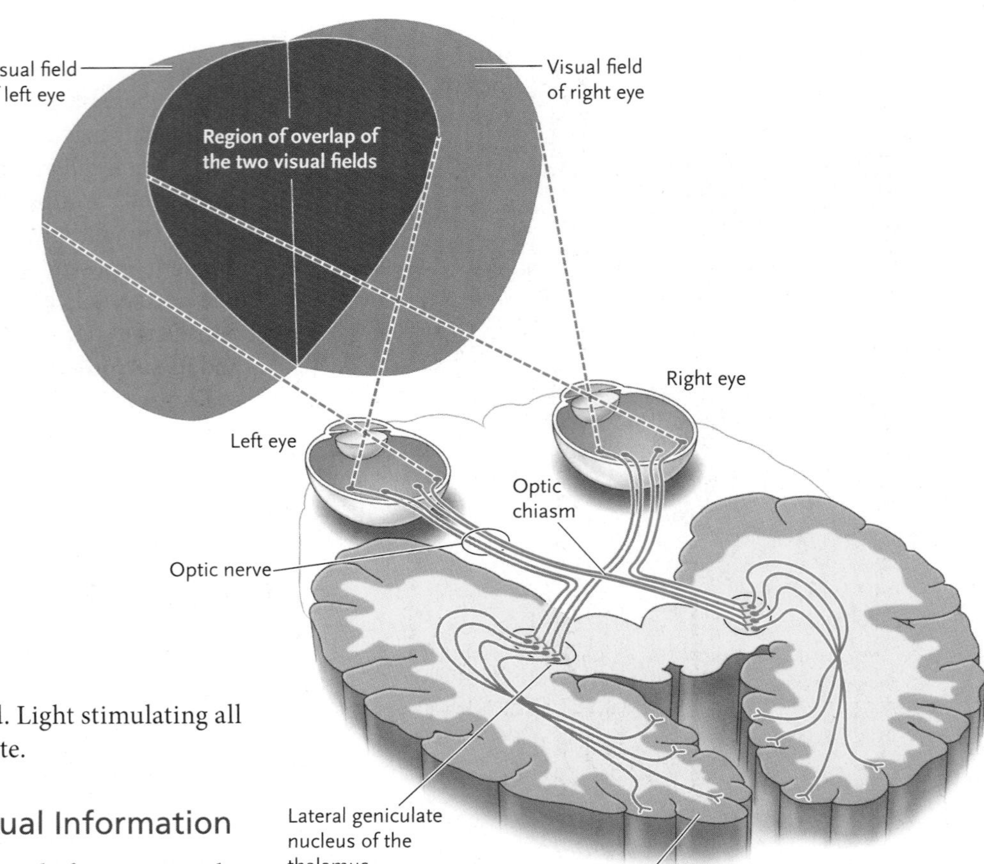
Visual field of left eye
Visual field of right eye
Region of overlap of the two visual fields
Right eye
Left eye
Optic chiasm
Optic nerve
Lateral geniculate nucleus of the thalamus
Visual cortex

FIGURE 39.17
Neural pathways for vision. Because half of the axons carried by the optic nerves cross over in the optic chiasm, the left half of the field seen by both eyes is transmitted to the visual cortex in the right cerebral hemisphere. The right half of the field seen by both eyes is transmitted to the visual cortex in the left cerebral hemisphere. As a result, the right hemisphere of the brain sees objects to the left of the center of vision, and the left hemisphere sees objects to the right of the center of vision.

ing to the particular wavelength absorbed. Light stimulating all three receptor types equally is seen as white.

The Visual Cortex Processes Visual Information

Just behind the eyes, the optic nerves converge before entering the base of the brain. A portion of each optic nerve crosses over to the opposite side, forming the **optic chiasm** (*chiasma* = crossing place). Most of the axons enter the **lateral geniculate nuclei** in the thalamus, where they make synapses with interneurons leading to the visual cortex **(Figure 39.17).**

Because of the optic chiasm, the left half of the image seen by both eyes is transmitted to the visual cortex in the right cerebral hemisphere, and the right half of the image is transmitted to the left cerebral hemisphere. The right hemisphere thus sees objects to the left of the center of vision, and the left hemisphere sees objects to the right of the center of vision. Communication between the right and left hemispheres integrates this information into a perception of the entire visual field seen by the two eyes.

If you look at a nearby object with one eye and then the other, you will notice that the point of view is slightly different. Integration of the visual field by the brain creates a single picture with a sense of distance and depth. The greater the difference between the images seen by the two eyes, the closer the object appears to the viewer.

The two optic nerves together contain more than a million axons, more than all other afferent neurons of the body put together. Almost one-third of the cerebral cortex in humans is devoted to visual information. These numbers give some idea of the complexity of the information integrated into the visual image formed by the brain.

STUDY BREAK 39.4 <

For vertebrate photoreception, define: (a) photopigment; (b)photoreceptor; (c) receptive field.

39.5 Chemoreceptors

Chemoreceptors form the basis of taste (gustation) and smell (olfaction), and measure the levels of internal body molecules such as oxygen, carbon dioxide, and hydrogen ions.

Invertebrates Have Either the Same or Different Receptors for Taste and Smell

In many invertebrates the same receptors serve for the senses of smell and taste. These receptors may be confined to certain locations or distributed over the body surface. For example, the cnidarian *Hydra* has chemoreceptor cells around its mouth that respond to glutathione, a chemical released from prey organisms ensnared in the cnidarian's tentacles. Stimulation of the chemoreceptors by glutathione causes the tentacles to retract, resulting in ingestion of the prey. By contrast, earthworms have taste/smell receptors distributed over their entire body surface.

Some terrestrial invertebrates, particularly insects, have clearly differentiated taste and smell receptors. In insects, taste receptors occur inside hollow sensory bristles called *sensilla* (singular, *sensillum*), which may be located on the antennae, mouthparts, or feet **(Figure 39.18).** Pores in the sensilla admit molecules from potential food to the chemoreceptors, which are specialized to detect sugars, salts, amino acids, or other chemicals. Many female insects have chemoreceptors on their ovipositors, which allow them to lay their eggs on food appropriate for the hatching larvae.

Insect olfactory receptors detect airborne molecules. Some insects use odor as a means of communication, such as the pher-

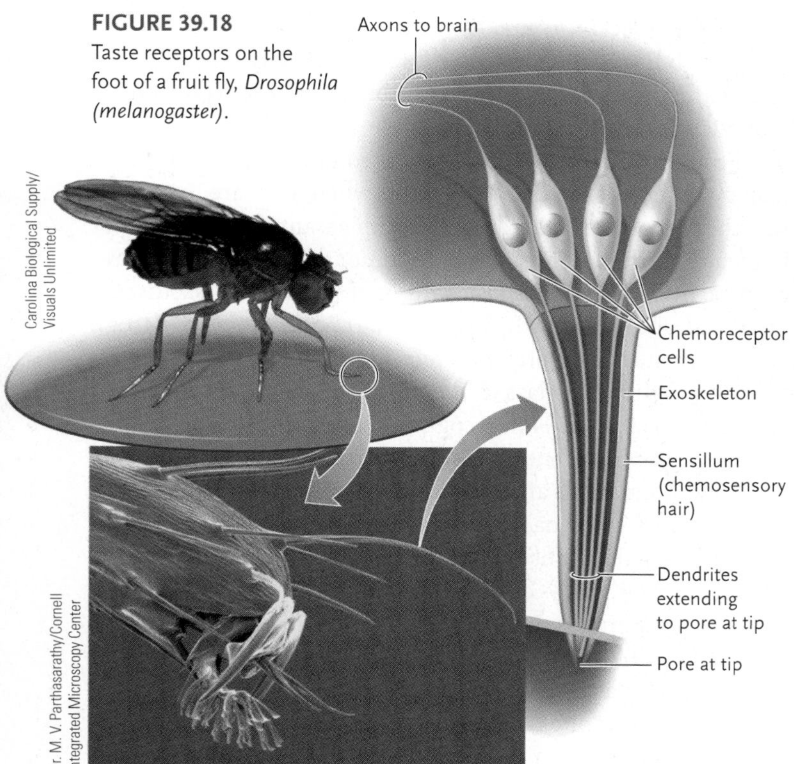

Axons to brain
Chemoreceptor cells
Exoskeleton
Sensillum (chemosensory hair)
Dendrites extending to pore at tip
Pore at tip
Carolina Biological Supply/ Visuals Unlimited
Dr. M. V. Parthasarathy/Cornell Integrated Microscopy Center

FIGURE 39.18 Taste receptors on the foot of a fruit fly, *Drosophila (melanogaster)*.

© A. Shay/OSF/Animals Animals—Earth Scenes
Louisa Howard/Dartmouth College
25 μm

FIGURE 39.19 The brushlike antennae of a male silkworm moth, *Bombyx mori*. Fine sensory bristles containing olfactory receptor cells cover the filaments of the antennae.

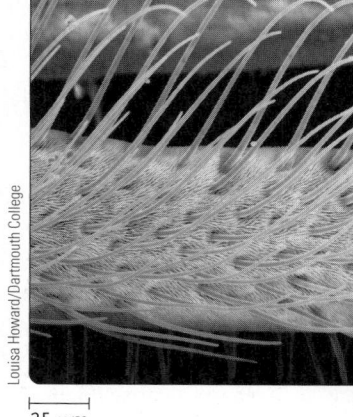

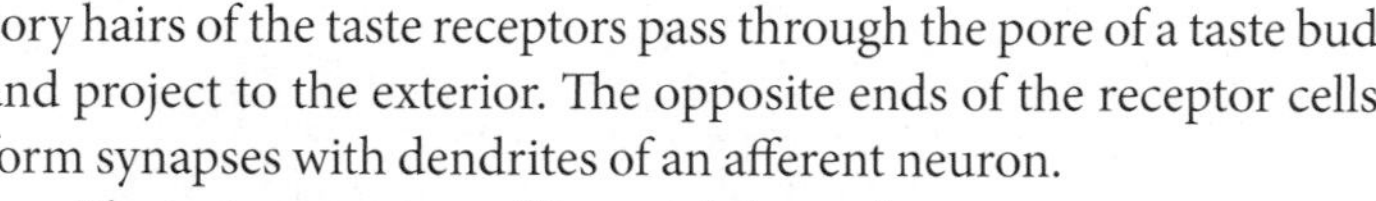

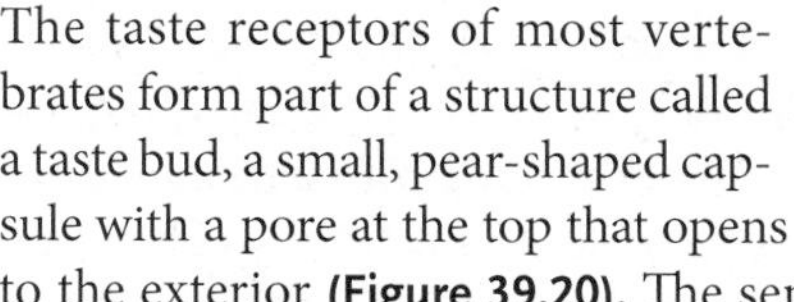

omones released into the air as sexual attractants by female moths. Olfactory receptors in the bristles of male silkworm moth antennae **(Figure 39.19)** have been shown experimentally to be able to detect pheromones released by a female of the same species in concentrations as low as one attractant molecule per 10^{17} air molecules; when as few as 40 of the 20,000 receptor cells on its antennae have been stimulated by pheromone molecules, the male moth responds to attract the female's attention. Ants, bees, and wasps identify members of the same hive or nest and communicate by means of odor molecules.

Taste and Smell Receptors Are Differentiated in Terrestrial Vertebrates

In terrestrial animals, taste involves the detection of potential food molecules in objects that are touched, while smell involves the detection of airborne molecules. Although both taste and smell receptors have hairlike extensions containing the proteins that bind environmental molecules, the hairs of taste receptors are derived from microvilli and contain microfilaments, while the hairs of smell receptors are derived from cilia and contain microtubules. Another significant difference between taste and smell in vertebrates is that information from taste receptors is typically processed in the parietal lobes, while information from smell receptors is processed in the olfactory bulbs and the temporal lobes.

Taste Receptors Are Located in Taste Buds

The taste receptors of most vertebrates form part of a structure called a taste bud, a small, pear-shaped capsule with a pore at the top that opens to the exterior **(Figure 39.20)**. The sensory hairs of the taste receptors pass through the pore of a taste bud and project to the exterior. The opposite ends of the receptor cells form synapses with dendrites of an afferent neuron.

The taste receptors of terrestrial vertebrates are concentrated in the mouth. Humans have about 10,000 taste buds scattered over the tongue, roof of the mouth, and throat. Those on the tongue are embedded in outgrowths called *papillae* (*papula* = pimple), which give the surface of the tongue its rough texture.

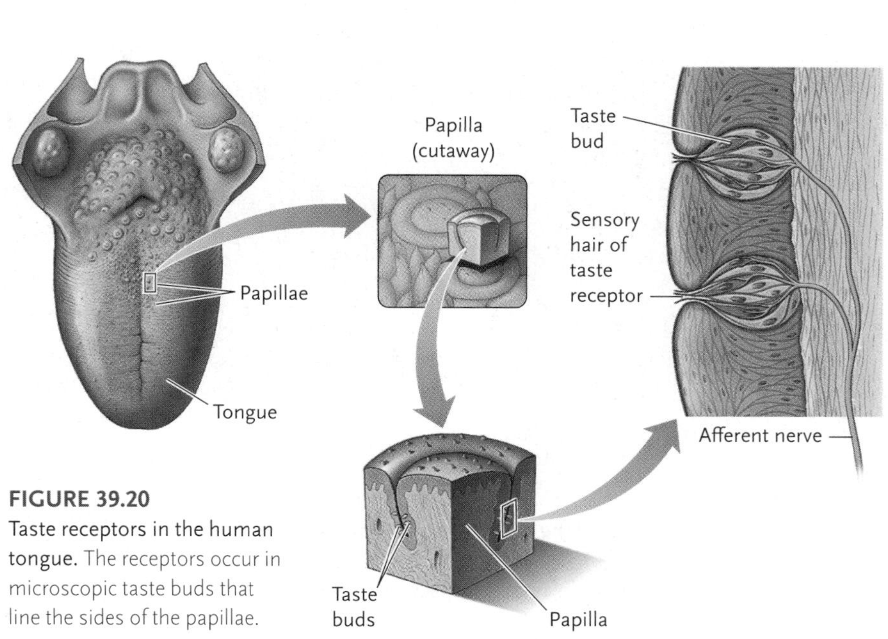

Papillae
Tongue
Papilla (cutaway)
Taste buds
Papilla
Taste bud
Sensory hair of taste receptor
Afferent nerve

FIGURE 39.20 Taste receptors in the human tongue. The receptors occur in microscopic taste buds that line the sides of the papillae.

Taste receptors on the human tongue are thought to respond to five basic tastes: sweet, sour, salty, bitter, and umami (savory). Some of the receptors for umami respond to the amino acid glutamate (familiar as monosodium glutamate, or MSG). Recent research indicates that the classes of receptors may all have many subtypes, each binding a specific molecule within that class.

Signals from the taste receptors are relayed to the thalamus. From there, some signals lead to gustatory centers in the cerebral cortex, which integrate them into the perception of taste. Other signals lead to the brain stem and limbic system, which links tastes to involuntary visceral and emotional responses. Through these connections, a pleasant taste may lead to salivation, secretion of digestive juices, sensations of pleasure, and even sexual arousal, whereas an unpleasant taste may produce revulsion, nausea, and even vomiting.

Olfactory Receptors Are Concentrated in the Nasal Cavities in Terrestrial Vertebrates

Receptors that detect odors are located in the nasal cavities. Bloodhounds have more than 200 million receptors in patches of olfactory epithelium in the upper nasal passages; humans have about 5 million olfactory receptors.

On one end, each olfactory receptor cell has 10 to 20 sensory hairs that project into a layer of mucus covering the olfactory area in the nose **(Figure 39.21)**. To be detected, airborne molecules must dissolve in the watery mucus solution. On the other end, the olfactory receptors make synapses with interneurons in the olfactory bulbs. Olfactory receptors are the only receptor cells that make direct connections with brain interneurons, rather than via afferent neurons.

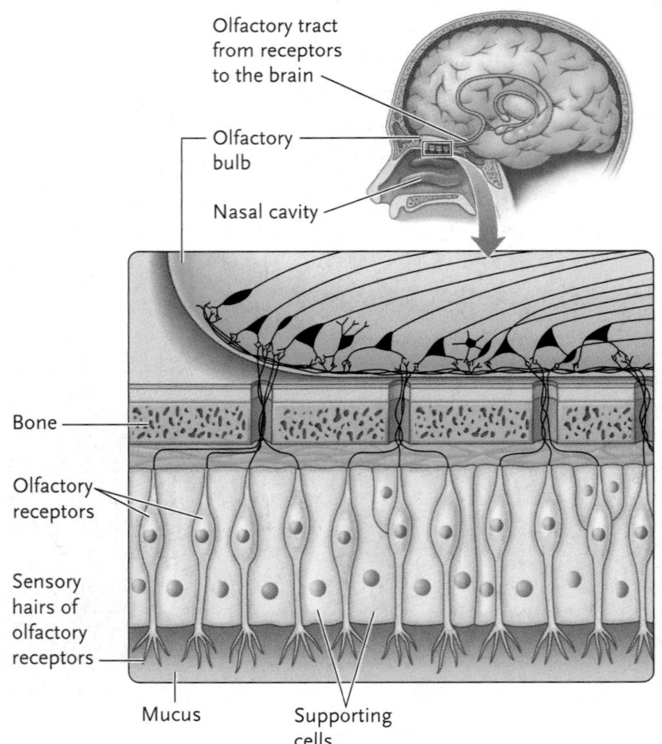
Olfactory tract from receptors to the brain
Olfactory bulb
Nasal cavity
Bone
Olfactory receptors
Sensory hairs of olfactory receptors
Mucus
Supporting cells

FIGURE 39.21
Olfactory receptors in the roof of the nasal passages in humans. Axons from these receptors pass through holes in the bone separating the nasal passages from the brain, where they make synapses with interneurons in the olfactory bulbs.

From the olfactory bulbs, nerves conduct signals to the olfactory centers of the cerebral cortex, where they are integrated into the perception of tantalizing or unpleasant odors. Most odor perceptions arise from a combination of different olfactory receptors. In the early 1990s, Richard Axel and Linda Buck discovered that about 1,000 different human genes give rise to an equivalent number of olfactory receptor types, each of which responds to a different class of chemicals. Axel and Buck received the Nobel Prize in 2004 in recognition of their research.

Olfaction contributes to the sense of taste because vaporized molecules from foods are conducted from the throat to the olfactory receptors in the nasal cavities. Olfactory input is the reason why anything that dulls your sense of smell—such as a head cold, or holding your nose—diminishes the apparent flavor of food.

Many mammals use odors as a means of communication. Individuals of the same family or colony are identified by their odor; odors are also used to attract mates and to mark territories and trails. Dogs, for example, use their urine to mark home territories with identifying odors. Humans use the fragrances of perfumes and colognes as artificial sexual attractants.

STUDY BREAK 39.5 <

1. **How do we distinguish different kinds of smells?**
2. **For terrestrial vertebrates, describe the pathway by which a signal generated by taste receptors leads to a response.**

39.6 Thermoreceptors and Nociceptors

Thermoreceptors detect changes in the surrounding temperature. Nociceptors respond to stimuli that may potentially damage tissues. Both types of receptors consist of free nerve endings formed by the dendrites of afferent neurons, with no specialized receptor structures surrounding them.

Thermoreceptors Can Detect Warm and Cold Temperatures and Temperature Changes

Most animals have thermoreceptors. Some invertebrates, such as mosquitoes and ticks, use thermoreceptors to locate their warm-blooded prey. Some snakes, including rattlesnakes and pythons, use thermoreceptors called *pit organs* to detect the body heat of warm-blooded prey animals **(Figure 39.22)**.

In mammals, current evidence suggests the existence of five thermoreceptor ranges. These begin at less than 8°C and extend to above 52°C. At the extremes, the neural signals are perceived as painful.

Some neurons in the hypothalamus of mammals also function as thermoreceptors, an ability that has only recently been in-

Pit organs
David Hosking/Frank Lane Picture Agency

FIGURE 39.22
The pit organs of an albino Western diamondback rattlesnake *(Crotalus atrox)*, located in depressions on both sides of the head below the eyes. These thermoreceptors detect infrared radiation emitted by warm-bodied prey animals such as mice.

vestigated. Not only do these neurons sense changes in brain temperature, but they also receive afferent thermal information. These neurons are highly sensitive to shifts from the normal body temperature, and trigger involuntary responses such as sweating, panting, or shivering, which restore normal body temperature.

Nociceptors Protect Animals from Potentially Damaging Stimuli

The signals from nociceptors—receptors in mammals and other vertebrates that detect damaging stimuli—are interpreted by the brain as pain. Pain is a protective mechanism; in humans, it prompts us to do something immediately to remove or decrease the damaging stimulus. Often pain elicits a reflex response—such as withdrawing the hand from a hot surface—that proceeds before we are even consciously aware of the sensation.

Various types of stimuli cause pain, including mechanical damage such as a cut or a blow to the body, and temperature extremes. Some nociceptors are specific for a particular type of damaging stimulus, whereas others respond to all kinds.

The axons that transmit pain are part of the somatic system of the PNS (see Section 38.2). They synapse with interneurons in the gray matter of the spinal cord, and activate neural pathways in the CNS by releasing the neurotransmitters glutamate or substance P (see Table 37.1). Glutamate-releasing axons produce sharp, prickling sensations that can be localized to a specific body part—the pain of stepping on a sharp stone, for example. Substance P–releasing axons produce dull, burning, or aching sensations, the location of which may not be easily identified—the pain of tissue damage such as stubbing your toe.

As part of their protective function, pain receptors adapt very little, if at all. Some pain receptors, in fact, gradually intensify the rate at which they send out action potentials if the stimulus continues at a constant level.

The CNS also has a pain-suppressing system. In response to stimuli such as exercise, sex, and stress, the brain releases *endorphins,* natural painkillers that bind to membrane receptors on substance P neurons, reducing the amount of neurotransmitter released.

Nociceptors contribute to the taste of some spicy foods, particularly those that contain hot peppers. In fact, researchers who study pain often use *capsaicin,* the organic compound that gives jalapeños and other peppers their hot taste, to identify nociceptors. To some, the burning sensation from capsaicin is addictive. Here is the reason. Nociceptors in the mouth, nose, and throat immediately transmit pain messages to the brain when they detect capsaicin. The brain responds by releasing endorphins, which act as a painkiller and create temporary euphoria. *Insights from the Molecular Revolution* describes a series of experiments investigating the molecular basis of the pain caused by capsaicin.

STUDY BREAK 39.6

What distinguishes thermoreceptors and nociceptors from the other types of sensory receptors discussed previously?

THINK OUTSIDE THE BOOK

You have now learned about various types of sensory receptors. On your own or collaboratively, investigate the role of breakdowns in sensory receptor pathways in causing human disease. Pick two diseases caused in this way and outline how the change in the sensory receptor pathway brings about disease symptoms.

39.7 Magnetoreceptors and Electroreceptors

Some animals have poorly developed visual systems but can gain information about their environment by sensing magnetic or electrical fields. In so doing, they directly sense stimuli that humans can detect only with scientific instruments.

Magnetoreceptors Are Used for Navigation

Some animals that navigate long distances, including migrating butterflies, beluga whales, sea turtles, homing pigeons, and foraging honeybees, have **magnetoreceptors** that allow them to detect and use Earth's magnetic field as a source of directional information (experiments with sea turtles are described in **Figure 39.23**).

The pattern of Earth's magnetic field differs from region to region yet remains almost constant over time, largely unaffected by changing weather and day and night. As a result, animals with magnetic receptors are able to monitor their location reliably. Although little is known about the receptors that detect magnetic fields, they may depend on the fact that moving a conductor, such as an electroreceptor cell, through a magnetic field generates an electric current.

Some magnetoreceptors may depend on the effect of Earth's magnetic field on the mineral *magnetite.*

Some Like It Hot: What is the nature of the capsaicin receptor?

HO, O, H, N, O

Capsaicin

Biting into a jalapeño pepper (a variety of *Capsicum annuum*) can produce a burning pain in your mouth strong enough to bring tears to your eyes. This painfully hot sensation is due primarily to *capsaicin,* a chemical that probably evolved in pepper plants as a defense against foraging animals. The defense is obviously ineffective against the humans who relish peppers and foods containing capsaicin (such as salsa). As you learned in this chapter, pain is an interpretation by the brain of action potentials generated by stimulation of nociceptors.

Research Question

What is the molecular nature of the capsaicin nociceptor?

Experiment

David Julius and his coworkers at the University of California, San Francisco, performed the key experiments that answered the question. Based on other studies, they knew that capsaicin activates an ion channel that leads to Ca^{2+} movement into the cell. Experimentally, an increase in Ca^{2+} within a cell can be detected by using a particular fluorescent dye and analysis under a microscope. Using that information, they set out to isolate and characterize the capsaicin receptor, following the steps shown in the figure.

Results

The researchers identified a transformant that showed capsaicin-stimulated Ca^{2+} uptake. They reasoned that the transformant had received a cDNA clone that expressed a capsaicin receptor. The group then carried out some follow-up experiments, injecting mRNA transcribed from the identified cDNA clone into both frog oocytes and cultured mammalian cells. Tests showed that both the oocytes and the cultured cells responded to capsaicin by admitting Ca^{2+}, providing further confirmation that the researchers had identified the capsaicin receptor cDNA.

Among the effects they noted when the receptor was introduced into oocytes was a response to heat. Increasing the temperature of the solution surrounding the oocytes from 22°C to about 48°C produced a strong calcium inflow.

Conclusion

The capsaicin nociceptor is activated both by capsaicin and heat in cells with the receptor. Therefore the feeling that your mouth is on fire when you eat a hot pepper probably results from the fact that, as far as your nociceptors and CNS are concerned, it *is* on fire.

Source: M. J. Caterina et al. 1997. The capsaicin receptor: a heat-activated ion channel in the pain pathway. *Nature* 389:816–824.

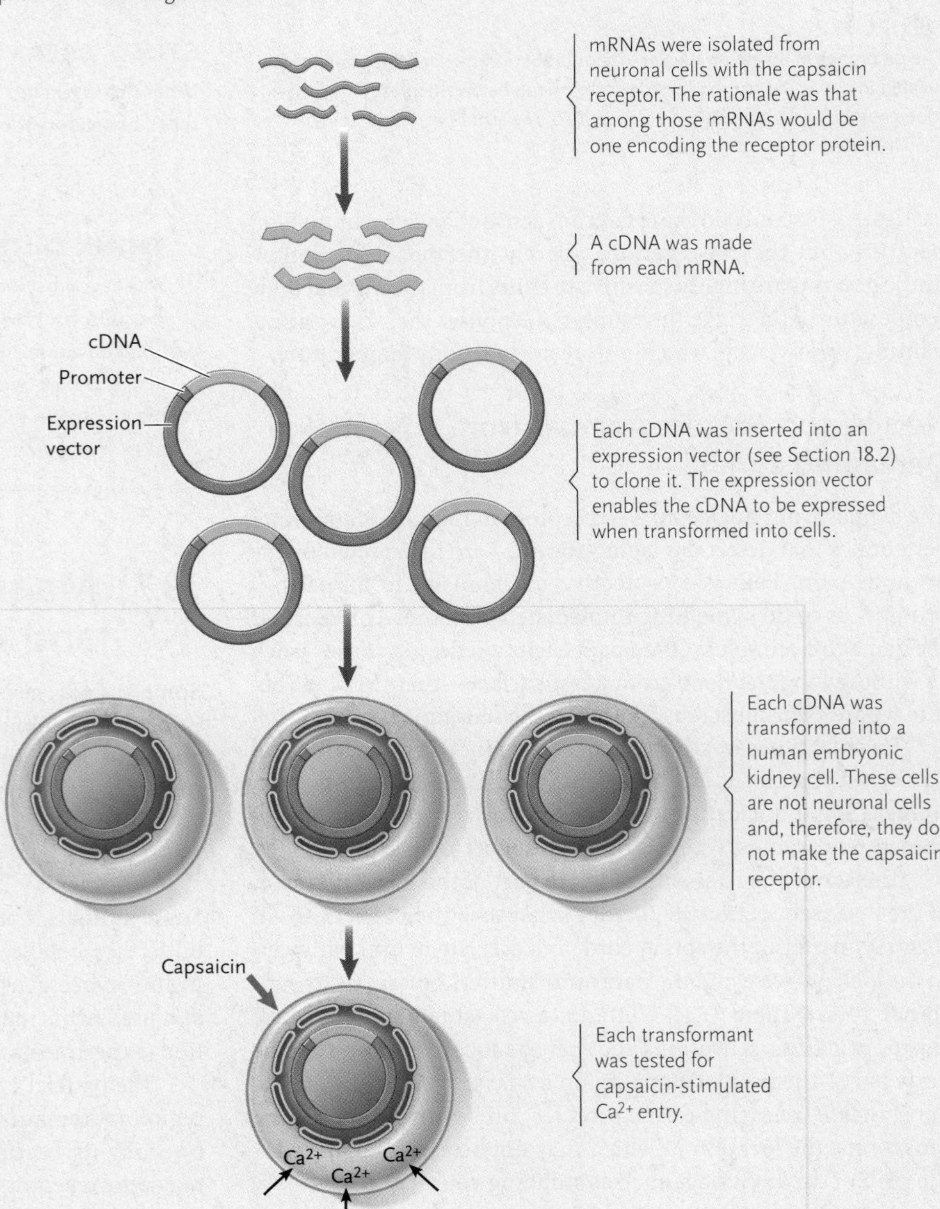

mRNAs were isolated from neuronal cells with the capsaicin receptor. The rationale was that among those mRNAs would be one encoding the receptor protein.
A cDNA was made from each mRNA.
cDNA
Promoter
Expression vector
Each cDNA was inserted into an expression vector (see Section 18.2) to clone it. The expression vector enables the cDNA to be expressed when transformed into cells.
Each cDNA was transformed into a human embryonic kidney cell. These cells are not neuronal cells and, therefore, they do not make the capsaicin receptor.
Capsaicin
Each transformant was tested for capsaicin-stimulated Ca2+ entry.
Ca2+
Ca2+
Ca2+

FIGURE 39.23 Experimental Research

Demonstration That Magnetoreceptors Play a Key Role in Loggerhead Sea Turtle Migration

Question: Do loggerhead sea turtles use a magnetoreceptor system for migration?

Experiment: Loggerhead sea turtles *(Caretta caretta)* that hatch along the east coast of Florida spend much of their lives traveling the North Atlantic current system around the Sargasso Sea, a pool of warm water with a unique seaweed system. Eventually and unerringly, the turtles return to their hatching beach for the mating season. Kenneth Lohmann of the University of North Carolina hypothesized that magnetoreception, likely involving magnetite, plays a central role in loggerhead migration. Lohmann tested his hypothesis using an experimental system in which the direction hatchling turtles swam was analyzed in different magnetic fields.

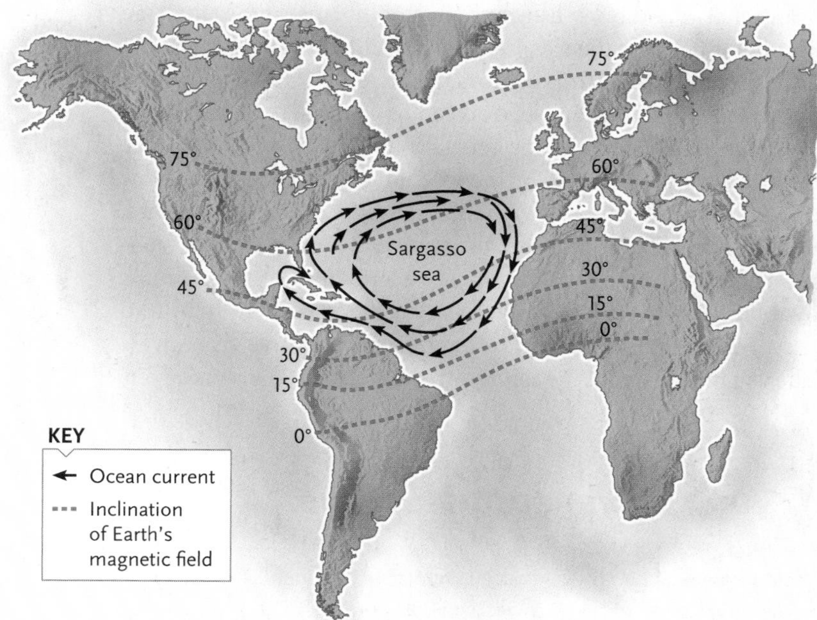
75°
60°
45°
30°
15°
0°
75°
60°
45°
30°
15°
0°
Sargasso sea
KEY
Ocean current
Inclination of Earth's magnetic field

1. Lohmann placed each turtle hatchling he tested in a harness and tethered it to a swiveling, electronic system in the center of a circular pool of water. The pool was surrounded by a large electromagnetic coil system that allowed the researchers to reverse the direction of the magnetic field. The direction the turtle swam was recorded by the tracking system and relayed to a computer.
2. Lohmann allowed the turtles to swim under two experimental conditions: half of the turtles swam in Earth's magnetic field, and the other half swam in a reversed magnetic field.

Kenneth Lohmann/University of North Carolina

Results: The turtle hatchlings tested in Earth's magnetic field swam in an east-to-northeast direction on average, mimicking the direction they follow normally when migrating at sea. The turtle hatchlings tested in the reversed magnetic field on average swam in a direction 180° opposite that of the hatchlings swimming in Earth's magnetic field.

Conclusion: The results indicate that loggerhead sea turtle hatchlings have the ability to detect Earth's magnetic field and use it as a way to orient their migration. Their direction of migration, east to northeast, matches the inclination of Earth's magnetic field in the Atlantic Ocean where they migrate (see map figure). Lohmann believes that the magnetoreception system in the turtles involves magnetite.

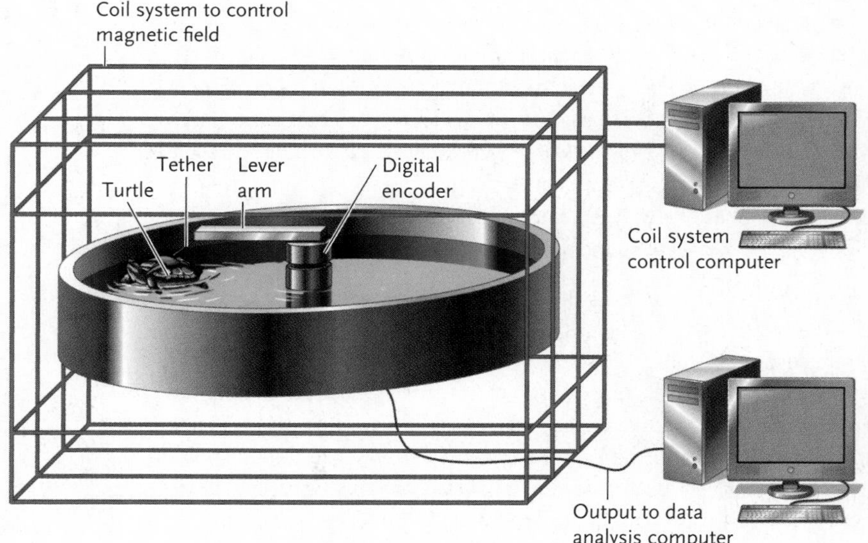
Coil system to control magnetic field
Turtle
Tether
Lever arm
Digital encoder
Coil system control computer
Output to data analysis computer

Source: K. J. Lohmann et al. 2001. Regional magnetic fields as navigational markers for sea turtles. *Science* 294:364–366.

Electroreceptors Are Used for Location of Prey or for Communication

Many sharks and bony fishes, some amphibians, and even some mammals (such as the star-nosed mole and duckbilled platypus) have specialized **electroreceptors** that detect electrical fields. The plasma membrane of an electroreceptor cell is depolarized by an electrical field, leading to the generation of action potentials. The electrical stimuli detected by the receptors are used to locate prey or navigate around obstacles in muddy water, or, by some fishes, to communicate. Some electroreception systems are passive—they detect electric fields in the environment, not the animal's own electric currents. Passive systems are used mainly to find prey. For example, the electroreceptors of sharks and rays can locate fish buried under the sand from the electrical currents generated by their prey's heartbeat or by the muscle contractions that move water over their gills.

Other electroreception systems are active—the animal emits and receives low voltage electrical signals, either to locate prey or to communicate with members of the same species. The electrical signals are generated by special electric organs. A few species, such as the electric eel and the electric catfish, produce discharges on the order of several hundred volts. These discharges are used to stun or kill prey. The voltage is high enough to stun, but not kill, a human.

STUDY BREAK 39.7

What are three ways electroreceptors are used in aquatic vertebrates?

UNANSWERED QUESTIONS

What happens when the senses get scrambled—when listening to music causes you to "see" colors, or when you "taste" certain words?

Synesthesia (joined senses) occurs when two senses, normally separate, are perceived together. For the most part, people with synesthesia—synesthetes—are born with it, and it tends to run in families. A recent study by Michael Esterman and his colleagues at the University of California, Berkeley, showed that the posterior parietal cortex, a region of the brain thought to be involved in sensory integration, appears to be crucial to sensory comingling. Some researchers think that this commingling is how the senses function early in development, when the nervous system is still immature. They believe that the senses normally separate from one another around four months after birth. In synesthetes, however, this separation is incomplete and two of their senses remain mingled.

London's Science Museum collaborated with Jamie Ward of University College London in an experiment that paired visual images and music. They wanted to determine if volunteers visiting the museum would prefer combinations of music and images designed by synesthetes over combinations that were designed by non-synesthetes. Interestingly, people found the synesthetic combinations more pleasing than the non-synesthetic ones. Thus, it is possible that everyone may have a built-in understanding of what sounds and colors go together.

In an evolutionary context, which "sense" developed first?

The descriptions of the senses in this chapter focus primarily on vertebrate sensory systems, but the nervous systems of many invertebrates can be quite complex. Indeed, squids, sea hares, leeches, horseshoe crabs, lobsters, and cockroaches have been quite instrumental in helping scientists understand the nervous system. As for senses, squids have sensitive eyes and accentuated smell and taste, and they respond to touch and vibration. Clearly, vertebrates are not the only multicellular organisms to develop senses.

However, is a nervous system necessary for organisms to have senses? Can single-celled organisms (which, of course, do not have nervous systems) respond to stimuli? Clearly, the answer is yes. Consider *Paramecium tetraurelia*, a single-celled organism covered with cilia that lives in water. It can detect substances in its environment and swim toward certain chemicals while avoiding others. It also responds to solid objects by turning when it runs into one. Thus, it has a "chemical sense" similar to taste or smell, and it responds to a type of "touch." We will probably never know which sense developed first, but some type of touch or chemical sense seems the most likely.

Think Critically

1. You serve your college roommate, who is very fond of scrambled eggs, scrambled eggs colored with green food coloring. You video your roommate eating the eggs, and show the video in your biology class. What sensory receptors are you hoping to stimulate most noticeably? If you were to suggest to your professor an imaging test for synesthetes, what imaging method might you choose (see Chapter 38)?
2. Does the fact that some prokaryotes contain a microbial rhodopsin suggest to you that the sense of vision has early evolutionary origins?

Rona Delay

Rona Delay is an associate professor in the Department of Biology at the University of Vermont. Her research centers on understanding how sensory receptors change or transduce information about the external world into a language the brain can understand. The focus of her research is the sense of smell. To learn more about Dr. Delay's work, go to http://www.uvm.edu/~biology/?Page=faculty/ronadelay.php&SM=facultysubmenu.html.

REVIEW KEY CONCEPTS

Go to **CENGAGENOW** at www.cengage.com/login to access quizzing, animations, exercises, articles, and personalized homework help.

39.1 Overview of Sensory Receptors and Pathways

- Sensory receptors are formed by the endings of afferent neurons, endings of afferent neurons enclosed in specialized structures, or specialized cells adjacent to the neurons. They detect stimuli such as mechanical pressure, sound waves, light, or specific chemicals. Action potentials generated by the receptors are carried by the axons of afferent neurons to pathways leading to specific parts of the brain, where signals are processed into sensory sensations (Figure 39.1).
- Receptors are specialized as mechanoreceptors, photoreceptors, chemoreceptors, thermoreceptors, and nociceptors. Some animals have receptors that detect electrical or magnetic fields.
- The routing of information from sensory receptors to particular regions of the brain identifies a specific stimulus as a sensation. The intensity of a stimulus is determined by the frequency of action potentials traveling along the neural pathways and the number of afferent neurons carrying action potentials.
- Many sensory systems show sensory adaptation, in which the frequency of action potentials decreases while a stimulus remains constant. Some sensory receptors, such as those related to pain, show no sensory adaptation.
- The brain uses signals from sensory receptors to generate an interpretation of the external and internal environments. That interpretation is our perception of the world. Each type of organism has a unique perception of the world.

Animation: Action potentials

39.2 Mechanoreceptors and the Tactile and Spatial Senses

- Mechanoreceptors detect touch, pressure, acceleration, or vibration. Touch and pressure receptors are free nerve endings or encapsulated nerve endings of sensory neurons (Figure 39.2).
- Mechanoreceptors called proprioceptors detect stimuli used by the CNS to monitor and maintain body and limb positions.
- Receptors in muscles, tendons, and joints of vertebrates detect changes in stretch and tension of body parts (Figure 39.3).
- Proprioceptors based on sensory hair cells generate action potentials when the hairs are moved (Figures 39.4–39.6).

Animation: Sensory receptors in the human skin

Animation: Dynamic equilibrium

39.3 Mechanoreceptors and Hearing

- Many invertebrates have mechanoreceptors in their skin or other surface structures that detect sound and other vibrations.
- Hearing relies on sensory hair cells in organs that respond to the vibrations of sound waves.
- In terrestrial vertebrates, the ear consists of three parts. The outer ear directs sound to the tympanic membrane (eardrum). Vibrations of the tympanic membrane are transmitted through one or more bones in the middle ear to the fluid-filled inner ear. In the inner ear, the vibrations are transmitted through membranes that bend the stereocilia of hair cells, leading to bursts of action potentials that are reflected in the frequency of the sound waves (Figure 39.8).

Animation: Ear structure and function

Animation: Properties of sound

39.4 Photoreceptors and Vision

- Invertebrates possess many forms of eyes, from the simplest, an ocellus, to single-lens eyes that are similar to vertebrate eyes (Figures 39.9–39.11).
- The photoreceptors of all animal eyes contain the pigment retinal, which absorbs the energy of light and uses it to generate changes in membrane potential.
- The transparent cornea admits light into the vertebrate eye. Behind the cornea, the iris controls the diameter of the pupil, regulating the amount of light that strikes the lens. The lens focuses an image on the retina lining the back of the eye, where photoreceptors and neurons carry out the initial integration of information detected by the photoreceptors (Figure 39.12).
- In terrestrial vertebrates, the lens is focused by adjusting its shape (Figure 39.13). The retina contains two types of photoreceptors, rods and cones. Rods are specialized for detecting light of low intensity; cones are specialized for detecting light of different wavelengths, which are perceived as colors.
- The light-absorbing pigment in photoreceptor cells consists of retinal combined with an opsin protein. When it absorbs light, retinal changes form, initiating reactions that alter the amount of neurotransmitter released by the photoreceptor cells (Figures 39.14 and 39.15).
- Rods and cones are linked to neurons in the retina that perform the initial processing of visual information. The processed signal is sent via the optic nerve through the lateral geniculate nuclei to the visual cortex (Figures 39.16 and 39.17).

Animation: Eye structure

Animation: Visual accommodation

Animation: Focusing problems

Animation: Organization of cells in the retina

Animation: Receptive fields

Animation: Path to visual cortex

Animation: Pathway to visual cortex

39.5 Chemoreceptors

- Chemoreceptors respond to the presence of specific molecules in the environment. In vertebrates, they form parts of receptor organs for taste (gustation) and smell (olfaction).
- Taste receptors detect molecules from food or other objects that come into direct contact with the receptor and are used primarily to identify foods (Figures 39.18 and 39.20).
- Olfactory receptors detect molecules from distant sources; besides identifying food, they are used to detect predators and prey, identify family and group members, locate trails and territories, and communicate (Figures 39.19 and 39.21).

Animation: Taste receptors

Animation: Olfactory pathway

39.6 Thermoreceptors and Nociceptors

- Thermoreceptors, which consist of free nerve endings located at the body surface and in limited numbers in the body interior, detect changes in body temperature.
- Nociceptors, located on both the body surface and interior, detect stimuli that can damage body tissues. Information from these receptors is integrated in the brain into the sensation of pain.

Animation: Referred pain

39.7 Magnetoreceptors and Electroreceptors

- Some vertebrates have electroreceptors that detect electrical currents and fields, or magnetoreceptors that detect magnetic fields (Figure 39.23).

UNDERSTAND AND APPLY

Test Your Knowledge

1. An ambulance siren in close proximity to a dog can cause the dog to howl in pain. Which receptors are responsible for this response?
   a. Thermoreceptors and chemoreceptors
   b. Photoreceptors and nociceptors
   c. Mechanoreceptors and nociceptors
   d. Chemoreceptors and mechanoreceptors
   e. Photoreceptors and chemoreceptors
2. The sensation of sheets lessens if you lie still in bed. The response is due to:
   a. sensory adaptation of mechanoreceptors.
   b. sensory adaptation of nociceptors.
   c. pH change receptors associated with sleep.
   d. the vestibular apparatus.
   e. vibration detecting systems.
3. The following situation is associated with movement and position in the human body:
   a. Statoliths in statocysts bend sensory hairs and trigger action potentials.
   b. If sensory hairs in the utricle are oriented horizontally and those in the saccule are oriented vertically, the person is lying down.
   c. When the head rotates, the endolymph in the semicircular canal pulls the cupula with it to activate sensory hair cells.
   d. Displacement of the utricle and saccule generates action potentials.
   e. If the body is spinning at a constant rate and direction, the cupula is displaced and action potentials are initiated.
4. Neuromast function is best described as:
   a. nonadapting pain receptors.
   b. stereocilia that bend and generate action potentials.
   c. statoliths that detect motion.
   d. motor axons that activate motion.
   e. cupulas that detect vibrations.
5. Structures in the vertebrate ear are activated by sound waves in the following order:
   a. oval window, tympanic membrane, semicircular canals, Golgi tendon organ, incus, malleus, stapes.
   b. organ of Corti, malleus, incus, stapes, auditory nerve, tympanic membrane.
   c. eustachian tube, round window, vestibular canal, tympanic canal, cochlear canal, oval window, pinna.
   d. basilar membrane, tectorial membrane, otoliths, utricle, saccule, malleus, cochlea.
   e. pinna, tympanic membrane, malleus, incus, stapes, oval window, cochlear duct.
6. The eyes of vertebrates and cephalopods are similar in structure and function. A difference between the vertebrate eye and the cephalopod eye is that the vertebrate eye has:
   a. an iris surrounding the pupil, whereas in cephalopods the pupil surrounds the iris.
   b. a lens that changes shape when focusing, whereas in cephalopods the lens moves back and forth to focus.
   c. a retina that moves in the socket when recording the image, whereas in cephalopods the retina changes shape when stimulated.
   d. a pupil that shrinks in size in bright light, whereas cephalopods have a pupil that enlarges in bright light.
   e. retinal synthesized from vitamin A, whereas cephalopods lack retinal.
7. Which of the following events does not occur during light absorption in the vertebrate eye?
   a. The retinal component of rhodopsin changes from *cis* to *trans* form.
   b. Rhodopsin, a G-protein–coupled receptor, triggers a signal transduction pathway to close Na^+ channels in the plasma membrane.
   c. The light stimulus passes from rods and cones to bipolar cells and horizontal cells and then to ganglion cells, whose axons compose the optic nerve.
   d. As light absorption increases, the rhodopsin response causes an increase in the release of neurotransmitters.
   e. When integrating information across the retina, horizontal cells connect the rods and cones, and amacrine cells join with the bipolar cells and ganglion cells.
8. The variety of color seen by humans is directly dependent upon the:
   a. activation of three different photopsins in cones.
   b. transmission of an image to separate brain hemispheres by the optic chiasm.
   c. transmission of impulses from rods across the lateral geniculate nuclei.
   d. lateral inhibition by amacrine cells.
   e. light stimulation of all photoreceptor types equally.
9. In terrestrial animals:
   a. the hairs of taste receptors are derived from cilia and contain microtubules.
   b. the hairs of smell receptors are derived from microvilli and contain microfilaments.
   c. signals from taste receptors are relayed to the temporal lobes.
   d. information from olfactory receptors is processed in the parietal lobes.
   e. connections from the olfactory bulbs lead to the limbic system.
10. In the human response to temperature or pain:
    a. all Ca^{2+} channels act as pain receptors.
    b. cold receptors are activated between 27°C and 37°C.
    c. pain receptors decrease the rate at which they send out action potentials if the pain is constant.
    d. nociceptors activated by capsaicin transmit pain messages to the brain.
    e. the CNS releases glutamate or substance P to dull the pain sensation.

Discuss the Concepts

1. Humans have about 200 million photoreceptors in two eyes, and about 32,000 sensory hair cells in two ears. About 3% of the somatosensory cortex is devoted to hearing, whereas roughly 30% of it is devoted to visual processing. Suggest an explanation for these differences from the perspective of natural selection and adaptation.
2. In owls and many other birds of prey, the fovea is located toward the top of the retina rather than at the center as in humans. This arrangement correlates with the birds' hunting behavior, in which they look down when they fly, scanning the ground for a meal. With this arrangement in mind, why do you think the standing owl in the picture is turning its head upside down?

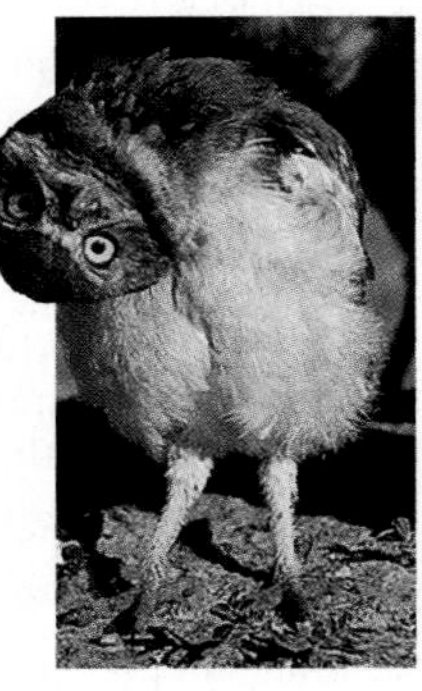
Chase Swift/Corbis

3. A patient made an appointment with her doctor because she was experiencing recurrent episodes of dizziness. Her doctor asked questions to distinguish whether she had sensations of lightheadedness, as if she were going to faint, or vertigo, as if she or objects near her were spinning around. Why was this clarification important in the evaluation of her condition?

Design an Experiment

The fruit fly *Drosophila melanogaster* can distinguish a large repertoire of odors in the environment. Their response may be to move toward food or away from danger. Moreover, particular odors play an important role in their mating behavior. The olfactory organs of a fruit fly are the antennae and an elongated bulge on the head called the maxillary pulp. Because of the ease with which fruit fly genes can be manipulated, identifying and studying their olfactory receptors likely would contribute significantly to our understanding of neural pathways of odor recognition more generally. How could you identify candidate fruit fly genes that encode components of olfactory receptors?

Interpret the Data

The graph in the right column plots relative olfactory bulb size for pairs of closely related birds (a through m), with each pair having a species that is more active at night (nocturnal) and one that is more active during the day (diurnal). The researchers hypothesized that nocturnal birds would have larger olfactory bulbs because they need to depend on senses other than sight.

1. From these data, which, if any, of the bird pairs supports the hypothesis of the researchers?
2. Do any of the pairs provide evidence against the hypothesis?
3. Looking at these data as a whole, can you draw any overall conclusions?

Be sure to explain your answers.

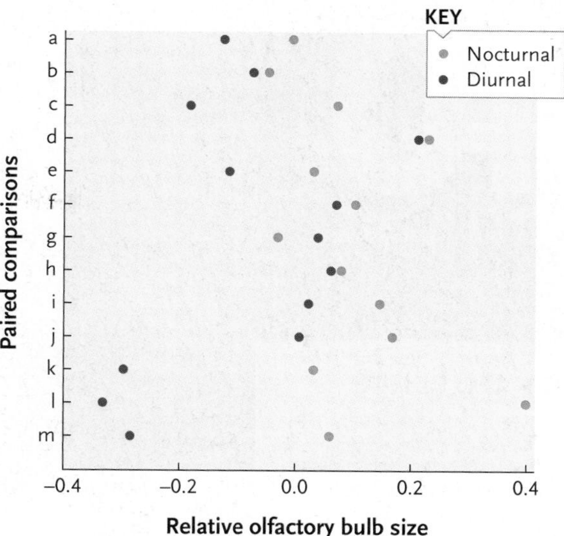
KEY
Nocturnal
Diurnal
Paired comparisons
a
b
c
d
e
f
g
h
i
j
k
l
m
−0.4
−0.2
0.0
0.2
0.4
Relative olfactory bulb size

Source: S. Healy and T. Guilford. 1990. Olfactory-bulb size and nocturnality in birds. *Evolution* 44:339–346.

Apply Evolutionary Thinking

In 2005, researchers took saliva and blood samples from six cats, including domestic cats, a tiger, and a cheetah, and found that all have a defective gene for one of the two chemoreceptor proteins needed to identify food as sweet. (The scientists conjecture that the lack of a sweet tooth may explain why cats are finicky eaters.) What are the evolutionary implications of the finding?

Express Your Opinion

Noise pollution from commercial shipping and other human activities generates low-frequency sounds that are believed to interfere with the acoustical signals that whales use for navigation, location of food, and communication. To what extent should we limit these activities to protect whales against potential harm? Would you support banning activities that exceeded a certain noise level from U.S. territorial waters? If so, how would you get other nations to do the same? Go to www.cengage.com/login to investigate both sides of the issue and then vote.

40

STUDY OUTLINE

Two North American bull elks contesting for cows in Yellowstone National Park. The shorter days of autumn trigger hormone production, battling, and reproductive behavior.

The Endocrine System

Why It Matters. . . Every September, as the days grow shorter and autumn approaches, bull elks *(Cervus canadensis)* begin to strut their stuff. Although they have grazed peacefully together at high mountain elevations from the Yukon to Arizona, they now become testy with each other. They also rasp at tree branches and plow the ground with their antlers. Soon, they descend to lower elevations, where the cow elks have been feeding in large nursery groups with their calves and yearlings.

The bulls move in among the cows and chase away the male yearlings. As part of the mating ritual, the bulls bugle, square off, strut, and circle; then they clash their antlers together, attempting to drive each other from the cows. The winning males claim harems of about 10 females each, a major prize.

After the mating season ends, tranquility returns. The cows again graze in herds; the males form now-friendly bachelor groups that also feed quietly in the meadows. Eating is their major occupation, storing nutrients in preparation for the snowy winter. The young will be born eight to nine months later, when summer returns.

The next year, the shortening days of late summer and fall again trigger the transition to mating behavior. Detected by the eyes and registered in the brain, reduced day length initiates changes in the secretion of long-distance signaling molecules called **hormones** (*horme* = to excite). Hormones released from one group of cells are transported through the circulatory system to other cells, their target cells, whose activities they change. Among the changes will be a rise in the concentration of hormones responsible for mating behavior.

We too are driven by our hormones. They control our day-to-day sexual behavior—often as outlandish as that of a bugling bull elk—as well as a host of other functions, from the concentration of salt in our blood, to body growth, to the secretion of digestive juices. Along with the central nervous system, hormones coordinate the activities of multicellular life.

The best-known hormones are secreted by cells of the **endocrine system** (*endo* = within; *krinein* = to separate), although hormones are actually produced by almost all organ systems in the body. The endocrine system, like the nervous system, regulates and coordinates distant organs. The two systems are structurally, chemically, and functionally related, but they control different types of activities. The nervous system, through its high-speed electrical signals, enables an organism to interact rapidly with the external environment, while the endocrine system mainly controls activities that involve slower, longer-acting responses. Typical responses to hormones may persist for hours, weeks, months, or even years.

The mechanisms and functions of the endocrine system are the subjects of this chapter. As in other chapters of this unit, we pay particular attention to the endocrine system of humans and other mammals. <

40.1 Hormones and Their Secretion

Cells signal other cells using neurotransmitters, hormones, and local regulators. Recall from Chapters 37, 38, and 39 that a neurotransmitter is a chemical released by an axon terminal at a synapse that affects the activity of a postsynaptic cell. Our focus in this chapter is hormones and local regulators.

The Endocrine System Includes Four Major Types of Cell Signaling

Four types of cell signaling occur in the endocrine system: classical endocrine signaling, neuroendocrine signaling, paracrine regulation, and autocrine regulation. In *classical endocrine signaling,* hormones are secreted into the extracellular fluid (ECF) by the cells of ductless secretory organs called **endocrine glands (Figure 40.1A).** (In contrast, *exocrine glands,* such as the sweat and salivary glands, release their secretions into ducts that lead outside the body or into the cavities of the digestive tract, as described in Section 36.2.) The hormones are circulated throughout the body in the blood and, as a result, most body cells are constantly exposed to a wide variety of hormones. (The cells of the central nervous system are sequestered from the general circulatory system by the blood–brain barrier, described in Section 38.3.) Only *target cells* of a hormone, those with *receptor proteins* recognizing and binding that hormone, respond to it. Through these responses, hormones control such vital functions as digestion, osmotic balance, metabolism, cell division, reproduction, and development. The action of hormones may either speed up or inhibit these cellular processes. For example, growth hormone stimulates cell division,

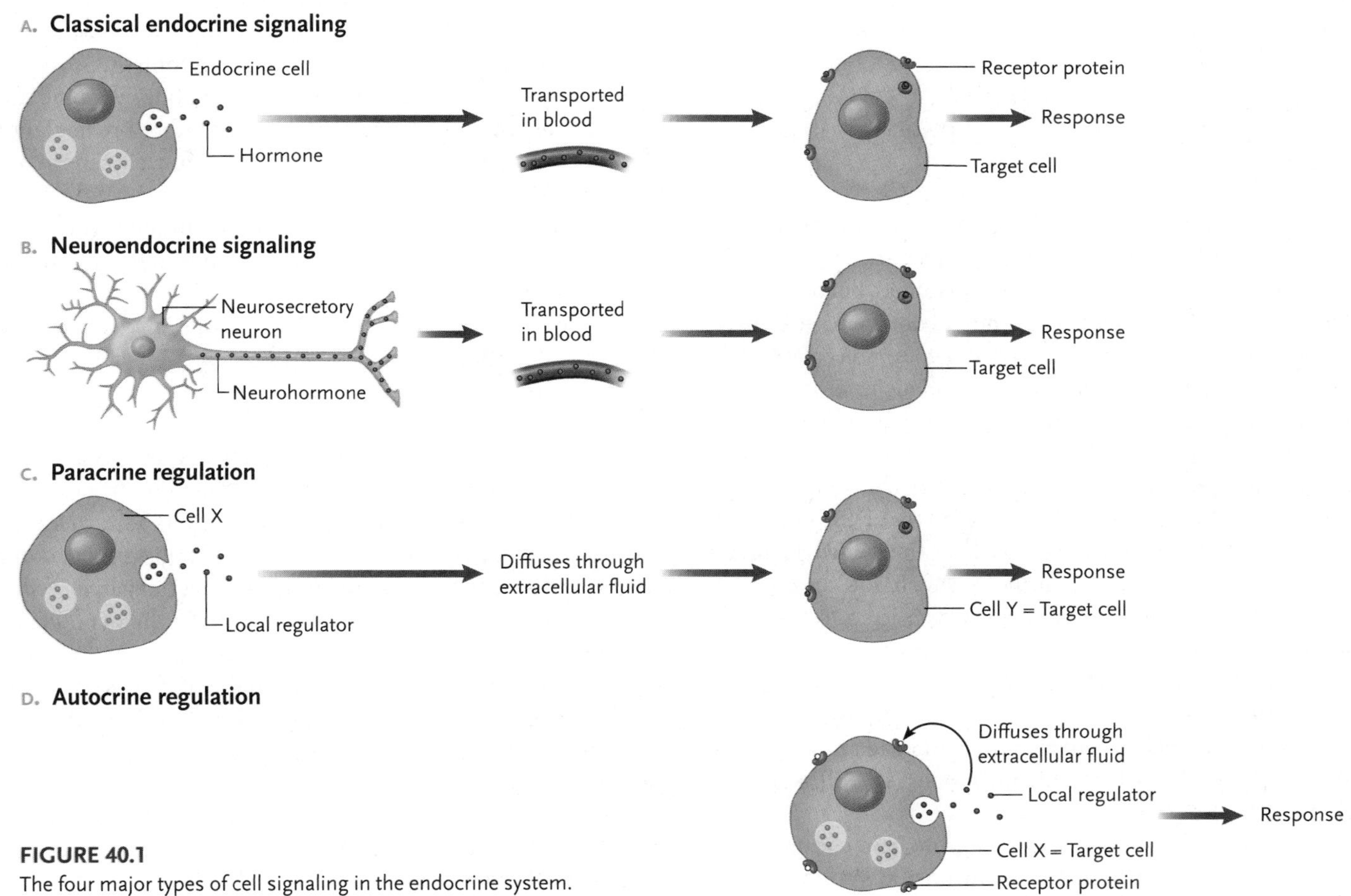
A. Classical endocrine signaling
Endocrine cell
Hormone
Transported in blood
Receptor protein
Response
Target cell
B. Neuroendocrine signaling
Neurosecretory neuron
Neurohormone
Transported in blood
Response
Target cell
C. Paracrine regulation
Cell X
Local regulator
Diffuses through extracellular fluid
Response
Cell Y = Target cell
D. Autocrine regulation
Diffuses through extracellular fluid
Local regulator
Response
Cell X = Target cell
Receptor protein

FIGURE 40.1
The four major types of cell signaling in the endocrine system.

whereas glucocorticoids inhibit glucose uptake by most cells in the body.

Hormones are cleared from the body by enzymatic breakdown, mainly in the liver and kidneys, but also in target cells themselves. Breakdown products are excreted in urine and feces. Depending on the hormone, the breakdown takes minutes to days.

In *neuroendocrine signaling,* specialized neurons called **neurosecretory neurons** release a hormone called a *neurohormone* into the circulatory system when appropriately stimulated **(Figure 40.1B).** The neurohormone is distributed by the circulatory system and elicits a response in target cells that have receptors for the hormone. Note that both neurohormones and neurotransmitters are secreted by neurons. An important distinction between neurohormones and neurotransmitters is that neurohormones are carried to target cells by the blood, whereas neurotransmitters act across a synaptic cleft (see Figure 37.6B). However, both neurohormones and neurotransmitters function in the same way—they cause cellular responses by interacting with specific receptors on target cells. For instance, gonadotropin-releasing hormone, a neurohormone secreted by the hypothalamus, controls the release of luteinizing hormone from the pituitary.

In *paracrine regulation,* a cell releases a signaling molecule that diffuses through the ECF and acts on nearby cells—regulation is *local* rather than at a distance, as is the case with hormones and neurohormones **(Figure 40.1C).** In some cases the local regulator acts on the same cells that produced it; this is called *autocrine regulation* **(Figure 40.1D).** For example, many of the growth factors that regulate cell division and differentiation act in both a paracrine and autocrine fashion.

Hormones and Local Regulators Can Be Grouped into Four Classes Based on Their Chemical Structure

More than 60 hormones and local regulators have been identified in humans. Many human hormones are either identical or very similar in structure and function to those in other animals, but other vertebrates as well as invertebrates have hormones not found in humans. Most hormones and local regulators fall into four molecular classes: amine, peptide, steroid, and fatty acid-derived molecules.

Amine hormones are involved in classical endocrine and neuroendocrine signaling. Most amine hormones are based on tyrosine. With one major exception, they are hydrophilic molecules, which diffuse readily into the blood and ECF. On reaching a target cell, they bind to receptors at the cell surface. The amine hormones include dopamine, epinephrine, and norepinephrine, already familiar as neurotransmitters released by some neurons (see Section 37.3). The exception is thyroxine, a hydrophobic amine hormone secreted by the thyroid gland. This hormone, which is based on a pair of tyrosines, passes freely through the plasma membrane and binds to a receptor inside the target cell, as do steroid hormones (see following discussion and Section 7.5).

The *peptide hormones* consist of amino acid chains, ranging in length from as few as three amino acids to more than 200. Some have carbohydrate groups attached to the amino acid chain. They are involved in classical endocrine signaling and neuroendocrine signaling. Peptide hormones are released into the ECF, and from there they enter the blood. One large group of peptide hormones, the **growth factors,** regulates the division and differentiation of many cell types in the body. Many growth factors act in both paracrine and autocrine manners as well as in classical endocrine signaling. Because they can switch cell division on or off, growth factors are an important focus of cancer research.

Steroid hormones are involved in classical endocrine signaling. All are hydrophobic molecules derived from cholesterol. They combine with hydrophilic carrier proteins to form water-soluble complexes that can diffuse through the ECF and enter the bloodstream. On contacting a cell, the hormone is released from its carrier protein, passes through the plasma membrane of the target cell, and binds to internal receptors in the nucleus or cytoplasm. Steroid hormones include aldosterone, cortisol, and the sex hormones. Steroid hormones may vary little in structure, but produce very different effects. For example, testosterone and estradiol, two major sex hormones responsible for the development of mammalian male and female characteristics, respectively, differ in little more than the presence or absence of a methyl group (see Figure 3.16).

Fatty acid-derived molecules are involved in paracrine and autocrine regulation. **Prostaglandins,** for example, are important as local regulators. Virtually every cell can secrete prostaglandins, and they are present at essentially all times. In semen, they enhance the transport of sperm through the female reproductive tract by increasing the contractions of smooth muscle cells, particularly in the uterus. During childbirth, prostaglandins secreted by the placenta work with a peptide hormone called oxytocin to stimulate labor contractions. Other prostaglandins induce contraction or relaxation of smooth muscle cells in many parts of the body, including blood vessels and air passages in the lungs. When released as a product of membrane breakdown in injured cells, prostaglandins may also intensify pain and inflammation.

Many Hormones Are Regulated by Feedback Pathways

The secretion of many hormones is regulated by feedback pathways. Most of these pathways are controlled by negative feedback—that is, a product of the pathway inhibits an earlier step in the pathway (see Section 36.4). For example, in some mammals, secretion by the thyroid gland is regulated by a negative feedback loop **(Figure 40.2).** Neurosecretory neurons in the hypothalamus secrete thyroid-releasing hormone (TRH) into the ECF and capillaries connecting the hypothalamus to the pituitary gland. In response, the pituitary releases thyroid-stimulating hormone (TSH) into the blood, which stimulates the thyroid gland to release thyroid hormones. As the thyroid hormone concentration in the blood increases, it begins to inhibit TSH secretion by the pituitary; this action is the negative feedback step. As a result, secretion of the thyroid hormones is reduced.

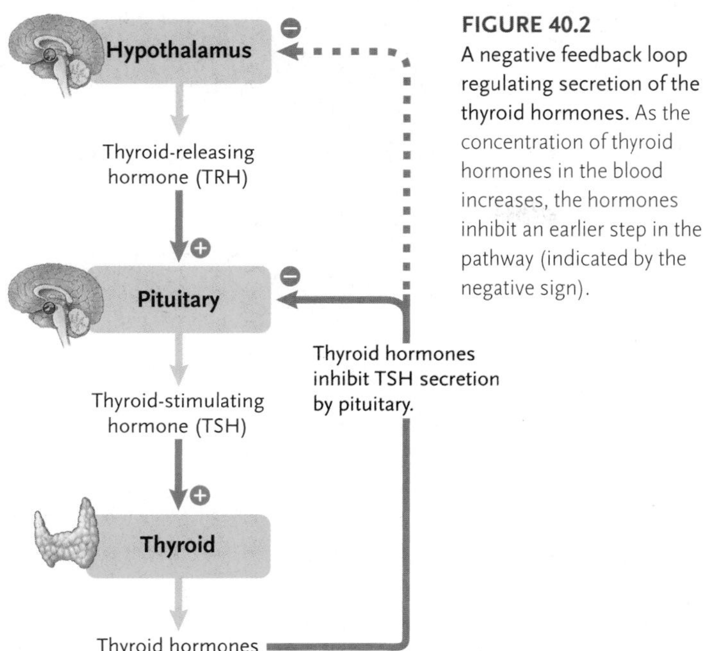
Hypothalamus
Thyroid-releasing hormone (TRH)
Pituitary
Thyroid hormones inhibit TSH secretion by pituitary.
Thyroid-stimulating hormone (TSH)
Thyroid
Thyroid hormones

FIGURE 40.2 A negative feedback loop regulating secretion of the thyroid hormones. As the concentration of thyroid hormones in the blood increases, the hormones inhibit an earlier step in the pathway (indicated by the negative sign).

Body Processes Are Regulated by Coordinated Hormone Secretion

Although we will talk mostly about individual hormones in the remainder of the chapter, body processes are affected by more than one hormone. For example, the blood concentrations of glucose, fatty acids, and ions such as Ca^{2+}, K^+, and Na^+ are regulated by the coordinated activities of several hormones secreted by different glands. Similarly, body processes such as oxidative metabolism, digestion, growth, sexual development, and reactions to stress are all controlled by multiple hormones.

In many of these systems, negative feedback loops adjust the levels of secretion of hormones that act in antagonistic (opposing) ways, creating a balance in their effects that maintains homeostasis. For example, consider the regulation of fuel molecules such as glucose, fatty acids, and amino acids in the blood. We usually eat three meals a day and fast to some extent between meals. During these periods of eating and fasting, four hormone systems act in coordinated fashion to keep the fuel levels in balance: (1) insulin and glucagon, secreted by the pancreas; (2) growth hormone, secreted by the anterior pituitary; (3) epinephrine and norepinephrine, released by the sympathetic nervous system and the adrenal medulla; and (4) glucocorticoid hormones, released by the adrenal cortex.

The entire system of hormones regulating fuel metabolism resembles the failsafe mechanisms designed by human engineers, in which redundancy, overlapping controls, feedback loops, and multiple safety valves ensure that vital functions are maintained at constant levels in the face of changing and even extreme circumstances.

STUDY BREAK 40.1 <

1. **Distinguish between a hormone and an exocrine secretion.**
2. **How is endocrine signaling different from autocrine signaling?**

40.2 Mechanisms of Hormone Action

Hormones control cell functions by binding to receptor molecules on or in their target cells. Small quantities of hormones can typically produce profound effects in cells and body functions as a result of **amplification** (see Figure 7.5). In amplification, binding of a hormone to a receptor triggers activation of many proteins, each of which then activates an even larger number of proteins for the next step in the cellular reaction pathway, and so on, increasing in magnitude for each step in the pathway. It has been estimated that, by amplification, a single molecule of epinephrine acting on a liver cell will liberate 10^6 molecules of glucose from stored glycogen.

Hydrophilic Hormones Bind to Surface Receptors, Activating Protein Kinases Inside Cells

Hormones that bind to receptor molecules in the plasma membrane—primarily hydrophilic amine and peptide hormones—produce their responses through signal transduction pathways (see Section 7.2). In brief, when a surface receptor binds a hormone, the receptor is activated and transmits the signal through the plasma membrane. Within the cell, the signal is transduced, changed into a form that causes the cellular response **(Figure 40.3A).** Typically, the reactions of signal transduction pathways involve protein kinases, enzymes that add phosphate groups to proteins. Adding a phosphate group to a protein may activate it or inhibit it, depending on the protein and the reaction. The particular response produced by a hormone depends on the kinds of protein kinases activated, the type of cell that can respond, and the types of target proteins they phosphorylate.

Two major types of surface receptors bind hydrophilic hormones: receptor tyrosine kinases and G-protein–coupled receptors. Receptor tyrosine kinases have a built-in protein kinase on the cytoplasmic side of the receptor (Figure 40.3A illustrates this type of receptor). Binding of a signal molecule to this type of receptor turns on the receptor's protein kinase, which activates the receptor. The activated receptor then initiates a signaling cascade within the cell, typically involving protein kinases, which phosphorylate target proteins, resulting in a cellular response. The cytoplasmic reactions receptor tyrosine kinases control when they are activated are described in detail in Section 7.3. Insulin, a peptide hormone that lowers glucose concentration in the blood, elicits a cellular response using a signal transduction pathway initiated by activating a receptor tyrosine kinase. Insulin acts mainly by binding to receptors on liver cells, adipose tissue (fat), and skeletal muscle cells to stimulate glucose transport into cells, conversion of glucose to glycogen, and other metabolic activities.

G-protein–coupled receptors lack a built-in protein kinase. This type of receptor responds to the binding of a hormone by activating a G protein associated with the cytoplasmic end of the receptor. Activated G protein then activates an effector molecule, which then generates a second messenger molecule. Second messengers activate protein kinases in the cell, which elicit the cellular response by phosphorylating target proteins. The cytoplasmic

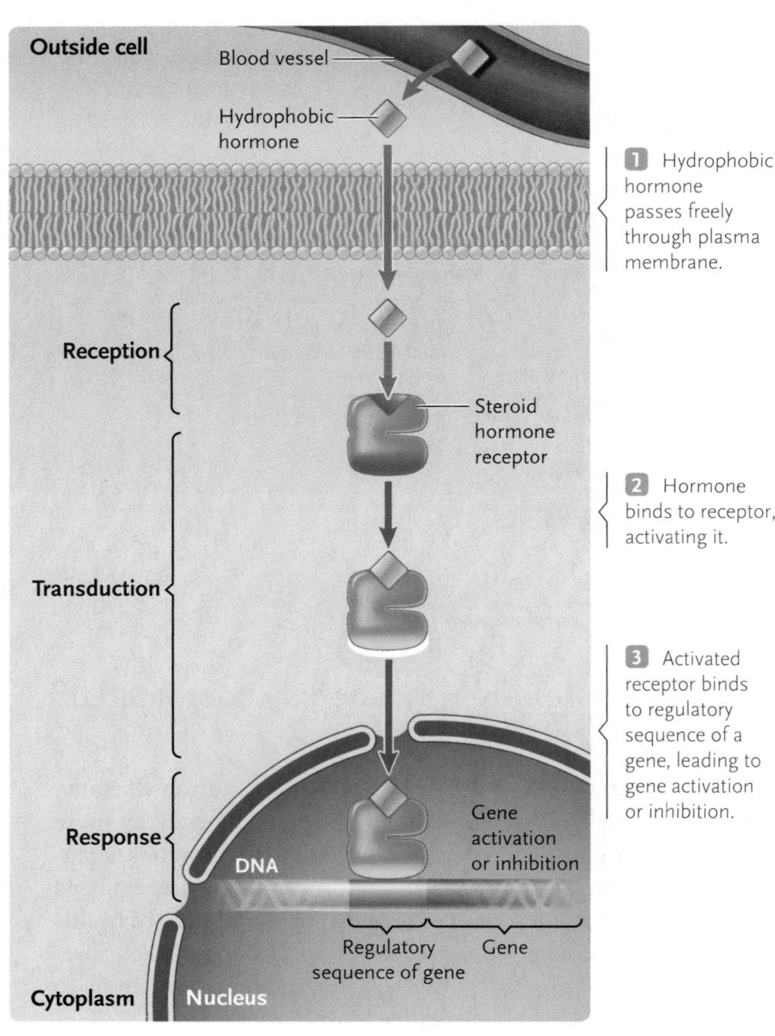
A. Hormone binding to receptor in the plasma membrane
Outside cell
Hydrophobic hormone
Blood vessel
Signal
Reception
1 Hormone binds to surface receptor and activates it.
Cytoplasmic end of receptor
Activation
Pathway molecule A
Activation
Transduction
2 Activated receptor triggers a signal transduction pathway.
Pathway molecule B
Activation
Pathway molecule C
Molecule that brings about response
Response
3 Transduction of the signal leads to cellular response.
Change in cell
Cytoplasm
B. Hormone binding to receptor inside the cell
Outside cell
Blood vessel
Hydrophobic hormone
1 Hydrophobic hormone passes freely through plasma membrane.
Reception
Steroid hormone receptor
2 Hormone binds to receptor, activating it.
Transduction
3 Activated receptor binds to regulatory sequence of a gene, leading to gene activation or inhibition.
Response
DNA
Gene activation or inhibition
Regulatory sequence of gene
Gene
Cytoplasm
Nucleus

FIGURE 40.3
Outline of the reaction pathways activated by hormones that bind to receptor proteins in the plasma membrane **(A)** or inside cells **(B)**.
In both mechanisms, the signal—the binding of the hormone to its receptor—is transduced to produce the cellular response.

reactions that G-protein–coupled receptors control when activated are described in detail in Section 7.4. Glucagon, a peptide hormone that raises glucose concentration in the blood, elicits a cellular response using a signal transduction pathway initiated by activating a G-protein–coupled receptor. When glucagon binds to surface receptors on liver cells, it triggers the breakdown of the glycogen stored in those cells into glucose. The glucose then is released into the circulatory system.

Hydrophobic Hormones Bind to Receptors Inside Cells, Activating or Inhibiting Genetic Regulatory Proteins

After passing through the plasma membrane, the hydrophobic steroid and thyroid hormones bind to internal receptors in the nucleus or cytoplasm (**Figure 40.3B,** and described in detail in Section 7.5). Binding of the hormone activates the receptor, which then binds to a regulatory sequence of specific genes. Depending on the gene, binding the regulatory sequence either activates or inhibits its transcription, leading to changes in protein synthesis that accomplish the cellular response. The characteristics of the response depend on the specific genes controlled by the activated receptors, and on the presence of other proteins that modify the activity of the receptor.

One of the actions of the steroid hormone aldosterone illustrates the mechanisms triggered by internal receptors **(Figure 40.4).** If blood pressure falls below optimal levels, aldosterone is secreted by the adrenal glands. The hormone affects only kidney cells that contain the aldosterone receptor in their cytoplasm. When activated by aldosterone, the receptor binds to the regulatory sequence of a gene, leading to the synthesis of proteins that increase reabsorption of Na^+ by the kidney cells. The resulting increase in Na^+ concentration in body fluids increases water retention and, with it, blood volume and pressure.

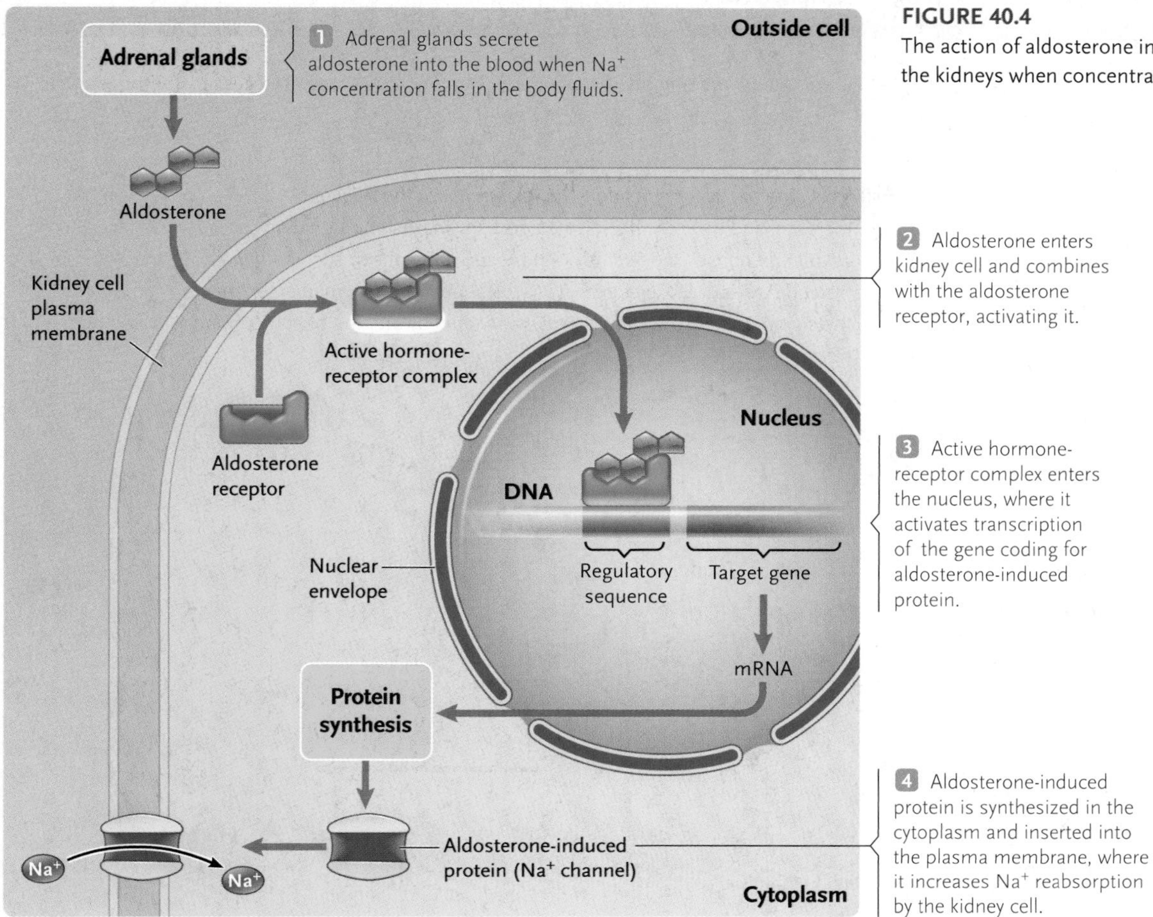
Adrenal glands
1 Adrenal glands secrete aldosterone into the blood when Na+ concentration falls in the body fluids.
Outside cell
Aldosterone
Kidney cell plasma membrane
Active hormone-receptor complex
Aldosterone receptor
Nucleus
DNA
Nuclear envelope
Regulatory sequence
Target gene
mRNA
Protein synthesis
Aldosterone-induced protein (Na+ channel)
Na+
Na+
Cytoplasm
2 Aldosterone enters kidney cell and combines with the aldosterone receptor, activating it.
3 Active hormone-receptor complex enters the nucleus, where it activates transcription of the gene coding for aldosterone-induced protein.
4 Aldosterone-induced protein is synthesized in the cytoplasm and inserted into the plasma membrane, where it increases Na+ reabsorption by the kidney cell.

FIGURE 40.4
The action of aldosterone in increasing Na^+ reabsorption in the kidneys when concentration of the ion falls in the blood.

Target Cells May Respond to More Than One Hormone, and Different Target Cells May Respond Differently to the Same Hormone

A single target cell may have receptors for several hormones and respond differently to each hormone. For example, vertebrate liver cells have receptors for the pancreatic hormones insulin and glucagon. Insulin increases glucose uptake and conversion to glycogen, which decreases blood glucose levels, while glucagon stimulates the breakdown of glycogen into glucose, which increases blood glucose levels.

Conversely, particular hormones interact with different types of receptors in or on a range of target cells. Different responses are then triggered in each target cell type because the receptors trigger different transduction pathways. For example, the amine hormone epinephrine secreted by the adrenal medulla prepares the body for handling stress (including dangerous situations) and physical activity. (Epinephrine is discussed in more detail in Section 40.4.) In mammals, epinephrine can bind to several different G-protein–coupled receptors (see Section 7.4) known as **adrenergic receptors** because of the adrenal origin of the hormones that bind to them. These receptors are categorized into two types: α-adrenergic receptors and β-adrenergic receptors. (The experimental demonstration that binding of epinephrine to a specific adrenergic receptor triggers a cellular response is described in **Figure 40.5**.) When epinephrine binds to an adrenergic receptor on a smooth muscle cell, such as that of a blood vessel, it triggers a response pathway that causes the cells to constrict, cutting off circulation to peripheral organs. When epinephrine binds to β_1-adrenergic receptors on heart muscle cells, the contraction rate of the cells increases, which in turn enhances blood supply. When epinephrine binds to β_2-adrenergic receptors on liver cells, it stimulates the breakdown of glycogen to glucose, which is released from the cell. The overall effect of these, and a number of other, responses to epinephrine secretion is to supply energy to the major muscles responsible for locomotion—the body is now prepared for handling stress or for physical activity.

Different receptors binding hydrophobic hormones also may generate diverse responses. *Insights from the Molecular Revolution* (pp. 926–927) describes an investigation that tested the cellular responses produced by different receptors binding the same steroid hormone.

In summary, the mechanisms by which hormones work have four major features:

1. Only the cells that contain surface or internal receptors for a particular hormone respond to that hormone.
2. Once bound by their receptors, hormones may produce a response that involves stimulation or inhibition of cellular processes through the specific types of internal molecules triggered by the hormone action.
3. Because of the amplification that occurs in both the surface and internal receptor mechanisms, hormones are effective in very small concentrations.
4. The response to a hormone differs among target organs.

FIGURE 40.5 **Experimental Research**

Demonstration That Binding of Epinephrine to β-Adrenergic Receptors Triggers a Signal Transduction Pathway within Cells

Question: Is binding of epinephrine to β-adrenergic receptors necessary for triggering a signal transduction pathway within cells?

Experiment: It was known that epinephrine triggers a signal transduction pathway within cells. First, activation of adenylyl cyclase causes the level of the second messenger cAMP to increase, and then cAMP activates protein kinases in a signaling cascade that generates a cellular response (see Section 7.4 for specific details of such pathways). Richard Cerione and his colleagues at Duke University Medical Center performed experiments to show whether the signal transduction pathway is stimulated by binding of epinephrine to β-adrenergic receptors.

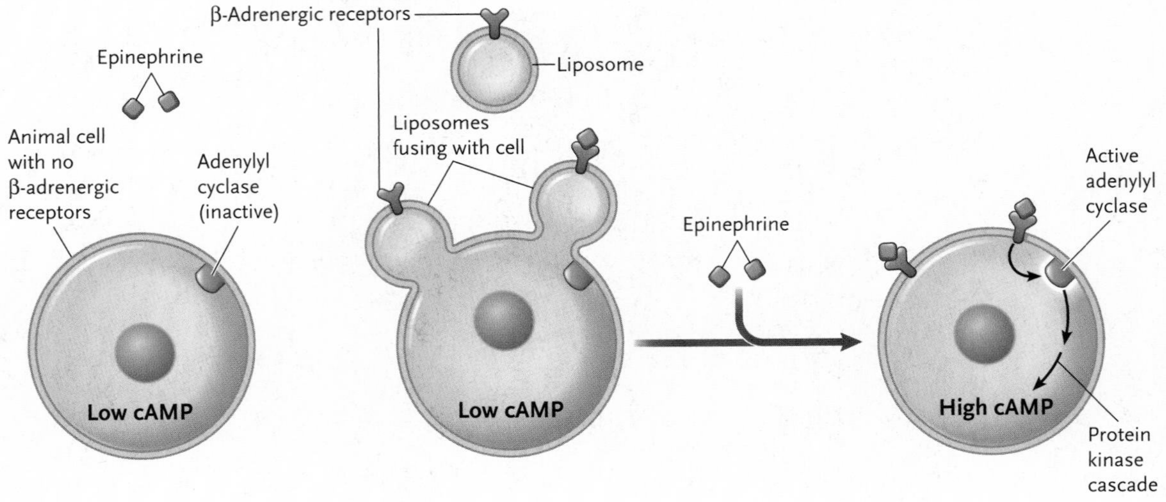
β-Adrenergic receptors
Liposome
Epinephrine
Animal cell with no β-adrenergic receptors
Adenylyl cyclase (inactive)
Low cAMP
Liposomes fusing with cell
Low cAMP
Epinephrine
Active adenylyl cyclase
High cAMP
Protein kinase cascade

1. Epinephrine was added to animal cells lacking β-adrenergic receptors.

Result: No change occurred to the low level of cAMP in those cells. This result demonstrated that epinephrine alone was not able to trigger an increase in cAMP.

2. Liposomes—artificial spherical phospholipid membranes—containing purified β-adrenergic receptors were fused with the animal cells, and then epinephrine was added.

Result: When the liposomes fused with the animal cells, β-adrenergic receptors became part of the fused cell's plasma membrane. Then, adding epinephrine triggered synthesis of cAMP, resulting in high levels of cAMP in the cells. This result demonstrated that β-adrenergic receptors must be present in the membrane for epinephrine to trigger an increase in cAMP in the cell. The simplest interpretation was that epinephrine bound to the β-adrenergic receptors, activating adenylyl cyclase within the cell.

Conclusion: The cellular response depended upon binding of the hydrophilic hormone to a specific plasma membrane-embedded receptor.

Source: R. A. Cerione et al. 1983. Reconstitution of the β-adrenergic receptors in lipid vesicles: Affinity chromatography-purified receptors confer catecholamine responsiveness on a heterologous adenylate cyclase system. *Proceedings of the National Academy of Sciences USA* 80:4899–4903.

In the next two sections we discuss the major endocrine cells and glands of vertebrates. The locations of these cells and glands in the human body and their functions are summarized in **Figure 40.6** and **Table 40.1.** Peptide hormones secreted by other body regions, including the stomach and small intestine, the thymus gland, the kidneys, and the heart will be described in the chapters in which these tissues and organs are discussed.

STUDY BREAK 40.2 <

1. **Compare and contrast the mechanisms by which glucagon and aldosterone cause their specific responses.**
2. **Explain how one type of target cell could respond to different hormones, and how the same hormone could produce different effects in different cells.**

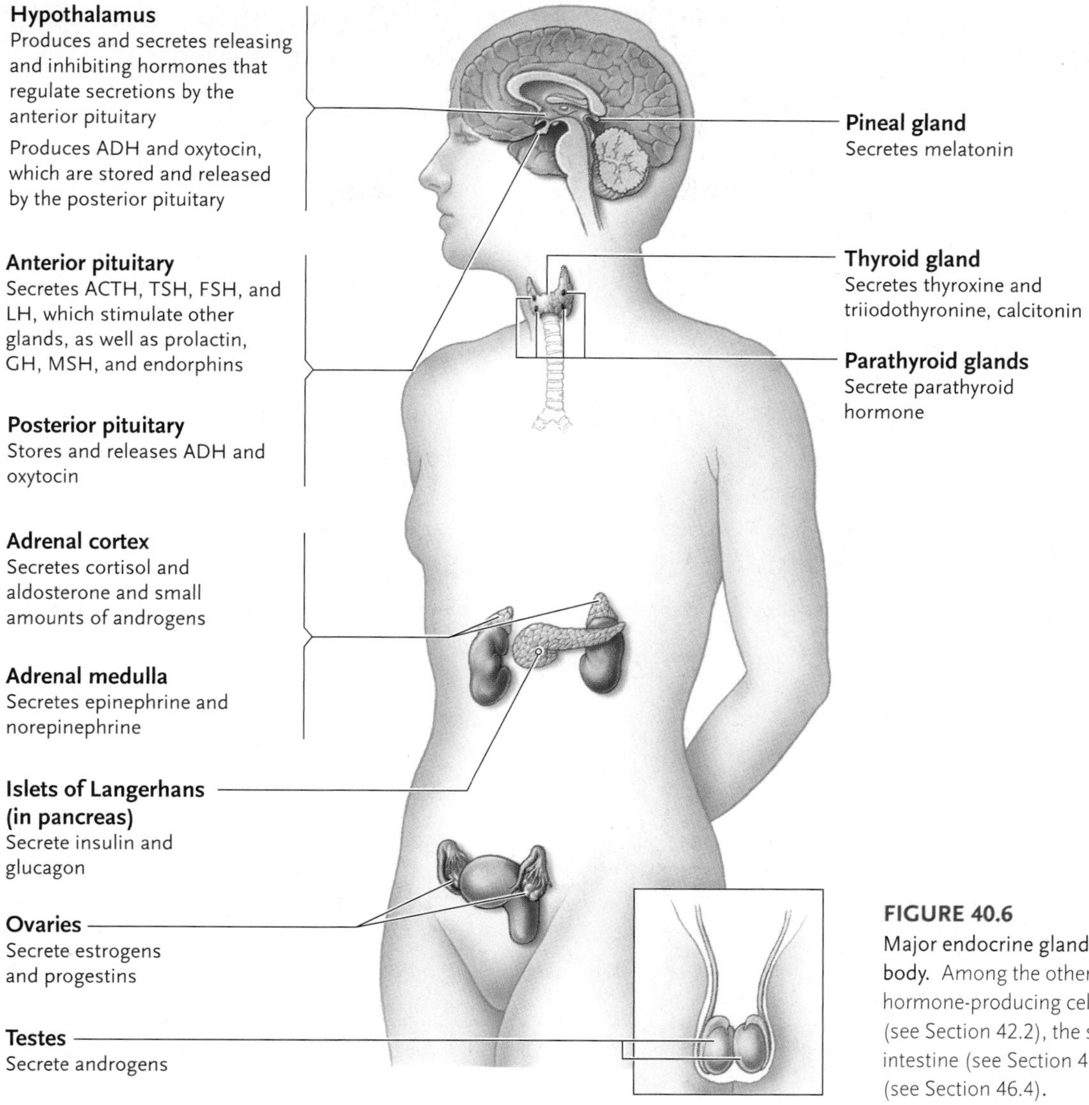

Hypothalamus
Produces and secretes releasing and inhibiting hormones that regulate secretions by the anterior pituitary
Produces ADH and oxytocin, which are stored and released by the posterior pituitary

Anterior pituitary
Secretes ACTH, TSH, FSH, and LH, which stimulate other glands, as well as prolactin, GH, MSH, and endorphins

Posterior pituitary
Stores and releases ADH and oxytocin

Adrenal cortex
Secretes cortisol and aldosterone and small amounts of androgens

Adrenal medulla
Secretes epinephrine and norepinephrine

Islets of Langerhans (in pancreas)
Secrete insulin and glucagon

Ovaries
Secrete estrogens and progestins

Testes
Secrete androgens

Pineal gland
Secretes melatonin

Thyroid gland
Secretes thyroxine and triiodothyronine, calcitonin

Parathyroid glands
Secrete parathyroid hormone

FIGURE 40.6
Major endocrine glands of the human body. Among the other organs that contain hormone-producing cells are the kidneys (see Section 42.2), the stomach and small intestine (see Section 45.4), and the heart (see Section 46.4).

40.3 The Hypothalamus and Pituitary

The hormones of vertebrates work in coordination with the nervous system. The action of several hormones is closely coordinated by the hypothalamus and the connected pituitary.

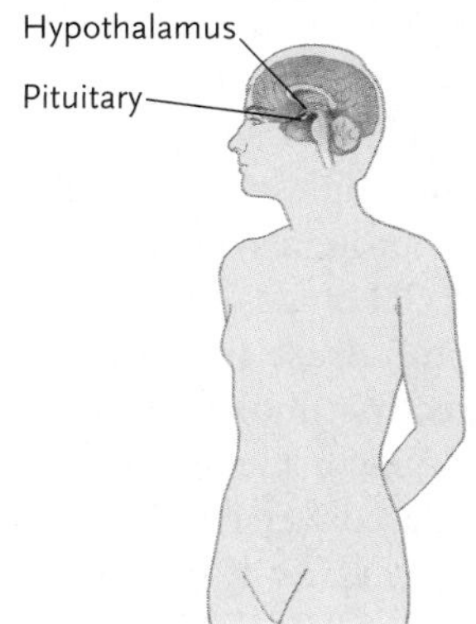
Hypothalamus
Pituitary

The hypothalamus is a region of the brain located in the floor of the cerebrum (see Section 38.3). The **pituitary gland,** consisting mostly of two fused lobes, is suspended just below it by a slender stalk of tissue that contains both neurons and blood vessels **(Figure 40.7).** The **posterior pituitary** contains axons and nerve endings of neurosecretory neurons that originate in the hypothalamus. The **anterior pituitary** contains nonneuronal endocrine cells that form a distinct gland. The two lobes are separate in structure and embryonic origins.

Under Regulatory Control by the Hypothalamus, the Anterior Pituitary Secretes Eight Hormones

The secretion of hormones from the anterior pituitary is controlled by peptide neurohormones called **releasing hormones (RHs)** and **inhibiting hormones (IHs),** which are released by the hypothalamus. These neurohormones are carried in the blood from the hypothalamus to the anterior pituitary in a *portal vein,* a special vein that connects the capillaries of the two glands. The portal vein provides a critical link between the brain and the endocrine system, ensuring that most of the blood reaching the anterior pituitary first passes through the hypothalamus.

RHs and IHs regulate hormone secretion by another endocrine gland. RHs and IHs regulate the anterior pituitary's secretion of another group of hormones; those hormones in turn control many other endocrine glands of the body, and also control some body processes directly.

Secretion of hypothalamic RHs is controlled by neurons containing receptors that monitor the blood to detect changes in body

TABLE 40.1 The Major Human Endocrine Glands and Hormones

Secretory Tissue or Gland	Hormones	Molecular Class	Target Tissue	Principal Actions
Hypothalamus	Releasing and inhibiting hormones	Peptide	Anterior pituitary	Regulate secretion of anterior pituitary hormones
Anterior pituitary	Thyroid-stimulating hormone (TSH)	Peptide	Thyroid gland	Stimulates secretion of thyroid hormones and growth of thyroid gland
	Adrenocorticotropic hormone (ACTH)	Peptide	Adrenal cortex	Stimulates secretion of glucocorticoids by adrenal cortex
	Follicle-stimulating hormone (FSH)	Peptide	Ovaries in females, testes in males	Stimulates egg growth and development and secretion of sex hormones in females; stimulates sperm production in males
	Luteinizing hormone (LH)	Peptide	Ovaries in females, testes in males	Regulates ovulation in females and secretion of sex hormones in males
	Prolactin (PRL)	Peptide	Mammary glands	Stimulates breast development and milk secretion
	Growth hormone (GH)	Peptide	Bone, soft tissue	Stimulates growth of bones and soft tissues; helps control metabolism of glucose and other fuel molecules
	Melanocyte-stimulating hormone (MSH)	Peptide	Melanocytes in skin of some vertebrates	Promotes darkening of the skin
	Endorphins	Peptide	Pain pathways of PNS	Inhibit perception of pain
Posterior pituitary	Antidiuretic hormone (ADH)	Peptide	Kidneys	Raises blood volume and pressure by increasing water reabsorption in kidneys
	Oxytocin	Peptide	Uterus, mammary glands	Promotes uterine contractions; stimulates milk ejection from breasts
Thyroid gland	Calcitonin	Peptide	Bone	Lowers calcium concentration in blood
	Thyroxine and triiodothyronine	Amine	Most cells	Increase metabolic rate; essential for normal body growth
Parathyroid glands	Parathyroid hormone (PTH)	Peptide	Bone, kidneys, intestine	Raises calcium concentration in blood; stimulates vitamin D activation

chemistry and temperature. For example, TRH is secreted in response to a drop in body temperature. Input to the hypothalamus also comes through numerous connections from control centers elsewhere in the brain, including the brain stem and limbic system. Negative feedback pathways regulate secretion of the releasing hormones, such as the pathway regulating TRH secretion.

Under the control of the hypothalamic RHs, the anterior pituitary secretes six major hormones into the bloodstream: prolactin, growth hormone, thyroid-stimulating hormone, adrenocorticotropic hormone, follicle-stimulating hormone, and luteinizing hormone, and two other hormones, melanocyte-stimulating hormone (MSH) and endorphins. **Prolactin (PRL)** influences reproductive activities and parental care in vertebrates. In mammals, PRL stimulates development of the secretory cells of mammary glands during late pregnancy, and stimulates milk synthesis after a female mammal gives birth. Stimulation of the mammary glands and the nipples, as occurs during suckling, leads to PRL release.

Growth hormone (GH) stimulates cell division, protein synthesis, and bone growth in children and adolescents, thereby causing body growth. GH also stimulates protein synthesis and cell division in adults. For these actions, GH binds to target tissues, mostly liver cells, causing them to release **insulin-like growth factor (IGF),** a peptide that directly stimulates growth processes. GH also controls a number of major metabolic processes in mammals of all ages, including the conversion of glycogen to glucose and fats to fatty acids as a means of regulating their levels in the blood. In addition, GH stimulates body cells to take up fatty acids and amino acids and limits the rate at which muscle cells take up glucose. These actions help maintain the

Secretory Tissue or Gland	Hormones	Molecular Class	Target Tissue	Principal Actions
Adrenal medulla	Epinephrine and norepinephrine	Amine	Sympathetic receptor sites throughout body	Reinforce sympathetic nervous system; contribute to responses to stress
Adrenal cortex	Aldosterone (mineralocorticoid)	Steroid	Kidney tubules	Helps control body's salt–water balance by increasing Na^+ reabsorption and K^+ excretion in kidneys
	Cortisol (glucocorticoid)	Steroid	Most body cells, particularly muscle, liver, and adipose cells	Increases blood glucose by promoting breakdown of proteins and fats
Testes	Androgens, such as testosterone*	Steroid	Various tissues	Control male reproductive system development and maintenance; most androgens are made by the testes
	Oxytocin	Peptide	Uterus	Promotes uterine contractions when seminal fluid is ejaculated into vagina during sexual intercourse
Ovaries	Estrogens, such as estradiol**	Steroid	Breast, uterus, other tissues	Stimulate maturation of sex organs at puberty, and development of secondary sexual characteristics
	Progestins, such as progesterone**	Steroid	Uterus	Prepare and maintain uterus for implantation of fertilized egg and the growth and development of embryo
Pancreas (islets of Langerhans)	Glucagon (alpha cells)	Peptide	Liver cells	Raises glucose concentration in blood; promotes release of glucose from glycogen stores and production from noncarbohydrates
	Insulin (beta cells)	Peptide	Most cells	Lowers glucose concentration in blood; promotes storage of glucose, fatty acids, and amino acids
Pineal gland	Melatonin	Amine	Brain, anterior pituitary, reproductive organs, immune system, possibly others	Helps synchronize body's biological clock with day length; may inhibit gonadotropins and initiation of puberty
Many cell types	Growth factors	Peptide	Most cells	Regulate cell division and differentiation
	Prostaglandins	Fatty acid	Various tissues	Have many diverse roles

*Small amounts secreted by ovaries and adrenal cortex.
**Small amounts secreted by testes.

availability of glucose and fatty acids to tissues and organs between feedings; this is particularly important for the brain. In humans, deficiencies in GH secretion during childhood produce *pituitary dwarfs,* who remain small in stature **(Figure 40.8).** Overproduction of GH during childhood or adolescence, often due to a tumor of the anterior pituitary, produces *pituitary giants,* who may grow above 2.4 m (8 ft) in height.

The other four major hormones secreted by the anterior pituitary control endocrine glands elsewhere in the body. **Thyroid-stimulating hormone (TSH)** stimulates the thyroid gland to grow in size and secrete thyroid hormones. **Adrenocorticotropic hormone (ACTH)** triggers hormone secretion by cells in the adrenal cortex. **Follicle-stimulating hormone (FSH)** controls egg development and the secretion of sex hormones in female mammals, and sperm production in males. **Luteinizing hormone (LH)** regulates part of the menstrual cycle in human females and the secretion of sex hormones in males. FSH and LH are grouped together as **gonadotropins** because they regulate the activity of the gonads (ovaries and testes). The roles of the gonadotropins and sex hormones in the reproductive cycle are described in Chapter 47.

Melanocyte-stimulating hormone (MSH) and **endorphins** are two other hormones produced by the anterior pituitary. MSH is named because of its effect in some vertebrates on melanocytes, skin cells that contain the black pigment melanin. For example, an increase in secretion of MSH produces a marked darkening of the skin of fishes, amphibians, and reptiles. The darkening is produced by a redistribution of melanin from the centers of the melanocytes throughout the cells. In adult humans, the pituitary secretes little or no MSH.

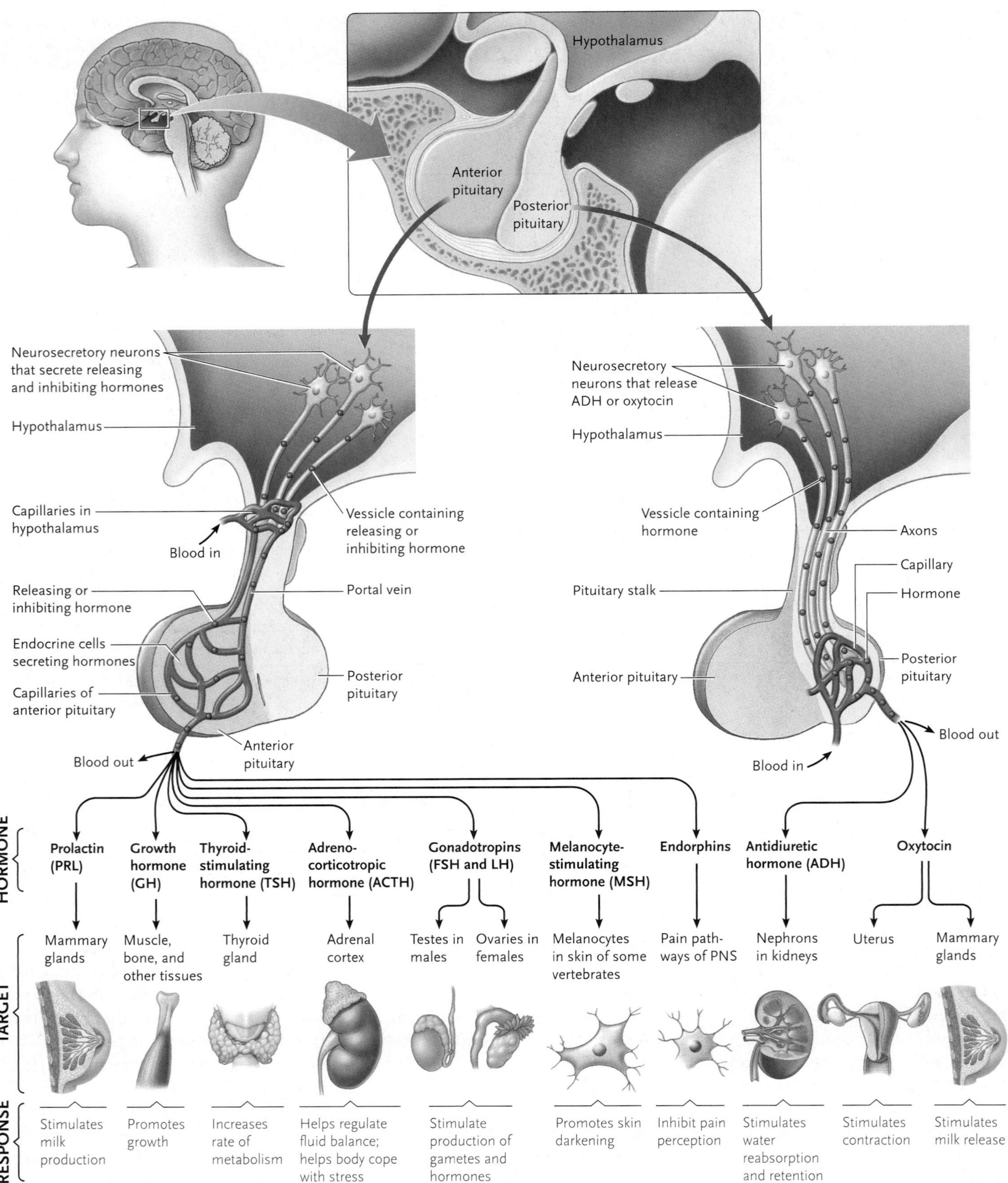

Hypothalamus
Anterior pituitary
Posterior pituitary
Neurosecretory neurons that secrete releasing and inhibiting hormones
Hypothalamus
Capillaries in hypothalamus
Blood in
Vessicle containing releasing or inhibiting hormone
Releasing or inhibiting hormone
Portal vein
Endocrine cells secreting hormones
Capillaries of anterior pituitary
Posterior pituitary
Anterior pituitary
Blood out
Neurosecretory neurons that release ADH or oxytocin
Hypothalamus
Vessicle containing hormone
Axons
Capillary
Pituitary stalk
Hormone
Anterior pituitary
Posterior pituitary
Blood out
Blood in
HORMONE
Prolactin (PRL)
Growth hormone (GH)
Thyroid-stimulating hormone (TSH)
Adreno-corticotropic hormone (ACTH)
Gonadotropins (FSH and LH)
Melanocyte-stimulating hormone (MSH)
Endorphins
Antidiuretic hormone (ADH)
Oxytocin
TARGET
Mammary glands
Muscle, bone, and other tissues
Thyroid gland
Adrenal cortex
Testes in males
Ovaries in females
Melanocytes in skin of some vertebrates
Pain pathways of PNS
Nephrons in kidneys
Uterus
Mammary glands
RESPONSE
Stimulates milk production
Promotes growth
Increases rate of metabolism
Helps regulate fluid balance; helps body cope with stress
Stimulate production of gametes and hormones
Promotes skin darkening
Inhibit pain perception
Stimulates water reabsorption and retention
Stimulates contraction
Stimulates milk release

FIGURE 40.7

The hypothalamus and pituitary. Hormones secreted by the anterior and posterior pituitary are controlled by neurohormones released in the hypothalamus.

Syndication International Ltd., 1986

FIGURE 40.8
The results of overproduction and underproduction of growth hormone by the anterior pituitary. The man on the left is of normal height. The man in the center is a pituitary giant, whose pituitary produced excess GH during childhood and adolescence. The man on the right is a pituitary dwarf, whose pituitary produced too little GH.

Endorphins, peptide hormones produced by the hypothalamus and pituitary, are also released by the anterior pituitary (see Table 40.1). In the peripheral nervous system (PNS), endorphins act as neurotransmitters in pathways that control pain, thereby inhibiting the perception of pain.

The Posterior Pituitary Secretes Two Hormones into the Circulatory System

The neurosecretory neurons in the posterior pituitary secrete two peptide hormones, antidiuretic hormone and oxytocin, directly into the circulatory system (see Figure 40.7).

Antidiuretic hormone (ADH) stimulates kidney cells to absorb more water from urine, thereby increasing the volume of the blood. The hormone is released when sensory receptor cells of the hypothalamus detect an increase in the blood's osmotic pressure during periods of body dehydration or after a salty meal. Ethyl alcohol reduces ADH secretion, explaining in part why alcoholic drinks increase the volume of urine excreted. Nicotine and emotional stress, in contrast, stimulate ADH secretion and water retention. After severe stress is relieved, the return to normal ADH secretion often makes a trip to the bathroom among our most pressing needs. The posterior pituitary also releases a flood of ADH when an injury results in heavy blood loss or some other event triggers a severe drop in blood pressure. ADH helps maintain blood pressure by reducing water loss and also by causing small blood vessels in some tissues to constrict.

Hormones with structure and action similar to ADH are also secreted in fishes, amphibians, reptiles, and birds. In amphibians, these ADH-like hormones increase the amount of water entering the body through the skin and from the urinary bladder.

We have noted that **oxytocin** stimulates the ejection of milk from the mammary glands of a nursing mother. Stimulation of the nipples in suckling sends neuronal signals to the hypothalamus, and leads to release of oxytocin from the posterior pituitary. The released oxytocin stimulates more oxytocin secretion by a positive feedback mechanism. Oxytocin causes the smooth muscle cells surrounding the individual alveoli of the mammary glands to contract, forcibly expelling the milk through the nipples. The entire cycle, from the onset of suckling to milk ejection, takes less than a minute in mammals. Oxytocin also plays a key role in childbirth, as discussed in Section 36.4.

In males, oxytocin is secreted into the seminal fluid by the testes. Like prostaglandins, when the seminal fluid is ejaculated into the vagina during sexual intercourse, oxytocin stimulates contractions of the uterus that aid movement of sperm through the female reproductive tract.

STUDY BREAK 40.3 <

1. **Summarize the functional relationship between the hypothalamus and the anterior pituitary gland.**
2. **What, in addition to the hormones they secrete, distinguishes the anterior and posterior pituitary glands?**

40.4 Other Major Endocrine Glands of Vertebrates

Besides the hypothalamus and pituitary, the body has seven major endocrine glands or tissues, many of them regulated by the hypothalamus–pituitary connection. These glands are the thyroid gland, parathyroid glands, adrenal medulla, adrenal cortex, gonads, pancreas, and pineal gland (shown in Figure 40.6 and summarized in Table 40.1).

The Thyroid Hormones Stimulate Metabolism, Development, and Maturation

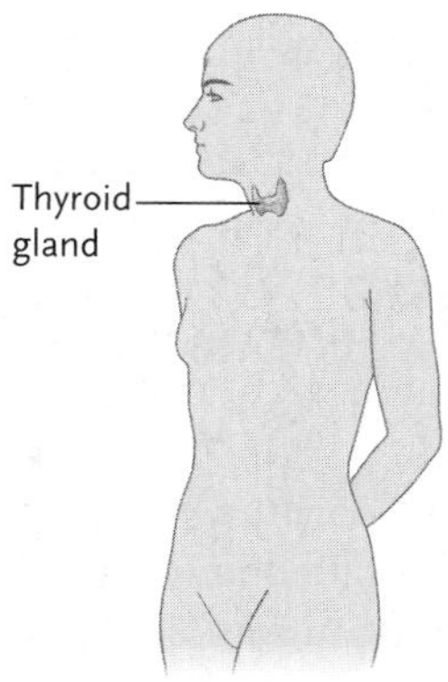
Thyroid gland

The **thyroid gland,** which is located in the front of the throat in humans, has a shape similar to that of a bowtie. It secretes the same hormones in all vertebrates. The primary circulating thyroid hormone, **thyroxine,** is known as T_4 because it contains four iodine atoms. The thyroid also secretes smaller amounts of a closely related hormone, **triiodothyronine** or T_3, which contains three iodine atoms. A supply of iodine in the diet is necessary for production of these hormones. Normally, their concentrations are kept at finely balanced levels in the blood by negative feedback loops such as that described in Figure 40.2.

Both T_4 and T_3 enter cells; however, once inside the cell, most of the T_4 is converted to T_3, the form that combines with internal receptors. Binding of T_3 to receptors alters gene expression, which brings about the hormone's effects.

The thyroid hormones are vital to growth, development, maturation, and metabolism in all vertebrates. They interact with GH for their effects on growth and development. Thyroid hormones also increase the sensitivity of many body cells to the

INSIGHTS FROM THE Molecular Revolution

Two Receptors for Estrogens

Estrogens have many different effects on female sexual development, behavior, and the menstrual cycle. One negative effect is to stimulate the growth of tumors in breast and uterine cancer. This cancer-enhancing effect can be reduced by administering *antiestrogens,* estrogen-like chemicals that bind competitively to estrogen receptors and block the sites that would normally be bound by the hormone. The antiestrogen *tamoxifen,* for example, inhibits the growth of breast tumors by blocking the activity of estrogen in breast tissue, but patients receiving it are at increased risk of developing uterine cancer.

Research Question

How can tamoxifen have opposite effects in two different tissues?

Experiment

A group of investigators led by Thomas Scanlan and Peter Kushner at the University of California, San Francisco, joined by others at the Karolinska Institute and the Karo Bio Company in Sweden, had discovered that humans have two highly similar estrogen receptors, ERα and ERβ. Could differences between them account for the opposing effects of tamoxifen in breast and uterine tissues?

To find out, the researchers constructed two pairs of recombinant DNA plasmids **(Figure 1).** One pair consisted of either the ERα receptor gene or the ERβ receptor gene, adjacent to a promoter for continuous transcription of the receptor gene in human tissue culture cells (Figure 1A). The other pair consisted of the firefly luciferase gene, which catalyzes a light-producing reaction, adjacent to one of two gene regulatory sequences, ERE or AP1, which act in estrogen-regulated systems (Figure 1B). In this experiment, the luciferase gene acts as a reporter gene, meaning that it provides a sensitive measure of the events occurring with the estrogen receptors and the gene regulatory sequences—luciferase has nothing to do with estrogen or estrogen activity.

The researchers introduced one receptor gene plasmid and one reporter gene plasmid in four possible combinations into human cell lines that do not normally make estrogen receptors. Estrogen receptors are produced in all four of the resulting cell lines because the gene for the receptor is transcribed from the introduced plasmid. Two of the cell lines make the ERα receptor and two make the ERβ receptor.

FIGURE 1 Schematic of the receptor genes and reporter genes on the recombinant DNA plasmids.

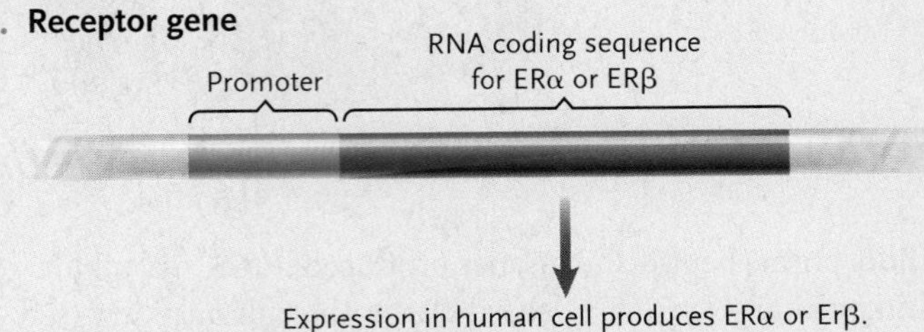
A. Receptor gene
Promoter
RNA coding sequence for ERα or ERβ
Expression in human cell produces ERα or Erβ.

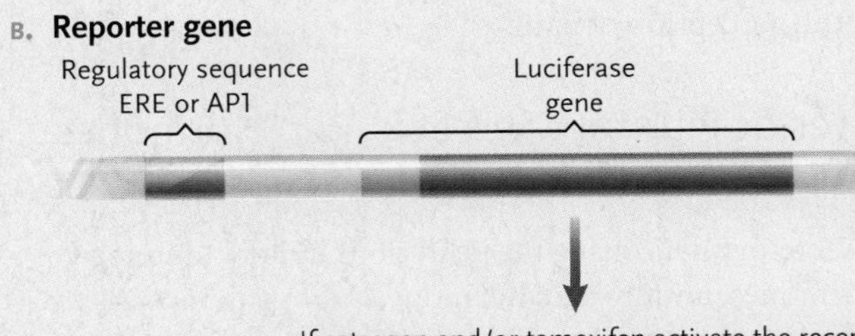
B. Reporter gene
Regulatory sequence ERE or AP1
Luciferase gene
If estrogen and/or tamoxifen activate the receptor, and if the activated receptor can bind to the regulatory sequence, then the luciferase gene is expressed. Luciferase is assayed in cell extracts; the reaction produces light, which is measured in a special instrument.

effects of epinephrine and norepinephrine, hormones released by the adrenal medulla as part of the "fight or flight response" (discussed further later in this chapter).

In amphibians such as frogs, thyroid hormones trigger **metamorphosis,** or change in body form from tadpole to adult **(Figure 40.9).** Thyroid hormones also contribute to seasonal changes in the plumage of birds and coat color in mammals.

In human adults, low thyroid output, *hypothyroidism,* causes affected individuals to be sluggish mentally and physically; they have a slow heart rate and weak pulse, and often feel confused and depressed. Hypothyroidism in infants and children leads to cretinism, that is, stunted growth and diminished intelligence. Overproduction of thyroid hormones in human adults,

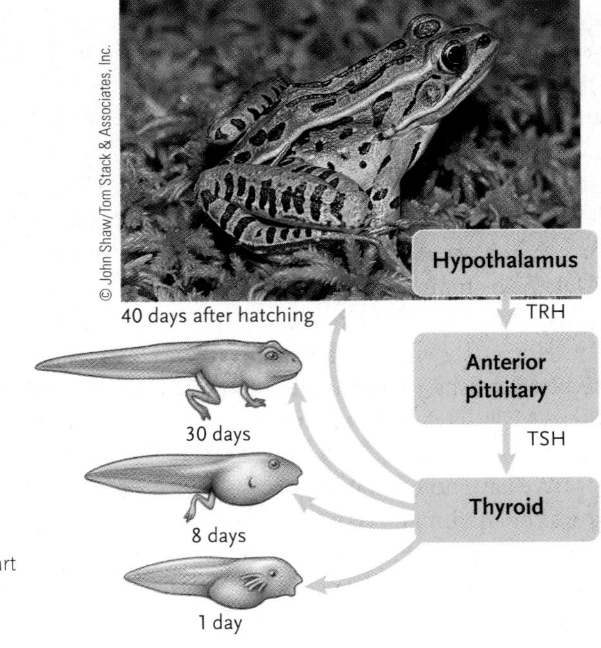
© John Shaw/Tom Stack & Associates, Inc.
Hypothalamus
TRH
Anterior pituitary
TSH
Thyroid
40 days after hatching
30 days
8 days
1 day

FIGURE 40.9
Metamorphosis of a tadpole into an adult frog, under the control of thyroid hormones. As a part of the metamorphosis, changes in gene activity lead to a change from an aquatic to a terrestrial habitat. TRH, thyroid-releasing hormone; TSH, thyroid-stimulating hormone.

With this experimental design, the researchers could test whether the ERα or ERβ receptor is activated by binding estrogen and/or tamoxifen, and also whether the activated receptor would bind to the luciferase plasmid. If these two conditions were met, luciferase would be synthesized, and its activity could be measured.

Results

The results are shown in **Figure 2.**

Conclusion

The experiments indicate that the previously baffling and opposing effects of tamoxifen on breast and uterine tissues occur because different estrogen receptors—ERα or ERβ—are present, and act on genes controlled by either the ERE or AP1 regulatory sequences. The results emphasize the fact that hormones can have distinct effects in different cell types depending on the types of receptors present.

Source: K. Paech et al. 1997. Differential ligand activation of estrogen receptors ERα and ERβ at AP1 sites. *Science* 277:1508–1510.

A. ERα gene and ERE regulatory sequence

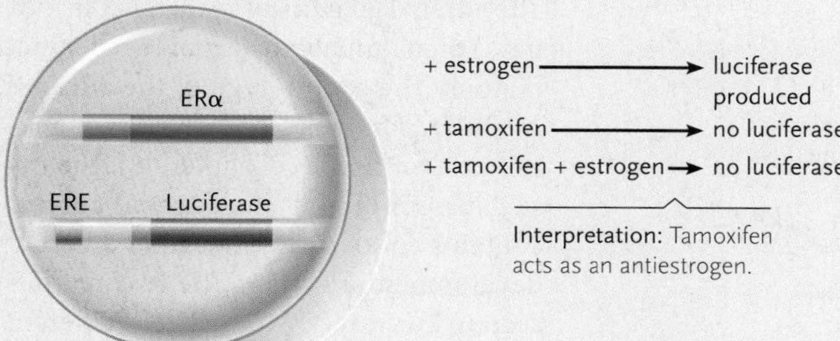

ERα
ERE
Luciferase
+ estrogen → luciferase produced
+ tamoxifen → no luciferase
+ tamoxifen + estrogen → no luciferase
Interpretation: Tamoxifen acts as an antiestrogen.

C. ERα gene and AP1 regulatory sequence

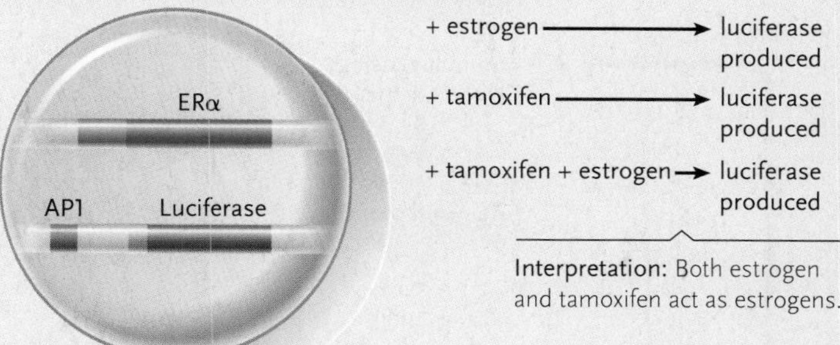

ERα
AP1
Luciferase
+ estrogen → luciferase produced
+ tamoxifen → luciferase produced
+ tamoxifen + estrogen → luciferase produced
Interpretation: Both estrogen and tamoxifen act as estrogens.

B. ERβ gene and ERE regulatory sequence

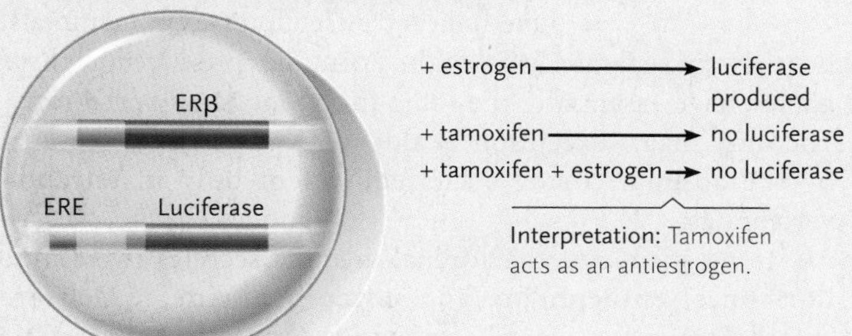

ERβ
ERE
Luciferase
+ estrogen → luciferase produced
+ tamoxifen → no luciferase
+ tamoxifen + estrogen → no luciferase
Interpretation: Tamoxifen acts as an antiestrogen.

D. ERβ gene and AP1 regulatory sequence

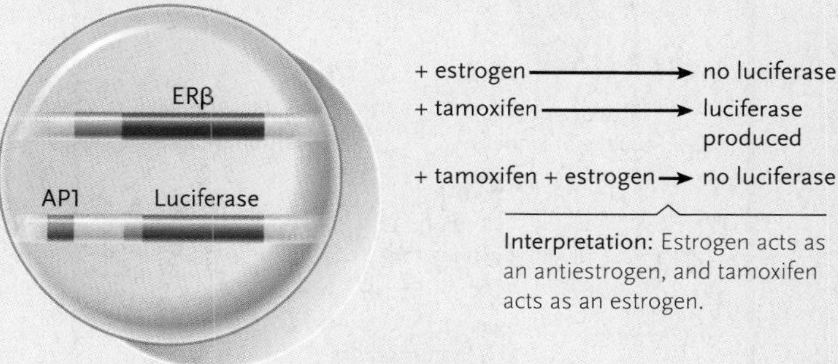

ERβ
AP1
Luciferase
+ estrogen → no luciferase
+ tamoxifen → luciferase produced
+ tamoxifen + estrogen → no luciferase
Interpretation: Estrogen acts as an antiestrogen, and tamoxifen acts as an estrogen.

FIGURE 2

Effects of estrogen and/or tamoxifen on luciferase expression in human cell lines containing a receptor gene plasmid and a reporter gene plasmid.

hyperthyroidism, produces nervousness and emotional instability, irritability, insomnia, weight loss, and a rapid, often irregular heartbeat. The most common form of hyperthyroidism is *Graves' disease,* characterized by inflamed, protruding eyes in addition to the other symptoms mentioned.

Insufficient iodine in the diet can cause *goiter,* enlargement of the thyroid. Without iodine, the thyroid cannot make T_3 and T_4 in response to stimulation by TSH. Because the thyroid hormone concentration remains low in the blood, TSH continues to be secreted, and the thyroid grows in size. Dietary iodine deficiency has been eliminated in developed regions of the world by the addition of iodine to table salt.

In mammals, the thyroid also has specialized cells that secrete the peptide hormone **calcitonin.** The hormone lowers the level of Ca^{2+} in the blood by inhibiting the ongoing dissolution of calcium from bone. Calcitonin secretion is stimulated when Ca^{2+} levels in blood rise above the normal range and inhibited when Ca^{2+} levels fall below the normal range.

The Parathyroid Glands Regulate Ca^{2+} Level in the Blood

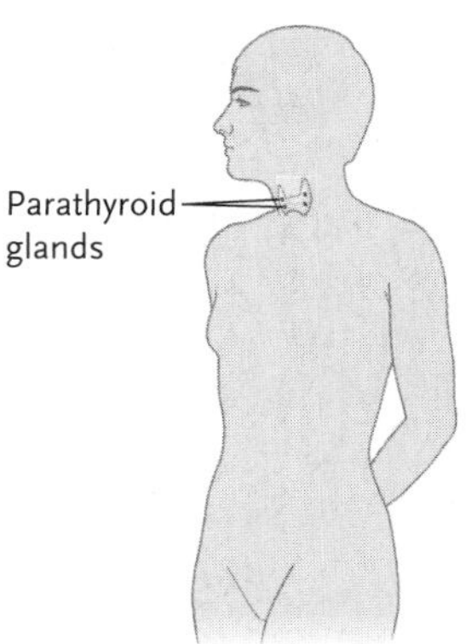

Parathyroid glands

The **parathyroid glands** occur only in tetrapod vertebrates—amphibians, reptiles, birds, and mammals. Each is a spherical structure about the size of a pea. Mammals have four parathyroids located on the posterior surface of the thyroid gland, two on each side. The single hormone they produce, called **parathyroid hormone (PTH),** is secreted in response to a fall in blood Ca^{2+} levels. PTH stimulates bone cells to dissolve the mineral matter of bone tissues, releasing both calcium and phosphate ions into the blood. The released Ca^{2+} is then available for enzyme activation, conduction of nerve signals across synapses, muscle contraction, blood clotting, and other uses. How blood Ca^{2+} levels control PTH and calcitonin secretion is shown in **Figure 40.10.**

PTH also stimulates enzymes in the kidneys that convert **vitamin D,** a steroidlike molecule, into its fully active form in the body. The activated vitamin D increases the absorption of Ca^{2+} and phosphates from ingested food by promoting the synthesis of a calcium-binding protein in the intestine; it also increases the release of Ca^{2+} from bone in response to PTH.

PTH underproduction causes Ca^{2+} concentration to fall steadily in the blood, disturbing nerve and muscle function—the muscles twitch and contract uncontrollably, and convulsions and cramps occur. Without treatment, the condition is usually fatal, because the severe muscular contractions interfere with breathing. Overproduction of PTH results in the loss of so much calcium from the bones that they become thin and fragile. At the same time, the elevated Ca^{2+} concentration in the blood causes calcium deposits to form in soft tissues, especially in the lungs, arteries, and kidneys (where the deposits form kidney stones).

FIGURE 40.10
Negative feedback control of PTH and calcitonin secretion by blood Ca^{2+} levels.

The Adrenal Medulla Releases Two "Fight or Flight" Hormones

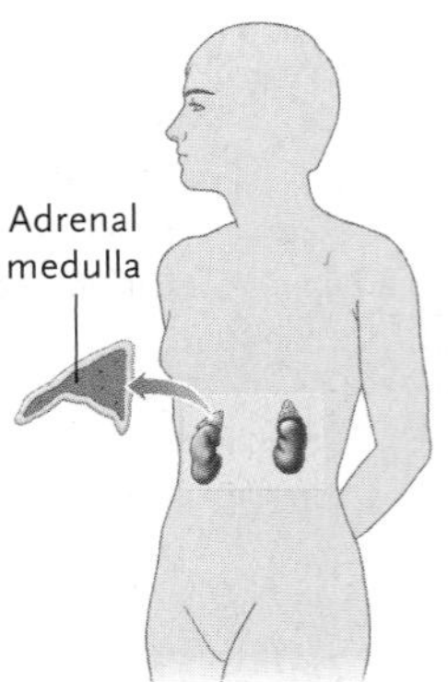

Adrenal medulla

The adrenal glands (*ad* = to; *renes* = kidneys) of mammals consist of two distinct regions. The central region, the **adrenal medulla,** contains secretory cells of neural crest origin (see Chapter 48); the tissue surrounding it, the **adrenal cortex,** contains endocrine cells with a different developmental origin. The two regions secrete hormones with entirely different functions. Nonmammalian vertebrates have glands equivalent to the adrenal medulla and adrenal cortex of mammals, but they are separate. Most of the hormones produced by these glands have essentially the same functions in all vertebrates. The only major exception is aldosterone, which is secreted by the adrenal cortex or its equivalent only in tetrapod vertebrates.

In most species, the adrenal medulla secretes two amine hormones, **epinephrine** and **norepinephrine,** which are **catecholamines,** chemical compounds derived from the amino acid tyrosine that circulate in the bloodstream. They bind to receptors in the plasma membranes of their target cells. (Epinephrine and norepinephrine are also secreted by some cells of the CNS and neurons of the sympathetic nervous system. In these cases, epinephrine and norepinephrine function as neurotransmitters between interneurons involved in a diversity of brain and body functions; see Section 37.3.)

Epinephrine and norepinephrine are secreted from sympathetic neurons and the adrenal medulla when the body encounters stresses such as emotional excitement, danger (fight-or-flight situations), anger, fear, infections, injury, even midterm and final exams. Epinephrine in particular prepares the body for handling stress or physical activity. The heart rate increases. Glycogen and fats break down, releasing glucose and fatty acids into the blood as fuel molecules. In the heart, skeletal muscles, and lungs, the blood vessels dilate to increase blood flow. Elsewhere in the body, the blood vessels constrict, raising blood pressure, reducing blood flow to the intestines and kidneys, and inhibiting smooth muscle contractions, which reduces water loss and slows down the digestive system. Airways in the lungs also dilate, helping to increase the flow of air.

The effects of norepinephrine on heart rate, blood pressure, and blood flow to the heart muscle are similar to those of epinephrine. However, in contrast to epinephrine, norepinephrine causes blood vessels in skeletal muscles to constrict. This antagonistic effect is largely canceled out because epinephrine is secreted in much greater quantities.

No known human diseases are caused by underproduction of the hormones of the adrenal medulla, as long as the sympathetic nervous system is intact. Overproduction of epinephrine and norepinephrine, which can occur if there is a tumor in the adrenal medulla, leads to symptoms duplicating a stress response.

The Adrenal Cortex Secretes Two Groups of Steroid Hormones That Are Essential for Survival

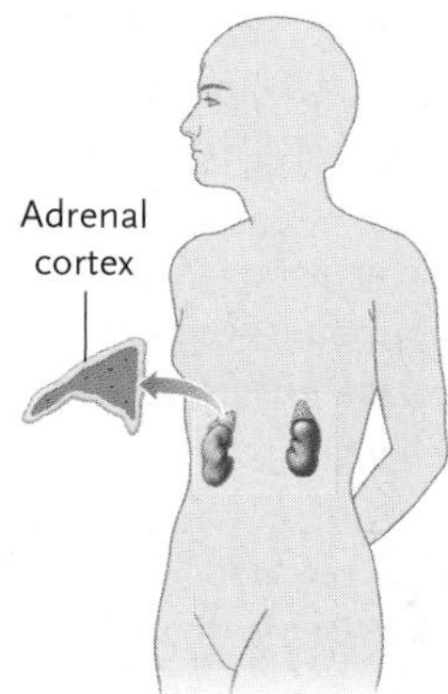
Adrenal
cortex

The adrenal cortex of mammals secretes two major types of steroid hormones: glucocorticoids help maintain the blood concentration of glucose and other fuel molecules, and mineralocorticoids regulate the levels of Na^+ and K^+ ions in the blood and ECF.

THE GLUCOCORTICOIDS The glucocorticoids help maintain glucose levels in the blood by three major mechanisms: (1) stimulating the synthesis of glucose from noncarbohydrate sources such as fats and proteins; (2) reducing glucose uptake by body cells except those in the central nervous system; and (3) promoting the breakdown of fats and proteins, which releases fatty acids and amino acids into the blood as alternative fuels when glucose supplies are low. The absence of down-regulation of glucose uptake to the CNS keeps the brain well supplied with glucose between meals and during periods of extended fasting. **Cortisol** is the major glucocorticoid secreted by the adrenal cortex.

Secretion of glucocorticoids is ultimately under control of the hypothalamus **(Figure 40.11).** Low glucose concentrations in the blood or elevated levels of epinephrine secreted by the adrenal medulla in response to stress are detected in the hypothalamus, leading to secretion of ACTH by the anterior pituitary. ACTH promotes the secretion of glucocorticoids by the adrenal cortex.

Overproduction of glucocorticoids makes blood glucose rise and increases fat deposition in adipose tissue and protein breakdown in muscles and bones. The loss of proteins from muscles causes weakness and fatigue; loss of proteins from bone, particularly collagens, makes the bones fragile and susceptible to breakage. Underproduction of glucocorticoids causes blood glucose concentration to fall below normal levels and diminishes tolerance to stress.

Glucocorticoids have anti-inflammatory properties and, consequently, they are used clinically to treat conditions such as arthritis or dermatitis. They also suppress the immune system and are used in the treatment of autoimmune diseases such as rheumatoid arthritis.

THE MINERALOCORTICOIDS In tetrapods, the mineralocorticoids, primarily **aldosterone,** increase the amount of Na^+ reabsorbed from the fluids processed by the kidneys and absorbed from foods in the intestine. They also reduce the amount of Na^+ secreted by salivary and sweat glands and increase the rate of K^+ excretion by the kidneys. The net effect is to keep Na^+ and K^+ balanced at the levels required for normal cellular functions, including those of the nervous system. Secretion of aldosterone is tightly linked to blood volume and indirectly linked to blood pressure (see Section 40.2 and Chapter 46).

Moderate overproduction of aldosterone causes excessive water retention in the body, so that tissues swell and blood pressure rises. Conversely, moderate underproduction can lead to excessive water loss and dehydration. Severe underproduction is rapidly fatal unless mineralocorticoids are supplied by injection or other means.

The adrenal cortex also secretes small amounts of androgens, steroid sex hormones responsible for maintenance of male characteristics, which are synthesized primarily by the gonads. These hormones have significant effects only if they are overproduced, as can occur with some tumors in the adrenal cortex. The result is altered development of the primary or secondary sex characteristics.

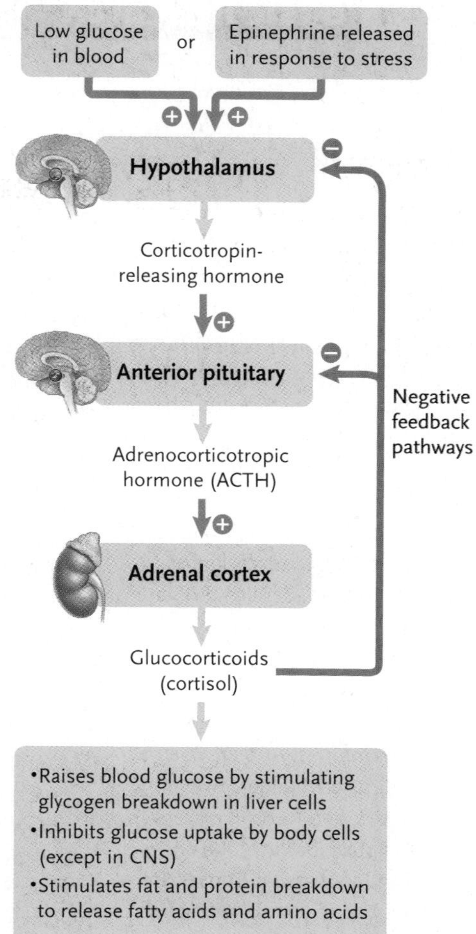
Low glucose in blood
or
Epinephrine released in response to stress
Hypothalamus
Corticotropin-releasing hormone
Anterior pituitary
Negative feedback pathways
Adrenocorticotropic hormone (ACTH)
Adrenal cortex
Glucocorticoids (cortisol)
•Raises blood glucose by stimulating glycogen breakdown in liver cells
•Inhibits glucose uptake by body cells (except in CNS)
•Stimulates fat and protein breakdown to release fatty acids and amino acids

FIGURE 40.11
Pathways linking secretion of glucocorticoids to low blood sugar and epinephrine secretion in response to stress.

FOCUS ON **Basic Research**

Neuroendocrine and Behavioral Effects of Anabolic-Androgenic Steroids in Humans

Anabolic-androgenic steroids (AAS) are synthetic derivatives of the natural steroid hormone testosterone. They were designed to have potent anabolic (tissue building) activity and low androgenic (masculinizing) activity in therapeutic doses. Overall, there are about 60 AAS that vary in chemical structure and, therefore, in their physiological effects.

AAS are used for treating conditions such as delayed puberty and subnormal growth in children, as well as for therapy in chronic conditions such as cancer, AIDS, severe burns, liver and kidney failure, and anemias. AAS are not used exclusively for medical purposes, however. Because of their anabolic effects, which include an increase in muscle mass, strength, and endurance, as well as acceleration of recovery from injuries, AAS are used by some athletes such as bodybuilders, weight lifters, baseball players, and football players. This use is actually abuse, because the doses typically administered for these purposes are far higher than therapeutic doses. AAS abuse is significant: in the early 1990s, about 1 million Americans had used or were using AAS to increase strength, muscle mass, or athletic ability. While originally limited to elite athletes, their use has trickled down to average athletes, including adolescents. It is estimated that perhaps 4% of high school students have used AAS. The greatest increase in AAS abuse over the past decade has been by adolescent girls.

Are AAS harmful at high doses? When researchers gave rodents doses of AAS comparable to those associated with human AAS abuse, they observed significant increases in aggression, anxiety, and sexual behaviors. These changes occur as a result of alterations in the neurotransmitters and other signaling molecules associated with those behaviors. All of these changes have been hypothesized to occur in human AAS abusers.

To study the effect of high doses of AAS on the human endocrine system, R. C. Daly and colleagues at the National Institute of Mental Health, in Bethesda, Maryland, administered the AAS methyltestosterone (MT) to normal (medication-free) human volunteers over a period of time in an in-patient clinic. The subjects were examined for the effects of MT on pituitary–gonadal, pituitary–thyroid, and pituitary–adrenal hormones, and the researchers attempted to correlate endocrine changes with psychological symptoms caused by the MT.

The researchers found, for instance, that high doses of MT caused a significant decrease in the levels of gonadotropins and gonadal steroid hormones in the blood. At the same time, thyroxine and TSH levels increased. No significant increases were seen in pituitary–adrenal hormones.

The decrease in testosterone levels correlated significantly with cognitive problems, such as increased distractibility and forgetfulness. The increase in thyroxine correlated significantly with a rise in aggressive behavior, notably anger, irritability, and violent feelings. There were no changes in activities associated with pituitary–adrenal hormones—energy, disturbed sleep, and sexual arousal—as was expected by the lack of change in those hormones.

In sum, behavioral changes associated with high doses of an AAS suggest that AAS-induced hormonal changes may well contribute to the adverse behavioral and mood changes that occur during AAS abuse. Clearly, there is every reason to believe that taking high doses of AAS for athletic gain alters the normal hormonal balance in humans, as it does in rodents.

The Gonadal Sex Hormones Regulate the Development of Reproductive Systems, Sexual Characteristics, and Mating Behavior

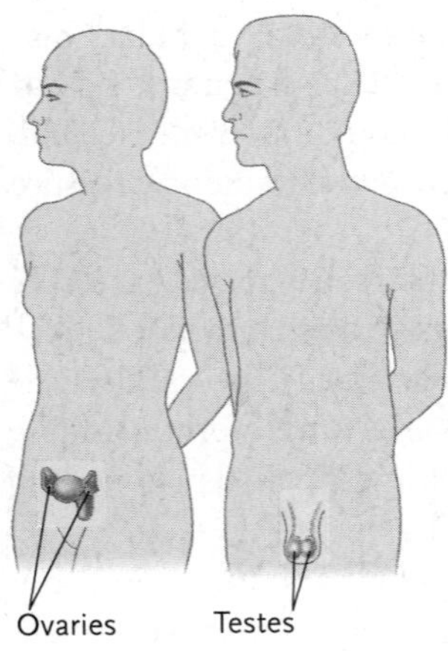
Ovaries
Testes

The **gonads,** the testes and ovaries, are the primary source of sex hormones in vertebrates. The steroid hormones they produce, the **androgens, estrogens,** and **progestins,** have similar functions in regulating the development of male and female reproductive systems, sexual characteristics, and mating behavior. Both males and females produce all three types of hormones, but in different proportions. Androgen production is predominant in males, while estrogen and progestin production is predominant in females. An outline of the actions of these hormones is presented here, and a more complete picture is given in Chapter 47.

The **testes** of male vertebrates secrete androgens, steroid hormones that stimulate and control the development and maintenance of male reproductive systems and masculine characteristics. The principal androgen is **testosterone,** the male sex hormone. In young adult males, a jump in testosterone levels stimulates puberty and the development of secondary sexual characteristics, including the growth of facial and body hair, muscle development, changes in vocal cord morphology, and development of normal sex drive. The synthesis and secretion of testosterone by cells in the testes is controlled by the release of luteinizing hormone (LH) from the anterior pituitary, which in turn is controlled by **gonadotropin releasing hormone (GnRH)** secreted by the hypothalamus.

A large number of synthetic derivatives of androgens, known as **anabolic-androgenic steroids (AAS)** (also referred to as **anabolic steroids**) mimic the effects of androgens. They have been in the news over the years because of their use by bodybuilders and other athletes in sports in which muscular strength is important. *Focus on Basic Research* discusses the potential adverse effects of anabolic-androgenic steroids.

The **ovaries,** under the stimulatory influence of follicle-stimulating hormone (FSH), produce estrogens, steroid hormones that stimulate and control the development and maintenance of female reproductive systems. The principal estrogen is **estradiol,** which stimulates maturation of sex organs at puberty and the development of secondary sexual characteristics. Ovaries also produce progestins, principally **progesterone,** the steroid hormone that prepares and maintains the uterus for implantation of a fertilized egg and the subsequent growth and development of an embryo. The synthesis and secretion of progesterone by cells in the ovaries is controlled by the release of LH from the anterior pituitary, which in turn is controlled by the same GnRH as in males.

The Pancreatic Islet of Langerhans Hormones Regulate Glucose Metabolism

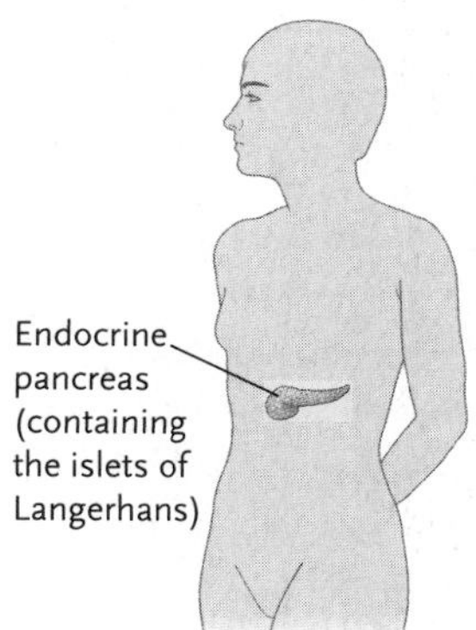
Endocrine pancreas (containing the islets of Langerhans)

Most of the pancreas, a relatively large gland located just behind the stomach, forms an exocrine gland that secretes digestive enzymes into the small intestine (see Chapter 45). However, about 2% of the cells in the pancreas are endocrine cells that form the islets of Langerhans. Found in all vertebrates, the islets secrete the peptide hormones insulin and glucagon into the bloodstream.

Insulin and glucagon regulate the metabolism of fuel substances in the body. **Insulin,** secreted by *beta cells* in the islets, acts mainly on cells of skeletal muscles, liver cells, and adipose tissue (fat). (Brain cells do not require insulin for glucose uptake.) Insulin lowers blood glucose, fatty acid, and amino acid levels and promotes their storage. That is, the actions of insulin include stimulation of glucose transport into cells, glycogen synthesis from glucose, uptake of fatty acids by adipose tissue cells, fat synthesis from fatty acids, and protein synthesis from amino acids. Insulin also inhibits glycogen degradation to glucose, fat degradation to fatty acids, and protein degradation to amino acids.

Glucagon, secreted by *alpha cells* in the islets, has effects opposite to those of insulin: it stimulates glycogen, fat, and protein degradation. Glucagon also uses amino acids and other noncarbohydrates as the input for glucose synthesis; this aspect of glucagon function operates during fasting. Negative feedback mechanisms that are keyed to the concentration of glucose in the blood control secretion of both insulin and glucagon to maintain glucose homeostasis **(Figure 40.12).**

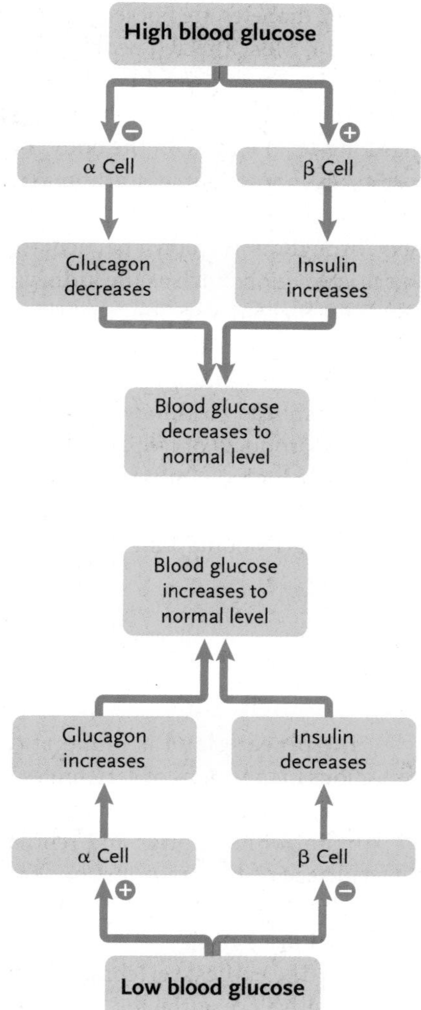
High blood glucose
α Cell
β Cell
Glucagon decreases
Insulin increases
Blood glucose decreases to normal level
Blood glucose increases to normal level
Glucagon increases
Insulin decreases
α Cell
β Cell
Low blood glucose

FIGURE 40.12
The action of insulin and glucagon in maintaining the concentration of blood glucose at an optimal level.

Diabetes mellitus, a disease that afflicts more than 14 million people in the United States and at least 200 million people worldwide, results from problems with insulin production or action. The three classic diabetes symptoms are frequent urination, increased thirst (and consequently increased fluid intake), and increased glucose in the blood (hyperglycemia). Frequent urination occurs because without insulin, body cells are not stimulated to take up glucose, leading to abnormally high glucose concentration in the blood. Excretion of the excess glucose in the urine requires water to carry it, which causes increased fluid loss from the blood and frequent trips to the bathroom. Food intake is necessary to offset the negative energy balance resulting from the decreased glucose uptake by cells or else weight loss will occur. Two of these classic symptoms gave the disease its name: *diabetes* is derived from a Greek word meaning "siphon," referring to the frequent urination, and *mellitus,* a Latin word meaning "sweetened with honey," refers to the sweet taste of a diabetic's urine. (Before modern blood or urine tests were developed, physicians tasted a patient's urine to detect the disease.)

The disease occurs in two major forms called *type 1* and *type 2.* Type 1 diabetes, which occurs in about 10% of diabetics, results from insufficient insulin secretion by the pancreas. This type of diabetes is usually caused by an autoimmune reaction that destroys pancreatic beta cells. To survive, type 1 diabetics must receive regular insulin injections (typically, a genetically engineered human insulin called Humulin); careful dieting and exercise also have beneficial effects, because active skeletal muscles do not require insulin to take up and utilize glucose.

In type 2 diabetes, insulin is usually secreted at or above normal levels, but the target cells of affected people have significantly reduced responsiveness to the hormone compared with the cells of normal people. About 90% of patients in the developed world with type 2 diabetes are obese. A genetic predisposition can also be a factor. Most affected people can lead a normal life by controlling their diet and weight, exercising, and taking drugs that enhance insulin action or secretion.

Diabetes has long-term effects on the body. Its cells, unable to utilize glucose as an energy source, start breaking down proteins and fats to generate energy. The protein breakdown weakens blood vessels throughout the body, particularly in the arms and legs and in critical regions such as the kidneys and retina of the eye. The circulation becomes so poor that tissues degenerate in the arms, legs, and feet. At advanced stages of the disease, bleeding in the retina causes blindness. The breakdown of circulation in the kidneys can lead to kidney failure. In addition, in type 1 diabetes, acidic products of fat breakdown (ketones) are produced in abnormally high quantities and accumulate in the blood. The resulting lowering of blood pH can disrupt heart and brain function, leading to coma and death if the disease is untreated.

The Pineal Gland Regulates Some Biological Rhythms

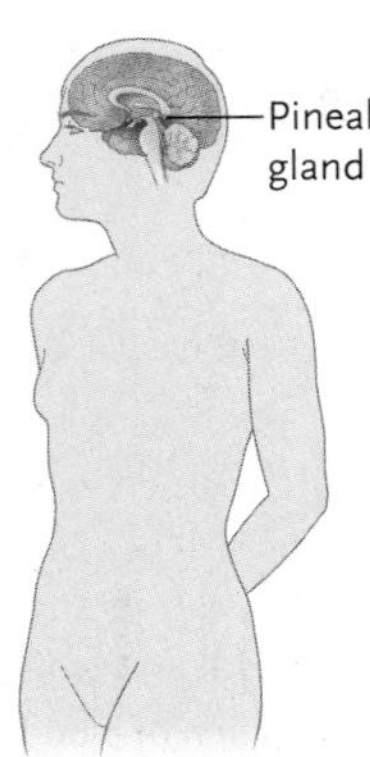
Pineal gland

The **pineal gland** is found at different locations in the brains of vertebrates—for example, in mammals, it is at roughly the center of the brain, while in birds and reptiles, it is on the surface of the brain just under the skull. The pineal gland regulates some biological rhythms.

The earliest vertebrates had a third, light-sensitive eye at the top of the head, and some species, such as lizards and tuataras (New Zealand reptiles), still have an eyelike structure in this location. In most vertebrates, the third eye became modified into a pineal gland, which in many groups retains some degree of photosensitivity. In mammals it is too deeply buried in the brain to be affected directly by light; nonetheless, specialized photoreceptors in the eyes make connections to the pineal gland.

In mammals, the pineal gland secretes a peptide hormone, **melatonin,** which helps to maintain daily biorhythms. Secretion of melatonin is regulated by an inhibitory pathway. Light hitting the eyes generates signals that inhibit melatonin secretion; consequently, the hormone is secreted most actively during periods of darkness. Melatonin targets a part of the hypothalamus called the *suprachiasmatic nucleus,* which is the primary biological clock coordinating body activity to a daily cycle. The nightly release of melatonin may help synchronize the biological clock with daily cycles of light and darkness. The physical and mental discomfort associated with jet lag may reflect the time required for melatonin secretion to reset a traveler's daily biological clock to match the period of daylight in a new time zone.

Melatonin also plays a role in other vertebrates. In some fishes, amphibians, and reptiles, melatonin and other hormones produce changes in skin color through their effects on *melanophores,* the pigment-containing cells of the skin. Skin color may vary with the season, the animal's breeding status, or the color of the background.

STUDY BREAK 40.4

1. **What effect does parathyroid hormone have on the body?**
2. **What hormones are secreted by the adrenal medulla, and what are their functions?**
3. **What are the two types of hormones secreted by the adrenal cortex, and what are their functions?**
4. **To what molecular class of hormones do estradiol and progesterone belong, and where do they act in the body?**

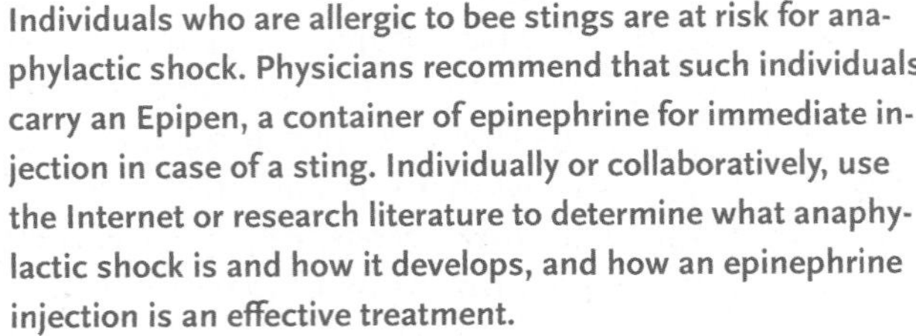

THINK OUTSIDE THE BOOK

Individuals who are allergic to bee stings are at risk for anaphylactic shock. Physicians recommend that such individuals carry an Epipen, a container of epinephrine for immediate injection in case of a sting. Individually or collaboratively, use the Internet or research literature to determine what anaphylactic shock is and how it develops, and how an epinephrine injection is an effective treatment.

40.5 Endocrine Systems in Invertebrates

In even the simplest animals, such as the cnidarian *Hydra,* hormones produced by neurosecretory neurons control reproduction, growth, and development of some body features. In annelids, arthropods, and mollusks, endocrine cells and glands produce hormones that regulate reproduction, water balance, heart rate, and sugar levels.

Some hormones occur in related forms in invertebrates and vertebrates. For example, fruit flies, mollusks, and humans have insulin-like hormones and receptors, even though molecular studies suggest that their last common ancestor existed more than 800 million years ago. Both invertebrates and vertebrates secrete peptide and steroid hormones, but most of the hormones' structures differ between the two groups, and therefore most have no effect when injected into members of the other group. However, the reaction pathways stimulated by the hormones are the same in both groups, suggesting that these regulatory mechanisms appeared very early in animal evolution.

Hormones have been studied in detail in only a few invertebrate groups, with the most extensive studies focusing on regulation of metamorphosis in insects. Butterflies, moths, and flies undergo the most dramatic changes as they mature into adults. They hatch from the egg as a caterpillar-like *larva.* During the larval stage, growth is accompanied by one or more *molts,* in which an old exoskeleton is shed and a new one forms. The insect then enters a typically nonmotile stage, the *pupa,* in which the body forms a thick, resistant coating, and finally it transforms into an adult.

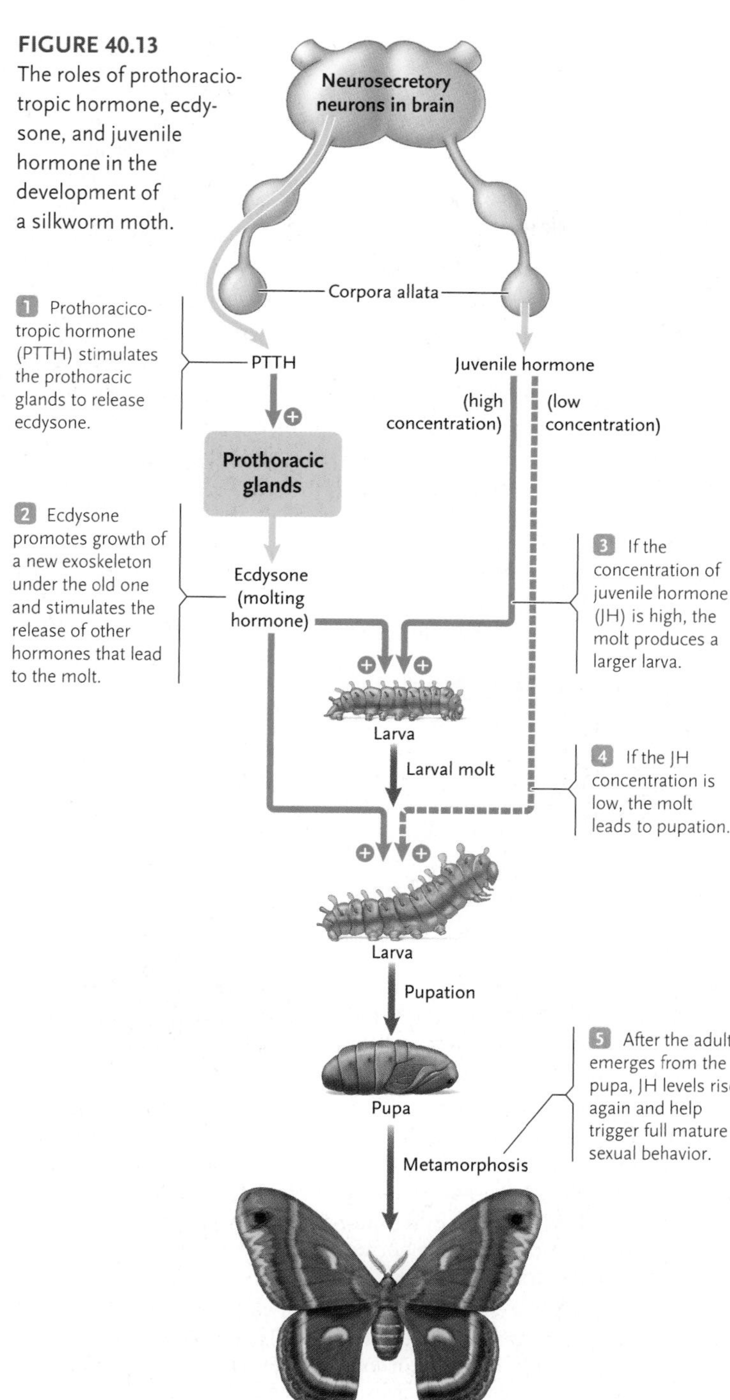
Neurosecretory neurons in brain
Corpora allata
1 Prothoracicotropic hormone (PTTH) stimulates the prothoracic glands to release ecdysone.
PTTH
Juvenile hormone
(high concentration)
(low concentration)
Prothoracic glands
2 Ecdysone promotes growth of a new exoskeleton under the old one and stimulates the release of other hormones that lead to the molt.
Ecdysone (molting hormone)
3 If the concentration of juvenile hormone (JH) is high, the molt produces a larger larva.
Larva
Larval molt
4 If the JH concentration is low, the molt leads to pupation.
Larva
Pupation
5 After the adult emerges from the pupa, JH levels rise again and help trigger full mature sexual behavior.
Pupa
Metamorphosis
Adult

FIGURE 40.13
The roles of prothoracicotropic hormone, ecdysone, and juvenile hormone in the development of a silkworm moth.

FIGURE 40.14
Control of molting by molt-inhibiting hormone (MIH), which is secreted by a gland in the eye stalks of crustaceans such as this crab.

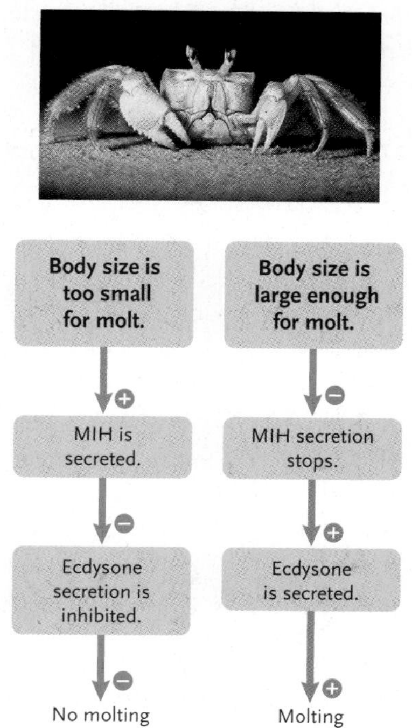
Body size is too small for molt.
MIH is secreted.
Ecdysone secretion is inhibited.
No molting
Body size is large enough for molt.
MIH secretion stops.
Ecdysone is secreted.
Molting

Three major hormones regulate molting and metamorphosis in insects: **prothoracicotropic hormone (PTTH)**, a peptide hormone secreted by neurosecretory neurons in the brain; **ecdysone** (*ekdysis* = getting out), a steroid hormone secreted by the *prothoracic glands;* and **juvenile hormone (JH)**, a terpenoid (a type of unsaturated hydrocarbon) secreted by the *corpora allata,* a pair of glands just behind the brain **(Figure 40.13).** The outcome of the molt depends on the level of JH. If it is high, the molt produces a larger larva; if it is low, the molt leads to pupation and the emergence of the adult.

Hormones that control molting have also been detected in crustaceans, including lobsters, crabs, and crayfish. Before growth reaches the stage at which the exoskeleton is shed, **molt-inhibiting hormone (MIH),** a peptide neurohormone secreted by a gland in the eye stalks, inhibits ecdysone secretion **(Figure 40.14).** As body size increases to the point requiring a molt, MIH secretion is inhibited, ecdysone secretion increases, and the molt is initiated.

In the next chapter we discuss the structure and functions of muscles, and their interactions with the skeletal system to cause movement. Muscle function depends primarily on the action of the nervous system, but the endocrine system plays a role in the control of smooth muscle contraction.

STUDY BREAK 40.5 <

How do hormones compare structurally and functionally in invertebrates and vertebrates?

UNANSWERED QUESTIONS

What are the cellular mechanisms for insulin resistance in patients with type 2 diabetes?

Insulin resistance is the condition in which the normal physiological levels of insulin are inadequate to produce a normal insulin response in the body. It plays an important role in the development of type 2 diabetes. Gerald Shulman and his research group at Yale Medical School have a long-term goal of elucidating the cellular mechanisms of insulin resistance. Once the mechanisms are known, therapeutic agents can be developed to reverse insulin resistance in patients with this type of diabetes. In their research, Shulman's group studies patients with type 2 diabetes as well as transgenic mouse models of insulin resistance.

Recall from the chapter that one of the effects of insulin is the conversion of glucose to glycogen. In one set of experiments, Shulman's group studied the rate of glucose incorporation into muscle glycogen. They discovered that muscle glycogen synthesis plays a major role in causing insulin resistance in patients with type 2 diabetes. More detailed studies showed that defects in insulin-stimulated glucose transport and glucose phosphorylation activity in muscles correlate with the early stages in the onset of type 2 diabetes.

Could the defect in glucose transport and phosphorylation activity be reversed? Shulman's group answered this question in a study of lean offspring of parents with type 2 diabetes. The offspring examined were insulin-resistant and synthesized insulin-stimulated muscle glycogen at a level only 50% that of normal individuals but, in contrast with their parents, they showed normal blood glucose levels. The potential for these individuals to develop type 2 diabetes later in life is high; that is, they are considered to be prediabetic. After six weeks of following a four-times-a-week aerobic exercise regime on a StairMaster, their insulin-stimulated muscle glycogen synthesis rates returned to normal due to correction of the glucose transport and glucose phosphorylation defects. Thus, the results suggest that regular aerobic exercise potentially could be useful in reversing insulin resistance in prediabetic individuals such as these offspring and, hence, that it might prevent the development of type 2 diabetes. More research is needed to see if that is the case.

Think Critically

1. Why did the researchers not use obese people or obese people with type 2 diabetes?
2. What type of muscle would you assay for this type of research?

Peter J. Russell

Peter J. Russell

REVIEW KEY CONCEPTS

Go to **CENGAGENOW** at www.cengage.com/login to access quizzing, animations, exercises, articles, and personalized homework help.

40.1 Hormones and Their Secretion

- Hormones are substances secreted by cells that control the activities of cells elsewhere in the body. The cells that respond to a hormone are its target cells. The best-known hormones are secreted by the endocrine system.
- The endocrine system includes four major types of cell signaling: classical endocrine signaling, in which endocrine glands secrete hormones; neuroendocrine signaling, in which neurosecretory neurons release neurohormones into the circulatory system; paracrine regulation, in which cells release local regulators that diffuse through the ECF to regulate nearby cells; and autocrine regulation, in which cells release local regulators that regulate the same cells that produced them (Figure 40.1).
- Most hormones and local regulators fall into one of four molecular classes: amines, peptides, steroids, and fatty acids.
- Many hormones are controlled by negative feedback mechanisms (Figure 40.2).

40.2 Mechanisms of Hormone Action

- Hormones typically are effective in very low concentrations in the body fluids because of amplification.
- Hydrophilic hormones bind to receptor proteins embedded in the plasma membrane, activating them. The activated receptors transmit a signal through the plasma membrane, triggering signal transduction pathways that cause a cellular response. Hydrophobic hormones bind to receptors in the cytoplasm or nucleus, activating them. The activated receptors control the expression of specific genes, the products of which cause the cellular response (Figure 40.3).
- As a result of the types of receptors they have, target cells may respond to more than one hormone, or they may respond differently to the same hormone.
- The major endocrine cells and glands of vertebrates are the hypothalamus, pituitary gland, thyroid gland, parathyroid glands, adrenal medulla, adrenal cortex, testes, ovaries, islets of Langerhans of the endocrine pancreas, and pineal gland. Hormones are also secreted by endocrine cells in the stomach and intestine, thymus gland, kidneys, and heart. Most body cells are capable of releasing prostaglandins (Figure 40.6).

Animation: Hormones and target cell receptors

Animation: Major human endocrine glands

40.3 The Hypothalamus and Pituitary

- The hypothalamus and pituitary together regulate many other endocrine cells and glands in the body (Figure 40.7).
- The hypothalamus produces hormones (releasing hormones and inhibiting hormones) that control the secretion of eight hormones by the anterior pituitary: prolactin (PRL), growth hormone (GH), thyroid-stimulating hormone (TSH), adrenocorticotropic hormone (ACTH), follicle-stimulating hormone (FSH), luteinizing hormone (LH), melanocyte-stimulating hormone (MSH), and endorphins.

- The posterior pituitary secretes antidiuretic hormone (ADH), which regulates body water balance, and oxytocin, which stimulates the contraction of smooth muscle in the uterus as a part of childbirth and triggers milk release from the mammary glands during suckling of the young.

Animation: Posterior pituitary function

Animation: Anterior pituitary function

40.4 Other Major Endocrine Glands of Vertebrates

- The thyroid gland secretes the thyroid hormones and, in mammals, calcitonin. The thyroid hormones stimulate the oxidation of carbohydrates and lipids, and coordinate with growth hormone to stimulate body growth and development. Calcitonin lowers the Ca^{2+} level in the blood by inhibiting the release of Ca^{2+} from bone. In amphibians, such as the frog, thyroid hormones trigger metamorphosis (Figure 40.9).
- The parathyroid glands secrete parathyroid hormone, which stimulates bone cells to release Ca^{2+} into the blood. PTH also stimulates the activation of vitamin D, which promotes Ca^{2+} absorption into the blood from the small intestine (Figure 40.10).
- The adrenal medulla secretes epinephrine and norepinephrine, which reinforce the sympathetic nervous system in responding to stress. The adrenal cortex secretes glucocorticoids, which help maintain glucose at normal levels in the blood, and mineralocorticoids, which regulate Na^+ balance and ECF volume. The adrenal cortex also secretes small amounts of androgens (Figure 40.11).
- The gonadal sex hormones—androgens, estrogen, and progestins—play a major role in regulating the development of reproductive systems, sexual characteristics, and mating behavior.
- The islet of Langerhans cells of the endocrine pancreas secrete insulin and glucagon, which together regulate the concentration of fuel substances in the blood. Insulin lowers the concentration of glucose in the blood and inhibits the conversion of noncarbohydrate molecules into glucose. Glucagon raises blood glucose by stimulating glycogen, fat, and protein degradation (Figure 40.12).
- The pineal gland secretes melatonin, which interacts with the hypothalamus to set the body's daily rhythms.

Animation: Parathyroid hormone action

Animation: Hormones and glucose metabolism

40.5 Endocrine Systems in Invertebrates

- Hormones control development and function of the gonads, manage salt and water balance in the body fluids, and control molting in insects and crustaceans.
- Three major hormones—prothoracicotropic hormone (PTTH), ecdysone, and juvenile hormone (JH)—control molting and metamorphosis in insects. Hormones that control molting are also present in crustaceans (Figures 40.13 and 40.14).

UNDERSTAND AND APPLY

Test Your Knowledge

1. Amine hormones are usually:
   a. hydrophilic when secreted by the thyroid gland.
   b. based on tyrosine.
   c. paracrine but not autocrine.
   d. not transported by the blood.
   e. repelled by the plasma membrane.
2. Prostaglandins would be best described as inducers of:
   a. male and female characteristics.
   b. cell division.
   c. nerve transmission.
   d. smooth muscle contractions.
   e. cell differentiation.
3. When the concentration of thyroid hormone in the blood increases, it:
   a. inhibits TSH secretion by the pituitary.
   b. stimulates TRH secretion by the hypothalamus.
   c. stimulates the pituitary to secrete TRH.
   d. stimulates the pituitary to secrete TSH.
   e. activates a positive feedback loop.
4. Which of the following statements about endocrine targeting and reception is correct?
   a. The idea that one hormone affects one type of tissue is illustrated when epinephrine binds to smooth muscle cells in blood vessels as well as to beta cells in heart muscle.
   b. The idea that one hormone affects one type of tissue is illustrated when epinephrine cannot activate both the receptors on liver cells and the beta receptors of heart muscle.
   c. The idea that a target cell can respond to more than one hormone is illustrated by a vertebrate liver cell responding to insulin and glucagon.
   d. The idea that a minute concentration of hormone can cause widespread effects demonstrates the specificity of cells for certain hormones.
   e. The idea that the response to a hormone is the same among different target cells is shown when different liver cells are activated by insulin.
5. The posterior pituitary secretes:
   a. hormones that control the hypothalamus.
   b. IGF, which simulates cell division and protein synthesis.
   c. ADH, which increases water absorption in the kidneys.
   d. oxytocin, which controls egg and sperm development.
   e. prolactin, which stimulates milk synthesis.
6. Blood levels of calcium are regulated directly by:
   a. insulin synthesized by the alpha cells of the pancreas.
   b. PTH made by the pituitary.
   c. vitamin D activated in the liver.
   d. prolactin synthesized by the anterior pituitary.
   e. calcitonin secreted by specialized thyroid cells.
7. If the human body is stressed, glucocorticoids:
   a. promote the breakdown of proteins in the muscles and bones.
   b. increase the amount of sodium reabsorbed from urine in the kidneys.
   c. decrease potassium secretion from the kidneys.
   d. decrease glucose uptake by cells in the nervous system.
   e. inhibit the synthesis of glucose from noncarbohydrate sources.
8. When blood glucose rises:
   a. alpha cells increase glucagon secretion.
   b. beta cells increase insulin secretion.
   c. urination decreases in a person with type 1 diabetes who has not recently received an insulin injection.
   d. glucagon stimulates the breakdown of amino acids into glycogen.
   e. target cells decrease their insulin receptors.

9. In mammals:
   a. the suprachiasmatic nucleus of the pineal gland controls both male and female reproductive systems.
   b. estradiol is produced by the hypothalamus to control ovulation.
   c. melatonin controls anabolic steroid production.
   d. GnRH stimulates LH to control testosterone production.
   e. progesterone increases the secretion of LH from the posterior pituitary.
10. Insect development is regulated by:
   a. ecdysone, a peptide secreted by the brain.
   b. juvenile hormone, a terpenoid secreted by the corpora allata near the brain.
   c. molt-inhibiting hormone, a steroid secreted by the prothoracic glands.
   d. prothoracicotropic hormone, a steroid secreted by the hypothalamus.
   e. melatonin, a peptide secreted by the brain in the larval stage.

Discuss the Concepts

1. A physician sees a patient whose symptoms include sluggishness, depression, and intolerance to cold. What disorder do these symptoms suggest?
2. Cushing's syndrome occurs when an individual overproduces cortisol; this rare disorder is also known as hypercortisolism. In children and teenagers, symptoms include extreme weight gain, retarded growth, excess hair growth, acne, high blood pressure, tiredness and weakness, and either very early or late puberty. Adults with the disease may also exhibit extreme weight gain, excess hair growth, and high blood pressure, and in addition may show muscle and bone weakness, moodiness or depression, sleep disorders, and reproductive disorders. Propose some hypotheses for the overproduction of cortisol in individuals with Cushing's syndrome.
3. A 20-year-old woman with a malignant brain tumor has her pineal gland removed. What kinds of side effects might this loss have?
4. In integrated pest management, a farmer uses a variety of tools to combat unwanted insects. These include applications of either hormones or hormone-inhibiting compounds to prevent insects from reproducing successfully. How might each of these hormone-based approaches disrupt reproduction?

Design an Experiment

The Environmental Protection Agency (EPA) defines endocrine disruptors as chemical substances that can "interfere with the synthesis, secretion, transport, binding, action, or elimination of natural hormones in the body that are responsible for the maintenance of homeostasis (normal cell metabolism), reproduction, development, and/or behavior." The chemicals, sometimes called environmental estrogens, come from both natural and man-made sources. A simple hypothesis is that endocrine disruptors act by mimicking hormones in the body. Many endocrine disruptors affect sex hormone function and, therefore, reproduction.

Examples of endocrine disruptors are the synthetic chemicals DDT (a pesticide), dioxins, and natural chemicals such as phytoestrogens (estrogen-like molecules in plants), which are found in high levels in soybeans, carrots, oats, onions, beer, and coffee.

Design an experiment to investigate whether a new synthetic chemical (pick your own interesting scenario) is an endocrine disruptor. (Hint: You probably want to work with a model organism.)

Interpret the Data

Researchers measured prolactin and growth hormone in the blood plasma of six subjects at regular intervals over a 24-hour period. The investigators were interested in determining if the levels of these two hormones cycle over a 24-hour period. The graph below presents the results of their experiments, with the y axis values showing the amount of each hormone expressed as a percentage of the 24-hour mean value, and each point showing the average of the six subjects' levels.

1. How did the level of growth hormone change over the 24-hour period and, in particular, how did it relate to the sleep period?
2. How did the level of prolactin change over the 24-hour period and, in particular, how did it relate to the sleep period?
3. Is there any consistent relationship between the patterns of prolactin and growth hormone in the plasma in the 24-hour period?
4. Do the results support the hypothesis that the changes in prolactin level over a 24-hour period are associated with the sleep–wake cycle?

Source: J. F. Sassin et al. 1972. Human prolactin: 24-hour pattern with increased release during sleep. *Science* 177:1205–1207.

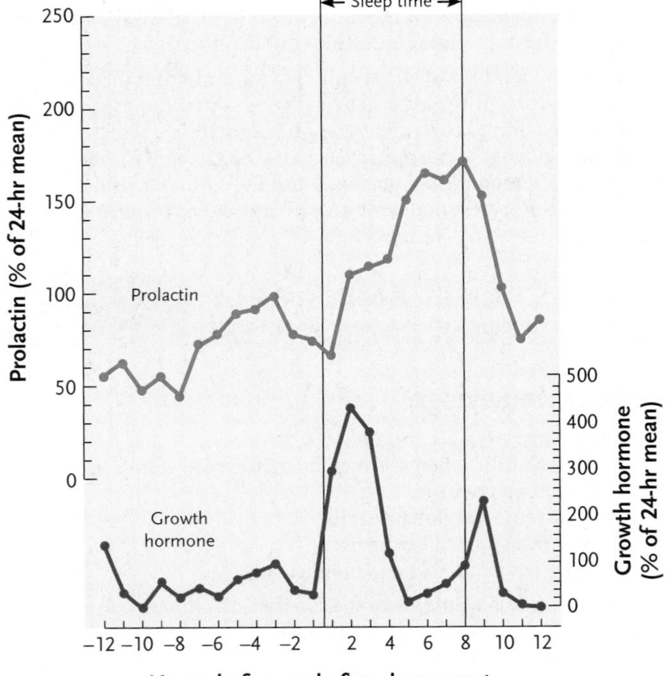
Sleep time
250
200
150
100
50
0
Prolactin (% of 24-hr mean)
Prolactin
Growth hormone
500
400
300
200
100
0
Growth hormone (% of 24-hr mean)
−12 −10 −8 −6 −4 −2 2 4 6 8 10 12
Hours before and after sleep onset

Mean concentrations of prolactin and growth hormone in plasma, averaged for six subjects and expressed as a percentage of the average concentrations measured for the 24-hr period.

Apply Evolutionary Thinking

Which endocrine system evolved earlier, endocrine glands or neurosecretory neurons? Support your conclusion with information obtained from online research.

Express Your Opinion

Crop yields that sustain the human population currently depend on agricultural pesticides, some of which may disrupt hormone function in frogs and other untargeted species. Should chemicals that may cause problems remain in use while researchers investigate them? Go to www.cengage.com/login to investigate both sides of the issue and then vote.

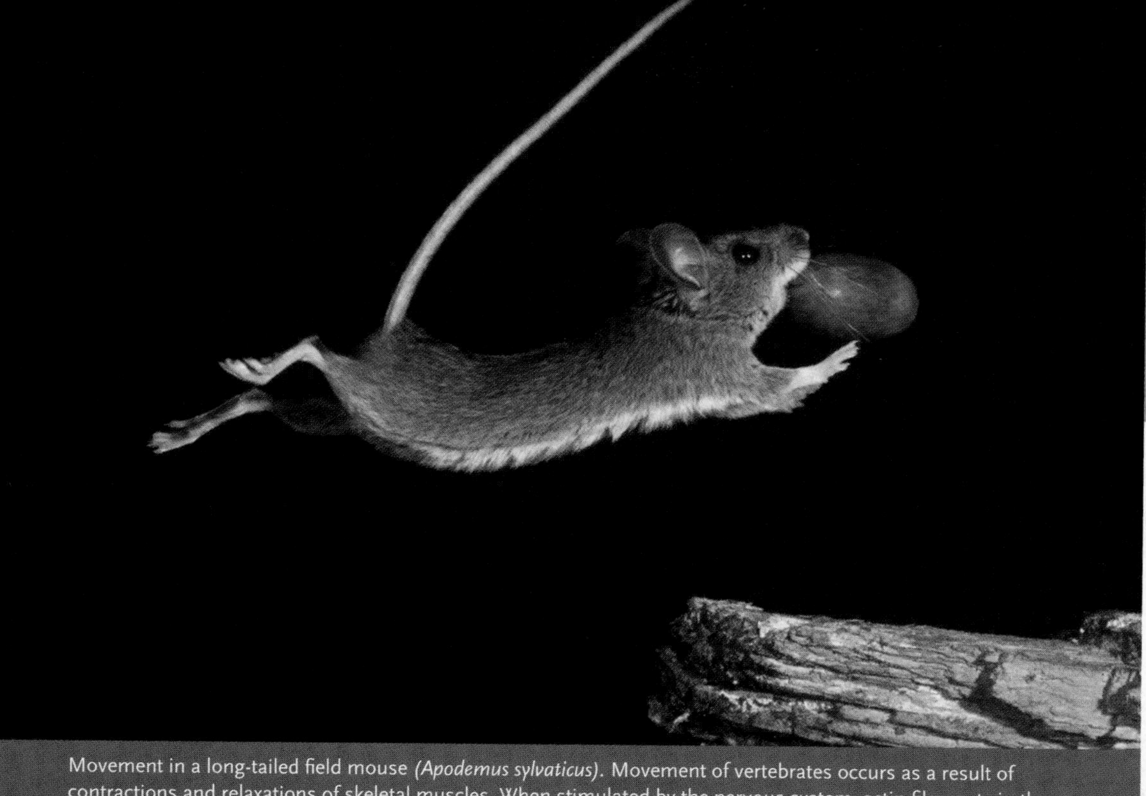

Movement in a long-tailed field mouse *(Apodemus sylvaticus)*. Movement of vertebrates occurs as a result of contractions and relaxations of skeletal muscles. When stimulated by the nervous system, actin filaments in the muscles slide over myosin filaments to cause muscle contractions.

G. Delpho/Peter Arnold, Inc.

41

STUDY OUTLINE

Muscles, Bones, and Body Movements

Why It Matters. . . A Mexican leaf frog *(Pachymedusa dacnicolor)* sits motionless, its eyes staring into space. When the frog detects an approaching cricket, it lunges forward at just the right moment, thrusts out its sticky tongue, and captures the prey. From the beginning of the movement until the frog's mouth closes, sealing the cricket's fate, requires only 260 msec—about one quarter of a second. How does the frog move so swiftly, and so surely?

As its prey draws near, neuronal signals travel from the frog's brain to the muscles that extend the frog's hind legs, causing those muscles to contract and propel the frog forward on its forelimbs toward the cricket. Within 50 msec after the jump begins, other signals contract the muscles of the lower jaw, opening the mouth. Then, a muscle on the upper surface of the tongue contracts, which raises the tongue and flips it out of the mouth **(Figure 41.1)**. As the tongue shoots forward, muscle contractions along the ventral side of the trunk arch the body and direct the head downward toward the prey. Within 80 msec after the lunge begins, the tip of the frog's tongue contacts the cricket. Completion of the lunge folds the tongue—and the cricket—into the frog's mouth, aided by contraction of a muscle on the bottom of the tongue. After the mouth closes, further muscle contractions pull the legs forward and fold them under the body.

In Section 36.2 you learned about the three types of muscle tissue: skeletal, cardiac, and smooth. Skeletal muscle is so named because most muscles of this type are attached by tendons to the skeleton of vertebrates.

116
128
196

Kiisa Nishikawa/Northern Arizona University

FIGURE 41.1
A Mexican leaf frog *(Pachymedusa dacnicolor)* **capturing a grasshopper.** The frog's movements were captured using a high-speed video camera linked to a millisecond timer, with a grid in the background that allowed precise measurement of the distances body parts traveled during the capture.

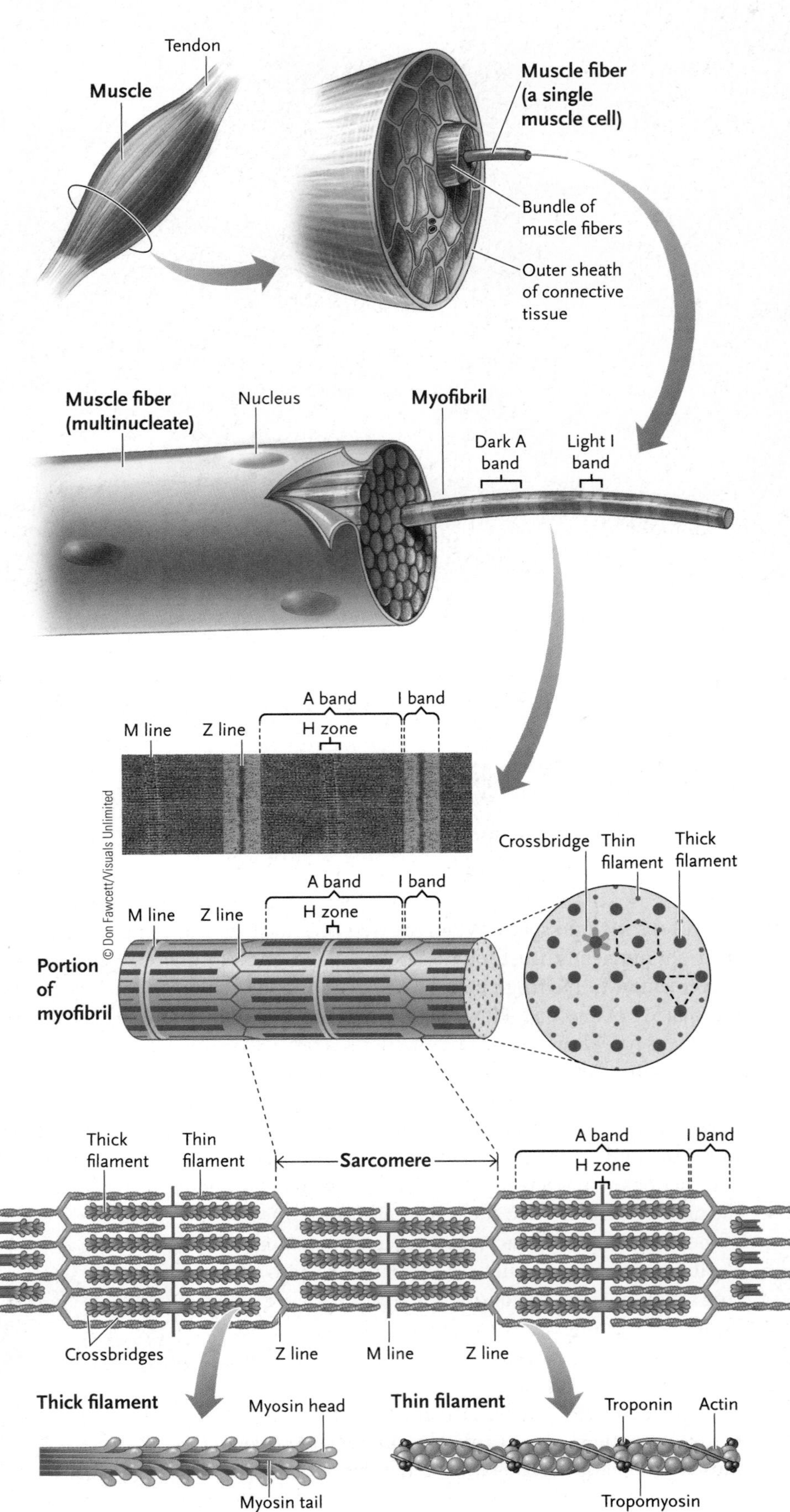

Tendon
Muscle
Muscle fiber (a single muscle cell)
Bundle of muscle fibers
Outer sheath of connective tissue
Muscle fiber (multinucleate)
Nucleus
Myofibril
Dark A band
Light I band
A band
I band
M line
Z line
H zone
© Don Fawcett/Visuals Unlimited
Crossbridge
Thin filament
Thick filament
Portion of myofibril
Thick filament
Thin filament
Sarcomere
Crossbridges
Z line
M line
Z line
Thick filament
Myosin head
Myosin tail
Thin filament
Troponin
Actin
Tropomyosin

FIGURE 41.2

Skeletal muscle structure. Muscles are composed of bundles of multinucleate cells called muscle fibers; within each muscle fiber are longitudinal bundles of myofibrils. The unit of contraction within a myofibril, the sarcomere, consists of overlapping myosin thick filaments and actin thin filaments. The myosin molecules in the thick filaments each consist of two subunits organized into a head and a double-helical tail. The actin subunits in the thin filaments form twisted, double helices, with tropomyosin molecules arranged head-to-tail in the groove of the helix and troponin bound to the tropomyosin at intervals along the thin filaments.

Cardiac muscle is the contractile muscle of the heart, and smooth muscle is found in the walls of tubes and cavities in the body, including blood vessels and intestines. In this chapter we describe the structure and function of skeletal muscles and the skeletal systems of invertebrates and vertebrates, and how muscles bring about movement. <

41.1 Vertebrate Skeletal Muscle: Structure and Function

Vertebrate **skeletal muscles** connect to bones of the skeleton. The cells forming skeletal muscles are typically long and cylindrical and contain many nuclei (shown in Figure 36.5A). Skeletal muscle is controlled by the somatic nervous system.

Most skeletal muscles in humans and other vertebrates are attached at both ends to bones of the skeleton and across a joint. (Some, such as those that move the lips, are attached to other muscles or connective tissues under skin.) Depending on its points of attachment, contraction of a single skeletal muscle may extend or bend body parts or may rotate one body part with respect to another. The human body has more than 600 skeletal muscles, ranging in size from the small muscles that move the eyeballs to the large muscles that move the legs.

Skeletal muscles are attached to bones by cords of connective tissue called *tendons* (see Section 36.2). Tendons vary in length from a few millimeters to some, such as those that connect the muscles of the forearm to the bones of the fingers, that are 20 to 30 cm long.

The Striated Appearance of Skeletal Muscle Fibers Results from a Highly Organized Internal Structure

A skeletal muscle consists of bundles of elongated, cylindrical cells called **muscle fibers,** which are 10 to 100 mm in diameter and run the entire length of the muscle **(Figure 41.2).** Muscle fibers are multinucleate—they contain many nuclei—reflecting their development by fusion of smaller cells. Some very small muscles, such as some of the muscles of the face, contain only a few hundred muscle fibers; others, such as the larger leg muscles, contain hundreds of thousands. In both cases, the muscle fibers are held in parallel bundles by sheaths of connective tissue that surround them in the muscle and merge with the tendons that connect muscles to bones or other structures. Muscle fibers are richly supplied with nutrients and oxygen by an extensive network of blood vessels that penetrates the muscle tissue.

Muscle fibers are packed with **myofibrils,** cylindrical contractile elements about 1 μm in diameter that run lengthwise inside the cells. Each myofibril consists of a regular arrangement of **thick filaments** (13–18 nm in diameter) and **thin filaments** (5–8 nm in diameter) (see Figure 41.2). The thick and thin filaments alternate with one another in a stacked set.

The thick filaments are parallel bundles of myosin molecules; each myosin molecule consists of two protein subunits that together form a *head* connected to a long double helix forming a *tail.* The head is bent toward the adjacent thin filament to form a *crossbridge.* In vertebrates, each thick filament contains some 200 to 300 myosin molecules and forms as many crossbridges. The thin filaments consist mostly of two linear chains of actin molecules twisted into a double helix, which creates a groove running the length of the molecule. Bound to the actin are *tropomyosin* and *troponin* proteins. Tropomyosin molecules are elongated fibrous proteins that are organized end to end next to the groove of the actin double helix. Troponin is a three-subunit globular protein that binds to tropomyosin at intervals along the thin filaments.

The arrangement of thick and thin filaments forms a pattern of alternating dark bands and light bands, giving skeletal muscle a striated appearance under the microscope (see Figure 41.2). The dark bands, called *A bands,* consist of stacked thick filaments along with the parts of thin filaments that overlap both ends. The lighter-appearing middle region of an A band, which contains only thick filaments, is the *H zone.* In the center of the H zone is a disc of proteins called the *M line,* which holds the stack of thick filaments together. The light bands, called *I bands,* consist of the parts of the thin filaments not in the A band. In the center of each I band is a thin *Z line,* a disc to which the thin filaments are anchored. The region between two adjacent Z lines is a **sarcomere** (*sarco-* = flesh, *meros* = segment); sarcomeres are the basic units of contraction in a myofibril.

At each junction of an A band and an I band, the plasma membrane folds into the muscle fiber to form a **T (transverse) tubule (Figure 41.3).** Encircling the sarcomeres is the **sarcoplasmic reticulum,** a complex system of vesicles modified from the smooth endoplasmic reticulum. Segments of the sarcoplasmic reticulum are wrapped around each A band and I band and are separated from the T tubules in those regions by small gaps.

An axon of an efferent neuron leads to each muscle fiber. The axon terminal makes a single, broad synapse with a muscle fiber called a **neuromuscular junction** (see Figure 41.3). The neuromuscular junction, T tubules, and sarcoplasmic reticulum are key components in the pathway for stimulating skeletal muscle contraction.

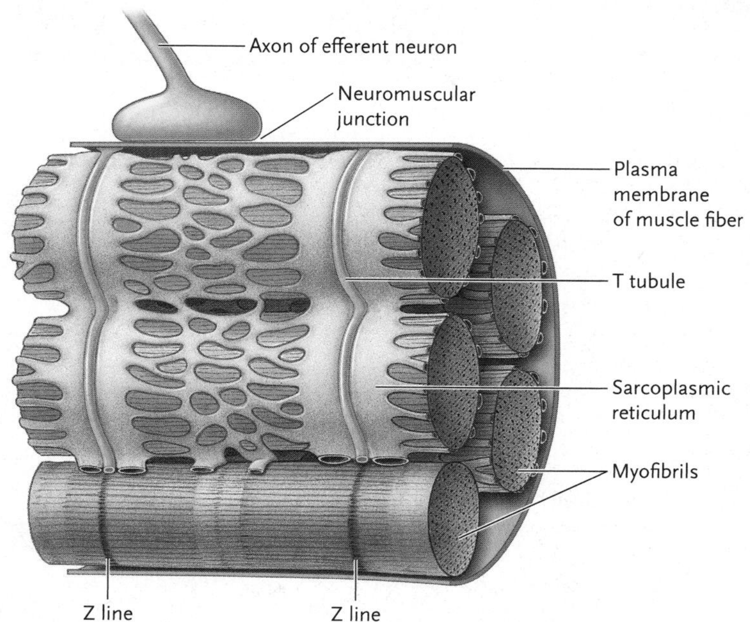

Axon of efferent neuron
Neuromuscular junction
Plasma membrane of muscle fiber
T tubule
Sarcoplasmic reticulum
Myofibrils
Z line
Z line

FIGURE 41.3
Components in the pathway for the stimulation of skeletal muscle contraction by neural signals. T (transverse) tubules are infoldings of the plasma membrane into the muscle fiber originating at each A band–I band junction in a sarcomere. The sarcoplasmic reticulum encircles the sarcomeres (it is cut away from the bottom myofibril in the figure), and segments of it end in close proximity to the T tubules.

During Muscle Contraction, Thin Filaments on Each Side of a Sarcomere Slide over Thick Filaments

The precise control of body motions depends on an equally precise control of muscle contraction by a signaling pathway that starts with action potentials traveling down an efferent neuron. These neural signals carry information from nerves to muscle fibers and trigger contraction.

MECHANISM OF MUSCLE FIBER CONTRACTION Contraction of a muscle fiber involves a process in which the thin filaments on each side of a sarcomere slide over the thick filaments toward the center of the A band, which brings the Z lines closer together, shortening the sarcomeres and contracting the muscle **(Figure 41.4).** This **sliding filament model** of muscle contraction depends on dynamic interactions between actin and myosin proteins in the two filament types. **Figure 41.5** presents a molecular model showing how an action potential arriving at the neuromuscular junction leads to the cyclic actin–myosin interactions that are the basis for muscle fiber contraction. That is, like neurons, skeletal muscle fibers are *excitable,* meaning that the electrical potential of their plasma membrane can change in response to a stimulus. The action potential leads to an increase in the concentration of Ca^{2+} in the cytosol of the muscle fiber. The increase in Ca^{2+} triggers the crossbridge cycle, in which repeated power strokes pull the actin thin filament over the myosin thick filament (see Figure 41.5, steps 4–6). When action potentials stop arriving at the neuromuscular junction, the muscle fiber stops contracting and relaxes.

FIGURE 41.4 **Experimental Research**

The Sliding Filament Model of Muscle Contraction

Question: What is the mechanism of muscle contraction?

Experiment: By 1954, researchers had established the locations and arrangements of actin and myosin in striated muscle. In that year, two independent teams—Andrew Huxley and Ralph Niedergerke of the University of Cambridge, UK, and Hugh Huxley and Jean Hanson of Massachusetts Institute of Technology—used high-resolution light microscopy techniques to study how the actin and myosin arrangements changed during muscle contraction.

Results: Their micrographs provided important evidence for the sliding of filaments. The stages of muscle contraction they outlined are shown in the figure.

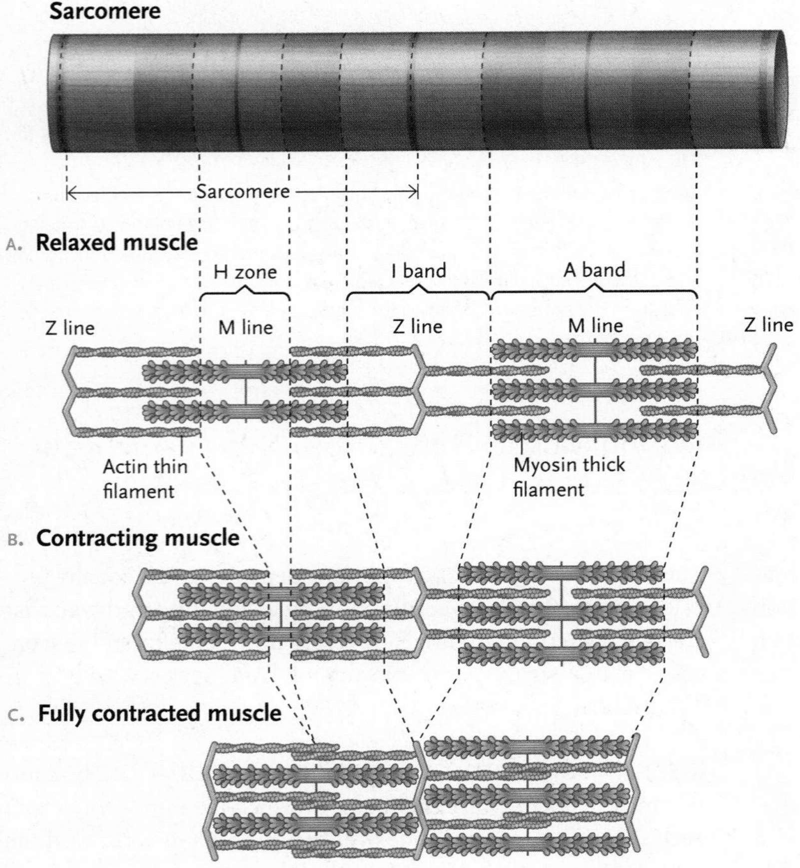

Sarcomere
Sarcomere
A. Relaxed muscle
H zone
I band
A band
Z line
M line
Z line
M line
Z line
Actin thin filament
Myosin thick filament
B. Contracting muscle
C. Fully contracted muscle

The figure illustrates a muscle fiber in a completely relaxed state; the muscle fiber is then shown as contracting, and finally as completely contracted. Note how the H zone, I band, and A band relate to one another at each stage. As the muscle fiber contracted, the key observations the researchers made were that: (1) the I band and the H zone each decrease in length in proportion to the shortening of the sarcomere; and (2) the A band remains constant in length.

Conclusion: The light microscopy evidence supported a model—the sliding filament model—in which muscle shortening (sarcomere shortening) results from increased overlap of thick and thin filaments not from any change in length of those filaments. This model was revolutionary in the field of muscle physiology, as models for muscle contraction at the time all proposed folding or coiling of the protein molecules in the filaments.

Sources: A. F. Huxley and R. Niedergerke. 1954. Structural changes in muscle during contraction. *Nature* 173:971–973; H. Huxley and J. Hanson. 1954. Changes in the cross-striations of muscle during contraction and stretch and their structural interpretation. *Nature* 173:973–976.

Crossbridge cycles based on actin and myosin power movements in all living organisms, from cytoplasmic streaming in plant cells and amoebae to muscle contractions in animals.

Although the force produced by a single myosin crossbridge is comparatively small, it is multiplied by the hundreds of crossbridges acting in a single thick filament, and by the billions of thin filaments sliding in a contracting sarcomere. The force, multiplied further by the many sarcomeres and myofibrils in a muscle fiber, is transmitted to the plasma membrane of a muscle fiber by the attachment of myofibrils to elements of the cytoskeleton. From the plasma membrane, it is transmitted to bones and other body parts by the connective tissue sheaths surrounding the muscle fibers and by the tendons.

Several gene mutations affecting muscle and nerve tissues interrupt the transmission of force and cause severe disabilities. Duchenne muscular dystrophy (DMD), for example, is caused by a mutation that weakens the cytoskeleton of the muscle fiber, causing the cells to rupture when contractile forces are generated. *Insights from the Molecular Revolution* describes experiments that may lead to a cure for this debilitating disease.

DEADLY INTERRUPTIONS OF THE CROSSBRIDGE CYCLE The mechanism controlling vertebrate muscle contraction can be blocked by several toxins and poisons. For example, the bacterium *Clostridium botulinum*, which grows in improperly preserved food, produces botulinum toxin, which blocks acetylcholine release in neuromuscular junctions. Affected body muscles are unable to contract, including the diaphragm, the muscle that is essential for inflating the lungs. When the diaphragm becomes paralyzed, the victim dies from respiratory failure. The toxin is so poisonous that 0.0000001 g is enough to kill a human; 600 g could wipe out the entire human population. A very low concentration of the same toxin, under the brand name Botox, is injected under the skin as a cosmetic treatment to remove or reduce wrinkles—if muscles cannot contract, then wrinkles cannot form.

Curare, extracted from the bark and sap of some South American trees, blocks acetylcholine from binding to its receptors in muscle fibers. The body muscles, including the diaphragm, become paralyzed and the victim dies of respiratory failure. Some indigenous peoples in South America took advantage of these effects by using curare as an arrow and dart poison in animal hunts.

In a natural process, within a few hours after an animal dies, Ca^{2+} diffuses into the cytoplasm of muscle cells and initiates the crossbridge cycle,

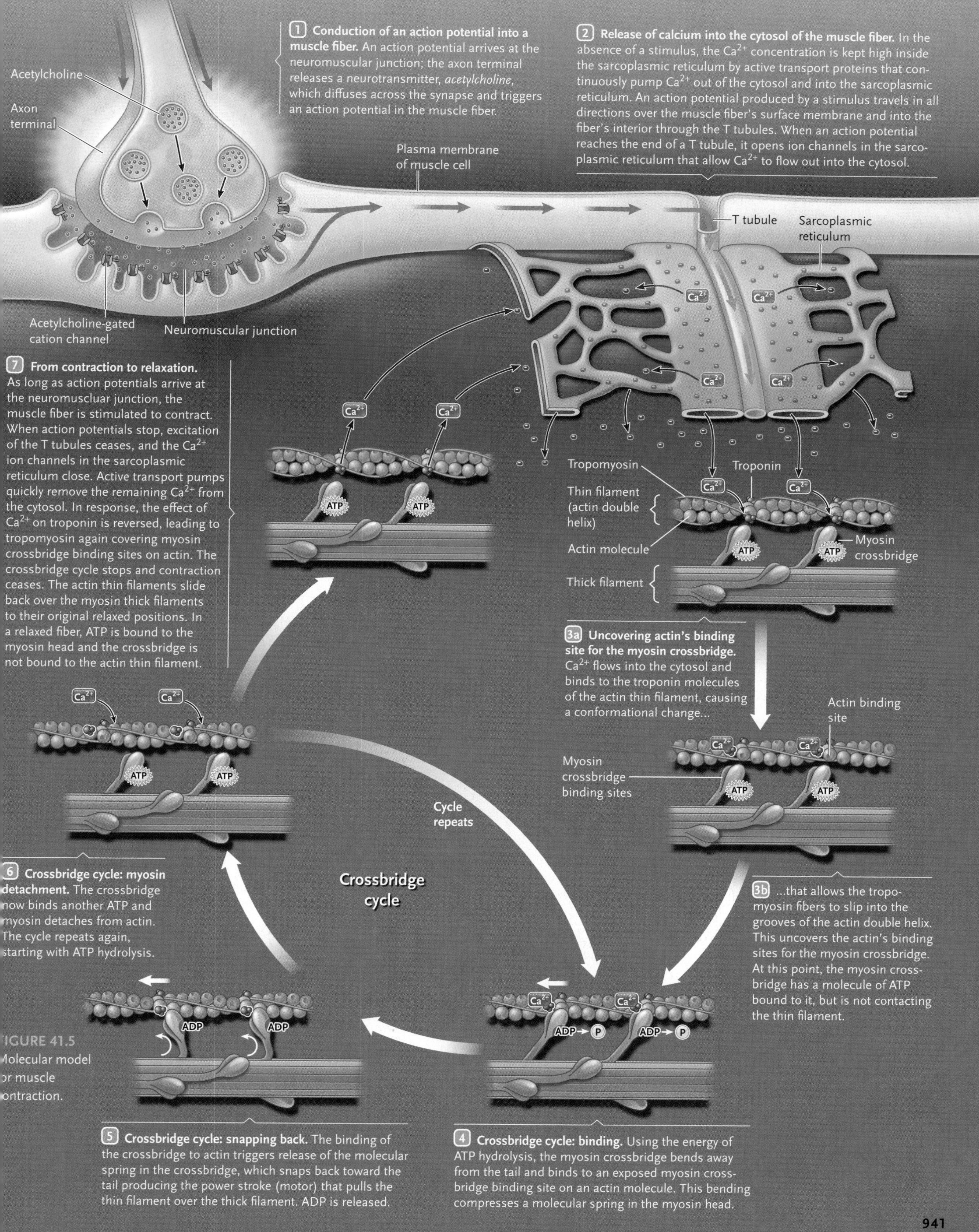
Acetylcholine
Axon terminal
Acetylcholine-gated cation channel
Neuromuscular junction
Plasma membrane of muscle cell
T tubule
Sarcoplasmic reticulum
1 Conduction of an action potential into a muscle fiber. An action potential arrives at the neuromuscular junction; the axon terminal releases a neurotransmitter, acetylcholine, which diffuses across the synapse and triggers an action potential in the muscle fiber.
2 Release of calcium into the cytosol of the muscle fiber. In the absence of a stimulus, the Ca2+ concentration is kept high inside the sarcoplasmic reticulum by active transport proteins that continuously pump Ca2+ out of the cytosol and into the sarcoplasmic reticulum. An action potential produced by a stimulus travels in all directions over the muscle fiber's surface membrane and into the fiber's interior through the T tubules. When an action potential reaches the end of a T tubule, it opens ion channels in the sarcoplasmic reticulum that allow Ca2+ to flow out into the cytosol.
Ca2+
Tropomyosin
Troponin
Thin filament (actin double helix)
Actin molecule
Myosin crossbridge
ATP
Thick filament
3a Uncovering actin's binding site for the myosin crossbridge. Ca2+ flows into the cytosol and binds to the troponin molecules of the actin thin filament, causing a conformational change...
Actin binding site
Myosin crossbridge binding sites
3b ...that allows the tropomyosin fibers to slip into the grooves of the actin double helix. This uncovers the actin's binding sites for the myosin crossbridge. At this point, the myosin crossbridge has a molecule of ATP bound to it, but is not contacting the thin filament.
ADP → P
4 Crossbridge cycle: binding. Using the energy of ATP hydrolysis, the myosin crossbridge bends away from the tail and binds to an exposed myosin crossbridge binding site on an actin molecule. This bending compresses a molecular spring in the myosin head.
ADP
5 Crossbridge cycle: snapping back. The binding of the crossbridge to actin triggers release of the molecular spring in the crossbridge, which snaps back toward the tail producing the power stroke (motor) that pulls the thin filament over the thick filament. ADP is released.
6 Crossbridge cycle: myosin detachment. The crossbridge now binds another ATP and myosin detaches from actin. The cycle repeats again, starting with ATP hydrolysis.
Crossbridge cycle
Cycle repeats
7 From contraction to relaxation. As long as action potentials arrive at the neuromusclar junction, the muscle fiber is stimulated to contract. When action potentials stop, excitation of the T tubules ceases, and the Ca2+ ion channels in the sarcoplasmic reticulum close. Active transport pumps quickly remove the remaining Ca2+ from the cytosol. In response, the effect of Ca2+ on troponin is reversed, leading to tropomyosin again covering myosin crossbridge binding sites on actin. The crossbridge cycle stops and contraction ceases. The actin thin filaments slide back over the myosin thick filaments to their original relaxed positions. In a relaxed fiber, ATP is bound to the myosin head and the crossbridge is not bound to the actin thin filament.

IGURE 41.5 Molecular model or muscle ontraction.

INSIGHTS FROM THE **Molecular Revolution**

A Substitute Player That May Be a Big Winner in Muscular Dystrophy

Duchenne muscular dystrophy (DMD) is an X-linked recessive disease (see Section 13.2), characterized by progressive muscle weakness, that primarily affects males—about 1 out of every 3,500 males is born with the disease. When DMD patients are 3 to 5 years old, their muscle tissue begins to break down, and by the time they are in their teens most can walk only with braces. Patients usually die of complications from degeneration of the heart and diaphragm muscle by their early twenties. Currently, there is no effective treatment for DMD.

The DMD gene was cloned in 1985. In its normal form, the gene encodes the protein *dystrophin*, which anchors a glycoprotein complex in the plasma membrane of a muscle fiber to the underlying actin cytoskeleton (see Section 5.3). In most people with DMD, segments of DNA are missing from the coding sequence of the gene, so the protein is incomplete and does not function. Without functional dystrophin, the plasma membrane of the muscle fibers is susceptible to tearing during contraction, which leads to muscle destruction. Creatine kinase (CK), an enzyme found predominantly in muscles and in the brain, leaks out of the damaged muscles and accumulates in the blood, which normally contains little CK. Elevated CK in the blood is diagnostic of muscle damage such as that found in DMD.

Research Question

Can DMD be cured by gene therapy?

Experiment

Kay E. Davies and her colleagues at Oxford University in England identified a protein that is structurally similar to dystrophin and appears to have a highly similar function. That protein, called *utrophin*, is made in small quantities in muscle fibers and normally functions only in neuromuscular junctions. The utrophin gene and its protein function normally in DMD patients.

The Davies team reasoned that utrophin might be able to substitute for the missing dystrophin in DMD patients if a means could be found to increase its quantity in muscle cells. For their research, they used *mdx* mice, a strain that has the dystrophin gene deleted and is therefore a mouse model of human DMD. **Figure 1** shows the researchers' experimental approach. By molecular techniques, they created a transgenic mouse strain carrying an engineered utrophin gene that was under the control of a strong promoter. (The construction of transgenic animals is described in Section 18.2.) Then by crossing a male of that strain with a female *mdx* mouse, they generated male offspring all of which were *mdx* (deleted dystrophin gene), and some of which (utrophin–*mdx* mice) had inherited the engineered utrophin gene. Those with the engineered utrophin gene expressed that gene at a level higher than that from the normal utrophin gene in the mouse because of the strong promoter controlling the gene. They assayed the two types of male offspring for CK in the blood and for evidence of muscle damage.

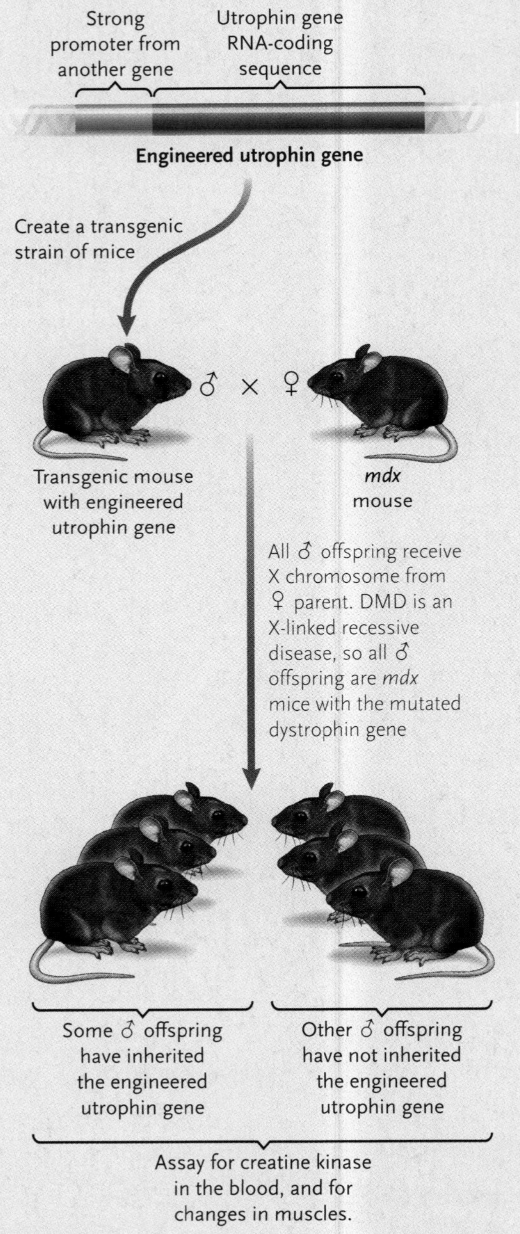

Strong promoter from another gene
Utrophin gene RNA-coding sequence
Engineered utrophin gene
Create a transgenic strain of mice
Transgenic mouse with engineered utrophin gene
♂ × ♀
mdx mouse
All ♂ offspring receive X chromosome from ♀ parent. DMD is an X-linked recessive disease, so all ♂ offspring are *mdx* mice with the mutated dystrophin gene
Some ♂ offspring have inherited the engineered utrophin gene
Other ♂ offspring have not inherited the engineered utrophin gene
Assay for creatine kinase in the blood, and for changes in muscles.

FIGURE 1
Experimental approach for testing the effect of increased utrophin gene expression on DMD.

Results

The researchers were excited to find that CK levels in the blood of the utrophin–*mdx* mice were reduced to 25% of the level in *mdx* mice, indicating that muscle damage was markedly decreased. This was confirmed by microscopic examination of the muscles. Other techniques showed that utrophin, instead of being concentrated in neuromuscular junctions as it is normally, was now distributed throughout the muscle plasma membranes.

Conclusion

The experiments showed that an elevated level of utrophin was able to substitute for dystrophin in the *mdx* mouse model of DMD, and decreased significantly the onset of disease symptoms. Moreover, no deleterious side effects from the overproduction of utrophin could be detected in the utrophin–*mdx* mice. This was the first study providing evidence to support the use of utrophin as a therapeutic strategy for DMD.

Promising as these results are, such germline gene therapy is not allowed for human patients. Instead, research groups have focused on developing a mechanism to upregulate the natural cellular utrophin genes; that is, to increase their levels of expression so as to increase the amount of utrophin protein produced. In early 2010, BioMarin Pharmaceuticals began testing the effects of a utrophin upregulator molecule on the progression of DMD.

Source: J. M. Tinsley et al. 1996. Amelioration of the dystrophic phenotype of *mdx* mice using a truncated utrophin transgene. *Nature* 384:349–353.

producing *rigor mortis,* a strong tension of essentially all the skeletal muscles that stiffens the entire body. The crossbridges become locked to the thin filaments because ATP production stops (remember that ATP is required to release the crossbridges from actin). The stiffness reverses as actin and myosin are degraded.

The Response of a Muscle Fiber to Action Potentials Ranges from Twitches to Tetanus

A single action potential arriving at a neuromuscular junction usually causes a single, weak contraction of a muscle fiber called a **muscle twitch (Figure 41.6A).** After a muscle twitch begins, the tension of the muscle fiber increases in magnitude for about 30 to 40 msec, and then peaks as the action potential runs its course through the T tubules and the Ca^{2+} channels begin to close. Tension then decreases as the Ca^{2+} ions are pumped back into the sarcoplasmic reticulum, falling to zero in about 50 msec after the peak.

If a muscle fiber is restimulated after it has relaxed completely, a new twitch identical to the first is generated (see Figure 41.6A). However, if a muscle fiber is restimulated before it has relaxed completely, the second twitch is added to the first, producing what is called *twitch summation,* a summed, stronger contraction **(Figure 41.6B).** If action potentials arrive so rapidly (about 25 msec apart) that the fiber cannot relax at all between stimuli, the Ca^{2+} channels remain open continuously and twitch summation produces a peak level of continuous contraction called **tetanus (Figure 41.6C).** (This is not to be confused with the disease of the same name, in which a bacterial toxin causes uncontrolled and continuous muscle contraction.) Contractile activity then decreases if either the stimuli cease or the muscle fatigues.

Tetanus (or tetanic contraction) is an essential part of muscle fiber function. If you lift a moderately heavy weight, for example, many of the muscle fibers in your arms enter tetanus and remain in that state until the weight is released. Even body movements that require relatively little effort, such as standing still but in balance, involve tetanic contractions of some muscle fibers.

Muscle Fibers Differ in Their Rate of Contraction and Resistance to Fatigue

Muscle fibers differ in their rate of contraction and resistance to fatigue, and thus can be classified as slow fibers, fast aerobic fibers, and fast anaerobic fibers. Their properties are summarized in **Table 41.1.** The proportions of the three types of muscle fibers tailor the contractile characteristics of each muscle to suit its function within the body.

Slow muscle fibers contract relatively slowly and the intensity of contraction is low because their myosin crossbridges hydrolyze ATP relatively slowly. They can remain contracted for relatively long periods without fatiguing. Slow muscle fibers typically contain many mitochondria and make most of their ATP by oxidative phosphorylation (cellular respiration; see Chapter 8). They have a low capacity to make ATP by anaerobic glycolysis. They also contain high concentrations of the oxygen-storing protein **myoglobin,** which greatly enhances their oxygen supplies. Myoglobin is closely related to hemoglobin, the oxygen-carrying protein of red blood cells. Myoglobin gives slow muscle fibers, such as those in the legs of ground birds such as chickens and ostriches, a deep red color. In sharks and bony fishes, strips of slow muscles concentrated in a band on either

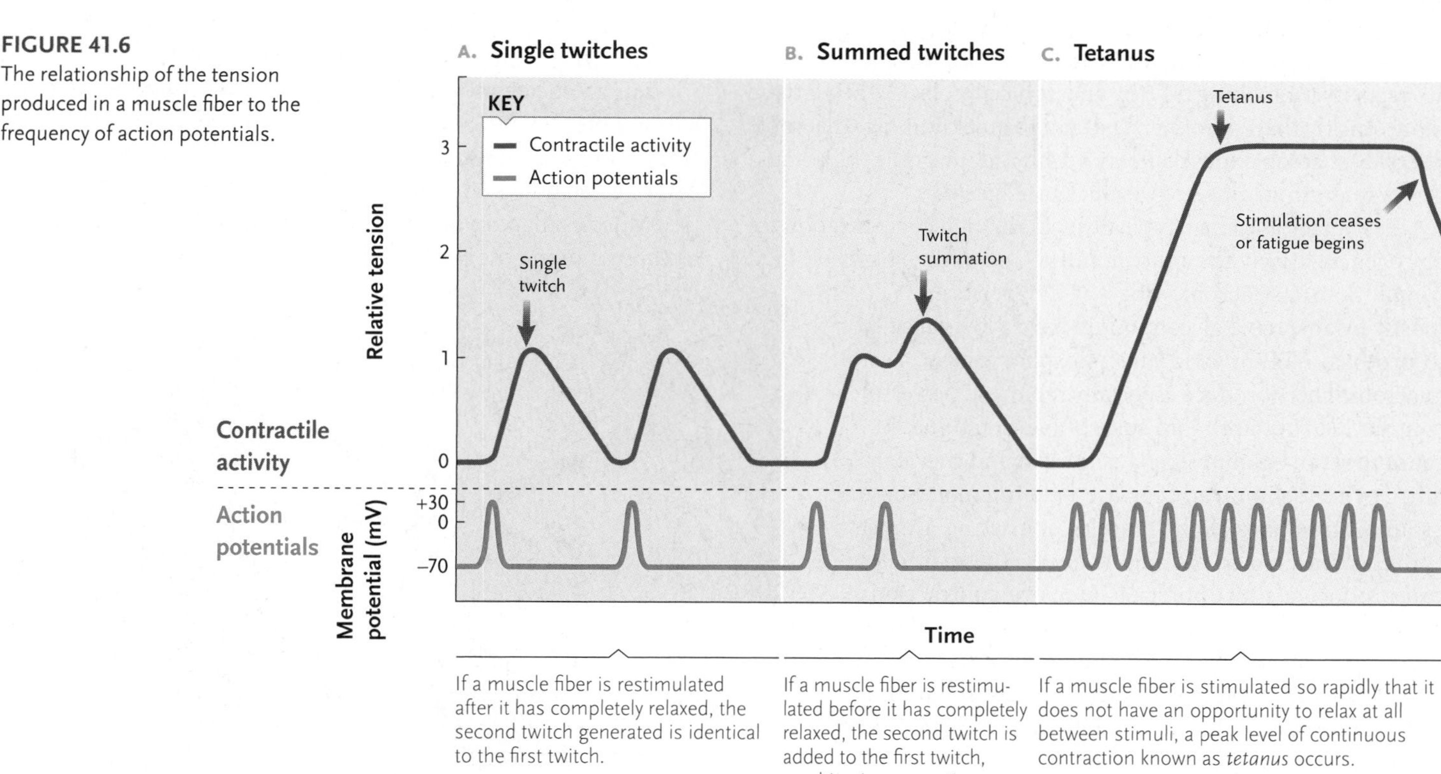
A. Single twitches
B. Summed twitches
C. Tetanus
KEY
Contractile activity
Action potentials
Tetanus
Stimulation ceases or fatigue begins
Twitch summation
Single twitch
Relative tension
3
2
1
0
Contractile activity
Action potentials
Membrane potential (mV)
+30
0
–70
Time
If a muscle fiber is restimulated after it has completely relaxed, the second twitch generated is identical to the first twitch.
If a muscle fiber is restimulated before it has completely relaxed, the second twitch is added to the first twitch, resulting in summation, essentially a summed, stronger contraction.
If a muscle fiber is stimulated so rapidly that it does not have an opportunity to relax at all between stimuli, a peak level of continuous contraction known as *tetanus* occurs.

FIGURE 41.6
The relationship of the tension produced in a muscle fiber to the frequency of action potentials.

TABLE 41.1	Characteristics of Slow and Fast Muscle Fibers in Skeletal Muscle		
	Fiber Type		
Property	Slow	Fast Aerobic	Fast Anaerobic
Contraction speed	Slow	Fast	Fast
Contraction intensity	Low	Intermediate	High
Fatigue resistance	High	Intermediate	Low
Myosin–ATPase activity	Low	High	High
Oxidative phosphorylation capacity	High	High	Low
Enzymes for anaerobic glycolysis	Low	Intermediate	High
Mitochondria	Many	Many	Few
Myoglobin content	High	High	Low
Fiber color	Red	Red	White
Glycogen content	Low	Intermediate	High

side of the body are used for slow, continuous swimming and maintaining body position.

Fast muscle fibers contract relatively quickly and powerfully because their myosin crossbridges hydrolyze ATP faster than those of slow muscle fibers. Fast aerobic fibers have abundant mitochondria, a rich blood supply, and a high concentration of myoglobin, which makes them red in color. They have a high capacity for making ATP by oxidative phosphorylation, and an intermediate capacity for making ATP by anaerobic glycolysis. They fatigue more quickly than slow fibers, but not as quickly as fast anaerobic fibers. Fast aerobic muscle fibers are abundant in the flight muscles of migrating birds such as ducks and geese.

Fast anaerobic fibers typically contain high concentrations of glycogen, relatively few mitochondria, and a more limited blood supply than fast aerobic fibers. They generate ATP mostly by anaerobic glycolysis and have a low capacity to produce ATP by oxidative phosphorylation. Fast anaerobic fibers produce especially rapid and powerful contractions but are more susceptible to fatigue. Because their myoglobin supply is limited and they contain few mitochondria, they are pale in color. Some ground birds have flight muscles consisting almost entirely of fast anaerobic muscle fibers. These muscles can produce a short burst of intensive contractions allowing the bird to escape a predator, but they cannot produce sustained flight. Most muscles of lampreys, sharks, fishes, amphibians, and reptiles also contain fast anaerobic muscle fibers, allowing the animals to move quickly to capture prey and avoid danger.

The muscles of humans and other mammals are mixed, and contain different proportions of slow and fast muscle fibers, depending on their functions. Muscles specialized for prolonged, slow contractions, such as the postural muscles of the back, have a high proportion of slow fibers and are a deep red color. The muscles of the forearm that move the fingers have a higher proportion of fast fibers and are a paler red than the back muscles. These muscles can contract rapidly and powerfully, but they fatigue much more rapidly than the back muscles.

The number and proportions of slow and fast muscle fibers in individuals are inherited characteristics. However, particular types of exercises can convert some fast muscle fibers between aerobic and anaerobic types. Endurance training, such as long-distance running, converts fast muscle fibers from the anaerobic to the aerobic type, and regimes such as weight lifting induce the reverse conversion. If the training regimes stop, most of the fast muscle fibers revert to their original types.

Skeletal Muscle Control Is Divided among Motor Units

The control of muscle contraction extends beyond the simple ability to turn the crossbridge cycle on and off. We can adjust a handshake from a gentle squeeze to a strong grasp, or exactly balance a feather or dumbbell in the hand. How are entire muscles controlled in this way? The answer lies in activation of the muscle fibers in blocks called **motor units.**

The muscle fibers in each motor unit are controlled by branches of the axon of a single efferent neuron **(Figure 41.7).** As a result, all those fibers contract each time the neuron fires an action potential. All the muscle fibers in a motor unit are of the same type—either slow, fast aerobic, or fast anaerobic. When a motor unit contracts, its force is distributed throughout the entire muscle because the fibers are dispersed throughout the muscle rather than being concentrated in one segment.

For a delicate movement, only a few efferent neurons carry action potentials to a muscle, and only a few motor units contract. For more powerful movements, more efferent neurons carry action potentials, and more motor units contract.

Muscles that can be precisely and delicately controlled, such as those moving the fingers in humans, have many motor units

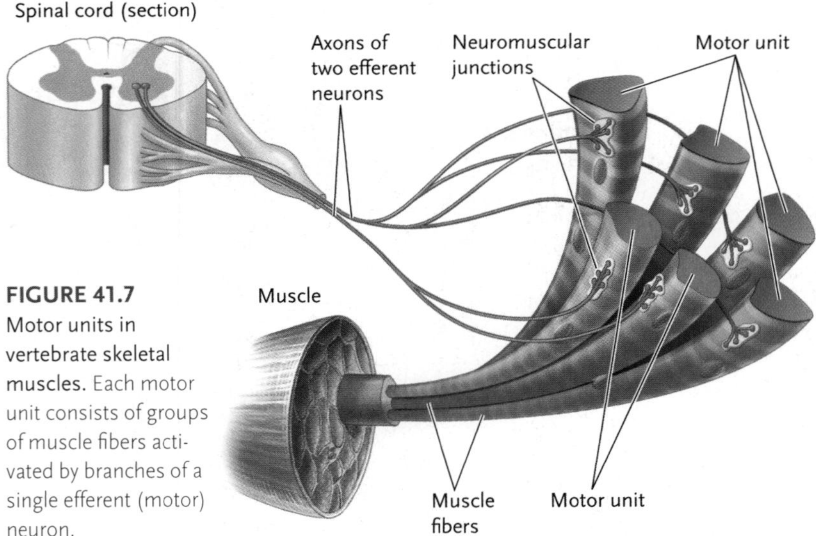
Spinal cord (section)
Axons of two efferent neurons
Neuromuscular junctions
Motor unit
Muscle
Muscle fibers
Motor unit

FIGURE 41.7 Motor units in vertebrate skeletal muscles. Each motor unit consists of groups of muscle fibers activated by branches of a single efferent (motor) neuron.

in a small area, with only a few muscle fibers—about 10 or so—in each unit. Muscles that produce grosser body movements, such as those moving the legs, have fewer motor units in the same volume of muscle but thousands of muscle fibers in each unit. In the calf muscle that raises the heel, for example, most motor units contain nearly 2,000 muscle fibers. Other skeletal muscles fall between these extremes, with an average of about 200 muscle fibers per motor unit.

Invertebrates Move Using a Variety of Striated Muscles

Invertebrates also have muscle cells in which actin-based thin filaments and myosin-based thick filaments produce movements by the same sliding mechanism as in vertebrates. Muscles that are clearly striated, which occur in virtually all invertebrates except sponges, have thick and thin filaments arranged in sarcomeres remarkably similar to those of vertebrates, except for variations in sarcomere length and the ratio of thin to thick filaments.

In invertebrates, an entire muscle is typically controlled by one or a few motor neurons. Nevertheless, invertebrate muscles are capable of finely graded contractions because individual neurons make large numbers of synapses with the muscle cells. As action potentials arrive more frequently at the synapses, more Ca^{2+} is released into the cells, and the muscles contract more strongly.

STUDY BREAK 41.1 <

1. **What is the mechanism by which muscle contraction occurs in response to a stimulus from the nervous system?**
2. **Outline the molecular events that take place in the sliding filament model of muscle contraction.**
3. **With respect to contraction intensity, fatigue resistance, oxidative phosphorylation capacity, number of mitochondria, and myoglobin content, what are the differences between slow, fast aerobic, and fast anaerobic muscle fibers?**

>

THINK OUTSIDE THE BOOK

You learned in Chapter 36 that smooth muscle is found in the walls of tubes and cavities of the body, including blood vessels and the urinary bladder. Smooth muscle cells have actin and myosin molecules arranged in a loose network rather than in bundles. Individually or collaboratively, research the mechanism of smooth muscle contraction and outline how it differs from the skeletal muscle contraction mechanism.

41.2 Skeletal Systems

Animal skeletal systems provide physical support for the body and protection for the soft tissues. They also act as a framework against which muscles work to move parts of the body or the entire organism. There are three main types of skeletons found in both invertebrates and vertebrates: hydrostatic skeletons, exoskeletons, and endoskeletons.

A Hydrostatic Skeleton Consists of Muscles and Fluid

A **hydrostatic skeleton** (*hydro-* = water, *statikos* = causing to stand) is a structure consisting of muscles and fluid that, by themselves, provide support for the animal or part of the animal; no rigid support, like a bone, is involved. A hydrostatic skeleton consists of a body compartment or compartments filled with water or body fluids, which are incompressible liquids. When the muscular walls of the compartment contract, they pressurize the contained fluid. If muscles in one part of the compartment are contracted while muscles in another part are relaxed, the pressurized fluid will move to the relaxed part of the compartment, distending it. In short, the contractions and relaxations of the muscles surrounding the compartments change the shape of the animal.

Hydrostatic skeletons are the primary support systems of cnidarians, flatworms, roundworms, and annelids. In all these animals, compartments containing fluids under pressure make the body semirigid and provide a mechanical support on which muscles act. For example, sea anemones have a hydrostatic skeleton consisting of several fluid-filled body cavities. The body wall contains longitudinal (muscles that run in the length dimension of the animal) and circular muscles (muscles that go around the body of the animal) that work against that skeleton. Between meals, longitudinal muscles are contracted (shortened), while the circular ones are relaxed, and the animal looks short and squat **(Figure 41.8A)**. The animal lengthens into its upright feeding position by contracting the circular muscles and relaxing the longitudinal ones **(Figure 41.8B)**.

In flatworms, roundworms, and annelids, striated muscles in the body wall act on the hydrostatic skeleton to produce creeping, burrowing, or swimming movements. Among these animals, annelids have the most highly developed musculoskeletal systems, with an outer layer of circular muscles surrounding the body, and an inner layer of longitudinal muscles running its

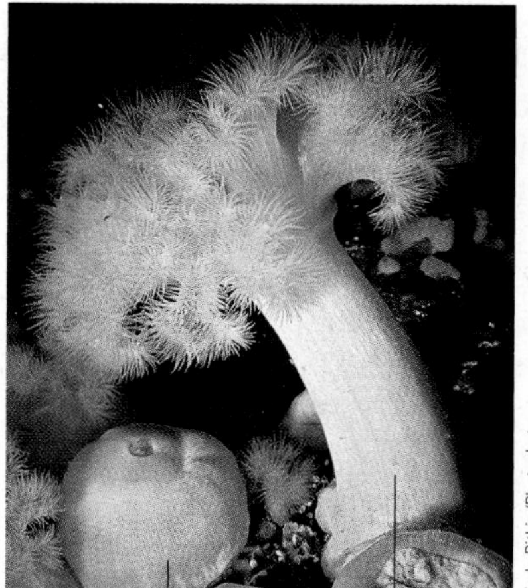
Linda Pitkin/Photoshot
A. Resting position
B. Feeding position

FIGURE 41.8
Sea anemones in **(A)** the resting and **(B)** the feeding position. In **(A),** longitudinal muscles in the body wall are contracted, and circular muscles are relaxed. In **(B),** the longitudinal muscles are relaxed, and the circular muscles are contracted. Both sets of muscles work against a hydrostatic skeleton.

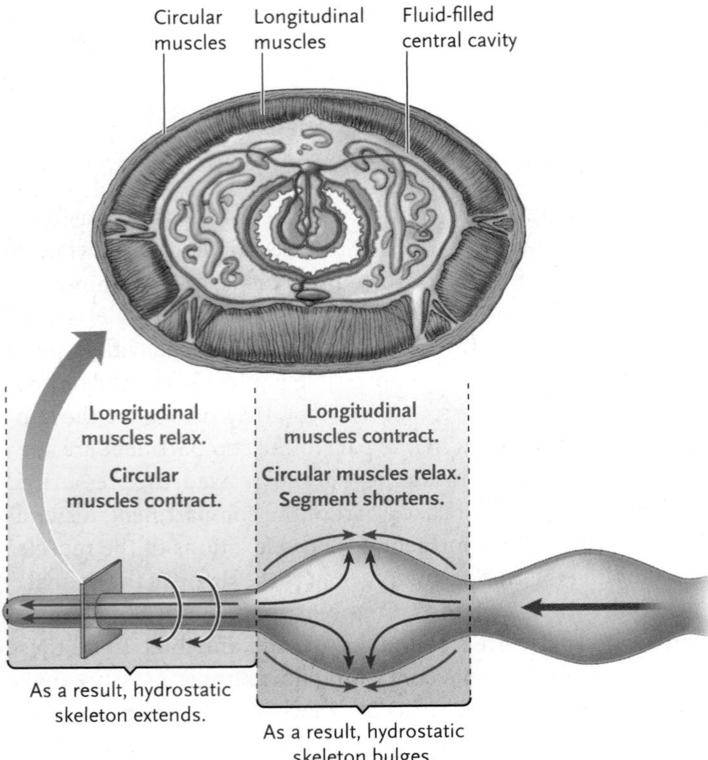
Circular muscles
Longitudinal muscles
Fluid-filled central cavity
Longitudinal muscles relax.
Circular muscles contract.
Longitudinal muscles contract.
Circular muscles relax. Segment shortens.
As a result, hydrostatic skeleton extends.
As a result, hydrostatic skeleton bulges.

FIGURE 41.9
Movement of an earthworm, showing how muscles in the body wall act on its hydrostatic skeleton. Contraction of the circular muscles reduces body diameter and increases body length, while contraction of the longitudinal muscles decreases body length and increases body diameter.

length **(Figure 41.9)**. Contractions of the circular muscles reduce the diameter of the body and increase the length of the worm; contractions of the longitudinal muscles shorten the body and increase its diameter. Annelids move along a surface or burrow by means of alternating waves of contraction of the two muscle layers that pass along the body, working against the fluid-filled body compartments of the hydrostatic skeleton.

Many arthropods have hydrostatic skeletal elements. In the larvae of flying insects, internal fluids held under pressure by the muscular body wall provide some body support. In spiders, the legs are extended from the bent position by muscles exerting pressure against body fluids.

Some structures of echinoderms are supported by hydrostatic skeletons. The tube feet of sea stars and sea urchins, for example, have muscular walls enclosing the fluid of the water vascular system (see Figure 30.2).

In vertebrates, the erectile tissue of the penis is a fluid-filled hydrostatic skeletal structure.

An Exoskeleton Is a Rigid External Body Covering

An **exoskeleton** (*exo* = outside) is a rigid external body covering, such as a shell, that provides support. In an exoskeleton, the force of muscle contraction is applied against that covering. An exoskeleton also protects delicate internal tissues such as the brain and respiratory organs.

Many mollusks, such as clams and oysters, have an exoskeleton consisting of a hard calcium carbonate shell secreted by glands in the mantle. Arthropods, such as insects, spiders, and crustaceans, have an external skeleton in the form of a chitinous cuticle, secreted by underlying tissue that covers the outside surfaces of the animals. Like a suit of armor, the arthropod exoskeleton has movable joints, flexed and extended by muscles that extend across the inside surfaces of the joints **(Figure 41.10)**. The exoskeleton protects against dehydration, serves as armor against predators, and provides the levers against which muscles work. In many flying insects, elastic flexing of the exoskeleton contributes to the movements of the wings.

In vertebrates, the shell of a turtle or tortoise is an exoskeletal structure, as are the bony plates, abdominal ribs, collar bones, and most of the skull of the American alligator.

An Endoskeleton Consists of Supportive Internal Body Structures Such as Bones

An **endoskeleton** (*endon* = within) consists of internal body structures, such as bones, that provide support. In an endoskeleton, the force of contraction is applied against those structures. Like exoskeletons, endoskeletons also protect delicate internal tissues such as the brain and internal organs.

Echinoderms have an endoskeleton consisting of *ossicles* (*ossiculum* = little bone), formed from calcium carbonate crystals. The shells of sand dollars and sea urchins are the endoskeletons of these animals.

The endoskeleton is the primary skeletal system of vertebrates. An adult human, for example, has an endoskeleton consisting of 206 bones **(Figure 41.11)**.

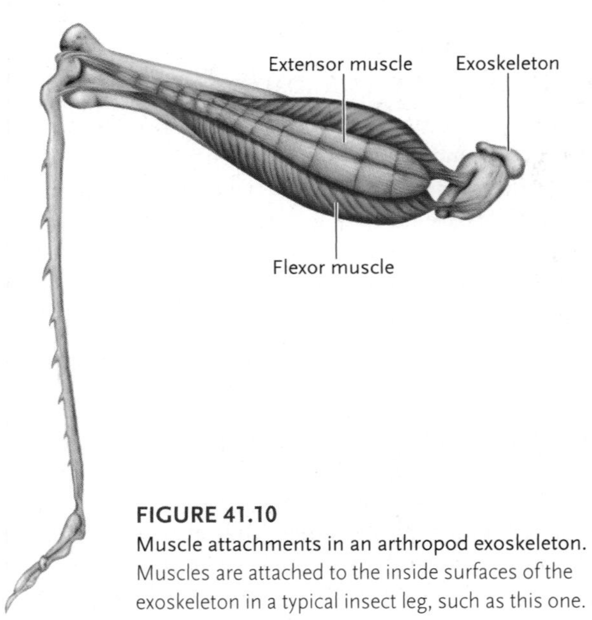
Extensor muscle
Exoskeleton
Flexor muscle

FIGURE 41.10
Muscle attachments in an arthropod exoskeleton. Muscles are attached to the inside surfaces of the exoskeleton in a typical insect leg, such as this one.

FIGURE 41.11
Major bones of the human body. The inset shows the structure of a limb bone, with the location of red and yellow marrow. The internal spaces lighten the bone's structure. The cartilage layer forms a smooth, slippery cushion between bones in a joint.

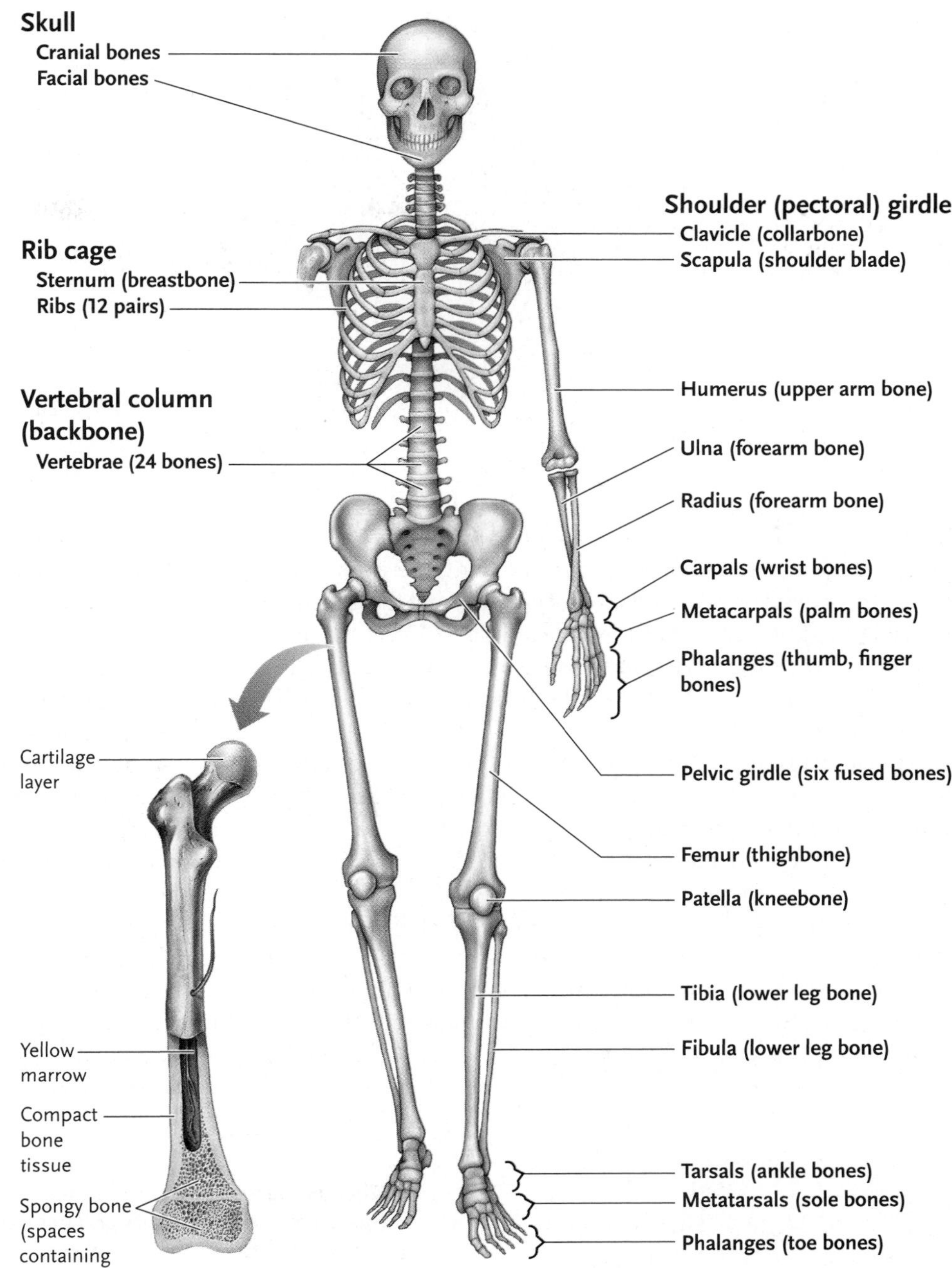

Skull
Cranial bones
Facial bones
Rib cage
Sternum (breastbone)
Ribs (12 pairs)
Vertebral column (backbone)
Vertebrae (24 bones)
Shoulder (pectoral) girdle
Clavicle (collarbone)
Scapula (shoulder blade)
Humerus (upper arm bone)
Ulna (forearm bone)
Radius (forearm bone)
Carpals (wrist bones)
Metacarpals (palm bones)
Phalanges (thumb, finger bones)
Pelvic girdle (six fused bones)
Femur (thighbone)
Patella (kneebone)
Tibia (lower leg bone)
Fibula (lower leg bone)
Tarsals (ankle bones)
Metatarsals (sole bones)
Phalanges (toe bones)
Cartilage layer
Yellow marrow
Compact bone tissue
Spongy bone (spaces containing red marrow)

Bones of the Vertebrate Endoskeleton Are Organs with Several Functions

The vertebrate endoskeleton supports and maintains the overall shape of the body and protects key internal organs. In addition, the skeleton is a storehouse for calcium and phosphate ions, releasing them as required to maintain optimal levels of these ions in body fluids. Bones are also sites where new blood cells form.

Bones are complex organs built up from multiple tissues, including bone tissue with cells of several kinds, blood vessels, nerves, and, in some, stores of adipose tissue. Bone tissue is distributed between dense, compact bone regions, which have essentially no spaces other than the microscopic canals of the osteons (see Figure 36.4D), and spongy bone regions, which are opened by larger spaces (see Figure 41.11). Compact bone tissue generally forms the outer surfaces of bones, and spongy bone tissue the interior. The interior of some flat bones, such as the hip bones and the ribs, are filled with *red marrow,* a tissue that is the primary source of new red blood cells in mammals and birds. The shaft of long bones such as the femur is opened by a large central canal filled with adipose tissue called *yellow marrow,* which is a source of some white blood cells.

Throughout the life of a vertebrate, calcium and phosphate ions are constantly deposited and withdrawn from bones. Hormonal controls maintain the concentration of Ca^{2+} ions at optimal levels in the blood and extracellular fluids (see Figure 40.10), ensuring that calcium is available for proper functioning of the nervous system, muscular system, and other physiological processes.

STUDY BREAK 41.2

1. **How do hydrostatic skeletons, exoskeletons, and endoskeletons provide support to the body? Give an example of each of these types in echinoderms and vertebrates.**
2. **What are the functions of the bones of the vertebrate endoskeleton?**

41.3 Vertebrate Movement: The Interactions between Muscles and Bones

The skeletal systems of all animals act as a framework against which muscles work to move parts of the body or the entire organism. In this section, the muscle–bone interactions that are responsible for the movement of vertebrates are described.

Joints of the Vertebrate Endoskeleton Allow Bones to Move and Rotate

The bones of the vertebrate skeleton are connected by joints. *Synovial joints,* the most-movable joints, consist of the ends of two bones enclosed by a fluid-filled capsule of connective tissue

FIGURE 41.12
A synovial joint, the human knee.

A. **Synovial joint cross section**

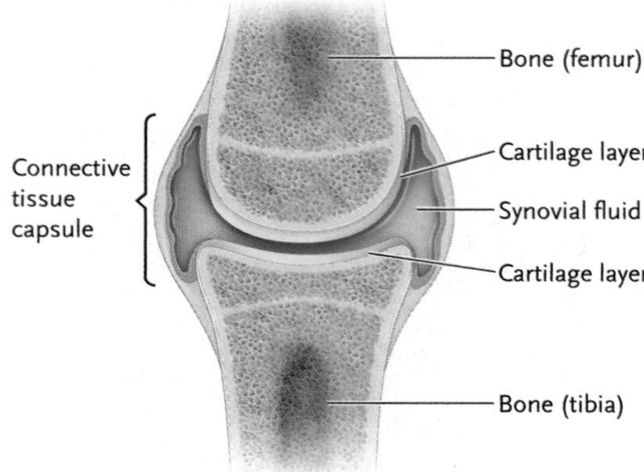
Bone (femur)
Connective tissue capsule
Cartilage layer
Synovial fluid
Cartilage layer
Bone (tibia)

B. **Ligaments reinforcing the knee joint**

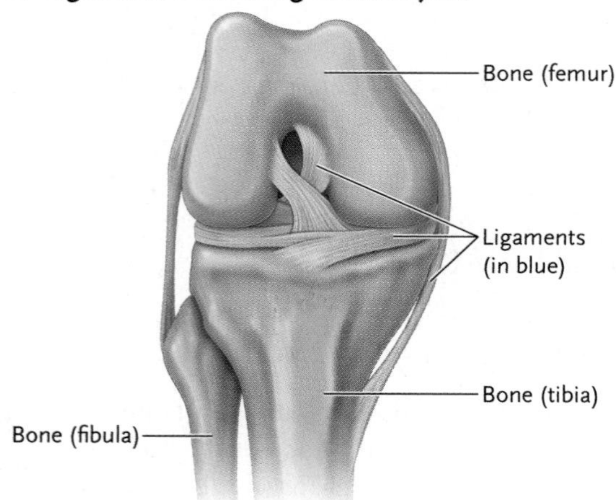
Bone (femur)
Ligaments (in blue)
Bone (tibia)
Bone (fibula)

(Figure 41.12A). Examples are the shoulders, elbows, wrists, fingers, knees, ankles, and toes. Within the joint, the ends of the bones are covered by a smooth layer of cartilage and lubricated by synovial fluid, which makes the bones slide easily as the joint moves. Synovial joints are held together by straps of connective tissue called *ligaments,* which extend across the joints outside the capsule **(Figure 41.12B).**

In **cartilaginous joints,** which are less movable, the ends of bones are covered with layers of cartilage, but have no fluid-filled capsule surrounding them. Fibrous connective tissue covers and connects the bones of these joints. They occur between the vertebrae and some rib bones.

The bones connected by movable joints work like levers. A lever is a rigid structure that can move around a pivot point known as a *fulcrum.* Levers differ with respect to where the fulcrum is along the lever and where the force is applied. The most common type of lever system in the body—exemplified by the elbow joint—has the fulcrum at one end, the load at the opposite end, and the force applied at a point between the ends **(Figure 41.13).** For this lever, the force applied must be much greater than the load, but it increases the distance the load moves as compared with the distance over which the force is applied. This allows small muscle movements to produce large body movements, and also allows movements such as running or throwing to be carried out at high speed.

At a joint, a muscle that causes movement in the joint when it contracts is called an **agonist.** Most of the bones of vertebrate skeletons are moved by muscles arranged in **antagonistic pairs: extensor muscles** extend the joint, meaning increasing the angle between the two bones, while **flexor muscles** do the opposite. (Antagonistic muscles are also used in invertebrates for movement of body parts—for example, the limbs of insects and arthropods.) In humans, one such pair is formed by the biceps brachii muscle at the front of the upper arm and the triceps brachii muscle at the back of the upper arm **(Figure 41.14).**

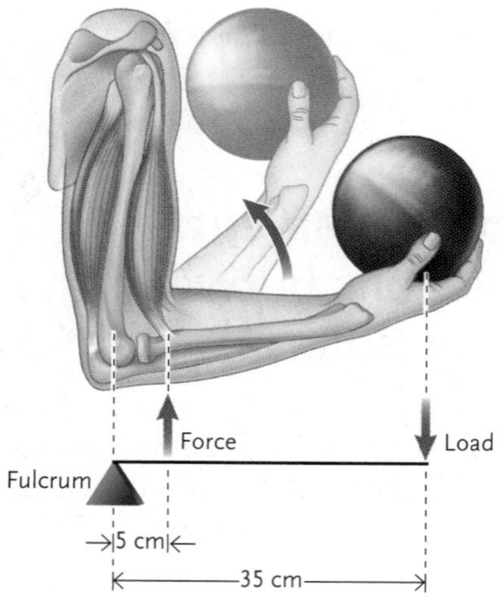
Force
Load
Fulcrum
5 cm
35 cm

FIGURE 41.13
A body lever: The lever formed by the bones of the forearm. The fulcrum (the hinge or joint) is at one end of the lever, the load is placed on the opposite end, and the force is exerted at a point on the lever between the fulcrum and the load.

A. **When the biceps muscle contracts and raises the forearm, its antagonistic partner, the triceps muscle relaxes.**

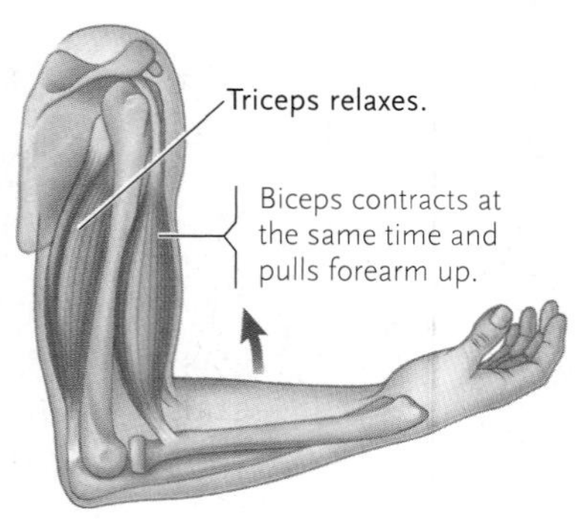
Triceps relaxes.
Biceps contracts at the same time and pulls forearm up.

B. **When the triceps muscle contracts and extends the forearm, the biceps muscle relaxes.**

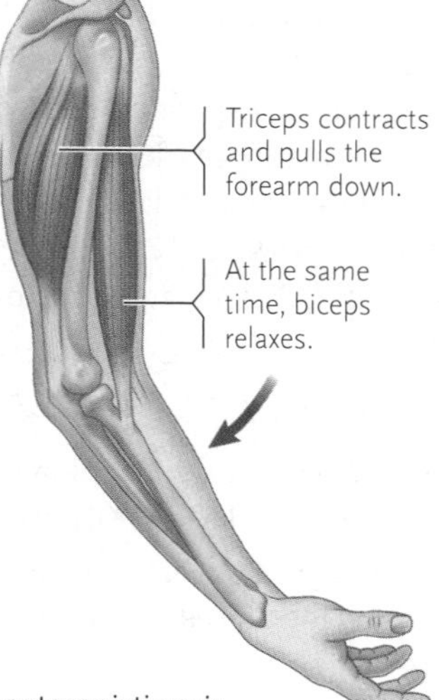
Triceps contracts and pulls the forearm down.
At the same time, biceps relaxes.

FIGURE 41.14
The arrangement and function of skeletal muscles in an antagonistic pair.

Vertebrates Have Muscle–Bone Interactions Optimized for Specific Movements

Vertebrates differ widely in the patterns by which muscles connect to bones, and in the length and mechanical advantage of the levers produced by these connections. These differences produce limbs and other body parts that are adapted for either power or speed, or the most advantageous compromise between these characteristics. Among burrowing mammals such as the mole, for example, the limb bones are short, thick, and heavy, and the points at which muscles attach produce levers that are slow to move but that need to apply smaller forces to move a load when compared with a human biceps. In contrast, a mammal such as the deer has relatively light and thin bones with muscle attachments producing levers that can generate rapid movement, moving the body easily over the ground.

STUDY BREAK 41.3 <

1. **What are the features of synovial joints and cartilaginous joints?**
2. **What are antagonistic muscle pairs?**

UNANSWERED QUESTIONS

How can muscle growth processes be controlled to improve the clinical treatment of muscular dystrophy and related disorders?

The mechanisms by which different muscle types develop and their impact on organismal metabolism are not well known. However, we do know that one of the proteins produced in muscle cells, myostatin, is a growth factor that inhibits skeletal muscle growth and development. Myostatin is therefore a potential therapeutic target for treating Duchenne muscular dystrophy, a degenerative and sometimes fatal disease associated with the progressive loss of skeletal muscle mass. Animals with nonfunctioning and mutant forms of the myostatin gene (for example, the Belgian Blue and Piedmontese cattle breeds) or myostatin knockout mice, in which the gene has been removed experimentally (see Section 18.2 and *Focus on Model Research Organisms* in Chapter 43), have significantly enhanced musculature that is commonly referred to as *double muscling*. Such mutations and enhanced skeletal muscle mass have also been described in a racing dog breed, the whippet, and recently in a young boy.

Our laboratory develops novel technologies to inhibit myostatin activity—in essence, inhibiting the inhibitor—and thereby stimulate skeletal muscle growth in both clinical and agricultural settings. We have recently determined that myostatin can also negatively regulate cardiac muscle growth. Thus, disrupting myostatin production or availability may help heart attack patients. Replacing damaged skeletal and cardiac muscle using adult or embryonic stem cells engineered to match either tissue type is another highly promising technique for treating these disorders. This technique could also be improved by using "antimyostatin" technologies that enhance growth of the transplanted cells.

How much do the metabolic processes of skeletal muscle specifically contribute to energy storage and whole body form?

Complications associated with obesity, particularly diabetes mellitus type 2 (see Section 40.4), have reached near-epidemic proportions worldwide. Type 2 diabetes is caused not by a lack of the pancreatic hormone insulin (as in type 1 diabetes) but by insulin resistance, in which tissue responses to insulin are inadequate. In both types of diabetes, however, the body is unable to properly process and store metabolites, mostly glucose. Type 2 diabetes can be a debilitating and fatal disease if poorly managed and often aggravates other diseases as well. Scientists now recognize that growth and metabolic processes are integrated and controlled by the same hormones, growth factors, and cytokines. Indeed, skeletal muscle is the largest consumer of metabolites and has the greatest potential to impact their circulating levels. Recent studies suggest that even small increases in muscle mass can significantly reduce fat mass and improve insulin responsiveness. Enhancing skeletal muscle growth in obese patients with type 2 diabetes could therefore improve treatments for both. The same antimyostatin technologies used to treat muscle growth disorders could therefore be used to treat severe obesity and type 2 diabetes with the goals of increasing muscle mass, decreasing fat mass, and improving insulin sensitivity.

Think Critically

Why do you think changes in skeletal muscle mass ultimately influence fat mass?

Buel (Dan) Rodgers

Buel (Dan) Rodgers is an associate professor in the Department of Animal Sciences at Washington State University, studying molecular endocrinology and animal genomics, specifically skeletal muscle growth and development. To learn more about Rodgers' research, visit http://www.ansci.wsu.edu/People/rodgers/faculty.asp.

REVIEW KEY CONCEPTS

Go to **CENGAGENOW** at www.cengage.com/login to access quizzing, animations, exercises, articles, and personalized homework help.

41.1 Vertebrate Skeletal Muscle: Structure and Function

- Skeletal muscles move the joints of the body. They are formed from long, cylindrical cells called muscle fibers, which are packed with myofibrils, contractile elements consisting of myosin thick filaments and actin thin filaments. The two types of filaments are arranged in an overlapping pattern of contractile units called sarcomeres (Figure 41.2).
- Infoldings of the plasma membrane of the muscle fiber form T tubules. The sarcomeres are encircled by the sarcoplasmic reticulum, a system of vesicles with segments separated from T tubules by small gaps (Figure 41.3).
- In the sliding filament mechanism of muscle contraction, the simultaneous sliding of thin filaments on each side of sarcomeres over the thick filaments shortens the sarcomeres and the muscle fibers, producing the force that contracts the muscle (Figure 41.4).
- The sliding motion of thin and thick filaments is produced in response to an action potential arriving at the neuromuscular junction. The action potential causes the release of acetylcholine, which triggers an action potential in the muscle fiber that spreads over its plasma membrane and stimulates the sarcoplasmic reticulum to release Ca^{2+} into the cytosol. The Ca^{2+} combines with troponin, inducing a conformational change that moves tropomyosin away from the myosin-binding sites on thin filaments. Exposure of the sites allows myosin crossbridges to bind and initiate the crossbridge cycle in which the myosin heads of thick filaments attach to a thin filament, pull, and release in cyclic reactions powered by ATP hydrolysis (Figure 41.5).
- When action potentials stop, Ca^{2+} is pumped back into the sarcoplasmic reticulum, leading to Ca^{2+} release from troponin, which allows tropomyosin to cover the myosin-binding sites in the thin filaments, thereby stopping the crossbridge cycle (Figure 41.5).
- A single action potential arriving at a neuromuscular junction causes a muscle twitch. Restimulation of a muscle fiber before it has relaxed completely causes a second twitch, which is added to the first, causing a summed, stronger contraction. Rapid arrival of action potentials causes the twitches to sum to a peak level of contraction called tetanus. Normally, muscles contract in a tetanic mode (Figure 41.6).
- Muscle fibers occur in three types. Slow muscle fibers contract relatively slowly, but do not fatigue rapidly. Fast aerobic fibers contract relatively quickly and powerfully, and fatigue more quickly than slow fibers. Fast anaerobic fibers can contract more rapidly and powerfully than fast aerobic fibers, but fatigue more rapidly. The fibers differ in their number of mitochondria and capacity to produce ATP (Table 41.1).
- Skeletal muscles are divided into motor units, consisting of a group of muscle fibers activated by branches of a single motor neuron. The total force produced by a skeletal muscle is determined by the number of motor units that are activated in the muscle (Figure 41.7).
- Invertebrate muscles contain thin and thick filaments arranged in sarcomeres, and contract by the same sliding filament mechanism that operates in vertebrates.

Animation: Structure of skeletal muscle

Animation: Sliding filament model

Animation: Troponin and tropomyosin

Animation: Energy sources for contraction

Animation: Types of contractions

Animation: Nervous system and muscle contraction

41.2 Skeletal Systems

- A hydrostatic skeleton is a structure consisting of a muscle-surrounded compartment or compartments filled with fluid under pressure. Contraction and relaxation of the muscles change the shape of the animal (Figures 41.8 and 41.9).
- In an exoskeleton, a rigid external covering provides support for the body. The force of muscle contraction is applied against the covering. An exoskeleton can also protect delicate internal tissues (Figure 41.10).
- In an endoskeleton, the body is supported by rigid structures within the body, such as bones. The force of muscle contraction is applied against those structures. Endoskeletons also protect delicate internal tissues. In vertebrates, the endoskeleton is the primary skeletal system (Figure 41.11).
- Bone tissue is distributed between compact bone, with no spaces except the microscopic canals of the osteons, and spongy bone tissue, which has spaces filled by red or yellow marrow (Figure 41.11).
- Calcium and phosphate ions are constantly exchanged between the blood and bone tissues. The turnover keeps the Ca^{2+} concentration balanced at optimal levels in body fluids.

Animation: Vertebrate skeletons

Animation: Human skeletal system

Animation: Structure of a femur

Animation: Long bone formation

41.3 Vertebrate Movement: The Interactions between Muscles and Bones

- The bones of a skeleton are connected by joints. A synovial joint, the most movable type, consists of a fluid-filled capsule surrounding the ends of the bones forming the joint. A cartilaginous joint, which is less movable, has smooth layers of cartilage between the bones with no surrounding capsule (Figure 41.12).
- The bones moved by skeletal muscles act as levers, with a joint at one end forming the fulcrum of the lever, the load at the opposite end, and the force applied by attachment of a muscle at a point between the ends (Figure 41.13).
- At a joint, an agonist muscle, perhaps assisted by other muscles, causes movement. Most skeletal muscles are arranged in antagonistic pairs, in which the members of a pair pull a bone in opposite directions. When one member of the pair contracts, the other member relaxes and is stretched (Figure 41.14).
- Vertebrates have a variety of patterns in which muscles connect to bones, giving different properties to the levers produced. Those properties are specialized for the activities of the animal.

Animation: Opposing muscle action

Animation: Human skeletal muscles

UNDERSTAND AND APPLY

Test Your Knowledge

1. Vertebrate skeletal muscle:
   a. is attached to bone by means of ligaments.
   b. may bend but not extend body parts.
   c. may rotate one body part with respect to another.
   d. is found in the walls of blood vessels and intestines.
   e. consists of cells that contain no nuclei.
2. In muscle fiber in a completely relaxed state:
   a. sarcomeres are regions between two H zones.
   b. discs of M line proteins called the A band separate the thick filaments.
   c. I bands are composed of the same thick filaments seen in the A bands.
   d. Z lines are adjacent to H zones, which attach thick filaments.
   e. dark A bands contain overlapping thick and thin filaments with a central thin H zone composed only of thick filaments.
3. The sliding filament contractile mechanism:
   a. causes thin filaments on each side of a sarcomere to slide over thick filaments toward the center of the A band, bringing the Z lines closer together.
   b. is inhibited by the influx of Ca^{2+} into the muscle fiber cytosol.
   c. lengthens the sarcomere to separate the I regions.
   d. depends on the isolation of actin and myosin until a contraction is completed.
   e. uses myosin crossbridges to stimulate delivery of Ca^{2+} to the muscle fiber.
4. During contraction of skeletal muscle:
   a. ATP stimulates Ca^{2+} to move out of the cytosol, which allows tropomyosin to bind myosin causing contraction of the thin filament.
   b. myosin crossbridges use ATP to relax the molecular spring in the myosin head, which pulls the thick filaments away from the thin actin filaments.
   c. actin binds ATP, allowing troponin in the thick filaments to form the myosin crossbridge.
   d. action potentials cause the release of Ca^{2+} allowing tropomyosin fibers to uncover the actin binding sites needed for the myosin crossbridge.
   e. botulinum toxin could increase the release of acetylcholine at the contracting muscle site.
5. When marathoners are running a race, most likely their:
   a. muscles have low concentrations of myoglobin.
   b. slow muscle fibers will do most of the work for the run.
   c. slow muscle fibers will remain in constant tetanus over the length of the run.
   d. fast muscle fibers will be employed in the middle of the run.
   e. slow muscle fibers are using ATP obtained primarily by anaerobic respiration.
6. Which description is characteristic of a motor unit?
   a. A single motor unit's muscle fibers vary among the slow/fast aerobic and slow/fast anaerobic forms.
   b. When receiving an action potential, a motor unit is controlled by a single efferent axon that causes all its fibers to contract.
   c. When a motor unit contracts, certain sections of the muscle as a whole remain relaxed.
   d. If a motor unit controls walking, it is found in large numbers in the same volume of muscle.
   e. If a motor unit controls finger movement, it contains a large number of muscle fibers that are stimulated over a large area.
7. Which of the following is *not* an example of a hydrostatic skeletal structure?
   a. the tube feet of sea urchins
   b. the body wall of annelids
   c. the mantle of squids
   d. the body wall of cnidarians
   e. the penis of mammals
8. Endoskeletons:
   a. protect internal organs and provide structures against which the force of muscle contraction can work.
   b. differ from exoskeletons in that endoskeletons do not support the external body.
   c. cannot be found in echinoderms.
   d. are composed of cartilaginous structures that form the skull.
   e. compose the arms and legs.
9. Connecting the bones of the vertebrate skeleton are:
   a. nonmovable synovial joints.
   b. ligaments holding together connective tissue of fibrous joints.
   c. cartilaginous joints held together by fluid-filled capsules.
   d. synovial joints lubricated by synovial fluid.
   e. fibrous joints that move around a fulcrum allowing small muscle movements to produce large body movements.
10. The movement of vertebrate muscles is:
    a. controlled primarily by cartilaginous joints with agonist muscles.
    b. antagonistic when it causes movement in the joint.
    c. caused by extensor muscles that flex the joint.
    d. caused by flexor muscles that extend the joint.
    e. exemplified by the biceps and triceps contracting or relaxing simultaneously.

Discuss the Concepts

1. A coach must train young athletes for the 100-meter sprint. They need muscles specialized for speed and strength, rather than for endurance. What kinds of muscle characteristics would the training regimen aim to develop? How would it be altered to train marathoners?
2. What kind of exercise program might the coach in question 1 recommend to an older person developing osteoporosis? Why?
3. Based on material in this chapter and in Chapter 40 on endocrine controls, outline some possible causes and physiological effects of calcium deficiency in an active adult.

Design an Experiment

Design an experiment using rats to determine whether endurance training alters the proportion of slow, fast aerobic, and fast anaerobic muscle fibers.

Interpret the Data

Osteogenesis imperfecta (OI) is a genetic disorder caused by a mutation in a gene for collagen. As bones develop, collagen forms a scaffold for deposition of mineralized bone tissue. In children with OI, the scaffold forms improperly, causing them to be born with multiple fractures in their fragile bones and to require many corrective surgeries. The tables show the results of an experimental test of a new drug. Treated children with OI, all less than 2 years old, were compared to similarly affected children of the same age who were not treated with the drug.

TABLE 1 Effect of Drug on Bone Properties in Children with OI

Treated Child	Vertebral Area, in cm² (Before Treatment)	Vertebral Area, in cm² (After Treatment)	Fractures Per Year
1	14.7	16.7	1
2	15.5	16.9	1
3	6.7	16.5	6
4	7.3	11.8	0
5	13.6	14.6	6
6	9.3	15.6	1
7	15.3	15.9	0
8	9.9	13.0	4
9	10.5	13.4	4
Mean	11.4	14.9	2.6

TABLE 2 Bone Properties in Children with OI Who Received No Drug Treatment

Control Child	Vertebral Area, in cm² (Initial)	Vertebral Area, in cm² (Final)	Fractures Per Year
1	18.2	13.7	4
2	16.5	12.9	7
3	16.4	11.3	8
4	13.5	7.7	5
5	16.2	16.1	8
6	18.9	17.0	6
Mean	16.6	13.1	6.3

1. An increase in the vertebral area during the 12-month period of the study indicates bone growth. How many of the treated children showed an increase?
2. How many of the untreated children showed an increase?
3. How did the rate of fractures in the two groups compare?
4. Do the results shown support the hypothesis that giving young children who have OI this drug, which slows bone breakdown, can increase bone growth and reduce fractures?

Apply Evolutionary Thinking

What characteristics of vertebrate muscle suggest that the genes for muscle structure were inherited from invertebrate ancestors?

Express Your Opinion

Dietary supplements are largely unregulated. Should they be placed under the jurisdiction of the Food and Drug Administration, which could subject them to more stringent testing for effectiveness and safety? Go to www.cengage.com/login to investigate both sides of the issue and then vote.

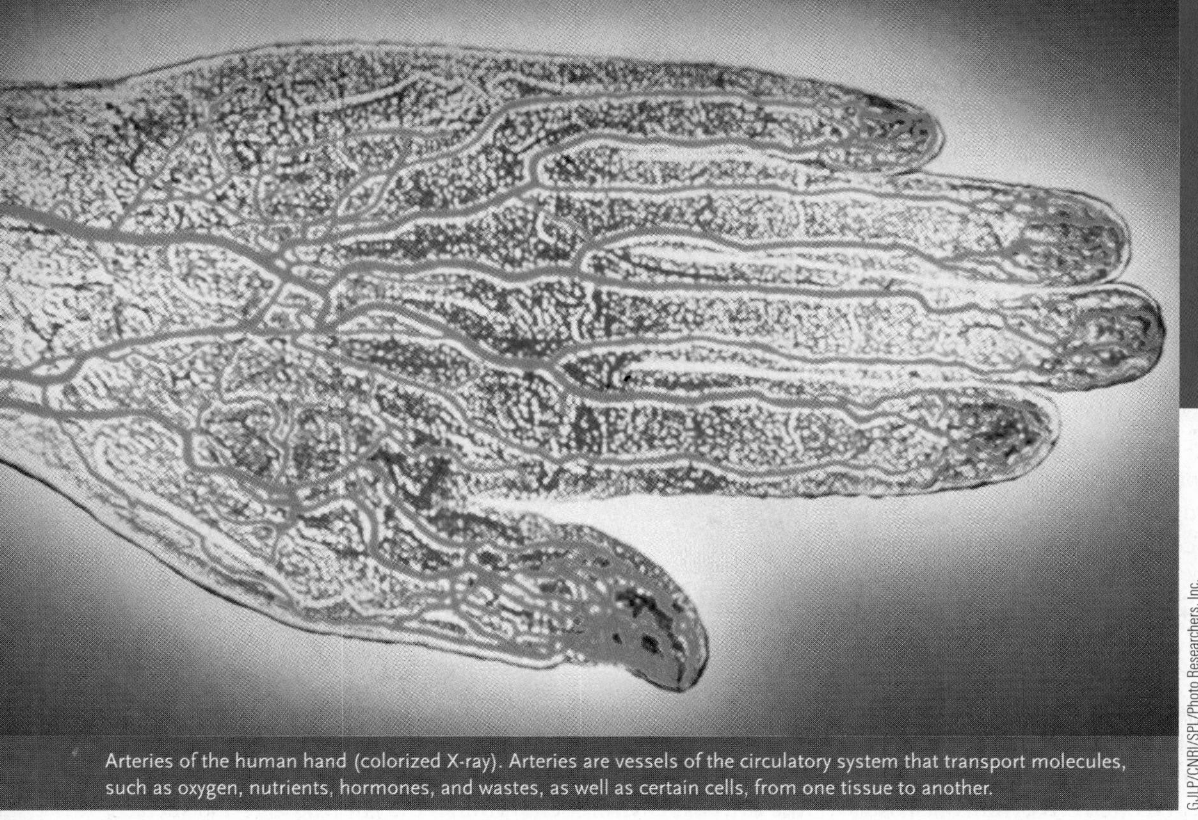

Arteries of the human hand (colorized X-ray). Arteries are vessels of the circulatory system that transport molecules, such as oxygen, nutrients, hormones, and wastes, as well as certain cells, from one tissue to another.

GJLP/CNRI/SPL/Photo Researchers, Inc.

42

STUDY OUTLINE

The Circulatory System

Why It Matters. . . Jimmie the bulldog stood on the stage of a demonstration laboratory at a meeting of the Royal Society in London in 1909, with one front paw and one rear paw in laboratory jars containing salt water **(Figure 42.1).** Wires leading from the jars were connected to a galvanometer, a device that can detect electrical currents. Jimmie's master, Dr. Augustus Waller, a physician at St. Mary's Hospital, was relating his experiments in the emerging field of *electrophysiology.* Among other discoveries, Waller had found that his apparatus detected the electrical currents produced each time the dog's heart beat.

Waller originally made his finding by experimenting on himself. He already knew that the heart creates an electrical current as it beats; other scientists had found this out by attaching electrodes directly to the heart of experimental animals. Looking for a painless alternative to that pro-

A. **Jimmie the bulldog**

From A. D. Waller, Physiology, The Servant of Medicine, Hitchcock Lectures, University of London Press, 1910

B. **Electrocardiogram**

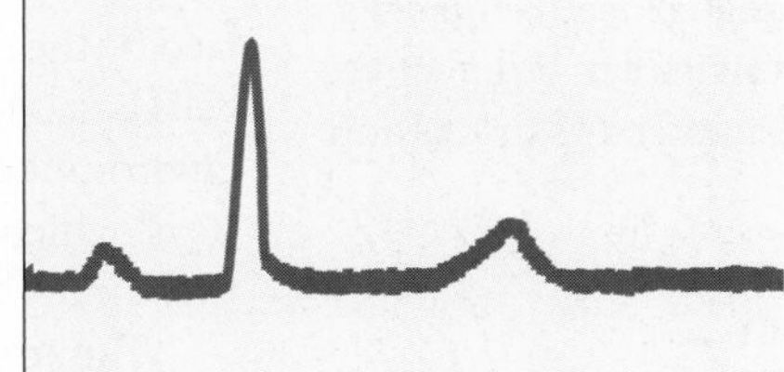

FIGURE 42.1
The first electrocardiograms.
(A) Jimmie the bulldog standing in laboratory jars containing saltwater, with wires leading to a galvanometer that recorded the electrical currents produced by his heartbeat. **(B)** One of Waller's early electrocardiograms.

cedure, Waller reasoned that because the human body can conduct electricity, his arms and legs might conduct the currents generated by the heart if they were connected to a galvanometer. Accordingly, Waller set up two metal pans containing saltwater and connected wires from the pans to a galvanometer. He put his bare left foot in one of the pans, and his right hand in the other one. The technique worked; the indicator of the galvanometer jumped each time his heart beat. And, it worked with Jimmie.

Waller also invented a method for recording the changes in current, which became the first electrocardiogram (ECG). He constructed a galvanometer by placing a column of mercury in a fine glass tube, with a conducting salt solution layered above the mercury. Changes in the current passing through the tube caused corresponding changes in the surface tension of the mercury, which produced movements that could be detected by reflecting a beam of light from the mercury surface. By placing a moving photographic plate behind the mercury tube, Waller could record the movements of the reflected light on the plate (Figure 42.1B shows one of his records).

The beating of Jimmie's heart, recorded as an electrical trace by Augustus Waller, is part of the actions of the **circulatory system,** an organ system consisting of a fluid, a heart, and vessels for moving important molecules, and often cells, from one tissue to another. Examples of transported molecules are oxygen (O_2), nutrients, hormones, carbon dioxide (CO_2), and other wastes.

We study the circulatory system in this chapter, with emphasis on the circulation of humans and other mammals. We also discuss the **lymphatic system,** an accessory system of vessels and organs that helps balance the fluid content of the blood and surrounding tissues and participates in the body's defenses against invading disease organisms. <

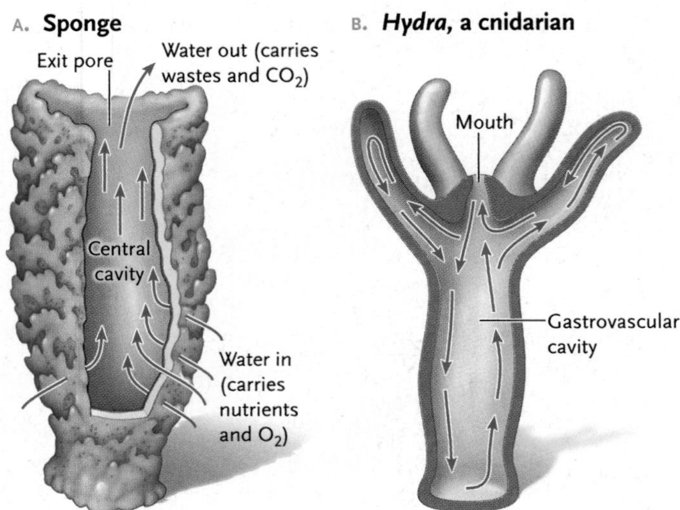
A. Sponge
Exit pore
Water out (carries wastes and CO_2)
Central cavity
Water in (carries nutrients and O_2)
B. Hydra, a cnidarian
Mouth
Gastrovascular cavity

FIGURE 42.2
Invertebrates with no circulatory system.

42.1 Animal Circulatory Systems: An Introduction

As you learned in the discussion of homeostasis in Section 36.4, the cells of all organisms must take in nutrients and O_2 from the external environment, and eliminate wastes such as CO_2 to that same environment. For very small or very thin animals, movement of such molecules occurs primarily by diffusion (see Section 6.2). Diffusion is slow over distances of even a few millimeters, however, making this mechanism unsuitable for larger and more complex animals. Circulatory systems overcome the problem by bringing fluids for gas and nutrient exchange close to the cells, where diffusion and other transport mechanisms can be effective. You will learn more about gas exchange in Chapter 44's discussion of the respiratory system.

Simple Animals Have No Distinct Circulatory System

The least complex animals, including sponges, cnidarians, and flatworms, function with no distinct circulatory system. These animals are aquatic or, like parasitic flatworms, live surrounded by the body fluids of a host animal. Their bodies are structured as thin sheets of cells that lie close to the fluids of the surrounding external environment. Substances diffuse between the cells and the external environment through the animal's external surface, or through the surfaces of internal channels and cavities that are open to the environment.

In sponges, water carrying nutrients and O_2 is pumped by surface cells with beating flagella through hundreds of pores in the body wall surrounding a central cavity; it passes through the cavity and leaves through a large exit pore, now carrying CO_2 and wastes **(Figure 42.2A).** Hydras, jellyfish, sea anemones, and other cnidarians have a central **gastrovascular cavity** with a mouth that opens to the outside and extensions that radiate into the tentacles and all other body regions **(Figure 42.2B).** Water enters and leaves through the mouth, serving both digestion and circulation.

Circulatory Systems of Complex Animals Share Basic Elements

In larger and more complex animals, most cells lie in cell layers too deep within the body to exchange substances directly with the external environment via diffusion. Instead, the animals have a set of tissues and organs—a circulatory system—that conducts O_2, CO_2, nutrients, and wastes between body cells and specialized regions of the animal where substances are exchanged with the external environment. In terrestrial vertebrates, for example, oxygen from the environment is absorbed in the lungs and is carried by the blood to all parts of the body; CO_2 released from body cells is carried by the blood to the lungs, where it is released to the environment. Wastes are conducted from body cells to the kidneys, which remove the wastes from the circulation and excrete them into the environment.

The animal circulatory systems carrying out these roles have three basic components:

1. **Fluid:** A specialized fluid medium carries O_2, CO_2, nutrients, and wastes, as well as cells, and plays a major role in homeostasis (see Section 36.4).

A. **Open circulatory system: no distinction between hemolymph and interstitial fluid**

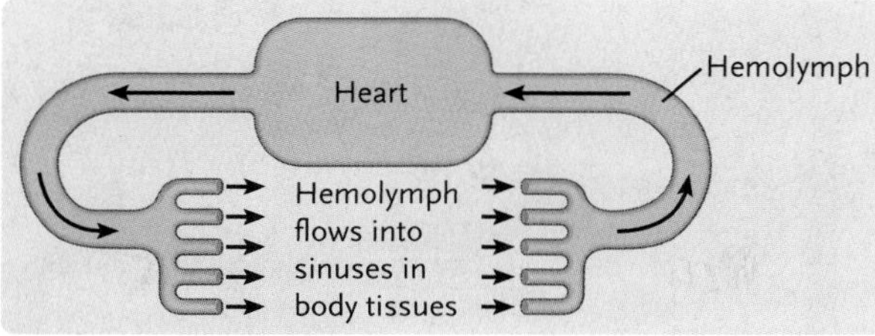
Heart
Hemolymph
Hemolymph
flows into
sinuses in
body tissues

B. **Closed circulatory system: blood separated from interstitial fluid**

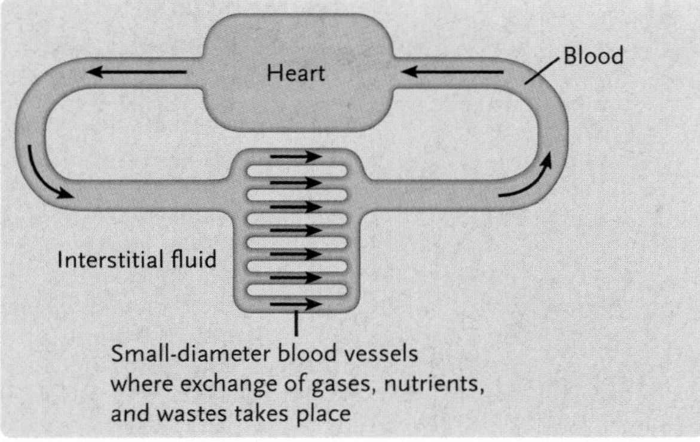
Heart
Blood
Interstitial fluid
Small-diameter blood vessels
where exchange of gases, nutrients,
and wastes takes place

FIGURE 42.3
Open and closed circulatory systems.

2. **Heart:** A muscular heart pumps the fluid through the circulatory system.
3. **Vessels:** Tubular vessels serve as conduits to distribute the fluid pumped by the heart.

Animal circulatory systems take one of two forms, either *open* or *closed* **(Figure 42.3)**. In an **open circulatory system,** vessels leaving the heart release bloodlike fluid termed **hemolymph** directly into body spaces, called **sinuses,** that surround organs. Thus, there is no distinction between hemolymph and *interstitial fluid,* the fluid immediately surrounding body cells (see Section 36.4). After flowing through the sinuses, the hemolymph reenters the heart through valves in the heart wall. In a **closed circulatory system,** the fluid—blood—is confined to blood vessels and is distinct from the interstitial fluid. Substances are exchanged between the blood and the interstitial fluid and then between the interstitial fluid and cells.

Most Invertebrates Have Open Circulatory Systems

Arthropods and most mollusks have open circulatory systems with one or more muscular hearts **(Figure 42.4)**. In an open system, most of the fluid pressure generated by the heart dissipates when the hemolymph is released into the sinuses. As a result, hemolymph flows relatively slowly. Open systems operate efficiently in these animals because they lead relatively sedentary lives; as a consequence, their tissues do not require O_2 and nutrients at the rate and quantities required by more active species. Among highly mobile and active species with open systems, such as insects and crustaceans, other

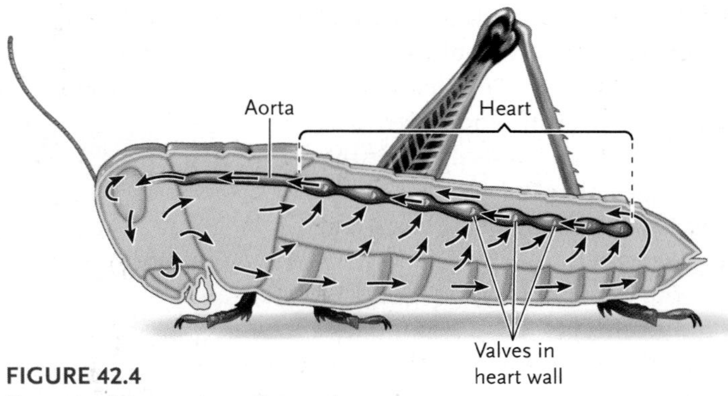
Aorta
Heart
Valves in
heart wall

FIGURE 42.4
Open circulatory system of a grasshopper.

adaptations compensate for the relatively slow distribution of hemolymph. In insects, for example, O_2 and CO_2 are exchanged efficiently with the environment by specialized air passages that branch throughout the body rather than by the hemolymph (these air passages, called *tracheae,* are discussed in Chapter 44).

Some Invertebrates and All Vertebrates Have Closed Circulatory Systems

Annelids, cephalopod mollusks such as squids and octopuses, and all vertebrates have closed circulatory systems. In these systems, vessels called **arteries** conduct blood away from the heart at relatively high pressure. From the arteries, the blood enters highly branched networks of microscopic, thin-walled vessels called **capillaries** that are well adapted for diffusion of substances. Nutrients and wastes are exchanged between the blood and body tissues as the blood moves through the capillaries. The blood then flows at relatively low pressure from the capillaries to larger vessels, the **veins,** which carry the blood back to the heart.

In many animals, but particularly the vertebrates, closed systems allow precise control of the distribution and rate of blood flow to different body regions by means of muscles that contract or relax to adjust the diameter of the blood vessels.

Vertebrate Circulatory Systems Have Evolved from Single to Double Blood Circuits

Several evolutionary trends accompanied the invasion of the different vertebrate groups into terrestrial habitats. Among the most significant is a trend from an effectively single circuit of blood pumped by the heart in sharks and bony fishes to the two completely separate circuits of birds and mammals. As part of this development, the heart evolved from one series of chambers to a double pump acting in two parallel series. In other words, depending on the vertebrate, the heart may consist of one or two **atria** (singular, *atrium*), the chambers that receive blood returning to the heart, and one or two **ventricles,** the chambers that pump blood from the heart.

THE SINGLE BLOOD CIRCUIT OF FISHES In fishes, the heart consists of two chambers arranged in a single line **(Figure 42.5A)**. The ventricle of the heart pumps blood into arteries leading to capillary networks in the gills, where the blood releases CO_2 and picks up O_2. The oxygenated blood is delivered to capillary networks in other body tissues where it delivers O_2 and picks up CO_2. The deoxygenated blood enters veins that carry it back to the atrium of the heart, and the single circuit is repeated.

The circulatory system of fishes is highly suited to the environments in which these animals live. Their bodies, supported by water and adapted for swimming, do not use as much energy in locomotion as do animals of an equivalent size moving on the land or in the air. Hence, fishes require less O_2 for their activities than terrestrial vertebrates do, and their relatively simple circulatory systems fully meet their O_2 requirements.

DOUBLE BLOOD CIRCUITS Vertebrate hearts changed significantly with the evolution of the first air-breathing fishes, such as the lungfish. In these fishes, the lung evolved as a respiratory organ in addition to gills. The lung necessitated a separate circuit because it is an additional organ for oxygenating the blood and, unlike other organs, does not receive oxygenated blood from the gills.

In amphibians, the separation into two circuits was accomplished by division of the atrium into two parallel chambers, the left and right atria, to produce a three-chambered heart, and by adaptations that keep oxygenated and deoxygenated blood partially separate as they are pumped by the single ventricle **(Figure 42.5B)**. In water, amphibians obtain O_2 primarily through the skin; the lungs are used only when the animal is in air. Oxygenated blood from lungs and skin enters veins that lead to the left atrium, whereas deoxygenated blood from the rest of the body enters the right atrium. Simultaneous contraction of the atria moves the two types of blood into the single ventricle, which pumps them—largely separately—into two circuits. Most of the oxygenated blood enters the **systemic circuit,** which supplies the blood for most of the tissues and cells of the body. Most of the deoxygenated blood enters the **pulmocutaneous circuit,** which leads to the lungs and

A. Circulatory system of fishes

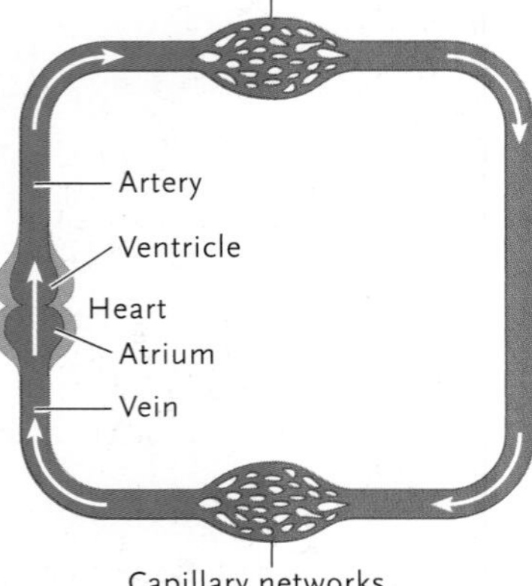

Capillary networks of gills
Artery
Ventricle
Heart
Atrium
Vein
Capillary networks in other body tissues

In fishes, the heart consists of a series of two chambers and pumps blood into one circuit. The ventricle pumps blood into arteries that lead to the capillary networks of the gills, where the blood releases CO_2 and picks up O_2. The oxygenated blood flows through other arteries to capillary networks in other body tissues where it releases O_2 and picks up CO_2. The deoxygenated blood enters veins that carry it to the atrium of the heart.

B. Circulatory system of amphibians

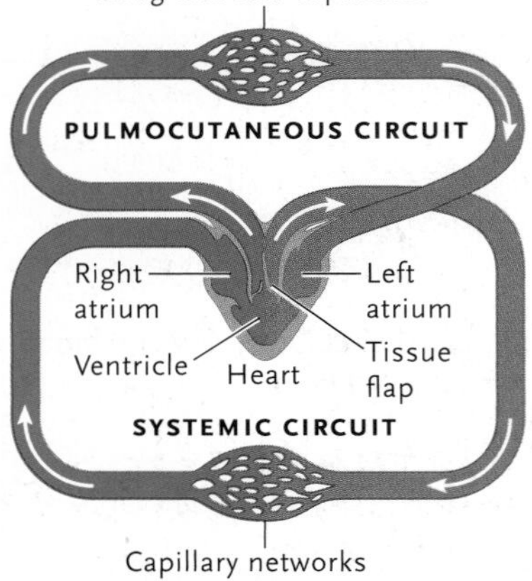

Lung and skin capillaries
PULMOCUTANEOUS CIRCUIT
Right atrium
Left atrium
Ventricle
Heart
Tissue flap
SYSTEMIC CIRCUIT
Capillary networks in other body tissues

In amphibians, the three-chambered heart pumps blood through two circuits. Oxygenated blood from the lungs and skin enters veins that lead to the left atrium, while deoxygenated blood from the rest of the body enters the right atrium. The atria contract simultaneously, pumping oxygenated and deoxygenated blood in the single ventricle. The two types of blood remain largely (90%) separate because of a smooth pattern of flow and a small tissue flap. Contraction of the ventricle moves most of the oxygenated blood into the systemic circuit, which delivers blood to most tissues and cells of the body. Deoxygenated blood moves into the pulmocutaneous circuit, which leads to the lungs and skin where CO_2 is released and O_2 is picked up.

C. Circulatory system of turtles, lizards, and snakes

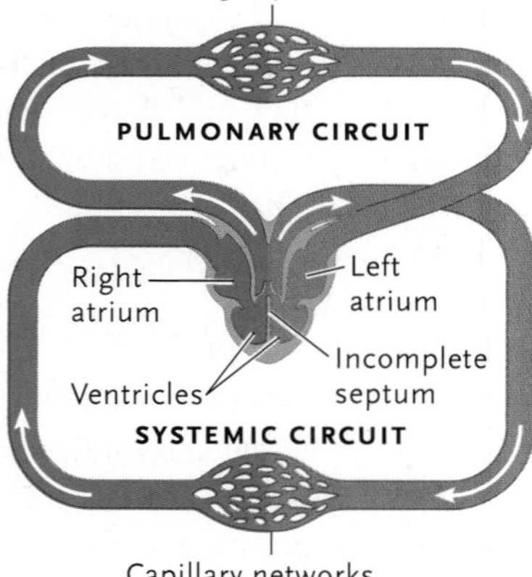

Lung capillaries
PULMONARY CIRCUIT
Right atrium
Left atrium
Ventricles
Incomplete septum
SYSTEMIC CIRCUIT
Capillary networks in other body tissues

In turtles, lizards, and snakes, an incomplete septum improves the separation of oxygenated blood from the lungs and deoxygenated blood from the rest of the body in the single ventricle.

D. Circulatory system of crocodilians, birds, and mammals

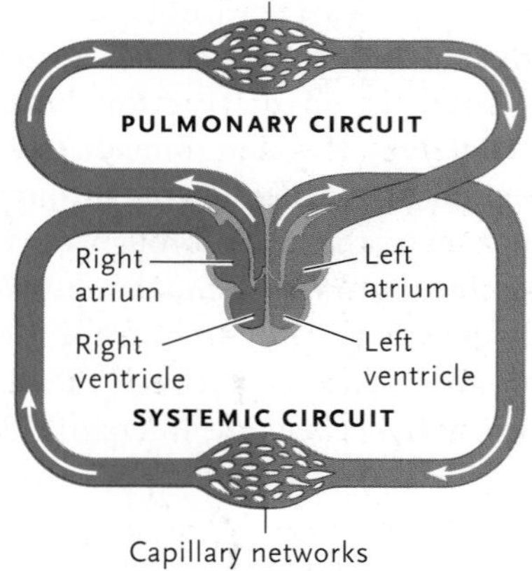

Lung capillaries
PULMONARY CIRCUIT
Right atrium
Left atrium
Right ventricle
Left ventricle
SYSTEMIC CIRCUIT
Capillary networks in other body tissues

In the four-chambered heart of crocodilians, birds, and mammals, a complete septum forms two ventricles and keeps the flow of oxygenated blood from the lungs and deoxygenated blood entirely separate from the rest of the body.

KEY

- Deoxygenated blood
- Oxygenated blood

FIGURE 42.5
Evolutionary developments in the heart and circulatory system of major vertebrate groups.

skin. Because the blood flows through separate systemic and pulmocutaneous circuits, the blood leaving the heart flows through only one capillary network in each circuit before returning to the heart. This separation greatly increases the blood pressure and flow in the systemic circuit as compared with that of fishes.

Turtles, lizards, and snakes also have two atria and a single ventricle. The ventricle is divided into right and left halves by a flap of connective tissue called the *septum,* which keeps the flow of oxygenated and deoxygenated blood almost completely separate in a systemic circuit and a **pulmonary circuit,** a branch leading to the lungs **(Figure 42.5C)**. The septum is incomplete in turtles, lizards, and snakes, allowing some mixing of oxygenated and deoxygenated blood. In crocodilians (crocodiles and alligators), the septum is complete **(Figure 42.5D)**.

Birds (which share ancestry with crocodilians) and mammals have a double heart consisting of two atria and two ventricles (as in Figure 42.5D). In effect, each half of the heart operates as a separate pump, restricting the blood circulation to completely separate pulmonary and systemic circuits. Blood is pumped by a ventricle in each circuit, so that both operate at relatively high pressure.

STUDY BREAK 42.1 <

1. **What are the three basic features of animal circulatory systems?**
2. **Distinguish between an open circulatory system and a closed circulatory system. Why do you think humans could not function with an open circulatory system?**
3. **Distinguish among atria, ventricles, arteries, and veins.**
4. **Distinguish among the systemic, pulmocutaneous, and pulmonary circuits.**

42.2 Blood and Its Components

In vertebrates, blood is a complex tissue containing a variety of cells suspended in a liquid matrix called the **plasma.** In addition to transporting molecules, blood stabilizes the pH and salt composition of body fluids and serves as a conduit for cells of the immune system. It also helps regulate body temperature by transferring heat between warmer and cooler body regions, and between the body and the external environment (the role of the blood in temperature regulation is discussed in Chapter 46).

For an average-sized adult human, the total blood volume is about 4 to 5 L—more than a gallon—and makes up about 8% of body weight. The plasma, a clear, straw-colored fluid, makes up about 55% of the volume of blood in human males and 58% in human females on average. Suspended in the plasma are three main types of blood cells—*erythrocytes, leukocytes,* and *platelets*—which constitute the hematocrit, the remainder of the blood volume. The typical components of human blood are shown in **Figure 42.6.**

In humans, blood cells develop in red bone marrow primarily in the vertebrae, sternum (breastbone), ribs, and pelvis. Blood cells originate from cells called *pluripotent* (*plura* = multiple; *potens* = power) *stem cells,* which retain the embryonic capacity to divide **(Figure 42.7)**. Pluripotent stem cells differentiate into two other types of cells, myeloid stem cells and lymphoid stem cells. Myeloid stem cells give rise to erythrocytes, platelets, and four types of leukocytes: the neutrophils, basophils, eosinophils, and monocytes/macrophages. Lymphoid stem cells give rise to two other types of leukocytes that function in the immune system: the B lymphocytes and T lymphocytes.

Plasma Is an Aqueous Solution of Proteins, Ions, Nutrient Molecules, and Gases

Plasma is so complex that its complete composition is unknown. Among its known components are water (91%–92% of its volume), glucose and other sugars, amino acids, plasma proteins, dissolved gases (mostly O_2, CO_2, and nitrogen), ions, lipids, vitamins, hormones and other signal molecules, and metabolic wastes, including urea and uric acid.

The plasma proteins fall into three classes, the *albumins,* the *globulins,* and *fibrinogen.* The **albumins,** the most abundant proteins of the plasma, are important for osmotic balance and pH buffering. They also transport a wide variety of substances through the circulatory system, including hormones, therapeutic drugs, and metabolic wastes. The **globulins** transport lipids (including cholesterol) and fat-soluble vitamins; a specialized subgroup of globulins, the *immunoglobulins,* constitute antibodies and other molecules contributing to the immune response. Some globulins are also enzymes. **Fibrinogen** plays a central role in the blood-clotting mechanism. All plasma proteins except immunoglobulins are synthesized in the liver; the immunoglobulins are synthesized by B and T lymphocytes (see Chapter 43).

The ions of the plasma include Na^+, K^+, Ca^{2+}, Cl^-, and HCO_3^- (bicarbonate). The Na^+ and Cl^- ions—the components of common table salt—are the most abundant. Some ions, particularly the bicarbonate ion, help maintain arterial blood at its characteristic pH, which in humans is slightly on the basic side at pH 7.4 (the bicarbonate ion and its role in pH balance are discussed further in Chapter 44).

Erythrocytes Are the Oxygen Carriers of the Blood

Erythrocytes—the red blood cells—carry O_2 from the lungs to body tissues. Each microliter of human blood contains about 5 million erythrocytes, which are small, flattened, and disclike. Indentations on each flattened surface make them *biconcave*—thinner in the middle than at the edges (see Figure 42.6). They are about 7 μm in diameter and 2 μm thick. Erythrocytes are highly flexible cells, able to squeeze through narrow capillaries.

As they mature, mammalian erythrocytes lose their nucleus, cytoplasmic organelles, and ribosomes, thereby limiting their metabolic capabilities and life span. The remaining cytoplasm contains enzymes that carry out glycolysis and large quantities of *hemoglobin,* the O_2-carrying protein of the blood. Most of the ATP produced by glycolysis is used to power active transport mechanisms that move ions in and out of erythrocytes.

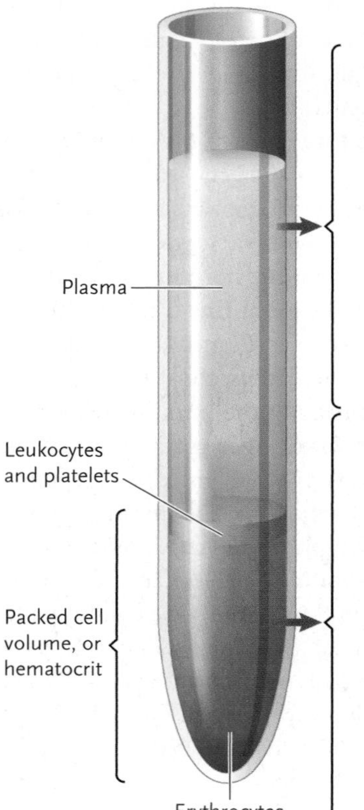
Erythrocyte (red blood cell)
Leukocyte (white blood cell)
Platelet
Plasma
Leukocytes and platelets
Packed cell volume, or hematocrit
Erythrocytes

FIGURE 42.6

Typical components of human blood. The colorized scanning electron micrograph shows the three major cellular components. The sketch of the test tube shows what happens when you centrifuge a blood sample. The blood separates into three layers: a thick layer of straw-colored plasma on top, a thin layer containing leukocytes and platelets, and a thick layer of erythrocytes. The table shows the relative amounts and functions of the various components of blood.

Plasma Portion: 55% (males)–58% (females) of total volume

Component	Percentage of Plasma Volume	Functions
1. Water	91–92	Solvent
2. Plasma proteins (albumin, globulins, fibrinogen, etc.)	7–8	Defense, clotting, lipid transport, roles in extracellular fluid volume, etc.
3. Ions, sugars, lipids, amino acids, hormones, vitamins, dissolved gases	1–2	Roles in extracellular fluid volume, pH, etc.

Cellular Portion (Hematocrit): 45% (males)–42% (females) of total volume

Component	Cells per Microliter	Functions
1. Erythrocytes (red blood cells)	4,800,000–5,400,000	Oxygen, carbon dioxide transport
2. Leukocytes (white blood cells)		
Neutrophils	3,000–6,750	Phagocytosis during inflammation
Lymphocytes	1,000–2,700	Immune response
Monocytes/macrophages	150–720	Phagocytosis in all defense responses
Eosinophils	100–360	Defense against parasitic worms
Basophils	25–90	Secrete substances for inflammatory response and for fat removal from blood
3. Platelets	250,000–300,000	Roles in clotting

Hemoglobin, the protein that gives erythrocytes and the blood its red color, consists of four polypeptides, each linked to a nonprotein *heme* group that contains an iron atom in its center (see Figure 3.20). The iron atom binds O_2 molecules as blood circulates through the lungs and releases the O_2 as blood flows through other body tissues. Chapter 44 describes the mechanisms of gas exchange and transport, including the process by which hemoglobin transports O_2.

Hemoglobin can also combine with carbon monoxide (CO), a byproduct of combustion. The combination, which is essentially irreversible, blocks hemoglobin's ability to combine with O_2, and can quickly lead to death. CO is also present in cigarette smoke.

Some 2 million to 3 million erythrocytes are produced in the average human each *second*. The life span of an erythrocyte in the circulatory system is about 120 days. At the end of their useful life, erythrocytes are engulfed and destroyed by *macrophages* (*macros* = big; *phagein* = to eat) (see Figure 42.7), a type of leukocyte, in the spleen, liver, and bone marrow.

A negative feedback mechanism keyed to the blood's O_2 content stabilizes the number of erythrocytes. If the O_2 content drops below the normal level, the kidneys release **erythropoietin** (EPO), a hormone that stimulates stem cells in bone marrow to increase erythrocyte production. EPO is also secreted after blood loss and when mammals move to higher altitudes. As new red blood cells enter the bloodstream, the O_2-carrying capacity of the blood rises. If the O_2 content of the blood rises above normal levels, EPO production falls in the kidneys and red blood cell production drops. The gene encoding human EPO has been cloned, allowing researchers to produce this protein

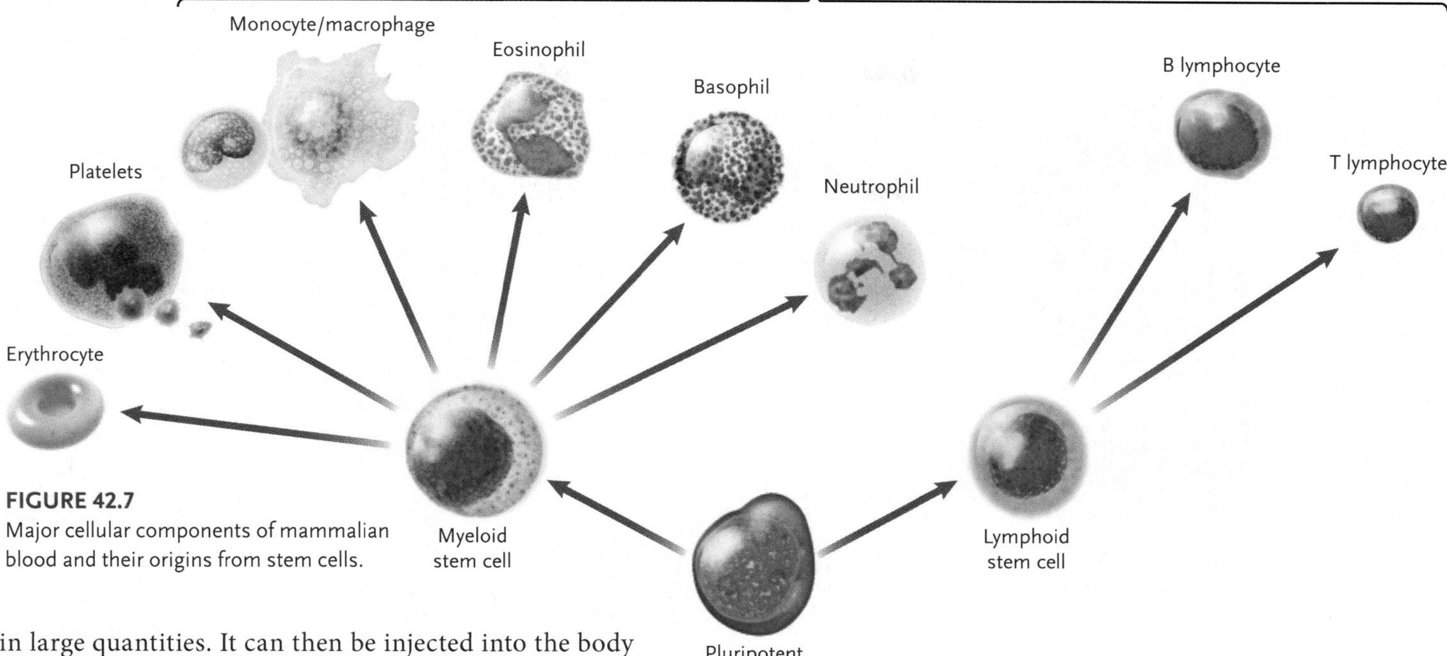
Leukocytes
Monocyte/macrophage
Eosinophil
Basophil
Neutrophil
B lymphocyte
T lymphocyte
Platelets
Erythrocyte
Myeloid stem cell
Pluripotent stem cell
Lymphoid stem cell

FIGURE 42.7
Major cellular components of mammalian blood and their origins from stem cells.

in large quantities. It can then be injected into the body to stimulate erythrocyte production, for example, in patients with anemia (lower-than-normal hemoglobin levels) caused by kidney failure or chemotherapy. It can also supplement or even replace blood transfusions. Some endurance athletes such as triathletes, bicycle racers, marathon runners, and cross-country skiers have used EPO to increase their erythrocyte levels to enhance performance. Such blood doping is deemed illegal by the governing organizations of most endurance sports and, as a result, many athletes have been sanctioned or banned in recent years.

Human blood groups are determined by antigens, the carbohydrate portions of particular glycoproteins on the surfaces of erythrocytes. Section 12.2 described the antigens of the human ABO blood group and how they are important in transfusions.

Disorders of erythrocytes are responsible for a number of human disabilities and diseases. The *anemias,* which result from too few or malfunctioning erythrocytes, prevent O_2 from reaching body tissues in sufficient amounts. Shortness of breath, fatigue, and chills are common symptoms of anemia. Anemia can be produced, for example, by blood loss from a wound, by certain infections, or by dietary insufficiencies.

Leukocytes Provide the Body's Front Line of Defense against Disease

Leukocytes eliminate dead and dying cells from the body, remove cellular debris, and defend against invading organisms. They are called white cells because they are colorless, in contrast to the strongly pigmented red blood cells. Also unlike red blood cells, leukocytes retain their nuclei, cytoplasmic organelles, and ribosomes as they mature, and hence are fully functional cells.

Some types of leukocytes are capable of continued division in the blood and body tissues. The specific types of leukocytes and their functions in the immune system are discussed in Chapter 43.

Platelets Induce Blood Clots That Seal Breaks in the Circulatory System

Blood **platelets** are oval or rounded vesicles about 2 to 4 μm in diameter that contain enzymes and other factors that take part in blood clotting. When blood vessels are damaged, collagen fibers in the extracellular matrix are exposed to the leaking blood. Platelets in the blood then stick to the collagen fibers and release signaling molecules that induce additional platelets to stick to them. The process continues, forming a plug that helps seal the damaged site. As the plug forms, the platelets release other factors that convert the soluble plasma protein fibrinogen into long, insoluble threads of **fibrin.** Cross-links between the fibrin threads form a meshlike network that traps blood cells and platelets and further seals the damaged area **(Figure 42.8).** The entire mass is a blood clot.

Mutations or diseases that interfere with the enzymes and factors taking part in the clotting mechanism can have serious effects and lead to uncontrolled bleeding. In the most common form of hemophilia, for example, a mutation in a single protein (called *clotting factor VIII*) interferes with the clotting re-

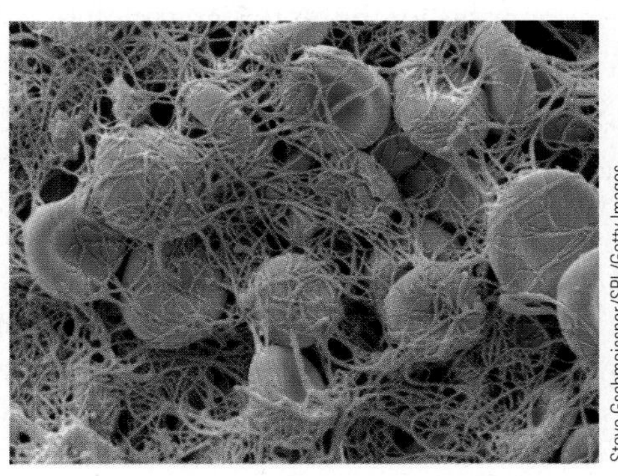

FIGURE 42.8
Red blood cells caught in a meshlike network of fibrin threads during formation of a blood clot.

Steve Gschmeissner/SPL/Getty Images

FIGURE 42.9
Cutaway view of the human heart showing its internal organization.

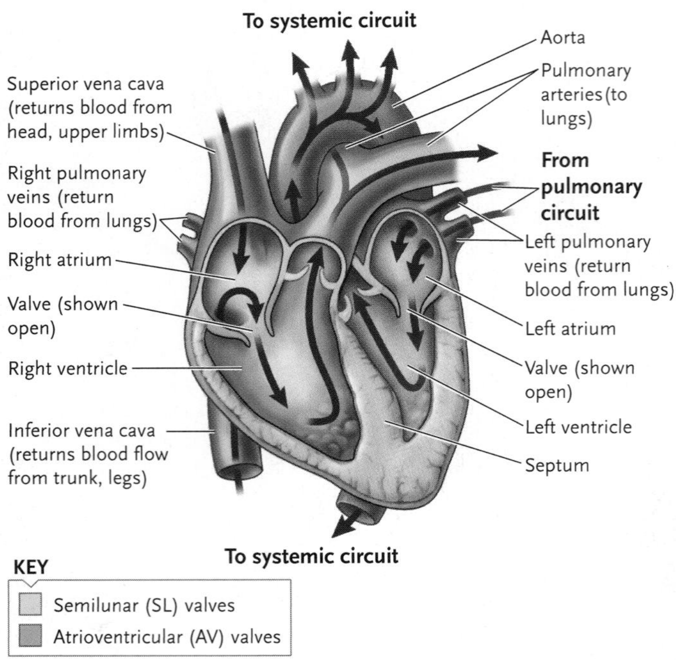

To systemic circuit
Aorta
Superior vena cava (returns blood from head, upper limbs)
Pulmonary arteries (to lungs)
From pulmonary circuit
Right pulmonary veins (return blood from lungs)
Left pulmonary veins (return blood from lungs)
Right atrium
Left atrium
Valve (shown open)
Valve (shown open)
Right ventricle
Left ventricle
Inferior vena cava (returns blood flow from trunk, legs)
Septum
To systemic circuit
KEY
Semilunar (SL) valves
Atrioventricular (AV) valves

action. Bleeding is uncontrolled in afflicted individuals; even small cuts and bruises can cause life-threatening blood loss.

STUDY BREAK 42.2 <

1. **Outline the life cycle of an erythrocyte.**
2. **How does the body compensate for a lower-than-normal level of oxygen in the blood?**
3. **What are the roles of leukocytes and platelets?**

42.3 The Heart

In mammals, the heart is structured from cardiac muscle cells (see Figure 36.6) forming a four-chambered pump, with two atria at the top of the heart pumping blood into two ventricles at the bottom of the heart **(Figure 42.9).** The powerful contractions of the ventricles push the blood at relatively high pressure into arteries leaving the heart. This arterial pressure is responsible for the blood circulation. Valves in the heart keep the blood from moving backward. Between each atrium and ventricle is an atrioventricular valve (AV valve), and between the ventricles and the arteries leaving the heart (the aorta and pulmonary arteries) are seminlunar valves (SL valves).

The heart of mammals pumps the blood through two completely separate circuits of blood vessels: the systemic circuit and the pulmonary circuit **(Figure 42.10).** The right atrium (toward the right side of the body) receives blood returning to the heart in vessels coming from the entire body, except for the lungs: the *superior vena cava* conveys blood from the head and forelimbs,

FIGURE 42.10
The pulmonary and systemic circuits of mammals. The right half of the heart pumps blood into the pulmonary circuit, and the left half of the heart pumps blood into the systemic circuit.

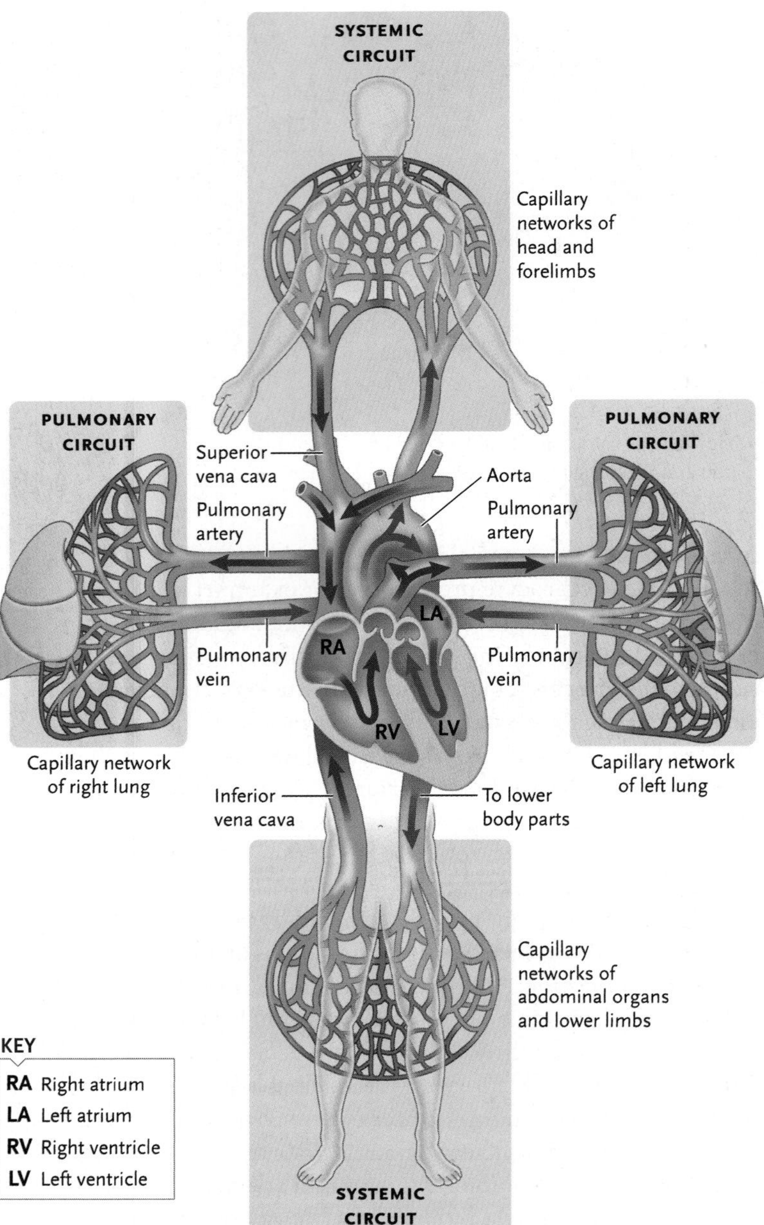

SYSTEMIC CIRCUIT
Capillary networks of head and forelimbs
PULMONARY CIRCUIT
PULMONARY CIRCUIT
Superior vena cava
Aorta
Pulmonary artery
Pulmonary artery
Pulmonary vein
Pulmonary vein
RA
LA
RV
LV
Capillary network of right lung
Capillary network of left lung
Inferior vena cava
To lower body parts
Capillary networks of abdominal organs and lower limbs
KEY
RA Right atrium
LA Left atrium
RV Right ventricle
LV Left ventricle
SYSTEMIC CIRCUIT

and the *inferior vena cava* conveys blood from the abdominal organs and the hind limbs. This blood is depleted of O_2 and has a high CO_2 content. The right atrium pumps blood into the right ventricle, which contracts to push the blood into the **pulmonary arteries** leading to the lungs. In the capillaries of the lungs, the blood releases CO_2 and picks up O_2. The oxygenated blood completes this pulmonary circuit by returning in **pulmonary veins** to the heart.

Blood returning from the pulmonary circuit enters the left atrium, which pumps the blood into the left ventricle. This ventricle, the most thick-walled and powerful of the heart's chambers, contracts to send the oxygenated blood coursing into a large artery, the **aorta,** which branches into arteries leading to

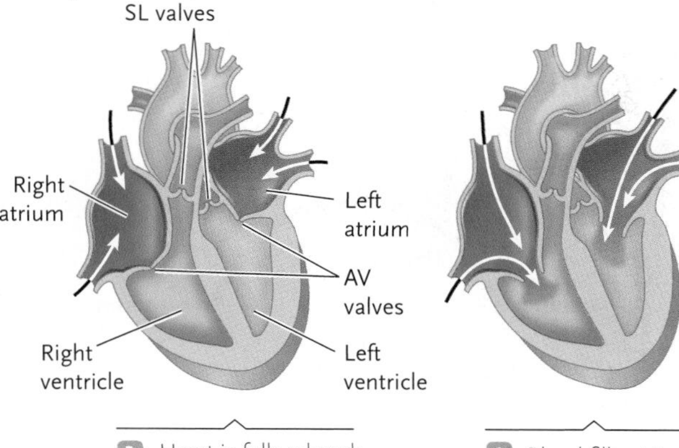
Diastole
SL valves
Right atrium
Left atrium
AV valves
Right ventricle
Left ventricle

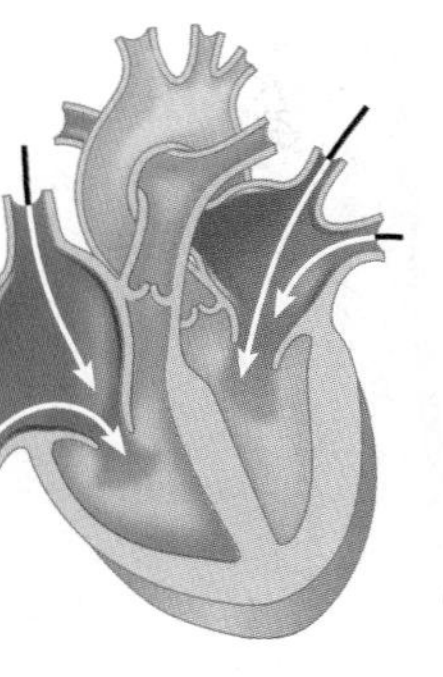

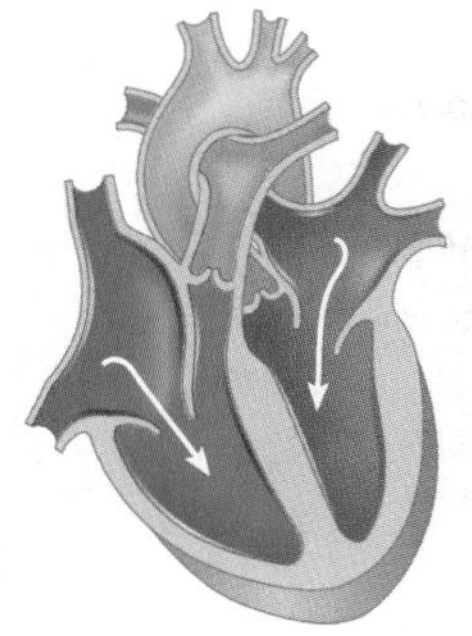

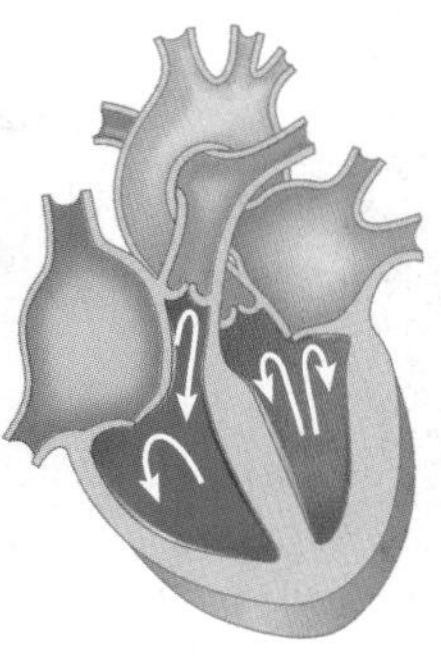
Systole

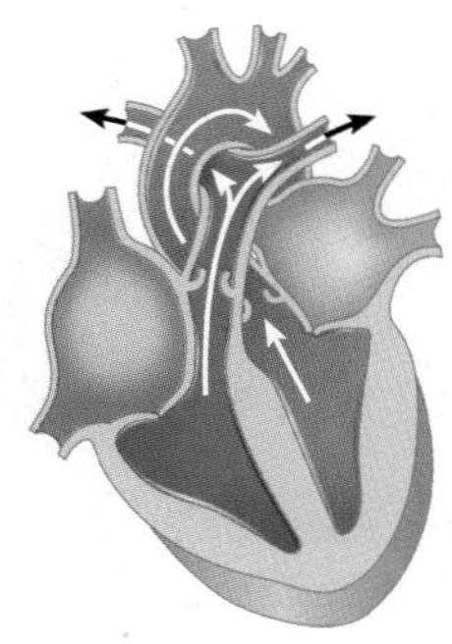

1 Heart is fully relaxed; atria begin to fill with blood; AV and SL valves are closed.

2 Blood fills atria; pressure pushes AV valves open; ventricles begin to fill.

3 When ventricles are about 80% full, atria contract and completely fill ventricles.

4 Ventricles begin to contract, forcing AV valves closed; SL valves remain closed.

5 As ventricles continue to contract, pressure in ventricle chambers rises above that in arteries leading from heart, forcing SL valves open and ejecting blood into aorta and pulmonary arteries. About two-thirds of blood in ventricles enters arteries. Ventricles now relax, lowering pressure in ventricle chambers below that in arteries. Blood pressure in ventricles now closes SL valves. The cycle begins again.

FIGURE 42.11
The cardiac cycle. The figure shows one cardiac cycle beginning at the start of diastole.

all body regions except the lungs. In the capillary networks of these body regions, the blood releases O_2 and picks up CO_2. The O_2-depleted blood collects in veins, which complete the systemic circuit. The blood from the veins enters the right atrium. The amount of blood pumped by the two halves of the heart is normally balanced so that neither side pumps more than the other. More details of gas exchange in the pulmonary circuit are presented in Chapter 44.

The heart also has its own circulation, called the *coronary circulation.* The aorta gives off two **coronary arteries** that course over the heart. The coronary arteries branch extensively, leading to dense capillary beds that serve the cardiac muscle cells. The blood from the capillary networks collects into veins that empty into the right atrium.

The Heartbeat Is Produced by a Cycle of Contraction and Relaxation of the Atria and Ventricles

Average heart rates vary among vertebrates, depending on body size and the overall level of metabolic activity. A human heart beats 72 times each minute, on average. The resting heart rate of a trained endurance athlete may be only 60% of this rate. The heart of a flying bat may beat 1,200 times a minute, whereas that of an elephant beats only 30 times a minute. The period of ventricular contraction and emptying of the heart is called the **systole** and the period of relaxation and filling of the heart between contractions is called the **diastole.** The systole–diastole sequence of the heart is called the **cardiac cycle (Figure 42.11).** At an average heart rate, the cardiac cycle takes a little over 0.8 sec.

The heart valves make a "lub-dub" sound when the heart beats, which you can hear by listening to the heart through a stethoscope. The "lub" sound occurs when the AV valves are pushed shut by the contraction of the ventricles; the "dub" sound is made when the SL valves are forced shut as the ventricles relax. *Heart murmurs* are abnormal sounds produced by turbulence created in the blood when one or more of the valves fails to open or close completely and blood flows backward.

In an adult human at rest, each ventricle pumps roughly 5 L of blood per minute—an amount roughly equivalent to the entire volume of blood in the body. At maximum rate and strength the human heart pumps about five times the resting amount.

The Cardiac Cycle Is Initiated within the Heart

Contraction of cardiac muscle cells is triggered by action potentials that spread along the muscle cell membranes. Some crustaceans, such as crabs and lobsters, have **neurogenic hearts,** that is, hearts that beat under the control of signals from the nervous system. Other animals, including all insects and all vertebrates, have **myogenic hearts,** that is, hearts that maintain their contraction rhythm autonomously. Indeed, it is the ability to sustain contractile rhythmicity in the mammalian heart that makes heart transplants possible.

Figure 42.12 illustrates the electrical control of the cardiac cycle. Two nodes, the sinoatrial node and atrioventricular node, generate the electrical signals. The **sinoatrial node (SA node)** controls the rate and timing of cardiac contraction in mammalian myogenic hearts by coordinating the contractions of individual cardiac muscle cells. The SA node consists of **pacemaker cells,** which are specialized cardiac muscle cells in the upper wall of the right atrium near the entry point for blood from the systemic circuit. Ion channels in these cells open and close in a cyclic, self-sustaining pattern that alternately depolarizes and repolarizes their plasma membranes. The regularly timed depolarizations initiate waves of contraction that travel over the heart.

The **atrioventricular node (AV node)** is located in the heart wall between the right atrium and right ventricle, just above the insulating layer between the atria and the ventricles. Cells of the AV

FIGURE 42.12
The electrical control of the cardiac cycle. The top part of the figure shows how a signal originating at the SA node leads to ventricular contraction. The bottom part of the figure shows the electrical activity for each of the stages as seen in an ECG. The colors in the hearts show the location of the signal at each step and correspond to the colors in the ECG.

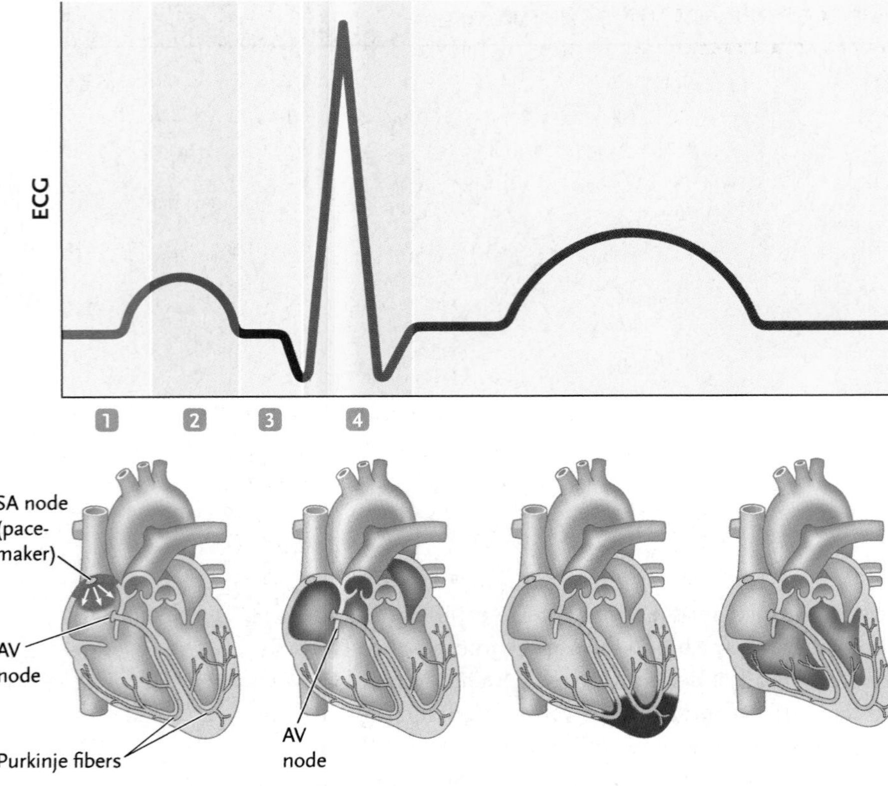

ECG
1
2
3
4
SA node (pace-maker)
AV node
Purkinje fibers
AV node

1 Depolarization signal. Pacemaker cells of SA node generate rhythmic depolarization signals initiating waves of contractions that travel over the heart. The waves cause cells of the atria to contract in unison and fill the ventricles with blood.

2 Contraction signal. A layer of connective tissue that separates and insulates the atria from the ventricles prevents a contraction signal from the SA node from spreading directly to the ventricles.

3 Signal from AV node. AV node cells excited by the atrial wave of contraction produce a signal that travels to the bottom of the heart via Purkinje fibers, which travel through the insulating layer to the bottom of the heart, where they branch into the walls of the ventricles.

4 Wave of contraction. The signal carried by the Purkinje fibers induces a wave of contraction that begins at the bottom of the heart and proceeds upward, squeezing the blood from the ventricles into the aorta and pulmonary arteries.

node are excited by the atrial wave of contraction, generating a signal that travels to the bottom of the heart via **Purkinje fibers.** The signal then induces a wave of contraction that begins at the bottom of the heart and proceeds upward. The transmission of a signal from the AV node to the ventricles takes about 0.1 sec; this delay gives the atria time to finish their contraction before the ventricles contract.

As Augustus Waller found, the electrical signals passing through the heart can be detected by attaching electrodes to different points on the surface of the body. The signals change in a regular pattern corresponding to the electrical signals that trigger the cardiac cycle, producing what is known as an **electrocardiogram** (**ECG;** also **EKG,** from German *elektrokardiogramm*). Many malfunctions of the heart alter the ECG pattern in characteristic ways, providing clues to the location and type of heart disease.

Arterial Blood Pressure Cycles between a High Systolic and a Low Diastolic Pressure

The pressure that a fluid in a confined space exerts is called *hydrostatic pressure.* That is, fluid in a container exerts some pressure on the wall of the container. Blood vessels are essentially tubular containers that are part of a closed system filled with fluid. Hence, the blood in vessels exerts hydrostatic pressure against the walls of the vessels. **Blood pressure** is the measurement of that hydrostatic pressure on the walls of the arteries as the heart pumps blood through the body. Blood pressure is determined by the force and amount of blood pumped by the heart and the size and flexibility of the arteries. In any person, blood pressure changes continually in response to activity, temperature, body position, emotional state, diet, and medications being taken.

In clinical medicine, blood pressure is measured with a *sphygmomanometer,* which consists of an inflatable cuff that wraps around the upper arm connected to an air pump and a mercury column. Likely you have had your blood pressure taken several times in your life. Two blood pressure measurements are typically recorded, representing different parts of the cardiac cycle. As the ventricles contract, a surge of high-pressure blood moves outward through the arteries leading from the heart. This peak of high pressure, called the **systolic blood pressure,** can be felt as a **pulse** by pressing a finger against an artery that lies near the skin, such as the arteries of the neck or the artery that runs along the inside of the wrist. Between ventricular contractions, the arterial blood pressure reaches a low point called the **diastolic blood pressure.**

For most healthy adults at rest, the systolic pressure measured at the upper arm is between 90 and 120 mm Hg, and the diastolic pressure is between 60 and 80 mm Hg. The numbers, written in the form 120/80 mm Hg and stated verbally as "120 over 80," refer to the height of a column of mercury in millimeters that would be required to balance the pressure exactly. The systolic and diastolic blood pressures in the pulmonary arteries are typically much lower, about 24/8 mm Hg.

The blood pressure in the systemic and pulmonary circuits is highest in the arteries leaving the heart, and drops as the blood passes from the arteries into the capillaries. By the time the blood returns to the heart, its pressure has dropped to 2 to 5 mm Hg, with no differentiation between systolic and diastolic pressures. The reduction in pressure occurs because the blood encounters resistance as it moves through the vessels, produced primarily by the friction created when blood cells and plasma proteins move over each other and over vessel walls.

Some people have **hypertension,** commonly called high blood pressure, a medical condition in which blood pressure is chronically elevated above normal values; that is, at least 140/90. In some cases, no specific medical cause can be found to explain the hypertension. In other cases, the hypertension results from another medical condition, such as kidney disease or disorders affecting the adrenal cortex. Hypertension can also be caused by certain medications, such as ibuprofen and steroids. Age is also a contributor to hypertension because over time the walls of blood vessels become stiffer as more collagen fibers are added, decreasing the elasticity of the arteries. During systole, these arteries cannot expand as much as they once could, and this results in a higher arterial blood pressure.

Hypertension is rarely severe enough to cause symptoms. However, in the long term, the increased pressure in the arteries can cause damage to organs. Hence, hypertension is treated because of the correlation with an increased risk for a number of medical conditions, including myocardial infarction (heart attack), cardiovascular accident (stroke), chronic renal failure, and retinal damage. Treatments to reduce hypertension typically involve lifestyle changes, such as weight loss and regular exercise; in the case of moderate to severe hypertension, drugs are also prescribed.

What happens to blood pressure during exercise? The answer depends on the type of exercise. During static exercise involving a sustained contraction of a muscle group or groups, such as weight lifting, both systolic and diastolic pressure increase. In elite weight lifters, for instance, blood pressure during lifts can reach 300/150 mm Hg. However, during dynamic exercise involving intermittent and rhythmical muscle contractions, such as running, bicycling, and swimming, only the systolic pressure increases.

STUDY BREAK 42.3 <

1. **What is the role of each of the four chambers of the mammalian heart in blood circulation?**
2. **Distinguish between systole and diastole.**
3. **How do neurogenic and myogenic hearts differ?**
4. **Describe the electrical events that occur during the cardiac cycle in a mammalian heart.**

THINK OUTSIDE THE BOOK >

Arrhythmia is any variation from the normal heartbeat rhythm. On your own or collaboratively, research two examples of arrhythmia and outline the physiological reasons for them.

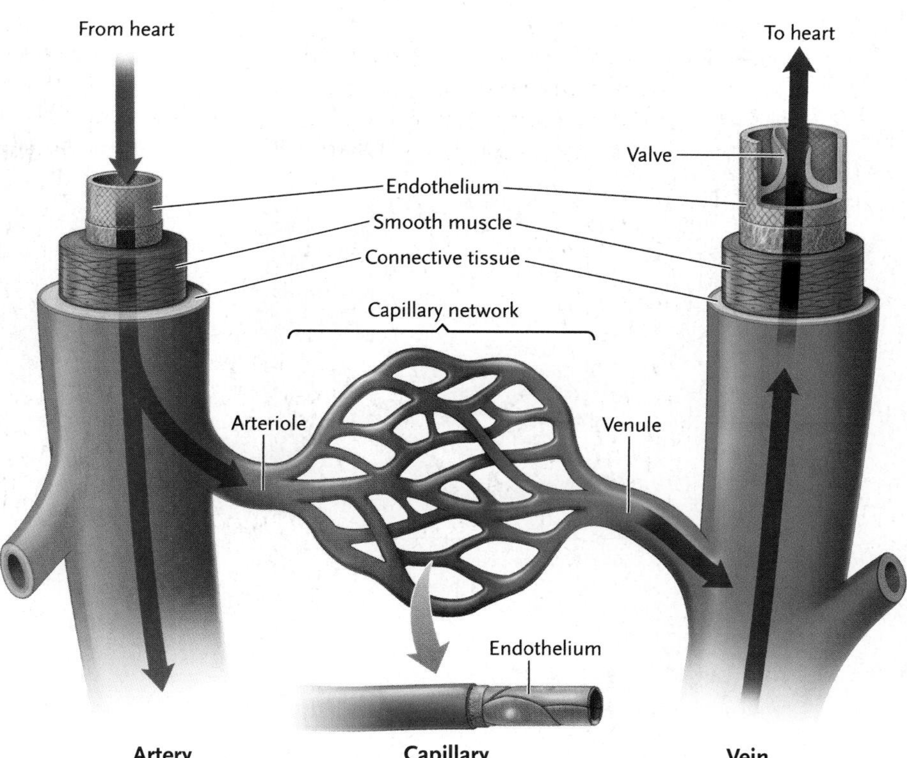
From heart
To heart
Valve
Endothelium
Smooth muscle
Connective tissue
Capillary network
Arteriole
Venule
Endothelium
Artery
Capillary
Vein

FIGURE 42.13
The structure of arteries, capillaries, and veins and their relationship in blood circuits.

42.4 Blood Vessels of the Circulatory System

Both the systemic and pulmonary circuits consist of a continuum of different blood vessel types that begin and end at the heart **(Figure 42.13).** From the heart, large arteries carry blood and branch into progressively smaller arteries that deliver the blood to the various parts of the body. When a small artery reaches the organ it supplies, it branches into yet smaller vessels, the **arterioles.** Within the organ, arterioles branch into capillaries, the smallest vessels of the circulatory system. The capillaries form a network in the organ that is used to exchange substances between the blood and the surrounding cells. Capillaries rejoin to form small **venules,** which merge into the small veins that leave the organ. The small veins progressively join to form larger veins that eventually become the large veins that enter the heart.

Arteries Transport Blood Rapidly to the Tissues and Serve as a Pressure Reservoir

Arteries have relatively large diameters and, therefore, provide little resistance to blood flow. Structurally, they are adapted to the relatively high pressure of the blood passing through them. The walls of arteries consist of three major tissue layers: (1) an outer layer of connective tissue containing collagen fibers mixed with fibers of the protein elastin, which gives the vessel recoil ability; (2) a relatively thick middle layer of vascular smooth muscle cells also mixed with elastin fibers; and (3) a one-cell-thick inner layer of flattened cells that forms an *endothelium,* a

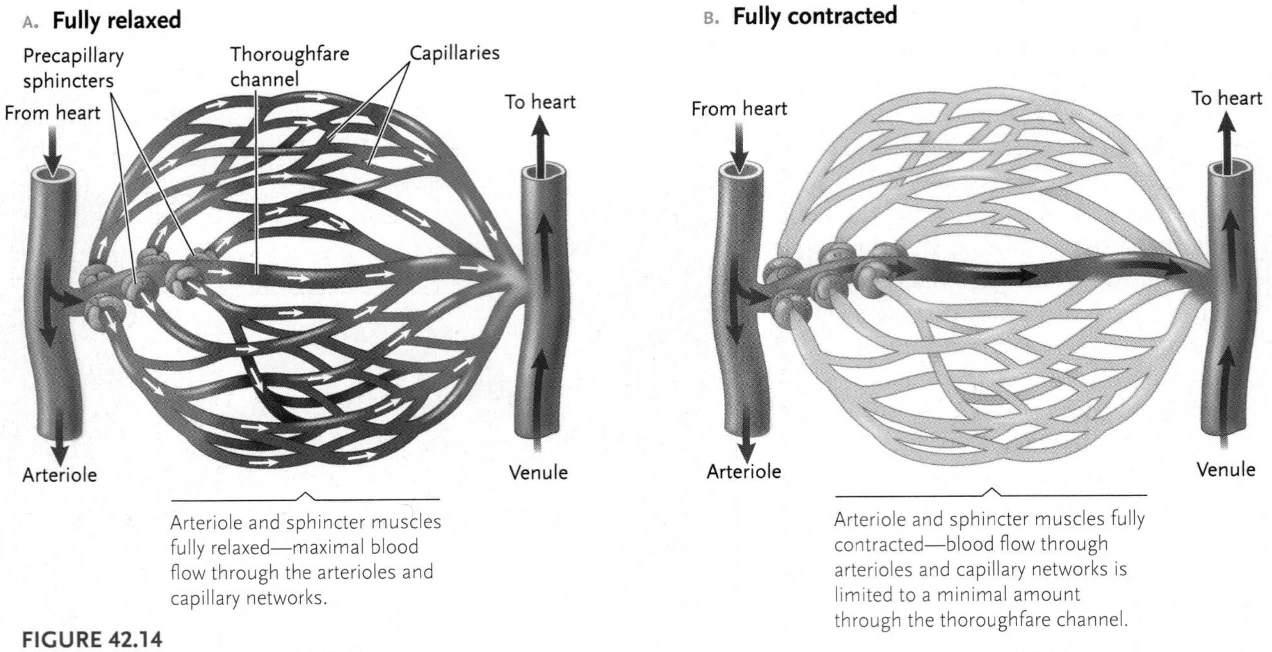

A. **Fully relaxed**

Arteriole and sphincter muscles fully relaxed—maximal blood flow through the arterioles and capillary networks.

B. **Fully contracted**

Arteriole and sphincter muscles fully contracted—blood flow through arterioles and capillary networks is limited to a minimal amount through the thoroughfare channel.

FIGURE 42.14
Control of blood flow through arterioles and capillary networks when arteriole and sphincter muscles are: (A) fully relaxed; and (B) fully contracted.

specialized type of epithelial tissue that lines the entire circulatory system (see Figure 42.13).

In addition to being the conduits for blood traveling to the tissues, arteries also act as a pressure reservoir to generate the force for blood movement when the heart is relaxing. When contraction of the ventricles pumps blood into the arteries, the amount of blood flowing into the arteries is greater than the amount flowing out into the smaller vessels downstream because of the higher resistance to blood flow in those smaller vessels. The arteries can accommodate the excess volume of blood because their elastic walls allow the arteries to expand in diameter. When the heart then relaxes and blood is no longer being pumped into the arteries, the arterial walls recoil passively back to their original state. The recoil pushes the excess blood from the arteries into the smaller downstream vessels. As a result, blood flow to tissues is continuous during systole and diastole.

Capillaries Are the Sites of Exchange between the Blood and Interstitial Fluid

Capillaries, which thread through nearly every tissue in the body, are arranged in networks that bring them within 10 μm of most body cells. They form an estimated 2,600 km^2 of total surface area for the exchange of gases, nutrients, and wastes with the interstitial fluid. Capillary walls consist of a single layer of endothelial cells.

CONTROL OF BLOOD FLOW THROUGH CAPILLARIES Blood flow through the capillary networks is controlled by contraction of smooth muscle in the arterioles **(Figure 42.14).** The capillaries themselves do not have smooth muscle but, in many cases, a small ring of smooth muscle called a *precapillary sphincter* is present at the junction between an arteriole and a capillary. Variation in the contraction of arteriole and sphincter smooth muscles adjusts the rate of flow through the capillary networks between the two flow limits. For example, during exercise, flow of blood through the capillary networks is increased severalfold over the resting state by relaxation of the precapillary sphincters.

CONTROL OF BLOOD VOLUME TO CAPILLARIES BY ARTERIOLES
The volume of blood flowing through an organ is adjusted by regulating the internal diameter of the arterioles of the organ. The blood leaving the arterioles enters the capillaries that branch from them. Although their total surface area is astoundingly large, the diameter of individual capillaries is so small that red blood cells must squeeze through most of them in single file. As a result, each capillary presents a high resistance to blood flow, so blood slows considerably as it moves through the capillaries **(Figure 42.15).** This is analogous to the slowing that occurs when a narrow, swiftly running stream widens into a broad pool. The slow movement of blood through the capillaries maximizes the time for exchange of substances between blood and tissues. As they leave the tissues, the capillaries rejoin to form veins. Veins have a reduced total cross-sectional area compared with capillaries, so the rate of flow increases as blood returns to the heart (see Figure 42.15).

EXCHANGE OF SUBSTANCES ACROSS CAPILLARY WALLS In most body tissues, narrow spaces between the capillary endothelial cells allow water, ions, and small molecules such as glucose to pass freely between the blood and interstitial fluid. Erythrocytes, platelets, and most plasma proteins are too large to pass between the cells and are retained inside the capillaries, except for molecules that are transported through the epithelial cells by specific carriers. Leukocytes, however, are able to squeeze actively between the cells and pass from the blood to the interstitial fluid.

There are exceptions to these general properties in some tissues; in the brain, for example, the capillary endothelial cells are

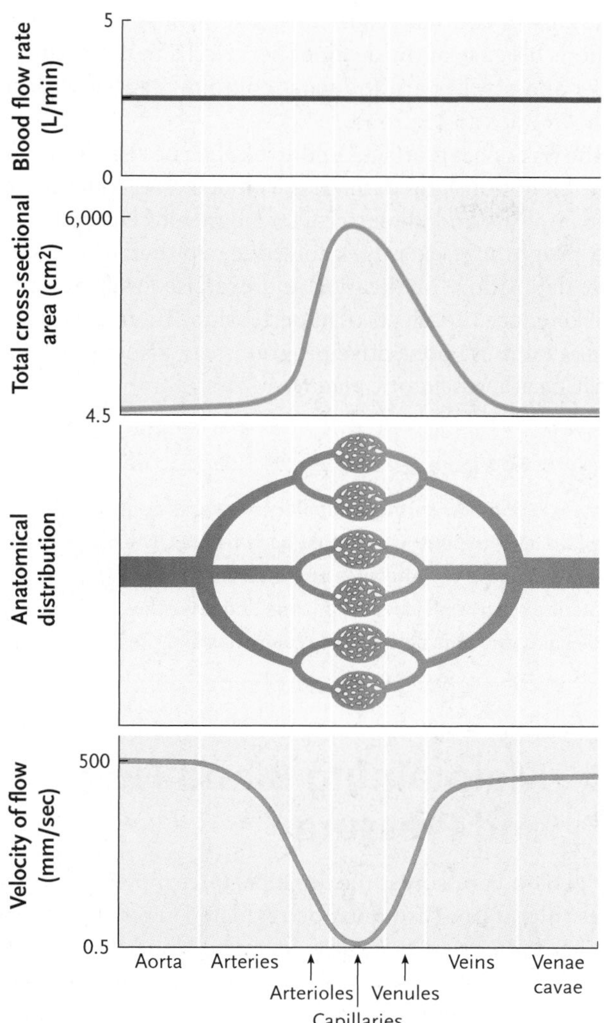
Blood flow rate (L/min)
5
0
Total cross-sectional area (cm²)
6,000
4.5
Anatomical distribution
Velocity of flow (mm/sec)
500
0.5
Aorta
Arteries
Arterioles
Capillaries
Venules
Veins
Venae cavae

FIGURE 42.15
Blood flow rate and velocity of flow in relation to total cross-sectional area of the blood vessels. The blood flow rate is identical throughout the circulatory system and is equal to the cardiac output. The velocity of flow in the different types of blood vessels is inversely related to the total cross-sectional area of all the vessels of a particular type: for example, the velocity is highest in the aorta, which has the smallest cross-sectional area, and lowest in the capillaries, which collectively have the largest total cross-sectional area.

tightly sealed together, preventing essentially all molecules and even ions from passing between them. The tight seals set up the **blood–brain barrier,** which limits the exchange between capillaries and brain tissues to molecules and ions that are specifically transported through the capillary endothelial cells (see Section 38.3 for additional discussion of the blood–brain barrier).

FORCES DRIVING THE EXCHANGE Two major mechanisms drive the exchange of molecules and ions between the capillaries and the interstitial fluid: (1) diffusion along concentration gradients; and (2) bulk flow.

Diffusion along concentration gradients occurs both through the spaces between the capillary endothelial cells and through their plasma membranes. Ions or molecules such as O_2 and glucose, which are more concentrated inside the capillaries, diffuse outward along their gradients; other ions or molecules that are more concentrated outside, such as CO_2, diffuse inward. Some molecules, such as O_2 and CO_2, diffuse directly through the lipid bilayer of the endothelial cell plasma membranes; others, such as glucose, pass by facilitated diffusion through transport proteins. The total diffusion is greatest at the ends of capillaries nearest the arterioles, where the concentration differences between blood plasma and interstitial fluid are highest, and least at the ends nearest the venules, where the inside/outside concentrations of diffusible substances are almost equal.

Bulk flow, which carries water, ions, and molecules out of the capillaries, occurs through the spaces between capillary endothelial cells. The flow is driven by the pressure of the blood, which is higher than the pressure of the interstitial fluid. Like diffusion, bulk flow is greatest in the ends of the capillaries nearest the arterioles, where the pressure difference is highest, and drops off steadily as the blood moves through the capillaries and the pressure difference becomes smaller.

Venules and Veins Serve as Blood Reservoirs in Addition to Conduits to the Heart

The walls of venules and veins are thinner than those of arteries and contain little elastin. Many veins have flaps of connective tissue that extend inward from their walls. These flaps form one-way valves that keep blood flowing toward the heart (see Figure 42.13).

Rather than stretching and contracting elastically, like arteries, the relatively thin walls of venules and veins can expand and contract over a relatively wide range, allowing them to act as blood reservoirs as well as conduits. At times, the venules and veins may contain from 60% to 80% of the total blood volume of the body. The stored volume is adjusted by skeletal muscle contraction and by the valves, in response to metabolic conditions and signals carried by hormones and neurotransmitters **(Figure 42.16).**

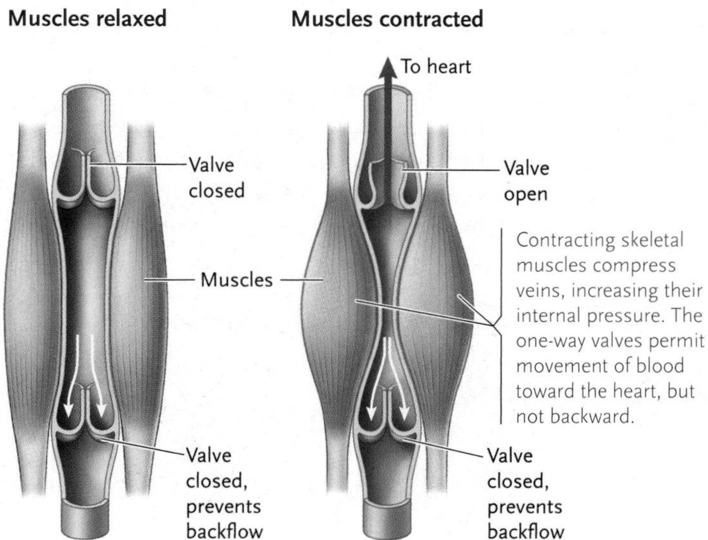
Muscles relaxed
Muscles contracted
To heart
Valve closed
Valve open
Muscles
Contracting skeletal muscles compress veins, increasing their internal pressure. The one-way valves permit movement of blood toward the heart, but not backward.
Valve closed, prevents backflow
Valve closed, prevents backflow

FIGURE 42.16
How skeletal muscle contraction, and the valves inside veins, helps move blood toward the heart. Contracting skeletal muscles compress veins, increasing their internal pressure. The one-way valves permit movement of blood toward the heart, but not backward.

When you sit without moving for long periods of time, as you might during an airline flight, the lack of skeletal muscular activity greatly reduces the return of venous blood to the heart. As a result, the blood pools in the veins of the body below the heart, making the hands, legs, and feet swell. The motionless blood can also form clots, particularly in the veins of the legs, a condition called **deep vein thrombosis.** Deep vein thrombosis often does not cause symptoms, but can cause serious medical problems if a clot breaks loose and moves elsewhere in the body, such as to the lungs. Raising the arms and getting up at intervals to exercise or contracting and relaxing the leg muscles as you sit can relieve this condition.

Disorders of the Circulatory System Are Major Sources of Human Disease

The layer of endothelial cells lining the arteries and veins is normally smooth and does not impede blood flow. However, several conditions, including bacterial and viral infections, chronic hypertension, smoking, and a diet high in fats, can damage the endothelial cells, exposing the underlying smooth muscle tissue, which begins a cycle of injury and repair leading to lesions. (*Focus on Applied Research* in Chapter 3 discussed the relationship between fats and cholesterol and coronary artery disease, and the *Unanswered Question* in Chapter 3 discussed how the enzymes that synthesize fatty acids and cholesterol are regulated in cells.) Thickened deposits of material called *atherosclerotic plaques* may form at the damaged sites **(Figure 42.17).** The plaques, which consist of cholesterol-rich fatty substances, smooth muscle cells, and collagen deposits, reduce the diameter of the blood vessel and impede blood flow. Worse, the damaged endothelial lining may stimulate platelets to adhere and trigger the formation of blood clots. The clots further reduce the vessel diameter and flow and may break loose, along with segments of plaque material, to block finer vessels in other regions of the body.

Atherosclerosis has its most serious effects in the smaller arteries of the body, particularly in the fine coronary arteries that serve the heart muscle. Here, the plaques and clots reduce or block the flow of blood to the heart muscle cells. Serious blockage can cause a heart attack—the death of cardiac muscle cells deprived of blood flow. The blockage of arteries in the brain by plaque material or blood clots released from atherosclerotic arteries is also a common cause of stroke—a loss of critical brain functions because of the death of nerve cells in the brain. Heart attacks and strokes are the most common causes of death in North America and Europe.

The risk of heart attacks and stroke can be reduced by avoiding the conditions that damage the blood vessel endothelium. A diet low in fats and cholesterol, avoidance of cigarette smoke, and a program of exercise can reduce epithelial damage and plaque deposition. Medication and exercise, or exercise alone, can also reduce the effects of hypertension. There are good indications that these preventive programs can also reduce the size of existing atherosclerotic plaques.

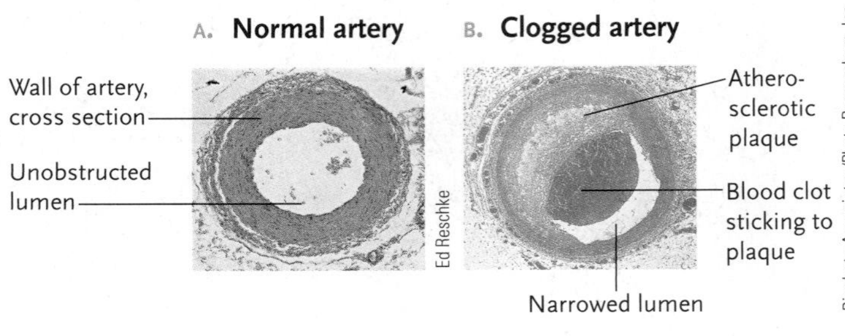
A. Normal artery
B. Clogged artery
Wall of artery, cross section
Unobstructed lumen
Athero-sclerotic plaque
Blood clot sticking to plaque
Narrowed lumen
Ed Reschke
Biophoto Associates/Photo Researchers, Inc.

FIGURE 42.17
Atherosclerosis. **(A)** A normal coronary artery. **(B)** A coronary artery that is partially clogged by an atherosclerotic plaque.

STUDY BREAK 42.4

1. How is blood flow through capillary networks controlled?
2. Explain how, in contrast to most body tissues, the brain does not allow exchange of molecules and ions with blood.
3. Describe the two major mechanisms that drive the exchange of molecules and ions between the capillaries and the interstitial fluid.

42.5 Maintaining Blood Flow and Pressure

Arterial blood pressure is the principal force moving blood to the tissues. Blood pressure must be regulated carefully so that the brain and other tissues receive adequate blood flow, but not so high that the heart is overburdened, risking damage to blood vessels. The three main mechanisms for regulating blood pressure are controlling *cardiac output* (the pressure and amount of blood pumped by the left and right ventricles), the degree of constriction of the blood vessels (primarily the arterioles), and the total blood volume. The sympathetic division of the autonomic nervous system and the endocrine system interact to coordinate these mechanisms.

Cardiac Output Is Controlled by Regulating the Rate and Strength of the Heartbeat

Regulation of the strength and rate of the heartbeat starts at stretch receptors called *baroreceptors* (a type of mechanoreceptor; see Section 39.2), located in the walls of blood vessels. By detecting the amount of stretch of the vessel walls, baroreceptors constantly provide information about blood pressure. The baroreceptors in the cardiac muscle, the aorta, and the carotid arteries (which supply blood to the brain) are the most crucial. Signals sent by the baroreceptors go to the medulla within the brain stem. In response, the brain stem sends signals via the autonomic nervous system that adjust the rate and force of the heartbeat: the heart beats more slowly and contracts less forcefully when arterial pressure is above normal levels, and it beats faster and contracts more forcefully when arterial pressure is below normal levels.

INSIGHTS FROM THE Molecular Revolution

Identifying the Role of a Hormone Receptor in Blood Pressure Regulation Using Knockout Mice

When the body experiences stress, the sympathetic division of the autonomic nervous system stimulates the adrenal medulla to release the hormones epinephrine and norepinephrine. These hormones increase the strength and rate of the heartbeat and change arteriole diameter, causing a temporary increase in blood pressure. The hormones exert their effects by binding to specific membrane-embedded receptors on target cells, activating them and triggering a cellular response via a signal transduction pathway. These G-protein–coupled receptors (see Section 7.4) are known as *adrenergic receptors* because of the adrenal origin of the hormones that bind to them.

Different types of adrenergic receptors are responsible for different responses to epinephrine and norepinephrine. For blood pressure regulation, the α_1-adrenergic receptor is a key receptor. Binding of norepinephrine, and to a lesser extent, epinephrine, to α_1 receptors on arteriolar smooth muscle causes vasoconstriction of arterioles, thereby contributing to an increase in blood pressure.

Researchers have identified three subtypes of α_1-adrenergic receptors: α_{1A}, α_{1B}, and α_{1D}. Some experimental results using drugs suggested that the α_{1D}-adrenergic receptor plays an important role in the control of blood pressure. Because the drugs used could not selectively affect just one subtype of receptor, confirmation of that conclusion was needed from more directed studies.

Research Question

What is the role of the α_{1D}-adrenergic receptor in the regulation of blood pressure by vasoconstriction?

Experiments

Gozoh Tsujimoto and colleagues at the Tokyo University of Pharmacy and Life Sciences, and the National Children's Medical Research Center, Tokyo, Japan, answered the question by making and studying mice with a knockout (deletion) of each of the two copies of the α_{1D} gene encoding the α_{1D} receptor (the technique for making a gene knockout is described in Section 18.2). They compared the cardiovascular functions of the $\alpha_{1D}^-/\alpha_{1D}^-$ knockout mice with normal mice, with the results shown in the figure.

$\alpha_{1D}^-/\alpha_{1D}^-$ knockout mouse

Cardiovascular functions of the knockout mice:

1. Modestly hypotensive (slightly lower-than-normal blood pressure) but a normal heart rate.
2. Normal levels of circulating norepinephrine and epinephrine.
3. Increasing doses of norepinephrine progressively increased blood pressure, but the response was markedly reduced versus normal mice.
4. The response to the vasoconstrictor norepinephrine is considerably reduced. This was shown in experiments in which the reseachers measured the effects of norepinephrine on contraction of segments of the aorta (see also Figure 42.19). Far less contraction was seen than in normal mice. This result showed that the α_{1D} receptor is directly involved in vascular smooth muscle contraction.

Conclusion

The researchers concluded that the α_{1D}-adrenergic receptor participates directly in sympathetic nervous system–driven regulation of blood pressure by vasoconstriction.

Source: A. Tanoue et al. 2002. The α_{1D}-adrenergic receptor directly regulates arterial blood pressure via vasoconstriction. *The Journal of Clinical Investigation* 109:765–775.

The O_2 content of the blood, detected by chemoreceptors in the aorta and carotid arteries, also influences cardiac output. If the O_2 concentration falls below normal levels, the brain stem integrates this information with the signals sent from baroreceptors and issues signals that increase the rate and force of the heartbeat. Too much O_2 in the blood has the opposite effect, reducing the cardiac output.

Hormones Regulate both Cardiac Output and Arteriole Diameter

Hormones secreted by several glands contribute to the regulation of blood pressure and flow. As part of the stress response, the adrenal medulla reinforces the action of the sympathetic nervous system by secreting epinephrine and norepinephrine into the bloodstream (see Section 40.4). Epinephrine in particular raises the blood pressure by increasing the strength and rate of the heartbeat and stimulating vasoconstriction of arterioles in some parts of the body, including the skin, gut, and kidneys. At the same time, by inducing the vasodilation of arterioles that deliver blood to the heart, skeletal muscles, and lungs, epinephrine increases the blood flow to these structures. *Insights from the Molecular Revolution* shows how gene manipulation experiments illuminated the role of a receptor for these hormones in regulating blood pressure.

The adrenal cortex and the posterior pituitary release hormones that regulate blood pressure. Those hormones and their effects are described in Chapter 46.

Local Controls Regulate Arteriole Diameter

Several automated mechanisms operate locally to increase the flow of blood to body regions engaged in increased metabolic activity, such as the muscles of your legs during an extended uphill bike ride. Low O_2 and high CO_2 concentrations, produced by the increased oxidation of glucose and other fuels, induce vasodilation of the arterioles serving muscles. Vasodilation, which increases the flow of blood and the O_2 supply, occurs when a signal

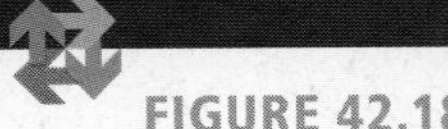

FIGURE 42.18 **Experimental Research**

Demonstration of a Vasodilatory Signal Molecule

Question: What causes arterial vasodilation?

Experiment: In live mammals, the neurotransmitter acetylcholine (see Chapter 37) is a potent vasodilator. Furchgott conducted in vitro experiments with strips of rabbit aorta to determine how acetylcholine-caused vasodilation occurred. He attached the aorta strips to an apparatus that recorded the tension of the strips. Contraction increases tension whereas relaxation (which is equivalent to vasodilation in the intact blood vessel) decreases tension. To the strips he first added norepinephrine to cause them to contract, and then added acetylcholine. He compared two key experimental set ups: (1) aorta strips from which the endothelial cells had been removed by rubbing; and (2) gently handled aorta strips with intact endothelium.

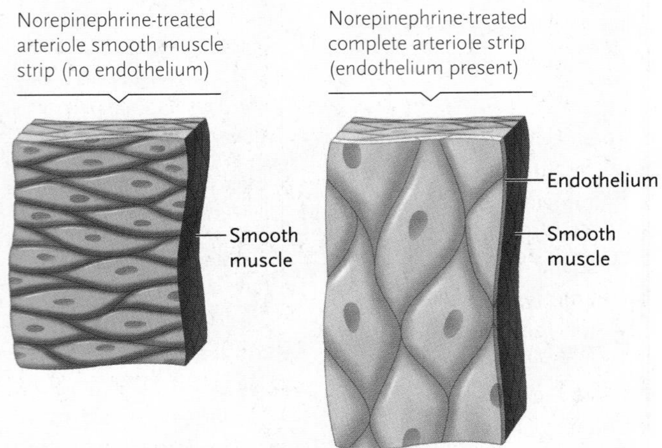
Norepinephrine-treated arteriole smooth muscle strip (no endothelium)
Norepinephrine-treated complete arteriole strip (endothelium present)
Endothelium
Smooth muscle
Smooth muscle
Acetylcholine:
No relaxation
Relaxation

Results: Furchgott's results are shown in the figure. Relaxation occurred on acetylcholine treatment only in the strips with intact endothelium. Furchgott hypothesized that the endothelial cells produced the vasodilatory signal molecule, which he named *endothelium derived relaxing factor (EDRF)*. He obtained more evidence that the endothelium released EDRF by performing a sandwich experiment. In this experiment he reconstituted arteriole tissue by combining isolated smooth muscle and layers of endothelium. The sandwich tissue also relaxed in Furchgott's experimental system, supporting his hypothesis.

Conclusion: Vasodilation of arterioles occurs as a result of release of a signal molecule from the endothelial cells of the arterioles. Subsequently, Furchgott's EDRF signal molecule was determined to be nitric oxide (NO). As you learned in Section 37.3, the impotency drug Viagra aids the erection of the penis by inhibiting an enzyme that normally reduces NO concentration in the penis. The now-higher level of NO results in vasodilation, increasing blood flow to the penis.

Source: R. F. Furchgott and J. V. Zawadzki. 1980. The obligatory role of endothelial cells in the relaxation of arterial smooth muscle by acetylcholine. *Nature* 288:373–376.

molecule is released from arteriole endothelial cells in body regions engaged in increased metabolic activity. In 1980 Robert Furchgott of the State University of New York, Health Science Center at Brooklyn, New York, reported the first evidence for a vasodilatory signal molecule produced by arterial endothelial cells **(Figure 42.18).** Subsequently, Louis Ignarro of the UCLA School of Medicine, and Ferid Murad of the University of Texas Medical School at Houston determined that Furchgott's signal molecule was the gas nitric oxide (NO). The three researchers received the Nobel Prize in 1998 for their "discoveries concerning nitric oxide as a signaling molecule in the cardiovascular system." NO is quickly broken down enzymatically after its release, ensuring that its effects are local.

STUDY BREAK 42.5

1. **Why is it important to regulate arterial blood pressure?**
2. **What are the three main mechanisms for regulating blood pressure?**
3. **How does epinephrine affect blood pressure?**

42.6 The Lymphatic System

Under normal conditions, a little more fluid from the blood plasma in the capillaries enters the tissues than is reabsorbed from the interstitial fluid into the plasma. The **lymphatic system** is an extensive network of vessels that collect the excess interstitial fluid and return it to the venous blood **(Figure 42.19).** The interstitial fluid picked up by the lymphatic system is called **lymph.** This system also collects fats that have been absorbed from the small intestine and delivers them to the blood circulation (see Chapter 45). The lymphatic system is a key component of the immune system.

Vessels of the Lymphatic System Extend throughout Most of the Body

Vessels of the lymphatic system collect the lymph and transport it to *lymph ducts* that empty into veins of the circulatory system. The *lymph capillaries,* the smallest vessels of the lymphatic system, are distributed throughout the body, intermixed intimately with the capillaries of the circulatory system. Although they are several times larger in diameter than the blood capillaries, the walls of lymph capillaries also consist of a single layer of endothelial cells surrounded by a thin network of collagen fibers. Interstitial fluid enters the lymph capillaries—becoming lymph—at sites in their walls where the endothelial cells overlap, forming a flap that is forced open by the higher pressure of the interstitial fluid. The openings are wide enough to admit all components of the interstitial fluid, including infecting bacteria, damaged cells, cellular debris, and lymphocytes.

The lymph capillaries merge into *lymph vessels* containing one-way valves that prevent the lymph from flowing backward. The lymph vessels lead to the thoracic duct and the right lymphatic

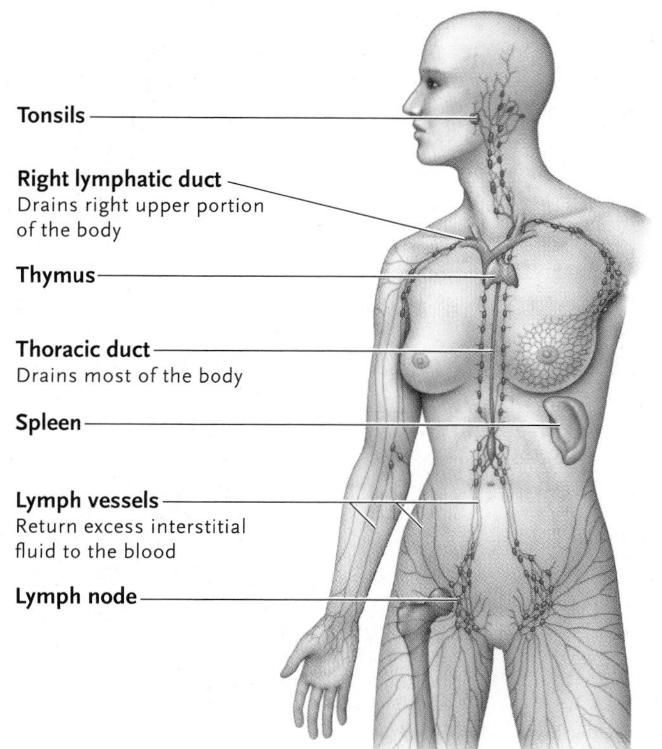
Tonsils
Right lymphatic duct
Drains right upper portion of the body
Thymus
Thoracic duct
Drains most of the body
Spleen
Lymph vessels
Return excess interstitial fluid to the blood
Lymph node

FIGURE 42.19
The human lymphatic system. Patches of lymphoid tissue in the small intestine and in the appendix also are part of the lymphatic system.

duct (see Figure 42.19), which empty the lymph into a vein beneath the clavicles (collarbones), adding it to the plasma in the vein.

Movements of the skeletal muscles adjacent to the lymph vessels and breathing movements help move the lymph through the vessels, just as they help move the blood through veins. Over a day's time, the lymphatic system returns about 3 to 4 L of fluid to the bloodstream.

Lymphoid Tissues and Organs Act as Filters and Participate in the Immune Response

The tissues and organs of the lymphatic system include the *lymph nodes*, the *spleen*, the *thymus*, and the *tonsils*. They play primary roles in filtering viruses, bacteria, damaged cells, and cellular debris from the lymph and bloodstream, and in defending the body against infection and cancer. Patches of lymphoid tissue are also scattered in other regions of the body, such as the small intestine and the appendix.

The **lymph nodes** are small, bean-shaped organs spaced along the lymph vessels and clustered along the sides of the neck, in the armpits and groin, and in the center of the abdomen and chest cavity (see Figure 42.19). Spaces in the nodes contain macrophages, a type of leukocyte that engulfs and destroys cellular debris and infecting bacteria and viruses in the lymph. The lymph nodes also contain other leukocytes, which produce antibodies that aid in the destruction of invading pathogens (discussed more in Chapter 43). Cancer cells that lodge in the nodes may be destroyed or may remain to grow and divide, forming new tumors within the nodes. Therefore, to reduce the risk of cancer spread, lymph nodes near a tumor typically are inspected and may be removed during tumor surgery.

The lymph nodes may become enlarged and painful if large numbers of bacteria or viruses carried by the lymph become trapped inside them. A physician usually checks for swollen nodes, particularly in the neck, armpits, and groin, as indicators of an infection in the region of the body served by the nodes.

STUDY BREAK 42.6 <

What is lymph, and how does it enter the lymph capillaries?

UNANSWERED QUESTIONS

How did vertebrate blood clotting evolve?

Vertebrate blood clotting is a complex process involving about two dozen different proteins found in the blood plasma. The system has the properties of a biochemical amplifier in that the exposure of a tiny amount of tissue initiates a series of proteolytic events, one protease activating another successively, with the climax being a large amount of localized thrombin that transforms fibrinogen into a fibrin clot.

Many years ago when I was a graduate student working in a laboratory devoted to blood proteins, I asked myself the question, How could blood clotting ever have evolved? The process seemed much too complicated to have been concocted in one fell swoop, so, I reasoned, it must have begun in a simpler fashion and gradually become more complicated. Certainly, other people had been asking similar questions about complex organs—the evolution of the eye, for example, had been considered by many scientists, including Charles Darwin—but facts about the evolution of individual proteins were just beginning to emerge.

In particular, Vernon Ingram, a scientist working at MIT, had just determined the amino acid sequences of the alpha and beta chains of hemoglobin and found that these two proteins were about 45% identical. They must be the products of a gene duplication, he reported. Given his observation, it seemed to me that the proteases involved in blood clotting ought also to be the products of gene duplication. At the time, none of their amino acid sequences was known, but several had unique properties that distinguished them from other proteases, so it stood to reason that they were related. Yet, some nonproteolytic proteins were also involved, and they must have been added to the process independently.

So where to start? I decided to compare the blood clotting process in a wide range of animals. As it happens, most animals, invertebrate and vertebrate alike, have a kind of blood (see Section 42.2), and in most cases it can be coagulated by various stressful events. But the vertebrate process looked unique to me and must have evolved independently of the system that occurs in lobsters (Crustacea), for example. Among the vertebrates, the most primitive (early diverging) creature I could get my hands on was the lamprey (see Section 30.4, where its place in the vertebrate lineage is discussed).

Because I was hoping to find a simple, predecessor scheme, I was a little disappointed to find that lampreys have a rather sophisticated coagulation process that involved many of the proteins observed in mammals. Certainly, a small amount of tissue factor provoked a thrombin generation that converted fibrinogen to fibrin, just as in humans, at least in a general way. At the time there was no way to determine whether lampreys use the equivalent of *all* the clotting factors on the way to generating thrombin. Some of the most important proteins in the human system occur in minute amounts, and there was no possibility of ever isolating them from the lamprey, short of collecting several barrels of lamprey blood.

Many years later, after the sequences for many of the clotting factors had been reported for humans, my colleague Da-Fei Feng and I were able to align the sequences with a computer and make a phylogenetic tree. It fit the notion of a series of gene duplications very well. Moreover, it implied that most of the duplications had occurred a long time ago, at the very dawn of the vertebrates.

In 2003, the complete genomic sequence of a modern bony fish, the pufferfish, was determined, and one could scan through it and see what genes it has. It has all but a few of the more peripheral clotting proteins found in humans, the central theme being the same. However, the amount of sequence difference between human and pufferfish proteins compared with the degree of difference observed between the duplicated genes suggested some of the gene duplications had occurred not long before the appearance of bony fish. Why not look at the lamprey genome, which diverged 50 million to 100 million years before the appearance of bony fish, to see if the preduplication genes were there? The reason is that the lamprey genome has not been totally sequenced.

Recently, various genome centers around the world have begun maintaining "trace databases." These are uncurated collections of raw DNA sequences determined by random shotgun methods and robotized sequencers. The data are not assembled in any way, and each entry is at best a fragment of a gene. Several hundred organisms, from bacteria to monkeys, are being logged automatically. Among them is the lamprey! So now one can get a glimpse of what genes the lamprey has. We have been scrutinizing this database, even while recognizing its limitations. The reason is that I am getting on in years and can't wait for some mammoth operation with the resources to assemble all the lamprey DNA fragments into a complete genome.

My initial searching of the lamprey trace database showed that the lamprey may lack at least two of the mainline clotting factors (Factors VIII and IX), each of which in other vertebrates is the result of a gene duplication. Now, very recently, a draft assembly of the lamprey genome sequence has been posted, and it bears out these conclusions completely. Furthermore, a Japanese team has been trying to clone clotting factors from another species of lamprey, and the ones they have found match perfectly with the ones found in the database. The missing ones are still missing!

Although the step-by-step evolution of vertebrate blood clotting factors is still technically an "unanswered question," I'm hoping these efforts will spur others to go for a more convincing, fully assembled lamprey genome that will cement the case.

Think Critically

1. Will studying lamprey clotting factors have much potential in leading to a genetic cure for hemophilia in humans?
2. Deep venous thrombosis (DVT) is a condition in which a blood clot forms in a vein deep inside the body (mainly in the lower extremities), often after major surgery or extended bed rest. (Surgery patients are given heparin, an anticoagulant, to forestall clot formation.) Explain very generally how a DVT occurs (i.e., how a blood clot forms), and speculate on the use of lampreys to research DVT.

Russell Doolittle

Russell Doolittle is professor emeritus at the University of California at San Diego. Finding out how blood clotting evolved is one of his major research interests. To learn more about Dr. Doolittle, go to http://www-biology.ucsd.edu/faculty/doolittle.html.

REVIEW KEY CONCEPTS

Go to **CENGAGENOW** at www.cengage.com/login to access quizzing, animations, exercises, articles, and personalized homework help.

42.1 Animal Circulatory Systems: An Introduction

- Only the simplest invertebrates—the sponges, cnidarians, and flatworms—have no circulatory systems (Figure 42.2).
- Animals with circulatory systems have a muscular heart that pumps a specialized fluid, such as blood, from one body region to another through tubular vessels. The blood carries O_2 and nutrients to body tissues and carries away CO_2 and wastes.
- Most invertebrates have an open circulatory system, in which the heart pumps hemolymph into vessels that empty into body spaces called sinuses before returning to the heart. Some invertebrates and all vertebrates have a closed system, in which the blood is confined in blood vessels throughout the body and does not mix directly with the interstitial fluid (Figure 42.3).
- In invertebrates, open circulatory systems occur in arthropods and most mollusks, whereas closed circulatory systems occur in annelids and in mollusks such as squids and octopuses. In vertebrates, the circulatory system has evolved from a heart with a single series of chambers, pumping blood through a single circuit, to a double heart that pumps blood through separate pulmonary and systemic circuits (Figures 42.4 and 42.5).

Animation: Types of circulatory systems

Animation: Circulatory systems

42.2 Blood and Its Components

- Mammalian blood is a fluid connective tissue consisting of erythrocytes, leukocytes, and platelets, suspended in a fluid matrix, the plasma (Figure 42.6).
- Plasma contains water, ions, dissolved gases, glucose, amino acids, lipids, vitamins, hormones, and plasma proteins. The plasma proteins include albumins, globulins, and fibrinogen.
- Erythrocytes contain hemoglobin, which transports O_2 between the lungs and all body regions (Figure 42.7).
- Leukocytes defend the body against infecting pathogens.
- Platelets are functional cell fragments that trigger clotting reactions at sites of damage to the circulatory system.

Animation: White blood cells

Animation: ABO compatibilities

Animation: Rh factor and pregnancy

42.3 The Heart

- The mammalian heart is a four-chambered pump. Two atria at the top of the heart pump the blood into two ventricles at the bottom of the heart, which pump blood into two separate pulmonary and systemic circuits of blood vessels (Figures 42.9 and 42.10).
- In both circuits, the blood leaves the heart in large arteries, which branch into smaller arteries, the arterioles. The arterioles deliver the blood to capillary networks, where substances are exchanged between the blood and the interstitial fluid. Blood is collected from the capillaries in small veins, the venules, which join into larger veins that return the blood to the heart (Figure 42.10).
- Contraction of the ventricles pushes blood into the arteries at a peak (systolic) pressure. Between contractions, the blood pressure in the arteries falls to a minimum (diastolic) pressure. The systole–diastole sequence is the cardiac cycle (Figure 42.11).
- Contraction of the atria and ventricles is initiated by signals from the SA node (pacemaker) of the heart (Figure 42.12).

Animation: The human heart

Animation: Human blood circulation

Animation: Cardiac cycle

Animation: Cardiac conduction

Animation: Examples of ECGs

42.4 Blood Vessels of the Circulatory System

- Blood is carried from the heart to body tissues in arteries; small branches of arteries, the arterioles, deliver blood to the capillaries, where substances are exchanged with the interstitial fluid. The blood is collected from the capillaries in venules and then returned to the heart in veins (Figure 42.13).
- The walls of arteries consist of an inner endothelial layer, a middle layer of smooth muscle, and an outer layer of elastic fibers. The smallest arteries, the arterioles, constrict and dilate to regulate blood flow and pressure into the capillaries.
- Capillary walls consist of a single layer of endothelial cells. Blood flow through capillaries is controlled by variation in contraction of the smooth muscles of arterioles and precapillary sphincters (Figure 42.14).
- In the capillary networks, the rate of blood flow is considerably slower than that in arteries and veins. This maximizes the time for exchange of substances between blood and tissues. Diffusion along concentration gradients and bulk flow drive the exchange of substances (Figure 42.15).
- Venules and veins have thinner walls than arteries, allowing the vessels to expand and contract over a wide range. As a result, they act as blood reservoirs as well as conduits.
- The return of blood to the heart is aided by pressure exerted on the veins when surrounding skeletal muscles contract and by respiratory movements. One-way valves in the veins prevent the blood from flowing backward (Figure 42.16).

Animation: Major human blood vessels

Animation: Capillary forces

Animation: Vessel anatomy

Animation: Vein function

Animation: Hemostasis

42.5 Maintaining Blood Flow and Pressure

- Blood pressure and flow are regulated by controlling cardiac output, the degree of blood vessel constriction (primarily arterioles), and the total blood volume. The autonomic nervous system and the endocrine system coordinate these mechanisms.
- Regulation of cardiac output starts with baroreceptors, which detect blood pressure changes and send signals to the medulla. In response, the brain stem sends signals via the autonomic nervous system that alter the rate and force of the heartbeat.
- Hormones secreted by several glands contribute to the regulation of blood pressure and flow.
- Local controls respond primarily to O_2 and CO_2 concentrations in tissues. Low O_2 and high CO_2 concentration causes dilation of arteriole walls, increasing the arteriole diameter and blood flow. High O_2 and low CO_2 concentrations have the opposite effects. NO released by arterial endothelial cells acts locally to increase arteriole diameter and blood flow (Figure 42.18).

Animation: Measuring blood pressure

42.6 The Lymphatic System

- The lymphatic system, a key component of the immune system, is an extensive network of vessels that collect excess interstitial fluid—which becomes lymph—and returns it to the venous blood (Figure 42.19).
- The tissues and organs of the lymphatic system include the lymph nodes, the spleen, the thymus, and the tonsils. They remove viruses, bacteria, damaged cells, and cellular debris from the lymph and bloodstream, and defend the body against infection and cancer.

Animation: Human lymphatic system

Animation: Lymph vascular system

UNDERSTAND AND APPLY

Test Your Knowledge

1. Compared with vertebrates, most invertebrates:
   a. lead more mobile lives.
   b. require a higher level of oxygen.
   c. have more complex layers of cells.
   d. have blood and interstitial fluid mixing directly in body spaces.
   e. require faster delivery and greater quantities of nutrients.
2. Which circulatory system description best matches the group(s) of animals?
   a. Cephalopod mollusks such as squids and octopuses have open circulatory systems with ventricles that pump blood away from the heart.
   b. Fishes have a single-chambered heart with an atrium that pumps blood through gills for oxygen exchange.
   c. Amphibians have the most oxygenated blood in the pulmocutaneous circuit and the most deoxygenated blood in the systemic circuit.
   d. Amphibians and reptiles use a two-chambered heart to separate oxygenated and deoxygenated blood.
   e. Birds and mammals pump blood to separate pulmonary and systemic systems from two separate ventricles in a four-chambered heart.
3. A healthy student from the coastal city of Boston enrolls at a college in Boulder, Colorado, a mile above sea level. An analysis of her blood in her first months at college would show:
   a. decreased macrophage activity.
   b. increased secretion of erythropoietin by the kidneys.
   c. increased signaling ability of platelets.
   d. anemia caused by malfunctioning erythrocytes.
   e. increased mitosis of leukocytes.

4. A characteristic of blood circulation through or to the heart is that:
   a. the superior vena cava conveys blood to the head.
   b. the inferior vena cava conveys blood to the right atrium.
   c. the pulmonary arteries convey blood from the lungs to the left atrium.
   d. the pulmonary veins convey blood into the left ventricle.
   e. the aorta branches into two coronary arteries that convey blood from heart muscle.
5. The heartbeat includes:
   a. the systole when the heart relaxes and fills.
   b. the diastole when the heart contracts and empties.
   c. pressure that causes the AV valves to open, filling the ventricles.
   d. rising pressure in the ventricles to open the AV valves and close the SL valves.
   e. the "lub" sound when the SL valves open and the "dub" sound when the AV valves close.
6. Keeping the mammalian cardiac cycle balanced is/are:
   a. an AV node between the right atria atrium and right ventricle, which induces a wave of contraction.
   b. pacemaker cells, which compose the AV node and signal the SA node.
   c. an insulating layer that isolates the SA node from the right atrium.
   d. ion channels in pacemaker cells, which close to depolarize their plasma membranes.
   e. neurogenic stimuli from the nervous system.
7. Hydrostatic pressure is best described as:
   a. the uncoordinated contractions that occur during heart attacks.
   b. a premature ventricular contraction that signifies a skipped beat.
   c. a high point of pressure called diastolic blood pressure.
   d. an arterial vasodilation caused by carbon dioxide.
   e. the pressure of blood on the walls of arteries.
8. Characteristics of veins and venules are:
   a. thick walls.
   b. large muscle mass in the walls.
   c. a large quantity of elastin in the walls.
   d. low blood volume compared with arteries.
   e. one-way valves to prevent backflow of blood.
9. When capillaries exchange substances:
   a. red blood cells move through the capillary lumen in double file.
   b. blood flow resistance is lower than it is in arteries and veins.
   c. water, ions, glucose, and erythrocytes pass freely between blood and tissues.
   d. diffusion along a concentration gradient and bulk flow are operating.
   e. diffusion is greatest closest to the arterioles.
10. How do lymphatic vessels and capillaries differ from other vessels and capillaries in the circulatory system?
   a. Lymph capillaries are larger in diameter than blood capillaries.
   b. Lymph capillary walls contain several layers of endothelial cells, which allows for efficient filtering of lymph.
   c. Lymph vessels are only distributed in specific regions of the body.
   d. Movement of lymph does requires skeletal muscle movement.
   e. The lymphatic system moves up to 8 L of fluid through the bloodstream daily.

Discuss the Concepts

1. *Aplastic anemia* develops when certain drugs or radiation destroy red bone marrow, including the stem cells that give rise to erythrocytes, leukocytes, and platelets. Predict some symptoms a person with aplastic anemia would be likely to develop. Include at least one symptom related to each type of blood cell.
2. In addition to the engine exhaust of boats and cars, carbon monoxide is also a component of cigarette smoke. What might be the impact of this phenomenon on a smoker's health?
3. In some people, the pressure of the blood pooling in the legs leads to a condition called *varicose veins,* in which the veins stand out like swollen, purple knots. Explain why this might happen, and why veins closer to the leg surface are more susceptible to the condition than those in deeper leg tissues.

Design and Experiment

Mice in which the apolipoprotein E gene has been knocked out (deleted) by genetic engineering methods have high levels of plasma cholesterol and readily develop atherosclerosis, particularly on diets high in cholesterol. The immunosuppressant drug rapamycin is being touted also to be a drug that can affect atherosclerosis. Design an experiment to determine whether and at what dose rapamycin is effective in reducing atherosclerosis caused by dietary cholesterol.

Interpret the Data

Nifedipine is an antihypertension medication that is also used to reduce the workload on the heart. The study below compares the effects of two different types of capsules used to administer the drug to patients (GITS—red line, and Cotracten X—blue line). Each graph shows changes with time following initial administration of the capsules. Figure A compares the levels of nifedipine in the blood with each capsule type. Figures B and C show the ability of the different capsules to reduce workload on the heart by altering blood pressure (Figure B) or heart rate (Figure C).

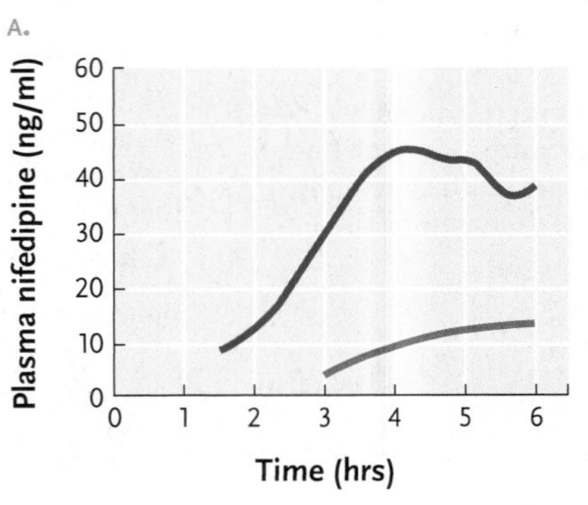
A.
Plasma nifedipine (ng/ml)
60
50
40
30
20
10
0
0 1 2 3 4 5 6
Time (hrs)

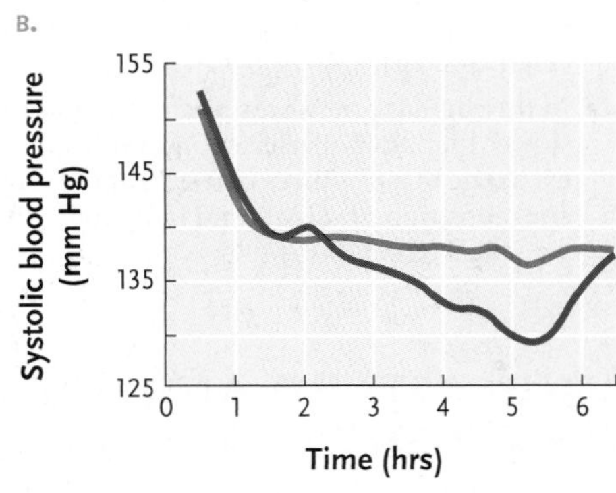
B.
Systolic blood pressure (mm Hg)
155
145
135
125
0 1 2 3 4 5 6
Time (hrs)

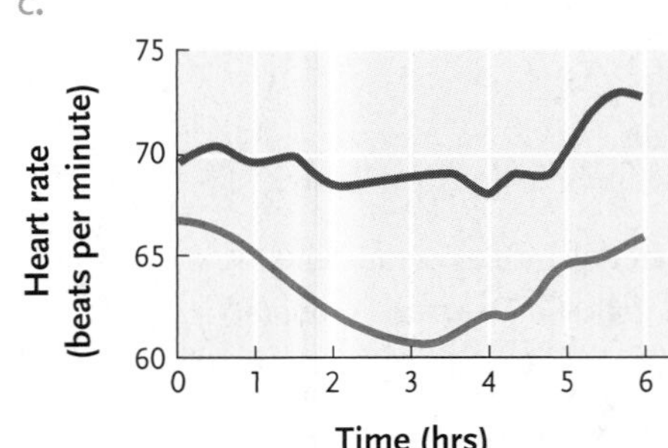
C.
Heart rate (beats per minute)
75
70
65
60
0 1 2 3 4 5 6
Time (hrs)

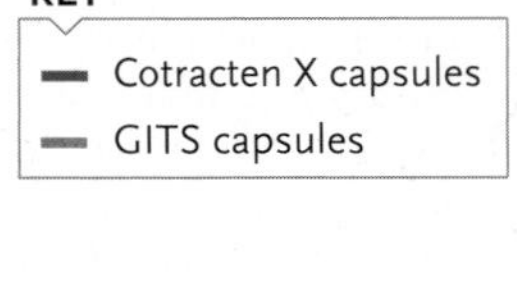
KEY
Cotracten X capsules
GITS capsules

1. Are elevated plasma levels of nifedipine (above 10 ng/mL) required for the drug to perform its actions?
2. Do both types of capsule appear to cause a reduction in the workload of the heart?
3. Which type of capsule would you expect to have fewer side-effects?

Source: M. J. Brown and C. B. Toal. 2008. Formulation of long-acting nifedipine tablets influences the heart rate and sympathetic nervous system response in hypertensive patients. *British Journal of Clinical Pharmacology* 65:646-652.

Apply Evolutionary Thinking

What is the evolutionary advantage of closure of the septum between the two ventricles to create a double circulatory system?

Express Your Opinion

Cardiopulmonary resuscitation (CPR) can make the difference between life and death after a cardiac arrest or a heart attack. Should public high schools in your state require all students to take a course in CPR? Is such a course worth diverting time and resources from the basic curriculum? Go to www.cengage.com/login to investigate both sides of the issue and then vote.

43

STUDY OUTLINE

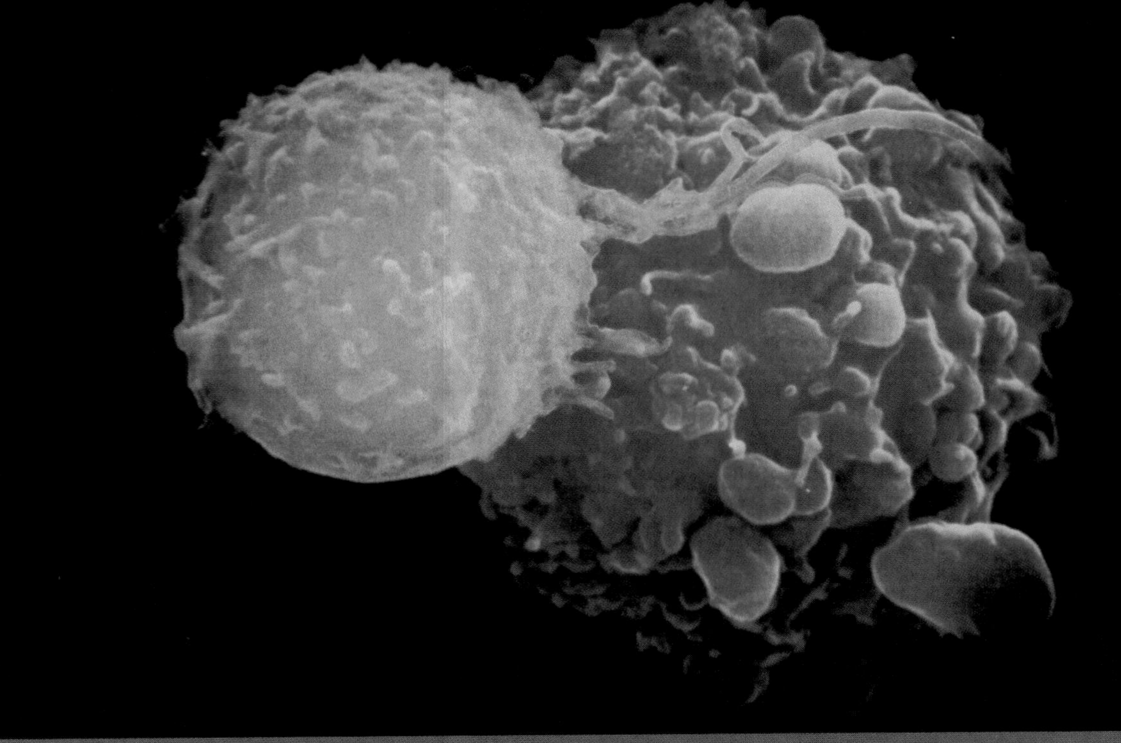

Death of a cancer cell. A cytotoxic T cell (orange) induces a cancer cell (mauve) to undergo apoptosis (programmed cell death). Cytotoxic T cells are part of the body's immune response system programmed to seek out, attach themselves, and kill cancer cells and pathogen-infected host cells.

Defenses against Disease

Why It Matters. . . Acquired immune deficiency syndrome (AIDS), which was first identified in the early 1980s, now infects about 33 million people worldwide, and continues to spread. Thousands of health-care workers, physicians, and researchers have joined the effort to control AIDS and develop effective treatments. One important aim is to develop an anti-AIDS *vaccine*—a substance that, when swallowed or injected, provides protection against infection by HIV (human immunodeficiency virus), which causes the disease.

The development of vaccines began with efforts to control smallpox, a dangerous and disfiguring infectious viral disease that once infected millions of people worldwide. As early as the twelfth century, healthy individuals in China sought out people who were recovering from mild smallpox infections, ground up scabs from their lesions, and inhaled the powder or pushed it into their skin. Variations on this treatment were effective in protecting many people against smallpox infection.

In 1796, an English country doctor, Edward Jenner, used a more scientific approach. He knew that milkmaids never got smallpox if they had contracted cowpox, a similar but mild disease of cows that can be transmitted to humans. Jenner decided to see if a deliberate infection with cowpox would protect humans from smallpox. He scratched material from a cowpox sore into a boy's arm, and 6 weeks later, after the cowpox infection had subsided, he scratched fluid from human smallpox sores into the boy's skin. (Jenner's use of the boy as an experimental subject would now be considered unethical.) The boy remained free from smallpox. Jenner carried out additional, carefully documented case studies with other patients with the same results. His technique became the basis for worldwide **vaccination** (*vacca* = cow) against smallpox. With improved vaccines, smallpox has now been eradicated from the human population.

Vaccination takes advantage of the **immune system** (*immunis* = exempt), the natural protection that is our main defense against infectious disease. This chapter focuses on the immune system and other defenses against infection, such as the skin. Our description emphasizes human and other mammalian systems, in which most of the scientific discoveries revealing the structure and function of the immune system have been made. At the end of the chapter we compare mammalian systems with the protective systems of nonmammalian vertebrates and invertebrates. Recognize that natural protection systems are not confined to animals. That is, all organisms have some ability to defend themselves against infectious agents such as viruses. For example, restriction enzymes (see Section 18.1), the molecular biologist's tool for cloning and other DNA manipulations, are a bacterium's natural defense against infecting bacteriophages. Plants also have defense mechanisms; they are described in Chapter 35. <

43.1 Three Lines of Defense against Pathogens

Every organism is constantly exposed to *pathogens,* disease-causing viruses or organisms such as infectious bacteria, protists, fungi, and parasitic worms. Humans and other mammals have three lines of defense against these threats:

1. **Physical barriers;** they are not part of the immune system.
2. The **innate immune system,** the inherited mechanisms that protect the body from many kinds of pathogens in a nonspecific way.
3. The **adaptive immune system,** the inherited mechanisms leading to the synthesis of molecules that target pathogens in a specific way.

Reaction to an infection takes minutes in the case of the innate immune system versus days for the adaptive immune system. Note that the three lines of defense are not either–or systems, they are cumulative actions of the body to defend against pathogens, meaning that physical barriers and the innate immune system still function against a specific pathogen even when the adaptive immune system is operational.

Epithelial Surfaces Are Anatomical Barriers That Help Prevent Infection

The body surface—the skin covering the body exterior and the epithelial surfaces covering internal body cavities and ducts, such as the lungs and intestinal tract—is the first line of defense. The tight junctions between epithelial cells of the body surface keep most pathogens (as well as toxic substances) from entering the body.

Many epithelial surfaces are coated with a mucus layer secreted by the epithelial cells that protects against pathogens as well as toxins and other chemicals. In the respiratory tract, ciliated cells constantly sweep the mucus, with its trapped bacteria and other foreign matter, into the throat, where it is coughed out or swallowed.

Many of the body cavities lined by mucous membranes have environments that are hostile to pathogens. For example, the strongly acidic environment of the stomach kills most bacteria and destroys many viruses that are carried there. Most of the pathogens that survive the stomach acid are destroyed by the digestive enzymes and bile secreted into the small intestine. The vagina, too, is acidic, which prevents many pathogens from surviving there. The mucus coating in some locations contains the enzyme lysozyme, which was secreted by the epithelial cells. Lysozyme breaks down the walls of some bacteria, causing them to lyse.

Two Immune Systems Protect the Body from Pathogens That Have Crossed External Barriers

The body's second line of defense is a series of generalized internal chemical, physical, and cellular reactions that attack pathogens that have breached the first line. These defenses include inflammation, which creates internal conditions that inhibit or kill many pathogens, and specialized cells that engulf or kill pathogens or infected body cells.

Innate immunity is the term for this initial response by the body to eliminate cellular pathogens, such as bacteria and viruses, and prevent infection. You are born with an innate immune system. Innate immunity provides an immediate, *nonspecific* response; that is, it targets any invading pathogen and has no memory of prior exposure to the pathogen. ("Memory" here means immunological memory, in which the body has a cellular record that it has been exposed previously.) It provides some protection against invading pathogens while a more powerful, specific response system is mobilized.

The third and most effective line of defense, **adaptive** (also called **acquired**) **immunity,** is *specific:* it recognizes individual pathogens and mounts an attack that directly neutralizes or eliminates them. It is so named because it is stimulated and shaped by the presence of a specific pathogen or foreign molecule. This mechanism takes several days to become protective. Adaptive immunity is triggered by specific molecules on pathogens that are recognized as being foreign to the body. The body retains a cellular memory of its first exposure to a foreign molecule, which enables it to respond more quickly if the pathogen is encountered again in the future.

Innate immunity and adaptive immunity together constitute the immune system, and the defensive reactions of the system are termed the **immune response.** Functionally, the two components of the immune system interconnect and communicate at the chemical and cellular levels. The immune system is the product of a long evolutionary history of compensating adaptations by both pathogens and their targets. Over millions of years of vertebrate history, the mechanisms by which pathogens attack and invade have become more efficient, but the defenses of animals against the invaders have kept pace.

White Blood Cells Are Key Participants in Both Immune Systems

White blood cells—leukocytes—and their derivatives, along with several types of plasma proteins, are responsible for the activities of the two immune systems. **Table 43.1** lists the major types of leukocytes and their functions. Most types of leukocytes are present in the blood only transiently as they travel to the tissues for their roles in the immune systems. Note that several of the types of white blood cells are **phagocytes.** Recall from Section 5.3 that phagocytes are cells that engulf bacteria or other cellular debris; the engulfment process is called **phagocytosis.**

TABLE 43.1 Major Types of Leukocytes and Their Functions

Type of Leukocyte	Function
Monocyte 	Differentiates into a macrophage when released from blood into damaged tissue
Macrophage 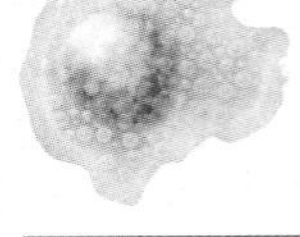	Phagocyte that engulfs infected cells, pathogens, and cellular debris in damaged tissues; helps activate lymphocytes carrying out immune response
Neutrophil 	Phagocyte that engulfs pathogens and tissue debris in damaged tissues
Eosinophil 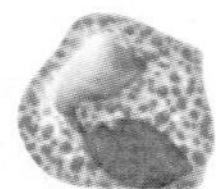	Secretes substances that kill eukaryotic parasites such as worms
Lymphocyte 	Main subtypes involved in innate and adaptive immunity are natural killer (NK) cells, B cells, plasma cells, helper T cells, and cytotoxic T cells. NK cells function as part of innate immunity to kill virus-infected cells and some cancerous cells of the host. The other cell types function as part of adaptive immunity: they produce antibodies, destroy infected and cancerous body cells, and stimulate macrophages and other leukocyte types to engulf infected cells, pathogens, and cellular debris
Basophil 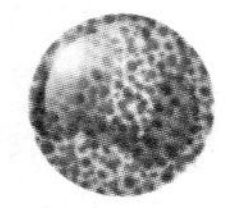	Located in blood, responds to IgE antibodies in an allergic response by secreting histamine, which stimulates inflammation

Most leukocytes originate from stem cells in the bone marrow, from where they are released into the blood; the exception is lymphocytes (see Section 42.2 and Figure 42.7), which originate from stem cells in the lymphoid tissues. Lymphoid tissues are part of the lymphatic system, which was described in Section 42.6. As a brief recap, the lymphatic system is an extensive network of vessels that collects excess interstitial fluid and returns it to the venous blood (see Figure 42.20). The interstitial fluid picked up is called lymph when it is within the lymphatic system. The lymphatic system, which includes the lymph nodes, the spleen, the thymus, and the tonsils, has a function in immunity. Potential pathogens that gain access to the lymph are filtered through the lymph nodes where white blood cells act to inactivate them. The spleen, the largest lymphoid tissue, performs a similar function to inactivate pathogens in blood coming through it.

STUDY BREAK 43.1

1. **What features of epithelial surfaces protect against pathogens?**
2. **What are the key differences between innate immunity and adaptive immunity?**

43.2 Innate Immunity: Nonspecific Defenses

In most cases, the body needs 7 to 10 days to develop a fully effective immune response against a pathogen that is invading the body for the first time. Innate immunity holds off invading pathogens in the meantime, killing or containing them until adaptive immunity comes fully into play. We have already discussed the body's anatomical barriers. Now let us look at the internal mechanisms of innate immunity: secreted molecules and cellular components. As you will see, cellular pathogens (such as bacteria) and viral pathogens elicit different responses.

Innate Immunity Provides an Immediate, General Defense against Invading Cellular Pathogens

Cellular pathogens—typically microorganisms—usually enter the body when injuries break the skin or epithelial surfaces. How does the host body recognize the pathogen as foreign? The answer is that the host has mechanisms to distinguish self from nonself. The innate immune system recognizes particular molecules that are common to many pathogens but absent in the host. For example, when the host body detects the lipopolysaccharide of gram-negative bacteria, it responds immediately to combat the pathogen.

Several types of specific cell-surface receptors in the host recognize the various types of molecules on microbial pathogens. The response depends on the receptor. For some receptors, the response is secretion of *antimicrobial peptides,* which kill the microbial pathogen. Other receptors trigger the host cell to engulf and destroy the pathogen, initiating inflammation. Still others activate the *complement system.*

ANTIMICROBIAL PEPTIDES All of our epithelial surfaces, namely skin, the lining of the gastrointestinal tract, the lining of the nasal passages and lungs, and the lining of the genitourinary tracts, are protected by antimicrobial peptides called *defensins.* Epithelial cells of those surfaces secrete defensins when attacked by a microbial pathogen. The defensins attack the plasma membranes of the pathogens, eventually disrupting them, thereby killing the cells. In particular, defensins play a significant role in innate immunity of the mammalian intestinal tract.

INFLAMMATION A tissue's rapid response to injury, including infection by most pathogens, involves **inflammation** (*inflammare* = to set on fire), the heat, pain, redness, and swelling that initially, or exclusively occur at the site of an infection. Inflammation at the site of an infection is called **local inflammation.** This inflammatory response is highly similar regardless of the triggering event, be it pathogen, mechanical damage, or chemical injury. Overall, the response operates to isolate the injured area and stop the spread of damage.

Several interconnecting mechanisms initiate local inflammation **(Figure 43.1).** Let us consider bacteria entering a tissue as a result of a wound. Cell-surface receptors on **macrophages** ("big eaters"; see Table 43.1)—a type of phagocyte—in the area of the wound recognize and bind to surface molecules on the pathogen, activating the macrophage to phagocytize (engulf) the pathogen (see Figure 43.1, step 1). Activated macrophages also secrete **cytokines,** molecules that bind to receptors on other host cells and, through signal transduction pathways, trigger a response. Usually, too few macrophages are present in the area of a bacterial infection to remove all of the bacteria.

The tissue damage also activates **mast cells** which then release **histamine,** an inflammatory signaling molecule (step 2). The histamine, along with cytokines from activated macrophages, dilates local blood vessels around the infection site and

FIGURE 43.1
The steps producing inflammation. The colorized micrograph on the left shows a macrophage engulfing a yeast cell.

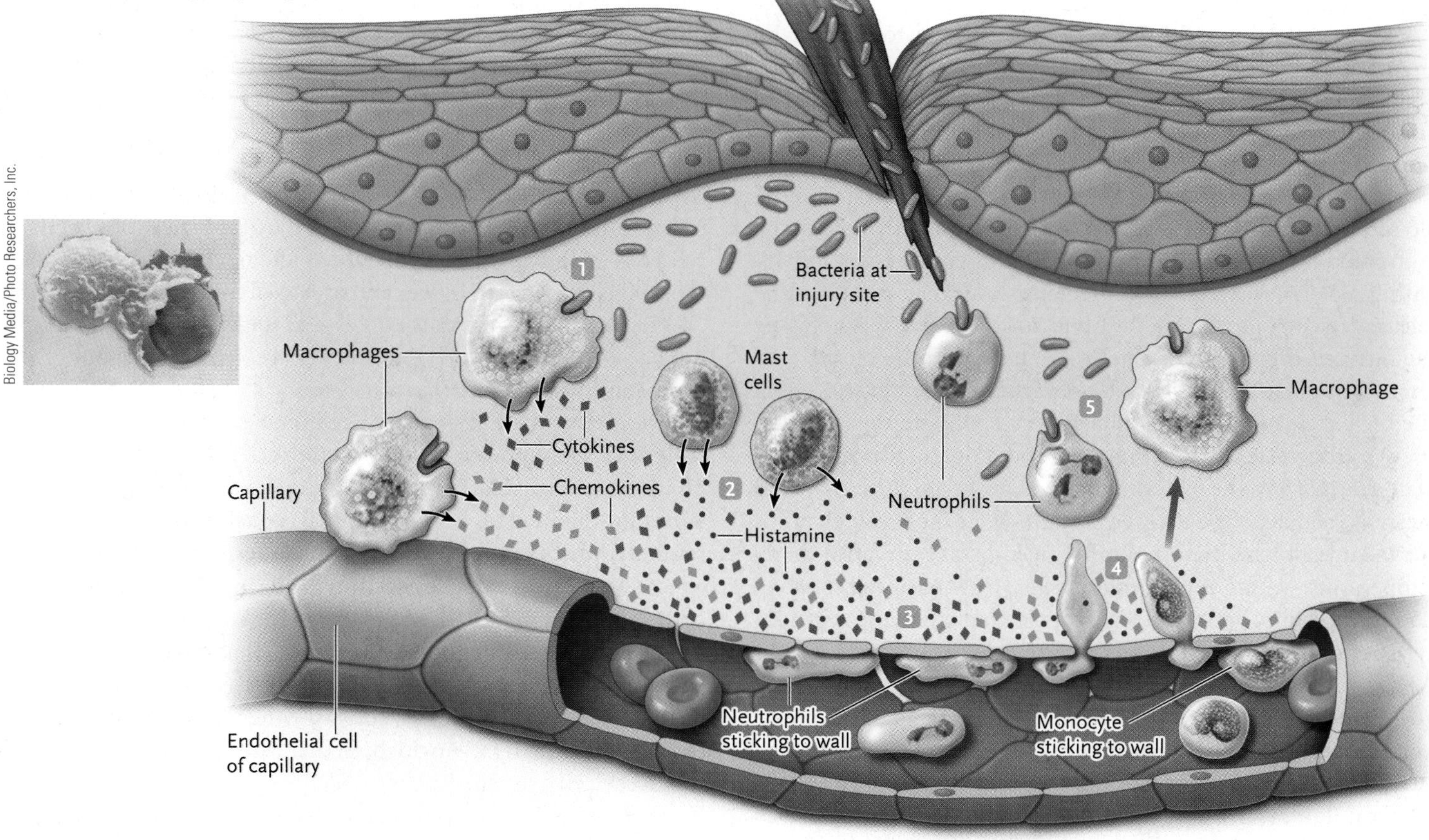

Biology Media/Photo Researchers, Inc.
Bacteria at injury site
Macrophages
Mast cells
Macrophage
Cytokines
Capillary
Chemokines
Neutrophils
Histamine
Neutrophils sticking to wall
Monocyte sticking to wall
Endothelial cell of capillary
1
2
3
4
5
1 A break in the skin introduces bacteria, which reproduce at the wound site. Activated resident macrophages engulf the pathogens and secrete cytokines and chemokines.
2 Mast cells in the area are activated by the tissue damage and release histamine.
3 Histamine and cytokines dilate local blood vessels and increase their permeability. The cytokines also make the blood vessel wall sticky, causing neutrophils and monocytes to attach.
4 Chemokines attract neutrophils and monocytes, which squeeze out between cells of the blood vessel wall and migrate to the infection site.
5 Monocytes differentiate into macrophages. Neutrophils and macrophages engulf the pathogens and destroy them.

increases their permeability, which increases blood flow and leakage of fluid from the vessels into body tissues (step 3). The response initiated by cytokines directly causes the heat, redness, and swelling of inflammation.

Cytokines also make the endothelial cells of the blood vessel wall stickier, causing circulating **neutrophils** and **monocytes** (see Table 43.1) to attach to it in massive numbers (step 3). From there, the neutrophils and monocytes are attracted to the infection site by **chemokines,** proteins also secreted by activated macrophages (step 4). Once in the damaged tissue, the monocytes differentiate into macrophages. Both the neutrophils (which, like macrophages, have cell-surface receptors that enable them to recognize pathogens) and the new macrophages engulf the pathogens (step 5).

Once a neutrophil or macrophage has engulfed a pathogen, it uses a variety of mechanisms to destroy it. These mechanisms include attacks by enzymes and defensins located in lysosomes and the production of toxic chemicals. The harshness of these attacks usually kills the neutrophils as well, while macrophages usually survive to continue their pathogen-scavenging activities. Dead and dying neutrophils, in fact, are a major component of the pus formed at infection sites. The pain of inflammation is caused by the migration of macrophages and neutrophils to the infection site and their activities there.

Some pathogens, such as parasitic worms, are too large to be engulfed by macrophages or neutrophils. In that case, macrophages, neutrophils, and **eosinophils** (see Table 43.1) cluster around the pathogen and secrete lysosomal enzymes and defensins in amounts that are often sufficient to kill the pathogen.

If tissue damage is extensive, or the infection spreads to the blood, a **systemic inflammation**—inflammation throughout the body—likely will occur. In systemic inflammation, chemical signaling molecules released by infected tissues may stimulate the release of neutrophils from the bone marrow, thereby increasing the number of white blood cells circulating in the blood. Systemic inflammation may also involve the onset of **fever.** Several chemicals released by macrophages in response to the infection act as **pyrogens** (*pyro* = fire; *gen* = production). Pyrogens stimulate release from the hypothalamus of *prostaglandins,* chemical messengers that act locally (see Section 40.1). The prostaglandins act to turn up the hypothalamic thermostat regulating body temperature, which produces the fever. How fever fights infection is uncertain. Some models propose that it enhances phagocytosis or interferes with bacterial propagation.

THE COMPLEMENT SYSTEM Another component of the innate immune system is the **complement system,** a group of more than 30 interacting plasma proteins that circulate in the blood and interstitial fluid. (As you will see, the complement system is also used in the adaptive immune system.) Complement proteins are activated when they recognize molecules on the surfaces of pathogens. Then, some of the proteins assemble into **membrane attack complexes,** which insert into the plasma membrane of many types of bacterial cells and create pores that allow ions and small molecules to pass through. As a result, the bacteria can no longer maintain osmotic balance, and they swell and lyse.

Combating Viral Pathogens Requires a Different Innate Immune Response

You have learned that for cellular pathogens such as bacteria recognition of specific nonself molecules is key to initiating innate immune responses. For viral pathogens, however, the innate immune system is unable to distinguish effectively between surface molecules of the virus and those of the host. Therefore, the host must use other strategies to provide some immediate protection against viral infections until the adaptive immune system, which can discriminate between viral and host proteins, is effective. Two main strategies involve interferon and natural killer cells.

INTERFERON When a virus infects a cell, the cell synthesizes and releases cytokines called **interferons** in response to being exposed to the viral genome. Interferons act both on the infected cell that produces them, an autocrine effect, and on neighboring uninfected cells, a paracrine effect (see Section 40.1). They work by binding to cell-surface receptors, triggering a signal transduction pathway that changes the gene expression pattern of the cells. The key changes include activation of a ribonuclease enzyme that degrades most cellular RNA and inactivation of a protein required for protein synthesis, thereby inhibiting most protein synthesis in the cell. These effects on RNA and protein synthesis inhibit replication of the viral genome, while putting the cell in a weakened state from which it may or may not recover.

NATURAL KILLER CELLS Cells that have been infected with virus must be destroyed. That is the role of **natural killer (NK) cells.** NK cells are a type of **lymphocyte,** a leukocyte that carries out most of its activities in tissues and organs of the lymphatic system (see Figure 42.20). NK cells circulate in the blood and kill target host cells—not only cells that are infected with virus, but also some cells that have become cancerous.

NK cells can be activated by cell-surface receptors or by interferons secreted by virus-infected cells. NK cells are not phagocytes; instead, they secrete granules containing *perforin,* a protein that creates pores in the target cell's membrane. Unregulated diffusion of ions and molecules through the pores causes osmotic imbalance, swelling, and rupture of the infected cell. NK cells also kill target cells indirectly by secreting *proteases* (protein-degrading enzymes) that pass through the pores. The proteases trigger **apoptosis** (programmed cell death). That is, the proteases activate other enzymes that cause the degradation of DNA which, in turn, induces pathways leading to the cell's death.

How does an NK cell distinguish a target cell from a normal cell? The surfaces of most vertebrate cells have particular *major histocompatibility complex (MHC) proteins* on them. You will learn about the role of these proteins in adaptive immunity in the next section; for now, just consider them to be tags on the cell surface. NK cells monitor the level of MHC proteins and respond differently depending on their level. An appropriately high level, as on normal cells, inhibits the killing activity of NK cells.

Viruses often inhibit the synthesis of MHC proteins in the cells they infect, lowering the levels of those proteins and identifying them to NK cells. Cancer cells also have low or, in some cases, no MHC proteins on their surfaces, which makes them a target for destruction by NK cells as well.

STUDY BREAK 43.2

1. **What are the usual characteristics of the inflammatory response?**
2. **What processes specifically cause each characteristic of the inflammatory response?**
3. **What is the complement system?**
4. **Why does combating viral pathogens require a different response by the innate immune system than combating bacterial pathogens? What are the three main strategies a host uses to protect against viral infections?**

43.3 Adaptive Immunity: Specific Defenses

Adaptive immunity is a defense mechanism that recognizes specific molecules as being foreign and clears those molecules from the body. The foreign molecules recognized may be free, as in the case of toxins, or they may be on the surface of viruses or cells such as pathogenic bacteria, cancer cells, pollen, or cells of transplanted tissues and organs.

Adaptive immunity develops only after the body is exposed to the foreign molecules and, hence, takes several days to become effective. This would be a significant problem in the case of invading pathogens were it not for the innate immune system, which combats the invaders in a nonspecific way within minutes after they enter the body. The two key distinctions between innate and adaptive immunity are:

1. Innate immunity is nonspecific whereas adaptive immunity is specific;
2. Innate immunity retains no memory of exposure to the pathogen whereas adaptive immunity retains a memory of the foreign molecule that triggered the response, thereby enabling a rapid, more powerful response if that pathogen is encountered again at a later time.

In Adaptive Immunity, Antigens Are Cleared from the Body by B Cells or T Cells

A foreign molecule that triggers an adaptive immune response is called an **antigen** (meaning "*anti*body *gen*erator"). Antigens are macromolecules; most are large proteins (including glycoproteins and lipoproteins) or polysaccharides (including lipopolysaccharides). Some nucleic acids can also act as antigens, as can various large, artificially synthesized molecules.

Some antigens enter the body from the environment; they include antigens on pathogens introduced beneath the skin, antigens in vaccinations, and inhaled or ingested macromolecules, such as toxins. Other antigens are produced within the body; they include proteins encoded by viruses that have infected cells and altered proteins produced by mutated genes, such as those in cancer cells.

Antigens are recognized in the body by two types of lymphocytes, B cells and T cells. **B cells** differentiate from stem cells in the bone marrow (see Section 42.2). It is easy to remember this as "B for bone." However, the "B" actually refers to the *bursa of Fabricus,* a lymphatic organ found only in birds; B cells were first discovered there. After their differentiation, B cells are released into the blood and carried to capillary beds serving the tissues and organs of the lymphatic system. Like B cells, **T cells** are produced by the division of stem cells in the bone marrow. Then, they are released into the blood and carried to the **thymus,** an organ of the lymphatic system (the "T" in "T cell" refers to the thymus). As you will see, two types of T cells—*helper T cells* and *cytotoxic T cells*—are involved in adaptive immunity.

The role of lymphocytes in adaptive immunity was demonstrated by experiments in which all of the leukocytes in mice were killed by irradiation with X-rays. These mice were then unable to develop adaptive immunity. Injecting lymphocytes from normal mice into the irradiated mice restored the response; other body cells extracted from normal mice could not restore the response. (For more on the use of mice as an experimental organism in biology, see *Focus on Model Research Organisms.*)

There are two types of adaptive immune responses: **antibody-mediated immunity** (also called **humoral immunity**) and **cell-mediated immunity.** In antibody-mediated immunity, B-cell derivatives called **plasma cells** secrete **antibodies,** highly specific protein molecules that circulate in the blood and lymph recognizing and binding to antigens and clearing them from the body. In cell-mediated immunity, a particular type of T cell becomes activated and, with other cells of the immune system, attacks foreign cells directly and kills them.

The body's reaction to an infection is not an either–or choice between antibody-mediated immunity and cell-mediated immunity. Rather, both responses typically occur and, depending on the infection, the relative strength of the two responses varies. Complicated chemical signaling controls how the two responses develop.

The steps of the adaptive immune response are similar for antibody-mediated immunity and cell-mediated immunity:

1. *Antigen encounter and recognition:* Lymphocytes encounter and recognize an antigen.
2. *Lymphocyte activation:* The lymphocytes are activated by binding to the antigen and proliferate by cell division to produce many clones of identical cells.
3. *Antigen clearance:* The many clones of activated lymphocytes clear the antigen from the body.
4. *Development of immunological memory:* Some activated lymphocytes differentiate into **memory cells** that circulate in the blood and lymph, ready to initiate a rapid immune response upon a later exposure to the same antigen.

The following discussions of antibody-mediated immunity and cell-mediated immunity expand on these steps.

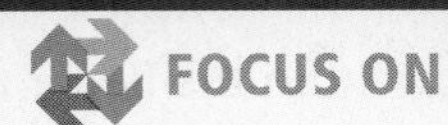

FOCUS ON Model Research Organisms

The Mighty Mouse

© Peter Skinner/Photo Researchers, Inc.

The mouse *(Mus musculus)* and its cells have been used to great advantage as models for research on mammalian developmental genetics, immunology, and cancer. The availability of the mouse as a research tool enables scientists to carry out experiments with a mammal that would not be practical or ethical with humans.

Mice are grown by the millions in laboratories all over the world. Its small size makes the mouse relatively inexpensive and easy to maintain in the laboratory, and its short generation time, compared with most other mammals, allows genetic crosses to be carried out within a reasonable time span. Mice can be mated when they are 10 weeks old; within 18 to 22 days the female gives birth to a litter of about 5 to 10 offspring. A female may be rebred within little more than a day after giving birth.

Mice have a long and highly productive history as experimental animals. Gregor Mendel, the founder of genetics, is known to have kept mice as part of his studies. The first example of a lethal allele was also found in mice, and pioneering experiments on the transplantation of tissues between individuals were conducted with mice. During the 1920s, Fred Griffith laid the groundwork for the research showing that DNA is the hereditary molecule in his work with pneumonia-causing bacteria in mice (see Section 14.1). More recently, genetic experiments with mice have revealed more than 500 mutants that cause hereditary diseases, immunological defects, and cancer in mammals including humans.

The mouse has also been the mammal of choice for experiments that introduce and modify genes through genetic engineering. One of the most spectacular results of this research was the production of giant mice by introducing a human growth hormone gene into a line of dwarfed mice that were deficient for this hormone (see Section 18.2 and Figure 18.12).

Genetic engineering has also produced "knockout" mice, in which a gene of interest is completely nonfunctional (see Section 18.2). The effects of the lack of function of a gene in the knockout mice often allow investigators to determine the role of the normal form of the gene. Some knockout mice have been developed to be defective in genes homologous to human genes that cause serious diseases, such as cystic fibrosis. This allows researchers to study the disease in mice with the goal of developing cures or therapies.

The revelations in developmental genetics from studies with the mouse have been of great interest and importance in their own right. But more and more, as we find that much of what applies to the mouse also applies to humans, the findings in mice have shed new light on human development and opened pathways to the possible cure of human genetic diseases.

The mouse has also been the main experimental choice for studies of immune responses. The results obtained have provided valuable insights into how the human immune systems work. However, mice and humans are separated by 65 million years of evolution and, as a result, the two organisms have significant differences in immune system mechanisms. These differences lead researchers to be cautious in relating mouse studies directly to humans when using mice in clinical studies involving the immune systems. For instance, the mouse model of multiple sclerosis has some differences from the human disease.

In 2002 the sequence of the mouse genome was reported. The sequence is enabling researchers to refine and expand their use of the mouse as a model organism for studies of mammalian biology and mammalian diseases.

Antibody-Mediated Immunity Involves Activation of B Cells, Their Differentiation into Plasma Cells, and the Secretion of Antibodies

An adaptive immune response begins as soon as an antigen is encountered and recognized in the body.

ANTIGEN ENCOUNTER AND RECOGNITION BY LYMPHOCYTES Antigens from the environment are encountered by B cells and T cells in the lymphatic system. Each B cell and each T cell is specific for a particular antigen, meaning that the cell can bind to only one particular molecular structure. The binding is so specific because the plasma membrane of each B cell and T cell is studded with thousands of identical receptors for the antigen; **B-cell receptors** on B cells and **T-cell receptors** on T cells **(Figure 43.2).**

B-cell receptors and T-cell receptors are encoded by different genes and thus have different structures. The B-cell receptor on a B cell (Figure 43.2A) corresponds to the antibody secreted by that cell when it is activated and differentiates into a plasma cell. Like its corresponding antibody, a B-cell receptor is a protein consisting of four polypeptide chains. At one end, the protein has two identical *antigen-binding sites,* regions that bind to a specific antigen. At the opposite end from the antigen-binding sites are *transmembrane domains,* which embed in the plasma membrane. T-cell receptors (Figure 43.2B) consist of a protein made up of two different polypeptides. T-cell receptors also have an antigen-binding site at one end and transmembrane domains at the other end.

Considering the entire populations of B cells and T cells in the body, there are multiple cells that can recognize each antigen but, most importantly, the populations (in normal persons) contain cells capable of recognizing any antigen. For example, each of us has about 10 trillion B cells that collectively have about 100 million different kinds of B-cell receptors. Importantly, these cells are present *before* the body has encountered the antigens.

The binding between antigen and receptor is an interaction between two molecules that fit together like an enzyme and its substrate. A given B-cell receptor or T-cell receptor does not bind to the whole antigen molecule, but to small re-

FIGURE 43.2
Antigen-binding receptors on B cells and T cells. The B-cell receptor corresponds to the antibody secreted by the cell when it is activated and differentiates into a plasma cell.

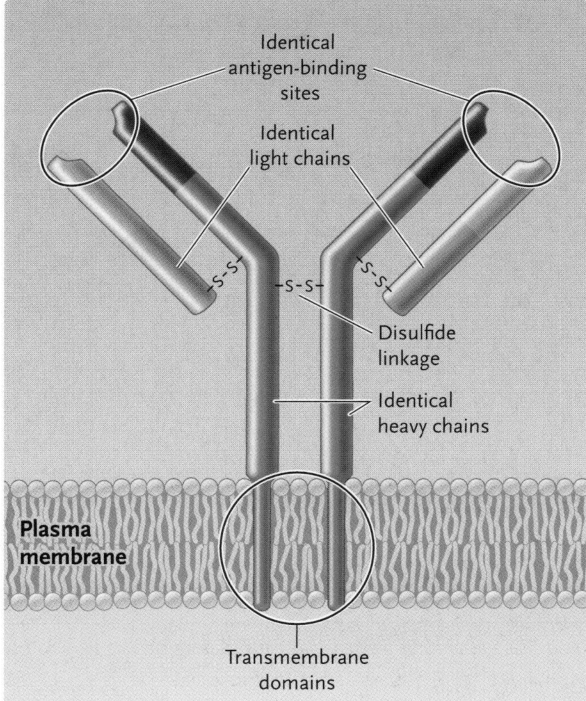

A. B-cell receptor
Identical antigen-binding sites
Identical light chains
S-S
Disulfide linkage
Identical heavy chains
Plasma membrane
Transmembrane domains

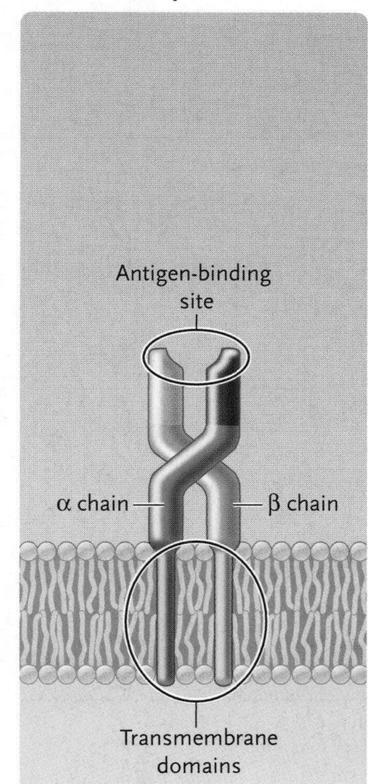

B. T-cell receptor
Antigen-binding site
α chain
β chain
Transmembrane domains

gions of it called **epitopes** or *antigenic determinants*. Therefore, several different B cells and T cells each with different receptors may bind to the population of a particular antigen encountered in the lymphatic system.

ANTIBODIES Antibodies are the core molecules of antibody-mediated immunity. Antibodies are large, complex molecules that belong to a class of proteins known as **immunoglobulins** (Ig). Each antibody molecule consists of four polypeptide chains: two identical **light chains** and two identical **heavy chains** about twice or more the size of the light chain **(Figure 43.3).**

Each polypeptide chain of an antibody molecule has a **constant (C) region** and a **variable (V) region.** The constant region of each antibody type has the same amino acid sequence for that part of the heavy chain, and likewise for that part of the light chain. The variable region of both the heavy and light chains, by contrast, has a different amino acid sequence for each antibody molecule in a population. The variable regions are the top halves of the polypeptides in the arms of the Y-shaped molecule; the three-dimensional folding of the variable regions of each arm creates the antigen-binding site. The antigen-binding site is the same for the two arms of an antibody molecule because its two heavy chains are identical, as are its two light chains, as mentioned earlier. However, the antigen-binding sites are different from antibody molecule to antibody molecule because of the amino acid differences in the variable regions of the two chain types.

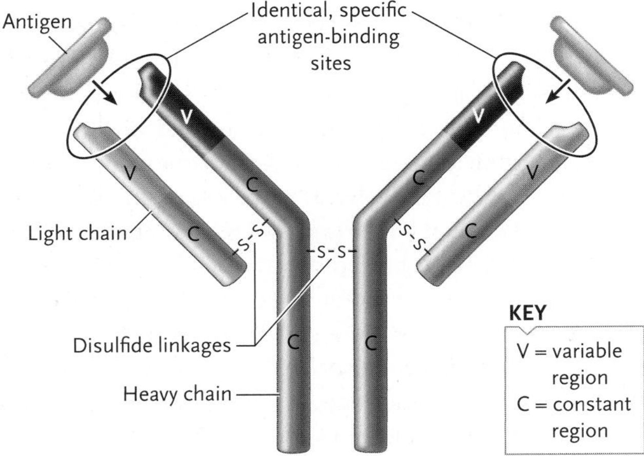

Antigen
Identical, specific antigen-binding sites
V
C
Light chain
S-S
Disulfide linkages
Heavy chain
KEY
V = variable region
C = constant region

FIGURE 43.3
The arrangement of light and heavy polypeptide chains in an antibody molecule. Disulfide (—S—S—) linkages hold the chains together, and the whole molecule folds into a Y-shape. The bonds between the two arms of the Y form a hinge that allows the arms to flex independently of one another. As shown, two sites, one at the tip of each arm of the Y, bind the same antigen.

The constant regions of the heavy chains in the tail part of the Y-shaped structure determine the **antibody class.** Humans have five different classes of antibodies—*IgM, IgG, IgA, IgE,* and *IgD* **(Table 43.2).** Due to differences in their heavy chain constant regions, they have specific structural and functional differences.

IgM remains bound to the cells that make it due to a region at the end opposite from the antigen-binding end that inserts into the plasma membrane of the cell. B-cell receptors on B cells are IgM molecules.

IgG is the most abundant antibody circulating in the blood and lymphatic system. IgG is produced in large amounts when the body is exposed a second time to the same antigen.

IgA is found mainly in secretions at particular locations in the body. In these locations, the antibodies bind to surface groups on pathogens and block their attachment to body surfaces. Breast milk transfers IgA antibodies and thus immunity to a nursing infant.

IgE is secreted by plasma cells of the skin and tissues lining the gastrointestinal tract and respiratory tract. IgE binds to basophils and mast cells, triggering release of histamine, which causes an inflammatory response. In this way, IgE mediates many allergic responses, such as hay fever, asthma, and hives. IgE also contributes to mechanisms that combat infection by parasitic worms.

IgD occurs with IgM as a receptor on the surfaces of B cells; its function is uncertain.

TABLE 43.2 Five Classes of Antibodies

Class	Structure (Secreted Form)	Location	Functions
IgM		Surfaces of unstimulated B cells (as monomer); free in circulation (as pentamer)	First antibodies to be secreted by B cells in primary response. When bound to antigen, promotes agglutination reaction, activates complement system, and stimulates phagocytic activity of macrophages.
IgG		Blood and lymphatic circulation	Most abundant antibody in primary and secondary responses. Stimulates phagocytosis and activates complement system. Crosses placenta, conferring passive immunity to fetus.
IgA		Body secretions such as tears, breast milk, saliva, and mucus	Blocks attachment of pathogens to mucous membranes; confers passive immunity for breastfed infants.
IgE		Skin and tissues lining gastro-intestinal and respiratory tracts (secreted by plasma cells)	Stimulates mast cells and basophils to release histamine; triggers allergic responses.
IgD		Surface of unstimulated B cells	Membrane receptor for mature B cells; probably important in B-cell activation (clonal selection).

THE GENERATION OF ANTIBODY DIVERSITY The human genome has approximately 20,000 genes, far fewer than necessary to encode 100 million different antibodies if two genes encoded one antibody, one gene for the heavy chain and one for the light chain. Instead, antibody diversity is generated during B-cell differentiation by a process involving three rearrangements of DNA segments that encode parts of the light and heavy chains. **Figure 43.4** shows the process that produces light-chain genes for an IgM antibody, the B-cell receptor; the production of IgM heavy-chain genes, and the light-chain and heavy-chain genes for the other classes of antibodies, is similar. The genes for the two different subunits of the T-cell receptor undergo similar rearrangements to produce the great diversity in antigen-binding capability of those receptors.

Light-chain genes have one C segment, but heavy-chain genes have five types of C segments, each of which encodes one of the constant regions of IgM, IgD, IgG, IgE, and IgA. In the assembly of functional heavy-chain genes, the inclusion of one of the five C segment types therefore specifies the class of antibody that will be made by the cell.

LYMPHOCYTE ACTIVATION Let us now follow the development of an antibody-mediated immune response for a bacterial pathogen in the circulation or in a tissue. Circulating viruses are dealt with in the same way. We follow the steps of recognition of an antigen by lymphocytes, activation of lymphocytes by antigen binding, and production of antibodies **(Figure 43.5).** Two phases are involved: in the first, particular T cells are activated; in the second, those T cells help B cells become activated and secrete antibodies. The numbered steps in the following description correspond to the events depicted in Figure 43.5:

1. **Engulfment of bacterium.** A type of phagocyte called a **dendritic cell** engulfs a bacterium in the infected tissue by phagocytosis. Dendritic cells, which have the same origin as leukocytes, are so named because they have many surface projections resembling dendrites of neurons. They recognize a bacterium as foreign through the same recognition mechanism used by macrophages in the innate immune system.
2. **Degradation of bacterium and release of antigens.** Engulfment of a bacterium activates the dendritic cell; the cell now migrates to a nearby lymph node. Then, within the dendritic cell, the vesicle containing the engulfed bacterium fuses with a lysosome. In the lysosome, the bacterium's proteins are degraded into short peptides, which function as antigens.
3. **Presentation of antigens on dendritic cell surface.** Within the dendritic cell, the antigens bind intracellularly to **class II major histocompatibility complex (MHC)** proteins.

   These proteins are named for a large cluster of 128 genes encoding them, called the **major histocompatibility complex.** Each individual of each vertebrate species has a unique combination of MHC proteins on almost all body cells, meaning

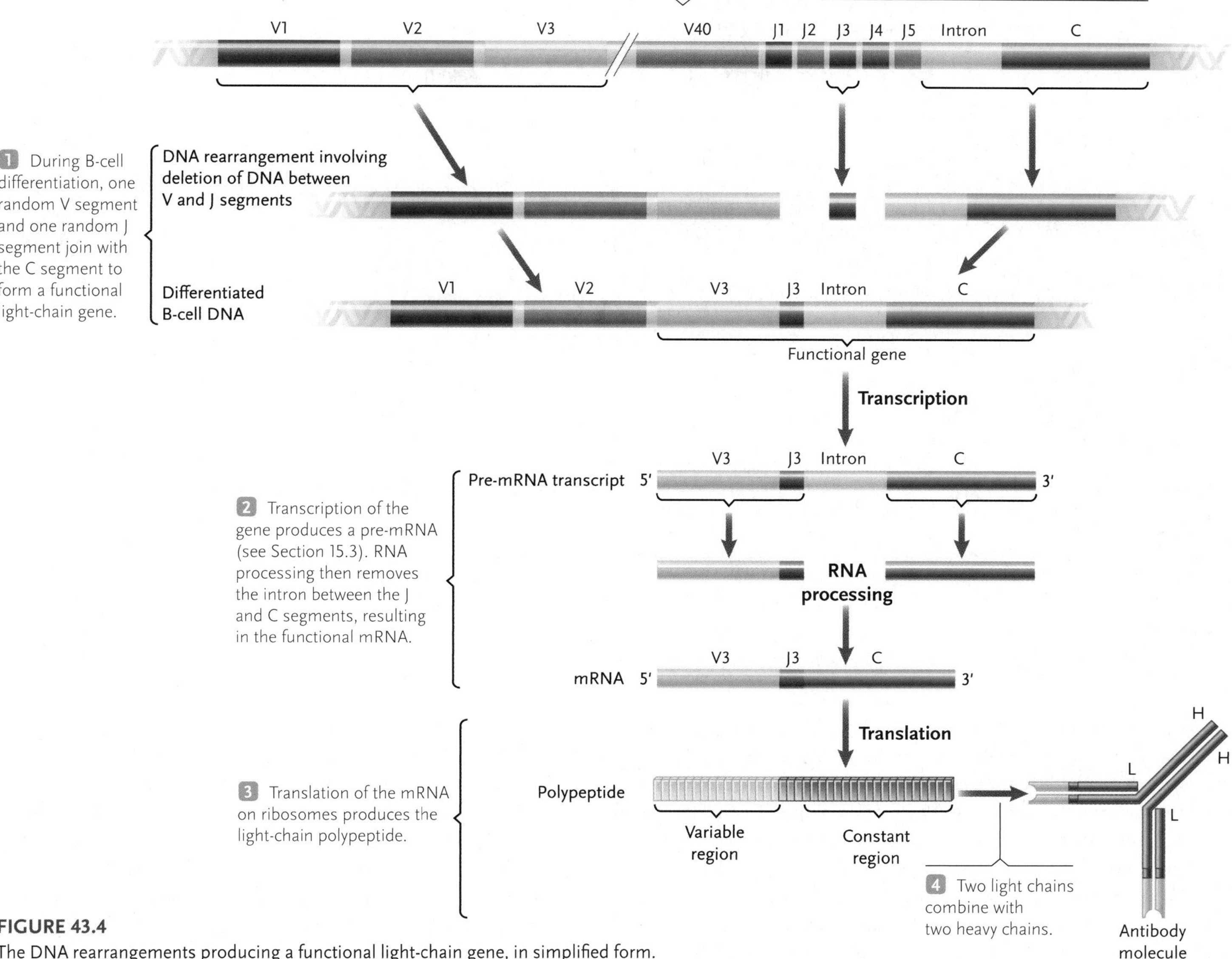
Light-chain gene containing multiple V (variable) and J (joining) segments and a single C (constant) segment in an undifferentiated B cell. One of each segment type is needed for a functional light-chain gene. Humans have 40 different V segments and 5 different J segments.
V1
V2
V3
V40
J1
J2
J3
J4
J5
Intron
C
1 During B-cell differentiation, one random V segment and one random J segment join with the C segment to form a functional light-chain gene.
DNA rearrangement involving deletion of DNA between V and J segments
Differentiated B-cell DNA
V1
V2
V3
J3
Intron
C
Functional gene
Transcription
V3
J3
Intron
C
Pre-mRNA transcript 5′
3′
2 Transcription of the gene produces a pre-mRNA (see Section 15.3). RNA processing then removes the intron between the J and C segments, resulting in the functional mRNA.
RNA processing
V3
J3
C
mRNA 5′
3′
Translation
3 Translation of the mRNA on ribosomes produces the light-chain polypeptide.
Polypeptide
Variable region
Constant region
4 Two light chains combine with two heavy chains.
H
H
L
L
Antibody molecule

FIGURE 43.4
The DNA rearrangements producing a functional light-chain gene, in simplified form.

that no two individuals of a species except identical siblings are likely to have exactly the same MHC proteins on their cells. The two classes of MHC proteins, class I and class II, have different functions in adaptive immunity, as you will see.

Once the antigen and class II MHC protein are bound, they migrate to the cell surface where the antigen is displayed. These steps, which occur in the dendritic cell after it has migrated to the lymph node, convert the cell into an **antigen-presenting cell (APC),** which can present the antigen to T cells.

4. **Interaction of antigen-presenting cell with lymphocyte.** The key function of an antigen-presenting cell is to present the antigen to a lymphocyte, which for antibody-mediated immunity is to a **$CD4^+$ T cell,** so called because it has a CD4 receptor on its surface. A specific $CD4^+$ T cell having a T-cell receptor with an antigen-binding site that recognizes an epitope of a bacterial antigen binds to the antigen on the antigen-presenting cell, and the two cells become linked together.
5. **Activation of T cell.** When the antigen-presenting cell binds to the $CD4^+$ T cell, the antigen-presenting cell secretes an *interleukin* (meaning "between leukocytes"), a type of cytokine that activates the associated T cell.
6. **Production of helper T cells.** The activated T cell secretes other interleukins, which act in an autocrine manner (see Section 40.1) to stimulate **clonal expansion,** the proliferation of the activated $CD4^+$ T cell by cell division to produce a clone of cells. These clonal cells differentiate into **helper T cells,** so named because they assist with the activation of other lymphocytes, in this case B cells. A helper T cell is an example of an **effector T cell,** meaning that it is involved in effecting—bringing about—the specific immune response to the antigen.

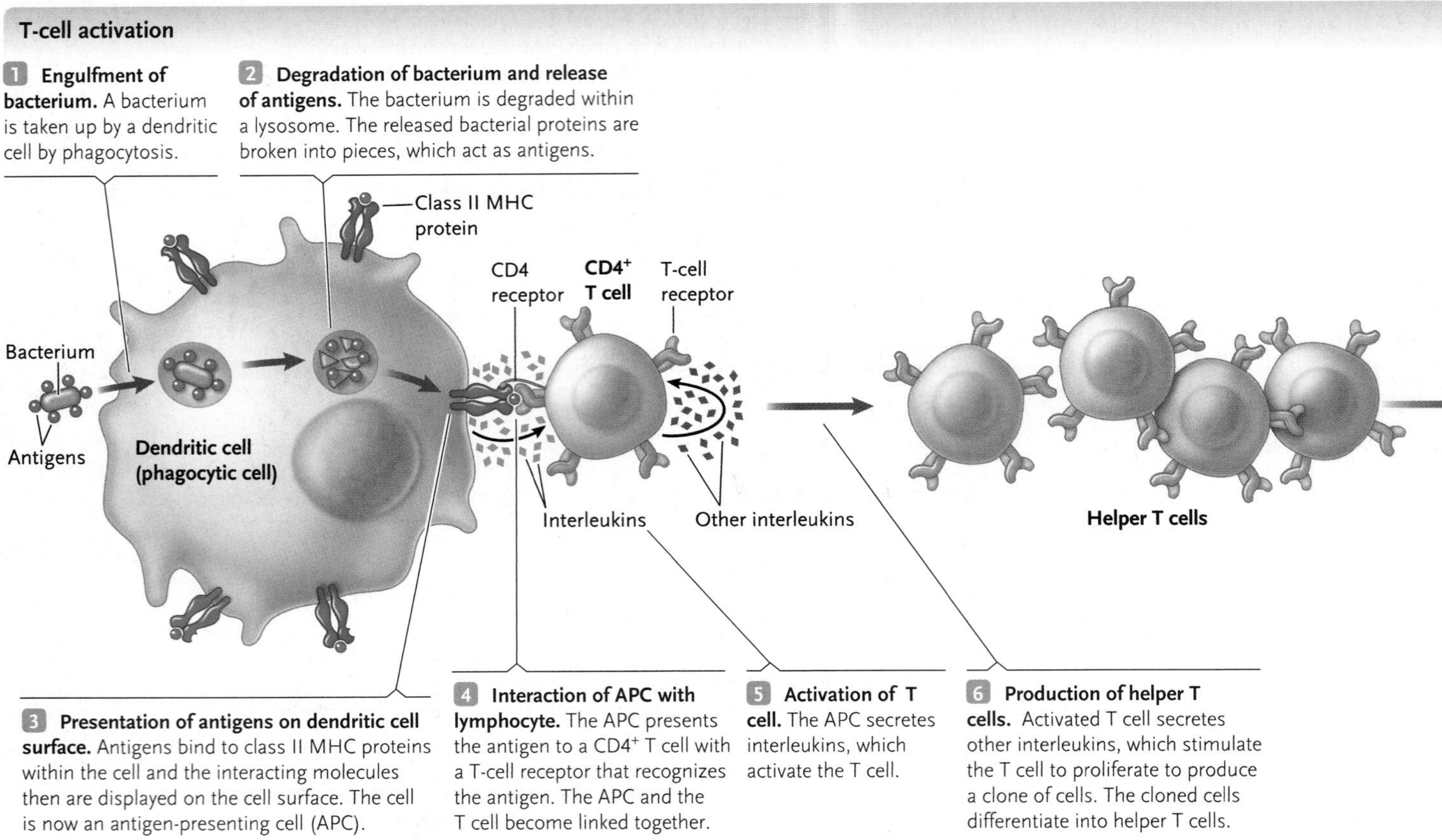

Antibody-mediated immune response
T-cell activation
1 Engulfment of bacterium. A bacterium is taken up by a dendritic cell by phagocytosis.
2 Degradation of bacterium and release of antigens. The bacterium is degraded within a lysosome. The released bacterial proteins are broken into pieces, which act as antigens.
Class II MHC protein
CD4 receptor
CD4+ T cell
T-cell receptor
Bacterium
Antigens
Dendritic cell (phagocytic cell)
Interleukins
Other interleukins
Helper T cells
3 Presentation of antigens on dendritic cell surface. Antigens bind to class II MHC proteins within the cell and the interacting molecules then are displayed on the cell surface. The cell is now an antigen-presenting cell (APC).
4 Interaction of APC with lymphocyte. The APC presents the antigen to a CD4+ T cell with a T-cell receptor that recognizes the antigen. The APC and the T cell become linked together.
5 Activation of T cell. The APC secretes interleukins, which activate the T cell.
6 Production of helper T cells. Activated T cell secretes other interleukins, which stimulate the T cell to proliferate to produce a clone of cells. The cloned cells differentiate into helper T cells.

FIGURE 43.5
The antibody-mediated immune response, illustrated for a bacterial pathogen.

7. **Presentation of antigens on B cell surface.** Antibodies are produced in and secreted by activated B cells. The activation of a B cell requires the B cell to present the antigen on its surface, and then to link with a helper T cell that has differentiated as a result of encountering and recognizing the *same* antigen. Antigen presentation on a B cell surface begins when B-cell receptors on the B cell bind to antigens on the surface of the bacterium it encounters. The bacterium plus B-cell receptor is then taken into the cell where the antigen is processed in the same way as in dendritic cells (see steps 2 and 3). The result is the presentation of antigen pieces on the B-cell surface in a complex with class II MHC proteins.
8. **Interaction of B cell with helper T cell.** When a B cell encounters a helper T cell (from step 6) displaying the same antigen, usually in a lymph node or in the spleen, the two cells become tightly linked together.
9. **Activation of B cell.** The linkage between cells stimulates the helper T cell to secrete interleukins that activate the B cell and then stimulate the B cell to proliferate, producing a clone of those B cells with identical B-cell receptors.
10. **Production of plasma cells and memory B cells.** Many of the cloned cells differentiate into relatively short-lived **plasma cells,** which now secrete the same antibody that was displayed on the parental B cell's surface to circulate in lymph and blood to attach the pathogen. The other cloned cells differentiate into **memory B cells,** which are long-lived cells that set the stage for a much more rapid response should the same antigen be encountered later.

Remember that there is an enormous diversity of randomly generated lymphocytes in the body, each with a particular receptor that may potentially recognize a particular antigen. **Clonal selection** is the term used for the process by which a particular lymphocyte is specifically selected for cloning when it recognizes a particular foreign antigen (**Figure 43.6;** this figure is an expansion of step 10 of Figure 43.5). The process of clonal selection was proposed in the 1950s by several scientists, most notably F. Macfarlane Burnet, Niels Jerne, and David Talmage. Burnet received the Nobel Prize in 1960 for his research in immunology.

CLEARING THE BODY OF FOREIGN ANTIGENS How do the antibodies produced in an antibody-mediated immune response clear foreign antigens from the body? Let us consider some examples concerning bacteria and viruses.

Two important mechanisms to clear foreign antigens from the body are neutralization and agglutination **(Figure 43.7).** In **neutralization,** toxins produced by invading bacteria, such as tetanus toxin, can be *neutralized* by antibodies (Figure 43.7A). The antibodies bind to the toxin molecules, preventing them from carrying out their damaging action. In **agglutination** (clumping), the pathogen is immobilized by the antibodies (Figure 43.7B). For instance, for intact bacteria at an infection site or

B-cell activation and antibody production

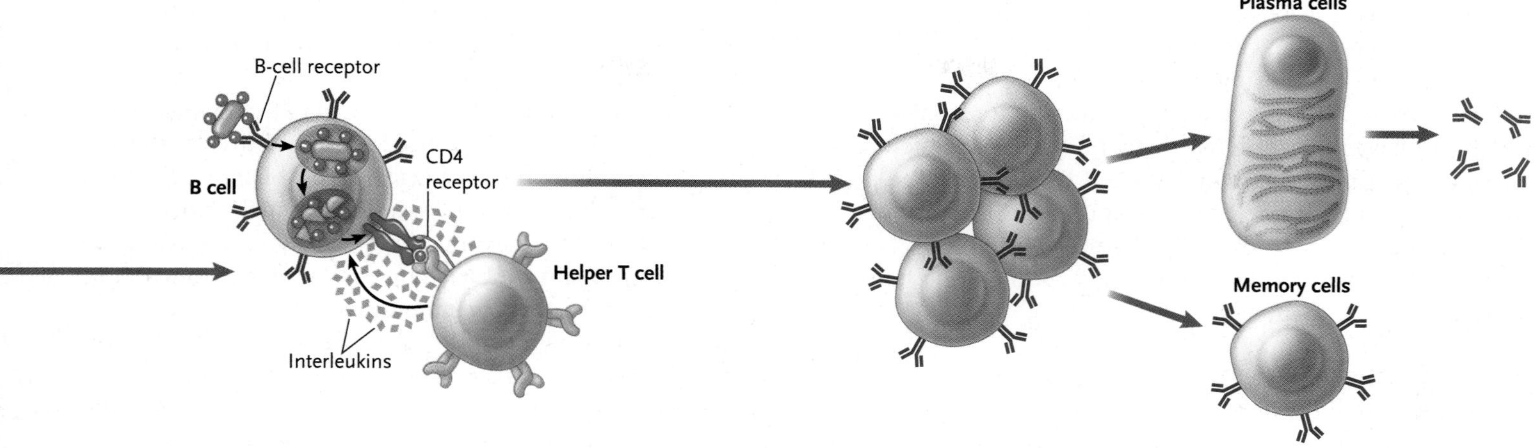

B-cell receptor
B cell
CD4 receptor
Helper T cell
Interleukins
Plasma cells
Memory cells

7 Presentation of antigens on B-cell surface. The B-cell receptor binds to antigen on the bacterium. Bacterium is engulfed and its macromolecules degraded. The antigens produced are displayed on cell surface bound to class II MHC proteins.

8 Interaction of B cell with helper T cell. The T-cell receptor of a helper T cell recognizes the antigen on the B cell and links the two cells together.

9 Activation of B cell. Interleukins secreted by the helper T cell activate the B cell and then stimulate B-cell proliferation to produce a clone of cells.

10 Production of plasma cells and memory B cells. Some cloned B cells differentiate into plasma cells, which secrete antibodies, while a few differentiate into memory B cells.

in the circulatory system, antibodies will bind to antigens on their surfaces. Because the two arms of an antibody molecule bind to different copies of the antigen molecule, an antibody molecule may bind to two bacteria with the same antigen. A population of antibodies against the bacterium, then, link many bacteria together into a lattice causing *agglutination* of the bacteria. Agglutination immobilizes the bacteria, preventing them from infecting cells. Antibodies can also agglutinate viruses, thereby preventing them from infecting cells.

More importantly, antibodies aid the innate immune response triggered by the pathogens. That is, antibodies bound to antigens stimulate the complement system. Membrane attack complexes are formed and insert themselves into the plasma membranes of the bacteria, leading to their lysis and death. In the case of viral infections, membrane attack complexes can insert themselves into the membranes surrounding enveloped viruses, which disrupts the membrane and prevents the viruses from infecting cells.

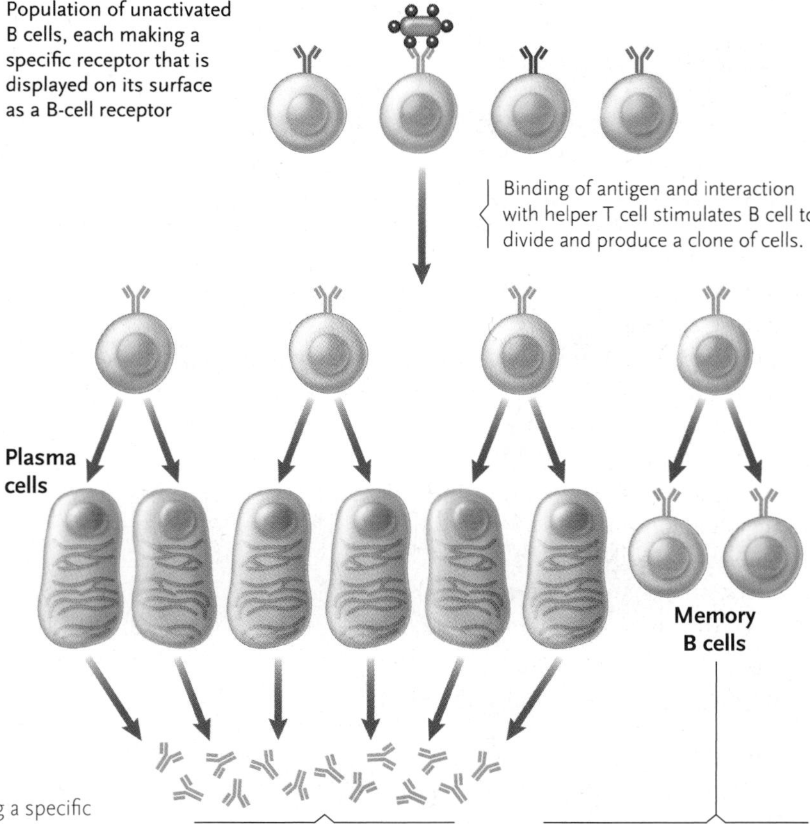

Population of unactivated B cells, each making a specific receptor that is displayed on its surface as a B-cell receptor
Binding of antigen and interaction with helper T cell stimulates B cell to divide and produce a clone of cells.
Plasma cells
Memory B cells
Some of the B-cell clones differentiate into plasma cells, which secrete antibodies.
A few B-cell clones differentiate into memory B cells, which respond to a later encounter with the same antigen.

FIGURE 43.6
Clonal selection. The binding of an antigen to a B cell already displaying a specific antibody to that antigen and interaction with helper T cells stimulates the B cell to divide and differentiate into plasma cells, which secrete the antibody, and memory cells, the cells that remain in the circulation ready to mount a response against the antigen at a later time.

Antibodies also enhance phagocytosis of bacteria and viruses. Phagocytic cells have receptors on their surfaces that recognize the heavy-chain end of antibodies (the end of the molecule opposite the antigen-binding sites). Antibodies bound to bacteria or viruses therefore bind to phagocytic cells, which then engulf the pathogens and destroy them.

For simplicity, the adaptive immune response has been described here in terms of a single antigen. Recognize that pathogens have many different types of antigens on their surfaces, which means that many different B cells are stimulated to proliferate and many different antibodies are produced. Pathogens therefore are attacked by many different types of antibodies, each targeted to one antigen type on the pathogen's surface.

IMMUNOLOGICAL MEMORY Once an immune reaction has run its course and the invading pathogen or toxic molecule has been eliminated from the body, division of the plasma cells and helper T-cell clones stops. Most or all of the clones die and are eliminated from the bloodstream and other body fluids. However, long-lived memory B cells and **memory helper T cells** (which differentiate from helper T cells), derived from encountering the same antigen, remain in an inactive state in the lymphatic system. Their persistence provides an **immunological memory** of the foreign antigen.

Immunological memory is illustrated in **Figure 43.8.** When exposed to a foreign antigen (labeled A in the figure) for the first time, a **primary immune response** results, following the steps already described. The first antibodies appear in the blood in 3 to 14 days and, by week 4, the primary response has essentially gone away. IgM is the main antibody type produced in a primary immune response. The primary immune response curve is followed whenever a new foreign antigen enters the body.

When a foreign antigen (here, antigen A) enters the body for a second or subsequent time, a **secondary immune response** results, while any new antigen (here, antigen B) introduced at the same time produces a primary response (see Figure 43.8). The secondary response is more rapid than a primary response because it involves the memory B cells and memory T cells that have been stored in the meantime, rather than having to initiate the clonal selection of a new B cell and T cell. Moreover, less antigen is needed to elicit a secondary response than a primary response, and many more antibodies are produced. The predominant antibody produced in a secondary immune response is IgG; the switch occurs at the gene level in the memory B cells.

Immunological memory forms the basis of vaccinations, in which antigens in the form of living or dead pathogens or antigenic molecules themselves are introduced into the body. After the primary immune response, memory B cells and memory T cells remaining in the body can mount an immediate and intense immune reaction against similar antigens in the dangerous pathogen. In Edward Jenner's technique, for example, introducing the cowpox virus, a related, less virulent form of

FIGURE 43.7
Neutralization **(A)** and agglutination **(B)**, two examples of mechanisms for clearing antigens from the body.

A. **Neutralization**

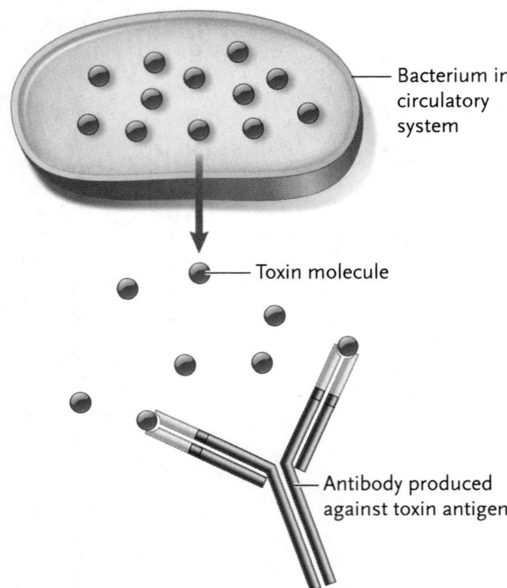
Bacterium in
circulatory
system
Toxin molecule
Antibody produced
against toxin antigen

B. **Agglutination**

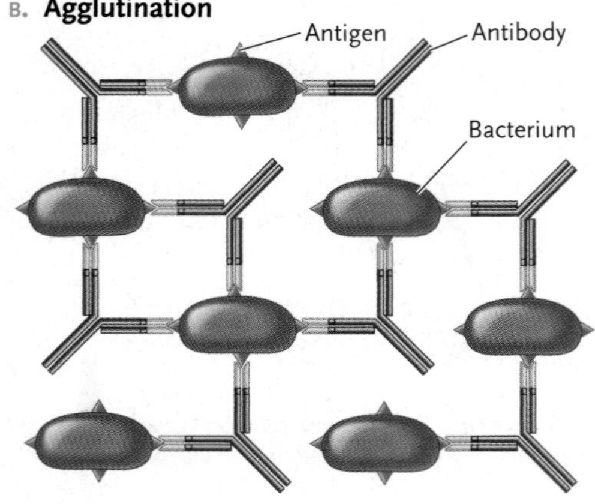
Antigen
Antibody
Bacterium

FIGURE 43.8
Immunological memory: primary and secondary responses to the same antigen.

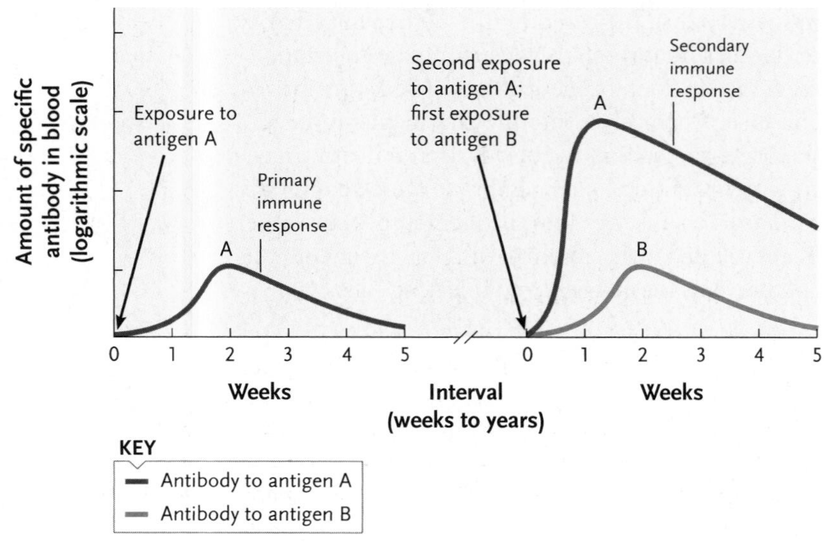
Amount of specific
antibody in blood
(logarithmic scale)
Exposure to
antigen A
Primary
immune
response
A
Second exposure
to antigen A;
first exposure
to antigen B
Secondary
immune
response
A
B
0
1
2
3
4
5
0
1
2
3
4
5
Weeks
Interval
(weeks to years)
Weeks
KEY
Antibody to antigen A
Antibody to antigen B

the smallpox virus, into healthy individuals initiated a primary immune response. After the response ran its course, a bank of memory B and T cells remained in the body, able to recognize quickly the similar antigens of the smallpox virus and initiate a secondary immune response. Similarly, a polio vaccine developed by Jonas Salk uses polio viruses that have been inactivated by exposing them to formaldehyde. Although the viruses are inactive, their surface groups can still act as antigens. The antigens trigger an immune response, leaving memory B and T cells able to mount an intense immune response against active polio viruses.

ACTIVE AND PASSIVE IMMUNITY **Active immunity** is the production of antibodies in the body in response to exposure to a foreign antigen—the process that has been described up until now. **Passive immunity** is the acquisition of antibodies as a result of direct transfer from another person. This form of immunity provides immediate protection against antigens that the antibodies recognize without the person receiving the antibodies having developed an immune response. Examples of passive immunity include the transfer of IgG antibodies from mother to fetus through the placenta and the transfer of IgA antibodies in the first breast milk fed from the mother to the baby. Compared with active immunity, passive immunity is a short-lived phenomenon with no memory, in that the antibodies typically break down within a month. However, in that time, the protection plays an important role. For example, a breast-fed baby is protected until it is able to mount an immune response itself, an ability that is not present until about a month after birth.

DRUG EFFECTS ON ANTIBODY-MEDIATED IMMUNITY Several drugs used to reduce the rejection of transplanted organs target helper T cells. Cyclosporin A, used routinely after organ transplants, blocks the activation of helper T cells and, in turn, the activation of B cells. Unfortunately, cyclosporin and other immunosuppressive drugs also leave the treated individual more susceptible to infection by pathogens.

Antibodies Have Many Uses in Research

The ability of the antibody-mediated immune system to generate antibodies against essentially any antigen provides an invaluable research tool to scientists, who can use antibodies to identify biological molecules and determine their locations and functions in cells. To obtain the antibodies, a molecule of interest is injected into a test animal such as a mouse, rabbit, goat, or sheep. In response, the animal develops antibodies capable of binding to the molecule. The antibodies are then extracted and purified from a blood sample.

To identify the cellular location of a molecule, antibodies made against the molecule are combined with a visible marker such as a dye molecule or heavy metal atom. When added to a tissue sample, the marked antibodies can be seen in the light or electron microscope localized to cellular structures such as membranes, ribosomes, or chromosomes, showing that the molecule forms part of the structure.

Antibodies can also be used to "grab" a molecule of interest from a preparation containing a mixture of all kinds of cellular molecules. For such studies, the antibodies are often attached to plastic beads that are packed into a glass column. When the mixture passes through the column, the molecule is trapped by attachment to the antibody and remains in the column. It is then released from the column in purified form by adding a reagent that breaks the antigen–antibody bonds.

Injecting a molecule of interest into a test animal typically produces a wide spectrum of antibodies that react with different parts of the antigen. Some of the antibodies also cross-react with other, similar antigens, producing false results that can complicate the research. These problems have been solved by producing **monoclonal antibodies,** each of which reacts only against the same segment (epitope) of a single antigen.

Georges Köhler and Cesar Milstein pioneered the production of monoclonal antibodies in 1975. In their technique, a test animal (usually a mouse) is injected with a molecule of interest **(Figure 43.9).** After the animal has developed an immune response, fully activated B cells are extracted from the spleen and placed in a cell culture medium. Because B cells normally stop dividing and die within a week when cultured, they are induced to fuse with cancerous lymphocytes called *myeloma cells,* forming single, composite cells called **hybridomas.** Hybridomas combine the desired characteristics of the two cell types—they produce antibodies like fully activated B cells, and they divide continuously and rapidly like myeloma cells.

Single hybridoma cells are then separated from the culture and used to start clones. Because all the cells of a clone are descended from a single hybridoma cell, they all make the same highly specific antibody, able to bind the same part of a single antigen. In addition to their use in scientific research, monoclonal antibodies are also widely used in medical applications such as pregnancy tests, screening for prostate cancer, and testing for AIDS and other sexually transmitted diseases.

In Cell-Mediated Immunity, Cytotoxic T Cells Expose "Hidden" Pathogens to Antibodies by Destroying Infected Body Cells

In **cell-mediated immunity,** cytotoxic T cells directly destroy host cells infected by pathogens, particularly those infected by a virus **(Figure 43.10).** The numbered steps in the following description correspond to the events depicted in the figure:

1. **Presentation of antigens on cell surface.** The killing process begins when some of the pathogens break down inside infected host cells, releasing antigens that are fragmented by enzymes in the cytoplasm. The antigen fragments bind to class I MHC proteins, which are delivered to the cell surface by essentially the same mechanisms as in B cells (see Figure 43.6). At the surface, the antigen fragments are displayed by the class I MHC protein, and the cell begins to function as an antigen-presenting cell.

FIGURE 43.9 **Research Method**

Production of Monoclonal Antibodies

Purpose: Injecting an antigen into an animal produces a collection of different antibodies that react against different parts of the antigen. Monoclonal antibodies are produced to provide antibodies that all react against the same epitope of a single antigen.

Protocol:

1. Inject antigen into mouse.
2. Extract activated B cells from spleen.

Activated B cells

Myeloma (cancer) cells

3. Fuse antibody-producing B cells with cancer cells to form fast-growing hybridoma cells.

Hybridoma cell

4. Grow clone from single hybridoma cells; test antibodies produced by clone for reaction against antigen.
5. Grow clone producing antibodies against antigen to large size.
6. Extract and purify antibodies.

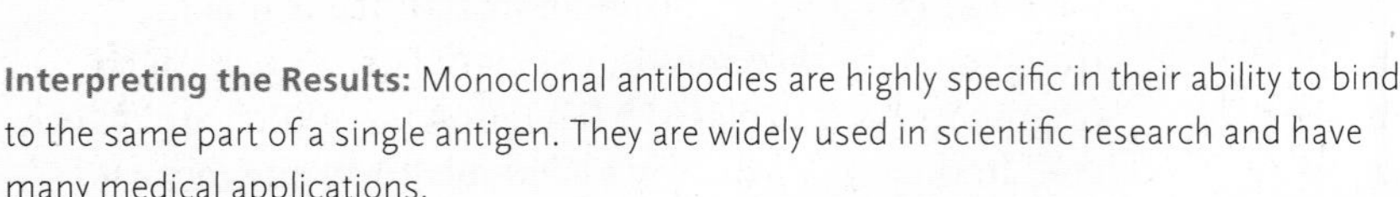

Interpreting the Results: Monoclonal antibodies are highly specific in their ability to bind to the same part of a single antigen. They are widely used in scientific research and have many medical applications.

Source: G. Köhler and C. Milstein. 1975. Continuous cultures of fused cells secreting antibody of predefined specificity. *Nature* 256:495–497.

2. **Activation of T cell.** The antigen-presenting cell binds to a type of T cell in the lymphatic system called a **$CD8^+$ T cell,** named because it has receptors named **CD8** on its surface in addition to T-cell receptors. The binding of the two cells depends on a specific $CD8^+$ T cell having a T-cell receptor with an antigen-binding site that recognizes the antigen fragment on the antigen-presenting cell. The interaction between the antigen-presenting cell and the $CD8^+$ T cell is one of two signals required to activate the T cell. The other activation signal is the presence of cytokines secreted by helper T cells generated in the antibody-mediated immune system in response to the same infection (see Figure 43.6, and recall that an infection triggers both antibody-mediated and cell-mediated immune responses).
3. **Production of cytotoxic T cells.** The activated $CD8^+$ T cell proliferates to form a clone. Some of the cells in the clone differentiate into **cytotoxic T cells,** whereas a few differentiate into **memory cytotoxic T cells.** Cytotoxic T cells are another type of effector T cell.
4. **Attack of infected cell by cytotoxic T cell.** T-cell receptors on the cytotoxic T cells again recognize the antigen fragment bound to class I MHC proteins on the infected cells (the antigen-presenting cells). The cytotoxic T cell then releases perforins, the molecules that will cause the death of the infected cell.
5. **Destruction of infected cell.** The infected cell is destroyed by mechanisms similar to those used by NK cells. The perforins released by the cytotoxic T cells create pores in the membrane of the infected cell. The leakage of ions and other molecules through the pores causes the infected cell to rupture. The cytotoxic T cell also secretes proteases that enter infected cells through the newly created pores and cause them to self-destruct by apoptosis. Rupture of dead, infected cells releases the pathogens to the interstitial fluid, where they are open to attack by antibodies and phagocytes.

Cytotoxic T cells can also kill cancer cells if their class I MHC molecules display fragments of altered cellular proteins that do not normally occur in the body. Another mechanism used by cytotoxic T cells to kill cells, and a process used by some cancer cells to defeat the mechanism, is described in *Insights from the Molecular Revolution*.

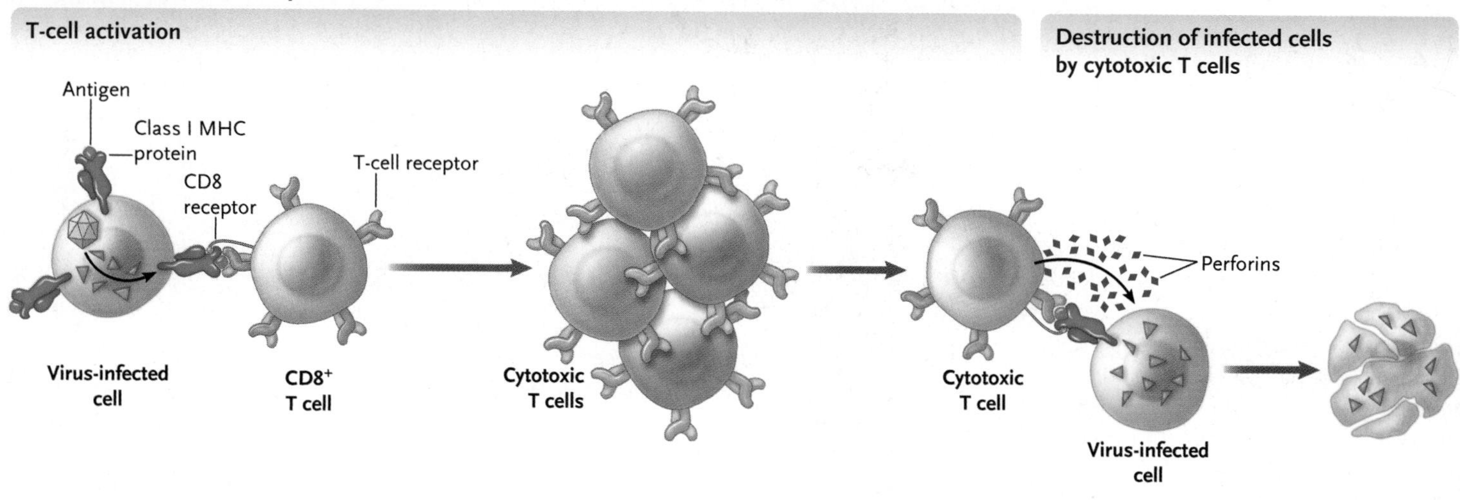
Cell-mediated immune response
T-cell activation
Destruction of infected cells by cytotoxic T cells
Antigen
Class I MHC protein
CD8 receptor
T-cell receptor
Perforins
Virus-infected cell
CD8⁺ T cell
Cytotoxic T cells
Cytotoxic T cell
Virus-infected cell

1 Presentation of antigens on cell surface. Viral proteins are degraded into fragments that act as antigens. The antigens are displayed on the cell surface bound to class I MHC proteins.

2 Activation of T cell. A T-cell receptor on a $CD8^+$ T cell recognizes an antigen bound to a class I MHC protein on an infected cell, and the two cells link together. The interaction along with cytokines from helper T cells activates the T cell.

3 Production of cytotoxic T cells. The activated $CD8^+$ T cell proliferates and forms a clone. The cloned cells differentiate into cytotoxic T cells and memory cytotoxic T cells.

4 Attack of infected cell by cytotoxic T cell. A T-cell receptor on a cytotoxic T cell recognizes the antigen bound to a class I MHC protein on the infected cell. The T cell releases perforins.

5 Destruction of infected cell. The perforins insert into the membrane of the infected cell, forming pores. Leakage of ions and other molecules (along with other events) causes the cell to lyse.

FIGURE 43.10
The cell-mediated immune response.

STUDY BREAK 43.3

1. **How, in general, do antibody-mediated and cell-mediated immune responses help clear the body of antigens?**
2. **Describe the general structure of an antibody molecule.**
3. **What are the principles of the mechanism used for generating antibody diversity?**
4. **What is clonal selection?**
5. **How does immunological memory work?**

THINK OUTSIDE THE BOOK

In this section you learned about the DNA rearrangements that occur to produce functional light-chain and heavy-chain genes. Susumu Tonegawa received a Nobel Prize in 1987 for his discovery of this mechanism of antibody diversity. He published the research in: N. Hozumi and S. Tonegawa. 1976. Evidence for somatic rearrangement of immunoglobulin genes coding for variable and constant regions. *Proceedings of the National Academy of Sciences USA* 73:3628–3632. Read the paper and make a flow chart outlining the experiments the investigators used in their research.

43.4 Malfunctions and Failures of the Immune System

The immune system is highly effective, but it is not foolproof. Some malfunctions of the immune system cause the body to react against its own proteins or cells, producing **autoimmune disease.** In addition, some viruses and other pathogens have evolved means to avoid destruction by the immune system. A number of these pathogens, including the AIDS virus, even use parts of the immune response to promote infection. Another malfunction causes the *allergic reactions* that most of us experience from time to time.

An Individual's Own Molecules Are Normally Protected against Attack by the Immune System

B cells and T cells are involved in the development of **immunological tolerance,** which protects the body's own molecules from attack by the immune system. Although the process is not completely understood, molecules present in an individual from birth are not recognized as foreign by circulating B and T cells, and do not elicit an immune response. The process of excluding self-reactive B and T cells goes on throughout the life of an individual.

Evidence that immunological tolerance is established early in life comes from experiments with mice. For example, if a foreign protein is injected into a mouse at birth, during the period in which tolerance is established, the mouse will not develop antibodies against the protein if it is injected later in life. Similarly, if mutant mice are produced that lack a given complement protein, so that the protein is absent during embryonic development, they will produce antibodies against that protein if it is injected during adult life. Normal mice do not produce antibodies if the protein is injected.

Autoimmune Disease Occurs When Immunological Tolerance Fails

The mechanisms setting up immunological tolerance sometimes fail, leading to an **autoimmune reaction**—the production of antibodies against molecules of the body. In most cases, the

INSIGHTS FROM THE Molecular Revolution

Some Cancer Cells Kill Cytotoxic T Cells to Defeat the Immune System

Among the arsenal of weapons employed by cytotoxic T cells to eliminate infected and cancerous body cells is the *Fas–FasL* system. Fas is a receptor that occurs on the surfaces of many body cells; FasL is a signal molecule that is displayed on the surfaces of some cell types, including cytotoxic T cells. If a cell with the Fas receptor contacts a cytotoxic T cell displaying the FasL signal, the effect for the Fas-bearing cell is something like stepping on a mine. When FasL is bound by the Fas receptor, a cascade of internal reactions initiates apoptosis and kills the cell with the Fas receptor.

Surprisingly, cytotoxic T cells also carry the Fas receptor, so they can kill each other by displaying the FasL signal. This mutual killing plays an important role in reducing the level of an immune reaction after a pathogen has been eliminated. In addition, cells in some regions of the body, such as the eye, the nervous system, and the testis, can be severely damaged by inflammation; these cells can make and display FasL, which kills cytotoxic T cells and reduces the severity of inflammation.

Research Question

Can cancer cells use the FasL signal to defeat the immune system?

Experiments

A group of Swiss researchers at the Universities of Lausanne and Geneva observed that many patients with malignant melanoma, a dangerous skin cancer, had a breakdown product associated with FasL in their bloodstream. This led them to hypothesize that some cancer cells survive elimination by the immune system by making and displaying the FasL signal, and thus killing any cytotoxic T cells that attack the tumor. Four experiments supported their hypothesis:

Experiment 1: The investigators extracted proteins from melanoma cells and tested whether FasL was present by using antibodies made against FasL.

Result: The test was positive, showing that FasL was in the tumor cells.

Experiment 2: The investigators used a molecule probe to see if the mRNA encoding FasL was present in the tumor cells.

Result: FasL mRNA was identified in the cells, indicating that the gene for FasL was active in those cells.

Experiment 3: The researchers tagged antibodies against the FasL protein with a dye molecule to make them visible under the light microscope and used the antibodies to see if they could detect FasL on cells in sections of melanoma tissue from patients.

Result: The investigators saw intense staining of the cells, showing that FasL was present.

Experiment 4: The researchers tested for the presence of the Fas receptor on the surface of the tumor cells.

Result: No Fas receptors were detected, indicating that Fas receptor synthesis was turned off in the melanoma cells.

Conclusion

The results show that the FasL of melanoma cells kills cytotoxic T cells that invade the tumor (see **Figure**). At the same time, the absence of Fas receptors ensures that the tumor cells do not kill each other. The presence of FasL and absence of the Fas receptor may explain why melanomas are rarely destroyed by an immune reaction, and also why many other types of cancer also escape immune destruction.

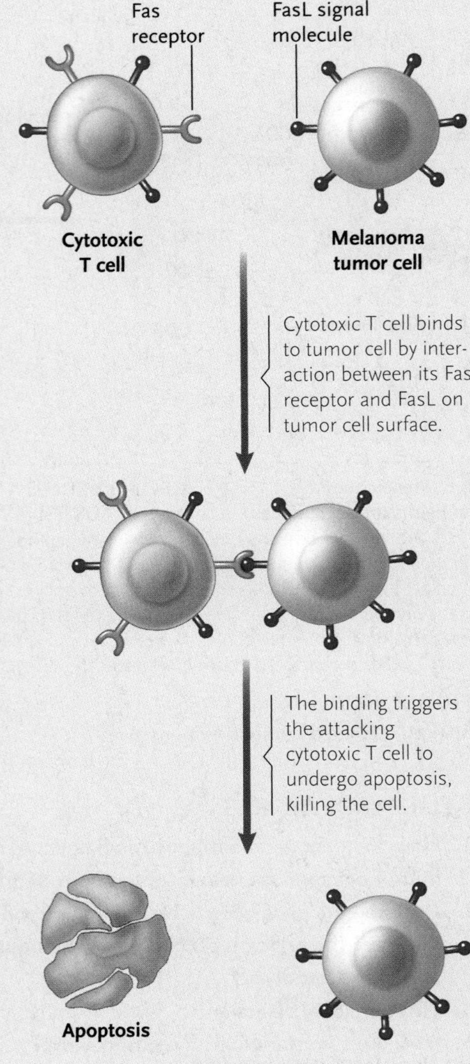

Fas receptor
FasL signal molecule
Cytotoxic T cell
Melanoma tumor cell
Cytotoxic T cell binds to tumor cell by interaction between its Fas receptor and FasL on tumor cell surface.
The binding triggers the attacking cytotoxic T cell to undergo apoptosis, killing the cell.
Apoptosis

Melanoma cells originate from pigment cells in the skin called *melanocytes*. Normal melanocytes do not contain FasL, indicating that synthesis of the protein is turned on as a part of the changes transforming normal melanocytes into cancer cells.

Source: M. Hahne et al. 1996. Melanoma cell expression of Fas(Apo-1/CD95) ligand: implications for tumor immune escape. *Science* 274:1363–1366.

effects of such anti-self antibodies are not serious enough to produce recognizable disease. However, in some individuals, about 5% to 10% of the human population, anti-self antibodies cause serious problems.

For example, type 1 diabetes (see Section 40.4) is an autoimmune reaction against the pancreatic beta cells producing insulin. The anti-self antibodies gradually eliminate the beta cells until the individual is incapable of producing insulin. **Systemic lupus erythematosus (lupus)** is caused by production of a wide variety of anti-self antibodies against blood cells, blood platelets, and internal cell structures and molecules such as mitochondria and proteins associated with DNA in the cell nucleus. People with lupus often become anemic and have problems with blood circulation and kidney function because antibodies, combined with body molecules, accumulate and clog capillaries and the microscopic filtering tubules of the kidneys. Lupus patients may also develop anti-self antibodies against the heart and kidneys. **Rheumatoid arthritis** is caused by a self-attack on connective tissues, particularly in the joints, causing pain and inflammation. **Multiple sclerosis** results from an autoimmune attack against a protein of the myelin sheaths insulating the surfaces of neurons. Multiple sclerosis can seriously disrupt nervous function, producing such symptoms as muscle weakness and paralysis, impaired coordination, and pain.

The causes of most autoimmune diseases are unknown. In some cases, an autoimmune reaction can be traced to injuries that expose body cells or proteins that are normally inaccessible to the immune system, such as the lens protein of the eye, to B and T cells. For a number of autoimmune diseases, genetic causes (for example, particular mutations, or combinations of alleles in the body) and/or environmental causes (for example, chemicals or drugs) are considered likely.

Some Pathogens Have Evolved Mechanisms That Defeat the Immune Response

Several pathogens regularly change their surface groups to avoid destruction by the immune system. By the time the immune system has developed antibodies against one version of the surface proteins, the pathogens have switched to different surface proteins that the antibodies do not match. These new proteins take another week or so to stimulate the production of specific antibodies; by this time, the surface groups change again. The changes continue indefinitely, always keeping the pathogens one step ahead of the immune system. Pathogens that use these mechanisms to sidestep the immune system include the protozoan causing African sleeping sickness, the bacterium causing gonorrhea, and the viruses causing influenza, the common cold, and AIDS.

Some viruses, such as HIV, the virus that causes AIDS, use parts of the immune system to get a free ride to the cell interior. The structure and life cycle of HIV were described in Section 17.2. For example, HIV has a surface molecule that is recognized and bound by the CD4 receptor on the surface of helper T cells. Binding to CD4 locks the virus to the cell surface and stimulates the membrane covering the virus to fuse with the plasma membrane of the helper T cell. (The protein coat of the virus is wrapped in a membrane derived from the plasma membrane of the host cell in which it was produced.) The fusion introduces the virus into the cell, initiating the infection and leading to destruction and death of the helper T cell. This reduces the population of helper T cells needed to stimulate the proliferation of B cells and cytotoxic T cells. HIV also infects and kills macrophages. The assault on lymphocytes cripples the immune system and makes the body highly vulnerable to infections and to development of otherwise rare forms of cancer. (Further details of HIV and AIDS are presented in *Focus on Applied Research.*)

Allergies Are Produced by Overactivity of the Immune System

The substances responsible for allergic reactions form a distinct class of antigens called **allergens,** which induce B cells to secrete an overabundance of IgE antibodies **(Figure 43.11).** The IgE antibodies, in turn, bind to receptors on mast cells in connective tissue and on **basophils,** a type of leukocyte in blood (see Table 43.1), inducing them to secrete histamine, which produces a severe inflammation. Most of the inflammation occurs in tissues directly exposed to the allergen, such as the surfaces of the eyes, the lining of the nasal passages, and the air passages of the lungs. Signal molecules released by the activated mast cells also stimulate mucosal cells to secrete floods of mucus and cause smooth muscle in air-

FIGURE 43.11

The response of the body to allergens. **(A)** The steps in sensitization after initial exposure to an allergen. **(B)** Production of an allergic response by further exposures to the allergen.

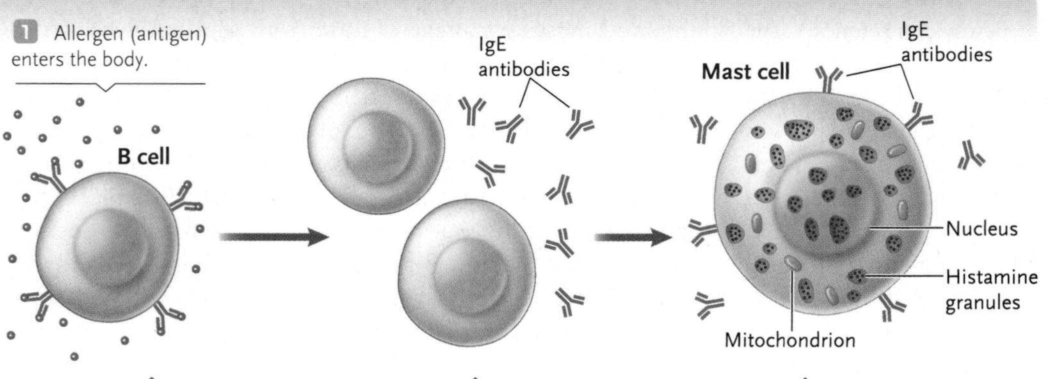
A. Initial exposure to allergen
1 Allergen (antigen) enters the body.
B cell
IgE antibodies
Mast cell
IgE antibodies
Nucleus
Histamine granules
Mitochondrion
2 Allergen binds B-cell surface antibodies; the B cell now processes the allergen and, with stimulation by a helper T cell (not shown), proceeds through the steps leading to cell division and antibody production.
3 Activated B-cell clone produces and secretes IgE antibodies active against the allergen.
4 IgE antibodies attach to mast cells in tissues, which have granules containing histamine molecules. Memory B and T cells capable of recognizing the allergen are also produced.

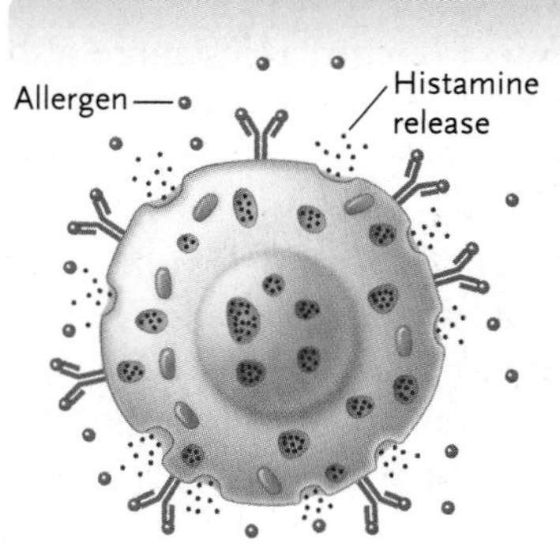
B. Further exposures to allergen
Allergen
Histamine release
5 After the first exposure, when the allergen enters the body, it binds with IgE antibodies on mast cells; binding stimulates the mast cell to release histamine and other substances.

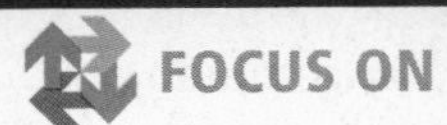 FOCUS ON **Applied Research**

HIV and AIDS

Acquired immune deficiency syndrome (AIDS) is a constellation of disorders that follows infection by the **human immunodeficiency virus, HIV** (see Figure 17.11). First reported in various countries in the late 1970s, HIV now infects more than 33 million people worldwide, 64% of them in Africa. AIDS is a potentially lethal disease, although drug therapy has reduced the death rate for HIV-infected individuals in many countries, including the United States.

HIV is transmitted when an infected person's body fluids, especially blood or semen, enter the blood or tissue fluids of another person's body. The entry may occur during vaginal, anal, or oral intercourse, or via contaminated needles shared by intravenous drug users. HIV can also be transmitted from infected mothers to their infants during pregnancy, birth, and nursing. AIDS is rarely transmitted through casual contact, food, or body products such as saliva, tears, urine, or feces.

As indicated in the main chapter text, HIV's destruction of macrophages and helper T cells cripples the immune system, which opens the door to infections and other diseases that are characteristic of AIDS. HIV enters helper T cells by binding of the *gp120* glycoprotein of the viral coat to the CD4 receptor of the T cell, followed by a viral protein-triggered fusion of the viral envelope with the T cell plasma membrane. Within the cell, the viral-encoded reverse transcriptase makes a DNA copy of the viral RNA genome. The DNA copy integrates into the host cell's chromosome where it is replicated as the cell multiplies. In this integrated state, the virus is protected from attack by the immune system. When the helper T cell is stimulated by an antigen, the integrated HIV genome is expressed to produce new viral RNA genomes and mRNAs that direct host cell ribosomes to make viral proteins, and new HIV virus particles are generated. Once released from the cell (which kills the cell), the viral particles may infect more body cells or another person.

As described in Section 17.2, symptoms after an initial infection are mild and then subside as antibodies against HIV viral proteins are generated in the body and viral particles are cleared from the circulation. However, by this time, the HIV genome has already been integrated into the genomes of infected helper T cells where it eludes the immune system. An infected person may then remain apparently healthy for years, yet can infect others. Both the transmitter and recipient of the virus may be unaware that the disease is present, making it difficult to control the spread of HIV infections.

With time, more and more helper T cells and macrophages are destroyed, eventually wiping out the body's immune response. The infected person becomes susceptible to opportunistic, secondary infections, such as a pneumonia caused by a fungus *(Pneumocystis carinii)*; drug-resistant tuberculosis; persistent yeast *(Candida albicans)* infections of the mouth, throat, rectum, or vagina; and infection by many common bacteria and viruses that rarely infect healthy humans. These secondary infections signal the appearance of full-blown AIDS. Steady debilitation and death typically follow within a period of years in untreated persons.

As yet, there is no cure for HIV infection and no vaccine that can protect against infection. The genes for the HIV coat proteins mutate constantly, making a vaccine developed against one form of the virus useless when the next form appears. Most of these mutations occur during replication of the virus, when reverse transcriptase makes a DNA copy of the viral RNA.

The development of AIDS can be greatly slowed by drugs that interfere with reverse transcription of the viral genomic RNA into the DNA copy that integrates into the host's genome. Treatment with a "cocktail" of drugs called *reverse transcriptase inhibitors (RTIs)* inhibits viral reproduction and destruction of helper T cells and extends the lives of people with the AIDS virus. The inhibiting cocktails are not a cure for AIDS, however, because the virus is still present in dormant form in helper T cells. If the therapy is stopped, the virus again replicates and the T-cell population drops.

Presently, the only certain way to avoid HIV infection is to refrain from unprotected sex with people whose HIV status is unknown, and from the use of contaminated needles of the type used to administer drugs intravenously.

---

ways to constrict (histamine also causes airway constriction). The resulting allergic reaction can vary in severity from a mild irritation to serious and even life-threatening debilitation. **Asthma** is a severe response to allergens involving constriction of airways in the lungs. **Antihistamines** (substances that block histamine receptors) are usually effective in countering the effects of the histamine released by mast cells.

An individual is *sensitized* by a first exposure to an allergen, which may produce only mild allergic symptoms or no reaction at all (see Figure 43.11A). However, the sensitization produces memory B and T cells; at the next and subsequent exposures, the system is poised to produce a greatly intensified allergic response (see Figure 43.11B).

In some persons, inflammation stimulated by an allergen is so severe that the reaction brings on a life-threatening condition called **anaphylactic shock.** Among other symptoms, extreme constriction of air passages in the lungs interferes with breathing, and massive leakage of fluid from capillaries causes the blood pressure to drop precipitously. Death may result in minutes if the condition is not treated promptly. In persons who have become sensitized to the venom of wasps and bees, for example, a single sting may bring on anaphylactic shock within minutes. Allergies developed against drugs such as penicillin and certain foods can have the same drastic effects. Anaphylactic shock can be controlled by immediate injection of epinephrine (adrenaline), which reverses the condition by constricting blood vessels and dilating air passages in the lungs.

STUDY BREAK 43.4

1. **What is immunological tolerance?**
2. **Explain how a failure in the immune system can result in an allergy.**

> **THINK OUTSIDE THE BOOK**
>
> A number of so-called new-generation drugs have been developed to treat patients with rheumatoid arthritis. One of them is abatacept (Orencia). Use the Internet or research literature to develop an outline of how abatacept works with respect to the adaptive immune system described in this section.

43.5 Defenses in Other Animals

This chapter has emphasized the mammalian immune system, the focus of most immunology research. We know relatively little about defenses against infections in most other vertebrate groups. Yet like all other physiological systems, mammalian defenses against pathogens are the result of evolution, and evidence of their functions can be seen in other vertebrate groups and also in invertebrates.

For example, molecular studies in sharks and rays have revealed DNA sequences that are clearly related to the sequences coding for antibodies in mammals. If injected with an antigen, sharks produce antibodies, formed from light- and heavy-chain polypeptides, capable of recognizing and binding the antigen. Although embryonic gene segments for the two polypeptides are arranged differently in sharks than they are in mammals, antibody diversity is produced by the same kinds of genetic rearrangements in both. Sharks also mount highly efficient nonspecific defenses, including production of a steroid that appears to kill bacteria and neutralize viruses.

Invertebrates lack specific immune defenses equivalent to antibodies and the activities of B and T cells, so their reactions to invading pathogens most closely resemble the nonspecific defenses of humans and other vertebrates. However, all invertebrates have phagocytic cells, which patrol tissues and engulf pathogens and other invaders. Some of the signal molecules that stimulate phagocytic activity, such as interleukins, appear to be similar in invertebrates and vertebrates.

Antibodies do not occur in invertebrates, but proteins of the immunoglobulin family are widely distributed. In at least some invertebrates, these Ig proteins have a protective function. In moths, for example, an Ig-family protein called *hemolin* binds to the surfaces of pathogens and marks them for removal by phagocytes.

Many invertebrates produce antimicrobial proteins such as lysozyme that are able to kill bacteria and other invading cells. Insects, for example, secrete lysozyme in response to bacterial infections.

> **STUDY BREAK 43.5**
>
> 1. Compare invertebrate and mammalian immune defenses.

UNANSWERED QUESTIONS

How does HIV evade the adaptive immunity system?

In cell-mediated immunity, a pathogen-infected antigen-presenting cell presents an antigen fragment bound to a class I major histocompatibility complex (MHC) protein to a $CD8^+$ T cell, stimulating the T cell to differentiate into cytotoxic T cells. The cytotoxic T cells then bind to the infected antigen-presenting cells and destroy them. Cytotoxic T cells act particularly against host cells infected by viral pathogens. HIV infects host cells but, rather than being eliminated by the host, this virus establishes a chronic infection that leads to the development of AIDS. That is, HIV evades the adaptive immunity system.

Kathleen Collins and her group at the University of Michigan Medical School have investigated the mechanism of this evasion. They have learned that the virus down-regulates the display of class I MHC proteins on the surfaces of HIV-infected antigen-presenting cells, which thereby limits the presentation of viral antigens by those cells. The down-regulation occurs by the action of the HIV Nef (*ne*gative *f*actor) protein. Nef binds to class I MHC molecules and inhibits them from moving through the Golgi complex to the cell surface. Without class I MHC molecules on the cell surface to present antigens, the immune response is compromised. The action of Nef, therefore, enhances the ability of HIV to induce AIDS.

A major research question addressed by Collins's lab is how does Nef disrupt the movement of class MHC molecules from the ER through the Golgi complex to the cell surface? Recent experiments have shown that the disruption results from an interaction of Nef with two cellular proteins, adaptor protein 1 and β-COP. The normal function of these cellular proteins is to help target other cellular proteins to their correct locations within the cell. Linking the two cellular proteins to Nef leads to trafficking of the class I MHC proteins to lysosomes (where they are degraded), rather than to the cell surface. The team is now working to develop pharmaceutical reagents aimed at blocking Nef action as a means of reducing HIV's attack on the immune system.

How does gene expression in a bacterial pathogen change during infection?

Characterizing the genetic events involved in the interactions between pathogens and hosts is important for understanding the development of an infectious disease in the host and for producing effective therapeutic treatments. You learned in this chapter that pathogenic bacteria are first combated by the innate immunity system. James Musser and his research group at The Methodist Research Institute, Houston, Texas, with collaborators at several other institutions, have characterized some aspects of the changes in gene expression for a bacterial pathogen, group A *Streptococcus* (GAS), during infection of a mammal.

GAS is a Gram-positive bacterium that is the cause of pharyngitis ("strep throat"; 2 million cases annually in the United States) and various other infections, including rheumatic heart disease. No vaccine is available to treat the infections. The researchers studied how gene expression in GAS changed during an 86-day period following an infection that caused pharyngitis in macaque monkeys. They are an excellent model for the study because the progression of pharyngitis caused by GAS in these monkeys is highly similar to that seen in human infections. There are three distinct phases, during which GAS can be detected by culturing throat swabs. In the first phase, colonization, GAS establishes infection of host cells, producing only mild pharyngitis. In the second phase, the acute phase, pharyngitis symptoms peak, as does the number of GAS bacteria. In the third phase, the asymptomatic phase, symptoms decrease and disappear along with a decrease in GAS bacteria.

Experimentally, the researchers analyzed gene expression from the entire genome of GAS in the three phases of pharyngitis using DNA microarrays (see Section 18.3). They found that the pattern of GAS gene expression changed over the course of the disease. Significantly, they saw characteristic gene expression patterns for each of the three phases of pathogen–host interaction. These results indicate that GAS regulates expression of its genes extensively as it establishes an infection, and as the host mounts an innate immune response against it.

This work likely will help direct future research efforts to control infections caused by GAS. More broadly, genomic studies of this kind are likely to provide insights into genetic events contributing to pathogenesis in other pathogen–host interactions. Currently, Musser's group and his collaborators are investigating the role of extracellular GAS proteins in host–pathogen interactions and the molecular steps involved in pathogenesis of human–GAS interactions.

Think Critically

How might research such as that by Musser's group on group A *Streptococcus* aid in the development of highly effective, novel antibiotics to which bacteria do not quickly evolve resistance?

Peter J. Russell

Peter J. Russell

REVIEW KEY CONCEPTS

Go to **CENGAGENOW** at www.cengage.com/login to access quizzing, animations, exercises, articles, and personalized homework help.

43.1 Three Lines of Defense against Pathogens

- Humans and other vertebrates have three lines of defense against pathogens. The first, which is nonspecific, is the barrier set up by the skin and mucous membranes.
- The second line of defense, also nonspecific, is innate immunity, an inborn system that defends the body against pathogens and toxins penetrating the first line.
- The third line of defense, adaptive immunity, is specific: it recognizes and eliminates particular pathogens and retains a memory of that exposure so as to respond rapidly if the pathogen is encountered again. The response is carried out by lymphocytes, a specialized group of leukocytes.

43.2 Innate Immunity: Nonspecific Defenses

- In the innate immune system, molecules on the surfaces of pathogens are recognized as foreign by receptors on host cells. The pathogen is then combated by the inflammation and complement systems.
- Epithelial surfaces secrete defensins, a type of antimicrobial peptide, in response to attack by a microbial pathogen. Defensins disrupt the plasma membranes of pathogens, killing them.
- Inflammation is characterized by heat, pain, redness, and swelling at the infection site. Several interconnecting mechanisms initiate inflammation, including pathogen engulfment, histamine secretion, cytokine release, and local blood vessel dilation and permeability increase (Figure 43.1).
- Large arrays of complement proteins are activated when they recognize molecules on the surfaces of pathogens. Some complement proteins form membrane attack complexes, which insert into the plasma membrane of many types of bacteria and cause their lysis. Fragments of other complement proteins coat pathogens, stimulating phagocytes to engulf them.
- Two nonspecific defenses are used to combat viral pathogens: interferons and natural killer cells.

Animation: Innate defenses

Animation: Inflammatory response

Animation: Complement proteins

Animation: Immune responses

Animation: Human lymphatic system

43.3 Adaptive Immunity: Specific Defenses

- Adaptive immunity, which is carried out by B and T cells, targets particular pathogens or toxin molecules.
- Antibodies consist of two light and two heavy polypeptide chains, each with variable and constant regions. The variable regions of the chains combine to form the specific antigen-binding site (Figure 43.3).
- Antibodies occur in five different classes: IgM, IgD, IgG, IgA, and IgE. Each class is determined by its constant region (Table 43.2).
- Antibody diversity is produced by genetic rearrangements in developing B cells that combine gene segments into intact genes encoding the light and heavy chains. The rearrangements producing heavy-chain genes and T-cell receptor genes are similar. The light- and heavy-chain genes are transcribed into pre-mRNAs, which are processed into finished mRNAs, which are translated on ribosomes into the antibody polypeptides (Figure 43.4).
- The antibody-mediated immune response has two general phases: T-cell activation and B-cell activation and antibody production. T-cell activation begins when a dendritic cell engulfs a pathogen and produces antigens, making the cell an antigen-presenting cell. The antigen-presenting cell secretes interleukins, which activate the T cell. The T cell then secretes other interleukins, which stimulate the T cell to proliferate, producing a clone of cells. The clonal cells differentiate into helper T cells (Figure 43.5).

- B-cell receptors on B cells recognize antigens on a pathogen and engulf it. The B cells then display the antigens. The T-cell receptor on a helper T cell activated by the same antigen binds to the antigen on the B cell. Interleukins from the T cell stimulate the B cell to produce a clone of cells with identical B-cell receptors. The clonal cells differentiate into plasma cells, which secrete antibodies specific for the antigen, and memory B cells, which provide immunological memory of the antigen encounter (Figures 43.5).
- Clonal expansion is the process of selecting a lymphocyte specifically for cloning when it encounters an antigen from among a randomly generated, large population of lymphocytes with receptors that specifically recognize the antigen (Figure 43.6).
- Antibodies clear the body of antigens by neutralizing or agglutinating them, or by aiding the innate immune response (Figure 43.7).
- In immunological memory, the first encounter of an antigen elicits a primary immune response and later exposure to the same antigen elicits a rapid secondary response with a greater production of antibodies (Figure 43.8).
- Active immunity is the production of antibodies in the body in response to an antigen. Passive immunity is the acquisition of antibodies by direct transfer from another person.
- Antibodies are widely used in research to identify, locate, and determine the functions of molecules in biological systems.
- Monoclonal antibodies are made by isolating fully active B cells from a test animal, fusing them with cancer cells to produce hybridomas, and using single hybridomas to start clones of cells, all of which make highly specific antibodies against the same epitope of an antigen (Figure 43.9).
- In cell-mediated immunity, cytotoxic T cells recognize and bind to antigens displayed on the surfaces of infected body cells, or to cancer cells. They then kill the infected body cell (Figure 43.10).

Animation: Antibody structure

Animation: Gene rearrangements

Animation: Antibody-mediated response

Animation: Clonal selection of a B cell

Animation: Cell-mediated response

43.4 Malfunctions and Failures of the Immune System

- In immunological tolerance, molecules present in an individual at birth normally do not elicit an immune response.
- In some people, the immune system malfunctions and reacts against the body's own proteins or cells, producing autoimmune disease.
- The first exposure to an allergen sensitizes an individual by leading to the production of memory B and T cells, which cause a greatly intensified response at the next and subsequent exposures.
- Most allergies result when antigens act as allergens by stimulating B cells to produce IgE antibodies, which leads to the release of histamine. Histamine produces the symptoms characteristic of allergies (Figure 43.11).

43.5 Defenses in Other Animals

- Antibodies, complement proteins, and other molecules with defensive functions have been identified in all vertebrates.
- Invertebrates rely on nonspecific defenses, including surface barriers, phagocytes, and antimicrobial molecules.

UNDERSTAND AND APPLY

Test Your Knowledge

1. Which of the following most accurately describes mammal defenses against disease-causing viruses or organisms?
   a. The three lines of defense work independently of each other in defending against a particular pathogen.
   b. Physical barriers are part of the immune system.
   c. The adaptive immune system reacts faster to pathogens than the innate immune system.
   d. Once the adaptive immune system is activated in response to a specific pathogen, the innate immune system stops functioning against that specific pathogen.
   e. White blood cells are key participants in adaptive immunity but not innate immunity.
2. Components of the inflammatory response include all *except:*
   a. macrophages.
   b. neutrophils.
   c. B cells.
   d. mast cells.
   e. eosinophils.
3. When a person's immune system resists infection by a pathogen after being vaccinated against it, this is the result of:
   a. innate immunity.
   b. immunological memory.
   c. a response with defensins.
   d. an autoimmune reaction.
   e. systemic inflammation.
4. One characteristic of a B cell is that it:
   a. has the same structure in both invertebrates and vertebrates.
   b. recognizes antigens held on class I major histocompatibility complex proteins.
   c. binds viral infected cells and directly kills them.
   d. makes many different B-cell receptors on its surface.
   e. has a B-cell receptor on its surface, which is the IgM molecule.
5. Antibodies:
   a. are each composed of four heavy and four light chains.
   b. display a variable end, which determines the antibody's location in the body.
   c. belonging to the IgE group are the major antibody class in the blood.
   d. found in large numbers in the mucous membranes belong to class IgG.
   e. function primarily to identify and bind antigens free in body fluids.
6. Place the following steps of recognition of an antigen by lymphocytes in the order in which they would occur:
   1. Bacteria degraded
   2. Dendritic cell engulfs bacteria
   3. T cells activated
   4. Antigens bind to class II MHC proteins
   5. Plasma cells and memory cells produced

   a. 1, 2, 3, 4, 5.
   b. 3, 2, 1, 4, 5.
   c. 4, 3, 1, 2, 5.
   d. 2, 1, 4, 3, 5.
   e. 5, 4, 3, 2, 1.

7. An antigen-presenting cell:
   a. can be a $CD8^+$ T cell.
   b. derives from a phagocytic cell and exposes an antigen to a lymphocyte.
   c. secretes antibodies.
   d. cannot be a B cell.
   e. cannot stimulate helper T cells.
8. Antibodies function to:
   a. deactivate the complement system.
   b. neutralize natural killer cells.
   c. clump bacteria and viruses for easy phagocytosis by macrophages.
   d. eliminate the chance for a secondary response.
   e. kill viruses inside of cells.
9. After Sally punctured her hand with a dirty nail she received both a vaccine and someone else's antibodies against tetanus toxin. The immunity conferred here is:
   a. both active and passive.
   b. active only.
   c. passive only.
   d. first active; later passive.
   e. innate.
10. Drugs are administered to patients to enhance the immune response when treating:
    a. organ transplant recipients.
    b. anaphylactic shock.
    c. rheumatoid arthritis.
    d. HIV infection.
    e. type I diabetes.

Discuss the Concepts

1. HIV wreaks havoc with the immune system by attacking helper T cells and macrophages. Would the impact be altered if the virus attacked only macrophages? Explain.
2. Given what you know about how foreign invaders trigger immune responses, explain why mutated forms of viruses, which have altered surface proteins, pose a monitoring problem for memory cells.
3. Cats, dogs, and humans may develop myasthenia gravis, an autoimmune disease in which antibodies develop against acetylcholine receptors in the synapses between neurons and skeletal muscle fibers. Based on what you know of the biochemistry of muscle contraction (see Section 41.1), explain why people with this disease typically experience severe fatigue with even small levels of exertion, drooping of facial muscles, and trouble keeping their eyelids open.

Design an Experiment

Space, the final frontier! Indeed, but being in space has some problems. Astronauts in space show a decline in their ability to mount an immune response and, consequently, develop a decreased resistance to infection. Two potentially important differences in physiology in space versus on Earth are more fluid flowing to the head and a lack of weight-bearing on the lower limbs. Could they be involved somehow in the deleterious effect on the immune system? Design an experiment to be done on Earth to answer this question.

Interpret the Data

In 2003, Michelle Khan and her coworkers published their findings on a 10-year study in which they followed cervical cancer incidence and HPV status in 20,514 women. All women who participated in the study were free of cervical cancer when the test began. Pap tests were taken at regular intervals, and the researchers used a DNA probe hybridization test to detect the presence of specific types of HPV in the women's cervical cells.

The results are shown as a graph of the incidence rate of cervical cancer by HPV type. Women who are HPV positive are often infected by more than one type, so the data were sorted into groups based on the women's HPV status ranked by type: either positive for HPV16; or negative for HPV16 and positive for HPV 18; or negative for HPV16/18 and positive for any other cancer-causing HPV; or negative for all cancer-causing HPV.

1. At 110 months into the study, what percentage of women who were not infected with any type of cancer-causing HPV had cervical cancer? What percentage of women who were infected with HPV16 also had cervical cancer?
2. In which group would women infected with both HPV16 and HPV18 fall?
3. Is it possible to estimate from this graph the overall risk of cervical cancer associated with infection of cancer-causing HPV of any type?
4. Do these data support the conclusion that being infected with HPV16 or HPV18 raises the risk of cervical cancer?

Cumulative incidence rate of cervical cancer correlated with HPV status in 20,514 women aged 16 years and older.

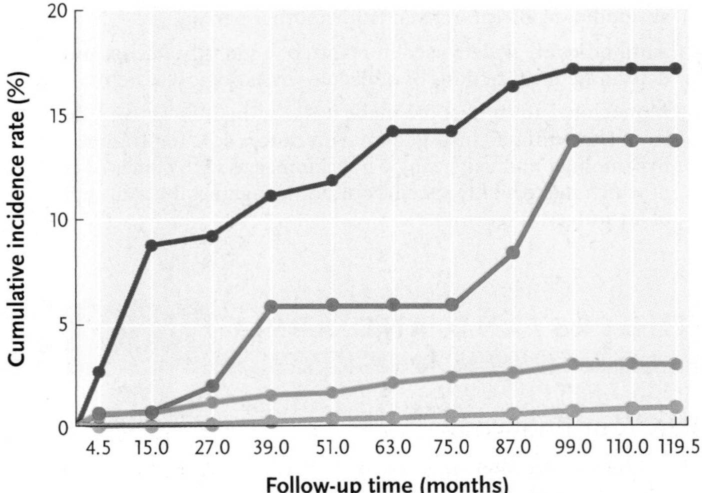
Cumulative incidence rate (%)
20
15
10
5
0
4.5
15.0
27.0
39.0
51.0
63.0
75.0
87.0
99.0
110.0
119.5
Follow-up time (months)

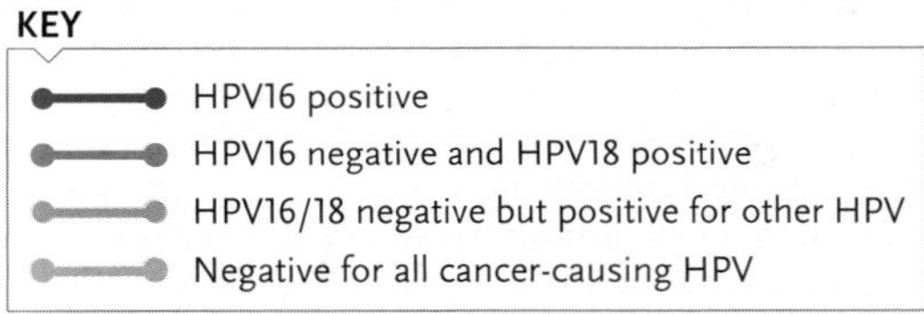
KEY
HPV16 positive
HPV16 negative and HPV18 positive
HPV16/18 negative but positive for other HPV
Negative for all cancer-causing HPV

Source: M. J. Khan et al. 2005. The elevated 10-year risk of cervical neoplasia in women with human papillomavirus (HPV) type 16 or 18 and the possible utility of type-specific HPV testing in clinical practice. *Journal of the National Cancer Institute* 97:1072–1079.

Apply Evolutionary Thinking

Defensins are found in a wide range of organisms, including plants as well as animals. What are the evolutionary implications of this observation?

Express Your Opinion

Drugs are available that can extend the life of patients with AIDS, but their high cost is more than people in most developing countries can afford to pay. Should the federal government offer incentives to companies to discount the drugs for developing countries? What about AIDS patients at home? Who should pay for their drugs? Go to www.cengage.com/login to investigate both sides of the issue and then vote.

Lining of the trachea (windpipe) shown in a colorized SEM, with mucus-secreting cells (white) and epithelial cells with cilia (pink). The trachea is positioned between the larynx and the lungs, providing a conduit for air entering and leaving the body.

SPL/Photo Researchers, Inc.

44

STUDY OUTLINE

Gas Exchange: The Respiratory System

Why It Matters. . . On October 25, 1999, at 9:19 A.M., the captain lined up Learjet N47BA on the runway at Orlando International Airport and opened the throttles. Within seconds, the corporate jet was airborne and climbing; 2 minutes later, at 9:21 EDT, the pilots reported passing through 9,500 feet.

As the jet continued its climb, the pressure of the outside air dropped steadily and with it the availability of the oxygen (O_2) that all animal life requires, including the two pilots and three passengers on the jet. Normally, in aircraft, the cabin pressure is maintained at a level equivalent to an altitude of 8,000 feet, more than sufficient to keep O_2 available to all on board. But, unknown to the pilots, the pressurization system was not functioning normally.

At 9:27 EDT, the controller at the Jacksonville Control Center instructed the jet to climb to 39,000 feet. The first officer's acknowledgment was the last radio transmission anyone was to hear from N47BA.

When humans experience increasingly higher altitudes, each breath brings less O_2 into the body. Of all the cells affected by reduced O_2, the ones most sensitive are those of the eyes and brain. Without an O_2 supply at 25,000 feet, most people progress from fully alert to unconscious in about 3 minutes; at 40,000 feet, the progression takes only 15 seconds.

The jet continued its climb, eventually reaching an altitude of 46,000 feet. When the pilots stopped responding to communications, military jets were sent to investigate. The military pilots could see no movement in the Learjet cabin and there was no response to their transmissions. The forward windshields of the Learjet were frosted over, indicating that warm air from the engines was not ventilating the cabin correctly. Evidently, the aircraft was maintaining its course through the autopilot, without conscious human direction.

Many hours later, at 12:11 P.M. CDT, one of the two engines failed: the aircraft, now unbalanced, rolled over and dived to a shattering impact in a field near Aberdeen, South Dakota. The subsequent investigation pointed to faulty operation of a valve controlling cabin pressurization as a likely cause of the accident. This tragic loss of life emphasizes the vital importance of O_2 to the survival of humans and other animals. In this chapter we discuss the respiratory system, the system that allows an animal to exchange CO_2 produced in the body for O_2 from the surroundings. The respiratory systems of animals reflect the environmental conditions under which they live, and this general principle has resulted in a truly remarkable array of adaptations. <

44.1 The Function of Gas Exchange

Physiological respiration is the process by which animals exchange gases with their surroundings—how they take in O_2 from the outside environment and deliver it to body cells, and remove CO_2 from body cells and deliver it to the environment **(Figure 44.1)**. The absorbed O_2 is used as the final electron acceptor for the oxidative reactions that produce ATP in mitochondria (see Section 8.4). The CO_2 released to the environment is a product of those oxidative reactions. Because they use O_2 and release CO_2, these ATP-producing reactions are called *cellular respiration.*

How gas exchange occurs in an animal depends on its respiratory medium—air or water—and the nature of its respiratory surface. The **respiratory medium** is the environmental source of O_2 and the "sink" for released CO_2. For aquatic animals the respiratory medium is water; for terrestrial animals, it is air. Amphibians and some fishes use both water and air as respiratory media. The exchange of gases with the respiratory medium by animals is called **breathing,** whether the medium is air or water.

The **respiratory surface,** formed by a layer of epithelial cells, provides the interface between the body and the respiratory medium. Oxygen is absorbed across the respiratory surface, and CO_2 is released. In all animals, the exchange of gases across the respiratory surface occurs by simple diffusion, movement of molecules from a region of higher concentration to a region of lower concentration (see Section 6.2).

Generally, the concentration of O_2 is higher in the respiratory medium than on the internal side of the respiratory surface, and thus the net diffusion of O_2 is inward. Carbon dioxide moves in the opposite direction because the CO_2 concentration is higher on the internal side of the respiratory surface than in the respiratory medium.

Respiratory surfaces typically have two structural properties that favor a high rate of diffusion: they are thin, and they have large surface areas. The rate of diffusion is inversely proportional to the square of the distance over which the diffusion occurs; diffusion rates are therefore higher through thin surfaces such as the single layer of epithelial cells forming many respiratory surfaces. And, the rate of diffusion is directly proportional to the surface area across which diffusion occurs, meaning that large surface areas allow for higher rates of gas exchange than small surface areas. In addition, the rate of diffusion becomes higher with larger concentration gradients and with increasing temperature.

In some relatively small animals, such as sponges, ctenophores, roundworms, flatworms, and some annelids, the entire body surface serves as the respiratory surface. All these animals are invertebrates that live in aquatic or moist environments.

In larger animals, specialized structures, *gills* and *lungs,* form the primary respiratory surface for exchanging gases with water and air, respectively. In insects, a **tracheal system,** an extensive system of branching tubes, channels air from the outside to the internal organs and most individual cells of the animal.

Because gases must dissolve in water to enter and leave epithelial cells, the respiratory surface must be wetted to function in gas exchange, either directly by the respiratory medium or by a thin film of water. For this reason, in water-breathing animals, **gills** are *evaginations* of the body: they extend outward into the respiratory medium. In terrestrial animals, **lungs** are typically pockets or *invaginations* of the body surface, buried deeply in the body interior where they are less susceptible to drying out. Terrestrial animals also have adaptations that moisten dry air before it reaches the respiratory surface. For example, in humans and other mammals, moisture is added to air as it passes through the mouth, nasal passages, throat, and air passages leading to the lungs.

The organ system responsible for gas exchange is termed the **respiratory system.** The respiratory system consists of all the parts of the body involved in exchanging air between the external environment and the blood. In mammals, this includes the airways leading to and into the lungs, the lungs themselves, and the structures of the chest used to move air through the airways into and out of the lungs.

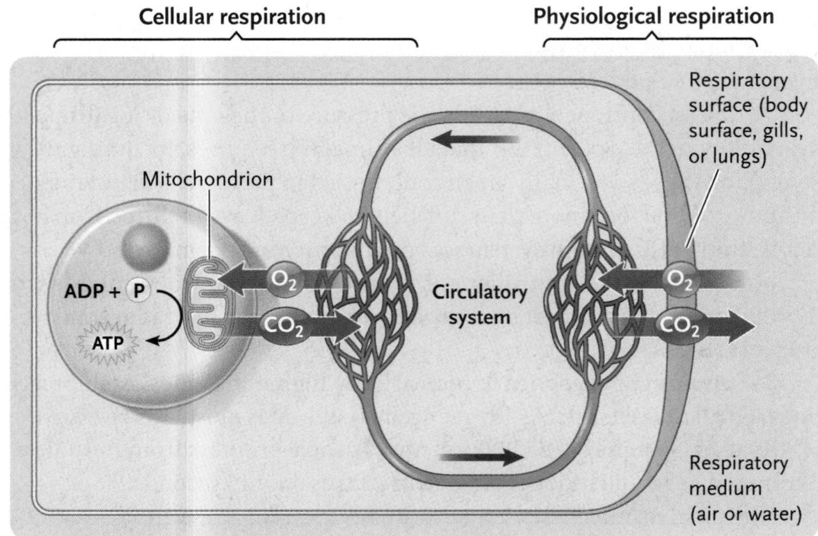
Cellular respiration
Physiological respiration
Respiratory surface (body surface, gills, or lungs)
Mitochondrion
ADP + P
ATP
O2
CO2
Circulatory system
O2
CO2
Respiratory medium (air or water)

FIGURE 44.1
The relationship between cellular respiration and physiological respiration.

Adaptations That Increase Ventilation and Perfusion of the Respiratory Surface Maximize the Rate of Gas Exchange

Two primary adaptations help animals maintain the difference in concentration between gases outside and inside the respiratory surface, thereby keeping the rate of gas exchange at maximal levels:

1. **Ventilation,** the flow of the respiratory medium (air or water, depending on the animal) over the external side of the respiratory surface.
2. **Perfusion,** the flow of blood or other body fluids on the internal side of the respiratory surface.

VENTILATION As they respire, animals remove O_2 from the respiratory medium and replace it with CO_2. Without ventilation, the concentration of O_2 would fall in the respiratory medium close to the respiratory surface, and the concentration of CO_2 would rise, gradually reducing the concentration gradients and dropping the rate of gas exchange below the minimum level required to sustain life. Examples of ventilation include the one-way flow of water over the gills in fish and many other aquatic animals and the in-and-out flow of air in the lungs of most vertebrates and in the tracheal system of insects.

PERFUSION The constant replacement of blood or another fluid on the internal side of the respiratory surface helps to keep the inside/outside concentration differences of O_2 and CO_2 at a maximum. In animals without a circulatory system, such as roundworms and flatworms, body movements help circulate body fluids beneath the skin. Most animals without a circulatory system are small or have thin, greatly flattened bodies, because all body cells must be located close to the respiratory surface to exchange O_2 and CO_2 adequately. In animals with a circulatory system, the circulatory system brings blood to the internal side of the respiratory surface, transporting CO_2 from all cells of the body—no matter how far they are from the respiratory surface—to exchange for O_2, which is then taken to all cells of the body.

Adaptations That Increase the Area of the Respiratory Surface Maximize the Quantity of Gases Exchanged

Most animals have adaptations that increase the quantity of gases exchanged by increasing the area of the respiratory surface. In animals whose skin serves as the respiratory surface, an elongated or flattened body form increases the area of the respiratory surface **(Figure 44.2A)**.

In animals with gills, the respiratory surface is increased by highly branched structures that include many fingerlike or platelike projections **(Figure 44.2B)**. Similarly, in animals with lungs or tracheae, the respiratory surface is increased by a multitude of branched tubes, folds, or pockets **(Figure 44.2C)**.

Water and Air Have Advantages and Disadvantages as Respiratory Media

Because their respiratory surfaces are exposed directly to the environment, water breathers have no problem keeping the respiratory surface wetted. However, aquatic animals face two main challenges in obtaining O_2 from water compared with terrestrial animals. First, water contains approximately one-thirtieth as much O_2 as air does (at 15°C). Therefore, to obtain the same amount of O_2, an aquatic animal must process 30 times as much of its respiratory medium as a terrestrial animal does. Second, water is about 1,000 times as dense as air and about 50 times as viscous. Therefore, it takes significantly more energy to move water than air over a respiratory surface. For this reason, ventilation in most aquatic animals takes place in a one-way direction. In bony fishes, for instance, water enters the mouth, flows over the gills, and exits through the gill covers, all in one direction.

In addition, temperature and solutes affect the O_2 content of water. That is, as either the temperature or the amount of solutes increases, the amount of gas that can dissolve in water decreases. Therefore, for obtaining O_2, aquatic animals that live in warm water are at a disadvantage compared with those that live in cold water. And, because levels of solutes (such as sodium chloride) are higher in seawater than in freshwater, aquatic animals living in seawater are at a disadvantage.

The relatively high O_2 content, low density, and low viscosity of air greatly reduce the energy required to ventilate the respiratory surface. These advantages allow animals with lungs to breathe in and out, reversing the direction of flow of the respiratory medium, without a large energy penalty. As you will see later, reversing the direction of flow decreases the efficiency of gas exchange.

A. **Extended body surface: flatworm**

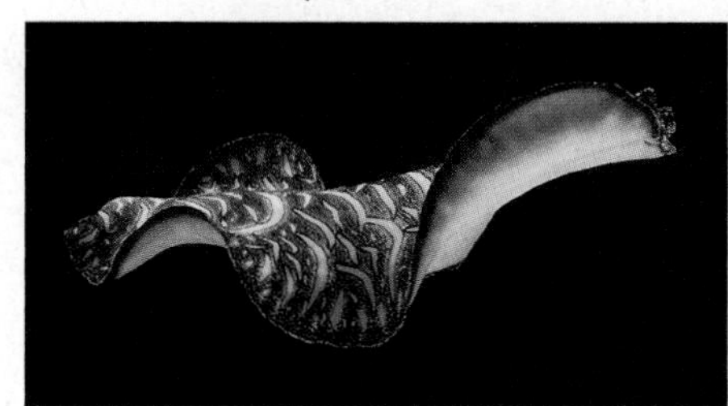

Peter Parks/OSF/Animals Animals–Earth Scenes

B. **External gills: mudpuppy**

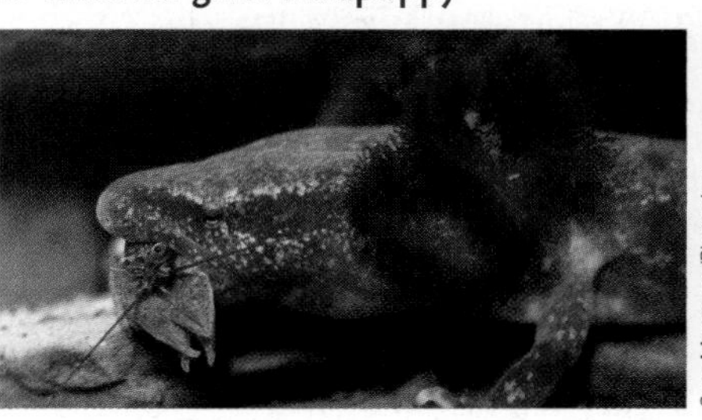

Gary Meszaros/Photoshot

C. **Lungs: human**

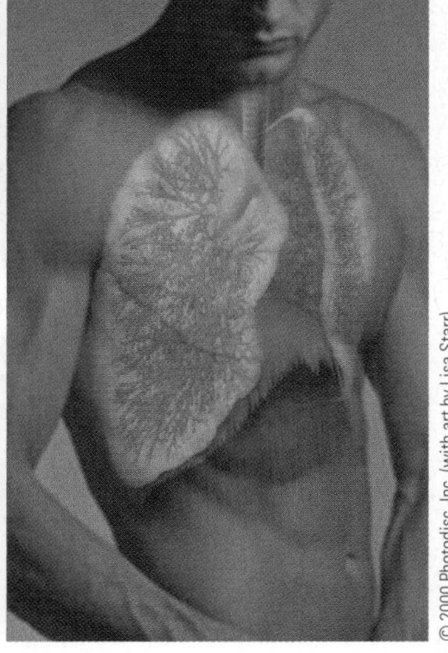

© 2000 Photodisc, Inc. (with art by Lisa Starr)

FIGURE 44.2
Adaptations increasing the area of the respiratory surface. (A) The flattened and elongated body surface of a flatworm. **(B)** The highly branched, feathery structure of the external gills in an amphibian, the mudpuppy *(Necturus)*. **(C)** The many branches and pockets expanding the respiratory surface in the human lung.

Another advantage of breathing air is that gas molecules diffuse nearly 10,000 times faster through air than through water. This increases the rate at which molecules of the gases at the respiratory surface exchange with those located farther away in the air and reduces the requirement for ventilation as compared with water.

A major disadvantage of air is that it constantly evaporates water from the respiratory surface unless the air is saturated with water vapor. Therefore, except in an environment with 100% humidity, animals lose water by evaporation during breathing and must replace the water to keep the respiratory surface from drying and causing the death of the surface cells.

We next turn to the adaptations that allow water-breathing and air-breathing animals to obtain O_2 and release CO_2 in aquatic and terrestrial environments. These adaptations allow animals to exploit the advantages and circumvent the disadvantages of water and air as respiratory media.

STUDY BREAK 44.1 <

1. **Distinguish between the roles of the respiratory medium and the respiratory surface in respiratory systems.**
2. **What is an advantage of water over air as a respiratory medium? What are two key advantages of air over water as a respiratory medium?**

44.2 Adaptations for Respiration

Although most animals that live in water exchange gases through the skin or gills, some, such as whales, seals, and dolphins, exchange gases through lungs (which originally evolved in aquatic creatures). And although most animals that live on land exchange gases through lungs, some, such as sow bugs and land crabs, exchange gases through gills, and others, such as insects, exchange gases using a tracheal system.

Aquatic Gill Breathers Exchange Gases More Efficiently Than Skin Breathers

Gills provide water-breathing animals, and a few air-breathers, with more efficient gas exchange than skin breathers have. In combination with the organized circulatory system common to these animals, gills also allow animals to live in more diverse habitats, and to achieve greater body mass, than animals that breathe primarily or exclusively through the skin.

EXTERNAL AND INTERNAL GILLS Gills are respiratory surfaces that are branched and folded evaginations of the body surface. **External gills** are gills that do not have protective coverings; they extend out from the body and are in direct contact with the water. With no protective coverings, external gills are exposed to mechanical damage and must be immersed in water

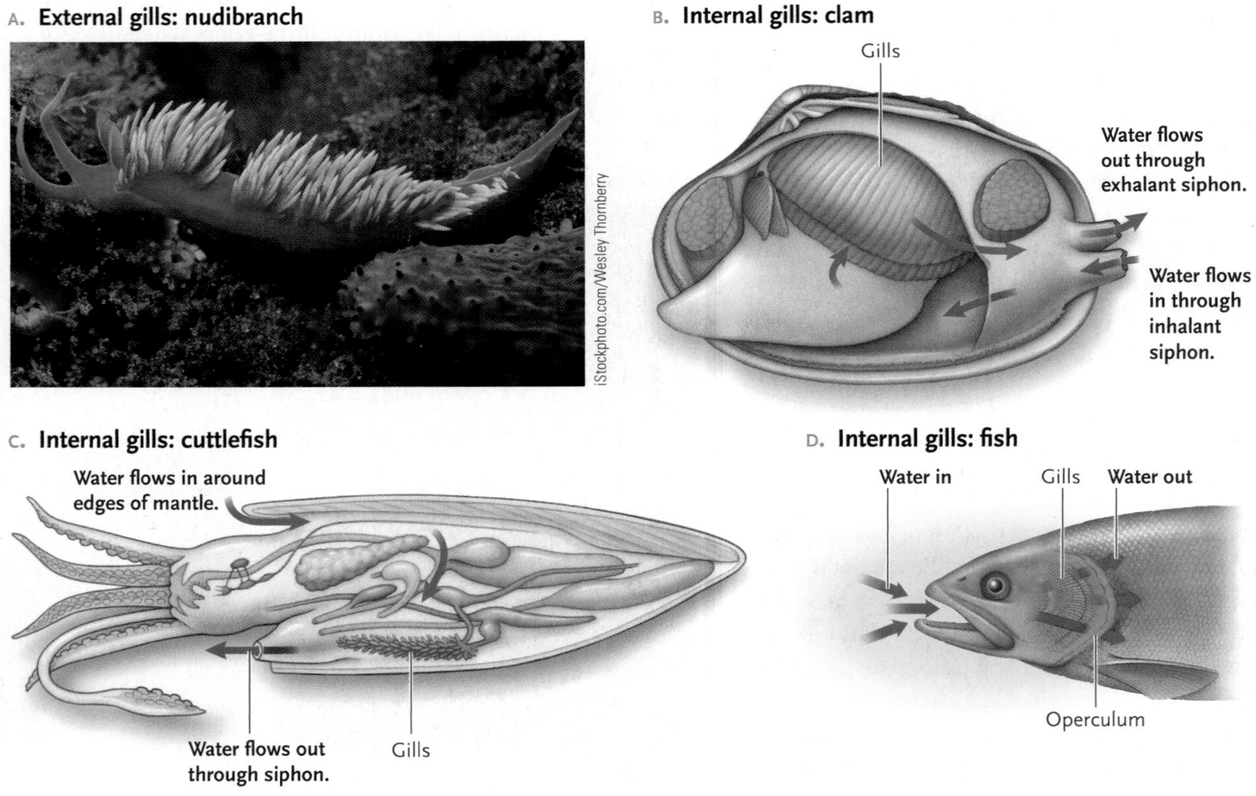

A. External gills: nudibranch
iStockphoto.com/Wesley Thornberry
B. Internal gills: clam
Gills
Water flows out through exhalant siphon.
Water flows in through inhalant siphon.
C. Internal gills: cuttlefish
Water flows in around edges of mantle.
Water flows out through siphon.
Gills
D. Internal gills: fish
Water in
Gills
Water out
Operculum

FIGURE 44.3
External and internal gills. **(A)** The external gills of a nudibranch *(Flabellina iodinea)*. **(B)** The internal gills in a clam. **(C)** The internal gills in a cuttlefish. **(D)** Internal gills of a bony fish. Water enters through the mouth and passes over the filaments of the gills before exiting through an opening at the edges of the flaplike protective covering, the operculum.

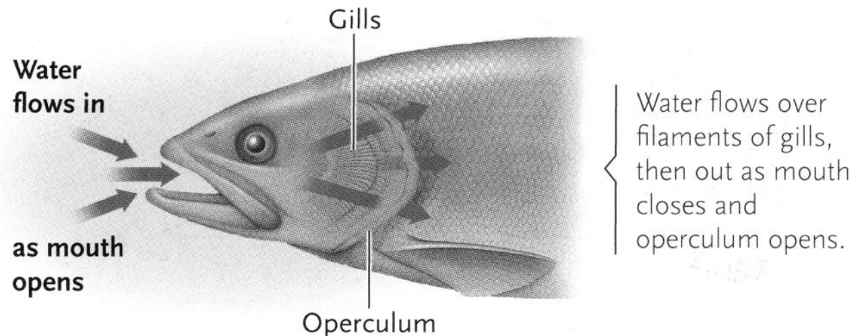

Gills
Water flows in
as mouth opens
Operculum
Water flows over filaments of gills, then out as mouth closes and operculum opens.

A. The flow of water around the gill filaments

B. Countercurrent flow in fish gills, in which the blood and water move in opposite directions

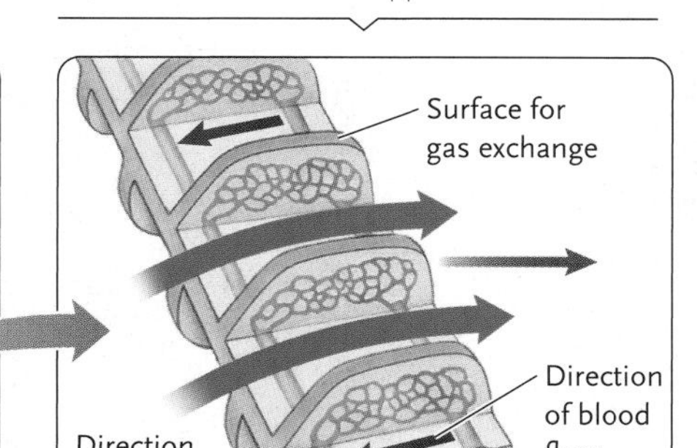

Gill arch
Filament of gill
Surface for gas exchange
Direction of blood flow
Direction of water flow
Oxygenated blood flows out of filament.
Deoxygenated blood flows into filament.

C. In countercurrent exchange, blood leaving the capillaries has the same O_2 content as fully oxygenated water entering the gills

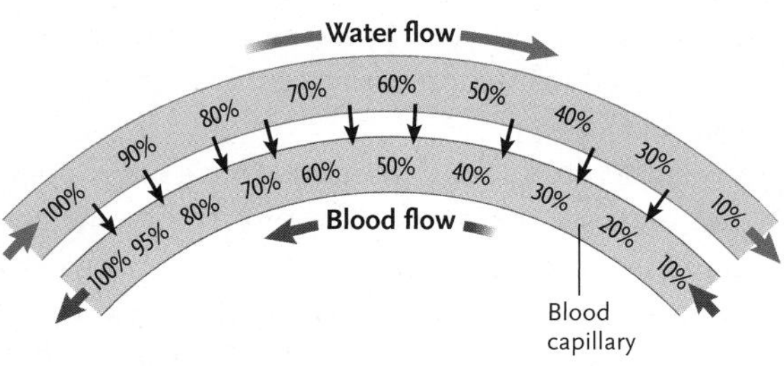

Water flow
100% 90% 80% 70% 60% 50% 40% 30% 10%
100% 95% 80% 70% 60% 50% 40% 30% 20% 10%
Blood flow
Blood capillary

FIGURE 44.4

Ventilation and countercurrent exchange in bony fishes. (A) Water flows around the gill filaments. **(B)** Water and blood flow in opposite directions through the gill filaments. **(C)** Countercurrent exchange: oxygen from the water diffuses into the blood, raising its oxygen content. The percentages indicate the degree of oxygenation of water (blue) and blood (red)

to keep them from collapsing or drying. For these reasons, animals with external gills, including some annelids and mollusks **(Figure 44.3A)**, aquatic insects, the larval forms of some bony fishes, and some amphibians, are limited to relatively protected aquatic environments.

Internal gills, by contrast, are located within chambers of the body that have a cover providing physical protection for the gills and protecting them from drying. Water must be brought to internal gills. Covered internal gills allow animals to live in highly diverse aquatic habitats and even in moist terrestrial habitats. Most crustaceans, mollusks, sharks, and bony fishes have internal gills. Some invertebrates, such as clams and oysters, use beating cilia to circulate water over their internal gills **(Figure 44.3B)**. Others, such as the cuttlefish, use contractions of the muscular mantle to pump water over their gills **(Figure 44.3C)**. In adult bony fishes, the gills extend into a chamber covered by gill flaps or *opercula* (singular, *operculum* = little lid) on either side of the head. The operculum also serves as part of a one-way pumping system that ventilates the gills **(Figure 44.3D)**.

Many Animals with Internal Gills Use Countercurrent Flow to Maximize Gas Exchange

Sharks, fishes, and some crabs take advantage of one-way flow of water over the gills to maximize the amounts of O_2 and CO_2 exchanged with water. In this mechanism, called **countercurrent exchange,** the water flowing over the gills moves in a direction opposite to the flow of blood under the respiratory surface.

Figure 44.4 illustrates countercurrent exchange in the uptake of O_2. At the point where fully oxygenated water first passes over a gill filament in countercurrent flow, the blood flowing beneath it in the opposite direction is also almost fully oxygenated. However, O_2 concentration is still higher in the water than in the blood, and the gas diffuses from the water into the blood, raising the concentration of O_2 in the blood almost to the level of the fully oxygenated water. At the opposite end of the filament, much of the O_2 has been removed from the water, but the blood flowing under the filament, which has just arrived from body tissues and is fully deoxygenated, contains even less O_2. As a result, O_2 also diffuses from the water to the blood at this end of the filament. All along the gill filament, the same relationship exists, so that at any point, the water is more highly oxygenated than the blood, and O_2 diffuses from the water into the blood across the respiratory surface.

The overall effect of countercurrent exchange is the removal of 80 to 90% of the O_2 content of water as it flows over the gills. In comparison, by breathing in and out and constantly reversing the direction of air flow, mammals manage to remove only about 25% of the O_2 content of air. Efficient removal of O_2 from water is important because of the much lower O_2 content of water compared with air.

Insects Use a Tracheal System for Gas Exchange

Insects breathe air by a respiratory system consisting of air-conducting tubes called **tracheae** (singular, *trachea*). The tracheae are invaginations of the outer epidermis of the animal, reinforced by rings of chitin, the material of the insect exoskeleton. They lead from the body surface and branch so extensively inside the animal that almost every cell is served by a microscopic branch **(Figure 44.5)**. Some branches even penetrate inside larger cells, such as those of insect flight muscles. The finest branches of the tracheae, called *tracheoles,* form the respiratory surface of the insect system.

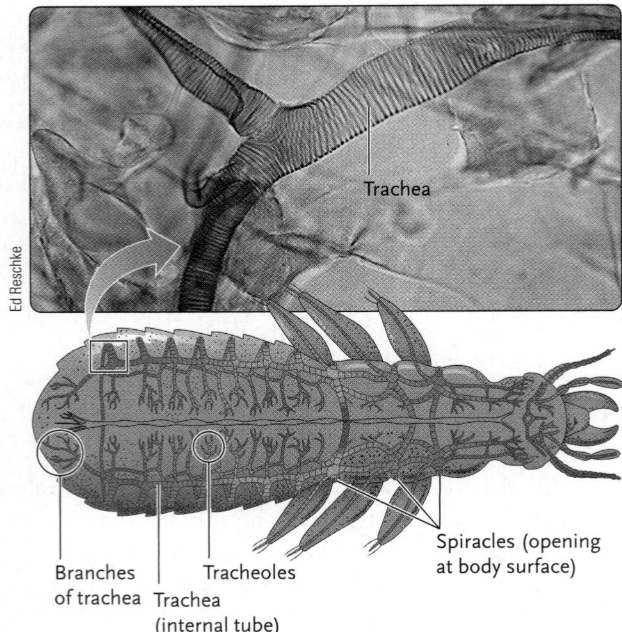
Trachea
Ed Reschke
Branches of trachea
Tracheoles
Trachea (internal tube)
Spiracles (opening at body surface)

FIGURE 44.5
The tracheal system of insects. Chitin rings, visible in the photomicrograph, reinforce many of the tracheae.

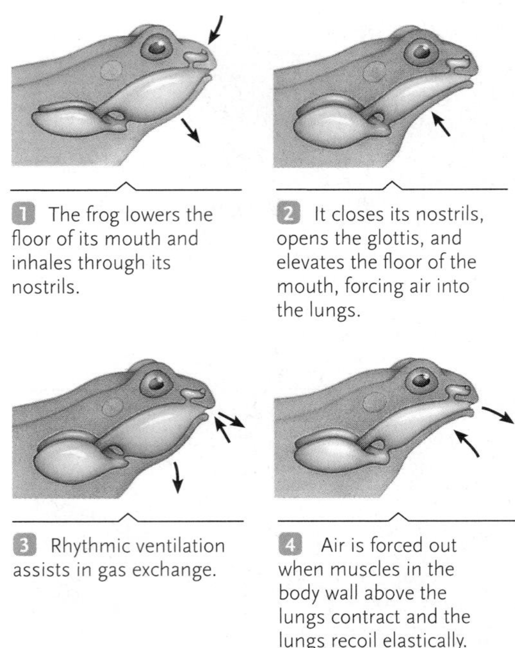
1 The frog lowers the floor of its mouth and inhales through its nostrils.
2 It closes its nostrils, opens the glottis, and elevates the floor of the mouth, forcing air into the lungs.
3 Rhythmic ventilation assists in gas exchange.
4 Air is forced out when muscles in the body wall above the lungs contract and the lungs recoil elastically.

FIGURE 44.6
Positive pressure breathing in an amphibian (a frog).

Tracheoles are dead-end tubes with very small fluid-filled tips that are in contact with cells of the body. Air is transported by the tracheal system to those tips, and gas exchange occurs directly across the plasma membranes of the body cells in contact with the tips. At places within the body, the tracheae expand into internal air sacs that act as reservoirs to increase the volume of air in the system.

Air enters and leaves the tracheal system at openings in the insect's chitinous exoskeleton called **spiracles** (*spiraculum* = airhole). In adult insects, the spiracles are located in a row on either side of the thorax and abdomen. The spiracles open and close in coordination with body movements to compress and expand the air sacs and pump air in and out of the tracheae. During insect flight, alternating compression and expansion of the thorax by the flight muscles also pump air through the tracheal system.

Lungs Allow Animals to Live in Completely Terrestrial Environments

Lungs are one of the primary adaptations that allowed animals to fully invade terrestrial environments. Some fishes and amphibians have lungs, as do all reptiles, birds, and mammals. All lungs are invaginated structures located internally in the body.

In some fishes, such as lungfishes, lungs and air breathing evolved as adaptations to survive in oxygen-poor water or temporarily in air when the water level dropped and exposed them. The lungs of these fishes consist of thin-walled sacs, which branch off from the mouth, pharynx, or parts of the digestive system; air is obtained by **positive pressure breathing,** a gulping or swallowing motion that forces air into the lungs.

The lungs of mature amphibians such as frogs and salamanders are also thin-walled sacs with relatively little folding or pocketing. Amphibians also fill their lungs by positive pressure breathing, in this case using a rhythmic motion of the floor of the mouth as the pump, in coordination with opening and closing of the nostrils **(Figure 44.6).**

The lungs of reptiles, birds, and mammals have many pockets and folds that increase the area of the respiratory surface, which contains dense, highly branched capillary networks. Mammalian lungs consist of millions of tiny air pockets, the **alveoli,** each surrounded by dense capillary networks. Reptiles and mammals fill their lungs by **negative pressure breathing**—by muscular contractions that expand the lungs, lowering the pressure of the air in the lungs and causing air to be pulled inward. (Mammalian negative pressure breathing is described in more detail in the next section.)

In birds, a countercurrent exchange system provides the most complex and efficient vertebrate lungs **(Figure 44.7).** In addition to paired lungs, birds have nine pairs of air sacs that branch off the respiratory tract. The air sacs, which collectively contain several times as much air as the lungs, set up a pathway that allows air to flow in one direction through the lungs, rather than in and out as in other vertebrates. Within the lungs, air flows through an array of fine, parallel tubes that are surrounded by a capillary network. The blood flows in the direction opposite to the air flow, setting up a countercurrent exchange. The countercurrent exchange allows bird lungs to extract about one-third of the O_2 from the air as compared with about one-fourth in the lungs of mammals.

A. Lungs and air sacs of a bird

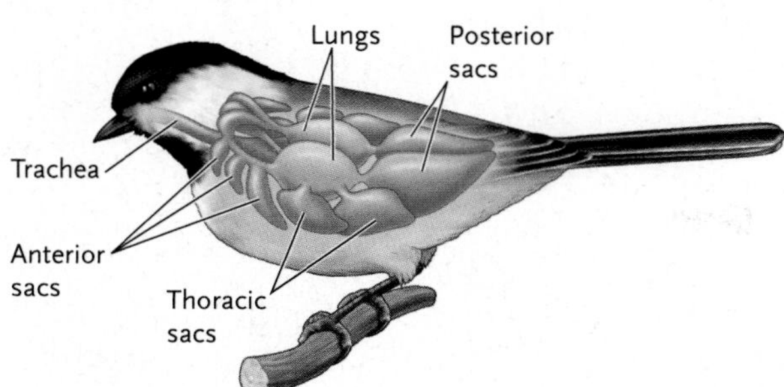

Lungs
Posterior sacs
Trachea
Anterior sacs
Thoracic sacs

B. Countercurrent exchange

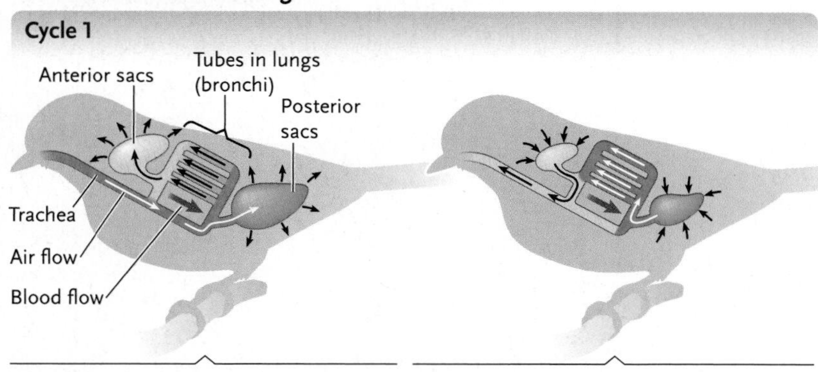

Cycle 1
Anterior sacs
Tubes in lungs (bronchi)
Posterior sacs
Trachea
Air flow
Blood flow

1 During the first inhalation, most of the oxygen flows directly to the posterior air sacs. The anterior air sacs also expand but do not receive any of the newly inhaled oxygen.

2 During the following exhalation, both anterior and posterior air sacs contract. Oxygen from the posterior sacs flows into the gas-exchanging tubes (bronchi) of the lungs.

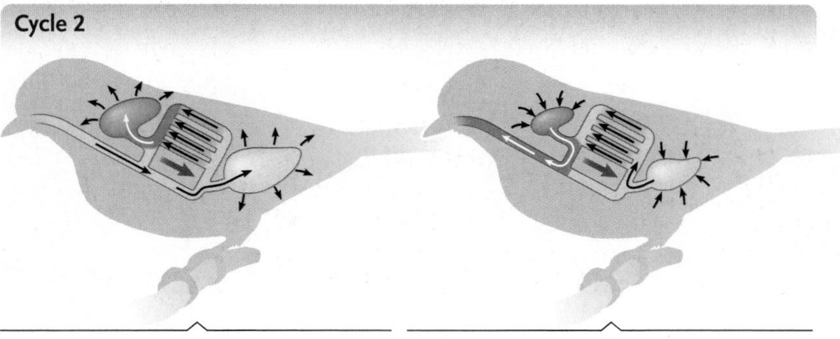

Cycle 2

1 During the next inhalation, air from the lung (now deoxygenated) moves into the anterior air sacs.

2 In the second exhalation, air from anterior sacs is expelled to the outside through the trachea.

FIGURE 44.7

Countercurrent exchange in bird lungs. **(A)** Unlike mammalian lungs, bird lungs do not expand and contract. Changes in pressure in the expandable air sacs move air in and out. **(B)** Air flows in one direction through the tubes of the lungs; blood flows in the opposite direction in the surrounding capillary network. Two cycles of inhalation and exhalation are needed to move a specific volume of air through the bird respiratory system.

44.3 The Mammalian Respiratory System

All mammals have a pair of lungs and a muscular diaphragm in the chest cavity that plays an important role in negative pressure breathing. (Birds and reptiles lack muscular diaphragms, illustrating their separate evolutionary lineage.) Rapid ventilation of the respiratory surface and perfusion by blood flow through dense capillary networks maximizes gas exchange.

The Airways Leading from the Exterior to the Lungs Filter, Moisten, and Warm the Entering Air

The human respiratory system is typical for a terrestrial mammal **(Figure 44.8).** Air enters and leaves the respiratory system through the nostrils and mouth. Hairs in the nostrils and mucus covering the surface of the airways filter out and trap dust and other large particles. Inhaled air is moistened and warmed as it moves through the mouth and nasal passages.

Next, air moves into the throat, or **pharynx,** which forms a common pathway for air entering the **larynx** (or "voice box") and food entering the esophagus, the tube that leads to the stomach. The airway through the larynx is open except during swallowing.

From the larynx, air moves into the trachea (or windpipe), which branches into two airways, the **bronchi** (singular, *bronchus*). The bronchi lead to the two elastic, cone-shaped lungs, one on each side of the chest cavity. Inside the lungs, the bronchi narrow and branch repeatedly, becoming progressively narrower and more numerous. The terminal airways, the **bronchioles,** lead into cup-shaped pockets, the alveoli (singular, *alveolus;* shown in Figure 44.8 insets).

Each of the 150 million alveoli in each lung is surrounded by a dense network of capillaries. By the time inhaled air reaches the alveoli, it has been moistened to the saturation point and brought to body temperature. The many alveoli provide an enormous area for gas exchange. If the alveoli of an adult human were flattened out in a single layer, they would cover an area approaching 100 square meters, about the size of a tennis court!

STUDY BREAK 44.2

1. **What advantages do gills confer on a water-breathing animal over skin breathing?**
2. **What is countercurrent exchange, and how is it beneficial for gas exchange?**
3. **How does the tracheal system of insects facilitate gas exchange with the cells of the body?**
4. **Distinguish between positive pressure breathing and negative pressure breathing in animals with lungs.**

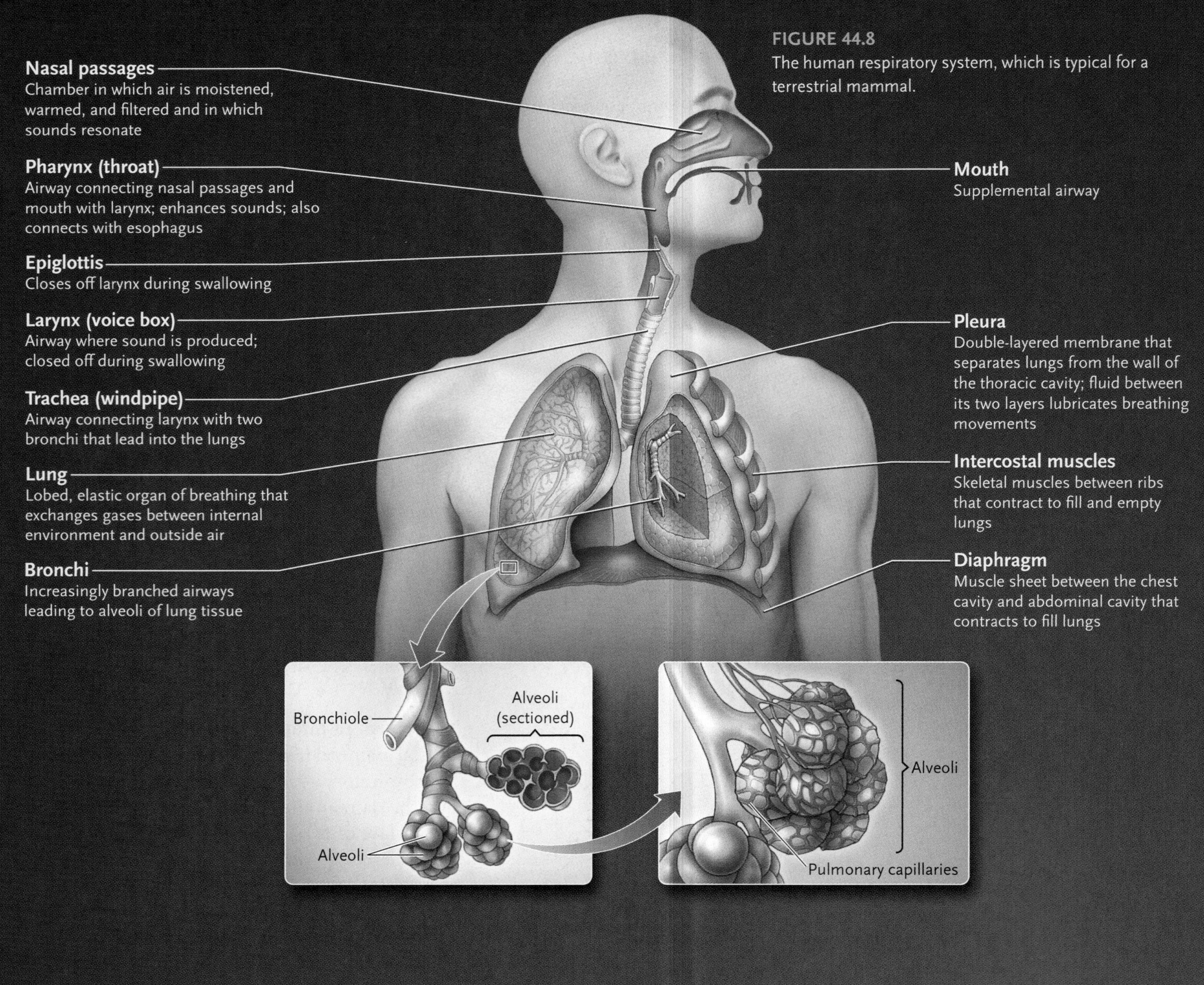
Nasal passages
Chamber in which air is moistened, warmed, and filtered and in which sounds resonate
Pharynx (throat)
Airway connecting nasal passages and mouth with larynx; enhances sounds; also connects with esophagus
Epiglottis
Closes off larynx during swallowing
Larynx (voice box)
Airway where sound is produced; closed off during swallowing
Trachea (windpipe)
Airway connecting larynx with two bronchi that lead into the lungs
Lung
Lobed, elastic organ of breathing that exchanges gases between internal environment and outside air
Bronchi
Increasingly branched airways leading to alveoli of lung tissue
Mouth
Supplemental airway
Pleura
Double-layered membrane that separates lungs from the wall of the thoracic cavity; fluid between its two layers lubricates breathing movements
Intercostal muscles
Skeletal muscles between ribs that contract to fill and empty lungs
Diaphragm
Muscle sheet between the chest cavity and abdominal cavity that contracts to fill lungs
Bronchiole
Alveoli (sectioned)
Alveoli
Alveoli
Pulmonary capillaries

FIGURE 44.8
The human respiratory system, which is typical for a terrestrial mammal.

The larynx, trachea, and larger bronchi are nonmuscular tubes encircled by rings of cartilage that prevent the tubes from compressing. The largest of the rings, which reinforces the larynx, stands out at the front of the throat as the Adam's apple; smaller supporting rings can be felt at the front of the throat just below the larynx. The walls of the smaller bronchi and the bronchioles contain smooth muscle cells that contract or relax to control the diameter of these passages, and with it, the amount of air flowing to and from the alveoli.

The epithelium lining each bronchus contains cilia and mucus-secreting cells. Bacteria and airborne particles such as dust and pollen are trapped in the mucus and then moved upward and into the throat by the beating of the cilia lining the airways. Infection-fighting macrophages also patrol the respiratory epithelium.

Tobacco smoke, by paralyzing the cilia lining the respiratory tract, interferes with the processes that clear bacteria and airborne particles from the lungs. The bacteria and foreign matter persisting in the lungs can cause infections and smoker's cough.

Contractions of the Diaphragm and Muscles between the Ribs Ventilate the Lungs

The lungs are located in the rib cage above the *diaphragm,* a dome-shaped sheet of skeletal muscle separating the chest cavity from the abdominal cavity. The lungs are covered by a double layer of epithelial tissue called the **pleura.** The inner pleural layer is attached to the surface of the lungs, and the outer layer is attached to the surface of the chest cavity. A narrow space between the inner and outer layers is filled with slippery fluid, which allows the lungs to move within the chest cavity without rubbing or abrasion as they expand and contract.

Contraction of muscles between the ribs and the diaphragm brings air into the lungs by a negative pressure mechanism. As an inhalation begins, the diaphragm contracts and flattens, and one set of muscles between the ribs, the *external intercostal* muscles, contracts, pulling the ribs upward and outward **(Figure 44.9).** These movements expand the chest cavity and lungs, lowering the air pressure in the lungs below that of the atmosphere. As a result, air is drawn into the lungs, expanding and filling them.

The expansion of the lungs is much like filling two rubber balloons. Like balloons, the lungs are elastic, and resist stretching as they are filled. Also like balloons, the stretching stores energy, which can be released to expel air from the lungs. When a person at rest exhales, the diaphragm and muscles between the ribs relax, and the elastic recoil of the lungs expels the air.

When physical activity increases the body's demand for O_2, other muscles help expel the air by forcefully reducing the volume of the chest cavity. Contractions of abdominal wall muscles increase abdominal pressure, exerting an upward-directed force on the diaphragm and thus pushing it upward. Contractions of *internal intercostal* muscles pull the chest wall inward and downward, causing it to flatten. As a result, the dimensions of the chest cavity decrease.

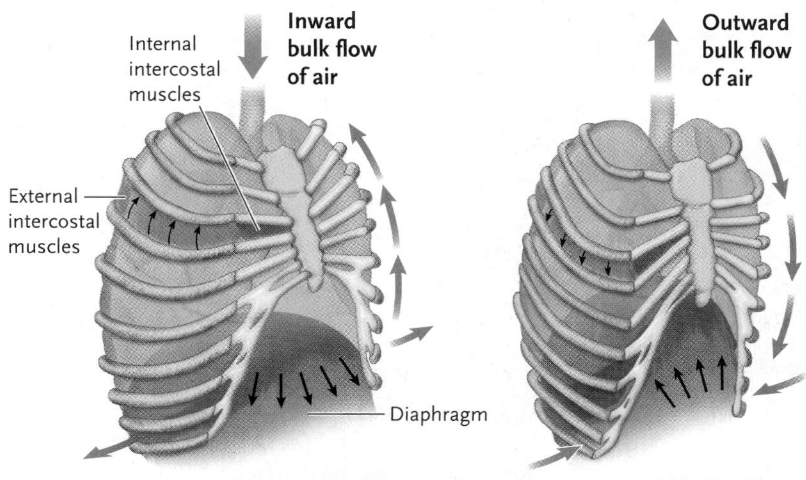
Internal intercostal muscles
Inward bulk flow of air
Outward bulk flow of air
External intercostal muscles
Diaphragm

Inhalation. Diaphragm contracts and moves down. The external intercostal muscles contract and lift rib cage upward and outward. The lung volume expands.

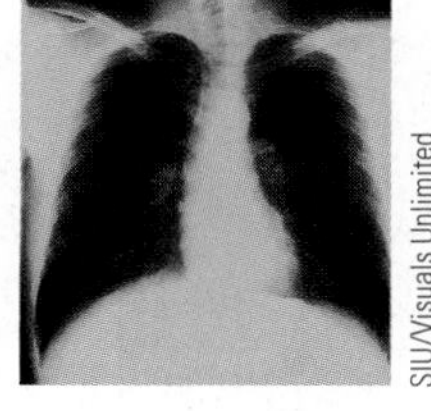
SIU/Visuals Unlimited

Exhalation during breathing or rest. Diaphragm and external intercostal muscles return to the resting positions. Rib cage moves down. Lungs recoil passively.

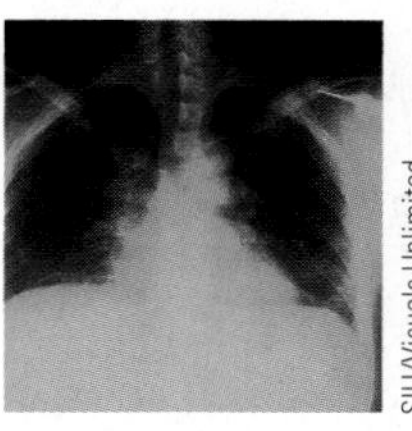
SIU/Visuals Unlimited

FIGURE 44.9

The respiratory movements of humans during breathing at rest. The movements of the rib cage and diaphragm fill and empty the lungs. Inhalation is powered by contractions of the external intercostal muscles and diaphragm, and exhalation is passive. During exercise or other activities characterized by deeper and more rapid breathing, contractions of the internal intercostal muscles and the abdominal muscles add force to exhalation. The X-ray images show how the volume of the lungs increases and decreases during inhalation and exhalation.

The Volume of Inhaled and Exhaled Air Varies over Wide Limits

The volume of air entering and leaving the lungs during inhalation and exhalation is called the **tidal volume.** In a person at rest, the tidal volume amounts to about 500 mL. As physical activity increases, the tidal volume increases to match the body's needs for O_2; at maximal levels, the tidal volume reaches about 3,400 mL in females and 4,800 mL in males. The maximum tidal volume is called the **vital capacity** of an individual.

Even after the most forceful exhalation, about 1,000 mL of air remains in the lungs in females, and about 1,200 mL in males; this is the **residual volume** of the lungs. In fact, the lungs cannot be deflated completely because small airways collapse during forced exhalation, blocking further outflow of air. Because air cannot be removed from the lungs completely, some gas exchange can always occur between blood flowing through the lungs and the air in the alveoli.

The Centers That Control Breathing Are Located in the Brain Stem

Breathing is controlled by centers in the medulla and pons, which form part of the brain stem **(Figure 44.10).** Groups of interneurons in the centers regulate the rate and depth of breathing, ranging from shallow, slow breathing when the body is at rest to the deep and rapid breathing of intense physical exercise, excitement, or fear. Over these extremes, the air entering and leaving lungs of a human male varies from as little as 5 to 6 L per minute to (for a brief time only) as much as 150 L per minute.

INTERNEURONS THAT REGULATE BREATHING Signals from interneurons in the medulla carried by efferent (motor) neurons of the autonomic system produce the breathing movements. Signals from a dorsal group of interneurons act as the primary stimulator of inhalation by causing the diaphragm and the external intercostal muscles to contract, which expands the chest cavity and produces an inhalation. In a person at rest, the signals are switched off as the lungs become moderately full: the rib muscles and the diaphragm relax, and a passive exhalation occurs. These signals act as the primary generator of breathing rhythm.

A ventral group of interneurons in the medulla can send signals for both inhalation and exhalation. These neurons become active only during physical exercise, fear, or other situations that require more oxygen when active rather than passive exhalation is needed. In that case, some of the ventral neurons send signals that stimulate the abdominal and internal intercostal muscles to contract, thereby causing active exhalation. Other neurons in the ventral group become stimulated by signals from the dorsal group, and then help increase inhalation activity when faster and deeper breathing is required.

Two interneuron groups in the pons modulate the signals originating from the medulla, fine-tuning and smoothing the muscle contractions so that inhalations and exhalations are gradual and controlled rather than sudden and abrupt. Signals sent from higher brain centers in the cerebrum can override the control of respiratory rate and depth by the brain stem. For example, as you speak or sing, or hold your breath, you can consciously alter or stop breath-

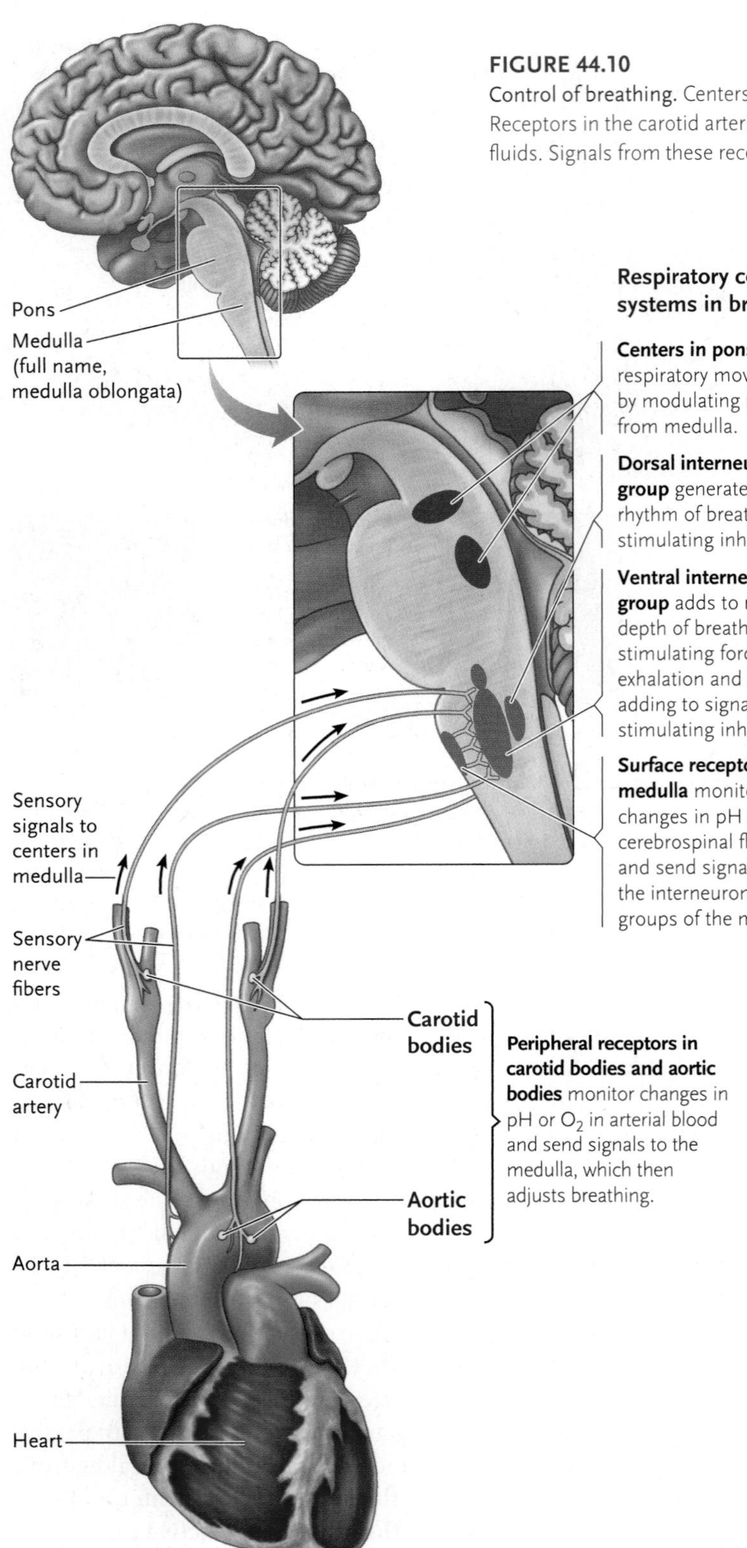
Pons
Medulla (full name, medulla oblongata)
Respiratory control systems in brain stem
Centers in pons smooth respiratory movements by modulating signals from medulla.
Dorsal interneuron group generates rhythm of breathing by stimulating inhalation.
Ventral interneuron group adds to rate and depth of breathing by stimulating forceful exhalation and by adding to signals stimulating inhalation.
Surface receptors of medulla monitor changes in pH of cerebrospinal fluid and send signals to the interneuron groups of the medulla.
Sensory signals to centers in medulla
Sensory nerve fibers
Carotid bodies
Carotid artery
Peripheral receptors in carotid bodies and aortic bodies monitor changes in pH or O_2 in arterial blood and send signals to the medulla, which then adjusts breathing.
Aortic bodies
Aorta
Heart

FIGURE 44.10
Control of breathing. Centers in the pons and medulla control the rhythm, rate, and depth of breathing. Receptors in the carotid arteries and aorta detect changes in the levels of O_2 and CO_2 in blood and body fluids. Signals from these receptors are integrated in the respiratory centers of the medulla and pons.

ing to match the demands of these activities. Breathing rate and depth are also modified by emotional states, controlled by centers in the limbic system of the brain (see Section 38.3). Thus breathing is altered as you laugh, gasp, groan, cry, and sigh.

RECEPTORS THAT SEND INFORMATION TO THE BRAIN CENTERS The brain centers controlling the rate and depth of breathing integrate sensory information sent by chemoreceptors that monitor O_2 and CO_2 levels in the blood and body fluids. The integration of sensory information serves to match breathing rate to the metabolic demands of the body. The chemoreceptors are located centrally on the surface of the medulla, and peripherally in **carotid bodies** in the carotid arteries leading to the brain and in **aortic bodies** in the large arteries leaving the heart (see Figure 44.10).

Surface receptors of the medulla detect changes in pH in the cerebrospinal fluid; the pH is determined mostly by the CO_2 concentration in the blood. (Remember that pH decreases as CO_2 levels increase.) The receptors in the carotid and aortic bodies detect changes in CO_2 and O_2 concentrations in the blood.

The CO_2 receptors in the medulla have the greatest effects on breathing. If increased body activities cause the CO_2 concentration to rise in the blood, the medulla receptors trigger interneuron groups in the medulla that increase the rate and depth of breathing. If CO_2 concentration falls, the receptors send signals to the medulla that lead to a slowing of the rate and depth of breathing.

The peripheral receptors in the carotid and aortic bodies detect changes in pH or O_2 concentration in arterial blood. When these receptors detect a decrease in blood pH they send signals to the medulla that cause the medulla to increase the rate and depth of breathing. Although the receptors in the carotid and aortic bodies also detect the O_2 level in arterial blood, the receptors do not respond until blood O_2 level falls below 60% of normal. This reaction makes the O_2 receptors act as a backup system that comes into play only when blood O_2 concentration falls to critically low levels.

Thus, the level of CO_2 in the blood and body fluids is much more closely monitored, and has a much greater effect on breathing, than the O_2 level. This reflects the fact that small fluctuations in blood pH have much greater effects on the ability of hemoglobin to carry oxygen, and on enzyme activity in the blood and interstitial fluid, than fluctuations in the O_2 level.

LOCAL CONTROLS Other, automated controls within the lungs match the rates of ventilation and perfusion by responding to O_2 concentrations in the blood. If air flow lags behind capillary blood flow, so that the O_2 level falls in the blood, the reduced O_2 concentration causes smooth muscles in the walls of arterioles in the lungs to contract. This reduces the flow of blood, thereby giving it more time to pick up O_2. Conversely, if blood flow lags behind, the rising blood O_2 concentration causes the smooth muscle cells in arteriole walls to relax, dilating the arterioles and increasing the rate of blood flow through lung capillaries. These local controls, in combination with the neural controls that regulate rate and depth of breathing, ensure that the respiratory system meets the body's varying need to obtain O_2 and release CO_2.

STUDY BREAK 44.3

1. **Explain how inhalation and exhalation occur in a mammal at rest.**
2. **You can consciously initiate and sustain an exhalation. What is going on muscularly in this case?**
3. **What is the most important feedback stimulus for breathing?**
4. **What is the role of the chemoreceptors in the medulla?**

THINK OUTSIDE THE BOOK

Individually or collaboratively, use the Internet or the research literature to answer the following questions: What is asthma, and what causes it? How do anti-inflammatory drugs and bronchodilators operate to relieve the symptoms of asthma? How is asthma similar to and different from COPD (chronic obstructive pulmonary disease)?

44.4 Mechanisms of Gas Exchange and Transport

In both the lungs and body tissues, gas exchange occurs when the gas diffuses from an area of higher concentration to an area of lower concentration. In this section, we consider the mechanics of gas exchange between air and the blood in mammals, and the means by which gases are transported between the lungs and other body tissues. A major part of this story involves hemoglobin, the vertebrate respiratory pigment.

The Proportion of a Gas in a Mixture Determines Its Partial Pressure

For gases, it is often more accurate and convenient to consider concentration differences as differences in pressure. When gases are present in a mixture, the pressure of each individual gas, called its **partial pressure,** is determined by its proportion in the mixture. Air, water, and blood all contain mixtures of gases, including oxygen, carbon dioxide, nitrogen, and other gases, so each gas exerts only a part of the total gas pressure. For example, the proportion of O_2 in dry air is about 21%, or 21/100. In dry air at sea level, the total atmospheric pressure under standard conditions is 760 mm Hg. The partial pressure of O_2, written as P_{O_2}, is equivalent to $760 \times 21/100$, or about 160 mm Hg. The proportion of CO_2 in dry air is about 0.04%, so its partial pressure, P_{CO_2}, is equivalent to $760 \times 0.04/100$, or about 0.3 mm Hg. For O_2 to diffuse inward across a respiratory surface, its partial pressure outside the surface must be greater than inside; for CO_2 to diffuse outward, its partial pressure inside must be greater inside than outside.

In the lungs, even though the P_{O_2} is reduced by mixing with the air in the residual volume, it is still much higher than the P_{O_2} in deoxygenated blood entering the network of capillaries in the lungs **(Figure 44.11).** As a result, O_2 readily diffuses passively down its partial pressure gradient from the alveolar air into the plasma solution in the pulmonary capillaries. Conversely, the P_{CO_2} in the lungs is much lower than the P_{CO_2} in the tissues. As a result, CO_2 readily diffuses passively down its partial pressure gradient from the pulmonary capillaries into the alveolar air.

Hemoglobin Greatly Increases the O_2-Carrying Capacity of the Blood

After entering the plasma, O_2 diffuses into erythrocytes, where it combines with hemoglobin. The combination with hemoglobin removes O_2 from the plasma, lowering the P_{O_2} of the plasma and allowing additional O_2 molecules to diffuse from alveolar air to the blood.

Recall from Section 42.2 that a mammalian hemoglobin molecule has four heme groups, each containing an iron atom that can combine reversibly with an O_2 molecule. A hemoglobin molecule can therefore bind a total of four molecules of O_2. The combination of O_2 with hemoglobin allows blood to carry about

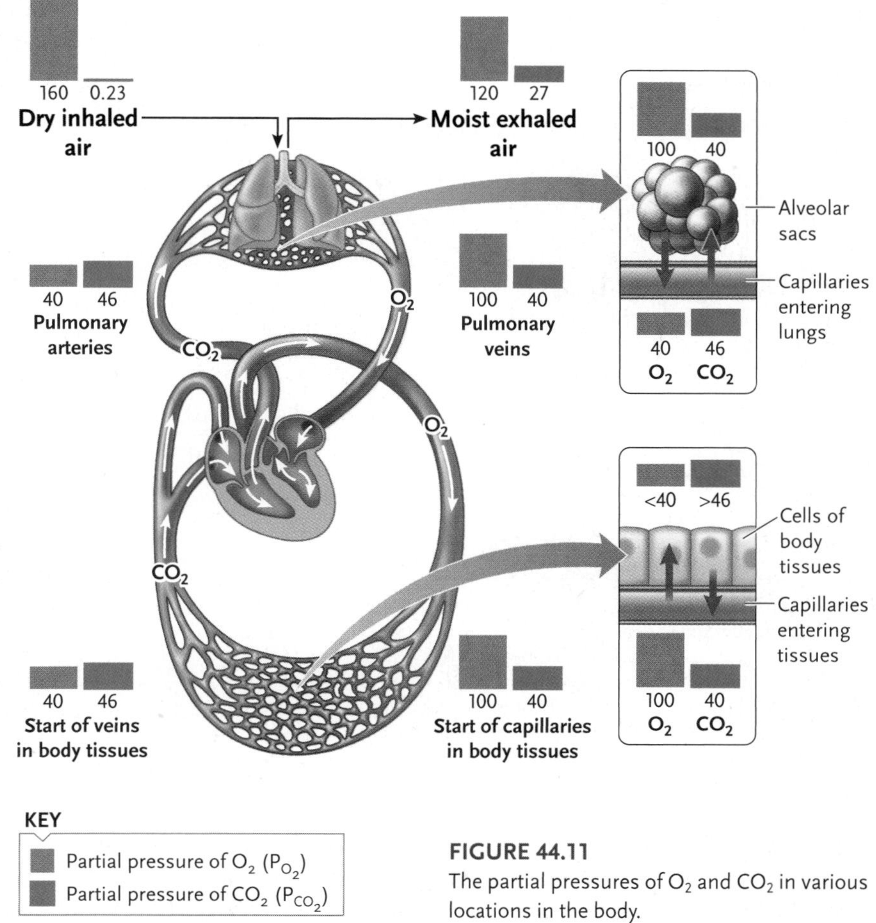

160 0.23
Dry inhaled air
120 27
Moist exhaled air
100 40
Alveolar sacs
Capillaries entering lungs
40 46
O_2 CO_2
40 46
Pulmonary arteries
100 40
Pulmonary veins
O_2
CO_2
O_2
CO_2
<40 >46
Cells of body tissues
Capillaries entering tissues
100 40
O_2 CO_2
40 46
Start of veins in body tissues
100 40
Start of capillaries in body tissues
KEY
Partial pressure of O_2 (P_{O_2})
Partial pressure of CO_2 (P_{CO_2})

FIGURE 44.11
The partial pressures of O_2 and CO_2 in various locations in the body.

60 times more O_2 (about 200 mL per liter) than it could if the O_2 simply dissolved in the plasma (about 3 mL per liter). About 98.5% of the O_2 in blood is carried by hemoglobin and about 1.5% is carried in solution in the blood plasma.

The reversible combination of hemoglobin with O_2 is related to the partial pressure of O_2 in a pattern shown by the *hemoglobin–O_2* dissociation curve in **Figure 44.12A.** (The curve is generated by measuring the amount of hemoglobin saturated at a given P_{O_2}.) The curve is S-shaped with a plateau region, rather than linear. The top, plateau part of the curve above 60 mm Hg is in the blood P_{O_2} range found in the pulmonary capillaries where O_2 is binding to hemoglobin. For this part of the curve, the blood remains highly saturated with O_2 over a relatively large range of P_{O_2}. Even at P_{O_2} levels much higher than shown on the graph (P_{O_2} theoretically can go up to 760 mm Hg), only a small extra amount of O_2 will bind to hemoglobin. The steep part of the curve between 0 and 60 mm Hg is in the blood P_{O_2} range found in the capillaries in the rest of the body. For this part of the curve, small changes in P_{O_2} result in a large change in the amount of O_2 bound to hemoglobin.

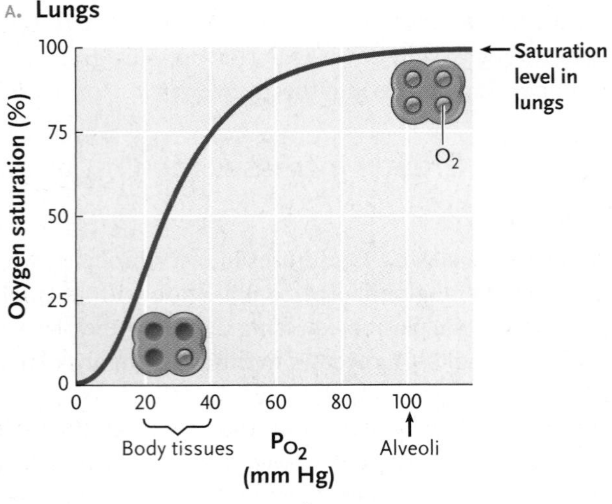
A. Lungs
Oxygen saturation (%)
100
75
50
25
0
Saturation level in lungs
O_2
0
20
40
60
80
100
Body tissues
P_{O_2} (mm Hg)
Alveoli

In the alveoli, in which the P_{O_2} is about 100 mm Hg and the pH is 7.4, most hemoglobin molecules are 100% saturated, meaning that almost all have bound four O_2 molecules.

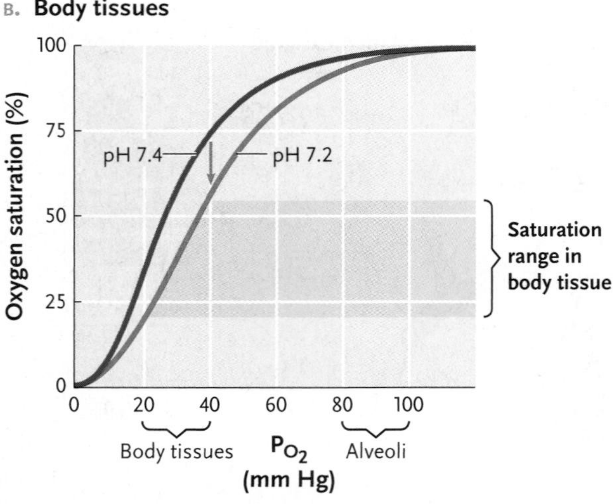
B. Body tissues
Oxygen saturation (%)
100
75
50
25
0
pH 7.4
pH 7.2
Saturation range in body tissue
0
20
40
60
80
100
Body tissues
P_{O_2} (mm Hg)
Alveoli

In the capillaries of body tissues, where the P_{O_2} varies between about 20 and 40 mm Hg depending on the level of metabolic activity and the pH is about 7.2, hemoglobin can hold less O_2. As a result, most hemoglobin molecules release two or three of their O_2 molecules to become between 25% and 50% saturated. Note that the drop in pH to 7.2 (red line) in active body tissues reduces the amount of O_2 hemoglobin can hold as compared with pH 7.4. The reduction in binding affinity at lower pH increases the amount of O_2 released in active tissues.

FIGURE 44.12
Hemoglobin–O_2 dissociation curves, which show the degree to which hemoglobin is saturated with O_2 at increasing P_{O_2}, in lungs (A) and body tissues (B).

The P_{O_2} of alveolar air is about 100 mg Hg (see Figure 44.11). As a result, most of the hemoglobin molecules are fully saturated in the blood leaving the alveolar networks, meaning that most of the hemoglobin molecules are bound to four O_2 molecules (see Figure 44.12A). The P_{O_2} of blood plasma has risen to approximately the same level as in the alveolar air, about 100 mm Hg. The blood has also changed color, reflecting the bright red color of oxygenated hemoglobin as compared with the darker red color of deoxygenated hemoglobin.

The oxygenated blood exiting from the alveoli collects in venules, which merge into the pulmonary veins leaving the lungs. These veins carry the blood to the heart, which pumps the blood through the systemic circulation to all parts of the body.

As the oxygenated blood enters the capillary networks of body tissues **(Figure 44.12B),** it encounters regions in which the P_{O_2} in the interstitial fluid and body cells is lower than that in the blood, ranging from about 40 mm Hg downward to 20 mm Hg or less (see Figure 44.11). As a result, O_2 diffuses from the blood plasma into the interstitial fluid, and from the fluid into body cells. As O_2 diffuses from the blood plasma into body tissues, it is replaced by O_2 released from hemoglobin.

Several factors contribute to the release of O_2 from hemoglobin, including increased acidity (lower pH) in active tissues. The acidity increases because oxidative reactions release CO_2, which combines with water to form carbonic acid (H_2CO_3). The lowered pH alters hemoglobin's conformation, reducing its affinity for O_2, a phenomenon called the **Bohr effect.** As a result of the reduced affinity, O_2 is released and used in cellular respiration.

The net diffusion of O_2 from blood to body cells continues until, by the time the blood leaves the capillary networks in the body tissues, much of the O_2 has been removed from hemoglobin. The blood, now with a P_{O_2} of 40 mm Hg or less (see Figures 44.11 and 44.12B), returns in veins to the heart, which pumps it through the pulmonary arteries to the lungs for another cycle of oxygenation.

Carbon Dioxide Diffuses down Concentration Gradients from Body Tissues into the Blood and Alveolar Air

The CO_2 produced by cellular oxidations diffuses from active cells into the interstitial fluid, where it reaches a partial pressure of about 46 mm Hg. Because this P_{CO_2} is higher than the 40 mm Hg P_{CO_2} in the blood entering the capillary networks of body tissues

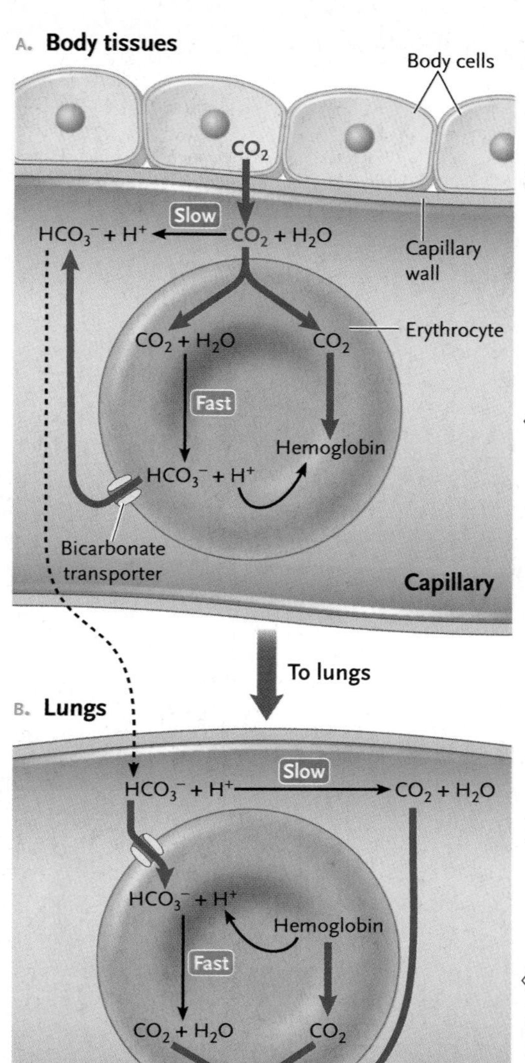
A. Body tissues
Body cells
CO_2
Slow
$HCO_3^- + H^+$
$CO_2 + H_2O$
Capillary wall
Erythrocyte
$CO_2 + H_2O$
CO_2
Fast
Hemoglobin
$HCO_3^- + H^+$
Bicarbonate transporter
Capillary
To lungs
B. Lungs
Slow
$HCO_3^- + H^+$
$CO_2 + H_2O$
$HCO_3^- + H^+$
Hemoglobin
Fast
$CO_2 + H_2O$
CO_2
CO_2
Capillary wall
Alveolar air
CO_2
Alveolar wall

FIGURE 44.13
The reactions occurring during the transfer of CO_2 from (A) body tissues (B) to alveolar air.

In body tissues, some of the CO_2 released into the blood combines with water in the blood plasma to form HCO_3^- and H^+. However, most of the CO_2 diffuses into erythrocytes, where some combines directly with hemoglobin and some combines with water to form HCO_3^- and H^+. The H^+ formed by this reaction combines with hemoglobin; the HCO_3^- is transported out of erythrocytes to add to the HCO_3^- in the blood plasma.

In the lungs, the reactions are reversed. Some of the HCO_3^- in the blood plasma combines with H^+ to form CO_2 and water. However, most of the HCO_3^- is transported into erythrocytes, where it combines with H^+ released from hemoglobin to form CO_2 and water. CO_2 is released from hemoglobin. The CO_2 diffuses from the erythrocytes and, with the CO_2 in the blood plasma, diffuses from the blood into the alveolar air.

(see Figure 44.11), CO_2 diffuses from the interstitial fluid into the blood plasma **(Figure 44.13A)**.

Some of the CO_2 remains in solution as a gas in the plasma. However, most of the CO_2, about 70%, combines with water to produce carbonic acid (H_2CO_3), which dissociates into bicarbonate (HCO_3^-) and H^+ ions. The reaction takes place both in the blood plasma and inside erythrocytes, where an enzyme, *carbonic anhydrase*, greatly speeds the reaction.

Most of the H^+ ions produced by the dissociation of carbonic acid combine with hemoglobin or with proteins in the plasma. The combination, by removing excess H^+ from the blood solution, *buffers* the blood pH, helping to maintain it at its near-neutral set point of 7.4. (Buffers are discussed in Section 2.5. The discussion includes the carbonic acid–bicarbonate buffer system used to buffer blood pH in humans and many other animals.) pH homeostasis is important because many enzymes are sensitive to minor changes in pH. The combined pathways absorbing CO_2 in the blood—solution in the plasma, conversion to bicarbonate, and combination with hemoglobin—help maintain the concentration gradient for gaseous CO_2 and keep its diffusion from the interstitial fluid into the blood at optimal levels.

The blood leaving the capillary networks of body tissues is collected in venules and veins and returned to the heart, which pumps it through the pulmonary arteries into the lungs. As the blood enters the capillary networks surrounding the alveoli, the entire process of CO_2 uptake is reversed **(Figure 44.13B)**. The P_{CO_2} in the blood, now about 46 mm Hg, is higher than the P_{CO_2} in the alveolar air, about 40 mm Hg (see Figure 44.11). As a result, CO_2 diffuses from the blood into the air. The diminishing CO_2 concentrations in the plasma, along with the lower pH encountered in the lungs, promote the release of CO_2 from hemoglobin. As CO_2 diffuses away, bicarbonate ions in the blood combine with H^+ ions, forming carbonic acid molecules that break down into water and additional CO_2. This CO_2 adds to the quantities diffusing from the blood into the alveolar air. By the time the blood leaves the capillary networks in the lungs, its P_{CO_2} has been reduced to the same level as that of the alveolar air, about 40 mm Hg (see Figure 44.11).

Carbon monoxide (CO), a colorless, odorless gas produced when fuels are incompletely burned, as in automobile exhaust or in a faulty furnace, gas appliance, or space heater, also binds to hemoglobin if it is inhaled into the lungs. It binds so strongly that it displaces O_2 from hemoglobin and drastically reduces the amount of O_2 carried to body tissues. If CO is inhaled in high quantity for even a few minutes, the reduction in oxygen delivered to the brain can lead to unconsciousness and brain damage. Sustained exposure leads to death by hypoxia (lack of oxygen). Because the brain regulates breathing based on CO_2 levels in blood rather than on O_2 levels, victims breathing CO can die from hypoxia without noticing anything amiss up to the point of unconsciousness. Interestingly, the combination of CO with hemoglobin, carboxyhemoglobin, is bright red. This has led to a myth often seen in textbooks that victims of CO poisoning turn a "classic cherry red" in color. However, this actually occurs in less than 2% of cases. *Insights from the Molecular Revolution* describes recent research testing the molecular basis for the binding of CO to hemoglobin and to *myoglobin*, a muscle protein with structure and properties similar to hemoglobin (myoglobin is discussed in Section 41.1).

STUDY BREAK 44.4

1. **Explain the role of hemoglobin in gas exchange.**
2. **Why is carbon monoxide potentially lethal?**

INSIGHTS FROM THE Molecular Revolution

Giving Hemoglobin and Myoglobin Air

Carbon monoxide, like O_2, combines directly with the heme group in the proteins myoglobin and hemoglobin. The heme group by itself, unassociated with the polypeptide chain of hemoglobin or myoglobin, has an affinity for CO some 10,000 times greater than for O_2. Combination of the heme group with the proteins reduces its affinity for CO to only 250 times greater than O_2 for hemoglobin and 30 times greater for myoglobin.

Research Question

How does combination with the proteins reduce the affinity of the heme group for CO so effectively?

Experiment

One hypothesis is that the reduction in affinity depends on a stabilizing hydrogen bond between O_2 and an amino acid, histidine, located in a protein pocket formed by a fold in the amino acid chain near the iron atom of the heme group. According to this proposal, the hydrogen bond allows O_2 to displace a water molecule that occupies the pocket when neither O_2 nor CO is bound. Because carbon monoxide cannot form this hydrogen bond, it displaces the water molecule less readily than O_2 does.

Research with myoglobin, conducted by John S. Olson and George Phillips and their colleagues at Rice University, supports the hydrogen-bond hypothesis. For their study, the team used a myoglobin gene isolated from a sperm whale and cloned in a plasmid cloning vector (see Section 18.1). They mutated the DNA of the myoglobin gene so that one of seven other amino acids was substituted for the histidine in the pocket of the encoded protein **(figure).** A highly active bacterial promoter was added to the altered genes, which were then introduced one at a time into *Escherichia coli* bacteria. The bacteria expressed the genes, producing the altered myoglobin molecules in quantity, thereby providing the researchers with seven different forms of myoglobin to test for binding affinity for O_2 (see the figure).

The seven amino acids substituted for the histidine—glycine, alanine, leucine, phenylalanine, threonine, valine, and glutamine—are all nonpolar, or uncharged (histidine is positively charged), and thus unlikely to form a hydrogen bond with O_2. Further, none of these amino acids except glycine and glutamine should be able to hold a water molecule stably in the binding pocket. To support the hydrogen-bond hypothesis, the affinity of O_2 for the altered myoglobin should be greatly reduced in the mutant forms of the molecule.

Results

Binding tests showed that the substitutions indeed reduced the affinity for O_2 by a factor that varied between 10 and 100 times. The greatest reduction was produced by the most nonpolar of the amino acids, leucine and phenylalanine. The smallest reduction was observed for glycine and glutamine.

Conclusion

The results strongly support the hydrogen-bond hypothesis.

Source: M. L. Quillin et al. 1993. High-resolution crystal structures of distal histidine mutants of sperm whale myoglobin. *Journal of Molecular Biology* 234:140–155.

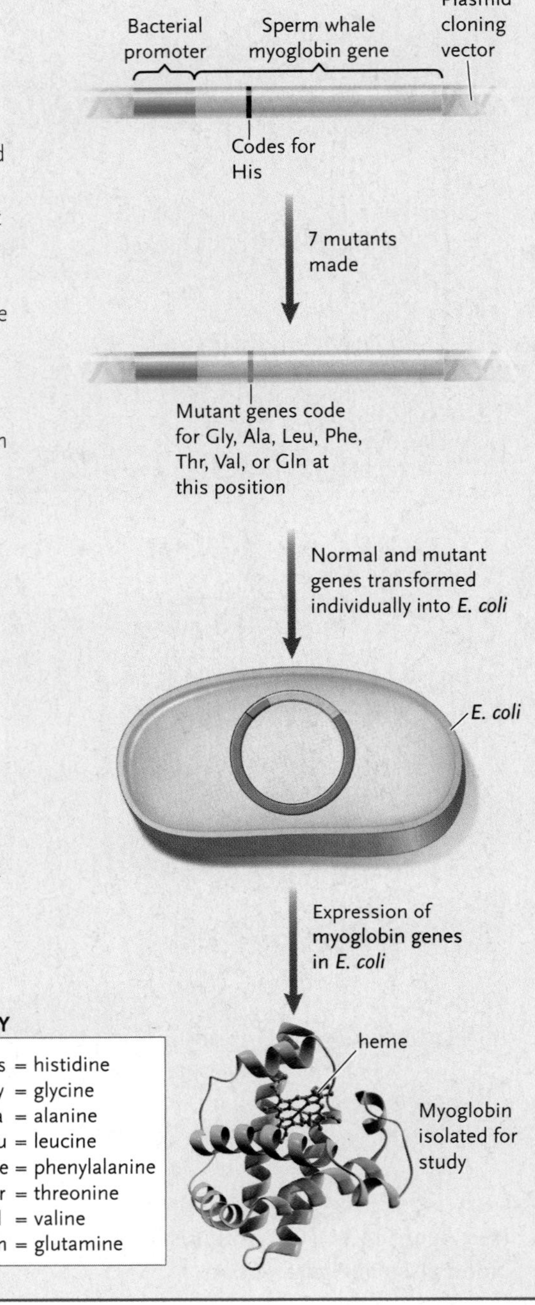

Bacterial promoter
Sperm whale myoglobin gene
Plasmid cloning vector
Codes for His
7 mutants made
Mutant genes code for Gly, Ala, Leu, Phe, Thr, Val, or Gln at this position
Normal and mutant genes transformed individually into E. coli
E. coli
Expression of myoglobin genes in E. coli
heme
Myoglobin isolated for study

KEY
His = histidine
Gly = glycine
Ala = alanine
Leu = leucine
Phe = phenylalanine
Thr = threonine
Val = valine
Gln = glutamine

44.5 Respiration at High Altitudes and in Ocean Depths

This chapter's introduction described some challenges to respiration that arise when humans travel to high altitude. In this concluding section we look at the effects of high altitude on respiration, along with the effects of increased pressure when humans and other mammals dive underwater.

High Altitudes Reduce the P_{O_2} of Air Entering the Lungs

As altitude increases, atmospheric pressure decreases, and with it, the P_{O_2} of alveolar air and the concentration gradient of O_2 across the respiratory surface. At 6,100 m (20,000 feet), where most people become unconscious unless they have supplemental O_2, the dry air pressure is about 380 mm Hg and the P_{O_2} is only $380 \times 21/100 =$ about 80 mm Hg, half that at sea level.

FIGURE 44.14 **Experimental Research**

Demonstration of a Molecular Basis for High-Altitude Adaptation in Deer Mice

Question: Do mutations in hemoglobin genes underlie high-altitude adaptation in deer mice?

Experiment: Hemoglobin of adult mammals consists of two α-globin and two β-globin polypeptides. Hemoglobin binds oxygen and transports it in the blood. In the low-oxygen environment of high altitudes, mammals can suffer hypoxia, a condition that results when insufficient oxygen is carried to the body tissues in arterial blood. Some mammals, such as deer mice *(Peromyscus maniculatus)*, live and thrive in a wide variety of environments, including at high altitude. What is the functional basis of this ability?

Deer mice have two gene loci each for the α-globin and β-globin polypeptides. Jay Storz and his colleagues at the University of Nebraska-Lincoln, and collaborators at the University of Kansas, the University of Porto in Portugal, and the University of Aarhus in Denmark, tested the hypothesis that, compared with their lowland relatives, deer mice living at high altitude have mutations in their α-globin and β-globin genes that lead to hemoglobin molecules adapted to carrying oxygen in low-oxygen environments.

The experimenters isolated DNA from deer mice living at high altitude and at low altitude and compared the sequences of the two α-globin and two β-globin genes.

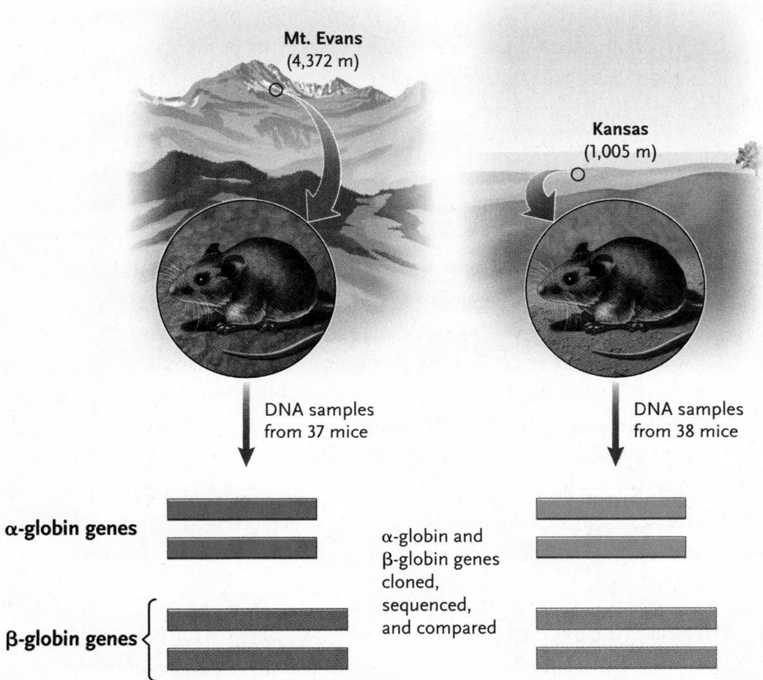
Mt. Evans
(4,372 m)
Kansas
(1,005 m)
DNA samples from 37 mice
DNA samples from 38 mice
α-globin genes
β-globin genes
α-globin and β-globin genes cloned, sequenced, and compared

Results: The investigators identified a number of mutations in the four genes of the high-altitude mice. Based on the biochemistry of hemoglobin, the researchers determined that the mutations increase the oxygen-binding affinity of hemoglobin compared with the low-altitude hemoglobin. This increased binding affinity in turn would increase the O_2 concentration in arterial blood, thereby enabling the mice to tolerate chronic hypoxia. None of the low-altitude control mice had the same mutations.

Conclusion: The research identified specific mutations presumably involved in the evolutionary adaptation of deer mice to high altitude.

Interestingly, the hemoglobin of high-altitude deer mice is functionally similar to fetal hemoglobins in humans and other placental mammals. Fetal hemoglobin is a special form of hemoglobin with two α-globin and two γ-globin polypeptides (the γ-globin polypeptide is related to the β-globin polypeptide). The sequence difference from adult hemoglobin results in the production of hemoglobin molecules that carry O_2 efficiently in the low-O_2 intrauterine environment.

Source: J. Storz et al. 2009. Evolutionary and functional insights into the mechanism underlying high-altitude adaptation of deer mouse hemoglobin. *Proceedings of the National Academy of Sciences USA* 106:14450–14455.

Humans who travel from sea level to elevations of 1,800 m (about 6,000 feet) or more often experience one or more unpleasant symptoms, including headache, blurred vision, dizziness, nausea, and fatigue. However, after a few weeks at higher elevation, the body adjusts by increasing the number of erythrocytes in the blood. The increase in erythrocyte production is stimulated primarily by a hormone, *erythropoietin* (EPO), which the kidneys secrete in greater quantities in response to a drop in blood O_2. Erythrocyte production slows when people return to lower altitudes. However, the erythrocyte count remains high for several weeks after high-altitude exposure. Athletes often train at high altitudes to increase their erythrocyte count, with the idea that it will improve their stamina and endurance at lower altitudes.

People who live at high altitudes from childhood develop more permanent changes, including an increase in the number of alveoli and more extensive capillary networks in the lungs. These anatomical characteristics are retained if they move to lower altitudes.

Some mammals evolutionarily adapted to high altitudes show genetically determined changes that are present throughout life. For example, deer mice that live at above 4,372 m (about 14,343 feet) have hemoglobin molecules with greater affinity for O_2 than the hemoglobin molecules of their lower-altitude relatives **(Figure 44.14).** As a result, hemoglobin becomes saturated with O_2 at the lower partial pressures typical of high altitudes. Increased binding affinity of hemoglobin for O_2 is also seen in birds adapted to life at high altitudes, such as the bar-headed goose *(Anser indicus)*. These birds have been observed flying over the peaks of the Great Himalayas, which have altitudes greater than 6,000 m.

Diving Mammals Are Adapted to Survive the High Partial Pressures of Gases at Extreme Depths

As a mammal such as a seal or whale dives from the surface, each additional 10 m of depth increases the partial pressure of dissolved gases by about 1 atmosphere. Below about 25 m or so, the pressure becomes so great that the lungs collapse and cease to function. Adaptations of diving mammals such as seals and whales allow these animals to survive the extreme pressure and lack of lung function, in some species for over an hour at ocean depths of more than a mile.

Among these adaptations are more blood per unit of body weight and more red blood cells, which are stored in the spleen and released during a dive. In addition, the muscles of these animals contain much greater quantities of the O_2-binding protein myoglobin than the muscles of land-dwelling mammals do. In all, the adaptations pack about twice as much O_2 per kilogram of body weight into a seal, for example, than into a human.

Other adaptations decrease O_2 consumption during a deep and prolonged dive. The heart rate slows by about 80% to 90% and the circulation of blood to internal organs and muscles is cut by as much as 95%, leaving only the brain with its normal blood supply. Even though most of the blood supply to muscles is cut off, the muscles continue to work by shifting to anaerobic oxidation. The lactic acid produced by anaerobic respiration in the muscles is not released into the blood until the animal returns to the surface.

These combined adaptations give seals and whales an amazing ability to dive to great depths and remain underwater for extended periods. Although average dives are on the order of 10 to 20 minutes, some sperm whales, tracked by sonar, have reached depths of 2,250 m (more than 7,000 feet) and remained underwater for as long as 82 minutes.

STUDY BREAK 44.5 <

1. **What are the key adaptations that diving mammals use to survive at significant ocean depths?**

UNANSWERED QUESTIONS

Does prenatal nicotine exposure alter development of respiratory neurons in the brainstem?

In this chapter you learned that the muscles of breathing are controlled in the brainstem, by groups of interneurons in the medulla oblongata and pons. The control of respiratory muscles (and thus breathing) by these neurons is called "central ventilatory control." Neonatal mammals that are exposed to nicotine *in utero* show various breathing abnormalities, such as reduced ventilatory output, increased frequency and duration of apneas (suspension of breathing), and delayed arousal in response to hypoxia (reduced blood oxygen levels) during sleep. One or several of these abnormalities may underlie sudden infant death syndrome (SIDS), also called "crib death" because victims are typically found dead in their crib. Clinical studies have shown that exposure to tobacco smoke is the number one risk factor for SIDS. Accordingly, laboratories, such as our own, are using animal models to examine how nicotine exposure alters development of central ventilatory control.

Several research methods can be used, depending on the particular question being addressed. In all of our experiments, neonatal rodents are exposed to nicotine *in utero* by implanting a small osmotic pump under the skin of a female rat that is 4–5 days pregnant. The pump releases nicotine at a prescribed rate, and the developing neonates are exposed to nicotine as the drug passes from mother to neonate via the placenta. When the neonates are born (on the 21st day of pregnancy), we study their breathing responses while they are awake or asleep using a device called a plethysmograph. This device senses the tiny pressure changes that accompany breathing in these small animals, and by adjusting chamber size, animals can be studied from birth to adulthood.

We can also dissect the brainstem, spinal cord, and rib cage from a neonatal animal, and place the preparation in a chamber for in vitro studies. Remarkably, this preparation is able to maintain rhythmic firing of respiratory neurons for up to 6 hours, allowing us to apply drugs and neurotransmitters to brainstem respiratory neurons while recording the electrical activity of neurons and respiratory muscle nerves.

Finally, we can prepare a brainstem slice, containing the most important central respiratory neurons and the hypoglossal nerve (this nerve innervates the tongue muscles, and it contains axons with rhythmic, respiratory-related activity), for detailed electrophysiological studies using the patch clamp technique (see Section 37.2). This preparation allows us to examine how prenatal nicotine exposure influences the membrane potential and firing properties of respiratory neurons.

To date, our studies have shown abnormal breathing in awake neonates, as well as an increase in inhibitory neurotransmission in respiratory neurons. Recently, we have shown that the intrinsic biophysical properties of respiratory neurons are altered in the nicotine-exposed pups; this means that when the cells are subjected to natural synaptic input from neighboring neurons (for example, from chemoreceptor neurons), their response is altered because the change in intrinsic properties is because of changes in the cell membrane. Current and future studies are directed at understanding the detailed cellular mechanisms that lead to these changes in intrinsic membrane properties. Understanding these mechanisms will hopefully lead to the development of drugs that can counteract nicotine's impact on the brain, as well as to an increased awareness and acceptance of the link between prenatal nicotine exposure and breathing abnormalities, resulting in more aggressive smoking prevention strategies.

Think Critically

1. If nicotine affects surface receptors of the medulla oblongata, is an infant's breathing rate being altered by carbon dioxide level or oxygen level?
2. Given your understanding of the peripheral nervous system and control of heart rate (from Section 38.2), is research investigating the effects of nicotine on SIDS and breathing focused on the somatic system or the autonomic system, and if the autonomic system, which division?

Ralph Fregosi

Ralph Fregosi is professor of physiology and neurobiology at the University of Arizona at Tucson. He does research on the neural control of breathing and teaches physiology to undergraduate and graduate students. Learn more about his research at http://www.physiology.arizona.edu/labs/rnlab/.

REVIEW KEY CONCEPTS

Go to **CENGAGENOW** at www.cengage.com/login to access quizzing, animations, exercises, articles, and personalized homework help.

44.1 The Function of Gas Exchange

- Physiological respiration is the process by which animals exchange O_2 and CO_2 with the environment (Figure 44.1).
- The two primary operating features of gas exchange are the respiratory medium, either air or water, and the respiratory surface, a wetted epithelium over which gas exchange takes place.
- In some invertebrates, the skin serves as the respiratory surface. In other invertebrates and all vertebrates, gills or lungs provide the primary respiratory surface (Figure 44.2).
- Simple diffusion of molecules from regions of higher concentration to regions of lower concentration drives the exchange of gases across the respiratory surface. The area of the respiratory surface determines the total quantity of gases exchanged by diffusion.
- The concentration gradients of O_2 and CO_2 across the respiratory surface are kept at optimal levels by ventilation and perfusion.

44.2 Adaptations for Respiration

- Animals breathing water keep the respiratory surface wetted by direct exposure to the environment. The high density and viscosity of water, and its relatively low O_2 content as compared with air, require water-breathing animals to expend significant energy to keep their respiratory surface ventilated.
- Air is high in O_2 content, allowing air-breathing animals to maintain higher metabolic levels than water breathers. The low density and viscosity of air as compared with water allow air breathers to ventilate the respiratory surface with relatively little energy. To accommodate water loss by evaporation, lungs typically are invaginations of the body surface, allowing air to become saturated with water before it reaches the respiratory surface.
- Gills are evaginations of the body surface. Water moves over the gills by the beating of cilia or is pumped over the gills by contractions of body muscles (Figures 44.2B and 44.3A–D).
- Water moves in a one-way direction over the gills of sharks, bony fishes, and some crabs, allowing these animals to use countercurrent exchange to maximize the exchange of gases over the respiratory surface (Figure 44.4).
- Insects breathe by means of tracheae, air-conducting tubes that lead from the body surface and send branches to essentially every cell in the body. Gas exchange takes place in the fluid-filled tips at the ends of the branches (Figure 44.5).
- Lungs consist of an invaginated system of branches, folds, and pockets. They may be filled by positive pressure breathing, in which air is forced into the lungs by muscle contractions, or by negative pressure breathing, in which muscle contractions expand the lungs, lowering the air pressure inside them and allowing air to be pulled into the lungs (Figures 44.6 and 44.9).

Animation: Examples of respiratory surfaces

Animation: Bony fish respiration

Animation: Frog respiration

Animation: Vertebrate lungs

Animation: Bird respiration

44.3 The Mammalian Respiratory System

- Air enters the respiratory system through the nose and mouth and passes through the pharynx, larynx, and trachea. The trachea divides into two bronchi, which lead to the lungs. Within the lungs, the bronchi branch into bronchioles, which lead into the alveoli, which are surrounded by dense networks of blood capillaries (Figure 44.8).
- Mammals inhale by a negative pressure mechanism. Air is exhaled passively by relaxation of the diaphragm and the external intercostal muscles between the ribs, and elastic recoil of the lungs. During deep and rapid breathing, the expulsion of air is forceful, driven by contraction of the internal intercostal muscles (Figure 44.9).
- The tidal volume of the lungs is the air moved in and out of the lungs during an inhalation and exhalation. The vital capacity is the total volume of air a person can inhale and exhale by breathing as deeply as possible. The air remaining in the lungs after as much air as possible is exhaled is the residual volume of the lungs.
- Breathing is controlled by a combination of local chemical controls and regulation by centers in the brain stem. These controls match the rate of air and blood flow in the lungs, and link the rate and depth of breathing to the body's requirements for O_2 uptake and CO_2 release (Figure 44.10).
- The basic rhythm of breathing is produced by interneurons in the medulla. When more rapid breathing is required, another group of interneurons in the medulla sends signals reinforcing inhalation and producing forceful exhalation. Two interneuron groups in the pons smooth and fine-tune breathing by stimulating or inhibiting the inhalation center in the medulla.
- Sensory receptors in the medulla, the carotid bodies, and the aortic bodies detect changes in the levels of O_2 and CO_2 in the blood and body fluids. The control centers in the medulla and pons adjust the rate and depth of breathing to compensate for changes in the blood gases.

Animation: Human respiratory system

Animation: Structure of an alveolus

Animation: Respiratory cycle

Animation: Changes in lung volume and pressure

Animation: Pressure-gradient changes during respiration

Animation: Heimlich maneuver

Animation: Vocal cords

44.4 Mechanisms of Gas Exchange and Transport

- The partial pressure of O_2 is higher in the alveolar air than in the blood in the capillary networks surrounding the alveoli causing O_2 to diffuse from the alveolar air into the blood. Most of the O_2 entering the blood combines with hemoglobin inside erythrocytes (Figure 44.11).
- A hemoglobin molecule can combine with four O_2 molecules. The large quantities of O_2 that combine with hemoglobin maintain a large gradient in partial pressure between O_2 in the alveolar air and in the blood (Figure 44.12).

- In body tissues outside the lungs, the O_2 concentration in the interstitial fluid and body cells is lower than in the blood plasma. As a result, O_2 diffuses from the blood into the interstitial fluid, and from the fluid into body cells.
- The partial pressure of CO_2 is higher in the tissues than in the blood. About 10% of this CO_2 dissolves in the blood plasma; 70% is converted into H^+ and HCO_3^- (bicarbonate) ions. The remaining 20% combines with hemoglobin (Figures 44.11 and 44.13A).
- In the lungs, the partial pressure of CO_2 is higher in the blood than in the alveolar air. As a result, the reactions packing CO_2 into the blood are reversed, and the CO_2 is released from the blood into the alveolar air (Figure 44.13C).

Animation: Partial pressure gradients

Animation: Globin and hemoglobin structure

44.5 Respiration at High Altitudes and in Ocean Depths

- In mammals that move to high altitudes, the number of red blood cells and the amount of hemoglobin per cell increase. These changes are reversed if the animals return to lower altitudes.
- Humans living at higher altitudes from birth develop more alveoli and capillary networks in the lungs.
- Some mammals and birds adapted to high altitudes have forms of hemoglobin with greater affinity for O_2, allowing saturation at the lower P_{O_2} typical of high altitudes (Figure 44.14).
- Marine mammals adapted to deep diving have a greater blood volume per unit of body weight, and their blood contains more red blood cells, with a higher hemoglobin content, than other mammals. Their muscles also contain more myoglobin than those of land mammals, allowing more O_2 to be stored in muscle tissues. During a dive, the heartbeat slows, and circulation is reduced to all parts of the body except the brain.

UNDERSTAND AND APPLY

Test Your Knowledge

1. Which of the following describes a respiratory medium?
   a. The liver of an amphibian, in which the rate of diffusion is high
   b. Neurons in the human brain, where CO_2 moves from the neurons to the blood
   c. The O_2 in the blood of humans
   d. Epithelial cells of fish that are in contact with the air
   e. Air and water
2. Which of the following describes a respiratory surface?
   a. A surface consisting of multiple layers of epithelial cells
   b. The exoskeleton of an insect
   c. The nasal passages of a mammal
   d. Thin surface consisting of a single layer of epithelial cells
   e. The outer membrane of a mitochondrion
3. A geothermal HVAC (heating ventilating and air conditioning) system air-conditions your house by passing warm air pulled from the house past a continuous flow of cool water pulled from the ground. How this heat exchanger works is analogous to:
   a. countercurrent exchange of gases in fish gills or bird lungs.
   b. diffusion of O_2 from blood to cells in shark tissues.
   c. diffusion of CO_2 from cells to blood in crabs.
   d. use of O_2 in cells in insects.
   e. excretion of CO_2 from mammalian cells.
4. Tracheal systems are characterized by:
   a. closed circulatory tubes that move gases.
   b. spiracles that move gases between cells and body fluids.
   c. body movements that compress and expand air sacs to pump air.
   d. positive pressure breathing, which swallows air into the body.
   e. negative pressure breathing, which lowers air pressure at the respiratory surfaces.
5. The partial pressure of O_2 in the atmosphere is 160 mm Hg, but when O_2 moves into the blood its partial pressure is one third lower than the atmospheric partial pressure. Which structure is it moving through at this partial pressure?
   a. Alveoli
   b. Bronchi
   c. Bronchioles
   d. Tracheae
   e. Pharynges
6. As a speed skater finishes the last lap of a race:
   a. his diaphragm and rib muscles contract when he exhales.
   b. positive pressure brings air into his lungs.
   c. his lungs undergo an elastic recoil when he inhales.
   d. his tidal volume is at vital capacity.
   e. his residual volume momentarily reaches zero.
7. A teenager is frightened when she is about to step onto the stage but then remembers to breathe deeply and slowly as she faces the audience. What is occurring here?
   a. Interneurons in the medulla cause the rib muscles to relax, followed later by stimulation and contraction of the intercostal muscles.
   b. Signals from the pons override the initial brain stem stimuli.
   c. The limbic system stabilized her emotional state, so there is no change in the mechanical movement of air.
   d. The brain signals the aortic bodies in the carotid arteries to adjust the breathing rate.
   e. Initial low CO_2 blood levels causing high pH are followed by increased CO_2 levels that lower pH.
8. Oxygen enters the blood in the lungs because relative to alveolar air:
   a. the CO_2 concentration in the blood is high.
   b. the CO_2 concentration in the blood is low.
   c. the O_2 concentration in the blood is high.
   d. the O_2 concentration in the blood is low.
   e. the process is independent of gas concentrations in the blood.

9. A hemoglobin O_2 dissociation curve:
   a. demonstrates that hemoglobin is about 50% saturated in the alveoli.
   b. shifts to the left when pH rises.
   c. demonstrates that hemoglobin holds less O_2 when the pH is higher.
   d. illustrates that oxygen saturation is not dependent on CO_2 levels.
   e. demonstrates why hemoglobin can bind O_2 at high pH in the lungs and release it at lower pH in the tissues.
10. The majority of CO_2 in the blood:
   a. is in the form of carbonic acid and bicarbonate ions.
   b. dissociates to add H^+ to the blood to raise its pH to 7.4.
   c. has a lower P_{CO_2} than the P_{CO_2} in the alveolar air.
   d. increases in the lung capillaries, which have a higher pH than the tissue capillaries.
   e. can be displaced on the hemoglobin molecule by CO if CO is inhaled.

Discuss the Concepts

1. Smoking has traditionally been considered to reduce the ability of athletes to run without becoming exhausted. Why might this be true?
2. People are occasionally found unconscious from breathing too much CO_2 (as from a charcoal heater placed indoors) or too much CO (as from auto exhaust in a closed garage). Would it be more advantageous to give pure O_2 to a person breathing too much CO_2 than simply moving the person to fresh air? Why? Which—pure O_2 or fresh air—would be best for a person unconscious from breathing CO? Why?
3. Hyperventilation, or overbreathing, is breathing faster or deeper than necessary to meet the body's needs. Hyperventilation reduces the CO_2 content of blood, but does not significantly increase the amount of O_2 available to tissues. Why might this be so?

Design an Experiment

Propose a hypothesis for the effect of zero gravity on respiration, and design an experiment to test the hypothesis.

Interpret the Data

Lycopene, which is abundant in tomatoes, is thought to have a potential benefit in protecting against many types of cancers. In the study below mice were exposed to tobacco smoke or normal air. The amount of lung tissue damage caused by each exposure was measured in mice that were given tomato juice (+) or water (−). To estimate the destruction of the alveolar wall, a destructive index (DI) was calculated from histological slides of the lung. All lungs will show some level of damage, but a DI above 10% is considered significant destruction.

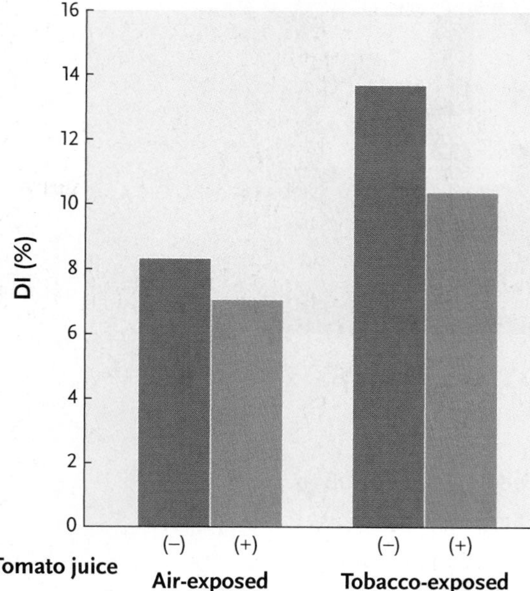
16
14
12
10
8
6
4
2
0
DI (%)
Tomato juice
(−)
(+)
(−)
(+)
Air-exposed
Tobacco-exposed

1. What is the effect of tobacco smoke on the lung tissue in these mice?
2. Do these data suggest that lycopene has a protective effect against tobacco's effects?

Source: S. Kasagi et al. 2006. Tomato juice prevents senescence-accelerated mouse P1 strain from developing emphysema induced by chronic exposure to tobacco smoke. *American Journal of Physiology-Lung Cellular and Molecular Physiology* 290:L396–L404.

Apply Evolutionary Thinking

From what you have learned in this chapter and in Chapter 30, do you think lungs evolved once, or on several occasions? Justify your answer.

Express Your Opinion

Tobacco is a worldwide threat to health and a profitable product for American companies. As tobacco use by its citizens declines, should the United States encourage international efforts to reduce tobacco use around the globe? Go to www.cengage.com/login to investigate both sides of the issue and then vote.

45

STUDY OUTLINE

A bald eagle *(Haliaeetus leucocephalus)*, snatching a fish out of the water with its talons.

Animal Nutrition

FIGURE 45.1
A deep sea anglerfish *(Melanocetus johnsoni)*, with its rod and lure lit and ready to attract prey.

Why It Matters. . . Invisible in the inky darkness, a deep-sea anglerfish (a member of the order Lophiiformes) lies in wait, its gaping mouth lined with sharp teeth. Just above the mouth dangles a glowing lure suspended from a fishing-rod-like spine that projects from the fish's dorsal fin **(Figure 45.1).** The lure resembles a tiny fish; it even wiggles back and forth in imitation of swimming movements. Its glow is produced by bioluminescent bacteria that live symbiotically in the lure's tissues.

A hapless fish is attracted to the lure. As it comes within range, the mouth of the anglerfish expands suddenly, creating a powerful suction that whips the prey in. The backward-angling fangs keep the prey from escaping. The strike takes only 6 ms (milliseconds), among the fastest of any known fishes.

Contractions of throat muscles send the prey to the stomach of the anglerfish, which can expand to accommodate a meal as large as the anglerfish itself. In the fish's digestive tract, acids and enzymes dissolve the body of the prey, gradually breaking it into molecules small enough to be absorbed into the bloodstream. In this function, the digestive system of the anglerfish is the same as that of any other vertebrate, including humans—it provides nutrients that allow the animal to live. And the adaptations of the anglerfish for feeding, although bizarre to human sensibilities, are no more remarkable than those of many other animals.

Animal **nutrition**—which includes the processes by which food is ingested, digested, and absorbed into body cells and fluids—is the subject of this chapter. Our discussion begins with the basic categories of animal foods and **ingestion,** the feeding methods used to take food into the digestive cavity. Then we examine the process of **digestion,** which is the splitting of carbohydrates, proteins, lipids, and nucleic acids in foods into chemical subunits small enough to be absorbed into an animal's body fluids and cells. The chapter also presents the main structural and functional features of digestive systems, with special emphasis on humans and other mammals. The adaptations animals use to obtain and digest food are among their most strongly defining anatomical and functional characteristics. <

45.1 Feeding and Nutrition

All organisms require sources of matter and energy for metabolism, homeostasis (maintaining their internal environment in a stable state; see Section 36.4), growth, and reproduction. For animals, meeting these nutritional requirements involves *feeding,* the uptake of food from the surroundings. Animals employ various feeding methods ranging from the ingestion of molecules in liquid solutions to eating entire organisms in one gulp. Once the food is ingested, digestive processes convert its molecules into absorbable subunits. In this section, we survey animal nutritional requirements and feeding methods as an introduction to animal digestive processes.

Animals Require Both Organic and Inorganic Molecules for Nutrition

Plants and other photosynthesizers need only sunlight as an energy source and a supply of simple inorganic precursors such as water, carbon dioxide, and minerals to make all the organic molecules they require. In contrast, animals require a constant diet of organic molecules as a source of both energy and nutrients that they cannot make for themselves.

Animals are classified according to their sources of organic molecules. **Herbivores** such as antelopes, horses, bison, giraffes, kangaroos, manatees, and grasshoppers obtain organic molecules primarily by eating plants. **Carnivores**—cats, Tasmanian devils, penguins, sharks, and spiders, for example—primarily eat other animals. We say "primarily" because many herbivores eat animal matter at times, and a number of carnivores occasionally eat plant material. An antelope will eat insects as it grazes, and a grizzly bear, although primarily carnivorous, also eats berries. **Omnivores,** such as crows, cockroaches, and humans, eat both plants and animals and, digestive enzymes allowing, may consume any source of organic matter.

Organic molecules are the basis for two of the most fundamental processes of life: they act as fuels for oxidative reactions which supply energy, and they are the building blocks for making complex biological molecules.

Energy supplies and requirements are usually described in terms of calories. A *calorie* (with a lowercase c) is the amount of energy required to raise the temperature of 1 g of pure water by 1°C, from 14.5°C to 15.5°C. In animal nutrition, calories are usually considered in units of 1,000 as kilocalories (kcal; the units listed on food packages in the United States [1 kcal = 4.2 kilojoules]) or Calories (with an uppercase C). One Calorie thus equals 1,000 calories. Carbohydrates contain about 4.2 kcal per gram, fats about 9.5 kcal per gram, and proteins about 4.1 kcal per gram.

Carbohydrates and fats are the primary organic molecules used as fuels. Animals whose intake of organic fuels is inadequate, or whose assimilation of such fuels is abnormal, suffer from **undernutrition.** Undernutrition is a form of **malnutrition,** which is a condition resulting from an improper diet. **Overnutrition,** the condition caused by excessive intake of specific nutrients, is another main type of malnutrition.

An animal suffering from undernutrition essentially is starving for one or more nutrients, taking in fewer calories than needed for daily activities. Animals with chronic undernutrition lose weight because they have to use energy-providing molecules of their own bodies as fuels. Mammals use stored fats and glycogen (animal starch) first. Once those stores have been used up, proteins are metabolized as fuels. The use of proteins as fuels leads to muscle wastage and, in the long term, to organ and brain damage, which eventually leads to death.

Organic molecules also serve as building blocks for carbohydrates, lipids, proteins, and nucleic acids. Animals can synthesize many of the organic molecules that they do not obtain directly in the diet by converting one type of building block into another. Typically, however, they cannot make certain amino acids and fatty acids from other organic molecules. These required organic building blocks are called **essential amino acids** and **essential fatty acids** because they must be obtained in the diet. If they are not obtained in the diet over a period of time, the animal may have serious health consequences. For instance, protein synthesis cannot continue unless all 20 amino acids are present. In the absence of essential amino acids in the diet, the animal would have to break down its own proteins to provide them for new protein synthesis.

Animals must also take in **vitamins,** organic molecules required in small quantities that the animal cannot synthesize for itself. Many vitamins are *coenzymes,* nonprotein organic subunits associated with enzymes that assist in enzymatic catalysis (see Section 4.4).

Individual species differ in the vitamins and essential amino acids and fatty acids they require. Various species also have differing dietary requirements for inorganic elements such as calcium, iron, and magnesium. These required inorganic elements are known collectively as **essential minerals.**

The essential amino acids, fatty acids, vitamins, and minerals are known collectively as an animal's **essential nutrients.** The list of essential nutrients differs from animal to animal. For domesticated animals, this means that specific feed formulations must be given to each type of animal. For instance, the essential nutrients for cats and dogs are different, which is why there are specific cat foods and dog foods, and why it is does not make good sense, nutritionally speaking, to feed cats and dogs human food.

Animals Obtain Nutrients in Fluid, Particle, or Bulk Form

All animals display adaptations that allow them to obtain the food they need in particular environments. Although these adaptations are amazingly varied, animals can be classified into one of four groups according to overall feeding methods and the physical state of the organic molecules they consume. These four groups are fluid feeders, suspension feeders, deposit feeders, and bulk feeders **(Figure 45.2).**

Fluid feeders obtain nourishment by ingesting liquids that contain organic molecules in solution. Among the invertebrates, aphids, mosquitoes, leeches, and spiders are examples of fluid feeders. Vertebrate fluid feeders include birds such as hummingbirds (Figure 45.2A), which feed on flower nectar; parasitic fishes such as lampreys, which feed on body fluids of their hosts; and some bats, which feed on nectar or blood. Many fluid feeders have mouthparts specialized to reach the source of their nourishment. For example, mosquitoes, bedbugs, and aphids have needlelike mouthparts that pierce body surfaces. Nectar-feeding birds and bats have long tongues that can extend deep within flowers. Some fluid feeders use enzymes or other chemicals to liquefy their food or to keep it liquid during feeding. For example, spiders inject digestive enzymes that liquefy tissues inside their victim and then suck up the liquid. The saliva of mosquitoes, leeches, and vampire bats includes an anticoagulant that keeps blood in liquid form during a feeding by inhibiting the clotting reaction.

Suspension feeders ingest small organisms that are suspended in water, such as bacteria, protozoa, algae, and small crustaceans, or fragments of these organisms. Among the suspension feeders are aquatic invertebrates such as clams, mussels, barnacles, many fishes, and even some birds and whales. These animals strain food particles suspended in water through a body structure covered with sticky mucus or through a filtering network of bristles, hairs, or other body parts. The trapped particles are then funneled into the animal's mouth. Bits of organic matter are trapped by the gills of bivalves such as clams and oysters, and plankton is filtered from water by the sievelike fringes of horny fiber hanging in the mouths of baleen whales (Figure 45.2B).

Deposit feeders pick up or scrape particles of organic matter from solid material they live in or on. Earthworms are deposit

A. **Fluid feeder**

B. **Suspension feeder**

Baleen

C. **Deposit feeder**

D. **Bulk feeder**

FIGURE 45.2
Adaptations for feeding. (A) A fluid feeder obtains nourishment by ingesting liquids containing nutrients. For example, a juvenile ruby-throated hummingbird *(Archilochus colubris)* obtains nectar from deep within a flower using its long bill and tongue. **(B)** A suspension feeder obtains nutrients by ingestion of small organisms in water. For example, a gray whale *(Eschrichtius robustus)* gulps water and mud from the sea bottom into its mouth, strains them out through its fringed baleen, and eats the crustaceans and other invertebrates that remain. **(C)** A deposit feeder obtains nutrients by picking up or scraping particles of organic matter from solid material. For example, a southwestern Atlantic fiddler crab *(Uca uruguayensis)* sifts edible material from the sediment it takes into its mouth. **(D)** A bulk feeder obtains nutrients by ingesting sizeable food items whole or in large pieces. This great blue heron *(Ardea herodias)* is feeding on a brook trout *(Salvelinus fontinalis)*.

feeders that eat their way through soil, taking the soil into their mouth and digesting and absorbing any organic material it contains. Some burrowing mollusks and tube-dwelling polychaete worms use body appendages to gather organic deposits from the sand or mud around them. Mucus on the appendages traps the organic material, and cilia move it to the mouth. The fiddler crab (*Uca* species) is also a deposit feeder (see Figure 45.2C). This animal has claws of markedly different sizes. The small claw picks up sediment and moves it to the mouth where the contents are sifted. The edible parts of the sediment are ingested, and the rest is put back on the sediment as a small ball. The feeding-related movement of the small claw over the larger claw looks like the crab is playing the large claw like a fiddle and hence gives the crab its name.

Bulk feeders are animals that consume sizeable food items whole or in large chunks. Most mammals eat this way, as do reptiles, most birds and fishes, and adult amphibians. Depending on the animal, adaptations for bulk feeding include teeth for tearing or chewing, and claws and beaks for holding large food items (Figure 45.2D), and jaws that are hinged or otherwise modified to permit a food mass to enter the mouth.

We now take up the processes by which animals, having fed, undertake the mechanical and chemical breakdown of food into absorbable molecular subunits.

STUDY BREAK 45.1 <

1. **What are carnivores, herbivores, and omnivores?**
2. **What are essential nutrients, and are they the same for all animals?**
3. **What is the difference between deposit feeders and suspension feeders?**

45.2 Digestive Processes

Digestive processes break food molecules into molecular subunits that can be absorbed into body fluids and cells. The breakdown occurs by **enzymatic hydrolysis,** in which chemical bonds are broken by the addition of H^+ and OH^- which are the components of a molecule of water (see Figure 3.5). Specific enzymes speed these reactions: *amylases* catalyze the hydrolysis of starches, *lipases* break down fats and other lipids, *proteases* hydrolyze proteins, and *nucleases* digest nucleic acids. Depending on the animal, the enzymatic hydrolysis of food molecules may take place inside or outside the body cells.

Intracellular Digestion Takes Place within Cells; Extracellular Digestion Occurs in an Internal Pouch or Tube

In **intracellular digestion,** cells take in food particles by endocytosis (described in Section 6.5). Inside the cell, the endocytic vesicle containing the food particles fuses with a lysosome, a vesicle containing hydrolytic enzymes. The molecular subunits produced by the hydrolysis pass from the vesicle to the cytosol. Any undigested material remaining in the vesicle is released to the outside of the cell by exocytosis (also discussed in Section 6.5). Only a few animals, primarily sponges and some cnidarians, break down food exclusively by intracellular digestion. In sponges, water containing particles of organic matter and microorganisms enters the animal's saclike body through pores in the body wall (see Figure 29.8B). In the body cavity, individual *choanocytes* (collar cells) lining the body wall trap the food particles, take them in by endocytosis, and transport them to amoeboid cells, which digest them intracellularly (see Figure 29.8C).

Extracellular digestion takes place outside body cells, in a pouch or tube that is enclosed within the body. Epithelial cells lining the pouch or tube secrete enzymes that digest the food. Processing food in specialized compartments in this way prevents the animal from digesting its own body tissues.

Most invertebrates and all vertebrates digest food primarily by extracellular digestion. From an adaptive standpoint, extracellular digestion greatly expands the range of available food sources by allowing animals to digest much larger food items than single cells can take in. Extracellular digestion also allows animals to eat large batches of food, which can be stored and digested while the animal continues other activities.

Saclike Digestive Systems Have a Single Opening through Which Food Enters and Undigested Matter Exits

Some animals, including flatworms and cnidarians such as hydras, corals, and sea anemones, have a saclike digestive system with a single opening, a mouth, which serves both as the entrance for food and the exit for undigested material. Since these animals lack a separate vascular system, water taken into the digestive chamber also serves to circulate nutrients and other materials through the various tissue layers. Because the digestive chamber contributes both to digestion and to circulation, it is called a **gastrovascular cavity.** In the flatworm *Dugesia,* for example, food is brought to the mouth by a protrusible **pharynx** (a throat that can be stuck out) and then enters the gastrovascular cavity **(Figure 45.3);** glands in the cavity wall secrete enzymes that begin the digestive process. Cells lining the cavity then take up the partially digested material by endocytosis and complete

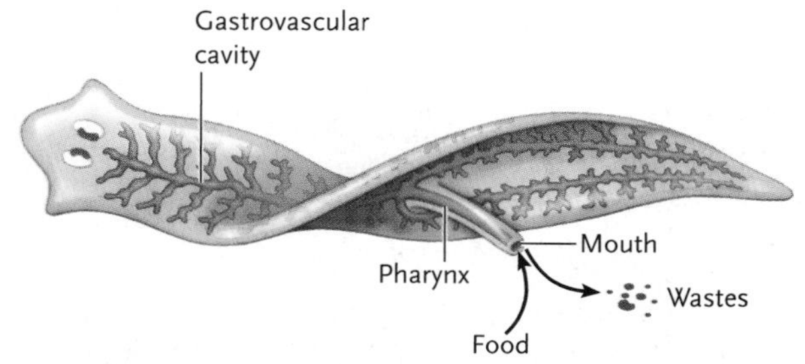
Gastrovascular cavity
Mouth
Pharynx
Wastes
Food

FIGURE 45.3
The digestive system of the flatworm *Dugesia.* The gastrovascular cavity (in blue) is a blind sac, with one opening to the exterior through which food is ingested and wastes are expelled.

digestion intracellularly. Undigested matter is released to the outside through the pharynx and mouth.

Digestive Tubes Typically Process Nutrients in Five Successive Steps

Most invertebrates and all vertebrates have a tubelike digestive system with two openings that form a separate mouth and anus; the digestive contents move in one direction through specialized regions of the tube, from the mouth to the anus. This type of digestive system is called a **digestive tube,** *gut, alimentary canal, digestive tract,* or *gastrointestinal (GI) tract.* Structurally, the inside of the digestive tube—called the **lumen**—is external to all body tissues. In other words, although entirely enclosed, the lumen is, in a functional sense, *outside* of the body.

In most animals with a digestive tube, digestion occurs in five successive steps, with each step taking place in a specialized region of the tube. The tube thus acts as a sort of biological disassembly line, with food entering at one end and passing through as many as five areas in which food processing occurs.

1. **Mechanical processing:** Chewing, grinding, and tearing break food chunks into smaller pieces, increasing their mobility and the surface area exposed to digestive enzymes.
2. **Secretion of enzymes and other digestive aids:** Enzymes and other substances that aid the process of digestion, such as acids, emulsifiers, and lubricating mucus, are released into the tube.
3. **Enzymatic hydrolysis:** Food molecules are broken down through enzyme-catalyzed reactions into absorbable molecular subunits.
4. **Absorption:** The molecular subunits are absorbed from the digestive contents into body fluids and cells.
5. **Elimination:** Undigested materials are expelled through the anus.

The material being digested is pushed along by muscular contractions of the wall of the digestive tube. During its progress through the tube, the digestive contents may be stored temporarily at one or more locations. The storage allows animals to take in larger quantities of food than they can process immediately, so that feedings can be spaced in time rather than continuous.

DIGESTION IN AN ANNELID The earthworm (genus *Lumbricus,* **Figure 45.4A**) is a deposit feeder. As it burrows, it pushes soil particles into its mouth. The particles pass from the mouth into the pharynx. Muscular activity then moves the particles through a connecting passage, the **esophagus,** into the **crop,** an enlargement of the digestive tube where the contents are stored and mixed with lubricating mucus. This mixture enters the **gizzard,** which contains grains of sand, and is ground into fine particles by muscular contractions of the wall. The pulverized mixture then enters a long **intestine,** where the organic matter is hydrolyzed by enzymes secreted into the digestive tube. As muscular contractions of the intestinal wall move the mixture along, cells lining the intestine absorb the molecular subunits produced by digestion. At the end of the intestine, the undigested residue is expelled through the anus.

DIGESTION IN AN INSECT Herbivorous insects such as the grasshopper **(Figure 45.4B)** tear leaves and other plant parts into small particles with hard external mouth parts. From the mouth, the food particles pass through the pharynx, where salivary secretions

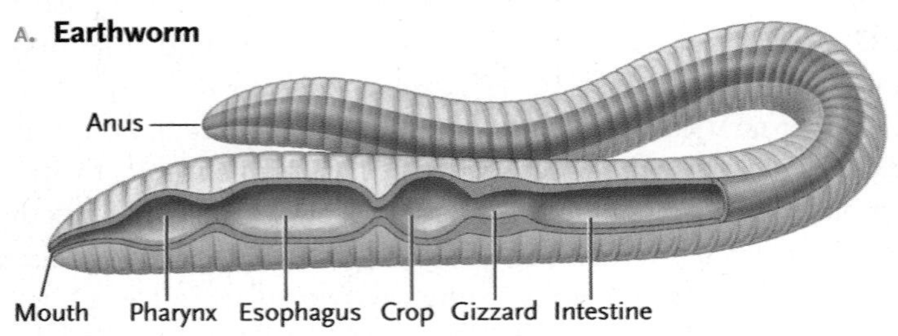
A. Earthworm
Anus
Mouth
Pharynx
Esophagus
Crop
Gizzard
Intestine

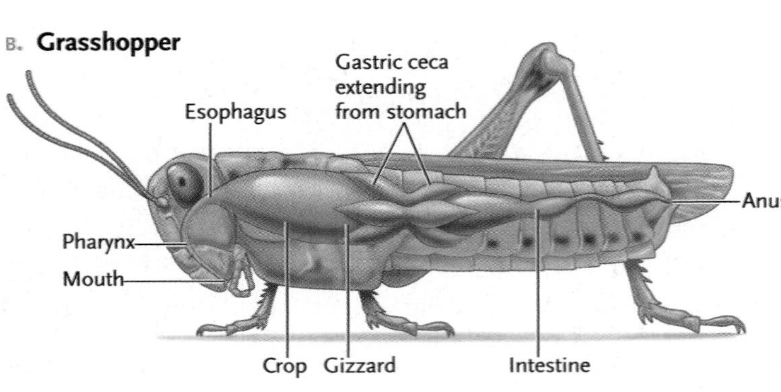
B. Grasshopper
Esophagus
Gastric ceca extending from stomach
Anus
Pharynx
Mouth
Crop
Gizzard
Intestine

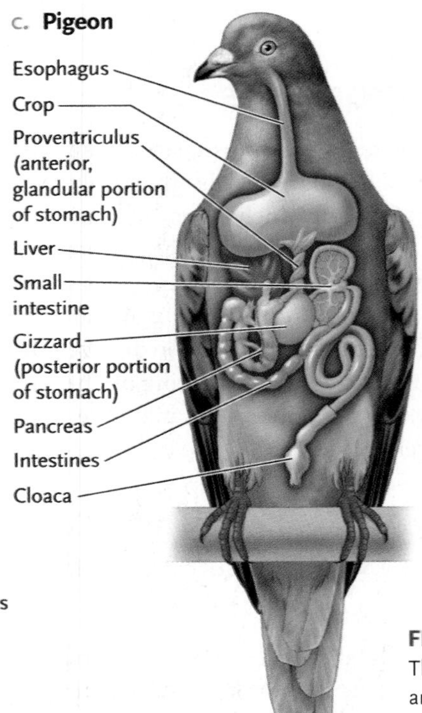
C. Pigeon
Esophagus
Crop
Proventriculus (anterior, glandular portion of stomach)
Liver
Small intestine
Gizzard (posterior portion of stomach)
Pancreas
Intestines
Cloaca

FIGURE 45.4
The digestive systems of an annelid, the earthworm **(A)**, an insect, the grasshopper **(B)**, and a bird, the pigeon *(Columba)* **(C)**.

moisten the mixture before it enters the esophagus and passes into the crop. These secretions begin the process of chemical digestion. From the crop, the food mass enters the gizzard, which grinds it into smaller pieces. These food particles enter the **stomach,** in which food is stored and digestion begins. Insect stomachs have saclike outgrowths, the *gastric ceca* (*caecus* = blind), where enzymes hydrolyze the digestive contents; the products of digestion are absorbed through the walls of the ceca. The undigested contents then move into the intestine for further digestion and absorption. At the end of the intestine, water is absorbed from the undigested matter and the remnants are expelled through the anus. The digestive systems of other arthropods are similar to the insect system.

DIGESTION IN A BIRD A pigeon **(Figure 45.4C)** picks up seeds with its bill. The bird's tongue moves the seeds into its mouth, where they are moistened by mucus-filled saliva and swallowed whole (birds have no teeth). The seeds then pass through the pharynx into the esophagus. (In some cases, birds crack open seeds with their bills and ingest seed kernels in a similar fashion.) The anterior end of the esophagus is tubelike; at the posterior end is the pouchlike crop, in which the bird can store large quantities of food. From the crop, the food passes into the anterior glandular portion of the stomach, called the *proventriculus,* which secretes digestive enzymes and acids. The posterior end is the gizzard, in which the seeds are ground into fine particles, aided by ingested bits of sand and rock. The food particles are released into the intestine, where the liver secretes bile and the pancreas adds digestive enzymes. The molecular subunits produced by enzymatic digestion are absorbed as the mixture passes along the intestine, and the undigested residues are expelled through the anus.

Many of the structures of the pigeon's digestive system, including the mouth, pharynx, esophagus, stomach, intestine, liver, and pancreas, occur in almost all vertebrates.

STUDY BREAK 45.2 <

1. **Distinguish between intracellular digestion and extracellular digestion.**
2. **What are the five steps of food processing in a digestive tube?**

>

THINK OUTSIDE THE BOOK

Use the Internet or a research paper search tool such as PubMed (http://www.ncbi.nlm.nih.gov/PubMed/) to list the research question being addressed in three projects involving bird digestive systems.

45.3 Digestion in Humans and Other Mammals

Mammals digest foods using the same five steps as other animals with a digestive tube: mechanical processing, secretion of enzymes and other digestive aids, enzymatic hydrolysis, absorption of molecular subunits, and elimination. The mammalian digestive system is a series of specialized digestive regions that perform these steps, including the mouth, pharynx, esophagus, stomach, small and large intestines, rectum, and anus **(Figure 45.5).** These regions are under the control of the nervous and endocrine systems. The digestive system is not identical in all mammals. Specific adaptations have produced variants of the digestive system such as occurs in humans. An example is the digestive system of ruminants (discussed in Section 45.5).

Humans Require Specific Essential Amino Acids, Fatty Acids, Vitamins, and Minerals in Their Diet

The human digestive system meets our basic needs for fuel molecules and for a wide range of nutrients, including the molecular building blocks of carbohydrates, lipids, proteins, and nucleic acids. If the diet is adequate, the digestive system also absorbs the essential nutrients—the amino acids, fatty acids, vitamins, and minerals that cannot be synthesized within our bodies.

ESSENTIAL AMINO ACIDS AND FATTY ACIDS There are eight essential amino acids for adult humans: lysine, tryptophan, phenylalanine, threonine, valine, methionine, leucine, and isoleucine. Infants and young children also require histidine. The proteins in fish, meat, egg whites, milk, and cheese supply all the essential amino acids, provided those foods are eaten in adequate quantities. In contrast, the proteins of many plants are deficient in one or more of the essential amino acids. Corn, for example, contains inadequate amounts of lysine, and beans contain little methionine. Vegetarians, and especially vegans who eat a diet with no animal-derived nutrients, must choose their foods carefully to obtain all of the essential amino acids **(Figure 45.6).** Such diets typically include combinations of foods, each of which provides some amino acids, and that together contain all of the essential amino acids. An example is including in the diet rice or corn or other grains (low in lysine but high in methionine) with lentils or soybeans (perhaps in the form of tofu) or other legumes (low in methionine but high in lysine).

If the diet lacks one or more essential amino acids, many enzymes and other proteins cannot be synthesized in sufficient quantities. The resulting protein deficiency is most damaging to the young, who must rapidly synthesize proteins for development and growth. Even mild protein starvation during pregnancy or for some months after birth can retard a child's mental and physical development.

Only two fatty acids, linoleic acid and linolenic acid, are essential in the human diet. Both are required for synthesis of phospholipids forming parts of biological membranes and certain hormones. Because almost all foods contain these fatty acids, most people have no problem obtaining them. However, people on a low-fat diet that is deficient in linoleic acid and linolenic acid are at serious risk for developing coronary heart disease. That is, there is an inverse correlation between the concentration of these essential fatty acids in the diet and the incidence of coronary heart disease. This is illustrated in the case of Hindu

FIGURE 45.5
The human digestive system.

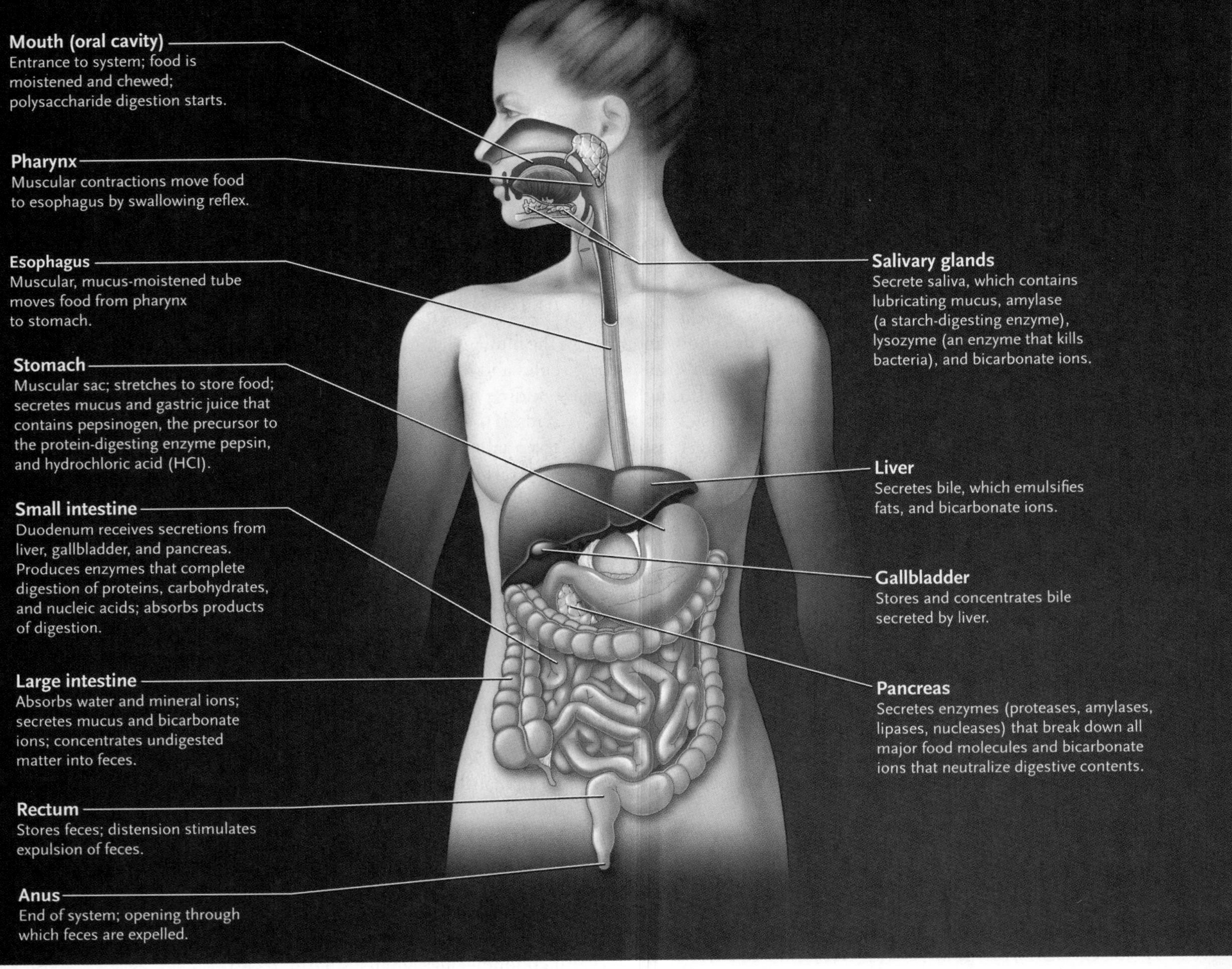

Mouth (oral cavity)
Entrance to system; food is moistened and chewed; polysaccharide digestion starts.

Pharynx
Muscular contractions move food to esophagus by swallowing reflex.

Esophagus
Muscular, mucus-moistened tube moves food from pharynx to stomach.

Stomach
Muscular sac; stretches to store food; secretes mucus and gastric juice that contains pepsinogen, the precursor to the protein-digesting enzyme pepsin, and hydrochloric acid (HCl).

Small intestine
Duodenum receives secretions from liver, gallbladder, and pancreas. Produces enzymes that complete digestion of proteins, carbohydrates, and nucleic acids; absorbs products of digestion.

Large intestine
Absorbs water and mineral ions; secretes mucus and bicarbonate ions; concentrates undigested matter into feces.

Rectum
Stores feces; distension stimulates expulsion of feces.

Anus
End of system; opening through which feces are expelled.

Salivary glands
Secrete saliva, which contains lubricating mucus, amylase (a starch-digesting enzyme), lysozyme (an enzyme that kills bacteria), and bicarbonate ions.

Liver
Secretes bile, which emulsifies fats, and bicarbonate ions.

Gallbladder
Stores and concentrates bile secreted by liver.

Pancreas
Secretes enzymes (proteases, amylases, lipases, nucleases) that break down all major food molecules and bicarbonate ions that neutralize digestive contents.

vegetarians from India. Their diet consists mainly of low-fat grains and legumes—clearly a low-fat diet—yet their rate of coronary heart disease is higher than that in the United States and Europe, where dietary fat content is higher.

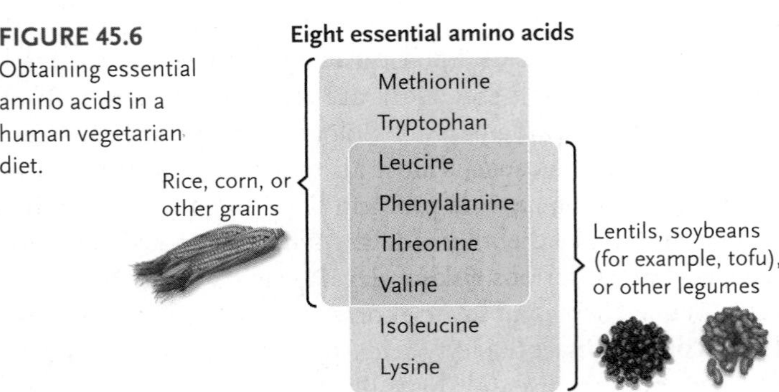

Eight essential amino acids
Methionine
Tryptophan
Leucine
Phenylalanine
Threonine
Valine
Isoleucine
Lysine
Rice, corn, or other grains
Lentils, soybeans (for example, tofu), or other legumes

FIGURE 45.6 Obtaining essential amino acids in a human vegetarian diet.

VITAMINS Humans require 13 known vitamins in their diet. Many metabolic reactions depend on vitamins, and the absence of one vitamin can affect the functions of the others. These essential nutrients fall into two classes: **water-soluble** (hydrophilic) **vitamins** and **fat-soluble** (hydrophobic) **vitamins** (summarized in **Table 45.1**). The body stores excess fat-soluble vitamins in adipose tissues, but any amount of water-soluble vitamins above daily nutritional requirements is excreted in the urine. Thus, meeting the daily minimum requirements of water-soluble vitamins is critical for good health. The body can tap its stores of fat-soluble vitamins to meet daily requirements, however, these stores are quickly depleted so prolonged deficiencies of the fat-soluble vitamins also negatively affect health.

Most of you get all the vitamins you need through a normal and varied diet that includes meats, fish, eggs, cheese, and vegetables. Vitamin supplements are usually necessary only for strict

TABLE 45.1 Vitamins: Sources, Functions, and Effects of Deficiencies in Humans

Vitamin	Common Sources	Main Functions	Selected Effects of Chronic Deficiency
Fat-Soluble Vitamins			
A (retinol)	Yellow fruits, yellow or green leafy vegetables; also in fortified milk, egg yolk, fish liver	Used in synthesis of visual pigments, and in growth of bone and teeth; maintains epithelial tissues	Dry, scaly skin; lowered resistance to infections; night blindness
D (calciferol)	Fish liver oil, egg yolk, fortified milk; manufactured when body exposed to sunshine	Promotes bone growth and mineralization; enhances calcium absorption from gut	Bone deformities (rickets) in children; bone softening in adults
E (tocopherol)	Whole grains, leafy green vegetables, vegetable oils	Antioxidant; helps maintain cell membrane and red blood cells	Lysis of red blood cells; nerve damage
K (naphthoquinone)	Intestinal bacteria; also in green leafy vegetables, cabbage	Promotes synthesis of blood-clotting protein by liver	Abnormal blood clotting, severe bleeding (hemorrhaging)
Water-Soluble Vitamins			
B_1 (thiamine)	Whole grains, green leafy vegetables, legumes, lean meat, eggs, nuts	Connective tissue formation; needed for folate utilization; coenzyme forming part of enzyme in oxidative reactions	Mental confusion; tingling sensations; muscular wasting; poor coordination; pain; vomiting; rapid heart rate; enlarged heart; congestive heart failure
B_2 (riboflavin)	Whole grains, poultry, fish, egg white, milk, lean meat	Coenzyme	Skin lesions
Niacin	Green leafy vegetables, potatoes, peanuts, poultry, fish, pork, beef	Coenzyme of oxidative phosphorylation	Sensitivity to light; damage to skin, gut, and nervous system, leading to, respectively, dermatitis, diarrhea, and dementia
B_6 (pyridoxine)	Spinach, whole grains, tomatoes, potatoes, meat	Coenzyme in amino acid and fatty acid metabolism	Skin, muscle, and nerve damage
Pantothenic acid	In many foods (meat, yeast, egg yolk especially)	Coenzyme in carbohydrate and fat oxidation; required for fatty acid and steroid synthesis	Fatigue; tingling in hands; headaches, nausea
Folic acid	Dark green vegetables, whole grains, yeast, lean meat; intestinal bacteria produce some folate	Coenzyme in nucleic acid and amino acid metabolism; promotes red blood cell formation	Anemia; inflamed tongue; diarrhea; impaired growth; mental disorders; neural tube defects and low birth weight in newborns
B_{12} (cobalamin)	Poultry, fish, eggs, red meat, dairy foods (not butter)	Coenzyme in nucleic acid metabolism; necessary for red blood cell formation	Anemia; nerve damage that causes tingling of hands and feet, loss of balance, depression, and, in severe cases, dementia
Biotin	Legumes, egg yolk; colon bacteria produce some	Coenzyme in fat and glycogen formation and amino acid metabolism	Scaly skin (dermatitis); sore tongue; brittle hair; depression; weakness
C (ascorbic acid)	Fruits and vegetables, especially citrus, berries, cantaloupe, cabbage, broccoli, green pepper	Vital for collagen synthesis; antioxidant	Scurvy (weakness, anemia, gum disease, and skin problems); delayed wound healing; impaired immunity

vegetarians, newborns, the elderly, and individuals who are taking medication that affects the body's uptake of nutrients.

Vitamin D (calciferol) differs from other essential vitamins because humans can synthesize it themselves, through the action of ultraviolet light on lipids in the skin. However, many people are not exposed to enough sunlight to make sufficient quantities of the vitamin, and so must rely on dietary sources. And, although humans cannot make vitamin K, much of the requirement for this vitamin is supplied through the metabolic activity of bacteria living in the large intestine. Vitamin K deficiency, therefore, is exceedingly rare in healthy persons. Vitamin K plays a role in blood clotting, so individuals with vitamin K deficiency will bruise easily and show increased blood clotting times. Vitamin K deficiency can be caused in persons on long-term antibiotic therapy because the antibiotics kill intestinal bacteria.

Other mammals have essentially the same vitamin requirements as humans, with some differences. For example, most mammals, with the exception of primates, guinea pigs, and fruit bats, can synthesize vitamin C. So far as is known, no animal can synthesize B vitamins, but ruminants such as cattle and deer are supplied with those vitamins by microorganisms that live in the digestive tract (see Section 45.5).

MINERALS Many minerals are essential in the human diet (**Table 45.2**). Some of them, called **macronutrients,** are required

TABLE 45.2 Major Minerals: Sources, Functions, and Effects of Deficiencies in Humans

Mineral	Sources	Functions	Selected Effects of Deficiency
Calcium (Ca)	Dairy products, leafy green vegetables, legumes, whole grains, nuts	Bone and tooth formation; blood clotting; neural and muscle action	Stunted growth; diminished bone mass (osteoporosis)
Chlorine (Cl)	Table salt, meat, eggs, dairy products	HCl formation in stomach; contributes to body's acid–base balance; necessary for neural function and water balance	Muscle cramps; impaired growth; poor appetite
Chromium (Cr)*	Meat, liver, cheese, whole grains, brewer's yeast, peanuts	Roles in carbohydrate metabolism	Impaired response to insulin; increased risk of type 2 diabetes mellitus
Cobalt (Co)*	Meat, liver, fish, milk	Constituent of vitamin B_{12} (required for red blood cell maturation)	Same as for vitamin B_{12} (see Table 45.1)
Copper (Cu)*	Nuts, legumes, seafood, drinking water, whole grains	Used in synthesis of melanin, hemoglobin, and some electron transfer system components in mitochondria	Anemia; changes in bone and blood vessels
Fluorine (F)*	Fluoridated water, tea, seafood	Bone and tooth maintenance	Tooth decay
Iodine (I)*	Marine fish, shellfish, iodized salt	Thyroid hormone formation	Goiter (enlarged thyroid), with metabolic disorders
Iron (Fe)	Liver, whole grains, green leafy vegetables, legumes, nuts, eggs, lean meat, molasses, dried fruit, shellfish	Component of hemoglobin, cytochrome, and myoglobin	Anemia
Magnesium (Mg)	Whole grains, green vegetables, legumes, nuts, dairy products	Required for action of many enzymes; roles in muscle and nerve function	Weak, sore muscles; impaired neural function
Manganese (Mn)*	Whole grains, nuts, legumes, many fruits	Activates many enzymes, including ones with roles in synthesis of urea and fatty acids	Abnormal bone and cartilage
Molybdenum (Mo)*	Dairy products, whole grains, green vegetables, legumes	Component of some enzymes	Impaired nitrogen excretion
Phosphorus (P)	Whole grains, legumes, poultry, red meat, dairy products	Component of bones and teeth, nucleic acids, ATP, and phospholipids	Muscular weakness; loss of minerals from bone
Potassium (K)	Meat, milk, many fruits, vegetables	Muscle and neural function; roles in protein synthesis	Muscular weakness
Selenium (Se)*	Meat, seafood, cereal grains, poultry, garlic	Constituent of several enzymes; antioxidant	Muscle pain
Sodium (Na)	Table salt, dairy products, meat, eggs	Acid–base balance; water balance; roles in muscle and neural function	Muscle cramps
Sulfur (S)	Meat, eggs, dairy products	Component of body proteins	Same as protein deficiencies
Zinc (Zn)*	Whole grains, legumes, nuts, meat, seafood	Component of digestive enzymes and transcription factors; roles in normal growth, wound healing, sperm formation, taste and smell	Impaired growth, scaly skin, impaired immune function

*Required in trace amounts in diet

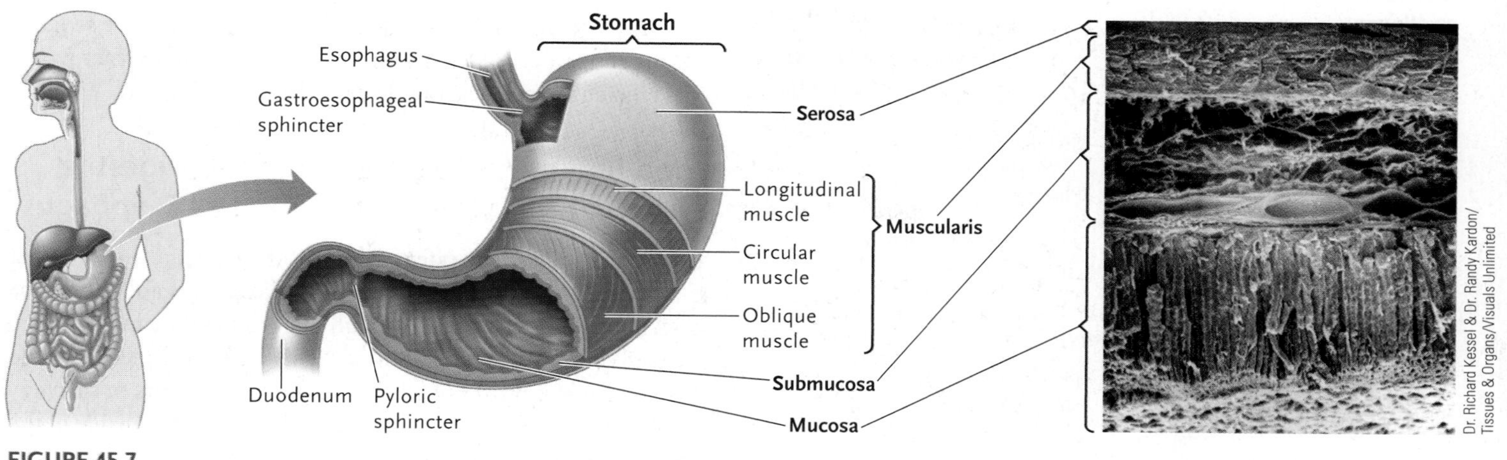
Stomach
Esophagus
Gastroesophageal sphincter
Serosa
Longitudinal muscle
Circular muscle
Oblique muscle
Muscularis
Submucosa
Mucosa
Duodenum
Pyloric sphincter
Dr. Richard Kessel & Dr. Randy Kardon/ Tissues & Organs/Visuals Unlimited

FIGURE 45.7
Layers of the gut wall in vertebrates, as seen in the stomach wall.

in amounts ranging from 50 mg to more than a gram per day; others, such as zinc, are **micronutrients,** or **trace elements,** required only in small amounts, some less than 1 mg per day. All of the minerals, although listed as elements, are ingested as compounds or as ions in solution.

A normal and varied diet supplies adequate amounts of the essential minerals. Supplements may be required for those on a strict vegetarian diet, the very young, and the aged. Overdoses of some minerals can cause problems; ingesting excess iron, for example, has been linked to liver, heart, and blood vessel damage; too much sodium can lead to elevated blood pressure and excess water retention in tissues.

We now turn to the structures that extract nutrients from ingested foods. We begin with a survey of digestive structures common to all vertebrates.

Four Major Layers of the Gut Each Have Specialized Functions in Digestion

The wall of the gut in mammals and other vertebrates contains four major layers, each with specialized functions. These layers are shown for the stomach in **Figure 45.7.**

1. The **mucosa,** which contains epithelial and glandular cells, lines the inside of the gut. The epithelial cells, which absorb digested nutrients, seal off the digestive contents from body fluids. The glandular cells secrete enzymes, substances such as lubricating mucus that aid digestion, and substances that adjust the pH of the digestive contents.
2. The **submucosa** is a thick layer of elastic connective tissue that contains neuron networks and blood and lymph vessels. The neuron networks provide local control of digestive activity and carry signals between the gut and the central nervous system. The lymph vessels carry absorbed lipids to other parts of the body.
3. In most regions of the gut, the **muscularis** is formed by two smooth muscle layers, a *circular layer* and a *longitudinal layer.* In a manner analogous to how similar layers of muscles alter the shape of an earthworm (see Figure 41.9), contraction of the circular muscles along with relaxation of the longitudinal muscles constricts the diameter of the gut and lengthens it, whereas contraction of the longitudinal muscles along with relaxation of the circular muscles causes the gut to assume a larger diameter and shortens it. The stomach also has an *oblique layer* running diagonally around its wall. The circular and longitudinal muscle layers of the muscularis coordinate their activities to push the digestive contents through the gut **(Figure 45.8).** In this mechanism, called **peristalsis,** the circular muscle layer contracts in a wave that passes along the gut, constricting the gut and pushing the digestive contents onward. Just in front of the advancing constriction, the longitudinal layer contracts, shortening and expanding the tube and making space for the contents to advance.
4. The outermost gut layer, the **serosa,** consists of connective tissue that secretes an aqueous, slippery fluid. The fluid lubricates the areas between the digestive organs and other organs, reducing friction between them as they move together as a result of muscle movement. Along much of the length of the digestive system, the serosa is continuous with the *mesentery,* a tissue that suspends the digestive system from the inner wall of the abdominal cavity.

Powerful rings of smooth muscle called **sphincters** form valves between major regions of the digestive tract. By contracting and relaxing, sphincters control the passage of the digestive contents from one region to the next, and ultimately through the anus.

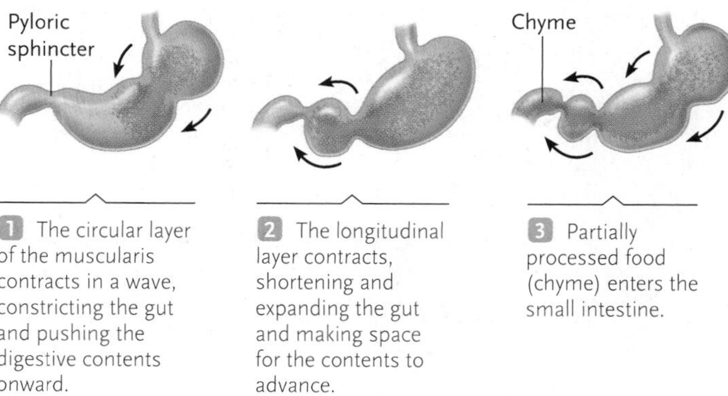
Pyloric sphincter
Chyme
1 The circular layer of the muscularis contracts in a wave, constricting the gut and pushing the digestive contents onward.
2 The longitudinal layer contracts, shortening and expanding the gut and making space for the contents to advance.
3 Partially processed food (chyme) enters the small intestine.

FIGURE 45.8
The waves of peristaltic contractions moving food through the stomach.

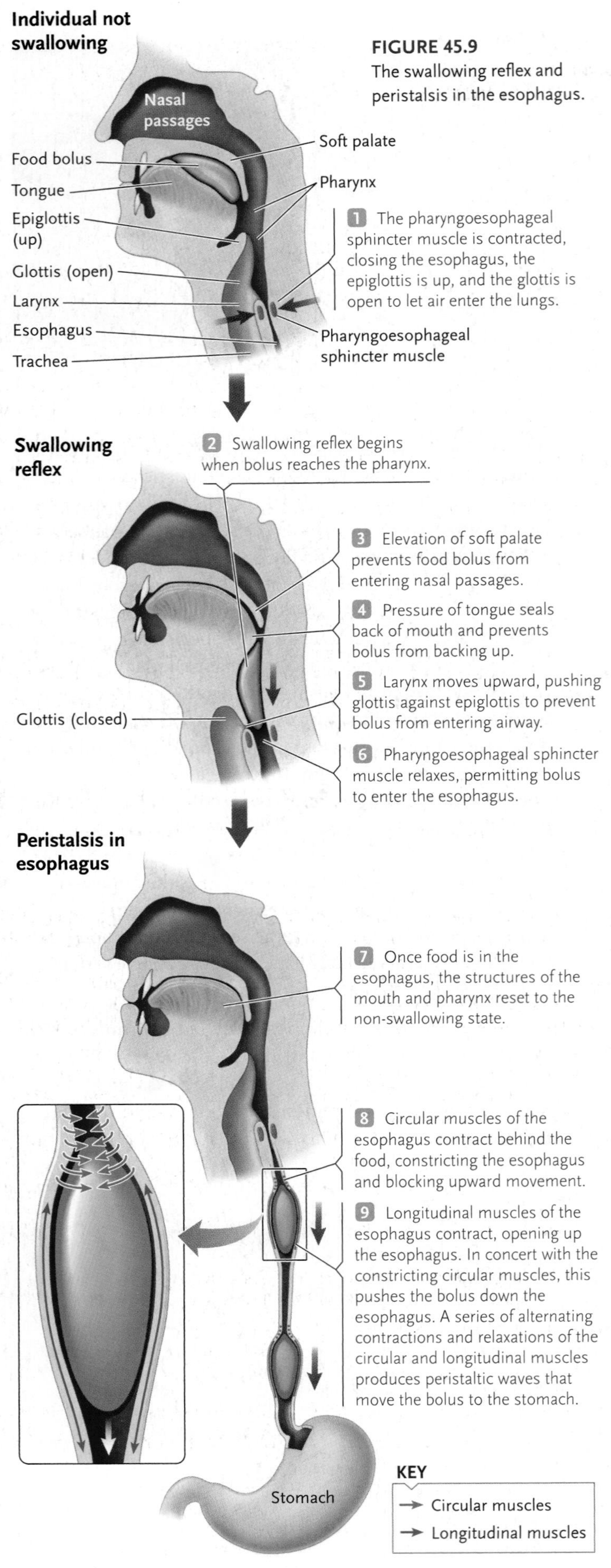
Individual not swallowing
Nasal passages
Food bolus
Tongue
Epiglottis (up)
Glottis (open)
Larynx
Esophagus
Trachea
Soft palate
Pharynx
1 The pharyngoesophageal sphincter muscle is contracted, closing the esophagus, the epiglottis is up, and the glottis is open to let air enter the lungs.
Pharyngoesophageal sphincter muscle
Swallowing reflex
2 Swallowing reflex begins when bolus reaches the pharynx.
3 Elevation of soft palate prevents food bolus from entering nasal passages.
4 Pressure of tongue seals back of mouth and prevents bolus from backing up.
5 Larynx moves upward, pushing glottis against epiglottis to prevent bolus from entering airway.
Glottis (closed)
6 Pharyngoesophageal sphincter muscle relaxes, permitting bolus to enter the esophagus.
Peristalsis in esophagus
7 Once food is in the esophagus, the structures of the mouth and pharynx reset to the non-swallowing state.
8 Circular muscles of the esophagus contract behind the food, constricting the esophagus and blocking upward movement.
9 Longitudinal muscles of the esophagus contract, opening up the esophagus. In concert with the constricting circular muscles, this pushes the bolus down the esophagus. A series of alternating contractions and relaxations of the circular and longitudinal muscles produces peristaltic waves that move the bolus to the stomach.
Stomach
KEY
Circular muscles
Longitudinal muscles

FIGURE 45.9
The swallowing reflex and peristalsis in the esophagus.

The specialized regions of the gut that perform the sequential processes of digestion in humans allow us to extract nutrients efficiently from the highly varied foods we ingest.

Food Begins Its Travel through the Digestive System in the Mouth, Pharynx, and Esophagus

The human digestive system in its normal contracted state in a living adult is about 4.5 m long. Fully relaxed, it is about twice as long. Food begins its travel through this tract in the mouth, where the teeth cut, tear, and crush food items into small pieces. While this mechanical processing is in progress, three pairs of **salivary glands** secrete saliva through ducts that open on the inside of the cheeks and under the tongue.

Saliva, which is more than 99% water, moistens the food. Saliva contains **salivary amylase,** which hydrolyzes starches to the disaccharide maltose. It also contains mucus, which lubricates the food mass, facilitating the formation of a **bolus** in preparation for swallowing. Bicarbonate ions (HCO_3^-) in saliva neutralize acids in the food and keep the pH of the mouth between 6.5 and 7.5, which is the optimal range for salivary amylase to function. Another component of saliva is *lysozyme,* an enzyme that kills bacteria by breaking open their cell walls. Some 1 to 2 L of saliva are secreted into the mouth each day.

After a suitable period of chewing, the bolus is pushed by the tongue to the back of the mouth, where touch receptors detect the pressure and trigger the **swallowing reflex (Figure 45.9).** The reflex is an involuntary action produced by contractions of muscles in the walls of the pharynx that direct food into the esophagus. Peristaltic contractions of the esophagus, aided by mucus secreted by the esophagus, propel the bolus toward the stomach. The passage of a bolus down the esophagus stimulates the **gastroesophageal sphincter** at the junction between the esophagus and the stomach (see Figure 45.7) to open and admit the bolus to the stomach. After the bolus enters the stomach, the sphincter closes tightly. If the closure is imperfect, the acidic stomach contents can enter the esophagus and produce the irritation and pain we recognize as *acid reflux* or heartburn.

We can consciously initiate the swallowing reflex. However, once the swallowing reflex has begun, we cannot voluntarily stop it, as you might have noticed when you get that feeling of a piece of food or a pill being stuck in the throat or chest. This is because the muscles of the pharynx and upper esophagus are skeletal muscles, which you can control, while the muscles below are smooth muscles, which you cannot control.

Involuntary movements of the tongue and soft palate at the back of the mouth prevent food from backing into the mouth or nasal cavities. Entry into the trachea (the airway to the lungs) is blocked by closure of the *glottis* (the space between the vocal cords) and an upward movement of the *larynx* (the voice box) at the top of the trachea, which closes against a flaplike valve, the **epiglottis.** You can feel the larynx and the front of the epiglottis bob upward if you place your hand on your throat while you swallow. If these blocking mechanisms fail, touch receptors in the nasal passages and larynx trigger coughing and sneezing reflexes that clear these passages.

The Stomach Stores Food and Continues Digestion

The stomach is a muscular, elastic sac that stores food and adds secretions that further the process of digestion. The mucosal layer of the stomach is an epithelium covered with tiny **gastric pits** that are entrances to millions of **gastric glands.** These glands extend deep into the stomach wall and contain cells that secrete some of the products needed to digest food.

The entry of food into the stomach activates stretch receptors in its wall. Signals from the stretch receptors stimulate the secretion of **gastric juice (Figure 45.10),** which contains **pepsinogen,** the precursor for the digestive enzyme **pepsin,** hydrochloric acid (HCl), and lubricating mucus. The stomach secretes about 2 L of gastric juice each day.

Pepsinogen is secreted by **chief cells** in the gastric glands and is converted to pepsin by the highly acid conditions of the stomach (see Figure 45.10). Once produced, pepsin itself catalyzes the reaction that converts more pepsinogen to pepsin. Pepsin begins the digestion of proteins by introducing breaks in polypeptide chains. The activation of pepsinogen illustrates a common theme in the digestive system: powerful hydrolytic enzymes that would be dangerous to the cells secreting them are synthesized in the form of inactive precursors and are not converted into active form until they are exposed to the digestive contents.

Parietal cells (see Figure 45.10) secrete H^+ and Cl^-, which combine to form HCl in the lumen of the stomach. The HCl lowers the pH of the digestive contents to pH 2 or lower, the level at which pepsin reaches optimal activity. To put this pH in perspective, lemon juice is pH 2.4, and sulfuric acid or battery acid is approximately pH 1. The acidity of the stomach also helps break up food particles and causes proteins in the digestive contents to unfold, which exposes their peptide linkages to hydrolysis by pepsin. The acid also kills most of the bacteria that reach the stomach and stops the action of salivary amylase.

A thick coating of alkaline mucus, secreted by **mucous cells** (see Figure 45.10), protects the mucosal layer of the stomach from attack by pepsin and HCl. Underneath the mucous barrier, tight junctions between cells prevent gastric juice from seeping into the stomach wall. Even so, some breakdown of the mucosal layer does occur. However, the damage is normally repaired quickly by the rapid division of mucosal cells.

Most bacteria cannot survive the highly acid environment of the stomach, but one, *Helicobacter pylori,* thrives there. In some people, the bacterium breaks down the mucous barrier, exposing the stomach wall to attack by HCl and pepsin. The resulting lesion, known as a **peptic** or **stomach ulcer,** causes stomach bleeding and pain. If untreated, an ulcer can become so deep that it perforates the stomach wall, with potentially fatal consequences. Ulcers are treated by taking an antibiotic that kills

FIGURE 45.10
Cells that secrete mucus, pepsin, and HCl in the stomach lining.

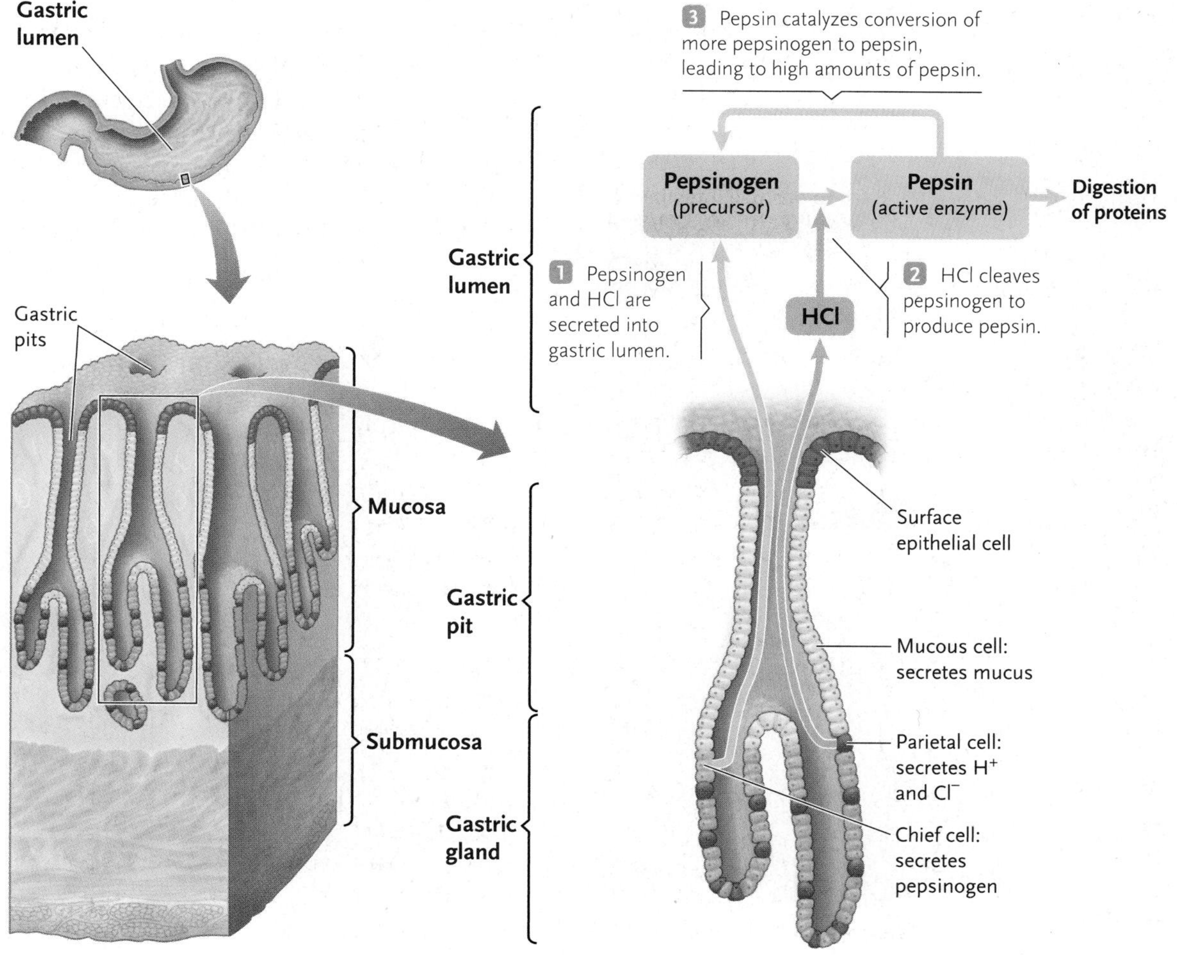

Gastric lumen
Gastric pits
Mucosa
Submucosa
Gastric lumen
Gastric pit
Gastric gland
3 Pepsin catalyzes conversion of more pepsinogen to pepsin, leading to high amounts of pepsin.
Pepsinogen (precursor)
Pepsin (active enzyme)
Digestion of proteins
1 Pepsinogen and HCl are secreted into gastric lumen.
HCl
2 HCl cleaves pepsinogen to produce pepsin.
Surface epithelial cell
Mucous cell: secretes mucus
Parietal cell: secretes H+ and Cl−
Chief cell: secretes pepsinogen

H. pylori. Barry J. Marshall of the University of Western Australia and J. Robin Warren of Royal Perth Hospital received a 2005 Nobel Prize for their discovery that a bacterium is responsible for most human ulcers (described in Figure 25.12).

As part of the digestive process, contractions of the stomach walls continually mix and churn the contents, which can amount to as much as 2 L when the stomach is full. Peristaltic contractions of the stomach wall move the digestive contents toward the **pyloric sphincter** (*pylorus* = gatekeeper) at the junction between the stomach and small intestine. The arrival of a strong stomach contraction relaxes and opens the valve briefly, releasing a pulse of the stomach contents, now called **chyme,** into the small intestine (see Figure 45.8).

Depending on the volume and composition of the stomach contents, it can take from 1 to 6 hours for the stomach to empty after a meal. Feedback controls that regulate the rate of gastric emptying tend to match it to the rate of digestion, so that food is not moved along faster than it can be chemically processed. In particular, when chyme with high fat content and high acidity enters the first part of the small intestine, it stimulates the mucosal layer to secrete hormones that slow stomach emptying. Fat is digested more slowly than other nutrients, and it is digested only in the lumen of the small intestine. Therefore, further emptying of the stomach is prevented until the processing of fat has been completed in the small intestine. This is why a fatty meal, such as a greasy pizza, feels so heavy in the stomach. Highly acidic chyme must be neutralized by bicarbonate in the small intestine. Unneutralized stomach acid would inactivate digestive enzymes secreted in the small intestine and, therefore, such acid would inhibit further emptying of the stomach until it is neutralized.

The Small Intestine Completes Digestion and Begins the Absorption of Nutrients

No absorption of nutrients occurs in the mouth, pharynx, or esophagus; with the exception of a few substances, such as alcohol, aspirin, caffeine, and water, little absorption occurs in the stomach. Most absorption begins, and digestion is completed, as the contents move through the small intestine.

The "small" in the small intestine refers to its diameter, about 3 cm. It is roughly 6 m long and complexly coiled within the abdominal cavity. The lining of the small intestine folds into ridges that are densely covered by microscopic, fingerlike extensions, the **intestinal villi** (singular, *villus*). In addition, the epithelial cells covering the villi have a **brush border** consisting of fingerlike projections of the plasma membrane called **microvilli (Figure 45.11).** The intestinal villi and microvilli are estimated to increase the absorptive surface area of the small intestine to as much as 300 m^2, which is about the size of a doubles tennis court.

SECRETIONS OF THE PANCREAS, LIVER, AND INTESTINAL MUCOSA Digestion in the small intestine depends on enzymes and other substances secreted by the intestine itself and by the pancreas and liver. The secretions from the pancreas and liver enter a common duct that empties into the lumen of the first segment of the small intestine, a short region about 20 cm long called the **duodenum (Figure 45.12).**

FIGURE 45.11

The structure of villi in the small intestine. The plasma membrane of individual epithelial cells of the villi extends into fingerlike microvilli, which greatly expand the absorptive surface of the small intestine. Collectively, the microvilli form the brush border of an epithelial cell of the intestinal mucosa.

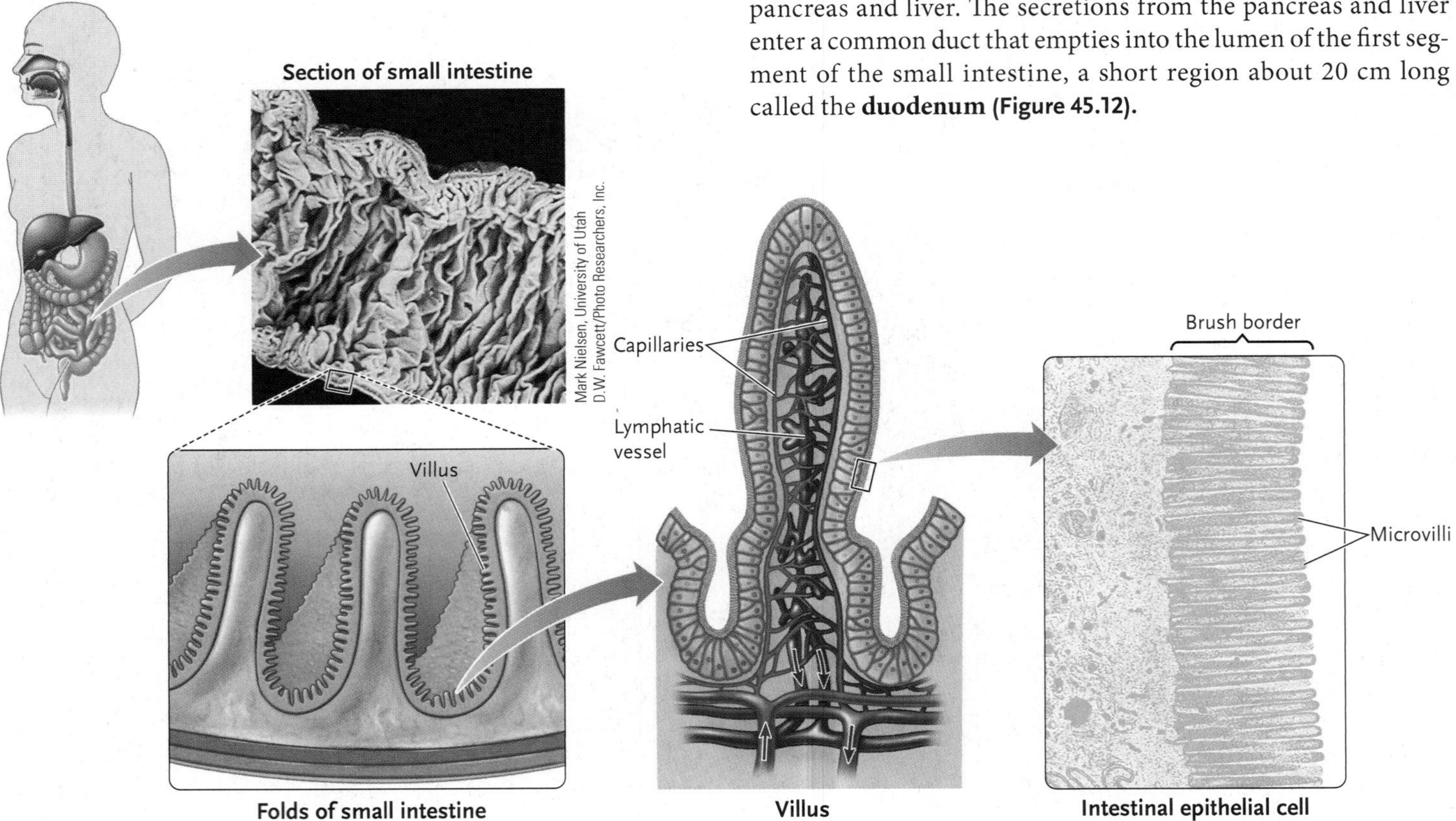
Section of small intestine
Mark Nielsen, University of Utah
D.W. Fawcett/Photo Researchers, Inc.
Villus
Folds of small intestine
Capillaries
Lymphatic vessel
Villus
Brush border
Microvilli
Intestinal epithelial cell

FIGURE 45.12
The ducts that deliver bile and pancreatic juice to the duodenum of the small intestine.

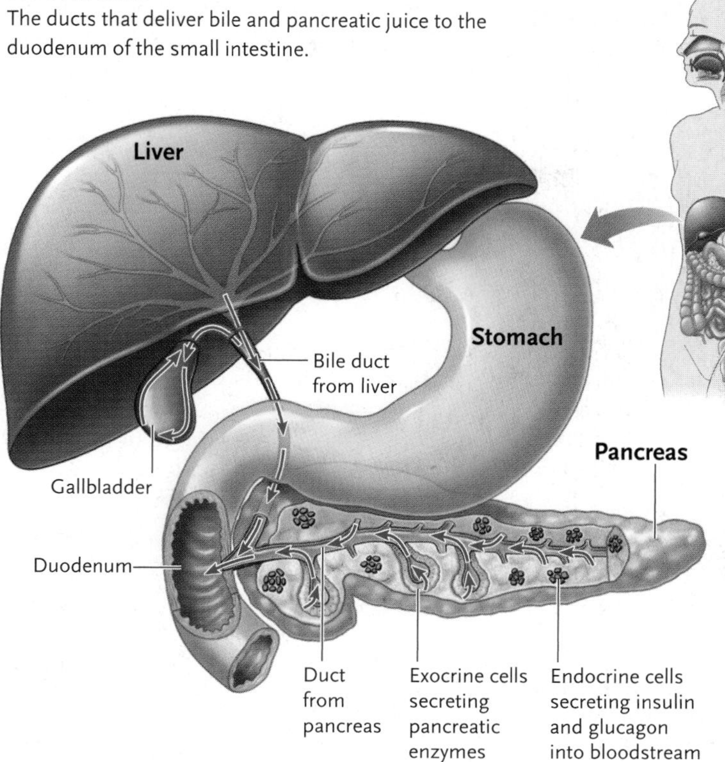
Liver
Stomach
Bile duct from liver
Pancreas
Gallbladder
Duodenum
Duct from pancreas
Exocrine cells secreting pancreatic enzymes
Endocrine cells secreting insulin and glucagon into bloodstream

In all, about 7 to 9 L of fluid from the stomach, liver, pancreas, and intestinal glandular cells enter the small intestine each day. About 95% of this amount is reabsorbed as water and nutrients as the digestive contents travel along the small intestine. Movement of the contents from the duodenum to the end of the small intestine takes about 3 to 5 hours.

In humans, the **pancreas** is an elongated, flattened gland located between the stomach and duodenum (see Figures 45.5 and 45.12). Exocrine cells in the pancreas secrete bicarbonate ions ($H_2CO_3^-$) and digestive enzymes—**pancreatic enzymes**—into ducts that empty into the lumen of the duodenum. The bicarbonate ions neutralize the acid in the chyme, bringing the digestive contents to a slightly alkaline pH. The alkaline pH allows optimal activity of the enzymes secreted by the pancreas, which include proteases, an amylase, nucleases, and lipases. All of these enzymes act in the lumen of the small intestine. Like pepsin, the proteases released by the pancreas are secreted in an inactive precursor form; contact with the digestive solution activates them. Among the active forms of these enzymes are **trypsin,** which hydrolyzes bonds within polypeptide chains, and **carboxypeptidase,** which cuts amino acids from polypeptide chains one at a time.

The **liver** secretes bicarbonate ions and **bile,** a mixture of substances including **bile salts,** cholesterol, and **bilirubin.** Bile salts are derivatives of cholesterol and amino acids that aid fat digestion through their detergent action. They form a hydrophilic coating around fats and other lipids, which allows the churning motions of the small intestine to *emulsify* the fats—break them into tiny droplets called **micelles,** as in the mixing of oil and vinegar in making a salad dressing. **Lipase,** a pancreatic enzyme, can then hydrolyze the fats in the micelles to produce monoglycerides and free fatty acids. Bilirubin, a waste product derived from worn-out red blood cells, is the yellow pigment that gives bile its color. Bacterial enzymes in the intestines modify bilirubin, resulting in the characteristic brown color of feces.

The liver secretes bile continuously. Between meals, when no digestion is occurring, bile is stored in the **gallbladder,** where it is concentrated by the removal of water. After a meal, entry of chyme into the small intestine stimulates the gallbladder to release the stored bile into the small intestine.

Brush-border epithelial cells on the villi of the small intestine secrete water and mucus into the intestinal contents. They also produce enzymes that complete the digestion of carbohydrates, proteins, and nucleic acids. Substrates for those enzymes are breakdown products produced by enzyme activity elsewhere in the digestive system. The resulting disaccharides, large peptides, dipeptides, and nucleotides are transported across the plasma membranes of the brush-border epithelial cells **(Figure 45.13).** Different **disaccharidases** break maltose, lactose, and sucrose into individual monosaccharides. Two proteases complete protein digestion: an *aminopeptidase* cuts amino acids from the end of a polypeptide, and a **dipeptidase** splits dipeptides into individual amino acids. Nucleases and other enzymes digest the nucleic acids: **nucleotidases** break them down into nucleosides, and **nucleosidases** convert the nucleosides to nitrogenous bases, five-carbon sugars, and phosphates.

Many adults lose the capacity to synthesize lactase, the enzyme that breaks down the milk sugar lactose. The lactose remaining in the intestine is broken down by bacteria, producing excess methane and CO_2. The accumulating gases distend the large intestine, producing pain, discomfort, and other symptoms of **lactose intolerance.** For many people lactose intolerance can be relieved by taking tablets containing lactase before eating milk products. One estimate is that 70% of the world's population is lactose intolerant. In the United States, between 30 and 50 million people are lactose intolerant, with some ethnic and racial groups affected more than others. For instance, over 80% of Native Americans, up to 80% of African Americans, and over 90% of Asian Americans are lactose intolerant. Generally, lactose intolerance is far less common among people of northern European descent.

ABSORPTION BY THE BRUSH-BORDER CELLS OF THE INTESTINAL MUCOSA The water-soluble products of digestion enter the intestinal mucosa cells by active transport or facilitated diffusion **(Figure 45.14A),** and water follows by osmosis. The nutrients are then transported from the mucosal cells into the extracellular fluids, from where they enter the bloodstream in the capillary networks of the submucosa.

The micelles formed by bile salts assist in the absorption of fatty acids, monoglycerides, fat-soluble vitamins, and cholesterol and other products of lipid breakdown by lipase **(Figure 45.14B).** When a micelle contacts the plasma membrane of a mucosal cell, the hydrophobic molecules within the droplet penetrate through the membrane and enter the cytoplasm.

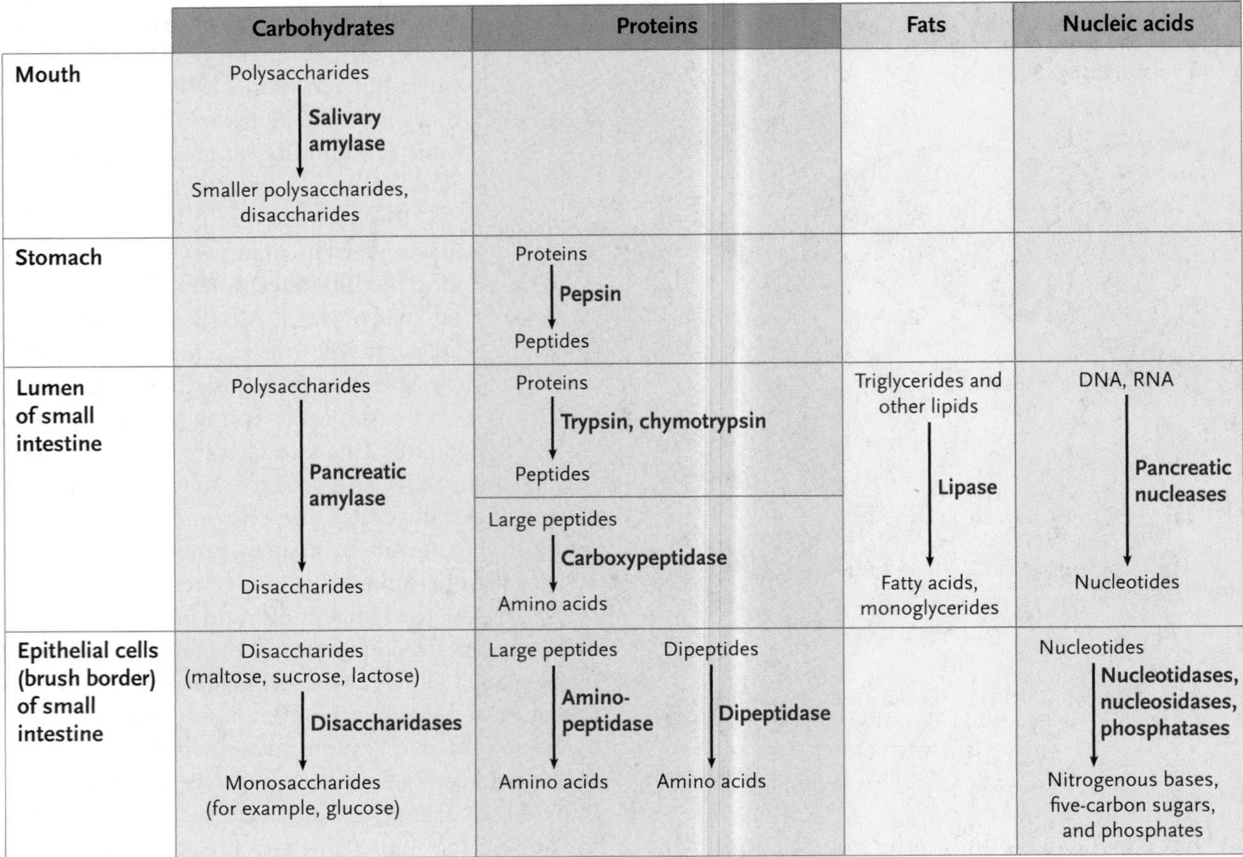
Carbohydrates
Proteins
Fats
Nucleic acids
Mouth
Polysaccharides
Salivary amylase
Smaller polysaccharides, disaccharides
Stomach
Proteins
Pepsin
Peptides
Lumen of small intestine
Polysaccharides
Pancreatic amylase
Disaccharides
Proteins
Trypsin, chymotrypsin
Peptides
Large peptides
Carboxypeptidase
Amino acids
Triglycerides and other lipids
Lipase
Fatty acids, monoglycerides
DNA, RNA
Pancreatic nucleases
Nucleotides
Epithelial cells (brush border) of small intestine
Disaccharides (maltose, sucrose, lactose)
Disaccharidases
Monosaccharides (for example, glucose)
Large peptides
Amino-peptidase
Amino acids
Dipeptides
Dipeptidase
Amino acids
Nucleotides
Nucleotidases, nucleosidases, phosphatases
Nitrogenous bases, five-carbon sugars, and phosphates

FIGURE 45.13
Enzymatic digestion of carbohydrates, proteins, fats, and nucleic acids in the human digestive system.

In the mucosal cells, the fatty acids and monoglycerides are combined into fats (specifically, triglycerides) and packaged into **chylomicrons,** small droplets covered by a protein coat. Cholesterol absorbed in the small intestine is also packed into the chylomicrons. The protein coat of the chylomicrons provides a hydrophilic surface that keeps the droplets suspended in the cytosol. After traveling across the mucosal cells, the chylomicrons are secreted into the interstitial fluid of the submucosa, where they are taken up by lymph vessels. Eventually, they are transferred with the lymph into the blood circulation.

Processing in the Liver

The capillaries absorbing nutrient molecules in the small intestine collect into veins that join to form a larger blood vessel, the **hepatic portal vein,** which leads to capillary networks in the liver. There, some of the nutrients leave the bloodstream and enter liver cells for chemical processing. Among the reactions taking place in the liver is the combination of excess glucose units into glycogen, which is stored in liver cells. This reaction reduces the glucose concentration in the blood exiting the liver to about 0.1%. If the glucose concentration in the blood entering the liver falls below 0.1% during a period of fasting between meals, the reaction reverses. The reversal adds glucose to return the blood concentration to the 0.1% level before it exits the liver. The hormonal activities controlling blood glucose levels in the blood are described in Section 40.4.

The liver also synthesizes the lipoproteins that transport cholesterol and fats in the bloodstream, detoxifies ethyl alcohol and other toxic molecules, and inactivates steroid hormones and many types of drugs.

As a result of the liver's activities, the blood leaving the liver has a markedly different concentration of nutrients than the blood carried into the liver by the hepatic portal vein. From the liver, blood is carried to the heart and then pumped by the heart to deliver nutrients to all parts of the body.

The Large Intestine Primarily Absorbs Water and Mineral Ions from Digestive Residues

The small intestine reabsorbs all but about 1 L of the 7 to 9 L of fluid released from the stomach. The remaining contents of the small intestine then move on to the large intestine. By this point, almost all nutrients have been hydrolyzed and absorbed.

A sphincter at the junction between the small and large intestines controls the passage of material between the two and prevents backward movement of the contents. The large intestine has an average diameter of 7.6 cm, more than twice that of the small intestine, but it is relatively short, about 1.5 m long in humans, as compared with the 6 m length of the small intestine. The inner surface of the large intestine is relatively smooth and contains no villi.

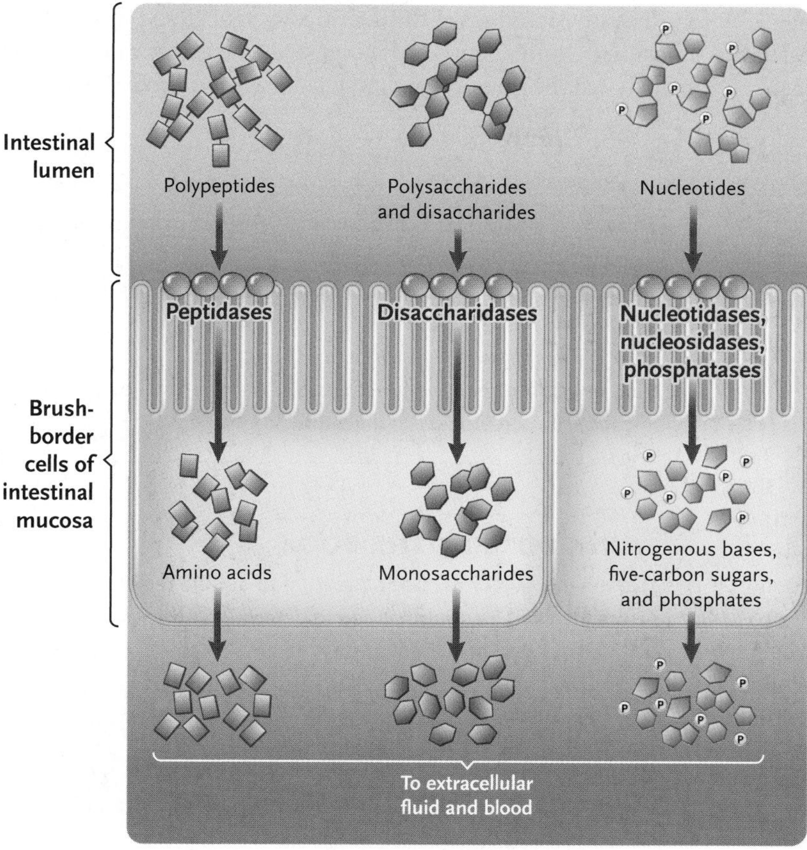
A. Absorption of water-soluble products of digestion by intestinal mucosa cells
Intestinal lumen
Brush-border cells of intestinal mucosa
Polypeptides
Polysaccharides and disaccharides
Nucleotides
Peptidases
Disaccharidases
Nucleotidases, nucleosidases, phosphatases
Amino acids
Monosaccharides
Nitrogenous bases, five-carbon sugars, and phosphates
To extracellular fluid and blood

Water-soluble molecules are broken into absorbable subunits at brush borders of mucosal cells and transported inside; the subunits are transported on the other side to extracellular fluid and blood.

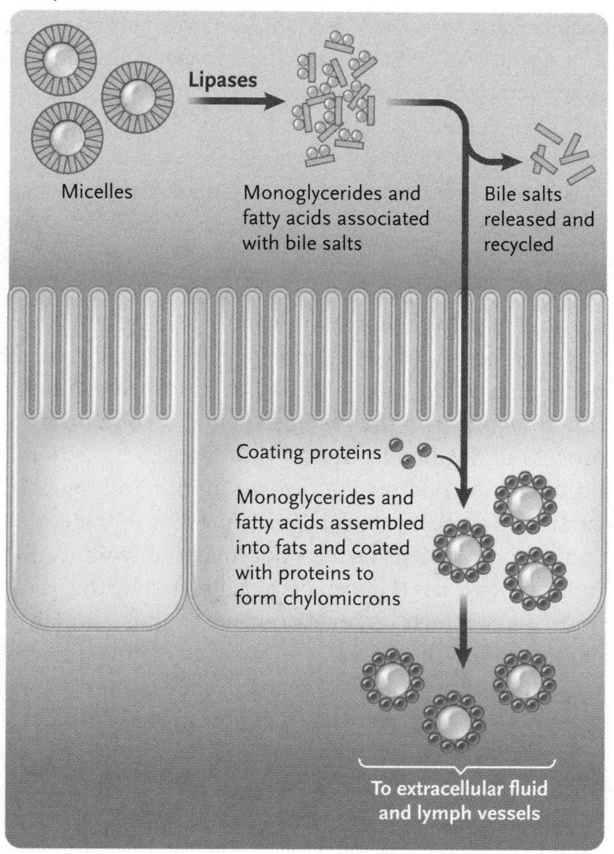
B. Absorption of fat-soluble products of digestion by intestinal mucosa cells
Lipases
Micelles
Monoglycerides and fatty acids associated with bile salts
Bile salts released and recycled
Coating proteins
Monoglycerides and fatty acids assembled into fats and coated with proteins to form chylomicrons
To extracellular fluid and lymph vessels

Micelles (fats coated with bile salts) are digested to monoglycerides and fatty acids, which penetrate into cells and are assembled into fats. The fats are coated with proteins to form chylomicrons, which are released by exocytosis to extracellular fluids, where they are picked up by lymph vessels.

FIGURE 45.14
Absorption of digestive products by the brush-border cells of the intestinal mucosa.

The large intestine has several distinct regions **(Figure 45.15).** At the junction with the small intestine, a part of the large intestine forms a blind pouch called the **cecum.** A fingerlike sac, the **appendix,** extends from the cecum. The appendix is on average 100 mm long and 7 mm in diameter; it is a vestigial structure with no known function in digestion. Since it contains patches of lymphoid tissue, it is presumed to have some role as part of the immune system. The cecum merges with the **colon,** the main part of the large intestine, which forms an inverted U. At its distal end, the colon connects with the final segment of the large intestine, the **rectum.**

The large intestine secretes mucus and bicarbonate ions and absorbs water and other ions, primarily sodium and chloride. The absorption of water condenses and compacts the digestive contents into solid masses, the **feces.** Normally, by the time the fecal matter reaches the rectum, it contains less than 200 mL of the fluid that enters the digestive tract each day. Diarrhea, by contrast, is an abnormal condition in which the fecal matter is highly fluid. The most common cause of diarrhea is a higher-than-normal rate of movement of materials through the small intestine, which does not leave adequate time

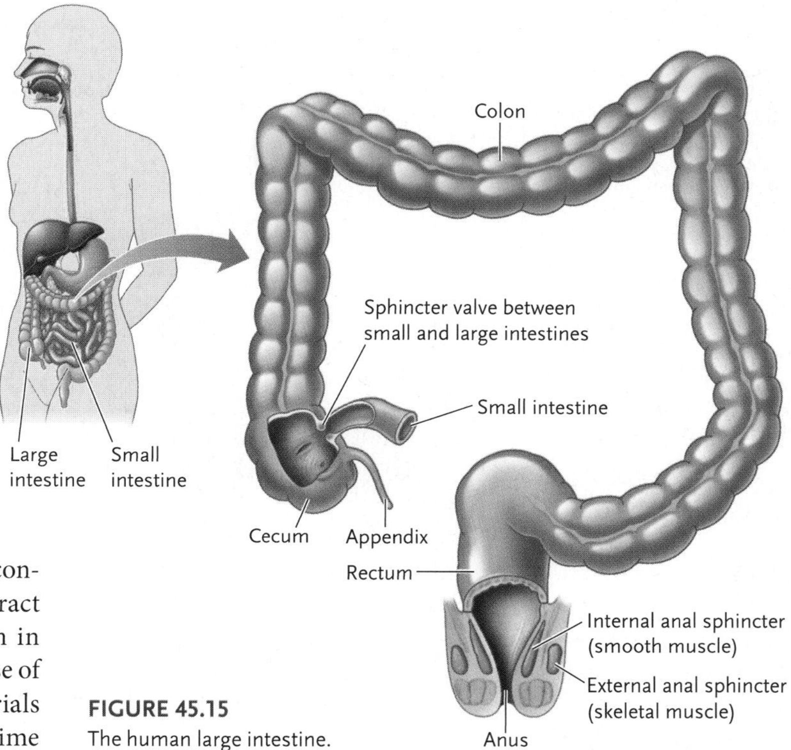
Large intestine
Small intestine
Colon
Sphincter valve between small and large intestines
Small intestine
Cecum
Appendix
Rectum
Internal anal sphincter (smooth muscle)
External anal sphincter (skeletal muscle)
Anus

FIGURE 45.15
The human large intestine.

for absorption of water to occur. The higher rate of movement can occur as a result of irritation of the small intestine wall caused by bacterial and viral infections, or because of emotional stress.

About 30% to 50% of the dry matter of feces in humans and other vertebrates consists of more than 500 species of bacteria that live as essentially permanent residents in the large intestine. Most common is the bacterium *Escherichia coli,* which also lives in the intestine of many other mammals. Intestinal bacteria metabolize sugars and other nutrients remaining in the digestive residue, and produce useful fatty acids and vitamins (such as vitamin K and the B vitamins folic acid and biotin), some of which are absorbed in the large intestine. Their activity also produces large quantities of gas—*flatus*—primarily CO_2, methane, and hydrogen sulfide. Most of the gas is absorbed through the intestinal mucosa, and the rest is expelled through the anus in the process of **flatulence.** The amount and composition of the gas produced depend on the type of food being ingested and the particular population of bacteria present in the large intestine. Some foods, such as beans, contain carbohydrates that humans cannot digest but that can be metabolized by the gas-producing intestinal bacteria. After eating such foods, humans may produce more gas than usual, and flatulence is more likely to occur.

When feces enter the rectum, they stretch the rectal wall. The stretching triggers a **defecation reflex** that opens the **anal sphincter** and expels the feces through the anus. Because the anal sphincter contains rings of voluntary skeletal muscle as well as involuntary smooth muscle (see Figure 45.15), we can resist the defecation reflex by voluntarily tightening the striated muscle ring.

Having completed the journey of ingested food through the digestive tract, we now turn our attention to the mechanisms that regulate the activities of the digestive system. These mechanisms coordinate one region of the digestive tract with another, and help match the production of nutrients with the body's needs.

STUDY BREAK 45.3

1. **What are the two classes of vitamins? Which of the two types is more critical in the diet and why?**
2. **What are the four layers of the mammalian gut? Which layer is responsible for peristalsis?**
3. **How is pepsin produced? What is its function?**
4. **Distinguish between the functions of the small and large intestines in the process of digestion.**

THINK OUTSIDE THE BOOK

Use the Internet or research literature to outline the mechanisms involved in vomiting, and to determine how vomiting is distinguished from regurgitation.

FIGURE 45.16
Control of digestion by receptors and hormones in the digestive system.

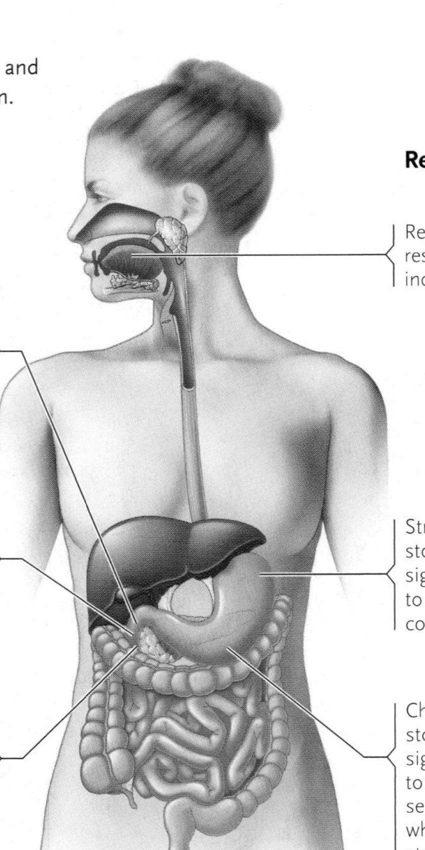

Hormonal controls
Acidic chyme stimulates release into the bloodstream of the hormone **secretin** from glandular cells in the small intestine. Secretin inhibits gastric emptying and gastric secretion and stimulates HCO_3^- secretion into the duodenum.
Fat (mostly) in chyme stimulates release of the hormone **cholecystokinin (CCK).** CCK inhibits gastric activity and stimulates secretion of pancreatic enzymes.
A meal entering the digestive tract stimulates **GIP** (glucose-dependent insulinotropic peptide) secretion, which triggers **insulin** release. Insulin stimulates the uptake and storage of glucose from the digested food.
Receptor controls
Receptors in the mouth respond to food by increasing salivary secretion.
Stretch receptors in the stomach respond to food, signaling neuron networks to increase stomach contractions.
Chemoreceptors in the stomach respond to food, signaling neuron networks to stimulate the stomach to secrete the hormone **gastrin,** which in turn stimulates the stomach to secrete HCl and pepsinogen.

45.4 Regulation of the Digestive Process

The digestive process is regulated and coordinated at many steps by controls that are largely automated. The autonomic nervous system, local neuron networks in the gut wall, and endocrine glands interact in these controls, in response to sensory information gathered by receptors in the digestive tract. The integration of the controls speeds up or slows down digestion to produce maximum efficiency in the breakdown of food molecules and the absorption of the nutrients.

Much of the control of the digestive system originates in the neuron networks of the submucosa. Other controls, particularly those regulating appetite and oxidative metabolism, originate in the brain, in control centers forming part of the hypothalamus.

The Digestive Tract Itself Has a Number of Control Systems

The movement of food through the digestive system is controlled by receptors in and hormones secreted by various parts of the system **(Figure 45.16).** Control starts with the mouth. Saliva is secreted constantly into the mouth. The presence of food activates receptors that increase the rate of salivary secretion by as much as 10-fold.

Swallowed food expands the stomach and sets off signals from stretch receptors in the stomach walls that increase the rate and strength of stomach contractions. At the same time, chemoreceptors in the stomach respond to the presence of food molecules, particularly proteins, with secretion of a hormone, **gastrin.** After traveling through the circulatory system, gastrin returns to the stomach, where it stimulates the secretion of HCl and pepsinogen (see Figure 45.10). These molecules are then used in the digestion of the protein that, as part of the swallowed meal, was responsible for their secretion. Gastrin also stimulates stomach and intestinal contractions, activities that keep the digestive contents moving through the digestive system when a new meal arrives.

Three hormones secreted when food is present in the duodenum also participate in regulating the digestive processes:

1. **Secretin.** When chyme is emptied into the duodenum, its acidic nature stimulates the release of the hormone **secretin** from glandular cells in the small intestine; the hormone enters the bloodstream. Secretin inhibits further gastric emptying to prevent more acid from entering the duodenum until the newly arrived chyme is neutralized. It also inhibits gastric secretion to reduce acid production in the stomach, and it stimulates HCO_3^- secretion into the lumen of the duodenum to neutralize the acid. If the acid is not neutralized, the duodenal wall will become damaged.
2. **Cholecystokinin (CCK).** Fat, and to a lesser extent protein, in the chyme that enters the duodenum stimulates the release of the hormone **cholecystokinin (CCK).** CCK inhibits gastric activity, thereby allowing time for nutrients in the duodenum to be digested and absorbed. It also stimulates the secretion of pancreatic enzymes, used to digest the macromolecules in the chyme.
3. **Glucose-dependent insulinotropic peptide (GIP).** The hormone **glucose-dependent insulinotropic peptide (GIP)** acts primarily to stimulate insulin release by the pancreas. When a meal is ingested, the body must change its metabolic state to use and store the new nutrients absorbed. Those activities are mostly under the control of insulin. Therefore, when a meal enters the digestive tract, GIP secretion is stimulated to trigger the release of insulin. Insulin is particularly important in stimulating the uptake and storage of glucose and so, not surprisingly, glucose in the blood draining the duodenum is directly detected as well by pancreatic β cells, the cells from which insulin is secreted (see Section 40.4).

The Hypothalamus Exerts Overall Controls

The hypothalamus contains two interneuron centers that work in opposition to control appetite and oxidative metabolism. One center stimulates appetite and reduces oxidative metabolism, and the other center stimulates the release of a peptide hormone called **α-melanocyte-stimulating hormone (α-MSH),** which inhibits appetite.

A major link in the control pathways is the peptide hormone *leptin* (*leptos* = thin), discovered in mice by Jeffrey Friedman and his coworkers at the Rockefeller University. Fat-storing cells secrete leptin when the deposition of fat increases in the body. Leptin travels in the bloodstream and binds to receptors in both centers in the hypothalamus. Binding stimulates the center that reduces appetite and inhibits the center that stimulates appetite. At the same time, leptin binds to receptors on body cells, triggering reactions that oxidize fatty acids rather than converting them into fats. When fat storage is reduced, leptin secretion drops off, and signals from other pathways activate the appetite-stimulating center in the hypothalamus and turn off the appetite-inhibiting center.

These controls closely match the activity of the digestive system to the amount and types of foods that are ingested, and they coordinate appetite and oxidative metabolism with the body's needs for stored fats. *Insights from the Molecular Revolution* describes recent investigations identifying signal molecules and receptors that regulate appetite and feeding behavior in mammals.

STUDY BREAK 45.4 <

1. **What differentiates the consequences of secretion of gastrin and cholecystokinin (CCK)?**
2. **How does the hypothalamus regulate the digestive process?**

45.5 Digestive Specializations in Vertebrates

Natural selection has modified the basic vertebrate digestive system into a multitude of structural and functional variations. The most common modifications are in the form of the mouth, teeth, and jaws; the structure and function of the esophagus, stomach, and cecum; and the length of the digestive tract. Vertebrates also vary in the types of enzymes secreted by the digestive system. For example, humans secrete an amylase into the saliva, but cats and pigs secrete a salivary lipase.

Teeth Are Adapted to Feeding Methods

To anthropologists and paleontologists, dentition (the number, kind, and arrangement of teeth) opens a window to an animal's diet and feeding method—and hence reveals a great deal about its habitat and lifestyle. For example, snakes have sharp, pointed teeth that curve backward into the mouth, which helps to ensure that prey (dead or living) does not slip out of the animal's mouth as muscles contract to swallow. The dentition is combined with specializations in jaw structure; many snakes have jaws with elastic connections that allow them to open wide enough to swallow prey whole.

Tooth specialization is especially evident among mammals **(Figure 45.17).** Typically, mammals have four types of upper and lower teeth. **Incisors,** located at the front of the mouth, are flat-

Food for Thought on the Feeding Response

A number of small proteins called *neuropeptides* regulate various physiological responses in humans and other mammals, including appetite and feeding, pain reception, and blood pressure regulation. They exert their effects by binding to receptors on the surfaces of neurons and other cells. For example, neuropeptide Y (NPY) strongly stimulates appetite and food uptake when it binds to neurons in the hypothalamus and other locations in the brain.

Research Question

What receptor in the hypothalamus binds NPY?

Experiments

A group of researchers at the Synaptic Pharmaceutical Corporation in New Jersey and Ciba-Geigy in Switzerland was one of several teams that tried to identify the receptor binding NPY in the hypothalamus. Their first step was to clone the gene for NPY using the steps shown in the **figure.** The cloned gene was identified by its expression: the encoded receptor inserted into the plasma membrane of a transformed mammalian cell and was detected by the binding of NPY.

The investigators then sequenced the cloned gene they had identified and showed that it encoded a previously unknown neuropeptide receptor, which they called Y5. An equivalent gene for this receptor was then identified in the human genome.

Next, the cloned cDNA was used in a Northern blotting experiment (see Section 18.2) to identify which tissues express the Y5 gene. The researchers probed several different rat tissues for the presence of Y5 mRNA.

Result

Their result showed that the Y5 gene is expressed only in the brain, with the strongest expression in the hypothalamus and the amygdala, a center associated with emotional responses.

Conclusion

The researchers had isolated a novel neuropeptide Y receptor gene, Y5. After testing the ability of various peptides to bind the Y5 receptor, and the effects of the binding on feeding behavior, they proposed that the binding of NPY initiates signals that stimulate hunger and the feeding response. Binding to receptors in the amygdala may add an emotional dimension to the craving for food. The researchers conjectured that studies of Y5 could help further our understanding of eating disorders and perhaps aid in the development of drugs to combat those disorders.

Indeed, the Y5 receptor became a focus of considerable drug discovery efforts. However, a recent study has burst the bubble—drugs designed to inhibit the receptor are not going to be effective in combating obesity. In this study, Andrew Turnbull and his colleagues at AstraZeneca in the United Kingdom tested whether one such drug affected feeding in rats. Unfortunately, the drug had no significant effect on the increase in food intake induced by NPY itself either in normal or genetically obese rats. Further, the drug had no effect on food intake or body weight in normal rats or in rats that were obese due to their diet. In short, the Y5 receptor is not a significant regulator of feeding behavior. The difference between these results and those of the previous study may reflect the experimental design—the compounds used in the earlier study may well have had other activities that were responsible for their effects on feeding behavior. Neuropeptide Y continues to be of great interest for research.

Source: C. Gerald et al. 1996. A receptor subtype involved in neuropeptide-Y-induced food intake. *Nature* 382:168–171; A. V. Turnbull. 2002. Selective antagonism of the NPY Y5 receptor does not have a major effect on feeding in rats. *Diabetes* 51:2441–2449.

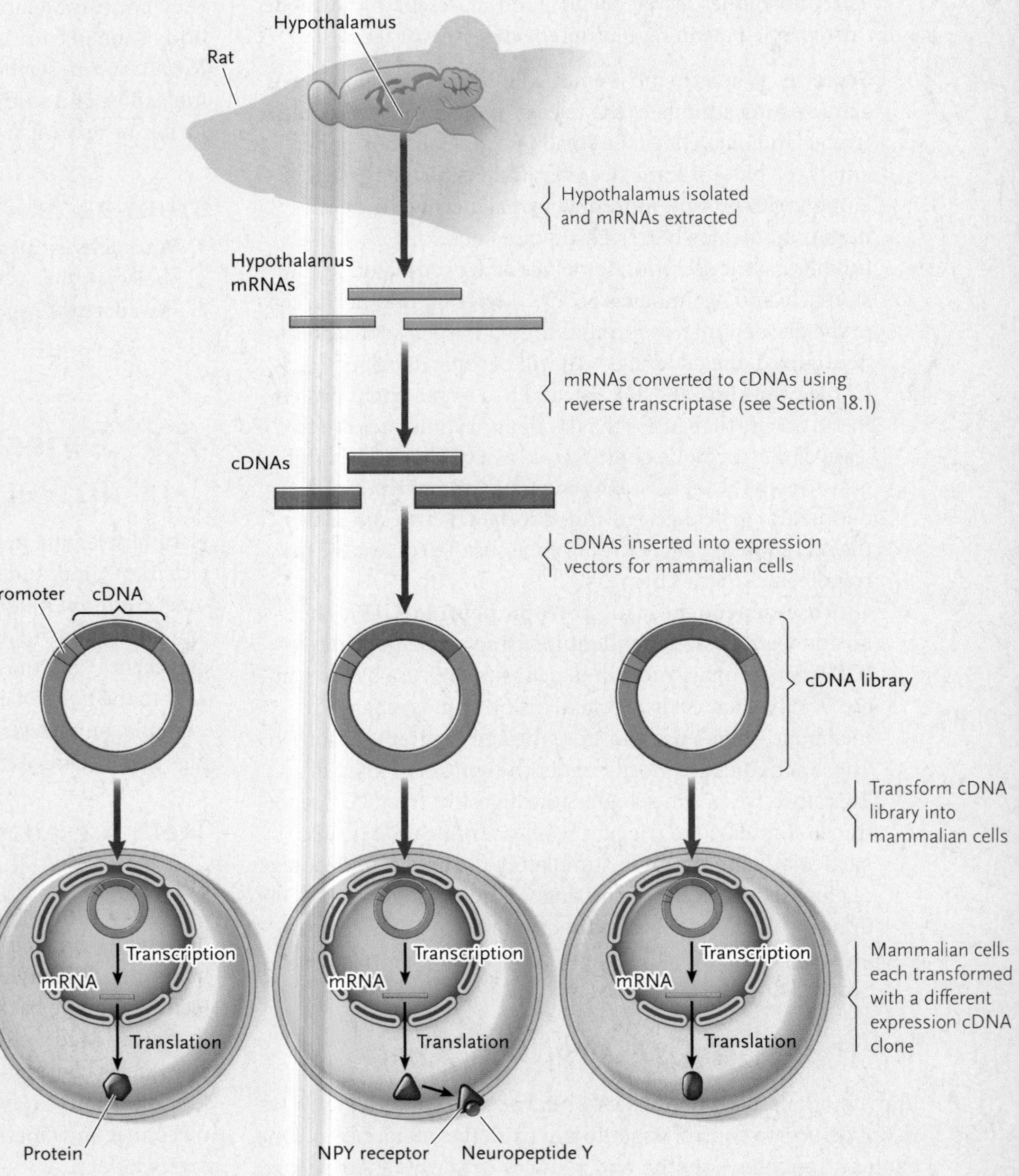
Hypothalamus
Rat
Hypothalamus isolated and mRNAs extracted
Hypothalamus mRNAs
mRNAs converted to cDNAs using reverse transcriptase (see Section 18.1)
cDNAs
cDNAs inserted into expression vectors for mammalian cells
Promoter
cDNA
cDNA library
Transform cDNA library into mammalian cells
Transcription
mRNA
Translation
Mammalian cells each transformed with a different expression cDNA clone
Protein
NPY receptor
Neuropeptide Y

Transformed cells screened for a cell that could bind neuropeptide Y (here, the central cell). That cell must have expressed the NPY receptor and contains the cDNA encoding the receptor.

Generalized mammalian dentition

Carnivore

Herbivore

Human

KEY

Molars | Canines | Premolars | Incisors

FIGURE 45.17
Mammalian dentition.

tened, chisel-shaped teeth used to nip or cut food. Horses use their prominent incisors to clip off blades of grasses. Pointed **canines** at the sides of the incisors are specialized for biting and piercing. Carnivores such as wolves and tigers use their long, sharp canines to pierce and kill prey, but the canines are minimally developed or absent in many herbivores. The blocky teeth at the sides of the mouth, the **premolars** and **molars,** have surface bumps, or *cusps,* that are used in crushing, grinding, and shearing food. Large premolars and molars with a ridged surface are characteristic of animals, such as deer, that consume fibrous plant material. The premolars and molars of some carnivores, such as cats, have sharp shearing surfaces that can slice meat efficiently. All four types of teeth are typically well developed in omnivores, such as humans.

The Length of the Intestine and Specializations of the Digestive Tract Reflect Feeding Patterns

There is a strong correlation between diet and the length of the digestive system **(Figure 45.18).** Vertebrates that feed primarily on nutrient-rich foods such as meat, blood, nectar, or insects, including carnivores (such as the dog), generally have a relatively short intestine. In contrast, herbivores (such as the rabbit) have a long intestinal tract and specializations of the esophagus, stomach, and cecum, or other structures that can store large volumes of plant material. Both the longer intestinal tract and greater storage capacity allow an herbivore to extract more nutrients from plant matter, which is relatively difficult to digest. Both types of intestine appear during the life cycle of frogs: a frog tadpole, which primarily eats algae, has a relatively long, coiled intestine, but after metamorphosis, the adult frog, which primarily eats insects, has a short intestine.

Carnivore (a dog)

Herbivore (a rabbit)

Fred Bruemmer

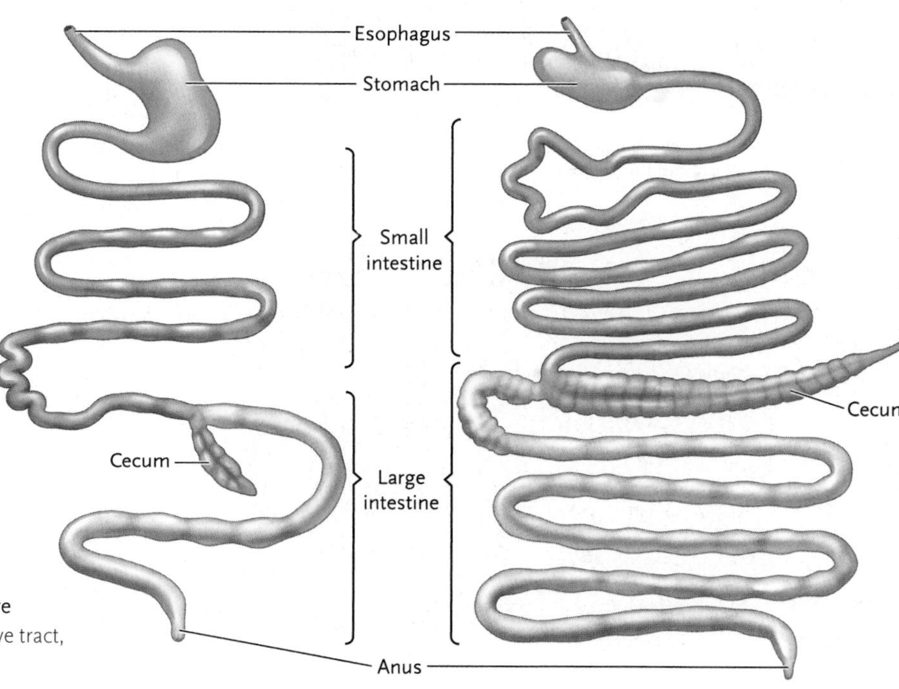
Esophagus
Stomach
Small intestine
Cecum
Large intestine
Cecum
Anus

FIGURE 45.18
Comparison of the length of the digestive tract in a carnivore and an herbivore. The carnivore has a relatively short digestive tract, while the herbivore's digestive tract is much longer.

Symbiotic Microorganisms Aid Digestion in a Number of Organisms

Many herbivores use the hydrolytic capabilities of microorganisms such as bacteria, protists, and fungi to aid digestion of plant material, housing them in specialized structures of the esophagus, stomach, or cecum. Unlike vertebrates, the microorganisms can synthesize *cellulase,* the enzyme that hydrolyzes the cellulose of plant cell walls into glucose subunits. This arrangement is a classic example of *mutualism* (a type of symbiosis in which both interacting organisms benefit); the herbivores benefit from the digestive capabilities of the microorganisms, and the microorganisms benefit from an ideal habitat and an abundant supply of nutrients.

The most remarkable adaptations for mutualistic digestion of plant matter among vertebrates occur in the **ruminants,** which include cattle, deer, goats, sheep, and antelopes. These animals have a complex, four-chambered stomach **(Figure 45.19).** The first three chambers are derived from the esophagus. After the ruminant's teeth tear, cut, and grind plant matter, boluses are swallowed. Most of them go to the **rumen** (the largest chamber), and the rest go to the **reticulum** (step 1). In these two chambers, symbiotic microorganisms hydrolyze cellulose in the plant matter into fuels for fermentation reactions (the oxygen level in the chambers is too low to support mitochondrial reactions). The fermentations generate various products, including alcohols, amino acids, and fatty acids, which are used as nutrients by the ruminants. Methane, another product, collects in the fermentation chambers. Ruminants belch the gas in huge quantities; a cow potentially can release more than 400 L of methane per day. In fact, cattle are estimated to contribute 20% of the methane polluting our atmosphere.

As part of the digestive process, a ruminant "chews its cud"—it regurgitates material from the rumen and reticulum, rechews it, and swallows it again (step 2). This process crushes the plants into smaller fragments, exposing more surface area to the microbial enzymes, and gives the enzymes more time to act. As a result, the food material is broken down further.

Reswallowed cud, consisting of matter that has been digested and liquefied by the microorganisms, bypasses the rumen and reticulum and instead goes to the **omasum,** where water is absorbed from the mass (step 3).

Matter then moves to the *abomasum* (the ruminant's gastric stomach) (step 4). There, the addition of acids and pepsin to the food mass kills the microorganisms and starts the process of typical vertebrate digestion. As the food mass moves to the small intestine, the dead microorganisms, which are a rich source of proteins, vitamins, and other nutrients, are digested and absorbed along with other hydrolyzable molecules in the digestive contents.

Although the ruminant digestive system is uniquely specialized, many other vertebrate species also have esophageal or gastric chambers containing plant-digesting symbiotic microorganisms. These include the camel, sloth, and langur monkey, and marsupials such as kangaroos and wallabies. One bird, the South American hoatzin *(Opisthocomus hoazin),* is known to have a crop in which microorganisms break down plant matter.

Many vertebrates house symbiotic, plant-digesting microorganisms in the cecum. Horses, elephants, rhinos, rabbits, koalas, many rodents, and some reptiles and birds, including the iguana and chicken, are all examples. However, because the cecum and the remainder of the large intestine have little capacity to absorb nutrients, microbial digestion in the cecum is not as productive as microbial digestion in the stomach.

Mutualistic microorganisms that synthesize essential amino acids and vitamins, and digest particular components of food that otherwise are indigestible, are found in all mammals. Recent research has demonstrated the importance of these microorganisms to nutrition. For example, **Figure 45.20** describes research experiments that correlate the particular array of bacteria in the human intestine with obesity.

We have seen that animals use various strategies to extract the available nutrients from foods. These nutritional strategies involve multiple steps combining both mechanical and

FIGURE 45.19
A ruminant, the pronghorn *(Antilocapra americanus),* and its four-chambered system that digests plant matter with the aid of mutualistic microorganisms.

Image copyright Koekeloer, 2010. Used under license from Shutterstock.com

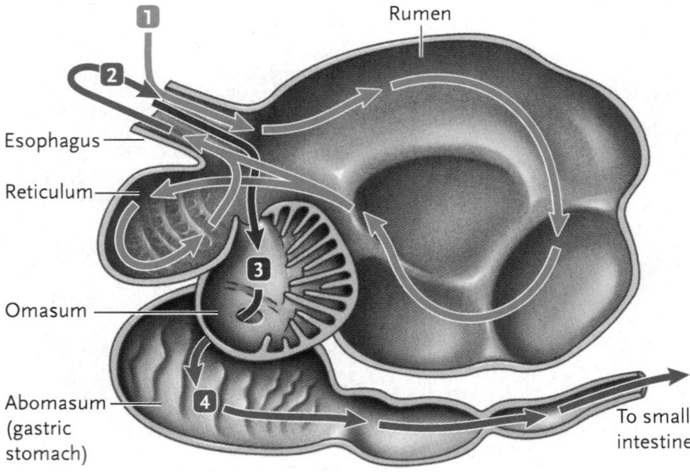
Chewing, swallowing, regurgitation, rechewing, and reswallowing of food through esophagus
1
2
Rumen
Esophagus
Reticulum
3
Omasum
4
Abomasum (gastric stomach)
To small intestine

1 Swallowed boluses go to rumen and reticulum where fermentation reactions by symbiotic microorganisms begin digesting the plant matter.

2 The animal chews its cud by regurgitating material, rechewing it, and swallowing it again.

3 Reswallowed cud goes to the omasum where water is absorbed.

4 Matter then moves to the abomasum where typical gastric digestion occurs.

FIGURE 45.20 **Experimental Research**

Association of Intestinal Bacteria Populations with Obesity in Humans

Question: Are particular populations of intestinal bacteria associated with obesity in humans?

Experiment: The human intestine contains 10^{13} to 10^{14} microorganisms, a number that is greater by an order of magnitude than the number of human somatic and germ cells. The mutualistic organisms in this intestinal flora synthesize essential amino acids and vitamins, and digest otherwise indigestible components of food, such as particular plant macromolecules.

Intestinal microorganisms consist largely of bacteria, with two groups predominating, the Firmicutes (a phylum of Gram-positive bacteria) and the Bacteroidetes (a phylum of Gram-negative bacteria). Research by Jeffrey Gordon and his group at Washington University School of Medicine, St. Louis, has shown that genetically obese mice (genotype *ob*/*ob*) have about 50% fewer Bacteroidetes, and correspondingly more Firmicutes, than do normal, lean, siblings (genotype ob^+/ob^+), leading to the conclusion that particular distributions of bacteria in the intestine correlate with obesity.

Gordon's group therefore decided to study the relationship between intestinal populations of the two types of bacteria and body fat in humans. They worked with 12 people classified as being obese according to the standard criterion of having a body mass index (BMI) of $\geq$30. BMI is calculated from the formula, BMI = weight (kg)/[height (m)]^2. BMI does not directly measure body fat, although it correlates well with the amount of body fat found in non-athletes. Figure A shows the experimental design of Gordon's research.

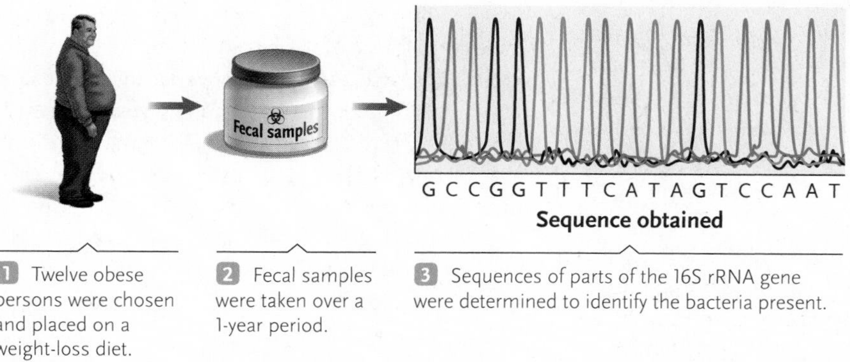

Fecal samples
G C C G G T T T C A T A G T C C A A T
Sequence obtained
1 Twelve obese persons were chosen and placed on a weight-loss diet.
2 Fecal samples were taken over a 1-year period.
3 Sequences of parts of the 16S rRNA gene were determined to identify the bacteria present.

FIGURE A

Sequencing specific regions of the 16S rRNA gene (which encodes an rRNA found only in the ribosome of prokaryotes) enabled the researchers to identify the bacteria in the fecal sample. The regions sequenced vary significantly among species, permitting the identification of species present in a mixed sample without having to culture the organisms. Analyzing these sequences is one of the methods used in molecular phylogenetics (see Section 23.3).

Results: The results showed that most of the bacteria present in the obese individuals and in two lean individuals (controls) were Firmicutes or Bacteroidetes. As shown in Figure B:

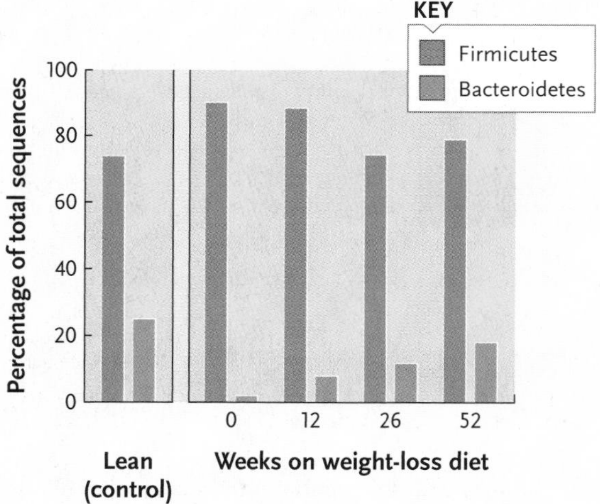

KEY
Firmicutes
Bacteroidetes
Percentage of total sequences
100
80
60
40
20
0
0
12
26
52
Lean (control)
Weeks on weight-loss diet

FIGURE B

1. Before embarking on the diet, obese individuals had more Fimicutes and fewer Bacteroidetes than lean individuals.
2. Over the course of the diet, as weight loss occurred, the relative abundance of Firmicutes decreased, and the relative abundance of Bacteroidetes increased.

Conclusion: Obese and lean individuals showed significant differences in their intestinal content of Firmicutes and Bacteroidetes. Moreover, the data from the effects of the weight-loss diet indicate a dynamic relationship between the intestinal populations of the two groups of bacteria and body fat content. Taken with the results of the mouse study, manipulation of intestinal microbial communities could be a potential way to control or treat obesity.

Sources: R. E. Ley et al. 2006. Human gut microbes associated with obesity. *Nature* 444:1022–1023; P. J. Turnbaugh et al. 2006. An obesity-associated gut microbiome with increased capacity for energy harvest. *Nature* 444:1027–1031.

chemical processing, which convert complex foodstuffs into the absorbable subunits that animals need to sustain life. Obtaining food is costly in terms of energy and risk, and animals have evolved many ways to make the most of it.

STUDY BREAK 45.5 <

1. How are the different types of teeth used in feeding?
2. What roles do symbiotic microorganisms play in digestion?

UNANSWERED QUESTIONS

How is energy partitioned in animals?

In this chapter you learned that ingested food molecules are broken down into smaller units that can be readily absorbed into body fluids and cells. The ingested energy is assimilated into the organism and partitioned into four main categories: maintenance metabolism, growth, reproduction, and storage. The top priority for energy allocation is maintenance metabolism—the energy required to seek and digest food and to support essential life processes. Researchers are studying the regulation of energy partitioning among the other categories when available energy exceeds that required for maintenance. The findings indicate that priorities for energy allocation vary between species as well as during an individual's life history. In the case of growth, for example, some species grow for a short period *(determinant growth)*, while other species grow throughout their lives *(indeterminant growth)*. Animal growth is controlled by genetic, environmental, and nutritional factors, but how such cues are integrated with each other and with other processes, such as reproduction and storage, remain to be determined. For example, the cues used to shut down growth and reproduction during fasting are not fully known. (In humans, malnourished juveniles become growth-retarded and adult females stop menstruating.) The extent to which severe energy restriction may result in metabolic adaptation (reduced metabolic rate) also needs to be examined.

How are foraging and feeding regulated?

In this chapter you also learned about the control of appetite and oxidative metabolism. While some of the players are known—among them hunger centers in the brain, appetite-stimulating hormones such as neuropeptide Y, and appetite-suppressing or satiety hormones such as leptin—a number of questions remain. For example, how are various sensory inputs, such as smell, integrated to initiate feeding? Similarly, how are other inputs, such as gut contents or increasing nutrient concentration in the blood, integrated to terminate feeding? Also, how is an animal's feeding strategy matched to nutrient quantity and quality in the environment and to the animal's metabolic pattern; for example, sluggish or active (tortoise or hare)? Inevitably, a complex of integrating chemicals—produced by the brain, gut, fats cells, and other tissues—will prove to be involved. Knowledge of this complex system will provide insight not only into obesity but into eating disorders such as anorexia as well.

How are lipids taken up and utilized?

Adipose and other cells take up lipids (mostly triglycerides) that have been hydrolyzed from chylomicrons and very-low-density lipoprotein (VLDL) by the enzyme lipoprotein lipase (LPL). Low-density lipoproteins (LDLs) are VLDL remnants and are the major means of delivering cholesterol to tissues. Normally, LDLs are taken up by receptor-mediated endocytosis. Cholesterol can be scavenged by high-density lipoproteins (HDLs) through the action of the enzyme lecithin cholesterol acyltransferase (LCAT), taken up by the liver, and removed from the body. Researchers have found that defects in LPL, LCAT, and LDL disrupt normal lipid metabolism and lead to severe health risks. For example, genetic defects in the LDL receptor are associated with familial hypercholesterolemia, in which excess cholesterol is left in the blood and causes heart disease and other abnormalities. Continued research on lipid and lipoprotein metabolism will be important for understanding obesity, type 2 diabetes, and cardiovascular disease.

Think Critically

1. In the most general terms, how is allocation to maintenance metabolism different for tortoises and hares (hint: think temperature regulation)?
2. Given what you learned about lipids in the chapter on biological molecules, what is the importance of triglycerides in the diet of animals?

Mark Sheridan

Mark Sheridan is the James A. Meier Professor of Biological Sciences at North Dakota State University, where he studies the control of growth, development, and metabolism in vertebrates. His current research focuses on somatostatin signaling, growth regulation, and environmental endocrinology. Find out more about Dr. Sheridan's research at http://www.ndsu.nodak.edu/ndsu/msherida/research.interests.

REVIEW KEY CONCEPTS

Go to **CENGAGENOW** at www.cengage.com/login to access quizzing, animations, exercises, articles, and personalized homework help.

45.1 Feeding and Nutrition

- Animals obtain organic molecules by eating other organisms. Herbivores primarily eat plants, carnivores primarily eat other animals, and omnivores eat animals, plants, and other sources of organic nutrients. The organic molecules are used as fuels for oxidative reactions providing energy and as building blocks for making complex biological molecules.
- Animals require essential substances in their diets—amino acids, fatty acids, vitamins, and minerals—that they cannot make for themselves.
- Animals may be classified with respect to feeding methods and the physical state of the organic molecules they eat. Fluid feeders ingest liquids containing organic molecules in solution. Suspension feeders eat small particles of organic matter or small organisms in suspension in fluids. Deposit feeders ingest small organic particles or organisms that are part of solid matter that the feeders live in or on. Bulk feeders consume large pieces of organisms, or entire large organisms (Figure 45.2).

45.2 Digestive Processes

- Digestion is the process of mechanical and chemical breakdown of food into molecular subunits small enough to be absorbed into body fluids and cells.
- Digestion may be intracellular or extracellular. Extracellular digestion allows food to be eaten in large batches, stored, and broken down while the animal carries out other activities.
- In animals with extracellular digestion, the digestive processes take place in an internal body cavity that is either a pouch or sac with one opening that serves as both mouth and anus, or a tube with two openings forming a mouth on one end and an anus on the other end (Figures 45.3 and 45.4).
- In animals with a digestive tube, digestion occurs in five stages: (1) mechanical processing, including chewing and grinding of food; (2) secretion of enzymes and other digestive aids into the digestive tract; (3) enzymatic hydrolysis of food molecules into molecular subunits; (4) absorption of the molecular subunits across cell membranes; and (5) elimination of undigested matter.
- Food particles and molecules are pushed through the digestive tube by muscular contractions of its wall. Storage of food at various locations in the tube allows animals to digest food while engaged in other activities.

Animation: Examples of digestive systems

45.3 Digestion in Humans and Other Mammals

- Adult humans require eight essential amino acids, 13 vitamins (Table 45.1) and a large number of essential minerals (Table 45.2).
- The mouth, pharynx, esophagus, stomach, intestine, and anus are common to the digestive system of mammals, including humans, and most vertebrates (Figure 45.5).
- The wall of the vertebrate gut is formed from four layers of tissues: the mucosa, the submucosa, the muscularis, and the serosa (Figure 45.7).
- Coordinated contractions of the circular and smooth muscles produce peristaltic waves that move the digestive contents from the mouth to the anus (Figure 45.8).
- Digestion begins in the mouth, where the teeth break the food into smaller bits. Salivary amylase, an enzyme that digests starch, is secreted into the food in the mouth. After chewing, the food is swallowed and travels through the pharynx and esophagus to reach the stomach (Figure 45.9).
- In the stomach, hydrochloric acid, the protein-digesting enzyme pepsin, and mucus are added to the food mass. The stomach churns the acid contents into chyme, which is released in pulses into the small intestine (Figure 45.10).
- Absorption of nutrients begins in the small intestine. Specializations of the small intestine to optimize absorption are the intestinal villi and microvilli (Figure 45.11).
- In the small intestine, digestive juices from the pancreas and liver add enzymes and digestive aids to the food mass (Figure 45.12). The pancreatic juice contains digestive enzymes and bicarbonate ions that neutralize the acidity of the digestive contents. The liver secretion, bile, contains bile salts, which emulsify fats, cholesterol, bilirubin, and additional bicarbonate ions.
- The small intestine secretes enzymes that complete most digestion. The mucosal cells of the small intestine absorb the molecular subunits created by digestion (Figures 45.13 and 45.14).
- Absorbed nutrients are delivered to the liver, where excess glucose is converted into glycogen and fats, and some of the amino acids are converted into plasma proteins or sugars. The liver also synthesizes cholesterol from lipids, carbohydrates, and other substances.
- The large intestine absorbs water and mineral ions from the digestive contents. At the end of the large intestine the undigested remnants, the feces, are expelled from the anus (Figure 45.15).

Animation: Human digestive system

Animation: Vitamins

Animation: Structure of the small intestine

Animation: Peristalsis

Animation: Structure of the large intestine

45.4 Regulation of the Digestive Process

- Digestion is regulated by signals from the autonomic nervous system, by the activity of neuron networks in the digestive tube wall, and by hormones secreted by the digestive system. The regulatory mechanisms operate in response to signals from sensory receptors that monitor the volume and composition of the digestive contents (Figure 45.16).

Animation: Body mass index

Animation: Caloric requirements

Animation: Chronology of leptin research

45.5 Digestive Specializations in Vertebrates

- Common variations in vertebrate digestive systems include modifications of the teeth, length of the digestive tract, and structure and function of the stomach.
- Mammals have four basic types of teeth—incisors for cutting, canines for piercing, and premolars and molars for cutting, grinding, and smashing food (Figure 45.17).
- Carnivores and vertebrates that eat other nutrient-rich foods have a relatively short digestive tract. Herbivores, which eat nutrient-poor foods, typically have a relatively long digestive tract that includes extensive storage regions (Figure 45.18).
- Many herbivores have digestive chambers in which symbiotic microorganisms digest plant matter into molecules that can be absorbed by the host (Figure 45.19).

Animation: Human teeth

Animation: Ruminant stomach function

UNDERSTAND AND APPLY

Test Your Knowledge

1. Required molecules that animals cannot synthesize are called:
   a. nutrients.
   b. essential nutrients.
   c. enzymes.
   d. proteins.
   e. carbohydrates.

2. Which of the following accurately describes a feeding style?
   a. Deposit feeders obtain nutrients from organic molecules in solution.
   b. Deposit feeders scrape organic matter from solid material on which they live.
   c. Fluid feeders digest organisms suspended in water.
   d. Fluid feeders strain food with networks of mucus or bristles and hairs.
   e. Suspension feeders consume sizable food whole or in chunks.

3. The order of successive steps in digestion is:
   a. absorption follows enzymatic hydrolysis.
   b. secretion of enzymes follows absorption of digestive material.
   c. mechanical processing follows enzyme secretion.
   d. mechanical processing follows enzymatic hydrolysis.
   e. enzymatic hydrolysis precedes secretion of digestive aids.
4. The esophagus, crop, gizzard, and intestine are found in:
   a. birds and mammals.
   b. insects and mammals.
   c. flatworms and birds.
   d. earthworms and birds.
   e. sponges and cnidarians.
5. All of the following are essential nutrients in humans *except:*
   a. vitamin B.
   b. calcium.
   c. glycogen.
   d. linoleic acid.
   e. vitamin K.
6. A specialized region of the gut is/are the:
   a. submucosa formed by circular and longitudinal layers.
   b. serosa lining the gut for absorption.
   c. mucosa composed of thick elastic connective tissue for movement.
   d. muscularis, an outer layer that secretes a slippery material to prevent friction with other organs.
   e. sphincters, which form valves between major digestive organs.
7. If the fat in whole milk is ingested:
   a. the stomach, with its high pH, will stimulate cells of the duodenum to hasten stomach emptying.
   b. parietal cells in the stomach will absorb it.
   c. in the small intestine, bile salts emulsify the fats and then lipase hydrolyzes them.
   d. lactase deficiency in the small intestine would prevent its digestion.
   e. microvilli will absorb the fat in the form of chylomicrons directly into the blood of the hepatic portal vein.
8. The role of the liver in digestion is to:
   a. synthesize aminopeptidase and dipeptidase to digest polypeptides.
   b. synthesize lipase to form free fatty acids.
   c. secrete trypsin to break the bonds in polypeptides.
   d. secrete bile and bicarbonate ions to help emulsify fats.
   e. store bile between meals.
9. Which of the following best describes regulation of digestion?
   a. GIP inhibits insulin release from the pancreas.
   b. Gastrin stimulates pancreatic secretion of HCl and pepsinogen.
   c. Secretin stimulates gastric emptying into the duodenum.
   d. CCK stimulates gastric activity to activate the duodenum.
   e. Leptin binds different hypothalamic receptors to stimulate or inhibit appetite.
10. An example of a digestive specialization is seen in:
   a. the long intestines characteristic of herbivores.
   b. the incisors being the dominant teeth in wolves.
   c. the canine teeth being the dominant teeth in deer.
   d. salivary lipase being made by humans.
   e. cellulose being made by humans.

Discuss the Concepts

1. As a person ages, the number of cells in the body steadily decreases and their energy needs decline. If you were planning a diet for an older person, what kind(s) of nutrients would you emphasize, and why? Which ones would you recommend the person consume less of? Include vitamins and minerals in your answer.
2. Formulate a healthy diet for a young, actively growing 7-year-old child, and explain why you have included each part of the diet. Refer to Question 1, above, for some issues to consider.
3. A baby develops symptoms of protein deficiency, and the attending physician suggests the cause is a genetic defect leading to a nonfunctional enzyme associated with digestion. Name at least three enzymes that might be likely suspects, and for each one explain how the defect would result in a protein deficiency.

Design an Experiment

Design experiments to test whether cigarette smoke affects the functioning of the various parts of the digestive system.

Interpret the Data

The human *AMY-1* gene encodes salivary amylase, an enzyme that breaks down starch. The number of copies of this gene varies, and people who have more copies generally make more of the enzyme. In addition, the average number of *AMY-1* copies differs among cultural groups.

George Perry and his colleagues hypothesized that duplications of the *AMY-1* gene would confer a selective advantage in cultures in which starch is a large part of the diet. To test this hypothesis, the scientists compared the number of copies of the *AMY-1* gene among members of seven cultural groups that differed in their traditional diets. The figure shows their results.

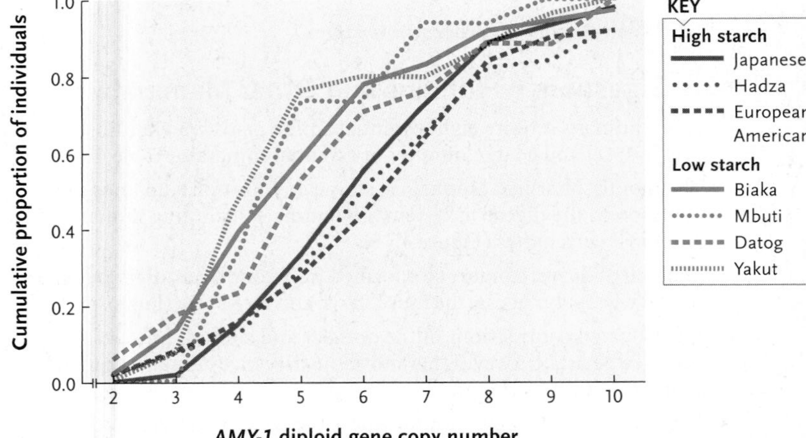
Cumulative proportion of individuals
1.0
0.8
0.6
0.4
0.2
0.0
2
3
4
5
6
7
8
9
10
AMY-1 diploid gene copy number
KEY
High starch
Japanese
Hadza
European American
Low starch
Biaka
Mbuti
Datog
Yakut

Number of copies of the *AMY-1* gene among members of cultures with traditional high-starch or low-starch diets. The Hadza, Biaka, Mbuti, and Datog are tribes in Africa. The Yakut live in Siberia.

1. Starchy tubers are a mainstay of Hadza hunter–gatherers in Africa, whereas fishing sustains Siberia's Yakut. Almost 60% of Yakut had fewer than 5 copies of the *AMY-1* gene. What percentage of the Hadza had fewer than 5 copies?
2. None of the Mbuti (rain-forest hunter–gatherers) had more than 10 copies of *AMY-1*. Did any European Americans have more than 10 copies of *AMY-1*?
3. Do these data support the hypothesis that a starchy diet favors duplications of the *AMY-1* gene?

Source: G. Perry et al. 2007. Diet and the evolution of human amylase gene copy number variation. *Nature Genetics* 39:1256–1260.

Apply Evolutionary Thinking

What is the advantage of a tubelike digestive system over a saclike digestive system?

Express Your Opinion

Many nutritionists suspect that increasing consumption of "fast foods" is contributing to rising levels of obesity. Should fast-food labels carry consumer warnings, as alcohol and cigarette labels do? Go to www.cengage.com/login to investigate both sides of the issue and then vote.

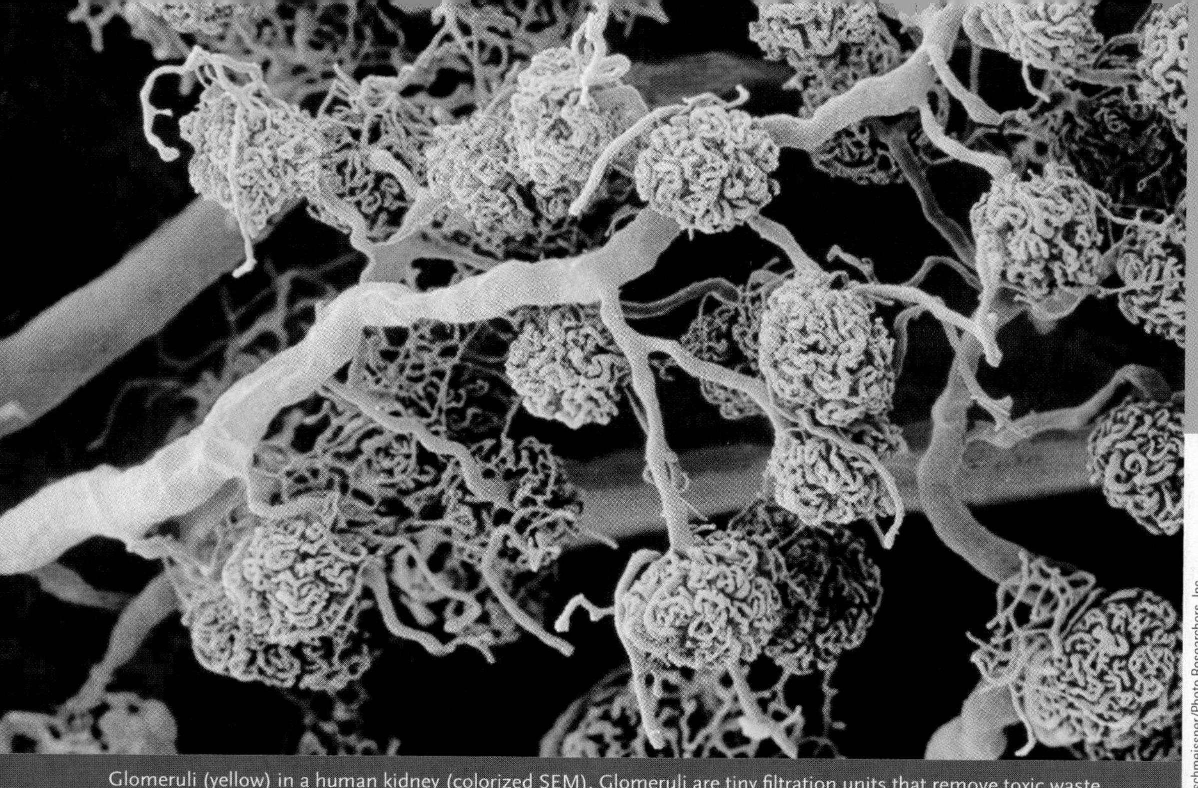

Glomeruli (yellow) in a human kidney (colorized SEM). Glomeruli are tiny filtration units that remove toxic waste products from the blood. The filtered fluid drains to the bladder as urine.

Steve Gschmeissner/Photo Researchers, Inc.

46

Regulating the Internal Environment

STUDY OUTLINE

Why It Matters. . . The crew of the World War II bomber *Lady Be Good* was assigned to fly a night mission to Naples, Italy, from a base on the North African coast on August 4, 1943. But trouble dogged the mission, forcing the crew to turn back before reaching their target. Navigational errors and a cloud layer led them to miss their home base and continue south over the hostile Sahara Desert. Some 440 miles from the coastline, with the fuel running out, the nine crew members parachuted from the aircraft. The bomber remained airborne for a few more minutes and then crash-landed, leaving its crew miles behind.

The eight men who survived began a northward trek with only half a canteen of water among them, in desert heat that reached 130°F during the day. In a testimony to the physiological mechanisms that conserve water and cool the body, they continued onward for eight days. But then, one by one, they died as prolonged dehydration and the continual merciless heat overwhelmed their homeostatic control systems for *water balance* (the equilibrium in inward and outward flow of water) and temperature regulation, and hyperthermia exceeded their capacity to survive. Rescue teams searched the desert for weeks after their disappearance, but no trace was found of the crew or their airplane.

The fate of the *Lady Be Good* remained unknown until 1958, when an oil exploration team flying over the desert spotted the aircraft, sitting largely intact in the desert sands. A two-year search finally led to the remains of the crew, some of them more than a hundred miles north of the downed bomber. Diaries found among the scattered effects told the poignant story of the flight and the futile struggle against the fiery desert environment.

This story illustrates only too clearly the trials of animal life under changing environmental conditions. Water and required nutrients may become more or less abundant. Temperatures may rise or fall. Through their homeostatic control systems, animals have an astounding capacity to compensate for fluctuating external conditions and to maintain the internal environment of their bodies within the relatively narrow limits that cells can tolerate (homeostasis is discussed in Chapter 36).

These limits, and the compensating mechanisms that maintain them, are the subjects of this chapter. First we examine **osmoregulation,** the control of water and ion balance, and the closely related topic of **excretion,** which helps maintain the body's water and ion balance while ridding the body of metabolic wastes. We then consider **thermoregulation,** the control of body temperature. <

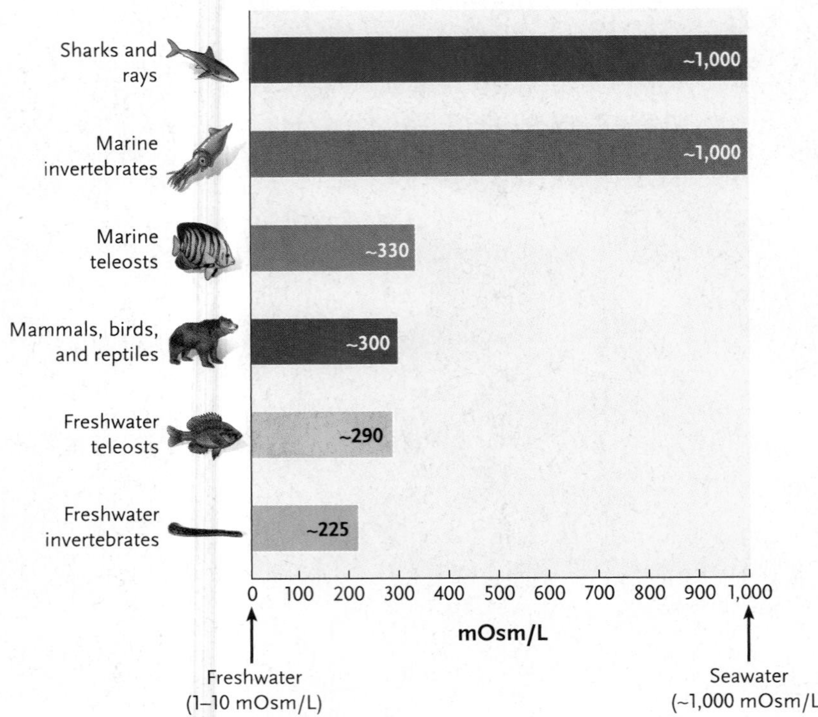

Sharks and rays
~1,000
Marine invertebrates
~1,000
Marine teleosts
~330
Mammals, birds, and reptiles
~300
Freshwater teleosts
~290
Freshwater invertebrates
~225
0 100 200 300 400 500 600 700 800 900 1,000
mOsm/L
Freshwater (1–10 mOsm/L)
Seawater (~1,000 mOsm/L)

FIGURE 46.1
Osmolarity of body fluids in some animal groups. Reptiles include turtles, lizards, and snakes.

46.1 Introduction to Osmoregulation and Excretion

Living cells contain water, are surrounded by water, and constantly exchange water with their environment. For the simplest animals, the water of the external environment directly surrounds cells. For more complex animals, an aqueous extracellular fluid (ECF) surrounds the cells, and is separated from the external environment by a body covering. In animals with a circulatory system, the ECF includes both the interstitial fluid immediately surrounding cells and the plasma, the fluid portion of the blood, or other circulated fluid.

In this section, we review the mechanisms cells use to exchange water and solutes with the surrounding fluid through osmosis and diffusion.

Osmosis Is a Form of Passive Diffusion

In osmosis (see Section 6.3 and Figures 6.9 and 6.10), water molecules move across a selectively permeable membrane from a region where they are more highly concentrated to a region where they are less highly concentrated. The difference in water concentration is produced by differing numbers of solute molecules or ions on the two sides of the membrane. The side of the membrane with a *lower* solute concentration has a *higher* concentration of water molecules, so water will move osmotically to the other side, where water concentration is *lower.*

A key factor in osmosis for helping maintain differences in solute concentration on either side of biological membranes is the selective permeability of those membranes. Proteins are among the most important solutes in establishing the conditions that produce osmosis.

The total solute concentration of a solution, called its **osmolarity,** is measured in *osmoles*—the number of solute molecules and ions (in moles)—per liter of solution. Because the total solute concentration in the body fluids of most animals is less than 1 osmole, osmolarity is usually expressed in thousandths of an osmole, or *milliosmoles* (mOsm). As shown in **Figure 46.1,** the osmolarity of body fluids in humans and other mammals is about 300 mOsm/L; marine fishes and marine invertebrates have higher values and freshwater fishes and freshwater invertebrates have lower values.

Considering solutions on either side of a selectively permeable membrane, a solution of higher osmolarity is said to be *hyperosmotic* to a solution of lower osmolarity, and a solution of lower osmolarity is said to be *hypoosmotic* to a solution of higher osmolarity. If the solutions on either side of a membrane have the same osmolarity, they are said to be *isoosmotic.* Water moves across the membrane between solutions that differ in osmolarity (see Figure 6.9), whereas when two solutions are isoosmotic, no net water movement occurs.

Animals Use Different Approaches to Keep Osmosis from Swelling or Shrinking Their Cells

For metabolic stability, animals must keep their cellular fluids and ECFs isoosmotic. In some animals, called **osmoconformers,** the osmolarity of the cellular and extracellular solutions simply matches the osmolarity of the environment. Most marine invertebrates are osmoconformers. Other animals, called **osmoregulators,** use control mechanisms to keep the osmolarity of cellular and extracellular fluids the same, but at levels that may differ from the osmolarity of the surroundings. Most freshwater and terrestrial invertebrates, and almost all vertebrates, are osmoregulators.

For terrestrial animals, one of the greatest challenges to osmoregulation is the limited supply of water in the environment—if the crew of the *Lady Be Good* had had an adequate supply of water, for example, they could probably have reached safety at the North African coast even without food.

Excretion Is Closely Tied to Osmoregulation

Control over osmolarity is partly maintained by removing certain molecules and ions from cells and body fluids and releasing them into the environment; thus, excretion is closely related to osmoregulation. Animals excrete H^+ ions to keep the pH of body fluids near the neutral levels required by cells for survival. They also excrete toxic products of metabolism, such as nitrogenous (nitrogen-containing) compounds resulting from the breakdown of proteins and nucleic acids, and breakdown products of poisons and toxins. Excretion of ions and metabolic products is accompanied by water excretion because water serves as a solvent for those molecules. Animals that take in large amounts of water may also excrete water to maintain osmolarity.

Microscopic Tubules Form the Basis of Excretion in Most Animals

Except in the simplest animals, minute tubular structures carry out osmoregulation and excretion **(Figure 46.2)**. The tubules are immersed in body fluids at one end (called the *proximal end* of the tubules), and open directly or indirectly to the body exterior at the other end (called the *distal end* of the tubules). The tubules are formed from a **transport epithelium**—a layer of cells with specialized transport proteins in their plasma membranes. The transport proteins move specific molecules and ions into and out of the tubule by either active or passive transport, depending on the particular substance and its concentration gradient.

Typically, the tubules function in a four-step process:

1. **Filtration.** Filtration is the nonselective movement of water and a number of solutes—ions and small molecules, but not large molecules such as proteins—into the proximal end of the tubules through spaces between cells. In animals with an open circulatory system, the water and solutes come from body fluids, with movement into the tubules driven by the higher pressure of the body fluids compared with the fluid inside the tubule. In animals with a closed circulatory system, such as humans, the water and solutes come from the blood in capillaries that surround the tubules, with the movement into the tubules similarly driven by hydrostatic pressure. (Open and closed circulatory systems are described in Section 42.1.)
2. **Reabsorption.** In reabsorption, some molecules (for example, glucose and amino acids) and ions are transported by the transport epithelium back into the ECF and eventually into the blood (in animals with closed circulatory systems) as the filtered solution moves through the excretory tubule.
3. **Secretion.** Secretion is a selective process in which specific small molecules and ions are transported from the ECF and blood into the tubules. Secretion is the second and more important route for eliminating particular substances from the body fluid or blood, filtration being the first. The difference between the two processes is that filtration is nonselective whereas secretion is selective for substances transported.
4. **Release.** The fluid containing waste materials—urine—is released into the environment from the distal end of the tubule. In some animals the fluid is stored in a bladder; in others, it is concentrated into a solid or semisolid form.

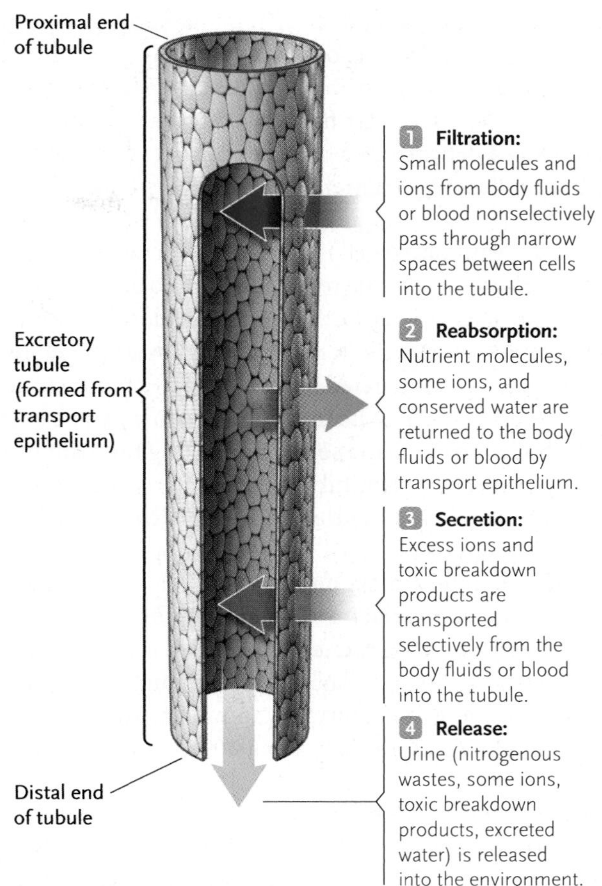
Proximal end of tubule
Excretory tubule (formed from transport epithelium)
Distal end of tubule
1 Filtration: Small molecules and ions from body fluids or blood nonselectively pass through narrow spaces between cells into the tubule.
2 Reabsorption: Nutrient molecules, some ions, and conserved water are returned to the body fluids or blood by transport epithelium.
3 Secretion: Excess ions and toxic breakdown products are transported selectively from the body fluids or blood into the tubule.
4 Release: Urine (nitrogenous wastes, some ions, toxic breakdown products, excreted water) is released into the environment.

FIGURE 46.2
Common structures and operations of the tubules carrying out osmoregulation and excretion in animals. The tubules are typically formed from a single layer of cells with transport functions.

In all vertebrates and many invertebrates, the excretory tubules are concentrated in specialized organs, the *kidneys*, which are discussed in later sections.

Animals Excrete Nitrogen Compounds as Metabolic Wastes

The metabolism of ingested food is a source of both energy and molecules for the biosynthetic activities of an animal. Importantly, metabolism of ingested food produces water—called *metabolic water*—that is used in chemical reactions and is involved in physiological processes such as the excretion of wastes.

The nitrogenous products of the breakdown of proteins, amino acids, and nucleic acids are excreted by most animals as *ammonia, urea,* or *uric acid,* or a combination of these substances **(Figure 46.3).** The particular molecule or combination of molecules produced depends on a balance among toxicity, water conservation, and energy requirements.

AMMONIA Ammonia (NH_3) is the result of a series of biochemical steps beginning with the removal of amino groups ($—NH_3^+$) from amino acids as a part of protein breakdown. Ammonia is readily soluble in water, but it is also highly toxic. Therefore, ammonia must either be excreted or converted to a nontoxic derivative. However, because of its toxicity, ammonia can be excreted from the body only in dilute solutions, making this path possible only in animals with a plentiful supply of water. Those animals include aquatic invertebrates, teleosts, and larval amphibians.

UREA All mammals, most amphibians, turtles, some marine fishes, and some terrestrial invertebrates combine ammonia with HCO_3^- and convert the product in a series of steps to *urea,* a soluble and relatively nontoxic substance. Although producing urea requires more energy than forming ammonia, excreting urea instead of ammonia requires only about 10% as much water.

URIC ACID Water is conserved further in some animals, including terrestrial invertebrates, lizards and snakes, and birds, by the formation of uric acid instead of ammonia or urea. Uric acid is nontoxic, and so insoluble that it precipitates in water as a crystal. (The white substance in bird droppings is uric acid.) The embryos of reptiles and birds also conserve water by forming uric acid, which is stored as a waste product.

Although making uric acid requires even more energy than urea, molecule for molecule it contains four times as much nitrogen as ammonia. And, because uric acid precipitates from water, it can be excreted as a concentrated paste. These factors conserve about 99% of the water that would be required to excrete an equivalent amount of nitrogen as ammonia.

We have now covered the basics of osmoregulation and excretion. In the sections that follow, we look at the specifics of these processes in different animal groups, beginning with the invertebrates.

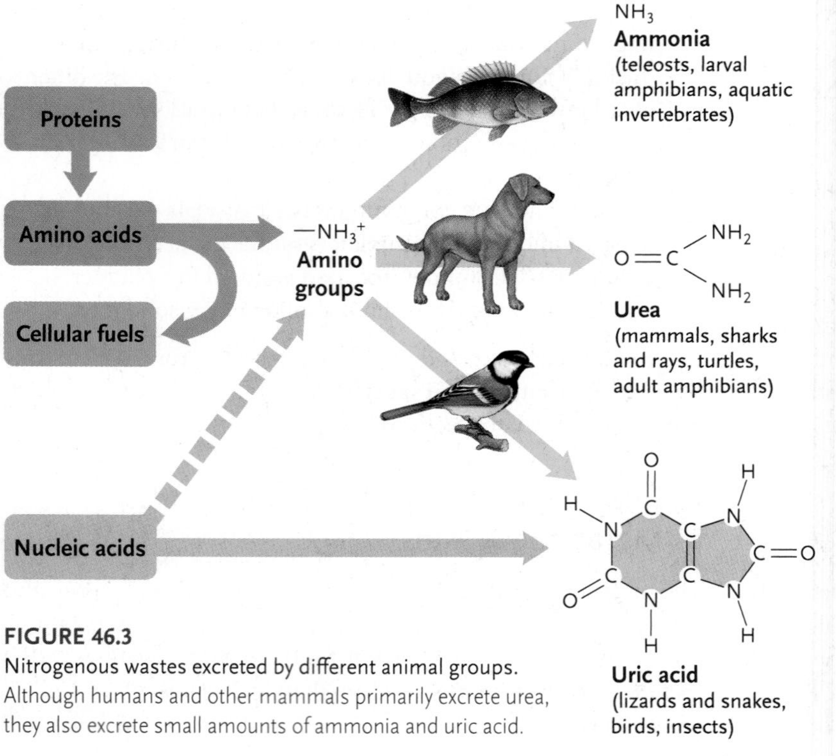

Proteins
Amino acids
Cellular fuels
Nucleic acids
$—NH_3^+$
Amino groups
NH_3
Ammonia
(teleosts, larval amphibians, aquatic invertebrates)
$O{=}C(NH_2)_2$
Urea
(mammals, sharks and rays, turtles, adult amphibians)
Uric acid
(lizards and snakes, birds, insects)

FIGURE 46.3
Nitrogenous wastes excreted by different animal groups.
Although humans and other mammals primarily excrete urea, they also excrete small amounts of ammonia and uric acid.

STUDY BREAK 46.1

Define the terms osmosis, osmolarity, hypoosmotic, osmoregulator, and transport epithelium.

46.2 Osmoregulation and Excretion in Invertebrates

Both osmoconformers and osmoregulators occur among the invertebrates. Except for the simplest groups, most invertebrates, whether osmoconformers or osmoregulators, carry out excretion by specialized excretory tubules.

Most Marine Invertebrates Are Osmoconformers; All Freshwater and Terrestrial Invertebrates Are Osmoregulators

Most marine invertebrates are osmoconformers. They release water, certain ions, and nitrogenous wastes—usually in the form of ammonia—directly from body cells to the surrounding seawater. The cells of these animals do not swell or shrink because the osmolarity of their intracellular and extracellular fluids and the surrounding seawater is the same, about 1,000 mOsm/L. Therefore, they do not have to expend energy to maintain their osmolarity. However, osmoconformers do expend energy to keep some ions, such as Na^+, at lower concentrations inside cells than in the surroundings.

By contrast, all freshwater invertebrates are osmoregulators because their cells could not survive if their internal ion concentrations were reduced to freshwater levels. These animals must expend energy to keep their internal fluids hyperosmotic to their surroundings. Although osmoregulation is energetically expensive, these invertebrates can live in more varied habitats than osmoconformers can.

Freshwater osmoregulators such as flatworms and mussels have cellular and extracellular fluids with an osmolarity higher than that of the external environment. This condition causes water to move constantly from the surroundings into their bodies. This excess water must be excreted, at a considerable cost in energy, to maintain the hyperosmotic state of their body fluids. These animals must also obtain the salts required to keep their body fluids hyperosmotic to freshwater. The salts are obtained from foods, and by actively transporting salt ions from the water into their bodies (even fresh water contains some dissolved salts). This active ion transport occurs through the skin or gills.

Terrestrial annelids (earthworms), arthropods (insects, spiders and mites, millipedes, and centipedes), and mollusks (land snails and slugs) must obtain salts from their surroundings, usually in their foods. Although they do not have to excrete water entering by osmosis, they must constantly replace water lost from their bodies.

In Invertebrate Osmoregulators, Specialized Excretory Tubules Participate in Osmoregulation and Carry Out Excretion

Invertebrate osmoregulators typically use specialized tubules for carrying out excretion. Three common types of these specialized tubules are *protonephridia,* found in flatworms and larval mollusks; *metanephridia,* found in annelids and most adult mollusks; and *Malpighian tubules,* found in insects and other arthropods. Each processes body fluids in a different way.

PROTONEPHRIDIA The flatworm *Dugesia* provides an example of the simplest form of invertebrate excretory tubule, the **protonephridium** (*protos* = first; *nephros* = kidney). In *Dugesia,* two branching networks of protonephridia run the length of the body **(Figure 46.4).** The proximal branches of the tubule network end with a *flame cell* containing a bundle of cilia that extend into the tubule and beat to move fluid through the tubule. When the hemolymph, the invertebrate equivalent to blood, passes through the protonephridia, some molecules and ions are reabsorbed and others, including nitrogenous wastes, are secreted into the tubules. The urine resulting from this filtration system is released through pores at the distal ends of the tubules where they reach the body surface.

FIGURE 46.4
The protonephridia of the flatworm *Dugesia,* showing a flame cell.

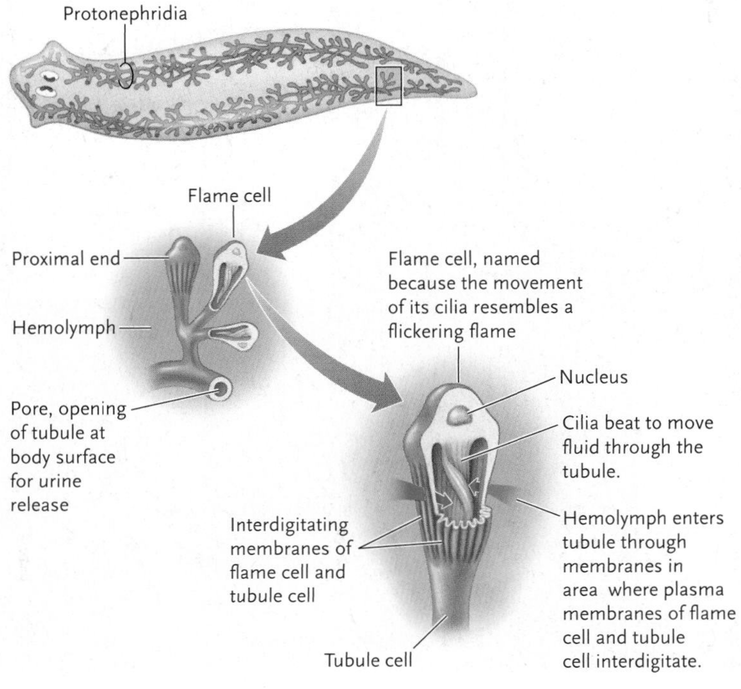

Protonephridia
Flame cell
Proximal end
Hemolymph
Pore, opening of tubule at body surface for urine release
Flame cell, named because the movement of its cilia resembles a flickering flame
Nucleus
Cilia beat to move fluid through the tubule.
Hemolymph enters tubule through membranes in area where plasma membranes of flame cell and tubule cell interdigitate.
Interdigitating membranes of flame cell and tubule cell
Tubule cell

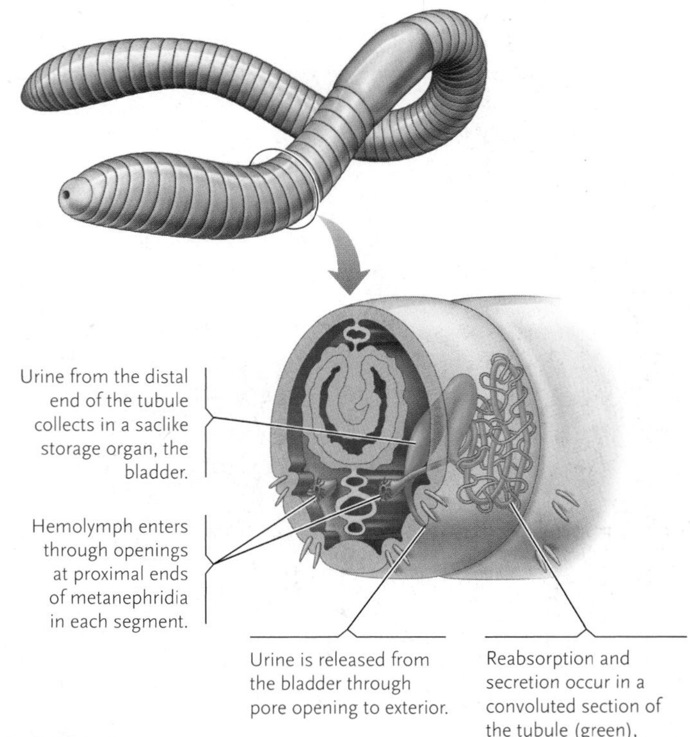

Urine from the distal end of the tubule collects in a saclike storage organ, the bladder.
Hemolymph enters through openings at proximal ends of metanephridia in each segment.
Urine is released from the bladder through pore opening to exterior.
Reabsorption and secretion occur in a convoluted section of the tubule (green), which is surrounded by a capillary network (red).

FIGURE 46.5
The metanephridium of an earthworm.

METANEPHRIDIA The excretory tubule of most annelids and adult mollusks, the **metanephridium** (*meta* = between), has a funnel-like proximal end surrounded with cilia that admits hemolymph. Like protonephridia, metanephridia are filtration systems. As hemolymph moves through the tubule, some molecules and ions are reabsorbed and other ions and nitrogenous wastes are secreted into the tubule and excreted from the distal end at the body surface.

Figure 46.5 shows the arrangement and operation of metanephridia in an earthworm. The proximal ends of a pair of metanephridia are located in each body segment, one on either side of the animal. Each tubule of the pair extends into the following segment, where it bends and folds into a convoluted arrangement surrounded by a network of blood vessels.

MALPIGHIAN TUBULES The excretory tubule of insects, the Malpighian tubule, has a closed proximal end that is immersed in the hemolymph **(Figure 46.6).** The distal ends of the tubules empty into the gut. In contrast to other excretory systems, which use pressure for the filtration step, Malpighian tubules use a different mechanism. They secrete K^+ into the lumen of the proximal segment, which makes the tubule fluid more positively charged and draws in Cl^- ions from the hemolymph surrounding the tubule. The accumulation of KCl makes the tubule fluid more concentrated, causing water from the hemolymph to move in by osmosis. Now organic wastes, in particular uric acid, are secreted into the tubule. The tubule empties into the gut. When the fluid reaches the hindgut and rectum, K^+ and Cl^- are reabsorbed, followed by

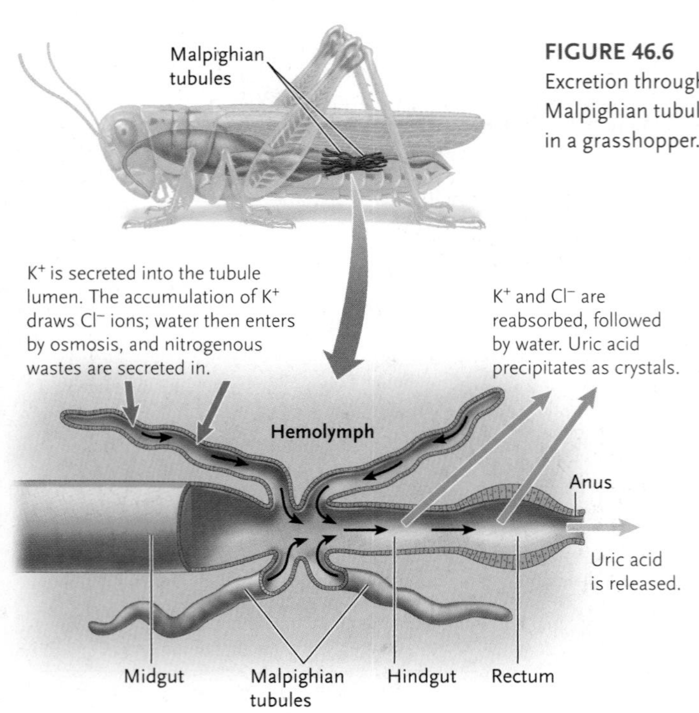
Malpighian tubules
K^+ is secreted into the tubule lumen. The accumulation of K^+ draws Cl^- ions; water then enters by osmosis, and nitrogenous wastes are secreted in.
K^+ and Cl^- are reabsorbed, followed by water. Uric acid precipitates as crystals.
Hemolymph
Anus
Uric acid is released.
Midgut
Malpighian tubules
Hindgut
Rectum

FIGURE 46.6 Excretion through Malpighian tubules in a grasshopper.

water. As the water is removed, uric acid precipitates as crystals, which mix with the undigested matter in the rectum and are released with the feces. In this way, nitrogenous waste excretion is accomplished with minimal water loss.

STUDY BREAK 46.2

How are protonephridia, metanephridia, and Malpighian tubules different? In which animal groups are each of these excretory tubules found?

46.3 Osmoregulation and Excretion in Mammals

In all vertebrates, specialized excretory tubules contribute to osmoregulation and carry out excretion. The excretory tubules, called **nephrons,** are located in a specialized organ, the kidney. We begin our survey of vertebrate osmoregulation and excretion with a description of the structure and function of the mammalian kidney.

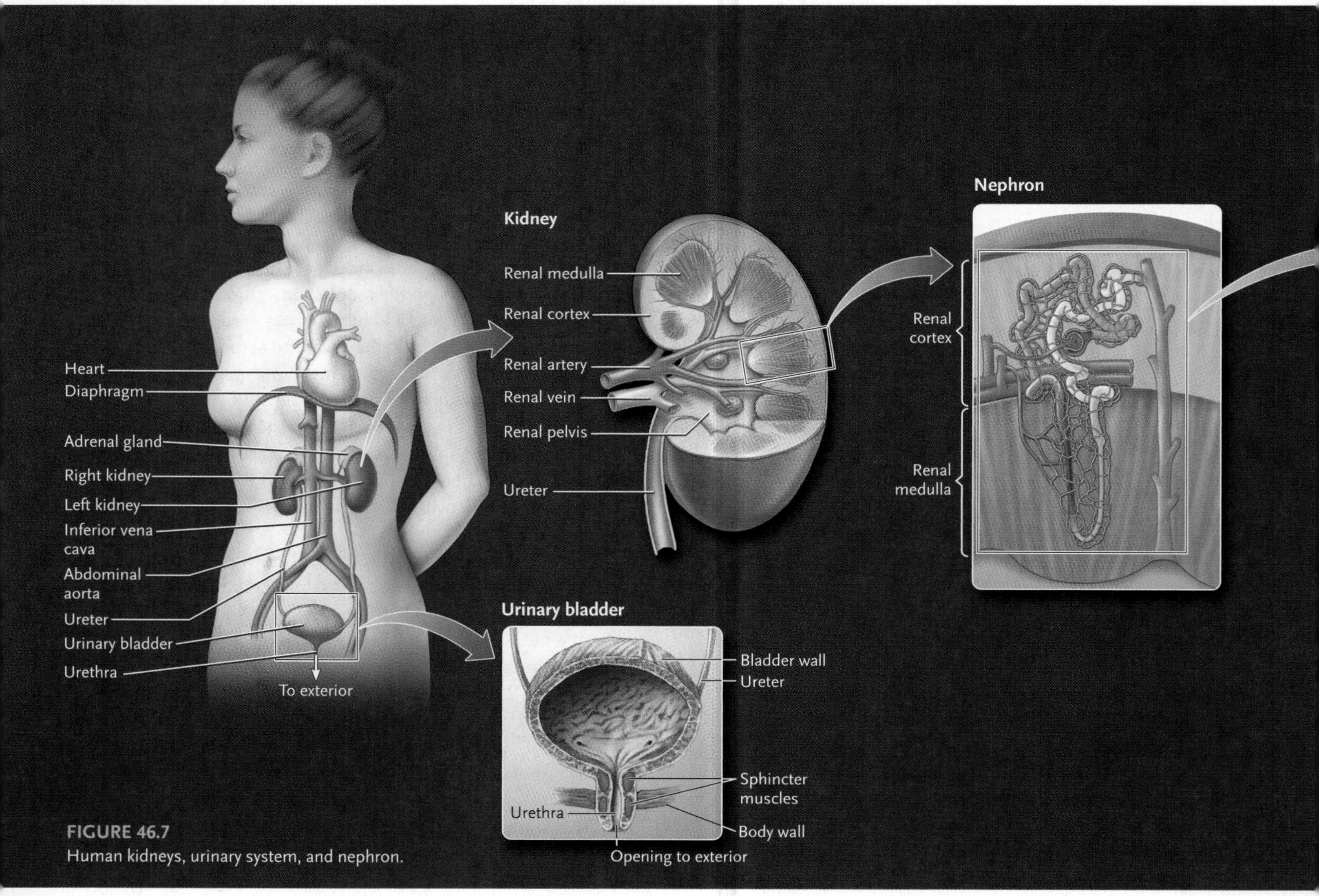

FIGURE 46.7 Human kidneys, urinary system, and nephron.

The Kidneys and Ureters, the Bladder, and the Urethra Form the Urinary System

Mammals have a pair of kidneys, located on either side of the vertebral column at the back of the abdominal cavity **(Figure 46.7)**. Internally, the mammalian kidney is divided into an outer **renal cortex** surrounding a central region, the **renal medulla.**

The **renal artery** carries blood into the kidney, where metabolic wastes and excess ions are moved into urine. The filtered blood is routed away from the kidney by the **renal vein.** The urine leaving individual nephrons is processed further in **collecting ducts** and then drains into a central cavity in the kidney called the **renal pelvis.**

From the renal pelvis, the urine flows through a duct called the **ureter** to the **urinary bladder,** a storage sac located outside the kidneys. Urine leaves the bladder through the **urethra,** a tube that (in most mammals) opens to the outside. In human females, the opening of the urethra is just in front of the vagina; in males, the urethra opens at the tip of the penis. The two kidneys and ureters, the urinary bladder, and the urethra constitute the mammalian urinary system.

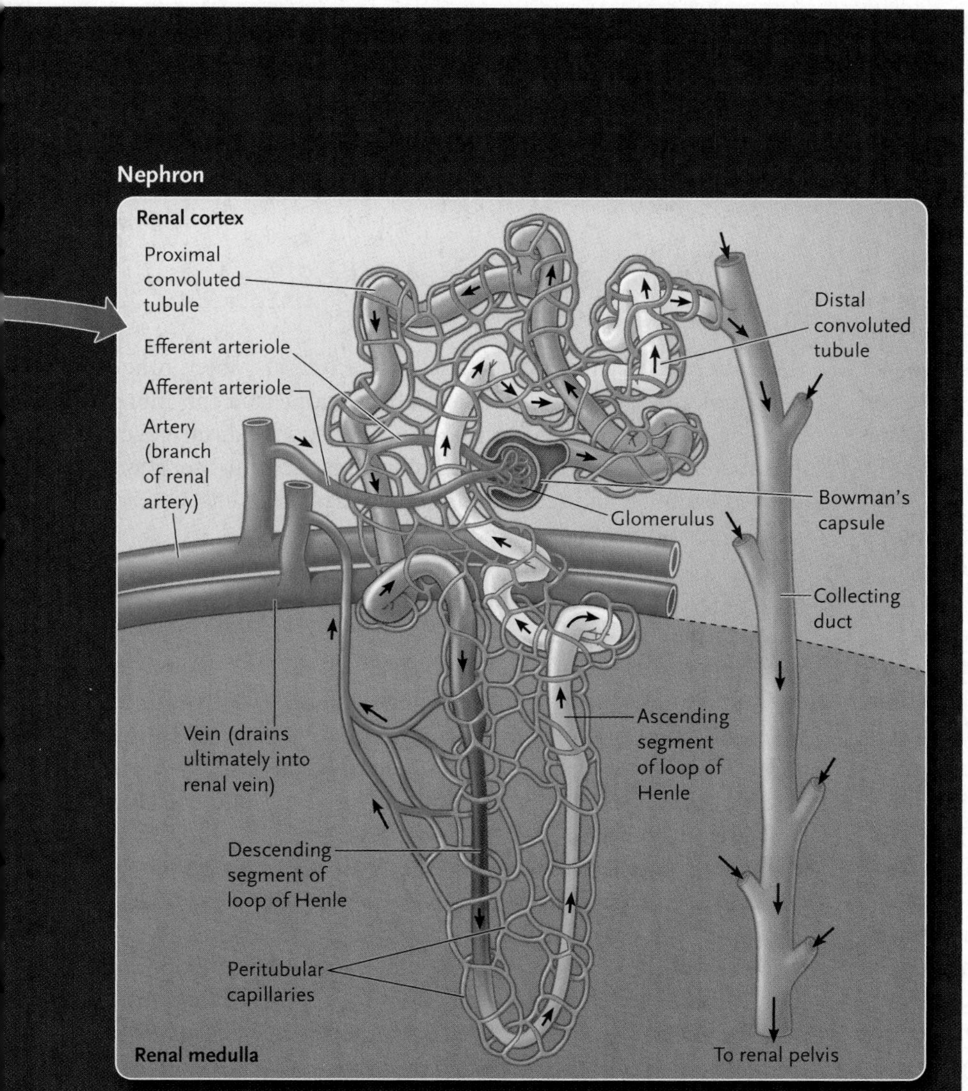

Two sphincter muscles control the flow of urine from the bladder to the urethra. In human infants, urination is an autonomic reflex triggered by stretch receptors in the bladder wall. When the bladder becomes full, the sphincters relax, smooth muscles in the bladder wall contract, and the urine is forced to the exterior. At about the age of two years, children learn to override the autonomic reflex by consciously keeping the striated sphincter contracted until urination is convenient.

Mammalian Nephrons Are Differentiated into Regions with Specialized Functions

Each human kidney has more than a million nephrons. Nephrons are differentiated into regions that perform successive steps in excretion. At its proximal end, a human nephron forms the **Bowman's capsule,** an infolded region that cups around a ball of arterial capillaries called the **glomerulus** (see Figure 46.7). The capsule and glomerulus are located in the renal cortex. Filtration in Bowman's capsule begins the process of excretion.

Following Bowman's capsule, the nephron forms a **proximal convoluted tubule** that descends into the renal medulla in a U-shaped bend called the **loop of Henle** and then ascends again to form a **distal convoluted tubule.** The distal tubule drains the urine into a collecting duct that leads to the renal pelvis. As many as eight nephrons may drain into a single collecting duct. The combined activities of the proximal convoluted tubule, the loop of Henle, the distal convoluted tubule, and the collecting duct convert the filtrate entering the nephron into urine.

Unlike most capillaries in the body, the capillaries in the glomerulus do not lead directly to venules. Instead, they form another arteriole that branches into a second capillary network called the **peritubular capillaries.** These capillaries thread around the proximal and distal convoluted tubules and the loop of Henle. Molecules and ions that are reabsorbed from the filtrate are transferred between the nephron and the peritubular capillaries. However, this transfer is not direct. Instead, the molecules or ions pass through the wall of the tubule, which is one cell layer thick; diffuse through the interstitial fluid; and then pass into the capillary through its wall, also one cell thick.

Mammalian Nephrons Interact with Surrounding Kidney Structures to Produce Hyperosmotic Urine

In mammals, urine is hyperosmotic to body fluids. All other vertebrates except for a few aquatic bird species produce urine that is hypoosmotic (or isoosmotic) to body fluids. Production of hyperosmotic urine is a water-conserving adaptation involving an interaction between the activities of the mammalian

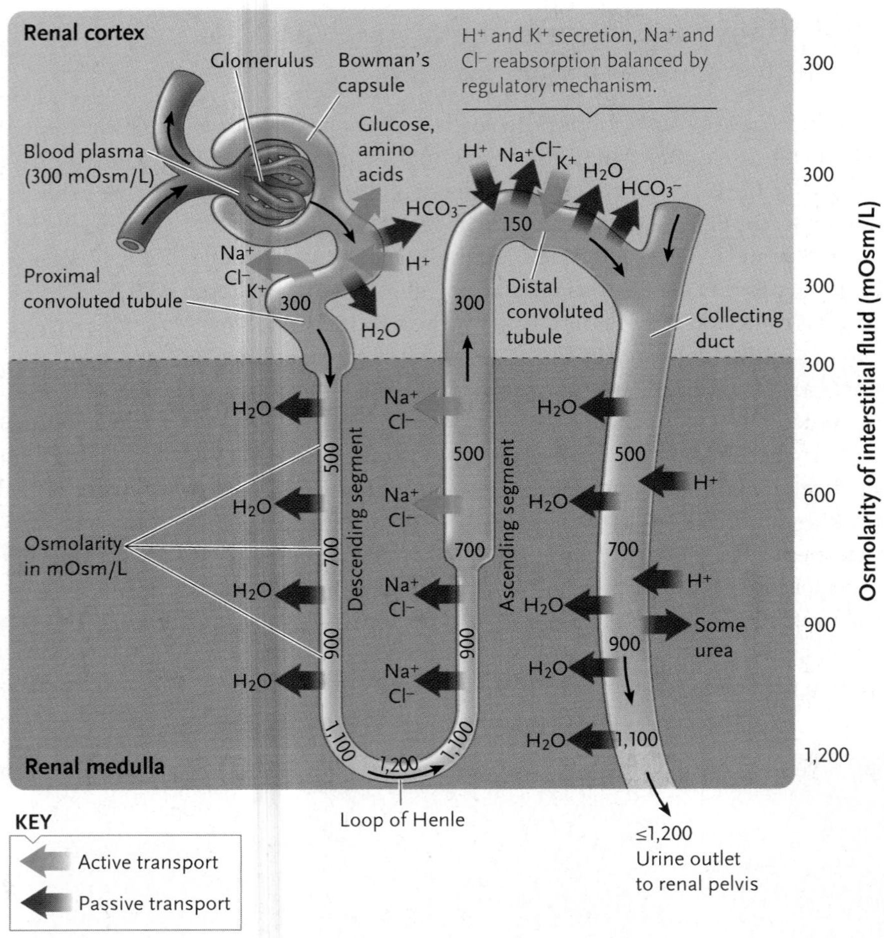

Renal cortex
Glomerulus
Bowman's capsule
H+ and K+ secretion, Na+ and Cl− reabsorption balanced by regulatory mechanism.
Blood plasma (300 mOsm/L)
Glucose, amino acids
H+
Na+
Cl−
K+
H2O
HCO3−
150
Proximal convoluted tubule
300
Distal convoluted tubule
Collecting duct
500
700
900
1,100
1,200
Descending segment
Ascending segment
Osmolarity in mOsm/L
Some urea
Renal medulla
Loop of Henle
≤1,200
Urine outlet to renal pelvis
KEY
Active transport
Passive transport
300
300
300
300
600
900
1,200
Osmolarity of interstitial fluid (mOsm/L)

FIGURE 46.8
The movement of ions, water, and other molecules to and from nephrons and collecting tubules in the human kidney. Nephrons in other mammals and in birds work in similar fashion. The numbers are osmolarity values in mOsm/L.

nephrons and the highly ordered structure of the mammalian kidney. Three features underlie this interaction:

1. The arrangement of the loop of Henle, which descends through the medulla and returns to the cortex again.
2. Differences in the permeability of successive regions of the nephron, established by a specific group of membrane transport proteins in each region.
3. A gradient in the concentration of molecules and ions in the interstitial fluid of the kidney, which increases gradually from the renal cortex to the deepest levels of the renal medulla.

These features interact to conserve nutrients and water, balance salts, and concentrate wastes for excretion from the body.

Filtration in Bowman's Capsule Begins the Process of Excretion

The mechanisms of excretion (shown in **Figure 46.8** and summarized in **Table 46.1**) begin in Bowman's capsule. The single layer of epithelial cells forming the glomerular capillaries is perforated by many pores that make the vessels much more permeable to water and solutes than other capillaries in the body. The higher pressure of the blood drives fluid containing ions, small nutrient molecules such as glucose and amino acids, and nitrogenous waste molecules (primarily urea) through the pores of the capillaries into the capsule. Blood cells and plasma proteins are too large to pass and are retained inside the capillaries.

The pressure driving fluid into Bowman's capsule is maintained because the diameter of the **afferent arteriole** delivering blood to the glomerulus is larger than that of the **efferent arteriole** that takes blood away from the glomerulus. That is, because more blood can enter the glomerulus through the afferent arteriole than can leave through the efferent arteriole, glomerular capillary pressure is maintained at a high level as a result of blood damming up in the capillaries.

In humans, Bowman's capsules collectively filter about 180 L (47.5 gallons) of fluid each day, from a daily total of 1,400 L (369.5 gallons) of blood that pass through the kidneys. The human body contains only about 2.75 L of blood plasma, meaning that the kidneys filter a fluid volume equivalent to 65 times the volume of the blood plasma each day. On average, more than 99% of the filtrate, mostly water, is reabsorbed in the nephrons, leaving about 1.5 L to be excreted daily as urine.

Reabsorption and Secretion Take Place in the Remainder of the Nephron

The fluid filtered into Bowman's capsule contains water, other small molecules, and ions at essentially the same concentrations as the blood plasma. By the time the fluid reaches the distal end of the tubules and passes through the collecting ducts, reabsorption out of the tubules and secretion into them have markedly altered the concentrations of all components of the filtrate.

THE PROXIMAL CONVOLUTED TUBULE Reabsorption of water, ions, and nutrients into the interstitial fluid is the main function of the proximal convoluted tubule. Na^+/K^+ pumps in the epithelium of the proximal convoluted tubule move Na^+ and K^+ from the filtrate into the interstitial fluid surrounding the tubule (see Figure 46.8). The movement of positive charges sets up a voltage gradient that causes Cl^- ions to be reabsorbed from within the tubule with the positive ions. Specific active transport proteins reabsorb essentially all the glucose, amino acids, and other nutrient molecules from the filtrate into the interstitial fluid, making the filtrate hypoosmotic to the interstitial fluid surrounding the tubule. As a result, water moves from the tubule into the interstitial fluid by osmosis. The osmotic movement is aided by *aquaporins,* transport proteins that form passages for water molecules in the transport epithelium of the tubule cells. (To review aquaporins, see Section 6.2 and

TABLE 46.1 Filtration, Reabsorption, and Secretion in Nephrons and Collecting Ducts

Segment	Location	Permeability and Movement	Osmolarity of Filtrate and Urine	Result of Passage
Bowman's capsule	Cortex	Water, ions, small nutrients, and nitrogenous wastes move through spaces between epithelia	300 mOsm/L, same as surrounding interstitial fluid	Water and small substances, but not proteins, pass into nephron
Proximal convoluted tubule	Cortex	Na^+ and K^+ actively reabsorbed, Cl^- follows; water leaves through aquaporins; H^+ actively secreted; HCO_3^- reabsorbed into plasma of peritubular capillaries; glucose, amino acids, and other nutrients actively reabsorbed	300 mOsm/L	67% of ions, 65% of water, 50% of urea, and all nutrients return to interstitial fluid; pH maintained
Descending segment of loop of Henle	Cortex into medulla	Water leaves through aquaporins; no movement of ions or urea	From 300 mOsm/L at top to 1,200 mOsm/L at bottom of loop	Additional water returned to interstitial fluid
Ascending segment of loop of Henle	Medulla into cortex	Na^+ and Cl^- actively transported out; no entry of water; no movement of urea	From 1,200 mOsm/L at bottom to 150 mOsm/L at top of loop	Additional ions returned to interstitial fluid
Distal convoluted tubule	Cortex	K^+ and Na^+ secreted via active transport into urine; Na^+ and Cl^- reabsorbed; water moves into urine through aquaporins; HCO_3^- reabsorbed into plasma of peritubular capillaries	From 150 mOsm/L at beginning to 300 mOsm/L at junction with collecting duct	Ion balance, pH balance
Collecting ducts	Cortex through medulla, empties into renal pelvis	Water moves out via aquaporins; no movement of ions; some urea leaves at bottom of duct	From 300 mOsm/L to 1,200 mOsm/L at junction with renal pelvis	More water and some urea returned to interstitial fluid; some H^+ added to urine

Unanswered Questions for Chapter 6.) The nutrients and water that entered the interstitial fluid move into the capillaries of the peritubular network.

Some substances are also transferred into the tubule. These are primarily H^+ ions by active transport and the products of detoxified poisons by passive diffusion (detoxification takes place in the liver). The secretion of H^+ ions into the filtrate helps balance the acidity constantly generated in the body by metabolic reactions. H^+ secretion is coupled with HCO_3^- reabsorption from the filtrate in the tubule to the plasma in the peritubular capillaries. Small amounts of ammonia are also secreted into the tubule.

In all, the proximal convoluted tubule reabsorbs about 67% of the Na^+, K^+, and Cl^- ions, 65% of the water, 50% of the urea, and essentially all the glucose, amino acids, and other nutrient molecules in the filtrate. The ions, nutrients, and water reabsorbed by the tubule are transported into the interstitial fluid, and then into capillaries of the peritubular network. Although half of the urea is reabsorbed, the constant flow of filtrate through the tubules keeps the concentration of nitrogenous wastes low in body fluids.

The proximal convoluted tubule has structural specializations that fit its function. The epithelial cells that make up its walls are carpeted on their inner surface by a brush border of microvilli. Like the brush border of epithelial cells in the small intestine (see Section 45.3), these microvilli greatly increase the surface area available for reabsorption and secretion.

THE DESCENDING SEGMENT OF THE LOOP OF HENLE The loop of Henle has dual functions: water is recaptured in the descending segment, and salts are reabsorbed in the ascending segment.

The filtrate leaving the proximal convoluted tubule enters the descending segment of the loop of Henle, where water is reabsorbed. As this segment descends, it passes through regions of increasingly higher solute concentrations in the interstitial fluid of the medulla (see Figure 46.8). (The generation of this concentration gradient is described later.) As a result, more water moves out of the tubule by osmosis as the fluid travels through the descending segment.

Aquaporins in the descending segment allow the rapid transport of water. The outward movement of water concentrates the molecules and ions inside the tubule, gradually increasing the osmolarity of the fluid to a peak of about 1,200 mOsm/L at the bottom of the loop. This is the same as the osmolarity of the interstitial fluid at the bottom of the medulla.

INSIGHTS FROM THE Molecular Revolution

An Ore Spells Relief for Osmotic Stress

Almost all cells respond to osmotic stress—osmotic imbalance with the surroundings—by adjusting the cytoplasmic concentration of *osmolytes*. When cells are surrounded by a hyperosmotic solution, for example, osmolytes accumulate in the cytoplasm, raising its osmolarity to match that of the surroundings.

In humans, cells in the renal medulla are regularly exposed to high solute concentrations in the interstitial fluid. These cells would quickly die from osmotic water loss if they were not protected by osmolytes. In these cells, as in many other types of mammalian cells, one of the predominant osmolytes is *sorbitol*, made from glucose in a reaction catalyzed by the enzyme *aldose reductase*. Placing cells in a hyperosmotic medium induces transcription of the aldose reductase gene.

Research Question

How does hyperosmotic stress activate the aldose reductase gene?

Experiments

The research of Joan D. Ferraris and her colleagues at the National Institutes of Health in Bethesda, Maryland, provided some molecular answers to the question. First, the researchers cloned a DNA segment containing the aldose reductase (AR) gene and its upstream promoter and regulatory sequences from a rabbit genomic library (see Chapter 18) **(Figure A).** (Recall from Section 16.2 that the promoter, the sequence immediately upstream of the gene's RNA coding sequence, plays a basic role in initiating transcription of the gene, but does not contain regulatory sequences. which are located upstream of the promoter.)

The researchers were particularly interested in identifying the regulatory sequence required for activating transcription in cells exposed to a hyperosmotic medium. They constructed a composite gene by fusing the upstream segment of the AR gene with the RNA coding sequence of the firefly luciferase gene **(Figure B).** The luciferase enzyme catalyzes a reaction that emits light as a product. By using light production as a measure of luciferase activity, the construct enabled the researchers to easily study the AR upstream region to identify the regulatory sequence.

The researchers transformed the composite gene into cultured rabbit renal cells, and then divided the cells into two groups. One group was maintained in an isoosmotic medium (300 mOsm/L); the other group was placed in a medium made hyperosmotic (500 mOsm/L) by added NaCl. After a period of incubation, the cells were collected and lysed, and the luciferase activity was measured. The cells in the hyperosmotic solution had several-fold higher luciferase activity than the cells in the isoosmotic solution, showing that the luciferase gene had been turned on under the high salt condition because of the presence of the regulatory sequence derived from the AR gene.

The next step was to isolate the particular regulatory sequence. First, they identified the promoter of the AR gene by adding various lengths of fragments immediately upstream of the AR coding sequence to the luciferase gene. They found that a 208-bp

THE ASCENDING SEGMENT OF THE LOOP OF HENLE The fluid then moves into the ascending segment of the loop of Henle, where Na^+ and Cl^- are reabsorbed into the interstitial fluid. As this segment ascends, it passes through regions of gradually lessening osmolarity in the medulla (see Figure 46.8). The ascending segment has membrane proteins that transport salt ions, but no aquaporins. Because water is trapped in the ascending segment, the osmolarity of the urine is reduced as the salt ions move out of the tubule.

In the part of the ascending segment immediately following the loop, the ion concentrations in the tubule filtrate are still high enough to move Na^+ and Cl^- out of the tubule by passive transport. Toward the top of the segment, they are moved out by active transport. Besides reducing the osmolarity of the filtrate in the ascending segment, the reabsorption of salt ions from the tubule into the interstitial fluid helps establish the concentration gradient of the medulla, high near the renal pelvis and low near the renal cortex. The energy required to transport NaCl from higher levels of the ascending segment makes the kidneys one of the major ATP-consuming organs of the body.

By the time the fluid reaches the cortex at the top of the ascending loop, its osmolarity has dropped to about 150 mOsm/L. During the travel of fluid around the entire loop of Henle, water, nutrients, and ions have been conserved and returned to body fluids, and the total volume of the filtrate in the nephron has been greatly reduced. Urea and other nitrogenous wastes have been concentrated in the filtrate. Little secretion occurs in either the descending or ascending segments of the loop of Henle.

THE DISTAL CONVOLUTED TUBULE Additional water is recovered by osmosis from the fluid in the distal convoluted tubule. In response to hormones triggered by changes in the body's salt concentrations (described in Section 46.4), varying amounts of K^+ and H^+ ions are secreted into the fluid, and varying amounts of Na^+ and Cl^- ions are reabsorbed. Bicarbonate ions are reabsorbed from the filtrate as in the proximal tubule.

The amounts of urea and other nitrogenous wastes remain the same. By the time the fluid—now urine—enters the collecting ducts at the end of the nephron, it is isoosmotic with blood plasma (about 300 mOsm/L) but very different in composition.

THE COLLECTING DUCTS The collecting ducts concentrate the urine. These ducts, which are permeable to water but not salt ions, descend from the cortex through the medulla of the kidney. As the ducts descend, they encounter the gradient of increasing solute concentration in the interstitial fluid of the medulla. This increase makes water move osmotically out of the ducts and greatly increases the concentration of the urine, which can become as sequence gave promoter activity **(Figure C)**; that is, they saw a low level of transcription of luciferase in isoosmotic conditions, but that level of transcription did not change in hyperosmotic conditions. Next, the researchers broke the DNA upstream of the promoter into fragments, attached them one at a time to the promoter–luciferase gene construct, and tested them for luciferase expression in isoosmotic and hyperosmotic conditions. Eventually, they identified the minimal sequence that activates the luciferase gene in cells placed in hyperosmotic medium; it proved to be the sequence 5′-CGGAAAATCAC-3′, beginning 1,105 bp upstream of the site where transcription begins. The investigators termed the activating sequence an *osmotic response element (ORE)*.

Conclusion

Molecular experiments fusing various parts of the sequence upstream of the AR gene to the luciferase gene illuminated for the first time the regulatory sequence responsible for activating the AR gene under hyperosmotic conditions. The result was a first step in elucidating the molecular mechanism for how hyperosmotic stress regulates the expression of genes involved in responding to that stress.

Sources: J. D. Ferraris et al. 1994. Cloning, genomic organization, and osmotic response of the aldose reductase gene. *Proceedings of the National Academy of Sciences, USA* 91:10742–10746; J. D. Ferraris et al. 1996. ORE, a eukaryotic minimal essential osmotic response element. *Journal of Biological Chemistry* 271:18318–18321.

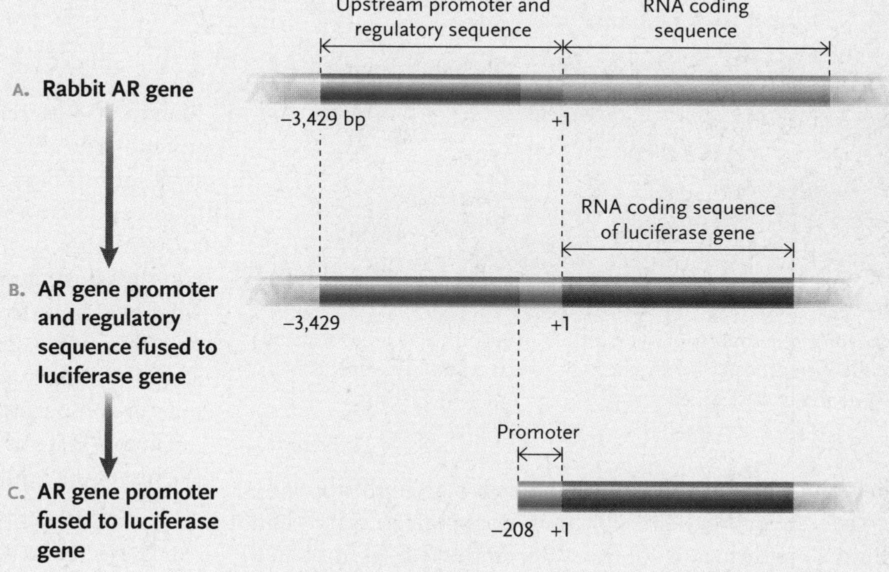
Upstream promoter and regulatory sequence
RNA coding sequence
A. Rabbit AR gene
–3,429 bp
+1
RNA coding sequence of luciferase gene
B. AR gene promoter and regulatory sequence fused to luciferase gene
–3,429
+1
Promoter
C. AR gene promoter fused to luciferase gene
–208
+1

high as 1,200 mOsm/L at the bottom of the medulla. Near the bottom of the medulla, the walls of the collecting ducts contain passive urea transporters that allow a portion of this nitrogenous waste to pass from the duct into the interstitial fluid. This urea adds significantly to the concentration gradient of solutes in the medulla.

In addition to these mechanisms, H^+ ions are actively secreted into the fluid by the same mechanism as in the proximal and distal convoluted tubules. The balance of the H^+ and bicarbonate ions established in the urine, interstitial fluid, and blood, achieved by secretion of H^+ into the urine by the nephrons and collecting ducts, is important for regulating the pH of blood and body fluids. The kidneys thus provide a safety valve if the acidity of body fluids rises beyond levels that can be controlled by the blood's buffer system (see Section 44.4).

At its maximum value of 1,200 mOsm/L, reached when water conservation is at its maximum, the urine at the bottom of the collecting ducts is about four times more concentrated than body fluids. It can also be as low as 50 to 70 mOsm/L, when very dilute urine is produced in response to conditions such as excessive water intake.

The high osmolarity of the interstitial fluid toward the bottom of the medulla would damage the medulla cells if they were not protected against osmotic water loss. The protection comes from high concentrations of otherwise inert organic molecules called *osmolytes* in these cells. The osmolytes, primarily a sugar alcohol called *sorbitol,* raise the osmolarity of the cells to match that of the surrounding interstitial fluid. *Insights from the Molecular Revolution* describes research that identified the genetic controls leading to sorbitol production in kidney medulla cells and other cells subjected to osmotic stress.

Urine flows from the end of the collecting ducts into the renal pelvis, and then through the ureters into the urinary bladder where it is stored. From the bladder, urine exits through the urethra to the outside.

Terrestrial Mammals Have Additional Water-Conserving Adaptations

Terrestrial mammals have other adaptations that complement the water-conserving activities of the kidneys. One is the location of the lungs deep inside the body, which reduces water loss by evaporation during breathing (see Section 44.1). Another is a body covering of keratinized skin. Skin is so nearly impermeable that it almost eliminates water loss by evaporation, except for the controlled loss through evaporation of sweat in mammals with sweat glands.

Water-conserving adaptations reach their greatest efficiency in desert rodents such as the kangaroo rat **(Figure 46.9)**. The proportion of nephrons with long loops extending deep

	Kangaroo rat	Human
Water gain (milliliters)		
From ingesting food	6.0	850
From drinking liquids	0.0	1,400
By metabolism	54.0	350
	60.0	2,600
Water loss (milliliters)		
In urine	13.5	1,500
In feces	2.6	200
By evaporation	43.9	900
	60.0	2,600

Claude Steelman/Wildshots, Inc.

FIGURE 46.9
A comparison of the sources of water for a human and a kangaroo rat (genus *Dipodymus*). Water conservation in the kangaroo rat is so efficient that the animal never has to drink water.

into the kidney medulla of kangaroo rats is very high, allowing them to excrete urine that is 20 times more concentrated than body fluids. Further, most of the water in the feces is absorbed in the large intestine and rectum. Lacking sweat glands, kangaroo rats lose little water by evaporation from the body surface. Much of the moisture in their breath is condensed and recycled by specialized passages in the nasal cavities. They stay in burrows during daytime, and come out to feed only at night.

About 90% of the kangaroo rat's daily water supply is generated from oxidative reactions in its cells. (Humans, by contrast, can make up only about 12% of their daily water needs from this source.) The remaining 10% of the kangaroo rat's water comes from its food. These structural and behavioral adaptations are so effective that a kangaroo rat can survive in the desert without ever drinking water.

Marine mammals, including whales, seals, and manatees, eat foods that are high in salt content and never drink fresh water. They are able to survive the high salt intake because they produce urine that is more concentrated than seawater. As a result, they are easily able to excrete all the excess salt they ingest in their diet.

STUDY BREAK 46.3

1. **Describe the structure of a human nephron from the proximal end to the distal end.**
2. **The urine entering the collecting ducts at the end of the nephron has an osmolarity essentially the same as that of fluids in other parts of the body. How is the urine subsequently made more concentrated?**

THINK OUTSIDE THE BOOK

On your own or collaboratively, research the symptoms and possible causes of chronic renal disease (also called chronic kidney disease). Then outline the hypothesis and research approach for a study of chronic renal disease that uses the rat as the experimental organism.

46.4 Regulation of Mammalian Kidney Function

Mammalian excretory functions are integrated into overall body functions by three primary control systems, which link kidney functions to blood pressure, to the osmolarity and pH of body fluids, and to the body's water balance. An autoregulation system located entirely within the kidney keeps glomerular filtration constant during relatively small variations in blood pressure, as when we move from sitting to standing. Two other systems involve hormonal controls that compensate for excessive loss of salt and body fluids and that adjust the rate of water uptake in the kidneys to compensate for excessive water intake or loss. These two hormonal systems regulate interactions between the kidneys and the rest of the body.

Autoregulation Involves Interactions between the Glomerulus and the Nephron

The autoregulation system responds almost instantly to keep the filtration rate constant during small variations in blood pressure. The system depends on signals from receptors in the **juxtaglomerular apparatus** (*juxta* = near), which is located at a point where the distal convoluted tubule contacts the afferent arteriole carrying blood to the glomerulus **(Figure 46.10)**. The receptors, located in the tubule wall, monitor the pressure and flow of fluid through the distal tubule. If a rise in blood pressure increases the filtration rate, the receptors release chemical signals that trigger constriction of the afferent arteriole. The constriction reduces blood flow through the glomerulus and lowers the filtration rate. A drop in filtration rate because of a fall in blood pressure has the opposite effect: signals from the receptors cause the arterioles to dilate, blood flow increases in the glomerulus, and the filtration rate rises.

The RAAS Responds to Na^+ by Triggering Na^+ Reabsorption

Major changes in blood volume and pressure occur when the body loses or gains Na^+ in excessive amounts. Excessive Na^+ loss may result from prolonged and heavy sweating, repeated vomiting, severe diarrhea, or insufficient Na^+ uptake in the diet. The Na^+ loss reduces the osmolarity of body fluids, which causes less water to be reabsorbed in the kidneys. The water loss reduces the volume of blood and interstitial fluid and causes the blood pressure to drop. Excessive Na^+ intake in salty foods may have the opposite effects. The body must compensate for significant changes in Na^+.

The **renin-angiotensin-aldosterone system (RAAS)** is the most important hormonal system involved in regulating Na^+ (see Figure 46.10). At normal body salt concentrations, the RAAS allows about 10 g of salt to be excreted in the urine each day. If excessive Na^+ is excreted, blood pressure and body fluid volume drop, and the glomerular filtration rate falls below levels that can be restored by the juxtaglomerular apparatus autoregulation. In response to this condition, the following events take place (see Figure 46.10):

- Cells in the juxtaglomerular apparatus secrete the enzyme **renin** into the bloodstream. (The RAAS also is activated to secrete renin when blood pressure or blood volume decreases independently of Na^+ levels, as in the case of a hemorrhage.)
- Renin cleaves the plasma protein *angiotensinogen* to produce **angiotensin I.**
- **Angiotensin-converting enzyme (ACE)** converts angiotensin I to **angiotensin II.**

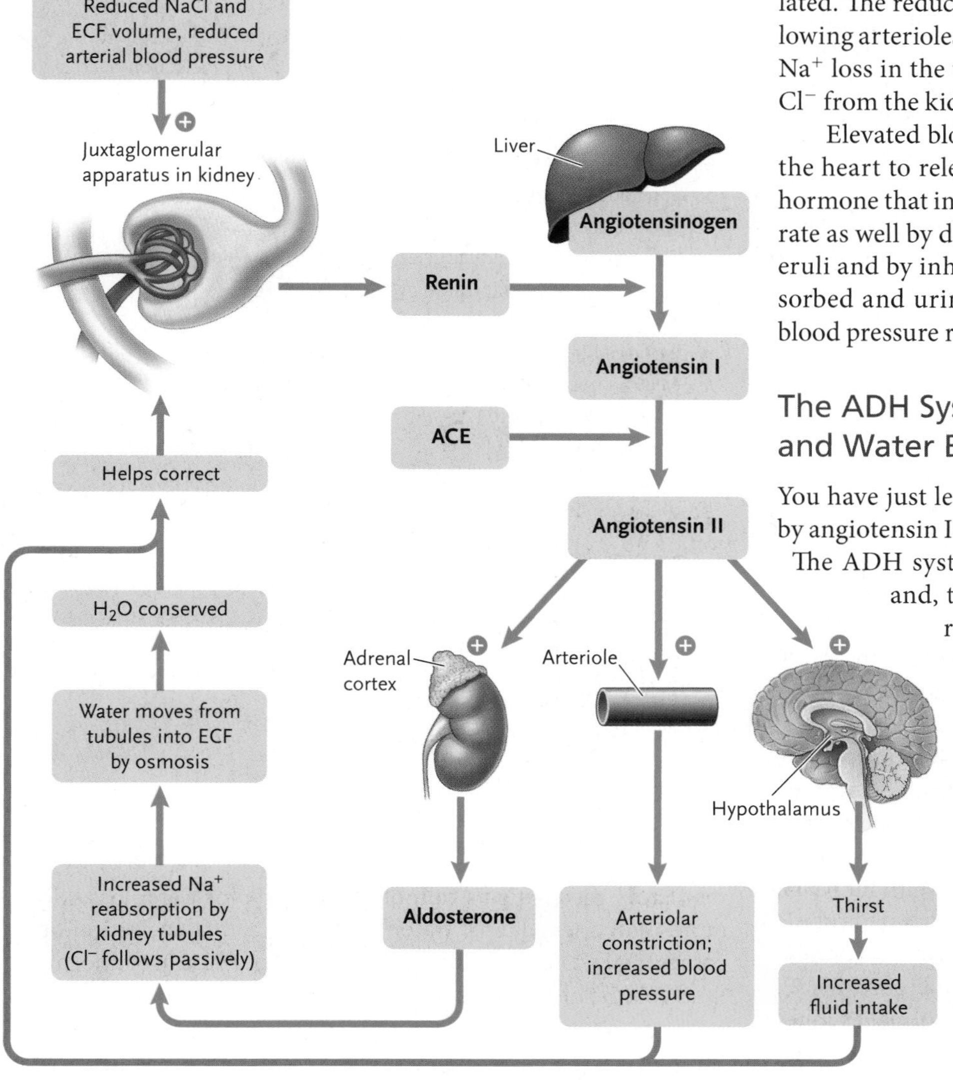

Reduced NaCl and ECF volume, reduced arterial blood pressure
Juxtaglomerular apparatus in kidney
Liver
Angiotensinogen
Renin
Angiotensin I
ACE
Angiotensin II
Helps correct
H₂O conserved
Adrenal cortex
Arteriole
Water moves from tubules into ECF by osmosis
Hypothalamus
Increased Na⁺ reabsorption by kidney tubules (Cl⁻ follows passively)
Aldosterone
Arteriolar constriction; increased blood pressure
Thirst
Increased fluid intake
KEY
ACE = angiotensin-converting enzyme
ECF = extracellular fluids (plasma and interstitial fluid)

FIGURE 46.10
The RAAS regulatory mechanism, which compensates for a fall in the salt concentration of the extracellular fluids and the reduced fluid volume and blood pressure that result from the lowered salt concentration.

Angiotensin II has three effects: (1) It quickly raises blood pressure by constricting arterioles in most parts of the body; (2) it stimulates synthesis of the steroid hormone **aldosterone** and its secretion from the adrenal cortex; and (3) it stimulates thirst so that more water is brought into the body. The aldosterone increases Na^+ reabsorption in the kidneys, which raises the osmolarity of body fluids. As a result, water moves from the tubules into the interstitial fluid, which conserves water. Angiotensin II may also stimulate secretion of **antidiuretic hormone (ADH)** (also called *vasopressin*) by the posterior pituitary (antidiuretic means "against urine output") (not shown in the figure). ADH increases water absorption in the kidneys. Overall, the combined effects of angiotensin act to return the blood pressure to normal.

In the opposite situation, when salt intake is too high, both body fluid volume and blood pressure rise above normal. Under these conditions, renin secretion is inhibited and, as a result, angiotensin production and aldosterone synthesis are not stimulated. The reduction in angiotensin lowers blood pressure by allowing arterioles to dilate; the reduction in aldosterone increases Na^+ loss in the urine by retarding the reabsorption of Na^+ and Cl^- from the kidney tubules.

Elevated blood pressure also stimulates specialized cells in the heart to release **atrial natriuretic factor (ANF),** a peptide hormone that inhibits renin release. ANF increases the filtration rate as well by dilating the arterioles that deliver blood to glomeruli and by inhibiting aldosterone release. As less Na^+ is reabsorbed and urine volume increases, both plasma volume and blood pressure return to normal.

The ADH System Also Regulates Osmolarity and Water Balance

You have just learned that the ADH system may be stimulated by angiotensin II of the RAAS in response to an increase in Na^+. The ADH system regulates osmolarity and water balance—and, therefore, urinary output—by increasing water reabsorption in the kidneys without changing the usual excretion of salt **(Figure 46.11).** Independently of being stimulated by angiotensin, ADH-secreting neurons in the hypothalamus act as **osmoreceptors.** When the osmolarity of the interstitial fluid bathing them increases, these neurons secrete more ADH from their endings in the posterior pituitary. In addition, they affect nearby neural circuits in the hypothalamus to increase thirst. The resulting increase in water ingestion helps compensate for water loss.

The action of ADH increases water reabsorption in the distal convoluted tubule and collecting ducts by promoting the insertion of more aquaporins into the epithelial membranes **(Figure 46.12).** As a result, urinary output is reduced and water is conserved. By stimulating thirst and water reabsorption in the kidneys,

FIGURE 46.11
The ADH regulatory system, which stimulates water reabsorption to compensate for a loss in the fluid volume of the extracellular fluids because of excessive water loss from the body.

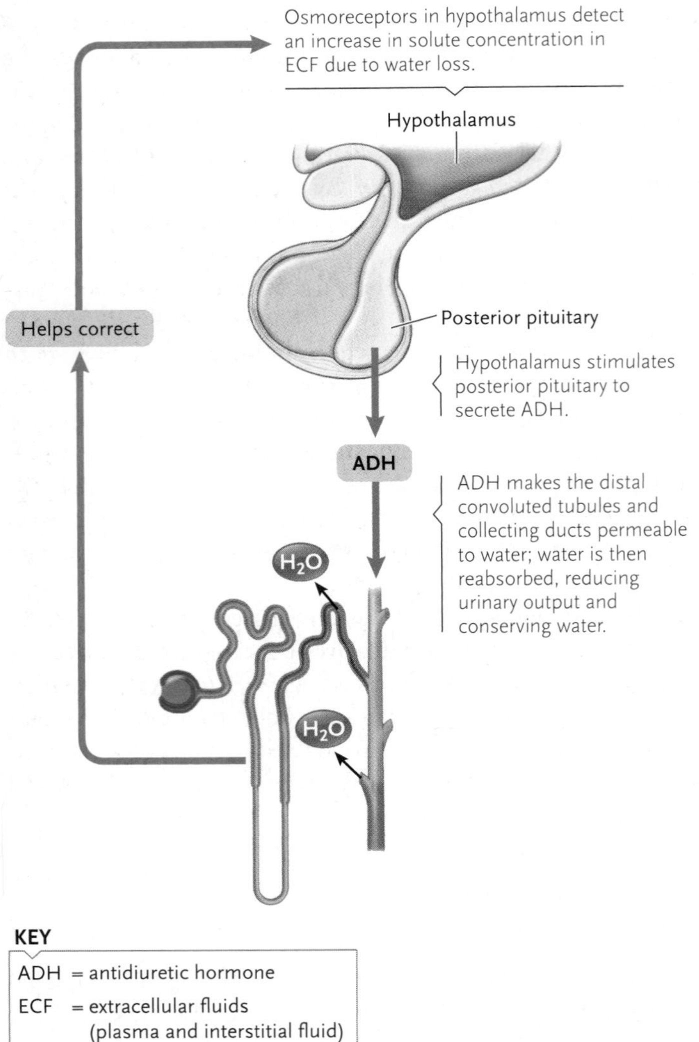

Osmoreceptors in hypothalamus detect an increase in solute concentration in ECF due to water loss.
Hypothalamus
Posterior pituitary
Helps correct
Hypothalamus stimulates posterior pituitary to secrete ADH.
ADH
ADH makes the distal convoluted tubules and collecting ducts permeable to water; water is then reabsorbed, reducing urinary output and conserving water.
H_2O
H_2O

KEY

ADH = antidiuretic hormone
ECF = extracellular fluids (plasma and interstitial fluid)

the body's depleted stock of water is restored. The newly added water dilutes the solutes in the body fluids to normal concentrations.

In the opposite condition, when there is a water excess in extracellular fluids, the osmolarity of those fluids drops below normal levels. Here, there is no stimulation of the osmoreceptors in the hypothalamus. Consequently, there is no sensation of thirst, and no ADH release from the posterior pituitary. Without ADH, water channels are retrieved from the distal convoluted tubule and collecting duct epithelial membranes (endocytosis of plasma membrane to produce intracellular vesicles; see Figure 46.12), and water recovery is significantly reduced. The animal excretes large volumes of dilute urine until the osmolarity of the extracellular fluids returns to normal. Alcohol also causes frequent urination by inhibiting ADH release.

Although the RAAS and ADH systems interact to regulate the body's water balance over a wide range of conditions, their regulatory mechanisms cannot compensate for water losses for more than a few days if water is unavailable. Dehydration becomes fatal when water loss amounts to about 12% of the normal fluid volume of the body.

FIGURE 46.12 **Experimental Research**

ADH-Stimulated Water Reabsorption in the Kidney Collecting Duct

Question: How does ADH cause water reabsorption in the kidney collecting ducts?

Experiment: Action of the peptide hormone ADH causes the reabsorption of water in the kidney by increasing the number of aquaporin2 (AQP2) water channels in the plasma membrane of the collecting duct cells. AQP2 is the only aquaporin expressed exclusively in the collecting duct. Mark Knepper and his colleagues at the National Institutes of Health, Bethesda, Maryland, and the University of Aarhus, Denmark, investigated the mechanism by which ADH causes AQP2 to appear in the plasma membrane. They tested the "shuttle" hypothesis, which states that ADH induces the movement of AQP2 from intracellular vesicles (IVs) to the plasma membrane of cells on the lumen side of the collecting duct.

The researchers first attached sections of collecting duct from rat kidneys to a pipet and treated them by passing an ADH solution through the duct **(Figure 1).** They then used immunoelectron microscopy to count the AQP2 in the IVs and in the plasma membrane of the lumen cells. That is, they cut treated ducts into thin sections and added an antibody that specifically recognizes AQP2. The antibody has gold particles attached, and when the antibody binds to AQP2, the gold particles bind too. Under the electron microscope, the gold is electron dense and is easily visualized. Counting the gold particles allows quantification of the AQP2.

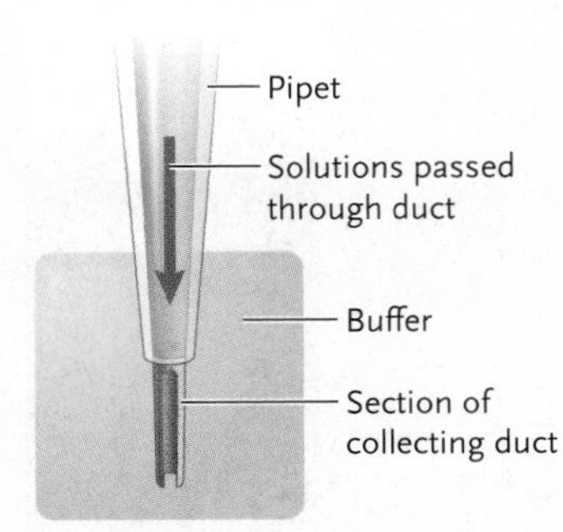

Pipet
Solutions passed through duct
Buffer
Section of collecting duct

FIGURE 1

Results: The researchers measured the distribution of AQP2 in the absence of ADH, in the presence of ADH, and then after ADH had been washed away. Their results are expressed as the ratio of AQP2 in the plasma membrane of lumen cells to AQP2 in IVs:

Before ADH treatment:	0.32 ± 0.03
In presence of ADH:	1.38 ± 0.14
After ADH removal:	0.35 ± 0.04

STUDY BREAK 46.4 <

Outline the roles of the RAAS and ADH system in regulating mammalian kidney function.

Conclusion: The results showed that ADH induces a change in distribution of AQP2, consistent with the shuttle hypothesis of AQP2 channels moving from IVs to the plasma membrane of lumen cells. The redistribution of AQP2 channels is reversible, as shown by the results after ADH removal.

Figure 2 shows the current molecular model for the ADH-induced redistribution of AQP2 channels in collecting duct epithelial cells.

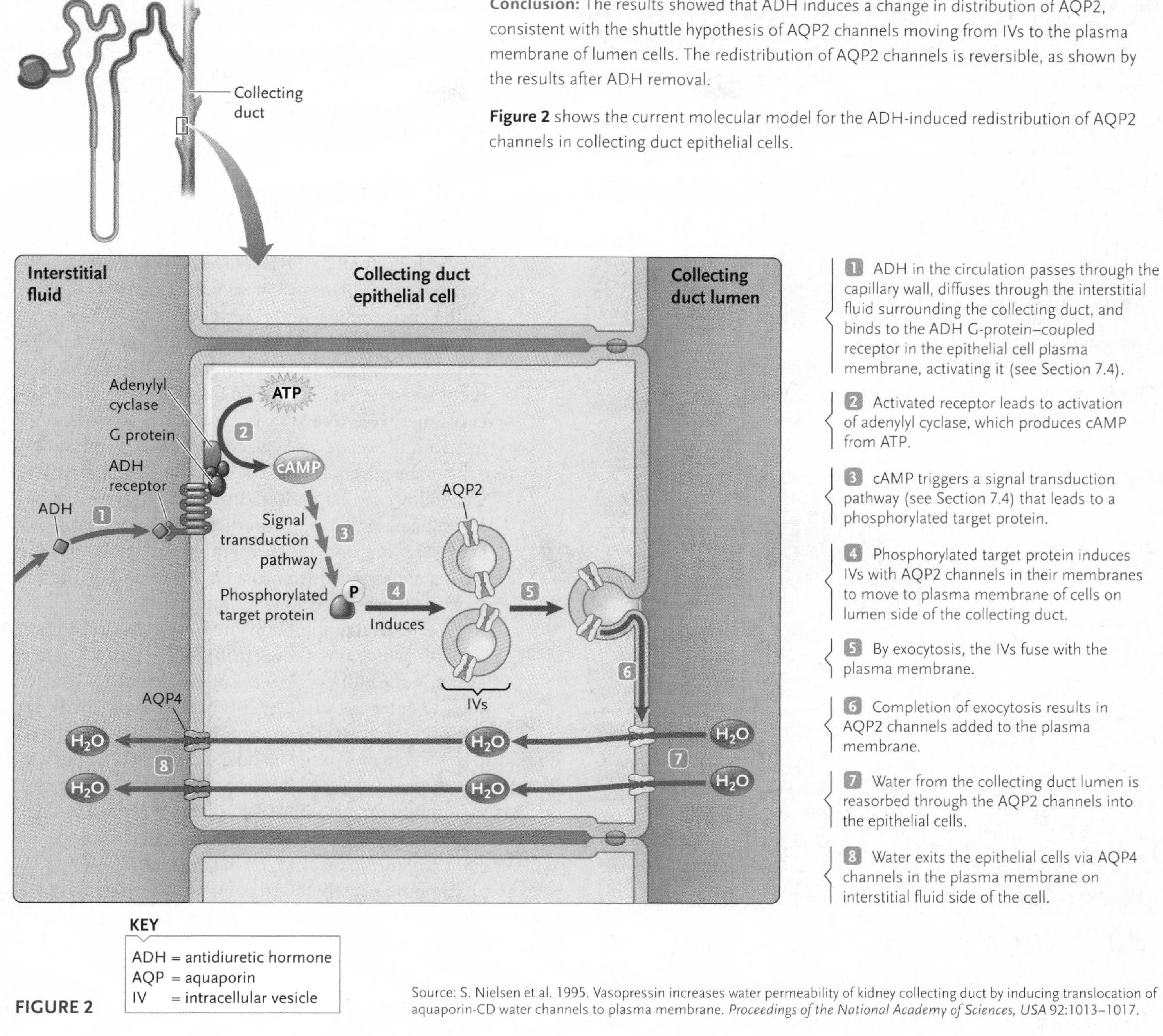

Collecting duct
Interstitial fluid
Collecting duct epithelial cell
Collecting duct lumen
Adenylyl cyclase
G protein
ADH receptor
ADH
ATP
cAMP
Signal transduction pathway
Phosphorylated target protein
P
Induces
AQP2
IVs
AQP4
H_2O
1 ADH in the circulation passes through the capillary wall, diffuses through the interstitial fluid surrounding the collecting duct, and binds to the ADH G-protein–coupled receptor in the epithelial cell plasma membrane, activating it (see Section 7.4).
2 Activated receptor leads to activation of adenylyl cyclase, which produces cAMP from ATP.
3 cAMP triggers a signal transduction pathway (see Section 7.4) that leads to a phosphorylated target protein.
4 Phosphorylated target protein induces IVs with AQP2 channels in their membranes to move to plasma membrane of cells on lumen side of the collecting duct.
5 By exocytosis, the IVs fuse with the plasma membrane.
6 Completion of exocytosis results in AQP2 channels added to the plasma membrane.
7 Water from the collecting duct lumen is reasorbed through the AQP2 channels into the epithelial cells.
8 Water exits the epithelial cells via AQP4 channels in the plasma membrane on interstitial fluid side of the cell.
KEY
ADH = antidiuretic hormone
AQP = aquaporin
IV = intracellular vesicle

FIGURE 2

Source: S. Nielsen et al. 1995. Vasopressin increases water permeability of kidney collecting duct by inducing translocation of aquaporin-CD water channels to plasma membrane. *Proceedings of the National Academy of Sciences, USA* 92:1013–1017.

46.5 Kidney Function in Nonmammalian Vertebrates

Most nonmammalian vertebrates secrete hypoosmotic urine; only a few species of aquatic birds produce urine that is hyperosmotic to body fluids. The particular adaptations that maintain osmolarity and water balance among these animals vary depending on whether retention of water or salts is the major issue.

Marine Fishes Conserve Water and Excrete Salts

Marine teleosts live in seawater, which is strongly hyperosmotic to their body fluids. As a result, they continually lose water to their environment by osmosis and must replace it by continual drinking. The kidneys of marine teleosts play little role in regulating salt in their body fluids because they cannot produce hyperosmotic urine that would both remove salt and conserve water. In-

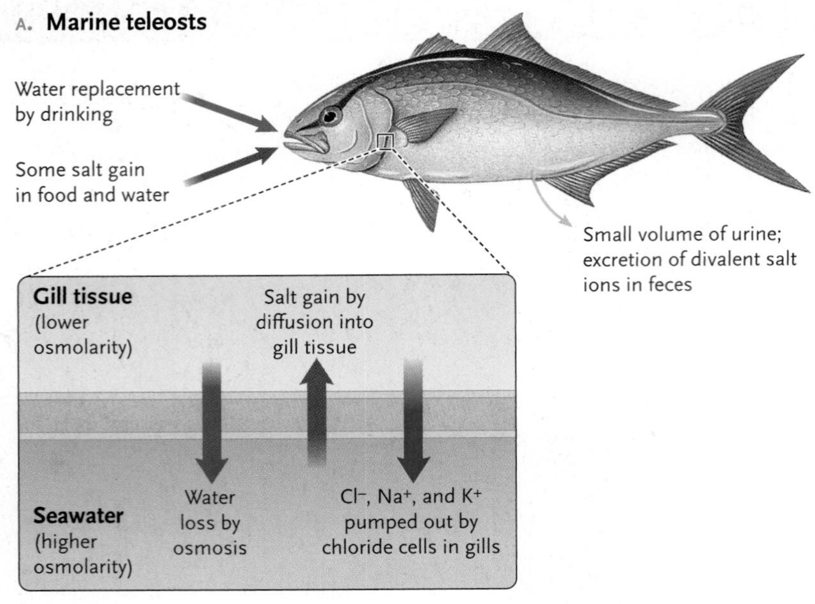
A. Marine teleosts
Water replacement by drinking
Some salt gain in food and water
Small volume of urine; excretion of divalent salt ions in feces
Gill tissue (lower osmolarity)
Salt gain by diffusion into gill tissue
Seawater (higher osmolarity)
Water loss by osmosis
Cl−, Na+, and K+ pumped out by chloride cells in gills

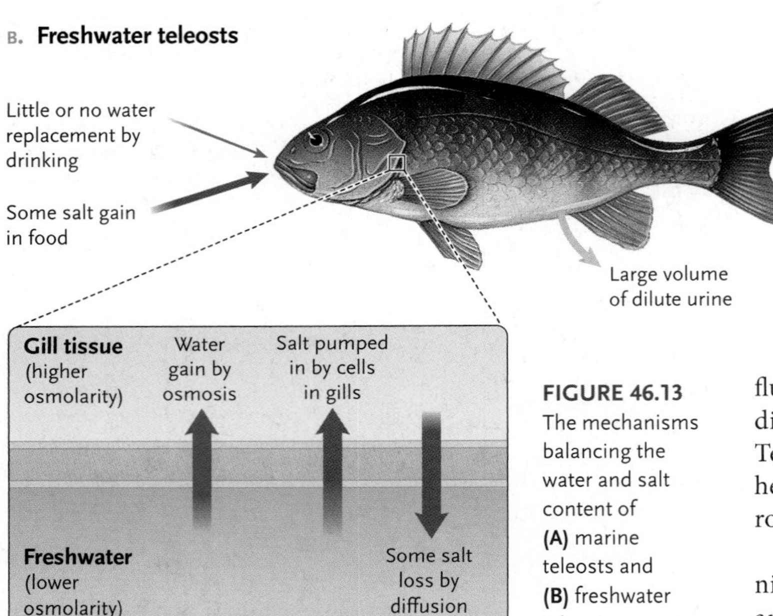
B. Freshwater teleosts
Little or no water replacement by drinking
Some salt gain in food
Large volume of dilute urine
Gill tissue (higher osmolarity)
Water gain by osmosis
Salt pumped in by cells in gills
Freshwater (lower osmolarity)
Some salt loss by diffusion

FIGURE 46.13 The mechanisms balancing the water and salt content of **(A)** marine teleosts and **(B)** freshwater teleosts.

stead, excess Na^+, K^+, and Cl^- ions are eliminated from the body by specialized cells in the gills, called *chloride cells,* which actively transport Cl^- into the surrounding seawater; the Na^+ and K^+ ions are also actively transported to maintain electrical neutrality **(Figure 46.13A).** Certain other ions in the ingested seawater, such as Ca^{2+} and Mg^{2+}, are removed by the kidneys in an isoosmotic urine. On balance, a marine teleost is able to retain most of the water it drinks and eliminate most of the salt, allowing its body fluids to remain hypoosmotic to the surrounding water with no need to secrete hyperosmotic urine. Nitrogenous wastes are released from the gills, primarily as ammonia, by simple diffusion. The kidneys play little role in nitrogenous-waste removal.

Sharks and rays have a different adaptation to seawater—the osmolarity of their body fluids is maintained close to that of seawater by retaining high levels of urea in body fluids, along with another nitrogenous waste, *trimethylamine oxide (TMAO).* The match in osmolarity keeps sharks and rays from losing water to the surrounding sea by osmosis, and they do not have to drink seawater continually to maintain their water balance. Excess salts ingested with food are excreted in the kidney and by specialized secretory cells in a *rectal salt gland* located near the anal opening.

Freshwater Fishes and Amphibians Excrete Water and Conserve Salts

The body fluids of freshwater fishes and aquatic amphibians (no amphibians live in seawater) are hyperosmotic to the surrounding water, which usually ranges from about 1 to 10 mOsm/L. Water therefore moves osmotically into their tissues. Such animals rarely drink, and they excrete large volumes of dilute urine to get rid of excess water **(Figure 46.13B).** In freshwater fishes, salt ions lost with the urine are replaced by salt in foods and by active transport of Na^+ and K^+ into the body by the gills; Cl^- follows to maintain electrical neutrality. Aquatic amphibians obtain salt in the diet and by active transport across the skin from the surrounding water. Nitrogenous wastes are excreted from the gills as ammonia in both freshwater fishes and aquatic amphibians.

Terrestrial amphibians must conserve both water and salt, which is obtained primarily in foods. In these animals, the kidneys secrete salt into the urine, causing water to enter the urine by osmosis. In the bladder, the salt is reclaimed by active transport and returned to body fluids. The water remains in the bladder, making the urine very dilute; during times of drought, it is reabsorbed as a water source. Terrestrial amphibians also have behavioral adaptations that help minimize water loss, such as seeking shaded, moist environments and remaining inactive during the day.

Larval amphibians, which are completely aquatic, excrete nitrogenous wastes from their gills as ammonia. Adult amphibians excrete nitrogenous wastes through their kidneys as urea.

Reptiles and Birds Excrete Uric Acid to Conserve Water

Terrestrial reptiles (lizards and snakes) conserve water by secreting nitrogenous wastes in the form of an almost water-free paste of uric acid crystals. Further water conservation occurs as the epithelial cells of the cloaca, the common exit for the digestive and excretory systems, absorb water from feces and urine before those wastes are excreted. Most birds conserve water by the same processes—they excrete nitrogenous wastes as uric acid and absorb water from the urine and feces in the cloaca. In reptiles, the scales covering the skin allow almost no water to escape through the body surface.

Reptiles and birds that live in or around seawater, including reptiles such as crocodilians, sea snakes, and sea turtles and birds such as seagulls, penguins, and pelicans, take in large quantities of salt with their food and rarely or never drink fresh water. These

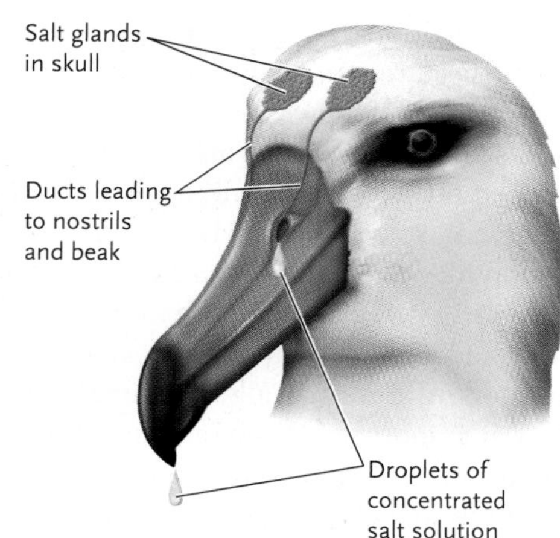
Salt glands in skull
Ducts leading to nostrils and beak
Droplets of concentrated salt solution

FIGURE 46.14 Salt glands in a bird living on a seacoast.

animals typically excrete excess salt through specialized *salt glands* located in the head **(Figure 46.14)**, which remove salts from the blood by active transport. The salts are secreted to the environment as a water solution in which salts are two to three times more concentrated than in body fluids. The secretion exits through the nostrils of birds and lizards, through the mouth of marine snakes, and as salty tears from the eye sockets of sea turtles and crocodilians. Neural and hormonal controls, essentially the same as those regulating osmolarity in mammals, control the rate of fluid secretion and its salt concentration.

In sum, the adaptations described in this section permit excretion of toxic wastes and allow animals to maintain the salt concentration of body fluids at levels that keep cells from swelling or shrinking.

STUDY BREAK 46.5 <

1. **How do marine and freshwater teleosts differ in water, salt, and nitrogenous-waste regulation?**
2. **Reptiles and birds excrete nitrogenous wastes in the form of uric acid. Is there an advantage to doing this as opposed to the mammalian process of excreting nitrogenous wastes as urea?**

46.6 Introduction to Thermoregulation

Another vital challenge for animals is maintaining their internal environment at temperatures that can be tolerated by body cells. Environmental temperatures vary enormously across Earth's surface. However, animal cells can survive only within a temperature range from about 0°C to 45°C (32°F to 113°F). Not far below 0°C, the lipid bilayer of a biological membrane changes from a fluid to a frozen gel, which disrupts vital cell functions, and ice crystals destroy the cell's organelles. At the other extreme, as temperatures approach 45°C, the kinetic motions of molecules become so great that most proteins and nucleic acids unfold from their functional form. Either condition leads quickly to cell death. Animals therefore usually maintain internal body temperatures somewhere within the 0°C to 45°C limits.

Temperature regulation—thermoregulation—is based on negative feedback pathways in which temperature receptors *(thermoreceptors)* detect changes from a temperature *set point.* Signals from the receptors trigger physiological and behavioral responses that return the temperature to the set point (thermoreceptors are discussed in Section 39.6; negative feedback mechanisms and set points are discussed in Section 36.4). All of the responses triggered by negative feedback mechanisms involve adjustments in the rate of heat-generating oxidative reactions within the body, coupled with adjustments in the rate of heat gain or loss at the body surface. The particular adaptations accomplishing these responses vary widely among species, however. And, although body temperature is closely regulated around a set point in all endotherms, the set point itself may vary over the course of a day and between seasons.

Thermoregulation Allows Animals to Reach Optimal Physiological Performance

Within the 0°C to 45°C range of tolerable temperatures, an animal's *organismal performance*—the rate and efficiency of its biochemical, physiological, and whole-body processes—varies greatly. For example, the speed at which the Middle Eastern lizard *Agama stellio* can sprint is low when the animal's body temperature is cold, rises smoothly with body temperature until it levels to a fairly broad plateau, and then drops off dramatically with further increases in body temperature **(Figure 46.15)**. The temperature range that provides good organismal performance varies from one species to another, however.

Animals Exchange Heat with Their Environments by Conduction, Convection, Radiation, and Evaporation

As part of thermoregulation, animals exchange heat with their environments. As with all physical objects, heat flows into animals if they are cooler than their surroundings and flows outward if they are warmer. This heat exchange occurs by four mechanisms: *conduction, convection, radiation,* and *evaporation* **(Figure 46.16)**.

Conduction is the flow of heat between atoms or molecules in direct contact. An animal loses heat by conduction when it contacts a cooler object and gains heat when it contacts an object that is warmer. **Convection** is the transfer of heat from a body to a fluid, such as air or water, that passes over its surface. The movement maximizes heat transfer by replacing fluid that has ab-

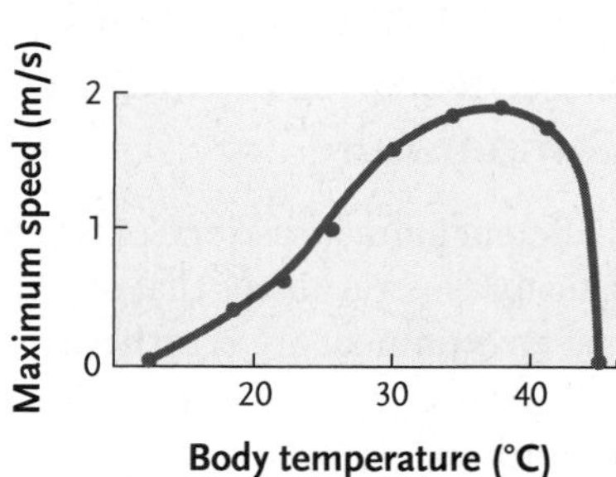
Maximum speed (m/s)
0
1
2
20
30
40
Body temperature (°C)

FIGURE 46.15 Body temperature and organismal performance. The maximum sprint speed of a lizard *(Agama stellio)* changes dramatically with body temperature.

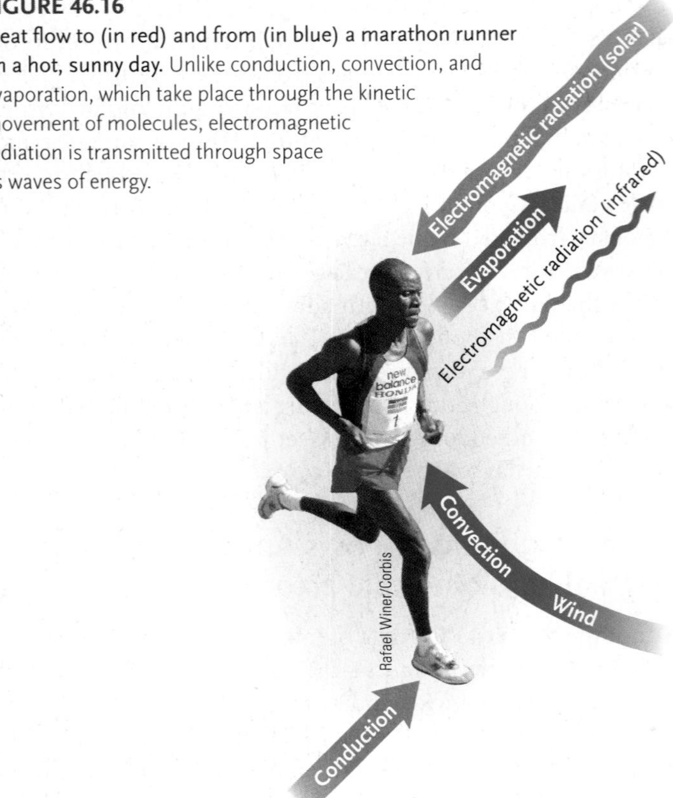
Electromagnetic radiation (solar)
Evaporation
Electromagnetic radiation (infrared)
Convection
Wind
Conduction
Rafael Winer/Corbis

FIGURE 46.16
Heat flow to (in red) and from (in blue) a marathon runner on a hot, sunny day. Unlike conduction, convection, and evaporation, which take place through the kinetic movement of molecules, electromagnetic radiation is transmitted through space as waves of energy.

sorbed or released heat with fluid at the original temperature. **Radiation** is the transfer of heat energy as electromagnetic radiation. Any object warmer than absolute zero (−273°C) radiates heat; as the object's temperature rises, the amount of heat it loses as radiation increases as well. Animals also gain heat through radiation, particularly by absorbing radiation from the sun. **Evaporation** is heat transfer through the energy required to change a liquid to a gas. Evaporation of water from a surface is an efficient way to transfer heat; when the water in sweat evaporates from the body surface, the body cools down because heat is being transferred to the evaporated water in the surrounding air.

All animals gain or lose heat by a combination of these four mechanisms. A marathon runner struggling with the heat on a sunny summer day, for example, loses heat by the evaporation of sweat from the skin, by convection as air flows over the skin, and by outward infrared radiation. The runner gains heat from internal biochemical reactions (especially oxidations), by absorbing infrared and solar radiation, and by conduction as the feet contact the hot ground. To maintain a constant body temperature, the heat gained and lost through these pathways must balance.

Ectothermic and Endothermic Animals Rely on Different Heat Sources to Maintain Body Temperature

Different animals use one of two major strategies to balance heat gain and loss. Animals that obtain heat primarily from the external environment are **ectotherms** (*ecto* = outside); those obtaining most of their heat from internal physiological sources are **endotherms** (*endo* = inside). All ectotherms generate at least some heat from internal reactions, however, and endotherms can obtain heat from the environment under some circumstances.

Virtually all invertebrates, fishes, amphibians, and lizards and snakes are ectotherms. These animals are popularly described as cold-blooded, although the body temperature of some, such as an active lizard, may be as high as or higher than ours on a sunny day. Ectotherms regulate body temperature by controlling the rate of heat exchange with the environment. Through behavioral and physiological mechanisms, they adjust body temperature toward a level that allows optimal physiological performance. However, most ectotherms are unable to maintain optimal body temperature when the temperature of their surroundings departs too far from that optimum, particularly when environmental temperatures fall. As a result, the body temperatures of ectotherms fluctuate with environmental temperatures, and they typically are less active when it is cold. Nevertheless, ectotherms are highly successful, particularly in warm environments.

The endotherms—birds, mammals, some fishes, sea turtles, and some invertebrates—keep their bodies at an optimal temperature by regulating two processes: (1) the amount of heat generated by internal oxidative reactions; and (2) the amount of heat exchanged with the environment. Because endotherms use internal heat sources to maintain body temperature at optimal levels, they can remain active over a broader range of environmental temperatures than ectotherms, and thus inhabit a wider range of habitats. However, endotherms require a nearly constant supply of energy to maintain their body temperatures. Because that energy is provided by food, endotherms typically consume much more food than ectotherms of equivalent size.

The difference between ectotherms and endotherms is reflected in their metabolic responses to environmental temperature **(Figure 46.17)**. For example, the metabolic rate of a resting mouse *increases* steadily as the environmental temperature falls from 25°C to 10°C (77°F to 50°F). This increase reflects the fact that to maintain a constant body temperature in a colder environment, endotherms must process progressively more food and generate more heat to compensate for their increased rate of heat loss. In this respect, an endotherm can be likened to a house in winter. To maintain a constant internal temperature, the homeowner must burn more oil or gas on a cold day than on a warm day.

By contrast, the metabolic rate of a resting lizard typically *decreases* steadily over the same temperature range. Because ectotherms do not maintain a constant body temperature, their biochemical and physiological functions, including oxidative reactions, slow down as environmental and body temperatures decrease. Thus, an ectotherm consumes and uses less energy when it is cold than when it is warm. This difference between ectotherms and endotherms is so fundamental that even samples of living tissue extracted from an ectotherm consume energy more slowly than equivalent samples from an endotherm.

Ectothermy and endothermy represent different strategies for coping with the variations in environmental temperature that all animals encounter; neither strategy is inherently superior to the other. Endotherms can remain fully active over a wide temperature range. Cold weather does not prevent them from foraging, mating, or escaping from predators, but it does increase

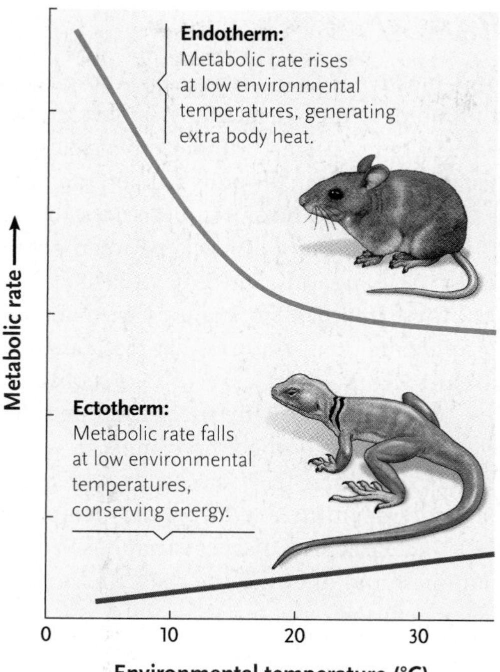
Endotherm:
Metabolic rate rises at low environmental temperatures, generating extra body heat.
Metabolic rate
Ectotherm:
Metabolic rate falls at low environmental temperatures, conserving energy.
0
10
20
30
Environmental temperature (°C)

FIGURE 46.17
Metabolic responses of ectotherms and endotherms to cooling environmental temperatures. At any temperature, the metabolic rates of endotherms are always higher than those of ectotherms of comparable size.

their energy and food needs—and, to satisfy their need for food, they may not have the option of staying curled up safely in a warm burrow. Ectotherms do not have the capacity to be active when environmental temperatures drop too low; they move sluggishly and are unable to capture food or escape from predators. However, because their metabolic rates are lower under such circumstances, so are their food needs, and they do not have to look for food and expose themselves to danger to the extent that endotherms do.

In the next section, we begin a more detailed examination of how animals actually regulate their body temperatures within these overall strategies.

STUDY BREAK 46.6

Distinguish between ectothermy and endothermy. Give one advantage and one disadvantage for each form of thermoregulation.

46.7 Ectothermy

Ectotherms vary widely in their ability to regulate internal body temperatures. For example, most aquatic invertebrates have such limited ability to thermoregulate that their body temperatures closely match those of the surrounding environment. These species live in or seek warm or temperate environments, where temperatures fall within a range that produces optimal physiological performance. Ectotherms with a greater ability to thermoregulate may occupy more varied habitats.

Ectotherms Are Found in All Invertebrate Groups

Most aquatic invertebrates are limited thermoregulators whose body temperature closely follows the temperature of their surroundings. However, even among these animals, some use behavioral responses to regulate body temperature. For example, a South American intertidal mollusk, *Echinolittorina peruviana,* is longer than it is wide. Researchers have shown that this animal orients itself as a means of thermoregulation. On sunny, summer days, it faces the sun, offering a smaller surface area for the sun's rays. On overcast summer days, or during the winter, it orients itself with a lateral side—which has the larger surface area—toward the sun's rays.

Invertebrates living in terrestrial habitats regulate body temperatures more closely. Many also use behavioral responses, such as moving between shaded and sunny regions, to regulate body temperature. Some winged arthropods, including bees, moths, butterflies, and dragonflies, use a combination of behavioral and heat-generating physiological mechanisms for thermoregulation. For example, in cool weather, these animals warm up before taking flight by rapidly vibrating the large flight muscles in the thorax, in a mechanism similar to shivering in humans. The tobacco hawk moth *(Manduca sexta)* vibrates its flight muscles until its thoracic temperature reaches about 36°C before flying. During flight, metabolic heat generated by the flight muscles sustains the elevated thoracic temperature, so much so that this moth produces more heat per gram of body weight than many mammals.

Most Fishes, Amphibians, and Reptiles Are Ectotherms

Vertebrate ectotherms—fishes, amphibians, and reptiles (turtles, lizards, and snakes)—also vary widely in their ability to thermoregulate. Most aquatic species have a more limited thermoregulatory capacity than that found among terrestrial species, particularly the reptiles. Some fishes, however, are highly capable thermoregulators.

FISHES The body temperatures of most fishes remain within one or two degrees of their aquatic environment. However, many fishes use behavioral mechanisms to keep body temperatures at levels allowing good physiological performance. Freshwater species, for example, may use opportunities provided by the thermal stratification of lakes and ponds (see Figure 49.24). During hot summer days they remain in deep, cool water, moving to the shallows to feed only during early morning and late evening when air and water temperatures fall.

AMPHIBIANS AND REPTILES The body temperatures of most amphibians also closely match environmental temperatures. Some terrestrial amphibians bask in the sun to raise their body temperature and seek shade to lower body temperature. However, basking can be dangerous to amphibians because they lose water rapidly through their permeable skin.

Thermoregulation is more pronounced among terrestrial reptiles. Some lizard species can maintain temperatures that are nearly as constant as those of endotherms. For small lizards, the most common behavioral thermoregulatory mechanism is shuttling between sunny (warmer) and shady (cooler) regions; in the deserts, lizards and other reptiles retreat into burrows during the hottest part of summer days. Some, such as the desert iguana *(Dipsosaurus dorsalis),* lose excess heat by *panting*—rapidly moving air in and out of the airways. The air movement increases heat loss by convection and by evaporation of water from the respiratory tract.

Lizards also frequently adjust their posture to foster heat exchange with the environment and control the angle of their body relative to the rays of the sun. Snakes and lizards can often be found on large rocks and on roads on chilly nights, taking advantage of the heat retained by the stone or concrete.

Ectotherms Can Compensate for Seasonal Variations in Environmental Temperature

Many ectotherms undergo physiological changes, called **thermal acclimatization,** in response to seasonal shifts in environmental temperature. These changes allow the animals to attain good physiological performance at both winter and summer temperatures.

For example, in the summer a bullhead catfish (*Ameiurus* species) can survive water temperatures as high as 36°C (97°F), but it cannot tolerate temperatures below 8°C (46°F). In the winter, however, the bullhead cannot survive water temperatures above 28°C (82°F), but can tolerate temperatures near 0°C (32°F).

Another acclimatizing change involves the phospholipids of biological membranes (see *Focus on Basic Research* in Chapter 6). For example, membrane phospholipids have higher proportions of double bonds in carp living in colder environments than in carp living in warmer environments. The higher proportion of double bonds makes it harder for the membrane to freeze. A higher proportion of cholesterol also protects membranes from freezing.

When seasonal temperatures fall below 0°C, some ectotherms add molecules to their body fluids that act as antifreeze molecules to depress their freezing point and retard ice crystal formation. For example, glycerol added to the cellular and extracellular fluids of a parasitic wasp (*Bracon* species) keeps the insect from freezing at temperatures as low as −45°C (−49°F). Similarly, antifreeze proteins allow fishes such as the winter flounder to remain active in seawater as cold as −1.8°C (29°F) (see *Insights from the Molecular Revolution* in Chapter 49).

Ectotherms thus control body temperature primarily by regulating heat exchange with the environment; internal-heat generating mechanisms contribute to the control mechanisms in some species, but are never the primary source of body heat. The opposite conditions occur among endotherms: although these animals also regulate heat exchange with the environment, their primary sources of body heat are internal.

STUDY BREAK 46.7

1. **Describe two mechanisms an ectothermic animal can use to regulate its temperature.**
2. **What is thermal acclimatization?**

46.8 Endothermy

Endotherms—mostly birds and mammals—have elaborate and extensive thermoregulatory adaptations. Highly specialized features of body structure interact with both physiological and behavioral mechanisms to keep the body temperature within a narrow range. Typically, the body temperatures of fully active individuals are held constant at levels between about 39°C to 42°C (102°F to 108°F) in birds, and 36°C to 39°C (97°F to 102°F) in mammals. These internal temperatures are maintained in the face of environmental temperatures that may range over much greater extremes, from as low as −42°C to as high as +48°C (−45°F to +120°F). Some highly specialized endotherms can even survive temperatures beyond these limits.

We begin by describing the basic feedback mechanisms that maintain body temperature, with primary emphasis on the human system. Later sections discuss variations in the responses of other mammals and of birds, and daily and seasonal variations in the temperature set point.

Information from Thermoreceptors Located in the Skin and Internal Structures Is Integrated in the Hypothalamus

Thermoreceptors (discussed in Section 39.6) are found in various locations in the human body, including the **integument** (skin; introduced in Chapter 36), spinal cord, and hypothalamus. Current evidence suggests the existence of five thermoreceptor types, each active in specific temperature ranges, beginning at less than 8°C and extending to above 52°C. Signals from the thermoreceptors are integrated in the hypothalamus and other regions of the brain to bring about compensating physiological and behavioral responses **(Figure 46.18).** The responses keep body temperature close to the set point, which varies normally in humans between 35.5°C and 37.7°C (96.0°F to 99.9°F) for the head and trunk. The appendages typically vary more widely in temperature; in freezing weather, for example, our arms, hands, legs, and feet are typically cooler in temperature than the body core—and the ears and nose especially so.

The hypothalamus was identified as a major thermoreceptor and response integrator in mammals by experiments in which various regions of the brain were heated or cooled with a temperature probe. Within the brain, only the hypothalamus produced thermoregulatory responses such as shivering or panting.

RESPONSES WHEN CORE TEMPERATURE FALLS BELOW THE SET POINT When thermoreceptors signal a fall in core temperature below the set point, the hypothalamus triggers compensating responses by sending signals through the autonomic nervous system. Among the immediate responses is constriction of the arterioles in the skin (vasoconstriction), which reduces the flow of blood to capillary networks in the skin. The reduced flow cuts down the amount of heat delivered to the skin and lost from the body surface. The reduction in flow is most pronounced in the skin covering the extremities, where

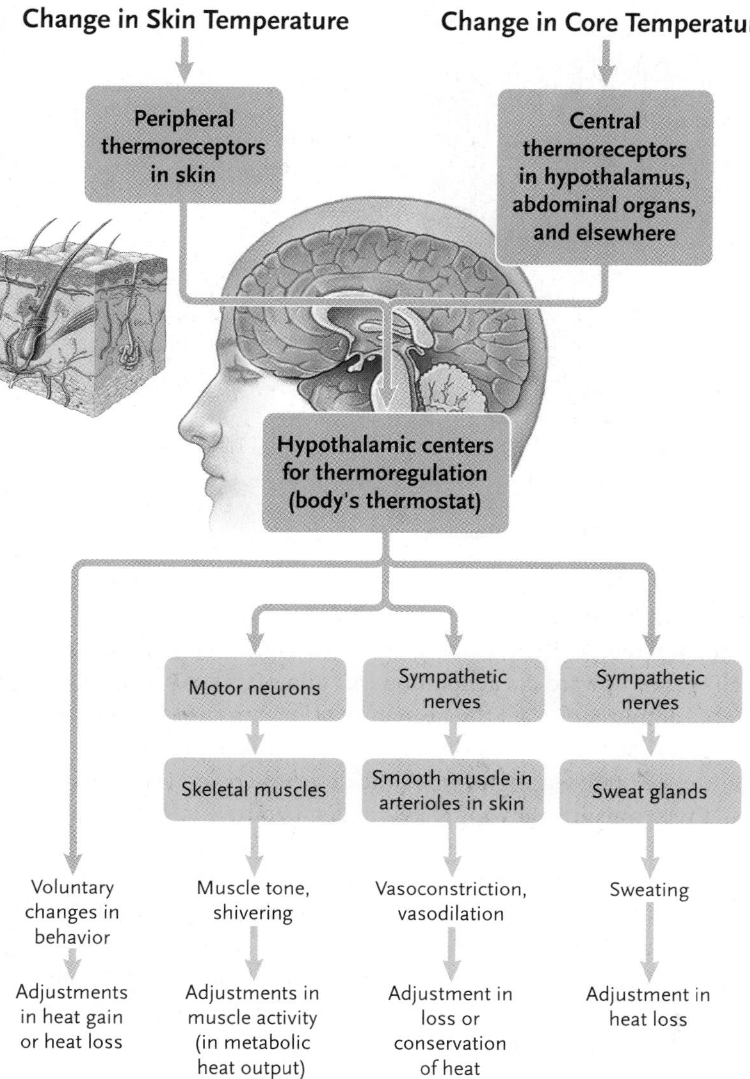
Change in Skin Temperature
Change in Core Temperature
Peripheral thermoreceptors in skin
Central thermoreceptors in hypothalamus, abdominal organs, and elsewhere
Hypothalamic centers for thermoregulation (body's thermostat)
Motor neurons
Sympathetic nerves
Sympathetic nerves
Skeletal muscles
Smooth muscle in arterioles in skin
Sweat glands
Voluntary changes in behavior
Muscle tone, shivering
Vasoconstriction, vasodilation
Sweating
Adjustments in heat gain or heat loss
Adjustments in muscle activity (in metabolic heat output)
Adjustment in loss or conservation of heat
Adjustment in heat loss

FIGURE 46.18
The physiological and behavioral responses of humans and other mammals to changes in skin and core temperature.

blood flow may be reduced by as much as 99% when core temperature falls.

Another immediate response is contraction of the smooth muscles erecting the hair shafts in mammals and feather shafts in birds, which traps air in pockets over the skin, reducing convective heat loss. The response is minimally effective in humans because hair is sparse on most parts of the body—it produces the goose bumps you experience when the weather gets chilly. However, in mammals with fur coats or in birds, erection of the hair or feather shafts significantly increases the thickness of the insulating layer that covers the skin.

Immediate behavioral responses triggered by a reduction in skin temperature also help reduce heat loss from the body. Mammals may reduce heat loss by moving to a warmer locale, curling into a ball, or huddling together.

If these immediate responses do not return body temperature to the set point, the hypothalamus triggers further responses, most notably the rhythmic tremors of skeletal muscle we know as shivering. The heat released by the muscle contractions and the oxidative reactions powering them can raise the total heat production of the body substantially. At the same time, the hypothalamus triggers secretion of epinephrine (from the adrenal medulla) and thyroid hormones (see Section 40.4). These hormones increase heat production by stimulating the oxidation of fats and other fuels. The generation of heat by oxidative mechanisms in nonmuscle tissue throughout the body is termed **nonshivering thermogenesis.**

In human newborn babies and many other mammals, the most intense heat generation by nonshivering thermogenesis takes place in a specialized **brown adipose tissue** (also called **brown fat**) that can produce heat rapidly. Heat is generated by a mechanism that uncouples electron transport from ATP production in mitochondria (see Section 8.4); the heat is transferred throughout the body by the blood. Animals that hibernate or are active in cold regions, as well as the young of many others, contain brown adipose tissue. In most mammals, brown adipose tissue is concentrated between the shoulders in the back and around the neck. In human newborn babies, this tissue accounts for about 5% of body weight. Typically the tissue shrinks as humans age, until it is absent or essentially so in most adults. However, if exposure to cold is ongoing, as is the case with male Finlanders who work outside during the year, significant amounts of the tissue remain.

If none of these responses succeeds in raising body temperature to the set point, the result is **hypothermia,** a condition in which the core temperature falls below normal for a prolonged period. In humans, a drop in core temperature of only a few degrees affects brain function and leads to confusion; continued hypothermia can lead to coma and death (see *Why It Matters* in Chapter 36).

RESPONSES WHEN CORE TEMPERATURE RISES ABOVE THE SET POINT When core temperature rises above the set point, the hypothalamus sends signals through the autonomic system that trigger responses lowering body temperature. As an immediate response, the signals relax smooth muscles of arterioles in the skin (vasodilation), increasing blood flow and with it, the heat lost from the body surface. In addition, in humans and other mammals with sweat glands, signals from the hypothalamus trigger the secretion of sweat, which absorbs heat as it evaporates from the surface of the skin.

Some endotherms, including dogs (which have sweat glands only on their feet) and many birds (which have no sweat glands), use panting as a major way to release heat. These physiological changes are reinforced by behavioral responses such as seeking shade or a cool burrow, plunging into cold water, or taking a cold drink.

When the heat gain of the body is too great to be counteracted by these responses, **hyperthermia** results. An increase of only a few degrees above normal for a prolonged period is enough to disrupt vital biochemical reactions and damage brain cells. Most adult humans become unconscious if their body temperature reaches 41°C (106°F) and die if it goes above 43°C (110°F) for more than a few minutes.

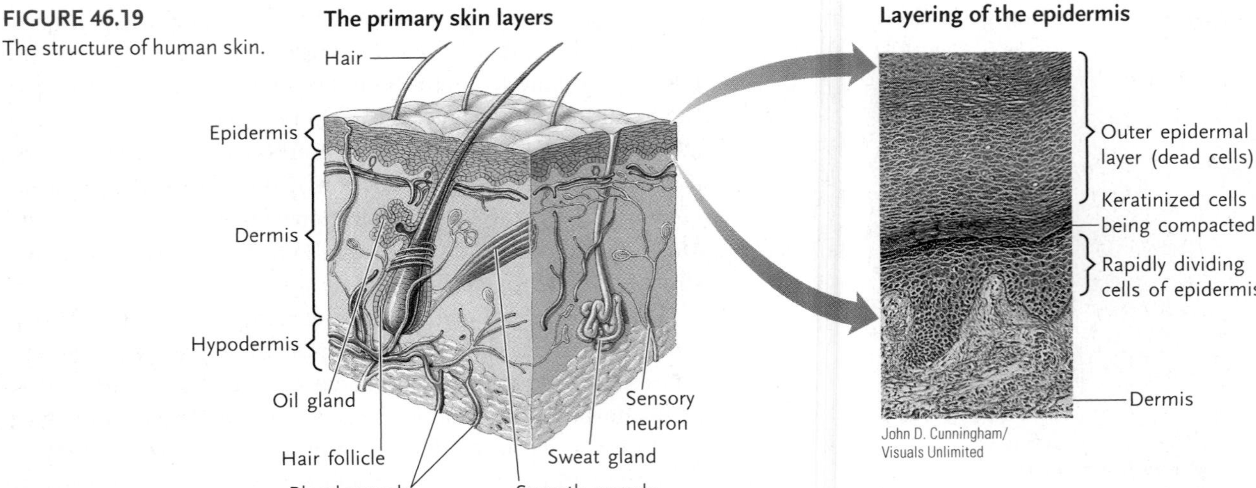
The primary skin layers
Hair
Epidermis
Dermis
Hypodermis
Oil gland
Hair follicle
Blood vessels
Sensory neuron
Sweat gland
Smooth muscle
Layering of the epidermis
Outer epidermal layer (dead cells)
Keratinized cells being compacted
Rapidly dividing cells of epidermis
Dermis
John D. Cunningham/ Visuals Unlimited

FIGURE 46.19 The structure of human skin.

The Skin Is Highly Adapted to Control Heat Transfer with the Environment

The skin of birds and mammals is an organ of heat transfer. The arterioles delivering blood to the capillary networks of the skin constrict or dilate to control blood flow and, with it, the amount of heat transferred from the body core to the surface.

The outermost living tissue of human skin, the **epidermis,** consists of cells that grow and divide rapidly **(Figure 46.19),** becoming packed with fibers of a highly insoluble protein, *keratin* (see Section 5.3). When fully formed, the epidermal cells die and become compacted into a tough, nearly impermeable layer that limits water loss primarily to evaporation of the fluids secreted by the sweat glands.

The sweat glands and hair follicles are embedded in the layer below the epidermis. Called the **dermis,** it is packed with connective tissue fibers such as collagen, which resist compression, tearing, or puncture of the skin. The dermis also contains thermoreceptors and the dense networks of arterioles, capillaries, and venules that transfer heat between the skin and the environment.

The innermost layer of the skin, the **hypodermis,** contains larger blood vessels and additional reinforcing connective tissue. The hypodermis also contains an insulating layer of fatty tissue below the dermal capillary network, which ensures that heat flows between the body core and the surface primarily through the blood. The insulating layer is thickest in mammals that live in cold environments, such as whales, seals, walruses, and polar bears, in which it is known as *blubber.*

The Set Point Varies in Daily and Seasonal Rhythms in Many Birds and Mammals

The temperature set point in many birds and mammals varies in a regular cycle during the day. In some, the daily variations are relatively small and not obviously keyed to changes in environmental temperature. In others, larger variations are correlated with daily or seasonal temperature changes.

Camels undergo a daily variation of as much as 7°C (13°F) in set point temperature. During the day, a camel's set point gradually resets upward, an adaptation that allows its body to absorb a large amount of heat. The heat absorption conserves water that would otherwise be lost by evaporation to keep the body at a lower set point. At night, when the desert is cooler, the thermostat resets again, allowing the body temperature to cool several degrees, releasing the excess heat absorbed during the day.

When the environmental temperature is cool, having a lowered temperature set point greatly reduces the energy required to maintain body temperature. In many animals, the lowered set point is accompanied by reductions in metabolic, nervous, and physical activity (including slower respiration and heartbeat), producing a sleeplike state known as **torpor.**

Many small animals and birds enter into **daily torpor,** a period of inactivity keyed to variations in daily temperature. These animals typically expend more energy per unit of body weight to keep warm than larger animals, because the ratio of body surface to volume increases as body size decreases. Hummingbirds, for example, feed actively during the daytime, when their set point is close to 40°C (104°F). During the cool of night, however, the set point drops to as low as 13°C (55°F); this allows the birds to conserve enough energy to survive overnight without feeding. Some nocturnal animals, including bats and small rodents such as deer mice, become torpid in cool locations during daylight hours when they do not actively feed. At night, their temperature set point rises and they become fully active **(Figure 46.20).**

Many animals enter a prolonged state of torpor tied to the seasons, triggered in most cases by a change in day length that signals the transition between summer and winter. The importance of day length has been shown by laboratory experiments in which animals have been induced to enter seasonal torpor by changing the period of artificial light to match the winter or summer day length.

Extended torpor during winter, called **hibernation** (*hibernus* = relating to winter), greatly reduces metabolic expenditures when food is unobtainable. Typically, hibernators must store large quantities of fats to serve as energy reserves.

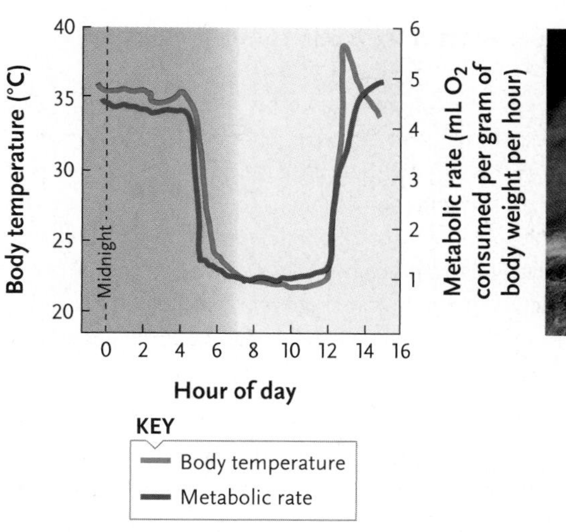

Body temperature (°C)
40
35
30
25
20
Midnight
Metabolic rate (mL O_2 consumed per gram of body weight per hour)
6
5
4
3
2
1
0 2 4 6 8 10 12 14 16
Hour of day
KEY
Body temperature
Metabolic rate

Inga Spence/Visuals Unlimited

FIGURE 46.20
Cycle of daily torpor in a deer mouse *(Peromyscus maniculatus)*.

The drop in body temperature during hibernation varies with the mammal. In some, such as hedgehogs, woodchucks, and squirrels, body temperature may fall by 20°C (36°F) or more. Body temperature even drops to near 0°C in some small hibernating mammals and, in the Arctic ground squirrel, the body supercools (goes to a below-freezing, unfrozen state) during hibernation, with body temperature dropping to about −3°C. Some ectotherms, including amphibians and reptiles living in northern latitudes and even some insects, also become torpid during winter.

Some mammals enter seasonal torpor during summer, called **estivation** (*aestivus* = relating to summer), when environmental temperatures are high and water is scarce. Some ground squirrels, for example, remain inactive in the cooler temperatures of their burrows during extreme summer heat. Many ectotherms, among them land snails, lungfishes, many toads and frogs, and some desert-living lizards, weather such climates by digging into the soil and entering a state of estivation that lasts throughout the hot dry season.

Some Animals Use a Form of Endothermy That Does Not Heat All of Their Cores

In contrast to birds and mammals, some animals show a form of endothermy that does not heat all of their cores. For example, some cold-water marine teleosts (such as tunas and mackerels) and some sharks (such as the great white) use endothermy in their aerobic swimming muscles to maintain a body core temperature as much as 10°C to 12°C warmer than their surroundings. These animals have in common the fact that they migrate over long distances, swimming continuously and, therefore, generating constant heat with the swimming muscles. That heat is insufficient to heat the entire body because too much heat is lost at the gill–water interface. These animals have evolved a *countercurrent heat exchanger* system between the swimming muscles and the gills to prevent most of the loss.

Recall from Section 44.2 that many animals with internal gills use countercurrent exchange to maximize the amounts of O_2 and CO_2 exchanged with water. A countercurrent exchange system can also be used for heat exchange. Cold blood from the gills is first routed through arteries under the skin, which is at the same temperature as the water. The blood enters the body core through small arteries that form the countercurrent heat exchanger along with small veins that bring warm blood from the swimming muscles in the core. Countercurrent exchange warms the arterial blood, which returns to the heart in veins and is then pumped around the body in arteries, including to the core and the gills. The overall result is that heat is retained within the muscles, so that the core body temperature can remain significantly higher than that of the surrounding water.

STUDY BREAK 46.8 <

Describe how thermoreceptors and negative feedback pathways achieve temperature regulation in endotherms.

UNANSWERED QUESTIONS

How can we use genetics to understand how the kidney works in humans?

As you have learned, the kidney is a complex and highly metabolically active organ. Before genetic tools were widely available, physiologists had worked out much of the basic physiology of kidney function at a macroscopic scale, but the specific molecules responsible for the functional specificity of different cell types within the kidney were not known.

A variety of disorders of kidney function can be inherited as Mendelian traits. In fact, the first human disorder clearly identified as following recessive inheritance, alkaptonuria, is characterized by a renal phenotype: black-colored urine. The disorder of renal growth known as polycystic kidney disease was recognized as a familial phenotype well before the structure of DNA was defined. Many other complex and common renal phenotypes (for example, blood pressure, and susceptibility to the development of decreased kidney filtration function) also have genetic components.

The use of molecular cloning methods in the 1980s and 1990s led to enormous growth in our understanding of the specific transport proteins used by the kidney to fine tune the extracellular environment. These discoveries were quickly followed by the elucidation of a large number of rare inherited disorders caused by mutations in the genes encoding many of these proteins. Careful observation of these human phenotypes, coupled with a biochemical understanding of the effects of these mutations, has in turn led to an improved understanding of basic renal physiology.

In the urine produced by a healthy kidney, protein is normally present only in very small amounts; little protein gets through the glomerular filtration barrier, and most of what does get through is reabsorbed in the renal tubule. By contrast, a large fraction of human kidney diseases are characterized by protein in the urine, a condition called proteinuria. Diabetic kidney disease, for example, the most common form of kidney disease in the western world, is characterized by increased urine protein and the abnormal accumulation of extracellular protein in the renal glomerulus, leading to a gradual decline in kidney function.

For the past decade, most of my own laboratory's research has focused on inherited diseases of the renal glomerulus, using methods of molecular cloning (see Section 18.1) to identify disease genes. Mutations in the actin-regulatory proteins α-actinin-4 and INF2 both cause late onset proteinuria and slowly progressive kidney dysfunction. Based on our studies and the research of other labs, we believe that fine regulation of the actin cytoskeleton is critical to maintaining the complex architecture of glomerular podocytes, a specialized epithelial cell that forms the final part of the glomerular filter. When this architecture is disturbed, the podocytes become more easily injured.

Large amounts of protein (on the order of 3 g/day) are the central feature of a serious medical condition known as the nephrotic syndrome. By studying very early onset forms of the nephrotic syndrome, proteins that are unique to the podocyte cell–cell junction have been identified as mutated in these disorders. For example, the cell surface protein nephrin, identified by these purely "reverse genetic" methods, has been shown to play important structural and signaling functions. Despite the growing list of genes known to be important to normal filtering of the blood, there is a tremendous amount of work to be done before we can answer the question of how the products of these genes interact to produce a functioning kidney.

Humans have many disadvantages as experimental organisms. We cannot do invasive experiments on them, we cannot do planned matings, and they take a long time to grow! But, because humans are intelligent creatures, they pay careful attention to phenotypes in themselves and others. The careful observation of renal phenotypes in humans followed by genetic analysis has led to a much better understanding of kidney disease and kidney function. It is important both for medicine and for understanding the kidney to use the information gained through genetic studies to inform biology.

Think Critically

Kidney transplantation is a standard form of treatment for kidney failure. In some instances, the kidney disease that affected the original kidneys returns to involve the transplanted kidney. What does this say about the mechanism of the disease? What would this say about the mechanism of an inherited form of kidney disease?

Martin Pollak

Martin Pollak is Chief of Nephrology at the Beth Israel Deaconess Medical Center in Boston and Associate Professor of Medicine at Harvard Medical School. The focus of his laboratory is to learn more about the causes of kidney disease in patients and families by studying genetics. To learn more visit: http://www.fsgs.bwh.harvard.edu/index.html.

REVIEW KEY CONCEPTS

Go to **CENGAGENOW** at www.cengage.com/login to access quizzing, animations, exercises, articles, and personalized homework help.

46.1 Introduction to Osmoregulation and Excretion

- Solute concentration is measured as osmolarity in milliosmoles per liter of solution (mOsm/L). A solution can be comparatively hyperosmotic, hypoosmotic, or isoosmotic to another solution. Water moving from a region of higher osmolarity to a region of lower osmolarity across a selectively permeable membrane is known as osmosis.
- Osmoregulators keep the osmolarity of body fluids different from that of the environment. Osmoconformers allow the osmolarity of their body fluids to match that of the environment.
- Molecules and ions must be removed from the body to keep cellular and extracellular fluids isoosmotic. In most animals, ECF is filtered through tubules formed from a transport epithelium and released to the exterior of the animal as urine (Figure 46.2).
- Nitrogenous wastes are excreted as ammonia, urea, or uric acid, or as a combination of these substances (Figure 46.3).

Animation: Diffusion, osmosis, and countercurrent systems

46.2 Osmoregulation and Excretion in Invertebrates

- Most marine invertebrates are osmoconformers. Because their body fluids are isoosmotic to seawater, they expend little or no energy on maintaining water balance.
- Freshwater and terrestrial invertebrates are osmoregulators, with body fluids that are hyperosmotic to their surroundings. They must expend energy to excrete water that moves into their cells by osmosis.
- The cells of the simplest marine invertebrates exchange water and solutes directly with the surrounding seawater. More complex invertebrates have specialized excretory tubules (Figures 46.4–46.6).

46.3 Osmoregulation and Excretion in Mammals

- In mammals and other vertebrates, excretory tubules are concentrated in the kidney.
- The mammalian excretory tubule, the nephron, has a proximal end at which filtration takes place, a middle region in which reabsorption and secretion occur, and a distal end that releases urine. A network of capillaries surrounding the nephron takes up ions and water and other molecules absorbed by the nephron. The urine leaving individual nephrons is processed further in collecting ducts and then pools in the renal pelvis. From there it flows through the ureter to the urinary bladder, and through the urethra to the exterior of the animal (Figure 46.7).
- At its proximal end, the nephron forms a cuplike Bowman's capsule around a ball of capillaries, the glomerulus. A filtrate consisting of water, other small molecules, and ions is forced from the glomerulus into Bowman's capsule, from which it travels through the nephron and drains into the collecting ducts and renal pelvis. The proximal convoluted tubule of the nephron secretes H^+ into the filtrate and reabsorbs Na^+, Cl^-, and K^+ along with water, HCO_3^-, and nutrients. In the descending segment of the loop of Henle, water is reabsorbed by osmosis. In the ascending segment of the loop, Na^+ and Cl^- are reabsorbed. In the distal convoluted tubule, the concentrations of H^+ and salts are balanced between the urine and the interstitial fluid surrounding the nephron. In the collecting ducts, additional H^+ is secreted into the urine and water is reab-

sorbed; some urea is also reabsorbed at the bottom of the ducts (Figure 46.8 and Table 46.1).

Animation: Human urinary system

Animation: Human kidney

Animation: Urine formation

Animation: Tubular reabsorption

46.4 Regulation of Mammalian Kidney Function

- The kidney's autoregulation system is activated by receptors in the juxtaglomerular apparatus. The receptors trigger constriction or dilation of the afferent arteriole to keep blood flow and filtration constant during small variations in blood pressure.
- When blood volume and blood pressure drop, the hormones of the renin-angiotensin-aldosterone system (RAAS) raise blood pressure by stimulating arteriole constriction and increasing NaCl reabsorption in the kidneys (Figure 46.10).
- ADH, which increases water reabsorption, is released from the pituitary when osmoreceptors detect an increase in the osmolarity of body fluids (Figures 46.11 and 46.12).

Animation: Structure of the glomerulus

46.5 Kidney Function in Nonmammalian Vertebrates

- Marine teleosts continually drink seawater to replace body water lost by osmosis to their hyperosmotic environment. Excess salts and nitrogenous wastes are excreted by the gills (Figure 46.13A).
- The body fluids of sharks and rays are isoosmotic with seawater. They do not lose water by osmosis, and do not drink seawater. Excess salts are excreted in the kidney and by a rectal salt gland.
- Body fluids of freshwater fishes and amphibians are hyperosmotic to their environment, and these animals must excrete the excess water that enters by osmosis. Body salts are obtained from food and, in fishes, through the gills (Figure 46.13B). Nitrogenous wastes are excreted from the gills of fishes and larval amphibians as ammonia and through the kidneys of adult amphibians as urea.
- Reptiles and birds conserve water by secreting nitrogenous wastes as uric acid and by absorbing water from urine and feces in the cloaca.

Animation: Water and solute balance

46.6 Introduction to Thermoregulation

- Animals must maintain body temperature at a level that provides optimal physiological performance. Heat flows between animals and their environment by conduction, convection, radiation, and evaporation (Figures 46.15 and 46.16).
- Ectothermic animals obtain heat energy primarily from the environment; endothermic animals obtain heat energy primarily from internal reactions (Figure 46.17).

Animation: Endotherms and ectotherms

46.7 Ectothermy

- Ectotherms obtain heat energy externally and control body temperature primarily by physiological or behavioral methods of regulating heat exchange with the environment.
- Many animals undergo thermal acclimatization, a structural or metabolic change in the limits of tolerable temperatures as the environment alternates between warm and cool seasons.

46.8 Endothermy

- Endotherms obtain heat energy primarily from internal reactions and maintain body temperature over a narrow range by balancing internal heat production against heat loss from the body surface.
- Internal heat production is controlled by negative feedback pathways triggered by thermoreceptors. When deviations from the temperature set point occur, signals from the receptors bring about compensating responses such as changes in blood flow to the body surface, sweating or panting, and behavioral modifications (Figure 46.18).
- The skin of endotherms is water-impermeable, reducing heat lost by direct evaporation of body fluids. The blood vessels of the skin regulate heat loss by constricting or dilating. A layer of insulating fatty tissue under the vessels limits losses to the heat carried by the blood. The hair of mammals and feathers of birds also insulate the skin. Erection of the hair or feathers reduces heat loss by thickening the insulating layer (Figure 46.19).
- The temperature set point in many birds and mammals varies in daily and seasonal patterns. During cooler conditions, a lowered set point is accompanied by torpor (Figure 46.20).
- Some animals show a form of endothermy in which part of their core is maintained at a temperature significantly higher than the surrounding environment.

Animation: Human thermoregulation

UNDERSTAND AND APPLY

Test Your Knowledge

1. Which of the following statements about osmoregulation is true?
   a. In freshwater invertebrates, salts move out of the body into the water because the animal is hypoosmotic to the water.
   b. A marine teleost has to fight gaining water because it is isoosmotic to the sea.
   c. Most land animals are osmoconformers.
   d. Vertebrates are usually osmoregulators.
   e. Terrestrial animals can regulate their osmolarity without expending energy.
2. One role of tubules in excretion is to:
   a. absorb H^+ ions to buffer body fluids.
   b. transport proteins across transport epithelium.
   c. reabsorb glucose and amino acids.
   d. move toxic substances from the filtrate into the cells composing the transport tubules.
   e. filter by maintaining a lower pressure in the fluid outside the tubule than inside it.
3. Products of metabolism in humans, as in:
   a. terrestrial amphibians, can include urea, which requires more energy to produce than ammonia.
   b. birds and reptiles, can include uric acid, which is nontoxic and excreted as a paste.
   c. sharks, are primarily excreted as ammonia.
   d. hydra, must be isoosmotic with the water ingested.
   e. other mammals, cannot be water as water comes only from what they drink.
4. Filtration and/or excretion can be performed by:
   a. ciliated metanephridia in insects.
   b. protonephridia containing flame cells in flatworms.
   c. a nephron and bladder in insects.
   d. Malpighian tubules on the segments of earthworms.
   e. the hindgut, which reabsorbs Na^+ and K^+ into the hemolymph of earthworms.

5. A mammalian nephron contains the:
   a. Bowman's capsule, which delivers the filtrate to the glomerulus.
   b. Bowman's capsule, which filters fluids, 99% of which will be excreted.
   c. proximal convoluted tubule, which moves Na^+ and K^+ into the filtrate of the interstitial fluids.
   d. proximal convoluted tubule, which reabsorbs K^+, Na^+, Cl^-, H_2O, and urea.
   e. proximal convoluted tubule, which lacks microvilli to ease fluid movement through it.
6. Which of the following correctly describes a part of kidney function?
   a. Collecting ducts dilute urine because they are permeable to salt but not water.
   b. In the ascending loop of Henle, Na^+ and Cl^- move into the tubules because the osmolarity of the filtrate is increased.
   c. The descending loop of Henle receives filtrate from the ascending loop.
   d. The distal convoluted tubule pumps water into the tubule by active transport.
   e. The renal pelvis receives urine from the collecting ducts and carries it to the ureters.
7. Which of the following is an example of autoregulation of kidney function?
   a. The RAAS regulates Na^+ by secreting renin when blood pressure or blood volume decreases.
   b. The ADH system regulates water balance by decreasing water reabsorption and increasing excretion of salt.
   c. Receptors in the juxtaglomerular apparatus of the distal convoluted tubule detect drops in blood pressure and cause a higher filtration rate.
   d. ANF is released by the kidney to increase renin release.
   e. Angiotensin II lowers blood pressure by constricting arterioles.
8. Deficient water levels in humans are prevented by:
   a. osmoreceptors in the hypothalamus that detect decreases in salt concentrations.
   b. the hypothalamus stimulating the posterior pituitary to secrete a hormone that allows the collecting ducts and distal convoluted tubules to be permeable to water.
   c. inhibiting ADH, which causes a rise in osmolarity of the ECF.
   d. producing dilute urine.
   e. drinking alcohol, which stimulates aldosterone to raise the osmolarity of body fluids.
9. Which best exemplifies ectothermy?
   a. The metabolic rate increases as the temperature decreases.
   b. Body temperature remains constant when environmental temperatures change.
   c. Food demand increases when temperatures drop.
   d. Virtually all invertebrate groups are ectotherms.
   e. No vertebrate groups are ectotherms.
10. Unique to endotherms is:
   a. torpor.
   b. thermal acclimatization.
   c. a nonchanging body temperature.
   d. response to seasonal temperature changes.
   e. thermoregulation by a hypothalamus.

Discuss the Concepts

1. A urinalysis reveals glucose, urea, hemoglobin, and sodium. Which of these substances are abnormal in urine, and why?
2. As a person ages, nephron tubules lose some of their ability to concentrate urine. What is the effect of this change?
3. Shivering increases air movement over the body surface. What effect does this air movement have on heat conservation in the shivering animal?
4. What heat transfer processes might account for the change in body temperature when a mammal's body temperature undergoes daily variations?

Design an Experiment

Design experiments to show the role of fluid consumption in thermoregulation during endurance exercise.

Interpret the Data

Products labeled as "organic" fill an increasing amount of space on supermarket shelves. What does that label mean? A food that carries the USDA's organic label must be produced without pesticides such as malathion and chlorpyrifos that conventional farmers typically use on fruits, vegetables, and grains.

Does eating organic food significantly affect the level of pesticide residues in a child's body? Chensheng Lu of Emory University used urine testing to find out. For 15 days, the urine of twenty-three children (aged 3 to 11) was monitored for breakdown products of pesticides. During the first 5 days (phase 1), children ate their standard, nonorganic diet. For the next 5 days (phase 2), they ate organic versions of the same types of foods and drinks. Then, for the final 5 days (phase 3), the children returned to their nonorganic diet.

Levels of metabolites (breakdown products) of malathion and chlorpyrifos in the urine of children taking part in a study of effects of an organic diet*

Study phase	No. of samples	Malathion metabolite Mean (μg/L)	Malathion metabolite Maximum (μg/L)	Chlorpyrifos metabolite Mean (μg/L)	Chlorpyrifos metabolite Maximum (μg/L)
1. Nonorganic	87	2.9	96.5	7.2	31.1
2. Organic	116	0.3	7.4	1.7	17.1
3. Nonorganic	156	4.4	263.1	5.8	25.3

*The difference in the mean level of metabolites in the organic and inorganic phases of the study was statistically significant.

1. During which phase of the experiment did the children's urine contain the lowest level of the malathion metabolite?
2. During which phase of the experiment was the maximum level of the chlorpyrifos metabolite detected?
3. Did switching to an organic diet lower the amount of pesticide excreted by children?

Source: C. Lu et al. 2006. Organic diets significantly lower children's dietary exposure to organophosphorus pesticides. *Environmental Health Perspectives* 114:260–263.

Apply Evolutionary Thinking

Humans produce urea as an excretion product, whereas reptiles and birds produce uric acid. Indeed, human kidneys are not as efficient as those of reptiles and birds. What does this mean in an evolutionary sense?

Express Your Opinion

Many companies use urine testing to screen for drug and alcohol use among prospective employees. Some people say this is an invasion of privacy. Do you think employers should be allowed to require a person to undergo urine testing before being hired? Go to www.cengage.com/login to investigate both sides of the issue and then vote.

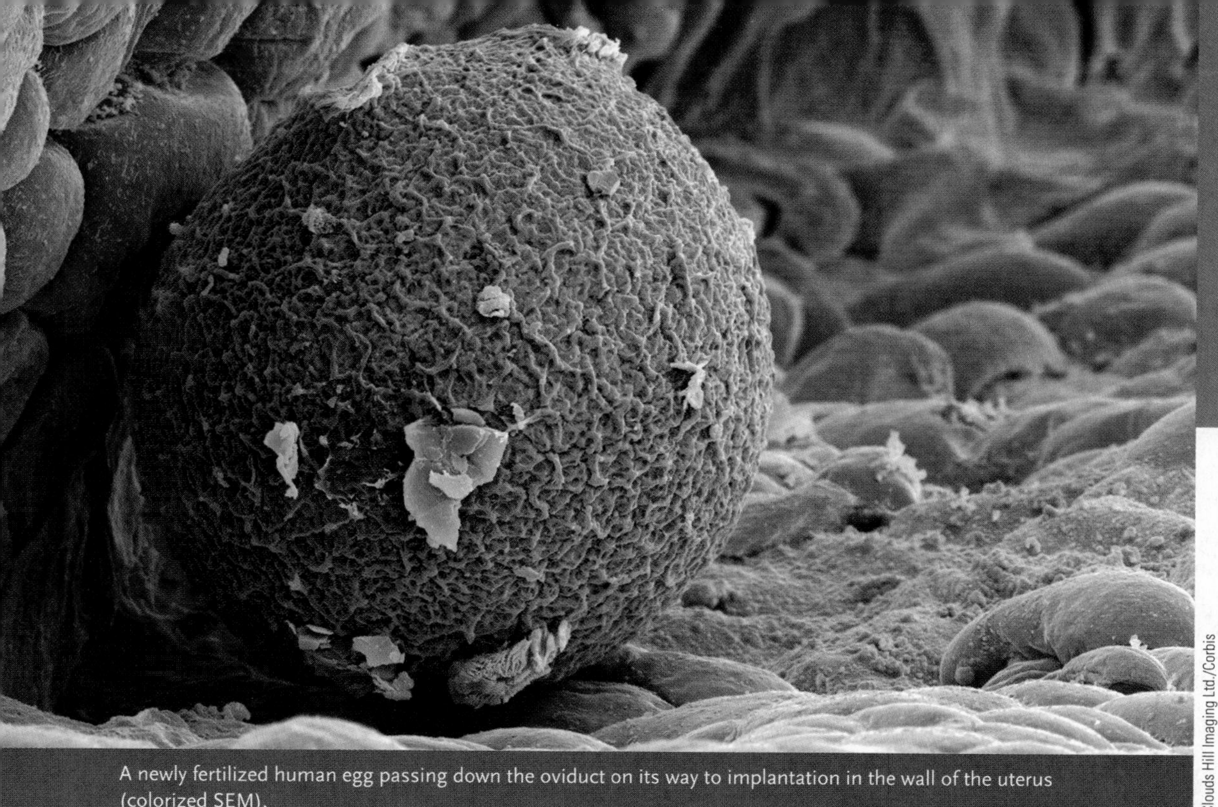

A newly fertilized human egg passing down the oviduct on its way to implantation in the wall of the uterus (colorized SEM).

47

STUDY OUTLINE

Animal Reproduction

Why It Matters. . . It is seven days after the October full moon and night is falling. All the inhabitants of the Samoan island of Tutuila who have access to a boat are gathered on the island's large lagoon. Some hold lanterns and look into the water; others have nets at the ready. They are awaiting the palolo worm *(Eunice viridis)* **(Figure 47.1)**, which has appeared in the water as the moon rises on this same night of the lunar year for as long as the islanders can remember.

The moon peeks over the horizon and the excitement of the crowd rises. Then, there they are, thousands of blue and green worms, squirming in the water like animated spaghetti. The boaters scoop up the worms by the netful and dump them into buckets. When the buckets are full, the islanders glide toward the shore where steaming pots are waiting, for palolo worms are a delicacy that the islanders savor only once a year. For the islanders, a night of feasting, singing, and dancing follows as they cook and eat the worms.

The worms that squirm to the surface to delight the islanders are actually not complete individuals. They are tail sections about 10 to 20 cm long that break from adults after they become filled with eggs or sperm (see Figure 47.1). The adults are polychaete annelids that live in burrows in coral reefs of the Samoan and Fiji islands. These annelids develop tail segments once a year, just after the October full moon. On the seventh night following the full moon, the tails break off and swim to the surface, where—if Samoan gourmets do not net them first—they disintegrate and release eggs and sperm by the millions, turning the water of the lagoon milky. The an-

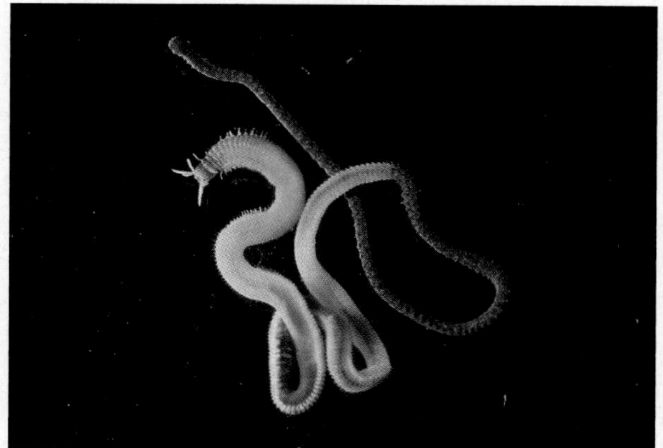

FIGURE 47.1
The palolo worm. Gametes are packed into segments of the tail section (in blue).

terior ends of the worms, safe in their burrows, will survive to produce tails for next year's mating frenzy. A biological clock in the worms, timed by periods of moonlight, precisely sets both the appearance of the mating swarm and indirectly, the appearance of the islanders with their boats.

The swarm of the palolo worms is only one of many adaptations that accomplish mating in animals. For animals that reproduce by eggs and sperm, the adaptations are as diverse as the number of species on Earth. This diversity allows individuals of the same species to find each other and unite eggs and sperm. Within the diversity, however, are underlying patterns that are shared by all animals.

Both the underlying patterns and the diversity of animal reproduction are the subjects of this chapter. We also discuss the development of eggs and sperm, and the union of egg and sperm that begins the development of a new individual. The next chapter continues with the events of development after eggs and sperm have united. <

47.1 Animal Reproductive Modes: Asexual and Sexual Reproduction

Reproduction is part of a life cycle in which individuals grow, develop, and reproduce according to instructions encoded in DNA. Rather than survival of the individual, reproduction is the means of passing on the genes of an individual to new generations of the species. As such, it is among the most vital functions of living organisms.

Two basic modes of reproduction operate in the animal kingdom. In **asexual reproduction,** a single individual gives rise to offspring without fusion of **gametes** (egg and sperm), that is, there is no genetic input from another individual. In **sexual reproduction,** male and female parents produce offspring through the union of egg and sperm generated by meiosis (meiosis is discussed in Chapter 11).

Asexual Reproduction Produces Offspring with Genes from Only One Individual

Many aquatic invertebrates and some terrestrial annelids and insects reproduce asexually. Asexual reproduction is rare among vertebrates. In asexual reproduction, one or many cells of a parent develop directly into a new individual. In a few animals that undergo asexual reproduction, the cells taking part are genetically varied products of meiosis, but in most they are the products of mitosis. The offspring therefore are genetically identical to one another and to the parent; in other words, they are genetic clones of the parent. For this reason, asexual reproduction of this kind is also called **clonal reproduction.**

Genetic uniformity of offspring can be advantageous in environments that remain stable and uniform. Asexual reproduction tends to preserve gene combinations, producing individuals that are successful in such environments. Further, individuals do not have to expend energy to produce gametes or find a mate. Asexual reproduction can also bring reproductive advantages to individuals living in sparsely populated areas, or to sessile animals, which cannot move from place to place.

Asexual reproduction involving mitosis occurs in animals by three basic mechanisms: *fission, budding,* and *fragmentation:*

1. **Fission.** In **fission,** the parent separates into two or more offspring of approximately equal size. Planarians (flatworms), for instance, reproduce asexually by fission; depending on the species, they may divide by transverse or longitudinal fission.
2. **Budding.** In **budding,** a new individual grows and develops while attached to the parent. Sponges, tunicates, and some cnidarians reproduce asexually by this mechanism. The offspring may break free from the parent, or remain attached to form a *colony.* In the cnidarian *Hydra,* for example, an offspring buds and grows from one side of the parent's body and then detaches to become a separate individual **(Figure 47.2).**
3. **Fragmentation.** In **fragmentation,** pieces separate from the body of a parent and develop *(regenerate)* into new individuals. Many species of cnidarians, flatworms, annelids, and some echinoderms can reproduce by fragmentation.

Some animals produce offspring by the growth and development of an egg without fertilization. The offspring may be haploid or diploid depending on the species. This form of asexual reproduction is called **parthenogenesis** (*parthenos* = virgin, *genesis* = origin). Because the egg from which a parthenogenetic offspring is produced derives from meiosis in the female parent, the offspring are not genetically identical to the parent or to each other. (How chromosome segregation and genetic recombination during meiosis produce gametes with gene combinations different from the parent is described shortly.)

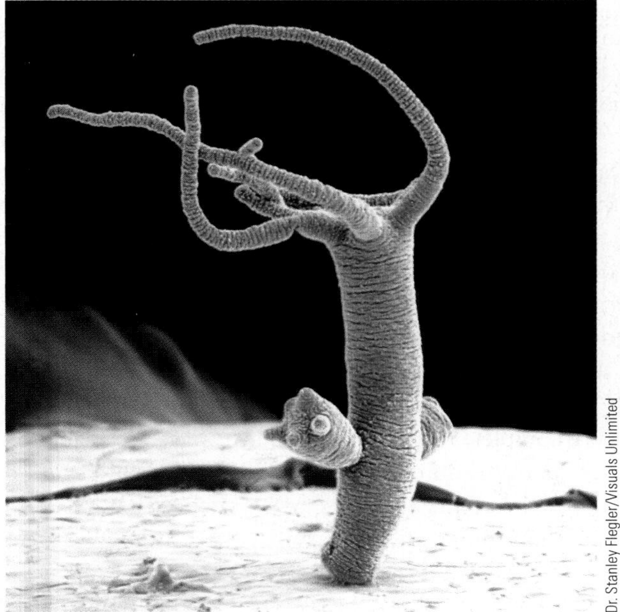
Dr. Stanley Flegler/Visuals Unlimited

FIGURE 47.2
Asexual reproduction by budding in *Hydra* (colorized SEM).

Parthenogenesis occurs in some invertebrates, including certain aphids, water fleas, bees, and crustaceans. In bees, for instance, haploid male drones are produced parthenogenetically from unfertilized eggs produced by reproductive females (queens) while new queens and sterile workers develop from fertilized eggs. Parthenogenesis also occurs in some vertebrates, for example, in certain fish, salamanders, amphibians, lizards, and turkeys. In these animals, an egg, produced by meiosis, typically doubles its chromosomes to produce a diploid cell that begins development. In single-sex species where females have two identical sex chromosomes, the offspring are female, whereas in single-sex species where males have two identical sex chromosomes, the offspring are male. For instance, all whiptail lizards (*Cnemidophorus* species) are females, produced solely by parthenogenesis. These females go through the motions of mating and copulation with each other.

Sexual Reproduction Generates Diversity among Offspring

Animals reproduce sexually by the union of sperm and eggs produced by meiosis. The overriding advantage of sexual reproduction is the generation of genetic diversity among offspring. This diversity increases the chance that, in a changing environment, at least some offspring will grow and reproduce successfully. Diversity also increases the chance that offspring may be able to live and reproduce in environments previously unoccupied by the species.

Two mechanisms that are part of meiosis give rise to the genetic diversity in eggs and sperm: *genetic recombination* (see Section 11.2) and the *independent assortment* of chromosomes of maternal and paternal origin (see Section 12.1). Genetic recombination mixes the alleles of parents into new combinations within chromosomes; independent assortment results in random combinations of maternal and paternal chromosomes in gamete nuclei. Additional variability is generated at fertilization when eggs and sperm from genetically different individuals fuse together at random to initiate the development of new individuals. To these sources of variability are added random DNA mutations, which are the ultimate source of variability for both sexual and asexual reproduction.

The disadvantages of sexual reproduction include the expenditure of energy and raw materials in producing gametes and finding mates. The need to find mates can also expose animals to predation and takes time from finding food and shelter and caring for existing offspring.

With these advantages and disadvantages in mind, we now turn to the mechanisms of sexual reproduction, which include both cellular and whole-organism activities. We begin with the cellular mechanisms in the next section.

STUDY BREAK 47.1

What are the advantages and disadvantages of asexual reproduction? Of sexual reproduction?

47.2 Cellular Mechanisms of Sexual Reproduction

The cellular mechanisms of sexual reproduction are **gametogenesis,** the formation of male and female gametes, and **fertilization,** the union of gametes that initiates development of a new individual. The pairing of a male and a female for the purpose of sexual reproduction is **mating.**

Gametogenesis Involves the Coordinated Events of Meiosis and Sperm and Egg Development

Gametes in most animals form from **germ cells,** a cell line that is set aside early in embryonic development and remains distinct from the other, **somatic cells** of the body. During development, the germ cells collect in specialized gamete-producing organs, the **gonads**—the **testes** (singular, *testis*) in males and **ovaries** in females. Mitotic divisions of the germ cells produce **spermatogonia** in males and **oogonia** in females; these are the cells that enter meiosis to give rise to gametes by **spermatogenesis** in males and **oogenesis** in females **(Figure 47.3).** In humans, each gamete has only one of 2^{23} possible combinations of parental chromosomes. In some animals, the germ cells also give rise to families of cells that assist gamete development.

Meiosis reduces the number of chromosomes from the diploid level characteristic of somatic cells of the species, in which there are two copies of each chromosome, to the haploid level of gametes, in which there is one copy of each chromosome. The fusion of a haploid sperm and egg during fertilization restores the diploid number of chromosomes and produces a **zygote,** the first cell of a new individual.

SPERMATOGENESIS Spermatogenesis produces mature, haploid sperm cells, also called **spermatozoa** (singular, *spermatozoon*) or simply *sperm* **(Figure 47.4).** The sperm of most animal species are motile cells, driven through a watery medium by the whiplike beating of a flagellum that extends from the posterior end of the cell. During maturation from spermatid to sperm, most of the cytoplasm is lost, except for mitochondria, which surround the base of a flagellum. These mitochondria produce the ATP used as the energy source for flagellar beating. At the head of the sperm, a specialized secretory vesicle, the **acrosome,** forms a cap over the nucleus. The acrosome contains enzymes and other proteins that help the sperm attach to and penetrate the surface coatings of an egg of the same species.

OOGENESIS Oogenesis produces mature, haploid egg cells, also called **ova** (singular, *ovum*) or simply *eggs.* The eggs of all animals are nonmotile cells, typically much larger than sperm of the same species. In spermatogenesis, all four products of meiosis develop into functional sperm. However, in oogenesis, only one of the cell products of meiosis develops into a functional egg, which retains almost all of the cytoplasm of the parent cell **(Figure 47.5).** The other products form nonfunctional cells called **polar bodies** (see

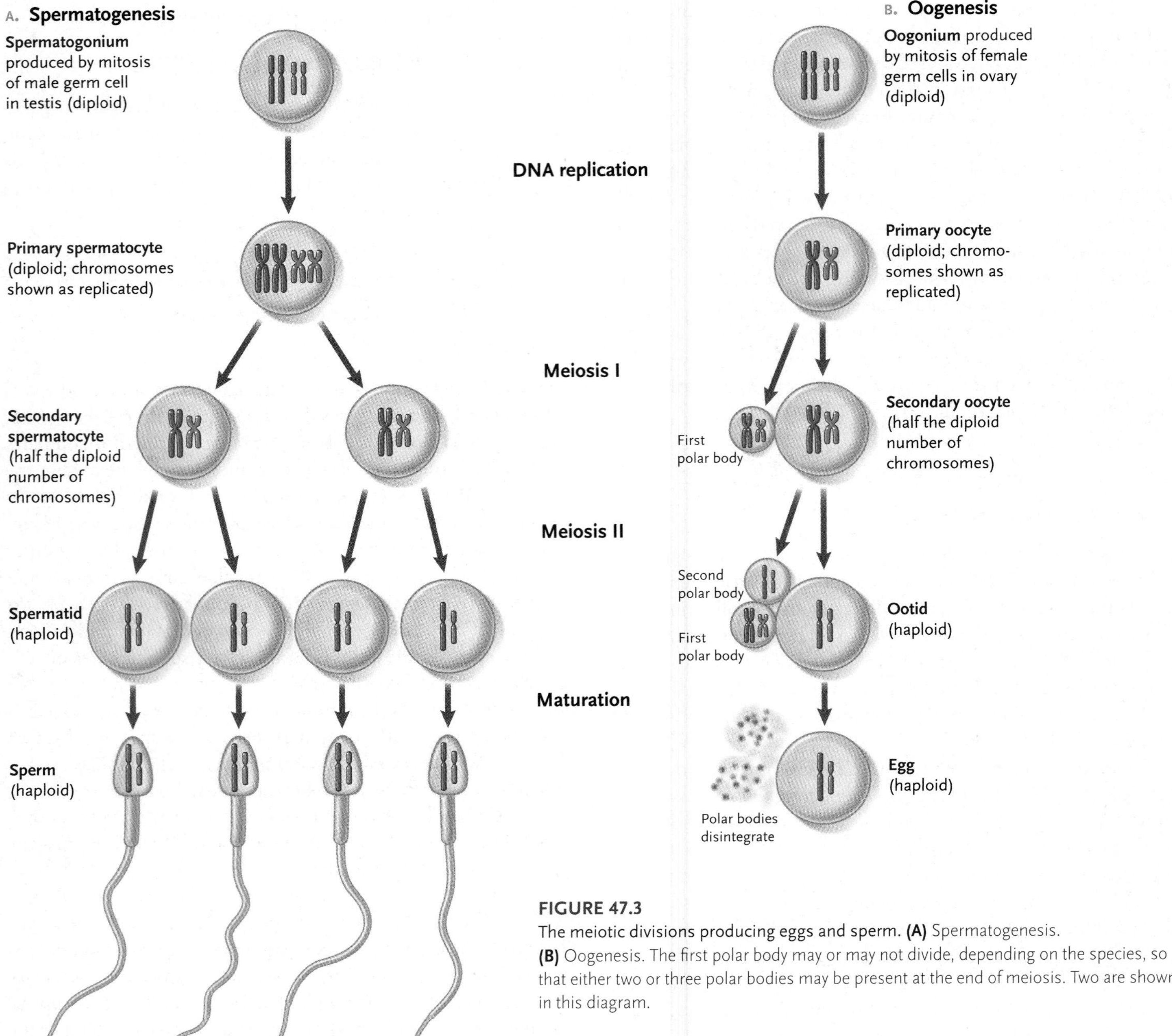
A. Spermatogenesis
Spermatogonium produced by mitosis of male germ cell in testis (diploid)
Primary spermatocyte (diploid; chromosomes shown as replicated)
Secondary spermatocyte (half the diploid number of chromosomes)
Spermatid (haploid)
Sperm (haploid)
DNA replication
Meiosis I
Meiosis II
Maturation
B. Oogenesis
Oogonium produced by mitosis of female germ cells in ovary (diploid)
Primary oocyte (diploid; chromosomes shown as replicated)
First polar body
Secondary oocyte (half the diploid number of chromosomes)
Second polar body
First polar body
Ootid (haploid)
Polar bodies disintegrate
Egg (haploid)

FIGURE 47.3
The meiotic divisions producing eggs and sperm. **(A)** Spermatogenesis. **(B)** Oogenesis. The first polar body may or may not divide, depending on the species, so that either two or three polar bodies may be present at the end of meiosis. Two are shown in this diagram.

Figure 47.3B). The unequal cytoplasmic divisions concentrate nutrients and other molecules required for development in the egg. In most species, the polar bodies eventually disintegrate and do not contribute to fertilization or embryonic development.

The oocytes of most animals do not complete meiosis until fertilization. In mammals, for example, oocytes stop developing at the end of the first meiotic prophase within a few weeks after a female is born. The oocytes remain in the ovary at this stage of development until the female is sexually mature. In humans, some oocytes may remain in prophase of the first meiotic division for perhaps 50 years. Then, one to several oocytes advance to the metaphase of the second meiotic division and are released from the ovary at intervals ranging from days to months, or at certain seasons, depending on the species. As in other animals, meiosis is completed at fertilization to produce the fully mature egg. Mature eggs are the largest cell type of an animal species.

An egg typically has specialized features, which include: (1) stored nutrients required for at least the early stages of embryonic development; (2) egg coats of one or more kinds which protect the egg from mechanical injury and infection and, in some species, protect the embryo after fertilization; and (3) mechanisms that prevent the egg from being fertilized by more than one sperm cell (discussed shortly).

The amount of stored nutrients in an egg varies with the animal. Mammalian eggs are microscopic, containing few stored nutrients. In mammals, the embryo develops inside the mother and is supplied with nutrients by the mother's body. In contrast, the relatively huge eggs of birds and reptiles contain all the nutrients required for complete embryonic development: the "yolk"

A. Human sperm

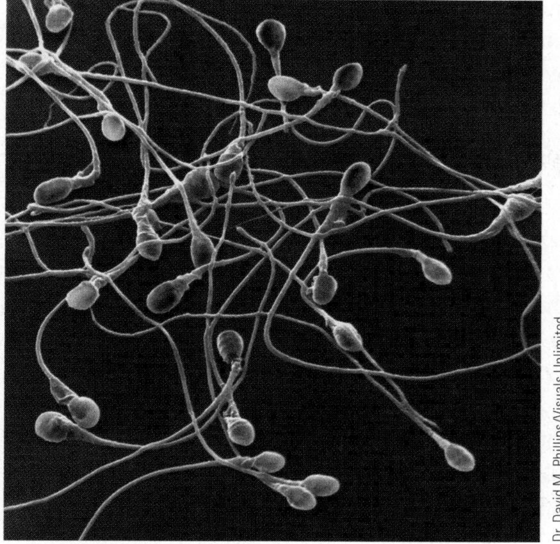
Dr. David M. Phillips/Visuals Unlimited

B. Sperm structure

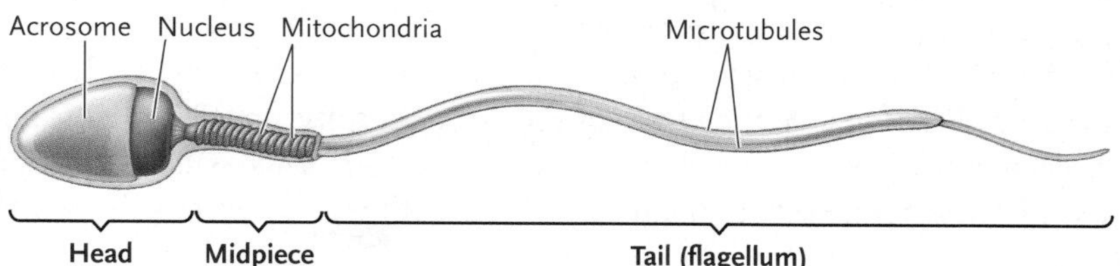
Acrosome
Nucleus
Mitochondria
Microtubules
Head
Midpiece
Tail (flagellum)

FIGURE 47.4
Spermatozoa. **(A)** Photomicrograph of human sperm. **(B)** Structure of a sperm.

contains the egg cell, and the "white" contains the nutrients. No matter what the size of an animal egg, however, most of the volume is cytoplasm, and the egg nucleus is microscopic or nearly so in all species.

Egg coats are surface layers added during oocyte development or fertilization in many species. The **vitelline coat,** called the **zona pellucida** in mammals (see Figure 47.5) is a gel-like matrix of proteins, glycoproteins, or polysaccharides located immediately outside the plasma membrane of the egg cell. Insect eggs have additional outer protein coats that form a hard, water-impermeable layer for preventing desiccation. Amphibians and some echinoderms have an additional outer egg jelly layer instead of a tough protein coat that protects the egg from drying out (see Figure 47.8).

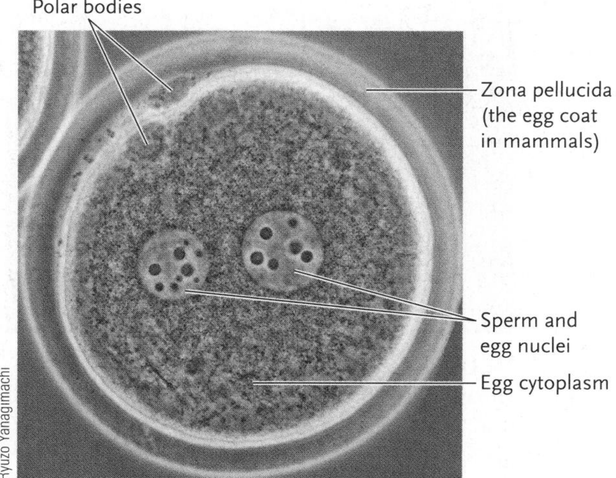
Polar bodies
Zona pellucida (the egg coat in mammals)
Sperm and egg nuclei
Egg cytoplasm
Ryuzo Yanagimachi

FIGURE 47.5
A mature hamster egg that has been fertilized.

In birds, reptiles, and one group of egg-laying mammals, the **monotremes,** the egg white, a thick solution of proteins, surrounds the vitelline coat. Outside the white is the *shell* of the egg, flexible and leathery in reptiles and mineralized and brittle in birds. Both the egg white and the shell are added while the egg—fertilized or not—is in transit through the **oviduct,** the tube through which the egg moves from the ovary to the outside of the body. In mammals, the egg is surrounded by **follicle cells** during its development. These cells, which grow from ovarian tissue, nourish the developing egg. They also make up part of the zona pellucida while the egg is in the ovary, and remain as a protective layer after it is released.

Fertilization Requires an Internal or External Aquatic Medium

Eggs and sperm are delivered from the ovaries and testes to the site of fertilization by oviducts in females and by sperm ducts in males; in many species, external accessory sex organs participate in the delivery. **Figure 47.6** shows examples of invertebrate and vertebrate reproductive systems. The nonmotile eggs move through the oviducts on currents generated by the beating of cilia lining the oviducts, or by contractions of the oviducts or the body wall.

Depending on the species, fertilization may be *external,* taking place in a watery medium outside the body of both parents, or *internal,* taking place in a watery fluid inside the body of the female. In **external fertilization,** which occurs in most aquatic invertebrates, bony fishes, and amphibians, sperm and eggs are shed into the surrounding water. The sperm swim until they collide with an egg of the same species. The process is helped by synchronization of female and male gamete release, and by the enormous quantities of gametes released. In some animals, such as sea urchins and amphibians, the sperm are attracted to the egg by diffusible attractant molecules released by the egg.

Most amphibians, even terrestrial species such as toads, mate in an aquatic environment. Frogs typically mate by a reflex response called *amplexus,* in which the male clasps the female tightly around the body with his forelimbs **(Figure 47.7).** The embrace stimulates the female to shed a mass of eggs into the water through the **cloaca**—the cavity in reptiles, birds, amphibians, and many fishes into which both the intestinal and genital tracts empty. As the eggs are released, they are fertilized by sperm released by the male.

Internal fertilization takes place in invertebrates such as annelids, some arthropods, and some mollusks, and in vertebrates such as reptiles, birds, mammals, some fishes, and some salamanders. In these animals, the sperm are released by the male close to or inside the entrance of the reproductive tract of the female. The sperm swim through fluids in the reproductive tract until they reach and fertilize each egg. In some species, molecules released by the egg attract the sperm to its outer coats. The physical act involving the introduction of the male's accessory sex organ (for example, penis) into a female's accessory sex organ (for example, vagina) to accomplish internal fertilization is known as **copulation.** Internal fertilization makes terrestrial life possible by providing the aquatic medium required for fertilization inside the female's body without the danger of gametes drying by exposure to the air.

Male reptiles, birds, and mammals have accessory sex organs that place sperm directly inside the reproductive tract of females, where fertilization takes place. In reptiles and birds, sperm fertilize eggs as they are released from the ovary and travel through the oviducts, before the shell is added. In mammals, the male's penis delivers sperm into the female's vagina. Unlike the cloaca of the reptiles and birds, which has both sexual and excretory functions, the vagina is specialized for reproduction. The introduced sperm swim into the tubular oviducts containing the eggs, and fertilization takes place.

A. **Insect (the fruit fly, *Drosophila*)**

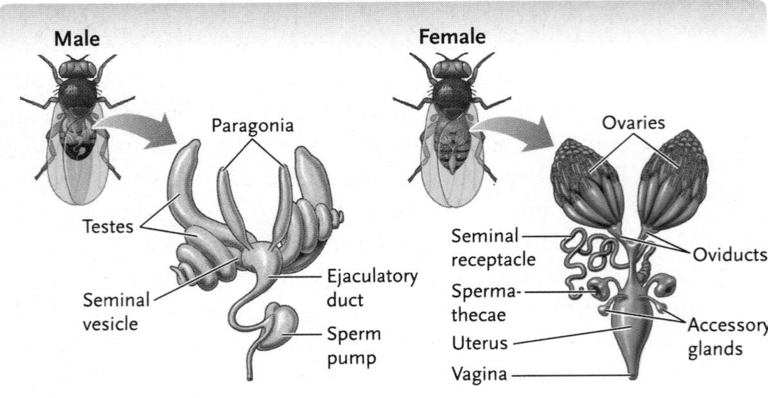
Male
Female
Paragonia
Testes
Seminal vesicle
Ejaculatory duct
Sperm pump
Ovaries
Seminal receptacle
Spermathecae
Uterus
Vagina
Oviducts
Accessory glands

B. **Amphibian (frog)**

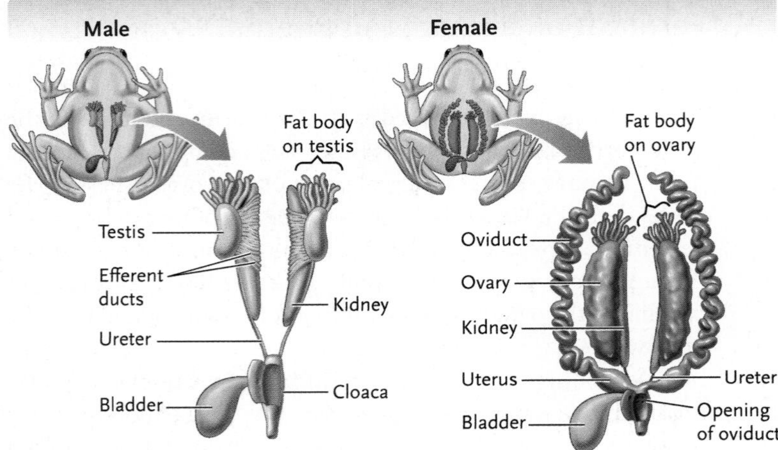
Male
Female
Fat body on testis
Testis
Efferent ducts
Ureter
Bladder
Kidney
Cloaca
Fat body on ovary
Oviduct
Ovary
Kidney
Uterus
Bladder
Ureter
Opening of oviduct

C. **Mammal (cat)**

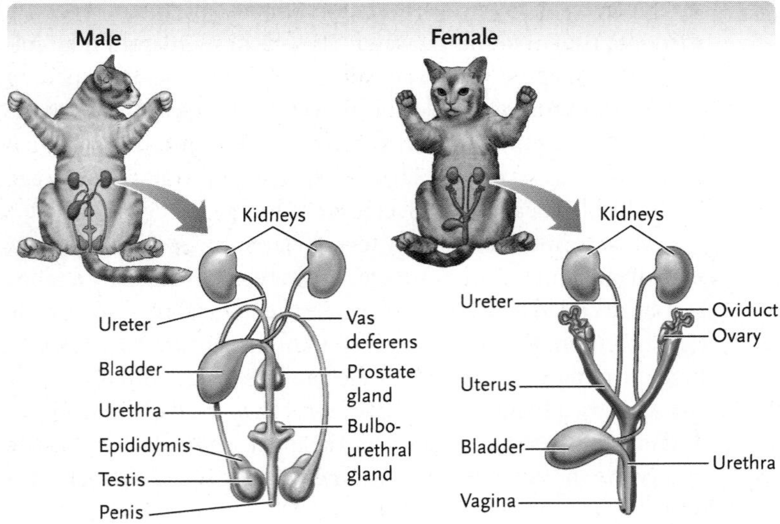
Male
Female
Kidneys
Ureter
Bladder
Urethra
Epididymis
Testis
Penis
Vas deferens
Prostate gland
Bulbourethral gland
Kidneys
Ureter
Uterus
Bladder
Vagina
Oviduct
Ovary
Urethra

FIGURE 47.6
Some reproductive systems. **(A)** An insect, *Drosophila* (fruit fly). **(B)** An amphibian, a frog. **(C)** A mammal, a cat. Female systems are shown in blue, and male systems in yellow.

Fertilization Involves Fusion of a Sperm and an Egg, Which Activates the Egg for Development

Once a sperm touches the outer surface of an egg of the same species **(Figure 47.8A),** receptor proteins in the sperm plasma membrane bind the sperm to the vitelline coat or zona pellucida. In most animals, only a sperm from the same species as the egg can recognize and bind to the egg surface.

Eggs
Hans Pfletschinger/Peter Arnold

FIGURE 47.7
A male leopard frog *(Rana pipiens)* clasping a female in a mating embrace known as amplexus. The tight squeeze by the male frog stimulates the female to release her eggs, which can be seen streaming from her body, embedded in a mass of egg jelly. Sperm released by the male fertilize the eggs as they pass from the female.

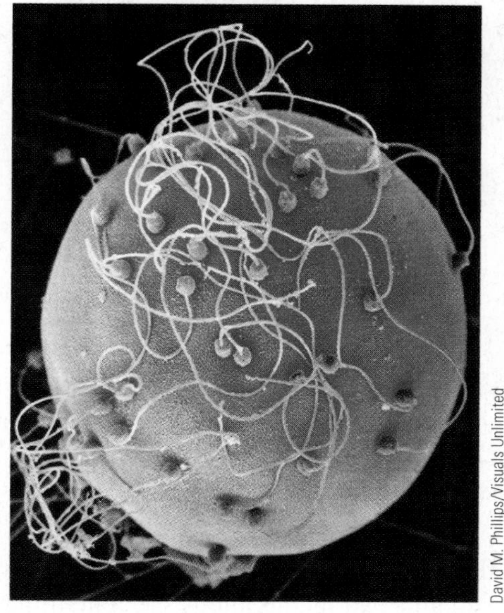
A. Sperm adhering to egg
David M. Phillips/Visuals Unlimited

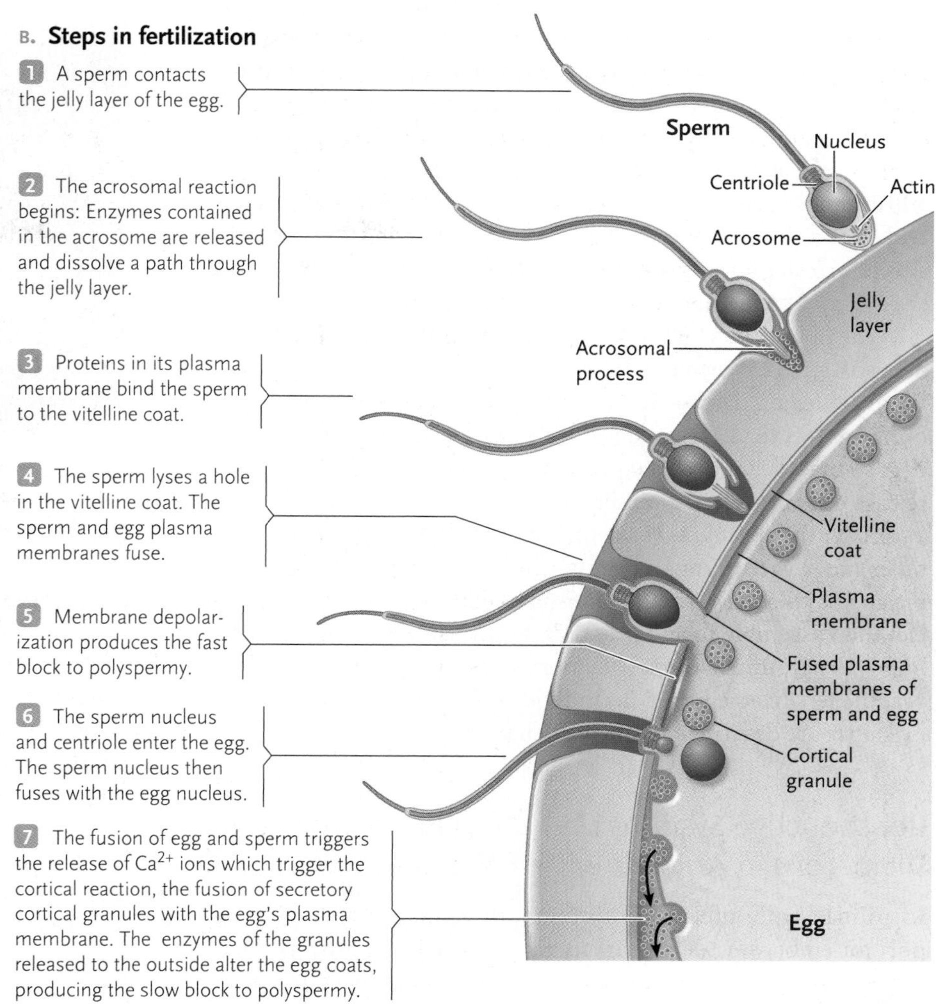
B. Steps in fertilization
1 A sperm contacts the jelly layer of the egg.
2 The acrosomal reaction begins: Enzymes contained in the acrosome are released and dissolve a path through the jelly layer.
3 Proteins in its plasma membrane bind the sperm to the vitelline coat.
4 The sperm lyses a hole in the vitelline coat. The sperm and egg plasma membranes fuse.
5 Membrane depolarization produces the fast block to polyspermy.
6 The sperm nucleus and centriole enter the egg. The sperm nucleus then fuses with the egg nucleus.
7 The fusion of egg and sperm triggers the release of Ca2+ ions which trigger the cortical reaction, the fusion of secretory cortical granules with the egg's plasma membrane. The enzymes of the granules released to the outside alter the egg coats, producing the slow block to polyspermy.
Sperm
Nucleus
Centriole
Actin
Acrosome
Jelly layer
Acrosomal process
Vitelline coat
Plasma membrane
Fused plasma membranes of sperm and egg
Cortical granule
Egg

FIGURE 47.8
Fertilization. **(A)** Sperm adhering to the surface coat of a sea urchin egg. Of the many sperm that may initially adhere to the outer surface of an egg, usually only one accomplishes fertilization. **(B)** Steps of fertilization in a sea urchin.

Species recognition is highly important in animals that carry out external fertilization, because the water surrounding the egg may contain sperm of many different species. It is less important in internal fertilization, because structural adaptations and behavioral patterns of mating usually limit sperm transfer from males to females of the same species.

FERTILIZATION After the initial attachment of sperm to egg, the events of fertilization proceed in rapid succession. **Figure 47.8B** shows the steps of fertilization in the sea urchin, a well-studied organism for invertebrate development and reproduction, and an often-used model for human fertility research. The steps are similar in other organisms. The attachment of sperm to egg triggers the **acrosomal reaction,** in which enzymes in the acrosome are released from the sperm and digest a path through the egg coats. The sperm, with its tail still beating, follows the path until its plasma membrane touches and fuses with the plasma membrane of the egg. Fusion introduces the sperm nucleus into the egg cytoplasm and activates the egg to complete meiosis and begin development.

EGG ACTIVATION AND BLOCKS TO POLYSPERMY Two mechanisms can prevent more than one sperm from fertilizing the egg: a *fast block* within seconds of fertilization, and a *slow block* within minutes.

In many invertebrate species, such as the sea urchin, the fusion of egg and sperm opens ion channels in the plasma membrane of the egg, spreading a wave of electrical depolarization over the egg surface, much like the nerve impulse traveling along a neuron. The depolarization alters the egg plasma membrane's potential from negative to positive so that it cannot fuse with any additional sperm, thereby eliminating the possibility that more than one set of paternal chromosomes enters the egg. Because it occurs within a few seconds after fertilization, the barrier set up by the wave of depolarization is called the **fast block to polyspermy** (Figure 47.8B, step 5).

In vertebrates, the wave of membrane depolarization following sperm–egg fusion is not as pronounced, and does not prevent additional sperm from fusing with the egg. However, any additional sperm nuclei entering the egg cytoplasm usually break down and disappear, so that only the first sperm nucleus to enter fuses with the egg nucleus.

In both invertebrates and vertebrates, fusion of egg and sperm triggers the release of stored calcium (Ca^{2+}) ions from the endoplasmic reticulum of the egg into the cytosol. The Ca^{2+} ions activate control proteins and enzymes that initiate intense metabolic activity in the fertilized egg, including a rapid increase in cellular oxidations and synthesis of proteins and other molecules.

The Ca^{2+} ions also trigger the **cortical reaction,** in which **cortical granules,** secretory vesicles just under the plasma membrane of the egg, fuse with the membrane and release their contents to the outside by exocytosis (see Figure 47.8B, step 7). Enzymes released from the cortical granules alter the egg coats within minutes after fertilization, so that no further sperm can attach and penetrate to the egg. Once this barrier, termed the **slow block to polyspermy,** is set up, no further sperm can reach the egg plasma membrane in any animal species.

The importance of Ca^{2+} to cortical granule release has been demonstrated experimentally. If Ca^{2+} is added to the cytoplasm, the granules are released in unfertilized eggs; conversely, if Ca^{2+}-binding chemicals are added to the cytoplasm of unfertilized eggs so that the Ca^{2+} concentration cannot rise, cortical granule release does not occur after fertilization.

After the sperm nucleus enters the egg cytoplasm, microtubules move the sperm and egg nuclei together in the egg cytoplasm and they fuse. The chromosomes of the egg and sperm nuclei then assemble together and enter mitosis. The subsequent, highly programmed events of embryonic development that convert the fertilized egg into an individual capable of independent existence are described in the next chapter.

Reproductive Systems May Be Oviparous or Viviparous in Animals with Internal Fertilization

In animals with internal fertilization, three major types of support for embryonic development have evolved: *oviparity,* meaning egg laying; *viviparity,* meaning giving birth to live offspring; and *ovoviparity,* meaning giving birth to live offspring that first hatch internally from eggs.

- **Oviparous** animals (*ovum* = egg, *parere* = to bring forth, to bear) lay eggs that contain the nutrients needed for development of the embryo outside the mother's body. Examples are insects, spiders, most reptiles, and birds. The only oviparous mammals are the *monotremes:* the echidnas and the duck-billed platypus *(Ornithorhynchus anatinus),* both of which inhabit Australia.
- **Viviparous** animals (*vivus* = alive) retain the embryo within the mother's body and nourish it during at least early embryo development. All mammals except the monotremes are viviparous. Viviparity is seen also in all other vertebrate groups except for the crocodiles, turtles, and birds. In viviparous animals, development of the embryo takes place in a specialized saclike organ, the **uterus** *(womb).* Among mammals, one group, called the *placental mammals* or *eutherians,* has a specialized temporary structure, the **placenta,** which connects the embryo to the uterus. The placenta facilitates the transfer of nutrients from the blood of the mother to the embryo and the movement of wastes in the opposite direction. Humans are placental mammals. The other group of mammals, the *marsupials* or *metatherians,* originally were called nonplacental mammals because of a belief that they lacked a placenta. In fact, they do have a placenta, but it derives from a different tissue than that of eutherians and does not connect the embryo and the uterus. Instead it provides nutrients to the embryo from an attached membranous sac containing yolk for only the early stages of its development. In many metatherians, the embryo is then born at an early stage and crawls over the mother's fur to reach the **marsupium,** an abdominal pouch within which it attaches to nipples and continue its development **(Figure 47.9).** Kangaroos, koalas, wombats, and opossums are marsupials.
- **Ovoviviparous** animals retain fertilized eggs within the body and the embryo develops using the nutrients provided by the egg. There is no uterus or placenta involved. When development is complete, the eggs hatch inside the mother and the offspring are released to the exterior. Ovoviviparity is seen in some fishes, lizards, and amphibians, many snakes, and many invertebrates.

John Cancalosi/Peter Arnold, Inc.

FIGURE 47.9
Developing offspring of a marsupial mammal, an opossum, attached to nipples in the marsupium (pouch) of the mother.

Hermaphroditism Is a Variation on Sexual Reproduction

Some animals have evolved modified mechanisms that they use as their normal sexual reproduction process. One of these mechanisms is **hermaphroditism** (from *Hermaphroditos,* the son of *Hermes* and *Aphrodite,* a Greek god and goddess), in which both mature egg-producing and mature sperm-producing tissue is present in the same individual. That is, hermaphroditic individuals are able to produce both eggs and sperm. Most flatworms, earthworms, land snails, and numerous other invertebrates are hermaphroditic. In humans and other mammals there are rare cases of the abnormal condition of individuals who have both testicular and ovarian tissues. They are not true hermaphrodites because they are not both male and female; rather, they are either male or female with ambiguous genitalia. Hence, they are called pseudohermaphrodites rather than true hermaphrodites.

Most hermaphroditic animals do not fertilize themselves. In those animals, self-fertilization is prevented by anatomical barriers that prevent individuals from introducing sperm into their own body, or by mechanisms in which the egg and sperm mature at different times. The prevention of self-fertilization maintains the genetic variability of sexual reproduction.

Hermaphroditism takes two forms: **simultaneous hermaphroditism,** in which individuals develop functional ovaries and testes at the same time, and **sequential hermaphroditism,** in which individuals change from one sex to the other. The two earthworms shown in **Figure 47.10** provide a common example of simultaneous hermaphroditism. The only known vertebrate simultaneous hermaphrodites are hamlets (genus *Hypoplectrus*), a group of predatory sea basses. Sequential hermaphroditism is seen among a number of invertebrates (for example, the slipper shell *Crepidula fornicata,* a gastropod), and some ectothermic vertebrates, notably fishes (for example, the clownfish, genus *Amphiprion*). In some species the initial sex is male (as with the slipper shell and the clownfish), and in others it is female.

STUDY BREAK 47.2

1. **What are egg coats, and what is their function? What egg coats do mammalian and bird eggs have?**
2. **How is the slow block to polyspermy brought about?**

THINK OUTSIDE THE BOOK

Use the Internet or research literature to outline a model for how hermaphroditism may have evolved.

A. **Mating earthworms**

Robin Chittenden; Frank Lane Picture Agency/Corbis

B. **Sex organs**

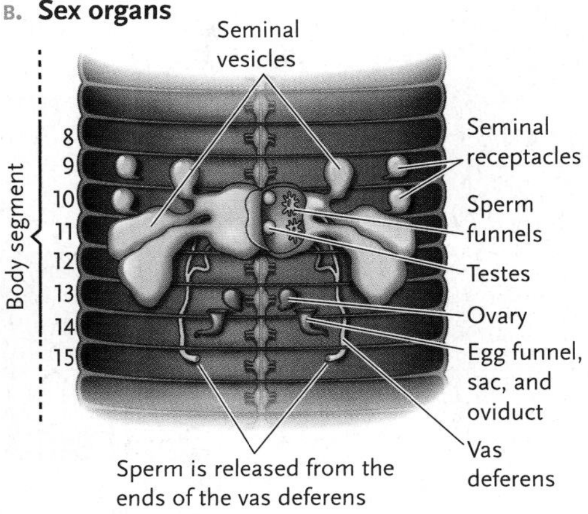
Seminal vesicles
Body segment
8
9
10
11
12
13
14
15
Seminal receptacles
Sperm funnels
Testes
Ovary
Egg funnel, sac, and oviduct
Vas deferens
Sperm is released from the ends of the vas deferens

FIGURE 47.10
Simultaneous hermaphroditism in the earthworm. **(A)** Copulation by a mating pair of earthworms, in which each earthworm releases sperm that fertilize eggs in its partner. **(B)** Sex organs in the earthworm.

47.3 Sexual Reproduction in Humans

Except for structural details, human reproduction is typical of that of eutherian (placental) mammals. Internally, these mammals have a pair of gonads, either ovaries or testes. The gonads have a dual function in mammals, as they do in all vertebrates: they both produce gametes and secrete hormones that are responsible for sexual development and mating behavior (see Section 40.4). Males have ducts that carry sperm from the testes to the exterior. Females have an oviduct that leads from each ovary to the uterus, in which fertilized eggs implant and proceed through embryonic development. Nutrients from the mother and wastes from the embryo are exchanged through the placenta. After birth, the newborn offspring is nourished with milk secreted by the mother's mammary glands.

In this section we survey reproductive structures and functions in humans as representative of eutherian mammals. Our story of human development continues in the next chapter, which traces the process from fertilization to birth.

Human Female Sexual Organs Function in Oocyte Production, Fertilization, and Embryonic Development

Human females have a pair of ovaries suspended in the abdominal cavity **(Figure 47.11).** An oviduct leads from each ovary to the uterus, which is a hollow, saclike organ with walls containing smooth muscle. The uterus is lined by the endometrium, which is formed by layers of connective tissue with embedded glands and is richly supplied with blood vessels. If an egg is fertilized and begins development, it must implant in the endometrium to continue developing. The lower end of the uterus, the **cervix,** opens

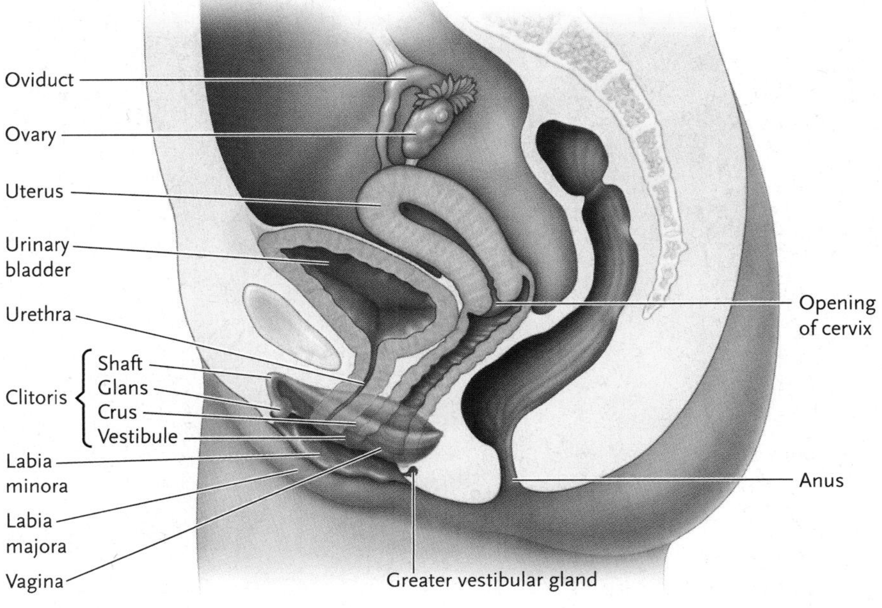
Oviduct
Ovary
Uterus
Urinary bladder
Urethra
Clitoris
Shaft
Glans
Crus
Vestibule
Labia minora
Labia majora
Vagina
Greater vestibular gland
Opening of cervix
Anus

FIGURE 47.11
The reproductive organs of a human female.

into a muscular canal, the **vagina,** which leads to the exterior. Sperm enter the female reproductive tract via the vagina and, at birth, the baby passes from the uterus to the outside through the vagina.

At the birth of a female, each of her ovaries contains about 1 million oocytes, arrested at the end of the first meiotic prophase. Of these oocytes, about 200,000 to 400,000 survive until a female becomes sexually mature; about 400 are **ovulated**—released into the oviducts as immature eggs—during the lifetime of a woman. The egg is released into the abdominal cavity and is pulled into the nearby oviduct by the current produced by the beating of the cilia lining the oviduct. The cilia also propel the egg through the oviduct and into the uterus. Fertilization of the egg occurs in the oviduct.

The external female sex organs, collectively called the **vulva,** surround the opening of the vagina. Two folds of tissue, the **labia minora,** run from front to rear on either side of the opening to the vagina. These folds are partially covered by a pair of fleshy, fat-padded folds, the **labia majora,** which also run from front to rear on either side of the vagina. At the anterior end of the vulva, the labia minora join to partly cover the head of the **clitoris.** The rest of the clitoris is within the body. The clitoris contains erectile tissue and has the same embryonic origins as the penis. A pair of **greater vestibular glands,** with openings near the entrance to the vagina, secretes a mucus-rich fluid that lubricates the vulva. The opening of the urethra, which conducts urine from the bladder, is located between the clitoris and the vaginal opening. Most nerve endings associated with erotic sensations are concentrated in the clitoris, in the labia minora, and around the opening of the vagina. When a human female is born, a thin flap of tissue, the **hymen,** partially covers the opening of the vagina. This membrane, if it has not already been ruptured by physical exercise or other disturbances, is broken by the first sexual intercourse.

Ovulation in Human Females Occurs in a Monthly Cycle

Reproduction in human females is under neuroendocrine control, involving complex interactions between the hypothalamus, pituitary, ovaries, and uterus. Under this control, approximately every 28 days from puberty to menopause, a female releases an egg from one of her ovaries. The cyclic events in the ovary leading to ovulation are known as the **ovarian cycle.** This cycle is coordinated with the **uterine cycle,** or **menstrual cycle** (*menstruus* = monthly), events in the uterus that prepare it to receive the egg if fertilization occurs. While the cycle is commonly considered a monthly one, there is a lot of variation among human females with a range of about 15 to 45 days, and an average is about 28 to 29 days.

THE OVARIAN CYCLE The ovarian cycle produces a mature egg **(Figure 47.12).** The starting point for the cycle is a primary oocyte in prophase of meiosis division I. The beginning of the cycle is triggered by an increase in the release of **gonadotropin-releasing hormone (GnRH)** by the hypothalamus. This hormone stimulates the pituitary to release **follicle-stimulating hormone (FSH)**

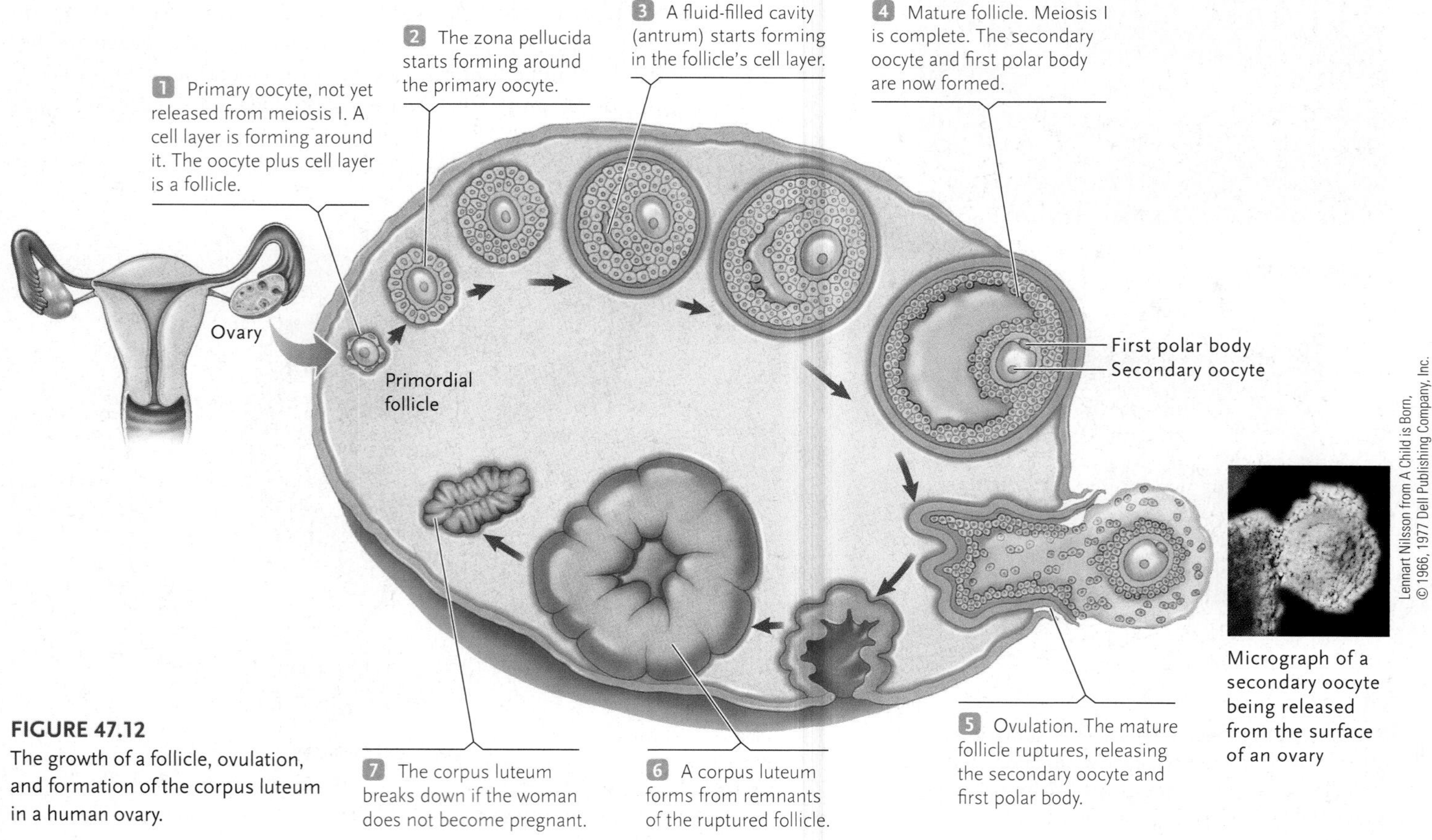
1 Primary oocyte, not yet released from meiosis I. A cell layer is forming around it. The oocyte plus cell layer is a follicle.
2 The zona pellucida starts forming around the primary oocyte.
3 A fluid-filled cavity (antrum) starts forming in the follicle's cell layer.
4 Mature follicle. Meiosis I is complete. The secondary oocyte and first polar body are now formed.
Ovary
Primordial follicle
First polar body
Secondary oocyte
5 Ovulation. The mature follicle ruptures, releasing the secondary oocyte and first polar body.
6 A corpus luteum forms from remnants of the ruptured follicle.
7 The corpus luteum breaks down if the woman does not become pregnant.
Micrograph of a secondary oocyte being released from the surface of an ovary
Lennart Nilsson from A Child is Born, © 1966, 1977 Dell Publishing Company, Inc.

FIGURE 47.12
The growth of a follicle, ovulation, and formation of the corpus luteum in a human ovary.

and **luteinizing hormone (LH)** into the bloodstream **(Figure 47.13A).** FSH stimulates six to 20 primary oocytes in the ovaries to be released from prophase of meiosis I and continue through the meiotic divisions. As the primary oocytes develop into secondary oocytes—which arrest in metaphase of meiosis II—they become surrounded by cells that form a **follicle** (day 2 of the cycle; **Figure 47.13B**). During this follicular phase, the follicle grows and develops and, at its largest size, becomes filled with fluid and may reach 12 to 15 mm in diameter. Usually only one follicle develops to maturity with release of the egg (secondary oocyte) by ovulation. If two or more follicles develop and their eggs are ovulated, multiple births can result.

As the follicle enlarges, FSH and LH interact to stimulate the follicular cells to secrete **estrogens** (female sex hormones), primarily **estradiol** (see Section 40.4) **(Figure 47.13C).** Initially, the estrogens are secreted in low amounts; at this level, the estrogens have a negative feedback effect on the pituitary, inhibiting its secretion of FSH. As a result, FSH secretion declines briefly. However, estrogen secretion increases steadily, and its level peaks at about 12 days after follicle development begins (day 14 of cycle). The high estrogen level now has a positive feedback effect on the hypothalamus and pituitary, increasing the release of GnRH and stimulating the pituitary to release a burst of FSH and LH. The increased estrogen levels also convert the mucus secreted by the uterus to a thin and watery consistency, making it easier for sperm to swim through the uterus.

The burst in LH secretion stimulates the follicle cells to release enzymes that digest away the wall of the follicle, causing it to burst and release the egg (see Figure 47.12); this is ovulation. LH also initiates the last phase of the menstrual cycle, the *luteal phase.* That is, LH causes the follicle cells remaining at the surface of the ovary to grow into an enlarged, yellowish structure, the **corpus luteum** (*corpus* = body, *luteum* = yellow) (see Figure 47.12). Acting as an endocrine gland, the corpus luteum secretes several hormones: estrogens, large quantities of **progesterone,** and **inhibin.** Progesterone, a female sex hormone, stimulates growth of the uterine lining and inhibits contractions of the uterus. Both progesterone and inhibin have a negative feedback effect on the hypothalamus and pituitary. Progesterone inhibits the secretion of GnRH. Without GnRH,

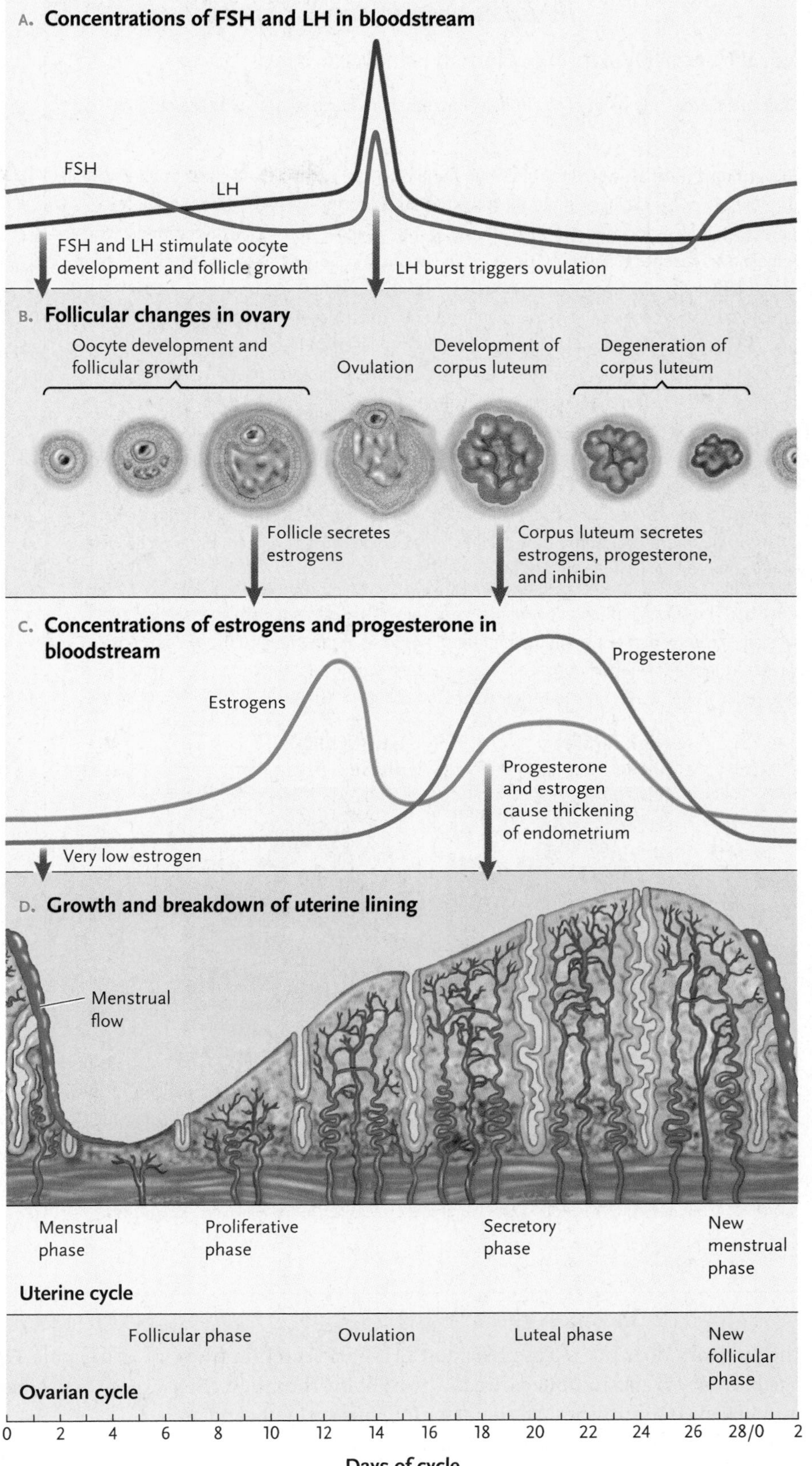

A. Concentrations of FSH and LH in bloodstream
FSH
LH
FSH and LH stimulate oocyte development and follicle growth
LH burst triggers ovulation
B. Follicular changes in ovary
Oocyte development and follicular growth
Ovulation
Development of corpus luteum
Degeneration of corpus luteum
Follicle secretes estrogens
Corpus luteum secretes estrogens, progesterone, and inhibin
C. Concentrations of estrogens and progesterone in bloodstream
Progesterone
Estrogens
Progesterone and estrogen cause thickening of endometrium
Very low estrogen
D. Growth and breakdown of uterine lining
Menstrual flow
Menstrual phase
Proliferative phase
Secretory phase
New menstrual phase
Uterine cycle
Follicular phase
Ovulation
Luteal phase
New follicular phase
Ovarian cycle
0 2 4 6 8 10 12 14 16 18 20 22 24 26 28/0 2
Days of cycle

FIGURE 47.13

The ovarian and uterine (menstrual) cycles of a human female. (A) The changing concentrations of FSH and LH in the bloodstream, triggered by GnRH secretion by the hypothalamus. **(B)** The cycle of follicle development, ovulation, and formation of the corpus luteum in the ovary. **(C)** The concentrations of estrogens and progesterone in the bloodstream. **(D)** The growth and breakdown of the uterine lining. The days of the monthly cycle are given in the scale at the bottom of the diagram.

FIGURE 47.14 **Experimental Research**

Vocal Cues of Ovulation in Human Females

Question: Does the pitch of the female voice change in association with ovulation in human females?

Experiment: Primate females other than human females display well-characterized visual or olfactory cues that signal their reproductive state. Some research has documented particular cues associated with ovulation in humans; those cues relate to femininity and female attractiveness. Based on those observations, Gregory Bryant and Martie Haselton of the University of California, Los Angeles, hypothesized that vocal cues associated with female attractiveness would increase in frequency (pitch) over the menstrual cycle. A higher voice pitch is a signal associated with higher levels of female sex hormones, and correlates with being younger, both of which correlate with high fertility. To test their hypothesis, the researchers recruited 69 women with normal menstrual cycles and collected two sets of vocal samples from them saying "Hi, I'm a student at UCLA." One set was taken during a high-fertility phase of the cycle (at a time near ovulation) and the other set was taken during a low-fertility phase (at the luteal phase). The phase of the menstrual cycle at the time of vocal sampling was confirmed directly by hormonal tests. The samples were analyzed for the pitches of the voices and significant differences were determined using statistical methods.

Results: The average pitch of women's voices when the subjects said a simple sentence was significantly higher during the high-fertility phase of the menstrual cycle compared with during the low-fertility phase. The researchers confirmed in blind studies that the difference in frequency is readily detectable.

High-fertility phase

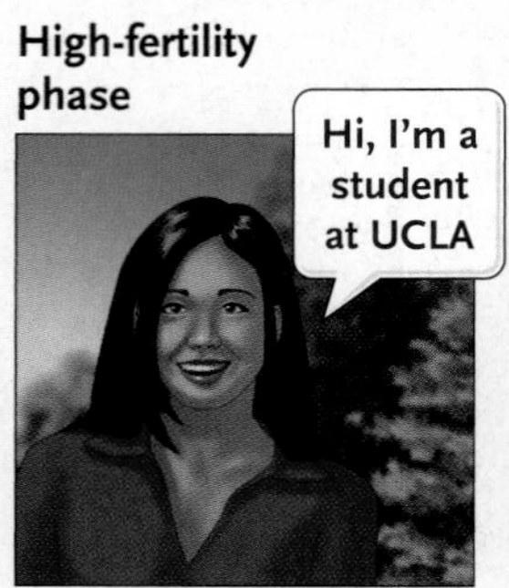
Hi, I'm a
student
at UCLA

Average pitch = 211 Hz

Low-fertility phase

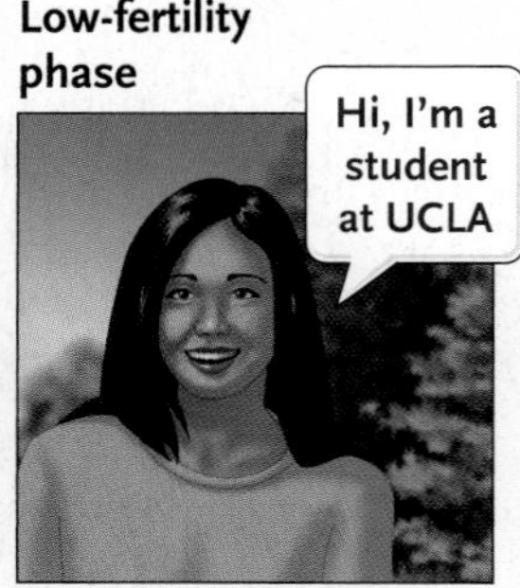
Hi, I'm a
student
at UCLA

Average pitch = 206 Hz

Conclusion: The change in vocal pitch seen for high-fertility (approaching ovulation) versus low-fertility phases of the menstrual cycle is seen as evidence for a cyclic fertility cue in the female human voice. Potentially the cue signals to males that females are in a high-fertility state.

Source: G. A. Bryant and M. G. Haselton. 2009. Vocal cues of ovulation in human females. *Biology Letters* 5:12–15.

the pituitary does not release FSH and LH. FSH secretion from the pituitary is also inhibited directly by inhibin. The fall in FSH and LH levels diminishes the signal for follicular growth, and no new follicles begin to grow in the ovary.

If fertilization does not occur, the corpus luteum gradually degenerates as cells are phagocytized and blood supply is cut off. By about 10 days after ovulation, little tissue remains, meaning that estrogen, progesterone, and inhibin are no longer secreted. In the absence of progesterone, *menstruation* begins (described in the next section). As progesterone and inhibin levels decrease, FSH and LH secretion is no longer inhibited, and a new monthly cycle begins.

THE UTERINE (MENSTRUAL) CYCLE The hormones that control the ovarian cycle also control the uterine (menstrual) cycle **(Figure 47.13D)**, keeping the processes connected physiologically. Day 0 of the monthly cycle in the figure is the beginning of follicular development in the ovary (see Figure 47.13B); in the uterus, this correlates with the time at which menstrual flow begins.

Menstrual flow results from the breakdown of the thickened endometrium. Blood and tissue breakdown products are released from the uterus to the outside through the vagina. When the flow ceases, at day 4 to 5 of the cycle, the endometrium begins to grow again; this is the proliferative phase. As the endometrium gradually thickens, the oocytes in both ovaries begin to develop further, eventually leading to ovulation at about 14 days after the beginning of the cycle, as already described. If fertilization does not take place, the uterine lining continues to grow for another 14 days after ovulation; this is the secretory phase. At the end of that time, the absence of progesterone results in the contraction of arteries supplying blood to the uterine lining, shutting down the blood supply and causing the lining to disintegrate. The menstrual flow begins. Contractions of the uterus, no longer inhibited by progesterone, help expel the debris. Prostaglandins released by the degenerating endometrium add to the uterine contractions, making them severe enough to be felt as the pain of "cramps," and also sometimes causing other effects such as nausea, vomiting, and headaches.

Menstruation—the menstrual flow—occurs only in human females and our closest primate relatives, gorillas and chimpanzees. In other mammals, the uterine lining is completely reabsorbed if a fertilized egg does not implant during the period of reproductive activity. The uterine cycle in those mammals is called the *estrous* cycle.

Hormonal tests can show where a woman is in her menstrual cycle. In non-human primates various cues inform the male of the reproductive state of a female. **Figure 47.14** describes a research study demonstrating that vocal cues may signal the fertility state of human females.

Human Male Sexual Organs Function in Sperm Production and Delivery

Organs that produce and deliver sperm make up the male reproductive system **(Figure 47.15)**. The testes are located outside the abdominal cavity. Sperm, which are produced by the testes, pass through tubules that enter the abdominal cavity and join with

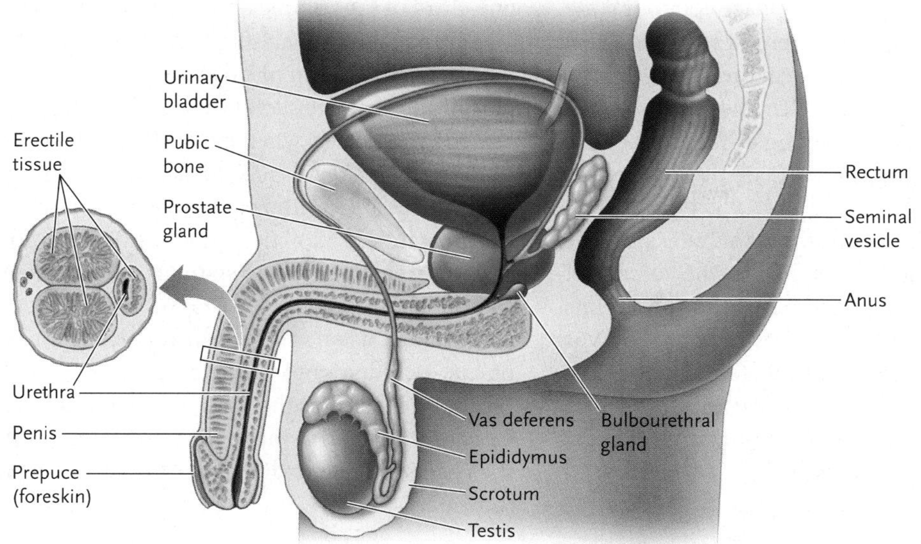
Urinary bladder
Erectile tissue
Pubic bone
Prostate gland
Rectum
Seminal vesicle
Anus
Urethra
Penis
Prepuce (foreskin)
Vas deferens
Epididymus
Scrotum
Testis
Bulbourethral gland

FIGURE 47.15
The reproductive organs of a human male.

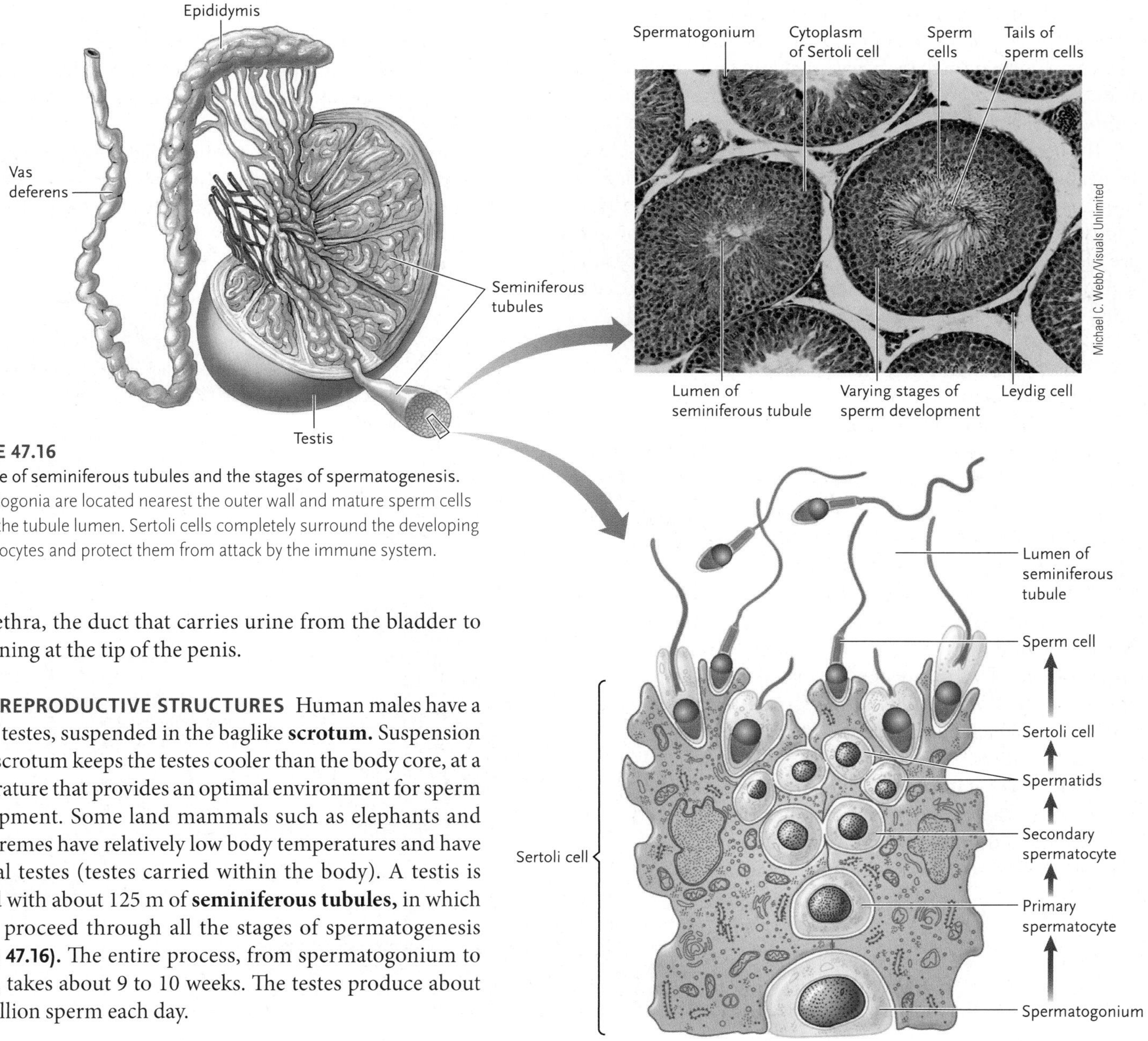

Epididymis
Vas deferens
Seminiferous tubules
Testis
Spermatogonium
Cytoplasm of Sertoli cell
Sperm cells
Tails of sperm cells
Lumen of seminiferous tubule
Varying stages of sperm development
Leydig cell
Michael C. Webb/Visuals Unlimited
Lumen of seminiferous tubule
Sperm cell
Sertoli cell
Spermatids
Secondary spermatocyte
Primary spermatocyte
Spermatogonium
Sertoli cell

FIGURE 47.16
Structure of seminiferous tubules and the stages of spermatogenesis. Spermatogonia are located nearest the outer wall and mature sperm cells nearest the tubule lumen. Sertoli cells completely surround the developing spermatocytes and protect them from attack by the immune system.

the urethra, the duct that carries urine from the bladder to an opening at the tip of the penis.

MALE REPRODUCTIVE STRUCTURES Human males have a pair of testes, suspended in the baglike **scrotum.** Suspension in the scrotum keeps the testes cooler than the body core, at a temperature that provides an optimal environment for sperm development. Some land mammals such as elephants and monotremes have relatively low body temperatures and have internal testes (testes carried within the body). A testis is packed with about 125 m of **seminiferous tubules,** in which sperm proceed through all the stages of spermatogenesis **(Figure 47.16).** The entire process, from spermatogonium to sperm, takes about 9 to 10 weeks. The testes produce about 130 million sperm each day.

Supportive cells called **Sertoli cells** completely surround the developing spermatocytes in the seminiferous tubules. They supply nutrients to the spermatocytes and seal them off from the body's blood supply. Other cells located in the tissue surrounding the developing spermatocytes, the **Leydig cells,** produce the male sex hormones, known as **androgens,** particularly **testosterone** (see Figure 47.16).

Mature sperm flow from the seminiferous tubules into the **epididymis,** a coiled storage tubule attached to the surface of each testis. Rhythmic muscular contractions of the epididymis move the sperm into a thick-walled, muscular tube, the **vas deferens** (plural, *vasa deferentia*), which leads into the abdominal cavity. Just below the bladder, the vasa deferentia empty into the urethra. During ejaculation, muscular contractions force the sperm into the urethra and out of the penis. The sperm are activated and become motile as they come in contact with alkaline secretions added to the ejaculated fluid by accessory glands.

Most of the interior of the penis is filled with three cylinders of spongelike tissue that become filled with blood and cause erection during sexual arousal. Although the human penis depends solely on engorgement of spongy tissue for erection, the males of many mammalian species, including bats, rodents, walruses, and most other primates, have a bone in the penis, the *baculum,* which helps maintain the penis in an erect state.

The penis ends in a soft, caplike structure, the **glans.** Most of the nerve endings producing erotic sensations are crowded into the glans and the region of the penile shaft just behind the glans. A loose fold of skin, the **prepuce** or **foreskin,** covers the glans (see Figure 47.15). In many cultures the prepuce is removed for hygienic, religious, or other ritualistic reasons by the procedure called **circumcision** (*circum* = around, *caedere* = cut).

ACCESSORY GLANDS AND THE SEMEN

About 150 million to 350 million sperm are released in a single ejaculation. Before they leave the body, these cells are mixed with the secretions of several accessory glands, forming the fluid known as **semen.** In humans, about two-thirds of the volume is produced by a pair of **seminal vesicles,** which secrete a thick, viscous liquid, the **seminal fluid,** into the vasa deferentia near the point where they join with the urethra. The seminal fluid contains prostaglandins that, when ejaculated into the female, trigger contractions of the female reproductive tract which help move the sperm into and through the uterus.

The large **prostate gland,** which surrounds the region where the vasa deferentia empty into the urethra, adds a thin, milky fluid to the semen. The alkaline prostate secretion, which makes up about one-third of the volume of the semen, raises the pH of the semen, and of the vagina, to about pH 6, the level of acidity best tolerated by sperm. The raised pH also activates motility of the sperm. As part of the prostate secretion, a fast-acting enzyme converts the semen to a thick gel when it is first ejaculated. The thickened consistency helps keep the semen from draining from the vagina when the penis is withdrawn. A second, slower-acting enzyme in the prostate secretion then gradually breaks down the semen clot and releases the sperm to swim freely in the female reproductive tract.

Finally, a pair of **bulbourethral glands** secretes a clear, mucus-rich fluid into the urethra before and during ejaculation. This fluid lubricates the tip of the penis and neutralizes the acidity of any residual urine in the urethra. In total, the secretions of the accessory glands make up more than 95% of the volume of semen; less than 5% is sperm.

Hormones Also Regulate Male Reproductive Functions

Many of the hormones regulating the menstrual cycle, including GnRH, FSH, LH, and inhibin, also regulate male reproductive functions. Testosterone, secreted by the Leydig cells in the testes, also plays a key role **(Figure 47.17).**

In sexually mature males, the hypothalamus secretes GnRH in brief pulses every 1 to 2 hours. The GnRH, in turn, stimulates the pituitary to secrete LH and FSH. LH stimulates the Leydig

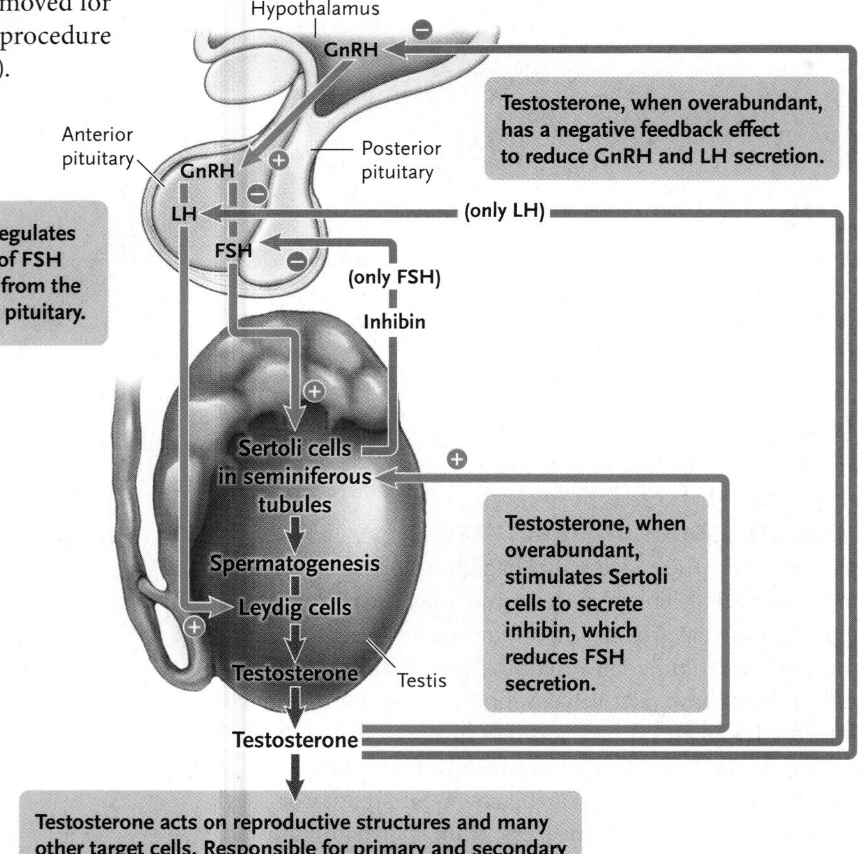
Hypothalamus
GnRH
Testosterone, when overabundant, has a negative feedback effect to reduce GnRH and LH secretion.
Anterior pituitary
Posterior pituitary
GnRH
LH
(only LH)
FSH
GnRH regulates release of FSH and LH from the anterior pituitary.
(only FSH)
Inhibin
Sertoli cells in seminiferous tubules
Spermatogenesis
Leydig cells
Testosterone
Testis
Testosterone, when overabundant, stimulates Sertoli cells to secrete inhibin, which reduces FSH secretion.
Testosterone
Testosterone acts on reproductive structures and many other target cells. Responsible for primary and secondary characteristics of the male, including sperm production.

FIGURE 47.17
Hormonal regulation of reproduction in the male, and the negative feedback systems controlling hormone levels.

cells to secrete testosterone, which stimulates sperm production and controls the growth and function of male reproductive structures. FSH stimulates Sertoli cells to secrete a protein and other molecules that are required for spermatogenesis.

The concentrations of these hormones are maintained by feedback mechanisms. If the concentration of testosterone falls in the bloodstream, the hypothalamus responds by increasing GnRH secretion. If the concentration of testosterone rises too high, negative feedback mechanisms operate to return its concentration to an optimal level in the bloodstream (see Figure 47.17).

Human Copulation Follows a Typical Mammalian Pattern

When the male is sexually aroused, the sphincter muscles controlling the flow of blood to the spongy erectile tissue of the penis relax, allowing the tissue to become engorged with blood. (The penis is a hydrostatic skeleton structure; see Section 41.2.) As the spongy tissue swells, it maintains the pressure by compressing and almost shutting off the veins draining blood from the penis. The engorgement produces an erection in which the penis lengthens, stiffens, and enlarges. During continued sexual arousal, lubricating fluid secreted by the bulbourethral glands may be released from the tip of the penis.

Female sexual arousal results in enlargement and erection of the clitoris, in a process analogous to erection of the penis in males. The labia minora also become engorged with blood and swell in size, and lubricating fluid is secreted onto the surfaces of the vulva by the greater vestibular glands. In addition to these changes, the nipples become erect by contraction of smooth muscle cells, and the breasts swell due to engorgement with blood.

Insertion of the penis into the vagina and the thrusting movements of copulation lead to the reflex actions of ejaculation, including spasmodic contractions of muscles surrounding the vasa deferentia, accessory glands, and urethra. During ejaculation, the sphincter muscles controlling the exit from the bladder close tightly, preventing urine from being released from the bladder and mixing with the ejaculate. Ejaculation is usually accompanied by *orgasm,* a sensation of intense physical pleasure that is the peak—climax—of excitement for sexual intercourse, followed by feelings of relaxation and gratification.

The motions of copulation stretch the vagina and stimulate the clitoris. The stretching and stimulation can also induce orgasm in females. The vaginal stretching also stimulates the hypothalamus to secrete oxytocin, which induces contractions of the uterus. The contractions keep the sperm in suspension and aid their movement through the reproductive tract. Uterine contractions are also induced by the prostaglandins in the semen.

Sperm reach the site of fertilization in the oviducts within 30 minutes after their ejaculation into the vagina. Of the millions of sperm released in a single ejaculation, only a few hundred actually reach the oviducts. After orgasm, the penis, clitoris, and labia minora gradually return to their unstimulated size. Females can experience additional orgasms within minutes or even seconds of a first orgasm, but most males enter a *refractory period* that lasts for 15 minutes or longer before they can regain an erection and have another orgasm.

A Human Egg Can Be Fertilized Only in the Oviduct

A human egg can be fertilized only during its passage through the third of the oviduct nearest the ovary. If the egg is not fertilized during the 12- to 24-hour period that it is in this location, it disintegrates and dies. However, sperm do not swim randomly for a chance encounter with the egg. Rather, they first swim up the cervical canal to reach the oviduct, and then are propelled up the oviduct by contractions of its smooth muscles. Further, researchers have found evidence that eggs release chemical attractant molecules that the sperm recognize, causing them to swim directly toward the egg. (*Insights from the Molecular Revolution* describes some of this research.)

To reach the egg, the fertilizing sperm uses enzymes in its plasma membranes to penetrate the layer of follicle cells surrounding the egg, and then by the acrosomal reaction dissolves a path through the zona pellucida coating the egg surface **(Figure 47.18).** As soon as the first sperm cell reaches the egg, the sperm and egg plasma membranes fuse, and the sperm cell enters the cytoplasm of the egg. Although only one sperm fertilizes the egg, the combined release of acrosomal enzymes from many sperm greatly increases the chance that a complete channel will be opened through the zona pellucida. Partly for this reason, a low sperm count is often a source of male infertility. Low sperm count has a number of causes, including infection, heat, frequent intercourse, smoking, and excess alcohol consumption.

The membrane fusion activates the egg. The sperm that has entered the egg releases nitric oxide, which stimulates the release

FIGURE 47.18 Fertilization in mammals.

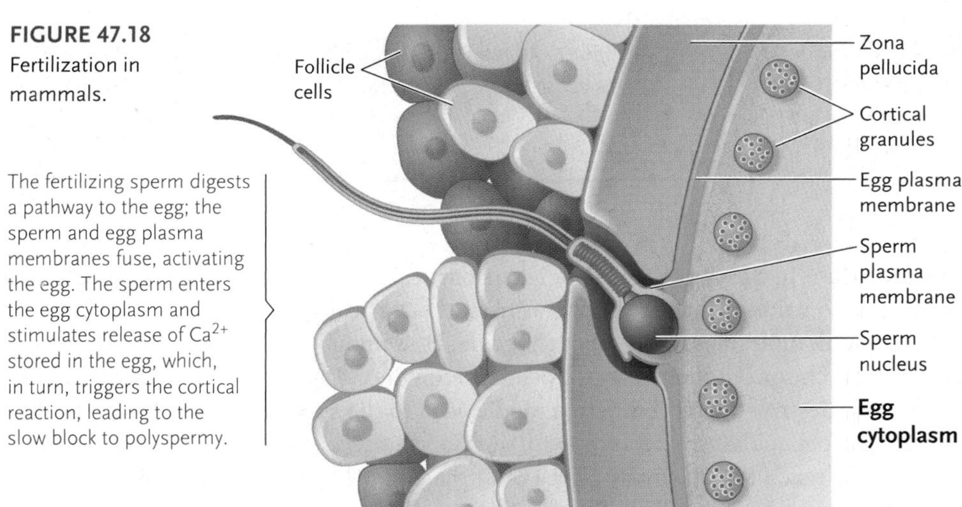
The fertilizing sperm digests a pathway to the egg; the sperm and egg plasma membranes fuse, activating the egg. The sperm enters the egg cytoplasm and stimulates release of Ca^{2+} stored in the egg, which, in turn, triggers the cortical reaction, leading to the slow block to polyspermy.
Follicle cells
Zona pellucida
Cortical granules
Egg plasma membrane
Sperm plasma membrane
Sperm nucleus
Egg cytoplasm

INSIGHTS FROM THE Molecular Revolution

Egging on the Sperm

Whether human eggs release attractants to draw sperm near has long been a subject of speculation and research. Some investigators had found that human sperm cells have receptors able to bind to chemical substances classified as odorants, aroma molecules that can be specifically recognized ("smelled"). In vertebrates, more than a thousand genes encode odorant receptors, most of them olfactory receptors associated with the senses of smell and taste. However, odorant receptors are also located on cell types that do not function in taste and smell, including sperm.

Research Question

Do sperm odorant receptors function in sperm–egg attraction?

Experiments

Marc Spehr and his colleagues at Ruhr University Bochum performed a number of molecular experiments to answer the question.

Experiment 1: The researchers tested testicular tissue for odorant receptor gene activity by using Northern blotting to probe for mRNAs with sequences typical of odorant receptor genes. (Northern blotting is described in Section 18.2.)
Result: Only two active odorant receptor genes were found in the testes: hOR17-2 (hOR = *h*uman *o*lfactory *r*eceptor), which had been detected previously by others; and hOR17-4, which was not previously known to be active.

Experiment 2: Spehr's group molecularly cloned the hOR17-4 gene and introduced the gene into cultured human embryonic kidney (HEK) cells to determine if the receptor encoded by the gene could respond to an odorant **(figure).** Previous work had shown that when an odorant receptor combines with the chemical it recognizes, it activates the receptor and triggers a cellular response, specifically reactions that lead to Ca^{2+} release in the cytoplasm (the IP3 pathway, described in Section 7.4), a property that can be measured readily. The investigators then tested a number of chemicals to see if any caused a cellular response.

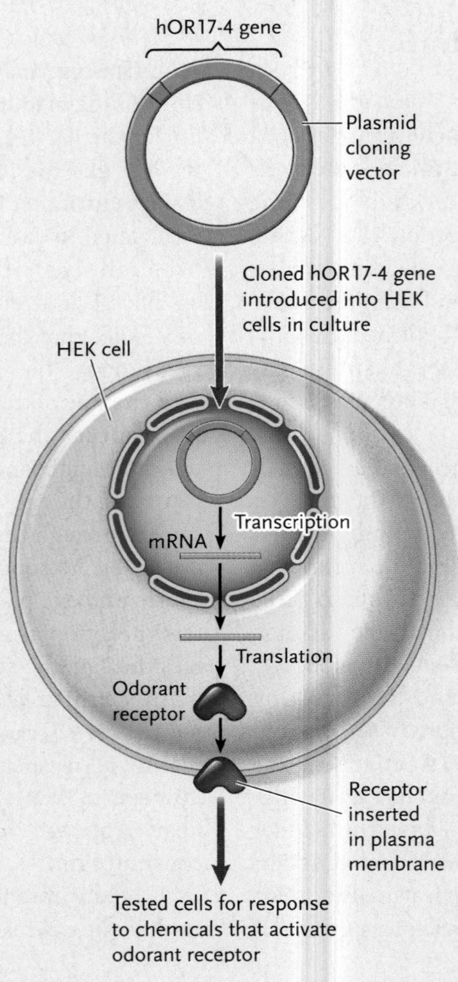
hOR17-4 gene
Plasmid cloning vector
Cloned hOR17-4 gene introduced into HEK cells in culture
HEK cell
Transcription
mRNA
Translation
Odorant receptor
Receptor inserted in plasma membrane
Tested cells for response to chemicals that activate odorant receptor

Result: Spehr's group identified a chemical, *bourgeonal,* that triggers a strong cellular response in the genetically modified cells.

Control HEK cells that lacked the cloned gene did not respond to bourgeonal.

Experiment 3: Because the HEK cells are not the natural cells involved in fertilization, the investigators next tested human sperm cells to see if they too would respond to bourgeonal.
Result: The sperm cells responded to the chemical by an increase in cytoplasmic Ca^{2+} concentration, indicating that the sperm cells have the hOR17-4 odorant receptor on their surfaces and, therefore, that the hOR17-4 gene was active during spermatogenesis.

Experiment 4: In a final experiment, the researchers exposed human sperm cells to gradients of bourgeonal solutions in micropipettes.
Result: The sperm swam consistently toward the regions of highest concentration, and swam faster and more directly as the concentration increased. This experiment showed that human sperm can detect and respond to chemical attractants by swimming toward the source of the attractant.

Conclusion

The results show that an odorant receptor is on a sperm cell's surface and that sperm will swim towards a chemical known to activate that receptor. Whether human eggs actually release such chemicals that act as sperm attractants remains to be determined. If so, the egg attractant detected by the hOR17-4 receptor is likely to resemble bourgeonal in chemical structure, because odorant receptors are highly specific in their responses.

As part of their research, the Spehr team found that another chemical, *undecanal,* strongly inhibits the binding of the hOR17-4 receptor to bourgeonal. With chemicals at hand that can both stimulate and eliminate sperm attraction, the system might provide a method for either contraception or procreation.

---

of stored Ca^{2+} in the egg. The Ca^{2+} triggers cortical granule release to the outside of the egg. Enzymes from the cortical granules crosslink molecules in the zona pellucida, hardening it and sealing the channels opened by acrosomal enzymes. The enzymes also destroy the receptors that bind sperm to the surface of the zona pellucida. As a result, no further sperm can bind to the zona or reach the plasma membrane of the egg; this is the slow block to polyspermy. The Ca^{2+} also triggers the completion of meiosis of the egg (recall that, up to that point, it is a secondary oocyte arrested in meiosis II). The sperm and egg nuclei then fuse and the cell is now a zygote. Mitotic divisions of the zygote soon initiate embryonic development.

The first cell divisions of embryonic development take place while the fertilized egg is still in the oviduct. By about 7 days after ovulation, the embryo passes from the oviduct and implants in the uterine lining. During and after implantation, cells associated with the embryo secrete **human chorionic gonadotropin (hCG),** a hormone that keeps the corpus luteum in the ovary from breaking down. Excess hCG is excreted in the urine; its presence in urine or blood provides the basis of pregnancy tests.

The continued activity of the corpus luteum keeps estrogen and progesterone secretion at high levels, maintaining the uterine lining and preventing menstruation. The high progesterone level also thickens the mucus secreted by the uterus, forming a plug that seals the opening of the cervix from the vagina. The plug keeps bacteria, viruses, and sperm cells from further copulation from entering the uterus.

Later in development, about 10 weeks after implantation, the placenta takes over the secretion of progesterone, hCG secretion drops off, and the corpus luteum regresses. However, the corpus luteum continues to secrete the hormone *relaxin,* which inhibits contraction of the uterus until the time of birth is near.

Infertility Has Many Possible Causes

About 10–15% of couples in the U.S. are infertile. **Infertility** is defined as the inability for the female of the couple to get pregnant after 12 months of frequent, contraceptive-free intercourse. Infertility may result from a cause in either member of the couple, or a combination of factors involving both members of the couple.

MALE INFERTILITY Male infertility may result from a problem with the sperm or with the delivery of the sperm into the vagina. Problems with the sperm include low sperm concentration or altered sperm shape that affects mobility. For example, testosterone deficiency can lead to a reduced sperm count, and sexually transmitted diseases may temporarily affect sperm motility. In many cases, however, the root cause of low sperm count is not known. Problems with sperm delivery include sexual issues such as erectile dysfunction and blockage of the epididymis or ejaculatory ducts. Environmental factors such as a person's health or lifestyle can also affect male fertility. Such factors include obesity, being underweight, malnutrition, emotional stress, alcohol or drug dependency, exposure to pesticides or other chemicals, and tobacco smoking.

FEMALE INFERTILITY Female infertility may result from any of a variety of causes including physical or hormonal changes. For example, inflammation of the Fallopian tubes (most frequently caused by *Chlamydia* infection) can cause damage, thereby affecting movement of the egg through the tube. Hormonal defects such as low levels of the hypothalamus–pituitary controlled LH and FSH, or certain medications, can block ovulation. Hormone deficiencies such as this can result from various perturbations of normal activity of the hypothalamus and pituitary, including tumors, injury, and excessive exercise. Many of the environmental factors that cause female infertility are the same as those that cause male infertility. In addition, female athletes with vigorous training regimes, such as cycling and running, may experience dysfunctions of their menstrual cycles, a condition called *athletic menstrual cycle irregularities.* The dysfunctions include amenorrhea (no menstrual periods), and cycles of normal length but without ovulation.

STUDY BREAK 47.3

Outline the roles of follicle-stimulating hormone (FSH) and luteinizing hormone (LH) in the ovarian cycle of a human female.

THINK OUTSIDE THE BOOK

The concept of athletic menstrual cycle irregularities was just introduced. Use the Internet or research literature to learn about possible hormonal differences in female athletes and to determine whether they might be involved in athletic menstrual cycle irregularities.

47.4 Methods for Preventing Pregnancy: Contraception

In human society, an unwanted pregnancy can be inconvenient at the least, or at the worst can have serious physical and social repercussions, particularly for the mother. Many methods exist for achieving **contraception**—the prevention of pregnancy—some old and others relatively new.

The oldest method of contraception is total abstinence from sex. Unfortunately, millions of years of animal evolution have stacked the cards against total abstinence by making the sex drive among the most powerful of compulsions. Millions of unwanted children attest to the failures of this method. Other methods of preventing pregnancy include techniques for (1) preventing the sperm from reaching the site of fertilization; (2) preventing ovulation; or (3) interfering with implantation if fertilization does occur. **Table 47.1** lists the most common contraceptive techniques and their reliability, based on one year of use. You should refer to the table as we discuss each of the methods of contraception in the following sections. Note that, while we discuss each method individually, combinations of particular methods may be used to reduce further the chance of pregnancy.

Vasectomy and Tubal Ligation Are the Most Effective Methods for Preventing Fertilization

A natural technique for preventing fertilization is the *rhythm method,* which consists of avoiding intercourse during the time of the month when the egg can be fertilized. Because sperm can survive for as long as five days in the female reproductive tract, intercourse should be avoided from five days before ovulation and, for safety's sake, for another four or five days after ovulation. The method is difficult to apply because of the typical unpredictability of the time of ovulation (and the power of the sex drive).

Another natural method to prevent fertilization is *withdrawal*—starting sexual intercourse, but withdrawing the penis before ejaculation. Unfortunately, once ejaculation begins, it proceeds as a series of reflexes that is extremely difficult to interrupt; in addition, some sperm may be present in lubrication produced prior to ejaculation.

TABLE 47.1	**Pregnancy Rates for Birth Control Methods**	
Method	**Lowest Expected Rate of Pregnancy**[a]	**Typical-Use Rate of Pregnancy**[b]
Rhythm method	1%–9%	25%
Withdrawal	4%	19%
Condom (male)	3%	14%
Condom (female)	5%	21%
Diaphragm and spermicidal jelly	6%	20%
Vasectomy (male sterilization)	0.1%	0.15%
Tubal ligation (female sterilization)	0.5%	0.5%
Contraceptive pill (combination estrogen/progestin)	0.1%	5%
Contraceptive pill (progestin only)	0.5%	5%
Implant (progestin)	0.09%	0.09%
Intrauterine device (IUD) (copper T)	0.6%	0.8%

[a]Rate of pregnancy when the birth control method was used correctly every time.
[b]Rate of pregnancy when the method was used typically, meaning that it may not have been always used correctly every time.
Source: U.S. Food and Drug Administration, http://www.fda.gov/fdac/features/1997/conceptbl.html. Data reported in 1997 for effectiveness of methods in a 1-year period.

The *condom,* a thin, close-fitting sheath of latex, lambskin, or polyurethane worn over the penis, is one of the traditional methods of preventing ejaculated sperm from entering the vagina. Condoms made from latex may also provide a barrier to the transmission of disease between sexual partners (condoms made from natural skin do not block viruses such as HIV). Pouchlike "female condoms," inserted into the vagina, prevent ejaculated sperm from entering the uterus.

The *diaphragm* is a cuplike rubber device that blocks the cervix in females. (The similar *cervical cap* is smaller and fits more closely over the cervix.) Typically a spermicidal jelly or cream is also used. To be most effective, a diaphragm and the spermicidal jelly must be inserted no more than an hour before intercourse, and left in place for the recommended time afterward.

The *intrauterine device (IUD)* is a small plastic or copper device inserted into the uterus just inside the cervix. The IUD remains in place as a long-term preventive measure; depending on the type, a single IUD is approved for five to 10 years of use. It is not completely clear how an IUD works; it is thought to cause a mild inflammation of the uterine lining that interferes with sperm function so that fertilization is less likely to occur. IUDs may also interfere with implantation of a fertilized egg but this may be a smaller contribution to their contraceptive effect. The IUD is effective as long as it is not deflected from its correct position in the uterus; unfortunately, this may happen without warning or the user's awareness. A few women also experience unpleasant side effects from the IUD such as cramps, uterine infections, or excessive menstrual bleeding.

Fertilization can also be prevented surgically, by cutting and closing off either the vasa deferentia in males or the oviducts in females. In *vasectomy,* the procedure for males, an incision is made in the scrotum and each vas deferens is severed and tied off. After vasectomy, the seminal fluid is still produced and ejaculated, but it does not contain sperm. In *tubal ligation,* the procedure for females, the oviducts are cut and tied off, or seared with heat (cauterized) to close them. The ligation prevents eggs from being fertilized or reaching the uterus. Neither vasectomy nor tubal ligation interferes with the production of sex hormones by the ovaries or testes, or results in any change in sexual behavior. Both operations are highly effective in preventing pregnancy. Although they can be reversed, the procedures are difficult and not always successful.

The Oral Contraceptive Pill Is the Most Effective Method for Preventing Ovulation

The primary method used to prevent ovulation is the *oral contraceptive pill,* or simply "the pill," containing a combination of estrogen and *progestin* (a synthetic form of progesterone) or progestin alone. In this highly effective method, the pill is taken daily for 20 to 21 days after the end of the menstrual flow and then stopped (placebo pills are taken for the remaining days of the cycle to maintain the routine of pill taking) to allow menstruation, and then the next month's course is begun. If pregnancy is desired, the pill is simply not taken after the menstrual flow.

The pill works by inhibiting the secretion of FSH and LH by the pituitary; without these hormones, ovulation does not occur. When the pill is stopped after 20 to 21 days, the resulting drop in progestin concentration causes the uterine lining to break down and initiates the menstrual flow. Since ovulation does not occur, fertilization and pregnancy are not possible.

Most pregnancies among women taking the pill result from failure to take it on schedule—often simply by forgetting to take the pill for a day or two at the wrong time of the month. Some women, about one in four, experience unpleasant side effects, such as nausea, tenderness of the breasts, irritability, nervousness, or changes in skin color or texture. Modern versions of the pill have almost eliminated the more serious side effects, such as increased incidence of breast cancer and formation of blood clots. However, cigarette smoking significantly increases the risk of heart attacks and strokes for women taking the pill. This risk increases with age and with the number of cigarettes smoked per day.

The Morning-After Pill Blocks Ovulation or Fertilization

Whatever the method of birth control, its effectiveness is improved if sex partners are highly motivated to avoid pregnancy, and careful in its use. The effectiveness of condoms, for example, is greatly improved if the penis is withdrawn immediately after ejaculation (before the semen has time to spread under the condom and leak into the vagina).

Another method used to prevent pregnancy is the so-called *emergency contraception pill,* commonly referred to as the "morning-after pill." These pills are administered after intercourse has occurred as a means to prevent pregnancy. A high-dosage synthetic progestin emergency contraception pill called Plan B is available in the United States without prescription to women who are 18 or older. This pill is highly effective if taken within 72 hours after unprotected sexual intercourse. Pregnancy tests do not work until significantly after this time. Research data show that Plan B works by blocking ovulation or fertilization. It may also inhibit implantation of a fertilized egg by altering the endometrium, but it has no effect if the process of implantation has begun.

Another emergency contraception pill is *mifepristone (RU-486),* which contains a molecule that binds to and blocks progesterone receptors in the uterine lining. The blockage prevents the lining from responding to progesterone and causes it to break down (that is, a menstrual period is initiated), taking with it any embryo that may have implanted. Mifepristone is approved in the United States for terminating pregnancies up to 49 days post-conception; the time period is longer in some foreign countries. It is available only by prescription.

In this chapter we have focused on animal reproduction up to the point of the fertilized egg. In the next chapter, we address the final stage of reproduction in sexually reproducing organisms, the development of a new individual from the fertilized egg.

STUDY BREAK 47.4 <

How does the oral contraceptive pill prevent pregnancy?

UNANSWERED QUESTIONS

What molecules are responsible for sperm recognition of egg coats?

Although fertilization is necessary for sexual reproduction, how sperm recognize eggs, particularly in mammals, has been an enigma. Fertilization success or failure may be attributed to how well complementary molecules on each gamete interact with each other. Furthermore, as discussed in Section 21.2, one of the prezygotic species isolating mechanisms relies on the fact that gamete receptors from one species do not recognize receptors from other species. However, in artificial environments (such as experimental situations), gametes from different species sometimes do recognize each other and form hybrids. These hybrids are usually explained by structural similarity of the molecules responsible for gamete recognition.

During fertilization in mammals, sperm must penetrate through several layers that surround the egg. Using motility and their own surface hyaluronidase, the sperm first move through the cumulus mass, a remnant of follicle cells (referred to as *cumulus cells*) and the sticky hyaluronate-containing matrix that cumulus cells produce. Next, the sperm bind to the zona pellucida, a tough acellular glycoprotein coat. In mammals, this coat is composed of three to four proteins, often referred to as ZP1, ZP2, ZP3, and ZP4. Mice, the mammal in which fertilization has been studied the most, produce ZP1, ZP2, and ZP3.

Starting in the 1980s, researchers have studied the function of zona pellucida glycoproteins. Early experiments studied the function of individual proteins using a competition assay. For this assay, zona pellucida proteins were dissolved and purified. Then, each was added individually to sperm to determine if it could occupy putative receptors on the plasma membrane of sperm cells and prevent the sperm from binding to cumulus-free eggs. Only ZP3 did this.

ZP3 also induces the acrosome reaction in sperm, which is necessary for sperm to penetrate the zona pellucida (discussed in Section 47.2). Subsequent research found that ZP3 bound to a form of the enzyme β-1,4-galactosyltransferase found on the sperm plasma membrane. Curiously, another ZP3 binding protein, sp56, was also found within the acrosome. These proteins may act sequentially to allow sperm to adhere to the zona pellucida (before and during or after the acrosome reaction).

Recent mouse genetic studies suggest that a more complex assembly of zona pellucida proteins is responsible for binding sperm. Other genetic studies indicate that additional receptors on sperm also contribute to sperm–zona pellucida adhesion. Further work is necessary to resolve the early biochemical data and the more recent results from genetic experiments. It is also important to clarify the role of other receptors that have been identified more recently. It seems likely that fertilization includes a number or perhaps a sequence of molecular adhesive steps that must function properly for normal fertility and that there may be molecular redundancy in such a fundamental process.

How is membrane fusion regulated during the sperm acrosome reaction?

To release the acrosome, the outer acrosomal membrane fuses at hundreds of points with the plasma membrane overlying it. These fusion points expand, forming vesicles that are released, freeing the acrosomal contents. Acrosomal proteins are released in two phases. First, the more soluble acrosomal proteins are released. Then, acrosomal proteins that initially remained bound to a protein matrix within the acrosome are released. Regulated membrane fusion during secretion has been studied in most detail in neurons. In neurons, each secretory vesicle forms a single fusion point with the plasma membrane. Although membrane fusion during the acrosome reaction involves hundreds of fusion points, and secretory vesicles in somatic cells form just one, we found that proteins promoting membrane fusion in neuronal secretion at the synapse were also involved in the sperm acrosome reaction.

Proteins called SNAREs (from Soluble N-ethylmaleimide-sensitive factor Attachment protein REceptor) promote membrane fusion in neuronal secretion. So-called "neuronal" SNAREs are also found in sperm near the acrosome. We have studied mice deficient in a protein called complexin that regulates SNARE function, and have found that sperm from these mice have defective acrosome reactions as well as markedly reduced fertility. Complexin appears to promote SNARE complex formation and then suspend the SNARE complex in an intermediate state in which membrane fusion can be completed readily by a Ca^{2+} influx into sperm.

Despite some molecular similarity to the neuronal process, membrane fusion in sperm is clearly unique. It is important to identify the important steps that prepare sperm for membrane fusion and to determine how SNARE function is regulated during the acrosome reaction. This should help understand sperm malfunctions that result in defects in the acrosome reaction, a frequent cause of male infertility.

Think Critically

One of the procedures used in fertility clinics is called intracytoplasmic sperm injection (ICSI). A sperm is aspirated into a pipette, and the pipette is driven through the zona pellucida and egg membrane. The sperm is then injected directly into an egg, and the egg develops into an embryo. Predict the types of fertility problems for which this procedure could be used to produce offspring.

David Miller

David Miller is Associate Professor in the Department of Animal Sciences at University of Illinois, Urbana–Champaign. His laboratory studies the molecular underpinnings of sperm maturation and fertilization. You can learn more about Miller's research by visiting: http://labs.ansci.illinois.edu/millerlab/.

REVIEW KEY CONCEPTS

Go to **CENGAGENOW** at www.cengage.com/login to access quizzing, animations, exercises, articles, and personalized homework help.

47.1 Animal Reproductive Modes: Asexual and Sexual Reproduction

- In asexual reproduction, a single parent gives rise to offspring without genetic input from another individual. In sexual reproduction, offspring are produced by the union of gametes—eggs and sperm—from two parents.
- Asexual reproduction involving mitosis occurs in animals by fission, budding, or fragmentation (Figure 47.2). In parthenogenesis, a form of asexual reproduction, females produce eggs that develop without being fertilized.
- In sexual reproduction, genetic variability is produced by the meiotic processes of genetic recombination and independent assortment.

47.2 Cellular Mechanisms of Sexual Reproduction

- Sexual reproduction includes two cellular processes, gametogenesis and fertilization, and a whole-organism process, mating. Gametogenesis is the formation of male and female gametes by meiotic cell division, followed by differentiation of the gametes; fertilization is the union of gametes that initiates development of new individuals (Figure 47.3).
- Gametogenesis takes place in the testes of males and in the ovaries of females. Sperm and eggs are delivered to the site of fertilization by sperm ducts in males and oviducts in females. External reproductive structures aid the delivery in many species.
- In male gametogenesis—spermatogenesis—each cell entering meiosis produces four haploid motile sperm cells. In female gametogenesis—oogenesis—each cell entering meiosis produces one haploid egg cell. The meiotic divisions of oogenesis concentrate almost all the cytoplasm in the single egg cell; the other division products are nonfunctional polar bodies (Figure 47.3).
- The egg contains stored nutrients and information required for at least the early stages of embryonic development. It is covered by one or more protective coats, and it has a mechanism that blocks additional sperm from entering after fertilization (Figure 47.5).
- Fertilization, which follows mating in most animals, may be external or internal. In external fertilization, sperm and eggs are shed into the surrounding water. In internal fertilization, sperm are released close to or inside the female reproductive ducts via copulation (Figure 47.6).
- When a sperm and egg touch during fertilization, their plasma membranes fuse, introducing the sperm nucleus into the egg cytoplasm. The sperm and egg nuclei then fuse to form a diploid zygote nucleus and initiate embryonic development (Figure 47.8).
- Oviparous animals lay eggs in which development of new individuals takes place outside the female's body. In viviparous animals, development takes place inside the female's body. In ovoviviparous animals, fertilized eggs are retained within the body while the embryo develops, the eggs hatch within the mother, and the offspring are then released from the body.
- In hermaphroditism, single individuals produce both mature egg-producing tissue and mature sperm-producing tissue (Figure 47.10).

47.3 Sexual Reproduction in Humans

- In females, eggs released from the ovaries travel through the oviducts to the uterus. The uterus opens into the vagina, the entrance for sperm and the exit for offspring during birth (Figure 47.11).
- The ovarian cycle produces an egg. The cycle begins with the release of GnRH by the hypothalamus, which stimulates the release of FSH and LH from the anterior pituitary. FSH stimulates oocytes in the ovaries to begin meiosis. One oocyte typically develops to maturity and is surrounded by cells that form a follicle (Figures 47.12 and 47.13).
- The enlarging follicle secretes estrogens, causing a burst in FSH and LH blood concentrations; at about day 14 of the cycle, the LH stimulates ovulation, the bursting of the follicle and the release of the egg. The remainder of the follicle forms the corpus luteum, which secretes estrogens, progesterone, and inhibin (Figures 47.12 and 47.13).
- Day 0 of the monthly uterine (menstrual) cycle correlates with the beginning of follicular development in the ovary and the beginning of the menstrual flow. Secretion of estrogen from the developing follicle stimulates the growth of a new endometrium. If fertilization does not occur, progesterone and inhibin maintain the endometrium until day 28 of the cycle, when the corpus luteum regresses. Without progesterone, the endometrium breaks down and is released as the menstrual flow (Figure 47.13).
- In males, sperm develop in seminiferous tubules in the testes and are released into the epididymis. When a male ejaculates, sperm travel from the epididymis to the vas deferens, and then through the urethra and the penis. The seminal vesicles, prostate gland, and bulbourethral glands add fluids to the sperm traveling to the outside (Figures 47.15 and 47.16).

- Sperm production in males is also controlled by LH and FSH. LH stimulates Leydig cells in the testes to secrete testosterone, which stimulates sperm production. FSH stimulates Sertoli cells in the testes to secrete molecules needed for spermatogenesis (Figure 47.17).
- During copulation, sperm are ejaculated into the vagina of the female. The sperm then swim through the female reproductive tract, aided by contractions of the oviduct and guided by molecules released by the egg. Upon contact with the egg in the oviduct, the acrosomes of sperm release enzymes that digest a path through the coats of the egg. When the fertilizing sperm contacts the egg, the sperm and egg plasma membranes fuse, releasing the sperm nucleus into the egg cytoplasm and activating the egg. The egg completes meiosis, and the sperm and egg nuclei fuse, producing the zygote (Figure 47.18).
- As the embryo implants, the hormone hCG sustains the corpus luteum, which continues to secrete estrogen and progesterone at high levels. These hormones maintain the uterine lining and prevent menstruation.
- Infertility has many causes, including physical problems, hormonal changes, or environmental factors.

Animation: Female reproductive system

Animation: Ovarian function

Animation: Menstrual cycle summary

Animation: Male reproductive system

Animation: Spermatogenesis

Animation: Route sperm travel

Animation: Hormonal control of sperm production

Animation: Hormones and the menstrual cycle

Animation: Fertilization

47.4 Methods for Preventing Pregnancy: Contraception

- Methods of contraception work by preventing sperm from reaching the site of fertilization, by preventing ovulation, or by interfering with implantation (Table 47.1).
- Methods for preventing fertilization include the rhythm method, the condom, the diaphragm or cervical cap, the IUD, and a vasectomy or tubal ligation.
- The oral contraceptive pill prevents ovulation. It contains a combination of estrogen and the progesterone-like progestin, which inhibiting the secretion of FSH and LH and follicle formation.
- The morning-after pill blocks ovulation or fertilization, and may also inhibit implantation.

UNDERSTAND AND APPLY

Test Your Knowledge

1. Asexual reproduction is most successful in:
   a. changing environments.
   b. sessile animals.
   c. densely settled populations.
   d. land animals.
   e. genetically varied individuals.
2. Which of the following processes does not increase genetic diversity?
   a. Parthenogenesis
   b. Random DNA mutations
   c. Genetic recombination
   d. Independent assortment
   e. Random combinations of paternal and maternal chromosomes
3. Gametogenesis has parallel stages in egg and sperm formation. The stage in eggs that is equivalent to spermatids is the:
   a. primary oocyte.
   b. oogonium.
   c. ovum.
   d. ootid and polar bodies.
   e. secondary oocyte and polar body.
4. External fertilization provides a male with relative certainty of paternity and is found in which of the following animal groups?
   a. Amphibians
   b. Birds
   c. Sharks
   d. Reptiles
   e. Mammals
5. The slow block to polyspermy:
   a. is caused by a change in membrane potential from negative to positive.
   b. triggers the movement of Ca^{2+} from the cytosol to the endoplasmic reticulum.
   c. triggers a decrease in egg oxidation and protein synthesis.
   d. describes the fusion of egg and sperm nuclei.
   e. includes the fusion of cortical granules with the egg's plasma membrane.
6. Some placental animals provide nutrients to their embryos from an attached membranous yolk-containing sac. They are called:
   a. oviparous animals.
   b. ovoviviparous animals.
   c. metatherians.
   d. eutherians.
   e. mammals.
7. Which activity is a step in the ovarian cycle?
   a. FSH stimulates the pituitary to release GnRH.
   b. When FSH and LH levels fall, the corpus luteum shrinks and the uterine lining breaks down.
   c. Luteinizing hormone stimulates the uterus to make progesterone.
   d. Estrogen levels initially have a positive feedback effect on the pituitary, which is followed by higher estrogen levels causing negative feedback.
   e. A fully developed corpus luteum inhibits uterine lining growth.
8. During spermatogenesis in mammals, sperm travels from the:
   a. Sertoli cells past the epididymis and urethra, through the vas deferens to the prepuce.
   b. seminal vesicles past the prostate gland, through the glans and prepuce to the bulbourethral glands.
   c. vestibular glands past the Leydig cells, through the accessory glands and epididymis to the vas deferens.
   d. labia past the bulbourethral glands, through the vas deferens and urethra to the epididymis.
   e. seminiferous tubules past the Sertoli cells, through the epididymis and vas deferens to the urethra.

9. The secondary oocyte in humans is fertilized in the:
   a. uterus.
   b. vagina.
   c. oviduct.
   d. cervical canal.
   e. ovary.
10. The most effective method to prevent fertilization is:
    a. the oral contraceptive.
    b. the IUD.
    c. the morning-after pill.
    d. vasectomy in men and tubal ligation in women.
    e. the rhythm method.

Discuss the Concepts

1. Currently under development is an "anti-pregnancy vaccine" that stimulates a woman's immune system to develop antibodies against human chorionic gonadotropin (hCG). How would this method prevent pregnancy?
2. Men sometimes have reduced fertility because of *testicular varioceles,* varicose veins in the testes in which blood pools. Based on what you now know of the conditions under which sperm develop properly, how do you think this condition might impair sperm development?
3. Spermatogenesis produces four sperm for each primary spermatocyte, but oogenesis produces only one egg for each primary oocyte. Why might these different outcomes be adaptive?
4. Sertoli cells protect spermatocytes from attack by antibodies during their development in the human male. What structures might protect the oocyte and egg from attack by antibodies in the human female?
5. Compare the advantages and disadvantages of sexual and asexual reproduction for an aphid and a parasitic worm.
6. It may be possible to develop a birth control drug that would prevent conception by interfering with fertilization. Outline the design for such a drug and explain exactly how it would work. (There may be more than one design that, in theory, would be effective.)

Design an Experiment

Design experiments to determine if, and at what dose, vitamin E can decrease menstrual cramping significantly.

Interpret the Data

Contamination of water by agricultural chemicals affects the reproductive function of some animals. Are there effects on humans? Epidemiologist Shanna Swan and her colleagues studied sperm collected from men in four cities in the United States (see the table). The men were partners of women who had become pregnant and were visiting a prenatal clinic, so all were fertile. Of the four cities, Columbia, Missouri, is located in the county with the most farmlands. New York City represents an area with no agriculture.

Data from a study of sperm collected from men who were partners of pregnant women who visited prenatal health clinics in one of four cities.

	Location of clinic			
	Columbia, Missouri	Los Angeles, California	Minneapolis, Minnesota	New York, New York
Average age	30.7	29.8	32.2	36.1
Percent nonsmokers	79.5	70.5	85.8	81.6
Percent with history of STD*	11.4	12.9	13.6	15.8
Sperm count (million/ml)	58.7	80.8	98.6	102.9
Percent motile sperm	48.2	54.5	52.1	56.4

*STD stands for sexually transmitted disease.

1. Where did researchers record the highest and lowest sperm counts?
2. In which cities did samples show the highest and lowest sperm motility (ability to move)?
3. Aging, smoking, and STDs adversely affect sperm. Could differences in any of these variables explain the regional differences in sperm count?
4. Do these data support the hypothesis that living near farmlands can adversely affect male reproductive function?

Source: S. Swan et al. 2003. Geographic differences in semen quality of fertile US males. *Environmental Health Perspectives* 111:414–420.

Apply Evolutionary Thinking

The nematode species *Caenorhabditis elegans* and *C. briggsae* are both hermaphroditic. Phylogenetic evidence indicates that the last common ancestor of these two species had a normal male–female mechanism of reproduction. What does this evidence suggest about their hermaphroditism?

Express Your Opinion

Fertility drugs induce multiple ovulations at the same time and increase the likelihood of high-risk multiple pregnancies. Should the use of such drugs be restricted to conditions that limit the number of embryos formed? Go to www.cengage.com/login to investigate both sides of the issue and then vote.

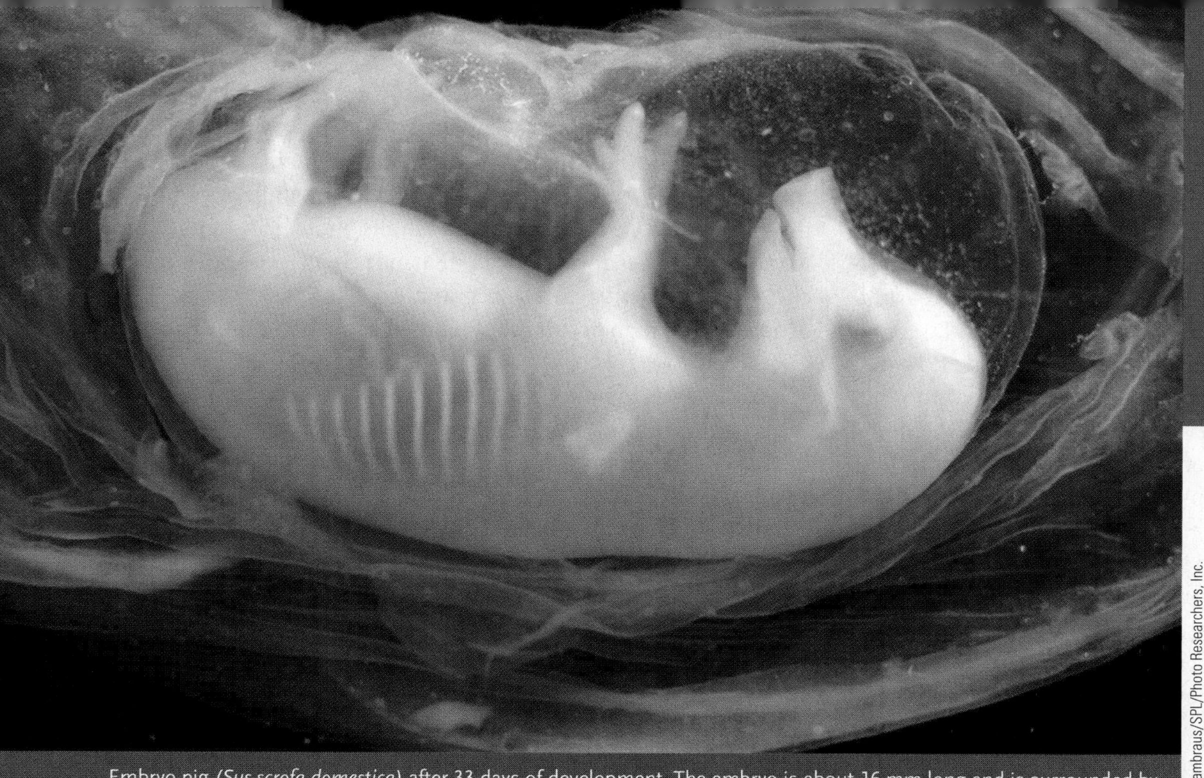

Embryo pig (*Sus scrofa domestica*) after 33 days of development. The embryo is about 16 mm long and is surrounded by several membranous sacs, including the fluid-filled amnion (closest to the embryo), which cushions and protects it.

Daniel Sambraus/SPL/Photo Researchers, Inc.

48

STUDY OUTLINE

Animal Development

Why It Matters. . . The uterine contractions announcing birth are taking place at shorter intervals and with greater intensity. The mother-to-be endures the discomfort and apprehension with the knowledge that the child that has been growing in her body will soon come into the world. It formed from a fertilized egg, about the size of a period on this page, and grew through a program of cell divisions, complex cell movements, and molecular interactions. She was unaware of these complexities except for movements of the fetus that became apparent about 14 weeks after she became pregnant.

Her baby's development required no conscious attention on her part: human development, like that of all animals, is programmed to proceed inexorably from fertilized egg to free-living offspring. Even childbirth is the result of programmed events that, once started, normally move to conclusion without requiring deliberate input from the mother.

Over the course of its development, the baby's body formed all the organ systems required for independent existence and, at its birth, they are already working to sustain its life. Most astonishing, perhaps, is the baby's brain. It formed from a tube of nerve tissue that bulged outward and enlarged, continually adding nerve cells and connecting them into circuits until it attained what may well be some of the most complexly organized matter in the universe—all as part of the automated events of development. Still, the human brain is unique only in the degree of its complexity and integrative capacity; the brains of other mammals are basically similar and develop through the same embryonic pathways.

The baby enters the outside world passing head first through the cervix, and then the vagina. Soon the rest of the body slips through, aided and lubricated in its passage by release of the fluid that surrounded and cushioned it in the uterus. In the first indignity of life, the baby is briefly held upside down to drain fluid from its lungs. This action triggers its first breath, followed by a satisfyingly loud cry.

The baby is proudly displayed to the mother, who greets it with love, relief, joy, and realization of the responsibilities the baby will bring. It is a girl, who with further luck and good care will continue developing through childhood, puberty, adult life, and old age, all through programs built into her hereditary molecules. As part of these passages, she may bring her own child into the world.

People have tried since ancient times to understand how development and birth take place. The scientific quest began with Aristotle, who observed chick development and correctly interpreted the functions of the placenta and umbilical cord in humans. The investigators who followed Aristotle concentrated on describing developmental changes in **morphology,** which is the form or shape of an organism, or of a part of an organism. More recently, investigators began to trace the molecular underpinnings of the morphological events.

In this chapter we survey the results of these investigations. We take up the story of animal development where the previous chapter left off, with the fertilized egg. We continue with the early events leading from the fertilized egg to the primary tissues of the embryo, and then trace the development of organs from these tissues. Next, we describe human development as representative of the process in mammals. Then, we survey the cellular and molecular bases of these mechanisms. At the cellular level, the development of an adult animal from a fertilized egg involves cell division, in which more cells are produced by mitosis; **cell differentiation,** in which changes in gene expression establish cells with specialized structure and function; and **morphogenesis** ("form creation"), the generation of the body form of the animal as differentiated cells end up in their appropriate sites. <

48.1 Mechanisms of Embryonic Development

Fertilization of an egg by a sperm cell produces a zygote. Embryonic development begins at this point and ultimately produces a free-living organism. All the instructions required for development are packed into the fertilized egg.

Developmental Information and Components for Growth Are Stored in the Fertilized Egg

Once the zygote is formed when egg and sperm nuclei fuse; mitotic divisions begin the developmental activity (see Section 47.2).

INFORMATION STORAGE IN THE EGG As introduced in Section 16.4, the initiation of development depends primarily on the DNA in the zygote nucleus, and on *cytoplasmic determinants,* that is, mRNA and protein molecules stored in the cytoplasm. Most of the mRNA and protein molecules are maternal in origin because the fertilizing sperm contribute essentially no cytoplasm to the zygote. The mRNAs and proteins direct the first stages of development up until the genes of the zygote become active.

OTHER COMPONENTS OF THE EGG In addition to cytoplasmic determinants, the egg cytoplasm also contains ribosomes and other cytoplasmic components required for protein synthesis and the early cell divisions of embryonic development. For example, the egg cytoplasm contains all the tubulin molecules required to form the spindles for early cell divisions. It also contains mitochondria, nutrients stored in granules in the yolk and in lipid droplets, and, in many animals, pigments that color the egg or regions of it.

Yolk contains nutrients. The eggs of insects, reptiles, and birds contain large amounts of yolk which supplies all of the nutrients for development of the embryo. The eggs of placental mammals, in contrast, contain very little yolk, which supplies nutrients only for the earliest stages of development.

Depending on the species, the yolk may be concentrated at one end or in the center of the egg, or distributed evenly throughout the cytoplasm. Its distribution influences the rate and location of cell division during early embryonic development. Typically, cell division proceeds more slowly in the region of the egg containing the yolk. In the large, yolky eggs of birds and reptiles, cell division takes place only in a small, yolk-free patch at the surface of the egg.

Unequal distribution of yolk and other components in a mature egg is termed **polarity.** For example, in most species the egg nucleus is located toward one end of the egg. This end of the egg, called the **animal pole,** typically gives rise to surface structures and the anterior end of the embryo. The opposite end of the egg, the **vegetal pole,** typically gives rise to internal structures such as the gut and the posterior end of the embryo. Yolk, when unequally distributed in the egg cytoplasm, is most frequently concentrated in the vegetal half of the egg.

Fertilization of the egg launches the early events of development. As you follow these events, keep in mind that the developmental processes involved work together to produce the body plan of the adult. For a bilaterally symmetrical animal such as a human or a dog, for example, the adult must develop three body axes: the anterior–posterior axis, the dorsal–ventral (back–front) axis, and the left–right axis **(Figure 48.1).**

FIGURE 48.1
Body axes: anterior–posterior, dorsal–ventral, and left–right.

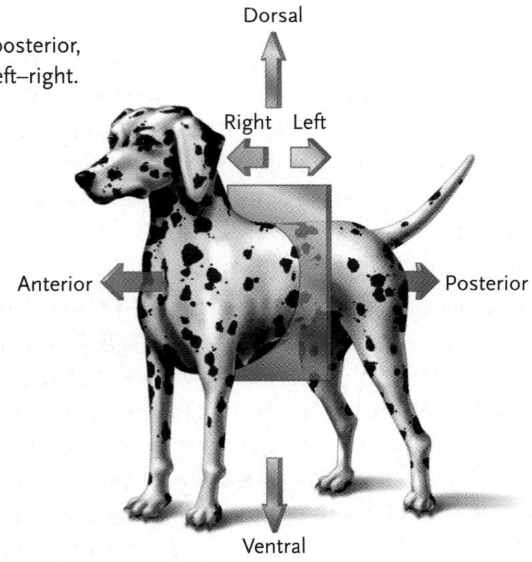

Dorsal
Right
Left
Anterior
Posterior
Ventral

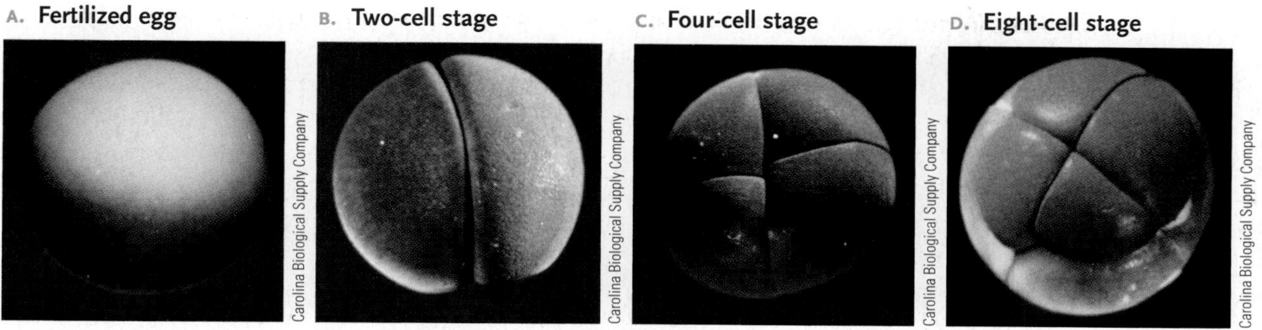
A. Fertilized egg
B. Two-cell stage
C. Four-cell stage
D. Eight-cell stage
Carolina Biological Supply Company
Carolina Biological Supply Company
Carolina Biological Supply Company
Carolina Biological Supply Company

FIGURE 48.2
The first three cleavage divisions of a frog embryo, which convert the fertilized egg into the eight-cell stage. Note that the cleavage divisions cut the volume of the fertilized egg into successively smaller cells, called blastomeres.

Cleavage, Gastrulation, and Organogenesis Are Early Events in Development

Soon after fertilization, the zygote begins **cleavage,** a series of mitotic divisions in which cycles of DNA replication and division occur without the production of new cytoplasm. As a result, the cytoplasm of the egg is partitioned into successively smaller cells without increasing the overall size or mass of the embryo **(Figure 48.2).** These cells are called **blastomeres** (*blastos* = bud or offshoot; *meros* = part or division). In the frog *Xenopus laevis,* for example, twelve cleavage divisions produce an embryo of about 4,000 cells.

Cleavage is the first of three major developmental processes that, with modifications, are common to the early development of most animals (described in detail for particular animals in the next section). Following cleavage, the second major process, **gastrulation,** produces an embryo with three distinct primary tissue layers. The third process, **organogenesis,** accomplishes the development of the major organ systems. At the end of organogenesis, the embryo has the body organization characteristic of its species. Gastrulation and organogenesis both involve the processes of cell division, cell movements, and cell rearrangements. **Figure 48.3** outlines these stages in the life cycle of a frog.

The cleavage divisions lead to three successive developmental stages that are common to the early development of most animals. The first stage, called a **morula** (*morula* = mulberry), is a solid ball or layer of blastomeres. As cleavage divisions continue, the ball or layer hollows out to form the second stage, the **blastula** (*ula* = small), in which the blastomeres enclose a fluid-filled cavity, the **blastocoel** (*koilos* = hollow) (see Figure 48.3).

Once cleavage is complete, the blastomeres undergo extensive cellular rearrangements. This morphogenetic process is called **gastrulation** and during the process the embryo is termed a **gastrula** (*gaster* = gut or belly) (see Figure 48.3). The details of gastrulation differ among animal groups, but the result is the same for all: some surface cells of the blastula move to an interior position, and three primary cell layers of the embryo are produced. The three layers are the **germ layers:** the outer **ectoderm** (*ecto* = outside; *derma* = skin), the inner **endoderm** (*endo* = inside), and the **mesoderm** (*meso* = middle) between the ectoderm and the endoderm. Gastrulation establishes body pattern; that is, each tissue and organ of the adult animal originates in one of the three primary cell layers of the gastrula **(Table 48.1).**

As gastrulation proceeds, embryonic cells begin to differentiate: they become recognizably different in biochemistry, structure, and function. The developmental potential of the cells also becomes more limited than that of the fertilized egg from which they originated. That is, although a fertilized egg is **totipotent,** meaning that it is capable of producing all the various types of cells of the adult, progressively the cells produced become more specialized. That is, totipotent cells give rise to **pluripotent** cells, which can give rise to most adult cell types, and then pluripotent cells give rise to **multipotent** cells which give rise to cells with particular functions. Thus, for example, a multipotent mesoderm cell may develop into muscle or bone but not normally into outside skin or brain.

The restriction of developmental potential does not occur, as was once thought, because the cells have lost all their genes except those for the structure and function of the cell type they will become. Rather, except in rare instances, the differentiating cells

TABLE 48.1 Origins of Adult Tissues and Organs in the Three Primary Tissue Layers

Primary Tissue Layer	Adult Tissues and Organs
Ectoderm	Skin and its elaborations, including hair, feathers, scales, and nails; nervous system, including brain, spinal cord, and peripheral nerves; lens, retina, and cornea of eye; lining of mouth and anus; sweat glands, mammary glands, adrenal medulla, and tooth enamel
Mesoderm	Muscles; most of skeletal system, including bones and cartilage; circulatory system, including heart, blood vessels, and blood cells; internal reproductive organs; kidneys and outer walls of digestive tract
Endoderm	Lining of digestive tract, liver, pancreas, lining of respiratory tract, thyroid gland, lining of urethra, and urinary bladder

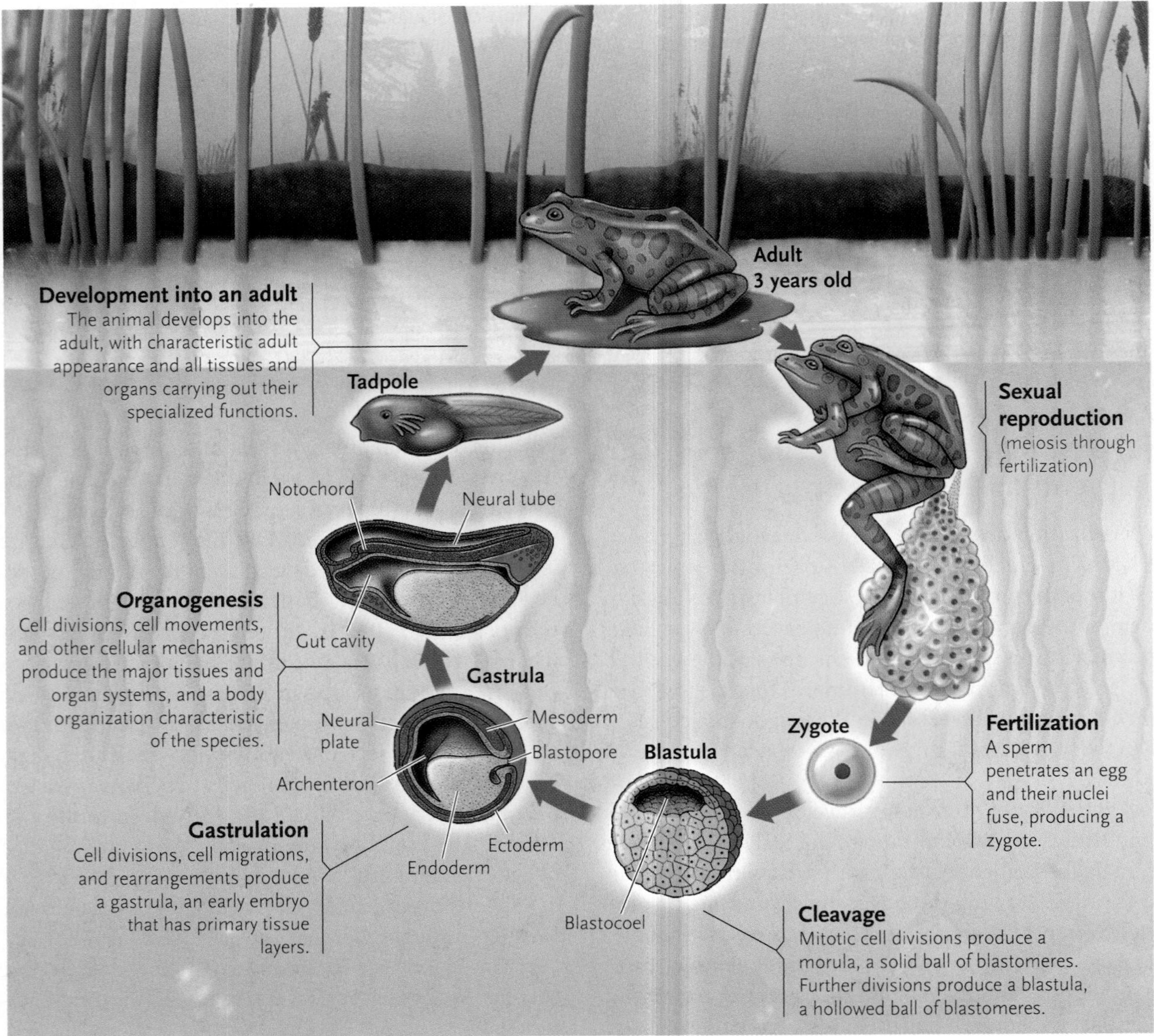
Adult
3 years old
Development into an adult
The animal develops into the adult, with characteristic adult appearance and all tissues and organs carrying out their specialized functions.
Tadpole
Sexual reproduction
(meiosis through fertilization)
Notochord
Neural tube
Organogenesis
Cell divisions, cell movements, and other cellular mechanisms produce the major tissues and organ systems, and a body organization characteristic of the species.
Gut cavity
Gastrula
Neural plate
Mesoderm
Blastopore
Blastula
Zygote
Fertilization
A sperm penetrates an egg and their nuclei fuse, producing a zygote.
Archenteron
Gastrulation
Cell divisions, cell migrations, and rearrangements produce a gastrula, an early embryo that has primary tissue layers.
Ectoderm
Endoderm
Blastocoel
Cleavage
Mitotic cell divisions produce a morula, a solid ball of blastomeres. Further divisions produce a blastula, a hollowed ball of blastomeres.

FIGURE 48.3
Stages of animal development shown in a frog.

actually all contain complete genomes of the organism, but each type of cell has a different program of gene expression. Several definitive experiments supporting this conclusion were carried out several decades ago by Robert Briggs and Thomas King of Lankenau Hospital Research Institute in Philadelphia (now Fox Chase Cancer Center) and extended by John B. Gurdon of the University of Cambridge, United Kingdom. In a typical experiment, the nucleus of a fertilized frog egg was destroyed by ultraviolet light. A micropipette was then used to transfer a nucleus from a fully differentiated tissue, intestinal epithelium, to the enucleated egg. Some of the eggs receiving the transplanted nuclei subsequently developed into normal tadpoles and adult frogs. This outcome was possible only if the differentiated intestinal cells still retained their full complement of genes. This conclusion was extended to mammals in 1997 when Ian Wilmut and his colleagues successfully cloned a sheep—Dolly—starting with an adult cell nucleus (see Section 18.2). The constancy of the genome during development was also discussed in Section 16.4.

A Number of Processes Are Responsible for Development

Development in all animals is accomplished by a number of processes that are under genetic control but are influenced to some extent by the environment (for example, temperature affects the rate of cell division). The processes are

1. Mitotic cell divisions.
2. Cell movements.
3. **Selective cell adhesions,** in which cells make and break specific connections to other cells or to the extracellular matrix.
4. **Induction,** in which one group of cells (the inducer cells) causes or influences another nearby group of cells (the responder cells) to follow a particular developmental pathway (see Section 16.4).
5. **Determination,** in which the developmental fate of a cell is set, and it is committed to becoming a particular cell type. Typically, determination is the result of induction.

6. **Differentiation,** which follows determination, establishes a cell-specific developmental program in cells. Differentiation results in cell types with clearly defined structures and functions.
7. **Apoptosis,** programmed cell death, in which tissues no longer required for continued development of the organism are removed.

Examples of these mechanisms in the examples of development are discussed in the following three sections.

STUDY BREAK 48.1

1. **How do cleavage divisions differ from cell division in an adult organism?**
2. **What are the primary cell layers of the embryo, and what process is responsible for producing them?**

48.2 Major Patterns of Cleavage and Gastrulation

With the principles of early embryonic development established, we describe cleavage and gastrulation in three animal groups that have been models in *embryology* (the study of embryos and their development): sea urchins, amphibians, and birds. Later in the chapter, we describe cleavage and gastrulation in humans and other mammals, which resemble the pattern in birds.

Sea Urchin Gastrulation Follows a Symmetrical Pattern That Reflects an Even Distribution of Yolk

The sea urchin is an echinoderm, a type of deuterostome (see Section 30.1). Cleavage divisions proceed at approximately the same rate in all regions of a sea urchin embryo (**Figure 48.4,** step 1), reflecting the uniform distribution of yolk in the sea urchin egg. These divisions continue until a blastula containing about a thousand cells is formed (step 2).

Gastrulation begins at the vegetal pole of the blastula. As a result of induction, some cells in the middle of that region become elongated and cylindrical, causing the region to flatten and thicken. Then, some cells break loose and migrate into the blastocoel (step 3). These cells, called *primary mesenchyme cells* (mesenchyme means "middle juice"), move around inside the blastocoel, making and breaking adhesions, until eventually they attach along the ventral sides of the blastocoel. These cells eventually become mesoderm cells (see step 7), from which the larval skeleton is produced. Next, the flattened vegetal pole of the blastula invaginates, pushing gradually into the interior (steps 4 and 5). The cells that invaginate are future endoderm cells. *Secondary mesenchyme cells* form at the top of the archenteron. These cells also eventually become mesoderm cells.

The inward movement of cells, in effect much like pushing in the side of a hollow rubber ball, generates a new cavity within

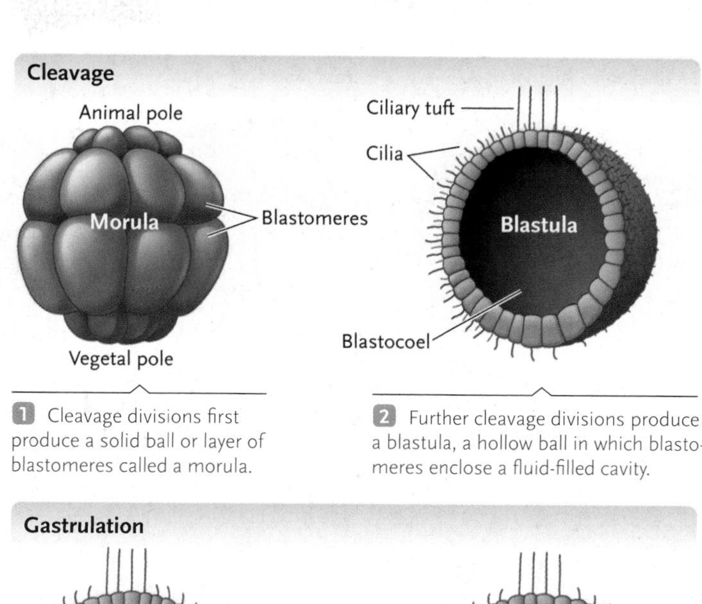

Cleavage
Animal pole
Morula
Blastomeres
Vegetal pole
Ciliary tuft
Cilia
Blastula
Blastocoel

1 Cleavage divisions first produce a solid ball or layer of blastomeres called a morula.

2 Further cleavage divisions produce a blastula, a hollow ball in which blastomeres enclose a fluid-filled cavity.

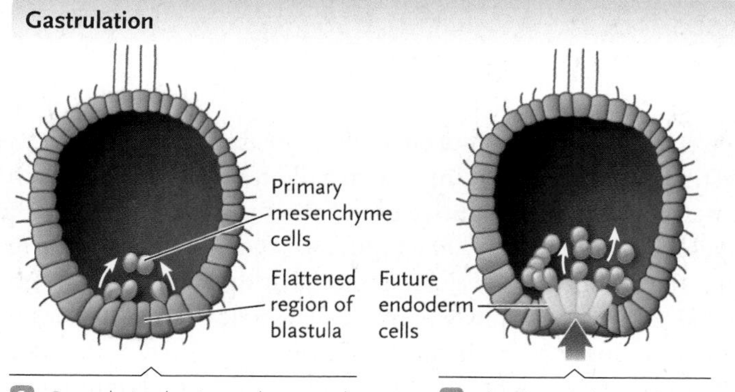

Gastrulation
Primary mesenchyme cells
Flattened region of blastula
Future endoderm cells

3 Gastrulation begins at the vegetal pole. Some cells are induced to change shape to produce a flattened region. Primary mesenchyme cells (future mesoderm) then break loose and migrate into the blastocoel.

4 The flattened vegetal pole consisting of future endoderm cells begins to invaginate.

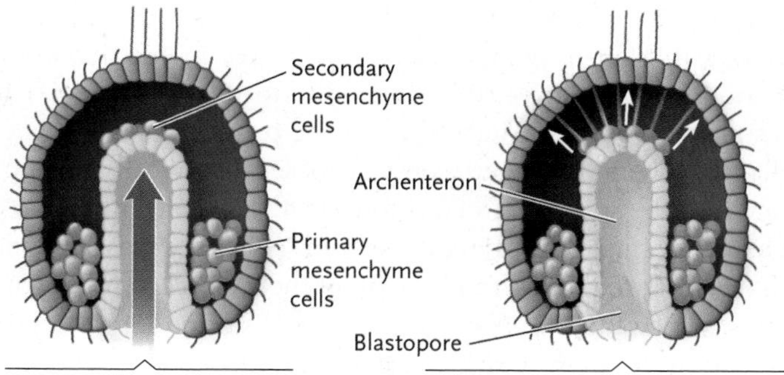

Secondary mesenchyme cells
Archenteron
Primary mesenchyme cells
Blastopore

5 Invagination continues. Secondary mesenchyme cells (also future mesoderm) form at the top of the archenteron.

6 Archenteron forms; secondary mesenchyme cells send out extensions that stretch across the blastocoel and adhere to the ectoderm.

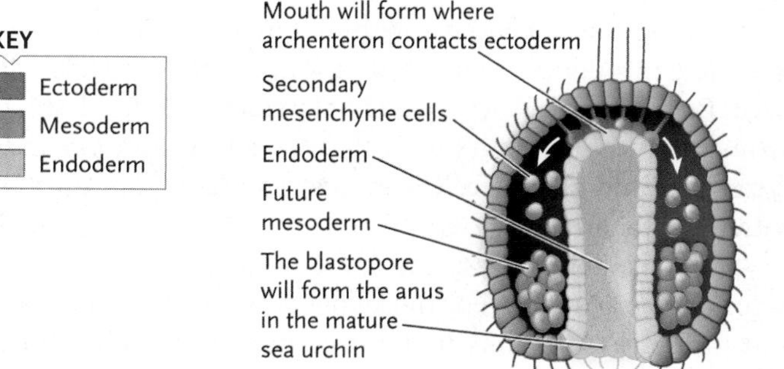

KEY
Ectoderm
Mesoderm
Endoderm
Mouth will form where archenteron contacts ectoderm
Secondary mesenchyme cells
Endoderm
Future mesoderm
The blastopore will form the anus in the mature sea urchin

7 Ectoderm and endoderm layers have formed; mesoderm cells are between them, some derived from primary mescenchyme cells and others from secondary mesenchyme cells.

FIGURE 48.4 Cleavage and gastrulation in the sea urchin.

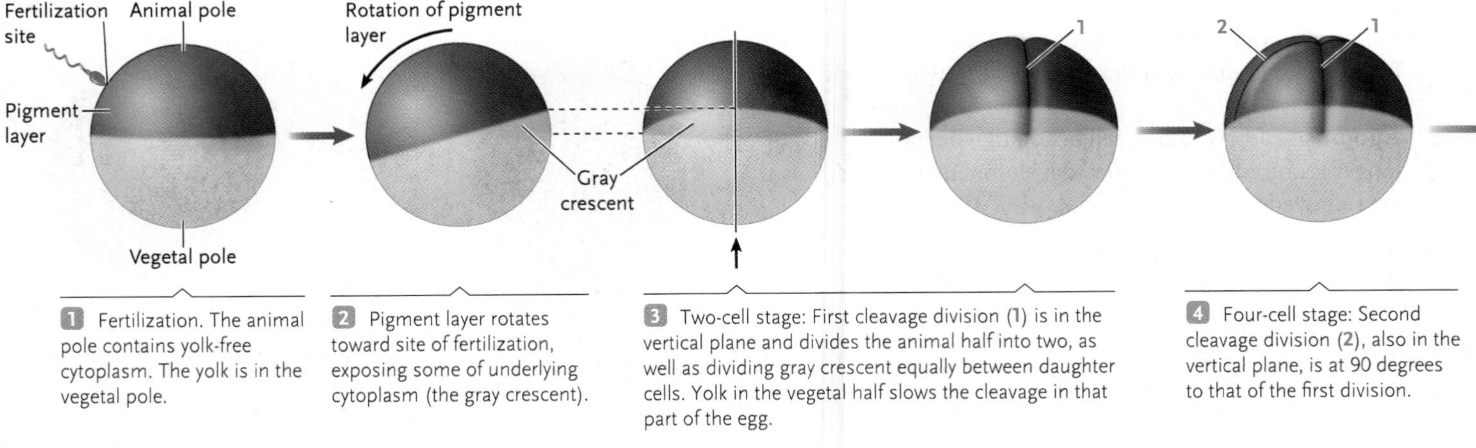
Fertilization site
Animal pole
Pigment layer
Vegetal pole
Rotation of pigment layer
Gray crescent
1
2
1 Fertilization. The animal pole contains yolk-free cytoplasm. The yolk is in the vegetal pole.
2 Pigment layer rotates toward site of fertilization, exposing some of underlying cytoplasm (the gray crescent).
3 Two-cell stage: First cleavage division (1) is in the vertical plane and divides the animal half into two, as well as dividing gray crescent equally between daughter cells. Yolk in the vegetal half slows the cleavage in that part of the egg.
4 Four-cell stage: Second cleavage division (2), also in the vertical plane, is at 90 degrees to that of the first division.

FIGURE 48.5
Fertilization and cleavage of a frog egg. The red numbers designate the cleavage divisions.

the embryo, the **archenteron** (*arche* = beginning; *enteron* = intestine or gut), which is lined with endoderm (step 6). The archenteron forms the primitive gut of the sea urchin embryo; the opening at one end is the **blastopore.** As the archenteron is forming, the secondary mesenchyme cells at the top of the invaginated cell layer send out extensions that stretch across the blastocoel and contact the inside of the ectoderm (step 6). These extensions make tight adhesions and then contract, pulling the invaginated cell layer inward and thereby eliminating most of the blastocoel.

At this point the embryo has two complete cell layers. The outer layer remaining from the original blastula surface makes up the ectoderm of the embryo. The second, inner layer, derived from the cells forming the archenteron, makes up the endoderm. Mesodermal cells are also beginning to form a third layer, the mesoderm. Some are derived from the primary mesenchyme cells and others from the secondary mesenchyme cells that migrate into the space between ectoderm and endoderm (step 7). The mesodermal cells originating from secondary mesenchyme cells give rise to mesodermal organs of the sea urchin. When the mesoderm layer is complete, the embryo has three complete layers: ectoderm, mesoderm, and endoderm. At this point, cells within each layer begin to differentiate, as evidenced by the synthesis of different proteins in each layer.

As the ectoderm, mesoderm, and endoderm develop, the embryo lengthens into an ellipsoidal shape with the blastopore marking its posterior end. From this point on, organ systems differentiate through further cell division, cell movements, selective cell adhesions, induction, and differentiation. The blastopore forms the anus; a mouth forms at the opposite, anterior end of the embryonic gut where the archenteron contacts and fuses with the ectoderm.

An archenteron and a blastopore are not present in the embryos of all organisms. For those that do have them, the blastopore gives rise to the anus or mouth of the embryo, depending on the animal group (see Section 29.2). In deuterostomes, which include echinoderms and chordates, the blastopore develops into the anus and the mouth forms at the opposite end of the embryonic gut formed by the archenteron. In the protostomes, which include annelids, arthropods, and mollusks, the blastopore develops into the mouth, and the anus forms at the opposite end of the embryonic gut.

Amphibian Cleavage and Gastrulation Are Influenced by an Unequal Distribution of Yolk

In amphibian eggs, such as those of frogs, yolk is concentrated in the vegetal half, which gives it a pale color. The animal half is darkly colored by a layer of pigment granules just below the surface. **Figure 48.5** shows the steps of fertilization and cleavage of a frog egg. The sperm typically fertilizes the egg in the animal half (step 1). After fertilization, the pigmented layer of cytoplasm rotates toward the site of sperm entry, exposing a crescent-shaped region of the underlying cytoplasm at the side opposite the point of sperm entry (step 2). This region, called the **gray crescent,** establishes the dorsal–ventral axis of the embryo, with the gray crescent marking the future dorsal side. The first cleavage division runs perpendicular to the long axis of the gray crescent and divides the crescent equally between the resulting two cells (step 3).

If one of the first two blastomeres does not receive gray crescent material, and the cells are separated experimentally, the cell without gray crescent divides to produce a disordered mass that stops developing. The cell receiving the gray crescent produces a normal embryo. Thus cytoplasmic material localized in the gray crescent is essential to normal development in frog embryos.

The second cleavage division is also in the vertical plane and at 90 degrees to the first cleavage division, producing the four-cell stage (step 4). At this point, the yolk in the vegetal half of the egg has slowed the progression of the first two cleavage furrows from going through that part of the egg. The third cleavage division is in the horizontal plane in the animal half of the egg and produces the eight-cell stage (step 5). By now the cleavage furrows of the first two cleavage divisions have continued all the way through the vegetal half of the egg. Embryo development continues with more divisions, first producing a morula (defined as a 16- to 64-cell stage in amphibians), and then a blastula (defined as at least 128 cells in amphibians) (step 6).

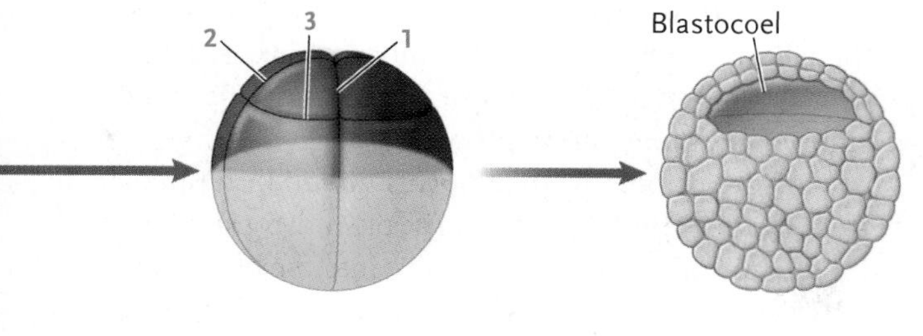
2
3
1
Blastocoel

5 Eight-cell stage: Third cleavage division (3) is in the horizontal plane within the animal half of the cell. Divisions I and II have now continued through the vegetal half of the egg.

6 Blastula stage (cross section): More cleavage divisions produce the blastula, which contains at least 128 cells and has a fluid-filled blastocoel.

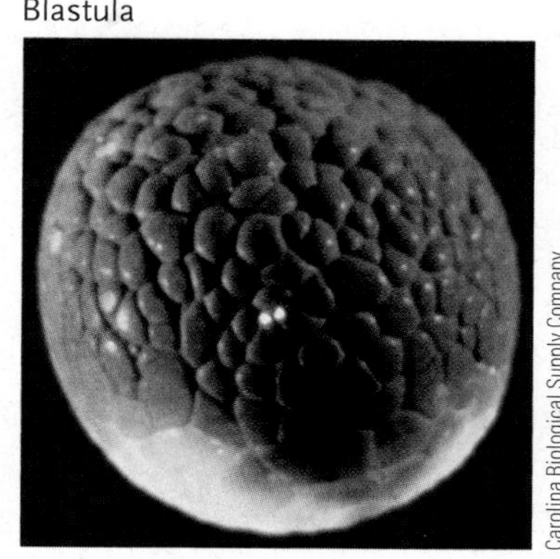
Blastula
Carolina Biological Supply Company

Gastrulation begins when cells from the animal pole move across the embryo surface and reach the region derived from the gray crescent (**Figure 48.6**, steps 1–2). This site is marked by a crescent-shaped depression rotated clockwise 90° called the **dorsal lip of the blastopore.** Cells changing shape and pushing inward from the surface in a process called **invagination** produce the depression. With continued inward movement of additional cells, the depression eventually forms a complete circle which is the blastopore.

Cells now migrate into the blastopore by **involution,** a process in which the cells entering from the outer layer of the embryo spread over the internal surfaces of the remaining exterior cells (step 3). A consequence of involution is that the pigmented cell layer of the animal half expands to cover the entire surface of the embryo. The cells of the vegetal half are enclosed by the movement and show on the outside as a yolk plug in the blastopore (step 4). In amphibians, the blastopore gives rise to the anus.

Continued involution moves cells into the interior and upward (see steps 3 and 4), forming two layers that line the inside top half of the embryo. The uppermost of these layers is induced to become the dorsal mesoderm (see step 4). The layer beneath it, which contains cells originating from both the outer surface of the embryo and the yolky interior, becomes the endoderm (shown in yellow). The pigmented cells remaining at the surface of the embryo form the ectoderm (shown in blue). The ventral mesoderm begins to be induced near the vegetal pole.

As the mesoderm and endoderm form, the depression created by the inward cell movements gradually deepens and extends inward as the archenteron (see steps 3 and 4), which displaces the blastocoel. The cells of the three primary cell layers continue to increase in number by further movements and divisions as development proceeds.

During frog gastrulation, cells of the dorsal lip of the blastopore are inducer cells that control blastopore formation; if the cells in the dorsal lip are removed and transplanted elsewhere in the egg, they cause a second blastopore—and a second embryo—to form in this region (see Section 48.5).

The events of gastrulation in frogs thus include the same developmental processes as in sea urchins—cell divisions, cell movements, selective adhesions, induction, and differentiation.

FIGURE 48.6
Gastrulation in a frog embryo.

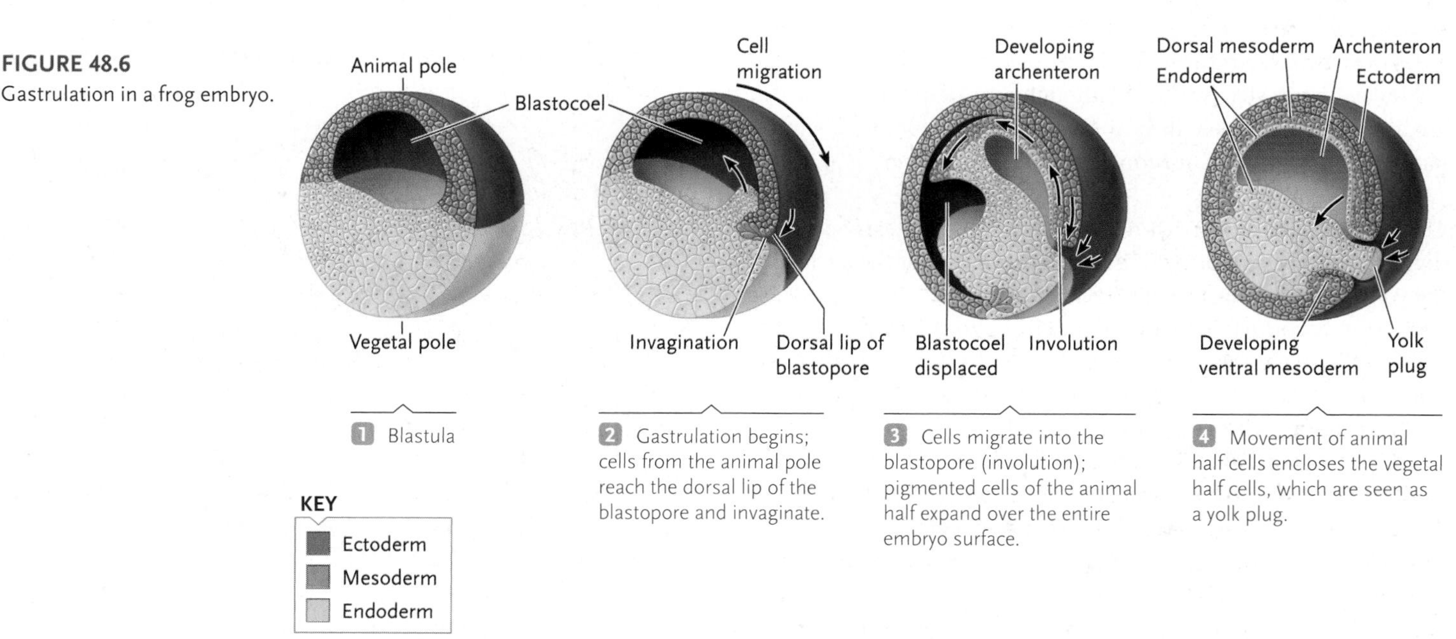
Animal pole
Blastocoel
Vegetal pole
Cell migration
Invagination
Dorsal lip of blastopore
Developing archenteron
Blastocoel displaced
Involution
Dorsal mesoderm
Archenteron
Endoderm
Ectoderm
Developing ventral mesoderm
Yolk plug
KEY
Ectoderm
Mesoderm
Endoderm

1 Blastula

2 Gastrulation begins; cells from the animal pole reach the dorsal lip of the blastopore and invaginate.

3 Cells migrate into the blastopore (involution); pigmented cells of the animal half expand over the entire embryo surface.

4 Movement of animal half cells encloses the vegetal half cells, which are seen as a yolk plug.

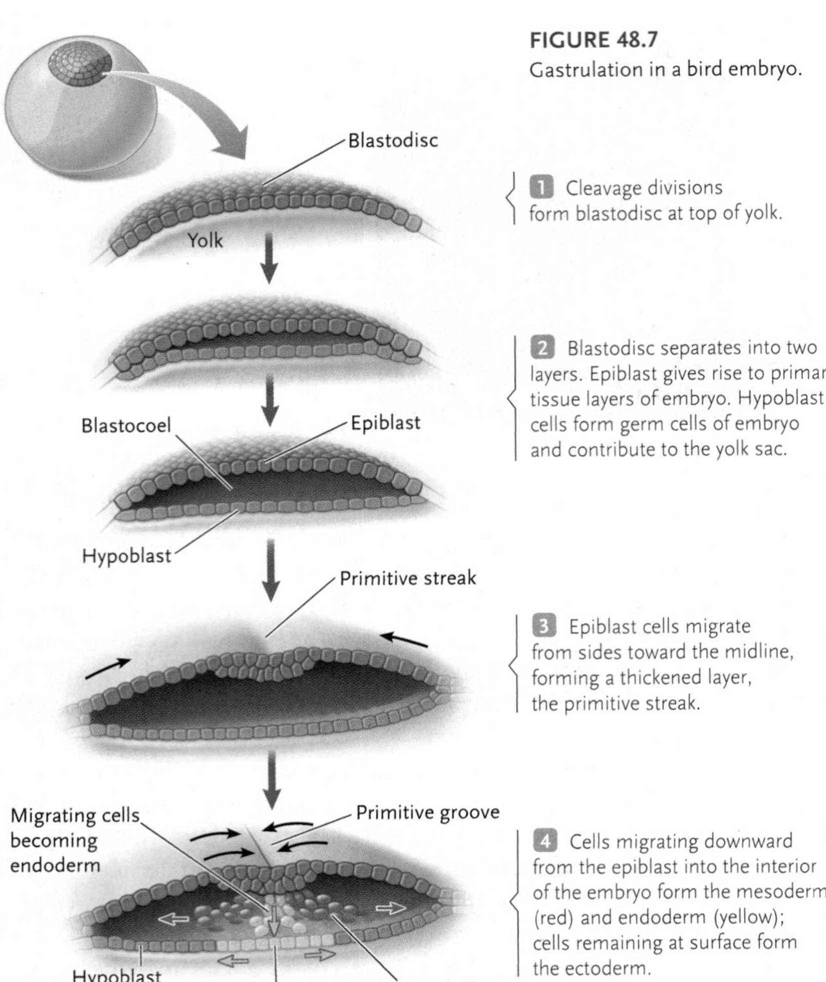
Blastodisc
Yolk
1 Cleavage divisions form blastodisc at top of yolk.
2 Blastodisc separates into two layers. Epiblast gives rise to primary tissue layers of embryo. Hypoblast cells form germ cells of embryo and contribute to the yolk sac.
Blastocoel
Epiblast
Hypoblast
Primitive streak
3 Epiblast cells migrate from sides toward the midline, forming a thickened layer, the primitive streak.
Migrating cells becoming endoderm
Primitive groove
4 Cells migrating downward from the epiblast into the interior of the embryo form the mesoderm (red) and endoderm (yellow); cells remaining at surface form the ectoderm.
Hypoblast
Endoderm
Migrating cells becoming mesoderm

FIGURE 48.7
Gastrulation in a bird embryo.

Gastrulation in Birds Proceeds at One Side of the Yolk

Gastrulation in amniotes (see Section 30.7) such as birds and reptiles is modified by the distribution of yolk, which occupies almost the entire volume of the egg. The portion of the cytoplasm that divides to give rise to the primary tissues of the embryo is confined to a thin layer at the egg surface. Although eggs of mammals (which are also amniotes) have little yolk, gastrulation follows a similar pattern, as discussed in Section 48.4.

CLEAVAGE AND GASTRULATION IN BIRDS The early cleavage divisions in birds produce a disclike layer of cells at the surface of the yolk called the **blastodisc** (**Figure 48.7,** step 1). When blastodisc formation is complete, the layer contains about 20,000 cells. The cells of the blastodisc then separate into two layers, called the **epiblast** (top layer) and **hypoblast** (bottom layer). The flattened cavity between them is the blastocoel (step 2).

Gastrulation begins as cells in the epiblast stream toward the midline of the blastodisc, thickening the epiblast in this region. The thickened layer—the **primitive streak**—begins forming in the posterior end of the embryo and extends toward the anterior end as more cells of the epiblast move into it (step 3). The primitive streak initially designates the future posterior end of the embryo, and by the time it has elongated fully, it has established the left and right sides of the embryo. The primitive streak forms on what will become the dorsal side of the embryo, with the ventral side below.

As the primitive streak forms, its midline sinks, forming the **primitive groove.** The primitive groove is a conduit for migrating cells to move into the blastocoel. The first cells to migrate through the primitive groove are epiblast cells (step 4), which will form the endoderm. Cells migrating laterally between the epiblast and the endoderm form the mesoderm. The epiblast cells left at the surface of the blastodisc form the ectoderm (see step 4). Thus all three of the primary tissue layers of the chick embryo arise from the epiblast. Once the three layers are formed, gastrulation is complete.

Of the cells in the hypoblast, only a few, near the posterior end of the embryo, contribute directly to the embryo. These hypoblast cells form the *germ cells* that, later in development, migrate to the developing gonads and establish the cell line leading to eggs and sperm (see Section 47.2).

FORMATION OF EXTRAEMBRYONIC MEMBRANES Each primary tissue layer of a bird embryo extends outside the embryo to form four **extraembryonic membranes (Figure 48.8),** which conduct nutrients from the yolk to the embryo, exchange gases with the environment outside the egg, and store metabolic wastes removed from the embryo. The **yolk sac** consists of extensions of mesoderm and endoderm that enclose the yolk. Although the yolk sac remains connected to the gut of the embryo by a stalk, yolk does not directly enter the embryo by this route. Instead, it is absorbed by blood vessels in the membrane, which transport the nutrients to the embryo. The **chorion,** produced from ectoderm and mesoderm, is the outermost membrane, which surrounds the embryo and yolk sac completely, and lines the inside of the egg shell. This membrane exchanges oxygen and car-

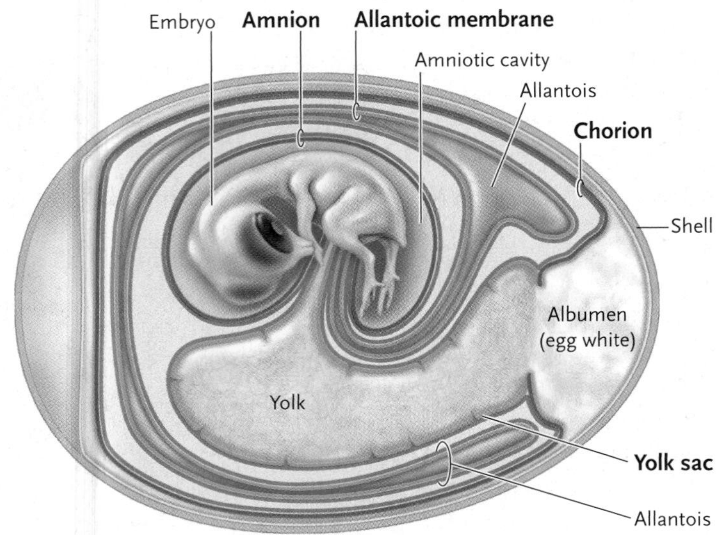
Embryo
Amnion
Allantoic membrane
Amniotic cavity
Allantois
Chorion
Shell
Albumen (egg white)
Yolk
Yolk sac
Allantois

FIGURE 48.8
The four extraembryonic membranes in a bird embryo (in bold).

bon dioxide with the environment through the shell of the egg. The **amnion** is the innermost membrane, which closes over the embryo to form the *amniotic cavity.* The cells of the amnion secrete *amniotic fluid* into the cavity, which bathes the embryo and provides an aquatic environment in which it can develop. Reptilian and mammalian embryos are also surrounded by an amnion and amniotic fluid. The adaptation of providing the embryo with an aquatic environment made the development of fully terrestrial vertebrates possible. The evolutionary importance of the amnion to the fully terrestrial vertebrates is recognized by classifying them together as amniotes (see Section 30.7). A membrane derived from mesoderm and endoderm that has bulged outward from the gut forms a sac called the **allantois.** This sac closely lines the chorion and fills much of the space between the chorion and the yolk sac. The allantois stores nitrogenous wastes (primarily uric acid) removed from the embryo. In addition, the part of the allantoic membrane that lines the chorion forms a rich bed of blood capillaries that is connected to the embryo by arteries and veins. This circulatory system delivers carbon dioxide to the chorion and picks up the oxygen that is absorbed through the shell and chorion.

STUDY BREAK 48.2

1. **What is the role of the gray crescent in amphibian development?**
2. **What evidence indicates that cells of the dorsal lip of the blastopore act as inducer cells?**
3. **What are the extraembryonic membranes in birds, and what are their functions?**

48.3 From Gastrulation to Adult Body Structures: Organogenesis

Following gastrulation, organogenesis—the process by which the ectoderm, mesoderm, and endoderm develop into organs—gives rise to an individual with the body organization characteristic of its species. Organogenesis involves the same mechanisms used in gastrulation—cell division, cell movements, selective cell adhesion, induction, and differentiation—plus an additional mechanism, *apoptosis,* in which certain cells are programmed to die (apoptosis is also discussed in Section 43.2 and later in this section). As with other aspects of development, the details of organogenesis differ among animal groups. The frog and the chick are used here as examples.

The Nervous System Develops from Ectoderm

In vertebrates, organogenesis begins with development of the nervous system from ectoderm, a process called **neurulation.** As a preliminary to neurulation, cells of the dorsal mesoderm form a solid rod of tissue, the **notochord,** which extends the length of the embryo under the dorsal ectoderm. Dorsal mesoderm cells under the ectoderm then induce the ectoderm cells above them to thicken and flatten into a longitudinal band called the **neural plate** (**Figure 48.9,** step 1).

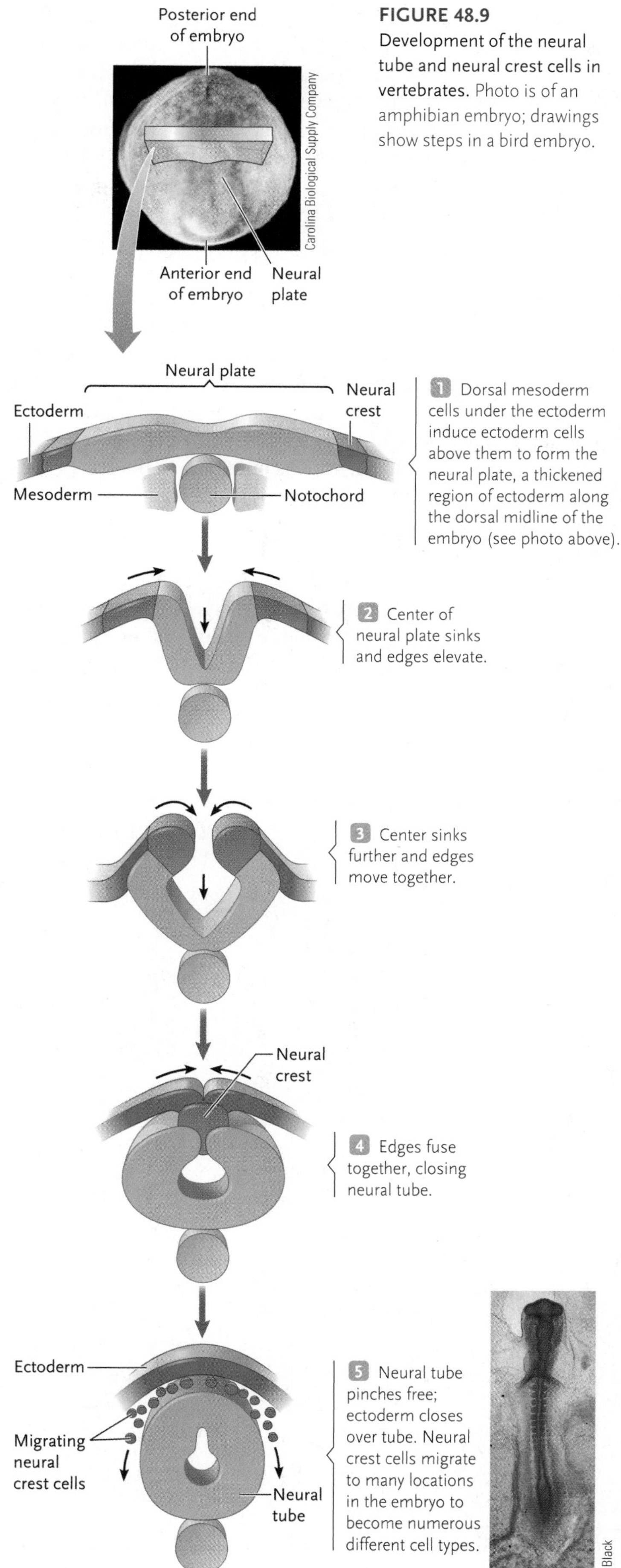
Posterior end of embryo
Carolina Biological Supply Company
Anterior end of embryo
Neural plate
Neural plate
Ectoderm
Neural crest
Mesoderm
Notochord
1 Dorsal mesoderm cells under the ectoderm induce ectoderm cells above them to form the neural plate, a thickened region of ectoderm along the dorsal midline of the embryo (see photo above).
2 Center of neural plate sinks and edges elevate.
3 Center sinks further and edges move together.
Neural crest
4 Edges fuse together, closing neural tube.
Ectoderm
Migrating neural crest cells
Neural tube
5 Neural tube pinches free; ectoderm closes over tube. Neural crest cells migrate to many locations in the embryo to become numerous different cell types.
S. Black

FIGURE 48.9 Development of the neural tube and neural crest cells in vertebrates. Photo is of an amphibian embryo; drawings show steps in a bird embryo.

Next, the neural plate sinks downward along its midline (step 2), creating a deep longitudinal groove. At the same time, the edges elevate along the sides of the neural plate. The groove becomes deeper and the edges move together (step 3). Next, the edges fuse together and close over the center of the groove, converting the neural plate into a **neural tube** that runs the length of the embryo (step 4). The neural tube then pinches off from the overlying ectoderm, which closes over the tube (step 5). The central nervous system, including the brain and spinal cord, develops directly from the neural tube.

During formation of the neural tube, cells of the **neural crest**—the region where the neural tube pinches off from the ectoderm—migrate to many locations in the developing embryo and become numerous different types of cells which contribute to a variety of organ systems. (The neural crest is one of the defining features of vertebrates.) Some cells develop into cranial nerves in the head; others contribute to the bones of the inner ear and skull, the cartilage of facial structures, and the teeth. Still others form ganglia of the autonomic nervous system, peripheral nerves leading from the spinal cord to body structures, and nerves of the developing gut. Neural crest cells also move to the skin, where they form pigment cells, and to the adrenal glands, where they form the medulla of the kidney. The migration of neural crest cells contributes to the development of all vertebrates.

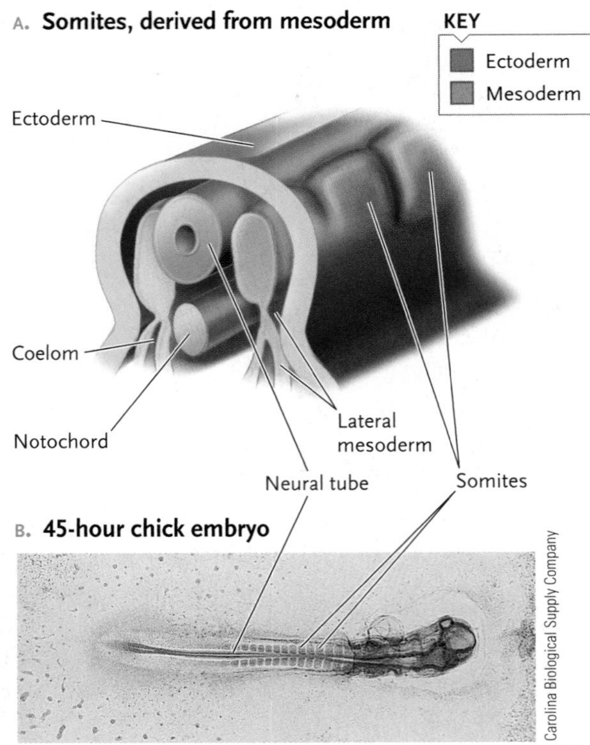
A. Somites, derived from mesoderm
KEY
Ectoderm
Mesoderm
Ectoderm
Coelom
Notochord
Lateral mesoderm
Neural tube
Somites
B. 45-hour chick embryo
Carolina Biological Supply Company

FIGURE 48.10
Later development of the mesoderm. **(A)** The somites develop into segmented structures such as the vertebrae, the ribs, and the musculature between the ribs. The lateral mesoderm gives rise to other structures, such as the heart and blood vessels and the linings of internal body cavities. **(B)** The somites in a 45-hour chick embryo.

Other structures differentiate in the embryo while the neural tube is forming. On either side of the notochord, the mesoderm separates into blocks of cells called **somites,** spaced one after the other along both sides of the notochord **(Figure 48.10).** The somites give rise to the vertebral column, the ribs, the repeating sets of muscles associated with the ribs and vertebral column, and muscles of the limbs. The mesoderm outside the somites, which extends around the primitive gut (lateral mesoderm in Figure 48.10), splits into two layers, one covering the surface of the gut, and the other lining the body wall. The space between the layers is the coelom of the adult (see Section 29.2).

Sequential Inductions and Differentiation Are Central to Eye Development

We now take up the development of the eye, to show how cellular mechanisms interact in organogenesis. Eyes develop by the same pathway in all vertebrates.

The brain forms at the anterior end of the neural tube from a cluster of hollow vesicles that swell outward from the neural tube (**Figure 48.11,** step 1). One paired set of vesicles, the *optic vesicles,* develop into the eyes. The figure depicts the optic vesicles in the brain of a frog embryo; the morphology of the forebrain, midbrain, and hindbrain in embryos differs among vertebrates.

The optic vesicles grow outward until they contact the overlying ectoderm, inducing a series of developmental responses in both tissues. The outer surface of the optic vesicle thickens and flattens at the region of contact and then pushes inward, transforming the optic vesicle into a double-walled *optic cup,* which ultimately becomes the retina. The optic cup induces the overlying ectoderm to thicken into a disclike swelling, the *lens placode* (step 2). The center of the lens placode sinks inward toward the optic cup, and its edges eventually fuse together, forming a ball of cells, the *lens vesicle* (step 3).

Ectoderm now closes over the lens vesicle, which then detaches from the surface ectoderm (step 4). The lens vesicle becomes the developing lens, the cells of which begin to synthesize *crystallin,* a fibrous protein that collects into clear, glassy deposits. The lens cells finally lose their nuclei and form the elastic, crystal-clear lens.

As the lens develops, neural crest cells migrate into the space between the lens and the surface epithelium. These neural crest cells form several layers of cells which eventually mature into the clear cornea. The developing cornea joins with the edges of the optic cup to complete the primary structures of the eye. Other cells contribute to accessory structures of the eye; for example, mesoderm and neural crest cells contribute to the reinforcing tissues in the wall of the eye and the muscles that move the eye. Figure 48.11, step 5, shows a fully developed vertebrate eye.

Many experiments have shown that the initial induction by the optic vesicle is necessary for development of the eye. For example, if an optic vesicle is removed before lens formation, the ectoderm fails to develop a lens placode and vesicle. Moreover, placing a removed optic vesicle under the ectoderm in other re-

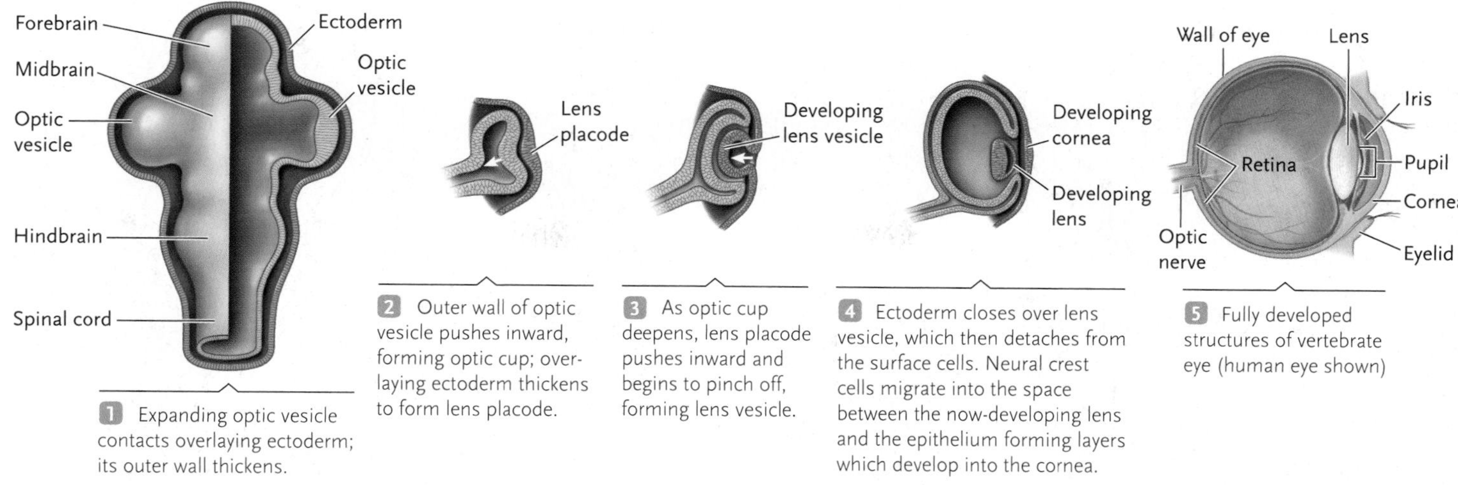
Forebrain
Midbrain
Optic vesicle
Hindbrain
Spinal cord
Ectoderm
Optic vesicle
Lens placode
Developing lens vesicle
Developing cornea
Developing lens
Wall of eye
Lens
Iris
Retina
Pupil
Cornea
Eyelid
Optic nerve
1 Expanding optic vesicle contacts overlaying ectoderm; its outer wall thickens.
2 Outer wall of optic vesicle pushes inward, forming optic cup; overlaying ectoderm thickens to form lens placode.
3 As optic cup deepens, lens placode pushes inward and begins to pinch off, forming lens vesicle.
4 Ectoderm closes over lens vesicle, which then detaches from the surface cells. Neural crest cells migrate into the space between the now-developing lens and the epithelium forming layers which develop into the cornea.
5 Fully developed structures of vertebrate eye (human eye shown)

FIGURE 48.11
Stages in the formation of the vertebrate eye from the optic vesicle of the brain and the overlying ectoderm.

gions of the head causes a lens to form in the new location. Or, if the ectoderm over an optic vesicle is removed and ectoderm from elsewhere in the embryo is grafted in its place, a normal lens will develop in the grafted ectoderm, even though in its former location it would not differentiate into lens tissue.

Eye development also demonstrates differentiation. Ectoderm cells that are induced to form the lens of the eye synthesize crystallin; in other locations, ectoderm cells typically synthesize a different protein, *keratin,* as their predominant cell product. Keratin is a component of surface structures such as skin, hair, feathers, scales, and horns. In other words, as a response to induction by the optic vesicle, the genes of the ectoderm cells coding for crystallin are activated, whereas genes coding for keratin are not expressed.

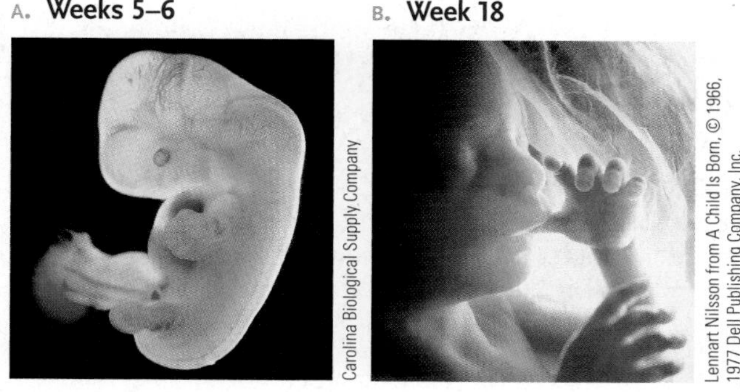
A. Weeks 5–6
B. Week 18
Carolina Biological Supply Company
Lennart Nilsson from A Child Is Born, © 1966, 1977 Dell Publishing Company, Inc.

FIGURE 48.12
An illustration of apoptosis in humans: the removal of tissue between developing fingers and toes produces the free fingers and toes seen later in development.

Apoptosis Eliminates Tissues That Are No Longer Required

Induction and differentiation build complex, specialized organs from the three fundamental tissue types. Complementing these processes is *apoptosis,* programmed cell death, which in this case removes tissues present during development but not in the fully formed organ. Apoptosis plays an important role in the development of animals, both vertebrates and invertebrates, for example:

- In frog development apoptosis occurs during metamorphosis, in which the tadpole changes into an adult frog. The tail of the tadpole becomes progressively smaller and finally disappears because its cells disintegrate and their components are absorbed and recycled by other cells. Cells that are eliminated by apoptosis, like those of a tadpole's tail, are typically parts of structures required at one stage of development but not for later stages.
- In humans, the developing fingers and toes are initially connected by tissue, forming a paddle-shaped structure; later in development, cells of the tissue die by apoptosis resulting in separated fingers and toes **(Figure 48.12).**
- Kittens and puppies, like many other mammals, are born with their eyes sealed shut by an unbroken layer of skin. Just after birth, cells die in a thin line across the middle of each eyelid, freeing the eyelids to open.
- During the pupation stage that converts a caterpillar to a butterfly, many tissues of the larva break down by apoptosis and are replaced by newly formed adult tissues.

Apoptosis results from gene activation, in response to molecular signals received by receptors at the surfaces of the marked cells. In effect, the signals amount to a death notice, delivered at a specific time during embryonic development. For example, in the nematode *C. elegans,* division of the fertilized egg leads to a total of 1,090 cells. Of these, exactly 131 die at prescribed times to produce a total of 959 cells in the adult hermaphrodite.

The molecular basis of apoptosis in *C. elegans* involves a molecule that can be considered a death signal binding to a receptor in the plasma membrane of a target cell for apoptosis. The receptor is activated, leading to activation of proteins that kill the cell. In the absence of the death signal, the killing proteins remain inactive.

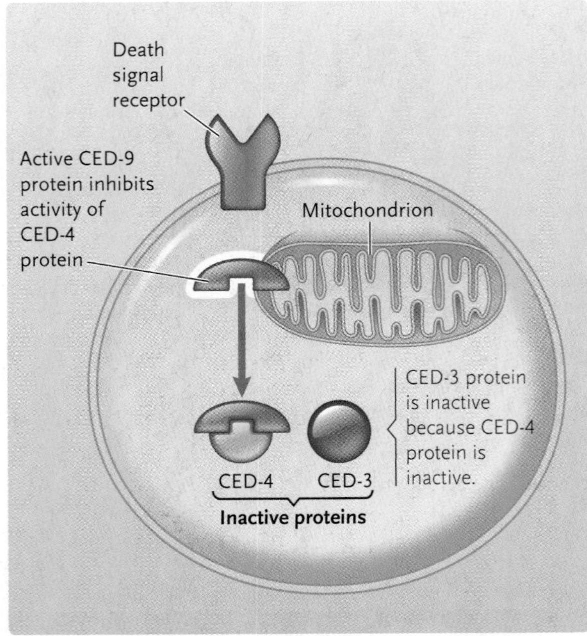
A. No death signal
Death signal receptor
Active CED-9 protein inhibits activity of CED-4 protein
Mitochondrion
CED-3 protein is inactive because CED-4 protein is inactive.
CED-4
CED-3
Inactive proteins

Apoptosis is inhibited as long as CED-9 protein is active; cell remains alive.

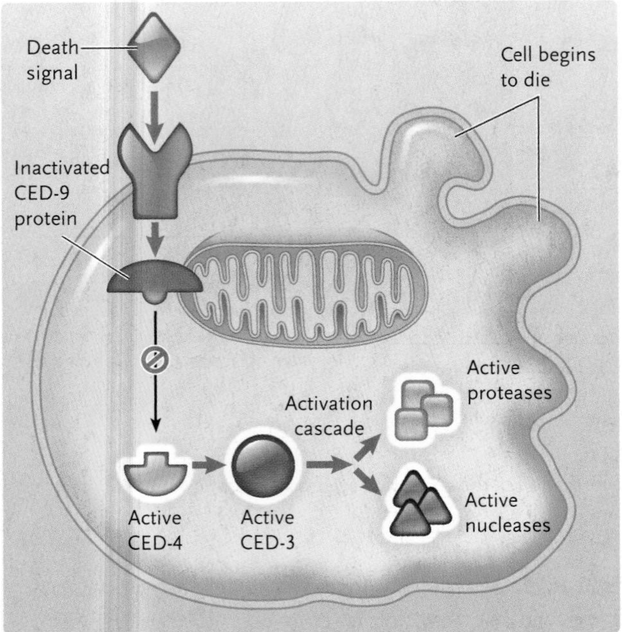
B. Death signal
Death signal
Cell begins to die
Inactivated CED-9 protein
Activation cascade
Active proteases
Active nucleases
Active CED-4
Active CED-3

When a death signal binds to the death signal receptor, it activates the receptor, which leads to inactivation of CED-9 protein. As a result, CED-4 protein is no longer inhibited and becomes active, activating CED-3 protein. Active CED-3 triggers a cascade of activations producing active proteases and nucleases, which cause the changes seen in apoptotic cells and lead eventually to cell death.

FIGURE 48.13
The molecular basis of apoptosis in the nematode, *C. elegans.* **(A)** In the absence of a death signal, no apoptosis occurs. **(B)** In the presence of a death signal, activation of CED-4 and CED-3 proteins triggers a pathway that leads to death of the cell.

Let us walk through the two situations:

- In the absence of a death signal, the membrane receptor is inactive **(Figure 48.13A).** This allows a protein associated with the outer mitochondrial membrane, CED-9 (encoded by the *ced-9 c*ell *d*eath gene) to inhibit CED-4 (encoded by the *ced-4* gene) and CED-3 (encoded by the *ced-3* gene), the two proteins that are needed to turn on the cell death program. Cells in this situation, with the *ced-9* gene being expressed and its product CED-9 being active, are those that normally survive to form the adult nematode.
- If a death signal binds to the receptor, the receptor becomes activated and the events that follow are typical of signal transduction pathways **(Figure 48.13B)** (see Sections 7.1–7.4). In this case, the activated receptor leads to inactivation of CED-9. Because CED-9 no longer is inhibiting CED-4, CED-4 is activated and, in turn, it activates CED-3. Activated CED-3 triggers a cascade of reactions, including the activation of proteases and nucleases that degrade cell structures and chromosomes as part of the cell death program.

STUDY BREAK 48.3 <

1. **What is the outcome of organogenesis?**
2. **What tissues or organs develop from the neural tube and neural crest cells?**

48.4 Embryonic Development of Humans and Other Mammals

The embryonic development of humans is representative of the placental mammals. In the uterus, the embryo is nourished by the placenta, which supplies oxygen and nutrients and carries away carbon dioxide and nitrogenous wastes.

Pregnancy or **gestation,** the period of mammalian development in the uterus, varies in different species. Larger mammals bearing larger young generally have longer gestation periods; for example, gestation takes 600 days in elephants, about 1 year in blue whales, and 21 days in hamsters.

In humans, gestation takes an average of 266 days, about 38 weeks, which amounts to about 40 weeks from the beginning of the menstrual cycle in which fertilization takes place. Human gestation is divided into three **trimesters,** each approximately 3 months long.

The major developmental events in human gestation—cleavage, gastrulation, and organogenesis—take place during the first trimester. By the fourth week, the embryo's heart is beating, and by the end of the eighth week, the major organs and organ systems have formed. From this point until birth, the developing human is called a **fetus.** Only 5 cm long by the end of the first trimester, the fetus grows during the second and third trimesters to an average length of 50 cm and an average mass of 3.5 kg (or about 19.7 inches and 7.7 pounds). The period of gestation ends with birth.

Cleavage and Implantation Occupy the First 2 Weeks of Development

We noted in Section 47.3 that human fertilization occurs when the egg is in the first third of the oviduct leading from the ovary to the uterus. After fertilization, cleavage divisions take place during passage of the developing embryo down the oviduct (also called the fallopian tube in mammals) and while it is still enclosed in the zona pellucida—the original coat of the egg (**Figure 48.14**).

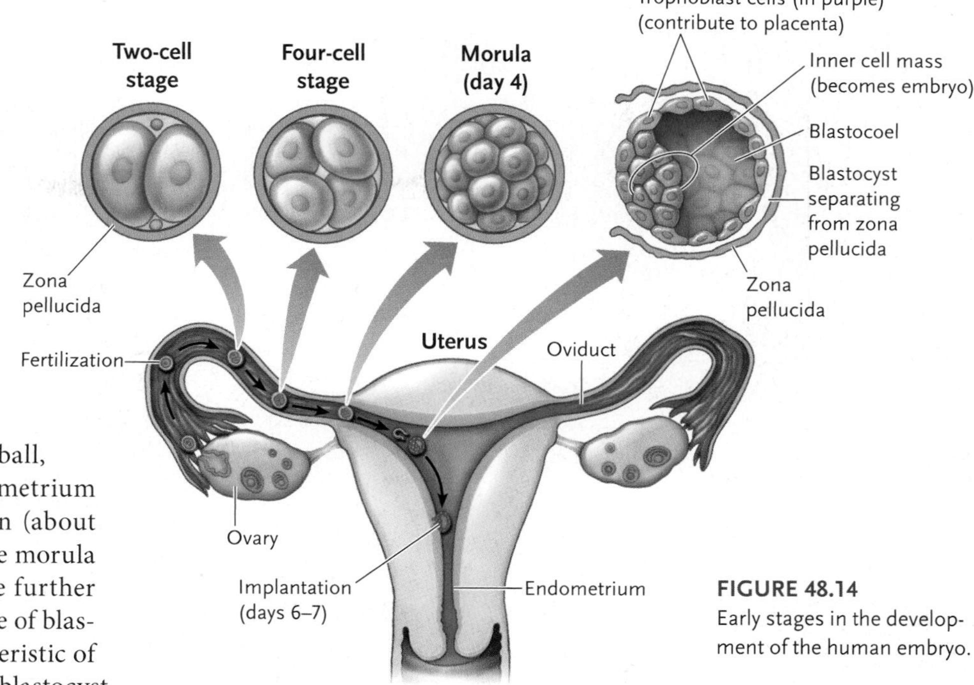

Two-cell stage
Four-cell stage
Morula (day 4)
Trophoblast cells (in purple) (contribute to placenta)
Inner cell mass (becomes embryo)
Blastocoel
Blastocyst separating from zona pellucida
Zona pellucida
Zona pellucida
Fertilization
Uterus
Oviduct
Ovary
Implantation (days 6–7)
Endometrium

FIGURE 48.14
Early stages in the development of the human embryo.

By day 4, the morula, a 16- or 32-cell ball, has been produced. By the time the endometrium (uterine lining) is ready for implantation (about 7 days after ovulation; see Section 47.3), the morula has reached the uterus and has undergone further cell divisions and differentiation into a type of blastula called a **blastocyst,** which is a characteristic of mammalian development. At this time, the blastocyst is a single-cell-layered hollow ball that is made up of about 70–100 cells, and contains a fluid-filled cavity. The blastocoel, the fluid-filled cavity, has a dense mass of cells localized to one side called the **inner cell mass.** This inner cell mass gives rise to the embryo as well as the yolk sac, the allantois, and the amnion. The rest of the blastocyst becomes tissues that support the development of the embryo in the uterus. The outer single layer of cells of the blastocyst is the **trophoblast.**

Next, when the blastocyst has increased to an appropriate size, it breaks out of the zona pellucida and sticks to the endometrium on the side with the inner cell mass (**Figure 48.15,** step 1). Implantation begins when the trophoblast secretes proteases that digest pathways between endometrial cells. The trophoblast then extends into the digested spaces, appearing like fingerlike projections into the endometrium. These cells continue to digest the nutrient-rich endometrial cells, serving both to produce a hole in the endometrium for the blastocyst and to release nutrients that the developing embryo can use.

Although the blastocyst is burrowing into the endometrium, the inner cell mass separates into the *embryonic disc,* which consists of two distinct cell layers (see Figure 48.15, step 1). The layer farther from the blastocoel is the epiblast, which gives rise to the embryo proper, and the layer nearer the blastocoel is the hypoblast, which gives rise to part of the extraembryonic membranes. When implantation is complete, the blastocyst has completely burrowed into the endometrium and is covered by a layer of endometrial cells (step 2).

Mammalian Gastrulation and Neurulation Resemble the Reptilian–Bird Pattern

Gastrulation proceeds as in birds (see Figure 48.7), with the formation of a primitive streak in the epiblast. Some epiblast cells remain in place, becoming the ectoderm, whereas others enter the streak to form the endoderm and mesoderm. The ectoderm, mesoderm, and endoderm are located initially in three layers; from this initial arrangement, the endoderm folds to form the primitive gut, and becomes surrounded with ectoderm and mesoderm. Neurulation in human and other mammalian embryos takes place essentially as in birds (see Figure 48.9).

Extraembryonic Membranes Give Rise to the Amnion and Part of the Placenta

Soon after the inner cell mass separates into the epiblast and hypoblast, a layer of cells separates from the epiblast along its top margin (see Figure 48.15, step 2). The fluid-filled space created by the separation becomes the amniotic cavity, and the layer of ectodermal cells forming its roof becomes the amnion, the extraembryonic membrane surrounding the cavity. The amnion expands until eventually it completely surrounds the embryo and suspends it in amniotic fluid. As in birds, the hypoblast develops into the yolk sac. However, in mammals, the mesoderm of the yolk sac gives rise to the blood vessels in the embryonic portion of the placenta.

Although the amnion is expanding around the embryonic disc, blood-filled spaces form in maternal tissue, and trophoblast cells grow rapidly around both the embryo and amnion to form the chorion (step 3). Next, a connecting stalk forms between the embryonic disc and the chorion, while the chorion begins to grow into the endometrium as fingerlike or treelike extensions called **chorionic villi** (step 4). When complete, the chorionic villi greatly increase the surface area of the chorion. Where these villi grow into the endometrium is the area of the future placenta. As the chorion develops, mesodermal cells of the yolk sac grow into it and form a rich network of blood vessels, the embryonic circulation of the placenta. The expanding chorion stimulates the

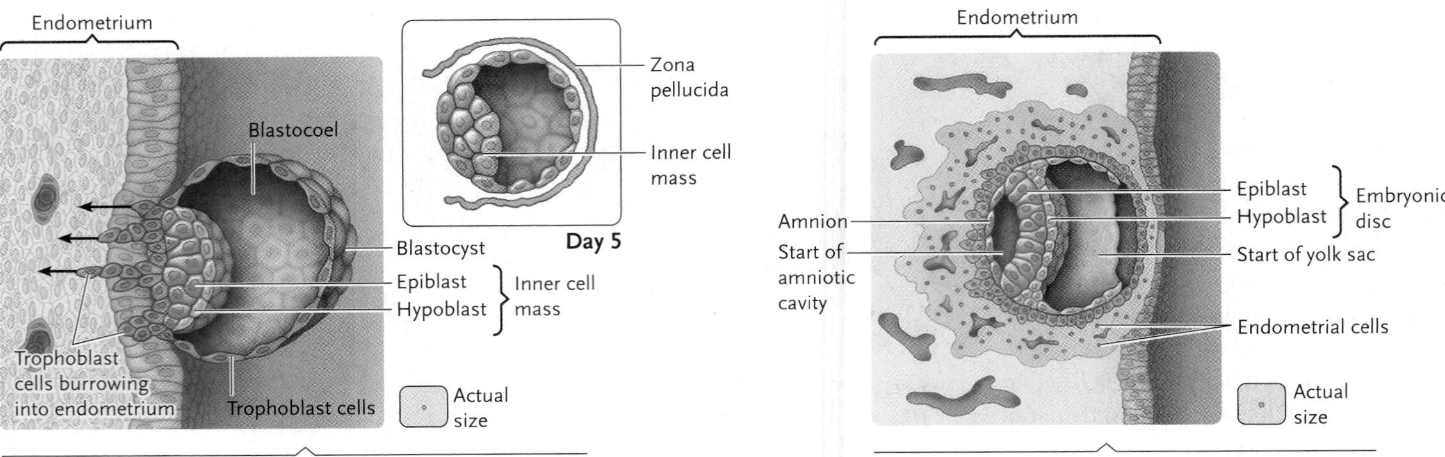
Endometrium
Blastocoel
Zona pellucida
Inner cell mass
Day 5
Blastocyst
Epiblast
Hypoblast
Inner cell mass
Trophoblast cells burrowing into endometrium
Trophoblast cells
Actual size
Endometrium
Amnion
Start of amniotic cavity
Epiblast
Hypoblast
Embryonic disc
Start of yolk sac
Endometrial cells
Actual size

1 Days 6–7: Surface cells of the blastocyst attach to the endometrium and start to burrow into it. Implantation is under way.

2 Days 10–11: A layer of epiblast cells separates, producing the amniotic cavity. The cells above the cavity become the amnion, which eventually surrounds the embryo. The hypoblast begins to form around the yolk sac.

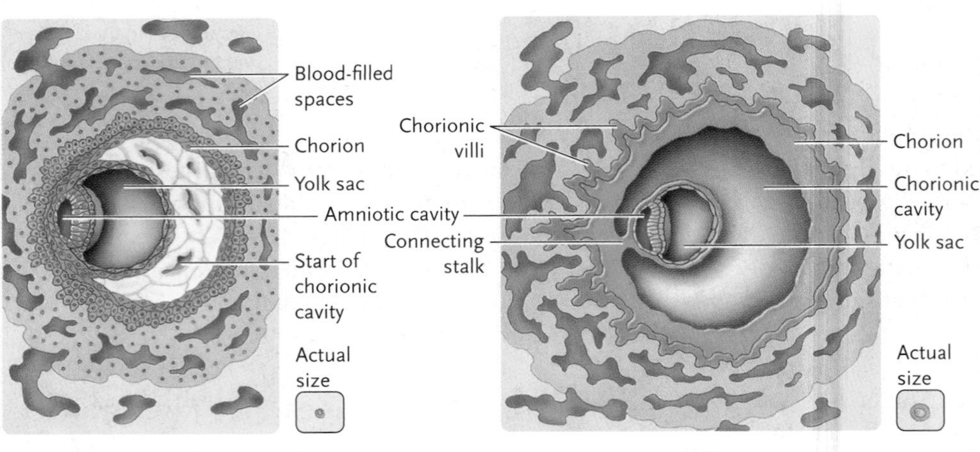
Blood-filled spaces
Chorion
Yolk sac
Amniotic cavity
Start of chorionic cavity
Actual size
Chorionic villi
Connecting stalk
Chorion
Chorionic cavity
Yolk sac
Actual size

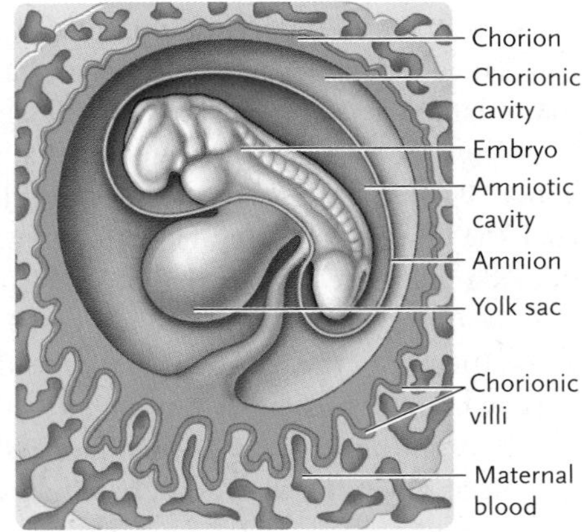
Chorion
Chorionic cavity
Embryo
Amniotic cavity
Amnion
Yolk sac
Chorionic villi
Maternal blood

3 Day 12: Blood-filled spaces form in maternal tissue. The chorion forms, derived from trophoblast cells, and encloses the chorionic cavity.

4 Day 14: A connecting stalk has formed between the embryonic disk and chorion. Chorionic villi, which will be features of a placenta, start to form.

5 Day 25: The chorion continues to grow into the endometrium, producing the chorionic villi. The chorion growth stimulates blood vessels of the endometrium to grow into the maternal circulation of the placenta.

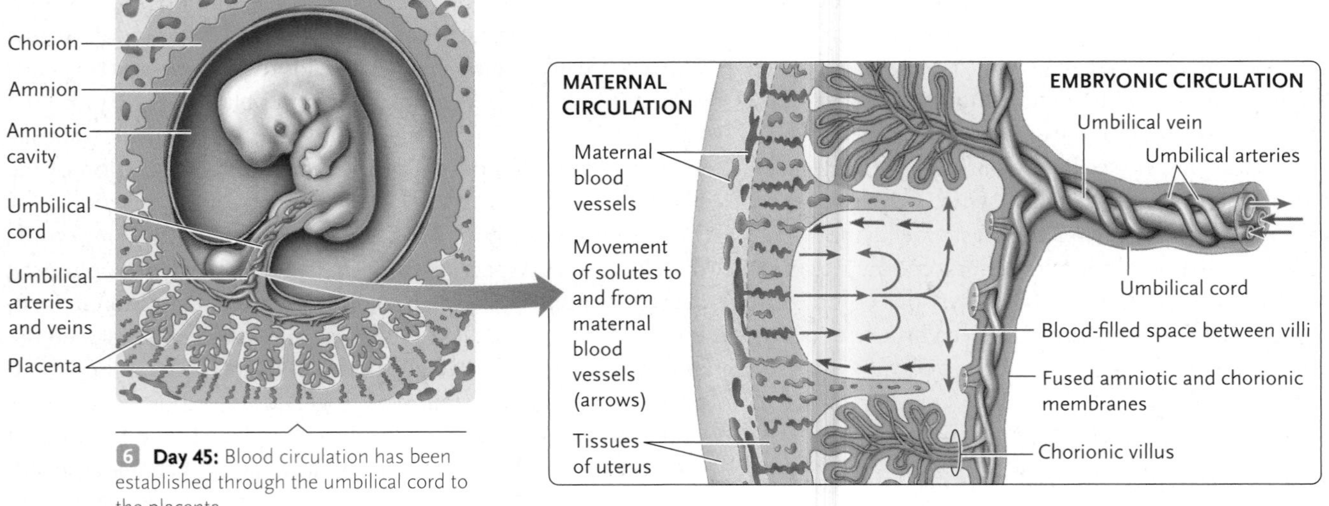
Chorion
Amnion
Amniotic cavity
Umbilical cord
Umbilical arteries and veins
Placenta
MATERNAL CIRCULATION
EMBRYONIC CIRCULATION
Maternal blood vessels
Movement of solutes to and from maternal blood vessels (arrows)
Tissues of uterus
Umbilical vein
Umbilical arteries
Umbilical cord
Blood-filled space between villi
Fused amniotic and chorionic membranes
Chorionic villus

6 Day 45: Blood circulation has been established through the umbilical cord to the placenta.

FIGURE 48.15
Implantation of a human blastocyst in the endometrium of the uterus and the establishment of the placenta.

blood vessels of the endometrium to grow into the maternal circulation of the placenta (step 5).

Within the placenta of humans, apes, monkeys, and rodents, the maternal circulation opens into spaces in which the maternal blood bathes the trophoblast layers of the placenta (step 6). The fetal capillaries are positioned next to the trophoblast so that there can be as few as two cell layers separating the maternal and fetal blood. (Different types of placentas are found in other mammals.) The embryonic circulation remains closed, however, so that the embryonic and maternal blood do not mix directly. This prevents the mother from developing an immune reaction against cells of the embryo, which may be recognized as foreign by the mother's immune system. Eventually, the placenta and its blood circulation grow to cover about a quarter of the inner surface of the enlarged uterus and reach the size of a dinner plate.

As the embryonic blood circulation develops, this connecting stalk between the embryo and placenta develops into the **umbilical cord,** a long tissue with blood vessels linking the embryo and the placenta. The vessels in the umbilical cord are derived from the extraembryonic membrane, the allantois. They conduct blood between the embryo and the placenta (shown in the inset for Figure 48.15, step 6).

Within the placenta, nutrients and oxygen pass from the mother's circulation into the circulation of the embryo. Besides nutrients and oxygen, many other substances taken in by the mother—including alcohol, caffeine, drugs, and toxins in cigarette smoke—can pass from mother to embryo. Carbon dioxide and nitrogenous wastes pass from the embryo to the mother, and are disposed of by the mother's lungs and kidneys.

If the presence of a genetic disease such as cystic fibrosis or Down syndrome is suspected, tests can be carried out on cells removed from the embryonic portion of the placenta or from the amniotic fluid, which contains cells derived from the embryo. The test using cells of the placenta is called *chorionic villus sampling;* the test using cells derived from the amniotic fluid is called *amniocentesis* (see Section 13.4). Chorionic villus sampling can be carried out as early as the eighth week, compared with 14 weeks for amniocentesis. Both tests carry some degree of risk to the embryo.

Further Growth of the Fetus Culminates in Birth

By the end of its fourth week, a human embryo is 3–5 mm long, which is 250–500 times the size of the zygote **(Figure 48.16A).** It has a tail and **pharyngeal arches,** which are embryonic features of all vertebrates. The pharyngeal arches contribute to the formation of the face, neck, mouth, larynx, and pharynx. After 5 to 6 weeks, most of the tail has disappeared and the embryo is beginning to take on recognizable human form **(Figure 48.16B).** At 8 weeks, the embryo, now a fetus, is about 2.5 cm long **(Figure 48.16C).** Its organ systems have formed, and its limbs, with fingers or toes at their ends, have developed **(Figure 48.16D).**

Figure 48.17 shows the hormonal events and associated physical events of birth. As the period of fetal growth comes to a close, the fetus typically turns so that its head is downward, pressed against the cervix. A steep rise in the levels of estrogen secreted by the placenta at this time causes cells of the uterus to express the gene for the receptor of the hormone *oxytocin.* The receptors become inserted into the plasma membranes of those cells. Oxytocin—which is secreted by the pituitary gland—

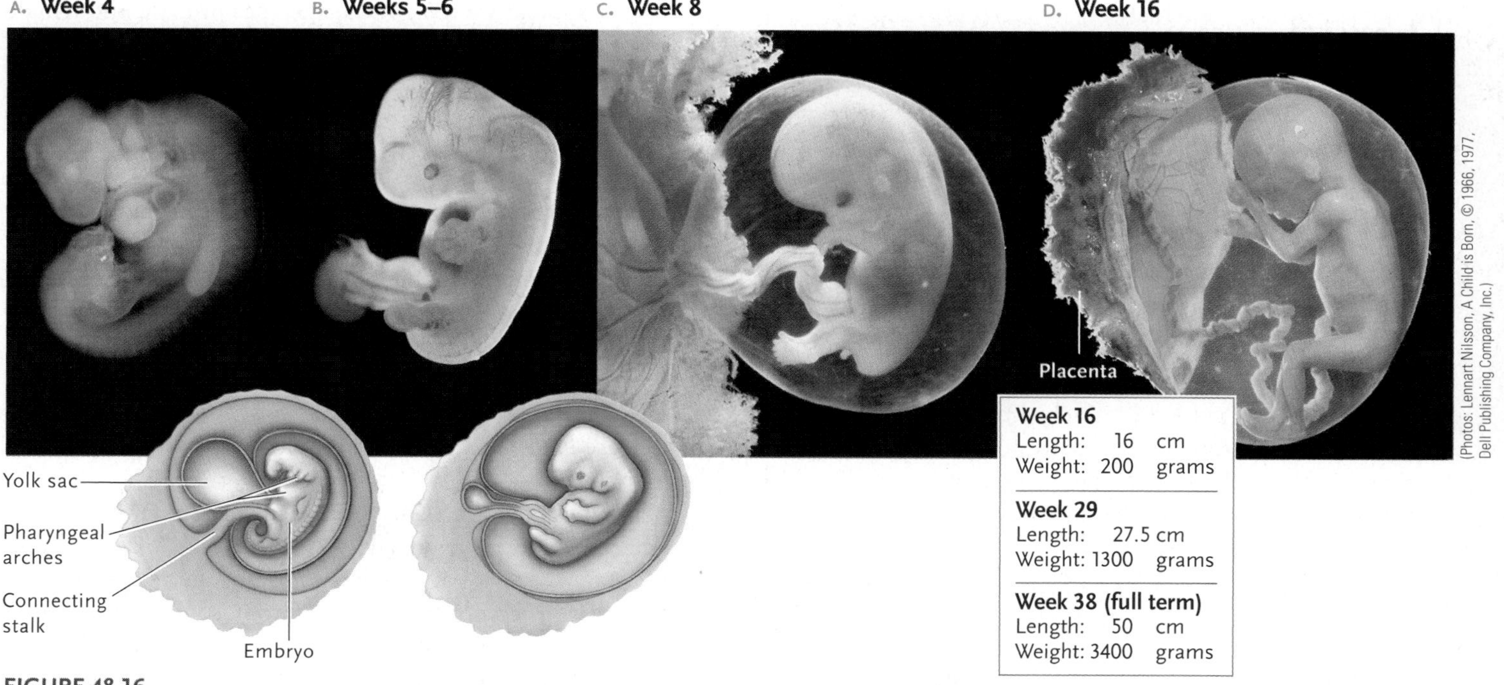
A. Week 4
B. Weeks 5–6
C. Week 8
D. Week 16
Placenta
Week 16
Length: 16 cm
Weight: 200 grams
Week 29
Length: 27.5 cm
Weight: 1300 grams
Week 38 (full term)
Length: 50 cm
Weight: 3400 grams
Yolk sac
Pharyngeal arches
Connecting stalk
Embryo
(Photos: Lennart Nilsson, A Child is Born, © 1966, 1977, Dell Publishing Company, Inc.)

FIGURE 48.16
The human embryo at various stages of development, beginning at week 4. The chorion has been pulled aside to reveal the embryo in the amnion at week 8 and week 16. By week 16, movements begin as nerves make functional connections with the forming muscles.

binds to its receptor, triggering the smooth muscle cells of the uterine wall to contract and begin the rhythmic contractions of labor. These contractions mark the beginning of **parturition** (*parturire* = to be in labor), the process of giving birth.

The contractions push the fetus further against the cervix and stretch its walls (step 1). In response, stretch receptors in the walls send nerve signals to the hypothalamus, which responds by stimulating the pituitary to secrete more oxytocin. In turn, the oxytocin stimulates more forceful contractions of the uterus, pressing the fetus more strongly against the cervix, and further stretching its walls. The positive feedback cycle continues, steadily increasing the strength of the uterine contractions.

As the contractions force the head of the fetus through the cervix (step 2), the amniotic membrane bursts, releasing the amniotic fluid. Usually, within 12 to 15 hours after the onset of uterine contractions, the head passes entirely through the cervix. Once the head is through, the rest of the body follows quickly and the entire fetus is forced through the vagina to the exterior, still connected to the placenta by the umbilical cord (step 3).

Umbilical cord
Vagina
Uterus
Partially dilated cervix

1 Contractions of the uterus press the head against the cervix, stretching the cervical opening.

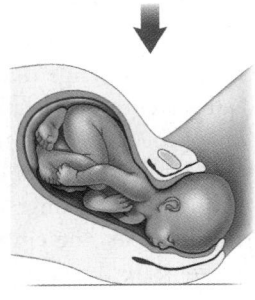

2 The head of the fetus begins to pass through the cervix and vagina.

Placenta
Uterus
Umbilical cord

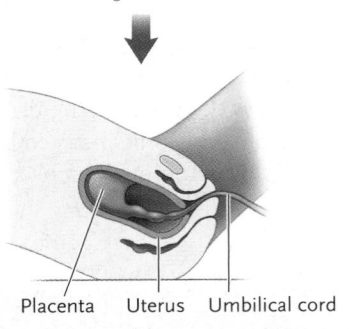

3 The placenta and umbilical cord will be forced out of the uterus as the "afterbirth."

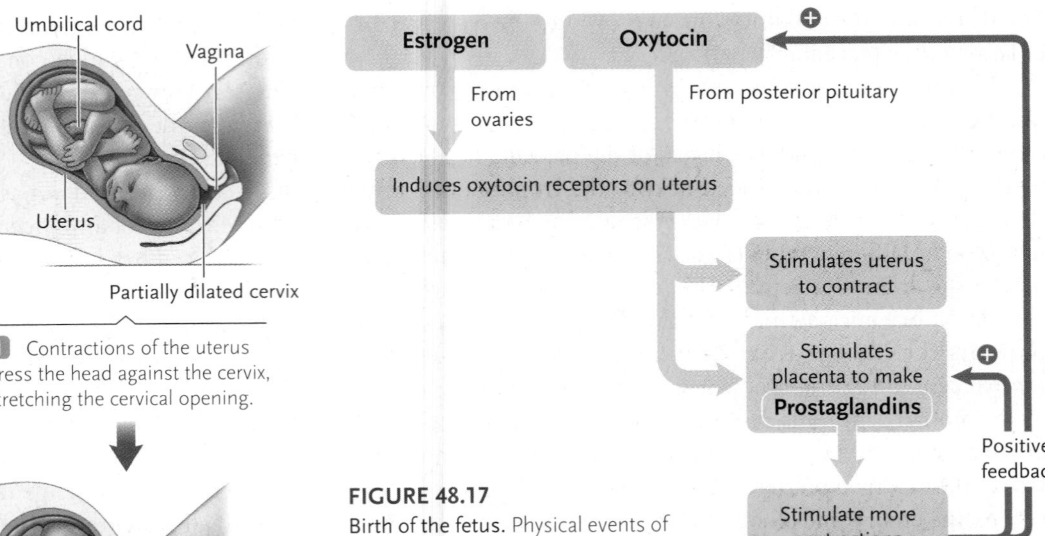

Estrogen
Oxytocin
From ovaries
From posterior pituitary
Induces oxytocin receptors on uterus
Stimulates uterus to contract
Stimulates placenta to make
Prostaglandins
Stimulate more contractions of uterus
Positive feedback

FIGURE 48.17
Birth of the fetus. Physical events of birth are shown to the left, and hormonal events of birth are shown above.

After the baby takes its first breath the umbilical cord is cut and tied off by the birth attendant. Contractions of the uterus continue, expelling the placenta and any remnants of the umbilical cord and embryonic membranes as the afterbirth, usually within 15 minutes to an hour after the infant's birth. The short length of umbilical cord still attached to the infant dries and shrivels within a few days. Eventually, it separates entirely and leaves a scar, the **umbilicus** or navel, to mark its former site of attachment during embryonic development.

The Mother's Mammary Glands Become Active after Birth

Before birth of the fetus, estrogen and progesterone secreted by the placenta stimulate the growth of the mammary glands in the mother's breasts. However, the high levels of these hormones prevent the mammary glands from responding to *prolactin,* the hormone secreted by the pituitary that stimulates the glands to produce milk. After birth of the fetus and release of the placenta, the levels of estrogen and progesterone fall steeply in the mother's bloodstream, and the breasts begin to produce milk (stimulated by prolactin) and secrete it (stimulated by oxytocin).

Continued milk secretion depends on whether the infant is suckled by the mother. If the infant is suckled, stimulation of the nipples sends nerve impulses to the hypothalamus, which responds by signaling the pituitary to release a burst of prolactin and oxytocin. Hormonal stimulation of milk production and secretion continues as long as the infant is breastfed.

So far, we have followed the development of a generic human, but certain aspects of development differ depending on the offspring's sex. Next we look at the specifics of male and female development.

A Gene on the Y Chromosome Determines the Development of Male or Female Sex Organs

The gonads and their ducts begin to develop during the fourth week of gestation. Until the seventh week, male and female embryos have the same set of internal structures derived from mesoderm, including a pair of gonads **(Figure 48.18A).** Each gonad is associated with two primitive ducts, the **Wolffian duct** (also called the *mesonephric duct*) and the **Müllerian duct** (also called the *paramesonephric duct*), which lead to a cloaca. At this time, the gonads are *bipotential:* they can develop into either male or female gonads.

The presence or absence of a Y chromosome determines whether the bipotential gonads and internal ducts develop into male or female gonads and their associated internal ducts. If the fetus has the XY combination of sex chromosomes, a single gene on the Y chromosome, *SRY (Sex-determining Region of the Y),* becomes active in the seventh week. The protein encoded by the gene sets a molecular switch that causes the primitive gonads to develop into testes. The fetal testes then secrete two hormones, testosterone and the *anti-Müllerian hormone (AMH).* The testosterone stimulates development of the Wolffian ducts into the male reproductive tract, including the epididymis and vas deferens **(Figure 48.18B)**, and seminal vesicles (see Figure 47.15). AMH causes the Müllerian ducts to degenerate and disappear. (*Insights from the Molecular Revolution* describes experiments that traced the activity of the *SRY* gene and its encoded protein in male development.) Testosterone additionally stimulates the development of the male genitalia.

If the fetus has the XX combination of chromosomes, no SRY protein is produced and the primitive gonads, under the influence of the estrogens and progesterone secreted by the placenta, develop into ovaries. The Müllerian ducts develop into the oviducts, uterus, and part of the vagina, and the Wolffian ducts degenerate and disappear **(Figure 48.18C)**. The female sex hormones additionally stimulate the development of the female external genitalia.

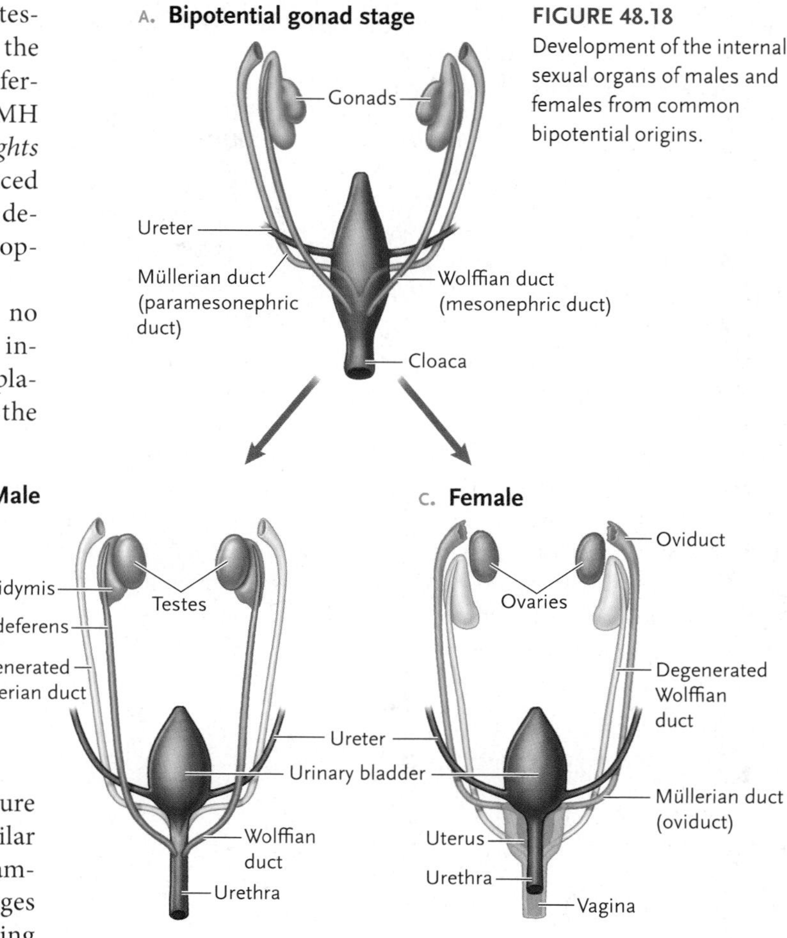

A. Bipotential gonad stage
Gonads
Ureter
Müllerian duct (paramesonephric duct)
Wolffian duct (mesonephric duct)
Cloaca
B. Male
Epididymis
Vas deferens
Degenerated Müllerian duct
Testes
Ureter
Urinary bladder
Wolffian duct
Urethra
C. Female
Oviduct
Ovaries
Degenerated Wolffian duct
Müllerian duct (oviduct)
Uterus
Urethra
Vagina

FIGURE 48.18 Development of the internal sexual organs of males and females from common bipotential origins.

Development Continues after Birth

Once fetal development is over, humans and other mammals, and indeed most other animals, follow a prescribed course of further growth and development that leads to the adult, the sexually mature form of the species. In humans, the internal and external sexual organs mature and secondary sexual characteristics appear at puberty. Similar changes occur in most mammals. There are, in fact, many examples among different animal groups of developmental changes that take place after hatching or birth. In some cases, offspring hatch that are distinctly different in structure from the adult. Examples among invertebrates include insects such as *Drosophila* and butterflies, in which eggs hatch to produce larva that undergo metamorphosis into the adult. Frogs similarly hatch as tadpoles, which undergo metamorphosis to produce the adult.

We have now described embryonic development in animals from a morphological perspective. In the rest of the chapter, we focus on the cellular mechanisms that underlie development.

STUDY BREAK 48.4 <

1. **Distinguish between the roles of the trophoblast and inner cell mass of the blastocyst in mammalian development.**
2. **What hormone would you use to induce labor in a pregnant woman?**

48.5 The Cellular Basis of Development

In the preceding sections, you learned about the processes of development from a mainly structural point of view. Underlying those developmental processes are specific cellular and molecular events. In this section, you will learn about some of the cellular events that underlie the stages of development. The genetic and molecular regulation of development was discussed in Section 16.4.

Cell Division Varies in Orientation and Rate during Embryonic Development

Recall from earlier in the chapter that *morphogenesis* is the generation of the body form as differentiated cells end up in their appropriate positions. In animals, morphogenesis occurs by changes in cell shape, cell position, and cell adhesion. For embryo morphogenesis, the *orientation* and *rate* of mitotic cell division have special significance. Regulation of these two features of mitotic cell division occurs at all stages of development.

The orientation of cell division refers to the angles at which daughter cells are added to older cells as development proceeds. Orientation is determined by the location of a furrow that separates the cytoplasm after mitotic division of the nucleus (furrowing is discussed in Section 10.2). The furrow forms in alignment with the spindle midpoint. Therefore, when the spindle is centrally positioned in the cell, the furrow leads to symmetric division of the cell. However, when the spindle is displaced to one end of the cell, the furrow leads to asymmetric division of the cell into a smaller and a larger cell. Little is known about how spindle positioning is regulated.

The rate of cell division primarily reflects the time spent in the G_1 period of interphase (see Section 10.2); once DNA replication begins, the rest of the cell cycle is usually of uniform length in all cells of the same species. As an embryo develops and cells

INSIGHTS FROM THE Molecular Revolution

Turning on Male Development

The switch to male development in mammalian embryos is triggered by the SRY protein encoded by the *SRY* gene on the Y chromosome. Individuals with a mutation in the *SRY* gene which results in an inactive protein develop into females, even though they have the XY combination of sex chromosomes.

Molecular studies revealed that the mutant SRY proteins have changes in single amino acids or have a missing segment. All the single amino acid changes are concentrated in a region of the SRY protein known as the *HMG box,* which can bind to DNA. This discovery suggested that SRY is a regulatory protein that binds to the control regions of genes such as *AMH,* which encodes the anti-Müllerian hormone, and turns on expression of those genes.

Research Question

Does SRY directly regulate the expression of the *AMH*?

Experiment

A group of investigators led by Michael Weiss at the Harvard Medical School and Massachusetts General Hospital in Boston carried out molecular studies testing whether SRY directly turns on the *AMH* gene.

For their experiments, the researchers attached the regulatory sequences that control transcription of the *AMH* gene to the amino acid-coding sequence of a gene encoding firefly luciferase (see **figure**). Firefly luciferase is an enzyme that catalyzes a reaction with the substrate luciferin to produce light. In this experiment, luciferase was used to detect the possible interaction between SRY and the *AMH* gene regulatory sequences. That is, if SRY binds to the *AMH* gene regulatory sequence and activates transcription, translation of the resulting mRNA produces luciferase. Luciferase activity is measured in cell extracts; the quantification of the amount of light produced in the reaction indicates the degree of activation of transcription caused by SRY. The data are then extrapolated to the effects of SRY on the complete *AMH* gene.

The researchers introduced the composite *AMH*–luciferase gene into embryonic cells removed from the developing gonads of XY rat embryos. A normal human *SRY* gene was then introduced into the gonad cells containing the artificial gene, and luciferase activity was measured.

Results

High luciferase activity was seen, confirming that the normal SRY protein activates the *AMH* gene. The experiment was repeated with a mutant *SRY* gene isolated from a human patient who had developed into a female even though she had the XY combination of sex chromosomes. In her case, the mutation resulted from a change of a single amino acid in the HMG box of the SRY protein. When her *SRY* gene was added to the embryonic rat gonad cells, there was no luciferase activity, indicating that her altered SRY protein could not turn on the *AMH* gene.

Conclusions

Adding a normal SRY protein to the *AMH* gene in a test tube showed that the protein binds directly to the gene. Tests with DNA-digesting enzymes showed that combination with SRY protects a segment of the control region of the *AMH* gene from attack by the enzymes. This protection indicated that SRY binds in this region, as expected for a regulatory protein.

Source: C. M. Haqq et al. 1994. Molecular basis of mammalian sexual determination: activation of Müllerian inhibiting substance gene expression by SRY. *Science* 266:1494–1500.

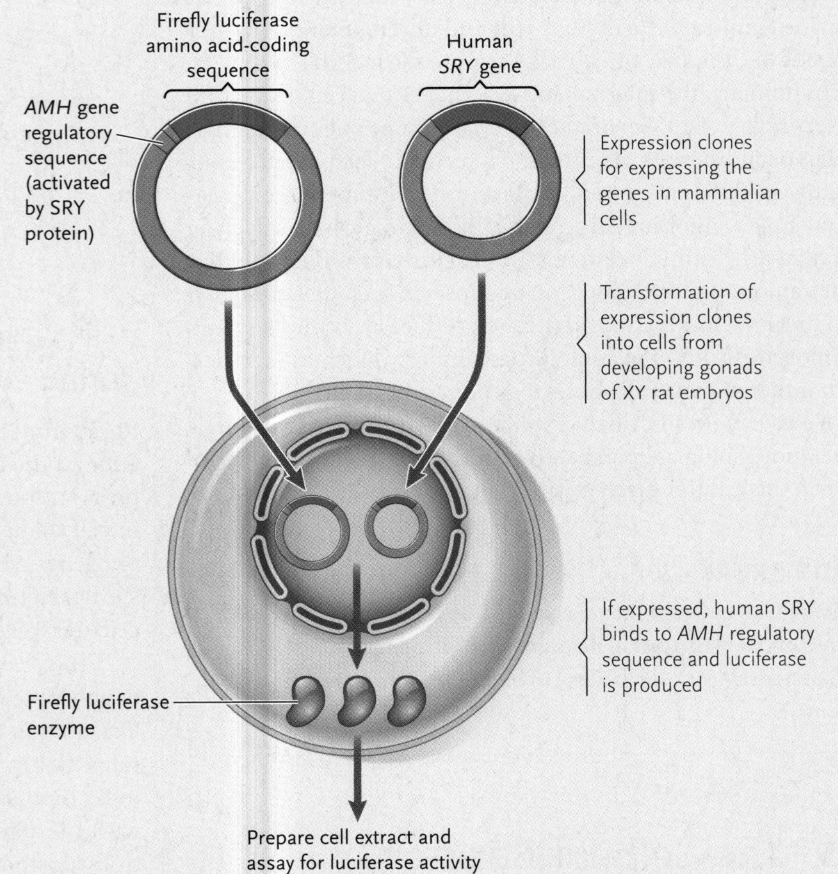
Firefly luciferase amino acid-coding sequence
Human *SRY* gene
AMH gene regulatory sequence (activated by SRY protein)
Expression clones for expressing the genes in mammalian cells
Transformation of expression clones into cells from developing gonads of XY rat embryos
If expressed, human SRY binds to *AMH* regulatory sequence and luciferase is produced
Firefly luciferase enzyme
Prepare cell extract and assay for luciferase activity

differentiate, the time spent in interphase increases and varies in length in different cell types. Therefore, different cell types proliferate at various rates, giving rise to tissues and organs with different cell numbers. Ultimately, the rate of cell division is under genetic control.

Frog egg cleavage provides examples of how both changes in orientation and rate of mitotic division affect development. The first two cleavages start at the animal pole and extend to the vegetal pole, producing four equal blastomeres (see Figure 48.2). The third cleavage occurs equatorially. However, because there is yolk in the vegetal region of the embryo, this cleavage furrow forms not at the equator but up higher toward the animal pole. The result is an eight-cell embryo with four small blastomeres in the animal region of the embryo, and four large blastomeres in the vegetal region. The blastomeres in the animal region of the embryo proceed to divide rapidly, while the blastomeres in the vegetal part of the embryo divide more slowly because division is inhibited by yolk. As a result, the morula produced consists of an animal region with many small cells, and a vegetal region with relatively few blastomeres.

Cell-Shape Changes and Cell Movements Depend on Microtubules and Microfilaments

We have seen that embryonic cells undergo changes in shape that generate movements, such as the infolding of surface layers to produce endoderm or mesoderm. Entire cells also move during the embryonic growth of animals, both singly and in groups. Both the shape changes and the whole-cell movements are produced by microtubules (powered by dyneins and kinesins) and microfilaments (powered by myosins) (see Section 5.3). Movements are also produced by changes in the rate of growth or by the breakdown of microtubules and microfilaments. Generally speaking, changes in both cell shape and cell movement play important roles in cleavage, gastrulation, and organogenesis.

CHANGES IN CELL SHAPE Changes in cell shape typically result from reorganization of the cytoskeleton. For example, during the development of the neural plate in frogs, the ectoderm flattens and thickens; that is, microtubules within cells in the ectoderm layer lengthen and slide farther apart, causing the cells to transition from a cubelike to a columnar shape **(Figure 48.19A)**.

Once formed, the neural plate sinks downward along its midline as a result of a change in cell shape from columnar to wedgelike **(Figure 48.19B)**. As the tops of the cells narrow, the entire cell layer is forced inward—it invaginates. How does this occur? Each wedge-shaped cell contains a group of microfilaments arranged in a circle at its top. Research suggests that the microfilaments slide over each other, tightening the ring like a drawstring and narrowing the top of the cell. This mechanism is supported by experiments in which cytochalasin, a chemical that interferes with microfilament assembly, was added to the cells. As a result, the microfilament circle was dispersed, and no invagination of the ectoderm occurred.

WHOLE-CELL MOVEMENTS Among the most striking examples of whole-cell movements in embryonic development are the cell movements during gastrulation and the often long-distance migrations of neural crest cells. These whole-cell movements involve the coordinated activity of microtubules and microfilaments. The typical pattern of movement is a repeating cycle of steps that resemble how an amoeba moves (see Section 5.3 and Figure 6.15) **(Figure 48.20)**.

How do the cells know where to go? Typically, cells migrate over the surface of stationary cells in one of the embryo's layers. In many developmental systems, migrating cells follow tracks formed by molecules of the extracellular matrix (ECM), secreted by the cells along the route over which they travel. An important track molecule is *fibronectin,* a fibrous, elongated protein of the

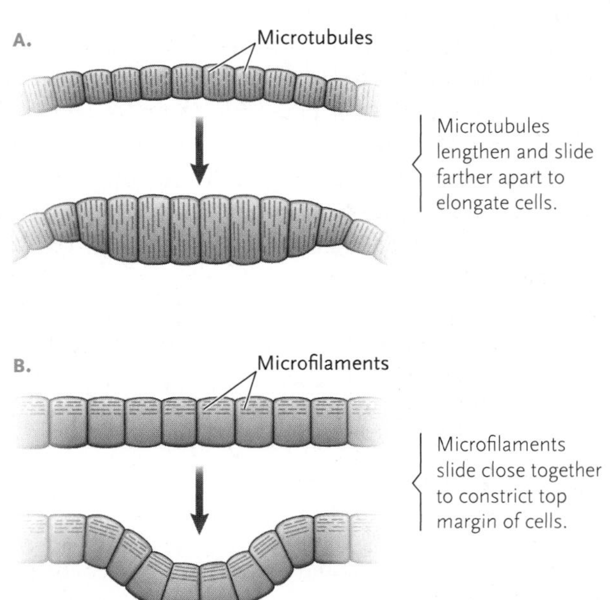
A.
Microtubules
Microtubules lengthen and slide farther apart to elongate cells.
B.
Microfilaments
Microfilaments slide close together to constrict top margin of cells.

FIGURE 48.19
The roles of microtubules **(A)** and microfilaments **(B)** in the changes in cell shape that produce developmental movements.

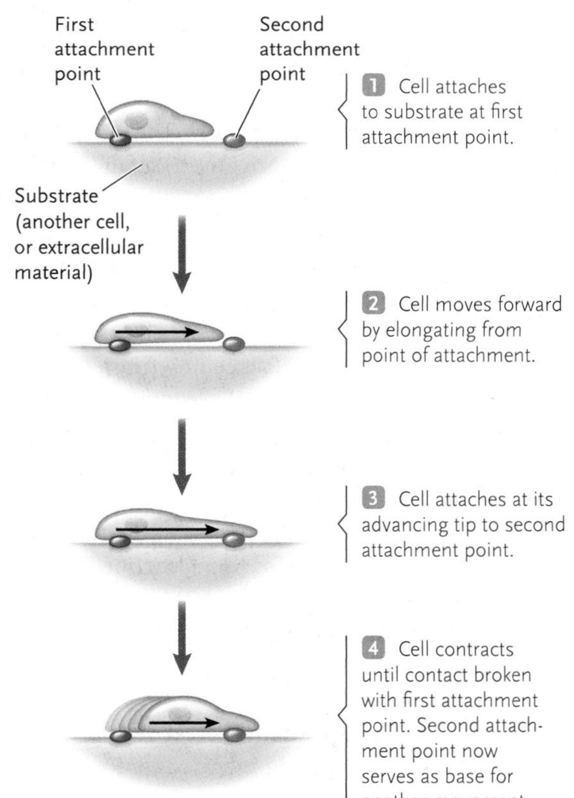
First attachment point
Second attachment point
1 Cell attaches to substrate at first attachment point.
Substrate (another cell, or extracellular material)
2 Cell moves forward by elongating from point of attachment.
3 Cell attaches at its advancing tip to second attachment point.
4 Cell contracts until contact broken with first attachment point. Second attachment point now serves as base for another movement.

FIGURE 48.20
The cycle of attachments, stretching, and contraction by which cells move over other cells or extracellular materials in embryos.

ECM. Migrating cells recognize and adhere to the fibronectin; in response, internal changes in the cells trigger movement in a direction based on the alignment of the fibronectin molecules.

Some migrating cells follow concentration gradients instead of molecular tracks. The gradients are created by the diffusion of molecules (often proteins) released by cells in one part of an embryo. Cells with receptors for the diffusing molecule follow the gradient toward its source, or move away from the source.

Selective Cell Adhesions Underlie Cell Movements

Selective cell adhesion, the ability of an embryonic cell to make and break specific connections to other cells, is closely related to cell movement. As development proceeds, many cells break their initial adhesions and move, forming new adhesions in different locations. Final cell adhesions hold the embryo in its correct shape and form. Junctions of various kinds, including tight, anchoring, and gap junctions, reinforce the final adhesions (see Section 5.5).

The selective nature of cell adhesions was first demonstrated in a classic experiment by Johannes Holtfreter of the University of Rochester and his student P. L. Townes. The researchers removed pieces of ectoderm, mesoderm, and endoderm from living amphibian embryos in the neurulation stage, separated them into individual cells, and added the cells in various combinations to a culture medium. Initially the cells clumped together at random into a ball. After a few hours, they sorted themselves out and moved into arrangements resembling their normal locations in the gastrula **(Figure 48.21).**

Further research has identified many cell surface proteins responsible for selective cell adhesions, including **cell adhesion molecules** (CAMs; see Section 5.5) and **cadherins** *(calcium-dependent adhesion molecules).* The cadherins are so named because they require calcium ions to set up adhesions. As cells develop, different types of CAMs or cadherins may appear or disappear from their surfaces as they make and break cell adhesions. The changes reflect alterations in gene activity, often in response to molecular signals arriving from other cells. For example, in the neural plate stage of neurulation in the frog, N-cadherin is on neural plate cells, keeping those cells together, whereas E-cadherin is on the adjacent ectodermal cells, keeping those cells together. The neural tube is produced when the neural plate cells separate from the ectodermal cells, whereas both cell types retain their respective cadherin type. The neural crest cells have neither cadherin bound to them, so they do not bind to each other and they disperse (as described earlier). However, if N-cadherin is expressed in the ectodermal cells through experimental manipulation, the forming neural tube does not separate from the flanking ectodermal cells because all of the cells are held together by N-cadherin.

As they begin to disperse, neural crest cells produce *integrins,* which insert into the plasma membrane. Recall from Section 5.5 that integrins are receptor proteins which span the plasma membrane. On the cytoplasmic side, integrins bind to microfilaments of the cytoskeleton. Integrins function to integrate changes outside and inside the cell by communicating changes in the extracellular matrix to the cytoskeleton. In the example here, the integrin receptors of the neural crest cells interact with proteins in the extracellular matrix in a way that causes cytoskeletal changes. The cytoskeletal changes move the cells along their migration pathways. More generally, integrins are involved in many key developmental processes.

The development of a new organism involves the cell shape changes, cell movements, and cell adhesions we have just discussed and, in addition, the programmed emergence of particular types of cells, or cell lineages, as we will see next.

Fate Mapping Maps Adult Structures onto Regions of the Embryos from Which They Developed

From the early days of studying development, embryologists have focused on describing not only how embryos form and develop, but exactly how adult tissues and organs are produced from the cells of the embryo. Thus, an important goal in embryology is to trace cell lineages from embryo to adult. For most organisms it is not possible to trace lineages at the individual cell level, primarily because of the complexity of the developmental process and the typical opacity of embryos. However, it has been possible to map adult or larval structures onto the region of the embryo from which each structure developed. This type of study is called *fate mapping,* and the result is called a **fate map.** Experimentally, fate mapping is done by following development of living embryos under the microscope, either using species in which the embryo is transparent, or by marking cells so they can be followed. Cells may be marked with vital dyes (dyes that do not kill cells), fluorescent dyes, or radioactive labels. Fate maps have been produced for a number of organisms, including the chick, *Xenopus,* and *Drosophila.*

In most cases a fate map is not detailed enough to relate how particular cells in the embryo gave rise to cells of the adult. The exception is the fate map of the nematode *Caenorhabditis elegans,* an organism that has a fixed, reproducible developmental pattern. This animal has a transparent body, and scientists have been able to map the fate—trace the **cell lineage**—of every somatic and germ-line cell as the zygote divides and the resulting embryo differentiates into the 959-cell adult hermaphrodite or the 1,031-cell adult male **(Figure 48.22).** All somatic cells of the adult can be traced from five somatic *founder cells* produced during early development. Knowing the cell lineages of *C. elegans* has been a valuable tool for research into the genetic and molecular control of development in this organism, for mutants affecting development have an easily visible effect.

Induction Depends on Molecular Signals Made by Inducing Cells

Recall from Section 48.1 that induction is the process by which a group of cells (the inducer cells) causes or influences a nearby group of cells (the responder cells) to follow a particular developmental pathway. Recall also that induction is the major process responsible for determination, in which the developmental fate of

FIGURE 48.21 **Experimental Research**

Demonstrating the Selective Adhesion Properties of Cells

Question: Do cells make specific connections to other cells?

Experiment: Johannes Holtfreter and P. L. Townes demonstrated that cells make specific connections to other cells, that is, that cells have selective adhesion properties.

1. Holtfreter and Townes separated ectoderm, mesoderm, and endoderm tissue from amphibian embryos soon after the neural tube had formed. They used embryos from amphibian species that had cells of different colors and sizes, so they could follow under the microscope where each cell type ended up. (The colors shown here are for illustrative purposes only.)

2. The researchers placed the tissues individually in alkaline solutions, which caused the tissues to break down into single cells.

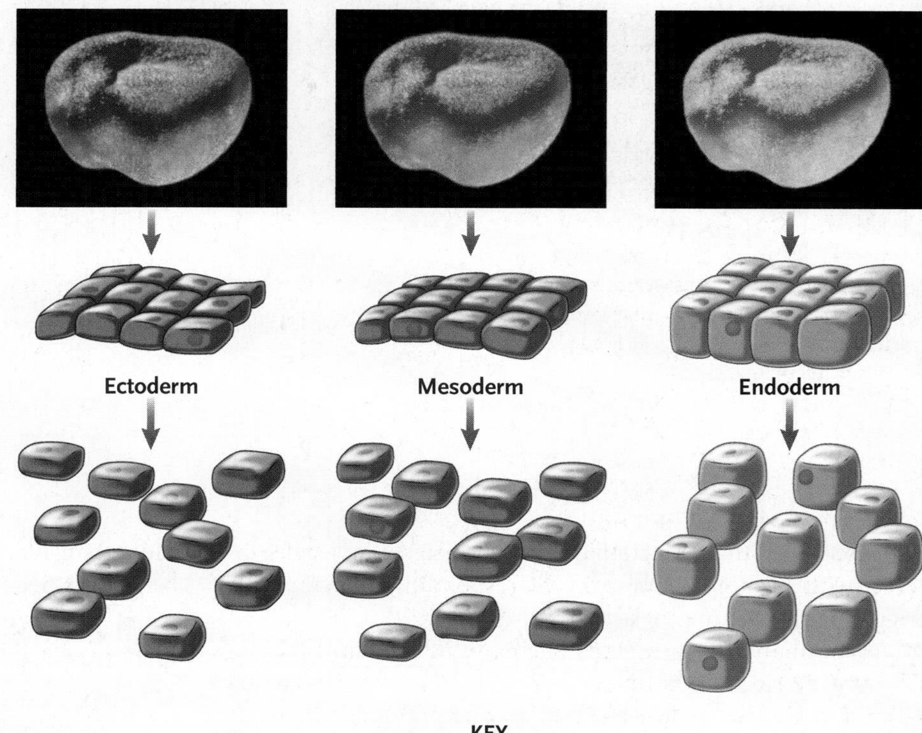

Amphibian embryos of different species
Ectoderm
Mesoderm
Endoderm

3. Holtfreter and Townes then combined suspensions of single cells in various ways. Shown here are ectoderm + mesoderm, and ectoderm + mesoderm + endoderm. When the pH was returned to neutrality, the cells formed aggregates. Through a microscope, the researchers followed what happened to the aggregates on agar-filled Petri dishes.

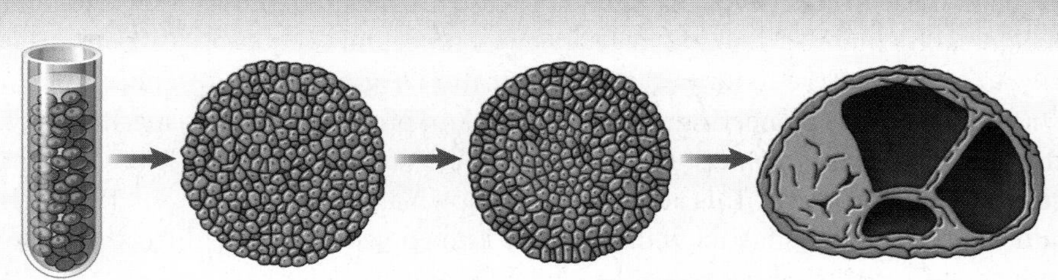

KEY
Ectoderm
Mesoderm
Endoderm
Ectoderm + Mesoderm

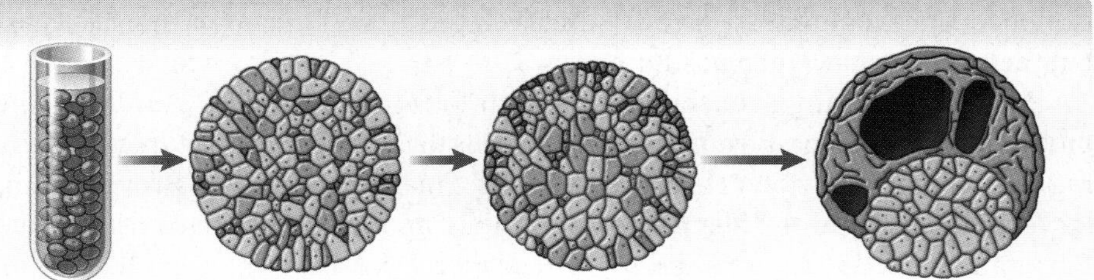

Ectoderm + Mesoderm + Endoderm

Results: In time the reaggregated cells sorted themselves with respect to cell type; that is, instead of the cell types remaining mixed, each cell type became separated spatially. That is, in the ectoderm + mesoderm mixture, the ectoderm moved to the periphery of the aggregate, surrounding mesoderm cells in the center. In no case did the two cell types remain randomly mixed. The ectoderm + mesoderm + endoderm aggregate showed further that cell sorting in the aggregates generated cell positions reflecting the positions of the cell types in the embryo. That is, the endoderm cells separated from the ectoderm and mesoderm cells and became surrounded by them. In the end, the ectoderm cells were located on the periphery, the endoderm cells were internal, and the mesoderm cells were between the other two cell types.

Conclusion: Holtfreter interpreted the results to mean that cells have selective affinity for each other; that is, cells have selective adhesion properties. Specifically, he proposed that ectoderm cells have positive affinity for mesoderm cells but negative affinity for endoderm cells, whereas mesoderm cells have positive affinity for both ectoderm cells and endoderm cells. In modern terms, these properties result from cell surface molecules that give cells specific adhesion properties.

Source: P. L. Townes and J. Holtfreter. 1955. Directed movements and selective adhesion of embryonic amphibian cells. *Journal of Experimental Zoology* 128:53–120.

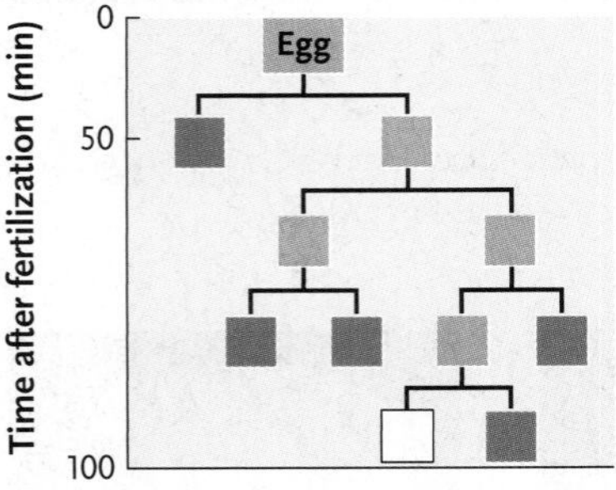
A. Founder cells
Time after fertilization (min)
0
50
100
Egg

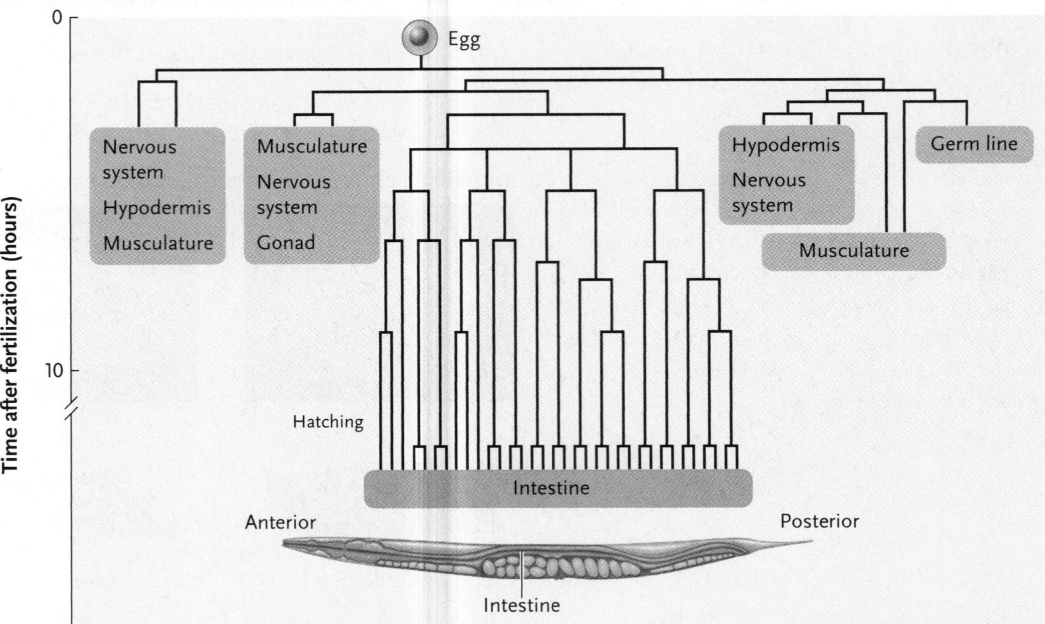
B. Cell lineage for intestinal cells
Time after fertilization (hours)
0
10
Egg
Nervous system
Hypodermis
Musculature
Musculature
Nervous system
Gonad
Hypodermis
Nervous system
Germ line
Musculature
Hatching
Intestine
Anterior
Posterior
Intestine

FIGURE 48.22
Cell lineages of *C. elegans.* **(A)** The founder cells (blue) produced in early cell divisions from which all adult somatic cells are produced. The cell in white gives rise to germ-line cells. **(B)** The cell lineage for cells that form the intestine. The detailed lineages for the other parts of the adult are not shown.

a cell is set, and that induction occurs when signal molecules interact with surface receptors on the responding cells, typically triggering changes in gene activity.

A German scientist, Hans Spemann of the University of Freiburg, carried out the first experiments identifying induction in embryos in the 1920s. He and his doctoral student, Hilde Mangold, found that if the dorsal lip of a newt embryo was removed and grafted into a different position on another newt embryo, on the ventral side for instance, cells moving inward from the dorsal lip induced a neural plate, a neural tube, and eventually an entire embryo to form in the new location **(Figure 48.23).** On the basis of his pioneering research, Spemann proposed that the dorsal lip is an *organizer,* acting on other cells to alter their course of development. This action is now known as *induction,* and the cells responsible for induction are known generally as inducer cells. Spemann received the Nobel Prize in 1935 for his research. (In 1924, the year their research paper was published, Mangold died in an accident when her kitchen gasoline heater exploded. She would likely have also received the Nobel Prize, but they are never awarded posthumously.)

Spemann's findings touched off a search for the inducing molecules that must pass from the inducing cells to the responding cells. It took many years to achieve success. Finally, molecular techniques led to the identification of inducing molecules. For example, in 1992, researchers constructed a DNA library from *Xenopus* gastrulas by isolating and cloning the cellular DNA in gene-size pieces. They made mRNA transcripts of the cloned genes and injected them into early *Xenopus* embryos in which the inducing ability of the mesoderm had been destroyed by exposure to ultraviolet light. Some of the injected mRNAs, translated into proteins in the embryos, were able to induce formation of a neural plate and tube and lead to a normal embryo. More than 10 other proteins acting as inducing molecules have been identified in the *Xenopus* system.

Differentiation Produces Specialized Cells without Loss of Genes

Differentiation is the process by which cells that have committed to a particular developmental fate by the determination process (see Section 48.1) now develop into specialized cell types with distinct structures and functions. As part of differentiation, cells concentrate on the production of molecules characteristic of the specific types. For example, 80% to 90% of the total protein that lens cells synthesize is crystallin.

Research into differentiation confirmed that as cells specialize, they retain all the genes of the original egg cell; except in rare instances, differentiation does not occur through selective gene loss. Several definitive experiments supporting this conclusion were carried out several decades ago by Robert Briggs and Thomas King of Lankenau Hospital Research Institute in Philadelphia (now Fox Chase Cancer Center), and extended by John B. Gurdon of the University of Cambridge, United Kingdom. In a typical experiment, the nucleus of a fertilized frog egg was destroyed by ultraviolet light. A micropipette was then used to transfer a nucleus from a fully differentiated tissue, intestinal epithelium, to the enucleated egg. Some of the eggs receiving the transplanted nuclei subsequently developed into normal tadpoles and adult frogs. This outcome was possible only if the differentiated intestinal cells still retained their full complement of genes. This conclusion was extended to mammals in 1997 when Ian Wilmut and his colleagues successfully cloned a sheep—Dolly—starting with an adult cell nucleus. (This experiment is described in Section 18.2.)

FIGURE 48.23 **Experimental Research**

Spemann and Mangold's Experiment Demonstrating Induction in Embryos

Question: Does induction occur in embryonic development?

Experiment: Hans Spemann and Hilde Mangold performed transplantation experiments with newt embryos, the results of which demonstrated that specific induction of development occurs in the embryos. The researchers removed the dorsal lip of the blastopore from one newt embryo and grafted it onto a different position—the ventral side—of another embryo. The two embryos were from different newt species that differed in pigmentation, allowing them to follow the fate of the tissue easily. The embryo with the transplant was allowed to develop.

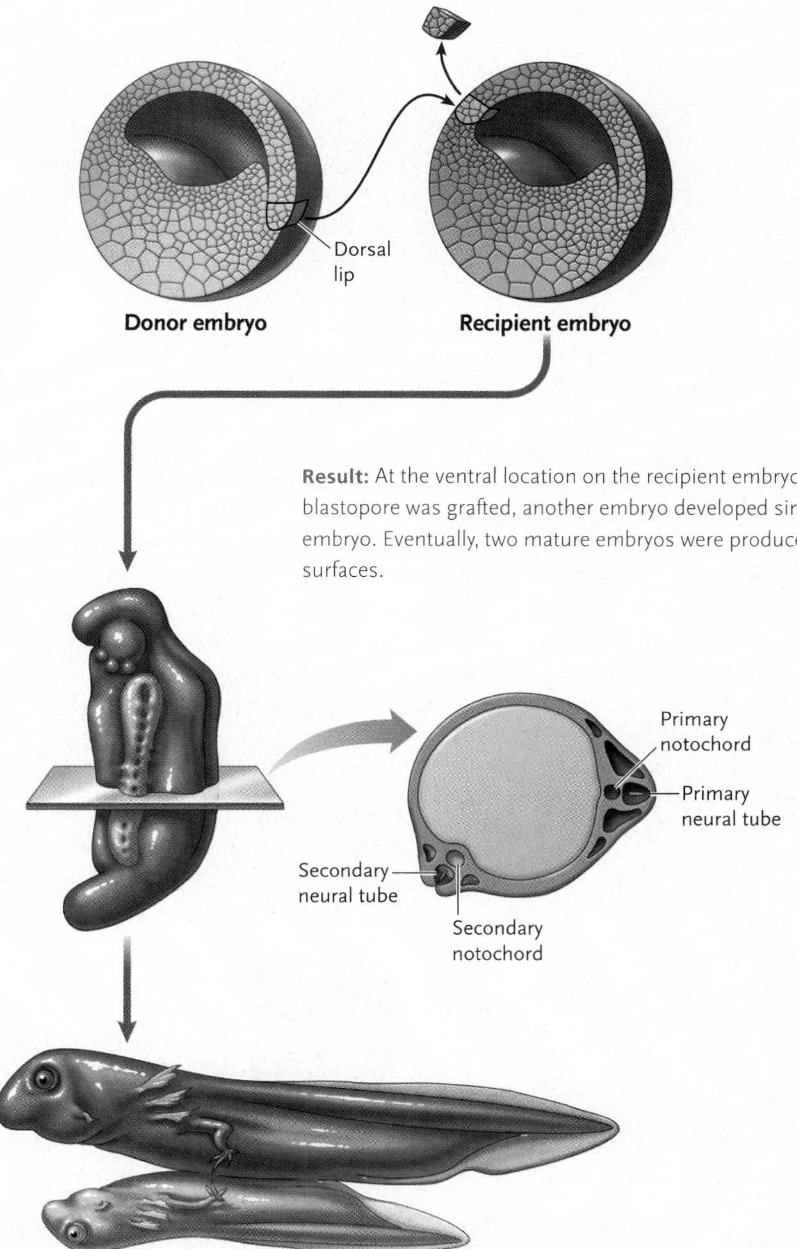
Dorsal
lip
Donor embryo
Recipient embryo
Primary
notochord
Primary
neural tube
Secondary
neural tube
Secondary
notochord

Result: At the ventral location on the recipient embryo where the dorsal lip of the blastopore was grafted, another embryo developed simultaneously with the recipient embryo. Eventually, two mature embryos were produced attached on their ventral surfaces.

Conclusion: The grafted dorsal lip of the blastopore induced a second gastrulation and subsequent development in the ventral region of the recipient embryo. The result demonstrated the ability of particular cells to induce the development of other cells.

Source: H. Spemann and H. Mangold. 1924. Über induktion von Embryonalanlagen durch implantation artfremder Organisatoren. *Roux's Archives of Developmental Biology* 100:599–638. [Reprinted as English translation by V. Hamburger: Induction of embryonic primordia by implantation of organizers from a different species. *International Journal of Developmental Biology* 45:13–38 (2001).]

STUDY BREAK 48.5

1. What are the key cellular events that contribute to morphogenesis in animals?
2. What is induction? What molecules are involved?

THINK OUTSIDE THE BOOK

Use the Internet or research literature to outline evidence associating an increase in the level of N-cadherin with breast cancer.

UNANSWERED QUESTIONS

How do cells that comprise animals know what sex an animal should be as it grows up?

This should be an easy question to answer, because we know that in mammals it is the presence of the Y chromosome (containing the gene *SRY* discussed earlier in this chapter) that determines the development of testes in males. Testes then secrete testosterone and anti-Müllerian hormone (AMH) which stimulate Wolffian duct development and Müllerian duct regression. In females, ovaries develop because of the absence of SRY protein and the Müllerian ducts form. Steroid hormones secreted from the gonads then act to differentiate other tissues, including the brain. However, in nature there are many reports of animals that are gynandromorphic; that is, they have both female and male characteristics. These animals can be either bilaterally symmetrical or mosaics (containing a mix of male and female cells). Gynandromorphs have been seen in chickens, song birds, blue crabs, spiders, and insects including butterflies and carpenter bees plus others. Some gynandromorphs that have been well studied include a song bird, the gynandromorphic Australian zebra finch, which had a testis and male plumage on the right side of the body and an ovary and female plumage on the left side of the body. Avian sex determination is chromosomal, and birds have a *ZW* sex-determination system which is reversed compared to the mammalian *XY* system. Female birds have heterogametic chromosomes *(ZW)*, and males have homogametic chromosomes *(ZZ)* and the sex-determining gene in birds is the Z-linked DMRT1 gene. The gynandromorphic zebra finch was studied by Dr. Arthur P. Arnold at the University of California, Los Angeles, where he and a team of researchers determined that the right side of the brain of the bird contained no *W* sex chromosomes while the left side contained both *Z* and *W* chromosomes. The fact that these animals develop naturally indicates that there are still many unanswered questions about sexual differentiation. What is not yet known is if the sex imposed on a developing cell comes entirely from hormonal signals originating from the outside of the cell, or if there are also signals coming from within the cells from the cell's own genome. Also in vertebrates, how do sex chromosomes, acting on individual cells, play a role in the differences that develop between male and female tissues, including the brain?

What initiates childbirth?

We learned in this chapter that there are a series of hormonal and physical events that are associated with childbirth. There is an increase in the secretion of oxytocin from the posterior pituitary gland and oxytocin receptors are found in the smooth muscle cells of the uterine wall. Oxytocin stimulates the muscle to begin the rhythmic contractions of labor marking the beginning of birth or parturition. This process has been well studied, but the initiation of labor is a complex process that has yet to be fully explained. What fetal-derived signals lead to the initiation of labor? How do these signals play a role in preterm labor? What roles do signals, such as prostaglandins, from the placenta play in regulating the timing of birth? What roles do the fetal hypothalamus and adrenal glands play in initiating childbirth? What hormones, such as estradiol from the mother, or cortisol and corticotropin-releasing factor from the fetus, play in sending the signals to begin the process of parturition? One researcher who is making strides in this area is Louis J. Muglia M.D., Ph.D. from Washington University School of Medicine in St. Louis, Missouri. Dr. Muglia has been researching the timing of birth and the challenges to prevent premature births. His research has demonstrated that prostaglandins are essential for the initiation of parturition and he has determined that the regulation of COX-1 is important for the initiation of parturition in mice. COX-1 (cyclooxygenase-1) is a protein that acts as an enzyme to speed up the production of prostaglandins in the stomach and the expression of the *COX-1* gene increases in the uterus during pregnancy. Identifying regulatory factors, such as COX-1, may provide critical information into timing the onset of parturition.

Think Critically

Based on this unanswered question, do you think there are mammals that are gynandromophs? Does your answer to this question alter the significance of the research?

Laura Carruth

Laura Carruth, an assistant professor of biology at Georgia State University, studies the genetic and hormonal factors that lead to sex differences in brain development. Her work focuses on model systems in songbirds (the Australian zebra finch) and mice. To learn more about Dr. Carruth's research, go to http://www2.gsu.edu/~wwwllc/.

REVIEW KEY CONCEPTS

Go to **CENGAGENOW** at www.cengage.com/login to access quizzing, animations, exercises, articles, and personalized homework help.

48.1 Mechanisms of Embryonic Development

- Developmental information is stored in both the nucleus and cytoplasm of the fertilized egg. The mRNA and protein molecules that direct the first stages of development are the cytoplasmic determinants.
- The unequal distribution of yolk and other components makes eggs polar. The animal pole typically gives rise to surface structures and the anterior end of the embryo, whereas the vegetal pole typically gives rise to internal structures of the embryo such as the gut (Figure 48.2).

- Following fertilization, cleavage divisions produce the morula. The morula hollows out to form the blastula, which develops into the gastrula, the stage in which rearrangements of cells produce the ectoderm, mesoderm, and endoderm. Gastrulation establishes the body pattern, in that the organs and other structures of embryo arise from these three tissue layers (Figure 48.3; Table 48.1).
- Development proceeds as a result of cell division, cell movements, selective adhesions, induction, determination, and differentiation.

Animation: Where embryos develop

Animation: Stages of development

48.2 Major Patterns of Cleavage and Gastrulation

- In sea urchin eggs, yolk is distributed evenly. As a result, cleavage divisions take place at the same rate in all regions of the embryo, and gastrulation follows a symmetrical pattern (Figure 48.4).
- In amphibian eggs, yolk is distributed unequally, with most in the vegetal pole. As a result, the rate of cell division is more rapid in the animal pole, and gastrulation shows an asymmetric pattern (Figures 48.5 and 48.6).
- In bird and reptile embryos, the cleavage divisions give rise to a flat disc of cells at the top of the yolk, which divides into the epiblast and the hypoblast. In gastrulation, cells of the epiblast migrate to the interior to form the endoderm and the mesoderm. The epiblast cells left at the surface form the ectoderm (Figure 48.7).
- In birds and reptiles, the yolk sac, chorion, amnion, and allantoic membrane form from extensions of the primary tissue layers. These extraembryonic membranes conduct nutrients from the yolk to the embryo, exchange gases with the environment, and store metabolic wastes (Figure 48.8).

Animation: Cytoplasmic localization

Animation: Process of gastrulation

48.3 From Gastrulation to Adult Body Structures: Organogenesis

- In organogenesis, the three primary tissues give rise to the tissues and organs of the embryo. Organogenesis begins with neurulation, the development of the nervous system from ectoderm (Figures 48.9 and 48.10).
- The mesoderm splits into somites, which give rise to the vertebral column and to the muscles of the ribs, vertebral column, and limbs (Figure 48.10).
- Development of the eye from optic vesicles is illustrative of the inductions and differentiations common to organogenesis in vertebrates (Figure 48.11).
- Apoptosis—programmed cell death—plays an important role in development by removing tissues present during development but not in the adult organ (Figures 48.12 and 48.13).

Animation: Neural tube formation

Animation: AER transplant

48.4 Embryonic Development of Humans and Other Mammals

- In humans, as in other placental mammals, cleavage divisions produce a morula that differentiates into a blastocyst. The blastocyst implants into the endometrium of the uterus, and its inner cell mass separates into the epiblast and hypoblast. The epiblast produces the ectoderm, mesoderm, and endoderm of the embryo (Figures 48.14 and 48.15A, B).
- Gastrulation, neurulation, differentiation of cell layers, and formation of extraembryonic membranes occur by mechanisms similar to those of bird and reptile embryos. Differentiation of ectoderm, mesoderm, and endoderm into their final tissues and organs is also similar in birds and reptiles.
- Extraembryonic membranes form in mammals by processes that are also similar to the reptilian–bird pattern. However, some of the membranes have altered functions, reflecting the minimal amount of yolk in mammalian embryos, and maintenance of the embryo by the placenta (Figure 48.15C–E).
- The placenta is connected to the embryo by the umbilical cord, which conducts blood between the embryo and the placenta (Figure 48.15F).
- Fetal growth proceeds until birth, when the fetus is forced from the uterine cavity and through the vagina by contractions of the uterus, stimulated by oxytocin (Figures 48.16 and 48.17).
- The mother's mammary glands secrete milk once the offspring is born. Suckling by the offspring stimulates prolactin and oxytocin release from the pituitary, which stimulates milk production and secretion from the glands, respectively.
- Embryos develop internal male or female sex organs from the same primitive structures. The presence or absence of a Y chromosome, which carries the key *SRY* gene, determines whether the internal structures develop into male or female sexual organs (Figure 48.18).
- Most animals continue development after hatching or birth, leading to the adult, the sexually mature form of the species.

Animation: Fertilization

Animation: Cleavage and implantation

Animation: First 2 weeks of development

Animation: Weeks 3 to 4 of development

Animation: Proportional changes during development

Animation: Structure of the placenta

Animation: Fetal development

Animation: Birth

Animation: Anatomy of the breast

48.5 The Cellular Basis of Development

- Development in animals involves the regulation of specific cellular events, including cell division, cell movement, and cell adhesion.
- Cell division in development varies in orientation and rate.
- Cell movements in development occur through changes in cell shape or the migrations of entire cells. Shape changes are produced by microtubules or microfilaments. In cell migrations, cells follow tracks in the embryo or move in response to gradients of signal molecules (Figures 48.19 and 48.20).
- Selective cell adhesions, which depend on surface glycoproteins including CAMs and cadherins, underlie many cell movements. The final cell adhesions hold the embryo in its correct shape and form (Figure 48.21).
- For some organisms, the origins of adult or larval structures have been mapped to regions of the embryo from which each structure derived (Figure 48.22).
- Induction results from the effects of signaling molecules of the inducing cells on the responding cells (Figure 48.23).
- In differentiation, cells change from embryonic form to specialized types with distinct structures and functions. Differentiation occurs by differential gene activation.

Animation: Embryonic induction

UNDERSTAND AND APPLY

Test Your Knowledge

1. Major contributors to the axes of the animal body are the:
   a. sperm and egg cytoplasm.
   b. sperm and egg chromosomes.
   c. ribosomes and mitochondria.
   d. egg nucleus and yolk.
   e. pigments.
2. The process by which cells undergo mitosis without a corresponding increase in cytoplasm is called:
   a. polarity.
   b. cleavage.
   c. gastrulation.
   d. organogenesis.
   e. induction.
3. Which of the following mechanisms does *not* contribute to zygote development?
   a. meiosis
   b. mitosis
   c. selective cell adhesions
   d. determination
   e. induction
4. A major event during gastrulation is:
   a. the outward movement of cells at the dorsal lip of the blastopore.
   b. the displacement of the archenteron by the blastocoel.
   c. the formation of the coelom from the endoderm.
   d. the extension of ectoderm and endoderm to form the yolk sac.
   e. the development of ectoderm to form epidermal and neural tissues.
5. The presence of the gray crescent in a frog embryo:
   a. corresponds to the location of the future mouth.
   b. corresponds to the location of sperm entry into the egg.
   c. will cause the formation of a disordered mass of cells.
   d. contains concentrated amounts of darkly colored pigment granules from the animal pole.
   e. indicates that the egg has been fertilized.
6. To contribute to the formation of a nervous system:
   a. the neural crest develops into motor neurons.
   b. the neural tube is converted into a neural plate.
   c. the notochord induces the overlying ectoderm to become a neural plate.
   d. the roof of the archenteron induces the formation of the neural tube.
   e. somites give rise to the autonomic nervous system.
7. In mammalian development:
   a. the morula develops into a trophoblast.
   b. the chorionic villi allow the blastocyst to move down the oviduct.
   c. the allantois takes over the work of the amnion.
   d. the pharyngeal arches transform into the pharynx, larynx, and nasal cavities.
   e. prolactin stimulates parturition.
8. In the development of the female sex organs:
   a. all ducts in the 7-week embryo become Wolffian ducts.
   b. the *SRY* gene is activated.
   c. anti-Müllerian hormone is secreted.
   d. the Müllerian ducts develop into oviducts.
   e. the mother secretes oxytocin.
9. In the embryonic development of the eye:
   a. the optic vesicle cells permanently adhere to each other to prevent movement, whereas the optic cup cells are very motile.
   b. signals from the optic cup trigger surface receptors on the lens placode.
   c. gradients determine that the ectoderm overlying the lens vesicle develops into the optic vesicle.
   d. microtubules powered by myosins and microfilaments powered by dyneins move the eye components around in the head region.
   e. cadherins function in the presence of calcium to allow the lens placode and optic cup to break apart.
10. Which of the following statements about development is incorrect?
    a. As microtubules lengthen and slide apart, the shape of cells within the ectoderm changes.
    b. The extracellular matrix molecule fibronectin guides the movement of cells during development.
    c. The furrow formed during mitosis of cells in the developing embryo is centrally located to ensure all cells divide symmetrically.
    d. Cell surface cadherins regulate the ability of developing cells to adhere to one another.
    e. Integrins are responsible for transmitting information from the extracellular matrix to the cytoskeleton.

Discuss the Concepts

1. Experimentally, it is possible to divide an amphibian egg so that the gray crescent is wholly within one of the two cells formed. If the two cells are separated, only the cell with the gray crescent will form an embryo with a long axis, notochord, nerve cord, and back musculature. The other forms a shapeless mass of immature gut and blood cells. Propose an explanation of these outcomes.
2. Developmental biologist Lewis Wolpert once observed that "it is not birth, marriage or death, but gastrulation which is truly the most important time in your life." In what sense is he correct?
3. Arguably, in sexually reproducing animals development begins when eggs and sperm form in the parents. In a paragraph, explain the rationale for this idea.
4. Investigators discovered a *Drosophila* protein that triggers development of the nerve cord on the ventral side of the embryos. When an mRNA encoding the protein was injected into cells on the ventral side of *Xenopus* embryos, dorsal structures were formed on the ventral side, including incomplete heads. What do these findings suggest about the evolution of embryonic development?

Design an Experiment

As you have learned in this chapter, embryogenesis in *Drosophila* has been well described. In Chapter 18 you also learned that the genome of *Drosophila* has been completely sequenced, allowing each gene in the genome to be cataloged. Design an experiment to identify all the genes that are activated during embryogenesis, and when they are activated.

Interpret the Data

People considering fertility treatments should be aware that such treatments raise the risk of multiple births and that multiple pregnancies are associated with an increased risk of some birth defects.

The table below shows the results of Yiwei Tang's study of birth defects reported in Florida from 1996 to 2000. Tang compared the incidence of various defects among single and multiple births. She calculated the relative risk for each type of defect based on type of birth, and corrected for other differences that might increase risk: maternal age, income, race, previous adverse pregnancy experience, education, Medicaid participation during pregnancy, and the infant's sex and number of siblings. A relative risk of less than 1 means a defect occurs less often with multiple births than single births. A relative risk of greater than 1 means that multiples are more likely to have a defect.

Prevalence, per 10,000 Live Births, of Various Types of Birth Defects among Multiple and Single Births

	Prevalence of defect		
	Multiples	**Singles**	**Relative Risk**
Total birth defects	358.50	250.54	1.46
Central nervous system defects	40.75	18.89	2.23
Chromosomal defects	15.15	14.20	0.93
Gastrointestinal defects	28.13	23.44	1.37
Genital/urinary defects	72.85	58.16	1.31
Heart defects	189.71	113.89	1.65
Musculoskeletal defects	20.92	25.87	0.92
Fetal alcohol syndrome	4.33	3.63	1.03
Oral defects	19.84	15.48	1.29

1. What was the most common type of birth defect in the single-birth group?
2. Was that defect more or less common in the multiple-birth group than among single births?
3. Tang found that multiples have more than twice the risk of single newborns for one type of defect. Which type?
4. Does a multiple pregnancy increase the relative risk of chromosomal defects in offspring?

Source: Y. Tang et al. 2006.The risk of birth defects in multiple births: A population-based study. *Maternal and Child Health Journal* 10:75–81.

Apply Evolutionary Thinking

In this chapter, you have learned that cleavage and gastrulation differ among vertebrates that have different amounts of yolk in their eggs. What are some features of early development that evolution has conserved across vertebrate groups?

Express Your Opinion

Sanitation and medical advances have greatly extended the average human life span, especially in developed countries. Some researchers are now looking for ways to extend the human life span even further. Do you think research into life extension should be supported by federal research funding? Go to www.cengage.com/login to investigate both sides of the issue and then vote.

49

STUDY OUTLINE

National Oceanic and Atmospheric Administration (NOAA)

Stormy weather in the biosphere. Atmospheric disturbances like Hurricane Katrina, seen here in a satellite photograph just before it arrival in New Orleans on August 28, 2005, often have a dramatic impact on living systems.

Ecology and the Biosphere

Why It Matters. . . The winter of 1997–1998 was one for the books. Record rainfall caused mudslides in California and flooding along the normally arid coast of Ecuador and Peru. But the annual rains never arrived in Asia and Australia, and fires consumed tropical rain forests in Indonesia and Malaysia. What caused these major climatic dislocations? Every 3 to 7 years, interactions between the upper layers of the Pacific Ocean and the atmosphere produce El Niño, a climatic event with global consequences **(Figure 49.1).** The 1997–1998 El Niño altered weather patterns worldwide, killing more than 2,000 people and causing at least $30 billion in property damage.

In most years, air flows from a high pressure system over the eastern Pacific toward a low pressure system over the western Pacific. These winds move surface water from east to west and bring heavy rains to parts of Asia and Australia. Winds also usually blow from the poles toward the equator along the western sides of continents, and Earth's rotation causes these winds to push ocean surface water westward, away from the coast. The displaced surface water is replaced by cold, deep, nutrient-rich water carried by vertical currents called *upwellings* (Figure 49.1A). The nutrients support complex marine food webs in the shallow water above the continental shelf. For example, the Peru Current along the west coast of South America once supported a rich anchovy fishery.

Ocean currents vary seasonally, however, and in late December or early January, a warm, nutrient-poor current flows eastward along the equator and then north and south along the coastlines of Central and South America. Peruvian fishermen call this warm current El Niño (Spanish for "the child"), because it reaches their coast around Christmas. It usually persists for only a few weeks.

In strong El Niño years, however, atmospheric pressure systems change over the Pacific, altering the prevailing winds and ocean currents. Equatorial winds weaken; surface currents reverse di-

A. **Usual pattern of Pacific Ocean currents**

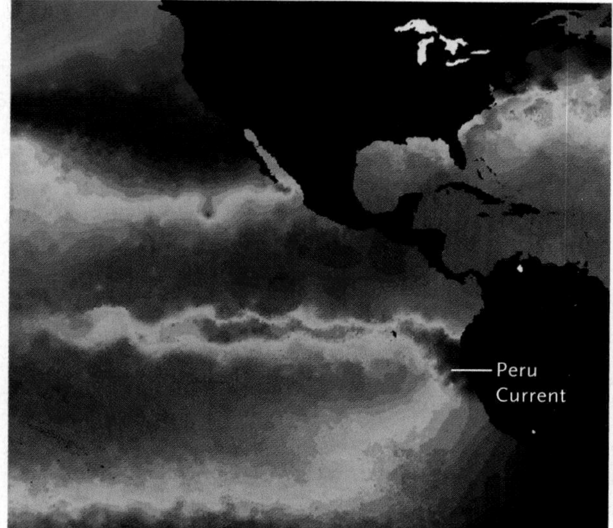
Peru
Current

In most years, the powerful Peru Current carries cold water from the ocean bottom to the surface off the west coast of South America. The cold surface water then flows westward along the equator toward a large pool of warm water in the western Pacific Ocean. In this satellite photo taken on May 31, 1988, dark red indicates the warmest water and dark green the coldest upwelled water.

B. **Pacific Ocean currents in an El Niño year**

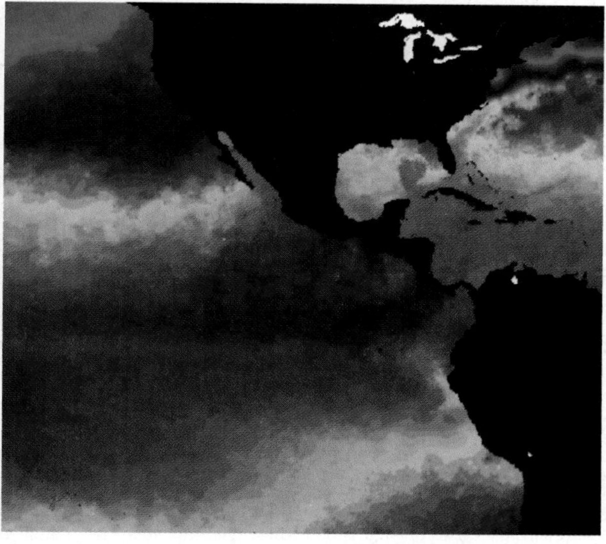
NASA

During an El Niño event, equatorial winds reverse directions and warm surface water flows eastward along the equator from the western Pacific Ocean toward South America. In this satellite photo taken on May 13, 1992, the warm water (orange) spreads up and down the west coast of North and South America, suppressing the upwelling of cold water by the Peru Current.

FIGURE 49.1 El Niño and Pacific Ocean currents.

rection, flowing from west to east; and a huge pool of warm ocean water accumulates in the eastern Pacific. During these shifts, the heavy rain that usually falls on Asia and Australia is instead delivered to the central and eastern Pacific. Thus, in strong El Niño years, Asia and Australia receive less rain than usual, and the west coasts of the Americas receive more. In the United States, winter temperatures are unusually high in the north central states and unusually low in the southern states.

El Niño episodes also alter sea surface temperature. When the warm current flowing from west to east reaches the continental shelf, it displaces the cold water of the Peru Current and prevents the usual upwelling (Figure 49.1B). These changes in ocean currents have catastrophic effects on marine life. Lacking sufficient nutrients, phytoplankton die, followed by fishes that eat phytoplankton, and seabirds that eat fishes. In combination with overfishing, the El Niño of 1972 drove the Peruvian anchovy population to the brink of extinction.

Some El Niño years are followed by a weather pattern called La Niña: the low pressure system over the western Pacific is accentuated, pulling air and ocean surface water from east to west. Low ocean surface temperatures extend from the coast of South America to Samoa. La Niña's effect on winter weather is opposite that of El Niño: parts of Asia and Australia are unusually wet; the northern United States experiences periods of cold, wet weather, whereas the southern region is unusually warm and dry.

El Niño and La Niña are two extremes of a global climate cycle called the El Niño Southern Oscillation, or ENSO (the name refers to fluctuations in air pressure over the tropical Pacific). ENSO, a product of large-scale interactions between the ocean and atmosphere, has a major impact on the **biosphere,** which comprises all organisms on Earth and the places where they live.

The devastation induced by severe El Niño and La Niña events introduces our unit on **ecology,** the study of interactions between organisms and their environments. All environments have both **biotic** (biological) and **abiotic** (nonbiological) components. The biotic environment includes all the organisms found in a particular place; the abiotic environment includes temperature, moisture, soil chemistry, and other physical factors. In addition to living organisms, the biosphere includes three abiotic components, which surround Earth's geological bulk like a skin. The **hydrosphere** encompasses all the water, including oceans and polar ice caps. The **lithosphere** includes the rocks, sediments, and soils of the crust. Finally, the **atmosphere** includes gases and airborne particles that envelop the planet.

In this chapter, we survey the biosphere with a wide-angle lens. After first describing how ecologists study the relationships between organisms and their environments, we examine Earth's environmental diversity as well as some mechanisms—the products of evolution—that allow organisms to cope with environmental variations in space and time. We then consider how variations in the physical environment influence the large-scale distributions of organisms on land, in fresh water, and in the sea. <

49.1 The Science of Ecology

The subject matter of ecology is so vast and so diverse that ecological research is often linked to work in genetics, physiology, anatomy, behavior, paleontology, and evolution as well as geology, geography, and environmental science. The connections between research in ecology and evolution are especially strong, because organisms exhibit evolutionary responses to their ecological relationships, and because ecological relationships change as organisms evolve. Moreover, many ecological phenomena occur over huge areas and long time spans. Thus, ecologists must devise clever ways to determine how environments influence organisms, how organisms respond to environmental factors, and how organisms change the environments in which they live.

Today, the science of ecology encompasses two related disciplines. The major research questions of *basic ecology* relate to the distribution and abundance of species and how they interact with each other and with the physical environment. Using these data as a baseline, workers in *applied ecology* develop conservation plans and amelioration programs to limit, repair, and mitigate ecological damage caused by human activities. Research in applied ecology overlaps to some extent with that in environmental science, but the latter discipline also encompasses perspectives from the social sciences, such as economics and political science.

Ecologists Study Levels of Organization Ranging from Individual Organisms to the Biosphere

Ecology can be divided into five increasingly complex and inclusive levels of organization. In **organismal ecology** researchers study the genetic, biochemical, physiological, morphological, and behavioral adaptations of organisms to the abiotic environment. We have described many such adaptations in Units V and VI; we describe the evolution of animal behavior in Chapter 55.

Population ecology, the subject of Chapter 50, focuses on **populations,** groups of individuals of the same species that live together. Population ecologists study how the size and other characteristics of populations change in space and time. Research in **community ecology,** discussed in Chapter 51, examines groups of populations that occur together in one area. Community ecologists study interactions between species, analyzing how predation, competition, and environmental disturbances influence a community's development, organization, and structure. Ecologists studying **ecosystem ecology,** considered in Chapter 52, explore the cycling of nutrients and the flow of energy between the biotic components of an ecological community and the abiotic environment. Finally, the largest-scale ecological studies focus on the biosphere, which we consider in this chapter. We discuss biodiversity and conservation biology in Chapter 53.

Ecologists Test Hypotheses with Observational and Experimental Data

Ecology has its roots in descriptive natural-history studies that date back to the ancient Greeks. Modern ecology was born in 1870 when the German biologist Ernst Haeckel coined the term *Oekologie* (*oikos* = house). Contemporary researchers still gather descriptive information about ecological relationships, but these observations are only a starting point for more rigorous studies.

Most ecologists create hypotheses about ecological relationships and how they change through time or differ from place to place. Like other scientists, some ecologists formalize these ideas in mathematical models that express clearly defined, but hypothetical, relationships among important variables in a system. Manipulation of a model, usually with the help of a computer, can allow researchers to ask what would happen if some of the variables or their relationships change. Thus, researchers can simulate natural events and large-scale experiments before investing time, energy, and money in field and laboratory work. Bear in mind, however, that mathematical models are no better than the ideas and assumptions they embody, and useful models are constructed only after basic observations define the relevant variables.

Ecologists often conduct field or laboratory studies to test the predictions of their hypotheses. In controlled experiments, researchers compare data from an experimental treatment (in which one or more variables are artificially manipulated) with data from a control (in which nothing is changed). Sometimes the distributions of species create "natural experiments," eliminating the need to manipulate variables. Studies of how islands of different sizes harbor different numbers of bird species (described in Section 51.7) provide an example of a natural experiment.

STUDY BREAK 49.1

1. **Why are studies of ecosystems more "inclusive" than studies of populations?**
2. **In what ways are mathematical models useful in ecological research?**

49.2 Environmental Diversity of the Biosphere

Numerous abiotic factors—sunlight, temperature, humidity, wind speed, cloud cover, and rainfall—contribute to a region's **climate,** the weather conditions prevailing over an extended period of time. Climates vary on global, regional, and local scales, and they undergo seasonal changes almost everywhere. As you will discover as you read this chapter, variations in climate over space and time influence the lives of all organisms.

Variations in Incoming Solar Radiation Create Global Climate Patterns

A global pattern of environmental diversity results from latitudinal variation in incoming solar radiation, Earth's rotation on its axis, and its orbit around the sun.

SOLAR RADIATION Earth's spherical shape causes the intensity of incoming solar radiation to vary from the equator to the poles **(Figure 49.2).** When sunlight strikes Earth directly at a 90° angle, as it does near the equator, it travels the shortest possible distance through the radiation-absorbing atmosphere and falls on the smallest possible surface area (Figure 49.2A). When sunlight ar-

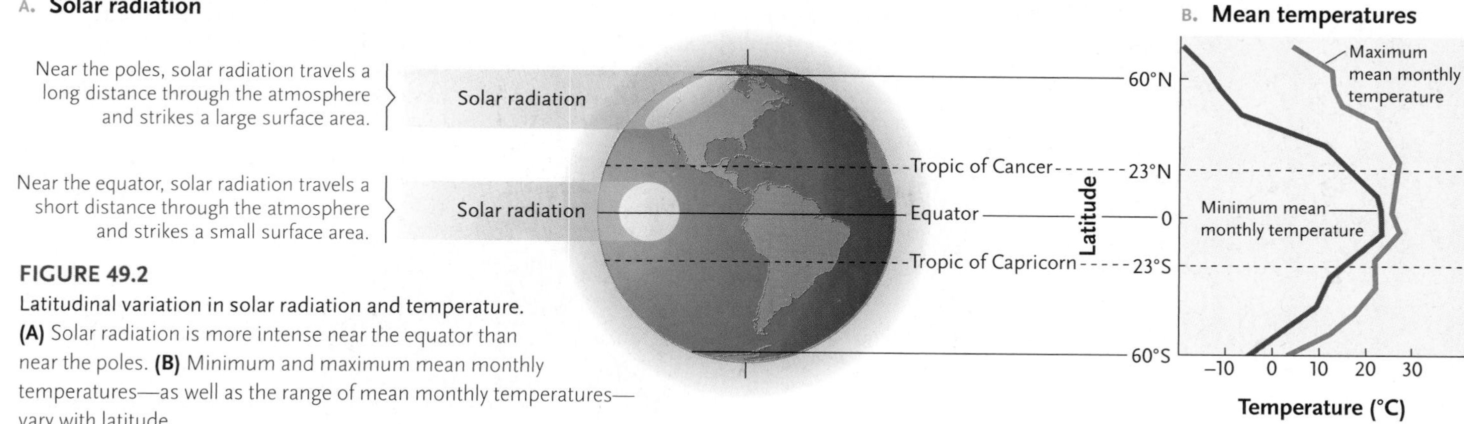
A. Solar radiation
Near the poles, solar radiation travels a long distance through the atmosphere and strikes a large surface area.
Near the equator, solar radiation travels a short distance through the atmosphere and strikes a small surface area.
Solar radiation
Solar radiation
Tropic of Cancer
Equator
Tropic of Capricorn
60°N
23°N
0
23°S
60°S
Latitude
B. Mean temperatures
Maximum mean monthly temperature
Minimum mean monthly temperature
−10
0
10
20
30
Temperature (°C)

FIGURE 49.2
Latitudinal variation in solar radiation and temperature. **(A)** Solar radiation is more intense near the equator than near the poles. **(B)** Minimum and maximum mean monthly temperatures—as well as the range of mean monthly temperatures—vary with latitude.

rives at an oblique angle, as it does near the poles, it travels a longer distance through the atmosphere and shines on a larger area. Thus, solar radiation is more concentrated near the equator than it is at higher latitudes, causing latitudinal variation in temperature (Figure 49.2B).

SEASONALITY Earth is tilted on its axis at a fixed position of 23.5° from the perpendicular to the plane on which it orbits the sun **(Figure 49.3).** This tilt produces seasonal variation in the duration and intensity of incoming solar radiation. The Northern Hemisphere receives its maximum illumination—and the Southern Hemisphere its minimum—on the June solstice (around June 22), when the sun shines directly over the Tropic of Cancer (23.5° N latitude). The reverse is true on the December solstice (around December 22), when the sun shines directly over the Tropic of Capricorn (23.5° S latitude). Twice each year, on the vernal and autumnal equinoxes (around March 21 and September 23, respectively), the sun shines directly over the equator.

Earth's tilt is permanent, and only the **tropics**—the latitudes between the Tropics of Cancer and Capricorn—ever receive solar radiation from directly overhead. Tropical regions experience only small seasonal changes in temperature and day length: environmental temperature is high, and days last approximately 12 hours throughout the year. (Tropical seasonality is reflected in the alternation of wet and dry periods rather than warm and cold seasons.) Seasonal variation in temperature and day length increases steadily toward the poles. Polar winters are long and cold with periods of continuous darkness, and polar summers are short with periods of continuous light.

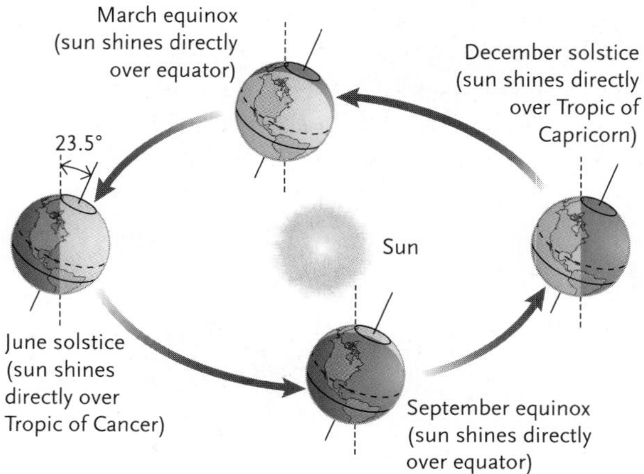
March equinox
(sun shines directly
over equator)
December solstice
(sun shines directly
over Tropic of
Capricorn)
23.5°
Sun
June solstice
(sun shines
directly over
Tropic of Cancer)
September equinox
(sun shines directly
over equator)

FIGURE 49.3
Seasonal variation in solar radiation. Earth's fixed tilt on its axis causes the Northern Hemisphere to receive more sunlight in June and the Southern Hemisphere to receive more in December. These differences are reflected in seasonal variations in day length and temperature, which are more pronounced at the poles than at the equator.

AIR CIRCULATION Sunlight warms air masses, causing them to expand, lose pressure, and rise in the atmosphere. The unequal heating of air at different latitudes initiates global air movements, producing three circulation cells in each hemisphere **(Figure 49.4).** Warm equatorial air masses rise to high altitude before spreading north and south. They eventually sink back to Earth at about 30° N and S latitude. At low altitude, some air masses flow back toward the equator, completing low-latitude circulation cells. Others flow toward the poles, rise at 60° latitude, and divide at high altitude. Some air flows toward the equator, completing the pair of middle-latitude circulation cells. The rest moves toward the poles, where it descends and flows toward the equator, forming the polar circulation cells.

The flow of air masses at low altitude creates winds near the planet's surface. But the surface rotates beneath the atmosphere, moving rapidly near the equator, where Earth's diameter is greatest, and slowly near the poles. Latitudinal variation in the speed of rotation deflects the movement of the rising and sinking air masses from a strictly north–south path into belts of easterly and westerly winds (see the right side of Figure 49.4); this deflection is called the Coriolis Effect. Winds near the equator are called the trade winds; those further from the equator are the temperate westerlies and easterlies, named for the direction from which they blow.

PRECIPITATION Differences in solar radiation and global air circulation create latitudinal variations in rainfall **(Figure 49.5).** Warm air holds more water vapor than cool air does. As air near the equator heats up, it absorbs water, primarily from the oceans. However, the warm air masses expand as they rise, and their heat energy is distributed over a larger volume, causing their tempera-

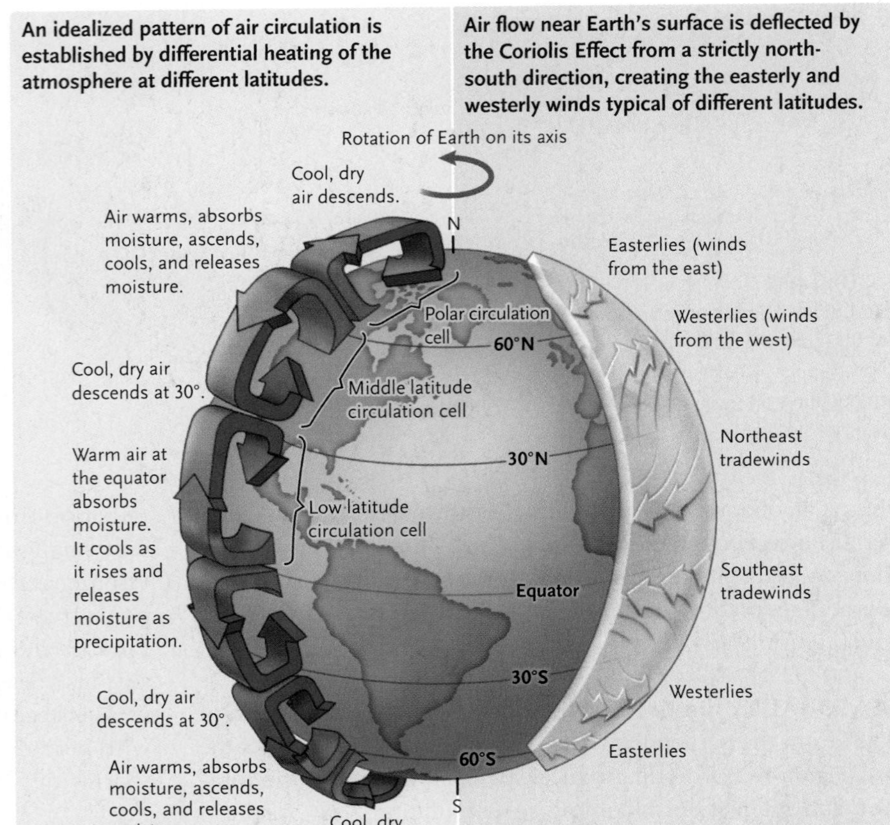

An idealized pattern of air circulation is established by differential heating of the atmosphere at different latitudes.
Air flow near Earth's surface is deflected by the Coriolis Effect from a strictly north-south direction, creating the easterly and westerly winds typical of different latitudes.
Rotation of Earth on its axis
Cool, dry air descends.
Air warms, absorbs moisture, ascends, cools, and releases moisture.
N
Easterlies (winds from the east)
Polar circulation cell
60°N
Westerlies (winds from the west)
Cool, dry air descends at 30°.
Middle latitude circulation cell
Northeast tradewinds
Warm air at the equator absorbs moisture. It cools as it rises and releases moisture as precipitation.
30°N
Low latitude circulation cell
Equator
Southeast tradewinds
30°S
Cool, dry air descends at 30°.
Westerlies
60°S
Easterlies
Air warms, absorbs moisture, ascends, cools, and releases moisture.
S
Cool, dry air descends.

FIGURE 49.4

Global air circulation. Latitudinal variations in the intensity of solar radiation cause equatorial air masses to warm and rise, initiating a global pattern of air movement in three circulation cells in each hemisphere. Air masses moving near Earth's surface are deflected from a strictly north–south flow by the planet's rotation, creating easterly and westerly winds.

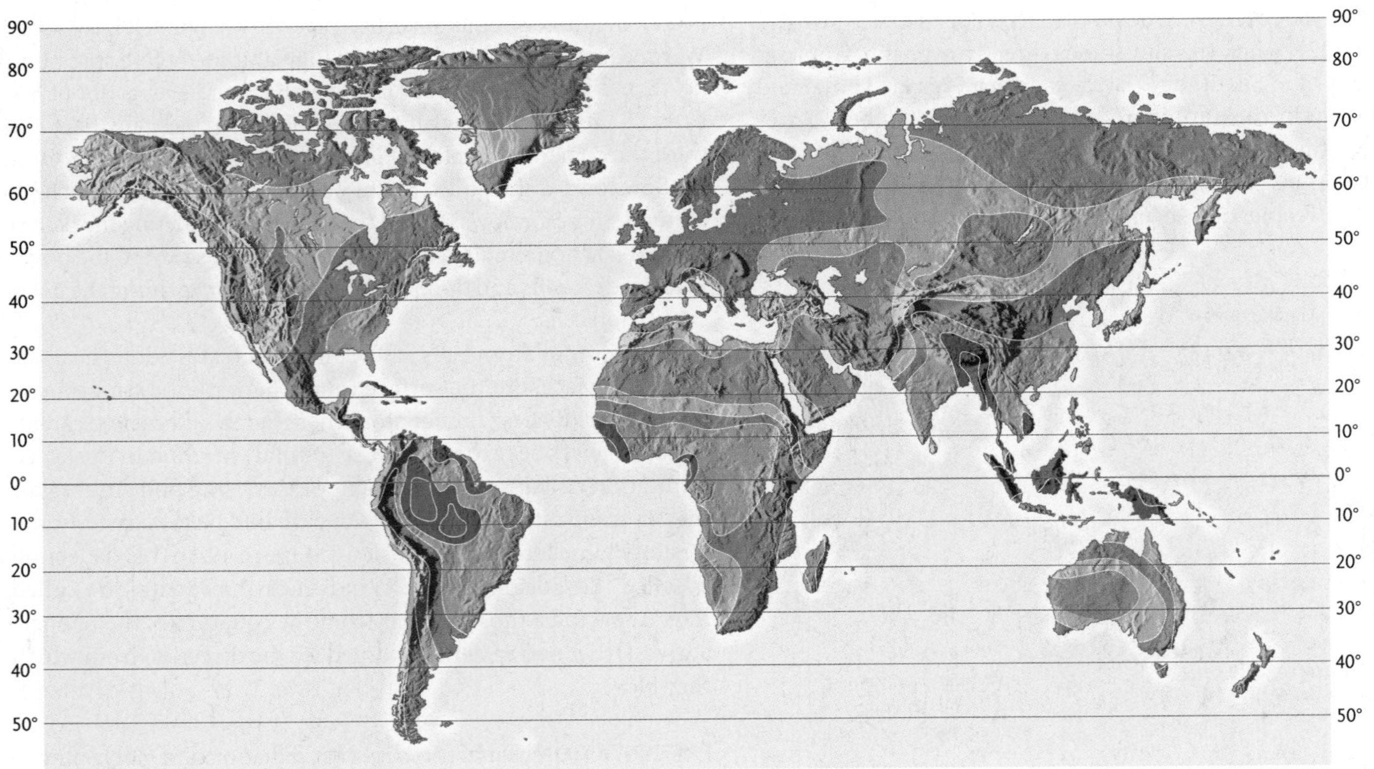

90°
80°
70°
60°
50°
40°
30°
20°
10°
0°
10°
20°
30°
40°
50°

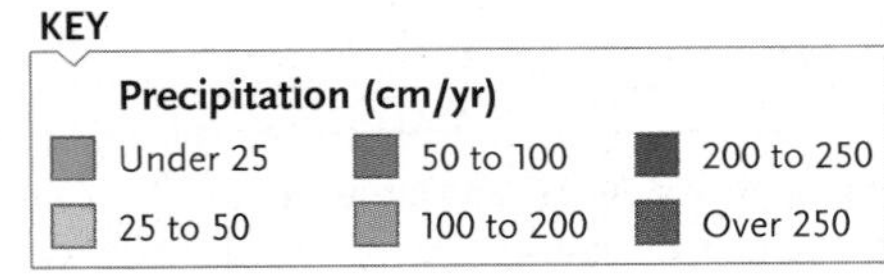

KEY
Precipitation (cm/yr)
Under 25
25 to 50
50 to 100
100 to 200
200 to 250
Over 250

FIGURE 49.5

Variations in precipitation. The tropics receive high annual rainfall, whereas regions near 30° latitude are usually dry. Local topographic features and ocean currents also influence precipitation patterns.

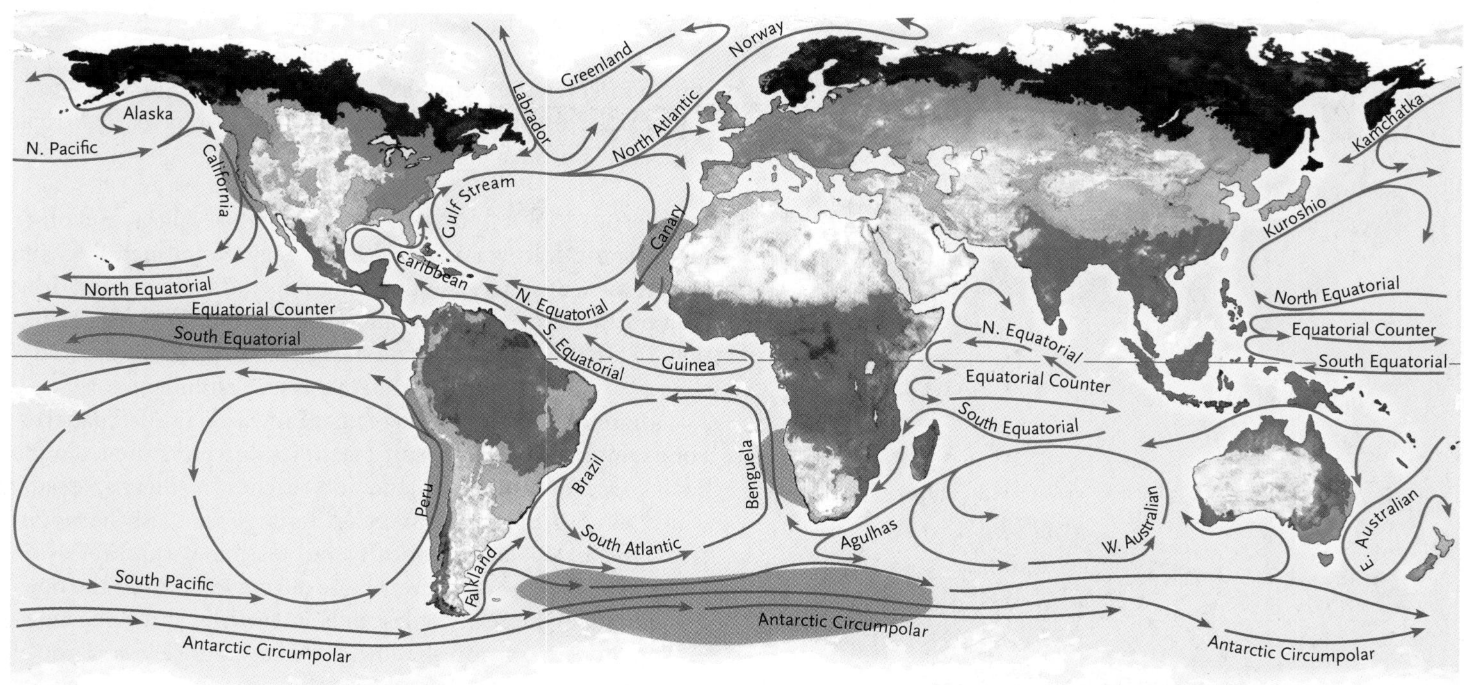

Alaska
N. Pacific
California
Labrador
Greenland
Norway
North Atlantic
Gulf Stream
Canary
Kamchatka
Kuroshio
Caribbean
North Equatorial
Equatorial Counter
South Equatorial
N. Equatorial
S. Equatorial
Guinea
N. Equatorial
Equatorial Counter
South Equatorial
North Equatorial
Equatorial Counter
South Equatorial
Peru
Brazil
Benguela
South Atlantic
Agulhas
W. Australian
E. Australian
Falkland
South Pacific
Antarctic Circumpolar
Antarctic Circumpolar
Antarctic Circumpolar

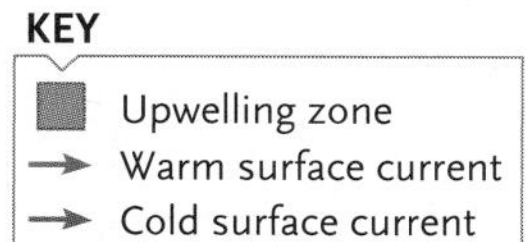

KEY
Upwelling zone
Warm surface current
Cold surface current

FIGURE 49.6
Ocean currents. Prevailing winds, Earth's rotation, gravity, the shape of ocean basins, and the positions of landmasses establish the direction and intensity of surface currents in the oceans. In general, warm currents flow away from the equator, and cold currents flow toward it.

ture to drop. A decrease in temperature without the actual *loss* of heat energy is called **adiabatic cooling.** After cooling adiabatically, the rising air masses release moisture as rain (see labels on the left side of Figure 49.4). Torrential rainfall is characteristic of warm equatorial regions, where rising, moisture-laden air masses cool as they reach high altitude.

As cool, dry air masses descend at 30° latitude, increased air pressure at low altitude compresses them, concentrating their heat energy, raising their temperature, and increasing their capacity to hold moisture. The descending air masses absorb water from the land, so these latitudes are typically dry. Some air masses continue moving poleward in the lower atmosphere. When they rise at 60° latitude, they cool adiabatically and release precipitation (see Figure 49.4), creating moist habitats in the northern and southern temperate zones.

OCEAN CURRENTS Latitudinal variations in solar radiation also warm the oceans' surface water unevenly. Because the volume of water increases as it warms, sea level is about 8 cm higher at the equator than at the poles. The volume of water associated with this "slope" is enough to cause surface water to move in response to gravity. The trade winds and temperate westerlies also contribute to the mass flow of water at the ocean surface. Thus, surface water flows in the direction of prevailing winds, forming major currents. Earth's rotation, the positions of landmasses, and the shapes of ocean basins also influence their movement.

Oceanic circulation is generally clockwise in the Northern Hemisphere and counterclockwise in the Southern **(Figure 49.6).** The trade winds push surface water toward the equator and westward until it contacts the eastern edge of a continent. Swift, narrow, and deep currents of warm, nutrient-poor water run toward the poles, parallel to the east coasts of continents. For example, the Gulf Stream flows northward along the east coast of North America, carrying warm water toward northwestern Europe. Cold water returns from the poles toward the equator in slow, broad, and shallow currents, such as the Peru Current, that parallel the west coasts of continents.

Regional and Local Effects Overlay Global Climate Patterns

Although global and seasonal patterns determine a site's climate, regional and local effects also influence abiotic conditions.

PROXIMITY TO THE OCEAN Currents running along seacoasts exchange heat with air masses flowing above them, moderating the temperature over nearby land. Breezes often blow from the sea toward the land during the day and in the opposite direction at night **(Figure 49.7).** These local effects sometimes override latitudinal variations in temperature. For example, the climate in London is much milder than that in Minneapolis, even though Minneapolis is slightly further south. Minneapolis has a **continental climate** that is not moderated by the distant ocean, but London has a **maritime climate,** tempered by winds that cross the nearby North Atlantic Current.

Ocean currents also affect moisture conditions in coastal habitats. For example, air masses absorb water as they move from west to east across the Pacific Ocean. They cool as they

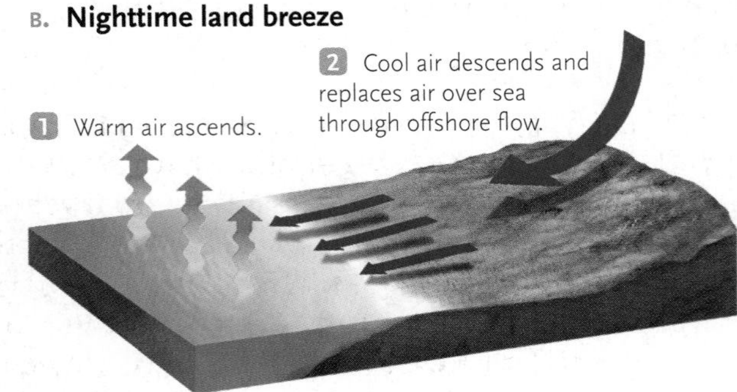
A. Daytime sea breeze
1 Warm air ascends.
2 Cool air descends and replaces air over land through onshore flow.
B. Nighttime land breeze
2 Cool air descends and replaces air over sea through offshore flow.
1 Warm air ascends.

FIGURE 49.7
Sea breezes and land breezes. **(A)** On a summer afternoon, when the land is warmer than the adjacent ocean, warm air rises over the land, and a cool sea breeze blows inland from the ocean. **(B)** At night, when the ocean is warmer than the land, the pattern is reversed.

cross the cold California Current, and when they reach land in northern California and Oregon during winter, their water vapor condenses into heavy fog and rain. During summer, however, land is warmer than the adjacent ocean. The air masses heat up as they cross the land, and they accumulate water, creating dry conditions.

Some regions experience **monsoon cycles** caused by seasonal reversals of wind direction. In the North American southwest, for example, summer heat causes air masses over land to rise, creating a zone of low pressure. Moist air from the nearby Gulf of California flows inland, where it rises and cools adiabatically, releasing substantial precipitation. Summer monsoon rains deliver one-third to one-half of the annual rainfall in Arizona and New Mexico. During the winter, when land is cooler than the nearby ocean, low-pressure systems form over the ocean, and winds blow from the land to the sea; thus, winters in the southwest are generally dry. Seasonal monsoon cycles also deliver torrential rainfall to parts of Africa, Asia, and South America.

THE EFFECTS OF TOPOGRAPHY Mountains, valleys, and other topographic features also influence regional climates. In the Northern Hemisphere, south-facing slopes are warmer and drier than north-facing slopes because they receive more solar radiation. In addition, adiabatic cooling causes air temperature to decline 3° to 6°C for every 1,000 m increase in altitude.

Mountains also establish regional and local rainfall patterns. For example, after a warm air mass picks up moisture from the Pacific Ocean, it moves inland and reaches the Sierra Nevada, which parallels the California coast. As it rises to cross the mountains, the air cools adiabatically and loses moisture, releasing heavy rainfall on the windward side **(Figure 49.8)**. After the now-dry air crosses the peaks, it descends and warms, absorbing moisture and forming a **rain shadow.** Habitats on the leeward side of mountains, such as the Great Basin Desert in western North America, are typically drier than those on the windward side.

MICROCLIMATE Although climate influences the overall distributions of organisms, the abiotic conditions that immediately surround them—the **microclimate**—have the greatest effect on survival and reproduction. For example, a fallen log on the forest floor creates a microclimate in the underlying soil that is shadier, cooler, and moister than surrounding soil exposed to sun and wind. Many animals, including some insects, worms, salamanders, and snakes, occupy these sheltered sites and avoid the effects of prolonged exposure to the elements.

STUDY BREAK 49.2

1. **How does Earth's spherical shape influence temperature and air movements at different latitudes?**
2. **What causes seasonality of the climate in the temperate zone?**
3. **Why do dry conditions occur at 30° N and S latitude?**
4. **Briefly describe how mountains influence local precipitation.**

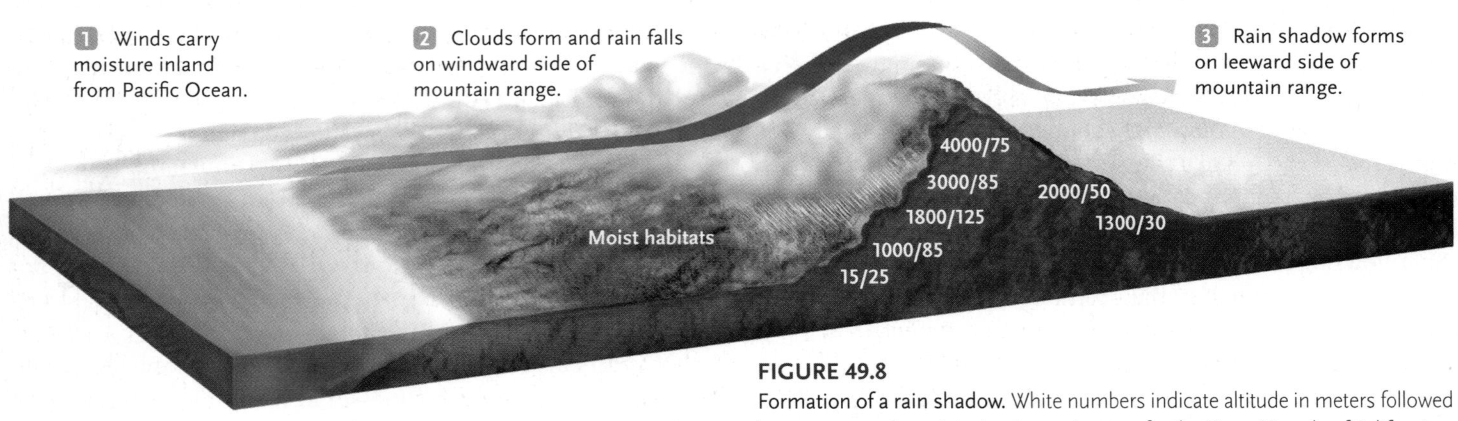
1 Winds carry moisture inland from Pacific Ocean.
2 Clouds form and rain falls on windward side of mountain range.
3 Rain shadow forms on leeward side of mountain range.
4000/75
3000/85
2000/50
1800/125
1300/30
Moist habitats
1000/85
15/25

FIGURE 49.8
Formation of a rain shadow. White numbers indicate altitude in meters followed by mean annual precipitation in centimeters for the Sierra Nevada of California.

INSIGHTS FROM THE

Molecular Revolution

Frozen Fish? Not If They Have Antifreeze Proteins

Polar-dwelling fishes, such as winter flounder, Alaskan plaice, and Arctic sculpin, have "antifreeze proteins" that prevent their bodies from freezing into solid ice at the extremely low environmental temperatures they encounter. As ice crystals begin to form within a fish's cells and tissues, the antifreeze proteins bind to the crystals and cover them with a protein coat that prevents further crystal growth and fusion. As a result, the fishes freeze only to an ultrafine slush that allows continued activity, including movement and feeding. The antifreeze proteins are small molecules containing between 30 and 50 amino acids.

Research Question

How do antifreeze proteins bind to ice crystals?

Experiment

Frank Sicheri and D. S. C. Yang at McMaster University in Hamilton, Ontario, and the Bio-Crystallography Laboratory of the VA Medical Center in Pittsburgh, Pennsylvania, sought a molecular answer to the question. They used X-ray crystallography (a technique like X-ray diffraction, described in Section 14.2, but applied to crystals) to work out the molecular structure of the antifreeze protein from the winter flounder *(Pseudopleuronectes americanus)*. They grew protein crystals in a solution at 4°C and then examined them by X-ray crystallography at 4°C and −180°C.

Conclusion

The X-ray crystallography data indicated that the 37 amino acids of the antifreeze protein wind into a single, linear alpha helix (see Section 3.5). Along one side of the helix, side groups of two polar amino acids, threonine and asparagine, extend from the surface at four evenly spaced locations in a flat plane, one at either end of the molecule and two within the helix. Sicheri and Yang propose that these locations, which are spaced 1.65 nm apart, are *ice-binding motifs*. The four motifs would fit nicely to the tips of ridges formed by water molecules on the surface of an ice crystal, which are spaced at intervals of 1.67 nm **(Figure).** Hydrogen bonds between the polar amino acid side groups and water molecules along the ridges would hold the proteins tightly to the surface of the ice crystal. The tight fit would prevent more water molecules from adding to the ice surface and thereby prevent further crystal growth.

Understanding how the fish antifreeze proteins work is not just a fascinating scientific issue. The description and characterization of the antifreeze proteins could lead to medical and industrial applications in situations where procedures and equipment must tolerate freezing conditions. In fact, the U.S. Food and Drug Administration has approved the use of an antifreeze protein—originally discovered in an arctic fish, but now produced by genetically engineered yeast—as an ingredient in ice cream to enhance its creamy texture.

Source: F. Sicheri and D. S. C. Yang. 1995. Ice-binding structure and mechanism of an antifreeze protein from winter flounder. *Nature* 375:427–431.

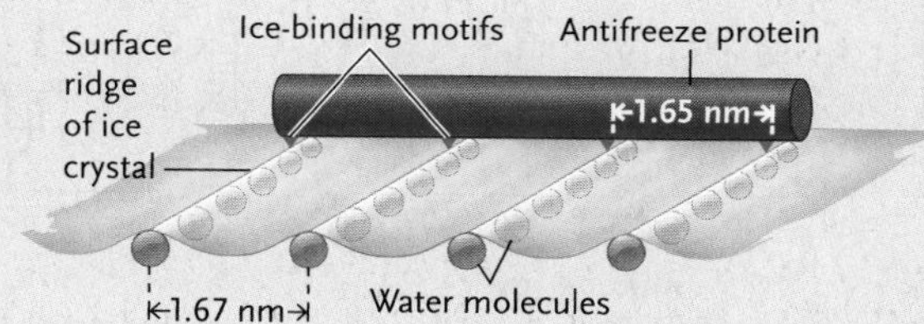
Surface ridge of ice crystal
Ice-binding motifs
Antifreeze protein
1.65 nm
1.67 nm
Water molecules

How winter flounder antifreeze protein may bind to the surface ridges of an ice crystal. Only the water molecules forming the tips of the ridges are shown.

THINK OUTSIDE THE BOOK

Using weather records for your community, characterize the climate in the area where you live. How do seasonal variations in temperature, precipitation, and day length compare to those observed at the poles and at the equator?

49.3 Organismal Responses to Environmental Variation and Climate Change

Daily and seasonal variations in physical factors have profound effects on the biology of individual organisms. Moreover, large-scale variations in environmental conditions, including those caused by global climate change, often influence the distributions of populations.

Organisms Use Homeostatic Responses to Cope with Environmental Variation

Many animals exhibit diverse homeostatic responses—biochemical, behavioral, physiological, and morphological—that enable them to maintain relatively constant conditions within their cells and tissues. Although the ability to use these responses almost certainly has a genetic basis, only some responses to environmental variation are *obligate* (that is, must always be used). *Insights from the Molecular Revolution* describes one such evolutionary response at the biochemical level. Many behavioral and physiological responses are *facultative*. In other words, animals may use them or not, as their immediate conditions demand. Here we provide two brief examples of facultative behavioral and physiological responses to variations in environmental temperature through space and time.

Like many ectothermic animals, lizards often use behaviors to regulate body temperature (see Figure 1.15 and Section 46.6). They commonly bask in sunny spots to raise body temperature

FIGURE 49.9 **Observational Research**

How Do Lizards Compensate for Altitudinal Variation in Environmental Temperature?

Hypothesis: Lizards living at high altitude can use behaviors to compensate for the low environmental temperatures they encounter.

Prediction: The percentage of lizards observed basking in the sun will increase with altitude, and mean air temperatures will vary more than mean lizard body temperatures among study sites distributed along an altitudinal gradient.

Method: Hertz and Huey measured the basking behavior as well as air temperatures and body temperatures of two closely related species of *Anolis* lizards distributed along an altitudinal gradient in the Dominican Republic. They surveyed populations of lizards at sea level, 550 m, 1,100 m, and 2,200 m elevation. They then compared the percentages of lizards basking and the mean air and lizard body temperatures at the four study sites.

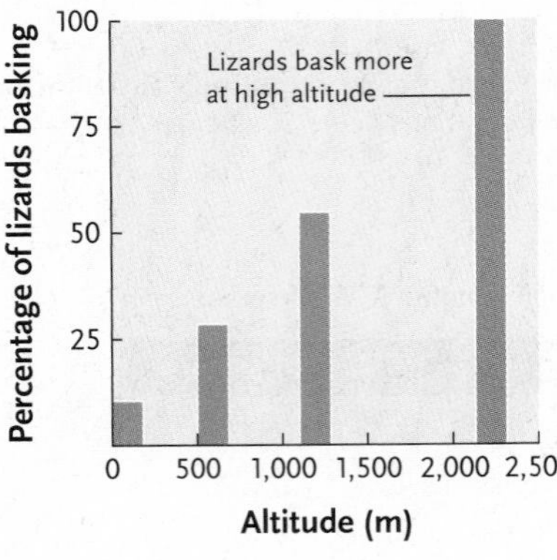

Percentage of lizards basking
Lizards bask more at high altitude
Altitude (m)
0
500
1,000
1,500
2,000
2,500
100
75
50
25

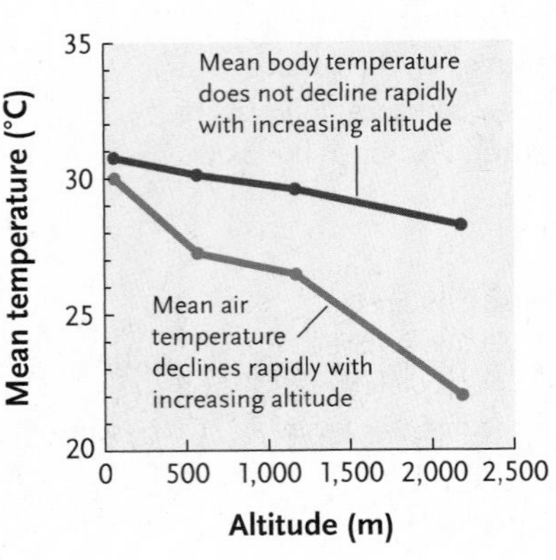

Mean temperature (°C)
Mean body temperature does not decline rapidly with increasing altitude
Mean air temperature declines rapidly with increasing altitude
Altitude (m)
35
30
25
20
0
500
1,000
1,500
2,000
2,500

Result: The percentage of lizards basking increased steadily with altitude. Mean air temperature differed by as much as 8°C among study sites, but mean body temperature differed by only 2°C.

Conclusion: Lizards living at high altitude bask in patches of sun more frequently, partially compensating for the low environmental temperatures in their habitats.

Source: P. E. Hertz and R. B. Huey. 1981. Compensation for altitudinal variation in the thermal environment by some *Anolis* lizards on Hispaniola. *Ecology* 62:515–521.

and seek shaded places to cool off. Many *Anolis* lizard species (see *Focus on Research Organisms* in Chapter 30) are distributed over broad altitudinal ranges, and populations living at high altitude encounter cooler environments than do those at low altitude. While they were graduate students at Harvard University, Paul E. Hertz, now of Barnard College, and Raymond B. Huey, now of the University of Washington, hypothesized that *Anolis* populations living at cool, high altitudes would bask more frequently than those living at warm, low altitudes. Hertz and Huey tested their hypothesis by observing *Anolis cybotes* and its close relative *Anolis shrevei* along an altitudinal gradient in the Dominican Republic **(Figure 49.9)**. Their results indicate that basking frequency increases steadily with altitude. Moreover, the body temperatures of the lizards vary much less with altitude than do air temperatures at the same localities. The researchers therefore concluded that increased basking frequency by lizards at high altitude partially compensates for the lower environmental temperatures they encounter.

The state of extreme physiological sluggishness called *torpor* is a facultative response to daily variations in environmental temperature. Endothermic animals use the heat generated by the metabolic breakdown of food to maintain high body temperature (see Section 46.6). However, small endotherms, such as hummingbirds, have a large relative surface area through which they lose body heat. When environmental temperature is low, they may lose heat faster than they can generate it, risking the total depletion of their energy reserves and death by starvation. The problem is particularly acute at night, when hummingbirds cannot feed to replenish their energy stores. F. Reed Hainsworth and Larry Wolf of Syracuse University discovered that the purple-throated carib *(Eulampis jugularis)*, a West Indian hummingbird, often becomes torpid at night, lowering its body temperature from 40° to 20°C. Because torpor reduces the temperature difference between their bodies and the environment, torpid birds lose heat less rapidly. At the nighttime environmental temperatures they usually encounter, the torpid hummingbirds may use 80% less energy than they would if they had not entered a temporarily dormant state.

Global Climate Change Affects the Ecology of Many Organisms

As described in Section 52.4, virtually all scientists agree that Earth's atmosphere and oceans are getting steadily warmer. What effect will rapid global climate change have on biological systems? Biologists hypothesize that, on the spatial scale of the biosphere, rising temperatures will affect the geographical distributions of populations, species, and communities. Models of climate change predict that the distributions of polar species will contract to even higher latitudes, the ranges of temperate and tropical species will expand or shift toward the poles, and lowland species will move to higher elevations. The models also predict that global warming will change the timing of important biological events. For example, plants whose flowering is triggered by warm springtime temperatures will flower earlier in the season; similarly, migratory animals will return from their wintering grounds and begin reproducing earlier in the year.

Camille Parmesan of the University of Texas at Austin and Gary Yohe of Wesleyan University tested these predictions with a massive literature review. They surveyed studies of changes in the geographical distributions and timing of springtime activities in a wide variety of herbaceous plants, trees, invertebrates, and vertebrates over roughly the past 100 years. Their analysis, published in 2003, suggested that the geographical ranges of 99 species of butterflies, birds, and alpine herbs in the Northern Hemisphere have shifted dramatically into habitats that had previously been too cold for them. Some species have expanded their distributions northward an average of 6.1 km per decade. Other species have shifted their distributions to higher altitude, an average of 6.1 m per decade. Their analysis also indicated that for 172 diverse species (including plants, butterflies, amphibians, and birds) springtime growth and reproduction have occurred on average 2.3 days earlier per decade. If these trends continue at the same rate, spring flowering and animal reproduction will occur one full month earlier in the year 2130 than it did in 2000.

A parallel analysis of less detailed data on 677 species of plants and animals suggested that 62% of the species surveyed showed trends toward earlier flowering, breeding, or growth. And for 434 species in which researchers documented a change in geographical distribution, 80% of the shifts were in the direction predicted by climate change models.

Climate warming is also changing the combinations of species that occur together within ecological communities. For example, in a paper published in 2008, Craig Moritz of the University of California, Berkeley, and his colleagues from Berkeley and Colorado State University compared the elevational distributions of small mammals in Yosemite National Park to data compiled a century earlier by naturalist Joseph Grinnell. The minimum temperatures recorded at Yosemite have increased about 3°C over the last century, reflecting the worldwide trend of global warming. The distributions of half of the 28 species they monitored had shifted upward an average of 500 m. Some species that had occupied low elevation sites expanded their ranges to include higher elevations. By contrast, some species that had lived only at middle and high elevation sites no longer live at middle elevations; their distributions are now restricted to higher elevations. As a result of these shifts in species distributions, different combinations of small mammals now occupy middle and high elevation sites in the park, and some species that live at high elevation may face extinction as environmental temperatures continue to rise and other species expand upward into their habitats **(Figure 49.10)**.

The effects of global climate change are not restricted to terrestrial environments. Ocean temperatures are also rising, and these changes are reflected in the distributions of marine species. For example, among invertebrates and fishes on the California coast, cold-adapted species have become less abundant and warm-adapted species more abundant. Comparable changes have been noted in other communities from Antarctica to the Arctic.

The geographical distributions of species and ecological communities have often changed with climate shifts over evolutionary time, but the rate of global warming has accelerated in your lifetime. As you will discover as you read Chapters 51 and 52, the factors that govern the structures of communities and ecosystems are complex, and scientists are far from being able to predict all of the consequences of these changes in detail. In the next section, we describe how today's climate affects species and community distributions on a biosphere-wide scale. You can be certain that biology texts in the twenty-second century will paint a very different portrait of these large-scale associations.

FIGURE 49.10
Endangered by climate change. The elevational range of the alpine chipmunk *(Tamius alpinus)* has shifted more than 600 m upward in the last 100 years, confining its populations to relatively small areas high in the Sierra Nevada of California.

Dr. Mark Chappell

STUDY BREAK 49.3

1. **How does the behavior of *Anolis* lizards in the Dominican Republic change over altitude?**
2. **What effect is global warming likely to have on the geographical distributions of organisms?**

49.4 Terrestrial Biomes

In Section 22.3 we described how convergent evolution produces morphological and physiological similarities in species that occupy similar environments. Early in the twentieth century, two American ecologists, Frederic Clements of the Carnegie Institution in Washington and Victor Shelford of the University of Illinois, generalized this observation by defining the **biome** as a vegetation type plus its associated microorganisms, fungi, and animals. Although vegetation is superficially similar throughout a biome, the particular species within it vary from place to place. For example, in eastern North America, the temperate deciduous forest biome includes beech-maple forests in the north and oak-hickory forests in the south. Before surveying eight major terrestrial biomes—*tropical forests, savannas, deserts, chaparral, temperate grasslands, temperate deciduous forests, evergreen coniferous forests,* and *tundra*—we consider how environmental factors influence their overall distribution.

Environmental Variation Governs the Distribution of Terrestrial Biomes

Because organisms—and the communities they form—are sensitive to abiotic factors, climate is the main determinant of biome distribution. A **climograph** portrays the particular combination of temperature and rainfall conditions where each terrestrial biome occurs **(Figure 49.11)**. For example, some deserts, grasslands, savannas, and tropical forests occur in areas that have comparable mean annual temperatures but vastly different rainfall. Conversely, some biomes, such as boreal forests, temperate deciduous forests, and savannas, are found under similar moisture conditions but different temperature regimes.

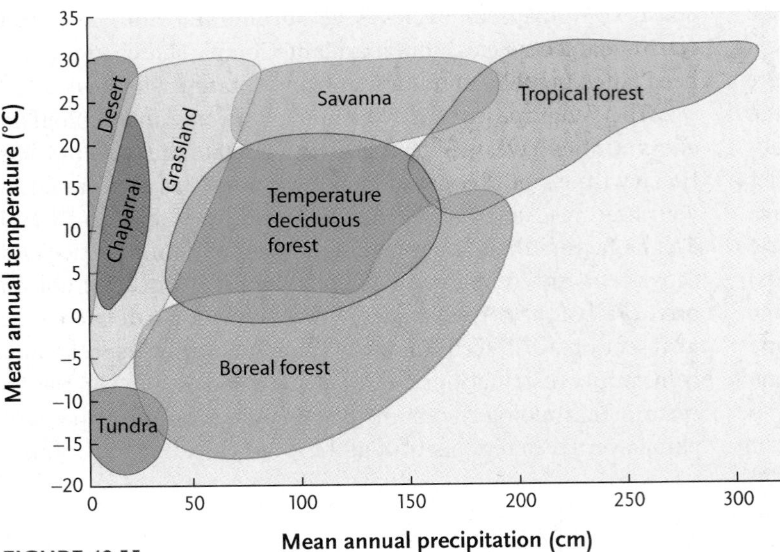

35
30
25
20
15
10
5
0
−5
−10
−15
−20
Mean annual temperature (°C)
Desert
Grassland
Chaparral
Savanna
Tropical forest
Temperature
deciduous
forest
Boreal forest
Tundra
0
50
100
150
200
250
300
Mean annual precipitation (cm)

FIGURE 49.11
Climograph. Each of the major terrestrial biomes occupies a characteristic combination of temperature and moisture conditions.

FIGURE 49.12
Terrestrial biomes. Climate governs the distributions of the world's major terrestrial biomes. High mountain ranges are indicated by stripes because different biomes occur at different elevations.

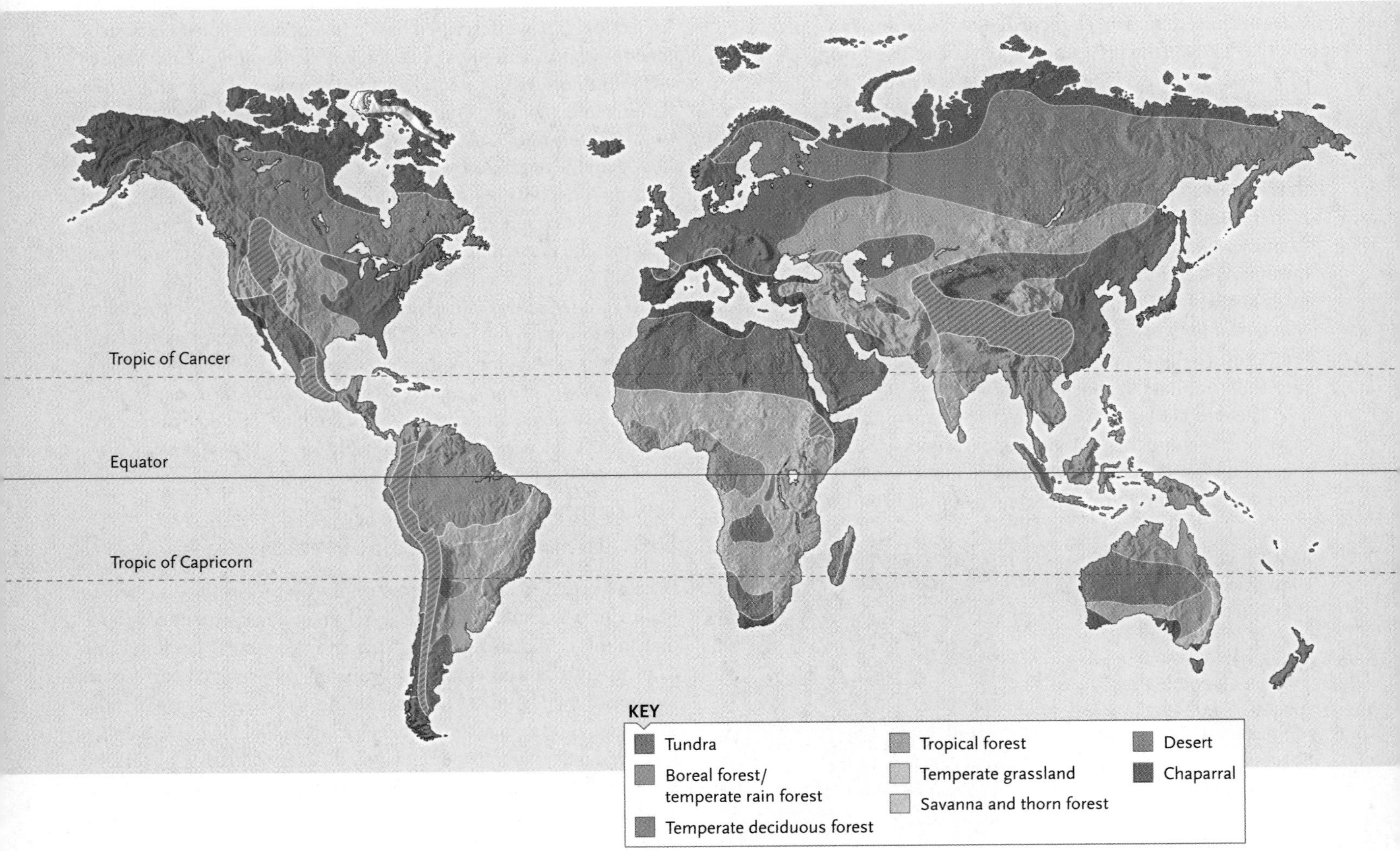

Although the climograph provides a general portrait of the temperature and moisture conditions where the different biomes occur, it does not address the details of environmental variation. For example, the climograph includes only mean annual temperature and rainfall, not seasonal variation in these factors. Two regions may have the same mean temperature even though one experiences blazingly hot summers and bitterly cold winters and the other has moderate temperature throughout the year; we would expect them to harbor different organisms. Moreover, the distributions of communities are also influenced by nonclimatic factors, such as regional variations in soil structure and mineral composition (see Section 33.2).

Because temperature and rainfall exhibit latitudinal patterns (displayed in Figures 49.2 and 49.5), the distributions of some terrestrial biomes appear as bands on a world map **(Figure 49.12).** But regional and local climatic variations influence these broad patterns. For example, chaparral is common in certain coastal habitats, whereas grasslands occur further inland at similar latitudes. Comparable bands of distinct vegetation form on mountainsides because temperature and moisture conditions also change with altitude.

Tropic of Cancer

Equator

Tropic of Capricorn

KEY

- Tundra
- Boreal forest/ temperate rain forest
- Temperate deciduous forest
- Tropical forest
- Temperate grassland
- Savanna and thorn forest
- Desert
- Chaparral

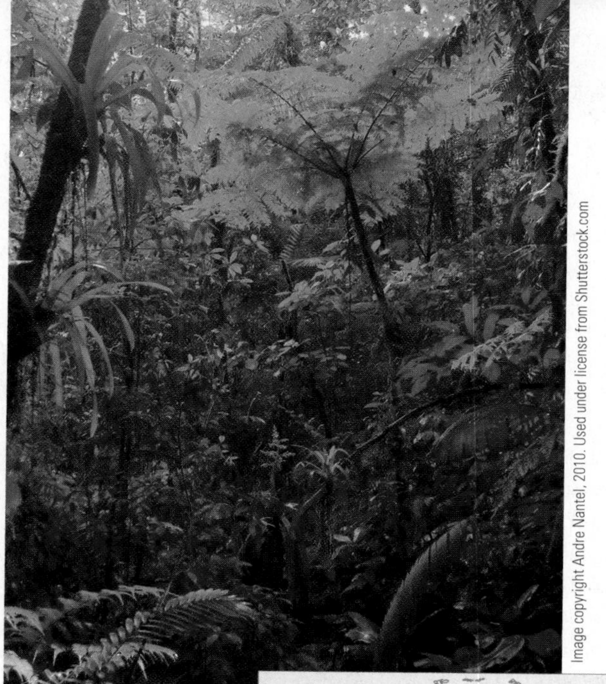

Image copyright Andre Nantel, 2010. Used under license from Shutterstock.com

FIGURE 49.13
Tropical forest. Many tropical rain forest trees, like these at Chachagua, Costa Rica, are covered with lianas and epiphytes.

Tropical Forests Include Earth's Most Species-Rich Communities

Three types of **tropical forests**—rain forest, deciduous forest, and montane forest—sweep across the parts of Africa, Asia, Australia, and Central and South America that receive intense solar radiation and heavy rainfall.

Tropical rain forests grow where some rain falls every month, mean annual rainfall exceeds 250 cm, mean annual temperature is at least 25°C, and humidity is above 80%. Limited by neither temperature nor water, the productivity of a tropical rain forest is exceptionally high (see Section 52.1). Trees replace their leaves throughout the year, producing a continuous rain of detritus that ants, land crabs, and other scavengers quickly consume. Decomposition is rapid in the hot, moist environment, and almost no litter accumulates on the ground. Because nutrients released by decomposition are promptly absorbed by vegetation or leached by rain, soil in tropical rain forests is nutrient-poor, with low humus content (see Section 33.2).

Tropical rain forests are usually layered (see Figure 51.19). The crowns of tall trees form a dense, tangled canopy that intercepts most incoming sunlight 40 to 45 m above the ground **(Figure 49.13).** Even the largest trees grow only shallow roots in the thin soil, but many have wide *buttresses,* woody lateral extensions of their trunks, that stabilize them in the ground. Shade-tolerant shrubs and small trees form understory layers below the canopy. The woody stems of lianas climb through both layers, and epiphytes, such as bromeliads and orchids, cover the trunks and branches of trees, especially in sunlit openings. In mature rain forests, the ground is surprisingly bare of leafy vegetation, because very little sunlight reaches the forest floor.

Tropical rain forests probably harbor more plant and animal species than all other terrestrial biomes combined. Ecologists have proposed numerous hypotheses to explain both the evolution and maintenance of the large number of species living in these communities (see Section 51.7), but no single hypothesis explains the pattern adequately. In fact, we do not even have a complete species list for any rain forest community, largely because most animals live in the highly productive canopy, which ecologists have only recently begun to study in detail (see *Focus on Basic Research*). The most extensive tracts of tropical rain forest occur in South America, central and western Africa, and Southeast Asia. Unfortunately, they are being cleared at an alarming rate (see Chapter 53); some experts predict that this biome will all but disappear before the middle of the twenty-first century.

Habitats centered at 20° north and south of the equator experience a pronounced summer rainy season and winter dry season. **Tropical deciduous forests** occur where winter drought reduces photosynthesis, and most trees drop their leaves. For example, the monsoon forests of Southeast Asia, which harbor teak and other tropical hardwoods, are as lush as tropical rain forests in the rainy season; but many trees are bare in the dry season.

High altitudes in the tropics support distinctive **tropical montane forests,** or "cloud forests," which are frequently enveloped in mist. The trees, often no more than 3 m tall, are densely covered with epiphytes, which thrive in the moisture-laden air. Cloud-forest plants grow slowly because productivity is limited by low temperatures, high humidity (making transpiration difficult), and sunlight-blocking clouds.

Savannas Grow Where Moderate Rainfall Is Highly Seasonal

Grasslands with few trees, the biome called **savanna,** grow in areas adjacent to tropical deciduous forests **(Figure 49.14).** Seasonality in tropical

iStockphoto.com/oversnap

FIGURE 49.14
Savanna. The African savanna, a warm grassland with scattered stands of shrubby trees, has an enormous concentration of large ungulates (hoofed, herbivorous mammals), such as these wildebeests *(Connochaetes taurinus)*.

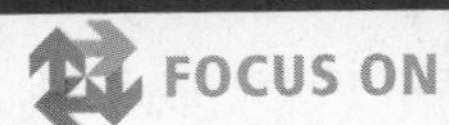

FOCUS ON **Basic Research**

Exploring the Rain Forest Canopy

Biological diversity in tropical rain forests has fascinated naturalists for centuries. Sadly, most of its organisms live beyond our reach. The forest canopy extends from 9 or 10 m above the ground to heights as great as 45 m, making the canopy inaccessible and largely unexplored. Early ecologists were able to study canopy-dwelling species only when they found a fallen tree or followed loggers into the forest. In the 1930s, a clever botanist trained monkeys to retrieve plants from the canopy, but these efforts provided little data about the ecological interactions that govern life in the treetops.

Many ecologists still study canopy-dwelling organisms from the safety of the ground. Binoculars provide a good view of fairly large vertebrates. And a hike along a ridgetop can provide a canopy-level view of trees growing in an adjacent valley or ravine. Some researchers use ropes to hoist nets or traps into the canopy, lowering them periodically to see what they have caught. Others spray a fog of insecticide into the canopy to kill small invertebrates, which then rain down onto plastic sheets spread below the trees. These ground-based techniques have led to the discovery of hundreds—perhaps thousands—of new arthropod species. Ecologists now collect huge samples of arthropods to study the species composition and structure of communities and to monitor changes in these communities over time. But distant observations and mass sampling techniques don't provide detailed data about which insects are feeding on a tree, how often hummingbirds pollinate a flower, or when a tiny lizard hunts its prey.

Today many ecologists routinely risk life and limb to collect detailed ecological data in the rain forest canopy. They climb trees and crawl along stout branches. Many build stable observation decks with walkways, allowing study on either side of the "trail."

What does this newfound access to the rain forest canopy add to our knowledge of organisms that live there? Researchers can measure the physical environment of the canopy and observe the physiological and behavioral adaptations of its plants and animals. For example, researchers are gathering data on the feeding habits and behavior of small animals that never venture to the ground, such as fruit-eating bats and birds. When coupled with information about the movement patterns of these animals, the data provide insight into the dispersal of seeds in the fruits. And an understanding of seed dispersal provides information for studies of the population ecology of rain forest trees.

Canopy ecologists have also discovered fascinating relationships between plants and their animal pollinators. For example, Donald Perry, a freelance biologist, discovered that birds are attracted to the sweet nectar of the vine *Norantea sessilis.* Feeding birds step on the vine's sturdy flowers; their feet become covered with the plant's pollen, which is embedded in a gummy substance. When the birds visit another vine of the same species, they transfer the pollen to that plant's flowers, providing cross-pollination, which appears to be necessary for the vine's reproduction.

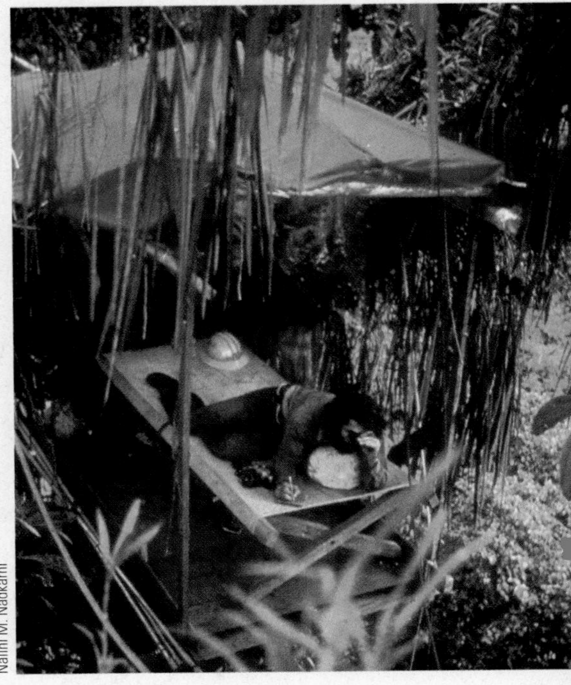

Nalini M. Nadkarni

A platform in the canopy of a tropical rain forest in Costa Rica provides a comfortable perch for Donald Perry to survey the pollinating activity of birds and bees.

Research in the tropical rain forest canopy promises exciting discoveries about ecological relationships in this unique biome, which is the most threatened on Earth (see Chapter 53). Such research is essential for developing a public appreciation of tropical forests and for creating conservation plans to preserve them.

and subtropical savannas is determined by the availability of water; although annual rainfall averages 90 to 150 cm, droughts typically last for months. Grasses are successful in semiarid conditions because their shallow roots harvest water efficiently. With the onset of seasonal rains, they grow quickly, reaching a height of 2 to 3 m. During the dry season, grasses die back and frequently burn, but their underground parts remain alive and resprout when water again becomes available. Shrubby trees outcompete grasses in moist, low-lying areas or on rocky ground, but periodic fires and grazing mammals eliminate most trees as seedlings.

The largest savannas stretch across eastern and southern Africa; smaller patches occur in India, Australia, and South America. African savannas are home to large herbivorous mammals, including antelopes, zebras, giraffes, and elephants, some of which fall prey to savanna predators, such as lions, leopards, cheetahs, and wild dogs. Grazing mammals follow the seasonal cycle of grasses, migrating away from dry areas to greener pastures.

Thorn forests grow at the arid borders of true savanna, where large mammals are less abundant. Grasses and other plants that store energy in large underground root systems grow among scrubby trees. Thorn forests are also highly seasonal, growing dramatically in the rainy season and dying back during the annual dry season, which may last for 8 to 9 months.

Deserts Develop Where Little Precipitation Falls

Deserts form where rainfall averages less than 25 cm per year. The hot deserts of the American Southwest, northern Chile, Australia, northern and southern Africa, and Arabia occur near 30° latitude, where descending air masses create very dry conditions.

FIGURE 49.15
Desert. The warm Sonoran Desert in Saguaro National Park near Tucson, Arizona, is home to columnar saguaro cacti *(Carnegiea gigantea)* and other drought-adapted plants.

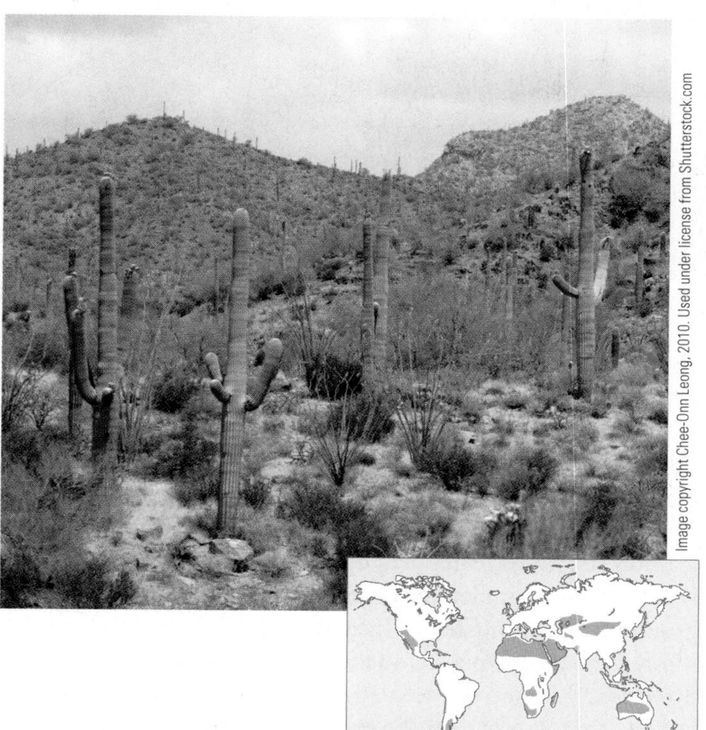

Image copyright Chee-Onn Leong, 2010. Used under license from Shutterstock.com

Cool deserts, such as the Gobi and Kyzyl-Kum of Asia and the Great Basin of North America, form in massive rain shadows at higher latitudes.

Desert conditions are often extreme. Rainfall arrives infrequently in heavy, brief pulses; and sudden runoff erodes topsoil, which often has high mineral content but little organic matter. Dry air and scant cloud cover allow most sunlight to reach the ground, raising daytime air and ground temperatures as high as 45°C and 70°C, respectively. At night, the surface loses heat quickly; in some deserts, temperatures drop below freezing in winter.

Desert vegetation is always sparse because arid environments do not favor large, leafy plants. Some deserts, such as the Namib of Africa and the Atacama-Sechura of South America, receive so little rainfall that large areas are practically devoid of vegetation. By contrast, the hot Sonoran Desert in northern Mexico, southeastern California, and southern Arizona harbors a diverse flora, including deep-rooted shrubs and shallow-rooted cacti **(Figure 49.15).** Mesquite and cottonwood trees grow long taproots into the permanent water supply below streambeds. Perennial plants often protect their tissues from herbivores with spines or toxic chemicals, and many use CAM photosynthesis to conserve water (see Section 9.4). After seasonal rains, annual plants germinate, mature, flower, and produce seeds before brutally dry conditions resume.

Deserts also support abundant animals, most of them fairly small. Ants, birds, and rodents often subsist on seeds. Some seed-eating mammals survive on the water they extract from food. Insects, some lizards, and mammals consume the sparse vegetation. Scorpions, lizards, and birds feed primarily on insects; and snakes, owls, and foxes prey on other animals. Most desert animals avoid the midday heat and dehydrating conditions; many retreat into underground burrows, where water vapor from their respiration cools and moistens the air. Many species are nocturnal or active only in the early morning and late afternoon.

Chaparral Grows Where Winters Are Cool and Wet and Summers Are Hot and Dry

A scrubby mix of short trees and low shrubs called **chaparral** dominates narrow sections of coastal land between 30° and 40° latitude, where winters are cool and wet and summers hot and dry. Seasonal rainfall averages only 25 to 60 cm per year. Chaparral occurs in central and southern California, central Chile, southwestern Australia, southern Africa, and the Mediterranean region.

Chaparral shrubs are dense, with hard, tough, evergreen leaves **(Figure 49.16).** They grow woody stems above ground and large root systems in the soil. Many species, such as sages (genus *Salvia*), produce toxic, aromatic compounds that inhibit the germination and growth of other plants. Just after the winter rains, the shrubs are covered with new leaves and flowers, and the vegetation teems with insects and breeding birds. During the hot, dry summers, however, most plants are dormant, and lightning sparks frequent fires. The aromatic oils and resins of many species, such as eucalyptus, make them highly flammable. Their aboveground parts burn swiftly, but they quickly resprout from large root crowns. Other species release seeds from fire-resistant cones or pods, and their seedlings grow in ash-enriched soil.

John C. Cunningham/Visuals Unlimited

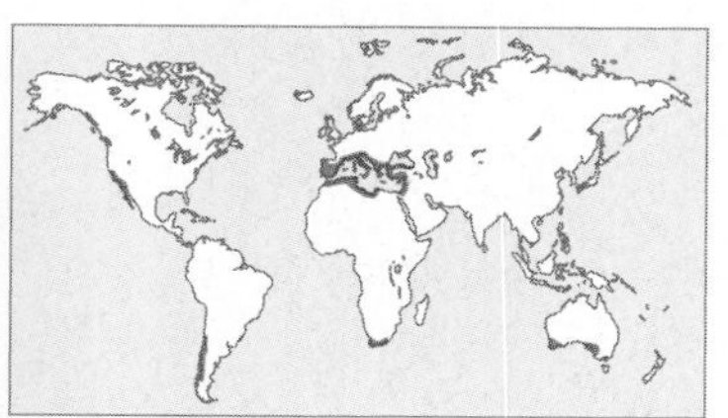

FIGURE 49.16
Chaparral. Chaparral covers broad expanses of hills in coastal areas of central and southern California.

Temperate Grasslands Are Subject to Periodic Disturbance

Temperate grasslands include the prairies of North America, the steppes of central Asia, the pampas of South America, and the veldt of southern Africa. They stretch across the interiors of

A. **Shortgrass prairie**

iStockphoto.com/YinYang

B. **Tallgrass prairie**

© Ray Wagner/Save the Tall Grass Prairie, Inc.

FIGURE 49.17

Temperate grassland. (A) The western plains of North America were once covered with shortgrass prairie, as shown in Custer State Park, South Dakota. Bison *(Bison bison)* were the dominant large herbivores. **(B)** Tallgrass prairie, like this lush patch in eastern Kansas, once covered the eastern plains.

continents, where winters are cold and snowy and summers are warm and fairly dry. Only 25 to 100 cm of rain falls unevenly through the year. Temperate grasslands are in a near-constant state of flux: seasonal drought, periodic fires, and grazing by mammals destroy the seedlings of shrubs and trees, preventing them from displacing perennial grasses and herbaceous plants (see Section 51.6). Grassland soil is rich in organic matter because the aboveground parts of most plants die and decompose annually.

In North America shortgrass prairie **(Figure 49.17A)** covers much of the west, where winds are strong, rainfall light and infrequent, and evaporation rapid. Drought-tolerant perennials have deep roots, and their underground rhizomes, which store energy, resprout quickly after a fire. Tallgrass prairie **(Figure 49.17B)** once occupied moister regions to the east of the shortgrass prairie. It boasted an abundance of legumes and sunflowers, often 3 m tall, but most of it was converted to farmland long ago; small patches still exist in nature preserves and in glades within eastern deciduous forests.

North American grasslands are still occupied by large grazing mammals, including pronghorns and bison, which once numbered in the millions. The most familiar burrowing mammal is the prairie dog, a rodent, but pocket gophers, ground squirrels, and jackrabbits are also common. Wolves were the primary large predators until they were hunted nearly to extinction. Coyotes, foxes, ferrets, hawks, and owls still take small prey today.

Temperate Deciduous Forests Experience Seasonal Dormancy

At temperate latitudes, with warm summers, cold winters, and annual precipitation between 75 and 250 cm, **temperate deciduous forests** grow at low to middle altitudes. In winter, low temperatures reduce photosynthetic rates, and snow and ice can damage leaves. Thus, most plants shed their leaves and grow new ones in spring **(Figure 49.18).** The thick layer of leaf litter, which releases mineral nutrients as it decomposes, enriches the soil. Decomposition is slow, however, because the growing season is only about 7 months long.

Temperate deciduous forests harbor fewer species than tropical forests. Trees form a canopy 10 to 35 m high, and woody shrubs form an understory below it. Herbaceous plants and a ground layer of mosses or liverworts grow below the shrubs. Many herbaceous plants, including some terrestrial orchids, flower early in spring, before trees produce sunlight-blocking leaves; others flower near the end of the growing season.

Forests of ash, beech, birch, chestnut, elm, and oak stretched unbroken across eastern North America, Europe, and eastern Asia before farmers cleared the land. In North America, introduced diseases and insects have nearly eliminated the once dominant species, such as American chestnut and American elm. Today, beech, birch, and maple predominate in the Northeast;

Summer

Thomas E. Hemmerly

Winter

Thomas E. Hemmerly

FIGURE 49.18

Temperate deciduous forest. Seasonal variations in temperature and water change the character of this forest south of Nashville, Tennessee.

iStockphoto.com/Grant Dougall

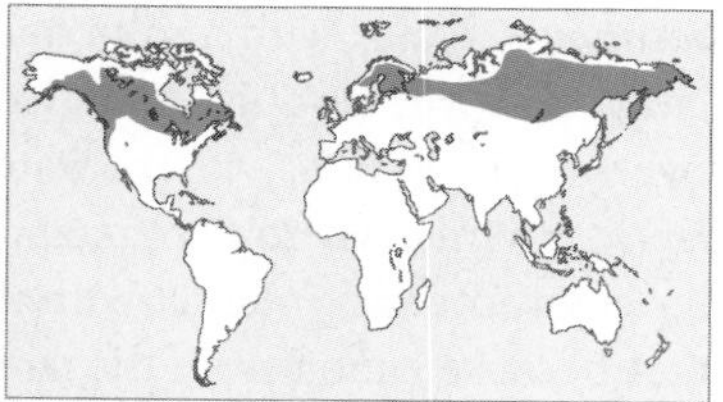

FIGURE 49.19

Boreal forest. Single-species stands of tall conifers dominate this boreal forest, the predominant forest at high latitudes in the Northern Hemisphere.

oak-hickory forests dominate farther south and west; and oak woodlands merge into tallgrass prairie to the west. Before the arrival of Europeans, deer, bison, bears, and pumas roamed the forests with many smaller species of animals. Today, small mammals such as voles, mice, chipmunks, squirrels, rabbits, opossums, and raccoons predominate, although deer and bears have recently surged in abundance.

Evergreen Coniferous Forests Predominate at High Northern Latitudes

The **boreal forest,** or **taiga** (Russian for "swamp forest"), is a circumpolar expanse of evergreen coniferous trees in Europe, Asia, and North America **(Figure 49.19).** Snow blankets the ground during long and extremely cold winters, but most precipitation falls during the short summer. In the northernmost taiga, plants grow quickly during long (18-hour) summer days.

Stands of white spruce and balsam fir dominate North America's boreal forest. Their needle-shaped leaves have a thick cuticle and recessed stomata that conserve water during winter, when groundwater is frozen. Fallen needles acidify the thin soil, which speeds the leaching of most nutrients, and few shrubs and herbaceous plants grow beneath the conifers. Lightning-sparked fires are common; some deciduous trees grow in areas opened by fire, but conifers eventually replace them. Cold streams, marshes, ponds, and lakes often dot the landscape; at flat, poorly drained sites, peat mosses, shrubs, and stunted trees dominate acidic bogs, called muskegs.

Most taiga is relatively undisturbed by humans, and it still harbors its native animals. Moose, elk, and deer are the dominant large herbivores. Hare as well as squirrels, porcupines, and other rodents also feed on plants. Some small animals are active all winter in runways they dig beneath the snow. Wolves, lynx, and wolverines prey on herbivores. Grizzly bears and black bears roam the forest, devouring seeds, berries, fishes, and small animals. Mosquitoes, black flies, and gnats are superabundant near bogs and lakes in summer.

Other types of coniferous forests grow in more southerly coastal lowlands where winters are mild and wet and the summers are cool. For example, a **temperate rain forest,** supported by heavy rain and fog, parallels the North American coast from Alaska into northern California. In western Washington State, the rain forest on the Olympic Peninsula receives 500 cm of rainfall per year, as much as some tropical forests. This temperate rain forest harbors some of the world's tallest trees, including Douglas fir and Sitka spruce to the north and coast redwoods to the south.

Tundra Comprises a Vast, Treeless Plain in the Northernmost Habitats

The treeless **arctic tundra** stretches from the boreal forests to the polar ice cap in Europe, Asia, and North America. Covering almost 5% of the land, this biome is windswept and wet. Winter temperatures are consistently below freezing. The 2-month summer is so cool that only the topmost layer of soil ever thaws, leaving the ground below perpetually frozen; in some areas, this **permafrost** is more than 500 m thick. Although less than 25 cm of precipitation falls each year, evaporation is slow, and permafrost is impermeable; thus, low-lying soil remains permanently waterlogged, forming bogs **(Figure 49.20A).** Anaerobic conditions and low temperatures retard decomposition, and soggy masses of detritus accumulate.

A. **Arctic tundra**

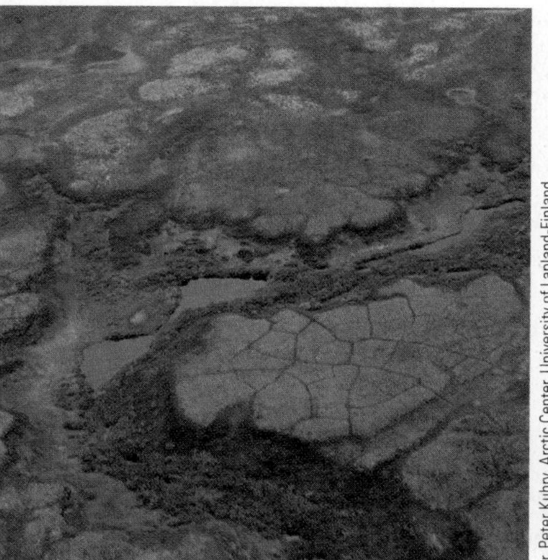

Dr. Peter Kuhry, Arctic Center, University of Lapland-Finland

B. **Alpine tundra**

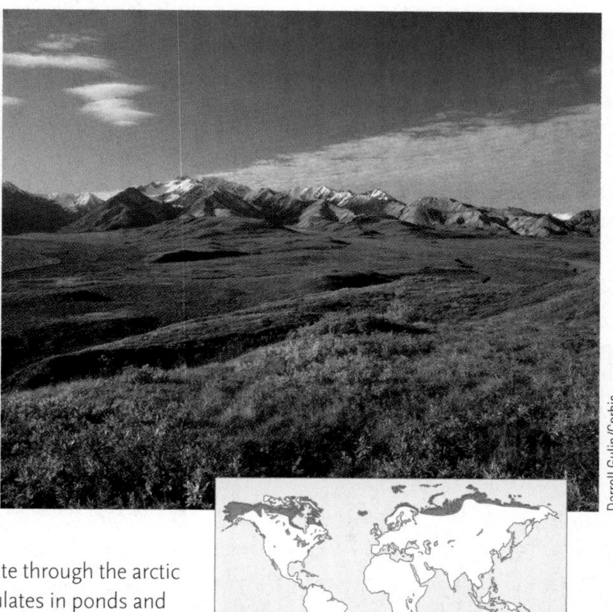

Darrell Gulin/Corbis

FIGURE 49.20

Tundra. (A) Rain and snowmelt cannot percolate through the arctic tundra's permafrost. In summer water accumulates in ponds and bogs, as shown in this aerial photograph of the tundra in northern Russia. **(B)** Compact, short plants form the alpine tundra, which occurs on mountaintops at temperate latitudes, such as those in Alaska's Denali National Park.

Plants in the tundra are short because the weak sunlight and minimal growing season provide barely enough energy and warmth for photosynthetic activity; moreover, strong winter winds shred any plants with a high profile. The vegetation consists of low-growing lichens, mosses, grasses, perennial herbs, dwarf shrubs, and a few stunted trees, usually less than 1 m tall. During summer's nearly continuous sunlight, plants flower profusely, and their fruits ripen fast.

Some animals, including herbivorous arctic hares, lemmings, and willow ptarmigans as well as predatory snowy owls, wolves, foxes, and lynx, are permanent tundra residents. In summer, herds of herbivorous musk oxen, caribou, and reindeer migrate there from boreal forests, and migratory shorebirds and waterfowl arrive to breed. Flying insects abound in summer, especially mosquitoes and black flies, which reproduce in boggy habitats.

A similar biome, called **alpine tundra,** occurs on high mountaintops throughout the world **(Figure 49.20B).** Dominant plants form cushions and mats that withstand the buffeting of strong winds. Winter temperatures are well below freezing, and shaded patches of snow persist even in summer. The thin, fast-draining soil is nutrient-poor, and photosynthetic activity is low.

STUDY BREAK 49.4 <

1. Which terrestrial biomes occur in habitats that receive the most rainfall?
2. Which terrestrial biomes are renewed by periodic fires?
3. Which terrestrial biomes have the tallest vegetation? Which ones have the shortest?
4. In which terrestrial biomes are the trees usually evergreen?

>

THINK OUTSIDE THE BOOK

What biome surrounds the community where you live? If you live in a city, how might you define an urban biome? How far would you need to travel to be in a different biome?

49.5 Freshwater Environments

Biomes are traditionally defined for terrestrial environments. Their aquatic counterparts comprise several distinctive habitats in either freshwater or marine environments. In freshwater environments, water with a salt concentration below 0.5% accumulates or moves through a landscape. Ecologists distinguish between *lotic* systems, where water flows through channels, and *lentic* systems, where water stands in an open basin. All freshwater environments interact with the surrounding terrestrial habitats, because runoff from the land carries a nearly constant input of nutrients. Highly productive **wetlands** often occur at the borders of freshwater environments. These marshes and swamps may harbor an astounding array of microorganisms, algae, plants, invertebrates, and vertebrates.

Streams and Rivers Carry Water Downhill to a Lake or Sea

Flowing-water environments start as seeps on high ground. As the water flows downhill, it collects into narrow streams, which merge to form wide rivers **(Figure 49.21).** Streams and rivers include three habitats. *Riffles* are shallow, fast-moving, turbulent stretches over a rough bottom of pebbles or rocks. *Pools* are deep, slow-moving areas with a smooth sand or mud bottom. *Runs* are deep, fast-moving stretches over smooth bedrock or sand. Streams generally have high flow rate, low volume, and lots of riffles and pools. As they merge into rivers, flow rate declines, but flow volume increases, and runs and pools predominate. Flow rate and volume also vary seasonally with the rate of water input from rainfall and snowmelt and geographically with altitude and topography.

Physical factors change over the length of a flowing-water system. The concentration of suspended particulate material is low in streams, but high in rivers, which are often turbid with silt. Temperature also rises as water flows downstream to warmer lowland habitats. Because oxygen is more soluble in cold water than in warm water, dissolved oxygen is usually higher in streams than in rivers. Erosion of the streambed and surrounding land provides the solute content of flowing water. Today, agricultural runoff and industrial and municipal wastes provide major input. In unpolluted streams, organic detritus provides more than 95% of the nutrients and energy entering aquatic food webs. This input is particularly important in streams flowing through dense forests, where vegetation blocks the sunlight necessary for photosynthesis.

FIGURE 49.21
Stream and river habitats. **(A)** In streams, such as this one in Virginia, water flows quickly through narrow channels, often with a rocky bottom. **(B)** In rivers, like the Rio Napo in Ecuador, water flows more slowly through broad channels, and suspended sediments often make the water murky.

A. **A stream**

E. F. Benfield, Virginia Tech

B. **A river**

iStockphoto.com/Steven Miric

FIGURE 49.22
Lakes. Lago Di Piani, a lake in the Dolomites region of Italy.

The flow of water affects every aspect of life in streams and rivers. In swift-moving riffles, primary producers cling permanently to fixed substrates, because phytoplankton are swept away by the current. Insect larvae and other invertebrates attach to the undersides of rocks; many species have a flattened shape to maintain a low profile in the current. By contrast, large rivers have dense populations of algae and cyanobacteria, which attach to rocks and other substrates, and rooted aquatic plants at the river's edge.

Lakes Are Bodies of Standing Water That Accumulates in Basins

Lakes and other standing-water biomes are generally fed by rainfall and by streams and rivers that carry water from surrounding lands **(Figure 49.22).** Because the availability of light affects photosynthesis by a lake's phytoplankton and plants, ecologists often distinguish between the **photic zone** of a lake, the surface water that sunlight penetrates, and the deeper **aphotic zone,** which is darker.

LAKE ZONATION Every lake includes zones, defined by depth and distance from the shore, that provide distinctive environments **(Figure 49.23).** In the **littoral zone,** the shallow water near the shore, sunlight penetrates to the bottom. Enriched by nutrients made available by decomposers and runoff, the littoral zone has high photosynthetic activity and is occupied by many species. Rooted aquatic plants, such as cattails and water lilies, grow above the surface, and "floating aquatics," such as duckweed, are common. Submerged vegetation harbors a rich community of microorganisms, epiphytes, and invertebrates. Numerous animals—insects, worms, snails, crayfish, fishes, frogs, turtles, and water birds—use the littoral zone to feed and reproduce.

The **limnetic zone,** the sunlit water beyond the littoral, supports plankton communities: the primary photosynthesizers are phytoplankton—cyanobacteria, diatoms, and green algae; they are eaten by zooplankton—rotifers, copepods, and other tiny heterotrophs. Small fishes, which feed on plankton, are themselves consumed by larger fishes, such as bass.

Photosynthesis is impossible, however, in the **profundal zone,** the perpetually dark water below the limnetic zone. Nevertheless, a constant rain of detritus from the limnetic zone supports a community of bacterial decomposers and animals that feed on dead or dying material, including worms, clams, insect larvae, and catfish.

SEASONAL CHANGES IN TEMPERATE LAKES In temperate areas, seasonal temperature variations induce changes in the vertical zonation of lakes **(Figure 49.24).** Like other liquids, water gets denser as it cools. But water has a unique property: it reaches maximum density at 4°C, with the density declining as it gets colder. Thus, water at 4°C sinks below water that is either warmer or colder; ice floats because it is less dense than very cold water.

During winter, ice forms on the surface of temperate zone lakes. Water temperature varies from near freezing just below the ice to 4°C at the bottom. Differences in the density of water at 0° and 4°C maintain this thermal stratification. In spring, as the ice melts, the warmer, denser water sinks; and the surface temperature gradually rises to 4°C. For a brief time, the temperature is uniform at all depths. Winds blowing across the lake create vertical currents that cause a **spring overturn,** mixing surface water with deep water. Oxygen at the surface moves to the bottom, and nutrients from the bottom move to the surface.

By midsummer, sunlight heats the top layer of the limnetic zone, called the **epilimnion,** to temperatures above 4°C. In large lakes, the epilimnion may be more than 10 m deep. In the deep water of the lake's profundal zone, called the **hypolimnion,** the temperature remains near 4°C. However, at the boundary between the epilimnion and the hypolimnion, water temperature changes abruptly over a narrow depth range, called the **thermocline.** The thermocline prevents vertical mixing because warm surface water floats above the thermocline, and cool deep water stays below it. During summer, nutrient-rich detritus sinks to the bottom of the lake, where decomposition depletes the oxygen dissolved in the hypolimnion. In autumn, declining sunlight and winds cause the epilimnion to

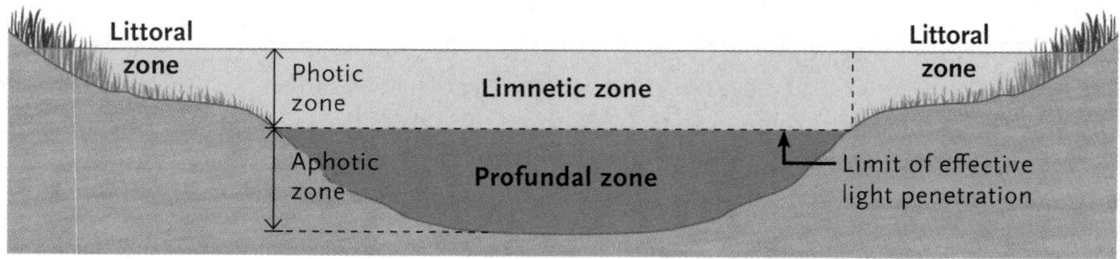
Littoral zone
Photic zone
Limnetic zone
Littoral zone
Aphotic zone
Profundal zone
Limit of effective light penetration

FIGURE 49.23
Lake zonation. The zonation in a lake is based on the water's depth and its distance from shore.

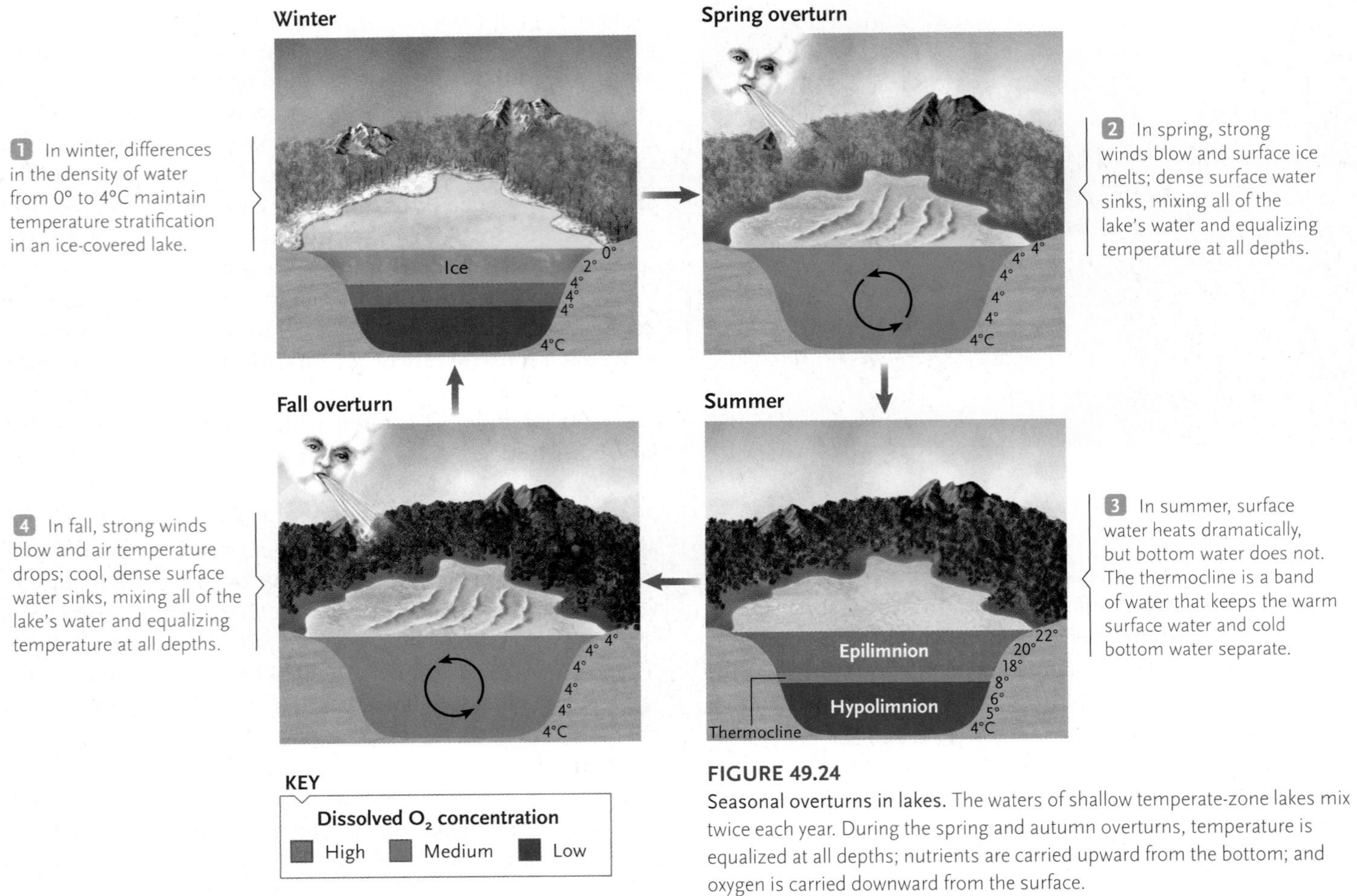
Winter
Spring overturn
1 In winter, differences in the density of water from 0° to 4°C maintain temperature stratification in an ice-covered lake.
Ice
0°
2°
4°
4°
4°
4°C
2 In spring, strong winds blow and surface ice melts; dense surface water sinks, mixing all of the lake's water and equalizing temperature at all depths.
4°
4°
4°
4°
4°
4°C
Fall overturn
Summer
4 In fall, strong winds blow and air temperature drops; cool, dense surface water sinks, mixing all of the lake's water and equalizing temperature at all depths.
4°
4°
4°
4°
4°
4°C
3 In summer, surface water heats dramatically, but bottom water does not. The thermocline is a band of water that keeps the warm surface water and cold bottom water separate.
Epilimnion
Hypolimnion
Thermocline
22°
20°
18°
8°
6°
5°
4°C
KEY
Dissolved O₂ concentration
High
Medium
Low

FIGURE 49.24
Seasonal overturns in lakes. The waters of shallow temperate-zone lakes mix twice each year. During the spring and autumn overturns, temperature is equalized at all depths; nutrients are carried upward from the bottom; and oxygen is carried downward from the surface.

cool, and as the water becomes denser, it sinks, eliminating the thermocline. Winds then mix the water vertically once again during an **autumn overturn,** and dissolved gases and nutrients are equalized at all depths.

Photosynthetic activity in the limnetic zone varies with the seasonal overturns. In spring, increased sunlight, warm temperatures, and the sudden availability of nutrients induce a bloom of photosynthesis and growth. As the season progresses and the thermocline prevents vertical mixing, nutrient levels dwindle in the epilimnion, and photosynthetic activity declines. By late summer, nutrient shortages limit photosynthesis. After the autumn overturn, nutrient cycling drives a short burst of productivity. But as days get shorter and temperature declines, productivity remains low until spring.

TROPHIC NATURE OF LAKES Ecologists classify lakes by their nutrient content and rates of photosynthetic activity. **Oligotrophic lakes** are poor in nutrients and organic matter, but rich in oxygen. Their low productivity keeps the water crystal clear, making them popular recreational sites. By contrast, **eutrophic lakes** are rich in nutrients and organic matter. The decomposition of organic matter depletes oxygen in the hypolimnion when the lake is stratified, and high productivity in the epilimnion often chokes the water with seasonal blooms of cyanobacteria and filamentous algae. Eutrophic lakes are often thick and "soupy," making them unattractive for recreation. Over long periods of time, as sediments accumulate, lakes naturally change from oligotrophic to eutrophic; their basins eventually fill with sediments, and terrestrial plants invade.

The addition of nutrients to a lake often disrupts its trophic condition (see *Why It Matters* for Chapter 52). In a classic experiment conducted in the late 1960s, David Schindler and his colleagues at The Experimental Lakes Project in Ontario, Canada, experimentally separated the two basins of a lake with a plastic curtain. The researchers added phosphates to one basin and used the other basin as a control. Within 2 months, the artificially enriched basin sported a bloom of cyanobacteria, a sign of eutrophication; the control basin remained oligotrophic and crystal clear **(Figure 49.25).**

STUDY BREAK 49.5 <

1. **How does the availability of dissolved oxygen vary from the headwaters of a stream to the mouth of a river?**
2. **What factors cause the seasonal overturns in lakes?**
3. **Why are oligotrophic lakes better for recreational purposes than eutrophic lakes?**

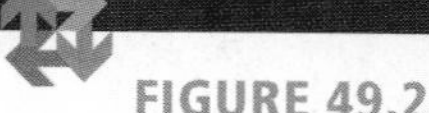

FIGURE 49.25 Experimental Research

What Causes Lake Eutrophication?

Question: Does the addition of excess phosphorus to a lake encourage the growth of phytoplankton, such as cyanobacteria?

Experiment: Schindler and his colleagues experimentally separated the two basins of a lake in Ontario, Canada, with a plastic curtain. The researchers added phosphates to one basin and used the other basin as a control.

Results: Within two months, the artificially enriched basin (in the upper left of the photo) sported a pale green bloom of cyanobacteria, a sure sign of eutrophication; the control basin remained oligotrophic and crystal clear.

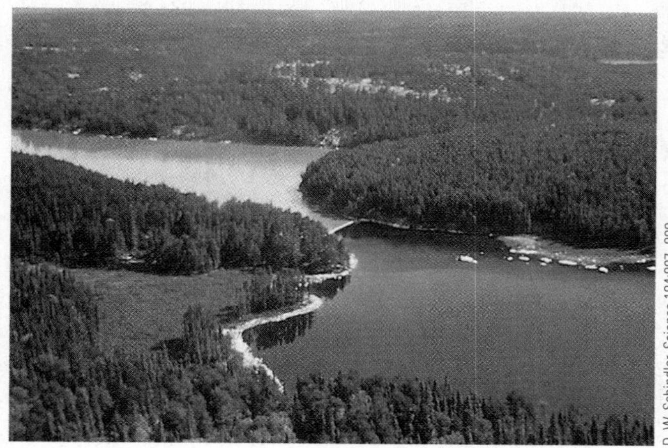
D. W. Schindler, Science 184:897–899

Conclusion: The addition of excess phosphorus to a lake encourages blooms of cyanobacteria, causing the lake to change from oligotrophic to eutrophic.

Source: D. W. Schindler. 1974. Eutrophication and recovery in experimental lakes: Implications for lake management. *Science* 184:897–899.

49.6 Marine Environments

Marine environments, in which salinity (salt concentration) averages about 3%, cover nearly three-fourths of Earth's surface and account for a large fraction of its photosynthetic activity. They also mediate important global processes: marine phytoplankton process large amounts of carbon dioxide, generating oxygen and moderating a major cause of global climate change (see Chapter 52).

As with standing freshwater environments, depth and distance from shore govern the physical characteristics of marine habitats. Ecologists describe ocean zonation in several ways **(Figure 49.26)**, including the distinction between the photic and aphotic zones. Another major distinction is between the **pelagic province,** the water, and the **benthic province,** the bottom sediments. The pelagic province includes the **neritic zone,** the shallow water above the continental shelves, and the **oceanic zone,** the deep water beyond them. The benthic province is divided into the **intertidal zone,** the shoreline that is alternately submerged and exposed by tides, and the **abyssal zone,** the bottom sediments that lie permanently below deeper water. Here we describe five marine environments—*estuaries, rocky and sandy coasts, continental shelves and oceanic banks, open ocean,* and *benthic regions*—that represent particular associations of organisms occupying different marine zones and provinces.

Estuaries Form Where Rivers Meet the Sea

Estuaries are coastal regions where seawater mixes with fresh water from rivers, streams, and runoff **(Figure 49.27).** Salinity is low where fresh water enters the estuary and high on the tidal side. After heavy rainfall, fresh water floods into the habitat, reducing salinity and raising water temperature. At high tide, cold, salty water flows in from the sea. All estuarine organisms must tolerate these variable conditions.

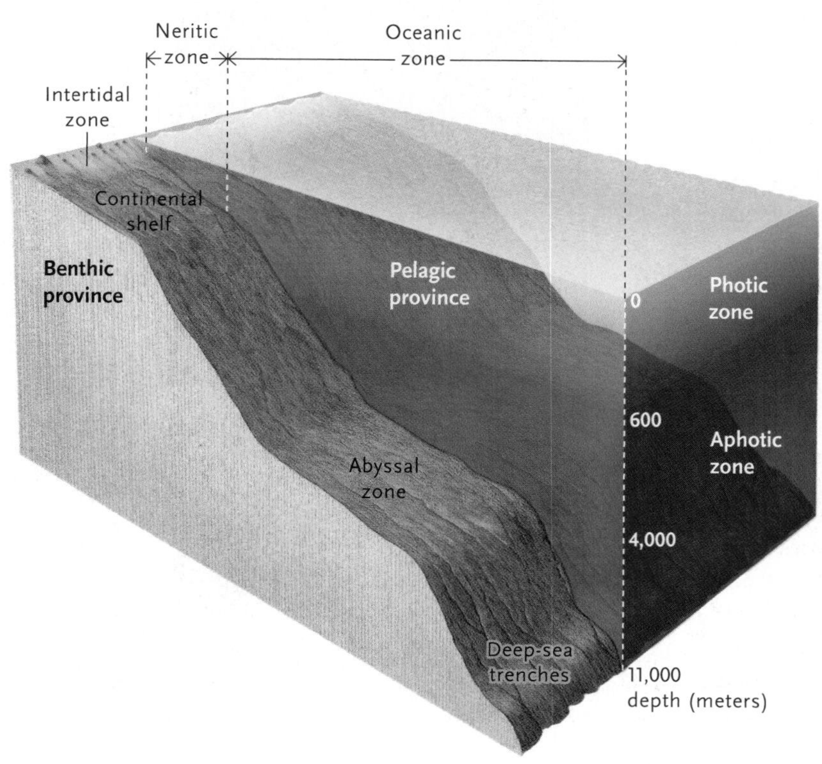
Neritic zone
Oceanic zone
Intertidal zone
Continental shelf
Benthic province
Pelagic province
Photic zone
Aphotic zone
Abyssal zone
Deep-sea trenches
0
600
4,000
11,000
depth (meters)

FIGURE 49.26
Oceanic zonation. Ecologists divide the ocean into the pelagic province (the water) and the benthic province (the ocean bottom). Zones are defined according to the depth of water (photic versus aphotic zones) and distance from shore (neritic versus oceanic zones in the pelagic province, intertidal versus abyssal zones in the benthic province). The different zones are not drawn to scale.

A. **Salt marsh grasses**

B. **Mangroves**

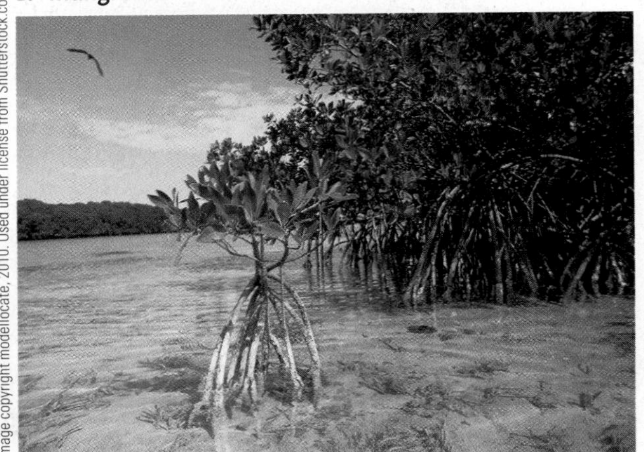

Image copyright modellocate, 2010. Used under license from Shutterstock.com

Douglas Peebles/Corbis

FIGURE 49.27
Estuaries. **(A)** The salt marsh grass (*Spartina* species) is the most common plant in a South Carolina estuary. **(B)** Red mangroves *(Rhizophora mangle)* are abundant in The Florida Keys.

Variations in local topography influence an estuary's physical features. Chesapeake Bay in Maryland, Mobile Bay in Alabama, and San Francisco Bay in California are broad, shallow estuaries. The estuaries in Alaska and British Columbia are narrow and deep, as are Norway's fjords. Many estuaries are bordered by **salt marshes,** tidal wetlands dominated by emergent grasses and reeds (Figure 49.27A). In tropical estuaries, the roots of densely packed mangrove trees penetrate the muddy bottom, accumulating sediments and slowly adding land to the shoreline (Figure 49.27B).

The constant input of nutrients and removal of wastes by the tides contribute to exceptionally high productivity in estuaries. The most common photosynthetic organisms include phytoplankton, salt-tolerant grasses and reeds that can withstand submergence at high tide, and algae that grow in mud and on plant surfaces. Roots and stems trap organic matter, which decomposes. The detritus (and bacteria clinging to it) supports nematodes, snails, crabs, and fishes; suspension-feeding mollusks and arthropods capture edible particles in the slowly moving water. Many marine arthropods and fishes breed in calm, shallow estuaries, where their young find abundant food and refuge from predators in the complex vegetation. Migratory birds use estuaries as rest stops, and shore birds and waterfowl use their muddy bottoms as rich feeding grounds, particularly at low tide.

Rocky and Sandy Coasts Experience Cyclic Periods of Exposure and Submergence

The intertidal zone, the area between low and high tide marks, is one of the most stressful habitats on Earth. On rocky shores, residents are battered by waves and floating debris. Sessile species, such as mussels and barnacles, attach to substrates with special structures or cement. Motile species, such as limpets and sea stars, simply hang on to rocks. Organisms that live high on the shore dry out at low tide, freeze in winter, and bake in summer. Exposed animals often seal themselves inside shells, and intertidal algae have thick polysaccharide coats that adsorb water and prevent dehydration.

Biotic interactions also take their toll. Organisms throughout the intertidal zone compete for attachment sites to avoid being washed away (see Figure 51.12). At low tide, predatory birds and mammals attack from above; at high tide, predatory fishes move in from the sea. Because the tides often scour detritus from the rocky intertidal, communities on rocky shores are largely supported by the photosynthetic algae and phytoplankton.

Rocky shores often have three zones **(Figure 49.28).** The *upper intertidal* is submerged only during the highest tide of the lunar

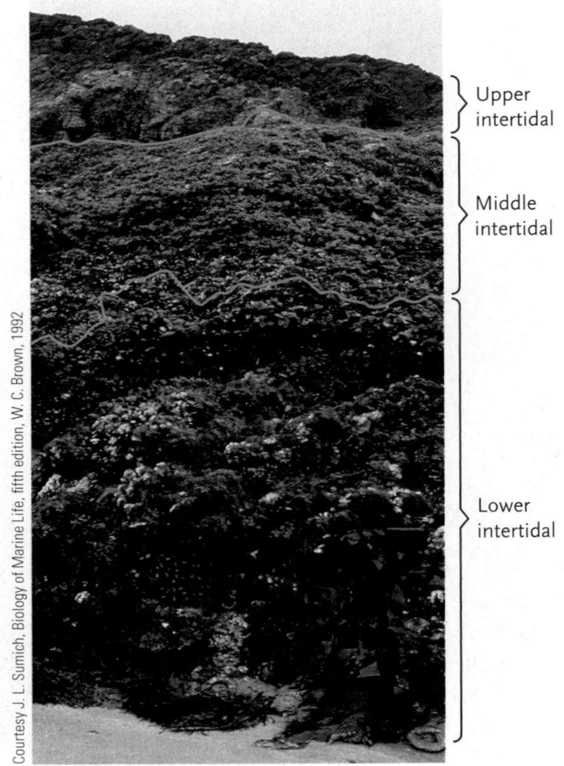
Upper intertidal
Middle intertidal
Lower intertidal

Courtesy J. L. Sumich, Biology of Marine Life, fifth edition, W. C. Brown, 1992

FIGURE 49.28
Vertical zonation in the intertidal. A rocky shore in the Pacific Northwest clearly exhibits vertical zonation. The distance between low and high tide marks on this rocky shore is about 3 m.

cycle. It is sparsely populated by barnacles, sturdy algae, and grazing and predatory snails. The *middle intertidal* is submerged daily during the highest regular tide and exposed during the lowest. Its tide pools are occupied by red, brown, and green algae, grazing and predatory mollusks, sponges, sea anemones, worms from several phyla, hermit crabs, echinoderms, and small fishes. Biodiversity is greatest in the *lower intertidal,* which is exposed only during the lowest tide of the lunar cycle. It is occupied by dense beds of algae, tunicates, echinoderms, other invertebrates, and fishes.

Sandy shores are composed of loose sediments that waves and currents constantly rearrange. Large plants and algae cannot grow on such unstable substrates, but organic debris imported from off shore or from nearby land supports animals that feed on detritus. Animals live in burrows, which they must frequently repair as the substrate shifts. Crabs and shorebirds live as scavengers or predators above the high tide mark. At night, beach hoppers and ghost crabs leave their burrows, seeking food. Marine worms, clams, crabs, and other invertebrates live in the sand between the high and low tide marks.

Light Penetrates the Shallow Water over Continental Shelves and Oceanic Banks

The neritic zone includes the shallow water over continental shelves and oceanic banks, underwater landmasses that rise to within 300 m of the surface. Although small in area, the neritic zone is highly productive and species-rich **(Figure 49.29).** Runoff from the land brings a steady inflow of nutrients, and upwelling (that is, vertical currents) and waves circulate nutrients from the bottom to the photic zone.

In temperate regions, giant kelp forests, which are among the most productive ecosystems, occupy some continental shelves and banks (Figure 49.29A). Kelp are enormous algae that attach to the bottom with giant holdfasts; their stipes ("stems") reach upward with fronds fanning out into the water. Sea anemones, snails, echinoderms, lobsters, and other invertebrates live in the kelp, where fishes and other predators consume them. Even where kelp does not grow, continental shelves and banks teem with life. Most of the important fisheries in the temperate zone occur there.

In the tropics, the warm but nutrient-poor water above continental shelves is often occupied by **coral reefs** (Figure 49.29B). Sunlight penetrates the clear water all the way to the bottom. Photosynthetic dinoflagellates, living as endosymbionts of the coral animals (see Section 26.2), and coralline algae are largely responsible for the photosynthesis that takes place there. Coral animals also feed on microscopic organisms and suspended particles. The reefs are the remains of corals, algae, and other organisms, and their structural complexity rivals that of tropical rain forests. Tides and currents carve ledges and caverns. Storms frequently disturb the reefs, creating openings in which new coral colonies can grow (see Section 51.5). A reef may be festooned with as many as 750 species of corals and a dizzying variety of algae. The diversity of coral skeletons provides a complex structure that is used by invertebrates from nearly every phylum and by a host of herbivorous and carnivorous fishes.

In the Open Ocean, Photosynthesis Occurs Only in the Sunlit Upper Layers

The oceanic zone lies beyond the continental shelves. Though generally low in nutrients, it is locally enriched by runoff from land and by upwelling bottom waters. The open ocean is typically cold, except in the tropics. The surface water is illuminated by

A. **Kelp forest**

B. **Coral reef**

FIGURE 49.29
Neritic zone. (A) Kelp forests, such as this one off the coast of California, often grow in the neritic zone along the coast at temperate latitudes. **(B)** This coral reef in Indonesia illustrates the structural complexity and biological diversity found in reef communities.

iStockphoto.com/Rod Kaye

FIGURE 49.30
Open ocean. A humpback whale *(Megaptera novaeangliae)* breaches (leaps out of the water).

sunlight, which warms it somewhat and allows photosynthesis. Most photosynthesis is undertaken in the top 50 m, however, because seawater filters light. Photosynthetic activity varies seasonally, as it does on land.

"Pastures" of phytoplankton are eaten by zooplankton, including copepods, shrimplike krill, small worms, cnidarians, and the larvae of invertebrates and fishes. Consumers that can actively swim against the currents, such as squids, fishes, marine turtles, and whales, are called **nekton (Figure 49.30).** Some consumers feed on plankton, and some prey on other nekton. Low light levels in water between about 50 and 600 m allow little photosynthesis, but many fishes and some mobile invertebrates are active at these depths, traveling into the sunlit zone to feed on organisms near the surface.

No sunlight ever penetrates the deepest part of the oceanic zone, below 600 m. Some of these abyssal regions, such as the Marianas Trench, are more than 9 km below the surface. Scientists have explored the deepest water in the ocean only during the past few decades, but we know that it is a cold (2° to 3°C), dark environment, where organisms live under tremendous pressure from the ocean above. Abyssal communities are surprisingly diverse, although population densities tend to be low. The denizens of the abyssal zone include invertebrates, bony fishes, and sharks. Some fishes and invertebrates are bioluminescent, producing spots of light that may serve for communication or as lures to entice prey within reach of their large jaws **(Figure 49.31).**

FIGURE 49.31
Deep sea. A deep-sea anglerfish *(Himantolophus groenlandicus)* uses a bioluminescent lure to attract prey to its formidable jaws.

Peter David/Getty Images

The Benthic Province Includes the Rocks and Sediments of the Ocean Bottom

The benthic province extends from the intertidal zone to the deep-sea trenches. In the oceanic zone, bottom sediments are composed of soft mud, fine particles of silt, detritus, and the shells of dead microscopic organisms. Species living in and on the bottom are collectively called **benthos.** Sunlight never strikes the benthic province of the open ocean, which is inhabited by bacteria, fungi, and a variety of animals. Sessile invertebrates, such as sponges, sea anemones, and clams, live amidst the sediments, and many motile animals, including worms, mollusks, crustaceans, echinoderms, and fishes, feed on the organic remains that sink from pelagic communities.

In 1976, researchers found communities thriving near hydrothermal vents at a depth of 3,000 m near the Galápagos Rift, a volcanically active boundary between two crustal plates. Near-freezing water seeps into fissures where it is heated to temperatures of 350°C or higher. Pressure forces the heated water upward, and minerals are leached from porous rocks as the water spews out through vents in the seafloor. This hydrothermal outpouring releases hydrogen sulfide, which chemosynthetic bacteria use as an energy source to fix carbon dioxide, just as plants and algae use sunlight for photosynthesis.

Some of these bacteria live as endosymbionts of giant clams and tube-dwelling worms **(Figure 49.32).** Worms in the genus *Riftia* lack digestive systems, subsisting entirely on the organic compounds produced by their chemosynthetic partners, which are so numerous that they account for up to half of the worms' body weight. The worms' blood contains both hemoglobin and sulfide-binding proteins, which deliver oxygen and sulfides to the bacteria.

Deep-sea communities also include sea anemones, crustaceans, and fishes. Researchers have located hydrothermal vent ecosystems in the South Pacific, near Easter Island; the North

Robert Vrijenhoek/MBARI

FIGURE 49.32
Deep benthos. Giant tube-dwelling worms *(Riftia)* and their chemosynthetic bacterial endosymbionts are common in hydrothermal vent communities on the deep ocean floor.

Pacific, off the coast of British Columbia; the Gulf of California, about 150 miles south of the tip of Baja California, Mexico; and the Atlantic.

Recent research in the deepest reaches of the ocean reveals that communities also exist in areas far from hydrothermal vents. These "cold seep" communities thrive on broad expanses of the seafloor, where extremely salty water percolates upward from the underlying rocks and sediment, carrying abundant minerals, hydrogen sulfide, and methane to areas that are accessible to organisms. Chemosynthetic bacteria, which grow in large mats, can metabolize these molecules, serving as a food source for animal communities that include sponges, worms, and bivalve mollusks.

STUDY BREAK 49.6 <

1. **What is the difference between the benthic and pelagic provinces of the ocean?**
2. **Which marine environments experience the largest fluctuations in salinity (salt concentration) over time?**
3. **Which marine regions receive abundant energy input from sunlight?**
4. **What is the source of nutrients and energy for the benthos of the oceanic zone?**
5. **What organisms are responsible for the synthesis of organic compounds in hydrothermal vent and cold seep communities, and how do they differ from those in the photic zone?**

UNANSWERED QUESTIONS

How will biomes change in response to anthropogenic (human-induced) global warming?

Will biomes remain largely intact and simply shift their geographical distributions northward or upward (to higher altitudes)? Or will the biomes we recognize today become disrupted, and new biomes arise as species associate in different combinations? Parmesan and Yohe estimated that 59% of wild species around the world have already shown some change in their geographical distributions in response to the relatively small level of global warming—a 0.7°C rise in average temperature—over the past 100 years. Documented responses to global warming vary from species to species, however. For example, only 20% of butterfly species in Spain, France, and North Africa have shifted their southern range boundaries northward, but 70% of butterfly species in the United Kingdom and Scandinavia have expanded their northern range borders further northward, sometimes by as much as 300 km over the past 30 years.

Thus, although the distributions of some species appear stable, the ranges of others are showing strong responses to global warming. At least two hypotheses may explain these patterns. First, some species may be stressed by rising temperatures but have not yet shown a measurable response. Second, the geographical distributions of some species may not be governed primarily by climate. Whatever the reason, the fact that we observe large variation in the response of different species suggests that not all species in a community are moving together. Thus, the existing communities of birds, butterflies, and trees are being disrupted—with some species moving and others not. Biologists have also noted differences in the response of different taxonomic groups. Butterflies in Europe and North America seem to be shifting their distributions northward and upward at about the same rate that temperatures are changing, but plants appear to lag behind. Alpine herbs in Switzerland, for example, have shifted their distributions upward at about half the rate that one might expect from the rate of regional warming, and it was not until 30 years after warming began that tree seedlings in Sweden started to colonize alpine habitats at higher elevations, shifting the treeline upward.

Even bigger questions remain. Will the vegetation that currently lives in the tropics expand into what is now the temperate zone and cover more of the planet? Some studies suggest that tropical lowland trees are already at their physiological limit—already showing signs of stress by shutting down photosynthesis on the hottest and driest days. Furthermore, climate model projections from a 2007 report of the Intergovernmental Panel on Climate Change consistently show substantial drying as well as warming in midlatitudes. If this projection is correct, many plants and animals now living in the wet tropics will be unable to shift northward into what may become an extreme desert climate. And what will happen to the arctic tundra? Researchers have already collected strong evidence that shrubs and trees are encroaching northward into the tundra of Alaska and Canada. The permafrost is melting, and the soil is drying. How do these observed changes in plant and animal distributions relate to the future? Human activity has already caused Earth's mean annual temperature to rise by 0.7°C in the past 100 years. Climate model projections suggest that further increases between 1.8°C and 4.0°C are likely; some models suggest the rise will be over 6.0°C. Can the tundra biome survive even the lowest projections—more than twice the warming it has already experienced?

How will evolution shape the ways that wild species respond to climate change? Which groups of organisms are likely to adapt, and which are likely to become extinct?

Populations are evolving all the time in response to changing selection regimes. Global warming is one of many human-driven environmental changes that could foster genetic change. Biologists have known for decades that organisms are locally adapted to the climatic conditions under which they routinely live. Scientists have documented local genetic changes toward more warm-adapted genotypes in fruit flies, mosquitoes, and the algal symbionts of corals. Do the observed genetic changes suggest that these species are adapting to anthropogenic global warming? Will other species follow suit? The fossil record suggests that during the Pleistocene glaciations, when Earth's temperature shifted between glacial periods (4°–8°C colder than now) and interglacial periods (today's temperatures), very few species became extinct and few experienced substantial morphological evolution. But, before the Pleistocene, Earth was much hotter than it is today, and the atmosphere had higher levels of CO_2. During the transition from these very warm, high CO_2 conditions to the colder, low CO_2 conditions of the Pleistocene, a large proportion of species became extinct. To how much climate change can organisms adapt? At what point is climate change extreme enough that species come to the limit of their genetic variation, can no longer adapt, and become extinct?

Camille Parmesan

Camille Parmesan is an associate professor in the Section of Integrative Biology at the University of Texas at Austin. Her recent work has focused on current impacts of climate change on wildlife, and especially on butterfly range shifts. To learn more about Dr. Parmesan's research, go to http://cluster3.biosci.utexas.edu/research/parmesanLab.

REVIEW KEY CONCEPTS

Go to **CENGAGENOW** at www.cengage.com/login to access quizzing, animations, exercises, articles, and personalized homework help.

49.1 The Science of Ecology

- Ecology is the study of the interactions between organisms and their environments. Basic ecology focuses on undisturbed natural systems, whereas applied ecology considers the effects of human disturbance.
- Ecologists conduct research at five levels of organization: organisms, populations, communities, ecosystems, and the biosphere.
- Ecologists test hypotheses about ecological relationships with experimental or observational data. They sometimes frame hypotheses in mathematical models.

49.2 Environmental Diversity of the Biosphere

- The biosphere encompasses all the regions on Earth where organisms live, including the atmosphere, hydrosphere, and lithosphere.
- Latitudinal variations in solar radiation establish global climate patterns (Figure 49.2). Earth's tilt on its axis causes seasonal variation in solar radiation and climate (Figure 49.3). Seasonal variations in day length and temperature increase steadily from tropical latitudes toward the poles.
- Unequal heating of the atmosphere causes air masses to flow in circulation cells that create worldwide wind and precipitation patterns (Figures 49.4 and 49.5). Ocean currents generally flow clockwise in the Northern Hemisphere and counterclockwise in the Southern Hemisphere (Figure 49.6).
- The oceans and local topographical features influence regional and local climates. Proximity to the ocean has a moderating effect on terrestrial climates (Figure 49.7). Habitats are generally wetter on the windward sides of mountains than on the leeward sides (Figure 49.8).

Animation: Air circulation and climate

Animation: Global air circulation patterns

Animation: Major climate zones and ocean currents

Animation: Rain shadow effect

Animation: El Niño Southern Oscillation

49.3 Organismal Responses to Environmental Variation and Climate Change

- Organisms use homeostatic responses to cope with environmental variation. Animals often use facultative behavioral and physiological mechanisms to respond to environmental temperature (Figure 49.9).
- Global climate change is affecting the ecology of many organisms. Many species are experiencing changes in their geographical ranges or in the timing of their reproduction (Figure 49.10).

49.4 Terrestrial Biomes

- Biomes are general types of vegetation and other associated organisms. Climate is the major determinant of terrestrial biome distributions (Figures 49.11 and 49.12).
- Tropical forest occurs at low latitudes where seasonality is determined by variations in rainfall rather than by day length and temperature (Figure 49.13). Tropical rain forests are the most species-rich terrestrial biome, but they grow on nutrient-poor soils.
- Savanna is tropical and subtropical grassland with scattered trees (Figure 49.14). Long dry seasons, fires, and grazing by large mammals prevent trees from replacing perennial grasses.
- Deserts form in arid regions where precipitation is low and temperature varies widely on a daily and seasonal basis (Figure 49.15).
- Chaparral is a coastal biome dominated by dense, woody shrubs and trees that resprout after periodic fires (Figure 49.16). Chaparral occurs where winters are mild and wet and summers hot and dry.
- Temperate grassland grows where winters are cold, summers are warm, and rainfall is moderate (Figure 49.17). Tree seedlings are eliminated from grasslands by droughts, periodic fires, and grazing by mammals. Grassland soils are rich and deep.
- Temperate deciduous forest flourishes at middle latitudes with abundant rainfall. The seasonality of the climate is reflected in the annual loss and regrowth of leaves (Figure 49.18).
- The boreal forest, or taiga, includes dense stands of coniferous trees at high latitudes, where winters are long and cold (Figure 49.19).
- Tundra is the northernmost biome, where plants grow in shallow topsoil over a layer of permafrost (Figure 49.20). The brief growing season and winter winds cause tundra plants to be very short.

Animation: Terrestrial biomes

49.5 Freshwater Environments

- Freshwater environments include both flowing-water and standing water systems.
- The physical characteristics of flowing-water environments change from the headwaters of a stream to the mouth of a river (Figure 49.21).
- The physical characteristics of standing-water environments change with the depth of water and distance from shore (Figures 49.22 and 49.23). Lakes exhibit marked vertical zonation and, in the temperate zone, undergo a seasonal mixing of their waters (Figure 49.24). Lakes are generally classified by their nutrient status and productivity (Figure 49.25).

Animation: Lake zonation

Animation: Lake turnover

Animation: Trophic nature of lakes

49.6 Marine Environments

- The oceans exhibit marked zonation based on water depth and distance from shore (Figure 49.26).
- Estuaries are highly productive tidal environments where rivers provide a constant input of nutrients and freshwater, and the tides carry away wastes (Figure 49.27).
- The intertidal zone is a stressful environment that is alternately submerged and exposed (Figure 49.28).
- Highly productive and diverse shallow-water communities thrive on continental shelves and oceanic banks. Kelp forests predominate at high latitudes, whereas coral reefs occur in the tropics (Figure 49.29).
- The open ocean is highly stratified because photosynthesis is possible only in the uppermost 50 m of water. Plankton are the primary producers in the uppermost layers (Figure 49.30). The deep sea includes many predatory species (Figure 49.31).
- Organisms of the sea floor occupy the benthic province. Falling detritus supports most benthic communities, but chemosynthetic bacteria support communities near deep-sea hydrothermal vents (Figure 49.32) and cold seeps.

Animation: Oceanic zones

Animation: Coastal upwelling

Animation: Rocky intertidal zones

Animation: Three types of reefs

Animation: Hydrothermal vent community

UNDERSTAND AND APPLY

Test Your Knowledge

1. The lithosphere includes all:
   a. oceans.
   b. ice caps.
   c. rocks, soils, and sediments.
   d. gases and airborne particles.
   e. places where organisms live.
2. Earth's 23.5° tilt on its axis directly causes:
   a. latitudinal variation in average annual rainfall.
   b. ocean currents to rotate clockwise in the Northern Hemisphere.
   c. microclimates to vary dramatically over short distances.
   d. low rainfall on the leeward side of mountain ranges.
   e. seasonal variation in the amount of solar radiation.
3. Adiabatic cooling causes rising air masses to:
   a. absorb moisture from Earth's surface.
   b. release precipitation.
   c. change the direction of the El Niño current.
   d. flow toward the equator from the poles.
   e. be deflected from a strictly northward or southward flow.
4. The term "rain shadow" describes the:
   a. low rainfall that is typical on the leeward side of mountains.
   b. low rainfall that is typical at 30° latitude.
   c. high rainfall that is typical on the windward side of mountains.
   d. blocking of rain by vegetation in dense tropical forests.
   e. low rainfall that is typical in the interior of continents.
5. The major climatic factors that govern the distributions of terrestrial biomes are:
   a. temperature only.
   b. rainfall only.
   c. wind speed only.
   d. temperature and rainfall.
   e. temperature, rainfall, and wind speed.
6. Which biome experiences the highest annual rainfall?
   a. tropical rain forest
   b. tropical savanna
   c. chaparral
   d. temperature grassland
   e. arctic tundra
7. From which biome are trees excluded by periodic fires and grazing herbivores?
   a. tropical rain forest
   b. thorn forest
   c. chaparral
   d. temperate grassland
   e. arctic tundra
8. The major source of nutrients in the headwaters of a small stream is from:
   a. dead leaves and other organic matter from adjacent land.
   b. photosynthesis by phytoplankton.
   c. photosynthesis by floating aquatic plants.
   d. the activity of chemoautotrophic bacteria.
   e. minerals from the underlying bedrock.
9. During the spring overturn in a temperate zone lake:
   a. oxygen is carried from the surface to the bottom, and nutrients are carried from the bottom to the surface.
   b. nutrients are carried from the surface to the bottom, and oxygen is carried from the bottom to the surface.
   c. nutrients and oxygen are carried from the bottom waters to the surface waters.
   d. nutrients and oxygen are carried from the surface waters to the bottom waters.
   e. oxygen concentration remains constant at all depths, and nutrients sink to the bottom.
10. In which habitat must organisms adjust regularly to changing salinity?
    a. salt marsh
    b. coral reef
    c. benthic province
    d. estuary
    e. riffle

Discuss the Concepts

1. Temperate grassland and chaparral often burn in lightning-induced fires, which stimulate the germination of seeds and regrowth of existing vegetation. Do you think that companies or the government should sell fire insurance to people who build expensive homes in places where periodic fires are virtually inevitable?
2. Boreal forests generally harbor many fewer species of trees than tropical forests do. Develop three hypotheses to explain this pattern. What data would you collect to test your hypotheses?
3. Many regions have been developed for agriculture, industry, and human habitation. Have our activities created new biomes? What physical environments are created by development, and what plants and animals occupy developed areas?

Design an Experiment

Design an experiment to test the hypothesis that streams receive much of their nutrients and energy from material that falls into them from overhanging vegetation.

Apply Evolutionary Thinking

If the geographical ranges of species change in response to global warming, what new selection pressures will organisms face as they move into ecological communities where they have not previously occurred? Your answer should address the effects of novel species interactions as well as the effects of encountering different physical environments.

Express Your Opinion

We cannot stop an El Niño from happening, but we might be able to minimize its environmental, social, and economic impacts. Would you support the use of taxpayer dollars to fund research into the causes and effects of El Niño? Go to www.cengage.com/login to investigate both sides of the issue and then vote.

Interpret the Data

Climate scientists at NOAA, the National Oceanographic and Atmospheric Administration, collect data about sea surface temperature (SST) and atmospheric conditions to predict El Niño and La Niña events before they occur. Researchers then calculate SST anomalies, the differences between current SSTs and historical averages. *Positive* anomalies of 0.5°C or more indicate that surface waters are much warmer than usual, a sign that El Niño is coming. *Negative* anomalies of 0.5°C or more indicate that surface waters are much cooler than usual, a sign of an upcoming La Niña. Anomalies that are within 0.5°C of average SSTs are identified as neutral conditions. The accompanying figure presents SST anomalies in the eastern Pacific Ocean from 1950 through 2008.

1. How many times have SSTs predicted a likely El Niño event or a likely La Niña event over the time represented in the graph?
2. Are all SST anomalies of equal magnitude and duration?
3. Is there a fixed pattern in the distribution of El Niño and La Niña events? In other words, does one type of event *always* follow the other?

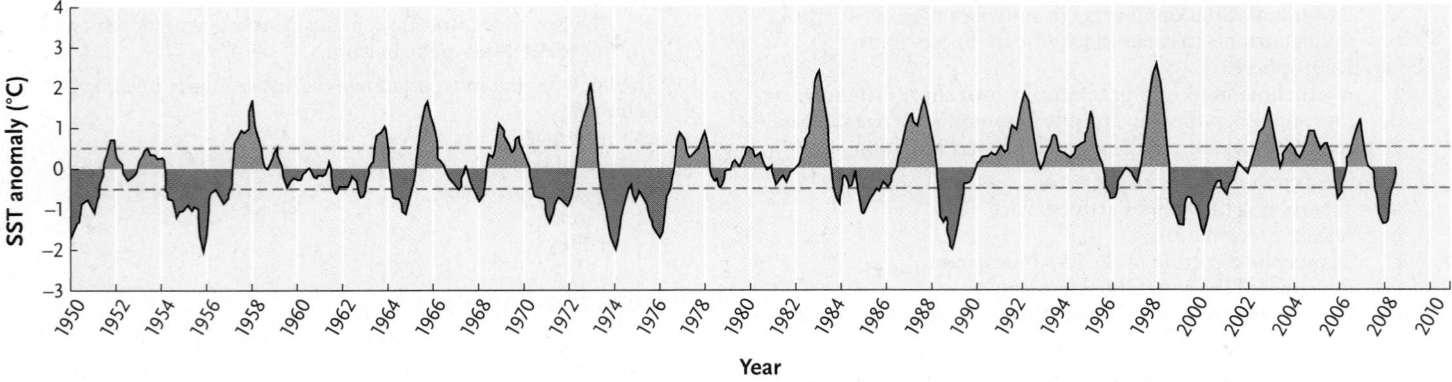
SST anomaly (°C)
4
3
2
1
0
−1
−2
−3
1950
1952
1954
1956
1958
1960
1962
1964
1966
1968
1970
1972
1974
1976
1978
1980
1982
1984
1986
1988
1990
1992
1994
1996
1998
2000
2002
2004
2006
2008
2010
Year

A population of Caribbean flamingos *(Phaenicopterus ruber)*. Each pair of flamingos in this breeding colony incubates a single egg in a mud nest.

© Gerry Ellis/Minden Pictures

50

STUDY OUTLINE

Population Ecology

Why It Matters. . . When humans immigrate to new places, they often transport familiar plants and animals from home, introducing them into their new gardens, fields, and forests. Some organisms fail to survive in the new environments. But other species—like the European starlings *(Sturnus vulgaris)* and house sparrows *(Passer domesticus)* that are now so common in North America—flourish, and sometimes become pests.

In 1859, an Australian rancher released a few pairs of European rabbits *(Oryctolagus cuniculus)* for sport hunting in the state of Victoria. The rabbits bred rapidly, sometimes producing litters of four or five offspring every month. They had no natural predators in Australia, and by 1900, an estimated 20 million rabbits had overrun much of the continent. Their advance was limited only by extreme climates, clay soil, and lack of food or water. The rabbits destroyed natural vegetation and the pastures that supported a large sheep industry. The government tried in vain to poison the rabbits. Ranchers introduced predators, hoping that they would eat rabbits faster than the rabbits could reproduce. But the rabbits continued to multiply. Eventually, the government built a "rabbit-proof fence" that stretched more than 3,200 km (2,000 miles) to keep the rabbits out of the rich pasture lands in Western Australia **(Figure 50.1).**

In 1950, scientists tackled the devastating problems caused by the introduced rabbits. Biologists collected myxoma virus (a relative of smallpox) from infected rabbits in South America and released it among the European rabbits in Australia. The virus was lethal to European rabbits, which had never evolved resistance to it. The first epidemic of myxomatosis killed more than 99% of infected rabbits. But in the following season, the virus killed only 90% of infected rabbits, and within a few years, the virus was killing only half the rabbits it infected. Clearly, some rabbits were becoming more resistant to the virus. Resistant rabbits survived and reproduced, comprising a larger percent-

FIGURE 50.1
Introduced organisms. European rabbits multiplied so rapidly and destroyed so much vegetation in Australia that the government built a fence across the country to prevent their spread.

age of the population over time (see Section 20.3 to review natural selection). Subsequent research showed that the virus had also become less virulent. Today, wildlife-control agents develop and release more deadly viruses to control the rabbit population.

This brief history of an introduced population identifies several questions about **population dynamics**—how the characteristics of populations change through time and vary from place to place—that we consider in this chapter. For example, why do some populations, like the rabbits in Australia, grow explosively, whereas others maintain reasonably stable numbers over long periods of time? How do interactions with abiotic and biotic factors in the environment influence the characteristics of populations? Finally, how do populations respond evolutionarily to their interactions with the environment? <

50.1 Population Characteristics

Populations have characteristics that transcend those of the individuals they comprise. For example, every population has a **geographical range,** the overall spatial boundaries within which it lives. Geographical ranges vary enormously. A population of snails might inhabit a small tidepool, whereas a population of marine phytoplankton might occupy an area that is orders of magnitude larger. Every population also occupies a **habitat,** the specific environment in which it lives, as characterized by its biotic and abiotic features. Ecologists also measure other population characteristics, such as size, distribution in space, and age structure.

A Population's Size and Density Determine the Amount of Resources It Uses

Population size is simply the number of individuals in a population at a specified time. **Population density** is the number of individuals per unit area or per unit volume of habitat. Species with large body sizes generally have lower population densities than species with smaller body sizes **(Figure 50.2).** Although population size and density are related measures, knowing a population's density provides more information about its relationship to the resources it uses. For example, if a population of 200 oak trees occupies 1 hectare (10,000 m^2), its population density is 200/10,000 m^2 or one tree per 50 m^2. But if a population of 200 oaks is spread over 5 hectares, its density is one tree per 250 m^2. Clearly, the second population is less dense than the first, and its members will have greater access to sunlight, water, and other resources.

Ecologists measure population size and density to monitor and manage populations of endangered species, economically important species, and agricultural pests. For large-bodied species, a simple head count provides accurate information. For example, ecologists survey the size and density of African bush elephant *(Loxodonta africana)* populations by flying over herds and counting individuals. Researchers use a variation on that technique to estimate population size in tiny organisms that live at high population densities. To estimate the density of aquatic phytoplankton, for example, you might collect water samples of known volume from representative areas in a lake and use a microscope to count the organisms; you could then extrapolate their population size and density based on the estimated volume of the entire lake. In other cases, researchers use the mark-release-recapture sampling technique **(Figure 50.3).**

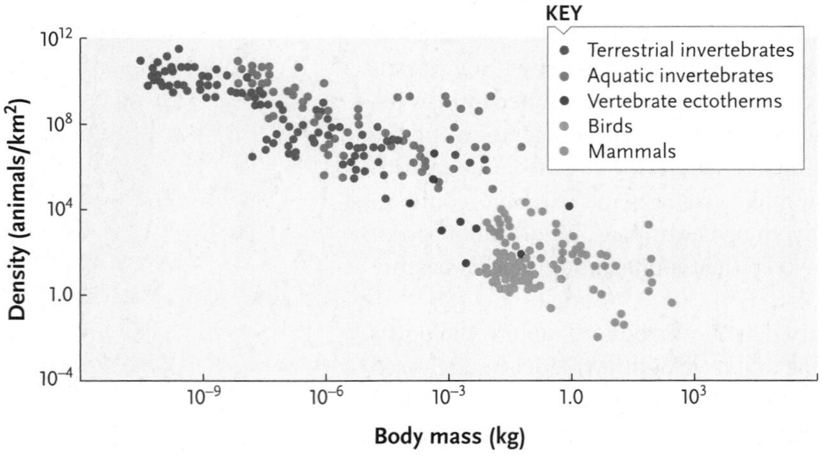

KEY
Terrestrial invertebrates
Aquatic invertebrates
Vertebrate ectotherms
Birds
Mammals

Density (animals/km²): 10^{12}, 10^{8}, 10^{4}, 1.0, 10^{-4}

Body mass (kg): 10^{-9}, 10^{-6}, 10^{-3}, 1.0, 10^{3}

FIGURE 50.2
Population density and body size. Population density generally declines with increasing body size among animal species. Similar trends exist for other types of organisms.

Populations Differ in How They Are Distributed in Space

Populations also vary in their **dispersion,** the spatial distribution of individuals within the geographical range. Ecologists define three theoretical patterns of dispersion: *random, clumped,* and *uniform* **(Figure 50.4).**

For some populations, environmental conditions don't vary much within a habitat, and individuals are neither attracted to nor repelled by others of their species. These populations exhibit **random dispersion,** which has a formal statistical definition that serves as a theoretical baseline for assessing whether organisms are clumped or uniformly distributed. In cases of random dispersion, individuals are distributed unpredictably. Some spiders, burrowing clams, and rainforest trees exhibit random dispersion.

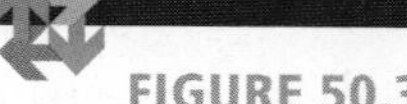

FIGURE 50.3 **Research Method**

Using Mark-Release-Recapture to Estimate Population Size

Purpose: Ecologists use the mark-release-recapture technique to estimate the population size of mobile animals that live within a restricted geographic range.

Protocol: A sample of organisms is captured, marked in some permanent but harmless way, and released. Insects and reptiles are marked with ink or paint, birds with rings on their legs, and mammals with ear tags or collars. Some time later, a second sample of organisms is captured, and the researcher notes what proportion of the second sample carries the mark. That proportion tells us what percentage of the total population was captured and marked at the first sampling. The total population size is estimated as (number marked) × (number in the second sample/number of marked recaptures).

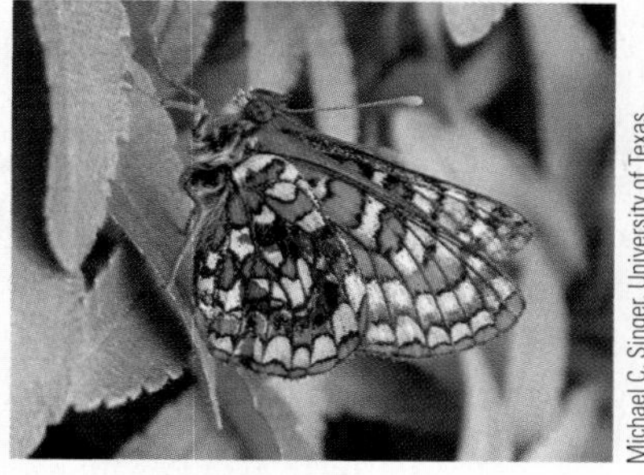

Michael C. Singer, University of Texas

Interpreting the Results: Imagine that you capture 120 butterflies, mark each with a black spot on its wing, and release them. A week later, you capture a second sample of 150 butterflies, and find that 30 of them have the black mark. Thus, you had marked one out of every five butterflies (30/150) on your first field trip. Because you captured 120 individuals on that first excursion, you would estimate that the total population size is 120 × (150/30) = 600 butterflies.

The technique is based on several assumptions that are critical to its accuracy: (1) that being marked has no effect on survival; (2) that marked and unmarked animals mix randomly in the population; (3) that no migration into or out of the population takes place during the estimating period; and (4) that marked individuals are just as likely to be captured as unmarked individuals. (Sometimes animals become "trap shy" or "trap happy," a violation of the fourth assumption.)

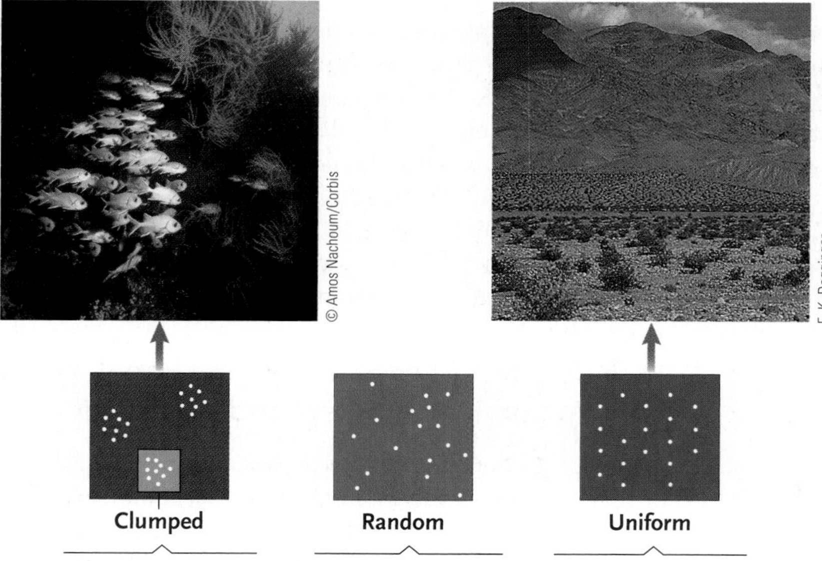
© Amos Nachoum/Corbis
E. K. Degginger
Clumped
A clumped dispersion pattern is one in which individuals are grouped more closely to each other than if they are randomly dispersed.
Random
A random dispersion pattern, in which organisms are distributed independently of each other, serves as a statistical yardstick for evaluating other dispersion patterns.
Uniform
A uniform dispersion pattern is one in which individuals are more widely separated from each other than they are if they are randomly dispersed.

FIGURE 50.4
Dispersion patterns. Schooling fishes, like these sabre squirrelfish *(Sargocentron spiniferum)* from the Maldives in the Indian Ocean, exhibit a clumped pattern of dispersion. A random pattern of dispersion, which is fairly rare in nature, occurs in organisms that are neither attracted to nor repelled by each other. Creosote bushes *(Larrea tridentata)* near Death Valley, California, exhibit uniform dispersion.

Three reasons explain why a **clumped dispersion**—with individuals grouped together—is extremely common in nature. First, suitable conditions often have a patchy distribution. For example, certain pasture plants may be clumped in small, scattered areas where cowpats fell months before, locally enriching the soil. Second, some animals live in social groups (see Section 55.5). Mates are easy to locate within groups, and individuals may cooperate in rearing offspring, feeding, or defending themselves from predators. Third, some organisms are clumped because of their reproductive pattern. Plants and animals that produce asexual clones, such as aspen trees and sea anemones, often occur in large aggregations (see Chapters 34 and 47). In other species, seeds, eggs, or larvae lack dispersal mechanisms, and offspring grow near their parents.

When the individuals in a population repel each other because resources are in short supply, they tend to be evenly spaced in their habitat, a pattern called **uniform dispersion.** For example, creosote bushes *(Larrea tridentata)* are uniformly distributed in the dry scrub deserts of the American Southwest. Mature bushes deplete the surrounding soil of water and secrete toxic chemicals, making it impossible for seedlings to grow. Moreover, seed-eating ants and rodents that live at the base of mature bushes consume any seeds that fall nearby. Territorial behavior, the defense of an area and its resources, produces uniform dispersion in animals (see Section 55.2).

Whether the spatial distribution of a population appears to be clumped, uniform, or random depends partly on how large an area an ecologist studies. Oak seedlings may be randomly dispersed on a spatial scale of a few square meters, but over an entire mixed hardwood forest, they are clumped under the parent trees.

In addition, the dispersion of animal populations often varies through time in response to natural environmental rhythms. Few habitats provide a constant supply of resources throughout the year, and many animals move from one habitat to another on a seasonal cycle. For example, tropical birds and mammals are often widely dispersed in deciduous forests during the wet season. But during the dry season, they crowd into narrow "gallery forests" along watercourses where evergreen trees provide food and shelter.

A Population's Age Structure, Generation Time, and Sex Ratio Influence How Quickly It Will Grow

All populations have an **age structure,** a statistical description of the relative numbers of individuals in each age class (see Section 50.6). Individuals can be roughly categorized as prereproductive (younger than the age of sexual maturity), reproductive, or postreproductive (older than the maximum age of reproduction). A population's age structure reflects its recent growth history and predicts its future growth potential. Populations that include many prereproductive individuals grew rapidly in the recent past and will continue to grow larger as the young individuals mature and reproduce.

Another characteristic that influences a population's growth is its **generation time,** the average time between the birth of an organism and the birth of its offspring. Generation time is usually short in species that reach sexual maturity at a small body size **(Figure 50.5).** Their populations often grow rapidly because of the speedy accumulation of reproductive individuals.

Populations also vary in their **sex ratio,** the relative proportions of males and females. In general, the number of females in a population has a bigger impact on population growth than the number of males because only females actually produce offspring. Moreover, in many species, one male can mate with several females, and the number of males may have little effect on the population's reproductive output. In northern elephant seals *(Mirounga angustirostris),* for example, mature bulls fight for dominance on the beaches where the seals mate, and only a few males may ultimately inseminate a hundred or more females. Thus, the presence of other males in the group has little effect on the size of future generations. However, in animals that form lifelong pair bonds, such as geese and swans, the number of males does influence reproduction in the population.

Population ecologists often try to determine the proportion of individuals in a population that are reproducing. This issue is particularly relevant to the conservation of any species in which individuals are rare or widely dispersed in the habitat (see Section 53.4). As *Insights from the Molecular Revolution* describes, ecologists now use DNA analysis to address this question. In the next section we consider factors that influence the age structure of a population and its potential for future growth.

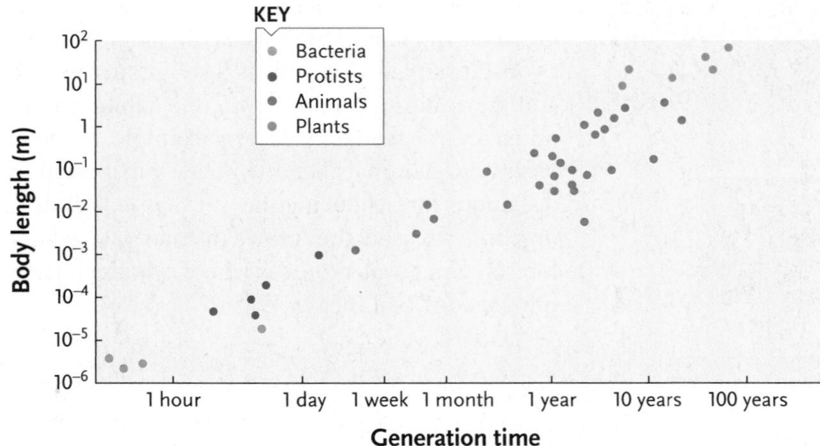

KEY
Bacteria
Protists
Animals
Plants
Body length (m)
10^2
10^1
1
10^{-1}
10^{-2}
10^{-3}
10^{-4}
10^{-5}
10^{-6}
1 hour
1 day
1 week
1 month
1 year
10 years
100 years
Generation time

FIGURE 50.5
Generation time and body size. Generation time increases with body size among bacteria, protists, plants, and animals. The logarithmic scale on both axes compresses the data into a straight line.

STUDY BREAK 50.1

1. **What is the difference between a population's size and its density?**
2. **What do the three patterns of dispersion imply about the relationships between individuals in a population?**

THINK OUTSIDE THE BOOK

Try to estimate the size of a population of plants or animals that live near your home or campus. Start by defining the boundaries of the population, identify the method you would use, and then make the estimate. How does the type of organism you are studying affect the accuracy of your estimate?

50.2 Demography

Populations grow larger through the birth of individuals and the **immigration** (movement into the population) of organisms from neighboring populations. Conversely, death and **emigration** (movement out of the population) reduce population size. **Demography** is the statistical study of the processes that change a population's size and density through time.

Ecologists use demographic analysis to predict a population's future growth. For human populations, these data help governments anticipate the need for social services such as schools and hospitals. Demographic data also allow conservation ecologists to develop plans to protect endangered species. For example, demographic data on the northern spotted owl *(Strix occidentalis caurina)* helped convince the courts to restrict logging in the owl's primary habitat, the old growth forests of the Pacific Northwest. *Life tables* and *survivorship curves* are among the tools ecologists use to analyze demographic data.

Life Tables Summarize a Population's Survival and Reproductive Rates

Although every species has a characteristic life span, few individuals survive to the maximum age possible. Mortality results from starvation, disease, accidents, predation, or the inability to find a suitable habitat. Life insurance companies first developed techniques for measuring mortality rates, but ecologists adapted these approaches to the study of nonhuman populations.

A **life table** summarizes the demographic characteristics of a population **(Table 50.1).** To collect life-table data for short-lived organisms, demographers

INSIGHTS FROM THE Molecular Revolution

Asocial Armadillos: Can armadillo reproductive patterns explain the expansion of their range?

Relatively little is known about the behavior of nine-banded armadillos *(Dasypus novemcinctus)*. They are slow moving and solitary animals, yet they have spread from Mexico and southern Texas through most of the southern United States in only 100 years.

Research Question

Given their almost completely asocial behavior, what proportion of an armadillo population successfully mates? And, given their slow movements, how did they expand their range?

Experiment

Paulo A. Prodöhl and his colleagues at the University of Georgia, University of Washington, and Valdosta State University used molecular techniques in an attempt to answer these questions. Their previous work had identified seven short tandem repeat (STR) loci in the armadillo genome. An STR locus consists of a segment of a chromosome with a short sequence repeated in series (see Section 18.2). In this case, the loci are approximately 200 bp long with 2-bp and 4-bp repeated sequences. Alleles of each locus vary in length because they differ in the number of copies of the repeated sequences. Because alleles of STR loci are inherited in the same way as alleles of genes, the researchers could trace parentage and migration patterns by comparing the allelic variations of the STR loci, in a manner analogous to human DNA fingerprinting.

Researchers collected small tissue samples from the ears of 290 armadillos living in the Tall Timbers Research Station near Tallahassee, Florida. They extracted genomic DNA from the tissue sample and used the polymerase chain reaction (PCR) to amplify alleles for each of the seven STR loci. They used gel electrophoresis to determine the sizes of the fragments amplified by PCR. The molecular techniques involved are discussed in Chapter 18.

Results

The investigators determined the relatedness of juveniles and adults using statistical methods to compare the gel patterns produced by their DNA **(Figure).** Adult males and females with gel patterns most similar to those of a given juvenile were considered to be its parents. When the data identified more than two possible parents, the male and female living closest to a juvenile were scored as the most likely candidates. These techniques allowed the investigators to assign parents to 69 juveniles. Only seven juveniles could not be assigned parents, possibly because the parents had died, avoided capture, or emigrated from the population.

Conclusion

The results from four years of study suggest that 36% to 46% of adult armadillos reproduced at least once, despite their asocial habits—a moderately successful reproductive rate. In general, parents and offspring lived between 800 and 1,500 m of each other, and individuals were usually recaptured within a 200-m radius of the same spot from season to season and from one year to the next. Thus, migration appears to be very limited, leaving the basis of their rapid spread unexplained.

Source: P. A. Prodöhl et al. 1998. Genetic maternity and paternity in a local population of armadillos assessed by microsatellite DNA markers and field data. *American Naturalist* 151:7–19.

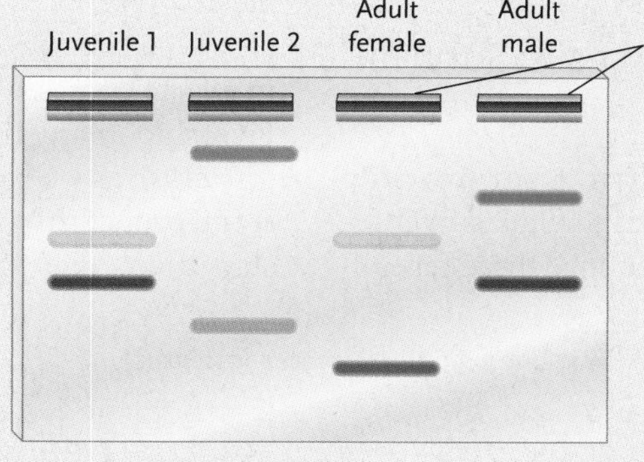
Juvenile 1
Juvenile 2
Adult female
Adult male
Wells for placing PCR samples

Theoretical gel electrophoresis result for STR alleles at one locus for two juveniles and an adult female and male. Juvenile 1 could be the offspring of the two adults because its two alleles match alleles present in the two parents. However, Juvenile 2 could not be the offspring of these two parents because neither of its alleles could have come from those two adults.

typically mark a **cohort,** a group of individuals of similar age, at birth and monitor their survival until all members of the cohort die. For organisms that live more than a few years, a researcher might sample the population for one or two years, recording the ages at which individuals die, and then extrapolate those results over the species' life span.

In any life table, the life span of the organisms is divided into age intervals of convenient length: days, weeks, or months for short-lived species; years or groups of years for longer-lived species. Mortality can be expressed in two complementary ways. **Age-specific mortality** is the proportion of individuals alive at the start of an age interval that died during that age interval. Its more cheerful reflection, **age-specific survivorship,** is the proportion of individuals alive at the start of an age interval that survived until the start of the next age interval. Thus, in Table 50.1, the age-specific mortality rate during the three-to-six-month age interval is 195/722 = 0.270, and the age-specific survivorship rate is 527/722 = 0.730. For any age interval, the sum of age-specific mortality and age-specific survivorship always equals 1. Life tables also summarize the proportion of the cohort that survived to a particular age, a statistic that identifies the probability that any randomly selected newborn will still be alive at that age. For the 3-to-6-month age interval in Table 50.1, this probability is 722/843 = 0.856.

TABLE 50.1 Life Table for a Cohort of 843 Individuals of the Grass *Poa annua* (Annual Bluegrass)

Age Interval (in months)	Number Alive at Start of Age Interval	Number Dying During Age Interval	Age-Specific Mortality Rate	Age-Specific Survivorship Rate	Proportion of Original Cohort Alive at Start of Age Interval	Age-Specific Fecundity (Seed Production)
0–3	843	121	0.144	0.856	1.000	0
3–6	722	195	0.270	0.730	0.856	300
6–9	527	211	0.400	0.600	0.625	620
9–12	316	172	0.544	0.456	0.375	430
12–15	144	90	0.625	0.375	0.171	210
15–18	54	39	0.722	0.278	0.064	60
18–21	15	12	0.800	0.200	0.018	30
21–24	3	3	1.000	0.000	0.004	10
24–	0	—	—	—	—	—

Source: M. Begon and M. Mortimer. *Population Ecology*. Sunderland, MA: Sinauer Associates, 1981. Adapted from R. Law. 1975.

Life tables also include data on **age-specific fecundity,** the average number of offspring produced by surviving females during each age interval. Table 50.1 shows, for example, that plants in the 3-to-6-month age interval each produced an average of 300 seeds. In some species, including humans, fecundity is highest in individuals of intermediate age. Younger individuals have not yet reached sexual maturity, and older individuals are past their reproductive prime. However, in some plants and animals fecundity increases steadily with age.

Survivorship Curves Depict Changes in Survival Rate over the Life Span

Survivorship data are depicted graphically in a **survivorship curve,** which displays the rate of survival for individuals over the species' average life span. Ecologists have identified three generalized survivorship curves (blue lines in **Figure 50.6**), although most organisms exhibit survivorship patterns that fall between these idealized patterns.

Type I curves reflect high survivorship until late in life, when mortality takes a great toll. Type I curves are typical of large animals that produce few young and provide them with extended care, which reduces juvenile mortality. For example, large mammals, such as the Dall's sheep *(Ovis dalli),* produce only one or two offspring at a time and nurture them through their vulnerable first year.

Type II curves reflect a relatively constant rate of mortality in all age classes, a pattern that produces steadily declining survivorship. Many lizards, such as the five-lined skink *(Eumeces fasciatus),* as well as songbirds and small mammals, face a constant probability of mortality from predation, disease, and starvation.

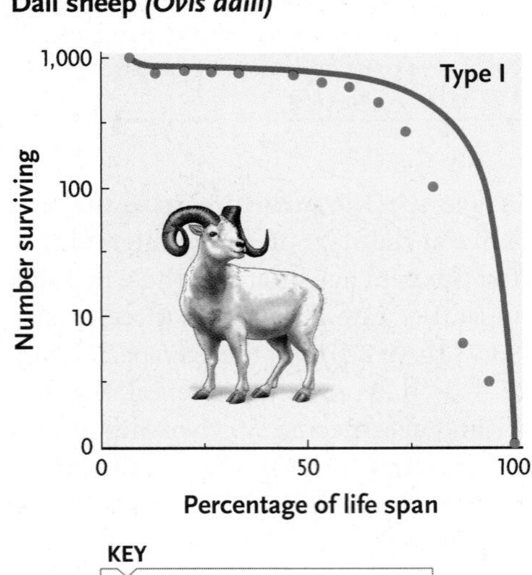

Dall sheep (Ovis dalli)
Type I
1,000
100
10
0
Number surviving
0
50
100
Percentage of life span
KEY
Theoretical
Empirical

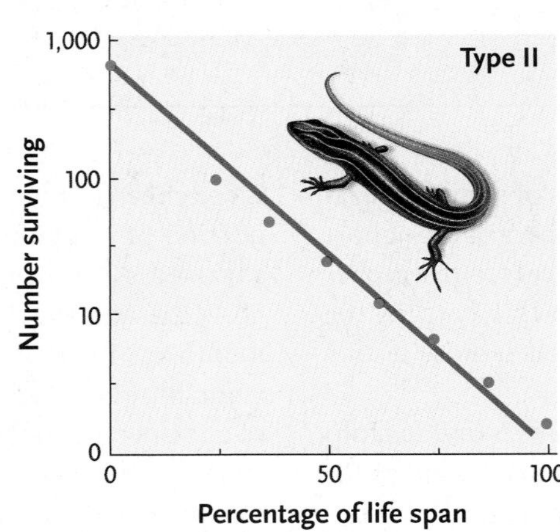

Five-lined skink (Eumeces fasciatus)
Type II
1,000
100
10
0
Number surviving
0
50
100
Percentage of life span

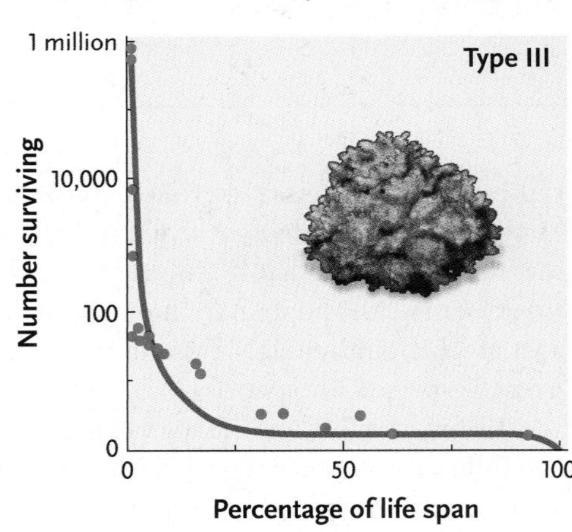

Perennial desert shrub (Cleome droserifolia)
Type III
1 million
10,000
100
0
Number surviving
0
50
100
Percentage of life span

FIGURE 50.6
Survivorship curves. The survivorship curves of many organisms (pink data points) roughly match one of three idealized patterns (blue curves).

Type III curves reflect high juvenile mortality, followed by a period of low mortality once the offspring reach a critical age and size. For example, *Cleome droserifolia,* a desert shrub from the Middle East, experiences extraordinarily high mortality in its seed and seedling stages. Researchers estimate that for every million seeds produced, fewer than 1,000 germinate, and only about 40 individuals survive their first year. Once a plant becomes established, however, its likelihood of future survival is higher, and the survivorship curve flattens out. Many plants, insects, marine invertebrates, and fishes exhibit type III survivorship.

STUDY BREAK 50.2 <

1. **What statistics are usually included in a life table?**
2. **Which type of survivorship curve is characteristic of humans in industrialized countries? Explain your answer.**

50.3 The Evolution of Life Histories

The analysis of life tables reveals how natural selection has produced different **life histories**—the lifetime patterns of growth, maturation, and reproduction—that maximize the number of surviving offspring an individual produces.

Organisms Face Trade-Offs in Their Allocation of Resources

Every organism is constrained by a finite **energy budget,** the total amount of energy that it can accumulate and use to fuel its activities. An organism's energy budget is like a savings account. When the individual accumulates more energy than it needs, it makes deposits to this account—energy is stored as starch, glycogen, or fat. When it expends more energy than it harvests, it makes withdrawals from its energy stores. But unlike a bank account, an organism's energy budget cannot be overdrawn, and no loans against future "earnings" are possible.

Organisms use the energy they harvest for three broadly defined functions: maintenance (the preservation of good physiological condition), growth, and reproduction. And when an organism devotes energy to any one of these functions, the balance in its energy budget is reduced, leaving less energy for the other functions.

Life History Patterns Vary Dramatically among Species

A fish, a deciduous tree, and a mammal illustrate the dramatic variations that exist in life history patterns **(Figure 50.7).** Larval coho salmon *(Oncorhynchus kisutch)* hatch in the headwaters of a stream, where they feed and grow for about a year before assuming their adult body form. After swimming downstream to the ocean, they remain at sea for a year or two, feeding voraciously and growing rapidly. Eventually, salmon use sun-compass, geomagnetic, and chemical cues to return to the rivers and streams where they hatched. The fishes swim upstream, and each female lays hundreds or thousands of relatively small eggs. After spending all of their energy reserves on the upstream journey and reproduction, their condition deteriorates and they die.

Most deciduous trees in the temperate zone, such as oaks (genus *Quercus*), begin their lives as seeds in late summer. The seeds remain metabolically inactive until the following spring or a later year. After germinating, trees collect nutrients and energy and continue to grow throughout their lives. Once they achieve a critical size, they may produce thousands of seeds annually for many years. Thus, growth and reproduction occur simultaneously through much of the trees' life.

European red deer *(Cervus elaphus)* are born in spring, and young remain with their mothers for an extended period, nursing and growing rapidly. After weaning, they feed on their own. Female red deer begin to breed after reaching adult size in their third year, producing one or two offspring annually until they are about 16 years old, when they reach their maximum life span and die.

How can we summarize the similarities and differences in the life histories of these organisms? All three species harvest

Coho salmon *(Oncorhynchus kisutch)*

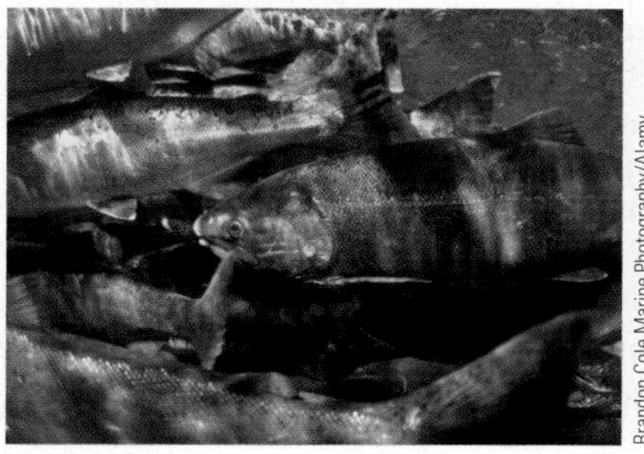
Brandon Cole Marine Photography/Alamy

White oak *(Quercus alba)*

Harry Rogers/Photo Researchers, Inc.

Red deer *(Cervus elaphus)*

Ron Perkins/Getty Images

FIGURE 50.7
Three organisms with very different life histories.

The Evolution of Life History Traits in Guppies

Some years ago, drenched with sweat and with fishnets in hand, two ecologists were engaged in fieldwork on the Caribbean island of Trinidad **(Figure A).** They were after guppies *(Poecilia reticulata)*—small fish that bear live young in shallow mountain streams. John Endler and David Reznick, then of the University of California, Santa Barbara, were studying the environmental variables that influence the evolution of life history patterns in guppies.

Male guppies are easy to distinguish from females. Males, which stop growing at sexual maturity, are smaller, and their scales have bright colors that serve as visual signals in intricate courtship displays. The drably colored females continue to grow larger throughout their lives.

In the mountains of Trinidad, guppies living in different streams—and even in different parts of the same stream—are eaten by one of two other fish species **(Figure B).** In some streams, a large pike-cichlid *(Crenicichla alta)* prefers mature guppies and tends not to spend time hunting small, immature ones. In other streams, a small killifish *(Rivulus hartii)* preys on immature guppies but does not have much success with the larger adults.

Reznick and Endler found that the life history patterns of guppies vary among streams with different predators. In streams with pike-cichlids, both male and female guppies mature faster and begin to reproduce at a smaller size and a younger age than their counterparts in streams where killifish live **(Figure C).** In addition, female guppies from pike-cichlid streams reproduce more often and produce smaller and more numerous young **(Figure D).** These differences allow guppies to avoid some predation. Those in pike-cichlid streams begin to reproduce when they are smaller than the size preferred by that predator. And those from killifish streams grow quickly to a size that is too large to be consumed by killifish.

David Reznick/University of California, Riverside

FIGURE A
David Reznick surveys a shallow stream in the mountains of Trinidad.

Male guppy (right) that shared a stream with pike-cichlids (below)
Male guppy (right) that shared a stream with killifish (below)

(Guppy photos: David Reznick/University of California, Riverside: computer enhanced by Lisa Starr; predator photos: Hippocampus Bildarchiv.)

FIGURE B
Male guppies from streams where pike-cichlids live (top) are smaller, more streamlined, and have duller colors than those from streams where killifish live (bottom). The pike-cichlid prefers to eat large guppies, and the killifish feeds on small guppies. Guppies are shown approximately life-size; adult pike-cichlids grow to 16 cm in length, and adult killifish grow to 10 cm.

energy throughout their lives. Salmon and deciduous trees continue to grow until old age, whereas deer reach adult size fairly early in life. Salmon produce many offspring in a single reproductive episode, whereas deciduous trees and deer reproduce repeatedly. However, most trees produce thousands of seeds annually, whereas deer produce only one or two young each spring.

What factors have produced these variations in life history patterns? Life history traits—like all population characteristics—are modified by natural selection. Thus, organisms exhibit evolutionary adaptations that increase the fitness of individuals. Each species' life history is, in fact, a highly integrated "strategy"—not in the human sense of planning ahead, but as a suite of selection-driven adaptations.

Ecologists Analyze the Individual Components of Life Histories

In analyzing life histories, ecologists often compare the number of offspring produced with the amount of care provided to each. They also consider the number of reproductive episodes in the organism's lifetime, and the age at which it first reproduces. Because these characteristics evolve together, a change in one trait is likely to influence the success of the others.

FECUNDITY VERSUS PARENTAL CARE If a female has a fixed amount of energy for reproduction, she can package that energy in various ways. By way of illustration, a female duck with 1,000 units of energy for reproduction might lay 10 eggs that each contain 100 units of energy per egg. A salmon, which has higher fecundity, might lay 1,000 eggs, each endowed with 1 unit of energy. The amount of energy invested in each offspring *before* it is born represents the **passive parental care** that the female provides. Passive parental care is provided through yolk in an egg, endosperm in a seed, or, in mammals, nutrients that cross the placenta.

Many animals, especially birds and mammals, also provide **active parental care** to offspring *after* their birth. In general, species that produce many offspring in a reproductive episode—such as the coho salmon—provide relatively little active parental care *to each offspring.* In fact, female coho salmon, which produce 2,400 to 4,500 eggs, die before their eggs even hatch. Conversely, species that produce only a few offspring at a time—such as the European red deer—provide a lot of care to each. A red deer doe nurses its single fawn for up to 8 months before weaning it.

Although these life history differences were correlated with the distributions of the two predatory fishes, they might result from some other, unknown differences between the streams. Endler and Reznick investigated this possibility with controlled laboratory experiments. They shipped groups of guppies to California, where they bred guppies from each kind of stream for two generations. Both experimental populations were raised under identical conditions in the absence of predation. Even when predators were absent, the two experimental populations retained their life history differences. These results provided evidence of a heritable genetic basis for the observed life history differences.

Endler and Reznick also examined the role of predators in the *evolution* of the size differences. They raised guppies for many generations in the laboratory under three experimental conditions—some alone, some with killifish, and some with pike-cichlids. As predicted, the guppy lineage that was subjected to predation by killifish became larger at maturity. Individuals that were small at maturity were frequently eaten, and their reproduction was limited. The lineage that was raised with pike-cichlids showed a trend toward earlier maturity. Individuals that matured at a larger size faced a greater likelihood of being eaten before they had reproduced.

Finally, when they first visited Trinidad, Endler and Reznick had introduced guppies from a pike-cichlid stream into another stream that contained killifish but no pike-cichlids or guppies. Eleven years later, the introduced guppy population had changed. As the researchers predicted, the guppies had become larger in size and reproduced more slowly, characteristics that are typical of natural guppy populations that live and die with killifish.

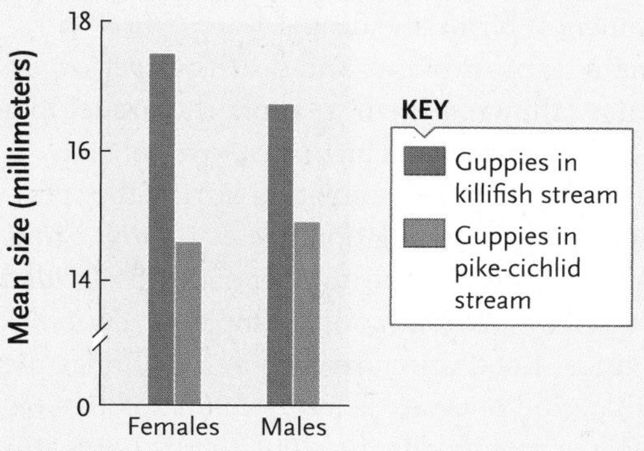
Mean size (millimeters)
18
16
14
0
Females
Males
KEY
Guppies in killifish stream
Guppies in pike-cichlid stream

FIGURE C
Guppies in streams occupied by pike-cichlids are smaller than those in streams occupied by killifish.

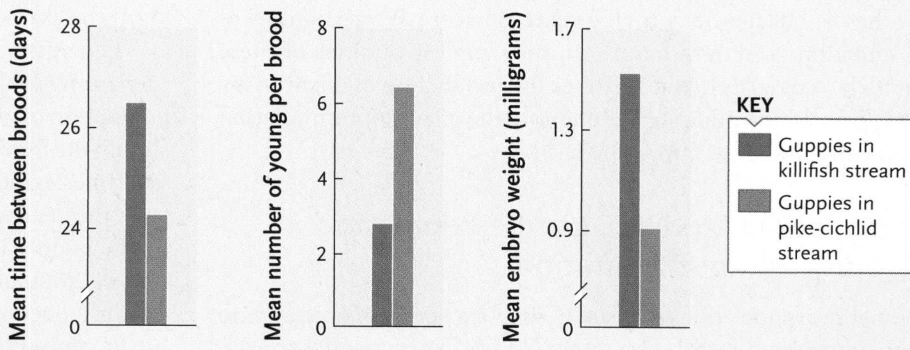
Mean time between broods (days)
28
26
24
0
Mean number of young per brood
8
6
4
2
0
Mean embryo weight (milligrams)
1.7
1.3
0.9
0
KEY
Guppies in killifish stream
Guppies in pike-cichlid stream

FIGURE D
Female guppies from streams occupied by pike-cichlids reproduce more often (shorter time between broods) and produce more young per brood and smaller young (lower embryo weight) than females living in streams occupied by killifish.

ONE REPRODUCTIVE EPISODE VERSUS SEVERAL A second life history characteristic adjusted by natural selection is the number of reproductive episodes in an organism's lifetime. Some organisms, like the coho salmon, devote all of their stored energy to a single reproductive event. Any adult that survives the upstream migration is likely to leave some surviving offspring. Other species, such as deciduous trees and red deer, reproduce multiple times. In contrast to salmon, individuals of these species devote only some of their energy budget to reproduction at any time, with the balance allocated to maintenance and growth. Moreover, in some plants, invertebrates, fishes, and reptiles, larger individuals produce more offspring than small ones do. Thus, one advantage of using only part of the energy budget for reproduction is that continued growth may result in greater fecundity at a later age. However, if an organism does not survive until the next breeding season, the potential advantage of putting energy into maintenance and growth is lost.

EARLY REPRODUCTION VERSUS LATE REPRODUCTION Individuals that first reproduce at the earliest possible age may stand a good chance of leaving some surviving offspring. But the energy devoted to reproduction is no longer available for maintenance and growth. Thus, early reproducers may be smaller and less healthy than individuals that delay reproduction in favor of these other functions. Conversely, an individual that delays reproduction may increase its chance of survival and its future fecundity by becoming larger or more experienced. But there is always some chance that it will die before the next breeding season, leaving no offspring at all. Thus, a finite energy budget and the risk of mortality establish a trade-off in the timing of first reproduction. Mathematical models suggest that delayed reproduction will be favored by natural selection if a sexually mature individual has a good chance of surviving to an older age, if organisms grow larger as they age, and if larger organisms have higher fecundity. Early reproduction will be favored if adult survival rates are low, if animals don't grow larger as they age, or if larger size does not increase fecundity.

Life history characteristics not only vary from one species to another, but they also vary among populations of a single species. *Focus on Basic Research* describes how predation influences life history characteristics in natural populations of guppies *(Poecilia reticulata)* in Trinidad.

STUDY BREAK 50.3

1. To what two broad categories of activities do children devote their energy budget?
2. Why do fecundity and the amount of parental care devoted to each offspring exhibit an inverse relationship?

50.4 Models of Population Growth

We now examine mathematical models of population growth that describe very different responses to changes in a population's density. *Exponential* models apply when populations experience unlimited growth. The *logistic* model applies when population growth is limited, often because available resources are finite. These simple models are tools that help ecologists refine their hypotheses, but neither provides entirely accurate predictions of population growth in nature. In the simplest versions of these models, ecologists define births as the production of offspring by any form of reproduction, and ignore the effects of immigration and emigration.

Exponential Models Describe Population Growth without Limitation

Sometimes populations increase in size for a period of time with no apparent limits on their growth. In models of exponential growth, population size increases steadily by a constant ratio **(Figure 50.8).** Bacterial populations provide the most obvious examples, but multicellular organisms also sometimes exhibit exponential population growth.

BACTERIAL POPULATION GROWTH Bacteria reproduce by binary fission. A parent cell divides in half, producing two daughter cells, which each divide to produce two granddaughter cells. Generation time in a bacterial population is simply the time between successive cell divisions. And if no bacteria in the population die, the population doubles in size each generation.

Bacterial populations grow quickly under ideal temperatures and with unlimited space and food. Consider a population of the human intestinal bacterium *Escherichia coli,* for which the generation time can be as short as 20 minutes. If we start with a population of one bacterium, the population doubles to two cells after one generation, to four cells after two generations, and to eight cells after three generations (Figure 50.8A). After only eight hours, or 24 generations, the population will number more than 16 million. And after a single day, or 72 generations, the population will number nearly 5×10^{21} cells. Although other bacteria grow more slowly than *E. coli,* it is no wonder that pathogenic bacteria, such as those causing cholera or plague, can quickly overtake the defenses of an infected animal.

EXPONENTIAL POPULATION GROWTH IN OTHER ORGANISMS By contrast to bacteria, many plants and animals live side-by-side with their offspring. In these populations, births increase a population's size and deaths decrease it. Over a given time period:

change in population size = number of births − number of deaths

We express this relationship mathematically by defining N as the population size; ΔN (pronounced "delta N") as the change in population size; Δt as the time period during which the change occurs; and B and D as the numbers of births and deaths, respectively, *during that time period.* Thus, $\Delta N/\Delta t$ symbolizes the change in population size over time, and

$$\Delta N/\Delta t = B - D$$

The preceding equation applies to any population for which we know the exact numbers of births and deaths.

Ecologists usually express births and deaths as *per capita* (per individual) rates, allowing them to apply the model to a population of any size. The per capita birth rate, symbolized b, is simply the number of births in the population during the specified time period divided by the population size: $b = (B/N)$. Similarly, the per capita death rate, d, is the number of deaths divided by the population size: $d = (D/N)$. If, for example, in a population of 2,000 field mice, 1,000 mice are born and 200 mice die during a one month period of time, then $b = 1{,}000/2{,}000 = 0.5$ births per individual per month, and $d = 200/2{,}000 = 0.1$ deaths

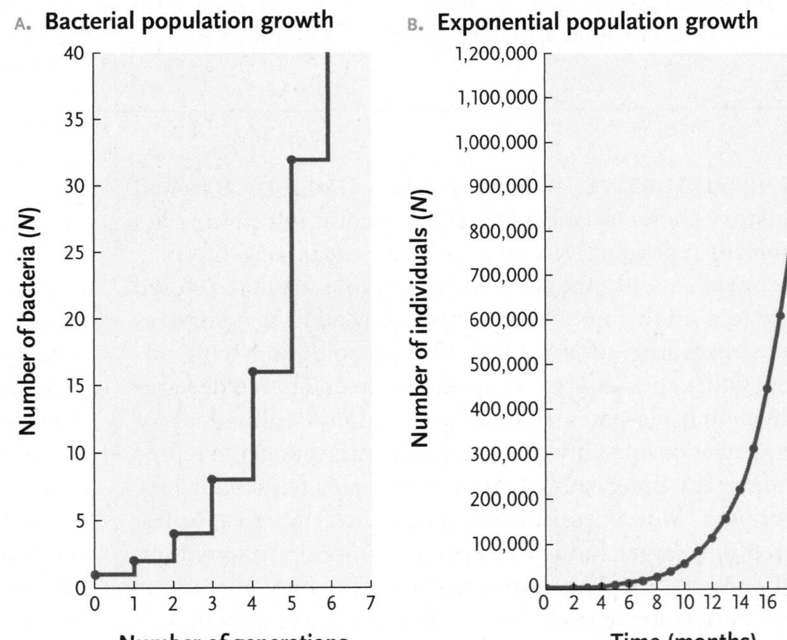
A. Bacterial population growth
Number of bacteria (N)
Number of generations
B. Exponential population growth
Number of individuals (N)
Time (months)

FIGURE 50.8
Changes in population size predicted by two models of unlimited population growth. **(A)** If all members of a bacterial population divide simultaneously, a plot of population size over time forms a stair-stepped curve in which the steps get larger as the number of dividing cells increases. **(B)** Exponential population growth produces a J-shaped curve of population size plotted against time. Although the per capita growth rate r remains constant, the increase in population size gets larger every month because more individuals are reproducing.

per individual per month. Of course, no mouse can give birth to half an offspring, and no individual can die one-tenth of a death. But these rates tell us the per capita birth and death rates *averaged over all mice in the population.* Per capita birth and death rates are always expressed over a specified time period. For long-lived organisms, such as humans, time is measured in years; for short-lived organisms, such as fruit flies, time is measured in days. We can calculate per capita birth and death rates from data in a life table.

We can now revise the population growth equation to use per capita birth and death rates instead of the actual numbers of births and deaths. The change in a population's size during a given time period ($\Delta N/\Delta t$) depends on the per capita birth and death rates, as well as on the number of individuals in the population. Mathematically, we can write

$$\Delta N/\Delta t = B - D = bN - dN = (b - d)N$$

or, in the notation of calculus,

$$dN/dt = (b - d)N$$

This equation describes the **exponential model of population growth.** (Note that in calculus, dN/dt is the notation for the population growth rate; the "d" in dN/dt is *not* the same "d" that we use to symbolize the per capita death rate.)

The difference between the per capita birth rate and the per capita death rate, $b - d$, is the **per capita growth rate** of the population, symbolized by r. Like b and d, r is always expressed per individual per unit time. Using the per capita growth rate, r, in place of $(b - d)$, the exponential growth equation is written

$$dN/dt = rN$$

If the birth rate exceeds the death rate, r has a positive value ($r > 0$), and the population is growing. In our example with field mice, $r = 0.5 - 0.1 = 0.4$ mice per mouse per month. If the birth rate is lower than the death rate, however, r has a negative value ($r < 0$), and the population is getting smaller. In populations where the birth rate equals the death rate, $r = 0$, and the population's size is not changing—a situation known as **zero population growth,** or ZPG. Even under conditions of ZPG, births and deaths still occur, but the numbers of births and deaths cancel out.

As long as a population's per capita growth rate is positive ($r > 0$), the population will increase in size. In our hypothetical population of field mice, we started with $N = 2{,}000$ mice, and calculated a per capita growth rate of 0.4 mice per individual per month. In the first month, the population grows by $0.4 \times 2{,}000 = 800$ mice. At the start of the second month, $N = 2{,}800$ and r still $= 0.4$. Thus, in the second month, the population grows by $0.4 \times 2{,}800 = 1{,}120$ mice. Notice that even though r remains constant, the *increase* in population size gets larger each month simply because more individuals are reproducing. In less than two years, the mouse population will grow to more than 1 million! A graph of exponential population growth has a characteristic J shape, getting steeper through time (Figure 50.8B). The population grows at an ever-increasing pace because the change in a population's size depends on the number of individuals in the population as well as its per capita growth rate.

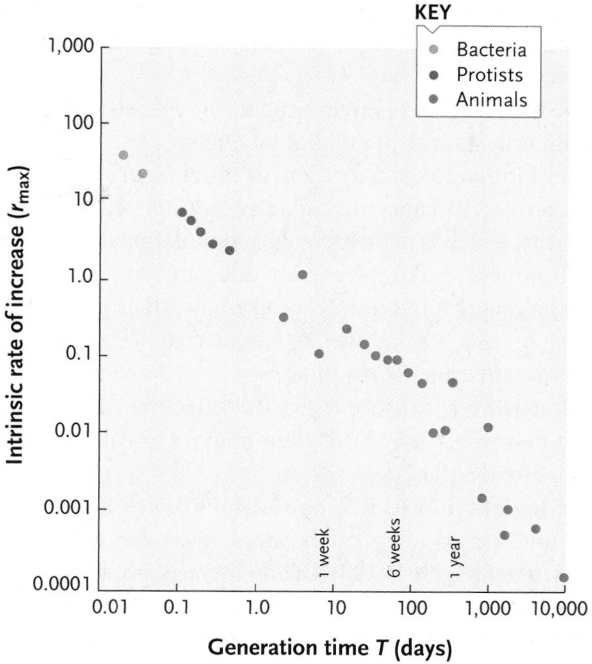
KEY
Bacteria
Protists
Animals
Intrinsic rate of increase (r_{max})
1,000
100
10
1.0
0.1
0.01
0.001
0.0001
1 week
6 weeks
1 year
0.01
0.1
1.0
10
100
1,000
10,000
Generation time T (days)

FIGURE 50.9
Generation time and r_{max}. The intrinsic rate of increase (r_{max}) is high for bacteria, protists, and animals with short generation time and low for those with long generation time. The logarithmic scale on both axes compresses the data into a straight line.

POPULATION GROWTH UNDER IDEAL CONDITIONS Imagine a hypothetical population living in an ideal environment—one with unlimited food and shelter, no predators, parasites, or disease, and a comfortable abiotic environment. Under such circumstances, which are admittedly unrealistic, the per capita birth rate is very high, the per capita death rate is very low, and the per capita growth rate, r, is as high as it can possibly be. This maximum per capita growth rate, symbolized r_{max}, is the population's **intrinsic rate of increase.** Under these ideal conditions, the exponential growth equation is

$$dN/dt = r_{max}N$$

When populations are growing at their intrinsic rate of increase, population size increases very rapidly. Across a wide variety of bacteria, protists, and animals, r_{max} varies inversely with generation time: species with short generation time (that is, those that mature quickly) have higher intrinsic rates of increase than those with long generation time **(Figure 50.9).**

The Logistic Model Describes Population Growth When Resources Are Limited

The exponential model predicts unlimited population growth. But we know from even casual observations that the population sizes of most species are somehow limited—we are not knee-deep in bacteria, rosebushes, or garter snakes. What factors limit the growth of populations? As a population gets larger, it uses more vital resources, and a shortage of resources may eventually develop. As a result, individuals may have less energy available for maintenance and reproduction, causing per capita birth rates to decrease and per capita death rates to increase. Changes in these rates reduce the population's per capita growth rate, causing population growth to slow down or to stop altogether.

THE LOGISTIC MODEL Environments provide enough resources to sustain only a finite population of any species. The maximum number of individuals that an environment can support indefinitely is termed its **carrying capacity,** symbolized *K*. The carrying capacity, which is defined for each population, is a property of the environment, and it varies from one habitat to another and in a single habitat through time. For example, the spring and summer flush of insects in temperate habitats supports large populations of insectivorous birds. But fewer insects are available in autumn and winter, causing a seasonal decline in the carrying capacity for birds. Many birds then migrate to habitats that provide more food and better weather.

The **logistic model of population growth** assumes that a population's per capita growth rate, *r*, decreases as the population gets larger **(Figure 50.10A).** In other words, population growth slows as the population size approaches the carrying capacity. The mathematical expression $(K - N)$ tells us how many individuals can be added to a population before it reaches carrying capacity. And the expression $(K - N)/K$ indicates what *percentage* of the carrying capacity is still available.

To create the logistic model, we factor the impact of carrying capacity into the exponential model by letting $r = r_{max}(K - N)/K$. This calculation reduces the per capita growth rate *r* from its maximum value (r_{max}) as *N* increases:

$$dN/dt = r_{max}N(K - N)/K$$

The calculation of how *r* varies with population size is straightforward **(Table 50.2).** In a very small population (*N* is much smaller than *K*), plenty of resources are still available; the value of $(K - N)/K$ is close to 1, and the per capita growth rate *r* is therefore close to the maximum possible (r_{max}). Under these conditions, population growth is close to exponential. If a population is large (*N* is close to *K*), few additional resources are available, the value of $(K - N)/K$ is small, and the per capita growth rate *r* is very low. When the size of the population exactly equals the carrying capacity, $(K - N)/K$ becomes zero, and so does the population growth rate—the situation defined as ZPG.

TABLE 50.2 The Effect of N on r and ΔN* in a Hypothetical Population Exhibiting Logistic Growth in which $K = 2{,}000$ and $r_{max} = 0.04$ per capita per year

N (population size)	$(K - N)/K$ (% of K available)	$r = r_{max}(K - N/K)$ (per capita growth rate)	$\Delta N = rN$ (change in N)
50	0.990	0.0396	2
100	0.950	0.0380	4
250	0.875	0.0350	9
500	0.750	0.0300	15
750	0.625	0.0250	19
1,000	0.500	0.0200	20
1,250	0.375	0.0150	19
1,500	0.250	0.0100	15
1,750	0.125	0.0050	9
1,900	0.050	0.0020	4
1,950	0.025	0.0010	2
2,000	0.000	0.0000	0

*ΔN rounded to the nearest whole number.

The logistic model of population growth predicts an S-shaped graph of population size over time, with the population slowly approaching its carrying capacity and remaining at that level **(Figure 50.10B).** According to this model, the population grows slowly when the population size is small, because there are few individuals reproducing. It also grows slowly when the popula-

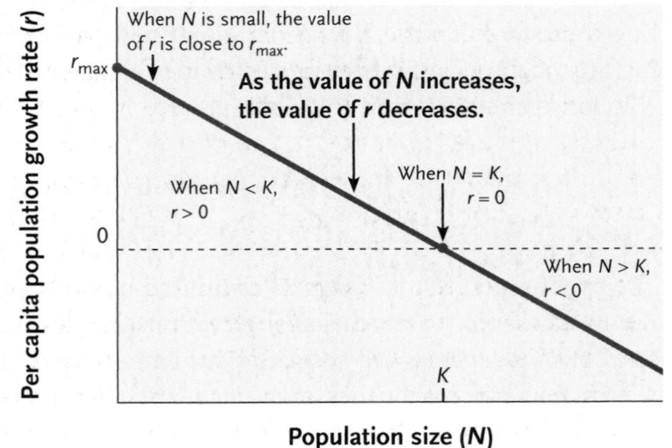
A. The predicted effect of N on r
Per capita population growth rate (r)
r max
When N is small, the value of r is close to r max.
As the value of N increases, the value of r decreases.
When N < K, r > 0
When N = K, r = 0
0
When N > K, r < 0
K
Population size (N)

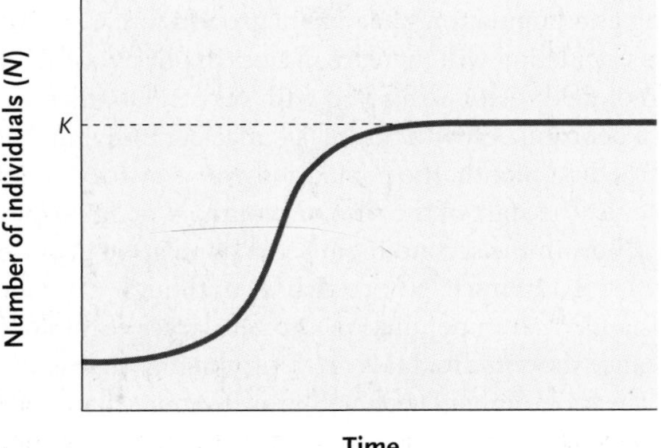
B. Population size through time
Number of individuals (N)
K
Time

FIGURE 50.10
The logistic model of population growth. **(A)** The logistic model assumes that the per capita population growth rate *r* decreases linearly as population size (*N*) increases. **(B)** The logistic model predicts that population size increases quickly at first, but then slowly approaches the carrying capacity (*K*).

tion size is large because the per capita population growth rate is low. The population grows quickly (dN/dt is highest) at intermediate population sizes, when a sizable number of individuals are breeding and the per capita population growth rate r is still fairly high (see Table 50.2).

INTRASPECIFIC COMPETITION The logistic model assumes that vital resources become increasingly limited as a population grows larger. Thus, the model is a mathematical portrait of **intraspecific** (within species) **competition,** the dependence of two or more individuals in a population on the same limiting resource. For animals, limiting resources can be food, water, nesting sites, refuges from predators, and, for sessile species (those permanently attached to a surface), space. For plants, sunlight, water, inorganic nutrients, and growing space can be limiting. The pattern of uniform dispersion described earlier often reflects intraspecific competition for limited resources.

In some very dense populations, the accumulation of poisonous waste products may also reduce survivorship and reproduction. Most natural populations live in open systems where wastes are consumed by other organisms or flushed away. But the buildup of toxic wastes is common in laboratory cultures of microorganisms. For example, yeast cells ferment sugar and produce ethanol as a waste product. Thus, the alcohol content of wine rarely exceeds 13% by volume, the ethanol concentration that poisons winemaking yeasts.

LOGISTIC GROWTH IN THE LABORATORY AND IN NATURE How well do species conform to the predictions of the logistic model? In simple laboratory cultures, relatively small organisms, such as *Paramecium,* some crustaceans, and flour beetles, often show an S-shaped pattern of population growth. Moreover, large animals that have been introduced into new environments sometimes exhibit a pattern of population growth that matches the predictions of the logistic model **(Figure 50.11).**

Nevertheless, some assumptions of the logistic model are unrealistic. For example, the model predicts that survivorship and fecundity respond immediately to changes in a population's density. But many organisms exhibit a delayed response, called a **time lag.** Some time lags occur because fecundity is usually determined by the availability of resources at some time in the past, when individuals were adding yolk to eggs or endosperm to seeds. Moreover, when food resources become scarce, individuals may use stored energy reserves to survive and reproduce, and the effects of crowding may not be felt until those reserves are depleted. As a result, the population size may temporarily overshoot its carrying capacity (Figure 50.11, center). Deaths may then outnumber births, causing the population size to drop below the carrying capacity, at least temporarily. Time lags can cause a population to oscillate around its carrying capacity.

Another unrealistic assumption of the logistic model is that the addition of new individuals to a population always decreases survivorship and fecundity, no matter how small the population is. But in small populations, modest population growth probably doesn't have much effect on these processes. In fact, most organisms probably require a minimum population density to survive and reproduce. For example, some plants flourish in small clumps that buffer them from physical stresses, whereas a single individual living in the open would suffer adverse effects. And in some animal populations, a minimum population density is necessary for individuals to find mates—an important issue in conservation biology (see Chapter 53).

STUDY BREAK 50.4

1. **How does the prediction of the exponential model of population growth differ from that of the logistic model?**
2. **What is carrying capacity? Is it a property of a habitat or of a population?**
3. **What is a time lag?**

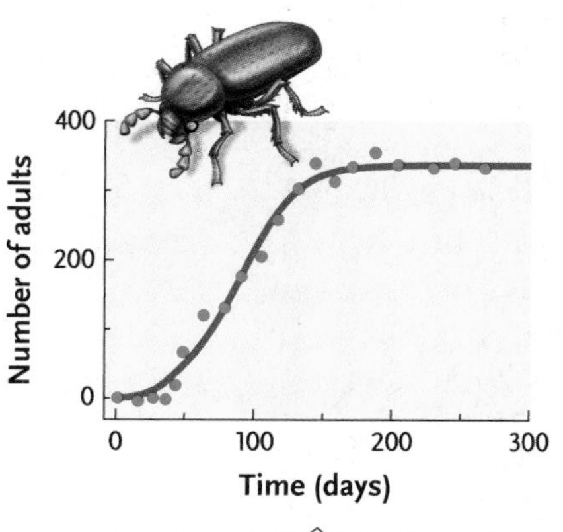

Number of adults
400
200
0
0
100
200
300
Time (days)

A laboratory population of the grain borer beetle *Rhizopertha dominica* showed logistic growth when its food was replenished weekly.

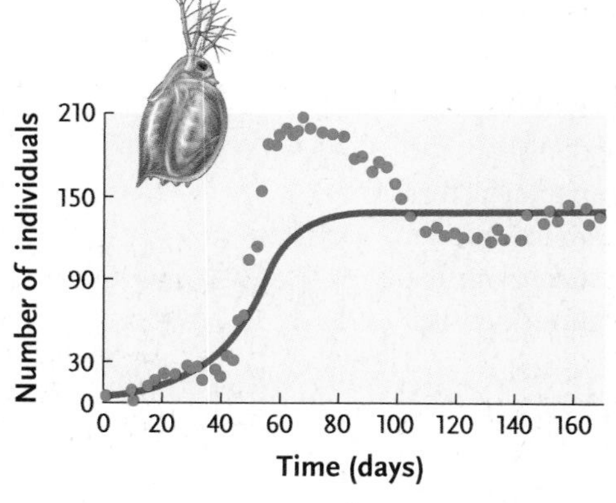

Number of individuals
210
150
90
30
0
0 20 40 60 80 100 120 140 160
Time (days)

A laboratory population of the water flea *Daphnia magna* overshot its carrying capacity; when population density increased, individuals relied upon stored energy reserves, causing a time lag in the appearance of density-dependent effects.

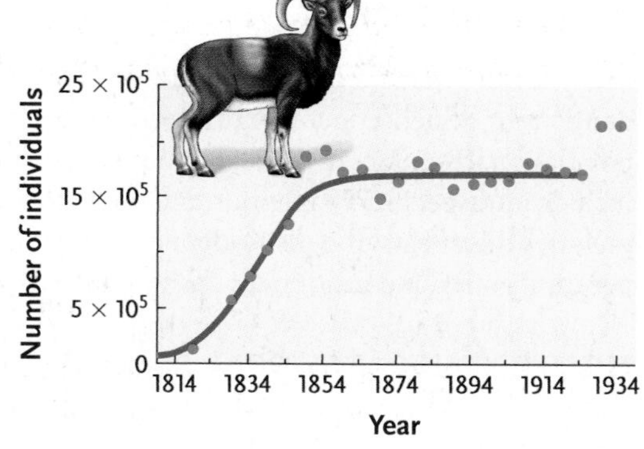

Number of individuals
25 × 10^5
15 × 10^5
5 × 10^5
0
1814 1834 1854 1874 1894 1914 1934
Year

European mouflon sheep *(Ovis musimon)* introduced into Tasmania exhibited logistic population growth; these data represent 5-year averages, smoothing out annual fluctuations in population size.

KEY

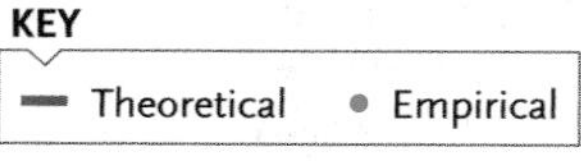

Theoretical
Empirical

FIGURE 50.11
Examples of logistic population growth.

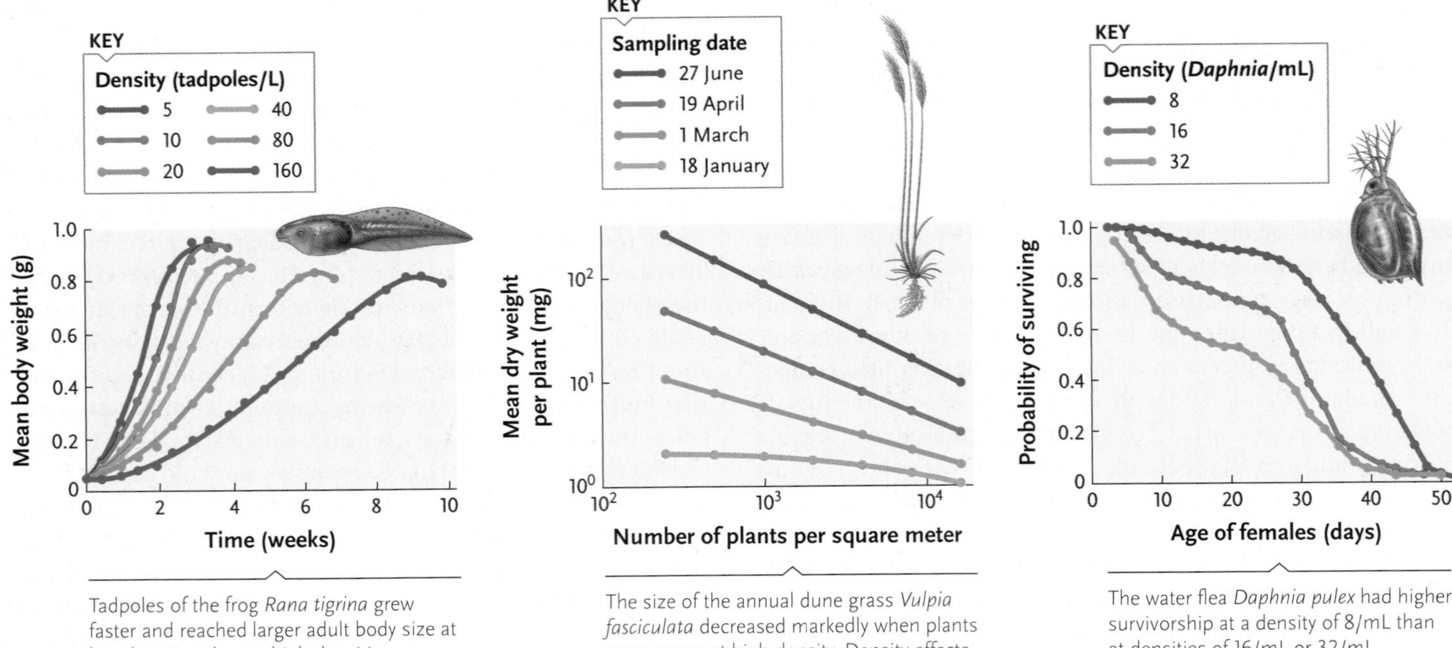

KEY
Density (tadpoles/L)
5
10
20
40
80
160
Mean body weight (g)
1.0
0.8
0.6
0.4
0.2
0
0
2
4
6
8
10
Time (weeks)
KEY
Sampling date
27 June
19 April
1 March
18 January
Mean dry weight per plant (mg)
10^2
10^1
10^0
10^2
10^3
10^4
Number of plants per square meter
KEY
Density (*Daphnia*/mL)
8
16
32
Probability of surviving
1.0
0.8
0.6
0.4
0.2
0
0
10
20
30
40
50
Age of females (days)

Tadpoles of the frog *Rana tigrina* grew faster and reached larger adult body size at low densities than at high densities.

The size of the annual dune grass *Vulpia fasciculata* decreased markedly when plants were grown at high density. Density effects became more accentuated through time, as the plants grew larger (indicated by the progressively steeper slopes of the lines).

The water flea *Daphnia pulex* had higher survivorship at a density of 8/mL than at densities of 16/mL or 32/mL.

FIGURE 50.12
Effects of crowding on individual growth, size, and survival.

50.5 Population Dynamics

Long-term studies on many species have shown that the size, density, age structure, and geographical ranges of most populations change somewhat from season to season and from year to year. As you have seen, some populations experience dramatic variations in size and other characteristics through time, whereas other populations appear to be much more stable. What environmental factors influence these aspects of population dynamics? Why do these characteristics fluctuate more in some populations than in others?

Density-Dependent Factors Often Regulate Population Size

Many factors that affect population dynamics are **density-dependent:** their influence increases or decreases with the density of the population. Examples of density-dependent environmental factors include intraspecific competition and predation. The logistic model includes the effects of density-dependence in its assumption that per capita birth and death rates change with a population's density.

THE EFFECTS OF CROWDING Numerous laboratory and field studies show that crowding (high population density) decreases the individual growth rate, adult size, and survival of plants and animals **(Figure 50.12).** Organisms living in extremely dense populations are unable to harvest enough resources; they grow slowly and tend to be small, weak, and less likely to survive. Gardeners understand this relationship, thinning their plants to a density that maximizes the number of vigorous individuals.

Crowding also has a negative effect on reproduction **(Figure 50.13).** When resources are in short supply, each individual

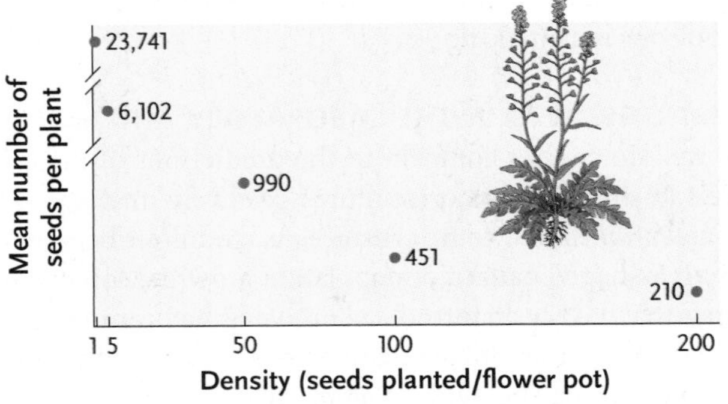

Mean number of seeds per plant
23,741
6,102
990
451
210
1
5
50
100
200
Density (seeds planted/flower pot)

The number of seeds produced by shepherd's purse *(Capsella bursa-pastoris)* decreased dramatically with increasing density in experimental pots.

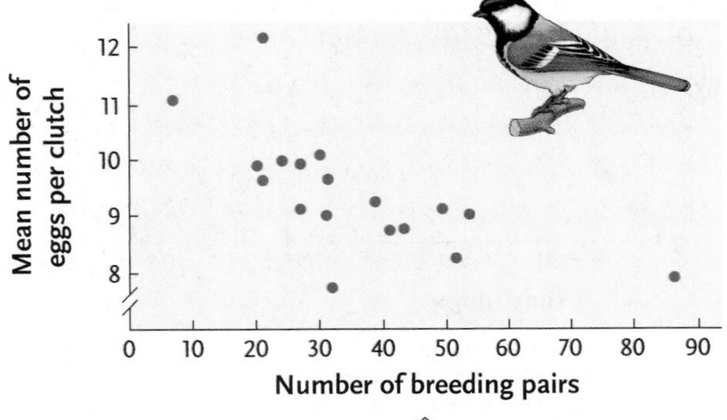

Mean number of eggs per clutch
12
11
10
9
8
0
10
20
30
40
50
60
70
80
90
Number of breeding pairs

The mean number of eggs produced by the great tit *(Parus major)*, a woodland bird, declined as the number of breeding pairs in Marley Wood increased.

FIGURE 50.13
Effects of crowding on fecundity.

© Alain Beignet BIOS/Peter Arnold, Inc.

FIGURE 50.14
A swarm of locusts. Migratory locusts *(Locusta migratoria)*, moving across an African landscape, can devour their own weight in plant material every day.

has less energy available for reproduction after meeting its basic maintenance needs. Hence, females in crowded populations produce either fewer offspring or smaller offspring that are less likely to survive.

In some species, crowding stimulates developmental and behavioral changes that may influence the density of a population. For example, migratory locusts *(Locusta migratoria)* can develop into either solitary or migratory forms in the same population. Migratory individuals have longer wings and more body fat, characteristics that allow them to disperse great distances. High population density increases the frequency of the migratory form, and huge numbers of locusts move away from the area of high density **(Figure 50.14)**, reducing the size of the original population.

These studies confirm the assumptions of the logistic equation, but they don't prove that natural populations are regulated by density-dependent factors. A convincing demonstration requires experimental evidence that an increase in population density causes population size to decrease, and that a decrease in density causes it to increase. In one study conducted in the 1960s, Robert Eisenberg of the University of Michigan experimentally increased the numbers of aquatic snails in some ponds, decreased them in others, and maintained natural densities in control ponds. Although adult survivorship did not differ between experimental and control treatments, snails in the high-density ponds produced fewer eggs, and those in the low-density ponds produced more eggs than those living at the control density. In addition, the survival rates of young snails declined as density increased. After four months, the densities in the two experimental groups converged on those in the control, providing strong evidence of density-dependent population regulation.

OTHER DENSITY-DEPENDENT FACTORS Our discussion of the logistic equation described intraspecific competition as the primary density-dependent factor regulating population size. Competition between populations of different species also exerts density-dependent effects on population growth, a topic we consider in Section 51.1.

Predation can also cause density-dependent population regulation. As a particular prey species becomes more numerous, predators may consume more of it because it is easier to find and catch. Once a prey species has exceeded a threshold density, predators may consume a *larger percentage* of the prey population, which is a density-dependent effect (see Figure 20.16). For example, on rocky shores in California, sea stars concentrate their feeding on the most abundant of several invertebrate species. When one prey species becomes common, predators feed on it disproportionately, drastically reducing its numbers.

Like predation, parasitism and disease cause density-dependent regulation of plant and animal populations. Infectious microorganisms spread quickly in a crowded population. In addition, if individuals crowded together are weak or malnourished, they are more susceptible to infection and may die from diseases that healthy organisms would survive.

Global Climate Change Is Increasing the Impact of Density-Independent Factors

Some populations are affected by **density-independent** factors, which reduce population size regardless of its density. If an insect population is not physiologically adapted to high temperature, a sudden hot spell may kill 80% of the insects whether they number 100 or 100,000. Fires, earthquakes, storms, and other natural disturbances may contribute directly or indirectly to density-independent mortality. But because such factors do not cause a population to fluctuate around its carrying capacity, density-independent factors do not *regulate* population size, although they may reduce it.

For many years, ecologists have recognized the strong effects of density-independent factors on populations of small-bodied species that cannot buffer themselves against environmental change. Their populations grow exponentially for a time, but shifts in climate or random events cause high mortality before populations reach a size at which density-dependent factors regulate their numbers. When conditions improve, populations grow exponentially—at least until another density-independent factor causes them to crash again. For example, a small Australian insect, *Thrips imaginis,* feeds on pollen and flowers of plants in the rose family; they are frequently abundant enough to damage the blooms. *Thrips* populations grow exponentially in spring, when many flowers are available and the weather is warm and moist **(Figure 50.15)**. But populations crash predictably during summer because *Thrips* do not tolerate extremely hot and dry conditions. After the crash, a few individuals survive in remaining flowers, forming the stock from which the population grows exponentially the following spring.

Recent research on the effects of global climate change suggests that the very rapid warming observed since the middle of the twentieth century (see Chapter 52) has also had a signifi-

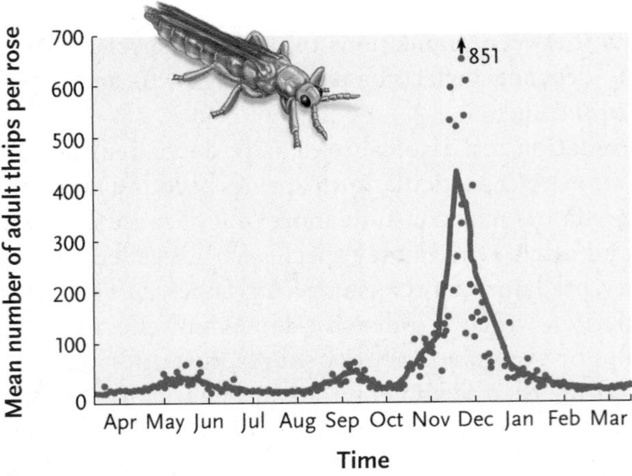
Mean number of adult thrips per rose
700
600
500
400
300
200
100
0
851
Apr May Jun Jul Aug Sep Oct Nov Dec Jan Feb Mar
Time

FIGURE 50.15

Booms and busts in a *Thrips* population. Populations of the Australian insect *Thrips imaginis* grow exponentially when conditions are favorable during spring (which begins in September in the southern hemisphere). The populations crash in summer, however, when hot and dry conditions cause high mortality.

cant impact on the population ecology of large, long-lived organisms. For example, in 2009, Phillip J. van Mantgem of the U.S. Geological Survey and colleagues from seven other institutions published an analysis of data collected between 1955 and 2007 in undisturbed coniferous forests of western North America. Their survey, which included 76 stands of trees between 200 and 1,000 years old, revealed that average mortality rates had increased 4% per year; the increase was most apparent in the Pacific Northwest, where mortality rates doubled in as little as 17 years **(Figure 50.16).** Increased mortality was evident at all elevations and in trees of all species, sizes, and ages.

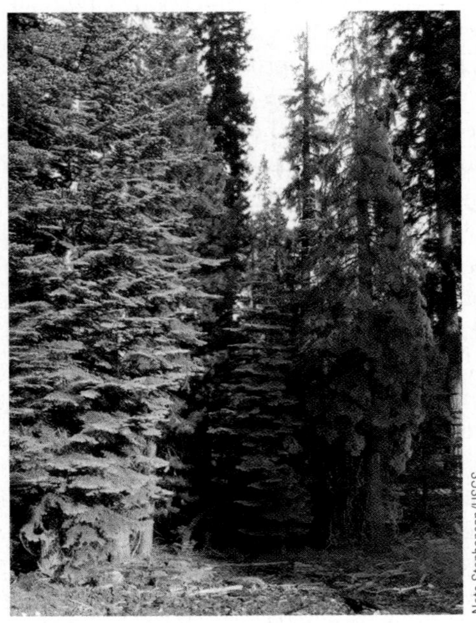
Nate Stephenson/USGS

FIGURE 50.16

Global climate change and mortality in trees. Like other conifers in western North America, red fir trees *(Abies magnifica)* in Sequoia National Park, California, have experienced increased mortality, almost certainly as a result of climate warming.

After considering several possible causes of the increased mortality, the researchers concluded that climate warming—about 0.5°C per decade in western North America—and the concomitant lengthening of summer droughts was the major contributing factor. The extended droughts not only prevent tree seedlings from establishing themselves and renewing the forests, but they also limit the photosynthetic activity of established trees and leave them more vulnerable to infestation by insect and fungal pests. These environmental stresses may lead to the large-scale death of western North American forests.

Interacting Environmental Factors Influence Population Dynamics in Complex Ways

Sometimes several density-dependent factors influence a population at the same time. For example, on small islands in the West Indies, the spider *Metepeira datona* is rare wherever lizards (*Ameiva festiva, Anolis carolinensis,* and *Anolis sagrei*) are abundant, but common where the lizards are rare or absent. To test whether the presence of lizards limits the abundance of spiders, David Spiller and Tom Schoener of the University of California, Davis, built screened fences around plots on islands where these species occur. They eliminated lizards from experimental plots, but left them in control plots. After two years, spider populations in some experimental plots were five times denser than those in control plots **(Figure 50.17).** In this case, lizards had two density-dependent effects on spider populations: they preyed upon spiders, and they competed with them for food.

Density-dependent factors can also interact with density-independent factors, limiting population growth. For example, food shortage caused by high population density (a density-dependent factor) may lead to malnourishment; in turn, malnourished individuals may be more likely to succumb to the stress of extreme weather (a density-independent factor).

Populations can also be affected by density-independent factors in a density-dependent manner. For example, animals often retreat into shelters to escape environmental stresses, such as floods or severe heat. If a population is small, most individuals can fit into a limited number of available refuges. But if a population is large, only a small proportion will find suitable shelter; the larger the population is, the greater the percentage of individuals that will experience the stress. Thus, although the density-independent effects of weather limit *Thrips* populations, it is the availability of flowers in summer—clearly a density-dependent factor—that regulates the size of the *Thrips* stock from which the population grows the following spring. Hence, both types of factors influence the size of *Thrips* populations.

Life History Characteristics Govern Fluctuations in Population Size through Time

Even casual observation reveals tremendous variation in how rapidly population size changes in different species. For example, new weeds often appear in a vegetable garden overnight, whereas the number of oak trees in a forest may remain relatively stable for years. Why do some species have the potential for explosive pop-

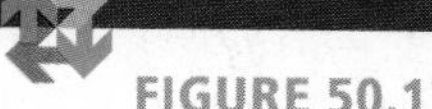

FIGURE 50.17 Experimental Research

Evaluating Density-Dependent Interactions between Species

Question: Does the population density of lizards on Caribbean islands have any effect on the population density of spiders?

Experiment: Spiller and Schoener built fences around a series of study plots on a small island in the Bahamas. They excluded all individuals of three lizard species from the experimental plots, but left resident lizards undisturbed in the control plots. They then made monthly measurements of population densities of the web-building spider *Metepeira datona* in both experimental plots and control plots.

Results: Over the 20-month course of the experiment, spider densities were as much as five times higher in the experimental plots than in the control plots.

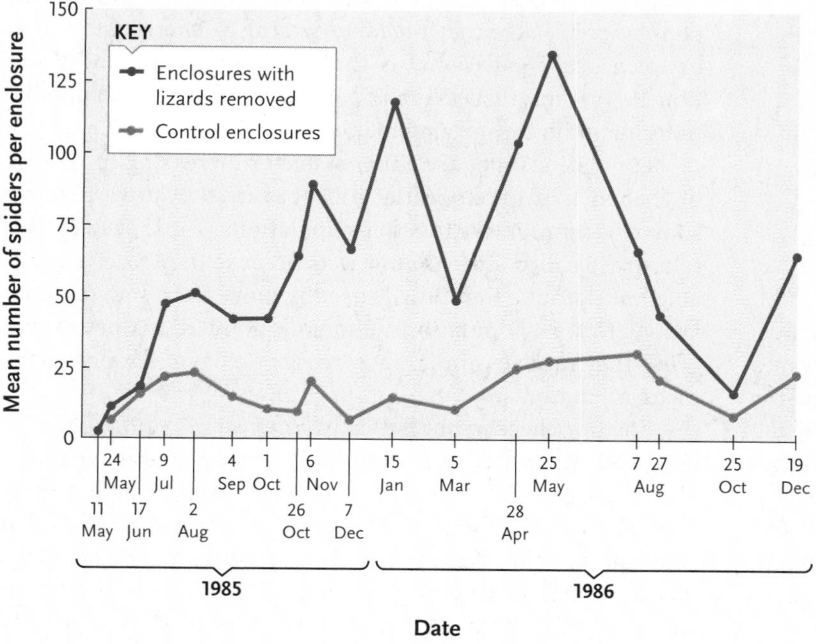

KEY
Enclosures with lizards removed
Control enclosures
Mean number of spiders per enclosure
150
125
100
75
50
25
0
24 May
9 Jul
4 Sep
1 Oct
6 Nov
15 Jan
5 Mar
25 May
7 Aug
27 Aug
25 Oct
19 Dec
11 May
17 Jun
2 Aug
26 Oct
7 Dec
28 Apr
1985
1986
Date

Conclusion: Spiller and Schoener concluded that the presence of lizards has a large impact on spider populations. The lizards not only compete with the spiders for insect food, but they also appear to prey on the spiders.

Source: D. A. Spiller and T. W. Schoener. 1988. An experimental study of the effects of lizards on web-spider communities. *Ecological Monographs* 58:57–77.

A. **An *r*-selected species**

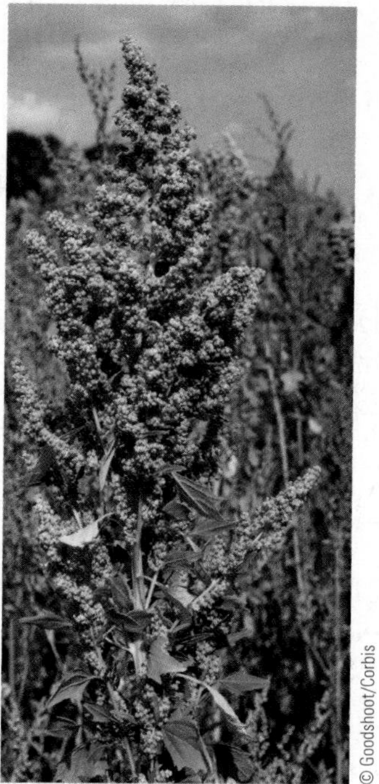
© Goodshoot/Corbis

B. **A *K*-selected species**

Nigel Cattlin/Holt/Holt Studios International, Ltd.

FIGURE 50.18
Life history differences. (A) An *r*-selected species, like quinoa *(Chenopodium quinoa)*, matures in one growing season and produces many tiny seeds, which were a traditional food staple for the indigenous people of North and South America. **(B)** A *K*-selected species, like the coconut palm *(Cocos nucifera)*, grows slowly and produces a few large seeds repeatedly during its long life.

ulation growth, but others do not? The answer lies in how natural selection has molded life history strategies that are adapted to different ecological conditions. Ecologists describe two divergent life history patterns—***r*-selected** species and ***K*-selected** species—with very different characteristics **(Table 50.3, Figure 50.18).** These strategies represent extremes on a continuum of possible patterns, and the life histories of most species actually fall somewhere between them.

Species with an *r*-selected life history are adapted to function well in rapidly changing environments. They are generally small, have short generation times, and produce numerous, tiny offspring, often in a single reproductive event. The offspring receive little or no parental care of any kind. Because species with short generation times tend to have high r_{max} (see Figure 50.9), their populations grow exponentially when environmental conditions are favorable—hence the name *r*-selected. Although their numerous offspring disperse and rapidly colonize available habitats, most die before reaching sexual maturity (Type III survivorship). Thus, the success of an *r*-selected life history depends on flooding the environment with a *large quantity* of young, only a few of which may be successful.

Because they have small body size, *r*-selected species lack physiological mechanisms to buffer them from environmental variation. Thus, as described earlier for the Australian thrips liv-

TABLE 50.3 Characteristics of *r*-Selected and *K*-Selected Species

Characteristic	*r*-Selected Species	*K*-Selected Species
Maturation time	Short	Long
Life span	Short	Long
Mortality rate	Usually high	Usually low
Reproductive episodes	Usually one	Usually several
Time of first reproduction	Early	Late
Clutch or brood size	Usually large	Usually small
Size of offspring	Small	Large
Active parental care	Little or none	Often extensive
Population size	Fluctuating	Relatively stable
Tolerance of environmental change	Generally poor	Generally good

ing in roses, survivorship and fecundity are often greatly influenced by density-independent factors, and population size fluctuates markedly. In good years, survivorship and fecundity may be high, and the population explodes. In bad years, survivorship and fecundity may be low, and the population crashes. Populations of *r*-selected species are often so greatly reduced by changes in abiotic environmental factors, such as temperature or moisture, that they never grow large enough to face a shortage of limiting resources. Thus, the carrying capacity for the species cannot be estimated, and changes in their population size cannot be described by the logistic model of population growth.

By contrast, *K*-selected species thrive in more stable environments. They are generally large, have long generation times, and produce offspring repeatedly during their lifetimes. Their offspring receive substantial parental care, either as energy reserves in an egg or seed or as active care, ensuring that most survive the early stages of life (Type I or Type II survivorship). Because *K*-selected species typically have a low r_{max}, their populations often grow slowly. The success of a *K*-selected life history therefore depends on the production of a relatively small number of *high quality* offspring that join an already well-established population.

The large body size of *K*-selected species allows them to use behavioral and physiological mechanisms to buffer themselves against environmental change, so that survivorship and fecundity do not fluctuate wildly in response to environmental variations. Instead, their populations are often affected by density-dependent factors, which regulate population size near the carrying capacity for the species—hence the name *K*-selected. For these species, natural selection has favored life history characteristics that result in stable population sizes: the production of relatively few offspring, extensive parental care, good competitive ability, a long life span, and repeated reproductions. Many large terrestrial vertebrates are examples of *K*-selected species.

Metapopulation Structure Allows Local Populations to Exchange Individuals

Although the models of population growth discussed earlier ignore the effects of immigration and emigration, individuals frequently disperse from one local population to another. To describe the dynamics of such movements, ecologists define a **metapopulation** as a group of neighboring populations that exchange individuals. All local populations within a metapopulation are not equal; they often differ in size, population growth rates, the suitability of their habitats, their exposure to predators, and other factors. Moreover, some may decline steadily (even to extinction), while others may increase.

Under favorable circumstances, a population may produce numerous offspring, some of which emigrate and join nearby populations, where they breed, providing a genetic connection between local populations (see the discussion of gene flow in Section 20.3). Thus, dispersal and gene flow between local populations maintain the metapopulation.

Populations that are either stable or increasing in size are described as **source populations** because they are a possible source of immigrants to other populations. Those that decline in size are called **sink populations** because they receive available immigrants. Individuals usually move from source populations to sink populations, and sink populations persist because they receive immigrants from source populations in the metapopulation.

The bay checkerspot butterfly, *Euphydryas editha bayensis* **(Figure 50.19),** provides an example of metapopulation dynamics. This species is restricted to serpentine grassland in the San Francisco Bay area (see Figure 51.18) because its larvae eat plants that grow only in that community. Human disturbance has fragmented much of the butterfly's natural habitat into patches of varying size, each of which may support a local butterfly population. The life cycle of these butterflies is always a race against time, because the larvae must feed and mature before dry summer weather kills their food plants. Populations in small patches often become extinct, but those occupying larger patches, where food plants stay alive longer, generally survive the seasonal drought. (Global climate warming has exacerbated the problem in recent years, causing food plants to die earlier in the season—and pushing the butterfly to the brink of extinction.) Butterfly populations in larger habitat patches still serve as source populations for emigrants that repopulate small habitat patches the following year. But the bay checkerspot is a poor flyer, and it cannot disperse long distances. Thus, small patches of suitable habitat harbor bay checkerspots only if they are close to a larger patch that serves as a source.

Some Species Exhibit Regular Cycles in Population Size

The population densities of many insects, birds, and mammals in the northern hemisphere fluctuate between species-specific lows and highs in a multiyear cycle. Arctic populations of small rodents vary in size over a four-year cycle, whereas snowshoe hares,

FIGURE 50.19 **Observational Research**

Do Immigrants from Source Populations Prevent Extinction of Sink Populations of the Bay Checkerspot Butterfly?

Paul Ehrlich

Map of serpentine habitat patches near Morgan Hill

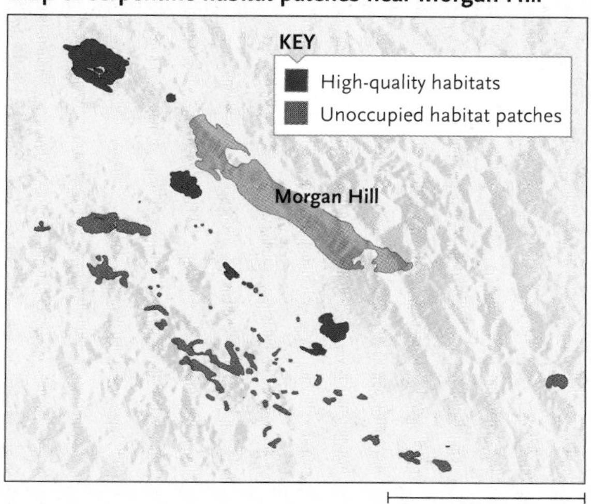

KEY
High-quality habitats
Unoccupied habitat patches
Morgan Hill
10 km

Hypothesis: Populations of the bay checkerspot butterfly *(Euphydryas editha bayensis)* living on small patches of suitable habitat are "sink" populations that frequently become extinct. Populations in large habitat patches can serve as a "source" of individuals to recolonize small habitat patches nearby.

Prediction: Because the bay checkerspot butterfly is a weak flyer, small patches of suitable habitat that are close to a large source population will be recolonized frequently. Patches of suitable habitat that are far from a large source population will be recolonized only rarely.

Method: Susan Harrison, Dennis D. Murphy, and Paul R. Ehrlich of Stanford University surveyed 59 small patches of serpentine grassland near San Jose, California, in 1986 and 1987. They estimated the "quality" of each patch based on the presence or absence of food plants on which bay checkerspots depend and on aspects of the physical environment that are important to these butterflies. They also measured the distance of each patch from Morgan Hill, a very large patch of suitable habitat that had sustained a bay checkerspot population for many years. In patches where they found butterflies, they estimated bay checkerspot population sizes.

Results: A complex statistical analysis revealed that both distance from the Morgan Hill population and habitat patch quality were important factors in determining whether bay checkerspots would be present or absent in small habitat patches. The authors noted that only the nine high-quality habitat patches near Morgan Hill (red on the map) were occupied by bay checkerspots. Of 50 unoccupied habitat patches, 6 were near Morgan Hill but of low quality; 18 were of high quality but far from Morgan Hill; and 26 were too far from Morgan Hill and of too low quality to support a population of bay checkerspots.

Conclusion: Populations of bay checkerspot butterflies that occupy large patches of suitable habitat serve as source populations for individuals that recolonize small patches of suitable habitat where butterfly populations frequently become extinct. However, because the bay checkerspot is a weak flyer, it recolonizes small patches of suitable habitat only if they are close to a source population.

Source: S. Harrison et al. 1988. Distribution of the bay checkerspot butterfly, *Euphydryas editha bayensis:* Evidence for a metapopulations model. *The American Naturalist* 132:360–382.

ruffed grouse, and lynxes have 10-year cycles. Ecologists documented such cyclic fluctuations more than a century ago, but none of the general hypotheses so far proposed explains the cycles in all species. The availability and quality of food, the abundance of predators, the prevalence of disease-causing microorganisms, and variations in weather may influence population growth. Furthermore, a cycling population's food supply and predators are themselves influenced by the population's size.

Theories of *intrinsic control* suggest that as an animal population grows, individuals undergo hormonal changes that increase aggressiveness, reduce reproduction, and foster dispersal to other areas. The dispersal phase of the cycle may be dramatic. For example, when populations of the Norway lemming *(Lemmus lemmus),* a rodent that lives in the Scandinavian arctic, reach their peak density, aggressive interactions drive younger and weaker individuals away from their place of birth. The exodus of many thousands of lemmings, scrambling over rocks and even cliffs, was sometimes incorrectly portrayed in nature films as a suicidal mass migration. Researchers do not yet know how widespread these hormonal and behavioral changes are among different species or exactly what regulates them.

Other explanations focus on *extrinsic control,* such as the relationship between a cycling species and its food or predators. A dense population may exhaust its food supply, increasing mortality and decreasing reproduction. But experimental food supplementation does not always prevent a decline in mammal populations, indicating that other factors are also at work.

Some mathematical models as well as some laboratory experiments on protists or small arthropods suggest that the cycles of predators and their prey are induced by time lags in each population's response to changes in density of the other **(Figure 50.20).** In the past, the 10-year cycles of snowshoe hares *(Lepus americanus)* and their feline predators, Canada lynxes *(Lynx canadensis),* were often cited as a classic example of such an interaction. But ecological relationships are often complex, and recent research has cast doubt on this straightforward explanation. Hare populations exhibit a 10-year fluctuation even on islands where lynxes are absent. Thus, the lynx cannot be solely responsible for the hare's cycle, although cycles in the hare populations may trigger cycles in populations of their predators.

Charles Krebs and his colleagues at the University of British Columbia studied hare and lynx interactions with a large-scale,

Tom Brakefield/Getty Images

FIGURE 50.20

The predator–prey model. Predator–prey interactions may contribute to density-dependent regulation of both populations. **(A)** A mathematical model predicts cycles in the numbers of predators and prey because of time lags in each species' responses to changes in the density of the other. (Predator population size is exaggerated in this graph: predators are usually less common than prey.) **(B)** The interaction between the Canada lynx *(Lynx canadensis)* and the snowshoe hare *(Lepus americanus)* was often described as a cyclic predator–prey interaction. The abundances of lynx and hare are based on counts of pelts that trappers sold to Hudson's Bay Company over a 90-year period. Recent research has shown that population cycles in snowshoe hares are caused by complex interactions between the hare, its food plants, and its predators.

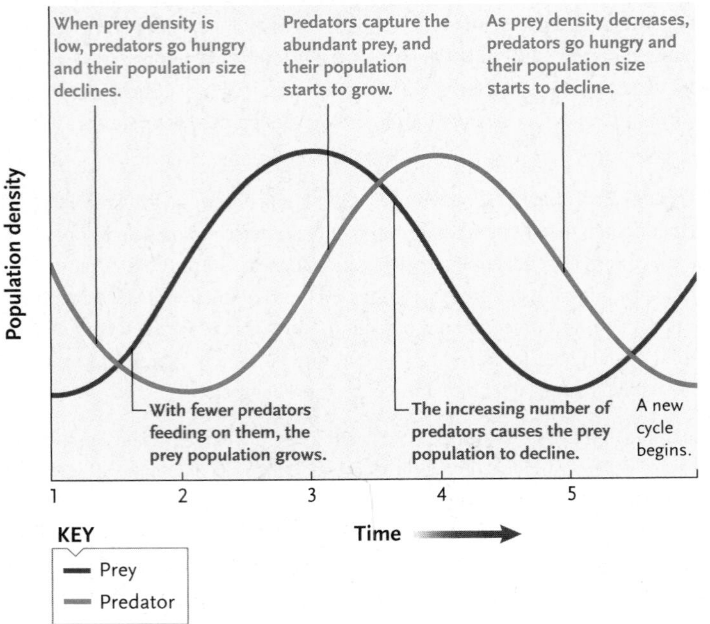

A. Predictions of a predator–prey model
When prey density is low, predators go hungry and their population size declines.
Predators capture the abundant prey, and their population starts to grow.
As prey density decreases, predators go hungry and their population size starts to decline.
With fewer predators feeding on them, the prey population grows.
The increasing number of predators causes the prey population to decline.
A new cycle begins.
Population density
1
2
3
4
5
Time
KEY
Prey
Predator

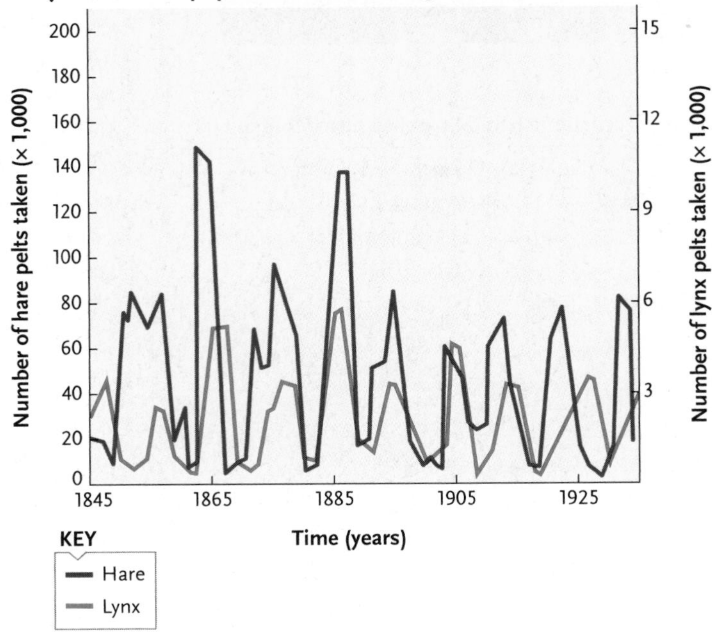

B. Lynx and hare population sizes through time
Number of hare pelts taken (× 1,000)
0
20
40
60
80
100
120
140
160
180
200
Number of lynx pelts taken (× 1,000)
3
6
9
12
15
1845
1865
1885
1905
1925
Time (years)
KEY
Hare
Lynx

multiyear experiment in the southern Yukon. They fenced experimental areas where they added food for the hares, excluded mammalian predators, or applied both experimental treatments; unmanipulated plots served as controls. Where mammalian predators were excluded, hare densities approximately doubled relative to the controls. Where food was added, hare densities tripled. But in plots where predators were excluded *and* food was added, the hare densities increased 11-fold. Krebs and his colleagues concluded that neither food availability nor predation alone is solely responsible for arctic hare population cycles; instead, complex interactions between the hares, their food plants, and their predators create the cyclic fluctuations in hare population size.

STUDY BREAK 50.5

1. **How can you tell whether an environmental factor causes density-dependent or density-independent effects on a population?**
2. **Are the effects of infectious diseases on populations more likely to be density-dependent or density-independent?**

50.6 Human Population Growth

How do human populations compare with those of other species we have studied? The worldwide human population surpassed 6 billion on October 12, 1999; it is projected to reach 7 billion by 2011 and 9 billion by 2050. Like many other species, most humans live in discrete populations, which vary in their demographic traits and access to resources. Although many of us live comfortably, at least a billion people are malnourished or starving, lack clean drinking water, and live without adequate shelter or health care. Even if it were possible to double the food supply, increased agricultural production would inevitably increase pollution and contribute to spoiled croplands, deforestation, and desertification, which are described in Chapter 53.

Human Populations Have Sidestepped the Usual Density-Dependent Controls

For most of human history, our population grew slowly; but over the past two centuries, the worldwide human population has grown exponentially **(Figure 50.21)**. Demographers have identified three ways in which humans have avoided the effects of density-dependent regulating factors.

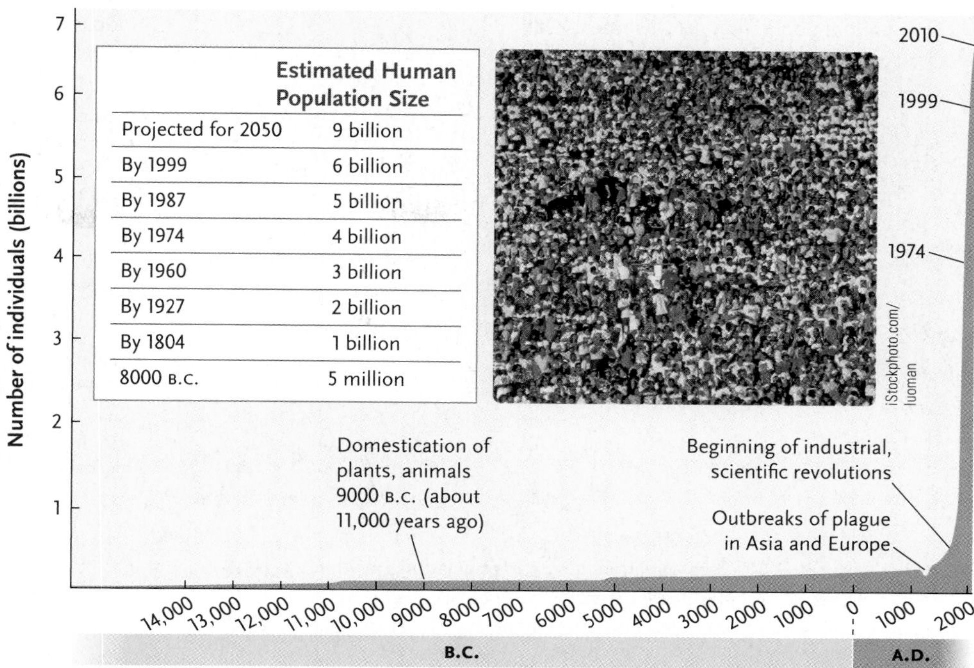

Estimated Human Population Size	
Projected for 2050	9 billion
By 1999	6 billion
By 1987	5 billion
By 1974	4 billion
By 1960	3 billion
By 1927	2 billion
By 1804	1 billion
8000 B.C.	5 million

Number of individuals (billions)
7
6
5
4
3
2
1
2010
1999
1974
iStockphoto.com/luoman
Domestication of plants, animals 9000 B.C. (about 11,000 years ago)
Beginning of industrial, scientific revolutions
Outbreaks of plague in Asia and Europe
14,000
13,000
12,000
11,000
10,000
9000
8000
7000
6000
5000
4000
3000
2000
1000
0
1000
2000
B.C.
A.D.

FIGURE 50.21
Human population growth. The worldwide human population grew slowly until 200 years ago, when it began to increase explosively. The dip in the mid-fourteenth century represents the death of 60 million Asians and Europeans from the bubonic plague. The table shows the years when the human population reached each additional billion people.

First, humans have expanded their geographical range into virtually every terrestrial habitat. Our early ancestors lived in tropical and subtropical grasslands, but by 40,000 years ago, they had dispersed through much of the world (see Section 30.13). Their success resulted from their ability to solve ecological problems by building fires, assembling shelters, making clothing and tools, and planning community hunts. Vital survival skills spread from generation to generation and from one population to another because language allowed the communication of complex ideas and knowledge.

Second, humans have increased the carrying capacities of habitats they occupy. About 11,000 years ago, many populations shifted from hunting and gathering to agriculture. They cultivated wild grasses, diverted water to irrigate crops, and used domesticated animals for food and labor. Such innovations increased the availability of food, raising both the carrying capacity and the population growth rate. In the mid-eighteenth century, people harnessed the energy in fossil fuels, and industrialization began in Western Europe and North America. Food supplies and the carrying capacity increased again, at least in the industrialized countries, through the use of synthetic fertilizers, pesticides, and efficient methods of transportation and food distribution.

Third, advances in public health have reduced the effects of critical population-limiting factors such as malnutrition, contagious diseases, and poor hygiene. Over the past 300 years, modern plumbing and sewage treatment, improvements in food handling and processing, and medical discoveries have reduced death rates sharply. Births now greatly exceed deaths, especially in less industrialized countries, resulting in rapid population growth.

Age Structure and Economic Development May Now Control Our Population Growth

Where have our migrations and technological developments taken us? It took almost 3 *million* years for the human population to reach 1 billion, 123 years to reach the second billion, and only 13 years to jump from 5 billion to 6 billion (see the inset table in Figure 50.21). Rapid population growth may now be an inevitable consequence of our age structure and economic development.

POPULATION GROWTH AND AGE STRUCTURE On a worldwide scale, the annual growth rate for the human population averaged roughly 1.2% ($r = 0.012$ new individuals per individual per year) between 2000 and 2009. Population experts expect that rate to decline, but even so, the human population will continue to grow at least until 2050.

The annual population growth rates of individual nations vary widely, however, ranging from slightly below 0% (that is, the population is getting smaller) to 3.9% in 2009 **(Figure 50.22A)**. The industrialized countries of Western Europe have achieved nearly zero population growth, and in some cases negative growth, but other countries—notably those in Africa, Latin America, and Asia—will experience huge increases over the next 15 to 40 years **(Figure 50.22B)**.

For all long-lived species, differences in age structure are a major determinant of differences in population growth rates **(Figure 50.23)**. The uniform age structure of countries with zero growth—with approximately equal numbers of people of reproductive and prereproductive ages—suggests that individuals have just been replacing themselves and that these populations will not experience a growth spurt when today's children mature. By contrast, the narrow-based age structure of countries with negative growth illustrates a continuing decrease in population size. Reproductives have been producing very few offspring, and the small group of prereproductives may not even replace themselves. Countries with rapid growth have a broad-based age structure, with many youngsters born during the previous 15 years. Worldwide, more than one-quarter of the human population falls within this prereproductive base. This age class will soon reach sexual maturity. Even if each woman produces only two offspring, populations will continue to grow rapidly because so many individuals are reproducing.

The age structure of the United States falls between those for countries with zero growth and countries with rapid growth.

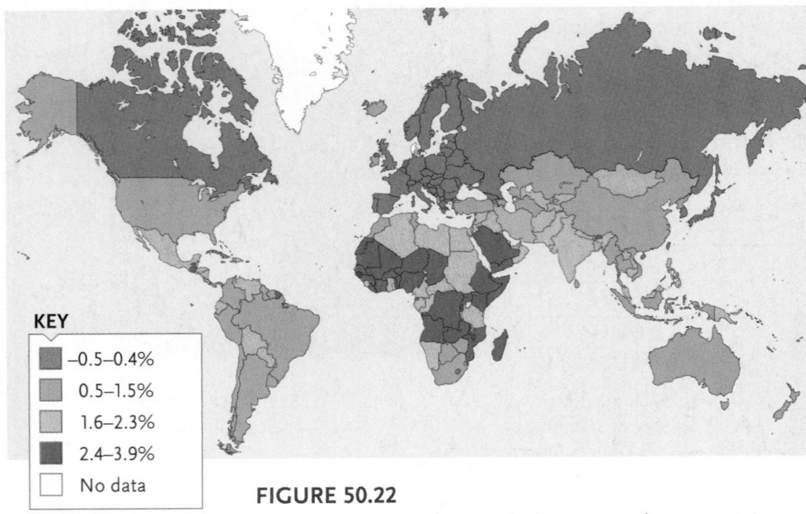
A. Mean annual population growth rates, 2009
KEY
−0.5–0.4%
0.5–1.5%
1.6–2.3%
2.4–3.9%
No data

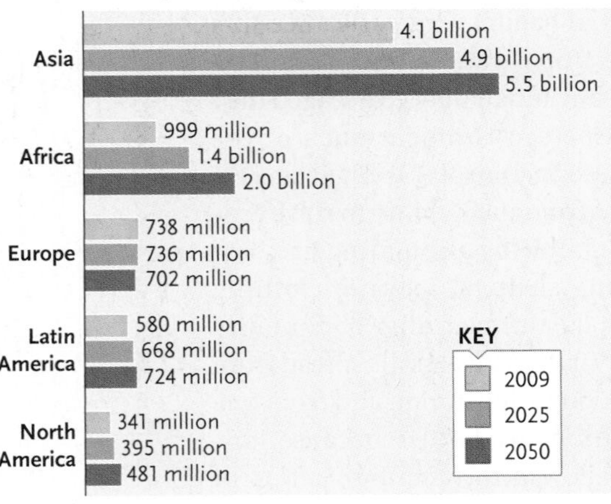
B. Actual and projected population sizes for major world regions
Asia
4.1 billion
4.9 billion
5.5 billion
Africa
999 million
1.4 billion
2.0 billion
Europe
738 million
736 million
702 million
Latin America
580 million
668 million
724 million
North America
341 million
395 million
481 million
KEY
2009
2025
2050

FIGURE 50.22
Local variation in human population growth rates. **(A)** Average annual population growth rates varied among countries in 2009. **(B)** In some regions, the population is projected to increase greatly by 2025 and even more by 2050; the population of Europe will likely decline. Source: Population Reference Bureau, Washington, D.C.

FIGURE 50.23
Age-structure diagrams. **(A)** Hypothetical age-structure diagrams differ for countries with zero, negative, and rapid population growth rates. The width of each bar represents the proportion of the population in each age class. **(B)** Age-structure diagrams for the United States and Bangladesh in 2009 (measured in millions of people) suggest that these countries will experience different population growth rates in the near future. Source: U.S. Census Bureau, International Data Base.

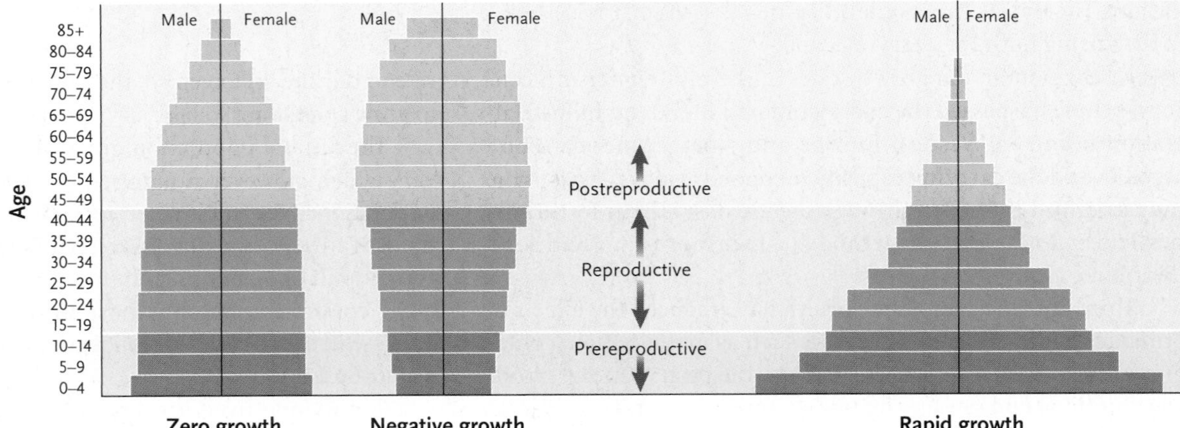
A. Hypothetical age distributions for populations with different growth rates
Age
85+
80–84
75–79
70–74
65–69
60–64
55–59
50–54
45–49
40–44
35–39
30–34
25–29
20–24
15–19
10–14
5–9
0–4
Male
Female
Postreproductive
Reproductive
Prereproductive
Zero growth
Negative growth
Rapid growth

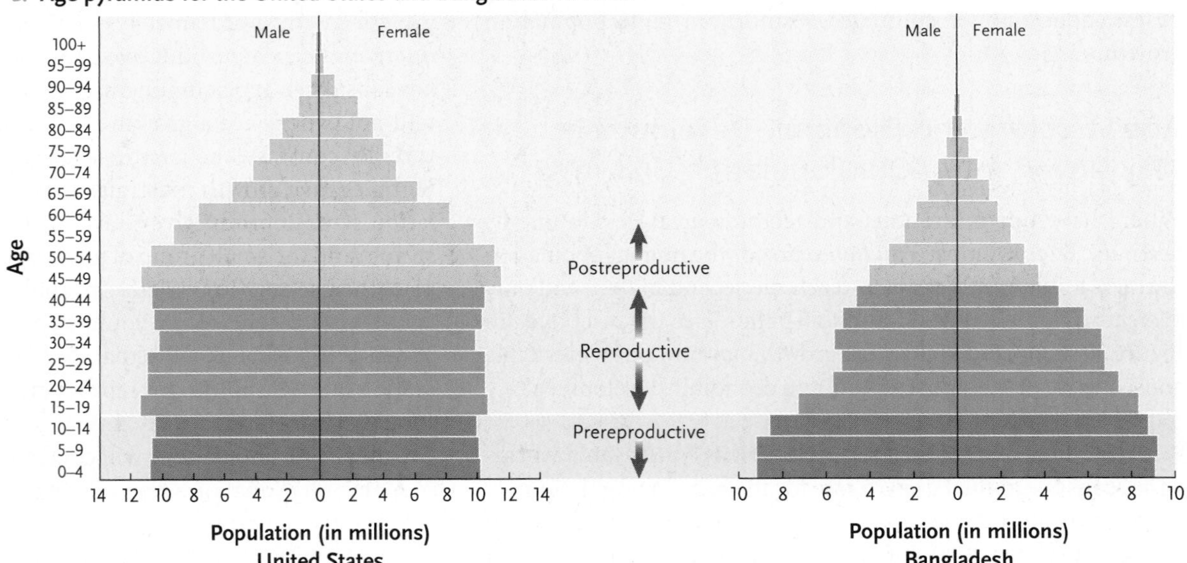
B. Age pyramids for the United States and Bangladesh in 2009
Age
100+
95–99
90–94
85–89
80–84
75–79
70–74
65–69
60–64
55–59
50–54
45–49
40–44
35–39
30–34
25–29
20–24
15–19
10–14
5–9
0–4
Male
Female
Postreproductive
Reproductive
Prereproductive
14 12 10 8 6 4 2 0 2 4 6 8 10 12 14
10 8 6 4 2 0 2 4 6 8 10
Population (in millions)
United States
Population (in millions)
Bangladesh

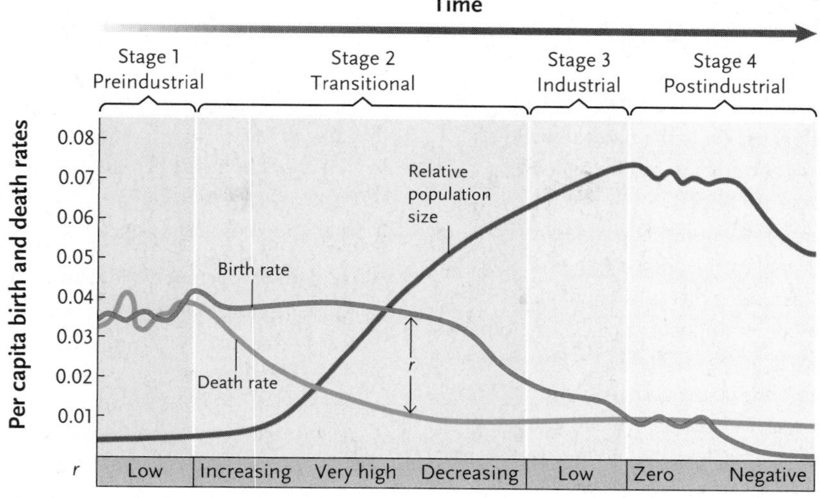
Time
Stage 1
Preindustrial
Stage 2
Transitional
Stage 3
Industrial
Stage 4
Postindustrial
Per capita birth and death rates
0.08
0.07
0.06
0.05
0.04
0.03
0.02
0.01
Relative
population
size
Birth rate
Death rate
r
r
Low
Increasing
Very high
Decreasing
Low
Zero
Negative

FIGURE 50.24

The demographic transition. The demographic transition model describes changes in the birth and death rates and relative population size as a country passes through four stages of economic development. The bottom bar describes the net population growth rate, *r*.

The average number of children per family has declined to the two that are necessary to replace their parents in the population. Nevertheless, the U.S. population will continue to grow slowly for the next couple of generations largely because of continued immigration.

POPULATION GROWTH AND ECONOMIC DEVELOPMENT The relationship between a country's population growth and its economic development can be depicted by the **demographic transition model (Figure 50.24).** This model describes historical changes in demographic patterns in the industrialized countries of Western Europe; we do not know if it accurately predicts the future for developing nations today.

According to this model, during a country's *preindustrial* stage, birth and death rates are high, and the population grows slowly. Industrialization begins a *transitional* stage, when food production rises, and health care and sanitation improve. The death rate declines, resulting in an increased rate of population growth. Later, as living conditions improve, the birth rate also declines, causing the population growth rate to drop. When the *industrial* stage is in full swing, population growth slows dramatically. People move from the countryside to cities, and urban couples often choose to accumulate material goods instead of having large families. Zero population growth is reached in the *postindustrial* stage. Eventually, the birth rate falls below the death rate, *r* falls below zero, and population size begins to decrease.

Today, the United States, Canada, Australia, Japan, Russia, and most of Western Europe are in the industrial stage. Their growth rates have already decreased or are slowly decreasing. In some Western European countries, birth rates are lower than death rates, and populations are getting smaller, indicating their entry into the postindustrial stage. Many less industrialized countries are in the transitional stage, but they may not have enough skilled workers or enough capital to make the transition to an industrialized economy. Thus, many poorer nations may be stuck in the transitional stage.

LIMITING POPULATION GROWTH Most governments realize that increased population size is now a major factor causing resource depletion, excessive pollution, and an overall decline in the quality of life. The principles of population ecology demonstrate that a slowing of population growth—or an actual decline in population size—can be achieved only by decreasing the birth rate or increasing the death rate. And because increasing mortality is neither a rational nor humane means of population control, most governments are attempting to lower birth rates with **family planning programs.** These programs educate people about ways to produce an optimal family size on an economically feasible schedule. Programs vary in their details, but all provide information on methods of birth control (see Section 47.4). When thoughtfully developed and carefully administered, family planning programs cause birth rates to decline significantly.

All species face limits to their population growth. We have postponed the action of most factors that limit population growth, but no amount of invention can expand the ultimate limits set by resource depletion and a damaged environment. We now face two options for limiting human population growth: we can make a global effort to limit our population growth, or we can wait until the environment does it for us.

STUDY BREAK 50.6 <

1. **How have humans sidestepped the controls that regulate populations of other organisms?**
2. **How does the age structure of a population influence its future population growth?**

UNANSWERED QUESTIONS

Are there universal governing principles in population ecology, similar to the laws of physical sciences? Or is the natural world so complex that each population must be considered individually, leaving us with just a series of case studies?

These types of broad questions motivated the founders of modern ecological studies, such as G. Evelyn Hutchinson and Robert MacArthur. Many ecologists have attempted to codify aspects of population ecology in terms of specific principles, sometimes imposing artificial dichotomies in the process. For example, this chapter considered whether or not natural populations are subject to either density-dependent or density-independent regulation. Ecologists have also attempted to uncover basic patterns in community ecology, which are described in the next chapter. We do know that some general principles are often important in governing the structure of populations or natural communities but, as yet, we cannot apply any of them to a specific system without also including a detailed study of that system.

What is the importance of scale in ecology?

Although an individual population can be a meaningful object of study, ecologists often collect data from multiple populations of the same species to compare results among the "replicates." However, although the separate populations may appear to be replicates, they are often quite different in appearance, age structure, life history, or other characteristics. Ecologists confronted with such variation might seek explanations in the differences between the populations' environments, including both abiotic and biotic factors. However, such local variation may also be attributable to the larger context, such as the landscape or surrounding communities.

One important manifestation of this question applies to how the populations of a species are distributed in space. In many cases, discrete populations are widely separated from one another. Such separation is easy to imagine in terms of fish that live in lakes or organisms that live on islands, but it also applies in a diversity of other organisms. For example, many plants and animals, like the Bay checkerspot butterflies discussed in this chapter, are found only on chemically distinct patches of soil that are distributed like islands across a terrestrial environment. Many lizards live in rock outcrops that dot the landscape. Other organisms live on cool, wet mountaintops surrounded by desert. Such isolation is often exaggerated by human modification of the landscape, which progressively fragments and isolates habitable environments from one another. How does such subdivision change the dynamics of the individual populations? How does it change the way they evolve? These are questions that have challenged population and evolutionary biologists for decades. The effects of humans on the environment are making the answers to these questions more than a theoretical concern.

What is the importance of evolution in ecological interactions?

Most research in ecology, ranging from formal models of population growth and regulation to empirical studies, treats populations as if they were unchanging—as if they were not evolving. This implicit perspective does not deny that evolution is happening, but treats it as if it happens on such a long time scale that it need not be considered in contemporary studies. However, many recent studies have shown that populations may evolve quickly, often on a year-to-year basis. If this observation is generally true, then ecological studies that do not include evolutionary change may be compromised. For example, the monitoring and management of commercially exploited fish populations are based entirely on models of population growth, demography, and life histories similar to those considered in this chapter. Commercial fisheries often capture a large proportion of a population every year, focusing on the largest adults. Although research has clearly shown that these practices are likely to select for earlier maturity at a smaller size, these findings have not yet been incorporated into fisheries-management policy. More generally, ecologists have not yet included sufficient emphasis on the interaction between evolution as it occurs on the scale of our day-to-day existence and the modeling and empirical study of ecological processes.

Think Critically

1. Why do you think the mathematical models describing population growth introduced in this chapter are not considered biological laws?
2. Considering life table data, why do you think population ecologists often "ignore" evolutionary change? How can molecular biology be used to incorporate evolution into population level studies?

David Reznick

David Reznick is Professor of Biology at the University of California, Riverside. He studies natural selection both from an experimental perspective and by testing evolutionary theory in natural populations. He works primarily with guppies on the island of Trinidad. Learn more about Dr. Reznick's work at http://www.biology.ucr.edu/people/faculty/Reznick.html.

REVIEW KEY CONCEPTS

Go to **CENGAGENOW** at www.cengage.com/login to access quizzing, animations, exercises, articles, and personalized homework help.

50.1 Population Characteristics

- A population's size and density can be measured directly or with sampling techniques (Figures 50.2 and 50.3).
- Organisms within a population may be clumped, uniformly distributed, or randomly distributed within their habitat (Figure 50.4). Clumped dispersion is the most common, but animals may change their dispersion pattern seasonally.
- The relative numbers of individuals of different ages determines a population's age structure. Generation time is the average time between an individual's birth and the birth of its offspring (Figure 50.5). A population's sex ratio is the relative proportion of males and females.

Animation: Mark-recapture method

Animation: Distribution patterns

50.2 Demography

- Demography is the study of the survivorship, reproduction, immigration, and emigration patterns that influence population characteristics.
- Life tables summarize age-specific mortality, survivorship, and age-specific fecundity of surviving individuals (Table 50.1).
- Survivorship curves depict a population's survival pattern over its life span. Ecologists define three general patterns of survivorship: high survivorship until late in life, a constant mortality level at all ages, and high juvenile mortality (Figure 50.6).

50.3 The Evolution of Life Histories

- An organism's energy budget mandates trade-offs in the allocation of energy to maintenance, growth, and reproduction (Figure 50.7).
- Natural selection has molded several interacting components of life history variation based upon the allocation of resources to growth, maintenance, and reproduction: the trade-off between fecundity and parental care; whether to reproduce once versus multiple times; and the age of first reproduction.

Animation: Life history patterns

Animation: Guppy characteristics

50.4 Models of Population Growth

- Bacteria reproduce by binary fission, and their populations double in size each generation (Figure 50.8A).
- The exponential growth model, $dN/dt = rN$, describes unlimited population growth. A graph of exponential growth is J-shaped (Figure 50.8B).
- The maximum population growth rates of populations, symbolized r_{max}, vary inversely with their generation times (Figure 50.9).
- The logistic model, $dN/dt = r_{max}N(K - N/K)$, includes the effects of resource limitation. The carrying capacity, K, is the maximum population size that an environment can sustain. The per capita population growth rate, r, decreases as N approaches K. A graph of logistic growth is S-shaped (Figure 50.10, Table 50.2).
- Some populations exhibit logistic growth in the laboratory and in nature, but time lags in response to increased density may cause N to oscillate around K (Figure 50.11).

Animation: Exponential growth

Animation: Effect of death on growth

Practice: Comparison of exponential and logistic population growth

50.5 Population Dynamics

- Density-dependent factors regulate population size by reducing individual growth rates, adult size, survivorship, and fecundity (Figures 50.12 and 50.13). Competition within populations or between species, predator–prey interactions, parasites, and infectious diseases can cause density-dependent population regulation (Figure 50.14).
- Abiotic environmental factors, which affect a population regardless of its size, cause density-independent limitation of population size (Figure 50.15).
- Global climate change is exaggerating the effects of density-independent factors on populations of large, long-lived organisms (Figure 50.16).
- Interactions between density-dependent and density-independent factors often influence population size (Figure 50.17).
- The life history patterns of most organisms fall between two extremes: r-selected species and K-selected species (Figure 50.18). They differ in many life history characteristics (Table 50.3).
- Within metapopulations, sink populations are often replenished by individuals dispersing from a source population (Figure 50.19).
- Some animal populations exhibit cyclic fluctuations in size (Figure 50.20). No general model has successfully explained all population cycles.

50.6 Human Population Growth

- Human populations have sidestepped density-dependent population regulation by expanding into most terrestrial habitats, increasing carrying capacity, and reducing death rates with improved medical care and sanitation (Figures 50.21 and 50.22).
- Age structure may now control human population growth rates (Figure 50.23). In countries with large numbers of young people, populations will continue to grow rapidly as they reach sexual maturity. The populations of countries with a uniform age structure will not experience much growth in the foreseeable future.
- The demographic transition model describes the influence of economic development on population growth (Figure 50.24).
- Many governments encourage population control through family planning programs.

Animation: Current and projected population sizes by region

Animation: Age structure diagrams

Animation: U.S. age structure

Animation: Demographic transition model

UNDERSTAND AND APPLY

Test Your Knowledge

1. Ecologists sometimes use mathematical models to:
   a. avoid conducting laboratory studies or field work.
   b. simulate natural events before conducting detailed field studies.
   c. make basic observations about ecological relationships in nature.
   d. collect survivorship and fecundity data to construct life tables.
   e. determine the geographical ranges of populations.
2. The number of individuals per unit area or volume of habitat is called the population's:
   a. geographical range.
   b. dispersion pattern.
   c. density.
   d. size.
   e. age structure.
3. One day you caught and marked 90 butterflies in a population. A week later, you returned to the population and caught 80 butterflies, including 16 that had been marked previously. What is the size of the butterfly population?
   a. 170
   b. 450
   c. 154
   d. 186
   e. 106
4. A uniform dispersion pattern implies that members of a population:
   a. cooperate in rearing their offspring.
   b. work together to escape from predators.
   c. use resources that are patchily distributed.
   d. may experience intraspecific competition for vital resources.
   e. have no ecological interactions with each other.

5. The model of exponential population growth predicts that the per capita population growth rate *r*:
   a. does not change as a population gets larger.
   b. gets larger as a population gets larger.
   c. gets smaller as a population gets larger.
   d. is always at its maximum level (r_{max}).
   e. fluctuates on a regular cycle.
6. A population of 1,000 individuals experiences 462 births and 380 deaths in 1 year. What is the value of *r* for this population?
   a. 0.842/individual/year
   b. 0.462/individual/year
   c. 0.380/individual/year
   d. 0.820/individual/year
   e. 0.082/individual/year
7. According to the logistic model of population growth, the absolute number of individuals by which a population grows during a given time period:
   a. gets steadily larger as the population size increases.
   b. gets steadily smaller as the population size increases.
   c. remains constant as the population size increases.
   d. is highest when the population is at an intermediate size.
   e. fluctuates on a regular cycle.
8. Which example might reflect density-dependent regulation of population size?
   a. An exterminator uses a pesticide to eliminate carpenter ants from a home.
   b. Mosquitoes disappear from an area after the first frost.
   c. The lawn dies after a month-long drought.
   d. Storms blow over and kill all the willow trees along a lake.
   e. The size of a clam population declines as the number of predatory herring gulls explodes.
9. A *K*-selected species is likely to exhibit:
   a. a Type I survivorship curve and a short generation time.
   b. a Type II survivorship curve and a short generation time.
   c. a Type III survivorship curve and a short generation time.
   d. a Type I survivorship curve and a long generation time.
   e. a Type II survivorship curve and a long generation time.
10. One reason why human populations have been able to sidestep the factors that usually control population growth is that:
    a. the carrying capacity for humans has remained constant since humans first evolved.
    b. agriculture and industrialization have increased the carrying capacity for our species.
    c. the population growth rate (*r*) for the human population has always been small.
    d. the age structure of human populations has no impact on its population growth.
    e. plagues have killed off large numbers of humans at certain times in the past.

Discuss the Concepts

1. Choose an animal or plant species that lives in your environment and identify the density-dependent and density-independent factors that might influence its population size. How could you demonstrate conclusively that the factors are either density-dependent or density-independent?
2. Many city-dwellers have noted that the density of cockroaches in apartment kitchens appears to vary with the habits of the occupants: people who wrap food carefully and clean their kitchen frequently tend to have fewer arthropod roommates than those who leave food on kitchen counters and clean less often. Interpret these observations from the viewpoint of a population ecologist.
3. How could you define the worldwide carrying capacity for humans? What factors would you have to take into account?

Design an Experiment

Design an experiment using fruit flies or some other small laboratory animal to test the hypothesis that delaying the age of first reproduction will decrease a population's per capita birth rate. Your experimental design should include experimental and control groups as well as details about your experimental methods and the data you would collect.

Interpret the Data

Gregory M. Erickson of Florida State University and colleagues from Florida State and the University of Alberta analyzed a fossil assemblage of 22 individuals of *Albertosaurus sarcophagus*, a relative of *Tyrannosaurus rex*. All 22 had died at roughly the same time—but at different ages—perhaps because of a drought or starvation. They used growth lines present in the fossilized leg bones to determine the age at death of each specimen. They had no data for very young individuals, and assumed a 60% mortality rate between birth and age two. The researchers constructed the following table extrapolated to an initial cohort of 1,000 individuals:

Age (years)	Number Surviving	Age (years)	Number Surviving
0	1,000	15	218
1	—	16	182
2	400	17	164
3	—	18	127
4	382	19	109
5	—	20	73
6	364	21	56
7	—	22	—
8	345	23	36
9	327	24	—
10	—	25	—
11	309	26	—
12	291	27	—
13	273	28	18
14	255		

Did this population of animals exhibit a Type I, Type II, or Type III survivorship pattern—or something intermediate? What does the life table tell you about the age at which *Albertosaurus sarcophagus* was most vulnerable to the factors causing mortality?

Source: After G. M. Erickson et al. 2006. Tyrannosaur life tables: An example of nonavian dinosaur population biology. *Science* 313:213–217.

Apply Evolutionary Thinking

Many animals, including humans and other primates, live long beyond their reproductive years. Develop an evolutionary hypothesis to explain this observation, and design a study that might test it.

Express Your Opinion

Some people oppose any deer hunting, whereas others see hunters as a logical substitute for an absence of natural predators. Do you support encouraging hunting in areas where the presence of too many deer is harming the habitat? Go to www.cengage.com/login to investigate both sides of the issue and then vote.

Three interacting populations. Ladybird beetles *(Coccinella septempunctata)* feed on aphids (order Hemiptera), which consume the sap of plants.

51

Population Interactions and Community Ecology

STUDY OUTLINE

Why It Matters. . . In some open woodlands in Central America, flocks of chestnut-headed oropendolas *(Psarocolius wagleri),* members of the blackbird family, build hanging nests in isolated trees **(Figure 51.1)**. Female giant cowbirds *(Molothrus oryzivorus)* often bully their way into a colony, laying an egg or two in each oropendola nest. Cowbirds are *brood parasites* on oropendolas, tricking them into caring for cowbird young. The cowbird chicks grow faster than oropendola chicks, and they consume much of the food that the oropendolas bring to their own offspring. Because cowbird chicks take food away from their oropendola nest mates, we might expect adult oropendolas to eject cowbird eggs and chicks from their nests—but often they don't.

Why do some oropendolas care for offspring that are not their own? In an ingenious study conducted in the 1960s, Neal Smith of the Smithsonian Tropical Research Institute determined that cowbird chicks could actually increase the number of offspring that some oropendolas raise. Oropendola chicks are frequently parasitized by botfly larvae, which feed on their flesh. The aggressive cowbird chicks snap at adult botflies and pick fly larvae off their nest mates. Although cowbird chicks eat food meant for oropendola chicks, they also protect them from potentially lethal parasites; twice as many young oropendolas survive in nests with cowbird chicks as in nests without them.

FIGURE 51.1
Potential victims of brood parasitism. Chestnut-headed oropendolas *(Psarocolius wagleri)* rear their young in elaborate hanging nests. Some populations of oropendolas are subject to brood parasitism by giant cowbirds *(Molothrus oryzivorus)*.

In other areas of Central America, oropendolas build nests near the hives of bees or wasps. These oropendolas chase cowbirds from their colonies, and when a cowbird does manage to sneak an egg into one of their nests, the oropendolas frequently eject it. Why do oropendolas in these colonies reject cowbird eggs? Smith determined that the swarms of bees and wasps keep botflies away from the oropendola colonies. At these sites, twice as many oropendola chicks survive in nests without cowbirds as in those that include them. Thus the oropendolas derive no benefit from having cowbird chicks in their nests, and natural selection has favored discriminating behavior in oropendolas that nest near bees and wasps.

The story of the oropendolas, cowbirds, botflies, bees, and wasps provides an example of the population interactions that characterize life in an **ecological community,** an assemblage of species living in the same place. And as this story reveals, the presence or absence of certain species may alter the effects of such interactions in almost unimaginably complex ways. We begin this chapter with a description of some of the many ways that populations in a community interact. We then examine how population interactions and other factors, such as the kinds of species present and the relative numbers of each species, influence a community's characteristics. <

51.1 Population Interactions

Population interactions usually provide benefits or cause harm to the organisms engaged in the interaction **(Table 51.1).** And because interactions with other species often affect the survival and reproduction of individuals, many of the relationships that we witness today are the products of long-term evolutionary modification. Before examining several general types of population interactions, we briefly consider how natural selection has shaped the relationships between interacting species.

TABLE 51.1 Population Interactions and Their Effects

Interaction	Effects on Interacting Populations	
Predation	+/−	Predators gain nutrients and energy; prey are killed or injured.
Herbivory	+/−	Herbivores gain nutrients and energy; plants are killed or injured.
Competition	−/−	Both competing populations lose access to some resources.
Commensalism	+/0	One population benefits; the other population is unaffected.
Mutualism	+/+	Both populations benefit.
Parasitism	+/−	Parasites gain nutrients and energy; hosts are injured or killed.

Coevolution Produces Reciprocal Adaptations in Species That Interact Ecologically

Population interactions change constantly. New adaptations that evolve in one species exert selection pressure on another, which then evolves adaptations that exert selection pressure on the first. The evolution of genetically based, reciprocal adaptations in two or more interacting species is described as **coevolution.**

Some coevolutionary relationships are straightforward. For example, ecologists describe the coevolutionary interactions between some predators and their prey as a race in which each species evolves adaptations that temporarily allow it to outpace the other. When antelope populations suffer predation by cheetahs, natural selection fosters the evolution of faster speed in the antelopes. Cheetahs then experience selection for increased speed so that they can overtake and capture antelopes. Other coevolved interactions provide benefits to both partners. For example, the flower structures of different monkey-flower species have evolved characteristics that allow them to be visited by either bees or hummingbirds (see Figure 21.7).

Although one can hypothesize a coevolutionary relationship between any two interacting species, documenting the evolution of reciprocal adaptations is difficult. As our introductory story about oropendolas and their parasites illustrated, coevolutionary interactions often involve more than two species. Indeed, most organisms experience complex interactions with numerous other species in their communities, and the simple portrayal of coevolution as taking place between two species rarely does justice to the complexity of these relationships.

Predation and Herbivory Define Many Relationships in Ecological Communities

Because animals acquire nutrients and energy by consuming other organisms, **predation** (the interaction between predatory animals and the animal prey they consume) and **herbivory** (the interaction between herbivorous animals and the plants they eat) are often the most conspicuous relationships in ecological communities.

ADAPTATIONS FOR FEEDING Both predators and herbivores have evolved remarkable characteristics that allow them to feed effectively. Carnivores use sensory systems to locate animal prey and specialized behaviors and anatomical structures to capture and consume it. For example, a rattlesnake (genus *Crotalus*) uses heat sensors on its head (see Figure 39.22) and chemical sensors in the roof of its mouth to find rats or other endothermic prey. Its hollow fangs inject toxins that kill the prey and begin to digest its tissues even before the snake consumes it. And elastic ligaments connecting the bones of its jaws and skull allow a snake to swallow prey that is larger than its head. Herbivores have comparable adaptations for locating and processing their food plants. Insects use chemical sensors on their legs and heads to identify edible plants and sharp mandibles or sucking mouthparts to consume plant tissues or sap. Herbivorous mammals have specialized teeth to harvest and grind tough vegetation (see Section 45.5).

All animals must select their diets from a variety of potential food items. Some species, described as *specialists,* feed on one or just a few types of food. Among birds, for example, the Everglades kite *(Rostrhamus sociabilis)* consumes just one prey species, the apple snail *(Pomacea paludosa).* Other species, described as *generalists,* have broader tastes. Crows (genus *Corvus*) consume food ranging from grains to insects to carrion.

How does an animal select what type of food to eat? Some mathematical models, collectively described as **optimal foraging theory,** predict that an animal's diet is a compromise between the costs and benefits associated with different types of food. Assuming that animals try to maximize their energy intake in a given feeding time, their diets should be determined by the time and energy it takes to pursue, capture, and consume a particular kind of food compared with the energy that food provides. For example, a cougar *(Puma concolor)* will invest more time and energy hunting a mountain goat *(Oreamnos americanus)* than a jackrabbit *(Lepus townsendii),* but the payoff for the cougar is a bigger meal.

Food abundance also affects food choice. When prey are scarce, animals often take what they can get, settling for food that has a low benefit-to-cost ratio. But when food is abundant, they may specialize, selecting types that provide the largest energetic return. Bluegill sunfish *(Lepomis macrochirus),* for example, feed on *Daphnia* and other small crustaceans. When crustacean density is high, the fish hunt mostly large *Daphnia,* which provide more energy for their effort; but when prey density is low, bluegills feed on *Daphnia* of all sizes **(Figure 51.2).**

Density of prey

Low | Medium | High

Proportions offered

Proportions encountered

Proportions eaten

When prey density is low, bluegills eat equal proportions of all prey sizes.

When prey density is high, bluegills eat proportionately more large prey than medium or small prey.

KEY
Large prey
Medium prey
Small prey

FIGURE 51.2
An experiment demonstrating that prey density affects predator food choice. Researchers tested the food size preferences of captive bluegill sunfish *(Lepomis macrochirus)* by offering them equal numbers of small, medium, and large-sized prey *(Daphnia magna)* at three different prey densities. Because large prey are the easiest to find, bluegills encountered them more frequently than small or medium-sized prey, especially at the highest prey density. The bluegills' selection of prey varied with prey density; they strongly preferred large prey when prey of all sizes were abundant.

DEFENSES AGAINST HERBIVORY AND PREDATION Because herbivory and predation have a negative impact on the organisms being consumed, plants and animals have evolved mechanisms to avoid being eaten. Some plants use spines, thorns, and irritating hairs to protect themselves from herbivores. Many plant tissues also contain poisonous chemicals that deter herbivores from feeding. For example, plants in the milkweed family (Asclepiadaceae) exude a milky, irritating sap that contains cardiac glycosides, even small amounts of which are toxic to vertebrate heart muscle. Other compounds mimic the structure of insect hormones, disrupting the development of insects that consume them. Most of these poisonous compounds are volatile, giving plants their typical aromas; some herbivores have coevolved the ability to recognize these odors and avoid the toxic plants. Recent research indicates that some plants increase their production of toxic compounds in response to herbivore feeding. For example, potato and tomato plants that have been damaged by herbivores produce higher levels of protease-inhibiting chemicals; these compounds prevent herbivores from digesting proteins they have just consumed, reducing the food value of these plant tissues.

Many animals have evolved an appearance that provides a passive defense against predation **(Figure 51.3).** Caterpillars that look like bird droppings, for example, may not attract much attention from a hungry predator. And as you learned in Chapter 1 (see Figure 1.9), **cryptic coloration** helps some prey (as well as some predators) to blend in with their surroundings.

Once discovered by a predator, many animals first try to run away. When cornered, they may try to startle or intimidate the predator with a display that increases their apparent size or ferocity **(Figure 51.4).** Such a display might confuse the predator just long enough to allow the potential victim to escape. Other species

A. **Bird dropping mimic (*Papilio cresphontes*)**

Edward S. Ross

B. **Damaged leaf mimic (*Mimetica* species)**

Frans Lanting/Minden Pictures

FIGURE 51.3
Hiding in plain sight. Some animals, such as **(A)** giant swallowtail butterfly larvae which resemble bird droppings and **(B)** some katydids that resemble insect-damaged leaves, do not attract the attention of predators.

Photo Researchers, Inc.

FIGURE 51.4
Startle defenses. A short-eared owl *(Asio flammeus)* increases its apparent size when threatened by a predator.

seek shelter in protected sites. For example, flexible-shelled African pancake tortoises *(Malacochersus tornieri)* retreat into rocky crevices and puff themselves up with air, becoming so tightly wedged between rocks that predators cannot extract them.

Other animals defend themselves actively. North American porcupines (genus *Erethizon*) release hairs modified into sharp, barbed quills that stick in a predator's mouth, causing severe pain and swelling. Other species fight back by biting, charging, or kicking an attacking predator. Chemical defenses also provide effective protection. Skunks release a noxious spray when threatened, and some frogs and toads produce neurotoxic skin secretions that paralyze and kill mammals. Some insects even protect themselves with poisons acquired from plants. The caterpillars of monarch butterflies *(Danaus plexippus)* are immune to the cardiac glycosides in the milkweed leaves they eat. They store these chemicals at high concentration, even through metamorphosis, making adult monarchs poisonous to vertebrate predators.

Courtesy of Ken Nemuras

FIGURE 51.5
Aposematic coloration. Poisonous animals, like the harlequin toad *(Atelopus varius)* from Central America, often have bright warning coloration.

Poisonous or repellant species often advertise their unpalatability with bright, contrasting patterns, called **aposematic coloration (Figure 51.5).** Although a predator might attack a black-and-white skunk, a yellow-banded wasp, or an orange monarch butterfly once, it quickly learns to associate the gaudy color pattern with pain, illness, or severe indigestion—and rarely attacks these easily recognized animals again.

Mimicry, in which one species evolves an appearance resembling that of another **(Figure 51.6),** is also a form of defense. In **Batesian mimicry,** named for English naturalist Henry W. Bates, a palatable or harmless species, the **mimic,** resembles an unpalatable or poisonous one, the **model.** Any predator that eats the poisonous model will subsequently avoid other organisms that resemble it. In **Müllerian mimicry,** named for German zoologist Fritz Müller, two or more unpalatable species share a similar appearance, which reinforces the lesson learned by a predator that attacks any species in the mimicry complex.

Despite the effectiveness of many antipredator defenses, coevolution has often molded the responses of predators to overcome them. For exam-

A. Batesian mimicry

Nigel Cattlin/Frank Lane Picture Agency

Drone fly *(Eristalis tenax)*, the mimic

© Simon Colmer and Abby Rex/Alamy

Honeybee *(Apis mellifera)*, the model

B. Müllerian mimicry

Robert Lubeck/Animals, Animals-Earth Scenes

Heliconius erato

© Patti Murray/Animals, Animals-Earth Scenes

Heliconius melpomene

FIGURE 51.6
Mimicry. (A) Batesian mimics are harmless animals that mimic a dangerous one. The harmless drone fly *(Eristalis tenax)* is a Batesian mimic of the stinging European honeybee *(Apis mellifera)*. **(B)** Müllerian mimics are poisonous species that share a similar appearance. Two distantly related species of butterfly, *Heliconius erato* and *Heliconius melpomene*, have nearly identical patterns on their wings.

A. Pinacate beetle *(Eleodes longicollis)*

Thomas Eisner, Cornell University

B. Grasshopper mouse *(Onychomys* species)

Thomas Eisner, Cornell University

FIGURE 51.7
Coevolution of predators and prey. (A) When disturbed by a predator, the pinacate beetle sprays a noxious chemical from its posterior end. **(B)** Grasshopper mice overcome this defense by shoving a beetle's rear end into the soil and dining on it headfirst.

FIGURE 51.8 **Experimental Research**

Gause's Experiments on Interspecific Competition in *Paramecium*

Question: Can two species of *Paramecium* coexist in a simple laboratory environment?

Experiment: Gause grew populations of two species, *Paramecium aurelia* and *Paramecium caudatum*, alone (single species cultures) or together (mixed cultures) in small bottles in his laboratory. To determine whether the growth of these populations followed the predictions of the logistic equation, Gause had to maintain a reasonably constant carrying capacity in each culture. Thus, he fed the cultures a broth of bacteria, and he eliminated their waste products (by centrifuging the cultures and removing some of the culture medium) on a regular schedule. He then monitored their population sizes through time.

Results: When grown separately, *P. caudatum* **(A)** and *P. aurelia* **(B)** each exhibited logistic population growth. But when the two species were grown together in a mixed culture **(C)**, *P. aurelia* persisted and *P. caudatum* was nearly eliminated from the culture.

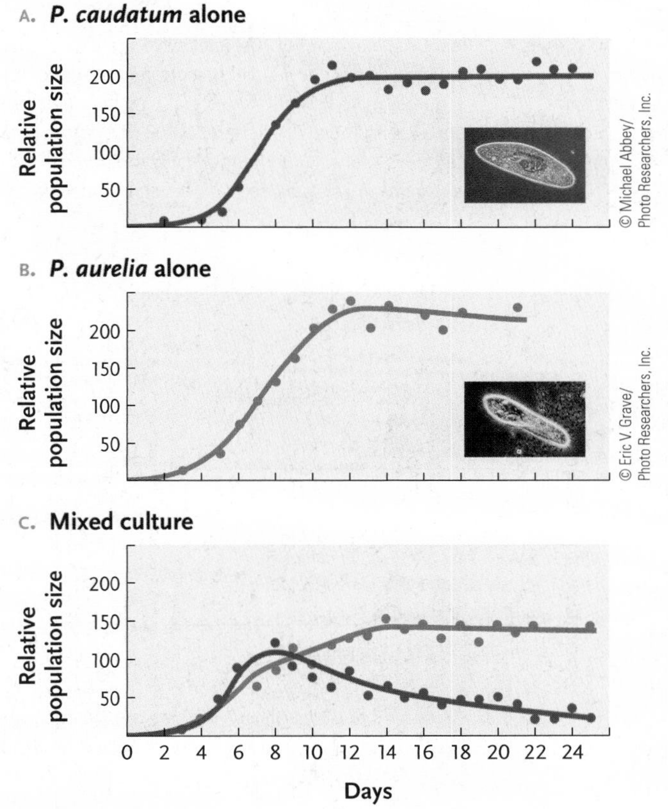
A. P. caudatum alone
Relative population size
200
150
100
50
© Michael Abbey/ Photo Researchers, Inc.
B. P. aurelia alone
Relative population size
200
150
100
50
© Eric V. Grave/ Photo Researchers, Inc.
C. Mixed culture
Relative population size
200
150
100
50
0 2 4 6 8 10 12 14 16 18 20 22 24
Days

Conclusion: Because one species was almost always eliminated from mixed species cultures, Gause formulated the competitive exclusion principle: populations of two or more species cannot coexist indefinitely if they rely on the same limiting resources and exploit them in the same way.

Source: G. F. Gause. 1934. *The Struggle for Existence.* Williams & Wilkins Company, London.

ple, when threatened by a predator, the pinacate beetle *(Eleodes longicollis)* raises its rear end and sprays a noxious chemical from a gland at the tip of its abdomen. Although this behavior deters many would-be predators, grasshopper mice (genus *Onychomys*) of the American southwest circumvent this defense: they grab the beetles and shove their abdomens into the ground, rendering the beetle's spray ineffective **(Figure 51.7)**.

Interspecific Competition Occurs When Different Species Depend on the Same Limiting Resources

Populations of different species often use the same limiting resources, causing **interspecific competition** (competition *between* species). The competing populations may experience increased mortality and decreased reproduction, responses that are similar to the effects of intraspecific competition (see Section 50.4). Interspecific competition reduces the size and population growth rate of one or more of the competing populations.

Community ecologists identify two main forms of interspecific competition. In **interference competition,** individuals of one species harm individuals of another species directly. Animals may fight for access to resources, as when lions chase smaller scavengers like hyenas and jackals from their kills. Similarly, many plant species, including creosote bushes (see Figure 50.4), release toxic chemicals that prevent other plants from growing nearby. In **exploitative competition,** two or more populations use ("exploit") the same limiting resource. The presence of one species reduces resource availability for the others, even in the absence of snout-to-snout or root-to-root confrontations. For example, in the deserts of the American Southwest, many bird and ant species feed largely on seeds. Thus, each seed-eating species may deplete the food supply available to others.

COMPETITIVE EXCLUSION AND THE NICHE CONCEPT In the 1920s, the Russian mathematician Alfred J. Lotka and the Italian biologist Vito Volterra independently proposed a model of interspecific competition, modifying the logistic equation (see Section 50.4) to describe the effects of competition between two species. In their model, an increase in the size of one population reduces the population growth rate (r) of the other.

A Russian biologist, G. F. Gause, tested the model experimentally in the 1930s. He grew cultures of two *Paramecium* species (ciliate protists) under constant laboratory conditions, regularly renewing food and removing wastes. Both species fed on bacteria suspended in the culture medium. When grown alone, each species exhibited logistic growth. When grown together in the same dish, however, *Paramecium aurelia* persisted at high density, but *Paramecium caudatum* was nearly eliminated **(Figure 51.8)**. These results inspired Gause to define the **competitive exclusion principle:** populations of two or more species cannot coexist indefinitely if they rely on the same limiting resources and exploit them in the same way. One species is inevitably more successful, harvesting resources more efficiently and producing more offspring than the other.

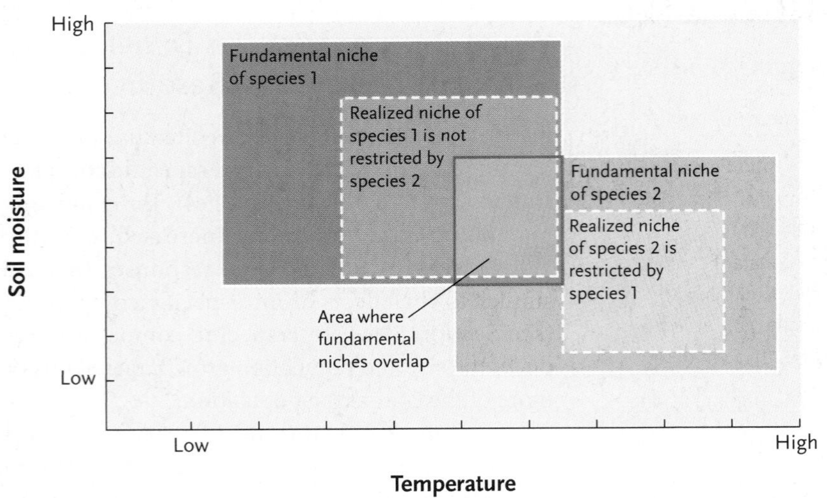
High
Low
Soil moisture
Fundamental niche of species 1
Realized niche of species 1 is not restricted by species 2
Fundamental niche of species 2
Realized niche of species 2 is restricted by species 1
Area where fundamental niches overlap
Low
High
Temperature

FIGURE 51.9
Fundamental versus realized niches. In this hypothetical example, both species 1 and species 2 can survive intermediate temperature and soil moisture conditions, as indicated by the green shading where their fundamental niches overlap. Because species 1 actually occupies virtually all of this overlap zone, its realized niche is not affected by the presence of species 2. By contrast, the realized niche of species 2 is restricted by the presence of species 1; species 2 occupies only the warmer and dryer parts of the habitat.

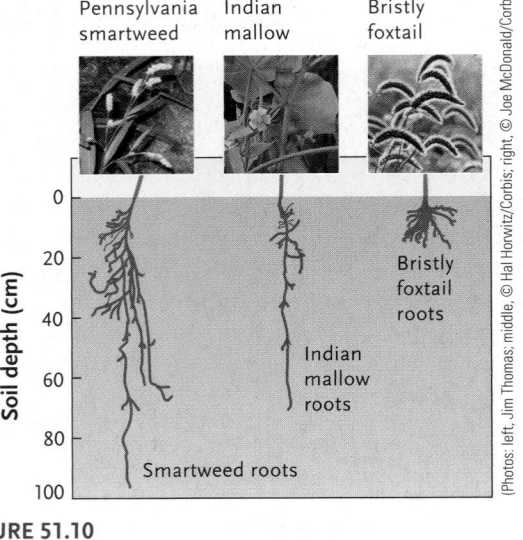

Pennsylvania smartweed
Indian mallow
Bristly foxtail
Soil depth (cm)
0
20
40
60
80
100
Bristly foxtail roots
Indian mallow roots
Smartweed roots
(Photos: left, Jim Thomas; middle, © Hal Horwitz/Corbis; right, © Joe McDonald/Corbis.)

FIGURE 51.10
Resource partitioning. The root systems of three plant species that grow in abandoned fields partition water and nutrient resources in soil. Bristly foxtail grass *(Setaria faberi)* has a shallow root system; Indian mallow *(Abutilon theophrasti)* has a moderately deep taproot; and Pennsylvania smartweed *(Polygonum pensylvanicum)* has a deep taproot that branches at many depths.

Ecologists developed the concept of the **ecological niche** as a tool for visualizing resource use and the potential for interspecific competition in nature. We define a population's niche by the resources it uses and the environmental conditions it requires over its lifetime. In this context, the niche includes food, shelter, and nutrients as well as abiotic conditions, such as light intensity and temperature, which cannot be depleted. In theory, one could identify an almost infinite variety of conditions and resources that contribute to a population's niche. In practice, ecologists usually analyze a few critical resources for which populations might compete. Sunlight, soil moisture, and inorganic nutrients are important resources for plants. Food type, food size, and nesting sites are important for animals.

Ecologists distinguish the **fundamental niche** of a population, the range of conditions and resources that it can possibly tolerate and use, from its **realized niche,** the range of conditions and resources that it actually uses in nature. Realized niches are smaller than fundamental niches, partly because all tolerable conditions are not always present in a habitat, and partly because some resources are used by other species. We can visualize competition between two populations by plotting their fundamental and realized niches with respect to one or more resources **(Figure 51.9).** If the fundamental niches of two populations overlap, they *might* compete in nature.

EVALUATING COMPETITION IN NATURE The observation that several populations use the same resource does not demonstrate that competition occurs. For example, all terrestrial animals consume oxygen, but they don't compete for oxygen because it is usually plentiful. Nevertheless, two general observations pro-

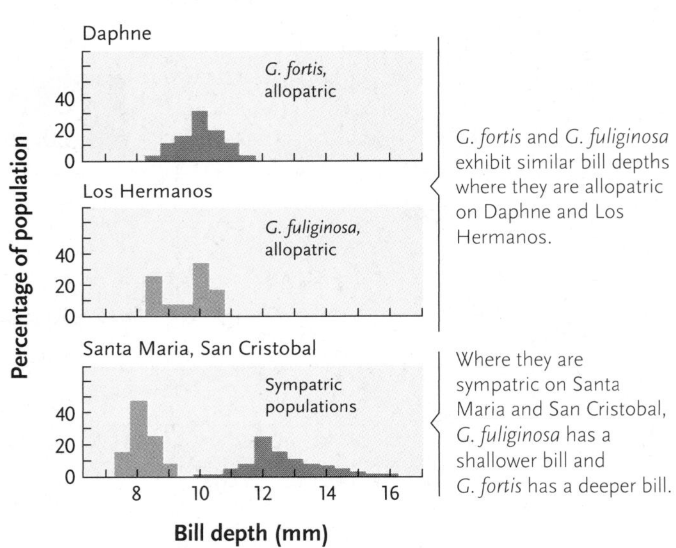
Daphne
G. fortis, allopatric
Los Hermanos
G. fuliginosa, allopatric
Santa Maria, San Cristobal
Sympatric populations
Percentage of population
0
20
40
8
10
12
14
16
Bill depth (mm)
G. fortis and G. fuliginosa exhibit similar bill depths where they are allopatric on Daphne and Los Hermanos.
Where they are sympatric on Santa Maria and San Cristobal, G. fuliginosa has a shallower bill and G. fortis has a deeper bill.

G. fortis
G. fuliginosa
Bill depth
Sam Fried/VIREO
© A. & J. Binns/VIREO

FIGURE 51.11
Character displacement. *Geospiza fortis* and *Geospiza fuliginosa* exhibit character displacement in the depth of their bills, a trait that is correlated with the sizes of seeds they eat.

FIGURE 51.12 **Experimental Research**

Demonstration of Competition between Two Species of Barnacles

Question: Do two barnacle species limit one another's realized niches in habitats where they coexist?

Experiment: Connell observed a difference in the distributions of two barnacle species on a rocky coast: *Chthamalus stellatus* occupies shallow water, and *Balanus balanoides* lives in deeper water. He then determined the fundamental niche of each species by removing either *Chthamalus* or *Balanus* from rocks and monitoring the distribution of each species in the absence of the other.

Results: When Connell removed *Balanus* from rocks in deep water, larval *Chthamalus* colonized the area and produced a flourishing population of adults. By contrast, the removal of *Chthamalus* from rocks in shallow water did not result in colonization by *Balanus*.

Control: No treatment
Results: *Chthamalus* occupies only shallow water and *Balanus* occupies only deep water.

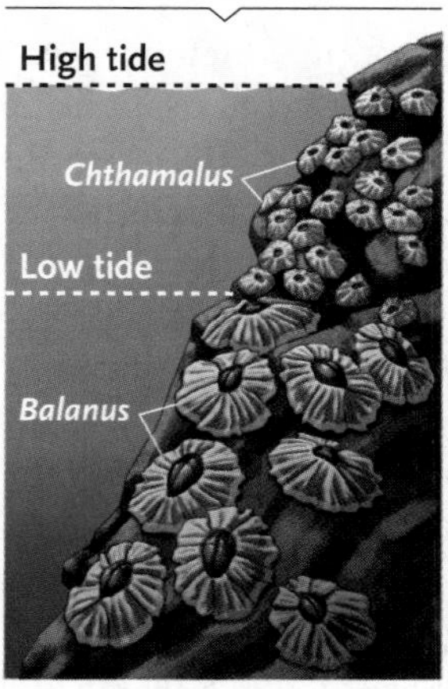
High tide
Chthamalus
Low tide
Balanus

Treatment 1: Remove *Balanus*
Results: In the absence of *Balanus*, *Chthamalus* occupies both shallow water and deep water.

Fundamental niche of Chthamalus

Treatment 2: Remove *Chthamalus*
Results: In the absence of *Chthamalus*, *Balanus* still occupies only deep water.

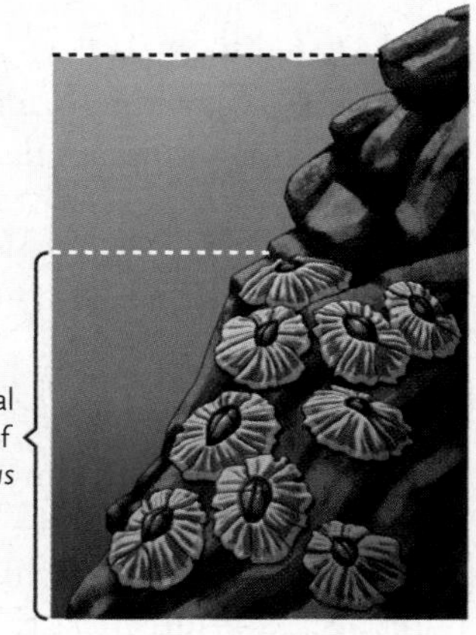
Fundamental niche of Balanus

Conclusion: In habitats where *Balanus* and *Chthamalus* coexist, the realized niche of *Chthamalus* is smaller than its fundamental niche because of competition from *Balanus*. The realized niche of *Balanus* is similar to its fundamental niche because it is not affected by the competitive interaction.

Source: Joseph H. Connell. 1961. The influence of interspecific competition and other factors on the distribution of the barnacle *Chthamalus stellatus*. *Ecology* 42:710–723.

vide *indirect* evidence that interspecific competition may have important effects. The first is the extremely common observation of **resource partitioning,** the use of different resources or the use of resources in different ways, by species living in the same place. For example, weedy plants might compete for water and dissolved nutrients in abandoned fields. But they avoid competition by partitioning these resources, collecting them from different depths in the soil **(Figure 51.10)**.

A second phenomenon that suggests the importance of competition is observed in comparisons of species that are sometimes sympatric (that is, living in the same place) and sometimes allopatric (that is, living in different places). In several studies of animals, researchers have documented **character displacement:** allopatric populations are morphologically similar and use similar resources, but sympatric populations are morphologically different and use different resources. The differences between the sympatric populations allow them to coexist without competing. Differences in bill size among sympatric finch species on the Galápagos Islands (see Sections 19.2 and 20.3) may be the product of character displacement **(Figure 51.11)**.

Data on resource partitioning and character displacement merely suggest the possible importance of interspecific competition in nature. To demonstrate *conclusively* that interspecific competition limits natural populations, one must show that the presence of one population reduces the population size or distribution of its presumed competitor. In a classic field experiment, Joseph Connell of the University of California, Santa Barbara, determined that competition between two barnacle species caused the realized niche of one species to be smaller than its fundamental niche **(Figure 51.12)**.

Connell first observed the distributions of barnacles in undisturbed habitats. *Chthamalus stellatus* is generally found in shallow water on rocky coasts, where it is periodically exposed to air. *Balanus balanoides* typically lives in deeper water, where it is usually submerged.

Connell determined the fundamental niche of each species by removing either *Chthamalus* or *Balanus* from rocks and monitoring the distribution of each species in the absence of the other. When Connell removed *Balanus* from rocks in deep water, larval *Chthamalus* colonized the area and produced a flourishing population of adults. Connell had observed that *Balanus* physically displaced *Chthamalus* from these rocks. Thus, interference competition from *Balanus* prevents *Chthamalus* from occupying areas where it would otherwise live. By contrast, the removal of *Chthamalus* from rocks in shallow water did not result in colonization by *Balanus*. *Balanus* is apparently unable to live in habitats that are frequently exposed to air. Connell therefore concluded that competition from *Chthamalus* does not affect the distribution of *Balanus*. Thus, the competitive interaction between these two species is asymmetrical: *Balanus* has a substantial effect on *Chthamalus*, but *Chthamalus* has virtually no effect on *Balanus*.

In Symbiotic Associations, the Lives of Two or More Species Are Closely Intertwined

Some species have a physically close ecological association called **symbiosis** (*sym* = together, *bio* = life, *sis* = condition). Biologists define three types of symbiotic interactions—*commensalism, mutualism,* and *parasitism*—that differ in their effects.

Commensalism, in which one species benefits and the other is unaffected, is rare in nature, because few species are unaffected by their interactions with another. One possible example is the relationship between cattle egrets *(Bubulcus ibis),* birds in the heron family, and the large grazing mammals with which they associate **(Figure 51.13).** Cattle egrets feed on insects and other small animals that their commensal partners flush from grass. Feeding rates of egrets are higher when they associate with large grazers than when they do not. The birds clearly benefit from this interaction, but the presence of birds has no apparent positive or negative impact on the mammals.

Fritz Polking/Frank Lane Picture Agency

FIGURE 51.13
Commensalism. Cattle egrets *(Bubulcus ibis)* feed on insects and other small animals flushed by the movements of large grazing mammals, like this African buffalo *(Syncerus caffer)*.

Mutualism, in which both partners benefit, is extremely common. The coevolved relationships between flowering plants and animal pollinators are largely mutualistic (see Figure 27.32). Animals that feed on a plant's nectar or pollen carry its gametes from one flower to another **(Figure 51.14).** Similarly, animals that eat the fruits of flowering plants disperse the seeds, "planting" them in a pile of nutrient-rich feces. These mutualistic relationships between plants and animals do not require active cooperation. Each species simply exploits the other for its own benefit.

Some associations between either bacteria or fungi and plants are also mutualistic. For example, mycorrhizae are fungi that grow alongside the roots of many plant species. These fungi facilitate the plants' uptake of nitrogen and phosphorus from the soil, and the plants provide the fungi with carbohydrates in return. Another important mutualism is the close relationship between the nitrogen-fixing bacterium *Rhizobium* and leguminous plants, such as peas, beans, and clover (see Section 33.3).

Mutualistic relationships between animal species are also common. For example, some small marine fishes feed on parasites that attach to the mouths and gills of large predatory fishes **(Figure 51.15).** Parasitized fishes hover motionless while the "cleaners" scour their tissues. The relationship is mutualistic because the cleaner fishes get a meal, and the larger fishes are relieved of parasites.

A. Flowering yucca plant

Harlo H. Hadow

B. Female yucca moth

Bob and Miriam Francis

A female yucca moth uses highly modified mouthparts to gather the sticky yucca pollen and roll it into a ball. She carries the pollen to another flower, and after piercing its ovary wall, she lays her eggs. She then places the pollen ball into the opening of the stigma.

C. Yucca moth larva

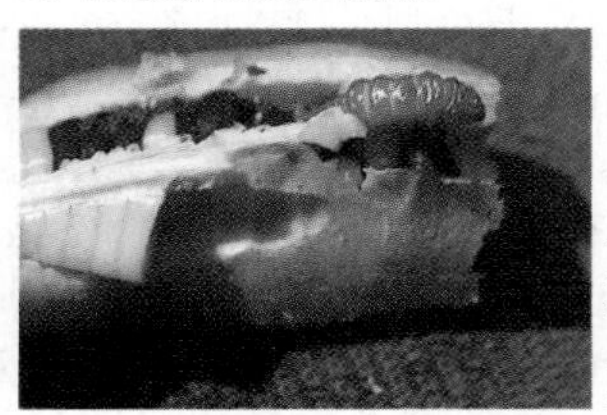

Harlow H. Hadow

When moth larvae hatch from the eggs, they eat some of the yucca seeds and gnaw their way out of the ovary to complete their life cycle. Enough seeds remain undamaged to produce a new generation of yuccas.

FIGURE 51.14
Mutualism between plants and animals. Several species of yucca plants (*Yucca* species) are each pollinated exclusively by one species of yucca moth (*Tegeticula* species). The adult stage of each moth appears at the time of year when its yucca plant flowers. These species are so mutually interdependent that the larvae of each moth species can feed on only one type of yucca, and the flowers of each yucca can be fertilized by only one species of moth. Most plant–pollinator mutualisms are much less specific.

Cleaner
wrasse
© Erik Schlogl

FIGURE 51.15

Mutualism between animal species. A large potato cod *(Epinephelus tukula)* from the Great Barrier Reef in Australia remains nearly motionless in the water while a striped cleaner wrasse *(Labroides dimidiatus)* carefully removes and eats ectoparasites attached to its lip. The potato cod is a predator, and the striped cleaner wrasse is a potential prey—but their mutualistic interaction supersedes a possible predator–prey interaction.

The relationship between the bullhorn acacia tree *(Acacia cornigera)* of Central America and a small ant species *(Pseudomyrmex ferruginea)* is one of the most highly coevolved mutualisms known **(Figure 51.16).** Each acacia is inhabited by an ant colony that lives in the tree's swollen thorns. The ants swarm out of the thorns to sting—and sometimes kill—herbivores that touch the tree. The ants also clip any vegetation that grows nearby. Thus, acacia trees that are colonized by ants grow in a space free of herbivores and competitors, and occupied trees grow faster and produce more seeds than unoccupied trees. In return, the plants produce sugar-rich nectar consumed by adult ants and protein-rich structures that the ants feed to their larvae. Ecologists describe the coevolved mutualism between these species as *obligatory,* at least for the ants; they cannot subsist on any other food sources.

Parasitism is a type of interaction in which one species, the **parasite,** uses another, the **host,** in a way that is harmful to the host. Parasite–host relationships are like predator–prey relationships: one population of organisms feeds on another. But parasites rarely kill their hosts quickly because a dead host is useless as a continuing source of nourishment.

Tapeworms and other parasites that live *within* a host are **endoparasites.** Many endoparasites acquire their hosts passively, when a host accidentally ingests the parasite's eggs or larvae (see *Focus on Applied Research,* Chapter 29). Endoparasites generally complete their life cycle in one or two host individuals. By contrast, leeches, aphids, mosquitoes, and other parasites that feed on the *exterior* of a host are **ectoparasites.** Most animal ectoparasites have elaborate sensory and behavioral mechanisms that allow them to locate specific hosts, and they feed on numerous host individuals during their lifetimes. Some plants, such as mistletoes (genus *Phoradendron*), live as ectoparasites on the trunks and branches of trees; their roots penetrate the host's xylem and extract water and nutrients.

Not all parasites feed directly on a host's tissues. The giant cowbirds described earlier are brood parasites, as are other species of cowbirds and cuckoos. Although oropendolas sometimes benefit from the presence of cowbirds, most brood parasites have negative effects on their hosts. For example, brood parasitism by the brown-headed cowbird *(Molothrus ater)* has played a large role in the near-extinction of Kirtland's warbler *(Dendroica kirtlandii).*

The feeding habits of some insects, called **parasitoids,** fall somewhere between true parasitism and predation. A female parasitoid lays eggs in the larva or pupa of another insect species, and her young consume the tissues of the living host. Because the hosts chosen by most parasitoids are highly specific, agricultural ecologists often release parasitoids to control populations of insect pests.

STUDY BREAK 51.1 <

1. **Why are some carnivores willing to spend more time and energy capturing large prey than small prey?**
2. **What are the differences between cryptic coloration, aposematic coloration, and mimicry? Can a mimic ever have aposematic coloration?**
3. **How can field experiments demonstrate conclusively that two species compete for limiting resources?**

THINK OUTSIDE THE BOOK >

Using the terms and concepts introduced in this chapter, describe the interactions that humans have with ten other species. Try to pick at least eight species that we do not eat.

A. **Ants patrolling an acacia**

Marie Read Natural History Photography

B. **Cleared area around an acacia**

David Thompson/OSF/Animals, Animals—Earth Scenes

FIGURE 51.16

A highly coevolved mutualism. (A) Bullhorn acacia trees *(Acacia cornigera)* provide colonies of small ants *(Pseudomyrmex ferruginea)* with homes in hollow enlarged thorns as well as other resources. Although individual ants are small, they are numerous and aggressive. **(B)** Because the ants attack herbivores and remove vegetation near their tree, acacias occupied by ants grow in a space that is free of herbivores and competitors.

51.2 The Nature of Ecological Communities

Ecologists have often debated the nature of ecological communities, asking if they have emergent properties that transcend the interactions among the populations they contain.

Most Ecological Communities Blend into Neighboring Communities

How do complex population interactions affect the organization and functioning of ecological communities? In the 1920s, ecologists in the United States developed two extreme hypotheses about the nature of ecological communities. Frederic Clements of the University of Minnesota championed an *interactive* view of communities. He described communities as "superorganisms," assemblages of species bound together by complex population interactions. According to this view, each species in a community requires interactions with a set of ecologically different species, just as every cell in an organism requires services that other types of cells provide. Clements believed that once a mature community was established, its **species composition**—the particular combination of species that occupy the site—was at *equilibrium*. If a fire or some other environmental factor disturbed the community, it would return to its predisturbance state.

Henry A. Gleason of the University of Michigan proposed an alternative, *individualistic* view of ecological communities. He believed that population interactions do not always determine species composition. Instead, a community is just an assemblage of species that are individually adapted to similar environmental conditions. According to Gleason's hypothesis, communities do not achieve equilibrium; rather, they constantly change in response to disturbance and environmental variation.

In the 1960s, Robert Whittaker of Cornell University suggested that ecologists could determine which hypothesis was correct by analyzing communities along environmental gradients, such as temperature or moisture **(Figure 51.17)**. According to Clements' interactive hypothesis, species that typically occupy the same communities should always occur together. Thus, their distributions along the gradient would be clustered in discrete groups with sharp boundaries between groups (see Figure 51.17A). According to Gleason's individualistic hypothesis, each species is distributed over the section of an environmental gradient to which it is adapted. Different species would have unique distributions, and species composition would change continuously along the gradient. In other words, communities would not be separated by sharp boundaries (see Figure 51.17B).

A. **Interactive hypothesis**

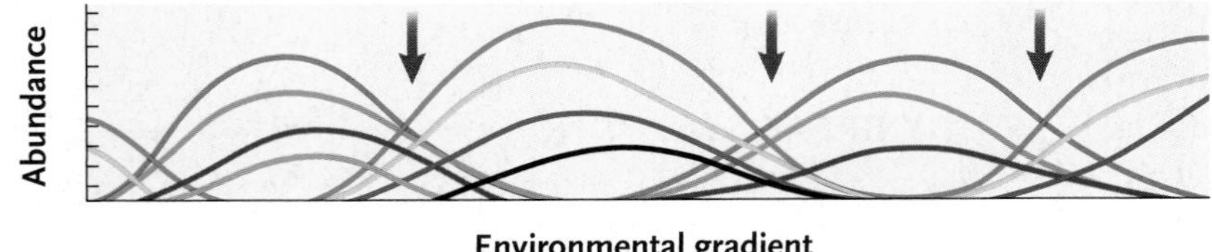
Abundance
Environmental gradient

The interactive hypothesis predicts that species within communities exhibit similar distributions along environmental gradients (indicated by the close alignment of several curves over each section of the gradient) and that boundaries between communities (indicated by arrows) are sharp.

B. **Individualistic hypothesis**

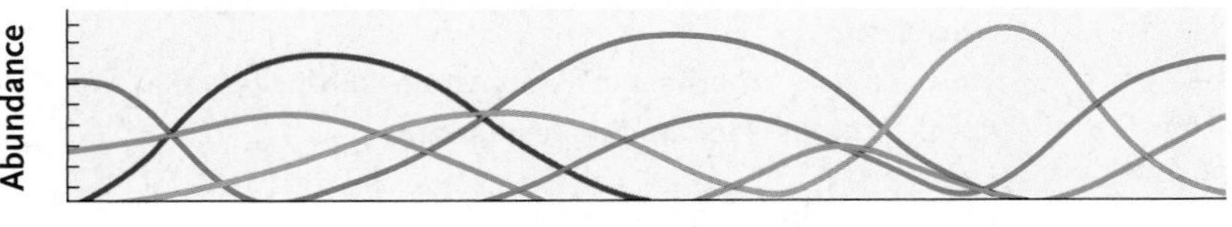
Abundance
Environmental gradient

The individualistic hypothesis predicts that species distributions along the gradient are independent (indicated by the lack of alignment of the curves) and that sharp boundaries do not separate communities.

C. **Siskiyou Mountains**

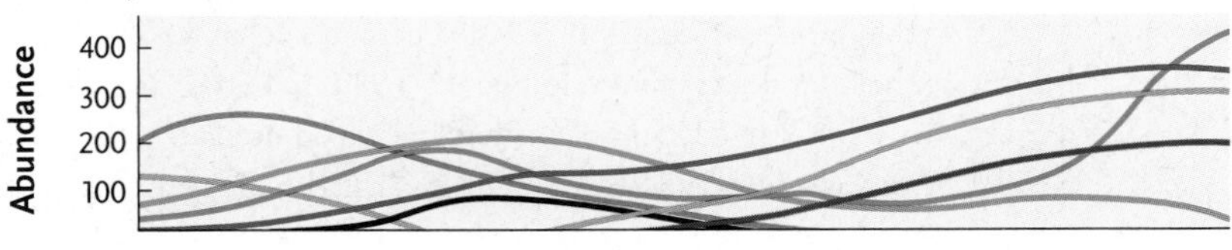
Abundance
400
300
200
100
Moisture gradient

D. **Santa Catalina Mountains**

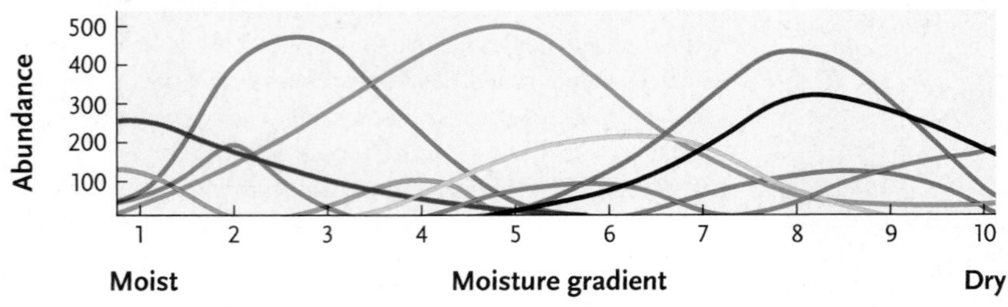
Abundance
500
400
300
200
100
1
2
3
4
5
6
7
8
9
10
Moist
Moisture gradient
Dry

Most gradient analyses support the individualistic hypothesis, as illustrated by distributions of tree species along moisture gradients in Oregon's Siskiyou Mountains and Arizona's Santa Catalina Mountains.

FIGURE 51.17
Two views of ecological communities. Each colored line represents the abundance of one plant species over the environmental gradient.

Most gradient analyses support Gleason's individualistic view of ecological communities. Environmental conditions vary continuously in space, and most plant distributions match these patterns (see Figure 51.17C, D). Species occur together in assemblages because they are adapted to similar conditions, and the species compositions of the assemblages change gradually across environmental gradients.

Nevertheless, the individualistic view does not fully explain all patterns observed in nature. Ecologists recognize certain assemblages of species as distinctive communities and name them accordingly—redwood forests and coral reefs are good examples. But the borders between adjacent communities are often wide transition zones, called **ecotones.** Ecotones are generally rich with species because they include plants and animals from both neighboring communities as well as some species that thrive only under transitional conditions. In some places, however, a discontinuity in a critical resource or some important abiotic factor produces a sharp community boundary. For example, chemical differences between soils derived from serpentine rock and sandstone establish sharp boundaries between communities of native California wildflowers and introduced European grasses **(Figure 51.18).**

STUDY BREAK 51.2 <

1. **Which view of communities suggests that they are just chance assemblages of species that happen to be adapted to similar abiotic environmental conditions?**
2. **Why would you often find more species living in an ecotone than you would in the communities on either side of it?**

Jasper Ridge Biological Preserve

FIGURE 51.18
Sharp community boundaries. Soils derived from serpentine rock have high magnesium and heavy metal content, which many plants cannot tolerate. Although native California wildflowers (bright yellow in this photograph) thrive on serpentine soil at the Jasper Ridge Preserve of Stanford University, introduced European grasses (green in this photograph) competitively exclude them from adjacent soils derived from sandstone.

51.3 Community Characteristics

Although the species composition of an ecological community may vary somewhat over geographical gradients, every community has certain characteristics that define its overall appearance and structure.

The Growth Forms of Plants Establish a Community's Overall Appearance

The growth forms of plants—their sizes and shapes—vary markedly in different environments. Warm, moist environments support complex vegetation with multiple vertical layers. For example, tropical forests include a canopy, formed by the tallest trees; an understory of shorter trees and shrubs; an herb layer under openings in the canopy; vinelike lianas; and epiphytes, which grow on the trunks and branches of trees **(Figure 51.19).** By contrast, physically harsh environments are occupied by low vegetation with simple structure. For example, trees on mountaintops buffeted by cold winds are short, and the plants below them cling to rocks and soil. Other environments support growth forms between these extremes (see Chapter 49).

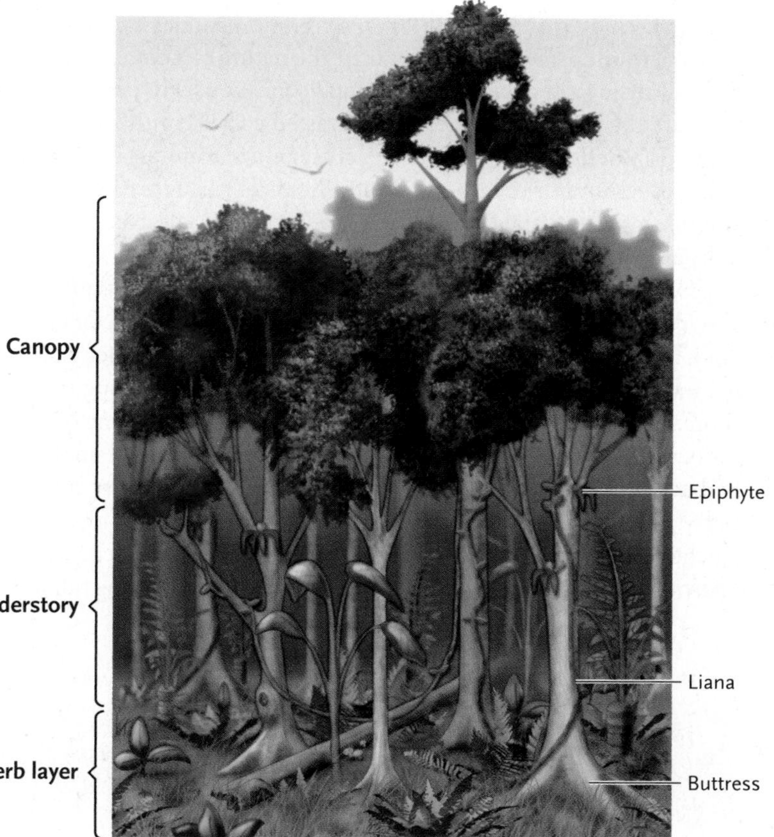
Canopy
Understory
Herb layer
Epiphyte
Liana
Buttress

FIGURE 51.19
Layered forests. Tropical forests include a canopy of tall trees and an understory of short trees and shrubs. Huge vines (lianas) climb through the trees, eventually reaching sunlight in the canopy, and epiphytic plants grow on trunks and branches, increasing the structural complexity of the habitat.

A. **Saltmarsh cordgrass *(Spartina alterniflora)***

Courtesy of Dr. John Bruno, University of North Carolina at Chapel Hill

B. **The effect of sawgrass patch length on the distribution of herbaceous plants**

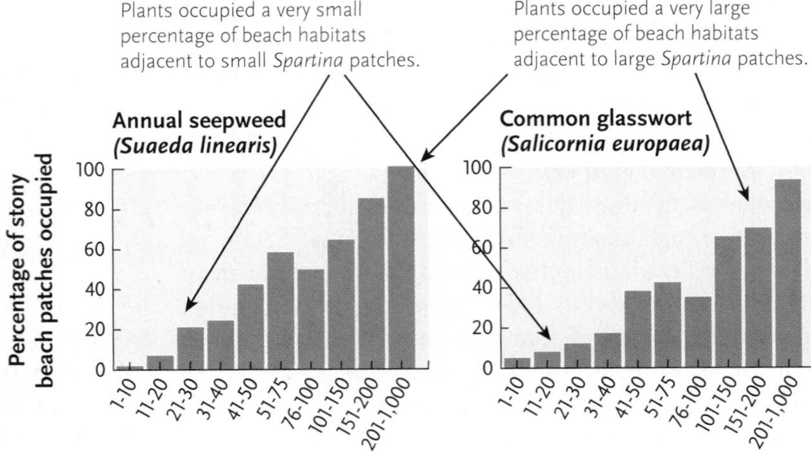

Plants occupied a very small percentage of beach habitats adjacent to small *Spartina* patches.
Plants occupied a very large percentage of beach habitats adjacent to large *Spartina* patches.
Annual seepweed (*Suaeda linearis*)
Common glasswort (*Salicornia europaea*)
Percentage of stony beach patches occupied
100
80
60
40
20
0
1-10
11-20
21-30
31-40
41-50
51-75
76-100
101-150
151-200
201-1,000
Length of adjacent *Spartina* patch (m)

FIGURE 51.20
Foundation species. (A) Saltmarsh sawgrass, the foundation species at the edges of Narragansett Bay, Rhode Island. **(B)** The presence or absence of herbaceous plants on stony beaches is strongly influenced by the length of the adjacent patch of sawgrass. The plants occupy very few beaches next to small sawgrass patches, but nearly all of the beaches next to large ones.

Foundation Species Moderate the Abiotic Environment within a Community

In many ecological communities, one common species can function as a **foundation species,** defining the nature of a community by creating locally stable environmental conditions. For example, trees are the foundation species in forested ecosystems, because their form defines the physical structure of the community, and their leaves and branches moderate short-term fluctuations in abiotic environmental factors like temperature, runoff from rainfall, and wind speed.

Saltmarsh cordgrass *(Spartina alterniflora)* is a foundation species in the wetlands surrounding Narragansett Bay, Rhode Island, because patches of this meter-high grass slow the velocity of the incoming tide and stabilize the stony beach habitat along the shore. In the absence of *Spartina,* tidal surges move the stones on the beach, disrupting the germination and growth of several small, herbaceous plant species. John F. Bruno of Brown University surveyed the plants growing adjacent to more than 350 *Spartina* patches of varying size. His research revealed that large patches of *Spartina* are more effective than small patches in moderating tidal effects. Not surprisingly, the percentage of a stony beach occupied by herbaceous plants was directly proportional to the length of the *Spartina* patch that bordered the beach **(Figure 51.20).**

Communities Differ in Species Richness and the Relative Abundance of Species

Communities differ greatly in their **species richness,** the number of species that live within them. For example, the harsh environment on a low desert island may support just a few species of microorganisms, fungi, algae (photosynthetic protists), plants, and arthropods. By contrast, tropical forests, which grow under milder physical conditions, include many thousands of species. Ecologists have studied global patterns of species richness (described below in Section 51.7) for decades. Today, as human disturbance of natural communities has already reached a tipping point, conservation biologists focus on such studies to determine which regions of Earth are most in need of preservation (see Chapter 53).

Within every community, populations differ in their commonness or the **relative abundance** of individuals. Some communities have just one or two **dominant species,** which represent a majority of the individuals present, as well as a number of rare species, each represented by just a few individuals. In other communities, species are represented by more equal numbers of individuals. For example, in a temperate deciduous forest in

Forest A: moderate species diversity

Forest B: high species diversity

Forest C: low species diversity

FIGURE 51.21
Species diversity. In this hypothetical example, each of three forests contains 50 trees. Forest A and forest B each include 10 tree species, but forest C includes only two tree species. Because forest A is dominated by one tree species, but forest B is not, ecologists would say that forest B is more diverse. Forest C, with only two tree species, is less diverse than the others.

West Virginia, tulip poplar *(Liriodendron tulipifera)* and sassafras *(Sassafras albidum)* are dominant, accounting for nearly 85% of the trees. By contrast, a tropical forest in Costa Rica may include more than 200 tree species, each making up only a small percentage of the total.

Species richness and relative abundance together contribute to a community characteristic that ecologists call **species diversity.** To demonstrate the concept of species diversity, we will compare three hypothetical forest communities **(Figure 51.21).** Two of the communities include 50 trees distributed among 10 species. In Forest A, the dominant species is represented by 39 individuals, two species by two individuals each, and seven species by one individual each. In Forest B, each of the 10 species is represented by five individuals. Although both communities have the same species richness (10 species), Forest A is less diverse than Forest B, because more than three-quarters of its trees are of the same species. The third forest has only two tree species (Forest C in Figure 51.21); it is less diverse than either of the others.

Feeding Relationships within a Community Determine Its Trophic Structure

All ecological communities, regardless of their species richness, also have a trophic structure (*trophe* = nourishment) that comprises all of the plant–herbivore, predator–prey, host–parasite, and potential competitive interactions **(Figure 51.22).**

TROPHIC LEVELS We can visualize the trophic structure of a community as a hierarchy of **trophic levels,** defined by the feeding relationships among its species (see Figure 51.22A). Photosynthetic organisms are the **primary producers,** the first trophic level. Primary producers are often described as **autotrophs** (*auto* = self) because they capture sunlight and convert it into chemical energy, using simple *inorganic* molecules acquired from the environment to build larger *organic* molecules that other organisms can use. Plants are the dominant primary producers in terrestrial communities. Multicellular algae (macroalgae) and plants are the major primary producers in shallow freshwater and marine environments. Photosynthetic or chemosynthetic bacteria, cyanobacteria, and "protists" are the primary producers in deep, open water.

Animals, by contrast, are **consumers.** Herbivores, which feed directly on producers, form the second trophic level, the **primary consumers.** Carnivores that feed on herbivores are the third trophic level, or **secondary consumers,** and carnivores that feed on other carnivores form the fourth trophic level, the **tertiary consumers.** For example, songbirds feeding on herbivorous insects are secondary consumers, and falcons feeding on songbirds are tertiary consumers. Some organisms, like humans and some bears, are **omnivores,** feeding at several trophic levels simultaneously.

A separate and distinct trophic level includes organisms that extract energy from the organic detritus (refuse) produced at other trophic levels. Scavengers, or **detritivores,** are animals such as earthworms and vultures that ingest dead organisms, digestive wastes, and cast-off body parts such as leaves and exoskeletons. **Decomposers** are small organisms, such as bacteria and fungi, that feed on dead or dying organic material. As described in Chapter 52, detritivores and decomposers serve a critical ecological function because their activity reduces organic material to small inorganic molecules that producers can assimilate.

All of the consumers in a community—the animals, fungi, and diverse microorganisms—are described as **heterotrophs** (*hetero* = other) because they acquire energy and nutrients by eating other organisms or their remains.

FOOD CHAINS AND WEBS Ecologists depict the trophic structure of a community in a **food chain,** a portrait of who eats whom. Each link in a food chain is represented by an arrow pointing from the food to the consumer. Simple, straight-line food chains are rare in nature because most consumers feed on more than one type of food, and because most organisms are eaten by more than one type of consumer. These complex relationships are portrayed as a **food web,** a set of interconnected food chains with multiple links.

In the food web for the waters off the coast of Antarctica (see Figure 51.22B), most organisms at the bottom of the food web are tiny, and they occur in vast numbers. Huge pastures of phytoplankton (microscopic algae and diatoms) are responsible for most photosynthesis. They are consumed by herbivorous zooplankton (some protists, copepods, and shrimplike krill), which are in turn eaten by larger species, such as carnivorous zooplankton, squids, fishes, and suspension-feeding baleen whales. Some of these secondary consumers are themselves eaten by birds and mammals at higher trophic levels. The top carnivore in this ecosystem, the orca *(Orcinus orca),* feeds on carnivorous birds and mammals. As you can see in Figure 51.22B, ecological relationships within a food web are complex because many species feed at more than one trophic level.

FOOD-WEB ANALYSIS In the late 1950s, Robert MacArthur of Princeton University pioneered the analysis of food webs to determine how the many links between trophic levels may contribute to a community's **stability**—its ability to maintain its species composition and relative abundances when environmental disturbances eliminate some species from the community. MacArthur hypothesized that in species-rich communities, where animals feed on many food sources, the absence of one or two species would have only minor effects on the structure and stability of the community as a whole. He therefore proposed a connection between species diversity, food-web complexity, and community stability.

Recent research has confirmed MacArthur's reasoning. For example, the average number of links per species generally increases with increasing species richness. Comparative food-web analyses also reveal that the relative proportions of species at the highest, middle, and lowest trophic levels are reasonably constant across communities. When researchers compared the number of prey species to the number of predator species in food webs from 92 communities of freshwater invertebrates, they discovered that, regardless of species richness, a community includes between two and three prey species for every predator species.

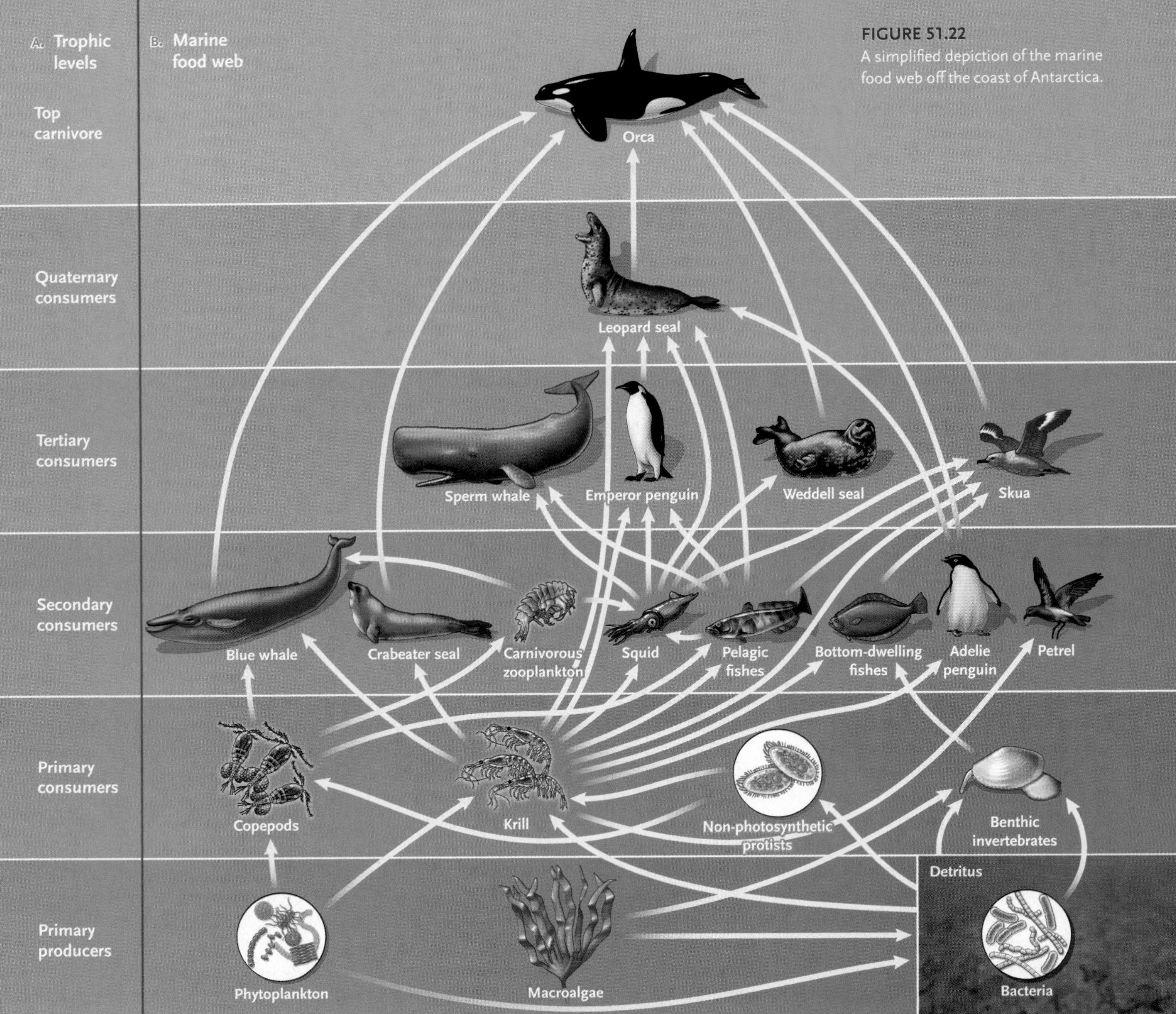

A. Trophic levels
B. Marine food web
Top carnivore
Quaternary consumers
Tertiary consumers
Secondary consumers
Primary consumers
Primary producers
Orca
Leopard seal
Sperm whale
Emperor penguin
Weddell seal
Skua
Blue whale
Crabeater seal
Carnivorous zooplankton
Squid
Pelagic fishes
Bottom-dwelling fishes
Adelie penguin
Petrel
Copepods
Krill
Non-photosynthetic protists
Benthic invertebrates
Detritus
Phytoplankton
Macroalgae
Bacteria

FIGURE 51.22 A simplified depiction of the marine food web off the coast of Antarctica.

Interactions among species in a food web are often complex, indirect, and hard to unravel. In desert communities of the American Southwest, for example, rodents and ants potentially compete for seeds, their main food source. And the plants that produce the seeds compete for water, nutrients, and space. Rodents generally prefer to eat large seeds, but ants prefer small seeds. Thus, feeding by rodents reduces the potential population sizes of plants that produce large seeds. As a result, the population sizes of plants that produce small seeds may increase, ultimately providing more food for ants.

Some analyses of food webs focus on interactions in which predators or prey have a significant influence on the growth rates and sizes of other populations in the community; these *strong interactions* can affect overall community structure. In the next section we provide examples of strong interactions when we describe how consumers influence the competitive interactions among populations of their prey.

STUDY BREAK 51.3

1. What plant growth forms are common in tropical forests?
2. What is the difference between species richness and relative abundance?
3. Peregrine falcons are predatory birds that have been introduced into many North American cities, where they feed primarily on pigeons. The pigeons eat mostly vegetable matter. To what trophic level do pigeons and peregrine falcons belong?

51.4 Effects of Population Interactions on Community Characteristics

Numerous studies have shown that interspecific competition and predation can influence a community's species composition.

Interspecific Competition Can Reduce Species Richness within Communities

Interspecific competition can cause the local extinction of species or prevent new species from becoming established in a community, thus reducing its species richness. During the 1960s and early 1970s, ecologists emphasized competition as the primary factor structuring communities. Observations of resource partitioning and character displacement suggested that some process had fostered differences in resource use among coexisting species, and competition provided the most straightforward explanation of these patterns.

Seeking to uncover direct evidence of competition, ecologists undertook many field experiments on competition in natural populations. The experiment on barnacles depicted in Figure 51.12 is typical of this approach, in which researchers determine whether adding or removing a species changes the distribution or population size of its presumed competitors. In the early 1980s, two independent reviews of the literature on these field experiments, one by Joseph Connell and the other by Thomas W. Schoener of the University of California, Davis, suggested that competition is sometimes a potent force. Connell's survey, which included 527 published experiments on 215 species, identified competition in roughly 40% of the experiments and more than 50% of the species. Schoener's review, which used different criteria to evaluate 164 experiments on approximately 400 species, found that competition affected more than 75% of the species.

Although these reviews confirm the importance of competition, the ecological literature upon which they were based probably contains several significant biases. First, ecologists who set out to study competition are more likely to study interactions in which they think competition occurs, and they are more likely to publish research that documents its importance. Accordingly, the literature includes more studies of competition in *K*-selected species than in *r*-selected species (review Section 50.5). Recall that populations of *r*-selected species, such as herbivorous insects, rarely reach carrying capacity, and competition may not limit their population sizes. Thus, the Connell and Schoener surveys may *overestimate* the importance of competition. Another bias, which Connell called "the ghost of competition past," *underestimates* the importance of competition. If, as many ecologists believe, resource partitioning and character displacement are the results of past competition, we are unlikely to witness much competition today, even though it was once important in structuring those population interactions.

Ecologists have still not reached consensus about whether interspecific competition strongly influences the species composition and structure of most communities. Plant ecologists and vertebrate ecologists, who often study *K*-selected species, generally believe that competition has a profound effect on species distributions and resource use. Insect ecologists and marine ecologists, who often study *r*-selected species, argue that competition is not the major force governing community structure, pointing instead to predation or parasitism and physical disturbance.

Predators Can Boost Species Richness by Stabilizing Competitive Interactions among Their Prey

Predators can influence the species richness and structure of communities by reducing the population sizes of their prey. On the rocky coast of the American Northwest, for example, algae and sessile invertebrates compete for attachment sites on rocks, a requirement for life on a wave-swept shore. California mussels *(Mytilus californianus)* are the strongest competitors for space, eliminating other species from the community. But at some sites, predatory sea stars *(Pisaster ochraceus)* preferentially feed on mussels, reducing their numbers and creating space for other species to grow. Because the interaction between *Pisaster* and *Mytilus* affects other species as well, it qualifies as a strong interaction.

In the 1960s, Robert Paine of the University of Washington conducted removal experiments to evaluate the effects of *Pisaster* predation **(Figure 51.23).** In predator-free experimental plots, mussels outcompeted barnacles, chitons, limpets, and other invertebrate herbivores, reducing species richness from 18 species to two or three. In control plots that contained predators, however, all 18 species persisted. Ecologists describe predators like *Pisaster* as **keystone species,** those that have a greater effect on community structure than their numbers might suggest.

Herbivores May Counteract or Reinforce Competition among Their Food Plants

Herbivores also exert complex effects on communities. In the 1970s, Jane Lubchenco, then of Harvard University, studied herbivory in a periwinkle snail *(Littorina littorea),* a keystone species on rocky shores in Massachusetts **(Figure 51.24).** Periwinkles preferentially graze on the tender green alga *Enteromorpha.* In tidepools, which are usually submerged, *Enteromorpha* outcompetes other algae. Moderate feeding by periwinkles, however, eliminates some *Enteromorpha,* allowing less competitive algal species to grow. Moderate herbivory by periwinkles therefore increases algal species richness in tidepools. But on high rocks, which are exposed to air during low tide, the dehydration-resistant red alga *Chondrus* is competitively dominant. Periwinkles don't eat the tough *Chondrus,* however, feeding instead on the less abundant and competitively inferior *Enteromorpha.* Thus, on exposed rocks, feeding by the snails reduces algal species richness.

STUDY BREAK 51.4 <

1. **How is the scientific literature on interspecific competition biased?**
2. **What are keystone species, and how do they influence species richness in communities?**

FIGURE 51.23 **Experimental Research**

Effect of a Predator on the Species Richness of Its Prey

Question: Does feeding by a predator influence the species richness and relative abundances of the species on which it feeds?

Experiment: The predatory sea star *Pisaster ochraceus* preferentially feeds on California mussels *(Mytilus californianus)*, which are the strongest competitors for space in rocky intertidal habitats in Washington State. Paine removed *Pisaster* from caged experimental study plots, but left control study plots undisturbed. He then monitored the species richness of *Pisaster's* invertebrate prey over many years.

Results: Paine documented an increase in California mussel populations in the experimental plots as well as complex changes in the feeding relationships among species in the intertidal food web. The overall effect of removing *Pisaster*, the top predator in this food web, was a rapid decrease in the species richness of invertebrates and algae. By contrast, control plots maintained their species richness over the course of the experiment.

Nancy Rotenberg/Animals, Animals-Earth Scenes

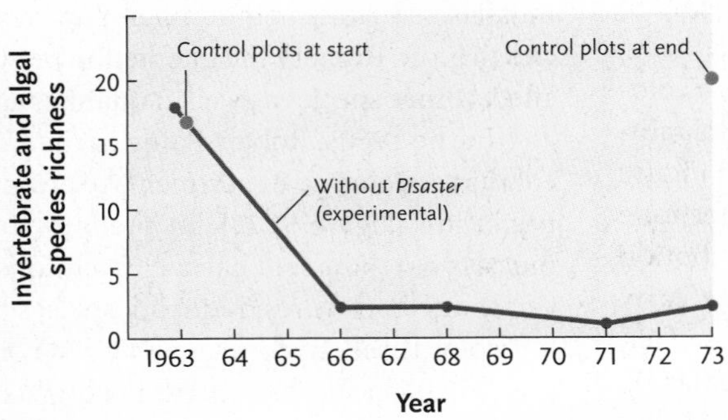
Control plots at start
Control plots at end
Without Pisaster (experimental)
Invertebrate and algal species richness
0
5
10
15
20
1963
64
65
66
67
68
69
70
71
72
73
Year

Conclusion: Predation by the sea star *Pisaster ochraceus* maintains the species richness of its prey by preventing mussels from outcompeting other invertebrates and algae on rocky shores.

Source: Robert T. Paine. 1974. Intertidal Community Structure. Experimental studies on the relationship between a dominant competitor and its principal predator. *Oecologia* 15:93–120.

51.5 Effects of Disturbance on Community Characteristics

Recent research tends to support the individualistic view that many communities are not in equilibrium and that their species composition changes frequently. Environmental disturbances—storms, landslides, fires, floods, and cold spells—often eliminate some species, providing opportunities for others to become established.

Frequent Disturbances Keep Some Communities in a Constant State of Flux

Physical disturbances are common in some environments. For example, lightning-induced fires commonly sweep through grasslands, powerful hurricanes routinely demolish patches of forest, and waves wash over communities that live at the edge of the sea.

Joseph Connell and his colleagues conducted an ambitious long-term study of the effects of disturbance on coral reefs, shallow tropical marine habitats that are among the most species-rich communities on Earth. In some parts of the world, reefs are routinely battered by violent storms, which wash corals off the substrate, creating bare patches in the reef. The scouring action of storms creates opportunities for coral larvae to settle on bare substrates and start new colonies; ecologists use the word *recruitment* to describe the process in which young individuals join a population.

From 1963 to 1992, Connell and his colleagues tracked the fate of the Heron Island Reef at the south end of Australia's Great Barrier Reef **(Figure 51.25).** The inner flat and protected crests of the reef are sheltered from severe wave action during storms, whereas some pools and crests are routinely exposed to physical disturbance. Because corals live in colonies of variable size, the researchers monitored coral abundance by measuring the percentage of the substrate (that is, the seafloor) that colonies covered. They revisited marked study plots at intervals, photographing and identifying individual coral colonies.

Five major cyclones crossed the reef during the 30-year study period. Coral communities in the exposed areas of the reef were in a nearly continual state of flux. In exposed pools, four of the five cyclones reduced the percentage of cover, often drastically. On exposed crests, the cyclone of 1972 eliminated virtually all of the corals, and subsequent storms slowed the recovery of these areas for more than 20 years. By contrast, corals in sheltered areas suffered much less storm damage. Nevertheless, their coverage also declined steadily during the study as a natural consequence of the corals' growth. As colonies grew taller and closer to the ocean's surface, their increased exposure to air resulted in substantial mortality.

Connell and his colleagues also documented recruitment, the growth of new colonies from settling larvae, in their study plots. They discovered that the rate at which new colonies developed was almost always higher in sheltered areas than in exposed areas. However, recruitment rates were extremely variable, depending in part on the amount of space that storms or coral growth had made available.

This long-term study of coral reefs illustrates that frequent disturbances prevent some communities from reaching an equilibrium determined by interspecific interactions. Changes in the coral reef community at Heron Island result from the combined effects of external disturbances that remove coral colonies from the reef and internal processes (growth and recruitment) that either eliminate colonies or establish new ones. In this community, growth and recruitment are slow processes, and disturbances are frequent. Thus, the community never attains equilibrium.

FIGURE 51.24 **Experimental Research**

The Complex Effects of an Herbivorous Snail on Algal Species Richness

Question: How does feeding by periwinkle snails *(Littorina littorea)* influence the species richness of algae in intertidal communities?

Experiment: Lubchenco manipulated the densities of periwinkle snails in tidepools and on exposed rocks in a rocky intertidal habitat by creating enclosures that prevented snails from either entering or leaving her study plots. She then monitored the species composition of algae in the study plots and graphed them against periwinkle density.

Results: The effects of periwinkle density on algal species richness varied dramatically between study plots in tidepools and on exposed rocks.

Periwinkle snails (*Littorina littorea*)

© Jane Burton/BruceColman, Ltd.

In tidepools

Algal species richness

12
10
8
6
4
2
0

0 100 200 300

Periwinkles per square meter

In tidepools, snails at low densities eat little algae and *Enteromorpha* competitively excludes other algal species, reducing species richness. At high snail densities, heavy feeding on all species reduces algal species richness. At intermediate snail densities, grazing eliminates some *Enteromorpha*, allowing other species to grow.

Enteromorpha **growing in tidepools**

© Jane Burton/BruceColman, Ltd.

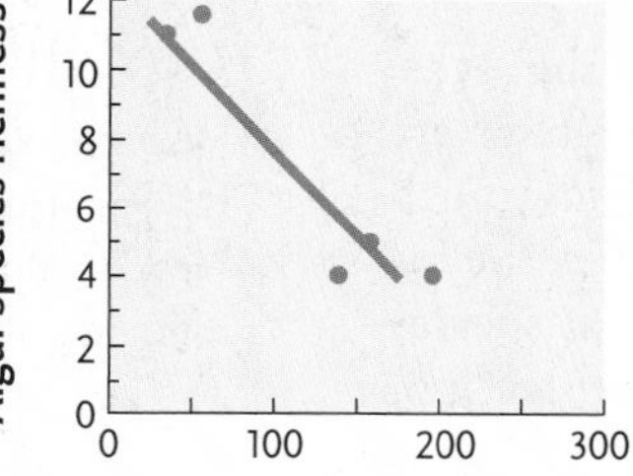
On exposed rocks
Algal species richness
12
10
8
6
4
2
0
0
100
200
300
Periwinkles per square meter

On exposed rocks, periwinkles never eat much *Chondrus*, but they consume the tender, less successful competitors. Thus, feeding by periwinkles reinforces the competitive superiority of *Chondrus*: as periwinkle density increases, algal species richness declines.

Chondrus **growing on exposed rocks**

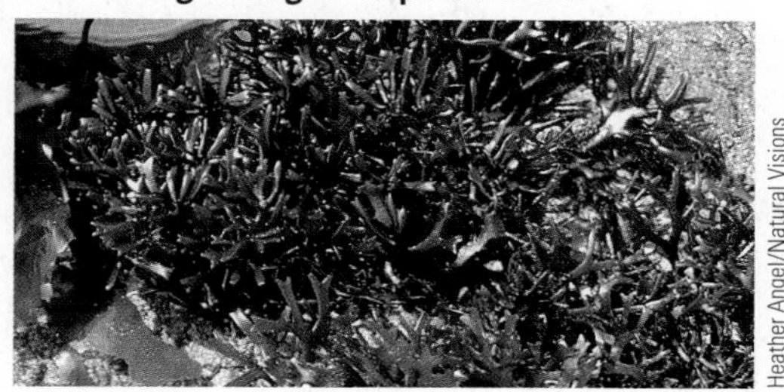

Heather Angel/Natural Visions

Conclusion: Grazing by periwinkle snails has complex effects on the species richness of competing algae. In tidepools, where periwinkle snails preferentially feed on *Enteromorpha*, the competitively dominant alga, snails at an intermediate density remove some *Enteromorpha*, which allows weakly competitive algae to grow, increasing species richness. Feeding by snails at either low or high densities reduces algal species richness. On exposed rocks, where periwinkle snails rarely eat the competitively dominant alga *Chondrus*, feeding by snails reduces algal species richness.

Source: Jane Lubchenco. 1978. Plant species diversity in a marine intertidal community: Importance of herbivore food preference and algal competitive abilities. *The American Naturalist* 112:23–39.

Moderate Levels of Disturbance May Foster High Species Richness

According to the **intermediate disturbance hypothesis,** proposed by Connell in 1978, species richness is greatest in communities that experience fairly frequent disturbances of moderate intensity. Moderate disturbances create some openings for *r*-selected species to arrive and join the community, but they allow *K*-selected species to survive. Thus, communities that experience intermediate levels of disturbance contain a rich mixture of species. Where disturbances are severe and frequent, communities include only *r*-selected species that complete their life cycles between catastrophes. Where disturbances are mild and rare, communities are dominated by long-lived *K*-selected species that competitively exclude other species from the community.

Several studies in diverse habitats have confirmed the predictions of the intermediate disturbance hypothesis. For example, Colin R. Townsend and his colleagues at the University of Otago studied the effects of disturbance at 54 stream sites in the Taieri River system in New Zealand. Disturbance occurs in these

FIGURE 51.25

The effects of storms on corals. Five tropical cyclones (marked by gray arrows) damaged corals on the Heron Island Reef during a 30-year period. Storms reduced the percentage cover of corals in exposed parts of the reef **(A)** much more than in sheltered parts of the reef **(B).**

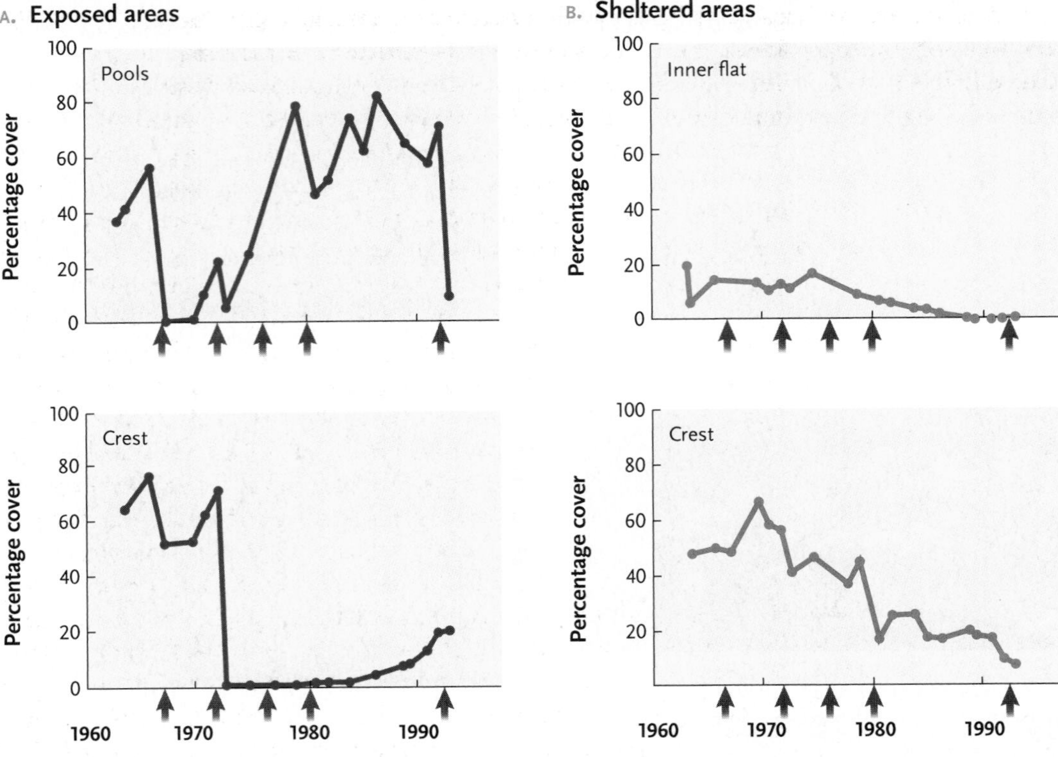

A. Exposed areas
B. Sheltered areas
Pools
Inner flat
Crest
Crest
Percentage cover
100
80
60
40
20
0
1960
1970
1980
1990

communities when water flow from heavy rains moves the rocks, soil, and sand in the streambed, disrupting the habitats where animals live. Townsend and his colleagues measured how much of the substrate moved in different streambeds to index the intensity of the disturbance. Their results indicate that species richness is highest in areas that experience intermediate levels of disturbance **(Figure 51.26).**

Some ecologists have also suggested that species-rich communities recover from disturbances more readily than do less diverse communities. For example, David Tilman and his colleagues at the University of Minnesota conducted large-scale experiments in midwestern grasslands on the relationship between species number and the ability of communities to recover from disturbance. Their results demonstrate that grassland plots with high species richness recover from drought faster than plots with fewer species.

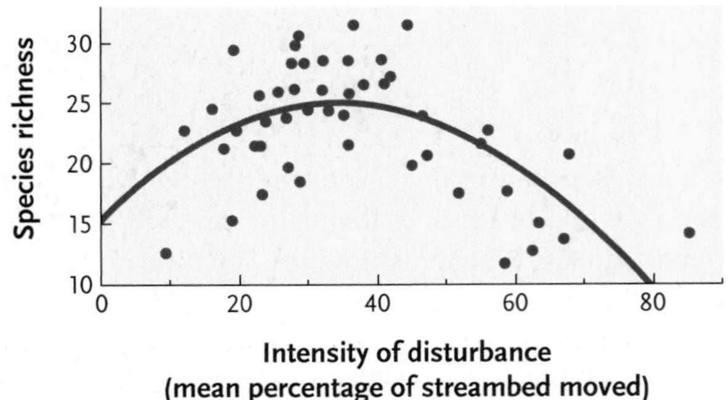

Species richness
30
25
20
15
10
0
20
40
60
80
Intensity of disturbance
(mean percentage of streambed moved)

FIGURE 51.26

An observational study that supports the intermediate disturbance hypothesis. In the Taieri River system in New Zealand, species richness was highest in stream communities that experienced an intermediate level of disturbance.

STUDY BREAK 51.5

1. How might disturbances from storms allow coral reefs to be rejuvenated by the recruitment of young individuals?
2. How do moderately severe and moderately frequent disturbances influence a community's species richness?

51.6 Ecological Succession: Responses to Disturbance

In response to disturbance, communities undergo **ecological succession,** a somewhat predictable series of changes in species composition over time.

Succession Begins after Disturbance Alters a Landscape or Changes the Species Composition of an Existing Community

Primary succession begins when organisms first colonize terrestrial habitats without soil, such as those created by erupting volcanoes and retreating glaciers **(Figure 51.27).** Lichens (see Section 28.3), which derive nutrients from rain and bare rock, are usually the first visible colonizers of such inhospitable habitats. They secrete mild acids that erode rock surfaces, initiating the slow development of soil, which is enriched by the organic material lichens produce. After lichens modify a site, mosses (see Section 27.2) colonize patches of soil and grow quickly.

As soil accumulates, hardy *r*-selected plants—grasses, ferns, and broad-leaved herbs—colonize the site from surrounding areas. Their roots break up rock, and as they die, their decaying

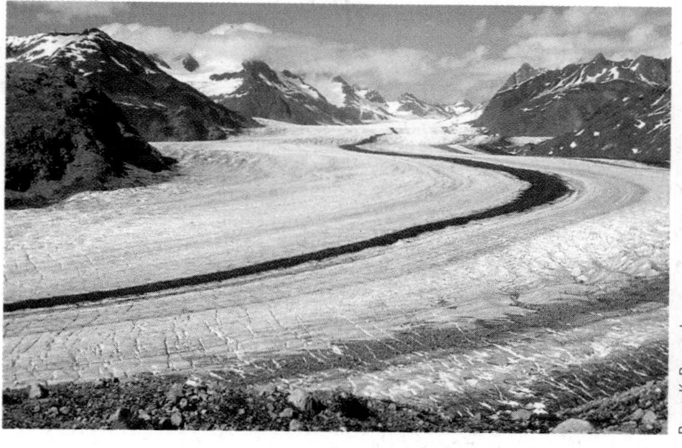

Roger K. Burnard

1 The glacier has retreated about 8 m per year since 1794.

Roger K. Burnard

2 This site was covered with ice less than 10 years before this photo was taken. When a glacier retreats, a constant flow of meltwater leaches minerals, especially nitrogen, from the newly exposed substrate.

Roger K. Burnard

3 Once lichens and mosses have established themselves, mountain avens (genus *Dryas*) grows on the nutrient-poor soil. This pioneer species benefits from the activity of mutualistic nitrogen-fixing bacteria, spreading rapidly over glacial till.

Roger K. Burnard

4 Within 20 years, shrubby willows (genus *Salix*), cottonwoods (genus *Populus*), and alders (genus *Alnus*) take hold in drainage channels. These species are also symbiotic with nitrogen-fixing microorganisms.

Roger K. Burnard

5 In time, young conifers, mostly hemlocks (genus *Tsuga*) and spruce (genus *Picea*), join the community.

Ed Degginger

6 After 80 to 100 years, dense forests of Sitka spruce *(Picea sichensis)* and western hemlock *(Tsuga heterophylla)* have crowded out the other species.

FIGURE 51.27

Primary succession following glacial retreat. The retreat of glaciers at Glacier Bay, Alaska, has allowed ecologists to document primary succession on newly exposed rocks and soil.

remains enrich the soil. Detritivores and decomposers facilitate these processes. As the soil gets deeper and richer, increased moisture and nutrients support bushes and, eventually, trees. Late successional stages are often dominated by *K*-selected species with woody trunks and branches that position leaves in sunlight and large root systems that acquire water and nutrients from soil.

In the classical view of ecological succession, long-lived species eventually dominate a community, and new species join it only rarely. This relatively stable, late successional stage is called a **climax community** because the dominant vegetation replaces itself and persists until an environmental disturbance eliminates it, allowing other species to invade. Local climate and soil conditions, the surrounding communities where colonizing species originate, and chance events determine the species composition of climax communities. However, recent research suggests that even "climax communities" change slowly in response to environmental fluctuations, as described below.

Secondary succession occurs after existing vegetation is destroyed or disrupted by an environmental disturbance, such as a fire, a storm, or human activity. The presence of soil makes the disturbed sites ripe for colonization. Moreover, the soil may contain numerous seeds that germinate after the disturbance. The early stages of secondary succession proceed rapidly, but later stages parallel those of primary succession.

Secondary succession in the north temperate zone is well studied in abandoned farms, called "old fields," where forests were cleared centuries ago. Because the transformation from old field back to forest takes at least a hundred years, ecologists use historical records to find the age of different stands of vegetation and reconstruct the successional sequence by comparing stands of different ages. In the Piedmont region of southeastern North America, an abandoned field is covered by crabgrass (genus *Digitaria*), an annual plant, during the first growing season. The following year, crabgrass is replaced by horseweed *(Conyza canadensis),* which cannot persist because it secretes substances that inhibit the germination of its own seeds. Ragweed *(Ambrosia artemisiifolia),* another annual, dominates during the third year, but it is gradually replaced by perennial asters (genus *Erigeron*) and broomsedges (genus *Andropogon*), which are, in turn, replaced by shrubs. Ten to fifteen years after the field was abandoned, pine (genus *Pinus*) seedlings germinate. Growing pines cast substantial shade and their fallen needles acidify the soil, making the site unsuitable for the plants from earlier successional stages. Because pines are intolerant of shade, pine seedlings don't flourish under mature pine trees. Thus, after 50 to 100 years, pines are replaced by a taller mixed hardwood forest of oaks (genus *Quercus*) and hickories (genus *Carya*), the seedlings of which are more shade tolerant than pines. The hardwood forest forms the climax community in the thick, moist soil after more than a century of successional change.

Similar climax communities sometimes arise from alternative successional sequences. For example, hardwood forests also develop in sites that were once ponds. During **aquatic succession,** debris from rivers and runoff accumulates in a body of water, causing it to fill in at its margins. The pond is transformed into a swamp, inhabited by plants adapted to a semisolid substrate. As larger plants get established, their high transpiration rates dry the soil, allowing other plant species to colonize. Given enough time, the site may become a meadow or forest, where an area of moist, low-lying ground is the only remnant of the original pond.

Community Characteristics Change during Succession

Several characteristics undergo directional change as succession proceeds. First, because *r*-selected species are short-lived and *K*-selected species long-lived, species composition changes rapidly in the early stages, but slowly in the late stages of succession. Second, species richness increases rapidly during the early stages because new species join the community faster than resident species become extinct; as succession proceeds, however, species richness stabilizes or may even decline. Third, in terrestrial communities that receive sufficient rainfall, the maximum height and total mass of the vegetation increase steadily as large species replace small ones, creating the complex structure of the climax.

Because plants influence the physical environment below them, the community itself increasingly moderates the microclimate. The shade cast by a forest canopy retains soil moisture and reduces temperature fluctuations. The trunks and canopy also reduce wind speed. By contrast, the short vegetation in an early successional stage does not effectively shelter the space below it.

Although ecologists usually describe succession in terms of vegetation, animals undergo succession, too. As the vegetation shifts, new resources become available, and animal species replace each other over time. Herbivorous insects, which often have strict food preferences, undergo succession along with their food plants. And as the herbivores change, so do their predators, parasites, and parasitoids. In old-field succession in eastern North America, different successional stages harbor a changing assortment of bird species **(Figure 51.28).**

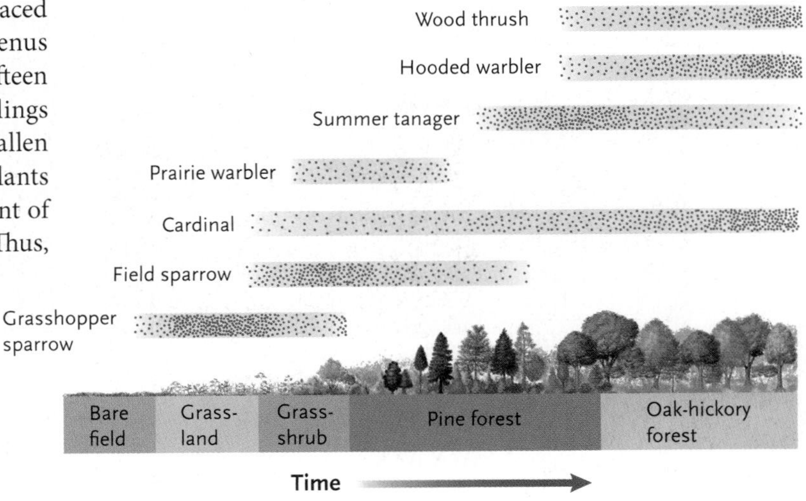
Wood thrush
Hooded warbler
Summer tanager
Prairie warbler
Cardinal
Field sparrow
Grasshopper sparrow
Bare field
Grass-land
Grass-shrub
Pine forest
Oak-hickory forest
Time

FIGURE 51.28
Succession in animals. Successional changes in bird species composition in an abandoned agricultural field in eastern North America parallel changes in plant species composition. Residence times of several representative species are illustrated. The density of stippling inside each bar illustrates the density of each species through time.

Several Hypotheses Help to Explain the Processes Underlying Succession

Differences in dispersal abilities, maturation rates, and life spans among species are at least partly responsible for ecological succession. Early successional stages harbor many *r*-selected species because they produce numerous small seeds that colonize open habitats and grow quickly. Mature successional stages are dominated by *K*-selected species because they are long-lived. Nevertheless, coexisting populations inevitably affect one another. Although the role of population interactions in succession is generally acknowledged, ecologists debate the relative importance of processes that either facilitate or inhibit the turnover of species in a community.

The **facilitation hypothesis** suggests that species modify the local environment in ways that make it less suitable for themselves but more suitable for colonization by species typical of the next successional stage. For example, when lichens first colonize bare rock, they produce a small quantity of soil, which is required by mosses and grasses that grow there later. According to this hypothesis, changes in species composition are both orderly and predictable because the presence of each stage facilitates the success of the next. Facilitation is very important in primary succession, but it may not be the best model of interactions that influence secondary succession.

The **inhibition hypothesis** suggests that new species are prevented from occupying a community by whatever species are already present. According to this hypothesis, succession is neither orderly nor predictable because each stage is dominated by whichever species happen to colonize the site first. Species replacements occur only when individuals of the dominant species die of old age or when an environmental disturbance reduces their numbers. Eventually, long-lived species replace short-lived species, but the precise species composition of a mature community is up for grabs.

Inhibition appears to play a role in some secondary successions that follow environmental disturbances. For example, rocky intertidal communities on sheltered shores in the Gulf of Maine include some habitat patches that are dominated by an alga *(Ascophyllum nodosum)* and other patches dominated by a mussel *(Mytilus edulis)*. Do these patches represent different alternative states for the mature community in this habitat? In winter, small patches in this habitat are sometimes scoured clean by sea-borne ice, after which the patch undergoes succession. In 2009, Peter Petraitis of the University of Pennsylvania and colleagues from several other institutions reported on the fate of habitat patches in which they had experimentally simulated "ice scour." If mussels colonized the cleared site first, they grew faster than the algae and eventually dominated the community. But if the alga colonized first, it provided cover for sea stars and other predators, which consumed mussels that subsequently grew there. Thus, the species composition of the mature community depended on which species arrived at the site first.

The **tolerance hypothesis** asserts that succession proceeds because competitively superior species replace competitively inferior ones. According to this model, early-stage species neither facilitate nor inhibit the growth of later-stage species. Instead, as more species arrive at a site and resources become limiting, competition eliminates species that cannot harvest scarce resources successfully. In the Piedmont region of North America, for example, hardwood trees are more tolerant of shade than pine trees are, and hardwoods gradually replace pines during succession. Thus, the climax community includes only strong competitors. Tolerance may explain the species composition of many transitional and mature communities.

At most sites, succession probably results from a combination of facilitation, inhibition, and tolerance, coupled with interspecific differences in dispersal, growth, and maturation rates. Moreover, within a community, the patchiness of abiotic factors also strongly influences plant distributions and species composition. In the deciduous forests of eastern North America, maples (genus *Acer*) predominate on wet, low-lying ground, but oaks (genus *Quercus*) are more abundant at higher and drier sites. Thus, a mature deciduous forest is more often a mosaic of species than a uniform stand of trees.

Disturbance and density-independent factors also play important roles, in some cases speeding successional change. In northern forests, for example, moose prefer to feed on deciduous shrubs, accelerating the rate at which conifers replace them. In other cases, disturbance inhibits successional change, establishing a *disturbance climax* or **disclimax community.** In many grassland communities (see Section 49.4), grazing by large mammals and periodic fires kill the seedlings of trees that would otherwise become established. Thus, disturbance prevents the succession from grassland to forest, and grassland persists as a disclimax community.

On a local scale, disturbances often destroy small patches of vegetation, returning them to an earlier successional stage. A hurricane may knock over a few trees in a forest, creating small, sunny patches of open ground. Locally occurring *r*-selected species take advantage of the resources that are suddenly available and quickly colonize the openings. These local patches then undergo succession that is out of step with the immediately surrounding forest. Thus, moderate disturbance, accompanied by succession in local patches, can increase species richness in many communities.

STUDY BREAK 51.6 <

1. **What is the difference between primary succession and secondary succession?**
2. **How does a climax community differ from early successional stages?**
3. **How do the three hypotheses about the causes of ecological succession view the role of population interactions in the successional process?**

51.7 Variations in Species Richness among Communities

Species richness often varies among communities according to a recognizable pattern. Two large-scale patterns of species richness—latitudinal trends and island patterns—have captured the attention of ecologists for more than a century.

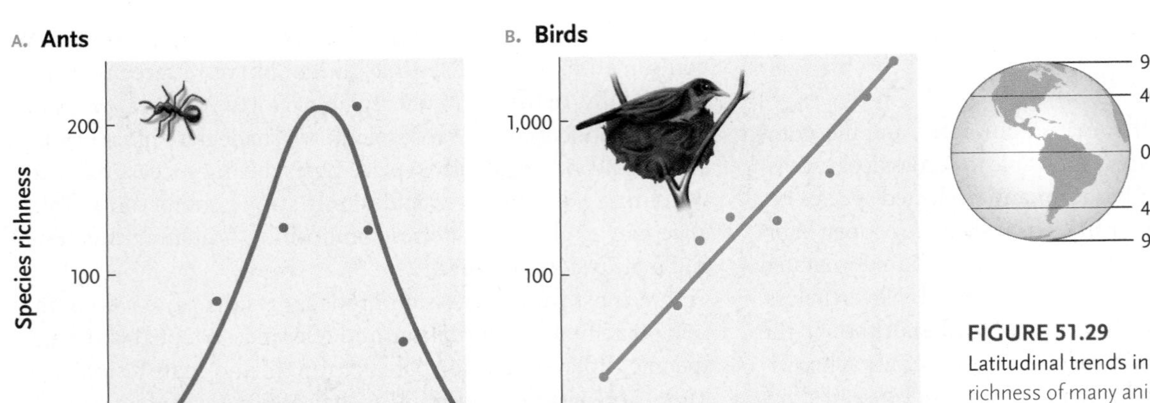
A. Ants
Species richness
200
100
0
90°N
60
30
0
30
60°S
Latitude
B. Birds
1,000
100
10
80°N
60
40
20
0°
Latitude

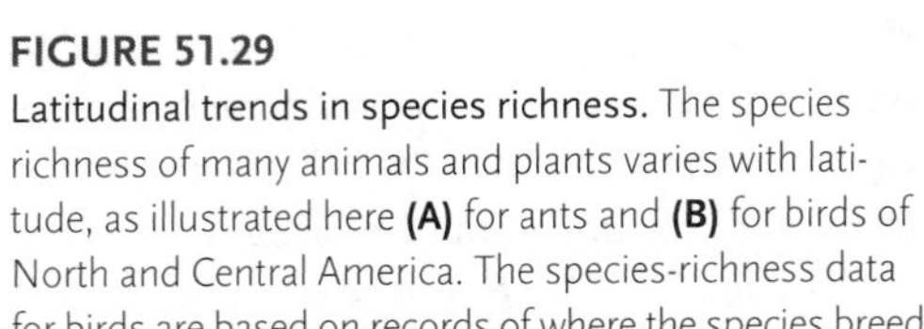
90°N
40°N
0°
40°S
90°S

FIGURE 51.29
Latitudinal trends in species richness. The species richness of many animals and plants varies with latitude, as illustrated here **(A)** for ants and **(B)** for birds of North and Central America. The species-richness data for birds are based on records of where the species breed.

Many Types of Organisms Exhibit Latitudinal Gradients in Species Richness

Ever since Darwin and Wallace traveled the globe (see Section 19.2), ecologists have recognized broad latitudinal trends in species richness. For many, but not all, plant and animal groups, species richness follows a latitudinal gradient, with the most species in the tropics and a steady decline in numbers toward the poles **(Figure 51.29).** Several general hypotheses may explain these striking patterns.

Some hypotheses propose historical explanations for the *origin* of high species richness in the tropics. The benign climate in tropical regions allows some tropical organisms to have more generations per year than their temperate counterparts. And, given the small seasonal changes in temperature, tropical species may be less likely than temperate species to migrate from one habitat to another, thus reducing gene flow between geographically isolated populations (see Section 21.3). These factors may have fostered higher speciation rates in the tropics, accelerating the accumulation of species. Tropical communities may also have experienced severe disturbance less often than communities at higher latitudes, where periodic glaciations have caused repeated extinctions. Thus, new species may have accumulated in the tropics over longer periods of time.

Other hypotheses focus on ecological explanations for the *maintenance* of high species richness in the tropics. Some resources are more abundant, predictable, and diverse in tropical communities. Tropical regions experience more intense sunlight, warmer temperatures in most months, and higher annual rainfall than temperate and polar regions (see Chapter 49). These factors provide a long and predictable growing season for the lush tropical vegetation, which supports a rich assemblage of herbivores, and through them many carnivores and parasites. Furthermore, the abundance, predictability, and year-round availability of resources allow some tropical animals to have specialized diets. For example, tropical forests support many species of fruit-eating bats and birds, which could not survive in temperate forests where fruits are not available year-round.

Species richness may therefore be a self-reinforcing phenomenon in tropical communities. Complex webs of population interactions and interdependency have coevolved in relatively stable and predictable tropical climates. Predator–prey, competitive, and symbiotic interactions may prevent individual species from dominating communities and reducing species richness.

The Theory of Island Biogeography Explains Variations in Species Richness

Although the species richness of communities may be stable over time, species composition is often in flux as new species join a community and others drop out. In the 1960s, Robert MacArthur of Princeton University and Edward O. Wilson of Harvard University addressed the question of why communities vary in species richness, using islands as model systems. Islands provide natural laboratories for studying ecological phenomena, just as they do for evolution (see *Focus on Basic Research* in Chapter 21). Island communities are often small, have well-defined boundaries, and are isolated from surrounding communities.

In developing the **equilibrium theory of island biogeography,** MacArthur and Wilson sought to explain variations in species richness on islands of different size and different levels of isolation from other landmasses **(Figure 51.30).** They hypothesized that the number of species on any island was governed by a give and take between two processes: the immigration of new species to the island and the extinction of species already there (Figure 51.30A).

According to the MacArthur–Wilson model, the mainland harbors a *species pool* from which species immigrate to offshore islands. Seeds and small arthropods are carried by wind or floating debris; some animals, such as birds, arrive under their own power. When few species are already on an island, the rate at which new species immigrate to the island is high. But as more species inhabit the island over time, the immigration rate declines because there are fewer species left in the mainland pool that can still arrive on the island as *new* colonizers.

Once a species immigrates to an island, its population grows and persists for some time. But as the number of species on the

A. **Immigration and extinction rates**

Immigration or extinction rate (species per unit time)

Immigration rate

Extinction rate

Equilibrium number

Number of species on island

The number of species on an island at equilibrium (indicated by the arrow) is determined by the rate at which new species immigrate and the rate at which species already on the island become extinct.

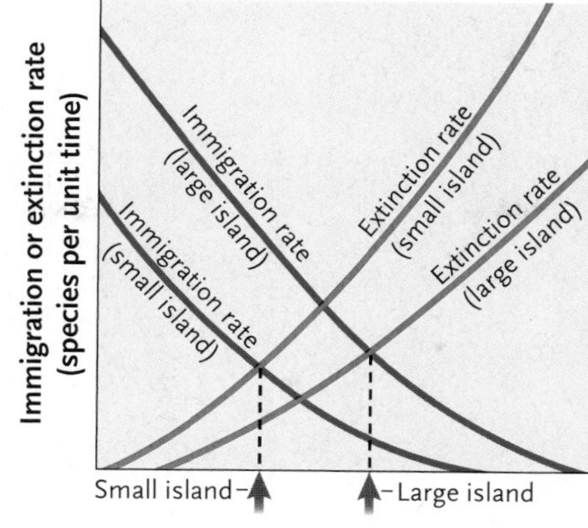
B. Effect of island size
Immigration or extinction rate (species per unit time)
Immigration rate (large island)
Immigration rate (small island)
Extinction rate (small island)
Extinction rate (large island)
Small island
Large island
Number of species on island

Immigration rates are higher and extinction rates lower on large islands than on small islands. Thus, at equilibrium, large islands have more species than small ones.

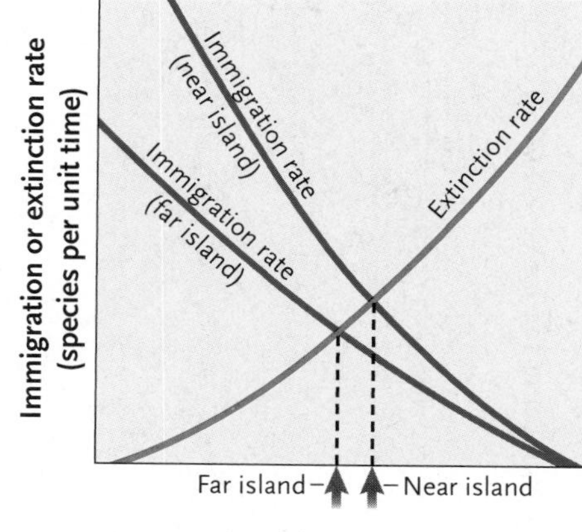
C. Effect of distance from mainland
Immigration or extinction rate (species per unit time)
Immigration rate (near island)
Immigration rate (far island)
Extinction rate
Far island
Near island
Number of species on island

Organisms leaving the mainland locate nearby islands more easily than distant islands, causing higher immigration rates on near islands. Thus, near islands support more species than far ones.

FIGURE 51.30
Predictions of the theory of island biogeography.

island increases, the rate at which those species go extinct also rises. The extinction rate increases through time partly because there are more species that can go extinct there. In addition, as the number of species on the island increases, competition and predator–prey interactions can reduce the population sizes of some species and drive them to extinction.

According to MacArthur and Wilson's theory, an equilibrium between immigration and extinction determines the number of species that ultimately occupy an island. In other words, once equilibrium is reached, the number of species remains relatively constant because one species already on the island goes extinct in about the same time it takes a new species to immigrate to the island. The model does not specify which species immigrate to the island or which ones already on the island go extinct. It simply predicts that the number of species on the island is in equilibrium, although species composition is not. The ongoing processes of immigration and extinction establish a constant turnover in the roster of species that live on any island.

The MacArthur–Wilson model explains why some islands harbor more species than others. Large islands have higher immigration rates than small islands do because they present a larger target for dispersing organisms. Moreover, large islands have lower extinction rates because they can support larger populations and provide a greater range of habitats and resources. Thus, at equilibrium, large islands have more species than small islands (Figure 51.30B). Similarly, islands near the mainland have higher immigration rates than distant islands do, because dispersing organisms are more likely to locate islands that are close to their point of departure. Distance does not affect extinction rates. Thus, at equilibrium, islands that lie closer to a mainland source have more species than more distant islands (Figure 51.30C).

The equilibrium theory's predictions about the effects of area and distance are generally supported by observational data on plants and animals **(Figure 51.31).** Daniel Simberloff, one of Wilson's graduate students at Harvard University, was the first person to test the theory's predictions experimentally; he monitored the immigration of arthropods to, and extinction of arthropods on, individual red mangrove trees in the Florida Keys **(Figure 51.32).** The trees, with canopies that spread from 11 to 18 m in diameter, grow in shallow water and are isolated from their neighbors; thus, each tree is an island that harbors an arthropod community. The species pool on the Florida mainland includes about 1,000 arthropod species, but each mangrove island contains no more than 40 species at one time.

After cataloging the species on each island, Simberloff and Wilson hired an extermination company to eliminate all arthropods on them (Figure 51.32A). Simberloff then monitored both the immigration of arthropods to the islands and the extinction of species that became established on them. He surveyed six islands regularly for two years and at intervals thereafter.

The results of this experiment confirm several predictions of MacArthur and Wilson's theory (Figure 51.32B). Arthropods recolonized the islands rapidly, and within eight or nine months the number of species living on each island had reached an equilibrium that was near the original species number. In addition, the island nearest to the mainland had more species than the most distant island. However, immigration and extinction were incredibly rapid, and Simberloff and Wilson suspected that some species went extinct even before they had noted their presence. The researchers also discovered that three years after the experimental treatments, the species composition of the islands was still changing constantly and did not remotely resemble the species composition in the islands before they were defaunated.

As described in *Insights from the Molecular Revolution,* the equilibrium view of species richness also applies to main-

A. **Distance effect**

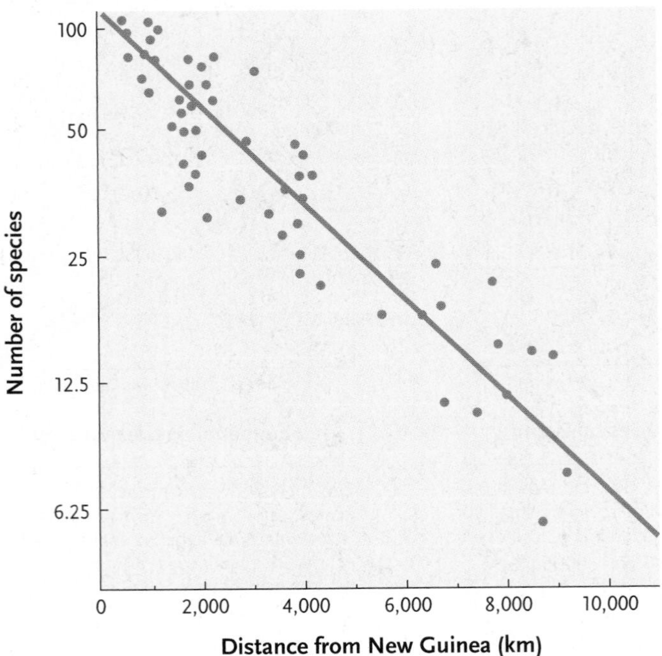

Number of species
100
50
25
12.5
6.25
0
2,000
4,000
6,000
8,000
10,000
Distance from New Guinea (km)

The number of lowland bird species on islands of the South Pacific declines with the islands' distance from the species source, the large island of New Guinea. Data in this graph were corrected for differences in the sizes of the islands. The number of bird species on each island is expressed as a percentage of the number of bird species on an island of equivalent size close to New Guinea.

B. **Area effect**

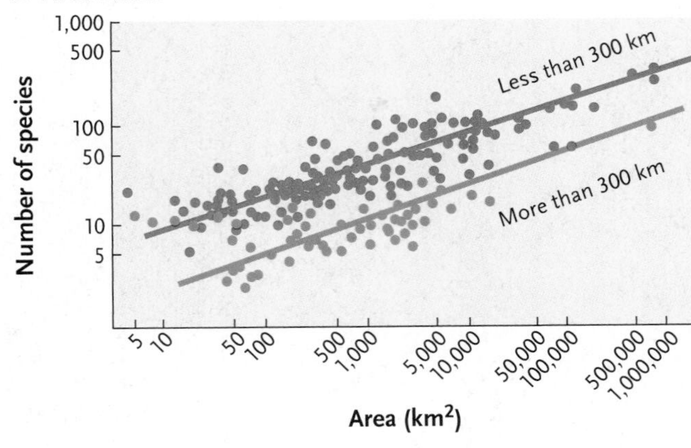

Number of species
1,000
500
100
50
10
5
Less than 300 km
More than 300 km
5
10
50
100
500
1,000
5,000
10,000
50,000
100,000
500,000
1,000,000
Area (km²)

The number of bird species on tropical and subtropical islands throughout the world increases dramatically with island area. The data for islands near to a mainland source and islands far from a mainland source are presented separately to minimize the effect of distance. Notice that the "distance effect" reduces the number of bird species on islands that are more than 300 km from a mainland source.

FIGURE 51.31
Factors that influence bird species richness on islands. (A) Fewer bird species occupy islands that are distant from the mainland source. **(B)** More bird species occupy large islands than small ones.

A. **The process of defaunation**

Daniel Simberloff, University of Tennessee

Daniel Simberloff, University of Tennessee

B. **The return of species richness over time**

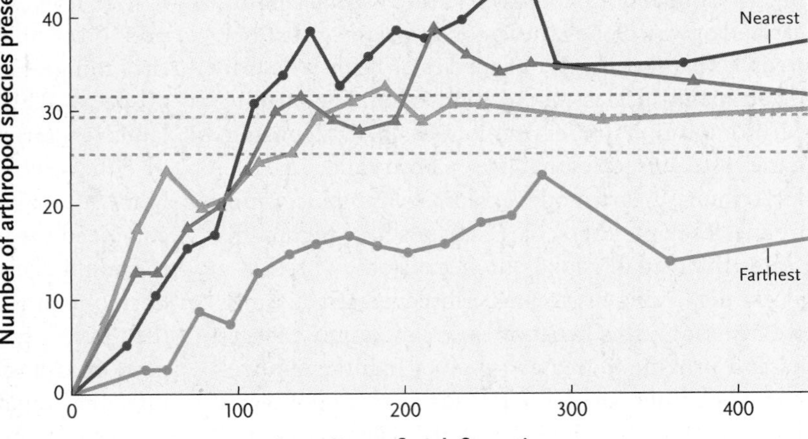

Number of arthropod species present
50
40
30
20
10
0
Nearest
Farthest
0
100
200
300
400
Days after defaunation

FIGURE 51.32
An experimental test of the theory of island biogeography. (A) After cataloguing the arthropods, Simberloff and Wilson hired an exterminating company to erect a tent over each mangrove island. Once the islands were fully covered, exterminators used methyl bromide—a pesticide that does not harm trees and leaves no residue—to eliminate all living arthropods. **(B)** On three of four islands, species richness gradually returned to the predefaunation level (indicated by color-coded dashed lines on the graph). The most distant island had not reached its predefaunation species richness after two years.

INSIGHTS FROM THE **Molecular Revolution**

The Species-Area Effect: Does bacterial species richness vary with "island" size?

Ecologists have long recognized that the number of animal or plant species inhabiting an island is directly proportional to the island's size. In fact, the "species-area effect" also applies to contiguous habitats that are not isolated from each other: larger areas generally harbor more species than smaller areas, at least in part because larger areas include a greater variety of distinctive resources that facilitate resource partitioning by resident species.

Nevertheless, the species-area effect is less dramatic in contiguous habitats than it is among islands of different sizes. In other words, a graph of species richness versus area for contiguous habitats has a lower slope than a graph of species richness versus area among isolated islands (see Figure 51.31B for graph of island data), probably because colonization rates from one habitat patch to another are higher and extinction rates within habitat patches are lower in contiguous habitats than on islands. Previous research had shown an especially limited species-area effect (that is, a low slope of the graph) for bacteria sampled in contiguous habitats; this result probably reflects the ubiquity of bacteria and the ease with which they colonize areas adjacent to those where they already occur. But researchers had not previously investigated the species-area effect for bacteria living in truly isolated island-like habitats.

A. Bacterial species richness as a function of treehole size

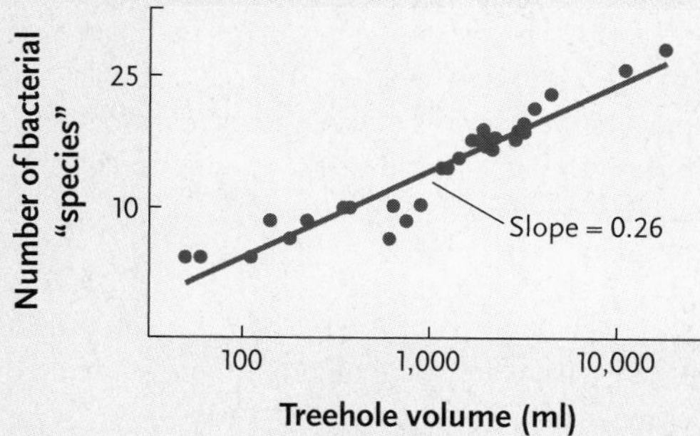
Number of bacterial "species"
25
10
Slope = 0.26
100
1,000
10,000
Treehole volume (ml)

B. Slope of species-area function for organisms in contiguous habitats and on islands

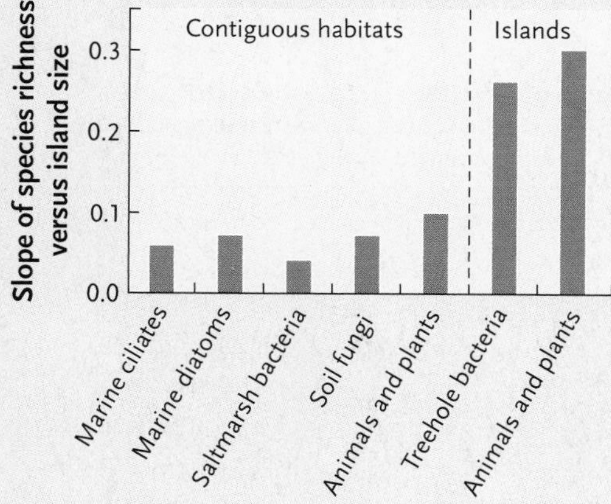
Slope of species richness versus island size
0.3
0.2
0.1
0.0
Contiguous habitats
Islands
Marine ciliates
Marine diatoms
Saltmarsh bacteria
Soil fungi
Animals and plants
Treehole bacteria
Animals and plants

Research Question

Do bacteria living on small islands show the dramatic species-area effect observed in plants and animals?

Experiment

The buttresses of large European beech trees *(Fagus sylvatica)* often form bark-lined, watertight basins that hold small, but permanent, bodies of water. Each of these treehole basins houses a miniature community—an "island" isolated from other such basins—that subsists on nutrients derived from decaying leaf litter. Thomas Bell and his colleagues at the University of Oxford, England, sampled 29 treeholes, measuring their volume (an index of island size) and estimating the bacterial species richness in each.

Instead of using laborious traditional culture techniques to identify the bacteria in each community, the researchers used a molecular approach. They transferred 50 mL of water and sediment from each treehole into sterile vials, and then extracted total DNA from subsamples. They separated out gene fragments for the 16S ribosomal subunit (16S rRNA), which is specific to bacteria, and then amplified them using PCR. The resulting product was analyzed with gel electrophoresis.

Results

Each band appearing on the gel was scored as a distinct bacterial "species." The number of bands identified in the sample from each community served as an estimate of its species richness. Species richness in these 29 bacterial communities was directly proportional to "island" size, as measured by the volume of each basin **(Figure A).**

Conclusion

The results of this study indicate that bacteria living in treehole islands are subject to species-area effects like those observed for animals and plants on true islands. Moreover, the slope of the line in Figure A is very similar to the slope of the species-area relationship for animals and plants on islands—and much higher than the slope of species-area effects for any organisms sampled in contiguous habitats **(Figure B).**

Source: T. Bell et al. 2005. Larger islands house more bacterial taxa. *Science* 308:1884.

land communities, which exist as islands in a metaphorical sea of dissimilar habitat. Lakes are "islands" in a "sea" of dry land, and mountaintops are habitat "islands" in a "sea" of low terrain. Species richness in these communities is partly governed by the immigration of new species from distant sources and the extinction of species already present. As human activities disrupt environments across the globe, undisturbed sites function as island-like refuges for threatened and endangered species. Conservation biologists now apply the general lessons of MacArthur and Wilson's theory to the design of nature preserves (see Chapter 53).

In the next chapter we examine ecosystems, which include ecological communities interacting with their abiotic environments, focusing on the movements of energy and nutrients.

STUDY BREAK 51.7

1. **What factors may foster the maintenance of high species richness in tropical communities?**
2. **According to the equilibrium theory of island biogeography, what are the effects of an island's size and its distance from the mainland on the number of species that can occupy it?**

UNANSWERED QUESTIONS

Do species interactions change predictably across environments?

As we learned in this chapter, the population interactions that occur between species range from mutualistic to parasitic. Some biologists have suggested that we should expect more competitive interactions between species in some kinds of environments, but more positive interactions in others. Community ecology will become a more quantitative and predictive discipline if researchers focus on how abiotic and biotic environmental factors—such as the presence of particular community members, environmental gradients, or global climate change—influence the strength of the interactions between species. For example, as physical environments become more stressful, the abundance and distribution of species should be determined less by resource limitation and more by the stress itself. Accordingly, plants tend to compete far less with each other in stressful environments than they do under ideal growing conditions. Scientists are now engaged in the intellectual feedback of theory development and experimental testing aimed at generating a predictive framework for particular types of interactions and their consequences for community structure.

What is the relative importance of positive versus negative interactions for community structure?

It was once suggested that ecologists in capitalist societies, like the United States, tend to more often study competition and predation, but ecologists in socialist societies tend to study mutualism. Although the truth of this anecdote is unclear, it is remarkable that ecologists still do not agree on the relative importance of positive interactions (for example, mutualism or commensalism) versus negative interactions (such as predation or competition) in generating community structure. Advances in this area of study may result from "factorial" experiments, in which two or more types of interactions are manipulated. For example, one might examine the relative effects of excluding pollinators versus excluding herbivores on the success of a plant population. In factorial experiments, the researcher can conclude that one factor has a bigger effect than the other, because all other factors were controlled. These sorts of experiments may eventually lead to an emerging picture of the relative importance of positive versus negative population interactions.

How does the evolutionary history of a species influence its ecology today?

The great evolutionary biologist Theodosius Dobzhansky once noted that "Nothing in biology makes sense except in the light of evolution." Although we know a great deal about both ecology and evolutionary biology, researchers are only beginning to explore the impact of an organism's evolutionary history on its ecology. This very active area of research includes the use of phylogenetic information (see Chapter 23), selection experiments (see Chapter 21), and a knowledge of the genetic basis of particular traits (see Chapter 12). For example, are closely related species more likely to compete with each other than more distantly related species are? Do organisms that are well adapted to particular environments fare poorly in other environments? Why do some organisms specialize in their resource use? Are the population dynamics that species experience shaped by past evolutionary events? These questions are currently being addressed, and the answers uncovered by researchers may unravel many current mysteries about the ecology of populations and communities.

Think Critically

1. Is mutualism really just reciprocal parasitism? Are the benefits a partner gains in a mutualism just "taken" from the other partner (or are benefits garnered by other means)?
2. Would you expect close relatives to share the same parasites? Why, and what might the consequence of this be for community structure of hosts and parasites?

Anurag Agrawal

Anurag Agrawal is an associate professor in the Departments of Entomology and Ecology and Environmental Biology at Cornell University. He studies the evolutionary and community ecology of plant–insect interactions. To learn more about Dr. Agrawal's research go to http://www.herbivory.com.

REVIEW KEY CONCEPTS

Go to **CENGAGENOW** at www.cengage.com/login to access quizzing, animations, exercises, articles, and personalized homework help.

51.1 Population Interactions

- Coevolution is the evolution of reciprocal adaptations in species that interact ecologically (Figure 51.1).
- Predators and herbivores use diverse adaptations to select, locate, capture, and ingest an appropriate diet (Figure 51.2). Plants have both structural and chemical defenses against herbivores. Animal prey may try to hide or escape from predators, defend themselves actively, or advertise their unpalatability (Figures 51.3–51.5). Some animal species mimic the appearance of poisonous species (Figure 51.6). Predators may evolve adaptations to counter prey defenses (Figure 51.7).
- Interspecific competition results if two or more populations use the same limiting resources; competition may lead to the extinction of one competitor (Figure 51.8). Ecologists use the ecological niche concept to visualize a population's resource use (Figure 51.9). Observations of resource partitioning (Figure 51.10) and character displacement (Figure 51.11) suggest that competition may be important, but only field experiments can demonstrate that competition occurs (Figure 51.12).

- Symbiosis is a close ecological association between species. In commensal interactions, one species benefits and the other is unaffected (Figure 51.13). In mutualistic interactions, both partners benefit (Figures 51.14–51.16). In parasitic interactions, one species benefits and the other is harmed.

Animation: Predator–prey interactions

Animation: Wasp and mimics

Animation: Competitive exclusion

Animation: Hairston's experiment

Animation: Resource partitioning

Practice: Understanding the major types of species interactions: competition, predation, parasitism, and mutualism

51.2 The Nature of Ecological Communities

- An interactive view suggests that species in a community are bound together in a complex web of necessary biotic interactions; an individualistic view recognizes communities as loose assemblages of organisms that have similar physical requirements (Figure 51.17).
- Ecotones occur where adjacent communities grade into one another; sharp boundaries occur between communities where a critical resource or an important abiotic factor is discontinuous (Figure 51.18).

51.3 Community Characteristics

- In warm, moist environments, vegetation is tall and has a complex physical structure (Figure 51.19). In harsh, cold environments, vegetation is short and has a simple physical structure.
- Foundation species moderate the physical environment within communities (Figure 51.20).
- Communities differ in species richness and the relative abundances of species. Both characteristics contribute to a community's species diversity (Figure 51.21).
- Organisms are classified as producers, consumers, detritivores, or decomposers. Ecologists depict the trophic structure (feeding relationships) of communities in food webs (Figure 51.22). Food-web analyses seek to identify generalities about trophic structure and its relationship to community stability.

Animation: Trophic levels in a simple food chain

Animation: Rain forest food web

51.4 Effects of Population Interactions on Community Characteristics

- Interspecific competition often affects the species composition and structure of communities.
- Predators may increase species richness by reducing the population size of the competitively most successful prey, thus allowing other prey species to occupy the community (Figure 51.23).
- Herbivores sometimes increase species richness and sometimes decrease it (Figure 51.24).

Animation: Effect of keystone species on diversity

51.5 Effects of Disturbance on Community Characteristics

- Environmental disturbances may eliminate populations from a community. Some communities, such as coral reefs, experience such frequent disturbance that their species composition is never at equilibrium (Figure 51.25).
- Disturbances of intermediate intensity and frequency allow both *r*-selected and *K*-selected species to occupy a site, increasing species richness (Figure 51.26).

51.6 Ecological Succession: Responses to Disturbance

- Ecological succession is a somewhat predictable change in species composition over time.
- Primary succession occurs on bare ground or rock (Figure 51.27). Secondary succession occurs where a community existed in the past (Figure 51.28).
- Species composition changes quickly and species richness rises rapidly during early successional stages. Early stages include short-lived *r*-selected species; later stages include long-lived *K*-selected species. Some communities eventually achieve a relatively stable climax state.
- Most communities include a mosaic of species that reflect patchiness in environmental conditions and the mixture of relatively undisturbed and recently disturbed sites.

Animation: Succession

51.7 Variations in Species Richness among Communities

- Communities near the equator have higher species richness than those near the poles (Figure 51.29). Explanations for this latitudinal gradient focus on either the origin or the maintenance of high species richness in the tropics.
- The equilibrium theory of island biogeography predicts that the number of species on an island represents a balance between the immigration of new species and the extinction of species already present (Figure 51.30). Studies show that large islands harbor more species than small islands, and islands near a mainland source have more species than distant islands (Figures 51.31 and 51.32).

Animation: Species diversity by latitude

Animation: Area and distance effects

UNDERSTAND AND APPLY

Test Your Knowledge

1. According to optimal foraging theory, predators:
   a. always feed on the largest prey possible.
   b. always feed on the prey that are easiest to catch.
   c. choose prey based on the costs of capturing and consuming it compared to the energy it provides.
   d. feed on plants when animal prey are scarce.
   e. have coevolved mechanisms to overcome prey defenses.
2. The use of the same limiting resource by two species is called:
   a. brood parasitism.
   b. interference competition.
   c. exploitative competition.
   d. mutualism.
   e. optimal foraging.

3. The range of resources that a population can possibly use is called:
   a. its fundamental niche.
   b. its realized niche.
   c. character displacement.
   d. resource partitioning.
   e. its relative abundance.
4. Differences in the bill sizes of finch species living on the same island in the Galápagos may be caused by:
   a. predation.
   b. character displacement.
   c. mimicry.
   d. interference competition.
   e. cryptic coloration.
5. Bacteria that live in the human intestine assist digestion and feed on nutrients the human consumed. This relationship might best be described as:
   a. commensalism.
   b. mutualism.
   c. endoparasitism.
   d. ectoparasitism.
   e. predation.
6. The table below shows how many individuals were recorded for each of five species in five separate communities (a–e). Which community has the highest species diversity?

Community	Species 1	Species 2	Species 3	Species 4	Species 5
a	90	10	0	0	0
b	80	10	10	0	0
c	25	25	25	25	0
d	2	4	6	8	80
e	20	20	20	20	20

7. A keystone species:
   a. is usually a primary producer.
   b. has a critically important role in determining the species composition of its community.
   c. is always a predator.
   d. usually reduces the species diversity in a community.
   e. usually exhibits aposematic coloration.
8. Species richness is often highest in communities where disturbances are:
   a. very frequent and severe.
   b. very frequent and of moderate intensity.
   c. very rare and severe.
   d. of intermediate frequency and moderate intensity.
   e. very rare and mild.
9. The change in the species composition of a community from bare and lifeless rock to climax vegetation is called:
   a. disturbance.
   b. competition.
   c. secondary succession.
   d. primary succession.
   e. facilitation.
10. The equilibrium theory of island biogeography predicts that the number of species found on an island:
    a. increases steadily until it equals the number in the mainland species pool.
    b. is greater on large islands than on small ones.
    c. is smaller on islands near the mainland than on distant islands.
    d. can never reach an equilibrium number.
    e. is greater for islands near the equator than for islands near the poles.

Discuss the Concepts

1. After reading about the two potential biases in the scientific literature on competition, describe how future studies of competition might avoid such biases.
2. How do human activities disrupt the process of succession in terrestrial communities? Would you describe most of our activities as mild disturbances, moderate disturbances, or severe disturbances?
3. Humans are destroying natural communities at an ever-increasing pace. Using the predictions of the theory of island biogeography, develop hypotheses about what might happen as patches of natural habitats get smaller and smaller. How would you test these hypotheses?

Design an Experiment

Chaparral, a community of woody shrubs that is fairly common in California, often grows adjacent to grassland. The two communities are consistently separated by a "bare zone," usually less than 1 m wide, where no vegetation of either type grows. Ecologists have proposed two possible explanations for this strip of bare soil: (1) that the leaves of chaparral shrubs release harmful, water-soluble chemicals that keep the grass seeds from germinating in the adjacent soil; and (2) that small mammals living in the dense cover provided by chaparral consume the grass seeds before they germinate; the animals don't venture very far from the shrubs because they would be easy targets for predatory hawks. Design a set of field experiments to test the two hypotheses.

Interpret the Data

The Mediterranean shrub *Hormathophylla spinosa* loses as much as 80% of its flowers and fruits to herbivorous mammals each year, and biologists interpret the spines on its flowering stems as an anti-herbivore adaptation. Jose M. Gomez and Regino Zamora of the University of Granada, Spain, conducted an exclosure experiment in which they used fences to protect some shrubs from feeding by herbivores and left other shrubs unprotected as controls. The accompanying graph illustrates the density of thorns on the experimental and control groups over a period of two years. How did the protected shrubs respond to the experimental reduction of feeding on the flowers and fruits? How did the unprotected shrubs respond to the control treatment? What benefits would unprotected shrubs derive from their response?

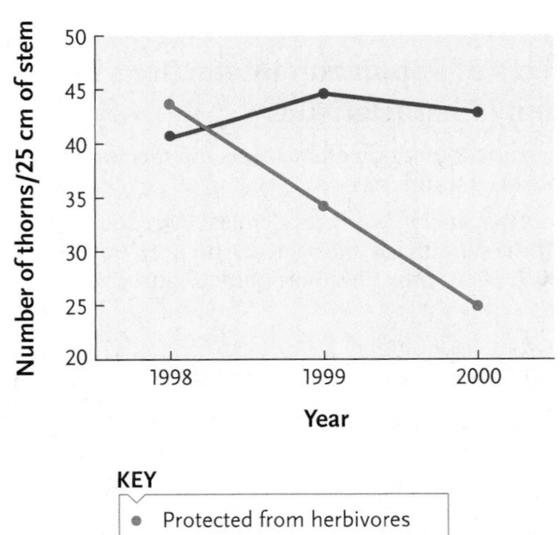
Number of thorns/25 cm of stem
50
45
40
35
30
25
20
1998
1999
2000
Year
KEY
Protected from herbivores
Unprotected from herbivores

Source: J. M. Gomez and R. Zamora. 2002. Thorns as induced mechanical defense in a long-lived shrub (*Hormathophylla spinosa*, Cruciferae). *Ecology* 83(4):885–890.

Apply Evolutionary Thinking

Five processes can foster microevolutionary change: gene flow, genetic drift, mutation, natural selection, and nonrandom mating (see Section 20.3). Which of those processes might contribute to the evolution of Batesian mimicry in two butterfly species? Would the same processes affect both the mimic and the model similarly? Which processes might have contributed to the evolution of the mutualistic relationship between ants and acacia trees, and how would their action on the two mutualists differ?

Express Your Opinion

Currently, only a fraction of the crates of foods and goods being imported into the United States are inspected for the inadvertent or deliberate presence of exotic species. Would the cost of added inspections be worth it? Go to www.cengage.com/login to investigate both sides of the issue and then vote.

52

STUDY OUTLINE

Silver Springs, Florida. This small river was the site of one of the earliest comprehensive studies of ecosystem structure and function.

Ecosystems

Why It Matters. . . Poor Lake Erie, the shallowest of the Great Lakes. Several major industrial cities, including Toledo, Cleveland, Erie, and Buffalo, sprawl along its shoreline. Most of its water comes from the Detroit River, which flows past Detroit; other rivers that flow into Lake Erie carry runoff from agricultural fields in Canada and the United States.

When Europeans first settled along its shores roughly 300 years ago, Lake Erie was a wetland paradise. Fishes and waterfowl reproduced in marshes and bays. Even after steel mills and oil refineries were built nearby in the 1860s and 1870s, the lake supported a busy fishing industry and was famous as a recreation area.

By 1970, wetlands had been filled for building, bays had been dredged for shipping lanes, and the shoreline had been converted to beaches. Worst of all, household sewage, industrial effluent, and agricultural runoff had so polluted the lake that it no longer supported the activities that had made it famous **(Figure 52.1).** The water was murky with algae and cyanobacteria; dead fishes washed up on the shore; local health departments closed beaches; and the fishing industry collapsed.

How can a vibrant natural resource become a foul-smelling dump? The answer lies in the human activities that disrupt an **ecosystem,** a biological community and the physical environment with which it interacts. Between the 1930s and the 1970s, Lake Erie's concentration of phosphorus, which had been a limiting nutrient, tripled, largely from household detergents and agricultural fertilizers. High phosphorus concentrations encouraged the growth of photosynthetic algae, changing the phytoplankton community. The density of coliform bacteria, which originate in the human gut and serve as indicators of organic pollution, also skyrocketed as a result of the surge in sewage and nutrients entering the lake.

Increased phytoplankton and bacterial populations depleted oxygen in the lake's waters, contributing to changes elsewhere in the lake. Mayflies (*Hexagenia* species), whose larvae live in well-

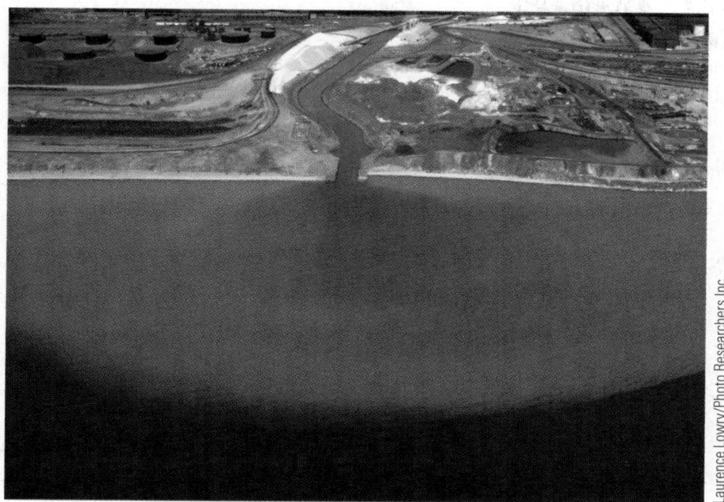

Laurence Lowry/Photo Researchers Inc.

FIGURE 52.1
Pollution of Lake Erie. A steel mill in Lackawanna, New York, discharged industrial wastes into Lake Erie until 1983, when the mill was closed.

oxygenated bottom sediments, had once been so abundant that their aerial breeding swarms were a public nuisance. But they became nearly extinct in the polluted lake, replaced by oligochaete worms, snails, and other invertebrates. Along with overfishing, changes in the bottom fauna shifted the composition of the fish community; the catch of desirable food fishes declined to almost zero by the mid-1960s.

In 1972, Canada and the United States began efforts to restore the lake. They spent billions of dollars to reduce the influx of phosphates and limited fishing of the most vulnerable native species. Nonnative salmon (*Onchorhynchus* species) and other predatory fishes were introduced in the hope that they could bring the lake back to its original condition. Even the accidental introduction of zebra mussels *(Dreissina polymorpha)*, an aquatic pest, inadvertently helped the effort because they feed on phytoplankton.

But, although somewhat improved, Lake Erie will never return to its former glory. Some native species are now extinct there, and the introduced species that replaced them function differently within the ecosystem. The lake still suffers periods of uncontrolled algal growth, fish kills, and high levels of harmful bacteria.

This story of an ecological disaster and partial recovery introduces ecosystem ecology, the branch of ecology that analyzes the flow of energy and the cycling of materials between an ecosystem's living and nonliving components. These processes make the resident organisms highly dependent on each other and on their physical surroundings. Ultimately, the Lake Erie ecosystem unraveled because human activities disrupted the flow of energy and the cycling of materials on which the organisms depended. <

52.1 Modeling Ecosystem Processes

All organisms require steady supplies of energy and nutrients for their maintenance, growth, and reproduction. Studies of ecosystems often focus on the inputs and outputs (that is, the gains and losses) of energy and nutrients to the ecosystem as a whole as well as the transfer of energy and nutrients within and between the ecosystem's biotic and abiotic components. Although the movements of energy and nutrients through an ecosystem are sometimes coupled, as when you eat a meal that contains both nutrients and calories, the inputs and outputs of energy and nutrients are fundamentally different (see Section 1.1). In virtually all ecosystems, sunlight constantly renews the supply of available energy, but, as dictated by the laws of thermodynamics (see Section 4.1), most of that energy is lost as heat in cellular respiration. By contrast, virtually all the nutrients that will ever be available for biological systems are already present on Earth, and they are constantly recycled between the abiotic and biotic components of ecosystems in what ecologists describe as **biogeochemical cycles.**

Researchers use several types of models to describe ecosystem processes. Food webs define the pathways through which energy and nutrients move within the biotic component of an ecosystem. Compartment models describe how nutrients move between living and nonliving nutrient reservoirs. Simulation models allow ecologists to predict how ecosystems will respond to perturbations of ecosystem processes.

Food Webs Illustrate the Transfer of Energy and Nutrients among Organisms

Food webs define the pathways by which energy and nutrients move through an ecosystem's biotic components (see Section 51.3). In most ecosystems, they move simultaneously through a *grazing food web* and a *detrital food web* **(Figure 52.2).** The grazing food web includes the producer, herbivore, and carnivore trophic levels. The detrital food web includes detritivores and decomposers. Because detritivores and decomposers subsist on the remains and waste products of organisms at every trophic level, the two food webs are closely interconnected. Detritivores also contribute to the grazing food web when carnivores eat them.

Compartment Models Track the Movement of Nutrients between Food Webs and Abiotic Reservoirs

Ecologists use a **compartment model** to describe nutrient cycling **(Figure 52.3).** Two criteria divide ecosystems into four compartments where nutrients accumulate. First, nutrient molecules and ions are described as either *available* or *unavailable*, depending on whether or not they can be assimilated by organisms. Second, nutrients are present either in *organic* material, the living or dead tissues of organisms, or in *inorganic* material, such as rocks and soil. For example, minerals in dead leaves on the forest floor are in the available organic compartment because they are in the remains of organisms that detritivores can eat. But calcium ions in limestone rocks are in the unavailable inorganic compartment because they exist in a nonbiological form that producers cannot assimilate.

Nutrients move rapidly within and between the available compartments. Living organisms are in the available organic compartment, and whenever heterotrophs consume food, they

FIGURE 52.2
Grazing and detrital food webs. Energy and nutrients move through two parallel food webs in most ecosystems. The grazing food web includes producers, herbivores, and carnivores. The detrital food web includes detritivores and decomposers. Each box in this diagram represents many species.

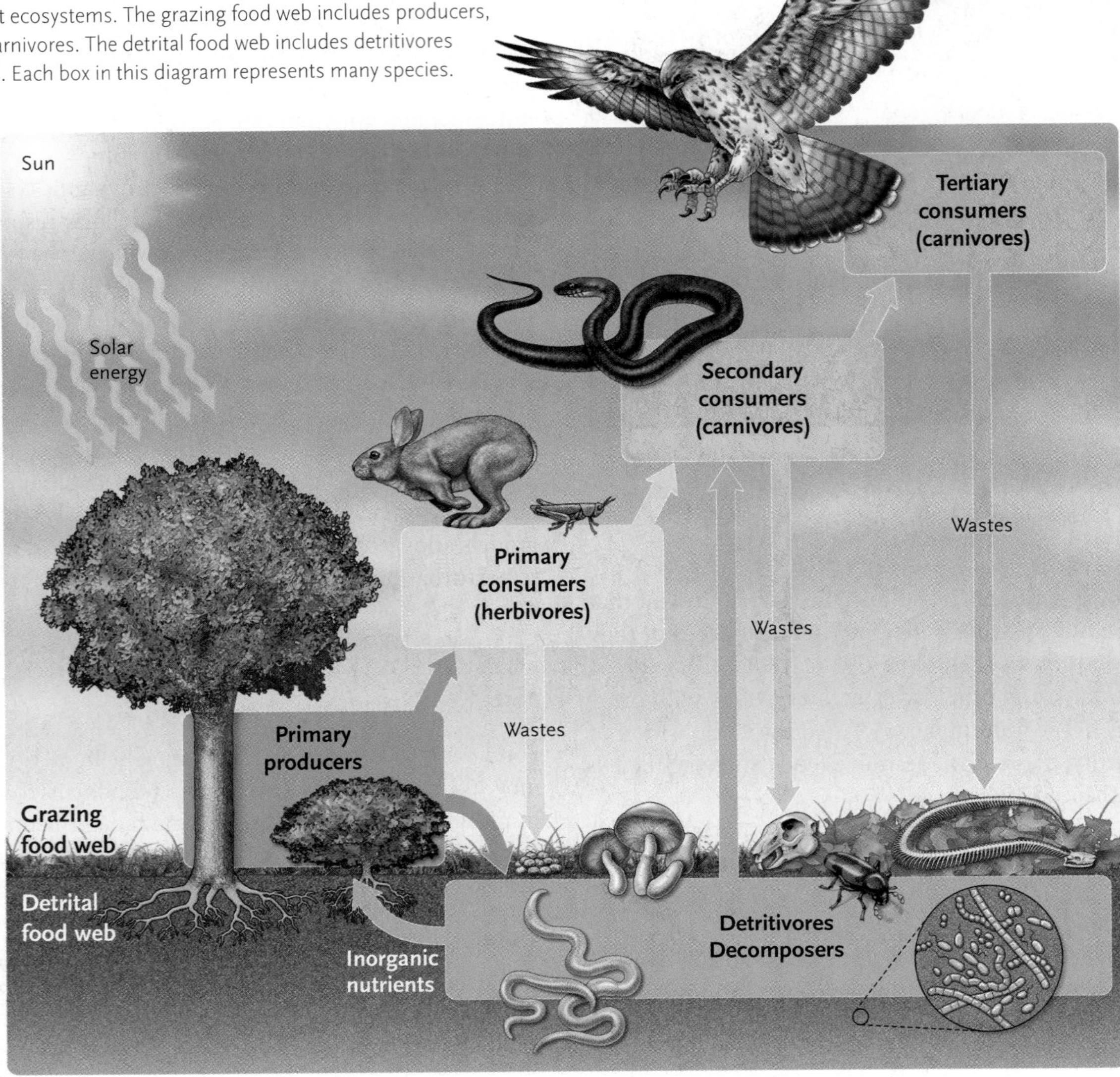
Sun
Solar energy
Tertiary consumers (carnivores)
Secondary consumers (carnivores)
Primary consumers (herbivores)
Primary producers
Wastes
Wastes
Wastes
Grazing food web
Detrital food web
Inorganic nutrients
Detritivores
Decomposers

recycle nutrients within that reservoir (indicated by the oval arrow in the upper left of Figure 52.3). Producers acquire nutrients from the air, soil, and water of the available inorganic compartment. Consumers also acquire nutrients from the available inorganic compartment when they drink water or absorb mineral ions through the body surface. Several processes routinely transfer nutrients from organisms to the available inorganic compartment. As one example, respiration releases carbon dioxide, moving both carbon and oxygen from the available organic compartment to the available inorganic compartment.

By contrast, the movement of materials into and out of the unavailable compartments is generally slow. Sedimentation, a long-term geological process, converts ions and particles of the available inorganic compartment into rocks of the unavailable inorganic compartment. Materials are gradually returned to the available inorganic compartment when rocks are uplifted and eroded or weathered. Similarly, over millions of years, the remains of organisms in the available organic compartment were converted into coal, oil, and peat of the unavailable organic compartment.

Except for the input of solar energy, we have described energy flow and nutrient cycling as though ecosystems were closed systems. In fact, most ecosystems exchange energy and nutrients with neighboring ecosystems. For example, rainfall carries nutrients into a forest ecosystem, and runoff carries nutrients from a forest into a lake or river. Ecologists have mapped the biogeochemical cycles of important elements, often by using radioactively labeled molecules that they can follow in the environment. As you study the details of the four biogeochemical cycles described in Section 52.3, try to understand them in terms of the generalized compartment model of nutrient cycling.

Simulation Models Predict the Effects of Perturbations on Ecosystem Processes

The compartment model described above is a *conceptual model* of how ecosystems function. In other words, it ignores the nuts-and-bolts details of exactly how a specific ecosystem functions in favor of a generalized portrait of how all ecosystems function. Although

FIGURE 52.3

A compartment model of nutrient cycling. Nutrients are recycled through four major compartments within ecosystems. Processes that move nutrients from one compartment to another are indicated on the arrows. The upper left quadrant includes the living organisms in food webs and their organic remains; the oval arrow represents animal predation on other animals. The pathways of energy flow within and from the food webs are also illustrated. Unlike nutrients, energy is not conserved, but is instead lost from the ecosystem.

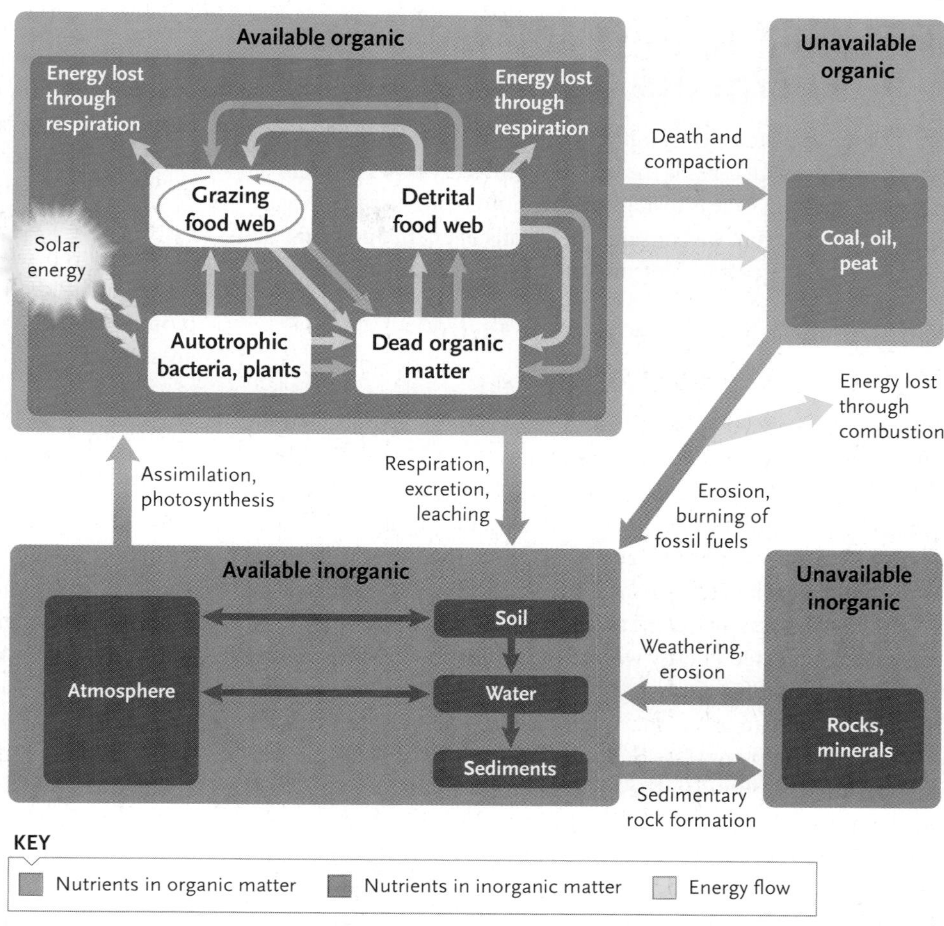

Available organic
Energy lost through respiration
Energy lost through respiration
Grazing food web
Detrital food web
Solar energy
Autotrophic bacteria, plants
Dead organic matter
Death and compaction
Unavailable organic
Coal, oil, peat
Energy lost through combustion
Assimilation, photosynthesis
Respiration, excretion, leaching
Erosion, burning of fossil fuels
Available inorganic
Atmosphere
Soil
Water
Sediments
Unavailable inorganic
Weathering, erosion
Rocks, minerals
Sedimentary rock formation
KEY
Nutrients in organic matter
Nutrients in inorganic matter
Energy flow

it is a useful tool, a conceptual model doesn't really help us predict what would happen, say, if we harvested 10 million tons of introduced salmon from Lake Erie every year. We could simply harvest the fishes and see what happens. But ecologists prefer less intrusive approaches to study the potential effects of disturbances.

To understand how an ecosystem will respond to specific changes in physical factors, energy flow, or nutrient availability, ecologists turn to **simulation modeling.** Researchers gather detailed information about a specific ecosystem and then create a series of mathematical equations that define its most important relationships. For example, one set of equations might describe how nutrient availability limits photosynthesis by autotrophs. Another might relate population growth of zooplankton to the abundance of phytoplankton. Other equations would relate the population dynamics of primary carnivores to the availability of their food, and still others would describe how the densities of primary carnivores influence reproduction in populations at both lower and higher trophic levels. Thus, a complete simulation model is a set of interlocking equations that collectively predict how changes in one feature of an ecosystem might influence others.

Creating a simulation model is no easy task, because the relationships within every ecosystem are complex. First, you would identify the important species, estimate their population sizes, and measure the average energy and nutrient content of each. Next, you would describe the food webs in which they participate, measure the quantity of food each species consumes, and estimate the growth and reproduction of individuals in each population. For the sake of completeness, you would also determine the ecosystem's energy and nutrient gains and losses caused by erosion, weathering, precipitation, and runoff. You would repeat these measurements seasonally to identify annual variation in the factors. Finally, you might repeat the measurements over several years to determine the effects of year-to-year variation in climate and chance events.

After collecting these data, you would write equations that quantify the relationships in the ecosystem, including information about how temperature and other abiotic factors influence the ecology of each species. At last, you could begin to predict—possibly in great detail—the effects of harvesting 10 million or even 50 million tons of salmon annually from Lake Erie. Of course, you would have to refine the model whenever new data became available.

As we attempt to understand larger and more complex ecosystems—and as we create larger and more complex environmental problems—modeling becomes an increasingly important tool. If a model is based on well-defined ecological relationships and good empirical data, it can allow us to make accurate predictions about ecosystem changes without the need for costly and environmentally damaging experiments. But like all ideas in science, a model is only as good as its assumptions, and models must constantly be adjusted to incorporate new ideas and recently discovered facts.

STUDY BREAK 52.1

1. **In the generalized compartment model of biogeochemical cycling, how are the compartments where nutrients accumulate classified?**
2. **What are the advantages and disadvantages of relying on conceptual models that describe ecosystem function?**
3. **What data must ecologists collect before constructing a simulation model of an ecosystem?**

52.2 Energy Flow and Ecosystem Energetics

Ecosystems receive a steady input of energy from an external source, which in almost all cases is the sun. But as energy flows through an ecosystem, much of it is lost as heat without being used by organisms. In this section, we consider the details of energy flow and the efficiency of energy transfer from one trophic level to another.

Sunlight Provides the Energy Input for Practically All Ecosystems

Every minute of every day, Earth's atmosphere intercepts roughly 19 kcal of solar energy per square meter. (Recall from Chapter 2 that 1 kcal = 1,000 calories.) About half that energy is absorbed, scattered, or reflected by gases, dust, water vapor, and clouds without ever reaching the planet's surface (see Chapter 49). Most energy that reaches the surface falls on bodies of water or bare ground, where it is absorbed as heat or reflected back into the atmosphere; reflected energy warms the atmosphere, as we discuss later in this chapter. Only a small percentage contacts primary producers, and most of that energy evaporates water, driving transpiration in plants (see Section 32.3).

Ultimately, photosynthesis converts less than 1% of the solar energy that arrives at Earth's surface into chemical energy. But primary producers capture enough energy to create an average of several kilograms of dry plant material per square meter per year. On a global scale, they produce more than 150 billion metric tons of new biological material annually. Some of the solar energy that producers convert into chemical energy is transferred to consumers at higher trophic levels.

The rate at which producers convert solar energy into chemical energy is an ecosystem's **gross primary productivity.** But like all other organisms, producers use energy for their own maintenance. After deducting the energy devoted to these functions, which are collectively called cellular respiration (see Section 8.1), whatever chemical energy remains is the ecosystem's **net primary productivity.** In most ecosystems, net primary productivity is between 50% and 90% of gross primary productivity. In other words, producers use between 10% and 50% of the energy they capture for their own respiration.

Ecologists generally measure primary productivity in units of energy captured (kcal/m^2/yr) or in units of biomass created (g/m^2/yr). *Biomass* is the dry weight of biological material per unit area or volume of habitat. (We measure biomass as the *dry* weight of organisms because their water content, which fluctuates with water uptake or loss, has no energetic or nutritional value.) You should not confuse an ecosystem's productivity with its **standing crop biomass,** the total dry weight of plants present at a given time. Net primary productivity is the *rate* at which the standing crop produces *new* biomass.

The energy captured by plants is stored in biological molecules—mostly carbohydrates, lipids, and proteins. Ecologists can convert units of biomass into units of energy or vice versa as long as they know how much carbohydrate, protein, and lipid a sample of biological material contains (4.2 kcal/g of carbohydrate; nearly 4.1 kcal/g of protein; and 9.5 kcal/g of lipid). Thus, net primary productivity is a measure of the rate at which producers accumulate energy as well as the rate at which new biomass is added to an ecosystem. Because it is far easier to measure biomass than energy content, ecologists usually measure changes in biomass to estimate productivity. New biomass takes several forms: the growth of existing producers; the creation of new producers by reproduction; and the storage of energy as carbohydrates or lipids. Because herbivores eat all three forms of new biomass, net primary productivity also measures how much new energy is available for primary consumers.

Primary Productivity Varies Greatly on Global and Local Scales

The potential rate of photosynthesis in any ecosystem is proportional to the intensity and the duration of sunlight, which vary geographically and seasonally (see Chapter 49). Sunlight is most intense and day length least variable near the equator. By contrast, light intensity is weakest and day length most variable near the poles. Thus, producers at the equator can photosynthesize nearly 12 hours a day, every day of the year. Near the poles, photosynthesis is virtually impossible during the long, dark winter; in summer, however, plants can photosynthesize around the clock.

Sunlight is not the only factor that influences the rate of primary productivity, however; temperature as well as the availability of water and nutrients also have big effects. For example, many of the world's deserts receive plenty of sunshine but have low rates of productivity because water is in short supply and the soil is nutrient-poor. Thus, mean annual net primary productivity varies greatly on a global scale **(Figure 52.4),** reflecting variations in these environmental factors (see Chapter 49).

On a finer geographical scale, within a particular terrestrial ecosystem, mean annual net productivity often increases with the availability of water **(Figure 52.5).** In systems with sufficient water, a shortage of mineral nutrients may be limiting. All plants need specific ratios of macronutrients and micronutrients for maintenance and photosynthesis (see Section 33.1). But plants withdraw nutrients from soil, and if nutrient concentration drops below a critical level, photosynthesis may decrease or stop altogether. In every ecosystem, one nutrient inevitably runs out before the supplies of other nutrients are exhausted. The element in short supply is called a **limiting nutrient** because its absence limits productivity. Productivity in agricultural fields is subject to the same constraints as productivity in natural ecosystems. Farmers increase productivity by irrigating (adding water to) and fertilizing (adding nutrients to) their crops.

In freshwater and marine ecosystems, where water is always readily available, the depth of the water and the combined availability of sunlight *and* nutrients govern the rate of primary productivity. Productivity is high in near-shore ecosystems where sunlight penetrates shallow, nutrient-rich waters. Kelp beds and coral reefs, for example, which occur along temperate and tropical coastlines respectively, are among the most productive ecosystems on Earth **(Table 52.1).** By contrast, productivity is low in the open waters of a large lake or ocean: sunlight penetrates only

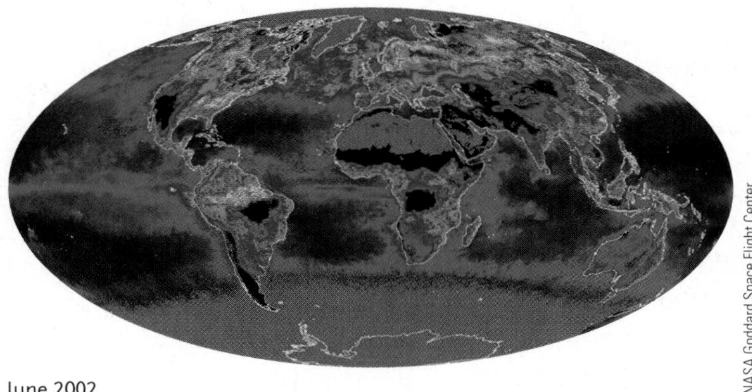
June 2002
NASA Goddard Space Flight Center

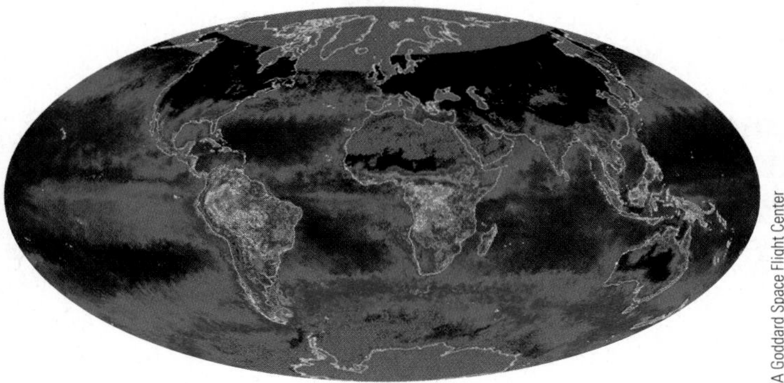
December 2002
NASA Goddard Space Flight Center
Net Primary Productivity (kg/m²/year)
0
1
2
3

FIGURE 52.4
Seasonal and global variation in net primary productivity. Satellite data for 2002 provide a visual portrait of net primary productivity across Earth's surface in June and in December. The key indicates productivity measured as kilograms of carbon fixed by photosynthesis per square meter per year.

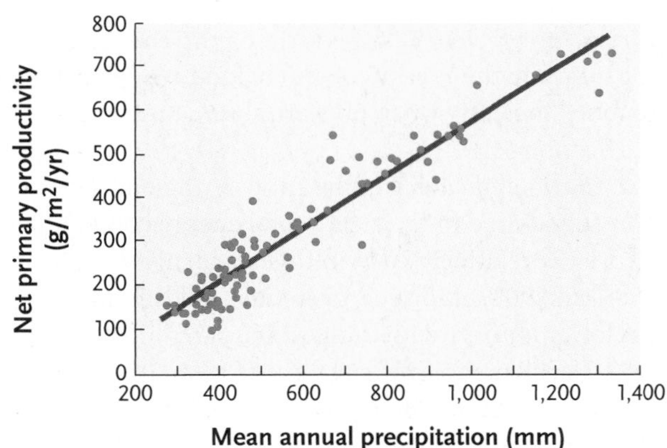
Net primary productivity (g/m²/yr)
800
700
600
500
400
300
200
100
0
200
400
600
800
1,000
1,200
1,400
Mean annual precipitation (mm)

FIGURE 52.5
Water and net primary productivity. Mean annual net primary productivity increases with mean annual precipitation among 100 sites in the Great Plains of North America. These data include only aboveground productivity.

TABLE 52.1 Standing Crop Biomass and Net Primary Productivity of Different Ecosystems

Ecosystem	Mean Standing Crop Biomass (g/m²)	Mean Net Primary Productivity (g/m²/yr)
Terrestrial Ecosystems		
Tropical rain forest	45,000	2,200
Tropical deciduous forest	35,000	1,600
Temperate rain forest	35,000	1,300
Temperate deciduous forest	30,000	1,200
Savanna	4,000	900
Boreal forest (taiga)	20,000	800
Woodland and shrubland	6,000	700
Cultivated land	1,000	650
Temperate grassland	1,600	600
Tundra and alpine tundra	600	140
Desert and thornwoods	700	90
Extreme desert, rock, sand, ice	20	3
Freshwater Ecosystems		
Swamp and marsh	15,000	2,000
Lake and stream	20	250
Marine Ecosystems		
Open ocean	3	125
Upwelling zones	20	500
Continental shelf	10	360
Kelp beds and reefs	2,000	2,500
Estuaries	1,500	1,500
World Average	**3,600**	**333**

From R. H. Whittaker. 1975. *Communities and Ecosystems*. 2nd ed. Macmillan.

the upper layers, and nutrients sink to the bottom. Thus, the two requirements for photosynthesis, sunlight and nutrients, are available in different places.

Although ecosystems vary in their net primary productivity, the differences are not always proportional to variations in their standing crop biomass (see Table 52.1). For example, biomass in temperate deciduous forests and temperate grasslands differs by a factor of 20, but the difference in their rates of net primary productivity is only twofold. Most biomass in trees is present in nonphotosynthetic tissues such as wood. As a result, their ratio of productivity to biomass is low ($1{,}200 \text{ g/m}^2 \div 30{,}000 \text{ g/m}^2 = 0.040$). By contrast, grasslands don't accumulate much biomass because annual mortality, herbivores, and fires remove plant material as it is produced; and their productivity to biomass ratio is much higher ($600 \text{ g/m}^2 \div 1{,}600 \text{ g/m}^2 = 0.375$).

Some ecosystems contribute more than others to overall net primary productivity **(Figure 52.6)**. Ecosystems that cover large areas make substantial contributions, even if their productivity is low. Conversely, geographically restricted ecosystems make

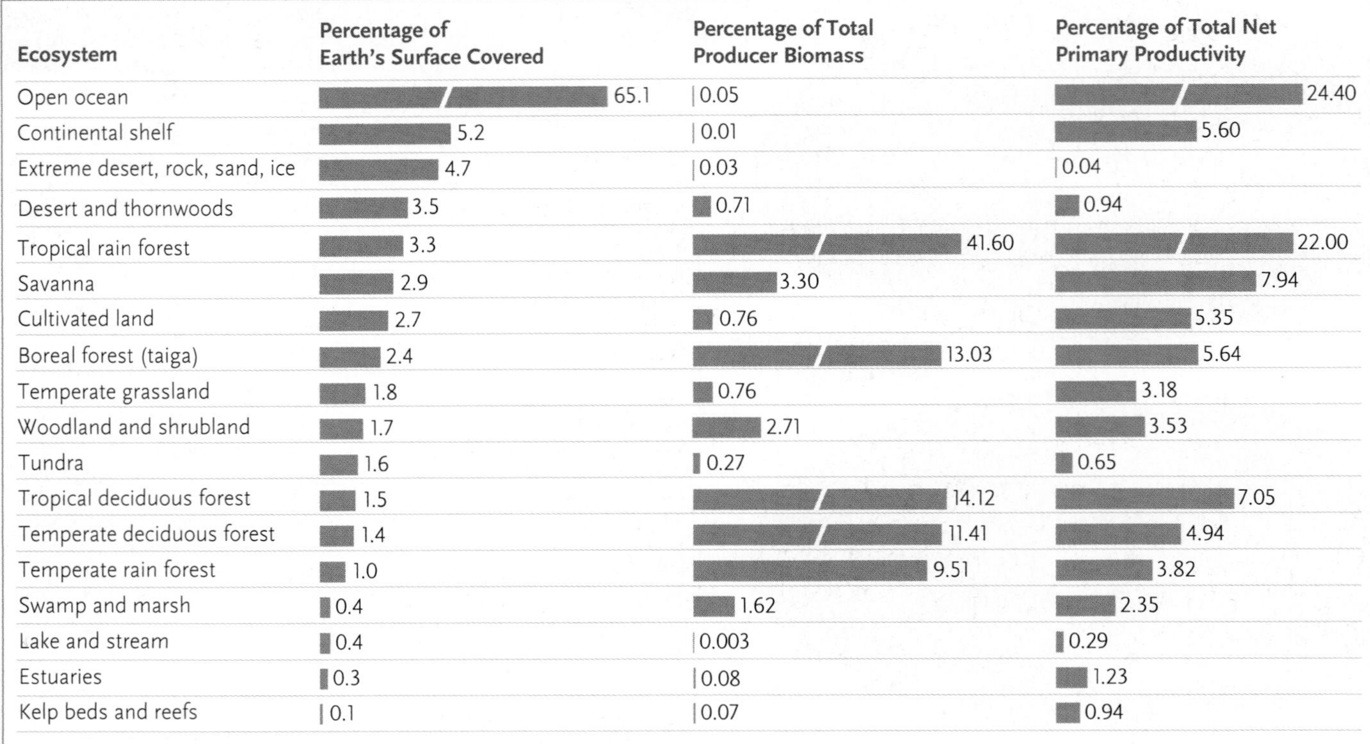

Ecosystem	Percentage of Earth's Surface Covered	Percentage of Total Producer Biomass	Percentage of Total Net Primary Productivity
Open ocean	65.1	0.05	24.40
Continental shelf	5.2	0.01	5.60
Extreme desert, rock, sand, ice	4.7	0.03	0.04
Desert and thornwoods	3.5	0.71	0.94
Tropical rain forest	3.3	41.60	22.00
Savanna	2.9	3.30	7.94
Cultivated land	2.7	0.76	5.35
Boreal forest (taiga)	2.4	13.03	5.64
Temperate grassland	1.8	0.76	3.18
Woodland and shrubland	1.7	2.71	3.53
Tundra	1.6	0.27	0.65
Tropical deciduous forest	1.5	14.12	7.05
Temperate deciduous forest	1.4	11.41	4.94
Temperate rain forest	1.0	9.51	3.82
Swamp and marsh	0.4	1.62	2.35
Lake and stream	0.4	0.003	0.29
Estuaries	0.3	0.08	1.23
Kelp beds and reefs	0.1	0.07	0.94

FIGURE 52.6
Biomass and net primary productivity. The percentage of Earth's surface that an ecosystem covers is not proportional to its contribution to the total biomass of producers or its contribution to the total net primary productivity.

large contributions if their productivity is high. For example, the open ocean and tropical rain forests contribute about equally to total global productivity, but for different reasons. Open oceans have low productivity, but they cover nearly two-thirds of Earth's surface. Tropical rain forests cover only a small area, but they are highly productive.

Some Energy Is Always Lost before It Is Transferred from One Trophic Level to the Next

Net primary productivity ultimately supports all the consumers in grazing and detrital food webs. Consumers in the grazing food web eat some of the biomass at every trophic level except the highest; uneaten biomass eventually dies and passes into detrital food webs. However, consumers assimilate only a portion of the material they ingest, and unassimilated material is passed as feces, which also supports detritivores and decomposers.

As energy is transferred from producers to consumers, some is stored in new consumer biomass, called **secondary productivity.** Nevertheless, two factors cause energy to be lost from the ecosystem every time it flows from one trophic level to another. First, animals use much of the energy they assimilate for maintenance or locomotion rather than the production of new biomass. Second, as dictated by the second law of thermodynamics, no biochemical reaction is 100% efficient; thus, some of the chemical energy liberated by cellular respiration is always converted to heat, which most organisms do not use.

Ecological efficiency is the ratio of net productivity at one trophic level to net productivity at the trophic level below it. For example, if the plants in an ecosystem have a net primary productivity of 100 $g/m^2/yr$ of new tissue and the herbivores that eat those plants produce 10 $g/m^2/yr$, the ecological efficiency of the herbivores is 10%. The efficiencies of three processes—harvesting food, assimilating ingested energy, and producing new biomass—determine the ecological efficiencies of consumers.

Harvesting efficiency is the ratio of the energy content of food consumed to the energy content of food available. Predators harvest food efficiently when prey are abundant and easy to capture (see Section 51.1).

Assimilation efficiency is the ratio of the energy absorbed from consumed food to the food's total energy content. Because animal prey is relatively easy to digest, carnivores absorb between 60% and 90% of the energy in their food; assimilation efficiency is lower for prey with indigestible parts like bones or exoskeletons. Herbivores assimilate only 15% to 80% of the energy they consume because cellulose is not very digestible.

Production efficiency is the ratio of the energy content of new tissue produced to the energy assimilated from food. Production efficiency varies with maintenance costs. For example, endothermic animals often use less than 10% of their assimilated energy for growth and reproduction, because they use energy to generate body heat (see Section 46.8). Ectothermic animals, by contrast, channel more than 50% of their assimilated energy into new biomass.

The overall ecological efficiency of most organisms is between 5% and 20%. As a rule of thumb, only about 10% of the energy accumulated at one trophic level is converted into biomass at the next higher trophic level, as illustrated by energy transfers at Silver Springs, Florida **(Figure 52.7).** Producers in

FIGURE 52.7 **Observational Research**

What Is the Pattern of Energy Flow within the Silver Springs Ecosystem?

Hypothesis: Only a small percentage of the energy present in a trophic level is transferred to the next higher trophic level in the ecosystem.

Prediction: The energy content of the organisms present in each trophic level will decline steadily from the lowest to highest trophic levels.

Method: Howard T. Odum and his team analyzed energy flow in an aquatic ecosystem at Silver Springs, Florida. The producers in this small spring are mostly aquatic plants. The herbivores include snails, shrimp, insects, fishes, and turtles. The carnivores include a variety of invertebrates and fishes. The top carnivores are large fish. Sunlight is available as an energy source all year round. After defining the food web in this ecosystem, the researchers estimated the biomass and energy content (kcal/g) of each trophic level. They then constructed a diagram that illustrates how much energy is present at each trophic level and how much energy is lost as it works its way through the food web.

20,810 To respiration and export
Detrital food web
Grazing food web
5,060 Respiration and export
To detrital food web
5 Top carnivore productivity
16 Respiration and export
21 kcal/m²/yr Consumed by top carnivores
90
21
111 carnivore productivity
272 Respiration and export
Detritivores and decomposers
To detrital food web
383 kcal/m²/yr Consumed by carnivores
98.8% of energy not harnessed
720
383
1,103 herbivore productivity
2,265 Respiration and export
To detrital food web
3,368 kcal/m²/yr Consumed by herbivores
Sun
To detrital food web
4,245
3,368
7,613 net primary productivity
13,197 Respiration and export
1.2% of energy harnessed
20,810 kcal/m²/yr Gross primary productivity

Results: The diagram illustrates annual energy flow for the spring ecosystem at Silver Springs, Florida. Numbers on the diagram indicate the quantity of energy (kcal/m^2/yr). Because the ecosystem is based on flowing water, small quantities of energy arrive from other ecosystems, and small quantities are exported in material carried away by stream flow.

Conclusion: The study supports the hypothesis that only a small proportion of the energy present at a trophic level is transferred to the next higher trophic level. Ultimately, all of the energy that passes through the grazing and detrital food webs is released as metabolically generated heat.

Source: H. T. Odum. 1957. Trophic structure and productivity of Silver Springs, Florida. *Ecological Monographs* 27:55–112.

FIGURE 52.8

Pyramids of energy. The pyramid of energy for Silver Springs, Florida, shows that the amount of energy (in kcal/m²/yr) passing through each trophic level decreases as it moves up the food web.

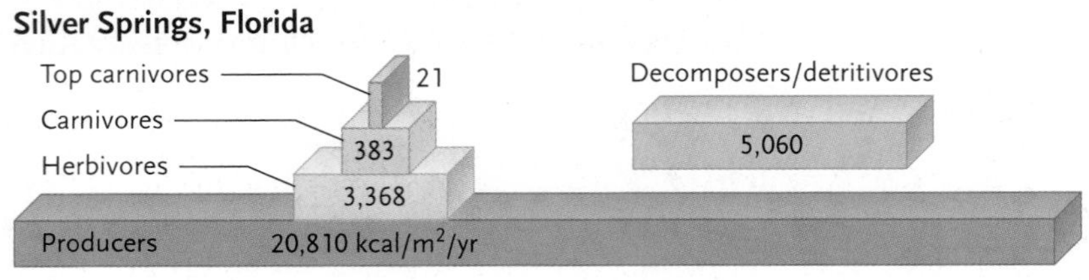
Silver Springs, Florida
Top carnivores
21
Decomposers/detritivores
Carnivores
383
5,060
Herbivores
3,368
Producers
20,810 kcal/m²/yr

the Silver Springs ecosystem convert 1.2% of the solar energy they intercept into chemical energy (represented by 20,810 kcal of gross primary productivity). However, they use about two-thirds of this energy for respiration, leaving only one-third to be included in new plant biomass, the net primary productivity. All consumers in the grazing food web (on the right in Figure 52.7) ultimately depend on this energy source, which dwindles with each transfer between trophic levels. Energy is lost to respiration and export (that is, the transport of energy-containing materials out of the ecosystem by flowing water) at each trophic level. In addition, substantial energy flows into the detrital food web (on the left in Figure 52.7) as organic wastes and uneaten biomass. To determine the ecological efficiency of any trophic level, we divide its productivity by the productivity of the level below it. For example, the ecological efficiency of midlevel carnivores at Silver Springs is 111 kcal/yr ÷ 1,103 kcal/yr = 10.06%.

The low ecological efficiencies that characterize most energy transfers illustrate one advantage of eating "lower on the food chain." Even though humans digest and assimilate meat more efficiently than vegetables, we might be able to feed more people if we all ate more vegetables directly instead of first passing these crops through another trophic level, such as cattle or chickens, to produce meat. The production of animal protein is costly because much of the energy fed to livestock is used for their own maintenance rather than the production of new biomass. But despite the economic—not to mention health-related—logic of a more vegetarian diet, a change in our eating habits alone won't eliminate food shortages or the frequency of malnutrition. Many regions of Africa, Australia, North America, and South America support vegetation that is suitable only for grazing by large herbivores. These areas could not produce significant quantities of edible grains and vegetables.

Ecological Pyramids Illustrate the Effects of Energy Losses

All organisms in a trophic level are the same number of energy transfers from the ecosystem's ultimate energy source. Plants are one energy transfer removed from sunlight; herbivores are two transfers away; carnivores feeding on herbivores are three transfers away; and carnivores feeding on other carnivores are four transfers away. As energy works its way up a food web, energy losses are multiplied in successive energy transfers, greatly reducing the energy available to support the highest trophic levels.

Consider a hypothetical example in which ecological efficiency is 10% for all consumers. Assume that the plants in a small field annually produce new tissues containing 100 kcal of energy. Because only 10% of that energy is transferred to new herbivore biomass, the 100 kcal in plants produces only 10 kcal of new herbivorous insects; only 1 kcal of new songbirds, which feed on insects; and only 0.1 kcal of new falcons, which feed on songbirds. Thus, after three energy transfers, only 0.1% of the energy from primary productivity remains at the highest trophic levels.

The inefficiency of energy transfer from one trophic level to the next has profound effects on ecosystem structure. Ecologists illustrate these effects in diagrams called **ecological pyramids.** Trophic levels are drawn as stacked blocks, with the size of each block proportional to the energy, biomass, or numbers of organisms present; primary producers are illustrated at the bottom of the pyramid and higher-level consumers at the top.

Pyramids of energy typically have wide bases and narrow tops because each trophic level contains on average only about 10% as much energy as the trophic level below it, as illustrated by the pyramid of energy for Silver Springs, Florida **(Figure 52.8).**

The progressive reduction in productivity at higher trophic levels, as illustrated in Figure 52.7, usually establishes a **pyramid of biomass (Figure 52.9).** The biomass at each trophic level is proportional to the chemical energy temporarily stored there. Thus,

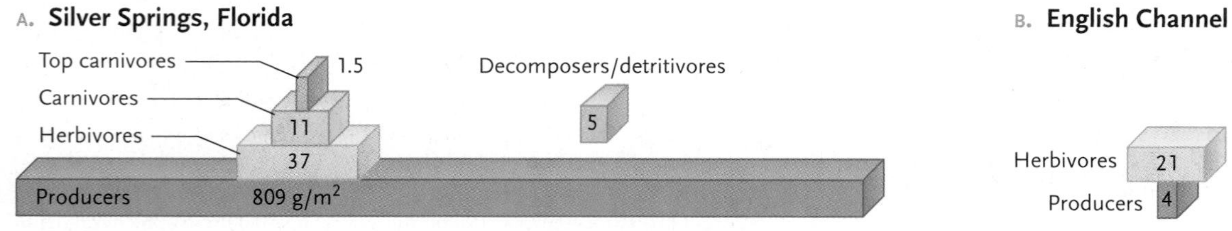
A. Silver Springs, Florida
Top carnivores
1.5
Decomposers/detritivores
Carnivores
11
5
Herbivores
37
Producers
809 g/m²
B. English Channel
Herbivores
21
Producers
4

FIGURE 52.9

Pyramids of biomass. (A) The pyramid of standing crop biomass for Silver Springs is bottom heavy, as it is for most ecosystems. **(B)** Some marine ecosystems, such as that in the English Channel, have an inverted pyramid of biomass because producers are quickly eaten by primary consumers. Only the producer and herbivore trophic levels are illustrated here. The data for both pyramids are given in grams of dry biomass per square meter.

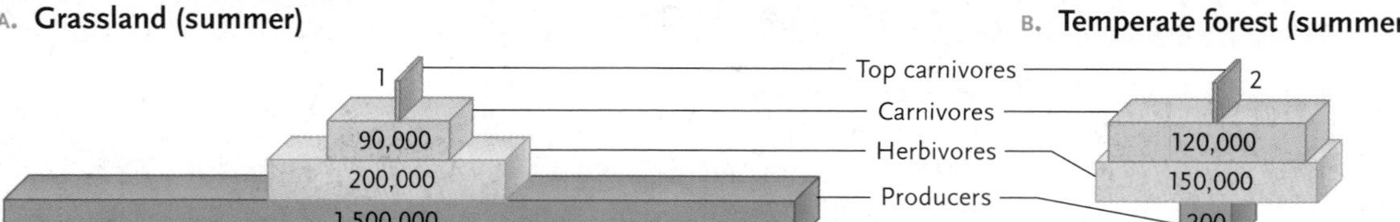
A. Grassland (summer)
B. Temperate forest (summer)
Top carnivores
Carnivores
Herbivores
Producers
1
90,000
200,000
1,500,000
2
120,000
150,000
200

FIGURE 52.10
Pyramids of numbers. **(A)** The pyramid of numbers (number of individuals per 1,000 m^2) for temperate grasslands is bottom-heavy because individual producers are small and very numerous. **(B)** The pyramid of numbers for forests may have a narrow base because herbivorous insects often outnumber the producers, which are large trees. Data for both pyramids were collected in summer. Detritivores and decomposers (soil animals and microorganisms) are not included because they are difficult to count.

in terrestrial ecosystems, the total mass of producers is generally greater than the total mass of herbivores, which is, in turn, greater than the total mass of predators (see Figure 52.9A). Populations of top predators—animals like mountain lions or alligators—contain too little biomass and energy to support another trophic level; thus, they have no nonhuman predators.

Freshwater and marine ecosystems sometimes exhibit inverted pyramids of biomass (see Figure 52.9B). In the open waters of a lake or ocean, primary consumers (zooplankton) eat the primary producers (phytoplankton) almost as soon as they appear. As a result, the standing crop of primary consumers at any moment in time is actually larger than the standing crop of primary producers. Food webs in these ecosystems are stable, however, because the producers have exceptionally high **turnover rates.** In other words, the phytoplankton divide and their populations grow so quickly that feeding by zooplankton doesn't endanger their populations or reduce their productivity. And on an annual basis, the *cumulative total* biomass of primary producers far outweighs that of primary consumers.

The reduction of energy and biomass also affects the population sizes of organisms at the top of a food web. Top predators are often relatively large animals, thus concentrating the limited biomass at the highest trophic levels in relatively few individuals **(Figure 52.10).** The extremely narrow top of this **pyramid of numbers** has grave implications for conservation biology. Because each top predator must patrol a large area to find sufficient food, the members of a population are often widely dispersed within their habitats. As a result, they are highly sensitive to hunting, habitat destruction, and random events, which can lead to local extinction (see Chapter 53). Top predators may also suffer from the accumulation of poisonous materials that move through food webs (see *Focus on Applied Research*). Even predators that feed below the top trophic level often suffer the ill effects of human activities.

Consumers Sometimes Regulate Ecosystem Processes in a Trophic Cascade

As you know from the preceding discussion, numerous abiotic factors—the intensity and duration of sunlight, rainfall, temperature, and the availability of nutrients—have significant effects on primary productivity. Primary productivity, in turn, has profound effects on populations of herbivores and the predators that feed on them. But what effect does feeding by these consumers have on primary productivity?

Recent research suggests that consumers may sometimes influence rates of primary productivity, especially in ecosystems with low species diversity and relatively few trophic levels. For example, food webs in North American salt marshes depend primarily on the productivity of salt marsh cordgrass *(Spartina alterniflora),* a foundation species that defines the nature of that coastal ecosystem (see Section 51.3). Cordgrass is consumed by herbivores, including the marsh periwinkle snail *(Littoraria irrorata).* In the southeastern United States, these herbivores are in turn consumed by blue crabs *(Callinectes sapidus),* mud crabs *(Eurytium limosum),* and terrapins *(Malaclemys terrapin).*

For many years, ecologists believed that the productivity of cordgrass was largely determined by the availability of nutrients in the salt marsh. However, research by Brian R. Silliman and Mark D. Bertness of Brown University showed that the productivity of cordgrass is actually regulated by what ecologists call a **trophic cascade**—predator–prey effects that reverberate through the population interactions at two or more trophic levels **(Figure 52.11).** Populations of the herbivorous periwinkle snails are controlled by the crabs and turtles that eat them. In the presence of these predators, snail populations are reduced in size, and the cordgrass grows luxuriantly. But when these predators are removed from the system, the snail populations grow rapidly, and feeding by the snails virtually eliminates the cordgrass, converting a highly productive ecosystem into a barren mudflat. Thus, cord grass productivity is *indirectly* controlled by the abundance of predators that eat the snails that eat the cordgrass.

STUDY BREAK 52.2 <

1. What is the difference between gross primary productivity and net primary productivity?
2. What environmental factors influence rates of primary productivity in terrestrial and aquatic ecosystems?
3. Why is energy lost from an ecosystem at every transfer from one trophic level to the trophic level above it?
4. How can the presence of predators influence an ecosystem's productivity?

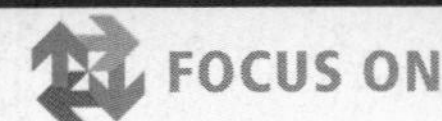

Applied Research

Biological Magnification and the Accumulation of Toxins at Higher Trophic Levels

The synthetic organic pesticide DDT (dichlorodiphenyl-trichloroethane) was first used widely during World War II. In the tropical Pacific, it killed mosquitoes that transmitted malarial parasites (*Plasmodium* species) to soldiers. In war-ravaged European cities, it controlled body lice that carried the bacteria causing typhus *(Rickettsia rickettsii)*. After the war, people started using DDT to kill agricultural pests, disease vectors, and insects in homes and gardens.

Although DDT is a stable hydrocarbon compound that is nearly insoluble in water, it is more mobile than its users expected. Winds carry it as a vapor, and water transports it as fine particles. DDT is also highly soluble in fats, accumulating in animal tissues—and it travels with animals wherever they go.

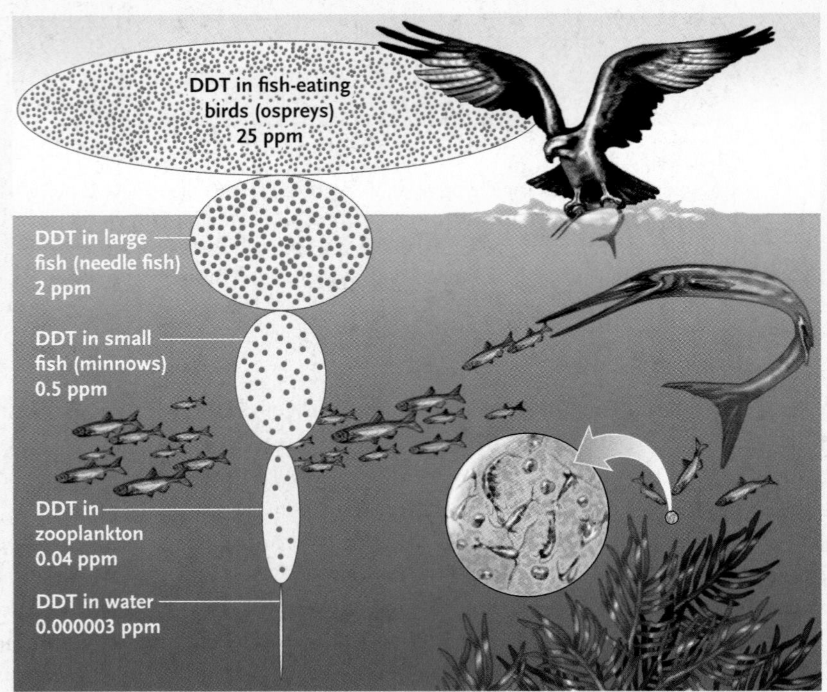

DDT in fish-eating birds (ospreys) 25 ppm
DDT in large fish (needle fish) 2 ppm
DDT in small fish (minnows) 0.5 ppm
DDT in zooplankton 0.04 ppm
DDT in water 0.000003 ppm

In this food web near Long Island Sound, New York, DDT concentration (measured in parts per million, ppm) was magnified nearly 10 million times between zooplankton and ospreys.

Unfortunately, consumers accumulate the DDT from all of the organisms they eat in their lifetimes. Primary consumers, like herbivorous insects, may ingest relatively small quantities. But a songbird that eats many insects will accumulate a moderate amount, and a predator that feeds on songbirds will accumulate even more. Thus, DDT and other nondegradable poisons become concentrated in organisms at higher trophic levels, a phenomenon called **biological magnification** (see **figure**). Although many organisms can partially metabolize DDT to other compounds, these products are also toxic or physiologically disruptive.

After the war, DDT moved rapidly through ecosystems, affecting organisms in ways that no one had predicted. In cities where DDT controlled Dutch elm disease, songbirds died after eating contaminated insects and seeds. In streams flowing through forests where DDT killed spruce budworms, salmon died because runoff carried the pesticide into their habitat. And in croplands around the world, new pests flourished because DDT indiscriminately killed the natural predators that had kept their populations in check.

Eventually, the effects of biological magnification began to show up in places far removed from the sites of DDT application. Top carnivores in some food webs were pushed to the brink of extinction. The reproduction of bald eagles, peregrine falcons, ospreys, and brown pelicans was disrupted because one DDT breakdown product interferes with the deposition of calcium in their eggshells. When birds tried to incubate their eggs, the shells cracked beneath the parents' weight. Even today, traces of DDT are found in the bodies of nearly all species, including in human fat and breast milk.

Since the 1970s, DDT has been banned in the United States, except for restricted applications to protect public health. Many hard-hit species have partially recovered, but some birds still lay thin-shelled eggs because they pick up DDT at their winter ranges in Latin America. As recently as 1990, the California State Department of Health recommended that a fishery off the coast of California be closed because DDT from industrial waste discharged 20 years earlier was still moving through that ecosystem. Moreover, DDT is still used in other countries, and some enters the United States on imported fruit and vegetables.

Biological magnification applies to many compounds that humans release into the environment. For example, polychlorinated biphenyls (PCBs), commonly used in the manufacture of plastics and electrical insulation, enter aquatic ecosystems in factory wastes. Their use has been banned in the United States since the 1970s. But these compounds break down very slowly, and vast deposits have accumulated in the bottom sediments of rivers and lakes. Once bottom-feeding organisms ingest them, the toxins work their way up food webs, accumulating at higher and higher concentrations in consumers. The effects on humans can be severe; pregnant women who regularly eat fish from the Great Lakes often give birth to children with below-average weight and neonatal behavioral problems. The pollution in some areas of New York State was so severe that the Department of Health advised people to avoid eating freshwater fish more than once a month. PCBs can be removed from aquatic ecosystems by dredging, but the dredging activity itself stirs up the polluted sediments, releasing the toxins into the water that flows above.

FIGURE 52.11 **Experimental Research**

A Trophic Cascade in Salt Marshes

A. **Herbivore damage and *Spartina* biomass**

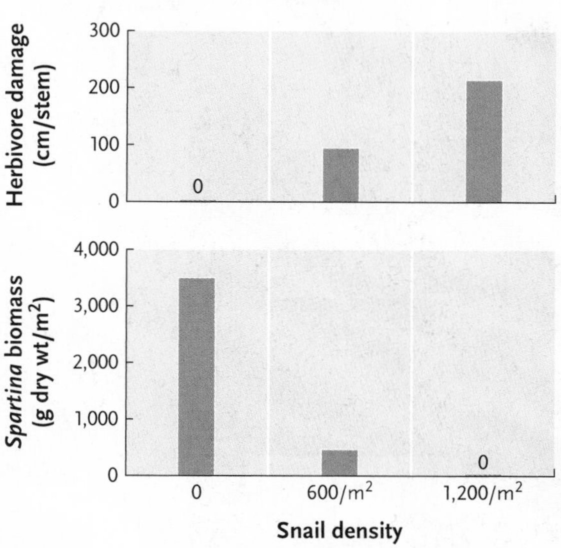
Herbivore damage (cm/stem)
300
200
100
0
0
Spartina biomass (g dry wt/m²)
4,000
3,000
2,000
1,000
0
0
0
600/m²
1,200/m²
Snail density

Question: How is the primary productivity of salt marsh grasses influenced by herbivores and their predators?

Experiment: Silliman and Bertness conducted a set of experiments to measure the effects of herbivores and their predators on the primary productivity of salt marsh cordgrass (*Spartina alterniflora*) along the coast of Sapelo Island, Georgia:

1. The researchers measured the effects on cordgrass of feeding by an herbivorous snail (*Littoraria irrorata*) by controlling snail densities within screened 1 m^2 enclosures (no snails, 600 snails, or 1,200 snails per enclosure) and measuring herbivore damage to cordgrass stems.
2. They created other enclosures that protected juvenile snails from their predators (crabs and terrapins) and compared the densities of juvenile snails that became established in protected *versus* unprotected areas.
3. They tethered snails to cordgrass stems at various locations in the marsh and monitored them to determine the rate at which they were consumed by predators.

All three experiments were replicated eight times in different parts of the marsh. The results reported here are from parts of the marsh where the cordgrass grew the tallest.

B. **Effect of snails on *Spartina***

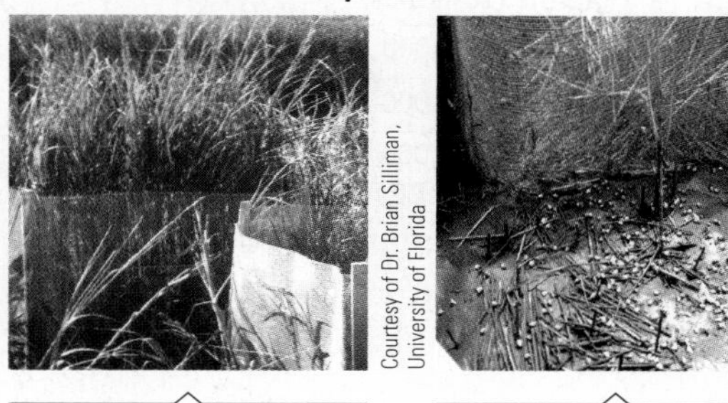
Courtesy of Dr. Brian Silliman, University of Florida
Courtesy of Dr. Brian Silliman, University of Florida

Spartina grew luxuriantly in an enclosure with no snails.

Spartina was virtually eliminated from an enclosure with snails at high density.

Results: The first experiment showed that herbivore damage increased with snail density, and cordgrass biomass decreased with snail density **(A).** The second experiment demonstrated that snail predators greatly reduce the density of juveniles that become established in the population: snail densities averaged 8.3 individuals/m^2 in unprotected enclosures versus 305/m^2 in protected enclosures. The third experiment revealed that 98% of the tethered snails were eaten by predators within 24 hours.

The researchers demonstrated that predators remove most herbivorous snails in this ecosystem, substantially reducing their impact on cordgrass. After predators are experimentally excluded from the system, snail densities increase dramatically, and high snail densities result in the near elimination of cordgrass **(B).**

C. **Trophic cascade**

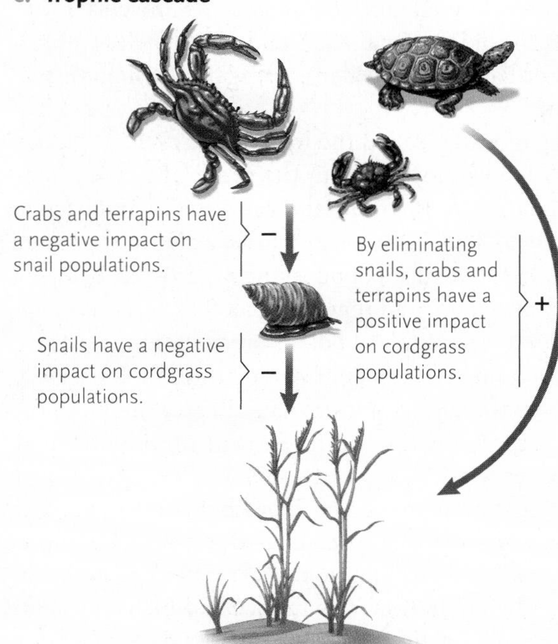
Crabs and terrapins have a negative impact on snail populations.
−
Snails have a negative impact on cordgrass populations.
−
By eliminating snails, crabs and terrapins have a positive impact on cordgrass populations.
+

Conclusion: The three experiments strongly suggest that cordgrass productivity is controlled by a top-down trophic cascade **(C).**

Blue crabs, a major predator on the snails, are a popular delicacy; the researchers note that so many blue crabs have been harvested from North American marshes that the crab populations are now imperiled. The elimination of this predator from marsh ecosystems is likely to result in a population explosion of snails, which in turn will virtually eliminate the cordgrass and thus the ecosystem itself.

Source: B. R. Silliman and M. D. Bertness. 2002. A trophic cascade regulates salt marsh primary production. *Proceedings of the National Academy of Sciences* 99:10500–10505.

A. **The water cycle**

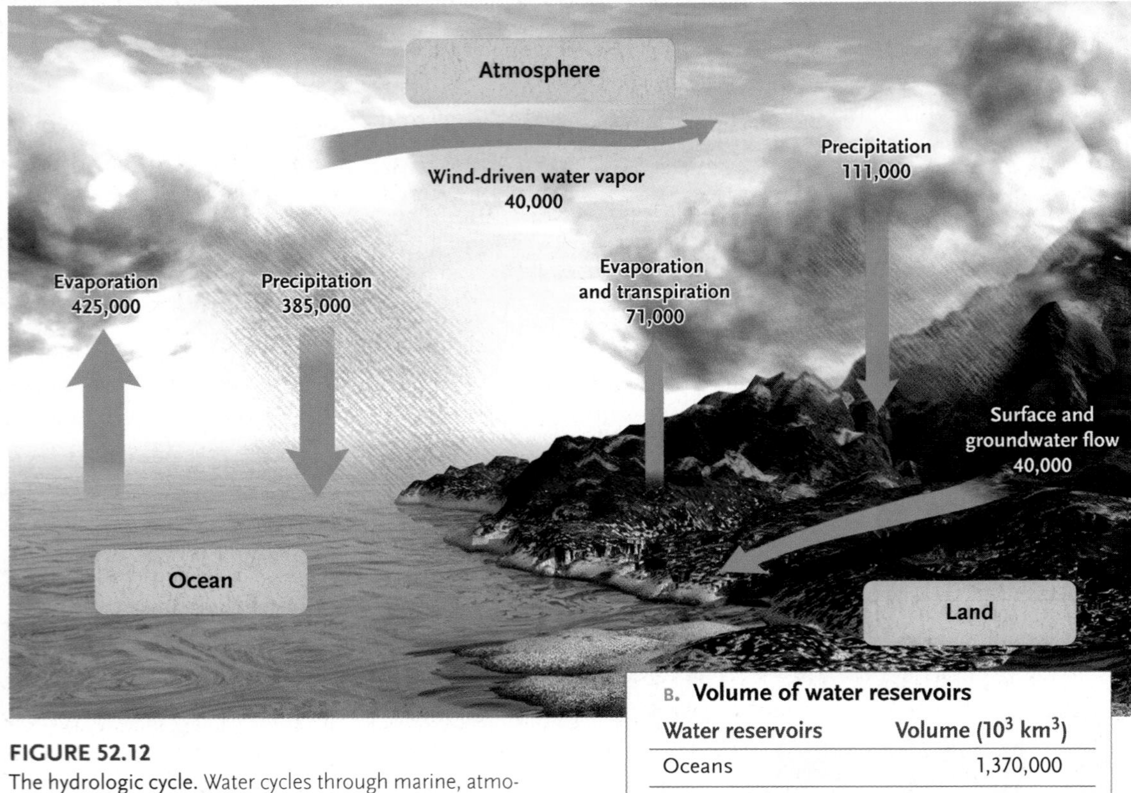

Atmosphere
Wind-driven water vapor
40,000
Precipitation
111,000
Evaporation
425,000
Precipitation
385,000
Evaporation
and transpiration
71,000
Surface and
groundwater flow
40,000
Ocean
Land

B. **Volume of water reservoirs**

Water reservoirs	Volume (10^3 km^3)
Oceans	1,370,000
Polar ice, glaciers	29,000
Groundwater	4,000
Lakes, rivers	230
Soil moisture	67
Atmosphere (water vapor)	14

FIGURE 52.12
The hydrologic cycle. Water cycles through marine, atmospheric, and terrestrial reservoirs. **(A)** Labels on the arrows list the amount of water (in km^3/yr) moved among reservoirs by various processes. **(B)** The oceans are by far the largest of the six major reservoirs of water on Earth.

52.3 Nutrient Cycling in Ecosystems

The availability of nutrients is as important to ecosystem function as the input of energy. Photosynthesis—the conversion of solar energy into chemical energy—requires carbon, hydrogen, and oxygen, which producers acquire from water and air. Producers also need nitrogen, phosphorus, and other minerals (see Table 33.1). A deficiency in any of these minerals can reduce primary productivity.

Earth is essentially a closed system with respect to matter. Thus, unlike energy, for which there is a constant cosmic input, Earth already contains virtually all the nutrients that will ever be available for biological systems. Biogeochemical cycles constantly circulate nutrient ions or molecules between the abiotic environment and living organisms. Unlike energy, which flows through ecosystems and is gradually lost as heat, matter is conserved in biogeochemical cycles. Although there may be local shortages of specific nutrients, Earth's overall supplies of these chemical elements are never depleted.

Nutrients take various forms as they pass through biogeochemical cycles. Some materials, such as carbon, nitrogen, and oxygen, form gases, which move through global *atmospheric cycles.* Geological processes move other materials, such as phosphorus, through local *sedimentary cycles,* carrying them between dry land and the seafloor. Rocks, soil, water, and air are the reservoirs where mineral nutrients accumulate, sometimes for many years.

The Hydrologic Cycle Recirculates All the Water on Earth

Although it is not a mineral nutrient, water is the universal intracellular solvent for biochemical reactions. Nevertheless, only a fraction of 1% of Earth's total water is present in biological systems at any time.

The cycling of water, called the **hydrologic cycle,** is global, with water molecules moving from the ocean into the atmosphere, to the land, through freshwater ecosystems, and back to the ocean **(Figure 52.12).** Solar energy causes water to evaporate from oceans, lakes, rivers, soil, and living organisms, entering the atmosphere as a vapor and remaining aloft as a gas, as droplets in clouds, or as ice crystals. It falls as precipitation, mostly in the form of rain and snow. When precipitation falls on land, water flows across the surface or percolates to great depth in the soil, eventually reentering the ocean reservoir through the flow of streams and rivers.

The hydrologic cycle maintains its global balance because the total amount of water that enters the atmosphere is equal to the amount that falls as precipitation. Most water enters the atmosphere through evaporation from the ocean, which represents the largest reservoir on the planet. A much smaller fraction

FOCUS ON Basic Research

Studies of the Hubbard Brook Watershed

Because water always flows downhill, local topography affects the movement of dissolved nutrients in terrestrial ecosystems. A **watershed** is an area of land from which precipitation drains into a stream or river system. Thus, each watershed represents a part of an ecosystem from which nutrients exit through a single outlet, much the way a bathtub empties through a single drain. When several streams join to form a river, the watershed drained by the river encompasses all of the smaller watersheds drained by the streams. For example, the Mississippi River watershed covers roughly one-third of the United States, and it includes watersheds drained by the Illinois, Missouri, and Tennessee Rivers as well as many other watersheds drained by smaller streams and rivers.

Because watersheds are relatively self-contained units, they are ideal for large-scale field experiments about nutrient flow in ecosystems. Herbert Bormann of Yale University and Gene Likens of Cornell University have conducted a classic experiment on this topic since the 1960s. Bormann and Likens manipulated small watersheds of temperate deciduous forest in the Hubbard Brook Experimental Forest in the White Mountain National Forest of New Hampshire. They measured precipitation and nutrient input into the watersheds, the uptake of nutrients by vegetation, and the amount of nutrients leaving the watershed via streamflow. Nutrients exported in streamflow were monitored in water samples collected from V-shaped concrete weirs built into bedrock below the streams that drained the watersheds **(Figure A).** Impermeable bedrock underlies the soil, preventing water from leaving the system by deep seepage.

After collecting several years of baseline data on six undisturbed watersheds, the researchers cut all the trees in one small watershed in 1965 and 1966. They also applied herbicides to prevent regrowth. After establishing this experimental treatment, they monitored the output of nutrients in streams that drained experimental and control watersheds.

Gene E. Likens from Gene E. Likens et al., Ecological Monographs, 40(1): 23–47, 1970

FIGURE A
Weir used to measure the volume and nutrient content of water leaving a watershed by streamflow.

They attributed differences in nutrient export between undisturbed watersheds (controls) and the clear-cut watershed (experimental treatment) to the effects of deforestation.

Bormann and Likens determined that vegetation absorbed substantial water and conserved nutrients in undisturbed watersheds. Plants used about 40% of the precipitation for transpiration. The rest contributed to runoff and groundwater. Control watersheds lost only about 8–10 kg of calcium per hectare each year, an amount that was replaced by the erosion of bedrock and input from rain. Moreover, control watersheds actually accumulated about 2 kg of nitrogen per hectare per year and slightly smaller amounts of potassium.

By contrast, the experimentally deforested watershed experienced a 40% annual increase in runoff. During a 4-month period in the summer, runoff increased 300%. Some mineral losses were similarly large. The net loss of calcium was 10 times higher than in the control watersheds **(Figure B)** and the loss of potassium 21 times higher.

Phosphorus losses did not increase; this mineral was apparently retained by the soil. However, the loss of nitrogen was an astronomical 120 kg per hectare per year. So much nitrogen entered the stream draining the experimental watershed that the stream became choked with algae and cyanobacteria. Thus, the results of the Hubbard Brook experiment suggest that deforestation increases flooding, decreases the fertility of ecosystems, and leads to nutrient enrichment of nearby aquatic ecosystems.

The Hubbard Brook Watershed project is an example of long-term ecological research (LTER), one of many in a network of such projects that provide remarkable insights into ecosystem processes. The network itself is funded by the National Science Foundation and described in detail at its web site: http://www.lternet.edu/.

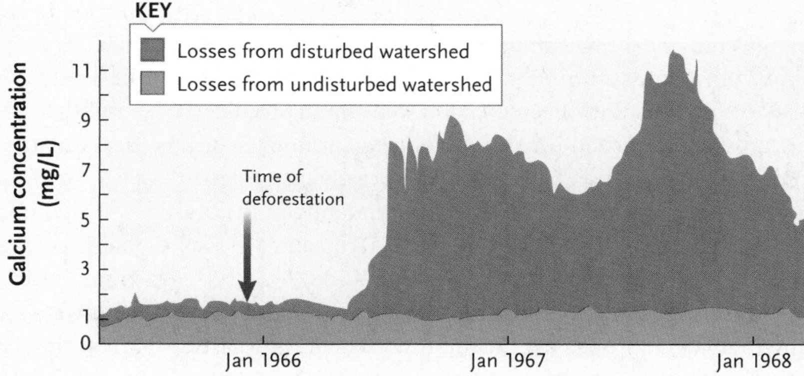

KEY
Losses from disturbed watershed
Losses from undisturbed watershed
Calcium concentration (mg/L)
0
1
3
5
7
9
11
Time of deforestation
Jan 1966
Jan 1967
Jan 1968

FIGURE B
Calcium losses from a deforested watershed were much greater than those from controls. The arrow indicates the time of deforestation in early winter. Mineral losses did not increase until after the ground thawed the following spring; increased runoff also caused large water losses from the watershed.

evaporates from terrestrial ecosystems, and most of that results from transpiration in green plants.

The constant recirculation provides fresh water to terrestrial organisms and maintains freshwater ecosystems such as lakes and rivers. Water also serves as a transport medium that moves nutrients within and between ecosystems, as demonstrated in a series of classic experiments in the Hubbard Brook Experimental Forest, described in *Focus on Basic Research*.

A. Amount of carbon in major reservoirs

Carbon reservoirs	Mass (10^{15} g)
Sediments and rocks	77,000,000
Ocean (dissolved forms)	39,700
Soil	1,500
Atmosphere	750
Biomass on land	715

B. Annual global carbon movement between reservoirs

Direction of movement	Mass (10^{15} g)
From atmosphere to plants (carbon fixation)	120
From atmosphere to ocean	107
To atmosphere from ocean	105
To atmosphere from plants	60
To atmosphere from soil	60
To atmosphere from burning fossil fuel	5
To atmosphere from burning plants	2
To ocean from runoff	0.4
Burial in ocean sediments	0.1

C. The global carbon cycle

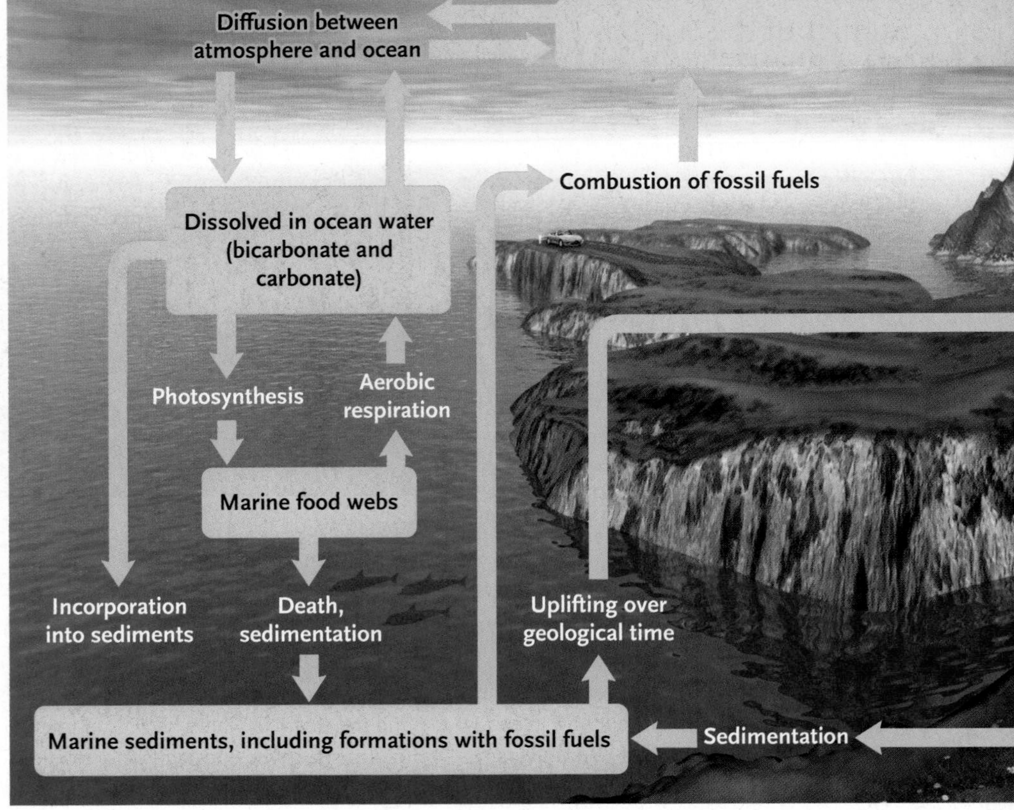

Diffusion between atmosphere and ocean
Combustion of fossil fuels
Dissolved in ocean water (bicarbonate and carbonate)
Photosynthesis
Aerobic respiration
Marine food webs
Incorporation into sediments
Death, sedimentation
Uplifting over geological time
Marine sediments, including formations with fossil fuels
Sedimentation

FIGURE 52.13

The carbon cycle. Marine and terrestrial components of the global carbon cycle are linked through an atmospheric reservoir of carbon dioxide. **(A)** By far, the largest amount of Earth's carbon is found in sediments and rocks. **(B)** Earth's atmosphere mediates most movements of carbon. **(C)** In this illustration of the carbon cycle, boxes identify major reservoirs, and labels on the arrows identify the processes that cause carbon to move between reservoirs.

The Carbon Cycle Includes a Large Atmospheric Reservoir

Carbon atoms provide the backbone of most biological molecules, and carbon compounds store the energy captured by photosynthesis (see Section 9.1). Carbon enters food webs when producers convert atmospheric carbon dioxide (CO_2) into carbohydrates. Heterotrophs acquire carbon by eating other organisms or detritus. Although carbon moves somewhat independently in the sea and on land, a common atmospheric pool of CO_2 creates a global **carbon cycle (Figure 52.13).**

The largest reservoir of carbon is sedimentary rock, such as limestone or marble. Rocks are in the unavailable inorganic compartment, and they exchange carbon with living organisms at an exceedingly slow pace. Most *available* carbon is present as dissolved bicarbonate ions (HCO_3^-) in the ocean. Soil, the atmosphere, and plant biomass form other significant, but much smaller, reservoirs of available carbon. Atmospheric carbon is mostly in the form of molecular CO_2, a product of aerobic respiration. Volcanic eruptions also release CO_2 into the atmosphere.

Sometimes carbon atoms leave the organic compartments for long periods of time. Some organisms in marine food webs build shells and other hard parts by incorporating dissolved carbon into calcium carbonate ($CaCO_3$) and other insoluble salts. When shelled organisms die, they sink to the bottom and are buried in sediments. The insoluble carbon that accumulates as rock in deep sediments may remain buried for millions of years before tectonic uplifting brings it to the surface, where erosion and weathering dissolve sedimentary rocks and return carbon to an available form.

Carbon atoms were also transferred to the unavailable organic compartment when soft-bodied organisms were buried in habitats where low oxygen concentration prevented decomposition. Under suitable geological conditions, these carbon-rich tissues were slowly converted to gas, petroleum, or coal, which humans now use as fossil fuels.

The Nitrogen Cycle Depends on the Activity of Diverse Microorganisms

All organisms require nitrogen to construct nucleic acids, proteins, and other biological molecules. Earth's atmosphere had a high nitrogen concentration long before life originated. Today, a global **nitrogen cycle** moves this element between the huge atmospheric pool of gaseous molecular nitrogen (N_2) and several much smaller pools of nitrogen-containing compounds in soils, marine and freshwater ecosystems, and living organisms **(Figure 52.14).**

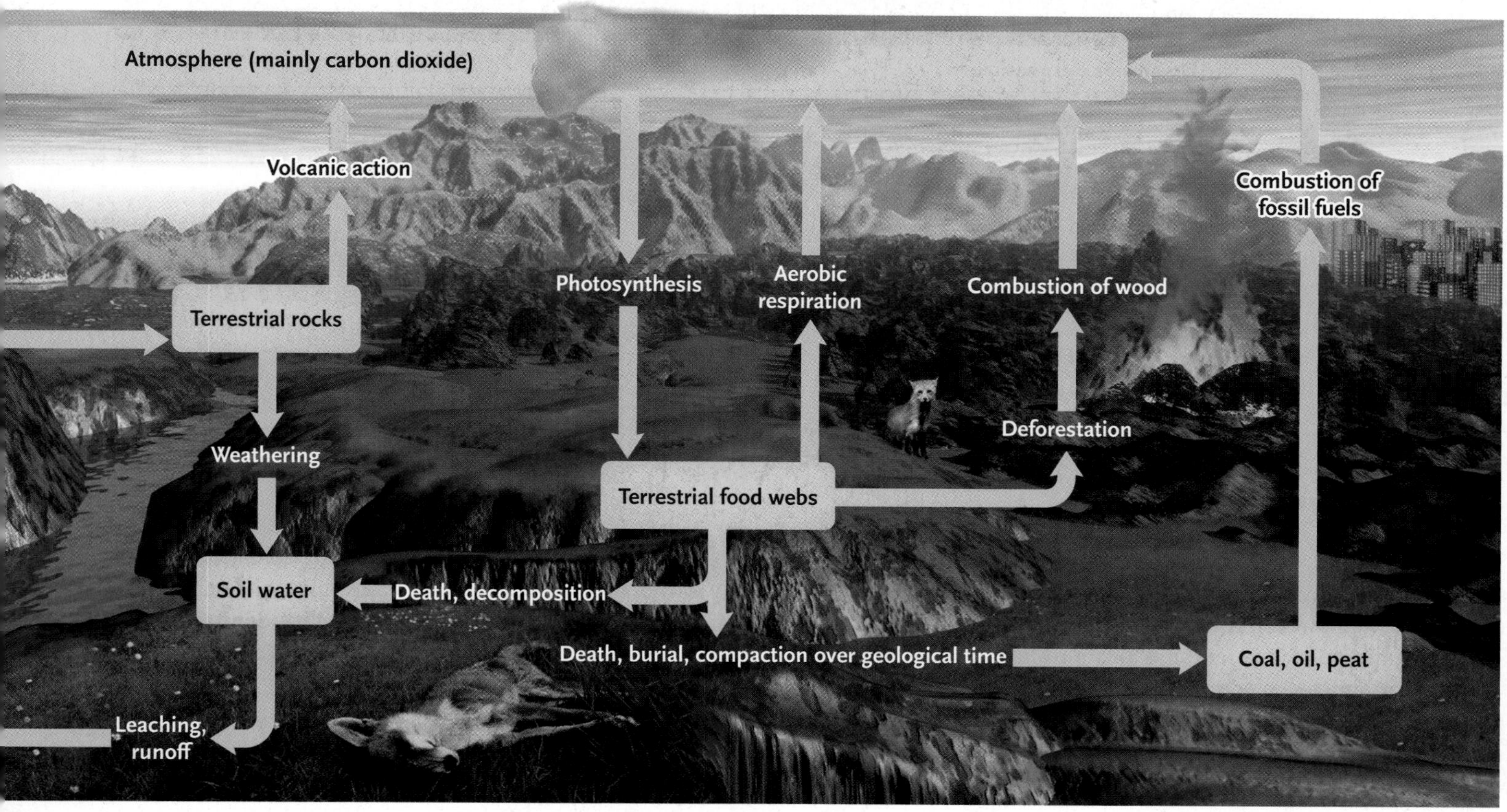

Atmosphere (mainly carbon dioxide)
Volcanic action
Combustion of fossil fuels
Photosynthesis
Aerobic respiration
Combustion of wood
Terrestrial rocks
Deforestation
Weathering
Terrestrial food webs
Soil water
Death, decomposition
Death, burial, compaction over geological time
Coal, oil, peat
Leaching, runoff

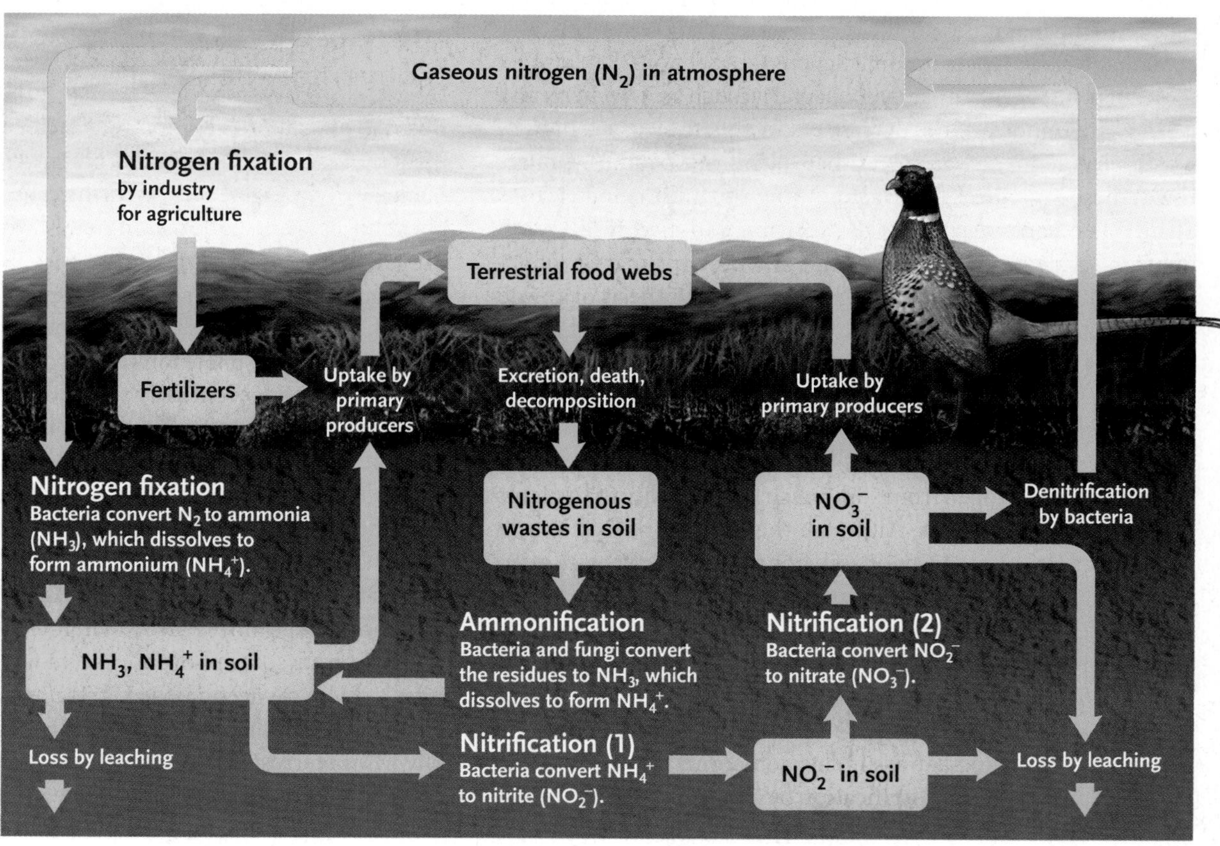

Gaseous nitrogen (N_2) in atmosphere
Nitrogen fixation
by industry
for agriculture
Terrestrial food webs
Fertilizers
Uptake by primary producers
Excretion, death, decomposition
Uptake by primary producers
Nitrogen fixation
Bacteria convert N_2 to ammonia (NH_3), which dissolves to form ammonium (NH_4^+).
Nitrogenous wastes in soil
NO_3^- in soil
Denitrification by bacteria
Ammonification
Bacteria and fungi convert the residues to NH_3, which dissolves to form NH_4^+.
Nitrification (2)
Bacteria convert NO_2^- to nitrate (NO_3^-).
NH_3, NH_4^+ in soil
Loss by leaching
Nitrification (1)
Bacteria convert NH_4^+ to nitrite (NO_2^-).
NO_2^- in soil
Loss by leaching

FIGURE 52.14
The nitrogen cycle in terrestrial ecosystems. Nitrogen cycles through terrestrial ecosystems when unavailable molecular nitrogen is made available through the action of nitrogen-fixing bacteria. Other bacteria recycle nitrogen within the available organic compartment through ammonification and two types of nitrification, converting organic wastes into ammonium ions and nitrates. Denitrification converts nitrate to molecular nitrogen, which returns to the atmosphere. Runoff carries various nitrogen compounds from terrestrial ecosystems into oceans, where it is recycled in marine food webs.

TABLE 52.2	Biochemical Processes That Influence Nitrogen Cycling in Ecosystems		
Process	**Organisms Responsible**	**Products**	**Outcome**
Nitrogen fixation	Bacteria: *Rhizobium, Azotobacter, Frankia* Cyanobacteria: *Anabaena, Nostoc*	Ammonia (NH_3), ammonium ions (NH_4^+)	Assimilated by primary producers
Ammonification of organic detritus	Soil bacteria and fungi	Ammonia (NH_3), ammonium ions (NH_4^+)	Assimilated by primary producers
Nitrification			
(1) Oxidation of NH_3	Bacteria: *Nitrosomonas, Nitrococcus*	Nitrite (NO_2^-)	Used by nitrifying bacteria
(2) Oxidation of NO_2^-	Bacteria: *Nitrobacter*	Nitrate (NO_3^-)	Assimilated by primary producers
Denitrification of NO_3^-	Soil bacteria	Nitrous oxide (N_2O), molecular nitrogen (N_2)	Released to atmosphere

Molecular nitrogen is abundant in the atmosphere, but triple covalent bonds bind its two atoms so tightly that most organisms cannot use it. However, three biochemical processes—nitrogen fixation, ammonification, and nitrification **(Table 52.2)**—convert nitrogen into nitrogen compounds that primary producers can incorporate into biological molecules such as proteins and nucleic acids. Secondary consumers obtain their nitrogen by consuming primary producers, thereby initiating the movement of nitrogen through the food webs of an ecosystem.

In **nitrogen fixation** (see Section 33.3), molecular nitrogen (N_2) is converted into ammonia (NH_3) and ammonium ions (NH_4^+). Certain bacteria, including *Azotobacter* and *Rhizobium,* which collect molecular nitrogen from the air between soil particles, are the major nitrogen fixers in terrestrial ecosystems. The cyanobacteria partners in some lichens (see Section 28.3) also fix molecular nitrogen. Other cyanobacteria, such as *Anabaena* and *Nostoc,* are important nitrogen fixers in aquatic ecosystems; the water fern (genus *Azolla*) plays that role in rice paddies. Collectively, these organisms fix an astounding 200 million metric tons of nitrogen each year; nitrogen fixation can also result from lightning and volcanic action. Plants and other primary producers assimilate and use this nitrogen in the biosynthesis of amino acids, proteins, and nucleic acids, which then circulate through food webs.

Some plants, including legumes (such as beans and clover), alders (*Alnus* species), and some members of the rose family (Rosaceae), are mutualists with nitrogen-fixing bacteria. These plants acquire nitrogen from soils much more readily than plants that lack such mutualists. Although these plants have the competitive edge in nitrogen-poor soil, nonmutualistic species often displace them in nitrogen-rich soil.

In addition to nitrogen fixation, several other biochemical processes make large quantities of nitrogen available to producers. **Ammonification** of detritus by bacteria and fungi converts organic nitrogen into ammonia (NH_3), which dissolves into ammonium ions (NH_4^+) that plants can assimilate; some ammonia escapes into the atmosphere as a gas. **Nitrification** by certain bacteria produces nitrites (NO_2^-), which are converted by other bacteria to usable nitrates (NO_3^-). All of these compounds are water-soluble, and water rapidly leaches them from soil into streams, lakes, and oceans.

Under conditions of low oxygen availability, **denitrification** by still other bacteria converts nitrites or nitrates into nitrous oxide (N_2O) and then into molecular nitrogen (N_2), which enters the atmosphere, completing the cycle. This action can deplete supplies of soil nitrogen in waterlogged or otherwise poorly aerated environments, such as bogs and swamps. In an interesting twist on the usual predator–prey relationships, several species of flowering plants that live in nitrogen-poor soils, such as Venus' fly trap *(Dionaea muscipula),* capture and digest small insects as their primary nitrogen source.

The Phosphorus Cycle Includes a Large Sedimentary Reservoir

Because phosphorus compounds lack a gaseous phase, this element moves between terrestrial and marine ecosystems in a sedimentary cycle **(Figure 52.15).** Earth's crust is the main reservoir of phosphorus, as it is for other minerals, such as calcium and potassium, that undergo sedimentary cycles.

Phosphorus is present in terrestrial rocks in the form of phosphates (PO_4^{3-}). In the **phosphorus cycle,** weathering and erosion carry phosphate ions from rocks to soil and into streams and rivers, which eventually transport them to the ocean. Once there, some phosphorus enters marine food webs, but most of it precipitates out of solution and accumulates for millions of years as insoluble deposits, mainly on continental shelves. When parts of the seafloor are uplifted and exposed, weathering releases the phosphates.

Plants absorb and assimilate dissolved phosphates directly, and phosphorus moves easily to higher trophic levels. All heterotrophs excrete some phosphorus as a waste product in urine and feces, which are decomposed, and producers readily absorb the phosphate ions that are released. Thus, phosphorus cycles rapidly *within* terrestrial communities.

Supplies of available phosphate are generally limited, however, and plants acquire it so efficiently that they reduce soil

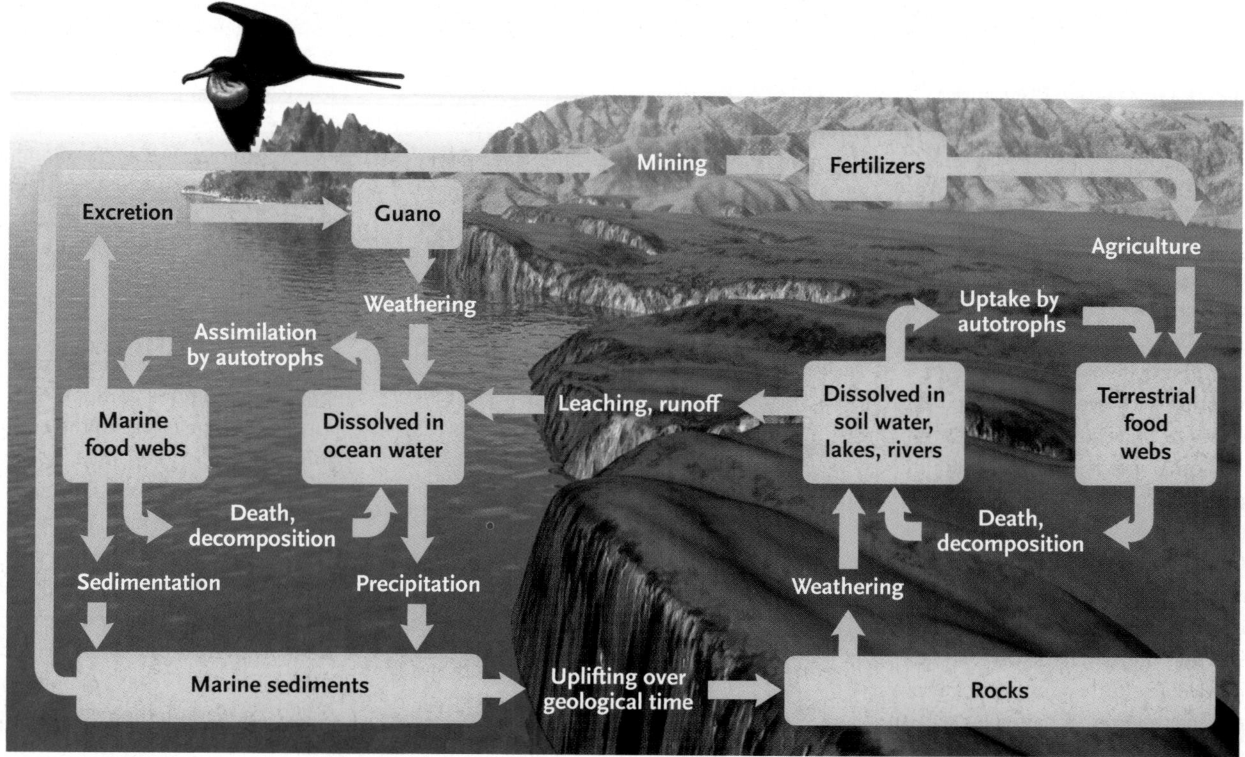
Mining
Fertilizers
Excretion
Guano
Agriculture
Weathering
Uptake by autotrophs
Assimilation by autotrophs
Leaching, runoff
Dissolved in soil water, lakes, rivers
Terrestrial food webs
Marine food webs
Dissolved in ocean water
Death, decomposition
Death, decomposition
Sedimentation
Precipitation
Weathering
Marine sediments
Uplifting over geological time
Rocks

FIGURE 52.15
The phosphorus cycle. Phosphorus becomes available to biological systems when wind and rainfall dissolve phosphates in rocks and carry them into adjacent soil and freshwater ecosystems. Runoff carries dissolved phosphorus into marine ecosystems, where it precipitates out of solution and is incorporated into marine sediments.

phosphate concentration to extremely low levels. Thus, like nitrogen, phosphorus is a common ingredient in agricultural fertilizers, and excess phosphates are pollutants of freshwater ecosystems. For many years, phosphate for fertilizers was obtained from *guano* (the droppings of seabirds that consume phosphorus-rich food), which was mined on small islands off the Pacific coast of South America. Most phosphate for fertilizer now comes from phosphate rock mined in Florida and other places with abundant marine deposits.

STUDY BREAK 52.3 <

1. **How does the global hydrologic cycle maintain its balance?**
2. **What processes move large quantities of carbon from an organic compartment to an inorganic compartment?**
3. **What microorganisms drive the global nitrogen cycle, and how do they do it?**
4. **What is Earth's main reservoir for phosphorus, and why is it recycled at such a slow rate from that reservoir?**

52.4 Human Disruption of Ecosystem Processes

Human activities—industrial processes, agricultural practices, and development—often disrupt biogeochemical cycles by moving materials from one nutrient compartment to another at unnaturally rapid rates. In this section we consider the consequences of human disruption of the carbon and nitrogen cycles.

Our Use of Combustible Energy Sources Has Disrupted the Carbon Cycle

The combustion of fossil fuels (oil, coal, and peat) and wood is transferring carbon from the reservoirs of organic nutrients to the atmospheric reservoir of inorganic nutrients at an unprecedented rate. Virtually all scientists agree that the resulting change in the worldwide distribution of carbon is having severe consequences for Earth's climate, including a general warming that will cause a significant rise in sea level.

Concentrations of certain gases in the lower atmosphere have a profound effect on global temperature, which in turn has enormous impact on global climate. Molecules of carbon dioxide (CO_2), water, ozone, methane, nitrous oxide, and other compounds collectively act like a pane of glass in a greenhouse (hence they are described as *greenhouse gases*). They allow the short wavelengths of visible light to reach Earth's surface; but they impede the escape of longer, infrared wavelengths back into space, trapping much of that energy as heat. In short, greenhouse gases foster the accumulation of heat in the lower atmosphere, a warming action known as the **greenhouse effect,** which prevents Earth from being a cold and lifeless planet.

Since the late 1950s, scientists have measured atmospheric concentrations of CO_2 and other greenhouse gases at several remote sampling sites, which are free of local contamination and

INSIGHTS FROM THE **Molecular Revolution**

Ocean Acidification: Does it affect gene expression?

Because CO_2 is soluble in water, the oceans readily absorb it from the atmosphere—so much so that ecologists describe the oceans as a "sink" for atmospheric CO_2. Some environmental scientists have even suggested that the transfer of CO_2 from the atmosphere to the oceans may mitigate rising atmospheric CO_2 concentration. However, increasing the concentration of dissolved CO_2 decreases oceanic pH; in other words, the oceans are becoming more acidic as they absorb more of this gas. If the CO_2 concentration of the atmosphere continues to increase at its present rate, scientists predict a decrease in ocean surface pH of 0.2–0.4 by the end of the twenty-first century. Researchers worry about the possible adverse effects of ocean acidification on ocean ecosystems. In particular, will marine organisms be able to adapt to such changes?

Ocean acidification is a potential stress to marine organisms. In the lab, scientists are using genomics tools to understand the molecular mechanisms that underlie physiological responses to environmental stresses such as changes in temperature, osmolarity, and oxygen availability. In one approach, researchers use DNA microarrays (see Chapter 18) to analyze changes in the expression of many protein-coding genes simultaneously. The set of genes expressed in a cell at any time is called a *transcriptome;* the study of transcriptomes is *transcriptomics.* Understanding how ocean acidification affects the transcriptomes of marine organisms would help scientists understand the physiological mechanism that organisms might use to tolerate a more acidic ocean in the future.

Research Question

Does the pH of seawater affect the transcriptome?

Experiment

To answer the question, Anne Todgham and Gretchen Hofmann of the University of California, Santa Barbara, performed DNA microarray experiments on sea urchin *(Strongylocentrotus purpuratus)* larvae under different pH conditions. They studied this species because its larval skeleton is altered by low pH conditions, and because its genome has been sequenced. To vary pH, the researchers bubbled different concentrations of CO_2 through the seawater in tanks with developing larvae. In the control, the CO_2 concentration was equivalent to current atmospheric conditions; it produced a pH of 8.01. Two experimental conditions reflected predictions (made by the Intergovernmental Panel on Climate Change in 2007) of atmospheric CO_2 concentrations in the year 2100. The moderate treatment reflected an optimistic value that assumed some success in limiting future increases in atmospheric CO_2 concentration; it produced a pH of 7.96. The high treatment assumed no success in limiting future increases in atmospheric CO_2 concentration; it produced a pH of 7.88. Todgham and Hofmann's experiment is outlined in the figure on the right.

Results

The researchers found subtle but statistically significant changes in the transcriptomes of the larvae growing in the moderate and high CO_2 conditions compared to the transcriptome of the larvae in the control condition:

- Moderate CO_2—83 of the 1,057 genes analyzed showed statistically significant decreases in expression; no genes showed increases in expression. Approximately half the genes with decreased expression are involved in biomineralization (that is, the biological process of building a mineralized skeleton), the cellular stress response, and energy metabolism.
- High CO_2—178 genes showed statistically significant changes in expression, with 160 of them showing decreases and 18 showing increases. Approximately 70% of the genes showing decreased expression are involved in apoptosis (programmed cell death), the cellular stress response, and energy metabolism.
- The transcriptomic responses to moderate and high CO_2 conditions were largely distinct, with only 10 genes showing altered expression under both treatments.

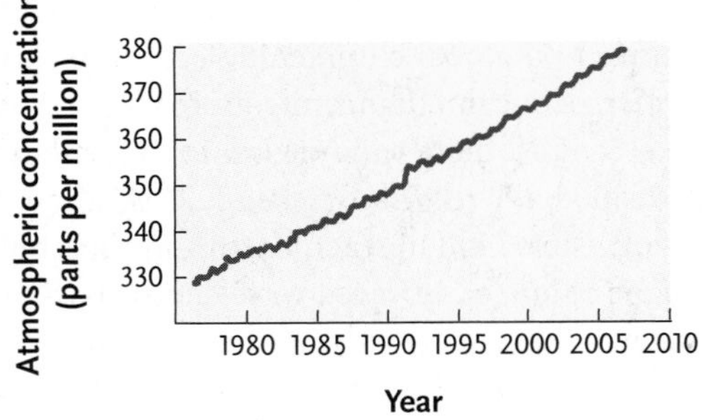

Atmospheric concentration (parts per million)
380
370
360
350
340
330
1980 1985 1990 1995 2000 2005 2010
Year

FIGURE 52.16
Increases in atmospheric concentration of carbon dioxide. Data have been collected at a remote monitoring station in Australia (Cape Grim, Tasmania) since 1976 and compiled by scientists at the Commonwealth Scientific and Industrial Research Organization, an agency of the Australian government.

reflect the average concentrations of these gases in the atmosphere. Results indicate that concentrations of greenhouse gases have increased steadily for as long as they have been monitored **(Figure 52.16).** The graph of atmospheric CO_2 concentration through time has a regular zigzag pattern that follows the annual cycle of plant growth in the northern hemisphere. Photosynthesis withdraws so much CO_2 from the atmosphere during the northern hemisphere summer that its concentration falls. The concentration is higher during the northern hemisphere winter, when aerobic respiration continues, returning carbon to the atmosphere, and photosynthesis slows. The zigs and the zags in the data for CO_2 represent seasonal highs and lows, but the midpoint of the annual peaks and troughs has increased steadily. Many

Conclusion

This transcriptomics study demonstrated that acidification affects the expression of numerous genes, many of which are involved in four major cellular processes: biomineralization, cellular stress response, metabolism, and apoptosis. Although one might have expected that various genes would be turned on to cope with the environmental stress, ocean acidification mostly decreased gene expression. Importantly, the study showed that pathways other than calcification, a process previously shown to be affected at the physiological level by pH changes, are altered by acidification. This result suggests that scientists must consider a wide range of physiological processes when assessing the possible effects of future ocean acidification on marine organisms.

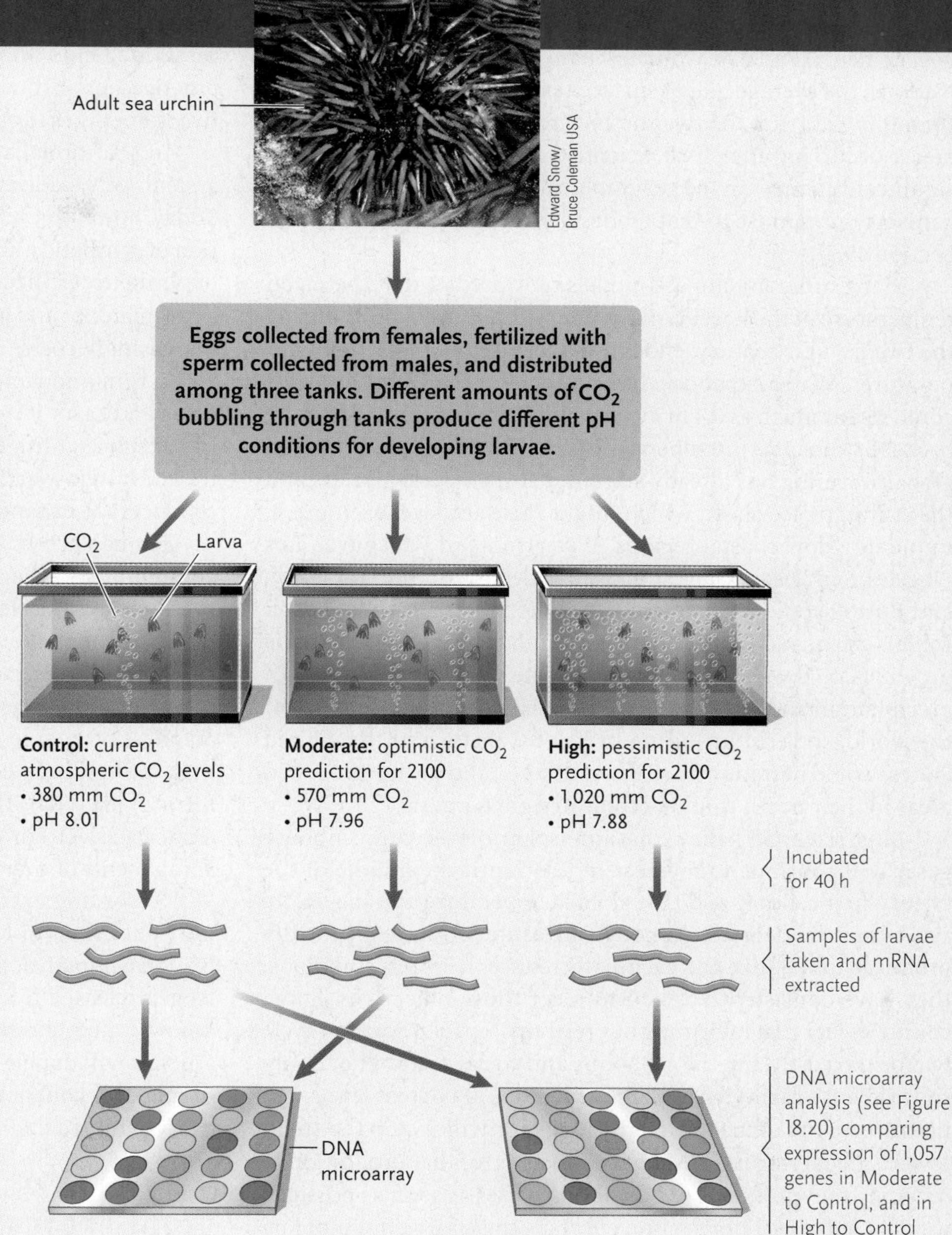

Adult sea urchin
Edward Snow/ Bruce Coleman USA
Eggs collected from females, fertilized with sperm collected from males, and distributed among three tanks. Different amounts of CO_2 bubbling through tanks produce different pH conditions for developing larvae.
CO_2
Larva
Control: current atmospheric CO_2 levels
• 380 mm CO_2
• pH 8.01
Moderate: optimistic CO_2 prediction for 2100
• 570 mm CO_2
• pH 7.96
High: pessimistic CO_2 prediction for 2100
• 1,020 mm CO_2
• pH 7.88
Incubated for 40 h
Samples of larvae taken and mRNA extracted
DNA microarray
DNA microarray analysis (see Figure 18.20) comparing expression of 1,057 genes in Moderate to Control, and in High to Control

Source: A. Todgham and G. E. Hofmann. 2009. Transcriptomic response of sea urchin larvae *Strongylocentrotus purpuratus* to CO_2-driven seawater acidification. *The Journal of Experimental Biology* 212:2579–2594.

scientists interpret these data as evidence of a rapid buildup of atmospheric CO_2, which represents a shift in the distribution of carbon in the major reservoirs on Earth. Scientists estimate that the atmospheric CO_2 concentration has increased 35% in the last 150 years and 15% in the last 30 years.

What has caused the increase in the atmospheric concentration of CO_2? Burning of fossil fuels and wood is the largest contributor, because CO_2 is a combustion product of this process. Today, humans burn more wood and fossil fuels than ever before. In a paper published in *Nature Geoscience* in December 2009, Corinne Le Quéré at the University of East Anglia, UK, and colleagues from other institutions estimated that CO_2 emissions from the combustion of fossil fuels increased 29% between 2000 and 2008. Although slightly more than half the CO_2 emissions are absorbed by forests and marine phytoplankton, vast tracts of tropical forests are being cleared and burned (see Section 53.2), reducing the biosphere's capacity to maintain the carbon cycle as it existed before recent human activities disrupted it.

Why is an increase in the atmospheric CO_2 concentration so alarming? As *Insights from the Molecular Revolution* describes, scientists are just beginning to discover many previously unforeseen effects of rising CO_2 levels in the atmosphere and in the oceans. Moreover, simulation models of the global climate suggest that increasing the concentration of any greenhouse gas is likely to intensify the greenhouse effect, contributing to a well-documented trend of global warming. According to the Intergovernmental Panel on Climate Change, an agency of the United Nations, the average global surface temperature increased more than 0.7°C during the twentieth century, with most of the increase occurring after 1980. Scientists have already documented significant changes in the geographical distributions of many organisms in response to that modest level of global warming (see Section 49.2).

Many different climate models now predict that the mean temperature of the lower atmosphere will rise 2.0°–4.5°C during the twenty-first century, enough to increase ocean surface temperatures. Water expands when heated and the global sea level could rise as much as 0.4 m just from this expansion. In addition, because atmospheric temperature is rising fastest near the poles, global warming has already started melting Arctic glaciers and the Antarctic ice sheet, which might raise sea level even more, inundating low coastal regions. Waterfronts in Vancouver, Los Angeles, San Diego, Galveston, New Orleans, Miami, New York, and Boston could be submerged. So might agricultural lands in India, China, and Bangladesh, where much of the world's rice is grown. Global warming could also disturb regional patterns of precipitation and temperature. Areas that now produce much of the world's grains, including parts of Canada and the United States, would become arid scrub or deserts, and the now-forested areas to their north would become dry grasslands.

Most scientists believe that atmospheric levels of greenhouse gases will continue to increase at least until the middle of the twenty-first century and that global temperature will inevitably rise by several degrees. Although governments have repeatedly promised to stabilize and eventually reduce their CO_2 emissions, they have consistently failed to meet those objectives. Some countries fear that reforming our relationship to the carbon cycle will be too costly, but the economic and social costs of not doing so will be far higher. Stabilizing emissions at current levels will not reverse the damage already done, nor will it stop the trend toward global warming. Many scientists agree that in addition to reducing our emissions of greenhouse gases as soon as possible, we must increase reforestation efforts because large tracts of forest can withdraw significant amounts of CO_2 from the atmosphere. We might also step up agricultural research to develop heat-resistant and drought-resistant crop plants, which may provide crucial food reserves in regions that will be subject to the most drastic climate change.

Human Activities Have Altered the Nitrogen Cycle

Human activities are also disrupting the nitrogen cycle, primarily through the use of nitrogen-containing fertilizers, the farming and processing of livestock, and the combustion of fossil fuels.

Of all nutrients required for primary production, nitrogen is often the least abundant. By definition, agriculture depletes soil nitrogen by removing nitrogen-containing plants from fields. Irrigation also fosters soil erosion and leaching, which remove even more. Traditionally, farmers rotated their crops, alternately cultivating nitrogen-fixing legumes and other types of plants in the same fields. In combination with other soil-conservation practices, crop rotation stabilized soils and kept them productive, sometimes for thousands of years.

In traditional agriculture, nearly all the nitrogen in living systems was made available by nitrogen-fixing microorganisms. Today, however, industrialized agriculture relies on the application of synthetic nitrogen-containing fertilizers. The production of synthetic fertilizers is expensive, and it uses fossil fuels both as a raw material and as an energy source; thus, fertilizer becomes increasingly costly as supplies of fossil fuels dwindle. Furthermore, rain and runoff leach excess fertilizer from agricultural fields and carry it into aquatic ecosystems. Like the phosphorus in Lake Erie, nitrogen has become a major pollutant of freshwater ecosystems, artificially enriching the waters and allowing producers to expand their populations (see Section 49.5).

Human activities also add an enormous volume of nitrogen compounds to the atmosphere. For example, large-scale livestock production leaves behind mountains of organic wastes, which, through the action of the anaerobic microorganisms that decompose it, releases nitrous oxide (NO_2), a greenhouse gas. Moreover, whenever we burn fossil fuels at high temperatures in combustion engines, molecular nitrogen (N_2), which is abundant in the atmosphere, combines with molecular oxygen (O_2) to form nitric oxide (NO). This molecule readily converts to nitrogen dioxide gas (NO_2) or nitric acid vapors (HNO_3), which are major components of smog and acid rain.

According to the Millennium Ecosystem Assessment, a report published in March 2005 with the support of the United Nations, human activities more than doubled the amount of nitrogen released from land into other nitrogen reservoirs in the second half of the twentieth century. The report projects that the amount will double again over the next 50 years. These transfers of material contribute significantly to the pollution of aquatic ecosystems, acid rain, and global climate change.

STUDY BREAK 52.4 <

1. **What is the greenhouse effect, and how does an increase in atmospheric CO_2 concentration affect it?**
2. **What human activities release the most CO_2 into the atmosphere?**
3. **What agricultural practices contribute to the disruption of the nitrogen cycle?**

>

THINK OUTSIDE THE BOOK

Examine the ingredients listed on the labels of three cleaning products that you use in your household or residence hall. Search the Internet or sources in your library for information about how the disposal of these compounds on land or in sewage might contribute to the disruption of a biogeochemical cycle.

UNANSWERED QUESTIONS

How does the carbon cycle of a forest respond to climate change and urbanization?

As you've read in this chapter, human influences on the environment can have dramatic unforeseen consequences for ecosystems, altering energy flow and nutrient cycling. Given the complexity of ecosystems—the myriad scales of influence and multiple interactions among the organisms, the physical environment, and climate change variables—the precise response of an ecosystem is difficult to predict, even with advanced ecosystem models. We do known with certainty, however, that the carbon, nitrogen, and water cycles of forested ecosystems in the northeastern United States are changing, and they are likely to continue to do so.

Carbon cycle research in forested ecosystems often entails building an ecosystem model from quantitative data on the various pools and fluxes of carbon in the ecosystem and how these change with time. Scientists then correlate these changes with the environmental conditions and derive a mechanistic understanding that they can use to make predictions about how the ecosystem will respond to future changes. In theory, *gross primary productivity* (GPP) should be predictable from a basic understanding of photosynthesis and a general description of the ambient environmental conditions. In practice, however, the complexity of canopy architecture and leaf positioning, the timing of recurring natural phenomena, and the effects of herbivory and leaf losses from abiotic factors all make accurate predictions more difficult. Furthermore, the problem is dynamic because age-related changes in stand structure, disturbance, invasion, drought, seasonality, and pests or pathogens all add spatial and temporal complexities. Scientists should be able to predict *net primary productivity* (NPP), a key parameter used by ecologists to classify the world's ecosystems, from measurements of the cellular respiration and the relative abundances of representative organisms from the ecosystem.

Quantifying GPP and NPP on a large spatial scale can be challenging, and discovering the underlying mechanisms that control ecosystem responses to changes in environmental conditions is difficult. For example, studies at Black Rock Forest, a deciduous-oak-dominated forest in New York State, revealed that temporal heterogeneity (seasonal variation in leaf and stem respiration) and spatial heterogeneity (variations in canopy and hill slope position) are important factors that must be included in models of canopy respiration. Nevertheless, some simplifications may be possible. For example, although the basal rate of respiration is quite variable and subject to acclimation, it may be predictable from basic plant properties such as their nitrogen concentrations. Furthermore, the temperature coefficient of respiration is relatively constant, greatly simplifying the construction of an ecosystem model. To consider the impact of tree respiration on ecosystem form and function fully, my research team experimented with models that explicitly consider physiological linkages between photosynthesis and respiration, as mediated by leaf carbohydrate pools. We found that when we included direct linkages to carbon gain in the analysis, the model correctly predicted a large (23%) decrease in the estimated nighttime canopy respiration during the growing season. This result emphasizes the need for a process-based modeling approach when estimating forest productivity.

Our research at Black Rock Forest has also demonstrated that human activities in New York City (60 miles to the south) may be influencing tree growth in both urban and rural areas, with significant changes in seedling size, biomass allocation, herbivory, stomatal densities, nutrient concentrations, efficiency of water use, and rates of key physiological processes such as photosynthesis and respiration. Urbanization has a clear effect on the land area developed, but current research is showing that human activities in urban areas also influence forested ecosystems in the surrounding rural areas. Understanding how human activity, climate change, and forest ecosystems interact is crucial if we are to make prudent and sustainable development decisions, preserving the health of the ecosystems and the services they provide.

Think Critically

Through what mechanisms might the presence of a large city influence primary productivity in a forest 60 miles away?

Kevin Griffin

Kevin Griffin is an associate professor at Columbia University's Lamont-Doherty Earth Observatory. His research centers on processes in plant and ecosystem ecology, the goal of which is to increase our understanding of both the role of vegetation in the global carbon cycle and the interactions between the carbon cycle and Earth's climate system. To learn more about Dr. Griffin's research, go to http://www.ldeo.columbia.edu/.

REVIEW KEY CONCEPTS

Go to **CENGAGENOW** at www.cengage.com/login to access quizzing, animations, exercises, articles, and personalized homework help.

52.1 Modeling Ecosystem Processes

- Ecosystems include biological communities and the abiotic environmental factors with which they interact (Figure 52.1).
- Interconnected grazing and detrital food webs define the pathways along which energy and nutrients move through the biological components of an ecosystem (Figure 52.2).
- Compartment models describe energy flow and nutrient cycling in ecosystems (Figure 52.3).
- Simulation models are interlocking mathematical equations that define the relationships between populations and between populations and the physical environment. They allow researchers to predict the effects of changes in ecosystem structure and function.

52.2 Energy Flow and Ecosystem Energetics

- Photosynthesis converts only a small portion of the solar energy that reaches Earth into chemical energy.
- An ecosystem's gross primary productivity is the rate at which producers convert solar energy into chemical energy. Producers use some energy for respiration; some is converted to heat; and some remains in the ecosystem as net primary productivity.

- Primary productivity is measured in units of energy captured or biomass produced per unit area per unit time. Net primary productivity indexes the energy available to support heterotrophs. Ecosystems vary in productivity and in their contributions to Earth's total productivity (Figure 52.4, Table 52.1).
- On land, primary productivity is limited by the availability of sunlight, water, and nutrients; temperature; and the amount of photosynthetic tissue present. In marine and aquatic ecosystems, primary productivity is limited when sunlight and nutrients are not available in the same place (Figures 52.5 and 52.6).
- Only a fraction of the energy at any trophic level is converted into biomass at higher trophic levels. As energy passes through a food web, an average of 90% is lost at each transfer between trophic levels, limiting the number of trophic levels that a food web can support (Figure 52.7).
- Ecological pyramids portray the effects of energy losses. For terrestrial ecosystems, pyramids of energy, biomass, and numbers generally have broad bases and narrow tops (Figures 52.8–52.10).
- Feeding by predators can influence primary productivity through a trophic cascade (Figure 52.11).

Animation: The role of organisms in an ecosystem

Animation: Food webs

Animation: Energy flow at Silver Springs

52.3 Nutrient Cycling in Ecosystems

- Earth is a closed system with respect to matter.
- Nutrients circulate in biogeochemical cycles between living organisms and nonliving reservoirs. Some biogeochemical cycles are atmospheric; others are sedimentary.
- Water circulates through the atmosphere, oceans, and terrestrial and freshwater ecosystems in a global hydrologic cycle. Water evaporates from the oceans and land and falls as precipitation. Runoff and streamflow return excess precipitation from the land to the oceans (Figure 52.12).
- The carbon cycles in terrestrial and aquatic ecosystems are linked through an atmospheric pool of CO_2, which primary producers assimilate. Respiration returns carbon to the atmosphere as CO_2. Earth's largest reservoir of carbon is unavailable in sedimentary rock. Other large reservoirs include coal, oil, and peat as well as dissolved bicarbonate and carbonate ions in seawater (Figure 52.13).
- Nitrogen is cycled between living organisms and an atmospheric pool of nitrogen gas. Bacteria and cyanobacteria make nitrogen available to the food web through the processes of nitrogen fixation, ammonification, and nitrification. Denitrification converts nitrogen compounds to molecular nitrogen, which enters the atmosphere (Figure 52.14, Table 52.2).
- Phosphorus undergoes a sedimentary cycle. Weathering and erosion of rock make phosphorus available; it is leached from soil and carried to the ocean. Dissolved phosphates precipitate out of seawater, forming insoluble deposits, which are eventually uplifted by tectonic processes (Figure 52.15).

Animation: Hydrologic cycle

Animation: Hubbard Brook experiment

Animation: Carbon cycle

Animation: Nitrogen cycle

Animation: Phosphorus cycle

52.4 Human Disruption of Ecosystem Processes

- Certain gases in the lower atmosphere create a greenhouse effect that traps heat near Earth's surface. Our reliance on the combustion of fossil fuels and wood has increased the atmospheric concentration of CO_2 substantially in recent decades (Figure 52.16), enhancing the greenhouse effect and raising the average surface temperature of the planet. Global warming has already had profound effects on the landscape and biological systems, and these effects are likely to increase for the foreseeable future.
- Human activities—particularly irrigation, the use of synthetic fertilizers, livestock farming, and the use of combustible fuels—have disrupted the nitrogen cycle. The consequences of shifting nitrogen among its natural reservoirs include the pollution of aquatic ecosystems, acid rain, and global climate change.

Animation: Greenhouse effect

Animation: Greenhouse gases

Animation: Carbon dioxide and temperature

UNDERSTAND AND APPLY

Test Your Knowledge

1. Which of the following events would move energy and material from a detrital food web into a grazing food web?
   a. A beetle eats the leaves of a living plant.
   b. An earthworm eats dead leaves on the forest floor.
   c. A robin catches and eats an earthworm.
   d. A crow eats a dead robin.
   e. A bacterium decomposes the feces of an earthworm.
2. The total dry weight of plant material in a forest is a measure of the forest's:
   a. gross primary productivity.
   b. net primary productivity.
   c. cellular respiration.
   d. standing crop biomass.
   e. ecological efficiency.
3. Which of the following ecosystems has the highest rate of net primary productivity?
   a. open ocean
   b. temperate deciduous forest
   c. tropical rain forest
   d. desert and thornwoods
   e. cultivated land
4. Endothermic animals exhibit a lower ecological efficiency than ectothermic animals because:
   a. endotherms are less successful hunters than ectotherms.
   b. endotherms eat more plant material than ectotherms.
   c. endotherms are larger than ectotherms.
   d. endotherms produce fewer offspring than ectotherms.
   e. endotherms use more energy to maintain body temperature than ectotherms.
5. The amount of energy available at the highest trophic level in an ecosystem is determined by:
   a. only the gross primary productivity of the ecosystem.
   b. only the net primary productivity of the ecosystem.
   c. the gross primary productivity and the standing crop biomass.
   d. the net primary productivity and the ecological efficiencies of herbivores.
   e. the net primary productivity and the ecological efficiencies at all lower trophic levels.

6. Some freshwater and marine ecosystems exhibit an inverted pyramid of:
   a. biomass.
   b. energy.
   c. numbers.
   d. turnover.
   e. ecological efficiency.
7. Which process moves nutrients from the available organic compartment to the available inorganic compartment?
   a. respiration
   b. erosion
   c. assimilation
   d. sedimentation
   e. photosynthesis
8. Which of the following materials has a sedimentary cycle?
   a. water
   b. oxygen
   c. nitrogen
   d. phosphorus
   e. carbon
9. Which of the following statements is supported by the results of studies at the Hubbard Brook Experimental Forest?
   a. Most of the energy captured by primary producers is lost before it reaches the highest trophic level in an ecosystem.
   b. Deforested watersheds experience significantly less runoff than undisturbed watersheds.
   c. Deforested watersheds lose more calcium and nitrogen in runoff than undisturbed watersheds.
   d. Nutrients generally move through biogeochemical cycles very quickly.
   e. Deforested watersheds generally receive more rainfall than undisturbed watersheds.
10. Nitrogen fixation converts:
   a. atmospheric molecular nitrogen to ammonia.
   b. nitrates to nitrites.
   c. ammonia to molecular nitrogen.
   d. ammonia to nitrates.
   e. nitrites to nitrates.

Discuss the Concepts

1. A lake near your home became overgrown with algae and pondweeds a few months after a new housing development was built nearby. What data would you collect to determine whether the housing development might be responsible for the changes in the lake?
2. Some politicians question whether recent increases in atmospheric temperature result from our release of greenhouse gases into the atmosphere. They argue that atmospheric temperature has fluctuated widely over Earth's history, and the changing temperature is just part of an historical trend. What information would allow you to refute or confirm their hypothesis? In addition, describe the pros and cons of reducing greenhouse gases as soon as possible versus taking a "wait and see" approach to this question.
3. If you could design the ideal farm animal—one that was grown as food for humans—from scratch, what characteristics would it have?
4. If you were growing a vegetable garden, identify the factors that might affect its primary productivity. How would you increase productivity? Identify some possible consequences of your gardening activities to nearby ecosystems.

Design an Experiment

Design an experiment to test the hypothesis that predators in an aquatic ecosystem regulate the ecosystem's primary productivity. Establish as many experimental ponds as you wish, and imagine stocking them with organisms at different trophic levels. If the hypothesis is correct, describe the results you would expect to record from each of your experimental treatments.

Interpret the Data

Amy Rosemond of Vanderbilt University and two colleagues at the Oak Ridge National Laboratory conducted a study of primary productivity in the algal community of Walker Branch, a stream in eastern Tennessee. In their experiment, they compared the productivity of algae that were eaten by snails (grazed) to the productivity of algae that were protected from herbivores (ungrazed) under three experimental treatments: (1) the addition of nitrogen to the stream; (2) the addition of phosphorus to the stream; and (3) the addition of both nitrogen and phosphorus to the stream. They compared the results of these treatments to control treatments of both grazed and ungrazed algae that received no nutrient enrichment. How would you interpret their results, presented in the accompanying graph? Was algal productivity controlled by the presence of herbivores, the addition of nutrients, or a combination of the two factors?

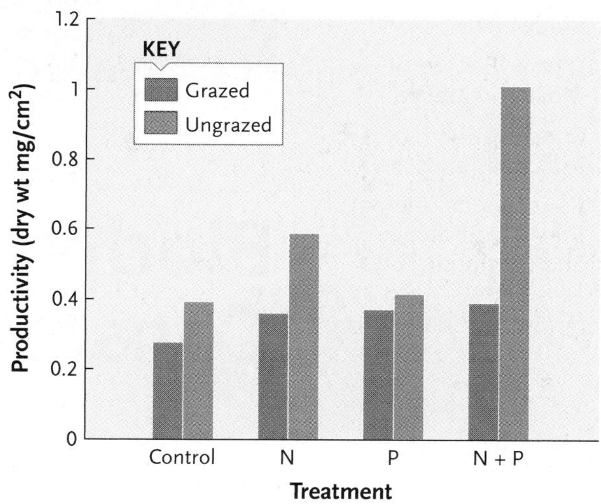
KEY
Grazed
Ungrazed
Productivity (dry wt mg/cm²)
1.2
1
0.8
0.6
0.4
0.2
0
Control
N
P
N + P
Treatment

Source: A. D. Rosemond et al. 1993. Top-down and bottom-up control of stream periphyton: effects of nutrients and herbivores. *Ecology* 74:1264–1280.

Apply Evolutionary Thinking

The dramatic loss of energy as it is transferred upward through a food web produces pyramids of energy, biomass, and numbers that are very narrow at the top. Which agent of evolution (Section 20.3) would you expect to influence allele frequencies in the relatively small populations that are typical of top predators? What are the long-term *evolutionary* effects of small population size?

Express Your Opinion

Emissions from motor vehicles are a major source of greenhouse gases. Many people buy large vehicles that use more fuel but are viewed as safer and more useful. Should such vehicles be taxed extra to discourage sales and offset their environmental costs? Can we expect the emergence of better fuels as well as more of the fuel-efficient, larger vehicles that are becoming available? Go to www.cengage.com/login to investigate both sides of the issue and then vote.

53

STUDY OUTLINE

JH Pete Carmichael/Getty Images Inc.

Florida panther *(Puma concolor coryi)*. Fewer than 100 individuals of this endangered subspecies survive.

Biodiversity and Conservation Biology

Why It Matters. . . Someone seems to have disappeared. Investigators thoroughly checked the missing subject's known haunts, but found no trace. They questioned others in the neighborhood, but came up with few leads. The subject was last seen alive in 1978. With so cold a trail to follow, investigators reluctantly marked the case file "Missing and Presumed Extinct."

The subject in this case was Miss Waldron's red colobus monkey, *Procolobus badius waldroni* **(Figure 53.1)**. Named for a traveling companion of the taxonomist who first described it in 1933, this distinctively colored subspecies lived in large and noisy social groups in a remote forest on the border between Ivory Coast and Ghana in West Africa.

John Oates of the City University of New York led a research team that tried to locate Miss Waldron's red colobus. They used every imaginable method, including visual and auditory censuses, searching for scat (dung) in natural habitats, interviewing local people, and looking in marketplaces where monkey meat is commonly traded. In 2000, more than 20 years after the last confirmed sighting, the researchers concluded that this monkey is probably extinct. A later search by a member of the team, William S. McGraw of Ohio State University, found the skin of one monkey that a hunter had shot six months before. But McGraw searched in vain for a living monkey, and he concluded that even if a few are still alive, the population is so small that continued hunting will surely eliminate it.

Procolobus badius waldroni may be the first primate subspecies to become extinct in more than 100 years—and only the second in the last 500 years.

Stephen D. Nash/Conservation International

FIGURE 53.1
Miss Waldron's red colobus. *Procolobus badius waldroni*, which weighed about 10 kg, may be the first primate subspecies to become extinct in more than 100 years.

1975
2001

NASA/Goddard Space Flight Center Scientific Visualization Studio

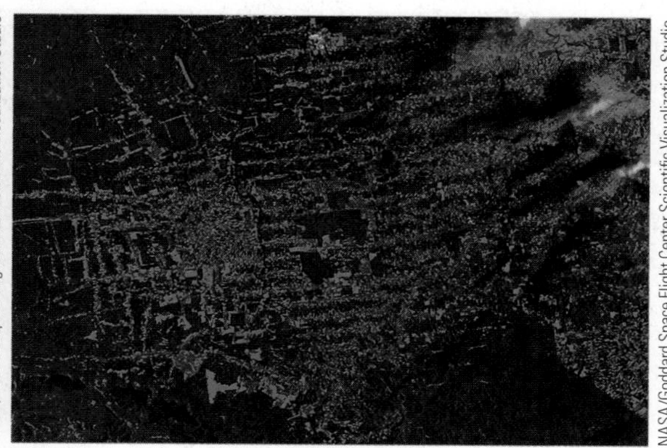

NASA/Goddard Space Flight Center Scientific Visualization Studio

FIGURE 53.2
Deforestation in the Amazon Basin. Satellite photos of Rondonia, in the Brazilian Amazon, show how much of the Amazon forest was cut (light green) between 1975 and 2001. Each photo illustrates an area approximately 60 by 85 km.

Monkeys and other primates are among the most closely monitored and protected species on Earth. Nonetheless, Oates and his colleagues concluded, these monkeys probably became extinct because they were hunted locally for food by a growing human population and because humans have destroyed their natural habitats.

Miss Waldron's red colobus is just one of many species driven to extinction every year. Current threats to biodiversity—all of which ultimately result from our disruption of natural populations, communities, and ecosystems—are massive. The likely loss of this monkey should warn us that many taxa are at risk, even those that are most rigorously protected.

When ecologists speak of **biodiversity,** they are referring to the richness of living systems. At the most fundamental level of biological organization, biodiversity encompasses the *genetic variation* that is raw material for adaptation, speciation, and evolutionary diversification (see Chapters 20 and 21). At a higher level of organization, biodiversity includes *species richness* within communities (see Section 51.3). The number and variety of species within a community influence its overall characteristics, population interactions, and trophic structure. Finally, biodiversity exists at the *ecosystem level.* Complex networks of interactions bind species in an ecosystem together, and because different ecosystems interact within the biosphere, damage to one ecosystem can reverberate through others.

In this chapter we first describe how human activities threaten biodiversity and reflect on why we should protect it. We then consider theoretical and practical approaches to conservation biology, the scientific discipline that focuses on preserving Earth's biological resources. <

53.1 The Biodiversity Crisis on Land, in the Sea, and in River Systems

Biodiversity is declining dramatically, perhaps faster than ever before in Earth's history. In this section, we describe the three broadest threats to biodiversity: the clearing of forests; the commercial overexploitation of marine fish populations; and hydrologic alterations of freshwater ecosystems. Bear in mind that these and other challenges are exacerbated by global climate change, which we have discussed in previous chapters.

Deforestation Disrupts the Carbon Cycle and May Lead to Desertification

Forests are among the habitats that humans most frequently clear and convert. According to the United Nations Forest Resources Assessment released in 2005, global deforestation occurs at a rate of 13 million hectares per year, or 25 hectares per minute. In other words, an area of forest equivalent to 46 football fields is cleared of all trees every minute of every day. An updated Forest Resources Assessment is being conducted in 2010, and experts expect its results to be even more alarming than those published in 2005.

Deforestation does not occur uniformly across the globe. Today, more than 90% of deforestation occurs in tropical regions, where many groups of organisms exhibit their highest diversity (described in Section 51.7). Forests are most often cut to clear land for grazing livestock; as a result, a few species of domesticated animals and the grasses they consume replace what had been a species-rich community. Brazil has experienced the most extensive recent damage, accounting for 25% of all deforestation during the late twentieth century **(Figure 53.2).** This assessment is particularly troubling because Brazil contains approximately 27% of the planet's total aboveground woody biomass.

Compounding the direct environmental damage, most tropical forests are burned as they are cleared, a process that adds CO_2 to the atmosphere, enhancing the greenhouse effect and increasing the rate of global warming (see Section 52.4). According to the Intergovernmental Panel on Climate Change, a Nobel Prize-winning agency of the United Nations, forest cutting now contributes nearly 20% of all greenhouse gases released into the atmosphere. Ironically, intact forests remove substantial quantities of carbon dioxide from the atmosphere, a capacity that is diminished with every tree felled.

Once a forest is cut, heavy grazing or farming drains nutrients from the soil. To remain productive, even the best agricultural or grazing lands require either the application of fertilizers or long periods during which the land lies fallow, allowing plants to replenish the soil naturally. Unfortunately, the soil where tropical forests grow is often of marginal value (for reasons described in Chapter 49), and it is rapidly degraded; it becomes

FIGURE 53.3
Desertification in the Sahel. **(A)** A satellite photo taken near the end of the dry season in June 2005 illustrates the severe desertification in parts of the Sahel region of Africa. Dark green areas are densely vegetated; light green areas are sparsely vegetated, and sand-colored areas are barren. **(B)** People who live in this region can barely eke out a living on the land.

A. **The Sahel region of Africa**

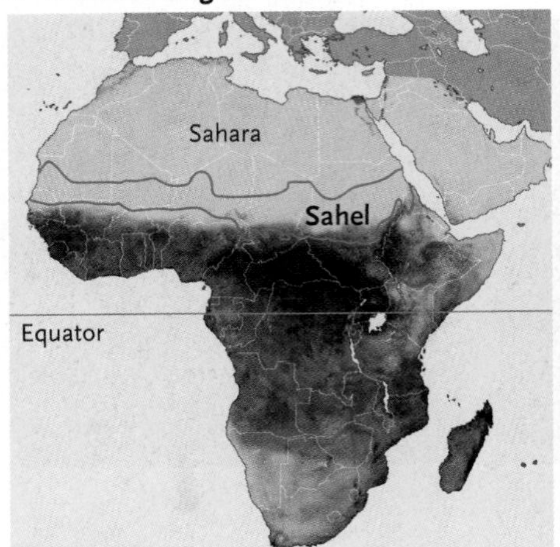
Sahara
Sahel
Equator

NASA's Earth Observatory

B. **Women preparing millet, a grain, in the Sahel**

Romano Cagnoni/Peter Arnold, Inc.

hard, even more nutrient-poor, unable to retain water, and likely to wash away.

When large tracts of subtropical forest are cleared and overused, the land often undergoes **desertification:** the groundwater table recedes to deeper levels; less surface water is available for plants; soil accumulates high concentrations of salts (a process called *salinization*); and topsoil is eroded by wind and water. In other words, the habitat is converted to desert.

Desertification speeds the loss of biodiversity locally, sometimes eliminating entire ecosystems. For example, desertification has decimated habitats in the Sahel region of Africa, just south of the Sahara **(Figure 53.3).** Excessive grazing of cattle and goats by an ever-expanding human population is the main reason for the Sahara's southward expansion at a rate of 5.5 to 8 km per year. Because the sand dunes of the expanding desert shift constantly, agriculture and grazing are nearly impossible, resulting in frequent famines among the inhabitants.

Desertification and salinization have also begun in the Everglades, a unique, shallow "river of grass" that covers much of southern Florida. The amount of fresh water flowing through South Florida to the Everglades has decreased approximately 70% since 1948, when an extensive network of canals and levees was built to reduce flooding. The rapidly growing human population in South Florida contributes directly to desertification, as groundwater is tapped for domestic use and to irrigate lawns, golf courses, and agricultural fields. Salt water from the Gulf of Mexico now intrudes into the water table, causing salinization of the soil. The Comprehensive Everglades Restoration Plan (CERP), approved by the U.S. Congress in 2000, seeks to restore the natural flow of the Everglades by 2030. This project may halt or reverse the desertification process.

Sadly, deforestation, desertification, and global warming reinforce each other in a positive feedback cycle (see *Focus on Applied Research* in Chapter 52). If scientists' projections are correct, desertification will lead to an increase in the average global temperature, speeding evaporation and the retreat of forests, which, in turn, will increase rates of desertification. If deforestation and desertification continue, we will soon lose a large proportion of Earth's forests and face a decrease in the area of habitable land.

Overexploitation Can Drive Species to Extinction

Many local extinctions result from **overexploitation,** the excessive harvesting of an animal or plant species. At a minimum, overexploitation leads to declining population sizes in the harvested species. In the most extreme cases, a species may be wiped out completely. Moreover, overexploitation can foster evolutionary changes in the exploited population, much the way guppies respond to natural predators in the streams of Trinidad (described in *Focus on Basic Research* in Chapter 50).

Humans have overexploited populations in every habitat we occupy. Today, overexploitation severely threatens marine ecosystems, the only environment from which we routinely harvest predators (such as tuna) as food. The fishery on the Grand Banks off the coast of Newfoundland, Canada, provides a sad example **(Figure 53.4).** For hundreds of years, fishermen used traditional line and small-net fishing to harvest a large but sustainable catch. During the twentieth century, however, new technology allowed them to locate and exploit schools of fishes more efficiently. As a result, roughly half the fish species harvested there are now overfished. Haddock *(Melanogrammus aeglefinus)* and yellowtail flounder *(Limanda ferruginea)* have been essentially eliminated from the Grand Banks; their populations will probably never recover. And because fishers preferentially harvest the oldest and largest individuals, which fetch a higher market price, Atlantic cod *(Gadus morhua)* now mature at a younger age (three years compared with five or six years) and smaller size.

As a consequence of overfishing, the average yield of the Grand Banks has declined to less than 10% of the highest historic levels. In the mid-1960s, Atlantic cod yielded a minimum of 350,000 tons per year. By the mid-1970s, the catch dropped to 50,000 tons per year. The Canadian government finally closed the fishery in 1993, after the cod catch fell below 20,000 tons for several consecutive years. But the damage had already been done: the most heavily exploited species are less marketable because of their smaller size, fish populations have decreased to dangerously low levels, and the fishing industry is itself imperiled. This sequence of events has been replicated in fisheries around the

A. **The Grand Banks**

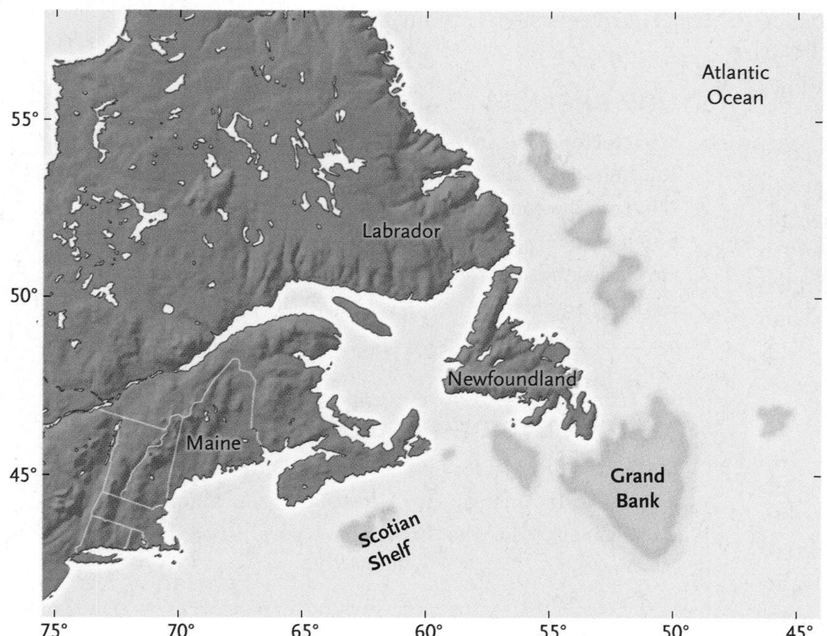

Atlantic Ocean
Labrador
Newfoundland
Maine
Grand Bank
Scotian Shelf
55°
50°
45°
75°
70°
65°
60°
55°
50°
45°

B. **Atlantic cod**

Labat J. M./Lanceau Y./Peter Arnold, Inc.

FIGURE 53.4
Overexploitation of North Atlantic fisheries. **(A)** The Grand Banks (sand-colored shading) were severely overfished in the late twentieth century, leading to the near extinction of many species, including the **(B)** Atlantic cod *(Gadus morhua)*.

world. Indeed, in a report published in 2003, Ransom A. Myers and Boris Worm of Dalhousie University in Nova Scotia estimated that modern fishing techniques have reduced the biomass of large predatory fishes by about 90% in marine ecosystems.

Hydrologic Alterations Endanger Freshwater and Wetland Ecosystems

Rivers have always played a key role in the development of human settlements because they provide a source of fresh water, a place to discard wastes, and a means to transport goods. Since ancient times, humans have also dammed rivers to capture reliable supplies of fresh water. When the human population—and the dams they built—were small, these **hydrologic alterations** (that is, changes to the pathways through which water moves in the hydrologic cycle) had primarily local effects. More recently, we have constructed massive dams that capture vast quantities of fresh water in reservoirs **(Figure 53.5)**. We distribute water from these reservoirs for agricultural, industrial, and domestic uses. The dams also generate hydroelectric power and allow us to control water flow to mitigate flooding of low-lying land. Today, the large scale and ubiquity of these hydrologic alterations have made freshwater ecosystems among the most endangered on Earth.

The damming of rivers and the diversion of their flow wreaks havoc on many interconnected ecosystems. As you learned in Section 49.5, the physical characteristics of rivers change predictably from the headwaters of streams to the estuaries where they empty into the sea. River-dwelling organisms are adapted to specific physical conditions—such as temperature, depth, and flow rate—that are characteristic of each section of the river system. And downstream, the flow of water supports distinct communities of organisms, each adapted to the different environments in floodplains, wetlands, and estuaries.

In 2002, Stuart E. Bunn and Angela H. Arthington of Griffith University in Australia identified four ways in which hydrologic alterations threaten freshwater biodiversity. First, the flow rate and volume of rivers are key determinants of their physical habitats, which have a major impact on the organisms that live there. For example, before it was dammed, the River Otra in Norway

1987
NASA

2006
NASA

FIGURE 53.5
Three Gorges Dam. In 2008, China completed the Three Gorges Dam, the world's largest hydroelectric dam, across the Yangtze River in Hubei Province. The dam is 2,300 m long and 185 m high. The reservoir behind it measures 660 km by 1.1 km and holds 39.3 km^3 of water. More than 1.25 million people had to be relocated from areas inundated by the reservoir.

experienced low winter water flows, but raging summer floods. These conditions established a regular pattern of disturbance that eliminated many rooted plants from the riverbed. Now dammed, the river has a more regulated flow regime that allows a huge accumulation of plant biomass.

Second, the life histories of aquatic species, which evolved in response to natural flow patterns, are disrupted by changes in river flows. For example, reproduction by many aquatic invertebrates and fishes is triggered by temperature and day-length cues. Because the water released through dams is often drawn from the depths of the reservoirs behind them, it is colder than the natural flow, changing the cues available to organisms. Researchers working in China discovered that the cold water released by dams delayed spawning by as much as 30 days in some fish species.

Third, dams reduce a river system's "connectivity" (that is, the continuity of flow through a river and its streams and tributaries). Reduced connectivity prevents fishes and other animals from migrating freely through a river system. For example, salmon undertake a spawning migration from the sea, swimming upstream into the tributaries and streams where they reproduce. Dams hamper this already difficult upstream journey. In the Pacific Northwest, more than 400 hydroelectric dams on the Columbia River system prevent many salmon from reaching their spawning grounds; they have reduced the breeding habitat for Chinook salmon *(Onchorhynchus tshawytscha)* by 75%. Similar problems have eliminated migratory fish species from rivers throughout the world.

Finally, dams and reservoirs facilitate the introduction and success of non-native species that thrive in disturbed habitats (discussed further below). For example, several species of large fishes collectively described as "Asian carp" have become established in the Mississippi River and its tributaries. These fishes feed voraciously on plankton. For years ecologists feared that they would spread into the Great Lakes, where they would outcompete native fish species that are the basis of a large fishing industry. In 2002 and 2004, the U.S. Army Corp of Engineers built two electric barriers in a canal that connected a tributary of the river to Lake Michigan, hoping to block the fishes' advance. But in late 2009 and early 2010, scientists detected genetic material from the carp on the far side of the fences. Some politicians have called for the closure of the canal as the only way to prevent the advance of these fishes into Lake Michigan; as of this writing, that proposal is the subject of several lawsuits.

Freshwater ecosystems are now under severe pressure. In the worst cases, human-induced hydrologic alterations have practically eliminated them: the Nile and the Colorado River now rarely discharge much water into the sea. Even in less dramatic cases, the effects of hydrologic alterations on biodiversity have been profound. Freshwater fish species have experienced marked declines in the last few decades. One 2006 estimate suggested that 56% of the freshwater fish species endemic to the Mediterranean region, more than 30% of the native species in North America, and 25% of those found in East Africa are now threatened with extinction. The status of freshwater invertebrates, though not as well documented, is probably comparable. Conservation biologists rank the restoration of natural flow patterns in river systems among their highest priorities.

STUDY BREAK 53.1

1. **What factors have increased the likelihood of desertification in southern Florida?**
2. **What are the consequences of the overexploitation of fish populations?**
3. **How does the construction of a dam disrupt the lives of river-dwelling organisms?**

THINK OUTSIDE THE BOOK

Search the Internet for updates about the spread of Asian carp in North America. Have they successfully invaded the Great Lakes? If so, what impact have they had on the native lake fishes?

53.2 Specific Threats to Biodiversity

Although the clearing of tropical forests, overexploitation of marine fisheries, and damming of rivers endanger entire ecosystems, many other human activities imperil natural populations. In this section, we briefly describe some of these threats.

Habitat Fragmentation Threatens Many Populations

When humans first colonize a pristine habitat, they build roads and then clear isolated areas for specific uses. Although this pattern of development initially affects only local populations, the negative effects spread rapidly to a regional scale. The remaining areas of *intact* habitat are inevitably reduced to small, isolated patches, a phenomenon that ecologists describe as **habitat fragmentation.**

Habitat fragmentation is a threat to biodiversity because small habitat patches can sustain only small populations of organisms. As you learned in Section 50.4, a habitat's *carrying capacity,* the maximum population size that it can support, varies with available resources. Populations that occupy small habitat patches inevitably experience low carrying capacities, a problem that is especially acute for species at the higher trophic levels (see Section 52.2). Furthermore, fragmented habitat patches are often separated by unsuitable habitat that organisms may be unable or unwilling to cross. As a result, individuals from one isolated population are unlikely to migrate into another, reducing gene flow between them. The combination of small population size and genetic isolation reduces genetic variability and fosters extinction (see Section 20.3).

In addition to reducing the amount of undisturbed habitat, habitat fragmentation jeopardizes the quality of the habitat that

FIGURE 53.6 Experimental Research

Predation on Songbird Nests in Forests and Forest Fragments

Question: Are songbird nests in small forest fragments more likely to be found by predators than nests in large forest patches?

Experiment: Wilcove placed between 13 and 50 artificial bird nests, each containing three quail eggs, in three habitat types: large areas of intact forest, rural forest fragments, and suburban forest fragments. He placed about half the nests at each study site on the ground at the base of a tree or shrub and half the nests 1 to 2 m above the ground in a sapling or shrub. He checked the nests after seven days to determine what proportion of the nests had been subjected to predation.

Result: Predators generally found a larger proportion of the artificial bird nests in small forest fragments than they did in large forest patches.

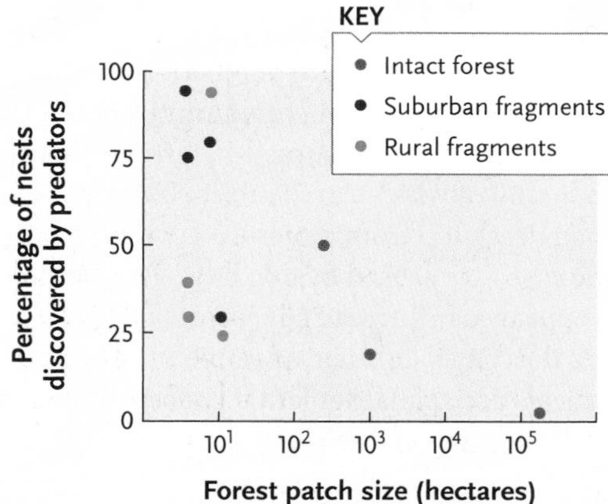
KEY
Intact forest
Suburban fragments
Rural fragments
Percentage of nests discovered by predators
100
75
50
25
0
10^1
10^2
10^3
10^4
10^5
Forest patch size (hectares)

Conclusion: Songbird nests are much more likely to suffer from predation in small forest fragments than they are in large patches of intact forest.

Source: D. S. Wilcove. 1985. Nest predation in forest tracts and the decline of migratory birds. *Ecology* 66:1211–1214.

remains. Human activities create noise and pollution that spread into nearby areas. The removal of natural vegetation disrupts the local physical environment, exposing the borders of the remaining habitat to additional sunlight, wind, and rainfall. Increased runoff compacts the soil and makes it waterlogged. These phenomena are collectively described as **edge effects.**

The effects of habitat fragmentation are often profound. For example, populations of forest-dwelling, migratory songbirds have declined markedly in eastern North America since the late 1940s, largely because of habitat fragmentation in their North American breeding grounds and in their Caribbean and South American wintering grounds.

In 1994, Scott K. Robinson of the Illinois Natural History Survey and David S. Wilcove of the Environmental Defense Fund identified three factors that decrease populations of migratory songbirds in fragmented breeding habitats. First, small forest patches often lack specific habitat types—such as streams, cool ravines, or dense ground cover—that many songbird species require. Second, songbirds breeding in forest patches are more likely to suffer from brood parasitism (described in the opening of Chapter 51) by brown-headed cowbirds *(Molothrus ater)* than are those breeding in intact forests. Parasitized songbirds rear fewer than half as many young as they might otherwise raise, and their populations decline accordingly.

The third factor that reduces songbird numbers in forest fragments is increased nest predation by blue jays *(Cyanocitta cristata),* American crows *(Corvus brachyrhynchos),* common grackles *(Quiscalus quiscula),* squirrels (genus *Sciurus*), raccoons *(Procyon lotor),* and domestic dogs and cats. These predators, which feed on songbird eggs and young, are now superabundant in rural and suburban areas, and they enter adjacent forest fragments in search of an easy meal. Wilcove tested the predation hypothesis experimentally by placing artificial nests with quail eggs in intact forests and in forest fragments. Although he did not observe predation directly, he found that predators discovered only 2% of the nests in the largest intact forest, but they often found 50% or more of the nests placed in small, suburban forest fragments **(Figure 53.6).**

Many Forms of Pollution Overwhelm Species and Ecosystems

The release of **pollutants**—materials or energy in forms or quantities that organisms do not usually encounter—poses another major threat to biodiversity.

Although chemical pollutants, the by-products or waste products of agriculture and industry, are released locally, many spread in water or air, sometimes on a continental or global scale. Within North America, for example, winds carry airborne pollutants from coal-burning power plants to the Northeast **(Figure 53.7).** Sulfur dioxide (SO_2), which dissolves in water vapor and forms sulfuric acid, falls as **acid precipitation,** acidifying soil and bodies of water. Many lakes in northeastern North America have experienced a precipitous drop in pH from historical readings near 6 to values that are now well below 5—a 10-fold increase in acidity. Although the lakes once harbored lush aquatic vegetation and teemed with fishes, they are now crystal clear and nearly devoid of life.

As residents of major cities and industrial areas know all too well, wastes produced by the combustion of fossil fuels in factories and automobile engines cause terrible local pollution, increasing rates of asthma and other respiratory ailments. Some airborne pollutants, notably CO_2, also join the general atmospheric circulation, where they contribute to the greenhouse effect and global warming.

Like air pollution, water pollution originates locally but has a much broader impact. Oil spills, for example, disrupt local ecosystems, killing most organisms near the spill. Because oil floats on water, it spreads rapidly to nearby areas. An explosion and fire on the Deepwater Horizon oil rig in April 2010 allowed many millions of gallons of oil to spill into the Gulf of Mexico. The un-

FIGURE 53.7

Acid precipitation. Coal-burning power plants (indicated by black dots) release air pollution that is carried northeast, where it falls as acid precipitation. The map shows the average pH of rainfall.

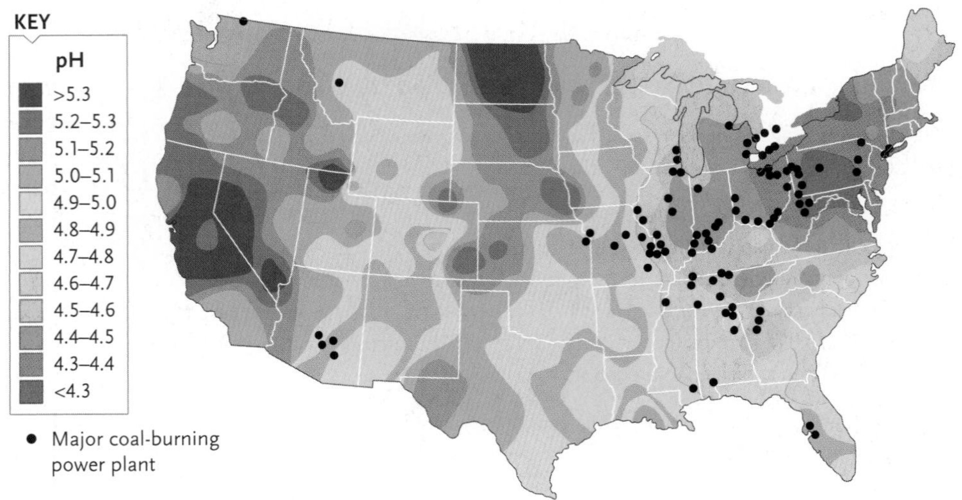
KEY
pH
>5.3
5.2–5.3
5.1–5.2
5.0–5.1
4.9–5.0
4.8–4.9
4.7–4.8
4.6–4.7
4.5–4.6
4.4–4.5
4.3–4.4
<4.3
Major coal-burning power plant

capped well continued to spew oil for three months, until it was partially plugged in July 2010. The oil spill had a devastating short-term effect on organisms in the Gulf and adjacent wetlands; scientists are still unable to predict its long-term effects on this delicate ecosystem.

Pollution can also have serious effects on terrestrial ecosystems. As a recent disaster in India, Nepal, and Pakistan illustrates, the application of synthetic compounds to agricultural fields or livestock can have dire and far-reaching consequences. For thousands of years, enormous populations of vultures (several *Gyps* species)—estimated at more than 40 million birds—consumed the abandoned carcasses of farm animals across South Asia. In the early 1990s, however, farmers began to administer diclofenac, a new and inexpensive anti-inflammatory drug, to injured livestock. Within a few years, vultures began to disappear; in 2006, scientists estimated that their populations had declined by more than 97%. Researchers determined that diclofenac, which causes fatal kidney failure in birds, was responsible for the deaths: vultures were ingesting substantial doses of the drug from the livestock carcasses they ate. All vulture species in South Asia are now on the verge of extinction, and although governments in the region have banned the sale of diclofenac, wildlife experts say that the vulture populations are unlikely to recover soon, if ever.

The decline in vulture populations has had a disastrous impact on urban and rural communities in South Asia. Livestock carcasses are now consumed by growing populations of feral dogs, many of which carry rabies. India has the world's highest human death toll from rabies—30,000 per year—and two-thirds of the cases are caused by dog bites. Populations of rats and flies also appear to be increasing. *Focus on Applied Research* in Chapter 52 describes another example of how pesticides and other chemicals accumulate at lethal concentrations in organisms living at higher trophic levels.

The Introduction of Exotic Species Often Eliminates Native Species

As humans travel from one habitat to another, we inevitably carry other species with us. Seeds cling to our legs, insects accompany us in our food and possessions, and some organisms hitch a ride on boats or cars. The introduction of nonnative organisms, called **exotic species,** into new habitats poses a serious threat to biodiversity.

Exotic species often prey upon, parasitize, or outcompete native species, leading to their extinction. Many have *r*-selected life

1970
1960
1950
1940
1930
1920
1900
1910

Denis Balibouse/X00870/Reuters/Corbis

FIGURE 53.8

Starling range expansion. After being introduced in New York City in 1890, European starlings *(Sturnus vulgaris)* increased their numbers and quickly extended their breeding range westward across North America. They reached the west coast by 1960 and Alaska by 1970.

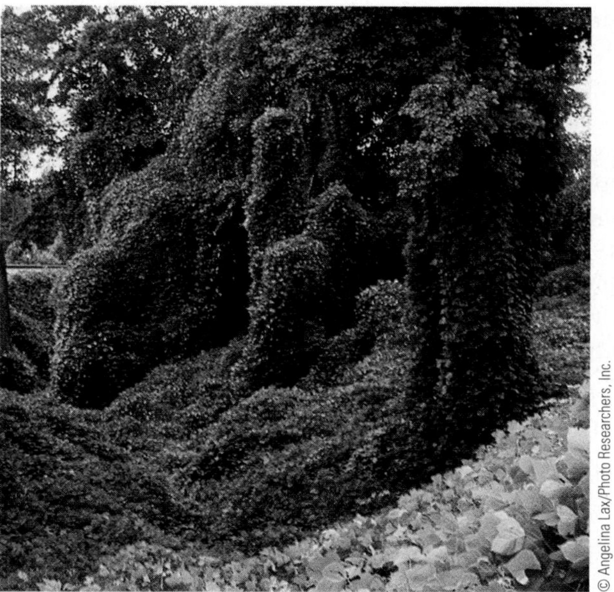
© Angelina Lax/Photo Researchers, Inc.

FIGURE 53.9
Kudzu, the vine that ate the South. Kudzu *(Pueraria lobata)*, an introduced vine, grows so quickly that it often covers living trees or even abandoned buildings.

histories; they mature quickly and reproduce prodigiously, and they thrive in the degraded habitats that humans so frequently create. In the absence of natural checks on population growth—such as competitors, predators, and parasites—exotics often experience exponential population growth (see Section 50.5).

The European starling *(Sturnus vulgaris)* provides an example of the explosive population growth and range expansion of an exotic species. These birds were released in North America in 1890 when a misguided individual, who wanted to introduce all of the bird species mentioned by Shakespeare into North America, imported them into Brooklyn, New York. Within 70 years, they had spread across the continent **(Figure 53.8)**; their population size is now estimated at 200 million. Starlings pose a serious threat to native birds, including several woodpecker species, because they successfully compete with them for nesting sites in natural cavities in trees.

Introduced plants often transform entire ecosystems. One of the best-known examples is kudzu *(Pueraria lobata)*, a fast-growing species from Asia. In the early 1900s, it was widely planted in the southeastern United States as a source of animal feed. Later, a government agency planted it to stabilize soils and decrease erosion on deforested hillsides. But when kudzu has access to abundant nutrients and water, it can grow up to 30 cm *per day*. It spread quickly across the South, literally overgrowing almost all native plants **(Figure 53.9)**.

Exotic insects often become pests of agricultural crops and native plants. The hemlock woolly adelgid *(Adelges tsugae)* was accidentally introduced into North America from Asia. The adelgid kills eastern hemlocks *(Tsuga canadensis)* by feeding on their sap. It now threatens the trees from North Carolina to Massachusetts **(Figure 53.10)**. But adelgids endanger far more than these evergreen trees. Hemlocks buffer the physical conditions below them: hemlock stands are cool in summer and warm in winter, sustaining a unique community of organisms that includes ruffed grouse *(Bonasa umbellus)*, turkey *(Meleagris gallopavo)*, white-tailed deer *(Odocoileus virginianus)*, and snowshoe hare *(Lepus americanus)*. Infested stands rarely survive more than a few years, and the communities established under pure stands of eastern hemlock will likely become extinct because of feeding by the adelgid.

The Spread of Disease-Causing Organisms Endangers Many Species

As exotic species become established in new habitats, they may carry disease-causing organisms with them. Because native species had no prior exposure to these pathogenic organisms, they never evolved resistance to them, leading to devastating outbreaks of disease. For example, amphibians have been in a worldwide decline since 1980. About one-third of the roughly 6,300 described species are now in danger of becoming extinct, and populations of nearly half of all amphibian species are declining in numbers. The sudden change in the status of these animals has sparked an enormous research effort aimed at identifying the pri-

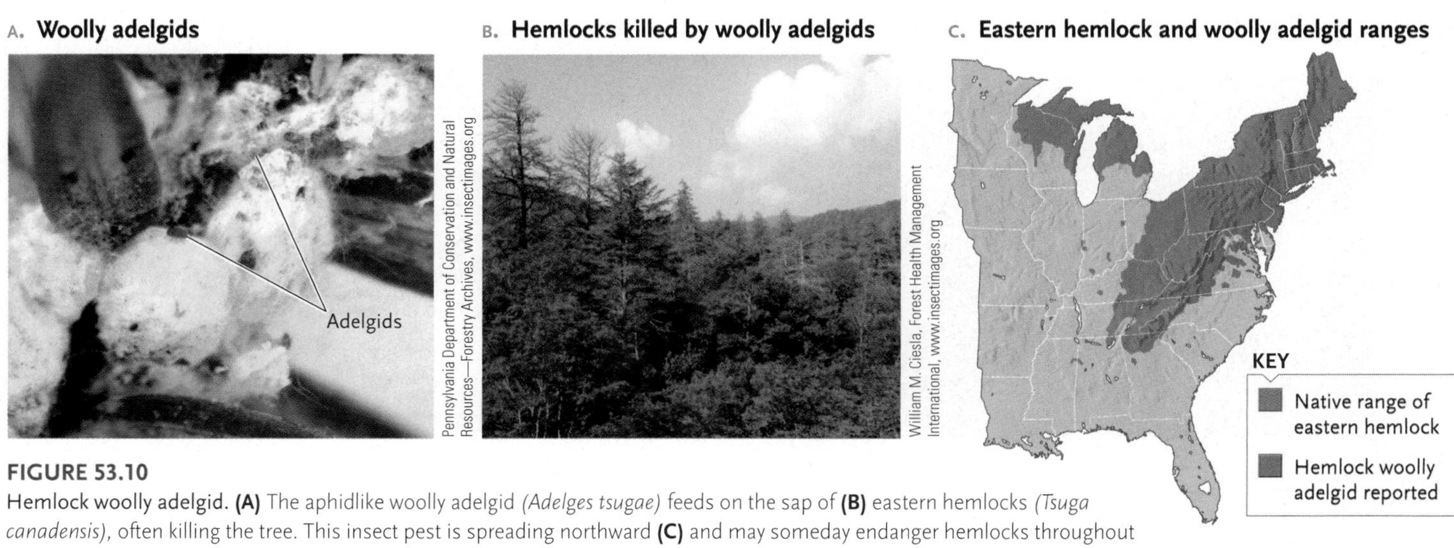
A. Woolly adelgids
Adelgids
Pennsylvania Department of Conservation and Natural Resources—Forestry Archives, www.insectimages.org
B. Hemlocks killed by woolly adelgids
William M. Ciesla, Forest Health Management International, www.insectimages.org
C. Eastern hemlock and woolly adelgid ranges
KEY
Native range of eastern hemlock
Hemlock woolly adelgid reported

FIGURE 53.10
Hemlock woolly adelgid. (A) The aphidlike woolly adelgid *(Adelges tsugae)* feeds on the sap of **(B)** eastern hemlocks *(Tsuga canadensis)*, often killing the tree. This insect pest is spreading northward **(C)** and may someday endanger hemlocks throughout their geographical range.

mary causal factors. Although habitat destruction and pollution have undoubtedly taken a great toll, scientists now attribute many recent amphibian declines and extinctions to infection by the chytrid fungus *Batrachochytrium dendrobatidis*. The fungus has been found in natural populations of more than 200 amphibian species; more than 1,000 others have been identified as susceptible to it **(Figure 53.11)**.

This pathogenic fungus was first described in 1999 by researchers investigating skin infections in amphibians from North America, Central America, and Australia. Scientists have since learned that the fungus is very strange indeed: it is the only species in its group known to infect vertebrates; it infects only amphibians, feeding on keratin in their skin and in the mouthparts of their tadpoles; and even though it has an aquatic life cycle, it can infect fully terrestrial amphibians that never enter standing water. Researchers believe that the fungus interferes with amphibians' oxygen acquisition and osmoregulation, two important physiological processes that are partly mediated by their skin. We still do not know how the fungus spreads among individuals or populations. It may be carried from place to place by amphibians and other vertebrates.

Where did the pathogen originate? In 2004, researchers found the infection in specimens of African clawed frogs *(Xenopus laevis)* that had been collected in 1938 and preserved in South African museums. Other researchers have found evidence of the infection in a Canadian population of bullfrogs *(Rana clamitans)* collected in 1961. Thus, scientists hypothesize that the infection originated in southern Africa and was confined there for decades. How did it spread so rapidly to other continents? Beginning in the 1930s, African clawed frogs were exported from South Africa to many countries for use in biological research and for the pet trade. Sadly, the frog's popularity among scientists and hobbyists has resulted in the spread of an infection likely to cause the extinction of many other amphibian species.

Why did the fungus start to devastate amphibian populations only in the 1980s? Some researchers suggest that even small increases in temperature and related changes in cloud cover and humidity—all the result of global climate warming—have favored the growth of the pathogenic fungus in some habitats with high amphibian diversity. Other researchers argue that climate warming and pollution stress amphibian populations, making them more susceptible to infection by many pathogens.

Biologists are working feverishly to learn more about the fungus and its role in amphibian declines before a majority of amphibian species become extinct. In a broader—and even more frightening—context, ecologists who study the dynamics of disease in natural population are just beginning to grapple with the likely effects of climate change and other consequences of human activity on the spread and success of pathogenic organisms, including those that infect humans.

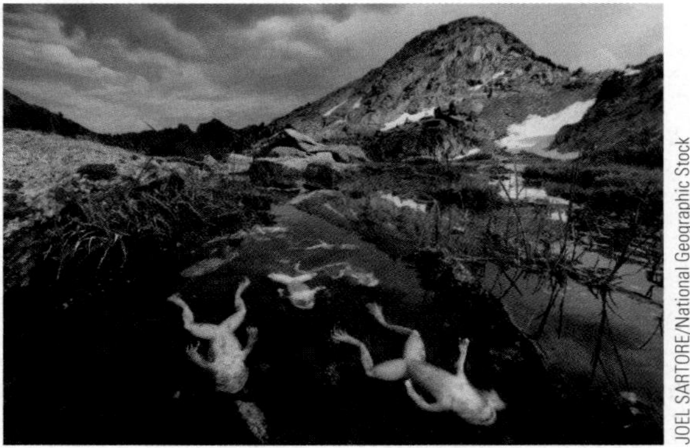
JOEL SARTORE/National Geographic Stock

FIGURE 53.11
Chytrid fungus infection. These mountain yellow-legged frogs *(Rana muscosa)* from King's Canyon National Park in the southern Sierra Nevada of California succumbed to chytrid fungus infection.

Human Activities Are Causing a Dramatic Increase in Extinction Rates

As you may remember from Section 22.4, extinction has been common in the history of life; roughly 10% of the species alive at any time in the past became extinct within 1 million years. These *background extinction rates* eliminated perhaps seven or eight species per year. Paleobiologists have also documented at least five *mass extinctions,* during which extinction rates increased greatly above the background rate for short periods of geological time (see Figure 22.15).

At present, Earth appears to be experiencing the greatest mass extinction of all time. According to Edward O. Wilson of Harvard University, extinction rates today may be 1,000 times the historical background rate, meaning that thousands of species are being driven to extinction each year. The vast majority of extinctions are a direct result of the destructive human activities discussed previously.

If humans are causing the current mass extinction, why didn't it begin long ago? The answer lies in our increased rate of population growth (see Section 50.6). During the nineteenth and twentieth centuries, improvements in food production, sanitation, and health care increased human life expectancy. Our ever-increasing population consumes resources and produces wastes at an escalating rate. And until we change the way we live in relation to the environment that we share with all other species, our negative impact will grow along with our global population.

STUDY BREAK 53.2

1. **How has habitat fragmentation affected breeding songbird populations in eastern North America?**
2. **What environmental factor has caused the demise of vulture populations throughout South Asia?**
3. **How do extinction rates today compare with the background extinction rate evident in the fossil record?**

THINK OUTSIDE THE BOOK

Search the Internet for updated information about the effects of the Deepwater Horizon oil spill in the Gulf of Mexico. Has the ecological impact of the spill become more evident over time?

53.3 The Value of Biodiversity

Given the many ways in which human activities are causing the dramatic decline in biodiversity, we should reflect on the present and future value of what we are destroying and contemplate why we might want to preserve it. Arguments for conserving biodiversity fall into three general groups: its direct benefit to humans, its indirect benefit to all living systems, and its intrinsic worth.

Biodiversity Benefits Humans Directly

Scientists constantly search for natural products that might provide humans with better food, clothing, or medicine. The development of a new medicine often begins when a scientist analyzes a traditional folk remedy or screens naturally occurring compounds for curative properties. Chemists then isolate and purify the active ingredient and devise a way to synthesize it in the laboratory. More than half of the 150 most commonly prescribed drugs were developed from natural products in this manner.

For example, *Taxol,* a drug treatment for breast and ovarian cancer, was isolated from the narrow strip of vascular cambium beneath the bark of the Pacific yew tree, *Taxus brevifolia* **(Figure 53.12).** Unfortunately, a fully grown, 100-year-old tree produces only a tiny amount of Taxol, and six trees must be destroyed to extract enough to treat one patient. Pacific yew trees are not abundant, and they grow slowly. Harvesting them for Taxol extraction could quickly lead to their extinction—and an end to the natural source of this life-saving compound. However, after much research, scientists now synthesize this widely used drug in the laboratory.

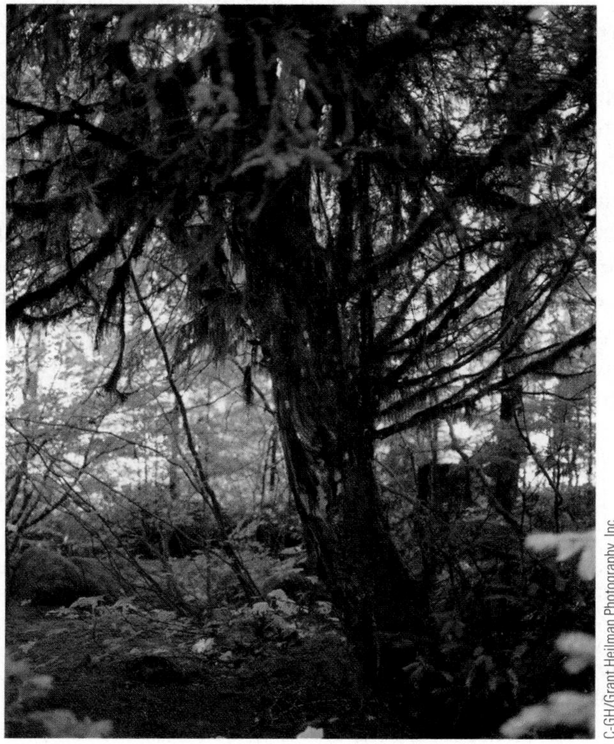

C-GH/Grant Heilman Photography, Inc.

FIGURE 53.12
The Pacific yew tree. The slow-growing Pacific yew *(Taxus brevifolia)* is the original source of Taxol, a compound that effectively fights several cancers.

Wild plants and animals also serve as sources of genetic traits that may improve agricultural crops and domesticated livestock. For example, corn *(Zea mays)* is an annual plant. Its cultivation requires yearly planting, a laborious activity that leads to the erosion of topsoil. Farmers would rather grow a perennial strain of corn, one that would produce grain for years after a single planting. In 1978, botanists discovered teosinte *(Zea diploperennis)* a perennial plant closely related to corn, in the mountains of western Mexico. Researchers crossed the two species, producing a perennial corn. If they can increase the yield of this hybrid, it may prove to be an economically valuable crop.

Today, many agricultural researchers use genetic engineering, the transfer of selected genes from one species into another (see Section 18.2), to alter crop plants more precisely than they can using hybridization. The transferred genes may be chosen to increase resistance to pests or environmental stress, promote faster growth, or increase shelf life after harvesting. However, many scientists and environmentalists fear that genetically modified crops may create environmental hazards that will inadvertently endanger biodiversity. For example, a genetically modified plant or animal that escaped into a natural habitat might compete with naturally occurring species. Or a genetically modified plant might poison harmless animals as well as insect pests.

Ecosystem Services Benefit All Forms of Life

Humans and other species derive indirect benefits when ecosystems perform the ecological processes on which all life depends. These **ecosystem services,** as they are called, include the decomposition of wastes, nutrient recycling, oxygen production, maintenance of fertile topsoil, and air and water purification.

Some ecosystem services can even mitigate environmental damage caused by humans. As you may recall from Section 52.4, the combustion of fossil fuels produces CO_2 and other waste products that accumulate in the atmosphere, increasing the greenhouse effect and fostering global warming. Photosynthetic organisms use CO_2 for essential metabolic processes; thus, the forests that we clear and, even more importantly, communities of marine phytoplankton, withdraw CO_2 from the atmosphere and incorporate it into living organisms (see Figure 52.13), in a phenomenon called *carbon sequestration.* Recent research indicates that these organisms are essential for limiting the damage caused by the burning of fossil fuels. In the long run, biodiversity's indirect benefits, provided in the form of ecosystem services, may be even more valuable to humans than the direct benefits.

Biodiversity Has Intrinsic Worth beyond Its Utility to Humans

Some ethicists argue that we should preserve biodiversity because it has intrinsic worth, independent of its direct or indirect value to humans. They note that humans are just one species among millions in the remarkable network of life. Countering this position is the view that our immediate needs should always rank above those of other species and that we should use them to maximize our own welfare. The latter view inevitably leads to the disruption of natural environments and the loss of biodiversity. Framed in this way, the debate lies more within the realms of philosophy and public policy than biology. Nevertheless, many people feel an emotional or spiritual connection to natural landscapes and the plants and animals they harbor. Thus, biodiversity enhances human existence in intangible ways.

STUDY BREAK 53.3 <

1. **How does biodiversity serve as a storehouse of genetic information that is potentially useful to humans?**
2. **What ecosystem services do naturally occurring organisms provide to humans?**

53.4 Where Biodiversity Is Most Threatened

To slow the current rate of extinction and loss of biodiversity, conservation biologists must first identify where species are likely to become extinct.

Conservation Biologists Identified Biodiversity Hotspots

If we are to limit the effects of human activities and preserve biodiversity, we must know how biodiversity is distributed. Although species richness within communities generally increases from the poles to the tropics (see Section 51.7), broad latitudinal surveys do not provide enough detail to be useful in this effort.

In a survey published in 2000, Norman Myers of Oxford University and his colleagues in England and the United States pinpointed 25 **biodiversity hotspots,** areas where biodiversity is both concentrated and endangered by human encroachment. To qualify as a biodiversity hotspot under Myers' criteria, an area must harbor at least 1,500 endemic plant species (those that are found nowhere else), and it must have already lost at least 70% of its natural vegetation. As human activity in natural habitats has increased, the number of terrestrial biodiversity hotspots has now grown to 34.

Myers used the number of endemic species as a criterion for identifying hotspots because endemics tend to have highly specific habitat or dietary requirements, low dispersal ability, and restricted geographical distributions. Indeed, locally distributed species account for much of Earth's biodiversity; and if the local habitats where these species occur are at risk of development, the species are also at risk. Although the 25 original hotspots occupy only 1.4% of Earth's land surface, they include the only remaining habitat for approximately 45% of all terrestrial plant species and 35% of all terrestrial vertebrate species.

Researchers Now Pinpoint Sites Where Extinctions Are Imminent

The identification of biodiversity hotspots tells us where biodiversity is both concentrated and threatened, but most of these areas are large and heavily populated. Conservation biologists need more detailed information to identify specific localities where their efforts will have the greatest impact.

Building on Myers' pioneering study, Taylor H. Ricketts of the World Wildlife Fund, working with 29 collaborators in the United States, Australia, and the United Kingdom, pinpointed sites where extinctions are imminent. In a paper published in 2005, they identified 585 locations in tropical forests, on islands, or in mountainous regions where 794 *trigger species*—highly endangered species of mammals, birds, reptiles, amphibians, and coniferous trees—are each confined to a single site **(Figure 53.13).** As defined by the Endangered Species Act, adopted by the U.S. Congress in 1973, an **endangered species** is one that is "in danger of extinction throughout all or a significant portion of its range."

The researchers used strict criteria for including a site in the list. First, it must harbor at least one species that has been officially designated as endangered by the World Conservation Union, an international organization. Second, it must contain at least 95% of the world population of that species. Third, it must have clearly definable boundaries within which habitats are distinct from those outside the boundary; examples of such bounded habitats include lakes, mountaintops, and forest fragments. The boundaries of the site thus define the area to be conserved. Although the 595 sites of imminent extinction are included within the biodiversity hotspots that Myers identified, the new approach has the practical advantage of pinpointing localized sites where conservation biologists can focus their efforts.

Ricketts and his colleagues noted that 794 trigger species are in danger of imminent extinction, compared with 245 species from the same taxonomic groups that are known to have become extinct in the last 500 years. Thus, the rate of extinction in these groups of organisms is accelerating rapidly. Their analysis also reveals that the proportion of extinctions in mainland habitats (as opposed to islands) is also growing: only 20% of historical extinctions occurred in low-lying mainland areas, whereas more than 60% of trigger species live on the mainland today. Moreover, their study detected a taxonomic shift in extinction: 53% of historical extinctions were of bird species, but 51% of the trigger species are amphibians. Finally, the data reveal a geographic shift in extinctions: whereas only 21% of historical extinctions were in the New World tropics, 50% of the trigger species live in Central America, South America, and on Caribbean islands. These results indicate that species living in the New World tropics, especially in wet forests, are in the greatest danger.

The data from the new analysis allow conservation biologists to target their efforts, but the task is daunting. Although one-third of the 585 sites lie completely within protected areas, more than 40% lack any protection at all. Most of the sites are small (median size approximately 12,000 hectares), suggesting that they might be easy to protect, but small sites are also the most vulnerable to human encroachment. Adding to the difficulty, the human population density in areas surrounding the sites is nearly triple the average density worldwide. Nevertheless, conservation biologists are optimistic that this new approach to pinpointing the areas most in need of their attention will allow them to develop appropriate strategies for preventing the extinction of the trigger species and others that live within these areas.

STUDY BREAK 53.4 <

1. What criteria do conservation biologists use to identify sites where extinctions are imminent?
2. Why are conservation biologists especially concerned about the rapid rate of deforestation in the New World tropics?

53.5 Conservation Biology: Principles and Theory

Conservation biology is an interdisciplinary science that focuses on the maintenance and preservation of biodiversity. In this section we describe how conservation biologists use theoretical concepts from systematics, population genetics, behavior, and ecology to develop ways to protect habitats and the endangered species that live within them. We introduce practical applications of conservation theory in the next section.

Systematics Organizes Our Knowledge of the Biological World

To develop a conservation plan for any habitat, scientists must start with an inventory of its species. Their primary tool is systematics, the branch of biology that discovers, describes, and organizes our knowledge of biodiversity (see Chapter 23). Cataloguing the diversity of life may be the most daunting task that biologists face. After more than 200 years of work, systematists have described and named approximately 1.6 million species. However, they realize that this number represents only a fraction of existing species.

In 1982, Terry Erwin of the Smithsonian Institution studied beetle biodiversity at the Tambopata National Reserve in Southern Peru. He sprayed biodegradable insecticide into the canopy of one large tree and collected 15,869 individual beetles, which he sorted into 3,429 species. More than 90% of the individual beetles he collected belonged to species that had not yet been described. Erwin used this astounding result and a complex mathematical model to predict that approximately 30 million species currently exist.

Nigel Stork of the Natural History Museum in London later questioned Erwin's conclusions. Using additional data and a modified set of assumptions, he estimated that the actual number of living species was closer to 100 million. If his figure is correct, more than 98% of species—most of them arthropods, nematodes, bacteria, and archaeans—are still unknown to science.

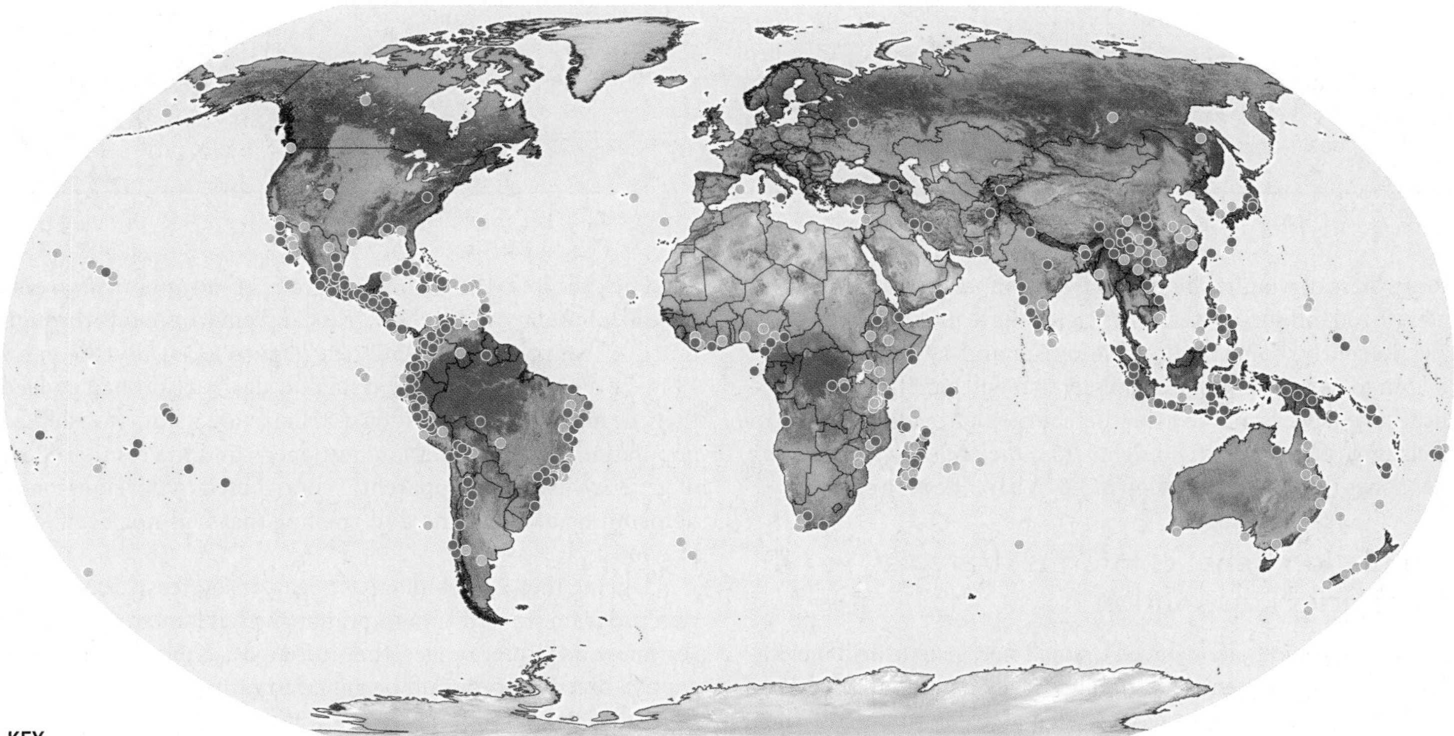

KEY
Unprotected or protection status unknown
Completely or partially protected

FIGURE 53.13
Pinpointing imminent extinctions. Taylor Ricketts and his colleagues identified 585 localities worldwide where at least one species of vertebrate or coniferous tree is in imminent danger of extinction. Sites marked in yellow are fully or partially protected. Those marked in red are unprotected or their protection status is unknown. The researchers mapped red localities over yellow ones to highlight areas in need of protection.

INSIGHTS FROM THE Molecular Revolution

Developing a DNA Barcode System

Everyone is familiar with the checkout scanners in stores: the cashier quickly passes an item's barcode over the scanner, and the register identifies it and records its price. The system works because the barcode on every item contains unique identifying information.

Research Question

In 2003, Paul Hebert, a population geneticist at the University of Guelph, Ontario, Canada, and his colleagues proposed an analogous method, called DNA barcoding, for identifying animal species quickly and accurately. The researchers envision using a handheld device to rapidly analyze DNA in the field; the resulting data would be sent to a database by cell phone, and minutes later an identification and a description of the species would appear on the instrument's screen.

Experiments

Hebert proposed using the first part of the *COI* (cytochrome oxidase 1) gene—a sequence of about 500 nucleotides—as the DNA barcode to distinguish animal species. This mitochondrial gene tends to vary greatly between animal species. Moreover, it appears to have no inserted or deleted DNA segments in most animals, making the alignment and comparison of sequences straightforward. Hebert's hope is that any *COI* gene sequence obtained in the field will provide a unique identifier for the species from which the DNA sample was obtained.

Results

Hebert's early tests of his barcode approach were promising and caught the attention of other researchers. For example, he and his collaborators first analyzed the *COI* gene sequence in the skipper butterflies of Costa Rica. Although adult skippers look pretty much alike, their caterpillars vary in appearance and in their food plant preferences, leading researchers to wonder if butterflies that had been assigned to one species *(Astraptes fulgerator)* might actually represent several. Analyses of the *COI* gene sequence sampled from 484 adults allowed Hebert and his colleagues to identify 10 distinctive DNA barcodes, suggesting that there are at least 10 species of skipper in Costa Rica rather than just one.

In early 2007, Hebert and his colleagues reported that they had used the DNA barcode to analyze 2,500 specimens of 643 North American bird species. The results were impressive: barcode differences between species were an order of magnitude greater than the differences within species, allowing the unambiguous identification of species from a short DNA sequence. Interestingly, the barcode analysis identified 15 probable new species that had not been previously identified and revealed that 8 supposed species of gull may be variants of just one species.

While the handheld analytical device is not yet ready for use in the field, Hebert's idea caught on quickly. In 2004 a consortium of major natural history museums and herbariums started the Barcode of Life Initiative, with the goal of creating a database of DNA barcodes linked to specimens already identified in their collections. The approach potentially could replace the traditional methods of systematic analysis using organismal and genetic characters to identify species.

Conclusion

Taken together, the results to date support using DNA barcodes, and using the *COI* gene sequence specifically for the barcode analysis, as a means of identifying animal species. Many researchers have joined the effort, and the Consortium for the Barcode of Life (http://barcoding.si.edu/) now includes 150 member organizations in the United States and Canada. They enter the results of their research into the Barcode of Life Data System (www.boldsystems.org), which, as of early 2010, includes entries for more than 800,000 specimens of nearly 68,000 species. Researchers are now seeking to internationalize the effort to encompass species from developing nations.

Source: P.D.N. Hebert et al. 2003. Biological identifications through DNA barcodes. *Proceedings of the Royal Society B* 270:313–321.

Regardless of whether biodiversity encompasses 30 million species or 100 million, systematists clearly have much work to do.

Recently, conservation biologists and systematists have begun to develop a new technology that will simplify the identification of species in the field, thereby facilitating the creation of a catalog of biodiversity. *Insights from the Molecular Revolution* describes the effort to develop a "DNA barcode scanner."

Population Genetics Informs Strategies for Species Preservation

When populations are reduced to small size, genetic drift inevitably reduces their genetic variability (see Section 20.3) and the evolutionary potential to adapt to changing environments. Thus, the loss of even a small fraction of a species' genetic diversity reduces its survival potential. To avoid this problem, conservationists strive not only to increase the population sizes of threatened or endangered species but to maintain or increase their genetic variation, both within and between populations.

For example, the whooping crane *(Grus americana)* was once an abundant bird in wet grassland environments through much of central North America **(Figure 53.14)**. By the early 1940s, excessive hunting and habitat destruction had caused their numbers to decline to just 21 individuals in two isolated populations. This population bottleneck and the resultant loss of genetic variability apparently contributed to developmental deformities of the spine and trachea that had not been seen previously.

During the 1970s, biologists began an aggressive conservation program. In addition to preserving habitats in the crane's summer and winter ranges, they initiated a carefully controlled captive breeding program designed to minimize the effects of inbreeding. Although more than 300 whooping cranes now survive in several wild and captive populations, recent research reveals that they still have a remarkably low level of genetic variability. As expected, the genetic effects of a severe population bottleneck may persist long after a population begins to increase in size.

Brian K. Miller/Animals, Animals—Earth Scenes

FIGURE 53.14
Whooping cranes. Endangered whooping cranes *(Grus americana)* winter in the Aransas National Wildlife Refuge in Corpus Christi, Texas.

Studies of Population Ecology and Behavior Are Essential to Conservation Plans

Conservation programs for animals also require data about target species' ecology and behavior, including their feeding habits, movement patterns, and rates of reproduction.

Sea otters *(Enhydra lutris)* are predatory marine mammals that live along the coastline of the North Pacific Ocean. In the early 1700s, they numbered approximately 300,000 individuals **(Figure 53.15)**, but commercial hunting reduced their numbers to about 3,000 individuals by the start of the twentieth century. Sea otters are keystone predators (see Section 51.4), and the destruction of sea otter populations had profound effects on the communities in which they lived. As the numbers of sea otters plummeted, populations of sea urchins, one of their favored prey, exploded; burgeoning sea urchin populations decimated local kelp beds, disrupting the communities of animals that live among these giant algae.

International treaties ended nearly all hunting of sea otters in 1911, and the populations subsequently recovered to about one-third of their original levels. Conservation biologists facilitated the recovery by reintroducing otters into southeastern Alaska, British Columbia, Washington, and California. Before deciding where otters should be reintroduced, scientists had to assess the resources available at different sites and determine how far individual otters would move, how rapidly they would reproduce, and how quickly their populations would spread. The reintroduction effort was successful at first. However, populations in California have experienced high mortality since the mid-1990s, and nearly half of those dying have been adults in their reproductive prime. Researchers have identified parasitic infections and heart disease as leading causes of death, suggesting that some coastal environments are so badly degraded that they may no longer support populations of this species.

Using complex mathematical models, conservation biologists often conduct a **population viability analysis** (PVA) to determine how large a population must be to ensure its long-term survival. PVAs evaluate phenomena that may influence the longevity of the population or species: habitat suitability, the likelihood of catastrophic events, and other factors that may cause fluctuations in demographics, population size, or genetic variability. When conducting a PVA, researchers must decide what level of risk is acceptable for a given survival time. For example, should a conservation plan attempt to ensure a 95% probability that the species will survive for 100 years, or should it specify a 99% survival probability? An increase in either the survival probability or the survival time requires an increase in the size of the population that must be conserved. The **minimum viable population size** identifies the smallest population that fits the desired specifications of the conservation plan. *Focus on Applied Research* describes how biologists used PVA in the conservation of an Australian marsupial, the yellow-bellied glider *(Petaurus australis).*

A. Sea otter

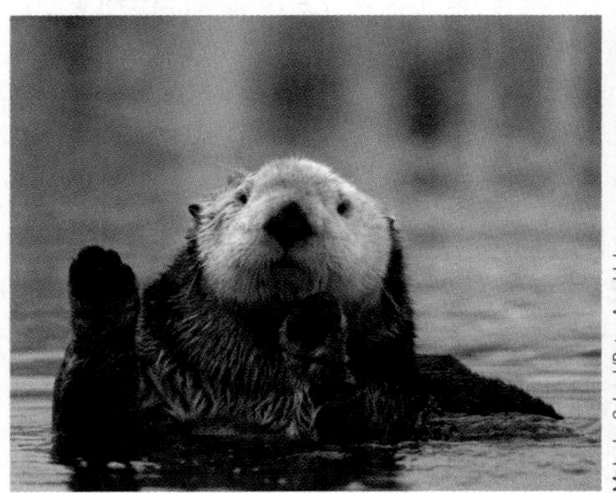
Malcolm Schuyl/Peter Arnold, Inc.

B. Geographical range of sea otters

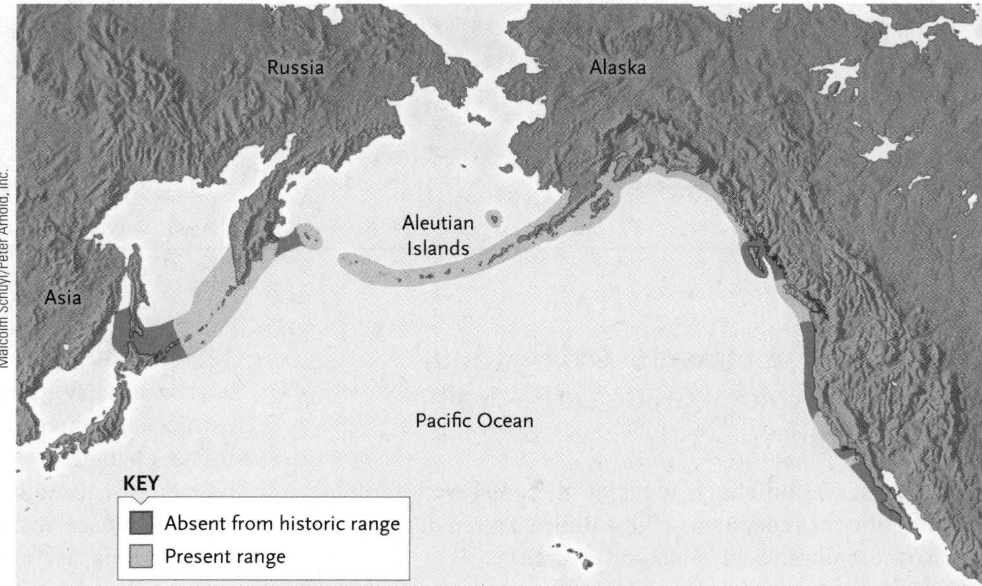
Russia
Alaska
Aleutian Islands
Asia
Pacific Ocean
KEY
Absent from historic range
Present range

FIGURE 53.15
Sea otters. After being hunted nearly to extinction, **(A)** sea otters *(Enhydra lutris)* have been reintroduced in many parts of their historical range **(B)**.

FOCUS ON Applied Research

Preserving the Yellow-Bellied Glider

Predicting the future is never easy, especially the future of a threatened species. But population viability analysis (PVA) allows conservation biologists to predict how a species will fare under a range of possible scenarios. An effective PVA for an animal species requires detailed information about its diet, predators, mating habits, habitat preferences, space requirements, demography, geographical distribution, responses to climatic fluctuations and human disturbances, and a host of other aspects of its biology.

The Australian yellow-bellied glider *(Petaurus australis)* provides an example of how PVA is essential for a conservation effort. This marsupial, about the size of a squirrel, lives in small family groups in undisturbed *Eucalyptus* forests along Australia's eastern coast. Each glider family maintains a home range (the area it uses for feeding and other activities) of 25 to 85 hectares; the home ranges of neighboring families do not overlap. As a result, the population density of gliders has never been high. But glider populations have declined precipitously as forests have been cleared, and the species is now considered threatened.

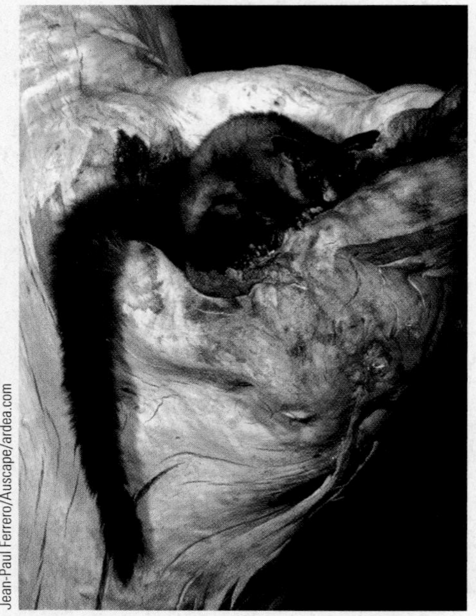
Jean-Paul Ferrero/Auscape/ardea.com

Using data from nearly 20 published papers, two Australian conservation biologists, Russ Goldingay of the University of Wollongong and Hugh Possingham of The University of Adelaide, conducted a PVA for this species. They estimated age distributions in glider populations as well as survival probabilities, litter sizes, sex ratios, lifespan, and home range sizes. They analyzed these data using a mathematical model that predicts the viability for populations of various sizes. In most PVAs, a population is considered viable if it has a 95% probability of surviving for 100 years. Goldingay and Possingham introduced additional complexity to their analysis by assessing the effects of unpredictable environmental events, such as drought, on breeding success. They also conducted sensitivity analyses to examine how changing the values of specific parameters—such as litter size, mortality rates of the different age classes, or the frequency and severity of droughts—might influence the general predictions of the viability model.

After many thousands of these calculations, the researchers concluded that a viable population of gliders would require at least 150 family groups. They also suggested that a population of that size would need approximately 18,000 hectares (roughly 70 square miles). Currently, only one of the 15 existing conservation reserves is that large.

Goldingay and Possingham did not factor some common environmental disturbances—fire, disease, or predation by introduced species—into their analyses. Such disturbances could decimate a small glider population in short order. Thus, the outlook for gliders may be bleaker than the researchers suggest, because their estimates of minimum viable population size and minimum necessary habitat size are almost certainly too low. Given only this information, we might predict that the glider will inevitably become extinct.

However, there is some hope for the yellow-bellied glider. Goldingay and Possingham assumed that gliders don't move between populations, a behavior that promotes gene flow, because they had no data on gene flow in this species. The movement of individuals between populations could reduce the required minimum viable population size by decreasing the likelihood of genetic drift and the extinction of local populations. Biologists may even be able to transplant gliders from one population to another, effectively creating source and sink populations. This procedure might increase population size and genetic diversity in the most endangered populations. If successful, such an approach could stave off extinction.

As a result of this PVA, conservation biologists can determine which of the remaining forest tracts are large enough to sustain a yellow-bellied glider population. Thus, they now know where to concentrate their limited resources to secure the future survival of this species. Although predicting the future is difficult, PVAs allow conservation biologists to make accurate and reliable recommendations for selective transplants that will contribute to the conservation of threatened species.

Source: R. Goldingay and H. Possingham. 1995. Area requirements for viable populations of the Australian gliding marsupial *Petaurus australis. Biological Conservation* 73:161–167.

Principles of Community Ecology and Landscape Ecology Guide Large-Scale Preservation Projects

Many conservation efforts focus on the preservation of entire communities or ecosystems. These projects often depend on the work of community and landscape ecologists.

SPECIES/AREA RELATIONSHIPS As you know from Chapter 51, community composition is dynamic: some species become extinct and others join the community through immigration. If we view patches of intact habitat as islands in a sea of unsuitable terrain, we can apply the predictions of the theory of island biogeography (see Section 51.7) to the design of protected areas. For example, we might expect that the number of species a

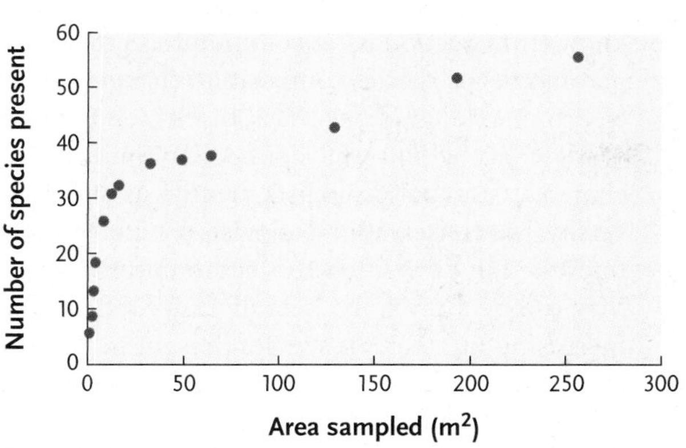
Number of species present
Area sampled (m²)
0
10
20
30
40
50
60
0
50
100
150
200
250
300

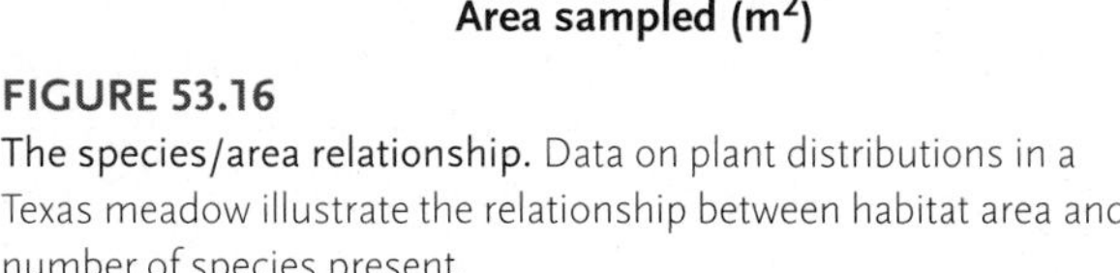

FIGURE 53.16
The species/area relationship. Data on plant distributions in a Texas meadow illustrate the relationship between habitat area and number of species present.

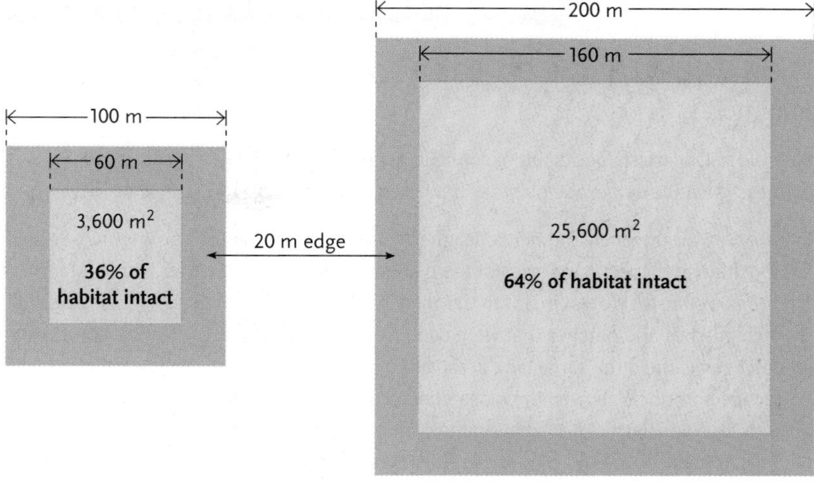
100 m
60 m
3,600 m²
36% of habitat intact
20 m edge
200 m
160 m
25,600 m²
64% of habitat intact

FIGURE 53.17
Edge effects and patch size. This hypothetical example illustrates how a 20-m-wide edge disrupts a larger fraction of a small habitat patch than a large habitat patch.

patch will support depends on its size and proximity to larger patches.

Indeed, ecologists recognized long ago that large habitat patches sustain more species than small patches do **(Figure 53.16)**. When plotted on an arithmetic scale, the relationship between species richness and habitat area increases sharply at first and then flattens. In other words, for relatively small habitat patches, even minor increases in area allow a large increase in the number of resident species; but as habitat patches get larger, the number of species present eventually levels off. You encountered an example of this relationship in our discussion of bird species richness on islands of different sizes (see Figure 51.31B).

As habitats become increasingly fragmented, edge effects exaggerate the species/area relationship in mainland habitat patches **(Figure 53.17)**. Consider two hypothetical patches of habitat: one is 100 m on a side, with a total area of 10,000 m^2; the other is 200 m on a side, with a total area of 40,000 m^2. Now, imagine that edge-effect disturbances penetrate 20 m into each patch from all directions. The small patch contains only 3,600 m^2 of intact habitat, but the large patch contains 25,600 m^2 of intact habitat. Although the large patch is only four times larger than the small patch, the large patch contains more than seven times as much *intact* habitat.

LANDSCAPE ECOLOGY Conservation biologists often use **landscape ecology** to design the size and geometry of nature reserves and other protected areas. Landscape ecology analyzes how large-scale ecological factors—such as the distribution of vegetation, topography, and human activity—influence local populations and communities.

When conservation biologists first applied concepts from landscape ecology to the design of protected areas, they debated whether nature preserves should comprise one large habitat patch or several smaller patches. Based on the species/area relationship, large patches should harbor more species than small patches; large patches would also experience proportionately smaller edge effects; and large patches would better support populations of large animals that need substantial resources. Nevertheless, some conservation biologists argued that clusters of physically separate preserves are more effective in maintaining metapopulations of endangered species (see Section 50.5), especially if the patches are interconnected by corridors of intact habitat. Individuals could move between preserves, reviving any local populations that experience a decline.

Some conservation biologists have worried that narrow landscape corridors, which by definition have large edges, might allow exotic species to invade protected areas. Ellen I. Damschen of North Carolina State University and several colleagues conducted an ambitious long-term field experiment to test the effects of landscape corridors on plant species richness **(Figure 53.18)**. Their results, published in 2006, suggest that habitat patches connected by corridors retain more native plant species than isolated patches do, and that corridors did not promote the entry of exotic species. Thus, corridors appear to be a useful feature in the design of nature preserves.

Landscape corridors are a key feature of efforts to prevent the extinction of the Florida panther (*Puma concolor coryi*, shown on page 1222). This subspecies is critically endangered: only 70 to 100 individuals remain from a population that once ranged throughout the southeastern United States. Other panther subspecies still inhabit the western states. Panthers are large predators, and each female requires nearly 20,000 hectares (more than 75 square miles) for hunting and breeding; males each require more than twice as much space.

Although the state and federal governments have set aside several panther conservation areas in Florida, 52% of the habitat panthers occupy is privately owned, and most of it is highly fragmented. Panthers frequently cross roads, and most panther

FIGURE 53.18 Experimental Research

Effect of Landscape Corridors on Plant Species Richness in Habitat Fragments

Question: Do landscape corridors connecting habitat patches influence the species richness of native and exotic plants within the habitat patches?

Experiment: Damschen and her colleagues studied changes in the community composition and species richness of the plants in open habitat patches within a longleaf pine *(Pinus palustris)* forest in South Carolina. Their experimental design included both isolated patches and patches that were connected to one another by a landscape corridor. All patches included the same land area, and their large size (1.375 ha each, including the landscape corridors) allowed the researchers to make a realistic assessment of the effects of landscape corridors. After creating the patches of open habitat within the forest in 2000, the researchers catalogued all plant species occurring in the patches through 2005, although they were unable to collect data in 2004.

Results: Over the course of the study, habitat patches that were connected by landscape corridors harbored increasingly more plant species than did isolated habitat patches. The researchers also noted that the difference in species richness between the two experimental treatments was caused by a difference in the number of native plant species present. The number of exotic species in connected and isolated habitat patches was similar.

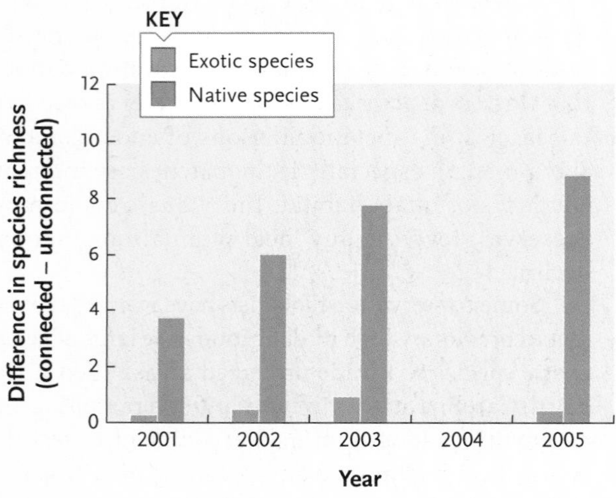

KEY
Exotic species
Native species
Difference in species richness (connected – unconnected)
0
2
4
6
8
10
12
2001
2002
2003
2004
2005
Year

Conclusion: Landscape corridors between patches of open habitat in longleaf pine forests increase the species richness of native species in open habitat patches, but they do not foster the entry of exotic species.

Source: E. I. Damschen et al. 2006. Corridors increase plant species richness at large scales. *Science* 313:1284–1286.

BETA-DIVERSITY Conservation biologists now often focus their efforts preserving assemblages of organisms rather than individual species. As you know from Section 51.2, communities grade into one another as species composition changes across environmental gradients. That discussion focused on diversity *within* well-defined communities, a characteristic that ecologists identify as alpha diversity. But conservation biologists are increasingly interested in diversity *across* communities, which they call beta diversity. Beta diversity reflects the increasing numbers of species present in an area that includes a wide variety of habitats, vegetation types, and small-scale environments. The concept of beta diversity is reflected in the slope of the species/area relationship: as the size of an area increases, so does the number of distinct environmental features it includes; and that environmental diversity supports a larger number of species.

By basing the design of nature preserves on the conservation of beta diversity, conservations biologists can establish reserves that will protect more species. An ideal reserve system might include several large areas, each including a diversity of small-scale environments suitable for species that do not disperse readily, and many small reserves interconnected by landscape corridors. A reserve system distributed over an important environmental gradient would include species that replace each other across it. And as the global climate continues to warm over the coming decades, reserve systems that include protected areas at different altitudes or latitudes might allow some species to migrate into cooler environments over time.

STUDY BREAK 53.5

1. **How does a population bottleneck increase the likelihood that a species will become extinct?**
2. **How does a population viability analysis assist in the development of a conservation plan for a species?**
3. **Would a single large nature preserve or several small preserves experience greater edge effects?**

deaths in Florida are caused by accidents with motor vehicles. Protected landscape corridors might enable panthers to move more safely between conservation areas. A preliminary study found that panthers already use such corridors, typically along wooded riverbanks, when they are available. The Florida Fish and Wildlife Service proposed the creation of an ambitious 6,100-hectare network of such corridors alongside the Caloosahatchee River to link several significant habitat fragments in neighboring counties.

53.6 Conservation Biology: Practical Strategies and Economic Tools

Conservation biology seeks to protect native species, communities, and ecosystems from the effects of human activity. Meeting that goal and reversing some of the existing damage requires the integration of biological research with economic and social realities.

FIGURE 53.19
The Albany Pine Bush habitat. **(A)** The Pine Bush lies entirely within the city limits of Albany, New York. It is home to about 50 threatened or endangered plant and animal species, including **(B)** the Karner Blue butterfly *(Lycaeides melissa samuelis)*.

A. **Albany Pine Bush**

Darlene Bordwell

B. **Karner Blue butterfly**

Ruth Cole/Animals, Animals—Earth Scenes

Conservation Efforts Aim to Preserve, Conserve, and Restore Habitats

Conservation groups often highlight efforts to preserve individual animal species, such as the giant panda *(Ailuropoda melanoleuca)* or California condor *(Gymnogyps californianus)*. The preservation of "charismatic megavertebrates," as these large animals are sometimes described, attracts substantial public support. Nonetheless, there is little point in trying to preserve natural populations of individual species if their habitats are in jeopardy. An alternative to species-based conservation focuses on the preservation of intact habitats; individual species are conserved as a consequence of preserving the habitats on which they depend. Conservation biologists approach this goal with a continuum of approaches, which fall into three general categories: *preservation, mixed-use conservation,* and *restoration.*

CONSERVATION THROUGH PRESERVATION In many countries, habitats are preserved when an individual or organization purchases them and enforces strict standards of land use. In sensitive habitats, people may be excluded altogether; in other cases, access is restricted and the exploitation of resources is controlled. This approach works well in countries with efficient law enforcement and a tradition of private land ownership. In the United States, for example, the Nature Conservancy has purchased tracts of land to preserve native species in every state.

The preservation approach has been successful in preserving portions of the Pine Bush habitat near Albany, New York **(Figure 53.19)**. This unique ecosystem arose approximately 11,000 years ago at the end of the last glacial period, when a massive deposit of sand was left near the western margin of Albany's current city limits. This sandy region formed an inland pine-barrens habitat in which pitch pine *(Pinus rigida)*, scrub oak *(Quercus ilicifolia)*, and dwarf chestnut oak *(Quercus prinoides)* are now the dominant vegetation. The Pine Bush is home to more than 50 plant and animal species that the state and federal government list as threatened or endangered. The habitat itself was once vulnerable because it lies within Albany's city limits; however, since 1988, the Pine Bush has been jointly owned and protected by New York state, local municipalities, and the Nature Conservancy.

MIXED-USE CONSERVATION When outright preservation is impractical, conservation biologists advocate mixed-use conservation, which combines the protection of some land parcels with the controlled development of others.

The Ngorongoro Conservation Area (NCA) in Tanzania provides an example of mixed-use conservation. The NCA covers 829,000 hectares of grassland and borders the Serengeti National Park. Because it houses a high concentration of wildlife, the NCA is one of the most visited tourist destinations in eastern Africa. For the past several hundred years, the Maasai people have herded cattle, goats, and sheep in the Serengeti and Ngorongoro **(Figure 53.20)**. The Maasai are nomadic pastoralists who frequently move their relatively small herds to new grazing areas in the region. As a result, their activities do not degrade the land or exclude native wildlife. In 1959 the Maasai agreed to vacate the Serengeti, which was converted into a national park, in return for retaining the rights to live and herd livestock within the NCA. The government of Tanzania helped create the necessary infrastructure within the NCA, including a constant water supply as well as social services. Under this agreement, 40,000 indigenous residents, most of them Maasai, live in this large and valuable conservation area.

Brian K. Miller/Animals, Animals—Earth Scenes

FIGURE 53.20
Mixed-use conservation. The Maasai use the Ngorongoro Conservation Area to graze cattle and goats.

CONSERVATION THROUGH RESTORATION Conservation biologists sometimes create restoration plans to reestablish the vitality of a previously disrupted community or ecosystem. This effort requires the removal of contaminants, impediments to the natural flow of water, and barriers to animal movement as well as the restoration of natural processes, such as periodic fires or floods. Most restoration projects also require replanting key plant communities and long-term management once restoration is complete.

Not all degraded habitats can be restored, and not all potential restoration projects are equally feasible. When making project decisions, restoration ecologists consider a number of factors: Will the restored habitat be suitable for rare or endangered species, and will its creation increase endemic biodiversity? Would the restoration reunite previously fragmented land parcels? Will the restored habitat experience the periodic disturbances, such as fires or floods, that are essential for its continued existence? What are the costs of implementing the plan and maintaining the area? Finally, would the restored land be valued by local residents, and will they support and maintain it?

A successful restoration project is currently underway in the Brazilian Atlantic Forest, sponsored by the Instituto de Pesquisas Ecológicas (IPÊ), a Brazilian nongovernmental organization. In western São Paulo state, near the Morro do Diabo state park, IPÊ is trying to recreate the natural Brazilian Atlantic Forest ecosystem by planting native trees in habitat corridors between remaining forest fragments. These corridors of native tree species should facilitate the preservation of species in those forest patches and supply valuable botanical resources for endemic wildlife and local residents.

Successful Conservation Plans Must Incorporate Economic Factors

Biologists can almost always develop a plan to conserve a species, community, or ecosystem. But to be successful, a plan must be economically feasible, and it must provide direct benefits to local residents whose lives it will affect.

LOCAL INVOLVEMENT Early conservation efforts simply set aside protected areas in which most human activities were banned. Local people were denied access to resources within the preserve—resources that were sometimes essential for their survival. Not surprisingly, these plans generated antipathy towards conservationists and the organisms they were trying to preserve.

For example, the northern spotted owl *(Strix occidentalis caurina)* lives only in old growth coniferous forests of the Pacific Northwest, where many local residents worked in forestry or supporting industries. The suggestion that the owl be listed as an endangered species triggered a bitter political battle between conservationists and local residents because the conservation plan for the owls required closing large tracts of forest to logging. Washington State listed the owl as an endangered species in 1988, but local residents, who lost jobs when logging was reduced, remain hostile to these conservation efforts.

Conservation plans are more successful if they provide local residents with benefits that depend on the existence of a preserve.

Ed Degginger/Color-Pic

FIGURE 53.21
Conservation and the local economy. Local residents support conservation efforts at the Royal Chitwan National Park, Nepal, because officials open the park for a grass harvest each year.

Chitwan National Park provides an excellent example. For more than 100 years, this area, located in south central Nepal near the northern border of India, was recognized as a hunting ground for local royalty. These activities decimated local populations of large mammals, especially the Bengal tiger *(Panthera tigris)* and one-horned rhinoceros *(Rhinoceros unicornis)*. The area was opened for settlement during the late 1950s, and immigrants swarmed into the fragile grassland. As the human population exploded, the populations of tigers and rhinos dwindled by the mid-1960s.

The area was converted into Royal Chitwan National Park in 1973. Today, humans are excluded from the park for most of the year. But each January, after the monsoon rains end and the grasses have dried, local residents are welcomed into the park to harvest the grass, which they use to thatch roofs, make mats, and feed domestic animals **(Figure 53.21)**. The area surrounding the Chitwan National Park has been designated as a buffer zone. Residents of the buffer zone receive 50% of the total revenue earned from park activities, including entrance fees. This income has changed the lifestyle of the people in the buffer zone, and they now value Chitwan and argue for its preservation. Today, more than 435 one-horned rhinos survive in Nepal, most of them in Chitwan National Park. And the Bengal tiger population of Chitwan has increased to 91 breeding adults.

ECOTOURISM In some preserves, governments enlist local residents in park development and operations, providing them with a viable livelihood. The most successful approach has been the development of **ecotourism,** in which visitors, often from wealthier countries, pay a fee to visit a nature preserve. Local people work as guides, cooks, and logistical and support staff.

Not everyone agrees that ecotourism is helpful. Critics note that increased human traffic may degrade habitats, and unregulated ecotourism can eventually lead to overdevelopment. For example, several million people visit national parks in the western United States annually. Traffic jams, automobile accidents, and long lines routinely plague visitors at the most popular sites.

Cranky ecotourists call for the construction of more roads and parking lots, which are inconsistent with the purpose of a national park because cars increase local air pollution and occasionally kill wildlife. In 2006, the government began charging a $20 fee for each automobile entering Yosemite National Park in California, hoping to limit the number of visitors arriving in private vehicles and to increase reliance on public transportation.

COUNTRYWIDE ECONOMIC APPROACHES In the mid-1990s, conservation biologists and economists developed the concept of **ecosystem valuation,** in which ecosystem services—such as carbon dioxide processing or water retention and purification, which are best provided by intact ecosystems—are assigned an economic value. These estimated values are used to negotiate contracts in which a private company or conservation organization pays a community, state, or country to maintain intact ecosystems. By one 1997 estimate, the gross global ecosystem valuation is roughly $18 trillion per year. If less obvious benefits provided by nature are tallied—soil formation, crop pollination, and nutrient cycling—the total value of ecosystem services rises to $33 trillion—almost twice the value of all goods produced by all humans on the planet!

The implementation of ecosystem valuation exchanges is determined on a case-by-case basis, depending on what ecosystem services the paying organization wants to preserve. Costa Rica is leading the way in this effort by creating valuation contracts with several corporations. For example, in 1998, the Monteverde Conservation League signed a contract with a local electrical company to ensure the continued flow of water from the Bosque Eterno de los Niños, a forest preserve. The company had plans to build a hydroelectric dam on the Rio Esperanza, and feared that deforestation upstream would disrupt water flow through the dam. The contract specifies that the electrical company will pay the people who live upstream to preserve their forests rather than cutting them. Thus, both the forests and water flow are preserved, maintaining the forest ecosystem and generating badly needed electricity.

Biodiversity is a precious resource that is disappearing rapidly throughout the world. It can still be conserved through a monumental effort to catalog the diversity of living organisms and develop an understanding of their ecological relationships. Perhaps the major challenges for conservation biologists are to educate the human population about the value of biodiversity and to develop conservation plans that will enlist the support of people who live among the threatened species.

STUDY BREAK 53.6

1. **Is the Pine Bush habitat in New York State an example of preservation, mixed-use conservation, or restoration?**
2. **How has the establishment of the Royal Chitwan National Park in Nepal been a successful conservation effort? How do conservation biologists measure its success?**
3. **How can the concept of ecosystem services be used to foster conservation of threatened habitats and species?**

UNANSWERED QUESTIONS

Conservation biology is a young science in a rapidly changing world. Our field is brimming with unanswered questions about scarcity, diversity, and extinction. Many of the questions proposed by conservation biologists are interesting, but only a subset may offer answers useful to policy makers, field practitioners, and nongovernmental organizations trying to find cost-effective ways to head off the global biodiversity crisis. I have selected two areas where basic conservation science can inform global policies and local efforts to prevent the current extinction crisis from dooming many species to oblivion.

How can we preserve species that are likely to become extinct?

Some estimates state that up to 50% of all species on Earth could disappear by the end of the century as a result of: (1) land conversion in the tropics, where more than 50% of all species occur, and (2) climate change. Professor E. O. Wilson of Harvard University has calculated that more than 100 species go extinct each day. So, ask the cynics, "What are they? Can you name them?"

To save species from imminent extinction, we need to know which species are highly endangered, where they live, what resources they require, and why they are threatened. Led by Taylor Ricketts, biologists from all major conservation organizations used published data to identify the "trigger species" of vertebrates and conifers most likely to go extinct in the next 20 years. We found 794 species in 585 locations for which the entire global population, ranked as endangered or critically endangered, is limited to a single site. Half of these sites contain rare, endemic amphibians, many of which are threatened by the deadly chytrid fungus. Can we find a way to stop the spread of this fungus, or reduce its virulence, to avoid a mass extinction of amphibians? Half of all threatened localities are on tropical mountains. Can we save these habitats as watersheds that provide vital ecosystem services to communities living downstream? What about the 350,000–400,000 species of vascular plants, perhaps 10% of which are known only from a single site? What about the millions of species of invertebrates, some so rare that their entire distribution may be limited to the crowns of a few tropical forest tree species? Can we protect all of these areas that are the last refuge for many species? Or is the task so overwhelming that we should try to save only what seems feasible?

Can we find the political will to protect global biodiversity?

Answers to this second question could provide a solution to those posed above. A new global treaty, REDD (Reduced Emissions from Deforestation and Degradation), is currently being negotiated. It offers the greatest conservation opportunity of our lifetimes. Can we seize it? If most of the world's endemic species live in the tropics and a high percentage occur in moist tropical forests, can we forge a new agreement that values these forests for the carbon they sequester, and in so doing, protect much of the world's biodiversity? Deforestation and degradation of moist tropical forests account for 20% of the greenhouse gas emissions recorded annually. These emissions could become much larger if peat swamp forests in Sumatra and other carbon-rich areas are burned, cleared, and converted to plantations. Can we harness the political will to create a carbon financing mechanism that rewards tropical countries that protect their carbon-rich and species-rich forests? Will REDD be a turning point in the race to secure a more stable climate and avoid biological catastrophe?

Think Critically

Write down 20 unanswered questions for which scientists must quickly find answers to save life on Earth. A recently published paper (W. J. Sutherland et al. 2009. One hundred questions of importance to the conservation of global biological diversity. *Conservation Biology* 23:557–567) offers a longer list. Does your list overlap with the most pressing 100 questions identified by biologists, conservationists, and policy-makers?

Eric Dinerstein

Eric Dinerstein is the World Wildlife Fund chief scientist and Vice President for Science. He is a co-architect and co-author of the Global 200 ecoregions, an analysis to identify the most biologically important ecoregions on Earth in the terrestrial, freshwater, and marine realms. He and his staff developed a framework for targeting priority areas for conserving tiger populations across Asia that is now widely adopted. To learn more about Dr. Dinerstein's work, go to: http://www.worldwildlife.org/who/experts/eric-dinerstein.html.

REVIEW KEY CONCEPTS

Go to **CENGAGENOW** at www.cengage.com/login to access quizzing, animations, exercises, articles, and personalized homework help.

53.1 The Biodiversity Crisis on Land, in the Sea, and in River Systems

- Deforestation is occurring at an alarming rate, especially in tropical regions (Figures 53.1 and 53.2). In addition to adding greenhouse gases to the atmosphere, deforestation may lead to desertification and the loss of entire ecosystems (Figure 53.3). Deforestation, desertification, and global warming reinforce each other in a positive feedback cycle.
- Overexploitation of natural populations reduces their sizes and may induce evolutionary responses in the exploited populations (Figure 53.4).
- Dams and other factors causing hydrologic alterations endanger freshwater ecosystems by changing the structure of riverine habitats, disrupting life histories of aquatic species, decreasing connectivity in river systems, and facilitating the establishment of exotic species (Figure 53.5).

Animation: Effects of deforestation

53.2 Specific Threats to Biodiversity

- Habitat fragmentation reduces the size of intact habitat patches, and edge effects diminish the quality of remaining habitat (Figure 53.6). Only small populations, which are subject to genetic drift and an increased likelihood of extinction, can inhabit small habitat patches.
- Although pollution is released locally, it often spreads regionally and globally, especially in bodies of water and the atmosphere (Figure 53.7).
- Exotic species often contribute to the extinction of native species through competition, predation, or parasitism (Figures 53.8–53.10). Humans frequently introduce exotics into communities either intentionally or inadvertently.
- The decline of amphibian species is at least partly caused by the spread of a pathogenic chytrid fungus (Figure 53.11). New and introduced disease-causing organisms pose serious threats to biodiversity.
- Although extinction has been common in the history of life, human activities have recently initiated what may be the greatest mass extinction of all time. Some biologists estimate that extinction rates today may be 1,000 times the background extinction rate.

Animation: Effect of air pollution in forests

Animation: Five major extinctions

53.3 The Value of Biodiversity

- Biodiversity provides direct benefits to humans because natural populations of organisms can be sources of useful natural products as well as genetic resources that can improve domesticated crops and animals (Figure 53.12).
- Biodiversity provides indirect benefits to humans by maintaining normal ecosystem processes, some of which help to counteract the harmful effects of human activities.
- Ethicists and environmentalists argue that biodiversity should be preserved simply because of its intrinsic worth.

53.4 Where Biodiversity Is Most Threatened

- Biodiversity hotspots harbor large numbers of endemic species and are threatened by human activities.
- Conservation biologists have pinpointed areas where several groups of vertebrates and coniferous plants are in imminent danger of extinction (Figure 53.13). An analysis of these data suggests that the extinction rate is accelerating and that species in the New World tropics are especially at risk.

Animation: Global crises by region and habitat

53.5 Conservation Biology: Principles and Theory

- Conservation biology draws its theoretical foundation from systematics, population genetics, population ecology, behavior, community ecology, and landscape ecology.
- Systematists provide taxonomic inventories of biodiversity that are helpful for establishing conservation priorities.
- Conservation biologists design breeding programs to maintain or increase the genetic variability of species being preserved (Figure 53.14).
- Conservation biologists study the population ecology and behavior of targeted species (Figure 53.15), and they may use population viability analyses to determine the minimum viable population size necessary to conserve threatened species.
- Studies in community ecology have established the generality of the species/area effect: large habitat patches harbor more species than small habitat patches do (Figure 53.16).
- From the perspective of landscape ecology, biologists have debated the advantages and disadvantages of establishing one large reserve versus several smaller ones that are connected by habitat corridors (Figures 53.17 and 53.18).

- Conservation biologists now include analyses of diversity across communities in their designs for nature preserves in an effort to preserve whole assemblages of organisms.

53.6 Conservation Biology: Practical Strategies and Economic Tools

- Efforts to conserve communities or ecosystems follow one of three general strategies. *Preservation* requires the restriction or prohibition of human access to the area (Figure 53.19). *Mixed-use conservation,* an approach that balances the conflicting demands of habitat preservation and development, allows local residents to use the protected area in limited ways (Figure 53.20). *Restoration* attempts to recreate natural communities and ecosystems in places that have already been degraded by human activities.
- Conservation plans must also incorporate economic and social factors to win local support. Most conservation plans now include the involvement of local residents to generate revenue for their communities (Figure 53.21). Ecosystem valuation also encourages the preservation of ecosystems by assigning them a significant economic value.

Animation: Sustainable resource management

UNDERSTAND AND APPLY

Test Your Knowledge

1. The greatest extinction in the history of life on Earth:
   a. occurred at the end of the Permian period.
   b. occurred at the end of the Cretaceous period.
   c. occurred at the end of the Ordovician period.
   d. occurred at the end of the Cambrian era.
   e. may be occurring now.
2. Which of the following activities is the most fundamental cause of the worldwide crisis in river ecosystems?
   a. overexploitation of predatory fishes
   b. damming of rivers
   c. pollution from power plants
   d. invasion by exotic species
   e. deforestation
3. Habitat fragmentation has damaged populations of breeding birds in North America because:
   a. the remaining habitat patches rarely contain enough food for birds to rear their offspring.
   b. the nests of birds in small habitat patches are frequently attacked by predators.
   c. pairs of breeding birds cannot easily move from one habitat patch to another.
   d. female birds cannot locate potential mates in small habitat patches.
   e. small habitat patches do not have enough edges to provide adequate hiding places.
4. Deforestation:
   a. is a problem only in the tropics.
   b. may speed desertification.
   c. is slowed by grazing and farming.
   d. permanently enriches the soil.
   e. leads to the formation of lush grasslands.
5. Chemical pollutants:
   a. can spread rapidly from the places they are released.
   b. do not appear to influence global climate change.
   c. have contributed to global mass extinctions.
   d. rarely affect natural bodies of water.
   e. rarely influence animals feeding at higher trophic levels.
6. Which of the following is most likely to be a biodiversity hotspot?
   a. a patch of forest in the middle of North America that is 500 km from the nearest big city
   b. a series of uninhabitable sand dunes in the Sahara Desert
   c. a botanical garden that houses representatives of 25,000 plant species
   d. a tropical island with many endemic species and a growing human population
   e. a suburban neighborhood where fields have been converted to backyards and playgrounds
7. Population viability analyses allow conservation biologists to:
   a. identify the source population from which an individual dispersed to a sink population.
   b. determine how large an area must be preserved for the protection of a threatened species.
   c. identify whether individuals of a threatened species are reproductively mature.
   d. predict the minimum population size of a threatened species that is likely to survive.
   e. predict whether a threatened species will use habitat corridors.
8. Beta diversity is a measure of:
   a. species diversity across community boundaries.
   b. the number of species within one community.
   c. the number of endemic species found in a particular place.
   d. the likelihood that a particular species will become extinct in the next 20 years.
   e. the minimum population size needed to conserve an endangered species.
9. For which of the following species has the use of habitat corridors been proposed as an important conservation tool?
   a. sea otters
   b. bay checkerspot butterflies
   c. Florida panthers
   d. whooping cranes
   e. Eastern hemlocks
10. The main goal of restoration ecology is the reestablishment of:
   a. natural patterns of water flow.
   b. the vitality of a degraded ecosystem.
   c. the historical corridors linking forest fragments.
   d. the natural barriers to animal movement.
   e. ecotourism.

Discuss the Concepts

1. National parks are often established in ecologically sensitive areas. In many places they have become so popular that visitors endanger the ecosystems the parks were originally designed to preserve. How can the goals of conservationists, who work to maintain intact ecosystems, be balanced with those of citizens who wish to visit intact ecosystems? In other words, how would you regulate domestic ecotourism?
2. How do concepts from population genetics, metapopulation dynamics, and beta-diversity apply to the design of nature preserves? Do they suggest different ideal designs for nature preserves?
3. Imagine that you are a conservation biologist who has been asked to develop a conservation plan for a species of lizard that lives in the deserts of the American Southwest. What sorts of data would you collect before developing a final plan?

Design an Experiment

Devise a field study to determine whether the species/area relationship applies to aquatic ecosystems, such as ponds and lakes, as it does to terrestrial habitats.

Interpret the Data

The accompanying table compares statistics on imminent extinctions versus historical extinctions in five groups of organisms. The data are divided into three geographic categories. In which of the three habitat categories are extinctions accelerating the most? In which group of organisms are extinctions accelerating the most? What do these data suggest about where conservation biologists should focus their efforts if the goal is to preserve as many species as possible?

Apply Evolutionary Thinking

Overexploitation of marine fish stocks has depleted natural populations and caused a reduction in the age and size at which many fish species become reproductively mature. What sort of government regulations of fishing might reverse the current trend toward smaller adult size? Explain your answer in terms of the selection pressures that fishing places on targeted species.

Express Your Opinion

Material goods can be manufactured in ways that protect biodiversity but often cost more than comparable goods produced without regard for the environment. As a consumer, are you willing to pay extra for the first kind? Go to www.cengage.com/login to investigate both sides of the issue and then vote.

Distribution of Species Facing Imminent Extinction (i.e., Trigger Species) and Historically Extinct Species among Taxa and Islands, Mountains, and Low Mainland Areas

	Islands*		Mountains†		Low mainland‡		Total	
Taxon	Trigger species	Extinct species	Trigger species	Extinct species	Trigger species	Extinct species	Trigger species	Extinct species
Mammals	80	49	35	5	16	19	131	73
Birds	128	121	51	1	38	7	217	129
Reptiles§	7	8	0	0	8	1	15	9
Amphibians	88	19	268	11	52	4	408	34
Conifers	9	0	12	0	2	0	23	0
Total	312	197	366	17	116	31	794	245

*Islands are landmasses smaller than Greenland and include mountainous sections of islands.
†Mountains are mountains on mainland landmasses (not on islands).
‡Low mainland regions are low-lying regions of continental mainlands.
§Reptiles include only turtles and tortoises, crocodilians, and iguanid lizards.

Source: T. H. Ricketts et al. 2005. Pinpointing and preventing imminent extinctions. *Proceedings of the National Academy of Sciences* 102:18497–18501.

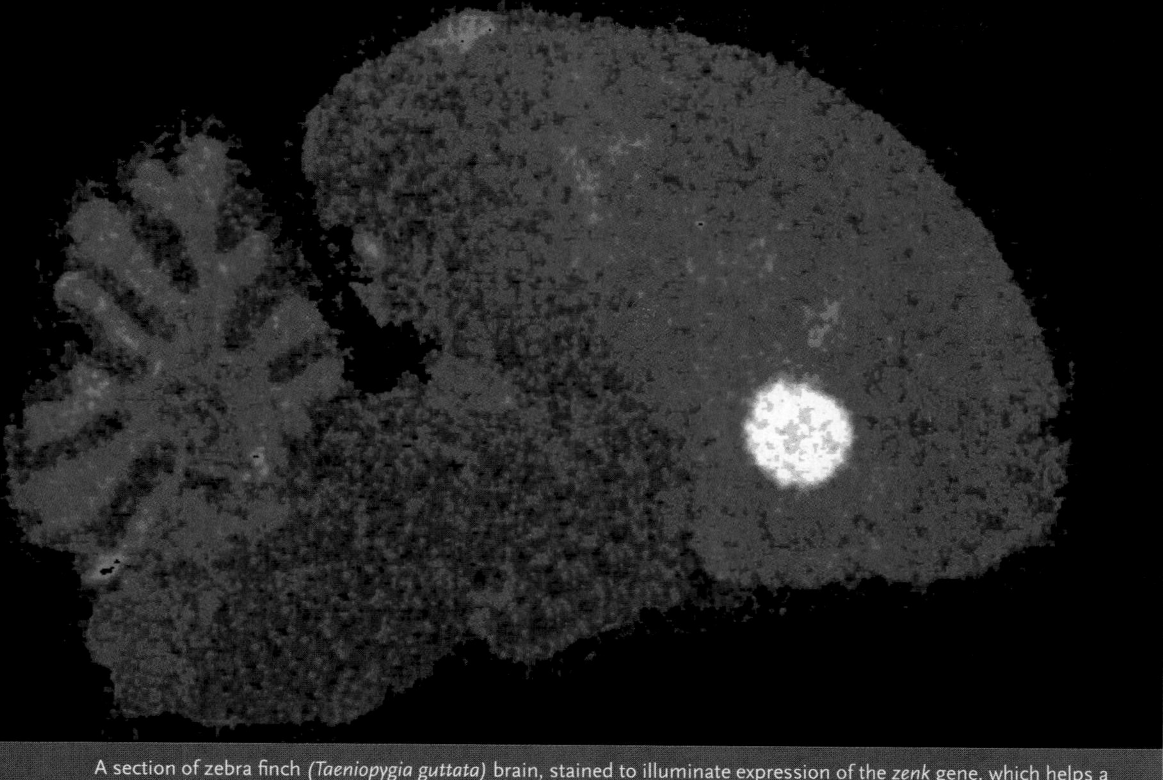

A section of zebra finch *(Taeniopygia guttata)* brain, stained to illuminate expression of the *zenk* gene, which helps a male bird reproduce his species' song.

54

STUDY OUTLINE

The Physiology and Genetics of Animal Behavior

Why It Matters. . . Male white-crowned sparrows *(Zonotrichia leucophrys)* are handsome birds with a song that birdwatchers describe as a "plaintive whistle" followed by a "husky trilled whistle." This distinctive song is a critical part of a male white-crown's **behavioral repertoire,** the set of actions that it can perform in response to stimuli in its environment. An adult male sparrow's song is one of the ways he struts his stuff. The song not only announces his presence to rival males, but it also signals to females that he is available as a potential mate. Experienced birders easily recognize this song, which differs from that of song sparrows *(Melospiza melodia)* and swamp sparrows *(Melospiza georgiana),* as sound spectrograms illustrate **(Figure 54.1).** In fact, every songbird species produces vocal signals that are characteristic of its species and its species alone.

The study of **animal behavior** involves discovering how animals respond to specific stimuli and why they respond in predictable and characteristic ways. A comprehensive approach to animal behavior studies first crystallized in the 1930s, when European researchers—notably Konrad Lorenz, Niko Tinbergen, and Karl von Frisch, who shared a Nobel Prize for their work in 1973—developed the discipline of **ethology,** which focuses on how animals behave in their natural environments. They analyzed how evolutionary processes shape inherited behaviors and the ways that animals respond to specific stimuli. Tinbergen identified four basic questions that any broad study of animal behavior should address: (1) What mechanisms trigger a specific behavioral response? (2) How does the expression of a behavior develop as an animal matures? (3) What is the behavior's function and how does it increase an animal's chances of surviving and reproducing? (4) How did the behavior evolve?

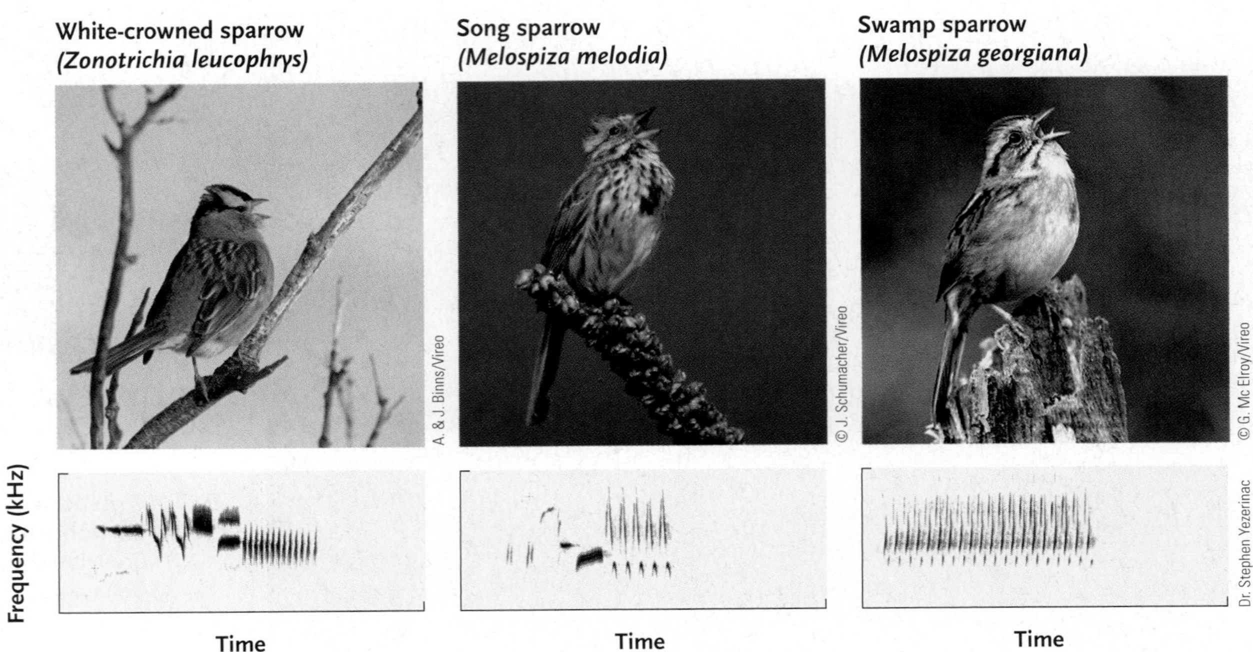
White-crowned sparrow
(Zonotrichia leucophrys)
Song sparrow
(Melospiza melodia)
Swamp sparrow
(Melospiza georgiana)
A. & J. Binns/Vireo
© J. Schumacher/Vireo
© G. Mc Elroy/Vireo
Dr. Stephen Yezerinac
Frequency (kHz)
Time
Time
Time

FIGURE 54.1
Songbirds and their songs. Sound spectrograms (visual representations of sound graphed as frequency versus time) illustrate differences in the songs of the white-crowned sparrow, the song sparrow, and the swamp sparrow.

Advances in **neuroscience**—the integrated study of the structure, function, and development of the nervous system—now allow researchers to explore the first and second questions in detail. Comparable advances in genetic analysis and evolutionary theory enable scientists to address the third and fourth questions. In this chapter, we examine the *proximate causes* of behavior—the genetic, cellular, physiological, and anatomical mechanisms that underlie an animal's ability to detect internal stimuli and environmental cues and react to them in species-specific ways. In Chapter 55, we consider the *ultimate causes* of animal behavior—its adaptive value and evolution. <

54.1 Genetic and Environmental Contributions to Behavior

For many years, animal behaviorists debated whether animals are born with the ability to perform most behaviors completely or whether experience is necessary to shape their actions. However, extensive research in neuroscience has demonstrated that no behavior is determined entirely by genetics or entirely by environmental factors. Instead, behaviors develop through complex gene–environment interactions. We illustrate such an interaction with a detailed description of the process through which male white-crowned sparrows learn their adult song.

Why do adult male white-crowned sparrows sing a song that no other species sings? One possible explanation is that they possess an innate (inborn) ability to produce their particular song—an ability so reliable that young males sing the "right" song the first time they try. According to this hypothesis, their distinctive song would be an example of an **instinctive behavior,** a genetically "programmed" response that appears in complete and functional form the first time it is used. An alternative hypothesis is that they acquire the song as a result of certain experiences, such as hearing the songs of adult male white-crowns that live nearby. In other words, this species' distinctive song might be an example of a **learned behavior,** one that is dependent upon having a particular kind of experience during development.

How can we determine which of these two hypotheses is correct? If the white-crowned sparrow's song is instinctive, isolated male nestlings that have never heard other members of their species should be able to sing their species' song when they mature. But if the learning hypothesis is correct, young birds deprived of certain essential experiences should not sing "properly" when they become adults.

In a set of pioneering experiments conducted at Rockefeller University, Peter Marler tested these alternative hypotheses. He took newly hatched white-crowns from nests in the wild and reared them individually in soundproof cages in his laboratory. Some of the chicks listened to recordings of a male white-crowned sparrow's song when they were 10 to 50 days old; others did not. The juvenile males in both groups first started to vocalize when they were about 150 days old. For many days, they produced whistles and twitters that only vaguely resembled the songs of adults. But gradually the young males that had listened to tapes of their species' song began to sing better and better approximations of that song. At about 200 days of age, they were right on target, producing a song that was nearly indistinguishable from the one they had heard months before. By contrast, males in the group that had not heard tape-recorded white-crown songs never came close to singing the way wild males do.

These results revealed that learning is essential for a young male white-crowned sparrow to acquire the full song of its spe-

TABLE 54.1	Instinctive Behaviors and Learned Behaviors	
	Behavior	**Description**
Instinctive behaviors (strong genetic basis)	Fixed action patterns	Stereotyped actions, often initiated by a sign stimulus
	Feeding behaviors	Innate food preferences and hunting tactics
	Defensive behaviors	Responses to predators
	Reproductive behaviors	Mating habits and parental care activities
Learned behaviors (strong experiential basis)	Imprinting	Affinity for caretaker species, developed during critical period
	Classical conditioning	Association between phenomena that are usually unrelated
	Operant conditioning	Trial-and-error learning
	Cognition	Insight learning in novel situations
	Habituation	Loss of responsiveness

cies. Although birds isolated as nestlings did sing instinctively, they needed the acoustical experience of listening to their species' song early in life if they were to reproduce it months later. We can therefore reject the hypothesis that white-crowned sparrows hatch from their eggs with the ability to produce the "right" song. Their species-specific song—and presumably those of other songbirds—has both instinctive and learned components.

Although early researchers generally classified behaviors as *either* instinctive *or* learned, we now know that most behaviors include *both* instinctive and learned components. Nevertheless, some behaviors have a strong instinctive basis, whereas others are mostly learned **(Table 54.1)**. Researchers use a variety of techniques to unravel the instinctive and learned components of specific behaviors. In some studies, observations and experiments based on simple environmental manipulation are sufficient. In others, researchers conduct more elaborate deprivation experiments (as described above), breeding studies, or experiments based on gene knockouts.

STUDY BREAK 54.1

1. **What is the difference between an instinctive behavior and a learned behavior?**
2. **How did the isolation of young male sparrows in soundproof cages allow Marler to conclude that learning was important to song acquisition?**

54.2 Instinctive Behaviors

Instinctive behaviors—which are often grouped into functional categories, such as feeding behaviors, defensive responses, mating behaviors, and parental care activities—can be performed without the benefit of prior experience. We therefore assume that they have a strong genetic basis and that natural selection has preserved them as adaptive behaviors.

Many Instinctive Behaviors Are Highly Stereotyped

Many instinctive behaviors are highly stereotyped; in other words, when triggered by a specific cue, they are performed over and over in almost exactly the same way. Such behaviors are called **fixed action patterns,** and the simple cues that trigger them are called **sign stimuli.** For example, sign stimuli and fixed action patterns govern the transfer of food from herring gull *(Larus argentatus)* parents to their offspring. Researchers found that very young chicks secure food from their parents through a begging response (the fixed action pattern), which is triggered when they see a red spot on the lower bill of an adult (the sign stimulus). This cue "releases" the begging behavior of hungry baby gulls, which peck at the spot on the parent's bill. In turn, the tactile stimulus delivered by the pecking chick serves as a sign stimulus that induces the adult bird to regurgitate food stored in its crop. The baby gulls then feed on the chunks of fish, clams, or other food that lie before them. We know that the spot on the parent's bill releases the begging response of the young gull because the same response is triggered by an artificial bill that looks only vaguely like a herring gull's bill, provided it has a dark contrasting spot near the tip **(Figure 54.2)**. Thus, even very simple cues can activate fixed action patterns.

Human infants often respond innately to the facial expressions of adults. For example, researchers can trigger smiling in even very young babies simply by moving a mask toward the infant, as long as the mask possesses two simple, diagrammatic eyes. Clearly the infant, like a nestling herring gull, is not reacting to every feature of a face; instead it focuses on simple cues, which function as sign stimuli that release a fixed behavioral response.

Natural selection has molded the behavior of some parasitic species to exploit the relationship between sign stimuli and fixed action patterns for their own benefit. For example, birds that are brood parasites lay their eggs in the nests of other species (see *Why It Matters* at the beginning of Chapter 51). When the brood parasite's egg hatches, the alien nestling mimics and even exaggerates sign stimuli that are ordinarily exhibited by its hosts' own chicks: opening its mouth, bobbing its head, and calling vigorously. These exaggerated behaviors elicit feeding by the foster parents, and the young brood parasite often receives more food than the hosts' own young **(Figure 54.3)**.

Although instinctive behaviors are often performed completely the first time an animal responds to a stimulus, they can be modified by an individual's experiences. For example, the fixed action patterns of a young herring gull change through

FIGURE 54.2 **Experimental Research**

The Role of Sign Stimuli in Parent–Offspring Interactions

Question: What feature of the parent's head triggers pecking behavior in young herring gulls?

Experiment: Niko Tinbergen and A. C. Perdeck tested the responses of young herring gull *(Larus argentatus)* chicks to cardboard cutouts of an adult herring gull's head and bill. They waved these models in front of the chicks and recorded how often a particular model elicited a pecking response from the chicks. One cutout included an entire gull's head with a red spot near the tip of the bill, another cutout included just the bill with the red spot, and the third cutout included the entire head but lacked the red spot.

Result: Young herring gulls pecked at the model of the bill with a red spot almost as often as they pecked at the model of an entire head with a red spot, but they pecked much less frequently at the model of an entire head that lacked a red spot.

Herring gulls *(Larus argentatus)*

Marie Read Natural History Photography

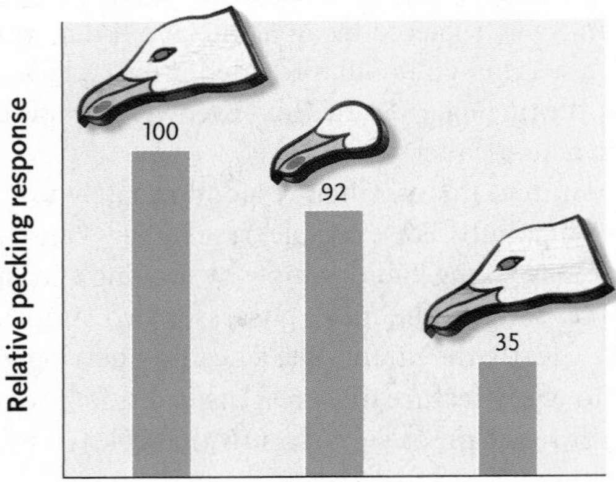
Relative pecking response
100
92
35
Model presented

Conclusion: Begging behavior by young herring gulls is triggered by a simple sign stimulus, the red spot on the parent's bill. Experimental tests revealed that herring gull chicks respond more to the presence of the contrasting spot than they do to the outline of an adult's head.

Source: N. Tinbergen and J. C. Perdeck. 1950. On the stimulus situation releasing the begging response in the newly hatched herring gull chick (*Larus argentatus argentatus* Pont.). *Behaviour* 3:1–39.

© Stephen Dalton/Photo Researchers, Inc.

FIGURE 54.3
Exploitation of a releaser. This young European cuckoo *(Cuculus canorus)*, a brood parasite, stimulates feeding behavior by its foster parent, a hedge sparrow *(Prunella modularis)*. It secures food by displaying exaggerated versions of the sign stimuli used by the host offspring to release feeding behavior by the parents.

time. Although the youngster initially begs by pecking at almost anything remotely similar to an adult gull's bill, it eventually learns to recognize the distinctive visual and vocal features associated with its parents. The chick uses this information to become increasingly selective about the stimuli that will elicit its begging behavior. Thus, instinctive behaviors can be modified in response to particular experiences during their early performances.

Behavioral Differences between Individuals May Reflect Underlying Genetic Differences

Because the performance of instinctive behaviors does not depend on prior experience, behavioral differences between individuals may reflect genetic differences between them. Stevan Arnold, then at the University of Chicago, tested that hypothesis by studying the innate responses of captive newborn garter snakes *(Thamnophis elegans)* to the olfactory stimuli provided by potential food items that they had never before encountered. Arnold measured the snakes' responses to cotton swabs that had been dipped in a smelly extract of banana slug *(Ariolimax columbianus)*, a shell-less mollusk. A snake "smells" by tongue-flicking, which draws volatile chemicals into a special sensory organ in the roof of its mouth. If the young snake had been born to a mother captured in coastal California, where adult garter snakes regularly eat banana slugs, it almost always began tongue-flicking at the slug-scented cotton swab **(Figure 54.4)**. By contrast, newborn snakes whose parents came from inland California, where banana slugs do not occur, rarely tongue-flicked at the swabs. Thus, although the coastal and inland snakes belong to the same species, their instinctive responses to the volatile chemicals associated with banana slugs were markedly different.

A. **Banana slug**

Image copyright Dee Golden, 2010. Used under license from Shutterstock.com

B. **Adult coastal garter snake eating a banana slug**

Stevan Arnold

C. **Newborn coastal garter snake "smelling" slug extract**

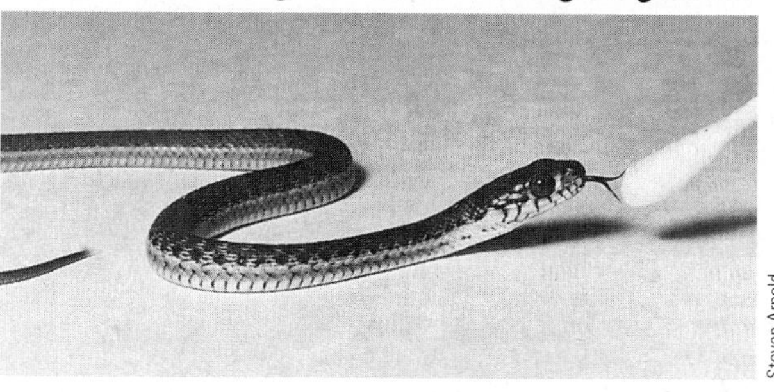
Stevan Arnold

FIGURE 54.4
Genetic control of food preference. **(A)** Banana slugs *(Ariolimax columbianus)* are a preferred food of **(B)** an adult garter snake *(Thamnophis elegans)* from coastal California. **(C)** A newborn garter snake from a coastal population flicks its tongue at a cotton swab drenched with tissue fluids from a banana slug.

In another experiment, Arnold tested whether newborn snakes would feed on bite-sized chunks of slug. After a brief flick of the tongue, 85% of the newborn snakes from a coastal population routinely struck at the slug and swallowed it, despite having had no prior experience with this prey. By contrast, only 17% of newborn snakes from the inland population ate slugs consistently, even when no other food was available. Arnold hypothesized that coastal and inland garter snakes possess different alleles at one or more gene loci controlling their odor-detection mechanisms, leading to differences in their behavior. To test this hypothesis, Arnold crossbred coastal and inland snakes. If genetic differences contribute to the different food preferences of the two snake populations, then hybrid offspring, which receive genetic information from each parent, should behave in an intermediate fashion. Results of the experiment confirmed his prediction: when presented with bite-sized chunks of slug, 29% of the newborn snakes of mixed parentage consumed them every time.

Many additional experiments have confirmed that genetic differences between individuals can translate into behavioral differences between them. *Insights from the Molecular Revolution* describes a striking example of a single gene that influences the grooming behavior of mice. Bear in mind that single genes do not control complex behavior patterns directly. Instead, the alleles present affect the kinds of enzymes that cells can produce, influencing the biochemical pathways involved in the development of an animal's nervous system. The resulting neurological differences can translate into a behavioral difference between individuals that have certain alleles and those that do not.

STUDY BREAK 54.2 <

1. **How do the chicks of brood parasites stimulate unwitting foster parents to feed them?**
2. **How did Arnold demonstrate that the receptiveness of garter snakes to a meal of banana slugs had a genetic basis?**

54.3 Learned Behaviors

Unlike instinctive behaviors, learned behaviors are not performed completely the first time an animal responds to a specific stimulus. Instead, they change in response to environmental stimuli that an individual experiences as it develops.

Behavioral scientists generally define **learning** as a process in which experiences change an animal's behavioral responses. Different types of learning occur under different environmental circumstances. In this section we consider *imprinting, classical conditioning, operant conditioning, cognition,* and *habituation.*

Some animals learn the identity of a caretaker or the key features of a suitable mate during a *critical period,* a restricted stage of development early in life. This type of learning is called **imprinting.** For example, newly hatched geese imprint on their mother's appearance and identity, staying near her for months. And when they reach sexual maturity, they try to mate with other geese, which exhibit the visual and behavioral stimuli on which they had imprinted as youngsters. When Konrad Lorenz, one of the founders of ethology, tended a group of newly hatched greylag geese *(Anser anser),* they imprinted on him instead of an adult of their own species **(Figure 54.5).** The male geese not only fol-

Thomas D. Mcavoy/Time Life Pictures/Getty Images

FIGURE 54.5
Imprinting. Having imprinted on him shortly after hatching, young greylag geese *(Anser anser)* frequently joined Konrad Lorenz for a swim.

INSIGHTS FROM THE Molecular Revolution

A Knockout by a Whisker: What is the function of disheveled genes in mice?

Almost all eukaryotic organisms share a series of developmental interactions called the *wingless/Wnt* pathway. The pathway was originally discovered in fruit flies, in which mutations in the pathway cause changes in the wings and other segmental structures.

Research Question

The mouse genome includes three genes closely related to *disheveled (dsh)*, one of the genes in the *Drosophila wingless/Wnt* pathway. No function has been identified for the proteins they encode, but they are highly active in both embryos and adults. Their function must be important, but what could it be?

Experiment

Nardos Lijam and his coworkers in several laboratories, including Case Western University, the Universities of Colorado and Maryland, and the National Institutes of Health in Bethesda, Maryland, answered this question by developing a line of mice that lacked *Dvl1*, one of the *disheveled* genes. First they constructed an artificial copy of *Dvl1* with the central section scrambled so that no functional proteins could be made from its encoded directions **(Figure A).** Next they introduced the artificial gene into embryonic mouse cells and injected the cells into very early mouse embryos **(Figure B).** Some mice grown from these embryos were heterozygotes, with one normal copy of the *Dvl1* gene and one nonfunctional copy. Matings between heterozygotes produced some individuals that carried two copies of the altered *Dvl1* gene and no normal copies. Individuals in which the normal gene is eliminated are called knockout mice for the missing gene **(Figure C).** (The procedure for making knockout mice is described in Section 18.2.)

Surprisingly, the knockout mice grew to maturity with no apparent morphological defects in any tissue examined, including the brain. Their motor skills, sensitivity to pain, cognition, and memory all appeared to be normal. However, their social behavior was a different story. When housed with normal mice, the knockouts failed to take part in common social activities: grooming, tail pulling, mounting, and sniffing. Rather than building nests and sleeping in huddled groups, as normal mice do, the knockouts tended to sleep alone in half-built nests. Heterozygotes for the *Dvl1* gene—that is, mice with one normal and one altered copy of the gene—behaved normally in all these social activities.

The knockout mice also jumped around wildly in response to abrupt, startling sounds while the response of normal mice was less extreme **(Figure D).** Because a specific neural circuit in the brain inhibits the startle response of normal mice, the reaction of the knockout mice suggested that this inhibitory circuit was probably altered. Humans with schizophrenia, obsessive-compulsive disorders, Huntington disease, and some other brain dysfunctions also show an intensified startle reflex similar to that of the *Dvl1* knockout mice.

Conclusion

The analysis revealed that the *Dvl1* gene modifies developmental pathways affecting complex social behavior in mice, and probably in other mammals. It was one of the first genes affecting mammalian behavior to be identified. The similarity in startle-reflex intensity between the knockout mice and humans with neurological or psychiatric disorders also suggests that mutations in the *Dvl* genes and the *wingless* developmental pathway may underlie some human mental conditions. If so, further studies of the *Dvl* genes may give us clues to the molecular basis of these ailments as well as a possible means to their cure.

Source: N. Lijam et al. 1997. Social interaction and sensorimotor gating abnormalities in mice lacking *Dvl1*. *Cell* 90:895–905.

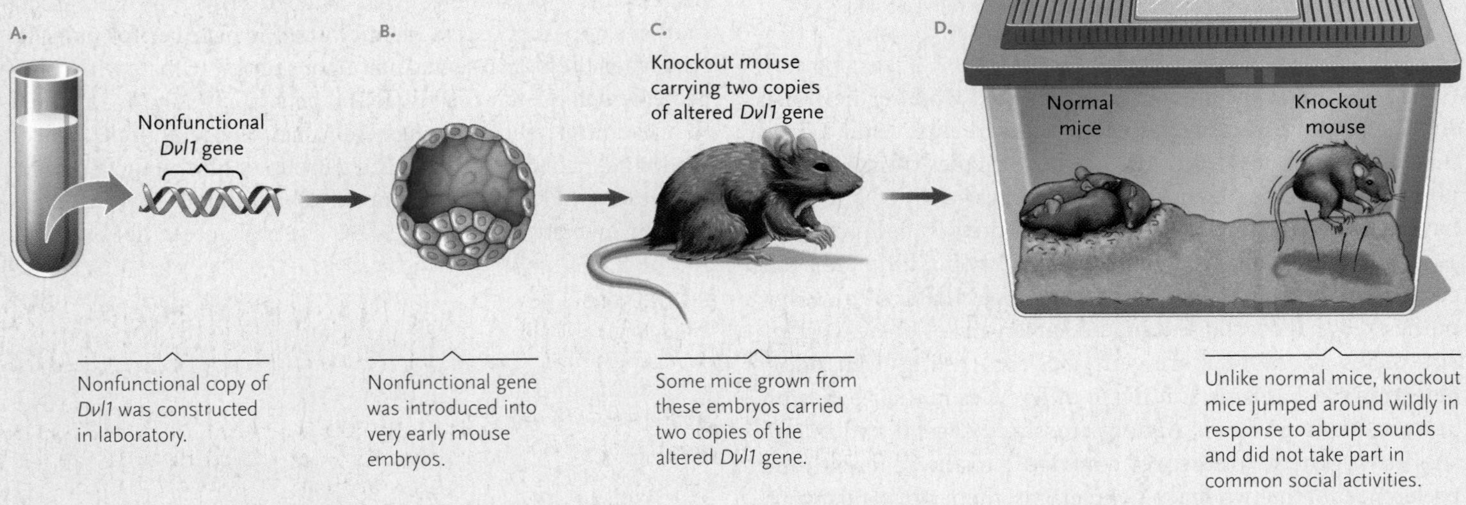
A.
Nonfunctional Dvl1 gene
B.
C.
Knockout mouse carrying two copies of altered Dvl1 gene
D.
Normal mice
Knockout mouse
Nonfunctional copy of Dvl1 was constructed in laboratory.
Nonfunctional gene was introduced into very early mouse embryos.
Some mice grown from these embryos carried two copies of the altered Dvl1 gene.
Unlike normal mice, knockout mice jumped around wildly in response to abrupt sounds and did not take part in common social activities.

lowed Lorenz about, but they also courted humans when they achieved sexual maturity.

Other forms of learning can occur throughout an animal's lifetime. Russian physiologist Ivan Pavlov's classic experiments with dogs explored **classical conditioning,** a type of learning in which animals develop a mental association between two phenomena that are usually unrelated. Dogs generally salivate when they eat. The food is called an *unconditioned stimulus* because the dogs respond to it instinctively; no learning is required for the stimulus (food) to elicit the response (salivation).

Courtesy of Bernd Heinrich, University of Vermont

FIGURE 54.6
Cognition in ravens *(Corvus corax)*. Ravens quickly figured out that they could retrieve food dangling on a string by repeating a series of actions in which they pulled up a loop of string and then held it in place.

In his experiment, Pavlov rang a bell just before offering food to dogs. After about 30 trials in which dogs received food immediately after the bell rang, the dogs associated the bell with feeding time, and they drooled profusely whenever it rang—even when no food was forthcoming. Thus, the bell became a *conditioned stimulus,* one that elicited a particular learned response. In classical conditioning, an animal learns to respond to a conditioned stimulus when it precedes an unconditioned stimulus that normally triggers the response. For example, your cat may become exceptionally friendly whenever she hears the sound of a can opener, another example of classical conditioning.

In another form of associative learning, called trial-and-error learning or **operant conditioning,** animals learn to link a voluntary activity, called an *operant,* with its favorable consequences, called a *reinforcement.* For example, a laboratory rat will explore a new cage randomly. If the cage is equipped with a bar that releases food when it is pressed, the rat will eventually lean on the bar by accident (the operant) and immediately receive a morsel of food (the reinforcement). After just a few such experiences, a hungry rat will learn to press the bar in its cage more frequently—as long as bar-pressing behavior is followed by access to food. Laboratory rats have also learned to press bars to turn off disturbing stimuli, such as bright lights.

Many researchers have wondered whether animals can solve a novel problem by using insight, "thinking up" a solution rather than making a series of trial-and-error attempts at resolving it. This process, called **cognition,** implies that an animal is aware of its circumstances, defines a specific goal, and then uses reasoning to achieve the goal. Bernd Heinrich of the University of Vermont demonstrated that common ravens *(Corvus corax)* used cognition to solve a problem that they (and presumably their ancestors) had never before encountered: how to capture a morsel of food that dangled below their perches on a string **(Figure 54.6).** After looking at the string and looking at the food, a raven started to pull on the string. It pulled up a loop of string, secured it against the perch with its foot, and then reached down again to pull up another loop. After repeating these actions six or eight times, the raven gained access to the food.

Animals typically lose their responsiveness to frequent stimuli that are not quickly followed by the usual reinforcement. This learned loss of responsiveness, called **habituation,** saves the animal the time and energy of responding to stimuli that are no longer important. For example, the sea hare *Aplysia,* a shell-less mollusk, typically responds to a touch on the side of its body by retracting its delicate gills, a response that helps protect it from approaching predators. But if an *Aplysia* is touched repeatedly over a short period of time with no harmful consequences, it stops retracting its gills.

STUDY BREAK 54.3

1. **Dogs typically wag their tails when they see their owners pick up a leash. What kind of learning does this demonstrate?**
2. **What type of learning allows you to sleep through your alarm clock when it rings to awaken you for biology class?**

THINK OUTSIDE THE BOOK

Observe a dog or a cat and analyze its behavior over a few days. Try to identify examples of behaviors that result from classical conditioning, operant conditioning, and habituation. Explain why you think those types of learning produced the behaviors you observed.

54.4 Neurophysiological Control of Behavior

Research in neuroscience has shown that all behavioral responses, even those that are either mostly instinctive or mostly learned, depend on an elaborate physiological foundation provided by the biochemistry and structure of neurons (nerve cells). The neurons that regulate an innate response as well as those that make it possible for an animal to learn something are products of a complex developmental process in which genetic information and environmental contributions are intertwined. Although the anatomical and physiological basis for some behaviors is present at birth, an individual's experiences alter cells of its nervous system in ways that produce particular patterns of behavior. In this section we use examples from research on the singing behavior of songbirds to explore general principles about the physiological basis of behavior that apply to many other kinds of animals.

Discrete Neural Circuits in Specific Brain Regions Control Singing Behavior in Songbirds

Marler's experiments (see Section 54.1) help explain the physiological underpinnings of singing behavior in male white-crowned sparrows. If acoustical experience shapes this behavior, a sparrow chick's brain must be able to acquire and store information present in the songs of other males. Then, months later, when the young male starts to sing, its nervous system must have special features that enable the bird to match its vocal output to the stored memory of the song that it had heard earlier. Eventually, when it achieves a good match, the sparrow's brain must "lock" on the now complete song and continue to produce it when the bird is singing.

Additional experiments have provided detailed information about the nature of the sparrow's nervous system. Young birds that did not hear taped song during their critical period, between 10 and 50 days old, never produced the full song of their species, even if they heard it later in life. In addition, young birds that heard recordings of *other* bird species' songs during the critical period never generated replicas of those songs as they matured. These and other findings suggested that certain neurons in the young male's brain are influenced only by appropriate stimuli, namely the acoustical signals from individuals of its own species, and only during the critical period. Neuroscientists have identified the neuron clusters, called *nuclei* (singular, *nucleus*), that make song learning and song production possible.

Moreover, every behavioral trait appears to have its own neural basis. For example, a male zebra finch, *Taeniopygia guttata* **(Figure 54.7)**, another songbird, can discriminate between the songs of strangers and the songs of established neighbors on adjacent **territories.** (In many bird species, territories are plots of land, defended by individual males or breeding pairs, within which the territory holders have exclusive access to food and other necessary resources. Territories are discussed further in Chapter 55.) The ability to discriminate between the songs of neighbors and those of strangers also involves a nucleus in the forebrain. Cells in this nucleus fire frequently the first time that the song of a new zebra finch is played to a test subject. But as the song is played again and again, these cells cease to respond, indicating that the bird becomes habituated to a now familiar song, although it still reacts to the songs of strangers. The neurophysiological networks that make this selective learning possible enable male zebra finches to behave differently toward familiar neighbors, which they largely ignore, and unfamiliar singers, which they attack and drive away.

© Juniors Bildarchiv/Alamy

FIGURE 54.7
Zebra finches. Native to Indonesia, zebra finches *(Taeniopygia guttata)*, have played an important role in studies of the physiological basis of song learning. The male has a striped throat.

The Activation of Specific Genes Fosters the Development of Nuclei That Regulate a Bird's Song

The role of genes in learning has been identified by research using new molecular and cellular techniques that reveal when a specific gene is active in neurons. When a bird is exposed to relevant acoustical stimuli, such as the songs of potential rivals of its own species, certain genes are "turned on" within neurons in the song-controlling nuclei of the bird's brain. For example, when a zebra finch hears the elements of its species' song, a gene called *zenk* becomes active in the brain, producing an enzyme that changes the structure and function of the neurons (see photo on p. 1245). In effect, the ZENK enzyme programs the neurons of the bird's brain to "anticipate" key acoustical events of potential biological importance. When these events occur, they trigger additional changes in the bird's brain that affect its actions. As a result, a territory owner habituates to (that is, learns to ignore) a singing neighbor with which it has already adjusted territorial boundaries; but it retains the ability to detect and repel new intruders of its own species, which represent a real threat to its continued control of its territory.

STUDY BREAK 54.4 <

1. **What research results suggest that certain neurons in a young male bird's brain are influenced only by acoustical signals from members of its own species and only during a critical period?**
2. **What happens to cells in a nucleus in the forebrain of a zebra finch after it hears a neighboring bird's song many times?**
3. **What is the role of the ZENK enzyme in song learning?**

54.5 Hormones and Behavior

Research on many animal species has revealed that hormones are the chemical signals triggering the performance of specific behaviors. They often accomplish this function by regulating the development of neurons and neural networks or by stimulating the cells within endocrine glands to release chemical signals.

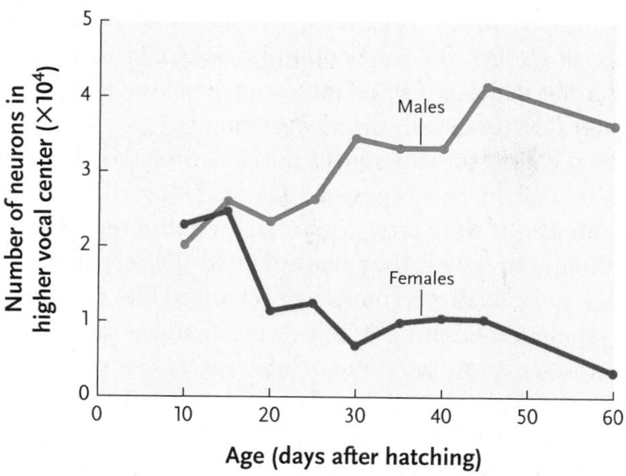

Number of neurons in higher vocal center ($\times 10^4$)
Males
Females
Age (days after hatching)
0
1
2
3
4
5
0
10
20
30
40
50
60

FIGURE 54.8
Hormonally induced changes in brain structure. The brains of young male zebra finches secrete estrogen, a hormone that stimulates the production of neurons in the higher vocal center; these changes contribute to song learning and singing behavior. Lacking this hormone, the brains of young female zebra finches lose neurons in the higher vocal center; females do not sing their species songs.

Hormones Regulate the Development of Cells and Networks That Form the Neural Basis of Behavior

How did the neurons in an adult zebra finch acquire the remarkable capacity to change in response to specific stimuli? In zebra finches, only males produce courtship songs. Very early in a male songbird's life, certain cells in its brain produce the hormone estrogen, which affects target neurons in an area of the developing brain called the *higher vocal center.* The presence of this hormone leads to a complex series of biochemical changes that result in the production of more neurons in parts of the brain that regulate singing. By contrast, the brains of developing females do not produce estrogen, and in the absence of this hormone, the number of neurons in the higher vocal center of females *declines* over time **(Figure 54.8).** Experiments have shown that when young female zebra finches are given estrogen, they produce more neurons in the higher vocal center; but the treated females do not sing later in life unless they are also treated with androgens (male hormones).

Thus, genetically induced hormone production contributes to song learning and singing behavior in male zebra finches by regulating the numbers and types of neurons in the brain centers that produce those behaviors. The development of these neurons primes them for additional changes in response to specific acoustical experiences during the bird's development. Moreover, specific stimuli, such as the songs of either familiar or unfamiliar males, can alter the genetic activity of the neurons that control the behavior of adult birds.

Changing Hormone Concentrations Alter the Behavior of Animals as They Mature

Just as estrogen influences the development of singing ability in zebra finches, other hormones mediate the development of the nervous system in other species. Indeed, a change in the concentration of a certain hormone is often the physiological trigger that induces important changes in an animal's behavior as it matures.

In honeybees *(Apis mellifera),* worker bees perform different tasks for the colony's welfare as they grow older: bees that are less than 15 days old tend to care for larvae and maintain the hive, whereas those that are more than 15 days old often make foraging excursions from the hive to collect the nectar and pollen that bees eat **(Figure 54.9).** These behavioral changes are induced by rising concentrations of juvenile hormone (see Section 40.5), which is released by a gland near the bee's brain. Despite its name, circulating levels of juvenile hormone actually increase as a honeybee gets older (see Figure 54.9).

Juvenile hormone may exert its effect on the bee's behavior by stimulating genes in certain brain cells to produce proteins that affect nervous system function. One such chemical, *octopamine,* stimulates neural transmissions and reinforces memories. It is concentrated in the antennal lobes, a part of the bee's brain that contributes to the analysis of chemical scents in the bee's external environment. Octopamine is present at higher concentrations in the older, foraging bees that have higher levels of juvenile hormone (see Figure 54.9). And when extra juvenile hormone is administered to bees experimentally, their production of octopamine increases. Thus, increased octopamine levels in the antennal lobes may help a foraging bee home in on the odors of flowers where it can collect nectar and pollen.

The honeybee example illustrates how genes and hormones interact in the development of behavior. Genes code for the production of hormones, which change the intracellular environment of assorted target cells. The hormones then directly or indirectly change the genetic activity and enzymatic biochemistry in their targets. If the cells in question are neurons, the changes in their biochemistry translate into changes in the animal's behavior.

Hormone Levels Affect Reproductive Activity in Many Animals

The African cichlid fish *(Haplochromis burtoni)* provides an example of how hormones regulate reproductive behavior. Some adult males maintain nesting territories on the bottom of Lake Tanganyika in East Africa. Territory holders are brightly colored, and they exhibit elaborate behavioral displays that attract egg-laden females to their territories. These males defend their real estate aggressively against neighboring territory holders and against incursions by males that have no territories of their own. By contrast, nonterritorial males are much less colorful and aggressive; they do not control a patch of suitable nesting habitat, and they make no effort to court females.

The behavioral differences between the two types of males are caused by differences in their levels of circulating sex hormones. Recall from Section 47.3 that gonadotropin-releasing hormone (GnRH) stimulates the testes to produce testosterone and sperm. When the circulating testosterone is carried to the brain, it modulates the activity of neurons that regulate sexual and aggressive behavior. In territorial fish the GnRH-producing neurons in the hypothalamus are large and biochemically active,

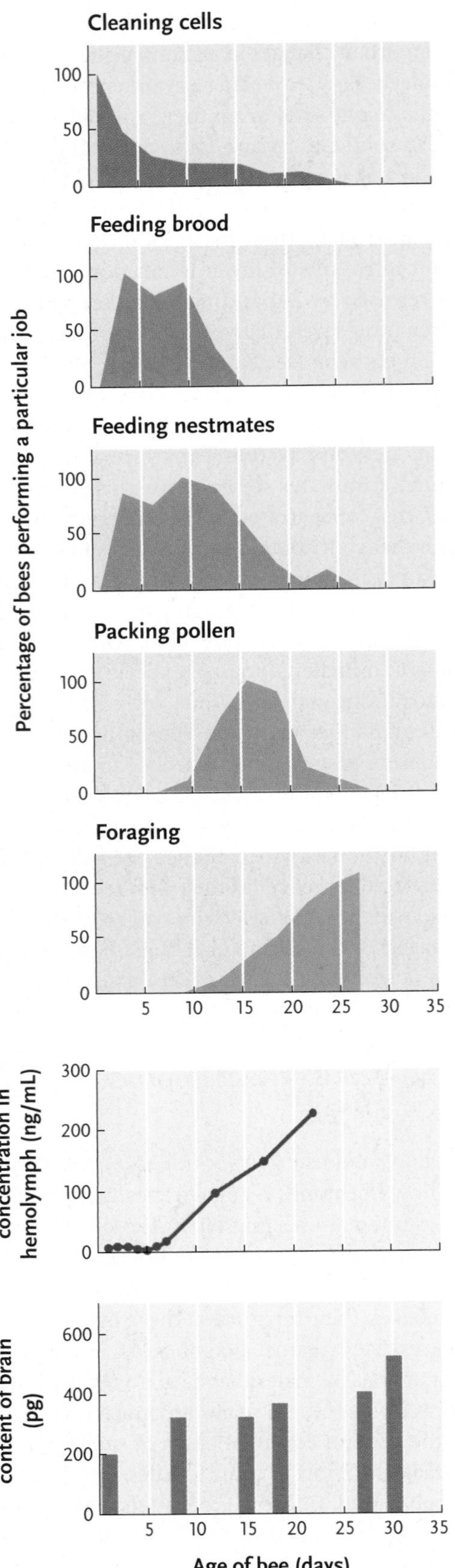
Cleaning cells
Feeding brood
Feeding nestmates
Packing pollen
Foraging
Percentage of bees performing a particular job
100
50
0
5
10
15
20
25
30
35
Mean juvenile hormone concentration in hemolymph (ng/mL)
300
200
100
0
Mean octopamine content of brain (pg)
600
400
200
0
Age of bee (days)

FIGURE 54.9
Age and task specialization in honeybee *(Apis mellifera)* workers. Young bees typically clean cells and feed the brood, and older workers leave the hive to forage for food. These behavioral changes are correlated with rising levels of juvenile hormone in their hemolymph and rising octopamine concentration in their brains.

but in nonterritorial fish they are small and inactive. In the absence of GnRH, the testes do not produce testosterone; the testosterone-deficient fish do not court females with sexual displays, nor do they usually attack other males.

What causes the differences in the neuronal and hormonal physiology of the two types of male fish? Russell Fernald and his students at Stanford University conducted laboratory experiments in which they manipulated the territorial status of males: some territorial males were changed into nonterritorial males; some nonterritorial males were changed into territorial males; and the territorial status of other males was left unchanged as a control **(Figure 54.10)**. Four weeks later, they compared the coloration and behavior as well as the size of the GnRH-producing cells in the brains of the experimental fishes with those of the control males that had retained their original status. Males that had held territories in the past, but had then been defeated by another male, quickly lost their bright colors and stopped being combative. Moreover, their GnRH-producing cells were smaller than those of the successful territory-holding controls. Conversely, males that gained a territory in the experiment quickly developed bright colors and displayed aggressive behaviors toward other males. And the GnRH-producing cells in their brains were larger than those of fishes that had maintained their status as non-territory-holding controls.

The neuronal, hormonal, and behavioral differences between the two experimental groups of males are therefore correlated with a key environmental variable: success or failure in the acquisition and maintenance of a territory. The fish can detect and store information about their aggressive interactions. The neurons that process this information transmit their input to the hypothalamus where it affects the size of the GnRH cells, which in turn dictates the hormonal state of the male. A decrease in GnRH production can turn a feisty territorial male into a subdued drifter, biding his time and building his energy reserves for a future attempt at defeating a weaker male and taking over his territory. If successful in regaining territorial status, the male's GnRH levels will increase again, and the once-peaceful male will revert to vigorous sexual and aggressive behavior.

Note the general similarity of these processes to those described for the white-crowned sparrow's song learning: the fish's brain possesses cells that can change their biochemistry, structure, and function in response to well-defined social stimuli. These physiological changes make it possible for the fish to modify its behavior, depending on its social circumstances. In the next section, we examine how the structure of the nervous system allows animals to respond to important environmental stimuli.

STUDY BREAK 54.5 <

1. **What is the effect of estrogen on the development of neurons in the higher vocal center of young zebra finches?**
2. **How might juvenile hormone production influence a bee's ability to recognize and locate appropriate food sources?**
3. **How does the loss of its territory change the brain chemistry of an African cichlid fish?**

FIGURE 54.10 **Experimental Research**

Effects of the Social Environment on Brain Anatomy and Chemistry

A. **African cichlid fish *(Haplochromis burtoni)***

Nonterritorial male
Territorial male
Russell Fernald, Stanford University

Question: How does the acquisition or loss of a territory affect the brain anatomy and chemistry of an African cichlid fish *(Haplochromis burtoni)*?

Experiment: Fernald and his students housed groups of male cichlids in aquariums in their laboratory. Males that established and maintained territories in the aquariums were brightly colored, whereas those that could not hold territories were pale and drab. The researchers then moved some small territorial males into tanks where larger males had already established territories. The newly introduced males could not establish and maintain territories under these experimental conditions, and therefore changed status from territorial to nonterritorial. The researchers also moved some large nonterritorial males into tanks with smaller territorial males. Under these experimental conditions, the newly introduced males quickly established and maintained territories, changing their status from nonterritorial to territorial. Other males, left in their original tanks so that their territorial status did not change, served as controls. Four weeks later, the researchers examined the brains of the experimental and control fish and measured the size of the neurons that produce GnRH, a hormone that stimulates bright coloration as well as aggressive behavior and mating behavior in males.

Result: The GnRH-producing cells in the brains of experimental males that had lost their territories were much smaller than those in the brains of control males that had maintained their territories. By contrast, the GnRH-producing cells in the brains of experimental males that had gained territories were much larger than those of control males that had never held territories.

B. **GnRH-secreting cells**

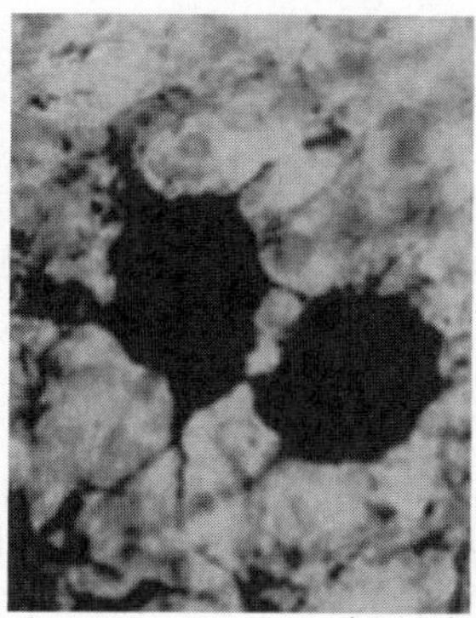

Territorial control 10 μm

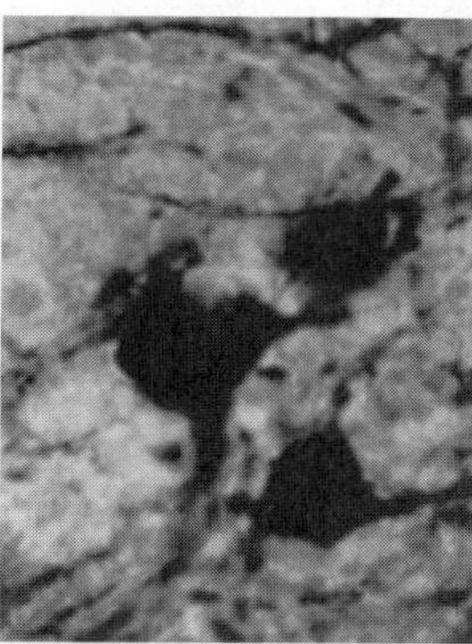

Territorial to nonterritorial experimentals

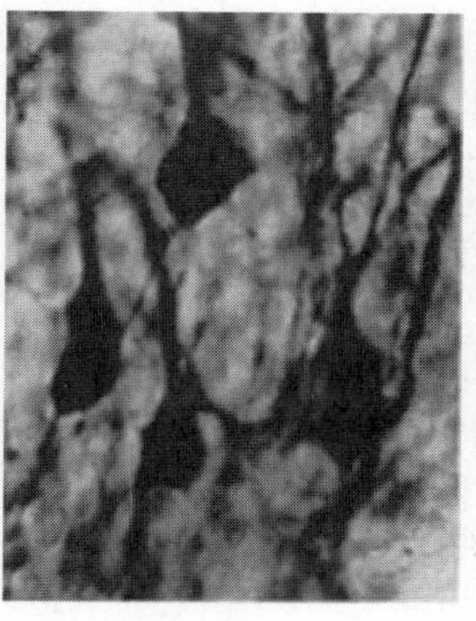

Nonterritorial control

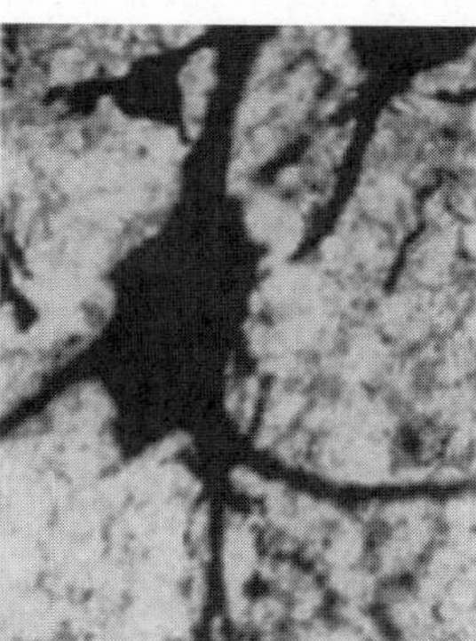

Nonterritorial to territorial experimentals

Conclusion: Changes in social status influence the size of brain cells producing hormones that influence the color and behavior of males.

Source: R. C. Francis et al. 1993. Social regulation of the brain–pituitary–gonadal axis. *Proceedings of the National Academy of Sciences* 90:7794–7798.

54.6 Nervous System Anatomy and Behavior

Although many behaviors result from gene–environment interactions and changes in hormone concentrations, some specific behaviors are produced by the anatomical structure of an animal's nervous system. Studies on a wide range of animal species demonstrate that the nervous systems of many animals provide rapid responses to key stimuli. In other species, sensory systems are structured to acquire a disproportionately large amount of information about those stimuli that are most important to survival and reproductive success.

Hard-Wired Connections between Sensory and Motor Systems Provide Rapid Behavioral Responses to Life-Threatening Stimuli

In some animals, important information acquired by the senses is relayed directly to motor neurons. Such a system provides crickets with a potentially lifesaving predator avoidance behavior.

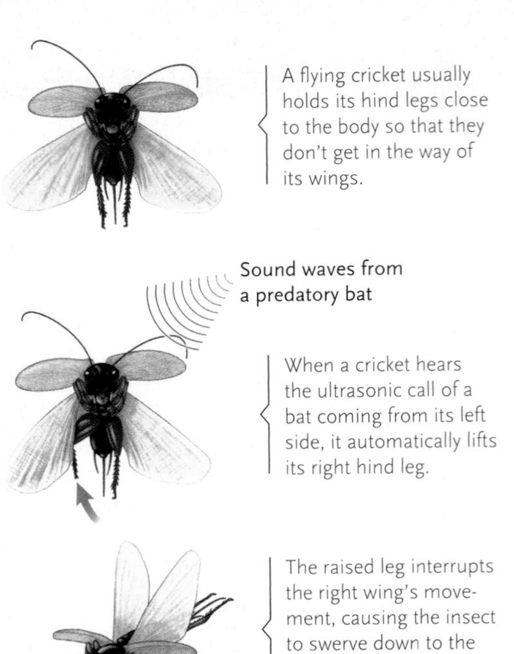
A flying cricket usually holds its hind legs close to the body so that they don't get in the way of its wings.
Sound waves from a predatory bat
When a cricket hears the ultrasonic call of a bat coming from its left side, it automatically lifts its right hind leg.
The raised leg interrupts the right wing's movement, causing the insect to swerve down to the right and away from the approaching predator.

FIGURE 54.11
A neural mechanism for escape behavior in the black field cricket *(Teleogryllus oceanicus)*.

Crickets and some other insects fly mainly at night, a behavior that allows them to avoid day-flying predatory birds. But flying crickets aren't safe even at night, because insect-eating bats can detect them in pitch darkness.

Bats detect potential prey by echolocation (see *Why It Matters* at the beginning of Chapter 39). They call almost continuously while flying at night, and the sound waves they produce bounce off items in their path, creating echoes that the bats hear and use to track their prey. Their vocalizations are of such high frequency (up to 100,000 hertz) that they lie outside the upper limit (20,000 hertz) of unaided human hearing. However, a bat's auditory apparatus and brain not only can hear ultrasound, as these high frequency sounds are called, but also can analyze ultrasonic echoes in a way that permits the bat to identify, approach, and capture flying insect prey. With enemies of this sort in its environment, a cricket flying at night is in real danger of being intercepted and eaten.

Crickets are not defenseless, however. Black field crickets *(Teleogryllus oceanicus)*, for example, hear ultrasound through ears in their front legs (see Figure 39.7). The approach of a calling bat causes sensory neurons connected to the ears to fire. However, to be of any use to the cricket, this information must be translated immediately into evasive action—and crickets have the anatomical and physiological equipment to do exactly that.

Imagine that a bat is zeroing in for the kill, rushing toward the left side of a flying cricket. The cricket's left ear will be bombarded with more intense ultrasound than its right ear, and the neurons that receive input from the ears will also be stimulated unequally. The cricket's nervous system is structured to relay incoming messages from the *left* ear to the motor neurons (muscle-regulating nerve cells) that control the *right* hind leg. Sufficient ultrasonic stimulation on the left side of the body will induce the motor neurons for the right hind leg to fire, causing muscle contractions in that leg. As the right hind leg jerks up, it blocks movement of the right hind wing, reducing the flight power generated on the right side of the cricket's body. The flying cricket then swerves sharply to the right and loses altitude, diving down and away from the approaching bat **(Figure 54.11)**. Thus, the anatomical structure of the cricket's nervous system produces a behavioral response that takes the cricket out of harm's way.

If all goes well for the cricket, it will gain the safety of foliage or leaf litter on the ground before the bat can reach it. Once there, echoes bouncing off the materials all around the cricket will mask any ultrasonic echoes coming from its body. The thwarted bat will be forced to look elsewhere for prey that responds less rapidly.

The Structure of Sensory Systems Allows Animals to Respond Appropriately to Different Stimuli

In some animals, the structure and neural connections of sensory systems allow them to distinguish potentially life-threatening stimuli from those that are more mundane. For example, fiddler crabs *(Uca pugilator)* live and feed on mud flats where they build burrows that provide safe refuge from predators, including crab-hunting shorebirds. But to use its burrow wisely, a crab must be able to distinguish between predatory gulls and its fellow fiddler crabs. Otherwise, it would dash for cover whenever anything moved in its field of vision.

Fiddler crabs possess long-stalked eyes that they hold above their carapace perpendicularly to the ground. John Layne, a neurophysiologist at Duke University, wondered whether a crab might distinguish between dangerous predators and fellow crabs by having a divided field of vision. A large predatory gull sailing in for the kill would stimulate receptors on the upper part of the eye, whereas a fellow crab, whose movements would be slightly below the midpoint of the eyes, would stimulate a lower set of visual receptors. If the receptors above and below the retinal equator relayed their signals to different groups of neurons, the crab's nervous system could be "wired" to provide different responses to the different stimuli.

Layne hypothesized that receptors above the midline of the eye activate neurons that control an escape response, so that stimulation from above would reliably trigger a dash for the burrow. By contrast, a moving stimulus at or below eye level, as when one crab approached another, would provide input to the neurons that allow a crab to behave appropriately to a male or female of its species.

Layne tested this hypothesis by placing crabs on an elevated platform in a glass jar. He then presented the same moving stimulus, a black square, to each crab at two heights; sometimes the stimulus circled the jar above the crab's eyes and sometimes below. Stimuli that activated the upper part of the

FIGURE 54.12 **Experimental Research**

Nervous System Structure and Appropriate Behavioral Responses

Question: Do fiddler crabs respond differently to stimuli that are presented above the midline of their visual field than they do to stimuli presented below it?

Experiment: Layne investigated this question by placing crabs on an elevated platform in a glass jar. He then presented the same moving stimulus, a black square, to each crab at varying heights; sometimes the stimulus circled the jar above the crab's eyes and sometimes below.

Results: Stimuli that activated the upper part of the retina did indeed induce escape behavior, but when the stimuli were below the retinal equator, the animal generally ignored the moving objects altogether.

Fiddler crab *(Uca pugilator)*

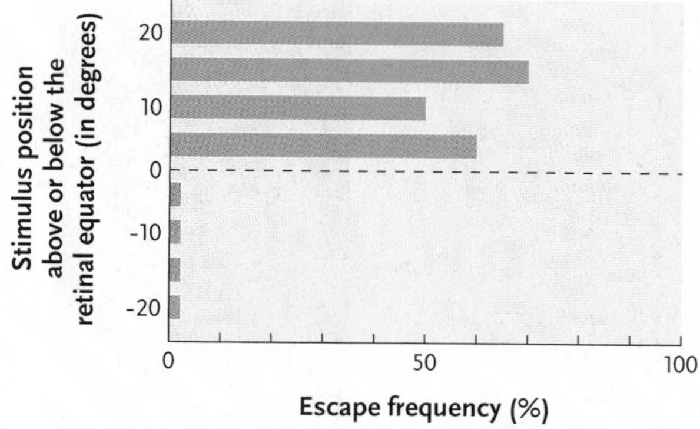
Stimulus position above or below the retinal equator (in degrees)
20
10
0
-10
-20
0
50
100
Escape frequency (%)

Conclusion: Specific nervous system connections between a fiddler crab's eyes and brain provide appropriate responses to different specific stimuli.

Source: J. E. Layne. 1998. Retinal location is the key to identifying predators in fiddler crabs *(Uca pugilator)*. *The Journal of Experimental Biology* 201:2253–2261.

retina did indeed induce escape behavior, but if the stimuli were below the retinal equator, the animal generally ignored the moving objects altogether **(Figure 54.12)**. Thus, specific nervous system connections between a fiddler crab's eyes and brain provide appropriate responses to different specific stimuli.

The Amount of Brain Tissue Devoted to Analyzing Sensory Information Varies from One Sensory System to Another

The match between the structure of an animal's nervous system and the real-world challenges it faces extends beyond the ability to avoid predators. For example, the star-nosed mole *(Condylura cristata)*, which lives in wet tunnels in North American marshlands, spends almost all of its life in complete darkness. Like nocturnal insect-eating bats, the mole must find food without benefit of visual cues; and, like the bats, it has a receptor-perceptual system that enables it to feed effectively. The star-nosed mole subsists largely on earthworms, and it uses its nose to locate them—but not by smell. Instead, as the mole proceeds down a tunnel, 22 fingerlike tentacles from its nose sweep the area directly ahead. These tentacles are covered with thousands of tactile (touch) receptors called Eimer's organs **(Figure 54.13A)**. Sensory nerve terminals in the Eimer's organs generate complex and detailed patterns of signals about the objects they contact. These messages are relayed by neurons to the cortex of the mole's brain, much of which is devoted to the analysis of information from the nose's tactile receptors.

The structural basis of the mole's sensory analysis is apparent when we consider that the amount of brain tissue responding to signals from the mole's nose contains many more cells than do the tissues that decode tactile signals from all other parts of the animal's body combined **(Figure 54.13B)**. Moreover, the brain does not treat inputs from all 22 of the mole's "fingers" equally. Instead, the brain devotes more cells to the tentacles closest to the mole's mouth, and fewer cells to analyzing messages from tentacles that are farther away.

The processing of tactile information in this species is clearly related to the importance of finding food in totally dark underground tunnels. Moreover, the extra attention given to signals from certain tentacles almost certainly helps the mole locate prey that are close to its mouth, allowing it to bite worms before they can move away after being touched.

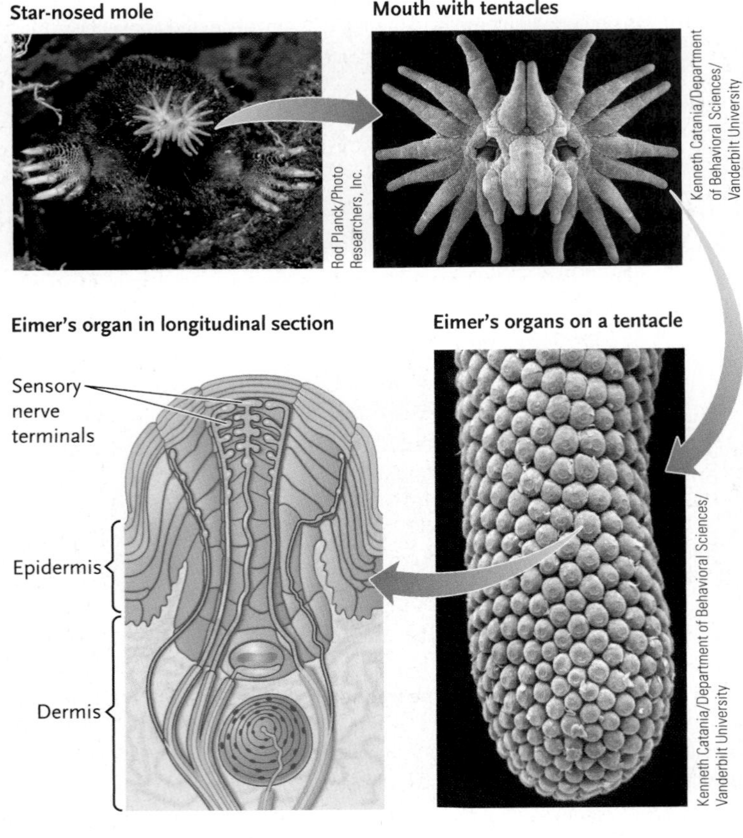
A. Sensory organs on the tentacle of a star-nosed mole
Star-nosed mole
Mouth with tentacles
Rod Planck/Photo Researchers, Inc.
Kenneth Catania/Department of Behavioral Sciences/ Vanderbilt University
Eimer's organ in longitudinal section
Eimer's organs on a tentacle
Sensory nerve terminals
Epidermis
Dermis
Kenneth Catania/Department of Behavioral Sciences/ Vanderbilt University

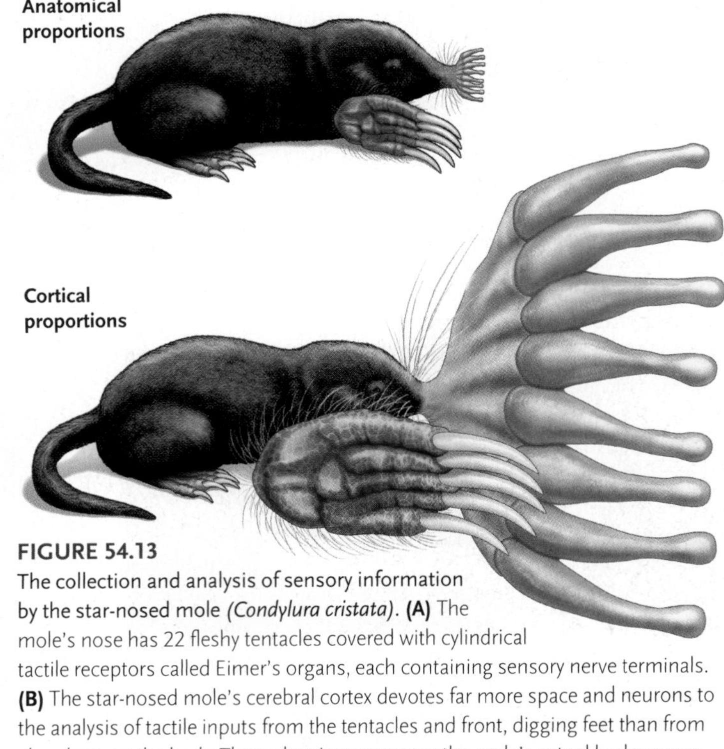
B. Comparison of anatomical proportions and cortical proportions
Anatomical proportions
Cortical proportions

FIGURE 54.13

The collection and analysis of sensory information by the star-nosed mole *(Condylura cristata)*. **(A)** The mole's nose has 22 fleshy tentacles covered with cylindrical tactile receptors called Eimer's organs, each containing sensory nerve terminals. **(B)** The star-nosed mole's cerebral cortex devotes far more space and neurons to the analysis of tactile inputs from the tentacles and front, digging feet than from elsewhere on the body. These drawings compare the mole's actual body proportions with the relative amount of cortical tissue that processes sensory information from the various parts of the body.

As these examples illustrate, animal nervous systems do not offer neutral and complete pictures of the environment. Instead, distorted and unbalanced perceptions of the world are advantageous because certain types of information are far more important for survival and reproductive success than others. In the next chapter, we examine the ecological circumstances and selective forces that have promoted the evolution of specific behaviors.

STUDY BREAK 54.6 <

1. **How does the anatomy of the cricket's nervous system help it avoid an approaching bat?**
2. **What behavioral response is elicited in a fiddler crab when its eyes detect movement above the midline of its visual field?**
3. **Explain the sensory mechanism that allows a star-nosed mole to locate earthworms in its tunnels.**

UNANSWERED QUESTIONS

How does an animal choose which behavior to perform?

Animals must perform many types of behaviors in their lifetimes. But what determines which behavior is most appropriate in any particular situation? In some cases, it appears that an animal performs whichever behavior it is most provoked to do at the time. For example, when the Eimer's organs of a star-nosed mole tell it that its tentacles have contacted an earthworm, the mole responds by biting into whatever is in front of its mouth. In other cases, an animal may perform whichever behavior will bring it the most reward. Researchers are studying how animals choose behaviors, using a combination of methods including theoretical modeling, neuroscience, and molecular biology. Male fruit flies often must choose between aggression and courtship, and molecular manipulations that result in the inability of a fruit fly to discriminate between males and females often result in the fly making the wrong choice. Understanding the behavioral choices that animals make will help us understand both the mechanisms and evolution of behavior.

How does experience change the brain to affect behavior?

Some behaviors develop only after an animal has had certain experiences. How does experience translate into a new or improved behavior? Researchers are tracing the paths from stimulus perception, to activation of specific neural circuits and molecular pathways in the brain, to changes in brain structure and chemistry that then lead to behavioral change. This research is helped enormously by the burgeoning field of molecular neuroscience. Knowing how experience translates into a new or improved behavior has broad implications for our understanding of learning and memory, brain plasticity, brain disease, and recovery from stroke.

How do genes influence behavior?

Genetic differences between individuals can translate into behavioral differences between them. But genes encode proteins, and the road between the transcription of a gene and the performance of a specific behavior is a "long and winding" one. Researchers are studying how genetic differences between individuals influence the biochemical pathways that shape the development of the nervous system or later modify its function. Insights into how genes influence behavior have profound implications for our understanding of brain function and perhaps even policy decisions about screening people for genes that may influence their behavior.

Critical Thinking

Imagine a male bird that is simultaneously confronted by a predator, a potential mate, and a tasty food item. In what order would you expect it to respond to these stimuli?

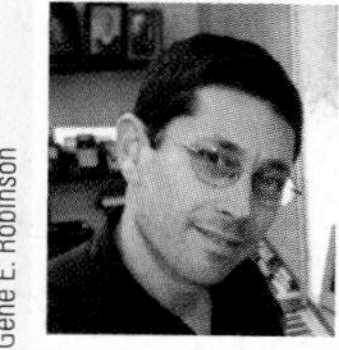
Gene E. Robinson

Gene E. Robinson is the G. William Arends Professor of Integrative Biology at the University of Illinois at Urbana–Champaign, where he studies social behavior and genomics using the honeybee. To learn more about Dr. Robinson's research, go to http://www.life.uiuc.edu/robinson.

REVIEW KEY CONCEPTS

Go to **CENGAGENOW** at www.cengage.com/login to access quizzing, animations, exercises, articles, and personalized homework help.

54.1 Genetic and Environmental Contributions to Behavior

- Although most behaviors have both instinctive and learned components, many are either largely instinctive or largely learned (Table 54.1). Some behaviors are produced only if the animal's nervous system acquires inputs from specific experiences during a critical stage of its development (Figure 54.1).

54.2 Instinctive Behaviors

- Instinctive behaviors are those that an animal performs completely the first time it is presented with a stimulus.
- Fixed action patterns are highly stereotyped behaviors that animals exhibit in response to simple cues called sign stimuli (Figures 54.2 and 54.3). Fixed action patterns often change through time in response to an animal's experiences.
- Behavioral differences between individuals often reflect underlying genetic differences. Research on garter snakes suggests that certain food preferences are genetically based (Figure 54.4).

Animation: Instinctive behavior in infants

Animation: Adaptive behavior in starlings

Animation: Snake taste preference

Animation: Cuckoo and foster parent

54.3 Learned Behaviors

- Learned behaviors develop only after an animal has had certain experiences in its environment. The different forms of learning include imprinting (Figure 54.5), classical conditioning, operant conditioning, and cognition (Figure 54.6). Habituation is a learned loss of responsiveness to specific stimuli.

54.4 Neurophysiological Control of Behavior

- Animal behavior requires an anatomical, physiological, and biochemical foundation based in the nervous system. An individual's experience alters cells of the nervous system in ways that produce particular patterns of behavior.
- The physiological basis of bird singing behavior resides in specific neuron clusters, called nuclei, that communicate with each other in the bird's brain.
- Bird song and some other behaviors develop only after specific genes are activated within the neurons that produce the behavior.

54.5 Hormones and Behavior

- Hormones can mediate the expression of specific behaviors by activating genes that change the biochemistry, morphology, and number of neurons in specific nuclei. Estrogen stimulates the production of neurons in the higher vocal center of male zebra finches (Figures 54.7 and 54.8).
- Age-related changes in hormone levels can alter the behavior of animals over the course of their lives. Changes in juvenile hormone concentration are correlated with changes in task specialization in honeybees (Figure 54.9).
- Behavioral interactions with other individuals can alter an animal's hormone levels, inducing changes in its behavior. Research on male cichlid fishes suggests that variations in coloration and aggressive behavior associated with territorial status and social interactions are mediated by the production of certain hormones (Figure 54.10).

Animation: Hormonal control of behavior

54.6 Nervous System Anatomy and Behavior

- Sensory information enables animals to make adaptive behavioral responses to their environments. In crickets and some other animals, sensory systems relay information to motor systems, inducing an almost instantaneous response (Figure 54.11).
- Nervous systems can generate prompt and effective responses to those environmental stimuli that may have a large impact on an animal's survival and reproductive success. Fiddler crabs respond differently to movements that occur either above or below the midline of their visual field (Figure 54.12).
- Sensory systems are often structured to provide more information about important environmental factors. A large fraction of a star-nosed mole's brain is devoted to analyzing input from tactile receptors on its nose (Figure 54.13).

UNDERSTAND AND APPLY

Test Your Knowledge

1. Marler concluded that white-crowned sparrows can learn their species' song only:
   a. after receiving hormone treatments.
   b. during a critical period of their development.
   c. under natural conditions.
   d. from their genetic father.
   e. if they are reared in isolation cages.
2. Instinctive behaviors are:
   a. performed completely the first time they are used.
   b. always modified by an animal's experiences.
   c. performed only by very young animals.
   d. often the product of habituation.
   e. sometimes called trial-and-error responses.
3. A stimulus that always causes an animal to behave in a highly stereotyped way is called:
   a. fixed action pattern.
   b. an instinct.
   c. habituation.
   d. a sign stimulus.
   e. a reinforcement.
4. Arnold's experiments on the feeding preferences of garter snakes demonstrated that food choice is largely governed by a snake's:
   a. early experiences.
   b. genetics.
   c. size and color.
   d. diet while it was developing inside its mother.
   e. trial-and-error learning.
5. Learning in which an animal associates two phenomena that it experiences at approximately the same time is called:
   a. imprinting.
   b. operant conditioning.
   c. classical conditioning.
   d. insight learning.
   e. habituation.
6. The development of the song system in male songbirds depends on:
   a. direct connections between sensory neurons and motor neurons.
   b. a decrease in the number of neurons in the song system.
   c. the behaviors of females, which stimulate hormone production.
   d. the successful defense of a territory.
   e. the production of estrogen early in life.
7. One of the functions of octopamine in foraging honeybees is to:
   a. increase the production of juvenile hormone.
   b. decrease the production of juvenile hormone.
   c. make the bees defend their territory more aggressively.
   d. stimulate neural transmissions and reinforce memories.
   e. increase the time they spend caring for larvae.
8. In cichlid fishes, high levels of the hormone GnRH:
   a. make females more receptive to male attention.
   b. cause males to be sexually aggressive but not territorial.
   c. stimulate a male to defend its territory.
   d. cause males to abandon their territories.
   e. cause males to lose their bright colors.
9. Sensory bias in the nervous system of a cricket ensures that ultrasound perceived on one side of the body will cause:
   a. a movement in a leg on the same side of the body.
   b. a movement in a leg on the opposite side of the body.
   c. the cricket to respond with a vocalization.
   d. the cricket to stop vocalizing.
   e. the cricket to fly toward the sound.
10. In the brain of a star-nosed mole, more cells decode:
    a. tactile information from its feet than from all other parts of its body.
    b. tactile information from the tentacles on its nose than from all other parts of its body.
    c. tactile information from its mouth than from all other parts of its body.
    d. visual information from the top part of its visual field than the bottom part.
    e. visual information from the bottom part of its visual field than the top part.

Discuss the Concepts

1. One day, while walking in the country, you see a rooster wade into a pond and begin to court a female mallard duck. What probably happened to the rooster early in life?
2. Using an example from your own experience, explain why habituation to a frequent stimulus might be beneficial. Also describe an example in which habituation might be harmful or even dangerous.
3. Is learning always superior to instinctive behavior? If you think so, why do so many animals react instinctively to certain stimuli? Are there some environmental circumstances in which being able to respond "correctly" the first time would have a big payoff?
4. Cockroaches have two small projections called *cerci* at the tip of the abdomen. You suspect that the cerci might be responsible for the insects' ability to detect predators, such as lizards, rushing toward them from behind. Under the microscope you see that each cercus is covered with fine hairs. What properties should these hairs have if they are part of a system that detects moving air pushed ahead by an approaching predator? How might the roach determine whether the danger was coming from the right or left side? How quickly should cercal information be processed compared with information about the chemicals in a food item?

Design an Experiment

You find that some fruit flies in your lab are quick to come to a dish containing citrus oils, but others are not as responsive. How could you test whether these behavioral differences are caused by genetic differences among the flies or environmental differences in their prior experience?

Apply Evolutionary Thinking

Some birds that are frequently kept as pets, such as parrots and myna birds, have the ability to imitate human speech faithfully. Develop a hypothesis that explains why the ability to be a good mimic might have evolved in these species. What features of the birds' brains might be involved in this behavior?

Interpret the Data

Carol B. Lynch of Wesleyan University wondered if differences she observed in the sizes of the nests built by house mice *(Mus musculus)* reflected genetically based differences in their behavior. She selectively bred mice that made especially large nests to create two "high nest weight lines," mice that made especially small nests to create two "low nest weight lines," and mice that built nests of average size as two "control lines." After 15 generations, she measured the amount of cotton that mice in each of the lines pulled into their cages to build nests. Her data are presented in the accompanying graph. Do her results suggest that variations in nest size among the different lines of mice have a genetic basis?

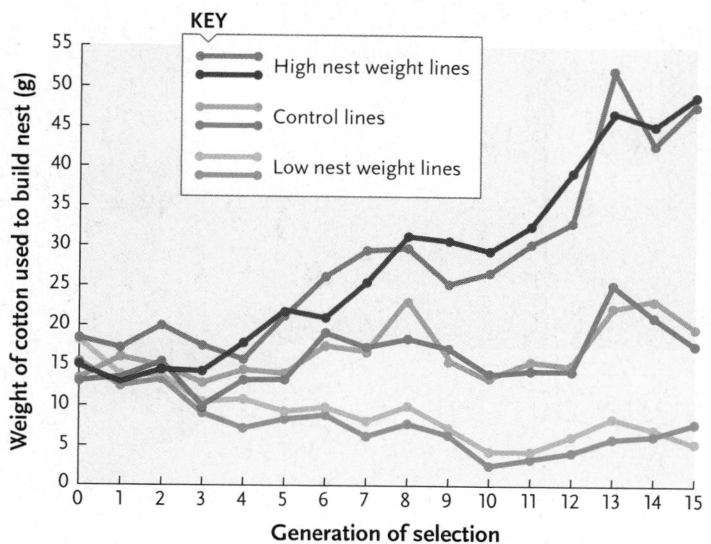
KEY
High nest weight lines
Control lines
Low nest weight lines
Weight of cotton used to build nest (g)
55
50
45
40
35
30
25
20
15
10
5
0
0 1 2 3 4 5 6 7 8 9 10 11 12 13 14 15
Generation of selection

Source: C. B. Lynch. 1980. Response to divergent selection for nesting behavior in *Mus musculus. Genetics* 96:757-765.

55

STUDY OUTLINE

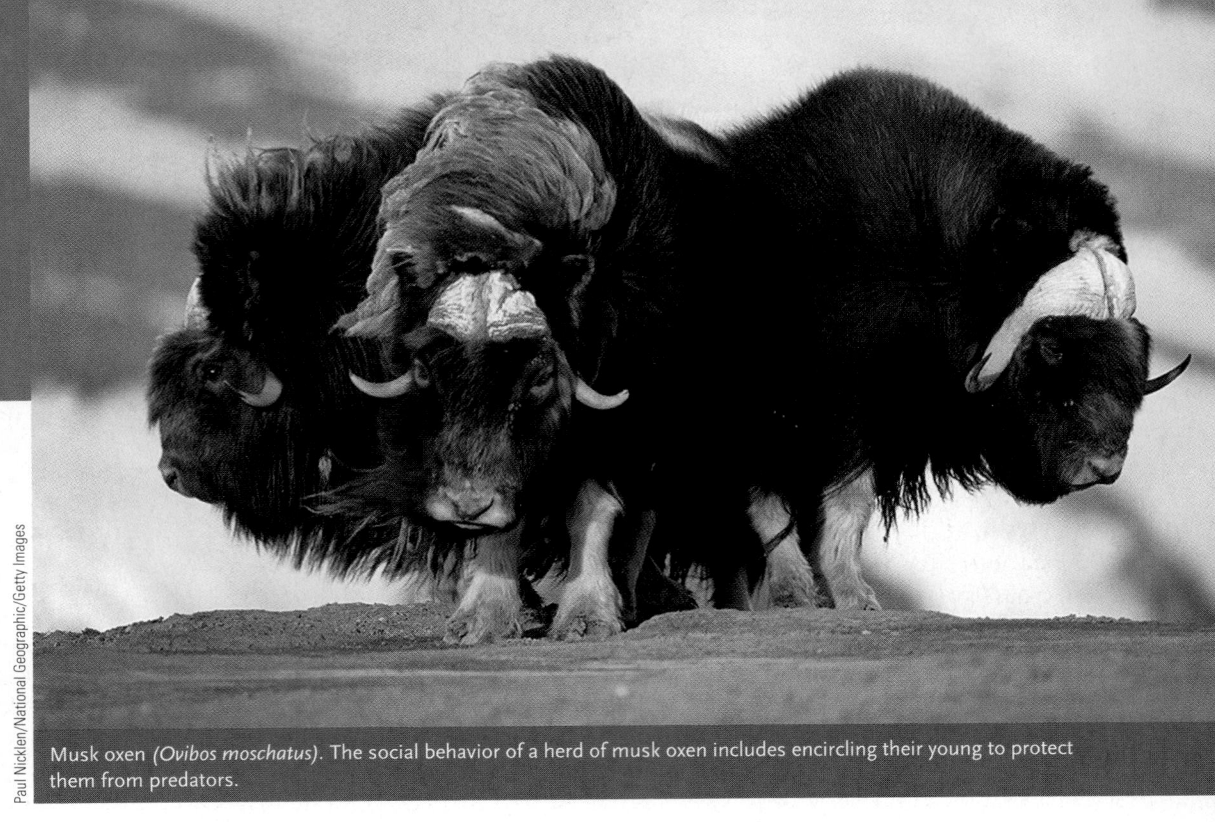

Paul Nicklen/National Geographic/Getty Images

Musk oxen *(Ovibos moschatus)*. The social behavior of a herd of musk oxen includes encircling their young to protect them from predators.

The Ecology and Evolution of Animal Behavior

© Edgar T. Jones/VIREO

FIGURE 55.1
Reproductive success. Parental care is just one of many behaviors required for successful reproduction in white-crowned sparrows *(Zonotrichia leucophrys)* and many other animal species. The number of surviving nestlings will determine the reproductive success of their parents.

Why It Matters. . . In early spring, male white-crowned sparrows *(Zonotrichia leucophrys)* leave their wintering grounds in Mexico and fly thousands of kilometers to their northern breeding range. There, they select patches of habitat that contain the resources necessary for breeding—suitable cover, potential nesting sites, and abundant food. Then, they start to sing and sing, repeating their song thousands of times every day. The songs are a form of communication through which males announce their presence to rival males and to females. Males also perform elaborate courtship behaviors. And once the young hatch, they communicate with their parents, eliciting the care they need before leaving the nest.

All of these behaviors carry significant costs and risks. For example, migration requires enormous energy expenditure; many migrating birds die before completing their trip. Moreover, singing males are conspicuous, and they may attract the attention of a hawk or some other predator. Given the costs and dangers associated with these behaviors, what benefits do the birds gain from performing them? The ultimate evolutionary benefit is obvious: with luck, individuals performing these complex and diverse behaviors may leave surviving offspring **(Figure 55.1)**.

Questions about ultimate benefits are fundamentally different from the questions we considered in the previous chapter, where we focused on *how* underlying physiological and genetic mechanisms enable animals to behave. In this chapter, we try to explain *why* animals behave as they do. Why do sparrows migrate to their breeding grounds, breed in certain habitats but not others, and expose themselves to predation when they sing? The behavior of animals is closely tied to ecological circumstances, and evolutionary biologists view most behaviors as an individual's responses to its environment. Moreover, like morphological traits, behaviors are subject to microevolutionary change (see Chapter 20). If particular alleles contribute even slightly to the development of a behavior that enhances an animal's fitness, natural selection will cause the frequency of those alleles to increase in the next generation.

Behavioral biologists apply ecological and evolutionary analyses to all forms of animal behavior, including those described above. In this chapter, we examine the ecology and evolution of several categories of animal behavior: orientation, navigation, and migration; habitat selection and territoriality; communication; reproductive behavior and mating systems; and social behavior, including behaviors described as altruistic. We close the chapter with a brief look at human behavior. <

FIGURE 55.2
Long-distance migration. Arctic terns *(Sterna paradisaea)* migrate from the high Arctic to Antarctica each year, a round-trip journey up to 70,000 km. This species' summer breeding range is shaded on the map.

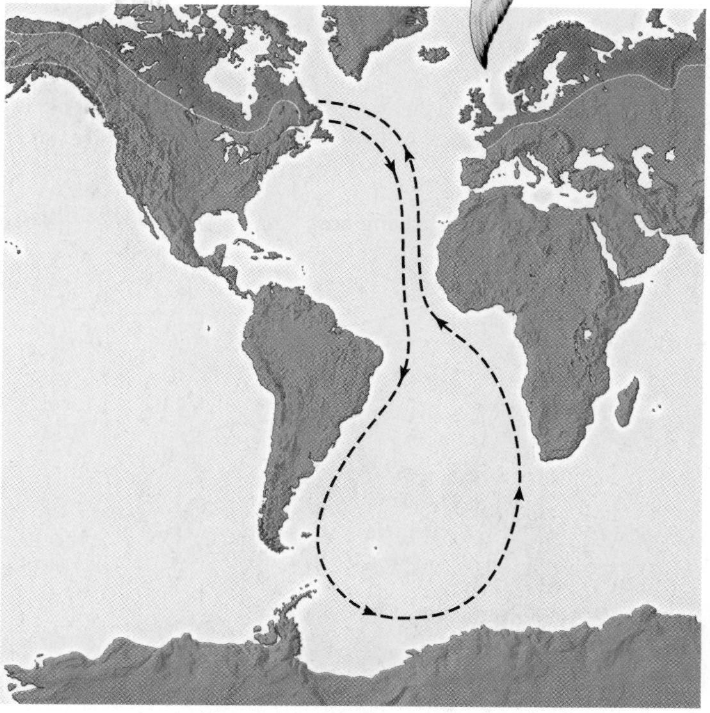

55.1 Migration and Wayfinding

Most animals move through their environments at some stage of their life cycles. Although some species move only short distances to find suitable environmental conditions, many others undertake large-scale movements on a seasonal schedule.

Migrating Animals Make Long Round-Trips on a Seasonal Cycle

Many animal species undertake a seasonal **migration,** traveling from the area where they were born to a distant and initially unfamiliar destination, and returning to their birth site later. The Arctic tern *(Sterna paradisaea),* a seabird, makes an annual round-trip migration of up to 70,000 km **(Figure 55.2).** Many other vertebrate species, including gray whales and salmon, also undertake long and predictable journeys. Even some arthropods migrate long distances. For example, spiny lobsters (*Panulirus* species) form long conga lines as they move between coral reefs and the open ocean floor on a seasonal cycle **(Figure 55.3).**

Animals Use Wayfinding Mechanisms to Guide Their Movements

Moving animals use various wayfinding mechanisms to arrive at their destination. Biologists group these mechanisms into three general categories: *piloting, compass orientation,* and *navigation.* Many species probably use a combination of these mechanisms to guide their movements.

The simplest wayfinding mechanism is **piloting,** in which animals use familiar landmarks to guide their journey. For example, gray whales *(Eschrichtius robustus)* migrate from Alaska to Baja California and back using visual cues provided by the Pacific coastline of North America.

Animals that do not undertake long migrations also use specific landmarks to identify their nest site or places where they have stashed food. In a famous experiment published in 1938, Niko Tinbergen showed that female digger wasps *(Philanthus triangulum),* which nest in soil, use visual landmarks to find

Howard Hall/Oxford Scientific/Index Stock

FIGURE 55.3
Migrating arthropods. Spiny lobsters *(Panulirus argus)* make seasonal migrations between coral reefs and the open ocean floor. As many as 50 individuals march in single file for several days.

FIGURE 55.4 **Experimental Research**

Using Landmarks to Find the Way Home

Question: How do female digger wasps *(Philanthus triangulum)* relocate their nests after flying off to search for food?

Experiment: Tinbergen arranged pinecones in a circle around the nest of a female digger wasp while she was still inside. After leaving the nest, she circled the area a few times, apparently noting nearby landmarks. Tinbergen then moved the circle of pinecones a short distance away.

Result: Each time the female returned, she searched for her nest within the pinecone circle. She was unable to find the nest unless Tinbergen replaced the pinecones in their original position.

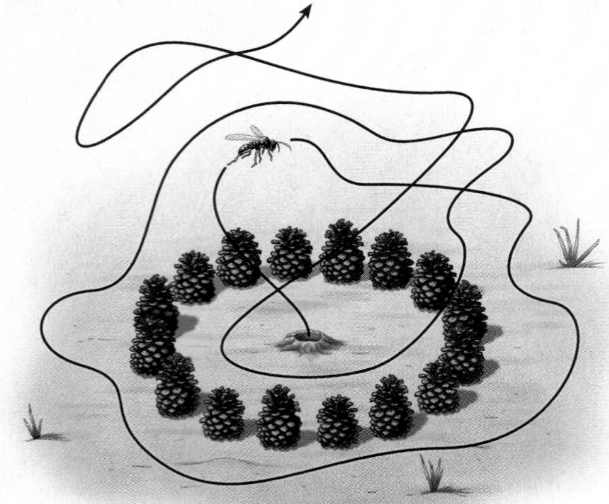
Wasp's flight pattern on leaving nest

Wasp's return, looking for nest
Nest

Conclusion: Female digger wasps use the location of local landmarks to find the entrances to their underground nests.

Source: N. Tinbergen. 1951. *The Study of Instinct.* Oxford University Press, New York.

their nests after flying off in search of food **(Figure 55.4).** Tinbergen arranged pinecones in a circle around one nest while the female was still inside. As she left, she flew around the area, apparently noting nearby landmarks. Tinbergen then moved the circle of pinecones a short distance away. Each time the female returned, she searched for her nest within the pinecone circle—but she never found it unless the pinecones were returned to their original position. In a follow-up study, Tinbergen rearranged the circle of pinecones into a triangle after females left their nests and added a ring of stones nearby. The returning females looked for their nest in the stone circle. Tinbergen concluded that digger wasps respond to the general outline or geometry of landmarks around their nests and not to the specific objects that create those landmarks.

In a more sophisticated wayfinding mechanism, **compass orientation,** animals move in a particular direction, often over a specific distance or for a prescribed length of time. Some day-flying migratory birds, for example, orient themselves using the sun's position in the sky in conjunction with an internal biological clock (see Section 40.4). The internal clock allows the bird to use the sun as a compass, compensating for changes in its position through the day; the clock may also allow some birds to estimate how far they have traveled since beginning their journey. Other migratory animals use polarized light or Earth's magnetic field as a compass.

Some birds that migrate at night use the positions of stars to determine their direction. The indigo bunting *(Passerina cyanea),* for example, flies about 3,500 km from the northeastern United States to the Caribbean or Central America each fall, and makes the return journey each spring. Stephen Emlen of Cornell University demonstrated that these birds use celestial cues to direct their migration **(Figure 55.5).** Emlen confined individual buntings in cone-shaped test cages. He lined the sides of the cages with blotting paper, placed inkpads on the bottom, and kept the cages in an outdoor enclosure so that the birds had a full view of the sky. Whenever a bird made a directed movement, its inky footprints indicated the direction in which it was trying to fly. Emlen found that on clear nights in fall, the footprints pointed to

FIGURE 55.5 **Experimental Research**

Experimental Analysis of the Indigo Bunting's Star Compass

Question: Do indigo buntings *(Passerina cyanea)* use the positions of stars in the night sky to orient their migrations?

Experiment: Emlen placed individual buntings in cone-shaped test cages. He lined the sides of the cages with blotting paper, placed inkpads on the bottom, and kept the cages in an outdoor enclosure so that the birds had a full view of the sky. Whenever a bird made a directed movement, its inky footprints indicated the direction in which it was trying to fly. Emlen predicted that the footprints would show the buntings' inclination to migrate south in autumn and north in spring.

Indigo bunting

R. & N. Bowers/VIREO

Blotting paper
Indigo bunting
Inkpad

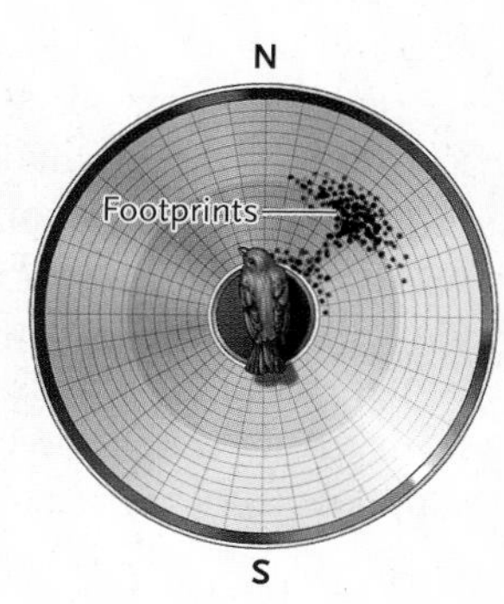
N
Footprints
S

Side (left) and overhead (right) views of the test cage with blotting paper on the sides and an inkpad on the bottom

Results: On clear nights in autumn, the footprints pointed to the south; on clear nights in spring, they pointed north. On cloudy nights, when buntings could not see the stars, their footprints were evenly distributed in all directions.

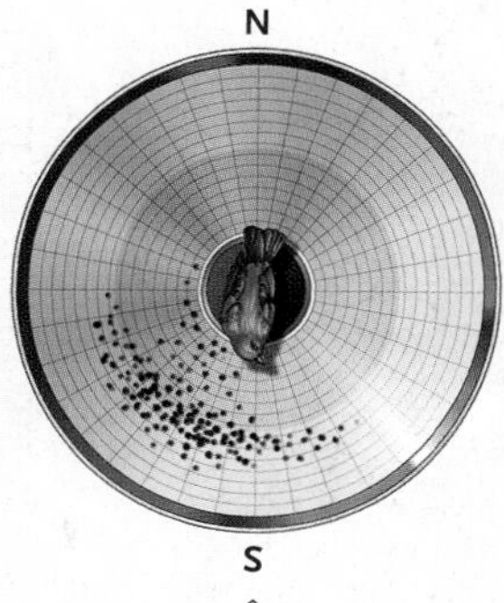
N
S

In autumn, the bunting footprints indicated that they were trying to fly south.

N
S

In spring, the bunting footprints indicated that they were trying to fly north.

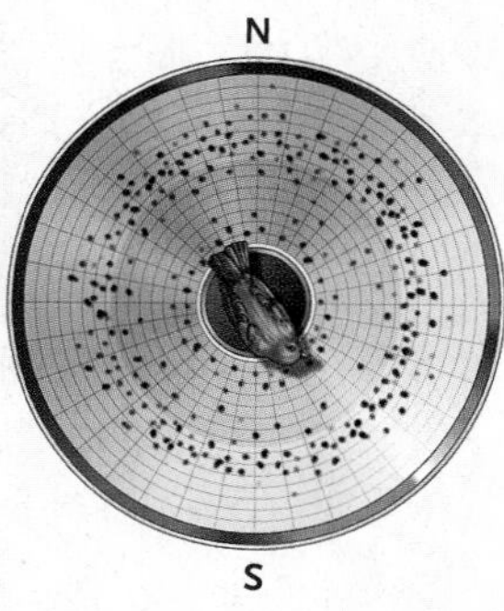
N
S

On cloudy nights, when buntings could not see the stars, their footprints indicated a random pattern of movement.

Conclusion: Indigo buntings use the positions of the stars to direct their seasonal migrations. When they could see the stars above their test cages, they moved in the predicted direction; but when clouds obscured their view of the stars, they moved in random directions.

Source: S. T. Emlen. 1967. Migratory orientation of the indigo bunting, *Passerina cyanea*. Part I: Evidence for use of celestial cues. *The Auk* 84:309–342.

the south; on clear nights in spring, they pointed north. On cloudy nights, when the buntings could not see the stars, their footprints were evenly distributed in all directions, indicating that their compass required a view of the stars.

The most complex wayfinding mechanism is **navigation,** in which an animal moves toward a specific destination, using both a compass and a "mental map" of where it is in relation to the destination. Human hikers in unfamiliar surroundings routinely use navigation to find their way home: they use a map to determine their current position and the necessary direction of movement and a compass to orient themselves in that direction. Scientists have documented true navigation in only a few animal species. Perhaps the most notable is the homing pigeon *(Columba livia)*, which can navigate to its home coop from any direction. Recent research suggests that homing pigeons probably use the sun's position as their compass and olfactory cues as their map.

Seasonal Variation in Food Supply May Explain the Evolution of Migratory Behavior

Migratory behavior entails obvious costs, such as the time and energy devoted to the journey and the risk of death from exhaustion or predator attack. Why then do some species migrate? What benefits accrue to an individual that undertakes a costly migration?

For migratory birds, the most widely accepted hypothesis focuses on seasonal changes in food supplies. The amount of insect food available in northern forests increases explosively during the warm spring and summer, providing abundant resources to produce eggs and rear offspring. Then, during the late fall and winter, insects all but disappear. A few bird species that forage on seeds and dormant insects do not head south. However, energy supplies are more predictably available in tropical overwintering grounds, and migratory birds may have a better chance of surviving there. The following spring they return north to exploit the food bonanza on their summer breeding grounds.

The two-way migratory journeys may provide other benefits as well. Avoiding the northern winter is probably adaptive because endotherms must increase their metabolic rates just to stay warm in cold climates (see Section 46.8). But in summer the days are longer at high latitudes than they are in the tropics (see Section 49.2), giving adult birds more time to collect enough food to rear a brood.

Seasonal changes in food supply also underlie the migration of monarch butterflies *(Danaus plexippus)*, which eat milkweed (*Asclepias* species) leaves as larvae and the nectar of milkweeds and other plants as adults. In eastern North America, milkweed plants grow only during spring and summer. Many adult monarchs head south in late summer, when milkweeds are beginning to die, migrating as much as 4,000 km from eastern and central North America to central Mexico, where they cluster in spectacular numbers **(Figure 55.6)**. Unlike migrant birds, these insects do not feed on their overwintering grounds. Instead, their meta-

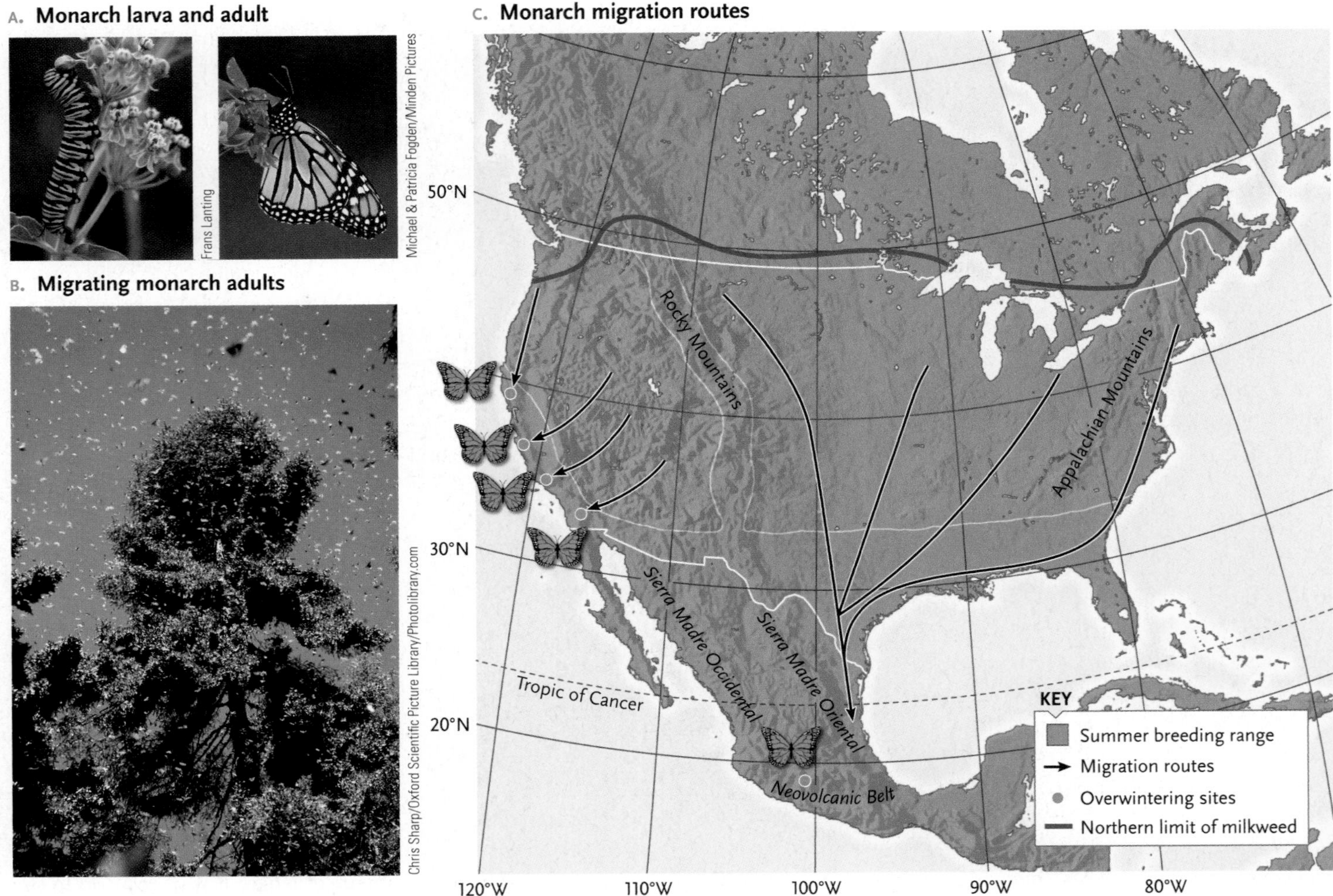
A. Monarch larva and adult
Frans Lanting
Michael & Patricia Fogden/Minden Pictures
B. Migrating monarch adults
Chris Sharp/Oxford Scientific Picture Library/Photolibrary.com
C. Monarch migration routes
50°N
30°N
20°N
120°W
110°W
100°W
90°W
80°W
Rocky Mountains
Appalachian Mountains
Sierra Madre Occidental
Sierra Madre Oriental
Tropic of Cancer
Neovolcanic Belt
KEY
Summer breeding range
Migration routes
Overwintering sites
Northern limit of milkweed

FIGURE 55.6
Monarch butterfly migrations. (A) Monarch butterflies *(Danaus plexippus)* feed primarily on milkweed plants. **(B)** When milkweed plants in their breeding range die back at the end of summer, millions of monarchs begin a southward migration. **(C)** Butterflies that live and breed east of the Rocky Mountains migrate to Mexico. After passing the winter in a semidormant state, they migrate northward the following spring. Monarchs living west of the Rocky Mountains winter in coastal California.

bolic rate decreases in the cool mountain air, and the butterflies become inactive for months, thereby conserving precious energy reserves. When spring arrives, the butterflies become active again and begin the return migration to northern breeding habitats. The northward migration is slow, however, and many individuals stop along the way to feed and lay eggs. But their offspring, and their offspring's offspring, continue the northward migration through the summer; some descendants eventually reach Canada for a final round of breeding. The summer's last generation then returns south to the spot where their ancestors, two to five generations removed, spent the previous winter.

For other animals, the migration to breeding grounds may provide special conditions necessary for reproduction. For example, gray whales migrate south to breeding grounds in quiet, shallow lagoons where predators are rare and warm water temperatures will not stress their calves.

STUDY BREAK 55.1 <

1. **What is the difference between piloting, compass orientation, and navigation?**
2. **What is the most probable selection pressure that has fostered seasonal migrations in birds?**

55.2 Habitat Selection and Territoriality

The geographical range of nearly every animal species includes a mosaic of habitat types. The breeding range of white-crowned sparrows, for example, encompasses forests, meadows, housing developments, and city dumps. An animal's choice of habitat is critically important because the habitat provides food, shelter, nesting sites, and the other organisms with which it interacts. If an animal chooses a habitat that does not provide appropriate resources, it will not survive and reproduce.

Animals Use Multiple Criteria for Selecting Habitats

On a large spatial scale, animals almost certainly use multiple criteria to select the habitats they occupy, but no research has yet established any general principles about how animals make these choices. When a migrating bird arrives at its breeding range, for example, it probably cues on large-scale geographical features, such as a pond or a patch of large trees. If it does not find the food or nesting resources it needs—or if other individuals have already depleted those resources—it may move to another habitat patch.

On a very fine spatial scale, basic responses to physical factors enable some animals to find suitable habitats. The simplest such mechanism is called a **kinesis** (*kinesis* = movement), a change in the rate of movement or the frequency of turning movements in response to environmental stimuli. For example, the terrestrial crustaceans known as wood lice (Isopoda) typically live under rocks and logs or in other damp places. Although these arthropods are not attracted to moisture *per se,* laboratory experiments have shown that when a wood louse encounters dry soil, it exhibits a kinesis, scrambling around and turning frequently; when it reaches a patch of moist soil, it moves much less. As a result, these animals accumulate in moist habitats. Biologists infer that this behavior is adaptive because wood lice exposed to dry soil quickly dehydrate and die. Other animals may exhibit a **taxis** (*taxis* = arrangement), a response that is directed either toward or away from a specific stimulus. For example, cockroaches (order Blattodea) exhibit negative phototaxis: they actively avoid light and seek darkness, which are behaviors that make them harder for visually oriented predators to detect.

Genetics and Learning Influence Habitat Selection

Biologists generally assume that habitat selection is adaptive and has been shaped by natural selection in most animal species. For example, some animals instinctively select habitats where they are well camouflaged, a means of avoiding detection by predators (see Figure 51.3); predators would discover and eliminate any individual that fails to select a matching background—along with any alleles responsible for the mismatch. Many insects have a genetically determined preference for the plants that they eat during their larval stage. Adults often restrict their mating and egg-laying activities to these food plants, effectively selecting the habitats where their offspring will live and feed, as described in the discussion of sympatric speciation (see Section 21.3).

Even vertebrates sometimes exhibit such innate preferences, as demonstrated by two closely related European bird species, blue tits *(Parus caeruleus)* and coal tits *(Parus ater).* Adult blue tits feed mostly in oak trees, whereas coal tits prefer to feed in pines. When researchers reared the young of both species in cages without any vegetation at all and then offered them a choice between oak branches and pine branches, coal tits immediately gravitated toward pines and blue tits toward oaks, strongly suggesting that the preference is innate **(Figure 55.7).** Further research demonstrated that each species feeds most successfully in the tree species it prefers. Thus, natural selection probably fostered these preferences.

Habitat preferences can be molded by experiences early in life, however. For example, the tadpoles of red-legged frogs *(Rana aurora)* usually live in aquatic habitats cluttered with sticks, strands of algae, and plant stems; when given a choice in the laboratory, they prefer striped backgrounds to plain ones. By contrast, tadpoles of the closely related cascade frog *(Rana cascadae)* live over gravel bottoms, and they prefer plain substrates over striped ones. However, when red-legged frogs are reared over plain substrates and cascade frogs over striped substrates, they no longer exhibit preferences for their usual substrates.

Animals Sometimes Defend Patches of Habitat for Their Exclusive Use

Under some circumstances, animals may defend a **territory** from other members of their species, retaining more or less exclusive use of the resources it contains. Territorial behavior occurs in all

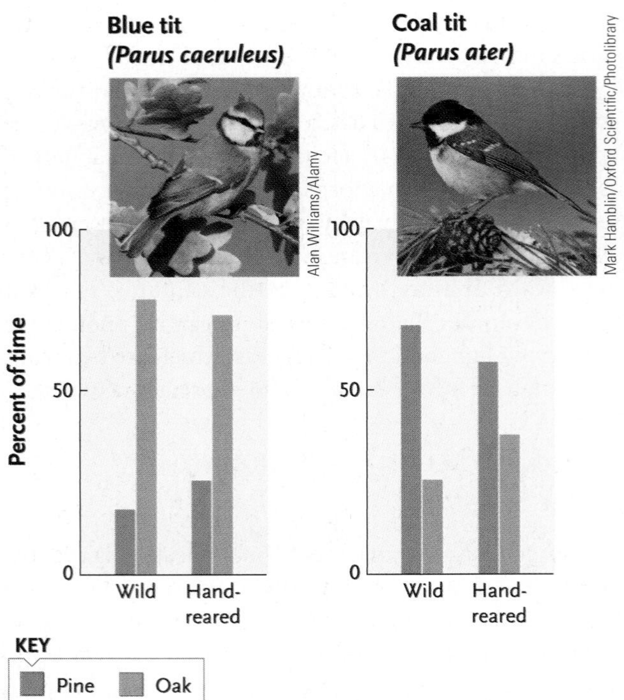
Blue tit
(Parus caeruleus)
Coal tit
(Parus ater)
Alan Williams/Alamy
Mark Hamblin/Oxford Scientific/Photolibrary
Percent of time
100
50
0
Wild
Hand-reared
KEY
Pine
Oak

FIGURE 55.7
Habitat selection by birds. Wild blue tits show a strong preference for oak trees, and coal tits show a strong preference for pine trees. Birds that were hand-reared in a vegetation-free environment showed identical, though slightly weaker, preferences.

major groups of vertebrates, many insects, and some other invertebrates, but it is by no means universal. In many organisms, territorial behavior occurs only during the breeding season.

Animals establish and defend territories only if some critical resource is in short supply. Moreover, the resource must be fixed in space so that the area around it can be defended. For example, during the breeding season, most songbirds defend a territory within which they build a nest and collect food for their young. By contrast, most sea birds, such as terns and penguins, do not defend feeding territories. Although they defend a tiny area around their nests, which they build on shore, they never attempt to defend the patches of ocean where they collect food; fishes come and go at will and thus do not constitute a defendable resource.

Territorial defense is always a costly activity. Patrolling territory borders, performing displays hundreds of times per day, and chasing intruders take time and energy. Moreover, territorial displays increase an animal's likelihood of being injured or captured by a predator.

Experiments conducted by Catherine Marler and Michael Moore of Arizona State University illustrate the cost of territorial behavior in Jarrow's spiny lizard *(Sceloporus jarrovi)*. Male lizards ordinarily exhibit strong territoriality only during the autumn mating season, when elevated blood levels of testosterone stimulate their aggressive behavior. The researchers implanted small doses of testosterone under the skin of experimental animals in June and July, during the *nonmating* season; controls received a placebo treatment. Testosterone-enhanced males were more active and displayed more often than control males. But experimental males spent less time feeding, even though they used about 30% more energy per day than control males. Over the course of about 7 weeks, a significantly higher percentage of experimental males died—a clear sign that engaging in territorial behavior is costly.

Maintaining a territory has definite benefits, however, such as having access to nesting sites, food supplies, and refuges from predators. For example, the surgeonfish *(Acanthurus lineatus)*, which lives in the coral reefs around American Samoa, may engage in as many as 1,900 chases per day to defend a small territory from other algae-eating fish species. But territory holders consume up to five times more food than nonterritory holders.

STUDY BREAK 55.2 <

1. **Why do wood lice tend to occur in moist parts of their habitat?**
2. **What are the costs of maintaining a territory, and what are the benefits?**

55.3 The Evolution of Communication

When resident animals advertise their presence in their territories, they are communicating information to nearby animals. In the formal language of animal behavior studies, all communication systems involve an interaction between a *signaler,* the animal that transmits information, and a *signal receiver,* the animal that intercepts the information and makes a behavioral response. Natural selection has adjusted the ability of signalers to transmit information and the ability of receivers to get the message.

Animal Signals Can Activate Different Sensory Receptors in Receivers

Biologists categorize animal signals according to the sensory receptors, or "channels," through which the signal acts: *acoustical, visual, chemical, tactile,* or *electrical.* Each channel has specific advantages.

Bird songs are examples of **acoustical signals;** a signaler produces a sound that is heard by a signal receiver. Many animals use the acoustical channel, including a host of nocturnal and burrow-dwelling insects and amphibians. These signals reach distant receivers, even at night and in cluttered environments where visual signals are less effective.

Because humans frequently use facial expressions and body language to send messages, **visual signals** are a familiar form of communication. In many animals, visual signals are *ritualized;* in other words, they have become exaggerated and stereotyped over evolutionary time, forming an easily recognized visual display **(Figure 55.8).** Visual displays can even be useful at night or in the darkness of the deep sea; some animals, such as fireflies and certain fishes, send bioluminescent signals to distant receivers.

© E. Mickleburgh/Ardea, London

FIGURE 55.8
Visual displays. The courtship display of a male wandering albatross *(Diomedea exulans)* includes ritualized postures and movements of the wings and body.

© Kenneth W. Fink/Photo Researchers, Inc.

FIGURE 55.10
Tactile signals. Grooming by hyacinth macaws *(Anodorhynchus hyacinthinus)* removes parasites and dirt from feathers. The close physical contact also promotes friendly relations between groomer and groomee.

Many species release **chemical signals,** which carry messages to signal receivers through the olfactory channel. Scent marking (spraying) by male dogs is an example. In particular, mammals and insects often communicate through **pheromones,** distinctive volatile chemicals released in minute amounts to influence the behavior of members of the same species. For example, a worker ant's body contains a battery of glands, each releasing a different pheromone **(Figure 55.9).** One set of pheromones recruits fellow workers to battle colony invaders; another set stimulates workers to collect food that has been discovered outside the colony. Other animals release pheromones to attract mates. Female silkworm moths *(Bombyx mori)* produce bombykol, a single molecule of which can generate a message in specialized receptors on the antennae of any male silkworm moth that is downwind (see Figure 39.19).

In many species, touch conveys important messages from a signaler to a receiver. **Tactile signals** operate only over very short distances, but for social animals living in close company, they play a significant role in the development of friendly bonds between individuals **(Figure 55.10).**

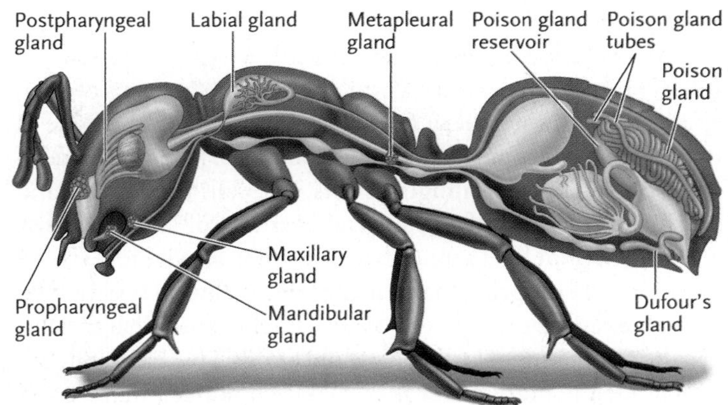
Postpharyngeal gland
Labial gland
Metapleural gland
Poison gland reservoir
Poison gland tubes
Poison gland
Maxillary gland
Mandibular gland
Propharyngeal gland
Dufour's gland

FIGURE 55.9
Chemical signals. An ant's body contains a host of pheromone-producing glands, each of which manufactures and releases its own volatile chemical or chemicals.

Some freshwater fish species, especially those that occupy murky tropical rivers where visual signals could not be seen, use weak **electrical signals** to communicate. These fishes have electric organs that can release charges of variable intensity, duration, and frequency, allowing substantial modulation of the message that a signaler sends. Among the New World knifefishes (order Gymnotiformes), including the electric eel *(Electrophorus electricus),* electrical discharges can signal threats, submission, or a readiness to breed.

Honey Bees Use Several Communication Channels to Transmit Complex Messages

When animals need to convey a complex message, they may use several channels of communication simultaneously. For example, as Karl von Frisch demonstrated, the dance of the honey bee *(Apis mellifera)* involves tactile, acoustical, and chemical communication **(Figure 55.11).** When a foraging honey bee discovers pollen or nectar, it returns to its colony and performs a complex dance on the vertical surface of the honeycomb in the complete darkness of the hive. The dancer moves in a circle, attracting a crowd of workers, some of which follow and maintain physical contact with the dancer. From the dance, they acquire information about the distance and direction they will need to fly to locate the food source.

When the food source is less than about 75 m from the hive, the bee performs a "round dance" (see Figure 55.11A). It moves in tight circles, swinging its abdomen back and forth. Bees surrounding the dancer produce a brief acoustical signal, which stimulates the dancer to regurgitate a sample of the food it discovered. The regurgitated sample serves as a chemical cue to other workers, which then leave the hive to search for that type of food.

If the food source is more distant, the forager performs what von Frisch described as the "waggle dance." The bee dances a half circle in one direction and then dances in a straight line while waggling its abdomen before dancing a half circle in the other

FIGURE 55.11
Dance communication by honey bees. Foraging honey bees *(Apis mellifera)* transmit information about the location of a food source by dancing on the vertical honeycomb. **(A)** If the food source is close to the hive, the forager performs a "round dance." **(B)** If the food source is more than about 75 m from the hive, the forager performs a "waggle dance" that indicates the distance to the food source. **(C)** The dancing bee indicates the direction to a distant food source by the angle of the waggle run.

A. **Round dance**

B. **Waggle dance**

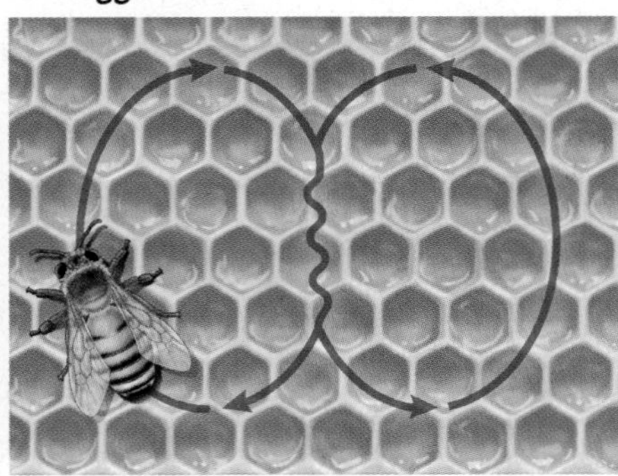

C. **Coding direction in the waggle dance**

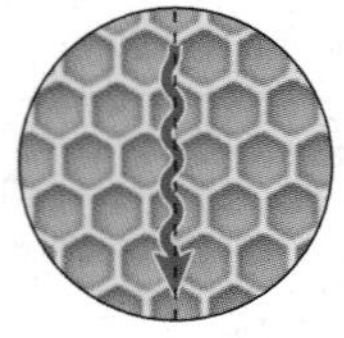

When the bee moves straight down the comb, other bees fly to the source directly away from the sun.

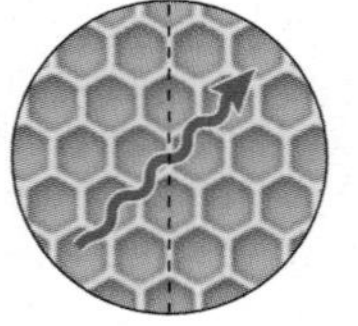

When the bee moves 45° to the right of vertical, other bees fly at a 45° angle to the right of the sun.

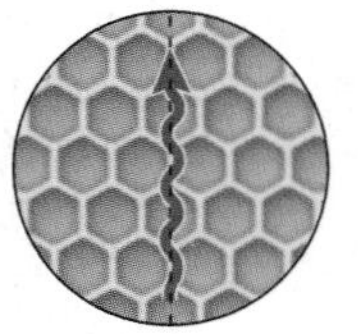

When the bee moves straight up the comb, other bees fly straight toward the sun.

direction (see Figure 55.11B). With each waggle, the dancer produces a brief buzzing sound. Von Frisch determined that the angle of the straight run relative to vertical on the honeycomb indicates the direction of the food source relative to the position of the sun (see Figure 55.11C). The duration of the waggles and buzzes that the bee makes on the straight run carries information about the distance to the food; the longer the time spent waggling and buzzing, the further the food is from the hive.

Biologists Use Evolutionary Hypotheses to Analyze Communication Systems

Signal receivers often respond to communication from signalers in predictable ways. For example, a male white-crowned sparrow generally avoids entering a neighboring territory simply because it hears the song of the resident male. Similarly, young male baboons often retreat without a fight when they see an older male's visual threat display **(Figure 55.12)**, even though they may lose the chance to mate with a female. Why do these receivers behave in ways that appear to be beneficial to their rivals, but not to themselves?

When biologists try to explain behavioral interactions, their hypotheses focus on how an animal's actions may allow it to contribute more offspring to the next generation. In our first example, the retreating male sparrow avoids wasting time and energy on a battle he is likely to lose. Moreover, ousting the current resident might be more tiring and risky than finding a suitable unoccupied breeding site. This hypothesis predicts that resident males should almost always win physical contests. In cases when an intruder wins a territory from a resident, it may do so only after a prolonged series of exhausting clashes. Observations of territorial species—whether birds, lizards, frogs, fish, or insects—generally support these predictions.

FIGURE 55.12
Threat displays. The threat display of a dominant male mandrill *(Mandrillus sphinx)*, used to drive away rival males, features exposed canines.

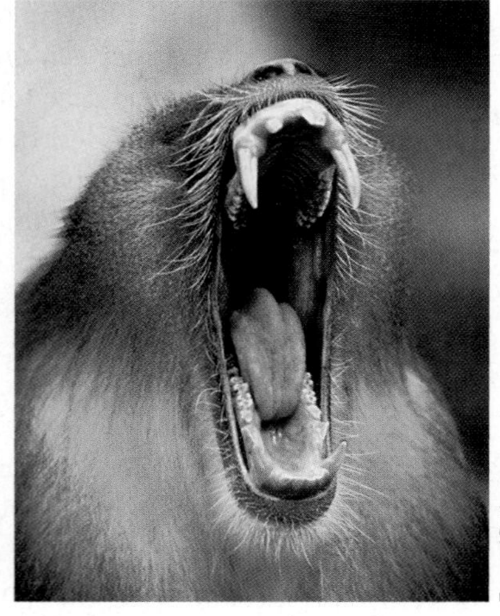

Tom and Pat Leeson

Applying a similar argument to competition among male baboons, we can predict that smaller or younger males will concede females to threatening older rivals without fighting. The signal receiver retreats after receiving the threat because he judges that he would be demolished in real combat—after all, a male baboon's canine teeth are not just for show. Evolutionary analyses therefore suggest that both the signaler and the signal receiver benefit from the transfer of information in their communication system.

An evolutionary analysis also helps to explain the strange yell of ravens *(Corvus corax)*, which scavenge carcasses of deer, elk, or moose in northern forests during winter. When one of these large birds comes across a food bonanza, it may call loudly, attracting a crowd of hungry ravens. The calling behavior and sharing of resources puzzled Bernd Heinrich of The University of Vermont. Wouldn't a quiet raven eat more, survive longer, and produce more offspring than a noisy bird? If natural selection favored the raven's calling behavior, we might expect that the cost of calling (in terms of lost food) would be offset by a reproductive benefit for the individual caller. Heinrich noticed that paired, territory-owning adults did not yell loudly when they found goat carcasses that he had hauled into the Maine woods; instead, they fed quietly. Only young, wandering ravens that happened upon a carcass in another bird's territory advertised their discovery. The signals of these birds attracted other nonterritorial ravens, which collectively overwhelmed the resident pair's attempts to defend their territory. Only then was a wanderer likely to have a chance to feed in an area that would otherwise be off-limits.

STUDY BREAK 55.3

1. Which channels do humans consciously use to communicate with each other?
2. How does a honey bee tell its hive-mates that it has discovered a distant food source?
3. Why do ravens sometimes announce their discovery of food?

THINK OUTSIDE THE BOOK

Observe a domesticated species—or an animal species that lives in a habitat near your home or university—and produce a catalog of the communication channels that it uses. What behaviors provide clues to the particular channels it uses to communicate?

55.4 The Evolution of Reproductive Behavior and Mating Systems

In many animal species, communication coordinates the reproductive activities of males and females and governs the interactions between parents and offspring. In this section, we examine how several elements of behavior contribute to the reproductive success of individuals.

Males and Females Use Different Reproductive Strategies

In sexually reproducing species, males and females often differ in their overall **reproductive strategies,** the set of behaviors that lead to reproductive success. This difference arises in part from a fundamental difference in the amount of **parental investment,** the time and energy devoted to the production and rearing of offspring, provided by the two sexes. Because eggs are much larger than sperm, females almost always contribute more energy than males to the production of a gamete.

A male might increase the number of offspring that carry his alleles simply by mating with multiple females, especially if he does not spend time and energy providing parental care to his offspring. Thus, in many animal species, males compete intensely for access to females, and any trait that increases a male's access or attractiveness to females has a big reproductive payoff.

Entirely different selection pressures operate on females. Their reproductive output is generally limited by the number of eggs they can produce, and mating with multiple males will not increase that number. But the success of her offspring may depend on the attributes of their father or the territory he holds. Thus, females of many species choose their mates carefully. In some cases, females mate with males whose territories include abundant resources, ensuring an ample food supply for their young. In other cases, females choose robust males that will contribute "good genes" (that is, alleles that confer a high likelihood of surviving and reproducing) to her offspring, increasing their chances of long-term success.

Male Competition for Females and Female Mate Choice Foster Sexual Selection

Male competition for access to females coupled with the females' choice of mates establishes a form of natural selection called **sexual selection,** that is, selection for mating success (see Section 20.3). As a result of sexual selection, males are larger than females in many species, and males have ornaments and weapons, such as horns and antlers, that are useful for attracting females as well as for butting, stabbing, or intimidating rival males. Males typically show off these elaborate structures in complex **courtship displays** to attract the attention of females. For example, male peafowl *(Pavo cristatus)* strut in front of females while spreading a gigantic fan of tail feathers, which they shake, rattle, and roll.

Why should females choose males that display exaggerated structures conspicuously? Biologists have developed several hypotheses to explain the attraction. First, a male's large size, bright feathers, or large horns might indicate that he is particularly healthy, that he can harvest resources efficiently, or simply that he has managed to survive to an advanced age. These traits are, in effect, signals of male quality, and if they reflect a male's genetic makeup, he is likely to fertilize a female's eggs with sperm containing successful alleles. In some cases, big, showy males hold large, rich territories, and females that choose them gain access to the resources their territories contain.

The degree to which females *actively* choose genetically superior mates varies among species. In the northern elephant seal *(Mirounga angustirostris),* for example, female choice is essentially passive. Large numbers of females gather on beaches to give birth to their pups before becoming sexually receptive again. Males locate these clusters of females and fight to keep other males away. The winners have exceptional reproductive success, but only after engaging in violent and relentless combat with rival males. In this kind of mating system, females are practically guaranteed to receive sperm from large and powerful males in superb physiological condition, attributes that may well be associated with alleles that will increase their offspring's chances of living long enough to reproduce.

In other species, females exercise more active mate choice, copulating only after inspecting a group of potential partners. Among birds, active female mate choice is most apparent at **leks,** display grounds where each male courts attentive females from a small territory. Male sage grouse *(Centrocercus urophasianus),* a lekking bird of western North America, gather in open areas among stands of sagebrush. Each male defends just a few square meters, where it struts in circles while emitting booming calls and showing off its elegant tail feathers and big neck pouches **(Figure 55.13).** Females wander among the displaying males, presumably analyzing the males' visual and acoustical displays. Eventually,

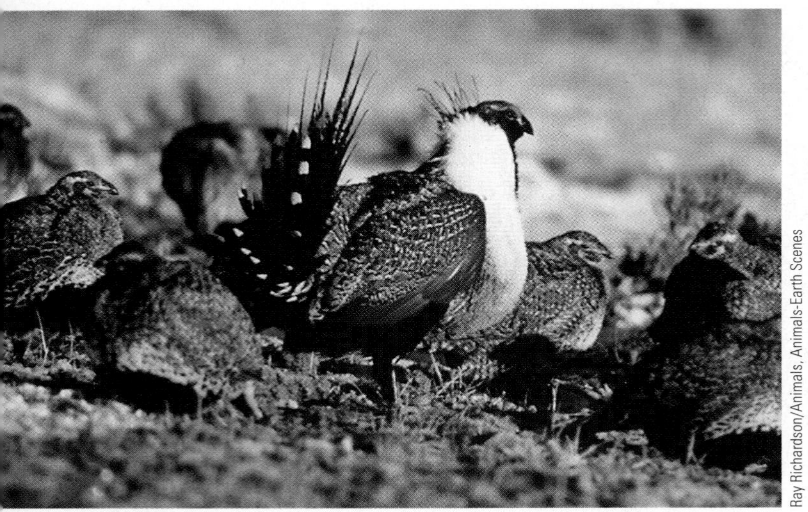
Ray Richardson/Animals Animals-Earth Scenes

FIGURE 55.13
Lekking behavior. Male sage grouse *(Centrocercus urophasianus)* use their ornamental feathers in visual courtship displays performed at a lek, where each male has his own small territory. The smaller brown females observe the prancing males before choosing a mate.

each female selects a mate from among the dozens of males present. Females repeatedly favor males that come to the lek daily, defend their small area vigorously, and display more frequently than the average lek participant. In other words, favored males can sustain their territorial defense and high display rate over long periods, an ability that may correlate with useful genetic traits.

Experimental studies of peafowl suggest that the top peacocks at a lek may indeed supply advantageous alleles to their offspring. In nature, peahens prefer males whose tails have many ornamental eyespots **(Figure 55.14).** In an experiment on captive birds, some females were mated to males with highly attractive tails, but others were paired with males whose tails were less impressive. The offspring of both groups were reared under uniform conditions for several months and then released into an English woodland. After three months on their own, the offspring of fathers with impressive tails survived better and weighed significantly more than did those whose fathers had less attractive tails. Apparently, a peahen's mate choice does provide her offspring with a survival advantage.

Ashley Cooper/Corbis

FIGURE 55.14
Sexual selection for ornamentation. The attractiveness of a peacock *(Pavo cristatus)* to females depends in part on the number of eyespots on his extraordinary tail. The offspring of males with elaborate tails are more successful than the offspring of males with plainer tails.

Another hypothesis argues that females select showy males even though their ornate structures may impede their locomotion or their elaborate displays may attract the attention of a predator. According to this hypothesis, any male that survives *despite* carrying such a handicap must have a very strong constitution indeed, and he will pass those successful alleles—as well as the alleles responsible for the ornamental handicap—to the female's offspring.

Patterns of Parental Care and Territoriality Influence Mating Systems

In the examples of mate choice just described, successful males inseminate many females, increasing their reproductive success dramatically. But one male mating with many females is only one of several **mating systems,** the ways in which males and females pair up. Some species are **promiscuous:** individuals do not form close pair bonds, and both males and females mate with multiple partners. Other species are **monogamous:** one male and one female form a long-term association. Finally, some species are **polygamous:** *either* males *or* females may have many mating partners. If one male mates with many females, the relationship is called **polygyny;** if one female mates with multiple males, it is described as **polyandry.**

Mating systems appear to have evolved to maximize reproductive success, partly in response to the amount of parental care that offspring require and partly in response to other aspects of a species' ecology. For example, the young of most songbird species, like the white-crowned sparrow, are helpless upon hatching; all they can do is open their mouths and peep, signaling to their parents that they are ready to be fed. These young require lots of parental care, and they are more likely to flourish if both parents bring food to the nest. As you might expect, nearly all songbirds are monogamous, and males and females team up to provide parental care to their offspring.

In some other bird species, such as red-winged blackbirds *(Agelaius phoeniceus),* males establish large, resource-filled territories, and females select mates largely by the quality of the real estate a male holds. Any male with an exceptionally fine territory will be desirable, even if another female has already established herself there. A second female may judge that more resources are available in his territory than in a neighboring one, despite competition with the other female. However, if many females have already settled in a male's territory, intense competition from them may make it less attractive. Given this pattern of habitat and mate choice by females, red-winged blackbirds have a polygynous mating system; males may fertilize the eggs of multiple females and provide little if any direct care to their offspring.

The Comparative Method Reveals How Nest-Building Behaviors Evolved in Swallows and Martins

Even before a pair of birds or other animals has mated, they often prepare a nest or den where they rear their young. The locations and structures of nests vary widely, sometimes even within a clade (that is, a monophyletic evolutionary lineage). For example, among the clade of birds that includes swallows and martins—two species of which are pictured at the opening of Chapter 21—some species excavate long nesting burrows in soil, others nest in pre-existing cavities in rocks or trees, and still others build mud nests of varying complexity on vertical substrates. Given the extraordinary variety of nesting habits in these birds, biologists long wondered about the evolution of their nest-building behavior. Which type of nest is ancestral within the group? Do closely related species share the same nesting habits?

David W. Winkler of Cornell University and Frederick H. Sheldon of the Academy of Natural Sciences in Philadelphia used "the comparative method," combining phylogenetic analysis with observations on behavior, to investigate these questions. Using data from DNA hybridization studies, they constructed a phylogenetic tree for 17 swallow and martin species; a more distantly related bird species, the tufted titmouse *(Parus bicolor)*, served as the outgroup **(Figure 55.15)**. (You may wish to review the

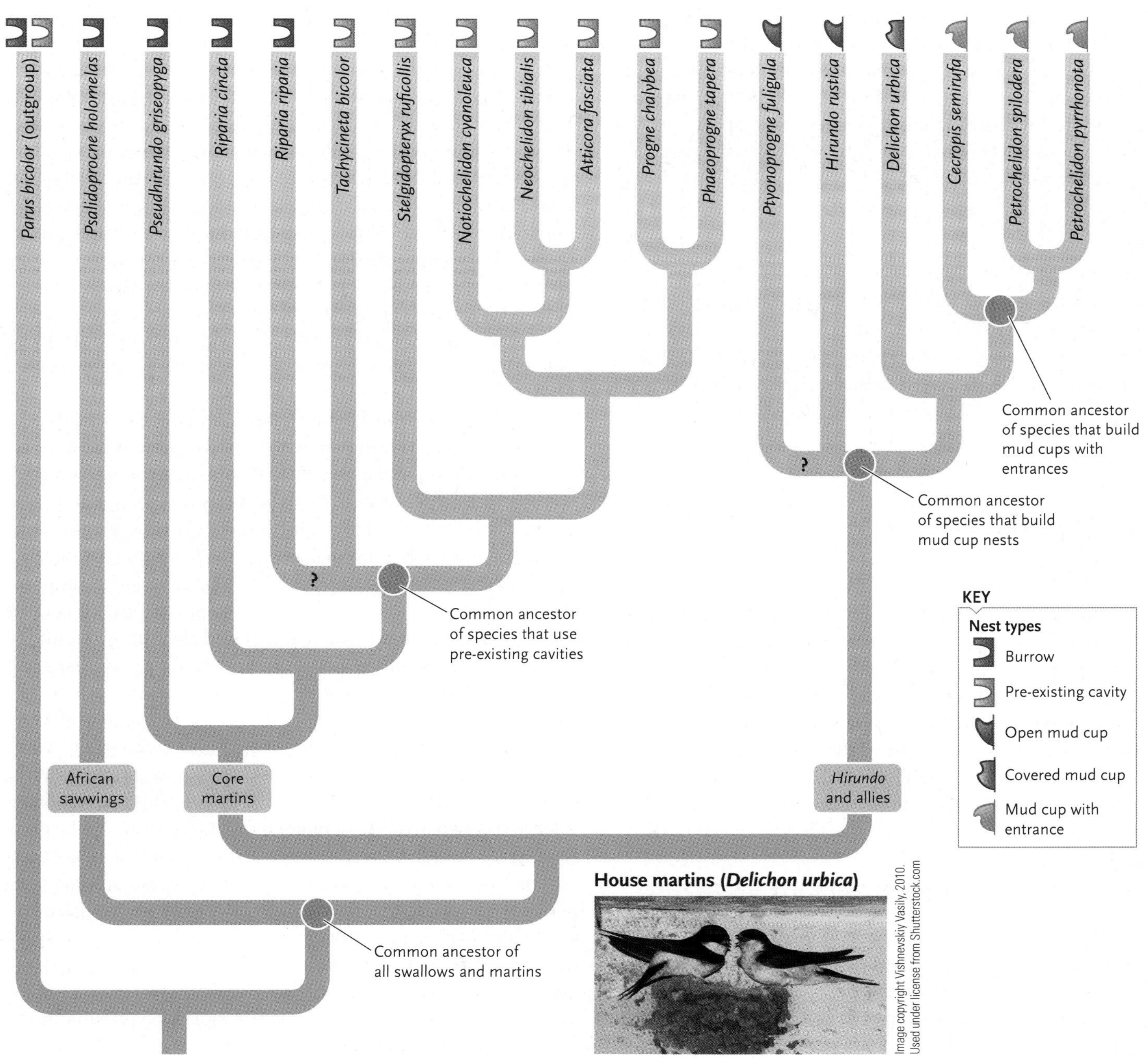

Parus bicolor (outgroup)
Psalidoprocne holomelas
Pseudhirundo griseopyga
Riparia cincta
Riparia riparia
Tachycineta bicolor
Stelgidopteryx ruficollis
Notiochelidon cyanoleuca
Neochelidon tibialis
Atticora fasciata
Progne chalybea
Phaeoprogne tapera
Ptyonoprogne fuligula
Hirundo rustica
Delichon urbica
Cecropis semirufa
Petrochelidon spilodera
Petrochelidon pyrrhonota
?
?
Common ancestor of species that build mud cups with entrances
Common ancestor of species that build mud cup nests
Common ancestor of species that use pre-existing cavities
KEY
Nest types
Burrow
Pre-existing cavity
Open mud cup
Covered mud cup
Mud cup with entrance
African sawwings
Core martins
Hirundo and allies
House martins (Delichon urbica)
Common ancestor of all swallows and martins

FIGURE 55.15

The comparative method, applied to nesting behavior in swallows and martins. Superimposing nest building behaviors on a phylogenetic tree for 17 species of swallows and martins allowed researchers to determine the pattern in which these diverse behaviors evolved. The question marks indicate phylogenetic relationships that the analysis could not resolve precisely.

construction of phylogenetic trees in Section 23.2.) The phylogenetic analysis revealed that an African species, the black sawwing *(Psalidoprocne holomelas),* branched off the tree first; thus, its behaviors probably represent the ancestral condition. The remaining species fall into one of two clades, the "core martins" and "*Hirundo* and its allies."

Winkler and Sheldon then superimposed data about nesting behaviors of the 18 species on the phylogenetic tree. Their results indicate that digging a nest burrow is the ancestral behavior because it is observed not only in the titmouse and African sawwings, but also in some "core martin" species that branched off the tree early in the group's evolution (on the left side of Figure 55.15). The analysis also shows that each of the other two nesting behaviors evolved once, presumably in the common ancestor of species that exhibit the behavior. Within the "core martins" clade, more recently derived species—all descended from one common ancestor—nest in pre-existing cavities. Within the "*Hirundo* and its allies" clade, all species build mud nests on vertical substrates. The phylogenetic tree also illustrates how the three different patterns of mud nest construction (open mud cup, covered mud cup, and mud cup with entrance) evolved within the latter clade. Thus, superimposing nesting behaviors on the phylogenetic tree allowed the researchers to identify when each of the observed nest-building behaviors first evolved.

STUDY BREAK 55.4 <

1. **For monogamous species, what characteristics of males should increase their attractiveness to females?**
2. **What activities do male and female sage grouse perform at a lek?**
3. **Why might a female red-winged blackbird settle on a territory that was already occupied by another female?**

55.5 The Evolution of Social Behavior

Social behavior, the interactions that animals have with other members of their species, has profound effects on an individual's reproductive success. Some animals are solitary, getting together only briefly to mate (rhinoceroses and leopards); others spend most of their lives in small family groups (gorillas); still others live in groups with thousands of relatives (termites and honey bees). Some species, such as some African antelopes and humans, live in large social units composed primarily of nonrelatives.

Group Living Carries both Benefits and Costs

Ecological factors have a large impact on the reproductive benefits and costs of social living. Groups of cooperating predators frequently capture prey more effectively than they would on their own. For example, white pelicans *(Pelecanus erythrorhynchos)* often encircle a school of fish before moving in for the kill. Conversely, prey that are subject to intense predation often gain safety in numbers. Those living in groups have more watchful eyes to detect an approaching predator. In addition, a predator may be confused when multiple prey scatter in many directions. Finally, few predators have the capacity to capture every individual in a prey cluster, so that some prey escape while the predator pursues others.

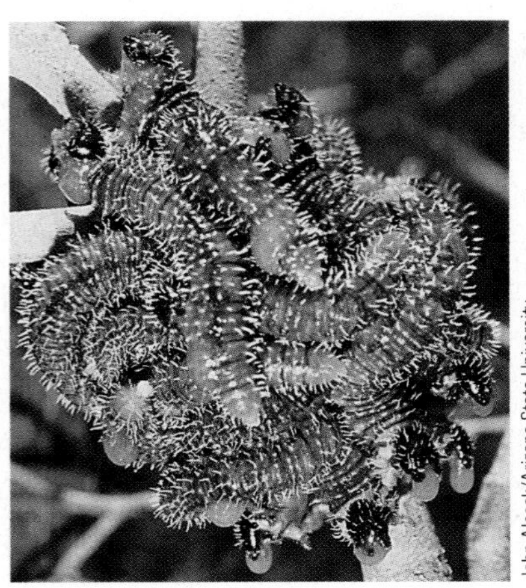
John Alcock/Arizona State University

FIGURE 55.16
Social defensive behavior. Australian sawfly *(Perga dorsalis)* caterpillars clump together on tree branches. These larvae each regurgitate yellow blobs of sticky, aromatic fluid. The accumulation of fluid from a large group of caterpillars successfully deters some predators.

Some prey species, such as musk oxen *(Ovibos moschatus),* join forces to defend themselves actively (see the photo that opens this chapter). Even some insects, such as Australian sawfly caterpillars *(Perga dorsalis),* exhibit cooperative defensive behavior **(Figure 55.16).** When predators disturb the caterpillars, all members of the group rear up and writhe about, regurgitating sticky, pungent oils that they have collected from the eucalyptus leaves they eat. Although the caterpillars can store these oils safely, they are toxic and repellent to bird predators.

A group of sawflies regurgitates more repellent eucalyptus oils than a single individual, which may explain why these insects form their simple societies. If this hypothesis is correct, solitary individuals should be at greater risk of being eaten than those that live communally. Birgitta Sillén-Tullberg of the University of Stockholm, Sweden, tested this prediction by offering sawfly caterpillars to young great tits *(Parus major),* a songbird species. Birds that received caterpillars one at a time consumed an average of 5.6, but those that received them in groups of 20 ate an average of only 4.1 caterpillars. As Sillén-Tullberg had predicted, the caterpillars were somewhat safer in a group than on their own.

In some environments, the costs of social clumping can be significant. These costs may include increased competition for food. For example, when thousands of royal penguins *(Eudyptes schlegeli)* crowd together in huge colonies **(Figure 55.17),** the pressure on the local food supplies is great, increasing the risk of starvation. Communal living also facilitates the spread of contagious diseases and parasites. Nestlings in large colonies of cliff swallows *(Petrochelidon pyrrhonota)* are often stunted in growth because their nests are swarming with blood-feeding parasites, which move easily from nest to nest under crowded conditions. Such costs are probably why the vast majority of animals do not live in large, complex societies.

FIGURE 55.17
Colonial living. Royal penguins *(Eudyptes schlegeli)* on Macquarie Island, between New Zealand and Antarctica, experience both benefits and costs from living together in huge colonies.

Anna Zuckerman-Vdovenko/Eye on the World Photography

Fitness Varies among the Members of a Dominance Hierarchy

Recognizing the costs as well as the benefits of social living, biologists have examined features of social living that appear to reduce the fitness of some individuals. For example, some animal species form **dominance hierarchies,** social systems in which each individual's behavior is governed by its place in a highly structured social ranking. In a typical dominance hierarchy, the dominant or *alpha* individual rules the roost; subordinate individuals typically concede valuable resources to more dominant animals without so much as a peep of protest.

Although dominant individuals gain first access to resources, they also incur costs. Frequent challenges from lower ranking individuals may induce a stress response in dominant animals, which must constantly defend their status. For example, in some primates, wild dogs, and other mammals, dominant males have higher blood levels of cortisol and other stress-related hormones (see Section 40.4) than do subordinates. Elevated cortisol levels may induce high blood pressure, the disruption of sugar metabolism, and other pathological conditions.

Why does a subordinate remain in the group when dominant companions reduce its chances for reproductive success? A possible explanation is that survival rates and reproductive success may be even lower for animals that live by themselves: a solitary baboon surely quickens the pulse of a passing leopard **(Figure 55.18).** A subordinate member of a group gains the benefits, such as protection against predators that come with being part of the group. Low-ranking males may even have the chance to mate with one of the group's females when dominant males are not watching, thus ensuring some representation of their alleles in the next generation. And if a low-ranking individual lives long enough, its social superiors may be toppled by predation, accidents, or old age, and a one-time subordinate may find itself high on the social register with food and mates galore.

FIGURE 55.18
The cost of living alone. A solitary olive baboon *(Papio anubis)* confronts a leopard *(Panthera pardus)* bravely, but without much chance of survival.

John Dominis/Time & Life Pictures/Getty Images

Kin Selection May Explain Altruistic Behavior in Some Animal Species

In some species, group members appear to sacrifice their own reproductive success to help individuals that are not their direct descendants; such behaviors are collectively called **altruism.** For example, subordinate members of a wolf pack do not reproduce, but they share captured prey with the dominant pair and that pair's offspring. Altruistic behavior, by its very definition, appears to contradict a basic premise of Darwinian evolutionary theory, namely that natural selection favors traits that increase an *individual's* relative fitness. Why don't subordinate wolves simply save the energy spent on helping, bide their time until they can become dominant, and then produce their own offspring?

Behavioral ecologist William D. Hamilton of University College, London, provided a solution to this puzzle. He recognized that alleles favoring altruism could be propagated indirectly if altruistic individuals sacrificed personal reproduction

Calculating Degrees of Relatedness

Purpose: The kin-selection hypothesis suggests that the extent of altruistic behavior exhibited by one individual to another is directly proportional to the percentage of alleles they share. The hypothesis therefore predicts that individuals are more likely to help close relatives because, by increasing a close relative's fitness, the individual is helping to propagate some of its own alleles. Researchers calculate the degree of relatedness between individuals to test this prediction.

Protocol: To calculate the degree of relatedness between any two individuals, we first draw a family tree that shows all of the genetic links between them. The alleles of a parent are shuffled by recombination and independent assortment in the gametes they produce, so we can calculate only the average percentage of a parent's alleles that offspring are likely to share.

Half siblings

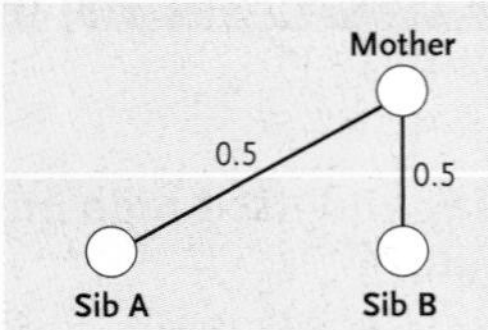

Mother
0.5
0.5
Sib A
Sib B

Relatedness = (0.5) (0.5) = 0.25

Full siblings

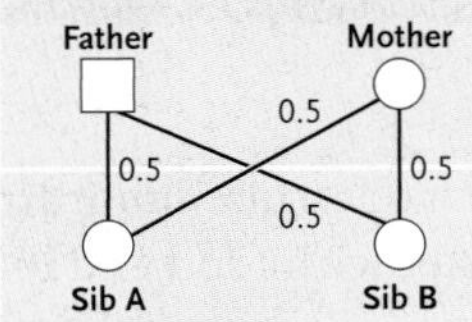

Father
Mother
0.5
0.5
0.5
0.5
Sib A
Sib B

Relatedness
Through mother = (0.5) (0.5) = 0.25
Through father = (0.5) (0.5) = 0.25
Total relatedness = 0.25 + 0.25 = 0.5

First cousins

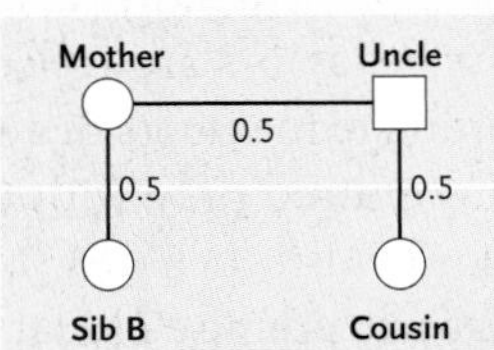

Mother
Uncle
0.5
0.5
0.5
Sib B
Cousin

Relatedness = (0.5) (0.5) (0.5) = 0.125

We start by considering *half* siblings, those who share only one genetic parent. Each sibling receives half of its alleles from its mother. Because a parent has only two alleles at each gene locus, the probability of sibling A getting a particular allele from its mother is 0.5 (decimal notation for 50%). Similarly, the probability of sibling B getting the same allele from its mother is also 0.5. Statistically, the probability that two independent events—in this case, the transfer of an allele to sibling A and the transfer of *the same* allele to sibling B—will both occur is the product of their separate probabilities. Thus, the likelihood that both siblings receive the same allele from their mother is $0.5 \times 0.5 = 0.25$.

Now consider two *full* siblings, who share the same genetic mother *and* father. They share 25% of their alleles through the mother *plus* 25% of their alleles through the father, for a total of 50% (half their alleles). In other words, the degree of relatedness for full siblings is 0.50.

Interpreting the Results: Each link drawn between a parent and an offspring or between full siblings indicates that those two individuals share, on average, 50% of their alleles. We can calculate the total relatedness between any two individuals by multiplying out the probabilities across all of the links between them. Thus, the degree of relatedness between a niece and an uncle is 0.25, and the degree of relatedness between first cousins is 0.125.

to help their relatives reproduce. Helping relatives in this way can propagate the helper's own genes because the family shares alleles inherited from their ancestors.

We can quantify the average percentage of alleles that relatives are likely to share by calculating their degree of relatedness **(Figure 55.19).** We start by considering half siblings who, by definition, share only one genetic parent. Half siblings share on average 25% of their alleles by inheritance from their shared parent, making their degree of relatedness 0.25. By contrast, full siblings, who share the same genetic mother *and* father, share 25% of their alleles through the mother *and* 25% of their alleles through the father, for a total, on average, of 25% + 25% = 50% of their alleles. In other words, the degree of relatedness for full siblings is 0.50. The degree of relatedness between a nephew or niece and an aunt or uncle is 0.25, and the degree of relatedness between first cousins is 0.125. Thus, individuals should be more likely to help close relatives because, by increasing a close relative's fitness, the individual is helping to propagate some of its own alleles.

If altruistic behavior reduces the reproductive success of an individual exhibiting that behavior, how could an allele that promotes altruistic behavior persist or even increase in frequency in a population? The answer depends on the overall number of offspring that carry the allele in the next generation. Altruistic behavior may increase the survival of the altruist's relatives—and they may share the allele in question. If the altruistic behavior allows the assisted relatives to produce proportionately more offspring than the altruist might have produced without helping them, the allele for altruism can increase in frequency in the population. This form of natural selection is aptly called **kin selection.**

For example, suppose a male wolf helps his parents rear four pups to adulthood, pups that would have died without the extra assistance provided by the altruist. Because the pups are his siblings, they share 50% of his genes; thus, on average the helper wolf has created "by proxy" two ($0.50 \times 4 = 2$) copies of any allele that contributed to his altruistic behavior. The costs of his altruism must be measured against this indirect reproductive success. If he had abstained from altruism, the helper wolf might have raised, say, two surviving offspring of his own. Each of his offspring would carry half of his alleles, preserving just one ($0.50 \times 2 = 1$) copy of a given allele. Under these hypothetical circumstances, reproducing on his own would have produced fewer copies of his alleles in the next generation than helping to raise his siblings.

Although our example of the altruistic wolf is hypothetical, biologists have observed sibling helpers in many bird and mammal species. The phenomenon is especially common among animals in which inexperienced young adults are unable to control sufficient resources to reproduce successfully on their own. Their altruistic behavior not only assists reproduction by their close relatives, but it may also provide useful practice for rearing their own future offspring.

Hamilton's kin-selection hypothesis explains altruistic behavior between closely related individuals, but behavioral biologists have also observed examples of altruism between nonrelatives. For example, the common vampire bat *(Desmodus rotundus)*, which feeds on the blood of sleeping mammals, must consume a meal every two days to avoid starving to death. Bats that have consumed a large meal often share their bounty with unrelated members of their group. Why would one bat share its resources with a nonrelative? Robert Trivers, then of Harvard University, proposed that individuals will help nonrelatives if they are likely to return the favor in the future. Trivers called this form of altruistic behavior **reciprocal altruism,** because each member of the partnership can potentially benefit from the relationship. Trivers hypothesized that reciprocal altruism would be favored by natural selection as long as individuals that do *not* reciprocate—called "cheaters" by behavioral biologists—are denied future aid. Observations of vampire bats and some other animals have confirmed Trivers' hypothesis: when a vampire bat accepts a "blood donation" from another bat, but then refuses to share food that it has collected, the other bats refuse to share their food with it in the future.

An Unusual Genetic System May Explain Altruism in Eusocial Insects

Hamilton's insights lead to a critical prediction about self-sacrificing behavior: altruism should usually be directed to close relatives. The evidence from many animal species overwhelmingly supports this prediction, but some species of ants, bees, wasps, and termites, those known as eusocial insects, provide a truly remarkable example. In **eusocial** insects, thousands of related individuals—a large percentage of them sterile female workers—live and work together in a colony for the reproductive benefit of a single queen and her mate(s). The workers may even die in defense of their colonies. How did this self-sacrificing social behavior evolve, and why does it persist over time? The failure of altruistic workers to reproduce should doom any alleles that promote altruism to early extinction.

For example, in a honey bee *(Apis mellifera)* colony, which may contain 30,000 to 50,000 related individuals, the only fertile female is the queen bee; all of the workers are her daughters **(Figure 55.20).** The queen's role in the colony is to reproduce. The workers perform all other tasks in maintaining the hive, from feeding the queen and her larvae to constructing new honeycomb and foraging for nectar and pollen. They also transfer food to one another and sometimes guard the entrance to the hive. Some pay the ultimate sacrifice when they sting intruders: this act of defense tears open the bee's abdomen, leaving the stinger and the poison sac behind in the intruder's skin, but killing the bee.

A. **Queen with sterile workers**

Kenneth Lorenzen

B. **Workers sharing food and passing pheromones**

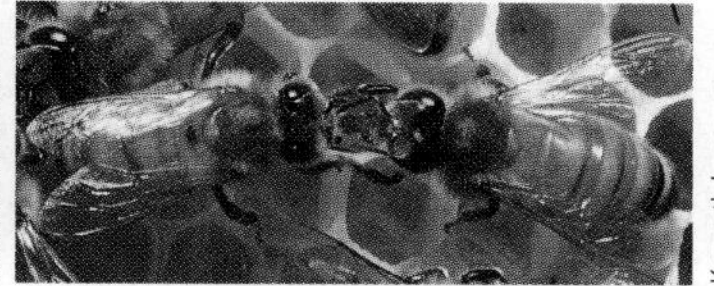
Kenneth Lorenzen

FIGURE 55.20
Life in a honey bee *(Apis mellifera)* colony. **(A)** A court of sterile worker daughters surrounds a queen bee, the only female of the colony that reproduces. **(B)** Worker bees routinely share food and transfer pheromones to one another.

Why do bees and other eusocial insects devote their entire lives to helping their mother produce hundreds of thousands of eggs? One answer may lie in a genetic phenomenon called **haplodiploidy,** an unusual pattern of sex determination in these insects **(Figure 55.21).** Like many other organisms, female bees are diploid, receiving one set of chromosomes from each parent. But

The drone (male parent) has only one set of chromosomes (symbolized by a red circle), which he contributes to the genome of every female worker. Thus, the workers are related to each other by 50% through their male parent.
Drone
♂
Queen
♀
The queen (female parent) has two sets of chromosomes (symbolized by a green triangle and a blue square). She contributes half of her alleles to each female offspring (either a triangle or a square) in this simplified presentation.

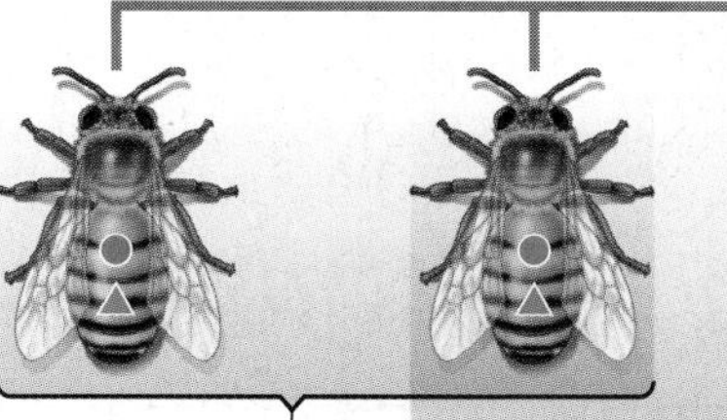

Workers that receive different alleles from the queen share no genetic relationship through their female parent. Thus, they are related to each other by 50% (the alleles inherited from their male parent).
Workers that receive the same alleles from the queen are related to each other by 50% through their male parent plus an additional 50% through their female parent.

FIGURE 55.21
Haplodiploidy. The genetic system of eusocial insects produces full siblings that have an exceptionally high degree of relatedness. Although this simplified model ignores recombination between the queen's two sets of chromosomes, it demonstrates how half the workers are related to each other by 50% and half are related to each other by 100%. Thus, the average degree of relatedness among workers is 75%. Including recombination would complicate the illustration, but the conclusion would be the same.

INSIGHTS FROM THE Molecular Revolution

Unadorned Truths about Naked Mole-Rat Workers

Naked mole-rats *(Heterocephalus glaber)* are sightless and essentially hairless burrowing mammals (see **Figure**) that live in mazes of subterranean tunnels in parts of Ethiopia, Somalia, and Kenya. Mole-rat colonies, which may include from 25 to several hundred individuals, contain a single "queen" and one to three males as the only breeding individuals. The remaining males and females are nonbreeding workers that, like the worker bees, ants, and termites of eusocial insect colonies, do all the labor: digging and defending the tunnels and caring for the queen and her mates.

Research Question

One of the many unanswered questions about these colonial mammals relates to the genetic structure of a colony. Does close kinship underlie the altruistic behavior of the workers? In other words, do they cooperate because they are all brothers and sisters?

© Gregory D. Dimijian/Photo Researchers, Inc.

Naked mole-rats *(Heterocephalus glaber)* live in colonies containing many workers that are effectively sterile.

Experiment

H. Kern Reeve and his colleagues at Cornell University investigated this question using molecular techniques resembling the DNA fingerprinting analysis often used to determine human kinship. The technique (see Section 18.2) relies on repeated DNA sequences that vary to a greater or lesser extent among individuals (that is, they are *polymorphic*). No two individuals (except identical twins) are likely to have exactly the same combination of sequences. Brothers and sisters with the same parents have the most closely related sequences; as relationships become more distant, the differences between sequences increase.

The researchers captured mole-rats living in four colonies in Kenya and housed individuals from each colony in a system of artificial tunnels. They collected samples of DNA from individual mole-rats and subjected them to fingerprinting analysis. First, the extracted DNA was fragmented by treating with a restriction endonuclease. The DNA fragments were separated by agarose gel electrophoresis and then transferred to a membrane filter by the Southern blot technique (see Figure 18.8). Next, the DNA fragments on the filter were hybridized independently with three radioactively labeled probes that identify three distinct groups of polymorphic sequences in the mole-rat DNA. The hybridization patterns were visualized by autoradiography. The pattern of bands, different for each individual (other than identical twins), is the DNA fingerprint.

The fingerprint of each mole-rat was compared with the fingerprints of individuals from the same and from different colonies. In the comparisons, bands that were the same in two individuals were scored as "hits." The number of hits was then analyzed to assign relatedness by noting which individuals shared the greatest number of bands.

Results

The comparisons revealed that individuals in the same mole-rat colony were indeed closely related—they shared an unusually high number of bands, higher than human siblings and approaching the band similarity of identical twins. The number of bands shared between individuals of different colonies was significantly lower, but still higher than that noted between unrelated individuals of other vertebrate species. The close relatedness of even separate colonies, as the investigators point out, may be due to similar selection pressures or to recent common ancestry among colonies in the same geographical region.

Conclusion

On the basis of their results, the researchers propose that the close genetic relatedness among individuals in a colony, which may increase the degree of altruistic behavior, is one of two major factors underlying the evolution and maintenance of the nonbreeding worker caste in the colonies. The second major factor, they propose, is the chance of survival, which is greater for mole-rats remaining in colonies than for those that attempt to live and breed on their own.

Source: H. K. Reeve et al. 1990. DNA "fingerprinting" reveals high levels of inbreeding in colonies of the eusocial naked mole-rat. *Proceedings of the National Academy of Sciences USA* 87:2496–2500.

male bees are haploid: they hatch from unfertilized eggs. When a queen bee mates with one drone (a male), all of the sperm he delivers are genetically identical because males have only one set of chromosomes. Thus, all workers inherit exactly the same set of alleles from their male parent, producing a 50% degree of relatedness among them. Like other diploid organisms, the workers are also related to each other by an average of 25% through their female parent. Adding these two components of relatedness, we see that workers are related to each other by an average of 75%, a higher degree of relatedness than they would have to any offspring they produced if they were fertile.

This extremely high degree of relatedness among the workers in some eusocial insect colonies may explain their exceptional level of cooperation. When Hamilton first worked out this explanation of social behavior in these insects, he suggested that the workers devote their lives to caring for their siblings—the queen's other offspring—because a few of those siblings, which carry 75% of the workers' alleles, may become future queens and produce enormous numbers of offspring themselves.

Nonbreeding workers also exist in a mammalian species, the naked mole-rat *(Heterocephalus glaber),* a small, almost hairless animal that lives in underground colonies of 70 to 80 individuals in eastern Africa. As described in *Insights from the Molecular Revolution,* recent studies have shown that the highly cooperative individuals occupying a colony share an exceptionally high proportion of their alleles.

FIGURE 55.22 Observational Research

An Evolutionary Analysis of Human Cruelty

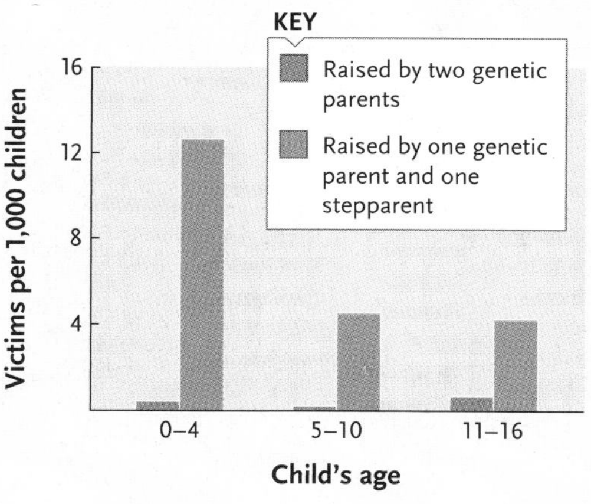
KEY
Raised by two genetic parents
Raised by one genetic parent and one stepparent
Victims per 1,000 children
16
12
8
4
0–4
5–10
11–16
Child's age

Hypothesis: Wilson and Daly hypothesized that child abuse would be more common in families in which parents and children were not genetically related than in families in which parents raise their biological offspring.

Prediction: Stepparents abuse their stepchildren more frequently than birth parents abuse their biological children.

Method: The researchers analyzed data on criminal child abuse within families that had been collected by the police department of a large Canadian city.

Result: The data indicated that children living with one stepparent and one genetic parent were 40 times more likely to suffer criminal abuse than children living with two genetic parents.

Conclusion: Children raised by one genetic parent and one stepparent are significantly more likely to suffer abuse than those raised by two genetic parents.

Source: M. Daly and M. Wilson. 1985. Child abuse and other risks of not living with both parents. *Ethology and Sociobiology* 6:197–210.

STUDY BREAK 55.5

1. **What do the social behaviors of musk oxen and sawfly larvae have in common?**
2. **Which animals in a dominance hierarchy are most likely to reproduce?**
3. **Why might the genetic system of many eusocial insects promote altruistic behavior?**

55.6 An Evolutionary View of Human Social Behavior

If we can analyze the evolutionary basis of the behavior of honey bees, naked mole-rats, and other animals, perhaps we can do the same for human behavior. According to Hamilton's kin selection hypothesis, we would expect human altruism toward nonrelatives to be rare. And it is true that *most* acts of human altruism are directed toward family members; huge sacrifices to help nonrelatives are relatively uncommon. But why, from an evolutionary perspective, do such charitable acts toward strangers occur at all?

Many behavioral biologists believe that reciprocal altruism can explain why humans have an evolved willingness to engage in low-cost acts of charity. Such behavior demonstrates their capacity for cooperation, and generosity is a socially approved trait that may confer benefits on those who exhibit it. This hypothesis yields the prediction that people who engage in charity will usually let others know about it. That prediction is supported by data showing that when organizers of blood drives offer small participation pins to donors, more people sign up to give blood.

Some researchers employ an evolutionary perspective to study difficult or painful societal issues, such as the occurrence of child abuse within families. Evolutionary theory leads us to predict that family members should generally help, not harm, one another. Margo Wilson and Martin Daly of McMaster University wondered if child abuse might be more common in reconstituted families, those with stepparents who are *not* genetically related to all the children in their care. To test this hypothesis, they examined data on criminal child abuse within families, made available by the police department of a Canadian city. In this city, the chance that a young child would be subject to criminal abuse was 40 times higher for children living with one stepparent and one genetic parent than for children who lived with both genetic parents **(Figure 55.22).**

This example illustrates the sort of insights that an evolutionary analysis of human behavior can provide. Wilson and Daly are not justifying or excusing child abusers. Neither are they claiming that abusive stepparenting is evolutionarily adaptive. Instead, their point is that humans may have some genetic characteristic that makes it more difficult to invest in children that they know are not their own, particularly if they also care for their own genetic children. These results are not just academic. Although a large majority of stepparents cope well with the difficulties of their role, a few do not. Knowledge of familial circumstances under which child abuse is more likely to occur may allow us to provide social assistance that would prevent some children from being abused in the future.

In recent years, the application of evolutionary thinking to human behavior has produced research on all sorts of questions. Sometimes the questions are interesting or even profound. Why do some tightly knit ethnic groups discourage intermarriage with members of other groups? At other times the issues may seem frivolous. Why do men often find women with certain physical characteristics attractive? Although evolutionary hypotheses about the adaptive value of behavior can be tested, helping us to understand why we behave as we do, they should never be used to *justify* behavior that is harmful to other individuals. Understanding why we get along or fail to get along with each other and the ability to make moral judgments about our behavior are uniquely human characteristics that set us apart from other animals.

STUDY BREAK 55.6

1. **How might evolutionary biologists explain altruistic behavior that people exhibit to non-relatives?**
2. **Why might stepparents provide fewer resources to their children than genetic parents do?**

UNANSWERED QUESTIONS

Who else is watching and listening?

Studies of communication have typically concentrated on the signaler and the intended receiver. But others are lurking in the background—eavesdroppers who are also attending to these signals. These third parties often use the signal to the signaler's detriment. Some flies, for example, listen for calling male crickets and deposit their larvae on the caller; the larvae eventually kill the cricket as they use him as a food source. Sometimes the signal that makes a male more attractive to females has the same effect on eavesdroppers. Male túngara frogs, for example, can add a "chuck" to the "whine" component of their call. These more complex calls make them more attractive to female frogs but also increase their risk of being captured by frog-eating bats or found by blood-sucking flies. Other animals circumvent this cruel bind by evolving signals in "private channels" to which intended receivers, but not predators, are privy. Ultraviolet signals in swordtail fishes and electrical signals in weakly electric fish are two examples. Increasingly, communication systems are being analyzed from the perspective of a complex communication network rather than the more simple two-way interactions.

Why do females prefer attractive mates?

One reason that females prefer attractive mates is that females can use the ornaments of males to judge their quality in terms of performance and survivorship. A superior male might provide better resources for the female and her offspring, or even pass on better genes to their young. Alternatively, male courtship traits must stand out against a sometimes chaotic background of environmental noise and signals from competing males. To stand out more than others, males might evolve signals that are better at stimulating the female's sensory, neural, and cognitive systems. If a species has evolved visual pigments that allow individuals to locate orange fruit, for example, males should evolve orange colors, because they will better stimulate the female's visual system. Such a scenario has been suggested for guppies. Researchers who study *sensory drive* try to understand how selection on sensory systems in one context, such as feeding, can influence functions in other contexts, such as mate choice.

To what degree is human behavior influenced by natural selection?

Cooperative and altruistic behavior sometimes evolve in animal societies. Since all societies face similar basic challenges, it is not surprising that cooperation can also be important in human societies. Strong evidence suggests that cooperation has evolved under selection in animal societies, but does logic demand that it has evolved under selection in humans as well? If not, what data would provide strong evidence for the evolution of human social behavior?

The field of evolutionary psychology, which poses these questions, is constrained because the invasive experimental approaches that have been so successful in animal biology cannot be applied to humans. Here the cross-cultural, comparative approach has made some important advances. These questions are being pursued by researchers in biology and psychology as well as anthropology and sociology. This mix of researchers with different approaches guarantees excitement and controversy about these very basic questions that ask why we are who we are.

Think Critically

Tamarins, a group of small-bodied New World monkeys, usually give birth to fraternal (non-identical) twins. Cells are sometimes exchanged between the two fetuses during development, resulting in chimeras. One male, for example, might have sperm cells which, according to their genetic constitution, are his brother's sperm and not his. This phenomenon explains the results of a laboratory study in which the genetic father of some offspring was really their uncle, who lived in a distant cage, and not the presumed "father," who was caged with the mother. Given what we know about how degrees of relatedness influence altruistic behavior, how might this phenomenon influence the social behavior of tamarins in the wild?

Michael J. Ryan

Michael J. Ryan is the Clark Hubbs Regents Professor in Zoology at The University of Texas at Austin, where he studies the evolution and function of animal behavior. Most of his work has addressed sexual selection and communication in frogs and fish. To learn more about Dr. Ryan's research, go to http://www.sbs.utexas.edu/ryan/.

REVIEW KEY CONCEPTS

Go to **CENGAGENOW** at www.cengage.com/login to access quizzing, animations, exercises, articles, and personalized homework help.

55.1 Migration and Wayfinding

- Some animals migrate seasonally, traveling from their birthplace to a distant locality and back again (Figures 55.2 and 55.3).
- Migrating animals use various behaviors to find their way. In piloting, animals use familiar landmarks to guide their journey (Figure 55.4). In compass orientation, animals use the position of the sun or stars, polarized light, or the Earth's magnetic field as a guide (Figure 55.5). In navigation, animals use mental maps of their position to find their destination.

- Biologists often interpret migratory behavior as an adaptive response to changing food supplies (Figure 55.6). Some animals breed in northern habitats when food is plentiful during the spring and summer. They generally head south to seasonally more productive habitats before the onset of winter.

55.2 Habitat Selection and Territoriality

- Animals use multiple criteria to select their habitats.
- Kineses and taxes help animals orient to appropriate portions of the habitats they occupy.
- Habitat selection often has a largely genetic basis, but learning and prior experience influence habitat selection in some species (Figure 55.7).
- Animals may maintain territories to gain exclusive use of defendable resources that are in short supply. The costs of territoriality include the time and energy devoted to territory defense, the risk of injury from fights, and exposure to predators.

55.3 The Evolution of Communication

- Animal communication occurs between a signaler, which sends a message, and a signal receiver, which receives and interprets the message.
- Animals communicate using acoustical, visual, chemical, tactile, or electrical signals (Figures 55.8–55.10). Each sensory channel provides specific advantages. Animals may use more than one channel simultaneously.
- Honey bees use a combination of tactile, acoustical, and chemical channels to share information about the location of food sources (Figure 55.11).

Animation: Honey bee dances

55.4 The Evolution of Reproductive Behavior and Mating Systems

- Males and females exhibit different reproductive strategies. Males can increase their reproductive success by inseminating the eggs of many females. Females generally seek mates that provide successful alleles to offspring, have access to abundant resources, or help care for young.
- Males often compete for access to females (Figure 55.12). Sexual selection has produced elaborate structures that males use for displays and for aggressive interactions with other males (Figures 55.13 and 55.14). Females may prefer males that have showy structures and great stamina, which are signs that they possess successful alleles.
- The type of mating system a species uses is tied to its pattern of territoriality and the amount of parental care the male parent provides.
- The comparative method allows biologists to trace the evolution of specific behaviors, such as nest building, by superimposing the behaviors on a phylogenetic tree for the lineage (Figure 55.15).

55.5 The Evolution of Social Behavior

- Social interactions between individuals have both benefits and costs. Group living may provide better protection from predators, more efficient feeding, and communal care of young (Figures 55.16 and 55.18). The costs of living in a group include increased competition for scarce resources and an increase in the spread of contagious diseases (Figure 55.17).
- Dominance hierarchies are highly structured societies in which some individuals have high status and first access to resources.
- Altruistic behavior appears to contradict Darwinian evolutionary theory, because altruistic individuals sacrifice their own fitness for the benefit of others. However, kin selection may ensure that individuals display altruistic behavior to close relatives that share some of their alleles (Figures 55.19).
- An unusual mechanism of sex determination, haplodiploidy, makes the workers in some eusocial insect colonies more closely related to each other than siblings in most species are (Figures 55.20 and 55.21). Haplodiploidy may have fostered the evolution of highly altruistic behavior.

Animation: Sawfly defense

55.6 An Evolutionary View of Human Social Behavior

- Although humans are more likely to provide assistance to close relatives than to nonrelatives, acts of charity to strangers are common, especially if altruists can advertise their generosity.
- An analysis of child abuse suggests that humans are more likely to abuse children to whom they are not genetically related than they are to abuse their biological children (Figure 55.22).

UNDERSTAND AND APPLY

Test Your Knowledge

1. Which of the following statements about animal migration is true?
   a. Piloting animals use the position of the sun to acquire information about their direction of travel.
   b. Animals migrating by compass orientation use mental maps of their position in space.
   c. Navigating animals use familiar landmarks to guide their journey.
   d. Navigating animals use a compass and a mental map of their position to reach a destination.
   e. Most migrating birds use olfactory cues to return to the place where they hatched from eggs.
2. In Marler and Moore's experiment with Jarrow's spiny lizard, what evidence from males that had received testosterone implants suggested that engaging in territorial behavior carries a heavy cost?
   a. They had to consume more water than control males.
   b. They mated with fewer females than control males.
   c. They ate more frequently than control males.
   d. They had higher death rates than control males.
   e. They weighed more than control males.
3. Which signal type would provide the fastest communication between bats flying in a dark forest?
   a. chemical signals
   b. acoustical signals
   c. visual signals
   d. tactile signals
   e. electrical signals

4. Squashing an ant on a picnic blanket often attracts many other ants to its "funeral." What kind of signal did squashing the ant likely produce?
   a. an electrical signal
   b. a visual signal
   c. an acoustical signal
   d. a chemical signal
   e. a tactile signal
5. Which of the following behaviors might have been produced by sexual selection?
   a. A male frog calls loudly and clearly from a pond during the breeding season.
   b. A young male goat bleats plaintively when left by its mother.
   c. A hen clucks to call its chicks closer when a predator approaches.
   d. A female lion ignores the sexual advances of a young male.
   e. A male dog is attracted to the odor of a female dog.
6. In comparison to males, the females of many animal species:
   a. compete for mates.
   b. choose mates that are well camouflaged in their habitats.
   c. choose to mate with many partners.
   d. are always monogamous.
   e. choose their mates carefully.
7. Social behavior:
   a. is exhibited *only* by animals that live in groups with close relatives.
   b. cannot evolve in animals that maintain territories.
   c. evolved because group living provides benefits to individuals in the group.
   d. is never observed in insects and other invertebrate animals.
   e. can be explained only by the hypothesis of kin selection.
8. Altruism is a behavior that:
   a. cannot evolve.
   b. has been observed only in insects.
   c. increases the number of offspring an individual produces.
   d. can indirectly spread the altruist's alleles.
   e. can evolve only in animals with a haplodiploid genetic system.
9. The degree of relatedness between a parent and its biological offspring:
   a. is the same as that between full siblings.
   b. is less than that between brother and sister.
   c. depends on how many siblings the parent has.
   d. promotes an individual's reproductive success.
   e. is the same as between first cousins.
10. The tendency for humans to be charitable to perfect strangers can be explained by the hypothesis of:
   a. sexual selection.
   b. kin selection.
   c. reciprocal altruism.
   d. polyandry.
   e. navigation.

Discuss the Concepts

1. In Chapter 53, you learned about some of the environmental changes associated with global warming. What effects might global warming have on animal species that undertake seasonal migrations?
2. The yellow-rumped whippersnapper, an imaginary species of songbird, always establishes breeding territories in forests where trees are interspersed among many small ponds. Design an experiment to determine what features of the environment this species uses to select its habitat.
3. Although females provide parental care far more often than males in the animal kingdom as a whole, exceptions exist, especially among birds and fishes. Develop three evolutionary hypotheses to explain why male birds are so likely to involve themselves in caring for their broods.

Interpret the Data

A female Seychelles warbler *(Acrocephalus sechellensis)* typically produces just one egg per nesting event. A male of this species will guard his mate from the attentions of neighboring males until she lays that egg, thereby increasing the likelihood that he was the one to father the offspring inside it. Jan Komdeur of the University of Groningen, The Netherlands, studied the behavior of male Seychelles warblers—before females had laid eggs—in relation to the number of other males that lived nearby. First, Komdeur watched focal males (that is, the ones being studied) for 30 s time periods and recorded the percentage of time periods in which they fed or guarded mates. Then, for focal males with more than three male neighbors, Komdeur experimentally reduced the number of neighbors to three and surveyed the behavior of the focal males again to see if it changed. His results are illustrated in the graphs reproduced below. (Because males sometimes performed both behaviors during the same observation period, some observation periods counted toward both behaviors; thus, the percentages of time periods spent foraging or mate guarding sum to more than 100%.) If a male warbler has many male neighbors, what cost is associated with ensuring his paternity of his mate's egg?

A. **Before removal of some male neighbors**

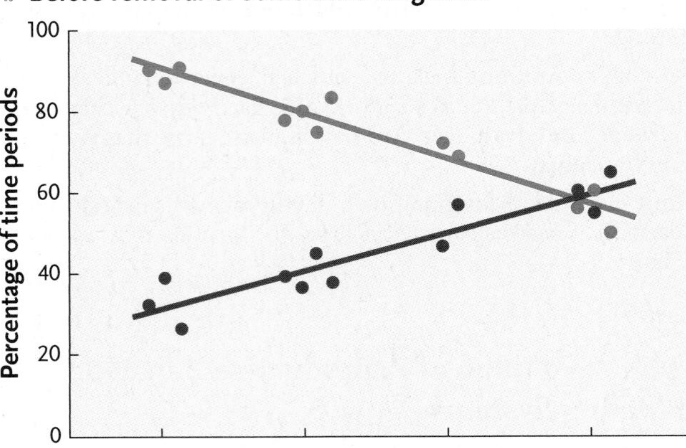
100
80
60
40
20
0
Percentage of time periods

B. **After removal of some neighbors**

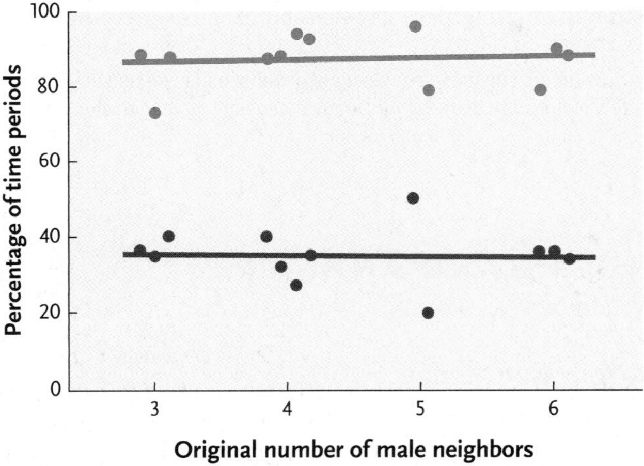
100
80
60
40
20
0
Percentage of time periods
3
4
5
6
Original number of male neighbors

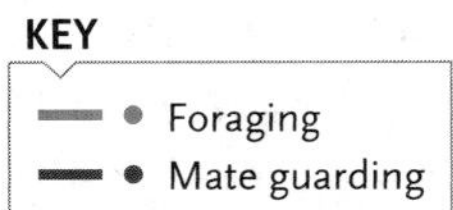
KEY
Foraging
Mate guarding

Source: J. Komdeur. 2001. Mate guarding in the Seychelles warbler is energetically costly and adjusted to paternity risk. *Proceedings of the Royal Society of London B* 268:2103–2111.

Design an Experiment

You discover that a particular butterfly species almost always lives in open meadows and almost never lives in nearby shaded forests. Design an experiment to test whether or not habitat selection by this species is adaptive.

Apply Evolutionary Thinking

African honeyguides (family Indicatoridae) are birds that call to humans and other mammals, leading them to honey bee colonies in woodlands. The mammals then open the hives to extract the honey, and the honeyguide feeds on the beeswax. How could a communication system between two species evolve?

Express Your Opinion

Africanized bees are slowly expanding their range in North America. Some researchers think the more we know about them, the better we will be able to protect ourselves. Should we fund more research into the genetic basis of their behavior? Go to www.cengage.com/login to investigate both sides of the issue and then vote.

Appendix A
Answers

Chapter 1

Study Break 1.1

1. The major levels in the hierarchy of life and some of their emergent properties are: cells—life; organisms—learning; populations—birth and death rates; communities, ecosystems, biosphere—diversity and stability.
2. Organisms use energy collected from the external environment for growth (including the production of new molecules and cells), maintenance and repair of body parts, and reproduction.
3. A life cycle is the series of structurally and functionally distinct developmental stages through which organisms pass.

Study Break 1.2

1. In artificial selection, humans selectively breed individuals with desirable heritable characteristics to enhance those traits in the next generation. In natural selection, genetically-based characteristics that increase survival and reproduction become more common in the next generation.
2. Random changes in DNA—mutations—may change the structure of proteins that contribute to the physical appearance and internal functions of an organism.
3. Being camouflaged may make an animal less likely to be noticed by a predator.

Study Break 1.3

1. In the cells of prokaryotic organisms, DNA is not separated from other parts of the cell. In the cells of eukaryotic organisms, DNA is enclosed within a nucleus.
2. Humans are classified in Domain Eukarya and Kingdom Animalia.

Study Break 1.4

1. A scientific hypothesis must be falsifiable. In other words, we must be able to imagine what sort of data would demonstrate that the hypothesis is incorrect.
2. The copper lizard models told the researchers how frequently lizards would perch in the sun just by chance and what the temperatures of non-thermoregulating lizards would be.
3. Model organisms are usually easy to maintain and study, and they have been so well studied that researchers already know a lot about their biology.
4. When scientists describe a set of ideas as a "theory," they recognize that the ideas have already withstood many scientific tests.

Test Your Knowledge

1. c, 2. b, 3. c, 4. d, 5. b, 6. d, 7. c, 8. a, 9. e, 10. d

Interpret the Data

Anolis gundlachi does not regulate its body temperature. The graphs show that lizards perch in patches of sun at about the same rate as randomly positioned models and that lizards and models exhibit similar distributions of body temperatures. These data suggest that this lizard species perches at random with respect to environmental factors that might influence its body temperature.

Chapter 2

Study Break 2.1

An element is a pure substance that consists of one type of atom. An atom is the smallest unit of an element that retains its chemical and physical properties. A molecule is a collection of atoms chemically combined in fixed numbers and ratios. Molecules can consist of the same atoms, as is seen for the two oxygen atoms in the molecule oxygen, or of different atoms, as in the combination of two hydrogen atoms and one oxygen atom in a molecule of water. Molecules with component atoms that are different, such as water, are compounds.

Study Break 2.2

1. Protons and neutrons are found in the nucleus of an atom. Electrons are found in orbitals located in energy levels (shells) that surround the nucleus.
2. Carbon-11 has six protons and five neutrons. Oxygen-15 has eight protons and seven neutrons.
3. The number of valence electrons—the electrons in the outermost shell of an atom—determines its chemical reactivity. If the outermost shell is not completely filled with electrons, the atom tends to be chemically reactive, whereas if that shell is completely filled, the atom is nonreactive.

Study Break 2.3

1. An ionic bond forms between atoms when those atoms gain or lose electrons completely. An example is the ionic bond in NaCl.
2. A covalent bond forms when atoms share a pair of valence electrons rather than gaining or losing them completely.
3. Electronegativity is a measure of an atom's attraction for the electrons it shares in a chemical bond with another atom. When electrons are shared equally, the atoms remain uncharged and the result is a nonpolar covalent bond. When electrons are shared unequally, one atom carries a partial negative charge and the other atom carries a partial positive charge. The molecule then has polarity, and the bond is a polar covalent bond.
4. In a chemical reaction, atoms or molecules interact to form new chemical bonds or break old ones. Atoms are added to or removed from molecules, or linkages of atoms in molecules are rearranged as a result of bond formation.

Study Break 2.4

1. Hydrogen bonds between neighboring water molecules produce a water lattice. The constant breakage and reformation of hydrogen bonds in the lattice allows water to flow easily. The polarity of water molecules also contributes to the properties of water. That is, in liquid water, the lattice resists invasion by other molecules unless the invading molecules also contain polar regions that can form competing attractions with water molecules. In that case, the water lattice opens, forming a cavity in which the polar or charged molecule can move. However, nonpolar molecules are unable to affect the water lattice. Hydrogen bonds also give water its unusual ability to resist changes in temperature by absorbing or releasing heat energy, its unusually high boiling point, and its unusually high internal cohesion and surface tension.
2. A solute is a dissolved substance. A solvent is a substance capable of dissolving another substance. A solution is a solute dissolved in a solvent. For example, salt (NaCl) is a solute that can dissolve in the solvent water.

Study Break 2.5

1. Acids are hydrogen ion (proton, H^+) donors; bases are proton acceptors. An acid dissociates in water to produce a hydrogen ion and an anion. Most bases dissociate in water to give hydroxide ions, which then accept protons to produce water.
2. Buffers act to control the pH of a solution. In living organisms, buffers keep the pH of body and cell fluids within a narrow range, enabling normal cell and body functions to occur. Outside of the normal pH range, the functions of proteins can be affected, thereby adversely affecting the functions of the organism.

Think Critically

Does the bacterium metabolize the arsenic? What actually is meant by the metaphor "eat" as used in this essay? If the bacterium does not metabolize the arsenic, what function does the arsenic play in the life of the bacterium?

Test Your Knowledge

1. d, 2. e, 3. a, 4. b, 5. a, 6. d, 7. d, 8. c, 9. e, 10. b

Interpret the Data

1. About 50 minutes
2. (a) Using a pH of 5.9, H^+ concentration is 0.0000012589 (1.2589×10^{-6}) *M* and OH^- concentration is 0.00000000789433 (7.9433×10^{-9}) *M*. (b) Using a pH of 2.5, H^+ concentration is 0.00316228 (3.16228×10^{-3}) *M* and OH^- concentration is 0.0000000000031623 (3.1623×10^{-12}) *M*.
3. Using a pH of 2.5 for peak reflux and pH of 4.0 for clinical reflux, the H^+ concentration is 0.003 *M* greater, and OH^- concentration is 0.0000000000968377 (9.68377×10^{-11}) *M* less than a pH of 4.0.

Chapter 3

Study Break 3.1

1. Organic molecules are molecules based on carbon. Hydrocarbons are a type of organic molecule that consists of carbon linked only to hydrogen atoms.
2. The maximum number of bonds that a carbon atom can form is four.
3. Carboxyl groups donate a hydrogen ion in water and therefore act as acids. Amino groups accept a hydrogen ion in water and therefore act as

bases. Phosphate groups donate hydrogen ions in water and therefore act as acids.

4. In a dehydration synthesis reaction, components of water (—H and —OH) are removed. In hydrolysis, the components of a water molecule are added to functional groups as molecules are broken down into smaller subunits.

Study Break 3.2

A monosaccharide is the structural unit of carbohydrate molecules. Monosaccharides are simple sugars such as trioses, pentoses, and hexoses. Glucose, galactose, and fructose are hexoses. A disaccharide is a molecule assembled from two monosaccharides linked by a dehydration synthesis reaction. Lactose, sucrose, and maltose are disaccharides. A polysaccharide is a polymer of monosaccharide subunits. The subunits are identical or different, depending on the particular polysaccharide. Glycogen, starch, cellulose, and chitin are polysaccharides.

Study Break 3.3

The three most common lipids found in living organisms are neutral lipids, phospholipids, and steroids. Most neutral lipids consist of a three-carbon backbone chain formed from glycerol, with each carbon linked to a fatty acid side chain. In the most common phospholipids, glycerol is the backbone, with two of its binding sites linked to fatty acids. The third binding site is linked to a polar phosphate group. Steroids have structures based on a framework of four carbon rings. Differences in side groups attached to the rings distinguish the different types of steroids.

Study Break 3.4

1. Differences in the side groups (R in the figures) give the amino acids their individual properties.
2. A peptide bond is the bond between the C of the carboxyl group of one amino acid and the N of the amino group of the adjacent amino acid (see Figure 3.19). The bond is formed in a dehydration synthesis reaction between an amino group of one amino acid and a carboxyl group of another amino acid.
3. Domains are distinct structural subdivisions in the final folded forms of proteins. They are the result of the amino acid sequence of the protein (the primary structure of the protein) and the secondary, tertiary, and quaternary (if more than one polypeptide is involved) structures of the protein.

Study Break 3.5

1. Nucleic acids are formed from nucleotide monomers. A nucleotide consists of a nitrogenous base, a five-carbon sugar, and a phosphate.
2. In DNA, the five-carbon sugar is deoxyribose; in RNA it is ribose. DNA has the pyrimidine nitrogenous base T (thymine), and RNA has U (uracil).

Think Critically

In Type 2 diabetes, SREBPs in the liver are activated, and this leads to overproduction of fatty acids and triglycerides. Inhibitors of SREBP processing might reduce plasma triglycerides and prevent their toxic effects on blood vessels.

Test Your Knowledge

1. a, 2. d, 3. c, 4. e, 5. c, 6. d, 7. b, 8. d, 9. b, 10. a

Interpret the Data

1. LDL was highest in the saturated fats diet group, and HDL was lowest in the *trans* fatty acids diet group.
2. The *trans* fatty acid group had the highest LDL-to-HDL ratio.
3. From best to worst diet: *cis* fatty acids, saturated fats, *trans* fatty acids

Chapter 4

Study Break 4.1

1. Kinetic energy is the energy of motion, whereas potential energy is stored energy.
2. An isolated system exchanges neither matter nor energy with its environment. A closed system can exchange energy, but not matter, with its environment. An open system can exchange both energy and matter with its environment.

Study Break 4.2

1. The change in energy content of a system, and its change in entropy. Reactions tend to be spontaneous when the products have less potential energy than the reactants, and when the products are less ordered than the reactants.
2. The greater the negative value of ΔG, the further a reaction will proceed toward completion and therefore the greater the concentration of product molecules versus reactant molecules.
3. An exergonic reaction releases free energy—ΔG is negative because the products contain less free energy than the reactants. An endergonic reaction requires free energy from the surroundings to run—ΔG is positive because the products contain more free energy than the reactants.

   In a catabolic reaction, energy is released during the breakdown of a complex reactant molecule to a simpler product molecule, whereas in an anabolic reaction, energy is used to create a product molecule that is more complex than the reactant molecule.

   Individual reactions may be exergonic or endergonic. When a series of reactions (each of which may have a positive or a negative ΔG value) forms a metabolic pathway, the pathway is catabolic if the overall sum of individual reaction ΔG values is negative, and it is anabolic if the overall sum of individual reaction ΔG values is positive.

Study Break 4.3

1. ATP contains the five-carbon sugar ribose, with the nitrogenous base adenine linked to one of the carbons and a chain of three phosphate bonds to another carbon. The phosphate groups are closely associated with each other and their negative charges strongly repel each other, making the bonding arrangements unstable and storing potential energy. Hydrolysis of ATP to remove one or two of the three phosphates is a spontaneous reaction that relieves the repulsion and releases large amounts of free energy.
2. Many individual reactions found in living cells are not spontaneous because they have a positive ΔG. By joining such a reaction to another reaction with a large negative ΔG, the reaction can be completed. The combined reaction is called a coupled reaction. ATP participates in coupled reactions in the enzyme-driven process of energy coupling. That is, ATP comes into close contact with a reactant molecule in an endergonic reaction. When ATP is hydrolyzed, the terminal phosphate group is transferred to the reactant molecule, which makes that molecule less stable so that the reaction continues spontaneously.

Study Break 4.4

1. Enzymes accelerate reactions by reducing the activation energy of a reaction, the initial input of energy required to start a reaction. Enzymes lower activation energy by rearranging the atoms and bonds of the reacting molecules into the transition state, an activated state that is highly unstable. With relatively little change in energy, the transition state can move forward toward products or backward toward reactants.
2. No.

Study Break 4.5

1. There is a maximum rate at which an enzyme can combine with substrates and release products. Beyond that point, increasing substrate concentration does not increase the reaction rate any further.
2. In competitive inhibition, the inhibitor competes with the normal substrate molecule for binding to the active site of the enzyme, whereas in noncompetitive inhibition, the inhibitor does not compete directly with the substrate for binding to the active site.
3. As the temperature increases, the increasing kinetic motions of the enzyme's amino acid chains eventually disrupt the enzyme's three-dimensional structure, causing it to unfold and denature. At that point, there is no enzyme activity.

Study Break 4.6

A ribozyme is an RNA molecule that accelerates the rate of a biological reaction. A ribozyme qualifies as an enzyme because it remains unchanged after the reaction is complete; that is, it is a true catalyst.

Think Critically

Variables that alter conformational changes, such as temperature, pH, and chemicals.

Test Your Knowledge

1. c, 2. d, 3. b, 4. c, 5. e, 6. a, 7. a, 8. b, 9. d, 10. e

Interpret the Data

1. ΔG is $-5.5 - 3.5 = -9$ kcal/mol for the mutated polypeptide and $-2.0 - 4.0 = -6.0$ kcal/mol for the parent polypeptide.
2. Binding of both of the polypeptides is exergonic and spontaneous, and the parent polypeptide would yield more free energy on binding.
3. No, because the ΔG is less negative, it would not be as energetically favorable.

Chapter 5

Study Break 5.1

The plasma membrane is a bilayer of lipid and suspended protein molecules that bounds the cytoplasm of a cell. The lipid bilayer is hydrophobic; therefore, it is a barrier to the passage of water-soluble substances. The membrane has protein channels through which selected water-soluble substances are able to pass.

Study Break 5.2

The DNA of a prokaryotic cell is located in the nucleoid region, a central area of the cell that has no membrane around it to separate it from the cytoplasm. In most prokaryotes, the DNA is a folded mass. In its unfolded state, it is a circular DNA molecule.

Study Break 5.3

1. Most of the DNA of a eukaryotic cell is found within a nucleus located roughly in the center of the cell. The nucleus is bounded by the nuclear envelope, a membrane that separates its contents from the cytoplasm. The DNA is complexed with proteins and organized into several linear chromosomes.
2. The nucleolus is an area within the nucleus. It forms around the genes for rRNA in the chromosomes and is the location where the information in those genes is copied into rRNA. The rRNA combines with proteins in the nucleolus to form the large and small ribosomal subunits. A large and a small ribosomal subunit function together in the cytoplasm to synthesize proteins.
3. The endomembrane system is a collection of organelles, membranous channels, and vesicles that form a major traffic network for the synthesis, distribution, and storage of proteins and other molecules. Its structure includes the endoplasmic reticulum (ER) and the Golgi complex. The rough ER has ribosomes on its outer surface. Proteins synthesized by those ribosomes enter the ER lumen, where they fold into their final shape and may be modified chemically. Then they are delivered to other regions of the cell within vesicles that pinch off from the rough ER. Most of the proteins made on rough ER go to the Golgi complex.

   The smooth ER consists of membranes that lack ribosomes. Functions of the smooth ER include synthesis of lipids that become parts of cell membranes.

   The Golgi complex is a stack of flattened, membranous sacs. The *cis* face of this organelle receives vesicles containing proteins released from the rough ER and continues their chemical modifications. The proteins are then sorted into vesicles that pinch off from the *trans* face of the Golgi complex, the side that faces the plasma membrane. Some of the released vesicles remain in the cytoplasm as storage vesicles of various types, whereas others, called secretory vesicles, release their contents to the outside of the cell.
4. A mitochondrion is an organelle enclosed by two membranes. The outer mitochondrial membrane is smooth and covers the outside of the organelle. The inner mitochondrial membrane is highly folded into cristae. Within the inner membrane is the mitochondrial matrix, which contains DNA and ribosomes. Most of the energy required for eukaryotic cellular activities is generated by reactions in the cristae and the matrix. Those reactions break down sugars, fats, and other fuel molecules into water and carbon dioxide, releasing energy mostly in the form of ATP.
5. The cytoskeleton is an internal cytoplasmic network of filaments and tubules composed mainly of actin and tubulin proteins. The function of the cytoskeleton is to maintain the shape of the cell, reinforce the plasma membrane, and organize internal structures. Changes in the cytoskeleton are responsible for movements of cell organelles, movements of parts of the cell, or movements of the whole cell.

Study Break 5.4

1. A chloroplast has two membranes: an outer boundary membrane and an inner boundary membrane. The latter, similar to the cristae of the mitochondrion, is highly folded. The two membranes enclose an inner compartment known as the stroma. Within the stroma is a membrane system composed of flattened, closed sacs called thylakoids. Chloroplasts are the sites of photosynthesis in plant cells. The thylakoid membranes contain chlorophyll and other molecules that absorb light energy and convert it into chemical energy. Enzymes in the stroma use the chemical energy to make carbohydrates and other complex organic molecules from water, carbon dioxide, and other simple inorganic precursors.
2. The tonoplast, the membrane surrounding the central vacuole, contains transport proteins that move substances into and out of the vacuole. Central vacuoles also store organic and inorganic salts, organic acids, sugars, storage proteins, pigments, and, in some cells, waste products. Chemical defense molecules are found in the central vacuoles of some plants.

Study Break 5.5

1. Anchoring junctions are spots or belts that run entirely around cells, effectively sticking adjacent cells together. Microfilaments (in adherens junctions) or intermediate filaments (in desmosomes) anchor the junction in the underlying cytoplasm. Tight junctions involve fusion of a network of junction proteins in the outer halves of the plasma membranes of adjacent cells, forming a tight seal that can keep even ions from moving between the cells. Gap junctions open direct channels between adjacent cells through which ions and small molecules can pass directly. The gap junctions are formed by aligned hollow protein cylinders in the plasma membranes of two cells.
2. The extracellular matrix (ECM) is a complex of proteins and polysaccharides secreted by the cells that it surrounds. Depending on the nature of the network of proteoglycans (carbohydrate-rich glycoproteins) in the ECM, the consistency ranges from soft and jellylike to hard and elastic.

Think Critically

Chapter 24 discusses the endosymbiotic theory. This is the theory that some organelles in eukaryotic cells, in particular mitochondria and chloroplasts, are derived from symbiotic relationships with prokaryotic cells. Evidence that supports this theory is discussed.

Test Your Knowledge

1. b, 2. e, 3. c, 4. a, 5. b, 6. c, 7. b, 8. b, 9. b, 10. d

Interpret the Data

Most of the cathepsin activity occurs in the fraction containing the lysosomal marker enzymes (fraction 5), thus localizing the worm cathepsin activity to the lysosome.

Chapter 6

Study Break 6.1

1. The fluid mosaic model proposes that the membrane consists of a fluid phospholipid bilayer in which proteins are embedded and float freely.
2. Integral proteins include transport proteins, receptor proteins, recognition proteins, and cell adhesion proteins. Peripheral proteins include microtubules, microfilaments, intermediate filaments, and proteins that link the cytoskeleton together.

Study Break 6.2

1. In passive transport, molecules and ions move across the membrane from the side with the higher concentration to the side with the lower concentration; that is, the difference in concentration provides the energy for passive transport. In active transport, molecules and ions move across the membrane from the side with the lower concentration to the side with the higher concentration, that is, against the concentration gradient. The energy for active transport comes from the hydrolysis of ATP.
2. The transport of substances through membranes based solely on molecular size and lipid solubility is simple diffusion. The diffusion of polar and charged molecules across membranes with the help of transport proteins is facilitated diffusion.

Study Break 6.3

1. Osmosis is the passive transport of water across a membrane. This movement follows concentration gradients. For osmosis to occur, there must be a selectively permeable membrane, that is, a membrane that will allow water molecules, but not molecules of the solute, to pass. As long as the solute is at different concentrations on the two sides of the membrane, water movement will occur; that is, it is not necessary for pure water to be present on one side of the membrane.
2. If animal cells are in a hypertonic solution, water molecules will move by osmosis from within the cells to the surrounding solution. If the outward movement of water exceeds the capacity of the cells to replace the lost water, the cells will shrink.

Study Break 6.4

1. Active transport is the movement of substances across membranes against their concentration gradients by pumps; the energy for active transport comes from ATP hydrolysis. In primary active transport, ATP hydrolysis directly drives the process; that is, the same protein that transports a substance also hydrolyzes ATP. In secondary active transport, ATP hydrolysis indirectly drives the process; that is, the transport proteins themselves do not hydrolyze ATP. Rather, the transporters use a favorable concentration gradient of ions, generated by primary active transport (where ATP hydrolysis is used), as their energy source for active transport of a different ion or molecule.
2. A membrane potential is a voltage difference across a membrane. Ion transport by membrane pumps contributes to this voltage difference. The sodium–potassium pump in the plasma membrane pushes three sodium ions out of the cell and two potassium ions into the cell with each turn of the pump. This leads to an accumulation of positive charges outside the membrane, causing the inside of the cell to become negatively charged with respect to the outside of the cell. In addition, an unequal distribution of ions across the membrane is created by passive transport. The electrical potential difference (voltage) across the plasma membrane is the membrane potential.

Study Break 6.5

1. In exocytosis, secretory vesicles in the cytoplasm contact and fuse with the plasma membrane, releasing their contents to the outside of the cell.
2. Endocytosis is a mechanism by which substances are brought into the cell from the exterior. The substances become trapped in pitlike depressions that bulge inward from the plasma membrane. The depression pinches off as an endocytic vesicle. In bulk-phase endocytosis, no binding by surface receptors is involved. Extracellular water is taken in together with any other molecules that are in solution in the water. This is the simplest form of endocytosis. In receptor-mediated endocytosis, molecules to be taken in become bound to the outer cell surface by receptor proteins. The receptor proteins are specific in that they recognize and bind only certain molecules from the solution that surrounds the cell. The molecules recognized are mostly proteins or other molecules carried by proteins. Once the receptors have bound their target molecules, the receptors collect into a coated pit, a depression in the plasma membrane. The pits, with the contained target molecules, pinch off from the plasma membrane to form endocytic vesicles.

Think Critically

Restricting the passage of protons aids in conservation of the membrane's electrochemical potential.

Test Your Knowledge

1. d, 2. b, 3. a, 4. c, 5. e, 6. d, 7. e, 8. b, 9. a, 10. c

Interpret the Data

1. Cell set B is displaying resistance because these cells fail to accumulate paclitaxel. They are using active transport to move the anticancer drug back out across the cell membrane.
2. The drug treatment imatinib has a slight but significant effect on increasing the accumulation of the labeled paclitaxel in set A, but a much more dramatic effect is seen using either of these drugs on cell set B. This suggests that in cell set B, the active transport mechanism is inhibited by the presence of these additional drugs.

Chapter 7

Study Break 7.1

Specificity of a cellular response depends on the signal–receptor interaction. Specificity starts with the signal molecule; that is, the specific signal molecule is the messenger that elicits a specific cellular response. For example, the hormone epinephrine causes glucose to be released into the bloodstream. Specificity also depends on the target cells; that is, only target cells respond to the signal molecule because they exclusively have receptors for the signal molecule.

Study Break 7.2

1. Protein kinases are enzymes that add phosphate groups to other proteins. The result of phosphorylation is that the protein will be either stimulated or inhibited in its activity. The cellular responses of signal transduction pathways are produced through the actions of protein kinases.
2. Amplification is the phenomenon of an increase in the magnitude of each step of a signal transduction pathway. Amplification typically occurs because the proteins conducting each step of the pathway are enzymes. That is, each enzyme, when it is becomes activated, activates large numbers of molecules entering the next step of the pathway.

Study Break 7.3

1. For a receptor tyrosine kinase to become activated, the signal molecule first binds to the receptor, which then assembles into a dimer. The receptor adds phosphate groups to tyrosines on the cytoplasmic side of itself, which activates the receptor.
2. A fully activated receptor tyrosine kinase has phosphorylated tyrosines on each of its two monomers. A signaling protein that recognizes a phosphorylated tyrosine as well as surrounding parts of the polypeptide binds to the receptor. Depending on the signaling protein, the binding itself may activate the signaling protein, or it is activated by tyrosine phosphorylation catalyzed by the receptor. In its activated form, the signaling protein initiates a transduction pathway leading to a cellular response. A given receptor can initiate different responses because different combinations of signaling proteins can bind to the receptor.

Study Break 7.4

1. The first messenger in a G-protein–coupled receptor–controlled pathway is the extracellular signal molecule. When it binds to the G-protein–coupled receptor, it activates a site on the cytoplasmic side of the receptor; the activated receptor, in turn, activates the G protein next to it.
2. The effector is activated by the G protein. The effector is a plasma-associated enzyme that generates a nonprotein signal molecule called the second messenger. The second messenger leads to the activation of protein kinases leading to the cellular responses triggered by the signal molecule.
3. A main way the pathway is turned off is by the conversion of cAMP to 5′-AMP by phosphodiesterase. As long as the receptor is bound by the signal molecule, cAMP is being generated by the activated effector. The continued synthesis of cAMP balances the degradation of cAMP by phosphodiesterase, ensuring that the pathway continues to run. However, if the signal molecule no longer is bound to the receptor, the effector again becomes inactive, cAMP therefore is not generated, and existing cAMP is rapidly degraded by phosphodiesterase. As a result, the protein kinase cascade is shut down and no cellular responses occur.

Study Break 7.5

1. The steroid receptor is within the cell, whereas the receptor tyrosine kinase and G-protein–coupled receptors are in the membrane or associated with the membrane. Also, the activated steroid receptor directly activates genes, whereas the other two receptors, when active, are just the first steps in pathways that may or may not activate genes.
2. A steroid hormone brings about a specific cellular response because whether a cell responds to a steroid hormone depends on whether it has the internal receptor for the hormone. Then within the cells with the receptor, the specific genes that are controlled are those with regulatory sequences that are recognized by the activated receptor.

Study Break 7.6

Signal transduction pathways, cellular response systems triggered by cell adhesion molecules, and communication pathways that involve gap junctions between adjacent cells might be integrated in a cross-talk network.

Think Critically

1. Perhaps the male's behavior induces the release of dopamine or another relevant neurotransmitter onto specific neurons in the brain of the female, which contain progestin receptors, and are involved in regulation of sexual behavior. The dopamine, in turn, may activate those receptors leading to neuronal changes resulting in the expression of sexual behavior.
2. To answer this question, you have to think of other situations in which the environment might cause release of a neurotransmitter that then via cross-talk activates a steroid receptor resulting in changes in behavior. Perhaps stimulation from pups activates neuronal steroid hormone receptors, resulting in changes in maternal behavior in a lactating mother. There are many other examples in which stimulation from the environment or another animal might, via neurotransmitter release, influence the function of a particular steroid hormone receptor (by activating it).

Test Your Knowledge

1. b, 2. a, 3. d, 4. c, 5. c, 6. c, 7. e, 8. b, 9. d, 10. e

Interpret the Data

1. In normal cells the least amount of CFTR was found in vesicles, while the greatest amount was found in endoplasmic reticulum. In cells with the CF mutation the least amount of CFTR was found in Golgi bodies, while the greatest amount was found in endoplasmic reticulum.
2. The CFTR protein is found in almost the same amount in the endoplasmic reticulum of normal and CF cells.
3. The CFTR protein is held up in the endoplasmic reticulum.

Chapter 8

Study Break 8.1

1. Oxidation is the removal of electrons from a substance; reduction is the addition of electrons to a substance.
2. Cellular respiration refers to the reactions in which oxygen is used as final electron acceptor; it includes the reactions that transfer electrons from organic molecules to oxygen and the reactions that make ATP. Oxidative phosphorylation is the process by which ATP is synthesized using the energy released by electrons as they are transferred to oxygen.

Study Break 8.2

1. The initial steps of glycolysis, which require 2 ATP, convert glucose to a phosphorylated derivative. The later steps, which release 4 ATP, remove electrons from the glucose derivatives and generate two molecules of pyruvate.
2. The redox reaction in glycolysis is the glyceraldehyde-3-phosphate (G3P) to 1,3-bisphosphoglycerate reaction (see Figure 8.7, step 6).
3. ATP is synthesized in glycolysis by substrate-level phosphorylation occurring in the 1,3-bisphosphoglycerate to 3-phosphoglycerate reaction (see Figure 8.7, step 7), and in the phosphoenolpyruvate (PEP) to pyruvate reaction (see Figure 8.7, step 10).
4. If excess ATP is present in the cytosol, it binds to and inhibits the activity of phosphofructokinase.

As a result, the concentration of fructose-1,6-bisphosphate, the product of the phosphofructokinase reaction, decreases, and the subsequent reactions of glycolysis are slowed or stopped. This is reversed when the ATP level in the cytosol decreases. In the end, this control mechanism helps prevent the needless oxidation of fuel molecules when the cell has an adequate supply of ATP.

Study Break 8.3

The three-carbon pyruvate molecules are transported from the cytosol into the mitochondria, where they are converted into two-carbon acetyl units through pyruvate oxidation. The citric acid cycle oxidizes the acetyl units completely to carbon dioxide with the transfer of electrons to NAD^+ or FAD.

Study Break 8.4

1. Each complex contains a unique combination of nonprotein carriers that pick up and release electrons.
2. The proton pumps push protons (H^+) from the mitochondrial matrix to the intermembrane compartment, increasing the proton concentration there. The resulting proton gradient produces an electrical gradient across the inner mitochondrial membrane with the matrix negatively charged with respect to the intermembrane compartment. The charge and proton concentration differences together provide energy for ATP synthesis in what is called proton-motive force.

Study Break 8.5

Fermentation occurs when oxygen is absent or limited. The electrons carried by the NADH produced in glycolysis are transferred to an organic molecule instead of the electron transfer system. In lactate fermentation, the end product of glycolysis, pyruvate, is converted to lactate. The lactate stores electrons temporarily, transferring them to the mitochondrial electron transfer system when the oxygen content of cells returns to normal. In alcoholic fermentation, pyruvate is converted into ethyl alcohol.

Think Critically

One possible hypothesis: There is a direct cause and effect relationship between altered regulation of mitochondrial expression and AD. If a significant number of mice whose genes were not genetically altered were to develop AD, this hypothesis would have to be rejected.

Test Your Knowledge

1. b, 2. a, 3. e, 4. c, 5. b, 6. e, 7. a, 8. d, 9. a, 10. c

Interpret the Data

1.

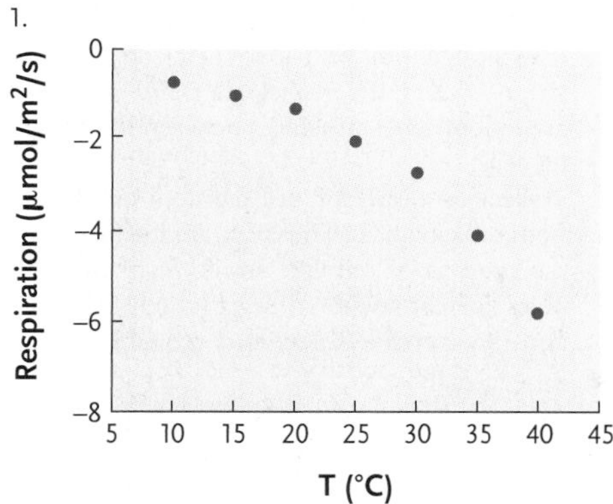

Respiration (μmol/m²/s)
0
−2
−4
−6
−8
5
10
15
20
25
30
35
40
45
T (°C)

2. 2
3. A curve represents the data best. A Q10 of 2 indicates a curve because the respiration rate is doubling for every 10°C increase in temperature. This type of relationship is an exponential increase.
4. Temperatures are predicted to increase, which would increase respiration according to these data. The data are insufficient to predict the actual increase in respiration since the plants would likely change their enzyme amounts or types (acclimate) in response to the rising temperature. Further, temperature is not the only factor that will likely change with rising CO_2. Other factors include precipitation, cloud cover, nutrient availability, etc.

Chapter 9

Study Break 9.1

1. The two stages of photosynthesis are the light-dependent reactions, in which the energy of sunlight is absorbed and converted into chemical energy in the form of ATP and NADPH, and the light-independent (dark) reactions, in which electrons carried by NADPH are used as a source of energy to convert carbon dioxide from inorganic to organic form.
2. In plants, photosynthesis takes place in the chloroplast. The light-dependent reactions are carried out on the thylakoid membranes and stromal lamellae. The light-independent reactions are carried out in the stroma.

Study Break 9.2

1. The chlorophyll *a* molecules in the antenna complexes are normal molecules of the pigment, consisting of a carbon ring structure with a magnesium atom bound at the center and an attached hydrophobic side chain. These chlorophyll *a* pigments absorb light. The chlorophyll *a* molecules in the reaction centers have modified light absorption properties that result from interactions with particular proteins of the photosystems. The two special chlorophyll *a* molecules of photosystem II are P680; those of photosystem I are P700. These pigment molecules receive light energy from normal pigments via inductive resonance. The energy is used to push electrons to an excited state.
2. The making of NADPH begins when electrons derived from water splitting are pushed to higher energies by light absorption in photosystem II. The high-energy electrons pass to a primary acceptor in photosystem II and then down an electron transfer system to P700 in photosystem I, losing energy along the way. Light energy absorbed by photosystem I again excites the electrons, which pass to different electron carriers, ending with ferredoxin. The ferredoxin transfers high-energy electrons to $NADP^+$, which is reduced to NADPH by $NADP^+$ reductase.
3. In the noncyclic electron flow pathway, electrons run through the entire set of photosystems and electron carriers, producing both NADPH and ATP. In the cyclic electron flow pathway, electrons flow cyclically around photosystem I; photosystem II is not involved. The cycle of electrons is through the cytochrome complex and plastocyanin to photosystem I, to ferredoxin, but then back to the cytochrome complex rather than on to $NADP^+$ reductase. Only ATP is produced by this pathway.

Study Break 9.3

1. Rubisco catalyzes a reaction combining carbon dioxide with RuBP to form two molecules of 3-phosphoglycerate (3PGA). Rubisco, an enzyme unique to photosynthetic organisms, is the key enzyme for producing the world's food because it is responsible for carbon dioxide fixation, a process that ultimately provides organic molecules for most of the world's organisms. Rubisco is the key regulatory site of the Calvin cycle for the following reason: During the daytime, sunlight powers the light-dependent reactions, and the NADPH and ATP produced by those reactions stimulate rubisco, which, in turn, keeps the Calvin cycle running. In darkness, however, NADPH and ATP levels are low, and as a result, rubisco's activity is inhibited and the Calvin cycle slows down or stops.
2. For each carbon atom that is released from the Calvin cycle in a carbohydrate molecule, one carbon dioxide molecule must enter the cycle. Therefore, to produce a molecule containing 12 carbon atoms, 12 molecules of carbon dioxide must enter the cycle.

Study Break 9.4

1. Photorespiration uses oxygen and releases CO_2. It occurs when oxygen concentrations are high relative to CO_2 concentrations. In that condition, rubisco acts as an oxygenase rather than a carboxylase, catalyzing the combination of RuBP with O_2 rather than CO_2. The toxic products formed by this reaction cannot be used in photosynthesis and are eliminated from the plant as CO_2. Photorespiration uses energy to salvage the carbons from phosphoglycolate, which greatly reduces the efficiency of energy use in photosynthesis. This can be seen in the reduced growth of plants grown under photorespiration conditions.
2. In the C_4 pathway, carbon fixation involves the reaction of CO_2 with phosphoenolpyruvate (PEP) to produce a four-carbon molecule, oxaloacetate. The oxaloacetate is reduced to malate by electrons transferred from NADPH, and malate then is oxidized to pyruvate in a reaction releasing CO_2, which is used in the rubisco-catalyzed first step of the Calvin cycle. In C_4 plants, carbon fixation and the Calvin cycle occur in different cell types: carbon fixation in mesophyll cells, and the Calvin cycle in bundle sheath cells. This alternative method of carbon fixation minimizes photorespiration.
3. In C_4 plants, carbon fixation and the Calvin cycle occur in different cell types, mesophyll cells and bundle sheath cells, respectively. In CAM plants, carbon fixation and the Calvin cycle occur at different times, at night and during the day, respectively.

Study Break 9.5

The reactions of photosynthesis and cellular respiration are essentially the reverse of one another, with CO_2 and H_2O the reactants of photosynthesis and the products of cellular respiration. Phosphorylation reactions involving electron transfer systems are part of each process, namely photophosphorylation in photosynthesis and oxidative phosphorylation in cellular respiration. G3P is an intermediate in both pathways: in photosynthesis it is a product of the Calvin cycle, and in cellular respiration it is generated in glycolysis in the conversion of glucose to pyruvate. In photosynthesis, G3P is used for the synthesis of sugars and other fuel molecules, and in cellular respiration, it is part of the catabolism of sugars to simpler organic molecules.

Think Critically

Imagine two types of desert plants. Some, like the wild watermelon plants of the Kalahari, grow very rapidly, with very robust photosynthesis, as soon as rain falls. They produce as many seeds as they can, quickly before severe drought sets in. This strategy requires rapid photosynthesis, even at the risk of photodamage. Others, desert scrub plants like sage, persist throughout the drought. They may grow slowly, with highly protected photosynthesis, even during times of rain. If they grew too fast, their large leaf surface areas would result in high water loss during drought. In some invasive plant species, aggressive photosynthesis and high growth rates have a selective advantage, allowing them to outcompete their rivals for resources like sunlight or growth space. In these cases, high rates of photosynthesis may be an advantage even as they risk photodamage. Other plants may invest for the longer term, building long-lived resilient structures (leaves, wood, etc.). The large investment may render risky photosynthetic strategies less viable, instead preferring more "conservative" down-regulatory strategies. Still other plants may have to contend with lack of key nutrients, where high photosynthetic rates or the need for rapid repair of the photosynthetic apparatus would be too "expensive" in terms of resources. Finally, humans have selected for traits in crop plants that are not related to photosynthetic yield, e.g. tasty fruits, disease resistance, short growing season. These traits often take precedence over photosynthetic efficiency.

Test Your Knowledge

1. d, 2. b, 3. a, 4. c, 5. e, 6. a, 7. b, 8. d, 9. d, 10. b

Interpret the Data

1. When PPFD is zero, no photosynthesis is occurring, but respiration continues to occur. Based on the definition of net photosynthesis here and respiration from Chapter 8 *Interpret the Data,* respiration alone would have a negative value of CO_2 flux.

   The temperature for a respiration rate of 0.2 $\mu mol/m^2/s$ is 25°C.

2. 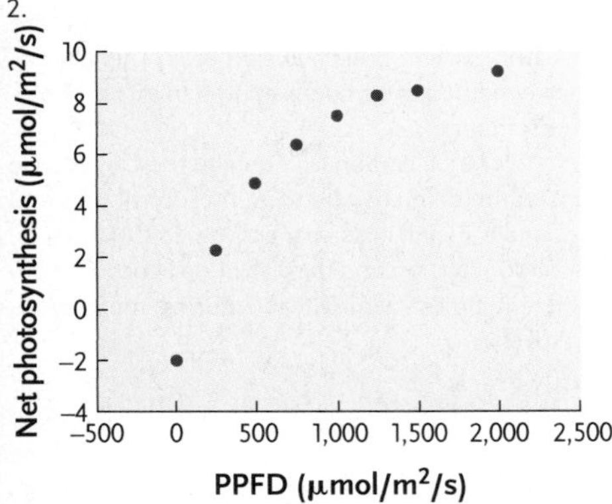

Net photosynthesis (μmol/m²/s)
10
8
6
4
2
0
−2
−4
−500
0
500
1,000
1,500
2,000
2,500
PPFD (μmol/m²/s)

3. A curve represents the data best. Net photosynthesis is saturating with increasing PPFD, indicating that some other factor (such as the amount of photosynthetic enzymes) is limiting photosynthesis.
4. Temperatures are predicted to increase, which would increase respiration according to the data and thus lower net photosynthesis. These data are insufficient to predict the actual impact of CO_2 concentrations on net photosynthesis because the plants would likely change their enzyme amounts or types (acclimate) in response to the rising temperature. Further, temperature is not the only factor that will likely change with rising CO_2. Other factors include precipitation, cloud cover, nutrient availability, etc.

Chapter 10

Study Break 10.1

1. (1) An elaborate master program of molecular checks and balances ensures an orderly and timely progression through the cell cycle. (2) The process of DNA synthesis replicates each DNA chromosome into two copies with almost perfect fidelity. (3) A structural and mechanical web of interwoven "cables" and "motors" of the mitotic cytoskeleton separates the DNA copies precisely into the daughter cells.
2. A linear DNA molecule complexed with proteins
3. In mitosis, DNA replication is followed by the equal separation of the replicated DNA molecules and their delivery to daughter cells. The process ensures that the two cell products of a division have the same DNA content and the same genetic information as the parent cell entering division has.

Study Break 10.2

1. In order, the stages of mitosis are prophase, prometaphase, metaphase, anaphase, and telophase.
2. Each eukaryotic chromosome has a specialized region known as a centromere. The centromere is where a complex of several proteins, called a kinetochore, forms. During mitosis, some spindle microtubules attach to each kinetochore. These connections determine the outcome of mitosis, because they attach the sister chromatids of each chromosome to microtubules leading to the opposite spindle poles; during anaphase, the spindle separates sister chromatids and pulls them to opposite spindle poles. In brief, the centromeres are key to chromosome segregation during mitosis. Although not mentioned in the chapter, this is also apparent when problems occur in which a chromosome fragment without a centromere breaks off from a chromosome. The fragment without a centromere cannot connect to the spindle and, hence, is not segregated properly.
3. Joined sister chromatid pairs attach to kinetochore microtubules during prometaphase and begin their migration to the metaphase plate. The spindle microtubules are also necessary for segregating the sister chromatids to opposite poles of the cell during anaphase. If colchicine is present, no spindle will form and no kinetochore microtubules are present to attach to the sister chromatids. Therefore, the cell will be stuck in mitosis with the condensed pairs of sister chromatids in an unorganized array.

Study Break 10.3

1. In animal cells, all of which have centrosomes, the spindle forms through division of the cell center. As the dividing centrosome separates into two parts, the microtubules of the spindle form between them. Plant cells lack centrosomes. In plant cells, the spindle microtubules simply assemble around the nucleus. In either case, the microtubules assemble in a parallel array that creates two poles in the dividing cell.
2. Chromosomes (sister chromatids) move apart during anaphase. During the anaphase movements, the kinetochores move along the kinetochore microtubules, which become shorter as anaphase progresses. The nonkinetochore microtubules slide over each other, decreasing the degree of overlap and pushing the poles farther apart. The total distance traveled by the chromosomes is the sum of the two movements.

Study Break 10.4

1. A Cdk can become active only once it has complexed with a cyclin protein. Each cyclin is present only during a particular segment of the cell cycle, controlled by when it is synthesized and degraded. Thus, the period in the cell cycle when a particular Cdk is active depends on when its activating cyclin is present.
2. When the kinase of the Cdk is activated upon binding to a cyclin, it phosphorylates target proteins in the cell, regulating their activities. Those proteins play roles in initiating or regulating key events of the cell cycle, namely DNA replication, mitosis, and cytokinesis. The progression through the cell cycle is regulated, then, by a succession of cyclin–Cdk complexes, each of which has specific regulatory effects.
3. An oncogene is an altered gene in an organism that contributes to the development of cancer—that is, uncontrolled cell division. Some of the genes that become oncogenes encode components of the cyclin–Cdk system that regulates cell division, whereas others encode proteins that regulate gene activity, form cell-surface receptors, or make up elements of the systems controlled by the receptors.
4. Metastasis is when cells break loose from a tumor, spread through the body, and grow into new tumors in other body regions.

Study Break 10.5

1. Prokaryotic cell division begins with replication of the bacterial chromosome, starting with duplication of the origin of replication. Once the origin of replication is duplicated, the two origins actively migrate to the two ends of the cells, a process that separates the two replicating chromosomes in the cell. Division of the cytoplasm then occurs by means of a partition of cell wall material that grows inward until the cell is separated into two parts. The cytoplasmic division divides the replicated DNA molecules and cytoplasmic structures between the daughter cells.
2. Present in eukaryotic cell division, but absent from prokaryotic cell division, are the following: the process of mitosis; any form of microtubules for chromosome segregation; a spindle apparatus; cyclin/CDK control proteins.

Think Critically

While *E. coli* is easy to culture, there is no known prokaryotic equivalent of mitosis; mammalian cells are hard to culture and cannot be kept in long-term culture (see *Focus on Basic Research*). *Focus on Model Organisms* provides abundant rationale for *Saccharomyces* as a model organism for this type of research. Ubiquitin is given this name because it is indeed ubiquitous—present in almost the same form in essentially all eukaryotes (see Chapter 16).

Test Your Knowledge

1. c, 2. a, 3. e, 4. b, 5. d, 6. d, 7. a, 8. b, 9. b, 10. c

Interpret the Data

1. Radium exposure does not have an immediate effect on mitosis. That is, for about two hours after initiating radium exposure, there is not a significant drop in the number of cells in the population undergoing mitosis. (The researchers interpreted this to mean that not only were cells completing mitosis, but other cells commenced mitosis.) However, after about two hours of radium exposure, mitosis in the cell population exhibits a sudden drop to zero. In other words, radium blocks mitosis after a long enough exposure. The corollary is that there was no evidence for radium stimulating mitosis.
2. The effect of radium exposure was not permanent. Mitosis resumed about two hours after radium was removed. (Importantly, this result also shows that the decrease in mitosis as a result of radium exposure was not because of death of the cells.)

Chapter 11

Study Break 11.1

1. Mitosis produces daughter cells that are genetically identical to the parent cell. Either a haploid or a diploid cell can undergo mitosis. Meiosis starts with a diploid cell. There is one round of DNA replication but two rounds of cell division, with the result that four haploid cells are produced from the parent diploid cell.
2. Recombination is the physical exchange of segments between the chromatids of homologous chromosomes. Recombination occurs in prophase I, when homologous chromosomes have each duplicated to produce sister chromatids and are aligned fully in an organization called a tetrad.
3. Meiosis II is the meiotic division that is similar to a mitotic division.

Study Break 11.2

1. There are three ways in which sexual reproduction generates genetic variability. First, recombination, which involves the physical exchange of segments between homologous chromatids in prophase I of meiosis, generates new combinations of alleles. Second, the random separation of homologous chromosomes during meiosis generates genetic variability. That is, in metaphase I, for each homologous pair of chromosomes, one chromosome makes spindle connections leading to one pole and the other chromosome connects to the opposite pole. This process operates independently for each homologous pair of chromosomes; thus, for each meiosis, random combinations of maternal and paternal chromosomes move to the poles during anaphase I. Third, the random joining of male and female gametes produces additional genetic variability.
2. The proportion of gametes that will have chromosomes that originate from the animal's female parent is: $(1/2)^6 = 1/64$.

Study Break 11.3

In animals, the diploid phase dominates the life cycle; mitotic divisions occur only in this phase. Meiosis in the diploid phase gives rise to products that develop directly into egg and sperm cells without undergoing mitosis.

In most plants, the life cycle alternates between haploid and diploid generations, both of which grow by mitotic divisions. Fertilization produces the diploid sporophyte generation; after growth by mitotic divisions, cells of the sporophyte undergo meiosis and produce haploid spores. The spores germinate and grow by mitotic divisions into the gametophyte generation. After growth of the gametophyte, cells develop directly into egg or sperm nuclei, which fuse in fertilization to produce the diploid sporophyte generation again.

Think Critically

Kinetochore structure because it mediates the attachment and movement of chromosomes. Chromosome segregation requires a physical connection between spindle microtubules and chromosomes; this attachment occurs at kinetochores.

Test Your Knowledge

1. b, 2. a, 3. d, 4. c, 5. b, 6. b, 7. b, 8. d, 9. a, 10. b

Interpret the Data

1. At 20°C, one mother produced 21 male offspring; assuming 100 offspring, this would be 21%. This was the highest. The lowest was 0 males produced by at least one mother at 20°C and 16-hr photoperiod. 15–20°C seems to result in more male production than either higher or lower temperatures. No mothers produced male offspring at 25°C; some mothers produced no male offspring at 11°C. Less than 10% males were produced at 25°C or 11°C or at 20°C and 16-hr photoperiod.
2. At the moderate temperatures, slightly more males are produced with 12-hr days rather than 16-hr days. However, there is a wide range. If time has an effect, the effect is much less than the effect temperature has on the production of males.

NOTE: The authors concluded that the difference seen for photoperiod was not statistically significant, meaning that the difference seen could be merely because of chance.

Chapter 12

Study Break 12.1

1. The genotypes of the two parents are *Aa Bb* × *Aa Bb*.
2. The genotypes of the parents are *Aa Bb* × *aa bb*. This is a testcross.

Study Break 12.2

1. The color pattern involved is an incompletely dominant trait.
2. The fur colors here involve multiple alleles of a single gene. The allele symbols are *C* for wild type, c^{ch} for chinchilla, c^h for Himalayan, and *c* for albino, with dominance in the order $C \rightarrow c^{ch} \rightarrow c^h \rightarrow c$. That is, the *C* allele is completely dominant to the c^{ch} allele, the c^{ch} allele is completely dominant to the c^h allele, and so on. Therefore, we have these genotypes and phenotypes:
   C with c^{ch}, c^h, or c = agouti (CC, Cc^{ch}, Cc^h, Cc)
   c^{ch} with c^{ch}, c^h, or c = chinchilla ($c^{ch}c^{ch}$, $c^{ch}c^h$, $c^{ch}c$)
   c^h with c^h or c = Himalayan (c^hc^h, c^hc)
   c with c = albino (cc)

Think Critically

If you sequence lots of species you will identify lots of regions of strong evolutionary constraint. If you look at many individuals within a species, you can look at the rate of fast-evolving alleles that might have responded to population-specific pressures or drift and start asking questions pertaining to the genome of the human. Ultimately, your choice will depend on the question you are trying to answer. If you are trying to understand developmental processes, cross-species comparisons are extremely useful, but if you are looking at immunity, maybe the within-species comparison is better.

Test Your Knowledge

1. (a) The *CC* parent produces all *C* gametes, and the *Cc* parent produces 1/2 *C* and 1/2 *c* gametes. All offspring would have colored seeds—half homozygous *CC* and half heterozygous *Cc*. (b) Both parents produce 1/2 *C* and 1/2 *c* gametes. Of the offspring, three-fourths would have colored seeds (1/4 *CC* + 1/2 *Cc*) and one-fourth would have colorless seeds (1/4 *cc*). (c) The *Cc* parent produces 1/2 *C* gametes and 1/2 *c* gametes, and the *cc* parent produces all *c* gametes. Half of the offspring are colored (1/2 *Cc*) and half are colorless (1/2 *cc*).
2. The genotypes of the parents are *Tt* and *tt*.
3. The taster parents could have a nontaster child, but nontaster parents are not expected to have a child who can taste PTC. The chance that they might have a taster child is 3/4. The chance of a nontaster child being born to the taster couple would be 1/4. Because each combination of gametes is an independent event, the chance of the couple having a second child, or any child, who cannot taste PTC is expected to be 1/4.
4. (a) All *A B*. (b) 1/2 *A b* + 1/2 *a b*. (c) 1/2 *A B* + 1/2 *a B*. (d) 1/4 *A B* + 1/4 *A b* + 1/4 *a B* + 1/4 *a b*.
5. (a) All *Aa BB*. (b) 1/4 *AA BB* + 1/4 *AA Bb* + 1/4 *Aa BB* + 1/4 *Aa Bb*. (c) 1/4 *Aa Bb* + 1/4 *Aa bb* + 1/4 *aa Bb* + 1/4 *aa bb*. (d) 1/4 *Aa Bb* + 1/8 *AA Bb* + 1/8 *Aa BB* + 1/8 *Aa bb* + 1/8 *aa Bb* + 1/16 *AA BB* + 1/16 *AA bb* + 1/16 *aa BB* + 1/16 *aa bb*.
6. (a) All *A B C*. (b) 1/2 *A B c* + 1/2 *a B c*. (c) 1/4 *A B C* + 1/4 *A B c* + 1/4 *a B C* + 1/4 *a B c*. (d) 1/8 *A B C* + 1/8 *A B c* + 1/8 *A b C* + 1/8 *A b c* + 1/8 *a B C* + 1/8 *a B c* + 1/8 *a b C* + 1/8 *a b c*.
7. Because the man can produce only 1 type of allele for each of the 10 genes, he can produce only 1 type of sperm cell with respect to these genes. The woman can produce 2 types of alleles for each of her 2 heterozygous genes, so she can produce $2 \times 2 = 4$ different types of eggs with respect to the 10 genes. In general, as the number of heterozygous genes increases, the number of possible types of gametes increases as 2^n, where n = the number of heterozygous genes.
8. Use a standard testcross; that is, cross the guinea pig with rough, black fur with a double recessive individual, *rr bb* (smooth, white fur). If your animal is homozygous *RR BB*, you would expect all the offspring to have rough, black fur.
9. One gene probably controls pod color. One allele, for green pods, is dominant; the other allele, for yellow pods, is recessive.

10. The cross $RR \times Rr$ will produce 1/2 *RR* and 1/2 *Rr* offspring. The cross $Rr \times Rr$ will produce 1/4 *RR*, 1/2 *Rr*, and 1/4 *rr* as combinations of alleles. However, the 1/4 *rr* combination is lethal, so it does not appear among the offspring. Therefore, the offspring will be born with only two types, *RR* and *Rr*, with twice as many *Rr* as *RR* in a 1:2 ratio (or 1/3 *RR* + 2/3 *Rr*).
11. The parental cross is *GG TT RR* × *gg tt rr*. All offspring of this cross are expected to be tall plants with green pods and round seeds, or *Gg Tt Rr*. When crossed, this heterozygous F_1 generation is expected to produce eight different phenotypes among the offspring: green–tall–round, green–dwarf–round, yellow–tall–round, green–tall–wrinkled, yellow–dwarf–round, green–dwarf–wrinkled, yellow–tall–wrinkled, yellow–dwarf–wrinkled, in a 27:9:9:9:3:3:3:1 ratio.
12. The genotypes are: bird 1, *Ff Pp;* bird 2, *FF PP;* bird 3, *Ff PP;* bird 4, *Ff Pp*.
13. Yes, it can be determined that the child is not hers, because the father must be AB to have both an A and B child with a type O wife; none of the woman's children could have type O blood with an AB father.
14. The cross is expected to produce white, tabby, and black kittens in a 12:3:1 ratio.
15. The mother is homozygous recessive for both genes, and the father must be heterozygous for both genes. The child is homozygous recessive for both genes. The chance of having a child with normal hands is 1/2, and that of having a child with woolly hair is 1/2. Using the product rule of probability, the probability of having a child with normal hands and woolly hair is 1/2 × 1/2 = 1/4.

Interpret the Data

Yes, the data support the hypothesis that the tolerance trait is dominant. Let us assign *T* as the symbol for the dominant allele for being tolerant, and *t* as the symbol for the recessive allele for being intolerant.

Cross 1: The crosses of true-breeding tolerant plants with true-breeding intolerant plants are *TT* × *tt*. The F_1 offspring are *Tt*. If the tolerance trait is dominant, the $F_1 \times F_1$ crosses are genotypically *Tt* × *Tt*, which produces an F_2 generation with a genotypic ratio of 1 *TT*:2 *Tt*:1 *tt*, which gives a phenotypic ratio of 3 alive (tolerant) : 1 dead (intolerant), which approximately matches the results.

Cross 2: As described for Cross 1, the F_1 plants have the *Tt* genotype. An F_1 × intolerant parent cross is *Tt* × *tt*. The offspring from this cross are expected to have a genotypic ratio of 1 *Tt*:1 *tt*, which gives a phenotypic ratio of 1 alive:1 dead, which approximately matches the results.

Cross 3: An F_1 × tolerant parent cross is *Tt* × *TT*. The offspring from this cross are expected to have a genotypic ratio of 1 *TT*:1 *Tt;* so all of the progeny are expected to be alive, which matches the results.

Chapter 13

Study Break 13.1

The cross to use is the testcross. Here, the testcross would be *Aa Bb* ×*aa bb*. A testcross is used so that you can follow the meiotic events in the dihybrid parent (including the consequences of crossing-over between linked genes), because all of the gametes from the testcross parent carry recessive alleles for the genes in the cross. A testcross shows linkage when the ratio of 1:1:1:1 for the four possible phenotypes is not seen. That is, the 1:1:1:1 ratio result occurs when two genes assort independently. However, if two genes are linked, there will be excess of the two parental classes of progeny compared with the two recombinant classes.

Study Break 13.2

The differences between sex-linked inheritance and autosomal inheritance are seen clearly when reciprocal crosses are made and followed through to the F_2 generation. If sex-linked inheritance is involved, a cross of miniature-winged female × normal-winged male flies will give an F_1 generation of all normal-winged female and all miniature-winged male flies. (This result would not be found with autosomal inheritance. Instead, all F_1 flies—both males and females—would have normal wings.) Selfing the F_1 flies will give an F_2 generation with 1:1 normal-winged:miniature-winged flies in both sexes. (For autosomal inheritance, you would see a 3:1 ratio of normal-winged:miniature-winged flies in both sexes.)

In the reciprocal cross of true-breeding normal-winged female × miniature-winged male, the F_1 flies will all have normal wings if sex-linked inheritance is involved. (This result is the same as for autosomal inheritance.) Selfing the F_1 flies will give an F_2 generation in which all females will have normal wings and the males will be 1/2 normal-winged and 1/2 miniature-winged. (For autosomal inheritance, you would see a 3:1 ratio of normal-winged:miniature-winged flies in both sexes.)

In summary, reciprocal crosses show different segregation patterns of phenotypes for sex-linked inheritance and autosomal inheritance. The two modes of inheritance are easiest to distinguish in a cross of a mutant female × wild-type male because then, in the F_1 generation, all males show the mutant phenotype when sex-linked inheritance is involved.

Study Break 13.3

(a) Duplication of a chromosome segment occurs when a segment breaks from one chromosome and is inserted into its homolog.

(b) A Down syndrome individual results when nondisjunction of chromosome 21 during meiosis occurs (usually in females), producing gametes with two copies of chromosome 21 and one copy of every other chromosome. When such a gamete fuses with a normal gamete, the result is a zygote with three copies of chromosome 21 and two copies of the other chromosomes. This individual will have Down syndrome. Aneuploidy is the term for the condition of extra or missing chromosomes.

(c) A translocation occurs when a broken segment of a chromosome becomes attached to a different, nonhomologous chromosome.

(d) Polyploidy means that there are more sets of chromosomes than the typical diploid set. Polyploidy may result if the spindle fails to function properly in mitosis of cell lines leading to gametes. Cells affected in this way will have twice the normal number of sets of chromosomes. When meiosis subsequently occurs, the gametes produced will have two sets of chromosomes. Fusion of these gametes with, for instance, a gamete with one set of chromosomes will produce a zygote with three sets of chromosomes—a triploid cell.

Study Break 13.4

1. Autosomal recessive inheritance: For a child to exhibit an autosomal recessive trait, he or she must inherit one recessive allele from each parent. For autosomal recessive inheritance to explain Simpson syndrome in the family, the father must be homozygous for the Simpson syndrome allele, *ss*, and the mother must be heterozygous, *Ss*. The expectation would be that 1/2 of the children would be *Ss* and 1/2 would be *ss*, regardless of sex, and that is what is found. Therefore, on the assumption that the mother is heterozygous, the syndrome could be an autosomal recessive trait. Sex-linked recessive inheritance: One characteristic of sex-linked recessive inheritance is that affected females pass on the trait to all their sons. Here, we start with an affected male, and he would have to be X^sY if it is a sex-linked recessive trait. To explain the children, we would have to assume that the mother is heterozygous, X^SX^s. The cross of $X^SX^s \times X^sY$ is expected to give 1/2 females with the syndrome and 1/2 males with the syndrome, which is what is described. Therefore, the syndrome could be a sex-linked recessive trait.
2. Autosomal recessive inheritance: In pedigrees of autosomal recessive traits, the appearance of progeny with a trait when both parents do not have the trait is one common feature. If we assume that both parents are heterozygous, *Ww*, then the children can be explained. That is, you can get both wiggly-eared children (*ww*, expected frequency 1/4) and non–wiggly-eared children (*WW* or *Ww*, combined expected frequency 3/4), regardless of sex. Therefore, wiggly ears could be an autosomal recessive trait based on this family. Sex-linked recessive inheritance: Because a male individual has only one X chromosome, it is not possible for two nonwiggler parents to produce a wiggler daughter. To get a wiggler daughter, the male would have to be X^wY (wiggler) and the female would have to be X^WX^w (nonwiggler), which is not the case here. Therefore, we cannot conclude that ear wiggling is a sex-linked recessive trait based on this family.

Study Break 13.5

A mutant trait that shows cytoplasmic inheritance is caused by an alteration in the DNA of an organelle, either the mitochondrion or the chloroplast. A key property of cytoplasmic inheritance is that a trait is transmitted by a parent to all offspring, regardless of sex. The most common form of this is maternal inheritance, in which the progeny inherit the trait from their mother, paralleling the inheritance of mitochondria and mitochondrial DNA from the female parent and not from the male parent. This pattern of inheritance would not be seen for genes on chromosomes in the nucleus (as explained in Chapter 11).

Think Critically

Not really. Genome instability in somatic cells, for example, *BRCA1* gene, is inherited. Therefore, the predisposition to cancer needs to be studied at both the level of germ-line cells and somatic cells. But, recombination and repair defects in a somatic cell are not passed on. (Note: Both germ-line and somatic mutations in *BRCA1* occur.)

Test Your Knowledge

1. All sons will be color-blind, but none of the daughters will be. However, all daughters will be heterozygous carriers of the trait.

2. The chance that her son will be color-blind is 1/2, regardless of whether she marries a normal or color-blind male.
3. All these questions can be answered from the pedigree. Polydactyly is caused by a dominant allele, and the trait is not sex-linked. The genotypes of each person are:

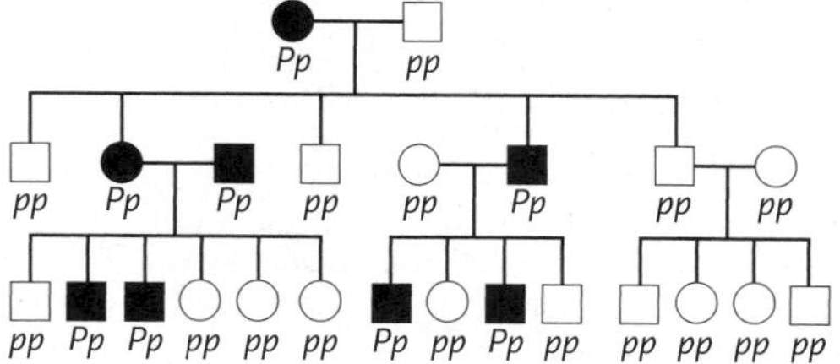

Pp
pp
pp
Pp
Pp
pp
pp
Pp
pp
pp
pp Pp Pp pp pp pp Pp pp Pp pp pp pp pp pp

4. The sequence of the genes is ADBC.
5. Let the allele for wild-type gray body color = b^+, and the allele for black body = b. Let the allele for wild-type red eye color = p^+, and the allele for purple eyes = p. Then the parents are:

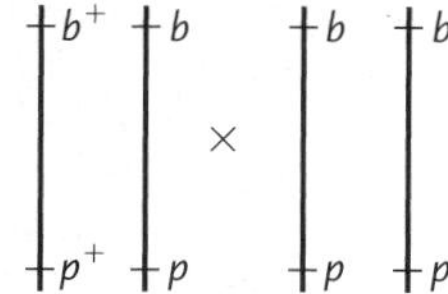

The F_1 flies with black bodies and red eyes are:

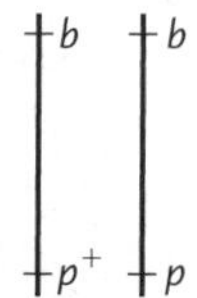

and the flies with gray bodies and purple eyes are:

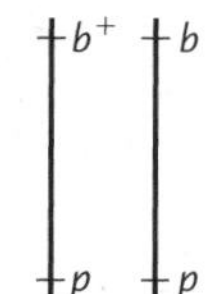

6. The genes are linked by their presence on the same chromosome (an autosome), but they are not sex-linked. Because the F_1 females must have produced 600 gametes to give these 600 progeny, and because 42 + 30 of these were recombinant, the percentage of recombinant gametes is 72/600, or 12%, which implies that 12 map units separate the two genes.
7. The initial cross is X^wX^w white female × $X^{w^+}Y$ red male. The F_1 females produced are $X^{w^+}X^w$, with red eyes, and the F_1 males are X^wY, with white eyes.

   The cross of an F_1 female with a parental male, therefore, is $X^{w^+}X^w$ (red) × $X^{w^+}Y$ (red). The offspring are $X^{w^+}X^{w^+}$ and $X^{w^+}X^w$ females, both of which have red eyes, and $X^{w^+}Y$ (red) and X^wY (white) males in equal proportions. Thus the phenotypic ratio for females is 2 red:0 white, and that for males is 1 red:1 white.

   The cross of an F_1 male with a parental female is X^wY (white) × X^wX^w (white). All progeny of both sexes will be white-eyed.
8. You might suspect that a recessive allele is sex-linked and is carried on one of the two X chromosomes of the female parent in the cross. When present on the single X of the male (or if present on both X chromosomes of a female) the gene is lethal.

Interpret the Data

1. Data are expressed as μg of DNA to control for the amount of sample analyzed from each individual.
2. Smoking promotes the formation of DNA adducts. The more the individual smoked the more adducts were observed.
3. Stopping smoking lowered the level of adduct formation.
4. The highest point in the data from the nonsmoker most probably reflects an exposure to high levels of passive or second-hand smoke in the nearby environment.

Chapter 14

Study Break 14.1

^{35}S-labeled phages in this scenario will have labeled protein coats and labeled DNA. When these phages infect bacteria, radioactivity enters the cell and is found in the progeny phages. In addition, radioactivity is found in the phage material removed by the blender. ^{32}P-labeled phages in this scenario are like the phages in Hershey and Chase's experiment; they have labeled DNA but unlabeled protein. When these phages infect bacteria, radioactivity enters the cell and is found in the progeny phages. No radioactivity is found in the phage material removed by the blender.

Study Break 14.2

1. Adenine and guanine are purines. Thymine and cytosine are pyrimidines.
2. Complementary base pairs are held together by hydrogen bonds. Each base is attached to the deoxyribose sugar by a covalent bond.
3. Watson and Crick described the right-handed double helix that consists of two sugar–phosphate backbones on the outside and complementary base pairs between the two backbones. A complementary base pair is a purine paired with a pyrimidine, more specifically, an A with a T, and a G with a C. The two strands of DNA are antiparallel. The key dimensions of the molecule are: diameter = 2 nm; 1 base pair = 0.34 nm; 1 turn of the helix = 10 base pairs = 3.4 nm.
4. The question focuses on the complementary base-pairing rules: A = T and G = C. If A = 20%, then T = 20%, giving 40% of the DNA as A-T base pairs. Therefore, 60% of the base pairs in this DNA molecule are G–C, and the percentage of C is 30%.

Study Break 14.3

1. Complementary base pairing ensures that the new DNA double helix is a faithful copy of the parental DNA double helix. For whatever base is exposed on the template strand, the DNA polymerase inserts the nucleotide with the complementary base.
2. DNA polymerases cannot initiate a DNA strand; they can add DNA nucleotides only to the 3′ end of an existing strand. The primer serves to provide a short stretch of nucleic acid that can be extended by DNA polymerase. The primer consists of RNA, rather than DNA, and is made by primase.
3. DNA polymerase III is the main DNA polymerase for replication in *E. coli*. This enzyme extends each primer that is synthesized on the lagging strand template, and synthesizes the leading strand. DNA polymerase I is used in lagging strand DNA synthesis. This enzyme replaces DNA polymerase when a new DNA fragment reaches the 5′ end of the Okazaki fragment that was made previously. With its 5′→3′ exonuclease activity, DNA polymerase I removes the RNA primer of that Okazaki fragment, and with its 5′→3′ polymerizing activity it replaces that primer with DNA nucleotides. In this way the Okazaki fragment is converted from an RNA–DNA hybrid into a DNA fragment.
4. Telomeres are buffers against the progressive loss of the ends of chromosomes by repeated rounds of replication. Only when the hundreds to thousands of copies of the telomere repeats have been lost are genes exposed. When those genes are lost by continued chromosome shortening and/or when chromosomes break down in the absence of telomeres, the cell is severely damaged.

Study Break 14.4

Proofreading prevents errors from being introduced into the DNA sequence. The DNA sequence in an organism's genome specifies everything about that organism—most notably, its function and reproduction. If significant errors occur during replication, gene sequences could be changed and the function of the organism could be adversely affected. Particularly if there is a high rate of errors, as there would be in the absence of proofreading, these errors would have potentially lethal consequences.

As part of the proofreading process, DNA polymerase reverses and removes the mispaired nucleotide. The enzyme then continues to move forward, inserting the correct nucleotide.

DNA repair mechanisms then look for and correct any errors that were not detected by proofreading. For example, repair enzymes remove a section of the newly synthesized DNA with the mismatch, and DNA polymerase synthesizes a replacement section with the correct base pairing.

Study Break 14.5

1. The nucleosome consists of two molecules each of histones H2A, H2B, H3, and H4 assembled into a nucleosome core particle wrapped with almost two turns of DNA. The diameter of the nucleosome is 10 nm.
2. Histone H1 is responsible for the next level of chromosome packing above the nucleosome. H1 binds to the exit/entry point of DNA on the nucleosome and to the linker DNA and brings about a coiling of the chromatin into the 30-nm chromatin fiber. The coiled structure is called the solenoid.

Think Critically

Cancer cells are stable if they grow and divide indefinitely. If a drug could target the telomerase, the system could be interrupted.

Test Your Knowledge

1. b, 2. d, 3. a, 4. a, 5. d, 6. c, 7. b, 8. a, 9. c, 10. b

Interpret the Data

1. Schedule B, CDDP (8h) followed by 5-FU, was the most effective treatment because it resulted in the lowest percentage of cell proliferation.
2. B was the only schedule in which cells were first treated with CDDP alone.

Chapter 15

Study Break 15.1

1. Although most enzymes are proteins, not all proteins are enzymes. And, some proteins consist of more than one polypeptide subunit. Each different polypeptide is encoded by a different gene, hence the one gene–one polypeptide hypothesis.
2. There are four different letters in the code (A, U, G, C), so a five-letter code would have 4^5 possible combinations 5 = 1,024 codons.

Study Break 15.2

1. 5′-GUUUAACCGAAUAAUGGCCUAC-3′
2. The promoter determines where transcription of a gene will begin. In prokaryotes, RNA polymerase binds to the nucleotide sequence of the promoter and orients in the correct way to transcribe the associated gene. In eukaryotes, transcription factors bind to the promoter and then recruit RNA polymerase, which then orients properly for transcription from the transcription start point.

Study Break 15.3

1. Both pre-mRNAs and mRNAs have a 5′ cap, exons, and a 3′ poly(A) tail. Only pre-mRNAs have introns, which are removed from pre-mRNAs to produce mRNAs.
2. Particular snRNPs bind to the ends of an intron using their contained RNAs to recognize the boundary sequences of the intron. Other snRNPs then bind, causing the intron to loop out, and completing the active spliceosome. Cleavage at each intron-exon junction, looping back of the intron on itself, and joining the two exons together completes the splicing event; the intron and snRNPs are then released.

Study Break 15.4

1. In eukaryotes, a complex of the small ribosomal subunit, initiator tRNA, initiation factors, and GTP binds to the 5′ cap of the mRNA and scans along the mRNA until it reaches the first AUG codon, which is the start codon. The anticodon of the initiator tRNA binds to the start codon, the large ribosomal subunit binds, and the initiation factors are released when GTP is hydrolyzed.

   In prokaryotes, a complex of the small ribosomal subunit, initiator tRNA, initiator factors, and GTP binds to the region of the mRNA where the AUG start codon is located, directed by a specific RNA sequence upstream of the start codon. The other steps are the same as those in eukaryotes.
2. The P site is where the tRNA with the growing polypeptide is located. Downstream of the P site is the A site. An incoming aminoacyl–tRNA enters the A site and its anticodon base-pairs with the codon of the mRNA in that site. When the polypeptide is transferred to the amino acid on the tRNA in the A site, the ribosome translocates one codon along the mRNA. As translocation takes place, the empty tRNA that was in the P site is moved to the E site. It remains there, blocking a new aminoacyl–tRNA entering the A site until translocation is finished. Then the empty tRNA is released from the ribosome.
3. Proteins found in the cytosol are made on free ribosomes.

   Proteins are sorted to the endomembrane system by cotranslational import. The proteins begin their synthesis on free ribosomes. These proteins have signal sequences at their N-terminal ends that direct them and the ribosome to dock with a receptor on the rough ER membrane. Continued translation inserts the growing polypeptide into the lumen of the ER, and the signal sequence is removed by signal peptidase. The proteins are then tagged to target them for sorting to their final destinations. Some proteins remain in the ER, whereas others are transported via the Golgi to vesicles for sorting to lysosomes, secreting them from the cell, or depositing them in the plasma membrane.

   Proteins are sorted to mitochondria, chloroplasts, microbodies, and the nucleus by posttranslational import. Proteins destined for the mitochondria, chloroplasts, and microbodies have short N-terminal transit sequences that target them to the organelle. A transit sequence interacts with a specific transport complexes on the organelle and the polypeptide is then taken into the organelle. The transit sequence is then removed. Proteins destined for the nucleus have a nuclear localization signal which is bound by a cytosolic protein. The complex then interacts with the nuclear pore complex and the polypeptide then is transported into the nucleus through the pore.

Study Break 15.5

1. A missense mutation involves a change from a sense codon to another sense codon that specifies a different amino acid. A silent mutation involves a change from one sense codon to another sense codon, but where both codons specify the same amino acid.
2. Genetic recombination occurs by crossing-over between two homologous sequences. TE transposition occurs by integration into a new location with which the TE has no sequence homology.
3. Both: (1) have inverted repeats at their ends; (2) contain a transposase gene; and (3) integrate into target sites and cause a duplication of the target site.
4. A transposon is a mobile genetic element that moves from one location to another in the genome as a DNA molecule.

   A retrotransposon is a mobile genetic element that moves from one location to another in the genome using an RNA intermediate. That is, the integrated DNA element is transcribed to produce an RNA copy. The RNA copy is reverse-transcribed into DNA, which then integrates at a new location in that genome.

Think Critically

1. The first ribosomes were made only of RNA.
2. Not to make proteins, but to make simple protein fragments (peptides) that bind to RNA to help RNA carry out its biological functions before the evolution of proteins.

Test Your Knowledge

1. b, 2. a, 3. e, 4. d, 5. b, 6. d, 7. b, 8. a, 9. e, 10. d

Interpret the Data

The codons specifying the amino acids are shown in the following figure:

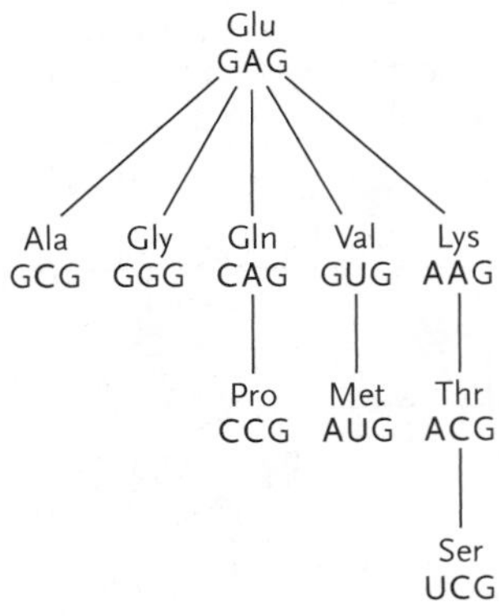
Glu
GAG
Ala
GCG
Gly
GGG
Gln
CAG
Val
GUG
Lys
AAG
Pro
CCG
Met
AUG
Thr
ACG
Ser
UCG

Chapter 16

Study Break 16.1

1. The Lac repressor is active when it is made. In normal cells, in the absence of lactose, the Lac repressor binds to the operator, blocking transcription. In the mutant, the Lac repressor is not made, so transcription can never be blocked because no repressor is available to bind to the operator. As a consequence, the lactose metabolizing enzymes will be made both in the presence and absence of lactose in the medium. The mutation involved is an example of a regulatory mutant. Mutations such as this were valuable to Jacob and Monod in developing the operon model for gene regulation.
2. The Trp repressor is inactive when it is made. In normal cells, in the presence of tryptophan, the Trp repressor is activated, binding to the operator and blocking transcription. In the mutant, the Trp repressor is not made, so the operon cannot be turned off when tryptophan is present. This means that the tryptophan biosynthesis enzymes will be produced both in the presence and absence of tryptophan.

Study Break 16.2

1. General transcription factors bind to the promoter and recruit RNA polymerase II, orienting the enzyme so that it will begin transcription at the beginning of the gene. Activators bind to regulatory sequences associated with genes and increase the rate of transcription. Activators that bind to regulatory sequences in the proximal promoter region interact directly with the general transcription factors at the promoter to exert their action. Activators that bind to regulatory sequences in the enhancer stimulate transcription indirectly. These latter activators bind to a coactivator that also binds to the complex of proteins at the promoter, and transcription then occurs at the maximal possible rate.
2. Histones are general negative regulators of gene expression. When DNA is complexed with histones in normal chromatin, gene promoters typically are not very accessible to the transcription machinery. By acetylating histones, the chromatin is remodeled, making the promoter now accessible to the transcription machinery. Acetylation of histones occurs in response to

the binding of an activator to a regulatory sequence associated with the gene.

Study Break 16.3

1. A microRNA (miRNA), in a complex with particular proteins, binds to an mRNA by complementary base pairing. Either the proteins cut the mRNA in the region of pairing, thereby destroying that molecule, or the double-stranded RNA region blocks translation.
2. Removal of the poly(A) tail would result in the mRNA not being translated.

Study Break 16.4

1. Determination is the process by which the developmental fate of a cell is set. Differentiation is the establishment of a cell-specific developmental program in cells. Determination and differentiation are under molecular control. Regulatory genes encode regulatory proteins that bind to promoters of the genes they control, switching the genes on or off depending on the interaction.
2. The segmentation genes subdivide the embryo progressively into regions, thereby determining the segments of the embryo and the adult. In essence, they organize the embryo into segments. The homeotic genes specify the identity of each segment with respect to the body part it will become.

Study Break 16.5

1. A tumor suppressor gene encodes a product that has an inhibitory role in the cell division cycle. If a tumor suppressor gene is mutated so that its product is nonfunctional or its function is significantly diminished, then the gene's inhibitory control of cell division is lost or reduced. As a result, the cell may progress toward division.
2. A proto-oncogene encodes a product that has a stimulatory role in cell division. Mutations that lead to increased levels of that product have converted a proto-oncogene to an oncogene. The increased amount of product is stimulatory to cell division.
3. Some miRNAs have been shown to regulate the expression of mRNA transcripts of particular tumor suppressor genes. Overexpression of those miRNAs can abnormally inhibit the activity of those tumor suppressor genes, thereby removing or decreasing inhibitory signals for cell proliferation. Other miRNAs have been shown to regulate the expression of mRNA transcripts of particular proto-oncogenes. Inactivation of those miRNAs means that the proto-oncogenes are expressed at higher than normal levels, which can stimulate cell proliferation.

Think Critically

Dr. Kay's lab uses mice to test RNAi therapies against hepatitis C virus. You could test various RNAi agents via an intravenous infusion to find a dose where the viral load (or level) is reduced substantially without causing any dangerous toxicities that have been determined by standard laboratory testing of blood samples. Kay's lab aims to reduce viral load by at least 100-fold.

Test Your Knowledge

1. d, 2. a, 3. b, 4. c, 5. e, 6. b, 7. a, 8. d, 9. a, 10. b

Interpret the Data

1. The woman would have an 8.6% chance of dying of cancer.
2. Her risk of dying of cancer is 18.0% if she carries a mutated *BRCA1* gene.
3. This study suggests a *BRCA1* mutation is more dangerous than a *BRCA2* mutation (in effect, *BRCA1* mutations are associated with a higher percentage of deaths). Sixteen out of 89 women with the *BRCA1* mutation died while only 1 out of 35 women with the *BRCA2* mutation died.
4. You would need to know the number of deaths of women who had a preventive mastectomy and the number of deaths of women who had a preventive oophorectomy.

Chapter 17

Study Break 17.1

1. An F^+ cell contains the F factor plasmid in addition to the chromosomal DNA. The F factor enables an F^+ cell to conjugate with an F^- cell, which lacks the F factor. By a special replication mechanism, a copy of the F factor is transferred to the recipient F^- cell in an $F^+ \times F^-$ conjugation, so the recipient is converted to an F^+ cell.

   An Hfr cell has the F factor integrated into the chromosomal DNA. When an Hfr cell conjugates with an F^- cell, the F factor begins its replicative transfer into the recipient as in an $F^+ \times F^-$ conjugation and, by that transfer mechanism brings in chromosomal genes from the Hfr donor. Those chromosomal genes can recombine with the genes in the recipient. Because replication of the F factor begins in the middle of the plasmid, the entire F factor cannot be transferred to the recipient unless the entire chromosome is transferred to the recipient, which occurs only rarely. Hence, the recipient remains F^- in this conjugation experiment.
2. In gene segregation in sexually reproducing organisms, genetic material moves from one generation to the next, basically by descent. In horizontal gene transfer, the movement of genetic material between organisms is other than by descent. For instance, in transformation, a DNA molecule in the environment is taken up by a recipient cell—the DNA has moved horizontally from a donor cell (from which the DNA was released) to the recipient cell. Genetic recombination in the recipient can alter the phenotype of that cell.

Study Break 17.2

1. A virulent phage always enters the lytic cycle when it infects a bacterial cell. The end result is the assembly of many progeny phages and their release into the surroundings when the cell breaks open.

   A temperate phage enters either the lytic cycle or the lysogenic cycle when it infects a cell. The lytic cycle is the same as that for a virulent phage. The lysogenic cycle involves integration of the phage's chromosome into the bacterial chromosome. In the integrated state the phage—now called the prophage—is inactive and replicates only when the bacterial chromosome replicates. In response to an adverse environmental signal to the cell, the phage chromosome can excise itself from the bacterial chromosome and enter the lytic cycle.
2. Animal cells: Viruses without an envelope bind by their recognition proteins to receptor proteins in the host cell's plasma membrane and are then taken into the cell by receptor-mediated endocytosis. For some enveloped viruses, the genome-containing capsid enters the cell when the envelope fuses with the host cell's plasma membrane. For other enveloped viruses, the complete virus, with the envelope, enters the cell by endocytosis.

   Plant cells: All plant viruses lack envelopes. They enter cells either through mechanical injuries to leaves and stems or by the action of biting and feeding insects.
3. Non-retrovirus RNA viruses have RNA genomes that are replicated in an RNA-to-RNA manner. Retroviruses have RNA genomes that are replicated via a DNA intermediate. That is, the RNA genome is copied to double-stranded DNA by reverse transcriptase and the DNA molecule integrates into the host cell's nuclear chromosomes, from which location new RNA viral genomes are transcribed.

Study Break 17.3

While all are infectious agents, viruses have protein coats, while viroids and prions do not. Viruses consist of a nucleic acid genome surrounded by a protein coat and, in some cases an envelope. Viroids are naked, single-stranded RNA circular molecules. Prions are proteins.

Think Critically

Researchers can further study the interactions between normal prion protein and prion aggregates and the effect of these interactions on neurons. One possible function for the normal prion protein is to act as a signaling molecule to regulate apoptosis. Interactions with prion particles may induce aberrant signaling through the normal prion protein. Since the normal prion protein also binds to many other molecules (for example, extracellular matrix components), one could also investigate the effects of the conversion process on the interactions of normal prion protein with other molecules.

Test Your Knowledge

1. e, 2. a, 3. b, 4. c, 5. d, 6. a, 7. c, 8. c, 9. a, 10. b

Interpret the Data

The answer is shown in the figure. The segments of donor chromosomes transferred by each Hfr strain are shown, with the arrowheads showing the beginning.

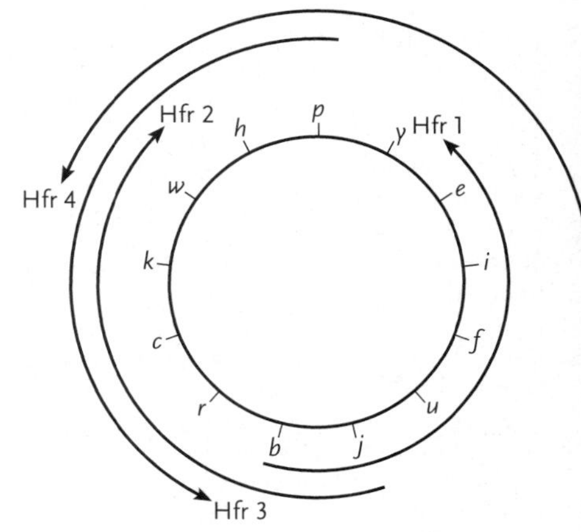

Hfr 2
h
p
y
Hfr 1
Hfr 4
w
e
k
i
c
f
r
u
b
j
Hfr 3

Chapter 18

Study Break 18.1

1. Each restriction enzyme recognizes a specific sequence in DNA, typically in the range of 4–6 bp, and cuts both strands of the DNA within the sequence. Restriction enzymes differ with respect to the DNA sequences—the restriction sites—they recognize and cut. The enzymes most useful for cloning produce sticky ends.
2. Replication origin so that the plasmid will replicate in *E. coli,* an antibiotic resistance gene to allow selection of bacteria containing the plasmid, a cluster of restriction sites (a multiple cloning site) to provide choices for inserting fragments of DNA, and the $lacZ^+$ gene to use in blue–white screening, in which colonies containing recombinant plasmids (white) are distinguished from colonies with vectors lacking inserted DNA (blue).
3. A cDNA library is a collection of clones in which DNAs that are complementary to the mRNAs in the cell have been inserted into a cloning vector. The array of particular clones in a cDNA library is directly related to the genes being expressed in the cells from which the mRNAs are isolated. Therefore, not all genes are represented in any given cDNA library. By contrast, a genomic library is a collection of clones containing all the sequences of an organism inserted into a cloning vector. A genomic library contains all of the genes of an organism, as well as noncoding sequences found between and within genes.
4. PCR is a method to amplify a specific segment of DNA. The amplification process depends on DNA replication and, therefore, requires primers. A limitation of PCR, then, is that DNA sequence information must be available for the segment of DNA to be amplified. Otherwise, the primers cannot be synthesized. The ingredients of a PCR are the template DNA containing the sequence to be amplified, the two primers, a buffer, and a DNA polymerase that is tolerant to the high temperature used to denature the DNA repeatedly during the cycles of the reaction.

Study Break 18.2

1. Each human (or other organism) has unique combinations and variations of DNA sequences. DNA fingerprinting exploits these combinations and variations to distinguish between different individuals (with the exception of identical twins). Using DNA technologies to analyze the particular regions of the genome showing sequence variation, it is possible to compare two DNA samples to see if they are from the same person or not. DNA fingerprinting is used in several ways, including forensics, paternity testing, and basic research.
2. A transgenic organism is one into which a gene or genes from an external source have been introduced as a means to modify the organism genetically.
3. Germ-line cells develop into reproductive cells, so modifying this cell type genetically will lead to the genetic modification being passed to offspring. In somatic cell gene therapy, germ-line cells and their products are not involved, so the inserted genes remain with the individual and are not passed to offspring.

Study Break 18.3

1. The three main areas of genome analysis are the following: (1) genome sequence determination and annotation; (2) functional genomics; and (3) comparative genomics.
2. In whole-genome shotgun genome sequencing, the entire genome is broken into thousands to millions of random, overlapping fragments, each fragment is cloned and sequenced, and then the genome sequence is assembled by computer on the basis of sequence overlaps between fragments.
3. To find protein-coding genes in a bacterium, computer algorithms search a genome sequence for open reading frames (ORFs), which are defined as the DNA equivalent of a start codon (ATG) in frame (a multiple of three away) from a stop codon (TAG, TAA, or TGA). Such a segment of DNA could potentially produce an mRNA that could be translated into a protein. Other analyses would be required to show if any given ORF identified in this way is an actual protein-coding gene.

   To find protein-coding genes in a mammal, one also needs to find ORFs. However, the presence of introns in eukaryotic protein-coding genes complicates matters. Therefore, more sophisticated computer algorithms are needed, in this case ones that attempt to locate exon-intron boundaries while they search for ORFs.
4. In a sequence similarity search, an input sequence (such as the DNA or amino acid sequence for a putative gene in a newly sequenced genome) is compared with all sequences in a database. Matches in the database can indicate the possible function of the gene of interest. That is, DNA sequences of two genes from different organisms will be similar if they are homologous genes with a common ancestor. Therefore, if the input gene sequence closely matches the sequence of a gene in the database with a known function, the function of the input gene can be inferred.
5. A genome-wide analysis of gene expression here could involve DNA microarray assays. mRNAs could be isolated from untreated tissue culture cells and from cells treated with a steroid hormone. Convert each mRNA preparation to cDNAs using reverse transcriptase, using nucleotide precursors with different fluorescent labels for the two batches; for example, green for the untreated (reference) sample and red for the treated (experimental) sample. Mix the two cDNAs and pump them through a DNA chip prepared to have spots with DNA representing every gene in the genome. Allow hybridization to occur, and analyze the hybridization by laser detection. The colors of the spots indicate which genes are affected by the steroid hormone. Purely red spots represent genes that are active only in hormone-treated cells. Purely green spots represent genes that are active only in untreated cells. Spots that are a mixture of green and red represent genes that are active in both types of cells. Based on controls, it would be possible to see if any of the genes have higher or lower expression levels in the treated cells.

Think Critically

Knowledge that an individual is genetically predisposed to developing a disease or is at greater risk for the aftereffects of disease based on genotype may cause emotional distress, and fear of discrimination by employers or insurance companies. Similar concerns may exist if genetic information suggests that disease treatment options for an individual are limited. The Genetic Information Nondiscrimination Act (GINA) was recently enacted to help address these concerns.

Test Your Knowledge

1. e, 2. c, 3. d, 4. a, 5. b, 6. b, 7. e, 8. a, 9. a, 10. d

Interpret the Data

An offspring must have one allele of each gene inherited from the male parent, and the other allele of each gene inherited from the female parent. The same goes for STR locus alleles.

Juvenile 1: Not the offspring of the two adults. The juvenile has one allele in common with the male parent (8 repeats = 72 bp), but its other allele came from neither adult (9 repeats = 81 bp). In other words, it is heterozygous for 8- and 9-repeat alleles.

Juvenile 2: Could be the offspring of the two adults. There was only one DNA fragment size amplified by PCR. The interpretation is that this juvenile is homozygous for a 6-repeat (54-bp) allele, which is why there is more DNA amplified (a thicker band). Each adult has a 6-repeat allele that could have passed to the juvenile.

Juvenile 3: Not the offspring of the two adults. This juvenile is heterozygous for 5-repeat (45-bp) and 11-repeat (99-bp) alleles. The 11-repeat allele could have been inherited from the female adult, but neither adult has the 5-repeat allele.

Juvenile 4: Could be the offspring. This juvenile is heterozygous for 6-repeat (54-bp) and 8-repeat (72-bp) alleles. It could have inherited the 6-repeat allele from either parent, but could have inherited the 8-repeat allele only from the male parent. Therefore, the 6-repeat allele would have had to come from the female parent.

Chapter 19

Study Break 19.1

1. Buffon did not understand how "anatomically perfect" animals could have useless structures.
2. Lamarck proposed that: all species change through time, the changes are inherited by the next generation, the changes arise in response to environmental conditions, and specific mechanisms caused the changes.
3. The concepts of gradualism and uniformitarianism suggest that Earth's major geological features were produced by the very slow action of geological processes observed today. Thus, it must have taken more than 6,000 years for these features to assume their present forms.

Study Break 19.2

1. Darwin observed that living organisms often resemble fossils found in the same area; that organisms found on South America resembled one another, even if they occupied different environments; and that many species found on the Galápagos Islands resembled species from the South American mainland.
2. Darwin realized that the effects of competition for resources in nature were similar to the action of a plant or animal breeder who used only certain individuals as parents to produce the next generation.

3. Darwin's theory relied on physical explanations for the origin of biodiversity, he recognized that evolutionary change takes place within a population rather than in individuals, he recognized that natural selection is a multistage process, and he emphasized the importance of environmental conditions to the process of natural selection.

Study Break 19.3

1. Two problems that slowed the acceptance of Darwin's theory are that Mendel's genetic studies focused on simple traits and that these traits often changed in just a few generations.
2. Microevolution refers to small, genetically based changes within populations. Macroevolution refers to larger-scale evolutionary changes observed in species and more inclusive groups. Current research suggests that microevolutionary change and macroevolutionary change result from the same evolutionary processes.
3. Evidence for evolution comes from studies of adaptation, the fossil record, historical biogeography, comparative morphology, and comparative molecular biology.

Think Critically

If climate change rapidly disrupts physical environments everywhere on Earth, Arctic species are unlikely to adapt to those changes. Given the frequency of extinction in the fossil record, one might assume that most of the species living today will become extinct eventually, and rapid climate change is likely to speed that process.

Test Your Knowledge

1. c, 2. e, 3. d, 4. b, 5. c, 6. b, 7. a, 8. b, 9. d, 10. d

Interpret the Data

The graph suggests that selective breeding of race horses did improve the speed of horses that won the Kentucky Derby between 1896 and 1960. However, the speed of winners has been stable since that time, suggesting that additional selection would not lead to future improvements in speed.

Chapter 20

Study Break 20.1

1. The variation in skunks is qualitative.
2. The researchers used artificial selection to change the activity levels of the mice.
3. Genetic variation, differing environmental effects on individuals, and interactions between genes and the environment affect phenotypic variation in a population.

Study Break 20.2

1. Genotype frequencies specify how alleles are combined in individuals, and allele frequencies specify how common the alleles are.
2. The Hardy–Weinberg principle is a null model because it identifies the conditions under which evolution will *not* occur.
3. If genotype frequencies are not already in equilibrium, they will stop changing after one generation of random mating.

Study Break 20.3

1. Mutation and gene flow tend to increase genetic variation within populations, and natural selection and genetic drift tend to decrease it.
2. Stabilizing selection increases the representation of the average phenotype in a population.
3. Sexual selection, like directional selection, favors extreme phenotypes.

Study Break 20.4

1. Diploidy protects harmful recessive alleles because dominant alleles mask their effects in heterozygotes.
2. A balanced polymorphism is one in which two or more phenotypes are maintained in fairly stable proportions over many generations.
3. The sickle-cell allele is rare in Northern Europe because, in the absence of the malarial parasite, it confers no advantage on individuals that carry it.

Study Break 20.5

1. The adaptive value of a trait can be evaluated by comparing closely related species that live in different environments.
2. Natural selection preserves traits that were useful when the organisms subject to selection were alive and reproducing.

Think Critically

If recombination induces mutations, then regions of high recombination should have many differences between species. This suggests a test for whether recombination rate and genetic variation are connected mechanistically vs. connected by the action of natural selection. Both mechanistic and selective explanations predict that regions of high recombination have lots of genetic variation *within* species—more than regions of low recombination—but only mechanistic explanations predict that genetic differences *between* species should be greater in regions of high recombination than in regions of low recombination.

Test Your Knowledge

1. c, 2. b, 3. c, 4. d, 5. b, 6. e, 7. a, 8. b, 9. c, 10. d

Interpret the Data

The graphs illustrate the action of directional selection favoring birds with deeper bills because the distribution of bill depths in the survivors (lower graph) is shifted slightly to the right of the distribution of bill depths in the population before the drought (upper graph).

Chapter 21

Study Break 21.1

1. The morphological species concept defines species based on morphological differences between them. The biological species concept defines species as populations that can successfully interbreed under natural conditions. The phylogenetic species concept defines a species as a cluster of populations with a recent shared evolutionary history.
2. Clinal variation is a pattern of smooth variation along a geographical gradient.

Study Break 21.2

1. Prezygotic isolating mechanisms either prevent individuals of different species from mating or prevent sperm of one species from fertilizing the eggs of another. Postzygotic isolating mechanisms limit the survivorship or reproductive capability of hybrid individuals.
2. The scenario illustrates a behavioral isolating mechanism.

Study Break 21.3

1. In the first stage of allopatric speciation, populations become geographically separated. In the second stage, they become reproductively isolated.
2. Some populations of bent grass survive better on unpolluted soil, whereas others survive better on polluted soil.
3. Insects from different host races spend most of their time on different host plant species. Thus, they rarely encounter each other and would be unlikely to mate under natural conditions.

Study Break 21.4

1. Natural selection cannot promote reproduction isolation in allopatric populations directly, but it can lead to genetic divergence that results in reproductive isolation.
2. Polyploidy changes the number of chromosome sets in the cells of an organism. Polyploid individuals are often reproductively isolated from their parent species because their gametes contain different numbers of chromosomes. When a gamete from a polyploid individual fuses with a gamete from a non-polyploid individual, the resulting offspring is usually sterile because its odd number of chromosomes cannot segregate properly during meiosis.

Think Critically

Most of the asexual organisms for which the biological species definition is not a good match are bacteria. Because these organisms do not exhibit as much morphological distinctness as many animals do, biologists rely on comparisons of their DNA or RNA sequences to analyze their evolutionary relationships. These data allow researchers to construct phylogenetic trees for the organisms they study, indicating that any new species definition that applies to them would probably resemble the phylogenetic species definition more than the morphological species definition.

Test Your Knowledge

1. a, 2. e, 3. c, 4. d, 5. b, 6. e, 7. b, 8. c, 9. a, 10. b

Interpret the Data

The graphs suggest that sympatric populations of the three frog species experience temporal prezygotic reproductive isolation. Populations that are allopatric exhibit substantial overlap in breeding times, especially in March, April, and May. By contrast, sympatric populations exhibit almost no overlap in breeding times.

Chapter 22

Study Break 22.1

1. Hard parts, such as the shells or bones of animals, are the materials most likely to fossilize.
2. The fossil record provides an incomplete portrait of life in the past because not all organisms are equally likely to form fossils; fossils do not form in all types of habitats; and fossils are often destroyed by geological processes and erosion.
3. The fossil record provides information about the morphology of ancient organisms; how structures changed over time; and the proliferation and extinction of evolutionary lineages. It also offers indirect evidence about the behavior, ecology, and physiology of organisms that lived in the past.

Study Break 22.2

1. Continental drift caused large-scale geographical separation of populations and lineages that subsequently evolved in isolation.

2. Sea levels fell whenever a large proportion of Earth's water was incorporated into glaciers.

Study Break 22.3

1. Continuous distribution requires no special explanations. Biologists infer that lineages with a continuous distribution simply occupy all or one part of their historical range.
2. Distantly related species that live in widely separated parts of the world may resemble each other because convergent evolution fosters similar adaptations to the environments they occupy.

Study Break 22.4

1. A population of organisms may occupy a new adaptive zone after the evolution of a key morphological innovation that allows it to use the environment in a unique way or after a once-successful group of organisms declines.
2. Huge volcanic eruptions triggered a chain of events that led to the mass extinction at the end of the Permian period. These eruptions warmed the atmosphere and oceans enough to melt frozen undersea methane reserves. The methane entering the atmosphere warmed the climate even further, creating a "runaway greenhouse" effect.
3. The first major adaptive radiation of animals took place in the Cambrian era.

Study Break 22.5

1. The horse lineage was highly branched, and it included some species that were larger and some that were smaller than their ancestors.
2. Phyletic gradualism predicts that morphological change is slow and steady, producing fossils with morphologies that are intermediate between the structures found in lower and higher strata. The punctuated equilibrium hypothesis predicts that morphological changes occur rapidly as new species form and that most species experience little morphological change for long periods of time.

Study Break 22.6

1. Allometry refers to the differential growth of body parts. Heterochrony refers to changes in the timing of developmental events.
2. Similar developmental control genes are present in a wide variety of animals, plants, fungi, and prokaryotes. Their widespread distribution suggests that they were present in the ancestor of all these organisms and have been conserved through countless generations.
3. The *Pitx1* gene is expressed in fin buds that later produce spines, and it is not expressed in those that fail to produce spines.

Think Critically

For genes that are very pleiotropic—that is, have many diverse functions—regulatory evolution may be more likely. Shifts in gene expression domains can produce rapid morphological change while leaving other functions intact. However, we can never rule out coding changes, especially if gene duplication has reduced pleiotropy.

Test Your Knowledge

1. a, 2. e, 3. b, 4. d, 5. c, 6. c, 7. a, 8. d, 9. e, 10. b

Interpret the Data

The data illustrate allometric growth in this evolutionary lineage. Brain size increased more than 200% over three million years, but body size increased only about 50%.

Chapter 23

Study Break 23.1

1. The system of binomial nomenclature avoids ambiguity in the naming of species because it assigns a unique two-part name to each species.
2. The taxonomic hierarchy helps biologists organize information about different species because it categorizes them into increasingly inclusive groups. Species that are included in a lower taxonomic category share many characteristics, whereas those included only in the same higher taxonomic category share fewer characteristics.

Study Break 23.2

1. A phylogenetic tree is a formal hypothesis about the evolutionary relationships among species. A classification is an arrangement of organisms into hierarchical groups that reflect their relatedness.
2. A monophyletic taxon contains an ancestor and all of its descendants. A polyphyletic taxon includes species from different clades. A paraphyletic taxon includes an ancestor and some, but not all, of its descendants.

Study Break 23.3

1. Systematists use homologous characters in their analyses because similarities in homologous characters indicate genetic relatedness and shared ancestry.
2. Morphological traits are often useful for tracing the evolutionary relationships within a group of organisms because they can be observed and measured in fossils as well as in living organisms.
3. Molecular characters provide several advantages in systematic analyses: (1) they provide abundant data; (2) molecular sequences can be compared between distantly related organisms that share no organismal characteristics or between closely related species with only minor morphological differences; and (3) proteins and nucleic acids are not directly affected by developmental or environmental factors that cause nongenetic morphological variation.

Study Break 23.4

Traditional systematics emphasizes the divergence of groups as well as branching evolution. Groups that have evolved very different morphology are sometimes placed in a taxon that is different from the one to which the rest of their clade is assigned.

Study Break 23.5

1. Outgroup comparison is a technique that compares the group under study, the ingroup, to more distantly related organisms, the outgroup, to identify the ancestral and derived versions of characters. Character states observed in the outgroup are considered ancestral.
2. In a cladistic analysis, organisms with synapomorphies (shared derived characters) are grouped together within a clade.
3. The principle of parsimony suggests that the cladogram or phylogenetic tree with the fewest number of hypothesized evolutionary changes is the best current hypothesis about the phylogenetic history of a group.

Study Break 23.6

1. An assumption that underlies the use of genetic sequence differences as molecular clocks is that mutations arise at a relatively constant rate.
2. Birds are more closely related to non-avian dinosaurs than they are to crocodilians.

Study Break 23.7

1. The gymnosperms are the sister clade of the angiosperms because the two groups branch off the same node.
2. Phylogenetic analyses of prokaryotes based on morphological data were not very successful because prokaryotes do not have many morphological features. Analyses based on molecular sequence data have been more successful because researchers can identify many molecular differences among the lineages of prokaryotes.

Think Critically

The Linnaean system of classification cannot easily incorporate the phenomenon of horizontal gene transfer because it pigeonholes species into a nested hierarchy of rigidly defined taxonomic categories. Although a phylogenetic view of biodiversity also depends on the vertical transmission of derived traits from one generation to the next, horizontal gene transfer can at least be illustrated on phylogenetic trees.

Test Your Knowledge

1. c, 2. b, 3. a, 4. b, 5. e, 6. a, 7. a, 8. e, 9. d, 10. c

Interpret the Data

The European wildcat is the domestic cat's closest relative. The clade that includes jaguars, lions, and leopards is the sister clade to tigers. Bobcats are more closely related to cougars than to ocelots because they share a more recent common ancestor with cougars.

Chapter 24

Study Break 24.1

1. Only in a reducing atmosphere can amino acids be produced from simpler chemicals and energy. Without amino acids there can be no life. Thus, any theory for the origin of life must consider the necessity of reducing conditions.
2. Current thinking is that early Earth's atmosphere *per se* was not reducing; scientists have looked for localized regions where conditions were of a reducing nature, such as ocean floor hydrothermal vents. Another hypothesis is that some amino acids had an extraterrestrial origin.

Study Break 24.2

Lipid bilayers in the form of membranes are present in all present-day cellular organisms. Therefore, it is more intuitive to conceive of a path from a simple lipid bilayer system to a membrane-bound cell than it is to conceive of a transition of a clay-based system to a cellular system.

Study Break 24.3

The basic tenet of the theory of endosymbiont origins for mitochondria and chloroplasts is that organelles such as mitochondria and chloroplasts originated from symbiotic relationships between two prokaryotic organisms. An anaerobic prokaryotic is proposed to have ingested an aerobic prokaryote, which persisted in the cytoplasm, continuing to respire aerobically. A gradual process of mutual adaptation transformed the cytoplasmic aerobes into mitochondria.

The same basic mechanism is believed to have led to the appearance of membrane-bound plastids (including chloroplasts) at a later time. In this case, nonphotosynthetic aerobic cells with mitochondria are proposed to have ingested photosynthetic prokaryotes that were perhaps similar to present-day cyanobacteria. Again, through mutual adaptation, the photosynthetic prokaryotes changed into plastids.

Think Critically

Two possible related explanations might be summarized as "division of labor" and "efficiency." In an "RNA world," RNA would have a dual function, being both the transcriptional unit (RNA copy of DNA) and the molecule that carries out the activity of catalyzing biochemical reactions. With a division of labor between proteins, the work unit, and RNA, the informational unit, RNA becomes more specialized and efficient and hence the whole process becomes more efficient. Proteins then evolve specialization to catalyze biochemical reactions.

Test Your Knowledge

1. d, 2. e, 3. d, 4. a, 5. b, 6. c, 7. e, 8. e, 9. d, 10. e

Interpret the Data

1. Asteroid impacts declined before levels of atmospheric oxygen began to rise.
2. Oxygen was much less abundant while carbon dioxide was much more abundant in the atmosphere when the first cells arose compared to levels today.
3. Oxygen (21%) is now more abundant in the atmosphere than carbon dioxide (currently 0.038%).

Chapter 25

Study Break 25.1

1. Prokaryotes have no major cytoplasmic organelles that are equivalent to the endoplasmic reticulum, the Golgi complex, mitochondria, or chloroplasts of eukaryotes, nor does it have lysosomes.

   The genetic material of a prokaryote generally is a single, circular DNA molecule localized in a non-membrane-bound central region of the cell called a nucleoid. By contrast, the genetic material of a eukaryote is distributed among a number of linear chromosomes, which consist of DNA complexed with basic proteins known as histones.

   Most prokaryotes are surrounded by a cell wall located outside the plasma membrane. Animal cells do not have a cell wall, but plants, fungi, and some other eukaryotes do. The compositions of the eukaryotic cell walls are chemically different from bacterial cell walls.
2. A chemoheterotroph oxidizes organic molecules as its energy source and obtains carbon in organic form. A photoautotroph uses light as its energy source and carbon dioxide as its carbon source.
3. Obligate anaerobes are poisoned by oxygen. They survive either by fermentation, or by a form of respiration in which inorganic molecules are used as final electron acceptors. Facultative anaerobes use oxygen when it is present but live by fermentation when conditions are anaerobic.
4. Nitrogen fixation is the reduction of atmospheric nitrogen (N_2) to ammonia (NH_3). Nitrification is the conversion of ammonium (NH_4^+) to nitrate (NO_3^-).

   Nitrogen fixation, an exclusively prokaryotic process, is the only means of replenishing the nitrogen sources used by most microorganisms, and by all animals and plants.
5. A biofilm is a complex aggregation of microorganisms, many or all of them prokaryotes, attached to a surface. Biofilms are used in a variety of beneficial applications, including bioremediation of toxic organic chemicals contaminating groundwater. On the other hand, biofilms can have adverse effects on human health. For instance, biofilms can result in antibiotic-resistant infections if they adhere to surgical materials, such as catheters and implants.

Study Break 25.2

1. The hypothesized evolutionary ancestor of present-day Proteobacteria is a purple, photosynthesizing bacterium.
2. Photosynthetic Proteobacteria carry out photosynthesis as either photoautotrophs (the purple sulfur bacteria) or photoheterotrophs (the purple nonsulfur bacteria). The photosynthesis process used does not use water as an electron donor and does not release oxygen as a byproduct of photosynthesis. Their photosynthetic pigment is a type of chlorophyll distinct from that of plants.

   Cyanobacteria are photoautotrophs that carry out photosynthesis by the same pathways as eukaryotic algae and plants do. They use the same chlorophyll as plants do for their main photosynthetic pigment, and they release oxygen as a by-product of photosynthesis.
3. An exotoxin is a toxic protein that is secreted from the bacterium that makes it, or is released when that bacterium lyses. An endotoxin is a normal lipopolysaccharide component of the outer membrane of Gram-negative bacteria; it is released when the bacteria die and lyse. An exoenzyme is an enzymatic protein that is released from cells.

   An exotoxin interferes with biochemical processes of body cells. An endotoxin overstimulates the immune system, often causing inflammation. Depending on the bacterium, the endotoxin release has different effects, which may include organ failure and death. An exoenzyme digests plasma membranes, causing cells of the infected host to rupture and die. Exoenzymes may also digest extracellular materials and red and white blood cells.

Study Break 25.3

1. See Table 25.1 for a comparison of the properties of organisms in each of the three domains. The classification of Archaea as a distinct domain was based on comparisons of DNA and rRNA sequences.
2. A methanogen lives in reducing environments, generating energy by converting substrates such as carbon dioxide, hydrogen gas, methanol or acetate into methane gas. All known methogens belong to the Euryarchaeota.
3. Extreme halophilic archaeans live in high-salt environments, requiring at least 1.5 M NaCl in order to live. Most of these organisms are aerobic chemoheterotrophs, obtaining energy from sugars, alcohols, and amino acids using pathways similar to those of bacteria. All known extreme halophilic archaeans belong to the Euryarchaeota.
4. Extreme thermophiles live in extremely hot environments such as thermal hot springs and hydrothermal ocean floor vents. Psychrophiles grow optimally at temperatures in the range −10 to −20°C, such as in the Antarctic and Arctic oceans.

Think Critically

One possibility is that the last universal common ancestor (LUCA) of all life on the planet had a genome composed not of DNA but of RNA. If the switch to DNA-based genomes occurred after LUCA, once in the lineage that led to Bacteria and once entirely independently in the lineage that led to Archaea and Eukarya, then the two machineries for replicating DNA need not be related to one another. A second possibility is that LUCA had a DNA-based genome but that the cellular replication machinery was displaced by that of a virus in either the bacterial or archaeal/eukaryal lineages.

Test Your Knowledge

1. d, 2. a, 3. c, 4. b, 5. e, 6. a, 7. d, 8. b, 9. c, 10. d

Interpret the Data

1. Freezing occurs at 0°C, and *Pseudomonas* will not grow at or near freezing temperatures.
2. Yes, the growth rate at 20°C (293 K) is 0.44 and at 4°C (277 K) it is 0.012. The growth rate is more than 36 times slower at refrigeration than room temperature, indicating that refrigeration is sufficient.

Chapter 26

Study Break 26.1

A protist is distinguished from a prokaryote by having typical eukaryotic cell features like a nuclear envelope surrounding its genetic material, and cell organelles such as mitochondria (in most protists), chloroplasts (in some protists), endoplasmic reticulum, Golgi complex, and so on.

Distinguishing protists from fungi, animals, and plants is more blurry. Fungi are nonmotile at all stages of their life cycles, while most protists are motile or have motile stages in their life cycles. Cell wall structure is also different from fungi, and from plants. Protists differ from both animals and plants by lacking highly differentiated structures and by not having complex developmental stages. Collagen, the extracellular support protein of animals, is absent in protists.

Study Break 26.2

1. Excavates' nuclear genomes contain genes that are of mitochondrial origin, arguing that they once had mitochondria.
2. The chloroplast will have two membranes: one derived from the plasma membrane of the engulfing eukaryote and the other from the plasma membrane of the cyanobacterium.

Think Critically

1. Evolution is a dynamic field of study utilizing many tools, from fossils to gene sequence data to complicated mathematical models. New ideas about the first eukaryotes and the significance of endosymbioses will continue to emerge by using the scientific method.
2. Anything is possible, but no current evidence suggests prokaryotes evolved from a single-celled eukaryote. Fossil evidence and evidence for the endosymbiotic theory support a prokaryote-first

world. Unlike *Giardia* that have mitochondrial genes, for example, prokaryotes do not have "eukaryotic" genes that appear to have no (or lost) function.

Test Your Knowledge

1. e, 2. d, 3. d, 4. b, 5. b, 6. c, 7. b, 8. a, 9. b, 10. c

Interpret the Data

1. The 50 μmol m^{-2} s^{1} irradiance treatment has a maximum at ~450 μmol m^{-2} s^{1}. The 1,200 μmol m^{-2} s^{1} irradiance treatment has a maximum above 1,200 μmol m^{-2} s^{1}. No, the maximum was different for the low irradiance treatment, because 50 μmol m^{-2} s^{1} was too low a level of light to maximize photosynthesis, so the diatoms increased their rates of photosynthesis in increased light levels. The 1,200 μmol m^{-2} s^{1} irradiance treatment reached its photosynthesis maximum at the treatment light level, indicating that this was an optimal light level.
2. The low irradiance treatment has a higher photosynthesis efficiency with light.

Chapter 27

Study Break 27.1

1. Evolution of a root system gave land plants access to minerals and water in soil and provided physical support for aerial parts. The evolving shoot system of land plants, including lignified tissues in stems, allowed vascular plants to grow taller and stay erect, thereby gaining better access to sunlight for photosynthesis. Reproductive structures borne on aerial stems (such as flowers) might serve as platforms for more efficient dispersal of spores from the parent plant. Vascular tissues were innovations for distributing water (xylem) and sugars (phloem) through the plant body.
2. Homosporous plants produce a single type of sexual spore and are in effect bisexual, with each gametophyte capable of producing both sperm and eggs. Heterosporous species, including angiosperms and gymnosperms, produce two types of spores, which develop into sexually different gametophytes that produce either sperm or eggs. Plant scientists associate the evolution of heterospory with several key reproductive innovations in land plant evolution, including the protection of male gametes inside pollen grains and the protection of plant embryos inside seeds.

Study Break 27.2

1. Like aquatic plants, bryophytes produce flagellated sperm that must swim through water to reach eggs, and they lack a complex vascular system (although some have a primitive type of conducting tissue). Bryophytes do have parts that are rootlike, stemlike, and leaflike, although the "roots" are rhizoids, and bryophyte "stems" and "leaves" did not evolve from the same structures that vascular plant stems and leaves did. Sporophytes of some species have a water-conserving cuticle and stomata. Like most plants, bryophytes also have both sexual and asexual reproductive modes.
2. In general, mosses are the bryophytes that most closely resemble vascular plants. Some species produce structurally complex gametophytes that have a central strand of primitive water-conducting tissue that resembles the xylem of vascular plants, and in a few species the water-conducting cells are surrounded by sugar-conducting tissue resembling the phloem of vascular plants.

Study Break 27.3

1. In bryophytes, the gametophyte is much larger than the sporophyte and obtains its nutrition independently. The comparatively tiny sporophyte remains attached to the gametophyte and depends on the gametophyte for much of its nutrition. In modern lycophytes (club mosses and their close relatives), the gametophyte is free-living—though it is nourished by mycorrhizae instead of carrying out photosynthesis—and it is smaller than the sporophyte, which is a photosynthetic autotroph.
2. Fern leaves often take the form of feathery fronds, and roots extend from underground stems called rhizomes. Whisk ferns lack true leaves and roots; instead, small leaflike scales dot an upright, green, branching stem, which arises from a horizontal rhizome system anchored by rhizoids. Horsetail sporophytes typically have underground rhizomes and roots that anchor the rhizome to the soil. The scalelike leaves are arranged in whorls about a photosynthetic stem.
3. In horsetails, the sporangia that produce spores are borne in strobili, and spores are carried away from the plant by air currents. In ferns, sporangia are produced on the lower surface or margin of leaves, and spores are forcefully dispersed from the parent plant when contraction of a beltlike annulus rips open the sporangium and ejects the spores.

Study Break 27.4

1. The four major reproductive adaptations that evolved in gymnosperms include pollen and the ovule, both of which shelter spores; pollination rather than dispersal of swimming sperm; and the seed as a "package" that protects and often nourishes the embryo.
2. The three basic parts of a seed are the embryo sporophyte, endosperm, and outer seed coat. The endosperm nourishes the embryo sporophyte, and the seed coat protects it.
3. Features that make conifer sporophytes structurally more complex than other gymnosperms include anatomically complex needlelike or scalelike leaves that are adapted to aridity and the production of resins as metabolic by-products.

Study Break 27.5

1. The lack of fossil early angiosperms, including obvious transitional forms, has made it difficult to trace a clear evolutionary path for flowering plants. As result plant scientists have proposed several, often conflicting, classification schemes for angiosperms.
2. Seeds leaves (cotyledons) and pollen morphology are two major features used to distinguish monocots and eudicots. Monocots have a single seed leaf and pollen grain with a single groove. Eudicots have two seed leaves and pollen grains with three grooves. Monocots (such as grasses, lilies, and palms) also generally have fibrous root systems, leaves with parallel veins, flower parts in multiples of three, and scattered vascular bundles in stems. Eudicots (most flowering trees and shrubs, roses, sunflowers, beans) usually have netlike leaf venation, a primary taproot, flower parts in fours or fives, and vascular tissues arranged in a ring.
3. Adaptations that have contributed to the evolutionary success of angiosperms include vascular tissue modifications that make transport of water and nutrients more efficient; double fertilization, which results in enhanced nutrition (endosperm) for embryos; physical protection of embryos within ovaries and seeds; and coevolution with animal pollinators, which increases the likelihood that pollination will occur.

Think Critically

1. Plants have been evolving continuously for millions of years and during that time have been able to adapt to a variety of conditions. Contemporary flowers have built on the framework established by the first flowers, evolving more diverse and refined methods of attracting pollinators, protecting developing seeds, etc.
2. Angiosperms have adapted to a wide variety of different habitats and life forms, ranging from long-lived forest trees to ephemeral opportunistic weeds of disturbed areas to drought-, salt-, or cold-resistant forms in extreme climates. This diversification has allowed flowering plants to colonize a broad range of particular environments. In addition, in parallel with the burst of speciation among insects, flowering plants have evolved to exploit a wide variety of pollination and dispersal mechanisms so that they can reliably reproduce and colonize new habitats.

Test Your Knowledge

1. b, 2. d, 3. e, 4. e, 5. a, 6. d, 7. a, 8. b, 9. a, 10. d

Interpret the Data

1. The data summarized in Figure A support the hypothesis that a single endosymbiosis event correlated with the evolution of all photosynthesizing organisms. This event corresponds with the node at which the clade including green plants, red algae, and glaucophytes diverged from the clade including alveolates and heterokonts (diatoms and their relatives).
2. Yes. The cladogram shows less evolutionary distance between green plants and glaucophytes than between green plants and red algae.

Chapter 28

Study Break 28.1

1. Some fungi are multicellular while others, the yeasts, are single cells. (Some species alternate between these two forms at different life cycle stages.) The cells of all fungi are surrounded by a hardened wall; in most cases the hardener is the polysaccharide chitin. The body of a multicellular fungus consists of a dense mesh of filaments called hyphae, which in some groups are separated into cell-like compartments by cross walls (septa). Aggregations of hyphae are the structural foundation for all other parts that develop as part of a multicellular fungus. For example, in some species modified hyphae form rhizoids that anchor the fungus to its substrate.
2. Fungal spores are microscopic, usually nonmotile reproductive cells in which haploid nuclei are surrounded by a tough outer wall. They are produced sexually or asexually. Sexual spores

are produced by genetically different parent fungi and may unite in a sexual process that gives rise to a diploid life stage. Asexual spores are genetically identical to the parent fungus and may give rise to a new, haploid individual.

3. Many fungal species have a life cycle stage called a dikaryon, which contains two haploid nuclei (a condition expressed as $n + n$). A dikaryon forms as the result of plasmogamy, a sexual stage in which the cytoplasms of two genetically different partners fuse. This fusion ensures genetic diversity in new individuals. At some point after a dikaryon forms, the nuclei fuse (karyogamy) to form a short-lived zygote. Meiosis in the zygote produces haploid nuclei that become packaged into sexual spores.

Study Break 28.2

1. The main phyla of fungi are the Chytridiomycota, Zygomycota, Glomeromycota, Ascomycota, and Basidiomycota. Chytrids are the only fungi that produce motile, flagellated spores. Zygomycetes often reproduce asexually, but sometimes reproduce sexually by way of hyphae that occur in + and − mating strains; haploid nuclei in the hyphae function as gametes. Following plasmogamy, further development produces zygospores in which karyogamy gives rise to diploid zygotes ($2n$ nuclei), which then undergo meiosis as sexual spores form. Glomeromycetes reproduce asexually, by way of spores that form at the tips of hyphae. In ascomycetes, chains of asexual spores called conidia, each containing a haploid nucleus, develop during asexual reproduction. Ascomycetes produce haploid sexual spores in pouchlike cells called asci. Most basidiomycetes reproduce only sexually: club-shaped basidia develop on a basidiocarp (for example, a "mushroom") and bear sexual spores on their outer surface. When dispersed, the spores may germinate and give rise to a haploid mycelium. Cytoplasmic fusion may occur between hyphae of two compatible mating strains, producing a dikaryotic mycelium from which basidiocarps may grow. Microsporidia, single-celled parasites that may be related to zygomycetes, resemble spores but lack mitochondria. Fungi for which no sexual life stage has been identified are placed in a convenience grouping called "conidial fungi."
2. Anatomically, the simplest fungi are chytrids, which are microscopic, and zygomycetes, which have aseptate hyphae. Ascomycetes, basidiomycetes, and glomeromycetes all form septate (walled) hyphae.
3. Most chytrids are aquatic, but some are parasites of insects, plants, and some animals. Others are symbiotic partners in the gut of cattle and some other herbivores.

   Many zygomycetes are saprobes in soil, feeding on plant detritus. Their metabolic activities release mineral nutrients that plant roots can take up. Some zygomycetes are parasites of insects or spoil stored grains, bread, fruits, and vegetables such as sweet potatoes. Others are used in manufacturing products such as industrial pigments and pharmaceuticals. Many more are highly destructive plant pathogens and several can be human pathogens, causing athlete's foot, ringworm infections, and more serious illnesses. The pink bread mold *Neurospora crassa* has been crucial in genetic research.

   Some basidiomycetes participate in vital mutualistic associations (mycorrhizae) with the roots of forest trees. Others produce prized edible mushrooms.

   Rusts and smuts are parasites that cause serious diseases in wheat, rice, and other plants.

   Glomeromycetes are all specialized to form mycorrhizae with plant roots.

Study Break 28.3

1. A lichen is a communal life form representing a symbiosis between a photosynthetic green alga or species of cyanobacteria (the photobiont) and a nonphotosynthetic fungus (the mycobiont). The algal cells supply the lichen's carbohydrates, most of which are absorbed by the fungus. In some cases, the alga is protected from desiccation or some other environmental threat.
2. A mycorrhiza is a symbiotic association between a fungus and plant roots. The fungal hyphae make mineral ions and sometimes water available to the plant's roots, and in exchange the fungus absorbs carbohydrates, amino acids, and possibly other growth-enhancing substances provided by the plant. Mycorrhizae greatly enhance the plant's ability to extract various nutrients, especially phosphorus and nitrogen, from soil, and they are crucial to the survival of many plant species.
3. In endomycorrhizae the hyphae of a fungus (typically a glomeromycete) enter plant roots, where their tips branch into clusters called arbuscules where exchanges of nutrients take place. In ectomycorrhizae the fungal partners are basidiomycetes. Their hyphae surround plant roots but do not penetrate them.

Think Critically

1. Ecological theory predicts that species with complete overlap in the way they use resources and interact with other species would not coexist in a community. Therefore, it is likely that ectomycorrhizal fungi that coexist on a single tree differ in factors such as how they gain nutrients from the tree, how they gain nutrients from the surrounding soil, how they interact with the tree (for example, what parts of the root system they colonize, what types and amounts of nutrients they provide, or what additional benefits they confer to the tree), where they occur in space (for example, at what soil depth they occur), or how they respond to different environmental conditions (such as rainfall and temperature).
2. Because mycorrhizal symbiosis may be controlled by the plant symbiont, the fungal symbiont, or both, the dominance of fungal symbionts within a community may result from interactions between different fungal symbionts (for example, direct competition between mycelia, or indirect competition for nutrients) or from differences in the way they interact with the tree (such as the "selection" by the tree of one symbiont over another).
3. Understanding the nature of interactions between plants, fungi, and the environment can be beneficial for protecting rare fungal or plant species, maintaining forest diversity, predicting the effects of climate or land use change, and improving the outcome of restoration efforts.

Test Your Knowledge

1. a,c, 2. b,e, 3. a, 4. d, 5. c,d, 6. e, 7. c, 8. a, 9. b,e, 10. d

Interpret the Data

1. In the dark, no. The number of fusion events in wild-type pairs showed the same pattern as matings between wild-type and mutant hyphae.
2. In white and blue light, the most fusion events occurred in crosses in which one or both hyphae were mutants, regardless of mating type. The fewest fusion events under these light conditions occurred in crosses in which one or both hyphae were the **a** mating type. In green and red light, the pattern shifted. The most fusion events occurred when the cross included one mutant and one wild-type hypha. The researchers hypothesized that in general light inhibits fusion of wild-type (+) strains of *C. neoformans.* These results support that hypothesis. Red light wavelengths predominate in shade, so it is not surprising that overall the most fusion events occurred in darkness and in red light conditions.

Chapter 29

Study Break 29.1

1. Several characteristics distinguish animals from plants: plant cells have cell walls, but animal cells do not; almost all plants are autotrophic, whereas all animals are heterotrophic; and plants are usually sessile, but animals are motile at some stage of their life cycle. Animals differ from fungi in that the cells of fungi have cell walls, and most fungi are sessile.
2. The ability of animals to move through the environment allows them to search for and pursue the food items that supply them with nutrients and energy.

Study Break 29.2

1. A tissue is a group of cells that share a common structure and function. The three primary tissue layers that contribute to the bodies of most animals are endoderm, mesoderm, and ectoderm.
2. Humans are bilaterally symmetrical.
3. The coelom is a space within which internal organs can move independently of the body wall muscles. The fluid within it provides protection for internal organs. In some animals the coelom functions as a hydrostatic skeleton.
4. Having a segmented body may allow an animal to survive damage to some parts of its body segments and may allow improved control over body movements.

Study Break 29.3

1. Molecular sequence studies have confirmed the distinctions between the Parazoa and the Eumetazoa, between the Radiata and the Bilateria, and between the Protostomia and the Deuterostomia.
2. A schizocoelom appears to be the ancestral body cavity among the protostomes.

Study Break 29.4

1. Sponges do not exhibit any kind of body symmetry.
2. A sponge gathers food from its environment by drawing water into its body through numerous small pores and harvesting particulate matter

from the water with its choanocytes, or collar cells.

Study Break 29.5

1. Cnidarians capture animal prey by stinging it with their nematocysts and using their tentacles to pull it into their mouths.
2. The anthozoans, including sea anemones and corals, have only a polyp stage in their life cycle.
3. Ctenophores capture microscopic plankton in sticky filaments on their two tentacles, which are then drawn across the mouth.

Study Break 29.6

1. Free-living flatworms have digestive, excretory, nervous, and reproductive systems. Tapeworms lack a digestive system.
2. Ectoprocts, brachiopods, and phoronid worms all have a circular or U-shaped feeding structure called a lophophore, a characteristic that reveals their close evolutionary relationship.
3. The anatomical and physiological systems that allow squids and other cephalopods to be more active than other types of mollusks include a closed circulatory system with accessory hearts and a complex nervous system with giant nerve fibers. Many cephalopods use their excurrent siphon to expel jets of water, allowing them to move rapidly through the environment.
4. The organ systems that exhibit segmentation in most annelid worms include respiratory surfaces; parts of the nervous, circulatory, and excretory systems; and the body wall and coelom.

Study Break 29.7

1. The cuticle protects a nematode from the digestive enzymes of its host.
2. Although the rigid exoskeletons of arthropods do not expand, these animals grow a new, soft exoskeleton inside the existing one. After shedding the old exoskeleton, they grow to a larger size by expanding the new exoskeleton with either water or air before it hardens.
3. The body regions of the four living subphyla of arthropods differ in how they have become fused. Chelicerates have a fused cephalothorax and an abdomen. Crustaceans show variable patterns, but many have a fused cephalothorax and an abdomen. Myriapods have a head and a trunk. Hexapods have a separate head, thorax, and abdomen.
4. Insects with incomplete metamorphosis hatch from their eggs as wingless nymphs, which vary in how closely they resemble adults; nymphs then undergo metamorphosis into the adult form. Insects with complete metamorphosis hatch from eggs as larvae, which are always very different from adults. After becoming a pupa, their cells and tissues are reorganized into the adult form.

Think Critically

1. Recent genetic comparisons of invertebrate taxa confirm the distinction between Parazoa and Eumetazoa and the distinction between Radiata and Bilateria.
2. The nucleotide sequences most likely to shed light on the diversity of living invertebrates are those in small subunit ribosomal RNA, mitochondrial DNA, and *Hox* genes of species within the protostome and deuterostome lineages.

Test Your Knowledge

1. b, 2. d, 3. c, 4. a, 5. c, 6. e, 7. e, 8. a, 9. d, 10. b

Interpret the Data

Leech therapy worked better than drug therapy to relieve the pain of arthritis in the thumbs of the patients tested. The effect of leech therapy persisted for at least two months after just one treatment.

Chapter 30

Study Break 30.1

1. Echinoderms have a water vascular system, which operates the tube feet that are used for locomotion and/or feeding.
2. Water enters the pharynx of a hemichordate through its mouth, and it exits the pharynx through the branchial slits. The animal extracts oxygen and particulate food from the water as it passes through the pharynx.

Study Break 30.2

1. The animal is not a chordate, because chordates have a dorsal nerve cord.
2. Vertebrates have an internal bony skeleton, including a cranium and vertebral column in most groups, as well as structures derived from neural crest cells.

Study Break 30.3

1. Vertebrates have multiple *Hox* gene complexes, which provide them with several copies of each *Hox* gene. Cephalochordates have just one *Hox* gene complex.
2. Of the three groups listed, Gnathostomata has the most species, and Amniota has the fewest.

Study Break 30.4

1. Hagfishes lack bone, paired fins, and scales in their skin. Hagfishes have neither a cranium nor a vertebral column, and lampreys have only rudimentary traces of vertebrae. These observations suggest that their lineages arose before these structures appeared in vertebrates.
2. The derived traits possessed by conodonts and ostracoderms include structures made of bone or a bonelike material and, in some ostracoderms, a brain divided into three regions.

Study Break 30.5

1. Sharks are more efficient predators than acanthodians or placoderms were because they have well-developed sensory systems to detect prey; their lightweight skeletons and absence of heavy body armor allow them to pursue prey rapidly; and they have numerous teeth that are replaced when damaged or worn, as well as a loosely attached upper jaw that permits them to suck in large chunks of food.
2. The air bladders of ray-finned bony fishes increase their locomotor abilities by allowing them to rise or sink easily in the water. Their fin rays allow them to engage in precise movements during locomotion.
3. The lungs of lungfish allow them to survive in environments with low oxygen content because they can acquire oxygen from the air.

Study Break 30.6

1. For the first tetrapods, the advantages of moving onto land included abundant food resources, the rarity of predators, and readily available oxygen. The disadvantages included the need for more skeletal support against gravity, mechanisms to prevent dehydration in air, and modifications of sensory systems so that they would function in air.
2. The parts of the amphibian life cycle that are most dependent on water are the egg and larval stages.

Study Break 30.7

1. The amniote egg freed amniotes from a dependence on standing water because it can survive on land. The shells of amniote eggs mediate gas exchange and water exchange with the environment.
2. Among the three amniote lineages, the Synapsida includes the mammals, the Anapsida includes turtles, and the Diapsida includes lizards, snakes, crocodilians, and birds.
3. Because lizards are lepidosaurs and both birds and crocodilians are archosaurs, crocodilians are more closely related to birds than they are to lizards.

Study Break 30.8

The overall structure of turtles differs from other amniotes in that their bodies are enclosed within a bony, keratin-covered shell.

Study Break 30.9

1. Besides their loss of legs, snakes differ from their lizard ancestors in having smaller skull bones, and the connections between them are more elastic.
2. Various species of lizards feed on vegetation, insects, or larger animals. Virtually all snakes are carnivores, and they swallow their prey whole.

Study Break 30.10

1. Several characteristics reveal the close evolutionary relationship of crocodilians and birds, including a four-chambered heart and maternal care of offspring.
2. The specific adaptations that allow birds to fly either reduce their weight or increase their muscle power. Weight-reducing adaptations include a lightweight skeleton, the absence of teeth and a urinary bladder, and the habit of laying an egg as soon as it has a shell. Power-promoting adaptations include large wing muscles, efficient digestive, respiratory, and circulatory systems, and a high metabolic rate.
3. The structure of a bird's bill reflects its diet. For example, hummingbirds, which drink nectar, have long thin bills, and parrots, which eat hard nuts, have stout sharp bills. Wings and feet are adapted to birds' flying habits and habitats. For example, ducks have webbed feet that allow them to paddle in water, and albatrosses have long thin wings that work efficiently for long distance flight.

Study Break 30.11

1. Most mammals were probably active at night during the Mesozoic era to avoid competition with and predation by dinosaurs, most of which were active during the day.
2. The key adaptations that allow mammals to be active under many types of environmental conditions include insulating fur and fat and a high metabolic rate that generates lots of body heat.
3. The three major groups of living mammals are distinguished on the basis of their reproductive habits. Monotremes lay eggs. Marsupials give birth to relatively undeveloped young after a short period of gestation. Placentals give birth to more developed young after a long period of gestation.

Study Break 30.12

1. The characteristics that allow many species of primates to spend a lot of time in trees include flexible shoulder and hip joints, grasping hands, and excellent depth perception.
2. The lowest taxonomic group that includes monkeys, apes, and humans is the Anthropoidea. The lowest taxonomic group that includes only apes and humans is the Hominoidea.
3. Gorillas, chimpanzees, and bonobos are the apes that spend the most time on the ground.

Study Break 30.13

1. Researchers usually use the criterion of bipedal locomotion to distinguish between humans and apes. Humans (that is, hominins) are bipedal, and apes are not.
2. The strongest evidence suggesting that Neanderthals and modern humans belong to different species comes from mtDNA sequence data: the differences between gene sequences of Neanderthals and humans are much greater than the differences between any two modern humans.
3. Independent genetic studies of mitochondrial DNA and the Y chromosome indicate that all human populations are descended from a common ancestor that originated in Africa and then migrated to various regions on Earth.

Think Critically

Viviparity could provide several benefits. By carrying her developing young, the mother can protect them from some detrimental environmental factors; she can regulate the temperature at which they develop by regulating her own body temperature; she can provide them with nutrients and energy throughout their development; and she can select a favorable environment into which she releases them when they are born.

Test Your Knowledge

1. a, 2. c, 3. d, 4. b, 5. e, 6. c, 7. e, 8. b, 9. d, 10. a

Interpret the Data

According to this analysis, turtles fall within the diapsid lineage, and they are more closely related to archosaurs (alligator and chicken) than to lepidosaurs (tuatara and iguana). The results of this analysis therefore contradict the traditional view.

Chapter 31

Study Break 31.1

1. A land plant's shoot system consists of its photosynthetic tissues and organs—stems, leaves, and buds. Stems are frameworks for upright growth and favorably position leaves for light exposure and flowers for pollination. Leaves increase a plant's surface area and thus its exposure to sunlight. Buds eventually extend the shoot or give rise to a new, branching shoot. The shoot system of a flowering plant also includes flowers and fruits. Parts of the shoot system store carbohydrates manufactured during photosynthesis.

   The root system usually grows below ground. It anchors the plant, and sometimes structurally supports its upright parts. It also absorbs water and dissolved minerals from soil and stores carbohydrates.
2. Meristem tissue is self-perpetuating embryonic tissue. Apical meristems, at the tips of shoots and roots, gives rise to a young plant's stems, buds, roots, and other primary tissues. In plants that show secondary growth, cylinders of lateral meristem tissue give rise to (often woody) secondary tissues that increase the diameter of older stems and roots.

Study Break 31.2

1. The ground tissue system makes up most of the plant body. It includes three types of structurally simple tissues—parenchyma, collenchyma, and sclerenchyma—each of which is composed mainly of one type of cell. Parenchyma makes up most of a plant's primary tissue and typically has air spaces between its cells, which are alive at maturity and can continue to divide. Subgroups of parenchyma cells are specialized for photosynthesis, secretion, and storage (of starch). Collenchyma is flexible ground tissue that contains cellulose. Its cells remain alive and metabolically active at maturity. They provide mechanical support for parenchyma and often collectively form strands or a sheathlike cylinder under the dermal tissue of growing shoot regions and leaf stalks. Cells of sclerenchyma are dead at maturity, but while alive they develop thick secondary walls that typically are lignified and provide additional support and protection in mature plant parts.
2. Xylem and phloem are the tissues of the vascular tissue system. The two types of xylem cells, called tracheids and vessel members, both develop thick, lignified secondary cell walls and die at maturity. The empty cell walls of abutting cells serve as pipelines for water and minerals. The conducting cells of phloem, called sieve tube members, form sieve tubes that conduct solutes, mainly sugars made during photosynthesis, throughout a plant.
3. The dermal tissue system serves as a skinlike protective covering for the plant body. Cells of the epidermis are tightly packed and cover the primary plant body. They secrete a cuticle that coats all plant parts except the very tips of the shoot and most absorptive parts of roots. Some epidermal cells become modified for specialized functions. Examples include guard cells, which form stomata; root hairs, which absorb water and minerals; and hairlike trichomes, which function in defense against herbivory or secrete sugars that attract pollinators.

Study Break 31.3

1. Stems have four main functions: (1) they provide mechanical support for body parts involved in growth, photosynthesis, and reproduction; (2) they house the vascular tissues (xylem and phloem), which transport products of photosynthesis, water and dissolved minerals, hormones, and other substances throughout the plant; (3) they often are modified to store water and food; and (4) they have specific stem regions that contain meristematic tissue, which gives rise to new cells of the shoot.

   A plant stem is divided into modules, each consisting of a node, where leaves are attached, and an internode, the space between nodes. New primary growth occurs in buds—a terminal bud at the apex of the main shoot, and lateral buds, which produce branches (lateral shoots), in the leaf axils. Meristem tissue in buds gives rise to leaves, flowers, or both.

   In eudicots, most primary growth in a stem's length occurs directly below the shoot apical meristem. When a meristematic cell divides, one of its daughter cells becomes an initial, a cell that remains as part of the meristem. The other daughter cell becomes a derivative, which typically divides once or twice and then enters on the path to differentiation. As derivatives differentiate, they give rise to three primary meristems: protoderm, procambium, and ground meristem. These primary meristems produce cells that differentiate into specialized cells and tissues. In eudicots, the primary meristems are also responsible for elongation of the plant body. Each primary meristem occupies a different position in the shoot tip. Outermost is protoderm, which gives rise to the stem's epidermis. Inward from the protoderm the ground meristem gives rise to ground tissue (mostly parenchyma). Procambium, which produces the primary vascular tissues, is sandwiched between ground meristem layers. In most plants, inner procambial cells give rise to xylem and outer procambial cells to phloem. The developing vascular tissues become organized into vascular bundles that are wrapped in sclerenchyma and thread lengthwise through the parenchyma. In the stems and roots of most eudicots and some conifers, the vascular bundles form a stele (vascular cylinder) that vertically divides the column of ground tissue into an outer cortex and an inner pith.
2. Leaves are organs specialized for photosynthesis. In both eudicots and monocots, the leaf blade provides a large surface area for absorbing sunlight and carbon dioxide. Many eudicot leaves have a broad, flat blade attached to the stem by a petiole. Unless a petiole is very short, it holds a leaf away from the stem and helps prevent individual leaves from shading one another. In most monocot leaves, such as those of rye grass or corn, the blade is longer and narrower and its base simply forms a sheath around the stem.
3. Leaves develop on the sides of the shoot apical meristem. Initially, meristem cells near the apex divide and their derivatives elongate. The resulting bulge enlarges into a thin, rudimentary leaf, or leaf primordium. As the plant grows and internodes elongate, the leaves become spaced at intervals along the length of the stem or its branches. Leaf tissues typically form several layers. Uppermost is epidermis, with cuticle covering its outer surface. Just beneath the epidermis is mesophyll, which is composed of loosely packed parenchyma cells that contain chloroplasts. Leaves of many plants, especially eudicots, contain two layers of mesophyll. Palisade mesophyll cells contain more chloroplasts and are arranged in compact columns with smaller air spaces between them, typically toward the upper leaf surface. Spongy mesophyll, which tends to be located toward the underside of a leaf, consists of irregularly arranged cells with a network of air spaces that enhance the uptake of carbon dioxide and release of oxygen during photosynthesis and account for 15% to 50% of a leaf's volume. Below the mesophyll is another cuticle-covered epidermal layer. Except in grasses and a few other plants, this layer contains most of the

stomata through which water vapor exits the leaf and gas exchange occurs. Vascular bundles form a network of veins throughout the leaf.

4. Plants that live for many years may spend part of their lives in a juvenile phase, then shift to a mature, or adult phase. The differences between juveniles and adults often are reflected in leaf size and shape, in the arrangement of leaves on the stem, or in a change from vegetative growth to a reproductive stage. Most woody plants must attain a certain size before their meristem tissue can respond to the hormonal signals that govern flower development.

Study Break 31.4

1. Most eudicots have a taproot system—a single main root, or taproot, that is adapted for storage and smaller branching lateral roots. As the main root grows downward, its diameter increases, and the lateral roots emerge along the length of its older, differentiated regions. Grasses and many other monocots develop a fibrous root system in which several main roots branch to form a dense mass of smaller roots. Fibrous root systems are adapted to absorb water and nutrients from the upper layers of soil, and tend to spread out laterally from the base of the stem.
2. The root apical meristem and the actively dividing cells behind it form the zone of cell division. Cells in the center of the root tip become the procambium; those just outside the procambium become ground meristem; and those on the periphery of the apical meristem become protoderm. The zone of cell division merges into the zone of elongation, where most of the increase in a root's length occurs. Above the zone of elongation, cells may differentiate further and take on specialized roles in the zone of maturation.
3. Primary root growth produces a system of vascular pipelines extending from root tip to shoot tip. The root procambium produces cells that mature into the root's xylem and phloem. Ground meristem gives rise to the root's cortex, its ground tissue of starch-storing parenchyma cells that surround the stele. In many flowering plants, the outer root cortex cells give rise to an exodermis, a thin band of cells that may limit water losses from roots and help regulate the absorption of ions. The innermost layer of the root cortex is the thin endodermis, which helps control the movement of water and dissolved minerals into the stele. Between the stele and the endodermis is the pericycle, which gives rise to lateral roots. In some cells in the developing root epidermis, the outer surface extends into root hairs.

Study Break 31.5

1. Secondary growth processes add girth to roots and stems over two or more growing seasons. In plant species that have secondary growth, older stems and roots become more massive and woody through the activity of two types of lateral meristems. One of these meristems, the vascular cambium, produces secondary xylem and phloem. The other, the cork cambium, produces cork, a secondary epidermis that is one element of bark.
2. Vascular cambium consists of two types of cells—fusiform initials and ray initials. Fusiform initials are derived from cambium inside the vascular bundles and give rise to secondary xylem and phloem cells. Secondary xylem forms on the inner face of the vascular cambium, and secondary phloem forms on the outer face. Ray initials are derived from the parenchyma cells between vascular bundles. Their descendants form spokelike rays of parenchyma cells—horizontal channels that carry water sideways through the stem. As the mass of secondary xylem inside the ring of vascular cambium increases, it forms hard tissue known as wood. Bark encompasses all the living and nonliving tissues between the vascular cambium and the stem surface. It includes the secondary phloem and the periderm, the outermost portion of bark that consists of cork, cork cambium, and secondary cortex.
3. Roots of some plant species undergo secondary growth, but the ring of vascular cambium develops differently than it does in stems. When their primary growth is complete, these roots have a layer of residual procambium between the xylem and phloem of the stele. The vascular cambium arises in part from this residual cambium, and in part from the pericycle. Eventually, the cambial tissues arising from the procambium and those arising from the pericycle merge into a complete cylinder of vascular cambium. As in stems, the vascular cambium gives rise to secondary xylem to the inside and secondary phloem to the outside. As secondary xylem accumulates, older roots can become thick and woody.

Think Critically

Despite producing progeny cells that adopt different fates, both shoot and root stem cells are pluripotent cells that share the properties of continuous division and self-renewal, and thus their molecular signatures are likely to be more similar than different.

Test Your Knowledge

1. d, 2. c, 3. b, 4. a, 5. c, 6. a, 7. d, 8. e, 9. c, 10. b

Interpret the Data

1. The data show wider than usual annual growth rings for the twentieth century. In their research report Dr. Jacoby and his coworkers concluded that "recent warming is unusual relative to temperatures of the past 450 years."
2. It was important for the study authors to begin by acknowledging that the width of rings formed by individual trees shows natural variations related to each tree's growth. Including a graph of the ring-width measurements from a single 500-year-old tree allows for a clear comparison of that pattern with the graph that charts aggregated data from many trees—which you will notice follows a quite similar pattern.
3. The lower line represents sample size in the study. It shows that relatively few very old trees (with rings formed before 1550) were available to be sampled, with a steady rise in the number of available trees that began forming rings in subsequent centuries. The line also reveals consistency in the number of core samples dating from about 1850, when the Industrial Revolution and more intensive burning of fossil fuels began. This consistency lends credence to the reliability of the study's findings.
4. Airborne smoke and ash reduce the amount of solar energy that reaches Earth's surface. Less available sunlight is likely to translate into reduced photosynthesis and plant growth, which in turn will be reflected in the width of tree rings that form during that period.

Chapter 32

Study Break 32.1

1. After an H^+ gradient is established (by proton pumping out of the cell), the resulting inward flow of H^+ down its concentration gradient provides the energy to actively transport other substances into the cell.
2. Water potential is potential energy stored in water. It is the driving force for osmosis, which in turn is responsible for the movement of water into and out of plant cells, including root cells.

Study Break 32.2

1. In the apoplastic pathway, water and dissolved substances do not pass through living root cells but instead move through the continuous network of adjoining cell walls and air spaces. When apoplastic water (and solutes) reach the endodermis, however, they must detour around the impermeable Casparian strip and pass through cells to move into the stele. The symplastic pathway passes through living cells. Water that diffuses into root cells moves in this pathway from cell to cell through plasmodesmata.
2. Epidermal cells of root hairs actively transport most mineral ions into root epidermal cells. These ions travel inward via the transmembrane pathway. Other ions may be dissolved in apoplastic water. They ultimately travel to the xylem in the symplast after crossing into and through endodermal cells of the Casparian strip. Once an ion reaches the stele it enters the xylem.

Study Break 32.3

1. In the cohesion–tension mechanism, water transport begins as water evaporates from the walls of mesophyll cells inside leaves and into the intercellular spaces. This water vapor escapes by transpiration through open stomata. As water molecules exit the leaf, they are replaced by others from the mesophyll cell cytoplasm. The water loss gradually reduces the water potential in a transpiring cell below the water potential in the leaf xylem. Water from the xylem in the leaf veins then follows the gradient into cells, replacing the water lost in transpiration.
2. Stomata open and close in response to changing environmental cues, such as light levels (detected via blue-light receptors), CO_2 concentration in the air spaces inside leaves, and the amount of water available to the plant. Stomata open when hydrogen ions are pumped out of guard cells, setting up the symport of H^+ and K^+ into the guard cells through ion channels. Water then follows by osmosis. Stomata close when H^+ pumping in guard cells ceases and K^+ moves out of guard cells, with water again following by osmosis. Through their ability to open and close, stomata help regulate water loss by plants and the uptake of carbon dioxide for photosynthesis.

Study Break 32.4

1. Translocation is the long-distance transport of substances in plants. The term generally applies to the transport of organic compounds, mainly

sucrose, in phloem. Transpiration is the evaporation of water from a plant's aerial parts, mainly leaves. This water moves from roots upward to aerial parts in the xylem.

2. The mechanism of pressure flow moves sucrose from a source (such as a leaf or stem) into sieve tubes. Pressure builds up at the source end of a sieve tube system as sucrose enters sieve tubes at sources and water follows by osmosis. Under high pressure, sucrose moves by bulk flow toward a sink (plant parts that take up sucrose as metabolic fuel), where the sugar is unloaded.

Think Critically

The results support a hypothesis that active transport may sometimes move substances through plasmodesmata. Active transport is unidirectional—a transported solute moves either into or out of a cell—and the tracer moved unidirectionally in the experiments. The halt in tracer movement when cell metabolism was inhibited by sodium azide also is consistent with an active transport mechanism, because active transport requires ATP produced by the cell's metabolism.

Test Your Knowledge

1. b, 2. d, 3. e, 4. a, 5. a, 6. b, 7. e, 8. c, 9. c, 10. c

Interpret the Data

1. In both graphs, the darker blue blocks represent the movement of deionized water through girdled branch segments over time. The lighter blue blocks represent the movement of a potassium chloride solution through girdled branch segments over time.
2. The results support the hypothesis because the values for the flow of pure water are consistently lower than the flow values of the KCl solution.
3. Results were substantially the same for both species, in that in both there was a marked increase in flow rates when the researchers switched from deionized water to the KCl solution. There was, however, a species-specific difference in the magnitude of the increased flow.

Chapter 33

Study Break 33.1

1. Plants require relatively large amounts of macronutrients such as nitrogen, sulfur, potassium, and calcium, and trace amounts of micronutrients such as iron, chlorine, zinc, nickel, and copper.
2. Plants vary in their nutritional requirements. For example, leafy plants require more nitrogen and magnesium than other plant types do, and alfalfa, a grass, requires significantly more potassium than lawn grasses do. An adequate amount of an essential element for one plant also may be toxic for another. For these reasons, the nutrient content of soils is an important factor determining which plants grow well in a given location.

Study Break 33.2

1. Humus is important in soil because it generally contains nutrient-rich organic material and because it absorbs water, which contributes to the water-holding capacity of soil.
2. The amount of water that is available in soil to be taken up by plant roots depends primarily on the relative proportions of different soil components. Water moves quickly through sandy soils, whereas soils rich in clay and humus tend to hold the most water.
3. A plant's ability to absorb soil minerals depends partly on cation exchange, in which one cation, usually H^+, replaces a soil cation. As H^+ enters the soil solution, it displaces adsorbed mineral cations attached to clay and humus, freeing them to move into roots. Anions in the soil solution, such as nitrate (NO_3^-), sulfate (SO_4^{2-}), and phosphate (PO_4^-), generally move more readily into root hairs. Soil pH also affects the availability of some mineral ions because chemical reactions in very acid ($pH < 5.5$) or very alkaline ($pH > 9.5$) soils can trigger chemical reactions that bind various mineral cations in compounds that are insoluble in soil water.

Study Break 33.3

1. As described in this chapter and in Section 28.3, a mycorrhiza is a symbiotic association between a fungus and plant roots. Most plants form mycorrhizal associations, which facilitate the plant's ability to extract soil nutrients such as nitrogen and phosphorus. As with plant roots, mineral ions enter fungal hyphae by way of transport proteins. Some of the plant's sugars and nitrogenous compounds nourish the fungus, and as the root grows, it takes up a portion of the minerals that the fungus has secured. In some types of mycorrhizae the fungus actually lives inside cells of the root cortex.
2. Nitrogen fixation refers to the incorporation of atmospheric nitrogen into compounds, especially nitrate (NO_3^-), that plants can readily take up. Ammonification is a process in which soil bacteria known as ammonifying bacteria break down decaying organic matter and convert it to ammonium (NH_4^+). In nitrification, nitrifying bacteria oxidize NH_4^+ to NO_3^-. Inside root cells, absorbed NO_3^- is converted by a multistep process back to NH_4^+. In this form, it is rapidly used to synthesize organic molecules, mainly amino acids.
3. Associations with bacteria supply nitrogen to certain types of plants, such as legumes. The host plant provides organic molecules that the bacteria use for cellular respiration, and the bacteria supply NH_4^+ that the plant uses to produce nitrogenous molecules. In legumes the nitrogen-fixing bacteria reside in root nodules. Usually, a single species of nitrogen-fixing bacteria colonizes a single legume species, drawn to the plant's roots by chemical attractants (mainly flavonoids) that the roots secrete. By way of exchanged molecular signals, bacteria then are able to penetrate a root hair and form a colony inside the root cortex. Each cell in a root nodule may contain several thousand bacteria (now called bacteroids). The plant takes up some of the nitrogen fixed by the bacteroids, and the bacteroids use some compounds produced by the plant.

Think Critically

1. Even after microorganisms satisfy their N requirements they continue to consume nitrogen-containing organic compounds because they need the carbon in these compounds to fuel metabolism and growth.
2. Proteins often become less soluble under low pH conditions. Also, chitin has a less complex structure than protein, because it is a polymer of a single monomer, whereas protein is a complex polymer comprising many different amino acids. Thus, chitin's simpler structure may require fewer enzymes to degrade.

Test Your Knowledge

1. e, 2. c, 3. d, 4. c, 5. b, 6. c, 7. a, 8. d, 9. a, 10. e

Interpret the Data

1. Overall, aspen leaf litter contained about twice as much nitrogen but the same amount of carbon as leaf litter of pines and spruces, so with aspens the ratio of nitrogen to carbon in leaf litter was about half the value as in the other species. Lodgepole pine litter (the pine needles) was significantly higher in both cellulose and lignin than were trembling aspen or Engelmann spruce litter. The study data showed no statistically different values among the species in either fine or coarse root biomass.
2. Although we see clear species-specific differences in the values, overall the species with the lowest litter nitrogen—the pines and spruces—have significantly higher ratios of lignin to nitrogen.
3. There was no significant difference in root biomass among the three tree species in the study plots.

Chapter 34

Study Break 34.1

1. The two alternating generations of plants are the sporophyte (spore-producing) and gametophyte (gamete-producing) generations.
2. Sporophytes produce spores that give rise to gametophytes. Gametophytes then may produce gametes; male gametophytes produce sperm, and female gametophytes produce eggs. In all seed plants the sporophyte is much larger and longer-lived than the gametophyte, and the gametophyte is protected within sporophyte tissues for all or part of its life. Gametophytes also are dependent upon the sporophyte for their nutrition.

Study Break 34.2

1. Flowers are specialized for reproduction. Before an angiosperm can produce a flower, biochemical signals (triggered in part by environmental cues such as day length and temperature) travel to the apical meristem of a shoot. In response, cells there change their activity: Instead of continuing vegetative growth, the shoot is modified into a floral shoot that will give rise to floral organs.
2. Pollen grains are the mature male gametophytes. They arise by the following steps:
   a. Spores that give rise to male gametophytes are produced in a flower bud's anthers.
   b. Diploid microsporocytes inside an anther's pollen sacs undergo meiosis; eventually each one produces four small haploid microspores.
   c. Microspores then divide by mitosis.
   d. One of the two resulting nuclei divides again by mitosis, yielding a three-celled immature gametophyte: two haploid sperm cells and a third cell that will control the development of a pollen tube after pollen lands on a receptive stigma.
   e. A mature male gametophyte consists of the pollen tube and sperm cells.

3. Female gametophytes develop inside ovules in a flower's carpels. They arise by the following steps:
   a. In an ovule, a diploid megasporocyte divides by meiosis, forming four haploid megaspores.
   b. Three (usually) of these megaspores disintegrate.
   c. The remaining megaspore undergoes three rounds of mitosis without cytokinesis. The result is a single large cell with eight nuclei arranged in two groups of four.
   d. One nucleus in each group migrates to the center of the cell.
   e. After the cell undergoes cytokinesis a cell wall forms around these two polar nuclei, forming a large "central cell."
   f. A wall also forms around each of the remaining nuclei, and three of them, including an egg cell, cluster near the micropyle.
   g. The result is an embryo sac containing seven cells and eight nuclei. This sac is the mature female gametophyte.

Study Break 34.3

1. A pollen grain that lands on a compatible stigma absorbs moisture and germinates a pollen tube, which burrows through the stigma and style toward an ovule. Chemical cues from the two synergid cells help guide the pollen tube toward the egg. Before or during these events, the pollen grain's haploid sperm-producing cell divides by mitosis, forming two haploid sperm. When the pollen tube reaches the ovule, it enters through the micropyle and an opening forms in its tip. The two sperm are released into the cytoplasm of a disintegrating synergid. Next double fertilization occurs: typically, one sperm nucleus fuses with the egg to form a diploid ($2n$) zygote. The other sperm nucleus fuses with the central cell, forming a cell with a triploid ($3n$) nucleus. Tissues derived from the $3n$ cell are called endosperm.
2. As a seed matures, the embryo inside it develops a root–shoot axis with root apical meristem at one end and shoot apical meristem at the other end. Depending on the plant group, one or two cotyledons also develop. In monocots a single large cotyledon develops and stores endosperm; protective tissues arise around the root and shoot apical meristems. In eudicots, two endosperm-storing cotyledons form. Near the micropyle the radicle (embryonic root) attaches to the cotyledon at a region called the hypocotyl. Beyond the hypocotyl is the epicotyl, which has the shoot apical meristem at its tip and which often bears a cluster of tiny foliage leaves, the plumule. At germination, when the root and shoot first elongate and emerge from the seed, the cotyledons are positioned at the first stem node with the epicotyl above them and the hypocotyl below them.
3. Imbibition causes the seed coat to split, and water and oxygen move more easily into the seed. Metabolism switches into high gear as cells divide and elongate to produce the seedling. Enzymes that were synthesized before dormancy become active; other enzymes are produced as the genes encoding them begin to be expressed. The increased gene activity and enzyme production mobilize the seed's food reserves in cotyledons or endosperm. Nutrients released by the enzymes sustain the developing seedling sporophyte until its root and shoot systems are established.

Study Break 34.4

1. Flowering plants may reproduce asexually (vegetatively) by fragmentation, in which cells in a piece of the parent plant dedifferentiate and then regenerate a whole plant; by apomixis, in which a diploid embryo develops from an unfertilized egg or from diploid cells in ovule tissue; or by the production of structures such as rhizomes or suckers from a nonreproductive plant part, typically meristem tissues in a bud on a root or stem.
2. Totipotency is the capacity of fully differentiated cells to dedifferentiate, return to an unspecialized embryonic state, and then develop into a fully functional mature plant. Plant tissue culture procedures trigger the development of a mass of dedifferentiated cells (a callus), some of which regain totipotency and develop into plantlets.

Study Break 34.5

1. A homeotic gene is a regulatory gene in the genome of an organism that encodes a transcription factor. In *A. thaliana,* homeotic genes govern the development of the root and shoot tissue systems, as well as of floral organs.
2. The two basic mechanisms of plant morphogenesis are oriented cell division and cell expansion. Oriented cell division establishes the general shape of a plant organ, and cell expansion enlarges the cells in a developing organ in particular directions. They increase in circumference (girth) when new cell walls form parallel to the nearest plant surface (such as the surface of a stem or tree trunk) or when cell walls form at right angles both to the nearest surface and to the transverse plane.
3. Leaves arise through a developmental program that begins with gene-regulated activity in meristematic tissue. Hormones or other signals may arrive at target cells via the stem's vascular tissue, activating genes that regulate development. Small phloem vessels penetrate a young leaf primordium almost immediately after it begins to bulge out from the underlying meristematic tissue, followed by xylem. A growing primordium becomes cone-shaped (wider at its base than at its tip). Rapid mitosis in a particular plane in cells along the flanks of the cone (perpendicular to the surface in eudicots and parallel to the surface in monocots) produce the leaf blade that is characteristic of the particular species. Leaf tip cells typically are the first to stop dividing. By the time a leaf has expanded to its mature size, mitosis has ended and the leaf is a fully functional photosynthetic organ.

Think Critically

1. Pollen tube guidance signals are likely to be species specific. Since they are involved in mating, it is likely that these signals ensure that successful pollination, and consequently fertilization, occur only within compatible species. It is also likely that the later the guidance event, the more specific is the guidance signal. For example, the guidance signal from an ovule is likely more selective and specific than those that mediate early interactions between a pollen tube and stigma.
2. If pollen tube growth and guidance are defective, seeds are not produced. 80% of the world's staple food is derived from seeds of crop plants. So studying this process is very important agriculturally. By understanding pollen tube guidance signals, we are equipping ourselves with knowledge using which we could (i) improve seed yield, (ii) regulate interspecies hybridizations and, thereby, generate novel plant hybrids, and (iii) contain pollen spreading from genetically modified crops—a possibility that will reassure concerned public and regulatory agencies.

Test Your Knowledge

1. c, 2. a, 3. b, 4. e, 5. d, 6. c, 7. b, 8. e, 9. c, 10. c

Interpret the Data

1. The variables were the concentration of gibberellin (10 μM GA_3 vs. 1 μM GA_3) and the effect of adding ABA to the test plants' growth medium.
2. Slightly more of the mutant seeds exposed to supplementary ABA germinated, but (as evidenced by the confidence intervals) the difference was not significant.
3. First, more seeds germinated at the higher concentration of GA_3 (10 μM). Second, regardless of the GA_3 concentration, supplemental ABA made no significant difference in the seed germination rate.

Chapter 35

Study Break 35.1

1. Auxins, gibberellins, cytokinins, and brassinosteroids all promote the growth of plant parts, and ethylene stimulates cell division in seedlings. Abscisic acid is the major growth-inhibiting plant hormone.
2. Ethylene is a good example of a hormone that can stimulate or inhibit growth at various stages of the plant life cycle. In seedlings, it simultaneously slows elongation of the stem and stimulates cell divisions that increase stem girth. In mature plants of deciduous species it governs senescence (including fruit ripening) and the abscission of flowers, fruits, and leaves. Studies of brassinosteroids have revealed that this family of steroid hormones has different effects in different tissues—for example, promoting the elongation of vascular tissue and pollen tubes, but inhibiting elongation in roots.

Study Break 35.2

1. General plant responses to attack include mobilization of jasmonates and salicylic acid, systemin (in tomato), the hypersensitive response, PR proteins, and secondary metabolites (phytoalexins). Gene-for-gene recognition is a pathogen-specific response.
2. Salicylic acid (SA) is considered to be a general systemic response to damage because experiments show that when a plant is wounded, soon thereafter SA can be detected in a variety of its tissues.
3. While the hypersensitive response is underway, SA also is synthesized and operates in other defensive chemical pathways in a plant. This effect includes the synthesis of PR (pathogenesis-related) proteins that attack pathogenic cells.

Study Break 35.3

1. Directional light of blue wavelengths is the direct stimulus for phototropism. The most widely accepted scientific explanation for gravitropism is the sinking of amyloplasts in cells surrounding vascular bundles in response to gravity. Sinking amyloplasts may provide a mechanical stimulus that triggers a gene-guided redistribution of IAA. The changing auxin gradient in turn adjusts a plant's growth pattern.
2. Unlike tropisms, nastic movements occur in response to nondirectional stimuli, such as mechanical pressure resulting from an insect brushing against hairlike sensory structures in the leaves of a Venus flytrap plant.

Study Break 35.4

1. Plant responses to changes in photoperiod rely on different chemical forms of the blue-green pigment phytochrome. Daylight converts the inactive phytochrome (P_r) to an active form, (P_{fr}). When light levels fall, P_{fr} reverts to P_r. This switching mechanism helps regulate light-related processes such as photosynthesis.
2. Photoperiod length is a factor in the seasonal flowering of many angiosperms. Spinach and irises are examples of long-day plants, which flower in spring when the period of daylight extends at least 9–16 hours. Chrysanthemums and potatoes are examples of short-day plants, which flower as day-length becomes shorter than some critical period, a condition that occurs naturally in the fall.
3. Dormancy is an adaptive response because it attunes a plant's growth to the most favorable environmental conditions for survival.

Study Break 35.5

1. Some response pathways may directly stimulate gene transcription whereas others might inhibit gene expression. Some pathways have an intermediary step, in which the original signal triggers the synthesis of regulatory proteins that in turn promote or inhibit the expression of still other genes.
2. Second messengers boost cellular responses to a hormonal signal by setting in motion a cascade of activated protein kinases, which in turn activate cell proteins that carry out the cell's response.
3. Although Chapter 7 focuses on signal transduction pathways in animals, in fact the basics of some transduction pathways are similar in animals and plants. In both groups we see examples of one-step signal transduction (in which a signal exerts its effect by binding a cell receptor) and of signaling cascades such as second messenger systems.

Think Critically

1. The genome reorganization occurs in a specific subset of the genome and reproducibly occurs under the same stress conditions. Therefore, it is not random, but the mechanism by which these regions of the genome are recognized is currently unknown. Most mutation systems are assumed to be random within the genome although there is evidence that transposons do have preferential sites.
2. The data indicate that the environment can direct higher rates of mutations to particular regions of the genome and, therefore, mutating and selecting plants simultaneously might provide a better frequency of adaptive mutations. Because plants are sessile, do they need mechanisms that enable the generation of diversity in the face of chronic stress growth conditions for evolution? If so, partitioning of the plant genome into regions that are protected from change (because they are essential) and variable regions (in which changes have phenotypic effects but are not lethal) is important to identify. Using the information might permit the more rapid breeding of stress tolerant crops.

Test Your Knowledge

1. e, 2. a, 3. c, 4. b, 5. b, 6. b,c, 7. a, 8. e, 9. c, 10. e

Interpret the Data

1. Overall, many more *Yucca* genes are up-regulated (expression enhanced) in response to BR and auxin than are down-regulated (expression suppressed).
2. The brassinosteroid (BR) down-regulated many more *Yucca* genes than did auxin. Also, a small number of genes that are up-regulated by auxin are down-regulated by BR. In an even smaller number of genes the reverse was true—that is, the genes were up-regulated by BR but down-regulated by auxin.
3. There is no overlap between the "up" and "down" bubbles of each hormone because a given hormone does not have contrasting effects on the same gene. Thus, with respect to each hormone there are two distinct groups of *Yucca* genes—those the hormone up-regulates and those it down-regulates.

Chapter 36

Study Break 36.1

1. Multicellularity made it possible for animals to create an internal fluid environment for fulfilling the nutrient supply, waste removal, and osmotic balance needs of individual cells. As a result, multicellular organisms could evolve to occupy a variety of habitats, including dry terrestrial environments, in which single cells cannot survive. Multicellularity also allowed major life functions to be distributed among specialized groups of cells, with each group having a single activity. The specialized groups of cells are typically organized into tissues, the tissues into organs, and the organs into organ systems.
2. A tissue is a group of cells with the same structure and function. Cells in the tissue work together to perform one or more activities. An organ integrates two or more different tissues into a structure that carries out a specific function. An organ system coordinates the activities of two or more organs to carry out a major body function.

Study Break 36.2

1. Exocrine and endocrine glands are formed by epithelia. An exocrine gland remains connected to the epithelium by a duct, whereas an endocrine gland is suspended in connective tissue underlying the epithelium, with no ducts leading to the epithelial surface.
2. Loose connective tissue, fibrous connective tissue, cartilage, bone, adipose tissue, and blood.
3. Skeletal, cardiac, and smooth.

Study Break 36.3

See Figure 36.10.

Study Break 36.4

A stimulus—a change in the external or internal environment—starts the homeostatic control system. The stimulus is detected by a *sensor*, an *integrator* compares the environmental change with a set point, and the *effector* or effectors become activated by the integrator and function to return the environmental parameter to the set point.

Think Critically

1. Some organisms needed to anticipate light in order to harvest its energy, whereas other organisms needed to avoid its damaging ultraviolet radiation. Thus, anticipation, via circadian rhythm, was and is crucial for survival.
2. Though this remains a question that will take years to answer, a circadian clock within blood vessels and their cells may help to adapt to changing temporal patterns of blood flow and pressure. Thus, it may be important for blood vessels to keep their own time, when or if the brain clock loses its own track of time.

Test Your Knowledge

1. b, 2. a, 3. e, 4. c, 5. b, 6. a, 7. c, 8. e, 9. e, 10. d

Interpret the Data

Yes, these data are consistent with the hypothesis. Taurine is known to inhibit vasopressin. The figure shows a method by which taurine could act as part of a negative feedback loop. Taurine release is directly stimulated by vasopressin; therefore, its levels are directly increased by vasopressin, which could lead to negative feedback of vasopressin release.

Chapter 37

Study Break 37.1

1. A dendrite receives signals and conducts them toward the cell body. An axon conducts signals away from the cell body toward another neuron or effector.
2. An afferent neuron conducts information from its sensory receptors to interneurons. Efferent neurons conduct signals from interneuron networks to effectors, the muscles and glands that carry out the response. The afferent and efferent neurons constitute the peripheral nervous system (PNS). Interneurons process information from afferent neurons and send a response to the efferent neurons. Interneurons form the brain and spinal cord, the central nervous system (CNS).
3. In an electrical synapse, the plasma membrane of the axon terminal of the presynaptic cell is in direct contact with the postsynaptic cell, allowing ions to pass directly between the cells when an electrical impulse arrives. In a chemical synapse, the presynaptic and postsynaptic cells are separated by a small gap. Neurotransmitters released from the presynaptic cell diffuse across the gap and bind to receptors in the plasma membrane of the postsynaptic cell. If enough neurotransmitter molecules bind to those receptors, the postsynaptic cell generates a new electrical impulse, which travels along its axon to the next neuron or effector in the circuit.

Study Break 37.2

1. At the peak of an action potential, the plasma membrane of the neuron enters a short refrac-

tory period, in which the threshold for generation of an action potential is much higher than normal. Only the region in front of the action potential can fire, meaning that the impulse can only move in one direction, that is, toward the axon tip. The refractory period remains in effect until the membrane again reaches the resting potential. By that time, the action potential has moved too far away to cause a second action potential in the same region.

2. Neurons insulated with myelin sheaths have gaps in the sheaths called nodes of Ranvier, where the axon membrane is exposed to extracellular fluids. The inward movement of sodium ions at a node produces depolarization and an action potential, but the adjacent myelin sheath prevents the excess positive ions from exiting through the membrane. Instead, they diffuse rapidly to the next node where they cause depolarization, inducing an action potential there. Continuation of this process allows the action potential to jump rapidly along the axon from node to node, at a faster rate than a nonmyelinated neuron of the same diameter.

Study Break 37.3

1. A neurotransmitter is synthesized in a neuron, is released into the synaptic cleft from a presynaptic axon terminal, and binds to receptors in the plasma membrane of the postsynaptic cell. Depending on the type of receptor, a neurotransmitter either stimulates or inhibits the generation of action potentials.
2. A neurotransmitter that binds to an ionotropic neurotransmitter receptor, like all neurotransmitters, is stored in synaptic vesicles in the cytoplasm at an axon terminal. When an action potential arrives at the terminal, the change in membrane potential opens voltage-gated Ca^{2+} channels in the axon terminal, allowing Ca^{2+} to flow back into the cytoplasm. The rise in Ca^{2+} concentration triggers the release of the neurotransmitters into the synaptic cleft by exocytosis. Neurotransmitters that bind to an ionotropic neurotransmitter receptor diffuse across the synaptic cleft and open or close ligand-gated ion channels in the postsynaptic neuron's membrane. Most of the channels regulate Na^+ or K^+ movement through the membrane, although some regulate Cl^-. Depending on the ion flow, action potentials are either stimulated or inhibited in the postsynaptic neuron.

Study Break 37.4

One way a postsynaptic neuron integrates signals is through summation of EPSPs and IPSPs that alter the neuron's membrane potential. EPSPs move the membrane potential toward the threshold for an action potential, while IPSPs move the membrane potential away from the threshold for an action potential. The final change in membrane potential depends on the particular array of EPSPs and IPSPs received. The patterns of synaptic connections made by a neuron also contribute to integration.

Think Critically

If we understood the biological nature of conscious thought, then perhaps we could design a more rational judicial system, for example.

Test Your Knowledge

1. d, 2. b, 3. a, 4. c, 5. e, 6. a, 7. c, 8. b, 9. e, 10. a

Interpret the Data

1. Cyanide inhibits the transport of Na^+ out of the squid axon, as evidenced by the decrease in the rate of ^{22}Na lost from the axon.
2. The addition of broken-down ATP has no effect on the cyanide-caused inhibition of Na^+ transport out of the axon.
3. The addition of ATP increases the rate of ^{22}Na transport out of the axon, somewhat reversing the effects of the cyanide. These results mean that cyanide likely is poisoning the Na^+/K^+ active transport pump, which depends upon ATP hydrolysis for its action. An injection of ATP overcomes the cyanide inhibition until that ATP is used up, and then the curve shows a decrease again in ^{22}Na transport out of the axon (at around the 3 hour point).

Chapter 38

Study Break 38.1

1. A nerve net is a loose meshwork of neurons organized in a radial pattern that reflects the radial symmetry of the animal in which it is found. Nerves are bundles of axons surrounded by connective tissue. Nerve cords are bundles of nerves.
2. Cephalization is the formation of a distinct head region containing ganglia, which form a major central control center or brain, and major sensory structures. Cephalization is found in more complex invertebrates and in all vertebrates.

Study Break 38.2

(a) Sympathetic nervous system; this is the classic "fight or flight" scenario for this autonomic nervous system division. (b) Parasympathetic nervous system.

Study Break 38.3

1. The blood–brain barrier is a distinct separation between the blood and the brain resulting from tight junctions between the epithelial cells forming the walls of the capillaries in the brain. The blood–brain barrier regulates exchanges between blood and brain, in particular preventing most dissolved substances from entering the cerebrospinal fluid, thereby protecting the brain and spinal cord from viruses, bacteria, and toxic substances that may be in the blood.
2. Gray matter consists of nerve cell bodies and dendrites; white matter consists of axons, many of them surrounded by myelin sheaths.
3. The brain stem regulates a number of the body's vital housekeeping functions without conscious involvement or control by the cerebrum, including heart and respiration rates, blood pressure, constriction and dilation of blood vessels, coughing, and reflex activities of the digestive system.
4. The cerebellum is an outgrowth of the pons, but it is structurally and functionally separate from the brain stem. The cerebellum has extensive connections with other parts of the brain, through which it receives sensory inputs from receptors in muscles and joints, from balance receptors in the inner ear, and from the receptors of touch, vision, and hearing. The cerebellum integrates the various sensory signals and compares them with signals from the cerebrum that control voluntary body movements. Information flow from the cerebellum to the cerebrum, brain stem, and spinal cord modifies and fine-tunes movements of the body. In humans, the cerebellum also is involved in the learning and memory of motor skills.

   The cerebral cortex is a thin, folded layer of gray matter that forms the surface of the cerebrum. The cerebral cortex contains sensory areas that receive and integrate sensory information of many kinds, including touch, pain, temperature, pressure, hearing, vision, smell, and taste. It also contains motor areas that are involved in controlling body movements and position, and association areas, which integrate information from the sensory areas and send responses to the motor area. Most higher functions of the human brain, including critical thinking, abstract thought, musical ability, and aspects of personality, involve activities of many regions of the cerebral cortex.

Study Break 38.4

Short-term memory involves transient changes in neurons. Short-term memory loss resulting from aging may be caused by a loss of control of the mechanisms involved, or, more likely, by a loss or degeneration of the neurons that constitute the short-term memory system.

Learning first involves storing memories. If the short-term memory system is faulty, perhaps due to loss or degeneration of neurons, then the transfer of short-term memories to the long-term memory system will be impaired. Another possibility is a loss or degeneration of the neurons responsible for long-term memory.

Think Critically

A good guess is that the generalist will show better learning. The reason is that it normally must find food across a wide variety of environmental conditions that might vary in unpredictability. Therefore, learning ability would be highly adaptive. The specialist could adapt quite well by "hard wiring" an attraction to all the specific chemical cues that are normally associated with its single food source. Of course, only an experimental test would verify this hypothesis.

Test Your Knowledge

1. d, 2. b, 3. b, 4. b, 5. e, 6. c, 7. a, 8. c, 9. c, 10. a

Interpret the Data

1. In the first 5-minute interval, the control rats were more active than the rats exposed prenatally to MDMA.
2. 76 photobeam breaks
3. The rats exposed to MDMA moved most during the last 5-minute interval.
4. Yes, because the two groups of rats had different patterns of movement. The control rats began to move around the cage almost immediately. The MDMA rats were less active to begin with but continued moving long after the control rats had settled down.

Chapter 39

Study Break 39.1

1. Sensory transduction is the conversion of a stimulus into a change in membrane potential.
2. The signals from particular sensory receptors are routed by afferent neurons to specific regions of the CNS. Processing of the incoming

signals by those regions gives the "sense" of the stimulus—a smell, pain, and so forth.

Study Break 39.2

1. Proprioceptors detect stimuli that are processed to provide the animal with information about movements and position of the body.
2. All proprioceptors are mechanoreceptors because they are stimulated by a mechanical force.

Study Break 39.3

1. Cephalopods have mechanoreceptors on their head and tentacles, similar to the lateral line of fishes. These mechanoreceptors detect vibrations in the water. Many insects have sensory receptors in the form of hairs or bristles that vibrate in response to sound waves, while some insects (for example, moths, grasshoppers, and crickets) have auditory organs on either side of the abdomen or on the first pair of walking legs.
2. In humans, vibrations representing sound frequencies are transmitted into the fluid-filled inner ear. They travel through the inner ear and cause the basilar membrane to vibrate in response, bending the sensory hair cells and stimulating them to release a neurotransmitter that triggers action potentials in afferent neurons leading from the inner ear. The vibrations from a particular sound frequency cause the basilar membrane to vibrate maximally at one particular location, stimulating the hair cells in that region to initiate action potentials. That information is sent to the brain, which integrates it into a perception of the sound stimulus.

Study Break 39.4

(a) A photopigment is an association of retinal with one of several different opsin proteins. (b) A photoreceptor is a receptor specialized for detection of colors (different wavelengths of light). (c) A ganglion cell's receptive field is the specific set of photoreceptors that sends signals to that cell. Receptive fields are usually circular and vary in size; the smaller the receptive field, the more precise the information sent to the brain, and the sharper the image.

Study Break 39.5

1. Most odor perceptions arise from a combination of different olfactory receptors, which are located in the nasal cavities in humans. We have about 1,000 different olfactory receptor types, each of which responds to a different class of chemicals. The stimulated olfactory receptors send signals via the olfactory bulbs to the olfactory centers of the cerebral cortex, where the signals are interpreted as particular smells.
2. Chemicals bind to taste receptors and generate signals. Signals from taste receptors are relayed to the thalamus. From there, some signals go to gustatory centers in the cerebral cortex where they are integrated to produce taste perception. Other signals go to the brain stem and limbic system, which links tastes to involuntary visceral and emotional responses, such as sensations of pleasure or revulsion.

Study Break 39.6

The other sensory receptors involve specialized receptor structures. Afferent neurons synapse with the receptors. When a receptor is stimulated, the change in membrane potential (sensory transduction) is transmitted to the afferent neuron, which transmits the signal to the interneuron networks of the CNS. By contrast, both thermoreceptors and nociceptors consist of free nerve endings formed by the dendrites of afferent neurons, with no specialized receptor structures involved.

Study Break 39.7

Electroreceptors are used in aquatic vertebrates for electrolocation (locating other animals such as prey), electrocommunication (communicating with other members of the same species), and killing prey (involving high-voltage discharge).

Think Critically

1. Photoreceptors (they will clearly "detect" that the color of the eggs is odd); chemoreceptors (while the taste of the eggs is not altered, the person eating the green eggs may make a facial expression that leads you to believe that he or she has detected an altered taste). Combining an MRI and a PET scan might be a good method for testing synesthesia.
2. Detection and response to light are logically a key sensory system for any living organism—from prokaryotes to plants to animals, but sensing light with photoreceptors does not to have to involve "vision." Yet, the genes for photoreceptor proteins would logically be of early evolutionary origin.

Test Your Knowledge

1. c, 2. a, 3. c, 4. b, 5. e, 6. b, 7. d, 8. a, 9. e, 10. d

Interpret the Data

1. Most of the bird pairs support the hypothesis, with k, l, and m providing the strongest data.
2. Pairs d and h show little difference between species, but pair g does not fit the hypothesis.
3. Overall the data appear to support the hypothesis.

Chapter 40

Study Break 40.1

1. A *hormone* is a signaling molecule secreted by one group of cells and transported through the circulatory system to other, target cells, whose activities they change. Specific target cells react to the hormone because they carry receptors for the hormone. The best-known hormones are secreted by cells of the endocrine system and elicit a response in target cells that have receptors for the hormone. An *exocrine secretion* is a molecule released from an exocrine gland. Such molecules are secreted into ducts that lead outside the body or into the cavities of the digestive tract.
2. In *endocrine signaling,* a hormone secreted by an endocrine gland into the ECF is transported in the blood and causes a response in a target cell that has a receptor for the hormone. In *autocrine signaling,* a local regulatory molecule is released from a cell and binds to cell surface receptors on the same cells that produce the molecule, triggering a response.

Study Break 40.2

1. Glucagon, a peptide hormone, triggers a response by binding to a surface receptor. Glucagon binding activates the receptor, which triggers a signal transduction pathway inside the cell, leading to phosphorylation of target proteins. The altered activities of the phosphorylated proteins produce responses in the target cells, in this case the breakdown of glycogen in liver cells to glucose.

   Aldosterone, a steroid hormone, passes though the plasma membrane and binds to an internal receptor in the cytoplasm or nucleus, activating it. The hormone-activated receptor complex binds to control sequences in the DNA that the receptor recognizes and either activates or inhibits transcription of the associated target genes. Through binding to a receptor, the hormone affects transcription of specific genes in target cells.
2. A target cell could respond to different hormones if it carries receptors for those different hormones. Turning this around, a target cell will respond only to one hormone if it just has the receptor for that hormone. The same hormone could produce different effects in different cells if there are different receptors for that hormone, each triggering a distinct response pathway.

Study Break 40.3

1. The hypothalamus and anterior pituitary gland are connected by nerve and vascular tissues. The hypothalamus releases peptide neurosecretory hormones into the linking blood vessels. These hormones regulate the secretion of peptide hormones by the anterior pituitary gland. The pituitary hormones regulate several key body systems.
2. Secretion of hormones from the anterior pituitary is controlled by releasing hormones (RHs) and inhibiting hormones (IHs), which are released by the hypothalamus. For the posterior pituitary, neurosecretory neurons directly secrete the hormones into the circulatory system.

Study Break 40.4

1. Parathyroid hormone stimulates the dissolution of calcium and phosphate ions from bone and their release into the bloodstream.
2. The adrenal medulla secretes epinephrine and norepinephrine. Epinephrine prepares the body for handling stress or physical activity by, among other actions, (1) increasing heart rate; (2) breaking down glycogen and fats, thereby releasing glucose and fatty acids into the blood for fuel; (3) dilating blood vessels in the heart, skeletal muscles, and lungs to increase blood flow; (4) constricting blood vessels elsewhere, thereby raising blood pressure, reducing blood flow to the intestine and kidneys, and inhibiting smooth muscle contraction, which reduces water loss and slows down the digestive system; and (5) dilating airways in the lungs, thereby increasing air flow.

   Norepinephrine has similar effects to epinephrine on heart rate, blood pressure, and blood flow to the heart muscle. In contrast to epinephrine, norepinephrine causes blood vessels in skeletal muscles to contract.
3. Glucocorticoids and mineralocorticoids are the two major classes of steroid hormones secreted by the adrenal cortex. Glucocorticoids help maintain the concentration of glucose and other fuel molecules in the blood, and mineralocorticoids regulate the levels of Na^+ and K^+ in blood and the ECF.
4. Estradiol is an estrogen and progesterone is a progestin; both are steroid hormones. Estradiol produced by the ovaries stimulates maturation of sex organs at puberty and development of secondary sexual characteristics. Progesterone, also produced by the ovaries, prepares and

maintains the uterus for implantation of a fertilized egg and the growth and development of an embryo.

Study Break 40.5

In general, invertebrates have fewer hormones regulating fewer body processes and responses than vertebrates do. Peptide and steroid hormones are produced in both invertebrates and vertebrates. However, most of those hormones are different in structure and molecular function in the two groups, even though the reaction pathways stimulated by the hormones are the same.

Think Critically

1. Not all obese people have type 2 diabetes, and not all people with type 2 diabetes are genetically predisposed. Note that being prediabetic often leads to obesity and development of type 2 diabetes.
2. Possibilities are skeletal, cardiac, or smooth muscle. Skeletal muscle would be least invasive to assay and probably most responsive to six weeks of exercise. Skeletal muscle is also the most abundant type of muscle tissue.

Test Your Knowledge

1. b, 2. d, 3. a, 4. c, 5. c, 6. e, 7. a, 8. b, 9. d, 10. b

Interpret the Data

1. The level of growth hormone was not constant over the 24-hour period. The level of this hormone increased markedly immediately after the onset of sleep and then declined in the middle of the sleep period. Another increase in growth hormone level occurred upon waking and then the level dropped to a level similar to that of presleep.
2. The level of prolactin also is not constant over the 24-hour period. The fluctuations in prolactin levels in the period before sleep are similar to those seen in that period for growth hormone. The level of prolactin increased in the sleep period, showing a rise immediately after sleep onset and continuing to a maximum level at the time of waking. The level of prolactin then decreased to a level similar to that of presleep.
3. Mostly there is no consistent relationship between the levels of the two hormones over the 24-hour period. The only exception is the initial rise in prolactin and in growth hormone that occurred with sleep onset.
4. Yes. The correlation between high level of prolactin with sleep and low level of prolactin with wakefulness supports the hypothesis. However, it cannot be determined from the data if the sleep-related rise in prolactin is directly dependent on sleep, or whether other factors triggered by sleep are involved.

Chapter 41

Study Break 41.1

1. An axon terminal of an efferent neuron makes a synapse called a neuromuscular junction with a muscle fiber. When an action potential arrives at that junction, the axon terminal releases acetylcholine, which triggers an action potential in the muscle fiber that moves in all directions over its surface and penetrates to the interior of the fiber through T tubules. When the action potential reaches the end of the T tubules, ion channels are opened in the sarcoplasmic reticulum that allow Ca^{2+} ions to flow from the sarcoplasmic reticulum into the cytosol. Troponin then binds the ion and undergoes a conformational change that allows tropomyosin to enter the grooves in the actin helix of the thin filaments. As a result, the myosin-binding sites are uncovered, and the crossbridge cycle is turned on, leading to muscle contraction.
2. In the sliding filament mechanism of muscle contraction, the thin filaments on each side of a sarcomere slide over the thick filaments toward the center of the A band, thereby bringing the Z lines closer together. A crossbridge cycle is responsible for the contraction. At the beginning of the cycle, the myosin crossbridge has an ATP bound to it and is not in contact with the actin of the thin filament. The ATP is hydrolyzed, causing the myosin crossbridge to bend away from the tail and bind to a myosin-binding site on the thin filament that was uncovered in response to the release of Ca^{2+} ions into the cytosol from the sarcoplasmic reticulum. When the myosin crossbridge binds to the actin, the crossbridge snaps back toward the myosin tail to produce the power stroke that pulls the thin filament over the thick filament. ADP and phosphate are released from the crossbridge in this step. A new molecule of ATP now binds to the crossbridge, causing the myosin to detach from the actin. The cycle then repeats.
3. Contraction intensity: low in slow aerobic fibers, intermediate in fast aerobic fibers, and high in fast anaerobic fibers.

   Fatigue resistance: high in slow aerobic fibers, intermediate in fast aerobic fibers, and low in fast anaerobic fibers.

   Oxidative phosphorylation capacity: high in slow aerobic fibers, high in fast aerobic fibers, and low in fast anaerobic fibers.

   Number of mitochondria: many in slow aerobic fibers, many in fast aerobic fibers, and few in fast anaerobic fibers.

   Myoglobin content: high in slow aerobic, high in fast aerobic, and low in fast anaerobic.

Study Break 41.2

1. A hydrostatic skeleton provides support to the body or body part through muscles acting on compartments filled with fluid under pressure. In vertebrates, the penis is a hydrostatic skeletal structure.

   An exoskeleton is a rigid external body covering. The force of muscle contraction against the covering provides support to the body or part of the body. In vertebrates, the shell of a turtle or tortoise and the bony plates in the skin, the abdominal ribs, the collarbones, and most of the bony skull of the American alligator are exoskeletal structures.

   An endoskeleton consists of internal body structures such as bones. The force of muscle contraction against the internal body structures provides support. In vertebrates, the endoskeleton is the primary skeletal system.
2. The bones of the vertebrate endoskeleton provide support for the body and body parts, protect key internal organs, store calcium and phosphate ions, and are the sites where new blood cells form.

Study Break 41.3

1. Synovial joints consist of the ends of two bones that are enclosed by a fluid-filled capsule of connective tissue. Fluid within the capsule and a smooth layer of cartilage over the ends of the bones enable the bones to slide easily as the joint moves. In these joints, ligaments extend across the joints across the capsule. The ligaments confine the motion of the joint and protect it to an extent from deleterious effects of heavy loads.

   Cartilaginous joints have no fluid-filled capsule surrounding them, and the bones involved are covered with cartilage. The bones of the joint are covered by and connected with fibrous connective tissue. Cartilaginous joints are less movable than synovial joints.
2. Antagonistic muscle pairs are muscles arranged so that bones can be extended, flexed, or rotated in opposite directions around a joint. For example, in humans, the biceps and triceps muscles are an antagonistic muscle pair.

Think Critically

Skeletal muscle consumes metabolites, including fatty acids and carbohydrates, both for movement and for building new muscle. Thus, maintaining and using new muscle could increase fat metabolism, its breakdown and oxidation for energy, and also divert carbohydrates, which are often converted to fat, away from fat cells and to muscle cells.

Test Your Knowledge

1. c, 2. e, 3. a, 4. d, 5. b, 6. b, 7. c, 8. a, 9. d, 10. b

Interpret the Data

1. Seven out of the nine treated children showed an increase in vertebral area during the study.
2. None of the six untreated control children showed an increase in vertebral area.
3. The treated children experienced a mean of 2.6 fractures per year, less than half of the mean of 6.3 fractures per year in the untreated children.
4. Yes

Chapter 42

Study Break 42.1

1. The three basic features of animal circulatory systems are (1) a specialized fluid medium for transporting molecules, exemplified by the blood of vertebrates; (2) a muscular heart for pumping the fluid; and (3) tubular vessels for distributing the fluid pumped by the heart.
2. In an open circulatory system, there is no distinction between blood and interstitial fluid. Vessels from the heart release hemolymph directly into body spaces and the fluid is subsequently collected and reenters the heart. In a closed circulatory system, blood is channeled in blood vessels leading to and from the heart and is distinct from the interstitial fluid.

   In an open circulatory system, most of the fluid pressure generated by the heart dissipates when the blood is released into the body spaces. Consequently, blood flows relatively slowly. Humans could not function as we do with such a system, because we would not be able to distribute oxygen efficiently throughout the body; nor would we be able to eliminate the wastes we produce.

3. Atria are chambers of the heart that receive blood returning to the heart. Ventricles are chambers of the heart that pump blood from the heart. Arteries are vessels of circulatory systems that conduct blood away from the heart at relatively high pressure. Veins are vessels of circulatory systems that carry blood back to the heart.
4. The systemic circuit of the circulatory system is the circuit from the heart to most of the tissues and cells of the body and back to the heart. The pulmocutaneous circuit goes from the heart to the skin and lungs or gills in amphibians and back to the heart. The pulmonary circuit goes from the heart to the lungs and back to the heart.

Study Break 42.2

1. Erythrocytes originate as pluripotent stem cells in the red bone marrow. In humans and other mammals, they lose their nucleus, cytoplasmic organelles, and ribosomes as they mature, becoming essentially a membrane-bound hemoglobin reservoir that is not capable of protein synthesis. At the end of their life span—about 4 months—erythrocytes are engulfed by macrophages, a type of leukocyte, in the spleen, liver, and bone marrow.
2. Low oxygen content triggers a negative feedback mechanism to increase erythrocytes in the blood. The kidneys are stimulated to synthesize the hormone erythropoietin (EPO), which stimulates stem cells in the bone marrow to increase erythrocyte production. EPO synthesis is stopped when the oxygen content of the blood rises above normal levels.
3. Leukocytes act as the first line of defense against invading organisms, eliminate dead and dying cells from the body, and remove cellular debris. Platelets assist in blood clotting. When blood vessels are damaged, platelets stick to the collagen fibers exposed to leaking blood and recruit other platelets to the site. Eventually a plug forms at the site, sealing off the damaged area.

Study Break 42.3

1. The right atrium receives blood in the systemic circuit returning to the heart in vessels coming from the entire body, with the exception of the lungs. The right atrium pumps blood into the right ventricle.

   The right ventricle receives blood from the right atrium, and pumps it into the pulmonary arteries going to the lungs, beginning the pulmonary circuit.

   The left atrium receives blood returning from the lungs in the pulmonary veins, completing the pulmonary circuit. The left atrium pumps this blood into the left ventricle.

   The left ventricle pumps the blood received from the left atrium into the aorta where the blood begins its path in the systemic circuit.
2. Systole is the period of contraction and emptying of the heart. Diastole is the period of relaxation and filling of the heart between contractions.
3. Neurogenic hearts beat under the control of signals from the nervous system. If the signals cease, this type of heart stops beating. Myogenic hearts maintain a contraction rhythm without signals from the nervous system. In the event of a serious trauma to the nervous system, this type of heart keeps beating.
4. Pacemaker cells of the SA node undergo regularly timed depolarizations, which initiate waves of contraction that travel over the heart. The waves stop before they reach the bottom part of the heart as they encounter an insulating layer of connective tissue. The contraction signals at this point excite cells of the AV node, and then pass to the bottom of the heart along Purkinje fibers. There they induce waves of contraction that begin at the bottom of the heart and move upwards, expelling blood from the ventricles into the aorta and pulmonary arteries.

Study Break 42.4

1. Blood flow through capillary networks is controlled by contraction of smooth muscles in arterioles and by contraction of precapillary sphincters at the junctions of capillaries and arterioles.
2. In most body tissues, there are small spaces between capillary endothelial cells, but in the brain the capillary endothelial cells are tightly sealed together, forming a blood–brain barrier.
3. The two major mechanisms are diffusion along concentration gradients and bulk flow. Diffusion along concentration gradients occurs both through the spaces between the capillary endothelial cells and through the plasma membranes. The direction of movement of the molecule or ion depends on the concentration gradient. Thus, oxygen and glucose, which are at higher concentrations in the capillaries, diffuse into the interstitial fluid and then into the cells of the tissues. Carbon dioxide, which is at a higher concentration in the interstitial fluid, diffuses into the capillaries.

   Bulk flow carries water, ions, and molecules out of the capillaries. Driven by the pressure of the blood, which is higher than the pressure of the interstitial fluid, bulk flow occurs through the spaces between capillary endothelial cells.

Study Break 42.5

1. Arterial blood pressure must be regulated within limits to provide sufficient blood flow for the brain and other tissues, and to prevent damage to blood vessels, tissues, and organs that would occur at high blood pressures.
2. The three main mechanisms for regulating blood pressure are: (1) controlling cardiac output (the pressure and amount of blood pumped by the ventricles); (2) controlling the degree of constriction of the blood vessels (mostly the arterioles); and (3) controlling the total blood volume.
3. The hormone epinephrine raises blood pressure by increasing the strength and rate of the heart rate, and by stimulating vasoconstriction of arterioles in certain parts of the body.

Study Break 42.6

Lymph is interstitial fluid, an aqueous solution containing molecules, ions, infecting bacteria, damaged cells, cellular debris, and lymphocytes. Because the lymph capillaries consist of a single layer of endothelial cells, the lymphatic fluid can enter them at sites where the endothelial cells overlap when the pressure of the interstitial fluid forces the flaps open. The openings produced are large enough for cells to enter.

Think Critically

1. Probably not. The most common form of hemophilia is a result of a mutation in clotting factor VIII (see Section 42.2), and lampreys lack that clotting factor.
2. When a blood vessel is injured during surgery or extended bed rest, the body uses platelets and fibrin to form a blood clot. The coagulation process is similar in lampreys and mammals, so the molecular basis of drugs used to prevent DVT could be elucidated by using DVT preventive drugs on lampreys.

Test Your Knowledge

1. d, 2. e, 3. b, 4. b, 5. c, 6. a, 7. e, 8. e, 9. d, 10. a

Interpret the Data

1. No. Nifedipine reduces blood pressure and heart rate prior to noticeable increases in plasma nifedipine (below 10 ng/mL).
2. Both capsules reduce blood pressure, but the GITS capsule seems to also reduce heart rate. Since reducing blood pressure or heart rate will decrease the workload on the heart, both will achieve this goal; however, GITS might be more effective.
3. Contracten XL causes greater elevation of nifedipine, which doesn't appear to be necessary for the beneficial effects. Based on these data alone, GITS capsules would be better at reducing workload with fewer side-effects.

Chapter 43

Study Break 43.1

1. Mucous layers are physical barriers to pathogens. Mucous layers may contain toxins and other chemicals that kill pathogens. The aqueous environment in contact with an epithelial surface may inhibit or kill pathogens by being acidic, or by containing enzymes or bile juices.
2. Innate immunity provides a nonspecific, immediate response to a pathogen. It is the first response system that comes into play when a pathogen is encountered, but it retains no memory of the encounter. The innate immune responses include inflammation and specialized cells that attack pathogens.

   Adaptive immunity, by contrast, provides a specific response to a pathogen, and it retains a memory of exposure to the pathogen so that it can respond more quickly to future attacks. The adaptive immune response takes several days to become protective, in contrast to minutes for an innate immune response.

Study Break 43.2

1. The inflammatory response is characterized by heat, pain, redness, and swelling at the infection site.
2. Inflammation-mediating molecules released at the infection site cause the dilation of local blood vessels and increase their permeability. As a result, blood flow increases and fluid leaks from the blood vessels. Heat, redness, and swelling are direct consequences of these effects.

   Pain is caused by the migration of macrophages and neutrophils to the infection site, and their activities at the site.
3. The complement system is a nonspecific defense mechanism in which more than 30 interacting plasma proteins are activated by molecules on the surface of pathogens. Activated complement system proteins produce membrane attack complexes that insert into the

plasma membrane of many types of bacterial cells and coat other types of cells with fragments of the complement proteins. Membrane attack complexes create pores that lead to loss of osmotic balance, which causes the pathogens to lyse. Pathogens coated with complement protein fragments are recognized by phagocytes that then engulf and destroy the pathogens.

4. Viral pathogens do not have surface molecules that are distinct from those of the host. The main strategies used by the host in innate immunity are interferon and natural killer cells.

Study Break 43.3

1. The antibody-mediated immune response uses antibodies secreted by plasma cells (differentiated from activated B cells) to target antigens in various body fluids. The cell-mediated immune response uses cytotoxic T cells to target and kill cells infected with a pathogen.
2. An antibody is a member of the immunoglobulin (Ig) family of proteins. It consists of four polypeptides, two of which are identical heavy chains and two of which are identical light chains. Structurally, an antibody looks like a Y. Each of the two arms of the Y involves pairing of a light chain with part of the heavy chain. The tail of the Y consists of the remaining parts of the heavy chains paired together. Each heavy chain and each light chain has a variable region and a constant region. The variable regions, which occur at the top half of the arms of the Y, form two identical binding sites for the antigen with which the antibody can react.
3. DNA rearrangements of genes during differentiation are responsible for generating diverse light-chain genes and heavy-chain genes for the B-cell receptor and for the T-cell receptor. In brief, there are many different V DNA segments, one C DNA segment, and a few different J (joining) DNA segments. During differentiation, a light-chain gene is assembled from one V segment, one J segment, and the C segment, creating a gene that can be transcribed. Transcription of the gene produces a pre-mRNA that is spliced to remove introns and generate the translatable mRNA. The various combinations of segments plus imprecision in the joining at the DNA level of the V and J segments produces many types of light-chain genes. A similar DNA rearrangement process occurs for heavy-chain genes. Together, this results in tremendous variability in the antibody molecules that can be made.
4. Clonal selection is the process by which an antigen stimulates the production of a clone of B-cell-derived plasma cells that secrete antibodies against that antigen. Clonal selection accounts for the rapid response, specific action, and diversity of antibodies seen for the adaptive immune system. It also accounts for immunological memory through the production of memory B cells and memory T cells.
5. Immunological memory is the aspect of adaptive immunity that allows for a rapid, intense immune reaction to develop if the body encounters an antigen it has seen before.

Study Break 43.4

1. Immunological tolerance, a feature of the adaptive immune system, protects the body's own molecules from attack by its immune system. B cells and T cells are involved in the development of immunological tolerance, but the exact process is not understood.
2. Basically, an allergy results from an overreaction of the immune system to a particular antigen. The substances responsible for allergic reactions are a class of antigens known as allergens. Allergens stimulate B cells to secrete IgE antibodies in high amounts. IgE binds to mast cells, which are then induced to oversecrete histamine. Histamine contributes to inflammation, and at high amounts, the histamine-induced inflammation can be severe and even life threatening.

Study Break 43.5

Invertebrates lack the adaptive immune system seen in mammals; they have no B or T cells, for instance. Invertebrate immune defenses, like the innate immune system of mammals, are nonspecific. For example, all invertebrates have phagocytic cells that engulf pathogens, and many invertebrates produce antimicrobial proteins such as lysozyme. Also, many invertebrates produce proteins of the immunoglobulin family. While these are not antibody molecules, in some invertebrates they provide a protective function against some pathogens.

Think Critically

The research might contribute to the development of antibiotics that target specific genes in specific pathogenic bacteria. The drugs could target specific events, and even specific phases of infection, rather than blocking protein synthesis in general or targeting the cell membrane as current antibiotics do.

Test Your Knowledge

1. a, 2. c, 3. b, 4. e, 5. e, 6. d, 7. b, 8. c, 9. a, 10. d

Interpret the Data

1. At 110 months, about 1% of the women who were not infected with any HPV type and about 17% of women who were infected with HPV16 had cervical cancer.
2. Women infected with both HPV16 and HPV18 must be in the HPV16 group.
3. Yes, by adding the percentages of individual groups.
4. Yes.

Chapter 44

Study Break 44.1

1. The respiratory medium is the environmental source of oxygen and the repository for released CO_2. For aquatic animals water is the respiratory medium, and for terrestrial animals it is air. The respiratory surface is the layer of epithelial cells between the body and the respiratory medium. Gas exchange occurs across the respiratory surface—O_2 in and CO_2 out of the body. In certain small animals, the body surface itself is the respiratory surface. Among larger animals, aquatic animals use gills, insects use tracheal systems, and terrestrial animals use lungs as the respiratory surface.
2. The advantage of water over air as a respiratory medium is that it readily enables the respiratory surface to remain wet at all times. Two key advantages of air over water as a respiratory medium are (1) there is much more oxygen in air than in water; and (2) air is much less dense and less viscous than water, so significantly less energy is needed to move air over the respiratory surface than to move water.

Study Break 44.2

1. The advantages of gills to water-breathing animals over skin breathing are greater efficiency of gas exchange, the ability to live in more diverse habitats, and the potential to achieve a greater body mass.
2. In countercurrent exchange, the respiratory medium flows in the opposite direction of the blood flow under the respiratory surface. Examples are the flow of water over the gills in sharks, fishes, and some crabs and the one-way flow of air through the lungs of birds; all these are opposite to the flow of blood. The advantage of this mechanism is that it maximizes the amounts of O_2 and CO_2 exchanged with the respiratory medium.
3. In the tracheal system, fine branches called tracheoles end in tips that are in contact with body cells. The tracheole tips are filled with fluid, and gas exchange occurs through the fluid and the plasma membrane of the body cells in contact with the tips. Air enters the tracheal system through spiracles on the body surface. Movement of air through the tracheal system occurs by muscle-driven contractions and expansions of air sacs within the system.
4. In positive pressure breathing, gulping, swallowing, or pumping action forces air into the lungs. In negative pressure breathing, muscular activity expands the lungs, lowering the pressure of air in the lungs and thereby causing air to be pulled inward.

Study Break 44.3

1. Contraction of the diaphragm and the external intercostal muscles pulls the ribs upward and outward, expanding the chest and lungs. By this negative pressure mechanism, the air pressure within the lungs is lower than outside of the body. The higher outside pressure drives air into the lungs, expanding and filling them. Relaxation of the diaphragm and the intercostal muscles reverses the pressure condition, and the elastic recoil of lungs expels the air.
2. For conscious exhalation, you contract the internal intercostal muscles between the ribs to pull the diaphragm down while simultaneously contracting abdominal muscles to compress the abdominal organs and force them upward against the diaphragm to expel air from the lungs.
3. The most important feedback stimulus for breathing is the level of CO_2 in the blood.
4. The chemoreceptors in the medulla respond to the CO_2 level in the blood. If it rises, the medulla triggers an increase in the rate and depth of breathing. If it falls, the medulla triggers a decrease in the rate and depth of breathing.

Study Break 44.4

1. Hemoglobin is present in red blood cells. O_2 diffuses from alveoli into the plasma solution in the capillaries and then into the red blood cells. In the red blood cells, the O_2 binds to hemoglobin, thereby lowering the partial pressure of O_2 in the plasma. This leads to more O_2 molecules diffusing down the oxygen concentration gradient from alveolar air to blood.

   The binding of O_2 to hemoglobin is reversible. Further, the affinity of hemoglobin for O_2 increases as the partial pressure of O_2 increases and vice versa. This property is important for determining the release of O_2 from hemoglobin keyed to tissue requirements.

2. CO has much greater affinity for hemoglobin than O_2 does, and so it displaces O_2 from hemoglobin, reducing the amount of O_2 carried in the blood. Because the brain does not monitor O_2 levels, and other receptors do not respond until blood O_2 levels are critically low, victims can easily lapse into unconsciousness and death.

Study Break 44.5

- More blood per unit body weight than land animals
- Additional red blood cells, many stored in the spleen and released during a dive
- More myoglobin in muscles than land animals
- Slowing of the heart by as much as 90%
- Reduction of blood circulation to internal organs and muscles by up to 95%
- Retention of lactic acid in the muscles with no release into the blood until the animal returns to the surface

Think Critically

1. Carbon dioxide level
2. Contraction of smooth muscle in the heart is under the control of the autonomic nervous system. Nicotine stimulates the postganglionic neurons of both the sympathetic (increases heart rate) and parasympathetic (decreases heart rate) divisions and disrupts heart rate rhythm. Therefore, research is focused on both divisions, as nicotine affects the opposing actions of these divisions.

Test Your Knowledge

1. e, 2. d, 3. a, 4. c, 5. a, 6. d, 7. b, 8. d, 9. e, 10. a

Interpret the Data

1. Tobacco smoke significantly increased the lung tissue damage in these mice.
2. The tobacco-exposed mice given tomato juice showed significantly less tissue damage than tobacco-exposed mice without tomato juice. This suggests that something in the tomato juice is protecting again lung tissue damage, which supports the hypothesis that lycopene has a protective effect.

Chapter 45

Study Break 45.1

1. Carnivores primarily eat other animals as their source of organic materials. Herbivores primarily eat plants as their source of organic materials. Omnivores obtain their organic materials from any digestible organic source.
2. Essential nutrients are the amino acids, fatty acids, vitamins, and minerals that the animal cannot make itself and must obtain from its diet. The list of essential nutrients varies among animal types.
3. Deposit feeders pick up or scrape particles of organic matter from solid material they live in or on. Suspension feeders ingest small organisms that are suspended in water. Depending on the animal, the ingested organisms may be bacteria, protozoa, algae, or small crustaceans, or fragments of those organisms.

Study Break 45.2

1. Intracellular digestion takes place within body cells, whereas extracellular digestion takes place outside the cells, either in a pouch or a tube enclosed by the body.
2. (1) Mechanical processing, to break up the food; (2) secretion of enzymes and other substances that aid digestion; (3) enzymatic hydrolysis of food molecules into simpler molecular subunits; (4) absorption of the molecular subunits into body fluids and cells; and (5) elimination of undigested materials.

Study Break 45.3

1. The two classes of vitamins are the water-soluble vitamins and the fat-soluble vitamins. Fat-soluble vitamins ingested in the diet in excess of bodily needs can be stored in adipose tissues. In contrast, excess water-soluble vitamins in the diet are excreted in the urine. Hence, it is critical that humans meet their daily requirements for the water-soluble vitamins.
2. The four layers of the gut are the mucosa, the submucosa, the muscularis, and the serosa. The muscularis layer, which is formed from two smooth muscle layers, is responsible for peristalsis.
3. Pepsin is secreted from chief cells into the stomach lumen as an inactive precursor molecule, pepsinogen. The pepsinogen is converted to pepsin by the highly acid conditions of the stomach. Pepsin is a digestive enzyme that begins the digestion of proteins by making breaks in polypeptide chains.
4. In the small intestine, the digestion of macromolecules into their molecular subunits occurs, and those subunits are absorbed. In the large intestine, water and mineral ions are absorbed from the remaining digestive contents, leaving the undigested remnants, the feces, which are expelled from the body.

Study Break 45.4

1. Gastrin stimulates the secretion of pepsinogen and HCl in the stomach. Pepsin generated by cleavage of pepsinogen digests protein in the swallowed food. Gastrin also stimulates contraction of the stomach and the intestines. Cholecystokinin (CCK), released in the duodenum in response largely to fat content in chyme, inhibits gastric activity. That is, CCK has the opposite effect of gastrin.
2. The hypothalamus has two interneuron centers that play a central role in regulating the digestive processes. One center stimulates appetite and reduces oxidative metabolism. The other center works in opposition to the first center by stimulating the release of a peptide hormone that inhibits appetite.

Study Break 45.5

1. Incisors nip or cut food. Canines bite and pierce food. Premolars and molars crush, grind, and shear food.
2. Symbiotic microorganisms aid digestion in many herbivores by assisting in the breakdown of plant material. The microorganisms synthesize cellulase, an enzyme vertebrates cannot make, which hydrolyzes the cellulose of plant cell walls into glucose subunits.

Think Critically

1. Tortoises are ectotherms and allocate energy to activity associated with maintaining body temperature; hares are endotherms and allocate energy to physiological processes to maintain body temperature.
2. They are used widely as stored energy. In fact, fats are an excellent source of energy in the diet.

Test Your Knowledge

1. b, 2. b, 3. a, 4. d, 5. c, 6. e, 7. c, 8. d, 9. e, 10. a

Interpret the Data

1. About 30% (0.30) of the Hazda had fewer than 5 copies of the *AMY-1* gene.
2. Yes, about 10% of the European-Americans had more than 10 copies of the *AMY-1* gene.
3. Yes, the data support the hypothesis that a starchy diet favors duplications of the *AMY-1* gene. Cultural groups with a high-starch diet have, proportionally, more copies of the *AMY-1* gene than cultural groups with low-starch diets.

Chapter 46

Study Break 46.1

Osmosis is a process in which water molecules move across a selectively permeable membrane from a region where they are more highly concentrated to one where they are less highly concentrated.

Osmolarity is the total solute concentration of a solution. Osmolarity is measured in osmoles per liter of solution, where an osmole is the number of solute molecules and ions (in moles).

A solution that is hypoosmotic has lower osmolarity than the solution on the other side of a selectively permeable membrane. The solution with the higher osmolarity is said to be hyperosmotic.

An osmoregulator is an animal that uses an active control mechanism to keep the osmolarity of cellular and extracellular fluids the same. This osmolarity value may differ from the osmolarity of the surroundings.

Tubules that carry out osmoregulation and excretion are formed from transport epithelium, a layer of cells with specialized transport proteins in their plasma membranes that move specific molecules and ions into and out of the tubule.

Study Break 46.2

Protonephridia are found in flatworms and larval mollusks. They are the simplest invertebrate excretory tubules. Body fluids enter the end of a protonephridium and are moved through the tubule by movement of cilia on the flame cells. As the fluids move through the protonephridium, some molecules and ions are reabsorbed, and others, including nitrogenous wastes, are secreted into the tubules. Excess fluid is released through pores connecting the network of protonephridia to the body surface.

Metanephridia are found in annelids and most adult mollusks. Body fluids enter the funnel-like proximal end, driven by cilia surrounding that end. Some molecules and ions are reabsorbed as the fluids move through the tubule, and other ions and nitrogenous wastes are secreted into the tubule and excreted from the body surface.

Malpighian tubules are found in insects and other arthropods. Body fluids enter the tubules through spaces between the tubule cells. The distal ends of the tubules empty into the gut. Pressure is not used in the filtration step in Malpighian tubules. The secretion of K^+ from the hemolymph into the lumen draws in Cl^-. The accumulation of KCl causes water to enter the tubule from the hemolymph by osmosis. Organic wastes now are secreted into the tubule. When the fluids reach the hindgut and rectum, K^+ and Cl^- are reabsorbed, followed by water. As water leaves, uric acid precipitates as crystals that are excreted from the rectum in the feces.

Study Break 46.3

1. At the proximal end, a human nephron forms a cuplike region called Bowman's capsule. Bowman's capsule surrounds the glomerulus, a complex of blood capillaries. The capsule and the glomerulus are located in the renal cortex. The proximal convoluted tubule descends in a U-shaped bend called the loop of Henle into the renal medulla, and ascends again to form the distal convoluted tubule. Up to eight distal convoluted tubules drain into one collecting duct.
2. The collecting ducts are permeable to water but not to salt ions. The ducts, which begin in the cortex and descend into the medulla of the kidney, become surrounded by an ever-increasing solute concentration. As the urine passes down the collecting ducts, water moves osmotically out of the ducts, causing the concentration of the urine to increase. At the bottom of the collecting ducts, the urine is about four times more concentrated than body fluids.

Study Break 46.4

The RAAS compensates for excessive loss of salt and body fluids. The ADH system compensates for excessive water intake or loss. Combined, the two systems play an important role in regulating the interactions between the kidneys and the rest of the body.

The RAAS is activated when the Na^+ concentration of body fluids falls, causing the volume of extracellular fluids and blood pressure also to drop. Cells in the juxtaglomerular apparatus secrete renin, which activates angiotensinogen to produce angiotensin I. Angiotensin-converting enzyme converts angiotensin I to angiotensin II, which has three effects. (1) It constricts arterioles around the body, thereby quickly increasing blood pressure. (2) It stimulates the secretion of aldosterone from the adrenal cortex, which increases Na^+ reabsorption in the kidneys and raises the osmolarity of body fluids; as a result, water moves from the tubules into the interstitial fluid, conserving water. (3) It stimulates thirst so that more water is brought into the body. The RAAS is suppressed if NaCl concentration in body fluids is higher than normal.

In the ADH system, ADH is released from the posterior pituitary when osmoreceptors in the hypothalamus detect an increase in the osmolarity of body fluids. ADH increases water reabsorption from the distal convoluted tubules and collecting ducts of the kidney; as a result, urinary output is reduced and water is conserved. In the case of a decrease in osmolarity, ADH release from the pituitary is inhibited, thereby decreasing water reabsorption in the kidney.

Study Break 46.5

1. Marine teleosts live in seawater, which is hyperosmotic to their body fluids. They drink seawater continuously to replace water lost to the environment by osmosis. Na^+, K^+, and Cl^- ions from the seawater they drink are excreted by the gills. Nitrogenous wastes are released from the gills, primarily as ammonia, by simple diffusion.

   Freshwater fishes live in fresh water, which is hypoosmotic to their body fluids. They take in water by osmosis and excrete excess water. Salts needed for bodily functions are obtained from food and by active transport through the gills from the water. Nitrogenous wastes are excreted from the gills as ammonia.
2. The excretion of urea in mammals involves expelling the urea in solution—urine. Thus, there is loss of water with excretion of nitrogenous waste as urea. Excretion of nitrogenous waste in the form of uric acid conserves water because uric acid crystals are almost water-free.

Study Break 46.6

Ectothermy applies to animals that obtain heat energy primarily from the environment; endothermy applies to animals that obtain heat energy primarily from internal reactions. Generally speaking, ectotherms are highly successful in warm environments, but their bodily functions slow down as the temperature drops. Endotherms remain active over a broader range of environmental temperatures than ectotherms do, but they require an almost constant supply of energy to maintain their body temperature.

Study Break 46.7

1. The thermoregulatory responses shown by ectotherms can be physiological, such as regulating blood flow to internal organs or to the skin, or behavioral, such as physically moving to a location in the environment suitable for their heat energy needs at the time.
2. Thermal acclimatization refers to the physiological changes that many ectotherms make to compensate for seasonal shifts in environmental temperature.

Study Break 46.8

Temperature regulation in endothermic animals involves mechanisms that balance internal heat production against heat loss from the body. Internal heat production is controlled by negative feedback pathways triggered by thermoreceptors in the skin, the hypothalamus, and the spinal cord. When a deviation from the set point occurs, this system operates to return the core temperature to that set point through changes in metabolic processes, behavioral activities, and control of heat loss at the skin surface.

Think Critically

Recurrence of kidney disease in a transplanted kidney suggests that the disease is not intrinsic to the kidney, but reflects factors made elsewhere that act on the kidney in some way. Recurrent disease is in general much less common in inherited forms of kidney disease which, for the most part, are defects in kidney-expressed proteins. One interesting clinical observation occurs in patients with Alport syndrome, a disease of the basement membrane protein type 4 collagen. Alport syndrome patients develop kidney disease, and can also have deafness and eye problems. Patients with kidney failure from Alport syndrome who receive kidney transplants sometimes develop new antibodies directed against normal type 4 collagen (which as absent previously). This can lead to a different form of kidney disease characterized by antibody-mediated inflammation of the glomeruli.

Test Your Knowledge

1. d, 2. c, 3. a, 4. b, 5. d, 6. e, 7. a, 8. b, 9. d, 10. e

Interpret the Data

1. Phase 2
2. Phase 1
3. Yes

Chapter 47

Study Break 47.1

Recall that asexual reproduction produces offspring with genes from only one parent. In most animals that undergo asexual reproduction, the offspring are produced by mitosis and, therefore, are genetically identical to one another and to the parent. Such genetic homogeneity can be advantageous in stable and uniform environments. Another advantage is that individuals do not need to use energy to produce gametes or to find and select a mate. Further, for individuals in sparse populations, or for sessile animals, asexual reproduction can be an advantage. A disadvantage is that a genetically homogeneous population may not adapt readily to new environments.

By contrast, sexual reproduction always generates genetic diversity among offspring. This provides a population the opportunity to adapt to changing environments, and perhaps to move to and colonize new environments. Disadvantages of sexual reproduction include the expenditure of energy and raw materials to produce gametes and to find and select mates.

Study Break 47.2

1. Egg coats are surface coats around the egg. They are added during oocyte development or fertilization. Egg coats provide protection against mechanical injury, infection by microorganisms, and, in some species, loss of water.

   Mammalian eggs have an egg coat called the zona pellucida immediately surrounding the egg. This gel-like coating is called the vitelline coat in other organisms. In birds, a thick solution of proteins—the egg white—surrounds a vitelline coat. Bounding the egg white is a hard shell.
2. The slow block to polyspermy occurs in many organisms, including mammals. The fusion of egg and sperm triggers an increase in calcium ions in the cytosol. The calcium ions activate proteins that initiate a high level of metabolic activity in the egg. The released calcium ions also cause the cortical granules to fuse with the plasma membrane of the egg and release their contents to the outside. Enzymes released from those granules alter the egg coats in a matter of minutes after fertilization, and this blocks further sperm from attaching and penetrating the egg.

Study Break 47.3

- FSH stimulates a number of oocytes to begin meiosis. A follicle forms around each oocyte.
- FSH and LH interact to stimulate follicle cells to secrete estrogens.
- Later in the cycle, an increased level of estrogen leads to a new burst of release of FSH and LH. This new LH stimulates follicle cells to release enzymes that digest away the follicle wall, leading to egg release.
- The new LH also causes the remaining follicle cells to grow into the corpus luteum.
- FSH and LH levels decrease. This removes the stimulatory signal for follicular growth, and no new follicles grow in the ovary.

Study Break 47.4

The oral contraceptive pill inhibits the secretion of FSH and LH by the pituitary. FSH and LH stimulate oocytes in the ovaries to begin meiosis. They also stimulate follicle cells that surround the oocytes to secrete estrogens. Later, LH causes ovulation—the release of an egg. Therefore, without FSH and LH, ovulation cannot occur.

Think Critically

It could be used in cases of a low sperm count (few sperm produced), poorly motile or immotile sperm, incomplete sperm maturation, or inability to produce sperm by ejaculation (sperm could be collected surgically from the epididymis).

Test Your Knowledge

1. b, 2. a, 3. d, 4. a, 5. e, 6. c, 7. b, 8. e, 9. c, 10. d

Interpret the Data

1. The lowest sperm count, 58.7 million/ml, was in Columbia, Missouri. The highest, 102.9 million/ml, was in New York, New York.
2. The lowest percent motile sperm was 48.2% in Columbia, Missouri, and the highest was 56.4% in New York, New York.
3. Based on these data, the adverse effects of age and STDs were unlikely explanations for the high sperm count in New York, New York, relative to the other locations, since New York had the highest average age and highest percentage of men with a history of STDs. The data do not show a clear correlation between smoking and sperm count, but it could have been a factor explaining the results.
4. The data support the hypothesis that living near farmland may adversely affect male reproductive function, but further study would be needed to eliminate other variables that could have contributed to the differences.

Chapter 48

Study Break 48.1

1. In cleavage divisions, cycles of DNA replication and division occur without the production of new cytoplasm. Therefore, the cytoplasm of the egg partitions into many cells without increasing in overall size or mass. In cell division in an adult organism, the mitotic cell cycle involves cycles of DNA replication and division interspersed with cell growth. That is, the cell grows, then divides, then the two progeny cells grow, then divide, and so on. New cytoplasm is produced during these cell cycles so that overall there is an increase in mass of the cells.
2. The three primary cell layers of the embryo are produced by gastrulation. They are ectoderm, mesoderm, and endoderm.

Study Break 48.2

1. The gray crescent establishes the dorsal–ventral axis, with the gray crescent marking the future dorsal side.
2. If cells of the dorsal lip of the blastopore are transplanted to another location in the egg, they cause a second blastopore, and subsequently a second embryo, to form in that region.
3. The extraembryonic membranes in birds and their functions are as follows:
   - Yolk sac: Surrounds the yolk, which provides nutrients to the embryo.
   - Chorion: Exchanges oxygen and carbon dioxide with the environment through the egg shell.
   - Amnion: Encloses the embryo and secretes amniotic fluid into the space that provides an aquatic environment in which the embryo can develop.
   - Allantois: Stores nitrogenous wastes, primarily in the form of uric acid, that are derived from the embryo. Part of the allantoic membrane forms a bed of capillaries that is connected to the embryo and that delivers carbon dioxide from the embryo to the chorion and picks up oxygen absorbed through the shell.

Study Break 48.3

1. The three primary tissues, ectoderm, mesoderm, and endoderm, give rise to the tissues and organs of the embryo.
2. The central nervous system, notably the brain and spinal cord, develops from the neural tube. Neural crest cells give rise to parts of the nervous system (including cranial nerves, ganglia of the autonomic nervous system, peripheral nerves from the spinal cord to body structures, and nerves of the developing gut) and contribute to a variety of other body structures (for example, bones of the inner ear and skull, cartilage of facial structures, teeth, pigment cells in the skin, and the adrenal medulla).

Study Break 48.4

1. Cells of the trophoblast are responsible for implantation of the blastocyst in the endometrium (uterine lining). When the blastocyst is ready for implantation, trophoblast cells secrete proteases that digest pathways between endometrial cells. Dividing trophoblast cells fill the spaces and, through continued digestion and division, the blastocyst burrows into the endometrium and eventually becomes covered by a layer of endometrial cells. From then on, cells originating in the trophoblast support the development of the embryo/fetus in the uterus through the production of the chorion.

   The inner cell mass becomes the embryo/fetus itself. During implantation, the inner cell mass separates into the epiblast and the hypoblast. The epiblast produces the ectoderm, mesoderm, and endoderm of the developing embryo. The hypoblast gives rise to part of the extraembryonic membranes.
2. Oxytocin. This hormone is responsible for stimulating contractions of the uterus.

Study Break 48.5

1. Cell division, cell movement, and cell adhesion.
2. Induction is the process whereby a region of the embryo acts on other cells to alter the course of development. Induction is brought about by proteins; hence, induction is under genetic control.

Think Critically

No (and, in fact, there are not any). This can be inferred from the description of sex determination in mammals and from the list of gynandromorph animals. The significance of the research is unchanged.

Much can be learned about sex determination in mammals by studying sex determination in other organisms, and one could argue that understanding why it is different in other animals only provides greater insight into sex determination in mammals.

Test Your Knowledge

1. d, 2. e, 3. b, 4. a, 5. e, 6. c, 7. d, 8. d, 9. b, 10. c

Interpret the Data

1. Heart defects were the most common birth defect in the single-birth group.
2. Heart defects were more common in the multiple-birth group than in the single-birth group.
3. Multiples had more than twice the risk of central nervous system defects than did single-birth infants.
4. No, the 0.93 relative risk of chromosomal defects indicates that multiple pregnancy decreases the relative risk of chromosomal defects in offspring.

Chapter 49

Study Break 49.1

1. Studies of ecosystems are more "inclusive" than studies of populations because ecosystems include the populations of many different species.
2. Mathematical models are useful in ecological research because they help scientists formalize hypotheses about the relationships between variables and because they allow researchers to simulate the effects of changing variables before investing time and resources in experiments or observational studies.

Study Break 49.2

1. Because of Earth's spherical shape, sunlight striking the planet's surface is more concentrated near the equator than at the poles. As a result, temperatures are higher at low latitudes. The concentrated sunlight near the equator heats the atmosphere, causing air masses near the equator to rise, establishing three circulation cells in the Northern Hemisphere and three in the Southern Hemisphere.
2. Earth's fixed tilt on its axis causes seasonal variation in the amount of sunlight striking the temperate zone as the planet orbits the sun.
3. Dry conditions prevail at 30° N and S latitudes because sinking air masses warm as they descend, causing them to absorb water from the land.
4. Mountains affect local precipitation because rising air masses on the windward side of a mountain cool adiabatically and release moisture. When the air masses descend on the leeward side of a mountain, they warm and absorb moisture, causing a rain shadow.

Study Break 49.3

1. *Anolis* lizards in the Dominican Republic bask more frequently at high elevation than they do at low elevation.
2. Global warming will likely cause the geographical distributions of species to shift or expand to higher latitudes and to higher elevations.

Study Break 49.4

1. Tropical rain forest and temperate rain forest are the terrestrial biomes that receive the most rainfall.
2. Savannas, chaparral, and temperate grasslands are renewed by periodic fires.
3. Tropical rain forests and temperate rain forests have the tallest vegetation. Arctic and alpine tundra have the shortest vegetation.
4. Trees are usually evergreen in tropical rain forest, temperate rain forest, and taiga.

Study Break 49.5

1. Dissolved oxygen concentration is usually high in the headwaters of a stream, gradually diminishing as water flows into a river.
2. The factors that cause seasonal overturns in lakes include seasonal changes in environmental temperature, variations in wind velocity, and the fact that water is densest at 4°C.
3. Oligotrophic lakes are better than eutrophic lakes for recreational purposes because the water in oligotrophic lakes is clear, whereas the water in eutrophic lakes is often clogged with strands of algae and cyanobacteria.

Study Break 49.6

1. The benthic province of the ocean includes all of the bottom sediments. The pelagic province includes all of the water.
2. Estuaries experience the largest fluctuations in salinity over time.
3. Estuaries, the intertidal zone, and the upper layer of the oceanic pelagic zone receive substantial energy inputs from sunlight.
4. The benthos of the oceanic zone receives nutrients and energy from the detritus sinking from the upper layers of water.
5. Chemoautotrophic bacteria are the primary producers of hydrothermal vent communities and cold seep communities. Unlike photosynthetic organisms of the photic zone, they use hydrogen sulfide and other molecules, instead of sunlight, as an energy source for their chemosynthetic activity.

Test Your Knowledge

1. c, 2. e, 3. b, 4. a, 5. d, 6. a, 7. d, 8. a, 9. a, 10. d

Interpret the Data

1. SSTs have predicted 19 El Niño events and 22 La Niña events from 1950 through 2008.
2. These events differ greatly in magnitude and duration.
3. There is no absolute pattern in the occurrence of the two types of events, although it appears that very pronounced events of one type are followed by an event of the other type.

Chapter 50

Study Break 50.1

1. A population's size is simply the number of individuals it contains. Its density is the number of individuals per area or volume of habitat occupied.
2. A clumped pattern of dispersion implies that individuals in the population help each other or that some vital resource in the environment also has a clumped distribution. A uniform pattern of dispersion implies that individuals in the population repel each other. A random pattern of dispersion does not imply either positive or negative interactions among individuals in the population.

Study Break 50.2

1. A life table usually summarizes statistics about the age-specific survival rates, age-specific mortality rates, and age-specific fecundity of a population.
2. Humans in the industrialized countries exhibit Type I survivorship curves because they provide lots of care to their offspring, thus reducing infant and childhood mortality to low levels.

Study Break 50.3

1. Children spend most of their energy on growth and maintenance; they devote energy to reproduction later in life.
2. Fecundity and the amount of parental care devoted to each offspring exhibit an inverse relationship because organisms that produce few offspring can devote substantial time and energy to each, whereas those that produce many offspring can devote only minimal time and energy to each.

Study Break 50.4

1. The model of exponential population growth predicts unlimited population growth through time, generating a J-shaped curve of population size versus time. The logistic model predicts that population growth slows down as the population approaches its carrying capacity, generating an S-shaped curve of population size versus time.
2. Carrying capacity is the maximum number of individuals in a population that an environment can support. The carrying capacity is thus a property of the environment with reference to a particular population.
3. A time lag is a delay in a population's response to a changing environment. It may cause a population's size to oscillate around its carrying capacity.

Study Break 50.5

1. The effects of density-dependent factors get stronger (that is, they affect a larger percentage of the individuals in the population) as the population's density increases. The effects of density-independent factors do not change (that is, they affect the same percentage of the individuals in a population) as the population's density changes.
2. The effects of infectious diseases are usually density-dependent because disease-causing pathogens spread more quickly through dense populations of the organisms they infect.

Study Break 50.6

1. Humans have sidestepped the controls that regulate the populations of other organisms by expanding their geographical range to include a wide variety of habitats, by increasing the carrying capacity through agricultural production, and by decreasing their death rate through the introduction of medical care and improved sanitation.
2. The age structure of a population influences its future population growth by determining how many individuals will reach reproductive age in the future. Populations with a bottom-heavy age structure (that is, with many young children) will experience a growth spurt when children alive today reach sexual maturity. Populations with a more even age structure will not experience a dramatic future increase in population size.

Think Critically

1. The "ideal" conditions under which the equations predict growth simply never exist, as is clear, for example, from the growth of populations illustrated in the chapter (see growth of *Thrips imaginis*). Therefore, unlike a law such as the law of gravity, population growth cannot simply be predicted by applying the logistic or exponential equation for population growth. The mathematical growth models are constructed after basic observations define the variables of the models.
2. The time scales for most population studies are too short to consider selection, genetic drift, or mutation. Molecular studies allow population biologists to measure changes in gene frequencies at the time scales of single generations and before those changes are manifested in noticeable changes in fecundity or survival (or morphology and behavior).

Test Your Knowledge

1. b, 2. c, 3. b, 4. d, 5. a, 6. e, 7. d, 8. e, 9. d, 10. b

Interpret the Data

The data in the *Albertosaurus* life table suggest that it had a survivorship pattern falling between Types I and II, though perhaps leaning more toward Type I. This species was most vulnerable to mortality between hatching and age 2.

Chapter 51

Study Break 51.1

1. Some carnivores spend more time and energy capturing large prey than small prey because large prey provide a larger return on their investment of time and energy in the hunt.
2. Cryptic coloration makes an organism inconspicuous, allowing it to blend in with its surroundings. Aposematic coloration makes an organism highly conspicuous, advertising its unpalatability. Mimicry allows one organism, the mimic, to resemble another species, the model; models are usually unpalatable or poisonous. A mimic will have aposematic coloration if it resembles an aposematic model.
3. Field experiments can demonstrate that two species are competing for limiting resources if the removal of one species increases population size or density in the other or if the addition of a potential competitor decreases the population size or density of the other.

Study Break 51.2

1. Gleason's individualistic view of communities suggests that they are just chance assemblages of species that happen to be adapted to similar abiotic environmental conditions.
2. Ecologists find more species living in an ecotone than in the communities on either side of it because ecotones contain species from both neighboring communities as well as species that are adapted to transitional environmental conditions.

Study Break 51.3

1. The plant growth forms found in tropical forests include a canopy of tall trees, an understory of shorter trees and shrubs, an herb layer, vinelike lianas, and epiphytes.
2. The species richness of a community is the number of species it contains. Relative abundance refers to the commonness or rarity of species in the community.
3. Pigeons, which eat grain and other vegetable matter, are included in the second trophic level, primary consumers. Peregrine falcons, which feed on pigeons and other birds, are in the third trophic level, secondary consumers.

Study Break 51.4

1. On the one hand, the ecological literature may overestimate the importance of competition be-

cause ecologists are more likely to study and publish papers on interactions in which competition is important than on interactions in which it is not. On the other hand, the literature may underestimate the importance of competition because, if strong competition between species cannot persist for long periods of time, we are unlikely to find populations competing strongly in nature.

2. Keystone species are those that have a substantial effect on community structure even if their populations are not very dense. Keystone species may either increase or decrease species richness in the communities they occupy.

Study Break 51.5

1. Strong storms allow coral communities to be rejuvenated through the recruitment of new individuals because they scour the seafloor, removing existing coral colonies from the community. These openings provide spaces where coral larvae may settle and initiate the growth of new colonies.
2. Moderately severe and moderately frequent disturbances increase a community's species richness by creating opportunities for *r*-selected species to colonize the habitat while allowing populations of *K*-selected species to persist.

Study Break 51.6

1. Primary succession occurs in places without soil; secondary succession occurs after a disturbance has destroyed vegetation.
2. A climax community differs from earlier successional stages in having taller, longer-lived vegetation, generally higher species richness, and a buffered physical environment under the vegetation.
3. Three hypotheses about the underlying causes of succession differ in how they view the role of population interactions. The facilitation hypothesis specifies no particular role for population interactions. The inhibition hypothesis suggests that the species already present prevent other species from joining a community. The tolerance hypothesis suggests that as environmental conditions within the community change during succession, only species that can compete strongly under the changing conditions will persist.

Study Break 51.7

1. Some explanations of the high species richness in the tropics suggest that the benign climate and historically low levels of severe disturbance have fostered more rapid rates of speciation. Other explanations suggest that the year-round availability of food resources and complex food webs allow more species to coexist in tropical regions.
2. According to the equilibrium theory of island biogeography, large islands will harbor more species than small islands, and islands that are close to the mainland will harbor more species than those that are farther away.

Think Critically

1. Many, but certainly not all, mutualisms involve partners "taking" resources from each other. In pollination, for example, bees take pollen, and plants take the bees' time and direct their movements.
2. Yes, because of shared traits among close relatives. This is one reason cannibalism may be rare.

Test Your Knowledge

1. c, 2. c, 3. a, 4. b, 5. b, 6. e, 7. b, 8. d, 9. d, 10. b

Interpret the Data

The protected shrubs responded to the exclusion of herbivores by steadily reducing the density of thorns on their flowering stems over two years. The unprotected shrubs continued to produce a high density of thorns on their flowering stems. The unprotected shrubs' production of thorns probably protected their flowers and fruits feeding by herbivores, allowing them to produce seeds.

Chapter 52

Study Break 52.1

1. In the generalized compartment model of biogeochemical cycling, nutrient pools are classified as available or unavailable and as organic or inorganic.
2. The advantage of using conceptual models of ecosystem function is that they are a simplification of the processes that determine ecosystem function in nature. The disadvantage of these models is that they do not include processes that carry nutrients and energy out of one ecosystem and into another; neither do they include the details of exactly how specific ecosystems function. Thus, conceptual models do not provide precise predictions about potential changes in ecosystem function.
3. Before constructing a simulation model of an ecosystem, ecologists must collect data about the population sizes of important species, the average energy and nutrient content of each, the food webs in which they participate, the quantity of food each species consumes, and the productivity of each population; the ecosystem's energy and nutrient gains and losses caused by erosion, weathering, precipitation, and runoff; and seasonal and annual variations in these factors.

Study Break 52.2

1. Gross primary productivity is a measure of the total amount of solar energy converted into chemical energy by the producers in an ecosystem. Net primary productivity is the amount of chemical energy that remains after deducting the producers' maintenance costs from the gross primary productivity.
2. In terrestrial ecosystems, primary productivity may be influenced by the availability of light, water, and nutrients and by the environmental temperature. In aquatic ecosystems, primary productivity is often limited by the joint availability of light and nutrients in the same place.
3. Energy is lost from an ecosystem at every transfer between trophic levels because some of the energy is not assimilated, because organisms use some of the energy they assimilate for maintenance costs, and because biological processes are never 100% efficient.
4. The presence of predators can influence an ecosystem's productivity by consuming herbivores and thereby changing the population dynamics of the herbivores and the plants they eat. These effects can reverberate through an ecosystem in a trophic cascade.

Study Break 52.3

1. The global hydrologic cycle maintains its balance because the amount of water returned to the atmosphere by evaporation and transpiration is equal to the amount that falls as precipitation. Runoff from the land maintains the balance between terrestrial and marine components of the cycle.
2. Respiration, excretion, leaching, and the burning of fossil fuels move large quantities of carbon from an organic compartment to an inorganic compartment of an ecosystem.
3. Bacteria, cyanobacteria, and fungi drive the global nitrogen cycle through their activities in nitrogen fixation, ammonification, nitrification, and denitrification.
4. Marine sediments are Earth's main reservoir for phosphorus, which is recycled slowly after geological uplifting and erosion make it available to producers.

Study Break 52.4

1. The greenhouse effect refers to the tendency of certain gases, called greenhouse gases, to foster the accumulation of heat in the lower atmosphere by hindering the escape of infrared heat into space. CO_2 enhances the greenhouse effect because it is one of the greenhouse gases.
2. The combustion of fossil fuels and wood is the human activity that releases the most CO_2 into the atmosphere.
3. Irrigation, the use of synthetic nitrogen-containing fertilizers, the harvesting of crops, and the rearing of livestock contribute to the disruption of the nitrogen cycle.

Think Critically

Large cities release substantial quantities of airborne pollutants that can be carried by prevailing winds even to distant forests. These materials can have a profound influence on the health and biological activity of primary producers in the forest.

Test Your Knowledge

1. c, 2. d, 3. c, 4. e, 5. e, 6. c, 7. a, 8. d, 9. c, 10. a

Interpret the Data

The productivity of the algal community in Walker Branch is controlled by both nutrients and herbivores. The graph illustrates that for grazed algae, the addition of nitrogen, phosphorus, or both nitrogen and phosphorus increased productivity over the level seen in the control treatment. Similarly, for ungrazed algae, the addition of nitrogen or the combination of nitrogen and phosphorus increased productivity over the level measured in the controls. The graph also illustrates the significant impact of grazing by herbivorous snails: in the control treatment and two of the three experimental treatments, ungrazed algae had much higher productivity than grazed algae.

Chapter 53

Study Break 53.1

1. The growing human population in south Florida has increased the likelihood of desertification there by withdrawing groundwater for agricultural, recreational, and residential uses faster than it is replenished.
2. Overexploitation of fish populations typically causes fishes to reach reproductive maturity at a smaller size and younger age, decreases population sizes, and sometimes leads to the extinction of populations.

3. The construction of dams changes the water flow in rivers, leading to changes in the physical structure of the river and the species that inhabit it; disrupting the environmental cues that trigger successful reproduction of organisms; inhibiting the free movement of migrating animals through all parts of the river system; and opening the habitat to invasion by species that originated elsewhere.

Study Break 53.2

1. Habitat fragmentation in eastern North America has affected breeding songbirds by reducing the variety of habitats available to them, increasing the frequency of brood parasitism of their nests, and increasing the rate of predation on their eggs and young.
2. The use of the anti-inflammatory drug diclofenac on livestock has decimated vulture populations in South Asia. Diclofenac is highly toxic to birds, and the vultures ingest it when they feed on the carcasses of livestock.
3. By one estimate, extinction rates today may be 1,000 times greater than the background extinction rate evident in the fossil record.

Study Break 53.3

1. Living systems are a storehouse of potentially useful genetic information because naturally occurring compounds may prove to be useful in the treatment of disease, in the manufacture of new products, or in agriculture.
2. Naturally occurring organisms provide many ecosystem services, such as the sequestration of carbon dioxide, fixation of nitrogen into forms that plants can absorb, recycling of nutrients with ecosystems, and the retention of water in ecosystems.

Study Break 53.4

1. Conservation biologists identify a site as one where extinction is imminent if it houses 95% or more of the individuals in an endangered species and it has definable boundaries that encompass distinctive habitats.
2. Conservation biologists are especially alarmed about deforestation in the New World tropics because these forests harbor many species in imminent danger of becoming extinct.

Study Break 53.5

1. Population bottlenecks—large, temporary reductions in a population's size—inevitably foster genetic drift, thereby reducing a population's genetic variability, which increases its likelihood of becoming extinct.
2. A population viability analysis allows a conservation biologist to identify the minimum population size that is likely to survive both predictable and unpredictable environmental change. It therefore specifies how many individuals must be conserved for the continued survival of the population and species.
3. Several small preserves would collectively experience more edge effects than one large preserve of the same total size.

Study Break 53.6

1. The Pine Bush habitat in New York state is an example of conservation through preservation.
2. The Royal Chitwan National Park has been judged a success because local residents benefit from the park's existence, and therefore support it, and because populations of many animals, including tigers and rhinoceroses, have increased within its borders.
3. Economists can determine the economic values of specific ecosystem services and convince local governments that it is economically beneficial to preserve ecosystems and the services they provide.

Think Critically

Student answers may vary. Because students are asked to come up with a list of unanswered questions on their own, we can't really provide an answer to this question.

Test Your Knowledge

1. e, 2. b, 3. b, 4. b, 5. a, 6. d, 7. d, 8. a, 9. c, 10. b

Interpret the Data

Extinctions in mountain habitats are accelerating the most, exhibiting a potential for more than a 20-fold (366/17) increase. Among groups, extinctions are accelerating the most among amphibians, which exhibit a potential 12-fold (408/34) increase. Conservation biologists should therefore focus their efforts on amphibians that live in mountainous regions.

Chapter 54

Study Break 54.1

1. An instinctive behavior is a genetically or developmentally programmed response that appears in complete and functional form the first time it is used. A learned behavior is one that is dependent on having a particular kind of experience during development.
2. Marler demonstrated that singing the correct species song is a learned behavior by isolating some young male sparrows in soundproof cages, thereby preventing them from hearing their species song. Because the isolated males never learned to sing the correct song, Marler concluded that the experience of hearing the species song was necessary for song learning.

Study Break 54.2

1. The chicks of brood parasites stimulate their foster parents to feed them by engaging in exaggerated behaviors—opening their mouths and begging and peeping vigorously—that serve as sign stimuli, triggering feeding behavior in the parents.
2. Arnold demonstrated that the receptiveness of garter snakes to a meal of banana slugs had a genetic basis by breeding snakes that almost always eat banana slugs with snakes that rarely eat them. The behavior of the hybrid offspring was intermediate between the behaviors of the two parent populations.

Study Break 54.3

1. Tail wagging by a dog when it sees its owner pick up a leash is an example of classical conditioning.
2. Sleeping through an alarm clock is an example of habituation.

Study Break 54.4

1. The conclusion that certain neurons in a young male bird's brain are influenced only by acoustical signals from its own species and only during a critical period is supported by two observations: (1) young birds that did not hear taped songs during the critical period never produced the full song of their species; and (2) young birds that heard recordings of *other* bird species' songs during the critical period never generated replicas of those songs as they matured.
2. After a zebra finch hears the song of a neighbor many times, cells in a nucleus in its forebrain habituate to that stimulus and stop responding to it.
3. The role of the ZENK enzyme in song learning is to program the nerve cells of the bird's brain to anticipate key acoustical events of potential biological importance.

Study Break 54.5

1. A high estrogen concentration in the brains of young male zebra finches stimulates the production of more neurons in the higher vocal center.
2. Juvenile hormone stimulates the production of octopamine, a protein that may help a foraging bee home in on the odors of flowers where it can collect nectar and pollen.
3. The loss of a territory by a male African cichlid fish causes its brain to produce less GnRH.

Study Break 54.6

1. The anatomy of the cricket's nervous system helps it avoid an approaching bat because sensory neurons on one side of the body send messages to motor neurons on the other side. When the cricket hears a bat approaching from the right, it automatically lifts a leg on the left side, blocking the left wing and causing the cricket to veer away from the oncoming bat's path.
2. When a fiddler crab detects movement above the midline of its visual field, it dashes into its burrow.
3. A star-nosed mole locates earthworms in its tunnels with touch receptors on the tentacles that sprout from its nose.

Think Critically

One might expect that the bird would first flee from the predator because it poses the greatest threat to the bird's survival and future fitness. In the absence of the predator, it might court the potential mate to insure its production of offspring, or it might pursue the food, depending on the season and its individual nutritional state.

Test Your Knowledge

1. b, 2. a, 3. d, 4. b, 5. c, 6. e, 7. d, 8. c, 9. b, 10. b

Interpret the Data

Lynch's results suggest differences in this behavior have a genetic basis, because only genetically based variation will respond to artificial selection. (See Section 20.1.)

Chapter 55

Study Break 55.1

1. When piloting, animals use familiar landmarks to guide their journey. In compass orientation, animals use external environmental cues such as the position of the sun or stars as a compass to move in a particular direction, often over a specific distance or for a prescribed length of time. When navigating, animals use a compass as well as a mental map of their position in relation to their destination.
2. Seasonal changes in temperature and food availability are the most likely selection pressures responsible for the evolution of migratory behavior in birds.

Study Break 55.2

1. Wood lice accumulate in moist habitats because they move much less in moist habitats than they do in dry habitats.
2. The costs of maintaining a territory include the time and energy needed to defend territory borders, the possibility of being injured or killed during a territorial encounter, and the increased likelihood of being noticed and captured by a predator. The benefits of holding a territory include access to all of the resources within the territory.

Study Break 55.3

1. Humans consciously use acoustical, visual, and tactile channels to communicate with each other.
2. A honey bee uses the waggle dance to communicate the location of a distant food source to its hive-mates.
3. A young wandering raven that discovers a food source in the territory of other ravens will call vigorously to attract other ravens from outside the territory. Collectively, the non-territory holders can overwhelm the defenses of the resident birds and then consume the food.

Study Break 55.4

1. For monogamous species, the males that are most attractive to females are those that can demonstrate their good genes with large showy morphological characteristics and elaborate behavioral displays and those that hold territories rich in resources.
2. Male sage grouse perform displays at a lek, trying to attract the attention of females. Female sage grouse go to a lek to evaluate the qualities of the males that are displaying there.
3. A female red-winged blackbird might settle in the territory of a male even in the presence of other resident females if the male's territory is very rich in resources.

Study Break 55.5

1. The social behavior of musk oxen and sawfly larvae includes cooperative defense of the group against predators.
2. Among the animals in a dominance hierarchy, the most dominant individuals are the most likely to reproduce.
3. The genetic system in eusocial insects promotes altruistic behavior because the workers in a colony are more closely related to each other than are siblings in most other animal species.

Study Break 55.6

1. People may exhibit altruistic behavior to nonrelatives because their acts of charity may induce others to be charitable toward them.
2. Stepparents might provide fewer resources to their children than genetic parents do because stepparents do not share as many genes with their stepchildren as genetic parents share with their biological children.

Think Critically

If the partner of a female is able to discriminate between offspring that are his own and those that are not, he should bias any paternal care in favor of his true son in preference to his nephew. The production of chimeras could also give rise to conflicts between a male and a female because the offspring would be more closely related to the female parent than to the female's partner (i.e., the offspring's uncle).

Test Your Knowledge

1. d, 2. d, 3. b, 4. d, 5. a, 6. e, 7. c, 8. d, 9. a, 10. c

Interpret the Data

The results suggest that the more male neighbors a Seychelles warbler has, the more time he spends guarding his mate; and the more time he spends guarding his mate, the less time he can devote to collecting food. When the number of neighboring males was experimentally reduced, the focal males spent more time feeding and less time guarding their mates.

Appendix B
Classification System

The classification system presented here is based on a combination of organismal and molecular characters and is a composite of several systems developed by microbiologists, botanists, and zoologists. This classification reflects current trends toward a phylogenetic approach to taxonomy, one that incorporates the ever more detailed information about the relationships of monophyletic lineages provided by new molecular sequence data. In keeping with these trends, we have omitted reference to the traditional taxonomic categories, such as "class" and "order." Instead, we present the major monophyletic lineages in each of the three domains, and we indicate their relationships within a nested hierarchy that parallels that of traditional Linnaean classification.

Although researchers generally agree on the identity of the major monophyletic lineages, the biologists who study different groups have not established universal criteria for identifying the somewhat arbitrary taxonomic categories included in the traditional Linnaean hierarchy. As a result, a "class" or an "order" of flowering plants may not be the equivalent of a "class" or an "order" of animals. In fact, as described in *Unanswered Questions* at the end of Chapter 23, systematic biologists are shifting toward a more phylogenetic approach to taxonomy and classification, such as the one represented here.

Bear in mind that we include this appendix to introduce the diversity of life and illustrate many of the evolutionary relationships that link monophyletic groups. Like all phylogenetic hypotheses, this classification is open to revision as new information becomes available. Moreover, the classification is incomplete because it includes only those lineages that are described in Unit Four.

Prokaryotes and Eukaryotes

Organisms fall into two groups, prokaryotes and eukaryotes, based on the organization of their cells. Prokaryotes consist of the Domains Bacteria and Archaea and are characterized by a central region, the nucleoid, which has no boundary membrane separating it from the cytoplasm, and by membranes typically limited to the plasma membrane. Most prokaryotes are single-celled, although some are found in simple associations. All other organisms are eukaryotes, which make up the Domain Eukarya. Eukaryotes are characterized by cells with a central, membrane-bound nucleus, and an extensive membrane system. Some eukaryotes are single-celled, while others are multicellular.

Domain Bacteria

The largest and most diverse group of prokaryotes. Includes photoautotrophs, chemoautotrophs, and heterotrophs.

PROTEOBACTERIA Purple sulfur bacteria, purple nonsulfur bacteria, and some chemoheterotrophs

SPIROCHETES Helically spiraled bacteria that move by twisting in a corkscrew pattern

CHLAMYDIAS Gram-negative intracellular parasites of animals, with cell walls that lack peptidoglycans

GRAM-POSITIVE BACTERIA Chemoheterotrophic bacteria with thick cell walls

CYANOBACTERIA Photoautotrophic Gram-negative bacteria that use the same chlorophyll as in plants

Domain Archaea

Prokaryotes that are evolutionarily between eukaryotic cells and the bacteria. Most are chemoautotrophs. None is photosynthetic. Originally discovered in extreme habitats, they are now known to be widely dispersed. Compared with bacteria, the Archaea have a distinctive cell wall structure and unique membrane lipids, ribosomes, and RNA sequences. Some are symbiotic with animals, but none is known to be pathogenic.

EURYARCHAEOTA Includes methanogens, extreme halophiles, and some extreme thermophiles

CRENARCHAEOTA Includes most of the extreme thermophiles, as well as psychrophiles; mesophilic species comprise a large part of plankton in cool, marine waters

KORARCHAEOTA Known only from DNA isolated from hydrothermal pools. As of this writing, none has been cultured and no species have been named.

Domain Eukarya

PROTISTS A collection of single-celled and multicelled lineages, which are almost certainly not a monophyletic group.

- **Excavata** (excavates) Single-celled animal parasites that have greatly reduced mitochondria, or organelles derived from mitochondria, and move by means of flagella; most have a scooped out (excavated) feeding apparatus on the ventral surface of the cell
  - **Metamonada** (metamonads)—consist of the Diplomonadida and the Parabasala
    - Diplomonadida (diplomonads)—cells have two nuclei; move by multiple freely beating flagella
    - Parabasala (parasabala)—move by freely beating flagella and an undulating membrane
  - **Euglenozoa** (euglenozoans)—mostly single-celled, highly motile cells that swim using flagella and have disc-shaped inner mitochondrial membranes
    - Euglenids—free-living with anterior flagella; most are photosynthetic autotrophs
    - Kinetoplastids—nonphotosynthetic, heterotrophs that live as animal parasites
- **Chromalveolata** (chromalveolates) Heterogeneous group with a range of forms and life styles
  - **Alveolata** (alveolates)—characterized by small membrane-bound vesicles called alveoli in a layer under the plasma membrane
    - Ciliophora (ciliates)—motile single-celled heterotrophs; swim by means of cilia
    - Dinoflagellata (dinoflagellates)—nonmotile single-celled marine heterotrophs or autotrophs; shell formed from cellulose plates
    - Apicomplexa (apicomplexans)—nonmotile parasites of animals with apical complex for attachment and invasion of host cells

Stramenopila (stramenopiles)—characterized by two different flagella

Oomycota (oomycetes)—water molds, white rusts, and mildews

Bacillariophyta (diatoms)—single-celled; covered by a glassy silica shell

Chrysophyta (golden algae)—colonial; each cell of the colony has a pair of flagella and is covered by a glassy shell consisting of plates or scales

Phaeophyta (brown algae)—photoautotrophic protists

Rhizaria (rhizarians) Amoebas with stiff, filamentous pseudopodia; some with outer shells

Radiolaria (radiolarians)—heterotrophic; glassy internal skeleton with projecting raylike strands of cytoplasm

Foraminifera (forams)—heterotrophic protists with shells consisting of organic matter reinforced by calcium carbonate

Chlorarachniophyta (chlorarachniophytes)—green, photosynthetic amoebas that also engulf food

Archaeplastida (archaeplastids) Red algae, green algae, and land plants, photosynthesizers with a common evolutionary origin

Rhodophyta (red algae)—marine seaweeds, typically multicellular, reddish in color; with plantlike bodies

Chlorophyta (green algae)—green photosynthetic single-celled, colonial, and multicellular protists that have the same photosynthetic pigments as plants; likely ancestor of land plants

Amoebozoa (amoebozoans) Includes most of the amoebas as well as the slime molds

Amoebas—single-celled; use non-stiffened pseudopods for locomotion and feeding

Cellular slime molds—heterotrophs; primarily individual cells; move by amoeboid motion, or as a multicellular mass

Plasmodial slime molds—heterotrophs; live as plasmodium, a large composite mass with nuclei in a common cytoplasm, that moves and feeds like a giant amoeba

Opisthokonta (opisthokonts) A single posterior flagellum at some stage in the life cycle; consist of the fungi, choanoflagellates, and animals

Choanoflagellata (choanoflagellates)—motile protists with a single flagellum surrounded by collar of closely packed microvilli; likely ancestor of animals and fungi

PLANTAE Multicellular autotrophs, mostly terrestrial, and most of which gain energy via photosynthesis; life cycle characterized by alternation of a gametophyte (gamete-producing) generation and sporophyte (spore-producing) generation

Hepatophyta (liverworts)—leafy or simple flattened thallus with rhizoids; no true leaves, stems, roots, or stomata (porelike openings for gas exchange); spores in capsules[1]

Anthocerophyta (hornworts)—simple flattened thallus, hornlike sporangia[1]

Bryophyta (mosses)—feathery or cushiony thallus; some with hydroids; spores in capsules[1]

Lycophyta (club mosses)—simple leaves, cuticle, stomata, true roots; most species have sporangia on sporophylls; fertilization by swimming sperm[2]

Pterophyta (ferns, whisk ferns, horsetails)—*Ferns*: Finely divided leaves; sporangia in sori. *Whisk ferns*: Branching stem from rhizomes; sporangia on stem scales. *Horsetails*: hollow stem, scale-like leaves, sporangia in strobili.[2]

Spermatophyta (seed plants)—vascular plants in which embryos develop within seeds

Gymnosperms—seeds born on stems, on leaves, or under scales

Cycadophyta (cycads)—shrubby or treelike with palmlike leaves; male and female strobili on separate plants

Ginkgophyta (ginkgoes)—lineage with a single living species (*Ginkgo biloba*); tree with deciduous, fan-shaped leaves; male, female reproductive structures on separate plants

Gnetophyta (gnetophytes)—shrubs or woody vinelike plants; male and female strobili on separate plants

Coniferophyta (conifers)—predominant extant gymnosperm group; mostly evergreen trees and shrubs with needlelike or scalelike leaves; male and female cones usually on the same plant

Anthophyta (angiosperms/flowering plants)—reproductive structures in flowers

Monocotyledones (monocots)—grasses, palms, lilies, orchids and their relatives; a single cotyledon (seed leaf); pollen grains have one groove

Eudicotyledones (eudicots)—roses, melons, beans, potatoes, most fruit trees, others; two cotyledons; pollen grains have three grooves

Other major angiosperm lineages: magnoliids (magnolias and relatives); Star anise (Family Illicium); water lilies (Family Nymphaeaceae); *Amborella* (Family Amborellaceae)

FUNGI Heterotrophic, mostly multicellular organisms with cell wall containing chitin and cell nuclei occurring in threadlike hyphae; life cycle typically includes both asexual and sexual phases, with sexual structures used as the basis for phylum-level classification. Single-celled species are known as yeasts.

Chytridiomycota (chytrids)—mostly aquatic; asexual reproduction by way of motile zoospores; sexual reproduction via gametes produced in gametangia; hyphae mostly aseptate

Zygomycota (zygomycetes)—terrestrial; asexual reproduction via nonmotile haploid spores formed in sporangia; sexual spores (zygospores) form in zygosporangia; aseptate hyphae

Glomeromycota (glomeromycetes)—terrestrial; asexual reproduction via spores at the tips of hyphae; form mycorrhizal associations with plant roots

Ascomycota (ascomycetes/sac fungi)—terrestrial and aquatic; sexual spores form in asci; asexual reproduction occurs via conidia (nonmotile spores); septate hyphae

Basidiomycota (basidiomycetes)—terrestrial; reproduction usually via sexual basidiospores produced by basidia; septate hyphae

Basidiomycetes: mushroom-forming fungi and relatives

Teliomycetes: rusts

Ustomycetes: smuts

Conidial fungi—not a true phylum but a convenience grouping of species for which no sexual phase is known

Microsporidia—single-celled sporelike parasites of animals, other groups; phylogeny uncertain

ANIMALIA Multicellular heterotrophs; nearly all with tissues, organs, and organ systems; motile during at least part of the life cycle; sexual reproduction in most; embryos develop through a series of stages; many with larval and adult stages in life cycle

Parazoa Animals lacking tissues and body symmetry

Porifera (sponges)—multicellular; extract oxygen and particulate food from water drawn into a central cavity

Eumetazoa Animals with tissues and either radial or bilateral symmetry

Radiata—acoelomate animals with radial symmetry and two tissue layers

Cnidaria (cnidarians)—two tissue layers; single opening into gastrovascular cavity; nerve net; nematocysts for defense and predation; some sessile, some motile; most are predatory, some with photosynthetic endosymbionts; freshwater and marine

Hydrozoa: hydrozoans

Scyphozoa: jellyfishes

Cubozoa: box jellyfishes

Anthozoa: sea anemones, corals

Ctenophora (comb jellies)—two (possibly three) tissue layers; feeding tentacles capture particulate food; beating cilia provide weak locomotion; marine

Bilateria—animals with bilateral symmetry and three tissue layers

PROTOSTOMIA—acoelomate, pseudocoelomate, or schizocoelomate; many with spiral, indeterminate cleavage; blastopore forms mouth; nervous system on ventral side

[1]Nonvascular plants (bryophytes)—A polyphyletic group of plants with no specialized structures for transporting water and nutrients; swimming sperm require liquid water for sexual reproduction

[2]Seedless vascular plants—A polyphyletic group of plants in which embryos are not housed inside seeds

Lophotrochozoa—many with either a lophophore for feeding and gas exchange or a trochophore larva

Ectoprocta (bryozoans)—coelomate; colonial; secrete hard covering over soft tissues; lophophore; sessile; particulate feeders; marine

Brachiopoda (lamp shells)—coelomate; dorsal and ventral shells; lophophore; sessile; particulate feeders; marine

Phoronida (phoronid worms)—coelomate; secrete tubes around soft tissues; lophophore; sessile; particulate feeders; marine

Platyhelminthes (flatworms)—acoelomate; dorsoventrally flattened; complex reproductive, excretory, and nervous systems; gastrovascular cavity in many; free-living or parasitic, often with multiple hosts; terrestrial, freshwater, and marine

Turbellaria: free-living flatworms

Trematoda: flukes

Monogenoidea: flukes

Cestoda: tapeworms

Rotifera (wheel animals)—pseudocoelomate; microscopic; complete digestive system; well-developed reproductive, excretory, and nervous systems; particulate feeders; major components of marine and freshwater plankton

Nemertea (ribbon worms)—schizocoelomate; proboscis housed within rhynchocoel; complete digestive tract; circulatory system; predatory; mostly marine

Mollusca (mollusks)—schizocoelomate; many with trochophore larva; many with shell secreted by mantle; body divided into head–foot, visceral mass, and mantle; well-developed organ systems; variable locomotion; herbivorous or predatory; terrestrial, freshwater, and marine

Polyplacophora: chitons

Gastropoda: snails, sea slugs, land slugs

Bivalvia: clams, mussels, scallops, oysters

Cephalopoda: squids, octopuses, cuttlefish, nautiluses

Annelida (segmented worms)—schizocoelomate; many with trochophore larva; segmented body and organ systems; well-developed organ systems; many use hydrostatic skeleton for locomotion; some predatory, some particulate feeders, some detritivores; terrestrial, freshwater, and marine

Polychaeta: marine worms

Oligochaeta: freshwater and terrestrial worms

Hirudinea: leeches

Ecdysozoa—cuticle or exoskeleton is shed periodically

Nematoda (roundworms)—pseudocoelomate; body covered with tough cuticle that is shed periodically; well-developed organ systems; thrashing locomotion; many are parasitic on plants or animals; mostly terrestrial

Onychophora (velvet worms)—schizocoelomate; segmented body covered with cuticle; locomotion by many unjointed legs; complex organ systems; predatory; terrestrial

Arthropoda (arthropods)—schizocoelomate; jointed exoskeleton made of chitin; segmented body, some with fusion of segments in head, thorax, or abdomen; complex organ systems; variable modes of locomotion, including flight; specialization of numerous appendages; herbivorous, predatory, or parasitic; terrestrial, freshwater, and marine

Trilobita: trilobites (extinct)

Chelicerata: horseshoe crabs, spiders, scorpions, ticks, mites

Crustacea: shrimps, crayfishes, lobsters, crabs, barnacles, copepods, isopods

Myriapoda: centipedes, millipedes

Hexapoda: springtails and insects

DEUTEROSTOMIA—enterocoelomate; many with radial, determinate cleavage; blastopore forms anus; nervous system on dorsal side in many

Echinodermata (echinoderms)—secondary radial symmetry, often organized around five radii; hard internal skeleton; unique water vascular system with tube feet; complete digestive system; simple nervous system; no circulatory or respiratory system; generally slow locomotion using tube feet; predatory, herbivorous, particulate feeders, detritivores; exclusively marine

Asteroidea: sea stars

Ophiuroidea: brittle stars

Echinoidea: sea urchins, sand dollars

Holothuroidea: sea cucumbers

Crinoidea: feather stars, sea lilies

Concentricycloidea: sea daisies

Hemichordata (acorn worms)—pharynx perforated with branchial slits; proboscis; complex organ systems; tube-dwelling in soft sediments; particulate or deposit feeders; exclusively marine

Chordata (chordates)—notochord; segmental body wall and tail muscles; dorsal hollow nerve chord; perforated pharynx; complex organ systems; variable modes of locomotion; extremely varied diets; terrestrial, freshwater, and marine

Cephalochordata: lancelets

Urochordata: tunicates, sea squirts

Vertebrata: vertebrates

Myxinoidea: hagfishes

Petromyzontoidea: lampreys

Placodermi: placoderms (extinct)

Chondrichthyes: sharks, skates, and rays

Acanthodii: acanthodians

Actinopterygii: ray-finned fishes

Sarcopterygii: fleshy-finned fishes

Amphibia: salamanders, frogs, caecelians

Synapsida: mammals

Anapsida: turtles

Diapsida: sphenodontids, lizards, snakes, crocodilians, birds

Glossary

3′ end The end of a polynucleotide chain at which a hydroxyl group is bonded to the 3′ carbon of a deoxyribose sugar.

3′ untranslated region (3′ UTR) The part of an mRNA between the stop codon and the 3′ end of the molecule; this region does not code for amino acids.

5′ cap In eukaryotes, a guanine-containing nucleotide attached in a reverse orientation to the 5′ end of pre-mRNA and retained in the mRNA produced from it. The 5′ cap on an mRNA is the site where ribosomes attach to initiate translation.

5′ end The end of a polynucleotide chain at which a phosphate group is bound to the 5′ carbon of a deoxyribose sugar.

5′ untranslated region (5′ UTR) The part of an mRNA between the 5′ end of the molecule and the start codon; this region does not code for amino acids.

10-nm chromatin fiber The most fundamental level of chromatin packing of a eukaryotic chromosome in which DNA winds for almost two turns around an eight-protein nucleosome core particle to form a nucleosome and linker DNA extends between adjacent nucleosomes. The result is a beads-on-a-string type of structure with a 10-nm diameter.

30-nm chromatin fiber Level of chromatin packing of a eukaryotic chromosome in which histone H1 binds to the 10-nm chromatin fiber causing it to package into a coiled structure about 30 nm in diameter and with about six nucleosomes per turn. Also referred to as a *solenoid.*

A site The site where the incoming aminoacyl–tRNA carrying the next amino acid to be added to the polypeptide chain binds to the mRNA.

abdomen The region of the body that contains much of the digestive tract and sometimes part of the reproductive system; in insects, the region behind the thorax.

abiotic Nonbiological, often in reference to physical factors in the environment.

abscisic acid (ABA) A plant hormone involved in the abscission of leaves, flowers, and fruits, dormancy of buds and seeds, and closing of stomata.

abscission In plants, the dropping of flowers, fruits, and leaves in response to environmental signals.

absorption Within the digestive tube, the process in which molecular subunits are absorbed from the digestive contents into body fluids and cells.

absorption spectrum Curve representing the amount of light absorbed at each wavelength.

abyssal zone The bottom sediments that lie permanently below deep ocean water.

accommodation A process by which the lens changes to enable the eye to focus on objects at different distances.

acid Proton donor that releases H^+ (and anions) when dissolved in water.

acid precipitation Rainfall with low pH, primarily created when gaseous sulfur dioxide (SO_2) dissolves in water vapor in the atmosphere, forming sulfuric acid.

acid-growth hypothesis A hypothesis to explain how the hormone auxin promotes growth of plant cells; it suggests that auxin stimulates H^+ pumps in the plasma membrane to move H^+ from the cell interior into the cell wall, which increases wall acidity, making the wall expandable.

acidity The concentration of H^+ in a water solution, as compared with the concentration of OH^-.

acoelomate A body plan of bilaterally symmetrical animals that lack a body cavity between the gut and the body wall.

acoustical signaling A means of animal communication in which a signaler produces a sound that is heard by a signal receiver.

acquired immune deficiency syndrome (AIDS) A constellation of disorders that follows infection by the HIV virus.

acquired immunity *See* adaptive (acquired) immunity.

acrosome A specialized secretory vesicle on the head of an animal sperm, which helps the sperm penetrate the egg.

acrosome reaction The process in which enzymes contained in the acrosome are released from an animal sperm and digest a path through the egg coats.

actin A protein that, in an interaction with the protein myosin, causes muscle contraction.

action potential The abrupt and transient change in membrane potential that occurs when a neuron conducts an electrical impulse.

action spectrum Graph produced by plotting the effectiveness of light at each wavelength in driving photosynthesis.

activation energy The initial input of energy required to start a reaction.

activator A regulatory protein that controls the expression of one or more genes.

active immunity The production of antibodies in the body in response to exposure to a foreign antigen.

active parental care Parents' investment of time and energy in caring for offspring after they are born or hatched.

active site The region of an enzyme to which substrate(s) bind and where catalysis occurs.

active transport The mechanism by which ions and molecules move against the concentration gradient across a membrane, from the side with the lower concentration to the side with the higher concentration.

adaptation Characteristic that helps an organism survive longer or reproduce more under a particular set of environmental conditions.

adaptation, evolutionary The accumulation of adaptive traits over time.

adaptation, sensory *See* sensory adaptation.

adaptive (acquired) immunity A specific line of defense against invasion of the body in which individual pathogens are recognized and attacked to neutralize and eliminate them.

adaptive immune system The third of three lines of defense that humans and other mammals have against threats of pathogens; the inherited mechanisms leading to the synthesis of molecules that target pathogens in a specific way.

adaptive radiation A cluster of closely related species that are each adaptively specialized to a specific habitat or food source.

adaptive trait A genetically based characteristic, preserved by natural selection, that increases an organism's likelihood of survival or its reproductive output.

adaptive zone An environment or part of an environment that may be occupied by a group of species exploiting resources in a similar manner.

adductor muscle A muscle that pulls inward toward the median line of the body; in bivalve mollusks, it pulls the shell closed.

adenine A purine that base-pairs with either thymine in DNA or uracil in RNA.

adherens junction Animal cell junction in which intermediate filaments are the anchoring cytoskeletal component.

adhesion The adherence of molecules to the walls of conducting tubes, as in plants.

adiabatic cooling A decrease in temperature without the actual loss of heat energy, occurring in air masses that expand as they rise in the atmosphere.

adipocytes Densely clustered cells within adipose tissue which are specialized for fat storage.

adipose tissue Connective tissue containing large, densely clustered cells called adipocytes that are specialized for fat storage.

adrenal cortex The outer region of the adrenal glands, which contains endocrine cells that secrete two major types of steroid hormones, the glucocorticoids and the mineralocorticoids.

adrenal medulla The central region of the adrenal glands, which contains neurosecretory neurons that secrete the catecholamine hormones epinephrine and norepinephrine.

adrenergic receptors G-protein–coupled receptors to which hormones originating in the adrenal gland can bind.

adrenocorticotropic hormone (ACTH) A hormone that triggers hormone secretion by cells in the adrenal cortex.

adventitious root A root that develops from the stem or leaves of a plant.

aerobe An organism that requires oxygen for cellular respiration.

aerobic respiration The form of cellular respiration found in eukaryotes and many prokaryotes in which oxygen is a reactant in the ATP-producing process.

afferent arteriole The vessel that delivers blood to the glomerulus of the kidney.

afferent neuron A neuron that transmits stimuli collected by a sensory receptor to an interneuron.

agar A gelatinous product extracted from certain red algae or seaweed used as a culture medium in the laboratory and as a gelling or stabilizing agent in foods.

agarose gel electrophoresis Technique by which DNA, RNA, or protein molecules are separated in an agarose gel subjected to an electric field.

age structure A statistical description or graph of the relative numbers of individuals in each age class in a population.

age-specific fecundity The average number of offspring produced by surviving females of a particular age.

age-specific mortality The proportion of individuals alive at the start of an age interval that died during that age interval.

age-specific survivorship The proportion of individuals alive at the start of an age interval that survived until the start of the next age interval.

agglutination One of two important mechanisms to clear foreign antigens from the body, the immobilization of pathogens by antibodies.

aggregate fruit A fruit that develops from multiple separate carpels of a single flower, such as a raspberry or strawberry.

agonist A muscle that causes movement in a joint when it contracts.

albumin The most abundant protein in blood plasma, important for osmotic balance and pH buffering; also, the portion of an egg that serves as the main source of nutrients and water for the embryo.

alcohol A molecule of the form R—OH in which R is a chain of one or more carbon atoms, each of which is linked to hydrogen atoms.

alcoholic fermentation Reaction in which pyruvate is converted into ethyl alcohol and CO_2 in a two-step series that also converts NADH into NAD^+.

aldehyde Molecule in which the carbonyl group is linked to a carbon atom at the end of a carbon chain, along with a hydrogen atom.

aldosterone A mineralocorticoid hormone released from the adrenal cortex that increases the amount of Na^+ reabsorbed from the urine in the kidneys and absorbed from foods in the intestine, reduces the amount of Na^+ secreted by salivary and sweat glands, and increases the rate of K^+ excretion by the kidneys, keeping Na^+ and K^+ balanced at the levels required for normal cellular function.

aleurone The thin layer of cells that separates the endosperm of a seed from the pericarp.

algin Alginic acid, found in the cell walls of brown algae.

allantois In an amniote egg, an extraembryonic membrane sac that fills much of the space between the chorion and the yolk sac and store's the embryo's nitrogenous wastes.

allele One of two or more versions of a gene.

allele frequency The abundance of one allele relative to others at the same gene locus in individuals of a population.

allergen A type of antigen responsible for allergic reactions, which induces B cells to secrete an overabundance of IgE antibodies.

allometric growth A pattern of postembryonic development in which parts of the same organism grow at different rates.

allopatric speciation The evolution of reproductive isolating mechanisms between two populations that are geographically separated.

allopolyploidy The genetic condition of having two or more complete sets of chromosomes from different parent species.

all-or-nothing principle The principle that an action potential is produced only if the stimulus is strong enough to cause depolarization to reach the threshold.

allosteric activator Molecule that converts an enzyme with an allosteric site, a regulatory site outside the active site, from the inactive form to the active form.

allosteric inhibitor Molecule that converts an enzyme with an allosteric site, a regulatory site outside the active site, from the active form to the inactive form.

allosteric regulation Specialized control mechanism for enzymes with an allosteric site, a regulatory site outside the active site, that may either slow or accelerate activity depending on the enzyme.

allosteric site A regulatory site outside the active site.

alpha (α) helix A type of secondary structure of a polypeptide in which the amino acid chain is twisted into a regular, right-hand spiral.

α-melanocyte-stimulating hormone (alpha-MSH) A peptide hormone released from the hypothalamus which inhibits appetite.

alpine tundra A biome that occurs on high mountaintops throughout the world, in which dominant plants form cushions and mats.

alternation of generations The regular alternation of mode of reproduction in the life cycle of an organism, such as the alternation between diploid (sporophyte) and haploid (gametophyte) phases in plants.

alternative hypothesis An explanation of an observed phenomenon that is different from the explanation being tested.

alternative splicing Mechanism by which a pre-mRNA in a eukaryotic cell is processed by reactions that join exons in different combinations to produce different mRNAs from a single gene.

altruism A behavioral phenomenon in which individuals appear to sacrifice their own reproductive success to help other individuals.

Alveolata (alveolates) A subgroup of the Chromalveolata protist evolutionary group; characterized by small, flattened, membrane-bound vesicles called alveoli in a layer just under the plasma membrane.

alveolus (plural, alveoli) One of the millions of tiny air pockets in mammalian lungs, each surrounded by dense capillary networks.

amacrine cell A type of neuron that forms lateral connections in the retina of the eye, connecting bipolar cells and ganglion cells.

amino acid A molecule that contains both an amino and a carboxyl group.

amino group Group that acts as an organic base, consisting of a nitrogen atom bonded on one side to two hydrogen atoms and on the other side to a carbon chain.

aminoacylation The process of adding an amino acid to a tRNA. Also referred to as *charging*.

aminoacyl–tRNA A tRNA linked to its "correct" amino acid, which is the finished product of aminoacylation.

aminoacyl–tRNA synthetase An enzyme that catalyzes aminoacylation.

ammonification A metabolic process in which bacteria convert organic nitrogen compounds into ammonia and ammonium ions; part of the nitrogen cycle.

amniocentesis Technique of prenatal diagnosis in which cells are obtained from the amniotic fluid for DNA testing, biochemical analysis, or for the presence of chromosomal mutations.

amnion In an amniote egg, an extraembryonic membrane that encloses the embryo, forming the amniotic cavity and secreting amniotic fluid, which provides an aquatic environment in which the embryo develops.

Amniota The monophyletic group of vertebrates that have an amnion during embryonic development.

amniote egg A shelled egg that can survive and develop on land.

amoeba A descriptive term for a single-celled protist that moves by means of temporary cellular projections called pseudopods.

Amoebozoa (amoebozoans) A protist evolutionary group that includes most of the amoebas as well as the cellular and plasmodial slime molds; characterized by the use of pseudopods for locomotion and feeding for all or part of their life cycles.

amplification An increase in the magnitude of each step as a signal transduction pathway proceeds.

amygdala A gray-matter center of the brain that works as a switchboard, routing information about experiences that have an emotional component through the limbic system.

amyloplast Colorless plastid that stores starch in plants.

anabolic-androgenic steroid (AAS) A synthetic derivative of androgens that mimics their effects. Also called an *anabolic steroid*.

anabolic pathway A metabolic pathway in which energy is used to build complicated molecules from simpler ones; also called a biosynthetic pathway. An individual reaction in an anabolic pathway is an anabolic reaction, also called a *biosynthetic reaction*.

anabolic reaction Metabolic reaction that requires energy to assemble simple substances into more complex molecules.

anabolic steroid *See* anabolic-androgenic steroid (AAS).

anaerobe An organism that does not require oxygen to live.

anaerobic respiration The form of cellular respiration found in some prokaryotes in which a molecule other than oxygen is used in the ATP-producing process.

anagenesis The slow accumulation of evolutionary changes in a lineage over time.

anal sphincter A muscular ring that controls the opening and closing of the anus.

anaphase The phase of mitosis during which the spindle separates sister chromatids and pulls them to opposite spindle poles.

anaphase-promoting complex (APC) An enzyme complex activated by M phase-promoting factor that controls the separation of sister chromatids and the onset of daughter chromosome separation in anaphase of mitosis.

anaphylactic shock A severe inflammation stimulated by an allergen, involving extreme swelling of air passages in the lungs that interferes with breathing, and massive leakage of fluid from capillaries that causes blood pressure to drop precipitously.

anapsida A lineage of amniotes that lack a bony temporal arch in the skull.

anatomy The study of the structures of organisms.

ancestral character state A trait that was present in a distant common ancestor.

anchoring junction Cell junction that forms belts that run entirely around cells, "welding" adjacent cells together.

androgen One of a family of hormones that promote the development and maintenance of sex characteristics.

aneuploid An individual with extra or missing chromosomes.

angiosperm A flowering plant. Its egg-containing ovules mature into seeds within protected chambers called ovaries.

angiotensin A peptide hormone that raises blood pressure quickly by constricting arterioles in most parts of the body; it also stimulates release of the steroid hormone aldosterone.

angiotensin I In the renin-angiotensin-aldosterone system (RAAS), the molecule produced by cleavage of the plasma protein angiotensinogen.

angiotensin II In the renin-angiotensin-aldosterone system (RAAS), the molecule converted from angiotensin I by angiotensin-converting enzyme (ACE); angiotensin II is a hormone that constricts arterioles to raise blood pressure, stimulates synthesis of aldosterone and its secretion from the adrenal cortex, and stimulates thirst.

angiotensin-converting enzyme (ACE) In the renin-angiotensin-aldosterone system (RAAS), the enzyme that converts angiotensin I to angiotensin II.

animal behavior The responses of animals to specific internal and external stimuli.

animal pole The end of the egg where the egg nucleus is located, which typically gives rise to surface structures and the anterior end of the embryo.

Animalia The taxonomic kingdom that includes all living and extinct animals.

anion A negatively charged ion.

annual An herbaceous plant that completes its life cycle in one growing season and then dies.

annulus In ferns, a ring of thick-walled cells that nearly encircles the sporangium and functions in spore release.

antagonistic pair Two skeletal muscles, one of which flexes as the other extends to move joints.

antenna A chemosensory appendage attached to the head of some adult arthropods.

antenna complex (light-harvesting complex) In photosystems, the sites at which light is absorbed and converted into chemical energy during photosynthesis, an aggregate of many chlorophyll pigments and a number of carotenoid pigments that serves as the primary site of absorbing light energy in the form of photons.

anterior Indicating the head end of an animal.

anterior pituitary The glandular part of the pituitary, composed of endocrine cells that synthesize and secrete several tropic and nontropic hormones.

anther The pollen-bearing part of a stamen.

antheridium (plural, antheridia) In plants, a structure in which sperm are produced.

Anthocerophyta The phylum comprising hornworts.

Anthophyta The phylum comprising flowering plants.

Anthropoidea The monophyletic lineage of primates comprising monkeys, apes, and humans.

antibody A highly specific soluble protein molecule that circulates in the blood and lymph, recognizing and binding to antigens and clearing them from the body.

antibody class The molecular type of an antibody determined by the constant regions of the heavy chains of the molecule.

antibody-mediated immunity Adaptive immune response in which plasma cells secrete antibodies.

anticodon The three-nucleotide segment in a tRNA that pairs with a codon in an mRNA.

antidiuretic hormone (ADH) A hormone secreted by the posterior pituitary that increases water absorption in the kidneys, thereby increasing the volume of the blood.

antigen A foreign molecule that triggers an adaptive immunity response.

antigen-presenting cell (APC) A cell that presents an antigen to T cells in antibody-mediated immunity and cell-mediated immunity.

antihistamines Substances that block histamine receptors thereby reducing symptoms of an allergic reaction.

antiparallel Strands of double-stranded DNA that run in opposite directions with the 3′ end of one strand opposite the 5′ end of the other strand.

antiport A secondary active transport mechanism in which a molecule moves through a membrane channel into a cell and powers the active transport of a second molecule out of the cell. Also referred to as *exchange diffusion*.

aorta A large artery from the heart that branches into arteries leading to all body regions except the lungs.

aortic body One of several small clusters of chemoreceptors, baroreceptors, and supporting cells located along the aortic arch, that measures changes in blood pressure and the composition of arterial blood flowing past it.

aphotic zone Deeper water of a lake or ocean where sunlight does not penetrate.

apical dominance Inhibition of the growth of lateral buds in plants due to auxin diffusing down a shoot tip from the terminal bud.

apical meristem A region of unspecialized dividing cells at shoot tips and root tips of a plant.

apical surface The outer surface of epithelial cells.

Apicomplexa (apicomplexans) A lineage within the Alveolata subgroup of the Chromalveolata protist evolutionary group; nonmotile parasites of animals characterized by the apical complex, a special grouping of fibrils, microtubules, and organelles at one end of the cell that functions in attachment and invasion of host cells.

apomixis In plants, the production of offspring without meiosis or formation of gametes.

apomorphy A derived character state.

apoplastic pathway The route followed by water moving through plant cell walls and intercellular spaces (the apoplast). *Compare* symplastic pathway.

apoptosis Programmed cell death.

aposematic coloration Bright, contrasting patterns that advertise the unpalatability of poisonous or repellant species.

appendicular skeleton The bones comprising the pectoral (shoulder) and pelvic (hip) girdles and limbs of a vertebrate.

appendix A fingerlike sac that extends from the cecum of the large intestine.

applied research Research conducted with the goal of solving specific practical problems.

aquaporin A specialized protein channel that facilitates diffusion of water through cell membranes.

aquatic succession A process in which debris from rivers and runoff accumulates in a body of fresh water, causing it to fill in at the margins.

aqueous humor A clear fluid that fills the space between the cornea and lens of the eye.

Archaea One of two domains of prokaryotes; archaeans have some unique molecular and biochemical traits, but they also share some traits with Bacteria and other traits with Eukarya.

Archaeplastida (archaeplastids) An evolutionary group consisting of the Rhodophyta and Chlorophyta protists, and the land plants; all are photosynthesizers.

archegonium The flask-shaped structure in which bryophyte eggs form.

archenteron The central endoderm-lined cavity of an embryo at the gastrula stage, which forms the primitive gut.

Archosauromorpha A diverse group of diapsids that comprises crocodilians, pterosaurs, and dinosaurs (including birds).

arctic tundra A treeless biome that stretches from the boreal forests to the polar ice cap in Europe, Asia, and North America.

arteriole A branch from a small artery at the point where it reaches the organ it supplies.

artery A vessel that conducts blood away from the heart at relatively high pressure.

artificial selection Selective breeding of organisms to ensure that certain desirable traits appear at higher frequency in successive generations.

ascending reticular formation Part of the reticular formation containing neurons that convey stimulatory signals via the thalamus to arouse and activate the cerebral cortex. It is responsible for the sleep–wake cycle. Also called the *reticular activating system*.

ascocarp In ascomycete (sac) fungi, a reproductive body that bears or contains asci.

ascus (plural, asci) A saclike cell in ascomycetes (sac fungi) in which meiosis gives rise to haploid sexual spores (meiospores).

asexual reproduction Any mode of reproduction in which a single individual gives rise to offspring without fusion of gametes; that is, without genetic input from another individual. See also *vegetative reproduction*.

association area One of several areas surrounding the sensory and motor areas of the cerebral cortex that integrate information from the sensory areas, formulate responses, and pass them on to the primary motor area.

aster Radiating array produced as microtubules extending from the centrosomes of cells grow in length and extent.

asthma In an allergic reaction, a severe response to allergens that involves constriction of airways in the lungs.

astrocyte A star-shaped glial cell that provides support to neurons in the vertebrate central nervous system.

asymmetrical Characterized by a lack of proportion in the spatial arrangement or placement of parts.

atmosphere The component of the biosphere that includes the gases and airborne particles enveloping the planet.

atom The smallest unit that retains the chemical and physical properties of an element.

atomic nucleus The nucleus of an atom, containing protons and neutrons.

atomic number The number of protons in the nucleus of an atom.

atomic weight The weight of an element in grams, equal to the mass number.

ATP (adenosine triphosphate) The primary agent that couples exergonic and endergonic reactions.

ATP/ADP cycle The continual hydrolysis and resynthesis of ATP in living cells.

ATP synthase A membrane-spanning protein complex that couples the energetically favorable transport of protons across a membrane to the synthesis of ATP.

atrial natriuretic factor (ANF) A peptide hormone that inhibits renin release and increases the filtration rate by dilating the arterioles that deliver blood to glomeruli and by inhibiting aldosterone release.

atrial siphon A tube through which invertebrate chordates expel digestive and metabolic wastes.

atriopore The hole in the body wall of a cephalochordate through which water is expelled from the body.

atrioventricular node (AV node) A region of the heart wall that receives signals from the sinoatrial node and conducts them to the ventricle.

atrium (plural, atria) A body cavity or chamber surrounding the perforated pharynx of invertebrate chordates; also, one of the chambers that receive blood returning to the heart.

autoimmune disease A malfunction of the immune system in which the body reacts against its own proteins or cells.

autoimmune reaction The production of antibodies against molecules of the body.

autonomic nervous system A subdivision of the peripheral nervous system that controls largely involuntary processes including digestion, secretion by sweat glands, circulation of the blood, many functions of the reproductive and excretory systems, and contraction of smooth muscles in all parts of the body.

autopolyploidy The genetic condition of having more than two sets of chromosomes from the same parent species.

autosomal dominant inheritance Pattern in which the allele that causes a trait is dominant, and only homozygous recessives are unaffected.

autosomal recessive inheritance Pattern in which individuals with a trait are homozygous for a recessive allele.

autosome Chromosome other than a sex chromosome.

autotroph An organism that produces its own food using CO_2 and other simple inorganic compounds from its environment and energy from the sun or from oxidation of inorganic substances.

autumn overturn A process in which winds mix the water in a lake vertically, equalizing the concentrations of dissolved gases and nutrients at all depths.

auxin Any of a family of plant hormones that stimulate growth by promoting cell elongation in stems and coleoptiles; inhibit abscission; govern responses to light and gravity, and have other developmental effects.

auxotroph A mutant strain that requires for its growth a nutrient supplement that is not needed by the wild-type strain.

Avogadro's number The number 6.022×10^{23}, derived by dividing the atomic weight of any element by the weight of an atom of that element.

***Avr* gene** A gene in certain plant pathogens that encodes a product triggering a defensive response in the plant.

axial skeleton The bones comprising the head and trunk of a vertebrate: the cranium, vertebral column, ribs, and sternum (breastbone).

axil The upper angle between the stem and an attached leaf.

axon The single elongated extension of a neuron that conducts signals away from the cell body to another neuron or an effector.

axon hillock A junction with the cell body of a neuron from which the axon arises.

axon terminal A branch at the tip of an axon that ends as a small, buttonlike swelling.

B cell A lymphocyte that recognizes antigens in the body.

Bacillariophyta (diatoms) A lineage of the Stramenopila subgroup of the Chromalveolata protist evolutionary group; single-celled organisms covered by an intricate, glassy silica shell.

bacillus (plural, bacilli) A cylindrical or rod-shaped prokaryote.

background extinction rate The average rate of extinction of taxa through time.

Bacteria One of the two domains of prokaryotes; collectively, bacteria are the most metabolically diverse organisms.

bacterial chromosome DNA molecule in bacteria in which hereditary information is encoded.

bacterial flagellum *See* flagellum.

bacteriophage A virus that infects bacteria. Also referred to as a *phage*.

bacteroid A rod-shaped or branched bacterium in the root nodules of nitrogen-fixing plants.

balanced polymorphism The maintenance of two or more phenotypes in fairly stable proportions over many generations.

bark The tough outer covering of woody stems and roots, composed of all the living and nonliving tissues between the vascular cambium and the stem surface.

Barr body The inactive, condensed X chromosome seen in the nucleus of female placental mammals.

basal angiosperm Any of the earliest branches of the flowering plant lineage; includes the star anise group and water lilies.

basal body Structure that anchors cilia and flagella to the surface of a cell.

basal lamina The membrane that fixes the epithelium to underlying tissues (also called the basement membrane).

basal nucleus One of several gray-matter centers that surround the thalamus on both sides of the brain and moderate voluntary movements directed by motor centers in the cerebrum.

basal surface The inner surface of epithelial cells.

base Proton acceptor that reduces the H^+ concentration of a solution.

base-pair mismatch An error in the assembly of a new nucleotide chain in which bases other than the correct ones pair together.

basic research Research conducted to search for explanations about natural phenomena in order to satisfy curiosity and to advance collective knowledge of living systems.

basidiocarp A fruiting body of a basidiomycete; mushrooms are examples.

basidiospore A haploid sexual spore produced by basidiomycete fungi.

basidium (plural, basidia) A small, club-shaped structure in which sexual spores of basidiomycetes arise.

basilar membrane A part of the floor of the cochlear duct, which anchors sensory hair cells in the organ of Conti and which vibrates in response to vibrations moving through the inner ear.

basophil A type of leukocyte that is induced to secrete histamine by allergens.

Batesian mimicry The form of defense in which a palatable or harmless species resembles an unpalatable or poisonous one.

Bayesian methods A statistical technique that evaluates a given phylogenetic tree by determining the probability that it is correct given the distribution of character states included in the analysis.

B-cell receptor (BCR) The receptor on B cells that is specific for a particular antigen.

behavioral isolation A prezygotic reproductive isolating mechanism in which two species do not mate because of differences in courtship behavior; also known as ethological isolation.

behavioral repertoire The set of actions that an animal can perform in response to stimuli in its environment.

benthic province The bottom sediments in the ocean.

benthos Species living in and on the bottom sediments of the ocean.

beta (β) sheet A type of primary structure in a polypeptide in which the amino acid chain zigzags in a flat plane to form a beta strand, and beta strands then align side by side in the same or opposite direction.

biennial A plant that completes its life cycle in two growing seasons and then dies; limited secondary growth occurs in some biennials.

Bikonta (bikonts) Eukaryotes with two flagella.

bilateral symmetry The body plan of animals in which the body can be divided into mirror image right and left halves by a plane passing through the midline of the body.

bilayer A membrane with two molecular layers.

bile A mixture of substances including bile salts, cholesterol, and bilirubin that is made in the liver, stored in the gallbladder, and used in the digestion of fats.

bile salts A component of bile, consisting of derivatives of cholesterol and amino acids, that aids fat digestion.

bilirubin A component of bile consisting of a waste product derived from worn-out red blood cells.

binomial Relating to or consisting of two names or terms.

binomial nomenclature The naming of species with a two part scientific name, the first indicating the genus and the second indicating the species.

biodiversity The richness of living systems as reflected in genetic variability within and among species, the number of species living on Earth, and the variety of communities and ecosystems.

biodiversity hot spot An area where biodiversity is both highly concentrated and endangered.

biofilm A microbial community consisting of a complex aggregation of microorganisms attached to a surface.

biogeochemical cycle Any of several global processes in which a nutrient circulates between the abiotic environment and living organisms.

biogeographical realm A major region of Earth that is occupied by distinct evolutionary lineages of plants and animals.

biogeography The study of the geographical distributions of plants and animals.

bioinformatics Field that fuses biology with mathematics and computer science that is used for the analysis of genome sequences.

biological clock An internal time-measuring mechanism that adapts an organism to recurring environmental changes.

biological evolution The process by which some individuals in a population experience changes in their DNA and pass those modified instructions to their offspring.

biological lineage An evolutionary sequence of ancestral organisms and their descendants.

biological magnification The increasing concentration of nondegradable poisons in the tissues of animals at higher trophic levels.

biological research The collective effort of individuals who have worked to understand how living systems function.

biological species concept The definition of species based upon the ability of populations to interbreed and produce fertile offspring.

bioluminescent An organism that glows or releases a flash of light, particularly when disturbed.

biomass The dry weight of biological material per unit area or volume of habitat.

biome A large scale vegetation type and its associated microorganisms, fungi, and animals.

bioremediation Applications of chemical and biological knowledge to decontaminate polluted environments.

biosignature Particular organic molecules in sedimentary rocks that could only have been formed by cellular activity.

biosphere All regions of Earth's crust, waters, and atmosphere that sustain life.

biosynthetic reaction An individual reaction in an anabolic pathway (biosynthetic pathway).

biota The total collection of organisms in a geographic region.

biotechnology The manipulation of living organisms to produce useful products.

biotic Biological, often in reference to living components of the environment.

bipedalism The habit in animals of walking upright on two legs.

bipolar cell A type of neuron in the retina of the eye that connects the rods and cones with the ganglion cells.

blade The expanded part of a leaf that provides a large surface area for absorbing sunlight and carbon dioxide.

blastocoel A fluid-filled cavity in the blastula embryo.

blastocyst An embryonic stage in mammals; a single-cell-layered hollow ball of about 120 cells with a fluid-filled blastocoel in which a dense mass of cells is localized to one side.

blastodisc A disclike layer of cells at the surface of the yolk produced by early cleavage divisions.

blastomere A small cell formed during cleavage of the embryo.

blastopore The opening at one end of the archenteron in the gastrula that gives rise to the mouth in protostomes and the anus in deuterostomes.

blastula The hollow ball of cells that is the result of cleavage divisions in an early embryo.

blending theory of inheritance Theory suggesting that hereditary traits blend evenly in offspring through mixing of the blood of the two parents.

blood A fluid connective tissue composed of blood cells suspended in a fluid extracellular matrix, plasma.

blood–brain barrier A specialized arrangement of capillaries in the brain that prevents most substances dissolved in the blood from entering the cerebrospinal fluid and thus protects the brain and spinal cord from viruses, bacteria, and toxic substances that may circulate in the blood.

blood pressure The measurement of the hydrostatic pressure on the walls of the arteries as the heart pumps blood through the body.

body system *See* organ system.

Bohr effect The reduction in affinity of hemoglobin for O_2 that results from a conformational change caused by lowered pH.

bolting Rapid formation of a floral shoot in plant species that form rosettes, such as lettuce.

bolus The food mass after chewing.

bone The densest form of connective tissue, in which living cells secrete the mineralized matrix of collagen and calcium salts that surrounds them; forms the skeleton.

bony temporal arches Bones that border the temporal fenestrae in the skulls of some amniotes.

book lungs Pocketlike respiratory organs found in some arachnids consisting of several parallel membrane folds arranged like the pages of a book.

boreal forest A biome that is a circumpolar expanse of evergreen coniferous trees in Europe, Asia, and North America.

Bowman's capsule An infolded region at the proximal end of a nephron that cups around the glomerulus and collects the water and solutes filtered out of the blood.

brachiation A pattern of locomotion among primates in which an individual swings below branches from one handhold to another.

brain A single, organized collection of nervous tissue in an organism's head that forms the control center of the nervous system and major sensory structures.

brain stem A stalklike structure formed by the pons and medulla, along with the midbrain, which connects the forebrain with the spinal cord.

branchial slits Openings in the walls of the pharynx that allow water to exit the pharynx.

brassinosteroid Any of a family of plant hormones that stimulate cell division and elongation and differentiation of vascular tissue.

breathing The exchange of gases with the respiratory medium by animals.

bronchiole One of the small, branching airways in the lungs that lead into the alveoli.

bronchus (plural, bronchi) An airway that leads from the trachea to the lungs.

brown adipose tissue A specialized tissue in which the most intense heat generation by nonshivering thermogenesis takes place.

brown algae *See* Phaeophyta.

brown fat *See* brown adipose tissue.

brush border The fingerlike projections of the plasma membrane of the epithelial cells covering the intestinal villi.

Bryophyta The phylum of nonvascular plants to which mosses are assigned.

bryophyte A general term for plants (such as mosses) that lack internal transport vessels.

budding A mode of asexual reproduction in which a new individual grows and develops while attached to the parent.

buffer Substance that compensates for pH changes by absorbing or releasing H^+.

bulbourethral gland One of two pea-sized glands on either side of the prostate gland, which secrete a mucous fluid that is added to semen.

bulk feeder An animal that consumes sizeable food items whole or in large chunks.

bulk flow The group movement of molecules in response to a difference in pressure between two locations.

bulk-phase endocytosis Mechanism by which extracellular water is taken into a cell together with any molecules that happen to be in solution in the water. Also referred to as *pinocytosis.*

C_3 pathway *See* light-independent reaction; also referred to as the *Calvin cycle.*

C_4 pathway In C_4 plants the pathway to fix CO_2 into oxaloacetate in mesophyll cells and then produce CO_2 for the Calvin cycle in bundle sheath cells.

Ca^{2+} pump (calcium pump) Pump that pushes Ca^{2+} from the cytoplasm to the cell exterior, and also from the cytosol into the vesicles of the endoplasmic reticulum.

cadherin A cell surface protein responsible for selective cell adhesions that require calcium ions to set up adhesions.

calcitonin A nontropic peptide hormone that lowers the level of Ca^{2+} in the blood by inhibiting the ongoing dissolution of calcium from bone.

callus An undifferentiated tissue that develops on or around a cut plant surface or in tissue culture.

calorie (cal) The amount of heat required to raise 1 g of water by 1°C, known as a "small" calorie; when capitalized, a unit equal to 1,000 small calories.

Calvin cycle *See* light-independent reaction.

calyx The outermost whorl of a flower, made up of sepals; early in the development of a flower, it encloses all the other parts, as in an unopened bud.

CAM pathway In CAM plants the pathway to fix CO_2 into oxaloacetate and then produce CO_2 for the Calvin cycle, both occurring in mesophyll cells, but separated by time of day. CAM stands for "crassulacean acid metabolism."

canines Pointed, conical teeth of a mammal, located between the incisors and the first premolars, that are specialized for biting and piercing.

CAP (catabolite activator protein) Key regulatory molecule involved in positive gene regulation of the *lac* operon.

CAP site Region in the promoter of the *lac* operon and in the promoters of a large number of other operons that control the catabolism of many sugars to which activated catabolite activator protein (CAP) binds, thereby enabling RNA polymerase to bind and transcribe the operon's structural genes.

capillary The smallest diameter blood vessel, with a wall that is one cell thick, which forms highly branched networks well adapted for diffusion of substances.

capsid The protective layer of protein that surrounds the nucleic acid core of a virus in free form; also known as a *coat.*

capsule An external layer of sticky or slimy polysaccharides coating the cell wall in many prokaryotes.

carapace A protective outer covering that extends backward behind the head on the dorsal side of an animal, such as the shell of a turtle or lobster.

carbon cycle The global circulation of carbon atoms, especially via the processes of photosynthesis and respiration.

carbonyl group The reactive part of aldehydes and ketones, consisting of an oxygen atom linked to a carbon atom by a double bond.

carboxyl group The characteristic functional group of organic acids, formed by the combination of carbonyl and hydroxyl groups.

carboxypeptidase An enzyme that cuts amino acids from polypeptide chains one at a time.

cardiac cycle The systole-diastole sequence of the heart.

cardiac muscle The contractile tissue of the heart.

carnivore An animal that primarily eats other animals.

carotenoid Molecule of yellow-orange pigment by which light is absorbed in photosynthesis.

carotid body A small cluster of chemoreceptors and supporting cells located near the bifurcation of the carotid artery that measures changes in the composition of arterial blood flowing through it.

carpel The reproductive organ of a flower that houses an ovule and its associated structures.

carrageenan A chemical extracted from the red alga *Eucheuma* that is used to thicken and stabilize paints, dairy products such as pudding and ice cream, and many other creams and emulsions.

carrier A heterozygote—an individual who carries a recessive mutant allele and could pass it on to offspring, but does not display its symptoms.

carrier protein Transport protein that binds a specific single solute and transports it across the lipid bilayer.

carrying capacity The maximum size of a population that an environment can support indefinitely.

Cartagena Protocol on Biosafety An international agreement that promotes biosafety as it relates to the handling and use of genetically modified organisms.

cartilage A tissue composed of sparsely distributed chondrocytes surrounded by networks of collagen fibers embedded in a tough but elastic matrix of the glycoprotein.

cartilaginous joint A joint between bones that is not very movable and in which the ends of the bones are covered with layers of cartilage but with no fluid-filled capsule surrounding them.

Casparian strip A thin, waxy impermeable band that seals abutting cell walls in roots; the strip helps control the type and amount of solutes that enter the stele by blocking the apoplastic pathway at the endodermis and forcing substances to pass through cells (the symplast).

catabolic pathway A metabolic pathway in which energy is released by the breakdown of complex molecules to simpler compounds. An individual reaction in a catabolic pathway is a catabolic reaction.

catabolic reaction Cellular reaction that breaks down complex molecules such as sugar to make their energy available for cellular work.

catalysis The process of accelerating a chemical reaction with a catalyst.

catalyst Substance with the ability to accelerate a spontaneous reaction without being changed by the reaction.

catastrophism The theory that Earth has been affected by sudden, violent events that were sometimes worldwide in scope.

catecholamine Any of a class of compounds derived from the amino acid tyrosine that circulates in the bloodstream, including epinephrine and norepinephrine.

cation A positively charged ion.

cation exchange Replacement of one cation with another, as on a soil particle.

$CD4^+$ T cell A type of T cell in the lymphatic system that has CD4 receptors on its surface. This type of T cell binds to an antigen-presenting cell in antibody-mediated immunity.

$CD8^+$ T cell A type of T cell in the lymphatic system that has CD8 receptors on its surface. This type of T cell binds to an antigen-presenting cell in cell-mediated immunity.

cDNA library The entire collection of cloned cDNAs made from the mRNAs isolated from a cell.

cecum A blind pouch formed at the junction of the large and small intestine.

cell Smallest unit with the capacity to live and reproduce.

cell adhesion molecule A cell surface protein responsible for selectively binding cells together.

cell adhesion protein Protein that binds cells together by recognizing and binding receptors or chemical groups on other cells or on the extracellular matrix.

cell body The portion of the neuron containing genetic material and cellular organelles.

cell center *See* centrosome.

cell culture Living cells growing in a growth medium in a laboratory vessel.

cell cycle The sequence of events during which a cell experiences a period of growth followed by nuclear division and cytokinesis.

cell differentiation A process in which changes in gene expression establish cells with specialized structure and function.

cell expansion A mechanism that enlarges the cells in specific directions in a developing organ.

cell fractionation Technique that divides cells into fractions containing a single cell component.

cell junction Junction that seals the spaces between cells and provides direct communication between cells.

cell lineage Cell derivation from the undifferentiated tissues of the embryo.

cell plate In cytokinesis in plants, a new cell wall that forms between the daughter nuclei and grows laterally until it divides the cytoplasm.

cell signaling The system of communication between cells through signaling pathways.

cell theory Three generalizations yielded by microscopic observations: all organisms are composed of one or more cells; the cell is the smallest unit that has the properties of life; and cells arise only from the growth and division of preexisting cells.

cell wall A rigid external layer of material surrounding the plasma membrane of cells in plants, fungi, bacteria, and some protists, providing cell protection and support.

cell-mediated immunity An adaptive immune response in which a subclass of T cells—cytotoxic T cells—becomes activated and, with other cells of the immune system, attacks host cells infected by pathogens, particularly those infected by a virus.

cellular respiration The process by which energy-rich molecules are broken down to produce energy in the form of ATP.

cellular slime mold Any of a variety of primitive organisms of the phylum Acrasiomycota, especially of the genus *Dictyostelium;* the life cycle is characterized by a slimelike amoeboid stage and a multicellular reproductive stage.

cellulose One of the primary constituents of plant cell walls, formed by chains of carbohydrate subunits.

centimorgan *See* map unit.

central canal The central portion of the vertebral column in which the spinal cord is found.

central dogma The name given by Francis Crick to the flow of information from DNA to RNA to protein.

central nervous system (CNS) One of the two major divisions of the nervous system containing the brain and spinal cord.

central vacuole A large, water-filled organelle in plant cells that maintains the turgor of the cell and controls movement of molecules between the cytosol and sap.

centriole A cylindrical structure consisting of nine triplets of microtubules in the centrosomes of most animal cells.

centromere A specialized chromosomal region that connects sister chromatids and attaches them to the mitotic spindle.

centrosome (cell center) The main microtubule organizing center of a cell, which organizes the microtubule cytoskeleton during interphase and positions many of the cytoplasmic organelles.

cephalothorax The anterior section of an arachnid, consisting of a fused head and thorax.

cerebellum The portion of the brain that receives sensory input from receptors in muscles and joints, from balance receptors in the inner ear, and from the receptors of touch, vision, and hearing.

cerebral cortex A thin outer shell of gray matter covering a thick core of white matter within each hemisphere of the brain; the part of the forebrain responsible for information processing and learning.

cerebrospinal fluid Fluid that circulates through the central canal of the spinal cord and the ventricles of the brain, cushioning the brain and spinal cord from jarring movements and impacts, as well as nourishing the CNS and protecting it from toxic substances.

cervix The lower end of the uterus.

channel protein Transport protein that forms a hydrophilic channel in a cell membrane through which water, ions, or other molecules can pass, depending on the protein.

chaparral A biome comprising a scrubby mix of short trees and shrubs that dominates coastal land between 30° and 40° latitude, where winters are cool and wet and summers hot and dry.

chaperone protein (chaperonin) "Guide" protein that binds temporarily with newly synthesized proteins, directing their conformation toward the correct tertiary structure and inhibiting incorrect arrangements as the new proteins fold.

character A specific heritable attribute or property of an organism.

character differences Alternative forms of characters. Also called a *trait*.

character displacement The phenomenon in which allopatric populations are morphologically similar and use similar resources, but sympatric populations are morphologically different and use different resources; may also apply to characters influencing mate choice.

character states One or more forms of a character used in a phylogenetic analysis.

charging *See* aminoacylation.

charophyte A member of the group of green algae most similar to the algal ancestors of land plants.

checkpoint Internal control of the cell cycle that prevents a critical phase from beginning until the previous phase is complete.

chelicerae The first pair of fanglike appendages near the mouth of an arachnid, used for biting prey and often modified for grasping and piercing.

chemical bond Link formed when atoms of reactive elements combine into molecules.

chemical equation A chemical reaction written in balanced form.

chemical reaction A reaction that occurs when atoms or molecules interact to form new chemical bonds or break old ones.

chemical signal Any secretion from one cell type that can alter the behavior of a different cell that bears a receptor for it; a means of cell communication.

chemical synapse A type of communicating connection between two neurons or a neuron and an effector cell in which an electrical impulse arriving at an axon terminal of the presynaptic cell triggers release of a neurotransmitter that crosses the gap and binds to a receptor on the postsynaptic cell, triggering an electrical impulse in that cell.

chemiosmotic hypothesis Model proposing that mitochondrial electron transfer produces an H^+ gradient and that the gradient powers ATP synthesis by ATP synthase.

chemoautotroph An organism that obtains energy by oxidizing inorganic substances such as hydrogen, iron, sulfur, ammonia, nitrites, and nitrates and uses carbon dioxide as a carbon source.

chemoheterotroph An organism that oxidizes organic molecules as an energy source and obtains carbon in organic form.

chemokine A protein secreted by activated macrophages that attracts other cells, such as neutrophils.

chemoreceptor A sensory receptor that detects specific molecules, or chemical conditions such as acidity.

chemotroph An organism that obtains energy by oxidizing inorganic or organic substances.

chiasmata *See* crossover.

chief cells Cells in the gastric glands of the stomach lining which secrete pepsinogen.

chitin A polysaccharide that contains nitrogen and is present in the cell walls of fungi and the exoskeletons of arthropods.

Chlorarachniophyta (chlorarachniophytes) A lineage within the Rhizaria protist evolutionary group; green, photosynthetic amoebas that also engulf food.

chlorophyll Molecule of green pigment that absorbs photons of light in photosynthesis.

Chlorophyta (green algae) A lineage within the Archaeplastida protist evolutionary group; autotrophs that carry out photosynthesis using the same pigments as plants.

chloroplast The site of photosynthesis in plant cells.

chlorosis An abnormal yellowing of plant tissues due to lack of chlorophyll; a sign of nutrient deficiency or infection by a pathogen.

choanocyte One of the inner layer of flagellated cells lining the body cavity of a sponge.

Choanoflagellata (choanoflagellates) A group of minute, single-celled protists found in water; the flask-shaped body has a collar of closely packed microvilli that surrounds the single flagellum by which it moves and takes in food.

cholecystokinin (CCK) Hormone released in response to fat, and to a lesser extent protein, in the chyme that enters the duodenum which inhibits gastric activity.

cholesterol The predominant sterol of animal cell membranes.

chondrocyte A cartilage-producing cell.

chorion In an amniote egg, an extraembryonic membrane that surrounds the embryo and yolk sac completely and exchanges oxygen and carbon dioxide with the environment; becomes part of the placenta in mammals.

chorionic villus (plural, villi) One of many treelike extensions from the chorion, which greatly increase the surface area of the chorion.

chorionic villus sampling Technique of prenatal diagnosis in which cells are obtained from portions of the placenta that develop from tissues of the embryo for DNA testing, biochemical analysis, or for the presence of chromosomal mutations.

Chromalveolata (chromalveolates) A protist evolutionary group consisting of the Alveolata and the Stramenopila.

chromatin Any assemblage of eukaryotic nuclear DNA molecules and their associated proteins.

chromatin remodeling Process in which the state of the chromatin is changed so that the proteins that initiate transcription can bind to a gene's promoter.

chromoplast Plastid containing red and yellow pigments.

chromosomal mutation A variation from the normal condition in chromosome structure or chromosome number.

chromosomal protein A histone and nonhistone protein associated with DNA in a eukaryotic nuclear chromosome.

chromosome In eukaryotic cells, a linear structure composed of a single DNA molecule complexed with protein. Each eukaryotic species has a characteristic number of chromosomes in the nucleus. Most prokaryotes have a single, usually circular chromosome with few or no associated proteins.

chromosome segregation The equal distribution of daughter chromosomes to each of the two cells that result from cell division.

chromosome theory of inheritance The principle that genes and their alleles are carried on the chromosomes.

Chrysophyta (golden algae) A lineage of the Stramenopila subgroup of the Chromalveolata protist evolutionary group; mostly colonial protists in which each cell of a colony has a pair of flagella and is surrounded by a glassy shell.

chylomicron A small triglyceride droplet covered by a protein coat.

chyme Digested content of the stomach released for further digestion in the small intestine.

ciliary body A fine ligament in the eye that anchors the lens to a surrounding layer of connective tissue and muscle.

Ciliophora (ciliates) A lineage of the Alveolata subgroup of the Chromalveolata protist evolutionary group.

cilium Motile structure, extending from a cell surface, that moves a cell through fluid or fluid over a cell.

circadian rhythm Any biological activity that is repeated in cycles, each about 24 hours long, independently of any shifts in environmental conditions.

circulatory system An organ system consisting of a fluid, a heart, and vessels for moving important molecules, and often cells, from one tissue to another.

circulatory vessel An element of the circulatory system through which fluid flows and carries nutrients and oxygen to tissues and remove wastes.

circumcision Removal of the prepuce for religious, cultural, or hygienic reasons.

cisternae Membranous channels and vesicles that make up the endoplasmic reticulum.

citric acid cycle Series of reactions in which acetyl groups are oxidized completely to carbon dioxide and some ATP molecules are synthesized. Also referred to as *Krebs cycle* and *tricarboxylic acid cycle.*

clade A monophyletic group of organisms that share homologous features derived from a common ancestor.

cladistics An approach to systematics that uses shared derived characters to infer the phylogenetic relationships and evolutionary history of groups of organisms.

cladogenesis The evolution of two or more descendant species from a common ancestor.

cladogram A branching diagram in which the endpoints of the branches represent different species of organisms, used to illustrate phylogenetic relationships.

claspers A pair of organs on the pelvic fins of male crustaceans and sharks, which help transfer sperm into the reproductive tract of the female.

class A Linnaean taxonomic category that ranks below a phylum and above an order.

class II major histocompatibility complex (MHC) A collection of proteins that present antigens on the cell surface of an antigen-presenting cell in an antibody-mediated immune response.

classical conditioning A type of learning in which an animal develops a mental association between two phenomena that are usually unrelated.

classification An arrangement of organisms into hierarchical groups that reflect their relatedness.

clathrin The network of proteins that coat and reinforce the cytoplasmic surface of cell membranes.

cleavage Mitotic cell divisions of the zygote that produce a blastula from a fertilized ovum.

climate The weather conditions prevailing over an extended period of time.

climax community A relatively stable, late successional stage in which the dominant vegetation replaces itself and persists until an environmental disturbance eliminates it, allowing other species to invade.

climograph A graph that portrays the particular combination of temperature and rainfall conditions where each terrestrial biome occurs.

cline A pattern of smooth variation in a characteristic along a geographical gradient.

clitoris The structure at the junction of the labia minora in front of the vulva, homologous to the penis in the male.

cloaca The cavity in reptiles, birds, amphibians, and many fishes into which both the intestinal and genital tracts empty.

clonal analysis A method of culturing meristematic tissue that contains a mutated embryonic cell having a readily observable trait, such as the absence of normal pigment.

clonal expansion The proliferation of the activated $CD4^+$ T cell by cell division to produce a clone of cells.

clonal reproduction The type of asexual reproduction in which the parent and offspring are genetically identical to one another.

clonal selection The process by which a lymphocyte is specifically selected for cloning when it encounters a foreign antigen from among a randomly generated, enormous diversity of lymphocytes with receptors that specifically recognize the antigen.

clone An individual genetically identical to an original cell from which it descended.

cloning vector DNA molecule into which a DNA fragment can be inserted to form a recombinant DNA molecule for the purpose of cloning.

closed circulatory system A circulatory system in which the fluid, blood, is confined in blood vessels and is distinct from the interstitial fluid.

clumped dispersion A pattern of distribution in which individuals in a population are grouped together.

cnidocyte A prey-capturing and defensive cell in the epidermis of cnidarians.

CO_2 fixation Process in which electrons are used as a source of energy to convert inorganic CO_2 to an organic form.

coactivator (mediator) In eukaryotes, a large multiprotein complex that bridges between activators at an enhancer and proteins at the promoter and promoter proximal region to stimulate transcription.

coat *See* capsid.

coated pit A depression in the plasma membrane that contains receptors for macromolecules to be taken up by endocytosis.

coccus (plural, cocci) A spherical prokaryote.

cochlea A snail-shaped structure in the inner ear containing the organ of hearing.

codominance Condition in which alleles have approximately equal effects in individuals, making the alleles equally detectable in heterozygotes.

codon Each three-letter word (triplet) of the genetic code.

coelom A fluid-filled body cavity in bilaterally symmetrical animals that is completely lined with derivatives of mesoderm.

coelomate A body plan of bilaterally symmetrical animals that have a coelom.

coenocytic Condition in which a single cell has many nuclei.

coenzymes Organic cofactors that include complex chemical groups of various kinds.

coevolution The evolution of genetically based, reciprocal adaptations in two or more species that interact closely in the same ecological setting.

cofactor An inorganic or organic nonprotein group that is necessary for catalysis to take place.

cognition A form of learning in which animals use insight to solve a novel problem; sometimes called trial-and-error learning.

cohesion The high resistance of water molecules to separation.

cohesion–tension mechanism of water transport A model of how water is transported from roots to leaves in vascular plants; the evaporation of water from leaves pulls water up in xylem by creating a continuous negative pressure (tension) that extends to roots.

cohort A group of individuals of similar age.

coleoptile A protective sheath that covers the shoot apical meristem and plumule of the embryo in monocots, such as grasses, as it pushes up through soil.

coleorhiza A sheath that encloses the radicle of an embryo until it breaks out of the seed coat and enters the soil as the primary root.

collagen Fibrous glycoprotein—very rich in carbohydrates—embedded in a network of proteoglycans.

collecting duct A location where urine leaving individual nephrons is processed further.

collenchyma A ground tissue that flexibly supports rapidly growing plant parts. Its elongated cells are alive at maturity and collectively often form strands or a sheath-like cylinder under the dermal tissue of growing shoot regions and leaf stalks.

colon The main part of the large intestine.

colony Multiple individual organisms of the same species living in a group.

combinatorial gene regulation The combining of a few regulatory proteins in particular ways so that the transcription of a wide array of genes can be controlled and a large number of cell types can be specified.

commaless The sequential nature of the words of the nucleic acid code, with no indicators such as commas or spaces to mark the end of one codon and the beginning of the next.

commensalism A symbiotic interaction in which one species benefits and the other is unaffected.

community Populations of all species that occupy the same area.

community ecology The ecological discipline that examines groups of populations occurring together in one area.

companion cell A specialized parenchyma cell that is connected to a mature sieve tube member by plasmodesmata and assists sieve tube members both with the uptake of sugars and with the unloading of sugars in tissues.

comparative genomics Comparison of the sequences of entire genomes (or extensive portions of them) to understand evolutionary relationships and the basic biological similarities and differences among species.

comparative morphology Analysis of the structure of living and extinct organisms.

compartment model A graphical depiction of the pathways through which nutrients and energy move between the living and non-living components of an ecosystem.

compass orientation A wayfinding mechanism that allows animals to move in a particular direction, often over a specific distance or for a prescribed length of time.

competitive exclusion principle The ecological principle stating that populations of two or more species cannot coexist indefinitely if they rely on the same limiting resources and exploit them in the same way.

competitive inhibition Inhibition of an enzyme reaction by an inhibitor molecule that resembles the normal substrate closely enough so that it fits into the active site of the enzyme.

complement system A nonspecific defense mechanism activated by invading pathogens, made up of more than 30 interacting soluble plasma proteins circulating in the blood and interstitial fluid.

complementary base pairing Feature of DNA in which the specific purine–pyrimidine base pairs A–T (adenine–thymine) and G–C (guanine–cytosine) occur to bridge the two sugar–phosphate backbones.

complementary DNA (cDNA) A DNA molecule that is complementary to an mRNA molecule, synthesized by reverse transcriptase.

complete digestive system A digestive system having a mouth at one end, through which food enters, and an anus at the other end, through which undigested waste is voided.

complete flower A flower in which all four whorls (sepals, petals, stamens, carpels) are present.

complete medium A growth medium containing a full complement of nutrient substances that a wild-type microorganism can make for itself.

complete metamorphosis The form of metamorphosis in which an insect passes through four separate stages of growth: egg, larva, pupa, and adult.

complex virus A bacteriophage with a DNA genome that has a tail attached at one side of a polyhedral head.

compound A molecule whose component atoms are different.

compound eye The eye of most insects and some crustaceans, composed of many faceted, light-sensitive units called ommatidia fitted closely together, each having its own refractive system and each forming a portion of an image.

concentration The number of molecules or ions of a substance in a unit volume of space.

concentration gradient The concentration difference that drives diffusion.

condensation reaction Reaction during which the components of a water molecule are removed, usually as part of the assembly of a larger molecule from smaller subunits. Also referred to as *dehydration synthesis reaction.*

conduction The flow of heat between atoms or molecules in direct contact.

cone In the vertebrate eye, a photoreceptor in the retina that is specialized for detection of different wavelengths (colors). In cone-bearing plants, a cluster of sporophylls.

conformation The overall three-dimensional shape of a protein.

conformational change Alteration in the three-dimensional shape of a protein.

conformers Animals having internal environments that change as the external environment changes.

conidiophore In ascomycete fungi, a modified hyphal branch that gives rise to conidia.

conidium (plural, conidia) An asexual spore produced by many species of ascomycetes.

Coniferophyta The major phylum of cone-bearing gymnosperms, most of which are substantial trees; includes pines, firs, and other conifers.

conjugation In bacteria, the process by which a copy of part of the DNA of a donor cell moves through the cytoplasmic bridge into the recipient cell where genetic recombination can occur. In ciliate protists, a process of sexual reproduction in which individuals of the same species temporarily couple and exchange genetic material.

connective tissue Tissue having cells scattered through an extracellular matrix; forms layers in and around body structures that support other body tissues, transmit mechanical and other forces, and in some cases act as filters.

conodont An abundant, bonelike fossil dating from the early Paleozoic era through the early Mesozoic era, now described as a feeding structure of some of the earliest vertebrates.

consciousness Awareness of oneself, identity, and surroundings, with understanding of the significance and likely consequences of events.

conservation biology An interdisciplinary science that focuses on the maintenance and preservation of biodiversity.

constant (C) region For the light and heavy polypeptides of a particular type of antibody molecule, the regions that have the same amino acid sequences for all molecules.

consumer An organism that consumes other organisms in a community or ecosystem.

contact inhibition The inhibition of movement or proliferation of normal cells that results from cell–cell contact.

continental climate Climate not moderated by the distant ocean.

continental drift The long-term movement of continents as a result of plate tectonics.

continuous distribution A geographical distribution in which a species lives in suitable habitats throughout a geographical area.

contraception The prevention of pregnancy.

contractile vacuole A specialized cytoplasmic organelle that pumps fluid in a cyclical manner from within the cell to the outside by alternately filling and then contracting to release its contents at various points on the surface of the cell.

control Treatment that tells what would be seen in the absence of the experimental manipulation.

convection The transfer of heat from a body to a fluid, such as air or water, that passes over its surface.

convergent evolution The evolution of similar adaptations in distantly related organisms that occupy similar environments.

copulation The physical act involving the introduction of the accessory sex organ of a male into the accessory sex organ of a female to accomplish internal fertilization.

coral reef A structure made from the hard skeletons of coral animals or polyps; found largely in tropical and subtropical marine environments.

corepressor In the regulation of gene expression in bacteria, a regulatory molecule that combines with a repressor to activate it and shut off an operon.

cork A nonliving, impermeable secondary tissue that is one element of bark.

cork cambium A lateral meristem in plants that forms periderm, which in turn produces cork.

cornea The transparent layer that forms the front wall of the eye, covering the iris.

corolla The structure formed collectively by the petals of a flower.

coronary arteries The arteries from the aorta that branch extensively over the heart, supplying blood to the cardiac muscle cells.

corpus callosum A structure formed of thick axon bundles that connect the two cerebral hemispheres and coordinate their functions.

corpus luteum Cells remaining at the surface of the ovary during the luteal phase; the structure acts as an endocrine gland, secreting several hormones: estrogens, large quantities of progesterone, and inhibin.

cortex Generally, an outer, rindlike layer. In mammals, the outer layer of the brain, the kidneys, or the adrenal glands. In plants, the outer region of tissue in a root or stem lying between the epidermis and the vascular tissue, composed mainly of parenchyma.

cortical granule A secretory vesicle just under the plasma membrane of an egg cell.

cortical reaction The reaction in which cortical granules fuse with the plasma membrane of the egg and release their contents to the outside.

cortisol The major glucocorticoid steroid hormone secreted by the adrenal cortex, which increases blood glucose by promoting breakdown of proteins and fats.

cotranslational import A mechanism by which a polypeptide being sorted via the endomembrane system in a eukaryotic cell begins its import into the endoplasmic reticulum simultaneously with translation of the mRNA encoding the polypeptide.

cotransport *See* symport.

countercurrent exchange A mechanism in which the water flowing over the gills moves in a direction opposite to the flow of blood under the respiratory surface.

coupled reaction Reaction that occurs when an exergonic reaction is joined to an endergonic reaction, producing an overall reaction that is exergonic.

courtship display A behavior performed by males to attract potential mates or to reinforce the bond between a male and a female.

covalent bond Bond formed by electron sharing between atoms.

cranial nerve A nerve that connects the brain directly to the head, neck, and body trunk.

cranium The part of the skull that encloses the brain.

crassulacean acid metabolism (CAM) A biochemical variation of photosynthesis that was discovered in a member of the plant family Crassulaceae. Carbon dioxide is taken up and stored during the night to allow the stomata to remain closed during the daytime, decreasing water loss.

Crenarchaeota A major group of the domain Archaea, separated from the other archaeans based mainly on rRNA sequences.

criss-cross inheritance The transmission pattern characteristic of an X-linked allele from a parent of one sex to a "child" of the opposite sex to a "grandchild" of the first sex.

crista Fold that expands the surface area of the inner mitochondrial membrane.

critical period A restricted stage of development early in life during which an animal has the capacity to respond to specific environmental stimuli.

crop Of birds, an enlargement of the digestive tube where the digestive contents are stored and mixed with lubricating mucus.

cross-fertilization Fertilization of one plant by a different plant.

crossing-over The recombination process in meiosis, in which chromatids exchange segments.

crossover Site of recombination during meiosis. Also referred to as a *chiasmata*.

cross-pollination *See* cross-fertilization.

cross-talk Interaction by which cell signaling pathways communicate with one another to integrate their responses to cellular signals.

cryptic coloration Coloration that allows an organism to match its background and hence become less vulnerable to predation or recognition by prey.

cryptochrome A light-absorbing protein that is sensitive to blue light and that may also be an important early step in various light-based growth responses.

C-terminal end The end of an amino acid chain with a —COO^- group.

cupula In certain mechanoceptors, a gelatinous structure with stereocilia extending into it that moves with pressure changes in the surrounding water; movement of the cupula bends the stereocilia, which triggers release of neurotransmitters.

cuticle The outer layer of plants and some animals, which helps prevent desiccation by slowing water loss.

Cycadophyta A phylum of palmlike gymnosperms known as cycads; the pollen-bearing and seed-bearing cones (strobili) occur on separate plants.

cyclic AMP (cAMP) In particular signal transduction pathways, a second messenger that activates protein kinases, which elicit the cellular response by adding phosphate groups to specific target proteins. cAMP functions in one of two major G-protein–coupled receptor–response pathways.

cyclic electron flow An electron transport pathway associated with photosystem I in photosynthesis that produces ATP without the synthesis of NADPH.

cyclin In eukaryotes, protein that regulates the activity of CDK (cyclin-dependent kinase) and controls progression through the cell cycle.

cyclin-dependent kinase (CDK) A protein kinase that controls the cell cycle in eukaryotes.

cytochrome Protein with a heme prosthetic group that contains an iron atom.

cytokine A molecule secreted by one cell type that binds to receptors on other cells and, through signal transduction pathways, triggers a response. In innate immunity, cytokines are secreted by activated macrophages.

cytokinesis Division of the cytoplasm into two daughter cells following nuclear division in mitosis or meiosis.

cytokinin A hormone that promotes and controls growth responses of plants.

cytoplasm All the parts of the cell that surround the central nucleus (eukaryotes) or nucleoid region (prokaryotes).

cytoplasmic determinants The mRNA and proteins stored in the egg cytoplasm that direct the first stages of animal development in the period before genes of the zygote become active.

cytoplasmic inheritance Pattern in which inheritance follows that of genes in the cytoplasmic organelles, mitochondria, or chloroplasts.

cytoplasmic streaming Intracellular movement of cytoplasm.

cytosine A pyrimidine that base-pairs with guanine in nucleic acids.

cytoskeleton The interconnected system of protein fibers and tubes that extends throughout the cytoplasm of a eukaryotic cell.

cytosol Aqueous solution in the cytoplasm containing ions and various organic molecules.

cytotoxic T cell A T lymphocyte that functions in cell-mediated immunity to kill body cells infected by viruses or transformed by cancer.

daily torpor A period of inactivity and lowered metabolic rate that allows an endotherm to conserve energy when environmental temperatures are low.

dalton A standard unit of mass, about 1.66×10^{-24} grams.

day-neutral plant A plant that flowers without regard to photoperiod.

decomposer A small organism, such as a bacterium or fungus, that feeds on the remains of dead organisms, breaking down complex biological molecules or structures into simpler raw materials.

deep vein thrombosis A medical condition that results when sitting for a long period of time causes blood to pool in the veins of the body below the heart and then clot, particularly in the legs.

defecation reflex The opening of the anal sphincter and expulsion of feces in response to feces entering the rectum and stretching the rectal wall.

degeneracy (redundancy) The feature of the genetic code in which, with two exceptions, more than one codon represents each amino acid.

dehydration synthesis reaction *See* condensation reaction.

deletion Chromosomal alteration that occurs if a broken segment is lost from a chromosome.

demographic transition model A graphical depiction of the historical relationship between a country's economic development and its birth and death rates.

demography The statistical study of the processes that change a population's size and density through time.

denaturation A loss of both the structure and function of a protein due to extreme conditions that unfold it from its normal conformation.

dendrite The branched extension of the nerve cell body that receives signals from other nerve cells.

dendritic cell A type of phagocyte, so called because it has many surface projections that resemble dendrites of neurons, which engulfs a bacterium in infected tissue by phagocytosis.

denitrification A metabolic process in which certain bacteria convert nitrites or nitrates into nitrous oxide and then into molecular nitrogen, which enters the atmosphere.

density-dependent Description of environmental factors for which the strength of their effect on a population varies with the population's density.

density-independent Description of environmental factors for which the strength of their effect on a population does not vary with the population's density.

deoxyribonucleic acid (DNA) The large, double-stranded, helical molecule that contains the genetic material of all living organisms.

deoxyribonucleotide Nucleotide containing deoxyribose as the sugar; deoxyribonucleotides are components of DNA.

deoxyribose A 5-carbon sugar to which a nitrogenous base and a phosphate group link covalently in a nucleotide of DNA.

depolarized State of the membrane (which was polarized at rest) as the membrane potential becomes less negative.

deposit feeder An animal that consumes particles of organic matter from the solid substrate on which it lives.

derivative One of the daughter cells produced when a plant cell divides; it typically divides once or twice and then enters on the path to differentiation.

derived character A new version of a trait found in the most recent common ancestor of a group.

dermal tissue system The plant tissue system that comprises the outer tissues of the plant body, including the epidermis and periderm; it serves as a protective covering for the plant body.

dermis The skin layer below the epidermis; it is packed with connective tissue fibers such as collagen, which resist compression, tearing, or puncture of the skin.

descending reticular formation Part of the reticular formation that receives information from the hypothalamus and connects with interneurons in the spinal cord that control skeletal muscle contraction, thereby controlling muscle movement and posture.

descent with modification Biological evolution.

desert A sparsely vegetated biome that forms where rainfall averages less than 25 cm per year.

desertification A process in which large tracts of subtropical forest are cleared and overused, the groundwater table recedes to deeper levels, less surface water is available for plants, soil accumulates high concentrations of salts, and topsoil is eroded by wind and water.

desmosome Anchoring junction for which microfilaments anchor the junction in the underlying cytoplasm.

determinate cleavage A type of cleavage in protosomes in which each cell's developmental path is determined as the cell is produced.

determinate growth The pattern of growth in most animals in which individuals grow to a certain size and then their growth slows dramatically or stops.

determination Mechanism in which the developmental fate of a cell is set.

detritivore An organism that extracts energy from the organic detritus (refuse) produced at other trophic levels.

development A series of programmed changes encoded in DNA, through which a fertilized egg divides into many cells that ultimately are transformed into an adult, which is itself capable of reproduction.

diabetes mellitus A disease that results from problems with insulin production or action.

diacylglycerol (DAG) In particular signal transduction pathways, a second messenger that activates protein kinases, which elicit the cellular response by adding phosphate groups to specific target proteins. DAG is involved in one of two major G-protein–coupled receptor–response pathways.

diapsid A member of a group within the amniote vertebrates having a skull with two temporal arches. Their living descendants include lizards and snakes, crocodilians, and birds.

diastole The period of relaxation and filling of the heart between contractions.

diastolic blood pressure The low point of the arterial blood pressure in the cardiac cycle that occurs between ventricular contractions.

diatoms *See* Bacillariophyta.

differentiation Process by which cells that have been committed to a particular developmental fate by the determination process now develop into specialized cell types with distinct structures and functions.

diffusion The net movement of ions or molecules from a region of higher concentration to a region of lower concentration.

digestion The splitting of carbohydrates, proteins, lipids, and nucleic acids in foods into chemical subunits small enough to be absorbed into the body fluids and cells of an animal.

digestive tube A tubelike digestive system with two openings that form a separate mouth and anus; the digestive contents move in one direction through specialized regions of the tube, from the mouth to the anus.

dihybrid A zygote produced from a cross that involves two characters.

dihybrid cross A cross between two individuals that are heterozygous for two pairs of alleles.

dikaryon The life stage in certain fungi in which a cell contains two genetically distinct haploid nuclei.

Dinoflagellata (dinoflagellates) A lineage of the Alveolata subgroup of the Chromalveolata protist evolutionary group; mostly single-celled marine phytoplankton that live as heterotrophs or autotrophs or sometimes using both modes of nutrition.

dioecious Having male flowers and female flowers on different plants of the same species.

dipeptidase An enzyme that splits dipeptides (two amino acids joined together) into individual amino acids.

diploblastic An animal body plan in which adult structures arise from only two cell layers, the ectoderm and the endoderm.

diploid An organism or cell with two copies of each type of chromosome in its nucleus.

directional selection A type of selection in which individuals near one end of the phenotypic spectrum have the highest relative fitness.

disaccharidase An enzyme that splits disaccharides (two sugar molecules joined together) into individual monosaccharides (single sugar molecules).

disclimax community An ecological community in which regular disturbance inhibits successional change.

discontinuous replication Replication in which a DNA strand is formed in short lengths that are synthesized in the direction opposite of DNA unwinding.

disjunct distribution A geographical distribution in which populations of the same species or closely related species live in widely separated locations.

dispersal The movement of organisms away from their place of origin.

dispersion The spatial distribution of individuals within a population's geographical range.

disruptive selection A type of natural selection in which extreme phenotypes have higher relative fitness than intermediate phenotypes.

dissociation The separation of water to produce hydrogen ions and hydroxide ions.

distal convoluted tubule The tubule in the human nephron that drains urine into a collecting duct that leads to the renal pelvis.

disulfide linkage Linkage that occurs when two sulfhydryl groups interact during a linking reaction.

DNA *See* deoxyribonucleic acid.

DNA chip *See* DNA microarray.

DNA fingerprinting Technique in which DNA samples are used to distinguish between individuals of the same species.

DNA helicase An enzyme that catalyzes the unwinding of DNA template strands.

DNA hybridization Technique in which a gene or sequence of interest is identified in a set of clones when it base pairs with a single-stranded DNA or RNA molecule called a nucleic acid probe.

DNA ligase In DNA replication, an enzyme that seals the nicks left after RNA primers are replaced with DNA.

DNA methylation Process in which a methyl group is added enzymatically to cytosine bases in the DNA.

DNA microarray A solid surface divided into a microscopic grid of thousands of spaces each containing thousands of copies of a DNA probe. DNA chips are used commonly for analysis of gene activity and for detecting differences between cell types. Also referred to as a *DNA chip.*

DNA polymerase I In *E. coli,* the replication enzyme that replaces the RNA primer at the start of a new DNA segment with DNA.

DNA polymerase III The principal replication polymerase in *E. coli* that synthesizes the majority of the new DNA.

DNA repair mechanism Mechanism to correct base-pair mismatches that escape proofreading.

DNA technologies Techniques to isolate, purify, analyze, and manipulate DNA sequences.

domain In protein structure, a distinct, large structural subdivision produced in many proteins by the folding of the amino acid chain. In systematics, the highest taxonomic category; a group of cellular organisms with characteristics that set it apart as a major branch of the evolutionary tree.

dominance The masking effect of one allele over another.

dominance hierarchy A social system in which the behavior of each individual is constrained by that individual's status in a highly structured social ranking.

dominant The allele expressed when paired with a recessive allele.

dominant species The species that is represented by a large proportion of the individuals present in an ecological community.

dormancy A period in the life cycle in which biological activity is suspended.

dorsal Indicating the back side of an animal.

dorsal lip of the blastopore A crescent-shaped depression rotated clockwise 90° on the embryo surface that marks the region derived from the gray crescent, to which cells from the animal pole move as gastrulation begins.

dosage compensation mechanism Mechanism in placental mammals by which the effects of most genes carried on the X chromosome in females are equalized in females (who have two X chromosomes) and males (who have one X chromosome).

double fertilization A characteristic feature of sexual reproduction in flowering plants. In the embryo sac, one sperm nucleus unites with the egg to form a diploid zygote from which the embryo develops, and another unites with two polar nuclei to form the primary endosperm nucleus.

double helix Two nucleotide chains wrapped around each other in a spiral.

double-helix model Model of DNA consisting of two polynucleotide strands twisted around each other.

downy mildews One of the subgroups of the Oomycota; parasites of plants.

duodenum A short region of the small intestine where secretions from the pancreas and liver enter a common duct.

duplication Chromosomal alteration that occurs if a segment is broken from one chromosome and inserted into its homolog.

E site The site where an exiting tRNA binds prior to its release from the ribosome in translation.

ecdysis Shedding of the cuticle, exoskeleton, or skin; molting.

ecdysone A steroid hormone secreted by the prothoracic glands of insects.

echolocation A technique for locating prey by making squeaking or clicking noises, and then listening for the echoes that bounce back from objects in their environment.

ecological community An assemblage of species living in the same place.

ecological efficiency The ratio of net productivity at one trophic level to net productivity at the trophic level below it.

ecological isolation A prezygotic reproductive isolating mechanism in which species that live in the same geographical region occupy different habitats.

ecological niche The resources a population uses and the environmental conditions it requires over its lifetime.

ecological pyramid A diagram illustrating the effects of energy transfer from one trophic level to the next.

ecological succession A somewhat predictable series of changes in the species composition of a community over time.

ecology The study of the interactions between organisms and their environments.

ecosystem Group of biological communities interacting with their shared physical environment.

ecosystem ecology A ecological discipline that explores the cycling of nutrients and the flow of energy between the biotic components of an ecological community and the abiotic environment.

ecosystem services The ecological processes on which all life depends, which include decomposition of wastes, nutrient recycling, oxygen production, maintenance of fertile topsoil, and air and water purification.

ecosystem valuation A process in which ecosystem services are assigned an economic value.

ecotone A wide transition zone between adjacent communities.

ecotourism An activity in which visitors, often from wealthy countries, pay a fee to visit a nature preserve.

ectoderm The outermost of the three primary germ layers of an embryo, which develops into epidermis and nervous tissue.

ectomycorrhiza A mycorrhiza that grows between and around the young roots of trees and shrubs but does not enter root cells.

ectoparasite A parasite that lives on the exterior of its host organism.

ectotherm An animal that obtains its body heat primarily from the external environment.

edge effect A phenomenon in which the removal of natural vegetation disrupts the local physical environment, exposing the borders of the remaining habitat to additional sunlight, wind, and rainfall.

effector In signal transduction, a plasma membrane-associated enzyme, activated by a G protein, that generates one or more second messengers. In homeostatic feedback, the system that returns the condition to the set point if it has strayed away.

effector T cell A cell involved in effecting—bringing about—the specific immune response to an antigen.

efferent arteriole The arteriole that receives blood from the glomerulus.

efferent neuron A neuron that carries the signals indicating a response away from the interneuron networks to the effectors.

egg cell The female reproductive cell.

elastin A rubbery protein in some connective tissues that adds elasticity to the extracellular matrix—it is able to return to its original shape after being stretched, bent, or compressed.

electrical signaling A means of animal communication in which a signaler emits an electric discharge that can be received by another individual.

electrical synapse A mechanical and electrically conductive link between two abutting neurons that is formed at the gap junction.

electrocardiogram (ECG) Graphic representation of the electrical activity within the heart, detected by electrodes placed on the body.

electrochemical gradient A difference in chemical concentration and electric potential across a membrane.

electromagnetic spectrum The range of wavelengths or frequencies of electromagnetic radiation extending from gamma rays to the longest radio waves and including visible light.

electron Negatively charged particle outside the nucleus of an atom.

electron microscope Microscope that uses electrons to illuminate the specimen.

electron transfer system Stage of cellular respiration in which high-energy electrons produced from glycolysis, pyruvate oxidation, and the citric acid cycle are delivered to oxygen by a sequence of electron carriers.

electronegativity The measure of an atom's attraction for the electrons it shares in a chemical bond with another atom.

electroreceptor A specialized sensory receptor that detects electrical fields.

element A pure substance that cannot be broken down into simpler substances by ordinary chemical or physical techniques.

elimination Referring to the digestive tube, the process in which undigested materials are expelled through the anus.

elongation factor (EF) A protein that aids in an elongation step of translation.

embryo An organism in its early stage of reproductive development, beginning in the first moments after fertilization.

embryo sac The female gametophyte of angiosperms, within which the embryo develops; it usually consists of seven cells: an egg cell, an endosperm mother cell, and five other cells with fleeting reproductive roles.

embryogenesis Stages of development from a fertilized egg to an embryo.

embryophyte Any plant in which the embryo is retained within maternal tissue.

emergent property Characteristic that depends on the level of organization of matter, but does not exist at lower levels of organization.

emigration The movement of individuals out of a population.

enantiomers Isomers that are mirror images of each other. Also referred to as *optical isomers.*

endangered species A species in danger of extinction throughout all or a significant portion of its range.

endemic species A species that occurs in only one place on Earth.

endergonic reaction Reaction that can proceed only if free energy is supplied.

endocrine gland Any of several ductless secretory organs that secrete hormones into the blood or extracellular fluid.

endocrine system The system of glands that release their secretions (hormones) directly into the circulatory system.

endocytic vesicle Vesicle that carries proteins and other molecules from the plasma membrane to destinations within the cell.

endocytosis In eukaryotes, the process by which molecules are brought into the cell from the exterior involving a bulging in of the plasma membrane that pinches off to form an endocytic vesicle.

endoderm The innermost of the three primary germ layers of an embryo, which develops into the gastrointestinal tract and, in some animals, the respiratory organs.

endodermis The innermost layer of the root cortex; a selectively permeable barrier that helps control the movement of water and dissolved minerals into the stele.

endomembrane system In eukaryotes, a collection of interrelated internal membranous sacs that divide a cell into functional and structural compartments.

endomycorrhiza A mycorrhiza in which the fungal hyphae penetrate into cells of the root.

endoparasite A parasite that lives in the internal organs of its host organism.

endoplasmic reticulum (ER) In eukaryotes, an extensive interconnected network of cisternae that is responsible for the synthesis, transport, and initial modification of proteins and lipids.

endorphin One of a group of small proteins occurring naturally in the brain and around nerve endings that bind to opiate receptors and thus can raise the pain threshold.

endoskeleton A supportive internal body structure, such as bones, that provides support.

endosperm Nutritive tissue inside the seeds of flowering plants.

endospore A small, metabolically inactive, asexual spore that develops within some bacterial cells when environmental conditions become unfavorable.

endosymbiont Organism that lives symbiotically within a host cell.

endosymbiont theory The proposal that the membranous organelles of eukaryotic cells (mitochondria and chloroplasts) may have originated from symbiotic relationships between two prokaryotic cells.

endotherm An animal that obtains most of its body heat from internal physiological sources.

endothermic Referring to a reaction that absorbs energy, that is, a reaction in which the products have more potential energy than the reactants.

endotoxin A lipopolysaccharide released from the outer membrane of the cell wall when a bacterium dies and lyses.

end-product inhibition *See* feedback inhibition.

energy The capacity to do work.

energy budget The total amount of energy that an organism can accumulate and use to fuel its activities.

energy coupling The process in living cells by which the hydrolysis of ATP is coupled to an endergonic reaction so that energy is not wasted as heat.

energy levels Regions of space within an atom where electrons are found. Also referred to as *shells.*

enhancer In eukaryotes, a region at a significant distance from the beginning of a gene containing regulatory sequences that determine whether the gene is transcribed at its maximum possible rate.

enterocoelom In deuterostomes, the body cavity pinched off by outpocketings of the archenteron.

enthalpy The potential energy in a system.

entropy Disorder, in thermodynamics.

enveloped virus A virus that has a surface membrane derived from its host cell.

enzymatic hydrolysis A process in which chemical bonds are broken by the addition of H^+ and OH^-, the components of a molecule of water.

enzyme Protein that accelerates the rate of a cellular reaction.

enzyme specificity The ability of an enzyme to catalyze the reaction of only a single type of molecule or group of closely related molecules.

eosinophil A type of leukocyte that targets extracellular parasites too large for phagocytosis in the inflammatory response.

epiblast The top layer of the blastodisc.

epicotyl The upper part of the axis of an early plant embryo, located between the cotyledons and the first true leaves.

epidermis A complex tissue that covers an organism's body in a single continuous layer or sometimes in multiple layers of tightly packed cells.

epididymis A coiled storage tubule attached to the surface of each testis.

epigenetics A phenomenon in which a change in gene expression does not involve a change in the DNA sequence of the gene or of the genome.

epiglottis A flaplike valve at the top of the trachea.

epilimnion The top layer of the limnetic zone in a lake.

epinephrine A nontropic amine hormone secreted by the adrenal medulla.

epiphyte A plant that grows independently on other plants and obtains nutrients and water from the air.

epistasis Interaction of genes, with one or more alleles of a gene at one locus inhibiting or masking the effects of one or more alleles of a gene at a different locus.

epithelial tissue Tissue formed of sheet-like layers of cells that are usually joined tightly together, with little extracellular matrix material between them. They protect body surfaces from invasion by bacteria and viruses, and secrete or absorb substances.

epitope The small region of an antigen molecule to which BCRs or TCRs bind.

equilibrium potential The electrical potential necessary to balance the diffusional potential of an ion at the plasma membrane of an axon.

equilibrium theory of island biogeography An hypothesis suggesting that the number of species on an island is governed by a give and take between the immigration of new species to the island and the extinction of species already there.

ER (endoplasmic reticulum) lumen The enclosed space surrounded by a cisterna.

erythrocyte A red blood cell, which contains hemoglobin, a protein that transports O_2 in blood.

erythropoietin (EPO) A hormone that stimulates stem cells in bone marrow to increase erythrocyte production.

esophagus A connecting passage of the digestive tube.

essential amino acid Any amino acid that is not made by the human body but must be taken in as part of the diet.

essential element Any of a number of elements required by living organisms to ensure normal reproduction, growth, development, and maintenance.

essential fatty acid Any fatty acid that the body cannot synthesize but needs for normal metabolism.

essential mineral Any inorganic element such as calcium, iron, or magnesium that is required in the diet of an animal.

essential nutrient Any of the essential amino acids, fatty acids, vitamins, and minerals required in the diet of an animal.

estivation Seasonal torpor in an animal that occurs in summer.

estradiol A form of estrogen.

estrogen Any of the group of female sex hormones.

estuary A coastal habitat where tidal seawater mixes with fresh water from rivers, streams, and runoff.

ethology A discipline that focuses on how animals behave in their natural environments.

ethylene A plant hormone that helps regulate seedling growth, stem elongation, the ripening of fruit, and the abscission of fruits, leaves, and flowers.

eudicot A plant belonging to the Eudicotyledones, one of the two major classes of angiosperms; their embryos generally have two seed leaves (cotyledons), and their pollen grains have three grooves.

Euglenids A lineage of the Euglenozoa subgroup of the Excavata protist evolutionary group; free-living with anterior flagella.

Euglenozoa (euglenozoans) A subgroup of the Excavata protist evolutionary group; single-celled, highly motile cells that swim by means of flagella and which contain mitochondria characterized by disc-shaped cristae.

Eukarya The domain that includes all eukaryotes, organisms that contain a membrane-bound nucleus within each of their cells; all protists, plants, fungi, and animals.

eukaryote Organism in which the DNA is enclosed in a nucleus.

eukaryotic chromosome A DNA molecule, with its associated proteins, in the nucleus of a eukaryotic cell.

euploid An individual with a normal set of chromosomes.

Euryarchaeota A major group of the domain Archaea, members of which are found in different extreme environments. They include methanogens, extreme halophiles, and some extreme thermophiles.

eusocial A form of social organization, observed in some insect species, in which numerous related individuals—a large percentage of them sterile female workers—live and work together in a colony for the reproductive benefit of a single queen and her mate(s).

eutrophic lake A lake that is rich in nutrients and organic matter.

evaporation Heat transfer through the energy required to change a liquid to a gas.

evolutionary developmental biology A field of biology that compares the genes controlling the developmental processes of different animals to determine the evolutionary origin of morphological novelties and developmental processes.

evolutionary divergence A process whereby natural selection or genetic drift causes populations to become more different over time.

exaptation A trait that is adaptive in a context different from the context in which it originally evolved.

Excavata A protist evolutionary group; single-celled animal parasites with greatly reduced mitochondria or organelles derived from mitochondria, that move by means of flagella, and most of which have a scooped out feeding apparatus on the ventral surface of the cell.

exchange diffusion *See* antiport.

excitatory postsynaptic potential (EPSP) The change in membrane potential caused when a neurotransmitter opens a ligand-gated Na^+ channel and Na^+ enters the cell, making it more likely that the postsynaptic neuron will generate an action potential.

excretion The process that helps maintain the body's water and ion balance while ridding the body of metabolic wastes.

exergonic reaction Reaction that has a negative ΔG because it releases free energy.

exocrine gland A gland that is connected to the epithelium by a duct and that empties its secretion at the epithelial surface.

exocytosis In eukaryotes, the process by which a secretory vesicle fuses with the plasma membrane and releases the vesicle contents to the exterior.

exodermis In the roots of some plants, an outer layer of root cortex that may limit water losses from roots and help regulate the absorption of ions.

exoenzyme An enzymatic protein released by a bacterium that digests plasma membranes and causes cells of the infected host to rupture and die.

exon An amino acid–coding sequence present in pre-mRNA that is retained in a spliced mRNA that is translated to produce a polypeptide.

exon shuffling Process by which existing amino acid–coding regions or domains are mixed into novel combinations to create new proteins.

exoskeleton A hard external covering of an animal's body that blocks the passage of water and provides support and protection.

exothermic Referring to a reaction that releases energy, that is, a reaction in which the products have less potential energy than the reactants.

exotic species A nonnative organism.

exotoxin A toxic protein that leaks from or is secreted from a bacterium and interferes with the biochemical processes of body cells in various ways.

experimental data Information that describes the result of a careful manipulation of the system under study.

experimental variable The variable in a scientific study that is manipulated by the experimenter.

exploitative competition Form of competition in which two or more individuals or populations use the same limiting resources.

exponential model of population growth Model that describes unlimited population growth.

expression vector Cloning vector that, in addition to normal features, contains regulatory sequences for transcription and translation of a gene.

extensor muscle A muscle that extends a joint, thereby increasing the angle between the two bones; with a flexor muscle constitutes the antagonistic pair of muscles that control a joint's movement.

external environment The environment outside of the bodies of multicellular organisms.

external fertilization The process in which sperm and eggs are shed into the surrounding water, occurring in most aquatic invertebrates, bony fishes, and amphibians.

external gill A gill that extends out from the body and lacks a protective covering.

extinction The death of the last individual in a species or the last species in a lineage.

extracellular digestion Digestion that takes place outside body cells, in a pouch or tube enclosed within the body.

extracellular fluid The fluid occupying the spaces between cells in multicellular animals.

extracellular matrix (ECM) A molecular system that supports and protects cells and provides mechanical linkages.

extraembryonic membrane A primary tissue layer extended outside the embryo that conducts nutrients from the yolk to the embryo, exchanges gases with the environment outside the egg, or stores metabolic wastes removed from the embryo.

F factor Plasmid in a donor bacterial cell that confers upon that cell the ability to conjugate with a recipient bacterial cell.

F pilus Structure on the cell surface that allows an F^+ donor bacterial cell (a cell containing an F factor) to attach to an F^- recipient bacterial cell (a cell lacking an F factor). Also referred to as a *sex pilus.*

F^- cell Recipient cell in conjugation between bacteria; it lacks an F factor.

F^+ cell Donor cell in conjugation between bacteria; it contains an F factor.

F_1 generation The first generation of offspring from a genetic cross.

F_2 generation The second generation of offspring from a genetic cross produced by interbreeding F_1 individuals.

facilitated diffusion Mechanism by which polar and charged molecules diffuse across membranes with the help of transport proteins.

facilitation hypothesis An hypothesis that explains ecological succession, suggesting that species modify the local environment in ways that make it less suitable for themselves but more suitable for colonization by species typical of the next successional stage.

facultative anaerobe An organism that can live in the presence or absence of oxygen, using oxygen when it is present and living by fermentation under anaerobic conditions.

familial (hereditary) cancer Cancer that runs in a family.

family A Linnaean taxonomic category that ranks below an order and above a genus.

family planning program A program that educates people about ways to produce an optimal family size on an economically feasible schedule.

fast block to polyspermy The barrier set up by the wave of depolarization triggered when sperm and egg fuse, making it impossible for other sperm to enter the egg.

fast muscle fiber A muscle fiber that contracts relatively quickly and powerfully.

fat Neutral lipid that is semisolid at biological temperatures.

fate map Mapping of adult or larval structures onto the region of the embryo from which each structure developed.

fat-soluble vitamin A vitamin that dissolves in liquid fat or fatty oils, in addition to water.

fatty acid One of two components of a neutral lipid, containing a single hydrocarbon chain with a carboxyl group linked at one end.

feather A sturdy, lightweight structure of birds, derived from scales in the skin of their ancestors.

feces Condensed and compacted digestive contents in the large intestine.

feedback inhibition In enzyme reactions, regulation in which the product of a reaction acts as a regulator of the reaction. Also referred to as *end-product inhibition.*

fermentation Process in which electrons carried by NADH are transferred to an organic acceptor molecule rather than to the electron transfer system.

fertilization The fusion of the nuclei of an egg and sperm cell, which initiates development of a new individual.

fetus A developing human from the eighth week of gestation onward, at which point the major organs and organ systems have formed.

fever A condition characterized by a rise in body temperature above the normal range.

fiber In sclerenchyma, an elongated, tapered, thick-walled cell that gives plant tissue its flexible strength.

fibrin A protein necessary for blood clotting; fibrin forms a web-like mesh that traps platelets and red blood cells and holds a clot together.

fibrinogen A plasma protein that plays a central role in the blood-clotting mechanism.

fibroblast The type of cell that secretes most of the collagen and other proteins in the loose connective tissue.

fibronectin A class of glycoproteins that aids in the attachment of cells to the extracellular matrix and helps hold the cells in position.

fibrous connective tissue Tissue in which fibroblasts are sparsely distributed among dense masses of collagen and elastin fibers that are lined up in highly ordered, parallel bundles, producing maximum tensile strength and elasticity.

fibrous root system A root system that consists of branching roots rather than a main taproot; roots tend to spread laterally from the base of the stem.

filament In flowers, the stalk of a stamen, which supports the anther.

filtration The nonselective movement of some water and a number of solutes—ions and small molecules, but not large molecules such as proteins—into the proximal end of the renal tubules through spaces between cells.

first law of thermodynamics The principle that energy can be transferred and transformed but it cannot be created or destroyed.

first messenger The extracellular signal molecule in signal transduction pathways controlled by G-protein–coupled receptors.

fission The mode of asexual reproduction in which the parent separates into two or more offspring of approximately equal size.

fixed action pattern A highly stereotyped instinctive behavior; when triggered by a specific cue, it is performed over and over in almost exactly the same way.

flagellum (plural, flagella) A long, threadlike, cellular appendage responsible for movement; found in both prokaryotes and eukaryotes, but with different structures and modes of locomotion.

flatulence Gas expelled through the anus.

flexor muscle A muscle that decreases the angle between the two bones at a joint, thereby performing the opposite action to an extensor muscle; with an extensor muscle constitutes the antagonistic pair of muscles that control a joint's movement.

flower The reproductive structure of angiosperms, consisting of floral parts grouped on a stem; the structure in which seeds develop.

fluid feeder An animal that obtains nourishment by ingesting liquids that contain organic molecules in solution.

fluid mosaic model Model proposing that the membrane consists of a fluid phospholipid bilayer in which proteins are embedded and float freely.

fMRI *See* functional magnetic resonance imaging.

follicle cell A cell that grows from ovarian tissue and nourishes the developing egg.

follicle-stimulating hormone (FSH) The pituitary hormone that stimulates oocytes in the ovaries to continue meiosis and become follicles. During follicle enlargement, FSH interacts with luteinizing hormone to stimulate follicular cells to secrete estrogens.

food chain A depiction of the trophic structure of a community; a portrait of who eats whom.

food vacuole A membrane-bound sac used for digestion.

food web A set of interconnected food chains with multiple links.

Foraminifera (forams) A lineage of the Rhizaria protist evolutionary group; heterotrophic marine organisms with shells consisting of organic matter reinforced by calcium carbonate.

forebrain The largest division of the brain, which includes the cerebral cortex and basal ganglia. It is credited with the highest intellectual functions.

foreskin A loose fold of skin that covers the glans of the penis.

formula The name of a molecule written in chemical shorthand.

fossil The remains or traces of an organism of a past geologic age embedded and preserved in Earth's crust.

foundation species A species that defines the nature of a community by creating locally stable environmental conditions.

founder effect An evolutionary phenomenon in which a population that was established by just a few colonizing individuals has only a fraction of the genetic diversity seen in the population from which it was derived.

fovea The small region of the retina around which cones are concentrated in mammals and birds with eyes specialized for daytime vision.

fragmentation A type of vegetative reproduction in plants in which cells or a piece of the parent break off, then develop into new individuals.

frameshift mutation Mutation in a protein-coding gene that causes the reading frame of an mRNA transcribed from the gene to be altered, resulting in the production of a different, and nonfunctional, amino acid sequence in the polypeptide.

free energy The energy in a system that is available to do work.

freeze-fracture technique Technique in which experimenters freeze a block of cells rapidly, then fracture the block to split the lipid bilayer and expose the hydrophobic membrane interior.

frequency-dependent selection A form of natural selection in which rare phenotypes have a selective advantage simply because they are rare.

fruit A mature ovary, often with accessory parts, from a flower.

fruiting body In some fungi, a stalked, spore-producing structure such as a mushroom.

functional genomics The study of the functions of genes and of other parts of the genome.

functional groups The atoms in reactive groups.

functional magnetic resonance imaging (fMRI) A scanning technique for studying activities of the cerebral cortex.

fundamental niche The range of conditions and resources that a population can possibly tolerate and use.

fungal spore In fungi, a haploid reproductive cell that can germinate and produce a mycelium.

furrow In cytokinesis, a groove that girdles the cell and gradually deepens until it cuts the cytoplasm into two parts.

G_0 phase The phase of the cell cycle in eukaryotes in which many cell types stop dividing.

G_1 phase The initial growth stage of the cell cycle in eukaryotes, during which the cell makes proteins and other types of cellular molecules, but not nuclear DNA.

G_2 phase The phase of the cell cycle in eukaryotes during which the cell continues to synthesize proteins and to grow, completing interphase.

gallbladder The organ that stores bile between meals, when no digestion is occurring.

gametangium In certain plants and fungi, a cell or organ in which gametes are produced.

gamete A haploid cell, an egg or sperm. Haploid cells fuse during sexual reproduction to form a diploid zygote.

gametic isolation A prezygotic reproductive isolating mechanism caused by incompatibility between the sperm of one species and the eggs of another; may prevent fertilization.

gametogenesis The formation of male and female gametes.

gametophyte In organisms in which alternation of generations occurs, notably plants and green algae, the multicellular haploid generation that produces gametes. Compare *sporophyte*.

ganglion A functional concentration of nervous system tissue composed principally of nerve-cell bodies, usually lying outside the central nervous system.

ganglion cell A type of neuron in the retina of the eye that receives visual information from photoreceptors via various intermediate cells such as bipolar cells, amacrine cells, and horizontal cells.

gap gene In *Drosophila* embryonic development, one of the first activated set of segmentation genes that progressively subdivide the embryo into regions, determining the segments of the embryo and the adult. Gap genes subdivide the embryo along the anterior–posterior axis into broad regions that later develop into several distinct segments.

gap junction Junction that opens direct channels allowing ions and small molecules to pass directly from one cell to another.

gastric glands Glands in the stomach wall that contain cells which secrete some of the products needed to digest food.

gastric juice A substance secreted by the stomach that contains the digestive enzyme pepsin.

gastric pits Entrances to gastric glands.

gastrin A hormone secreted in response to stimulation of chemoreceptors in the stomach by the presence of food molecules, particularly proteins.

gastroesophageal sphincter Ring of muscle at the junction of the esophagus and the stomach which controls the movement of a bolus of food into the stomach.

gastrovascular cavity A saclike body cavity with a single opening, a mouth, which serves both digestive and circulatory functions.

gastrula The developmental stage resulting when the cells of the blastula migrate and divide once cleavage is complete.

gastrulation The second major process of early development in most animals, which produces an embryo with three distinct primary tissue layers.

gated channel Ion transporter in a membrane that switches between open, closed, or intermediate states.

gemma (plural, gemmae) Small cell mass that forms in cuplike growths on a thallus.

gene A unit containing the code for a protein molecule or one of its parts, or for functioning RNA molecules such as tRNA and rRNA.

gene flow The transfer of genes from one population to another through the movement of individuals or their gametes.

gene pool The sum of all alleles at all gene loci in all individuals in a population.

gene therapy Correction of genetic disorders using genetic engineering techniques.

gene-for-gene recognition A mechanism in which plants can detect an attack by a specific pathogen; the product of a specific plant gene interacts with the product of a specific pathogen gene, triggering the plant's defensive response.

general transcription factor (basal transcription factor) In eukaryotes, a protein that binds to the promoter of a gene in the area of the TATA box and recruits and orients RNA polymerase II to initiate transcription at the correct place.

generalized compartment model A model used to describe nutrient cycling in which two criteria—organic *versus* inorganic nutrients and available *versus* unavailable nutrients—define four compartments where nutrients accumulate.

generalized transduction Transfer of bacterial genes between bacteria using virulent phages that have incorporated random DNA fragments of the bacterial genome.

generation time The average time between the birth of an organism and the birth of its offspring.

genetic code The nucleotide information that specifies the amino acid sequence of a polypeptide.

genetic counseling Counseling that allows prospective parents to assess the possibility that they might have a child affected by a genetic disorder.

genetic drift Random fluctuations in allele frequencies as a result of chance events; usually reduces genetic variation in a population.

genetic engineering The use of DNA technologies to alter genes for practical purposes.

genetic equilibrium The point at which neither the allele frequencies nor the genotype frequencies in a population change in succeeding generations.

genetic recombinants Nonparental combinations of alleles. In eukaryotes they result from crossing-over in meiosis.

genetic recombination The process by which the combinations of alleles for different genes in two parental individuals become shuffled into new combinations in offspring individuals.

genetic screening Biochemical or molecular tests for identifying inherited disorders after a child is born.

genetically modified organism (GMO) A transgenic organism.

genomic imprinting Pattern of inheritance in which the expression of a nuclear gene is based on whether an individual organism inherits the gene from the male or female parent.

genomic library A collection of clones that contains a copy of every DNA sequence in a genome.

genotype The genetic constitution of an organism in terms of its genes and alleles.

genotype frequency The percentage of individuals in a population possessing a particular genotype.

genus A Linnaean taxonomic category ranking below a family and above a species.

geographical range The overall spatial boundaries within which a population lives.

germ cell An animal cell that is set aside early in embryonic development and gives rise to the gametes.

germ layers The three primary cell layers of the embryo: the outer ectoderm, the inner endoderm, and the mesoderm between the ectoderm and the endoderm.

germ-line cells Cells that develop into sperm or eggs.

germ-line gene therapy Experiment in which a gene is introduced into germ-line cells of an animal to correct a genetic disorder.

gestation The period of mammalian development in which the embryo develops in the uterus of the mother.

gibberellin Any of a large family of plant hormones that regulate aspects of growth, including cell elongation.

gill A respiratory organ formed as evagination of the body that extends outward into the respiratory medium.

gill arch One of the series of curved supporting structures between the slits in the pharynx of a chordate.

gill slit One of the openings in the pharynx of a chordate through which water passes out of the pharynx.

Ginkgophyta A plant phylum with a single living species, the ginkgo (or maiden-hair) tree.

gizzard The part of the digestive tube that grinds ingested material into fine particles by muscular contractions of the wall.

gland A cell or group of cells that produces and releases substances nearby, in another part of the body, or to the outside.

glans A soft, caplike structure at the end of the penis, containing most of the nerve endings producing erotic sensations.

glial cell A nonneuronal cell contained in the nervous tissue that physically supports and provides nutrients to neurons, provides electrical insulation between them, and scavenges cellular debris and foreign matter.

globulin A plasma protein that transports lipids (including cholesterol) and fat-soluble vitamins; a specialized subgroup of globulins, the immunoglobulins, constitute antibodies and other molecules contributing to the immune response.

glomerulus A ball of blood capillaries surrounded by Bowman's capsule in the human nephron.

glucagon A pancreatic hormone with effects opposite to those of insulin: it stimulates glycogen, fat, and protein degradation.

glucocorticoid A steroid hormone secreted by the adrenal cortex that helps maintain the blood concentration of glucose and other fuel molecules.

glucose-dependent insulinotropic peptide (GIP) Hormone released in response to a meal entering the digestive tract which triggers insulin release from the pancreas.

glycocalyx A carbohydrate coat covering the cell surface.

glycogen Energy-providing carbohydrates stored in animal cells.

glycolipid A lipid molecule with carbohydrate groups attached.

glycolysis Stage of cellular respiration in which sugars such as glucose are partially oxidized and broken down into smaller molecules.

glycoprotein A protein with carbohydrate groups attached.

glycosidic bond Bond formed by the linkage of two α-glucose molecules with oxygen as a bridge between a carbon of the first glucose unit and a carbon of the second glucose unit.

Gnathostomata The group of vertebrates with moveable jaws.

golden algae *See* Chrysophyta.

Golgi complex In eukaryotes, the organelle responsible for the final modification, sorting, and distribution of proteins and lipids.

Golgi tendon organ A proprioceptor of tendons.

gonad A specialized gamete-producing organ in which the germ cells collect. Gonads are the primary source of sex hormones in vertebrates: ovaries in the female and testes in the male.

gonadotropin A hormone that regulates the activity of the gonads (ovaries and testes).

gonadotropin releasing hormone (GnRH) A tropic hormone secreted by the hypothalamus that causes the pituitary to make luteinizing hormone (LH) and follicle stimulating hormone (FSH).

G-protein–coupled receptor In signal transduction, a surface receptor that responds to a signal by activating a G protein.

graded potential A change in membrane potential that does not necessarily trigger an action potential.

gradualism The view that Earth and its living systems changed slowly over its history.

gradualist hypothesis The hypothesis that large changes in either geological features or biological lineages result from the slow, continuous accumulation of small changes over time.

Gram stain technique A technique of staining bacteria to distinguish between types of bacteria with different cell wall compositions.

Gram-negative Describing bacteria that do not retain the stain used in the Gram stain technique.

Gram-positive Describing bacteria that appear purple when stained using the Gram stain technique.

granum Structure in the chloroplasts of higher plants formed by thylakoids stacked one on top of another.

gravitropism A directional growth response to Earth's gravitational pull that is induced by mechanical and hormonal influences.

gray crescent A crescent-shaped region of the underlying cytoplasm at the side opposite the point of sperm entry exposed after fertilization when the pigmented layer of cytoplasm rotates toward the site of sperm entry.

gray matter Areas of densely packed nerve cell bodies and dendrites in the brain and spinal cord.

greater vestibular gland One of two glands located slightly below and to the left and right of the opening of the vagina in women. They secrete mucus that provides lubrication, especially when the woman is sexually aroused.

green algae *See* Chlorophyta.

greenhouse effect A phenomenon in which certain gases foster the accumulation of heat in the lower atmosphere, maintaining warm temperatures on Earth.

gross primary productivity The rate at which producers convert solar energy into chemical energy.

ground meristem The primary meristematic tissue in plants that gives rise to ground tissues, mostly parenchyma.

ground tissue system One of the three basic tissue systems in plants; includes all tissues other than dermal and vascular tissues.

growth factor Any of a large group of peptide hormones that regulates the division and differentiation of many cell types in the body.

growth hormone (GH) A hormone that stimulates cell division, protein synthesis, and bone growth in children and adolescents, thereby causing body growth.

guanine A purine that base-pairs with cytosine in nucleic acids.

guard cell Either of a pair of specialized crescent-shaped cells that control the opening and closing of stomata in plant tissue.

guttation The exudation of water from leaves as a result of strong root pressure.

gymnosperm A seed plant that produces "naked" seeds not enclosed in an ovary.

H^+ pump *See* proton pump.

habitat The specific environment in which a population lives, as characterized by its biotic and abiotic features.

habitat fragmentation A process in which remaining areas of intact habitat are reduced to small, isolated patches.

habituation The learned loss of responsiveness to stimuli.

half-life The time it takes for half of a given amount of a radioisotope to decay.

haplodiploidy A pattern of sex determination in insects in which females are diploid and males are haploid.

haploid An organism or cell with only one copy of each type of chromosome in its nuclei.

Hardy–Weinberg principle An evolutionary rule of thumb that specifies the conditions under which a population of diploid organisms achieves genetic equilibrium.

haustorium (plural, haustoria) The hyphal tip of a parasitic fungus that penetrates a host plant and absorbs nutrients from it; likewise in parasitic flowering plants, a root that can penetrate a host's tissues and absorb nutrients.

head The anteriormost part of an organism's body, containing the brain, sensory structures, and feeding apparatus. For a bacteriophage, the usually polyhedral part of the virus containing the genetic material.

head–foot In mollusks, the region of the body that provides the major means of locomotion and contains concentrations of nervous system tissues and sense organs.

heart The muscular organ that pumps fluid through the circulatory system.

heartwood The inner core of a woody stem; composed of dry tissue and nonliving cells that no longer transport water and solutes and may store resins, tannins, and other defensive compounds.

heat of vaporization The heat required to give water molecules enough energy of motion to break loose from liquid water and form a gas.

heat-shock protein (HSP) Any of a group of chaperone proteins that are present in all cells in all life forms. They are induced when a cell undergoes various types of environmental stresses like heat, cold, and oxygen deprivation.

heavy chain The heavier of the two types of polypeptide chains that are found in immunoglobulin and antibody molecules.

helical virus A virus in which the protein subunits of the coat assemble in a rodlike spiral around the genome.

helper T cell A clonal cell that assists with the activation of B cells.

hemolymph The circulatory fluid of invertebrates with open circulatory systems, including mollusks and arthropods.

hepatic portal vein The blood vessel that leads to capillary networks in the liver.

Hepatophyta The phylum that includes liverworts and their bryophyte relatives.

herbicide A compound that, at proper concentration, kills plants.

herbivore An animal that obtains energy and nutrients primarily by eating plants.

herbivory The process in which herbivores consume plants.

hermaphroditism The mechanism in which both mature egg-producing and mature sperm-producing tissue are present in the same individual.

heterochrony Changes in the relative rate of development of morphological characters.

heterogametic sex The sex that produces two types of gametes with respect to the sex chromosomes.

heterosporous Producing two types of spores, "male" microspores and "female" megaspores.

heterotroph An organism that acquires energy and nutrients by eating other organisms or their remains.

heterozygote An individual with two different alleles of a gene.

heterozygote advantage An evolutionary circumstance in which individuals that are heterozygous at a particular locus have higher relative fitness than either homozygote.

heterozygous State of possessing two different alleles of a gene.

Hfr cell A special donor cell that can transfer genes on a bacterial chromosome to a recipient bacterium.

hibernation Extended torpor during winter.

hindbrain The lower area of the brain that includes the brain stem, medulla oblongata, and pons.

hippocampus A gray-matter center that is involved in sending information.

histamine An inflammatory signaling molecule.

histone A small, positively charged (basic) protein that is complexed with DNA in the chromosomes of eukaryotes.

historical biogeography The study of the geographical distributions of plants and animals in relation to their evolutionary history.

homeobox A region of a homeotic gene that corresponds to an amino acid section of the homeodomain.

homeodomain An encoded transcription factor of each protein that binds to a region in the promoters of the genes whose transcription it regulates.

homeostasis A steady internal condition maintained by responses that compensate for changes in the external environment.

homeostatic control system The processes and activities responsible for homeostasis.

homeotic gene Any of the family of genes that determines overall body plan (the structure of body parts) during embryonic development.

hominin A member of a monophyletic group of primates, characterized by an erect bipedal stance, that includes modern humans and their recent ancestors.

Hominoidea The monophyletic group of primates that includes apes and humans.

homogametic sex The sex that produces only one type of gamete with respect to the sex chromosomes.

homologies Characteristics shared by a set of species because they inherited them from their common ancestor.

homologous chromosomes The two chromosomes of each pair in a diploid cell—one of the pair derives from the maternal parent and the other derives from the paternal parent. Homologous chromosomes have the same genes, in the same order, in their DNA.

homologous traits Characteristics that are similar in two species because they inherited the genetic basis of the trait from their common ancestor.

homoplasies Characteristics shared by a set of species, often because they live in similar environments, but not present in their common ancestor; often the product of convergent evolution.

homosporous Producing only one type of spore.

homozygote An individual with two copies of the same allele.

homozygous State of possessing two copies of the same allele.

horizon A noticeable layer of soil, such as topsoil, having a distinct texture and composition that varies with soil type.

horizontal cell A type of neuron that forms lateral connections among photoreceptor cells in the retina of the eye.

horizontal gene transfer Movement of genetic material between organisms other than by descent.

hormone A signaling molecule secreted by a cell that can alter the activities of any cell with receptors for it; in animals, typically a molecule produced by one tissue and transported via the bloodstream to another specific tissue to alter its physiological activity.

host A species that is fed upon by a parasite.

host race A population of insects that may be reproductively isolated from other populations of the same species as a consequence of their adaptation to feed on a specific host plant species.

***Hox* gene** A type of homeotic gene that controls an organism's overall body plan.

human chorionic gonadotropin (hCG) A hormone that keeps the corpus luteum in the ovary from breaking down.

human immunodeficiency virus (HIV) A retrovirus that causes acquired immunodeficiency syndrome (AIDS).

humoral immunity *See* antibody-mediated immunity.

humus The organic component of soil remaining after decomposition of plants and animals, animal droppings, and other organic matter.

hybrid breakdown A postzygotic reproductive isolating mechanism in which hybrids are capable of reproducing, but their offspring have either reduced fertility or reduced viability.

hybrid inviability A postzygotic reproductive isolating mechanism in which a hybrid individual has a low probability of survival to reproductive age.

hybrid sterility A postzygotic reproductive isolating mechanism in which hybrid offspring cannot form functional gametes.

hybrid zone A geographical area where the hybrid offspring of two divergent populations or species are common.

hybridoma A B cell that has been induced to fuse with a cancerous lymphocyte called a myeloma cell, forming a single, composite cell.

hydration layer A surface coat of water molecules that covers other polar and charged molecules and ions.

hydrocarbon Molecule consisting of carbon linked only to hydrogen atoms.

hydrogen bond Noncovalent bond formed by unequal electron sharing between hydrogen atoms and oxygen, nitrogen, or sulfur atoms.

hydrologic alterations Human-induced changes in the pathways through which water moves in the hydrologic cycle.

hydrologic cycle The global cycling of water between the ocean, the atmosphere, land, freshwater ecosystems, and living organisms.

hydrolysis Reaction in which the components of a water molecule are added to functional groups as molecules are broken into smaller subunits.

hydrophilic In chemistry and biology, referring to polar molecules that associate readily with water.

hydrophobic In chemistry and biology, referring to nonpolar substances that are excluded by water and other polar molecules.

hydroponic culture A method of growing plants not in soil but with the roots bathed in a solution that contains water and mineral nutrients.

hydrosphere The component of the biosphere that encompasses all the waters on Earth, including oceans, rivers, and polar ice caps.

hydrostatic skeleton A structure consisting of muscles and fluid that, by themselves, provide support for the animal or part of the animal; no rigid support, like a bone, is involved.

hydroxyl group Group consisting of an oxygen atom linked to a hydrogen atom on one side and to a carbon chain on the other side.

hymen A thin flap of tissue that partially covers the opening of the vagina.

hyperpolarized The condition of a neuron when its membrane potential is more negative than the resting value.

hypersensitive response A plant defense that physically cordons off an infection site by surrounding it with dead cells.

hypertension Commonly called high blood pressure, a medical condition in which blood pressure is chronically elevated above normal values.

hyperthermia The condition resulting when the heat gain of the body is too great to be counteracted.

hypertonic Solution containing dissolved substances at higher concentrations than the cells it surrounds.

hypha (plural, hyphae) Any of the thread-like filaments that form the mycelium of a fungus.

hypoblast The bottom layer of a blastodisc.

hypocotyl The region of a plant embryo's vertical axis between the cotyledons and the radicle.

hypodermis The innermost layer of the skin that contains larger blood vessels and additional reinforcing connective tissue.

hypolimnion The deep water of the profundal zone of a lake.

hypothalamus The portion of the brain that contains centers regulating basic homeostatic functions of the body and contributing to the release of hormones.

hypothermia A condition in which the core temperature falls below normal for a prolonged period.

hypothesis A tentative explanation for an observation, phenomenon, or scientific problem that can be tested by further investigation.

hypotonic Solution containing dissolved substances at lower concentrations than the cells it surrounds.

ice lattice A rigid, crystalline structure formed when a water molecule in ice forms four hydrogen bonds with neighboring molecules.

IgA The class of antibodies found mainly in secretions at particular locations in the body; IgA functions to bind to surface groups on pathogens and block their attachment to body surfaces.

IgD The class of antibodies that occurs with IgM as a receptor on the surfaces of B cells; the function of IgD is uncertain.

IgE The class of antibodies secreted by plasma cells of the skin and tissues lining the gastrointestinal tract and respiratory tract that binds to basophils and mast cells to trigger the release of histamine, thereby causing an inflammatory response.

IGF *See* insulin-like growth factor.

IgG The most abundant antibody type circulating in the blood and lymphatic system that is involved in primary and secondary immune responses.

IgM The first antibody type secreted by B cells in a primary immune response.

imbibition The movement of water into a seed as the water molecules are attracted to hydrophilic groups of stored proteins; the first step in germination.

immigration Movement of organisms into a population.

immune response The defensive reactions of the immune system.

immune system The body's system of defenses against disease, composed of certain white blood cells and antibodies.

immunoglobulin A specific protein substance produced by plasma cells to aid in fighting infection.

immunological memory The capacity of the immune system to respond more rapidly and vigorously to the second contact with a specific antigen than to the primary contact.

immunological tolerance The process that protects the body's own molecules from attack by the immune system.

imperfect flower A type of incomplete flower that has stamens or carpels, but not both.

imprinting The process of learning the identity of a caretaker and potential future mate during a critical period early in life.

inbreeding A special form of nonrandom mating in which genetically related individuals mate with each other.

incisors Flattened, chisel-shaped teeth of mammals, located at the front of the mouth, that are used to nip or cut food.

incomplete dominance Condition in which the effects of recessive alleles can be detected to some extent in heterozygotes.

incomplete flower A flower lacking one or more of the four floral whorls.

incomplete metamorphosis In certain insects, a life cycle characterized by the absence of a pupal stage between the immature and adult stages.

incurrent siphon A muscular tube that brings water containing oxygen and food into the body of an invertebrate.

incus The second of the three sound-conducting middle ear bones in vertebrates, located between the malleus and the stapes.

independent assortment Mendel's principle that the alleles of the genes that govern two characters assort independently during formation of gametes.

indeterminate cleavage A type of cleavage, observed in many deuterostomes, in which the developmental fates of the first few cells produced by mitosis are not determined as soon as cells are produced.

indeterminate growth Growth that is not limited by an organism's genetic program, so that the organism grows for as long as it lives; typical of many plants. *Compare* determinate growth.

inducer Concerning regulation of gene expression in bacteria, a molecule that turns on the transcription of the genes in an operon.

inducible operon Operon whose expression is increased by an inducer molecule.

induction A mechanism in which one group of cells (the inducer cells) causes or influences another nearby group of cells (the responder cells) to follow a particular developmental pathway.

infection thread In the formation of root nodules on nitrogen-fixing plants, the tube formed by the plasma membrane of root hair cells as bacteria enter the cell.

infertility The inability for the female of a couple to get pregnant after 12 months of frequent, contraceptive-free intercourse.

inflammation The heat, pain, redness, and swelling that occur at the site of an infection.

ingestion The feeding methods used to take food into the digestive cavity.

inheritance The transmission of DNA (that is, genetic information) from one generation to the next.

inhibin A peptide that, in females, is an inhibitor of FSH secretion from the pituitary thereby diminishing the signal for follicular growth. In males, inhibin inhibits FSH secretion from the pituitary, thereby decreasing spermatogenesis.

inhibiting hormone (IH) A hormone released by the hypothalamus that inhibits the secretion of a particular anterior pituitary hormone.

inhibition hypothesis An hypothesis suggesting that new species are prevented from occupying a community by whatever species are already present.

inhibitory postsynaptic potential (IPSP) A change in membrane potential caused when hyperpolarization occurs, pushing the neuron farther from threshold.

initial A plant cell that remains permanently as part of a meristem and gives rise to daughter cells that differentiate into specialized cell types.

initiation factor (IF) A protein that aids an initiation step of translation.

initiator codon *See* start codon.

innate immune system A nonspecific line of defense against pathogens that includes inflammation, which creates internal conditions that inhibit or kill many pathogens, and specialized cells that engulf or kill pathogens or infected body cells.

inner boundary membrane Membrane lying just inside the outer boundary membrane of a chloroplast, enclosing the stroma.

inner cell mass The dense mass of cells within the blastocyst that will become the embryo.

inner ear That part of the ear, particularly the cochlea, that converts mechanical vibrations (sound) into neural messages that are sent to the brain.

inner mitochondrial membrane Membrane surrounding the mitochondrial matrix.

inorganic molecule Molecule without carbon atoms in its structure.

inositol triphosphate (IP_3) In particular signal transduction pathways, a second messenger that activates transport proteins in the endoplasmic reticulum to release Ca^{2+} into the cytoplasm. IP_3 is involved in one of two major G-protein–coupled receptor–response pathways.

insertion sequence A transposable element that contains only genes for its transposition.

instinctive behavior A genetically "programmed" response that appears in complete and functional form the first time it is used.

insulin A hormone secreted by beta cells in the islets, acting mainly on cells of nonworking skeletal muscles, liver cells, and adipose tissue (fat) to lower blood glucose, fatty acid and amino acids levels, and promote the storage of those molecules.

insulin-like growth factor (IGF) A peptide released by target tissues in response to binding by growth hormone that directly stimulates growth processes.

integral protein Protein embedded in a phospholipid bilayer.

integration The sorting and interpretation of neural messages and the determination of the appropriate response(s).

integrator In homeostatic feedback, the control center that compares a detected environmental change with a set point.

integument Skin.

intercalated disk The joining point between two cells in cardiac muscle.

interference competition Form of competition in which individuals fight over resources or otherwise harm each other directly.

interferon A cytokine produced by infected host cells affected by viral dsRNA, which acts both on the infected cell that produces it, an autocrine effect, and on neighboring uninfected cells, a paracrine effect.

interkinesis A brief interphase separating the two meiotic divisions.

intermediate disturbance hypothesis Hypothesis proposing that species richness is greatest in communities that experience fairly frequent disturbances of moderate intensity.

intermediate filament A cytoskeletal filament about 10 nm in diameter that provides mechanical strength to cells in tissues.

intermediate-day plant A plant that flowers only when daylength falls between the values for long-day and short-day plants.

internal environment The fluid within an organism that supplies all the needs of individual cells.

internal fertilization The process in which sperm are released by the male close to or inside the entrance of the reproductive tract of the female.

internal gill A gill located within the body that has a cover providing physical protection for the gills. Water must be brought to internal gills.

interneuron A neuron that integrates information to formulate an appropriate response.

internode The region between two nodes on a plant stem.

interphase The first stage of the mitotic cell cycle, during which the cell grows and replicates its DNA before undergoing mitosis and cytokinesis.

interspecific competition The competition for resources between species.

interstitial fluid The fluid occupying the spaces between cells in multicellular animals.

intertidal zone The shoreline that is alternately submerged and exposed by tides.

intestinal villus A microscopic, fingerlike extension in the lining of the small intestine.

intestine The portion of digestive system where organic matter is hydrolyzed by enzymes secreted into the digestive tube. As muscular contractions of the intestinal wall move the mixture along, cells lining the intestine absorb the molecular subunits produced by digestion.

intracellular digestion The process in which cells take in food particles by endocytosis.

intraspecific competition The dependence of two or more individuals in a population on the same limiting resource.

intrinsic rate of increase The maximum possible per capita population growth rate in a population living under ideal conditions.

intron A non-protein-coding sequence that interrupts the protein-coding sequence in a eukaryotic gene. Introns are removed by splicing in the processing of pre-mRNA to mRNA.

invagination The process in which cells changing shape and pushing inward from the surface produce an indentation, such as the dorsal lip of the blastopore.

inversion Chromosomal alteration that occurs if a broken segment reattaches to the same chromosome from which it was lost, but in reversed orientation, so that the order of genes in the segment is reversed with respect to the other genes of the chromosome.

invertebrate An animal without a vertebral column.

involution The process by which cells migrate into the blastopore.

ion A positively or negatively charged atom.

ionic bond Bond that results from electrical attractions between atoms that have lost or gained electrons.

ionotropic neurotransmitter receptor A receptor in the postsynaptic membrane to which a neurotransmitter binds directly, causing an associated ion channel gate to open or close thereby altering the flow of a specific ion or ions in the postsynaptic cell.

iris Of the eye, the colored muscular membrane that lies behind the cornea and in front of the lens, which by opening or closing determines the size of the pupil and hence the amount of light entering the eye.

islets of Langerhans Endocrine cells that secrete the peptide hormones insulin and glucagon into the bloodstream.

isomers Two or more molecules with the same chemical formula but different molecular structures.

isotonic Equal concentration of water inside and outside cells.

isotope A distinct form of the atoms of an element, with the same number of protons but different number of neutrons.

jasmonate Any of a group of plant hormones that help regulate aspects of growth and responses to stress, including attacks by predators and pathogens.

juvenile hormone (JH) A peptide hormone secreted by the corpora allata, a pair of glands just behind the brain in insects.

juxtaglomerular apparatus A group of receptors that monitor the pressure and flow of fluid through the distal tubule of the kidney.

karyogamy In plants, the fusion of two sexually compatible haploid nuclei after cell fusion (plasmogamy).

karyotype A characteristic of a species consisting of the shapes and sizes of all the chromosomes at metaphase.

keeled sternum The ventrally extended breastbone of a bird to which the flight muscles attach.

ketone Molecule in which the carbonyl group is linked to a carbon atom in the interior of a carbon chain.

keystone species A species that has a greater effect on community structure than its numbers might suggest.

kilocalorie (kcal) The scientific unit equivalent to a Calorie and equal to 1,000 small calories.

kin selection An explanation for altruistic behavior to close relatives, allowing them to produce proportionately more surviving copies of the altruist's genes than the altruist might otherwise have produced on its own.

kinesis A change in the rate of movement or the frequency of turning movements in response to environmental stimuli.

kinetic energy The energy of motion.

kinetochore A specialized structure consisting of proteins attached to a centromere that mediates the attachment and movement of chromosomes along the mitotic spindle.

Kinetoplastids A lineage of the Euglenozoa subgroup of the Excavata protist evolutionary group; nonphotosynthetic, heterotrophic cells that live as animal parasites and that contain a single mitochondrion with a large DNA–protein deposit called a kinetoplast.

kingdom A Linnaean taxonomic category that ranks below a domain and above a phylum.

Kingdom Animalia The taxonomic kingdom that includes all living and extinct animals.

Kingdom Fungi The taxonomic kingdom that includes all living or extinct fungi.

Kingdom Plantae The taxonomic kingdom encompassing all living or extinct plants.

Korarchaeota A group of Archaea recognized solely on the basis of rRNA coding sequences in DNA taken from environmental samples.

Krebs cycle *See* citric acid cycle.

***K*-selected species** Long-lived species that thrive in more stable environments.

labia majora A pair of fleshy, fat-padded folds that partially cover the labia minora.

labia minora Two folds of tissue that run from front to rear on either side of the opening to the vagina.

lactate fermentation Reaction in which pyruvate is converted into lactate.

lactose intolerance The unpleasant intestinal symptoms that result when an individual loses the capacity to synthesize lactase, the enzyme that breaks down the milk sugar.

lagging strand The new DNA strand synthesized discontinuously during replication in the direction opposite to that of DNA unwinding.

lagging strand template The DNA template strand for the lagging strand.

landscape ecology The field that examines how large-scale ecological factors—such as the distribution of plants, topography, and human activity—influence local populations and communities.

larva A sexually immature stage in the life cycle of many animals that is morphologically distinct from the adult.

larynx The voice box.

latent phase The time during which a virus remains in the cell in an inactive form.

lateral bud A bud on the side of a plant stem from which a branch may grow.

lateral geniculate nuclei Clusters of neurons located in the thalamus that receive visual information from the optic nerves and send it on to the visual cortex.

lateral inhibition Visual processing in which lateral movement of signals from a rod or cone proceeds to a horizontal cell and continues to bipolar cells with which the horizontal cell makes inhibitory connections, serving both to sharpen the edges of objects and enhance contrast in an image.

lateral line system The complex of mechanoreceptors along the sides of some fishes and aquatic amphibians that detect vibrations in the water.

lateral meristem A plant meristem that gives rise to secondary tissue growth. *Compare* primary meristem.

lateral root A root that extends away from the main root (or taproot).

lateralization A phenomenon in which some brain functions are more localized in one of the two hemispheres.

lateral-line system The complex of organs and sensory receptors along the sides of many fishes and amphibians that detects vibrations in water.

leaching The process by which soluble materials in soil are washed into a lower layer of soil or are dissolved and carried away by water.

leading strand The new DNA strand synthesized during replication in the direction of DNA unwinding.

leading strand template The DNA template strand for the leading strand.

leaf primordium A lateral outgrowth from the apical meristem that develops into a young leaf.

learned behavior A response of an animal that is dependent upon having a particular kind of experience during development.

learning A process in which experiences stored in memory change the behavioral responses of an animal.

leghemoglobin An iron-containing, red-pigmented protein produced in root nodules during the symbiotic association between *Bradyrhizobium* or *Rhizobium* and legumes.

lek A display ground where males each possess a small territory from which they court attentive females.

lens The transparent, biconvex intraocular tissue that helps bring rays of light to a focus on the retina.

Lepidosauromorpha A monophyletic lineage of diapsids that includes both marine and terrestrial animals, represented today by sphenodontids, lizards, and snakes.

leukocyte A white blood cell, which eliminates dead and dying cells from the body, removes cellular debris, and participates in defending the body against invading organisms.

Leydig cell A cell that produces the male sex hormones.

lichen A single vegetative body that is the result of an association between a fungus and a photosynthetic partner, often an alga.

life cycle The sequential stages through which individuals develop, grow, maintain themselves, and reproduce.

life history The lifetime pattern of growth, maturation, and reproduction that is characteristic of a population or species.

life table A chart that summarizes the demographic characteristics of a population.

ligament A fibrous connective tissue that connects bones to each other at a joint.

light chain The lighter of the two types of polypeptide chains found in immunoglobulin and antibody molecules.

light microscope Microscope that uses light to illuminate the specimen.

light-dependent reaction The first stage of photosynthesis, in which the energy of sunlight is absorbed and converted into chemical energy in the form of ATP and NADPH.

light-independent reaction The second stage of photosynthesis, in which electrons are used as a source of energy to convert inorganic CO_2 to an organic form. Also referred to as the *Calvin cycle*.

lignification The deposition of lignin in plant cell walls; it anchors the cellulose fibers in the walls, making them stronger and more rigid, and protects the other wall components from physical or chemical damage.

lignin A tough, rather inert polymer that strengthens the secondary walls of various plant cells and thus helps vascular plants to grow taller and stay erect on land.

limbic system A functional network formed by parts of the thalamus, hypothalamus, and basal nuclei, along with other nearby gray-matter centers—the amygdala, hippocampus, and olfactory bulbs—sometimes called the "emotional brain."

limiting nutrient An element in short supply within an ecosystem, the shortage of which limits productivity.

limnetic zone The sunlit, open water in a lake, beyond the zone where plants rooted in the bottom can grow.

linkage The phenomenon of genes being located on the same chromosome.

linkage map Map of a chromosome showing the relative locations of genes based on recombination frequencies.

linked genes Genes on the same chromosome.

linker A short segment of DNA extending between one nucleosome and the next in a eukaryotic chromosome.

lipase A pancreatic enzyme that hydrolyzes fats.

lipopolysaccharide A large molecule that consists of a lipid and a carbohydrate joined by a covalent bond.

lithosphere The component of the biosphere that includes the rocks, sediments, and soils of the crust.

littoral zone The shallow, sunlit water near the shore of a lake or pond.

liver A large organ whose many functions include aiding in digestion, removing toxins from the body, and regulating the chemicals in the blood.

loam Any well-aerated soil composed of a mixture of sand, clay, silt, and organic matter.

local homeostatic controls Controls that operate only within an organ where a change in the internal environment needs to be addressed.

local inflammation Inflammation that occurs at the site of an infection.

locus The particular site on a chromosome at which a gene is located.

logistic model of population growth Model of population growth that assumes that a population's per capita growth rate decreases as the population gets larger.

long-day plant A plant that flowers in spring when dark periods become shorter and day length becomes longer.

long-term memory Memory that stores information from days to years or even for life.

long-term potentiation A long-lasting increase in the strength of synaptic connections in activated neural pathways following brief periods of repeated stimulation.

loop of Henle A U-shaped bend of the proximal convoluted tubule.

loose connective tissue A tissue formed of sparsely distributed cells surrounded by a more or less open network of collagen and other glycoprotein fibers.

lophophore The circular or U-shaped fold with one or two rows of hollow, ciliated tentacles that surrounds the mouth of brachiopods, bryozoans, and phoronids and is used to gather food.

lumen The inside of the digestive tube.

lung One of a pair of invaginated respiratory surfaces, buried in the body interior where they are less susceptible to drying out; the organs of respiration in mammals, birds, reptiles, and most amphibians.

luteinizing hormone (LH) A hormone secreted by the pituitary that stimulates the growth and maturation of eggs in females and the secretion of testosterone in males.

Lycophyta The plant phylum that includes club mosses and their close relatives.

lymph The interstitial fluid picked up by the lymphatic system.

lymph node One of many small, bean-shaped organs spaced along the lymph vessels that contain macrophages and other leukocytes that attack invading disease organisms.

lymphatic system An accessory system of vessels and organs that helps balance the fluid content of the blood and surrounding tissues and participates in the body's defenses against invading disease organisms.

lymphocyte A leukocyte that carries out most of its activities in tissues and organs of the lymphatic system. Lymphocytes play major roles in immune responses.

lysogenic cycle Cycle in which the DNA of the bacteriophage is integrated into the DNA of the host bacterial cell and may remain for many generations.

lysosome Membrane-bound vesicle containing hydrolytic enzymes for the digestion of many complex molecules.

lytic cycle The series of events from infection of one bacterial cell by a phage through the release of progeny phages from lysed cells.

M phase-promoting factor (MPF) A complex of M cyclin and cyclin-dependent kinase 1 (Cdk1). The complex initiates mitosis and orchestrates some of its key events.

macroevolution Large-scale evolutionary patterns in the history of life, producing major changes in species and higher taxonomic groups.

macronucleus In ciliophorans, a single large nucleus that develops from a micronucleus but loses all genes except those required for basic "housekeeping" functions of the cell and for ribosomal RNAs.

macronutrient In humans, a mineral required in amounts ranging from 50 mg to more than 1 gram per day. In plants, a nutrient needed in large amounts for the normal growth and development.

macrophage A phagocyte that takes part in nonspecific defenses and adaptive immunity.

magnetoreceptor A receptor found in some animals that navigate long distances which allows them to detect and use Earth's magnetic field as a source of directional information.

magnification The ratio of an object as viewed to its real size.

magnoliids An angiosperm group that includes magnolias, laurels, and avocados; they are more closely related to monocots than to eudicots.

major histocompatibility complex A large cluster of genes encoding the MHC proteins.

malleus The outermost of the sound-conducting bones of the middle ear in vertebrates.

malnutrition A condition resulting from a diet that lacks one or more essential nutrients.

Malpighian tubule The main organ of excretion and osmoregulation in insects, helping them to maintain water and electrolyte balance.

mammary glands Specialized organs of female mammals that produce energy-rich milk, a watery mixture of fats, sugars, proteins, vitamins, and minerals.

mantle One or two folds of the body wall that lines the shell and secretes the substance that forms the shell in mollusks.

map unit The unit of a linkage map, equivalent to a recombination frequency of 1%. Also referred to as a *centimorgan.*

maritime climate Climate tempered by ocean winds.

marsupium An external pouch on the abdomen of many female marsupials, containing the mammary glands, and within which the young continue to develop after birth.

mass The amount of matter in an object.

mass extinctions The disappearance of a large number of species in a relatively short period of geological time.

mass number The total number of protons and neutrons in the atomic nucleus.

mast cell A type of cell dispersed through connective tissue that releases histamine when activated by the death of cells, caused by a pathogen at an infection site.

maternal chromosome The chromosome derived from the female parent of an organism.

maternal inheritance A type of uniparental inheritance in which all progeny (both males and females) have the phenotype of the female parent.

maternal-effect gene One of a class of genes that regulate the expression of other genes expressed by the mother during oogenesis and that control the polarity of the egg and, therefore, of the embryo.

mating The pairing of a male and a female for the purpose of sexual reproduction.

mating systems The social systems describing how males and females pair up.

mating type A genetically defined strain of an organism (such as a fungus) that can only mate with an organism of the opposite mating type; mating types are often designated + and −.

matter Anything that occupies space and has mass.

maximum likelihood methods A statistical technique that compares alternative phylogenetic trees with specific models of evolutionary change.

mechanical isolation A prezygotic reproductive isolating mechanism caused by differences in the structure of reproductive organs or other body parts.

mechanical processing In the digestive tube, the process of chewing, grinding, and tearing food chunks into smaller pieces.

mechanoreceptor A sensory receptor that detects mechanical energy, such as changes in pressure, body position, or acceleration. The auditory receptors in the ears are examples of mechanoreceptors.

medusa The tentacled, usually bell-shaped, free-swimming sexual stage in the life cycle of a coelenterate.

megapascal A unit of pressure used to measure water potential.

megaspore A plant spore that develops into a female gametophyte; usually larger than a microspore.

meiosis The division of diploid cells to haploid progeny, consisting of two sequential rounds of nuclear and cellular division.

meiosis I The first division of the meiotic cell cycle in which homologous chromosomes pair and undergo an exchange of chromosome segments, and then the homologous chromosomes separate, resulting in two cells, each with the haploid number of chromosomes and with each chromosome still consisting of two chromatids.

meiosis II The second division of the meiotic cell cycle in which the sister chromatids in each of the two cells produced by meiosis I separate and segregate into different cells, resulting in four cells each with the haploid number of chromosomes.

melanocyte-stimulating hormone (MSH) A hormone secreted by the anterior pituitary that controls the degree of pigmentation in melanocytes.

melatonin A peptide hormone secreted by the pineal gland that helps maintain daily biorhythms.

membrane attack complexes (MAC) An abnormal activation of the complement (protein) portion of the blood, forming a cascade reaction that brings blood proteins together, binds them to the cell wall, and then inserts them through the cell membrane.

membrane potential An electrical voltage that measures the potential inside a cell membrane relative to the fluid just outside; it is negative under resting conditions and becomes positive during an action potential.

memory The storage and retrieval of a sensory or motor experience, or a thought.

memory B cell In antibody-mediated immunity, a long-lived cell expressing an antibody on its surface that can bind to a specific antigen. A memory B cell is activated the next time the antigen is encountered, producing a rapid secondary immune response.

memory cell An activated lymphocyte that circulates in the blood and lymph, ready to initiate a rapid immune response upon subsequent exposure to the same antigen.

memory cytotoxic T cells Cells differentiated from cytotoxic T cells that are stored for subsequent cell-mediated immunity involving the same antigens.

memory helper T cell In cell-mediated immunity, a long-lived cell differentiated from a helper T cell, which remains in an inactive state in the lymphatic system after an immune reaction has run its course and ready to be activated upon subsequent exposure to the same antigen.

meninges Three layers of connective tissue that surround and protect the spinal cord and brain.

menstrual cycle A cycle of approximately 1 month in the human female during which an egg is released from an ovary and the uterus is prepared to receive the fertilized egg; if fertilization does not occur, the endometrium breaks down, which releases blood and tissue breakdown products from the uterus to the outside through the vagina.

meristem An undifferentiated, permanently embryonic plant tissue that gives rise to new cells forming tissues and organs.

mesenteries Sheets of loose connective tissue, covered on both surfaces with epithelial cells, which suspend the abdominal organs in the coelom and provide lubricated, smooth surfaces that prevent chafing or abrasion between adjacent structures as the body moves.

mesoderm The middle layer of the three primary germ layers of an animal embryo, from which the muscular, skeletal, vascular, and connective tissues develop.

mesohyl The gelatinous middle layer of cells lining the body cavity of a sponge.

mesophyll The ground tissue located between the two outer leaf tissues, composed of loosely packed parenchyma cells that contain chloroplasts.

messenger RNA (mRNA) An RNA molecule that serves as a template for protein synthesis.

metabolism The biochemical reactions that allow a cell or organism to extract energy from its surroundings and use that energy to maintain itself, grow, and reproduce.

metabotropic neurotransmitter receptor A receptor in the postsynaptic membrane, typically of the G-protein–coupled receptor type, to which a neurotransmitter binds as a first messenger, activating the receptor and triggering generation of a second messenger that brings about changes in ion-conducting channels in the membrane.

metagenomics Analysis of DNA sequences of the genomes in entire communities of microbes that have been isolated from the environment.

Metamonada (metamonads) A subgroup of the Excavata protist evolutionary group containing the Diplomonadida and the Parabasala.

metamorphosis A reorganization of the form of certain animals during postembryonic development.

metanephridium The excretory tubule of most annelids and mollusks.

metaphase The phase of mitosis during which the spindle reaches its final form and the spindle microtubules move the chromosomes into alignment at the spindle midpoint.

metapopulation A group of neighboring populations that exchange individuals.

micelles The tiny droplets produced when fats are emulsified.

microbody Small, membrane-bound organelle that carries out vital reactions linking metabolic pathways.

microclimate The abiotic conditions immediately surrounding an organism.

microevolution Small-scale genetic changes within populations, often in response to shifting environmental circumstances or chance events.

microfilament A cytoskeletal filament composed of actin.

microfossil The remains of a cell that has decayed and been filled in by calcium carbonate or silica.

micronucleus In ciliophorans, one or more diploid nuclei that contains a complete complement of genes, functioning primarily in cellular reproduction.

micronutrient Any mineral required by an organism only in trace amounts.

micropyle A small opening at one end of an ovule through which the pollen tube passes prior to fertilization.

microRNA (miRNA) One of the major types of small regulatory RNAs in eukaryotes involved in RNA interference (RNAi).

microsatellite *See* short tandem repeat (STR).

microscope Instrument of microscopy with different magnifications and resolutions of specimens.

microscopy Technique for producing visible images of objects that are too small to be seen by the human eye.

microspore A plant spore from which a male gametophyte develops; usually smaller than a megaspore.

microsporidium (plural, microsporidia) A fungal parasite of animals; many mycologists believe they make up a possible sixth phylum within the Kingdom Fungi.

microtubule A cytoskeletal component formed by the polymerization of tubulin into rigid, hollow rods about 25 nm in diameter.

microtubule organizing center (MTOC) An anchoring point near the center of a eukaryotic cell from which most microtubules extend outward.

microvilli Fingerlike projections forming a brush border in epithelial cells that cover the villi.

midbrain The uppermost of the three segments of the brainstem, serving primarily as an intermediary between the rest of the brain and the spinal cord.

middle ear The air-filled cavity containing three small, interconnected bones: the malleus, incus, and stapes.

middle lamella Layer of gel-like polysaccharides that holds together walls of adjacent plant cells.

migration The predictable seasonal movement of animals from the area where they are born to a distant and initially unfamiliar destination, returning to their birth site later.

mimic The species in Batesian mimicry that resembles the model.

mimicry A form of defense in which one species evolves an appearance resembling that of another.

mineralocorticoid A steroid hormone secreted by the adrenal cortex that regulates the levels of Na^+ and K^+ in the blood and extracellular fluid.

minimal medium A growth medium containing the minimal ingredients that enable a nonmutant organism, such as *E. coli,* to grow.

minimum viable population size The smallest population size that is likely to survive both predictable and unpredictable environmental variation.

miRNA-induced silencing complex (miRISC) Protein complex containing an miRNA that binds to sequences in the 3′ UTRs of target mRNAs, resulting in either inhibition of translation of the mRNAs or their degradation.

mismatch repair Repair system that removes mismatched bases from newly synthesized DNA strands.

missense mutation A base-pair substitution mutation in a protein-coding gene that results in a different amino acid in the encoded polypeptide than the normal one.

mitochondrial electron transfer system Series of electron carriers that alternately pick up and release electrons, ultimately transferring them to their final acceptor, oxygen.

mitochondrial matrix The innermost compartment of the mitochondrion.

mitochondrion Membrane-bound organelle responsible for synthesis of most of the ATP in eukaryotic cells.

mitosis Nuclear division that produces daughter nuclei that are exact genetic copies of the parental nucleus.

model The species in Batesian mimicry that is resembled by the mimic.

model organism An organism with characteristics that make it a particularly useful subject of research because it is likely to produce results widely applicable to other organisms.

modern synthesis A unified theory of evolution developed in the middle of the twentieth century.

molarity (*M*) The number of moles of a substance dissolved in 1 L of solution.

molars Posteriormost teeth of mammals, with a broad chewing surface for grinding food.

mold Asexual, spore-producing stage of many multicellular fungi.

mole (mol) Amount of substance that contains as many atoms or molecules as there are atoms in exactly 12 g of carbon-12, which is 6.022×10^{23}.

molecular clock A technique for dating the time of divergence of two species or lineages, based upon the number of molecular sequence differences between them.

molecular phylogenetics Approach of using DNA or amino acid sequence comparisons to determine evolutionary relationships among organisms.

molecular weight The weight of a molecule in grams, equal to the total mass number of its atoms.

molecule A unit composed of atoms combined chemically in fixed numbers and ratios.

molt-inhibiting hormone (MIH) A peptide neurohormone secreted by a gland in the eye stalks of crustaceans that inhibits ecdysone secretion.

monoclonal antibody An antibody that reacts only against the same segment (epitope) of a single antigen.

monocot A plant belonging to the Monocotyledones, one of the two major classes of angiosperms; monocot embryos have a single seed leaf (cotyledon) and pollen grains with a single groove.

monocyte A type of leukocyte that enters damaged tissue from the bloodstream through the endothelial wall of the blood vessel.

monoecious Having both "male" flowers (which possess only stamens) and "female" flowers (which possess only carpels).

monogamy A mating system in which one male and one female form a long-term association.

monohybrid An F_1 heterozygote produced from a genetic cross that involves a single character.

monohybrid cross A genetic cross between two individuals that are each heterozygous for the same pair of alleles.

monophyletic taxon A group of organisms that includes a single ancestral species and all of its descendants.

monoploid An individual with one set of chromosomes instead of the usual two.

monosaccharides The smallest carbohydrates, containing three to seven carbon atoms.

monotreme A lineage of mammals that lay eggs instead of bearing live young.

monounsaturated Fatty acids with one double bond.

monsoon cycle A wind pattern that brings seasonally heavy rain to a region by blowing moisture-laden air from the sea to the land.

morphogenesis Orderly, genetically programmed changes in the size, shape, and proportion of body parts of an organism; the process by which specialized tissues and organs form.

morphological species concept The concept that all individuals of a species share measurable traits that distinguish them from individuals of other species.

morphology The form or shape of an organism, or of a part of an organism.

morula The first stage of animal development, a solid ball or layer of blastomeres.

motif A highly specialized region in a protein produced by the three-dimensional arrangement of amino acid chains within and between domains.

motile Capable of self-propelled movement.

motor neuron An efferent neuron that carries signals to skeletal muscle.

motor unit A block of muscle fibers that is controlled by branches of the axon of a single efferent neuron.

mRNA splicing Process that removes introns from pre-mRNAs and joins exons together.

mucosa The lining of the gut that contains epithelial and glandular cells.

mucous cells Cells of gastric pits that secrete alkaline mucus which protects the mucosal layer of the stomach.

Müllerian duct The bipotential primitive duct associated with the gonads that leads to a cloaca.

Müllerian mimicry A form of defense in which two or more unpalatable species share a similar appearance.

multicellular organism Individual consisting of interdependent cells.

multiple alleles More than two different alleles of a gene.

multiple sclerosis An autoimmune disease resulting from an attack against a protein of the myelin sheaths insulating the surfaces of neurons.

multipotent In development, referring to cells derived from pluripotent cells; multipotent cells give rise to cells with particular functions.

muscle fiber A bundle of elongated, cylindrical cells that make up skeletal muscle.

muscle tissue Cells that have the ability to contract (shorten) forcibly.

muscle twitch A single, weak contraction of a muscle fiber.

muscularis The muscular coat of a hollow organ or tubular structure.

mutation A spontaneous and heritable change in DNA.

mutualism A symbiotic interaction between species in which both partners benefit.

mycelium A network of branching hyphae that constitutes the body of a multicellular fungus.

mycobiont The fungal component of a lichen.

mycorrhiza A mutualistic symbiosis in which fungal hyphae associate intimately with plant roots.

myofibril A cylindrical contractile element about 1 mm in diameter that runs lengthwise inside the muscle fiber cell.

myogenic heart A heart that maintains its contraction rhythm with no requirement for signals from the nervous system.

myoglobin An oxygen-storing protein closely related to hemoglobin.

myosin A protein that interacts with the protein actin to cause muscle contraction.

Na^+/K^+-ATPase *See* Na^+/K^+ pump.

Na^+/K^+ pump Pump that pushes 3 Na^+ out of the cell and 2 K^+ into the cell in the same pumping cycle. Also referred to as the *sodium–potassium pump* or as *Na^+/K^+-ATPase*.

Nanoarchaeota A group of Archaea that was proposed based on rRNA sequence analysis of a thermophilic archaean found in a symbiotic relationship with another thermophilic archaean; most probably a subgroup of the Euryarchaeota.

nastic movement In plants, a reversible response to nondirectional stimuli, such as mechanical pressure or humidity.

natural history The branch of biology that examines the form and variety of organisms in their natural environments.

natural killer (NK) cell A type of lymphocyte that destroys virus-infected cells.

natural selection The evolutionary process by which alleles that increase the likelihood of survival and the reproductive output of the individuals that carry them become more common in subsequent generations.

natural theology A belief that knowledge of God may be acquired through the study of natural phenomena.

navigation A wayfinding mechanism in which an animal moves toward a specific destination, using both a compass and a "mental map" of where it is in relation to the destination.

negative feedback The primary mechanism of homeostasis, in which a stimulus—a change in the external or internal environment—triggers a response that compensates for the environmental change.

negative pressure breathing Muscular contractions that expand the lungs, lowering the pressure of the air in the lungs and causing air to be pulled inward.

nekton Animals that can actively swim against water currents.

nematocyst A coiled thread, encapsulated in a cnidocyte, that cnidarians fire at prey or predators, sometimes releasing a toxin through its tip.

nephron A specialized excretory tubule that contributes to osmoregulation and carries out excretion, found in all vertebrates.

neritic zone The shallow water of the oceans above the continental shelves.

nerve A bundle of axons enclosed in connective tissue and all following the same pathway.

nerve cord A bundle of nerves that extends from the central ganglia to the rest of the body, connected to smaller nerves.

nerve net A simple nervous system that coordinates responses to stimuli but has no central control organ, or brain.

nervous tissue Tissue that contains neurons, which serve as lines of communication and control between body parts.

net primary productivity The chemical energy remaining in an ecosystem after a producer's cellular respiration is deducted from the gross primary productivity.

neural circuit A connection between axon terminals of one neuron and the dendrites or cell body of a second neuron; typically a neural circuit consists of an afferent (sensory) neuron connected to one or more interneurons, and an efferent neuron.

neural crest A band of cells that arises early in the embryonic development of vertebrates near the region where the neural tube pinches off from the ectoderm; later, the cells migrate and develop into unique structures.

neural plate Ectoderm thickened and flattened into a longitudinal band, induced by notochord cells.

neural signaling The process by which an animal responds appropriately to a stimulus.

neural tube A hollow tube in vertebrate embryos that develops into the brain, spinal cord, spinal nerves, and spinal column.

neurogenic heart A heart that beats under the control of signals from the nervous system.

neuromuscular junction The junction between a nerve fiber and the muscle it supplies.

neuron An electrically active cell of the nervous system responsible for controlling behavior and body functions.

neuroscience The integrated study of the structure, function, and development of the nervous system.

neurosecretory neuron A neuron that releases a neurohormone into the circulatory system when appropriately stimulated.

neurotransmitter A chemical released by an axon terminal at a chemical synapse.

neurulation The process in vertebrates by which organogenesis begins with development of the nervous system from ectoderm.

neutral lipid Energy-storing molecule consisting of a glycerol backbone and three fatty acid chains.

neutral variation hypothesis An evolutionary hypothesis that some variation at gene loci coding for enzymes and other soluble proteins is neither favored nor eliminated by natural selection.

neutralization One of two important mechanisms that clear foreign antigens from the body; toxins produced by an invading pathogen are inactivated (neutralized) by antibodies.

neutron Uncharged particle in the nucleus of an atom.

neutrophil A type of phagocytic leukocyte that attaches to blood vessel walls in massive numbers when attracted to the infection site by chemokines.

nitrification A metabolic process in which certain soil bacteria convert ammonia or ammonium ions into nitrites that are then converted by other bacteria to nitrates, a form usable by plants.

nitrogen cycle A biogeochemical cycle that moves nitrogen between the huge atmospheric pool of gaseous molecular nitrogen and several much smaller pools of nitrogen-containing compounds in soils, marine and freshwater ecosystems, and living organisms.

nitrogen fixation A metabolic process in which certain bacteria and cyanobacteria convert molecular nitrogen into ammonia and ammonium ions, forms usable by plants.

nitrogenous base A nitrogen-containing molecule with the properties of a base.

nociceptor A sensory receptor that detects tissue damage or noxious chemicals; their activity registers as pain. In systematics, the point of intersection of two branches of a phylogenetic tree.

node In plants, the point on a stem where one or more leaves are attached.

node of Ranvier The gap between two Schwann cells, which exposes the axon membrane directly to extracellular fluids.

noncompetitive inhibition Inhibition of an enzyme reaction by an inhibitor molecule that binds to the enzyme at a site other than the active site and, therefore, does not compete directly with the substrate for binding to the active site.

noncyclic electron flow Pathway in photosynthesis in which electrons travel in a one-way direction from H_2O to $NADP^+$.

nondisjunction The failure of homologous pairs to separate during the first meiotic division or of chromatids to separate during the second meiotic division.

nonhistone protein All the proteins associated with DNA in a eukaryotic chromosome that are not histones.

nonpolar association Association that occurs when nonpolar molecules clump together.

nonpolar covalent bond Bond in which electrons are shared equally.

nonsense codon *See* stop codon.

nonsense mutation A base-pair substitution mutation in a gene in which the base-pair change results in a change from a sense codon to a nonsense codon in the mRNA. The polypeptide translated from the mRNA is shorter than the normal polypeptide because of the mutation.

nonshivering thermogenesis The generation of heat by oxidative mechanisms in non-muscle tissue throughout the body.

nonvascular plant *See* bryophyte.

norepinephrine A nontropic amine hormone secreted by the adrenal medulla.

Northern blot analysis Method of analysis in which RNA molecules extracted from cells or a tissue are separated by size using electrophoresis, transferred to a filter paper, and then probed to reveal the presence of particular RNAs.

notochord A flexible rodlike structure, constructed of fluid-filled cells surrounded by tough connective tissue, which supports a chordate embryo from head to tail.

N-terminal end The end of a polypeptide chain with an —NH_3^+ group.

nuclear envelope In eukaryotes, membranes separating the nucleus from the cytoplasm.

nuclear localization signal A short amino acid sequence in a protein that directs the protein to the nucleus.

nuclear pore complex A large, octagonally symmetrical, cylindrical structure that functions to exchange molecules between the nucleus and cytoplasm and prevents the transport of material not meant to cross the nuclear membrane. A nuclear pore—a channel through the complex—is the path for the exchange of molecules.

nucleoid The central region of a prokaryotic cell with no boundary membrane separating it from the cytoplasm, where DNA replication and RNA transcription occur.

nucleolus The nuclear site of rRNA transcription, processing, and ribosome assembly in eukaryotes.

nucleoplasm The liquid or semiliquid substance within the nucleus.

nucleoside Chemical structure containing only a nitrogenous base and a five-carbon sugar.

nucleosome The basic structural unit of chromatin in eukaryotes, consisting of DNA wrapped around a histone core.

nucleosome core particle An eight-protein particle formed by the combination of two molecules each of H2A, H2B, H3, and H4, around which DNA winds for almost two turns.

nucleosome remodeling complex Multiprotein complex that uses the energy of ATP hydrolysis to slide a nucleosome along the DNA of a eukaryotic chromosome to expose the promoter of a gene, or to restructure the nucleosome without moving it to allow transcription factors to bind.

nucleotide The monomer of nucleic acids consisting of a five-carbon sugar, a nitrogenous base, and a phosphate.

nucleus The central region of eukaryotic cells, separated by membranes from the surrounding cytoplasm, where DNA replication and messenger RNA transcription occur.

null hypothesis A statement of what would be seen if the hypothesis being tested were wrong.

null model A conceptual model that predicts what one would see if a particular factor had no effect.

nutrition The processes by which an organism takes in, digests, absorbs, and converts food into organic compounds.

obligate aerobe A microorganism that uses oxygen for cellular respiration and requires oxygen in its surroundings to support growth.

obligate anaerobe A microorganism that cannot use oxygen and can grow only in the absence of oxygen.

observational data Basic information on biological structures or the details of biological processes.

oceanic zone The deep ocean water beyond the continental shelves.

ocellus (plural, ocelli) The simplest eye, which detects light but does not form an image.

oil Neutral lipid that is liquid at biological temperatures.

Okazaki fragments The short lengths of lagging strand DNA produced by discontinuous replication.

olfactory bulb A gray-matter center that relay inputs from odor receptors to both the cerebral cortex and the limbic system.

oligodendrocyte A type of glial cell that populates the CNS and is responsible for producing myelin.

oligosaccharin A complex carbohydrate that in plants serves as a signaling molecule and as a defense against pathogens.

oligotrophic lake A lake that is poor in nutrients and organic matter, but rich in oxygen.

omasum One of the four stomach chambers of a ruminant; in this chamber, water is absorbed from the mass of digesting material.

ommatidium (plural, ommatidia) A faceted visual unit of a compound eye.

omnivore An animal that feeds at several trophic levels, consuming plants, animals, and other sources of organic matter.

oncogene A gene capable of inducing one or more characteristics of cancer cells.

one gene–one enzyme hypothesis Hypothesis showing the direct relationship between genes and enzymes.

one gene–one polypeptide hypothesis Restatement of the one gene–one enzyme hypothesis, taking into account that some proteins consist of more than one polypeptide and not all proteins are enzymes.

oocyte A developing gamete that becomes an ootid at the end of meiosis.

oogenesis The process of producing eggs.

oogonium A cell that enters meiosis and gives rise to gametes, produced by mitotic divisions of the germ cells in females.

Oomycota (water molds, white rusts, and downy mildews) A lineage of the Stramenopila subgroup of the Chromalveolata protist evolutionary group; funguslike organisms that lack chloroplasts and live as heterotrophs.

open circulatory system An arrangement of internal transport in some invertebrates in which the vascular fluid, hemolymph, is released into sinuses, bathing organs directly, and is not always retained within vessels.

open reading frame (ORF) Segment of a protein-coding gene or an mRNA transcribed from such a gene that involves a start codon separated by a multiple of three nucleotides from one of the stop codons. The ORF is a potential polypeptide-coding sequence.

operant conditioning A form of associative learning in which animals learn to link a voluntary activity, an operant, with its favorable consequences, the reinforcement.

operator A DNA regulatory sequence that controls transcription of an operon.

operculum A lid or flap of the bone serving as the gill cover in some fishes.

operon A cluster of prokaryotic genes and the DNA sequences involved in their regulation.

Opisthokonta (opisthokonts) A protist evolutionary group; a broad group of eukaryotes in which a single posterior flagellum is found at some stage in the life cycle.

opsin One of several different proteins that bond covalently with the light-absorbing pigment of rods and cones (retinal).

optic chiasm Location just behind the eyes where the optic nerves converge before entering the base of the brain, a portion of each optic nerve crossing over to the opposite side.

optical isomers *See* enantiomers.

optimal foraging theory A set of mathematical models that predict the diet choices of animals as they encounter a range of potential food items.

oral hood Soft fleshy structure at the anterior end of a cephalochordate that frames the opening of the mouth.

orbital The region of space where the electron "lives" most of the time.

order A Linnaean taxonomic category of organisms that ranks above a family and below a class.

organ Two or more different tissues integrated into a structure that carries out a specific function.

organ of Corti An organ within the cochlear duct that contains the sensory hair cells detecting sound vibrations transmitted to the inner ear.

organ system The coordinated activities of two or more organs to carry out a major body function such as movement, digestion, or reproduction.

organelles The nucleus and other specialized internal structures and compartments of eukaryotic cells.

organic acid (carboxylic acid) Acid for which the characteristic functional group is a carboxyl group (—COOH).

organic molecule Molecule based on carbon.

organismal ecology An ecological discipline in which researchers study the genetic, biochemical, physiological, morphological, and behavioral adaptations of organisms to their abiotic environments.

organogenesis The development of the major organ systems, giving rise to a free-living individual with the body organization characteristic of its species.

oriented cell division Cell division in different planes; establishes the overall shape of a plant organ.

origin of replication A specific region at which DNA replication commences. Bacterial chromosomes have single origins of replication (*ori*) whereas eukaryotic chromosomes have multiple origins.

orthogenesis An obsolete theory that evolution is goal oriented, striving to perfect organisms.

oscula One or more openings in a sponge through which water is expelled.

osmoconformer An animal in which the osmolarity of the cellular and extracellular solutions matches the osmolarity of the environment.

osmolarity The total solute concentration of a solution, measured in osmoles—the number of solute molecules and ions (in moles)—per liter of solution.

osmoreceptor A chemoreceptor in the hypothalamus that responds to changes in the osmolarity of the fluid surrounding it, which reflects the osmolarity generally of the body fluids.

osmoregulation The regulation of water and ion balance.

osmoregulator An animal that uses control mechanisms to keep the osmolarity of cellular and extracellular fluids the same, but at levels that may differ from the osmolarity of the surroundings.

osmosis The passive transport of water across a selectively permeable membrane in response to solute concentration gradients, a pressure gradient, or both.

osmotic pressure A state of dynamic equilibrium in which the pressure of the solution on one side of a selectively permeable membrane exactly balances the tendency of water molecules to diffuse passively from the other side of the membrane due to a concentration gradient.

osteoblast A cell that produces the collagen and mineral of bone.

osteoclast A cell that removes bone minerals and recycles them through the bloodstream.

osteocyte A mature bone cell.

osteon The structural unit of bone, consisting of a minute central canal surrounded by osteocytes embedded in concentric layers of mineral matter.

ostracoderm One of an assortment of extinct, jawless fishes that were covered with bony armor.

otolith One of many small crystals of calcium carbonate embedded in the otolithic membrane of the hair cells.

outer boundary membrane A smooth membrane that surrounds a chloroplast, enclosing the stroma.

outer ear The external structure of the ear, consisting of the pinna and meatus.

outer membrane In Gram-negative bacteria, an additional boundary membrane that covers the peptidoglycan layer of the cell wall.

outer mitochondrial membrane The smooth membrane covering the outside of a mitochondrion.

outgroup comparison A technique used to identify ancestral and derived characters by comparing the group under study to more distantly related species that are not otherwise included in the analysis.

ova *See* ovum.

oval window An opening in the bony wall that separates the middle ear from the inner ear.

ovarian cycle The cyclic events in the ovary leading to ovulation.

ovary In animals, the female gonad, which produces female gametes and reproductive hormones. In flowering plants, the enlarged base of a carpel in which one or more ovules develop into seeds.

overexploitation The excessive harvesting of an animal or plant species, potentially leading to its extinction.

overnutrition The condition caused by excessive intake of specific nutrients.

oviduct The tube through which the egg moves from the ovary to the outside of the body.

oviparous Referring to animals that lay eggs containing the nutrients needed for development of the embryo outside the mother's body.

ovoviviparous Referring to animals in which fertilized eggs are retained within the body and the embryo develops using nutrients provided by the egg; eggs hatch inside the mother.

ovulation The process in which oocytes are released into the oviducts as immature eggs.

ovule In plants, the structure in a carpel in which a female gametophyte develops and fertilization takes place.

ovum (plural, *ova*) A female sex cell, or egg.

oxidation The removal of electrons from a substance.

oxidative phosphorylation Synthesis of ATP in which ATP synthase uses an H^+ gradient built by the electron transfer system as the energy source to make the ATP.

oxidized Substance from which the electrons are removed during oxidation.

oxytocin A hormone that stimulates the ejection of milk from the mammary glands of a nursing mother.

P generation The parental individuals used in an initial genetic cross.

P site The site in the ribosome where the tRNA carrying the growing polypeptide chain is bound during translation.

pacemaker cell A specialized cardiac muscle cell in the upper wall of the right atrium that sets the rate of contraction in the heart.

paedomorphosis A common form of heterochrony in which juvenile characteristics are retained in a reproductive adult.

pairing Process in meiosis in which homologous chromosomes come together and pair. Also referred to as *synapsis*.

pair-rule gene In *Drosophila* embryonic development, one of the second activated set of segmentation genes that progressively subdivide the embryo into regions, determining the segments of the embryo and the adult. Pair-rule genes control the division of the embryo into units of two segments each.

paleobiology The study of ancient organisms.

pancreas A mixed gland composed of an exocrine portion that secretes digestive enzymes into the small intestine and an endocrine portion, the islets of Langerhans, that secretes insulin and glucagon.

pancreatic enzymes Digestive enzymes secreted by exocrine cells in the pancreas into ducts that empty into the lumen of the duodenum.

parapatric speciation Speciation between populations with adjacent geographical distributions.

paraphyletic taxon A group of organisms that includes an ancestral species and some, but not all, of its descendents.

parapodia Fleshy lateral extensions of the body wall of aquatic annelids, used for locomotion and gas exchange.

parasite An organism that feeds on the tissues of or otherwise exploits its host.

parasitism A symbiotic interaction in which one species, the parasite, uses another, the host, in a way that is harmful to the host.

parasitoid An insect species in which a female lays eggs in the larva or pupa of another insect species, and her young consume the tissues of the living host.

parasympathetic division The division of the autonomic nervous system that predominates during quiet, low-stress situations, such as while relaxing.

parathyroid gland One of a pair of glands that produce parathyroid hormone (PTH) (found only in tetrapod vertebrates).

parathyroid hormone (PTH) The hormone secreted by the parathyroid glands in response to a fall in blood Ca^{2+} levels.

parenchyma A ground tissue with cells having a thin primary wall, which is pliable and permeable to water. Parenchyma cells may be specialized for photosynthesis, storage, secretion, or other tasks.

parental Phenotypes identical to the original parental individuals.

parental investment The time and energy devoted to the production and rearing of offspring.

parietal cells Cells of gastric glands that secrete H^+ and Cl^-, which combine to form HCl in the lumen of the stomach.

parthenogenesis A mode of asexual reproduction in which animals produce offspring by the growth and development of an egg without fertilization.

partial diploid A condition in which part of the genome of a haploid organism is diploid. Recipients in bacterial conjugation between an Hfr and an F^- cell become partial diploids for part of the Hfr bacterial chromosome.

partial pressure In a mixture of gases, the pressure of each individual gas.

parturition The process of giving birth.

passive immunity The acquisition of antibodies as a result of direct transfer from another person.

passive parental care The amount of energy invested in offspring—in the form of the energy stored in eggs or seeds or energy transferred to developing young through a placenta—before they are born.

passive transport The transport of substances across cell membranes without expenditure of energy, as in diffusion.

paternal chromosome The chromosome derived from the male parent of an organism.

pathogenesis-related (PR) protein A hydrolytic enzyme that breaks down components of a pathogen's cell wall.

pattern formation The arrangement of organs and body structures in their proper three-dimensional relationships.

pectoral girdle A bony or cartilaginous structure in vertebrates that supports and is attached to the forelimbs.

pedicellariae Small pincers at the base of short spines in starfishes and sea urchins.

pedigree Chart that shows all parents and offspring for as many generations as possible, the sex of individuals in the different generations, and the presence or absence of a trait of interest.

pelagic province The open water in the oceans.

pellicle A layer of supportive protein fibers located inside the cell, just under the plasma membrane, providing strength and flexibility instead of a cell wall.

pelvic girdle A bony or cartilaginous structure in vertebrates that supports and is attached to the hind limbs.

pepsin An enzyme made in the stomach that breaks down proteins.

pepsinogen The inactive precursor molecule for pepsin.

peptic ulcer Lesion in the stomach wall resulting from attack by HCl and pepsin.

peptide bond A link formed by a dehydration synthesis reaction between the $—NH_2$ group of one amino acid and the —COOH group of a second.

peptidoglycan A polymeric substance formed from a polysaccharide backbone tied together by short polypeptides, which is the primary structural molecule of bacterial cell walls.

peptidyl transferase An enzyme that catalyzes the reaction in which an amino acid is cleaved from the tRNA in the P site of the ribosome and forms a peptide bond with the amino acid on the tRNA in the A site of the ribosome.

peptidyl–tRNA A tRNA linked to a growing polypeptide chain containing two or more amino acids.

per capita growth rate The difference between the per capita birth rate and the per capita death rate of a population.

perception The conscious awareness of our external and internal environments derived from the processing of sensory input.

perennial A plant in which vegetative growth and reproduction continue year after year.

perfect flower A flower that has both male (stamen) and female (carpel) sexual organs.

perfusion The flow of blood or other body fluids on the internal side of the respiratory surface.

pericarp The fruit wall.

pericycle A tissue of plant roots, located between the endodermis and the phloem, which gives rise to lateral roots.

periderm The outermost portion of bark; consists of cork, cork cambium, and secondary cortex.

peripheral nervous system (PNS) All nerve roots and nerves (motor and sensory) that supply the muscles of the body and transmit information about sensation (including pain) to the central nervous system.

peripheral protein Protein held to membrane surfaces by noncovalent bonds formed with the polar parts of integral membrane proteins or membrane lipids.

peristalsis The rippling motion of muscles in the intestine or other tubular organs characterized by the alternate contraction and relaxation of the muscles that propel the contents onward.

peritoneum The thin tissue derived from mesoderm that lines the abdominal wall and covers most of the organs in the abdomen.

peritubular capillary A capillary of the network surrounding the glomerulus.

permafrost Perpetually frozen ground below the topsoil.

peroxisome Microbody that produces hydrogen peroxide as a by-product.

PET *See* positron emission tomography.

petal Part of the corolla of a flower, often brightly colored.

petiole The stalk by which a leaf is attached to a stem.

pH scale The numerical scale used by scientists to measure acidity.

Phaeophyta (brown algae) A lineage of the Stramenopila subgroup of the Chromalveolata evolutionary subgroup; photosynthetic autotrophs that range from microscopic forms to giant kelps.

phage *See* bacteriophage.

phagocytosis Process in which some types of cells engulf bacteria or other cellular debris to break them down.

pharyngeal arches Embryonic features of all vertebrates; they contribute to the formation of the face, neck, mouth, larynx, and pharynx.

pharynx The throat. In some invertebrates, a protrusible tube used to bring food into the mouth for passage to the gastrovascular cavity; in mammals, the common pathway for air entering the larynx and food entering the esophagus.

phenotype The observable or measurable (biochemical, molecular) characteristics of an organism that are produced by an interaction between the genotype and the environment.

phenotypic variation Differences in appearance or function between individual organisms.

pheromone A distinctive volatile chemical released in minute amounts to influence the behavior of members of the same species.

phloem The food-conducting tissue of a vascular plant.

phloem sap The solution of water and organic compounds that flows rapidly through the sieve tubes of plants.

phosphate group Group consisting of a central phosphorus atom held in four linkages: two that bind —OH groups to the central phosphorus atom, a third that binds an oxygen atom to the central phosphorus atom, and a fourth that links the phosphate group to an oxygen atom.

phosphodiester bond The linkage of nucleotides in polynucleotide chains by a bridging phosphate group between the 5′ carbon of one sugar and the 3′ carbon of the next sugar in line.

phospholipid A phosphate-containing lipid.

phosphorus cycle A biogeochemcial cycle in which weathering and erosion carry phosphate ions from rocks to soil and into streams and rivers, which eventually transport them to the ocean, where they are slowly incorporated into rocks.

phosphorylation The addition of a phosphate group to a molecule.

photic zone Surface water of a lake or ocean that sunlight penetrates.

photoautotroph Photosynthetic organism that uses light as its energy source and carbon dioxide as its carbon source.

photobiont The photosynthetic component of a lichen.

photoheterotroph An organism that uses light as the ultimate energy source but obtains carbon in organic form rather than as carbon dioxide.

photoperiodism The response of plants to changes in the relative lengths of light and dark periods in their environment during each 24-hour period.

photophosphorylation The synthesis of ATP coupled to the transfer of electrons energized by photons of light.

photopigment Light-absorbing pigment.

photopsin One of three photopigments in which retinal is combined with different opsins.

photoreceptor A sensory receptor that detects the energy of light.

photorespiration A process that metabolizes a by-product of photosynthesis.

photosynthesis The conversion of light energy to chemical energy in the form of sugar and other organic molecules.

photosystem A large complex into which the light-absorbing pigments for photosynthesis are organized with proteins and other molecules.

photosystem I In photosynthesis, a protein complex in the thylakoid membrane that uses energy absorbed from sunlight to synthesize NADPH.

photosystem II In photosynthesis, a protein complex in the thylakoid membrane that uses energy absorbed from sunlight to synthesize ATP.

phototroph An organism that obtains energy from light.

phototropism The tendency of a plant shoot to bend toward a source of light.

phyletic gradualism hypothesis The hypothesis that most morphological change occurs gradually over long periods of time.

PhyloCode A formal set of rules governing phylogenetic nomenclature.

phylogenetic species concept A concept that seeks to delineate species as the smallest aggregate population that can be united by shared derived characters.

phylogenetic tree A branching diagram depicting the evolutionary relationships of groups of organisms.

phylogeny The evolutionary history of a group of organisms.

phylum (plural, phyla) A major Linnaean division of a kingdom, ranking above a class.

physical barriers The first of three lines of defense that humans and other mammals have against threats of pathogens; parts of the organism that physically block access to a pathogen. Examples are skin and epithelial surfaces covering internal body cavities and ducts.

physiological respiration The process by which animals exchange gases with their surroundings—how they take in oxygen from the outside environment and deliver it to body cells, and remove carbon dioxide from body cells and deliver it to the environment.

physiology The study of the functions of organisms—the physico-chemical processes of organisms.

phytoalexin A biochemical that functions as an antibiotic in plants.

phytochrome A blue-green pigmented plant chromoprotein involved in the regulation of light-dependent growth processes.

phytoplankton Microscopic, free-flowing aquatic plants and protists.

phytosterol A sterol that occurs in plant cell membranes.

piloting A wayfinding mechanism in which animals use familiar landmarks to guide their journey.

pilus (plural, pili) A hair or hairlike appendage on the surface of a prokaryote.

pinacoderm In sponges, an unstratified outer layer of cells.

pineal gland A light-sensitive, melatonin-secreting gland that regulates some biological rhythms.

pinna The external structure of the outer ear, which concentrates and focuses sound waves.

pinocytosis *See* bulk-phase endocytosis.

pith The soft, spongelike, central cylinder of plant stems, composed mainly of parenchyma.

pituitary A gland consisting mostly of two fused lobes suspended just below the hypothalamus by a slender stalk of tissue that contains both neurons and blood vessels; it interacts with the hypothalamus to control many physiological functions, including the activity of some other glands.

placenta A specialized temporary organ that connects the embryo and fetus with the uterus in mammals, mediating the delivery of oxygen and nutrients.

plasma The clear, yellowish fluid portion of the blood in which cells are suspended. Plasma consists of water, glucose and other sugars, amino acids, plasma proteins, dissolved gases, ions, lipids, vitamins, hormones and other signal molecules, and metabolic wastes.

plasma cell A large antibody-producing cell that develops from B cells.

plasma membrane The outer limit of the cytoplasm responsible for the regulation of substances moving into and out of cells.

plasmid A DNA molecule in the cytoplasm of certain prokaryotes, which often contains genes with functions that supplement those in the nucleoid and which can replicate independently of the nucleoid DNA and be passed along during cell division.

plasmodesma (plural, plasmodesmata) A minute channel that perforates a cell wall and contains extensions of the cytoplasm that directly connect adjacent plant cells.

plasmodial slime mold A slime mold of the class Myxomycetes.

plasmodium The composite mass of plasmodial slime molds consisting of individual nuclei suspended in a common cytoplasm surrounded by a single plasma membrane.

plasmogamy The sexual stage of fungi during which the cytoplasms of two genetically different partners fuse.

plasmolysis Condition due to outward osmotic movement of water, in which plant cells shrink so much that they retract from their walls.

plastids A family of plant organelles that includes chloroplasts, amyloplasts, and chromoplasts.

plastron The ventral part of the shell of a turtle.

plate tectonics The geological theory describing how Earth's crust is broken into irregularly shaped plates of rock that float on its semisolid mantle.

platelet An oval or rounded cell fragment enclosed in its own plasma membrane, which is found in the blood; they are produced in red bone marrow by the division of stem cells and contain enzymes and other factors that take part in blood clotting.

pleiotropy Condition in which single genes affect more than one character of an organism.

pleura The double layer of epithelial tissue covering the lungs.

ploidy The number of chromosome sets of a cell or species.

plumule The rudimentary terminal bud of a plant embryo located at the end of the hypocotyl, consisting of the epicotyl and a cluster of tiny foliage leaves.

pluripotent In development, referring to cells derived from totipotent cells; pluripotent cells can give rise to most adult cell types.

polar association Association that occurs when polar molecules attract and align themselves with other polar molecules and with charged ions and molecules.

polar body A nonfunctional cell produced in oogenesis.

polar covalent bond Bond in which electrons are shared unequally.

polar nucleus In the embryo sac of a flowering plant, one of two nuclei that migrate into the center of the sac, become housed in a central cell, and eventually give rise to endosperm.

polar transport Unidirectional movement of a substance from one end of a cell (or other structure) to the other.

polarity The unequal distribution of yolk and other components in a mature egg.

pollen grain The male gametophyte of a seed plant.

pollen sac The microsporangium of a seed plant, in which pollen develops.

pollen tube A tube that grows from a germinating pollen grain and carries sperm cells to the ovary.

pollination The transfer of pollen to a seed plant's female reproductive parts by air currents or on the bodies of animal pollinators.

pollutant Materials or energy in a form or quantity that organisms do not usually encounter.

poly(A) polymerase The enzyme in eukaryotes that adds a chain of adenine nucleotides—the poly(A) tail—to the cleaved 3′ end of a pre-mRNA.

poly(A) tail The string of A nucleotides added posttranscriptionally to the 3′ end of a cleaved pre-mRNA molecule and retained in the mRNA produced from it that enables the mRNA to be translated efficiently and protects it from attack by RNA-digesting enzymes in the cytoplasm.

polyadenylation signal Sequence near the 3′ end of a eukaryotic gene which, in the pre-mRNA transcript of the gene, specifies where the transcript should be cleaved. Once cleaved, a poly(A) tail is added to the 3′ end of the RNA.

polyandry A polygamous mating system in which one female mates with multiple males.

polygamy A mating system in which either males or females may have many mating partners.

polygenic inheritance Inheritance in which several to many different genes contribute to the same character.

polygyny A polygamous mating system in which one male mates with many females.

polyhedral virus A virus in which the coat proteins form triangular units that fit together like the parts of a geodesic sphere.

polymerase chain reaction (PCR) Process that amplifies a specific DNA sequence from a DNA mixture to an extremely large number of copies.

polymorphism The existence of discrete variants of a character among individuals in a population.

polyp The tentacled, usually sessile stage in the life cycle of a coelenterate.

polypeptide The chain of amino acids formed by sequential peptide bonds.

polyphyletic taxon A group of organisms that belong to different evolutionary lineages and do not share a recent common ancestor.

polyploid An individual with one or more extra copies of the entire haploid complement of chromosomes.

polyploidy The condition of having one or more extra copies of the entire haploid complement of chromosomes.

polysaccharide Chain with more than 10 linked monosaccharide subunits.

polysome The entire structure of an mRNA molecule and the multiple associated ribosomes that are translating it simultaneously.

polyunsaturated Fatty acid with more than one double bond.

population All the individuals of a single species that live together in the same place and time.

population bottleneck An evolutionary event that occurs when a stressful factor reduces population size greatly and eliminates some alleles from a population.

population density The number of individuals per unit area or per unit volume of habitat.

population dynamics Changes in the characteristics of populations through time or over space.

population ecology The ecological discipline that focuses on how a population's size and other characteristics change in space and time.

population genetics The branch of science that studies the prevalence and variation in genes among populations of individuals.

population size The number of individuals in a population at a specified time.

population viability analysis A mathematical analysis used by conservation biologists to determine the minimum viable population size for threatened or endangered species.

positive feedback A mechanism that intensifies or adds to a change in internal or external environmental condition.

positive pressure breathing A gulping or swallowing motion that forces air into the lungs.

positron emission tomography (PET) A scanning technique for studying activities of the cerebral cortex.

posterior Indicating the tail end of an animal.

posterior pituitary The neural portion of the pituitary, which stores and releases two hormones made by the hypothalamus, antidiuretic hormone and oxytocin.

postsynaptic cell The neuron or the surface of an effector after a synapse that receives the signal from the presynaptic cell.

posttranslational import A mechanism by which proteins are sorted to their final cellular locations in a eukaryotic cell after they have been made on free ribosomes in the cytosol.

postzygotic isolating mechanism A reproductive isolating mechanism that acts after zygote formation.

potential energy Stored energy.

predation The interaction between animals and the animal prey they consume.

prediction A statement about what the researcher expects to happen to one variable if another variable changes.

pregnancy The period of mammalian development in which the embryo develops in the uterus of the mother.

premolars Teeth located in pairs on each side of the upper and lower jaws of mammals, positioned behind the canines and in front of the molars.

pre-mRNA (precursor-mRNA) The primary transcript of a eukaryotic protein-coding gene, which is processed to form messenger RNA.

prenatal diagnosis Techniques in which cells derived from a developing embryo or its surrounding tissues or fluids are tested for the presence of mutant alleles or chromosomal alterations.

prepuce Foreskin; a loose fold of skin that covers the glans of the penis.

pressure flow mechanism In vascular plants, pressure that builds up at the source end of a sieve tube system and pushes solutes by bulk flow toward a sink, where they are removed.

presynaptic cell The neuron with an axon terminal on one side of the synapse that transmits the signal across the synapse to the dendrite or cell body of the postsynaptic cell.

prezygotic isolating mechanism A reproductive isolating mechanism that acts prior to the production of a zygote, or fertilized egg.

primary active transport Transport in which the same protein that transports a substance also hydrolyzes ATP to power the transport directly.

primary cell layers The ectoderm, mesoderm, and endoderm layers that form the embryonic tissues.

primary cell wall The initial cell wall laid down by a plant cell.

primary consumer An herbivore, a member of the second trophic level.

primary endosymbiosis In the model for the origin of plastids in eukaryotes, the first event in which a eukaryotic cell engulfed a photosynthetic cyanobacterium.

primary growth The growth of plant tissues derived from apical meristems. *Compare* secondary growth.

primary immune response The response of the immune system to the first challenge by an antigen.

primary meristem Root and shoot apical meristems, from which a plant's primary tissues develop. *Compare* lateral meristem.

primary motor area The area of the cerebral cortex that runs in a band just in front of the primary somatosensory area and is responsible for voluntary movement.

primary plant body The portion of a plant that is made up of primary tissues.

primary producer An autotroph, usually a photosynthetic organism, a member of the first trophic level.

primary somatosensory area The area of the cerebral cortex that runs in a band across the parietal lobes of the brain and registers information on touch, pain, temperature, and pressure.

primary structure The sequence of amino acids in a protein.

primary succession Predictable change in species composition of an ecological community that develops on bare ground.

primary tissue A plant tissue that develops from an apical meristem.

primase An enzyme that assembles the primer for a new DNA strand during DNA replication.

primer A short nucleotide chain made of RNA that is laid down as the first series of nucleotides in a new DNA strand, or made of DNA for use in the polymerase chain reaction (PCR).

primitive groove In development of birds, the sunken midline of the primitive streak that acts as a conduit for migrating cells to move into the blastocoel.

primitive streak In development of birds, the thickened region of the embryo produced by cells of the epiblast streaming toward the midline of the blastodisc.

Principle of Independent Assortment Mendel's principle that the alleles of the genes that govern two characters assort independently during formation of gametes.

principle of parsimony A principle of systematic biology that states that a particular trait is unlikely to evolve independently in separate evolutionary lineages.

Principle of Segregation Mendel's principle that the pairs of alleles that control a character segregate as gametes are formed, with half the gametes carrying one allele, and the other half carrying the other allele.

prion An infectious agent that contains only protein and does not include a nucleic acid molecule.

probability The possibility that an outcome will occur if it is a matter of chance.

procambium The primary meristem of a plant that develops into primary vascular tissue.

product An atom or molecule leaving a chemical reaction.

product rule Mathematical rule in which the final probability is found by multiplying individual probabilities.

profundal zone The perpetually dark layer below the limnetic zone in a lake.

progesterone A female sex hormone that stimulates growth of the uterine lining and inhibits contractions of the uterus.

progestin A class of sex hormones synthesized by the gonads of vertebrates and active predominantly in females.

prokaryote Organism in which the DNA is suspended in the cell interior without separation from other cellular components by a discrete membrane.

prokaryotic chromosome A single, typically circular DNA molecule.

prolactin (PRL) A peptide hormone secreted by the anterior pituitary that stimulates breast development and milk secretion in mammals.

prometaphase A transition period between prophase and metaphase during which the microtubules of the mitotic spindle attach to the kinetochores and the chromosomes shuffle until they align in the center of the cell.

promiscuity A mating system in which individuals do not form close pair bonds, and both males and females mate with multiple partners.

promoter The site to which RNA polymerase binds (prokaryotes) or to which general transcription factors bind and recruit RNA polymerase (eukaryotes) for initiating transcription of a gene.

promoter proximal element Regulatory sequence within the promoter proximal region, a region upstream of the promoter of a eukaryotic protein-coding gene. Regulatory proteins bind to promoter proximal elements.

promoter proximal region Upstream of a eukaryotic gene, a region containing regulatory sequences for transcription called promoter proximal elements.

proofreading mechanism Mechanism during DNA replication in which DNA polymerase backs up and removes a mispaired nucleotide from a newly synthesized DNA strand and then adds the correct nucleotide to the growing chain.

prophage A viral genome inserted in the host cell DNA.

prophase The beginning phase of mitosis during which the duplicated chromosomes within the nucleus condense from a greatly extended state into compact, rodlike structures.

proprioceptor A mechanoreceptor that detects stimuli used in the CNS to maintain body balance and equilibrium and to monitor the position of the head and limbs.

prostaglandin One of a group of local regulators derived from fatty acids that are involved in paracrine and autocrine regulation.

prostate gland An accessory sex gland in males that adds a thin, milky fluid to the semen and adjusts the pH of the semen to the level of acidity best tolerated by sperm.

proteasome Large cytoplasmic protein complex in eukaryotic cells that degrades ubiquitinylated proteins.

protein Molecules that carry out most of the activities of life, including the synthesis of all other biological molecules. A protein consists of one or more polypeptides depending on the protein.

protein chip *See* protein microarray.

protein kinase Enzyme that transfers a phosphate group from ATP to one or more sites on particular proteins.

protein microarray Similar in concept to a DNA microarray, a solid surface with a microscopic grid with thousands of spaces containing probes for analyzing the proteome, the complete set of proteins encoded by the genome of an organism. Also referred to as a *protein chip.*

protein phosphatase Enzyme that removes phosphate groups from target proteins.

protein-coding gene A gene encoding a protein.

proteome The complete set of proteins that can be expressed by the genome of an organism.

proteomics The study of the proteome.

prothoracicotropic hormone (PTTH) A peptide hormone secreted by neurosecretory neurons in the brains of insects that is one of the hormones regulating molting and metamorphosis.

protists A diverse and polyphyletic group of single-celled and multicellular eukaryotic species.

protocell A primitive cell-like structure that has some of the properties of life and that might have been the precursor of cells.

protoderm The primary meristem that will produce stem epidermis.

proton Positively charged particle in the nucleus of an atom.

proton pump Pump that moves hydrogen ions across membranes and pushes hydrogen ions across the plasma membrane from the cytoplasm to the cell exterior. Also referred to as *H^+ pump.*

protonema The structure that arises when a liverwort or moss spore germinates and eventually gives rise to a mature gametophyte.

protonephridium The simplest form of invertebrate excretory tubule.

proton-motive force Stored energy that contributes to ATP synthesis, as well as to the cotransport of substances to and from mitochondria.

proto-oncogene A gene that encodes various kinds of proteins that stimulate cell division. Mutated proto-oncogenes—oncogenes—contribute to the development of cancer.

protoplast The cytoplasm, organelles, and plasma membrane of a plant cell.

protoplast fusion A plant breeding process in which protoplasts are fused into a single cell.

provirus DNA copy of a retrovirus RNA genome that becomes inserted into host DNA.

proximal convoluted tubule The tubule between the Bowman's capsule and the loop of Henle in the nephron of the kidney, which carries and processes the filtrate.

pseudocoelom A fluid- or organ-filled body cavity between the gut (a derivative of endoderm) and the muscles of the body wall (a derivative of mesoderm).

pseudocoelomate A body plan of bilaterally symmetrical animals with a body cavity that lacks a complete lining derived from mesoderm.

pseudopod (plural, pseudopodia) A temporary cytoplasmic extension of a cell.

psychrophile An archaean or bacterium that grows optimally at temperatures in the range -10 to -20°C.

Pterophyta The plant phylum of ferns and their close relatives.

pulmocutaneous circuit In amphibians, the branch of a double blood circuit that receives deoxygenated blood and moves it to the skin and lungs or gills.

pulmonary arteries The arteries from the right ventricle of the heart that lead to the lungs.

pulmonary circuit The circuit of the cardiovascular system that supplies the lungs.

pulmonary veins The veins that carry oxygenated blood from the lungs to the right atrium of the heart.

pulse The measurable rate of ventricular contraction of the heart detected at the peak of high pressure in an artery.

pulvinus (plural, pulvini) A jointlike, thickened pad of tissue at the base of a leaf or petiole; flexes when the leaf makes nastic movements.

punctuated equilibrium hypothesis The evolutionary hypothesis that most morphological variation arises rapidly during speciation events in isolated populations at the edge of a species' geographical distribution.

Punnett square Method for determining the genotypes and phenotypes of offspring and their expected proportions by combining gametes and their probabilities of occurrence in a matrix table.

pupa The nonfeeding stage between the larva and adult in the complete metamorphosis of some insects, during which the larval tissues are completely reorganized within a protective cocoon or hardened case.

pupil The dark center in the middle of the iris through which light passes to the back of the eye.

purine A type of nitrogenous base with two carbon–nitrogen rings.

pyloric sphincter The muscular ring at the junction between the stomach and small intestine that controls the flow of materials from the stomach.

pyramid of biomass A diagram that illustrates differences in standing crop biomass in a series of trophic levels.

pyramid of energy A diagram that illustrates the amount of energy that flows through a series of trophic levels.

pyramid of numbers A diagram that illustrates the number of individual organisms present in a series of trophic levels.

pyrimidine A type of nitrogenous base with one carbon–nitrogen ring.

pyrogens Chemicals released by macrophages in response to infection that stimulate prostaglandin release from the hypothalamus thereby leading to an increase in body temperature (fever).

pyruvate oxidation (pyruvic acid oxidation) Stage of cellular respiration in which the three-carbon molecule pyruvate is converted into a two-carbon acetyl group that is completely oxidized to carbon dioxide.

qualitative variation Variation that exists in two or more discrete states, with intermediate forms often being absent.

quantitative trait A character that displays a continuous distribution of the phenotype involved, typically resulting from several to many contributing genes.

quantitative trait loci (QTLs) The individual genes that contribute to a quantitative trait.

quantitative variation Variation that is measured on a continuum (such as height in human beings) rather than in discrete units or categories.

quaternary structure The arrangement of polypeptide chains in a protein that contains more than one chain.

quiescent center A region in a root apical meristem where there is no cell division.

***R* gene** A resistance gene in a plant; dominant R alleles confer enhanced resistance to plant pathogens.

R plasmid A bacterial plasmid containing genes that provide resistance to unfavorable conditions.

radial cleavage A cleavage pattern in deuterostomes in which newly formed cells lie directly above and below other cells of the embryo.

radial symmetry A body plan of organisms in which structures are arranged regularly around a central axis, like spokes radiating out from the center of a wheel.

radiation The transfer of heat energy as electromagnetic radiation.

radicle The rudimentary root of a plant embryo.

radioactivity The giving off of particles of matter and energy by decaying nuclei.

radioisotope An unstable, radioactive isotope.

Radiolaria (radiolarians) A lineage of the Rhizaria protist evolutionary group; characterized by axopods, slender, raylike strands of cytoplasm supported internally by long bundles of microtubules.

radiometric dating A dating method that uses measurements of certain radioactive isotopes to calculate the absolute ages in years of rocks and minerals.

radula The tooth-lined "tongue" of mollusks that scrapes food into small particles or drills through the shells of prey.

rain shadow An area of reduced precipitation on the leeward side of a mountain.

random coil An arrangement of the amino acid chain providing flexible regions that allow sections of the chain to bend.

random dispersion A pattern of distribution in which the individuals in a population are distributed unpredictably in their habitat.

rapid eye movement (REM) sleep The period during deep sleep when the delta wave pattern is replaced by rapid, irregular beta waves characteristic of the waking state. The person's heartbeat and breathing rate increase, the limbs twitch, and the eyes move rapidly behind the closed eyelids.

reabsorption The process in which some molecules (for example, glucose and amino acids) and ions are transported by the transport epithelium back into the body fluid (animals with open circulatory systems) or into the blood in capillaries surrounding the tubules (animals with closed circulatory systems) as the filtered solution moves through the excretory tubule.

reactants The atoms or molecules entering a chemical reaction.

reaction center Part of photosystems I and II in chloroplasts of plants. In the light-dependent reactions of photosynthesis, the reaction center receives light energy absorbed by the antenna complex in the same photosystem.

reading frame The series of codons for a polypeptide encoded by the mRNA.

realized niche The range of resources and environmental conditions actually used by a population in nature.

receptacle The expanded tip of a flower stalk that bears floral organs.

reception In signal transduction, the binding of a signal molecule with a specific receptor in a target cell.

receptive field The region surrounding a receptor within which the receptor responds to a stimulus.

receptor potential The change in membrane potential exhibited by a sensory receptor in response to a stimulus in its receptive field.

receptor protein Protein that recognizes and binds molecules from other cells that act as chemical signals.

receptor tyrosine kinase In signal transduction, a surface receptor with built-in protein kinase activity.

receptor-mediated endocytosis The selective uptake of macromolecules that bind to cell surface receptors concentrated in clathrin-coated pits.

recessive An allele that is masked by a dominant allele.

reciprocal altruism Form of altruistic behavior in which individuals help nonrelatives if they are likely to return the favor in the future.

reciprocal cross A genetic cross in which the two parents are switched with respect to which trait is associated with each sex.

recognition protein Protein in the plasma membrane that identifies a cell as part of the same individual or as foreign.

recombinant Phenotype with a different combination of traits from those of the original parents.

recombinant chromosomes Chromosomes that contain nonparental combinations of alleles. In eukaryotes they are generated by crossing-over in meiosis.

recombinant DNA DNA from two or more different sources joined together.

recombination *See* genetic recombination.

recombination frequency In the construction of linkage maps of diploid eukaryotic organisms, the percentage of testcross progeny that are recombinants.

rectum The final segment of the large intestine.

red tide A growth in dinoflagellate populations that causes red, orange, or brown discoloration of coastal ocean waters.

redox reaction Coupled oxidation–reduction reaction in which electrons are removed from a donor molecule and simultaneously added to an acceptor molecule.

reduced Substance that receives electrons during reduction.

reduction The addition of electrons to a substance.

reflex A programmed movement that takes place without conscious effort, such as the sudden withdrawal of a hand from a hot surface.

refractory period A period that begins at the peak of an action potential and lasts a few milliseconds, during which the threshold required for generation of an action potential is much higher than normal.

regulators Animals that maintain factors of the internal environment in a relatively constant state.

regulatory gene Gene that encodes a protein that regulates the expression of a structural gene or genes.

regulatory protein Protein that binds to a regulatory sequence of a gene or genes to control the transcription of the gene or genes.

regulatory sequence DNA sequence involved in the regulation of a gene or genes to which a regulatory protein binds to control the transcription of the gene or genes.

reinforcement The enhancement of reproductive isolation that had begun to develop while populations were geographically separated.

relative abundance The relative commonness of populations within a community.

relative fitness The number of surviving offspring that an individual produces compared with the number left by others in the population.

release The process in which urine is released into the environment from the distal end of the excretory tubule.

release factor A protein that recognizes stop codons in the A site of a ribosome translating an mRNA and terminates translation. Also referred to as the *termination factor.*

releasing hormone (RH) A peptide neurohormone that control the secretion of hormones from the anterior pituitary.

renal artery An artery that carries bodily fluids into the kidney.

renal cortex The outer region of the mammalian kidney that surrounds the renal medulla.

renal medulla The inner region of the mammalian kidney.

renal pelvis The central cavity in the kidney where urine drains from collecting ducts.

renal vein The vein that routes filtered blood away from the kidney.

renaturation The reformation of a denatured protein into its folded, functional state.

renin An enzyme secreted by cells in the juxtaglomerular apparatus into the bloodstream that converts a blood protein into the peptide hormone angiotensin.

renin-angiotensin-aldosterone system (RAAS) The most important hormonal system involved in regulation of Na^+ in mammals.

replica plating Technique for identifying and counting genetic recombinants in conjugation, transformation, or transduction experiments in which the colony pattern on a plate containing solid growth medium is pressed onto sterile velveteen and transferred to other plates containing different combinations of nutrients.

replicates Multiple subjects that receive either the same experimental treatment or the same control treatment.

replication bubble The two Y-shaped replication forks joined together at the tops of the Ys after DNA is unwound at an origin of replication.

replication fork The region of DNA synthesis where the parental strands separate and two new daughter strands elongate.

repressible operon Operon whose expression is prevented by a repressor molecule.

repressor A regulatory protein that prevents the operon genes from being expressed.

reproduction The process in which parents produce offspring.

reproductive isolating mechanism A biological characteristic that prevents the gene pools of two species from mixing.

reproductive strategy One of a set of behaviors that lead to reproductive success.

residual volume The air that remains in lungs after exhalation.

resilin An elastic protein related to elastin, found only in insects and some crustaceans.

resolution The minimum distance two points in a specimen can be separated and still be seen as two points.

resource partitioning The use of different resources or the use of resources in different ways by species living in the same place.

respiratory medium The environmental source of O_2 and the "sink" for released CO_2. For aquatic animals, the respiratory medium is water; for terrestrial animals, it is air.

respiratory surface A layer of epithelial cells that provides the interface between the body and the respiratory medium.

respiratory system All the parts of the body involved in exchanging air between the external environment and the blood.

response In signal transduction, the last stage in which the transduced signal causes the cell to change according to the signal and to the receptors on the cell. In the nervous system, the output resulting from the integration of neural messages.

resting potential A steady negative membrane potential exhibited by the membrane of a neuron that is not stimulated—that is, not conducting an impulse.

restriction endonuclease (restriction enzyme) An enzyme that cuts DNA at a specific sequence.

restriction fragment A DNA fragment produced by cutting a long DNA molecule with a restriction enzyme.

restriction fragment length polymorphisms When comparing different individuals, restriction enzyme–generated DNA fragments of different lengths from the same region of the genome.

reticular activating system *See* ascending reticular formation.

reticular formation A complex network of interconnected neurons that runs through the length of the brain stem, connecting to the thalamus at the anterior end and to the spinal cord at the posterior end.

reticulum One of the four chambers of the stomach of a ruminant; in this chamber, fermentation reactions by symbiotic microorganisms begin digesting boluses of swallowed plant matter.

retina A light-sensitive membrane lining the posterior part of the inside of the eye.

retrotransposon A transposable element that transposes via an intermediate RNA copy of the transposable element.

retrovirus A virus with an RNA genome that replicates via a DNA intermediate.

reverse transcriptase An enzyme that uses RNA as a template to make a DNA copy of the retrotransposon. Reverse transcriptase is used to make DNA copies of RNA in test tube reactions.

reversible The term indicating that a reaction may go from left to right or from right to left, depending on conditions.

rheumatoid arthritis An autoimmune disease that results by a self-attack on connective tissues, particularly in the joints, causing pain and inflammation.

Rhizaria (rhizarians) A protist evolutionary group; amoebas with filamentous pseudopods.

rhizome A horizontal, modified stem that can penetrate a substrate and anchor the plant.

Rhodophyta (red algae) A lineage of the Archaeplastida protist evolutionary group; mostly free-living autotrophic marine seaweeds.

rhodopsin The retinal-opsin photopigment.

ribonucleic acid (RNA) A polymer assembled from repeating nucleotide monomers in which the five-carbon sugar is ribose. Cellular RNAs include mRNA (which is translated to produce a polypeptide), tRNA (which brings an amino acid to the ribosome for assembly into a polypeptide during translation), and rRNA (which is a structural component of ribosomes). The genetic material of some viruses is RNA.

ribonucleotide Nucleotide containing ribose as the sugar; ribonucleotides are components of RNA.

ribose A five-carbon sugar to which the nitrogenous bases in nucleotides link covalently.

ribosomal RNA (rRNA) The RNA component of ribosomes.

ribosome A ribonucleoprotein particle that carries out protein synthesis by translating mRNA into chains of amino acids.

ribosome binding site In translation initiation in prokaryotes, a sequence just upstream of the start codon which directs the small ribosomal subunit to bind and orient correctly for the complete ribosome to assemble and start translating in the correct spot.

ribozyme An RNA-based catalyst that is part of the biochemical machinery of all cells.

ring species A species with a geographic distribution that forms a ring around uninhabitable terrain.

RNA *See* ribonucleic acid.

RNA interference (RNAi) The phenomenon of silencing a gene posttranscriptionally by a small, single-stranded RNA that is complementary to part of an mRNA.

RNA polymerase An enzyme that catalyzes the assembly of ribonucleotides into an RNA strand.

RNA world model A model which states that the first genes and enzymes were RNA molecules.

rod In the vertebrate eye, a type of photoreceptor in the retina that is specialized for detection of light at low intensities.

root An anchoring structure in land plants that also absorbs water and nutrients and (in some plant species) stores food.

root cap A dome-shaped cell mass that forms a protective covering over the apical meristem in the tip of a plant root.

root hair A tubular outgrowth of the outer wall of a root epidermal cell; root hairs absorb much of a plant's water and minerals from the soil.

root nodule A localized swelling on a root in which symbiotic nitrogen-fixing bacteria reside.

root pressure The pressure that develops in plant roots as the result of osmosis, forcing xylem sap upward and out through leaves. *See also* guttation.

root primordium A rudimentary root.

root system An underground (or submerged) network of roots with a large surface area that favors the rapid uptake of soil water and dissolved mineral ions.

rough ER Endoplasmic reticulum with many ribosomes studding its outer surface.

round window A thin membrane that faces the middle ear.

***r*-selected species** A short-lived species adapted to function well in a rapidly changing environment.

RuBP carboxylase/oxygenase (rubisco) An enzyme that catalyzes the key reaction of the Calvin cycle, carbon fixation, in which CO_2 combines with RuBP (ribulose 1,5-bisphosphate) to form 3-phosphoglycerate.

rumen One of the four chambers of the stomach of a ruminant; in this chamber, fermentation reactions by symbiotic microorganisms begin digesting boluses of swallowed plant matter.

ruminant An animals that has a complex, four-chambered stomach.

S phase The phase of the eukaryotic cell cycle during which DNA replication occurs.

saccule A fluid-filled chamber in the vestibular apparatus that provides information about the position of the head with respect to gravity (up versus down), as well as changes in the rate of linear movement of the body.

salicylic acid (SA) In plants, a chemical synthesized following a wound that has multiple roles in plant defenses, including interaction with jasmonates in signaling cascades.

salivary amylase A substance that hydrolyzes starches to the disaccharide maltose.

salivary gland A gland that secretes saliva through a duct on the inside of the cheek or under the tongue; the saliva lubricates food and begins digestion.

salt marsh A tidal wetland dominated by emergent grasses and reeds.

saltatory conduction A mechanism that allows small-diameter axons to conduct impulses rapidly.

saprobe An organism nourished by dead or decaying organic matter.

sapwood The newly formed outer wood located between heartwood and the vascular cambium. Compared with heartwood, it is wet, lighter in color, and not as strong.

sarcomere The basic unit of contraction in a myofibril.

sarcoplasmic reticulum In vertebrate muscle fibers, a complex system of vesicles modified from the smooth endoplasmic reticulum that encircles the sarcomeres. The sarcoplasmic reticulum is part of the pathway for the stimulation of muscle contraction by neural signals.

saturated enzymes Enzymes for which increases in substrate concentration have no effect on the reaction rate.

saturated fatty acid Fatty acid with only single bonds linking the carbon atoms.

savanna A biome comprising grasslands with few trees, which grows in areas adjacent to tropical deciduous forests.

schizocoelom In protostomes, the body cavity that develops as inner and outer layers of mesoderm separate.

Schwann cell A type of glial cell in the PNS that wraps nerve fibers with myelin and also secretes regulatory factors.

scientific method An investigative approach in which scientists make observations about the natural world, develop working explanations about what they observe, and then test those explanations by collecting more information.

scientific name A two-part name identifying the genus to which a species belongs and designating a particular species within that genus.

scientific theory A broadly applicable idea or hypothesis that has been confirmed by every conceivable test.

sclereid A type of sclerenchyma cell; sclereids typically are short and have thick, lignified walls.

sclerenchyma A ground tissue in which cells develop thick secondary walls, which commonly are lignified and perforated by pits through which water can pass.

scrotum The baglike sac in which the testes are suspended in many mammals.

scutellum The shield-shaped cotyledon of a grass.

second law of thermodynamics Principle that for any process in which a system changes from an initial to a final state, the total disorder of the system and its surroundings always increases.

second messenger In particular signal transduction pathways, an internal, nonprotein signal molecule that directly or indirectly activates protein kinases, which elicit the cellular response.

secondary active transport Transport indirectly driven by ATP hydrolysis.

secondary cell wall A layer added to the cell wall of plants that is more rigid and may become many times thicker than the primary cell wall.

secondary consumer A carnivore that feeds on herbivores, a member of the third trophic level.

secondary endosymbiosis In the model for the origin of plastids in eukaryotes, the second event, in which a nonphotosynthetic eukaryote engulfed a photosynthetic eukaryote.

secondary growth Plant growth that originates at lateral meristems and increases the diameter of older roots and stems. *Compare* primary growth.

secondary immune response The rapid immune response that occurs during the second (and subsequent) encounters of the immune system of a mammal with a specific antigen.

secondary plant body The part of a plant made up of tissues that develop from lateral meristems.

secondary productivity Energy stored in new consumer biomass as energy is transferred from producers to consumers.

secondary structure Regions of alpha helix, beta strand, or random coil in a polypeptide chain.

secondary succession Predictable changes in species composition in an ecological community that develops after existing vegetation is destroyed or disrupted by an environmental disturbance.

secondary tissue In plants, the tissue that develops from lateral meristems.

secretin Hormone released into the bloodstream from glandular cells in the small intestine in response to stimulation by acidic chyme.

secretion A selective process in which specific small molecules and ions are transported from the body fluids (in animals with open circulatory systems) or blood (in animals with closed circulatory systems) into the excretory tubules.

secretory vesicle Vesicle that transports proteins to the plasma membrane.

seed The structure that forms when an ovule matures after a pollen grain reaches it and a sperm fertilizes the egg.

seed coat The outer protective covering of a seed.

segment polarity gene In *Drosophila* embryonic development, one of the third activated set of segmentation genes that progressively subdivide the embryo into regions, determining the segments of the embryo and the adult. Segment polarity genes set the boundaries and anterior–posterior axis of each segment in the embryo, thereby determining the regions that become segments of larvae and adults.

segmentation The production of body parts and some organ systems in repeating units.

segmentation genes Genes that work sequentially, progressively subdividing the embryo into regions, determining the segments of the embryo and the adult.

segregation The separation of the pairs of alleles that control a character as gametes are formed.

selective cell adhesion A mechanism in which cells make and break specific connections to other cells or to the extracellular matrix.

selectively neutral *See* neutral variation hypothesis.

selectively permeable Membranes that selectively allow, impede, or block the passage of atoms and molecules.

self-fertilization (self-pollination) Fertilization in which sperm nuclei in pollen fertilize egg cells of the same plant.

self-incompatibility In plants, the inability of a plant's pollen to fertilize ovules of the same plant.

semen The secretions of several accessory glands in which sperm are mixed prior to ejaculation.

semicircular canal A part of the vestibular apparatus that detects rotational (spinning) motions.

semiconservative replication The process of DNA replication in which the two parental strands separate and each serves as a template for the synthesis of new progeny double-stranded DNA molecules.

seminal fluid Fluid secreted by the seminal vesicles that contains prostaglandins, which when ejaculated into the female trigger contractions of the female reproductive tract that help move the sperm into and through the uterus.

seminal vesicle A vesicle that secretes seminal fluid.

seminiferous tubule One of the tiny tubes in the testes where sperm cells are produced, grow, and mature.

senescence The biologically complex process of aging in mature organisms that leads to the death of cells and eventually the whole organism.

sense codon A codon that specifies an amino acid.

sensitization Increased responsiveness to mild stimuli after experiencing a strong stimulus; one of the simplest forms of memory.

sensor A tissue or organ that detects a change in an external or internal factor such as pH, temperature, or the concentration of a molecule such as glucose.

sensory adaptation A condition in which the effect of a stimulus is reduced if it continues at a constant level.

sensory neuron A neuron that transmits stimuli collected by their sensory receptors to interneurons.

sensory receptor A receptor formed by the dendrites of afferent neurons, or by specialized receptor cells making synapses with afferent neurons that pick up information about the external and internal environments of the animal.

sensory transduction The conversion of a stimulus into a change in membrane potential.

sepal One of the separate, usually green parts forming the calyx of a flower.

septum (plural, septa) A thin partition or cross wall that separates body segments.

sequential hermaphroditism The form of hermaphroditism in which individuals change from one sex to the other.

serosa The serous membrane: a thin membrane lining the closed cavities of the body; has two layers with a space between that is filled with serous fluid.

Sertoli cell One of the supportive cells that completely surrounds developing spermatocytes in the seminiferous tubules. Follicle-stimulating hormone stimulates Sertoli cells to secrete a protein and other molecules that are required for spermatogenesis.

sessile Unable to move from one place to another.

set point The level at which the condition controlled by a homeostatic pathway is to be maintained.

seta (plural, setae) A chitin-reinforced bristle that protrudes outward from the body wall in some annelid worms.

sex chromosomes Chromosomes that are different in male and female individuals of the same species.

sex pilus *See* F pilus.

sex ratio The relative proportions of males and females in a population.

sex-linked gene Gene located on a sex chromosome.

sexual dimorphism Differences in the size or appearance of males and females.

sexual reproduction The mode of reproduction in which male and female parents produce offspring through the union of egg and sperm generated by meiosis.

sexual selection A form of natural selection established by male competition for access to females and by the females' choice of mates.

shells *See* energy levels.

shoot system The stems and leaves of a plant.

short interfering RNA (siRNA) One of the major types of small regulatory RNAs in eukaryotes involved in RNA interference (RNAi).

short tandem repeat (STR) Short 2–6-bp DNA sequence repeated in series.

short-day plant A plant that flowers in late summer or early autumn when dark periods become longer and light periods become shorter.

short-term memory Memory that stores information for seconds.

sieve tube A series of phloem cells joined end to end, forming a long tube through which nutrients are transported; a feature mainly of flowering plants.

sieve tube member Any of the main conducting cells of phloem that connect end to end, forming a sieve tube.

sign stimulus A simple cue that triggers a fixed action pattern.

signal peptide A short segment of amino acids to which the signal recognition particle binds, temporarily blocking further translation. A signal peptide is found on polypeptides that are sorted to the endoplasmic reticulum. Also referred to as *signal sequence.*

signal recognition particle (SRP) Protein-RNA complex that binds to signal sequences and targets polypeptide chains to the endoplasmic reticulum.

signal sequence *See* signal peptide.

signal transduction The series of events by which a signal molecule released from a controlling cell causes a response (affects the function) of target cells with receptors for the signal. Target cells process the signal in the three sequential steps of reception, transduction, and response.

silencing Phenomenon in which methylation of cytosines in eukaryotic promoters inhibits transcription and turns the genes off.

silent mutation A base-pair substitution mutation in a protein-coding gene that does not alter the amino acid specified by the gene.

simple diffusion Mechanism by which certain small substances diffuse through the lipid part of a biological membrane.

simple fruit A fruit that develops from a single ovary; in many of them at least one layer of the pericarp is fleshy.

simulation modeling An analytical method in which researchers gather detailed information about a system and then create a series of mathematical equations that predict how the components of the system interact and respond to change.

simultaneous hermaphroditism A form of hermaphroditism in which individuals develop functional ovaries and testes at the same time.

single-lens eye An eye type that works by changing the amount of light allowed to enter into the eye and by focusing this incoming light with a lens.

single-stranded binding protein (SSB) Protein that coats single-stranded segments of DNA, stabilizing the DNA for the replication process.

sink Any region of a plant where organic substances are being unloaded from the phloem and used or stored.

sink population In metapopulation analysis, a population that routinely declines in size after being replenished by immigrants from a source population.

sinoatrial node (SA node) The region of the heart that controls the rate and timing of cardiac muscle cell contraction.

sinus A body space that surrounds an organ.

siRNA-induced silencing complex (siRISC) Protein complex containing an siRNA that binds to a sequence in a target RNA resulting in cleavage of that RNA.

sister chromatid One of two exact copies of a chromosome duplicated during replication.

sister clades Two evolutionary lineages (that is, clades) that emerge from the same node in a phylogenetic tree.

sister species Two species that are descended from the same recent ancestral species.

skeletal muscle A muscle that connect to bones of the skeleton, typically made up of long and cylindrical cells that contain many nuclei.

sliding DNA clamp A protein that encircles the DNA and binds to the DNA polymerase to tether the enzyme to the template, thereby making replication more efficient.

slime layer A coat typically composed of polysaccharides that is loosely associated with bacterial cells.

slime molds Members of the Amoebozoa; heterotrophic protists that, at some stage of their life cycle, exist as individuals that move by amoeboid motion but that exist in more complex forms the remainder of the time.

slow block to polyspermy The process in which enzymes released from cortical granules alter the egg coats within minutes after fertilization, so that no other sperm can attach and penetrate to the egg.

slow muscle fiber A muscle fiber that contracts relatively slowly and with low intensity.

small interfering RNA (siRNA) A class of single-stranded RNAs that cause RNA interference.

small ribonucleoprotein particle A complex of RNA and proteins.

smooth ER Endoplasmic reticulum with no ribosomes attached to its membrane surfaces. Smooth ER has various functions, including synthesis of lipids that become part of cell membranes.

smooth muscle A relatively small and spindle-shaped muscle cell in which actin and myosin molecules are arranged in a loose network rather than in bundles.

social behavior The interactions that animals have with other members of their species.

sodium–potassium pump *See* Na^+/K^+ pump.

soil solution A combination of water and dissolved substances that coats soil particles and partially fills pore spaces.

solenoid *See* 30-nm chromatin fiber.

solute The molecules of a substance dissolved in water.

solution Substance formed when molecules and ions separate and are suspended individually, surrounded by water molecules.

solvent The water in a solution in which the hydration layer prevents polar molecules or ions from reassociating.

somaclonal selection A procedure in which somatic embryos derived from tissue culture are screened to identify those having desired characteristics, such as disease resistance.

somatic cell Any of the cells of an organism's body other than reproductive cells.

somatic embryo A plant embryo that is genetically identical to the parent because it arose through asexual means.

somatic gene therapy Gene therapy in which genes are introduced into somatic cells.

somatic nervous system A subdivision of the peripheral nervous system controlling body movements that are primarily conscious and voluntary.

somites Paired blocks of mesoderm cells along the vertebrate body axis that form during early vertebrate development and differentiate into dermal skin, bone, and muscle.

soredium (plural, soredia) A specialized cell cluster produced by lichens, consisting of a mass of algal cells surrounded by fungal hyphae; soredia function like reproductive spores and can give rise to a new lichen.

sorus (plural, sori) A cluster of sporangia on the underside of a fern frond; reproductive spores arise by meiosis inside each sporangium.

source In plants, any region (such as a leaf) where organic substances are being loaded into the sieve tube system of phloem.

source population In metapopulation analyses, a population that is either stable or increasing in size.

Southern blot analysis Technique in which labeled probes are used to detect specific DNA fragments that have been separated by gel electrophoresis.

spatial summation The summation of EPSPs produced by firing of different presynaptic neurons.

specialized transduction Transfer of bacterial genes between bacteria using temperate phages that have incorporated fragments of the bacterial genome as they make the transition from the lysogenic cycle to the lytic cycle.

speciation The process of species formation.

species A group of populations in which the individuals are so closely related in structure, biochemistry, and behavior that they can successfully interbreed.

species cluster A group of closely related species recently descended from a common ancestor.

species composition The particular combination of species that occupy a site.

species diversity A community characteristic defined by species richness and the relative abundance of species.

species richness The number of species that live within an ecological community.

specific epithet The species name in a binomial.

specific heat The amount of heat required to increase the temperature of a given quantity of water.

spermatocyte A developing gamete that becomes a spermatid at the end of meiosis.

spermatogenesis The process of producing sperm.

spermatogonium (plural, spermatogonia) A cell that enters meiosis and gives rise to gametes, produced by mitotic divisions of the germ cells in males.

spermatozoan Also called sperm; a haploid cell that develops into a mature sperm cell when meiosis is complete.

sphincter A powerful ring of smooth muscle that forms a valve between major regions of the digestive tract.

spinal cord A column of nervous tissue located within the vertebral column and directly connected to the brain.

spinal nerve A nerve that carries signals between the spinal cord and the body trunk and limbs.

spindle The structure that separates sister chromatids and moves them to opposite spindle poles.

spindle pole One of the pair of centrosomes in a cell undergoing mitosis from which bundles of microtubules radiate to form the part of the spindle from that pole.

spinneret A modified abdominal appendage from which spiders secrete silk threads.

spiracle An opening in the chitinous exoskeleton of an insect through which air enters and leaves the tracheal system.

spiral cleavage The cleavage pattern in many protostomes in which newly produced cells lie in the space between the two cells immediately below them.

spiral organ *See* organ of Corti.

spiral valve A corkscrew-shaped fold of mucous membrane in the digestive system of elasmobranchs, which slows the passage of material and increases the surface area available for digestion and absorption.

spirillum (plural, spirilla) Any flagellated aerobic bacterium twisted helically like a corkscrew.

spliceosome A complex formed between the pre-mRNA and small ribonucleoprotein particles, in which mRNA splicing takes place.

spongocoel The central cavity in a sponge.

spontaneous reaction Chemical or physical reaction that occurs without outside help.

sporadic (nonhereditary) cancer Cancer that is not inherited.

sporangium (plural, sporangia) A single-celled or multicellular structure in fungi and plants in which spores are produced.

spore A haploid reproductive structure, usually a single cell, that can develop into a new individual without fusing with another cell; found in plants, fungi, and certain protists.

sporophyll A specialized leaf that bears sporangia (spore-producing structures).

sporophyte An individual of the diploid generation produced through fertilization in organisms that undergo alternation of generations; it produces haploid spores.

sporopollenin A tough polymer in the walls of spores and pollen grains, the presence of which helps such structures resist decay.

spring overturn The mixing of surface water with deep water in a lake or pond, causing oxygen at the surface to move to the bottom, and nutrients from the bottom to move to the surface.

SRP (signal recognition particle) receptor A protein on the membrane of the endoplasmic reticulum that binds the signal recognition particle.

stability The ability of a community to maintain its species composition and relative abundances when environmental disturbances eliminate some species from the community.

stabilizing selection A type of natural selection in which individuals expressing intermediate phenotypes have the highest relative fitness.

stamen A "male" reproductive organ in flowers, consisting of an anther (pollen producer) and a slender filament.

standing crop biomass The total dry weight of plants present in an ecosystem at a given time.

stapes The smallest of three sound-conducting bones in the middle ear of tetrapod vertebrates.

starch A storage polysaccharide in plants consisting of branched or unbranched chains of glucose subunits.

start codon The first codon read in an mRNA in translation—AUG. Also referred to as the *initiator codon.*

statocyst A mechanoreceptor in invertebrates that senses gravity and motion using statoliths.

statolith A movable starch- or carbonate-containing stonelike body involved in sensing gravitational pull.

stele The central core of vascular tissue in roots and shoots of vascular plants; it consists of the xylem and phloem together with supporting tissues.

stem cells Cells capable of undergoing many divisions in an unspecialized, undifferentiated state and that also have the ability to differentiate into specialized cell types.

stereocilia Microvilli covering the surface of hair cells clustered in the base of neuromasts.

steroid A type of lipid derived from cholesterol.

steroid hormone receptor Internal receptor that turns on specific genes when it is activated by binding a signal molecule.

steroid hormone response element The DNA sequence to which the hormone-receptor complex binds.

sterol Steroid with a single polar —OH group linked to one end of the ring framework and a complex, nonpolar hydrocarbon chain at the other end.

sticky end End of a DNA fragment generated by digestion with a restriction enzyme, with a single-stranded structure that can form hydrogen bonds with a complementary sticky end on any other DNA molecule cut with the same enzyme.

stigma The receptive end of a carpel where deposited pollen germinates.

stimulus A component of a negative feedback control system maintaining homeostasis, specifically an environmental change that triggers a response.

stoma (plural, stomata) The opening between a pair of guard cells in the epidermis of a plant leaf or stem, through which gases and water vapor pass.

stomach The portion of the digestive system in which food is stored and digestion begins.

stomach ulcer *See* peptic ulcer.

stop codon A codon that does not specify amino acids. The three nonsense codons are UAG, UAA, and UGA. Also referred to as the *nonsense codon* and *termination codon.*

Stramenopila (stramenopiles) A subgroup of the Chromalveolata protist evolutionary group; protists with two different flagella, one with hollow tripartite projections, and one that is plain.

stratification Horizontal layering of sedimentary rocks beneath the soil surface.

strict aerobe Cell with an absolute requirement for oxygen to survive, unable to live solely by fermentations.

strict anaerobe Organism in which fermentation is the only source of ATP.

strobilus *See* cone.

structural gene Gene that encodes a protein that has a function other than gene regulation.

structural isomers Two molecules with the same chemical formula but atoms that are arranged in different ways.

style The slender stalk of a carpel situated between the ovary and the stigma in plants.

submucosa A thick layer of elastic connective tissue that contains neuron networks and blood and lymph vessels.

subsoil The region of soil beneath topsoil, which contains relatively little organic matter.

subspecies A taxonomic subdivision of a species.

substrate The particular reacting molecule or molecular group that an enzyme catalyzes.

substrate-level phosphorylation An enzyme-catalyzed reaction that transfers a phosphate group from a substrate to ADP.

sugar–phosphate backbone Structure in a polynucleotide chain that is formed when deoxyribose sugars (in DNA) or ribose sugars (in RNA) are linked by phosphate groups in an alternating sugar–phosphate–sugar–phosphate pattern.

sulfhydryl group Group that works as a molecular fastener, consisting of a sulfur atom linked on one side to a hydrogen atom and on the other side to a carbon chain.

sum rule Mathematical rule in which final probability is found by summing individual probabilities.

surface tension The force that places surface water molecules under tension, making them more resistant to separation than the underlying water molecules.

survivorship curve Graphic display of the rate of survival of individuals over a species' life span.

suspension feeder An animal that ingests small food items suspended in water.

suspensor In seed plants, a stalklike row of cells that develops from a zygote and helps position the embryo close to the nourishing endosperm.

swallowing reflex The involuntary action produced by contractions of muscles in the walls of the pharynx that direct food into the esophagus.

swim bladder A gas-filled internal organ that helps bony fishes maintain buoyancy.

symbiosis An interspecific interaction in which the ecological relations of two or more species are intimately tied together.

symmetry Exact correspondence of form and constituent configuration on opposite sides of a dividing line or plane.

sympathetic division Division of the autonomic nervous system that predominates in situations involving stress, danger, excitement, or strenuous physical activity.

sympatric speciation Speciation that occurs without the geographic isolation of populations.

symplastic pathway The route taken by water that moves through the cytoplasm of plant cells (the symplast). *Compare* apoplastic pathway.

symport The transport of two molecules in the same direction across a membrane. Also referred to as *cotransport.*

synapomorphy A derived character state found in two or more species.

synapse A site where a neuron makes a communicating connection with another neuron or an effector such as a muscle fiber or gland.

synapsida A lineage of amniotes that have one bony temporal arch near the back of the skull; includes mammals.

synapsis *See* pairing.

synaptic cleft A narrow gap that separates the plasma membranes of the presynaptic and postsynaptic cells.

synaptonemal complex A protein framework that tightly holds together homologous chromosomes as they pair.

systematics The branch of biology that studies the diversity of life and its evolutionary relationships.

systemic acquired resistance A plant defense response to microbial invasion; defensive chemicals including salicylic acid may spread throughout a plant, rendering healthy tissues less vulnerable to infection.

systemic circuit In amphibians, the branch of a double blood circuit that receives oxygenated blood and provides the blood supply for most of the tissues and cells of a body.

systemic homeostatic controls Controls that are initiated outside of an organ or organ system to control that organ's or organ system's activity.

systemic inflammation Inflammation that occurs throughout the body.

systemic lupus erythematosus (lupus) An autoimmune disease caused by production of a wide variety of anti-self antibodies against blood cells, blood platelets, and internal cell structures and molecules such as mitochondria and proteins associated with the DNA in the cell nucleus; characterized by anemia and problems with blood circulation and kidney function.

systemin A plant peptide hormone that functions in defense responses to wounds.

systems biology An area of biology that studies the organism as a whole to unravel the integrated and interacting network of genes, proteins, and biochemical reactions responsible for life.

systole The period of contraction and emptying of the heart.

systolic blood pressure The peak of high blood pressure in the cardiac cycle that occurs as the heart ventricles contract and move blood outward through the arteries.

T (transverse) tubule The tubule that passes in a transverse manner from the sarcolemma across a myofibril of striated muscle.

T cell A lymphocyte produced by the division of stem cells in the bone marrow and then released into the blood and carried to the thymus. T cells participate in adaptive immunity.

tactile signal A means of animal communication in which the signaler uses touch to convey a message to the signal receiver.

taiga *See* boreal forest.

taproot system A root system consisting of a single main root from which lateral roots can extend; often stores starch.

target site The location in a genome to which a transposable element moves when it transposes.

taxis A behavioral response that is directed either toward or away from a specific stimulus.

taxon (plural, taxa) A name designating a group of organisms included within a category in the Linnaean taxonomic hierarchy.

taxonomic hierarchy A system of classification based on arranging organisms into ever more inclusive categories.

taxonomy The science of the classification of organisms into an ordered system that indicates natural relationships.

T-cell receptor (TCR) A receptor that covers the plasma membrane of a T-cell, specific for a particular antigen.

tectorial membrane A membrane that extends the length of the cochlear canal of the inner ear in which the stereocilia of the sensory hair cells of the organ of Corti are embedded.

telomerase An enzyme that adds telomere repeats to chromosome ends.

telomeres Repeats of simple-sequence DNA that maintain the ends of linear chromosomes.

telophase The final phase of mitosis, during which the spindle disassembles, the chromosomes decondense, and the nuclei re-form.

temperate bacteriophage Bacteriophage that may enter an inactive phase (lysogenic cycle) in which the host cell replicates and passes on the bacteriophage DNA for generations before the phage becomes active and kills the host (lytic cycle).

temperate deciduous forest A forested biome found at low to middle altitudes at temperate latitudes, with warm summers, cold winters, and annual precipitation between 75 and 250 cm.

temperate grassland A non-forested biome that stretches across the interiors of most continents, where winters are cold and snowy and summers are warm and fairly dry.

temperate rain forest A coniferous forest biome supported by heavy rain and fog, which grows where winters are mild and wet and the summers are cool.

template A nucleotide chain used in DNA replication for the assembly of a complementary chain.

template strand The DNA strand that is copied into an RNA molecule during gene transcription.

temporal fenestrae Openings near the back of the skull in some amniote vertebrates.

temporal isolation A prezygotic reproductive isolating mechanism in which species live in the same habitat but breed at different times of day or different times of year.

temporal summation The summation of several EPSPs produced by successive firing of a single presynaptic neuron over a short period of time.

tendon A type of fibrous connective tissue that attaches muscles to bones.

terminal bud A bud that develops at the apex of a shoot.

termination codon *See* stop codon.

termination factor *See* release factor.

terminator Specific DNA sequence for a gene that signals the end of transcription of a gene. Terminators are common for prokaryotic genes.

territory A plot of habitat, defended by an individual male or a breeding pair of animals, within which the territory holders have exclusive access to food and other necessary resources.

tertiary consumer A carnivore that feeds on other carnivores, a member of the fourth trophic level.

tertiary structure The overall three-dimensional folding of a polypeptide chain.

testcross A genetic cross between an individual with the dominant phenotype and a homozygous recessive individual.

testis (plural, testes) The male gonad. In male vertebrates, they secrete androgens and steroid hormones that stimulate and control the development and maintenance of male reproductive systems.

testosterone A hormone produced by the testes, responsible for the development of male secondary sex characteristics and the functioning of the male reproductive organs.

tetanus A situation in which a muscle fiber cannot relax at all between stimuli, and twitch summation produces a peak level of continuous contraction.

tetrad Homologous pair of eukaryotic chromosomes during the first meiotic division consisting of four chromatids.

Tetrapoda A monophyletic lineage of vertebrates that includes animals having four feet, legs, or leglike appendages.

T-even bacteriophage Virulent bacteriophages, T2, T4, and T6, that have been valuable for genetic studies of bacteriophage structure and function.

thalamus A major switchboard of the brain that receives sensory information and relays it to the regions of the cerebral cortex concerned with motor responses to sensory information of that type.

thallus A plant body not differentiated into stems, roots, or leaves.

thermal acclimatization A set of physiological changes in ectotherms in response to seasonal shifts in environmental temperature, allowing the animals to attain good physiological performance at both winter and summer temperatures.

thermocline The narrow depth range in a lake where water temperature changes abruptly at the boundary between the epilimnion and the hypolimnion.

thermodynamics The study of the energy flow during chemical and physical reactions.

thermoreceptor A sensory receptor that detects the flow of heat energy.

thermoregulation The control of body temperature.

thick filament A type of filament in striated muscle composed of myosin molecules; they interact with thin filaments to shorten muscle fibers during contraction.

thigmomorphogenesis A plant response to a mechanical disturbance, such as frequent strong winds; includes inhibition of cellular elongation and production of thick-walled supportive tissue.

thigmotropism Growth in response to contact with a solid object.

thin filament A type of filament in striated muscle composed of actin, tropomyosin, and troponin molecules; they interact with thick filaments to shorten muscle fibers during contraction.

thorax The central part of an animal's body, between the head and the abdomen.

thorn forest A forested biome that grows at the arid borders of true savanna, where large mammals are less abundant.

threshold potential In signal conduction by neurons, the membrane potential at which the action potential fires.

thymine A pyrimidine that base-pairs with adenine.

thymus An organ of the lymphatic system that plays a role in filtering viruses, bacteria, damaged cells, and cellular debris from the lymph and bloodstream, and in defending the body against infection and cancer.

thyroid gland A gland located beneath the voice box (larynx) that secretes hormones regulating growth and metabolism.

thyroid-stimulating hormone (TSH) A hormone that stimulates the thyroid gland to grow in size and secrete thyroid hormones.

thyroxine (T_4) The main hormone of the thyroid gland, responsible for controlling the rate of metabolism in the body.

Ti (tumor-inducing) plasmid A plasmid used to make transgenic plants.

tidal volume The volume of air entering and leaving the lungs during inhalation and exhalation.

tight junction Region of tight connection between membranes of adjacent cells.

time lag The delayed response of organisms to changes in environmental conditions.

tissue A group of cells and intercellular substances with the same structure that function as a unit to carry out one or more specialized tasks.

tolerance hypothesis Hypothesis asserting that ecological succession proceeds because competitively superior species replace competitively inferior ones.

tonoplast The membrane that surrounds the central vacuole in a plant cell.

topoisomerase An enzyme that relieves the overtwisting and strain of DNA ahead of the replication fork.

topsoil The rich upper layer of soil where most plant roots are located; it generally consists of sand, clay particles, and humus.

torpor A sleeplike state produced when a lowered set point greatly reduces the energy required to maintain body temperature, accompanied by reductions in metabolic, nervous, and physical activity.

torsion The realignment of body parts in gastropod mollusks that is independent of shell coiling.

totipotent Having the capacity to produce cells that can develop into or generate a new organism or body part.

trace element An element that occurs in organisms in very small quantities ($<0.01\%$); in nutrition, a mineral required by organisms only in small amounts.

tracer Isotope used to label molecules so that they can be tracked as they pass through biochemical reactions.

trachea In insects, an extensively branched, air-conducting tube formed by invagination of the outer epidermis of the animal, and reinforced by rings of chitin. In vertebrates, the windpipe, which branches into the bronchi.

tracheal system A branching network of tubes that carries air from small openings in the exoskeleton of an insect to tissues throughout its body.

tracheid A conducting cell of xylem, usually elongated and tapered.

tracheophyte A plant with xylem, phloem, and usually well-developed roots, stems, and leaves.

traditional systematics An approach to systematics that uses phenotypic similarities and differences to infer evolutionary relationships, grouping together species that share both ancestral and derived characters.

trait One of the forms of a genetic character.

transcription The mechanism by which the information encoded in DNA is made into a complementary RNA copy.

transcription factor (TF) In eukaryotes, one of the group of proteins that bind to the promoter and then recruit RNA polymerase, the enzyme for transcription of the associated gene.

transcription initiation complex Combination of general transcription factors with RNA polymerase II.

transcription unit A region of DNA that transcribes a single primary transcript.

transcriptional regulation The processes that directly control gene activity.

transduction In cell signaling, the process of changing a signal into the form necessary to cause the cellular response. In prokaryotes, the process in which DNA is transferred from donor to recipient bacterial cells by an infecting bacteriophage.

transfer cell Any of the specialized cells that form when large amounts of solutes must be loaded or unloaded into the phloem; they facilitate the short-distance transport of organic solutes from the apoplast into the symplast.

transfer RNA (tRNA) The RNA that brings amino acids to the ribosome for addition to the polypeptide chain.

transformation The conversion of the hereditary type of a cell by the uptake of DNA released by the breakdown of another cell.

transgenic An organism that has been modified to contain genetic information from an external source.

transition state An intermediate arrangement of atoms and bonds that both the reactants and the products of a reaction can assume.

transit sequence Short amino acid sequence on a protein that serves to direct the protein to the appropriate organelle (other than the nucleus or ER) in a eukaryotic cell.

translation The use of the information encoded in mRNA to assemble amino acids into a polypeptide.

translocation In genetics, a chromosomal alteration that occurs if a broken segment is attached to a different, nonhomologous chromosome. In vascular plants, the long-distance transport of substances by xylem and phloem.

transmembrane pathway The path followed by water when it enters root cells by crossing the cells' plasma membranes.

transmission In neural signaling, the sending of a message along a neuron, and then to another neuron or to a muscle or gland.

transpiration The evaporation of water from a plant, principally from the leaves.

transport The controlled movement of ions and molecules from one side of a membrane to the other.

transport epithelium A layer of cells with specialized transport proteins in their plasma membranes.

transport protein A protein embedded in the cell membrane that forms a channel allowing selected polar molecules and ions to pass across the membrane.

transposable element (TE) A sequence of DNA that can move from one place to another within the genome of a cell.

transposase An enzyme that catalyzes some of the reactions inserting or removing the transposable element from the DNA.

transposition The movement of a transposable element from one site to another in a genome.

transposon A bacterial transposable element with an inverted repeat sequence at each end enclosing a central region with one or more genes.

tricarboxylic acid cycle *See* citric acid cycle.

trichocyst A dartlike protein thread that can be discharged from a surface organelle for defense or to capture prey.

trichome A single-celled or multicellular outgrowth of the plant epidermis; examples include root hairs and leaf trichomes that resemble hairs.

triglyceride A nonpolar compound produced when a fatty acid binds by a dehydration synthesis reaction at each of glycerol's three —OH-bearing sites.

triiodothyronine (T_3) A hormone secreted by the thyroid gland that regulates metabolism.

trimester A division of human gestation, three months in length.

triploblastic An animal body plan in which adult structures arise from three primary germ layers, endoderm, mesoderm, and ectoderm.

trochophore The small, free-swimming, ciliated aquatic larva of various invertebrates, including certain mollusks and annelids.

trophic cascade The effects of predator–prey interactions that reverberate through other population interactions at two or more trophic levels in an ecosystem.

trophic level A position in a food chain or web that defines the feeding habits of organisms.

trophoblast The outer single layer of cells of the blastocyst.

tropical deciduous forest A tropical forest biome that occurs where winter drought reduces photosynthesis, and most trees drop their leaves seasonally.

tropical forest Any forest that grows between the Tropics of Capricorn and Cancer, a region characterized by high temperature and rainfall and thin, nutrient-poor topsoil.

tropical montane forest A tropical forest biome of short trees, which are frequently enveloped in mist; also known as a "cloud forest."

tropical rain forest A dense tropical forest biome that grows where some rain falls every month, mean annual rainfall exceeds 250 cm, mean annual temperature is at least 25°C, and humidity is above 80%.

tropics The latitudes between 23.5° N and 23.5° S, the Tropics of Cancer and Capricorn.

tropism The turning or bending of an organism or one of its parts toward or away from an external stimulus, such as light, heat, or gravity.

true-breeding Individual that passes traits without change from one generation to the next.

trypsin A digestive enzyme that hydrolyzes bonds within polypeptide chains.

tumor-suppressor gene A gene that encodes proteins that inhibit cell division.

turgor pressure The internal hydrostatic pressure within plant cells.

turnover rate The rate at which one generation of producers in an ecosystem is replaced by the next.

tympanic membrane Also called the eardrum, a sheet of tissue forming the boundary between the outer ear and the middle ear which vibrates in response to sound waves that move through the auditory canal.

tympanum A thin membrane in the auditory canal that vibrates back and forth when struck by sound waves.

umbilical cord A long tissue with blood vessels linking the embryo and the placenta.

umbilicus Navel; the scar left when the short length of umbilical cord still attached to the infant after birth dries and shrivels within a few days.

undernutrition A condition in animals in which intake of organic fuels is inadequate, or whose assimilation of such fuels is abnormal.

undulating membrane In parabasalid protists, a finlike structure formed by a flagellum buried in a fold of the cytoplasm that facilitates movement through thick and viscous fluids.an expansion of the plasma membrane in some flagellates that is usually associated with a flagellum.

unicellular organism Individual consisting of a single cell.

uniform dispersion A pattern of distribution in which the individuals in a population are evenly spaced in their habitat.

uniformitarianism The concept that the geological processes that sculpted Earth's surface over long periods of time—such as volcanic eruptions, earthquakes, erosion, and the formation and movement of glaciers—are exactly the same as the processes observed today.

Unikonta (unikonts) Eukaryotes with a single flagellum.

uniparental inheritance A pattern of inheritance in which all progeny (both males and females) have the phenotype of only one of the parents.

universal A feature of the nucleic acid code, with the same codons specifying the same amino acids in all living organisms.

unreduced gamete A gamete that contains the same number of chromosomes as a somatic cell.

unsaturated fatty acid Fatty acid with one or more double bonds linking the carbons.

ureter The tube through which urine flows from the renal pelvis to the urinary bladder.

urethra The tube through which urine leaves the bladder. In most animals, the urethra opens to the outside.

urinary bladder A storage sac located outside the kidneys.

uterine cycle The menstrual cycle.

uterus A specialized saclike organ, in which the embryo develops in viviparous animals.

utricle A fluid-filled chamber of the vestibular apparatus that provides information about the position of the head with respect to gravity (up versus down), as well as changes in the rate of linear movement of the body.

vaccination The process of administering a weakened form of a disease to patients as a means of giving them immunity to a more serious form of the disease.

vagina The muscular canal that leads from the cervix to the exterior.

valence electron An electron in the outermost energy level of an atom.

Van der Waals forces Weak molecular attractions over short distances.

variable An environmental factor that may differ among places or an organismal characteristic that may differ among individuals.

variable (V) region For the light and heavy polypeptides of a particular type of antibody, the regions that have different amino acids sequences from molecule to molecule.

vas deferens The tube through which sperm travel from the epididymis to the urethra in the male reproductive system.

vascular bundle A cord of plant vascular tissue; often multistranded with both xylem and phloem.

vascular cambium A lateral meristem that produces secondary vascular tissues in plants.

vascular plant *See* tracheophyte.

vascular tissue system One of the three tissue systems in plants that provide the foundation for plant organs; it consists of transport tubes for water and nutrients.

vegetal pole The end of the egg opposite the animal pole, which typically gives rise to internal structures such as the gut and the posterior end of the embryo.

vegetative reproduction Asexual reproduction in plants by which new individuals arise (or are created) without seeds or spores; examples include fragmentation from the parent plant or the use of cuttings by gardeners.

vein In a plant, a vascular bundle that forms part of the branching network of conducting and supporting tissues in a leaf or other expanded plant organ. In an animal, a vessel that carries the blood back to the heart.

ventilation The flow of the respiratory medium (air or water, depending on the animal) over the respiratory surface.

ventral Indicating the lower or "belly" side of an animal.

ventricle In the brain, an irregularly shaped cavity containing cerebrospinal fluid. In the heart, a chamber that pumps blood out of the heart.

venule A capillary that merges into the small veins leaving an organ.

vernalization The stimulation of flowering by a period of low temperature.

vertebrae The series of bones that form the vertebral column of vertebrate animals.

vertebral column The series of vertebrae that surrounds and protects the dorsal nerve cord and forms the supporting axis of the body.

vertebrate A member of the monophyletic group of tetrapod animals that possess a vertebral column.

vesicle A small, membrane-bound compartment that transfers substances between parts of the endomembrane system.

vessel In plants, one of the tubular conducting structures of xylem, typically several centimeters long; most angiosperms and some other vascular plants have xylem vessels.

vessel member Any of the short cells joined end to end in tubelike columns in xylem.

vestibular apparatus The specialized sensory structure of the inner ear of most terrestrial vertebrates that is responsible for perceiving the position and motion of the head and, therefore, for maintaining equilibrium and for coordinating head and body movements.

vestigial structure An anatomical feature of living organisms that no longer retains its function.

vibrio Any of various short, motile, S-shaped or comma-shaped bacteria of the genus *Vibrio*.

vicariance The fragmentation of a continuous geographical distribution by nonbiological factors.

virion A complete virus particle.

viroid A plant pathogen that consists of strands or circles of RNA, smaller than any viral DNA or RNA molecule, that have no protein coat.

virulent bacteriophage Bacteriophage that kills its host bacterial cells during each cycle of infection.

virus An infectious agent that contains either DNA or RNA surrounded by a protein coat.

visceral mass In mollusks, the region of the body containing the internal organs.

visual signal A means of communication in which animals use facial expressions or body language to send messages to other individuals.

vital capacity The maximum tidal volume of air that an individual can inhale and exhale.

vitamin An organic molecule required in small quantities that the animal cannot synthesize for itself.

vitamin D A steroidlike molecule that increases the absorption of Ca^{2+} and phosphates from ingested food by promoting the synthesis of a calcium-binding protein in the intestine; it also increases the release of Ca^{2+} from bone in response to PTH.

vitelline coat A gel-like matrix of proteins, glycoproteins, or polysaccharides immediately outside the plasma membrane of an egg cell.

vitreous humor The jellylike substance that fills the main chamber of the eye, between the lens and the retina.

viviparous Referring to animals that retain the embryo within the mother's body and nourish it during at least early embryo development.

voltage-gated ion channel A membrane-embedded protein that opens and closes as the membrane potential changes.

vulva The external female sex organs.

water lattice An arrangement formed when a water molecule in liquid water establishes an average of 3.4 hydrogen bonds with its neighbors.

water molds *See* Oomycota.

water potential The potential energy of water, representing the difference in free energy between pure water and water in cells and solutions; it is the driving force for osmosis.

watershed An area of land from which precipitation drains into a single stream or river.

water-soluble vitamin A vitamin with a high proportion of oxygen and nitrogen able to form hydrogen bonds with water.

wax A substance insoluble in water that is formed when fatty acids combine with long-chain alcohols or hydrocarbon structures.

weight A measure of the pull of gravity on an object.

wetland A highly productive ecotone often at the border between a freshwater biome and a terrestrial biome.

white blood cell *See* leukocyte.

white matter The myelinated axons that surround the gray matter of the central nervous system.

white rusts *See* Oomycota.

wilting The drooping of leaves and stems caused by a loss of turgor.

wobble hypothesis Hypothesis stating that the complete set of 61 sense codons can be read by fewer than 61 distinct tRNAs because of particular pairing properties of the bases in the anticodons.

Wolffian duct A bipotential primitive duct associated with the gonads that leads to a cloaca.

wood The secondary xylem of trees and shrubs, lying under the bark and consisting largely of cellulose and lignin.

X chromosome Sex chromosome that occurs paired in female cells and single in male cells.

X-linked gene A gene on the X chromosome.

X-linked inheritance The pattern of inheritance of an X-linked gene.

X-linked recessive inheritance Pattern in which displayed traits are due to inheritance of recessive alleles carried on the X chromosome.

X-ray diffraction Method for deducing the position of atoms in a molecule.

xylem The plant vascular tissue that distributes water and nutrients.

xylem sap The dilute solution of water and solutes that flows in the xylem.

Y chromosome Sex chromosome that is paired with an X chromosome in male cells.

yeast A single-celled fungus that reproduces by budding or fission.

yolk The portion of an egg that serves as the main energy source for the embryo.

yolk sac In an amniote egg, an extraembryonic membrane that encloses the yolk.

zero population growth A circumstance in which the birth rate of a population equals the death rate.

zona pellucida A gel-like matrix of proteins, glycoproteins, or polysaccharides immediately outside the plasma membrane of the egg cell.

zone of cell division The region in a growing root that consists of the root apical meristem and the actively dividing cells behind it.

zone of elongation The region in a root where newly formed cells grow and elongate.

zone of maturation The region in a root above the zone of elongation where cells do not increase in length but may differentiate further and take on specialized roles.

zooplankton Small, usually microscopic, animals that float in aquatic habitats.

zygospore A multinucleate, thick-walled sexual spore formed in zygomycete fungi (Zygomycota).

zygote A fertilized egg.

Index

The letter i *designates illustration;* t *designates table;* **bold** *designates defined or introduced term.*